2003 IEEE International Conference on Robotics and Automation

September 14 - 19, 2003

The Grand Hotel, Taipei, Taiwan

Proceedings

Volume 3

Pages 2907 — 4450

Sponsored by

The IEEE Robotics and Automation Society
National Science Council, Taiwan, R.O.C.
Ministry of Education, Taiwan, R.O.C.

Hosted by

National Chung-Cheng University
   Automation Research Center
National Taiwan University
   Center for Information and Electronics Technology
   Center for Nano Science and Technology

In Cooperation with

Chinese Institute of Automation Engineers
Industrial Technology Research Institute
TECO Electric and Machinery Co., Ltd.
Metal Industry Research and Development Center

ICRA 2003 Proceedings (3 Volumes)

Additional copies may be ordered from:

IEEE Service Center
445 Hoes Lane
P.O. Box 1331
Piscataway, NJ 08855-11331, U.S.A.

IEEE Catalog Number:	03CH37422
ISBN:	0-7803-7736-2
Library of Congress Catalog Number:	90-640158
ISSN:	1050-4729
IEEE Catalog Number (CD-ROM):	03CH37422C
ISBN (CD-ROM):	0-7803-7737-0

Copyright and Reprint Permission:

Abstracting is permitting with credit to the source. Libraries are permitted to photocopy beyond the limit of U.S. copyright law for private use of patrons those articles in this volume that carry a code at the bottom of the first page, provided the pre-copy fee indicated in the code is paid through Copyright Clearance Center, 222 Rosewood Drive, Danvers, MA 09123, U.S.A. For other copying, reprint, or republication permission, write to IEEE Copyrights Manager, IEEE Operations Center, 445 Hoes Lane, P.O. Box 1331, Piscataway, NJ08855-1331, U.S.A. All rights reserved. Copyright ©2003 by the Institute of Electrical and Electronics Engineers, Inc.

Printed in Taiwan.

The Institute of Electrical and Electronics Engineers, Inc.

Foreword

The 2003 IEEE International Conference on Robotics and Automation (ICRA 2003) is being held in Taipei, the capital city of Taiwan, R.O.C. After Japan and Korea, this is the 3rd time in the 20 years history of IEEE Robotics and Automation Society (RAS) that this extremely prestigious annual conference is to be held in Asia again. With the last ICRA being held in Washington D.C., U.S.A., the conference location in Taiwan at almost opposite side of the globe simply provides the most wonderful opportunity for the R&A field scholars and practitioners to explore more in depth the academia, economy, high tech. industry, culture, etc. of a part in Asia – Taiwan, besides enhancement of sociality and experience sharing

A record number of papers, 1176, have been submitted to this conference Program Committee (PC) from 48 countries completely through the web on-line system developed locally, which demonstrates excellent capabilities in handling huge amount of paper submissions from all over the world and the subsequent review and registration matters. Special thanks should go to the Program Co-Chairs, Shigeo Hirose, C.S.G. Lee, Bruno Siciliano, and B.H. Lee and all PC members for their extraordinary work in reviewing such incredible amount of submitted papers and in selecting 714 papers for presentations over 14 parallel tracks per day in the three day main program. Topics of the accepted papers cover quite wide spectrum and are roughly grouped into 34 categories. To serve the conference participants with a technical program of greater variety, the sessions of paper presentation are made of several kinds besides the regular ones, including Organized/Invited Session, Industry Session, Video Session, and three plenary speeches as well as thee panel discussions. Such solid program will not be possible without tremendous help and professional support from members of the organizing committee and all other conference committees.

We wish you will have a fruitful conference and a very pleasant stay in Taiwan.

Ren C. Luo
General Chair

Li-Chen Fu
Program Chair

Conference Organization

Advisory Committee
T. J. Tarn, Washington Univ., St. Louis, U.S.A.
Toshio Fukuda, Nagoya Univ., Japan

National Advisory Committee

Chintay Shih	Ho-Ming Huang	Hong Sun	Jia-Ming Shyu
Powen Hsu	Stan Shih	Yuan T. Lee	

Organizing Committee

General Chair
Ren C. Luo
President
National Chung Cheng University
Taiwan

General Co-Chairs
S. H. Lu, Taiwan
G. Bekey, U.S.A.
F. Harashima, Japan
G. Giralt, France

Program Co-Chairs
C.S. George Lee, U.S.A
Bruno Siciliano, Italy
B. H. Lee, Korea
Shigeo Hirose, Japan

Awards Committee Chair Steve Hsia, U.S.A.
Awards Committee Co-Chair Antonio Bicchi, Italy

Program Chair
Li-Chen Fu
Professor
National Taiwan University
Taiwan

Publicity Chair Y. F. Zheng, U.S.A.
Publications Chair H.-P. Huang, Taiwan
Video Proceedings Chair Rajiv V. Dubey, U.S.A.
Finance Chair F.-T. Cheng, Taiwan
Finance Co-Chair Rong-Shine Lin, Taiwan
Workshop/Tutorial Chair Ken Goldberg, U.S.A.
Local Arrangement Chair Tsai-Yen Li, Taiwan
Exhibitions Chair MuDer Jeng, Taiwan
Registration Chair Yi-Ping Hung, Taiwan

Program Committee

Program Chair Li-Chen Fu, Taiwan
Co-Chairs C. S. George Lee, U.S.A.
Bruno Siciliano, Italy
B. H. Lee, Korea
Shigeo Sugano, Japan
Vice-Chairs Kuang-Chao Fan, Taiwan
Chang-Huan Liu, Taiwan
Tsu-Tian Lee, Taiwan

Awards Committee

Awards Committee Chair	Steve Hsia, U.S.A.
Co-Chair	Antonio Bicchi, Italy
Members	Raja Chatila, France
	Toshio Fukuda, Japan
	Ian Walker, U.S.A.
	Rajiv V. Dubey, U.S.A.

Publicity Committee

Publicity Committee Chair	Y. F. Zheng, U.S.A.
Members	Kazuhito Yokoi, Japan
	Shigeaki Yanai, Japan

Publications Committee

Publications Committee Chair	H.-P. Huang, Taiwan
Members	Argon Chen, Taiwan
	Chiou-Shann Fuh, Taiwan

Video Proceedings Committee

Video Proceedings Committee Chair	Rajiv V. Dubey, U.S.A.
Members	I-Ming Chen, Singapore
	T. Kesavadas, U.S.A.
	Oussama Khatib, U.S.A.
	Peter B. Luh, U.S.A.
	Shigeki Sugano, Japan

Workshop/Tutorial Committee

Workshop/Tutorial Committee Chair	Ken Goldberg, U.S.A.
Members	Ian Walker, U.S.A.
	Michael Wang, HongKong
	Simon DiMaio, Canada
	Wes Huang, U.S.A.
	Wolfram Burgard, U.S.A.

Local Arrangement Committee

Local Arrangement Committee Chair	Tsai-Yen Li, Taiwan
Members	Jyi-Shane Liu, Taiwan
	Maw-Kae Hor, Taiwan
	Tzuu-Hseng S. Li, Taiwan
	Jwu-Sheng Hu, Taiwan

Exhibitions Committee

Exhibitions Committee Chair	MuDer Jeng, Taiwan
Members	I-Ming Chen, Singapore
	Pau-Lo Hsu, Taiwan
	Da-Yin Liao, Taiwan
	Robin Qiu, U.S.A.

Technical Program Committee

Asia/Oceania

Akihito Sano	Argon Chen	Atsuo Takanishi
Boo Hee Nam	Chia-Ju Wu	Ching-Chih Tsai
Ching-Long Shih	Chiou-Shann Fuh	Der-Baau Perng
Er Meng Joo	Fan-Tien Cheng	Fumihito Arai
Fumitoshi Matsuno	Han-Pang Huang	Hideki Hashimoto
Hidenori Ishihara	Hugh Durrant-Whyte	I-Ming Chen
In-Joong Ha	James T. Lin	Jane Hsu
Jang-Myung Lee	Jen-Hwa Guo	Jin Young Choi
Jong Hyeon Park	Jongwon Kim	Jun Ota
Kai-Tai Song	Katsushi Ikeuchi	Kazuhiro Kosuge
Kazuihito Yokoi	Kazuo Kiguchi	Kazuo Tanie
Kiyoshi Komoriya	Koji Ito	Kuu-Young Young
Makoto Kaneko	Marcelo H. Ang Jr.	Masakatsu Fujie
Maw-Kae Hor	Max Q.-H. Meng	Michael Yu Wang
Minoru Asada	MuDer Jeng	N. Viswanadham
Sang-Rok Oh	Shi-Chung Chang	Shigeki Sugano
Shinichi Yuta	Suguru Arimoto	Suhkan Lee
Takafumi Matsumaru	Takashi Tsubouchi	Toshio Fukuda
Tsuneo Yoshikawa	Tzuu-Hseng S. Li	Woonchul Ham
Yangsheng Xu	Yasuyoshi Yokokohji	Yi-Ping Hung
Yoji Yamada	Yoshiaki Shirai	Yuechao Wang
Yugeng Xi	Zengqi Sun	Zexiang Li

Americas

Alan A. Desrochers	Andrew Goldenberg	Andrew Kusiak
Anthony A. Maciejewski	Antti J. Koivo	Arthur C. Sanderson
Bijoy K. Ghosh	Bradley J. Nelson	C. L. Philip Chen
Daniel Koditschek	David E. Orin	David J. Kriegman
Feiyue Wang	George Bekey	Gregory S. Chirikjian
Hami Kazerooni	Ian D. Walker	J. Y. S (John) Luh
Jay Lee	Jeffrey C. Trinkle	Jeremy Cooperstock
Jing Xiao	John Feddema	John M. Hollerbach

Jorge Angeles
Ken Goldberg
Lydia Kavraki
Matt Mason
Nancy Amato
Ning Xi
Peter Will
Ralph Hollis
Russell H. Taylor
Sunil K. Agrawal
Wayne J. Book
William R. Hamel
Yuan F. Zheng

Junku Yuh
Kok-Meng Lee
Lynne E. Parker
MengChu Zhou
Nikos Papanikolopoulos
Paul Oh
Pradeep K. Khosla
Richard M. Voyles
Seth Hutchinson
Vincent Hayward
Wesley E. Snyder
Xiaoping Yun

Kamal Gupta
Louis L. Whitcomb
Marjorie Skubic
Ming C. Lin
Nilanjan Sarkar
Peter B. Luh
Rajiv V. Dubey
Robin Murphy
Sooyong Lee
Vladimir Lumelsky
William A. Gruver
Yilong Chen

Europe/Mediterranean

Alessandro De Luca
Angel P. del Pobil
Benedetto Allotta
Christian Laugier
Davide Brugali
Eugenio Guglielmelli
Fabrizio Caccavale
François Pierrot
Gianantonio Magnani
Giuseppe Menga
Herman Bruyninckx
J. Norberto Pires
Jorge Dias
Kostas Kyriakopoulos
Manfred Hiller
Massimo Caccia
Radu Horaud
Rolf Johansson
Stefano Stramigioli

Alessandro Giua
Antonio Bicchi
Bernard Espiau
Claudio Melchiorri
Elzbieta Roszkowska
Evangelos Papadopoulos
Federico Thomas
Frans Groen
Giulio Sandini
Giuseppe Oriolo
Hong Qiao
Jadran Lenarcic
Joris De Schutter
Krzysztof Kozlowski
Maria Chiara Carrozza
Paolo Dario
Raja Chatila
Ruediger Dillmann
Vincenzo Parenti Castelli

Alicia Casals
Aydan M. Erkmen
Cecilia Laschi
David Lane
Erwin Prassler
Eve Coste-Maniere
Francesco Basile
Gerd Hirzinger
Giuseppe Casalino
Henrik Christensen
Imre Rudas
Jean-Pierre Merlet
Jose Neira
Luigi Villani
Martin Buss
Paolo Fiorini
Roland Siegwart
Stefano Chiaverini
Wisama Khalil

Table of Contents

Foreword i
Conference Organization iii
Technical Program Committee v
Table of Contents vii

TuA1: Mobile Robot Navigation (I)

Foliage Discrimination Using a Rotating Ladar 1
Andres Castano and Larry Matthies

Moving Obstacle Detection for a Skid-Steered Vehicle Endowed with a Single 2-D Laser Scanner 7
Kostas J. Kyriakopoulos and Nikos Skounakis

A Neural Network Based Torque Controller for Collision-Free Navigation of Mobile Robots 13
Simon X. Yang, Tiemin Hu, Xiaobu Yuan, Peter X Liu, and Max Meng

Studying the Feasibility of Energy Harvesting in a Mobile Sensor Network 19
Mohammad Rahimi, Hardik Shah, Gaurav S. Sukhatme, John Heidemann, and Deborah Estrin

A Robotic Walker that Provides Guidance 25
Aaron Morris, Raghavendra Donamukkala, Anuj Kapuria, Aaron Steinfeld, Judith T. Matthews, Jacqueline Dunbar-Jacobs, and Sebastian Thrun

TuA2: Control of Biped Robot (I)

An Order n Dynamic Simulator for a Humanoid Robot with a Virtual Spring-Damper Contact Model 31
Yoonkwon Hwang, Eiichi Inohira, Atsushi Konno, and Masaru Uchiyama

Towards a Factored Analysis of Legged Locomotion Models 37
Richard Altendorfer, Daniel E. Koditschek, and Philip Holmes

Frontal Plane Algorithms for Dynamic Bipedal Walking 45
Chee-Meng Chew and Gill A. Pratt

Contact Phase Invariant Control for Humanoid Robot Based on Variable Impedant Inverted Pendulum Model 51
Tomomichi Sugihara and Yoshihiko Nakamura

Aerial Posture Control for 3D Biped Running Using Compensator around Yaw Axis 57
Sang-Ho Hyon and Takashi Emura

TuA3: Mobile Robot Wheel Mechanisms

Development of Arm Equipped Single Wheel Rover: Effective Arm-Posture-Based Steering Method 63
Kazuhiro Motomura, Atsushi Kawakami, and Shigeo Hirose

Kinematic Modelling of Wheeled Mobile Manipulators 69
B. Bayle, J.-Y. Fourquet, and M. Renaud

Trail-Laying Robots for Robust Terrain Coverage 75
Jonas Svennebring and Sven Koenig

Curvature Based Point Stabilization for Compliant Framed Wheeled Modular Mobile
Robots 83
Brian W. Albiston and Mark A. Minor

Scout Robot Motion Model 90
Sascha A. Stoeter, Ian T. Burt, and Nikolaos Papanikolopoulos

TuA4: Agriculture and Off-Road Robotics

Control of a Heavy Material Handling Agricultural Manipulator Using µ-Synthesis and
Robust Gain Scheduling 96
Satoru Sakai, Koichi Osuka, Michihisa Iida, and Mikio Umeda

A Generalized Newton Method for Identification of Closed-Chain Excavator Arm
Parameters 103
Yahya H. Zweiri, Lakmal D. Seneviratne, and Kaspar Althoefer

Proprioceptive Control for a Robotic Vehicle over Geometric Obstacles 109
Kenneth J. Waldron, Ronald C. Arkin, Douglas Bakkum, Ernest Merrill, and Muhammad Abdallah

Adaptive Control for Car Like Vehicles Guidance Relying on RTK GPS:
Rejection of Sliding Effects in Agricultural Applications 115
R. Lenain, B. Thuilot, C. Cariou, and P. Martinet

On-Line Soil Property Estimation for Autonomous Excavator Vehicles 121
Choopar Tan, Yahya H Zweiri, Kaspar Althoefer, and Lakmal D Seneviratne

TuA5: 3D Vision (I)

Uncertainty-Driven Viewpoint Planning for 3D Object Measurements 127
Y.F. Li and Z.G. Liu

Overview of Coded Light Projection Techniques for Automatic 3D Profiling 133
Jordi Pagès, Joaquim Salvi, Rafael García, and Carles Matabosch

Registration and Segmentation for 3D Map Building: A Solution Based on Stereo Vision
and Inertial Sensors 139
Jorge Lobo, Luis Almeida, João Alves, and Jorge Dias

3D Modeling of Historic Sites Using Range and Image Data 145
Peter K. Allen, Ioannis Stamos, A. Troccoli, B. Smith, M. Leordeanu, and Y.C. Hsu

Detection of Classes of Features for Automated Robot Programming 151
Markus Vincze, Andreas Pichler, and Georg Biegelbauer

TuA6: Computer Aided Scheduling

A Polynomial Algorithm for a Two-Job Shop Scheduling Problem with Routing Flexibility 157
Yazid Mati and Xiaolan Xie

Efficient Selection of Scheduling Rule Combination by Combining Design of Experiment
and Ordinal Optimization-Based Simulation 163
Bo-Wei Hsieh, Shi-Chung Chang, and Chun-Hung Chen

Deadlock-Free Scheduling of Flexible Manufacturing Workcells Using Automata Theory 169
   Hamid R. Golmakani, James K. Mills, and Beno Benhabib

Utility and Stability Measures for Agent-Based Dynamic Scheduling of Steel Continuous Casting 175
   D. Ouelhadj, P.I. Cowling, and S. Petrovic

Platform-Based AS/RS for Container Storage 181
   Chuanyu Chen, Shell-Ying Huang, Wen-Jing Hsu, Ah Cheong Toh, and Chee Kit Loh

TuA7: Actuator Design

Development of a Hot Gas Actuator for Self-Powered Robots 188
   Michael Goldfarb, Eric J. Barth, Michael A. Gogola, and Joseph A. Wehrmeyer

Development of a Passively Operating Load-Responsive Transmission 194
   Hitoshi Maekawa and Kiyoshi Komoriya

Analytic and Experimental Study on Fast Response MR-Fluid Actuator 202
   Naoyuki Takesue, Junji Furusho, and Yuuki Kiyota

Dynamic Modeling and Analysis of a Transmission-Based Robot Servoactuator 208
   William R. Hamel, Sewoong Kim, Renbin Zhou, and Arnold Lumsdaine

Development of Isokinetic Exercise Machine Using ER Brake 214
   Takehito Kikuchi, Junji Furusho, and Kunihiko Oda

TuA8: Bio-Robotics

Automatic EMG Feature Evaluation for Controlling a Prosthetic Hand Using a Supervised Feature Mining Method: An Intelligent Approach 220
   Han-Pang Huang, Yi-Hung Liu, and Chun-Shin Wong

Rhythmic Movement by Neural Oscillator with Periodic Stimulus 226
   Hiroaki Hirai and Fumio Miyazaki

Visuo-Motor Coordination of a Humanoid Robot Head with Human-like Vision in Face Tracking 232
   Cecilia Laschi, Hiroyasu Miwa, Atsuo Takanishi, Eugenio Guglielmelli, and Paolo Dario

An Extensor Mechanism for an Anatomical Robotic Hand 238
   David D. Wilkinson, Michael Vande Weghe, and Yoky Matsuoka

A Clinical Jaw Movement Training Robot for Lateral Movement Training 244
   Akihisa Okino, Takahiro Inoue, Hideaki Takanobu, Atsuo Takanishi, Kayoko Ohtsuki, Masatoshi Ohnishi, and Yoshio Nakano

TuA9: Mobility and Manipulation

Mood and Task Coordination of Home Robots 250
   Myung-Jin Jung, Fumihito Arai, Yasuhisa Hasegawa, and Toshio Fukuda

Remote Book Browsing System Using a Mobile Manipulator 256
   Tetsuo Tomizawa, Akihisa Ohya, and Shin'ichi Yuta

Environment Identification by Comparing Maps of Landmarks 262
   Jens-Steffen Gutmann, Masaki Fukuchi, and Kohtaro Sabe

Behavior-Based Mobile Manipulation Inspired by the Human Example 268
B.J.W. Waarsing, M. Nuttin, and H. Van Brussel

Reliability Analysis of Mobile Robots 274
Jennifer Carlson and Robin R. Murphy

TuA10: Micro Robotics (I)

Wet Shape Memory Alloy Actuators for Active Vasculated Robotic Flesh 282
Stephen A. Mascaro and H. Harry Asada

A Thermally Actuated Polymer Micro Robotic Gripper for Manipulation of Biological Cells 288
Ho-Yin Chan and Wen J. Li

Actively Servoed Multi-Axis Microforce Sensors 294
Yu Sun, D. P. Potasek, D. Piyabongkarn, R. Rajamani, and B.J. Nelson

Pico-Newton Order Force Measurement Using a Calibrated Carbon Nanotube Probe by Electromechanical Resonance 300
Fumihito Arai, Masahiro Nakajima, Lixin Dong, and Toshio Fukuda

Novel Touch Sensor with Piezoelectric Thin Film for Microbial Separation 306
Fumihito Arai, Kouhei Motoo, Paul G.R. Kwon, Toshio Fukuda, Akihiko Ichikawa, and Tohoru Katsuragi

TuA11: Mechanism Design

A Multifunctional Hybrid Hip Joint for Improved Adaptability in Miniature Climbing Robots 312
Satya P. Krosuri and Mark A. Minor

Compliant Constant-Force Mechanism with a Variable Output for Micro/Macro Applications 318
Dhiraj R. Nahar and Thomas Sugar

Design and Foot Contact of a Leg Mechanism with a Flexible Gear System 324
K.H. Low and Aiqiang Yang

Conceptual Design and Kinematic Analysis of a 3-DOF Robot Wrist 330
Meng Li, Tian Huang, and Zhanxian Li

A Novel Mechanism for Implementing Multiple Collocated Spherical Joints 336
Paul Bosscher and Imme Ebert-Uphoff

TuA12: Geometry Issues in Robotics

A Branch-and-Prune Algorithm for Solving Systems of Distance Constraints 342
Josep Maria Porta, Federico Thomas, Lluís Ros, and Carme Torras

Exact Collision Detection of Two Moving Ellipsoids under Rational Motions 349
Yi-King Choi, Wenping Wang, and Myung-Soo Kim

Coordinate-Free Formulation of a 3-2-1 Wire-Based Tracking Device Using Cayley-Menger Determinants 355
Federico Thomas, Erika Ottaviano, Lluís Ros, and Marco Ceccarelli

Examination by Software Simulation on Preliminary-Announcement and Display of Mobile Robot's Following Action by Lamp or Blowouts 362
Takafumi Matsumaru, Hisashi Endo, and Tomotaka Ito

Float Arm V: Hyper-Redundant Manipulator with Wire-Driven Weight-Compensation Mechanism ... 368
Shigeo Hirose, Tomoyuki Ishii, and Atsuo Haishi

TuA13: Human Robot Interaction (I)

Coordination of Human and Mobile Manipulators Formation in a Perceptive Reference Frame ... 374
Jindong Tan, Ning Xi, Amit Goradia, and Weihua Sheng

Proposal and Evaluation of Natural Language Human-Robot Interface System Based on Conversation Theory ... 380
Yasushi Nakauchi, Piyawat Naphattalung, Takeshi Takahashi, Takashi Matsubara, and Eiichi Kashiwagi

Visual Human Machine Interface by Gestures ... 386
Manel Frigola, Josep Fernández, and Joan Aranda

Realizing Personality in Audio-Visually Triggered Non-Verbal Behaviors ... 392
Hiroshi G. Okuno, Kazuhiro Nakadai, and Hiroaki Kitano

Robot Recognizes Three Simultaneous Speech by Active Audition ... 398
Kazuhiro Nakadai, Hiroshi G. Okuno, and Hiroaki Kitano

TuA14: SLAM

Airborne Simultaneous Localisation and Map Building ... 406
Jong-Hyuk Kim and Salah Sukkarieh

Real Time Data Association for FastSLAM ... 412
Juan Nieto, Jose Guivant, Eduardo Nebot, and Sebastian Thrun

Outdoor Exploration and SLAM Using a Compressed Filter ... 419
John Folkesson and Henrik Christensen

Linear Time Vehicle Relocation in SLAM ... 427
José Neira, Juan D. Tards, and José A. Castellanos

A Genetic Algorithm for Simultaneous Localization and Mapping ... 434
Tom Duckett

TuM1: Mobile Robot Navigation (II)

3D-Odometry for Rough Terrain—Towards Real 3D Navigation ... 440
Pierre Lamon and Roland Siegwart

Smooth and Efficient Obstacle Avoidance for a Tour Guide Robot ... 446
Roland Philippsen and Roland Siegwart

Internal Posture Sensing for a Flexible Frame Modular Mobile Robot ... 452
Roy Merrell and Mark A. Minor

Mobile Robot Navigation Using Sensor Fusion ... 458
Fernando Lizarralde, Eduardo V.L. Nunes, Liu Hsu, and John T. Wen

Optimal Navigation and Object Finding without Geometric Maps or Localization ... 464
Benjamín Tovar, Steven M. LaValle, and Rafael Murrieta

TuM2: Control of Biped Robot (II)

A Small Biped Entertainment Robot Exploring Attractive Applications ···· 471
  Yoshihiro Kuroki, Masahiro Fujita, Tatsuzo Ishida, Ken'ichiro Nagasaka, and Jin'ichi Yamaguchi

Development of Walking Manipulator with Versatile Locomotion ···· 477
  Yusuke Ota, Tatsuya Tamaki, Kan Yoneda, and Shigeo Hirose

Sensor and Control Design of a Dynamically Stable Biped Robot ···· 484
  K. Löffler, M. Gienger, and F. Pfeiffer

Double Spherical Joint and Backlash Clutch for Lower Limbs of Humanoids ···· 491
  Masafumi Okada, Tetsuya Shinohara, Tatsuya Gotoh, Shigeki Ban, and Yoshihiko Nakamura

TuM3: Omnidirectional Vehicles

Two Wheels Caster Type Odometer for Omni-Directional Vehicles ···· 497
  Nobuhiro Ushimi, Motoji Yamamoto, and Akira Mohri

CVT Control of an Omnidirectional Mobile Robot with Steerable Omnidirectional Wheels for Energy Efficient Drive ···· 503
  Kyung-Seok Byun and Jae-Bok Song

Exponential Control Law for a Multi-Degree of Freedom Mobile Robot ···· 509
  Gabriel Ramirez and Saïd Zeghloul

A Common Reference Object Concept to Cooperative Transportation ···· 515
  Xin Yang, Keigo Watanabe, Kiyotaka Izumi, and Kazuo Kiguchi

The Dynamic Modeling and Analysis for an Omnidirectional Mobile Robot with Three Caster Wheels ···· 521
  Jae Heon Chung, Byung Ju Yi, Whee Kuk Kim, and Hogil Lee

TuM4: Helicopter/Air Vehicle

Autonomous Hovering Control and Test for Micro Air Vehicle ···· 528
  Huai-yu Wu, Zhao-ying Zhou, and Dong Sun

Micro Air Vehicle: Architecture and Implementation ···· 534
  Huai-yu Wu, Dong Sun, Zhao-ying Zhou, Shen-shu Xiong, and Xiao-hao Wang

Lateral Path-Following GPS-Based Control of a Small-Size Unmanned Blimp ···· 540
  Emmanuel Hygounenc and Philippe Souères

Low-Cost Flight Control System for a Small Autonomous Helicopter ···· 546
  Jonathan M. Roberts, Peter I. Corke, and Gregg Buskey

Integrated Modeling and Robust Control for Full-Envelope Flight of Robotic Helicopters ···· 552
  Marco La Civita, George Papageorgiou, William C. Messner, and Takeo Kanade

TuM5: Omnidirectional Vision

Omnidirectional Vision for an Autonomous Helicopter ···· 558
  Stefan Hrabar and Gaurav S. Sukhatme

Iconic Memory-Based Omnidirectional Route Panorama Navigation ···· 564
  Yasushi Yagi, Kousuke Imai, and Masahiko Yachida

Multibody Motion Estimation and Segmentation from Multiple Central Panoramic Views 571
Omid Shakernia, René Vidal, and Shankar Sastry

Visual Odometry from an Omnidirectional Vision System 577
Roland Bunschoten and Ben Kröse

Formation Control of Nonholonomic Mobile Robots with Omnidirectional Visual Servoing and Motion Segmentation 584
René Vidal, Omid Shakernia, and Shankar Sastry

TuM6: Diagnostics and Networked Manufacturing Systems

Web-Based Hardware-Neutral Sequential Controller 590
Chiaming Yen, Wujeng Li, and Jui Cheng Lin

Development of a Web-Services-Based e-Diagnostics Framework 596
Min-Hsiung Hung, Fan-Tien Cheng, and Sze-Chien Yeh

A Novel Approach to Fault Diagnostics and Prognostics 604
C. Kwan, X. Zhang, R. Xu, and L. Haynes

Intelligent Diagnosis Method of Multi-Fault State for Plant Machinery Using Wavelet Analysis, Genetic Programming and Possibility Theory 610
Peng Chen, Masatoshi Taniguchi, and Toshio Toyota

Distributed Modeling and Simulation of 300mm Fab Intrabay Automation Systems Using Distributed Agent Oriented Petri Nets 616
Chung-Hsien Kuo and Chien-Sheng Huang

TuM7: Actuators and Drivers

A New Approach to Compensate Friction in Robotic Actuators 622
Sebastião Cícero Pinheiro Gomes, and Vagner Santos Da Rosa

Modeling and Control of a Monopropellant-Based Pneumatic Actuation System 628
Eric J. Barth, Michael A. Gogola, and Michael Goldfarb

Actuator Failure Detection and Isolation Using Generalized Momenta 634
Alessandro De Luca and Raffaella Mattone

Development on Conveyance Module with New Power Drive Mechanism for Thin Wire Production System 640
Hidenori Ishihara, Kimihito Yukawa, and Atsutoshi Ikeda

Large-Scale Servo Control Using a Matrix Wire Network for Driving a Large Number of Actuators 646
Kyu-Jin Cho, Samuel Au, and H. Harry Asada

TuM8: Medical Diagnostic Robotics

Remote Actuation Mechanism for MR-Compatible Manipulator Using Leverage and Parallelogram—Workspace Analysis, Workspace Control, and Stiffness Evaluation 652
Yoshihiko Koseki, Noriho Koyachi, Tatsuo Arai, and Kiyoyuki Chinzei

The Development of a Bendable Colonoscopic Tip 658
G. Thomann, M. Bétemps, and T. Redarce

Touching Stomach by Air ⋯ 664
Makoto Kaneko, Tomohiro Kawahara, Satoshi Matsunaga, Toshio Tsuji, and Shinji Tanaka

An MR Compatible Robotic Technology ⋯ 670
R. Moser, R. Gassert, E. Burdet, L. Sache, H.R. Woodtli, J. Erni, W. Maeder, and H. Bleuler

Impedance Controller and Its Clinical Use of the Remote Ultrasound Diagnostic System ⋯ 676
Norihiro Koizumi, Shin'ichi Warisawa, Hiroyuki Hashizume, and Mamoru Mitsuishi

TuM9: Dexterous Hand and Control

DLR Hand II: Hard- and Software Architecture for Information Processing ⋯ 684
S. Haidacher, J. Butterfass, M. Fischer, M. Grebenstein, K. Joehl, K. Kunze, M. Nickl, N. Seitz, and G. Hirzinger

Design of a New Grasper Having XYZ Translational Motion ⋯ 690
Dong Yi, Byung-Ju Yi, and WheeKuk Kim

Under-Actuated Passive Adaptive Grasp Humanoid Robot Hand with Control of Grasping Force ⋯ 696
Wenzeng Zhang, Qiang Chen, Zhenguo Sun, and Dongbin Zhao

DLR Hand II: Experiments and Experiences with an Anthropomorphic Hand ⋯ 702
Ch. Borst, M. Fischer, S. Haidacher, H. Liu, and G. Hirzinger

Novel Fingertip Equipped with Soft Skin and Hard Nail for Dexterous Multi-Fingered Robotic Manipulation ⋯ 708
Kouji Murakami and Tsutomu Hasegawa

TuM10: Distributed Robotic Systems

Automatic Locomotion Pattern Generation for Modular Robots ⋯ 714
Akiya Kamimura, Haruhisa Kurokawa, Eiichi Yoshida, Kohji Tomita, Satoshi Murata, and Shigeru Kokaji

From Local to Global Behavior in Intelligent Self-Assembly ⋯ 721
Chris Jones and Maja J. Matarić

Functional Reactive Programming as a Hybrid System Framework ⋯ 727
Izzet Pembeci and Gregory Hager

Formation Constrained Multi-Robot System in Unknown Environments ⋯ 735
Zhiqiang Cao, Liangjun Xie, Bin Zhang, Shuo Wang, and Min Tan

Enveloping Obstacles with Hexagonal Metamorphic Robots ⋯ 741
Jennifer E. Walter, Elizabeth M. Tsai, and Nancy M. Amato

TuM11: Parallel Robot (I)

An In-Parallel Actuated Manipulator with Redundant Actuators for Gross and Fine Motions ⋯ 749
Yukio Takeda, Kazuki Ichikawa, Hiroaki Funabashi, and Kazuya Hirose

Type Synthesis of 4-DOF Parallel Manipulators ⋯ 755
Qinchuan Li and Zhen Huang

Singularity-Free Path Planning of Parallel Manipulators Using Clustering Algorithm and Line Geometry ⋯ 761
Anjan Kumar Dash, I-Ming Chen, Song Huat Yeo, and Guilin Yang

Singularity Analysis of the HALF Parallel Manipulator with Revolute Actuators · 767
Xin-Jun Liu, Jongwon Kim, and Kun-Ku Oh

The Management of Parallel-Manipulator Singularities Using Joint-Coupling · 773
I-Ming Chen, Jorge Angeles, Theingi, and Chuan Li

TuM12: Computational Intelligence (I)

Evolutionary Acquisition of Handstand from Backward Giant Circle by a Three-Link Rings Gymnastic Robot · 779
Takaaki Yamada, Keigo Watanabe, and Kazuo Kiguchi

Dexterous Manipulation from Pinching to Power Grasping—Strategy Selection According to Object Dimensions and Grasping Position · 785
Yasuhisa Hasegawa, Hayato Ioka, Toshio Fukuda, and Kensaku Kanada

Extend QDSEGA for Controlling Real Robots—Acquisition of Locomotion Patterns for Snake-like Robot · 791
Kazuyuki Ito, Tetsushi Kamegawa, and Fumitoshi Matsuno

A Study toward Cognitive Action with Environment Recognition by a Learning Space Robot · 797
Kei Senda, Tsutomu Matsumoto, and Yuzo Okano

Learning New Representations and Goals for Autonomous Robots · 803
Williams Paquier and Raja Chatila

TuM13: Haptic Interface (I)

Enabling Multi-Finger, Multi-Hand Virtualized Grasping · 809
Federico Barbagli, Kenneth Salisbry Jr., and Roman Devengenzo

Improvement of Passive Elements for Wearable Haptic Displays · 816
Sadao Kawamura, Katsuya Kanaoka, Yuichiro Nakayama, Jinwoo Jeon, and Daisuke Fujimoto

Sampled and Continuous Time Passivity and Stability of Virtual Environments · 822
Jee-Hwan Ryu, Yoon Sang Kim, and Blake Hannaford

The Haptic Scissors: Cutting in Virtual Environments · 828
A.M. Okamura, R.J. Webster III, J.T. Nolin, K.W. Johnson, and H. Jafry

Digital Clay: Architecture Designs for Shape-Generating Mechanisms · 834
Paul Bosscher and Imme Ebert-Uphoff

TuM14: Localization (I)

Online Simultaneous Localization and Mapping with Detection and Tracking of Moving Objects: Theory and Results from a Ground Vehicle in Crowded Urban Areas · 842
Chieh-Chih Wang, Charles Thorpe, and Sebastian Thrun

Using Visual Features to Build Topological Maps of Indoor Environments · 850
Paul E. Rybski, Franziska Zacharias, Jean-François Lett, Osama Masoud, Maria Gini, and Nikolaos Papanikolopoulos

Vision-Based Fast and Reactive Monte-Carlo Localization · 856
Thomas Röfer and Matthias Jüngel

Object-Based Localization and Mapping Using Loop Constraints and Geometric Prior Knowledge ... 862
Masahiro Tomono and Shin'ichi Yuta

Putting the 'I' in 'Team': An Ego-Centric Approach to Cooperative Localization ... 868
Andrew Howard, Maja J Matarić, and Gaurav S. Sukhatme

TuP1: Vision-Based Navigation (I)

Direct Plane Tracking in Stereo Images for Mobile Navigation ... 875
Jason Corso, Darius Burschka, and Gregory Hager

Location Estimation and Trajectory Prediction of Moving Lateral Vehicle Using Two Wheel Shapes Information in 2-D Lateral Vehicle Images by 3-D Computer Vision Techniques ... 881
Chih-Chiun Lai and Wen-Hsiang Tsai

Path-Dependent Gaze Control for Obstacle Avoidance in Vision Guided Humanoid Walking ... 887
Javier F. Seara, Klaus H. Strobl, and Günther Schmidt

Mobile Robot Navigation in Dynamic Environments Using Omnidirectional Stereo ... 893
Hiroshi Koyasu, Jun Miura, and Yoshiaki Shirai

Self-Positioning with an Omnidirectional Stereo System ... 899
Jyun-ichi Eino, Toshinobu Takashi, Jyun-ichi Takiguchi, and Takumi Hashizume

TuP2: Control of Biped Robot (III)

A Motion Planning Approach to Fast Parking Control of Mobile Robots ... 905
Ti-Chung Lee, Chi-Yi Tsai, and Kai-Tai Song

Online Humanoid Walking Control System and a Moving Goal Tracking Experiment ... 911
Koichi Nishiwaki, Satoshi Kagami, James J. Kuffner, Masayuki Inaba, and Hirochika Inoue

Autonomous Reactive Control for Simulated Humanoids ... 917
Petros Faloutsos, Michiel van de Panne, and Demetri Terzopoulos

The Sway Compensation Trajectory for a Biped Robot ... 925
Ryo Kurazume, Tsutomu Hasegawa, and Kan Yoneda

Online Footstep Planning for Humanoid Robots ... 932
James Kuffner, Satoshi Kagami, Koichi Nishiwaki, Masayuki Inaba, and Hirochika Inoue

TuP3: Multi-Mobile Robot System (I)

Decentralized Control of Multiple Mobile Manipulators Based on Virtual 3-D Caster Motion for Handling an Object in Cooperation with a Human ... 938
Yasuhisa Hirata, Youhei Kume, Zhi-Dong Wang, and Kazuhiro Kosuge

Time-Optimal Cooperative Control of Multiple Robot Vehicles ... 944
Tomonari Furukawa, Hugh F. Durrant-Whyte, and Gamini Dissanayake

Analysis of Formation Control of Cooperative Transportation of Mother Ship by SMC ... 951
Masaki Yamakita, Yasuaki Taniguchi, and Yuichirou Shukuya

Multi Mobile Robot Navigation Using Distributed Value Function Reinforcement Learning ... 957
Sharareh Babvey, Omid Momtahan, and Mohammad R. Meybodi

TuP4: Underwater Robotics

Trajectory Sonar Perception 963
Richard J. Rikoski and John J. Leonard

Vision-Based Localization of an Underwater Robot in a Structured Environment 971
M. Carreras, P. Ridao, R. Garcia, and T. Nicosevici

Vision-Based Linear Motion Estimation for Unmanned Underwater Vehicles 977
Massimo Caccia

Determining the Bodily Motion of a Biomimetic Underwater Vehicle under Oscillating
Propulsion 983
Jenhwa Guo, Forng-Chen Chiu, Chih-Chieh Chen, and Yueh Sheng Ho

Correction of Shading Effects in Vision-Based UUV Localization 989
R. Garcia, X. Cufí, M. Carreras, and P. Ridao

TuP5: 3D Vision (II)

Design of an Artificial Mark to Determine 3D Pose by Monocular Vision 995
Rie Katsuki, Jun Ota, Takahisa Mizuta, Tomomi Kito, Tamio Arai, Tsuyoshi Ueyama, and Tsuyoshi Nishiyama

Foveated Observation of Shape and Motion 1001
James Davis and Xing Chen

Automated Multisensor Polyhedral Model Acquisition 1007
D. Ortín, J.M.M. Montiel, and A. Zisserman

Structure and Pose from Single Images of Symmetric Objects with Applications to Robots
Navigation 1013
Allen Y. Yang, Wei Hong, and Yi Ma

A Comparison of Gaussian and Mean Curvatures Estimation Methods on Triangular
Meshes 1021
Tatiana Surazhsky, Evgeny Magid, Octavian Soldea, Gershon Elber, and Ehud Rivlin

TuP6: Semiconductor Factory Automation

Effective OHT Dispatching for Differentiated Material Handling Services in 300mm Wafer
Foundry 1027
Chia-Nan Wang and Da-Yin Liao

ERCN* Merged Nets for Modeling Degraded Behavior and Parallel Processes in
Semiconductor Manufacturing Systems 1033
MuDer Jeng, Xiaolan Xie, and Sheng-Luen Chung

Schedule Stabilization and Robust Timing Control for Time-Constrained Cluster Tools 1039
Ja-Hee Kim and Tae-Eog Lee

Structural Analysis of Resource Allocation Systems with Synchronization Constraints 1045
Spyros Reveliotis

An Overview of Semiconductor Fab Automation Systems 1050
Sheng-Luen Chung and MuDer Jeng

TuP7: Adaptive Control

Robustness of Image-Based Visual Servoing with Respect to Depth Distribution Errors 1056
Ezio Malis and Patrick Rives

Uniform Parametric Convergence in the Adaptive Control of Manipulators:
A Case Restudied 1062
Antonio Loría, Rafael Kelly, and Andrew R. Teel

Novel Approach in the Adaptive Control of Systems Having Strongly Nonlinear Coupling
between Their Unmodeled Internal Degrees of Freedom 1068
Imre J. Rudas, K. Kozlowski, József K. Tar, and Karel Jezernik

Adaptive Control of Robot Manipulators Using Multiple Neural Networks 1074
Choon-Young Lee and Ju-Jang Lee

Collision Detection System for Manipulator Based on Adaptive Impedance Control Law 1080
Shinya Morinaga and Kazuhiro Kosuge

TuP8: Endoluminal Surgery-Microendoscopy (I)

Review of Locomotion Techniques for Robotic Colonoscopy 1086
Irwan Kassim, Wan S. Ng, Gong Feng, Soo J. Phee, Paolo Dario, and Charles A. Mosse

Functional Colonoscope Robot System 1092
Byungkyu Kim, Younkoo Jeong, Hyun-young Lim, Jong-Oh Park, Arianna Menciassi, and Paolo Dario

Hyper Redundant Miniature Manipulator "Hyper Finger" for Remote Minimally Invasive
Surgery in Deep Area 1098
Koji Ikuta, Takahiko Hasegawa, and Shinichi Daifu

Development of Remote Microsurgery Robot and New Surgical Procedure for Deep and
Narrow Space 1103
Koji Ikuta, Keiichi Yamamoto, and Keiji Sasaki

Simulating Soft Tissue Cutting Using Finite Element Models 1109
C. Mendoza and C. Laugier

TuP9: Grasping Analysis

Dynamic Preshaping for a Robot Driven by a Single Wire 1115
Mitsuru Higashimori, Makoto Kaneko and Masatoshi Ishikawa

Capturing a Concave Polygon with Two Disc-Shaped Fingers 1121
Attawith Sudsang and Thanaphon Luewirawong

Optimization of Grasping by Using a Required External Force Set 1127
Tetsuyou Watanabe and Tsuneo Yoshikawa

Ranking Planar Grasp Configurations for a Three-Finger Hand 1133
Eris Chinellato, Robert B. Fisher, Antonio Morales, and Ángel P. del Pobil

On the Force Capability of Underactuated Fingers 1139
Lionel Birglen and Clément M. Gosselin

TuP10: Micro Systems

Biomimetic Sensor Suite for Flight Control of a Micromechanical Flying Insect:

Design and Experimental Results ... 1146
Wei-Chung Wu, Luca Schenato, Robert J. Wood, and Ronald S. Fearing

Model Identification and Attitude Control for a Micromechanical Flying Insect Including Thorax and Sensor Models ... 1152
Xinyan Deng, Luca Schenato, and Shankar S. Sastry

Dispersion Behaviors for a Team of Multiple Miniature Robots ... 1158
Janice L. Pearce, Paul E. Rybski, Sascha A. Stoeter, and Nikolaos Papanikolopoulos

Synthetic Gecko Foot-Hair Micro/Nano-Structures for Future Wall-Climbing Robots ... 1164
Metin Sitti and Ronald S. Fearing

Design and Implementaion of MARG Sensor for 3-DOF Orientation Measurement of Rigid Bodies ... 1171
Eric R. Bachmann, Xiaoping Yun, Doug McKinney, Robert B. McGhee, and Michael J. Zyda

TuP11: Parallel Robot (II)

Mobility Analysis of Lower-Mobility Parallel Manipulators Based on Screw Theory ... 1179
Qinchuan Li and Zhen Huang

Design and Control of a Novel 4-DOFs Parallel Robot H4 ... 1185
H.B. Choi, O. Company, F. Pierrot, A. Konno, T. Shibukawa, and M. Uchiyama

Vision-Based Kinematic Calibration of a H4 Parallel Mechanism ... 1191
Pierre Renaud, Nicolas Andreff, Frédéric Marquet, and Philippe Martinet

Determination of the Optimal Geometry of Modular Parallel Robots ... 1197
J-P. Merlet

Type Synthesis of 5-DOF Parallel Manipulators ... 1203
Qinchuan Li and Zhen Huang

TuP12: Computational Intelligence (II)

Planning and Control of UGV Formations in a Dynamic Environment: A Practical Framework with Experiments ... 1209
Yongxing Hao, Benjamin Laxton, Sunil K. Agrawal, Edward Lee, and Eric Benson

A New Approach for Structural Credit Assignment in Distributed Reinforcement Learning Systems ... 1215
Yu Zhong, Guochang Gu, and Rubo Zhang

Genetic Algorithm Based Path Planning for a Mobile Robot ... 1221
Jianping Tu and Simon X. Yang

Results for Outdoor-SLAM Using Sparse Extended Information Filters ... 1227
Yufeng Liu and Sebastian Thrun

Autonomous Feature-Based Exploration ... 1234
P. Newman, M. Bosse, and J. Leonard

TuP13: Haptic Interface (II)

Stable Haptic Interaction with Switched Virtual Environments ... 1241
Saurabh Mahapatra and Miloš Žefran

Telerobotic Haptic System to Assist the Performance of Occupational Therapy Tests by
Motion-Impaired Users ··········· 1247
Norali Pernalete, Wentao Yu, Rajiv Dubey, and Wilfrido A. Moreno

A Telemanipulation System for Psychophysical Investigation of Haptic Interaction ········· 1253
B.J. Unger, R.L. Klatzky, and R.L. Hollis

The FeTouch Project ··········· 1259
B. la Torre, D. Prattichizzo, F. Barbagli, and A. Vicino

Development of a System for Experiencing Tactile Sensation from a Robot Hand by
Electrically Stimulating Sensory Nerve Fiber ··········· 1264
Makoto Shimojo, Takafumi Suzuki, Akio Namiki, Takashi Saito, Masanari Kunimoto, Ryota Makino, Hironori Ogawa, Masatoshi Ishikawa, and Kunihiko Mabuchi

TuP14: Localization (II)

Mobile Robot Self-Localization Based on Global Visual Appearance Features ··········· 1271
Chao Zhou, Yucheng Wei, and Tieniu Tan

Fast and Accurate Vision-Based Pattern Detection and Identification ··········· 1277
James Bruce and Manuela Veloso

Ultrasonic Self-Localization and Pose Tracking of an Autonomous Mobile Robot via Fuzzy
Adaptive Extended Information Filtering ··········· 1283
Hung-Hsing Lin, Ching-Chih Tsai, Jui-Cheng Hsu, and Chih-Fu Chang

Probabilistic Visual Recognition of Artificial Landmarks for Simultaneous Localization and
Mapping ··········· 1291
David Prasser and Gordon Wyeth

3D Localization and Tracking in Unknown Environments ··········· 1297
P. Saeedi, D.G. Lowe, and P.D. Lawrence

TuE1: Vision-Based Navigation (II)

Weighted Line Fitting Algorithms for Mobile Robot Map Building and Efficient Data
Representation ··········· 1304
Samuel T. Pfister, Stergios I. Roumeliotis, and Joel W. Burdick

Learning Self-Organizing Maps for Navigation in Dynamic Worlds ··········· 1312
Rui Araújo, Gonçalo Gouveia, and Nuno Santos

BE-Viewer: Vision-Based Navigation System to Assist Motor-Impaired People in Docking
Their Mobility Aids ··········· 1318
Angelo M. Sabatini, Vincenzo Genovese, and Eliseo S. Maini

Using Learned Visual Landmarks for Intelligent Topological Navigation of Mobile Robots ········· 1324
M. Mata, J.M. Armingol, A. de la Escalera, and M.A. Salichs

Environment Modelling for Topological Navigation Using Visual Landmarks and
Range Data ··········· 1330
F. Lerasle, J. Carbajo, M. Devy, and J.B. Hayet

TuE2: New Challenges in Biped Locomotion

Running Pattern Generation and Its Evaluation Using a Realistic Humanoid Model ··········· 1336
Takashi Nagasaki, Shuuji Kajita, Kazuhito Yokoi, Kenji Kaneko, and Kazuo Tanie

Synthesis of Walking Primitive Databases for Biped Robots in 3D-Environments ······· 1343
J. Denk and G. Schmidt

Gait Transitions for Walking and Running of Biped Robots ······· 1350
Ohung Kwon and Jong Hyeon Park

Development and Control of Autonomous, Biped Locomotion Using Efficient Modeling, Simulation, and Optimization Techniques ······· 1356
Michael Hardt, Oskar von Stryk, Dirk Wollherr, and Martin Buss

Co-Evolution of Morphology and Walking Pattern of Biped Humanoid Robot Using Evolutionary Computation—Designing the Real Robot ······· 1362
Ken Endo, Takashi Maeno, and Hiroaki Kitano

TuE3: Dynamic Control of Multi-Legged Robot

On the Stable Passive Dynamics of Quadrupedal Running ······· 1368
Ioannis Poulakakis, Evangelos Papadopoulos, and Martin Buehler

Template Based Control of Hexapedal Running ······· 1374
Uluc Saranli and Daniel E. Koditschek

Stair Descent in the Simple Hexapod 'Rhex' ······· 1380
D. Campbell and M. Buehler

Towards a Dynamic Actuator Model for a Hexapod Robot ······· 1386
Dave McMordie, Chris Prahacs, and Martin Buehler

A Leg Configuration Sensory System for Dynamical Body State Estimates in a Hexapod Robot ······· 1391
Pei-Chun Lin, Haldun Komsuoğlu, and Daniel E. Koditschek

TuE5: Visual Sensing and Application (I)

A Truncated Least Squares Approach to the Detection of Specular Highlights in Color Images ······· 1397
Jae Byung Park and Avinash C. Kak

A Kite and Teleoperated Vision System for Acquiring Aerial Images ······· 1404
Paul Y. Oh and Bill Green

Appearance-Based Representation and Recognition of Human Motions ······· 1410
M. Masudur Rahman and Seiji Ishikawa

Automatic Detection and Response to Environmental Change ······· 1416
Scott Lenser and Manuela Veloso

Sky/Ground Modeling for Autonomous MAV Flight ······· 1422
Sinisa Todorovic, Michael C. Nechyba, and Peter G. Ifju

TuE6: Petri Nets in Automated Systems Design (I)

AGV Routing for Conflict Resolution in AGV Systems ······· 1428
Naiqi Wu and MengChu Zhou

Colored Timed Petri-Net and GA Based Approach to Modeling and Scheduling for Wafer Probe Center ······· 1434
Shun-Yu Lin, Li-Chen Fu, Tsung-Che Chiang, and Yi-Shiuan Shen

Petri Net Controllers to Enforce Disjunction of GMECs 1440
Francesco Basile, Ciro Carbone, and Pasquale Chiacchio

Property-Preserving Composition of Augmented Marked Graphs that Share Common Resources 1446
H.J. Huang, L. Jiao, and T.Y. Cheung

A Novel Siphon-Based Deadlock Control Method for FMS 1452
ZhiWu Li and MengChu Zhou

TuE7: Control Applications (I)

Globally Adaptive Decentralized Control of Time-Varying Robot Manipulators 1458
Su-Hau Hsu and Li-Chen Fu

Kinematic Control of a Platoon of Autonomous Vehicles 1464
Gianluca Antonelli and Stefano Chiaverini

Dynamic Modeling and Input Shaping of Thermal Bimorph MEMS Actuators 1470
Dan O. Popa, Byoung Hun Kang, John T. Wen, Harry E. Stephanou, George Skidmore, and Aaron Geisberger

A Switching Control Strategy for Nonlinear Dynamic Systems 1476
Mingjun Zhang and Tzyh-Jong Tarn

Control of Biomimetic Locomotion via Averaging Theory 1482
Patricio A. Vela and Joel W. Burdick

TuE8: Human Robot Interaction (II)

A Method for Preventing Accidents due to Human Action Slip Utilizing HMM-Based Dempster-Shafer Theory 1490
Yoji Yamada, Tetsuya Morizono, Yoji Umetani, and Yukitaka Sonohara

EMG Classification for Prehensile Postures Using Cascaded Architecture of Neural Networks with Self-Organizing Maps 1497
Han-Pang Huang, Yi-Hung Liu, Li-Wei Liu, and Chun-Shin Wong

Narrative Situation Assessment for Human-Robot Interaction 1503
Björn Jensen, Roland Philippsen, and Roland Siegwart

Configuring Sensors by User Learning for a Locomotion Aid Interface 1509
R. Thieffry, E. Monacelli, P. Henaff, and S. Delaplace

Face Direction-Based Human-Computer Interface Using Image Observation and EMG Signal for the Disabled 1515
Inhyuk Moon, Kyunghoon Kim, Jeicheong Ryu, and Museong Mun

TuE13: Sensor Application

Information-Theoretic Coordinated Control of Multiple Sensor Platforms 1521
Ben Grocholsky, Alexei Makarenko, and Hugh Durrant-Whyte

High-Precision Task-Space Sensing and Guidance for Autonomous Robotic Localization 1527
Goldie Nejat and Beno Benhabib

Geometry Design of an Elastic Finger-Shaped Sensor for Estimating Friction Coefficient by Pressing an Object 1533
Takashi Maeno and Tomoyuki Kawamura

Recognizing Surface Properties Using Impedance Perception 1539
Ryo Kikuuwe and Tsuneo Yoshikawa

Vision and Tactile Sensing for Real World Tasks 1545
D. Kragic, S. Crinier, D. Brunn, and H.I. Christensen

WA1: Map Building

Automatic Map Acquisition for Navigation in Domestic Environments 1551
Philipp Althaus and Henrik I. Christensen

Map Building with Mobile Robots in Dynamic Environments 1557
Dirk Hähnel, Rudolph Triebel, Wolfram Burgard, and Sebastian Thrun

An Evaluation of Sequential Monte Carlo Technique for Simultaneous Localisation and Map-Building 1564
David C.K. Yuen and Bruce A. MacDonald

Image-Based Localization with Depth-Enhanced Image Map 1570
Dana Cobzas, Hong Zhang, and Martin Jagersand

Temporal Landmark Validation in CML 1576
Juan Andrade-Cetto and Alberto Sanfeliu

WA2: Mobile Robot Design and Localization

TopBot: Automated Network Topology Detection with a Mobile Robot 1582
Paul Blaer and Peter K. Allen

A New Localization System for Autonomous Robots 1588
Sergio Hernández, Carlos A. Morales, Jesús M. Torres, and Leopoldo Acosta

Probabilistic Models of Dead-Reckoning Error in Nonholonomic Mobile Robots 1594
Yu Zhou and Gregory S. Chirikjian

MICAbot: A Robotic Platform for Large-Scale Distributed Robotics 1600
M. Brett McMickell, Bill Goodwine, and Luis Antonio Montestruque

Accurate Relative Localization Using Odometry 1606
Nakju Doh, Howie Choset, and Wan Kyun Chung

WA3: Humanoid Robotics Software Platform: OpenHRP

Whole Body Teleoperation of a Humanoid Robot—A Method of Integrating Operator's Intention and Robot's Autonomy 1613
Neo Ee Sian, Kazuhito Yokoi, Shuuji Kajita, Fumio Kanehiro, and Kazuo Tanie

Biped Walking Pattern Generation by Using Preview Control of Zero-Moment Point 1620
Shuuji Kajita, Fumio Kanehiro, Kenji Kaneko, Kiyoshi Fujiwara, Kensuke Harada, Kazuhito Yokoi, and Hirohisa Hirukawa

Pushing Manipulation by Humanoid Considering Two-Kinds of ZMPs 1627
Kensuke Harada, Shuuji Kajita, Kenji Kaneko, and Hirohisa Hirukawa

The First Humanoid Robot that Has the Same Size as a Human and that Can
Lie down and Get up 1633
*Fumio Kanehiro, Kenji Kaneko, Kiyoshi Fujiwara, Kensuke Harada, Shuuji Kajita, Kazuhito Yokoi,
Hirohisa Hirukawa, Kazuhiko Aakchi, and Takakatsu Isozumi*

Experimental Evaluation of the Dynamic Simulation of Biped Walking of Humanoid Robots 1640
*Hirohisa Hirukawa, Fumio Kanehiro, Shuuji Kajita, Kiyoshi Fujiwara, Kazuhito Yokoi, Kenji Kaneko, and
Kensuke Harada*

WA4: Network Robotics

Tele-Coordinated Control of Multi-Robot Systems via the Internet 1646
Imad Elhajj, Ning Xi, Amit Goradia, Chow Man Kit, Yun Hui Liu, and Toshio Fukuda

Connectivity-through-Time Protocols for Dynamic Wireless Networks to Support Mobile
Robot Teams 1653
Nageswara S.V. Rao, Qishi Wu, S. Sitharama Iyengar, and Arul Manickam

Control and Data Transmission for Internet Robots 1659
Peter Xiaoping Liu, Max Q-H Meng, Jason Gu, Simon X. Yang, and Chao Hu

A Scalable Approach to Human-Robot Interaction 1665
Ashley D. Tews, Maja J. Matarić, and Gaurav S. Sukhatme

A Service-Based Network Architecture for Wearable Robots 1671
Ka Keung Lee, Ping Zhang, and Yangsheng Xu

WA5: Stereo Vision and Visual Tracking

A Real-Time Robust Eye Tracking System for Autostereoscopic Displays Using
Stereo Cameras 1677
Chan-Hung Su, Yong-Sheng Chen, Yi-Ping Hung, Chu-Song Chen, and Jiun-Hung Chen

Real-Time Object Tracking Using Multi-Resolution Critical Points Filters 1682
Jérôme Durand and Seth Hutchinson

Correspondence-Free Stereo Vision for the Case of Arbitrarily-Positioned Cameras 1688
Ding Yuan and Ronald Chung

High Resolution Catadioptric Omni-Directional Stereo Sensor for Robot Vision 1694
Shih-Schön Lin and Ruzena Bajcsy

A Voting Stereo Matching Method for Real-Time Obstacle Detection 1700
Mohamed Hariti, Yassine Ruichek, and Abderrafiaa Koukam

WA6: Manufacturing Systems Architecture and Design (I)

Distributing 3D Manufacturing Simulations to Realize the Digital Plant 1705
E. Freund and D.H. Pensky

Flexible Material Handling System Using Smart-Carriers and Powerline Communication 1711
Eric Wade and H. Harry Asada

The Development of Distributed Web-Based Rapid Prototyping Manufacturing System 1717
Ren C. Luo and Jyh Hwa Tzou

Development of a Generic Tester for Distributed Object-Oriented Systems 1723
Fan-Tien Cheng, Chin-Hui Wang, and Yu-Chuan Su

Development of a STEP-Based Collaborative Product and Process Development System for Manufacturability Evaluation with Reverse Engineering 1731
Rong Shean Lee and Jo Peng Tsai

WA7: Supply Chain Design, Analysis and Optimization

Design of Six Sigma Supply Chains 1737
Dinesh Garg, Y. Narahari, and N. Viswanadham

Supply Chain Planning with Order/Setup Costs and Capacity Constraints: A New Lagrangian Relaxation Approach 1743
Haoxun Chen and Chengbin Chu

Supply Chain Performance Evaluation: A Simulation Study 1749
Yan Tu, Peter B. Luh, Weidong Feng, and Katsumi Narimatsu

An Optimization-Based Approach for Distributed Project Scheduling 1756
Ming Ni, Peter B. Luh, and Bryan Moser

Robust Supply Chain Design: A Strategic Approach for Exception Handling 1762
Roshan Gaonkar and N. Viswanadham

WA8: Robotic Surgery

A Wireless Temperature Measurement Guide Rod for Internal Bone Fixation Surgery 1768
Raymond H.W. Lam, Wen J. Li, and Ning Xi

Robotic Needle Insertion: Effects of Friction and Needle Geometry 1774
M.D. O'Leary, C. Simone, T. Washio, K. Yoshinaka, and A.M. Okamura

Design of All-Accelerometer Inertial Measurement Unit for Tremor Sensing in Hand-Held Microsurgical Instrument 1781
Wei Tech Ang, Pradeep K. Khosla, and Cameron N. Riviere

Force Control and Breakthrough Detection of a Bone Drilling System 1787
Wen-Yo Lee and Ching-Long Shih

Needle Insertion and Radioactive Seed Implantation in Human Tissues: Simulation and Sensitivity Analysis 1793
Ron Alterovitz, Ken Goldberg, Jean Pouliot, Richard Taschereau, and I-Chow Hsu

WA9: Grasping and Manipulation (I)

A New Algorithm for Three-Finger Force-Closure Grasp of Polygonal Objects 1800
Jia-Wei Li, Ming-He Jin, and Hong Liu

Estimating Finger Contact Location and Object Pose from Contact Measurements in 3-D Grasping 1805
S. Haidacher and G. Hirzinger

Simplified Generation Algorithm of Regrasping Motion—Performance Comparison of Online-Searching Approach with EP-Based Approach 1811
Yasuhisa Hasegawa, Masaki Higashiura, and Toshio Fukuda

A Polyhedral Bound on the Indeterminate Contact Forces in 2D Fixturing and Grasping Arrangements 1817
Elon Rimon, Joel W. Burdick, and Toru Omata

Automatic Grasp Planning Using Shape Primitives ... 1824
Andrew T. Miller, Steffen Knoop, Henrik I. Christensen, and Peter K. Allen

WA10: Micro Robotics (II)

Polymer-Based New Type of Micropump for Bio-Medical Application ... 1830
Shuxiang Guo and Kinji Asaka

A High Force Miniature Gripper Fabricated via Shape Deposition Manufacturing ... 1836
Cesare Stefanini, Mark R. Cutkosky, and Paolo Dario

Microrobotics Using Composite Materials: The Micromechanical Flying Insect Thorax ... 1842
R.J. Wood, S. Avadhanula, M. Menon, and R.S. Fearing

Torque Sensing of Finger Joint Using Strain-Deformation Expansion Mechanism ... 1850
Yong Yu, Takashi Ishitsuka, and Showzow Tsujio

Digital Polymer Motor for Robotic Applications ... 1857
H.R. Choi, K.M. Jung, J.W. Kwak, S.W. Lee, H.M. Kim, J.W. Jeon, and J.D. Nam

WA11: Parallel Robot (III)

Identifiablity of Geometric Parameters of 6-DOF PKM Systems Using a Minimum Set of Pose Error Data ... 1863
Tian Huang, Jinsong Wang, Derek G. Chetwynd, and David J. Whitehouse

Optimal Design of Parallel Manipulators via LMI Approach ... 1869
Y.J. Lou, G.F. Liu, and Z.X. Li

I4: A New Parallel Mechanism for Scara Motions ... 1875
Sébastien Krut, Olivier Company, Michel Benoit, Hiromichi Ota, and François Pierrot

Singular Loci Analysis of 3/6-Stewart Manipulator by Singularity-Equivalent Mechanism ... 1881
Yanwen Li, Zhen Huang, and Longhui Chen

Mobility Analysis of a 3-5R Parallel Mechanism Family ... 1887
Qinchuan Li and Zhen Huang

WA12: Computational Intelligence (III)

An Iterative Framework for Projection-Based Image Sequence Registration ... 1893
Joaquín Salas

An Atlas Framework for Scalable Mapping ... 1899
Michael Bosse, Paul Newman, John Leonard, Martin Soika, Wendelin Feiten, and Seth Teller

Probabilistic Cooperative Localization and Mapping in Practice ... 1907
Ioannis Rekleitis, Gregory Dudek, and Evangelos Milios

Local Exploration: Online Algorithms and a Probabilistic Framework ... 1913
Volkan Isler, Sampath Kannan, and Kostas Daniilidis

Pure Range-only Sub-Sea SLAM ... 1921
P. Newman and J. Leonard

WA13: Human Robot Interaction (III)

Dimensionality Reduction and Reproduction with Hierarchical NLPCA Neural Networks

—Extracting Common Space of Multiple Humanoid Motion Patterns ········ 1927
Koji Tatani and Yoshihiko Nakamura

Switching Control of Position/Torque Control for Human-Robot Cooperative Task
—Human-Robot Cooperative Carrying and Peg-in-Hole Task ········ 1933
Toru Tsumugiwa, Atsushi Sakamoto, Ryuichi Yokogawa, and Kei Hara

Behavior Developing Environment for the Large-DOF Muscle-Driven Humanoid Equipped with Numerous Sensors ········ 1940
Ikuo Mizuuchi, Tomoaki Yoshikai, Daisuke Sato, Shigenori Yoshida, Masayuki Inaba, and Hirochika Inoue

Wearable-Based Evaluation of Human-Robot Interactions in Robot Path-Planning ········ 1946
Ritsu Shikata, Takayuki Goto, Hiroshi Noborio, and Hiroshi Ishiguro

Spatial Motion Constraints: Theory and Demonstrations for Robot Guidance Using Virtual Fixtures ········ 1954
Panadda Marayong, Ming Li, Allison M. Okamura, and Gregory D. Hager

WA14: Sensor Localization and Mapping

Marco Polo Localization ········ 1960
Eric Beowulf Martinson and Frank Dellaert

Constrained Initialization for Bearing-only SLAM ········ 1966
Tim Bailey

Sound Localization Based on Mask Diffraction ········ 1972
S.S. Ge, A.P. Loh, and F. Guan

Outdoor Navigation of a Mobile Robot between Buildings Based on DGPS and Odometry Data Fusion ········ 1978
Kazunori Ohno, Takashi Tsubouchi, Bunji Shigematsu, Shoichi Maeyama, and Shin'ichi Yuta

Simultaneous Localization and Mapping with Unknown Data Association Using FastSLAM ········ 1985
Michael Montemerlo and Sebastian Thrun

WM1: Mobile Robot Navigation (III)

A Navigation Framework for Multiple Mobile Robots and Its Application at the Expo.02 Exhibition ········ 1992
Kai O. Arras, Roland Philippsen, Nicola Tomatis, Marc De Battista, Martin Schilt, and Roland Siegwart

Adapting Navigation Strategies Using Motions Patterns of People ········ 2000
Maren Bennewitz, Wolfram Burgard, and Sebastian Thrun

Navigation of Cleaning Robots Using Triangular-Cell Map for Complete Coverage ········ 2006
Joon Seop Oh, Yoon Ho Choi, Jin Bae Park, and Yuan F. Zheng

Integrating Terrain Maps into a Reactive Navigation Strategy ········ 2012
Ayanna Howard, Barry Werger, and Homayoun Seraji

μNAV: A Minimalist Approach to Navigation ········ 2018
Alessandro Scalzo, Antonio Sgorbissa, and Renato Zaccaria

WM2: Control of Quadruped Walking Robot

A Biologically Inspired Four Legged Walking Robot ········ 2024
S. Peng, C.P. Lam, and G.R. Cole

Firm Standing of Legged Mobile Manipulator 2031
Takashi Tagawa, Yasumichi Aiyama, and Hisashi Osumi

Adaptive Dynamic Walking of a Quadruped Robot 'Tekken' on Irregular Terrain Using a Neural System Model 2037
Y. Fukuoka, H. Kimura, Y. Hada, and K. Takase

Adaptive Running of a Quadruped Robot on Irregular Terrain Based on Biological Concepts 2043
Z.G. Zhang, Y. Fukuoka, and H. Kimura

Adaptive Gait for a Quadruped Robot on 3D Path Planning 2049
Hiroshi Igarashi and Masayoshi Kakikura

WM3: Snake-like Robots

Control of Locomotion and Head Configuration of 3D Snake Robot (SMA) 2055
Masaki Yamakita, Minoru Hashimoto, and Takeshi Yamada

Control of Redundant 3D Snake Robot Based on Kinematic Model 2061
Fumitoshi Matsuno and Kentaro Suenaga

Control of a 3-Dimensional Snake-like Robot 2067
Shugen Ma, Yoshihiro Ohmameuda, Kousuke Inoue, and Bin Li

Analysis of Creeping Locomotion of a Snake Robot on a Slope 2073
Shugen Ma, Naoki Tadokoro, Bin Li, and Kousuke Inoue

Dynamic Control for a Tentacle Manipulator with SMA Actuators 2079
Mircea Ivanescu, Nicu Bizdoaca, and Deniela Pana

WM4: Intelligent Transportation Systems

Locating Nearby Vehicles on Highway at Daytime Based on the Front Vision of a Moving Car 2085
Ming-Yang Chern and Bor-Yeu Shyr

Traffic Monitoring Based on Real-Time Image Tracking 2091
Ching-Po Lin, Jen-Chao Tai, and Kai-Tai Song

Driver Assistance: An Integration of Vehicle Monitoring and Control 2097
Lars Petersson, Nicholas Apostoloff, and Alexander Zelinsky

Using Bayesian Programming for Multi-Sensor Multi-Target Tracking in Automotive Applications 2104
C. Coué, Th. Fraichard, P. Bessière, and E. Mazer

The Lane Recognition and Vehicle Detection at Night for a Camera-Assisted Car on Highway 2110
Ming-Yang Chern and Ping-Cheng Hou

WM5: 3D Vision (III)

Viewpoint Selection for Object Reconstruction Using only Local Geometric Features 2116
K. Jonnalagadda, R. Lumia, G. Starr, and J. Wood

Robust Model-Based 3D Object Recognition by Combining Feature Matching with Tracking 2123
Sungho Kim, Inso Kweon, and Incheol Kim

Dynamically Reconfigurable Visual Sensing for 3D Perception ... 2129
S.Y. Chen and Y.F. Li

Simultaneous Shape and Motion Recovery: Geometry, Optimal Estimation, and Coordinate Descent Algorithms ... 2135
Joonhyuk Choi, F.C. Park, and Munsang Kim

Vision-Based 2.5D Terrain Modeling for Humanoid Locomotion ... 2141
Satoshi Kagami, Koichi Nishiwaki, James J. Kuffner, Kei Okada, Masayuki Inaba, and Hirochika Inoue

WM6: Manufacturing Systems Architecture and Design (II)

i-Fork: A Flexible AGV System Using Topological and Grid Maps ... 2147
H. Martínez Barberá, J.P. Canovas Quiñonero, M. Zamora Izquierdo, and A. Gomez Skarmeta

Functional Model Based Object-Oriented Development Framework for Mechatronic Systems ... 2153
Mingjun Zhang, William Fisher, Peter Webb, and Tzyh-Jong Tarn

Avoiding Unsafe States in Manufacturing Systems Based on Polynomial Digraph Algorithms ... 2159
Yin Wang and Zhiming Wu

The Application and Verification of Banker's Algorithm for Deadlock Avoidance in Flexible Manufacturing System with SPIN ... 2165
Gang Xu and Zhiming Wu

New Finishing System for Metalic Molds Using a Hybrid Motion/Force Control ... 2171
Fusaomi Nagata, Keigo Watanabe, Yukihiro Kusumoto, Kunihiro Tsuda, Kiminori Yasuda, Kazuhiko Yokoyama, and Naoki Mori

WM7: Control Applications (II)

Zero Power Control of 0.5KWh Class Flywheel System Using Magnetic Bearing with Gyroscopic Effect ... 2176
Ya-Chong Zhang, Guo-Ji Sun and Ya-Jun Zhang

Control Design for the Rotation of Crane Loads for Boom Cranes ... 2182
O. Sawodny, A. Hildebrandt, and K. Schneider

Application of Artificial Pneumatic Rubber Muscles to a Human Friendly Robot ... 2188
Toshiro Noritsugu, Daisuke Sasaki, and Masahiro Takaiwa

Autonomous Control of a Horizontally Configured Undulatory Flap Propelled Vehicle ... 2194
Stephen Hsu, Chris Mailey, Ethan Eade, and Jason Janét

Motion Planning of Aerial Robot Using Rapidly-Exploring Random Trees with Dynamic Constraints ... 2200
Jongwoo Kim and James P. Ostrowski

WM8: Rehabilitation Robotics (I)

Exoskeleton for Human Upper-Limb Motion Support ... 2206
Kazuo Kiguchi, Takakazu Tanaka, Keigo Watanabe, and Toshio Fukuda

Functional Assessment of Hand Orthopedic Disorders Using a Sensorised Glove: Preliminary Results ... 2212

Silvestro Micera, Ettore Cavallaro, Rossella Belli, Franco Zaccone, Eugenio Guglielmelli, Paolo Dario, Diego Collarini, Bruno Martinelli, Chiara Santin, and Renzo Marcovich

Modelling of the Human Paralysed Lower Limb under FES 2218
David Guiraud, Philippe Poignet, Pierre-Brice Wieber, Hassan El Makksoud, François Pierrot, Bernard Brogliato, Philippe Fraisse, Etienne Dombre, Jean-Louis Divoux, and Pierre Rabischong

Maneuvering a Bed Sheet for Repositioning a Bedridden Patient 2224
Binayak Roy, Arin Basmajian, and H. Harry Asada

Experimental Analysis of an Innovative Prosthetic Hand with Proprioceptive Sensors 2230
M.C. Carrozza, F. Vecchi, F. Sebastiani, G. Cappiello, S. Roccella, M. Zecca, R. Lazzarini, and P. Dario

WM9: Grasping and Manipulation (II)

Force Passivity in Fixturing and Grasping 2236
Michael Yu Wang and Yun-Hui Liu

Capturing a Convex Object with Three Discs 2242
Jeff Erickson, Shripad Thite, Fred Rothganger, and Jean Ponce

Application of "Generalized Attractive Region" in Orienting 3D Polyhedral Part 2248
Hong Qiao

Complementarity Formulation for Multi-Fingered Hand Manipulation with Rolling and Sliding Contacts 2255
Masahito Yashima and Hideya Yamaguchi

Implementation of Multi-Rigid-Body Dynamics within a Robotic Grasping Simulator 2262
Andrew T. Miller and Henrik I. Christensen

WM10: Multiple Robots Coordination

Cooperative Exploration of Mobile Robots Using Reaction-Diffusion Equation on a Graph 2269
Chomchana Trevai, Yusuke Fukazawa, Jun Ota, Hideo Yuasa, Tamio Arai, and Hajime Asama

A Strategy and a Fast Testing Algorithm for Object Caging by Multiple Cooperative Robots 2275
ZhiDong Wang, Vijay Kumar, Yasuhisa Hirata, and Kazuhiro Kosuge

Multi-Robot Team Response to a Multi-Robot Opponent Team 2281
James Bruce, Michael Bowling, Brett Browning, and Manuela Veloso

A Behaviour-Based Manipulator for Multi-Robot Transport Tasks 2287
Antonios K. Bouloubasis, Gerard T McKee, and Paul S. Schenker

Multi-Robot Task-Allocation through Vacancy Chains 2293
Torbjørn S. Dahl, Maja J. Matarić, and Gaurav S. Sukhatme

WM11: Reconfigurable Robot and Special Robot

Inverse Dynamics of Gel Robots Made of Electro-Active Polymer Gel 2299
Mihoko Otake, Yoshiharu Kagami, Yasuo Kuniyoshi, Masayuki Inaba, and Hirochika Inoue

Simple Self-Transfer Aid Robotic System 2305
Yoshihiko Takahashi, Go Manabe, Katsumi Takahashi, and Takuro Hatakeyama

Highly Compliant and Self-Tightening Docking Modules for Precise and

Fast Connection of Self-Reconfigurable Robots ············ 2311
Behrokh Khoshnevis, Peter Will, and Wei-Min Shen

Adaptability of Reconfigurable Robotic Systems ············ 2317
Z.M. Bi, W.A. Gruver, and W.J. Zhang

Design of a New Exoskeletal Mechanism for a Shoulder Joint of Wearable Robots: The Wearable HEXA Mechanism ············ 2323
Tetsuya Morizono, Yoji Yamada, Yoji Umetani, Takahisa Yamamoto, Tetsuji Yoshida, and Shigeru Aoki

WM12: Mathematical Optimization

GA-Based Robust H_2 Controller Design Approach for Active Suspension Systems ············ 2330
Chein-Chung Sun, Huan-Yuan Chung, and Wen-Jer Chang

Stability on a Manifold: Simultaneous Realization of Grasp and Orientation Control of an Object by a Pair of Robot Fingers ············ 2336
S. Arimoto, J.-H. Bae, and K. Tahara

Intrinsic Localization and Mapping with 2 Applications: Diffusion Mapping and Marco Polo Localization ············ 2344
Frank Dellaert, Fernando Alegre, and Eric Beowulf Martinson

Slack Variable Method for State Variable Constraint ············ 2350
Takeuchi Hiroki

Vector Quantization for State-Action Map Compression ············ 2356
Ryuichi Ueda, Takeshi Fukase, Yuichi Kobayashi, and Tamio Arai

WM13: Human Robot Interaction (IV)

Interaction among Human, Machine and Patient with Work State Transition ············ 2362
Takeshi Koyama, Takayuki Tanaka, Kazuo Tanaka, and Maria Q. Feng

Enabling Real-Time Full-Body Imitation: A Natural Way of Transferring Human Movement to Humanoids ············ 2368
Marcia Riley, Ales Ude, Keegan Wade, and Christopher G. Atkeson

Visual Transformations in Gesture Imitation: What You See is What You Do ············ 2375
Manuel Cabido-Lopes and José Santos-Victor

Affect-Sensitive Human-Robot Cooperation: Theory and Experiments ············ 2382
Pramila Rani, Nilanjan Sarkar, and Craig A. Smith

Multi-Robot Human-Interaction and Visitor Flow Management ············ 2388
B. Jensen, G. Froidevaux, X. Greppin, A. Lorotte, L. Mayor, M. Meisser, G. Ramel, and R. Siegwart

WM14: Sensor-Based Robotics (I)

A Multiagent Multisensor Based Real-Time Sensory Control System for Intelligent Security Robot ············ 2394
Ren C. Luo and Kuo L. Su

Robotic Catching Using a Direct Mapping from Visual Information to Motor Command ············ 2400
Akio Namiki and Masatoshi Ishikawa

Computing C-Space Entropy for View Planning with a Generic Range Sensor Model ············ 2406
Pengpeng Wang and Kamal Gupta

Towards a Haptic Black Box for Free-Hand Softness and Shape Discrimination ········· 2412
Enzo Pasquale Scilingo, Nicola Sgambelluri, Danilo De Rossi, and Antonio Bicchi

Development of a Smart Robotic Gripper for Shape and Vibration Control of Flexible
Payloads: Theory and Experiments ········· 2418
Edward J. Park, Gary Li, and James K. Mills

WP1: Mobile Robot Path Planning

Extracting Optimal Paths from Roadmaps for Motion Planning ········· 2424
Jinsuck Kim, Roger A. Pearce, and Nancy M. Amato

Towards Motion Autonomy of a Bi-Steerable Car: Experimental Issues from Map-Building
to Trajectory Execution ········· 2430
J. Hermosillo, C. Pradalier, S. Sekhavat, C. Laugier, and G. Baille

Simplified Navigation and Traverse Planning for a Long-Range Planetary Rover ········· 2436
David P. Miller, Li Tan, and Scott Swindell

An Adaptive Motion Prediction Model for Trajectory Planner Systems ········· 2442
A. Elnagar and A.M. Hussein

Region Exploration Path Planning for a Mobile Robot Expressing Working Environment by
Grid Points ········· 2448
Yusuke Fukazawa, Trevai Chomchana, Jun Ota, Hideo Yuasa, Tamio Arai, and Hajime Asama

WP2: Control of Biped Robot (IV)

Trajectory Planning for Smooth Transition of a Biped Robot ········· 2455
Zhe Tang, Changjiu Zhou, and Znqi Sun

Motion Control of Biped Robots Using a Single-Chip Drive ········· 2461
Sung-Nam Oh, Kab-Il Kim, and Seungchul Lim

Development of Dinosaur-like Robot TITRUS—The Efficacy of the Neck and Tail of
Miniature Dinosaur-like Robot TITRUS-III ········· 2466
Kensuke Takita, Toshio Katayama, and Shigeo Hirose

Cooperation of Dynamic Patterns and Sensory Reflex for Humanoid Walking ········· 2472
Guang Wang, Qiang Huang, Juhong Geng, Hongbin Deng, and Kejie Li

Analysis of Dynamics of Passive Walking from Storage Energy and Supply Rate ········· 2478
Akihito Sano, Yoshito Ikemata, and Hideo Fujimoto

WP3: Cooperative Control of Multi-Vehicle Systems

Vehicle Motion Planning Using Stream Functions ········· 2484
Stephen Waydo and Richard M. Murray

Obstacle Avoidance in Formation ········· 2492
Petter Ögren and Naomi Ehrich Leonard

Abstraction and Control for Groups of Fully-Actuated Planar Robots ········· 2498
Calin Belta and Vijay Kumar

Cooperative Task Planning of Multi-Robot Systems with Temporal Constraints ········· 2504
Feng-Li Lian and Richard Murray

Control of Small Formations Using Shape Coordinates 2510
Fumin Zhang, Michael Goldgeier, and P.S. Krishnaprasad

WP4: Space Robots

Self-Assembly in Space via Self-Reconfigurable Robots 2516
Wei-Min Shen, Peter Will, and Berok Khoshnevis

Trajectory Control of a Flexible Space Manipulator Utilizing a Macro-Micro Architecture 2522
T.W. Yang, Z.Q. Sun, S.K. Tso, and W.L. Xu

Automated Object Capturing with a Two-Arm Flexible Manipulator 2529
Tomohiro Miyabe, Atsushi Konno, and Masaru Uchiyama

Instrument Deployment for Mars Rovers 2535
L. Pedersen, M. Bualat, C. Kunz, S. Lee, R. Sargent, R. Washington, and A. Wright

Evolution of the NASA/DARPA Robonaut Control System 2543
M.A. Diftler, R. Platt Jr., C.J. Culbert, R.O. Ambrose, and W.J. Bluethmann

WP5: Visual Sensing and Application (II)

Real-Time Vision-Based Contour Following with Laser Pointer 2549
Wen-Chung Chang and Mong-Lu Chai

Modeling of Ultrasound Sensor for Pipe Inspection 2555
Francisco Gomez, Kaspar Althoefer, and Lakmal D. Seneviratne

Experiments Using a Laser-Based Transducer and Automated Analysis Techniques
for Pipe Inspection 2561
Olga Duran, Kaspar Althoefer, and Lakmal D. Seneviratne

Real-Time Estimation of Facial Expression Intensity 2567
Ka Keung Lee and Yangsheng Xu

Evaluation of Shadow Range Finder: SRF for Planetary Surface Exploration 2573
Yasuharu Kunii and Taeko Gotoh

WP6: Assembly Systems Design and Planning (I)

Experiments in Fixturing Mechanics 2579
Joel W. Burdick, Yongqiang Liang, and Elon Rimon

An Easily Reconfigurable Robotic Assembly System 2586
Yusuke Maeda, Haruka Kikuchi, Hidemitsu Izawa, Hiroki Ogawa, Masao Sugi, and Tamio Arai

Efficient Contact State Graph Generation for Assembly Applications 2592
Feng Pan and Joseph M. Schimmels

Active Sensing for the Identification of Geometrical Parameters during Autonomous
Compliant Motion 2599
Tine Lefebvre, Herman Bruyninckx, and Joris De Schutter

Haptic Modeling of Contact Formations and Compliant Motion 2605
Jing Xiao, Qi Luo, and Song You

WP7: Control Applications (III)

Sliding Control for Linear Uncertain Systems 2611
C.W. Tao, M.L. Chan, and W.Y. Wang

State Variance Constrained Fuzzy Controller Design for Nonlinear TORA Systems with Minimizing Control Input Energy 2616
Wen-Jer Chang and Sheng-Ming Wu

Robust Spatially Sampled Controller Design for Banding Reduction in Electrophotographic Process 2622
Cheng-Lun Chen, George T.C. Chiu, and Jan P. Allebach

Design of a Static Anti-Windup Compensator that Optimizes L_2 Performance: An LMI Based Approach 2628
Nobutaka Wada and Masami Saeki

Smart Neuro-Fuzzy Based Control of a Rotary Hammer Drill 2634
Chr. W. Frey, A. Jacubasch, H.-B. Kuntze, and R. Plietsch

WP8: Endoluminal Surgery-Microendoscopy (II)

Micro Hydrodynamic Actuated Multiple Segments Catheter for Safety Minimally Invasive Therapy 2640
Koji Ikuta, Hironobu Ichikawa, Katsuya Suzuki, and Takahiro Yamamoto

A Laparoscopic Robot with Intuitive Interface for Gynecological Laser Laparoscopy 2646
Hsiao-Wei Tang, Hendrik Van Brussel, Dominiek Reynaerts, Jos Vander Sloten, and Philippe R. Koninckx

Design of an Advanced Tool Guiding System for Robotic Surgery 2651
Jan Peirs, Dominiek Reynaerts, Hendrik Van Brussel, Gudrun De Gersem, and Hsiao-Wei Tang

A New Active Microendoscope for Exploring the Sub-Arachnoid Space in the Spinal Cord 2657
Luca Ascari, Cesare Stefanini, Arianna Menciassi, Sambit Sahoo, Pierre Rabischong, and Paolo Dario

Development of a Remote Minimally-Invasive Surgical System with Operational Environment Transmission Capability 2663
Mamoru Mitsuishi, Jumpei Arata, Katsuya Tanaka, Manabu Miyamoto, Takumi Yoshidome, Satoru Iwata, Shin'ichi Warisawa, and Makoto Hashizume

WP9: Grasping and Manipulation (III)

Regrasp Planning for a 4-Fingered Hand Manipulating a Polygon 2671
Attawith Sudsang and Thanathorn Phoka

Extending Fingertip Grasping to Whole Body Grasping 2677
Robert Platt Jr., Andrew H. Fagg, and Roderic A. Grupen

Convergence Analysis and Experimental Study of Geometric Algorithms for Real-Time Grasping Force Optimization 2683
G.F. Liu, J.J. Xu, and Z.X. Li

Randomized Manipulation Planning for a Multi-Fingered Hand by Switching Contact Modes 2689
Masahito Yashima, Yoshikazu Shiina, and Hideya Yamaguchi

A Comparative Study of Geometric Algorithms for Real-Time Grasping Force Optimization 2695
G.F. Liu, J.J. Xu, and Z.X. Li

WP10: Multi-Robot Systems (I)

Rules and Control Strategies of Multi-Robot Team Moving in Hierarchical Formation 2701
Tien-Sung Chio and Tzyh-Jong Tarn

Multi-Vehicle Pursuit-Evasion: An Agent-Based Framework 2707
Kingsley Fregene, Diane Kennedy, and David Wang

Efficient Exploration without Localization 2714
Maxim A. Batalin and Gaurav S. Sukhatme

Real Time Multi-UAV Simulator 2720
Ali Haydar Göktoğan, Eric Nettleton, Matthew Ridley, and Salah Sukkarieh

Learning to Role-Switch in Multi-Robot Systems 2727
Eric Martinson and Ronald C. Arkin

WP11: Redundant Robots

A Miniature Inspection Robot Negotiating Pipes of Widely Varying Diameter 2735
Koichi Suzumori, Shuichi Wakimoto, and Masanori Takata

Measurement Method for Compliance of Vertical-Multi-Articulated Robot
—Application to 7-DOF Robot PA-10 2741
Toru Tsumugiwa, Ryuichi Yokogawa, and Kei Hara

Obstacle Avoidance of Redundant Manipulators Using a Dual Neural Network 2747
Yunong Zhang and Jun Wang

A New Method for Motion Planning of Redundant Manipulators Using Singular Configurations 2753
J. Alfonso Pámanes G. and JoséLuis Zapata D.

Motion Analysis of a Kinematically Redundant Seven-DOF Manipulator under the Singularity-Consistent Method 2760
Dragomir N. Nenche and Yuichi Tsumaki

WP12: Control Architectures

The Real-Time Motion Control Core of the Orocos Project 2766
Herman Bruyninckx, Peter Soetens, and Bob Koninckx

An Object-Oriented Controller Architecture for Flexible Parts Feeding Systems 2772
Greg Causey

A Hierarchical Behavior-Based Approach to Manipulation Tasks 2780
Zbigniew Wasik and Alessandro Saffiotti

Embedded FPGA-Based Control of a Multifingered Robotic Hand 2786
Giuseppe Casalino, Fabio Giorgi, Alessio Turetta, and Andrea Caffaz

Tripodal Schematic Design of the Control Architecture for the Service Robot PSR 2792
Gunhee Kim, Woojin Chung, Munsang Kim, and Chongwon Lee

WP13: Telerobotics

Virtual Fixture Architectures for Telemanipulation 2798
Jake J. Abbott and Allison M. Okamura

Robonaut Task Learning through Teleoperation 2806
Richard Alan Peters II, Christina L. Campbell, William J. Bluethmann, and Eric Huber

Predictive Display Models for Tele-Manipulation from Uncalibrated Camera-Capture of Scene Geometry and Appearance 2812
Keith Yerex, Dana Cobzas, and Martin Jagersand

Kalman Filter Analysis for Quantitative Comparison of Sensory Schemes in Bilateral Teleoperation Systems 2818
M. Cenk Çavuşoğlu and Frank Tendick

Workspace Deformation Based Teleoperation for the Increase of Movement Precision 2824
A. Casals, L. Muñoz, and J. Amat

WP14: Mobile Robot Localization

A Reliable Position Estimation Method of the Service Robot by Map Matching 2830
Dongheui Lee, Woojin Chung, and Munsang Kim

Adaptive Real-Time Particle Filters for Robot Localization 2836
Cody Kwok, Dieter Fox, and Marina Meilă

Efficient Entropy-Based Action Selection for Appearance-Based Robot Localization 2842
J.M. Porta, B. Terwijn, and B. Kröse

Mobile Robot Localization with an Incomplete Map in Non-Stationary Environments 2848
Kanji Tanaka, Tsutomu Hasegawa, Hongbin Zha, Eiji Kondo, and Nobuhiro Okada

Feature-Based Localization Using Scannable Visibility Sectors 2854
Jinsuck Kim, Roger A. Pearce, and Nancy M. Amato

WE2: Industry Session (I) : Advanced Industrial Robot Systems

The Development of the Advanced Carrier System by IDC (Intelligent Data Career) 2860
M. Sameshima, N. Kawauchi, and T. Oomichi

An Autonomous Production System that Coexists Harmoniously with Human —Development of Autonomous Mobile Robot System 2865
Hitoshi Hibi

A Fast Collision-Free Path Planning Method for a General Robot Manipulator 2871
Shingo Ando

The Latest Robot Systems which Reinforce Manufacturing Sector 2878
Shinsuke Sakakibara

WE3: Industry Session (II) : Robotics and Automation in Biotechnology

On the Trajectory Formation of the Human Arm Constrained by the External Environment 2884
K. Ohta, M.M. Svinin, Z.W. Luo, and S. Hosoe

Micromachined Fluid Ejector Arrays for Biotechnological and Biomedical Applications 2892
Gökhan Perçin, Göksenin G. Yaralioglu, and Butrus T. Khuri-Yakub

DNA Microarray Manufacturing Factory Automation 2895
Mingjun Zhang, Karen Griswold, William Fisher, and Tzyh-Jong Tarn

An Auto-Teach/Re-Teach Implementation of Industrial Robots for Bio-Product
Manufacturing Automation 2901
WeiMin Tao, Bert Larson, and Clay Kim

ThA1: Motion Planning (I)

Path Planning Using Learned Constraints and Preferences 2907
Gregory Dudek and Saul Simhon

A Method for Handling Multiple Roadmaps and Its Use for Complex Manipulation Planning 2914
F. Gravot and R. Alami

Incremental Low-Discrepancy Lattice Methods for Motion Planning 2920
Stephen R. Lindemann and Steven M. LaValle

Online Motion Planning Using Incremental Construction of Medial Axis 2928
Ellips Masehian, M.R. Amin-Naseri, and S. Esmaeilzadeh Khadem

On Addressing the Run-Cost Variance in Randomized Motion Planners 2934
Pekka Isto, Martti Mäntylä, and Juha Tuominen

ThA2: Micro and Pipe Crawler Walking Robot

A Ciliary Based 8-Legged Walking Micro Robot Using Cast IPMC Actuators 2940
Byungkyu Kim, Jaewook Ryu, Younkoo Jeong, Younghun Tak, Byungmok Kim, and Jong-Oh Park

Motion Planning for a Three-Limbed Climbing Robot in Vertical Natural Terrain 2946
Timothy Bretl, Stephen Rock, and Jean-Claude Latombe

"MORITZ" a Pipe Crawler for Tube Junctions 2954
Andreas Zagler and Friedrich Pfeiffer

Microfabricated Thermally Actuated Microrobot 2960
Agnès Bonvilain and Nicolas Chaillet

PCG: A Foothold Selection Algorithm for Spider Robot Locomotion in 2D Tunnels 2966
Amir Shapiro and Elon Rimon

ThA3: Humanoid Robotics Project of METI

A Plant Maintenance Humanoid Robot System—Navigation System of Autonomous and
Tele-Operation Fusion Control 2973
Naoto Kawauchi, Shigetoshi Shiotani, Hiroyuki Kanazawa, Taku Sasaki, and Hiroshi Tsuji

Development of User Interface for Humanoid Service Robot System 2979
Takashi Nishiyama, Hiroshi Hoshino, Kazuya Sawada, Yoshihiko Tokunaga, Hirotatsu Shinomiya, Mitsunori Yoneda, Ikuo Takeuchi, Yukiko Ichige, Shizuko Hattori, and Atsuo Takanishi

Cooperative Works by a Human and a Humanoid Robot 2985
Kazuhiko Yokoyama, Hiroyuki Handa, Takakatsu Isozumi, Yutaro Fukase, Kenji Kaneko, Fumio Kanehiro, Yoshihiro Kawai, Fumiaki Tomita, and Hirohisa Hirukawa

Application of Humanoid Robots to Building and Home Management Services 2992
Naoyuki Sawasaki, Toshiya Nakajima, Atsushi Shiraishi, Shinya Nakamura, Kiyoshi Wakabayashi, and Yusuke Sugawara

A Tele-Operated Humanoid Robot Drives a Backhoe 2998
Hitoshi Hasunuma, Katsumi Nakashima, Masami Kobayashi, Fumisato Mifune, Yoshitaka Yanagihara, Takao Ueno, Kazuhisa Ohya, and Kazuhito Yokoi

ThA4: Complex Robotic Systems

A Synchronization Approach to the Multual Error Control of a Mobile Manipulator ... 3005
Dong Sun and Garry Feng

Mode Shape Compensator for Improving Robustness of Manipulator Mounted on Flexible Base ... 3011
Jun Ueda and Tsuneo Yoshikawa

Landing Control of Acrobat Robot (SMB) Satisfying Various Constraints ... 3017
Teruyoshi Sadahiro and Masaki Yamakita

Cable-Suspended Planar Parallel Robots with Redundant Cables: Controllers with Positive Cable Tensions ... 3023
So-Ryeok Oh and Sunil K. Agrawal

Optimal Motion Planning of Free-Flying Robots ... 3029
R. Lampariello, S. Agrawal, and G. Hirzinger

ThA5: Visual Servoing (I)

Positioning Control of the Arm of the Humanoid Robot by Linear Visual Servoing ... 3036
Kyota Namba and Noriaki Maru

Sliding PID Uncalibrated Visual Servoing for Finite-Time Tracking of Planar Robots ... 3042
V. Parra-Vega and J.D. Fierro-Rojas

Adaptive Sliding Mode Uncalibrated Visual Servoing for Finite-Time Tracking of 2D Robot ... 3048
V. Parra-Vega, J.D. Fierro-Rojas, and A. Espinosa-Romero

Visual Servoing Based on Dynamic Vision ... 3055
Ali Alhaj, Christophe Collewet, and François Chaumette

An Experimental Study of Hybrid Switched System Approaches to Visual Servoing ... 3061
Nicholas R. Gans and Seth A. Hutchinson

ThA6: Assembly Systems Design and Planning (II)

Error-Tolerant Execution of Complex Robot Tasks Based on Skill Primitives ... 3069
Ulrike Thomas, Bernd Finkemeyer, Torsten Kröger, and Friedrich M. Wahl

Strategies of Human-Robot Interactions for Automatic Microassembly ... 3076
Antoine Ferreira

Admittance Selection for Planar Force-Guided Assembly for Single-Point Contact with Friction ... 3082
Shuguang Huang and Joseph M. Schimmels

Dynamic Modeling of the Body Inversion for Automated Transfer of Live Birds ... 3089
Kok-Meng Lee and Chris Shumway

Sufficient Conditions for Admittance to Ensure Planar Force-Assembly in Multi-Point Frictionless Contact ... 3095
Shuguang Huang and Joseph M. Schimmels

ThA7: Flexible Manipulator Control and Estimation

Decoupling Based Cartesian Impedance Control of Flexible Joint Robots ... 3101
Christian Ott, Alin Albu-Schäffer, Andreas Kugi, and Gerd Hirzinger

Design and Simulation or Robust Composite Controllers for Flexible Joint Robots ... 3108
H.D. Taghirad and M.A. Khosravi

Design and Experimental Evaluation of a Single Robust Position/Force Controller for a Single Flexible Link (SFL) Manipulator in Collision ... 3114
Kamyar Ziaei and David W.L. Wang

Estimation of the Flexural States of a Macro-Micro Manipulator Using Acceleration Data ... 3120
K. Parsa, J. Angeles, and A.K. Misra

A New Impedance Control Concept for Elastic Joint Robots—A Case of 1 DOF Robot with Programmable Linear Passive Impedance ... 3126
R. Ozawa and H. Kobayashi

ThA8: Neuro-Robotics (I)

From Visuo-Motor Self Learning to Early Imitation—A Neural Architecture for Humanoid Learning ... 3132
Yasuo Kuniyoshi, Yasuaki Yorozu, Masayuki Inaba, and Hirochika Inoue

Learning about Objects through Action—Initial Steps towards Artificial Cognition ... 3140
Paul Fitzpatrick, Giorgio Metta, Lorenzo Natale, Sajit Rao, and Giulio Sandini

Motor Learning Model Using Reinforcement Learning with Neural Internal Model ... 3146
Jun Izawa, Toshiyuki Kondo, and Koji Ito

Statistical Analysis and Comparison of Questionnaire Results of Subjective Evaluations of Seal Robot in Japan and U.K. ... 3152
Takanori Shibata, Kazuyoshi Wada, and Kazuo Tanie

A Biologically Inspired Homeostatic Motion Controller for Autonomous Mobile Robots ... 3158
Do-Young Yoon, Sang-Rok Oh, Gwi-Tae Park, and Bum Jae You

ThA9: Grasping and Manipulation (IV)

The HIT/DLR Dexterous Hand: Work in Progress ... 3164
X.H. Gau, M.H. Jin, L. Jiang, Z.W. Xie, P. He, L. Yang, Y.W. Liu, R. Wei, H.G. Cai, H. Liu, J. Butterfass, M. Grebenstein, N. Seitz, and G. Hirzinger

Development of a Soft-Fingertip and Its Modeling Based on Force Distribution ... 3169
Kwi-Ho Park, Byoung-Ho Kim, and Shinichi Hirai

From Nominal to Robust Planning: The Plate-Ball Manipulation System ... 3175
Giuseppe Oriolo, Marilena Vendittelli, Alessia Marigo, and Antonio Bicchi

Motion Sensing for Robot Hands Using MIDS ... 3181
Alan H. F. Lam, Raymond H.W. Lam, Wen J. Li, Martin Y.Y. Leung, and Yunhui Liu

Mechatronic Design of Innovative Fingers for Anthropomorphic Robot Hands ... 3187
L. Biagiotti, F. Lotti, C. Melchiorri, and G. Vassura

ThA10: Micro Robotics (III)

Microassembly of 3-D MEMS Structures Utilizing a MEMS Microgripper with a Robotic Manipulator ... 3193
Nikolai Dechev, William L. Cleghorn, and James K. Mills

Micromanipulation Contact Transition Control by Selective Focusing and Microforce Control 3200
Ge Yang and Bradley J. Nelson

Two-Dimensional Signal Transmission Technology for Robotics 3207
Hiroyuki Shinoda, Naoya Asamura, Mitsuhiro Hakozaki, and Xinyu Wang

Micropeg Manipulation with a Compliant Microgripper 3213
Woo Ho Lee, Byoung Hun Kang, Young Seok Oh, Harry Stephanou, Arthur C. Sanderson, George Skidmore, and Matthew Ellis

Levitated Micro-Nano Force Sensor Using Diamagnetic Materials 3219
Mehdi Boukallel, Joël Abadie, and Emmanuel Piat

ThA11: Robot Design

Design and Modeling of Classes of Spatial Reactionless Manipulators 3225
Abbas Fattah and Sunil K. Agrawal

A Comparison of the Oxford and Manus Intelligent Hand Prostheses 3231
P.J. Kyberd and J.L. Pons

Mechatronics Design and Kinematic Modelling of a Singularityless Omni-Directional Wheeled Mobile Robot 3237
W.K. Loh, K.H. Low, and Y.P. Leow

Design of a 6DOF Haptic Master for Teleoperation of a Mobile Manipulator 3243
Dongseok Ryu, Changhyun Cho, Munsang Kim, and Jae-Bok Song

A Passive Robot System for Measuring Spacesuit Joint Damping Parameters 3249
H. Wang, X.H. Gao, M.H. Jin, L.B. Du, J.D. Zhao, H.Y. Hu, H.G. Cai, T.Q. Li, and H. Liu

ThA12: Identification and Control

Parameters Identification and Vibration Control for Modular Manipulators 3254
Yangmin Li, Yugang Liu, Xiaoping Liu, and Zhaoyang Peng

Experimental Identification and Evaluation of Performance of a 2DOF Haptic Display 3260
Antonio Frisoli and Massimo Bergamasco

A Robust Friction Control Scheme of Robot Manipulators 3266
Jeng-Shi Chen and Jyh-Ching Juang

Identification of the Dynamic Parameters of the Orthoglide 3272
Sylvain Guegan, Wisama Kahlil, and Philippe Lemoine

Experimental Dynamic Identification of a Fully Parallel Robot 3278
Andrès Vivas, Philippe Poignet, Frédéric Marquet, François Pierrot, and Maxime Gautier

ThA13: Teleoperation

Pattern-Based Arcdhitecture for Building Mobile Robotics Remote Laboratoires 3284
A. Khamis, D.M. Rivero, F. Rodríguez, and M. Salichs

Digital Passive Geometric Telemanipulation 3290
C. Secchi, S. Stramigioli, and C. Fantuzzi

Impedance Reflecting Rate Mode Teleoperation 3296
F. Mobasser, K. Hashtrudi-Zaad, and S.E. Salcudean

Laboratory Tools for Robotics and Automation Education ⋯ 3303
Claudio Cosma, Mirko Confente, Debora Botturi, and Paolo Fiorini

Passivity Analysis of Sampled-Data Interactive Systems ⋯ 3309
Ravi Hebbar and Wyatt S. Newman

ThA14: Sensor-Based Robotics (II)

Uncalibrated Visual Servoing Technique Using Large Residual ⋯ 3315
G.W. Kim, B.H. Lee, and M.S. Kim

STOMP: A Software Architecture for the Design and Simulation of UAV-Based Sensor Networks ⋯ 3321
Erik D. Jones, Randy S. Roberts, and T.C. Steve Hsia

Uncalibrated Robotic 3-D Hand-Eye Coordination Based on the Extended State Observer ⋯ 3327
Hongyu Ma and Jianbo Su

Robust Visual Tracking Using a Fixed Multi-Camera System ⋯ 3333
Vincenzo Lippiello, Bruno Siciliano, and Luigi Villani

Development of Piezoelectric Bending Actuators with Embedded Piezoelectric Sensors for Micromechanical Flapping Mechanisms ⋯ 3339
Domenico Campolo, Ranjana Sahai, and Ronald S. Fearing

ThM1: Motion Planning (II)

Online Motion Planning Using Laplace Potential Fields ⋯ 3347
Diego Alvarez, Juan C. Alvarez, and Rafael C. González

Potential-Based Path Planning for Robot Manipulators in 3-D Workspace ⋯ 3353
Chien-Chou Lin and Jen-Hui Chuang

Dual Dijkstra Search for Paths with Different Topologies ⋯ 3359
Yusuke Fujita, Yoshihiko Nakamura, and Zvi Shiller

A Novel Potential-Based Path Planning of 3-D Articulated Robots with Moving Bases ⋯ 3365
Chien-Chou Lin, Chi-Chun Pan, and Jen-Hui Chuang

Improved Analysis of D* ⋯ 3371
Craig Tovey, Sam Greenberg, and Sven Koenig

ThM2: Mobile Robot Control (I)

On the Nonlinear Controllability of a Quasiholonomic Mobile Robot ⋯ 3379
Alessio Salerno and Jorge Angeles

Exploiting Redundancy to Implement Multi-Objective Behavior ⋯ 3385
Yuandong Yang, Oliver Brock, and Roderic A. Grupen

Bilateral Time-Scaling for Control of Task Freedoms of a Constrained Nonholonomic System ⋯ 3391
Siddhartha S. Srinivasa, Michael A. Erdmann, and Matthew T. Mason

Development of a Mobile Robot for Visually Guided Handling of Material ⋯ 3397
T.I. James Tsay, M.S. Hsu, and R.X. Lin

Trajectory Planning of Mobile Manipulator with Stability Considerations ⋯ 3403
Seiji Furuno, Motoji Yamamoto, and Akira Mohri

ThM3: Mobile Robot Control (II)

Motion Control for Vehicle with Unknown Operating Properties
—On-Line Data Acquisition and Motion Planning ... 3409
Kazuya Okawa and Shin'ichi Yuta

Remote Control of a Mobile Robot Using Distance-Based Reflective Force ... 3415
J.B. Park, B.H. Lee, and M.S. Kim

Motion Planning for Humanoid Walking in a Layered Environment ... 3421
Tsai-Yen Li, Pei-Feng Chen, and Pei-Zhi Huang

Enhancing the Reactive Capabilities of Integrated Planning and Control with Cooperative Extended Kohonen Maps ... 3428
Kian Hsiang Low, Wee Kheng Leow, and Marcelo H. Ang Jr.

RoboDaemon—A Device Independent, Network-Oriented, Modular Mobile Robot Controller ... 3434
Gregory Dudek and Robert Sim

ThM4: New Robotics

A Vision-Based Haptic Exploration ... 3441
Hiromi T. Tanaka, Kiyotaka Kushihama, Naoki Ueda, and Shin'ichi Hirai

Vision Based Shape Estimation for Continuum Robots ... 3449
Michael Hannan and Ian Walker

Learning Human Control Strategy for Dynamically Stable Robots: Support Vector Machine Approach ... 3455
Yongsheng Ou and Yangsheng Xu

Perceptual Navigation Strategy: A Unified Approach to Interception of Ground Balls and Fly Balls ... 3461
Keshav Mundhra, Thomas G. Sugar, and Michael K. McBeath

Stereo Omnidirectional Vision for a Hopping Robot ... 3467
Mirko Confente, Paolo Fiorini, and Giovanni Bianco

ThM5: Visual Tracking

Real-Time Tracking and Pose Estimation for Industrial Objects Using Geometric Features ... 3473
Youngrock Yoon, Guilherme N. DeSouza, and Avinash C. Kak

Trajectory Generation for Constant Velocity Target Motion Estimation Using Monocular Vision ... 3479
Eric W. Frew and Stephen M. Rock

Confluence of Parameters in Model Based Tracking ... 3485
D. Kragic and H.I. Christensen

Visual Position Tracking Using Dual Quaternions with Hand-Eye Motion Constraints ... 3491
Tomas Olsson, Johan Bengtsson, Anders Robertsson, and Rolf Johansson

Fast 3D Tracking of Non-Rigid Objects ... 3497
Nobuhiro Okada and Martial Hebert

ThM6: Computer Aided Production Planning (I)

A General Framework for Automatic CAD-Guided Tool Planning for Surface Manufacturing 3504
Heping Chen, Ning Xi, Weihua Sheng, Yifan Chen, Allen Roche, and Jeffrey Dahl

Conflict-Free Routing of AGVs on the Mesh Topology Based on a Discrete-Time Model 3510
Jianyang Zeng and Wen-Jing Hsu

Development of an Automatic Mold Polishing System 3517
Ming J. Tsai, Jau-Lung Chang, and Jian-Feng Haung

Locating and Checking of a BGA Pin's Position Using Gray Level 3523
Chi-Wei Ruo and Ching-Long Shih

High Efficient Robotic De-Palletizing System for the Non-Flat Ceramic Industry 3529
J. Norberto Pires and Sérgio Paulo

ThM7: Enterprise-Level Modeling and Analysis

Multi-Objective Differential Evolution and Its Application to Enterprise Planning 3535
Feng Xue, Arthur C. Sanderson, and Robert J. Graves

Task Planning with Transportation Constraints: Approximation Bounds, Implementations and Experiments 3542
Ovidiu Daescu, Derek Soeder, and R.N. Uma

Improvement of Product Sustainability 3548
Meimei Gao, MengChu Zhou, and Fei-Yue Wang

A Data Mining Based Clustering Approach to Group Technology 3554
Mu-Chen Chen, Hsiao-Pin Wu, and Chia-Ping Lin

Modular Petri Net Based Modeling, Analysis and Synthesis of Dedicated Production Systems 3559
G.J. Tsinarakis, K.P. Valavanis, and N.C. Tsourveloudis

ThM8: Neuro-Robotics (II)

Robust Modeling of Dynamic Environment Based on Robot Embodiment 3565
Kuniaki Noda, Mototaka Suzuki, Naofumi Tsuchiya, Yuki Suga, Tetsuya Ogata, and Shigeki Sugano

Strategy Acquirement by Survival Robots in Outdoor Environment 3571
Pitoyo Hartono, Keishiro Tabe, Kenji Suzuki, and Shuji Hashimoto

A Bio-Inspired Approach for Regulating Visco-Elastic Properties of a Robot Arm 3576
L. Zollo, B. Siciliano, E. Guglielmelli, and P. Dario

Realization of Autonomous Search for Sound Blowing Parameters for an Anthropomorphic Flutist Robot 3582
Shuzo Isoda, Manabu Maeda, Yuji Hiramatsu, Yu Ogura, Hideaki Takanobu, Atsuo Takanishi, and Kunimitsu Wakamatsu

A New Mental Model for Humanoid Robots for Human Friendly Communication —Introduction of Learning System, Mood Vector and Second Order Equations of Emotion 3588
Hiroyasu Miwa, Tetsuya Okuchi, Kazuko Itoh, Hideaki Takanobu, and Atsuo Takanishi

ThM9: Manipulation

Planning Velocities of Free Sliding Objects for Dynamic Manipulation 3594
  Qingguo Li and Shahram Payandeh

Experiments in Nonsmooth Control of Distributed Manipulation 3600
  T.D. Murphey, J.W. Burdick, J. Burgess, and A. Homyk

Multi-Agent Cooperative Manipulation with Uncertainty: A Neural Net-Based Game
  Theoretic Approach 3607
  Qingguo Li and Shahram Payandeh

Cartop Manipulation 3613
  Wesley H. Huang, Kartik Babu, and Jonathan A. Bandlow

Smooth Feedback Control Algorithms for Distributed Manipulators 3619
  T.D. Murphey and J.W. Burdick

ThM10: Nano Robots and Manipulations

Nanotube Devices Fabricated in a Nano Laboratory 3624
  Lixin Dong, Fumihito Arai, Masahiro Nakajima, Pou Liu, and Toshio Fukuda

Nano/Micro Technologies for Single Molecule Manipulation and Detection 3630
  Tza-Huei Wang and Chih-Ming Ho

Platform Technology for Manipulation of Cells, Proteins and DNA 3636
  Gwo-Bin Lee and Long-Ming Fu

3-D Nanomanipulation Using Atomic Force Microscopes 3642
  Guangyong Li, Ning Xi, Mengmeng Yu, and Wai Keung Fung

Bundled Carbon Nanotubes as Electronic Circuit and Sensing Elements 3648
  Victor T.S. Wong and Wen J. Li

ThM11: Robot Design and Analysis

Stiffness Analysis of the Humanoid Robot WABIAN-RIV: Modelling 3654
  Giuseppe Carbone, Hun-ok Lim, Atsuo Takanishi, and Marco Ceccarelli

Auto-Calibration for a Parallel Manipulator with Sensor Redundancy 3660
  Y.K. Yiu, J. Meng, and Z.X. Li

Optimum Force Balancing with Mass Distribution and a Single Elastic Element for a
  Five-Bar Parallel Manipulator 3666
  Gürsel Alici and Bijan Shirinzadeh

Kinematics and Dynamics of a Cable-like Hyper-Flexible Manipulator 3672
  Hiromi Mochiyama and Takahiro Suzuki

Development of Parallel Manipulator "NINJA" with Ultra-High-Acceleration 3678
  Kiyoshi Nagai, Masaharu Matsumoto, Ken'ichiro Kimura, and Ban Masuhara

ThM12: Impedance Control

Robotic Force Control Using Observer-Based Strict Positive Real Impedance Control 3686
  Rolf Johansson and Anders Robertsson

Impact When Robots Act Wisely 3692
Eunjeong Lee, Juyi Park, Cheryl B. Schrader, and Pyung-Hun Chang

Stiffness Control of a Three-Link Redundant Planar Manipulator Using the Conservative Congruence Transformation (CCT) 3698
Yanmei Li and Imin Kao

Cartesian Impedance Control of Redundant Robots: Recent Results with the DLR-Light-Weight-Arms 3704
Alin Albu-Schäffer, Christian Ott, Udo Frese, and Gerd Hirzinger

The Passivity of Natural Admittance Control Implementations 3710
Mark Dohring and Wyatt Newman

ThM13: Virtual Reality

Interactive Rendering of Deformable Objects Based on a Filling Sphere Modeling Approach 3716
François Conti, Oussama Khatib, and Charles Baur

Passivity-Based High-Fidelity Haptic Rendering of Contact 3722
Mohsen Mahvash and Vincent Hayward

On the Calibration of Deformation Model of Rheology Object by a Modified Randomized Algorithm 3729
Hiroshi Noborio, Ryo Enoki, Shohei Nishimoto, and Takumi Tanemura

Constructing Rheologically Deformable Virtual Objects 3737
Masafumi Kimura, Yuuta Sugiyama, Seiji Tomokuni, and Shinichi Hirai

Post-Stabilization for Rigid Body Simulation with Contact and Constraints 3744
Michael B. Cline and Dinesh K. Pai

ThM14: Nonholonomic Path Planning

Point-to-Point Paths Generation for Wheeled Mobile Robots 3752
Diogo P.F. Pedrosa, Adelardo A.D. Medeiros, and Pablo J. Alsina

Optimization-Based Formation Reconfiguration Planning for Autonomous Vehicles 3758
Shannon Zelinski, T. John Koo, and Shankar Sastry

On the Use of Low-Discrepancy Sequences in Non-Holonomic Motion Planning 3764
Abraham Sánchez, René Zapata, and Claudio Lanzoni

Smooth Path Planning by Using Visibility Graph-like Method 3770
Tomomi Kito, Jun Ota, Rie Katsuki, Takahisa Mizuta, Tamio Arai, Tsuyoshi Ueyama, and Tsuyoshi Nishiyama

Implementation of Autonomous Fuzzy Garage-Parking Control by an FPGA-Based Car-like Mobile Robot Using Infrared Sensors 3776
Tzuu-Hseng S. Li, Shih-Jie Chang, and Yi-Xiang Chen

ThP1: Motion Planning (III)

On Energy-Minimizing Paths on Terrains for a Mobile Robot 3782
Zheng Sun and John Reif

Optimal Strategies to Track and Capture a Predictable Target ... 3789
Alon Efrat, Héctor H. González-Baños, Stephen G. Koburov, and Lingeshwaran Palaniappan

Planning Multi-Goal Tours for Robot Arms ... 3797
Mitul Saha, Gildardo Sánchez-Ante, and Jean-Claude Latombe

Trajectory Generation for Vehicle Moving with Constraints on a Complex Terrain ... 3804
Ken-Jui Tsao, Li-Sheng Wang, Po-Ting Kuo, and Fan-Ren Chang

Robot Motion Planning Using Adaptive Random Walks ... 3809
Stefano Carpin and Gianluigi Pillonetto

ThP2: Control of Multi-Legged and Multi-Joint Robot

FSW (Feasible Solution of Wrench) for Multi-Legged Robots ... 3815
Takao Saida, Yasuyoshi Yokokohji, and Tsuneo Yoshikawa

Intelligent Control of an Experimental Articulated Leg for a Galloping Machine ... 3821
Luther R. Palmer, David E. Orin, Duane W. Marhefka, James P. Schmiedeler, and Kenneth J. Waldron

Implementing Configuration Dependent Gaits in a Self-Reconfigurable Robot ... 3828
K. Støy, W.-M. Shen, and P. Will

Controlling a Marionette with Human Motion Capture Data ... 3834
Katsu Yamane, Jessica K. Hodgins, and H. Benjamin Brown

Achieving Periodic Leg Trajectories to Evolve a Quadruped Gallop ... 3842
Darren P. Krasny and David E. Orin

ThP3: Multi-Mobile Robot System (II)

Development of Omni-Directional Vehicle with Step-Climbing Ability ... 3849
Daisuke Chugo, Kuniaki Kawabata, Hayato Kaetsu, Hajime Asama, and Taketoshi Mishima

A Distributed Route Planning Method for Multiple Mobile Robots Using Lagrangian Decomposition Technique ... 3855
Tatsushi Nishi, Masakazu Ando, Masami Konishi, and Jun Imai

Multi-Robot Task Allocation: Analyzing the Complexity and Optimality of Key Architectures ... 3862
Brian P. Gerkey and Maja J Matarić

Explicit Communication in Designing Efficient Cooperative Mobile Robotic System ... 3869
Y.K. Lam, E.K. Wong, and C.K. Loo

A Hybrid-Systems Approach to Potential Field Navigation for a Multi-Robot Team ... 3875
Jing Ren and Kenneth A. McIsaac

ThP4: Robot Programming through Visual Observation and Model-Based Knowledge

Calculating Possible Local Displacement of Curve Objects Using Improved Screw Theory ... 3881
Jun Takamatsu, Koichi Ogawara, Hiroshi Kimura, and Katsushi Ikeuchi

Knot Planning from Observation ... 3887
Takuma Morita, Jun Takamatsu, Koichi Ogawara, Hiroshi Kimura, and Katsushi Ikeuchi

Estimation of Essential Interactions from Multiple Demonstrations ... 3893
Koichi Ogawara, Jun Takamatsu, Hiroshi Kimura, and Katsushi Ikeuchi

Synthesize Stylistic Human Motion from Examples 3899
Atsushi Nakazawa, Shinichiro Nakaoka, and Katsushi Ikeuchi

Generating Whole Body Motions for a Biped Humanoid Robot from Captured Human Dances 3905
Shinichiro Nakaoka, Atsushi Nakazawa, Kazuhito Yokoi, Hirohisa Hirukawa, and Katsushi Ikeuchi

ThP5: Visual Servoing (II)

Improving Camera Displacement Estimation in Eye-in-Hand Visual Servoing: A Simple Strategy 3911
Graziano Chesi and Koichi Hashimoto

Optimal Landmark Configuration for Vision-Based Control of Mobile Robots 3917
Darius Burschka, Jeremy Geiman, and Gregory Hager

Visual Navigation of an Autonomous Robot Using White Line Recognition 3923
Huaming Li, Changhai Xu, Qionglin Xiao, and Xinhe Xu

A Switching Control Law for Keeping Features in the Field of View in Eye-in-Hand Visual Servoing 3929
Graziano Chesi, Koichi Hashimoto, Domenico Prattichizzo, and Antonio Vicino

Visual Registration and Navigation Using Planar Features 3935
Gabriel A.D. Lopes and Daniel E. Koditschek

ThP6: Computer Aided Production Planning (II)

A Computer-Aided Probing Strategy for Workpiece Localization 3941
Zhenhua Xiong, Michael Yu Wang, and Zexiang Li

Structured Product Coding System (SPCS) for Product Cost Evaluation in a CAE/CAD/CAM Product (C3P) Environment 3947
Chi-haur Wu, Swee M. Mok, and Yujun Xie

"Unilateral" Fixturing of Sheet Metal Parts Using Modular Jaws with Plane-Cone Contacts 3953
K. Gopalakrishnan, Matthew Zaluzec, Rama Koganti, Patricia Deneszczuk, and Ken Goldberg

Realization of Fault Tolerant Manufacturing System and Its Scheduling Based on Hierarchical Petri Net Modeling 3959
YoungWoo Kim, Akio Inaba, Tatsuya Suzuki, and Shigeru Okuma

ThP7: Intelligent/Flexible Machine Control

A New Concept of Modular Parallel Mechanism for Machining Applications 3965
Damien Chablat and Philippe Wenger

An Error Restraining Method for Accurate Freeform Surface Cutting 3971
A. Jaganathan and Y.J. Lin

Robotic Metal Spinning—Shear Spinning Using Force Feedback Control 3977
Hirohiko Arai

Robot Trajectory Integration for Painting Automotive Parts with Multiple Patches 3984
Heping Chen, Ning Xi, Zhouhua Wei, Yifan Chen, and Jeffrey Dahl

A Novel 2-DOF Parallel Mechanism Based Design of a New 5-Axis Hybrid Machine Tool 3990
Xin-Jun Liu, Xiaoqiang Tang, and Jinsong Wang

ThP8: Rehabilitation Robotics (II)

Psychological and Social Effects of Robot Assisted Activity to Elderly People Who Stay at a Health Service Facility for the Aged 3996
Kazuyoshi Wada, Takanori Shibata, Tomoko Saito, and Kazuo Tanie

Therapy of Hemiparetic Walking by FES 4002
Markus Weber and Friedrich Pfeiffer

Assistance of Self-Transfer of Patients Using a Power-Assisting Device 4008
Kiyoshi Nagai, Isao Nakanishi, and Hideo Hanafusa

Development of Rehabilitation System for the Upper Limbs in a NEDO Project 4016
Ken'ichi Koyanagi, Junji Furusho, Ushio Ryu, and Akio Inoue

Implementation of a Path Planner to Improve the Usability of a Robot Dedicated to Severely Disabled People 4023
M. Mokhtari, B. Abdulrazak, R. Rodriguez, and B. Grandjean

ThP9: Contact

Modeling the Kinematics and Dynamics of Compliant Contact 4029
Vincent Duindam and Stefano Stramigioli

Inverse and Direct Dynamics of Constrained Multibody Systems Based on Orthononal Decomposition of Generalized Force 4035
Farhad Aghili

The 6 x 6 Stiffness Formulation and Transformation of Serial Manipulators via the CCT Theory 4042
Shih-Feng Chen

Dynamic Performance Analysis of Non-Redundant Robotic Manipulators in Contact 4048
Alan Bowling and ChangHwan Kim

Nonholonomic Dynamic Rolling Control of Reconfigurable 5R Closed Kinematic Chain Robot with Passive Joints 4054
Tasuku Yamawaki, Osamu Mori, and Toru Omata

ThP10: Multi-Robot Systems (II)

Layered Multi Agent Architecture with Dynamic Reconfigurability 4060
Eiichi Inohira, Atsushi Konno, and Masaru Uchiyama

Coordinating the Motions of Multiple Robots with Kinodynamic Constraints 4066
Jufeng Peng and Srinivas Akella

Scalability and Schedulability in Large, Coordinated, Distributed Robot Systems 4074
John D. Sweeney, Huan Li, Roderic A. Grupen, and Krithi Ramamritham

Real-Time Path Planning with Deadlock Avoidance of Multiple Cleaning Robots 4080
Chaomin Luo, Simon X. Yang, and Deborah A. Stacey

Hybrid Systems Modeling of Cooperative Robots 4086
Luiz Chaimowicz, Mario F.M. Campos, and Vijay Kumar

ThP11: Parallel Robot (IV)

Inverse Dynamics and Simulation of a 3-DOF Spatial Parallel Manipulator 4092
  Yu-Wen Li, Jin-Song Wang, Li-Ping Wang, and Xin-Jun Liu

Development of Force Displaying Device Using Pneumatic Parallel Manipulator and Application to Palpation Motion 4098
  Masahiro Takaiwa and Toshiro Noritsugu

Workspace and Dexterity Analyses of Hexaslide Machine Tools 4104
  A.B. Koteswara Rao, P.V.M. Rao, and S.K. Saha

Task Teaching to a Force-Controlled High-Speed Parallel Robot 4110
  Daisuke Sato, Takeshi Shitashimizu, and Masaru Uchiyama

Dynamic Analysis of Clavel's Delta Parallel Robot 4116
  Stefan Staicu and D.C. Carp-Ciocardia

ThP12: Learning Control

Learning Implicit Models during Target Pursuit 4122
  Chris Gaskett, Peter Brown, Gordon Cheng, and Alexander Zelinsky

A New Algorithm of Adaptive Iterative Learning Control for Uncertain Robotic Systems 4130
  Chun-Te Hsu, Chiang-Ju Chien, and Chia-Yu Yao

Learning to Optimize Mobile Robot Navigation Based on HTN Plans 4136
  Thorsten Belker, Martin Hammel, and Joachim Hertzberg

Design and Implementation of a Behavior-Based Control and Learning Architecture for Mobile Robots 4142
  Il Hong Suh, Sanghoon Lee, Bong Oh Kim, Byung Ju Yi, and Sang Rok Oh

On Learning Control with Limited Training Data 4148
  Yongsheng Ou and Yangsheng Xu

ThP13: Intelligent Environment

Ada—Intelligent Space: An artificial Creature for the Swiss Expo.02 4154
  Kynan Eng, Andreas Bäbler, Ulysses Bernardet, Mark Blanchard, Marcio Costa, Tobi Delbrück, Rodney J Douglas, Klaus Hepp, David Klein, Jonatas Manzolli, Matti Mintz, Fabian Roth, Ueli Rutishauser, Klaus Wassermann, Adrian M Whatley, Aaron Wittmann, Reto Wyss, and Paul F M J Verschure

Human Behavior Interpretation Systems Based on View and Motion-Based Aspect Models 4160
  Masayuki Furukawa, Yoshio Kanbara, Takashi Minato, and Hiroshi Ishiguro

Collaborative Capturing of Experiences with Ubiquitous Sensors and Communication Robots 4166
  Norihiro Hagita, Kiyoshi Kogure, Kenji Mase, and Yasuyuki Sumi

Self-Identification of Distributed Intelligent Networked Device in Intelligent Space 4172
  Hideki Hashimoto, Joo-Ho Lee, and Noriaki Ando

Expression Method of Human Locomotion Records for Path Planning and Control of Human-Symbiotic Robot System Based on Spacial Existence Probability Model of Humans 4178
  Rui Fukui, Hiroshi Morishita, and Tomomasa Sato

ThP14: Path Planning with Uncertainty

Safe Path Planning in an Uncertain-Configuration Space — 4185
Alain Lambert and Dominique Gruyer

On-Line Safe Path Planning in Unknown Environments — 4191
Weidong Chen, Changhong Fan, and Yugeng Xi

Robot Motion Decision-Making System in Unknown Environments — 4197
S. Boonphoapichart, S. Komada, T. Hori, and W.A. Gruver

Probability of Success and Uncertainty Analysis in Path Planning — 4203
Dapena Eladio and Moreno Luis

A Neural Network Model that Calculates Dynamic Distance Transform for Path Planning and Exploration in a Changing Environment — 4209
Dmitry V. Lebedev, Jochen J. Steil, and Helge Ritter

ThE1: Multi-Robot Motion Planning

Motion Planning for a Crowd of Robots — 4215
Tsai-Yen Li and Hsu-Chi Chou

Motion Planning for Multiple Mobile Robots Using Dynamic Networks — 4222
Christopher M. Clark, Stephen Rock, and Jean-Claude Latombe

Reduced Order Motion Planning for Nonlinear Symmetric Distributed Robotic Systems — 4228
M. Brett McMickell and Bill Goodwine

Evasion of Multiple, Intelligent Pursuers in a Stationary, Cluttered Environment Using a Poisson Potential Field — 4234
Ahmad A. Masoud

Closed Loop Navigatipon for Multiple Non-Holonomic Vehicles — 4240
Savvas G. Loizou and Kostas J. Kyriakopoulos

ThE2: Mobile Robot Systems

Designing a Secure and Robust Mobile Interacting Robot for the Long Term — 4246
N. Tomatis, G. Terrien, R. Piguet, D. Burnier, S. Bouabdallah, Kai O. Arras, and R. Siegwart

The Current Opinion on the Use of Robots for Landmine Detection — 4252
S. Rajasekharan and C. Kambhampati

Sensor-Based Motion Planning for Car-like Mobile Robots in Unknown Environments — 4258
Claudio Lanzoni, Abraham Sánchez, and René Zapata

Heterogeneous Implementation of an Adaptive Robotic Sensing Team — 4264
Bardley Kratochvil, Ian T. Burt, Andrew Drenner, Derek Goerke, Bennett Jackson, Colin McMillen, Christopher Olson, Nikolaos Papanikolopoulos, Adam Pfeifer, Sascha A. Stoeter, Kristen Stubbs, and David Waletzko

A System for Volumetric Robotic Mapping of Abandoned Mines — 4270
Sebastian Thrun, Dirk Hähnel, David Ferguson, Michael Montemerlo, Rudolph Triebel, Wolfram Burgard, Christopher Baker, Zachary Omohundro, Scott Thayer, and William Whittaker

ThE5: Vision Based Control

Application of Moment Invariants to Visual Servoing — 4276
Omar Tahri and François Chaumette

Modeling and Vision-Based Control of a Micro Catheter Head for Teleoperated In-Pipe Inspection — 4282
Saliha Boudjabi, Antoine Ferreira, and Alexandre Krupa

Visual Servoing of a Car-like Vehicle—An Application of Omnidirectional Vision — 4288
Kane Usher, Peter Ridley, and Peter Corke

Quadrotor Control Using Dual Camera Visual Feedback — 4294
Erdinç Altuğ, James P. Ostrowski, and Camillo J. Taylor

ThE7: Petri Nets in Automated Systems Design (II)

Production Cycle-Time Analysis Based on Sensor-Based Stage Petri Nets for Automated Manufacturing Systems — 4300
ShihSen Peng and MengChu Zhou

A Colored Timed Petri Net Model to Manage Resources in Complex Automated Manufacturing Systems — 4306
Maria Pia Fanti

Controller Synthesis via Mapping Task Sequence to Petri Nets in Multi-Agent Collaboration Applications — 4312
Wenbiao Han and Mohsen A. Jafari

Fuzzy Petri Nets for Monitoring and Recovery — 4318
Daniel Racoceanu, Eugenia Minca, and Noureddine Zerhouni

Information Systems as a Tool for Specification of Concurrent Systems — 4324
Zbigniew Suraj

ThE11: Parallel Robot (V)

Dynamic Modeling of a Parallel Robot: Application to a Surgical Simulator — 4330
N. Leroy, A.M. Kökösy, and W. Perruquetti

Path Trackability and Verification for Parallel Manipulators — 4336
C.K. Kevin Jui and Qiao Sun

Control and Experiments of a Multi-Purpose Bipedal Locomotor with Parallel Mechanism — 4342
Yusuke Sugahara, Tatsuro Endo, Hun-ok Lim, and Atsuo Takanishi

Design of a Redundantly Actuated Leg Mechanism — 4348
Byung Rok So, Byung-Ju Yi, Wheekuk Kim, Sang-Rok Oh, Jongil Park, and Young Soo Kim

Probabilistic Motion Planning for Parallel Mechanisms — 4354
J. Cortés and T. Siméon

ThE12: Robot Control

Robust Task-Space Control of Hydraulic Robots — 4360
O. Becker, I. Pietsch, and J. Hesselbach

Passivity Monitor and Software Limiter which Guarantee Asymptotic Stability of Robot Control Systems ··· 4366
Katsuya Kanaoka and Tsuneo Yoshikawa

Hierarchical Velocity Field Control for Robot Manipulators ··· 4374
Javier Moreno and Rafael Kelly

Better Robot Tracking Accuracy with Phase Lead Compensated ILC ··· 4380
Yongqiang Ye and Danwei Wang

Forcefree Control with Independent Compensation for Inertia, Friction and Gravity of Industrial Articulated Robot Arm ··· 4386
Satoru Goto, Masatoshi Nakamura, and Nobuhiro Kyura

ThE13: Remote Robotics

Challenges in VR-Based Robot Teleoperation ··· 4392
Cheng-Peng Kuan and Kuu-young Young

Adaptive Fusion of Sensor Signals Based on Mutual Information Maximization ··· 4398
Tetsushi Ikeda, Hiroshi Ishiguro, and Minoru Asada

Reinforcement Learning Congestion Controller for Multimedia Surveillance System ··· 4403
Ming-Chang Hsiao, Kao-Shing Hwang, Shun-Wen Tan, and Cheng-Shong Wu

Student Performance Evaluation in Web Based Access to Robot Supported Laboratories ··· 4408
H.E. Motuk, A.M. Erkmen, and I. Erkmen

Co-Operative Control of Internet Based Multi-Robot Systems with Force Reflection ··· 4414
Wang-tai Lo, Yun-Hui Liu, Imad Elhajj, Ning Xi, Yinghai Shi, and Yuechao Wang

ThE14: Probabilistic Roadmap

The Bridge Test for Sampling Narrow Passages with Probabilistic Roadmap Planners ··· 4420
David Hsu, Tingting Jiang, John Reif, and Zheng Sun

Improving the Connectivity of PRM Roadmaps ··· 4427
Marco Morales, Samuel Rodríguez, and Nancy M. Amato

HPRM: A Hierarchical PRM ··· 4433
Anne D. Collins, Pankaj K. Agarwal, and John L. Harer

A General Framework for Sampling on the Medial Axis of the Free Space ··· 4439
Jyh-Ming Lien, Shawna L. Thomas, and Nancy M. Amato

A General Framework for PRM Motion Planning ··· 4445
Guang Song, Shawna Thomas, and Nancy M. Amato

Authors' Index ··· A-1

Categories' Index ··· A-18

Path Planning Using Learned Constraints and Preferences

Gregory Dudek and Saul Simhon

Centre for Intelligent Machines, McGill University
3480 University St, Montréal, Québec, Canada H3A 2A7

Abstract—In this paper we present a novel method for robot path planning based on learning motion patterns. A motion pattern is defined as the path that results from applying a set of probabilistic constraints to a "raw" input path. For example, a user can sketch an approximate path for a robot without considered issues such as bounded radius of curvature and our system would then elaborate it to include such a constraint. In our approach, the constraints that generate a path are learned by capturing the statistical properties of a set of training examples using supervised learning. Each training example consists of a pair of paths: an unconstrained (raw) path and an associated preferred path. Using a Hidden Markov Model in combination with multi-scale methods, we compute a probability distribution for successive path segments as a function of their context within the path and the raw path that guides them. This learned distribution is then used to synthesize a preferred path from an arbitrary input path by choosing some mixture of the training set biases that produce the maximum likelihood estimate. We present our method and applications for robot control and non-holonomic path planning.

I. INTRODUCTION

Traditionally, motion constraints have been represented by analytic methods that constrain the differential geometry of the set of admissible paths. Path planning typically entails solving an optimization problem with respect to these constraint equations. In contrast, this paper presents a radically different approach to path planning. Constraints (or preferences) are expressed in terms of a set of examples that illustrate how the robot is permitted to move. Further, these examples indicate how to elaborate an input path from a user or high-level planner (which is typically not acceptable in itself) into a suitable acceptable output path. Informally, the examples say: "if a user asks you to do something like *this* than what you should actually perform is a maneuver like *that*".

Traditional constraint equations for motion control are complex relations that typically model the dynamics of a mobile robot based on its mechanical design. Our method can be used to simulate such constraints without having to explicitly model them. The goal is to learn from examples of constrained motions to properly generate novel paths.

Motion constraints are not only used for modeling a robot's mechanical configuration. In some applications, equations are constructed to model task specific motion requirements, such as a sweeping pattern for full floor coverage or a suitable behavior to scan the environment using a narrow-beam sensor. Specialized paths also occur various specialized contexts; in the classic 1979 film "The In-Laws" Peter Falk instructs Alan Arkin to run along a "serpentine" path while going elsewhere – our system could readily accommodate such a preference bias as well. Additionally, in applications such as obstacle avoidance, motions are not only related to the robot's pose but are also a function of the perceived environment. In all of these examples, the underlying core problem consists of finding a valid transformation between two components: 1) the idealized "raw" path that directs the robot to a goal without taking into account certain preferences or constraints, and 2) the refined path that attempts to reach the goal while also satisfying the system constraints.

Whatever the constraints, expressing them in a suitable formal framework is often challenging. Further, the processes of finding allowable solutions can be costly, particularly since the solution techniques are often engineered for a specific context. In our work, we develop a method to implicitly learn motion constraints. Given a set of sample *motion patterns*, each with an associated *control path*, the algorithm captures statistical properties over the length of the path and configures a Hidden Markov Model. Given a new input control path and the configured HMM, the algorithm generates a new path that is statistically consistent with the learned patterns, enforcing the desired local constraints. As such, the algorithm can be applied over a variety of training examples to generate a rich set of paths without having to explicitly model the constraints.

II. RELATED WORK

Path planning has been extensively examined by many authors. One of the key ideas in the area is the notion of path planning under non-holonomic constraints, specifically path planning using a bound on the radius of curvature on the vehicle [1]. Notable work in the field includes the landmark results of Dubins [2] and Reeds and Shepp [3] on optimal trajectories. Much of this work deals with the quest for an optimal path (or trajectory) under a motion constraint which is expressed analytically (for example a derivative constraint). Prevalent solution techniques include analytic solutions (or expressions regarding their bounds), search methods that seek to optimize a path, and planners that start with a path of one form and seek to refine it.

In particular, a classic approach to the application of non-holonomic constraints is to find an (optimal) unconstrained solution and then apply recursive constrained path refinement to the sub-regions to achieve an admissible plan [1]. This is also typical of probabilistic motion planning methods [4], [5]. Similarly, jerky paths are sometimes smoothed using energy minimization methods [6], [7].

This work shares that common spirit in that it takes an initial path as input and produces a refined path as its result. While traditional methods such as those cited above typically accomplish path refinement based on highly specialized constraints, typically in the domain of differential geometry, our methods learns from examples of acceptable paths. That is, the significant constraints or preferences are indicated by showing the appropriate refinements that should be applied in specific cases.

This idea of learning to generalize specific examples to a broad ensemble of cases is, of course, the crux of classical machine learning [8]. Learning using Markov models is a longstanding classic research area, although to our knowledge it has never been applied to problems like this one. Likewise, although there has been some prior work on the relationship between learning and planning, most of this has dealt with more traditional plan formulation problems [9] or on learning suitable cues that control or determine plan synthesis or execution [10], [11].

III. LEARNING MOTION PATTERNS

Our objective is to learn attributes from training examples in order to synthesize a constrained path given an arbitrary unconstrained path. Figure 1 shows some training examples that capture a non-holonomic constraint: a bound on the maximum radius of curvature. The examples show smooth right angle turns, wide D-shaped turns, narrow U-shaped turns and an example of a parallel-parking type motion with a direction reversal (note that at every path along these paths the orientation of the vehicle is also represented but not always shown in the figure). In each example, both the desired path and the associated control path are displayed.

We present a method to represents those significant attributes using a Hidden Markov Model. A Hidden Markov Model encodes the dependencies of successive elements of a set of *hidden* states along with their relationship to *observable* states. It is typically used in cases where a set of states, that exhibit the Markov property, are not directly measurable but only their effect is visible through other observable states. Formally, a Hidden Markov Model Λ is defined as follows:

$$\Lambda = \{M, B, \pi\} \quad (1)$$

where M is the transition matrix with transition probabilities of the hidden states, $p\{h_i(t) \mid h_j(t-1)\}$, B is the confusion matrix containing the probability that a hidden state h_j generates an observation o_i, $p\{o_i(t) \mid h_j(t)\}$, and π is the initial distribution of the hidden states.

There is an abundance of literature on Hidden Markov Models and the domain is frequently decomposed into 3 critical sub-problems:

- Evaluation, where the likelihood of an HMM is evaluated for a sequence of observations, $p\{o \mid \Lambda\}$.
- Decoding, where the maximum likelihood sequence of hidden states is predicted for a given HMM and an observation sequence, $\max_h p\{h \mid o, \Lambda\}$.
- Learning, where the transition probabilities, the confusion matrix and the initial distribution that best fit an observed set of examples are estimated.

Given only the observations, learning is most commonly performed by algorithms such as the Baum-Welch algorithm or generalized Expectation-Maximization methods. In our application, we have direct access to both the hidden and observable states. They consist of sample points from the desired motion patterns and the associated control paths respectively, which are readily available. Therefore, we can estimate an HMM by the the statistics of the training data, calculating probabilities of successive elements of the desired motions and their relationship to the control paths.

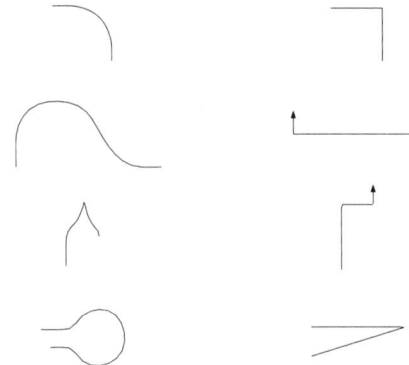

Fig. 1. Samples of a training set simulating non-holonomic motions. Paths on the left display the constrained motions while paths on the right display the associated unconstrained control paths. Where specified, arrows indicate additional constraints on the direction of motion to account for the orientation of the robot. The full set consists of the above at several orientations.

A. Hidden Paths

We represent a path by a curve over 2D space parametrized by the arc-length. Let α represent a parametric curve $(x(t), y(t))$ where t is the arc-length of the curve from $0 <= t <= T$. Since we can, in principle, encode a function using only its derivations, we assume our paths are suitably normalized and encode them as a discreet succession of tangent angles $\theta(t)$.

Consider a stochastic process Δ as the source for a family of paths. As such, each a path is considered to be a random signal with characteristics described by the probability density function of the process. Let α denote the curve representing the constrained motion pattern and $\theta(t)$ as the tangent angles of that path. We assume that the sequence of samples $\theta(t)$ from all constrained motion patterns exhibit an nth order Markov property, i.e. a Markov Process:

$$p\{\theta(t+1) \mid \theta(t), \theta(t-1), \ldots, \theta(t-n+1)\} =$$
$$p\{\theta(t+1) \mid \theta(t), \theta(t-1), \ldots, \theta(0)\}$$

This *locality* condition states that information from recent samples is sufficient to predict the next sample point. Further, the dependency is considered to be invariant, where relationships between successive points are *stationary* with respect to the arc-length.

Sample points of the motion patterns are represented by hidden states in the HMM. Given an ensemble of training examples, we estimate the transition probabilities by the statistics of successive elements in the set and construct the transition matrix M where:

$$P_\theta(t+1) = M\, P_\theta(t) \tag{2}$$

The transition matrix propagates the information embedded in the probability distribution $P_\theta(t)$ to predict the next distribution $P_\theta(t+1)$. We assume a uniform initial probability distribution $\pi = P_\theta(0)$, providing equal likelihoods to all paths at time zero.

B. Observable Paths

Sample points of the control paths are represented by observable states in the HMM, which characterizes the relationship to the constrained motion patterns. Based on this relationship, an input path can condition the distribution in equation 2 and bias the synthesis according to the prescribed characteristics. Because the input path can be any arbitrary shape, we assume that samples of the control paths are independent. For all t and k in the domain:

$$p\{\theta(t) \mid \theta(k)\} = p\{\theta(t)\}$$

That is, previous points generally do not provide information on what the next point may be. This assumption adheres to the HMM condition that the observable state sequence is independent over time.

Let β denote the curve representing the associated control path and $\phi(t)$ as the tangent angles of that path. Then:

$$\alpha = \Psi\, \beta \tag{3}$$

where Ψ is some mapping that transforms the control path to the constrained motion pattern. The mapping in essence encodes the constraint relationship between the coupled pair. Given a normalized ensemble of constrained/unconstrained curves, we estimate the probabilities of the confusion matrix B from the statistics of associated sample points $(\theta(t), \phi(t))$ and form the following relation:

$$P_\phi(t) = B\, P_\theta(t) \tag{4}$$

where the elements of the confusion matrix are the conditional probabilities $p(\phi_i | \theta_j)$ for all states i and j. This is analogous to the inverse relation of the mapping Ψ in equation 3. However, using Bayes law, one can show that solving the decoding problem for a HMM in a maximum likelihood sense is analogous to solving for the desired transformation Ψ. (It is in-essence solving the inverse problem in a maximum likelihood sense.)

IV. SYNTHESIS

Given a set of observations and an HMM trained with a family of path patters, we generate a new path pattern by solving for the maximum likelihood hidden state sequence

$$\max_{\theta_i \ldots \theta_n} p\{\theta(0), \theta(1), \ldots, \theta(T) \mid \phi(0), \phi(1), \ldots, \phi(T), \Lambda\}$$
$$or$$
$$\max_{\alpha} p\{\alpha \mid \beta, \Lambda\} \tag{5}$$

Also known as the decoding problem, the above problem can be solved iteratively. At each time interval, we propagate the underlying probability distribution as in equation 2 and maintain states with maximum consistency across successive elements. The resulting distribution is then conditioned by the current observation as in equation 4. (This method is analogous to the *Viterbi algorithm*.) We iterate up to time T to produce a sequence of probability distributions $\{P_\theta(0), P_\theta(1), \ldots, P_\theta(T)\}$. To instantiate a path, we can select states with maximum probability from each distribution. However, independently selecting states in a greedy fashion can result to an inconsistent sequence, it may break the continuity of valid links between successive elements. Rather, we instantiate the state with maximum probability at time T and then backtrack by choosing the previous most likely state that would generate the current one. Backtracking is essential for generating a consistent path as not only does it consider the links between successive states, but also propagates future information back to earlier points. Figure 7 shows an example path manually drawn and the resulting generated path.

A. Multi-Dimensional State Space

Implementation of a first order Markov Model is generally achievable by storing the transition probabilities in a memory array. However, preliminary empirical results showed that for most training examples, a first order Markov Model does not capture enough information to properly generate the paths. Further, higher order Markov Models increase the state space exponentially and storage of a transition matrix is not practical. To address this issue, we do not explicitly compute and store the transition matrix, rather, we only maintain a list of candidate states with strictly positive probabilities. The algorithm then performs a search comparing the training set and candidate list. When a match occurs, the probability of the proceeding state is calculated and added to the list of candidates for the next sample point.

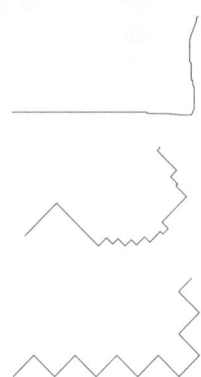

Fig. 3. Generating a sweeping style path pattern. Training data consists of the zig-zag patterns at several directions, each associated to straight line segments. The top curve shows the input path. The middle curve shows the synthesized path using a first order assumption. The bottom curve shows the resulting path using higher order states.

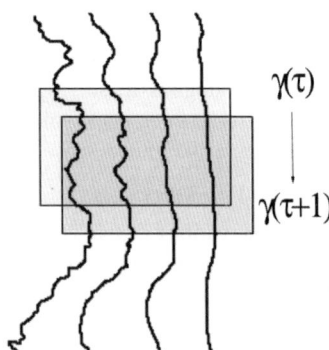

Fig. 2. Each multi-dimensional state consists of points within a fixed size window spanning the multi-scale curve model. When a match for state $\gamma(t)$ is found, the next state $\gamma(t+1)$ is added to the list.

To store high-order information, each state is represented by a multi-dimensional element γ_τ where each dimension corresponds to a sample point further in history. Without exaggerating the dimensionality of the state space, additional history can be attained by sampling over a multi-scale representation of the curve $\gamma_{\tau,s}$. A curve is filtered several times to produce lower scale versions. A single point on a lower scale curve represents a summary for the region of the high scale curve (see figure 2). Figure 3 shows an example where higher scale structures are important to capture. It is easy to see how the first order assumption does not capture enough information to generate the pattern while a synthesis using higher order states produces more consistent results.

Given the flexibility of the multi-dimensional model, additional relevant attributes that are required to further constrain the system can be easily incorporated. For example, we include an additional dimension to provide control over the *direction of motion* of the robot. Normally, the direction is implicitly defined in the input path, however, there are cases where we may need to further constrain the system in directions that are not necessarily along the path. For example, this can be used to provide control to align the robot axis orientation at particular points on the path. The training data must also contain this additional parameter in order to define the preferred behavior given both the path and the direction (see figure 1. Figure 4 shows an example where we condition the synthesis over various directions of motion in order to align the orientations of the robot's axis. At first, the initial direction of motion points vertically while the path progresses horizontally. The robot performs a D-shape turn for proper alignment. Later along the curve, the robot performs several other turns in order to align to the other specified directions.

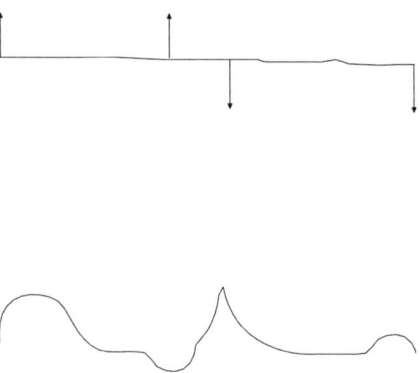

Fig. 4. The above was generated using the training example for non-holonomic constraints. The input path (top) is a hand-drawn path with several desired directions of motion (shown by arrows). The resulting path (bottom) consists of parts of a D-shape turn, a U-shape turn and a parallel parking style motion in order to end up in the right motion directions at the corresponding points.

B. State Blurring

The input path used for conditioning may not provide exact matches to the training data. The path may be generated by some noisy path planner or even manually drawn by a human operator. Further, even with perfect inputs, quantization errors are likely to occur in both the hidden states and the observations. Such errors can abruptly terminate the synthesis by conditioning or propagating all probabilities to zero. Blurring the probability matrices, or synonymously the probability vector, avoids this issue. The probability distribution is modeled as Gaussian mixture over the state space. The probability for state γ_i is given by:

$$p(\gamma_i) = \frac{1}{N}\sum_j p(\gamma_j) e^{\frac{-(\gamma_i - \gamma_j)^2}{v^2}} \quad (6)$$

For the multi-scale representation, the goodness of the match $(\gamma_i - \gamma_j)$ is based on a weighted difference over the scales. Such a blur may result in too many matches where every combination of states will produce strictly positive probabilities. This causes computational complexities and high dimensional models may not be solved in practical time. Therefore, we threshold over the tail of the Gaussian and normalize.

C. Coherency Measures

Since a Stationary Markov Process is assumed, there is no sense of progression or continuity of paths along the arc-length. The search is performed irrespective of the parametric position, choosing matches arbitrarily along the path. Further, the synthesis may get *stuck* at a state or a cycle through small set of states (know as absorbing states or irreducible communication classes). In such a situation, the propagated probabilities will model disjoint and self contained distributions. Conditioning over the observables and using a multi-scale model reduces the chance of this occurrence over long intervals. However, due to the nature of the training set, where there are many line segments that have few distinct features, such situations still often occurs. To address this issue, we define a measure for *coherency over arc-length* as a measure of the number of out-of-sequence states in a synthetic curve. To bias the synthesis for more coherent paths, we enforce a penalty on matches that are out-of-sequence. The probability is penalized by a factor of τ to help promote more coherent path.

While some degree of divergence is necessary to fulfill the desired motion pattern, we wish to avoid situations where the generated curve diverges too much from the input curve. Since the state space only represents the tangent angles as a function of arc-length, there is no indication of how close the generated curve is to the input curve. Therefore, we define a measure for *spatial coherency to input* as a measure of the average distance between the input curve and the generated curve over Cartesian space. To generate more spatially coherent paths, we include a magnetic force that biases the distributions to prefer points that are closer. At each sample point t the probability is updated by:

$$p(\gamma_i) = \frac{p(\gamma_i)}{N(1+kd^2)} \quad (7)$$

where d is the distance between the input sample point and the resulting sample point generated by the maximum likelihood path up to and including the candidate state, k is the influence factor and N is a normalization constant. Figure 5 shows an example comparing generated paths with and without the coherency conditions. (One can also formulate these measures as regularization terms in a variational calculus problem minimizing entropy.)

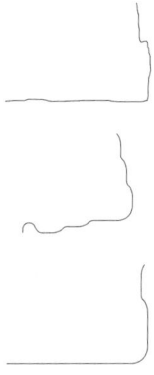

Fig. 5. The above example was generated using the training example for non-holonomic constraints. The input path (top) is a hand-drawn curve. Below it (middle) shows the generated path without any coherence factors and the bottom curve shows the generated path taking into account both spatial and arc-length coherence.

V. RESULTS

Experiments were performed using path styles for sweep patterns, curled patterns and bounded turning radius patterns. The input paths are arbitrary paths hand drawn by a user. When the direction of motion is not specified by the input curve, it is taken along the tangent of that path. All experiments were executed on a Linux PC with a 1GHz Pentium IV processor and 1GB of RAM. The results were generated in real time.

Figure 6 shows examples where a single input path is used to generate several path styles. It can be seen how the resulting paths form analogies to the input path with respect to the learned path styles. Each resulting pattern maintains the local consistency while following the general direction of the input.

Figure 7 shows an example synthesis with non-holonomic constraints. The generated path follows a

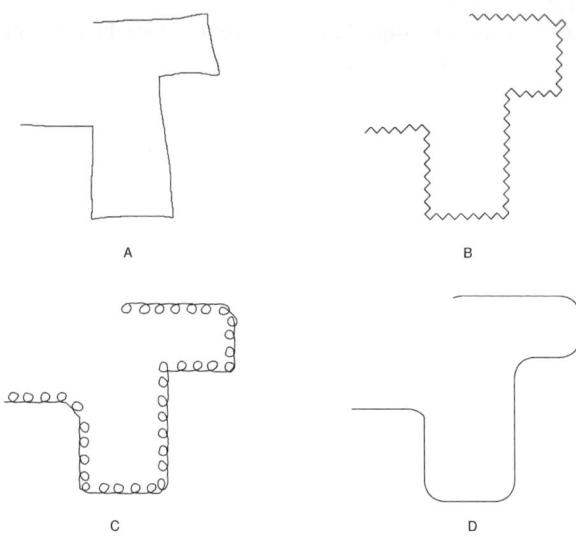

Fig. 6. The above example shows the input path (A) and the synthesized paths (B,C,D) using various training styles. The three training sets consist of a zig-zag patter for a sweep motion, a curl pattern for a narrow-beam sensor scan and bounded radius of curvature pattern.

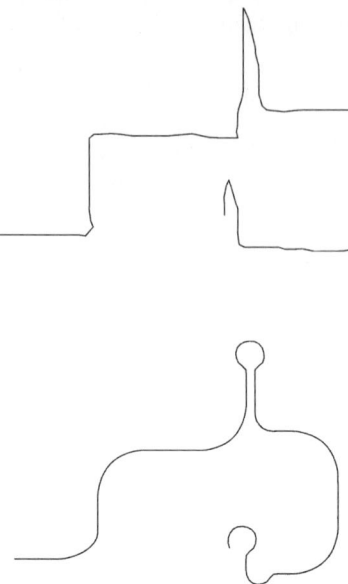

Fig. 7. The training examples are samples of motions with non-holonomic constraints. The input path (top) is a hand drawn (noisy) path. Below shows the generated path maintaining the bounded turning radius constraint for non-holonomic motions.

smooth curve outlining the overall shape of the noisy input path. It can bee seen that the bottom right turn was an extended loop about the corner rather than the typical smoothed out turn. This was due to backtracking effects where the proceeding segment consisted of second turn, immediately after the first. The limited room for turning forced the path to extend around the corner. Figure 8 shows another example with non-holonimic constraints. The input curve restricts the initial direction of motion as displayed by the arrow. This results in the motion pattern with a direction reversal at point (B).

VI. FUTURE WORK

We plan on extending this work to take into account obstacles. Conditioning the probabilities over some occupancy grid can be a possible direction of work. Further, we wish to investigate a method that would perform synthesis computations using measurements over local reference frames. This can help avoid the requirement that training data must span all the desired orientations. In addition, another direction for future work is to examine applications of the method using related multiple curve signals, such as control for multi-robot navigation or some other correlated attributes.

VII. CONCLUSION

We have presented an approach to path refinement: that is, to producing accepting paths for a robot given input curves that indicate roughly what kind of path is suitable. The method consists of using *a-priori* data to automatically learn constraints and preferred patterns to configure

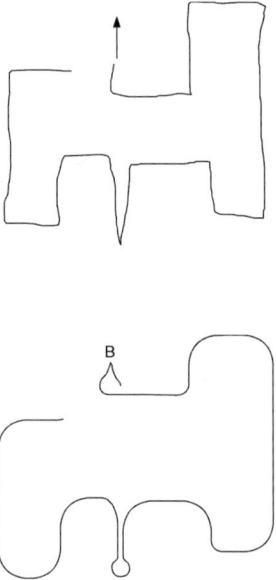

Fig. 8. The figure above was generated using training examples for non-holonomic constraints. The input path (top) is a hand drawn path with a restriction on the initial direction of motion. Below shows the generated path where point (B) marks a direction reversal.

a Hidden Markov Model. The learned probabilities are used to synthesize a new path given an arbitrary normally unconstrained path. Experimental results display how a synthesized path exhibits the constraint properties of the training data while following the overall shape of the input curve.

In this discussion we have assumed that when a path is generated, we know *a priori* which family of statistical biases we should apply. In practice, it may be that in one part of a path we want one style of locomotion and in another part we expect a different style. How to incorporate two different types of bias in the system and, further, how to make a transition between them remains a topic we are still investigating.

Since our approach is based in local refinements to an input curve, it is not readily able to apply large-scale reconfigurations to a path (it a possible, but leads to various technical problems). As such, this approach is best suited to problems where at least the homotopy class of the desired solution is clearly indicated. While this is a restriction, it is not apparent that any other mode of operation would be desirable.

VIII. REFERENCES

[1] J.-C. Latombe, *Robot Motion Planning*. Norwell, MA: Kluwer Academic Publishers, 1991.

[2] L. E. Dubins, "On curves of minimal length with a constraint on average curvature, and with prescribed initial and terminal positions and tangents," in *American Journal of Mathematics*, vol. 79, pp. 497–517, 1957.

[3] Reeds and L. Shepp, "Optimal paths for a car that goes both forwards and backwards," in *Pacific Journal of Mathematics*, vol. 145(2), pp. 367–393, 1990.

[4] L. K. J.-C. Latombe, "Randomized preprocessing of configuration space for fast planning," in *IEEE Internation Conference on Robotics and Automation*, 1994.

[5] L. K. J.-C. Latombe, "Probabilistic roadmaps for path planning in high-dimensional configuration spaces," in *IEEE Transactions on Robotics and Automation*, vol. 12, August 1996.

[6] S. Singh and M. C. Leu, "Optimal trajectory generation for robotic manipulators using dynamic programming," in *ASME Journal of Dynamic Systems, Measurement and Control*, vol. 109, 1989.

[7] J.-P. Laumond, P. E. Jacobs, M. Taix, and R. M. Murray, "A motion planner for nonholonomic mobile robots," in *IEEE Transactions on Robotics and Automation*, vol. 10, pp. 577–593, 1994.

[8] M. Learning, "Tom mitchell," in *McGraw Hill*, 1997.

[9] X. Wang, "Learning planning operators by observation and practice," in *Artificial Intelligence Planning Systems*, pp. 335–340, 1994.

[10] J. Miura, "Hierarchical vision-motion planning with uncertainty: Local path planning and global route selection."

[11] S. P. Engelson, "Learning robust plans for mobile robots from a single trial," in *AAAI/IAAI, Vol. 1*, pp. 869–874, 1996.

A method for handling multiple roadmaps and its use for complex manipulation planning.

F. Gravot, R. Alami

LAAS-CNRS, 7, Avenue du Colonel Roche,
31077 Toulouse CEDEX 04, France.
e-mail: Fabien.Gravot@laas.fr, Rachid.Alami@laas.fr

Abstract

We propose a resolution scheme that is aimed to solve a wide range of multi-robot planning problems in a 3D geometrical environment. This complements our efforts in developing manipulation planning algorithms [1, 15, 14] to deal with multiple robots and several movable objects problems.

While the elementary planning step relies on Probabilistic Roadmap Methods (PRMs), the main contribution here is a reasoning level that adapts its control over the construction and extension of a number of roadmaps. The coherence between roadmaps and the hierarchical search process are done through a new type of graph built for a subset of objects or robots in the environment and called Elementary Kinematic Graph.

This paper describes the main ingredients of the proposed framework, and its first results.

1 Introduction

The Problem. We address the manipulation planning problem with several objects and several robots. Indeed, we are willing to build a system that is able to produce plans for cooperative manipulation tasks where several robots participate. Some tasks will correspond to individual Pick&Place actions while some others will correspond to one robot passing an object to another one, or even to several robots handling simultaneously a (big) object. A multi-robots plan will be produced according to the robots capacities and accessibility.

We have developed a first resolution scheme that is able to build plans for such problems. The current implementation is already able to deal with a few robots (holonomic or not) with movable objects and continuous grasps and placements[15, 14]. The presence of objects that can only be moved by the robots (so called "movable" objects versus obstacles which are fixed objects) leads to complex instances of the motion planning problem. Indeed, the robot may be forced because of kinematic constraints or because of collision avoidance to find intermediate positions and re-grasping steps for the movable object. Besides, it may be necessary sometimes to displace objects that are directly concerned by the task, simply because they prevent the robots to reach some desired configurations.

Let us note here that we are interested in the planning phase. The execution step would need techniques based on distributed control as developed for example in [9, 4].

Motion planning [10] is a challenging problem that involves dealing with elaborate physical constraints and high-dimensional configuration spaces. The fastest existing deterministic planner has a complexity exponential in the number of degrees of freedom of the robot. On the other hand, a class of randomized planners, in particular, *Probabilistic Roadmap Methods* (PRMs) [8, 13], have been used successfully in high-dimensional configuration spaces. More recently, the use of several roadmaps to solve motion planning with kinematic constraints [7, 11] or manipulation tasks [12, 15, 14], have been proposed. In our approach, we develop a resolution scheme based on a number of methods for reasoning on roadmaps construction and handling.

In manipulation problems, one robot can perform an object *transfer* or move alone [12, 15, 14]. This leads to roadmaps for *transfer* or *transit* movement. Moreover, a robot grasping an object on the ground defines a closed kinematic chain. The probability to draw such valid configuration is null. So we also have a *grasp* roadmap, with specific methods [3], to represent these key events in manipulation problems. One object alone can only be put in a valid (stable) position, so we use the last main type of roadmaps, the *placement* roadmap.

The Key Ideas. The planner is based on a search through various roadmaps. Each roadmap type is built with specific random and local planning functions. In fact, when two or more robots are involved, several methods can be used to find a solution. One can use directly the *Probabilistic Roadmap Methods* on the union of all configurations. This (theoretically) allows to solve multi-robot problems, but is time consuming, even for simple tasks, and often gives inefficient solutions (all the robots are moving quite randomly). An other way is to plan for each robot and then to coordinate them [2, 5]. This solution is much faster and allows dynamic interactions, but there is no guaranty to find a solution if it exists. Our idea here is not to merge trajectories, but merge roadmaps.

For example, with two robots, the use of two different roadmaps gives the possibility to deal with the product of their nodes. Even if such product is not valid, only the collision between robots must be done to validate a search process. Thence it is possible to reuse roadmaps and to revalue their validity with several configurations of other robots. Moreover, searching a solution in separate roadmaps seems to lead to more realistic solutions than global planning, because the search is guided with individual solution.

2 Graphs Definition

2.1 The Probabilistic Roadmaps

To deal with 3D geometrical tasks, we use the well known *Probabilistic Roadmaps Methods* (PRMs) [8, 13]. It is one of the two types of graph used. These roadmaps are graphs that store configurations of the robots and edges to link them. They catch the free space domain topology. The edges are valid local-path allowing to go from one node to any connected node in the graph. We use $\mathcal{G}$ to denote these roadmaps. In those roadmaps, the sub-sets of nodes fully connected are the roadmap connect components $\mathcal{C}$.

The main idea of our method, is to divide the complexity by the separation of the problems. So we use roadmaps to catch the free-space for sub-problems. As we said before (§1), we want to reuse the information for each object. So we decide to build a roadmap for each one. Moreover, for all closed kinematic chains, it could be useful to have a specific roadmap [3, 7, 8, 14] with specific methods. For instance, To randomly choose a configuration of a robot that grabs a specified object is null in probability. So there is a need for specific methods[3].

In the same way, we define roadmaps for sub-set elements of objects or robots when there are linked between themselves. But we do not define roadmaps when there is no relation between objects. To represent two independent robots, we have two independent roadmaps.

Because those roadmaps are independents, they are not fully valid. We have to check the collisions with other objects to validate them. And we will do it only if it is necessary. Note that it is possible to build also heuristic graph with no-valid edges, but that could help the construction of other roadmaps [6]. In this paper we focus on *crossable roadmaps*, roadmaps that are used to extract a solution, all their edges are valid for the set of objects taken into account.

2.2 The Elementary Kinematic Graphs

This graph will be used for representing the link between roadmaps. Like the roadmaps, they store nodes (kinematic node $\mathcal{K}n$) and edges (kinematic link $\mathcal{K}l$) to link them. Into an *Elementary Kinematic Graphs* ($\mathcal{KG}$) the nodes connected to each other are grouped in component ($\mathcal{KC}$). These graphs are defined for a sub-set of the elements (movable objects and robots) taken into account in the problem. Let us consider the problem illustrated in the example section (§4). We have two robots (robotized forklifts) and two movable objects (a big box and a flat box that the user wants to displace). One $\mathcal{KG}$ can be built for example for one robot, one robot and one object, or all robots and objects.

The nodes of these $\mathcal{KG}$ do not represent a configuration of a robot, but a connected component of roadmaps. There are two node types:

The roadmap component: the $\mathcal{K}n$ represent a component of a roadmap that takes into account the same set of elements than the $\mathcal{KG}$. For example (table 1), a connected component of the placement roadmap of the flat box is a $\mathcal{K}n$.

The set of $\mathcal{KC}$: the $\mathcal{K}n$ represent a set of kinematic component. For instance (table 1), the set of one $\mathcal{KC}$ of one forklift moving alone and the $\mathcal{KC}$ of the flat box represents one $\mathcal{K}n$ of the forklift and flat box $\mathcal{KG}$.

Table 1 represents all the possible $\mathcal{K}n$ types for all $\mathcal{KG}$ that are built. Note that presently we cannot use the symmetries. So we have in fact one $\mathcal{KG}$ for forklift1 and flat box, and one for forklift2 and that flat box, even if the two forklifts are identical. So there are some redundancies between graphs that can be easily avoided with the $\mathcal{KG}$.

All these types of $\mathcal{KG}$ and $\mathcal{K}n$ are computed with a domain analysis. The number of $\mathcal{KG}$ used is less than the number of all element combination. In fact, we build a $\mathcal{KG}$ only when the elements interact strongly, i.e. there is a roadmap $\mathcal{G}$ that uses those elements. We also build the $\mathcal{KG}$ needed to represent the initial and final states of a problem ($\mathcal{KG}$(all) in table 1).

Table 1: Elementary Kinematic Graphs example

$\mathcal{KG}$	Type of kinematic node
flat box	$\mathcal{C}$(Placement)
big box	$\mathcal{C}$(Placement)
forklift	$\mathcal{C}$(Movement)
forklift & flat box	$\mathcal{C}$(Grasp) $\mathcal{C}$(Transfer) $\mathcal{KC}$(forklift, flat box)
2 forklifts & flat box	$\mathcal{C}$(Both Grasp) $\mathcal{KC}$(forklift1 & flat box, forklift2) $\mathcal{KC}$(forklift2 & flat box, forklift1)
forklift & big box	$\mathcal{C}$(Grasp) $\mathcal{C}$(Transfer) $\mathcal{KC}$(forklift, big box)
all	$\mathcal{KC}$(2 forklifts & flat box, big box) $\mathcal{KC}$(forklift1 & flat box, forklift2 & big box) $\mathcal{KC}$(forklift2 & flat box, forklift1 & big box)

Like the nodes, there are also two types of kinematic link $\mathcal{K}l$:

The link between same roadmaps: this $\mathcal{K}l$ links two $\mathcal{K}n$ based on $\mathcal{KC}$. It only show that the $\mathcal{K}n$ have the same decomposition of roadmaps, so with the same connected component for all elements it is possible to switch between those two $\mathcal{K}n$ with no cost (Figure 1). This link appears because we try to reduce the number of necessary $\mathcal{KG}$.

The link between different roadmaps: this $\mathcal{K}l$ links a $\mathcal{K}n$ based on a roadmap connected component $\mathcal{C}$ to another $\mathcal{K}n$. This links are given by the problem definition. This allows for instance (Figure 1) to one forklift not carrying anything (c) to grasp the flat box (b) then to carry it (a) for after putting it on the ground (b) and drop it (c). This explain how to combine the specific methods used in the roadmaps.

Figure 1 represents the type of $\mathcal{K}n$ and $\mathcal{K}l$ for two forklifts and a flat box. It shows how the domain analysis divides a problem with many objects and robots into sub-problems much simpler and how to link them. It also shows that the problem could be solve hierarchically, for instance (Figure 1) crossing a $\mathcal{KC}$ of two forklifts and one flat box may need to divide the problem to moving one forklift and moving the other with the flat box, this one can be also divide in simpler problems. We will see that in the next section §3.2.

3 Solver Protocol

For a given problem, we need a way to find a solution. For that we may increase the roadmaps size (§3.1) and find a way between those roadmaps (§3.2 & §3.3).

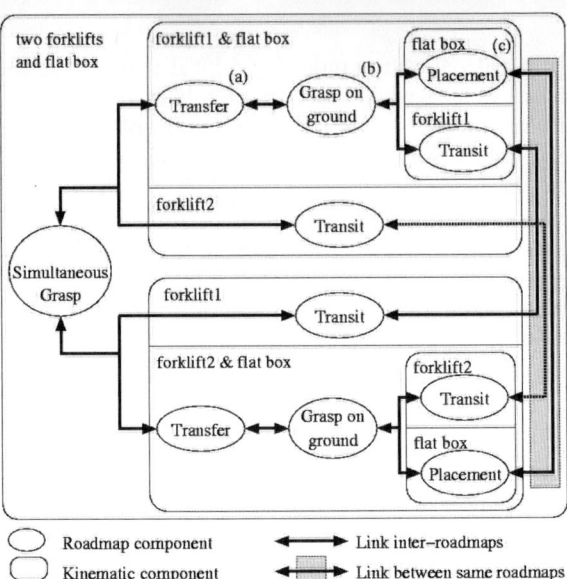

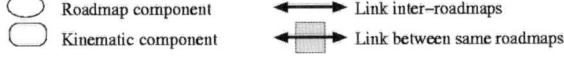

Figure 1: Elementary Kinematic Graphs definition

3.1 The Heuristic Rules

To increase the roadmaps sizes we use heuristic rules that analyze the roadmaps to choose which one must be increased and how to do it [6]. Those rules are able to combine nodes from other roadmaps to build new nodes in a given roadmap. For instance, nodes of the flat box position can be used to generate new nodes corresponding to the robot grasping the object on the ground. In an other hand, a node of grasping the object on the ground leads to several valid nodes: the object on the ground, the robot alone and the robot carrying the object. The $\mathcal{K}l$ between different roadmaps are based on those relations (Figure 1 (a) with (b) and (c) with (b)).

We will not detail further these heuristic rules. But note that they are presently independent of the search process. This will certainly change in the future versions of the planner in order to adapt the roadmap building process to the current search state.

3.2 The Hierarchical Search

As we already said in §2.2, the structure of the $\mathcal{KG}$ exhibits a hierarchical organization. Figure 2 gives an idea of the method used to search a solution through those roadmaps.

Note that we can partially define the goal (in our case only the final position of the flat box counts). But with the $\mathcal{KC}$ system it is easy to know all the $\mathcal{K}n$ that embed the goal. For example for the $\mathcal{KG}$ with all the robots and objects, the start node contains the goal solution, note that the other $\mathcal{K}n$ also, probably contain the goal solution.

But even if a $\mathcal{K}n$ contains the goal it does not mean that it can reach it by itself. In fact the kinematic

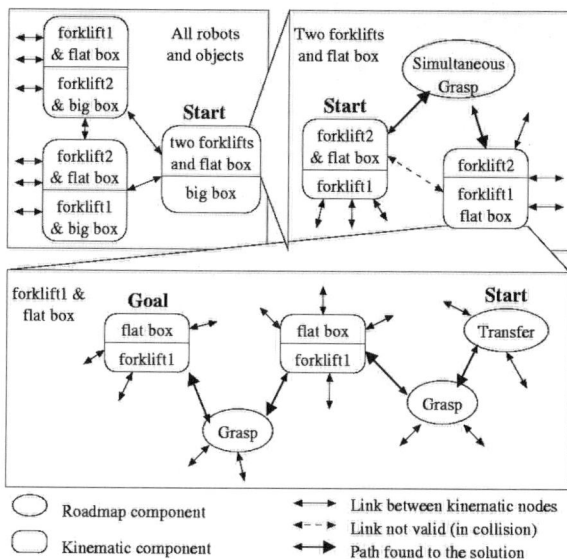

Figure 2: *Search through $\mathcal{KG}$*

nodes are build on independent roadmaps, the interaction between the robots and the objects are not fully tested. The search process needs to validate each edge used. Figure 2 shows one edge that is not valid. In fact, if the big box was not placed to block the robots path, it would be possible to directly link the $\mathcal{K}n$ in the $\mathcal{KG}$ of two forklifts with the flat box. Note that the start $\mathcal{K}n$ of this $\mathcal{KG}$ contains the goal solution, but the big box forbid the forklift 2 to carry alone the flat box to its final position.

The search of a solution and the edge validation is done in the same time. This avoids to test the combination of all roadmaps that correspond to a global planning on the configurations union. In our case we will do the minimal tests that the search process finds useful.

The search process consist in searching a path in one level of the $\mathcal{KG}$ and each time that it is necessary to switch nodes between $\mathcal{K}n_1$ and $\mathcal{K}n_2$, they need to test if it is possible. For that, it is necessary to search between the graphs that compose $\mathcal{K}n_1$ if a path exists between the current state of all robots and objects, and the position needed to reach a link to $\mathcal{K}n_2$. Figure 2 underlines the hierarchical research, each time there is a need to refine the search for a sub-set of elements.

This search process has to handle goal partially defined, or multiple goal definition. Indeed it is possible to have several $\mathcal{K}l$ between two $\mathcal{K}n$. Each like can precise a different state to switch of $\mathcal{K}n$, leading to several sub-goals definition.

Presently we use an $\mathcal{A}^*$ algorithm to search a path through each level. Even if we use an admissible heuristic, the hierarchic structure does not easily allows to extend the search front in the same way for each level. So we do not have a minimal solution. Moreover, the heuristic does not take into account the interaction between objects leading the search process to many not valid search branch. To avoid this we define an other type of goal.

3.3 Partial goals

In the search process the goals are partially defined. It could be a sub set of configuration, but also a sub set of $\mathcal{K}n$. For example a link between a group of same roadmaps (§2.2) does not need to pass through precise configurations, but only through defined connected component that are indeed $\mathcal{K}n$. In the same way, the goal of the problem can be, to put a object in a certain area that corresponds to a roadmap and then to $\mathcal{K}n$. In fact, a goal is represented by a tree that matches the hierarchy. The leafs of this tree define how precise the goal is. There is three possibilities: nothing specified, a $\mathcal{K}n$, or a configuration node (the most precise).

Then, as we said previously (§3.2), there is not only one possibility for each node in the tree definition. So for each level of the goal definition there is not an unique choice, but several. So we have a list of trees to define the goal.

The partial goal have three sources. The first, is the problem definition. When a $\mathcal{K}n$ contains this goal, the first thing to do is to check if it is possible to reach it without leaving the current $\mathcal{K}n$ (Figure 2).

The second is given by the link between $\mathcal{K}n$ as explained in the previous section (§3.2).

The last possibility, is the heuristic process. If we compute distance between configurations for the heuristic, we obtain an admissible heuristic that does not take into account the interaction between the robots and the movable objects. For example, the goal do not specify a final position for the big box, so its current position, even if it is a blocking position is valid for the goal. Hence, the big box is well placed for the configuration heuristic. Then the search process try nearly all the possibilities before moving it. To avoid this king of problem, we decide to detect such situation and to put new sub goals. In fact, when it is not possible to link two nodes in the same connected component, we score the elements that block the path. When an element is really annoying (score above a given limit), a new sub goal that avoid the collision detected is given. This allows to free a path for other robots, and to solve the big box problem.

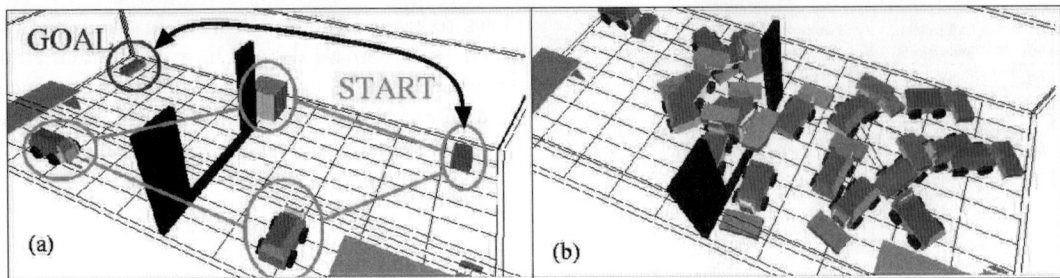

Figure 3: *Example of two robots forklift and two movable objects: The environment consists of two rooms connected by a single door. There is a "window" in the wall separating the rooms. The task consists in displacing the flat box (a). The goal position of the robots and of the big box is left unspecified. (b) shows one of the roadmap of on forklift carrying the flat box.*

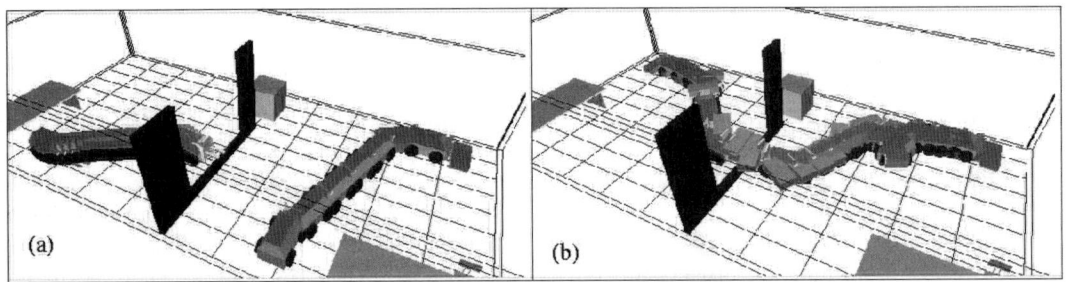

Figure 4: *The first solution found consists in passing the flat box from one robot to the other through the window.*

4 Examples

In this section we will detail the example used along this paper (Figure 3 4 4). In this example we have two mobile robots that can lift palette. They are non holonomic, following the Reed and Shepp curves. There are two palettes that can be carried, one with a flat box, and one with a big box.

The problem is defined with the initial state for both robots and movable objects (Figure 3 (a)). The goal is partially defined with only the desired position for the flat box (Figure 3 (a)). The problem also define continuous set of grasping position and placement [15, 14] for objects and robot even if here those degree of freedom are small to keep realistic object transfer (object well placed on the robot). The last information that must be given by the problem is the type of roadmaps needed and how they are linked. Presently these data are given by the user, but for multi-robot manipulation problem the same type of roadmaps are always used. So we hope to automate this definition.

Figure 3 (b) shows an example of roadmap. It is the forklift transfer roadmap. All the nodes where the forklift is carrying the flat box appear in this roadmap. We can see link to the goal position and to a position to give the flat box to the other robot.

Figures 4 or 5 show two possible solutions found by the planner. The planner do not give a unique solution. It is base on PRMs so random configurations are used. This can lead to different solution.

The first solution (Figure 4) do not move the big box. First one robot goes to take the flat box and the second comes near the big window in the wall to wait for the flat box. Then (Figure 4 (b)), the robot carrying the flat box give it to the other robot which can reach the goal.

In the second solution (Figure 5), one robot remove the big box and the other carry the flat box to its goal position.

If we increase the roadmaps enough. It is possible to choose between the possible solutions by changing the cost of roadmaps edges, the roadmaps transitions or the level of annoyance of obstacle before moving them. For example if the big box is heavy, the cost for carrying it, is high, so the first solution appear more often. If giving the flat box between robots is difficult, the transition to the roadmap of both robots carrying the flat box must have a high cost, so the second solution is often extracted.

5 Conclusions

We have presented a new approach to multi-robot manipulation planning that deals with a set of roadmaps by the use of a new type of graph: the *Elementary Kinematic Graphs*. This structure allows to divide the complexity of a problem with the defi-

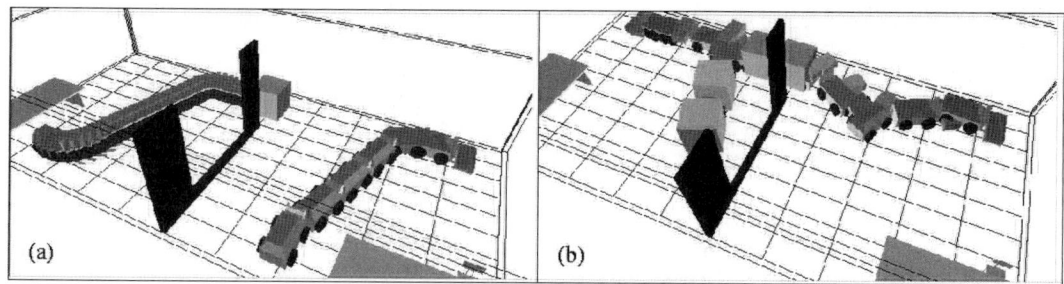

Figure 5: The second solution consists in displacing the big box which was blocking the door and then, allowing the second robot to reach directly the final placement position.

nition of sub-domains. It provides a mean to search a solution in a hierarchical way.

The first results are encouraging, but many improvement could be done in order to have the planner to be fully functional. Some ideas have been already underlined, for example to automate the roadmap definition, to use heuristic rules for the roadmap construction that depend of the state of the research process. The use of common roadmaps for identical objects or robots seems to be easily implementable in the kinematic structures, but leads to problems with less specific search.

In fact our last efforts have been put on the solver engine. A purely symbolic level as been added to guide a state based search process to avoid the limitation $\mathcal{A}^*$ when it deals with collision checking with several mobiles objects, but this feature is beyond the scope of this paper.

References

[1] R. Alami, J.P. Laumond and T. Siméon: "Two manipulation planning algorithms." In *Algorithmic Foundations of Robotics (WAFR94)*, K. Goldberg et al (Eds), AK Peters, 1995.

[2] R. Alami, F. Ingrand, and S. Qutub, "A Scheme for Coordinating Multi-robot Planning Activities and Plans Execution", in *ECAI*, 1998

[3] J. Cortès, T. Siméon and J.P. Laumond: "A Random Loop Generator for Planning the motions of closed kinematic chains with PRM methods". In *IEEE International Conference on Robotics & Automation*, Washington D.C., May 2002

[4] J.P. Desai, J. Ostrowski, and V. Kumar, "Controlling formations of multiple mobile robots". In *IEEE International Conference on Robotics and Automation*, May 1998.

[5] F. Gravot, R. Alami, "An extention of the Plan-Merging Paradigm for multi-robot coordination" In *IEEE International Conference on Robotics & Automation*, Seoul, May 2001

[6] F. Gravot, R. Alami, T. Siméon, "Playing with Several Roadmaps to Solve Manipulation Problems". In *IEEE International Conference on Intelligent Robots and Systems*, Lausanne, September 2002

[7] Li Han and N.M. Amato: "A Kinematics-Based Probabilistic Roadmap Method for Closed Chain Systems". In *Proceedings of the Workshop on Algorithmic Fundations of Robotic (WAFR '00)*, March 2000, pp. 233-246

[8] L. Kavraki and J.C. Latombe: "Ramdomized preprocessing of configuration space for fast path planning". In *IEEE International Conference on Robotics & Automation*, San Diego, May 1994

[9] O. Khatib, "Mobile manipulation: the robotic assistant". In *Robotics and Autonomous Systems*, (26):2175–183, 1999.

[10] J.C. Latombe: "Robot Motion Planning". Kluwer Academic Publishers, Boston, MA, 1991

[11] S.M. La Valle, J.H. Yakey and L.E. Kavraki: "A Probabilistic Roadmap Approach for Systems with Closed Kinematic Chains". In *IEEE International Conference on Robotics & Automation*, Detroit, Michigan, May 1999

[12] C. Nielsen and L.E. Kavraki: "A Two-Level Fuzzy PRM for Manipulation Planning". In *IEEE International Conference on Intelligent Robots and Systems*, Takamatsu, Japan, November, 2000

[13] C. Nissoux, T. Siméon and J.P. Laumond: "Visibility based probabilistic roadmaps". In *IEEE International Conference on Intelligent Robots & Systems*, Korea, October, 1999

[14] A. Sahbani, J. Cortès and T. Siméon: "A Probabilistic Algorithm for Manipulation Planning under Continuous Grasps and Placements". *IEEE International Conference on Intelligent Robots and Systems*, Lausanne, September 2002

[15] T. Siméon, J. Cortès, A. Sahbani and J.P. Laumond: "A Manipulation Planner for Pick and Place Operations under Continious Grasps and Placements". In *IEEE International Conference on Robotics & Automation*, Washington D.C., May 2002

Incremental Low-Discrepancy Lattice Methods for Motion Planning

Stephen R. Lindemann   Steven M. LaValle

Dept. of Computer Science
University of Illinois
Urbana, IL 61801 USA
{slindema, lavalle}@uiuc.edu

Abstract

We present deterministic sequences for use in sampling-based approaches to motion planning. They simultaneously combine the qualities found in many other sequences: i) the incremental and self-avoiding tendencies of pseudo-random sequences, ii) the lattice structure provided by multiresolution grids, and iii) low-discrepancy and low-dispersion measures of uniformity provided by quasi-random sequences. The resulting sequences can be considered as multiresolution grids in which points may be added one at a time, while satisfying the sampling qualities at each iteration. An efficient, recursive algorithm for generating the sequences is presented and implemented. Early experiments show promising performance by using the samples in search algorithms to solve motion planning problems.

1 Introduction

Sampling the configuration space has been one of the fundamental issues in developing practical motion planners. Some approaches use context-specific heuristics to concentrate samples in critical places [1, 4, 8, 19]. Classical grid-search approaches (see survey in [9]) and also recent "lazy" approaches [3, 2, 5, 14]) have focused more on how to sample the configuration space before taking obstacles into account. In this paper, we consider sampling issues from this perspective, and ask: What is the best way to sample the space?

Here are several desirable criteria that we will consider for an infinite sequence of samples over a bounded configuration space:

1. **Uniformity:** Good covering of the space is obtained without clumping or gaps. This can be formulation in terms of optimizing discrepancy or dispersion [12, 13].

2. **Lattice structure:** For any sample, the location of nearby samples can easily be determined.

3. **Incremental quality:** For any i, if the sequence is suddenly terminated, it has decent coverage. This is an advantage over a sequence that only provides high-quality coverage for a fixed n.

A simple grid generated by scanning has good lattice structure and uniformity, but fails to provide good incremental quality. Quality coverage is only obtained at values of i that yield a complete grid at some resolution. It is intuitive that *closed* sequences (in which the total number of samples is specified in advance) can achieve better coverage than infinite or *open* sequences (in which the total number is not specified), but this sacrifices incremental quality; for many applications, paying a small penalty in coverage to achieve incremental quality is a welcome exchange. A random sequence exhibits incremental quality, but at the expense of lattice structure and even uniformity (clumps and gap are prevalent with high probability [11, 13]). Thus, random sequences and grids appear to be quite complementary. In probabilistic roadmap (PRM) approaches (e.g., [1, 10, 15, 20]), one is usually willing to sacrifice the first two properties to obtain the last.

After considering the tradeoffs, we wondered whether it is possible to define sequences that provide all three qualities listed above. It turns out that this can be done, and the resulting sequences of samples are the primary contribution of this paper. We next provide formal definitions of discrepancy and dispersion, which our sequence will optimize.

2 Uniformity Measures

Uniform sampling criteria and techniques have been developed by numerous mathematicians over the past century. Excellent overviews of the subject include [12, 13]. Here we briefly introduce only the concepts needed for this paper. Let $X = [0,1]^d \subset \Re^d$ define a space over which to generate samples. Define a *range space*, $\mathcal{R}$, as a collection of subsets of X. Let $R \in \mathcal{R}$ denote one such subset. Reasonable choices for $\mathcal{R}$ include the set of all axis-aligned rectangles, the set of all balls, or the set of all convex subsets.

Let $\mu(R)$ denote the Lebesgue measure (or volume) of subset R. If the samples in P are uniform in some ideal sense, then it seems reasonable that the fraction of these samples that lie in any subset R should be roughly $\mu(R)$ (divided by $\mu(X)$, which is simply one). We define the *discrepancy* [18] to measure how far from ideal the point set P is:

$$D(P, \mathcal{R}) = \sup_{R \in \mathcal{R}} \left| \frac{|P \cap R|}{N} - \mu(R) \right| \qquad (1)$$

in which $|\cdot|$ applied to a finite set denotes its cardinality.

Whereas discrepancy is based on measure, a metric-based criterion, called *dispersion*, can be introduced:

$$\delta(P, \rho) = \sup_{q \in X} \min_{p \in P} \rho(q, p). \qquad (2)$$

Above ρ denotes any metric, such as Euclidean distance or ℓ^∞. Intuitively, this corresponds to the radius of the largest empty ball (assuming all ball centers lie in $[0,1]^d$).

3 One-Dimensional Sampling

To gain an understanding of the issues, it is helpful to first consider the case of sampling a one-dimensional space. In this case, a sequence introduced by van der Corput in 1935 achieves all three desired criteria with beautiful simplicity [17]. Consider a binary representation of points in $[0,1]$. A one-dimensional "grid" can be made by counting in binary. For example, if the resolution is 8, then samples are taken at: 0.000, 0.001, 0.010, 0.011, 0.100, etc. Of course, this scanning behavior of the sequence does not have incremental quality.

The *van der Corput* sequence simply takes the binary counting above and reverses the order of the bits. During the original scan, the least significant bit alternates in every step, but this only yields a small change in value. By reversing bit order, the change is maximized, causing the coverage to be nearly uniform at every point in the sequence. After bit reversal, the sequence is: 0.000, 0.100, 0.010, 0.110, 0.001, 0.101, 0.011, 0.111. An infinite sequence is constructed by using reversed-bit representations of higher binary numbers. The next eight samples are obtained by reversing binary representations of 8 through 15.

This deterministic sequence is ideal in many ways; it satisfies all three of the criteria from Section 1. It is asymptotically optimal in terms of discrepancy ($\mathcal{R}$ is a set of intervals), and also in terms of dispersion (note that this would not be achieved by a random sequence). It has a trivial lattice structure. Finally, the sequence is incremental because at any given time, the sequence can be stopped while still yielding low discrepancy and low dispersion. If the sequence is stopped at $i = 2^k$ for any integer k, then all samples are equally spaced, much as in a classical grid. More importantly, if the sequence is stopped elsewhere, the distribution of points is still good, which would not be the case of the resolution was simply improved by scanning.

4 Higher Dimensional Sampling

For use in motion planning, straightforward extensions of the van der Corput sequence to $[0,1]^d$ would be very useful; unfortunately, such sequences have not been found. Simply making a vector-valued sequence will only generate samples along a diagonal line. Halton used the bit-reversal technique to extend the sequence, using a different base for each dimension [7]. His method is as follows: choose d distinct primes $p_1, p_2, \ldots, p_d$ (usually the first d primes, $p_1 = 2$, $p_2 = 3$, $\ldots$). To construct the ith sample, consider the digits of the base p representation for i in the reverse order: $i = a_0 + pa_1 + p^2 a_2 + p^3 a_3 + \ldots$, in which $a_j \in \{0, 1, \ldots, p-1\}$. Define the following element of $[0,1]$:

$$r_p(i) = \frac{a_0}{p} + \frac{a_1}{p^2} + \frac{a_2}{p^3} + \frac{a_3}{p^4} + \cdots.$$

The ith sample in the Halton sequence is

$$(r_{p_1}(i), r_{p_2}(i), \ldots, r_{p_d}(i)), \qquad i = 0, 1, 2, \ldots.$$

This sequence is known to produce asymptotically-optimal discrepancy. It satisfies the first and last criteria from Section 1; therefore, it is a useful sequence. For virtually all randomized motion planning algorithms, one can replace a pseudo-random sequence with the deterministic Halton sequence because of the satisfaction of these properties. Recent experimental results in [6] show Halton points performing well versus other sampling techniques in the context of the PRM.

It is possible, though, to construct an alternative generalization of the van der Corput sequence, which is able to satisfy all three criteria? The neighborhood structure offered by a lattice is particularly useful in the context of motion planning. For example, in the probabilistic roadmap method, substantial time is invested in performing nearest-neighbor queries to build the roadmap. In a lattice, this information is already implicitly defined.

In addition to having grid structure, the van der Corput sequence naturally creates a *multiresolution* grid as it progressively fills in gaps in the unit interval. A generalization of the van der Corput sequence which has grid structure should have this property as well, for several reasons. First, for many problems it is impossible to know ahead of time what an appropriate resolution might be. In addition to this, it is intuitive that a multiresolution approach yields more incremental quality than a grid of fixed resolution. Finally, any open sequence (such as the van der Corput sequence) which also has grid structure *must* be multiresolution, because if the resolution is fixed, then the number of samples is fixed as well, which is a contradiction for an open sequence.

In summary, what is required is a multi-dimensional generalization of the van der Corput sequence: an open sequence generating a multiresolution grid and satisfying the criteria given at the beginning of this paper. Below, we present such a sequence.

5 A New Sequence

Before proceeding to describe our new sequence, several definitions will prove useful. Consider a classical grid in the d-dimensional unit cube, $[0,1]^d \subset \Re^d$; we define a multiresolution classical grid of resolution level l to be a grid with 2^{dl} points (i.e., 2^l points per axis). From this definition, it is apparent that a grid of resolution level l contains all the points from resolution level $(l-1)$, and that of all grids having this property, it is the one with the fewest points (assuming that all dimensions are required to have the same number of points per axis).

In addition to considering classical grids, it is worthwhile to examine Sukharev grids as well [11, 16]. Consider a grid in the d-dimensional unit cube, with k points per axis; the unit cube may then be divided into k^d regions. While the classical grid places a vertex at the origin of each region, the Sukharev grid places a vertex at the center of each region. This has the advantage of optimizing ℓ^∞ dispersion, defined in Section 2. While this difference may not seem very large, it is significant when the grids are taken to be multiresolution; while a multiresolution classical grid has 2^{dl} points for resolution level l, a multiresolution Sukharev grid has 3^{dl} points. Most of our ideas apply equally to both classical and Sukharev grids; while we will deal primarily with classical grids for the sake of brevity, we will note applications to the Sukharev case as well.

We describe the points of a classical grid of resolution l as follows:

$$P_l^n = \left\{ \left(\frac{i_1}{2^l}, \cdots, \frac{i_n}{2^l}\right) : i \in \mathbb{Z}, 0 \le i \le 2^l - 1 \right\}.$$

One may also define the *grid region* associated with point j at resolution level l as:

$$G_{j,l} = \left[j_1, j_1 + \frac{1}{2^l}\right) \times \cdots \times \left[j_n, j_n + \frac{1}{2^l}\right).$$

Similar definitions may be made for the Sukharev case.

With these definitions in mind, we may proceed to consider the sequence itself. To motivate the way our sequence is generated, consider a d-dimensional classical grid of resolution level 1, having 2^d total points. A first question is raised immediately: in what order should these points be placed? Two possible criteria for making a decision are dispersion and discrepancy. As seen above, these measure the uniformity of coverage of the space; therefore, they are natural criteria for choosing the optimal placement order. However, using dispersion as the decision criterion for a multiresolution grid often results in ties. In fact, in the case of a Sukharev grid or a classical grid on a toroidal manifold, the ℓ^∞ dispersion remains constant between complete resolution levels. For example, a Sukharev grid with i points, $3^{dl} \le i < 3^{d(l+1)}$ will have the same ℓ^∞ dispersion as the grid with 3^{dl} points (for a more-detailed explanation of the relationship between ℓ^∞ dispersions of classical and Sukharev grids, see [11]). Given this fact, it seems best to use discrepancy as the decision criterion.

From the discussion of discrepancy in Section 2, a range space $\mathcal{R}$ must be chosen over which to calculate the discrepancy. Preferably, we should choose one which is suitable for grids and grid regions. Hence, we define the set of *canonical rectangles*, similar to the b-ary canonical boxes in [12]: given positive integers n and m, let $\mathcal{Q}_m^n$ be the following family of n-dimensional canonical rectangles:

$$\mathcal{Q}_m^n = \left\{ \left[\frac{i_1}{2^m}, \frac{i_1 + j_1}{2^m}\right) \times \cdots \times \left[\frac{i_n}{2^m}, \frac{i_n + j_n}{2^m}\right) : \right.$$
$$\left. i, j \in \mathbb{Z}, 0 \le i \le 2^m - 1, 1 \le j \le \min(2^m - i, 2) \right\}.$$

This closely relates to the previous definitions regarding the points of a multiresolution grid and their associated grid regions. In fact, the rectangles of $\mathcal{Q}_m^n$ may be as wide in any dimension as a single grid region at resolution level $m-1$, or two grid regions at resolution level m. For the case of $m = 1$, visualize $\mathcal{Q}_1^n$ as the set of all convex unions of the 2^d grid regions of a unit cube. Finally, define $\widetilde{\mathcal{Q}}_m^n = \bigcup_{i=0}^{m} \mathcal{Q}_i^n$. Again, these definitions apply to the case of classical grids, but analogous formulations may be made for Sukharev grids.

Now, let discrepancy be taken over the set $\mathcal{Q}_1^d$ be the criterion for determining the optimal order of the points of resolution level 1. Using this criterion, an optimal ordering list L of the first 2^d grid points (the first resolution level) may be explicitly computed (or, if we are dealing with a Sukharev grid and use a suitable modified range space, we may compute the optimal ordering list L_s for the first 3^d points). Hence, from this point on assume that we have such an ordering. Note that this only yields the correct ordering for the first resolution level. How should we fill in the next resolution level?

To answer this question, recognize that L may be viewed as an ordering not only of the first 2^d samples, but of the grid regions of the unit cube. This identification can be made because the discrepancy is calculated over a range space consisting of unions of these grid regions. Since this is the case, to maintain optimality over $\mathcal{Q}_1^d$, any future samples must follow this ordering: each future group of 2^d points must iterate through the ordering L in the same way as the first 2^d points. Hence, sample i must fall into the region of the unit cube specified by $L[i \bmod 2^d]$. However, this solves only part of our problem. Suppose that we know that point i of the second resolution level must fall into $G_{j,1}$ for some j; however, where within $G_{j,1}$ should it be placed? Recognize that during the transition from resolution level l to $(l+1)$, each grid region is subdivided into 2^d subregions. Since our initial ordering scheme determined the optimal ordering of placement of 2^d points within a region, we may recursively apply it to each subregion (with an

```
GET_SAMPLE(n, L, origin, factor)
1   Point sample ← origin;
2   index ← n % |L|;  //remainder of integer division
3   nextN ← n/|L|;    //quotient of integer division
4   sample ← sample + (currentFactor × L[index]);
5   if (nextN = 0)
6       return sample;
7   else
8       f ← factor/2;
9       return GET_SAMPLE(nextN, L, sample, f);
```

Figure 1: Recursively generate a new sample from the vertices of a classical grid.

appropriate scaling factor, of course). An algorithm that implements this approach is given in Figure 1.

Therefore, define the infinite sequence for a classical grid as $S_d = \{s_0, s_1, \ldots : s_i = GET\_SAMPLE(i, L, \vec{0}, 1.0)\}$, in which the zero vector denotes the origin and L is the ordering list appropriate for dimension d. A brief examination of GET_SAMPLE will yield insight into the behavior of the sample sequence. At each recursion level, the integer remainder (which is less than 2^d) tells the function which grid region of the current resolution level it should go to. The function then updates the sample to be the origin of that grid region. The integer quotient tells the function how many times that grid region has previously been visited, which in turn specifies how the function behaves when it is called recursively on that region. Finally, the function returns when it determines that it has found the exact location of the sample.

It is important to note one particular feature. Suppose that point i of resolution level l is added to grid region $G_{j,(l-1)}$. Then, by the nature of the recursion, a corresponding point will have to be added to every other grid region $G_{k,(l-1)}, k \neq j$ before another point is added to $G_{j,(l-1)}$. This feature contributes to the quality of uniformity discussed in the introduction, and will contribute to the following proof, which shows that the sequence retains optimality under recursion.

Theorem 1 *Take the first i elements of the sampling sequence S_d.*

1. *This sequence is a multiresolution grid sampling sequence of length i.*

2. *From the set of multiresolution grid sampling sequences, it is discrepancy-optimal over $\widetilde{\mathcal{Q}}_l^n$, in which $l = \lceil \log_{2^d} i \rceil$, i.e., the current resolution level.*

Proof: (1) For this to be the case, the sequence must form a classical grid for every $i = 2^{dl}, l \in \mathbb{Z}$. We show this to be the case by induction on l. First, take the base case $l = 0$. The first point of the sequence is the origin, which is a classical grid of size 1. Now, assume that $i = 2^{dl}$, and that the sequence formed a classical grid for every $j = 2^{dm}, m \in \mathbb{Z}, 0 \leq m \leq l-1$. Now, $2^{dl} - 2^{d(l-1)} = 2^{d(l-1)}(2^d - 1)$; by the observation preceding this theorem, this implies that each grid region $G_{j,(l-1)}$ had $2^d - 1$ points added to the point already placed in it at the previous resolution level. Moreover, these were added according to the specification of the ordering list L, which places points on the 2^d grid vertices of a certain region. Therefore, each region contains the 2^d points of a classical grid. Since the union of two classical grids of uniform resolution results in a classical grid of the same resolution, at $i = 2^{dl}$ the samples form a classical grid of resolution l. Therefore, our inductive hypothesis is shown to be true, and part (1) is proven.

(2) We also show this by induction on the current resolution l. First, we note that resolution level 0 consists of only one point, which is placed on the origin, and it is trivially optimal over $\widetilde{\mathcal{Q}}_0^n$, which consists simply of the unit cube. Now, assume that i is such that the current resolution level is l, and that all sequences up to length $2^{d(l-1)}$ are optimal over $\widetilde{\mathcal{Q}}_{l-1}^n$. We will show that the sequence is optimal over $\widetilde{\mathcal{Q}}_l^n$, and the proof will be complete.

From the definition of $\widetilde{\mathcal{Q}}_l^n$, $\widetilde{\mathcal{Q}}_l^n = \widetilde{\mathcal{Q}}_{(l-1)}^n \cup \mathcal{Q}_l^n$. Also, we know that the first $2^{d(l-1)}$ points were added in the optimal order with respect to $\widetilde{\mathcal{Q}}_{(l-1)}^n$, by assumption; denote the grid region associated with the j-th sample of that complete grid as $G_{j,(l-1)}$ (the j-th sample is located at the origin of $G_{j,(l-1)}$). Each of the points of the current resolution level fall into one of the $G_{j,(l-1)}$; moreover, by the nature of the recursion, the order in which they fall into the $G_{j,(l-1)}$ is the same order that the original $2^{d(l-1)}$ points did. Consequently, all points of the current resolution level are optimal with respect to the set of rectangles $\widetilde{\mathcal{Q}}_{(l-1)}^n$.

Now, examine the rectangles which are part of $\mathcal{Q}_l^n$, which can be partitioned into the set of all rectangles which are completely contained within one of the $G_{j,(l-1)}$ above, and those which are not. For those which are completely contained within one of the $G_{j,(l-1)}$, optimality is clearly seen. Denote as p_j the subset of the sample sequence contained in $G_{j,(l-1)}$; by the definition of the recursion, the points $G_{j,(l-1)}$ are added to $G_{j,(l-1)}$ in precisely the optimal order defined in L; since this is the case for all $G_{j,(l-1)}$, the point sequence is optimal for all rectangles completely enclosed in some $G_{j,(l-1)}$.

Finally, we must consider the set of all rectangles in $\mathcal{Q}_l^n$ which are not enclosed in any $G_{j,(l-1)}$. First, note that for $\mathcal{Q}_l^n$ the maximum width of any rectangle in a single dimension is $1/2^{(l-1)}$, the size of each block $G_{j,(l-1)}$. Let $r \in \mathcal{Q}_l^n$ be a rectangle partially enclosed in gr_j; then, there are three possibilities for each dimension of r: first, it is entirely in $G_{j,(l-1)}$; second, it covers the top half of $G_{j,(l-1)}$ and the bottom half of some other block; or third, it covers the bottom half of $G_{j,(l-1)}$ and the top half of some other block. Denote by r_o the portion of r which is outside of $G_{j,(l-1)}$, and by r_i the portion which is inside. Now, one can take the reflection of r about r_i; then, each part of r_o is mapped to a place inside $G_{j,(l-1)}$. Define q to be a rectangle resulting from such a reflection, and note that $q \subseteq G_{j,(l-1)}$ and $q \in \mathcal{Q}_l^n$. Hence, we know that q is part of a set of rectangles for which optimal discrepancy has already been shown.

Let P_c be the set of points in the sequence for which analogous points can be found in every grid region $G_{j,(l-1)}$, and let P_i be the remaining points. Note that all points in P_i are analogous to each other by the observation immediately preceding this theorem; this implies that no rectangle in $\mathcal{Q}_l^n$ can contain more than one of these points. Now, take some rectangle r as described above. If this rectangle contains a point $p \in P_i$, then denote $G_{j,(l-1)}$ as the grid region containing this point; else, choose it to be any grid region containing a portion of r. Define r_o, r_i as above; then, we may once again take the reflection of r about r_i. The rectangle q obtained from this reflection has measure identical to rectangle r, and it contains the same number of points. We know this to be the case, because by assumption all points in r_o have analogues in $G_{j,(l-1)}$. This is the case since $p \in P_i$ is contained in r_i, by our choice of $G_{j,(l-1)}$; thus, all points in r_o are part of P_c and consequently have analogues in every grid region of resolution $(l-1)$. Thus, since r and q have identical measures and numbers of enclosed points, and q is part of the set for which optimal discrepancy has been shown, it is impossible for r to hurt the total discrepancy.

Since we have now shown that the sequence is optimal over $\widetilde{\mathcal{Q}}_{(l-1)}^n$ and $\mathcal{Q}_l^n$, we know that the sequence is optimal over $\widetilde{\mathcal{Q}}_l^n = \widetilde{\mathcal{Q}}_{(l-1)}^n \cup \mathcal{Q}_l^n$. Therefore, our inductive hypothesis is shown to be true and part (2) of the theorem is proven. ∎

6 Useful Properties for Motion Planning

Thus far we have defined a sample sequence which incrementally builds a multiresolution grid in an order which is discrepancy-optimal over an appropriately chosen range space. While this is of value on its own, we are particularly interested in using this sequence for motion planning applications, especially those which depend heavily on having a good sample set (e.g., the PRM). Hence, we now examine several properties of this sample sequence, to demonstrate the potential benefits of this sequence in motion planning applications.

A first consideration is the amount of time required to generate each sample. If it is computationally expensive to generate the sample sequence, this may offset time gained through the quality of the sequence. Hence, we give bounds on the time required to generate a particular sample. (In this and all future considerations, all scalar mathematical operations are considered to be constant time, since they depend on internal representation only. Vector operations are considered to be $O(d)$ time.)

Property 1 *The position of the i-th sample in the d-dimensional sampling sequence S_d can be generated in $O(\log i)$ time.*

Proof: The recursive function specified in Figure 1 may be written as $GL(i) = GL(i/2^d) + O(d)$. The solution to this recursion is $O(d \log_{2^d} i)$. Since $\log_{2^d} i = (\log i)/d$, the final result is $O(d(\log i)/d) = O(\log i)$. ∎

For purposes of comparison, pseudo-random samples usually require $O(d)$ time and Halton samples require $O(d \log i)$.

In the introduction, we stated that lattice structure is desirable because the location of neighbors can easily be determined. It is well-known that all points in a lattice can be specified in terms of a colloction of d linearly-independent basis vectors $b_1, \ldots, b_d$. In the case of a grid, the basis vectors are simply the columns of the $d \times d$ identity matrix. By adding (or subtracting) these basis vectors, the neighbors of a point can be found immediately. We define the i-*neighbors* of a point p as those points which may be reached by adding or subtracting i distinct basis vectors, $1 \leq i \leq n$.

However, our points are specified in terms of their index in the sequence; based on this index alone, it is unclear how to calculate the index of a neighbor in the sample space. It is possible to do so, however. The algorithm is too long to present here in its entirety; thus, we will sketch its operation.

For any element in the ordering list L, it is possible to store the order indices of all of the i-neighbors of each element. Since any i-neighbor may be found through a sequence of 1-neighbors, it suffices to store the order indices of each elements 1-neighbors, of which there are $2d$. The space required to do this is consequently $O(d 2^d)$, since $|L| = 2^d$.

Now, suppose we wish to find a particular 1-neighbor of sample i. As in the GET_SAMPLE function, we may execute a recursion, storing the sample index of each "ancestor" and the order index between ancestors. We then use this information along with the neighborhood information stored with each element in the ordering list L to find the sample index of the desired neighbor. Then, we perform a simple query to see if the sample corresponding to this index exists.

Property 2 *Let the number of samples taken so far be N. Then, a 1-neighbor of any of these samples can be found in $O((\log N)/d)$ time.*

Proof: Apply the algorithm described above. The function will recurse at most $O((\log N)/d)$ times and generate as many ancestors, similar to the GET_SAMPLE function. Note that this requires only $O((\log N)/d)$ rather than $O(\log N)$ time as in the analysis of GET_SAMPLE because the actual sample location is not being remembered; only indices are being calculated, which are dimension-independent. Now, to find the desired neighbor, the entire ancestor chain may have to be traversed (this is similar to binary addition, in which adding 1 may result in each bit needing to be changed). However, only a constant amount of work is done each time (applying some simple formulas to obtain new indices). After doing this, the sample index of the desired neighbor has been calculated; if this value is larger than N, then the neighbor does not exist. If a vector of pointers to previous samples is kept, simply indexing into this vector will allow one to determine the previously-calculated position of this sample. Therefore, the total time required is simply $O((\log N)/d)$. ∎

The scheme described in the proof above can easily be adapted to the case of motion planning, in which some samples of index less than N may not exist, due to being in collision with some obstacle. In this case, the corresponding entry in the vector is nil, and the query returns that the vertex does not exist. Also, the fact that we can calculate neighbors in this way suggests the potential for developing "lazy" planners that can search the space without allocating huge amounts of space for storing edge connections and neighborhood information.

This method is an improvement over naïve search (in which the proximity of every sample to the initial point is checked), and can be used to find 2-,...,d-neighbors in addition to 1-neighbors; however, there may be situations in which it is desired to determine the radius necessary to connect to the i-neighbors of a particular point at resolution level l, for use in naïve search. At resolution level l, the distance between 1-neighbors is $1/2^l$; hence, the distance between i-neighbors is $\sqrt{i(1/2^l)^2} = \sqrt{i}/2^l$. Thus, by setting the connection radius appropriately, one may use some other technique to connect neighboring grid points. In passing, it should be noted that to prevent a point from connecting with non-neighboring points (e.g., those of distance $2/2^l$ in a single direction), i must be 3 or less.

Finally, since the grid is multiresolution, it may be possible to reduce collision checks in resolution level l by remembering some results from resolution level $(l-1)$. We give a bound on the number of collision checks that may be saved in this way.

Property 3 *If samples are connected only to their 1-neighbors, then at most a fraction of $\frac{1}{2^{d-1}}(1 - \frac{1}{2^l-1})$ of the collision checks required at resolution level l may be saved.*

Proof: The total number of points in a grid of resolution level l is 2^{dl}. Assume for a moment that we are on a toroidal manifold, so that each point has $2d$ 1-neighbors. Then the total number of edges is $d2^{dl}$, since each edge is shared by two points. Now, to correct for being in $\Re^d$, we must remove some edges. For each dimension, a fraction of 2^l of the edges cross the boundary (since we may recall that the number of points per axis is 2^l). Therefore, we must remove $d2^{dl}/2^l = d2^{(d-1)l}$ edges, leading to a total of $d2^{dl} - d2^{(d-1)l}$ edges in resolution level l. Since the fraction of collision checks saved is the same as the fraction of new edges covered by edges of the previous resolution level, we find that we may save:

$$\frac{2(2^{d(l-1)} - 2^{(d-1)(l-1)})}{2^{dl} - 2^{(d-1)}} = 2 \cdot 2^{-d}\frac{2^{dl} - 2 \cdot 2^{(d-1)l}}{2^{dl} - 2^{(d-1)}} =$$
$$\frac{1}{2^{d-1}}\left(1 - \frac{2^{(d-1)l}}{2^{dl} - 2^{(d-1)}}\right) = \frac{1}{2^{d-1}}\left(1 - \frac{1}{2^l - 1}\right)$$

∎

From the equation above, it can be seen that while there may be some savings for low-dimensional applications, there will be only slight savings for higher-dimensional problems. Consequently, we expect the primary benefits of our sequence's lattice structure to be in its implicitly-defined neighbors, rather than in collision check savings.

7 Experimental Results

While extensive empirical testing is needed to conclusively determine the practical utility of our sequences, we have conducted several experiments using the classical grid-based sequence in a PRM-like planner. Our experiments indicate that dispersion and discrepancy are good measures of sample sequence quality and that grid-based sequences with high incremental quality can be acceptable replacements for random samples in roadmap planners. For comparison purposes, we tried three different sampling schemes: random sampling, multiresolution grid sampling in scanning order, and the discrepancy-optimal order discussed in this paper. The four experimental setups can be seen in Figures 2 and 3, and results in Figures 4 and 5. In these experiments, all grid sampling methods were configured to connect only to their neighbors in the current grid resolution level (although the algorithm described in Section 6 was not used). We used two different connection rules with the random samples: one uses a fixed radius, and the other attempts to connect to the k nearest neighbors. Since our grid sampling methods attempted at most $2d$ connections per new node, we set $k = 2d$.

In each experiment, we found that grid-sampling methods performed well when compared to random sampling. Discrepancy-optimal grid sampling outperformed random sampling in each experiment, and scanning-order grid sampling performed reasonably well. However, we believe that in general, scanning-order grid sampling will not perform as well as random sampling, because its incremental quality is so low. On the other hand, discrepancy-optimal grid sampling sequences have good incremental quality and we would therefore expect

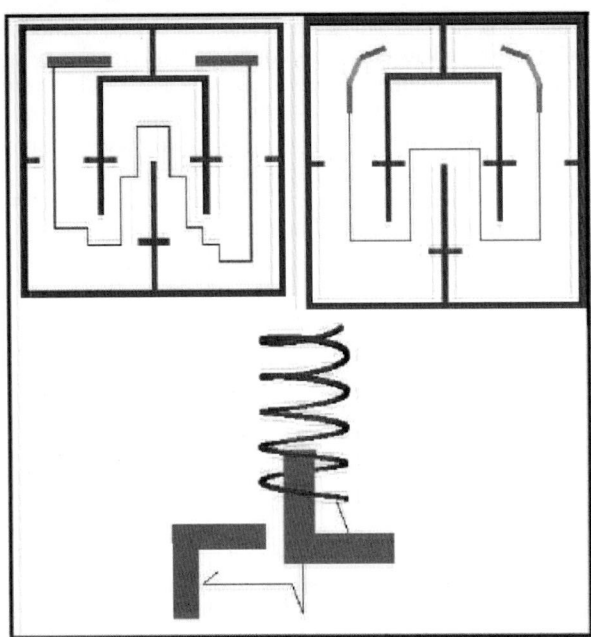

Figure 2: Preliminary experiments: top row, from the left: moving a rigid bar through a maze ($d = 3$), moving a rigid chain through a maze ($d = 5$); bottom row, removing an L-shaped robot from a spring ($d = 6$).

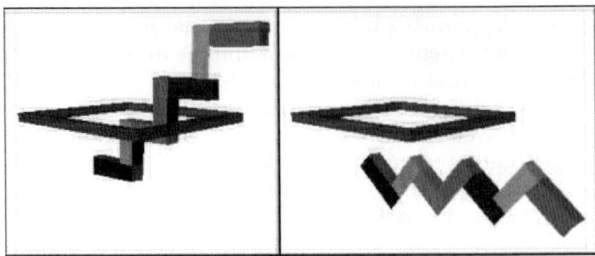

Figure 3: Moving a robot arm with a fixed base through a rectangle ($d = 6$). On the left is the initial state, on the right the goal state.

Prob.	Dim	R_Rad	R_KNear	Scan	Opt
Bar	3	440.52	922.44	684	877
Links	5	6674.04	2404.84	6785	4854
Elbow	6	4399.58	4518.04	3168	2413
Arm	6	148.58	1040.13	2704	1652

Figure 4: Comparisons of the number of nodes used in the experiments. R_Rad uses random samples and a fixed connection radius, R_KNear uses random samples and attempts k-nearest connections, Scan uses multiresolution grid samples in scanning order, and Opt uses the sequences introduced in this paper. Random sampling sequences are averaged over 50 trials.

them to perform well across a broad range of tests. In our experiments, we did not see the running time correspond precisely to the number of nodes in the roadmap; there are several reasons for this. Most importantly, the performance of the roadmap planners can depend heavily on the connection method chosen and its implementation. In particular, a fixed-radius connection technique with a large radius may result in few nodes but a large number of connection attempts, while a k-nearest approach or grid-based connection technique may have more nodes yet attempt connections more conservatively. After a connection method has been chosen, there are still several ways one can tune parameters or optimize performance for certain problems. Other researchers have recognized the difficulty of making good experimental comparisons for PRM-style planners, and work has been done to develop a more complete experimental analysis of different techniques [6]. While this degree of thoroughness is outside the scope of this paper, we believe that our experiments give a good estimate of expected performance.

8 Conclusions and Future Work

In conclusion, we have presented a new sample sequence, which satisfies all of the desirable criteria explained in Section 1 (uniformity, lattice structure, and incremental quality), and which is an arbitrary-dimensional generalization of the van der Corput sequence. As a low-discrepancy, low-dispersion sequence, it provides good coverage of the sample space; as a lattice, it has implicit neighborhood structure, which can be exploited in planning algorithms; and having incremental quality, it provides good coverage if terminated after any sample and can be easily interchanged with other incremental sampling techniques. These properties suggest that this sequence will be of benefit to the motion planning community. In particular, we believe that this sequence is a useful replacement for random sampling in PRM-style planners.

There are several directions for future work. In the previous section, we expressed the desire to do compre-

Prob.	Dim	R_Rad	R_KNear	Scan	Opt
Bar	3	1.98	1.81	1.08	1.48
Links	5	373.26	844.58	438.00	219.10
Elbow	6	71.11	777.08	63.20	29.56
Arm	6	44.28	69.93	33.91	13.94

Figure 5: Comparisons of the construction times (in seconds) corresponding to the results of the previous figure. The experiments were implemented in Gnu C++ on a 2.0GHz PC running Linux.

hensive experimental analysis; as part of this, we would like to determine limits on the dimensions for which this sequence is useful, and to gain insight into the relative merits of different types of grid sampling methods. Second, we would like to discover a more elegant way to describe and generate the sequence (such as the bit-reversal description appropriate for the van der Corput and Halton sequences); currently, we use a less-appealing recursive scheme based on an explicitly-calculated ordering for the first resolution level. Third, we would like to investigate the possibility of using other sets of rectangles for discrepancy calculations. Fourth, we have already mentioned that the extension to the Sukharev grid is fairly straightforward; similarly, an extension to an arbitrary lattice is not difficult. We would like to implement and test both of these extensions. Finally, we plan to continue to develop software generating and utilizing this sample sequence for use in our own planners, and to make the software available for use by the community (the latest versions of this software are available at http://msl.cs.uiuc.edu/).

Acknowledgments This work was funded in part by NSF Awards 9875304, 0118146, and 0208891. We also thank Robert Bohlin, Michael Branicky, Bruce Donald, Mike Erdmann, and Lydia Kavraki for helpful recent discussions.

References

[1] N. M. Amato and Y. Wu. A randomized roadmap method for path and manipulation planning. In *IEEE Int. Conf. Robot. & Autom.*, pages 113–120, 1996.

[2] R. Bohlin. Path planning in practice; lazy evaluation on a multi-resolution grid. In *IEEE/RSJ Int. Conf. on Intelligent Robots & Systems*, 2001.

[3] R. Bohlin and L. Kavraki. Path planning using lazy prm. In *IEEE Int. Conf. Robot. & Autom.*, 2000.

[4] V. Boor, N. H. Overmars, and A. F. van der Stappen. The gaussian sampling strategy for probabilistic roadmap planners. In *IEEE Int. Conf. Robot. & Autom.*, pages 1018–1023, 1999.

[5] M. Branicky, S. M. LaValle, K. Olsen, and L. Yang. Quasi-randomized path planning. In *Proc. IEEE Int'l Conf. on Robotics and Automation*, pages 1481–1487, 2001.

[6] R. Geraerts and M. H. Overmars. A comparative study of probabilistic roadmap planners. In *Proc. Workshop on the Algorithmic Foundations of Robotics (to appear)*, December 2002.

[7] J. H. Halton. On the efficiency of certain quasi-random sequences of points in evaluating multi-dimensional integrals. *Numer. Math.*, 2:84–90, 1960.

[8] C. Holleman and L. E. Kavraki. A framework for using the workspace medial axis in PRM planners. In *IEEE Int. Conf. Robot. & Autom.*, pages 1408–1413, 2000.

[9] Y. K. Hwang and N. Ahuja. Gross motion planning– A survey. *ACM Computing Surveys*, 24(3):219–291, September 1992.

[10] L. E. Kavraki, P. Svestka, J.-C. Latombe, and M. H. Overmars. Probabilistic roadmaps for path planning in high-dimensional configuration spaces. *IEEE Trans. Robot. & Autom.*, 12(4):566–580, June 1996.

[11] S. M. LaValle and M. S. Branicky. On the relationship between classical grid search and probabilistic roadmaps. In *Proc. Workshop on the Algorithmic Foundations of Robotics (to appear)*, December 2002.

[12] J. Matousek. *Geometric Discrepancy.* Springer-Verlag, Berlin, 1999.

[13] H. Niederreiter. *Random Number Generation and Quasi-Monte-Carlo Methods.* Society for Industrial and Applied Mathematics, Philadelphia, USA, 1992.

[14] G. Sánchez and J.-C. Latombe. A single-query bi-directional probabilistic roadmap planner with lazy collision checking. In *Int. Symp. Robotics Research*, 2001.

[15] T. Simeon, J.-P. Laumond., and C. Nissoux. Visibility based probabilistic roadmaps for motion planning. *Advanced Robotics Journal*, 14(6), 2000.

[16] A. G. Sukharev. Optimal strategies of the search for an extremum. *U.S.S.R. Computational Mathematics and Mathematical Physics*, 11(4), 1971. Translated from Russian, *Zh. Vychisl. Mat. i Mat. Fiz.*, 11, 4, 910-924, 1971.

[17] J. G. van der Corput. Verteilungsfunktionen I. *Akad. Wetensch.*, 38:813–821, 1935.

[18] H. Weyl. Über die Gleichverteilung von Zahlen mod Eins. *Math. Ann.*, 77:313–352, 1916.

[19] S. A. Wilmarth, N. M. Amato, and P. F. Stiller. Maprm: A probabilistic roadmap planner with sampling on the medial axis of the free space. In *IEEE Int. Conf. Robot. & Autom.*, pages 1024–1031, 1999.

[20] Y. Yu and K. Gupta. On sensor-based roadmap: A framework for motion planning for a manipulator arm in unknown environments. In *IEEE/RSJ Int. Conf. on Intelligent Robots & Systems*, pages 1919–1924, 1998.

Online Motion Planning Using Incremental Construction of Medial Axis

Ellips Masehian	M.R. Amin-Naseri	S. Esmaeilzadeh Khadem
masehian@modares.ac.ir	amin_nas@modares.ac.ir	khadem@modares.ac.ir
Industrial Engineering Dep.	Industrial Engineering Dep.	Mechanical Engineering Dep.
Tarbiat Modares University	Tarbiat Modares University	Tarbiat Modares University
Tehran, 14115-143	Tehran, 14115-143	Tehran, 14115-143
IRAN	IRAN	IRAN

Abstract — This paper deals with the online path planning of mobile robots. We first suggest a systematic method to incrementally construct the *Medial Axis* of the workspace. This is done by using sensor information for land-marking the nodes of medial axis, which will guide the robot to explore the unknown environment thoroughly. Next, this approach is implemented in an online motion planning algorithm. This method can be generalized to higher spaces, and uses only the line of sight information of sonar sensors. It is much simpler than HGVG method, and is *complete*. The simulations showed good results for different environments.

I. INTRODUCTION

Collision detection is the most important factor of path planning [8], [10]. A planner should guarantee a collision-free movement of the robot, otherwise the system will fail to function properly due to a breakdown of the hardware. This must be done automatically by the motion planner.

When no representation of the surrounding is available, a map of the environment has to be built incrementally. The motion planning based on information collected from its sensors is called *online* motion planning, or *sensor-based* motion planning. Hence, the mobile robot faces three fundamental questions, which are "Where am I?" "Where am I going?" and "How can I get there?". The first question, that of position estimation, is commonly referred to as *localization*. A naive approach to robot localization is to use odometers or accelerometers to measure the displacements of the robot. This approach, known as *dead reckoning*, is subject to errors due to external factors beyond the robot's control, such as wheel slippage, or collisions. More importantly, dead reckoning errors increase without bound unless the robot employs sensor feedback in order to recalibrate its position estimate.

A number of works for online motion planning exist in the literature. Lumelsky presented algorithms for a point robot to move from a source point to a destination point, using touch sensing in a planar terrain populated with arbitrary shaped obstacles. Cox and Yap developed algorithms to navigate a rod to a destination position in planar polygonal terrains. A good survey of these works on online path planning is provided in [13].

Specifically, some works take advantage of the maximum clearance property of the Voronoi diagram, and attempt to build the Voronoi diagram iteratively. [14] presents proofs for four basic properties of Voronoi diagrams: 1) finiteness, 2) connectivity, 3) local constructability, and 4) terrain visibility. Then they suggest an algorithm for the navigation of a circular robot in unknown terrains by iteratively visiting the Voronoi vertices. Rao then extended these results to generalized polygons in plane [15].

In [6], Choset has introduced a new concept, the *Hierarchical Generalized Voronoi Graph*, which takes advantage of some "bridge" edges (called GVG^2) to maintain the connectivity of the Generalized Voronoi Graph in higher dimensions than planar workspace. They take advantage of numerical continuation techniques to construct Generalized Voronoi Graph incrementally, relying on information acquired from ultrasonic sensors embedded in the periphery of a polygonal mobile robot. Since our work has points in common with this work, we will discuss this method later.

The next subsections introduce Voronoi diagrams and Medial axis transform, as foundations of our approach.

A. Voronoi Diagrams

Voronoi diagram is defined as the set of points that are equidistant from two or more Object features. The Voronoi diagram partitions the space into regions, where each region contains one feature. For each point in a region, this feature is the closest feature to the point than any other feature.

The Voronoi diagram has $O(n)$ edges and can be efficiently constructed in $\Omega(n \log n)$ time, where n is the number of features. There are different methods to compute Voronoi diagrams, mainly discussed in [1].

Another advantage of Voronoi methods is the fact that the object's initial connectedness is directly transferred to the diagram, whereas other methods have to restore connectedness artificially in a post-processing step [16].

In an $\Re^k$ space, the *k-equidistant face* is the set of points equidistant to objects $C_{i_1}, ..., C_{i_k}$ such that each point is closer to objects $C_{i_1}, ..., C_{i_k}$ than any other object. The *Generalized Voronoi Graph* (GVG) is the collection of *m*-equidistant faces (i.e. generalized Voronoi edges) and *m*+1-equidistant faces (i.e. generalized Voronoi vertices, or, *meet points*). The *Generalized Voronoi Diagram* (GVD) (or *Medial Axis Transform*) is the locus of points equidistant to *two* obstacles, whereas the GVG is the locus of points equidistant to *m* obstacles. Therefore, in $\Re^m$, the GVD is *m*–1-dimentional, and the GVG one-dimensional. In planar case, the GVG and GVD coincide.

B. Medial Axis Transform

As noted before, the Generalized Voronoi Diagram is also knows as *Medial Axis Transform* (*MAT*). This concept has first appeared in the literature in 1967, when H. Blum introduced the notion of a *skeleton* in his paper about MAT [2]. There he compared the *symmetric* or medial axis transform with a grass fire phenomenon, where the fire on the borders of a grass field advances toward the center. The fire fronts will meet and quench in some points which form the medial axis. Blum showed that these points are the centers of *Maximal Inscribed Discs (MID)*. To mathematically express the MID in 2D, we need to define some terms.

Definition 1. Let $\mathcal{W}$ stand for the workspace and $\mathcal{C}$ its configuration space. Then $\mathcal{C}_{free}$ represents the free configuration space, and $\mathcal{C}_{obs}$ denotes for the Cspace occupied by obstacles. Let the set of all possible distance values between any two elements in the $\mathcal{C}_{free}$ be called $\mathcal{D}$:

$$\mathcal{D} := \{\|x-y\| \mid x, y \in \mathcal{C}\} \quad (1)$$

The *Distance Transform* $DT : \mathcal{C}_{free} \rightarrow \mathcal{D}$ assigns to every $\mathcal{C}_{free}$ element the minimal distance to the $\mathcal{C}_{obs}$:

$$DT(x \in \mathcal{C}_{free}) := \min\{\|x-y\| \mid y \in \mathcal{C}_{obs}\}, \quad (2)$$

where $\|\cdot\|$ is some arbitrary metric. The *Distance Map* is the set of all $\mathcal{C}_{free}$ elements along with their associated distance values:

$$DM(\mathcal{C}_{free}) := \{x, DT(x) \mid x \in \mathcal{C}_{free}\}. \quad (3)$$

Definition 2. Since no boundary point is closer to x than $DT(x)$, every element $(x, DT(x))$ of the distance map defines the *Locally Maximal Disc* centered around x:

$$LMD(x) := \{y \mid \|x-y\| < DT(x)\} \quad (4)$$

describing the disc with maximal radius from among the values in $\mathcal{D}$ and centered around x which is completely contained in $\mathcal{C}_{free}$.

Definition 3. A *Maximal Inscribed Disc* (*MID*) is a locally maximal disc which is not completely contained in any other disc. The set of maximal inscribed discs in $\mathcal{C}_{free}$ is thus (5)

$$MID(\mathcal{C}_{free}) := \{LMD(x \in \mathcal{C}_{free}) \mid \forall y \in N(x) : LMD(x) \not\subset LMD(y)\}$$

where $N(x)$ denotes the neighborhood of x. A MID touches at least two boundary points of $\mathcal{C}_{free}$.

Definition 4. The locus of the centers of maximal inscribed discs comprises the *Medial Axis*, and the transformation of an object to its medial axis is called *Medial Axis Transform* (*MAT*).

Fig. 1 shows the Maximal Inscribed Discs and Medial Axis of an L-shaped environment. The Medial Axis is generally constructed in $O(n\log n)$ time [3].

It is important to note that Blum's concept can be extended to three-dimensional space in a straightforward way. The centers of *Maximal Inscribed Balls* (*MIB*) in a 3D object form a 2D medial axis, also called *Medial Surface*.

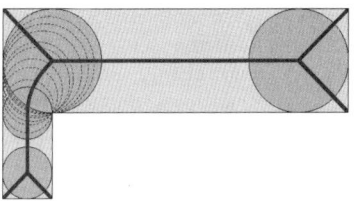

Fig. 1– *Medial axis* (middle line) and some *Maximal Inscribed Discs* for a simple 2D L-shape. The axis is a piecewise quadratic curve representing the local symmetry axes. (From [16]).

The next section of this paper proposes a new method for incremental construction of medial axis, and an online path planning is suggested based on that in section III. The discussions and conclusion come in sections IV and V.

II. INCREMENTAL CONSTRUCTION OF MEDIAL AXIS

Sensor based planning by incremental construction of medial axis (or GVG) encompasses three phases: 1) connecting the start point to GVG, 2) navigating through GVG roadmap and 3) constructing a path to the vicinity of the goal. The properties associated with each phase are called *accessibility*, *connectivity* and *departability* respectively, and are essential for a roadmap regarded for path planning.

Since our work has similarities to the incremental construction of Generalized Voronoi Diagram studied in [6] and [7], we first deal with their work.

The importance of their work lies in its completeness, and its applicability to higher dimensions than planar. The structure of their algorithm depends extensively to the mechanism of the robot in hand, and is tailored for discrete information acquired from sensor readings.

The main procedure for incrementally building the GVG edge involves mathematical computations, that is, numerical continuation techniques, as well as a need for implementing a correction step through Newton's recursive correction function, again a mathematical calculation tailored for meet points considered $m+1$-equidistant faces. Although this approach is precise, the usually limited number of sensors and their incomplete perception of the world overshadow this advantage and force the robot to 'guess' its next direction, as it is obvious in fig. 2(b).

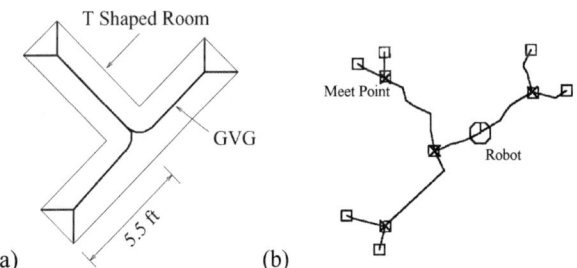

Fig. 2 – (a) The actual environment. (b) GVG computed by method proposed in [4].

The meet points (i.e. Voronoi vertices) are perceived by a comparative analysis of different sensor readings [5]; that is, by watching for an abrupt change in the direction of the (negated) gradients to the m closest obstacles (fig. 3(a)). They attribute a *meet point* as an $m+1$-equidistant point in $\Re^m$ space (e.g. 3 in planar case) and establish their calculations on this non-generic assumption. However this is not the case with many situations where more than $m+1$ Voronoi edges conjoin at a single meet point (fig 3(b)), or there are inaccuracies in defining the borders of obstacles.

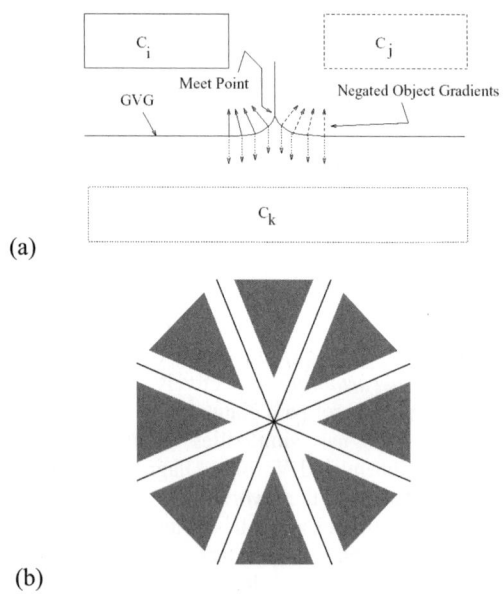

(a)

(b)

Fig. 3 – (a) Meet point detection according to [6]. (b) More than 3 Voronoi edges may coincide in a single meet point.

Moreover, they did not propose a specific method to determine the threshold beyond which the gradient variations are considered "meaningful". Also, it is not clear which gradient must be traced to identify an 'abrupt' change in direction. For instance, in the fig. 4, the robot tracing an edge will find the same gradients in meet point as on edge, and will not perceive a gradual or abrupt change in the vicinity of the meet point.

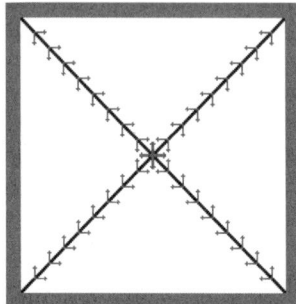

Fig. 4 – Voronoi edges and tangent vectors induced by surrounding obstacles.

This approach renders numerous problems like dead reckoning error, localization error, sharp corners problem, *weak meet points* and problems due to hyper-symmetrical environments, which are addressed in [9], [11] and [12]. Also they have not proposed a systematic way to visit meet points, i.e. they explore untraversed edges in no particular order.

We have tried to resolve these problems by adopting a *distributed* (discretized) representation of workspace. This representation enables the planner to reliably find and follow the direction of Voronoi edge, without performing complex computations for edge tracing and correction procedures. Also, taking advantage of the properties of Maximal Inscribed Disks (and their centers), the meet point detection is a simpler task.

Now we can propose our new algorithm for incremental construction of Generalized Voronoi Diagram based on the medial axis notation presented in sec. I. The outline of the algorithm is as follows:

- Starting from an arbitrary point, move to the GVG in the direction of negated gradient of MID $\cap$ OB (OB is Obstacle Boundary)
- From the first accessed point on GVG, extend the constructed graph in the direction of greatest MID at its neighborhood. Note that unlike the tangent step and correction method in [6], we explore the *neighborhood* of the recent point based on the resolution of discretization.
- When a meet point or an obstacle boundary point is reached, add this meet point to the available meet points, T. extend the GVG from a meet point whose MID radius is maximum in the meet point matrix T (this is different from HGVG).
- Repeat the above steps until no meet point is remained with unexplored edges, i.e. $T = \emptyset$.

The result of applying this algorithm through an experiment is depicted in figure 5. The dashed circle is the Maximal Inscribed Disc (MID), and the black solid line is the GVG. Obstacles are shown gray.

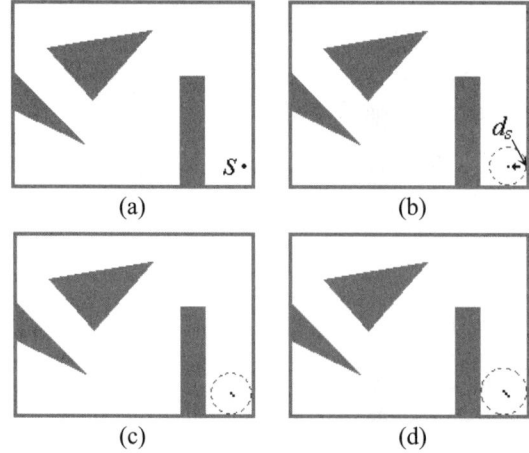

Fig. 5 – Incremental construction of GVD.

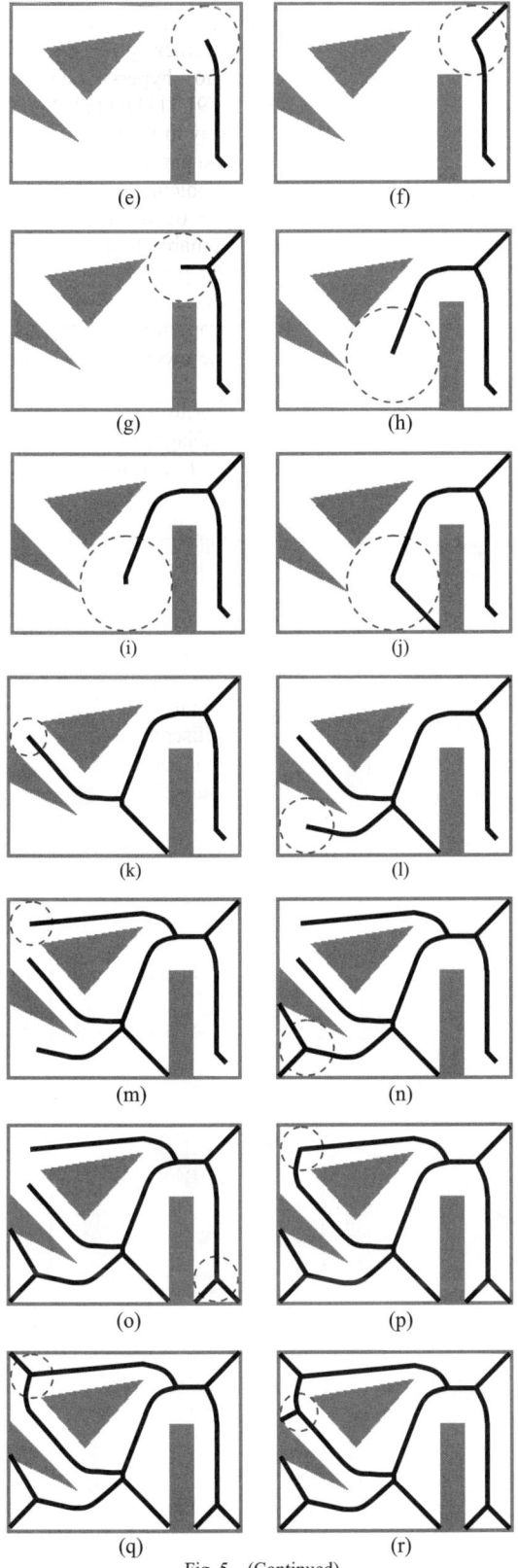

Fig. 5 – (Continued)

Figure 5(a) shows the start point. The first accessed point and its MID is drawn in fig.(b), which is the result of step 1 (of the algorithm explained in fig. 6). Fig. 5(c) and 5(d) show the result of step 2. These new points are those in neighborhood of previous point and have $j(x) \geq 2$. Note that $j(x)$ implicitly denotes the number of edges emanated from point x. The step 2 is recursively executed until a new meet point is reached (case 3.2) (fig.5(e)). The meet points are then sorted in T according to their MID radius (step 4). The algorithm picks up the last reached meet point, which has the greatest MID radius (step 5), and explores an arbitrarily selected edge through step 2 (fig.5(f)). Upon reaching a boundary point (case 3.1), the first element of T is checked for its number of $j(x) = \{\text{MID}(y) \cap OB\}$, which yields 2, hence case 5.2. So the step 2 is repeated and traces a new edge. This procedure is repeated until neither the first accessed point, nor a meet point remains unexplored.

Details of the algorithm are now presented in fig. 6.

STEP 1. Construct a Maximal Inscribed Disk (MID) centered on an arbitrary start point s, preferably near a corner. Find the intersection d_s of the MID of s (i.e. MID(s)) with the Obstacle Boundary, OB. Extend the line connecting s and d_s until it reaches a GVG point. Essentially, the MID of this GVG point intersects the OB in j points ($j \geq 2$). Mark this point x_i ($i = 1$), as the first accessed GVG point, and create the strings GVG and T, such that:

$GVG = \{x_i\}$, $T = \{(x_i, |j(x_i)|)\}$, $j(x_i) = \{\text{MID}(x_i) \cap OB\}$.

The GVG is the set of points on Generalized Voronoi Graph, $j(x_i)$ is the set of points on OB which are also located on Maximal Inscribed Disc centered on x_i, T is the set of pairs of meet points and the number of their respective $j(x_i)$, and $|\bullet|$ means the cardinality of the set.

STEP 2. For the first element (x_i) of T, search among its neighboring points (which are not on GVG) to find the point y with greatest MID radius. Mathematically, find y such that:

$\{y \,|\, \forall z \in N(x_i) \wedge \exists y \in N(x_i), r_{\text{MID}(y)} \geq r_{\text{MID}(z)}, j(y) \geq 2, y \notin GVG\}$

where $N(x_i)$ is the neighborhood of x_i, and $r_{\text{MID}(z)}$ is the radius of the maximal inscribed disk of point z. If y is not unique, select an arbitrary point among them. This step adds a new point to GVG. So set $GVG = GVG \cup \{y\}$.

STEP 3. If $j(y) = 2$ it means an edge is being traced. Repeat the step 2 until:

3.1. A $y \in N(x_i)$ is found such that $y \in OB$. This means that the edge has collided with the obstacle boundary. Go to step 5.

3.2. A $y \in N(x_i)$ is found such that $j(y) \geq 3$. This means that the edge has reached a meet point. Go to step 4.

Fig. 6 – Algorithm for incremental construction of Medial Axis.

STEP 4. Update the set $T = T \cup \{(y, |j(y)|)\}$, and sort its elements according to a descending order of the radius of MIDs for all points (i.e. $r_{MID(y_j)}$). Note that T contains the very beginning point and all visited meet points, so for $\forall x \in T, |j(x)| \geq 2$.

STEP 5. For the first element of T, e.g. $(y_1, |j(y_1)|)$, (which is either a meet point, or the first accessed point, and has the biggest disc radius), check:
5.1. If $|j(y)| > 2$, then set $j(y) = j(y) - 1$. This operation marks off the explored branches of the meet point.
5.2. If $|j(y)| = 2$, then set $j(y) = j(y) - 1$, $T = T - \{(y, j(y))\}$. This operation indicates that a meet point is explored fully.

STEP 6. If $T \neq \emptyset$, then there are still unexplored meet points, so go to step 2, else STOP.

Fig. 6 – (Continued).

Since there are $O(n)$ edges and vertices in GVG that are traversed iteratively, the proposed algorithm is supposed to build the GVG in $O(n)$ time for discretized environments, which is better than the known $O(n\log n)$. Also because the edges emanated from meet points are explored in the order of their biggest MID radii (which may not be adjacent or neighboring), there may be 'shifts' in exploration areas, and so the algorithm is devised to run in offline mode. Therefore, to implement this method in online contexts and handle real situations, the above algorithm is modified in the next section.

III. ONLINE PATH PLANNING

Based on the GVG construction algorithm introduced in previous section, we will present a new online path planning method. But for the sake of conciseness, we avoid explaining the whole procedure, which is predominantly the same algorithm explained in section II. Only the modifications are discussed.

In incremental construction of GVG, two items are selected arbitrarily: a) The selection of next edge (between $j(y)$ edges) to be explored, and b) The selection of next meet point to be explored, when two or more meet points have greatest MID radii.

For online path planning, these selective behaviors are directed toward the construction of a graph that converges more rapidly to the goal point. The principle is that whenever more than one alternative are available, the next point is one that is closest to the goal point. This can be attained by a simple distance check for candidate points. This will result in a graph (subset of the total GVG) that extends toward the goal greedily. So the probability of finding the goal increases.

It may happen that while following the more promising edge, the robot encounters obstacle boundaries, or local minima. Here, to avoid collision, we define a *safety radius*; that is, the robot considers an obstacle *bumping* if the distance to it is less than the safety radius. In such cases, the robot backtracks the last traversed edge, marks it "dead-end", and explores the other edge(s) of last meet point visited. If there are no more unexplored edges emanating from that meet point, the robot selects the penultimate meet point and explores its edges. This backtracking strategy can be attributed as *reverse depth-first search*.

The main steps of the online path planning algorithm remains the same as incremental GVG construction algorithm, except those mentioned above; that is, the edge and meet point selection is improved, and a backtracking step is added.

The termination condition is of course the access to goal point. This task requires an additional step, named *departure*. If the goal point is located on the roadmap, this step is redundant; otherwise, the robot must be able to leave the roadmap towards the goal. If the goal is within the robot's line of sight, a straight line connects GVG to goal. This task is sometimes facilitated by constructing a *virtual* GVG, which is a bounded set of voronoi edges surrounding the goal [4]. The virtual GVG is generally connected to the main GVG; therefore goal can be connected to the total GVG by a straight line. We applied our online algorithm to the example in fig. 7. Here the robot did not encounter dead-ends.

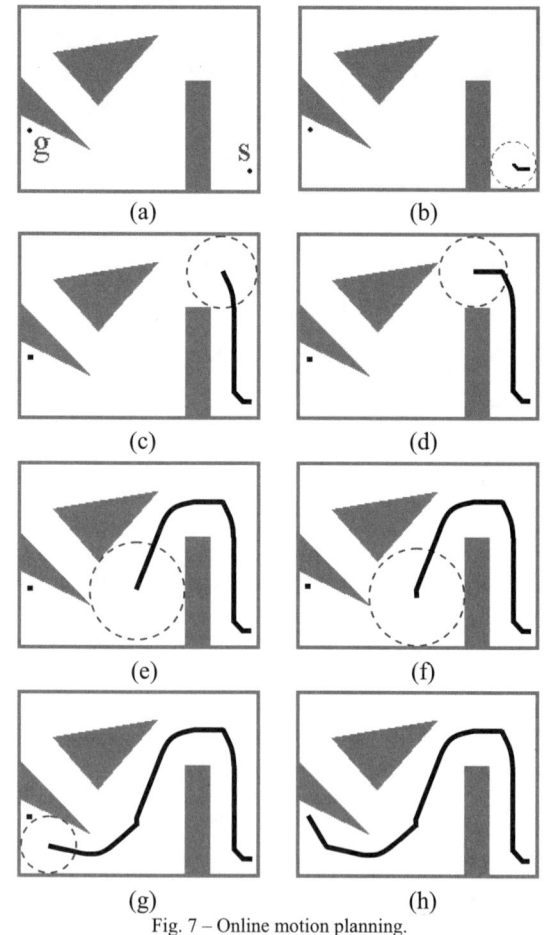

Fig. 7 – Online motion planning.

IV. DISCUSSION

For representing the workspace (or C-space) we utilize a grid-based representation. This yields facilities in approximating errors in sonar sensing readings, where each pixel is assigned a value indicating the likelihood that it overlaps an obstacle. Although this method may require memory storage, it obviates many computational techniques for reducing localization errors. Especially for two-dimensional workspaces this is a rational choice. This representation also enables a simple definition of obstacles, without limiting their convexity, or shape.

Regarding that all the visited meet points and the number of unexplored edges emanating from them are stored in the matrix T, the meet point locations are constantly being matched. Therefore the localization errors are dynamically corrected and do not accumulate.

An interesting property of our online planning method is its *completeness*, i.e. it has the ability to find the path to goal point if it is attainable, or report no such a path exists otherwise. Since the method utilizes a *retraction* roadmap of the environment (GVG), and the GVG is connected itself [14], any path that contains the roadmap plus links to start and goal points is also connected. Therefore our method will reach the goal if it lies in C_{free}. To this point, the algorithm is *exact*. Now the question is: Will the algorithm report that a path to the goal does not exist, if it is the case? To answer, recall the step 6 of incremental GVG construction algorithm: "If $T \neq \varnothing$ go to step 2 else STOP". For the online algorithm, if the situation $T = \varnothing$ is materialized *before* the goal is reached, then the algorithm will terminate and report appropriately.

Also, taking advantage of the GVG roadmap in high dimensional spaces, and extendibility of the MID to higher spaces (e.g. Maximal Inscribed Balls, MIB), the proposed method can be generalized to higher spaces, which is among our future research.

To make the algorithm more efficient, we also integrated the Potential Fields approach into this method 10]. Among the alternative edges or meet-points, the one with least potential is selected as the next and promising edge/meet-point to be explored. The repulsive potential of obstacles will discourage the robot to move towards the vicinity of their boundary. The attractive potential of goal also facilitates the departure phase.

V. CONCLUSION

In this paper, we have proposed a new algorithm for systematically constructing the medial axis using sensory data. This concept is then utilized to develop an online path planning method, which can be used to explore unknown environments, especially labyrinthine spaces. The algorithm is fast, simple, complete, and extendable to higher spaces.

REFERENCES

[1] Aurenhammer, F. and Klein, R., Voronoi diagrams. In J. Sack and G. Urrutia, editors, *Handbook of Computational Geometry*, Elsevier Science Publishing, pages 201-290, 2000.

[2] Blum, H., A Transformation for Extracting New Descriptors of Shape, In W. Waltendunn, editor, *Models for the Perception of Speech and Visual Form*, MIT Press, 1967.

[3] Choi, H., Choi, S. and Moon, H., Mathematical Theory of Medial Axis Transform, *Pacific Journal of Mathematics*, 1997, Vol. 181, No. 1, pp. 57-88.

[4] Choset, H. and Burdick, J., Sensor Based Planning, Part II: Incremental Construction of the Generalized Voronoi Graph, *IEEE / ICRA '95*, Nagoya, Japan, 1995.

[5] Choset, H., Konukseven, I., and Burdick, J., Mobile Robot Navigation: Issues in Implementation the Generalized Voronoi Graph in the Plane, *IEEE / MFI*, Washington DC, 1996.

[6] Choset, H. and Burdick, J., Sensor-Based Exploration: The Hierarchical Generalized Voronoi Graph, *International Journal of Robotics Research*, Vol. 19(2), 2000.

[7] Choset, H., Walker, Sean, Eiamsa-Ard, K. and Burdick, J., Sensor-Based Exploration: Incremental Construction of the Hierarchical Generalized Voronoi Graph, *International Journal of Robotics Research*, Vol. 19(2), 2000.

[8] Hwang, Y.K, and Ahuja, N., Gross Motion Planning - A Survey, *ACM Computing Surveys*, 24(3), 1992.

[9] Konukseven, I. and Choset, H., Mobile Robot Navigation: Implementing the GVG in the Presence of Sharp Corners, *IROS '97*, Grenoble, France, 1997.

[10] Latombe, J.C., *Robot motion planning*, Kluwer Academic publishers, London, 1991.

[11] Nagatani, K., Choset, H., and Thrun, S., Towards Exact Location without Explicit Localization, *IEEE / ICRA '98*, Lueven, Belgium, 1998.

[12] Nagatani, K. and Choset, H., Toward Robust Sensor Based Exploration by Constructing Reduced Generalized Voronoi Graph, *IROS '99*, Kyongju, Korea 1999.

[13] Rao, N.S.V., Kareti, S., Shi, W. and Iyenagar, S.S., *Robot Navigation in Unknown Terrains: Introductory Survey of Non-Heuristic Algorithms*, Oak Ridge National Laboratory Technical Report, ORNL/TM-12410, July 1993.

[14] Rao, N.S.V., Stolzfus, N. and Iyengar,S.S., A "Retraction" method for Learned Navigation in Unknown Terrains for a Circular Robot, *IEEE Transaction on Robotics and Automation*, 7(5), pp. 699-707, 1991.

[15] Rao, N.S.V., Robot navigation in unknown generalized polygonal terrains using vision sensors, *IEEE Transactions on Systems, Man and Cybernetics*, vol. 25, no. 6, 1995, pp. 947-962.

[16] Schirmacher, Hartmut, *Extracting Graphs from Three Dimensional Neuron Datesets*, Diploma Thesis, Inst. of Comp. Sci., Friedrich-Alexander Uni. Erlangen-Nürnberg, 1998.

On Addressing the Run-Cost Variance in Randomized Motion Planners

Pekka Isto
*Helsinki University of Technology,
Espoo, Finland
evp@cs.hut.fi*

Martti Mäntylä
*Helsinki Institute of Information
Technology, Helsinki, Finland
Martti.Mantyla@hiit.fi*

Juha Tuominen
*Helsinki University of Technology,
Espoo, Finland
jtu@cs.hut.fi*

Abstract

The decades of research in motion planning have resulted in numerous algorithms. Many of the most successful algorithms are randomized and can have widely differing run-times for the same problem instance from run to run. While this property is known to be undesirable from user's point of view, it has been largely ignored in past research. This paper introduces the large run-cost variance of randomized motion planners as a distinct issue to be addressed in future research. Run-cost variance is an important performance characteristic of an algorithm that should be studied together with the mean run-cost. As a positive example of possibilities for reducing the run-cost variance of a randomized motion planner, simple heuristic techniques are introduced and investigated empirically.

1. Introduction

Motion planning capability is an essential property of any autonomous robotic system [1]. When formulated as the classical "mover's" problem, it can be shown that the problem is PSPACE-hard [2]. The most basic variation of the problem assumes that all obstacles are known and static during the planning and execution of the motion for the movable device. A large number of motion planning algorithms have been presented for this and other variations of the problem [1][2][3]. Due to the complexity of the problem, practical planners must be based on heuristic and incomplete algorithms.

The complexity of the problem guarantees that for any algorithm there exists a problem instance that exhibits the intractable worst-case behavior. Heuristics can take advantage of assumed regularities in the problem, but when the heuristics is known, the construction of a "deceptive" problem instance exhibiting worst-case behavior is usually quite straightforward. Relaxing the completeness requirement has allowed the development of randomized motion planners that are complete only in probabilistic sense. Randomized planners use either some randomized search procedure, e.g. [4][5], or random sampling of the configuration space (*cspace*) for subgoals in combination of local search between the subgoals, e.g. [6][7][8][9].

The current motion planners that rely on configuration space samples for subgoals are very sensitive to the selection of the particular sampling sequence, be it a particular segment of a pseudo-random number sequence or a particular type of a quasi-random number sequence. Structured sampling techniques, e.g. [10][11][12], usually rely on assumptions on the task structure that are easily violated [13]. Since randomized algorithms tend to break the task structure, they are usually robust against "deceptive" inputs. However, this robustness comes at the expense of being able to reliably estimate the length of a particular run of the algorithm due to run-cost variance.

This paper introduces the large run-cost variance of randomized motion planners as a distinct issue to be addressed by the research community in the future work. Variance should be accounted for in the research methodology, especially if it is empirical [14]. If a new or variant randomized motion planner is presented, appropriate descriptive statistics should be presented to describe the run-cost variance of the proposed motion planner. Users are known to find run-cost variance distressing [15], therefore motion planners with low variance are preferred. The size of variance is also directly involved in comparisons between various planners or components of planners, since large variance can invalidate the statistical significance of an observed difference in average performance.

Furthermore, in order to demonstrate prospects for addressing the variance problem, simple techniques with variance reduction effect are presented. The techniques are heuristic in nature and their effects are verified empirically. The techniques are applicable to motion planners that use samples in the configuration space as subgoals to be connected with a local search procedure. In essence, they make the local search procedure more powerful when it is observed that a large number of samples or local planner calls are needed for the solution and thus the particular sampling sequence cannot be relied to be successful for the task at hand. Since the search procedure used here is deterministic, this can be seen as shifting the balance of computation away from the randomized sampling procedure to the deterministic search procedure.

The next section presents a brief overview of randomized motion planners. Section 3 introduces PRM motion planner variants with heuristics that improve performance both in the expected run-cost and the run-cost variance. Section 4 describes the empirical methodology used in this paper and section 5 presents the results. Finally, section 6 presents the conclusions and issues for future work.

2. Previous Work

Randomization techniques have been introduced into motion planning algorithms in order to avoid expending large amount of computation when a deterministic planner exhibits the worst-case behavior. Classical cell decomposition methods construct a *cspace* representation for the entire *cspace* although only a part may be necessary for the construction of the solution. Heuristic strategies can be used to restrict the construction of the *cspace* representation to promising areas, e.g. [16][17]. However, when the heuristics fails and guides the construction to some unimportant area, the complexity of the problem may again reveal itself with large costs in terms of time and space requirements. This renders planners utilizing straightforward best-first search strategies useless for practical problems, since the planners tend to get trapped examining dead-end regions of the *cspace*. This has been long known as the problem of local minima [1].

The same problem is also present when the guiding heuristics takes the form of a potential field over the workspace or the *cspace*. A theoretical result establishes that all global navigation functions have saddle points [18], and therefore, simple gradient procedures can terminate at some configuration other than the intended goal configuration of the problem. Again, the question becomes what to do once the (deterministic) primary technique ceases to make progress.

Randomization was introduced in order to solve the local minima problem. A randomized search procedure can be obtained by combining a heuristic search strategy with randomly generated subgoal configurations. The heuristic strategy is first attempted between the original start and the original goal configuration. If the heuristics guides the search to a local minimum, a random subgoal is generated and the heuristic strategy is attempted via the subgoal configuration. As more subgoals are generated and path segments are generated between them with the heuristic strategy, they will form a graph that approximates the connectivity of the *cspace* [6][19]. The heuristic strategy can be as simple as a straight-line interpolation between the subgoals [2]. This general idea has evolved into a randomized version of the classical roadmap approach [20][10][21] and it has been a topic for extensive research under the name of probabilistic roadmap (PRM) planning. Other similar motion planners use more complex heuristics for search and subgoal generation, e.g. [9][7].

An early and influential motion planner called Randomized Path Planner (RPP) combined the potential field approach with a random walk procedure to escape from the local minima [4]. Several variations of the planner were introduced, but it had difficulties in solving problems that required long walks against the potential [22]. A special random exploration procedure was later introduced to improve the efficiency of randomized search [5].

Recently, the use of quasi-random sampling has been proposed as an alternative for the more usual pseudo-random sampling [23]. While eliminating the variance caused by different seeds for the random number generator, it remains brittle with respect to small but critical changes in the workspace of the problem.

Very few user experiences with applications utilizing motion planning algorithms have been reported in the literature. Experiences with a maintainability study tool based on RPP indicate that users find the variance inherent in randomized algorithms very disturbing [15]. Although the run-cost variance has been known to be a serious deficiency of the sampling based randomized motion planners [24], no methods have been presented to directly address this problem. This observation has motivated the consideration of the amount of variance as a measure of the performance of an algorithm and the investigation in the possibilities of reducing the variance.

3. Heuristic Techniques for Variance Reduction

The general idea of the heuristic strategy presented in this paper is to increase the relative effort spent in deterministic local search procedure when it is observed that the randomized sampling procedure does not succeed rapidly in producing subgoals that cover the *cspace* adequately. Two variations of this general idea are derived with a simple PRM planner using bidirectional A^* search [25] based local planner and parameterized formulas for increasing the competence of the local planner.

Although A^* search algorithm makes a (resolution) complete motion planner when combined with one of the cell decomposition methods, it has an exponential memory consumption. The memory consumption makes it practically impossible to escape deep local minima, since the algorithm can exhaust any available central memory. Therefore, it alone is not suitable for motion planning, but must be combined with other methods, here with pseudo-random sample subgoals. The approach is to use A^* search local planner on a grid approximation of the *cspace* from one sample to the other, but to discontinue A^* search when it appears to be too costly to generate the path segment with A^*. The underlying assumption is that a sequence of subgoals can be found so that A^* search succeeds in generating all the path segments from the start to target configurations within a defined limit on the search efficiency on each segment.

The essential property of a search algorithm based local planner is that it allows simple evaluation and control of the difficulty of the path segments that the local planner is capable of yielding. A measure of search efficiency can be obtained from the ratio of the size of the examined search space to the length of the best available path candidate towards the target sample. Noting that the local planning is performed in discrete rectangular grid representation of the *cspace* lets us realize the above measure with simple counting operations. Let $F(C)$ be the total number of collision-free configurations examined by the local planner until the examination of configuration C. Furthermore, let $g(C)$ be the distance from the start sample to the configuration C currently examined by the local planner. Now, an efficiency measure $O(C)$ can be defined as

$$O(C) = \frac{F(C)}{g(C)}.$$

The competence of the local planner can be controlled by setting an upper limit O_{th} for $O(C)$ and discontinuing the search in the local planner, if the limit is exceeded. Thus, the local planner is only competent to solve problems in which the search algorithm does not at any point violate the given efficiency limit by requiring more than the set upper limit of grid point examinations for each step proceeded away from the start sample along the currently best solution candidate path.

Note that $O(C)$ measure is similar to Nilsson's penetrance [25]. The difference is that penetrance is evaluated once after the (optimal) path is found, but $O(C)$ is evaluated continuously during the search and used as a control during the search.

The efficiency of a heuristic search algorithm such as A^* is highly dependant on the guiding heuristic function. The usual form of the guiding function is $f(C) = A \times g(C) + B \times h(C)$, where $g(C)$ is as above, $h(C)$ is a heuristic estimate for the cost from the currently examined configuration C to the target sample, and A and B are constants. The guiding function used here is greedy with smaller $A=3$ and larger $B=5$. Manhattan distance in the discrete *cspace* is used for both $g(C)$ and $h(C)$. The heuristic estimate $h(C)$ has an additional tie breaking in favor of configurations that repeat a motion of a joint along the solution path candidate.

The sampling strategy used here is simple pseudo-random sampling until a roadmap connecting all the seed configurations of the given test task is obtained. Candidate sample pairs for the local planner are produced by selecting for the newly generated sample up to $k=10$ closest samples from each connected component of the roadmap at the sample generation time. Euclidean distance is used as the distance metric in the selection.

An interesting problem is the question of how to determine a threshold for $O(C)$. The majority of PRM planners have static local planners. For this study, static local planners with well-defined capability can be obtained by setting a fixed upper limit o for $O(C)$.

Since the competence of the local planner can be set separately for each call, information gained during the roadmap construction can be used to determine a suitable value for $O(C)$ limit with the goal of reducing the overall roadmap construction cost. Two such adaptation strategies are proposed here and studied experimentally in the following sections. The first strategy involves increasing the $O(C)$ limit linearly with the size of the roadmap. The intuition behind this strategy is that more difficult problems require larger roadmap and a more capable local planner to adequately capture the connectivity of the *cspace*. Furthermore, as more samples are added to the roadmap, the failure probability of the local planner decreases [26]. If the failure probability is lower, failures that are more expensive can be tolerated. This strategy has a global character in the sense that it determines a single increasing $O(C)$ limit for all the samples in the roadmap. The following parameterized formula is used to determine the threshold O_{th} for $O(C)$ during roadmap construction at a particular roadmap size of S:

$$O_{th}(s) = \frac{S}{s} \times 32. \quad (1)$$

A strategy of local character is defined by setting the O_{th} threshold separately for each sample in the roadmap. The strategy uses a measure of difficultiness of the *cspace* around a particular sample. For each sample v the fraction of successful calls of local planner is computed:

$$r_s(v) = \frac{N_s(v)+1}{N(v)+1},$$

where $N(v)$ is the total number of local planner calls with the configuration space sample v either as start or target and $N_s(v)$ is the number of calls that succeeded in producing a path segment to or from the sample v. This measure is very similar to failure ratio [8].

A value of O_{th} is computed for both start and target samples with the parameterized formula:

$$O_{th}(v,n) = 1 + \frac{n}{r_s(v)}, \quad (2)$$

and the maximum is used as the current threshold value. The intuition behind this strategy is to use an estimate of the difficultiness of the *cspace* region to control the capability of the local planner. A small r_s suggests that sample resides in a difficult region of the *cspace* and thus the local planner should be given opportunity to search the region more broadly.

Parameterized heuristics present a problem of selecting the values for the heuristic parameters. If the properties of the expected motion planning problems are known, then the parameters should of course be tuned for those problems using preliminary experiments. When tuning is not possible or desirable, then some on-line procedure can be used to select the value.

In this paper a metaplanner is used to select the values for parameters o, s and n. Perhaps unsurprisingly, the selection is done randomly from a set of reasonable values for each parameter. The metaplanner selects a parameter value uniformly from the set at the start of the execution of the PRM planner. Based on the preliminary experiments, the reasonable ranges of parameter values were determined and the sets defined to be *{2, 4, 8, 16, 32}* for o, *{300, 1000, 3000, 9000}* for s and *{0.01, 0.03, 0.1, 0.3}* for n. Thus, the planners compared in the experimental section of this paper are PRM variants with bidirectional A^* search between the samples. Planner PRM-C uses a static local planner at a competence level o selected by the metaplanner at the start of the run. PRM-G uses the global adaptation of the local planner according to the equation (1) with parameter value s selected by the metaplanner. Similarly, PRM-L uses the local adaptation strategy defined by the equation (2) and parameter n. The research question is whether the planners PRM-C, PRM-G and PRM-L differ in location or spread of their empirical run-cost distributions. The results are taken to indicate possible benefits of increasing the capability of the local A^* search during planning (PRM-G, PRM-L) against keeping it static (PRM-C).

4. On the Empirical Methodology

The run-cost of a Las Vegas -type randomized algorithm varies from run to run according to some distribution $\mathcal{D}$. Usually, deriving the true run-cost distribution of a randomized algorithm is a formidable task and no such results exists for any of the well-known motion planning algorithms. Sometimes the true distribution is sampled empirically and the results are presented in the form of a histogram [19]. More often the run-cost distribution is described by reporting estimates for the location and spread of the true distribution e.g. [9][27][28]. This implies a parametric model $\mathcal{D}(l,s)$ for the run-time distribution, where l describes the location of the distribution and s the spread of the distribution. Sample average and median are typically used for estimating the location and sample variance, standard deviation or range for estimating the spread.

Most often, however, only an estimate of the location parameter is presented by reporting the sample average of the run-cost over some number of runs. This case is unsatisfactory since it hides one of the most problematic characters of the randomized motion planning algorithms. Furthermore, it makes it impossible to evaluate the statistical significance of differences between algorithms or variations, e.g. when comparing the performance of one sampling strategy with another. An estimate of the standard deviation is needed for the most elementary test of statistical significance, namely the t-test. Of course, the outcome of the test or analysis of variance should be given as an indication of the statistical significance.

The following section presents descriptive statistics on the run-cost of the proposed PRM variants and results from the statistical tests of the significance of the differences between the variants. The test problems are two well-known benchmark problems proposed in the literature. The Hwang and Ahuja benchmark problem is a 5 degrees-of-freedom robotics motion planning problem for a SCARA-type robot [2]. The task was designed to represent a realistic but non-pathological problem for a manipulator. The task involves removing a hook from a wicket and a subsequent backtracking motion to avoid a large obstacle (see Fig. 1). No generally available geometric model for the task exists, but a difficult version of the problem was produced for this study.

The second test problem is the Alpha Puzzle benchmark problem proposed by Amato *et al.* [11]. The problem is intended to represent 6 degrees-of-freedom disassembly problems and it is designed to have a narrow passage that the movable object must pass. Several versions of the Alpha Puzzle exist with varying difficultiness. This study uses Alpha Puzzle version 1.2, which is of medium difficultiness. The original Alpha Puzzle problem involves separating the two intertwined α-shaped loops. The loops can be intertwined in two different ways with the prongs of the loops either in symmetric (first image in figure 2) or anti-symmetric (middle image in figure 2) orientations. Since the intention of this paper is to evaluate the performance of the various strategies in capturing the full connectivity of the test problems, both intertwined configurations together with a separated configuration (last image in figure 2) are inserted as seed configurations to the roadmap at the beginning of the construction.

The roadmap construction with the PRM variants is continued until a roadmap connecting all the seed configurations of the task is obtained and the number of performed collision checks is taken as the run-cost measure. The grid approximation of *cspace* for A^* search has resolution of 128 steps for each degree-of-freedom for the Hwang and Ahuja benchmark problem and 512 for

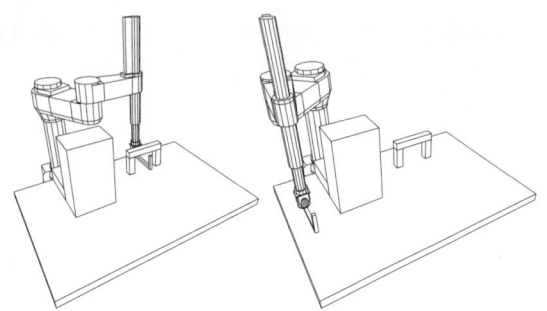

Figure 1: The seed configurations for a version of the benchmark problem proposed by Hwang and Ahuja.

Figure 2: The seed configurations for the Alpha Puzzle 1.2 benchmark task.

the Alpha Puzzle problem.

5. Empirical Results

Table 1 gives descriptive statistics for the experiments with the PRM variants. The table gives run-cost mean, standard deviation and the coefficient of variation for a sample of 240 runs. The coefficient of variation expresses standard deviation as a percentage of mean, so it can reveal if the standard deviation changes together with the mean.

As can be seen in table, PRM-G and PRM-L have considerably better performance than PRM-C both in terms of mean run-cost and the standard deviation of the run-cost for the version of the Hwang and Ahuja benchmark problem. PRM-L not only has the absolute standard deviation improved but also the coefficient of variation is smaller. This indicates that not only has the scale of the run-cost distribution changed but also its shape. Ryan-Einot-Gabriel-Welsch multiple comparison procedure [29] can declare the difference between static (PRM-C) and adaptive (PRM-G, PRM-L) planner variants statistically significant at $\alpha=0.01$ (experimentwise), but fails to detect statistically significant difference between the two proposed heuristics. Levene's test of homogeneity of variances [30] detects a very significant difference in the standard deviations (p=0.0071). For the Alpha Puzzle 1.2 problem the results are not as good. There is an improvement in standard deviation when using the proposed heuristics, but that difference is not statistically significant. Neither are the differences in the means.

The failure of the heuristics to yield statistically significant differences for the Alpha Puzzle 1.2 may be explained by the fact that it is very difficult to generate "good" subgoal samples for this task. Finding critical samples for this problem from a pseudo-random sampling sequence is a rare event and the increase in the capability of the local planner fails to make it sufficiently more frequent. It can be stated that the behavior of the planners on this test problem is determined by the "narrow passage" nature of the problem, as samples are required in a small bottleneck area in the *cspace*. Narrow passages are a well-known difficulty for all PRM type planners [31] and the proposed heuristics does not overcome this problem.

6. Conclusions And Future Work

This paper introduced the run-cost variance of randomized motion planners as a distinct research issue. In future research in randomized motion planners the run-cost variance should be considered as an important performance characteristic of the planner. Descriptive statistics and statistical tests used in this paper will provide researchers with tools to address the issue of variance and strengthen the research methodology.

As a positive example of possibilities for reducing the run-cost variance of a randomized motion planner, simple heuristic techniques were introduced and investigated empirically. A typical analysis of the run-cost means shows that the techniques can provide a statistically significant improvement. An analysis of the run-cost variance shows an additional benefit of the techniques and helps to select between two techniques that have similar expected run-cost. But like all heuristics, also the techniques presented here may fail to provide the expected benefit.

The heuristic techniques for variance reduction presented in this paper can also be used with other search procedures that have a means to control the extent of search in the local planner. Possible search procedures include randomized ones such as RPP [4] and RRT [5]. But due to the lack of a theoretical model of the phenomena, the effectiveness of the proposed techniques with other search procedures must be investigated empirically.

Acknowledgments

Alpha Puzzle was designed by Boris Yamrom, GE Corporate Research & Development Center. The model was provided by the DSMFT research group at Texas A&M University. Johannes Lehtinen provided the geometric model for the Hwang and Ahuja task. Janne Ravantti provided access to the cluster computer at the

Bamford Laboratory, University of Helsinki.

References

[1] J. C. Latombe, Robot Motion Planning, Kluwer Academic Publishers, Norwell, Mass. 1991.

[2] Y. K. Hwang, N. Ahuja, Gross Motion Planning - A Survey, ACM Computing Surveys, Vol. 24, No. 3, Sep. 1992, 219-291.

[3] K. Gupta, A. P. del Pobil (eds.), Practical Motion Planning in Robotics: Current Approaches and Future Directions, John Wiley & Sons, West Sussex, 1998.

[4] J. Barraquand, J.C. Latombe, Robot Motion Planning: A Distributed Representation Approach, Int. Journal of Robotics Research, vol. 10, no. 6, Dec. 1991, 628-649.

[5] S. M. LaValle, J. J. Kuffner, Jr., Rapidly-Exploring Random Trees: Progress and Prospects, Workshop on Algorithmic Foundations of Robotics, A K Peters, Wellesley, 2001, 293-308.

[6] B. Glavina, Solving Findpath by Combination of Goal-Directed and Randomized Search, Proc. of the 1990 IEEE Int. Conf. on Robotics and Automation, IEEE, 1990, 1718-1723.

[7] E. Mazer, J. Ahuactzin, G. Talbi, P. Bessiere, The Ariadne's Clew Algorithm, Journal of Artificial Intelligence Research, Vol. 9, 1998, 295-316.

[8] L. E. Kavraki, P. Svestka, J.C. Latombe, M. Overmars, Probabilistic Roadmaps for Path Planning in High-Dimensional Configuration Spaces, IEEE Transactions on Robotics and Automation, Vol. 12, No. 4, 1996, 566-580.

[9] P. Isto, A Two-level Search Algorithm for Motion Planning, Proc. of the 1997 IEEE Int. Conf. on Robotics and Automation, IEEE Press, 2025-2031.

[10] M. H. Overmars, P. Svestka, A Probabilistic Learning Approach to Motion Planning, Technical Report UU-CS-1994-03, Department of Computer Science, Utrecht University, The Netherlands, January 1994.

[11] N. M. Amato, O. B. Bayazit, L. K. Dale, C. Jones, D. Vallejo, OBPRM: An Obstacle-Based PRM for 3D Workspaces, Workshop on Algorithmic Foundations of Robotics, 1998, 155-168.

[12] C. Holleman, L. Kavraki, A Framework for Using the Workspace Medial Axis in PRM Planners, Proc. of the 2000 IEEE Int. Conf. on Robotics and Automation, IEEE.

[13] P. Isto, Constructing Probabilistic Roadmaps with Powerful Local Planning and Path Optimization. Proc. of the 2002 IEEE/RSJ Int. Conf. on Intelligent Robots and Systems, IEEE, 2323-2328.

[14] P. R. Cohen, Empirical Methods for Artificial Intelligence, The MIT Press, Cambridge, 1995.

[15] H. Chang, T.-Y. Lai, Assembly Maintainability Study with Motion Planning, Proc. of the 1995 IEEE Int. Conf. on Robotics and Automation, IEEE, Los Alamitos, CA, 1012-1019.

[16] B. R. Donald, A Search Algorithm for Motion Planning with Six Degrees of Freedom, Artificial Intelligence, Vol. 31, No. 3, March 1987, 295-353.

[17] K. Kondo, Motion Planning with Six Degrees of Freedom by Multistrategic Bidirectional Heuristic Free-Space Enumeration, IEEE Transactions on Robotics and Automation, Vol. 7, No. 3, June 1991, 267-277.

[18] D. E. Koditschek, Exact Robot Navigation by Means of Potential Functions: Some Topological Considerations, Proc. of the 1987 IEEE Int. Conf. on Robotics and Automation Conf., IEEE, 1-6.

[19] B. Glavina, A Fast Motion Planner for 6-DOF Manipulators in 3-D Environments. Proc. of the Fifth Int. Conf. on Advanced Robotics, IEEE Press, 1991, 1176-1181.

[20] L. Kavraki, J.-C. Latombe, Randomized Preprocessing of Configuration Space for Fast Path Planning, Proc. of the 1994 IEEE Int. Conf. on Robotics and Automation, IEEE, 2138-2145.

[21] Th. Horsch, F. Schwarz, H. Tolle, Motion Planning with Many Degrees of Freedom - Random Reflections at C-Space Obstacles, Proc. of the 1994 IEEE Int. Conf. on Robotics and Automation, IEEE, 3318-3323.

[22] X. Zhu, K. Gupta, On Local Minima and Random Search in Robot Motion Planning, Unpublished Manuscript, 1993.

[23] M. Branicky, S. M. LaValle, K. Olsen, L. Yang. Quasi-Randomized Path Planning. Proc. of the 2001 IEEE Int. Conf. on Robotics and Automation, IEEE, 1481-1487.

[24] F. Lamiraux, L. E. Kavraki, Planning Paths for Elastic Objects under Manipulation Constraints, Int. Journal of Robotics Research, Vol. 20, No. 3, 2001, 188-208.

[25] N. J. Nilsson, Principles of Artificial Intelligence, Springer-Verlag, Berlin, 1982.

[26] L. Kavraki, J.C. Latombe, R. Motwani, P. Raghavan, Randomized Query Processing in Robot Motion Planning, Proc. of the ACM SIGACT Symposium on Theory of Computing (STOC), ACM, Las Vegas, 1995, 353-362.

[27] D. J. Challou, D. Boley, M. Gini, V. Kumar, C. Olson, Parallel Search Algorithms for Robot Motion Planning, In [3], 115-131.

[28] S. Caselli, M. Reggiani, R. Sbravati, Parallel Path Planning with Multiple Evasion Strategies, Proc. of the 2001 IEEE Int. Conf. on Robotics and Automation, IEEE, 260-266.

[29] I. Einot, K. R. Gabriel, A Study of the Powers of Several Methods of Multiple Comparisons, Journal of the American Statistical Association, Vol. 70, No. 351, September 1975, 574-583.

[30] R. E. Glaser, Levene's Robust Test of Homogeneity of Variances, In Kotz & Johnson (eds.), Encyclopedia of Statistical Sciences, Vol 4, John Wiley & Sons, 1983, 608-610.

[31] J.-C. Latombe, Motion Planning: A Journey of Robots, Molecules, Digital Actors, and Other Artifacts, Int. Journal of Robotics Research, Vol. 18, No. 11, November 1999, 1119-1128.

	Hwang and Ahuja problem			Alpha Puzzle 1.2		
	Mean	Std. Dev.	Coeff. Var.	Mean	Std. Dev.	Coeff. Var.
PRM-C	2,021,367	3,308,323	164	1,431,266	1,061,356	74
PRM-G	804,865	1,357,438	169	1,473,386	924,775	62
PRM-L	816,762	800,089	98	1,357,358	921,015	68

Table 1: The mean and standard deviation of the run-cost in collision checks and the coefficient of variation for the problems of figure 1 and figure 2. The sample size is 240 runs.

A Ciliary Based 8-Legged Walking Micro Robot Using Cast IPMC Actuators

Byungkyu Kim[a], Jaewook Ryu[a], Younkoo Jeong[a], Younghun Tak[a], Byungmok Kim[a], Jong-Oh Park[b]
[a] Microsystem Research Center, Korea Institute of Science and Technology,
[b] Intelligent Microsystem Center, P.O. Box 131, Cheongryang, Seoul, 130-650, South Korea
E-mail bkim@kist.re.kr

Abstract

We have proposed a prototype model of walking micro robot using IPMC (Ionic Polymer Metal Composite) actuators. The stiffness of IPMC actuator is a key parameter to implement a walking robot. Therefore, the casting process is developed to increase the stiffness of the actuator by controlling thickness of ion-exchange polymer film. The process of fabricating a solid film from liquid state of ion-exchange polymer is difficult since any process parameter and handling material are not disclosed and has to be set by trial and error. The bending characteristics and generative tip force of IPMC actuator under variation of thickness and length of the actuator and voltage input are investigated. Also, mechanical model is derived to predict the generative tip force and displacement of IPMC actuator according to the variation of thickness. With cast film based IPMC actuators, a ciliary type 8-legged micro robot, which can be operated in aqueous surroundings like inside of human body, is constructed and tested. The robot shows good reliability and can reach up to 17 mm/min in speed.

1. INTRODUCTION

Recently, IPMC (Ionic Polymer Metal Composite) actuators have attracted attention due to its softness, lightness, flexibility and biocompatibility compared to other actuators such as piezo, electro active ceramic (EAC) and shape memory alloy (SMA). The IPMC actuator requires and consumes relatively low voltage and electric power. It can produce fairly large bending motion compared to piezos and SMA actuators. In addition, it can be actuated in a wet condition or even in water. Many applications using the IPMC actuator such as an underwater swimming microrobot[1], an active catheter[2], a capsule micropump[3] and a micro gripper[4] were realized. In walking robot area, Tadokoro et. al presented a new actuation device using a pair of IPMCs called Elliptic Friction Drive (EFD) actuator elements[5]. Tadokoro developed the bundle type distributed actuation device using EFD actuator elements[6]. In addition, methods to increase the generative force of the IPMC actuator has been studied by changing the counter-ions of the IPMC actuator[7] and the effect of cations on the stiffness of the IPMC actuator has been studied as well[8]. The typical IPMC actuator has the thickness between 100 and 300 μm because the IPMC actuator is fabricated based on a commercialized solid ion-exchange polymer film such as Nafion™ film that has the thickness of 100-300 μm. Therefore, the stiffness and generative force at the tip of the IPMC actuator is limited. However, using the casting method of the liquid ion-exchange polymer, the thickness of the ion-exchange polymer can be varied with ease. By the casting process, the thickness of the Nafion™ film could be increased up to several millimeters and be reduced down to less than 50 μm. The thickness could increase more a few millimeters. The **ca**st **Na**fion™ **fi**lm, so called "caNafi" in this paper is made from the liquid Nafion™ solution and used to fabricate the IPMC actuators. The stiffness of the caNafi based IPMC actuator increases dramatically. Therefore, we could build a ciliary motion based 8-legged walking robot by utilizing an IPMC actuator as a leg of the robot. The performance of the robot is evaluated through experiment and simulation.

2. EXPERIMENT

2.1 CaNafi based IPMC actuator

The methods to cast liquid Nafion™ into a solid Nafion™ film and to fabricate an IPMC are described below. The overall process is divided into casting procedures of liquid Nafion™ Solution and electroless plating process.

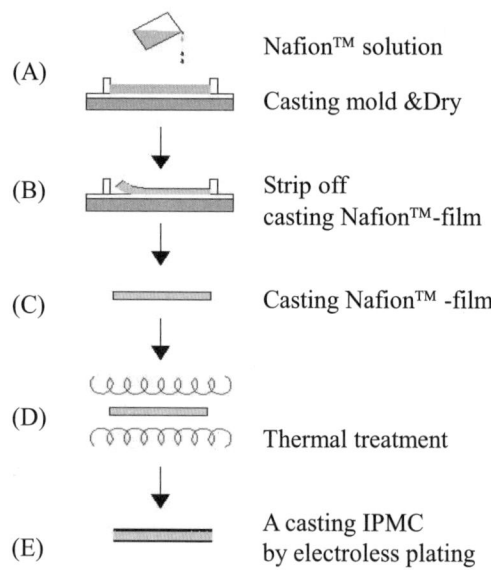

Figure 1. IPMC fabrication procedures based on CaNafi

(A) step: Pouring Nafion™ solution into a casting mold.
Pouring Nafion™ solution (EW1000, Dupont) into the mold carefully not to create bubbles. Drying a cast Nafion™ film. Solidifying Nafion™ solution (EW1000, Dupont) that is poured into the mold with protective shield is on. Normally solidification is processed under the atmospheric condition around 25°C for several days.
(B,C) step: Stripping off cast Nafion™ film.
Stripping cast Nafion™ film off the casting mold.
(D)step: Curing a cast Nafion™ film.
After stripping off a cast Nafion™ film from the mold, place the cast Nafion™ film into a vacuum oven at 1 bar and 100°C for an hour for thermal treatment.
(E) step: Electroless chemical plating process
For coating Pt layer on cast Nafion™ film, proper amount of precursor salts is required.

During Casting Procedures of Liquid Nafion™, liquid Nafion™ is transformed into a solid Nafion™ film and the thickness of the Nafion™ film is tailored. After the thermal treatment, the weak and brittle film becomes rigid. Electroless chemical plating process is used to fabricate the IPMC actuator.

2.2 Measurement Setup

The measurement setup was prepared to measure the bending displacement and generative force at the tip of the IPMC actuators.

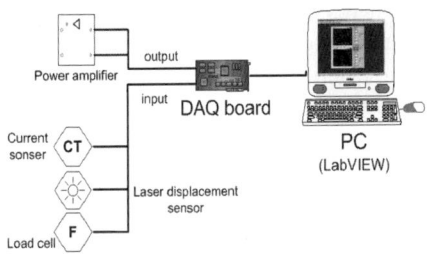

Figure 2. A schematic and real measurement setup of the IPMC actuators

In this experiment setup, LabVIEW by National Instrument combined with DAQ board is used to control the output signal to the actuators. To provide enough voltage to the IPMC actuator the output signal is amplified by a power amplifier (PA26 and OPA547 by Apex) and the output signal is directly connected with the IPMC actuator. LabVIEW is pre-programmed to control sequence, frequency, duty rate, phase and offset as well as voltage. A laser displacement sensor (LB-12 by KEYENCE), a force sensor (Transducer Techniques) and a current sensor (HFP2 series by HINODE) are connected with DAQ board. The laser displacement sensor is aimed perpendicular to the tip of the IPMC actuator for bending displacement and the load cell is installed horizontally to the IPMC actuator for the generative force measurement. All information is directly sent to the LabVIEW by DAQ board for the record and storage of data.

3. RESULTS

The bending displacements of the IPMC actuators were measured with respect to the different voltages as well as different lengths of the IPMC actuator. Figure 3 shows the maximum bending displacement related with applied voltage at the different actuator lengths. The maximum bending displacement becomes larger as the input voltage gets larger and the actuator length becomes longer. The width and the thickness of the IPMC actuator are kept to 4 mm and 1.15 mm respectively. The IPMC actuator in a cantilever was tested under the

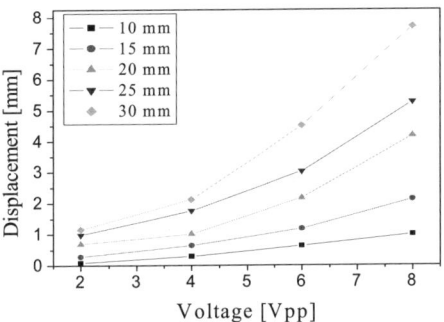

Figure 3. Maximum bending displacements of the caNafi based IPMC actuator vs. applying voltage

frequency of 0.5 Hz with square waveform. To be able to infer the proper relation between input voltages and the bending displacement, the displacement was divided by the square of the actuator length (L^2) of the actuator. The approximated equation, which is extracted from the polynomial curve fitting based on the relation between the input voltages and the bending displacements divided by L^2 is driven as follows.

$$\frac{\delta_{max}}{L^2} = 1.25 \times 10^{-5} V^3 - 2.5 \times 10^{-5} V^2 + 5 \times 10^{-4} V - 4 \times 10^{-4} \quad (1)$$

Using the equation (1), the maximum bending displacement at the tip of the IPMC actuator could be estimated with respect to different applied voltages and lengths of the actuator.

According to Tadokoro et. Al[9], the operation of the IPMC actuator results from the strain change in the IPMC actuator, which is generated by electric potential difference between the two electrodes of the IPMC actuator. Thus, the generative tip force of the IPMC actuator depends on the potential difference between the electrodes and also the potential difference is depending on the surface resistance of the IPMC actuator. According to Kim and Shahinpoor, the force generation by the IPMC actuator is varied depending on the surface resistance of the actuator[10]. The surface resistance of the IPMC actuator is shown in figure 4. In this graph, the surface resistance becomes larger as the actuator length increases. Therefore, the potential difference across the two electrodes will be varied depending on the distance from the base that is, in this case, the electrical source. As the length gets far away from the electrical source, resistance increases. Thus, the potential difference across the two electrodes recedes. Therefore, the tip force generated by the IPMC actuator becomes smaller as the length of the actuator increases. As shown in figure 5, the actual force measurements on the IPMC actuator at different points shows good agreement with the above presumption.

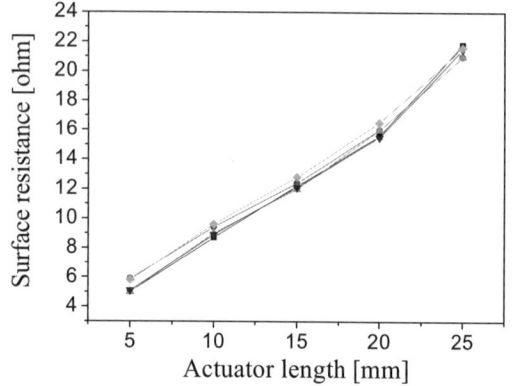

Figure 4. Surface resistances of the IPMC actuator in cantilever configuration

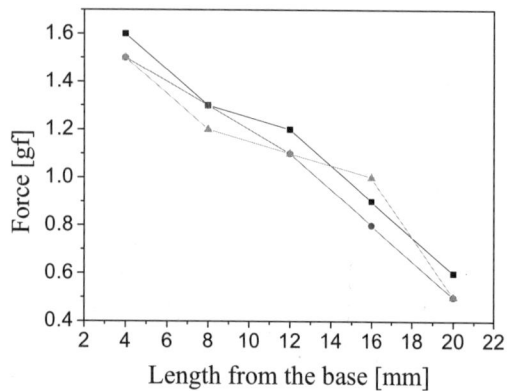

Figure 5. Measured distributed forces on the IPMC actuator in cantilever configuration

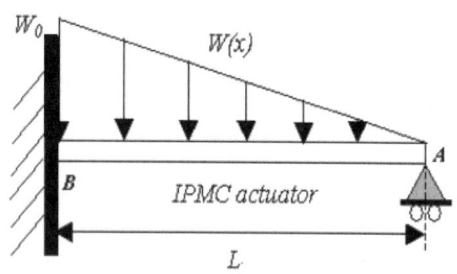

Figure 6. Actual cantilever modeling of the IPMC actuator

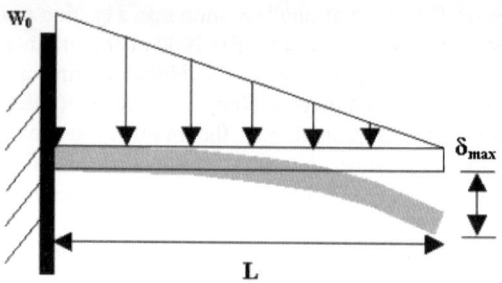

Figure 7. A selected cantilever model of the caNafi based IPMC actuator.

The actual cantilever model of the IPMC actuator is provided in figure 6. The load cell is represented as a support in this modeling, since the load cell provides the reacting force at the tip of the IPMC actuator during the measurements. Based on the proposed cantilever model of the IPMC actuator, the tip force generated by the IPMC actuator is found theoretically. By differential equation of the elastic curve, the reaction force at the tip of the IPMC actuator in cantilever configuration is,

$$R = \frac{1}{10} w_0 L \qquad (2)$$

In Figure 7, if there is no support at the tip of the IPMC actuator and it sustains the triangularly distributed load, then

$$w_0 = \frac{30 EI \delta_{max}}{L^4} \qquad (3)$$

Where δ_{max} is the maximum tip bending displacement by the triangularly distributed load on the IPMC actuator. Finally, the reaction at the tip of the IPMC actuator becomes as follows.

$$R = \frac{3 EI \delta_{max}}{L^3} \qquad (4)$$

Here, L is the total length of the IPMC actuator, EI is the stiffness of the IPMC actuator and δ_{max} is the measured maximum tip bending on the IPMC actuator.

As well as the maximum force generated by the IPMC actuator, the stiffness plays an important role in many practical applications. Increasing the thickness of the IPMC actuator by the casting of liquid ion-exchange polymer increases the stiffness as well as the generative tip force. Therefore, the stiffness (EI) of the caNafi based IPMC actuator is obtained based on equation (4) in different thickness after we measure the reaction force corresponding to the displacement of IPMC actuator. Also, the stiffness of the caNafi and the stiffness of the caNafi with "Na" cation are presented to provide the guideline to obtain the specific stiffness of the IPMC actuator. Therefore, we can estimate the stiffness of the caNafi based IPMC actuator from figure 8 just after obtain the thickness of the caNafi film. Also, we can estimate the generative tip force based on equation (1), (4) and figure 8 in case of caNafi thickness-1.15 mm under various voltage inputs. The comparison between the experimental result and the estimated result using equation (1), (4) and figure 8 shows good agreement as shown in figure 9. Based on above data and some additional experimental data at various thicknesses, it is possible to design the required specifications and the process parameters for the caNafi IPMC actuator at any specific applications.

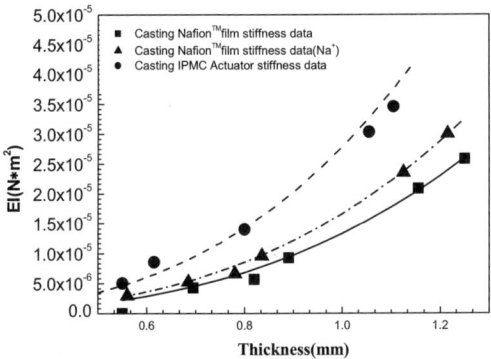

Figure 8. Stiffness (EI) of the IPMC actuator with various thickness (width=4 mm, length=30 mm)

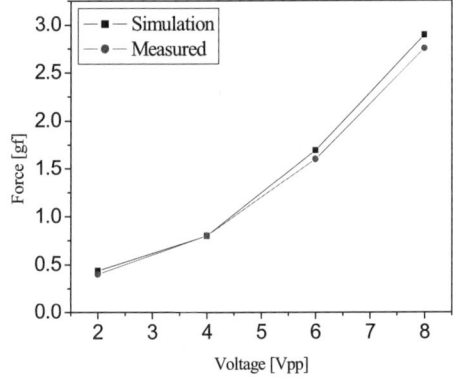

Figure 9. Tip displacement vs. generative tip force with variation of input voltage (width=4 mm, length=30 mm, thickness=1.15 mm)

4. APPLICATION OF THE CANAFI BASED IPMC ACTUATOR AS A WALKING ROBOT

4.1 Ciliary Motion Based Walking Robot

A ciliary-motioned 8-legged walking robot driven by the caNafi based IPMC actuator is constructed. The locomotion device is utilizing the bending motion of the IPMC actuators. Beneche and Riethmüller proposed and developed the concept a micro conveyor system based on ciliary motion[11]. Later Ebefors et, al developed a walking silicon micro-robot based on the ciliary motion[12] with polyimide joints. The walking principle and sequence for the proposed walking robot are provided below in figure 10. As the front leg is pushed downward at the second stage and the rear leg is folded upward at the fourth stage, the robot moves forward with a fixed distance (x) respectively. Therefore, the total distance that the robot can advance per cycle is $2x$. The robot has the total of eight legs, four legs on each side. As shown in figure 11, the legs are oriented in opposite direction, the dotted line (a) is in forward direction and the solid line (b) is in backward direction. The robot has the length and width of 6 cm and 3 cm respectively without the actuators. With the actuators, the length increases to 6.5 cm and the width increase to 4.2 cm. The height of the robot becomes 1.5 cm (see figure 12). The electric wires and other electric components are adequately embedded on the body of the robot.

The robot is composed of low-density material such as Styrofoam to reduce weight and 100 μm wire is used to reduce the wire stiffness. Stainless steel is used for the electrodes of the robot to prevent from corrosion. The weights of the robot with the actuators and without the actuators are 3.6 g and 4.4 g.

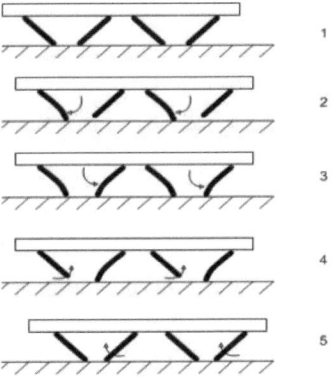

Figure 10. Walking principle and sequence of the ciliary motion based 8-legged walking device

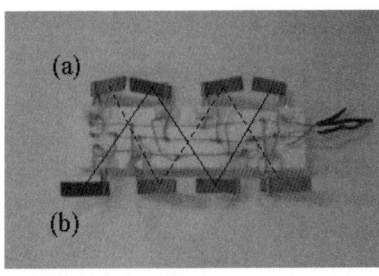

Figure 11. Legs orientation and layout of the ciliary motion based 8-legged robot.

The caNafi based IPMC actuators have enough stiffness and produce enough generative force for locomotion. The dimension of the actuators is 20 mm in length, 4 mm in width and 1.15 mm in thickness and the applied voltage is ±4 V. In this case, one caNafi based IPMC actuator can produce about 5 gf. Overall the robot can generate more than 40 gf at the tip of the IPMC actuators that is about 10 times greater than its own body weight.

4.2 Modeling of the IPMC driven robot

By the walking principle, the horizontal displacement plays important roles on the performance of the robot. Actually the displacement of the robot depends on the horizontal displacement of the IPMC actuators rather than the vertical displacement. The vertical displacement, however, is necessary to make non-contact actuators be bent freely and the robot move forward or backward as the result of it. The IPMC actuators are mounted with the angle of 45 degrees with respect to the ground. The modeling of the IPMC driven robot is provided below.

First, the coordinates of the IPMC actuator are defined as equation (5) when the actuator is tilted with the angle of θ as shown in figure 13.

$$P_x = [\delta_v], \quad P_y = [\delta_h - L] \quad (5)$$

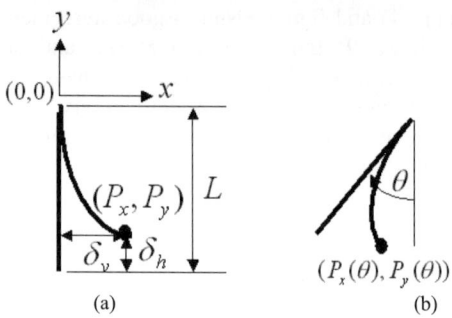

Figure 13. (a) Coordinate axis at normal state and (b) coordinate axis at tilted state

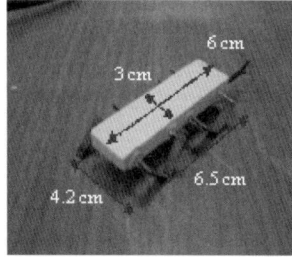 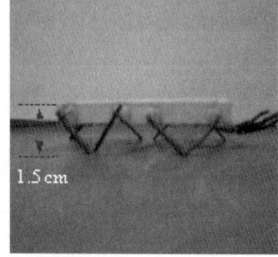

Figure 12. Dimension of the ciliary motion based 8-legged robot

If the angle θ is applied to the original pose, the coordinates of the end point of the IPMC actuator is changed at a given angle θ as equations (6) and (7).

$$P_x(\theta) = [-L \cdot \sin\theta + \delta_h \cdot \sin\theta + \delta_v \cdot \cos\theta] \quad (6)$$

$$P_y(\theta) = [-L \cdot \cos\theta + \delta_h \cdot \cos\theta - \delta_v \cdot \sin\theta] \quad (7)$$

Where, $\delta_h \cong 0$ (8)

$$\delta_v = L^2 \cdot [1.25 \times 10^{-5} \cdot V^3 - 2.5 \times 10^{-5} \cdot V^2 + 5 \times 10^{-4} \cdot V - 4 \times 10^{-4}] \quad (9)$$

According to the walking sequence of the ciliary-motioned 8-legged walking device, the movement change per cycle can be calculated as $2 \times \max(P_x(\theta))$ with the assumption of no slip between the IPMC actuator and the surface. Therefore, the moving speed of the proposed 8-legged walking device is calculated as equation (10) when the frequency of the control signal is defined as f and the value is low enough to ignore the magnitude change of the IPMC actuator with the frequency changes.

$$V = 2 \cdot \max(P_x(\theta)) \cdot f \quad (10)$$

The magnitude and the phase of the deflection δ_v at the end of the IPMC actuator change according to frequency changes of the input control signal. To evaluate the performance of the actuator based on the frequency changes, the response of the actuator with various frequencies was tested and the results were analyzed to

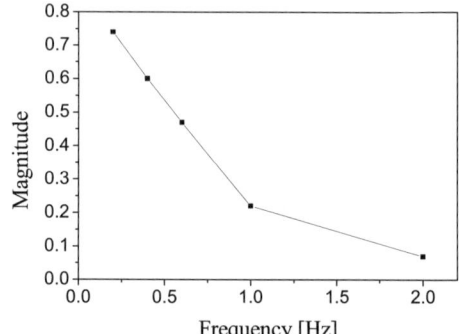

(a) Magnitude response with various frequencies.

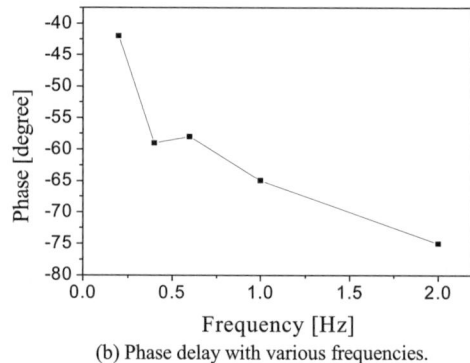

(b) Phase delay with various frequencies.

Figure 14. Frequency response of the IPMC actuator

$$V = 2 \cdot \max(P_x(\theta)) \cdot f \cdot mag(f) \quad (11)$$

understand them. The response analysis result with various control frequencies is shown in figure 14. Based on this analysis result the estimated speed of the locomotive robot is derived as equation (11) if the magnitude change of the actuator is defined as $mag(f)$. The estimated and measured moving speeds of the locomotive device with various control frequencies are plotted as shown in figure 14. According to the figure 15, it seems that the there is an optimized frequency range for the fastest movement for the proposed robot and the value is around 0.8 Hz. The speed of the IPMC robot increases as the frequency changes from 0.2 Hz to 1.0 Hz with the increment of 0.2 Hz. As shown in figure 15, the speed of the robot increases until the frequency reaches to 0.8 Hz. At 1.0 Hz, the speed decreases dramatically. This is because as the frequency increases the bending motion gets fast, but the magnitude of the bending displacement decreases as shown in Fig 14(a). It is believed that there is an optimal frequency around 0.8 Hz. The difference between the simulated and the measured is due to the weight of the robot and the slip between the IPMC actuator and the surface is not considered in the simulation. In addition, hysterisis of the IPMC actuator should be considered for the better result.

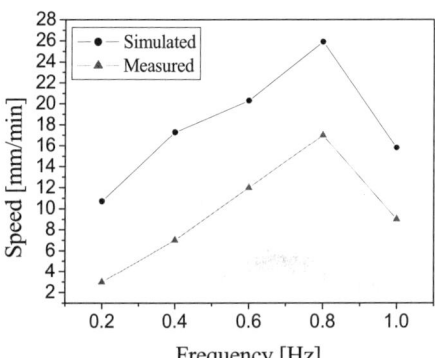

Figure 15. Comparison between estimated and measured speed of the IPMC actuator driven robot

5. CONCLUSION

As an actuator for robot application, caNafi (**ca**st **Nafi**on[TM] **fi**lm) based IPMC actuator is verified. To construct a walking robot, the casting process of liquid ion-exchange polymer is developed. Based on performance test, we also present the guideline to design the stiffness and generative force that the actuator produces. Therefore, the generative tip force can be estimated after the cast Nafion[TM] film of arbitrary thickness is obtained. With caNafi based IPMC actuator, a ciliary motion based walking robot is embodied. The robot produces about 10 times generative force of its own weight at the tip of the IPMC actuators. The robot runs up to 17 mm/min under operation frequency 0.8 Hz. Unlike to other actuators the IPMC powered robot can be operated in aqueous surroundings.

6. ACKNOWLEDGEMENT

This research was supported by the Intelligent Microsystem Center, Seoul, Korea, which is carrying out one of the 21st century's New Frontier R&D Projects sponsored by the Korea Ministry Of Science & Technology.

7. REFERENCES

[1] Shuxiang Guo, Toshio Fukuda, Norihiko Kato and Keisuke Oguro, "Development of underwater microrobot using ICPF actuator", in *Proceedings of the 1998 IEEE International Conference on Robotics & Automation*, Vol 2, pp.1829-1834, Leuven, Belgium.

[2] Shuxiang Guo, Toshio Fukuda, Tatsuya Nakamura, Fumihito Arai, Keisuke Oguro and Makoto Negoro, "Micro Active Guide Wire Catheter System-Characteristic Evaluation, Electrical Model and Operability Evaluation of Micro Active Catheter," in *Proceedings of the 1996 IEEE International Conference on Robotics and Automation*, Vol. 3, pp. 2226 –2231.

[3] Shuxiang Guo, Siji Hata, Koichi Sugumoto, Toshio Fukuda and Keisuke Oguro, "Development of a New Type of Capsule Micropump," in *Smart Structure and Materials, Proc. SPIE* 3669, pp. 322-329, 1999.

[4] R. Lumia and M. Shahinpoor, "Microgripper Design Using Electro-active Polymers," in *Proceedings of SPIE the 1999 IEEE International Conference on Robotics and Automation*, Vol. 3, pp. 2226 –2231.

[5] Satoshi Tadokoro, Takahiko Murakami, Satoshi Fuji, Ryu Kanno, Motofumi Hattori, Toshi Takamori, Keisuke Oguro, "An elliptic friction drive element using an ICPF actuator," in *IEEE Control Systems Magazine*, Vol. 17, Issue 3, pp. 60-68, June 1997.

[6] Satoshi Tadokoro, Satoshi Fuji, Mitsuaki Fushimi, Ryu Kanno, Tetsuya Kimura, Toshi Takamori, "Development of a distributed actuation device consisting of soft gel actuator elements", in *Proceedings of the 1998 IEEE International Conference on Robotics & Automation*, Vol. 3, 2155-2160, Leuven, Belgium.

[7] Mohsen Shahinpoor and Kwang J. Kim, "Effects of the Counter-ions on the Performance of IPMCs," in *Smart Structure and Materials, Proc. SPIE* 3669, 110-120, 2000.

[8] Yoseph Bar-Cohen, "Electroactive Polymer (EAP) Actuators as Artificial Muscles-Reality, Potential and Challenges," pp. 152-154.

[9] Satoshi Tadokoro, Masahiko Fukuhara, Yoseph Bar-Cohen, Keisuke Pguro, Toshi Takamori, "A CAE Approach in Application of Nafion-Pt Composite (ICPF) Actuators – Analysis for Surface Wipers of NASA MUSES_CN Nanorovers," in *Smart Structure and Materials, Proc. SPIE* 3669, 262-272, 2000.

[10] Kwang J. Kim and Mohsen Shahinpoor, "The effect of surface-electrode resistance on the performance of ionic polymer-metal composite (IPMC) artificial muscles", in *Smart Structure and Materials, Proc. SPIE* 3669, pp. 308-319, 1999.

[11] W. Beneche and W. Riethmüller, "Application of Silicon-Microactuators based on Bimorph Structures," *Proc. Of IEEE MEMS'89*, Salt Lake City, USA, Feb 20-22,1989, pp. 116-120.

[12] Thorbjörn Ebefors, Johan U. Mattsson, Edvard Kalvesten and Goran Stemme, "A Walking Silicon Micro-robot", Presented at the 10th Int. Conference on Solid-State Sensors and Actuators (TRANSDUCERS99), June 7-10, 1999, pp. 1202-1205, Sendai, Japan.

Motion Planning for a Three-Limbed Climbing Robot in Vertical Natural Terrain

Timothy Bretl and Stephen Rock

Aerospace Robotics Lab
Department of Aeronautics and Astronautics
Stanford University, Stanford, CA 94305
{tbretl, rock}@sun-valley.stanford.edu

Jean-Claude Latombe

Robotics Laboratory
Computer Science Department
Stanford University, Stanford, CA 94305
latombe@cs.stanford.edu

Abstract

This paper presents a general framework for planning the quasi-static motion of a three-limbed climbing robot in vertical natural terrain. The problem is to generate a sequence of continuous one-step motions between consecutive holds that will allow the robot to reach a particular goal hold. A detailed algorithm is presented to compute a one-step motion considering the equilibrium constraint only. The overall framework combines this local planner with a heuristic search technique to generate a complete plan. An online implementation of the algorithm is demonstrated in simulation.

1 Introduction

The work presented in this paper is part of an effort to develop critical technologies that will enable the design and implementation of an autonomous robot able to climb vertical natural terrain. To our knowledge, this capability has not been demonstrated previously for robotic systems. Prior approaches have dealt with artificial terrain, either using special "grasps" (e.g., pegs, magnets) adapted to the terrain's surface or exploiting specific properties or features of the terrain (e.g., ducts and pipes) [1-12].

Developing this capability will further our understanding of how humans perform such complex tasks as climbing and scrambling in rugged terrain. This may prove useful in the future development of sophisticated robotic systems that will either aid or replace humans in the performance of aggressive tasks in difficult terrain. Examples include robotic systems for such military and civilian uses as search-and-rescue, reconnaissance, and planetary exploration.

Many issues need to be addressed before real robots can climb real, vertical natural terrain. This paper presents preliminary work in the area of motion planning. A general framework for climbing robots is presented and this framework is instantiated to compute climbing motions of the three-limbed robot shown in Figure 1.

Fig 1. A three-limbed climbing robot moving vertically on natural surfaces.

1.1 Problem Statement

The robot of Figure 1 consists of three limbs. Each limb has two joints, one located at the center of the robot (called the pelvis) and one at the midpoint of the limb. Motion is assumed to be quasi-static (as is usually the case in human climbing) and to occur in a vertical plane, with gravity. The low complexity of this robot's kinematics makes it suitable for studying the planning of climbing motions.

The terrain is modeled as a vertical plane to which is attached a collection of small, angled, flat surfaces, called "holds," that are arbitrarily distributed. The endpoint of each robot limb can push or pull at a single point on each hold, exploiting friction to avoid sliding.

A climbing motion of the robot consists of successive steps. Between any two consecutive steps, all three limb endpoints achieve contact with distinct holds. During each step, one limb moves from one hold to another, while the other two endpoints remain fixed. The robot can use the degrees of freedom in the linkage formed by the corresponding two limbs to maintain quasi-static equilibrium and to avoid sliding on either of the two supporting holds. In addition, during a step, the torque at any joint should not exceed the actuator limits and the limbs should not collide with one another. These constraints define the feasible subset of the configuration space of the robot in each step. A path in this subset defines a one-step motion.

The overall planning problem is the following: given a model of the terrain, an initial robot configuration where it rests on a pair of holds, and a goal hold, generate a series of one-step motions that will allow the robot to move in quasi-static equilibrium from the initial configuration to an end configuration where one limb endpoint is in contact with the goal hold.

This paper presents an algorithm to compute a one-step motion considering the equilibrium constraint only. Adding the actuator-limit and self-collision constraints, though still undone, does not seem to raise major difficulties. The overall planner combines this "local planner" with a heuristic search technique to determine a sequence of holds from the initial configuration to the goal hold.

1.2 Related Work

The search space, which will be described in Section 3, is a hybrid space, involving both continuous and discrete actions. Many different methods are available for motion planning through continuous spaces, including cell decomposition, potential field, and roadmap algorithms [13]. Discrete actions can be included in these methods directly, such as at the level of node expansion in a roadmap algorithm, but this approach generally leads to a slow implementation that is specific to a particular system.

Previous work on motion planning for legged robots has developed tools for addressing these hybrid search spaces for some systems. This work can be categorized by whether or not the planning is done offline, in order to generate a reactive gait, or online, in order to allow non-gaited motion specific to a sensed environment.

Gaited planners generate a predefined walking pattern offline, assuming a regular environment. This pattern is used with a set of heuristics or behaviors to control the robot online based on current sensor input. Gaited planning was used by [2, 11], for example, to design patterns for climbing pipes and ducts. Other methods such as [14] are based on the notion of support triangles for maintaining equilibrium. Stability criteria such as the zero-moment-point have been used to design optimal walking gaits [15]. Dynamic gaiting and bounding also have been demonstrated [16-18]. Recent work [19, 20] has attempted to provide unifying mathematical tools for gait generation. Each of these planning algorithms would be very effective in portions of a natural climbing environment with a sustained feature such as a long vertical crack of nearly uniform width. However, something more is needed for irregular environments such as the one studied in this paper, where the surfaces on which the robot climbs are angled and placed arbitrarily.

Non-gaited planners use sensed information about the environment to create feasible motion plans online. Most previous work on non-gaited motion planning for legged robots has focused on a particular system model, the spider robot. The limbs of a spider robot are assumed to be massless, which leads to elegant representations of their free space for quasi-static motion based on support triangles [21-23]. These methods have been extended to planning dynamic motions over rough terrain [24, 25]. The analysis used in these methods breaks down, however, when considering robots that do not satisfy the spider-robot assumption. For example, additional techniques were necessary in [26, 27] to plan non-gaited walking motions for humanoids, which clearly do not satisfy this assumption. To address the high number of degrees of freedom and the high branching factor of the discrete search through possible footsteps, these techniques were based on heuristic discretization and search algorithms. This paper considers a robot with fewer degrees of freedom in a more structured search space where it is possible to achieve much better performance than with these heuristic methods. Similar issues were addressed by [28] in designing a motion-planning algorithm for character animation, although this algorithm was meant to create "realistic," rather than strictly feasible, motion.

There is also some similarity between non-gaited motion planning for legged locomotion and for grasping and robotic manipulation, particularly in the concept of a manipulation graph [29-32]. Both types of planning require making discrete and continuous choices.

1.3 Contribution

The major contribution of this paper is a detailed analysis of one-step motion for the three-limbed climbing robot.

First, the properties of the continuous configurations at which the robot is in equilibrium are established. These properties are used to define the feasible set of robot configurations at each pair of holds. In particular, it is shown that the connectivity of the four-dimensional continuous feasible space of the robot can be preserved when planning in a two-dimensional subspace. This result reduces the complexity of the one-step planning problem and leads to a fast, online implementation.

Then, an overall framework is presented for planning a sequence of one-step motions from a specific configuration on an initial pair of holds to a goal hold. Heuristic methods are used to guide this discrete search, based on observation of the way in which human climbers plan their motions.

2 Notation and Terminology

Figure 2 illustrates the notation and terms used in the rest of this paper to describe the three-limbed robot.

The robot consists of three identical limbs meeting at the *pelvis*, whose location is denoted by (x_C, y_C). Each limb consists of two segments and has two revolute joints, one located at the pelvis, the other between the two segments. For simplicity, the six limb segments are assumed to have equal mass and length L, but this assumption can be relaxed easily. In total, the robot's configuration space has six dimensions (two for each limb). It is assumed that the revolute joints are not limited by any internal mechanical stops.

In Figure 2, the endpoints of the two limbs of the robot are at *holds* 1 and 2 located at (x_1, y_1) and (x_2, y_2), while the third limb is moving. The two-limbed linkage between (x_1, y_1) and (x_2, y_2) is called the *contact chain* and the other limb is the *free limb*. The constraint that two limb endpoints be at (x_1, y_1) and (x_2, y_2) reduces the set of possible configurations of the robot to a four-dimensional subspace, which is denoted by $\mathbf{C}(i,j)$, since both the contact chain and the free limb now have two degrees of freedom. Any motion of the robot maintaining these two contacts will occur in $\mathbf{C}(i,j)$. The configuration of the robot in $\mathbf{C}(i,j)$ can be uniquely specified by the angles (θ_1, θ_2) of the free limb, the position (x_C, y_C) of the pelvis, and two additional binary variables identifying the direction of the knee bends in the contact chain (see Figure 6(a)).

The location of the robot's center of mass (CM), which is not shown, is (x_{CM}, y_{CM}). The location of the CM's of the contact chain and the free limb are $(x_{CM,chain}, y_{CM,chain})$ and $(x_{CM,free}, y_{CM,free})$, respectively.

Friction at each hold is modeled using Coulomb's law. More precisely, each hold (x_1, y_1) and (x_2, y_2) exerts a reaction force on the corresponding limb endpoint. According to Coulomb's law, this force must point into a *friction cone* to avoid slipping (see Figure 3). The orientation ϕ_1 and ϕ_2 of each cone is normal to each hold, and the half-angle $\Delta\phi_1$ and $\Delta\phi_2$ of each cone is determined by each hold's coefficient of friction. For the robot to be in quasi-static equilibrium, two forces, one in each cone, must exist that exactly compensate for the gravitational force (see Section 3.1). This condition will select a subset of $\mathbf{C}(i,j)$, which in this paper is called the *feasible* space and is denoted by $\mathbf{F}(i,j)$. Any motion of the robot maintaining both contact at the two holds and quasi-static equilibrium must occur in $\mathbf{F}(i,j)$.

Since the three limbs are identical, there is no need to identify them. In particular, the same configurations can be achieved by the robot while maintaining contact at two holds, independent of which two limbs form the contact chain. It is also assumed that the robot does not bring two limb endpoints to the same hold.

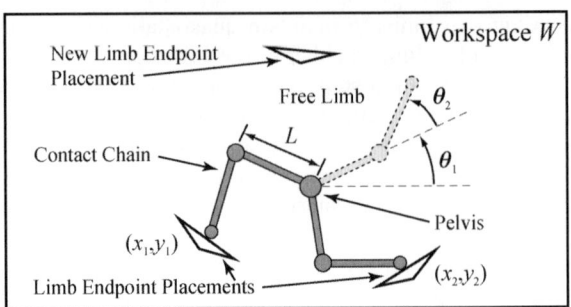

Fig. 2. The different components of the three-limbed climbing robot.

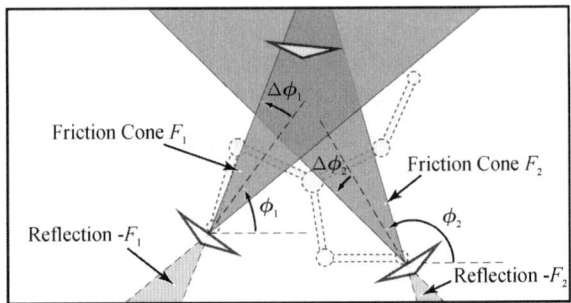

Fig. 3. The friction cones for two limb endpoint placements.

3 Equilibrium Analysis

In this section, it is assumed that the robot rests on two given holds i and j, as in Figures 2 and 3. This section establishes properties of the configurations in $\mathbf{C}(i,j)$ at which the robot is in equilibrium. These properties define the feasible subspace $\mathbf{F}(i,j)$.

3.1 Equilibrium Constraint

The only external forces acting on the robot are gravity and the reaction forces at the two holds. The gravitational force acts at the robot's center of mass, the position of which varies as the robot moves. The two reaction forces act at the endpoints of the contact chain, which have fixed positions. Therefore, the equilibrium constraint can be represented completely by a condition on the location of the center of mass.

The work in [33, 34] provides criteria for static equilibrium in a two-dimensional workspace. In particular, it notes that if a body acted upon by gravity and two external forces is in equilibrium, it will remain so with arbitrary vertical translation of its center of mass. This observation yields the following proposition when the external forces are subject to friction constraints:

Proposition 1. Consider an articulated body that is acted upon by a gravitational force $-g\hat{y}$ at (x_{CM}, y_{CM}) and that is in contact with two surfaces at points (x_1, y_1) and (x_2, y_2) with associated friction cones F_1 and F_2. Contact forces can be chosen to place the body in static equilibrium without slipping if and only if there exists $y \in \mathbf{R}$ and unit vectors $\hat{f}_1 \in F_1$, $\hat{f}_2 \in F_2$ such that the following conditions hold:

(1a) $(\hat{f}_1 + \hat{f}_2) \cdot \hat{y} > 0$
(1b) $(\hat{f}_1 \cdot \hat{x})(\hat{f}_2 \cdot \hat{x}) \leq 0$
(1c) Lines through (x_1, y_1) and (x_2, y_2) parallel to $\hat{f}_1$ and $\hat{f}_2$, respectively, intersect at (x_{CM}, y).

This proposition allows the equilibrium of the robot to be tested given the location of its center of mass. For example, Figure 4(a) shows a configuration of the climbing robot and illustrates graphically that all three conditions in the proposition are verified, so the robot is in equilibrium. However, the proposition does not specify the shape of the equilibrium region—the region in which the center of mass must lie—although it does indicate that this region must be a union of vertical columns in the workspace.

In fact, using Proposition 1 it can be shown that the equilibrium region is always a *single* vertical column, whose boundaries are easy to calculate. This result, stated as Proposition 2, can be explained intuitively. In two dimensions, the center of mass of a body resting at two points on horizontal supporting surfaces can only vary between these two points. Rotating the support surfaces can only lead to widening or narrowing the vertical column. However, the formal proof given below is more technical.

Proposition 2. Consider the articulated body of Proposition 1. The region over which (x_{CM}, y_{CM}) can vary while this body remains in static equilibrium is a single vertical column in the workspace.

Proof. From Proposition 1, it is clear that the equilibrium region is defined by the projection on the x-axis of the set of all points (x, y) for which unit vectors $\hat{f}_1 \in F_1$, $\hat{f}_2 \in F_2$ can be found satisfying Conditions 1a-1c. The problem is that this set of points is not convex, and in fact is not necessarily connected. However, it can be broken down into the union of convex sets, each of which projects to a connected segment on the x-axis. Further, it can be shown that each projected segment overlaps in such a way that the entire x-projection is a single connected segment, proving the result.

Assume without loss of generality that $x_2 > x_1$, that $x_1 \neq x_2$ ($x_1 = x_2$ is a degenerate case that can be handled separately), and that each friction cone has a half-angle $\Delta\phi < 90°$ (true for flat contact surfaces).

First, notice that a point (x, y) satisfies Condition 1c only if it lies in the intersection $(F_1 \cup -F_1) \cap (F_2 \cup -F_2)$. Call the set of points that additionally satisfy Conditions 1a-1b the set of *feasible* intersection points for cones F_1 and F_2. The equilibrium region is the x-projection of this set.

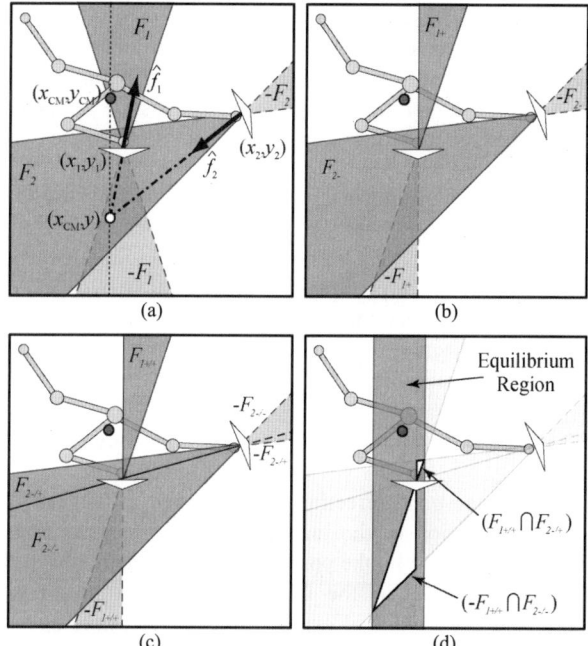

Fig. 4. An example calculation of the equilibrium region associated with a set of limb endpoint placements for the three-limbed robot.

Next, divide each friction cone F_1 and F_2 into two parts. Let F_{1+} be that part of F_1 containing points such that $x > x_1$, and F_{1-} be that part of F_1 containing points such that $x < x_1$. Likewise, divide F_2 into F_{2+} and F_{2-} using x_2. Since each friction cone has a half-angle $\Delta\phi < 90°$, each of F_{1+}, F_{1-}, F_{2+}, and F_{2-} must be either a single cone or empty.

Since Condition 1b indicates that force vectors must lie in opposite x-directions, the set of feasible intersection points for the cones F_1 and F_2 is equal to the union of the set of feasible intersection points for F_{1+} and F_{2-}, facing inward, and F_{1-} and F_{2+}, facing outward. For example, since F_{2+} is empty for the two friction cones shown in Figure 4(a), only the intersection of the inward-facing cones F_{1+} and F_{2-} needs to be considered, as shown in Figure 4(b).

For inward-facing cones, condition 1a can be used to show that the set of feasible intersection points (x, y) must satisfy $(y-y_1)(x-x_2) > (y-y_2)(x-x_1)$. Further divide F_{1+} into the cones $F_{1+/+}$, containing points such that $(y-y_1)(x_2-x_1) > (y_2-y_1)(x-x_1)$, and $F_{1+/-}$, containing points such that $(y-y_1)(x_2-x_1) < (y_2-y_1)(x-x_1)$. Analogously, divide F_{2-}. This procedure, shown in Figure 4(c), simply divides the friction cones into the part that points above the line connecting (x_1, y_1) and (x_2, y_2) and the part that points below this line. Again, since each friction cone has a half-angle $\Delta\phi < 90°$, each of the sub-cones must be either a single cone or empty.

Then from condition 1a, the set of feasible intersection points for the inward-facing cones is exactly the union of subsets $(-F_{1+/+} \cap F_{2-/-}) \cup (F_{1+/+} \cap F_{2-/+}) \cup (F_{1+/-} \cap -F_{2-/+})$, as shown in Figure 4(d).

Each of these three subsets is the intersection of two convex cones, so is convex with an x-projection that is a single connected segment. Further, it is easy to show that if any two subsets are nonempty, their x-projections must overlap at either x_1 or x_2, as is the case in Figure 4(d). Therefore, the x-projection of the set of feasible intersection points for inward-facing cones is a single connected segment.

An identical argument shows the same result for the set of outward-facing cones. In addition, it is easy to show that if the x-projections for both the inward- and outward-facing cones are nonempty, then they must overlap at either x_1 or x_2. Thus, the x-projection for the entire set of feasible intersection points is a single connected segment, proving Proposition 2. □

3.2 Feasible Space for a Given Pelvis Location

Given knee bend directions as described in Section 2, the configuration of the three-limbed climbing robot is uniquely specified by the position (x_C, y_C) of the pelvis and the angles (θ_1, θ_2) of the free limb. This section first establishes an analytical expression of the feasible space Θ_{FL} of the free limb given (x_C, y_C). Next, this expression is used to characterize the connectivity of Θ_{FL} in Proposition 3.

Since the location of the CM of the contact chain is fixed by the given (x_C, y_C), the equilibrium region shown in Section 3.1 to be a vertical column defined by some (x_{min}, x_{max}) can be transformed from a constraint on the location of the CM of the entire robot to a constraint on the CM of the free limb only. Under the geometry and mass assumptions made in Section 2, the following relationship holds:

$$x_{CM,free} = (3x_{CM} - 2x_{CM,chain}) \quad (1)$$

So, the center of mass of the free limb must be within the column $(x_{min,free}, x_{max,free})$ in the workspace where

$$x_{min,free} = (3x_{min} - 2x_{CM,chain})$$
$$x_{max,free} = (3x_{max} - 2x_{CM,chain}) \quad (2)$$

A pelvis location (x_C, y_C) is feasible with respect to the equilibrium constraint only if a configuration of the free limb exists such that $x_{CM,free} \in [x_{min,free}, x_{max,free}]$. The center of mass of the free limb is located at

$$x_{CM,free} = x_C + \frac{L}{4}(3\cos\theta_1 + \cos(\theta_1 + \theta_2)) \quad (3)$$

From Equations (1)-(3), a pelvis location is feasible only if

$$x_{min,free} \geq x_C - L$$
$$x_{max,free} \leq x_C + L \quad (4)$$
$$x_{min,free} \leq x_{max,free}$$

For any feasible pelvis location, the equilibrium region of the center of mass of the free limb can be cropped such that $[x_{min,free}, x_{max,free}] \subset [x_C - L, x_C + L]$, since values outside these bounds are unattainable.

The solutions of Equation 3 for a fixed value of $x_{CM,free}$ define a one-dimensional curve in the configuration space of the free limb. Curves for several values of $x_{CM,free}$ are

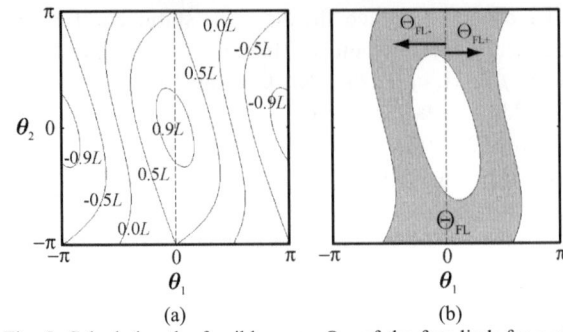

Fig. 5. Calculating the feasible space Θ_{FL} of the free limb for a given pelvis location.

shown in Figure 5(a). Since the mapping from (θ_1, θ_2) to $x_{CM,free}$ is single-valued, no two such curves intersect. The feasible space Θ_{FL} of the free limb is the region between the solution curves for $x_{CM,free} = x_{min,free}$ and $x_{CM,free} = x_{max,free}$, as shown in Figure 5(b) for $(x_{min,free}, x_{max,free}) = (x_C - 0.1L, x_C + 0.7L)$.

Since the feasible space Θ_{FL} depends on both x_C and $x_{CM,chain}$, which itself is a complicated function of (x_C, y_C), it is difficult to compute the four-dimensional feasible space of the robot. However, the following proposition characterizes the connectivity of this space:

Proposition 3. Partition Θ_{FL} as

$$\Theta_{FL-} = \Theta_{FL} \cap \{(\theta_1, \theta_2) | \theta_1 \leq 0\}$$
$$\Theta_{FL+} = \Theta_{FL} \cap \{(\theta_1, \theta_2) | \theta_1 \geq 0\} \quad (5)$$

Also, for any $x' \in [x_C - L, x_C + L]$, define

$$(\theta_1, \theta_2)_{x'-} = (-\cos^{-1}\frac{x'-x_C}{L}, 0)$$
$$(\theta_1, \theta_2)_{x'+} = (\cos^{-1}\frac{x'-x_C}{L}, 0) \quad (6)$$

Then the following results hold:

(3a) Let $\tilde{x} \in [x_{min,free}, x_{max,free}]$. Then $(\theta_1, \theta_2)_{\tilde{x}+} \in \Theta_{FL+}$ and $(\theta_1, \theta_2)_{\tilde{x}-} \in \Theta_{FL-}$.

(3b) Θ_{FL+} and Θ_{FL-} are both connected spaces.

(3c) Θ_{FL} is connected if and only if
$$x_{min,free} \notin [x_C - \tfrac{L}{2}, x_C + \tfrac{L}{2}] \text{ or}$$
$$x_{max,free} \notin [x_C - \tfrac{L}{2}, x_C + \tfrac{L}{2}].$$

Proof. Result 3a follows trivially, since $(\theta_1, \theta_2)_{x'-}$ and $(\theta_1, \theta_2)_{x'+}$ are solutions to Equation 3 for $x_{CM,free} = \tilde{x}$ such that $\theta_1 \leq 0$ and $\theta_1 \geq 0$, respectively. This result implies that any attainable value of $x_{CM,free}$ is attainable with $\theta_2 = 0$, and that continuous paths between two values of $x_{CM,free}$ for $\theta_2 = 0$ always exist in both Θ_{FL-} and Θ_{FL+}. Since the boundary of Θ_{FL+} is defined by curves of constant $x_{CM,free}$, and since, as mentioned above, these curves do not intersect, then a curve of constant $x_{CM,free} = \tilde{x}$ between $(\theta_1, \theta_2) \in \Theta_{FL+}$ and $(\theta_1, \theta_2)_{\tilde{x}+}$ that lies completely within Θ_{FL+} always exists. So a path between any two configurations $(\theta_1, \theta_2)_i \in \Theta_{FL+}$ and $(\theta_1, \theta_2)_f \in \Theta_{FL+}$ can always be generated by moving from $(\theta_1, \theta_2)_i$ along a curve of constant $x_{CM,free} = \tilde{x}_i$ to $(\theta_1, \theta_2)_{\tilde{x}_i+}$, moving along $\theta_2 = 0$ to $(\theta_1, \theta_2)_{\tilde{x}_{f+}}$, and moving along a curve of constant

$x_{CM,free} = \tilde{x}_f$ to $(\theta_1, \theta_2)_f$. Therefore, Θ_{FL+} is connected. The result for Θ_{FL-} follows identically, so Result 3b holds. To prove Result 3c, notice that Θ_{FL} is connected if and only if some $(\theta_1, \theta_2) \in \Theta_{FL}$ exists such that $\theta_1 = 0$ or $\theta_1 = \pm\pi$. From Equation 3, this is equivalent to saying that $x_{CM,free} \notin [x_C - \frac{L}{2}, x_C + \frac{L}{2}]$ at some $(\theta_1, \theta_2) \in \Theta_{FL}$, which from Result 3b can occur if and only if either $x_{min,free} \notin [x_C - \frac{L}{2}, x_C + \frac{L}{2}]$ or $x_{max,free} \notin [x_C - \frac{L}{2}, x_C + \frac{L}{2}]$. □

3.3 Implications

Proposition 3 implies that for any feasible pelvis location the feasible space of the free limb can be divided into two non-empty, connected components, Θ_{FL+} and Θ_{FL-}. Therefore, using Result 3a it is possible to extend any feasible continuous path of the pelvis to a feasible path of the entire robot, such that the configuration of the free limb remains in either Θ_{FL+} or Θ_{FL-}. This key result yields the continuous planning approach described in this section to compute one-step motions of the robot.

First, decompose the four-dimensional feasible space $F(i,j)$ into four subsets as illustrated in Figure 6(a). Each subset corresponds to a pair of knee bends in the limbs forming the contact chain. In each subset, the position of the CM of the contact chain is uniquely determined by the position of the pelvis. Therefore, the feasibility of a pelvis location in each subset is determined by Equation 4. Transitions between subsets can occur only within one-dimensional curves along their boundaries, which correspond to feasible positions of the pelvis in which one of the limbs is fully stretched out.

Further partition each subset into two parts according to the sign of the configuration parameter θ_1 of the free limb, as illustrated in Figure 6(b). In one subset ($\theta_1 \geq 0$), the first segment of the free limb points upward; in the other subset, it points downward. Notice that the sign of θ_1 also serves to distinguish Θ_{FL+} from Θ_{FL-}, so robot configurations in each of the two parts of each subset correspond to free limb configurations entirely in either Θ_{FL+} or Θ_{FL-}. Transitions between the two parts can occur only within two-dimensional regions where Θ_{FL} is connected, i.e. where the conditions of Result 3c are satisfied.

Suppose for a pair of holds (i,j) that in each of the four subsets shown in Figure 6(a) an explicit representation can be built of the two-dimensional region formed by the feasible positions of the pelvis. From Proposition 3, this region is identical in the two parts of each subset corresponding to Θ_{FL+} and Θ_{FL-}. Therefore, the connected components of each of the eight subsets shown in Figure 6(b) can be determined. Likewise, suppose that an explicit representation can be built of the one-dimensional transition curves between the subsets corresponding to different knee bends and of the two-dimensional transition regions between the parts of each of these subsets corresponding to Θ_{FL+} and Θ_{FL-}. Using these transition

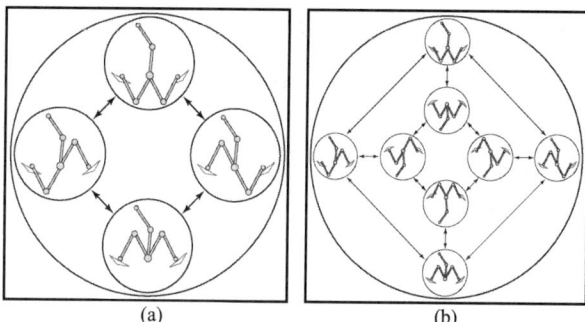

Fig. 6. Decompositions of $F_P(i,j)$ into four sub-spaces based on knee bends and into eight sub-spaces based additionally on whether the free limb is pointing up or down.

curves and regions, the connected components of each of the eight subsets can be linked to form a discrete graph.

Components of this graph are the connected components of the two-dimensional space of feasible pelvis positions of the robot. This space, which is denoted $F_P(i,j)$, is the projection of $F(i,j)$ onto (x_C, y_C). To plan a continuous path between any two points in a connected component of $F(i,j)$, it suffices to plan a path in the corresponding component of $F_P(i,j)$. This path can then be lifted to $F(i,j)$ using Proposition 3. Define a distinct state for each such connected component of $F_P(i,j)$, and denote it $(i,j)_A$, $(i,j)_B$, etc.

Finding analytic representations of $F_P(i,j)$ and the transition regions is impractical. In the current implementation, these regions are constructed using probabilistic roadmaps similar to those in [35]. For the three-limbed robot, a deterministic two-dimensional grid approximation would work as well. However, this approach might scale poorly to climbing robots with more than three limbs.

4 Overall Planning Framework

This section describes an overall framework for planning a sequence of one-step motions from a specific configuration on an initial pair of holds to a goal hold.

4.1 Search Space

The search space for the three-limbed climbing robot is a hybrid space, involving both continuous and discrete actions. Discrete actions correspond to placing a limb endpoint on a hold or removing it from the hold. They decompose the overall climbing motion into a sequence of steps. During each step, two limb endpoints are positioned at two given holds and the action is a continuous motion that brings the endpoint of the free limb from one hold to another. Therefore, motion planning can be divided into *discrete planning* and *continuous planning* operations, each with its own search space.

Section 3 described the continuous-planning search space of the robot, when its limbs are in contact with two holds i and j, and presented a method of generating the components of the feasible part of this space (e.g. $(i,j)_A$, $(i,j)_B$, etc.).

The discrete-planning search space is the collection of all of these components, for all pairs of holds (i,j) in the workspace. So, each state in this search space is a single component $(i,j)_M$ of $\mathbf{F}_P(i,j)$. Each possible successor of a state $(i,j)_M$ is another state of the form $(i,k)_N$ or $(j,k)_O$, which is a single component of $\mathbf{F}_P(i,k)$ or $\mathbf{F}_P(j,k)$ for a different pair of holds (i,k) or (j,k).

A link to a successor is a robot configuration with limbs in contact with holds i, j, and k, that satisfies two conditions. First, the position of the pelvis in this configuration must be common to both $(i,j)_M$ and $(i,k)_N$ (or $(j,k)_O$). Second, the four-dimensional representation of this configuration in $\mathbf{C}(i,j)$ and $\mathbf{C}(i,k)$ (or $\mathbf{C}(j,k)$) must be in $\mathbf{F}(i,j)$ and $\mathbf{F}(i,k)$ (or $\mathbf{F}(j,k)$), respectively. The second condition must be satisfied because the link defines a *specific* free-limb configuration, while the components $(i,j)_M$ and $(i,k)_N$ (or $(j,k)_O$) specify *compliant* free-limb configurations only. Note that this is the only point in the planning process at which $\mathbf{F}(i,j)$, rather than $\mathbf{F}_P(i,j)$, need be considered. In the current implementation, these link configurations are generated using a random sampling technique.

For example, consider the environment shown in Figure 7(a). The robot is initially located on holds (0,1) with a goal of reaching hold 4. The discrete-planning search space is shown in Figure 7(b). In this example, only $\mathbf{F}_P(2,3)$ has more than one component, $(2,3)_A$ and $(2,3)_B$.

4.2 Algorithm

In practice, it is too costly to compute the entire search space online for a reasonably sized environment. Instead, heuristic methods are used to guide the discrete search and the components of $\mathbf{F}_P(i,j)$ are only computed as each pair of holds (i,j) is explored.

For example, a necessary condition for a link between two pairs of holds (i,j) and (i,k) is that holds i and k be distant by less than $2L$. Likewise, the equilibrium regions (x_{min}, x_{max}) for both pairs of holds (see Section 3.1) must overlap. These simple conditions make it possible to quickly filter out many successor holds.

Another useful heuristic is to pre-compute rough discrete plans, without any continuous-planning exploration, using conservative approximations for the components $(i,j)_M$ of each $\mathbf{F}_P(i,j)$. In almost all cases, it has been found that each of the eight subsets illustrated in Figure 6(b) contains a *single* connected component. Using this decomposition, the entire discrete search space can be computed online. The resulting nominal plan is then used to guide a discrete search using the exact decomposition of every $\mathbf{F}_P(i,j)$.

The appropriateness of this approach is motivated by observation of the way in which human climbers plan

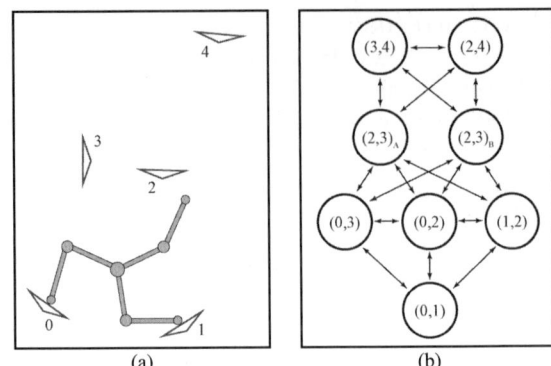

Fig. 7. An example environment for the three-limbed climbing robot and the corresponding connectivity graph between discrete states.

their motion. The resulting path, often called a *sequence*, consists of a series of *moves*, such as a back-step or high-step, between an ordered set of hand and foot placements (see [36, 37]). Each move does not specify a continuous path, but rather a discrete choice that is exactly analogous to the selection of knee-bend directions for the three-limbed robot.

Future observation of human climbers may suggest other useful heuristics, such as a consideration of the size of equilibrium regions.

5 Simulation

Figure 8 shows the result of applying the planning algorithm described in Sections 3 and 4 to move the three-limbed robot through the simulated vertical terrain illustrated in Figure 7(a). Notice that the center of mass of the robot stays within the equilibrium region, as required.

Other results, including animations of 3D-simulations, are available online at http://arl.stanford.edu/~tbretl/.

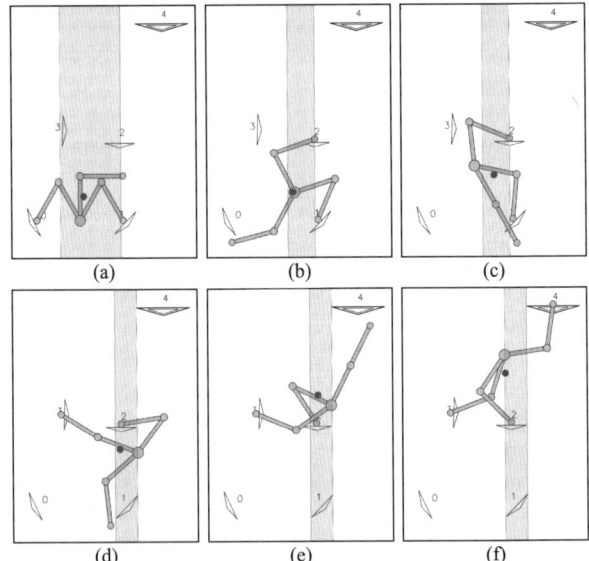

Fig. 8. Results of a simulation, shown as a time-sequence.

6 Future Work

This paper presented a framework for planning the motion of a three-limbed climbing robot in vertical terrain and showed the results of applying this framework in simulation. Current work concerns the application of the planning algorithm to experimental hardware. As part of this effort, the continuous-planning method described in Section 3 is being extended to handle additional motion constraints, more complicated robot geometries, imperfectly known environments, and three-dimensional terrain. Future work will address other fundamental issues such as sensing, control, hardware design, and grasping.

Acknowledgements T. Bretl is partially supported by an NDSEG fellowship through the ASEE and by a Herbert Kunzel Fellowship. The authors would also like to thank D. Halperin and T. Miller for their helpful contributions.

References

[1] P. Pirjanian, C. Leger, E. Mumm, B. Kennedy, M. Garrett, H. Aghazarian, S. Farritor, and P. Schenker, "Distributed Control for a Modular, Reconfigurable Cliff Robot," IEEE Int. Conf. on Robotics and Automation, 2002.

[2] A. Madhani and S. Dubowsky, "Motion Planning of Mobile Multi-Limb Robotic Systems Subject to Force and Friction Constraints," IEEE Int. Conf. on Robotics and Automation, 1992.

[3] S. Hirose, K. Yoneda, and H. Tsukagoshi, "Titan Vii: Quadruped Walking and Manipulating Robot on a Steep Slope," IEEE Int. Conf. on Robotics and Automation, 1997.

[4] M. Nilsson, "Snake Robot - Free Climbing," in *IEEE Control Systems Magazine*, vol. 18, Feb 1998, pp. 21-26.

[5] J. C. Grieco, M. Prieto, M. Armada, and P. G. d. Santos, "A Six-Legged Climbing Robot for High Payloads," IEEE Int. Conf. on Control Applications, 1998.

[6] H. Dulimarta and R. L. Tummala, "Design and Control of Miniature Climbing Robots with Nonholonomic Constraints," 4th World Congress on Intelligent Control and Automation, Jun 2002.

[7] S. W. Ryu, J. J. Park, S. M. Ryew, and H. R. Choi, "Self-Contained Wall-Climbing Robot with Closed Link Mechanism," IEEE/RSJ Int. Conf. on Intelligent Robots and Systems, 2001.

[8] W. Yan, L. Shuliang, X. Dianguo, Z. Yanzheng, S. Hao, and G. Xuesban, "Development & Application of Wall-Climbing Robots," IEEE Int. Conf. on Robotics and Automation, 1999.

[9] H. Amano, K. Osuka, and T.-J. Tarn, "Development of Vertically Moving Robot with Gripping Handrails for Fire Fighting," IEEE/RSJ Int. Conf. on Intelligent Robots and Systems, 2001.

[10] Z. M. Ripin, T. B. Soon, A. B. Abdullah, and Z. Samad, "Development of a Low-Cost Modular Pole Climbing Robot," TENCON, 2000.

[11] W. Neubauer, "A Spider-Like Robot That Climbs Vertically in Ducts or Pipes," IEEE/RSJ/GI Int. Conf. on Intelligent Robots and Systems, 1994.

[12] K. Iagnemma, A. Rzepniewski, S. Dubowsky, P. Pirjanian, T. Huntsberger, and P. Schenker, "Mobile Robot Kinematic Reconfigurability for Rough-Terrain," Sensor Fusion and Decentralized Control in Robotic Systems III, 2000.

[13] J.-C. Latombe, *Robot Motion Planning*. Boston, MA: Kluwer Academic Publishers, 1991.

[14] Y. Golubev and E. Selenskii, "The Locomotion of a Six-Legged Walking Robot in Horizontal Cylindrical Pipes with Viscous Friction," *J. of Computer and Systems Sciences Int.*, pp. 349-356, 2001.

[15] K. i. Nagasaka, H. Inoue, and M. Inaba, "Dynamic Walking Pattern Generation for a Humanoid Robot Based on Optimal Gradient Method," IEEE Int. Conf. on Systems, Man, and Cybernetics, 1999.

[16] M. Berkemeier, "Modeling the Dynamics of Quadrupedal Running," *Int. J. of Robotics Research*, vol. 17, Sep 1998.

[17] M. Buehler, U. Saranli, D. Papadopoulos, and D. Koditschek, "Dynamic Locomotion with Four and Six-Legged Robots," Int. Symp. on Adaptive Motion of Animals and Machines, 2000.

[18] M. F. Silva, J. A. T. Machado, and A. M. Lopes, "Performance Analysis of Multi-Legged Systems," IEEE Int. Conf. on Robotics and Automation, 2002.

[19] B. Goodwine and J. Burdick, "Motion Planning for Kinematic Stratified Systems with Application to Quasi-Static Legged Locomotion and Finger Gaiting," 4th Int. Workshop on Algorithmic Foundations of Robotics, Mar 2000.

[20] B. Goodwine and J. Burdick, "Controllability of Kinematic Control Systems on Stratified Configuration Spaces," *IEEE Tr. on Automatic Control*, vol. 46, pp. 358-368, 2001.

[21] J.-D. Boissonnat, O. Devillers, and S. Lazard, "Motion Planning of Legged Robots," *SIAM J. on Computing*, vol. 30, pp. 218-246, 2001.

[22] J.-D. Boissonnat, O. Devillers, L. Donati, and F. Preparata, "Motion Planning of Legged Robots: The Spider Robot Problem," *Int. J. of Computational Geometry and Applications*, vol. 5, pp. 3-20, 1995.

[23] J.-D. Boissonnat, O. Devillers, and S. Lazard, "Motion Planning of Legged Robots," *Rapport de Recherche INRIA*, vol. 3214, 1997.

[24] S. Kajita and K. Tani, "Study of Dynamic Biped Locomotion on Rugged Terrain," IEEE Int. Conf. on Robotics and Automation, 1991.

[25] S. Bai, K. H. Low, and M. Y. Teo, "Path Generation of Walking Machines in 3d Terrain," IEEE Int. Conf. on Robotics and Automation, 2002.

[26] J. Kuffner, Jr., S. Kagami, K. Nishiwaki, M. Inaba, and H. Inoue, "Dynamically-Stable Motion Planning for Humanoid Robots," *Autonomous Robots*, vol. 12, pp. 105-118, 2002.

[27] J. Kuffner, Jr., K. Nishiwaki, S. Kagami, M. Inaba, and H. Inoue, "Footstep Planning among Obstacles for Biped Robots," IEEE/RSJ Int. Conf. on Intelligent Robots and Systems, 2001.

[28] M. Kalisiak and M. v. d. Panne, "A Grasp-Based Motion Planning Algorithm for Character Animation," Eurographics Workshop on Computer Animation and Simulation, 2000.

[29] R. Alami, J. P. Laumond, and T. Simeon, "Two Manipulation Planning Algorithms," in *Algorithmic Foundations of Robotics*, K. Goldberg, D. Halperin, J.-C. Latombe, and R. Wilson, Eds. Wellesley, MA: A K Peters, 1995, pp. 109-125.

[30] J. Ponce, S. Sullivan, A. Sudsang, J.-D. Boissonnat, and J.-P. Merlet, "On Computing Four-Finger Equilibrium and Force-Closure Grasps of Polyhedral Objects," *Int. J. of Robotics Research*, vol. 16, pp. 11-35, Feb 1997.

[31] A. Bicchi and V. Kumar, "Robotic Grasping and Contact: A Review," IEEE Int. Conf. on Robotics and Automation, 2000.

[32] M. Yashima and H. Yamaguchi, "Dynamic Motion Planning Whole Arm Grasp Systems Based on Switching Contact Modes," IEEE Int. Conf. on Robotics and Automation, 2002.

[33] R. Mason, E. Rimon, and J. Burdick, "The Stability of Heavy Objects with Multiple Contacts," IEEE Int. Conf. on Robotics and Automation, 1995.

[34] R. Mason, E. Rimon, and J. Burdick, "Stable Poses of 3-Dimensional Objects," IEEE Int. Conf. on Robotics and Automation, 1997.

[35] L. E. Kavraki, P. Svetska, J.-C. Latombe, and M. Overmars, "Probabilistic Roadmaps for Path Planning in High-Dimensional Configuration Spaces," *IEEE Tr. on Robotics and Automation*, vol. 12, pp. 566-580, 1996.

[36] D. Graydon and K. Hanson, *Mountaineering: The Freedom of the Hills*, 6th Rev edition ed: Mountaineers Books, Oct 1997.

[37] J. Long, *How to Rock Climb!*: Chockstone Press, May 2000.

"MORITZ" a Pipe Crawler for Tube Junctions

Andreas Zagler

Applied Mechanics
Technical University of Munich
Garching, Germany 85748
zagler@amm.mw.tum.de

Friedrich Pfeiffer

Applied Mechanics
Technical University of Munich
Garching, Germany 85748
pfeiffer@amm.mw.tum.de

Abstract

This paper deals with the further development of the tube crawling robot "MORITZ", which was built at the Technical University of Munich. This robot can climb through pipes of different inclinations. It is improved in the recent research project, so that it also can manage tube junctions. After a brief introduction about problems concerning such a robot the used gait pattern is explained. Additionally some simulation results - showing the load at the robot during a motion through a tube crossing - are presented. Moreover the new developed joints in the central body as well as the sensors are described. The paper ends with a brief description of the crawler's control and its modifications for the new climbing maneuver in tube crossings.

1 Introduction

Leakages of pipe systems lead to a loss of the transported medium and may have a harmful effect on the environment. Therefore companies like the chemical, the power supply or the waste water industry require machines for the inspection and repair of their large pipe systems. Currently such robots are driven by wheels, chains or they float with the medium. Another possibility to realize the movement is to use a legged locomotion, which allows a higher flexibility than the common systems. Rossman [5, 6] built such a crawler (figure 1 left) at the Technical University of Munich. It is able to crawl in tubes of any inclination from horizontal up to vertical pipes. The robot also manages curved pipes with a diameter of $60 - 70\,cm$. To enable this the crawler is equipped with eight legs each with two driven joints, which can achieve torques up to $78\,Nm$. To move the crawler spreads four of its legs against the pipe wall to generate friction forces. These can carry its weight of $20\,kg$ and an additional load, while the other four legs swing to the next stance point. During this motion the robot control has to ensure permanently that there is sufficient friction force to bear the robot. However "MORITZ's" movement is limited to its working room limitations.

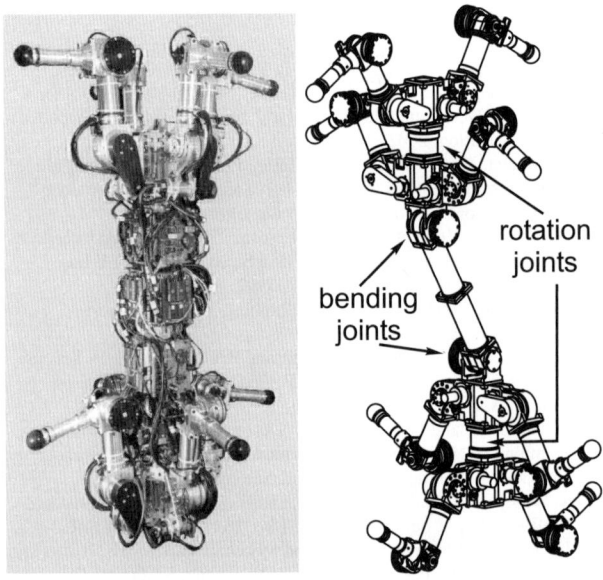

Figure 1: Pipe crawling robot "MORITZ"

The current research project aims to enhance the crawling capabilities of this robot to enable it to climb through pipe junctions. To make this possible the control has to find suitable stance points for the implementation of the movement. The analysis of the gait pattern, the joint angle limitation and the length of the leg showed that new joints (two bending joints and two rotation joints) have to be implemented in the central body (figure 1 right). The two bending joints divide the central body into three parts - the front and rear body, on which the legs are mounted, and the middle body. This configuration allows the robot to bend its central body and to maneuver in crossings, which are located in this bending plane. The other two joints (the rotation joints) separate each group of legs into two leg pairs to allow a rotational motion around the longitudinal axis of the crawler.

In section 2 the gait pattern will be described in more detail. Section 3 shows simulation results and in section 4 the design of the joints are explained. The used sensors as well as its redundancies is explained in section 5. Finally the control of the robot is described in section 6.

2 The gait pattern

To climb through the pipe the robot has to spread its legs against the tube wall to generate the necessary friction forces. Therefore two leg pairs, each consisting of two opposite legs, are needed to bear the robot (stance phase) while the other legs swing to the next stance point (air phase). Because of the limited degrees of freedom in each leg (only 2 degrees of freedom) those four legs are used to bear the robot, which lie in one leg plane (e.g. legs 1, 3, 5 and 7 in drawing plane of figure 2). When the legs, in the air phase, reach

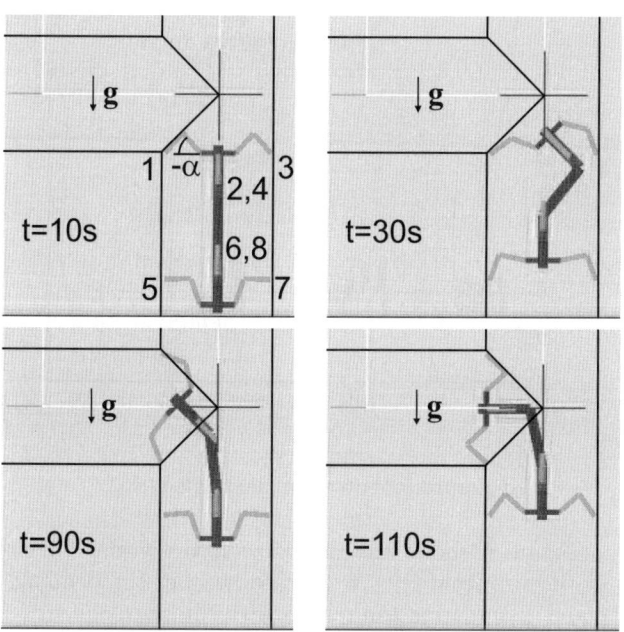

Figure 2: The gait pattern

their new stance points, their tasks are changing and they will have to bear the robot, while the other ones can change into the air phase and swing ahead. Due to this alternating bearing planes a three dimensional motion of the central body is realized.

To achieve a movement through a tube junction the following modifications are necessary. First of all the robot has to rotate around its longitudinal axis to get its bending plane parallel to the junction plane - both planes are parallel to the drawing plane in figure 2. Therefore the robot spreads in the pipe either with its outer or inner two leg pairs. Then the robot rotates the two leg pairs on the front and the rear body that are in the air phase to the new position. For this movement the robot requires the rotation joints. After that the robot changes its bearing plane and orientates its leg planes into normal position (figure 2, $t = 10\,s$). Now the bending joints allow to bend the central body as shown in figure 2 ($t = 30\,s$). At this position the bearing plane is alternated again (legs 2, 4, 6 and 8 are bearing the robot) and therefore the front and rear body can achieve a motion only in their bearing planes. That is the reason why the middle body is needed. The middle body decouples the forward motion of the front body from the rear body and thus avoids a drilling friction in the stance feet. In the following the bearing plane alternates again and the robot will be in position $t = 90\,s$. The robot finally bends its front body completely into the intersecting tube ($t = 110\,s$). Analogical to this the rear body will move into the intersecting tube.

3 Simulation results

To analyze the motion under different joint configurations and to determine the loads at the joints as well as the structure a multi-body simulation program was developed. The examination of these joint con-

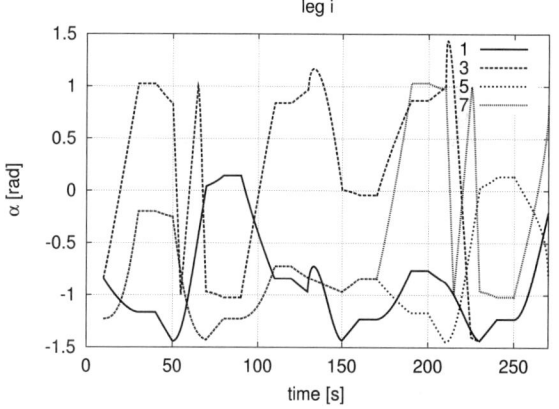

Figure 3: Leg joint angle α

figurations resulted in the concept with four new joints as described above. To simulate the motion of the robot a target path is given, which was transformed into minimal coordinates via a newton approximation. Figure 3 shows the angle α for the legs in the bending plane during the maneuver through a tube junction.

α describes the angle between the central body and the first leg segment, while $\alpha = \pm\frac{\pi}{2}$ means that the leg segment is parallel to the central body. The joint angle never exceeds the limitation of $\pm\frac{\pi}{2}$.

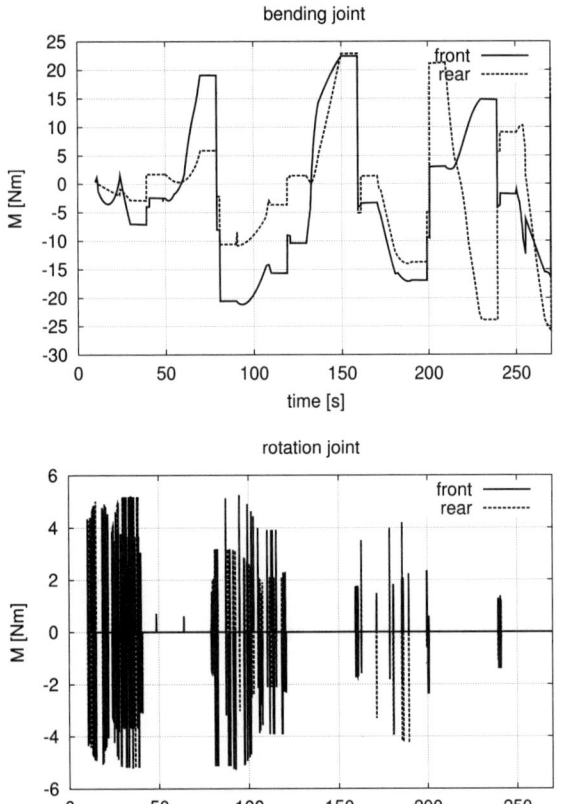

Figure 4: Torques of the central body joints

For the load determination an optimization of the feet forces is necessary. This optimization is realized with a linear programming under the quality function

$$\max_{\forall \text{ joints } i} \left(\frac{|M_i|}{M_{i,limit}} \right) \longrightarrow \text{MIN!}$$

and the following constraints:

- The target position has to be observed.
- The friction constraint must be fulfilled, which means that for the linear programming the friction cone has to be linearized to a friction pyramid.
- The limits of the drives $M_{i,limit}$ have to be observed.
- A maximal foot force of 350N must not be exceeded. This limitation is necessary due to a implemented force sensor.

The quality function was chosen to minimize the maximal joint load. For the solution of this min-max-problem, it can be transformed in a linear programming if all constraints are linear [3]. In figure 4 the computed joint moments of the central body are shown. The loads at the rotation joints result from the large lever arm (matching the pipe diameter) and small differences in the friction forces lateral to the leg plane.

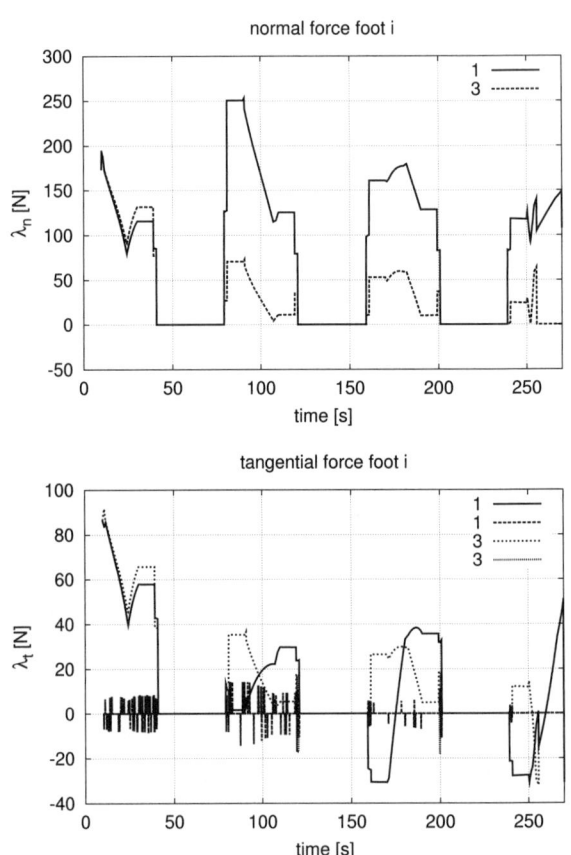

Figure 5: Forces of the front feet

Figure 5 shows the normal and tangential forces at the two front legs, which are lying in the bending plane. Comparing the results with the positions visualized in figure 2 it can be shown that the normal force in leg 3 decreases. This results from the direction of the gravitation vector **g** which causes leg 1 to bear the robot while the normal force in leg 3 generates an additional load. The remaining normal force in leg 3 is used to generate a friction force for the compensation of the tangential force. The tangential forces between $t = 10\,s$ and $t = 40\,s$ are used to bear the robot in the vertical pipe and thus have to be generated out of the normal force. The tangential force of leg one for the time period $t > 50\,s$ is smaller than the generated friction force.

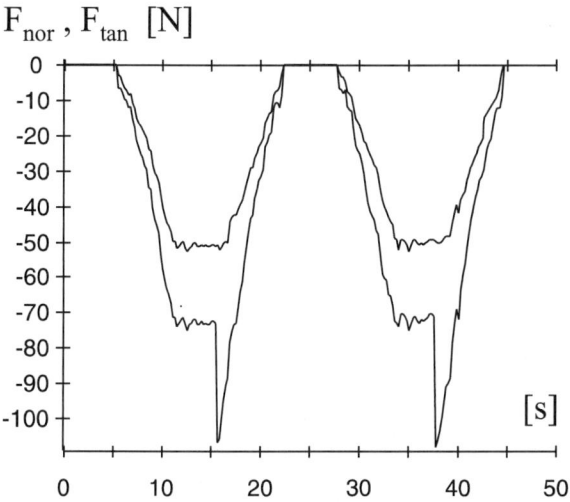

Figure 6: Measured leg forces for a motion in a vertical tube with friction coefficient $\mu = 0.75$

A further verification is done by comparing the simulated forces for the redesigned robot with the measured ones of the present robot (Figure 6). In order to get comparable results we consider a motion through a vertical pipe combined with a force optimization analog to the one described above. Due to the existing limitations in the gait pattern only the first time period can be compared. It can be seen that the tangential force is 30% greater than the measured force, which corresponds to the increased weight of the robot. The difference between the normal forces results from the reduced friction coefficient of $\mu = 0.5$ for the simulation. Based on this and further simulations of the climbing maneuver in different conditions the gait pattern as described is selected. Finally the simulated loads of the central body joints serve as estimation for the joint construction.

4 The central body joints

The great amount of joint as well as the demand for a low system weight, which is required for this kind of motion, claims for a light weight and space-saving design. Due to the existing power supply, DC motors in combination with Harmonic-Drive-gears have to be used. With respect to the simulation results ESCAP motors 35NT2R82, which have a good efficiency weight and a compact size, are selected. These motors admit a continuous torque of $0.115\,Nm$. That is the reason why gears with a reduction of $i \approx 480$ are needed for the bending joints. This gear reduction is generated by a bevel gear with $i = 3$ and a Harmonic-Drive-gear (HD-gear) HFUC 20 with $i = 161$. Figure 7 shows the design of the bending joint.

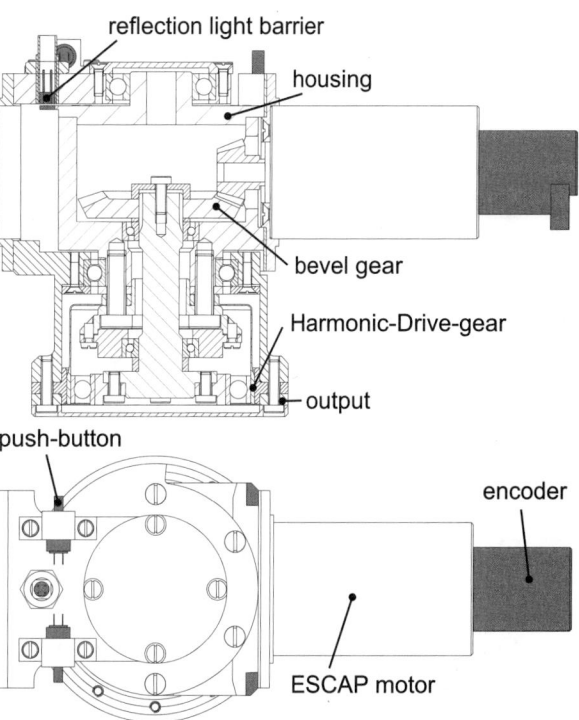

Figure 7: The bending joint

The motor actuates the drive shaft, which is fixed to the wave-generator of the HD-gear via the bevel gear. The drive shaft itself is bedded by two angular contact bearings in the housing, to which the middle body of the robot is connected. The flex spline is fixed to the housing and thus the output is connected to the circular spline of the HD-gear. This output is bedded in the housing with two ball bearings, because only small axial forces have to be transmitted from the legs to the central body.

For the design of the rotation joint the same motor is used, but due to the smaller loads and the different rotation axis another gear ratio as well as another mounting is selected. A HD-gear HFUC 17 with a reduction of $i = 121$ is sufficient and allows a high transmission accuracy as well as a high angular repeatability. This time the motor drives the shaft and thus the wave-generator of the HD-gear directly. The drive shaft is again bedded in the housing with two angular contact bearings, to which one of the inner leg pair is connected. The flex-spline is also fixed to the housing thus connecting the output again to the circular-spline. But this time the output is bedded via two angular contact bearings, due to the great bending torques, which have to be transmitted from the output to the housing. Figure 8 shows the design of the rotation joint.

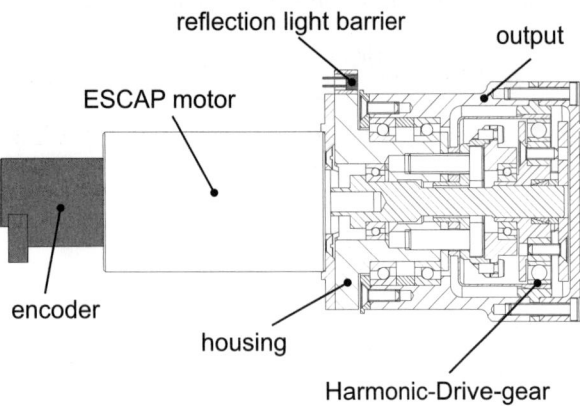

Figure 8: The rotation joint

5 The sensors of the crawler

Each leg has two actuated joints and thus two potentiometers for measuring the angles and two tachometer generators for measuring the angular velocity. This redundancy is used by the robot control to evaluate both sensor signals for integrity and therefore can halt the robot to avoid damage if one sensor fails.

Also a light weight force/torque sensor, which was developed for this robot [5], is integrated in the outer leg segment. Gálvez improved this sensor not only to measure the contact forces but also to allow the controller to determine the normal vector in the contact point of the foot [2, 7].

For the control of the central body joints sensors are needed to measure the angle and angular velocity. Both is done via encoders which are connected to the motors. However this kind of sensor would lack the sensor's redundancy as well as the possibility to determine the correct reference point. The reference signal which is sent by the encoder is repeated every revolution of the motor and thus wouldn't be unique. To solve this problem a reflection light barrier is implemented in the joints which divides the adjusting range into two parts, one with reflection and another one without. On the one hand this signal is used to find the correct rotation direction on startup, to get the transition point from the reflected area to the unreflected one. On the other hand this resulting transition point signals that the next encoder reference defines the authentic reference. Finally, to solve the redundancy problem two different solutions are realized. For the bending joint two push-buttons are mounted which shut off the power of the motors when the joint reaches the end of the adjusting range. The redundancy of the other joint is done with a second reflection light barrier which signals faulty angular positions to the controller which will shut off the power of the motors. Therefore the sensors are mounted in a way that ensures that only functional light sensors send a faultless signal. Both senor systems can be seen in figures 7 and 8.

The advantage of the encoder over the analog sensors is a higher accuracy of the determination of the angle as well as a reduction of computation time and measurement noise. In an experimental setup a resolution of 2000 signals per motor revolution is reached for all new joints. But due to the restricted computation power of the microcontroller rounding errors must be expected in the control computations.

6 Crawler control

For the control of the robot all sensor signals are evaluated by five Siemens C167CS microcontrollers, which are mounted on the central body. One controller acts as a central unit, while each of the remaining four units controls a leg pair. A CAN bus system is used for the communication between the microcontrollers. Thus it is possible to give simple control commands or to get information about the system via an external PC.

The implementation of a decentral control architecture similar to the one of the six legged walking machine [1, 4] enables to divide the whole control problem into different subproblems. On the one hand the controllers have to ensure that the legs are coordinated with each other and on the other hand they have to execute the operations. Each of the two task levels (the coordination level and the operating level) can be divided again into a central and a local part. Figure 9 shows this task distribution.

- The *central coordination level* coordinates the characteristic phases of the two leg planes. The controller decides which legs have to bear the load and which ones can swing ahead to the next stance point on their own. Furthermore, problems which can only be mastered by a reaction of the whole robot are solved on this level (e.g. the legs of one plane cannot find a contact point in their whole working area). Therefore this control level also can influence the local coordination level.

- The *local coordination level* controls the changes between the different leg phases (stance, protract,

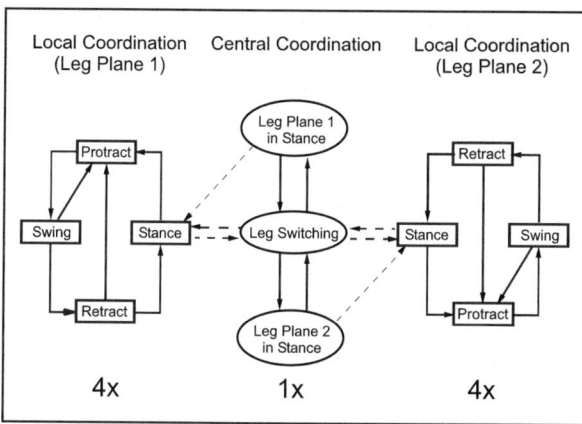

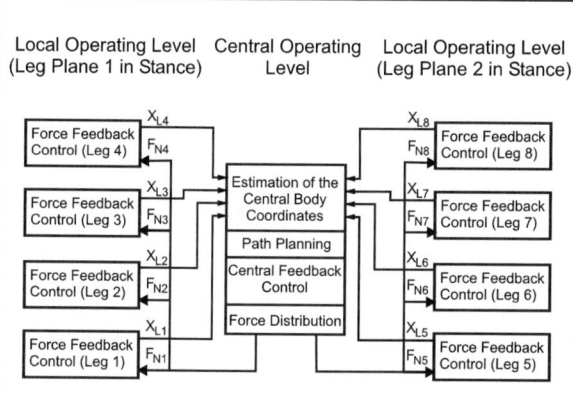

Figure 9: Control scheme for the crawler

responding communication between the leg controllers.

7 Conclusion

In this paper the tube crawling robot "MORITZ" as well as its improvements to climb through tube junctions are presented. "MORITZ" is able to move with a legged locomotion through pipes of any inclination and bears its weight only with friction forces. For the new features the robot gets four new joints in the central body which allows to rotate in the pipe and to bend its body. The gait pattern that will be used to walk through a pipe crossing is explained and a simulation describes this movement. The optimization, which is used to distribute the feet forces, is elucidated. Afterwards the design and the sensors for the new joints are explained. Finally the control concept, which is divided in decentralized versus centralized tasks and in coordination versus operating tasks, is described.

References

[1] ELTZE, J.: *Biologisch orientierte Entwicklung einer sechsbeinigen Laufmaschine*, Fortschrittberichte VDI, Reihe 17, Nr.110, VDI-Verlag, Düsseldorf, 1994.

[2] GÁLVEZ, J. A.; DE SANTOS, P. G.; PFEIFFER, F.: *Intrinsic Tactile Sensing for the Optimization of Force Ditstribution in a Pipe Crawling Robot*, IEEE/ASME Transactions on Mechatronics, (2001).

[3] GILL, P.; MURRAY, W.; WRIGHT, M.: *Practical Optimization*, Academic Press, London, 1988.

[4] PFEIFFER, F.; ELTZE, J.; WEIDEMANN, H.-J.: *The TUM-Walking Machine*, Intelligent Automation and Soft Computing, 1 (1995), pp. 307–323.

[5] ROSSMANN, T.: *Eine Laufmaschine für Rohre*, Fortschrittberichte VDI, Reihe 8, Nr. 732, VDI-Verlag, Düsseldorf, 1998.

[6] ROSSMANN, T.; PFEIFFER, F.: *Control of a Tube Crawling Robot*, Proc. of the European Mechanics Colloquium Euromech 375, Lehrstuhl B für Mechanik, ed., Munich, Germany, 1998.

[7] ZAGLER, A.; PFEIFFER, F.: *Joint Design and Sensors for a Pipe Crawler*, Robotik 2002, VDI, ed., 2002.

swing and retract). If a predefined event occurs the controller switches to the corresponding leg phase. For example a leg in swing phase is switched by the local coordination level to the retract phase, if the leg has passed a geometric limit in the front area. The central coordination may influence this process under certain conditions. This control level also reacts to disturbances like avoiding small obstacles or finding no contact.

- The *central operating level* controls the positions and the velocities of the central bodies. For this purpose an appropriate force distribution is computed to determine the joint torques similar to the one described in section 3. These joint torques are then sent to the local operating levels.

- The *local operating level* controls the applied forces during the contact phase and the motion during the different air phases of a single leg. While the motion in the air phase is really a local control problem, the leg forces in the stance phase are strongly coupled and therefore the force control cannot be implemented without a cor-

Microfabricated Thermally Actuated Microrobot

Agnès Bonvilain, Nicolas Chaillet

Laboratoire d'Automatique de Besançon
LAB UMR CNRS 6596 / IMFC FR W0067
25 rue Alain Savary
25000 Besançon
France
agnes.bonvilain@ens2m.fr , nicolas.chaillet@ens2m.fr

Abstract – The work presented in this paper concerns the study, the design, the fabrication and the experimentation of a legged microrobot. Legs were developed by integrating in each two thermal bimorph actuators, which give two degrees of freedom. After their fabrication and their experimentation, the legs were integrated on a body to realize the complete structure of the microrobot in a monolithic way. This microrobot was totally made by microfabrication in a clean room. The microrobot has a total volume of 6 mm x 3,5 mm x 0,5 mm and possesses six legs. Its experimentation allowed to determine its potential performances. This microrobot can find numerous applications in microrobotics, such as the inspection of stuffy environments, the micro-escorting or as a plug and play function in a system.

I. INTRODUCTION

The miniaturization of robots does not only implies a simple scale factor applied to existing macrotechnologies. It is often necessary to use new principles for actuation and fabrication. The fabrication of complete structures of microrobots in millimeter-length or in micrometric dimensions often needs the use of microtechnologies in clean room, originally comming from microelectronics technologies.

The objective of our work consists of the realization of a microrobot with legs, moving on a surface not participating in its locomotion, which we define by :
- a robot manufactured collectively by microfabrication,
- a robot whose volume is lower than a cube of 10 mm aside,
- a robot carrying out of the tasks in the microworld.

In our state of the art, we were interested in the microrobots concerned with this definition. At this day, seven microrobots of this type or closed to were already studied worldwide [1] to [7].

The state of the various works shows the difficulties met for the realization of such structures, notably because of technological barriers, but also because of the difficulties in integration of actuators in so small dimensions. For example, the microrobot described in [1] moves only on a surface which participates in its locomotion, it is the most smaller. In the microrobot described in [4], the actuation was not integrated yet. The microrobot described in [5] is under development, and exceeds the dimensions that we settled, but on the other hand, it will be totally wireless and autonomous. The microrobot reported in [6] moves only on the glass, and it exceeds the dimensions that we settled. Finally, the microrobot proposed in [7] has in our knowledge never be fabricated. Only the microrobots reported in [2] and [3] are achieved and correspond to the requirements that we settled for ours.

The paper is divided in three main parts. In the first one we present the thermal bimorph actuator which are integrated in the legs. The second part explains the structure and the functionning of the leg. Then in the third part we describe the complete structure of the microrobot, its experimentations and its potential performances.

II. THERMAL BIMORPH ACTUATOR

For the actuation of our microrobot, different actuators were studied, notably the piezoelectric and the thermal actuators. The actuator used for the movement of legs is the thermal bimorph. It was chosen for the importance of its deflection and of its density of energy [8].

A. Principle of functionning

In this work, several types of legs were studied, fabricated and experimented [8] to [11]. Only the last ones which have given the best results are described here. In this most recent version, the legs are actuated by thermal bimorphs constituted of a layer of SU-8 resin (60 µm) and a layer of silicon (20 µm) (see figure 1). Between these two layers, there is two very thin layers : one of insulating material (1,2 µm of oxide) and an other one of metal (1 - 2 µm of gold) . They are used for warming the actuator by Joule effect. Both active constituents (i.e. resin and silicon) were chosen, for the important difference of their coefficient of thermal expansion ($6,2.10^{-6}$ K^{-1} for silicon and 52.10^{-6} K^{-1} for SU-8 resin), as well as for the possibility of using them in microfabrication. When one heats the bimorph, the resin extends much more than the silicon and causes a flexion as shown in dotted lines on figure 1.

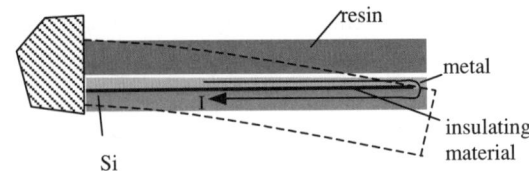

Figure 1 : plan of the thermal bimorph actuator with its deformed shape in dotted lines

The actuator is heated by Joule effect. The current crosses the silicon and the gold as indicated on figure 1. The SU-8 resin is heated by thermal conduction.

The deflection of a free embeded bimorph is given by the following equation [12] :

$$\delta = \frac{V.L^2.\Delta T}{2.e} \quad (1)$$

δ is the free deflection of the bimorph (m),
V is the specific curvature defined further (K⁻¹),
L is the length of the bimorph (m),
ΔT is the rise of temperature of the bimorph when it is heated (K),
e is the thickness of the bimorph (m).

The blocking force at the tip of a bimorph is given by the equation :

$$P = \frac{V.E.l.e^2.\Delta T}{8.L} \quad (2)$$

P is the external force applied at the tip of the bimorph which cancels the deflection for a given rise of temperature ΔT (N),
E is the mean Young modulus of the two constituents (Pa),
l is the width of the bimorph (m).

The specific curvature V is given by the following equation, in which the indication 1 is allocated to the least dilatable material (i.e. the silicon) and the indication 2 in the material the most dilatable (i.e. the SU-8 resin) :

$$V = \frac{2}{3} \cdot \frac{\alpha_2 - \alpha_1}{1 + \frac{(E_1.e_1^2 - E_2.e_2^2)^2}{4.E_1.e_1.E_2.e_2.e^2}} \quad (3)$$

V is the specific curvature (K⁻¹),
α_1, α_2 are the coefficients of thermal expansion of each material (K⁻¹),
E_1, E_2 are the Young modulus of both constituents at considered temperatures (Pa),
e_1, e_2 are the thickness of each material of the bimorph (m),
e is the thickness of the bimorph (m).

B. Experimentations

The bimorph has a length of 2 mm and a width of 100 µm. The fabrication is made by microfabrication. The experimentations allow us to measure the deflection according to the applied power, as it is presented on figure 2.

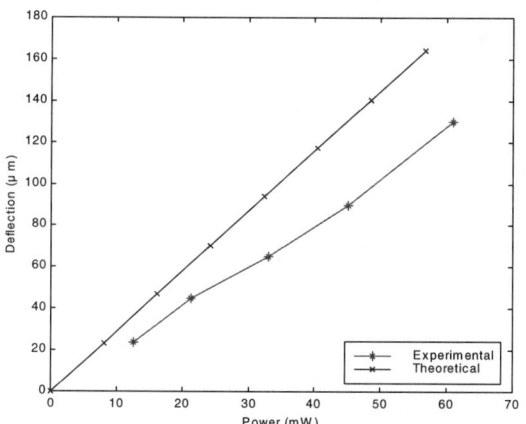

Figure 2 : theoretical and experimental free deflection of the actuator according to the applied electric power

The theoretical results are obtained with the following equation :

$$\Delta T = \frac{Pc}{S.h} \text{ so } \delta = \frac{V.L^2.Pc}{2.e.S.h} \quad (4)$$

Pc is the power consumption (W),
S is the convection surface of the actuator (m²),
h is the convection coefficient (W/m².°C).

The figure 3 present the blocking force versus the deflection for a fixed ΔT. We can say that it is the caracteristic of the bimorph actuator.

The results of the experimentations show the great magnitude of the free deflection of this type of actuator (130 µm), as well as the importance of the effort which can be generated (more than 6.5 mN for the blocking force). Previous actuators made of aluminium and silicon give a big power consumption which were divided by a factor of ten with this new SU-8resin and silicon actuator [8].

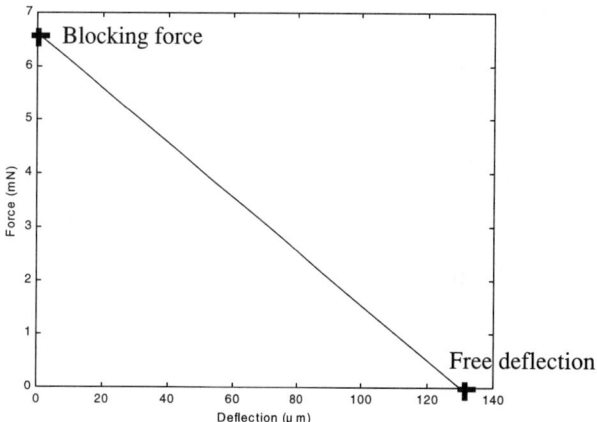

Figure 3 : blocking force at the end of the bimorph according to its free deflection (at a fixed ΔT ≈ 100 °C)

III. THE LEG

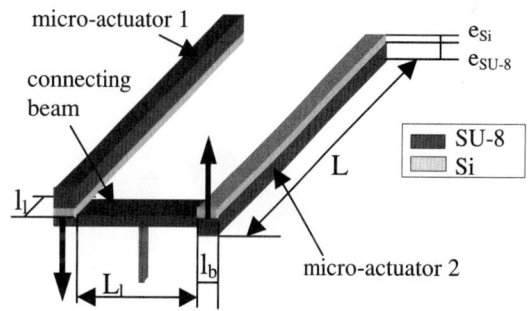

Figure 4 : plan of the leg, the arrows at the tip of the actuators indicate their direction of deflection

Two micro-actuators such as those presented above were then integrated to realize a leg with two degrees of freedom. The structure of the leg, actuated by two thermal bimorphs working in opposition, and connected by a connecting beam made of SU-8 resin (20 µm) is presented in figure 4. The dimensions are : L = 2 mm, L_1 = 500 µm, l_b = 100 µm, l_1 = 100 µm, e_{Si} = 20 µm and e_{SU-8} = 60 µm.

The operating cycle of the leg is presented in figure 5, the leg being seen at its extremity. On this figure, the currently actuated bimorph is represented by the letter "A", and arrows indicate the direction of the corresponding leg motion. On the other hand, the initial position of the microleg is drawn in dotted lines.

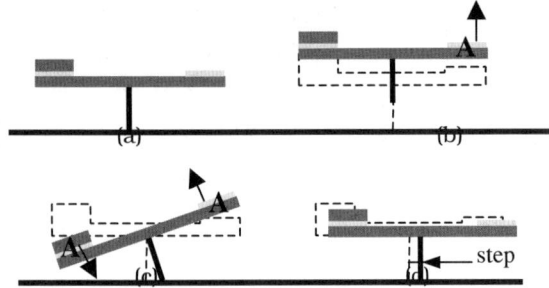

Figure 5 : plan of the operating cycle of the leg
(a) initial position of the leg,
(b) the bimorph which bend upward is actuated and the leg gets up, the rigidity of the connecting beam is sufficient to pull the complete leg,
(c) both bimorphs are actuated, the leg tilts, and the foot gets down on the ground,
(d) the actuation of bimorphs is stopped, and the leg resumes its initial shape by making a step.

Numeric simulations were made with Ansys 5.6, to verify the deformations of the leg, as well as to determine the temperature in the leg. This last one allowed to verify that the extremity of the foot stays permanently at ambient temperature, that allows the leg not to alter the surface on which it moves. The figure 6 (a) presents the deformations of the leg when both actuators are powered, and the figure 6 (b) present the distribution of the temperature along the leg when the both actuators are powered. The cooling of the actuators is done only by convection.

Legs were fabricated by microfabrication. Forty two legs were manufactured simultaneously on a three inches silicon wafer and required about fifty stages of manufacture and two working weeks in clean room. The figure 7 presents the result of this fabrication.

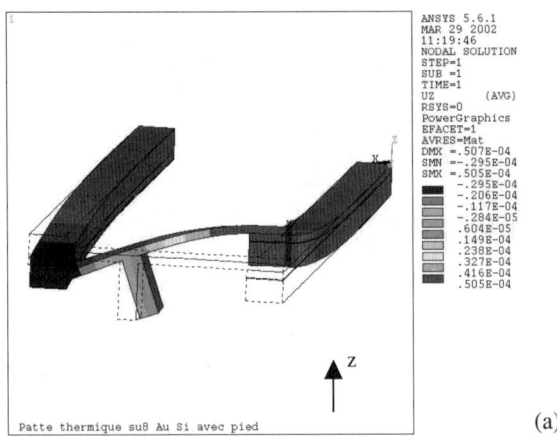

(a)

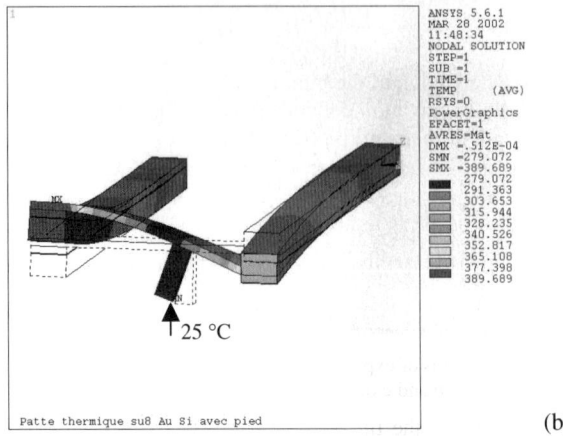

(b)

Figure 6 : Numeric simulation (a) of the deflections on z axis of the leg (b) of the distribution of temprature along the leg, when both actuators are powered

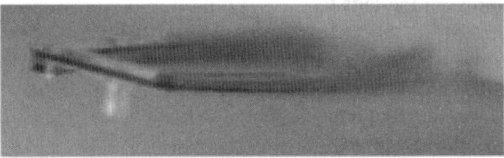

Figure 7 : photo of a leg (700µm width)

Several legs were experimented, and the figure 8 presents the photos of a leg in the positions a, b and c given on the figure 5. On the figure 8 (a), the connecting beam is not horizontal, this is due to residual constraints of fabrication.
On the figure 8, the ground and the initial position of the foot were represented by white lines. The results of the experimentations allow to give the performances of these legs.

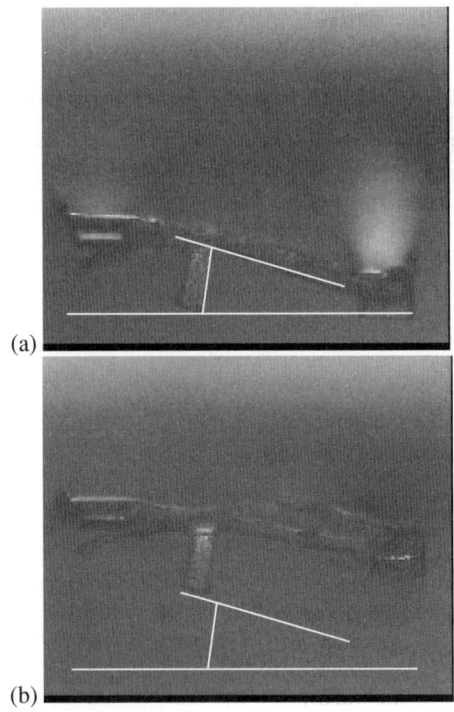

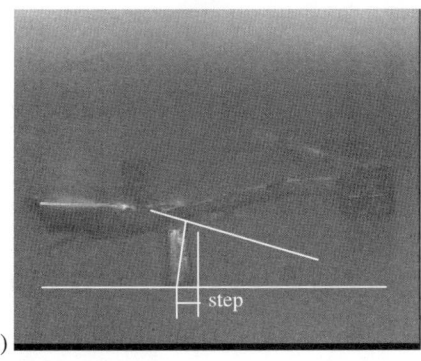

Figure 8 : photos of experimentation of the leg in the three positions a, b and c of the operating cycle of figure 5

Their deflection at the tip is 130 µm (to move the leg up). The weight of a leg is 54.22 µg, and the force which can be developed for no deflection is 6,6 mN, that is the equivalent of 0,673 g. Calculations made possible to show that this allows to intend the legs to carry a body, but also the energy and control necessary for its functioning. This calculations are not detailed here. Besides, the length of the step was measured at 60 µm, and a dynamic experimentation allowed to determine the frequency of functioning of the leg at 1.7 Hz. This allows to estimate the walking speed of the microrobot of about 100 µm/s.

IV. THE MICROROBOT

This section presents the complete mechanical structure of the microrobot integrating six of the previously described legs (see figure 9). It then gives a monolithic structure (without any assembling step) based on a silicon body.

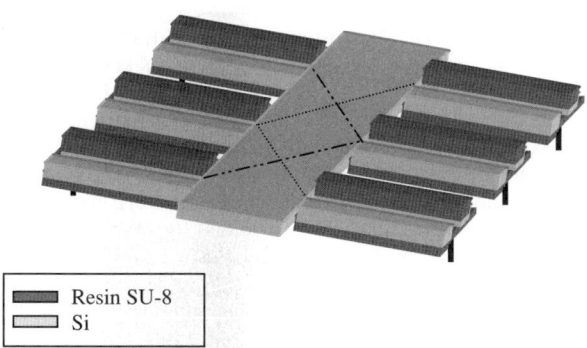

Figure 9 : the complete mechanical structure of the microrobot

Its manufacture requires approximately the same number of stage as for the legs and the same time. On a wafer of three inches silicon wafer, twenty four microrobots are manufactured simultaneously. The result of this fabrication is presented in figure 10.

The version of the microrobot presented in figure 10 could not be powered. Then for the first experimentations, the microrobots which are presented in this paper stay on the silicon wafer in which they are made, as it is shown on figure 11 (a). The connections of legs among them are made by bonding on the wafer (see figure 11 (b)). External needles allow to supply the legs of the microrobot (see figure 11 (c)).

The control is carried out using a Display Signal Processing (DSP) associated with an electronic card.

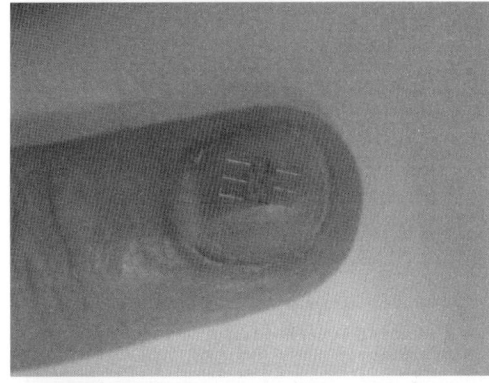

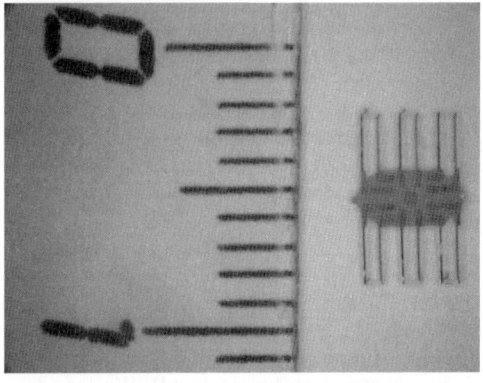

Figure 10 : photos of the microrobot on a finger nail and behind a ruler

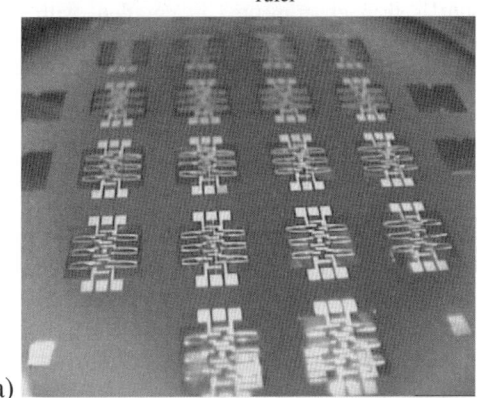
(a)

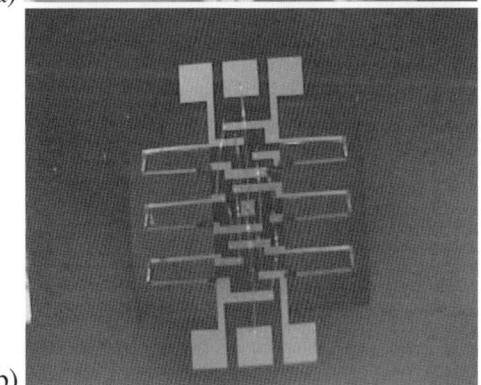
(b)

Figure 11 : photo of microrobots (a) united of the wafer (b) with the bounding (c) with the needles

The coordination of the legs motions are done as for the normal walking of real insects, i.e. the legs are actuated three by three : a group of three is done by the to extreme legs of one side and the central leg of the other side. A DSP and an I/O electronic card are used to generate synchronized signals to control the legs cycles. Various experimentations were made. The first experimentation consists in the motion of the six legs of the microrobot in a synchronized way. During this experimentation, the current is measured, it allows to calculate the electric power (437 mW) and the consumed energy (157 mJ) for one step. Tests are made from 0.2 to 2 Hz (the bandwidth of each leg is 1.7 Hz). Figure 12 presents the frequency response (experimental and identification results) of the actuator described in part II. We can identify in a first approximation with a first order.

The microrobot is then experimented in duration, with a frequency of 0,5 Hz during more than one hour, what corresponds to more than 2.10^3 operating cycles. This test entails no deterioration.

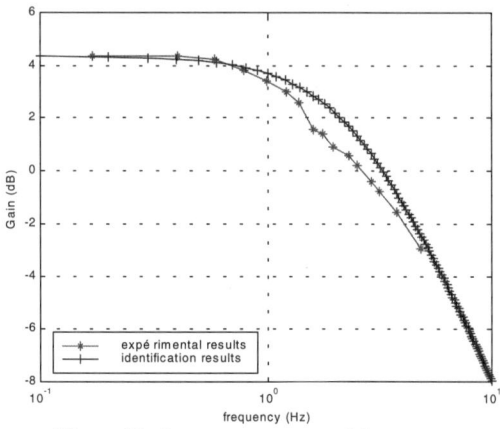

Figure 12 : frequency response of the actuator

An other experimentation is done, which consist in loading an immobile prototype of microrobot with a constant weight. Its shows that the microrobot can carry a mass of 2 grams without becoming deformed (figure 13 (a) and (b)). More significant masses could not be put on the microrobot for a problem of stability.

A last experimentation showed that the microrobot did not slip on a plan made of paper inclined until 20° compared to the horizontal (figure 14). The same experimentation gives 12° for a surface of polished glass.

Figure 13 (a) : profile of the microrobot without load

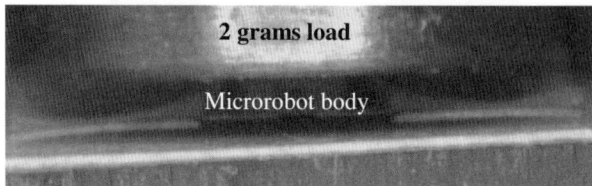

Figure 13 (b) : profile of the microrobot with 2 grams load

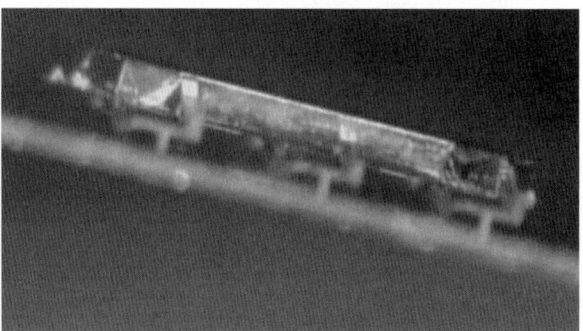

Figure 14 : plan out of paper inclined at 20°

The experimentations allowed to detemine the characteristics of the microrobot. The dimensions of the microrobot are : 6 mm in width, 3,5 mm in length and 0,5 mm in thickness. It is thus smaller than a cube of 10 mm aside. Figure 17 presents the congestion of the microrobot in a cube of 10 mm aside. The walking speed is calculated at 100 µm/s, which remains to verify. It can carry a load of 5,3 g for a weight of 2,33 mg. Knowing that at present, lithium microbatteries have a density of energy of the order of 500 to 600 J/g, knowing the energy consumption of the microrobot (157mJ / cycle), and knowing the load which it can carry (5g), it is possible to evaluate the approximate time of the autonomy of the microrobot, i.e. about 3 hours, without considering the weight of the control.

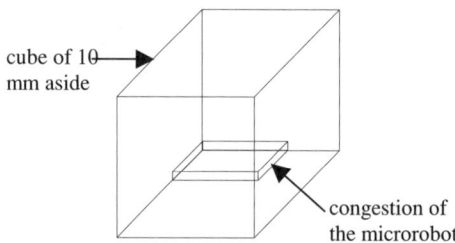

Figure 15 : congestion of the microrobot in a cube of 10 mm aside

V. CONCLUSION

If one makes a comparison with microrobots already studied worldwide, our microrobot seems to be the smallest to be able to move on a surface not participating in its locomotion, and it is the one which can carry the biggest load.

The obtained results are very encouraging, and allow us to envisage the re-design of the electrical connections and the fabrication of new prototypes, which will allow to experiment the microrobot in more real walking condition. Other experimentations can then be made on variable surfaces roughness, even the clearing obstacles. The load carrying can also be tested.

In a more distant future, its energy and control can be also integrated, as well as sensors and elements of telecommunication.

VI. ACKNOWLEDMEMENTS

We want to thank Pascal Blind from the CTM and Jean Claude Jeannot from the IMFC for their technical support during the manufacture of the different structures.

VII. REFERENCES

[1] Takashi Yasuda, Isao Shimoyama, Hirofumi Miura ; "Microrobot Actuated by a Vibration Energy Field" ; The 7th International Conference on Solid-State Sensors and Actuators ; pp 42-45 ; 1996.

[2] P.E. Kladitis, V.M. Bright ; "Prototype Microrobots for Micro Positionning and Micro Unmanned Vehicles" ; Sensors and Actuators 80 ; pp 132-137 ; 2000.

[3] Thorbjörn Ebefors, Johan Ulfstedt-Mattsson, Edvard Kalversten, Göran Stemme, "A Walking Silicon Micro-Robot", 10th Int Conference on Solid-State Sensors and Actuators (Transducers'99), 7-10 june 1999, Sendai, Japan, pp 1202-1205.

[4] Richard Yeh, Ezekiel J. J. Kruglick and Kristofer S. J. Pister ; "Surface-micromachined components for articulated microrobots" ; Journal of microelctromechanical systems ; Vol. 5 ; N°1 ; pp 10-17 ; march 1996.

[5] P Basset, A Kaiser, P Bigotte, D Collard, L Buchaillot ; "A large stepwise motion electrostatic actuator for a wireless microrobot" ; Proceeding of MEMS'02, pp 606-609.

[6] Matthew H. Mohebbi, Mason L. Terry, Karl F. Böhringer ; "Omnidirectional Walking Microrobot Realized by Thermal Microactuator Arrays" ; Proceeding of 2001 ASME International Mechanical Engineering Congress and Exposition ; New York ; November 11-16 200.

[7] Jan G. Smits ; "Is Micromecanics Becoming a New Subject for Academic Courses or the design of a Piezoelectric on Silicon microrobot" ; Industry Symposium, IEEE, 1989.

[8] J. Agnus, A. Bonvilain, G. Cabodevila, N. Chaillet, Y. Haddab, P. Rougeot "Study and Development of a Station for Manipulation Tasks in the Microworld", SPIE Conference on Microrobotics and Microassembly, Vol. 3834, pp 141-152, Boston, September 1999.

[9] Agnès Bonvilain, Nicolas Chaillet, "Some Prototypes of Silicon-based Thermal Actuated Microlegs for an Insect-like Microrobot", Mechatronics'01, Besancon France, septembre 2001.

[10] Agnès Bonvilain, Nicolas Chaillet, "Fabrication and Experimentation of Microlegs for an Insect-like Microrobot", International Conference on Microrobotics and Microassembly III, Spie Vol. 4568, pp 163-174, Boston USA, october 2001.

[11] Agnès Bonvilain, Jean René Coudevylle, Pascal Blind, Nicolas Chaillet, "Micromachined Thermal Actuated Microlegs for an Insect-like Microrobot", International Conference on Micromachining and Microfabrication process technologie VII, Spie Vol. 4557, pp 403-414, San Francisco USA, october 2001.

[12] Wen-Hwa CHU, Mehran MEHREGANY, Robert L MULLEN ; "Analysis of tip deflection and force of a bimetallic cantilever microactuator" ; Journal Micromechanic and Microengeneering ; N° 3 ; pp 4-7 ; 1993.

PCG: A Foothold Selection Algorithm for Spider Robot Locomotion in 2D Tunnels

Amir Shapiro and Elon Rimon
Technion, Israel Institute of Technology
amirs@tx.technion.ac.il, elon@robby.technion.ac.il

Abstract *This paper presents an algorithm, called* PCG, *for planning the foothold positions of spider-like robots in planar tunnels bounded by piecewise linear walls. The paper focuses on 3-limb robots, but the algorithm generalizes to robots with a higher number of limbs. The input to the PCG algorithm is a description of a tunnel having an arbitrary piecewise linear geometry, a lower bound on the amount of friction at the contacts, as well as start and target foothold positions. Using efficient convex programming techniques, the algorithm approximates the possible foothold positions as a collection of cubes in contact c-space. A graph structure induced by the cubes has the property that its edges represent feasible motion between neighboring sets of 3-limb postures. This motion is realized by lifting one limb while the other two limbs brace the robot against the tunnel walls. A shortest-path search along the graph yields a 3-2-3 gait pattern that moves the robot from start to target using a minimum number of foothold exchanges. Simulation results demonstrate the PCG algorithm in a tunnel environment.*

1 Introduction

Many robotic tasks are suited for legged robots that interact with the environment in order to achieve stable locomotion. For example, surveillance of collapsed structures for survivors [13], inspection and testing of complex pipe systems [9], and maintenance of hazardous structures such as nuclear reactors [11]. Our ultimate goal is to develop spider-like mechanisms that navigate quasistatically in such complex environments. A *spider-like robot* consists of k articulated limbs attached to a central body, such that each limb ends with a *footpad* (Figure 1). This paper presents an algorithm called PCG (short for Partitioned Cubes Gaiting), for planning the foothold positions of spider-like robots in planar tunnel environments.

In our setup, the robot moves by exerting forces on tunnel walls which are mounted on a horizontal plane. The robot is supported against gravity by frictionless contacts mounted under the mechanism (Figure 1). In general, a spider-like robot must have at least *three* limbs in order to move quasistatically in a planar tunnel[1]. At every instant the spider braces against the tunnel walls in static equilibrium using two or three limbs. During a 2-limb posture the spider moves its free limb to the next foothold position. During a 3-limb posture the spider changes its internal geometry in preparation for the next limb lifting. The PCG algorithm is presented in the context of such 3-limb robots, but we also discuss the generalization of the algorithm to higher number of limbs.

We make the following assumptions. First, we assume piecewise linear tunnel walls with known geometry. Second, the entire tunnel lies in a horizontal plane so that gravity is excluded. Third, each limb can only push against the environment, using its footpad. Fourth, each footpad contacts the tunnel walls via a frictional point contact, with a known lower bound on the coefficient of friction. The ith foothold position is parametrized by $s_i \in [0, L]$, where L is the total length of the tunnel walls. The footholds of the entire k-limb mechanism are parametrized by *contact c-space*, $(s_1, \ldots, s_k) \in [0, L]^k$ (Figure 2). Last, we lump the kinematic structure of the robot into a single parameter called the *robot radius*. This parameter, denoted R, is the length of a fully stretched limb. The algorithm uses R to ensure that the selected footholds can be reached from the robot's central base.

Relationship to prior work. The use of contact c-space is common in the grasp planning literature. In particular, Nguyen [8] and Ponce et al. [10] introduced the notion of *contact independent regions*. Given a k-finger grasp of a planar object, a contact independent region is a k-dimensional cube in contact c-space aligned with the coordinate axes. This cube represents k segments along the object's boundary, such that any placement of the k contacts inside these segments generates an equilibrium grasp. We use a similar notion in our representation of the feasible footholds as cubes in contact c-space. Each cube represents three segments along the tunnel walls, such that any placement of three footpads inside these seg-

---
[1] In quasistatic motion inertial effects due to moving parts of the robot are kept small relative to the forces of interaction between the robot and the environment.

Figure 1: Top view of a 3-limb spider robot moving in a planar tunnel environment.

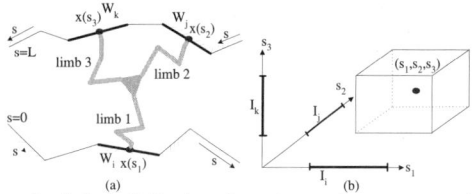

Figure 2: (a) A 3-limb robot in a planar tunnel, and (b) the parametrization of its contact c-space.

ments results in a feasible 3-limb equilibrium posture. Other relevant grasping papers discuss finger gaiting. Hong et al. [3] describe 3 and 4-finger gaits for planar objects. However, they assume that once an object is grasped, the fingers may not change their order along the object's boundary. In contrast, we impose no restriction on the order of the footpads along the tunnel walls. Goodwine et al. [15] investigate the stratification of the full configuration space associated with finger gaiting. While their approach is justifiable for the design of feedback control laws, motion planning can be carried out in lower dimensional spaces such as contact c-space.

In the multi-legged locomotion literature, Boissonnat et al. [1] discuss a motion planning algorithm for multi-legged robots that move in a gravitational field over a flat terrain. Much like our approach, they lump the kinematic structure of the robot into a reachability radius, and compute a sequence of stable stances from start to target. Our work differs from the work of Boissonnat et al. in two fundamental ways. First, we consider motions where the robot stably braces against tunnel walls rather than maintaining stable stances against gravity. Second, they allow the legs to contact only discrete point sites, while we allow arbitrary footpad placement along the tunnel walls. Other papers that consider motion planning for multi-legged robots are [2, 4, 6, 7, 14].

This paper focuses on the portion of the PCG algorithm that plans a sequence of footholds in contact c-space. In Section 2 we characterize the feasible 3-limb postures in contact c-space. These postures must be reachable, form stable equilibria, and satisfy a condition that allows their inclusion in a 3-2-3 gait pattern. In Section 3 we establish a key result, that the feasible 3-limb postures are a union of convex sets in contact c-space. It is also shown in this section that the approximation of a convex set by p maximal cubes is a convex optimization problem. In Section 4 we describe the PCG algorithm. The algorithm approximates the convex sets by contact independent cubes, then searches a graph induced by the cubes for the shortest sequence of footholds from start to target. In Section 5 we run the PCG algorithm on a simulated tunnel environment. Finally, in the concluding section we discuss the generalization of the algorithm to higher number of limbs.

2 The Feasible 3-Limb Postures

In this section we characterize the feasible 3-limb postures as inequality constraints in contact c-space. The feasible 3-limb postures must form stable equilibria, be reachable, and satisfy the following *gait feasibility* condition. This condition requires that a 3-limb posture will contain two distinct 2-limb postures—one for entering the 3-limb posture by establishing a new foothold, and one for leaving the 3-limb posture by releasing some other foothold.

Next we introduce notation that would allow us to write the above conditions as inequalities in contact c-space. Let $W_1, \ldots, W_n$ denote the tunnel walls, and let $I_1, \ldots, I_n$ be the partition of $[0, L]$ into intervals that parametrize the individual walls (Figure 2). Let t_i and n_i denote the unit tangent and unit normal to the ith wall. Points along W_i are given by $\boldsymbol{x}(s) = x_i + s t_i$, where x_i is the initial vertex of W_i and $s \in I_i$. Given a contact force $\boldsymbol{f}_i$, $f_i^t = \boldsymbol{f}_i \cdot t_i$ and $f_i^n = \boldsymbol{f}_i \cdot n_i$ are the tangent and normal components of $\boldsymbol{f}_i$. The friction cone at a contact along the ith wall, denoted FC_i, is given by $FC_i = \{\boldsymbol{f}_i : f_i^n \geq 0 \text{ and } -\mu f_i^n \leq f_i^t \leq \mu f_i^n\}$, where μ is the coefficient of friction.

Gait feasibility requires that a 3-limb will contain two distinct 2-limb postures. As a preparation, we review the conditions for equilibrium and stability of 2-limb postures. A 2-limb mechanism forms an equilibrium posture if the line segment connecting the two contacts lies inside the two friction cones [8]. As a stability criterion we use the notion of *force closure*—a posture where the mechanism can resist any perturbing wrench by suitable adjustment of its contact forces. In general, a planar equilibrium posture is force closure and hence stable if the contact forces of the unperturbed posture lie in the interior

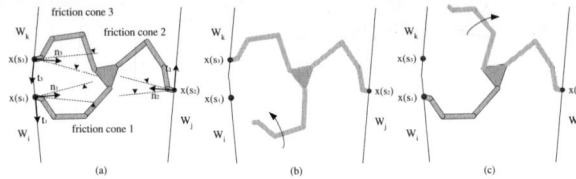

Figure 3: (a) A gait feasible 3-limb posture, (b)-(c) contains two distinct 2-limb postures.

of the respective friction cones.

Let two limbs with indices l and m contact the tunnel walls W_i and W_j. Then for a 2-limb stable equilibrium, the vector $x(s_m)-x(s_l)$ must lie in the interior of FC_i, while $-(x(s_m)-x(s_l))$ must lie in the interior of FC_j. This condition defines a set in the (s_l, s_m) plane, denoted $\mathcal{E}_{ij}^{lm}$, given by

$$\mathcal{E}_{ij}^{lm} = \{(s_l, s_m) \in I_i \times I_j : \\ \left|(x(s_m)-x(s_l))\cdot t_i\right| < \mu(x(s_m)-x(s_l))\cdot n_i, \\ \left|(x(s_l)-x(s_m))\cdot t_j\right| < \mu(x(s_l)-x(s_m))\cdot n_j\}.$$

It is important to note that the inequalities describing $\mathcal{E}_{ij}^{lm}$ are linear in s_l and s_m. Hence $\mathcal{E}_{ij}^{lm}$ is a *convex polygon* in the (s_l, s_m) plane. When $\mathcal{E}_{ij}^{lm}$ is considered as a subset of the contact c-space of a 3-limb mechanism, it becomes a three-dimensional prism orthogonal to the (s_l, s_m) plane. The prism is denoted with an $\times$ for the limb that does not participate in the 2-limb posture. The 2-limb equilibrium set $\mathcal{E}_{ij}^{12}$ thus becomes the prism $\mathcal{P}_{ij\times}$, the sets $\mathcal{E}_{ij}^{13}$ becomes $\mathcal{P}_{i\times j}$, and $\mathcal{E}_{ij}^{23}$ becomes $\mathcal{P}_{\times ij}$.

Reachability constraint of 3-limb postures. A 3-limb posture is reachable when its footholds lie within the robot's radius R. For each wall triplet W_i, W_j, W_k the reachability constraint is given by

$$\mathcal{R}_{ijk} = \{(s_1, s_2, s_3) \in I_i \times I_j \times I_k : \exists c \in \mathbb{R}^2 \\ \max\{\|x(s_1)-c\|, \|x(s_2)-c\|, \|x(s_3)-c\|\} \leq R\}, \quad (1)$$

The point c appearing in (1) can be interpreted as the center of a disc containing the three foothold positions, such that the disc radius is bounded by R. As discussed below, the elimination of the existential quantifier in (1) results in a set which is bounded by quadratic surfaces in contact c-space.

Gait feasibility of 3-limb postures. A 3-limb posture is gait feasible if it contains two distinct 2-limb equilibrium postures (Figure 3). Let us write this constraint in a cell $I_i \times I_j \times I_k$ of contact c-space. This cell corresponds to contact with the walls W_i, W_j, W_k, and gait feasibility is satisfied by intersection of pairs of 2-limb prisms associated with the three walls. There are three such pairs—$(\mathcal{P}_{ij\times}, \mathcal{P}_{i\times k})$, $(\mathcal{P}_{ij\times}, \mathcal{P}_{\times jk})$, and $(\mathcal{P}_{\times jk}, \mathcal{P}_{i\times k})$—and the resulting set of feasible 3-limb postures in the cell, denoted $\mathcal{F}_{ijk}$, is given by

$$\mathcal{F}_{ijk} = (\mathcal{P}_{ij\times} \cap \mathcal{P}_{i\times k} \cap \mathcal{R}_{ijk}) \cup (\mathcal{P}_{ij\times} \cap \mathcal{P}_{\times jk} \cap \mathcal{R}_{ijk}) \\ \cup (\mathcal{P}_{\times jk} \cap \mathcal{P}_{i\times k} \cap \mathcal{R}_{ijk}). \quad (2)$$

Note that the same three walls appear in *six* cells in contact c-space, each corresponding to a specific assignment of the limbs to the three walls. The entire collection of feasible 3-limb postures is the union of all such sets over all ordered wall triplets. We end with the following assertion [12]. *It is always possible to transfer forces between two 2-limb postures contained in a feasible 3-limb posture, while the mechanism is kept in static equilibrium with three fixed footholds.*

3 Convexity of Feasible 3-Limb Postures

In this section we establish two convexity results that will be used by the PCG algorithm. First we establish that the feasible 3-limb postures are a union of convex sets in contact c-space. Then we show that the approximation of a convex set by p maximal cubes is a convex optimization problem.

3.1 Convexity of the feasible postures

The set $\mathcal{F}_{ijk}$ of feasible 3-limb postures is specified in (2) as a union of three sets, each corresponding to a particular pair of 2-limb postures. The following lemma asserts that each of these sets is convex in contact c-space.

Lemma 3.1. *In each cell $I_i \times I_j \times I_k$ of contact c-space, the set $\mathcal{F}_{ijk}$ of feasible 3-limb postures is a union of three convex sets.*

Proof: Consider the set $\mathcal{P}_{ij\times} \cap \mathcal{P}_{\times jk} \cap \mathcal{R}_{ijk}$ in (2). The prisms $\mathcal{P}_{ij\times}$ and $\mathcal{P}_{\times jk}$ are defined by intersection of linear inequalities, and are therefore convex polytopes in contact c-space. Next consider the set $\mathcal{R}_{ijk}$. The existential quantifier in (1) acts on a set, denoted $\bar{\mathcal{R}}_{ijk}$, which is defined in the five-dimensional space (s_1, s_2, s_3, c): $\bar{\mathcal{R}}_{ijk} = \{(s_1, s_2, s_3, c) \in I_i \times I_j \times I_k \times \mathbb{R}^2 : \max\{\|x(s_1)-c\|, \|x(s_2)-c\|, \|x(s_3)-c\|\} \leq R\}$. The norm function $\|x-c\|$ is convex in (x, c) space, and each $x(s_i)$ is linear in s_i. Hence the functions $\|x(s_i)-c\|$ are convex in (s_1, s_2, s_3, c) space. The pointwise maximum of convex functions is a convex function. Hence $\bar{\mathcal{R}}_{ijk}$ is convex in (s_1, s_2, s_3, c) space. But $\mathcal{R}_{ijk}$ is the projection of $\bar{\mathcal{R}}_{ijk}$ onto contact c-space. Since projection preserves convexity, $\mathcal{R}_{ijk}$ is convex. Finally, the intersection of convex sets is convex, hence $\mathcal{P}_{ij\times} \cap \mathcal{P}_{\times jk} \cap \mathcal{R}_{ijk}$ is convex. □

To summarize, the set $\mathcal{F}_{ijk}$ is the union of three convex sets, each bounded by planar surfaces associated with the 2-limb prisms, and quadratic surfaces associated with the reachability constraint [12].

3.2 Convexity of Cube Approximation

We now discuss the approximation of the convex sets comprising $\mathcal{F}_{ijk}$ by maximal cubes. Consider the approximation of a three-dimensional convex set $\mathcal{S}$ by p cubes, where the cubes have arbitrary center and dimensions. We assume as input a desired relative configuration for the p cubes, where a *relative configuration* is a specification of an adjacency relation between the cubes in terms of a set of separating planes, such that no two cubes can possibly intersect. Each of the separating planes is defined in terms of the relative position of two cubes, and does not restrict the absolute position of the two cubes. The ith cube is parametrized by its center $c_i \in \mathbb{R}^3$, and its dimensions along the coordinate axes, $h_i \in \mathbb{R}^3$. The optimization therefore takes place in the $6p$-dimensional space whose coordinates are $(c_1, h_1, \ldots, c_p, h_p)$. Our objective is to maximize the total volume of the cubes. However, the sum of the cubes' volumes is not a convex function of the optimization variables. Rather, we use a normalized total volume function given by[2]

$$\phi(c_1, h_1, \ldots, c_p, h_p) = \sum_{i=1}^{p} (h_{i1} h_{i2} h_{i3})^{\frac{1}{3}}.$$

Next we list the constraints involved in the cube approximation problem. First we have the requirements that the cubes' dimensions be non-negative, $h_{ij} \geq 0$, and that their centers lie inside contact c-space, $0 \leq c_{ij} \leq L$ ($i = 1, \ldots, p, j = 1, 2, 3$). Second, the relative configuration of the cubes is specified by a list of separating planes, each involving the center and dimensions of two cubes separated by the plane. Last, we must ensure that the cubes lie inside the convex set $\mathcal{S}$. The following proposition asserts that the maximization of ϕ over p cubes contained in $\mathcal{S}$ is a convex optimization problem.

Proposition 3.2 ([12]). *The maximization of $\phi = \sum_{i=1}^{p} (h_{i1} h_{i2} h_{i3})^{\frac{1}{3}}$ over p cubes contained in a convex set $\mathcal{S}$ and satisfying a relative-configuration specification is a convex optimization problem.*

It is worth mentioning that convex optimization algorithms, for instance the ellipsoid algorithm used in our implementation, generate an ϵ-accurate solution in $O(m^2 l \log(1/\epsilon))$ time, where m is the number of optimization variables and l the number of steps required to evaluate the constraints.

[2]This function was suggested to us by Prof. A. Nemirovsky.

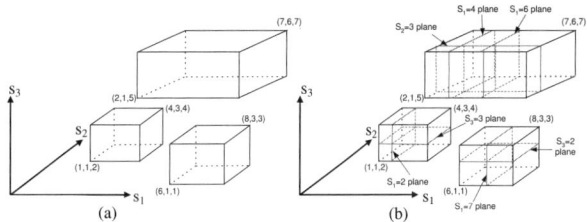

Figure 4: (a) Three cubes in contact c-space, and (b) their mutual partition into sub-cubes along the separating planes.

4 The PCG Algorithm

We begin with an overview of the algorithm. The set of feasible 3-limb postures in each contact c-space cell, $\mathcal{F}_{ijk}$, is a union of three convex sets. However, usually each cell contains at most one convex set, and we describe the algorithm under the assumption of a single convex set per cell. The algorithm first approximates each of the convex sets by p maximal cubes. The number of cubes and their relative configuration are user-specified inputs whose selection is discussed below. In order to describe the next stage of the algorithm we introduce the notion of cube orientation. A maximal cube parametrizes a set of feasible 3-limb postures, each containing two distinct 2-limb postures. The two 2-limb postures necessarily share a limb in common (Figure 3). However, this common limb *cannot be lifted,* since its lifting would destroy both 2-limb postures. By construction, all the 3-limb postures parametrized by a given maximal cube have the same common limb. Thus we associate with each maximal cube an *orientation vector,* which is aligned with the s_i-axis of the limb that cannot be lifted from the 3-limb postures parametrized by the cube.

In the second stage the algorithm partitions the maximal cubes as follows. The algorithm constructs an arrangement of all the separating planes of the cubes, where each separating plane contains one of the cubes' faces. Using this arrangement, the algorithm partitions the cubes as illustrated in Figure 4. The figure shows three cubes and their mutual partition along the separating planes into sub-cubes. During the partition process, each sub-cube inherits the orientation vector of its parent cube. The resulting sub-cubes have disjoint interiors, and they satisfy the following projection property. Any two sub-cubes either have precisely the same projection on one of the coordinate planes, or their projection on all three coordinate planes are disjoint. If two sub-cubes share a projection they are called *compatible*, and the s_i-axis aligned with the direction of projection is called the *direction of compatibility.*

In the third stage the algorithm constructs a graph called the *sub-cube graph*. The nodes of the graph are center points of the sub-cubes. The edges of the graph are assigned unit weights. Each edge connects compatible sub-cubes whose direction of compatibility is orthogonal to the orientation vector of the two sub-cubes. The meaning meaning of the orthogonality condition is discussed below. Finally, the start and target 3-limb postures, denoted S and T, are added as special nodes to the sub-cube graph. The construction of edges from S and T to the other nodes of the graph is described below.

In the last stage, the algorithm searches the sub-cube graph for the shortest path from S to T. The shortest path on the graph minimizes the number of foothold exchanges along the path from S to T. However, this minimality is relative to the cube approximation obtained in the first stage of the algorithm. A formal description of the algorithm follows.

PCG Algorithm:
Input: Geometrical description of an n-wall tunnel. A value for the coefficient of friction. Start and target 3-limb postures S and T. A value for the number of cubes p and their relative configuration.
1. Cube approximation:
   1.1 Determine which cells $I_i \times I_j \times I_k$ contain a non-empty set $\mathcal{F}_{ijk}$ of feasible 3-limb postures.
   1.2 Approximate each non-empty set $\mathcal{F}_{ijk}$ by p maximal cubes. Assign an orientation vector to each maximal cube.
2. Cube partition:
   2.1 Construct an arrangement of the separating planes of all maximal cubes.
   2.2 Subdivide each maximal cube into sub-cubes along the separating planes. Assign to each sub-cube the orientation vector of its parent cube.
3. Graph construction:
   3.1 Define a *sub-cube graph* as described above.
   3.2 Define S and T as special nodes and connect them to the graph as described below.
   3.3 Assign unit weight to all edges.
4. Graph search:
   Search for the shortest path along the sub-cube graph from S to T.

Let us discuss the meaning of the edges in the sub-cube graph. An edge represents lifting and replacement of a particular limb. The lifting of a limb must leave the robot with a stable 2-limb posture. The orientation vector of a sub-cube describes which limb my not be lifted from the 3-limb postures parametrized by the sub-cube. Hence all edges emanating from a node must be *orthogonal* to the orientation vector of the sub-cube associated with the node. Moreover, all edges of the sub-cube graph are *straight lines parallel to the s_i-axes in contact c-space* (Figure 6). Another aspect of the edges is reachability—motion of a limb between any two sub-cubes connected by an edge can always be executed such that reachability is maintained throughout the limb's motion [12].

Next consider the construction of edges from S and T to the other nodes of the sub-cube graph. Let S and T be feasible 3-limb postures with their own orientation vector. For S and T, compatibility with a sub-cube means that the projection of the sub-cube on one of the coordinate planes contains the corresponding projection of the node. Having defined orientation and compatibility for S and T, the edges connecting these nodes to the other nodes of the graph are constructed by the rule specified in step 3.1 of the algorithm. A second technical issue is the selection of a relative configuration for the p cubes. We specify in each cell a relative configuration that separates the p cubes along a coordinate axis which is *orthogonal* to the cell's orientation vector. Adjacent maximal cubes consequently overlap along the cell's allowed directions of motion, thereby preserving the connectivity of the set of feasible 3-limb postures in the cell.

Let us discuss some notable features of the algorithm. First, the uniform weight assignment reflects our desire to minimize the number of foothold exchanges along the path. Second, the algorithm treats the motion of a limb between walls and along a single wall in a uniform manner. Last, the size of the sub-cube graph increases with p. However, if an edge exists in the graph for low values of p, it would persist in the graph for larger values of p. Consequently, *the path from start to target only becomes shorter as p increases.* The computational complexity of the algorithm is analyzed in Ref. [12]. Under the reasonable assumption that the spider robot can reach from any given position only a small number of walls which is bounded by a constant, the algorithm runs in $O(np^6 \log(np))$ time, where n is the number of tunnel walls and p is the number of cubes per cell in contact c-space.

5 Simulation Results

In this section we run the PCG algorithm in the tunnel depicted in Figure 7. The tunnel consists of six walls whose lengths in cm are marked in the figure. The robot reachability radius is $R = 60$ cm, and the coefficient of friction is $\mu = 0.5$. Note that this simple tunnel already contains significant geometric features: the two walls at the bottom form a closing cone, the tunnel next turns leftward and becomes two parallel walls, and finally the two walls at the top form an opening cone. These geometric features are significant, since *the robot must use friction effects to*

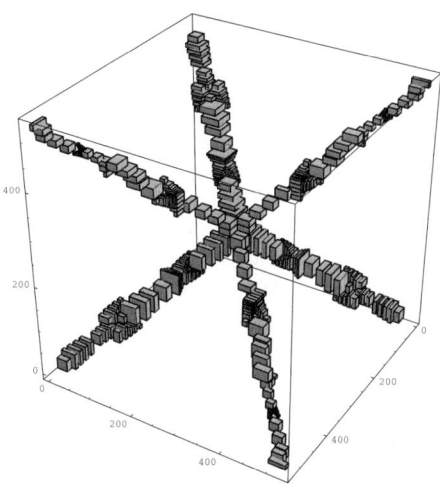

Figure 5: The collection of 270 maximal cubes approximating the feasible 3-limb postures.

traverse such features.

Let us now discuss the computation of the feasible 3-limb postures in contact c-space. The collection of feasible 3-limb postures has a *six-fold symmetry* consisting of six symmetric "arms:" every non-empty cell represents an assignment of the three limbs to a triplet of walls, and there are six permutations of the three limbs on the triplet of walls. The arms are roughly aligned with the diagonals of contact c-space, for the following reason. The coordinate projection of each arm covers the entire length of the tunnel. Each arm can therefore be visualized as "dragging" the 3-limb mechanism as a single rigid body along the entire length of the tunnel. There are nine non-empty cells in each arm, giving a total of 54 non-empty cells in the entire contact c-space.

Next consider the cube approximation of the feasible 3-limb postures. We use $p = 5$ cubes per cell and compute the maximal cubes using the ellipsoid algorithm. The $p = 5$ value preserves the connectivity of the set of feasible 3-limb postures, while still being sufficiently low to allow reasonable execution time. The result of running the ellipsoid algorithm on the non-empty cells of contact c-space appear in Figure 5. Since there are 54 non-empty cells, the resulting cube approximation of contact c-space contains $5 \cdot 54 = 270$ maximal cubes. The algorithm next partitions each of the maximal cubes along the separating planes of the other maximal cubes. The partitioning of the maximal cubes generated 28,299 sub-cubes in each of the six arms of contact c-space (the resulting sub-cubes are not shown).

The algorithm next constructs the sub-cube graph, and searches the graph for the shortest path from the start to target postures. The result of computing the shortest path is shown in Figure 6. Each segment in the figure is an edge of the sub-cube graph that represents one limb lifting and re-placement. Figure 7 shows the same path in physical space, where each foothold is marked by its index in the sequence of steps taken by the robot. Let us denote the sequence of 3-limb postures by (i_1, i_2, i_3), where i_j is the foothold position of limb j at the ith posture. Then the path computed by the algorithm consists of the 3-limb postures: $S = (1,2,3) \rightarrow (4,2,3) \rightarrow (4,5,3) \rightarrow (4,5,6) \rightarrow (7,5,6) \rightarrow (7,8,6) \rightarrow (7,8,9) \rightarrow (7,10,9) \rightarrow (11,10,9) \rightarrow (11,10,12) \rightarrow (13,10,12) \rightarrow (13,14,12) \rightarrow (13,14,15) \rightarrow (16,14,15) \rightarrow (16,17,15) \rightarrow (16,17,18) \rightarrow (19,17,18) \rightarrow (19,20,18) \rightarrow (19,20,21) \rightarrow (22,20,21) \rightarrow (22,20,23) \rightarrow (22,24,23) \rightarrow (25,24,23) \rightarrow (25,24,26) \rightarrow (25,27,26) \rightarrow (28,27,26) \rightarrow (28,27,29) \rightarrow T = (30,27,29)$. This sequence describes a 3-2-3 gait pattern, where successive 3-limb postures are interspersed by a 2-limb posture that allows motion of a limb between the two 3-limb postures. The path generated by the algorithm is minimal in terms of the number of foothold exchanges, where minimality is relative to the cube approximation of the feasible 3-limb postures (Figure 5).

6 Conclusion

We presented the PCG algorithm, for selecting the footholds of a 3-limb robot in planar tunnel environments with an arbitrary piecewise linear geometry. The algorithm assumes knowledge of the tunnel geometry and a lower bound on the amount of friction at the contacts. The algorithm approximates the collection of feasible 3-limb postures by p maximal cubes in each non-empty cell of contact c-space. Then it partitions the cubes and searches the sub-cube graph for the shortest 3-2-3 gait sequence from start to target. The algorithm's main strength is its emphasize on achieving contact independent foothold placement sequences. Each sub-cube parametrizes three contact independent wall segments, and each edge can be realized by limb lifting and re-placement between any two postures in the two sub-cubes connected by the edge. Thus a controller for the robot's limbs need only ensure footpad placement within the segments parametrized by the sub-cubes. The main weakness of the algorithm is the lack of a procedure for selecting the parameter p. This topic is under investigation.

Finally, it seems that the algorithm directly generalizes to k-limb mechanisms that move with a $k - (k-1) - k$ gait pattern. Contact c-space in this case is k-dimensional, and one must first establish that the feasible k-limb postures in this space are a union of convex sets. If this is the case, the algorithm can be directly applied to such mechanisms. However, the computational complexity of the algorithm would become $O(np^{k+3} \log(np))$. A second

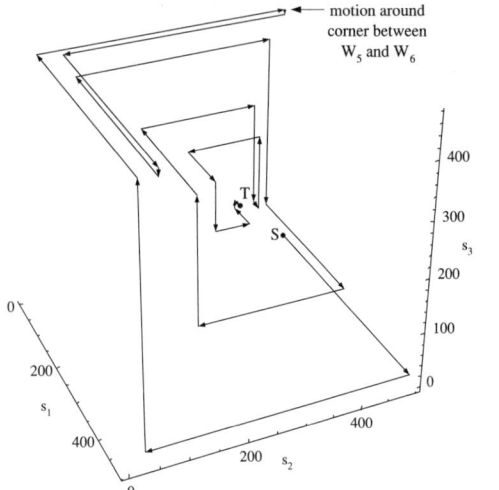

Figure 6: The shortest path from S to T along the edges of the sub-cube graph in contact c-space.

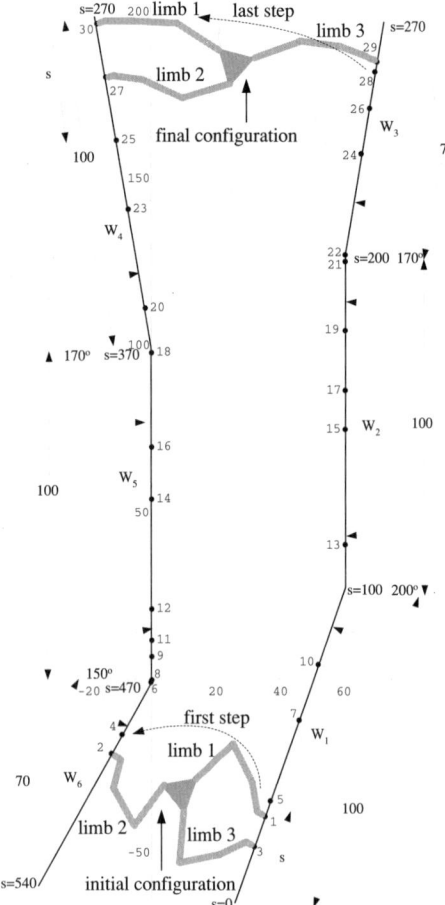

Figure 7: The tunnel environment used in the simulations, and the sequence of footholds generated by the PCG algorithm.

more challenging topic is how to plan the footholds of a k-limb mechanism using a variable gait pattern.

References

[1] J.-D. Boissonnat, O. Devillers, and S. Lazard. Motion planning of legged robots. *SIAM J. of Computing*, 30:218–246, 2000.

[2] S. Hirose and O. Kunieda. Generalized standard foot trajectory for a quadruped walking vehicle. *Int. J. of Robotics Research*, 10(1):2–13, 1991.

[3] J. Hong, G. Lafferriere, B. Mishra, and X. Tan. Fine manipulation with multifinger hands. *Icra*, 1568–1573, 1990.

[4] J. K. Lee and S. M. Song. Path planning and gait of walking machine in an obstacle-strewn environment. *J. Robotics Sys.*, 8:801–827, 1991.

[5] S. Leveroni and K. Salisbury. Reorienting objects with a robot hand using grasp gaits. *7th Int. Symp. on Robotics Research*, 2–15, 1995.

[6] A. Madhani and S. Dubowsky. Motion planning of mobile multi-limb robotic systems subject to force and friction constraints. *Icra*, 233–239, 1992.

[7] D. Marhefka and D. Orin. Gait planning for energy efficiency in walking machines. *Icra*, 474–480, 1997.

[8] V.-D. Nguyen. Constructing force closure grasps. *Int. J. of Robotics Research*, 7(3):3–16, 1988.

[9] F. Pfeiffer, T. Rossmann, N. Bolotnik, F. Chernousko, G. Kostin. Simulation and optimization of regular motions of a tube-crawling robot. *Multibody Sys. Dyn.*, 5:159–184, 2001.

[10] J. Ponce and B. Faverjon. On computing three-finger force closure grasps of polygonal objects. *IEEE Trans. on Robotics and Aut.*, 11(6):868–881, 1995.

[11] J. Savall, A. Avello, and L. Briones. Two compact robots for remote inspection of hazardous areas in nuclear power plants. *Icra*, 1993–1998, 1999.

[12] A. Shapiro and E. Rimon. Pcg: A foothold selection algorithm for spider robot locomotion in 2D tunnels. Tech. report, Dept. of ME, Technion, http://www.technion.ac.il/~robots, July 2002.

[13] T. J. Stone, D. S. Cook, B. L. Luk. Robug III—an 8-legged teleoperated walking and climbing robot for disordered hazardous environments. *Mech. Incorp. Engineer*, 7(2):37–41, 1995.

[14] K. van der Doel and D. K. Pai. Performance measures for locomotion robots. *J. of Robotic Systems*, 14(2):135–147, 1997.

[15] Y. Wei and B. Goodwine. Stratified motion planning on non-smooth domains with application to robotic legged locomotion and manipulation. *Icra*, 3546–3551, 2002.

A Plant Maintenance Humanoid Robot System
– Navigation system of Autonomous and Tele-operation Fusion Control –

Naoto Kawauchi*, Shigetoshi Shiotani*, Hiroyuki Kanazawa*, Taku Sasaki*, Hiroshi Tsuji**

*Takasago Research & Development Center, Mitsubishi Heavy Industries, Ltd.
2-1-1, Shinhama, Takasago, Hyogo 676-8686, Japan
** Kobe Ship Yard &Machinery Works, Mitsubishi Heavy Industries, Ltd.
1-1-1, Wadasaki-cho, Hyogo-ku, Kobe 652-8585, Japan

Abstract

In the Power Plant under operation, there are many important equipments which the operator must inspect in the non-operation state. Our milestone of the humanoid robot which can inspect for the man is moving around the Power Plant and inspection experiments by using navigation system of autonomous and tele-operation fusion control. We constructed and implemented a navigation system, which is controlled by Tag (RF-ID) and Remote teaching human interface. The robot just reads a tag with information on the plant maintenance, and the robot can act with the information. Therefore, switching of the program movement of the robot and the remote control become very easy.

This research and Development is done with the HRP (Humanoid Robot Project: 2000 ~ 2002) project of METI (Ministry of Economy, Trade and Industry, JAPAN).

1. Introduction

There have been various types of robots for plant maintenance in the past. [1][2] However, most of them have long crawler transfer system that need for climbing up steps and cannot go through curved narrow space. Hence, a plant maintenance system by a humanoid robot with the equivalent mobility as human is needed.

Figure 1 shows two plant maintenance missions.

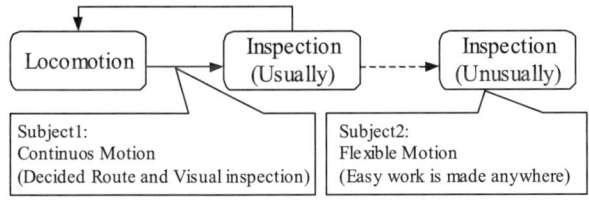

Figure 1:   Plant Maintenance Missions

(1) To move on steps, slopes, pipe-works and the restricted width of the passages along programmed route in Figure 2 and inspect of pressure gauges by CCD camera and measuring temperature of plant equipments and so on in Figure 3.

As the conventional navigation methods of robots, locomotion environment maps and the marks like white lines have been used. However, because a plant is large, it is important that the map can be made simply. We want to avoid the extreme power to add newly extra equipment because it isn't always built at the same time for the robot movement. Therefore, as for the navigation of the robot, remote control is mainly used.

Figure 2 :   A Plant Maintenance Humanoid Robot Image (Continuous Locomotion through the many kind of equipment.)

Figure 3:   A Plant Maintenance Humanoid Robot Image (The inspect of pressure meters by CCD camera and measuring temperature of plant equipments.)

(2) If an unexpected event occurs, robot must move there through un-programmed route and perform easy works such as opening valves. Furthermore, after such work is done, it should be necessary that it is returned to programmed route easily.

Figure 4 shows the technological problems for achieving continuous locomotion and inspection in the plant with obstacles listed above.

This paper shows a navigation system of autonomous and tele-operation fusion control by using Humanoid robot. An autonomous navigation is carried out by RF-ID (Radio Frequency Identification System) tag and accuracy positioning magnetic tapes. And a remote teaching interface supports remote control easy.

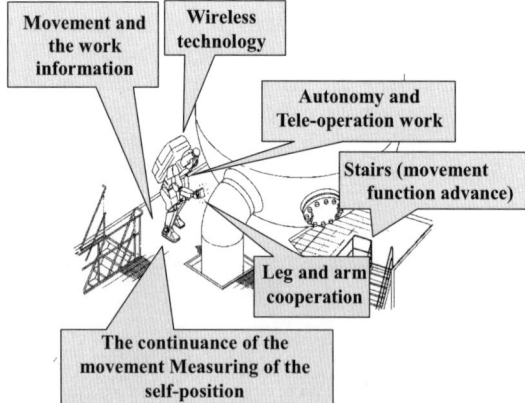

Figure 4:   Plant Maintenance Technology

2. Autonomous Navigation System with RF-ID Tags

2.1 Merits of RF-ID Tags

In comparison to the conventional methods such as the tape guidance for automatic guided vehicle, the plant maintenance robot needs simplification and reliability of the guide.

RF-ID tags can perform data-exchange with non-contact for navigation and maintenance such as present position and direction, next inspection position and task and so on. [3]

We arrange these tags on the floor and have controlled locomotion of autonomous guided vehicles in the carriage system of the factory. [4] By this navigation system, the following advantages may be expected. [5], [6]

- Guidance tapes and the external instrument for measuring the self-position of the humanoid robot are eliminated. (Infrastructure-less)
- In the case of a wireless control, the central control software is simplified.
- Sure data-exchange for navigation data improves reliabilities of control systems.

2.2 Problems to Be Solved for Navigation

The problems of the autonomous control are followings.

Probrem-1: Locating its position in the vast plant.
"Where is the humanoid robot in the plant exactly?"

Problem-2: Accurate positioning of the humanoid robot before climbing up and down the steps.
"What obstacle is ahead of the humanoid robot now? What should its position be?"

The problems of the returning a programmed route from tele-operating control are followings.

Problem-3: Reading and checking of the instruments that is not in the programmed route, and returning to the round route.
"When an unforeseen event occurs, how will it respond? Will it return to the round routine simply?"

2.3 Solutions

Our navigation method solves above problems by following method.

(1) When the humanoid robot reaches within wireless communication area of the tag, the robot reads its present position from the RF-ID tag as shown in Figure 5. However, accuracy of self-location depends on the wireless communication range of the tag. It is about $\phi 50$cm.

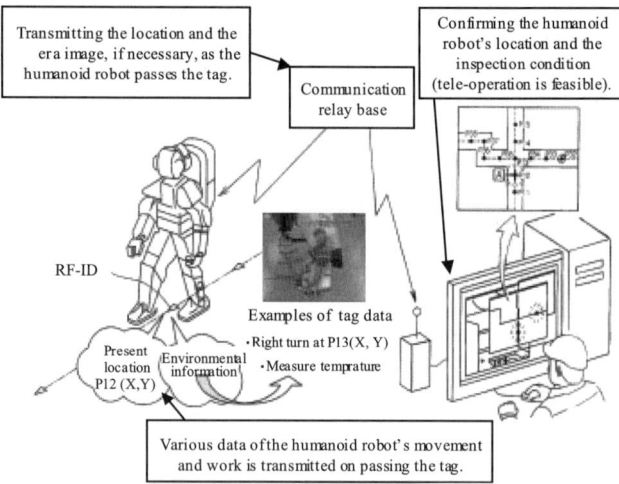

Figure 5:   RF-ID Tag Navigation System

(2) The accuracy positioning is performed using a magnetic tape on the floor and its reading sensors on both ankles of the robot as shown in Figure 6. Its accuracy is 2cm at the position and 3 degrees at the posture.

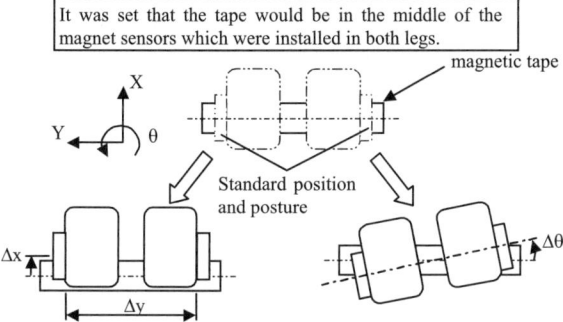

Figure 6: Accuracy positioning of both legs using magnetic sensors

(3) In the case returning from the tele-operation to autonomous control, the operator makes the robot cameras on the head turn toward a nearest RF-ID tag. The robot can calculate the position of the nearest tag by this operation and return to programmed route automatically. After the robot moves to the tag and reads desirable position and direction on the tag and next tag position from the tag as process (1) mentioned above, the robot adjusts self posture and goes toward next tag.

3. Construction of the Navigation System

Figure 7 shows the structure of navigation system with the RF-ID tag, which is composed, of RF-ID reader/writer, controller and AIO/DIO board as shown in Figure 8.

The configuration of the navigation program is shown in Figure 9. The navigation software (1) in Figure 9 makes the humanoid robot move or maintain the plant equipments by sending commands read from the tag using software (2) and (3) in Figure 9.

The specification of RF-ID tag is shown in Table 1. And information to be written in RF-ID tag is shown in Table 2. RF-ID tag memory has two partitions: Read and Write; the following is a simple description of its functions.

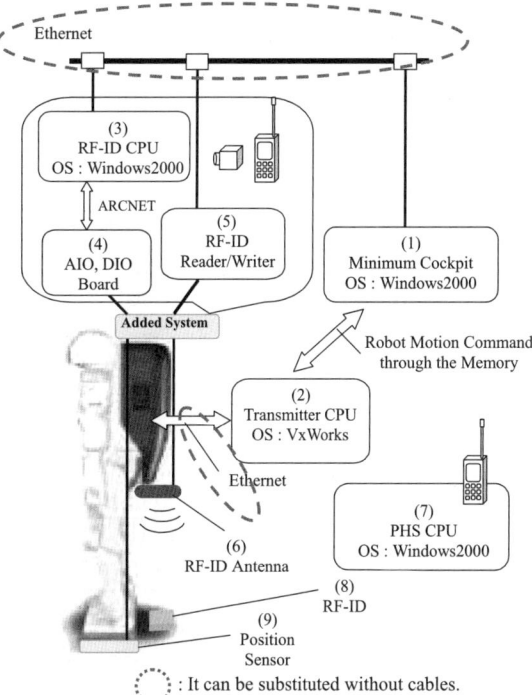

Figure 7: Structure of RF-ID Tag Navigation System

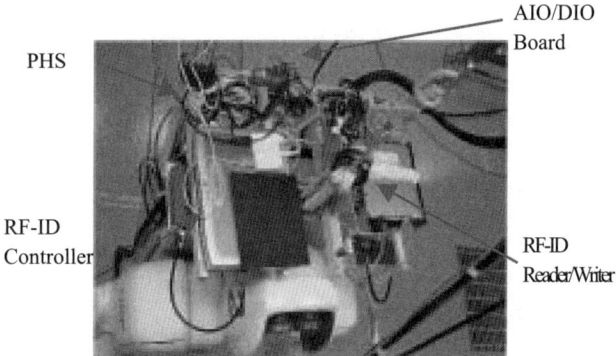

Figure 8: Added System of a Robot Platform

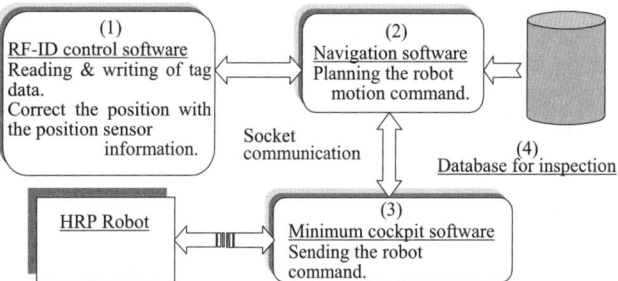

Figure 9: RF-ID Tag Navigation Software Construction

Table 1: RF-ID Tag Specification
(JAPN RF Solution Co., LTD.)

Item		Spec.
Dimension	Antenna	110 * 100 * 16 mm
	Tag Reader / Writer	233 * 337 * 98 mm (Include Cover +)
	Tag (RF-ID)	73 * 55 * 8 mm
Weight	Antenna	0.14 kg
	Tag Reader / Writer	3.3 kg (Include Cover +)
Memory Area		110 byte
Transport Rate	Read	165 ms
	Write	about 2.75 s
Transport Length	Read	1000 mm
	Write	700 mm
Transport Area		φ 500mm
Multi-Read		Possible
Environment (on the Steel)		Possible

+ : Possible to attached robot without the cover

Table 2: RF-ID Tag Data

Number	Robot Process	Data Definition	Size (Byte)	Remarks
Read-1	Recognizing the current position	Tag number Absolute (x, y, z) Absolute(θ)	14	It dose not resister (x, y, z).
Read-2	Past Record	Last passing date Inspection result	–	It is the same as write-1, 2 area.
Read-3	Tasks at the current position	Motion Command ・walking ・Side walking ・Rotation ・Steps(Up/Down) ・Inspection and so on	55	
Read-4	Locating the next tag	Relative(x, y, z, θ)	12	
Write-1	Date passed	Year/Month/Day, Hour/Minute/Second	20	
Write-2	Record	Inspection result	8	
		Total Size	109	

Read-1: This memory area is used for the robot to confirm the current location of the tag, that is the robot position in the plant.

However, it is not quite possible to control the tag location with x, y, z in such a vast plant, this section is not in use.

Read-2: This memory area is used for the robot to confirm the time when it passes on the tag, such as year, month, date and hour.

Read-3: This memory area has the task command of the robot such as walking, side-walking and turn there.

Read-4: Location of the next RF-ID tag based on the current position, and traveling direction.

Write-1: Time (year, month, date and time), when the humanoid robot passes the RF-ID tag, is written here.

Write-2: Section for writing sensing results such as inspections.

4. Basic Locomotion Ability of Humanoid Robot

To determine the specification of navigation system such as tag interval, magnetic tape interval and accuracy to adjust the legs using magnetic sensors and in order to confirm for the robot to walk on the environments described in Table 3.; a walking accuracy and stability we measured while the motion of arms of the humanoid robot.

Table 3: Required Specification for Moving within the Power Plant

Item	Required Specification
Floor gradient	Maximum 2/100
Floor undulation	±5mm within 1m square
Ramp / Step	150 to 250 mm (include stairs)
Passage	720 mm at airlock entrance

4.1 Locomotion Ability

1) Walking on the floor
   - Straight movement (3m): approximately 20mm
   - Rotating movement: approximately ±2 degrees
   - Width for passing: approximately 1300 mm
   - Slope: no problem seen up to 2/100
   - Ramp: no problem seen on the uneven surface up to 20mm

2) Ascent and descent of the staircases (Tested at the height of 160 mm)
   - Required positioning accuracy: less than ±25mm
   - Required posture accuracy: ± 2~3 deg
   - Acceptable ramp height: −30 to +10 mm.

4.2 Stability for Arm Performance

Humanoid robot is expected to perform some work with its arms. The movable range of its arms, required for stabilizing its legs, was measured to determine its standing position against the work site. The result showed that the humanoid robot, 1600 mm of height with arm length of 800mm, requires the distance of 500 mm, (measured vertically from its shoulders), which is considerably close, from the panel or the instruments.

5. Inspection Tools

Special tools have been developed for the humanoid robot to enable valve operation and non-contact measurement of temperature and humidity.

Special inspection tools are restricted to be compact and light weight for the humanoid robot to move about with them. With that in mind, following devices were developed:

- Wireless structure: a compact I/F board (Motor-drive is also available) controlled by ARCNET communication is installed.
- Sharing of various systems: a combined unit of the base tool of communication parts and the work tools for inspection work.

Temperature / humidity measuring tool based on the design concept mentioned above is shown in Figure 9.

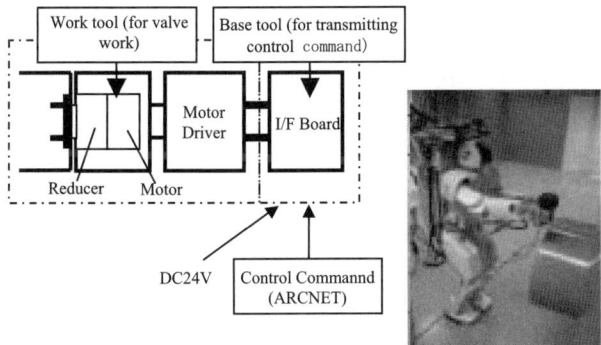

Figure 9: Special inspection tools design concept and Temperature/ humidity measuring tool

6. Construction of Experiment Site

Mock-up structure shown in Figure 10 was constructed to provide the experiment to verify the following performance by simulating the navigation within the plant.

① To move (walk) over the obstacles such as pipes on the passage
② To turn left or right at the corner
③ To up or down the stairs and to turn at the narrow landing
④ To move to inspect the gauge and instrument at the landing (access to high / low point)
⑤ To measure temperature and humidity by reading the gauge and to open / close valves
⑥ To return to the starting point via narrow walkway

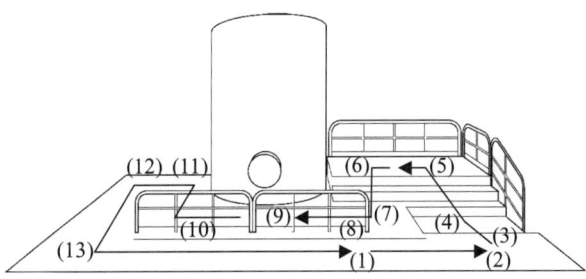

Figure 10: Plant Mock-up Simulation Structure

7. Locomotion and Inspection Experiments

We experiment the inspection by a humanoid robot in the plant mock-up shown in the figure 10 using the RF-ID tag navigation system. Figure 11 shows the processes. We have confirmed for the humanoid robot to go round there autonomously, inspection of temperature and humidity of the directed instruments by tele-operating control and to return from tele-operating control to autonomous control easily by teaching of nearest tag position using camera of the robot.

8. Remote Teaching Human Interface

It has been stated about the autonomous navigation that the inspection route, the stairs, the pipe, narrow passages were too complex to continue. But, as for the round inspection of the plant, it is necessary that correspondence to the matter outside the assumption such as a movement to the point that it is not in the programmed route and valve opening and closing can be done rapidly, too. So we have developed a remote control of the robot in the unexpected situation, and the interface of easy leading of the robot to the remote destination.

An operator inputs the characteristics point of the movement applicable thing to the camera image of the robot head indicated in the monitor, and this interface is taught with a mouse. The position of the applicable thing is measured with the image from this characteristics point. Then, movement route is formed automatically. That precision is ±10 - 50mm toward a measurement distance 500 - 3000mm.

It needed one or more cameras, collision sensors, and so on, to realize a narrow passage walk by the usual remote control. But this interface made the movement of the complex course easy. Figure 12 shows the remote teaching interface and a side walking of a narrow passage.

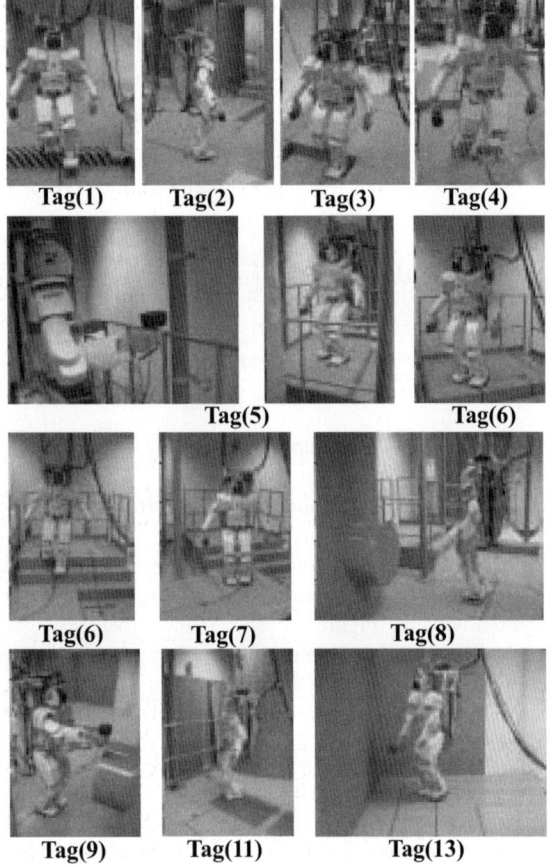

Figure 11: Autonomous Locomotion and Inspection in Plant Mock-up

9. Summaries and the Future Works

It was verified that the humanoid robot is capable of autonomous movement through the stairs, narrow passages and ramps of the mock-up of plant environment, and to perform non-contact inspection operation. And the interface for remote control to do a robot simply against the unexpected matter was developed, and we have confirmed that the movement of the narrow passage could be done easily. Future works are to verify its maintenance movement, with contact, such as opening and closing of the valves.

This work is done as a part of Technology Application Research and Development of Ministry of Economy, Trade and Industry via NEDO through MSTC.

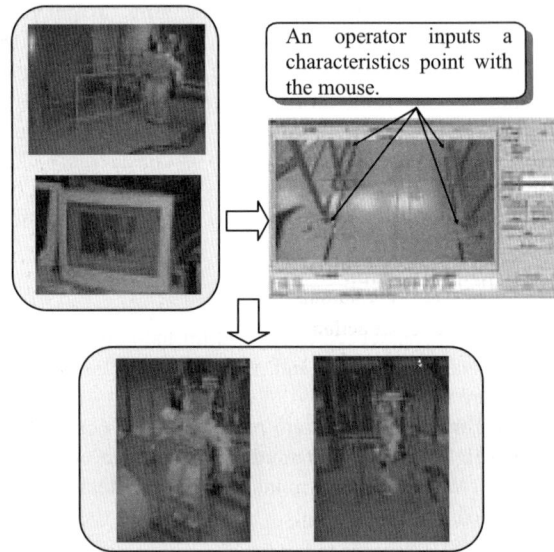

Figure 12: A remote teaching interface and a side walking

References

[1] T. Mano, and S. Hamada: Development of Robotic System for Nuclear Facility Emergency Preparedness, Robotic Society of Japan, Vol. 19 No. 6, pp.714-721, 2001

[2] S. Hamada, and T. Mano: Survey Report on Robotics System for Nuclear Facility Emergency Preparedness, Robotic Society of Japan, Vol. 19 No. 6, pp.678-684, 2001

[3] H. Asama, T. Fujii, H. Kaetsu, I. Endo, and T. Fujita: Distributed Task Processing by a multiple Autonomous Robot System Using an Intelligent Data Carrier System, Intelligent Automation and Soft Computing, An International Journal vol. 6, no. 3, pp. 215-224, 2000

[4] M. Samejima, N. Kawauchi, and T. Oomichi: Development of the Transport System of Coming Generation by IDC, JSME In Proc. of Robotics and Mechatronics, 2PI-J10,2001

[5] N. Kawauchi, T. Sasaki, R. Hiura, M. Tamura, K. Ohnishi, and T. Oomichi: The Plant Maintenance Humanoid Robot System (1st report) - About a Navigation System and Experimental Environment - , In Proc. RSJ Annual Conference, pp. 339-340, 2001

[6] H. Kanazawa, N. Kawauchi, T. Sasaki, S. Asano, and M. Tamura: The Plant Maintenance Humanoid Robot System (2nd report) - Navigation System of Autonomous and Teleoperation Fusion Control - , In Proc. RSJ Annual Conference, 3D14, 2001

Development of User Interface for Humanoid Service Robot System

Takashi Nishiyama*, Hiroshi Hoshino*, Kazuya Sawada*,
Yoshihiko Tokunaga*, Hirotatsu Shinomiya*, Mitsunori Yoneda*
Ikuo Takeuchi**, Yukiko Ichige**, Shizuko Hattori** and Atsuo Takanishi***

*Systems Technology Research Laboratory, Matsushita Electric Works, Ltd.
1048, Kadoma, Osaka, 571-8686, Japan,
**Mechanical Engineering Research Laboratory, Hitachi, Ltd.
502, Kandatsu, Tsuchiura, Ibaraki, 300-0013, Japan,
***Department of Mechanical Engineering, Waseda University
3-4-1, Ookubo, Shinjuku-ku, Tokyo, 169-8555, Japan

Abstract - METI has launched a national 5-year-project called Humanoid Robotics Project (HRP) since 1998 FY. Because humanoid robots have a human-like figure and can walk in a biped way and can take an action like a human being, we consider applying the humanoid robot platform in HRP to service fields of caring people such as the elderly, patients, etc. Taking the technical limitation of the current humanoid robot into account, we assume a situation where the robot serves such people. We regard the robot as users' avatar or another existence of users, and consider the class of users: a nurse, a patient, and a person in a remote site. This allows us to propose four types of user interfaces for the humanoid service robot system: an on/off-line user interface for a nurse who wants to control the robot as his avatar, a user interface for a patient who wants to control the robot as another existence, and a user interface for a person in a remote site who wishes to control the robot as his avatar. These user interfaces are described in this paper.

I. INTRODUCTION

In the Humanoid Robotics Project (HRP), we have been applying the humanoid robot platform to a field where the humanoid robot serves such people as the elderly, the patients and so on [1]. Taking the technical limitation of the current robot into account, we assume a situation or scenario where the robot serves such people. We regard the robot as users' avatar or another existence of users [2], and consider the class of users: a nurse, a patient, and a person in a remote site. This leads us to classify the user interfaces for the humanoid service robot system as follows:

- user interfaces for a nurse who wants to control the robot as his avatar
- user interfaces for a patient who wants to talk to the robot in order to have the robot to serve him
- user interfaces for a person in a remote site who wants to control the robot as his avatar

In the first and third case, the robot is regarded as users' avatar. In the second case, the robot is regarded as another existence of users. This paper clarifies the architecture of the humanoid service robot system embedding these user interfaces, and shows some examples illustrating that the humanoid service robot system is working in a real world.

II. CONCEPT OF HUMANOID SERVICE ROBOT SYSTEM

In the field of service robots, robots are required to operate and communicate with people in usual human-living environments. A humanoid biped robot having a human-like figure can be considered suitable for service fields, since people can easily regard and recognize the humanoid robot as their partners from the outlook and behavior of the humanoid robot. Therefore, we assume that a humanoid robot, situated in a hospital, is helping nurses and patients to do their activities and also assisting people in a remote site to communicate with people in the hospital shown in **Fig.1**. When we consider robots servicing users, robots can be regarded as two types: users' avatars or other existences of users. Taking this into account, we assume the following requirements of users for humanoid service robot systems.

req.1) A nurse wants to use a humanoid robot as his avatar. Also, if possible, he wants the robot to do a portion of his total task.

req.2) A patient wants to communicate and ask a humanoid robot to do a certain task, since he regards the robot as another existence.

req.3) A person in a remote site wants to use a humanoid robot as his avatar.

In this study, in order to satisfy req. 1), we set a remote control cockpit in a nurse station shown in Fig.1. This enables a nurse in a nurse station to control a humanoid robot in a hospital room in real-time. As the control cockpit, we propose to improve a minimum cockpit developed in the former phase of HRP from 1998 FY util 1999 FY [3]. We call this improved cockpit **modified minimum cockpit** [4].

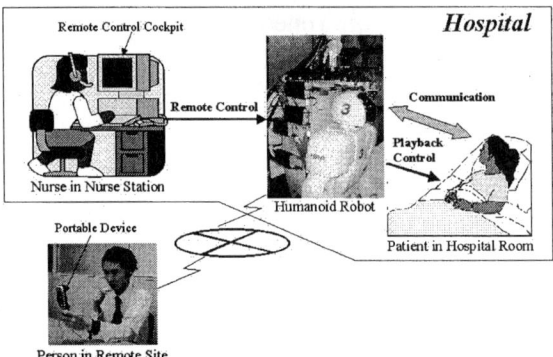

Fig.1 Concept of Humanoid Service Robot System

The modified minimum cockpit is utilized for the nurse to control the robot in real-time. However, the nurse cannot always control the robot. The nurse wants the robot to move or operate automatically to some degree. For instance, when the nurse controlled the robot to move around the environment in the past, if he controls the robot to move along the same path, then he desires the robot to move automatically along with that path. Also, the nurse wants the robot to manipulate an object in an automatic way, if he controlled the robot to manipulate the same object in the past. In this study, to satisfy these needs, we propose a **teaching-playback function** for the humanoid service robot system. That is, while the user (nurse) operates the robot system, he can save or store his operational sequence as an associated command to the system. When the user operates the robot in the same way as the past, he selects the associated command among the list of commands, to designate the proper command. This teaching-playback function is an on-line system to be embedded in the modified minimum cockpit. Also, we have been developing another teaching-playback system called '**extended task library system**'. This extended task library system is an off-line system, in which a user (nurse) can store an associated command allowing the robot to behave automatically. That is, this system is a Virtual Reality (VR) system where a virtual robot is operating in a virtual environment. Since both the virtual robot and virtual environment are the same as the real system, the user can make or generate his own command designating the robot's behavior, and can simulate and check whether that command is effective or not through the virtual environment.

In order to satisfy req.2), we propose a software user interface called '**robot avatar agent**', that enables the humanoid robot to behave and communicate with people in rich and intelligent way [5]. The robot avatar agent can talk to a patient and tries to understand what the patient wants the robot to do. We introduce a voice dialogue function into the agent system since we regard such function to be suitable for the patient or the elderly. Here, what the patient wants the robot to do is limited to the tasks executable for the current humanoid robot system. That is, executable tasks are commands or macro-commands, which are predefined ones by the extended task library system. Also, the robot avatar agent controls the gesture motions and facial expression of the robot. The facial expression is introduced to express the emotion of the robot.

In order to satisfy req.3), we propose a **network-based user interface for robot control**. As shown in Fig.1, a person in a remote site uses a portable device such as a PDA (portable data assistant) or a portable telephone to communicate with a patient in a hospital through the network. The proposed system monitors the conversation between the person in the remote site and the patient. And the system chooses a word told by the person in the remote site, which is associated to a command executable for the humanoid robot as the robot's behavior/gesture. This allows the person in the remote site to express his intent or emotion to the patient not only through his voice but also through the behaviors/gestures of the humanoid robot. Here, the behaviors/gestures of the robot are limited to the ones, which are predefined by the extended task library system. Since cameras and microphones are mounted on both the robot and the portable device, the audio-visual communication through the network is realized in a bi-directional way.

The modified minimum cockpit, the extended task library system, the robot avatar agent, and the network-based user interface for robot control are detailed in the following sections.

III. MODIFIED MINIMUM COCKPIT

When we improve the minimum cockpit developed in the former phase of HRP, we consider the following functions.

- In the minimum cockpit, when the operator controls the robot to walk, he only issues a discrete command designating a goal position and direction. Thus, a function enabling the operator to issue a continuous command is desired.

- In the minimum cockpit, when the operator controls the direction of the camera mounted on the robot, he uses buttons of the 3D mouse. This is not useful for the operator to control the camera direction.
- In the minimum cockpit, the operator cannot control the waist (back) position of the robot.
- In the minimum cockpit, the operator monitors images of the robot camera through the normal display.

In order to realize or modify these functions, we design as follows.

- A joystick issues a walking continuous command.
- A head-tracker worn by the user designates pan-tilt directions of the robot camera.
- A 3-dimensional mouse device is utilized to designate the waist height of the robot. 3D mouse is also used to control the position and posture of the robot hand.
- A head-mounted-display (HMD) provides the operator with images of robot camera.

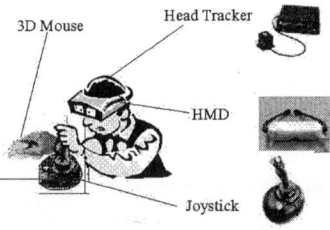

Fig.2 Overview of Modified Minimum Cockpit

Fig.2 shows an overview of the modified minimum cockpit. **Fig.3** illustrates an outlook of operating the cockpit. The operator (nurse) wears the HMD (PTV-K240DP, TekGear), and controls the joystick to move the humanoid robot. The joystick (Side Winder ForceFeedback 2, Microsoft), the head-tracker (IS-300 Pro, InterSense), and the 3D mouse (Magellan Plus 3D Controller, Logicad) are connected with a controlling PC. The controlling PC monitors the control information from these devices to send this information to the robot controller. Each unit walking of going forward, backward, rightward, leftward and rotating clockwise, counter-clockwise is previously generated as each associated command and stored in the robot controller. While the controlling PC detects the operation of joystick, the controlling PC continues to issue the associated command to the robot controller. In order to control the position and posture of the robot hand, instead of using 3D mouse, we developed a new device enabling the operator to designate 6 degrees of freedom (**Fig.4**), and compares the usability of this device with that of 3D mouse.

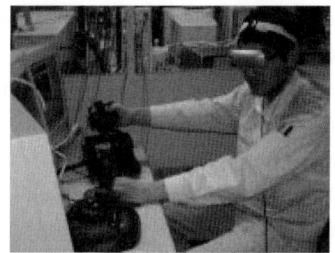

Fig.3 Outlook of Operating the Cockpit

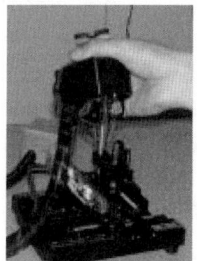

Fig.4 A Novel Operating Device

IV. DEVELOPMENRT OF THE EXTENDED TASK LIBRARY [6]

To build up a task library in the HRP, we first developed the basic motion library for three dimensional walk menus as walking straight, turning, going up/down the stairs, walking on the rugged ground. Moreover, we developed the extended task library. This library can execute a whole body of robot motion including a motion of arm. With the library programming functionality, complex motion can be realized as a sequence of the basic unit motion in the menu. The linking motion between the basic unit motion of the robot is also automatically generated for the continuous motion control. The control data for the three dimensional walk adaptive to the land shape generated by this library was tested and proved to be consistent with the mechanical dynamics using the developed dynamic simulator. The data consistency was also proved to control the test-bed hardware model, which is the scaled humanoid robot, to walk using the extended task library.

Also, we developed the test-bench of the 6 degrees of freedom leg-shaped mechanical elements to improve the accuracy of the control algorithm. The accuracy of the simulation was verified with the consideration of the delay of the joint response.

Also we developed the graphical user interface to use the developed library through the network. User can easily make the robot motion control sequence with built-in block programming function, generate the motion control data using the CORBA-connected network modules on-line, then visualize the robot motion using the VRML robot model on the display. The developed interface was made to help complex operation such as choosing the extended task library menus, specifying the control parameters, and testing the sequence of the motion through the distributed network. It also helps the dynamics analysis in automatic generation of the necessary scripts and files, the script specifying the simulation time for example, to use the dynamics simulator. This network interface provides the seamless system operation in generating the walking motion sequence, verifying with the mechanical dynamics simulation, and confirming the robot motion with the VRML. With the developed VRML Motion Simulator, user can make their own motion plan and simulate the assist task for human-care in the hospital as an example of a real application for humanoid. It is implemented on the Web Browser, with the application specific function menu. The outlook of the interface with the VRML model room objects is in **Fig.5**. With simple operation as selecting the menu, pushing the button, and indicating the control parameter, user can make a plan of the task motion of humanoid, and view the humanoid motion using VRML model through the network. As a future work, more complicated task will be implemented and proved to control real humanoid and other intelligent devices for human-care application.

Fig.5 The Service Task Motion Simulator for Human-care

V. ROBOT AVATAR AGENT

According to the requirements mentioned in section **2.**, we specify the functions as follows.

func.1) A voice dialogue system is embedded.

func.2) A facial expression is realized by Computer Graphics (CG), to be displayed as the robot face.

func.3) The robot avatar agent can control gestures/behaviors of the humanoid robot.

Concerning func.1), what we human beings are saying is usually various, vague and incomplete. Thus, when we ask another person to do a certain task, we cannot say whole conditions /words related to the task at a time. It is usual for us to communicate or interact with each other repeatedly. Therefore, a voice dialogue system is to be developed adaptive to this situation. Here, as an example, we take a robot task requiring three types of attributes: what to do (action), against what to do (object), and where to do (action place). We develop a voice dialogue system that can understand these three attributes through the conversation with a person (patient). In order to construct a voice dialogue system, we have to analyze and model a conversation process between persons [7], [8]. In this study, we model the process using a finite state diagram.

Concerning func.2), we mount a small LCD on the head of the robot, which shows facial expressions using cartoon faces shown in **Fig.6**. The facial expressions are introduced to show the internal states of the robot avatar agent to another person (e.g. patient). For instance, a patient wishes to check whether the agent understands what he asks the robot to do. Here, we utilize various cartoon expressions such as 'pleased', 'disappointed', 'angry', 'being at a loss', 'serious', to represent what the agent understands and recognizes.

Fig.6 Facial Expression of the Robot

It should be noted that when the user tele-operates the robot, he watches the environment around the robot through the binocular camera shown in **Fig.7**. On the other hand, when the user allows the robot to behave automatically and to display facial expressions on the LCD, the LCD should be faced to another person around the robot such as a patient. That is, two modes are needed: the binocular camera faced and the LCD faced. Thus, to switch these two modes alternately, we utilize the tilting motion of the camera platform. In order to allow another person (e.g. patient) to see the facial expressions on the small LCD, the camera platform tilts to the floor direction.

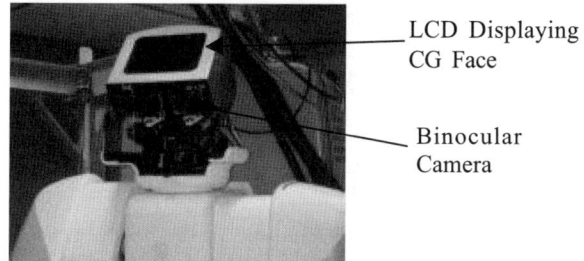

Fig.7 Outlook of the LCD Mounted on the Robot

Concerning func.3), since the robot has a human-like figure, it is possible for the robot to represent gestures such as indicating, shaking the head, stooping down, standing up and so on. In this study, we predefine and store (macro-) commands, which execute these gestures, using the extended task library system. These (macro-)commands are activated and executed if they are needed.

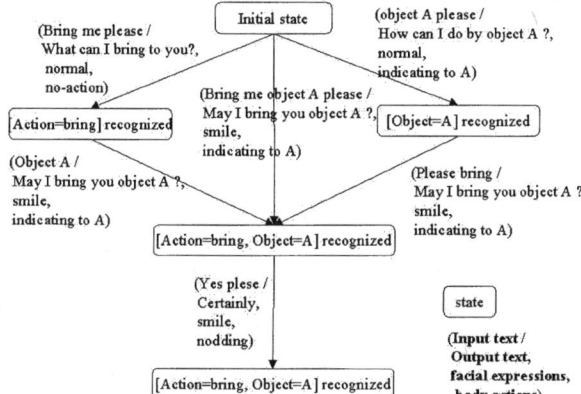

Fig.8 Finite State Diagram of Communication Process

The robot avatar agent integrates and controls these three functions mentioned above. In this study, we extend the finite state diagram modeling a voice dialogue system to represent the behaviors of the robot avatar agent. That is, we add the attributes representing the facial expressions and gestures of the robot on the arcs between the states shown in **Fig.8**. This figure shows an example illustrating the communication process between the robot avatar agent and the patient. In this conversation, the patient talks to the agent to ask the robot to bring the object to him. States represented by circles '○' indicate which attributes the agent recognizes among 'action' and 'object'. On the arcs represented by '→', (Input text / Output text, facial expressions, gestures) are indexed. 'Input text' designates a text recognized by speech recognition engine. 'Output text' designates a text synthesized by speech synthesis

engine.

Fig.9 shows the architecture of the robot avatar agent. As the speech recognition engine, we adopt the 'Julius'[9] that recognizes a large-scaled continuous speech. When the patient speaks to the agent, the speech recognition engine transforms the voice input to a text. The agent matches this input text with the finite state diagram to transit the recognized states. At the same time, the agent refers to the indexes on the arcs, and controls the subsystems of speech syntheses, facial expressions and robot's gestures. The output text is transformed to voices by speech synthesis engine. The robot's gestures are generated and executed by macro-commands.

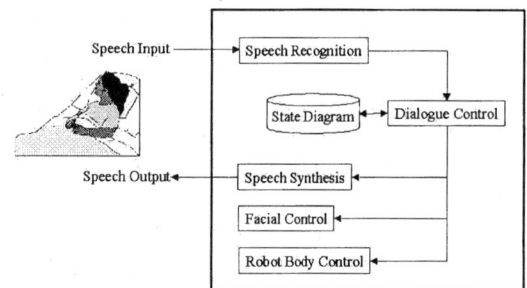

Fig.9 Architecture of Robot Avatar Agent

VI. NETWORK-BASED USER INTERFACE FOR ROBOT CONTROL

According to the requirements mentioned in section **2.**, we clarify the following functions of the network-based user interface.

func.1) The speech recognition is realized via the network.

func.2) The recognized words can be transformed into gestures of the robot.

func.3) A person in a remote can communicate with a patient through the robot. The both communicate with each other using audio and video in a bi-directional way.

Concerning func.1), we can construct a system configuration, in which a client subsystem such as a portable device recognizes speech spoken by a person in a remote site. In this case, only recognized text is sent to a server subsystem through the network. In this study, however, we construct another system configuration, in which a server subsystem recognizes speech spoken by a person in a remote site, since we consider that the current network speed allows speech to flow from the client to the server without delay.

Concerning func.2), the network-based user interface for robot control chooses and picks up a word among speech by a person in a remote site, which can be transformed and be executable for the robot system. For instance, greeting words such as 'Hello', 'Good-bye', or emotional words such as 'so pleased', 'very disappointed' are chosen among the conversation. These words are translated into gestures of the humanoid robot.

Concerning func.3), as an example of portable devices, we can select the portable telephone named 'FOMA', which has the small camera. The images captured by this camera are sent through the network to be presented on the small LCD mounted on the robot.

VII. OPERATION EXAMPLES

In order to operate the total system, we have to integrate the subsystems mentioned in sections **3.**, **4.**, **5.**, **6.**, with the humanoid robot. In this study, we utilize a shared memory device called reflective memory and an optical fiber to connect with each other. This enables each subsystem to refer to the states of the total system. **Fig.10** shows the block diagram of the total system.

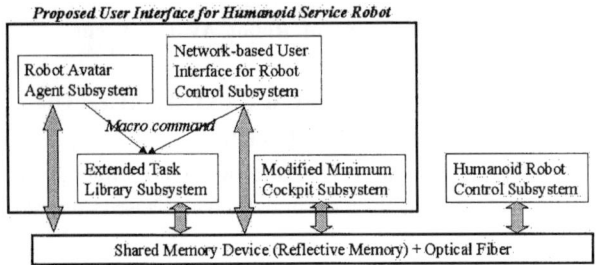

Fig.10 Block Diagram of the Total System

We demonstrated the working of the total system according to the following scenario. The nurse in the nurse station handed in a medicine to the robot, and told the robot to bring the medicine to the patient on the bed. The robot walked to the patient and handed in the medicine to him shown in **Fig.11**. These behaviors were executed by a sequence of commands (a macro-command) in an automatic way. Then the patient talked to the robot, and asked the robot to bring equipment on the table to him. The robot avatar agent controlled this conversation. After this conversation, the robot moved to the equipment by the automatic control through a macro-command, and picked up the equipment by the remote control through the nurse. The robot instructed physical exercises to the patient. These exercise behaviors were also predefined by extended task library system. The communication between the patient and the person in the remote site was also realized.

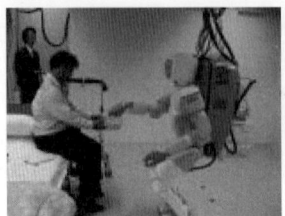

Fig.11 Photograph of Handing in the Medicine

VIII. CONCLUSIONS

In this paper, we described the user interface for the humanoid service robot system. The users' requirements for such user interface were followed by each user interface proposed and developed: the modified minimum cockpit, the extended task library system, the robot avatar agent and network-based user interface for robot control. The working scenario of the total humanoid service robot system was also mentioned.

ACKNOWLEDGEMENT

This study is being conducted by MSTC, Manufacturing Science and Technology Center as part of the "Humanoid Robotics Projects", for the NEDO, New Energy and Industrial Technology Development Organization, under the ISTF, Industrial Science and Technology Frontier Program of METI, Ministry of Economic Trade and Industry.

REFERENCES

[1] Research and Development of Humanoid Robotics System, Annual Report in 2001, MSTC, 2001 (in Japanese).

[2] S. Tachi, Introduction to Robotics, Chikuma Shobo, 2002 (in Japanese).

[3] S. Tachi, K. Komoriya, K. Sawada, T. Itoko, K. Inoue, Telexistence Control Cockpit System for HRP, *Journal of Robot Society of Japan,* Vol.19, No.1, pp.16-27, 2001 (in Japanese).

[4] H. Hoshino, T. Nishiyama, K. Sawada: Development of a Tele-operation System in the Personal Care Service Field, *Proc. of Annual Conference of Robot Society of Japan,* to appear, 2002 (in Japanese).

[5] T. Nishiyama, H. Hoshino, K. Sawada: Development of Communication Agent for Humanoid Service Robot System, *Proc. of SICE System Integration Division Annual Conference,* pp.455-456, 2001 (in Japanese).

[6] Y. Nemoto, I. Takauchi, K. Sawada, T. Nishiyama, M. Fujie, A. Takanishi, Development of a Basic Service Task Library for HRP Humanoid, *Proc. of 2001 IEEE/RSJ International Conference on Intelligent Robots and Systems,* 2001.

[7] N. M. Fraser, G. N. Gilbert, Simulating Speech Systems, *Computer Speech and Language,* Vol.5, pp.81-99, 1991.

[8] I. Katsuse, M. Takahashi, A. Teraoka, N. Kishida, K. Fukuda, T. Ngasako, Speech Interface for Remote-control of the Robot, *Journal of Human Interface Society,* Vol.3, No.2, pp.111-119, 2001 (in Japanese).

[9] K. Shikano, K. Itoh, T. Kawahara, K. Takeda, M. Yamamoto, Speech Recognition System, Ohmsha, 2001 (in Japanese).

Cooperative Works by a Human and a Humanoid Robot

Kazuhiko YOKOYAMA[*], Hiroyuki HANDA[*], Takakatsu ISOZUMI[**], Yutaro FUKASE[†], Kenji KANEKO[‡],
Fumio KANEHIRO[‡], Yoshihiro KAWAI[‡], Fumiaki TOMITA[‡] and Hirohisa HIRUKAWA[‡]

* Yaskawa Electric Corporation
  12-1 Ohtemachi, Kokura-Kita-ku, Kitakyushu, Fukuoka 803-8530, Japan
** Kawada Industries, Inc.
  122-1 Hagadai, Haga-machi, Haga-gun, Tochigi 321-3325, Japan
† Shimizu Corporation
  1-2-3 Shibaura, Miano-ku, Tokyo 105-8007, Japan
‡ National Institute of Advanced Industrial Science and Technology
  1-1-1 Umezono, Tukuba, Ibaraki 305-8568, Japan

Email: {yoko, hand}@yaskawa.co.jp, taka.isozumi@kawada.co.jp, fukase@shimz.co.jp,
{k.kaneko, f-kanehiro, y.kawai, f.tomita, hiro.hirukawa}@aist.go.jp

- *Abstract* --- **We have developed a humanoid robot HRP-2P with a biped locomotion controller, stereo vision software and aural human interface to realize cooperative works by a human and a humanoid robot. The robot can find a target object by the vision, and carry it cooperatively with a human by biped locomotion according to the voice commands by the human. A cooperative control is applied to the arms of the robot while it carries the object, and the walking direction of the robot is controlled by the interactive force and torque through the force/torque sensor on the wrists. The experimental results are presented in the paper.**

1. Introduction

Honda first demonstrated the capability of humanoid robot 1996 through the development of P2 [1] and based on that success, this field of research has been increasing and spreading in various ways. People has slowly become more inclined and interested to use this technology in their respective fields, for example, in personal service, high-risk maintenance task, construction industries, etc.

To cope with these social needs, Ministry of Economy, Trade and Industry of Japan has promoted the Humanoid Robotics Project (HRP) since 1998 for five years [2]. The project is divided into two phases. In the first phase, the robot platform, the remote control cockpit and the virtual robot platform were developed. In the second phase from 2000 to 2002, the applications of humanoid robots have been investigated using the platforms.

In the second phase, we have been enhancing an idea of introducing a humanoid robot to work cooperatively with a human in the outdoor [3]. Figure 1 is an illustration of such a cooperative work in a construction environment. In this environment, there are many chores that are carried out by a pair of an expert and a novice. The replacement of the novice with a humanoid robot enables us to reduce the number of workers and the cost, and to supply labors to counter the shortage of manpower in the aged society, which is one of the most serious problems in Japan.

We selected two typical works that are doing by the pair - one is carrying an external wall panel, and the other is mounting the panel on the frame of a house. To replace the novice with a humanoid robot, we have been developing four technologies, mainly.

1) Hardware/software to walk on uneven surfaces, and to prevent damage in the event of tipping
2) Sensors to position the wall panel and the grasping point
3) Communication between human being and the robot conveniently
4) Software to control the arms and the legs cooperatively

In the first half of second phase, we have combined these hardware and software into a new humanoid robot HRP-2P and demonstrated to carry a panel with a human on the flat floor. This paper outlines the results of these developments.

This paper is organized as follows. Section 2 describes the above building blocks implemented on HRP-2P. Section 3 presents the experimental results. Section 4 concludes the paper.

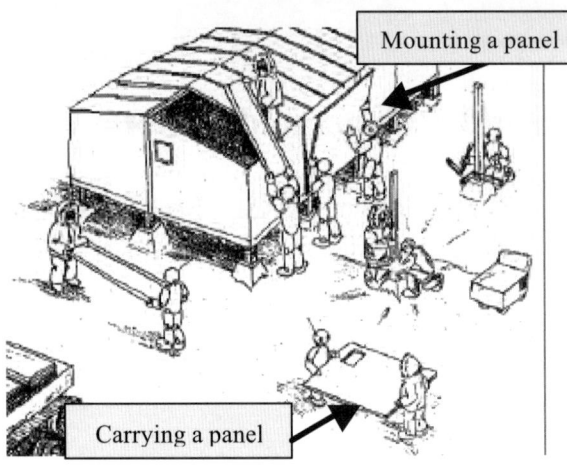

Figure 1. Outdoor Worksite

2. Building Blocks

2.1. Humanoid Robot HRP-2P

HRP-2 is a new humanoid robot platform, whose manufacturing process is in progress in phase two of HRP. The design concepts of HRP-2 are light, compact, but performable for application tasks like cooperative works in the open air shown in Fig. 1 [3]. As a result, HRP-2 is designed to be feminine size. Figure 2 shows the prototype of HRP-2 [4].

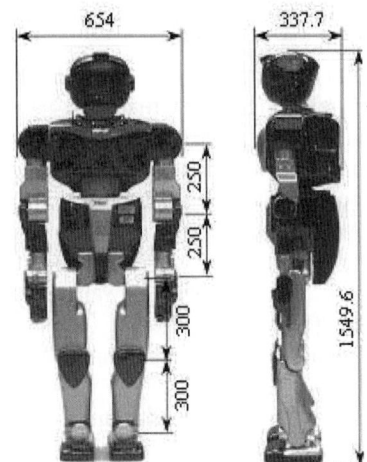

Figure 2. Humanoid Robot HRP-2P

As shown in Fig. 2, HRP-2P has unique configurations. One is that the hip joint of HRP-2P has a cantilever type structure as well as HRP-2L which is the leg module of HRP-2 [6]. The other is that HRP-2P has two waist joints.

The humanoid robot tends to tip over easily, since the area of the foot sole that supports the whole body is so small and limited. The motions of the body during tasks may easily make the humanoid robot lose its balance as well as those of the arms. From this observation, the mechanism for prevention of tipping over is a very important requisite to realize a really useful humanoid robot.

The tipping over easily occurs when the target ZMP is going to the outside of the support polygon made by supporting feet [5]. Since it is so hard to recover from a tipping over, our approach to prevent tipping over is to construct the mechanism, which easily enables to make the target ZMP to be inside of the support polygon.

A mechanism, which enables the robot to have a wide sphere of landing point for the swinging leg, would be one solution for our approach. The reason is that we can appropriately shape the support polygon for the phase of double supported legs by selecting the landing point of the swing leg. By shaping the support polygon for the phase of double supported legs immediately, the tipping over would be prevented, even if humanoid robot begins to tip over. Especially, crossing legs further can make the support polygon to be on the opposite side of supported leg. To realize a wide sphere of landing point for the swing leg, the hip joint of HRP-2P has a cantilever type structure as shown in Fig. 2. Because the cantilever type structure enables the robot to have less collision between both inside upper-limbs and also to cross legs [6].

The other factor throwing the humanoid robot off balance is caused from rolling motions of the gait. The mechanism, which makes the trajectory of the center of gravity (COG) of the upper body smooth with less rolling motion, is also effective in the prevention of tipping over. To reduce rolling motion of the gait, the cantilever type structure also plays an important role. Since the cantilever type structure can make the length between hip joints shorter, this structure can make the length between landing points of pitch axis shorter too [6].

From these discussions, we designed the cantilever type structure to achieve the mechanism for prevention of tipping over. This structure enables the robot to cross legs as well as to make a protector between legs for minimal damage in the event of tipping over.

HRP-2P would not be able to avoid tipping over, even though HRP-2P has the cantilever type structure as explained above. When HRP-2P tips over during cooperative works in open air, we request HRP-2P to get up by a humanoid robot's own self. To realize such a humanoid, a waist joint with 2 D.O.F. (pitch axis and yaw axis) is necessary for HRP-2P.

The waist joint brings several advantages. One is that the structure of HRP-2P can be lithe. The lither the upper body is, the smoother its gait is. Another is that the moment generated in the yaw axis of HRP-2P can be suppressed by using waist motion. This compensation will be done in the near future. Furthermore, the waist joint makes a working space of arm extended. Although HRP-2P has 6 D.O.F. in each arm and 1 D.O.F. in each hand, waist motion gives a redundancy to the arm motion.

2.2. Vision System on HRP-2P

HRP-2P has a stereo vision system composed of three cameras(Fig. 3). Two horizontal cameras are separated by 120 mm, and the third camera is 60mm upper than them. We adopt three cameras system since it is difficult for stereo vision composed of two cameras to detect the horizontal line. The weight of the vision system is less than 700 g. Relatively short focus lens is used in order to measure a position of an object, whose standoff is from 0.5m to 4 m. The camera's shutter speed is controlable by a computer to adapt various lightning condition. The image processing is based on VVV System [7]. VVV consists of several image processing modules and we can reconstruct 3D shape model, detect a object, measure position and track a object with it.

Figure 3. Vision System

2.2.1. Correction of Distortion

Through a plastic shield in front of the cameras, the object image is distorted. It is practically difficult to model the shield shape and the position of the camera correctly. So we make a conversion table between a distorted image to a image without the shield. On a calibration board, circle patterns are arranged at even intervals and a lager circle is arranged near the center of the board. We capture the board with and without the shield (Fig. 4) and detect the positions of each circle. The larger circle position on each image is used as positioning reference of all the other circles and gets the correspondence of all circle points between two images. Using an interpolation of this correspondence, we make each pixel conversion table. We also correct lens distortion using the same image without the shield. As a result, the distortion is reduced within 0.2 pixel.

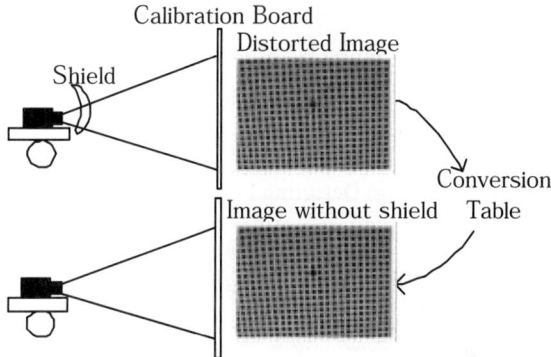

Figure 4. Correcting Shield Distortion

2.2.2. Coordinates of the Cameras and the Robot

The detected position of an object is represented by a vision coordinate system. In order to use this position data, we have to transform the vision coordinate system to the robot coordinate system.

A marker is put on the robot finger and HRP-2P moves the hand to several places in the filed of the vision (Fig. 5). HRP-2P captures image and gets the position of the finger at each place. Using the detected marker positions on the vision coordinate system and the finger positions on the robot coordinate system, we calculate the transform matrix from the vision coordinate system to the robot coordinate system.

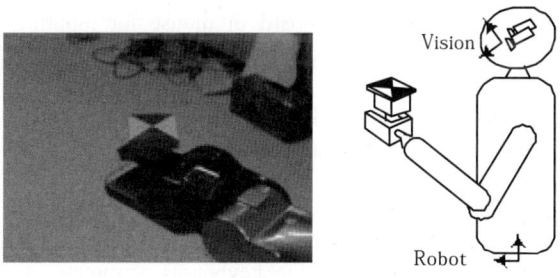

Figure 5. Calibrating Transform Matrix

2.2.3. Recognition of Panel

An object recognition function is necessary for HRP-2P to grasp and carry a panel with a human operator. A model based recognition system [8] is used, and the grasping point was calculated within 2 mm error from the results of recognizing the panel top edge. Figure 6 shows (a)HRP-2P detecting panel, (b) captured images and (c)a result of matching a panel model.

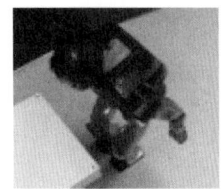

a) Detecting Panel

b) Captured Images

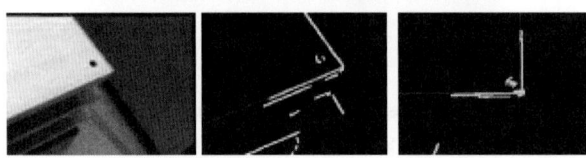

c) Matching panel model on image

Figure 6. Detecting Panel

2.3. Human Interface

In the cooperative works by a human and a robot, an user-friendly interface is important. The target work is that a human and a robot carry an exterior wall panel together as shown in Fig. 7. The human needs to use his/her both hands to carry the object. In this case, it is not suitable to use a keyboard or mouse for inputting commands and a joy-stick for operations. However, the human interface using voice has been gaining importance in practice with the advancement of speech recognition technology and the processing speed of recent CPUs. Also in the industrial robot, the direct-teaching by force sensor is already seen in practice [9]. We call this method as force sensing operation. Under these backgrounds, the keyboards or mouse and the joy-stick were replaced by the voice input method using speech recognition and the force sensing operation

method respectively. This section describes the developed voice input system while section 2.4 describes the force sensing operation in detail.

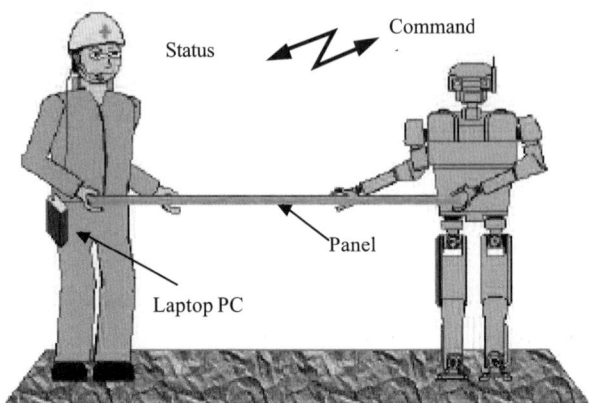

Figure 7. Work example

The voice instructions have been developed and installed on a laptop PC that has been carried by the human. The communication is taken place between PC and the robot. PC recognizes speech, converts it to generate command, and transmits to the robot. The robot executes command after interpretation. The human also gets a confirmation message from the PC through headphone that the command is sent to the robot. The human also sees the robot if it executes his command. However, the problem arises when a command has not been sent to robot due to the following failures.
1) Failure in speech recognitions
2) Failure at communications

Besides, all practical system, recognition and communication take some time period. Sometimes this time period may create a new problem. For example, if the time is long and the human is restless, the human may send the same command twice before the first command is executed by the robot. In that case, the robot will repeat the same action during the execution of second command. All these problems can be solved by developing a system that sends information at various stages as shown in the flow diagram of Fig. 8. The system tells the human using different sounds in each of the stages as shown below.
1) Speech recognition is started (Sound A)
2) Recognition is impossible (Sound B)
3) Out of commands (Sound C)
4) Under communication (Sound D)

With this development, it is possible for the human to

know the state of the system without using special devices and to transmit command to the robot with certainty.

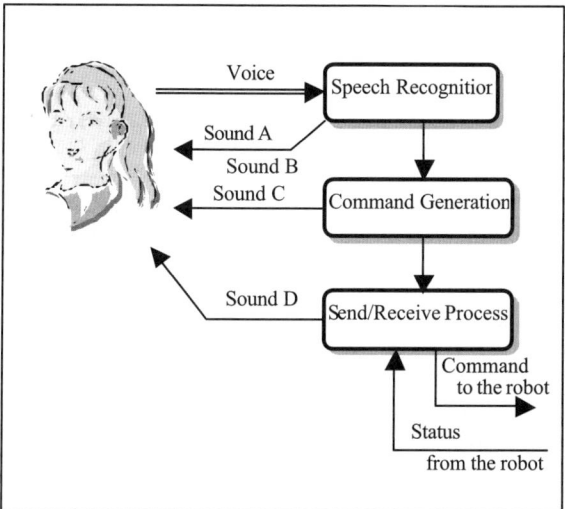

Figure 8. The flow of sound information

2.4 Force Sensing Operation System

When a human and a robot work on an object cooperatively, it is necessary to tell the robot the moving direction and speed continuously. For example, when operating a robot using voice instruction, the human must direct the robot continuously with words such as "Right", "Left", "Clockwise", "Anticlockwise", "Before", "Back", "Stop", etc. Furthermore, in addition to these commands, the speed information are also to be sent with words like "Slow", "Fast", "Medium", etc. However, the human has to speak continuously without any mistake then. This can be overcome by using force sensor. When the human moves the object, the force is transferred to the robot through the object. This force is sensed by force sensor. However, the biggest problem, using force as a prime mover to operate the robot, arises from other side. This transmitted force generates tipping moment on the robot that creates unstable movement. To overcome this difficulty some kind of cushion between the object and robot is necessary in order to use force sensing operation effectively. In this case, it is possible to apply an impedance control based on force sensing operation. Kosuge et al, adopted the impedance control to a dual-arm mobile robot, and confirmed the effectiveness [10],[11].

When dual-arm grasped one side of the object like a long panel and human moved the other side to right and left, it is difficult to recognize translation or rotation by the force sensors mounted on wrists of the robot. To cope with this problem, we took measures to change translation or rotation via voice instructions.

Hence, using voice instruction, impedance model on robot's arms, can be written by simple equation as

$$M_d \Delta \ddot{X} + D_d \Delta \dot{X} + K_d \Delta X = F,$$

where M_d is an inertia matrix, D_d is a viscous matrix, K_d is a stiffness matrix, and F is the force and moment vector. When F is applied, it will produce the displacement vector ΔX. And ΔX is added to target poses of robot's hands.

F applied from the human must include some factors of disturbance, like a jerk. On the other hand, the calculated ΔX from the impedance control changes continuously and smoothly. Therefore the walk velocity is expressed using the ΔX instead of F, and its vector v is determined by

$$v = G_L \Delta X,$$

where G_L is a coefficient matrix which changes the displacement vector ΔX into walk velocity vector v. The robot generates movement patterns, such as steps on case of walking robot in real time using velocity vector and moves along the direction of the force transmitted by the human.

Hence, using impedance control, the transmitted force on the robot by the human is absorbed smoothly avoiding any jerk created to tip over the robot. Therefore, applying force sensing operation, it is possible to operate the robot intuitively along the direction and speed without considering any command signals.

3. Experiments

A series of experiments have been carried out with the humanoid robot working cooperatively with a human (operator) to move an object, say panel, from the loading zone to some other place. The panel is 5kg in weight having dimension of 1.8 x 0.6 m.

The followings are the detail of experiments to demonstrate the capabilities.

a) The operator work while saying "Start work" using voice control. The robot moves in front of a panel as shown Fig. 9a.

b) The operator says "Get panel position" and the robot starts to search the panel (Fig. 9b). Based on the recognition of panel position using stereo vision, the robot automatically judges whether the current standing position is sufficient to grasp the panel or not. If not, the robot also automatically adjusts the standing position as shown in Fig. 10.

c) The operator says "Grasp panel". The robot grasp

and lifts the panel using the pose information acquired by the stereo vision. (Fig. 9c)

d) The operator lifts the panel at the same time and says "Move panel". The robot changes its mode from position control to impedance control and moves along the direction of the force applied by the operator. (Fig. 9d)
e) The operator guides the robot to move to the unloading zone and says "Put panel". (Fig. 9e)
f) The robot releases the panel after recognizing the voice command "Put panel", and finishes the work as shown in Fig. 9f.

Figure 11 shows an experiment of the cooperative work. The robot is not taught its destination, but it is possible to reach in front of a panel by the force sensing operation.

In our daily life, an interaction between humans is common and natural but that between a human and a robot through voice is far from real. This experiment shows a human, communicating with a humanoid robot, performs a work cooperatively. The work is not complex to human but complex enough to execute by a robot.

Our humanoid robot performs the task using biped locomotion, vision, cooperative handling control, and voice instructions.

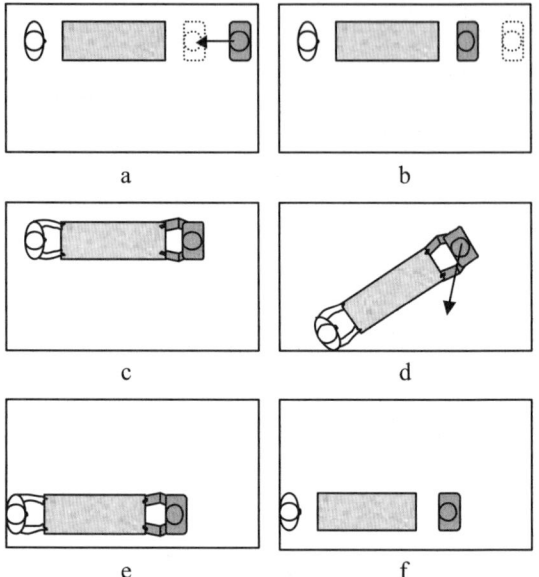

Figure 9. The scenario of an experiment

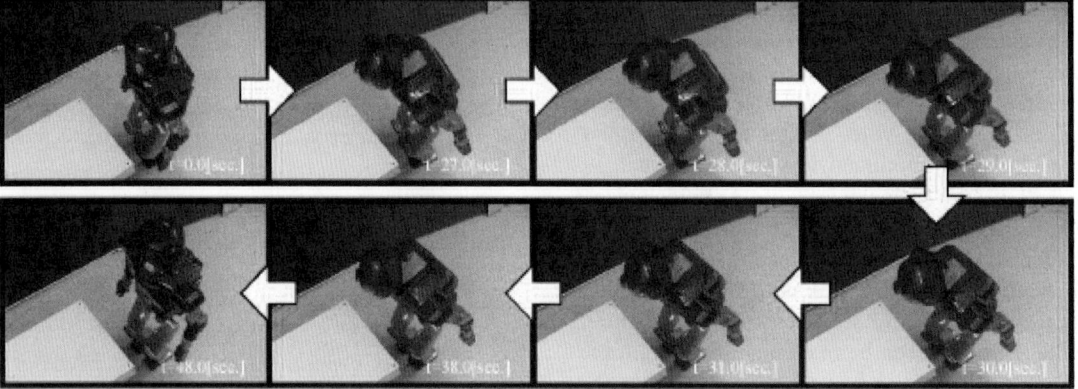

Figure 10. The recognition of panel position and the standing position adjustment

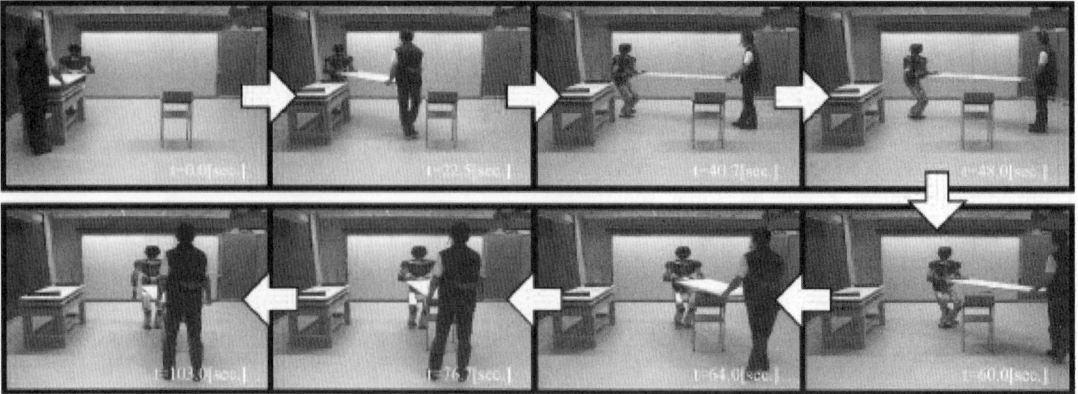

Figure 11. Cooperative work with human

4. Conclusions

In this paper, humanoid robot HRP-2P has been developed to realize the cooperative work between a human and the humanoid robot. This robot is equipped with the stereo vision system, human interface and force sensing operation system. HRP-2P is one of the first humanoid robots on which various functions are realized and can execute a significant task.

In the rest of the second phase, we plan to develop a final model HRP-2 which is designed based on the experimented results on HRP-2P. The software, for the final goal such as carrying the external wall panel on uneven surfaces, mounting it on the frame of the house, etc., has to be modified and improved accordingly. In addition, we would like to develop a robot that can minimize damage in an event of tipping and is capable of standing up, if tipped.

Acknowledgments

This research was supported by the Humanoid Robotics Project (HRP) of the Ministry of Economy, Trade and Industry (METI), through the New Energy and Industrial Technology Development Organization (NEDO) and the Manufacturing Science and Technology Center (MSTC). The authors would like to express sincere thanks to them for their financial supports.

This successful development of cooperative works by humanoid robot HRP-2P and a human would not be achieved without helpful discussions from our cooperative members. The authors would like to thank sincerely the member, Kazuhito Yokoi, Shuuji Kajita and Kiyoshi Fujiwara from AIST, Junichirou Maeda from Shimizu Co., Kazuhiko Akachi from Kawada Industries, Inc. and Kenichi Yasuda of Yaskawa Elec. Co.

References

[1] K. Hirai, "Current and Future Perspective of Honda Humanoid Robot," Proc. IEEE/RSJ Int. Conference on Intelligent Robots and Systems, pp.500-508, 1997.

[2] H. Inoue, S. Tachi, Y. Nakamura, K. Hirai, N. Ohyu, S. Hirai, K.Tanie, K. Yokoi, and H. Hirukawa, "Overview of Humanoid Robotics Project of METI," Proc. the 32nd Int. Symposium on Robotics, pp. 1478-1482, 2001.

[3] K. Yokoyama, J. Maeda, T. Isozumi, and K. Kaneko, "Application of Humanoid Robots for Cooperative Tasks in the Outdoors," Proc. Int. Conference on Intelligent Robots and Systems, Workshop2 (Oct. 29, 2001), 2001.

[4] K. Kaneko, F. Kanehiro, S. Kajita, K. Yokoyama, K. Akachi, T. Kawasaki, S. Ohta, and T. Isozumi, , "Design of Prototype Humanoid Robotics Platform for HRP," Proc. IEEE Int. Conference on Intelligent Robots and Systems, (to appear), 2002.

[5] M. Vukobratovic and D. Juricic, "Contribution to the Synthesis of Biped Gait," IEEE Tran. On Bio-Medical Engineering, Vol. 16, No. 1, pp. 1-6, 1969.

[6] K. Kaneko, S. Kajita, F. Kanehiro, K. Yokoi, K. Fujiwara, H. Hirukawa, T. Kawasaki, M. Hirata, and T. Isozumi, "Design of Advanced Leg Module for Humanoid Robotics Project of METI," Proc. IEEE Int. Conference on Robotics and Automation, pp. 38-45, 2002.

[7] Tomita, Yoshimi, Ueshiba, Kawai, Sumi, Matsushita, Ichimura, Sugimoto, Ishiyama: R&D of Versatile 3D Vision System VVV, Proc. IEEE Int'l Conf. on SMC'98, pp.4510-4516, 1998.

[8] Sumi, Kawai, Yoshimi, Tomita: 3D Object Recognition in Cluttered Environments by Segment-Based Stereo Vision, International Journal of Computer Vision, 46, 1, pp.5-23, 2002.

[9] Y. Inoue, K. Takaoka, "Teaching Method for Industrial Robot", Journal of Robotics Society of Japan, pp. 22-25, 1996.

[10] K. Kosuge, H. Kakuya, and Y. Hirata. "Control Algorithm of Dual Arms Mobile Robot for Cooperative Works with Human". Proc. of IEEE Int. Conf. On Systems, Man, and Cybernetics, pp. TA10-3, 2001

[11] R. Suda and K. Kosuge. "Handling of Object by Mobile Robot Helper in Cooperation with a Human". Proc. Of the 32th Int. symposium on Robotics, pp.550-555, 2001

Application of Humanoid Robots to Building and Home Management Services

Naoyuki Sawasaki[1] Toshiya Nakajima[1] Atsushi Shiraishi[1]
Shinya Nakamura[2] Kiyoshi Wakabayashi[2] Yusuke Sugawara[2]

[1]FUJITSU LIMITED; 4-1-1 Kamikodanaka; Nakahara-ku, Kawasaki 211-8588; Japan
[2]SOHGO SECURITY SERVICES CO.,LTD.; 2-14 Ishijima; Koto-ku, Tokyo 135-0014; Japan)

Abstract – This paper describes the development of applications for the building and home management service section of the "Humanoid Robotics Project (HRP)" supported by METI. The objective of this section is to develop a system that enables users to control the humanoid robots in their homes remotely via the Internet with a simple mobile terminal. To construct such a system, we focused on developing sensor-based autonomy sufficient to achieve task-level teleoperations. The system included view simulation based on CG technologies to help users perceive conditions at the robot's site and the client-server architecture necessary for remote task execution to reduce the load on the mobile terminal. Based on these technologies, we implemented a prototype system and conducted some experiments that illustrate that the system achieves the basic required functions for a building and home management service.

1. INTRODUCTION

Humanoid robots have been a principal topic in recent robotics research. After Honda demonstrated the capability of their prototype humanoid robot, Japan's Ministry of Economy, Trade and Industry (METI) launched a five-year project called the "Humanoid Robotics Project" (HRP) in FY1998[1]. In phase one (FY1998 to FY1999), a humanoid robot, a tele-existence cockpit to control the robot and an equivalent virtual robot were developed to serve as a common research and development platform. In phase two (FY2000 to FY2002), based on intensive investigation into future expected needs, development began on several application systems.

Fujitsu Limited and Sohgo Security Services Co., Ltd. collaborated to develop an application system for a building and home management service.

This paper describes the development of the building and home management service. The paper is organized as follows. Section 2 overviews the technical issues involved and describes the basic design concepts of the application system. Section 3 introduces our prototype system implemented on the platforms developed during phase one. Section 4 describes some experiments that illustrate the capabilities of the developed system. Section 5 is a summary of this paper.

2. OVERVIEW OF THE BUILDING AND HOME MANAGEMENT SERVICE

Many people think it could be quite useful for people to have a robot to look after their house in their absence.

The objective of the building and homes management service is to develop a system that enables users to remote control one or more humanoid robots in their houses remotely via the Internet with a simple mobile terminal, as

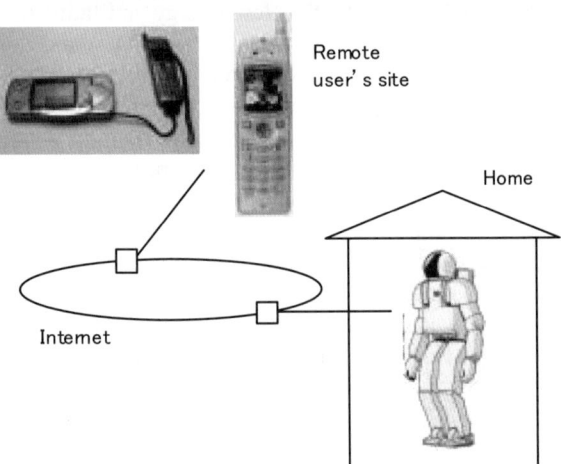

Fig. 1 Building and home management service

shown in Figure 1. In this configuration, however, it is impossible to apply the tele-existence cockpit developed in phase one because of the limited computing power and user interface of mobile terminals and the limited capacity of the communication paths. The main principal problem in this application is how to make it easier to facilitate remote control of a wide variety of complicated motions of a humanoid robot in spite of these restrictions.

To solve this problem, we focused on the following technologies:
· Sensor-based autonomy sufficient to implement task level teleoperation
· View simulations based on CG-related technologies to help users perceive the conditions at the robot's site.

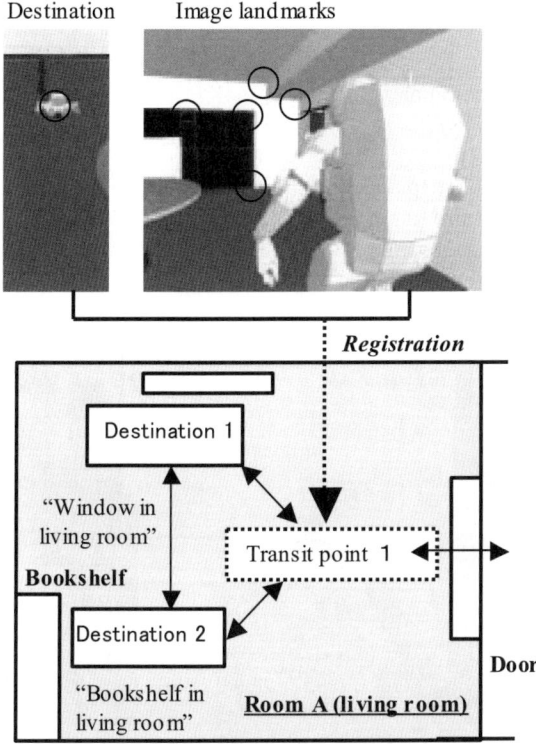

Fig. 2 Concept of sensor-based autonomy

Fig. 3 Examples of view simulation

- Remote control client-server architecture to reduce the load on the mobile terminal

2.1 Sensor-based Autonomy for Walking and Hand Operations

Although a humanoid robot can execute a wide variety of tasks almost anywhere in a house, it is quite difficult to define a desired task completely to the robot. In many cases, people just expect the robot to do simple tasks such as holding apparatuses to check the condition of the house and doing specific operations in their absence. From this point of view, it is still useful and practical that the user teach the desired locations and rather limited operations to be performed by the robot before they are actually needed. If the tasks are limited, the robot can execute them autonomously by using its sensors and the information registered by the user. This enables task-level teleoperation.

The scheme we developed is shown in Figure 2. In the teaching phase, the user guides the robot by the basic motion commands to the desired locations and registers the names of the locations and related landmarks. A landmark consists of a pair of image features and their corresponding positions in terms of world coordinates so as to be identified and measured by the vision sensor of the robot for self-localization. The user can also resister desired hand operations related to the desired objects at the desired location.

To avoid complicated procedures of registering specific hand operations, we developed a basic library of hand operations that consist of built-in motion sequence templates for each operation. By using this library, the user can register the desired hand operation by simply registering the name of the template, the name of the object, and the image of the object. The system then represents these names and images as a graph structure, as shown in Figure 2.

In the autonomous execution of a specified task, the proper path to the desired location is determined by graph search, and then the robot starts walking there. The robot might walk past registered locations before reaching the final destination. In these cases, it executes self-localization at each registered location to correct any error in position from the planned path. After reaching the specified location, the robot identifies and measures the registered objects by its vision sensor and executes the registered hand operations on them.

2.2 View Simulation

The image from the robot's camera must be presented to the user so the condition at the site of the robot can be known. Unlike the tele-existence cockpit developed in phase one, it is impossible to represent high-quality 3-D images in real time on a mobile terminal. In practice, only low-resolution still images from the single camera can be shown. Therefore, the user might find it difficult to get enough information to control the robot.

To solve this problem, we developed a view simulator based on CG technologies to represent simulated images from several angles, which are quite helpful for the user in perceiving the target site. Figure 3 is an example of a bird's eye view image, a plane view image, and a simulated camera image from the robot generated by the view simulator.

2.3 Remote Control Client-Server

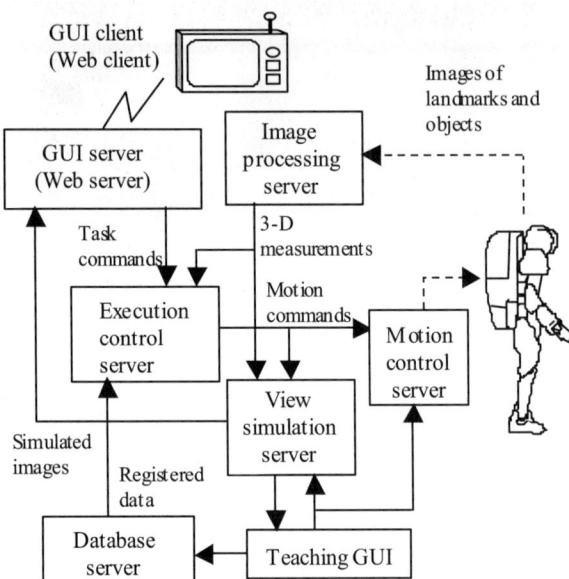

Fig. 4 Architecture of the prototype system

TABLE 1
SPECIFICATIONS OF ADOPTED TERMINALS

	Mobile phone	PDA
Communication speed (bps)	32K - 64K	32K - 64K
Weight (g)	< 100	180 - 250
Display size (pixel)	120 x 160	240 x 320
Display size (mm)	29 x 38	55 x 70
Web capability	Browser	Browser, Java

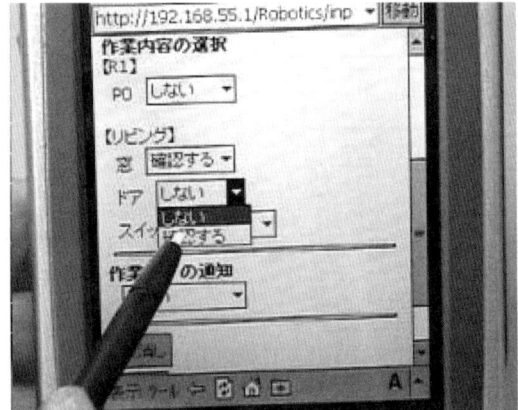

Fig. 5 Task specification menus on the PDA

The mobile terminals we adopted are small and lightweight thus very suitable for control terminals of this application. On the other hand, since computing power, displaying area and wireless data communication speed of the terminals are relatively smaller and slower than those of note PCs with LANs, for instance, we must make special efforts to develop a graphical user interface (GUI) on these terminals for robot control.

To resolve the restrictions described above, we developed the GUI system as a client-server application (GUI client and GUI server). The GUI client runs on a mobile terminal and is basically a web client that performs no heavy computation. It is customized to transmit robot control commands entered by the user to the GUI server and to display results of the commands. The GUI server runs on another PC in the house and serves as a broker, communicating with the terminal and the Execution control server. This client-server system enables us to define short-length task-level commands for robot control. These are sent from the client and each of them is converted on the server to a long sequence of basic motion commands. Thus we can reduce traffic on the relatively slow telephone connection and also improve usability of the GUI by inputting a few simple task-level commands.

3. IMPLEMENTATION OF PROTOTYPE SYSTEM

The architecture of our prototype system is shown in Figure 4. The execution control server organizes the image processing server, the view simulation server, the database server, and the motion control server. It receives task-level commands from the GUI server and executes the task autonomously by generating the commands for image processing, view simulation, and motion control based on the data registered in the database server. The details of implementing the GUI client-server, the image processing server, the view simulation server, and the teaching GUI are described in this section.

3.1 GUI Client-Server

Since the final goal is to realize robot control system that can easily be used by average people when they are away from home, we adopted inexpensive, widely used commercial mobile hardware for use as robot control terminals. The terminals are mobile telephones and personal digital assistants (PDAs), both equipped with web browsers and functions of wireless data communication via public phone lines. The specifications are listed in Table 1.

Figure 5 shows an example to input tasks that the robot should perform. Each task is predefined and has a form of "workplace" + "object" + "operation". All workplaces and objects are listed out on the display and the user selects an operation associated to object. For example, opening a window in a living room will consist of operation "open" for object "window" at workplace "living room". Note that the user selects *only* tasks and is not necessary to determine a path along which the robot walks to workplace.

The user can specify multiple tasks at one time, and the robot performs them automatically, one after another. Since it is convenient for the user to be able to check the progress of a task, the GUI indicates progress by displaying the status of each task that is one of "done", "processing" and "waiting". The user can also see two types of images, one

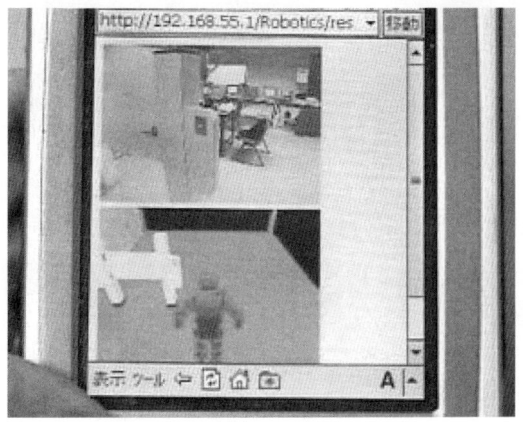

Fig. 6 Images of the task results

Fig. 8 Manual control GUI on the PDA

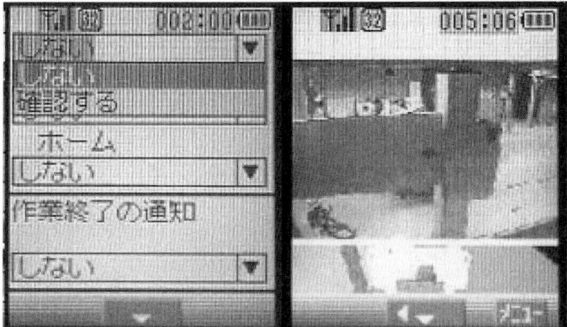

Fig. 7 Equivalent GUI on the mobile phone

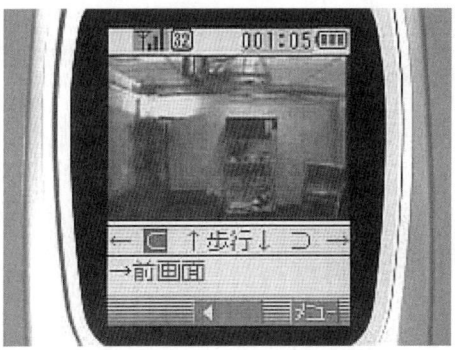

Fig. 9 Manual control GUI on the mobile phone

taken by the camera on the robot and the other generated by the View Simulator, to check conditions around the robot after each task was done (Figure 6). Though the GUI shown in Figure 5 and 6 are for the PDA, equivalent GUI is also possible for the mobile phone with smaller size of display (Figure 7).

In addition to this autonomous control function, the GUI also has a capability to control the robot manually that extends flexibility to various kinds of tasks. The user can instruct the robot to walk forward and backward, turn left and right, pan and tilt its camera, and so on. The actions are confirmed by images sent from the camera that is mounted on the robot, as shown in Figure 8 and 9. The images on the mobile phone are still and updated by the user's key operation, whereas the images on the PDA are streaming video generated by a Java program. The program that is downloaded from the GUI server employs the User Datagram Protocol and minimum buffering for quick response.

3.2 Image Processing Server

The image processing server performs the image processing required for the execution of autonomous tasks including identification of registered landmarks and objects, 3-D measurements of their positions in the robot coordinates by stereoscopic vision, and self-localization of the robot based on the measurements of the landmarks. Because the landmarks and the objects are different in each individual house, we developed a coarse-to-fine image matching algorithm for their identification and for detecting stereoscopic correspondence. We use an intensity-based area correlation as the matching criteria, so as to identify any specific kinds of image features. Although correlation-based image matching requires a huge amount of computation, we employ a dedicated image processor developed by Fujitsu[2] to accelerate the processing speed to a practical level.

3.3 View Simulation Server

The view simulator receives the same commands as the motion control server and generates simulated images of the robot's site from different perspectives. Because the robot platform developed in phase one does not output details of the internal status of the robot, the view simulator cannot precisely synchronize the status of the robot model with that of the actual robot.

We designed the view simulator to update the model robot based on its own timer according to the motion command and to synchronize its location with the real robot

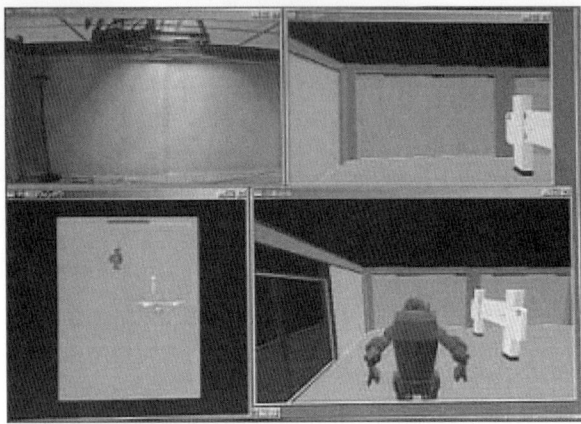

Fig. 10 Teaching GUI software

Fig. 12 Registration of landmarks

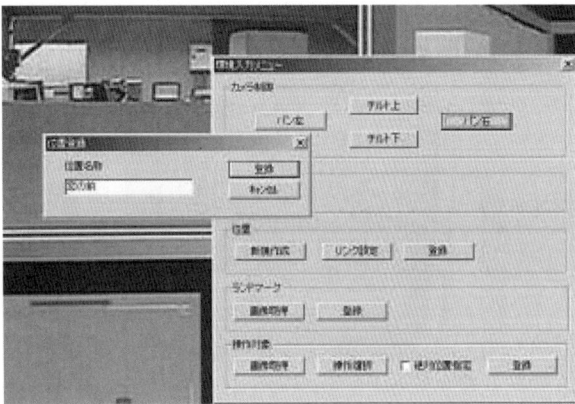

Fig. 11 GUI menus for registration

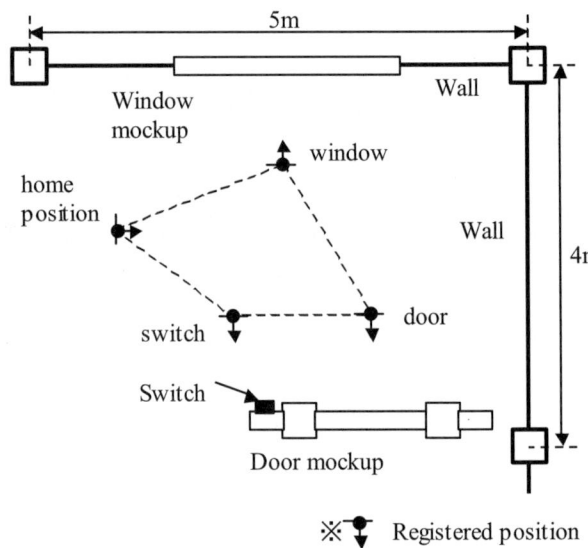

Fig. 13 Experimental setup

only when self-localization is performed by the image processing server.

3.4 Teaching GUI

The teaching GUI software provides applications for registering locations, objects, and operations. As described in Section 2, in order to teach the locations, the objects, and the operations to the robot, the user must guide the robot to the specific locations and register the images of the landmarks and objects at those locations.

To help the user do this, we developed GUI software that utilizes the CG technologies shown in Figure 10. The lower-left window is the plane view image generated by the view simulator. The upper-left window is the image from the camera mounted on the robot, while the upper-right window is the simulated CG image from the camera. The user can guide the robot to the desired location by pointing to the location in the plane view image with a pointing device. After guiding the robot to the desired location, the name of the location is registered on the GUI menu shown in Figure 11.

To register the landmark, the user must specify the rectangular region in the image shown in the upper-left window and the corresponding region in the image shown in the upper-right window, as shown in Figure 12. The software automatically registers the specified region as the image of the landmark and its corresponding position in the world coordinates calculated from the 3-D model.

To register an object for hand operations, the user can utilize the same GUI menu and select one of the built-in operations. This software dramatically simplifies the user registration procedure.

4. EXPERIMENTS OF REMOTE OPERATIONS

To demonstrate the usefulness of our system in practical situations, we conducted some experiments on remote operation in a mockup of a living room. First, the four locations shown in Figure 13, home position, window, door, and switch were registered using the teaching GUI software. Then, several tasks were specified by a PDA. Figures 14 and 15 are the images of the robot walking according to the

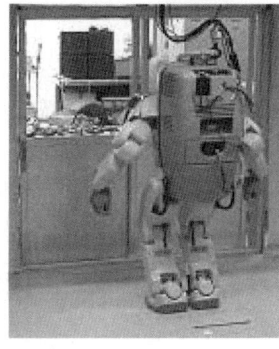

Fig. 14 Autonomous walking to the window

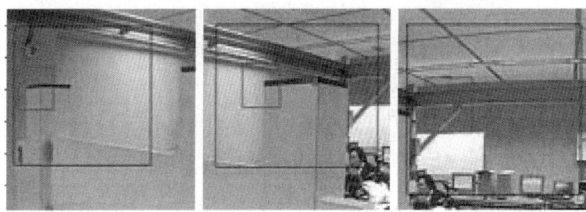

Fig. 16 Landmark detection for self-localization

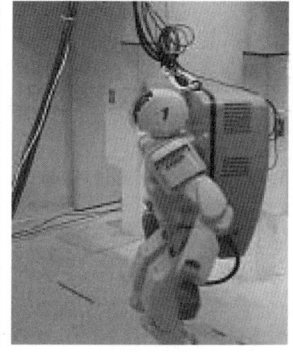
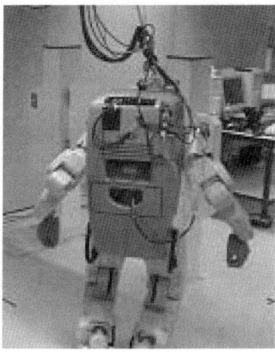

Fig. 15 Autonomous walking to the switch

Fig. 17 Autonomous execution of the hand operation

specified operations including get image at the window and turn on the switch. The robot first walked to the registered location of the window and got the image of the window, then walked to the registered location of the switch and turned the switch on. Figure 16 shows the results of image processing of identifying landmarks for self-localization after reaching the switch. By measuring the identified landmarks, the robot can estimate its current location that resulted from walking. Figure 17 shows execution of the turn on operation of the switch. The robot first detected the "switch" object by matching it with the registered image of the object and measured its position by stereoscopic vision, and then generated the motion command for turning on the switch based on the registered motion template. We also conducted other experiments, including the remote execution of manual walking a step at a time and task execution combining manual walking and autonomous task operations.

These experiments illustrated that our system was able to achieve the basic required functions of a practical system for a building and home management service.

5. CONCLUSION

This paper described the development of applications for the building and home management service section in HRP, which is supported by METI. The objective of this service is to develop a system that enables users to control humanoid robots in their homes remotely via the Internet with a simple mobile terminal. To produce such a system, we focused on the development of sensor-based autonomy, that is sufficient to achieve task-level teleoperation, view simulations based on CG technologies to help users perceive the condition at the robot's site, and client-server architecture for remote task execution to reduce the load on the mobile terminal.

Several experiments conducted with our prototype system demonstrated that our system achieved the basic required functions of a practical system for a building and home management service.

6. ACKNOWLEDGMENTS

The authors extend special thanks to Professor Hirochika Inoue of The University of Tokyo, the project leader, and also to Professor Masayuki Inaba of The University of Tokyo and Tetsuo Koutoku of AIST for their technical guidance.

7. REFERENCES

[1] H. Inoue et al., HRP: Humanoid Robotics Project of MITI, IEEE Humanoid 2000.

[2] T. Morita, "Tracking vision system for real-time motion analysis," Advanced Robotics, Vol. 12, No. 6, pp. 609-617, 1999.

A Tele-operated Humanoid Robot Drives a Backhoe

Hitoshi Hasunuma[‡], Katsumi Nakashima[‡], Masami Kobayashi[‡], Fumisato Mifune[‡],
Yoshitaka Yanagihara[††], Takao Ueno[††], Kazuhisa Ohya[††], and Kazuhito Yokoi[†]

[‡] System Technology Development Center, Kawasaki Heavy Industries, Ltd.
118 Futatsuzuka Noda, Chiba 278-8585, Japan
{hasunuma, nakasima, kobayasi, mifune}@tech.khi.co.jp

[††] Institute of Technology, Tokyu Construction Co., Ltd.
3062-1, Soneshita, Tana, Sagamihara, Kanagawa 229-1124, Japan
{yanagihara.yoshitaka, ueno.takao, ooya.kazuhisa}@tokyu-cnst.co.jp

[†] Intelligent Systems Institute,
National Institute of Advanced Industrial Science and Technology (AIST)
AIST Tsukuba Central 2, Tsukuba 305-8568 Japan
kazuhito.yokoi@aist.go.jp

Abstract

In this paper, we will describe our attempt for a tele-operated humanoid robot to drive a backhoe. It will be possible for a humanoid robot to drive an industrial vehicle instead of a human operator. If a humanoid robot can be operated by a human operator from a remote site, it enables to use a general type of vehicle safely in a dangerous field. We introduced a backhoe as the target vehicle to show the possibility of driving a vehicle in a sitting posture. The robot sits down with balancing on a cockpit of the backhoe and manipulates control levers for driving. We developed a portable remote control device and remote control methods to operate the humanoid robot. For the evaluation, it is tested to sit down and manipulate the levers on the driving cockpit of the real backhoe. To compare it with a human's work, the efficiency is close to practical use.

1 Introduction

In recent years, many universities [1, 9, 10, 13, 15] and some companies [4, 7] have produced humanoid robots. However the application area of the humanoid robots is still limited in the research, the advertisement, and the entertainment.

Since 1998 Japanese fiscal year (JFY), Ministry of Economy, Trade and Industry (METI) has promoted the research and development project of "Humanoid Robotics (HRP)." The aim of the projects is to find some suitable applications for humanoid robots. Five applications of humanoid robots are being developed in HRP [6]. For one of them, we are developing a tele-operated humanoid robot that drives an industrial vehicle.

A humanoid robot has substantial advantages when operating a machine that human beings usually use, because a humanoid robot can operate the machine without any previous modifications of it. Furthermore, by using other machines, a humanoid robot expands its ability as human does.

We have selected industrial vehicles as the machines, because there is the strong demand of an unmanned industrial vehicle that can be used at a hazardous field. If a humanoid robot can operate an industrial vehicle and it can be operated from a remote site, the industrial vehicle can work at a hazardous place without any danger for human operators.

In recent years, several researchers have contributed work concerned with automated or tele-operated industrial vehicles [2, 8, 12, 11, 16].

However there are some advantages for using a tele-operated humanoid robot to operate an industrial vehicle.

- It might be not so difficult to make a typical industrial vehicle tele-operated. However it is impossible to make all kinds of industrial vehicle tele-operated. If we can put a tele-operated humanoid robot on a cockpit of an industrial vehicle, it instantaneously becomes a tele-operated one.

- A tele-operated humanoid robot can not only operate a vehicle but also get out of the vehicle and do ancillary activities, such as checking the vehicle. Moreover, the usage of the tele-operated humanoid robot is not limited to the operation of the vehicle. We can introduce a humanoid robot that is engaged in other works on demand.

A lift truck was selected as the first target of an industrial vehicle [3]. We have succeeded in making the experiments as for the proxy drive of a lift truck in a standing posture by a tele-operated humanoid robot. The time to carry out the model tasks, that include the sequence of transferring a load on the pallet, is about 300 [s]. It took approximately three times longer than the human operator did.

For the next target, a backhoe, which is operated in a sitting posture, has been selected in order to demonstrate that the same humanoid robot can operate the different kinds of vehicles. Additionally, we take into account the fact that the frequency in use of a backhoe is highest in all of industrial vehicles which are used in the construction industry.

In this paper, we will describe our attempt for the tele-operated humanoid robot to drive a backhoe.

The rest of the paper is organized as follows: In Section 2, we present three main equipments for research and development; a backhoe, humanoid robot HRP-1S, a remote control device of HRP-1S. Some important items for the remote operation of a backhoe by a tele-operated humanoid robot such as avoidance of self-collision and vision assistance are described in Section 3. Section 4 shows results of some basic experiments. We conclude the paper in Section 5.

2 Equipments for Tele-operating a Backhoe

2.1 Backhoe

The backhoe shown in Fig. 1 was chosen as a model in which an evaluation experiment is possible. Its specifications are shown in Table 1.

The cockpit of backhoe was slightly reconstructed, because the present humanoid robot is a little bit fat and has very limited movable area of its four limbs.

Figure 1: Backhoe

Table 1: Specifications of the backhoe

Bucket Capacity $[m^3]$	0.08
Full Length $[mm]$	4475
Over All Width $[mm]$	1450
Over All Hight $[mm]$	2355
Weight of Machine $[kg]$	2400
Rated Engine Output $[kW]$	14.7

The points of reconstruction are flattening the floor for the robot getting into and moving operation levers to the position where the vision system can recognize and where the robot can make into operation within the joint angle limitation of both arms. The cockpit was also reconstructed because of the protection sheet installation. However, it is a premise not to give reconstruction that becomes advantageous to a robot about circumference environment in the future.

2.2 Humanoid Robot

We introduced the humanoid robot HRP-1S that has developed in HRP. HRP-1S has 1600 [mm] height, 600 [mm] width, and 117 [kg] weight including batteries. The mechanical hardware is identical with HRP-1[5]. It has 12 d.o.f. in two legs and 16 d.o.f in two arms including hands with 1 d.o.f. grippers.

Each joint is actuated by a brushless DC servo motor with a harmonic-drive reduction gear. Brushless DC servo amplifiers, a Ni-Zn battery, a wireless ethernet modem are embedded in the body.

The body is equipped with an inclination sensor that consists of gyroscopes and G-force sensors. Each foot and wrist is equipped with a force/torque sensor, respectively. Two video cameras are mounted in the head.

Figure 2 shows the control hardware system of

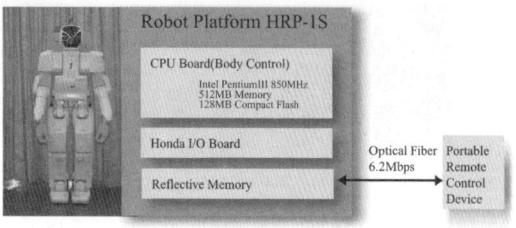

Figure 2: Humanoid Robot HRP-1S

Figure 3: Master Arm

HRP-1S. The real-time controller runs on the CPU board in the backpack of HRP-1S, whose operating system is ART-Linux. The whole body motion control software has been newly developed and installed [14].

2.3 Portable Remote Control Device

We developed a new portable remote control device for humanoid robot HRP-1S. The motion of arms, legs and head of HRP-1S can be operated remotely by using the control device. In some cases of applying a tele-operated humanoid robot to various tasks, we want to transport the remote control device near the robot. Therefore, the remote control device can be decomposed to the six units; two master arms, one set of master foot, two amplifier units, and one PC unit. Each unit is designed within 20 [kg] in order to be carried easily. In order to assemble and disassemble the separated units easily, it is important to reduce the number of connecting cables. A serial line, which is up to the IEEE1394 standard, is installed to connect between the PC unit and the amplifier units.

As a motion of arms of HRP-1S should be operated precisely for manipulating control levers of the backhoe, we introduced a master arm as the arm operation device. For commanding of walk such as "Walk forward five meters!" it is possible to preprogram the walk motion and execute it by using a switch. However, in some situations, we strongly want to tele-operate both arms and legs simultaneously. For this reason, we introduced a special device that can be operated by leg. We call it "master foot". The head of HRP-1S is operated with a small pointing device mounted on the master arm. The details of the master arm device and the master foot device are described in the next.

2.3.1 Master Arm Device

There are many remote control devises proposed [17]. However, some of them have only three d.o.f. force feedback and are not suitable for the tele-operation of a humanoid robot. We have also developed the master arm device at the phase one of HRP [18]. It has a good performance, but it is too heavy to transport it to a operation site. The six d.o.f. master arm device developed is shown in Fig.3.

Direct drive motors are introduced as drive motors of the master arm because of their low friction performance and counter weights are equipped with the link mechanism for gravity balancing. To operate the master arm easily, the pantograph mechanism, which clears out interaction between translation and orientation motion of the grip, is installed on the master arm.

The motion of the hand of HRP-1S is operated according to the position and orientation of the grip of the master arm. The force and moment on the wrist of HRP-1S are fed back to the operator through the master arm.

In order to tele-operate the robot arm with force feedback, it is important that the operator can feel force and moment correctly. In the case that a rotational axis of the wrist mechanism, which produces the feedback moment, doesn't pass through the operator's grasp point, it is difficult for the operator to recognize pure moment. This is because the operator feels the same directional forces at the both side edge of his palm.

On the other hand, in the case that all rotational axes of the wrist mechanism are cross at the operator's grasp point, the operator can easily recognize pure moment. This is because the operator feels the opposite directional forces at the both side edge of his palm.

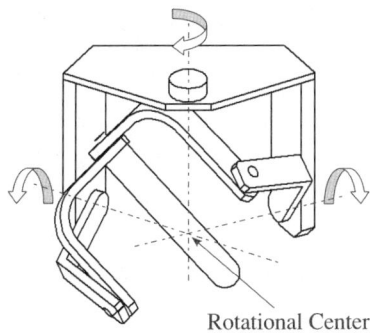

Figure 4: Rotational Parallel Mechanism

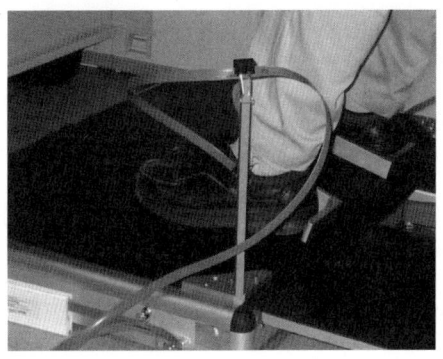

Figure 5: Master Foot

To realize such a structure, it is easy to think of using a universal joint mechanism. However, a universal joint mechanism is not enough small for the wrist mechanism of the portable master arm. Therefore we introduced a rotational parallel mechanism shown in Fig.4. Two d.o.f. rotational motion; pitch and roll, are realized in this mechanism and one rotational motion; yaw, are generated by turning whole mechanism around the vertical axis.

2.3.2 Master Foot Device

Two tape type sensors were introduced for the master foot as shown in Fig.5. Each sensor is attached on each ankle of the operator and detects the position of the ankle. The weight of the tape type sensor is so light that the operator feels free. It is easy to equip the master foot device rather than the convetional device [21].

The main purpose of the master foot is to operate the leg at the same time when the robot arm is operated with the master arm.

The leg of the robot is controlled mainly on a parallel with a ground. For this reason, we restricted the input from the master foot only to the position input. As the human operator feels force at the foot less precisely than at the hand, force feedback function is not installed on the master foot control.

3 Remote Control Technologies for Tele-operating a Backhoe

Some sophisticate concepts, such as "Supervisory Control [19]" and "Shared Autonomy [20]", have been proposed for a tele-operation system. When a humanoid robot is tele-operated in various kinds of situation, the operator often wants to change a tele-operation method to a suitable one for easy operation. We selected three operational modes, as shown below, according to the operation of the robot for getting on, sitting down and manipulating levers.

1. The playback mode: The robot is controlled according to a reference motion pattern that were generated by the program off-line. We use this mode for seating. Note that the seating motion pattern is reproduced easily by setting the height of the seat.

2. The manual mode: The robot is manually tele-operated by using the remote control device. Both the master arm device and the master foot device can be used. It applies the leg control after seating and the head control.

3. The semi-automatic mode: The operator gives the robot a partial command of a whole body motion and the rest of the motion is automatically calculated. The main purpose of this mode is to reduce the responsibility of the operator in the manual operation. We applied this mode for the tele-operation of the arm and walking to ride on the bachkhoe.

In the following sections, we describe two examples of the semi-automatic mode.

3.1 Avoidance of Self-collision of Arms

Even if a shape of a humanoid robot is similar to the human's one, it is difficult for the human operator to predict self-collisions between the arm and other parts of the humanoid robot. We introduced a semi-automatic function; the operator commands the position and orientation of the hand of the robot, and the control system of the robot automatically avoids a self-collision.

HRP-1S has 7 d.o.f. arm that has one d.o.f. redundancy. It can change its elbow position even the hand keeps the same position and orientation in the world coordinate system. The control system changes the elbow position in order to avoid a self-collision. A parameter, called "elbow angle", which represents the position of the elbow, would be utilized.

The value of the elbow angle is set to the rotational angle at which the elbow rotates around the line segment between the wrist joint and the shoulder joint. When the elbow locates below on the vertical plane for the body including the line segment, the value of the elbow angle is set to zero.

The operator commands the position and orientation of the each hand using the master arm and a default value of the each elbow angle using the switch on the master arm. In the semi-automatic mode, the elbow angle is updated automatically to avoid self-collisions between the arm and other parts of the robot.

3.2 Vision Assistance

In order to set a target position of the hand more easily, we introduced a vision assistance function.

The operator positions a cursor to an object on a stereo image getting from two cameras installed in the head of HRP-1S. The position of the object is detected by processing the image and then the robot moves its hand to a neighbor of the object in order to prevent the collision between the hand and the object caused by the error of the image processing.

4 Experiments

The cockpit of the backhoe from which its travel equipment and arm mechanism were removed was used in the experiments just to evaluate a performance of basic operations such as seating and operating levers.

4.1 Seating Experiment

HRP-1S sits down in two steps; 1) the robot puts its hip on the seat, 2) the robot puts its legs forward for increasing the sitting stability.

At the first step, the robot should take a seat with keeping its balance in order that its hip parts shall contact softly with the seat, because the parts are not strong enough to receive a big impact force.

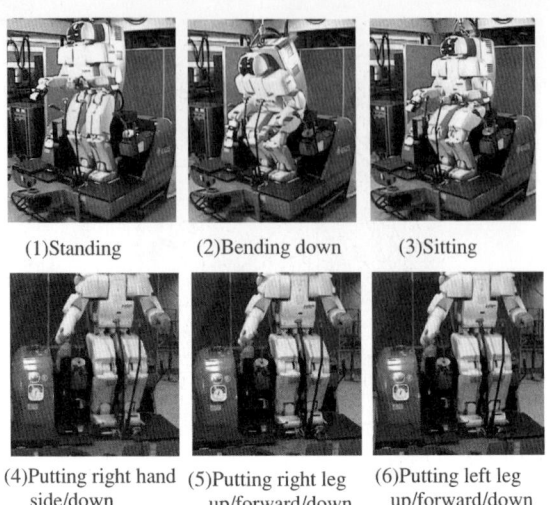

(1)Standing   (2)Bending down   (3)Sitting

(4)Putting right hand side/down   (5)Putting right leg up/forward/down   (6)Putting left leg up/forward/down

Figure 6: Seating sequence

In a standing posture, the hip of the robot is just above the feet just like human's. In the case that the robot takes a seat with keeping its upper body upright and its balance, the hip just goes right down and never goes out of the foot print. Therefore the robot bends its upper body forward and put its hip back and down. After contacting with the seat, the robot makes its upper body upright.

At the second step, HRP-1S puts its legs forward in order to enlarge the supporting area made from both feet and the hip. This makes the robot seat more stably.

To put the legs forward, the robot should lift it up from the ground because the sole of the foot is made with rubber whose friction is large. However, in the initial sitting posture just after seating, the robot cannot lift the foot up from the ground, because the center of gravity of the robot locates outside of the supporting triangle made from the other foot and the hip. Therefore HRP-1S put down its right hand onto a side bar to support some of its weight and extend the supporting area. After that, HRP-1S lifts up its right leg, puts it forward, and down it. When HRP-1S lifts up its left leg, it is no need to put down its left hand. Because the center of gravity of the robot is located in the supporting area made from its right leg and the hip.

The sequence of seating is shown in Fig.6.

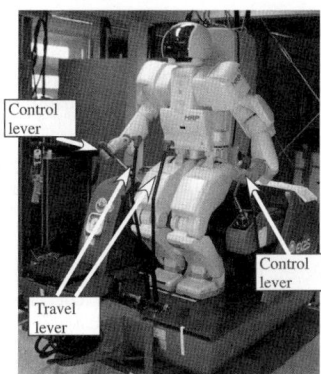

Figure 7: The cockpit of the real backhoe with HRP-1S

Table 2: Operational Time Comparisons: Travel Lever Operation Time [s]

Side + Direction	A: Man	B: Robot	B/A
1: Right + Front	10.9	15.4	1.4
2: Right + Rear	10.4	18.4	1.8
3: Left + Front	10.7	15.5	1.5
4: Left + Rear	10.8	20.5	1.9
1 and 3	11.9	19.0	1.6
1 and 4	11.6	19.7	1.7
2 and 3	11.8	20.7	1.7
2 and 4	12.3	18.8	1.5

Table 3: Operational Time Comparisons: Control Lever Operation Time [s]

Side + Direction	A: Man	B: Robot	B/A
1: Right + Front	11.3	19.3	1.7
2: Right + Rear	11.8	24.3	2.1
3: Left + Front	12.0	22.8	1.9
4: Left + Outside	11.2	15.9	1.4
5: Left + Inside	11.3	15.0	1.3
6: Right + Outside	12.6	16.7	1.3
7: Right + Inside	10.8	16.2	1.5
1 and 4	11.7	22.0	1.9
1 and 5	11.7	17.4	1.5
2 and 4	11.0	22.0	2.0
2 and 5	11.0	18.6	1.7

4.2 Operating Lever Experiments

Figure 7 shows HRP-1S seating on the cockpit of the real backhoe. At the cockpit of the backhoe, there are two types of lever: the travel lever and the control lever.

The operator can recognize and operate the travel levers easily by using HRP-1S, because they are located just in front of the robot. On the other hand, the control levers are located at the side of HRP-1S so that it is not easy to recognize and operate them.

In order to evaluate the workability, the time required to do lever operation through HRP-1S was compared with the time required for the human operator seated at the cockpit directly to do the same operation. Figure 8 shows scenes of the operating levers experiments. The results are shown in Table 2, 3. The time required for the experiments using HRP-1S is up to 2.1 times longer than it for the human operator.

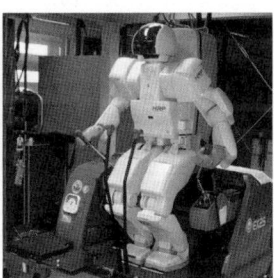

(a)Travel lever operation   (b)Control lever operation

Figure 8: Lever operations

5 Conclusions

This paper presented the current status of research and development on the application to tele-operate an industrial vehicle by a tele-operated humanoid robot undergoing on HRP in Japan.

We have newly developed a portable remote control device that can be used for the tele-operation of humanoid robot HRP-1S.

We have succeeded in seating the humanoid robot on the cockpit of the real backhoe and manipulating the control levers by tele-operation.

The operator could manipulate levers through the robot about twice as much time as directly.

To sit down on the cockpit, a special tactics for seating was needed because of mechanical limitation of the robot.

Now, we are trying to make it real that a tele-operated humanoid robot drives a real backhoe and excavates a yard in the open air. To realize this, we are doing research and development of the following items:

- A strategy for riding HRP-1S on the cockpit of the backhoe,
- A special suit for HRP-1S, which is dust-proofing and drip-proofing,
- A method to protect HRP-1S from vibration of a backhoe in a real operation.

Acknowledgements

We thank Manufacturing Science and Technology Center (MSTC), New Energy and Industrial Technology Development Organization (NEDO), Ministry of Economy, Trade and Industry (METI) for their entrusting development of the project "Humanoid Robotics", and the members cooperating in the project for their constructive support. We also thank Prof. Inoue of the University of Tokyo; Dr. Tanie, Dr. Hirukawa, and Dr. Kanehiro of AIST; and Prof. Yokokohji of the Kyoto University for their various instructive advices.

References

[1] T. Furuta, Y. Okomura, and K. Tomiyama. Design and Construction of a Series of Compact Humanoid Robots and Development of Biped Walk Control Strategies, *Proc. IEEE-RAS Int. Conf. Humanoid Robots*, 2000.

[2] E. Gambao and C. Balaguer. Robotics and Automation in Construction, *IEEE Robotics and Automation Magazine*, Vol.9, No.1, pp. 4–6, 2002.

[3] H. Hasunuma, M. Kobayashi, H. Moriyama, T. Itoko, Y. Yanagihara, T. Ueno, K. Ohya, and K. Yokoi. A Teleoperated Humanoid Robot Drives a Lift Truck, *Proc. IEEE Int. Conf. Robotics and Automation,* pp. 2246–2252, 2002.

[4] M. Hirose, Y. Haikawa, and T. Takenaka. Introduction of Honda Humanoid Robots Development, *Proc. Advanced Science Institute 2001*, No. 16, pp. 1–8, 2001.

[5] H. Inoue, S. Tachi, K. Tanie, K. Yokoi, S. Hirai, H. Hirukawa, K. Hirai, S. Nakayama, K. Sawada, T. Nishiyama, O. Miki, T. Itoko, H. Inaba, and M. Sudo. HRP: Humanoid Robotics Project of MITI, *Proc. IEEE-RAS Int. Conf. Humanoid Robots*, 2000.

[6] H. Inoue, S. Tachi, Y. Nakmura, K. Hirai, N. Ohyu, S. Hirai, K. Tanie, K. Yokoi, and H. Hirukawa. Overview of Humanoid Robotics Project of METI, *Proc. Int. Symp. Robotics*, pp. 1478–1482, 2001.

[7] T. Ishida, Y. Kuroki, J. Yamaguchi, M. Fujita, and T. Doi. Motion Entertainment by a Small Humanoid Robot based on Open-R, *Proc. IEEE/RSJ Int. Conf. Intelligent Robots and Sytems*, 2001.

[8] R. A. Jarvis. Sensor Rich Teleoperation of an Excavating Machine, *Proc. Field and Service Robotics*, pp. 238–243, 1999.

[9] S. Kagami, K. Nishiwaki, T. Sugihara, J. J. Kuffner, M. Inaba, and H. Inoue. Design and Implementation of Software Research Platform for Humanoid Robotics : H6 *Proc. IEEE Int. Conf. Robotics and Automation*, pp. 2431–2436, 2001.

[10] F. Kanehiro, M. Inaba, and H. Inoue. Development of a Two-Armed Bipedal Robot that can Walk and Carry Objects, *Proc. IEEE/RSJ Int. Conf. Intelligent Robot and Systems*, pp. 23–28, 1996.

[11] S. E. Salcudean, K. Hashtrudi-Zaad, S. Tafazoli, S. P. DiMaio and C. Reboulet. Bilateral Matched Impedance Teleoperation with Application to Excavator Control, *Proc. IEEE Int. Conf. Robotics and Automation*, pp. 133–139, 1998.

[12] C.E. Thorpe. Vision and Navigation The Carnegie Mellon Navlab, *Kluwer Academic Publishers,* 1990.

[13] J. Yamaguchi, E. Soga, S. Inoue, and A. Takanishi. Development of a Bipedal Humanoid Robot - Control Method of Whole Body Cooperative Dynamic Biped Walking, *Proc. IEEE Int. Conf. Robotics and Automation*, pp. 368–374, 1999.

[14] K. Yokoi, F. Kanehiro, K. Kaneko, K. Fujiwara, S. Kajita and H. Hirukawa. A Honda Humanoid Robot Controlled by AIST Software, *Proc. IEEE-RAS Int. Conf. Humanoid Robots 2001*, pp. 259–264, 2001.

[15] M. Gienger, K. Löffler, and F. Pfeiffer. Towards the Design of a Biped Jogging Robot, *Proc. IEEE Int. Conf. Robotics and Automation*, pp. 4140–4145, 2001.

[16] W. R. Hamel and P. Murray. Observations Concerning Internet-based Teleoperations For Hazardous Environments, *Proc. IEEE Int. Conf. Robotics and Automation*, pp. 638–643, 2001.

[17] G. C. Burdea. Force and Touch Feedback for Virtual Reality, *John Wiley & Sons*, 1996.

[18] H. Hasunuma, H. Kagaya, M. Koyama, J. Fujimori, F. Mifune, H. Moriyama, M. Kobayashi, T. Itoko, and S. Tachi. Teleoperation Master-arm with Gripping Operation Devices, *Proc. 9th Int. Conf. Machine Automation*, pp. 567–572, 2000.

[19] W. R. Ferell and T. B. Sheridan. Supervisory Control of Remote Manipulation, *IEEE Spectrum*, pp. 81–88, 1967.

[20] S. Hirai, T. Sato, M. Kakikura, and T. Matsui. Integration of a Task Knowledge Base and a Cooperative Manuvering System, *Proc. IEEE/RSJ Int. Conf. Intelligent Robots and Systems*, pp. 349–354, 1990.

[21] Graspy, *Meta Motion* (2002) http://www.metamotion.com/

A Synchronization Approach to the Mutual Error Control of A Mobile Manipulator *

Dong Sun and Garry Feng

Department of Manufacturing Engineering and Engineering Management

City University of Hong Kong

83 Tat Chee Avenue

Kowloon, Hong Kong

Tel: (852) 2788-8405, Fax: (852) 2788-8423

medsun@cityu.edu.hk

Abstract

Precise tracking control of a mobile manipulator is a challenging problem in which tracking errors of the vehicle and the manipulator merge and jointly affect the endpoint trajectory performance. A synchronization approach to minimize the mutual errors between the vehicle and the manipulator is reported in this paper. The basic idea is to utilize the cross-coupling concept to cooperate motions of the vehicle and the manipulator so that the both tracking errors are compensated each other. An adaptive synchronized controller is proposed to guarantee asymptotic convergence to zero of the position tracking error and the synchronization error of the mobile manipulator. The controller is in a decentralized architecture for easy implementation, and is able to address model uncertainty problem. Simulation results verify the effectiveness of the proposed approach.

1 Introduction

A number of researchers have studied on integrating locomotion and manipulation of mobile manipulators in the past years. One strategy is to specify the position of the endpoint of the manipulator for the reference point, and the mobile vehicle moves in such a way that the manipulator is brought into the preferred configuration [7]. The other strategy is to plan vehicle motion first and then plan manipulator motion, given a desired endpoint trajectory at an expected configuration [2]. A real-time approach for singularity avoidance by modeling the mobile manipulator as a redundant system and controlling the null motion, was recently reported in [4].

This paper discusses a precise tracking control of a mobile manipulator in a desired endpoint trajectory that is decomposed into a desired trajectory for the mobile vehicle and a desired trajectory for the manipulator, in a similar strategy to [2]. Since the vehicle and the manipulator track their own desired trajectories respectively and simultaneously, both tracking errors of the vehicle and the manipulator merge and jointly affect the accuracy of the endpoint trajectory tracking. Here, a challenging problem is that the vehicle and the manipulator, each separately controlled by its own controller, need to generate coordinate motions so that the mutual error between the vehicle and the manipulator is minimized.

Synchronized control provides a unique set of advantages and opportunities to such error compensation of the mobile manipulator. Specially, if the mobile vehicle generates a tracking error during tracking, the manipulator must have a synchronous response to generate the same tracking error but in opposite direction. As a result, the mutual error (or synchronization error) between the vehicle and the manipulator is zero, and the effect of these two tracking errors to the endpoint trajectory is eliminated. An effective synchronization strategy is to utilize the cross-coupling concept [3]. The majority of the previous work on cross-coupling control were proposed mainly for machine tools (i.e., [6] etc.) A few of applications of the cross-coupling control have also been founded in robotics [1]. Recently, the author utilized the cross-coupling concept to control two driving wheels of a mobile vehicle in a synchronous manner so that the differential position error of two wheels and its integral approach zero [5], which helps to maintain the vehicle in the desired trajectory path. This synchroniza-

---

*The work described in this paper was partially supported by a grant from the Research Grants Council of the Hong Kong Special Administrative Region, China [Project No. CityU 1085/01E], and a grant from City University of Hong Kong (Project No. 7001365).

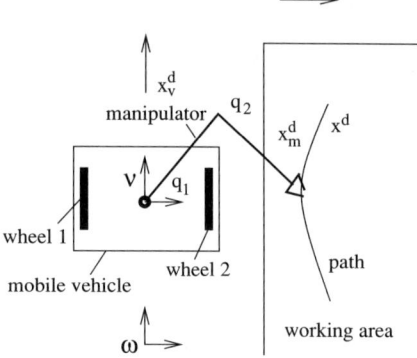

Figure 1: A mobile manipulator in trajectory following

tion approach is extended in this paper, by considering the synchronization of vehicle wheels [5] as well as the synchronization of the vehicle and the manipulator for error compensation between them. The proposed control algorithm guarantees asymptotic convergence to zero of position tracking and synchronization errors.

2 Control Strategy

Figure 1 illustrates a mobile manipulator comprised of a robotic arm with two joints $q_1(t)$ and $q_2(t)$, mounted on a differential mobile vehicle. Denote $x^d(t)$ as the desired endpoint trajectory in the world frame Σw, which satisfies

$$x^d(t) = x_v^d(t) + x_m^d(t) \qquad (1)$$

where $x_v^d(t)$ and $x_m^d(t)$ denote desired trajectories of the mobile vehicle and the manipulator, in the world frame Σw and the vehicle coordinate frame Σv, respectively. The desired motion trajectory $x_v^d(t)$ is designed in such a way that the vehicle maintains the preferred configuration of the manipulator when the manipulator follows $x^d(t)$. The preferred configuration of the manipulator is determined based on manipulability [8]. From [7], the manipulator measure is maximized for $q_2 = \pm 90°$ and arbitrary $q_1(t)$. Therefore, only one joint (q_1) is controlled to correspond to the vehicle motion to follow $x^d(t)$. The other joint (q_2) is used as redundancy to maintain the preferred configuration of the manipulator. For simplicity, in the following control design, the manipulator is treated as a 1-DOF lumped mass mounted on the mobile vehicle.

Consider that the vehicle with two independent driving wheels has the configuration specified by a vector $\begin{bmatrix} x_v(t) & \theta(t) \end{bmatrix}^T$, where $x_v(t)$ is a 2×1 vector representing coordinates at the center of mass of the vehicle platform, and $\theta(t)$ denotes the orientation or heading angle with respect to the frame Σw. It

is assumed that driving wheels are under pure rolling condition and no slip exists. Recently, the author [5] proposed a control strategy under which tracking displacements of two driving wheels, defined by $x_1(t)$ and $x_2(t)$, are directly controlled in a synchronous manner. The synchronization effort is aimed to regulate the differential displacement error of two driving wheels to zero and thus maintain the vehicle in the desired trajectory path. In this study, the previous approach [5] is extended by adding synchronous control of the manipulator mounted on the vehicle. For simplicity, the vehicle path considers the line segments only, as in [2].

Define the position synchronization error between two vehicle wheels as

$$\epsilon_v(t) = e_1(t) - e_2(t) \qquad (2)$$

where $e_i(t) = x_i^d(t) - x_i(t)$, represents the displacement error of the wheel. Keeping $\epsilon_v(t) = 0$ and $\int_0^t \epsilon_v(w)dw = 0$ can help the vehicle to maintain in the desired trajectory path.

In case of the mobile manipulator, the tracking displacement error of the center of mass of the platform, denoted by

$$e_v(t) = \frac{e_1(t) + e_2(t)}{2} \qquad (3)$$

coupled with the joint position error of the manipulator in the frame Σv, denoted in the joint space by

$$e_m(t) = q_1^d(t) - q_1(t)$$

may affect the endpoint trajectory following significantly. It is necessary to develop a method that accommodates errors $e_v(t)$ and $e_m(t)$. This can be achieved by synchronous control of the vehicle and the manipulator. Define the synchronization error between the vehicle and the manipulator as

$$\epsilon(t) = e_v(t) + \lambda(t) e_m(t) \qquad (4)$$

where $\lambda(t)$ is a bounded non-zero coupling parameter that demonstrates the kinematic constraint between the tracking displacement of the vehicle and the joint trajectory of the manipulator. The control goal is therefore to regulate $\begin{bmatrix} e_1(t) \\ e_2(t) \\ e_m(t) \end{bmatrix} \to 0$ and at the same time, 1) to synchronize motions of two wheels of the vehicle so that $\epsilon_v(t) \to 0$ and $\int_0^t \epsilon_v(w)dw \to 0$, which helps the vehicle to maintain in the desired trajectory path; and 2) to synchronize motions of the vehicle and the manipulator so that $\epsilon(t) \to 0$, which guarantees that the effects of $e_v(t)$ and $e_m(t)$ to the endpoint trajectory are cancelled each other.

3 Control Algorithm

The dynamics of the mobile vehicle with two wheel-subsystems, and the manipulator with consideration of one degree of freedom $(q_1(t))$, are respectively expressed by

$$H_i(x_i)\ddot{x}_i(t) + C_i(x_i,\dot{x}_i)\dot{x}_i(t) + F_i(x_i,\dot{x}_i)$$
$$= Y_i(x_i,\dot{x}_i,\ddot{x}_i)p_i = \tau_i \quad (i=1,2) \quad (5)$$
$$H_m(q_1)\ddot{q}_1(t) + C_m(q_1,\dot{q}_1)\dot{q}_1(t) + F_m(q_1,\dot{q}_1)$$
$$= Y_m(q_1,\dot{q}_1,\ddot{q}_1)p_m = \tau_m \quad (6)$$

where H_i and H_m are inertia terms, C_i and c_m are nonlinear terms, F_i and F_m are external force disturbances, Y_i and Y_m are regression matrices, p_i and p_m are vectors containing unknown dynamic parameters of the model, and τ_i and τ_m denote the input torques.

Introduce a new concept named *coupled position error* that contains the position tracking error and the synchronization error(s), expressed by

$$e_1^*(t) = e_1(t) + \beta_v \int_0^t \epsilon_v(w)dw + \beta \int_0^t \epsilon(w)dw \quad (7)$$
$$e_2^*(t) = e_2(t) - \beta_v \int_0^t \epsilon_v(w)dw + \beta \int_0^t \epsilon(w)dw \quad (8)$$
$$e_m^*(t) = \lambda(t)e_m(t) + \beta \int_0^t \epsilon(w)dw \quad (9)$$

where β_v and β are positive coupling parameters. Note that $e_1^*(t)$ and $e_2^*(t)$ contain both synchronization errors $\epsilon_v(t)$ and $\epsilon(t)$, while $e_m^*(t)$ contains $\epsilon(t)$ only. Define command vectors for each driving wheel of the vehicle and the manipulator as

$$u_1(t) = \dot{x}_1^d(t) + \beta_v \epsilon_v(t) + \beta \epsilon(t) + \Lambda e_1^*(t) \quad (10)$$
$$u_2(t) = \dot{x}_2^d(t) - \beta_v \epsilon_v(t) + \beta \epsilon(t) + \Lambda e_2^*(t) \quad (11)$$
$$u_m(t) = \lambda(t)\dot{q}_1^d(t) + \dot{\lambda}(t)q_1^d(t) + \beta \epsilon(t) + \Lambda e_m^*(t) \quad (12)$$

where λ is a control gain. We then have

$$r_1(t) = u_1(t) - \dot{x}_1(t) = \dot{e}_1^*(t) + \Lambda e_1^*(t) \quad (13)$$
$$r_2(t) = u_2(t) - \dot{x}_2(t) = \dot{e}_2^*(t) + \Lambda e_2^*(t) \quad (14)$$
$$r_m(t) = u_m(t) - \lambda(t)\dot{q}_1(t) = \dot{e}_m^*(t) + \Lambda e_m^*(t) \quad (15)$$

Design input torques as follows, in a decentralized architecture,

$$\tau_1 = \hat{H}_1(x_1)\dot{u}_1(t) + \hat{C}_1(x_1,\dot{x}_1)u_1(t) + \hat{F}_1(x_1,\dot{x}_1) +$$
$$k_r r_1(t) + k_v \epsilon_v(t) + \frac{1}{2}k_\epsilon \epsilon(t) \quad (16)$$
$$= Y_1(x_1,\dot{x}_1,u_1,\dot{u}_1)\hat{p}_1(t) + k_r r_1(t) + k_v \epsilon_v(t) + \frac{1}{2}k_\epsilon \epsilon(t)$$
$$\tau_2 = \hat{H}_2(x_2)\dot{u}_2(t) + \hat{C}_2(x_2,\dot{x}_2)u_2(t) + \hat{F}_2(x_2,\dot{x}_2) +$$
$$k_r r_2(t) - k_v \epsilon_v(t) + \frac{1}{2}k_\epsilon \epsilon(t) \quad (17)$$
$$= Y_2(x_2,\dot{x}_2,u_2,\dot{u}_2)\hat{p}_2(t) + k_r r_2(t) - k_v \epsilon_v(t) + \frac{1}{2}k_\epsilon \epsilon(t)$$
$$\tau_m = \hat{H}_m(q_1)\lambda^{-1}(\dot{u}_m(t) - \dot{\lambda}\dot{q}_1(t)) + \hat{C}_m(q_1,\dot{q}_1)\lambda^{-1}u_m(t)$$
$$+ \hat{F}_m(q_1,\dot{q}_1) + \lambda^{-1}k_r r_m(t) + \lambda k_\epsilon \epsilon(t) \quad (18)$$
$$= Y_m(q_1,\dot{q}_1,u_m,\dot{u}_m)\hat{p}_m(t) + \lambda^{-1}k_r r_m(t) + \lambda(t)k_\epsilon \epsilon(t)$$

where $\hat{(\cdot)}$ denotes the estimate of the true model dynamics, k_r, k_v, and k_ϵ are positive control gains. $\hat{p}_1(t)$, $\hat{p}_2(t)$ and $\hat{p}_m(t)$ are estimated parameter vectors, which are subject to the following adaptation laws

$$\dot{\hat{p}}_i(t) = \Gamma_i Y_i^T(x_i,\dot{x}_i,u_i,\dot{u}_i)r_i(t) \quad (i=1,2) \quad (19)$$
$$\dot{\hat{p}}_m(t) = \Gamma_m Y_m^T(q_1,\dot{q}_1,u_m,\dot{u}_m)\lambda^{-1}(t)r_m(t) \quad (20)$$

where Γ_i and Γ_m are positive control gains. Define the model estimation errors as

$$\tilde{p}_i(t) = p_i - \hat{p}_i(t) \quad (i=1,2) \quad (21)$$
$$\tilde{p}_m(t) = p_m - \hat{p}_m(t) \quad (22)$$

Then, the adaptation laws (19) and (20) can be rewritten in another forms

$$\dot{\tilde{p}}_i(t) = -\Gamma_i Y_i^T(x_i,\dot{x}_i,u_i,\dot{u}_i)r_i(t) \quad (i=1,2) \quad (23)$$
$$\dot{\tilde{p}}_m(t) = -\Gamma_i Y_m^T(q_1,\dot{q}_1,u_m,\dot{u}_m)\lambda^{-1}(t)r_m(t) \quad (24)$$

Substituting controllers (16) $\sim$ (18) into dynamic equations (5) and (6) leads to the following closed-loop dynamics

$$H_1(x_1)\dot{r}_1(t) + C_1(x_1,\dot{x}_1)r_1(t) + k_r(t)r_1(t) +$$
$$k_v \epsilon_v(t) + \frac{1}{2}k_\epsilon \epsilon(t) = Y_1 \tilde{p}_1(t) \quad (25)$$
$$H_2(x_2)\dot{r}_2(t) + C_2(x_2,\dot{x}_2)r_2(t) + k_r(t)r_2(t) -$$
$$k_v \epsilon_v(t) + \frac{1}{2}k_\epsilon \epsilon(t) = Y_2 \tilde{p}_2(t) \quad (26)$$
$$H_m(q_1)\lambda^{-1}\dot{r}_m(t) + C_m(q_1,\dot{q}_1)\lambda^{-1}r_m(t) +$$
$$\lambda^{-1}k_r(t)r_m(t) + \lambda k_\epsilon \epsilon(t) = Y_m \tilde{p}_m(t) \quad (27)$$

It will be shown in the following that the proposed adaptive synchronized controllers (16) $\sim$ (24) guarantee asymptotic convergence to zero of position tracking errors $e_i(t)$ and $e_m(t)$, and synchronization errors $\epsilon_v(t)$ and $\epsilon(t)$.

Theorem 1 *If the control gain k_r is large enough to satisfy*

$$k_r \geq H_m(q_1)\lambda(t)\dot{\lambda}^{-1}(t) \quad (28)$$

the proposed controllers (16) $\sim$ (24) guarantee asymptotic convergence to zero of position tracking errors and synchronization errors, i.e.,
$\begin{bmatrix} e_1(t) \\ e_2(t) \\ e_m(t) \end{bmatrix} \to 0$ *and*
$\begin{bmatrix} \epsilon_v(t) \\ \int_0^t \epsilon_v(w)dw \\ \epsilon(t) \end{bmatrix} \to 0$ *as $t \to \infty$.*

Proof: Define a positive definite function as

$$V(t) = \sum_{i=1}^{2}[\frac{1}{2}H_i r_i^2 + \frac{1}{2}\tilde{p}_i^T \Gamma_i^{-1} \tilde{p}_i] + [\frac{1}{2}H_m(\lambda^{-1} r_m)^2 + \frac{1}{2}\tilde{p}_m^T \Gamma_m^{-1} \tilde{p}_m] + \frac{1}{2}k_v \epsilon_v^2 + k_v \Lambda \beta_v (\int_0^t \epsilon_v(w) dw)^2 + \frac{1}{2}k_\epsilon \epsilon^2 + k_\epsilon \Lambda \beta (\int_0^t \epsilon(w) dw)^2 \quad (29)$$

Differentiating $V(t)$ with respect to time yields

$$\dot{V}(t) = \sum_{i=1}^{2}[H_i r_i \dot{r}_i + \frac{1}{2}\dot{H}_i r_i^2 + \tilde{p}_i^T \Gamma_i^{-1} \dot{\tilde{p}}_i] + [H_m \lambda^{-2} r_m \dot{r}_m + \frac{1}{2}\dot{H}_m(\lambda^{-1} r_m)^2 + H_m \lambda^{-1} \dot{\lambda}^{-1} r_m^2 + \tilde{p}_m^T \Gamma_m^{-1} \dot{\tilde{p}}_m]$$
$$+ k_v \epsilon_v \dot{\epsilon}_v + 2k_v \Lambda \beta_v \epsilon_v \int_0^t \epsilon_v(w) dw + k_\epsilon \epsilon \dot{\epsilon}$$
$$+ 2k_\epsilon \Lambda \beta \epsilon \int_0^t \epsilon(w) dw \quad (30)$$

Multiplying both sides of closed-loop equations (25) $\sim$ (26) by r_i, and (27) by $\lambda^{-1} r_m$, and then substituting resulting equations into (30) yields

$$\dot{V}(t) = -k_r(r_1^2 + r_2^2) - (k_r - H_m \lambda \dot{\lambda}^{-1})(\lambda^{-1} r_m)^2 - \quad (31)$$
$$(r_1 - r_2)k_v \epsilon_v - [\frac{1}{2}(r_1 + r_2) + r_m]k_\epsilon \epsilon + k_v \epsilon_v \dot{\epsilon}_v +$$
$$2k_v \Lambda \beta_v \epsilon_v \int_0^t \epsilon_v(w) dw + k_\epsilon \epsilon \dot{\epsilon} + 2k_\epsilon \Lambda \beta \epsilon \int_0^t \epsilon(w) dw$$

From (13) $\sim$ (15) and (7) $\sim$ (9), one obtains that

$$r_1 - r_2 = \dot{e}_1^* - \dot{e}_2^* + \Lambda(e_1^* - e_2^*) \quad (32)$$
$$= \dot{\epsilon}_v + (2\beta_v + \Lambda)\epsilon_v + 2\Lambda \beta_v \int_0^t \epsilon_v(w) dw$$

and

$$\frac{1}{2}(r_1 + r_2) + r_m$$
$$= \frac{1}{2}[\dot{e}_1^* + \dot{e}_2^* + \Lambda(e_1^* + e_2^*)] + (\dot{e}_m^* + \Lambda e_m^*)$$
$$= (\frac{\dot{e}_1 + \dot{e}_2}{2} + \lambda \dot{e}_m + \dot{\lambda} e_m) +$$
$$\Lambda(\frac{e_1 + e_2}{2} + \lambda(t)e_m) + 2\beta\epsilon + 2\Lambda\beta \int_0^t \epsilon(w) dw$$
$$= \dot{\epsilon} + (2\beta + \Lambda)\epsilon + 2\Lambda\beta \int_0^t \epsilon(w) dw \quad (33)$$

Substituting (32) and (33) into (31) yields

$$\dot{V}(t) = -k_r(r_1^2 + r_2^2) - (k_r - H_m \lambda \dot{\lambda}^{-1})(\lambda^{-1} r_m)^2 - (2\beta_v + \Lambda)k_v \epsilon_v^2 - (2\beta + \Lambda)k_\epsilon \epsilon^2 \quad (34)$$

If k_r is large enough to satisfy (28), $\dot{V}(t) \leq 0$. Then, r_i, $\lambda^{-1} r_m$, ϵ_v, and ϵ are bounded in terms of L_2 norm.

Since λ^{-1} is bounded, r_m is bounded. From (13) $\sim$ (15), e_i^*, $\dot{e}_i^*$, e_m^*, and $\dot{e}_m^*$ are also bounded, and so are $\dot{e}_i$ and $\dot{e}_m$ by differentiating (7) $\sim$ (9). $\dot{\epsilon}_v$ and $\dot{\epsilon}$ are further bounded by differentiating (2) and (4). From closed-loop dynamic equations (25) $\sim$ (27), $\dot{r}_i$ and $\dot{r}_m$ are bounded. Therefore, r_i, r_m, ϵ_v, and ϵ are uniformly continuous, since $\dot{r}_i$, $\dot{r}_m$, $\dot{\epsilon}_v$, and $\dot{\epsilon}$ are bounded. From Barbalat's lemma, it follows

$$r_i(t) \to 0, \ r_m(t) \to 0, \ \epsilon_v(t) \to 0, \ \epsilon(t) \to 0 \ as \ t \to \infty$$

From (13) $\sim$ (15), one obtains

$$e_i^*(t) \to 0 \ \dot{e}_i^*(t) \to 0 \ e_m^*(t) \to 0 \ \dot{e}_m^*(t) \to 0$$

From (4), one obtains

$$e_v + \lambda e_m = \epsilon \to 0 \quad (35)$$

Combining (7) $\sim$ (9) yields

$$\frac{e_1^* + e_2^*}{2} - e_m^* = e_v - \lambda e_m \to 0 \quad (36)$$

Combining (35) and (36) yields

$$\lambda e_m \to 0 \Rightarrow e_m \to 0$$
$$e_v \to 0 \Rightarrow e_1 + e_2 \to 0$$

Since $e_1 - e_2 = \epsilon_v \to 0$, one further obtains that $e_1 \to 0$ and $e_2 \to 0$. From (7) $\sim$ (8), one finally obtains $\int_0^t \epsilon_v(w) dw \to 0$. □

4 Case Studies

Simulations were performed to verify the effectiveness of the proposed method. As shown in Figure 1, the mobile vehicle moves along a straight line to aid the manipulator to follow a desired trajectory $x^d(t)$ for the assembly task. During the tracking, the manipulator maintains the preferred configuration with $q_2 = -90°$. The task coordinate $x_m(t)$ of the manipulator in the frame Σv satisfies

$$x_m(t) = \begin{bmatrix} l_1 c_1 + l_2 c_{12} \\ l_1 s_1 + l_2 s_{12} \end{bmatrix} = \begin{bmatrix} l_1 c_1 + l_2 s_1 \\ l_1 s_1 - l_2 c_1 \end{bmatrix} \quad (37)$$

where $c_1 = \cos q_1$, $s_1 = \sin q_1$, $c_{12} = \cos(q_1 - 90°) = s_1$, and $s_{12} = \sin(q_1 - 90°) = -c_1$. l_1 and l_2 are lengths of two links of the manipulator, respectively. In the simulation, one selected $l_1 = 1m$ and $l_2 = 0.6m$.

Suppose that the initial configuration of the mobile vehicle is $\begin{bmatrix} 0 & 0 & 90° \end{bmatrix}^T$. The initial configuration of the manipulator is $q_1(0) = 45°$ and $q_2(0) = -90°$. The initial endpoint position is $x(0) = x_m(0) = \begin{bmatrix} 1.1312 & 0.2828 \end{bmatrix}^T (m)$. The desired endpoint trajectory $x^d(t)$ consists of the desired vehicle trajectory $x_v^d(t)$ in the frame Σw and and the desired manipulator trajectory $x_m^d(t)$ in the frame Σv. Let $x_v^d(t)$ vary from the initial position $x_v^d(0) = \begin{bmatrix} 0 & 0 \end{bmatrix}^T$ (m) to the final position $x_v^d(f) = \begin{bmatrix} 0 & 1 \end{bmatrix}^T$ (m), in a straight line,

$$x_v^d(t) = x_v^d(0) + (x_v^d(f) - x_v^d(0))(1 - exp(-t)) \quad (38)$$

The variation of $x_m(t)$ is realized by the variation of $q_1(t)$. Let the desired $q_1^d(t)$ vary from $q_1^d(0) = 45°$ to $q_1^d(f) = 60°$, in a trajectory specified by

$$q_1^d(t) = q_1^d(0) + (q_1^d(f) - q_1^d(0))(1 - exp(-t)) \quad (39)$$

Then, $x_m^d(t)$ can be known by substituting (39) into (37). The kinematic relationship between $x_v(t)$ and $x_m(t)$ can be realized by the coupling parameter $\lambda(t)$. The endpoint position error $\Delta x_m(t)$ can be derived by differentiation of $x_m(t)$ in (37). Since the vehicle moves along the y direction, the error compensation in y direction only is required. Denote $\Delta x_m(t)$ in y direction as

$$\Delta x_m^y(t) = (l_1 c_1 + l_2 s_1)\Delta q_1 = \lambda e_m \quad (40)$$

where $\lambda = l_1 c_1 + l_2 s_1$, and $e_m = \Delta q_1$. To eliminate the effect caused by $e_v(t)$ of the vehicle and $\Delta x_m^y(t)$ of the manipulator, the following synchronization error is defined and controlled to be zero

$$\epsilon(t) = e_v(t) + \Delta x_m^y(t) = e_v(t) + \lambda e_m(t) \quad (41)$$

Assume that dynamic modeling parameters of the mobile manipulator are $H_1 = 10 kg \cdot m^2$, $H_2 = 9 kg \cdot m^2$, $H_m = 8 kg \cdot m^2$, $C_1 = 10 N \cdot m \cdot s/rad$, $C_2 = 9 N \cdot m \cdot s/rad$, $C_m = 8 N \cdot m \cdot s/rad$, and $F_1 = F_2 = F_m = 0 N \cdot m$. Let the estimated parameter vectors $\hat{p}_1(0) = \hat{p}_2(0) = \hat{p}_m(0) = 0$ at the initial time. The control gains were chosen to be: $\Lambda = 4 sec^{-1}$, $k_r = 300 N \cdot m \cdot sec/rad$, $\beta_v = \beta = 30 sec^{-1}$, $k_v = k_\epsilon = 10 N \cdot m/rad$, and $\Gamma = diag\{0.001\}$.

Figure 2 illustrates the tracking error of the center of mass of the vehicle platform $e_v(t)$, and the joint position error of the manipulator $e_m(t)$. Note that $e_v(t)$ and $e_m(t)$ are in opposite direction, which helps them to be compensated each other. The endpoint position error of the mobile manipulator is shown in Figure 3. Figure 4 illustrates synchronization errors $\epsilon_v(t)$ and $\epsilon(t)$, respectively. It can be seen that all above errors exhibit good convergence to zero.

For comparison, an adaptive control without motion synchronization between the vehicle and the manipulator, namely $\beta = 0$ and $k_\epsilon = 0$, was further applied to the mobile manipulator. (Note that the mobile vehicle still utilized the synchronized control to maintain the vehicle in the desired trajectory.) Figure 5 illustrates the tracking errors of the vehicle $e_v(t)$ and the joint position error of the manipulator $e_m(t)$, respectively. The two errors have the same signs and cannot be compensated each other. Figure 6 illustrates the endpoint position error of the mobile manipulator. Compared with that in Figure 3, the endpoint position error becomes greater because the synchronization error between the vehicle and the manipulator, $\epsilon(t)$, increases obviously under the control without synchronization, as shown in Figure 7.

5 Conclusions

This paper describes an approach to the control of the mutual errors of mobile manipulators, using a synchronization effort to compensate for the errors of the vehicle and the manipulator each other to ensure precise endpoint trajectory tracking. An adaptive synchronized controller has been proposed by incorporating cross-coupling technology into adaptive architecture. It is proven that the proposed controller guarantees asymptotic convergence to zero of position and synchronization errors. Simulation results verifies effectiveness of the proposed approach.

References

[1] L. Feng, Y. Koren, and J. Borenstein. "Cross-coupling motion controller for mobile robots". *IEEE Control Systems*, pages 35–43, December 1993.

[2] Q. Huang, K. Tanie, and S. Sugano. "Coordinated motion planning for a mobile manipulator considering stability and manipulation". *International Journal of Robotics Research*, 19(8):732–742, 2000.

[3] Y. Koren. "Cross-coupled biaxial computer controls for manufacturing systems". *ASME Journal of Dynamic Systems, Measurement, and Control*, 102:265–272, 1980.

[4] G. Marani, J. Kim, J. Yuh, and W. K. Chung. "A real-time approach for singularity avoidance in resolved motion rate control of robotic manipulators". *Proceedings of IEEE International Conference on Robotics and Automation*, pages 1973–1978, 2002.

[5] D. Sun, H. N. Dong, and S. K. Tso. "Tracking stabilization of differential mobile robots using adaptive synchronized control". *Proceedings of IEEE International Conference on Robotics and Automation*, pages 2638–2643, 2002.

[6] M. Tomizuka, J. S. Hu, and T. C. Chiu. "Synchronization of Two Motion Control Axes Under Adaptive Feedforward Control". *ASME Journal of Dynamic Systems, Measurement, and Control*, 114(6):196–203, 1992.

[7] Y. Yamamoto and X. Yun. "Coordinating locomotion and manipulation of a mobile manipulator". *IEEE Transactions on Automatic Control*, 39(6):1326–1332, 1994.

[8] T. Yoshikawa. *"Foundations of Robotics: Analysis and Control"*. Cambridge, MA: MIT Press, 1990.

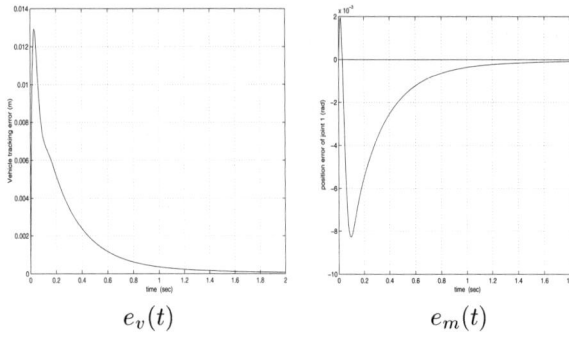

Figure 2: Position tracking errors of the vehicle and the manipulator

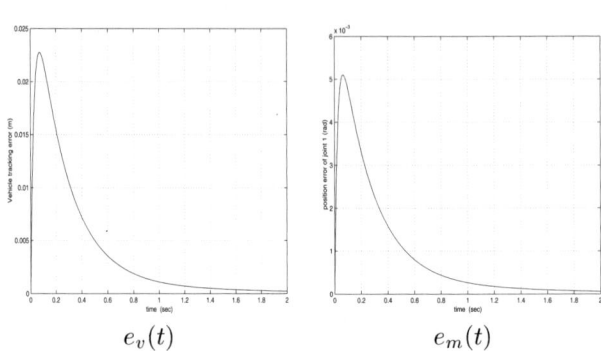

Figure 5: Position tracking errors of vehicle and manipulator without synchronization

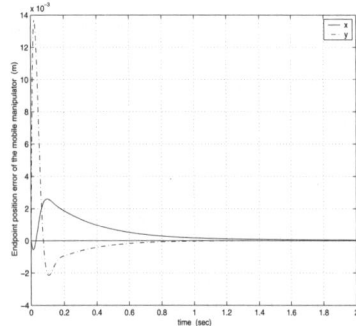

Figure 3: Endpoint position error of the mobile manipulator

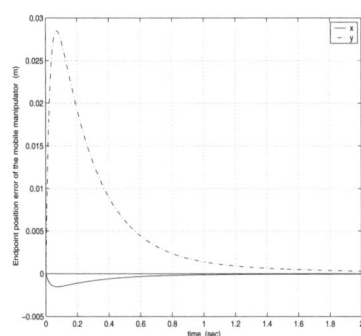

Figure 6: Endpoint position error without synchronization

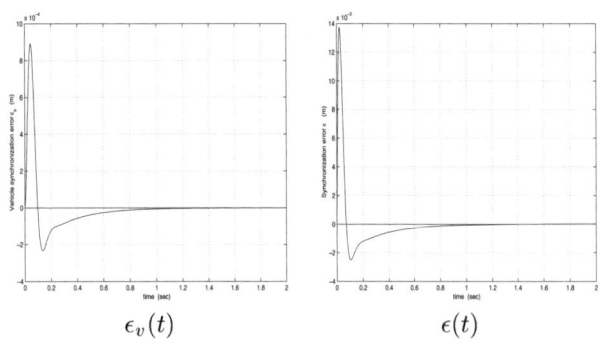

Figure 4: Synchronization errors

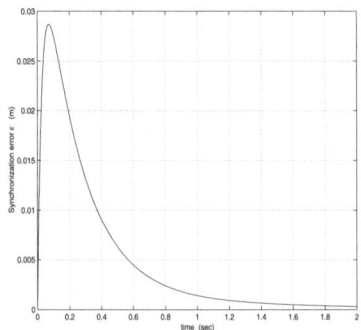

Figure 7: Synchronization error between the vehicle and the manipulator without synchronization

Mode Shape Compensator for Improving Robustness of Manipulator Mounted on Flexible Base

Jun Ueda* and Tsuneo Yoshikawa**
* Nara Institute of Science and Technology (NAIST), Nara, 630-0192, Japan,
E-mail:uedajun@is.aist-nara.ac.jp
** Department of Mechanical Engineering, Kyoto University, Kyoto, 606-8501, Japan,
E-mail:yoshi@mech.kyoto-u.ac.jp

Abstract— In this paper, the concept of the 'robust arm configuration' (RAC) is expanded using a mode shape compensator. This compensator improves the robustness of the arm configuration which is far out of the RAC. The compensator consists of a constant gain matrix and acceleration of each joint. The design method for this mode shaping matrix is presented based on the mode shaping algorithm. Effect upon the manipulability is also examined. It suggests that inclining of DME's principal axes, resulting in changing the rigid body dynamics, by the compensator leads to the improvement of the robustness. The validity is confirmed by a numerical example performed with a 2 DOF planar manipulator mounted on a 1 DOF flexible base. A high bandwidth settling is realized with obtained compensator.

Keywords— Flexible Robots, Flexible Base, Mechanical Resonance, Modal Analysis, Robust Control

I. INTRODUCTION

The demand of high speed and accurate assembling is increasing with the advance of industrial products. One of the primary factors of limiting the bandwidth is the flexibility of the base on which a tool or a manipulator is mounted. Although many researches has been presented for manipulators with link or joint flexibility[1][2][3], this problem due to the base flexibility has hardly been studied.

Simultaneous structure/control design has been studied[4][5][6][7][8] for this interaction problem. Recently, to overcome the ambiguity in mixing both design parameters, a passivity-based mechanical design [9][10][11] has been investigated. It is indicated that making the part of the frequency domain where passivity is satisfied as large as possible may improve the controllability and the robustness of the mechanical system itself.

We have examined the controllability of rigid manipulator mounted on a flexible base by applying this mechanical design criterion to the evaluation of arm configuration. It was shown that there exists a set of arm configurations where high robustness and controllability are obtained. We defined this configuration as the 'robust arm configuration' (RAC) where the linearized dynamics is positive real. A high-gain task-space control of the end effector can be applied in the neighborhood of the RAC based on the Lyapnov indirect method. However, there still remains an instability problem of the configuration which is far out of the RAC. A considerable solution is adding a viscosity with velocity feedback to the joint axes[12]. This method is valid for space-craft control. However, considering the application for robot manipulators, it deteriorates the manipulability to great extent.

In this paper, the concept of the 'robust arm configuration' (RAC)[13] is expanded by using a mode shape compensator. It is a novel dynamic compensator which improves robustness using acceleration of joint axes. The compensator makes the system close to positive real as the RAC. A numerical design method using the measure which indicates the distance from the RAC is also presented. The validity of this approach is confirmed by a numerical example. Effects upon the change of manipulability is also examined.

II. ROBUST ARM CONFIGURATION OF MANIPULATOR

A. Flexibility of the Base

A typical situation we are going to study is given in Fig. 1. A flexible assembly cell which consists of 2 degrees of freedom (DOF) planar rigid manipulator and two pallets on moving tracks is considered. As shown in Fig. 2, the parts are taken out from the parts feeder and are attached to the workpiece. Fig. 3 shows its dynamics model. The base driven by a ball screw on which the manipulator is mounted is considered as a passive visco-elastic joint. Since the control bandwidth of the ball screw is not high compared with the manipulator, the ball screw is positioned beforehand and is not driven during the assembly. During a motion, the root of the base twists due to the flexibility of the ball screw and the guide rods that support the base.

In general, it is very difficult to attach additional sensors to measure the base flexibility. Compensation is through a task-space feedback control. The positioning error between the pallet and the end effector is measured directly using a narrow visual sensor attached on the pallet or the end effector. If the servo gain of the task-space control is increased, the full-closed loop is easily destabilized.

B. Dynamic Equation

The system consists of an m DOF flexible base and an n DOF non redundant manipulator is considered, which includes the system of Fig. 3 as a special case.

$\boldsymbol{\tau}$ denotes joint torque, $\boldsymbol{q}$ the joint axis displacement vector. The dynamic equation is as follows.

$$\boldsymbol{\tau} = \boldsymbol{M}(\boldsymbol{q})\ddot{\boldsymbol{q}} + \boldsymbol{h}(\boldsymbol{q},\dot{\boldsymbol{q}}) \qquad (1)$$

$\boldsymbol{\tau} = [\boldsymbol{\tau}_p^T, \boldsymbol{\tau}_a^T]^T \in \Re^N$ ($N = m + n$) where $\boldsymbol{\tau}_p = [\tau_1 \ldots \tau_m]^T$ is torque generated by passive visco-elastic joints, $\boldsymbol{\tau}_a = [\ \tau_{m+1}\ \cdots\ \tau_{m+n}\ ]^T$ is actuator torque generated by motors. Similarly, $\boldsymbol{q} = [\boldsymbol{q}_p^T, \boldsymbol{q}_a^T]^T \in \Re^N$ where $\boldsymbol{q}_p = [\ q_1\ \cdots\ q_m\ ]^T$ and $\boldsymbol{q}_a = [\ q_{m+1}\ \cdots\ q_{m+n}\ ]^T$. The subscript p denotes the passive visco-elastic joints and a denotes the active joints. Note that $\boldsymbol{q}_p$ cannot be measured directly. $\boldsymbol{M}(\boldsymbol{q}) \in \Re^{N \times N}$ denotes the inertia matrix and $\boldsymbol{h}(\boldsymbol{q},\dot{\boldsymbol{q}})$ are the centrifugal and the Coriolis factor. The effect of the gravity is neglected as it is considered unimportant. A task-space control by measuring the positioning error is necessary to compensate the uncertainty of the inverse kinematics of a manipulator and uncertain position of the moving tracks. In this paper, we focus on a task-space feedback control of the end effector using Jacobian transpose. Let $\boldsymbol{r} = \boldsymbol{r}(\boldsymbol{q}) \in \Re^n$, the position of the end effector, $\boldsymbol{r}_d$ as the desired position of $\boldsymbol{r}$, and $\bar{\boldsymbol{q}}_p = [\bar{\theta}_1 \cdots \bar{\theta}_m]^T$ as the equilibrium points of $\boldsymbol{q}_p$. $\boldsymbol{r}_d$ is realized by $\boldsymbol{q}_d = [\bar{\boldsymbol{q}}_p^T, \boldsymbol{q}_{ad}^T]^T$ where $\bar{\boldsymbol{q}}_{ad} = [\theta_{(m+1)d} \cdots \theta_{(m+n)d}]^T$ and it satisfies $\boldsymbol{r}_d = \boldsymbol{r}(\boldsymbol{q}_d)$. Let $\boldsymbol{r}_v = \boldsymbol{r}_v(\boldsymbol{q}_a)$ be the position of the end effector when the displacement of the base is zero, so that $\boldsymbol{r}(\boldsymbol{q}_d) = \boldsymbol{r}_v(\boldsymbol{q}_{ad})$. It is assumed that the torque $\tau_1, \cdots, \tau_m$ generated by passive joints are working to settle to the equilibrium points $\bar{\theta}_1, \cdots, \bar{\theta}_m$:

$$\tau_i = k_{pi}(\bar{\theta}_i - \theta_i) - k_{vi}\dot{\theta}_i (i=1,\cdots,m) \quad (2)$$

where k_{pi} and k_{vi} are the stiffness and the viscous coefficient respectively.

Hereafter, we examine the local asymptotic stability of the nonlinear system (1) around $\boldsymbol{r}_d$, (around $\bar{\boldsymbol{q}}_p$, $\boldsymbol{q}_{ad}$ for joint space) by its linearized model. In this regard, the centrifugal and Coriolis factor can be ignored since these terms are not related to the stability. Substituting (2) for (1), we obtain

$$\begin{bmatrix} \boldsymbol{0} \\ \boldsymbol{\tau}_a \end{bmatrix} = \boldsymbol{M}\ddot{\boldsymbol{q}} + \boldsymbol{D}\dot{\boldsymbol{q}} + \boldsymbol{K}(\boldsymbol{q}-\boldsymbol{q}_d) \quad (3)$$

where $\boldsymbol{D} = \text{diag}(k_{v1},\cdots,k_{vm},0,\cdots,0)$, $\boldsymbol{K} = \text{diag}(k_{p1},\cdots,k_{pm},0,\cdots,0)$ are the stiffness and the viscous matrix respectively.

Replacing $\boldsymbol{x} = \left[\dot{\boldsymbol{q}}^T, (\boldsymbol{q}-\boldsymbol{q}_d)^T\right]^T$ and $\boldsymbol{y} = \boldsymbol{r}(\boldsymbol{q}) - \boldsymbol{r}(\boldsymbol{q}_d)$, n-inputs n-outputs state space representation in the work coordinate system is obtained:

$$\boldsymbol{P}(\boldsymbol{q}_d, s) = \boldsymbol{C}(s\boldsymbol{I}-\boldsymbol{A})^{-1}\boldsymbol{B} \quad (4)$$

where $\boldsymbol{A} = \begin{bmatrix} -\boldsymbol{M}^{-1}\boldsymbol{D} & -\boldsymbol{M}^{-1}\boldsymbol{K} \\ \boldsymbol{I} & \boldsymbol{O} \end{bmatrix}$, $\boldsymbol{B} = \begin{bmatrix} \boldsymbol{M}^{-1}\begin{bmatrix} \boldsymbol{O} \\ \boldsymbol{I} \end{bmatrix} \\ \boldsymbol{O} \end{bmatrix} \boldsymbol{J}_v^T$, $\boldsymbol{C} = [\ \boldsymbol{O}\ \boldsymbol{J}\ ]$. $\boldsymbol{J} = \partial \boldsymbol{r}/\partial \boldsymbol{q}^T \in \Re^{n \times (m+n)}$, $\boldsymbol{J}_v = \partial \boldsymbol{r}_v/\partial \boldsymbol{q}_a^T \in \Re^{n \times n}$ are Jacobian matrices.

We have assumed that this flexibility is not measured, then $\boldsymbol{\tau}_a$ is calculated as $\boldsymbol{\tau}_a = \boldsymbol{J}_v^T \boldsymbol{f}$, where $\boldsymbol{f}$ denotes the control input force in the taskspace calculated by a feedback controller $\boldsymbol{K}_c(s)$: $\boldsymbol{f} = -\boldsymbol{K}_c(s)\boldsymbol{y}$. Note that we use $\boldsymbol{P}(\boldsymbol{q}_d, s)$ when we emphasize that $\boldsymbol{P}(s)$ is a function of $\boldsymbol{q}_d$, otherwise we use $\boldsymbol{P}(s)$.

C. Modal Analysis

By modal analysis, $\boldsymbol{P}(s)$ can be expressed as a linear sum of the rigid mode and the m vibration modes. $\boldsymbol{A}$ has in total $2(n+m)$ poles, $2n$ of these poles are zero, corresponding to the rigid mode, $2m$ conjugate complex poles correspond to the vibration modes.

Let $\lambda_0 = 0, \lambda_i, \bar{\lambda}_i$ $(i=1\cdots,m)$ be $2m+1$ distinct eigenvalues of $\boldsymbol{A}$. λ_0, λ_i and $\bar{\lambda}_i$ correspond to the rigid mode and the ith vibration modes respectively. We define matrix $\boldsymbol{U}$ and $\boldsymbol{V}$ using $\boldsymbol{u}$ which satisfy: $\boldsymbol{A}\boldsymbol{u}_{2j-1} = \lambda_0 \boldsymbol{u}_{2j-1} = \boldsymbol{o}, (j=1,\cdots,n)$, $\boldsymbol{A}\boldsymbol{u}_{2j} = \boldsymbol{u}_{2j-1}$, $\boldsymbol{A}\boldsymbol{u}_{2(n+i)-1} = \lambda_i \boldsymbol{u}_{2(n+i)-1}, (i=1,\cdots,m)$, $\boldsymbol{A}\boldsymbol{u}_{2(n+i)} = \bar{\lambda}_i \boldsymbol{u}_{2(n+i)}$ where $\boldsymbol{U} \stackrel{\triangle}{=} [\ \boldsymbol{u}_1\ \cdots\ \boldsymbol{u}_{2n}\ \boldsymbol{u}_{2n+1}\ \cdots\ \boldsymbol{u}_{2(m+n)}\ ]$ and $\boldsymbol{V}^* \stackrel{\triangle}{=} \text{col}\left[\boldsymbol{v}_1^*, \boldsymbol{v}_2^*, \cdots, \boldsymbol{v}_{2(m+n)}^*\right] = \boldsymbol{U}^{-1}$. $\boldsymbol{v}^*$ denotes the complex conjugate transpose of $\boldsymbol{v}$. Applying $\boldsymbol{U}, \boldsymbol{V}$ to $\boldsymbol{A}$, the modal

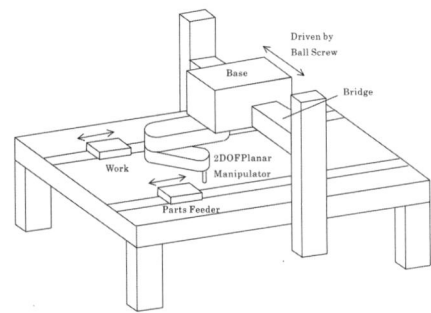

Fig. 1. Flexible Assembly System

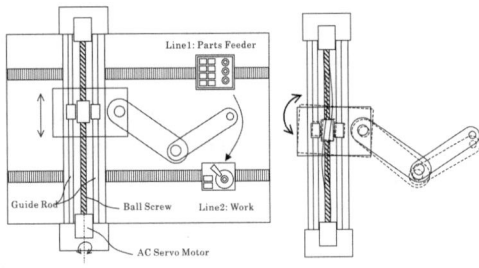

Fig. 2. Flexibility of the Base

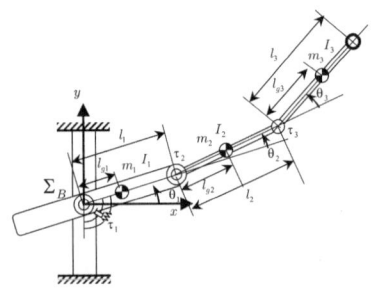

Fig. 3. 2 DOF Manipulator with 1 DOF Base

analyzed transfer function[14] of $P(s)$ is obtained as:

$$P(s) = \frac{1}{s^2}R_0 + \sum_{i=1}^{m} \frac{1}{s^2 + 2\zeta_i\hat{\omega}_i s + \hat{\omega}_i^2}R_i \quad (5)$$

$$R_0 = \sum_{j=1}^{n} Cu_{2j-1}v_{2j}^* B = J_v\hat{M}^{-1}J_v^T \quad (6)$$

$$R_i = -2\text{Re}\left(\bar{\lambda}_i(Cu_{2(n+i)-1})(v_{2(n+i)-1}^* B)\right) \quad (7)$$

R_0 corresponds to the rigid body mode and R_0 is positive semi-deninite. $\hat{M} = [M_{kl}] \in \Re^{n \times n}(m+1 \le k, l \le m+n)$ is a partial matrix of M. R_i is called residue matrix. Note that R_0 is positive-semi-definete and rank$(R_i) = 1$ at most from (7). In addition, $\hat{\omega}_i = |\lambda_i|, \zeta_i = -\text{Re}(\lambda_i)/|\lambda_i|$ is obtained where ζ_i and $\hat{\omega}_i$ represent the damping coefficient and the natural frequency respectively.

D. Definition of Robust Arm Configuration

Let $G(s) \triangleq sP(s)$ be the transfer function from the control input f to the velocity $\dot{y}$: We have shown that $G(s)$ is positive real if and only if $R_i = R_i^T > 0$[13].

Definition q_d which satisfies

$$\lambda_{min}(G(q_d, j\omega) + G^T(q_d, -j\omega)) = 0 \quad (8)$$

is called the 'robust arm configuration' (RAC) where $\lambda_{min}(G)$ denotes the minimum eigenvalue of G.

The RAC is a set of special configurations where the linearized system is positive real. Note that the dimension of the RAC is not equal to that of the taskspace. Lyapnov indirect method and Passivity theory suggest that a local asymptotic stability for the original nonlinear system can be guaranteed with a finite but high feedback gain within a small area including the RAC. Measuring the flexible part and the inverse dynamics solution are not necessary.

E. Robustness Measure of Arm Configuration

Recall that if the rank$(R_i + R_i^T) \le 2$, there exist only minimum and maximum eigenvalue. The system is positive real if rank$(R_i + R_i^T) = 1$ and $\lambda_{min}(R_i + R_i^T) \ge 0$. We have proposed the following robustness measure which indicates the distance from the RAC:

$$w_i(q) = \lambda_{min}(R_i + R_i^T) \quad (9)$$

The characteristic of this measure is that $w_i \le 0$ commonly holds and $w_i = 0$ holds only on the RAC for ith mode. The configuration where w_i is near 0 is preferable for better controllability.

The advantages of this measure are as follows. First, it is easily calculated by $n \times n$ numerical computation of eigenvalues. Second, it clarifies the optimality of the arm configuration based on the positive realness. Third, it does not depend on the underlying control law. Forth, it identifies the robustness for individual mode.

This measure is closely related to the robust stability margin based on the normalized coprime factorization [15]. It supports the optimality of the robust arm configuration in the case of considering a whole class of stable controller.

III. MODE SHAPE COMPENSATOR

A. Improving Robustness Using Acceleration of Joint Axes

In general, without the presence of additional sensors; strain gauges, accelerometers and so on, it is assumed that the displacement of the flexibility, $q_p = [\ q_1\ \ldots\ q_m\ ]^T$ cannot be measured. The displacement of the active joints $q_a = [\ q_{m+1}\ \ldots\ q_{m+n}\ ]^T$ can be measured if encoders are attached. We assume that $\ddot{q}_a = [\ \ddot{q}_{m+1}\ \ldots\ \ddot{q}_{m+n}\ ]^T$ can be ideally measured. Set the joint input torque as:

$$\tau_a = \tilde{\tau}_a + \hat{E}\ddot{q}_a \quad (10)$$

where $\tilde{\tau}_a$ is calculated by an error feedback scheme $K_c(s)$ shown in Fig. 4; $\tilde{\tau}_a = -J_v^T K_c(s)y$

Substituting (10) for (1), the following equation is obtained:

$$\begin{bmatrix} 0 \\ \tilde{\tau}_a \end{bmatrix} = (M - E)\ddot{q} + D\dot{q} + K(q - q_d) \quad (11)$$

where $E = \begin{bmatrix} O^{m \times m} & O^{m \times n} \\ O^{n \times m} & \hat{E}^{n \times n} \end{bmatrix}$. Equation (11) shows that the inertia matrix M is modified to $(M - E)$.

The second term of (10) modifies the vibration mode so that the the same property of the RAC, i.e., positive-realness, is obtained in a direct way. We named this type of compensator the 'mode shape compensator'. The calculation of the mode shape compensator is simple, since it consists of a constant gain matrix and acceleration of joint axes.

$\hat{E}$ is a constant gain matrix which is designed for each arm configuration. We call $\hat{E}$ the 'mode shaping matrix' corresponding to the mode shape compensator. It is clear that not all elements of M but only those who correspond to active joints can be modified. However, an appropriately designed $\hat{E}$ improves robustness drastically as shown in the next part.

B. Design of Mode Shaping Matrix

It is necessary that $(M - E) = (M - E)^T > 0$ is satisfied to avoid instability since $M = M^T > 0$ in original systems.

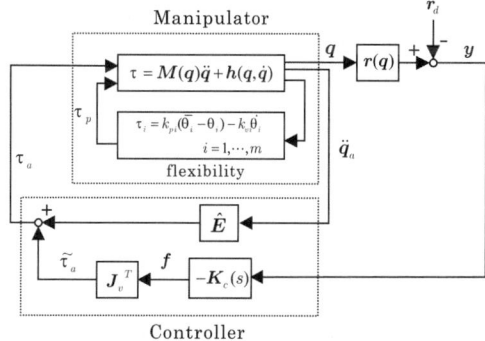

Fig. 4. Mode Shape Compensator and Controller

Since $w_i(\boldsymbol{q}_d) = 0$ holds on the RAC of the ith mode, $\hat{\boldsymbol{E}}$ should be designed aiming at $w_i(\boldsymbol{q}_d) \to -0$.

A mode shaping method[16] is applied for designing the mode shaping matrix where the natural frequency and the residue are originally used as representative parameters.

For simplicity, suppose that $m = 1$. The property of the first mode can be represented as $\boldsymbol{\phi}_1(\hat{\boldsymbol{E}}) \triangleq [\ w_1 \ \ \hat{\omega}_1\ ]^T$ where w_1 is the distance from the RAC and $\hat{\omega}_1$ is the natural frequency. We set the desired natural frequency as $\hat{\omega}_{1d}$. $\hat{\boldsymbol{E}}$ is designed aiming at

$$\boldsymbol{\phi}_1(\hat{\boldsymbol{E}}) \to \boldsymbol{\phi}_{1d}(\hat{\boldsymbol{E}}_d) = [\ 0 \ \ \hat{\omega}_{1d}\ ]^T \quad (12)$$

Recall that $\hat{\boldsymbol{E}} = \hat{\boldsymbol{E}}^T$ so that $n(n+1)/2$ elements (upper triangle part) are independent. A design vector is defined as $\boldsymbol{\rho} \triangleq [\hat{e}_{11}, \cdots, \hat{e}_{1n}, \hat{e}_{22}, \cdots, \hat{e}_{2n}, \cdots, \hat{e}_{nn}]^T \in \Re^{\frac{n(n+1)}{2} \times 1}$ where $\hat{\boldsymbol{E}} = [\hat{e}_{kl}] (1 \le k, l \le n)$. Let $\boldsymbol{\phi}_1^k$ and $\boldsymbol{\rho}^k$ be the representative parameters and the design vector after the kth iteration. The next design vector $\boldsymbol{\rho}^{k+1}$ is obtained by updating $\Delta\boldsymbol{\rho}^k$ as

$$\boldsymbol{\rho}^{k+1} = \boldsymbol{\rho}^k + \Delta\boldsymbol{\rho}^k \quad (13)$$

$$\Delta\boldsymbol{\rho}^k = g_1 \boldsymbol{H}_1^{k+}(\boldsymbol{\phi}_{1d} - \boldsymbol{\phi}_1^k) \quad (14)$$

where $g_1(>0)$ is a scalar gain and $\boldsymbol{H}_1^k = \partial \boldsymbol{\phi}_1^k / \partial \boldsymbol{\rho}^T$.

It is difficult to obtain $\boldsymbol{H}_1^k$ analytically, hence an estimation process is applied as follows. Let $\Delta\boldsymbol{\rho}_1, \Delta\boldsymbol{\rho}_2, \ldots, \Delta\boldsymbol{\rho}_l$ be small variations from $\boldsymbol{\rho}^k$ and the corresponding representative parameters $\Delta\boldsymbol{\phi}_1, \Delta\boldsymbol{\phi}_2, \ldots, \Delta\boldsymbol{\phi}_l$ from $\boldsymbol{\phi}_1$ where

$$\Delta\boldsymbol{\phi}_j = \boldsymbol{H}_1^k \Delta\boldsymbol{\rho}_j (j = 1, 2, \cdots, l \le \frac{n(n+1)}{2}) \quad (15)$$

The estimated $\boldsymbol{H}_1^k$ which minimizes the squared error of $|\Delta\boldsymbol{\phi}_j - \boldsymbol{H}_1^k \Delta\boldsymbol{\rho}_j|$ is given by:

$$\boldsymbol{H}_1^k = \boldsymbol{H}_1^{k-1} + (\boldsymbol{Y} - \boldsymbol{H}_1^{k-1}\boldsymbol{X})\boldsymbol{X}^+ \quad (16)$$

where $\boldsymbol{X} = [\Delta\boldsymbol{\rho}_1 \ \Delta\boldsymbol{\rho}_2 \ \ldots \ \Delta\boldsymbol{\rho}_l] \in \Re^{\frac{n(n+1)}{2} \times l}$, $\boldsymbol{Y} = [\Delta\boldsymbol{\phi}_1 \ \Delta\boldsymbol{\phi}_2 \ \ldots \ \Delta\boldsymbol{\phi}_l] \in \Re^{2 \times l}$.

It is not difficult to design the mode shaping matrix of the system which contains multiple ($m \ge 2$) modes. A task priority based mode shaping method [16] can be applied.

One might argue that $\boldsymbol{\phi}_i(\hat{\boldsymbol{E}}) \triangleq w_i$ is only appropriate for the purpose of improving robustness; a desired natural frequency should not be clearly given. However, in practical calculation, the natural frequency sometimes converges into an unrealistic frequency since the redundancy is too high. For this reason, a certain desired natural frequency is given.

IV. NUMERICAL DESIGN EXAMPLE

A. 2 DOF Planar Manipulator Mounted on Flexible Base

The system shown in Fig. 3 is considered. Table I shows the link parameters. θ_i is the joint angle of axis i, m_i is the mass, l_i and I_i are the length and the moment of inertia of link i and l_{gi} is the distance between the axis i and

TABLE I
LINK PARAMETERS

m_1	20.0(kg)	l_1	0.2(m)
m_2	7.0(kg)	l_2	0.3(m)
m_3	5.0(kg)	l_3	0.25(m)
I_1	0.066(kgm^2)	l_{g1}	0.02(m)
I_2	0.0525(kgm^2)	l_{g2}	0.15(m)
I_3	0.026(kgm^2)	l_{g3}	0.125(m)

the center of mass respectively. Assume that no singular configuration is passed while working. The working range is $-\pi < \theta_2 < \pi$, $0 < \theta_3 < \pi$. The stiffness and viscous constant are $k_{p1} = 3.0 \times 10^4$(Nm/rad), $k_{v1} = 10$(Nms/rad).

The compensator, for the end effector (0.33,-0.25) shown in Fig. 5, is designed. Fig. 6 shows the measure w_1 from the RAC without compensating action where the distance of the contour line is 0.015. We have shown that the robustness of this configuration is not high so that the closed loop system is destabilized [13].

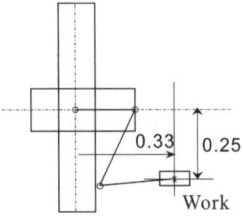

Fig. 5. Example Configuration

B. Design of Mode Shaping Matrix

Recall $\hat{\boldsymbol{E}} = \begin{bmatrix} \hat{e}_{11} & \hat{e}_{12} \\ \hat{e}_{21} & \hat{e}_{22} \end{bmatrix}$ where $\hat{e}_{12} = \hat{e}_{21}$. Set $\boldsymbol{\rho} \triangleq [\hat{e}_{11}\ \hat{e}_{12}\ \hat{e}_{22}]^T$, $\boldsymbol{\phi}_{1d}(\hat{\boldsymbol{E}}_d) = [\ 0\ \ 346.8\ ]^T$ and $g_1 = 0.02$. The modify of the natural frequency is done by the compensation of the inertia. Therefore, the desired natural frequency is set fractionally higher than the original value to avoid a large mode shaping matrix. After 800 recursive calculation loops, the mode shaping matrix is obtained as $\hat{\boldsymbol{E}} = \begin{bmatrix} 0.12 & 0.079 \\ 0.079 & -0.15 \end{bmatrix}$, for original $\hat{\boldsymbol{M}} = \begin{bmatrix} 0.58 & 0.013 \\ 0.013 & 0.104 \end{bmatrix}$. As a result, (1,1) element is canceled approximately 20 %, practically, it is considered in the range of safty. Fig. 7 shows the transition of w_1, $\hat{\omega}_1$ and $\lambda_{\min}(\boldsymbol{M} - \boldsymbol{E})$ respectively. It shows that w_1 and $\hat{\omega}_1$ are converged to the desired values, and simultaneously the robustness is improved. Additionally, $(\boldsymbol{M} - \boldsymbol{E})$ stays positive definite so that the system with the compensator maintains its stability. The residue matrix is modified and becomes close to positive semi-definite:

$$\boldsymbol{R}_1 = \begin{bmatrix} -0.032 & -0.075 \\ 0.037 & 0.085 \end{bmatrix} \to \begin{bmatrix} 0.036 & 0.048 \\ 0.05 & 0.066 \end{bmatrix} \quad (17)$$

Fig. 8 shows the proposed measure w_1 with the obtained compensator. In certain region, $(\boldsymbol{M} - \boldsymbol{E})$ is not positive

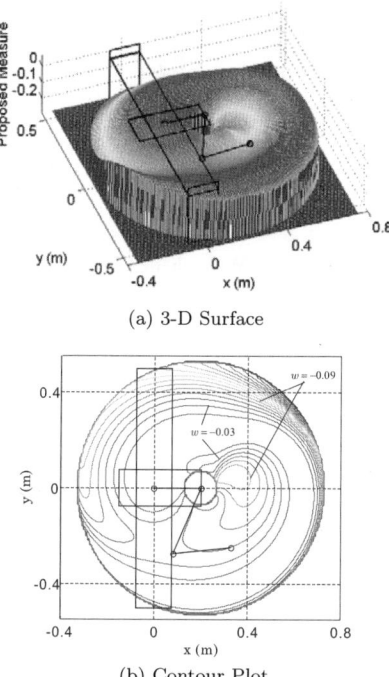

(a) 3-D Surface

(b) Contour Plot

Fig. 6. Proposed Measure of Primary Dynamics

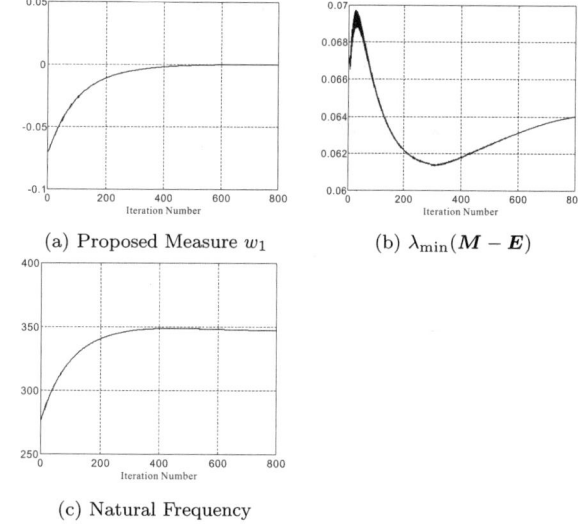

(a) Proposed Measure w_1

(b) $\lambda_{\min}(M - E)$

(c) Natural Frequency

Fig. 7. Design of Mode Shaping Matrix

definite, so that the system is not stable. However, in the neighborhood of the designed configuration, the robustness is successfully improved.

A simulation is performed; a point to point control from (0.33,0.25) to (0.33,-0.25). While the arm is moving, joint-level tracking control is applied to follow the generated path $[0 - 0.1]$(s). Then the controller is switched to the task-space feedback controller $[0.1 - 0.4]$(s) in the neighborhood of (0.33,-0.25). The applied settling controller is a task-space PD feedback: $\boldsymbol{K}_c(s) = \boldsymbol{K}_p + \boldsymbol{K}_v s$, then the actuator torque is calculated as:

$$\boldsymbol{\tau}_a = -\boldsymbol{J}_v^T(\boldsymbol{q}_a)(\boldsymbol{K}_p \boldsymbol{y} + \boldsymbol{K}_v \dot{\boldsymbol{y}}) \tag{18}$$

where $\boldsymbol{K}_p = \text{diag}(1.0 \times 10^5, 1.0 \times 10^5)$(N/m), $\boldsymbol{K}_v = \text{diag}(1.5 \times 10^3, 1.5 \times 10^3)$(Ns/m). Fig. 9 shows the positioning error for each control scheme. Without the compensator, the system is unstable and oscillates. However, with the compensator, it satisfactorily settles.

C. Effects upon Manipulability

The mode shape compensator modifies the dynamics of the rigid body so that the vibration mode is shaped. As a result of this mode shaping, the robustness is improved. It is necessary to examine how much the manipulability of the rigid body changes. This effect is estimated by evaluating the dynamic manipulability ellipsoid (DME) [17] of 2 DOF rigid body part without its flexible base.

Consider the relation from a control input $\boldsymbol{f}$ to a tip acceleration $\ddot{\boldsymbol{y}}$. Assuming that the displacement of the base is zero, a linearized input-output relation is obtained as:

$$\ddot{\boldsymbol{y}} = \boldsymbol{J}_v(\hat{\boldsymbol{M}} - \hat{\boldsymbol{E}})^{-1}\boldsymbol{J}_v^T \boldsymbol{f} \tag{19}$$

Then the DME within $\|\boldsymbol{f}\| \leq 1$ satisfies:

$$\ddot{\boldsymbol{y}}^T \boldsymbol{J}_v^{-T}(\hat{\boldsymbol{M}} - \hat{\boldsymbol{E}})^T \boldsymbol{J}_v^{-1} \boldsymbol{J}_v^{-T}(\hat{\boldsymbol{M}} - \hat{\boldsymbol{E}}) \boldsymbol{J}_v^{-1} \ddot{\boldsymbol{y}} \leq 1 \tag{20}$$

Fig. 10 shows these ellipsoids for the whole range of motion. The distance of the contour line is 0.15. The dashed line denotes the original DME and the solid line denotes the modified one. Note that in Fig. 10, the desired natural frequencies are determined as $\hat{\omega}_{1d} = \hat{\omega}_1 + 100$ rad/sec and executed the calculation for 800 times. Robustness is improved and the positioning error stably converges for each configuration. In Fig. 10, the contour plot of the measure w_1 without the compensator is superimposed.

Fig. 10 reveals that the size of the ellipsoid itself does not drastically change because the mode shaping matrix is relatively small to the inertia matrix. However, the principal axes of modified ellipsoids tend to incline compared to the original ones. Furthermore, the inclination of the principal axes is large where the robustness is not high, while it is not large in the neighborhood of the RAC. These results indicate feasible design criteria of the compensator. That is, inclining the principal axes by a compensator leads to a good design.

V. Conclusion

In this paper, the concept of the 'robust arm configuration' (RAC)[13] has been expanded. A mode shape compensator has been proposed which improves the robustness of the arm configuration which is far out of the RAC. The compensator makes the system close to positive real so that a high-gain feedback can be obtained just as in the neighborhood of the RAC. The compensator consists of a constant gain matrix and an acceleration of each joint. The

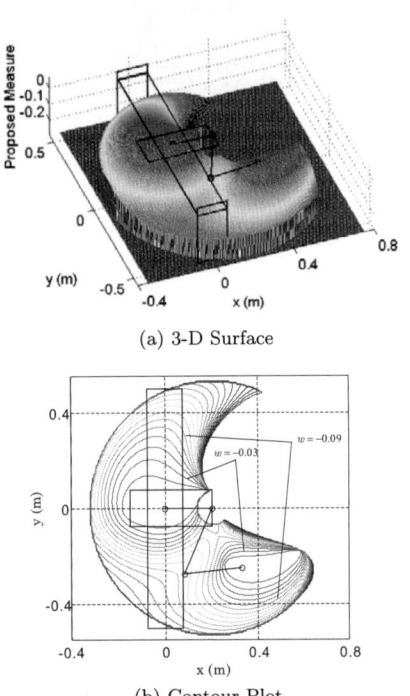

(a) 3-D Surface

(b) Contour Plot

Fig. 8. Proposed Measure with Obtained Mode Shape Compensator

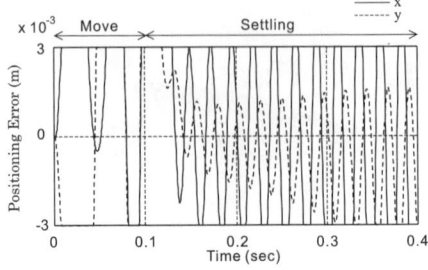

(a) Without Mode Shape Compensator

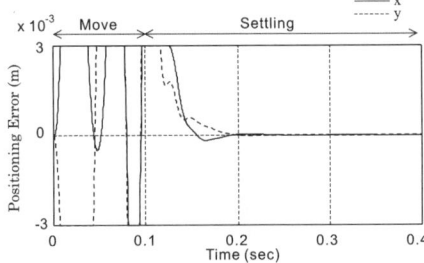

(b) With Mode Shape Compensator

Fig. 9. Settling Performance

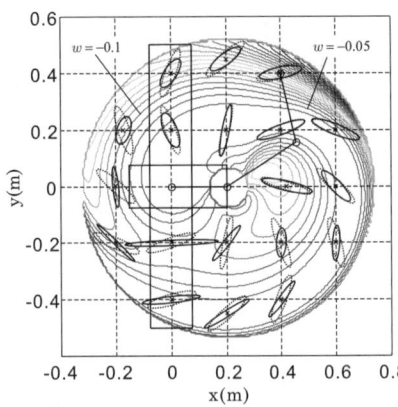

Fig. 10. Change of DME of Rigid Body (Whole Region)

structure is simple and easy to use. The method of designing the mode shaping matrix has been proposed based on the task priority mode shaping method. The validity of this approach was confirmed by a numerical example. Manipulability analysis suggests that inclining of DME's principal axes by the compensator leads to the improvement of the robustness.

References

[1] M. W. Spong, "Modeling and Control of Elastic Joint Robots", *ASME J. Dynamic Systems, Measurement and Control*, Vol. 109, pp.310–319, 1987.

[2] P. Tomei, "A Simple PD Controller for Robots with Elastic Joints", *IEEE Trans. Automatic Control*, Vol. 36, No. 10, pp.1208–1213, 1985.

[3] A. De Luca, "Feedforward/Feedback Laws for the Control of Flexible Robots", *Proc. IEEE Int. Conf. Robotics and Automation*, pp.233–240, 2000

[4] J. Onoda, R. T. Haftka, "An approach to structure/ control simultaneous optimization for large flexible spacecraft", *AIAA Journal*, Vol. 25, No. 8, pp.1133–1138 , 1987.

[5] A. M. A. Hamdan, A. H. Nayfeh, 'Measures of Modal Controllability and Observability for First-and Second-Order Linear Systems', *Trans. AIAA, Journal of Guidance and control*, vol. 12, No. 3, pp.421–428, 1989.

[6] D. S. Bodden, J. L. Junkins, "Eigenvalue Optimization Algorithms for Structure/Controller Design Iterations", *Trans. AIAA, J. Guidance and control*, Vol. 8, No. 6, pp.697–706. 1985.

[7] A. Mayzus, K. Grigoriadis, "Integrated Structural and Control Design for Structural Systems via LMIs", *Proc. IEEE Int. Conf. Control Applications*, pp.75–78, 1999.

[8] A. C. Pil, H. Asada, "Integrated Structure/ Control Design of Mechanical Systems Using a Recursive Experimental Optimization Method", *IEEE/ASME Trans. Mechatronics*, Vol. 1, No. 3, pp.191–203, 1996.

[9] S. Hara, "Integration of structure design and control system synthesis", *Proc. TITech COE/SMS Symposium*, pp.53–60, 1998.

[10] T. Iwasaki, "Integrated System Design by Separation", *Proc. IEEE Conf. Control Application*, pp.97–102, 1999.

[11] T. Iwasaki, S. Hara, H. Yamauchi, "Structure/control integration with finite frequency positive real property", *Proc. TITech COE/Super Mechano-System Symposium*, pp.126–135, 2000.

[12] S. Manabe, K. Tsuchiya, "Controller Design of Flexible Spacecraft Attitude Control", *9th IFAC World Congress*, III, pp.30–35, 1984.

[13] J. Ueda, T. Yoshikawa, "Robust Arm Configuration of Manipulator Mounted on Flexible Base", *Proc. IEEE Int. Conf. Robotics and Automation*, pp.1321–1326, 2002.

[14] C. T. Chen, C. A. Desser, "Controllability and Observability of Composite Systems", *Trans. IEEE Automatic Control*, Vol. 12, No. 4, pp.402–409, 1967.

[15] D. C. McFarlane, K. Glover, "Robust Controller Design Using Normalized Coprime Factor Plant Descriptions", Springer-Verlag.

[16] T. Yoshikawa, J. Ueda, "Task Priority Based Mode Shaping Method for In-phase Design of Flexible Structures Aiming at High Speed and Accurate Positioning", *Proc. IEEE Int. Conf. Robotics and Automation*, pp.1806–1812, 2001.

[17] T. Yoshikawa, "Dynamic Manipulability of Robotic Mechanism", *J. Robotic Systems*, Vol. 2 No. 1, pp.113–124, 1985.

Landing Control of Acrobat Robot (SMB) Satisfying Various Constraints

Teruyoshi SADAHIRO[1]

[1] Dept. of Mechanical and Control Engineering
Tokyo Institute of Technology
2-12-1 Ohokayma, Meguro-ku, Tokyo, Japan

Masaki YAMAKITA[1,2]

[2] Bio-Mimetic Control Research Center,
RIKEN
2271-130 Anagahora, Shidami, Moriyama-ku
Nagoya, Japan

Abstract— The C.O.E. acrobat robot called SMB (Super Mechano-Boy) is considered as a model of a gymnast with a horizontal bar. Our final objective is to perform skillful motions with the system same as a gymnast. In this paper, we consider the landing control after actions on the horizontal bar using a linear complementarity problem. We also consider a parameter estimation using the relationship between Lagrange multiplier and sensitivity of the constrained system.

I. INTRODUCTION

The Acrobot introduced by Spong[1], which is a simple example of two-link planar robot with a single actuator at the elbow, is a well-known under-actuated system. This system is similar to a human who suspends a horizontal bar. And there were proposed many control algorithms to realize acrobatic motions with a horizontal bar like as human gymnasts [2] [3] [4] [5].

Fig.1 shows the C.O.E. acrobat robot (SMB) to be considered in this paper. It consists of nine links and 8 actuators. C.O.E acrobat robot is also considered as a model of a gymnast with a horizontal bar. Our final objective is to perform skillful motions such as swinging from pendant state, forward upward circling, handstand on the horizontal bar, giant swing, fling off, somersault and landing.

In this paper, we consider to realize the landing in order to solve Linear Complementarity Problem(LCP) which minimize kinetic energy with various constraints. And we also consider to estimate an unknown friction coefficient. It is important to know it accurately in order to realize the landing. But, it is difficult to measure it accurately which is usually decided by the material of the landing point. An other point of view, if we can select the material freely, we have to know the necessary friction coefficient to realize landing. Thus, we propose a way of estimating the unknown friction coefficient using a feature of LCP that gives the Lagrange multiplier simultaneously and the relationship between Lagrange multiplier and sensitivity of the constrained system.

II. MODELING

Because the body weight is dominant, we constrain the links of legs as one link. And for simplicity, we also

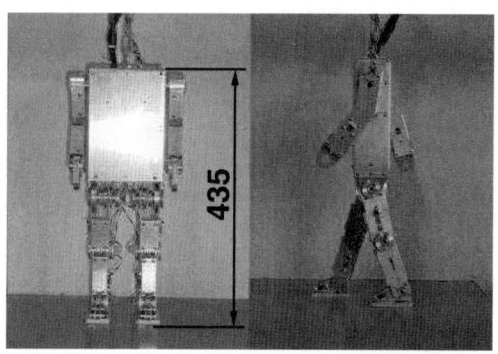

Fig. 1. C.O.E. acrobat robot (SMB)

constrain the couple of arms and legs each other. Therefore the robot can be considered a planar 3 links robot.(Fig.2)

In this figure, we shall denote by l_i its length, by r_i the distance between its center of mass and the joint of a previous link, by m_i its mass and by I_i its moment of inertia(i=1,2,3) and by (x, y) the position of the edge of the first link. The torques at the first and second joint are indicated by τ_1 and τ_2, respectively.

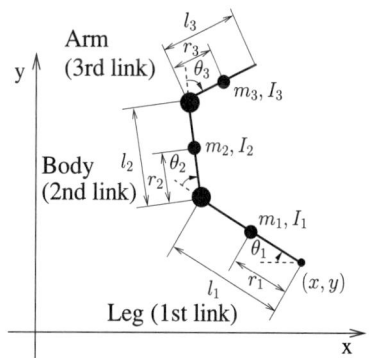

Fig. 2. Model of the robot

TABLE I
PARAMETERS OF THE ACROBAT ROBOT

i	$m_i[kg]$	$I_i[kg \cdot m^2]$	$l_i[m]$	$r_i[m]$
1	0.80	3.32 $\times 10^{-3}$	0.223	0.112
2	3.10	1.15 $\times 10^{-2}$	0.211	0.144
3	0.85	2.35 $\times 10^{-3}$	0.182	0.079

Then the dynamical equation for the system is given by

$$D(q)\ddot{q} + h(q,\dot{q}) = \begin{bmatrix} 0_{3\times 1} \\ \tau \end{bmatrix} \quad (1)$$

where $q = (x, y, \theta_1, \theta_2, \theta_3)^T$ is a generalized coordinates, $D(q) \in R^{5\times 5}$ is an inertia matrix, $h(q,\dot{q}) \in R^{5\times 1}$ is a colioris and gravity matrix, and $\tau = (\tau_1, \tau_2)^T$ is a control input.

Each parameters of the robot are shown in Table I which was obtained by the real robot parameters.

III. LANDING CONTROL

A. Optimized Input Using LCP

For realizing a soft landing, we consider a problem minimizing the kinetic energy of the robot just after the impact. It is given by

$$K_+ = \frac{1}{2}\dot{q}_+^T D(q) \dot{q}_+ \quad (2)$$

where $\dot{q}_+$ is the generalized velocities just after the impact. And we also consider that it is restricted with some appropriate constraints such as slipping conditions, impact conditions, angular momentum conditions and velocity conditions. In this problem, system input are generalized coordinates and generalized velocities just before the impact denoted as q and $\dot{q}_-$, respectively.

First, we assume that the initial velocity (V_{dx}, V_{dy}) and the initial angular momentum L_d are given. We additionally assume that the posture of the robot are given because of the beauty of the landing. So we can be also given the generalized coordinates q. We also assume that the impact is perfectly inelastic collision. If we let J_c denotes the vector of the velocity constraints at the landing point, F denotes the change of a generalized momentum, then

$$J_c \dot{q}_+ = 0. \quad (3)$$

Dynamic equation just before and after impact is given by

$$D\dot{q}_+ = D\dot{q}_- - J_c^T F. \quad (4)$$

we can get the new relationships solving the simultaneous equations Eq.(3) and Eq.(4) are given by

$$F = (J_c D^{-1} J_c^T)^{-1} J_c \dot{q}_- =: X\dot{q}_- \quad (5)$$
$$\dot{q}_+ = (I - D^{-1} J_c^T (J_c D^{-1} J_c^T)^{-1} J_c)\dot{q}_- =: Y\dot{q}_-. \quad (6)$$

Using Eq.(6), Eq.(2) can be rewritten as

$$K_+ = \frac{1}{2}\dot{q}_-^T Y^T D(q) Y \dot{q}_- \quad (7)$$

which denotes the quadratic equation of $\dot{q}_-$.

Now, we can formulate the problem.

> given $(V_{dx}, V_{dy}), L_d, q,$
> find $\dot{q}_-$ s.t.
>   minimize: K_+ Eq.(7)
>   sub. to: slipping condition   Eq.(12)
>   impact condition   Eq.(16)
>   angular momentum condition
>   Eq.(17)(19)
>   velocity constraints   Eq.(22)

In this study, LCP is used to solve the problem. Constraints in the problem are explained later.

LCP can solve the Convex Quadratic problems(CQP) and the Dual problem of CQP(DQP) simultaneously. And we let landing problem as CQP and let validity of constraints as DQP, then we can solve these simultaneously. The relationships between CQP, DQP and LCP is described at Appendix . Projection and Contraction(PC) method[8] is used for solving LCP.

B. Slipping Condition

If F_y is the magnitude of the normal force and (F_x, F_z) denote the magnitude of the tangential forces, from the Coulomb's law, the range of tangential forces which is not slipped at a contact point should satisfy

$$\sqrt{F_x^2 + F_z^2} \leq \mu F_y \quad (8)$$

where μ is the friction coefficient.[7]

In this paper, we consider the problem as planner one and so Eq.(8) is given by

$$|F_x| \leq \mu F_y. \quad (9)$$

From Eq.(5), these are given by

$$F_x = \begin{bmatrix} 1 & 0 \end{bmatrix} X \dot{q}_- \quad (10)$$
$$F_y = \begin{bmatrix} 0 & 1 \end{bmatrix} X \dot{q}_-. \quad (11)$$

By using Eq.(10), Eq.(11) and expanding Eq.(9), we can rewrite it by

$$\begin{bmatrix} \begin{bmatrix} 0 & 1 \end{bmatrix} X \\ \begin{bmatrix} 1 & \mu \end{bmatrix} X \\ \begin{bmatrix} -1 & \mu \end{bmatrix} X \end{bmatrix} \dot{q}_- \geq 0. \quad (12)$$

C. Impact Condition

Let ΔT_m denote the impulse of each links, and take it as the effect of the landing impact. It is given by

$$\Delta T_m = J_m J_{cdm}(\dot{q}_- - \dot{q}_+)$$
$$= J_m J_{cdm} D^{-1} J_c^T (J D^{-1} J_c^T)^{-1} J_c \dot{q}_-$$
$$= Z \dot{q}_- \quad (13)$$

where J_m is a transformation matrix of the motor displacement and J_{cdm} is the coupled drive matrix.[6]

We can consider that ΔT_m is caused by the tension of wire. Hence the necessary condition of the impulse of the landing impact is given by

$$|S \Delta T_m| \leq T_{limit} \quad (14)$$

where

$$S = \begin{bmatrix} 0 & 0 & 1 & 0 & 0 \\ 0 & 0 & 0 & 1 & 0 \end{bmatrix} \quad (15)$$

is a matrix to only deal with the links of driven by wires.

Thus the impact condition is formed by

$$\begin{bmatrix} -SZ \\ SZ \end{bmatrix} \dot{q}_- \geq \begin{bmatrix} -T_{limit} \\ -T_{limit} \end{bmatrix}. \quad (16)$$

D. Angular Momentum Condition

By the law of conservation of angular momentum, motions of links in the air have no effect on angular momentum, namely the angular momentum before the impact should satisfy

$$I_g(q) \dot{q}_- = L_d \quad (17)$$

where $I_g(q)$ is the inertia moment of the center of gravity(CG).

We also consider the angular momentum of the landing point after the impact to be small and clockwise rotation in order to obtain the much control time after the landing, namely it is given by

$$L_f \geq I_\ell(q) \dot{q}_+ \geq 0 \quad (18)$$

where $I_\ell(q)$ is the inertia moment of the landing point.

Thus the angular momentum condition is formed by

$$\begin{bmatrix} I_\ell(q) Y \\ -I_\ell(q) Y \end{bmatrix} \dot{q}_- \geq \begin{bmatrix} 0 \\ -L_f \end{bmatrix}. \quad (19)$$

E. Velocity Condition

If we don't consider the effect of the loss of the air friction and so on, actually these are much smaller than in this case, the velocity of CG at the landing (V_{x_l}, V_{y_l}) is given by

$$V_{x_l} = V_{dx} \quad (20)$$
$$V_{y_l} = -\sqrt{V_{dy}^2 + 2gh_g} \quad (21)$$

where g is acceleration of gravity and h_g is the distance of falling of CG.

By the derivative of CG, the relationship between the velocity and generalized coordinates just before the impact is derived as

$$\begin{bmatrix} W_x \\ W_y \end{bmatrix} \dot{q}_- = \begin{bmatrix} V_{x_l} \\ V_{y_l} \end{bmatrix} \quad (22)$$

where

$$W_x = \frac{1}{M} \begin{bmatrix} M & 0 & \alpha_x + \beta_x + \gamma_x & \beta_x + \gamma_x & \gamma_x \end{bmatrix} \quad (23)$$

$$W_y = \frac{1}{M} \begin{bmatrix} 0 & M & \alpha_y + \beta_y + \gamma_y & \beta_y + \gamma_y & \gamma_y \end{bmatrix} \quad (24)$$

$$M = m_1 + m_2 + m_3 \quad (25)$$
$$\alpha_x = (m_1 r_1 + m_2 l_1 + m_3 l_1) \sin \theta_1 \quad (26)$$
$$\alpha_y = (m_1 r_1 + m_2 l_1 + m_3 l_1) \cos \theta_1 \quad (27)$$
$$\beta_x = (m_2 r_2 + m_3 l_2) \sin(\theta_1 + \theta_2) \quad (28)$$
$$\beta_y = (m_2 r_2 + m_3 l_2) \cos(\theta_1 + \theta_2) \quad (29)$$
$$\gamma_x = m_3 r_3 \sin(\theta_1 + \theta_2 + \theta_3) \quad (30)$$
$$\gamma_y = m_3 r_3 \cos(\theta_1 + \theta_2 + \theta_3). \quad (31)$$

F. LCP Formulation

If we let $x_2 = \dot{q}_-$ and use the relationship of Appendix, then we can formulate the problem using Eq.(7), (12), (16), (17), (19), (22) as LCP given by

$$H_{22} = Y^T D(q) Y \quad (32)$$

$$A_{12} = \begin{bmatrix} \begin{bmatrix} 0 & 1 \end{bmatrix} X \\ \begin{bmatrix} 1 & \hat{\mu} \end{bmatrix} X \\ \begin{bmatrix} -1 & \hat{\mu} \end{bmatrix} X \\ -SZ \\ SZ \\ I_\ell(q) Y \\ -I_\ell(q) Y \end{bmatrix} \in R^{9 \times 5} \quad (33)$$

$$A_{22} = \begin{bmatrix} I_g(q) \\ W_x \\ W_y \end{bmatrix} \in R^{3 \times 5} \quad (34)$$

$$b_1 = \begin{bmatrix} 0 \\ 0 \\ 0 \\ -T_{limit} \\ T_{limit} \\ 0 \\ -L_f \end{bmatrix} \in R^{9 \times 1} \quad (35)$$

$$b_2 = \begin{bmatrix} L_{cg} \\ V_{x_l} \\ V_{y_l} \end{bmatrix} \in R^{3 \times 1}. \quad (36)$$

And the other matrices are obviously zero matrix.

IV. PARAMETER ESTIMATION

It is difficult to measure a friction coefficient accurately which is usually decided by the material of the landing place. And if we can select the material freely, we have to know the necessary friction coefficient to realize landing. In this study, we consider a way of estimating an unknown parameter $\hat{\mu}$ using a feature of LCP that gives the Lagrange multiplier simultaneously and the relationship between Lagrange multiplier and sensitivity of the constraints system.[9]

If μ is known, Eq. (12) must hold. That is

$$\begin{cases} -F_y \leq 0 \\ -F_x - \hat{\mu} F_y \leq 0 \\ F_x - \hat{\mu} F_y \leq 0. \end{cases} \tag{37}$$

These equations are formed by

$$g(x) \leq \phi \tag{38}$$

where

$$g(x) = [-F_y, \; -F_x - \hat{\mu} F_y, \; F_x - \hat{\mu} F_y]^T \tag{39}$$
$$\phi = [0,0,0]^T. \tag{40}$$

Now, let x^* denote the solution, x_i denote the candidate of the ith iterative solution, J_i denote the performance index of x_i, y_i denote the Lagrange multiplier that have a duality for the constraints, $\hat{\mu}_i$ denote the ith iterative estimation of μ. And x_i hold the constraints and is given by

$$g(x_i) \leq \theta. \tag{41}$$

We, furthermore, assume that $\hat{\mu}_i$ does not slip. Namely

$$\hat{\mu}_i \leq \mu. \tag{42}$$

In this setting, we consider to update $\hat{\mu}_i$. We also assume the situation that holds

$$\begin{bmatrix} 0 & 1 & 0 \\ 0 & 0 & 1 \end{bmatrix} \lambda_i \neq 0, \tag{43}$$

namely we assume that the second and/or third constraints make sense.

And we assume that (x_i, λ_i) is neighborhood of (x^*, λ^*) and sensitivity analysis is valid approximately. That is

$$\nabla_d f(x_i(c,d))|_{0,0} \doteq -\lambda_i. \tag{44}$$

Let J_{i+1} is the performance index value when we update x_i and keep $\hat{\mu}_i$. We then define the variation of the performance index value

$$\Delta J_i \triangleq J_{i+1} - J_i \leq 0. \tag{45}$$

If $\Delta \hat{\mu}_i > 0$ is a variation of $\hat{\mu}_i$, the the constraints are given by

$$\begin{cases} -F_{y_i} \leq 0 \\ -F_{x_i} - (\hat{\mu}_i + \Delta \hat{\mu}_i) F_{y_i} \leq 0 \\ F_{x_i} - (\hat{\mu}_i + \Delta \hat{\mu}_i) F_{y_i} \leq 0 \end{cases} \tag{46}$$

We can rewrite it by

$$g(x_i) \leq \begin{bmatrix} 0 \\ F_{y_i} \\ F_{y_i} \end{bmatrix} \Delta \hat{\mu}_i \tag{47}$$

If we let d used in sensitivity analysis be

$$d = \begin{bmatrix} 0 \\ F_{y_i} \\ F_{y_i} \end{bmatrix} \Delta \hat{\mu}_i \tag{48}$$

then, the expected value of J_i obtained by changing d is given by

$$\lambda_i^T d = \lambda_i^T \begin{bmatrix} 0 \\ F_{y_i} \\ F_{y_i} \end{bmatrix} \Delta \hat{\mu}_i \leq 0 \tag{49}$$

On the other hand, ΔJ_i was defined as an improvement of the performance index value when we keep $\hat{\mu}_i$. And so the conservative way of deciding $\Delta \hat{\mu}_i$ is given by

$$\Delta \hat{\mu}_i = \frac{\Delta J_i}{\lambda_i^T \begin{bmatrix} 0 \\ F_{y_i} \\ F_{y_i} \end{bmatrix}}. \tag{50}$$

After all, we iterate the simulation using updated estimation given by

$$\hat{\mu}_{i+1} = \hat{\mu}_i + \Delta \hat{\mu}_i. \tag{51}$$

And if the robot slips with $\hat{\mu}_i$, then $\hat{\mu}_i$ will not update and only x_i will be updated. An other way of updating when $\hat{\mu}_i$ is not proper is updating $\hat{\mu}_{i+1}$ using the binary searching.

V. NUMERICAL EXAMPLES

A. Landing Control

In this simulation, we give the posture of the robot $[\theta_1, \theta_2, \theta_3] = [60, 5, 90]$ (deg), initial velocity of CG $(0.840, 1.46)$, initial angular momentum of CG $L_d = 0.60$, and we iterate PC method with various friction coefficient μ until $i = 20000$. We also give the limit impulse of each wire as $\Delta T_{limit} = 10.0$.

Additionally, M is numerically unstable by its structure, it is modified as

$$D \rightarrow D + \epsilon I. \tag{52}$$

The results are shown by Fig.3. And the converged values are shown by Table.II.

TABLE II
CHANGE OF THE OPT. VELOCITIES W.R.T. FRICTIONS

μ	converge	$\dot{x}$	$\dot{y}$	$\dot{\theta}_1$	$\dot{\theta}_2$	$\dot{\theta}_3$
0.001	12618	2.93	3.92	-21.8	16.9	5.32
0.01	8505	2.94	3.92	-21.8	16.9	5.29
0.1	5904	2.93	3.84	-21.4	16.5	5.13
0.3	6013	2.98	3.77	-21.3	16.4	4.59
0.7	6940	3.04	3.68	-21.1	16.3	3.87

TABLE III
CONVERGED OPTIMAL VELOCITY

$\dot{x}$	$\dot{y}$	$\dot{\theta}_1$	$\dot{\theta}_2$	$\dot{\theta}_3$
3.30	4.01	-23.4	14.8	2.56

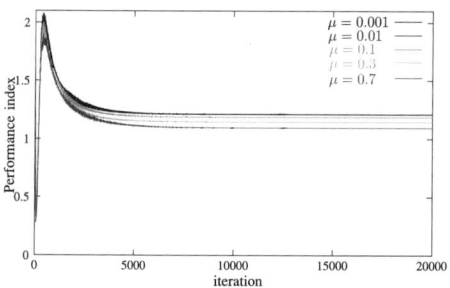

Fig. 3. Convergence of the performance index

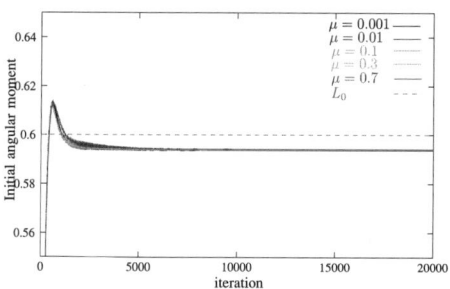

Fig. 4. Change of resultant angular moment w.r.t. μ

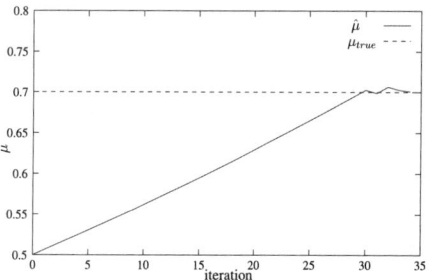

Fig. 5. Estimation of friction coefficient

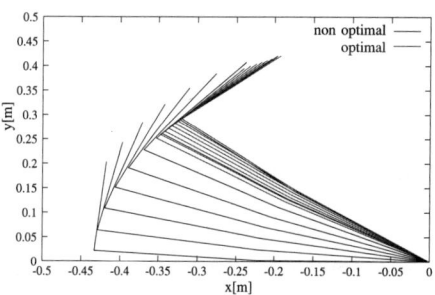

Fig. 6. Stick diagram of optimal landing

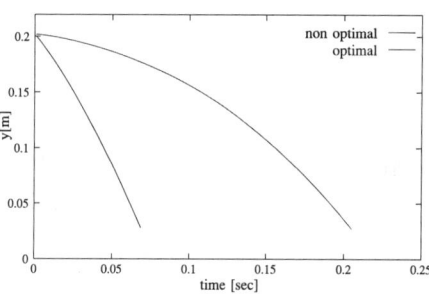

Fig. 7. Time response of the body height

It is apparent from the figure that kinetic energy generated by the landing impact can be small provided μ is large.

In the figure, in all cases, kinetic energy is grown at first then converge. But Fig.4 show that angular momentum condition is not satisfied in some steps, so these steps make no sense of optimized solutions. In actuality, it is obviously the iteration numbers when the simulation is converged as shown by Table.II. And the offset between the true value and iteration results in Fig.4 are the computational residual caused by Eq. (52).

B. Parameter Estimation

In this simulation, we assume that the true value of the friction coefficient $\mu_{true} = 0.7$, and we compute the estimation value by each iterations. The results is shown in Fig.5. It is shown that the estimation value is almost same the true value. And it is confirmed that this method can estimate the friction coefficient.

Additionally, we can expect the performance index value is improved using the method as much as updating by sensitivity. In actuality, performance index value is 1.0865 using the method, it is smaller than the value in the case without the optimization. The converged values are shown in Table.III.

In Fig.6 and Fig.7, we illustrate a simulation that shows the posture of the robot after landing impact using the optimized value that satisfies the constraints and an unoptimized value in order to compare the results.

It is apparent from the figures that in the optimized case the robot falls down slower than the unoptimized case.

VI. CONCLUSION

In this paper, we have proposed the motion design which minimizes kinetic energy with various constraints to solve LCP. We also propose a way of estimating the unkown friction coefficient using a feature of LCP that gives the Lagrange multiplier simultaneously and the relationship between Lagrange multiplier and sensitivity of the constraints system. The advantages of the method is what it accomplish optimizing the landing motion and estimating the unkown friction coefficient simultaneously.

Additionally, we can also consider some other possibility of the proposed simultaneous estimation method as follows:

- PC method is an algorithm that the norm of $\|u - u^*\|$ is monotonically decreasing, then the method can be applied in the case of that Ω^* have a little change. Namely, we can consider the method is robust for a model error.
- If the hardware has a mechanism that can tune the strength of wires, it is possible to optimize this mechanism in order to use our proposed method. Applying it to a human gymnast, we seems that he harden him muscle which needs to perform acrobatic motion until accomplishing it.

The future work is to confirm the efficiency of the method and to realize landing experimentally.

This work was supported in part by the Grant-in-Aid for C.O.E. Research #09CE2004 of the Ministry of Education, Science and Culture, Japan.

VII. REFERENCES

[1] M.Spong, "The Swing up control problem foe the Acrobot", *IEEE Control Systems Magazine*, 15-2, 1995, pp.49-55.

[2] G.Boone, "Minimum-time Control of the Acrobot", *Proc. of the 1997 IEEE Int.Conf. on Robotics and Automation*, pp.3281-3287.

[3] T.Nam, T.Mita, I.Pantelidis and M.Yamakita, "Swing-up Control and Singular Problem of a Acrobot System ", *Journal of Robotics Soc. Japan*, vol.20, no.1, Jan. 2002, pp.85-88.

[4] Y.Michitsuji, H.Sato,and M.Yamakita, "Giant swing via forward upward circling of the Acrobat-robot", *Proc. of 2001 American Control Conference*, pp.3262-3267.

[5] M.Yamakita and T.Yonemura, "Stabilization of Acrobat Robot in Upright Position on a Horizontal Bar", *Proc. of the 2002 IEEE Robotics and Automation*, pp.3093-3098

[6] Shigeo Hirose and Mikio Sato. "Coupled Drive of the Multi-DOF Robot", *Proc. of the 1989 IEEE Int.Conf. on Robotics and Automation*, pp.1610-1616.

[7] R.M. Murray, Z.Li and S.S.Sastry, *A Mathematical Introduction to ROBOTIC MANIPULATION*, CRC Press; 1993, Chapter 5.

[8] Bingsheng He, "A Projection and Contraction Method for a Class of Linear Complementarity Problems and Its Application in Convex Quadratic Programming", *Appl. Math. Optim.*,vol 25, 1992, pp.247-262.

[9] D.G.Luenberger, *Linear and Nonlinear Programming 2nd Edition*, Addison Wesley; 1989, Chapter 10.

APPENDIX

If x^* is a solution of CQP, then there exists a y^* such that $u^* = (x^*, y^*)$ is a solution of DQP. Let

$$\Omega^* = \{u^* = (x^*, y^*) | x^* \text{ solves (CQP)}, \\ (x^*, y^*) \text{ solves (DQP)}\} \quad (53)$$

Then a necessary and sufficient condition for $u = (x, y) \in \Omega^*$ is given by

$$\begin{cases} A_{11}x_1 + A_{12}x_2 \geq b_1, \\ A_{21}x_1 + A_{22}x_2 = b_2, \\ x_1 \geq 0, \\ H_{11}x_1 + H_{12}x_2 - A_{11}^T y_1 - A_{21}^T y_2 + c_1 \geq 0, \\ H_{12}x_1 + H_{22}x_2 - A_{12}^T y_1 - A_{22}^T y_2 + c_2 = 0, \\ y_1 \geq 0, \\ x_1^T(H_{11}x_1 + H_{12}x_2 - A_{11}^T y_1 - A_{21}^T y_2 + c_1) \\ \qquad = 0, \\ y_1^T(A_{11}x_1 + A_{12}x_2 - b_1) = 0. \end{cases}$$
(54)

Thus if

$$M = \begin{bmatrix} H & -A^T \\ A & 0 \end{bmatrix}, \quad q = \begin{bmatrix} c \\ -b \end{bmatrix} \quad u = \begin{bmatrix} x \\ y \end{bmatrix} \quad (55)$$

then the problem can be described as

$$u \geq 0, \quad Mu + q \geq 0, \quad u^T(Mu + q) = 0. \quad (56)$$

This is obviously a LCP.

Cable-Suspended Planar Parallel Robots with Redundant Cables: Controllers with Positive Cable Tensions

So-Ryeok Oh, Ph.D. Student
Sunil K. Agrawal, Ph.D., Professor

Mechanical Systems Laboratory, Department of Mechanical Engineering
University of Delaware, Newark, DE 19716, U.S.A.
oh, agrawal@me.udel.edu

Abstract — Cable-suspended robots are structurally similar to parallel actuated robots but with the fundamental difference that cables can only pull the end-effector but not push it. From a scientific point of view, this feature makes feedback control of cable-suspended robots lot more challenging than their counterpart parallel-actuated robots. In the case with redundant cables, feedback control laws can be designed to make all tensions positive while attaining desired control performance. This paper describes these approaches and their effectiveness is demostrated through simulations of a three degree-of-freedom cable suspended robots with four, five, and six cables.

1 Introduction

In the last decades, robots have made tremendous in-roads into industries for manufacturing and assembly. However, for long reach robotics such as inspection and repair in shipyards and airplane hangars, application of robotics is still in its infancy([1],[2]). Conventional robots with serial or parallel structures are impractical for these applications since the workspace requirements are higher by three to four orders of magnitude than what the conventional robots can provide. However, cables have the unique property - they carry loads in tension but not in compression[3].

This paper presents approaches to handle input tension constraints using the robot's redundancy([4],[5]). We use the tension in the redundant cables to satisfy the positive input constraints. At a point in the state space, positive tension constraints in the input can be represented by a set of linear inequalities in the input tensions. The feasible space enclosed by these linear inequalities can be characterized mathematically and sketched graphically for up to six cables. However, selection of input tensions point-wise in the state space may not always provide continuity of the tension trajectories. Due to this reason, we also propose a second method of trajectory planning which uses trajectory parameterization in conjunction with collocation to ensure smooth input tensions during the path [6].

The organization of this paper is as follows: Section 2 describes the kinematic and dynamic equations of the robot. Section 3 outlines a method for control based on feedback linearization. Section 4 provides the details of obtaining the feasible region for the cases with 4, 5, and 6 cables. In addition, the constraint problem is reformulated for linear and quadratic programming. Section 5 shows a method to find the globally continuous tensions by trajectory planning. The results of simulation are presented in Section 6.

We have fabricated an experiment test-bed of a planar cable suspended robot with four cables as shown in Fig. 1. In this cable robot, the suspension points of the cables in the ground frame and attachment points of the cable in the end-effector frame can be changed. Due to presence of four cables, it is a redundantly actuated system and will provide a setup for validation of the ideas described in this paper.

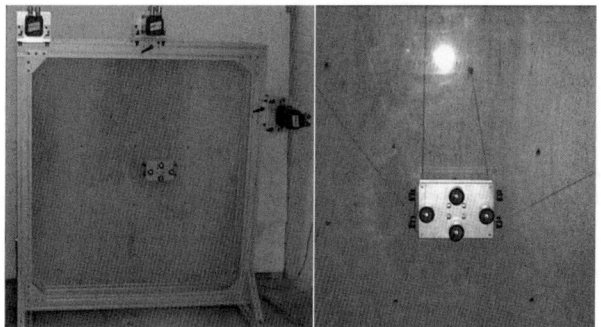

Figure 1: A prototype of a planar cable robot with four cables fabricated at University of Delaware. The fourth cable and the motor are located at the back and are not visible in the front view.

2 System Dynamic Model

Our model of a planar cable robot consists of a moving platform (MP) that is connected by n cables to a base platform shown in Fig. 2. The cable i is connected to MP as shown in Fig. 2. M is the center of the MP. The cable separation angles on MP is denoted by α_i. An inertial reference frame $F_0(XY)$ is located at 0 and a moving reference frame $F_M(xy)$ is located on MP at its center of mass M. The orientation of MP is specified by θ_e. The origin of F_M is given by a vector from 0 to M with x_e and y_e as its components.

2.1 Cable Kinematics and Statics

The position vector of point a_i with respect to F_M is written as

$$\begin{bmatrix} b_i c\alpha_i & b_i s\alpha_i \end{bmatrix}^T. \qquad (1)$$

where c and s stand for cos and sin, respectively and b_i is the distance between points M and a_i. The transformation matrix of frame F_M with respect to frame F_0 can be written as

$$^0T_M = \begin{bmatrix} c\theta_e & -s\theta_e & x_e \\ s\theta_e & c\theta_e & y_e \\ 0 & 0 & 1 \end{bmatrix}. \quad (2)$$

Therefore, the position vector of points a_i with respect to F_0 is

$$\begin{bmatrix} ^0r_i \\ 1 \end{bmatrix} = {}^0T_M \begin{bmatrix} ^Mr_i \\ 1 \end{bmatrix}, i = 1 \cdots n. \quad (3)$$

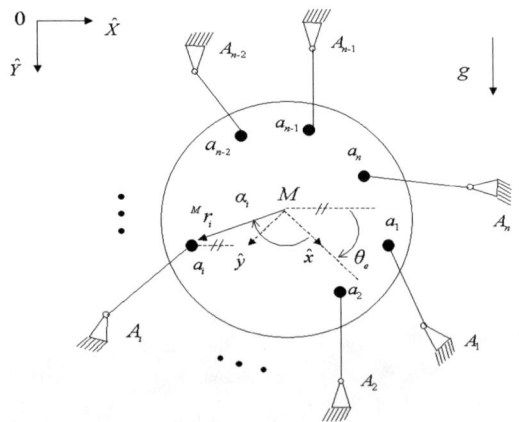

Figure 2: A sketch of the cable system along with geometric parameters for the robot with n cable.

Upon substitution of ^0T_M from Eq.(2) into Eq.(3), one leads to

$$^0r_i = \begin{bmatrix} x_e + b_i \cdot c\theta_e c\alpha_i - b_i \cdot s\theta_e s\alpha_i \\ y_e + b_i \cdot s\theta_e c\alpha_i + b_i \cdot c\theta_e s\alpha_i \end{bmatrix}, i = 1 \cdots n. \quad (4)$$

Moreover, the position vector of suspension point A_i of cable i with respect to reference point 0 is written as

$$^0p_i = \begin{bmatrix} d_i \\ h_i \end{bmatrix}, i = 1 \cdots n. \quad (5)$$

Hence, the vector $\overrightarrow{a_iA_i}$ for cable i is

$$l_i = {}^0p_i - {}^0r_i = \begin{bmatrix} l_{ix} \\ l_{iy} \end{bmatrix}$$

$$= \begin{bmatrix} d_i - x_e - b_i \cdot c\theta_e c\alpha_i + b_i \cdot s\theta_e s\alpha_i \\ h_i - y_e - b_i \cdot s\theta_e c\alpha_i - b_i \cdot c\theta_e s\alpha_i \end{bmatrix}$$

$$i = 1 \cdots n. \quad (6)$$

The static equilibrium equation of MP can be used to obtain the forces in the cables.

$$\begin{array}{ll} \sum F_x = 0 & \sum_1^n T_i c\theta_i = 0 \\ \sum F_y = 0 \Rightarrow & \sum_1^n T_i s\theta_i + \overline{m}g = 0 \\ \sum M_z = 0 & \sum_1^n T_i s_i = 0 \end{array} \quad (7)$$

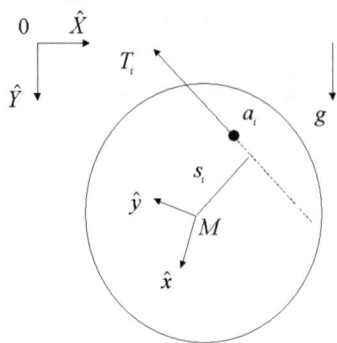

Figure 3: A sketch of parameter s_i.

where

$$c\theta_i = \frac{l_{ix}}{\| l_i \|}, s\theta_i = \frac{l_{iy}}{\| l_i \|}, i = 1 \cdots n$$

and s_i is the normal distance between M and the cable axis i and can be expressed using Fig. 3 as $s_i = b_i \cdot s(\theta_e + \alpha_i - \theta_i)$. Eqs.(7) can be written in matrix form as

$$Au = F \quad (8)$$

where

$$A = \begin{bmatrix} \frac{l_{1x}}{\|l_1\|} & \cdots & \frac{l_{nx}}{\|l_n\|} \\ \frac{l_{1y}}{\|l_1\|} & \cdots & \frac{l_{ny}}{\|l_n\|} \\ s_1 & \cdots & s_n \end{bmatrix} \quad (9)$$

$$u = \begin{bmatrix} T_1 & T_2 & \cdots & T_n \end{bmatrix}$$
$$F = \begin{bmatrix} F_x & F_y & M_z \end{bmatrix}. \quad (10)$$

2.2 System Dynamics

During the motion,

$$F = \begin{bmatrix} \overline{m}\ddot{x} \\ \overline{m}(\ddot{y}_e - g) \\ I_z\ddot{\theta}_e \end{bmatrix}, \quad (11)$$

where $\overline{m}$ is the mass and I_z is the moment of inertia of the end-effector about its center of mass along $\hat{Z}$. The equations of motion can be written alternatively in the following general form

$$D\ddot{x} + \mathcal{G} = A(x)u \quad (12)$$

where D is the inertia matrix for the system and g is the vector of gravity terms. Their expressions are

$$D = \begin{bmatrix} \overline{m} & 0 & 0 \\ 0 & \overline{m} & 0 \\ 0 & 0 & I_z \end{bmatrix}, \mathcal{G} = \begin{bmatrix} 0 \\ -\overline{m}g \\ 0 \end{bmatrix}$$

and $x = [x_e, y_e, \theta_e]^T$. The above dynamic model is valid only for $u \geq 0$, i.e., the cables are in tension. A positive tension implies that the cable is pulling the attachment point of the end-effector.

3 Feedback Controller

In this section, we consider controllers based on feedback linearization theory to asymptotically stabilize the system to $x_d(t)$, while satisfying the property $u(t) \geq 0$, $t > 0$. We define $\tilde{x}(t) = x(t) - x_d(t)$. For the system given in Eq. (12), with a feedback law of the form

$$A(x)u = Dv + \mathcal{G}, \quad (13)$$

the system dynamics becomes

$$\ddot{x} = v. \quad (14)$$

With the choice $v = \ddot{x}_d - K_p\tilde{x} - K_d\dot{\tilde{x}}$, where K_p and K_d are positive diagonal matrices, we can ensure that $x(t)$ will asymptotically track $x_d(t)$. The solution of Eq.(13) depends on the number of cables. With three cables, Eq.(13) has 3 equations in 3 unknowns. If the three equations are linearly independent, there is a unique solution for the problem. For more than three cables, Eq.(13) is an underdetermined system of equations and has many solutions if AA^T is invertible. In this case, the general solution for Eq.(13) can be written as

$$u = \bar{u} + N(A)m. \quad (15)$$

Here, $\bar{u}$ is the minimum norm solution of Eq.(13) derived using the pseudo inverse of matrix A and is given by

$$\bar{u} = A^T(AA^T)^{-1}\left[D(\ddot{x}_d - K_p\tilde{x} - K_d\dot{\tilde{x}}) + \mathcal{G}\right]. \quad (16)$$

Here, $N(A)$ is the null space or kernel of matrix A and m is a $(n-r)$ dimensional underdetermined vector, where r is the rank of matrix A and n is the number of cables.

4 Pointwise Feasible Region

On using the input constraint $u(t) \geq 0$ for all the time $t > 0$, the resulting condition is

$$\bar{u} + N(A)m \geq 0. \quad (17)$$

Since $A(x)$ is nonlinear in x, it is hard to get a solution that is globally valid in the state space. However, it is possible to get the solution at a specific point x in the state space. It is clear from Eq.(17) that a feasible solution at a specific x is characterized by a convex region bounded by n linear inequalities on the elements of $m_i, i = 1, \cdots, k$, where k is the number of linearly independent columns or rank of $N(A)$. In the following discussions, we assume that matrix A has full row rank of 3 and the null space has a dimension $(n-3)$.

4.1 One extra cable

With one extra cable, the four linear inequalities in m_1 become

$$\begin{bmatrix} \bar{u}_1 \\ \bar{u}_2 \\ \bar{u}_3 \\ \bar{u}_4 \end{bmatrix} + \begin{bmatrix} n_{11} \\ n_{21} \\ n_{31} \\ n_{41} \end{bmatrix} m_1 \geq 0. \quad (18)$$

So, the feasible region F_A of m_1 is described by the common interval bounded by four linear inequalities shown in Fig. 4(a). Here p_i is the solution point when each component of Eq.(18) is made to be an equality. If F_A is empty, the tension constraints can not be met.

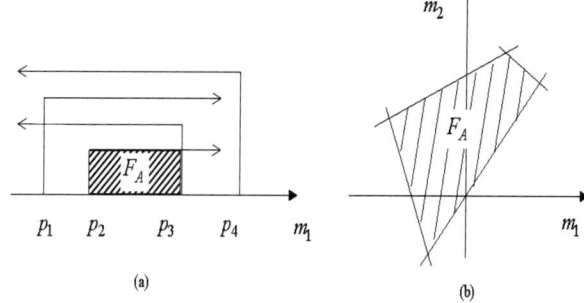

Figure 4: A sketch of feasible region for m with (a) 4 cables and (b) 5 cables.

4.2 Two extra cable

Two extra cables result in 2-dimensional vector m and five linear inequalities given by

$$\begin{bmatrix} \bar{u}_1 \\ \bar{u}_2 \\ \bar{u}_3 \\ \bar{u}_4 \\ \bar{u}_5 \end{bmatrix} + \begin{bmatrix} n_{11} & n_{12} \\ n_{21} & n_{22} \\ n_{31} & n_{32} \\ n_{41} & n_{42} \\ n_{51} & n_{52} \end{bmatrix} \begin{bmatrix} m_1 \\ m_2 \end{bmatrix} \geq 0. \quad (19)$$

To determine the feasible area, the main computation steps are: (i) find the intersection points p_i of every pair of two equations formed by converting the inequalities to equalities, (ii) check if the solution satisfies all the remaining inequalities. The computation of p_i requires solving sets of $^nC_2$ linear equations in two variables, followed by inequality checks to determine feasibility. A typical feasible region is shown in Fig. 4(b).

4.3 Three extra cable

Three extra cables result in 3-dimensional vector m and six linear inequalities given by

$$\begin{bmatrix} \bar{u}_1 \\ \bar{u}_2 \\ \bar{u}_3 \\ \bar{u}_4 \\ \bar{u}_5 \\ \bar{u}_6 \end{bmatrix} + \begin{bmatrix} n_{11} & n_{12} & n_{13} \\ n_{21} & n_{22} & n_{23} \\ n_{31} & n_{32} & n_{33} \\ n_{41} & n_{42} & n_{43} \\ n_{51} & n_{52} & n_{53} \\ n_{61} & n_{62} & n_{63} \end{bmatrix} \begin{bmatrix} m_1 \\ m_2 \\ m_3 \end{bmatrix} \geq 0. \quad (20)$$

It's feasible region is determined using a procedure similar to the case with two extra cables. One needs to find the intersection points p_i of every set of three equations formed by converting the inequalities to corresponding equalities and check if the solution satisfies all the remaining inequalities. The feasible domain is typically a volume in 3-dimensionsl space of (m_1, m_2, m_3).

One can generalize this method for more cables. As expected, the computations will increase with the number of extra cables. If all components of the minimum norm solution in Eq.(16) are greater than zero, one may not need to compute the feasible region. In case, the minimum norm solution results in some negative tensions, a criteria is needed to choose a point inside the feasible region. Several criteria are discussed in the upcoming sections based on linear programming and quadratic programming. In addition, we take the criteria of selecting a point m in the feasible area with minimum norm, i.e., minimum tensions in the extra cables that make all the tensions positive. We label this selection criteria FA.

4.4 Linear Programming

A Linear Programming(LP) can be described as follows:

$$minimize \quad cx \quad (21)$$

$$subject\ to \quad Ax \leq b$$
$$x_l \leq x \leq x_u \quad (22)$$

where x is the vector of decision variables, A is a matrix with constant coefficients, c and b are constant vectors. It is well known that the solution of a LP problem is one of the vertices of the feasible area. Geometrically, this solution can be obtained by shifting the level set $cx = k$ parallel to itself along the direction of decreasing k until the solution becomes infeasible. Essentially, this results in a vertex solution. The LP cost to be minimized can be taken as $m_1 + m_2 + \cdots + m_k$.

4.5 Quadratic Programming

The structure of constraints in a quadratic programming(QP) is similar to that of LP. The objective function is allowed to contain a quadratic term such as

$$minimize \quad x^T H x \quad (23)$$

where H is a positive definite matrix.

With the given structure of the QP problem, the optimal solution is obtained as the point where the level set, an ellipsoid $x^T H x = k$ centered about the origin touches the feasible region for the smallest value of k. The feasible solution may not lie on a vertex.

5 Globally Continuous Solution

The solution of Eq.(17), through characterization of the feasible space in Section 4, may result in tensions which are pointwise feasible but discontinuous in time. From an experimentation point of view, it is important for the solution to be continuous to avoid instability.

The objective is to develop trajectories $x(t)$ over $[0, t_f]$ to steer the system between given boundary conditions and satisfy the positive input constraints. One can choose $x(t)$ to have the following form:

$$x(t) = \Phi_0(t) + \sum_{i=1}^{M} P_i \phi_i(t). \quad (24)$$

The form of $x(t)$ is made admissible in the following way: (i) $\Phi_0(t)$ are 3-dimension vector functions that satisfy boundary conditions of $x(t)$ and its derivatives, (ii) $\phi_i(t)$ are scalar functions that satisfy the boundary conditions in their homogeneous form, and (iii) P_i are 3-dimension vectors of constant

coefficients. Here, $\phi_i(t)$ are chosen such that together with $\Phi_0(t)$, they span a complete set. When the form of $x(t)$ from Eq.(24) is substituted in Eq.(17), the inequalities becomes

$$u = \bar{u}(t, P_i) + N(A(t, P_i))m \geq 0. \quad (25)$$

Eq.(25) must be valid at all time during $[0, t_f]$, i.e., it represents an infinite number of constraints on the coefficients $P_1, \cdots, P_M, m_1, \cdots, m_k$. A number of schemes may be used to transform these infinite constraints into a finite number of constraints. We use a collocation grid in time to form a finite number of nonlinear inequality constraints in the coefficients $P_1, \cdots, P_M, m_1, \cdots, m_k$.

A finite collocation grid is selected within $[0, t_f]$. At each collocation point, the n constraints of Eq.(25) are satisfied. If needed, one can ensure satisfaction of the constraints between the collocation points by bounding a finite number of derivatives of the constraints at the collocation points. On choosing $N+2$ collocation points at $t_0, t_1, \cdots, t_N, t_f$ such that $t_0 < t_1 < \cdots < t_N < t_f$ we get a total of $(N+2)n$ inequalities on the mode coefficients $P_1, \cdots, P_M, m_1, \cdots, m_k$ which have the following form

$$u_j = \bar{u}(t_j, P_m) + N(A(t_j, P_m))m_l \geq 0$$
$$j = 1, \cdots, N+2, m = 1, \cdots, M, l = 1, \cdots, k. \quad (26)$$

The parameters $P_1, \cdots, P_M, m_1, \cdots, m_k$ can be solved using Matlab 6 routines such as $fmincon$, designed to solve nonlinear programming problems. This method allows computation of continuous trajectories for control.

6 Simulation

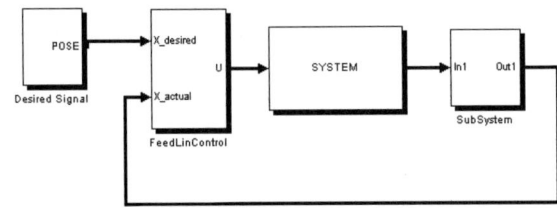

Figure 5: A Simulink block diagram for the cable system.

A simulation of the cable robot was developed in Matlab Simulink to verify the feedback controller concepts outlined in the previous sections. Fig. 5 shows the Simulink model for the controllers. The user specifies a desired position and orientation for the end-effector plate. The controller block implements the control law given in Eq.(15). The ouputs of the controller block are the cable tensions. If each tension is positive, it is fed to the system dynamic model. If at least one is negative, the inputs are modified to become positive using the null space contribution by computing the feasible solution using FA, LP, QP.

6.1 Four cable robot

We consider a specific move of the end-effector from $x_0 = [0.48, 0.24, 0°]^T$. Fig. 6 show the graphs for four control meth-

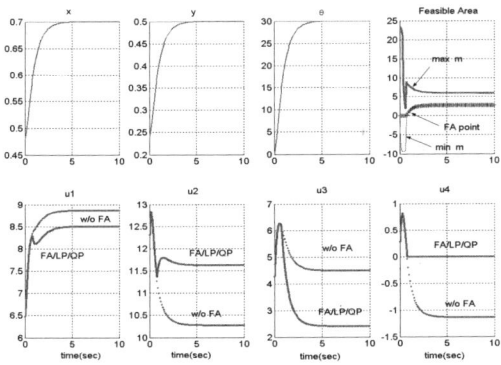
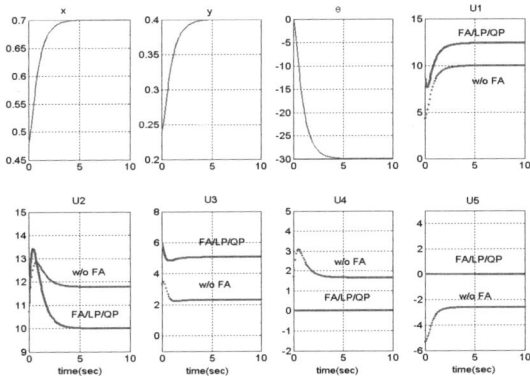

Figure 6: Plots of position, orientation, feasible area, and cable tensions for the 4-cable robot without and with null space components using FA, LP, and QP.

Figure 7: Plots of position, orientation, and cable tensions for the 5-cable robot without and with null space contributions using FA, LP, and QP.

ods: (i) No null space contribution, (ii) null space contribution using FA, i.e., minimum norm of the tension in extra cable, (iii) Linear Programming(LP), and (iv) Quadratic Programming(QP). The first three plots of Fig. 6 show that the motions of the cable robot in position and orientation are exactly the same in all four cases. This is due to the fact that the null space term does not affect the control trajectories and thereby the state trajectories. The fourth plot is the feasible area for m. The dotted line in this plot shows the feasible point selected by the algorithm. If Eq.(16) results in positive tension, the null space contribution is not invoked. Otherwise, the algorithm is invoked to make all tensions positive. The last four plots show the cable tensions. When not using the null space term, the tension in the fourth cable is negative after 0.8 seconds. However, FA, LP, and QP use the available feasible region of the null space to prevent the tension from becoming negative. The tension histories from the algorithm FA, LP, and QP turn out to be the same.

6.2 Five cable robot

For 5 cable robot, the position, orientation, and correspoding tension graphs are shown in Fig. 7. The results are similar to the case with 4 cables. Fig. 8 shows the shape of the feasible region, which has the shape of a triangle at any time instant. The figure also shows the points selected by FA method during the simulation.

6.3 Six cable robot

A 6 cable model was developed to check the effect of redundancy on the feasible region in the null space. As one will expect, the feasible region is a volume at each time instance. We can observe that last two tension graphs have negative values when the null space contribution were not used. Utilizing the null space term, we can satisfy positive input constraints. The plots 4-7 in Fig. 9 show that the tensions of QP are slightly different from those of FA and LP.

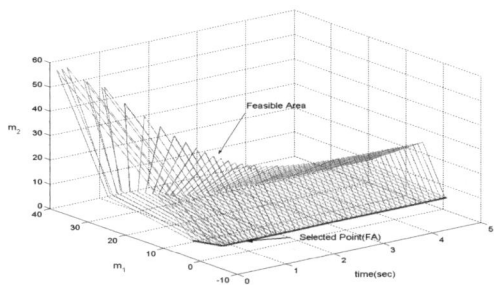

Figure 8: Plots of the feasible region for the null space vector of the 5-cable robot during the trajectory and selected point using FA.

6.4 Continuous Trajectory

The trajectory planning method is to prevent discontinuity in tensions illustrated by the example of a four cable robot. We select the feasible solution of $x(t)$ to have the following form.

$$x(t) = \begin{bmatrix} \Phi_{10} \\ \Phi_{20} \\ \Phi_{30} \end{bmatrix} + \begin{bmatrix} p_{11} \\ p_{12} \\ p_{13} \end{bmatrix} \phi_1 + \begin{bmatrix} p_{21} \\ p_{22} \\ p_{23} \end{bmatrix} \phi_2 \quad (27)$$

where

$$\Phi_{10} = a_0 + a_1 t + a_2 t^2 + a_3 t^3 + a_4 t^3 (t_f - t) + a_5 t^3 (t_f - t)^2$$
$$\Phi_{20} = b_0 + b_1 t + b_2 t^2 + b_3 t^3 + b_4 t^3 (t_f - t) + b_5 t^3 (t_f - t)^2$$
$$\Phi_{30} = c_0 + c_1 t + c_2 t^2 + c_3 t^3 + c_4 t^3 (t_f - t) + c_5 t^3 (t_f - t)^2$$

$$\begin{aligned} \phi_1 &= t^3 (t_f - t)^3 \\ \phi_2 &= t^4 (t_f - t)^3. \end{aligned} \quad (28)$$

The coefficients a_i, b_i, c_i are determined from eighteen conditions, $(x, \dot{x}, \ddot{x}, y, \dot{y}, \ddot{y}, \theta, \dot{\theta}, \ddot{\theta})_0 = (0.48, 0, 0, 0.24, 0, 0, 0°, 0, 0)$ and $(x, \dot{x}, \ddot{x}, y, \dot{y}, \ddot{y}, \theta, \dot{\theta}, \ddot{\theta})_6 = (0.6, 0, 0, 0.4, 0, 0, -20°, 0, 0)$ with $t \in [0, 6]$.

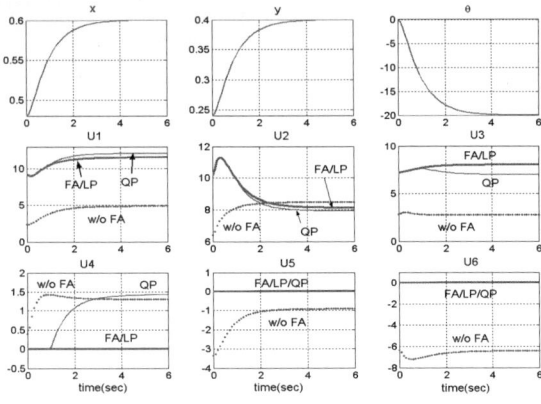

Figure 9: Plots of position, orientation, and cable tensions for the 6-cable robot without and with null space contribution using FA, LP, and QP.

Equally spaced five collocation points were used in the simulation and the objective function was defined by the sum of norms for $x(t)$ and its derivatives. The solutions of p_{ij}, m are $p_{11} = 0.01, p_{12} = -0.01, p_{13} = 0.02, p_{21} = 0.01, p_{22} = 0, p_{23} = 0.01$, and $m = 2.78$. The solid lines in plots 1-3 of Fig. 10 are the planned trajectories which are calculated by the proposed scheme mentioned and are used as desired trajectories of the planar robot. Dotted lines are the actual state trajectories while tracking the desired trajectories starting out from an initial error. Plot 4-8 in Fig. 10 are the tension histories, which show that this method satisfies not only the tension constraints but also results in smooth tensions.

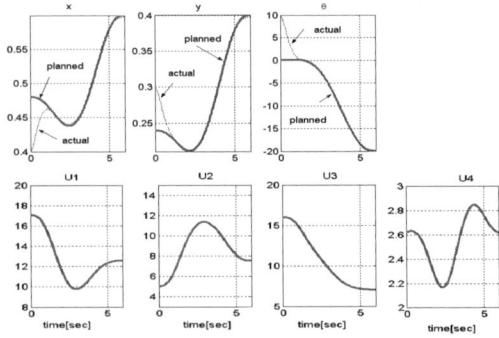

Figure 10: Plots of planned trajectory(solid line), actual states(dotted line) of the positions and orientation, and four tension.

7 Conclusions

This paper presented approaches for control of cable suspended robots with redundant cables to follow prescribed trajectories, while keeping all tensions positive during motion. This was achieved by adjusting the tensions in the extra cables by identifying a feasible space for these tensions. It was shown that this feasible space is a polytope at any given point in the state space and is described by a set of linear inequalities. This polytope was sketched for planar cable robots with one, two, and three extra cables. The feasible space is a line for one extra cable and increases in dimension by one, i.e., becomes an area and a volume as extra cables are added. The performance of the controllers was compared when null space contribution was added from the cables using simple heuristics, linear programming, and quadratic programming. These methods were classified as point-wise since they do not ensure continuity of the cable tensions as the trajectory evolves in time. An alternative method was proposed which uses a global continuous description of the trajectory while satisfying the constraints. Dynamic simulations were presented to show the effectiveness of all these methods with satisfactory results.

8 Acknowledgments

The authors appreciate financial supports of NSF award No. IIS-0117733, NIST MEL award No. 60NANB2D0137, and PTI/NIST award No. AGR20020506.

References

[1] Aghili, F., Buehler, and Hollerbach, J. M., "Dynamics and Control of Direct-Drive Robots with Positive Joint Torque Feedback", *Proceedings of International Conference on Robotics and Automation*, Albuquerque, New Mexico, 1156-1161, 1997.

[2] Albus, J., Bostelman, R., and Dagalakis, N., "The NIST Robocrane", *Journal of Research of National Institute of Science and Technology*, Vol. 97, No. 3, 373-385, 1992.

[3] Abdullah B. Alp, Sunil K. Agrawal, "Cable Suspended Robots: Design, Planning and Control", *Proceedings of International Conference on Robotics and Automation*, Washington, DC, 4275-4280, 2002.

[4] A.A. Maciejewski, C.A. Klein, "Obstacle avoidance for kinematically redundant manipulators in dynamically varying environments", *International Journal of Robotics Research*, Vol 5, No. 4, 109-117, 1985.

[5] Y. Nakamura, H. Hanafusa, "Inverse Kinematics Solutions with Singularity Robustness for Robot Manipulator Control", *Transaction of the ASME Journal of Dynamic System, Measurement, and Control*, Vol. 108, 163-171, 1986.

[6] Faiz, N. and Agrawal, S. K., Murray, R. M., "Trajectory Planning of Differentially Flat Systems with Dynamics and Inequalities", *AIAA Journal of Guidance, Control, ad Dynamics*, Vol. 24, No. 2, 219-227, 2001.

Optimal Motion Planning for Free-Flying Robots

R. Lampariello*, S. Agrawal**, G. Hirzinger*

*Institute of Robotics and Mechatronics
German Aerospace Center (DLR)
82234 Weßling, Germany
Roberto.Lampariello@dlr.de

**Department of Mechanical Engineering
University of Delaware
Newark, DE 19716
Agrawal@me.udel.edu

Abstract—This paper addresses the problem of motion planning for free-flying robots. Full state actuation is considered to allow for large displacements of the spacecraft. Motion planning is formulated as an optimization problem and kinematic as well as dynamic constraints are considered. The chosen optimization criteria are spacecraft actuation and final time. The proposed method allows solutions which do not require any spacecraft actuation for those end goals for which the robot motion is sufficient.

I. INTRODUCTION

This paper deals with the general problem of motion planning for free-flying robots. Much effort has already been dedicated to such a problem, with the dynamic interaction between the robot motion and the base (spacecraft). The non-integrability of the angular momentum conservation law has given rise to interesting path planning problems and solutions ([1], [2], [3] to mention a few). These problems however deal with a particular case of free-flying robot dynamics, for which it is assumed that no external actions are present. For this case, the robot is often termed free-floating as opposed to free-flying. It is however more generally true that a free-flying robot will be subject to spacecraft control actions, for large displacements, as well as to non-negligible orbital disturbances. The free-floating approximation can only be reasonably assumed in some specific situations, where the robot is engaged in *local motions* and *short operations*. Otherwise, the free-floating assumption is not valid. It is perhaps then useful to consider the path planning problem in a more general context.

The first point to note is that in space energy is indeed precious, at least in the form of thruster fuel. The motion planning strategy should then account for this and be in this sense optimal. Since the free-flying operational condition should not necessarily discard the free-floating one - the two should be complementary - the planner should allow solutions to be found for which no base control action is necessary, neglecting orbital disturbances. Furthermore, the final robot-spacecraft configuration should be judiciously chosen, in accordance with the optimality criterion and in relation to the given initial configuration. Of all the possible final robot-spacecraft configurations which result in the desired final state of the robot end-effector in inertial space, the path planning solution should converge to the dynamically optimal one.

The second major point considered here is the fact that the execution of a desired path will necessarily involve some actuator dynamics. A dynamic model of the robot is used together with an optimization routine such that an optimal path can be generated to account for the control bounds. This can ensure that, in the absence of disturbances and modelling simplifications, the desired path is dynamically feasible as well as optimal.

We also point out that a dynamic description of the robot allows the inclusion of models for external actions which are present in Earth orbit. The ETS-VII experiments have in fact shown that these actions (torques) can be significant, at least in Low Earth Orbit [4]. We, however, suggest that these are neglected for the path planning phase and are dealt with by a path tracker which is designed to compensate for them (see [5]).

The modelling of the robot as a fully actuated system also allows to choose between different operational strategies, free-flying (actuation on all spacecraft and robot states), free-floating (actuation on robot states) or attitude controlled (actuation on rotational states of spacecraft and on all robot states). The latter would be necessary, for example, if the attitude of the spacecraft had to be contained within a certain operational window to avoid communication loss with ground. These different strategies can be chosen by simply commanding the desired spacecraft behavior in the desired state variables definition.

The general problem of robot optimal motion planning has been addressed in [6], [7], [8]. These authors address real-time applicability as well as collision avoidance for fixed base robots, with various optimization strategies, including multiple shooting, semi-infinite parameter optimization and polytopic representation of collision constraints. The more specific case of free-flying robots is treated in [9], [10], [11]. The first two, however, only treat the free-floating case while the last addresses the collision avoidance problem in detail.

We present here a first step into the direction of developing a planner for the problem described above. A spatial dynamical model is derived for a free-flying robot with six-degree-of-freedom robot arm and with rheonomically driven joints. An initial and final robot end-effector state (position and orientation) are defined in Cartesian space. The motion planning is then solved as a single shooting problem, with inequality constraints on the joint kinematics and on the actuator dynamics and equality constraints on the final robot end-effector state. Finally, the motion planning is chosen to be done in joint space, rather than in task space, to avoid the problem of dynamic singularities. This does not allow to formulate a Boundary Value Problem, since the final robot configuration is unknown and should derive from the optimal solution.

II. Modelling and Equations of Motion

Consider the free-flying robot shown in figure 1, composed of rigid bodies connected by revolute joints. Every element of the system is characterised by a local frame of reference $\{O^i, \underline{e}^i\}$, placed at the joint connecting it with the previous element along the kinematic chain. If $i = 0$, the frame of reference is relative to the base body (satellite), placed in some arbitrary position within it, while if $i = e$, the frame of reference is relative to the end-effector. The inertial frame is expressed as $\{O^I, \underline{e}^I\}$.

The quantity $\underline{r}^i$, shown in figure 1, is the position vector of body i. The position of the ith revolute joint, whose rotation vector is $\underline{u}^i$, is described by the variable θ^i, which is measured relative to an arbitrary initial reference robot configuration. Vectors $\underline{c}^i$ and $\underline{d}^i$ represent the distance from the joint to the centre of mass and to the following joint of body i.

The equations of motion of the system with rheonomically driven joints, first described in [5] for the planar case, will now be briefly described for the spatial case. These can first be written in descriptor form using the Newtonian-Eulerian formulation, as follows:
Kinematics:

$$\underline{\dot{r}}^i = \underline{v}^i \qquad (1)$$

$$\dot{\mathbf{A}}^i := \tilde{\underline{\omega}}^i \mathbf{A}^i \qquad (2)$$

Dynamics:

$$\frac{d}{dt}\left(m^i \underline{v}^i\right) = \underline{f}^i + \underline{f}_c^i, \qquad (3)$$

$$\frac{d}{dt}\left(\underline{\underline{I}}^i \cdot \underline{\omega}^i\right) = \underline{t}^i + \underline{t}_c^i \qquad (4)$$

Constraints:

$$\underline{r}^{i+1} = \underline{r}^i + \underline{d}^i, \qquad (5)$$

$$\mathbf{A}^{i+1} = \mathbf{B}^{i+1} \mathbf{A}^i \qquad (6)$$

where $\underline{r}^i$ is the absolute position, $\mathbf{A}^i$ is the direction cosine matrix, $\underline{v}^i$ the translational velocity and $\underline{\omega}^i$ the angular velocity of body i relative to the inertial frame (expressed in inertial coordinates in Eq. (2)). $\mathbf{B}^i$ is the relative

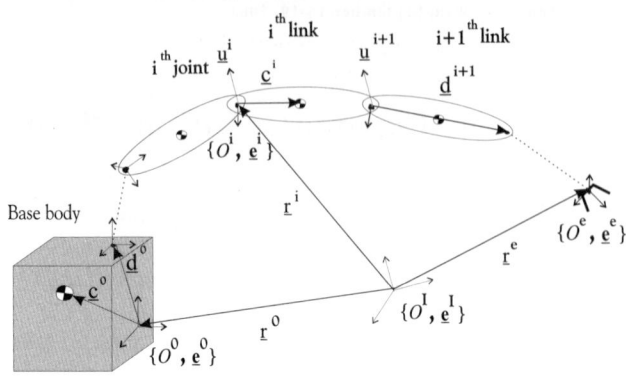

Fig. 1. Reference frames and geometrical quantities of the free-flying robot

rotation matrix between frames i and $i - 1$, function of θ^i. Furthermore, m^i is the mass, $\underline{\underline{I}}^i$ the inertia tensor referred to the centre of mass, $\underline{f}^i$ and $\underline{t}^i$ are the sums of the external forces and torques and $\underline{f}_c^i$ and $\underline{t}_c^i$ are the sums of the constraint forces and torques (arising from the revolute joints) on the ith body.

After choosing a suitable frame of reference in which to express the vector quantities and a suitable parameterisation of the angular position, as for example Euler angles ϕ^i, the equations can be written in matrix form. Choosing the inertial frame $\{O^I, \underline{e}^I\}$, the kinematic and dynamic equations become:

$$\dot{\mathbf{x}}_I = \mathbf{X}(\mathbf{x}_I)\mathbf{x}_{II} \qquad (7)$$

$$\mathbf{J}\dot{\mathbf{x}}_{II} = \mathbf{Q} + \mathbf{\Lambda} = \mathbf{Q} + \boldsymbol{\tau} + \boldsymbol{\tau}_c \qquad (8)$$

where

$$\mathbf{x}_I = \begin{bmatrix} [\mathbf{r}^i] \\ [\phi^i] \end{bmatrix} \qquad \mathbf{x}_{II} = \begin{bmatrix} [\mathbf{v}^i] \\ [\omega^i] \end{bmatrix} \qquad (9)$$

are vector arrays related by the matrix $\mathbf{X}$, which in turn can be determined by the standard formulation of the Euler angles parameterisation (see for example [12], p.83) and

$$\mathbf{J} = \begin{bmatrix} \mathbf{m} & \mathbf{0} \\ \mathbf{0} & \mathbf{I} \end{bmatrix} \quad \mathbf{Q} = \begin{bmatrix} \mathbf{0} \\ -[\tilde{\omega}^i \mathbf{I}^i \omega^i] \end{bmatrix} \qquad (10)$$

$$\boldsymbol{\tau} = \begin{bmatrix} [\mathbf{f}^i] \\ [\mathbf{t}^i] \end{bmatrix}.$$

Furthermore, $\mathbf{m} = \text{diag}(m^i \mathbf{E}_3)$, where $\mathbf{E}_3$ is a three dimensional identity matrix, and $\mathbf{I} = \text{diag}(\mathbf{I}^i)$.

The differential algebraic equations (7) - (8) with (5) - (6) can then be transformed into a set of ordinary differential equations. The Cartesian position variables of the system defined in Eq. (9) are replaced by the independent position state space variables which can be chosen to be

$$\mathbf{y}_I = [\mathbf{r}^0, \phi^0, \boldsymbol{\theta}]^T \quad \mathbf{y}_{II} = [\mathbf{v}^0, \boldsymbol{\omega}^0, \dot{\boldsymbol{\theta}}]^T \qquad (11)$$

where $\boldsymbol{\theta} = [\theta^i]$. The relationship between the variables $\mathbf{x}_{II}$ and $\mathbf{y}_{II}$ can be expressed as

$$\mathbf{x}_{II} = \boldsymbol{\psi}\,\mathbf{y}_{II} + \hat{\boldsymbol{\psi}}\,\mathbf{z}_{II} \qquad (12)$$

where $\boldsymbol{\psi}$ is termed the modal matrix. Furthermore, $\mathbf{z}_{II}$ includes all the locked and kinematically (or rheonomically) driven velocity state variables of the system (for further details refer to [12]).

Eq. (12) can then be differentiated in time and substituted into Eq. (8) to obtain:

$$\dot{\mathbf{y}}_I = \bar{\mathbf{X}}(\mathbf{y}_I)\,\mathbf{y}_{II} \qquad (13)$$
$$\bar{\mathbf{J}}\,\dot{\mathbf{y}}_{II} = \bar{\mathbf{Q}} + \bar{\boldsymbol{\tau}}. \qquad (14)$$

Note that the resultant equations are now free of the unknown constraint force vector $\boldsymbol{\tau}_c$, which results from the application of d'Alembert's principle.

A. Rheonomic joints

Noting then that Eq. (12), with rheonomically driven joints, is written as:

$$\mathbf{x}_{II} = \bar{\boldsymbol{\psi}}\,\bar{\mathbf{y}}_{II} + \hat{\bar{\boldsymbol{\psi}}}\,\bar{\mathbf{z}}_{II} + \hat{\bar{\boldsymbol{\psi}}}_l\,\bar{\mathbf{z}}_{II\,l}, \qquad (15)$$

where

$$\bar{\mathbf{y}}_{II} = [\mathbf{v}^0,\ \boldsymbol{\omega}^0]^T,\quad \bar{\mathbf{z}}_{II} = \dot{\boldsymbol{\theta}}, \qquad (16)$$

and $\bar{\mathbf{z}}_{II\,l}$ now contains only the locked velocity state variables, it follows, using the same procedure above, that the equations of motion become:

$$\dot{\bar{\mathbf{y}}}_I = \bar{\mathbf{X}}^0(\bar{\mathbf{y}}_I)\,\bar{\mathbf{y}}_{II} \qquad (17)$$
$$\bar{\mathbf{J}}^0\,\dot{\bar{\mathbf{y}}}_{II} = \bar{\mathbf{Q}}^0 + \bar{\boldsymbol{\tau}}^0 + \boldsymbol{\Upsilon}. \qquad (18)$$

In Eqs. (18) - (18) the superscript 0 relates to the case with rheonomic joints. Also, in the equation

$$\boldsymbol{\Upsilon} = \boldsymbol{\psi}^T\,\mathbf{J}\,\hat{\bar{\boldsymbol{\psi}}}\,\ddot{\boldsymbol{\theta}}^0, \qquad (19)$$

where now $\ddot{\boldsymbol{\theta}}^0$ is a prescribed function of time.

Note that the order of the equations of motion (18) is now only six, equal to the degrees of freedom of the base body.

B. Modelling summary

From the above derivation it follows that:

- the fully-actuated system is represented by Eqs. (14) - (14) - note that actuation on the spacecraft and robot is expressed by vector $\bar{\boldsymbol{\tau}}$;
- the fully-actuated system with rheonomically driven joints is represented by Eqs. (18) - (18) - note that actuation on the robot does not appear, as its motion is described by the prescribed time function $\ddot{\boldsymbol{\theta}}^0$. Actuation on the spacecraft is expressed by vector $\bar{\boldsymbol{\tau}}^0$;
- the free-floating system is described by Eqs. (14) - (14) or by Eqs. (18) - (18) for $\bar{\boldsymbol{\tau}}^0 = 0$.

III. TRAJECTORY PLANNING

The path planning problem is generally defined here as the point-to-point problem, i.e. that of determining the time history of the robot joints and spacecraft state (position and orientation) in order to move the end-effector of the robot from a given initial to a given final state in inertial space. As we are considering the trajectory planning problem, the equivalent time history for the actuation variables will also be determined.

Using the notation in figure 1, the initial and final end-effector position and orientation are defined as:

$$\underline{r}^e(t_0) = \underline{r}^e_0 \qquad \underline{\phi}^e(t_0) = \underline{\phi}^e_0 \qquad (20)$$
$$\underline{r}^e(t_f) = \underline{r}^e_{\text{des}} \qquad \underline{\phi}^e(t_f) = \underline{\phi}^e_{\text{des}}, \qquad (21)$$

where t_0 and t_f are the initial and final time of the maneuver and $\underline{\phi}^e$ is a set of Euler angles to represent the orientation of the frame of reference $\{O^e, \underline{\mathbf{e}}^e\}$

Furthermore, the path planning solution has to satisfy kinematic and dynamic constraints. The former can be expressed as

$$\theta^i_{\min} \leq \theta^i \leq \theta^i_{\max},\ 1 \leq i \leq n, \qquad (22)$$

where n is the number of joints of the robot (taken to be six).

The dynamic constraints are simply

$$\bar{\tau}^{0\,i}_{\min} \leq \bar{\tau}^{0\,i} \leq \bar{\tau}^{0\,i}_{\max},\ 1 \leq i \leq 6, \qquad (23)$$
$$\bar{\tau}^{i}_{\min} \leq \bar{\tau}^{i} \leq \bar{\tau}^{i}_{\max},\ 6 < i \leq n+6, \qquad (24)$$

where $\bar{\tau}^i$ is generally the actuation force for the state variable i and superscript 0 relates to the base body. Note that we are considering the case of full actuation, meaning that the states of the robot are taken to be the six degrees of freedom of the spacecraft ($1 \leq i \leq 6$) and the n degrees of freedom of the robot ($6 < i \leq 6+n$).

As we anticipated in the introduction, the path planning problem is taken as an optimization problem. The $6+n$ states of the robot-spacecraft system are parameterised in time while Eqs. (21) - (21) and (22) - (24) will be taken as equality and inequality constraints respectively. Regarding the cost function, we have initially chosen to optimize for the base actuation in order to minimise for the thruster fuel consumption. We have chosen to optimize for both translational and rotational spacecraft actuation as an example. The cost function Γ can then be expressed mathematically as

$$\Gamma = \sum_{i=1}^{6} \int_0^{t_f} \|\bar{\tau}^{0\,i}(t)\|\ dt, \qquad (25)$$

where Γ can be considered as the impulse, if the torques are assumed to be computed as the product of a force by a unit moment arm. Then the impulse can be related to an energy content because it is equal to the product of the necessary fuel mass times its specific impulse (which is a given constant).

A. Constraint equations

The left hand sides of the equality constraints (21) - (21) are simply expressed as

$$\underline{r}^e = \underline{r}^0 + \sum_{i=0}^{n} \underline{d}^i \quad (26)$$

$$\mathbf{A}(\boldsymbol{\phi}^e) = \mathbf{A}(\boldsymbol{\phi}^0)\mathbf{A}(\boldsymbol{\theta}) \quad (27)$$

where $\mathbf{A}$ is a rotation matrix, function of a set of rotation parameters ϕ or of joint angles θ.

For inequality constraints (22) - (24), the inverse dynamics problem has to be solved. Given a time history of the $6+n$ state variables $\mathbf{y}_I = [\,\mathbf{r}^0 \;\; \boldsymbol{\phi}^0 \;\; \boldsymbol{\theta}\,]^T$ and their first and second derivatives $\dot{\mathbf{y}}_I = [\,\mathbf{v}^0 \;\; \boldsymbol{\omega}^0 \;\; \dot{\boldsymbol{\theta}}\,]^T$ and $\ddot{\mathbf{y}}_I = [\,\dot{\mathbf{v}}^0 \;\; \dot{\boldsymbol{\omega}}^0 \;\; \ddot{\boldsymbol{\theta}}\,]^T$, then the vector of actuator actions $\bar{\boldsymbol{\tau}}$ is given by Eq. (14).

B. Use of the free-floating solution

The equations defined in the previous section are conceptually complete for the solution of the path planning problem we are wanting to solve. However, in order to improve the quality of the optimal solutions and the efficiency of the optimizing routine a further step is introduced. We first want to distinguish between maneuvers in what we define the *local workspace* - the workspace of the robot with no spacecraft actuation - and those in the *global workspace* - ideally all reachable free space. The distinction is useful here because the first kind of maneuver does not *require* any spacecraft actuation and the optimal solution in the sense we have defined should be exactly this one. This can be called the *free-floating solution*. The motion of the base for these solutions is complex or in any case would require a polynomial of high degree in order for it to be suitably approximated. A high degree polynomial means a high number of optimization parameters and a longer running time for the optimization algorithm.

An alternative approach is to solve for the free-floating motion for a given time evolution of the joints and add this to a parametric function of the spacecraft states, such that the latter can be simply set to zero if free-floating solutions are sought. Although one might argue that the overall running time is comparable to that of the previous approach, we have chosen to go this way. A free-floating solution can in this way be described exactly.

Therefore, equations of motion (18) - (18) for the free-flying robot with rheonomically driven joints are used. The integration of Eq. (18) supplies the solution $\bar{\mathbf{y}}_{II}^{ff} = [\mathbf{v}^{ff} \;\; \boldsymbol{\omega}^{ff}]^T(t)$, the free-floating solution. Note that this solution depends on the joint motion variables, which are parameterised in time. The free-floating solution can then be added to a time parameterised term to finally satisfy the equality constraints (21) - (21) (see section IV). The inequality constraints (24) - (24) can then be satisfied by use of Eq. (14).

IV. TRAJECTORY PLANNING SOLVER

The optimization problem is then:

$$\min_{\mathbf{p}} \Gamma \quad (28)$$

where $\mathbf{p}$ is the set of parameters appearing in the states parametric functions defined below. The cost function is given by Eq. (25) and Eqs. (21) - (21) and (22) - (24) are the equality and inequality constraints respectively.

Furthermore, the parameterisation of the state variables is as follows:

$$\mathbf{r}^0(t) = \mathbf{r}^{ff}(t) + f_t^5(t;\mathbf{p}) \quad (29)$$
$$\mathbf{v}^0(t) = \mathbf{v}^{ff}(t) + f_t^4(t;\mathbf{p}) \quad (30)$$
$$\dot{\mathbf{v}}^0(t) = \dot{\mathbf{v}}^{ff}(t) + f_t^3(t;\mathbf{p}) \quad (31)$$
$$\mathbf{A}(\boldsymbol{\phi}^0)(t) = \mathbf{A}(\boldsymbol{\phi}^0)(0) \int_0^t \tilde{\boldsymbol{\omega}}^0 \mathbf{A}(\boldsymbol{\phi}^0) \, d\bar{t} \quad (32)$$
$$\boldsymbol{\omega}^0(t) = \boldsymbol{\omega}^{ff}(t) + f_r^4(t;\mathbf{p}) \quad (33)$$
$$\dot{\boldsymbol{\omega}}^0(t) = \dot{\boldsymbol{\omega}}^{ff}(t) + f_r^3(t;\mathbf{p}) \quad (34)$$
$$\theta^i(t) = f_j^5(t;\mathbf{p}) \quad 1 \leq i \leq n \quad (35)$$
$$\dot{\theta}^i(t) = f_j^4(t;\mathbf{p}) \quad 1 \leq i \leq n \quad (36)$$
$$\ddot{\theta}^i(t) = f_j^3(t;\mathbf{p}) \quad 1 \leq i \leq n \quad (37)$$

where generally function f^n is a polynomial function of degree n and given a polynomial function of degree i, e.g. f_j^5, then the following polynomial functions of degree $i-1, i-2$ are the successive derivatives of the first. These functions have been parameterised with a polynomial of degree 5 in order to set the further desired conditions that:

- velocity and acceleration of parameterised variables are at initial and final time zero;
- the initial position is given by the definition of the initial conditions of the robot-spacecraft configuration.

It follows that parameter vector $\mathbf{p}$ is composed of the following quantities:

$$\mathbf{p} = [\theta^1 \; \theta^2 \; \theta^3 \; \theta^4 \; \theta^5 \; \theta^6 \; r^{01}{}_b \; r^{02}{}_b \\ r^{03}{}_b \; \phi^{01}{}_b \; \phi^{02}{}_b \; \phi^{03}{}_b]^T (t_f), \quad (38)$$

where $r^{0i}{}_b$ are the components of the second term on the right hand side of Eq. (30) and $\phi^{0i}{}_b$ are the equivalent for the rotational motion. The latter are such that the integral of the angular velocity $\boldsymbol{\omega}^0$ (Eq. (33)) satisfies the orientation constraint on the end-effector. Furthermore, each of the first terms on the right hand side of Eqs. (30) - (35) is taken from the solution (integration) of Eq. (18), which is in turn obtained with the aid of Eqs. (36) - (37). Eqs. (30) - (37) are then used with Eq. (14) to compute vector $\bar{\boldsymbol{\tau}}$.

Note that the chosen polynomial parameterisation for the state variables allows only one free parameter for each position variable. This is the minimum number of parameters possible and a broader range of functions could be used for the optimization problem by choosing higher degree polynomials or B-splines, for possibly better optimal solutions, but at the clear expense of computation time. As a first approach the simplest function representation was chosen here.

Furthermore, due to the boundedness of function $\theta^i(t)$ between $\theta^i(t_0)$ and $\theta^i(t_f)$, the kinematic inequality constraints only need to be checked at $t = t_f$.

The optimization method used for the resolution of the above problem is Sequential Quadratic Programming (SQP).

V. RESULTS

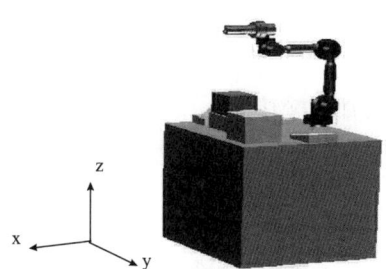

Fig. 2. Model example of free-flying robot. Initial configuration is shown.

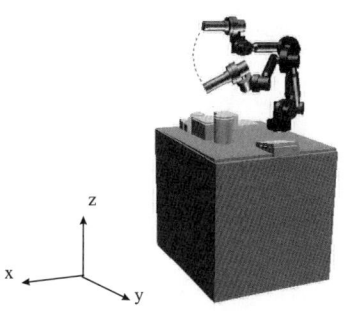

Fig. 3. Maneuver in local workspace - free-floating solution. Initial and final configurations are shown.

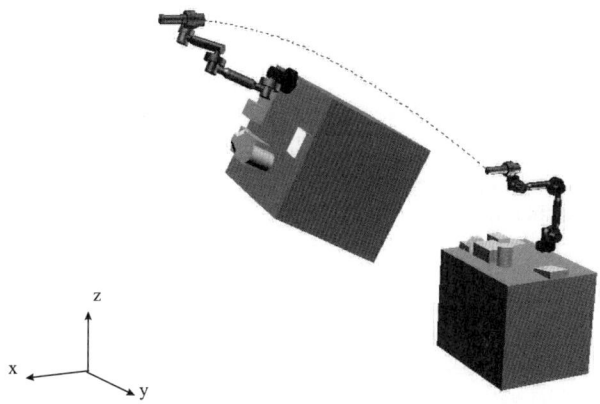

Fig. 4. Maneuver in global workspace. Initial and final configurations are shown.

Two paths are considered for the description of the proposed method: a path in the local workspace and a path in the global workspace. Consider the free-flying robot shown in figure 2. The inertial parameters of the robot where chosen to be those of the ETS-VII satellite. The initial state of the end-effector corresponding to the shown configuration of the robot is:

$$\underline{r}_0^e = [\,0.10\ -0.83\ 4.49\,]^T \quad \underline{\phi}_0^e = [\,0.0\ 0.0\ 0.0\,]^T.$$

For the first path, the final desired state of the end-effector was chosen to be

$$\underline{r}_{\text{des}}^e = [\,0.40\ 0.0\ 3.8\,]^T \quad \underline{\phi}_{\text{des}}^e = [\,-0.5\ 0.5\ 0.0\,]^T.$$

The initial guess was taken to be
$\mathbf{p}_0 = [\,0.1\ 0.1\ 0.1\ 0.1\ 0.1\ 0.1\ 0.0001\ ...\ 0.0001\,]^T.$
The solution was hence found to be
$\mathbf{p} = [\,0.85\ 0.62\ -0.20\ 0.40\ -0.94\ 0.12\ 0.0\ ...\ 0.0\,]^T.$
The cost function for this example was 8 e-3 kg m/s, indicating that no base actuation is present (this is also clear from the zero values of the last six parameters). The path is shown in Fig. 3. The computation time was about 130 seconds on a standard Sgi machine. For the second maneuver, the final desired state of the end-effector was chosen to be

$$\underline{r}_{\text{des}}^e = [\,6.0\ -0.83\ 7.00\,]^T \quad \underline{\phi}_{\text{des}}^e = [\,0.0\ 0.0\ 0.0\,]^T.$$

The initial guess was taken to be the same as for the previous case. The solution was hence found to be

$\mathbf{p} = [\,-1.35\ 0.3\ -0.94\ -0.88\ 0.45\ 1.37\ 1.97\ 0.42\ 3.94\ 0.03\ 1.09\ -0.07\,]^T.$

The cost function for this case was 4460 kg m/s, of which 1815 kg m/s from the translational forces. The computation time was about 165 seconds. The path is shown in Fig. 4. Kinematic constraints (22) were applied as

$$\theta_{\min}^i = -1.5, \quad \theta_{\max}^i = +1.5, \ 1 \leq i \leq n.$$

The spacecraft actuator's effort is shown in Fig. 5, for the spacecraft actuation torque τ^{04}. Note that the final time t_f is 50.0 seconds. The chosen parameterisation however does not allow to lower the control bounds below the found solution since the parameterisation function is only dependent on the final robot-spacecraft position. However, to avoid saturation of the actuators within the context of trajectory planning, a variable execution time is sufficient. The last maneuver is then repeated with bounds on the base body actuation forces, taken to be

$$\tau_{\min}^{0i} = -20.0, \quad \tau_{\max}^{0i} = +20.0, \ 1 \leq i \leq 6.$$

The results, also given in Fig. 5, are clearly showing the optimal use of the actuators to minimise the final time for the given dynamical limits of the robotic system. The

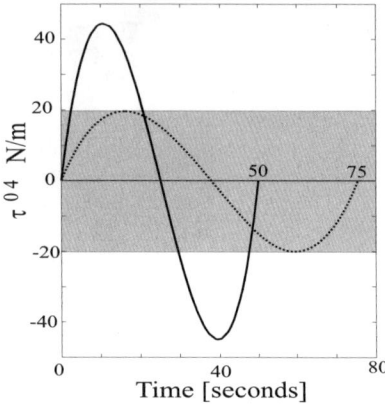

Fig. 5. Spacecraft actuation torque $\tau^{0\,4}$ for unbound (continuos line) and bounded solution (dotted line)

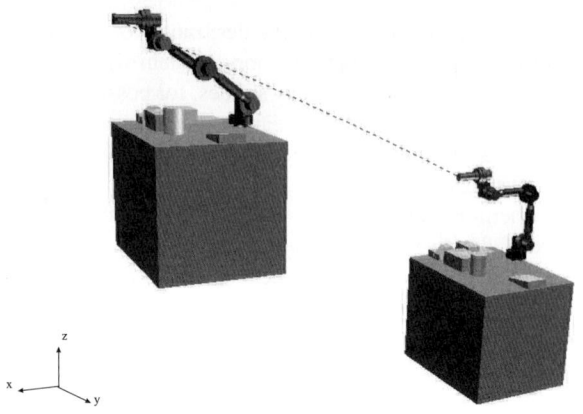

Fig. 6. Maneuver in global workspace - attitude controlled spacecraft for zero angular motion. Initial and final configurations are shown.

final time t_f is now 75.37 seconds. Furthermore, the cost function has been defined as $\Gamma = t_f$, as opposed to the conflicting function defined in Eq. (25).

A. Maneuvers for attitude controlled operational strategy

The second maneuver, shown in Fig. 4, is repeated for the attitude controlled case. This is achieved by simply fixing the desired rotational states of the base to their initial values and the angular velocity and acceleration to zero. The resulting maneuver is shown in Fig. 6. Note that the solution in this case is
$\mathbf{p} = [\,1.00\;\;1.50\;\;-1.42\;\;-0.08\;\;-1.00\;\;0.00\;\;5.46$
$-1.42\;\;3.30\;\;0.00\;\;0.00\;\;0.00\,]^T$
and the cost function is 5033 kg m/s, of which 2108 kg m/s from the translational forces. Comparing this result with that of the maneuver shown in Fig. 4, we find that the cost is in this case higher, due to the constraint imposed on the base motion.

B. Orbital disturbances

These could be included in the model to aid the optimal solution. However their effect leads to a monotonic acceleration of the system. Although this could still be of aid for part of the maneuver, we believe that they should be dealt with by the spacecraft control only, gaining in planning simplicity, and because the gain in including them in the dynamic model would be minimal. Only for very slow motions would their effect be of real relevance.

VI. CONCLUSION

An optimization method has been applied to the motion planning problem of free-flying robots. Solutions have been found for local and global motions, where for the latter the unnecessary spacecraft actuation has been shown to be efficiently avoided. Both kinematic and dynamic constraints have been satisfied allowing for minimum spacecraft actuation or minimum time of execution. The times of computation have also been shown to be promisingly short. Further work will address collision avoidance and the development of an initial guess criterion for optimization robustness.

VII. REFERENCES

[1] Y. Nakamura, R. Mukherjee: "Nonholonomic Path Planning of Space Robots via a Bi-Directional Approach", *Proc. of IEEE Int. Conf. on Robotics and Automation (ICRA)*, Cincinnati, OH, May 1990.

[2] Mukherjee, R., Zurowski, M.: "Reorientation of a structure in space using a three-link rigid manipulator", *Journal of Guidance, Control and Dynamics*, Vol.17, No.4, July-August 1994.

[3] Y. Nakamura, T. Suzuki: "Planning Spiral Motions of Nonholonomic Free-Flying Space Robots", *J. of Spacecraft and Rockets*, Vol.34, No.1, Jan.-Feb. 1997.

[4] K. Landzettel, B. Brunner, G. Hirzinger, R. Lampariello, G. Schreiber, B.-M. Steinmetz: "A Unified Groud Control and Programming Methodology for Space Robotics Applications - Demonstrations on ETS-VII", *International Symposium on Robotics (ISR 2000)*, Montreal, Canada, May 2000.

[5] R. Lampariello, G. Hirzinger: "Free-flying Robots - Inertial Parameter Identification and Control Strategies", *ESA Workshop on Advanced Space Technologies for Robotics and Automation*, Estec, Noordwijk, The Netherland, December 2000.

[6] M. C. Steinbach, H.G. Bock, G.V. Kostin, R.W. Longman: Mathematical Optimisation in Robotics: Towards Automated High Speed Motion Planning, *Surveys Math. Indust.* 7(4), 303-340, 1998.

[7] N. Faiz, S. K. Agrawal: "Trajectory Planning of Robots with Dynamics and Inequalities", *Proc. of the IEEE Int. Conf. on Robotics and Automation (ICRA)*, San Francisco, CA, April 2000.

[8] M. Schlemmer, G. Gruebel: "Real-Time Collision Free Trajectory Optimisation of Robot Manipulators via Semi-Infinite Parameter Optimisation", *International Journal of Robotics Research*, vol. 17, no.9, September 1998, pp.1013-1021.

[9] V.H. Schulz: "Reduced SQP Methods for Large-Scale Optimal Control Problems in DAE with Application to Path Planning Problems for Satellite Mounted Robots", Ph.D. thesis at the University of Heidelberg, 1996.

[10] S. Dubowsky, M. Torres: *Minimizing Attitude Control Fuel in Space Manipulator Systems*, Proc. of the Int. Symp. on Artificial Intelligence: Robotics and Automation in Space, Japan, Nov. 1990.

[11] O. Brock, L.E. Kavraki: *Towards Real-time Motion Planning in High-dimensional Space*, Proc. of the Int. Symp. on Robotics and Automation 2000.

[12] Roberson, E.R.; Schwertassek, R.: *Dynamics of Multibody Systems*, Springer-Verlag, 1988.

Positioning Control of the Arm of the Humanoid Robot by Linear Visual Servoing

Kyota Namba
Graduate School of Systems Engineering
Wakayama University
930 Sakaedani, Wakayama 640-8510
Japan

Noriaki Maru
Department of Systems Engineering
Wakayama University
930 Sakaedani, Wakayama 640-8510
Japan
maru@sys.wakayama-u.ac.jp

Abstract - This paper presents a positioning control of the arm of the humanoid robot by linear visual servoing. Linear visual servoing is based on the linear approximation between binocular visual space and joint space of the arm of the humanoid robot. It is very robust to calibration error, especially to camera angle errors and joint angle errors, because it uses neither camera angles nor joint angles to calculate feedback command. In this paper, we propose a method to expand work space of linear visual servoing by using neck joint. We obtain the linear approximation matrix in wide space and express it as a function of the neck angle by using the neural network. Some experimental results are presented to demonstrate the effectiveness of the proposed method.

1 INTRODUCTION

Visual feedback is indispensable for the intelligent robots that work in dynamic changing environment. Various kinds of mechanisms of visual feedback have been proposed and they are called *visual servoing* [1][2][3][4]. Mitsuda et al. proposed a simple visual servoing scheme called *linear visual servoing*[5][7] and showed the effectiveness in 2D positioning control. Linear visual servoing is based on the linear approximation between binocular visual space and joint space of the arm of the humanoid robot which has a similar kinematic structure to a human being. The relationship makes it possible to generate joint velocities from image data using a constant linear mapping. Linear visual servoing is very robust to calibration error, especially to camera angle errors and joint angle errors, because it uses neither camera angles nor joint angles to calculate feedback command. Hence, it is especially suitable for the humanoid robots which use *active stereo vision*[9][10]. That is, it is possible to turn cameras to facilitate visual processing, even if the arm is under control by visual servoing. Furthermore, the amount of the calculation is very small compared to the conventional visual servoing schemes, because it only needs both the time-invariant constant matrix and the image coordinates of the feature points.

Although we showed that 3D linear visual servoing is also realizable[8], the work space was limited to the front space of the robot. In this paper, we propose a method to expand work space of linear visual servoing by using neck joint. We calculate the linear approximation matrix in wide space and express it as a function of the neck angle by using the neural network. It is because the elements of linear approximation matrix in 3D linear visual servoing varies nonlinear to the vertical angle of the neck joint. Some experimental results are presented to demonstrate the effectiveness of the proposed method.

2 3D LINEAR VISUAL SERVOING

2.1 Model of the Hand-eye Coordination of the Humanoid robot

Fig.1 shows the hand-eye coordination of the Humanoid Robot which has a similar kinematic structure to a human being. The arm consists of two links and two joints. The elbow joint has 1 d.o.f. and the shoulder joint has 2 d.o.f. The shoulder joint is located at the origin of arm coordinate Σ_A. The two cameras are mounted on pan-tilt heads, and the heads are mounted on a base frame which turns horizontally and vertically round the neck joints. We show the parameters of the hand-eye coordination of the Humanoid robot in **Table 1**. These parameters are defined to be proportional to those of a human being. They are the most suitable values for the approximation of the transformation from binocular visual space to joint space as a linear mapping[6].

Table 1: Parameters of the model

Link Length	$L_1 = 450$, $L_2 = 710$
Camera Position	W=365, K=345, G=0
Baseline Length/2	E=50
Neck Length	H=165, M=60
Focal Length	f=4.5 (unit : mm)

2.2 Binocular Visual Space

The binocular visual space is defined as the vergence angle γ and the viewing directions θ, δ(see **Fig.2**). This s-

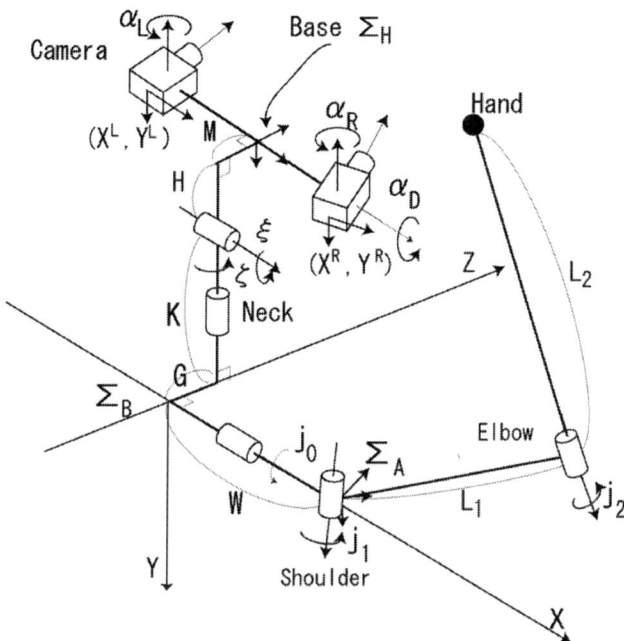

Fig. 1: Model of the hand-eye coordination of the humanoid robot

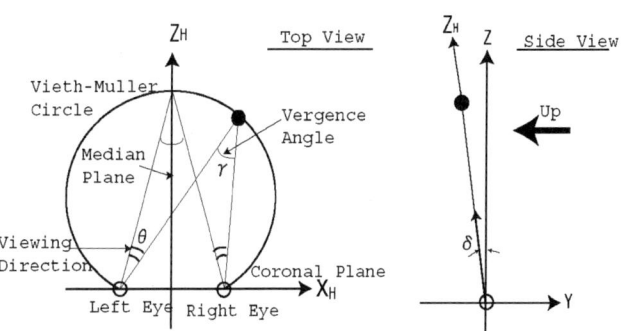

Fig. 2: Binocular visual space

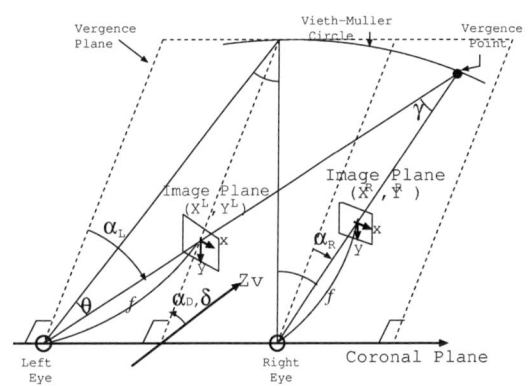

Fig. 3: Model of the active stereo cameras

pace has been employed by psychologists and physiologists as a model of binocularly-perceived space. The binocular visual coordinate of a fixation point is described as

$$\mathbf{V} = \begin{bmatrix} \gamma \\ \theta \\ \delta \end{bmatrix} = \begin{bmatrix} \alpha_L - \alpha_R \\ (\alpha_L + \alpha_R)/2 \\ \alpha_D \end{bmatrix}, \quad (1)$$

where α_L, α_R and α_D are the camera angles.

The binocular visual space has a close relation to the camera image. We show the stereo camera geometry in **Fig.3**.

The coordinates of a feature point projected on the camera image planes are transformed into binocular visual coordinates by

$$\mathbf{V} = \begin{bmatrix} \alpha_L - \alpha_R \\ (\alpha_L + \alpha_R)/2 \\ \alpha_D \end{bmatrix} + \begin{bmatrix} (X^L - X^R)/f \\ (X^L + X^R)/2f \\ (Y^L + Y^R)/2f \end{bmatrix}, \quad (2)$$

where (X^L, Y^L) and (X^R, Y^R) are the coordinates of a feature point in the left and right image respectively. Note that we use the approximation such as $\tan^{-1}(X^{L,R}/f) \simeq X^{L,R}/f$. This approximation is available around the fixation point. Camera angles and image data are transformed into binocular visual coordinates.

2.3 Linear Approximation

Fig.4 and **Fig.5** show the joint space of the arm projected onto Cartesian space and binocular visual space respectively when neck angle $\zeta, \xi = 0$. These figures indicate that the joint space projected onto binocular visual space is more linear than that onto Cartesian space.

We linearize the transformation between binocular visual space and joint space using the least-squares approximation in front space defined as $-20[\text{deg}] \leq j_0 \leq 20[\text{deg}], 20[\text{deg}] \leq j_1 \leq 60[\text{deg}], 60[\text{deg}] \leq j_2 \leq 100[\text{deg}]$. Then the transformation from binocular visual space to joint space is given by

$$\boldsymbol{j} = \boldsymbol{RV} + \boldsymbol{C}, \quad (3)$$

where $\boldsymbol{V} = (\gamma, \theta, \delta)^T$ and $\boldsymbol{j} = (j_0, j_1, j_2)^T$, $\mathbf{R}$: $matrix(3 \times 3), \mathbf{C} : vector(3 \times 1)$.

The least-squares approximation using binocular visual space results in

$$\boldsymbol{R} = \begin{pmatrix} -1.432 & -0.148 & -1.144 \\ -9.234 & -2.207 & 0.303 \\ 16.888 & 1.678 & -0.439 \end{pmatrix}, \quad (4)$$

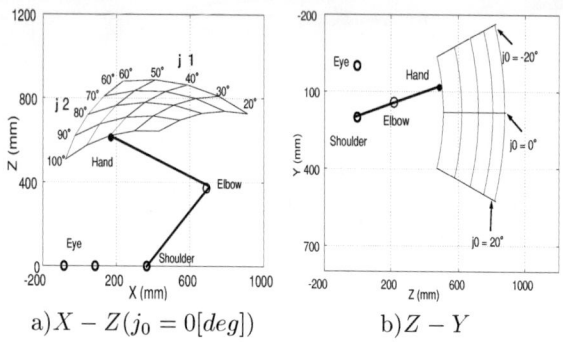

Fig. 4: Joint space projected onto Cartesian space

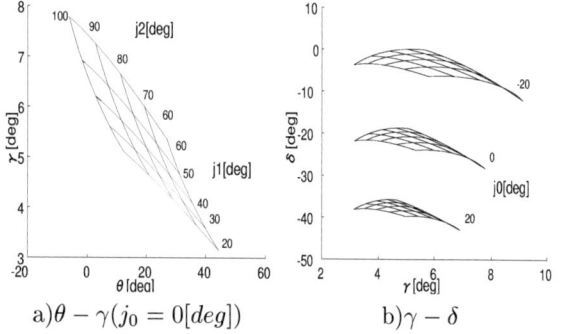

Fig. 5: Joint space projected onto binocular visual space

$$C = \begin{bmatrix} 3.086 \\ 162.869 \\ -99.663 \end{bmatrix}. \quad (5)$$

Fig.6 shows the joint space obtained by linear approximation.

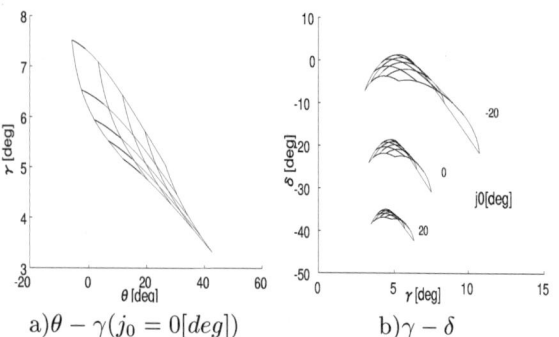

Fig. 6: Joint space obtained by linear approximation

2.4 Linear Visual Servoing

The 3D linear visual servoing using binocular visual space is given by

$$\begin{aligned} \boldsymbol{u} &= -\lambda \boldsymbol{R}(\boldsymbol{V} - \boldsymbol{V}_d), \\ &= -\lambda \boldsymbol{R} \begin{bmatrix} \{(X^L - X^R) - (X_d^L - X_d^R)\}/f \\ \{(X^L + X^R) - (X_d^L + X_d^R)\}/2f \\ \{(Y^L + Y^R) - (Y_d^L + Y_d^R)\}/2f \end{bmatrix}, \end{aligned}$$

Table 2: Region for approximation

$550 \leq r \leq 775$	(unit:mm)
$60 \leq \phi \leq 90$	(unit:deg)
$-15 \leq \varphi \leq 15$	(unit:deg)

$$\begin{aligned} &= -\lambda \boldsymbol{RT}(\boldsymbol{I} - \boldsymbol{I}_d), \quad (6) \\ \boldsymbol{T} &= \begin{pmatrix} 1/f & -1/f & 0 & 0 \\ 1/2f & 1/2f & 0 & 0 \\ 0 & 0 & 1/2f & 1/2f \end{pmatrix}, \\ \boldsymbol{I} &= (X^L, X^R, Y^L, Y^R)^T, \\ \boldsymbol{I}_d &= (X_d^L, X_d^R, Y_d^L, Y_d^R)^T, \end{aligned}$$

where $\boldsymbol{u}$ are control signals to joint velocity controllers, $\boldsymbol{V}$ is the binocular visual coordinates of the hand, $\boldsymbol{V}_d$ is the binocular visual coordinates of a target and λ is a scalar gain, $\boldsymbol{R}$ is the linear approximation matrix of the inverse kinematics obtained in the previous section.

Linear visual servoing is very robust to calibration error, especially to camera angle errors and joint angle errors, because the control law includes neither camera angles nor joint angles. Furthermore, the amount of the calculation is very small compared to the conventional visual servoing schemes.

3 EXPANSION OF WORK SPACE

3.1 Linear Approximation in Peripheral Region

Although we have shown that the 3D linear visual servoing is realizable, the workspace of the arm of the humanoid robot is narrow. It is because the approximation region is limited to its front space of the humanoid robot. One method to expand workspace is to expand the linear approximation region. But the expansion of linear approximation region lead to the increase of approximation error. In this paper, we expand workspace of linear visual servoing by using neck joint.

In this section, we use Spherical coordinate system (r,ϕ,φ) to describe the region of linear approximation. **Fig.7** and **Table 2** show the region of linear approximation when neck angle $\zeta, \xi = 0$. We rotate this region according to the neck angle and linearize the transformation from binocular visual space to joint space using 441 point in this region. We show the elements of $\boldsymbol{R}$ as the continuous lines in **Fig.8**$\sim$ **Fig.10**. In these figures, $A_i, B_i, C_i, (i = 0, 1, 2)$ denotes the coefficients of j_0, j_1, j_2 respectively. Although it is reported that the elements of $\boldsymbol{R}$ varies linear to the neck angle ζ, we can see that they varies nonlinear to the neck angle ξ from these figures.

Fig.11 shows the SSD error of linear approximation. This figure indicates that the error become large according to the rotation of the neck.

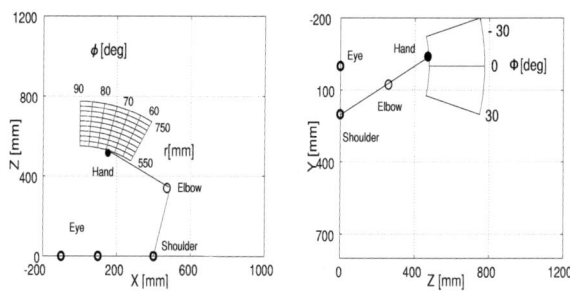

Fig. 7: Region for approximation($\zeta, \xi = 0$)

Table 3: Neural network	
input layer	2
hidden layer	330
output layer	12
learning pattern	3271

3.2 Approximation of R by Neural Network

We approximate the elements of $\boldsymbol{R}$ for linear visual servoing as a function to the neck angle ζ, ξ by using neural network. It is because that the data size to express $\boldsymbol{R}$ become large and interpolation or extrapolation is needed to obtain $\boldsymbol{R}$ for arbitrary neck angle. **Table 3** shows both the data for input, hidden and output layer and learning data for neural network. We use tanh for output function of neuron to produce plus and minus values. We show the elements of $\boldsymbol{R}$ which are approximated by neural network as the broken lines in **Fig.8**~**Fig.10**. We can see that the approximation of nonlinear function is almost realized.

3.3 Linear Visual Servoing using Neck Joint

Then the transformation from binocular visual space to joint space is given by

$$\boldsymbol{j} = \boldsymbol{R}(\zeta,\xi)\boldsymbol{V} + \boldsymbol{C}(\zeta,\xi), \quad (7)$$

where $\boldsymbol{R}(\zeta,\xi), \boldsymbol{C}(\zeta,\xi)$ are the function of ζ, ξ.

When the neck angle is kept constant, the linear visual servoing using neck joint is given by

$$\boldsymbol{u} = -\lambda \boldsymbol{R}(\zeta,\xi)(\boldsymbol{V} - \boldsymbol{V}_d) \quad (8)$$

When the neck rotates, the equation of linear visual servoing become a little complicated and camera angles are needed to calculate binocular visual space.

4 EXPERIMENT

4.1 Experimental System

Fig.12 shows the experimental system. We used PA-10(Mitsubishi Heavy Industry Co. Ltd.) as a 2 linkage

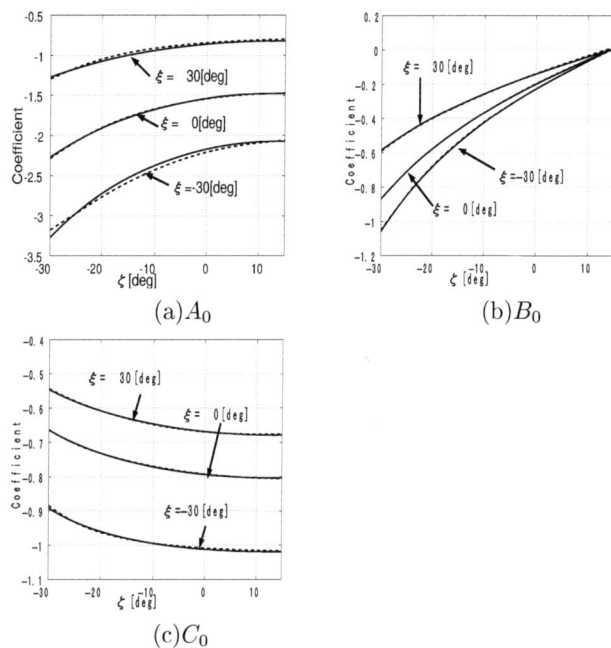

Fig. 8: Change of approximation coefficients(j_0)

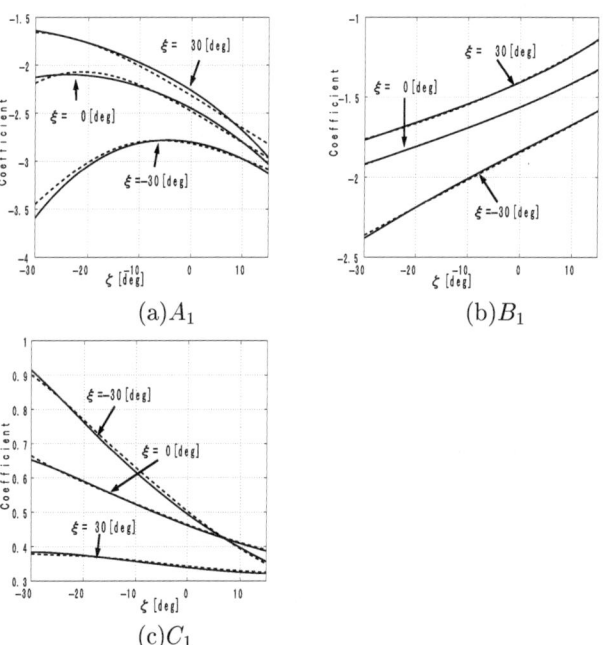

Fig. 9: Change of approximation coefficients(j_1)

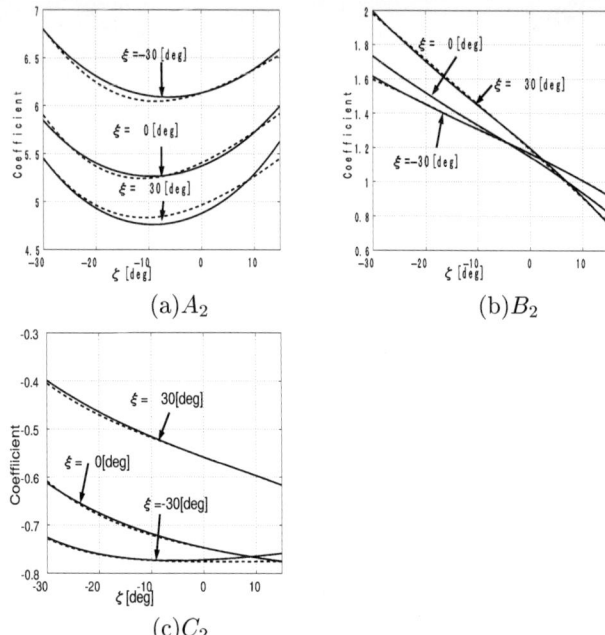

(a)A_2

(b)B_2

(c)C_2

Fig. 10: Change of approximation coefficients(j_2)

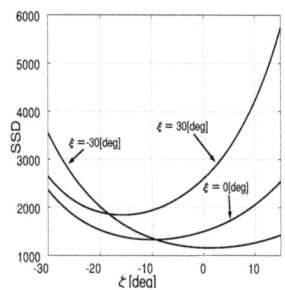

Fig. 11: Error of linear approximation

and 3 d.o.f. arm by fixing the other joints except 2 shoulder joints and 1 elbow joint, although it has 7 d.o.f.. We used two EVI-G20(SONY Co. Ltd.) as the stereo cameras which pan and tilt independently. The stereo cameras are mounted on the neck which pan and tilt by two DC servo motors(Harmonic Drive Systems Inc.). We used G6-300(Gateway Co. Ltd.) as a host computer to control neck joint and MaxPCI workstation(DataCube Co. Ltd.) as a image processing and a host computer to control the arm of the humanoid robot. We attached a yellow marker at the hand and a red object to simplify image processing. The stereo images are converted in an image using a picture separator(Video Device Co. Ltd.) and input to MaxPCI workstation. The stereo images are binarized by each color and the gravity centers are calculated to obtain the image coordinates of both the hand and a target. Then the feedback commands are calculated and sent to the motor driver of PA-10 using ARCNET. In this experiment, we used $\lambda = 150$ and the sampling time is 33[ms].

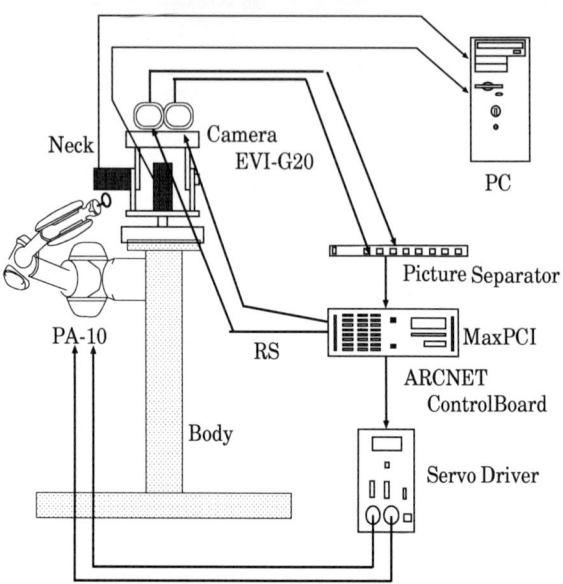

Fig. 12: Experimental system

4.2 Experimental Results

Fig.13 shows 3D trajectories of the hand when the neck angle $\zeta = \xi = 0$. In this figure, 'Without Presumption' and 'Presumption' means the data with law $\boldsymbol{R}$ and with approximated $\boldsymbol{R}$ by neural network. This figure shows that the approximation error of coefficients of matrix $\boldsymbol{R}$ by neural network does not affect the ability of linear visual servoing.

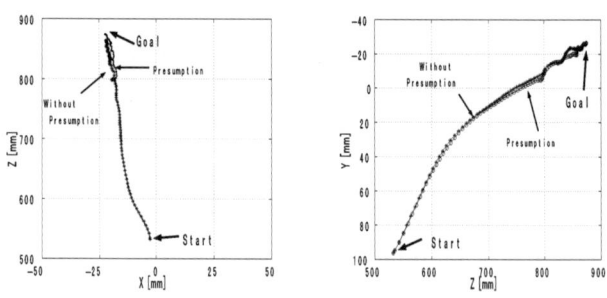

Fig. 13: Trajectories of the hand($\zeta = 0, \xi = 0$)

Fig.14$\sim$ **Fig.17** show the 3D trajectories of the hand when the neck rotates. In these figures, the circle denotes the hand in every 33ms. As can be seen from these figures, the expansion of work space of linear visual servoing is realized.

5 CONCLUSION

This paper has presented a positioning control of the arm of the humanoid robot by linear visual servoing. Linear visual servoing is based on the linear approximation

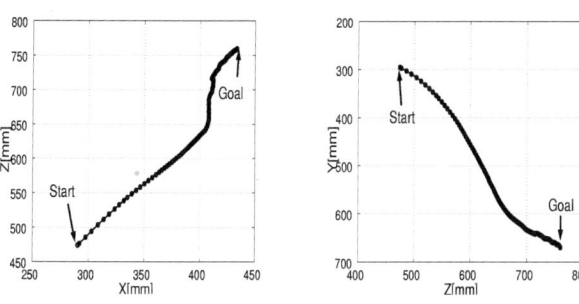

Fig. 14: Trajectories of the hand($\zeta = 30, \xi = -30$)

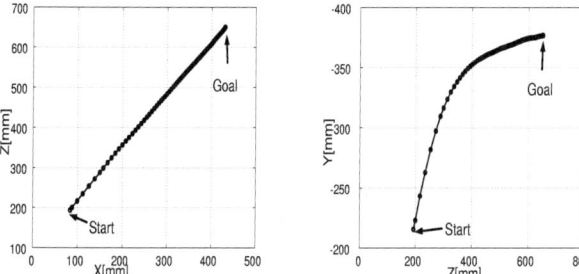

Fig. 15: Trajectories of the hand($\zeta = 30, \xi = 30$)

between binocular visual space and joint space in human-like hand-eye robot. It is very robust to calibration error, especially to camera angle errors and joint angle errors, because it uses neither camera angels nor joint angles to calculate feedback command. Although the method is effective in 3D positioning control, the work space was limited to its front space. In this paper, we have proposed a method to expand work space of linear visual servoing by using neck joint. We calculate the linear approximation matrix in wide space and express it as a function of the neck angle by using the neural network. It is because the elements of linear approximation matrix in 3D linear visual servoing varies nonlinear to the vertical angle of the neck joint. We have shown some experimental results to demonstrate the effectiveness of the proposed method. Although the neck angle is kept constant in this report, the neck can rotate during visual servoing. We are currently investigating about these situations. In this paper, we did not describe how to control general 6(or higher) d.o.f. arms. We will investigate how to control such arms

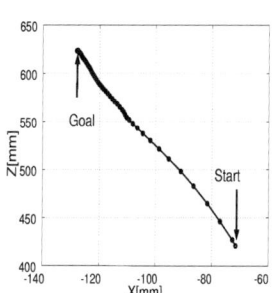 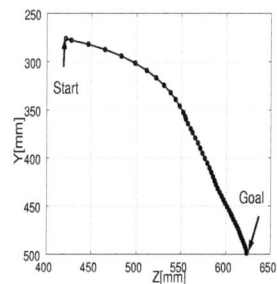

Fig. 16: Trajectories of the hand($\zeta = 15, \xi = -30$)

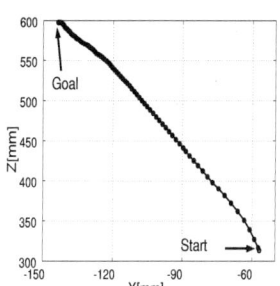 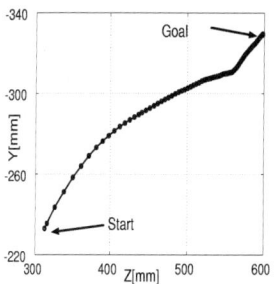

Fig. 17: Trajectories of the hand($\zeta = 15, \xi = 30$)

by using 3D linear visual servoing.

References

[1] L.E.Weiss, A.C.Sanderson and C.P.Neuman, "Dynamic sensor-based control of robots with visual feedback", *IEEE Trans. Robotics and Automation*, vol.RA-3, No.5, pp.404–417, 1987.

[2] B.Espiau, F.Chaumette and P.Rives, "A new approach to visual servoing in robotics", *IEEE Trans. Robotics and Automation*, vol.8, no.3, pp.313–326, 1992.

[3] K.Hashimoto, T.Ebine and H.Kimura, "Visual Servoing with Hand-Eye manipulator - Optimal Control Approach", *IEEE Trans. Robotics and Automation*, vol.12, no.5, pp.766–774, 1996.

[4] N.Maru, H.Kase, S.Yamada and F.Miyazaki "Manipulator control by visual servoing with the stereo vision", in *Proceedings of the IEEE/RSJ Int. Conf. on Intelligenet Robots and Systems*, vol.3, pp.1865–1870, 1993.

[5] T. Mitsuda, N.Maru, K.Fujikawa, F.Miyazaki: "Visal Servoing based on Linear Approximation of the Inverser Kinematics" Journal of the Robotics Society of Japan vol.14no.5pp.743-7501996. (in Japanese)

[6] T. MitsudaN.MaruK.FujikawaF.Miyazaki "Linear Approximation of the Inverser Kinematics by using a Binocular Visual Space", Journal of the Robotics Society of Japan, vol.14, no.8, pp.1145-1151, 1996. (in Japanese)

[7] T.Mitsuda, N.Maru and F.Miyazaki, "Binocular visual servoing based on linear time-invariant mapping", *Advanced Robotics*, Vol.11, No.5, pp.429–443, 1997.

[8] Kyota Namba and N.Maru, "3-D positioning control by linear visual servoing", in *Proceedings of Artificial Life and Robotics*, 2002.(CD-ROM)

[9] J.Aloimonos, I.Weiss and A.Aandyopadhyay "Active Vision", in *Proceedings of the First Int. Conf. on Computer Vision*, pp.552–573, 1987

[10] D.H.Ballard, C.M.Brown, "Principles of Animate Vision", *CVGIP:Image Understanding*, Vol. 56, No. 1, pp.3–21, 1992

Sliding PID Uncalibrated Visual Servoing for Finite-time Tracking of Planar Robots

V. Parra-Vega, and J. D. Fierro-Rojas
Mechatronics Division, CINVESTAV, México
Email: (vparra,jfierro)@mail.cinvestav.mx

Abstract— A decentralized state visual feedback control scheme is presented using an uncalibrated camera. Full nonlinear robot dynamics is considered only in the stability analysis. The closed loop equation renders local exponential tracking by means of the existence of a chattering-free second order sliding mode regime for all time. Additionally, a time-varying feedback gain induces a well-posed terminal attractor of visual tracking errors to prove finite-time tracking of image-based desired trajectories. The system is computationally very inexpensive in comparison to the few dynamic visual servoing controllers available in the literature since the regressor is not computed. Simulations results presented, which confirms the expected dynamic closed-loop performance.

I. INTRODUCTION

There have been much research on kinematic-based visual servoing for robot manipulators, in contrast to the very few schemes available for dynamic-based visual servoing. Despite the impressive achievements of kinematic-based visual servoing systems, these systems move very slowly, in part because of the latency on processing the visual information, and in part because of the controller is computed using kinematic information, while the robot is a complex highly coupled nonlinear system.

On the other hand, technological breakthrough of digital Firework cameras cameras (which are easy to implement because neither expensive dedicated hardware nor complex software development are required) have opened the doors of this field to non-computer scientists, who used to dominate this area. Now, more robot control oriented researcher are proposing dynamic-based visual servoing schemes, who in theory guarantee better performance than kinematic-based visual servoing schemes.

In the design of visual servoing controllers, two important assumption must be taken into account: first the camera parameters are hardly known, and uncalibrated camera must be assumed, and second, the robot parameters are also unknown. These constraints have lead to some researcher very recently to explore adaptive control techniques to deal with asymptotic stability in such circumstances [7], [1], [2], [5], [3].

There are, however, two important problems of dynamic-based visual servoing schemes. The first deals with the on-line computational complexity associated to the structure of the controller, which increases the latency of the closed-loop system of the above. The second one, dynamic-based visual servoing controller delivers a complex nonlinear controller, which can be interpreted only by control-oriented researcher, which reduces the scope of these controllers. In this paper, in the real of [6], and encouraged by [7] and [3], we develop a controller that does not deal with problem to guarantee tracking in image space.

A. Contribution

A decentralized nonlinear PID-like structure that induces a second order sliding mode dynamic-based visual feedback controller for planar manipulators. The robot dynamics are not used in the stage of implementation of the controller, though robot dynamics are explicitly included in the stability analysis, in contrast to kinematic-based visual servoing. Local exponential and finite time convergence of image-based tracking errors are guaranteed. Thus, our proposal stands as the first globally exponentially stable dynamic-based visual servoing tracking scheme supported by rigorous stability proof. Furthermore, it is the first controller that certainly guarantee finite-time convergence. The main advantage of this scheme over [7] [1], [2], [5] is that the regressor does not need to be computed, which leads to a very efficient coding. And it keeps a PID-like structure, very easy to tune, though conservative bound is required. Although Our controller is developed only for the planar case (including gravity), it can be extended to the spatial case if knowledge of jacobian matrix is assumed, and a proper stereovision camera model is considered. To illustrate the performance of the proposed controller we present some simulations that confirms the expected very fast convergence behavior of the trajectory errors.

II. ROBOT MODEL

The dynamics of a serial n−link rigid, non-redundant, fully actuated robot manipulator can be written as follows

$$H(q)\ddot{q} + C(q,\dot{q})\dot{q} + G(q) = \tau - f_r \quad (1)$$

where $q \in R^n$ is the vector of generalized joint displacements, with $\dot{q} \in R^n$ its velocity, $\tau \in R^{n\times 1}$ stands for the vector applied joint torques, $H(q) \in R^{n\times n}$ is the symmetric positive definite manipulator inertia matrix, $C(q,\dot{q})\dot{q} \in R^n$ stands for the vector of centripetal and Coriolis torques, $G(q) \in R^n$ is the vector of gravitational torques, and finally $f_r = B_0\dot{q} \in R^n$ stands for the viscous forces, wherein $B_0 \in R^{n\times n}$ is the coefficient of viscous forces. Two important properties of robot dynamics, useful for stability analysis, are the following.

Property 1: Skew simmetry The time derivative of the inertia matrix, and the centripetal and Coriolis matrix satisfy a skew-symmetric matrix

$$X^T \left[\frac{1}{2}\dot{H}(q) - C(q,\dot{q})\right] X = 0, \quad \forall X \in R^n \quad (2)$$

Property 2: Linear parametrization Robot dynamics are linearly parameterizable

$$H(q)\ddot{q} + C(q,\dot{q})\dot{q} + G(q) = Y_b \theta_b \quad (3)$$

in terms of a known regressor $Y_b = Y_b(q,\dot{q},\ddot{q}) \in R^{n \times p_1}$ and an unknown but constant vector $\theta_b \in R^{p_1}$ of robot parameters as follows

Property 3: Boundedness of robot dynamics There exists positive scalars β_i, where $i = 0, ..., 5$ such that

$$\begin{aligned}
\|H(q)\| &\geq \lambda_m(H(q)) > 0, \\
\|H(q)\| &\leq \lambda_M(H(q)) < \infty, \\
\|C(q,\dot{q})\| &\leq \beta_2 \|\dot{q}\|, \\
\|G(q)\| &\leq \beta_3 < \infty
\end{aligned} \quad (4)$$

where $\lambda_m(A) \geq \beta_0 > 0$, $\lambda_M(A) \leq \beta_1 < \infty$ stand for the minimum and maximum eigenvalues of an $A \in \Re^{n \times n}$ matrix, respectively. Norms $\|A\| = \sqrt{\lambda_M(A^T A)}$, and $\|b\|$ of vector $b \in R^n$ stand for the induced Frobenius and vector Euclidean norms, respectively. These constants can be computed from the state of the system, desired trajectories, feedback gains, and a conservative upper bounds of the dynamic model of the robot arm, assuming $q_d \in C^2$.

III. CAMERA MODEL

For planar robots, and using thin lens without aberration, [4] presented the following widely accepted fixed (static) camera configuration, whose basic mathematical description of this system consists of a composition of four transformations defined as follows

- Joint to Cartesian transformation
- Cartesian to Camera transformation
- Camera to CCD transformation
- CCD to Image transformation

Then, it arises the following.

A. Forward kinematics

The position $x \in R^2$ of the end-effector given in the screen coordinate frame can be calculated by

$$\begin{aligned}
x &= \alpha_0 \frac{\lambda_f}{\lambda_f - z} \begin{bmatrix} -1 & 0 \\ 0 & 1 \end{bmatrix} R_0 f(q) + \beta \\
&= \alpha R f(q) + \beta
\end{aligned} \quad (5)$$

where α_0 is the scaling parameter[1], $R_0 \in R^{2 \times 2}$ stands for the 2×2 upper square matrix of $R_\theta \in SO(3)$, that is $R_0 = \begin{bmatrix} -\cos(\theta) & \sin(\theta) \\ \sin(\theta) & \cos(\theta) \end{bmatrix}$, and

$$\alpha = \alpha_0 \frac{\lambda_f}{\lambda_f - z} < 0 \quad (6)$$

$$R = \begin{bmatrix} \cos(\theta) & -\sin(\theta) \\ \sin(\theta) & \cos(\theta) \end{bmatrix} \quad (7)$$

$$\beta = \alpha_0 \frac{\lambda_f}{\lambda_f - z} \begin{bmatrix} -{}^v O_{b1} \\ {}^v O_{b2} \end{bmatrix} + \begin{bmatrix} u_c \\ v_c \end{bmatrix} \quad (8)$$

[1]Without loss of generality, α_0 can be considered as a 2×2 scaling matrix.

When any intrinsic parameters, such as the depth of field z, camera position offset ${}^v O_b$, and/or focal length λ_f are unknown, it is said that the camera is not calibrated (that is α, θ, and β are unknown), which is a more relevant problem for visual servoing since usually these parameters are hardly known in any practical implementation.

B. Differential kinematics

The *standard* differential kinematics of a serial kinematic chain under consideration yields

$$^B\dot{x}_E = J(q)\dot{q} \quad (9)$$

where $^B\dot{x}_E$ is the velocity of the end-effector with respect to the base frame, $J(q)$ is the analytic Jacobian matrix of the manipulator, and $\dot{q}$ is the (generalized) joint velocities. By using equation (9), the differential kinematic model of (5) becomes

$$\dot{x} = \alpha R J \dot{q} \quad (10)$$

where $\dot{x} \in R^2$ denotes the velocity of the end-effector with respect to the image frame.

C. Inverse differential kinematics

According to the above equation, the following mapping appears

$$\dot{q} = J^{-1} R^{-1} \alpha^{-1} \dot{x} \equiv J^{-1} R_\alpha^{-1} \dot{x} \quad (11)$$

to establish an explicit dependence of joint velocity coordinates $\dot{q}$ in terms of image velocity vector $\dot{x}$, where $R_\alpha^{-1} = (\alpha R)^{-1}$.

D. Assumptions and a key proposition

we consider the following assumptions and propositions.

Assumptions 1: Image $(x, \dot{x})$ and joint $(q, \dot{q})$ coordinates are available.

Assumptions 2: calibration parameters $(\alpha, \theta, \beta$ are unknown.

Assumptions 3: Analytic jacobian matrix J is not available.

Assumptions 4: Inertial robot (θ_b) and camera (θ_v) parameters are unknown.

Proposition 1: [2], [7] For any vector $z \in R^2$ the product with $J^{-1} R_\alpha^{-1} \in R^{2 \times 2}$ can be represented in the following linear form

$$J^{-1} R_\alpha^{-1} z = Y_v(q, z) \theta_v \quad (12)$$

whose elements of $Y_v(q, z) \in R^{2 \times p_2}$ do not depend neither on the rotation matrix nor links length, and $\theta_v \in R^{p_2}$ is composed of parameters of the rotation matrix and parameters of the Jacobian matrix.

Remark 1: Proposition 1 is well-posed for the reachable workspace away from singularities of a 2 degrees of freedom (DoF) robot manipulator. Notice that such linear parameterization does not apply for an spatial robot manipulator of 3 or more DoF. This is the reason that our approach is limited for 2 DoF. However, our scheme can be easily extended to the 3 or more DoF if J is known by using a similar camera model for the spatial case. The reason of this claim is that formulation includes the full nonlinear dynamics of n DOF, and hence it remains to incorporate only the stereo vision model, similar to [7].

IV. Problem Statement

This can be stated as follows

Design a smooth joint torque robot control input for an uncalibrated camera and unknown full nonlinear robot dynamics (regressor is not available), such that the closed loop system guarantees fast image-based trajectory tracking.

The previous statement poses some new issues into the dynamic visual servoing problem: apart from kinematic-based visual servoing [4], wherein robot dynamics are ignored, dynamic-based visual servoing makes use of the model of the system through the regressor [1], [?], [3], [5], [7]. In this paper, we are concerned to dynamic-based visual servoing when the regressor is not available, which is usually the case for industrial robots, and anyway for research robots it is very hard to know exactly the regressor of the system because typically a simplified model of the friction is considered, or some gyroscopic terms are neglected. Before going into the details lets us sketch briefly our proposal.

A. Structure of the proposed controller

The fix camera is modelled as a static operator (5) that relates position and velocities of image and joint coordinates. Then, we are interested in designing an uncalibrated joint output error manifold $\bar{s}_q$ related to a measurable visual error manifold s_x, such that establishment of a passivity inequality for $\langle \bar{s}_q, \tau^* \rangle$ implies dissipativity for output s_x. This means that if we can find τ^* independently of the regressor, then s_x converges, and the controller will satisfy the problem given above. To this end, we need to derive the robot dynamics in $\bar{s}_q$ coordinates, such that the passivity inequality dictates the control structure as well as the storage function. Doing so will allow us to explore some known techniques in passivity-based robot control.

V. Joint and Visual Manifolds

Since passivity arises at the velocity level, then it is customary in passivity-based robot control to design velocity-level extended errors to stabilize the system around them. Then the extended error is given certain structure to finally obtain the desired stability properties. Now, consider the following visual error manifold s_x in image coordinates error

$$s_x = \dot{x} - \dot{x}_r \quad (13)$$

and the joint error manifold s_q in joint coordinates is given by

$$s_q = \dot{q} - \dot{q}_r \quad (14)$$

Now, lets design $\dot{x}_r$ and $\dot{q}_r$.

A. Definition of error manifolds

1) Visual error manifold: Consider the following nominal reference with respect to the image frame

$$\dot{x}_r = \dot{x}_d - \lambda \Delta x + s_d - K_i v \quad (15)$$
$$\dot{v} = sgn(s_\delta) \quad (16)$$

where x_d and $\dot{x}_d$ denote the desired position and velocity of the end-effector with respect to the image frame, respectively, and

$$s_\delta = s - s_d \quad (17)$$
$$s = \Delta \dot{x} + \lambda \Delta x \quad (18)$$
$$s_d = s(t_0) \exp^{-\kappa t} \quad (19)$$

with the integral feedback gain $K_i = K_i^T \in R_+^{n \times n}$ whose precise lower bound is to be defined yet; $\lambda > 0$; $\kappa > 0$; the $sgn(y)$ is the entrywise discontinuous $signum(y)$ function of $y \in R^n$; $\Delta x = x - x_d$ is the image-based end-effector position tracking error; $s_d \in C^1$ and $s_\delta(t_0) = 0$ $\forall\ t$. Substituting (15)-(16) into (13) yields the visual error manifold in image coordinates

$$s_x = s_\delta + K_i \int_{t_0}^{t} sgn(s_\delta)(\zeta) d\zeta \quad (20)$$

In this way, the derivative of (15) becomes

$$\ddot{x}_r = \ddot{x}_d - \lambda \Delta \dot{x} + \dot{s}_d - K_i sgn(s_\delta) \quad (21)$$

Now for the joint space.

2) Joint error manifold: Accordingly to (11), a nominal reference $\dot{q}_r$ in the joint space is defined as follows

$$\dot{q}_r = J^{-1} R_\alpha^{-1} \dot{x}_r \quad (22)$$

Thus, the calibrated joint error manifold s_q is given by

$$s_q = \dot{q} - \dot{q}_r \equiv J^{-1} R_\alpha^{-1} (\dot{x} - \dot{x}_r)$$
$$= J^{-1} R_\alpha^{-1} s_x \quad (23)$$

to establish a dipheomorphism between joint and visual errors manifolds at the velocity level. But $J^{-1} R_\alpha^{-1}$ is not available, and a new nominal reference $\dot{\bar{q}}_r \neq \dot{q}_r$ must be designed

Uncalibrated camera: Since by assumptions J and R_α are not available, then (22) cannot be used and (23) cannot be implemented. That is, the joint error manifold s_q needs all intrinsic and extrinsic camera parameters, as well as image jacobian, and analytical jacobian. This is quite restrictive since camera parameters are usually unknown. Therefore, s_q is not available, and an estimate of s_q must be proposed. Since s_x does not require calibration, then s_x remains the same.

Consider similar to (11)

$$\dot{\bar{q}}_r = \bar{J}^{-1} \bar{R}_\alpha^{-1} \dot{x}_r \quad (24)$$

Using proposition 1, equation (24) becomes

$$\dot{\bar{q}}_r = Y_v \bar{\theta}_v \quad (25)$$

where $Y_v = Y_v(q, \dot{x}_r) \in R^{n_1 \times p_2}$ is known and $\bar{\theta}_v \in R^{p_2}$ is tuned such that $\bar{J}^{-1} \bar{R}_\alpha^{-1}$ is well-posed. From equations (23), (24) and proposition 1, the uncalibrated joint error manifold $\bar{s}_q$ vector is given by

$$\bar{s}_q = \dot{q} - \dot{\bar{q}}_r \quad (26)$$

Notice that (27) can be written

$$\bar{s}_q = \dot{q} - \dot{\bar{q}}_r = \dot{q} - \dot{\bar{q}}_r \pm \dot{q}_r$$
$$= s_q - Y_v \bar{\theta}_v + Y_v \theta_v$$
$$= s_q - Y_v \Delta \theta_v \quad (27)$$

where $\Delta\theta_v = \bar{\theta}_v - \theta_v$. Thus $\ddot{\bar{q}}_r$ becomes, according to (24),

$$\ddot{\bar{q}}_r = \bar{J}^{-1}\bar{R}_\alpha^{-1}\ddot{x}_r + \dot{\bar{J}}^{-1}\bar{R}_\alpha^{-1}\dot{x}_r \qquad (28)$$

In this way, the following property arises

Property 4: Boundedness of nominal references There exists positive scalars β_i, where $i = 0,...,5$ such that

$$\|\dot{\bar{q}}_r\| \leq \sigma_3\sigma_9(\sigma_1 + \lambda_M(\lambda)\|\Delta\dot{x}\| + \sigma_5 + \lambda_M(K_i)\|v\|) \qquad (29)$$

$$\|\ddot{\bar{q}}_r\| \leq \sigma_8\|\dot{q}\| + \sigma_4\lambda M(\lambda)\|\dot{q}\|\|\Delta x\| + \sigma_4\lambda_M(K_i)\|\dot{q}\|\|v\| + \sigma_6\lambda_M(\lambda)\|\Delta\dot{x}\| + \sigma_6\lambda_M(K_i) + \sigma_9 \qquad (30)$$

where $\sigma_i > 0$, $i = 1,...,9$ can be computed from the image-based $x, \dot{x}$, joint state $q, \dot{q}$, desired trajectories $x_d, \dot{x}_d$, feedback gains as follows

$$\sigma_1 \geq \|\dot{x}_d\| \qquad (31)$$
$$\sigma_2 \geq \|\ddot{x}_d\| \qquad (32)$$
$$\sigma_3 \geq \|\frac{\delta \bar{J}^{-1}}{\delta q}\| \qquad (33)$$
$$\sigma_4 \geq \|\lambda_M(R_\alpha^{-1})\sigma_3\| \qquad (34)$$
$$\sigma_5 \geq \|S_d\| \qquad (35)$$
$$\sigma_6 \geq \|\lambda_M(\bar{J}^{-1})\lambda_M(R_\alpha^{-1})\| \qquad (36)$$
$$\sigma_7 \geq \|\kappa\sigma_1\| \qquad (37)$$
$$\sigma_8 \geq \sigma_4(\sigma_1 + \sigma_5) \qquad (38)$$
$$\sigma_9 \geq \sigma_6(\sigma_3 + \sigma_7) \qquad (39)$$

Property 5: Boundedness of the parametrization of $Y_r\theta$. By using property 2 in terms of (24) and (28), the following regressor arises

$$H(q)\ddot{\bar{q}}_r + C(q,\dot{q})\dot{\bar{q}}_r + G(q) = Y_{\bar{b}}\theta_b \qquad (40)$$

The $\mathcal{L}_2$ norm of the function $Y_{\bar{b}}\theta_b$ in (40) is upper bounded according to properties 2 and 3 in the following way

$$\|Y_r\Theta\| \leq \|H(q)\|\|\ddot{\bar{q}}_r\| + \|C(q,\dot{q})\|\|\dot{\bar{q}}_r\| + \|G(q)\|$$
$$\leq \lambda_M(H(q))(\sigma_8\|\dot{q}\| + \sigma_4\lambda_M(\lambda)\|\dot{q}\|\|\Delta x\| + \sigma_4\lambda_M(K_i)\|\dot{q}\|\|v\| + \sigma_6\lambda_M(\lambda)\|\Delta\dot{x}\| + \sigma_6\lambda_M(K_i)) + \sigma_9\beta_2\|\dot{q}\|(\sigma_3\sigma_9(\sigma_1 + \lambda_M(\lambda)\|\Delta\dot{x}\| + \sigma_5 + \lambda_M(K_i)\|v\|)) + \beta_3$$
$$\leq \eta(\Delta x, \Delta\dot{x}, \dot{q}, \sigma_i, \beta_i, K_i, \lambda) \qquad (41)$$

where $\bar{\beta}_3 = \beta_1\beta_5 + \beta_3$, and $\eta(\Delta x, \Delta\dot{x}, \dot{q}, \sigma_i, \beta_i)$ is a smooth scalar functional.

VI. CONTROL AND STABILITY ANALYSIS

Adding and subtracting (40) to (1) produces the open-loop error equation

$$H(q)\dot{\bar{s}}_q + C(q,\dot{q})\bar{s}_q = \tau - Y_{\bar{b}}\theta_b \qquad (42)$$

At this stage, the problem is dual: in one hand compute τ such that $\bar{s}_q$ be globally stable subject to unknown $Y_{\bar{b}}$, and unknown θ_b; on the other, hand prove that boundedness of $\bar{s}_q$ implies local convergence of $\Delta x, \Delta\dot{x}$.

Theorem 1: Consider a robot manipulator (1) in closed loop with the following decentralized sliding PID visual servoing scheme

$$\tau = -K_d\bar{s}_q \qquad (43)$$

Then, the closed-loop system yields locally exponentially $\lim_{t\to\infty}\Delta x = 0$, $\lim_{t\to\infty}\Delta\dot{x} = 0$ provided that K_d and K_i are tuned large enough, for small enough initial error conditions.

A. Stability analysis

The following closed-loop error equation between (1) and (43) arises

$$H(q)\dot{\bar{s}}_q = -(C(q,\dot{q}) + K_d)\bar{s}_q - Y_{\bar{b}}\theta_b + \tau^* \qquad (44)$$

where $\tau^* \equiv 0$ is useful only for stability analysis, to produce the following energy storage function from the passivity inequality $\langle \bar{s}_q, \tau^* \rangle$

$$V = \frac{1}{2}\bar{s}_q^T H(q)\bar{s}_q \qquad (45)$$

Derivative of (45) and property 4 give rise to

$$\dot{V} = -\bar{s}_q^T K_d \bar{s}_q - \bar{s}_q^T Y_{\bar{b}}\theta_b$$
$$\leq -K_{d1}\|\bar{s}_q\|^2 + \eta(\Delta x, \Delta\dot{x}, \dot{q}, \sigma_i, \beta_i, K_i, \lambda)\|\bar{s}_q\| \qquad (46)$$

where $K_d = K_{d1}^T K_{d1}$, and we have used properties 1 and 4. Since $\bar{s}_q$ is a function of $(\Delta x, \Delta\dot{x}, \sigma_i, K_i, \lambda)$, then for small $\Delta x(t_0), \Delta\dot{x}(t_0)$ there exists a large enough feedback gain K_d such that, according to the Barbalath Lemma, there arises stability of $\bar{s}_q$. That is $\bar{s}_q$ is bounded in the $\mathcal{L}_2$ sense. Thus, there exists positive constants $\rho_j, i = 1,...,5$ such that $|\bar{s}_q| \leq \rho_1$, and $|\dot{\bar{s}}_q| \leq \rho_2$, and since equation (23) $|s_x| \leq \rho_3, |\dot{s}_x| \leq \rho_4$. Also, since s_δ is function of s_x, then $\dot{s}_\delta \leq \rho_5$. Therefore, this chain of implications also implies that the state $q, \dot{q}$ of the robot be bounded since $\bar{s}_q, \dot{\bar{s}}_q|$ are bounded.

Now, lets proceed to prove that if K_i is large enough then there arises a sliding mode for $s_\delta = 0$. Consider the derivative of equation (23), gives rise to $\bar{s}_q = s_q - Y_v\Delta\theta_v \to \bar{s}_q = J^{-1}R_\alpha^{-1}s_x - Y_v\Delta\theta_v$. Multiplying the previous equation by $R_\alpha J$, becomes

$$R_\alpha J\bar{s}_q = s_x - R_\alpha J Y_v \Delta\theta_v \qquad (47)$$

Using (20), equation (47) becomes

$$R_\alpha J\bar{s}_q = s_\delta + K_i \int_{t_0}^{t} sgn(s_\delta)(\sigma)d\sigma - R_\alpha J Y_v\Delta\theta_v$$

After few manipulations, finally we arrive to

$$\dot{s}_\delta = -K_i sgn(s_\delta) + \dot{s}_x \qquad (48)$$

Now, in order to produce the sliding mode condition for s_δ, multiply (48) by s_δ^T to obtain

$$s_\delta^T \dot{s}_\delta = -K_i|s_\delta|) + s_\delta^T \dot{s}_x$$
$$\leq -K_i|s_\delta| + |s_\delta||\dot{s}_x|$$
$$\leq -K_i|s_\delta| + |s_\delta||\dot{s}_x|$$
$$\leq -K_i|s_\delta| + \rho_5|s_\delta|$$
$$\leq -\nu|s_\delta| \qquad (49)$$

where $\nu = K_i - \rho_5$. Thus, if $K_i > \rho_9$, the sliding mode condition is met, and a second order sliding mode regime is induced at $s_\delta = 0$ for some time $t = t_s \equiv \frac{|s_q(t_0)|}{\nu}$. Moreover, since $s_\delta(t_0) = 0 \forall t_0$, then $t_s = 0$, and a sliding mode regime is induced for all time at $s_\delta = 0$. This implies the local exponential convergence for image-based tracking errors, i.e. $\Delta x, \Delta \dot{x} \to \lim_{t \to \infty}$ with continuous control. $\diamondsuit$

VII. REMARKS

A. Two control feedback loops

The stability analysis suggest that a damping force equivalent to $K_d \bar{s}_q \; Nm$ is implemented to stabilize the system around the uncalibrated joint error manifold $\bar{s}_q = 0$. On the other hand, $\dot{v}(= sgn(s_\delta))$ switches over the visual error manifold $s_\delta = 0$ to induce precisely a sliding mode at $s_\delta = 0$, to finally converge toward its equilibrium $\Delta x = 0$, and $\Delta \dot{x} = 0$.

B. Continuous controller

The controller (43) depends only on continuous signals, so chattering is not an issue.

C. PID-like structure

The controller (43) can be written as follows

$$\begin{aligned} \tau &= -K_d \bar{s}_q \equiv -K_d \{s_q - Y_v \Delta \theta_v\} \\ &= -K_d (J^{-1} R_\alpha^{-1} \{\Delta \dot{x} + \lambda \Delta x - s(t_0)\exp^{-\kappa t} + \\ &\quad K_i \int_{t_0}^t sgn(s_\delta)\} - K_d Y_c \Delta_v \theta_v) \end{aligned} \quad (50)$$

notice that the feedback gains depend on q through the *real* jacobian and real calibration parameters, plus $\Sigma_1(t)$, which is useful for inducing a sliding mode for any initial condition and $\Sigma_2(t)$ compensates for uncalibrated visual information.

Finally, (50) exhibits a PID-like structure plus vanishing term $\Sigma_1(t) = K_d s(t_0)\exp^{-\kappa t}$ and a nonlinear compensation input $\Sigma_2(t) = K_d Y_c \Delta_v \theta_v$ as follows

$$\begin{aligned} \tau &= -K_p(t)\Delta x - K_v(t)\Delta \dot{x} - \bar{K}_i I + \Sigma_{12}(t) \\ \dot{I} &= sgn(s_\delta) \end{aligned} \quad (51)$$

where

$$\begin{aligned} K_p(t) &= K_d J^{-1} R_\alpha^{-1} \lambda \\ K_v(t) &= K_d J^{-1} R_\alpha^{-1} \\ \bar{K}_i &= K_d K_i \end{aligned}$$

and $\Sigma_{12}(t) = \Sigma_1(t) + \Sigma_2(t)$.

D. Control structure

It is evident the simple structure of this controller, only an estimate of the composition of the image jacobian, the rotation matrix and the scaling and depth parameters are required. Notice that using (43), rather (27) or than (50) or (51).

E. 3D visual servoing?

Our control scheme is valid only for the planar case, but if the jacobian is available, this algorithm can be extended to $n \geq 3$ degrees of freedom simply by using a proper camera projection model, and substituting the real jacobian by the estimate one in the controller.

VIII. FINITE-TIME CONVERGENCE

In our previous paper [6], a time-based generator is proposed to induce terminal attractors for the rigid model of a robot manipulator. It is show there that if a feedback gain is properly tuned according to a time base generator, then a time-varying feedback gain induces finite-time convergence of tracking errors, with smooth control effort. Similar idea is developed here to introduce a terminal attractor in the visual error manifold s_δ. Surprisingly, the only modification to be made to our controller (43) is to design a time varying feedback gainλ in equation (15). It is easy to see, according to the developments of proof of theorem 1, and [6] that the following holds.

Theorem 2: Consider a robot manipulator (1) subject to unknown robot dynamics and unknown camera parameters in closed loop with the sliding PID visual servoing scheme (43), with λ of (18) given by

$$\lambda(t) = \lambda_0 \frac{\dot{\xi}}{(1-\xi)+\delta} \quad (52)$$

where $\lambda_0 = 1 + \epsilon$, for small positive scalar ϵ such that λ_0 is close to 1, and $0 < \delta \ll 1$. The generator $\xi = \xi(t) \in C^2$ must be provided by the user so as to ξ goes from $0 \to 1$ in finite time t_g. The $\dot{\xi} = \dot{\xi}(t)$ is a bell shaped derivative of ξ such that $\dot{\xi}(t_0) = \dot{\xi}(t_g) \equiv 0$. Then, the closed-loop system yields finite-time convergence $\Delta x(t_g) = 0, \Delta \dot{x}(t_g) = 0$, for an arbitrary given finite time $t = t_s > 0$ provided that K_d and K_i are tuned large enough, for small enough initial error conditions.

Proof. The proof immediately follows theorem 1 and [6] since $s_\delta = 0$ holds also for $\lambda = \lambda(t)$. That is,

$$\begin{aligned} s_\delta = 0 \;\to\; & \Delta \dot{x} + \lambda(t)\Delta x = s(t_0)\exp^{-\kappa t} \to \\ & \Delta \dot{x} = -\lambda(t)\Delta x + s(t_0)\exp^{-\kappa t} \end{aligned} \quad (53)$$

According to [6], equation (53) yields a terminal attractor at $t = t_g > 0$ for image-based trajectories, that is in a user-defined time t_g it arises $\Delta x(t_g) = 0, \Delta \dot{x}(t_g) = 0$. $\diamondsuit$

Robot parameter	Value
Length link l_1, l_2	$0.4, 0.3 \; m$
Center of gravity 1,2 l_{c1}, l_{c2}	$0.1776, 0.1008 \; m$
Mass link m_1, m_2	$9.1, 2.5714 \; kg$
Inertia link I_1, I_2	$0.284, 0.0212 \; kgm^2$
Gravity acceleration g_z	$-9.8 \; m/sec^2$

TABLE I

ROBOT PARAMETERS.

Vision parameters	Value
Clock-wise rotation angle θ	$\frac{\pi}{8} \; rad$
Scale factor α_v	$77772 \; pixeles/m$
Depth field of view z	$1.5 \; m$
Camera offset ^vO_b	$[-0.2 \; -0.1]^T \; m$
Offset $\Sigma_I \; o_I$	$[0.0005 \; 0.0003]^T \; m$
Focal length λ_f	$0.008 \; m$

TABLE II

CAMERA PARAMETERS.

IX. SIMULATIONS

A two-rigid link, planar robot is considered, with a CCD camera, see parameters in tables 1 and 2. All inertial parameters of the robot arm and all parameters of the camera, including depth of field of view, are unknown. The endpoint of the manipulator is requested to draw a circle in image space defined with respect to the vision frame $y_d = (y_{d1}, y_{d2})^T = (0.1\cos\omega t + 0.25, 0.1\sin\omega t + 0.15)$, where $\omega = 0.5$ rad/sec. Simulations are carried in Matlab 6.0, with RungeKutta45 as the numerical solver, at 1 ms, so it was assumed that visual data arrives also at $1ms$, otherwise a technique to obtain $x, \dot{x}$ at $1ms$ should be implemented, for instance an observer or simple polynomial approximation running one step behind or a Kalman Filter to predict the following desired trajectory. Results presented in Figure 1 and 2 yield exponential and finite-time tracking capabilities of the controllers of theorems 1 and 1, with a remarkable smooth control effort.

X. CONCLUSIONS

A decentralized nonlinear PID plus a bias term continuous controller for dynamic visual servoing that guarantees tracking tasks in image space for dynamical robot manipulators and uncalibrated camera is proposed. The controller is coined as sliding PID control scheme because it induces a second order sliding mode regime at the visual manifold.

XI. REFERENCES

[1] Hsu, L., and Aquino, P., Adaptive Visual Tracking with Uncertain Manipulator Dynamics and Uncalibrated Camera, 1999 Proc. 38th IEEE Conference on Decision and Control, pp. 1248-1253.

[2] Bishop, B.E., and M.W. Spong, Adaptive calibration and control of 2D monocular visual servo systems, IFAC Symp. on Robot Control, France, 1997.

[3] V. Parra-Vega, J.D. Fierro-Rojas, A. Espinosa-Romero, Uncalibrated Sliding Mode Visual Servoing of Uncertain Robot Manipulators, IEEE/RSJ International Conference on Intelligent Robots and Systems, September 30 - October 4, 2002, Switzerland.

[4] Hutchinson, S., Hager, G.D., and Corke, P.I., A Tutorial on Visual Servo Control, Trans. on Robotics and Automation, 12:651-670, 1996.

[5] W.E. Dixon, E. zergeroglu, Y. fang, and D.M. Dawson, Object Tracking by a Robot Manipulator: Robust Cooperative Visual Servoing Approach, Proceeding of the 2002 IEEE International Conference on Robotics & Automation, 211-216.

[6] V. Parra-Vega, Second Order Sliding Mode Control for Robot Arms with Time Base Generators for Finite-time Tracking, Dynamics and Control, 11 (2): 175-186, April 2001.

[7] Y. Shen, G.Xiang, Y.H. Liu, K. Li, Uncalibrated visual Servoing of Planar Robots, Proceeding of the 2002 IEEE International Conference on Robotics & Automation, Washinton, D.C., 580-585.

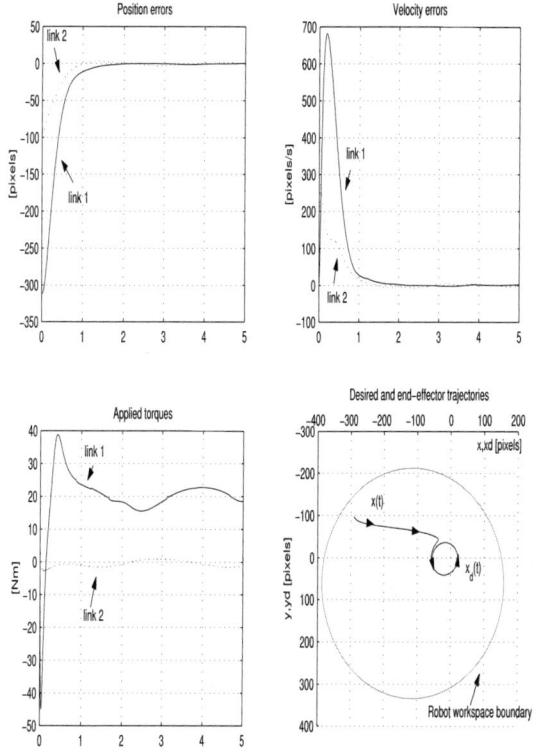

Fig. 1. Performance of Theorem 1: exponential convergence of image coordinates.

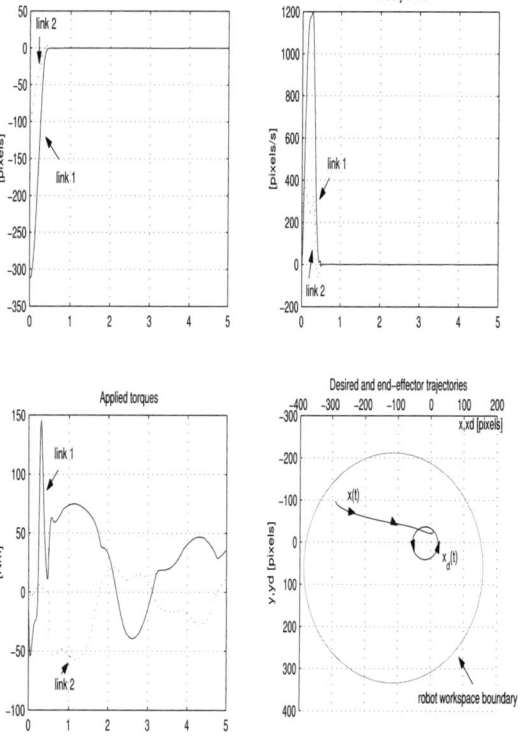

Fig. 2. Performance of Theorem 2: finite-time convergence at $t_g = 0.5\ s$, image coordinates.

Adaptive Sliding Mode Uncalibrated Visual Servoing for Finite-time Tracking of 2D Robot

V. Parra-Vega, J. D. Fierro-Rojas
(vparra,jfierro)@mail.cinvestav.mx
Mechatronics Division
CINVESTAV, México

A. Espinosa-Romero
arturoe@cic3.iimas.unam.mx
Institute for Applied Math. and Systems
UNAM, México

Abstract—A globally convergent visual feedback control scheme is proposed for the dynamic model of uncertain planar robot manipulators with uncalibrated camera. Additionally, a time-varying feedback gain induces a terminal attractor in the visual error manifold to guarantee finite-time convergence of image-based tracking errors. Simulations results of a two degrees-of-freedom manipulator with uncalibrated CCD camera are presented to illustrate the dynamic closed-loop performance.

I. INTRODUCTION

More than two decades of mature research on kinematic-based visual servoing have rendered certainly impressive visual servoing systems. However, since these systems ignore the dynamics of the robots, the real-time performance of are quite low, despite the outstanding achievements in a number of fields. Since camera parameters are hardly known, lot of research has gone into off-line kinematic calibration before going to real-time applications. Moreover, some on-line kinematic tracking schemes have been also proposed[1].

On the other hand, dynamic-based visual servoing schemes consider explicitly the robot dynamics, so as to compensate them to achieve better dynamic response. The obvious advantage of this scheme is the fact that on-line compensation of uncalibrated camera can be easily carried out along the explicit computation of the controller. Very recently, uncalibrated spatial visual servoing tasks have been proposed using adaptive control schemes for dynamic robot arms to guarantee local tracking subject to uncertainty on the parameters [7], [1], [2], [5]. These schemes exploits the fact that the rotation matrix is constant, and formal and rigorous stability analysis support these results. However, these papers assume knowledge of the analytic jacobian matrix, and furthermore, those are singular at rotation angle $\theta = \frac{\pi}{2}$. For planar uncalibrated visual servoing tasks, [7] propose a regulation scheme

[1]An important overlooked problem with kinematic-based visual servoing schemes is that the numerically computed output of these schemes stand for the desired joint velocity of a fast velocity robot control loop, and the desired position is obtained by integrating this output. Then, the output should be validated to assure that a desired joint position is in the visual field. In contrast, dynamic-based scheme do not deal with this problem, since joint torques are explicitly computed.

that removes the requirement of the image jacobian. For tracking, similar technique is followed by [3] using a discontinuous first order sliding mode controller.

A. Contribution

In this paper, inspired by [7], and in the realm of [3], we develop an adaptive second order sliding mode global tracking visual feedback controller for planar manipulators in a modified image-based approach under unknown parameters. Thus, our proposal stands as the first globally exponentially stable tracking scheme supported by rigorous stability proof based on passivity-based approach, with continuous control effort. Furthermore, it is the first controller that certainly guarantee finite-time convergence, which means not only faster response but also that zero tracking error in image space. No acceleration is required, and only the visual flow of three landmarks are computed. Our controller is developed only for the planar case (with gravity), however it can be extended to the spatial case if knowledge of jacobian matrix is assumed. To illustrate the performance of the proposed controller we present some simulations that confirms the expected very fast convergence behavior of the trajectory errors.

II. ROBOT MODEL

The dynamics of a serial $n-$link rigid, non-redundant, fully actuated robot manipulator can be written as follows

$$H(q)\ddot{q} + C(q,\dot{q})\dot{q} + G(q) = \tau - f_r \qquad (1)$$

where $q \in R^n$ is the vector of generalized joint displacements, with $\dot{q} \in R^n$ its velocity, $\tau \in R^{n \times 1}$ stands for the vector applied joint torques, $H(q) \in R^{n \times n}$ is the symmetric positive definite manipulator inertia matrix, $C(q,\dot{q})\dot{q} \in R^n$ stands for the vector of centripetal and Coriolis torques, $G(q) \in R^n$ is the vector of gravitational torques, and finally $f_r = B_0\dot{q} \in R^n$ stands for the viscous forces, wherein $B_0 \in R^{n \times n}$ is the coefficient of viscous forces[2]. Two important properties of robot dynamics, useful for stability analysis, are the following.

[2]Without loss of generality, our controller can be applied with similar results if we consider dynamic friction, for instance the LuGre model.

Property 1: The time derivative of the inertia matrix, and the centripetal and Coriolis matrix satisfy a skew-symmetric matrix

$$X^T \left[\frac{1}{2}\dot{H}(q) - C(q,\dot{q})\right] X = 0, \quad \forall X \in R^n \quad (2)$$

Property 2: Robot dynamics are linearly parameterizable in terms of a known regressor $Y_b = Y_b(q,\dot{q},\ddot{q}) \in R^{n \times p_1}$ and an unknown but constant vector $\theta_b \in R^{p_1}$ of robot parameters as follows

$$H(q)\ddot{q} + C(q,\dot{q})\dot{q} + G(q) = Y_b \theta_b \quad (3)$$

III. CAMERA MODEL

A. Camera Visual Forward, Inverse and Differential Kinematics

For planar robots, and using thin lens without aberration, [4] presented the following widely accepted static camera model, whose basic mathematical description of this system consists of a composition of four transformations defined as follows

- Joint to Cartesian transformation
- Cartesian to Camera transformation
- Camera to CCD transformation
- CCD to Image transformation

Then, it arises the following.

1) Forward kinematics: The position $x \in R^2$ of the end-effector given in the screen coordinate frame can be calculated by

$$\begin{aligned} x &= \alpha_0 \frac{\lambda_f}{\lambda_f - z} \begin{bmatrix} -1 & 0 \\ 0 & 1 \end{bmatrix} R_0 f(q) + \beta \\ &= \alpha R f(q) + \beta \end{aligned} \quad (4)$$

where $f(q)$ stands for the forward kinematic mapping, α_0 is the scaling parameter[3], $R_0 \in R^{2 \times 2}$ stands for the 2×2 upper square matrix of $R_\theta \in SO(3)$, that is $R_0 = \begin{bmatrix} -\cos(\theta) & \sin(\theta) \\ \sin(\theta) & \cos(\theta) \end{bmatrix}$, and

$$\alpha = \alpha_0 \frac{\lambda_f}{\lambda_f - z} < 0 \quad (5)$$

$$R = \begin{bmatrix} \cos(\theta) & -\sin(\theta) \\ \sin(\theta) & \cos(\theta) \end{bmatrix} \quad (6)$$

$$\beta = \alpha_0 \frac{\lambda_f}{\lambda_f - z} \begin{bmatrix} -^v O_{b1} \\ ^v O_{b2} \end{bmatrix} + \begin{bmatrix} u_c \\ v_c \end{bmatrix} \quad (7)$$

When any intrinsic parameters, such as the depth of field z, camera position offset $^v O_b$, and/or focal length λ_f are unknown, it is said that the camera is not calibrated (that is α, θ, and β are unknown), which is a more relevant problem for visual servoing since usually these parameters are hardly known in any practical implementation.

[3]Without loss of generality, α_0 can be considered as a 2×2 scaling matrix.

2) Differential kinematics: The *standard* differential kinematics of a serial kinematic chain under consideration yields

$$^B\dot{x}_E = J(q)\dot{q} \quad (8)$$

where $^B\dot{x}_E$ is the velocity of the end-effector with respect to the base frame, $J(q)$ is the analytic Jacobian matrix of the manipulator, and $\dot{q} \in R^n$ is the (generalized) joint velocities. By using equation (8), the differential kinematic model of (4) becomes

$$\dot{x} = \alpha R J \dot{q} \quad (9)$$

where $\dot{x} \in R^2$ denotes the velocity of the end-effector with respect to the image frame.

3) Inverse differential kinematics: According to the above equation, the following mapping appears

$$\dot{q} = J^{-1} R^{-1} \alpha^{-1} \dot{x} \equiv J^{-1} R_\alpha^{-1} \dot{x} \quad (10)$$

to establish an explicit dependence of joint velocity coordinates $\dot{q}$ in terms of image velocity vector $\dot{x}$, where $R_\alpha^{-1} = (\alpha R)^{-1}$.

B. Assumptions and a key proposition

Consider the following assumptions and propositions.

Assumptions 1: Image $(x, \dot{x})$ and joint $(q, \dot{q})$ coordinates are available. Three tracking landmarks are required to compute x, and numerical differentiation renders $\dot{x}$. No optical flow computation is needed.

Assumptions 2: calibration parameters (α, θ, β) are unknown.

Assumptions 3: Analytic jacobian matrix J is not available.

Assumptions 4: Inertial robot (θ_b) and camera (θ_v) parameters are unknown.

Proposition 1: For any vector $z \in R^2$ the product with $J^{-1} R_\alpha^{-1} \in R^{2 \times 2}$ can be represented in the following linear form

$$J^{-1} R_\alpha^{-1} z = Y_v(q, z) \theta_v \quad (11)$$

whose elements of $Y_v(q,z) \in R^{2 \times p_2}$ do not depend neither on the rotation matrix nor links length, and $\theta_v \in R^{p_2}$ is composed of parameters of the rotation matrix and parameters of the Jacobian matrix.

Remark 1: Proposition 1 is evident from the fact of the linear parametrization property of robot dynamics, and thus is well-posed for the reachable workspace away from singularities of a 2 degrees of freedom (DoF) robot manipulator. Notice that such linear parameterization does not apply for an spatial robot manipulator of 3 or more DoF. This is the reason that our approach is limited for 2 DoF. However, our scheme can be easily extended to the 3 or more DoF if J is known by using a similar camera model for the spatial case. The reason of this claim is that formulation includes the full nonlinear dynamics of n DoF, and hence it remains to incorporate only the stereo vision model, similar to [7].

C. Problem Statement

Design a smooth joint robot control input with uncalibrated camera and uncertain robot parameters of the full nonlinear robot model, such that the closed loop system guarantees finite time image-based tracking trajectories, without any off-line adaptation.

Before going into the details lets us scketch briefly our proposal.

To make clear the exposition of our controller, first the known parametric case (the camera is calibrated) is derived, and afterwards the unknown parametric case (the camera is no calibrated) that satisfies the problem above is presented.

IV. WITH CALIBRATED CAMERA

A. Definition of error manifolds

1) Visual error manifold: Consider the following nominal reference with respect to the image frame

$$\dot{x}_r = \dot{x}_d - \lambda \Delta x + s_d - K_i \int_{t_0}^{t} sgn(s_\delta)(\zeta) d\zeta \quad (12)$$

where x_d and $\dot{x}_d$ denote the desired position and velocity of the end-effector with respect to the image frame, respectively, and

$$s_\delta = s - s_d \quad (13)$$
$$s = \Delta \dot{x} + \lambda \Delta x \quad (14)$$
$$s_d = s(t_0) \exp^{-\kappa t} \quad (15)$$

with the integral feedback gain $K_i = K_i^T \in R_+^{n \times n}$ whose precise lower bound is to be defined yet; $\lambda > 0$; $\kappa > 0$; the $sgn(y)$ is the entrywise discontinuous $signum(y)$ function of $y \in R^n$; $\Delta x = x - x_d$ is the image-based end-effector position tracking error; $s_d \in C^1$ and $s_\delta(t_0) = 0$ $\forall\, t$. Substituting (12) into the visual error manifold $s_x = \dot{x} - \dot{x}_r$ yields

$$s_x = s_\delta + K_i \int_{t_0}^{t} sgn(s_\delta)(\zeta) d\zeta \quad (16)$$

In this way, the derivative of (12) becomes

$$\ddot{x}_r = \ddot{x}_{cont} + Z \quad (17)$$

where

$$\ddot{x}_{cont} = \ddot{x}_d - \lambda \Delta \dot{x} + \dot{s}_d - K_i \tanh(\beta s_\delta) \quad (18)$$
$$Z = \tanh(\beta s_\delta) - sgn(s_\delta) \quad (19)$$

with $\tanh(y)$ as the continuous entrywise hyperbolic tangent function of vector $y \in R^n$, $\sigma = \sigma^T \in R_+^{n \times n}$, $\beta = \beta^T \in R_+^{n \times n}$.

2) Joint error manifold: According to (10), a nominal reference $\dot{q}_r$ in the joint space is defined as follows

$$\dot{q}_r = J^{-1} R_\alpha^{-1} \dot{x}_r \quad (20)$$

Thus, the joint error manifold $s_q = \dot{q} - \dot{q}_r$ in joint space is given by

$$\begin{aligned} s_q &= \dot{q} - \dot{q}_r \equiv J^{-1} R_\alpha^{-1} (\dot{x} - \dot{x}_r) \\ &= J^{-1} R_\alpha^{-1} s_x \end{aligned} \quad (21)$$

Finally, notice that the corresponding $\ddot{q}_r$ is calculated by

$$\begin{aligned} \ddot{q}_r &= J^{-1} R_\alpha^{-1} \ddot{x}_r + \dot{J}^{-1} R_\alpha^{-1} \dot{x}_r \\ &= \mathcal{C} + \mathcal{D} \end{aligned} \quad (22)$$

where $Z = \tanh(\beta s_\delta) - sgn(s_\delta)$, $\mathcal{C} = J^{-1} R_\alpha^{-1} \ddot{x}_{cont}$ is composed of continuous signals, and $\mathcal{D} = J^{-1} R_\alpha^{-1} Z$ is discontinuous, but bounded. To design the controller we need to write the robot dynamics in terms of s_q coordinates. Furthermore, to avoid chattering, $\mathcal{D}$ should not be used.

B. Open-loop robot dynamics

By using property 2 in terms of (20) and (22), the following regressor arises

$$H(q)\ddot{q}_r + C(q,\dot{q})\dot{q}_r + G(q) = Y_c \theta_b + H(q)\mathcal{D} \quad (23)$$

where Y_c is continuous, as follows

$$Y_c \theta_b = H(q)\mathcal{C} + C(q,\dot{q})\dot{q}_r + G(q) \quad (24)$$

Adding and subtracting $Y_c \theta_b + H(q)\mathcal{D}$ to (1) produces

$$H(q)\dot{s}_q + C(q,\dot{q})s_q = \tau - Y_c \theta_b - H(q)\mathcal{D} \quad (25)$$

then the problem becomes in computing τ in (25) such that $s_q \to 0$ subject to unknown θ_b, and $\mathcal{D}$.

C. Control Law

Consider the following control law

$$\tau = -K_d s_q + Y_c \hat{\theta}_b \quad (26)$$
$$\dot{\hat{\theta}}_b = -\Gamma Y_c^T s_q \quad (27)$$

where $\Gamma = \Gamma^T \in R_+^{p1 \times p1}$, $K_d = K_d^T \in R_+^{p1 \times p1}$. We then have the following result.

Theorem 1: Consider a calibrated camera and the parametric uncertain robot manipulator (1) controlled by the adaptive sliding mode visual servoing scheme (26)-(27). Then, the robotic system yields second order sliding mode regime with global exponential tracking of image-based tracking errors, that is $\Delta x, \Delta \dot{x} \to 0$.

Proof. The closed-loop equation between (25) and (26)-(26) yields

$$H(q)\dot{s}_q = -C(q,\dot{q})s_q - Y_c \Delta \theta_b - H(q)\mathcal{D} - K_d s_q \quad (28)$$
$$\Delta \dot{\theta}_b = \Gamma Y_{br}^T s_q \quad (29)$$

where $\Delta\theta_b = \theta_b - \hat{\theta}_b$ Now, note that the passivity inequality $\langle s_q, \tau^* \rangle$, for τ^* a virtual control input, produces the following energy storage function

$$V = \frac{1}{2}\{s_q^T H(q) s_q + \Delta\theta^T \Gamma^{-1} \Delta\theta\} \quad (30)$$

whose, in turn, its rate of change yields

$$\dot{V} = -s_q^T K_d s_q - s_q^T H(q) J^{-1} R_\alpha^{-1} Z \quad (31)$$

Since $\|Z\|$ is bounded by unity, equation (31) becomes

$$\begin{aligned}\dot{V} &\leq -s_q^T K_d s_q + |s_q| \|H(q)\| |J^{-1} R_\alpha^{-1}| \\ &\leq -s_q^T K_d s_q + \rho_0 |s_q|\end{aligned} \quad (32)$$

where we have used Property 1, and $0 \geq \rho_0 \geq |H(q)||J^{-1}R_\alpha^{-1}|$. Existence of ρ_0 is assured provided that inertial matrix and $J^{-1}R_\alpha^{-1}$ are bounded in the whole workspace away from singularities. Then, there exists a large enough feedback gain K_d such that, according to the Barbalath Lemma, there arises stability of s_q, that is s_q is bounded, with $\mathcal{L}_\infty$ boundedness for $\Delta\theta$. Thus, there exists positive constants $\rho_i, i = 1, ..., 4$ such that $|s_q| \leq \rho_1, |\dot{s}_q| \leq \rho_2$, and since equation (21) $|s_x| \leq \rho_3, |\dot{s}_x| \leq \rho_4$. Therefore, this chain of implications also implies that the state $x, \dot{x}, q, \dot{q}$ are bounded. Now, multiplying the derivative of equation (21) ($\dot{s}_\delta = -K_i sgn(s_\delta) + R_\alpha J \dot{s}_q + R_\alpha \dot{J} s_q$), by using (16), by s_δ^T yields

$$\begin{aligned}s_\delta^T \dot{s}_\delta &= -s_\delta^T K_i sgn(s_\delta) + s_\delta^T R_\alpha J \dot{s}_q + s_\delta^T R_\alpha \dot{J} s_q \\ &\leq -K_i |s_\delta| + \epsilon_0 |s_\delta| |\dot{s}_q| + \epsilon_1 |s_\delta| |\dot{q}| \\ &\leq -\mu |s_\delta|, \quad \mu = K_i - \epsilon_4,\end{aligned} \quad (33)$$

where $\epsilon_0 \geq \rho_2 |R_\alpha J|$, $\epsilon_1 \geq \rho_1 |R_\alpha \frac{\partial J}{\partial q}|$, $\epsilon_2 \geq \epsilon_0 \rho_2$, $\epsilon_3 \geq \epsilon_1 |\dot{q}|$, $\epsilon_4 = \epsilon_2 + \epsilon_3$. Thus, if $K_i > \epsilon_4$, then the sliding mode condition is verified, and a second order sliding mode regime is induced at $s_\delta = 0$ for some time $t = t_s \equiv \frac{|s_q(t_0)|}{\mu}$. However, since $s_\delta(t_0) = 0 \forall t_0$, then $t_s = 0$, and thus, a sliding mode regime is induced for all time at $s_\delta = 0 \rightarrow s = s_d \Rightarrow \Delta\dot{x} = -\lambda\Delta x + s(t_0)exp^{-\kappa t} \Rightarrow x \rightarrow x_d, \dot{x} \rightarrow \dot{x}_d$ exponentially, with continuous control. $\Diamond$

V. WITH UNCALIBRATED CAMERA: EXPONENTIAL CONVERGENCE

A. Uncalibrated joint error manifold

When camera parameters are not known, then joint error manifold s_q is not available, and a new nominal reference $\dot{\bar{q}}_r \neq \dot{q}_r$ is proposed as follows

$$\begin{aligned}\dot{\bar{q}}_r &= \bar{J}^{-1} \bar{R}_\alpha^{-1} \dot{x}_r \\ &= Y_v \bar{\theta}_v\end{aligned} \quad (34)$$

where $Y_v = Y_v(q, \dot{x}_r)$ and $\bar{\theta}_v$ is tuned such that $\bar{J}^{-1}\bar{R}_\alpha^{-1}$ is well-posed. From equations (21), (34) and proposition 1, the uncalibrated joint error manifold $\bar{s}_q$ vector is given by

$$\begin{aligned}\bar{s}_q &= \dot{q} - \dot{\bar{q}}_r \\ &= \dot{q} - \dot{\bar{q}}_r \pm \dot{q}_r \\ &= s_q - Y_v \bar{\theta}_v + Y_v \theta_v \\ &= s_q - Y_v \Delta\theta_v\end{aligned} \quad (35)$$

where $\Delta\theta_v = \bar{\theta}_v - \theta_v$. Thus $\ddot{\bar{q}}_r$ becomes, according to (22) and (34),

$$\begin{aligned}\ddot{\bar{q}}_r &= \bar{J}^{-1} \bar{R}_\alpha^{-1} \ddot{x}_r + \dot{\bar{J}}^{-1} \bar{R}_\alpha^{-1} \dot{x}_r \\ &= \bar{J}^{-1} \bar{R}_\alpha^{-1} (\ddot{x}_{cont} + Z) + \dot{\bar{J}}^{-1} \bar{R}_\alpha^{-1} \dot{x}_r \\ &= \bar{C} + \bar{\mathcal{D}}\end{aligned} \quad (36)$$

where $\bar{C} = \bar{J}^{-1}\bar{R}_\alpha^{-1}\ddot{x}_{cont} + \dot{\bar{J}}^{-1}\bar{R}_\alpha^{-1}\dot{x}_r$ is composed of continuous signals, and $\bar{\mathcal{D}} = \bar{J}^{-1}\bar{R}_\alpha^{-1}Z$ is discontinuous, but bounded, and anyway $\bar{\mathcal{D}}$ cannot be used because it is discontinuous. In order to compensate the effects on robot dynamics due to definition of new nominal references ($\dot{\bar{q}}_r \neq \dot{q}_r, \ddot{\bar{q}}_r \neq \ddot{q}_r$, and therefore $\bar{s}_q \neq s_q$), it is convenient also to find the right expressions for $\dot{\bar{s}}_q$ in terms of $\dot{s}_q$ as follows

$$\begin{aligned}\dot{\bar{s}}_q &= \ddot{q} - \ddot{\bar{q}}_r \equiv \ddot{q} - \ddot{\bar{q}}_r \pm \ddot{q}_r \\ &= (\ddot{q} - \ddot{q}_r) - (\ddot{\bar{q}}_r - \ddot{q}_r) \\ &= \dot{s}_x - (\bar{C} - C) - (\bar{\mathcal{D}} - \mathcal{D}) \\ &= \dot{s}_q - \Delta C - \Delta \mathcal{D}\end{aligned} \quad (37)$$

where $\Delta C = (\bar{J}^{-1}\bar{R}_\alpha^{-1} - J^{-1}R_\alpha^{-1})\ddot{x}_c + (\dot{\bar{J}}^{-1}\bar{R}_\alpha^{-1} - \dot{J}^{-1}R_\alpha^{-1})\dot{x}_r$ that is $\Delta C = Y_{cont}\Delta\bar{\theta}_v + \dot{Y}_v\Delta\bar{\theta}_v$, and $\Delta \mathcal{D} = (\bar{J}^{-1}\bar{R}_\alpha^{-1} - J^{-1}R_\alpha^{-1})Z = Y_z\Delta\bar{\theta}_v$.

B. Open-loop robot dynamics

By using property 2 in terms of (34) and (36), the following regressor arises

$$H(q)\ddot{\bar{q}}_r + C(q,\dot{q})\dot{\bar{q}}_r + G(q) = Y_{\bar{c}}\theta_b + H(q)\bar{\mathcal{D}} \quad (38)$$

where $Y_{\bar{c}}\theta_b = H(q)\bar{C} + C(q,\dot{q})\dot{\bar{q}}_r + G(q)$, with $Y_{\bar{c}}$ as a continuous regressor. Adding and subtracting $(Y_{\bar{c}}\theta_b + H(q)\bar{\mathcal{D}})$ to (1) produces the open-loop error equation

$$H(q)\dot{\bar{s}}_q + C(q,\dot{q})\bar{s}_q = \tau - Y_{\bar{c}}\theta_b - H(q)\bar{\mathcal{D}} \quad (39)$$

At this stage, the problem is dual: in one hand in computing τ such that $\bar{s}_q \rightarrow 0$ be stable subject to unknown θ_b, and unknown $H(q)\bar{\mathcal{D}}$; on the other hand prove that boundedness of $\bar{s}_q$ implies convergence of $\Delta x, \Delta \dot{x}$.

C. Main Result

Consider the following control law

$$\tau = -K_d \bar{s}_q + Y_{\bar{c}} \hat{\theta}_b \quad (40)$$

$$\dot{\hat{\theta}}_b = -\Gamma Y_{\bar{c}}^T \bar{s}_q \quad (41)$$

We then have the following result.

Theorem 2: Consider a robot manipulator (1) with the second order sliding mode visual servoing scheme (40)-(41), subject to unknown robot and all camera parameters. Then, the closed-loop system yields globally exponentially $\lim_{t\to\infty} \Delta x = 0$, $\lim_{t\to\infty} \Delta \dot{x} = 0$.

Proof. The following closed-loop error equation between (1) and (40)-(41) arises

$$H(q)\dot{\bar{s}}_q = -(C(q,\dot{q}) + K_d)\bar{s}_q - Y_{\bar{c}}\Delta\theta_b - H(q)\bar{D}$$
$$\Delta\dot{\hat{\theta}}_b = \Gamma Y_{\bar{c}}^T \bar{s}_q \quad (42)$$

Similarly to theorem 1, the following energy storage function arises from the passivity inequality $\langle \bar{s}_q, \tau^* \rangle$

$$V = \frac{1}{2}\{\bar{s}_q^T H(q)\bar{s}_q + \Delta\theta^T \Gamma^{-1}\Delta\theta\} \quad (43)$$

whose, in turn, its rate of change yields

$$\begin{aligned}\dot{V} &= -\bar{s}_q^T K_d \bar{s}_q - \bar{s}_q^T H(q)\bar{D} \\ &= -\bar{s}_q^T K_d \bar{s}_q - \bar{s}_q^T H(q) \bar{J}^{-1}\bar{R}_\alpha^{-1} Z \\ &\leq -\bar{s}_q^T K_d \bar{s}_q + |\bar{s}_q| |H(q)| |\bar{J}^{-1}\bar{R}_\alpha^{-1}| \\ &\leq -\bar{s}_q^T K_d \bar{s}_q + \rho_5 |\bar{s}_q| \quad (44)\end{aligned}$$

where we have used Property 1, and $0 \geq \rho_5 \geq |H(q)||J^{-1}R_\alpha^{-1}|$. Then, there exists a large enough feedback gain K_d such that, according to the Barbalath Lemma, there arises stability of $\bar{s}_q$. That is $\bar{s}_q$ is bounded, with $\mathcal{L}_\infty$ boundedness for $\Delta\theta$. Thus, there exists positive constants $\rho_j, i = 6, ..., 9$ such that $|\bar{s}_q| \leq \rho_6, |\dot{\bar{s}}_q| \leq \rho_7$, and since equation (21) $|s_x| \leq \rho_8, |\dot{s}_x| \leq \rho_9$. Therefore, this chain of implications also implies that the state $q, \dot{q}$ are bounded. Now, the derivative of equation (21), gives rise to $\bar{s}_q = s_q - Y_v \Delta\theta_v \to \bar{s}_q = J^{-1}R_\alpha^{-1}s_x - Y_v \Delta\theta_v$. Multiplying the previous equation by $R_\alpha J$, becomes

$$R_\alpha J \bar{s}_q = s_x - R_\alpha J Y_v \Delta\theta_v \quad (45)$$

Using (16), equation (45) becomes

$$R_\alpha J \bar{s}_q = s_\delta + K_i \int_{t_0}^t \text{sgn}(s_\delta)(\sigma)d\sigma - R_\alpha J Y_v \Delta\theta_v$$

After few manipulations, finally we arrive to

$$\dot{s}_\delta = -K_i \text{sgn}(s_\delta) + \dot{s}_x \quad (46)$$

Now, in order to produce the sliding mode condition for s_δ, multiply (46) by s_δ^T to obtain

$$\begin{aligned}s_\delta^T \dot{s}_\delta &= -K_i |s_\delta| + s_\delta^T \dot{s}_x \\ &\leq -K_i |s_\delta| + |s_\delta||\dot{s}_x| \\ &\leq -K_i |s_\delta| + |s_\delta||\dot{s}_x| \\ &\leq -K_i |s_\delta| + \rho_9 |s_\delta| \\ &\leq -\nu |s_\delta| \quad (47)\end{aligned}$$

where $\nu = K_i - \rho_9$. Thus, if $K_i > \rho_9$, the sliding mode condition is met, and a second order sliding mode

regime is induced at $s_\delta = 0$ for some time $t = t_s \equiv \frac{|s_q(t_0)|}{\nu}$. Moreover, since $s_\delta(t_0) = 0 \forall t_0$, then $t_s = 0$, and a sliding mode regime is induced for all time at $s_\delta = 0$. This implies the global exponential convergence for image-based tracking errors, i.e. $\Delta x, \Delta \dot{x} \to \lim_{t\to\infty}$ with continuous control. $\diamond$

VI. WITH UNCALIBRATED CAMERA: FINITE-TIME CONVERGENCE

In our previous paper [6], a time-based generator is proposed to induce terminal attractors for the rigid model of a robot manipulator. It is show there that if a feedback gain is properly tuned according to a time base generator, then a time-varying feedback gain induces finite-time convergence of tracking errors, with smooth control effort. Similar idea is developed here to introduce a terminal attractor in the visual error manifold. Surprisingly, the only modification to be made to the controller of previous section is to design a time varying λ in the equation (12). It is easy to see, according to the developments of proof of theorem 1, and [6] that the following holds.

Theorem 3: Consider a robot manipulator (1) and subject to unknown robot and camera parameters in closed loop with the adaptive the second order sliding mode visual servoing scheme (40)-(41), with λ given by

$$\lambda(t) = \lambda_0 \frac{\dot{\xi}}{(1-\xi)+\delta} \quad (48)$$

where $\lambda_0 = 1+\epsilon$, for small positive scalar ϵ such that λ_0 is close to 1, and $0 < \delta \ll 1$. The generator $\xi = \xi(t) \in C^2$ must be provided by the user so as to ξ goes from $0 \to 1$ in finite time t_g. The $\dot{\xi} = \dot{\xi}(t)$ is a bell shaped derivative of ξ such that $\dot{\xi}(t_0) = \dot{\xi}(t_g) \equiv 0$. Then, the closed-loop system yields finite-time convergence $\Delta x(t_s) = 0, \Delta\dot{x}(t_s) = 0$, for an arbitrary given finite time $t = t_s > 0$.

Proof. The proof immediately follows theorem 2 and [6]. $\diamond$

Robot parameter	Value
Length link l_1, l_2	$0.4, 0.3\ m$
Center of gravity 1,2 l_{c1}, l_{c2}	$0.1776, 0.1008\ m$
Mass link m_1, m_2	$9.1, 2.5714\ kg$
Inertia link I_1, I_2	$0.284, 0.0212\ kgm^2$
Gravity acceleration g_z	$-9.8\ m/sec^2$

TABLE I

ROBOT PARAMETERS.

VII. DISCUSSION

The continuous control structure yields tracking of a smooth desired trajectory x_d given in image space by implementing two very well defined feedback control loops, each one with a defined task. The outer robot control loop (26)-(27) establishes passive interconnection,

Vision parameters	Value
Clock-wise rotation angle θ	$\frac{\pi}{8}\ rad$
Scale factor α_v	$77772\ pixeles/m$
Depth field of view z	$1.5\ m$
Camera offset ^vO_b	$[-0.2\ \ -0.1]^T\ m$
Offset $\Sigma_I\ o_I$	$[0.0005\ \ 0.0003]^T\ m$
Focal length λ_f	$0.008\ m$

TABLE II

CAMERA PARAMETERS, AND SET UP.

in terms of s_q, for the adaptive block to compensate for parametric uncertainty of robot parameters, while the inner visual control loop $\dot{v} = sgn(s_\delta)$ plays the role of a second order sliding mode controller for $\dot{s}_\delta$ dynamics.

Though the camera model is quite simple, and 2D visual servoing, and calibration for monocular 2D vision apparently seems to represent a not so hard problem, this result set a step further to propose a solid formal mathematical framework to deal with the largely ignored robot dynamics in the treatment of visual servoing for robot. In this way, this paper allows to include eventually more complicated camera models, with varying depth of field, for 3D case.

With respect to [7] our result proposes faster convergence, which is relevant for visual servoing, and with respect to [3] our paper guarantees a continuous control effort, which is important for any real time application.

Finally, the nontrivial finite time visual error convergence guarantees,in theory, zero tracking errors, despite the dynamics of the robot manipulator. This is the fastest convergence rate the robot may achieve.

VIII. SIMULATIONS

A two-rigid link, planar robot is considered, with a CCD camera, see parameters in tables 1 and 2. All inertial parameters of the robot arm and all parameters of the camera, including depth of field of view, are unknown. The endpoint of the manipulator is requested to draw a circle in image space defined with respect to the vision frame $y_d = (y_{d1}, y_{d2})^T = (0.1\cos\omega t + 0.25, 0.1\sin\omega t + 0.15)$, where $\omega = 0.5$ rad/sec. Simulations are carried in Matlab 6.0, with RungeKutta45 as the numerical solver, at 1 ms. Results are presented in Figure 1 and 2, wherein it can be visualized the exponential, and finite-time tracking capabilities of the controllers of theorems 2 and 3. The spikes observed in the finite-time controller require further tuning on the feedback gains, since there is not apparent reason of such spikes.

IX. CONCLUSIONS

A two servoloops controller for dynamic visual servoing that guarantees tracking tasks in image space for uncertain robot manipulators and uncalibrated camera is proposed. One control loop, an adaptive control loop, compensates for robot uncertain parameters using a joint error manifold, while a visual error manifold stands for a switching manifold for a visual second order sliding mode controller to compensate for camera uncertain parameters. This structure, derived using passivity considerations, allows to straighforwardly introduce a well posed terminal attractors to yield finite-time convergence of image-based tracking errors. The control is continuous, so chattering is not an issue. Our control scheme is valid only for the planar case, but if the jacobian is available, this algorithm can be extended to $n \geq 3$ degress of freedom by using a proper camera projection model. Finally, simulation allows to visualized the expected closed-loop properties of the three theorems.

X. REFERENCES

[1] Hsu, L., and Aquino, P., "Adaptive visual tracking with uncertain manipulator dynamics and uncalibrated camera" Proc. 38th IEEE Conference on Decision and Control, pp. 1248-1253, Phoenix, December 1999.

[2] Bishop, B.E., and M.W. Spong, Adaptive calibration and control of 2D monocular visual servo systems, IFAC Symp. on Robot Control, Nantes, France, 1997.

[3] V. Parra-Vega, J.D. Fierro-Rojas, A. Espinosa-Romero, Uncalibrated Sliding Mode Visual Servoing of Uncertain Robot Manipulators, IEEE/RSJ International Conference on Intelligent Robots and Systems, September 30 - October 4, 2002, Lausanne, Switzerland.

[4] Hutchinson, S., Hager, G.D., and Corke, P.I., A titorial on visual servo control, Trans. on Robotics and Automation, 12:651-670, 1996 .

[5] W.E. Dixon, E. zergeroglu, Y. fang, and D.M. Dawson, Object tracking by a Robot Manipulator: Robust Cooperative Visual Servoing Approach, Proceeding of the 2002 IEEE International Conference on Robotics & Automation, Washinton, D.C., 211-216.

[6] V. Parra-Vega, Second Order Sliding Mode Control for Robot Arms with Time Base Generators for Finite-time Tracking, Dynamics and Control, 11 (2): 175-186, April 2001.

[7] Y. Shen, G.Xiang, Y.H. Liu & K. Li, Uncalibrated visual Servoing of Planar Robots, Proceeding of the 2002 IEEE International Conference on Robotics & Automation, Washinton, D.C., 580-585.

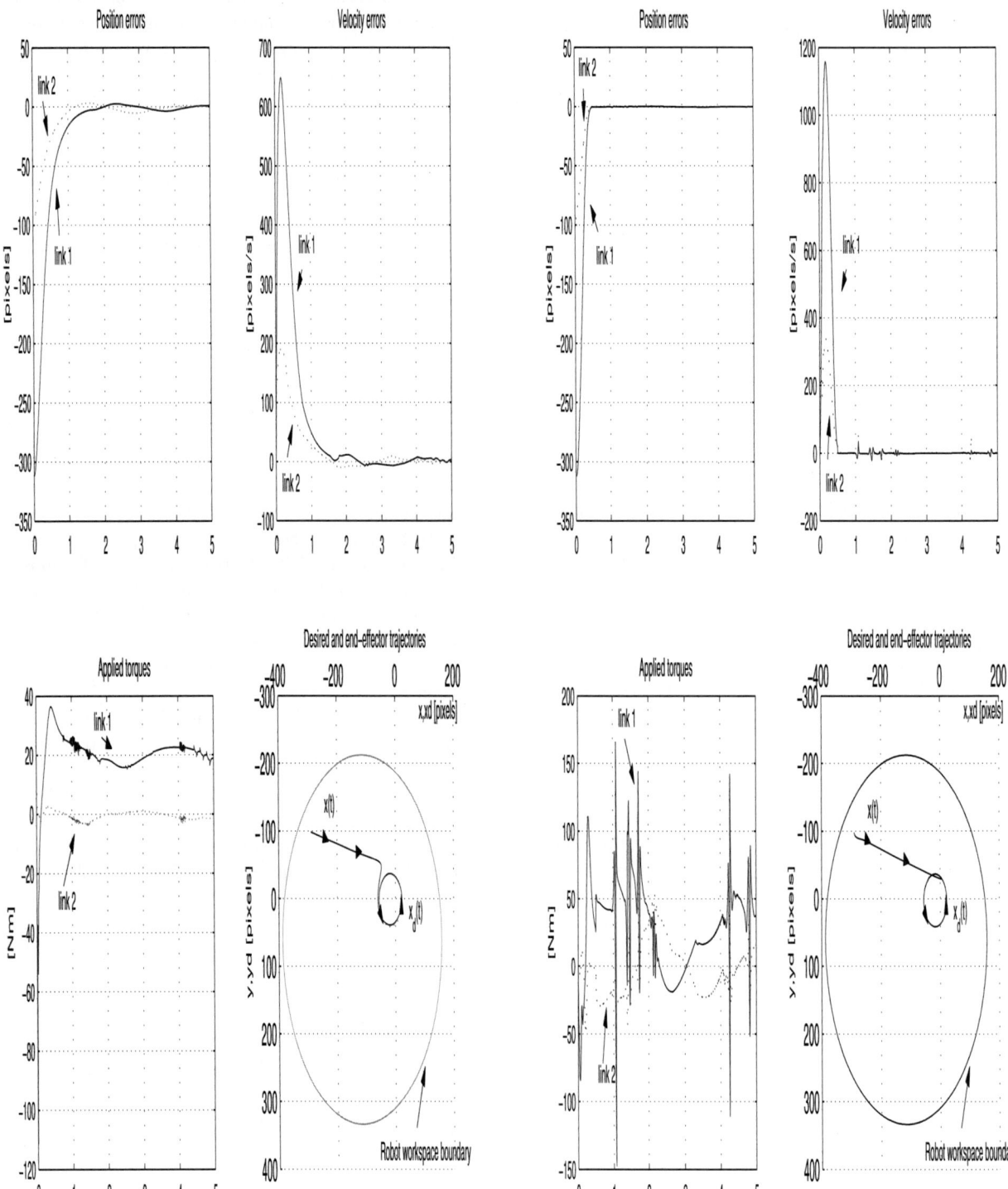

Fig. 1. Exponential tracking case of theorem 2.

Fig. 2. Finite time tracking case, theorem 3.

Visual Servoing Based on Dynamic Vision

Ali Alhaj

Christophe Collewet

François Chaumette

Cemagref
17 Avenue de Cucillé
35044 Rennes Cedex, France
ali.alhaj@cemagref.fr

Cemagref
17 Avenue de Cucillé
35044 Rennes Cedex, France
christophe.collewet@cemagref.fr

IRISA / INRIA Rennes
Campus Universitaire de Beaulieu
35042 Rennes Cedex, France
francois.chaumette@irisa.fr

Abstract

This paper deals with the way to achieve positioning tasks by visual servoing in the case of planar and motionless objects whose shape is unknown. In fact, we consider here complex objects like those one can encount in the agrifood industry. A 3D reconstruction phase is first considered yielding to the parameters of the plane. This computation is based on the measurement of the 2D velocity in a region of interest and on the measurement of the camera velocity. Once the parameters of the plane are sufficiently stable, a visual servoing scheme is used to control the orientation of the camera with respect to the object and to ensure that it remains in the camera field of view for any desired orientation. This 3D reconstruction phase is maintained active during the servoing to improve the accuracy of the parameters and, consequently, to obtain a small positioning error. Finally, experimental results validate the proposed approach as well as for its effectiveness than for the obtained accuracy.

1 Introduction

2D visual servoing allows the realization of robotic tasks directly from visual features acquired by a camera. These features are compared with the desired ones, extracted from the desired position of the camera with respect to the considered object[1]. Nevertheless, we still cannot achieve positioning tasks with regard to partially known objects. Indeed, except rigid manufactured goods for which a model often exists, we rarely have a precise description of the object or of the desired visual features, either because these objects can be subject to deformations or simply because of their natural variability. Such cases can appear for example in surgical domain, agrifood industry, agriculture or in unknown environments (underwater, space). In [2], the authors use a specific motion to perform an alignment task without a precise description of the desired visual features. Unfortunately, their study is restricted to planar motions. In [3], thanks to dynamic visual features, a positioning task consisting in moving the camera to a position parallel to a planar object of unknown shape is achieved. However, such an approach needs particular motion parameters estimation [4] currently leading to a high computation duration and, consequently, to a low control scheme rate. In addition, this approach cannot be used for any specified orientation of the camera. This case has been taken into account in [5], where geometric features are used. This approach is based on the maximization in the image of the surface of a triangle built from three feature points. To do that, three tasks have to be performed sequentially yielding, in some cases, to excessive durations of the task. Moreover, tracking the feature points can be difficult, depending, in the case of agrifood products, on their texture.

The approach described in this paper proposes to treat the same problem, that is the realization of positioning tasks with respect to a planar and motionless object of unknown shape for any specified orientation of the camera. Since the shape of the object is considered as unknown, a 3D reconstruction phase by *dynamic vision* is first performed. This computation is based, as in [3], on the measurement of the 2D motion in a region of interest, but here we use it to recover the structure of the object. This way to proceed allows, as will be shown later, more flexibility to synthetize the control law, in particular to ensure that the object remains in the camera field of view.

The paper is organized as follows: first, we present in Section 2 a brief review on previous works relevant to 3D reconstruction by dynamic vision. We formulate then the problem in Section 3 and describe how to obtain the structure of the object in Section 4. Section 5 details the way we synthetize the control law. Finally, experimental results concerning objects of unknown shape are presented in Section 6.

2 Previous works

Let us consider a point P of the object described by $\underline{P} = (X, Y, Z)^T$ in the camera frame, with the Z axis the camera optical axis. Assuming without loss of generality a unit focal length, this point projects in p, described by $\underline{p} = (x, y, 1)^T$, according to

$$\underline{p} = \frac{\underline{P}}{Z} \quad (1)$$

which yields to the well-known relation [6]

$$\begin{pmatrix} \dot{x} \\ \dot{y} \end{pmatrix} = \begin{pmatrix} -1/Z & 0 & x/Z & xy & -1-x^2 & y \\ 0 & 1/Z & y/Z & 1+y^2 & -xy & -x \end{pmatrix} T_c \quad (2)$$

where $T_c = (\underline{V}^T, \underline{\Omega}^T)^T$ is the camera velocity and $\underline{V} = (V_x, V_y, V_z)^T$ and $\underline{\Omega} = (\Omega_x, \Omega_y, \Omega_z)^T$ its translational and rotational components respectively. This relation can be rewritten as follows

$$\begin{cases} \dot{x} = \dfrac{-V_x + xV_z}{Z} + xy\Omega_x - (1+x^2)\Omega_y + y\Omega_z \\ \dot{y} = \dfrac{-V_y + yV_z}{Z} + (1+y^2)\Omega_x - xy\Omega_y - x\Omega_z. \end{cases} \quad (3)$$

In this equation, only the depth Z is unknown if $\underline{p}$, $\underline{\dot{p}}$ and T_c can be measured.

Various ways to estimate Z exist, they are based on different approaches to cope with $\underline{\dot{p}}$. The most immediate way is to approximate the velocities $\dot{x}$ ($\dot{y}$) by $\frac{\Delta x}{\Delta t}$ ($\frac{\Delta y}{\Delta t}$) [7]. However, this method does not provide accurate results because of errors introduced by the discretization. A way to greatly improve these results is to impose $\dot{x} = \dot{y} = 0$ [8]. This may be realized easily using visual servoing, this approach is then relevant to *active vision*. Another approach is based on the assumption that the brightness of p remains constant during the motion. This assumption leads to the well-known additional constraint [6]

$$\dot{x}I_x + \dot{y}I_y + I_t = 0 \quad (4)$$

where I_x, I_y and I_t represent the spatio-temporal derivatives of the intensity of the considered point in the image. By substituting $\dot{x}$ and $\dot{y}$ given by (3) in (4), an expression of Z can be obtained [9, 10] (note that these works treat the more general case where T_c is also supposed to be unknown). Such approaches, known as *direct approaches*, require accurate estimations of the derivatives I_x, I_y and I_t and therefore, are not very accurate in practice. Another way is to locally model the surface of the object in the neighborhood of P. That provides an expression of $1/Z$ in function of the chosen parameterization, which can be used in (3) to exhibit a parametric model of the 2D motion. On the other hand, these parameters can be obtained by a method of computation of the 2D motion, like in [4] for example. Finally, an expression of the structure of the object can be extracted [11] (here too, by considering a second point, the case where T_c is unknown is treated). These approaches are known as *indirect approaches* since they require an intermediate computation of the 2D motion.

The main benefit of our approach with regard to the previous works is that we explicitly use parameters obtained by 3D reconstruction in the control scheme. This allows us to compute the orientation error and to synthetize easily the control law, in particular to take into account any desired orientation of the camera.

3 Modeling

If we assume that the considered object is planar, or at least planar in a neighborhood of P, we have

$$Z = AX + BY + C \quad (5)$$

which, according to (1), can be rewritten in function of $\underline{p}$ as follows

$$\frac{1}{Z} = \alpha x + \beta y + \gamma \quad (6)$$

with $\alpha = -\frac{A}{C}$, $\beta = -\frac{B}{C}$ and $\gamma = \frac{1}{C}$. By substituting this expression in (3), one can show that the 2D motion is exactly modeled by a parametric model with 8 parameters [12]. By neglecting the second order terms, we obtain

$$\begin{pmatrix} \dot{x} \\ \dot{y} \end{pmatrix} = \begin{pmatrix} a_0 + a_1 x + a_2 y \\ b_0 + b_1 x + b_2 y \end{pmatrix} \quad (7)$$

where

$$\begin{cases} a_0 &=& -\Omega_y - \gamma V_x \\ a_1 &=& -\alpha V_x + \gamma V_z \\ a_2 &=& \Omega_z - \beta V_x \\ b_0 &=& \Omega_x - \gamma V_y \\ b_1 &=& -\Omega_z - \alpha V_y \\ b_2 &=& -\beta V_y + \gamma V_z \end{cases} \quad (8)$$

which can be rewritten under a matrix form as follows

$$M\underline{\Theta} = \underline{\Gamma} \quad (9)$$

with

$$M = \begin{pmatrix} 0 & 0 & -V_x \\ -V_x & 0 & V_z \\ 0 & -V_x & 0 \\ 0 & 0 & -V_y \\ -V_y & 0 & 0 \\ 0 & -V_y & V_z \end{pmatrix}, \quad \underline{\Gamma} = \begin{pmatrix} a_0 + \Omega_y \\ a_1 \\ a_2 - \Omega_z \\ b_0 - \Omega_x \\ b_1 + \Omega_z \\ b_2 \end{pmatrix} \quad (10)$$

and
$$\underline{\Theta} = \begin{pmatrix} \alpha & \beta & \gamma \end{pmatrix}^T. \quad (11)$$

Using a measure of the 2D motion parameters and the camera velocity, $\underline{\Theta}$ can be easily obtained by solving (9):
$$\underline{\Theta} = (M^T M)^{-1} M^T \underline{\Gamma}. \quad (12)$$

However, the solution is correct only if the matrix $M^T M$ is well conditioned, that is if
$$\min(\nu_i) > \nu_s, \ i = 1 \ldots 3 \quad (13)$$

where the ν_i's are the eigenvalues of $M^T M$ and ν_s a given threshold. We will use this property in Section 6.

Next Sections 4 and 5 show respectively how to compute the 2D motion parameters and to synthetize the control law to achieve the positioning task.

4 Estimation of the 2D motion parameters

The algorithms for the computation of 2D motion parameters given by (7) are relatively complex and not very easy to implement. In addition, they are expensive with regard to the computational cost. In fact, it is more judicious to investigate an approach based on the measurement of the displacement in the image rather than on the use of the velocity. This approach is similar to track the point p from an image to another. We review here the most classical approach [13] and we will see that its degree of complexity does not exceed that of the inversion of a 6x6 matrix.

Here too, we assume that the brightness of p remains unchanged during the motion, so we can write
$$I(x, y, t) = I(x + \dot{x}\Delta t, y + \dot{y}\Delta t, t + \Delta t) \quad (14)$$

where Δt represents the control scheme period, and $\dot{x}$ and $\dot{y}$ are modeled by (7).

Because of the noise, (14) is generally not satisfied. Therefore, a solution is to move the problem to an optimization one to find the parameters which have to minimize the following residue
$$\epsilon = \sum_{W} (I(x, y, t) - I(x + \dot{x}\Delta t, y + \dot{y}\Delta t, t + \Delta t))^2 \quad (15)$$

where W is a window of interest centered in p.

To carry out the optimization, we have to assume that Δt and the displacements are sufficiently small. If so, a first order Taylor expansion of $I(x + \dot{x}\Delta t, y + \dot{y}\Delta t, t + \Delta t)$ can be performed and substituted in (15) to obtain a linear system in function of the required parameters. Usually, this system is inverted by using an iterative Newton-Raphson style algorithm to take into account the error introduced by the Taylor expansion.

To ensure the convergence of the minimization process, p is selected from points of interest extracted from the first image. Moreover, we choose p as the best point in the sense that it will be correctly tracked during the motion. To do that, the method described in [14] has been used.

5 Control law

First, let us remember the task to achieve. The goal is to ensure a given final orientation of the camera with respect to plane π described by (5) and, also to ensure that P will still remain in the camera field of view.

Once $\underline{\Theta}$ is estimated, the unit normal $\underline{n}$ of plane π in P in the camera frame can be derived. However, in the case of any orientation we rather have to consider $\underline{n}^* = \mathcal{R}\underline{n}$ where $\mathcal{R}$ is the rotation matrix computed from the desired orientation (see Figure 1). Therefore, we have to move the camera so that $\underline{Z} = \underline{n}_c$ with $\underline{Z}$ the unit vector carried by the optical axis and $\underline{n}_c = -\underline{n}^*$. This rotation to perform can be expressed under the form $\underline{u}\theta$ where $\underline{u}$ represents the rotation axis and θ the rotation angle around this axis
$$\underline{u} = \frac{\underline{Z} \wedge \underline{n}_c}{\|\underline{Z} \wedge \underline{n}_c\|} \quad (16)$$

and
$$\theta = \arccos(\underline{Z}.\underline{n}_c) \quad (17)$$

The camera orientation being known, it is possible to compute the control law. We used the one described in [15]. Indeed, it ensures that P remains in the camera field of view since the trajectory of p is a straight line between the current position p and the desired position p^* (which has been chosen as the principal point of the image). We describe here briefly this approach known as *hybrid visual servoing*.

First, $\underline{p}_r$ is defined as follows
$$\underline{p}_r = \frac{1}{Z^*}\underline{P} = \frac{Z}{Z^*}\underline{p} \quad (18)$$

with Z^* the desired depth for P in final position. One can then show that
$$\underline{\dot{p}}_r = \left(-\frac{1}{Z^*}\mathbb{I}_3 \ [\underline{p}_r]_\times\right) T_c \quad (19)$$

where the notation $[\underline{v}]_\times$ denotes the antisymmetric matrix associated to $\underline{v}$.

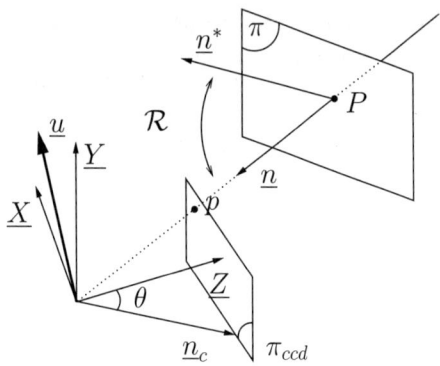

Figure 1: Rotation to perform by the camera.

The change of the orientation can be expressed as

$$\frac{d(\underline{u}\theta)}{dt} = \begin{pmatrix} 0 & L_\Omega \end{pmatrix} T_c \quad (20)$$

with

$$L_\Omega(\underline{u},\theta) = \mathbb{I}_3 - \frac{\theta}{2}[\underline{u}]_\times + \left(1 - \frac{\text{sinc}(\theta)}{\text{sinc}^2\left(\frac{\theta}{2}\right)}\right)[\underline{u}]_\times^2 \quad (21)$$

where $\text{sinc}(\theta) = \sin(\theta)/\theta$.

Expressions (19) and (20) can be merged as follows

$$\frac{d}{dt}\begin{pmatrix} \underline{p}_r \\ \underline{u}\theta \end{pmatrix} = \begin{pmatrix} -\frac{1}{Z^*}\mathbb{I}_3 & [\underline{p}_r]_\times \\ 0 & L_\Omega \end{pmatrix} T_c = L T_c \quad (22)$$

Finally, the desired positioning task can be expressed in term of regulation to zero of the following task function

$$\underline{e} = \begin{pmatrix} \underline{p}_r - \underline{p}_r^* \\ \underline{u}\theta \end{pmatrix} \quad (23)$$

yielding to the camera velocity

$$T_c = -\lambda \widehat{L}^{-1} \underline{e} \quad (24)$$

λ being a positive gain and $\widehat{L}$ an approximation of L.

Let us note that the values of Z and $\underline{p}$ required for the computation of $\underline{p}_r$ are obtained respectively thanks to (6) and by integration of (7).

6 Experimental results

In order to validate the proposed algorithm, we present here experimental results for three different desired orientations. The experimental system is described in [5].

The object consists of a photograph of a trout steak fixed on a planar support. To evaluate the positioning accuracy of our method, this support makes possible to express precisely the transformation matrix between the camera frame and the object one with the method used in [5]. This matrix is characterized by the Euler's angles denoted ϕ_X, ϕ_Y, ϕ_Z which respectively represent the angles of the X, Y and Z rotations.

Furthermore, since the object is motionless, one can improve the accuracy on $\underline{\Theta}$. Indeed, in a fixed frame, one can express a value $\underline{\Theta}^f$ that can be filtered since a fixed value has to be obtained. Thereafter, this value is expressed in the camera frame to be used in the control law. Moreover, proceeding this way allows to know when $\underline{\Theta}^f$ is stable enough to be used in the control law (typically fifteen acquisitions are sufficient). Thus, a preliminary phase with a constant velocity is required. We imposed $V_x = V_y = 2$ cm/s, $V_z = 0$ and $\underline{\Omega} = 0$.

Finally, the algorithm consists of three phases, a first phase at constant velocity, a second phase when both reconstruction and servoing are performed, and a last phase where only the servoing operates. This last phase occurs when the constraint given by (13) is no more satisfied (which occurs near convergence when the motion is very small).

The following values were used for all the experiments: $\lambda = 0.06$, $\nu_s = 1.10^{-6}$, $\Delta t = 800$ ms, $Z^* = 70$ cm and the size of W has been fixed to 55×55 pixels.

The first experiment consists in positioning the camera parallel to π. Figures 2.a depicts respectively the behavior of parameters A, B and C in a fixed frame; Figures 2.b the components of $\underline{n}$; Figures 2.c-d the behavior of the motion parameters, respectively a_0, b_0 and a_i, b_i ($i = 1,2$). The components of the camera velocity are represented on Figure 2.e, the normalized error on Figure 2.f (defined as $\frac{\|\underline{e}\|_{t=k\Delta t}}{\|\underline{e}\|_{t=0}}$), the estimated depth on Figure 2.g and finally, the initial and final images are reported respectively on Figures 2.h-i.

First, Figure 2.f confirms that the control law converges since the normalized error tends towards zero, as well as $\widehat{Z}$ towards Z^* (Figure 2.g). In the same way, as expected, p tends towards the principal point following a straight line (see Figure 2.i where the trajectory of p is drawn, the first segment corresponding to the motion relative to the preliminary phase). One can also remark on Figure 2.a that the reconstruction stops around the 120th iteration, while the stop condition relative to the normalized error is not reached before the 300th iteration approximately. For this experiment, the initial orientation of the camera was $\phi_X = 9.66°$, $\phi_Y = 23.77°$ and $\phi_Z = 9.46°$, the orientation after the servoing was $\phi_X = 2.44°$ and $\phi_Y = 1.39°$ (let us recall that ϕ_Z is not controlled).

Consequently, the positioning error is less than 2.5°.

The second experiment consists in positioning the camera so that $\phi_X = 0°$ and $\phi_Y = 20°$. Figures 3.a-i describe the behavior of the same variables as for the previous experiment. The same comments can be made, in particular concerning the convergence of the control law. However, as seen in Figure 3.i, the trajectory of p does not follow a straight line, probably, because of a relatively bad behavior of the point tracking algorithm. But this phenomenon has no bad consequence, since the positioning error remains small: the final orientation is $\phi_X = -0.65°$ and $\phi_Y = 20.16°$. For this experiment the initial orientation was $\phi_X = 16.88°$, $\phi_Y = 11.22°$ and $\phi_Z = 7.40°$.

For the last experiment the desired orientation was $\phi_X = -10°$ and $\phi_Y = 0°$. Figures 4.a-i describe the same variables as previously. Here again, the positioning error is small, we have $\phi_X = -8.90°$ and $\phi_Y = -0.06°$. The initial orientation was $\phi_X = 6.58°$, $\phi_Y = -10.94°$ and $\phi_Z = 2.83°$.

7 Conclusion and future works

We have presented in this paper a way to achieve positioning tasks by visual servoing when the desired image of the object cannot be precisely described and for any desired orientation of the camera. However, we have to assume the object to be planar and motionless. The approach is based on a 3D reconstruction which allows the estimation of the current orientation of the object with respect to the camera, and thereafter to the elaboration of the control law. Experimental results validated our algorithm, low positioning errors were observed ($\approx 2°$). However, we can regret that our method is quite sensitive to the calibration of the robot since it necessitates a measurement of the camera velocity.

In the future, an interesting prospect is to extend the algorithm to the case of nonplanar objects. The algorithm should not be too much modified by considering locally a parametric modeling around P.

References

[1] S. Hutchinson, G. D. Hager, and P. I. Corke, "A tutorial on visual servo control," *IEEE Trans. on Robotics and Automation*, vol. 12, no. 5, pp. 651–670, October 1996.

[2] B. Yoshimi and P. K. Allen, "Active uncalibrated visual servoing," in *IEEE Int. Conf. on Robotics and Automation, ICRA'94*, San Diego, May 1994, pp. 156–161.

[3] A. Crétual and F. Chaumette, "Visual servoing based on image motion," *Int. Journal of Robotics Research*, vol. 20, no. 11, pp. 857–877, November 2001.

[4] J.M. Odobez and P. Bouthemy, "Robust multiresolution estimation of parametric motion models," *Journal of Visual Communication and Image Representation*, vol. 6, no. 4, pp. 348–365, December 1995.

[5] C. Collewet and F. Chaumette, "Positioning a camera with respect to planar objects of unknown shape by coupling 2-d visual servoing and 3-d estimations," *IEEE Trans. on Robotics and Automation*, vol. 18, no. 3, pp. 322–333, June 2002.

[6] B.K.P. Horn and B.G. Schunck, "Determining optical flow," *Artificial Intelligence*, vol. 16, no. 1–3, pp. 185–203, August 1981.

[7] P. Rives and M. Xie, "Toward dynamic vision," in *Proc. IEEE Workshop on Interpretation of 3D scenes*, Austin, Texas, November 1989.

[8] F. Chaumette, S. Boukir, P. Bouthemy, and D. Juvin, "Structure from controlled motion," *IEEE Trans. on Pattern Analysis and Machine Intelligence*, vol. 18, no. 5, pp. 492–504, May 1996.

[9] B. K. P. Horn and E. J. Weldon, "Direct methods for recovering motion," *Int. Journal of Computer Vision*, vol. 2, no. 1, pp. 51–76, June 1988.

[10] S. Negahdaripour and B. K. P. Horn, "Direct passive navigation," *IEEE Trans. on Pattern Analysis and Machine Intelligence*, vol. 9, no. 1, pp. 168–176, January 1987.

[11] S. Negahdaripour and S. Lee, "Motion recovery from image sequences using only first order optical flow information," *Int. Journal of Computer Vision*, vol. 9, no. 3, pp. 163–184, 1992.

[12] G. Adiv, "Determining 3d motion and structure from optical flow generated by several moving objects," *IEEE Trans. on Pattern Analysis and Machine Intelligence*, vol. 7, no. 4, pp. 384–401, July 1985.

[13] J. Shi and C. Tomasi, "Good features to track," in *IEEE Int. Conf. on Computer Vision and Pattern Recognition, CVPR'94*, Seattle, June 1994, pp. 593–600.

[14] C. Kermad and C. Collewet, "Improving feature tracking by robust points of interest selection," in *6th International Fall Workshop on Vision, Modeling, and Visualization, VMV'2001*, Stuttgart, Germany, November 21-23, 2001, pp. 415–422.

[15] E. Malis and F. Chaumette, "Theoretical improvements in the stability analysis of a new class of model-free visual servoing methods," *IEEE Trans. on Robotics and Automation*, vol. 18, no. 2, pp. 176–186, April 2002.

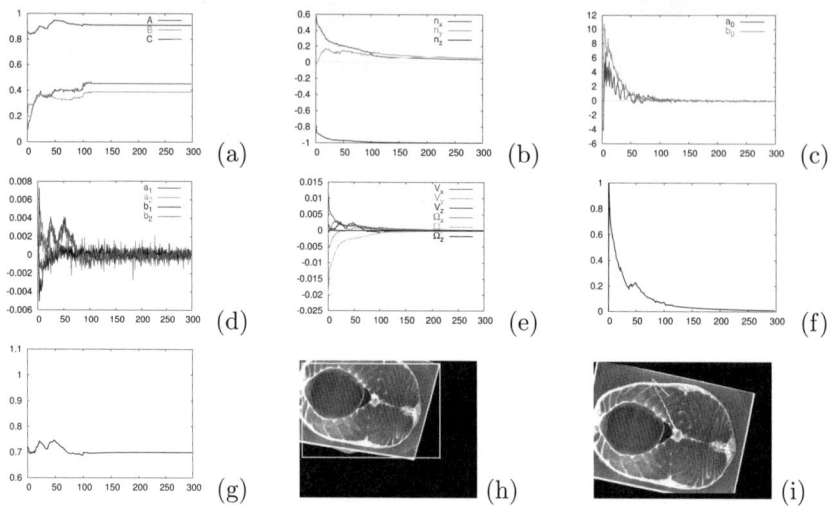

Figure 2: 1st experiment: (a) Parameters of the plane in a fixed frame. (b) Components of the normal $\underline{n}$ in the camera frame. (c) Parameters a_0, b_0. (d) Parameters a_1, b_1, a_2, b_2. (e) Kinematic screw. (f) Normalized error. (g) Estimated depth $\widehat{Z}$. (h) Initial image. (i) Final image.

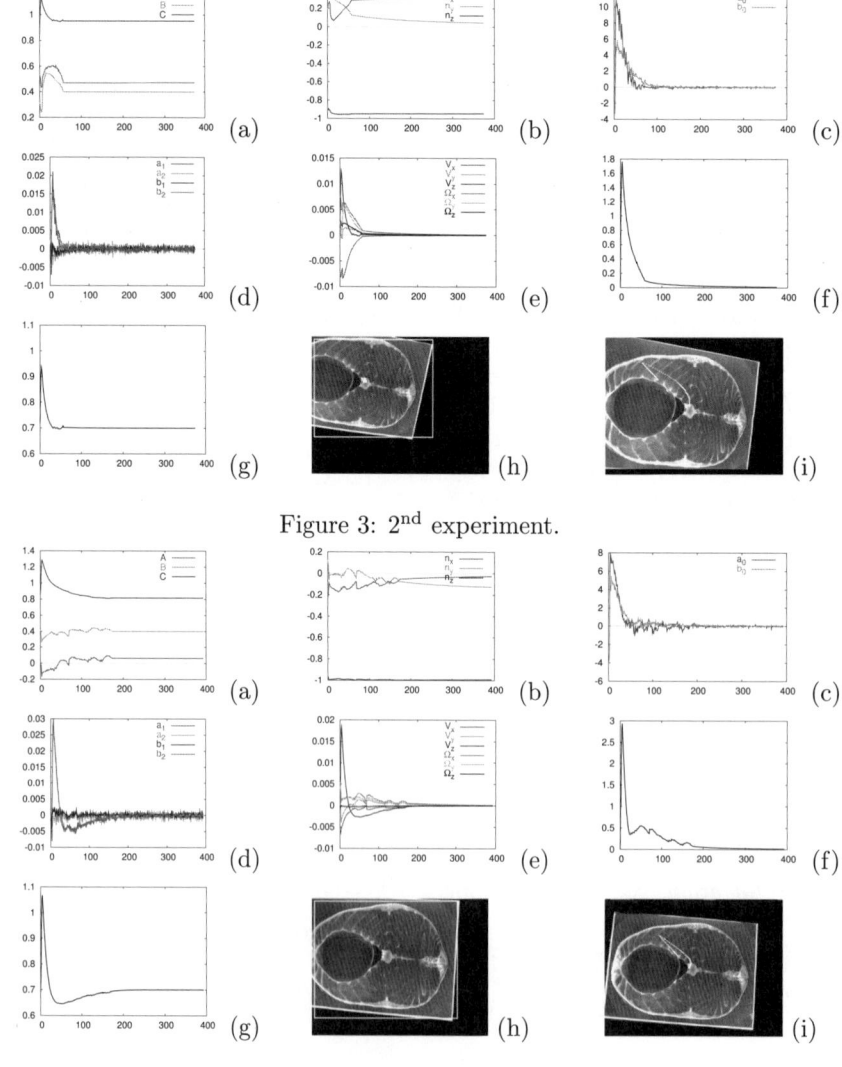

Figure 3: 2nd experiment.

Figure 4: 3rd experiment.

An Experimental Study of Hybrid Switched System Approaches to Visual Servoing

Nicholas R. Gans and Seth A. Hutchinson

ngans@uiuc.edu, seth@uiuc.edu
Dept. of Electrical and Computer Engineering
The Beckman Institute for Advanced Science and Technology
University of Illinois at Urbana-Champaign
Urbana, IL USA

Abstract—In the recent past, many researchers have developed control algorithms for visual servo applications. In this paper, we introduce a new hybrid switched system approach, in which a high-level decision maker selects between two visual servo controllers. We have evaluated our approach with simulations and experiments using three individual visual servo systems and three candidate switching rules. The proposed method is very promising for visual servo tasks in which there is a significant distance between the initial and goal configuration, or the task is one that can cause an individual visual servo system to fail.

I. Intro

Visual servoing has proven to be a highly effective means to control a robot manipulator through the use of visual data. It provides a high degree of accuracy using even simple camera systems and robustness in the face of signal error and uncertainty of system parameters.

Visual servo methods have classically been divided into two camps, Position Based Visual Servoing (PBVS) and Image Based Visual Servoing (IBVS). There are extensive resources detailing these methods [1–4]. In the late nineties, Chaummette outlined a number of problems that cannot be solved using the traditional local linearized approaches to visual servo control [5]. This resulted in a variety of partitioned visual servo systems which used the image Jacobian linearization of IBVS for specific degrees of freedom, and 3D techniques exemplified in PBVS for the remainder [6–10].

Rather than combining systems, another approach is the use of hybrid switched systems, i.e., systems comprised of a set of continuous subsystems along with a discrete switching control [11, 12]. Hybrid switched systems can offer an increased region of stability and increased rate of convergence, and there exists the potential to switch between unstable systems in a pattern that makes the total system stable.

Section II will provide an introduction to the theory behind hybrid switched systems. In Section III, we discuss our individual visual servo systems. In Sections IV and V we will present the two switched systems along with their simulated and experimental results.

II. Hybrid Switched System Control

The theory of hybrid switched control systems, i.e., systems that comprise a number of continuous subsystems and a discrete system that switches between them, has received notable attention in the control theory community [11–13]. In general, a hybrid switched system can be represented by the differential equation

$$\dot{x} = f_\sigma(x) : \sigma \in \{1..n\} \quad (1)$$

where f_σ is a collection of n distinct functions. For our purposes, it is convenient to explicitly note that the switching behavior directly affects the choice of the control input u

$$\dot{x} = f_\sigma(x, u_\sigma) : \sigma \in \{1..n\}. \quad (2)$$

A useful interpretation is to consider σ to be a discrete signal, switching among discrete values in 1..n. The value σ at time t determines which function $f(x, u_\sigma)$ is used. The signal σ is typically classified as state-dependent or time-dependent, depending on whether switches are caused by the state of x or the time t, although overlap does exist between these classes. In our research we explored state-dependent switching contingent on the state of the image plane or camera pose, a time-dependent switch induced by a random variable, and a combined method where a random variable influenced by the state determined switches.

The systems we present are each comprised of two visual servo controllers; each visual servo controller provides a velocity screw, $u = [T_x, T_y, T_z, \omega_x, \omega_y, \omega_z]^T$, and a switching rule determines which is used as the actual control input at each control cycle.

The stability of a switched system is not insured by the stability of the individual controllers. Indeed, a collection of stable systems can become unstable when inappropriately switched. As an illustration, Figure 1 (from [13]) shows trajectories for two asymptotically stable subsystems in (a) and (b). A set of switches resulting in a stable system is shown in (c), while a series of switches resulting in an unstable system are shown in (d).

Stability of switched systems can be proven using Lyapunov's direct method [13, 14]. Generally this requires establishing a common Lyapunov function that works for all subsystems. Alternately, one can establish a family of Lyapunov functions for the systems such that at each switch, the value of the function at the end of that interval is less than the value of the function of the interval that proceeded it, as illustrated for a one dimensional family of two functions in Figure 2.

Stability of a switched system can be extremely difficult to prove. However, we have performed extensive empirical evaluations that demonstrate the efficacy of our approach. We will turn our attention to establishing stability in the near future.

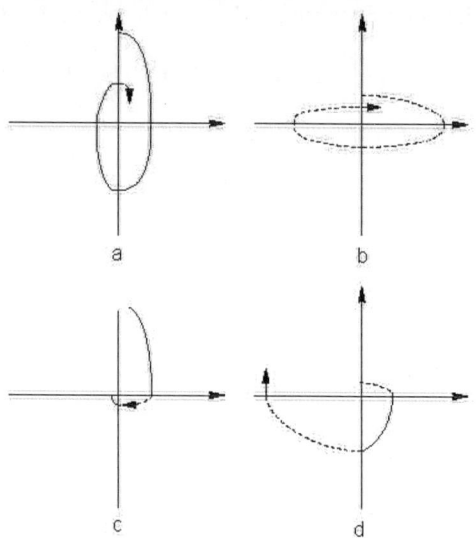

Fig. 1. trajectories of switched systems

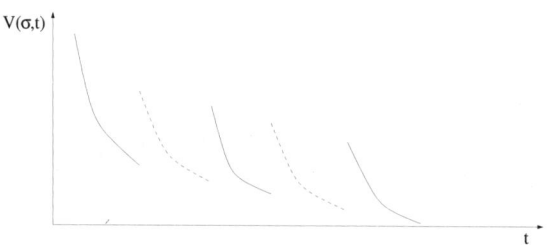

Fig. 2. stable family of Lyapunov function

III. THREE VISUAL SERVO CONTROLLERS

We present the three visual servo controllers used by our switched systems.

A. Homography Based Controller

The homography method exploits the epipolar constraints between two images of planar feature points. The homography matrix has been used previously for visual servoing in [6, 8] to control a restricted set of degrees of freedom. We however, use it to control all degrees of freedom.

Define $\mathbf{f}^*$, $\mathbf{f}$, as the homogeneous coordinates in two images of a set of 3D points lying on a plane π, where * denotes features in the goal image.. These are related by

$$\mathbf{f}^* = \mathbf{H}\mathbf{f} \quad (3)$$

where $\mathbf{H}$ is the calibrated homography matrix. As shown in [15, 16], $\mathbf{H}$ can be decomposed as

$$\mathbf{H} = \mathbf{R}(\mathbf{I}_3 - \frac{\mathbf{t}\mathbf{n}^T}{d}) \quad (4)$$

where $\mathbf{I}_3$ is a 3×3 identity matrix and $\mathbf{R}$ and $\mathbf{t}$ are the rotation matrix and translation vector, respectively, relating the two camera views. The parameter $\mathbf{n}$ is the the normal of the plane π and describes the orientation of π with respect to the current camera view; d is the distance from the current camera origin to the plane π. We calculate the vector $\mathbf{T} = [T_x, T_y, T_z]^T = \hat{d}\mathbf{t}$, where $\hat{d}$ is an estimate of d. Given knowledge of the geometry of the feature point locations it is possible to accurately estimate $\hat{d}$ and so determine $\mathbf{t}$ to the proper scale. From the rotation matrix $\mathbf{R}$, we extract the roll, pitch and yaw angles, $\omega_z, \omega_x, \omega_y$, obtaining the velocity screw $u = k[T_x, T_y, T_z, \omega_x, \omega_y, \omega_z]$ in which k is a scalar gain constant, or a 6×6 gain matrix..

Of the numerous methods to calculate $\mathbf{H}$, we have used a linear solution since visual servoing, in general, requires quicker calculations than iterative methods may provide. Decomposing the homography as in (4) is not a trivial exercise and generally cannot be solved to a unique solution. Additional information pertaining to the use of the homography in visual servoing can be found in [6]

Since this method provides rotation and translation vectors to completely realize the camera's goal position from its current position, it shares many of the performance characteristics of PBVS systems. Namely this system will perform optimally in Cartesian space. The end effector will typically follow the shortest path to the goal position. This, however, can lead to large motions of the features in the image space. This can cause the feature points to leave the field of view, resulting in system failure. We will define system failure as any time that a system cannot zero the error within 250 iterations.

B. Affine-Approximation Controller

For camera motions that do not involve rotation about the camera x- or y- axes, the initial and goal images will be related by an affine transformation. While this is a constrained set of motions, it is common in many situations such as aligning camera with a component on a conveyer belt.

Define $\tilde{\mathbf{f}}^*$, $\tilde{\mathbf{f}}$, as the calibrated pixel coordinates of two points in the image plane. Again, * indicates the features are in the goal image. Then these points are related by the affine transformation

$$\begin{aligned}\tilde{\mathbf{f}}^* &= \mathbf{A}\tilde{\mathbf{f}} + \mathbf{b} \\ &= \begin{bmatrix} 1 & a \\ 0 & 1 \end{bmatrix} \begin{bmatrix} s_1 & 0 \\ 0 & s_2 \end{bmatrix} \begin{bmatrix} C_\theta & -S_\theta \\ S_\theta & C_\theta \end{bmatrix} \begin{bmatrix} f_x \\ f_y \end{bmatrix} + \begin{bmatrix} t_x \\ t_y \end{bmatrix}\end{aligned} \quad (5)$$

in which C_θ and S_θ denote respectively $\cos\theta$ and $\sin\theta$; f_x and f_y are image point coordinates; a, s_i, and θ describe the skew, scale, and rotation respectively; and $\mathbf{b}$ is the translation. Both $\mathbf{A}$ and $\mathbf{b}$ can be obtained by solving a linear system of equations, and QR decomposition can then be used to determine a, s_i, and θ.

$$\tilde{\mathbf{f}}^* = \mathbf{R}\mathbf{Q}\mathbf{f} + \mathbf{b} = \begin{bmatrix} r_{11} & r_{12} \\ 0 & r_{22} \end{bmatrix} \begin{bmatrix} q_{21} & q_{12} \\ q_{11} & q_{22} \end{bmatrix} \begin{bmatrix} f_x \\ f_y \end{bmatrix} + \begin{bmatrix} t_x \\ t_y \end{bmatrix} \quad (6)$$

The $\mathbf{Q}$ matrix is a permutation of the rotation matrix in (5), and rotation θ about the camera z-axis equals $\arcsin(q_{21})$. During an affine transformation, the rotations about the $x-$ and $y-$ axes are, by definition, zero. The r_{11} and r_{22} of (6) respectively equal the s_1 and s_2 parameters of (5) and provide $z-$axis translation to scale. Translation along the $x-$ and $y-$ axes are defined to scale in the vector $\mathbf{b}$. Multiplying the scaled translations by a depth estimate will provide true values. Again, knowledge of the feature point geometry will allow for the depth estimate to be accurately derived.

Given a, s_i, θ, t_x, t_y we again have the position and orientation relating the initial and camera goal positions. This controller provides the velocity screw $u = k[t_x, t_y, s_2, 0, 0, \theta]$ where k is a gain constant or matrix. Note that if there is no rotation about the $x-$ or $y-$axes, we will have $s_1 = s_2$.

C. IBVS

There is a vast amount of literature regarding IBVS systems [4, 5, 17, 18]. In IBVS systems, the control exists in the image space. In the common case of a camera mounted on the robot end effector, the motion of a two-dimensional feature point $\mathbf{f} = [\mathbf{u}, \mathbf{v}]^T$ in the image is related to the velocity screw of the end effector $\dot{\mathbf{r}} = [T_x, T_y, T_z, \omega_x, \omega_y, \omega_z]^T$ by the relation

$$\dot{\mathbf{f}} = \mathbf{J_{im}}(\mathbf{u}, \mathbf{v}, \mathbf{z})\dot{\mathbf{r}}, \quad (7)$$

where $\mathbf{J_{im}}$ is the image Jacobian [1, 2]. Given at least three feature points, it is possible to use (7) to construct the control law

$$\mathbf{u} = k\mathbf{J_{im}^+}(\mathbf{r})\dot{\mathbf{f}} \quad (8)$$

where $\mathbf{u}$ will be the velocity screw and k is a scalar gain factor or a 6×6 gain matrix.

Under this control, feature points tend to move in straight lines to their goal positions. This provides desirable performance in the image space, but as first reported by Chaumette [5], it can lead to extraneous motions of the end effector in 3D Cartesian space. These motions can lead to singular positions for the robot or singularities in the image Jacobian, leading to task failure.

IV. A IBVS/HOMOGRAPHIC HYBRID SWITCHED SYSTEM

Our first switched system presented here uses the IBVS and homographic methods as sub-systems. A higher level decision maker determines which system to use a each iteration. This system was designed in hopes of maximizing the strengths IBVS and PBVS systems as discussed in Section III.

As noted in Section II, we explored three switching signals:

State-Dependent Switching. We attempt to avoid the weaknesses of both systems by switching when the current system is approaching a problematic state. We determine a threshold level for how far the feature points will be allowed to stray from the center of the image, as well as a threshold on the distance we will allow the camera to move from the feature points. At each iteration, we compare each switching parameter to its threshold. If we are using IBVS and move past the threshold distance from the feature point plane we switch to PBVS to bring us towards the goal position and end camera retreat. If we are using PBVS and the feature points move outside the threshold distance to the image center we switch to IBVS to bring the image points towards their goal configuration, which is commonly centered.

Random Switching. Random switching has been used in control systems for such tasks as task routing [19]. At each iteration, We use a binary random variable to select between the two systems with equal probability. This provides a strong test to stability under arbitrary switching, and can tell us whether an undesirable switching pattern may result in instability.

Biased Random Switching. The state-dependent levels discussed above are now inputs to a probabilistic function used to determine the next system used. The farther the camera is from the image plane, the more likely the system is to choose the PBVS method. Likewise, the farther the feature points are from the image center, the more likely the system is to choose to use IBVS. The probability will be unity at either threshold.

We first present a series of simulations to show performance under ideal conditions. Simulations were performed for an ideal camera with a 512×512 pixel array, with each pixel measuring $10\mu m \times 10\mu m$ and a focal length of 7.8mm. We allowed perfect depth estimation. Visual servoing was halted if the pixel error was reduced below 1 pixel, or had converged to steady state for ten iterations.

We simulate a goal image where the feature points are close to the image border, and an error image where the camera is rotated by $160°$ about the optical axis. This is an extremely difficult task for the individual subsystems. In our simulations, using only the PBVS method would result in a loss of the feature points, and using only the IBVS method induced a camera retreat of 10 meters. Either of these would likely cause failure in a physical system. All three methods of switching were successfully able to zero the error.

The first simulation was the state-dependent switching system. Figure 3 hows the feature point errors, the velocity screw, the value of our switching parameters and the feature point positions at each iteration. Tick marks at the bottom of the graphs show the system currently being used: black for IBVS, cyan for PBVS. The color of the position lines follow the same color scheme regarding which system is determining the motion.

The feature points begin far from the center of the image, so we begin in IBVS mode to bring the points towards the goal. Indeed, the maximum error decreases, along with a sharp increase in the distance of the camera from the feature points. We enforced a threshold of 1.75 meter for the camera distance, so the camera switches to PBVS when camera retreat reaches this distance. The camera retreat is corrected along with completion of the rotation.

Figure 4 shows results for the random switching method. The feature points are kept within the image, and the camera retreats less than under the state based switching method. This is due to the large number of switches; since the switching is entirely random, it is possible to select the IBVS method the majority time and experience extreme camera retreat, although this never happened during our simulations or experiments. It does take slightly longer in this case to zero the error that state based switching.

Our final simulation result is for probabilistic switching, shown in Figure 5. The feature point trajectories closely resemble those of the random method. The feature point excursion is kept low, and the camera distance is also lower than that experienced under either the state-dependent method or random method, and it is slightly faster than both other systems as well.

Our experiments were performed using a camera mounted on the end effector of a PUMA 560 robot. The camera is a Sony VFW-V500, which has a 640×480 color pixel display. The lens focal length is 14.4mm. The feature points consisted of four color dots on a black sheet. The image was thresholded in

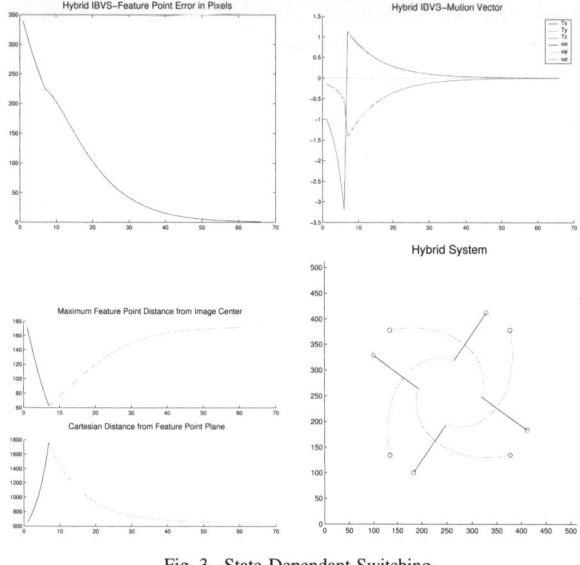

Fig. 3. State-Dependant Switching

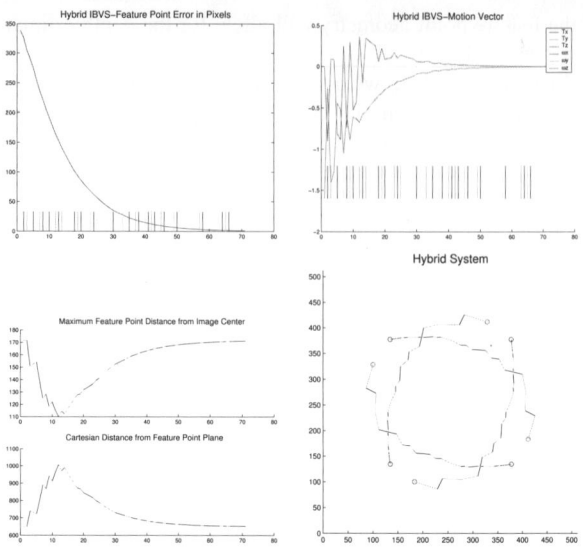

Fig. 5. Probabilistic Switching

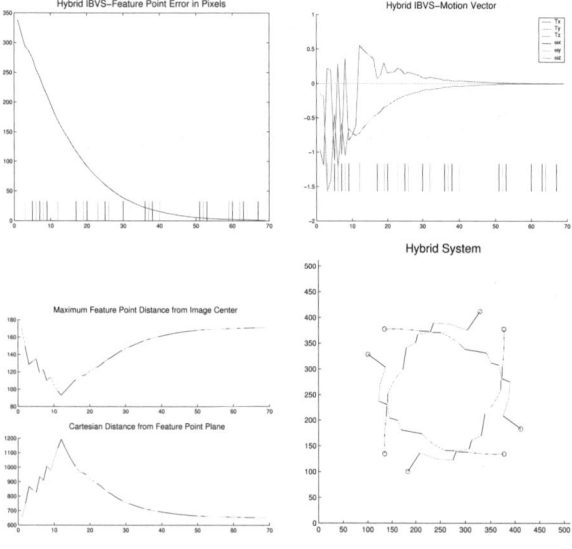

Fig. 4. Random Switching

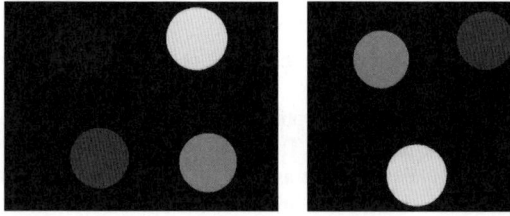

Fig. 6. Goal and Initial Image

RBG space to locate the center points of each dot. This provided 4 co-planar feature points.

The first experiments were very similar to the simulations. They involved a goal image with the camera very close to the feature points, and an initial offset consisting of a large rotation about the optical axis. Two such images can be seen in Figure 6.

Naturally, the individual systems did not perform ideally in our experiments due to the presence of camera calibration and depth estimation errors. For instance, the the feature point error did not strictly become smaller at every iteration under IBVS. The systems do well however, and did perform expected motions such as IBVS camera retreat during rotation. We feel that the fact that the systems worked well, even when the subsystems did not perform ideally, is a testament to the strength of the switched system.

During live experiments, both PBVS and IBVS systems failed this task if used individually. PBVS lost the feature points, and IBVS, experienced a great deal of camera retreat, losing focus of the image, and ultimately losing the feature points when making rotations. The figures show the same data we presented in the simulated results, with some minor changes. The graphs of both the feature point error and feature point trajectories have the color of the dot they correspond to, and trajectories with a black shadow indicate that PBVS was used to calculate that motion.

Figure 7 shows the results for the state-dependent switching method. Since the feature points are close to the image edge we are begin by using IBVS. As expected, the camera retreats rotates, bringing the the feature points towards the image center along with an increase in camera distance. Finally PBVS takes over and is able to reduce both. The system is unable to completely zero the feature point error, after 250 iterations when visual servoing was halted. Results for random selection are seen in Figure 8. The maximum feature distance from the image center is higher than seen in the state-dependent method, but the camera retreat is kept much lower. Due to the lower camera retreat, the system is able to zero the error faster than either the state-dependent or probabilistic methods. However, the maximum feature point distance is 250 pixels; clearly, the error must be along the horizontal image axis, or the feature points would have left the field of view and the system would have failed. This indicates a potential for system failure using the random method, though the system never failed during our experiments.

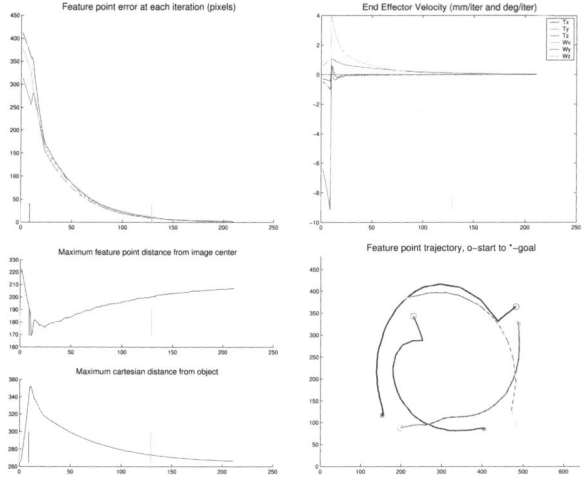

Fig. 7. State-Dependant Switching

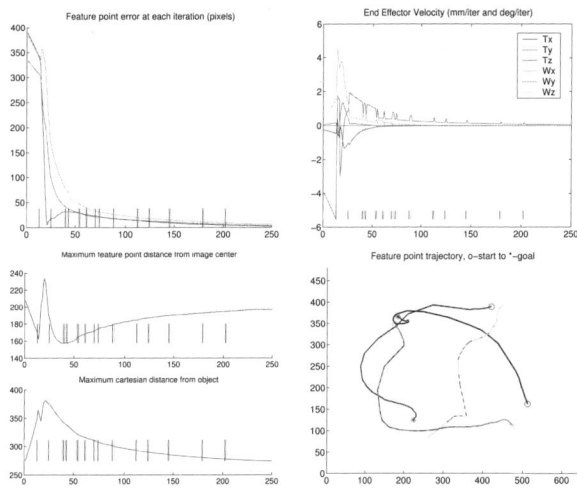

Fig. 9. Probabilistic Switching

Fig. 10. Goal and Initial Image

V. AN AFFINE/HOMOGRAPHIC HYBRID SWITCHED SYSTEM

The major strength of the homography-based controller in this system is that it is the only controller capable of handling general motions which include rotation about the x and $y-$axes. If the camera motion does not involve such rotations, the two approaches have similar performance. However, in the presence of noise, the affine method is much more accurate. We conducted a series of Monte Carlo tests in which both systems performed

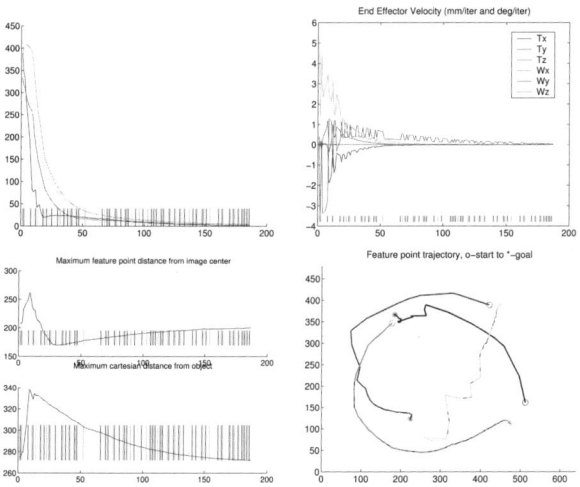

Fig. 8. Random Switching

The probabilistic method is presented in Figure 9. The method choices are identical to the state-dependent method for the first thirty iterations. After this point the switching does become fairly random, though the velocity screw shows much evidence of change due to switching. This system is also unable to completely zero the error after 250 iterations, due to remaining optical axis translation.

We repeated the experiments using an oblique view involving heavy rotation about the camera y-axis, moderate rotation about the camera z-axis, and translation along all axes. The goal and initial images are shown in figure 10.

For this task, both subsystems are capable of zeroing the feature point error, and camera retreat is not a dangerous factor here. Results can be seen in figures 11, 12 and 13. The state-dependent and probabilistic methods again perform very similarly. The random system is able to zero the error more quickly than the other two methods, but in general experiences a larger feature point error.

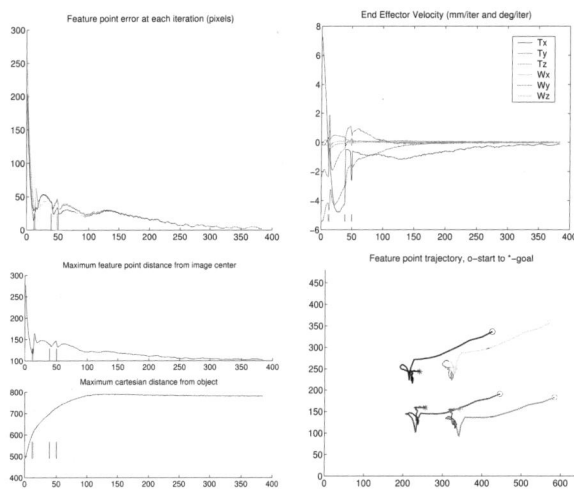

Fig. 11. State-Dependant Switching

3065

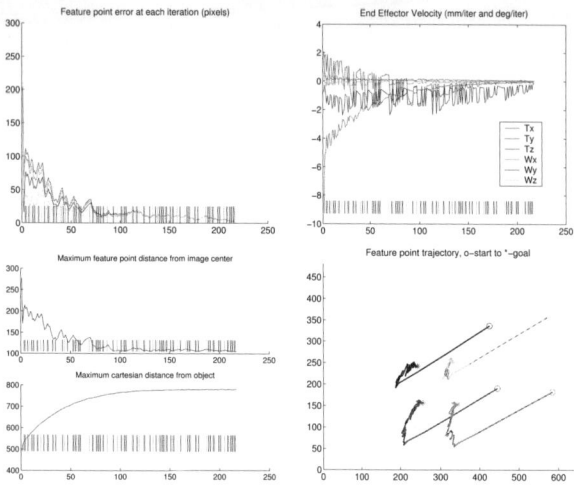

Fig. 12. Random Switching

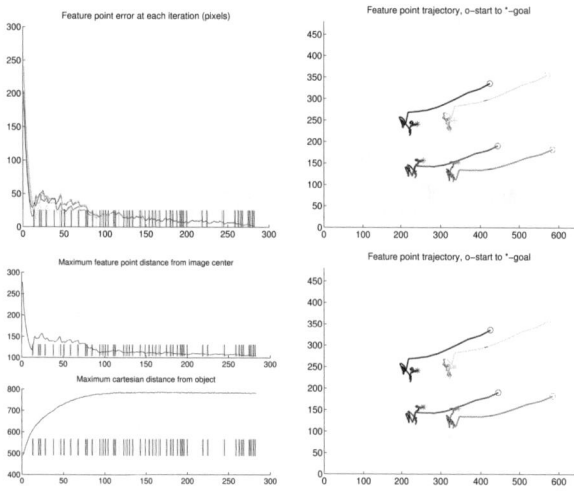

Fig. 13. Probabilistic Switching

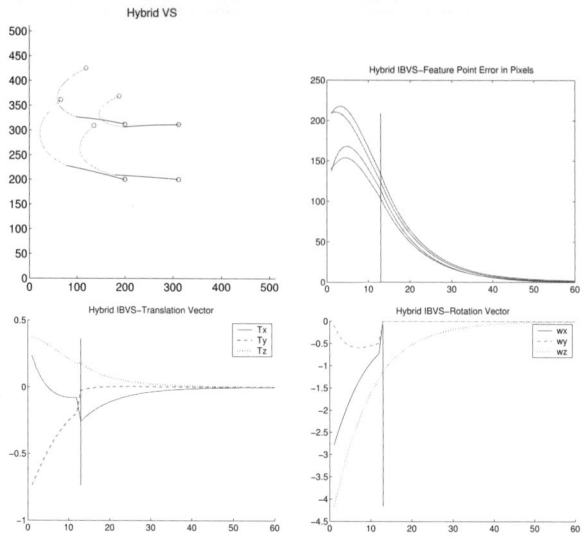

Fig. 14. General Motion, State-Dependant Switching

an identical affine motion under the effects of increasing white noise. Under large amounts of noise, the homography-based method typically had an error in the pose estimation that was fifty times greater than the affine approach, and error in the total rotation was almost fifteen times greater.

We again explored three switching rules:

State-Dependent Switching. After solving the homography, we take the RMS value of the rotations about $x-$ and $y-$ and compare it to a threshold value, if the amount of rotation is less than this value, we select the affine solution. We have found that the affine method can successfully zero the error for any motion with a rotation about y or x less then $0.5°$.

Random Switching. As discussed in Section IV.

Biased Random Switching. The current RMS of $x-$ and $y-$ rotation is used to determine the value of a random variable which selects the current system.

We performed simulations and experiments of a similar camera motion task, an oblique view of the image plane requiring motion along each degree of freedom to zero the error. The simulation and camera configurations are the same as discussed in Section IV.

We first present the simulations using the state-dependent switching method. For the results in Figure 14 a rotation of thirty degrees of the feature point plane about both the world y-axes is followed by a a moderate camera translation along all degrees of freedom and rotation about the optical axis. This causes a general motion involving all degrees of freedom.

The top left image shows the feature point trajectory; black line segments are motions induced by the affine method, while cyan portions are induced by the homographic method. We see the homographic method used for the first portion of the motion, with a switch to the affine method when the x and y rotation have been reduced. The upper right graph shows the pixel error for the four feature points. A black vertical line at the bottom of the graph indicates a switch to the affine method, a cyan line indicates a switch to the homographic system. Clearly a switch to the affine method occurred at about the twelfth iteration, and causes a slight incongruity in the velocities, though the error remains fairly smooth.

We repeated the previous test, but added a Gaussian random variable with variance 0.5 pixels to the feature point locations, simulating white noise. The results are seen in Figure 15. The system remains to the homographic method for almost all of the motion, briefly switching to the affine method at several points when the noise causes the $RMS(\omega_x, \omega_y)$ term to exceed the threshold. The system still zeros the error, though it takes longer than previous tests, over 75 iterations. The velocities are extremely rough in appearance due to the noise.

Figure 16 shows test results using the random switching method. The trajectory and motion vectors are irregular, though the feature point errors are smooth. It is worth noting that this system avoids the large feature point motions of the state-dependent switching system, which almost loses the feature points from the image plane.

We finally simulated the biased switching rule, shown in Figure 17. It has a similar appearance to the state-dependent switching system, remaining in the homography-based method for the

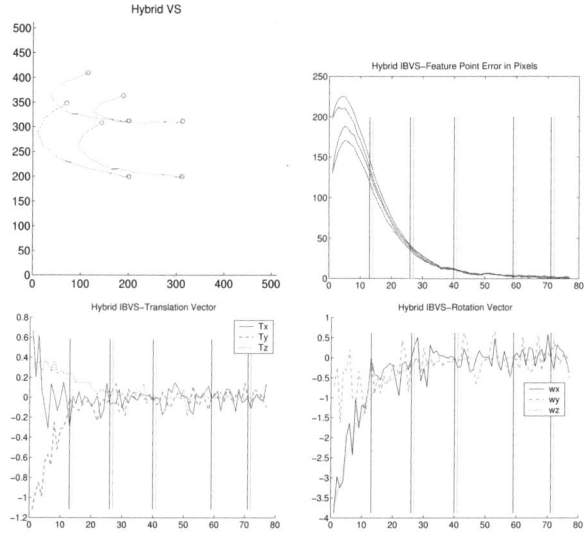

Fig. 15. General Motion with Noise, State-Dependant Switching

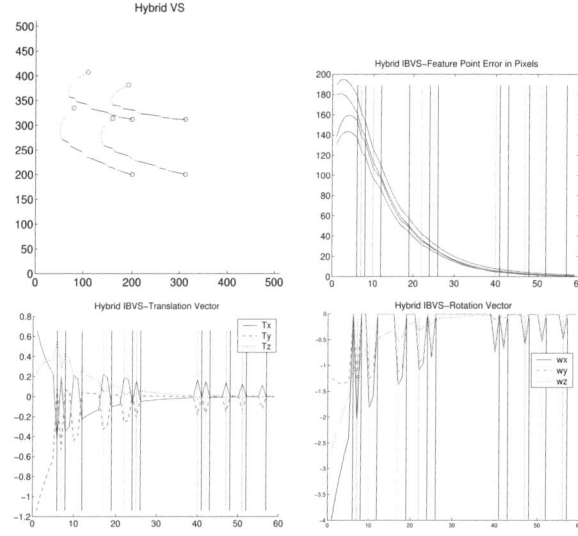

Fig. 17. General Motion, Biased Random Switching

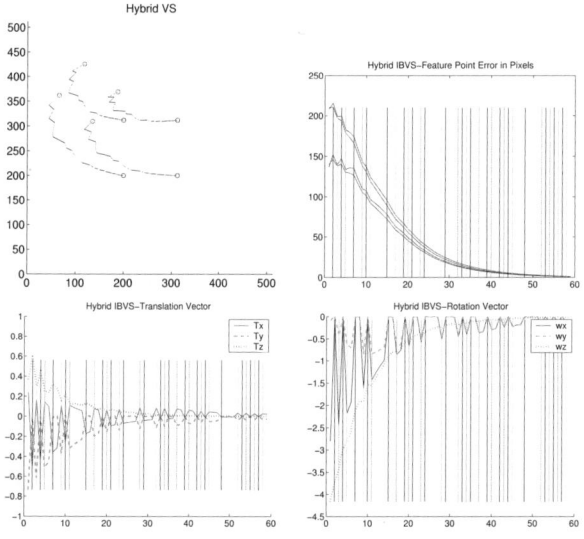

Fig. 16. General Motion, Random Switching

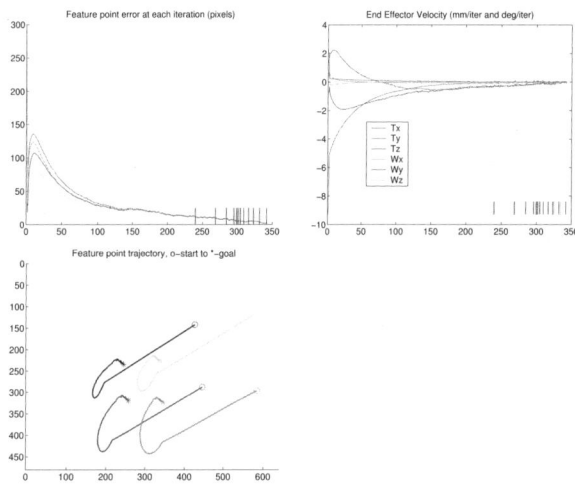

Fig. 18. Experiment Results State-Dependant Switching

first portion, while performing x and y rotations. At this point the probabilistic effects begin to surface as the controller oscillates between methods, with a slight preference for the affine method. The error is zeroed in just under 60 iterations, and the error plots and trajectories are quite smooth while the velocities are irregular.

Simulations provided a good idea of the performance characteristics of our switched system controller. We then conducted experiments also involving an oblique view, as seen in Figure 10. The experimental setup was the same as that described in Section IV.

We first explored the state-dependent switching method. Since there is a great deal of rotation about that the camera $y-$axis, we expect that it will use the homographic method for the majority of iterations, and switch to using the affine method when the y axis rotation has become very small. Figure 18 shows the feature point error, the velocity screw of the vector and the recorded feature point position for each iteration. In the first two graphs, small lines on the bottom of the graph indicate a switch has taken place; a black line indicates the homography method is being used for the following iteration, while a cyan line indicated the affine method will be used. We do see the homographic method used for almost two thirds of the iterations, at which point it switches between the affine solution and the homographic method as the amount of $y-$ axis rotation becomes negligible. The third graph shows the trajectory the point followed in the image plane. Portions of the lines with a black shadow indicate when the affine method is being used.

Figure 19 shows the results of random switching. All the measured values are much more chaotic. The feature point error tends to be greater, as does the magnitude of the velocity screw variables. However, the error is still zeroed in approximately the same amount of time and we also avoid the extremely large initial motion which the homographic method introduced in the

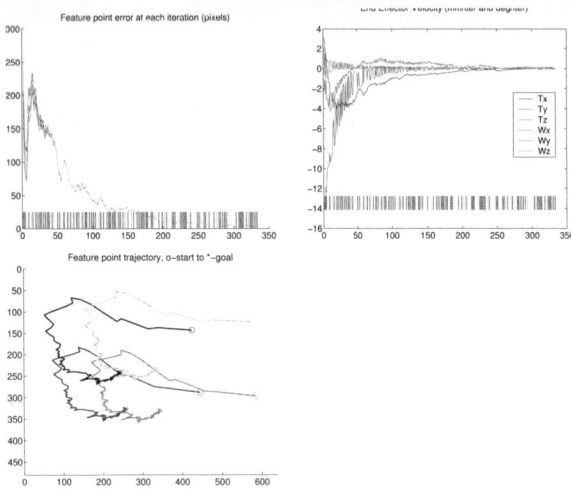

Fig. 19. Experiment Results Random Switching

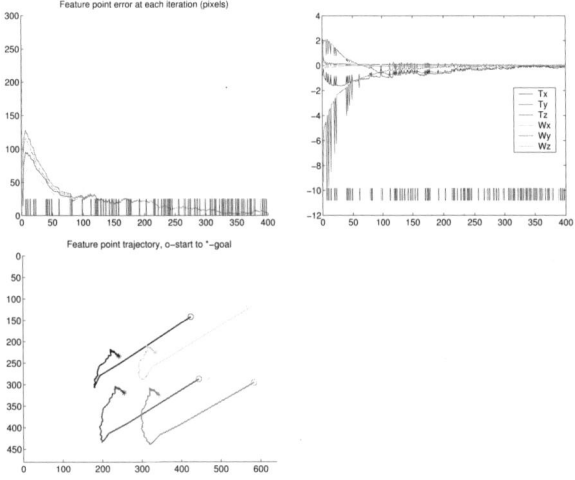

Fig. 20. Experiment Results Probabilistic Switching

state-dependent case.

Finally, Figure 20 shows the probabilistic system. The feature point error is similar to the state-dependent method, though it tends to be slightly smaller. Likewise the velocity screw tends to have similar shape and size to the state-dependent method. The system is, however, not able to zero the error as quickly as the other switching methods.

VI. CONCLUSION

We have presented two switched hybrid control visual servo systems. Each switched system is composed of two visual subsystems, and a decision maker that generates a signal to switch between them depending on the current state of the system, the time, or both. Simulation and experimental results are extremely promising. The IBVS/Homographic system showed a great deal of potential. It was ostensibly stable in all of our tests and successfully zeroed a task error that caused either individual subsystem to fail. The Affine/Homographic system also displayed stability in all of our tests, but the practicality of this system may be limited.

There remain future avenues to explore. There are more systems that could be integrated into a hybrid switched system framework, as well as the complication of including more than two continuous subsystems. There also remains the question of stability, which has not been resolutely established. Our experimental results are certainly compelling, and indicate this is a fruitful field for development.

REFERENCES

[1] L. E. Weiss, A. C. Sanderson, and C. P. Neuman, "Dynamic sensor-based control of robots with visual feedback," *IEEE Journal of Robotics and Automation*, vol. RA-3, pp. 404–417, Oct. 1987.

[2] J. Feddema and O. Mitchell, "Vision-guided servoing with feature-based trajectory generation," *IEEE Trans. on Robotics and Automation*, vol. 5, pp. 691–700, Oct. 1989.

[3] P. Martinet, J. Gallice, and D. Khadraoui, "Vision based control law using 3d visual features," 1996.

[4] S. Hutchinson, G. Hager, and P. Corke, "A tutorial on visual servo control," *IEEE Trans. on Robotics and Automation*, vol. 12, pp. 651–670, Oct. 1996.

[5] F. Chaumette, "Potential problems of stability and convergence in image-based and position-based visual servoing," in *The confluence of vision and control* (D. Kriegman, G. Hager, and S. Morse, eds.), vol. 237 of *Lecture Notes in Cont. and Info. Sci.*, pp. 66–78, Springer-Verlag, 1998.

[6] E. Malis, F. Chaumette, and S. Boudet, "2-1/2d visual servoing," *IEEE Trans. on Robotics and Automation*, vol. 15, pp. 238–250, Apr. 1999.

[7] G. Morel, T. Liebezeit, J. Szewczyk, S. Boudet, and J. Pot, "Explicit incorporation of 2d constraints in vision based control of robot manipulators," in *Experimental Robotics VI* (P. Corke and J. Trevelyan, eds.), vol. 250 of *Lecture Notes in Cont. and Info. Sci.*, pp. 99–108, Springer-Verlag, 2000.

[8] K. Deguchi, "Optimal motion control for image-based visual servoing by decoupling translation and rotation," in *Proc. Int. Conf. Intelligent Robots and Systems*, pp. 705–711, Oct. 1998.

[9] P. Corke and S. Hutchinson, "A new partitioned approach to image-based visual servo control," *IEEE Trans. on Robotics and Automation*, vol. 17, no. 4, pp. 507–515, 2001.

[10] N. R. Gans, P. I. Corke, and S. A. Hutchinson, "Performance tests of partitioned approaches to visual servo control," in *Proc. IEEE Int'l Conf. on Robotics and Automation*, 2002.

[11] M. Branicky, V. Borkar, and S. Mitter, "A unified framework for hybrid control," in *Proc. of the 33rd IEEE Conf. on Decision and Control*, 1994.

[12] R. W. Brockett, *Hybrid models for motion control systems*. 1993. H. L. Trentelman and J. C. Willems, Eds.

[13] D. Liberzon and A. Morse, "Basic problems in stability and design of switched systems," *IEEE Control Systems Magazine 19*, 1999.

[14] M. Branicky, "Multiple lyapunov functions and other analysis tools for switched and hybrid systems," in *IEEE Trans. Automat. Contr.*, 1998.

[15] O. Faugeras and F. Lustman, "Motion and structure from motion in a piecewise planar environment," *International Journal of Pattern Recognition and Artificial Intelligence*, vol. 2, no. 3, pp. 485–508, 1988.

[16] Z. Zhang and A. Hanson, "3d reconstruction based on homography mapping," in *ARPA Image Understanding workshop, Palm Springs, CA*, 1996.

[17] B. Espiau, F. Chaumette, and P. Rives, "A new approach to visual servoing in robotics," *IEEE Trans. on Robotics and Automation*, vol. 8, pp. 313–326, June 1992.

[18] R. Kelly, R. Carelli, O. Nasisi, B. Kuchen, and F. Reyes, "Stable visual servoing of camera-in-hand robotic systems," in *IEEE Trans. on Mechatronics*, 2000.

[19] R. Boel and J. van Schuppen, "Distributed routing for load balancing," in *Proceedings of the IEEE, vol.77, Iss.1, 1989*, pp. 210–221, jan 1989.

Error-Tolerant Execution of Complex Robot Tasks Based on Skill Primitives

Ulrike Thomas, Bernd Finkemeyer, Torsten Kröger and Friedrich M. Wahl

*Institute for Robotics and Process Control, Technical University of Braunschweig,
Mühlenpfordtstrasse 23, 38106 Braunschweig, Germany, {u.thomas,b.finkemeyer,t.kroeger}@tu-bs.de*

Abstract— This paper presents a general approach to specify and execute complex robot tasks considering uncertain environments. Robot tasks are defined by a precise definition of so-called skill primitive nets, which are based on Mason's hybrid force/velocity and position control concept, but it is not limited to force/velocity and position control. Two examples are given to illustrate the formally defined skill primitive nets. We evaluated the controller and the trajectory planner by several experiments. Skill primitives suite very well as interface to robot control systems. The presented hybrid control approach provides a modular, flexible, and robust system; stability is guaranteed, particularly at transitions of two skill primitives. With the interface explained here, the results of compliance motion planning become possible to be examined in real work cells. We have implemented an algorithm to search for mating directions in up to three-dimensional configuration-spaces. Thereby, on one hand we have released compliant motion control concepts and on the other hand we can provide solutions for fine motion and assembly planning. This paper shows, how these two fields can be combined by the general concept of skill primitive nets introduced here, in order to establish a powerful system, which is able to automatically execute prior calculated assembly plans based on CAD-data in uncertain environments.

I. INTRODUCTION

In this paper we present methods for error-tolerant execution of complex robot tasks using the very general approach of skill primitive nets. In the past decades, many assembly planning systems have been developed e.g. [1], [2], [3], [4]. In addition, control techniques for assembly have reached a mature state, e.g. [5], [6], [7], [8], [9]. Nevertheless, there still exists a gap between assembly planning and assembly execution. The big challenge is to obtain a system for assembly, which is able to execute prior calculated assembly plans. Executing an assembly plan according to precisely modelled objects means to deal with uncertainties. CAD data of objects usually are based on polyhedral models, but the real world is not polyhedral at all. To overcome these problems, the integration of sensors in the execution process is essential. Mason [10] has introduced a hybrid force/velocity and position control concept many years ago. The implementation of these concepts leads to a very general way to execute sensor based robot motion primitives in a consistent and integrative manner. We call these simple commands skill primitives, which provide a unique interface to our PC-based robot control system [11]. Task frame formalisms of hybrid control techniques have been investigated frequently in the past, e.g. [12], [13]. In contrast to these approaches, we have incorporated a sophisticated trajectory planner resolving contradictions during execution. For robot programming purposes, the techniques known from literature up to now are weakly integrated only. With our approach we unburden the programmer as well as an automatic planning system from all controlling details.

We consider a complex robot task as a network of skill primitives, where the next actions are chosen depending on sensor values. The skill primitive net represents all possible actions. Our approach provides the opportunity of sensor based fine motion planning and it also gives the flexibility of programming various robot tasks in an intuitive and practical manner incorporating uncertainty and error handling.

In [14], skill primitives have been applied to describe sensor based motions. Each robot task exists of a sequence of such motions, which are programmed off-line. In [15], [16], appropriate robot tasks are selected automatically, e.g. by classifying symbolic spatial relations as features [17]. In this paper, we give results on automatically planned and generated robot tasks in terms of skill primitive nets. Our aim is to demonstrate, that with a skill primitive based robot programming methodology, we obtain a flexible and general programming interface to describe complex robot tasks. Our approach satisfies the following claims:

- fast and robust execution of assembly tasks;
- generality to program serial, parallel or mobile robots with complex sensors;
- provision of a powerful interface between assembly planning and execution of assembly tasks, hiding all control and implementation details;
- possibility for handling deformable objects.

With the application of the methodology suggested here, assembly *planning* and assembly *execution* comes closer. In this paper, we demonstrate the potential of our approaches by executing the well-known robot tasks *object placing* and *bayonet bulb insertion*, which frequently occur in industrial applications. The second task can be considered as a more difficult peg-in-hole problem, because of the light bulb's pins, and the socket's spring, which has to be compressed during screwing.

II. A GENERAL APPROACH FOR ROBOT TASK EXECUTION: SKILL PRIMITIVE NETS

Robot tasks can be generated by our or some other assembly planner [16]. They also could be gained from a teaching/learning process [18] or they can be taken from human-linguistic instructions [19]. As will be shown here, skill primitive nets suit very well as interface between the planning and the execution process. In contrast to [20], we are not applying petri-nets, because we do not consider concurrency and, in addition, the semantics of petri-nets are overloaded for our purposes. Let us first define a skill primitive following Mason's concept:

Definition 1: A skill primitive $SP := \{\gamma, \mathbf{C}, \mathbf{CoC}, \lambda\}$ consists of a tuple where:

- γ $\in \Gamma$; Γ is a non empty set of control actions with $\Gamma := \{HybridMove, Close, Open, On, Off, ...\}$.
- $\mathbf{C}$ $\mathbf{C} \in \mathbb{R}^{6 \times 6}$ is the compliance frame, defining the motion relative to the center of compliance. $\mathbf{C}$ can be defined as function f depending on measured sensor values $S(t)$, $\mathbf{C} := f(S(t)) \to \mathbb{R}^{6 \times 6}$. Herewith, $\mathbf{C}$ can be set dynamically according to the specified function $f \in \mathcal{F}$.
- $\mathbf{CoC}$ $\in \mathbb{R}^{4 \times 4}$ the center of compliance relative to $\mathbf{TCP}$; the $\mathbf{CoC}$ may also change during execution, hence $\mathbf{CoC} := g(S(t)) \to \mathbb{R}^{4 \times 4}$, with $g \in \mathcal{F}$.
- λ $\in \Lambda$ is a boolean expression defined as function, $\lambda(S(t)) :\to \{true, false\}$; with $a \in S(t)$ and $b \in \mathbb{R}$, we define h as comparison function: $h(a \text{ eq } b) \to \{true, false\}$ $\forall eq \in \{\leq, \geq, =, <, >\}$, then λ is a conjunction $\lambda \to \{h \text{ op } \lambda \mid h\}$ with $op \in \{\vee, \wedge\}$.

Skill primitives are either hybrid motions or tool operations. To each of these, a compliance frame $\mathbf{C}$ is associated, which specifies desired values for a motion relative to the $\mathbf{CoC}$. Both $\mathbf{CoC}$ and $\mathbf{C}$ can be defined as functions, so that they can change throughout one skill primitive. This might be useful e.g. for a screwing operation, where the $\mathbf{CoC}$ changes according to the position of the screw within the thread. A skill primitive stops as soon as its boolean expression λ is evaluated to be true. Subsequently, execution of the next skill primitive is started.

A robot task RT consists of a skill primitive net, where each primitive is represented by a node and each branch represents a possible successor action with respect to the current state. We define skill primitive nets as follows:

Definition 2: A skill primitive net $SPN := \{\Sigma, \Pi, \Xi, \Upsilon, \Omega\}$ is a directed graph with attributed arcs and definite start, stop, and error states. With:

- Σ is a finite set of non empty $\{SP^*\}$.
- Π is a finite non empty set of start states.
- Ξ is a finite non empty set of stop states.
- Υ is a finite set of error states.
- Ω is a finite set of directed arcs, so that $\forall \omega_{ij} \in \Omega$ $\exists \sigma_i, \sigma_j \in \bigcup_{\Sigma, \Pi, \Xi, \Upsilon}$, with $\omega_{ij} = (\sigma_i \times \sigma_j)$, where $\sigma_i \neq \sigma_j$. Each arc is attributed with $\omega_{ij} := (\lambda_{ij}, g_{ij})$, where g_{ij} is used to determine the new center of compliance for the next skill primitive σ_j: $\mathbf{CoC}_j := g_{ij}(S_i(t)) \to \mathbb{R}^{4 \times 4}$, with $g_{ij} \in \mathcal{F}$. $\lambda_{ij} \in \Lambda$ is a boolean expression, as defined above, so that the attributes $\lambda_{i[1...n]}$ of all n succeeding arcs of σ_i, yield $\lambda_i = \bigvee_{k=1}^{n} \lambda_{ik}$. This means, the disjunction of all succeeding arcs defines the stop condition of the current skill primitive.

By this definition, we are able to describe any kind of robot task by means of skill primitive nets. The so defined nets give the flexibility to handle all interferences during execution. If an unknown situation occurs, the skill primitive execution is stopped; the next primitive in all cases is well-defined, because of the property, that the disjunction of all equations λ_{ij} attributing the outgoing arcs ω_{ij} of a skill primitive σ_i per definition is equal to the stop condition implemented in the current primitive. For each different contact formation during execution, the corresponding branch in the skill primitive net has to be selected. The robot tasks *object placing* and *bayonet bulb insertion* will illustrate the concepts introduced so far:

A. Object Placing

In Fig. 1, a skill primitive net of the object placing task is illustrated, assuming that the surface, on which to place the polyhedral object is not known precisely, and/or that the object might have been gripped inaccurately.

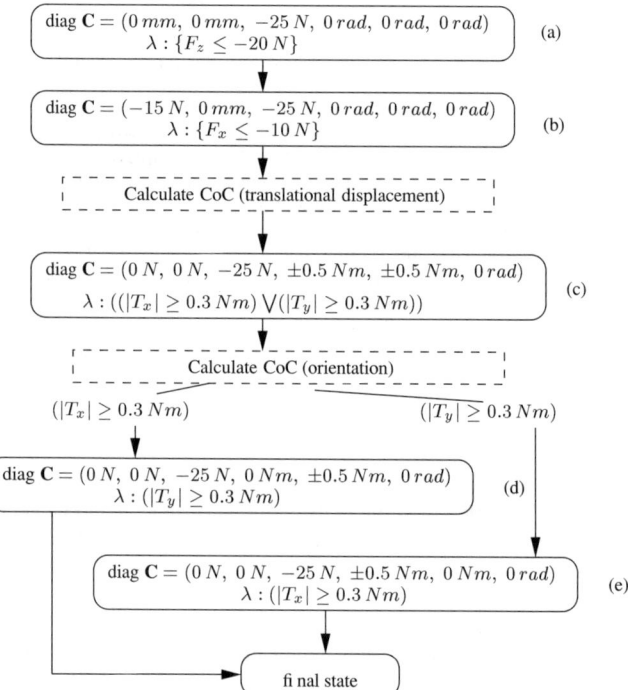

Fig. 1. A skill primitive net of the object placing task

The most probable situation is that the first contact is a vertex/face contact. To move the robot into that contact formation we set up the skill primitive (a) with:

$$\text{diag } \mathbf{C} = (0\,mm,\ 0\,mm,\ -25\,N,\ 0\,rad,\ 0\,rad,\ 0\,rad)^T,$$

with stop condition $F_z \leq -20N$. This skill primitive causes a movement in z-direction until contact is established. Other possible contact formations could be an edge/face or a face/face contact. Here, for simplification we assume a vertex/face contact. With the next skill primitive (b), the object is pressed in (x,z)-direction onto the surface in order to determine the contact point. The **CoC** is set to this point; by the following skill primitive (c), we rotate the object about a vector lying in the (x,y)-plane until one of the stop condition yields: $(((|T_x| \leq 0.3\,Nm) \vee (|T_y| \leq 0.3\,Nm))$. By this, we can determine the edge of the established edge/face contact; we reorientate the **CoC** so that its x, y-axes are collinear to the two edges of the object touching the surface. Depending on the measured torque (refer to the two different branches in the skill primitive net), we then rotate around the x-axis (e) or the y-axis (d) respectively, so that either equation $|T_x| \leq 0.3\,Nm$ or $|T_y| \leq 0.3\,Nm$ stops the robot motion. By this, the object is placed accurately and the final state is reached. All other unpredictable measurements result in error states.

B. Bayonet Bulb Insertion

Fig. 2 left shows the light bulb to be inserted into a bayonet socket. Fig. 2 right illustrates the configuration space (z, φ_z), where the light bulb as the moving part is considered as a moving point and the socket defines the configuration space obstacle. Subject to limited space, we are assuming, that the

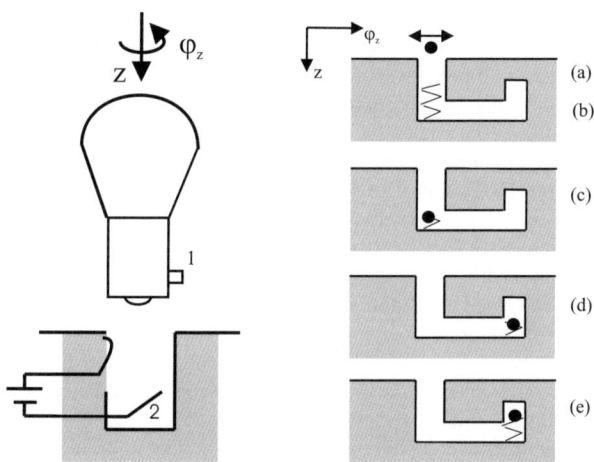

Fig. 2. The light bulb and the bayonet socket (left), the corresponding configuration space (z, φ_z) (right)

rotational axis of the bulb is fairly well aligned with the axis of the socket. The decomposition of the robot task leads to the skill primitive net shown in Fig. 3. Two traversals of the net are possible. The right branch will be chosen in case where the bulb is rotated perfectly around the z-axis, so that the light bulb's pin (1 in Fig. 2) fits into the notch. The left branch represents an error situation in z-orientation. First, we set up a skill primitive (a), so that the light bulb directly comes in contact with either the border of the socket or the spring (2 in Fig. 2). In both cases the second skill primitive (b) is set to

$$\text{diag } \mathbf{C} = (0\,N,\ 0\,N,\ -25\,N,\ 0\,rad,\ 0\,rad,\ -0.5\,Nm)^T,$$

which leads to a rotation around the z-axis, while a contact force of $-25N$ in z-direction is established. The difference between the two above mentioned cases is that the light bulb already is at the down position of the notch or that it is in front of the notch, which can be detected by either $((F_z \leq -15\,N) \wedge (T_z \leq -0.2\,Nm))$ or $((F_z \geq -5\,N) \wedge (P_z \geq 3mm) \wedge (T_z = 0))$. If the light bulb already has been turned in the notch to secure the bulb, a small movement in negative z-direction is necessary until a positive force in z-direction is measured (e). In case the light bulb is on top of the notch a new contact force in z-direction has to be applied until the spring is compressed (c). By the next skill primitive, the bulb is rotated around the z-axis until a high torque around the z-axis is produced (d). Now, the bulb properly can be secured as mentioned above (e). In all other cases defined error states are reached.

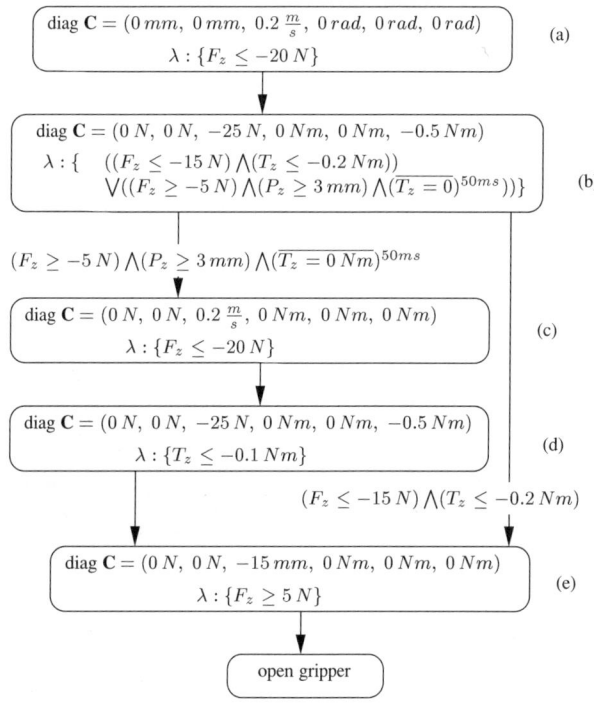

Fig. 3. A skill primitive net for a light bulb to be inserted into a bayonet socket

III. DECOMPOSITION OF ROBOT TASKS INTO SKILL PRIMITIVE NETS

For a complete automatic assembly process, i.e. from planning to execution, a suitable solution is, to generate assembly tasks in terms of skill primitive nets. In the past, we have introduced an algorithm for the automatical recognition of robot tasks contained in assembly sequence plans, which have been generated by an assembly planner [15], [16]. The classification of robot tasks was based on certain criteria, e.g. the type of depart space, symbolic spatial relations etc. [17]. In this approach, robot tasks had to be programmed off-line by humans. In this paper, we give first results on the automatic decomposition of robot tasks into skill primitive nets. The user may specify a complex robot task such as the bayonet insertion task by clicking on appropriate surfaces in a CAD environment, and by selecting symbolic spatial relations. To specify, that the bulb axis is collinear to the axis of the socket, a user might define: *Shaft of bulb fits hole of socket.*

Planning concepts for skill based execution have been introduced e.g. in [21], [22]. Usually it is necessary to find a path from free space to the specified goal position. Using the reverse order for solving this problem is well-known as assembly by disassembly. For this purpose, we calculate the (x, y, z)-C-space obstacle [23] in the goal position of the object to be mated (refer to Fig. 2 (e) and Fig. 4). Based on the object geometry we generate possible mating directions by testing all face normals and edge vectors of collision with the C-space obstacles. The largest vector points down the right vertical notch in Fig. 2 (d), (e). In a second iteration, because no translational DOF is remaining, no suitable vector in the (x, y, z)-C-space can be found. To obtain possible rotational movements of the object, we build up the C-spaces (z, φ_x), (z, φ_y), (z, φ_z), where the vector computed before coincides with the z-axis. A vector larger than a certain threshold can be found only in (z, φ_z)-space. In the next step, the (x, y, z)-C-space has to be calculated according to the intermediate orientation of the removable object. Finally, the mating direction shown in Fig. 4 can be calculated. Concatenating the

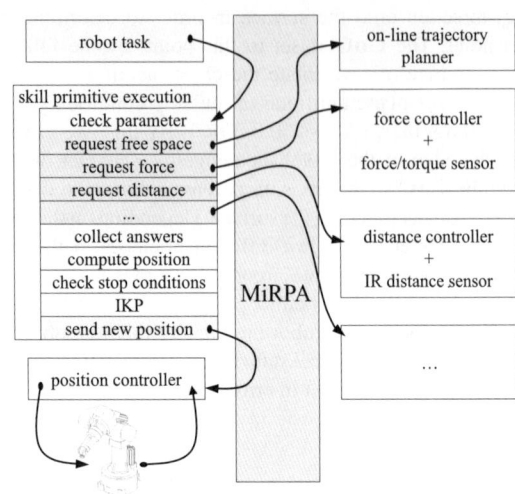

Fig. 5. Software architecture of the control system

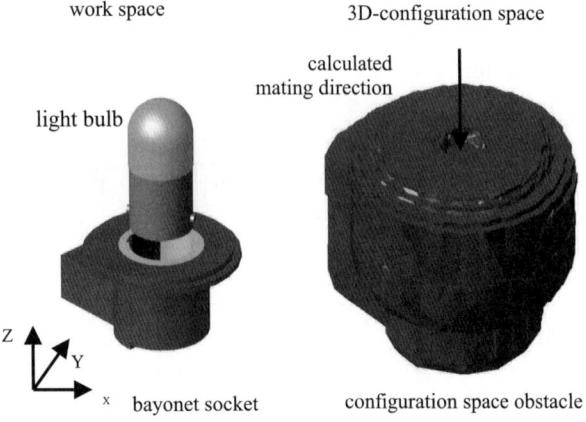

Fig. 4. The 3-dimensional configuration space of the light bulb and its corresponding bayonet socket, consisting of approximately 40 000 triangles.

calculated vectors we find a path from (a,b) to (c), (d) and (e) in Fig. 2. For each phase a compliance frame $\mathbf{C}$ has to be generated; e.g. from (c) to (d):

$$\operatorname{diag} \mathbf{C} = (0\,N,\ 0\,N,\ -25\,N,\ 0\,rad,\ 0\,rad,\ -0.5\,Nm)^T\ .$$

The DOFs are set according to the occurrence of contacts, one produced by the spring and the other produced by borders of the notch. For each step a new skill primitive has to be generated. As mentioned above, branches in the skill primitive net represent uncertainties, which can be modelled by contact formations.

The threshold values in the skill primitive net are obtained from a look-up-table, which contains possible contact forces and suitable velocities. The positions are calculated based on the CAD data.

For the object placing task, the (x, y, z)-C-space obstacle is a half space, for which we easily can obtain the mating direction. For the generation of contact formations, approaches have been developed already [24]. We will develop methods to map contact graphs to our skill primitive nets in the near future.

IV. SKILL PRIMITIVES AS INTERFACE TO ROBOT CONTROL SYSTEMS

A desired state and motion of robot manipulators can be defined by skill primitives introduced in the above sections. We have developed a new control architecture able to interpret skill primitives. The challenge of such a control system is that the control structure must be adapted to the current skill primitive. E.g., one DOF of the **CoC** is force controlled and all remaining DOFs are position or velocity controlled. The control loops vary for each DOF. They may depend on different physical variables. E.g. the DOFs, which have been defined by forces or torques, must be closed by feedback of a force/torque sensor. The DOFs with free motions are controlled by an on-line trajectory planner. If one DOF has been defined by a specified constant distance to some object in the environment, the control loop must be closed via a suitable distance sensor.

It is obvious that there are many possibilities to build combinations of the various control loops per DOF. As the user of skill primitives is not necessarily a control engineer, the control system automatically must handle different control loops. Of course, the involved control concepts are well-known from literature e.g. [5], [6], [7], [8], [9], but the problem of switching between different controllers is scarcely considered anywhere. The following question arises: How can the control system handle the skill primitive transitions? Imagine for example the transition from free space movements into compliant motion. The controller of the concerned DOF has to switch from position to force control. Stability and smooth movements of the robot must be guaranteed during all task states. By our new control system, we present a first approach to solve this problem. The software of our new control system is based on the middleware MiRPA (Middleware for Robotic and Process Control Applications) [11]. Its usage allows the implementation of very modular software. The control system easily can be adapted to different requirements by starting appropriate modules (even during runtime). The software architecture of our control system is shown in Fig. 5. Each block in the picture represents a software module. These modules either can be run on one computer only, or they can be distributed in a computer network. The block in the center of Fig. 5 represents the middleware. It manages the communication between the modules. All modules are able to communicate with each other via MiRPA. The block "robot task" contains the representation of the skill primitive net described in section II. It sends the parameterized skill primitives to the block "skill primitive execution". This block handles all involved control loops: It is the main module for the skill primitive execution. Its functionality will be explained below. It generates robot positions in joint space and sends them to the block "position controller". The remaining blocks contain the implementations of the different controller types.

The first step of the skill primitive execution is to interpret its parameters. Especially the compliance frame $\mathbf{C}$ must be analyzed. It is obvious that the physical units of its elements decide, which controller is involved to control the corresponding DOF. The installed controllers have been numbered from 0 to k. From $\mathbf{C}$ the 6×6 selection matrix $\mathbf{S_i}$ is generated for each installed controller. The index i is the number of the controller. A "1" in the diagonal of $\mathbf{S_i}$ selects the corresponding DOF to be controlled with the controller numbered with i. If $\mathbf{C}$ has been interpreted in a correct way, the following equation must be true: $\mathbf{I} = \sum_{i=0}^{k} \mathbf{S_i}$, where $\mathbf{I}$ is the unit matrix. This selection matrix is well-known from literature [5], but in our approach the definition is not limited to positions and forces. We can select for each DOF any other physical variable to be controlled.

$\mathbf{S_i}$ is used to select the controller. A request is sent to all controllers, whose selection matrix $\mathbf{S_i}$ is not equal to zero. The request includes the desired diag $\mathbf{C}$ and the $\mathbf{CoC}$. Its return value is a position difference vector within the $\mathbf{CoC}$. E.g., analyzing the following $\mathbf{C}$ matrix

$$\text{diag } \mathbf{C} = (5\ mm, 0\ mm, -40\ N, 0\ rad, 0\ rad, 0\ rad)^T$$

results in the selection matrices

$$\text{diag } \mathbf{S_0} = (1, 1, 0, 1, 1, 1)^T$$
$$\text{diag } \mathbf{S_1} = (0, 0, 1, 0, 0, 0)^T$$
$$\text{diag } \mathbf{S_2} = (0, 0, 0, 0, 0, 0)^T$$

$\mathbf{S_0}$ is the selection matrix for free space movements. If it is not equal to zero, a new position must be interpolated by the trajectory planner. To achieve this, a request is sent to the corresponding control-block "on-line trajectory planner" (see Fig. 5). It computes the next motion step depending on the desired $\mathbf{C}$ and considers current position, velocity and acceleration. As this is done on-line in each control cycle, the planned trajectory can be adapted ad hoc. This results in very smooth robot movements, even if the system switches from one skill primitive to the next one.

The matrix $\mathbf{S_1}$ selects the DOFs, which are force controlled. As it is also not equal to zero, a request is sent to the "force controller" block. The displacement in that DOF is determined by the measured force/torque values and the $\mathbf{C}$ matrix.

$\mathbf{S_2}$ selects the DOFs, which are controlled by the feedback of a distance sensor. As in this example, no distance has been specified in $\mathbf{C}$, no request is sent to the "distance controller" and $\mathbf{S_2}$ equals zero.

The results of the requested controllers are collected as follows: $\Delta \mathbf{x} = \sum_{i=0}^{k} \mathbf{S_i f_i}(\mathbf{C}, \mathbf{CoC})$. The $\mathbf{f_i}$ function represents the involved control block and its control law computing the displacement. As all calculations are computed with respect to the $\mathbf{CoC}$, the results must be transformed into Cartesian robot positions. Subsequently, the stop conditions are checked. If the result is logical true or if an error occurs, a message with the current robot state is sent to the robot task and $\mathbf{C}$ is set to safe values. That usually means, the robot is kept in the current state.

The inverse kinematic model (IKP) computes the new desired robot position in joint space. The result is sent to the "position controller". The described procedure is repeated every cycle. Up to now, we have realized a cycle time of 6 ms; but we are very optimistic to decrease this value to less than 1 ms.

The resulting control scheme of the skill primitive execution is a cascade of two control loops. The outer loop is closed by additional sensors like distance sensors, force/torque sensors or the trajectory planner. Control commands are positions, which are controlled by the inner loop. This position interface has the great advantage, to keep the robot stable. If an error occurs in the outer loop, e.g. the skill primitive net is erroneous, stability is guaranteed. It does not move the robot along/around the affected DOF. If we would use a velocity interface to the outer loop, the robot would keep on moving with the last velocity value. As many commercial manipulators contain a position interface in joint space and as in several systems the used power electronics can be commanded by positions, the presented concept easily can be adapted to these and other systems.

To avoid unstable robot movements it is necessary to ensure, that the controlled physical values of each DOF are linear independent.

The control architecture outlined above may be considered as complex realization of a cascaded controller, but it provides the postulated flexibility. In our approach the implementation has been kept very modular, i.e. sensors and their corresponding control laws easily can be integrated into the system or can be changed by the user even during runtime. Only the measurements of each sensor must be transformed into the $\mathbf{CoC}$.

Due to the usage of MiRPA it is possible to distribute the control algorithms to a net of computers. Hence, different control algorithms can work in parallel. Systems with intensive calculations, like a vision system, easily can be integrated into the skill primitive concept too. As it is also possible to start and to stop additional sensor controllers during runtime, it is feasible to change sensors during one assembly task. I.e., complex tools with integrated sensors can be changed by a tool changing system in a very simple way.

V. EXPERIMENTAL RESULTS

The relevant values occurring during the execution of the example skill primitive net shown in Fig. 3 (bayonet bulb insertion), F_z and T_z, are illustrated in Fig. 6. During the first phase (a) the robot has already gripped the light bulb and heads for contact with the socket. The z-axis of the $\mathbf{CoC}$ is coincident with the center axis of the light bulb. At the transition from phase (a) to phase (b) contact is detected, since the force F_z has reached the stop condition of $-20N$. The desired contact force

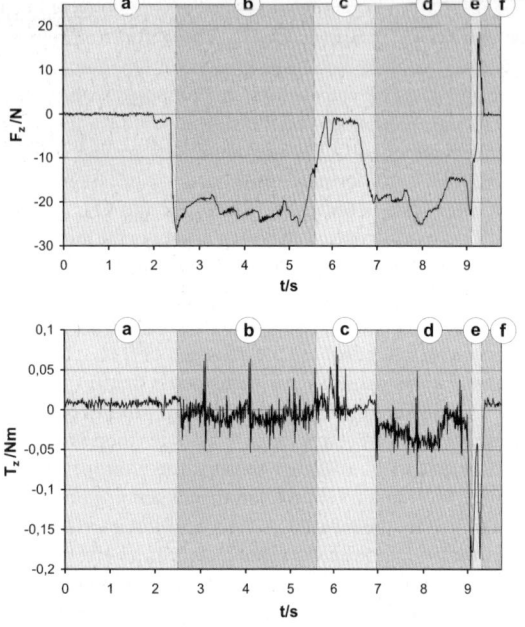

Fig. 6. Force F_z and torque T_z while inserting a light bulb into a bayonet socket

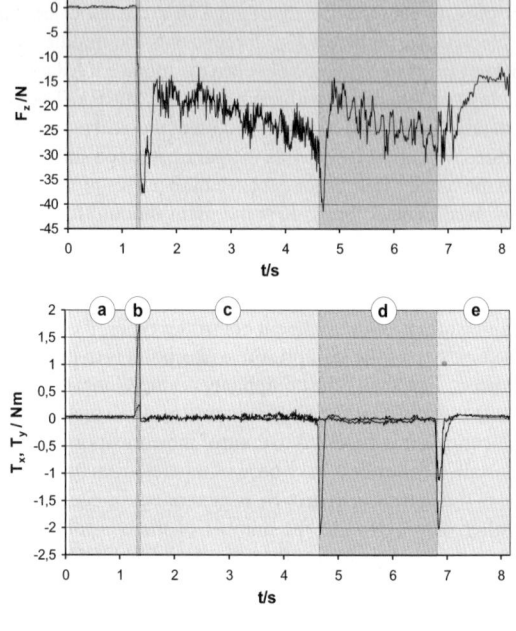

Fig. 7. Force F_z, torque T_x and torque T_y during the placing of a polyhedral object onto a plane surface

of the following skill primitive is set to $-25N$, and the robot starts to rotate the light bulb on the socket surface. The noise of T_z in phase (b) is caused by the roughness of the socket surface. When the bayonet pins are in the correct position at the transition from (b) to (c), the light bulb moves into the socket's hole, which is recognizable because of the zero-converging force. The increasing force at the end of phase (c) is caused by the socket spring, which completely is compressed at the changeover to (d), where the second rotation about the socket's axis starts. At the end of phase (d) the torque T_z increases, i.e. the light bulb pins are at the initial stop; subsequently the device is pushed upwards in phase (e) until the force F_z reaches a positive signed threshold value. The light bulb insertion is completed, i.e. the gripper opens in phase (f). Fig. 8 top raw illustrates this robot task execution. As second example, the placing of a polyhedral object onto a plane surface was chosen, since translational and rotational DOFs are not independent. The relevant magnitudes F_z, T_x and T_y are shown in Fig. 7. Starting with phase (a), the robot heads for contact with the surface. The z-axis of the **CoC** equals the direction of motion towards the surface. The increasing force at the transition from (a) to (b) indicates contact. A very short test movement is executed to establish a force that uses a moment arm, which is orthogonal to the surface. Depending on the measured torques and forces the point of contact is calculated and set as **CoC** at the beginning of phase (c). I.e. the leap to zero in the torque diagram is not caused by a motion, but by a translational displacement of the **CoC**. Subsequently, the object rotates about the x- and the y-axes of the **CoC** in phase (c) until a torque threshold value about one of these axes is exceeded. The contact 'edge' is calculated and the **CoC** is rotated about the z-axis, such that the x-axis or

the y-axis is coincident to the contact edge. This axis is used as rotation axis in phase (d). After a torque threshold value is exceeded, the object plane and the surface plane are coplanar and the final state is reached in phase (e). Three pictures of the sequence of this robot task are shown in Fig. 8 bottom raw.

Videos of both experiments can be downloaded from the internet at

http://www.cs.tu-bs.de/rob/download/welcome.htm.

VI. CONCLUSION AND FUTURE WORK

In this paper, we have presented a new approach filling the gap between assembly planning and plan execution by appropriate control. The approach is manipulator-independent providing a sophisticated and powerful interface between hybrid robot control and automatic assembly planning systems. Because of the modularity of the control architecture, embedding any kind of sensors as well as employment of further control algorithms becomes possible very easily. Based on these features, flexible and error-tolerant execution of complex robot tasks is feasible. This provides an enormous advantage in handling deformable objects with robots and uncertain robot work cell environments, which can be programmed automatically or manually using skill primitive nets.

We will conduct experiments along these lines with parallel manipulators designed in the German Collaborative Research Centre 562. Even the integration of sophisticated sensor systems, e.g. computer vision systems, does not require a system redesign and thus easily can be implemented. If necessary, complex algorithms required to solve these kind of problems can be distributed in a computer network to obtain the required

Fig. 8. Experiments - Top: Inserting a light bulb into a bayonet socket; bottom: Placing a polyhedral object on a flat surface

real-time performance in various application fields. We will test the described concepts by means of complete assemblies, e.g. gearboxes, headlights, engine blocks etc., to demonstrate a fully automated solution, reaching from assembly planning to plan execution.

REFERENCES

[1] L. Lieberman, M. Wesley, "AUTOPASS: An Automatic Programming System for Computer Controlled Mechanical Assembly", *IBM J. Res. Dev.* Vol. 21, No. 4 , 1977.

[2] T. Lozano-Pérez, P. H. Winston, "LAMA: A Language For Automatic Mechanical Assembly", *International Joint Conference. Artificial Intelligence*, 1987.

[3] S. G. Kaufmann, R. H. Wilson, R. E. Jones, T. L. Calton, "The archimedes 2 mechanical assembly planning system", *IEEE International Conference on Robotics and Automation*, pp. 3361-3368, 1996.

[4] R. E. Jones, R. H. Wilson, T. L. Claton, "On Contraints in Assembly Planning", *IEEE Transactions on Robotics and Automation*, Vol. 14, No. 6, pp. 849-863, December 1998.

[5] H. H. Raibert, J. J. Craig, "Hybrid Position/Force Control of Manipulators", *ASME Journal of Dynamic Systems, Measurement and Control*, Vol. 102, pp. 126-133, 1981.

[6] D. E. Whitney, "Force Feedback Control of Manipulator Fine Motions", *ASME J. of Dync. Sys. Meas. Contr.* 99, pp. 91-97, 1977.

[7] T. Yoshikawa, "Force Control of Robot Manipulators", *Proc. of the IEEE Inter. Conf. on Robotics and Automation*, San Francisco, pp. 220-226, 2000.

[8] S. P. Patarinski, R. G. Botev, "Robot Force Control: A Review", *Mechatronics* Vol.3, No.4, pp. 377-398, 1993.

[9] J. De Schutter, H. Van Brussel: "Compliant Robot Motion II. A Control Approach Based on External Control Loops", The International Jour. of Robotics Research, Vol. 7, No.4, pp. 18-33, August 1988.

[10] M. T. Mason, "Compliance and force control for computer controlled manipulators", *IEEE Transactions on Systems, Man, and Cybernetics*, Vol. 11, no. 6, pp. 418-432, June 1981.

[11] B. Finkemeyer, M. Borchard, F. M. Wahl, "A Robot Control Architecture Based on an Object Server," *IASTED International Conference Robotics and Manufacturing*, Cancunm Mexiko, pp. 46-40, 2001.

[12] H. Bruynincks, J. De Schutter, "Specification of Force-Controlled Actions in the Task Frame Formalism - A Synthesis", *IEEE Transaction on Robotics and Automation*, Vol. 12. No. 4, pp. 581-589, August 1996.

[13] J. De Schutter, J. Van Brussel, "Compliant Robot Motion I. A Formalism for Specifying Compliant Motion Tasks", *The International Journal of Robotics Research*, Vol. 7, No. 4, pp. 3-17, August 1988.

[14] T. Hasegawa, T. Suehiro, K. Takase, "A Model-Based Manipulation System with Skill-Based Execution", *IEEE Transactions on Robotics and Automation*, Vol. 8, No. 5, pp. 535-544, 1992.

[15] H. Mosemann, F. M. Wahl, "Automatic Decomposition of Planned Assembly Sequences Into Skill Primitives", *IEEE Transactions on Robotics and Automation* Vol. 17, No. 5, pp. 709-718, 2001.

[16] U. Thomas, F. M. Wahl, "A System for Automatic Planning, Evaluating and Execution of Assembly Sequences for Industrial Robots", *IEEE/RSJ International Conference on Intelligent Robots and Systems* , pp.1458-1465, 2001.

[17] A. P. Ambler, R. J. Popplestone, "Inferring the positions of bodies from specified spatial relationships", *Artificial Intelligence*, pp. 157-174, 1975.

[18] T. Fukuda, M. Nakaoka, T. Ueyama, Y. Hasegawa, "Direct teaching and error recovery method for assembly task based on a transition process of a constraint condition", *IEEE International Conference on Robotics and Automation*, Soul, Korea, pp. 1518-1523, May 2001.

[19] J. Zhang, Y. von Collani, A. Knoll, "Development of a robot agent for interactive assembly", 4th International Symposium on Distributed Autonomous Robotic Systems, Karlsruhe, 1998.

[20] B. J. McCarragher, "Task Primitives for the Discret Event Modelling and Control of 6-DOF Assembly Task", *IEEE Transactions on Robotics and Automation*, Vol. 12, No. 2. pp. 280-289, April 1996.

[21] A. Nakamura, T. Ogasawara, T. Suehiro, H. Tskune, "Skill Based Backprojection for Fine Motion Planning", *IEEE/ International Conference on Intelligent Robots and Systems*, pp. 526-583, 1996.

[22] M. Erdmann, "Using Backprojection for Fine Motion Planning", *International Journal of Robotics Research* Vol. 3, No. 1, pp. 19-45, 1986.

[23] D. Lozano-Perez, "Spatial Planning: A Configuration Space Approach, *IEEE Transaction on Robotics and Automation*, Vol. C32, No. 2, pp. 108–120, 1983.

[24] X. Ji, J. Xiao, "Planning Motions Compliant to Complexe Contact State", *The International Journal of Robotics Research*, Vol. 20, No. 6, pp. 446-465, 2001.

Strategies of Human-Robot Interaction for Automatic Microassembly

Antoine Ferreira

Laboratoire Vision et Robotique, ENSIB-University of Orléans
10, Bd. Lahitolle, 18020 Bourges, France,
antoine.ferreira@ensi-bourges.fr

Abstract : *In this paper new intelligent control strategies are presented allowing a cooperative work between a human operator and a microassembly platform working in a partially structured microworld. Human integration is essential in the microworld where autonomous control alone would not be successful. The allocation of the control among human and robot occurs dynamically according to the state of the task execution. In the following, different task-level control issues (shared, traded, cooperative and priority) are discussed and tested on a micro device assembly workcell.*

1. INTRODUCTION

At present, operations to manipulate micromechanical parts (MEMS, MOEMS, micro-electronics), and biological cells (embryo, molecules,…) are accomplish manually with the help of micromanipulation systems. As the remote microworld is not well understood by the operator, augmented human sensing, scaled manipulation, dexterous haptic interfaces are concentring most of the research efforts made on *direct teleoperation* [1-3]. In these cases, a human operator is inside the control loop and as all movements are executed directly by the manipulator, possible collisions due to fatigue, imprecision of scaled manipulation and lack of dexterity can destroy components in the microworld. Moreover, future mass production of 3-D MEMS will require fully automatic assembly of hybrid components with very limited human interaction in order to preserve potential economic benefits. A better solution for operating is to have the operator outside the control loop. The operator gives "high level" commands by depicting different tasks to reach a goal. In this case, we speak of *supervised teleoperation*. Different research projects are currently investigated [4,5,6] but the full automatic control approach in the microworld seems to be very challenging. The more autonomous the system is, the more structured the microworld should be, and more specific its tasks are. Small uncertainties in the microworld, such as, parameter variation between the world model and the real world, noisy visual sensing information or unpredictable dynamic effects (friction, adhesion) causes the microassembly system break down.

How to minimize microeffects on every level of planning and execution system? We do not try to eliminate the uncertainties in the environment nor increases the intelligence of the microrobotic system. The main idea is to compensate the uncertain and dynamic varying micro environments by the ability, decision-making, and intervention capacities of human behavior in order to intervene in supervising and controlling with different proficiency levels. The principle consists of sharing the control of the microrobotic workcell between the operator and the remote visual controller (RVC) according to the context of action, taking advantage of their unique strengths and helping each other in their areas of weaknesses in micromanipulation. Figure 1 illustrates such human-machine interaction during on-line execution of the programmed virtual tasks developed in this work.

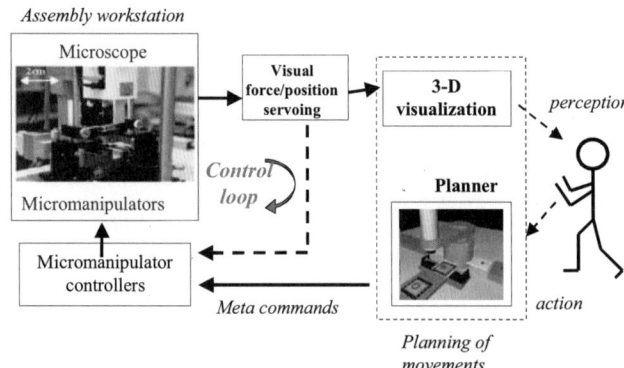

Fig. 1. Supervisory control of a microassembly workcell assisted by Virtual Reality.

This paper deals with the sharing of multilevel task-based control between a human operator and the microassembly workcell which makes use of the planning, monitoring and intervening supervisory control functions. In Section 2 we present the developed visually servoed microassembly system. In Section 3 an overview of the control-framework sharing human-robot interactions are modeled through a mathematical formalism. Application to typical *pick and place* tasks are analyzed and experimented following the shared control approach in Section 4. In Section 5 we present and discuss the obtained results.

2. TELEOPERATED MICROASSEMBLY WORKCELL

Currently, the microrobotic assembly system consists of two concentrated motion micromanipulators (main and sub) equipped with a microtool holder, a coarse motion worktable (driving with piezoelectric actuators, the movable ranges: 30×30×30mm), a multidegree of freedom fine positioning system (driving with piezo-actuators, the movable ranges: 100×100×100μm). The orientation of the end-effectors can also be changed by two accurate ultrasonic motors. A reflecting light-type optical microscope (OM) is used as a top-view vision sensor. A high-definition camera on the OM is connected to the Matrox Co. Frame-grabber which enables real-time image viewing of the microworld on the teleoperator interface.

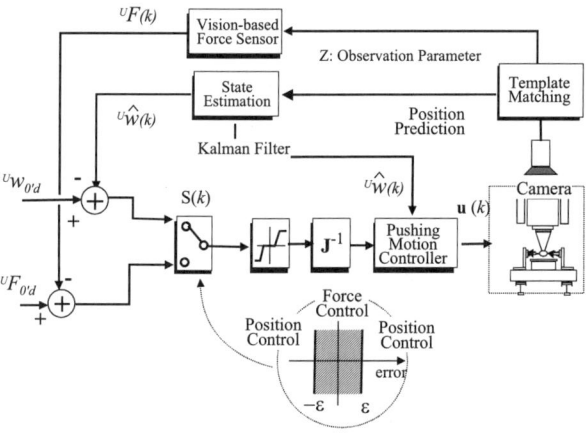

Fig. 2. Scheme of visually servoed position and force remote controller (RVC).

For task-oriented teleoperation, the workcell has to be controlled automatically. Figure 2 presents the strategy adopted for integrating visual force and position feedback of microassembly tasks performed by two micromanipulators. For more details, the reader may refer to [7,8]. The *carrier* micromanipulator allows to transport a microobject, using stepping pushing operation, to its final destination before to be handled. The object orientation is controlled in real time by a tracking system observing the scene change during motion through a microscope. In addition, we calculate predictions of pattern positions on-line for optimal trajectory control. This is done using an Extended Kalman Filter (EKF). A vision-based buckling type force sensor is used for estimation of object contact forces [9]. The two inputs are the desired state of visual features observed by the optical microscope, $^U w_{0'd}$, and the desired contact force at the object/tool interfaces, $^U F_{0'd}$. The desired feature state represents visual features of the object upon to initiate the microhandling operation. When the error vector in visual features $e_{0'} = [^{O'}p_e^T, \theta_e]^T$ is large, the switching controller S(k) uses pure vision feedback to control the position object. When the error vector is less than a threshold value ε, the handling operation can be initiated by switching the controller S(k) to the vision-based force control feedback. The object is handled and maintained in a stable way when the desired contact force $^U F_{0'd}$ is reached.

3. LIMITATIONS OF AUTOMATIC MICROASSEMBLY APPROACH

3.1. Task planning using Virtual Reality

The design issues of automatic microassembly include assembly planning (how the components can be put together?) and subtask planning (generation of instructions to the micromanipulators to carry out the assembly sequence). A typical *pick-and-place* task-based teleoperation approach has been modeled (positioning, handling and insertion) where the operator determines the elementary subtasks to be performed (Fig.3), and the visually servoed controller realizes these operations with the aid of a virtual reality (VR) guiding-system interface. The description of the task plan is given in Table 1.

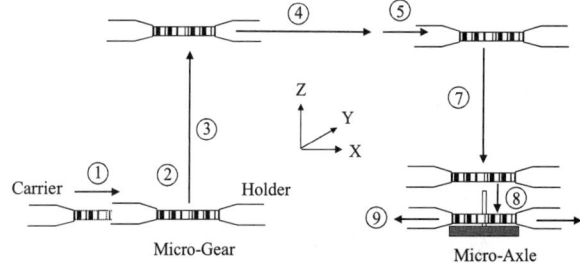

Fig. 3. Task plan decomposition of a microassembly task: insertion of a microgear into a microaxle.

Subtask	Definition
SE1 : ①	Pushing the object to its final position
SE2 : ②	Handling the object by both micromanipulators
SE3 : ③	Departing
SE4 : ④	Coarse motion of the manipulators holding the object
SE5 : ⑤	Fine motion for accurate positioning
SE6 : ⑥	Approaching the object to the microsystem
SE7 : ⑦	Establishing a defined alignment with the micro-axle
SE8 : ⑧	Microinsertion into the pin axle
SE9 : ⑨	Release of the microhandling operation

Table I. Planning task description.

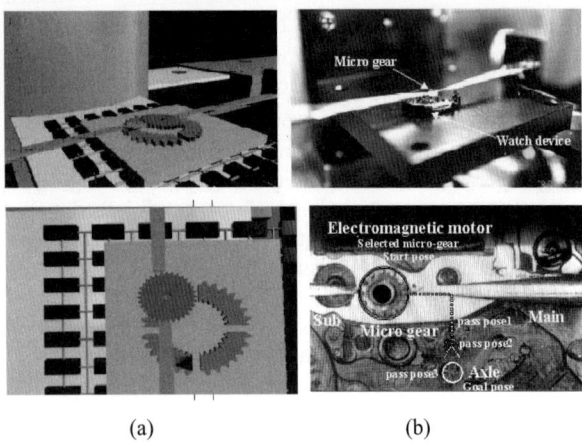

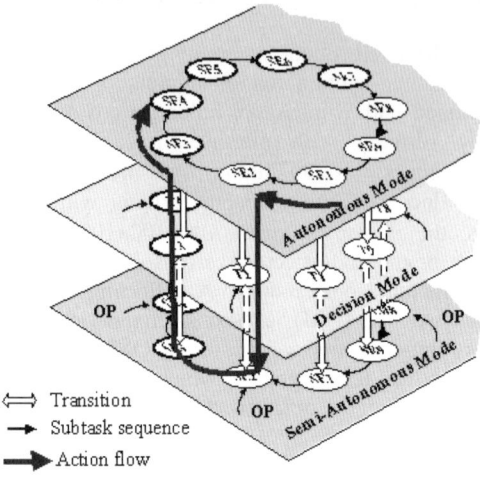

Fig. 4 : Planning-based VR (a) of assembly tasks, through off-line programming. On-line assembly (b) of a gear hole in a desired axle by decomposition of the subtasks, i.e., transportation (*pass pose1*), alignment (*pass pose2*), and insertion (*pass pose3*).

The combination of vision servoing techniques and VR-based simulation system allows to plan the manipulation tasks and to insure the guided-movements of the micromanipulators preventing any collisions and also increasing safety and reliability of the distributed assembly workcell. In our system, a VR-teleoperator interface giving a 3-D map of the scene is provided to the user for automatic positioning sequences in an absolute frame. We build our VR application in C using WTK (World tool kit, sense8^{TM}) and then we added our C application in our master control application [7]. As illustration, Fig.4 shows planned and on-line execution of subtasks depicted in Table1: ⑤ coarse and fine transportation of the handled object is controlled by vision feedback (*pass pose1*), ⑥ then, alignment operation by controlling accurately the cooperative movements of both piezoelectric manipulators through 3-D viewing (*pass pose2*), and finally ⑦ automatic insertion operation of the microgear into a microaxle (*pass pose3*).

3.2. Error Recovery

From the previous experiments with the microassembly workcell, small uncertainties in the microworld, such as parameter variation between the virtual world model and the real world, noisy visual sensing or unpredictable dynamic microeffects (i.e. friction forces during insertion, sticking object to the gripper) lead to degraded operating modes on every level of planning. If an operation fails, the incident is reported to the supervision which analyses the fault either requests manual assistance. Classical full automatic task-plan approach was unreliable. Supervisory control strategy demands for sensor surveillance during the assembly plan, requests

Fig. 5. Integration of the human-robot interaction into the task plan.

manual assistance during faulty micromanipulation, and requires support to solve the conflict between the internal model and the real microworld. Human integration into robot control is considered to compensate these limitations by the ability, the decision-making and the intervention capacities of human behavior.

4. HUMAN INTEGRATION IN ROBOT CONTROL

A method for integrating different levels of human-robot interactions into the task plan is given in the following which tolerates some degrees of uncertainty and gives some degree of freedom to the human. The integration of human in the "control loop" of the workcell needs to define, for each elementary subtask (SE) of the *pick-and-place task*, which is executed automatically, symmetrically semi-autonomous subtasks. These two levels of task execution, i.e., autonomous and semi-autonomous modes, are defined according to the degree of autonomy of the workcell facing to on-line execution problems. When the workcell is not able to accomplish the required task autonomously, it switches into the semi-automatic mode and gives then the operator the possibility to interact. Fig. 5 shows the scheme of integration of the human operator (OP) into the task plan. Contrarily to [10], we introduced a decision mode which activates the transition between the two levels represented by the bidirectional white arrows. The nature of the cooperative interaction between the operator and the assembly workcell changes according to the evolution of the task. To each SE corresponds a predetermined control allocation specified by the user during off-line programming. This control allocation is defined according to the capabilities of the robot and the human. A given degree of freedom is controlled by the partner of the human-robot team, who has the best sensing, actuating and cognitive capabilities to control it.

5. CONTROL ALLOCATION OF THE DECISION MODE

5.1. Basic Human-Robot Interactions

The question now is how do we combine basic behaviors to achieve a desirable supervisory control during the task plan? Let a human operator and the remote visual controller (RVC) defined by two vectors to cooperate for some task. Each process do the task by considering the contribution of the second one as shown in Fig.6a.

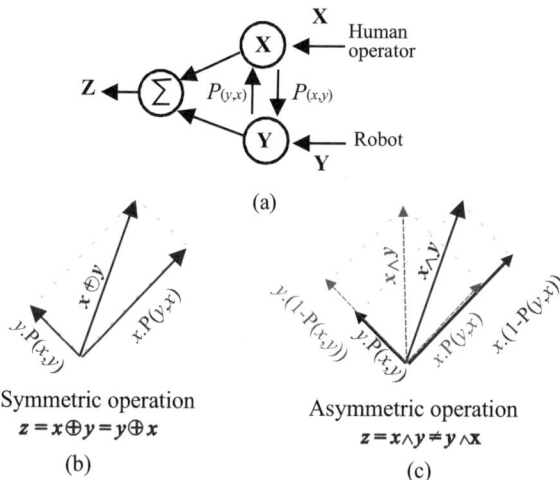

Figure 6. Decision making between human and robot: (a) basic flow of information and (b) vector representation of the symmetric and asymmetric human-robot interaction.

The behavior function $P(x,y)$ expresses the human-robot interaction;

$$P(x,y) = e^{-\left(\frac{ax^2+bxy+cy^2}{x^2+y^2}\right)} \text{ and } x,y \in \Re. \quad (1)$$

where the variables x and y are the contribution of each partner. The constants are settled to $a=1,02889$, $b=0,3574$ and $c=0$. If we note $P(x,y)$, it implies that the second process y acts by taking in consideration the contribution of the first process, x. On the contrary, the function can be reversed $P(y,x)$. From the definition of the behavior function $P(x,y)$, we can express the output behavior vector, z, resulting from the summation of the two input vectors (Fig.6a). The latter gives particular behavior outputs. In the following, several analogical gates illustrating the different human-robot interactions are defined. These gates are of two kinds: symmetric and asymmetric. The symmetric gates perform operations such as union (OR) and intersection (AND) while the asymmetric ones implement priority operations.

- *Cooperative Interaction.* We define the gate-OR as the result of a cooperation between two operators to perform some task. If x is the contribution of the first operator and y that of the second then the cooperation result is defined as :

$$z = x \oplus y = xP(y,x) + yP(x,y) \quad (2)$$

where $xP(y,x)$ is the contribution of the first operator by taking in consideration the contribution of the second one and $yP(x,y)$ means the reverse (see Fig.6b). Both operators contribute in relation to their magnitude. It is clear that this operation is symmetric. Now, we must impose properties to $P(y,x)$ of common sense :

(1): The cooperation of the same operator gives for both cases the same result, it means that : $P(x,x)=1/2$.

(2): The absence of cooperation of an operator does not act any more on the result. It means that $x \oplus 0 = x$ implying that for $x \neq 0$, $P(0,x) = 1$. In mathematical terms, zero is the neutral element of the $\oplus$ law.

(3): If one can define the opposite contribution $-x$ of x then $x \oplus -x = 0$.

(4): The contribution of two operators is superior than the smallest contribution of both processes and less than the largest one $\min(x,y) \leq x \oplus y \leq \max(x,y)$. This relationship is true if we assume that $0 < P(x,y) + P(y,x) \leq 1$.

These properties can be summarized as follows:

- $z = x \oplus y = y \oplus x$
- $z = cx \oplus cy = c(y \oplus x)$
- $x \oplus x = x$
- $x \oplus 0 = x$
- $x \oplus -x = 0$
- $\min(x,y) \leq x \oplus y \leq \max(x,y)$

- *Sustain Interaction.* It is pointed out that the gate-AND is the negation of the gate-OR and by using a "probabilistic" reasoning we can deduce :

$$z = x \otimes y = x(1 - P(y,x)) + y(1 - P(x,y)) \quad (3)$$

as negation of the OR function (2). The task is performed fastest when both processes simultaneously increase their performances. No task is accomplished if either process input is not activated. This operation is also symmetric as (2).

- $z = x \otimes y = y \otimes x$
- $z = cx \otimes cy = c(y \otimes x)$
- $x \otimes x = x$
- $x \otimes 0 = 0$
- $x \otimes -x = 0$
- $\min(x,y) \leq x \otimes y \leq \max(x,y)$

When the two operators are closely related, the asymmetric gates implement priority relationships such as.

-*Traded Interaction.* The *x*-port is assigned an exceptional prevalence over the *y*-port. The latter is put-through directly to the output as long as the former is absent. However, once the input is at the prevalent port it strongly dominates the output. It is expressed by

$$z = x \wedge y = xP(y,x) + yP(x^n, y). \quad (4)$$

where $n \in \Re$ given more importance to x-port. It is clear that this operation is asymmetric but verifies all the others properties as in (2). It means that in the absence of the first operator the second does not act.

- $z = x \wedge y \neq y \wedge x$
- $x \wedge 0 = x$ and $0 \wedge y = y$
- $x \wedge -y = -x \wedge y$
- $\min(x,y) \leq x \wedge y \leq \max(x,y)$

- *Shared Interaction*. As the x-input grows, the share of the y-input to the output is increased. The absence of x-input inhibits the output. In the absence of the y-input, the x-input is linearly passed to the output.

$$z = x!y = xP(y,x) + y(1 - P(x,y)). \quad (5)$$

Practically, it means that in the absence of the first operator the second does not act.

- $z = x!y = y!x$
- $x!0 = x$ and $0!y = 0$
- $z = cx!cy = c(y!x)$
- $x!-x = 0$ and $-x!x = 0$
- $x!x = x$
- $\min(x,y) \leq x!y \leq \max(x,y)$

6. Experimental Strategies of Human-Robot Interaction

Based on our approach, an experimental evaluation of the different strategies of human-robot interaction has been carried out for reliable and efficient 3-D microassembly.

6.1. Shared Control

In shared control, the teleoperator and the RVC control different subtasks of the workcell simultaneously. A common example which has been tested is the position and force control during the pickup and the transportation of the microobject. The respective shared tasks SE5–6 are described in Fig.8a considering several factors. Firstly, collisions between objectives, micromanipulators and MEMS device occur due to the limited working distance of the microscope objective. Secondly, to guide automatically transportation operation through the entire assembly scene, it needs to implement an efficient multi-view tracking system which is not feasible. Considering those limitations, the human operator handles the position control of both telemanipulators using "moment-to-

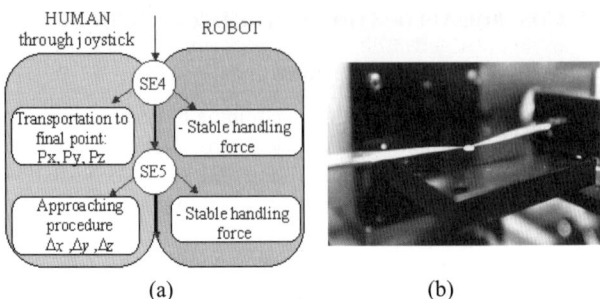

Fig.8: (a) Function allocation according to the shared control task SE5-SE6 during semi-autonomous mode and (b) on-line execution from a side-view.

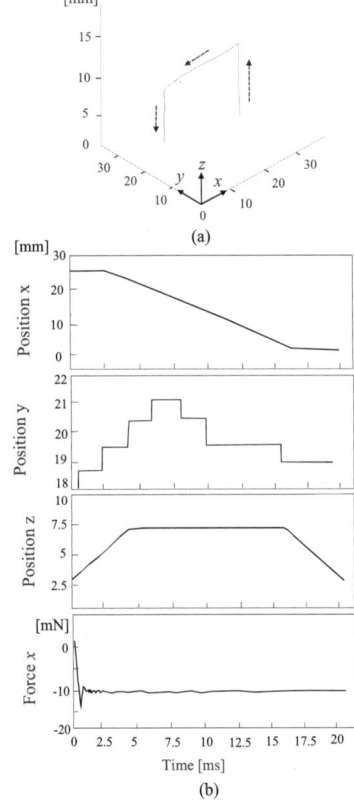

Fig.9: Experimental trajectory of the micro end-effector and position and force applied to the end-effector during execution.

moment" commands from a force-controlled joystick and controls visually the transportation through the multi viewing CCD sensors (Fig.8b). A 3-D view of the end-effector trajectory is shown in Fig.9a. The evolution of the position (P_x, P_y, P_z) of the end-effector controlled by the human is shown in Fig.9b. Meanwhile the RVC is responsible for microforce interaction with the handled object (force control). The transitions to switch to the semi-autonomous mode (SE3 to SE4) or to the autonomous mode (SE5 to SE6) are initiated automatically. Experiments have shown that shared control was well adapted for subtasks where there was some, i.e., weakness of visual sensing through the entire assembly scene, restricted motion and dexterity due to obstacles or need for complex manipulation skills.

5.3. Traded Control

In traded control, the operator and the remote controller (RVC) take turns to control the workcell according to the situation of successful control. At the operation level, the subtasks are temporally assigned to the operator and the RVC according their fitness to the the subtask execution. We tested a microinsertion subtask as planed in Fig.10(a).

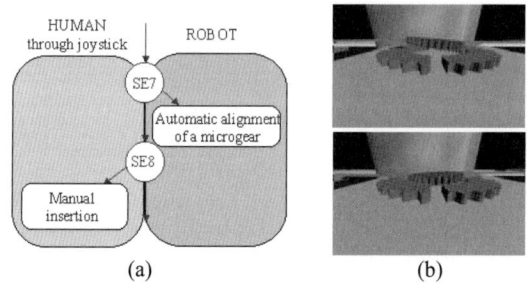

Fig. 10: (a) Function allocation according to the traded control task SE7-SE8 during semi-autonomous mode and (b) on-line execution.

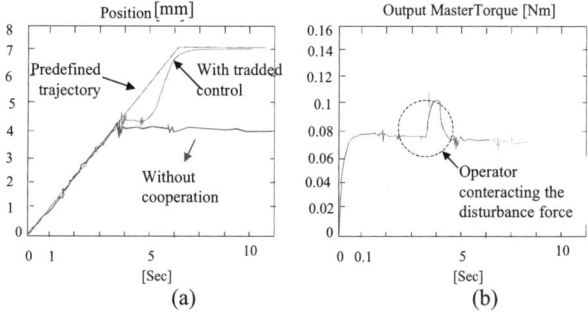

Fig. 11: (a) Position with and without traded control during desired motion and (b) on-line counteracting force applied by the operator through a joystick due to misalignment errors.

The telemanipulator is following the predefined trajectory, which is taken temporally by the local remote controller under normal conditions. Since the assembly clearance is small (Fig.10(b)), the part contact the MEMS substrate after it moves into the hole. As friction forces are important, handling operation becomes potentially unstable. The assembly can be more tolerant of alignment errors due to compliance if the microgripper can be controlled actively. When the operator notices some disturbance forces during execution, he takes over the control and helps the telemanipulator with the aid of the joystick interface. The transitions to switch to the semi-autonomous mode from SE7 to SE8 are initiated manually when the operator applies the corresponding forces on the joystick. All SEs are guarded motion, so that if the applied forces cross a given transition threshold, the transition is initiated. After that, the operator will release the control to the remote controller. Figure 11 shows the measured position/force control applied by the operator to the one-armed manipulator to counteract friction forces. Critical situations should be avoided with the aid of human manipulation capabilities.

5. CONCLUSION

In this paper, we have addressed the problem of new strategies of control using multilevel man-robot interactions for automatic microassembly. The main idea is to find a compromise between workcell autonomy and human interaction. We present a method which integrates the human into the task plan according to the state of the execution. The combination of these capabilities occurs through the definition of behavior functions. Their experimental integration into the vision-based microrobotic workcell has shown their efficiency and reliabilty during 3-D microassembly tasks. Their main advantages are, i.e., the error occurs on-line, the operator interacts directly with the workcell, direct force/position human feedback and right mixture of robot autonomy and human interaction adapted to the state of the microassembly plan.

6. REFERENCES

[1]: Zhou Y., Nelson B.J., Vikramaditya, "Fusing force and vision feedback for micromanipulations", IEEE ICRA, Leuven, Belg., pp.1220-1225 (1998).
[2]: Ferreira A., Cassier C., Haddab Y., Rougeot P., Chaillet N., "Macro-micro teleoperated system with 3-D visual and haptic feedback", SPIE Int. Symp. on Microrobots and Microassembly, Newton, MA,pp.112-123 (2001).
[3]: Li X., Zong G., Bi S., "Development of global vision system for biological automatic micromanipulation system", IEEE ICRA, Seoul, Korea, pp.127-132 (2001).
[4] : Fatikow S., Munassypov R., Rembold U., "Assembly planning and plan decomposition in an automated microrobot-based microassembly desktop station", Journal of Intelligent Manufacturing, **9**, pp. 73-92, 1998.
[5]: Yang G., Gaines J.A., Nelson B.J., "A flexible experimental workcell for efficient and reliable wafer-level 3D microassembly", IEEE ICRA, Seoul, South Korea, pp.133-138 (2001).
[6]: Thompson J.A., Fearing R.S., "Automating microassembly with ortho-tweezers and force sensing", IEEE/RSJ IROS, Hawaii, pp.1327-1334 (2001).
[7] Ferreira A., Hirai S., Fontaine J.G., "Automation of a teleoperated micro-assembly desktop station supervised by virtual reality", Trans. On Control, Automation, and Systems Engineering, Vol.4, No.1, 2002, pp.23-31.
[8] Cassier C., Ferreira A., Hirai S. "Combination of vision techniques and VR-based simulation for semi-autonomous microassembly workstation", IEEE ICRA, Washington DC, pp.2189-2196 (2002).
[9]: Ferreira A., Fontaine J-G., "New vision-based flexible force sensor for microteleoperation systems", XVI World Congress, sept.25-28, Vienna, Austria, pp.101-106 (2000).
[10] : Hoeniger T., "Dynamically shared control in human-robot teams through physical interactions", IEEE/RSJ IROS, Canada, 1998, pp.744-749.

Admittance Selection for Planar Force-Guided Assembly for Single-Point Contact with Friction

Shuguang Huang and Joseph M. Schimmels

Department of Mechanical and Industrial Engineering
Marquette University
Milwaukee, Wisconsin 53201-1881, USA
E-mail: {huangs, j.schimmels}@marquette.edu

Abstract— This paper identifies procedures for selecting the appropriate admittance to achieve reliable planar force-guided assembly for single-point frictional contact cases. A set of conditions that are imposed on the admittance matrix is presented. These conditions ensure that the motion that results from contact reduces part misalignment. We show that, for bounded misalignments, if an admittance satisfies the misalignment-reduction conditions at a finite number of contact configurations and a given coefficient of friction μ_M, then the admittance will also ensure that the conditions are satisfied at all intermediate configurations for all coefficients less than μ_M.

I. INTRODUCTION

Admittance control has been used in assembly tasks to provide force regulation and force guidance. In robotic assembly tasks, the admittance maps contact forces into changes in the velocity of the body held by the manipulator. To achieve reliable assembly, the manipulator admittance must be appropriate for the particular assembly task. In this paper, we identify procedures used to select the appropriate manipulator admittance for planar assembly with friction.

We consider a simple form of admittance, a linear admittance control law. For planar applications, this admittance behavior has the form:

$$\mathbf{v} = \mathbf{v}_0 + \mathbf{A}\mathbf{w} \qquad (1)$$

where $\mathbf{v}_0$ is the nominal twist (a 3-vector for planar cases), $\mathbf{w}$ is the contact wrench (force and torque) measured in the body frame (a 3-vector), and $\mathbf{A}$ is the admittance matrix (a 3×3 matrix).

Many researchers have addressed the use of admittance for force-guidance. Whitney [1], [2] proposed that the compliance of a manipulator be structured so that contact forces lead to decreasing errors. Peshkin [3] addressed the synthesis of an accommodation (inverse damping) matrix using least squared optimization. Asada [4] used a similar unconstrained optimization procedure for the design of an accommodation neural network.

A reliable admittance selection approach is to design the control law so that, at each possible part misalignment, the contact force always leads to a motion that reduces the existing misalignment. The approach is referred to as *force-assembly* and has been successful for workpart into fixture insertion when errors are infinitesimal [5], [6], [7].

For force-assembly, the motion resulting from contact must instantaneously reduce misalignment. Since the configuration space of a rigid body is non-Euclidian, there is no "natural" metric for finite spatial error. In [8], several body-specific *metrics* are established. One metric is based on the Euclidean distance between a single point on the body and its location when properly positioned.

Previously, we have considered sufficient conditions on the admittance to ensure planar force-assembly in *frictionless* single-point contact [9], [10]. In the study, we considered a *measure* of error based on the Euclidean distance between a single (fixed) point on the held body and its location when properly positioned. The misalignment reduction condition of force-assembly requires that, at each possible misalignment, the contact force yields a motion that reduces the misalignment. Using the point-based measure of misalignment discussed above, this condition can be expressed mathematically if we let $\mathbf{d}$ (a 3-vector for planar motion) be the line vector from the selected point at its proper mated position to its current position. Then, for error reducing motion, the condition is:

$$\mathbf{d}^T \mathbf{v} = \mathbf{d}^T (\mathbf{v}_0 + \mathbf{A}\mathbf{w}) < 0 \qquad (2)$$

which must be satisfied for all possible misalignments.

In this paper, we investigate single-point *frictional* contact using the same error measure. We show that, by identifying an admittance matrix that satisfies the error-reduction conditions at a *finite* number of extremal contact configurations and at a specified coefficient of friction, the error reduction requirements are also satisfied for *all* intermediate configurations and for all coefficients of friction less than the one specified. The friction model considered is "hard" point contact satisfying Coulomb's law [11].

Planar bodies in single-point contact have two types of stable contact states. One is referred to as "vertex-edge" contact ({v-e}, Fig. 1a); the other is referred to

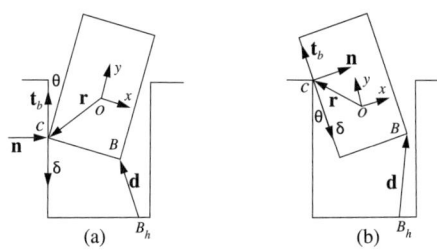

Fig. 1. Planar single-point contact. (a) Vertex-edge contact state. (b) Edge-vertex contact state.

as "edge-vertex" contact ($\{e\text{-}v\}$, Fig. 1b).

In this paper, the motion of a rigid body constrained by a frictional contact is derived in Section II. Sufficient conditions for error reduction for vertex-edge and edge-vertex contact states are obtained in Section III and Section IV, respectively. These conditions show that an admittance matrix satisfying the error reduction conditions at the boundaries of a set of contact configurations and the coefficients of friction, also satisfies the error-reduction conditions at all intermediate configurations for all intermediate coefficients of friction. A brief summary is presented in Section V.

II. Motion of a Rigid Body Constrained by a Frictional Contact

In this section, the planar motion a rigid body constrained by a single frictional surface is studied. First, the constrained compliant motion equation for frictional contact is derived. Then, the error-reduction function, describing the appropriate motion response in terms of the constraint and the admittance, is obtained.

A. Motion of a Constrained Rigid Body

Consider a rigid body interacting with a frictional surface as shown in Fig. (1). Let $\mathbf{n}$ (unit 2-vector) be the surface normal (pointing toward the held body) and let $\mathbf{t}_b$ (unit 2-vector) be the unit vector tangent to the surface at the contact point. Then, the direction of friction $\mathbf{t}$ must be along $\mathbf{t}_b$, i.e., $\mathbf{t} = \pm\mathbf{t}_b$.

The unit wrenches associated with the normal force and the friction force have the form:

$$\mathbf{w}_n = \begin{bmatrix} \mathbf{n} \\ (\mathbf{r} \times \mathbf{n}) \cdot \mathbf{k} \end{bmatrix}, \quad \mathbf{w}_t = \begin{bmatrix} \mathbf{t} \\ (\mathbf{r} \times \mathbf{t}) \cdot \mathbf{k} \end{bmatrix} \quad (3)$$

where $\mathbf{r}$ is the position vector from the origin of the coordinate frame to the point of contact, c, and $\mathbf{k}$ is the unit vector orthogonal to the plane.

Let ϕ be the magnitude of the normal contact force. Since we only consider sliding motion, the overall contact wrench is:

$$\mathbf{w} = (\mathbf{w}_n + \mu\mathbf{w}_t)\phi \quad (4)$$

where μ is the coefficient of friction.

By the control law (1), the motion of the body is:

$$\mathbf{v} = \mathbf{v}_0 + \mathbf{A}(\mathbf{w}_n + \mu\mathbf{w}_t)\phi. \quad (5)$$

Due to "hard" point contact, the motion of the rigid body cannot penetrate the surface. Thus, the reciprocal condition [12] must be satisfied:

$$\mathbf{w}_n^T \mathbf{v} = \mathbf{w}_n^T \mathbf{v}_0 + \mathbf{w}_n^T \mathbf{A}(\mathbf{w}_n + \mu\mathbf{w}_t)\phi = 0.$$

The magnitude ϕ is determined from:

$$\phi = \frac{-\mathbf{v}_0^T \mathbf{w}_n}{\mathbf{w}_n^T \mathbf{A}\mathbf{w}_n + \mu \mathbf{w}_n^T \mathbf{A}\mathbf{w}_t}. \quad (6)$$

Substituting (6) into (5) yields

$$\mathbf{v} = \frac{(\mathbf{v}_0\mathbf{w}_n^T - \mathbf{v}_0^T\mathbf{w}_n\mathbf{I})\mathbf{A}(\mathbf{w}_n + \mu\mathbf{w}_t)}{\mathbf{w}_n^T \mathbf{A}\mathbf{w}_n + \mu \mathbf{w}_n^T \mathbf{A}\mathbf{w}_t} \quad (7)$$

where $\mathbf{I}$ is the 3×3 identity matrix.

For planar motion, the normal $\mathbf{n}$ and tangent space base vector $\mathbf{t}_b$ at the contact point are known. The direction of the friction force ($\mathbf{t} = \mathbf{t}_b$ or $\mathbf{t} = -\mathbf{t}_b$) is uniquely determined by satisfying the following conditions: 1) ϕ in (6) is positive, and 2) $\mathbf{v}^T \mathbf{t} < 0$. Thus, $\mathbf{t}$ is known for a given contact point. The compliant motion can be determined by (7).

B. Error-Reduction Function

If the compliant motion is error-reducing, condition (2) must be satisfied for a given point. Thus,

$$E = \frac{\mathbf{d}^T(\mathbf{v}_0\mathbf{w}_n^T - \mathbf{v}_0^T\mathbf{w}_n\mathbf{I})\mathbf{A}(\mathbf{w}_n + \mu\mathbf{w}_t)}{\mathbf{w}_n^T \mathbf{A}\mathbf{w}_n + \mu \mathbf{w}_n^T \mathbf{A}\mathbf{w}_t} < 0. \quad (8)$$

To avoid singularity in (7), the denominator must have no root over the range considered. Since $\mathbf{A}$ is positive definite, $\mathbf{w}_n^T \mathbf{A}\mathbf{w}_n > 0$, the denominator is positive for $\mu = 0$. Thus, we assume that, for $\mu \in [0, \mu_M]$, the inequality

$$\mathbf{w}_n^T \mathbf{A}\mathbf{w}_n + \mu \mathbf{w}_n^T \mathbf{A}\mathbf{w}_t > 0 \quad (9)$$

is satisfied. Therefore, the error-reduction function can be expressed as:

$$F_{er} = \mathbf{d}^T(\mathbf{v}_0\mathbf{w}_n^T - \mathbf{v}_0^T\mathbf{w}_n\mathbf{I})\mathbf{A}(\mathbf{w}_n + \mu\mathbf{w}_t). \quad (10)$$

In the following two sections, error-reduction conditions are obtained for the two single-point contact states.

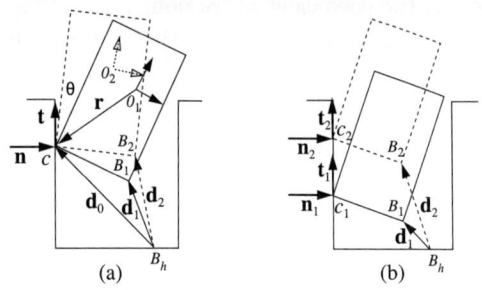

Fig. 2. Vertex-edge contact state. (a) Orientational variation. (b) Translational variation.

III. VERTEX-EDGE CONTACT STATE

In this section, vertex-edge contact is considered. As shown in Fig. 1a, the configuration of the body can be determined by the orientation of the body θ and the location of the contact point δ.

Suppose that θ varies within the range of $[-\theta_M, \theta_M]$, and δ varies within the range of $[-\delta_M, \delta_M]$. We prove that, if an admittance matrix $\mathbf{A}$ satisfies a set of conditions at a finite number of configurations for $\mu = 0, \mu_M$, then the $\mathbf{A}$ matrix ensures error-reducing motion for all configurations $\theta \in [-\theta_M, \theta_M]$, $\delta \in [-\delta_M, \delta_M]$, and all coefficients of friction $\mu \in [0, \mu_M]$.

Similar to the approach used for the frictionless case [9], we first consider orientational and translational variation separately. Then, by combining the two variation cases, sufficient conditions for the general case are obtained.

A. Orientational Variation

Consider only orientation variation as illustrated in Fig. 2a. In this case, both the direction of the error reduction vector $\mathbf{d}$ and the direction of the contact wrench $\mathbf{w}$ (in the body frame) are changed by changing the orientation. We prove that, for variation $\theta_M \leq \frac{\pi}{4}$, if $\mathbf{A}$ satisfies a set of conditions at orientation $\theta = 0$, then an error-reducing motion is ensured for all configurations $\theta \in [-\theta_M, \theta_M]$.

A.1 Error Reduction Function

Let $\mathbf{w}_0$ be the contact wrench, and $\mathbf{d}_0$ be the position vector associated with $\theta = 0$. Suppose that at $\theta = 0$, error-reducing motion is obtained, i.e.,

$$\mathbf{d}_0^T \mathbf{v}_0 + \mathbf{d}_0^T \mathbf{A} \mathbf{w}_0 < 0. \quad (11)$$

Consider a rotation given by an angle change θ. Let $\mathbf{n}_0$ and $\mathbf{t}_0$ be the unit vectors in the directions of the normal force and friction force respectively when $\theta = 0$, then in the body coordination frame, the two vectors associated with $\theta \in [-\theta_M, \theta_M]$ are:

$$\mathbf{n}_\theta = \mathbf{R}(\theta)\mathbf{n}_0, \quad \mathbf{t}_\theta = \mathbf{R}(\theta)\mathbf{t}_0, \quad (12)$$

where $\mathbf{R}$ is the rotation matrix associated with θ having the form:

$$\mathbf{R}(\theta) = \begin{bmatrix} \cos\theta & \sin\theta \\ -\sin\theta & \cos\theta \end{bmatrix}. \quad (13)$$

The unit contact normal and friction wrenches calculated using (3) are:

$$\mathbf{w}_n(\theta) = \begin{bmatrix} \mathbf{R}\mathbf{n}_0 \\ (\mathbf{r} \times \mathbf{R}\mathbf{n}_0) \cdot \mathbf{k} \end{bmatrix}, \quad (14)$$

$$\mathbf{w}_t(\theta) = \begin{bmatrix} \mathbf{R}\mathbf{t}_0 \\ (\mathbf{r} \times \mathbf{R}\mathbf{t}_0) \cdot \mathbf{k} \end{bmatrix}, \quad (15)$$

where $\mathbf{r}$ is the position vector from the origin of the body frame to the contact point (constant).

Since all configurations considered correspond to pure rotation about the contact point, the position vector of B for an intermediate configuration can be expressed in the body frame as:

$$\mathbf{d}'_\theta = \mathbf{R}\mathbf{d}'_0 + \mathbf{d}' \quad (16)$$

where $\mathbf{d}'_0$ is the position vector from B_h to the contact point c, $\mathbf{d}'$ is the position vector from c to point B_1. Note that $\mathbf{d}'_0$ is a constant in the global frame and $\mathbf{d}'$ is constant in the body frame. Then, the line vector of B relative to its properly mated position B_h (expressed in the body frame) is obtained:

$$\mathbf{d}_\theta = \begin{bmatrix} \mathbf{d}'_\theta \\ (\mathbf{r}_\theta \times \mathbf{d}'_\theta) \cdot \mathbf{k} \end{bmatrix} \quad (17)$$

where $\mathbf{r}_\theta$ is the vector from the body frame origin to point B.

Since $\mathbf{d}_\theta$, $\mathbf{w}_n$ and $\mathbf{w}_t$ each involve first order terms in $\sin\theta$ and $\cos\theta$, the error-reduction function (10) can be expressed as a third order polynomial in $\sin\theta$ and $\cos\theta$. Further, by the relation $\sin^2\theta = 1 - \cos^2\theta$, the function can be written in the form:

$$F_{er}(\theta) = c_1 \cos^3\theta + c_2 \sin\theta \cos^2\theta + c_3 \cos^2\theta + \\ c_4 \sin\theta \cos\theta + c_5 \sin\theta + c_6 \cos\theta + c_7 \quad (18)$$

where the c_i's are functions of the admittance matrix $\mathbf{A}$ and the friction coefficient μ having the form:

$$c_i = a_i + \mu b_i, \quad i = 1, ..., 7. \quad (19)$$

A.2 Error Reduction Conditions

To achieve error reduction at all other configurations and for any value of friction less than μ_M considered, $F_{er}(\theta)$ must be negative for all $\theta \in [-\theta_M, \theta_M]$ and $\mu \in [0, \mu_M]$. Now consider F_{er} as a function of (θ, μ), then $F_{er}(\theta, \mu)$ only contains a first order term in μ.

In the following, we first obtain error reduction conditions for $\theta \in [-\theta_M, \theta_M]$ for both $\mu = 0$ and $\mu = \mu_M$. Then, we prove that the conditions for the extremal friction coefficients ensure error-reducing motion for any intermediate $\mu \in [0, \mu_M]$.

By an appropriate rearrangement, (18) can be written as:

$$F_{er}(\theta, \mu) = (c_1 \cos^3 \theta + c_3 \cos^2 \theta + c_6 \cos \theta + c_7) + (c_2 \cos^2 \theta + c_4 \cos \theta + c_5) \sin \theta. \quad (20)$$

For $\mu = 0$, $c_i = a_i$. A conservative "more positive" function $F_0^+(\theta)$ for $\theta > 0$ is constructed based on (20) by the following
- If $a_i > 0$, replace the corresponding $\cos \theta$ with 1 (by setting $\theta = 0$);
- If $a_i < 0$, replace the corresponding $\cos \theta$ with $\cos \theta_M$.

As such, $F_0^+(\theta)$ has the form:

$$F_0^+(\theta) = a + a^+ \sin \theta. \quad (21)$$

It can be seen that for any $0 \leq \theta \leq \theta_M$,

$$F(\theta)|_{\mu=0} \leq F_0^+(\theta). \quad (22)$$

For $\theta < 0$, a conservative "more positive" function $F_0^-(\theta)$ is constructed based on (20) by the following:
- For the terms involving $\sin \theta$, if $a_i > 0$, replace the corresponding $\cos \theta$ with $\cos \theta_M$; if $a_i < 0$ replace the corresponding $\cos \theta$ with 1.
- For the terms involving only $\cos \theta$, if $a_i > 0$, replace the corresponding $\cos \theta$ with 1; if $a_i < 0$, replace the corresponding $\cos \theta$ with $\cos \theta_M$.

As such, F_0^- has the form:

$$F_0^-(\theta) = a + a^- \sin \theta. \quad (23)$$

It can be seen that for any $-\theta_M \leq \theta \leq 0$,

$$F(\theta)|_{\mu=0} \leq F_0^-(\theta). \quad (24)$$

Because $\sin \theta$ is a monotonic function over $[-\frac{\pi}{4}, \frac{\pi}{4}]$, $F_0^+(0) < 0$ and $F_0^+(\theta_M) < 0$ ensure that $F_0^+(\theta) < 0$ for all $\theta \in [0, \theta_M]$; and $F_0^-(0) < 0$ and $F_0^-(-\theta_M) < 0$ ensure that $F_0^-(\theta) < 0$ for all $\theta \in [-\theta_M, 0]$. Since $F_0^-(0) = F_0^+(0)$, the following set of 3 inequalities:

$$a < 0 \quad (25)$$
$$a + a^+ \sin \theta_M < 0 \quad (26)$$
$$a - a^- \sin \theta_M < 0 \quad (27)$$

ensures that $F(\theta)|_{\mu=0} < 0$ for all $\theta \in [-\theta_M, \theta_M]$.

Using the same procedure for $F(\theta)|_{\mu=\mu_M}$, two conservative "more positive" functions $F_{\mu_M}^-(\theta)$ and $F_{\mu_M}^+(\theta)$ are constructed:

$$F_{\mu_M}^+(\theta) = e + e^+ \sin \theta, \quad (28)$$
$$F_{\mu_M}^-(\theta) = e + e^- \sin \theta. \quad (29)$$

Thus, the following set of 3 inequalities:

$$e < 0 \quad (30)$$
$$e + e^+ \sin \theta_M < 0 \quad (31)$$
$$e - e^- \sin \theta_M < 0 \quad (32)$$

ensures that $F(\theta)|_{\mu=\mu_M} < 0$ for all $\theta \in [-\theta_M, \theta_M]$.

Although inequalities (25)-(27) and (30)-(32) are constructed for two friction coefficients $\mu = 0, \mu_M$, they are sufficient error reduction conditions for all $\mu \in [0, \mu_M]$. In fact, since the error-reduction function F contains only first order term of μ, then, for any $\theta \in [-\theta_M, \theta_M]$ and $\mu \in [0, \mu_M]$,

$$\min\{F_{er}(\theta, 0), F_{er}(\theta, \mu_M)\} \leq F_{er}(\theta, \mu) \quad (33)$$
$$\leq \max\{F_{er}(\theta, 0), F_{er}(\theta, \mu_M)\}.$$

Since the sets of inequalities (25)-(27) and (30)-(32) ensure $F_{er}(\theta, 0) < 0$ and $F_{er}(\theta, \mu_M) < 0$, thus, from (33), $F_{er}(\theta, \mu) < 0$ for $\forall \theta \in [-\theta_M, \theta_M], \forall \mu \in [0, \mu_M]$ is ensured by these inequalities.

B. Translational Variation

Now consider the translational variation of the contact configuration illustrated in Fig. 2b. In this case, only translation along the edge is allowed, and the contact force does not change in the body frame. The configuration of the body can be determined by a vector $\mathbf{d}$ (Fig. 2b).

Suppose that, at the two extremal configurations characterized by $\mathbf{d}_1$ and $\mathbf{d}_2$, the error reduction conditions are satisfied:

$$\mathbf{d}_1^T \mathbf{v}_0 + \mathbf{d}_1^T \mathbf{A} \mathbf{w}_1 < 0, \quad (34)$$
$$\mathbf{d}_2^T \mathbf{v}_0 + \mathbf{d}_2^T \mathbf{A} \mathbf{w}_2 < 0, \quad (35)$$

where $\mathbf{w}_1$ and $\mathbf{w}_2$ are total contact wrenches at the two locations c_1 and c_2.

For any $\alpha, \beta \geq 0$,

$$(\alpha \mathbf{d}_1 + \beta \mathbf{d}_2)^T \mathbf{v}_0 + (\alpha \mathbf{d}_1 + \beta \mathbf{d}_2)^T \mathbf{A} \mathbf{w} < 0. \quad (36)$$

At any intermediate configuration, the $\mathbf{d}$ vector is expressed as a convex combination of the vectors $\mathbf{d}_1$ and $\mathbf{d}_2$, i.e.,

$$\mathbf{d} = \alpha \mathbf{d}_1 + \beta \mathbf{d}_2 \quad (37)$$

where $\alpha, \beta \geq 0$ and $\alpha + \beta = 1$.

Since the contact wrench $\mathbf{w}$ is the same in the body frame for all contact configurations, $\mathbf{w} = \mathbf{w}_1 = \mathbf{w}_2$. Substituting (37) into (36) yields:

$$F_{er} = \mathbf{d}^T \mathbf{v}_0 + \mathbf{d}^T \mathbf{A} \mathbf{w} < 0.$$

Thus, for translational variation, if at two configurations the error reduction condition is satisfied, then the error reduction condition must be satisfied for all intermediate configurations bounded by these two configurations.

It is noted that the contact wrench $\mathbf{w}_i$'s in (34) and (35) include friction. Because the coefficient of friction μ is linear in F_{er}, satisfying the error-reduction conditions at $\mu = 0, \mu_M$ ensures that the same conditions are satisfied for all $\mu \in [0, \mu_M]$.

C. General Case

Because of the linear dependence of the error-reduction function on the boundary configurations for the translational-only variation, similar to the frictionless case presented in [9], the results presented in III-A and III-B can be generalized to the vertex-edge contact state involving both translational and orientational variations. Thus we have:

Proposition 1: For a vertex-edge contact state with variation of orientation $[-\theta_M, \theta_M]$ and variation of translation $[-\delta_M, \delta_M]$, if at the two configurations with different contact boundary locations $[-\delta_M, \delta_M]$ the admittance satisfies inequalities (25)-(27) and (30)-(32) for $\mu = 0$ and $\mu = \mu_M$, then the admittance will satisfy the error reduction condition for all configurations bounded by the four configurations, $[-\delta_M, -\theta_M]$, $[-\delta_M, \theta_M]$, $[\delta_M, -\theta_M]$, $[\delta_M, \theta_M]$, for all $\mu \in [0, \mu_M]$.

Therefore, for an edge-vertex contact state, to ensure that the motion response due to contact is error reducing for all configurations considered, function values at only two configuration extremals and two coefficients of friction need be tested.

IV. Edge-Vertex Contact State

Consider "edge-vertex" contact. As shown in Fig. 1b, the configuration of the body can be determined by two parameters, (δ, θ).

Suppose that θ varies within the range of $[-\theta_M, \theta_M]$, and δ varies within the range of $[-\delta_M, \delta_M]$. We prove that, if an admittance matrix $\mathbf{A}$ satisfies a set of conditions at the "boundary" points for $\mu = 0$ and μ_M, then the $\mathbf{A}$ matrix ensures error-reducing motion for all configurations $\theta \in [-\theta_M, \theta_M]$, $\delta \in [-\delta_M, \delta_M]$, and $\mu \in [0, \mu_M]$.

In this case, the error-reduction function does not linearly depend on the configuration parameter θ or δ when considering either orientational or translational variation separately. As a consequence, a somewhat more complicated evaluation is used in which the orientational and translational variation are considered simultaneously.

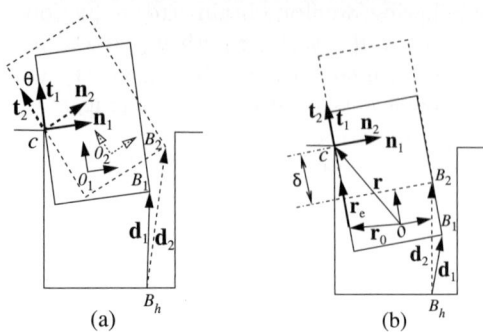

Fig. 3. Edge-vertex contact state. (a) Orientational variation: the contact wrench $\mathbf{w}$ is constant in the body frame while the error-measure vector $\mathbf{d}$ is a nonlinear function of θ. (b) Translational variation: both the contact wrench $\mathbf{w}$ and the error-measure vector $\mathbf{d}$ are functions of δ.

A. Error-Reduction Function

In order to obtain the error-reduction function, we first express the contact wrench and the error-measure vector $\mathbf{d}$ in terms of δ and θ.

For an edge-vertex contact state, as shown in Fig. 3a, when the held body rotates relative to the fixtured body about the contact point, the description of the contact wrench does not change in a body-based coordinate frame. When the held body translates relative to the fixtured body, the description of the contact wrench changes in a body-based coordinate frame as the contact point changes (although its direction is constant). Thus, the contact force is a function involving only the translational variable δ.

As shown in Fig. 3b, in the body frame, the direction of the surface normal is constant while the position vector of the contact point, $\mathbf{r}$, varies. For an arbitrary δ, $\mathbf{r}$ can be expressed as:

$$\mathbf{r}_\delta = \mathbf{r}_0 + \mathbf{r}_e \delta \tag{38}$$

where $\mathbf{r}_0$ is a vector from the body frame to a center point of the edge (constant) and $\mathbf{r}_e$ is the unit vector along the edge.

By (3), the unit wrench corresponding to the surface normal and friction are:

$$\mathbf{w}_n = \begin{bmatrix} \mathbf{n} \\ (\mathbf{r}_\delta \times \mathbf{n}) \cdot \mathbf{k} \end{bmatrix}, \quad \mathbf{w}_t = \begin{bmatrix} \mathbf{t} \\ (\mathbf{r}_\delta \times \mathbf{t}) \cdot \mathbf{k} \end{bmatrix}. \tag{39}$$

It can be seen that in the body frame, the directions of $\mathbf{w}_n$ and $\mathbf{w}_t$ are constant while the last components (the moment terms) are linear functions of δ.

Let $\mathbf{d}'_0$ be the error-measure 2-vector at $(\theta, \delta) = (0, 0)$, then for an arbitrary δ with $\theta = 0$, the error-measure vector $\mathbf{d}'$ is:

$$\mathbf{d}'_\delta = \mathbf{d}'_0 + \mathbf{r}_e \delta, \quad \delta \in [-\delta_M, \delta_M] \tag{40}$$

where $\mathbf{r}_e$ is a unit vector along the contacting edge. Note that $\mathbf{d}_0'$ is constant in the global coordinate frame while $\mathbf{r}_e$ is constant in body coordinate frame. Thus for an arbitrary orientation $\theta \in [-\theta_M, \theta_M]$ and $\delta \in [-\delta_M, \delta_M]$, the error-measure 2-vector $\mathbf{d}'$ is a function of δ and θ having the form:

$$\mathbf{d}'(\delta, \theta) = \mathbf{R}\mathbf{d}_0' + \mathbf{r}_e \delta, \tag{41}$$

where $\mathbf{R}$ is the rotation matrix having the form of (13).

The line vector associated with $\mathbf{d}'(\delta, \theta)$ can be calculated:

$$\mathbf{d}(\delta, \theta) = \begin{bmatrix} \mathbf{R}\mathbf{d}_0' \\ (\mathbf{r}_B \times \mathbf{R}\mathbf{d}_0') \cdot \mathbf{k} \end{bmatrix} + \delta \begin{bmatrix} \mathbf{r}_e \\ (\mathbf{r}_B \times \mathbf{r}_e) \cdot \mathbf{k} \end{bmatrix} \tag{42}$$

where $\mathbf{r}_B$ is the position vector from the body frame origin to point B.

Thus, for any intermediate configuration (δ, θ), because $\mathbf{w}_n$ and $\mathbf{w}_t$ in (39) each only contain first order terms in δ and $\mathbf{d}(\delta, \theta)$ in (42) only contains first order terms in $\sin\theta$, $\cos\theta$ and δ, the error-reduction function (10) can be expressed as a third order polynomial in δ in the form:

$$F_{er}(\delta, \theta) = f_3 \delta^3 + f_2 \delta^2 + f_1 \delta + f_0 \tag{43}$$

where the coefficients f_i's have the form:

$$f_i = a_i \cos\theta + b_i \sin\theta. \tag{44}$$

Also note that, μ appears in the coefficients of $\mathbf{w}_t$. Therefore, the coefficients a_i and b_i have the form:

$$a_i = (p_i + \mu p_i'), \quad b_i = (q_i + \mu q_i'), \tag{45}$$

where p_i, p_i', q_i and q_i' are functions of the admittance $\mathbf{A}$.

B. Sufficient Conditions for Error-Reduction

The error-reduction condition requires that the error-reduction function in (43) must be negative in the range of configurations considered. In order to obtain sufficient conditions, we construct two functions F_0 and F_M by replacing the $\cos\theta$ terms in (45) with 1 and $\cos\theta_M$ respectively:

$$F_0(\delta, \theta) = (a_3\delta^3 + a_2\delta^2 + a_1\delta + a_0) + (b_3\delta^3 + b_2\delta^2 + b_1\delta + b_0)\sin\theta, \tag{46}$$

$$F_M(\delta, \theta) = (a_3\delta^3 + a_2\delta^2 + a_1\delta + a_0)\cos\theta_M + (b_3\delta^3 + b_2\delta^2 + b_1\delta + b_0)\sin\theta. \tag{47}$$

For small θ (e.g., $\theta \leq \frac{\pi}{8}$), F_0 and F_M are close approximations of F_{er}, and for any (δ, θ) in the range considered,

$$\min\{F_0, F_M\} \leq F_{er} \leq \max\{F_0, F_M\}. \tag{48}$$

Thus, if both F_0 and F_M are negative over the range $\delta \in [-\delta_M, \delta_M]$ and $\theta \in [-\theta_M, \theta_M]$, error reducing motion is ensured.

For a given θ, both F_0 and F_M are third order polynomials in δ. To obtain conditions on F_0 and F_M, we first evaluate the bounds on the coefficients of these two polynomials.

By (46) and (47), the coefficients of δ^i in F_0 and F_M are:

$$f_i^0(\mu, \theta) = (p_i + p_i'\mu) + (q_i + q_i'\mu)\sin\theta, \tag{49}$$
$$f_i^M(\mu, \theta) = (p_i + p_i'\mu)\cos\theta_M + (q_i + q_i'\mu)\sin\theta. \tag{50}$$

If the range of μ is $[0, \mu_M]$, it can be proved that f_i^0 and f_i^M achieve their maximum and minimum values only at the boundary points $(0, \pm\theta_M)$ and $(\mu_M, \pm\theta_M)$. This can be verified by evaluating the Hessian matrices of f_i^0 and f_i^M. In fact, the Hessian matrix of f_i^0 with respect to (μ, θ) is:

$$\text{Hess}(f_i^0) = \begin{bmatrix} 0 & q_i'\cos\theta \\ q_i'\cos\theta & -(q_i + q_i'\mu)\sin\theta \end{bmatrix}.$$

Since for $|\theta| \leq \frac{\pi}{8}$, $\det(\text{Hess}) = -q_i'^2\cos\theta < 0$, the Hessian is indefinite and the function f_i^0 cannot have a maximum or minimum in the interior of the area $[0, \mu_M] \times [-\theta_M, \theta_M]$ [13]. Thus, the maximum (minimum) values of f_i^0 can be chosen from its four values at the 4 boundary points: $(0, \pm\theta_M)$ and $(\mu_M, \pm\theta_M)$. The same property holds true for f_i^M.

Denote

$$s_M = \max\{|f_i^0|, |f_i^M|, \ i = 1, 2, 3\}, \tag{51}$$
$$s_0 = \min\{|f_0^0|, |f_0^M|\}. \tag{52}$$

We prove that if

$$\frac{s_0}{s_M + s_0} > \delta_M, \tag{53}$$

then both F_0 and F_M have no root for all $\delta \in [-\delta_M, \delta_M]$, $\theta \in [-\theta_M, \theta_M]$ and $\mu \in [0, \mu_M]$.

Consider the function F_0 in (46). For an arbitrary $\theta_0 \in [-\theta_M, \theta_M]$ and an arbitrary $\mu_0 \in [0, \mu_M]$, F_0 is a third order polynomial in a single-variable δ:

$$F_0(\delta, \theta_0) = c_3\delta^3 + c_2\delta^2 + c_1\delta + c_0 \tag{54}$$

where

$$c_i = (p_i + \mu_0 p_i') + (q_i + \mu_0 q_i')\sin\theta_0. \tag{55}$$

Let

$$c_M = \max\{|c_1|, |c_2|, |c_3|\}, \tag{56}$$

then, as shown in [10], a root of F_0, ξ, must satisfy

$$|\xi| \geq \frac{|c_0|}{c_M + |c_0|}. \tag{57}$$

3087

Since $\theta_0 \in [-\theta_M, \theta_M]$ and $\mu_0 \in [0, \mu_M]$, by (51) and (52), we have:

$$c_M \leq s_M, \qquad |c_0| \geq s_0. \qquad (58)$$

Therefore,

$$\frac{s_M}{s_0} \geq \frac{c_M}{c_0} \qquad (59)$$

which leads to

$$|\xi| \geq \frac{|c_0|}{c_M + |c_0|} \geq \frac{s_0}{s_M + s_0} > \delta_M. \qquad (60)$$

Thus, F_0 has no root in $[-\delta_M, \delta_M]$ for all $\theta \in [-\theta_M, \theta_M]$ and $\mu \in [0, \mu_M]$. The same reasoning applies to F_M. Therefore, the functions F_0 and F_M do not change sign if inequality (53) is satisfied. By (48), F_{er} has no root in the same bounded area. Since the s_M in (51) and s_0 in (52) are functions of the admittance **A**, (53) imposes a constraint on **A**. In summary, we have:

Proposition 2: For an edge-vertex contact state, if: i) at the configuration $[\delta, \theta] = [0, 0]$, the admittance satisfies the error reduction condition (2), and ii) condition (53) is satisfied for the configuration boundary points $[\pm\delta, \pm\theta]$ and the maximum value of friction coefficient μ_M, then the admittance will satisfy the error reduction conditions for all configurations bounded by these four configurations and friction coefficient $\mu \leq \mu_M$.

Thus, for an edge-vertex contact state, to ensure that contact yields error-reducing motion for the body, only four configuration extremals at two extremal coefficients of friction need be tested.

V. SUMMARY

In this paper, the error reduction condition for a single point on the held body in frictional contact is considered. We have presented an approach for admittance selection of a planar rigid body motion for force-guided assembly with friction. We have shown that, for one point contact cases, the admittance control law can be selected based on their behavior at a *finite* number of configurations and at two extremal coefficients of friction of the contact. If the error reduction conditions are satisfied at these configurations with these two coefficients of friction, error reduction will be satisfied for all intermediate configurations and all intermediate coefficients of friction. Thus, for a given set of bounded misalignments, a single admittance control law that satisfies these conditions guarantees the proper assembly of a given pair of mating parts.

ACKNOWLEDGMENTS

This research was supported in part by a Ford Motor Company URP award and the National Science Foundation under grant IIS 0010017.

REFERENCES

[1] D. E. Whitney, "Force feedback control of manipulator fine motions," *ASME Journal of Dynamic Systems, Measurements and Control*, vol. 98, no. 2, 1977.

[2] D. E. Whitney, "Quasi-static assembly of compliantly supported rigid parts," *ASME Journal of Dynamic Systems, Measurements, and Control*, vol. 104, no. 1, pp. 65–77, 1982.

[3] M. A. Peshkin, "Programmed compliance for error-corrective manipulation," *IEEE Transactions on Robotics and Automation*, vol. 6, no. 4, pp. 473–482, 1990.

[4] H. Asada, "Teaching and learning of compliance using neural net," *IEEE Transactions on Robotics and Automation*, vol. 9, no. 6, pp. 863–867, 1993.

[5] J. M. Schimmels and M.A. Peshkin, "Admittance matrix design for force guided assembly," *IEEE Transactions on Robotics and Automation*, vol. 8, no. 2, pp. 213–227, 1992.

[6] J. M. Schimmels and M. A. Peshkin, "Force-assembly with friction," *IEEE Transactions on Robotics and Automation*, vol. 10, no. 4, pp. 465–479, 1994.

[7] J. M. Schimmels, "A linear space of admittance control laws that guarantees force-assembly with friction," *IEEE Transactions on Robotics and Automation*, vol. 13, no. 5, pp. 656–667, 1997.

[8] J. M. R. Martinez and J. Duffy, "On the metric of rigid body displacements for infinite and finite bodies," *ASME Journal of Mechanical Design*, vol. 117, no. 1, pp. 41–47, 1995.

[9] S. Huang and J. M. Schimmels, "Sufficient conditions used in admittance selection for planar force-guided assembly," in *Proceedings of the IEEE International Conference on Robotics and Automation*, Washington, D.C., May 2002, pp. 538–543.

[10] S. Huang and J. M. Schimmels, "Sufficient conditions used in admittance selection for force-guided assembly of polygonal parts," *IEEE Transactions on Robotics and Automation*, (submitted for publication).

[11] S. Huang and J. M. Schimmels, "The motion of a compliantly suspended rigid body constrained by a frictional surface," in *Proceedings of the ASME International Congress and Exposition*, New York, NY, November 2001, vol. 2.

[12] M. S. Ohwovoriole and B. Roth, "An extension of screw theory," *ASME Journal of Mechanical Design*, vol. 103, no. 4, pp. 725–735, 1981.

[13] G. A. Korn, *Mathematical Handbook for Scientists and Engineers*, McGraw-Hill Book, Inc., 1968.

Dynamic Modeling of the Body Inversion for Automated Transfer of Live Birds

Kok-Meng Lee and Chris Shumway
The George W. Woodruff School of Mechanical Engineering
Georgia Institute of Technology
Atlanta, Georgia 30332-0405
Correspondence email: kokmeng.lee@me.gatech.edu

ABSTRACT:- Body rotation under free fall along a desired trajectory can be found in many applications such as sports, entertainment, and manufacturing. An appropriately designed body path could lower the forces at the joints during inversion and thus minimizing potential injury. This paper presents a method of developing dynamic models that characterize the interaction between the body of a live object undergoing inversion and the mechanical system driving the rotation. The method offers an effective means to analyze the sensitivity of the design and operational parameters on the body rotation. The models have been validated experimentally. The simulated and experimental results offer significant insights to the joint forces and a means to improve the body dynamics. While the results have immediate application in inverting live birds for poultry meat processing, we expect the model will provide a basis for analyzing body rotational dynamics in other applications such as gymnastics and roller coasters.

Index terms: body rotation, live-bird handling automation, inversion, prototyping, design simulation

1. INTRODUCTION

Body rotation can be found in many applications such as sports, inversion therapy, entertainment, and manufacturing. An appropriate designed body path could lower the forces at the joints during inversion and thus minimize potential injuries. A good understanding of the interaction between the body undergoing inversion and the mechanical system driving the rotation can offer significant insights to the influences of the design parameters, the gravity, the initial momentum, and the external forces on the body trajectory and the joint forces.

Body rotation has been studied in many different fields. In the pole-vault event in track and field [Hubbard, 1980; Griner 1984], the design of the highly elastic pole must take into account the body rotation of the vaulter who applies a compressive load and a bending moment to the upper end of the pole during the vault. In therapy, the inversion of the human body has been used to relieve back and neck pain by gently stretching the vertebrae using the person's own body weight. In manufacturing of meat products, objects must be inverted for subsequent processing. In robotics, the nature of a gymnastic maneuver was originally examined for the purpose of programming a mechanism to execute a forward flip [Hodgins and Raibert, 1990] or computer graphic animation [Raibert *et al.*, 1993]. Saito and Fukuda [1994] studied the motion of a long-arm ape for designing Brachiation robots, much like a gymnast on a high bar. These early researches were motivated by the design and control of mobile robots, the dynamic models were generally based on relatively simple mechanisms. More recently, dynamic models were developed to help coaches teach novice gymnasts the kip [Nakawaki *et al.*, 1998], develop more sophisticated skilful motion on a high bar [Michitsuji *et al.* 2001], or better understand the skill required to perform backward giant circling on the rings [Yamada *et al.* 2002]. These investigations formulated the gymnast as a three-link pendulum system, focusing on the body rotation about a specific, stationary point under the influences of gravity.

Unlike most of published works in robotics, where the emphasis has been placed on the control of a mechanism aimed at animating a live subject, this paper focuses on developing models and algorithms to characterize the body dynamics under the influences of the mechanical system that drives the body inversion. The equations of body rotational motion are derived using the Lagrangian method, which are subject to constraints imposed by the track on which the body is transported and the motion limit of the limbs. Although a number of methods have been proposed for analyzing complex mechanisms, for example [Chen, 1998; Caputo 2001], no experimental verification was attempted in these publications. This paper offers the following:

(1) The formulation of an inversion system, which takes into account the dynamics of both the body undergoing inversion and the mechanical system driving the rotation, is presented:

Using a set of well-defined system parameters and a set of redundant generalized coordinates, we offer an effective method to analyze the sensitivity of the system/body parameters on the dynamic loading of the driving mechanism as well as the resulting forces/torques at the joints of the rotating body.

(2) The model has been experimentally validated:

The dynamic models have been validated by comparing the simulations against results obtained experimentally. The error is estimated by calculating the residuals of the constraint equations, which should equal to zero if the constraints are satisfied. The maximum error was found within 0.3mm.

(3) A simulation algorithm for assessing the effects of the design changes has been demonstrated:

The validated model has an immediate application in automating live-bird inversion process for poultry meat process, which has been used to analyze the effects of inversion track design on the joint forces and improve the body dynamics during rotation.

It is expected that the models presented here will provide a basis for analyzing body rotational dynamics in other applications such as gymnastics and roller coasters.

2. SYSTEM OVERVIEW

Figure 1 shows the inverter which consists of the chain-conveyor and the trolley on which the pallet is mounted. The trolley is controlled by a motor-gear-chain mechanism via a link connecting the chain and the trolley by means of pin joints at P_g and A_1. The reference coordinate system XY is attached at the driving shaft as shown in Figure 1. The rollers, P_f and P_b, follow the pre-defined elliptical inversion path:

$$\frac{X^2}{e_X^2} + \frac{Y^2}{e_Y^2} = 1$$

where e_X and e_Y are the characteristic dimensions of the inversion path along the X and Y-axes respectively. The pallet is constrained to move perpendicularly to the trolley through a spring-damper system. The trolley can be characterized by the following parameter vector:

$$D = [e_X \quad e_Y \quad r_g \quad d_1 \quad d_2 \quad d_3 \quad h_o \quad \varepsilon]^T$$

where d_1 is the length of the connecting link; d_2 and d_3 are the distances of the front and back rollers from A_1 respectively; h_o is the nominal length of the spring-damper system; r_g is the gear radius; and ε is the distance between J_1 and A_3. For a given chain velocity v_g, the pallet has only one DOF (the displacement of A_3 with respect to A_2 from its nominal position.

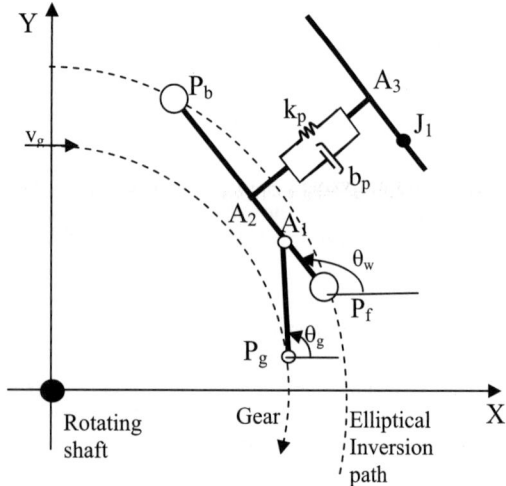

Figure 1 Schematics of the gear-chain-trolley mechanism

The object is modeled as a 2D, 4-limb serial mechanism as shown in Figure 2, where J_1, J_2, J_3, J_4 and J_c are the foot, ankle, knee, hip joints and the body center respectively. The variation of the object sizes can be characterized by the vector S:

$$S = [\ell_1 \quad \ell_2 \quad \ell_3 \quad \ell_4 \quad \varphi \quad \eta \quad \lambda \quad m]^T$$

where ℓ_1, ℓ_2, and ℓ_3 represents the lengths of the foot, shank, and thigh respectively; ℓ_4 denotes the distance between J_4 and J_c; φ is the angle between ℓ_4 and the x-axis as shown in Figure 2; η and λ represent the characteristic dimensions of the body (or link 4); and m is the mass of the body. The 2D mechanism has 3-DOF (X_c, Y_c, φ_c) but 4 degrees of mobility (φ_1, φ_2, φ_3, φ_4) and thus, it has one kinematical redundancy. However, since J_1 is fixed on the pallet surface and φ_2 and φ_3 rotate between 0° and 180°, the motions of J_2 and J_3 have finite ranges.

3. ANALYTICAL MODEL

The model of the body inversion is formulated using Lagrange dynamics. The generalized coordinates characterizing the motion are defined by the vector

$$q = [q_p \quad q_b]^T$$

where $q_p = [P_{gX} \quad P_{gY} \quad P_{fX} \quad P_{fY} \quad P_{bX} \quad P_{bY} \quad \theta_g \quad \theta_w \quad h]^T$

$$q_b = [J_{1X} \quad J_{1Y} \quad \varphi_1 \quad \varphi_2 \quad \varphi_3 \quad \varphi_4 \quad X_c \quad Y_c \quad \varphi_c]^T,$$

and X_c and Y_c are the position coordinates of J_c. Since the pallet has 1-DOF and the object has 3-DOF with one kinematical redundancy, the system is subject to 13 constraints given by Equations (1)-(10), which relate the 18 generalized coordinates.

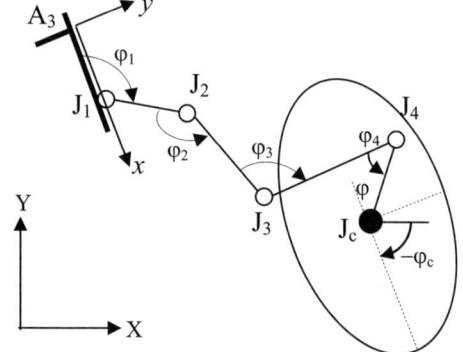

Figure 2: Schematic illustrating the object kinematics

3.1 Kinematical Constraints

For a specified angular rotation θ_r, the position and orientation of the object can be described by Equations (1) and (2) respectively:

$$\begin{bmatrix} X_c \\ Y_c \end{bmatrix} = \underline{J}_1 + \ell_1 \begin{bmatrix} \cos\phi_1 \\ \sin\phi_1 \end{bmatrix} - \ell_2 \begin{bmatrix} \cos\phi_2 \\ \sin\phi_2 \end{bmatrix} + \ell_3 \begin{bmatrix} \cos\phi_3 \\ \sin\phi_3 \end{bmatrix} - \ell_4 \begin{bmatrix} \cos(\varphi_5 - \varphi_c) \\ \sin(\varphi_5 - \varphi_c) \end{bmatrix} \quad (1)$$

$$\varphi_c = \phi_4 + \theta_w - \pi + \varphi \quad (2)$$

where

$$\begin{bmatrix} J_{1X} \\ J_{1Y} \end{bmatrix} = \underline{P}_g + d_1 \begin{bmatrix} \cos\theta_g \\ \sin\theta_g \end{bmatrix} + \left(\frac{d_3 - d_2}{2} - \varepsilon\right) \begin{bmatrix} \cos\theta_w \\ \sin\theta_w \end{bmatrix} + (h_0 + h) \begin{bmatrix} \sin\theta_w \\ -\cos\theta_w \end{bmatrix} \quad (3)$$

$$\begin{bmatrix} P_{gX} \\ P_{gY} \end{bmatrix} = \begin{bmatrix} r_g \cos\theta_r \\ r_g \sin\theta_r \end{bmatrix} \quad (4)$$

and

$$\phi_n = \theta_w + \sum_{m=1}^{n} (-1)^m \varphi_m$$

Since the trolley is constrained to follow the inversion path, the roller positions, (P_{fX}, P_{fY}) and (P_{bX}, P_{bY}), must be solved from the following set of non-linear constraint equations in order to determine the angular displacements θ_g and θ_w:

$$\frac{P_{fX}^2}{e_X^2} + \frac{P_{fY}^2}{e_Y^2} = 1 \quad (5)$$

$$\frac{P_{bX}^2}{e_X^2} + \frac{P_{bY}^2}{e_Y^2} = 1 \quad (6)$$

$$(P_{bX} - P_{fX})^2 + (P_{bY} - P_{fY})^2 = (d_2 + d_3)^2 \quad (7)$$

$$(P_{fX} + d_2 \cos\theta_w - P_{gX})^2 + (P_{gX} + d_2 \sin\theta_w - P_{gY})^2 = d_1^2 \quad (8)$$

Once the locations of the rollers are known with respect to the XY frame, the angular displacements, θ_g and θ_w, can be determined from Equations (9) and (10):

$$\theta_g = \tan^{-1}\left(\frac{P_{fY} + d_2 \sin\theta_w - P_{gY}}{P_{fX} + d_2 \cos\theta_w - P_{gX}}\right) \quad (9)$$

$$\theta_w = \tan^{-1}\left(\frac{P_{bY} - P_{fY}}{P_{bX} - P_{fX}}\right) \qquad (10)$$

3.2 Additional Constraints at Contact Area

Figure 3 illustrates the forces at the contact area between the object and the pallet, where f_n, and f_t are the normal and tangential components of the reaction acting on the object, and τ is the corresponding moment about the object center. Based on the conditions at the contact area, the inversion process can be divided into three regions.

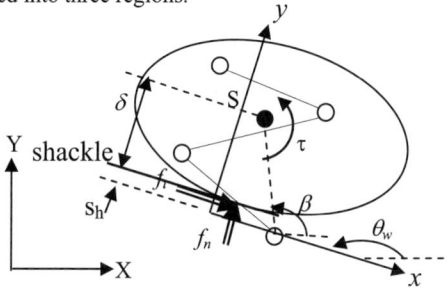

3(a) Forces and moment at the contact

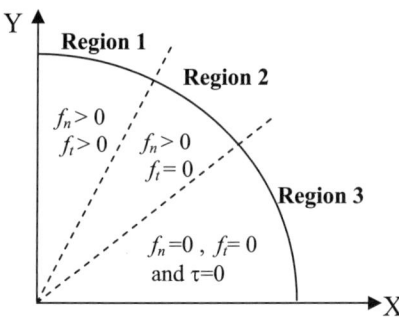

3(b) Conditions for transition

Figure 3 Constraints imposed at contact area

Region 1:
The body is in contact with the surface and is prevented to move backward (which could arise when the mechanical system starts from rest). Thus, $f_n > 0$ and $f_t > 0$, and the bird center is fixed with respect to the toe joint J_1.

$$\ell_T \sin(\theta_w - \beta) = s_h + \delta \qquad (11)$$

where
$$\ell_T = \sqrt{(J_{cX} - J_{1X})^2 + (J_{cY} - J_{1Y})^2}$$
$$\beta = \tan^{-1}\left(\frac{J_{cY} - J_{1Y}}{J_{cX} - J_{1X}}\right)$$

and
$$\Delta^2 = (J_{cX} - S_X)^2 + (J_{cY} - S_Y)^2 \qquad (12)$$

where
$$\begin{bmatrix} S_X \\ S_Y \end{bmatrix} = \underline{F}_w + d_2 \begin{bmatrix} \cos(\theta_w) \\ \sin(\theta_w) \end{bmatrix} + (h + s_h + \lambda) \begin{bmatrix} \sin(\theta_w) \\ -\cos(\theta_w) \end{bmatrix}$$

The underside of the object flattens which allows sliding but not rolling, causing the object orientation to remain parallel to the shackle surface.

$$\varphi_c = \theta_w - \pi \qquad (13)$$

Region 2:
When the body slides forward but it remains in contact with the surface, $f_t = 0$ and $f_n > 0$. Equation (12) no longer applies.

Region 3:
The object body is no longer in contact with the surface. Equations (11)-(13) are removed since $f_n = 0$ and $f_t = 0$.

3.3 Dynamic Model

The time derivates of Equations (1) through (13) can be written in the following form:

$$[a]\dot{\underline{q}} = \underline{0} \qquad (14)$$

where $[a]$ is a nxN Jacobian matrix for N generalized coordinates and n constraint equations. The Lagrange equations of motion can be written as

$$\frac{d}{dt}\left(\frac{\partial T(\dot{q},q)}{\partial \dot{q}_j}\right) - \frac{\partial T(\dot{q},q)}{\partial q_j} + \frac{\partial D(\dot{q})}{\partial \dot{q}_j} + \frac{\partial U(q)}{\partial q_j} = Q_j + \sum_{i=1}^{n}\lambda_i a_{ij}(q) \qquad (15)$$

where the kinetic energy

$$T(\dot{q},q) = \frac{1}{2}\sum_{i=1}^{7}(m_i V_i^2 + I_i \omega_i^2) ; \qquad (16)$$

the Rayleigh dissipation function

$$D(\dot{q}) = \frac{1}{2}b_P y^2 + \frac{1}{2}\sum_{i=1}^{4}b_i \dot{\varphi}_i^2 ; \qquad (17)$$

the potential energy

$$U(q) = \sum_{i=1}^{7} m_i g P_{iY} + \frac{1}{2}k_P y^2 + \frac{1}{2}\sum_{i=1}^{4} k_i \delta_i^2 ; \qquad (18)$$

Q_j is the external forces applied on the system; and the last term accounts for the constraints through the Lagrange multiplier λ_i. In Equations (16) and (18), m_i and I_i ($I = 1, \dots, 7$) are the masses and the moments of inertia of the connecting link, the trolley, the pallet, the four limbs of the bird respectively; P_{iY} and V_i are the Y-component position and the absolute velocity at the center of the ith mass; and ω_I is the component of the vector defined by

$$\underline{\omega} = \begin{bmatrix} \dot{\theta}_G & \dot{\theta}_w & \dot{\theta}_w & \dot{\phi}_1 & \dot{\phi}_2 & \dot{\phi}_3 & \dot{\phi}_4 \end{bmatrix}^T \qquad (19)$$

Due to the space limitation, the detailed derivation is omitted, which can be shown that Equations (15) has the form:

$$[M(q)]\ddot{q} + [C(\dot{q},q)]\dot{q} + D_V(\dot{q}) + G(q) = Q + [a(q)]^T \lambda \qquad (20)$$

where
$$[M(q)] = \left[\frac{\partial^2 T(\dot{q},q)}{\partial \dot{q}_1^2} \cdots \frac{\partial^2 T(\dot{q},q)}{\partial \dot{q}_j^2} \cdots \frac{\partial^2 T(\dot{q},q)}{\partial \dot{q}_n^2}\right]^T \qquad (21)$$

$$[C(\dot{q},q)]\dot{q} = \left[\cdots \left(\frac{\partial^2 T(\dot{q},q)}{\partial q_j \partial \dot{q}_j}\dot{q} - \frac{\partial T(\dot{q},q)}{\partial q_j}\right)\cdots\right]^T \qquad (22)$$

$$D_V(\dot{q}) = \frac{\partial D(\dot{q})}{\partial \dot{q}_j} \qquad (23)$$

$$G(q) = \frac{\partial U(q)}{\partial q_j} \qquad (24)$$

$$[a(q)]^T \lambda = \sum_{i=1}^{n} \lambda_i a(q)_{i,j} \qquad (25)$$

Equation (20) is in derivative form while its constraint equations are algebraic. In order to convert the coupled differential-algebraic equation into a form more appropriate for computation, the time derivative of the velocity constraint Equation (14) is used:

$$[a]\ddot{q} + [\dot{a}]\dot{q} = \underline{0} \qquad (26)$$

Equations (20) and (26) are augmented, which result in the following form:

$$\begin{bmatrix} [M(q)] & -[a]^T \\ -[a] & [0] \end{bmatrix} \begin{Bmatrix} \ddot{q} \\ \lambda \end{Bmatrix} = \begin{Bmatrix} Q - [F(q,\dot{q})]\dot{q} \\ [\dot{a}]\dot{q} \end{Bmatrix} \qquad (27)$$

where $F(\dot{q},q) = [C(\dot{q},q)]\dot{q} + D_V(\dot{q}) + G(q)$.

At any time instant, the numerical values for the generalized position vector $\underline{q}$ and velocity vector, $\underline{\dot{q}}$, are known; thus, the numerical values of $[M(\underline{q})]$, $\underline{F}(\dot{q},q)$, $[a(\underline{q})]$, and $[\dot{a}(\underline{q})]$ can be computed. To explicitly solve $\ddot{q}$ and $\underline{\lambda}$, Equation (27) is rewritten as

$$\begin{Bmatrix} \ddot{q} \\ \underline{\lambda} \end{Bmatrix} = \begin{bmatrix} [M(q)] & -[a(q)]^T \\ -[a(q)] & [0] \end{bmatrix}^{-1} \begin{Bmatrix} Q - F(\dot{q},q) \\ [\dot{a}(q)]\dot{q} \end{Bmatrix} \qquad (28)$$

With appropriate initial conditions, the above equation can be solved numerically. The forces/torques acting at the toe, ankle, knee and hip joints corresponding to $\varphi_1, \varphi_2, \varphi_3,$ and φ_4 respectively can then be calculated from $[a(q)]^T \lambda$:

$$\tau_1 = \lambda_1 \sum_{i=1}^{3} \ell_I s_i (-1)^{i-1} - \lambda_2 \sum_{i=1}^{3} \ell_I c_i (-1)^{i-1} - \lambda_3 \qquad (30a)$$

$$\tau_2 = \lambda_1 \sum_{i=2}^{3} \ell_I s_i (-1)^{i} - \lambda_2 \sum_{i=2}^{3} \ell_I c_i (-1)^{i} + \lambda_3 \qquad (30b)$$

$$\tau_3 = \ell_3 s_3 \lambda_1 - \ell_3 c_3 \lambda_2 - \lambda_3 \qquad (30c)$$

$$\tau_4 = \lambda_3 \qquad (30d)$$

where λ_1, λ_2 and λ_3 are the Lagrange multipliers associated with the velocity constraints derived from the X and Y components of Equation (1) and (2) respectively.

4. SIMULATION RESULTS

Since the model has an immediate application in live bird handling [Lee, 2000], the object chosen is a model bird mechanism designed with dimensions from a typical broiler so that the dynamic models presented above can be experimentally verified before testing with a live bird. Simulations were performed based on the design of a live-bird inverting system developed at Georgia Tech as shown in Figure 4. The values of the design and object parameters used in the simulation are given as follows:

$$D = [111 \ \ 148 \ \ 147 \ \ 100 \ \ 65 \ \ 65 \ \ 140 \ \ 30]^T$$
$$S = [70 \ \ 90 \ \ 80 \ \ 60 \ \ 37 \ \ 57 \ \ 97 \ \ 1.6]^T$$

where lengths are in mm; angles in degrees; and mass in kilograms (kg). The masses for the connecting rod, the trolley and the pallet are 0.1, 1.1, and 3.4 kg respectively. The masses for the limbs ℓ_1, ℓ_2 and ℓ_3 and ℓ_4 are 0.03, 0.08, and 0.1 kg. The masses are assumed at the mid points of the respective components (or the limbs). The mass of the body is 1.6 kg.

In the simulation, the non-conservative external force vector Q is assumed to be empty and s_h is 2.5mm. The coefficients for the spring and damper in the pallet suspension are 4.68kN/m and 11.1 Ns/m respectively. Before the inversion, the bird moves with the pallet at a chain speed of 0.457m/s, and the velocities in the Y direction are assumed to be zero. Other initial conditions are given as follows:

$$q_p(0) = [-20 \ \ -147 \ \ -57 \ \ 148 \ \ -184 \ \ 148 \ \ 180 \ \ 180 \ \ 126]^T$$
$$q_b(0) = [-90 \ \ 273 \ \ 5 \ \ 40 \ \ 58 \ \ -14 \ \ -108 \ \ 333 \ \ 0]^T$$

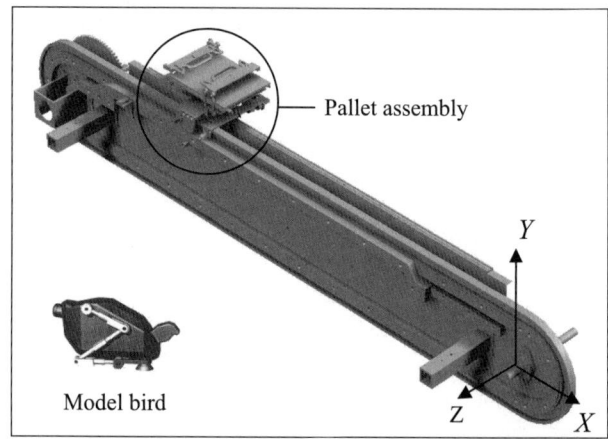

Figure 4 CAD model o the bird and the inverter

Figure 5 shows the trajectory of the body as it inverts, which the dashed lines divide the three regions as described in Figure 3(b). Still shot 1 is in Region #1 where $f_n > 0$ and $f_t > 0$. Still shot 2 is in Region #2 where the ellipse is sliding down the shackle but remains in contact with the shackle, where $f_n > 0$ and $f_t = 0$. Still shots 3 through 7 are in Region #3 where $f_n = f_t = 0$, where still shot 3 shows the bird body just leaving contact with the shackle surface, while still shot 7 shows the end of the simulation.

In the numerical computation, the transition from one region to the next is determined by the reaction forces acting on the bird by the mechanical system, f_n and f_t. Since only the algebraic sign of these constraint forces are needed the Lagrange multipliers associated with the velocity constraints derived from Equations (10) and (11) can be used to avoid the complexity of computing. Figure 6 is a plot showing the Lagrange multipliers that represent the constraint forces, f_n and f_t, and thus the progression from region to region.

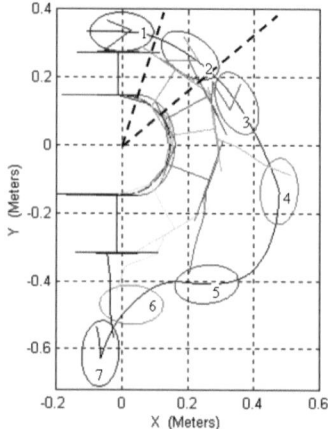

Figure 5 Simulated body trajectory

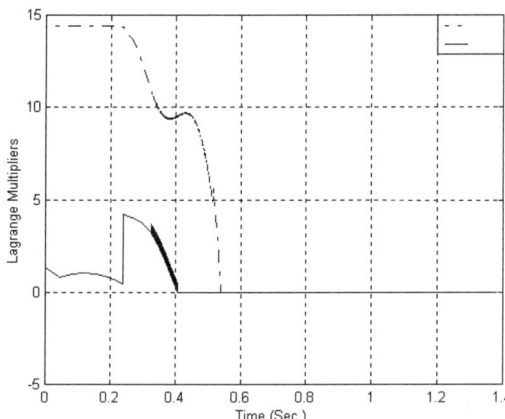

Figure 6 Lagrange multipliers representing f_n and f_t

Figure 7 shows the angular velocity of the orientation experiencing a sudden increase at $t = 0.72$ second, which corresponds to the still shot 4 in Figure 5. The body approaches $-90°$ and then rotates back to above horizontal (or $0°$). This is undesired since once the body is inverted it should kept at the orientation $\varphi_c = -90°$. The velocity change, as seen in still shot 4, is contributed to the bird limb angles reaching their respective maximums. In other words, the limbs have been stretched out to a singularity configuration, and the bird body rotates about point J_4 as illustrated in Figure 8. To accomplish J_4 rotation, an internal force from the stretched limbs is applied onto the bird body, which also causes the angular velocity and thus the bird angular momentum to rapidly change.

The simulation error can be estimated by calculating the residuals of the constraint equations, which should equal to zero if the constraints are satisfied. Figure 9 graphs the constraints with the greatest errors, which is the constraint of the front wheel. The maximum value of the error is less than 0.3mm.

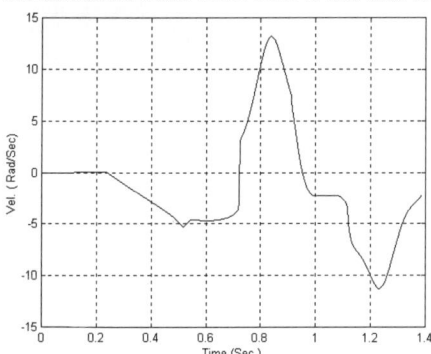

Figure 7 Angular velocity of the elliptical body

5. EXPERIMENTAL VERIFICATION

In order to validate the analytical model, an experiment was performed using the test bed developed at Georgia Tech. The two toe-joints J_1 of a model bird mechanism were fixed on the pallet by means of suction cups. The motion trajectory was recorded using a video camera and digitized for off-line analysis. Figure 10 shows the experimentally recorded trajectory of the model bird, which agrees well with the simulated centers. Similar comparison of the object orientation is graphed in Figure 11. These comparisons show the experimental results agree well with the trends predicted by simulation, where some discrepancies are expected since friction occurred in the physical joints were neglected in the model. At the instant when the bird body is at its singularity configuration which results in a drastic change in body orientation from $-90°$ and then rotates back to approximately $0°$ as shown in Figures 12 (a) and (b) respectively.

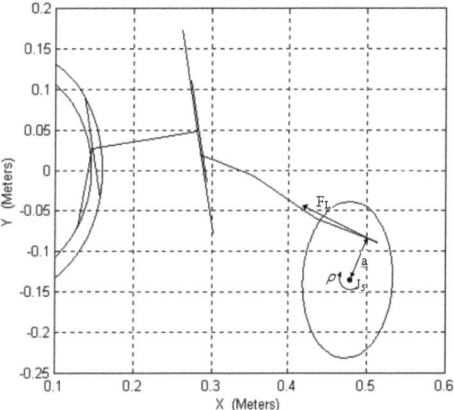

Figure 8 Configuration at the signularity

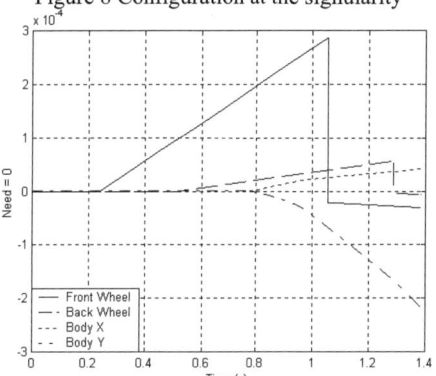

Figure 9 residual of the kinematical constraints

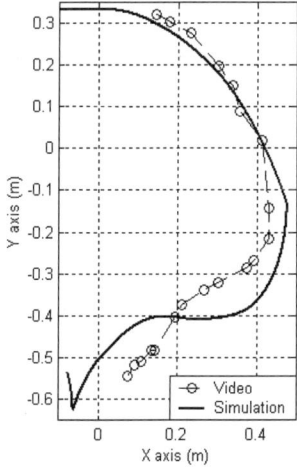

Figure 10 Comparison between experimental results and simulations of bird center

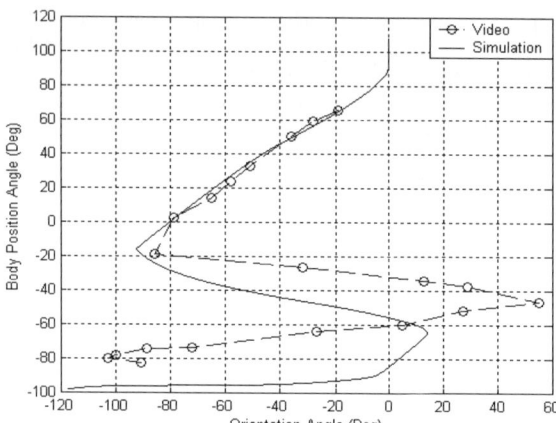

Figure 11 Comparison between experimental results and simulations of bird orientation

6. CONCLUSIONS

A dynamic model that characterizes the dynamics of the body undergoing inversion along a predetermined track has been developed. The model derived using constrained Lagrange dynamics offers an effective means to analyze the sensitivity of the design and operational parameters on the dynamics of the body and the forces and moments at the joints.

The model has been validated by comparing the simulated trajectory of the body against the results obtained experimentally. Comparisons show the experimental results agree well with the trends predicted by simulation.

Although the results presented here have an immediate application in meat processing industry, it is expected that the model will also provide a basis for analyzing body rotational dynamics in other applications such as sports, inversion therapy, entertainment, and manufacturing.

REFERENCE

Caputo, M.R., 2001, "Further Results on Lagrange Multipliers with Several Binding Constraints," *Economic Letters* 70, 335-340.

Chen, Y.H. 1998, Second-order Constraints for Equations of Motion of Constrained Systems, *IEEE/ASME Trans. on Mechatronics*, Vol 3. No. 3, September, 240-248.

Griner, G. M. 1984, "A Parametric Solution to the Elastic Pole Vaulting Pole Problem," *ASME Journal of Applied Mechanics*, Vol. 106.

Hodgins, J. K. and M. H. Raibert, 1990, "Biped Gymnastics," *International Journal of Robotics Research*, Vol. 9, No. 2 April, 115-132.

Habbard, M., "Dynamics of the Pole Vault," *Journal of Biomechanics*, Vol. 13, No. 11, 1980, 965-976.

Lee, K-M., 2000, "Design Criteria for Developing an Automated Live Bird Transfer System," *Proc. of the 2000 IEEE ICRA*, April 24-28, San Francisco, CA, Vol. 2, pp. 1138-1143; also in *IEEE Trans. on Robotics and Automation*, Vol. 17, Issue 4, Aug. 2001 pp. 483-490.

Micitsuji, Y., H. Sato, and M. Yamakita, 2001, "Giant Swing via Forward Upward Circling of the Acrobat-Robot," *Proc. of ACC*, Arlington, VA, 3262-3267.

Nakawaki, D., S. Joo, and F. Miyazaki, 1998, "Dynamic Modeling Approach to Gymnastic Coaching," *Proc. of IEEE ICRA*, Leuven, Belgium 1069-1076.

Raibert, H. R., J. K. Hodgins, R.P. Playter, and R. P. Ringrose, 1993, "Animation of Legged Maneuvers: Jumps, Somersaults, and Gait Transitions," *JRSJ* Vol. 11, No.3, 333-342.

Saito, F., T. Fukuda, and F. Arai, 1994, "Swing and Locomotion Control for a Two-Link Brachiating Robot," *IEEE Control Systems*, February, 5-11.

Yamada, T., K. Watanabe, K. Kiguchi, and K. Izumi, 2002, "Acquiring Performance Skill of Backward Giant Circle by Rings Gymnastic Robot," *Proc. of IEEE ICRA*, Washington DC, 1565-1570.

ACKNOWLEDGEMENT

The Agriculture Technology Research Program (ATRP) and the U.S. Poultry and Eggs Association have supported this project. The test bed was developed with helps from the ATRP's staff.

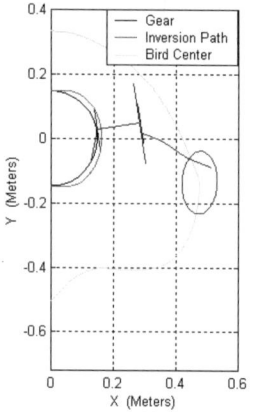

(a) singularity occurring at about 80°

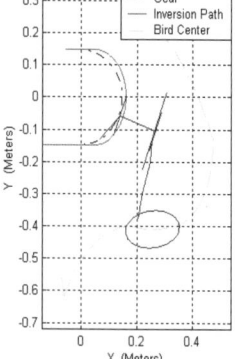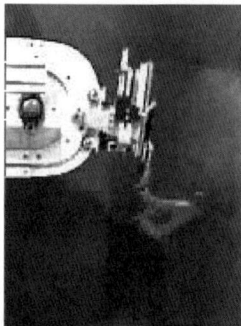

(b) 110°

Figure 12 Comparison between experimental results and simulations of bird configuration at singularity

Sufficient Conditions for Admittance to Ensure Planar Force-Assembly in Multi-Point Frictionless Contact

Shuguang Huang and Joseph M. Schimmels

Department of Mechanical and Industrial Engineering
Marquette University
Milwaukee, Wisconsin 53201-1881, USA
E-mail: {huangs, j.schimmels}@marquette.edu

Abstract— An important issue in the development of force guidance assembly strategies is the specification of an appropriate admittance control law. This paper identifies procedures for selecting the appropriate admittance to achieve reliable planar force-guided assembly for multi-point contact cases. Conditions that restrict the admittance behavior for each of the various types of two-point contact are presented. These conditions ensure that the motion that results from contact reduces part misalignment for each case. We show that, for bounded misalignments, if the conditions are satisfied for a finite number of contact configurations, the conditions ensure that force guidance is achieved for all configurations within the specified bounds.

I. INTRODUCTION

Admittance control has been used in assembly tasks to provide force regulation and force guidance. In these tasks, the admittance maps contact forces into changes in the velocity of the held body. To achieve reliable assembly through force-guidance, an appropriate admittance must be selected. For linear admittance behavior, the planar control law has the form:

$$\mathbf{v} = \mathbf{v}_0 + \mathbf{A}\mathbf{w} \qquad (1)$$

where $\mathbf{v}_0$ is the nominal velocity (a 3-vector), $\mathbf{w}$ is the contact wrench measured in the body frame (a 3-vector), and $\mathbf{A}$ is the admittance matrix (a 3×3 matrix).

Many researchers have addressed the use of an admittance for force-guidance. Whitney [1], [2] proposed that the compliance of a manipulator be structured so that contact forces lead to decreasing errors. Peshkin [3] addressed the synthesis of an accommodation (inverse damping) matrix by specifying the desired force/motion relation at a sampled set of positional errors for a planar assembly task. An unconstrained optimization was then used to obtain an accommodation matrix. Asada [4] used a similar optimization procedure for the design of an accommodation neural network rather than an accommodation matrix. Others [5], [6] provided synthesis procedures based on spatial intuitive reasoning. None of these approaches, however, ensures that the admittance selected will, in fact, be reliable.

A reliable admittance selection approach is to design the control law so that, at each possible part misalignment, the contact force always leads to a motion that reduces the existing misalignment. The approach is referred to as *force-assembly* [7], [8], [9]. A condition for force-assembly is that the admittance matrix $\mathbf{A}$ is positive semidefinite [7].

For force-assembly, the motion resulting from contact must instantaneously reduce misalignment. Since the configuration space of a rigid body is non-Euclidian, there is no natural metric for finite error. In [10], several body-specific rigid body metrics were identified. These metrics are based on the Euclidean distance between one (or more) point on the body and its corresponding location when properly positioned.

Previously, we have considered sufficient conditions on the admittance to ensure planar force-assembly in frictionless *single-point* contact [11], [12]. In the study, we considered a *measure* of error based on the Euclidean distance between a single (fixed) point on the held body and its location when properly positioned. The error reduction conditions developed require that, at each possible misalignment, the contact force yields a motion that reduces this distance. A set of sufficient conditions was identified that ensures planar force-assembly in single-point contact.

This paper considers planar rigid body assembly in *multi-point* frictionless contact. The mathematical description of error-reducing motion for two-point contact is derived in Section II. The solution strategy is presented in Section III. The strategy ensures that conditions imposed at a finite number of contact configurations guarantee that the conditions are also satisfied at the infinite number of contact configurations within a specified range. Sections IV addresses the sufficient conditions for each of the various types of multi-point contact. A brief summary is presented in Sections V.

II. ERROR-REDUCING MOTION

In this section, the error-reducing motion of a constrained rigid body in two-point contact is analyzed. First, the equation describing the constrained motion of a rigid body is derived. Then, the error-reduction function for general two-point contact is obtained.

A. Constrained Rigid Body Motion

Consider planar motion of a rigid body in 2-point contact with another part. Let $\mathbf{n}_i$ be the surface normal at contact point i and let $\phi_i \geq 0$ be the magnitude of the normal force. Then, for two-point contact, the total contact wrench is

$$\mathbf{w} = \mathbf{w}_1 \phi_1 + \mathbf{w}_2 \phi_2 \qquad (2)$$

where

$$\mathbf{w}_i = \begin{bmatrix} \mathbf{n}_i \\ (\mathbf{r}_i \times \mathbf{n}_i) \cdot \mathbf{k} \end{bmatrix}$$

and where $\mathbf{r}_i$ is the position vector from the origin of the coordinate frame to the contact point and $\mathbf{k}$ is the unit vector orthogonal to the plane.

Alternately, if we denote

$$\mathbf{W} = [\mathbf{w}_1, \mathbf{w}_2] \in \mathbb{R}^{3 \times 2}, \quad \boldsymbol{\phi} = [\phi_1, \phi_2]^T \in \mathbb{R}^2,$$

then the total contact wrench is: $\mathbf{w} = \mathbf{W}\boldsymbol{\phi}$.

By the control law (1), the motion of the rigid body is given by:

$$\mathbf{v} = \mathbf{v}_0 + \mathbf{A}\mathbf{W}\boldsymbol{\phi}. \qquad (3)$$

To maintain contact [13], the reciprocal condition requires:

$$\mathbf{w}_i^T \mathbf{v} = 0 \implies \mathbf{W}^T \mathbf{v} = \mathbf{0}.$$

Substituting this into (3) and solving for $\boldsymbol{\phi}$, we have:

$$\boldsymbol{\phi} = -[\mathbf{W}^T \mathbf{A} \mathbf{W}]^{-1} \mathbf{W}^T \mathbf{v}_0. \qquad (4)$$

Thus the equation for constrained motion is obtained:

$$\mathbf{v} = \mathbf{v}_0 - \mathbf{A}\mathbf{W}[\mathbf{W}^T \mathbf{A} \mathbf{W}]^{-1} \mathbf{W}^T \mathbf{v}_0. \qquad (5)$$

B. Error-Reduction Function

For planar motion of a rigid body with two point contact, if the contact is maintained, the body has only one degree of freedom (DOF). The instantaneous motion of the body is a rotation about the body's instantaneous center.

If the instantaneous center is at infinity, the motion of the body is a pure translation. This is the simplest case: the error-reduction at any given configuration within a contact state ensures the error-reduction for all configurations within that contact state.

If the instantaneous center is not at infinity (generic case), it is uniquely determined by the geometry of the contact for each configuration. This paper addresses the contact states of this case.

As stated previously, assembly error-reduction requires that, at any instant, the motion of the body must be toward its properly mated position. Consider

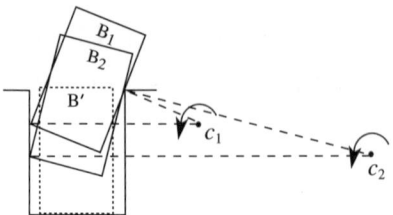

Fig. 1. Error-Reducing Motion: the angular motion of the rigid body must be along a specific direction for all configurations of a given two-point contact state.

the two-point contact state shown in Fig. 1. For error-reduction, the direction of rotation of the body about the instantaneous center c_i must cause the body to move toward the properly mated position B'. Since error-reduction must hold for any configuration, the angular motion of the body must be along a specific direction for all configurations within the same contact state. Thus, the error-reducing motion for two-point contact is solely indicated by the angular velocity of the constrained body.

Now consider the angular motion in (5). Let $\mathbf{e}_3 = [0, 0, 1]^T$ and $\mathbf{v}_0 = [v_{x0}, v_{y0}, \omega_0]^T$, then, the orientational component in (5) is:

$$\omega = \mathbf{e}_3^T \mathbf{v} = \omega_0 - \mathbf{e}_3^T \mathbf{A}\mathbf{W}[\mathbf{W}^T \mathbf{A}\mathbf{W}]^{-1} \mathbf{W}^T \mathbf{v}_0. \qquad (6)$$

Let $[\mathbf{W}^T \mathbf{A}\mathbf{W}]^*$ be the adjugate of $[\mathbf{W}^T \mathbf{A}\mathbf{W}]$. Then (6) can be written as:

$$\omega = \frac{\det(\mathbf{W}^T \mathbf{A}\mathbf{W})\omega_0 - \mathbf{e}_3^T \mathbf{A}\mathbf{W}[\mathbf{W}^T \mathbf{A}\mathbf{W}]^* \mathbf{W}^T \mathbf{v}_0}{\det(\mathbf{W}^T \mathbf{A}\mathbf{W})}. \qquad (7)$$

Since $\mathbf{A}$ is positive definite, $\det(\mathbf{W}^T \mathbf{A}\mathbf{W}) > 0$ and we only need to consider the following function:

$$F_{er} = \det(\mathbf{W}^T \mathbf{A}\mathbf{W})\omega_0 - \mathbf{e}_3^T \mathbf{A}\mathbf{W}[\mathbf{W}^T \mathbf{A}\mathbf{W}]^* \mathbf{W}^T \mathbf{v}_0$$

which can be expressed as:

$$\begin{aligned}
F_{er} = &\, [(\mathbf{w}_1^T \mathbf{A}\mathbf{w}_1)(\mathbf{w}_2^T \mathbf{A}\mathbf{w}_2) - (\mathbf{w}_1^T \mathbf{A}\mathbf{w}_2)^2]\omega_0 \\
&- (\mathbf{w}_1^T \mathbf{v}_0)(\mathbf{a}_3^T \mathbf{w}_1)(\mathbf{w}_2^T \mathbf{A}\mathbf{w}_2) \\
&+ (\mathbf{w}_1^T \mathbf{v}_0)(\mathbf{a}_3^T \mathbf{w}_2)(\mathbf{w}_1^T \mathbf{A}\mathbf{w}_2) \\
&+ (\mathbf{w}_2^T \mathbf{v}_0)(\mathbf{a}_3^T \mathbf{w}_1)(\mathbf{w}_1^T \mathbf{A}\mathbf{w}_2) \\
&- (\mathbf{w}_2^T \mathbf{v}_0)(\mathbf{a}_3^T \mathbf{w}_2)(\mathbf{w}_1^T \mathbf{A}\mathbf{w}_1)
\end{aligned} \qquad (8)$$

where $\mathbf{a}_3$ is the 3rd column of the admittance matrix $\mathbf{A}$.

Since the function F_{er} in (8) indicates the sign of the orientational motion for the body, it is used as the error-reduction function for the two-point contact case. If, error-reducing motion is achieved at one configuration and the angular velocity of the body (indicated by F_{er}) does not change sign within the range of configurations, then error-reduction is ensured for all configurations within the contact state.

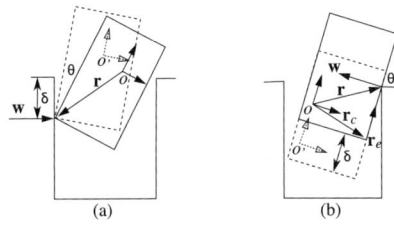

Fig. 2. Single-point Contact State. (a) $\{v\text{-}e\}$: vertex-edge contact state. (b) $\{e\text{-}v\}$: edge-vertex contact state.

III. SOLUTION STRATEGY

In general, the error-reduction function F_{er} in (8) depends on the geometries of the parts in contact. In this section, a solution strategy to obtain sufficient conditions for error-reduction is presented. With this strategy, a set of sufficient conditions can be obtained for bounded configurations without having to explicitly describe the variation in part configuration within a given contact state.

A. Contact States

Polygonal planar bodies in single-point contact have two basic types of contact. One is referred to as "vertex-edge" contact ($\{v\text{-}e\}$); the other is referred to as "edge-vertex" contact ($\{e\text{-}v\}$). In "vertex-edge" contact, one vertex of the held body is in contact with one edge of its mating part (Fig. 2a). In "edge-vertex" contact, one edge of the held body is in contact with one vertex of the mating fixtured part (Fig. 2b).

The basic types of two-point contact are the various combinations of two single-point contacts. There are three types of two-point contact: 1) one $\{e\text{-}v\}$ and one $\{v\text{-}e\}$ contact ($\{e\text{-}v,v\text{-}e\}$), 2) two $\{e\text{-}v\}$ contact ($\{e\text{-}v,e\text{-}v\}$), and 3) two $\{v\text{-}e\}$ contact ($\{v\text{-}e,v\text{-}e\}$).

B. Contact Wrenches

In the following, the contact wrenches for the two basic types of single-point contact are presented. We show that, although the body configuration is determined by two variables (δ, θ), the unit contact wrench for each type of contact only depends on one of them. As a consequence, two-point contact is readily expressed in terms of two variables.

B.1 Vertex-Edge Contact

Consider the case for which one vertex of the body is in contact with one edge of its mating part ($\{v\text{-}e\}$ contact state). As shown in Fig. 2a, the direction of the contact wrench $\mathbf{w}$ is constant in the global coordinate frame and the relative position of the contact is constant in the body frame. Suppose that the relative body orientation is given by angle θ, then the direction of the contact force also changes in θ in the body frame. Thus, the unit contact wrench in the body frame can be expressed as:

$$\mathbf{w} = \begin{bmatrix} \mathbf{Rn} \\ (\mathbf{r} \times \mathbf{Rn}) \cdot \mathbf{k} \end{bmatrix} \quad (9)$$

where $\mathbf{n}$ is the surface normal for a chosen configuration and $\mathbf{R}$ is the rotation matrix associated with θ having the form:

$$\mathbf{R} = \begin{bmatrix} \cos\theta & -\sin\theta \\ \sin\theta & \cos\theta \end{bmatrix}. \quad (10)$$

Therefore, in the body frame, $\mathbf{w}$ is a single-variable function in θ for a specified contact state.

B.2 Edge-Vertex Contact

Consider the case for which one edge of the body is in contact with one vertex of its mating part ($\{e\text{-}v\}$ contact state). As shown in Fig. 2b, the direction of the contact wrench is constant in the body frame but the location of the contact varies, thus the unit contact wrench can be expressed as:

$$\mathbf{w} = \begin{bmatrix} \mathbf{n} \\ [(\mathbf{r}_c + \mathbf{r}_e \delta) \times \mathbf{n}] \cdot \mathbf{k} \end{bmatrix} \quad (11)$$

where $\mathbf{r}_c$ identifies a location on the edge and $\mathbf{r}_e$ is the unit vector along the edge (both are constant in the body frame). Thus, $\mathbf{w}$ is a single-variable function in δ for a given contact state.

B.3 Multi-Point Contact

Since two-point contact is a combination of the two single-point contact cases, the contact wrench for two-point contact is a combination of the two corresponding single-point contact wrenches.

The error-reduction function for two-point contact [calculated using (8)] involves the two contact wrenches $\mathbf{w}_1$ and $\mathbf{w}_2$. Since each unit wrench in [(9) or (11)] is a function of δ or θ, in general, the error-reduction function can always be expressed as a function of two variables, i.e.,

$$F_{er} = F_{er}(\xi, \eta) \quad (12)$$

where ξ and η are θ or δ depending on the contact state. For example, for $\{e\text{-}v,e\text{-}v\}$ contact, both ξ and η involve the two displacement variables along the corresponding edges, δ_1 and δ_2.

C. Mathematical Requirement

If the parts remain in contact, the planar motion of the rigid body has only 1 DOF. Therefore, the two parameters ξ and η in (12) must be related by the geometry of the parts. For example, for a given geometry, ξ can be expressed as a function of η (or vice

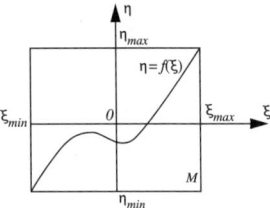

Fig. 3. Two-point Contact State: in the bounded area ξ and η are treated as two independent variables regardless of their relation $\eta = f(\xi)$.

versa). The function, however, could be complicated (highly nonlinear) for a specific geometry.

Note that δ represents a relative position along an edge of the held body and θ represents a relative orientation between the two parts. The ranges for δ and θ can be determined from bounds on relative misalignment or by bounds determined by the contact state. Thus the range of the two parameters ξ and η are readily determined.

For a given geometry and contact state, ξ and η are related by a function $\eta = f(\xi)$. The error-reduction condition requires that for all configurations on the curve $\eta = f(\xi)$, the orientational error of the body is monotonically reduced by contact. This means that the error-reduction function F_{er} has the appropriate sign along the curve $\eta = f(\xi)$.

Since the function is geometry specific and difficult to determine, we consider a set of more conservative conditions based on the range of the two variables.

Suppose that the range of ξ and η are $[\xi_{min}, \xi_{max}]$ and $[\eta_{min}, \eta_{max}]$, respectively. Consider the rectangular area M bounded by $[\xi_{min}, \xi_{max}]$ and $[\eta_{min}, \eta_{max}]$ as shown in Fig. 3. If in the bounded area the error-reduction condition is satisfied, then for any configuration considered, the error-reduction condition must be satisfied. Mathematically, this condition can be imposed on the function F_{er} as:

1. For one point $(\xi_0, \eta_0) \in M$, the error-reduction condition is satisfied, i.e.,

$$F_{er} = F_{er}(\xi_0, \eta_0) < 0. \quad (13)$$

2. For all points in M, F_{er} does not change sign.

As such, sufficient conditions for error-reduction motion are established. This approach enables us to treat the parameters (ξ, η) as two independent variables regardless of the geometrical relationship between them.

For the variables (ξ, η), F_{er} can be written as a polynomial in ξ with the coefficients being functions of η. Since both variables ξ and η are bounded, the extremals of the coefficients can be determined. Based on these extremals, a single-variable polynomial is constructed. Using an approach similar to that used for the single-point contact case [12], sufficient conditions for (13) are obtained. Since the conditions impose constrains on the admittance matrix, a set of sufficient conditions for an admittance to ensure force-assembly is identified.

IV. CONDITIONS FOR TWO-POINT CONTACT

In this section, the sufficient conditions are obtained for each type of contact state. Below, for each type of contact, 1) the error-reduction function F_{er} is specified, 2) bounds of the coefficients in F_{er} are identified, and 3) specific conditions for satisfying error-reduction are presented.

In each case, the range for each of the variables can be transformed to be centered about zero, e.g., $[\delta_{min}, \delta_{max}] \Rightarrow [-\delta_M, \delta_M]$ to facilitate subsequent analysis.

A. Conditions for $\{e\text{-}v, v\text{-}e\}$ Contact

In this case, the two-point contact wrench is a combination of the two corresponding single-point contact wrenches.

Using the notation developed in Section III-B, the contact wrenches for the $\{v\text{-}e\}$ and $\{e\text{-}v\}$ contact are obtained by (9) and (11) respectively,

$$\mathbf{w}_1 = \mathbf{w}_1(\theta), \qquad \mathbf{w}_2 = \mathbf{w}_2(\delta). \quad (14)$$

Since $\mathbf{w}_1$ in (9) contains only first order terms in $\cos\theta$ and $\sin\theta$, and $\mathbf{w}_2$ in (11) contains only a linear term in δ, (8) can be expressed as a function of (θ, δ) in the form:

$$F_{er}(\theta, \delta) = q_2 \delta^2 + q_1 \delta + q_0 \quad (15)$$

where q_i's are functions of θ having the forms:

$$q_i = a_i \cos^2\theta + b_i \cos\theta \sin\theta + c_i \cos\theta + d_i \sin\theta + e_i. \quad (16)$$

If at a given configuration error-reduction is satisfied: $F_{er} < 0$, then in order for all θ and δ to satisfy the condition, we need to obtain conditions such that F_{er} has no root in the range of consideration.

A.1 Bounds on the Coefficients

In order to analyze the root of the function F_{er}, we evaluate the bounds on the coefficients q_i's in (16).

First, consider the two terms involving $\cos\theta$ alone $(a_i \cos^2\theta + c_i \cos\theta)$ in (16). If we denote:

$$p_{i1}^+ = \max\{(a_i \cos^2\theta_M + c_i), (a_i \cos^2\theta_M + c_i \cos\theta_M),$$
$$(a_i + c_i \cos\theta_M), (a_i + c_i)\}, \quad (17)$$

$$p_{i1}^- = \min\{(a_i \cos^2\theta_M + c_i), (a_i \cos^2\theta_M + c_i \cos\theta_M),$$
$$(a_i + c_i \cos\theta_M), (a_i + c_i)\}, \quad (18)$$

then for $\forall \theta \in [-\theta_M, \theta_M]$ and $i = 0, 1, 2$,

$$p_{i1}^- \leq (a_i \cos^2\theta + c_i \cos\theta) \leq p_{i1}^+. \quad (19)$$

Consider the two terms involving $\sin\theta$ (i.e., $b_i \cos\theta \sin\theta + d_i \sin\theta$) in (16). If we denote

$$p_{i2}^+ = \max\{0, (b_i + d_i)\sin\theta_M, (b_i \cos\theta_M + d_i)\sin\theta_M\},$$
$$p_{i2}^- = \min\{0, (b_i + d_i)\sin\theta_M, (b_i \cos\theta_M + d_i)\sin\theta_M\}.$$

Then, for $\forall \theta \in [-\theta_M, \theta_M]$ and $i = 0, 1, 2$,

$$p_{i2}^- \leq (b_i \cos\theta \sin\theta + d_i \sin\theta) \leq p_{i2}^+. \quad (20)$$

Thus,

$$(p_{i1}^- + p_{i2}^- + e_i) \leq q_i \leq (p_{i1}^+ + p_{i2}^+ + e_i). \quad (21)$$

If we denote:

$$q_{Mi} = p_{i1}^+ + p_{i2}^+ + e_i, \quad (22)$$
$$q_{mi} = p_{i1}^- + p_{i2}^- + e_i, \quad (23)$$

then, the bounds for q_i's are determined:

$$q_{mi} \leq q_i \leq q_{Mi}, \quad i = 0, 1, 2, \quad (24)$$

where all q_{mi}'s and q_{Mi}'s are functions of the admittance matrix (independent of the configuration).

A.2 Sufficient Conditions for Error-Reduction

Since the bounds of q_i's are determined, a single-variable polynomial is constructed for which the method used for single-point contact case [12] is applied.

First, the error-reduction condition must be satisfied at one configuration in the range considered (say, at $[\theta, \delta] = [0, 0]$).

To consider all configurations, we construct a polynomial by:

$$F_p(\delta) = q_{M2}\delta^2 + q_{M1}\delta + q_{m0} \quad (25)$$

where the coefficients are constants defined in (22) and (23). Denote

$$q_M = \max\{|q_{M2}|, |q_{M1}|\}. \quad (26)$$

It is proved [12] that, if

$$\frac{|q_{m0}|}{q_M + |q_{m0}|} > \delta_M, \quad (27)$$

then, $F_p(\delta)$ has no root in $[-\delta_M, \delta_M]$. Since the coefficients of $F_p(\delta)$ are extremal values of q_i's in the range considered, the condition (27) ensures that the function $F_{er}(\theta, \delta)$ in (15) has no root in $[-\delta_M, \delta_M]$ for any given $\theta \in [-\theta_M, \theta_M]$. In fact, for a given $\theta \in [-\theta_M, \theta_M]$, $F_{er}(\theta, \delta)$ is a polynomial in δ. As shown in [12], a root of $F_{er}(\theta, \delta)$, δ_θ, must satisfy

$$\delta_\theta \geq \max\{\frac{|q_0|}{|q_1| + |q_0|}, \frac{|q_0|}{|q_2| + |q_0|}\} \geq \frac{|q_{m0}|}{q_M + |q_{m0}|} > \delta_M$$

which ensures that $F_{er}(\theta', \delta)$ has no root in $[-\delta_M, \delta_M]$. Therefore, constraints (13) and (27) are a set of sufficient conditions for error-reduction for all configurations in a contact state.

B. Conditions for $\{e\text{-}v, e\text{-}v\}$ Contact State

In this case, the two-point contact wrench is a combination of the two $\{e\text{-}v\}$ contact wrenches.

Using the notation developed in Section III-B, the contact wrenches for the two $\{e\text{-}v\}$ contacts are obtained by (11):

$$\mathbf{w}_i = \begin{bmatrix} \mathbf{n}_i \\ [(\mathbf{r}_{ci} + \mathbf{r}_{ei}\delta_i) \times \mathbf{n}_i] \cdot \mathbf{k} \end{bmatrix}, \quad i = 1, 2 \quad (28)$$

where $\mathbf{r}_{ci}$ and $\mathbf{r}_{ei}$ are constant vectors associated with edge i.

By (8), the error reduction function can be expressed in terms of two variables δ_1 and δ_2:

$$F(\delta_1, \delta_2) = (a_2\delta_1^2 + b_2\delta_1 + c_2)\delta_2^2 + (a_1\delta_1^2 + b_1\delta_1 + c_1)\delta_2 + (a_0\delta_1^2 + b_0\delta_1 + c_0) \quad (29)$$

where a_i's, b_i's and c_i's are all functions of the admittance matrix $\mathbf{A}$.

Denote

$$q_i(\delta_1) = a_i\delta_1^2 + b_i\delta_1 + c_i, \quad i = 0, 1, 2. \quad (30)$$

Since the q_i's are quadratic functions, it is not difficult to determine their extreme values for $\delta_1 \in [-\delta_{1M}, \delta_{1M}]$. In fact, if we denote

$$q_{mi} = \min\{q_i(-\delta_{1M}), q_i(\delta_{1M}), q_i(\frac{b_i}{2a_i})\}, \quad (31)$$

$$q_{Mi} = \max\{q_i(-\delta_{1M}), q_i(\delta_{1M}), q_i(\frac{b_i}{2a_i})\}, \quad (32)$$

$$q_M = \max\{q_{M1}, q_{M2}\}, \quad (33)$$

then, for all $\delta_1 \in [-\delta_{1M}, \delta_{1M}]$.

$$q_{mi} \leq q_i \leq q_{Mi}. \quad (34)$$

By the same reasoning used for the $\{e\text{-}v, v\text{-}e\}$ contact case in Section V-A2, a similar condition is obtained

$$\frac{|q_{m0}|}{q_M + |q_{m0}|} > \delta_M. \quad (35)$$

This condition ensures that $F_{er}(\delta_1, \delta_2)$ has no root for all $\delta_1 \in [-\delta_{1M}, \delta_{1M}]$ and $\delta_2 \in [-\delta_{2M}, \delta_{2M}]$. Therefore, constraints (13) and (35) are a set of sufficient conditions for error-reduction for all configurations in a contact state.

C. Conditions for {v-e,v-e} Contact State

In this case, the two-point contact wrench is a combination of the two {v-e} contact wrenches.

Using the notation developed in Section III-B, the contact wrenches for the two {v-e} contact are obtained by (9):

$$\mathbf{w}_i = \begin{bmatrix} \mathbf{R}\mathbf{n}_i \\ (\mathbf{r}_i \times \mathbf{R}\mathbf{n}_i) \cdot \mathbf{k} \end{bmatrix}, \quad i = 1, 2. \quad (36)$$

The error-reduction function (8) can be used directly. Since the wrenches involve only one variable θ, the error-reduction function is a single-variable function in the form:

$$F_{er}(\theta) = a_1 \cos^4 \theta + a_2 \cos^3 \theta \sin \theta + a_3 \cos^2 \theta \sin \theta$$
$$+ a_4 \cos \theta \sin \theta + a_5 \cos \theta + a_6 \sin \theta + a_0, \quad (37)$$

where a_i's are functions of the admittance matrix $\mathbf{A}$.

If we denote

$$p_1(\theta) = a_1 \cos^4 \theta + a_5 \cos \theta + a_0, \quad (38)$$
$$p_2(\theta) = a_2 \cos^3 \theta + a_3 \cos^2 \theta + a_4 \cos \theta + a_6, \quad (39)$$

then $F_{er}(\theta)$ can be expressed as:

$$F_{er}(\theta) = p_1(\theta) + p_2(\theta) \sin \theta. \quad (40)$$

Since $|\theta| \leq \theta_M \ll \frac{\pi}{4}$, the bounds for the single-variable functions $p_1(\theta)$ and $p_2(\theta)$ can be obtained by the approach used for the single-point contact case [12].

Let p_{mi} and p_{Mi} be the bounds of p_i, i.e.,

$$p_{m1} \leq p_1 \leq p_{M1}, \quad (41)$$
$$p_{m2} \leq p_2 \leq p_{M2}, \quad (42)$$

and

$$p_M = \max\{|p_{2m}|, |p_{2M}|\}. \quad (43)$$

Then, it can be proved that, the conditions

$$p_{1m} - p_M \sin \theta_M < 0 \quad (44)$$
$$p_{1m} + p_M \sin \theta_M < 0 \quad (45)$$

ensure that for all $\theta \in [-\theta_M, \theta_M]$, $F_{er}(\theta) < 0$. Therefore, constraints (44)-(45) combined with (13) are a set of sufficient conditions for error-reduction for all configurations in a contact state.

V. Summary

In this paper, we identified procedures for selecting the appropriate admittance to achieve reliable planar force-guided assembly for multi-point contact cases. Conditions imposed on the admittance matrix for each of the various types of two-point contact are presented. We show that, for bounded misalignments, if the conditions are satisfied for a finite number of contact configurations, the conditions ensure that force guidance is achieved for all configurations within the specified bounds.

In this study, two variables are used to describe the contact wrenches for a rigid body (with 1-DOF) in two-point contact. With this approach, the error-reduction function is expressed in two independent variables regardless of the geometry of the contact. For each type of contact, by evaluating the bounds on the coefficients in the error-reduction function, specific conditions for satisfying error-reduction are obtained.

Acknowledgment

This research was supported in part by a Ford Motor Company URP award, and the National Science Foundation under grant IIS 0010017.

References

[1] D. E. Whitney, "Force feedback control of manipulator fine motions," *ASME Journal of Dynamic Systems, Measurements and Control*, vol. 98, no. 2, 1977.

[2] D. E. Whitney, "Quasi-static assembly of compliantly supported rigid parts," *ASME Journal of Dynamic Systems, Measurements, and Control*, vol. 104, no. 1, pp. 65–77, 1982.

[3] M. A. Peshkin, "Programmed compliance for error-corrective manipulation," *IEEE Transactions on Robotics and Automation*, vol. 6, no. 4, pp. 473–482, 1990.

[4] H. Asada, "Teaching and learning of compliance using neural net," *IEEE Transactions on Robotics and Automation*, vol. 9, no. 6, pp. 863–867, 1993.

[5] E. D. Fasse and J. F. Broenink, "A spatial impedance controller for robotic manipulation," *IEEE Transactions on Robotics and Automation*, vol. 13, no. 4, pp. 546–556, 1997.

[6] Jr. Marcelo H. Ang and Gerry B. Andeen, "Specifying and achieving passive compliance based on manipulator structure," *IEEE Transactions on Robotics and Automation*, vol. 11, no. 4, pp. 504–515, 1995.

[7] J. M. Schimmels and M.A. Peshkin, "Admittance matrix design for force guided assembly," *IEEE Transactions on Robotics and Automation*, vol. 8, no. 2, pp. 213–227, 1992.

[8] J. M. Schimmels and M. A. Peshkin, "Force-assembly with friction," *IEEE Transactions on Robotics and Automation*, vol. 10, no. 4, pp. 465–479, 1994.

[9] J. M. Schimmels, "A linear space of admittance control laws that guarantees force-assembly with friction," *IEEE Transactions on Robotics and Automation*, vol. 13, no. 5, pp. 656–667, 1997.

[10] J. M. R. Martinez and J. Duffy, "On the metric of rigid body displacements for infinite and finite bodies," *ASME Journal of Mechanical Design*, vol. 117, no. 1, pp. 41–47, 1995.

[11] S. Huang and J. M. Schimmels, "Sufficient conditions used in admittance selection for planar force-guided assembly," in *Proceedings of the IEEE International Conference on Robotics and Automation*, Washington, D.C., May 2002, pp. 538–543.

[12] S. Huang and J. M. Schimmels, "Sufficient conditions used in admittance selection for force-guided assembly of polygonal parts," *IEEE Transactions on Robotics and Automation*, (submitted for publication).

[13] M. S. Ohwovoriole and B. Roth, "An extension of screw theory," *ASME Journal of Mechanical Design*, vol. 103, no. 4, pp. 725–735, 1981.

Decoupling Based Cartesian Impedance Control of Flexible Joint Robots

Christian Ott*, Alin Albu-Schäffer*, Andreas Kugi** and Gerd Hirzinger*

*German Aerospace Center (DLR) - Institute of Robotics and Mechatronics
**Saarland University - Chair of System Theory and Automatic Control
christian.ott@dlr.de

Abstract—This paper addresses the impedance control problem for flexible joint manipulators. An impedance controller structure is proposed, which is based on an exact decoupling of the torque dynamics from the link dynamics. A formal stability analysis of the proposed controller is presented for the general tracking case. Preliminary experimental results are given for a single flexible joint.

I. INTRODUCTION

Whenever a robotic manipulator is supposed to get in contact with its environment in order to perform some manipulation tasks, a compliant behaviour of the manipulator is desired. The achievement of such a compliant behaviour by control therefore got a classical problem in robotics research [3]. For the case of a manipulator with rigid joints, various approaches to this problem have been studied in the literature and led to control techniques such as impedance control, admittance control or stiffness control. Compared to this, only little work has been spent on the compliant control problem for robotic manipulators with flexible joints.

Maybe the most obvious approach to the impedance control problem for a robot with flexible joints is based on a singular perturbation analysis of the flexible joint model. From this perspective an impedance controller may be designed in the same manner as for a robot with rigid joints. The flexibility of the joints is then treated in a sufficiently fast inner torque control loop. While this approach is very attractive at a first glance due to the simplicity of the resulting controllers, it has the conceptual problem that the singular perturbation approach does not admit a formal stability analysis for the complete model of the flexible joint robot without referring to an approximate consideration.

In contrast to this approach, in this paper an impedance controller is presented which fully accounts for the flexibility of the joints. The proposed controller structure is based on an internal torque controller which decouples the torque dynamics from the link dynamics exactly and thus leads to a cascaded structure. The desired torque for this inner loop results from a standard impedance control law. Stability results from the theory of cascaded systems [8] can then be used to prove the stability of the overall closed loop system.

Notice that the proposed combination of a decoupling based torque controller with an outer control law for the link positions is strongly related to the works of Lin and Goldenberg. While their design idea in [6] and [7] is similar to the one followed in this paper, their focus lies merely on the position control problem and leads to different controllers. Consequently, their stability analysis in [6] and [7] cannot be applied to the impedance control problem in a straightforward manner.

The paper is organized as follows. First, the considered model is given in Section II. In Section III the proposed torque controller is presented and compared to a simpler singular perturbation based controller. Next, the impedance control law is given in Section IV. Section V contains the stability proof of the decoupling based torque controller from Section III in combination with the outer impedance controller from Section IV. Finally, an experimental comparison of the proposed controller to a singular perturbation based controller is presented for a single flexible joint in Section VI.

II. CONSIDERED MODEL

In this paper a simplified model of a robot with n flexible joints is considered as proposed in [12]

$$M(q)\ddot{q} + C(q,\dot{q})\dot{q} + g(q) = \tau + \tau_{ext}, \quad (1)$$

$$B\ddot{\theta} + \tau = \tau_m, \quad (2)$$

$$\tau = K(\theta - q). \quad (3)$$

Herein, $q \in \Re^n$ is the vector of link positions and $\theta \in \Re^n$ the vector of motor positions. The vector of transmission torques is denoted by τ. Equation (1) contains the symmetric and positive definite mass matrix $M(q)$, the vector of Coriolis and centripetal torques $C(q,\dot{q})\dot{q}$ and the vector of gravitational torques $g(q)$. B and K are diagonal matrices containing the motor inertias and the stiffnesses for the individual joints. τ_m is the vector of motor torques which will serve as the control input and τ_{ext} is a vector of external torques which are exhibited by the manipulator's environment.

Herein it is furtheron assumed that the external torques τ_{ext} can be measured. This can be realized at least in applications where these torques at joint level result from forces and torques at the manipulator's endeffector and can therefore be measured e.g. by a 6DOF force/torque-sensor. For the further analysis, the model (1)-(2) may be rewritten by choosing $(q^T, \dot{q}^T, \tau^T, \dot{\tau}^T)^T$ as state variables

$$M(q)\ddot{q} + C(q,\dot{q})\dot{q} + g(q) = \tau + \tau_{ext}, \quad (4)$$
$$BK^{-1}\ddot{\tau} + \tau = \tau_m - BM(q)^{-1}(\tau + \tau_{ext} - C(q,\dot{q})\dot{q} - g(q)). \quad (5)$$

Based on this model a cascaded control design procedure will be presented in the next sections. An outer loop impedance controller (treated in Section IV) generates a desired torque vector τ_d for an inner torque control loop. The design of the torque controller is treated in the next section.

III. TORQUE CONTROLLER

Obviously, some undesired terms of the torque dynamics equation (5) may be easily compensated with a feedback compensation of the form

$$\tau_m = u + BM(q)^{-1}(\tau + \tau_{ext} - C(q,\dot{q})\dot{q} - g(q)), \quad (6)$$

with the new input variables u. This leads to the system

$$M(q)\ddot{q} + C(q,\dot{q})\dot{q} + g(q) = \tau + \tau_{ext}, \quad (7)$$
$$BK^{-1}\ddot{\tau} + \tau = u. \quad (8)$$

From replacing the torque variables[1] by introducing a *desired torque*-variable τ_d and a torque error variable z

$$z = \tau - \tau_d, \quad (9)$$

we obtain

$$M(q)\ddot{q} + C(q,\dot{q})\dot{q} + g(q) = z + \tau_d + \tau_{ext}$$
$$BK^{-1}(\ddot{z} + \ddot{\tau}_d) + z + \tau_d = u.$$

Based on this system two different torque controllers are given in the following. The first controller results from a singular perturbation approach and shall be used as a reference for the second controller later on in the experiments. The second controller achieves an exact decoupling of the torque dynamics from the link dynamics. This exact decoupling has the conceptual advantage that it admits, in combination with the impedance controller from the next section, a stability analysis for the complete flexible model without the need to refer to an approximate consideration as in the case of the singular perturbation based torque controller.

[1] which means shifting the steady state to 0

A. Singular Perturbation Based Controller

It shall not be in the scope of this paper to treat the singular perturbation analysis of flexible joint robots in detail. Instead we refer to [4] for a comprehensive treatment of the theoretical basis. In a singular perturbation analysis of (4)-(5), the flexible joint model is virtually split up into a fast and a slow subsystem for the joint torques τ and the link positions q respectively. From these two subsystems it is then possible to design an inner loop controller for τ and an outer loop controller for q separately.
For this study a singular perturbation based controller similar to the one in [9] is considered

$$u = \tau_d + BK^{-1}(-K_s\dot{\tau} - K_t z), \quad (10)$$

which is (under a singular perturbation consideration) sufficient to stabilize the joint torque dynamics around the equilibrium point $\tau = \tau_d$. The matrices K_s and K_t herein are some positive definite controller gain matrices. Notice that[2], the controller (10) results in the following link dynamics:

$$M(q)\ddot{q} + C(q,\dot{q})\dot{q} + g(q) = \tau_d + \tau_{ext}.$$

In the following the controller of (10) is extended by some additional terms, which achieve an exact decoupling of the torque dynamics but are not necessary from a singular perturbation point of view.

B. Decoupling Based Torque Controller

It is easy to see that an exact feedback decoupling of the torque dynamics may be obtained by a feedback law of the form

$$u = \tau_d + BK^{-1}(\ddot{\tau}_d - K_s\dot{z} - K_t z). \quad (11)$$

Again, the matrices K_s and K_t are chosen as some positive definite matrices. This leads to a system in cascaded form

$$M(q)\ddot{q} + C(q,\dot{q})\dot{q} + g(q) = z + \tau_d + \tau_{ext}, \quad (12)$$
$$\ddot{z} + K_s\dot{z} + (K_t + KB^{-1})z = 0. \quad (13)$$

Notice that compared to (10) the controller in (11) also contains the time derivatives of the desired torque τ_d up to the second order.
In the next section, the design of τ_d is treated such that a desired impedance behaviour is achieved.

IV. TASK SPACE IMPEDANCE CONTROLLER

In this section an impedance controller is presented which can be used in combination with the two torque controllers from the last section.
It is assumed that the desired behaviour of the manipulator can be described in task-space coordinates[3] $x \in \Re^m$ and

[2] again under a singular perturbation consideration
[3] e.g. describing the endeffector movement

the mapping between these coordinates and the link angles is known $x = f(q)$. The relevant coordinate mappings for the first and the second derivatives can then be computed via the Jacobian $J(q) = \frac{\partial f(q)}{\partial q}$ as

$$\dot{x} = J(q)\dot{q}, \quad (14)$$
$$\ddot{x} = J(q)\ddot{q} + \dot{J}(q)\dot{q}. \quad (15)$$

For reasons of simplicity only the nonredundant and nonsingular case shall be treated herein, thus it is assumed that $m = n$ and that the Jacobian $J(q)$ has full rank in the considered region of the workspace. A description of an appropriate singularity treatment can be found in [5]. It is furtheron assumed that in the considered workspace the vector function $f(q)$ is a one-to-one mapping. Under these assumptions the coordinates x completely describe the rigid-body-behaviour of the robot and can be used as generalized coordinates.

Notice that, while the above-mentioned assumptions on the coordinates x are trivially fulfilled for a joint space consideration (with $x = q$) in the whole workspace, in the case of a desired impedance behaviour in Cartesian coordinates it is generally not possible to find coordinates which fulfill the assumptions globally [13].

However, the analysis in this paper focuses on globally valid statements and is therefore formally based on a globally valid set of coordinates x.

With these assumptions, the external torques τ_{ext} can also be written in task space coordinates as F_{ext} via the well known relationship

$$\tau_{ext} = J(q)^T F_{ext} \quad (16)$$

and equation (12) may also be rewritten as[4]

$$\Lambda(x)\ddot{x} + \mu(x,\dot{x})\dot{x} + p(x) = \quad (17)$$
$$J(q)^{-T}(z + \tau_d) + F_{ext},$$

with the equivalent task space mass matrix

$$\Lambda(x) = J(q)^{-T} M(q) J(q)^{-1}, \quad (18)$$
$$\mu(x,\dot{x}) = J(q)^{-T} C(q,\dot{q}) J(q)^{-1} - \Lambda(q) \dot{J}(q) J(q)^{-1} \quad (19)$$

and $p(x) = J(q)^{-T} g(q)$.

It is well known that in the case of a rigid robot an arbitrary second order impedance behaviour can be (at least theoretically) realized by feedback. It can be shown that for a flexible joint robot this is not possible exactly, therefore only an approximation can be expected here. By comparing the structure of equation (17) to the dynamical equations of a robot with rigid joints one can see that the only difference is the occurrence of the torque error term $J(q)^{-T} z$ in (17). If for the design of the impedance controller this term is neglected, then τ_d can be chosen according to an impedance controller for a robot with rigid joints.

In our previous experiments with singular perturbation based controllers the realization of a desired second order impedance behaviour with an arbitrary inertia turned out to be very difficult [1]. Also, in many applications the main focus merely lies on the realization of a desired stiffness and damping behaviour. Consequently, it is also considered herein that the desired behaviour in the task space can be characterized by a positive definite damping matrix D_d and a positive definite stiffness matrix K_d, while the manipulator's mass matrix shall be maintained. Thus, for a given trajectory $x_d(t)$, the desired behaviour of the manipulator with respect to an external force F_{ext} is given by

$$\Lambda(x)\ddot{e}_x + (\mu(x,\dot{x}) + D_d)\dot{e}_x + K_d e_x = F_{ext} \quad (20)$$
$$e_x = x - x_d.$$

The desired trajectory $x_d(t)$ is assumed to be continuously differentiable up to the order two. For the desired impedance behaviour, and under the above assumption on $J(q)$ and $x_d(t)$, three important properties shall be mentioned:

Property 1: For $F_{ext} = 0$, the system (20) with the positive definite matrices K_d and D_d is uniformly globally asymptotically stable.

Property 2: For $\dot{x}_d(t) = 0$, the system (20) with the positive definite matrices K_d and D_d gets time-invariant and represents a passive mapping from the external force F_{ext} to the velocity error $\dot{e}_x$.

Property 3: The matrix $\dot{\Lambda}(x) - 2\mu(x,\dot{x})$ is skew symmetric.

The proof of property 1 can be found, e.g., in [10]. Although it is drawn therein only for the case of joint space coordinates ($x = q$), it is obviously also valid for general coordinates x under the above-mentioned assumptions. Property 2 can be shown easily with the storage function $V_s = \frac{1}{2}\dot{e}_x^T \Lambda(x) \dot{e}_x + \frac{1}{2} e_x^T K_d e_x$.

Notice that property 2 is very important from a practical point of view. For cases when the robot is to be expected to get in contact with an unknown environment, a common assumption is that the environment can be represented by a passive[5] system which is in feedback interconnection with the robot. Then the above passivity property is sufficient in order to preserve the stability of the whole system. Obviously, for the case of a rigid joint robot (with $z = 0$), the desired behaviour (20) can be achieved for the system (17) by the following control law

$$\tau_d = g(q) + J(q)^T (\Lambda(x)\ddot{x}_d + \mu(x,\dot{x})\dot{x}_d - D_d \dot{e}_x - K_d e_x). \quad (21)$$

---

[4] The substitution $q = f^{-1}(x)$ may be obmitted in the following.

[5] with respect to the input/output-pair $(\dot{e}_x, -F_{ext})$

In the case of the flexible model, the controller (21) in combination with the torque controller (11) leads to the following closed loop dynamics

$$\Lambda(x)\ddot{e}_x + (\mu(x,\dot{x}) + D_d)\dot{e}_x + K_d e_x = \quad (22)$$
$$F_{ext} + J(q)^{-T}z ,$$
$$\ddot{z} + K_s\dot{z} + (K_t + KB^{-1})z = 0 . \quad (23)$$

While the desired impedance characteristics is realized only approximately for the flexible joint robot with the described controller, as one can see from (22), property 1 and property 2 still hold for the system (22)-(23). This is shown in detail in the next section.

V. STABILITY ANALYSIS

First we formulate the main result of this paper in form of two propositions.

Proposition 1: *For $F_{ext} = 0$, the system (22)-(23) with the positive definite matrices K_s, K_t, K_d and D_d is uniformly globally asymptotically stable.*

Proposition 2: *For $\dot{x}_d(t) = 0$, the system (22)-(23) with the positive definite matrices K_s, K_t, K_d and D_d gets time-invariant and represents a passive mapping from the external force F_{ext} to the velocity error $\dot{e}_x$.*

A. Proof of Prop. 1

For the stability analysis of the system (22)-(23) it is important to notice that the system is time-variant due to the occurence of $x_d(t)$ in the equations of motion. In order to rewrite the system (22)-(23) for $F_{ext} = 0$ in the state variables $e_x, \dot{e}_x, z, \dot{z}$ only, it is convenient to make the following substitutions: $J(e_x,t) = J(f^{-1}(x))$, $\Lambda(e_x,t) = \Lambda(x)$ and $\mu(e_x,\dot{e}_x,t) = \mu(x,\dot{x})$. Also for the linear part of the system the substitutions $w = (w_1^T, w_2^T)^T = (z^T, \dot{z}^T)^T$ and

$$A = \begin{bmatrix} 0 & I \\ -K_s & -(K_t + KB^{-1}) \end{bmatrix}$$

are made. This leads to the system:

$$\Lambda(e_x,t)\ddot{e}_x + (\mu(e_x,\dot{e}_x,t) + D_d)\dot{e}_x + K_d e_x = J(e_x,t)^{-T}w_1 ,$$
$$\dot{w} = Aw .$$

Notice that this system has a cascaded structure because the linear system $\dot{w} = Aw$ does not depend on the state variables e_x and $\dot{e}_x$. For a nonlinear time-invariant system in such a cascaded form to be asymptotically stable it is necessary to show that all solutions of the coupled system remain bounded and the uncoupled subsystems are asymptotically stable [11]. Loria extended this result to the time-variant case in [8]. In order to apply this result to the system (22)-(23), the following theorem of [8] is reproduced:

Theorem 1: *Consider the system*

$$\dot{y}_1 = f_1(y_1,t) + h(y,t)y_2 \quad (24)$$
$$\dot{y}_2 = f_2(y_2,t) \quad (25)$$

with $y = (y_1^T, y_2^T)^T$. The functions $f_1(y_1,t)$, $f_2(y_2,t)$ and $h(y,t)$ are continuous in their arguments, locally Lipschitz in y, uniformly in t, and $f_1(y_1,t)$ is continuously differentiable in both arguments. This system is uniformly globally asymptotically stable if and only if the following assumptions hold:

- *There exists a nondecreasing function $H(\cdot)$ such that*

$$\|h(y,t)\| \leq H(\|y\|) . \quad (26)$$

- *The systems*

$$\dot{y}_1 = f_1(y_1,t)$$
$$\dot{y}_2 = f_2(y_2,t)$$

are uniformly globally asymptotically stable
- *The solutions of (24)-(25) are uniformly globally bounded.*

The proof of this theorem can be found in [8].
Notice that for the system (22)-(23) the existence of a nondecreasing function $H(\cdot)$ for which (26) holds is fulfilled due to the assumption that the Jacobian $J(q)$ is nonsingular. Thus, there exists a $\delta \in \Re$, $0 < \delta < \infty$, such that

$$\|J(e_x,t)^{-T}\| \leq \sup_{t \in [0,\infty[} \sqrt{\lambda_{max}(J(e_x,t)^{-1}J(e_x,t)^{-T})}$$
$$< \delta$$

with $\lambda_{max}(A(t))$ as the maximum eigenvalue of $A(t)$ at the time t.
Uniformly globally asymptotic stability of the two uncoupled subsystems is given by property 1 and the fact that the linear system $\dot{w} = Aw$ is even globally exponentially stable for positive definite matrices K_s and K_t.
Hence it is sufficient to show that all solutions of the coupled system are uniformly globally bounded. Before this is shown, two well known matrix lemmas, which will be used in the following, shall be given without proof ([6], [4]).

Lemma 1: *Suppose that a symmetric matrix A is partitioned as*

$$A = \begin{bmatrix} A_{1,1} & A_{1,2} \\ A_{1,2}^T & A_{2,2} \end{bmatrix} \quad (27)$$

where $A_{1,1}$ and $A_{2,2}$ are square. Then the matrix A is positive definite if and only if $A_{1,1}$ is positive definite and $A_{2,2} > A_{1,2}^T A_{1,1}^{-1} A_{1,2}$.

Lemma 2: *Given an arbitrary positive definite matrix Q, one can find a unique positive definite solution P of the Lyapunov equation $A^T P + PA = -Q$ if and only if the matrix A is Hurwitz.*

Consider the positive definite function[6]

$$V_c = \frac{1}{2}\dot{e}_x^T \Lambda(e_x,t)\dot{e}_x + \frac{1}{2}e_x^T K_d e_x + \frac{1}{2}w^T P w \quad (28)$$

with a positive definite matrix P. Under consideration of the well known skew symmetry property 3 the time derivative of V_c along the solutions of (22)-(23) is given by

$$\dot{V}_c = -\dot{e}_x^T D_d \dot{e}_x - \frac{1}{2}w^T Q w + \dot{e}_x^T J(e_x,t)^{-T} w_1$$

where $Q = -(PA + A^T P)$ can be an arbitrary positive definite matrix, because A is Hurwitz for positive definite matrices K_s and K_t and the matrix P in V_c is positive definite (see Lemma 2).

Obviously, $\dot{V}_c$ can be written in matrix form

$$\dot{V}_c = - \begin{bmatrix} \dot{e}_x \\ w_1 \\ w_2 \end{bmatrix}^T N \begin{bmatrix} \dot{e}_x \\ w_1 \\ w_2 \end{bmatrix}$$

with

$$N = \begin{bmatrix} D_d & \begin{bmatrix} -\frac{1}{2}J(e_x,t)^{-T} & 0 \end{bmatrix} \\ \begin{bmatrix} -\frac{1}{2}J^{-1}(e_x,t) \\ 0 \end{bmatrix} & Q \end{bmatrix}$$

From Lemma 1 it follows that a necessary and sufficient condition for N to be positive definite[7] is

$$J^T(e_x,t)D_d J(e_x,t) > \frac{1}{4}Q^{-1} ,$$

which can be fulfilled for every positive definite matrix D_d, because by assumption $J(e_x,t)$ does not get singular and the matrix Q is some positive definite matrix which may be chosen arbitrarily. Hence, one can conclude that

$$\dot{V}_c(\dot{e}_x, e_x, w, t) \leq 0 .$$

At this point it is worth mentioning that V_c is bounded from above and below by some time-invariant, radially unbounded and positive definite functions $W_1(\dot{e}_x, e_x, w)$ and $W_2(\dot{e}_x, e_x, w)$

$$W_1(\dot{e}_x, e_x, w) \leq V_c(\dot{e}_x, e_x, w, t) \leq W_2(\dot{e}_x, e_x, w)$$
$$W_1(\dot{e}_x, e_x, w) = \frac{1}{2}\lambda_1 \|\dot{e}_x\|_2^2 + \frac{1}{2}e_x^T K_d e_x + \frac{1}{2}w^T P w$$
$$W_2(\dot{e}_x, e_x, w) = \frac{1}{2}\lambda_2 \|\dot{e}_x\|_2^2 + \frac{1}{2}e_x^T K_d e_x + \frac{1}{2}w^T P w$$

where

$$0 < \lambda_1 < \inf_{t \in [0,\infty[} \lambda_{min}(\Lambda(e_x,t)) <$$
$$\sup_{t \in [0,\infty[} \lambda_{max}(\Lambda(e_x,t)) < \lambda_2 < \infty$$

[6] Notice that the positive definiteness of V_c is ensured by the fact that the eigenvalues of the matrix $\Lambda(e_x,t)$ are bounded from above and below by some positive constants for all $t \in \Re$ and all $e_x \in \Re^n$

[7] Notice also that, in addition to Lemma 1, it is also possible to show that all eigenvalues of N are bounded from above and below by some positive constants, because Q can be chosen arbitrarily and the matrix $J(e_x,t)$ is nonsingular.

with $\lambda_{min}(A(t))$ and $\lambda_{max}(A(t))$ as the minumum and maximum eigenvalue of $A(t)$ at the time t.

From this property and the fact that $\dot{V}_c \leq 0$, it can be shown that the solutions of (22)-(23) are globally uniformly bounded. Proposition 1 follows then from Theorem 1.

Notice that the need to refer to Theorem 1 in this stability proof results from the facts that on the one hand the considered system is time-varying and on the other hand the time derivative of the chosen function V_c is not negative definite but only negative semidefinite. This fact, together with the remark that Q can be arbitrarily chosen, are the most important differences to the proofs in [6] and [7].

B. Proof of Prop. 2

Choosing V_c as the considered storage function yields for $\dot{V}_c$ in the case of $F_{ext} \neq 0$

$$\dot{V}_c = - \begin{bmatrix} \dot{e}_x \\ w_1 \\ w_2 \end{bmatrix}^T N \begin{bmatrix} \dot{e}_x \\ w_1 \\ w_2 \end{bmatrix} + \dot{e}_x^T F_{ext} , \quad (29)$$

The matrix N has already be shown to be positive definite. From this one can conclude the passivity property from Proposition 2 easily.

VI. EXPERIMENTAL RESULTS

In order to show the advantage of the proposed controller compared to a simpler singular perturbation based controller an experimental comparison with a single flexible joint is presented. The chosen hardware setup is shown

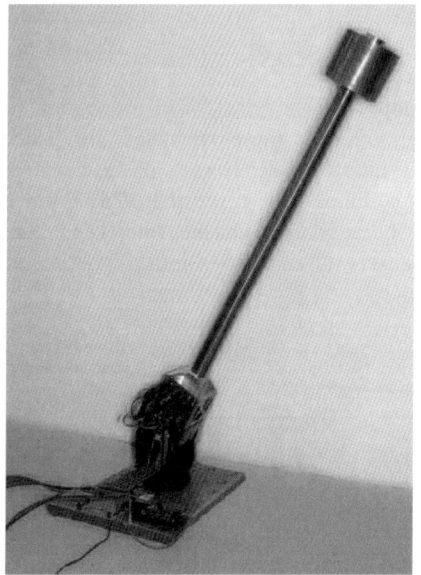

Fig. 1. Hardware setup

in figure 1 and consists of a joint as used in the DLR light weight robots [2] with a mass of approx. 10 kg attached to it as a load. These joints have torque sensors in addition to the common motor position sensors. In order to get a full state measurement, these signals are differentiated numerically. The link position q and its first derivative are then computed from the motor position θ and the joint torque τ via the known joint stiffness k:

$$q = \theta - \tau/k \quad (30)$$
$$\dot{q} = \dot{\theta} - \dot{\tau}/k \quad (31)$$

Notice that the second and third order derivatives of q, which are necessary for the implementation of the impedance controller with the decoupling based torque controller, can be computed from (1) if the external torque τ_{ext} can be measured.

In the experiment only the regulation case shall be considered, and the desired link position is given by q_d. Notice that for a single flexible joint with a constant link side inertia m the only nonlinearity results from the effects of gravity. Therefore, in this experiment the desired stiffness and damping values for the desired behaviour of the link position with respect to external forces can be characterized with a chosen cutoff frequency $w_{bd,q}$ and a damping factor ξ_q. The desired behaviour, from which the controller gains k_d and d_d can be computed, is then given by the linear system:

$$m\ddot{q} + m(2\xi_q w_{bd,q})\dot{q} + mw_{bd,q}^2(q - q_d) = \tau_{ext}$$

Also for the design of the torque controller gains k_s and k_t the desired torque dynamics are characterized by a linear behaviour of the form

$$\ddot{z} + 2\xi_z w_{bd,z}\dot{z} + w_{bd,z}^2 z = 0$$

with a cutoff frequency $w_{bd,z}$ and a damping factor ξ_z. The chosen values of these parameters are given in table I. While the cutoff frequency $w_{bd,z}$ of the torque dynamics has been chosen as high as possible, only a low damping factor has been chosen. On the other hand, the desired impedance is well damped. In the experiment, a step

TABLE I
CHOSEN PARAMETERS FOR THE EXPERIMENT

$w_{bd,q}$	$2\pi 3.5$ rad/s
ξ_q	0.7
$w_{bd,z}$	$2\pi 18$ rad/s
ξ_z	0.2

for the desired position q_d from 9 to 10 degrees was commanded in the absence of external torques. The same impedance controller was used in combination with the two torque controllers from Section III. Figure 2 shows the comparison of the ideal step response to the measured result with the decoupling based torque controller. In Fig. 3

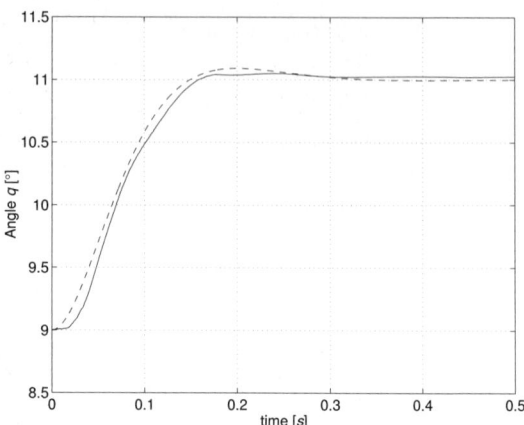

Fig. 2. Step response for the decoupling based controller: ideal (dashed) and measured (solid) response

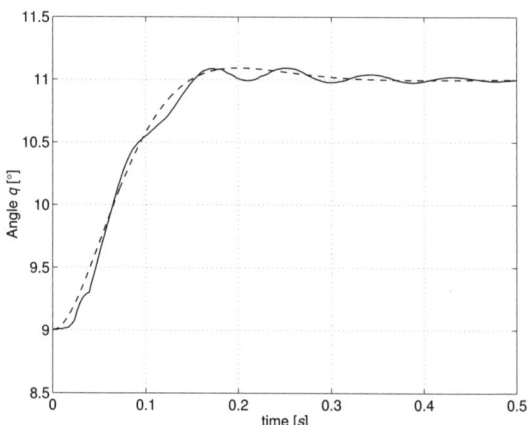

Fig. 3. Step response for the singular perturbation based controller: ideal (dashed) and measured (solid) response

the result with the singular perturbation based torque controller is given. Notice that in this experiment both $w_{bd,q}$ and $w_{bd,z}$ were chosen quite high. Clearly, for impedance controllers with a considerably lower bandwidth and better damped torque control loop, the difference of the step responses of the two controllers would be smaller. But in this experiment it is shown that, for cases where the singular perturbation based controller reaches its limit, the proposed decoupling based controller behaves much better due to the exact decoupling of the torque dynamics.

The same can be seen from the comparison of the torque error. This comparison is given in Fig. 4. Here the inital torque error due to the step of the desired position q_0 is diminished considerably faster in case of the decoupling based controller.

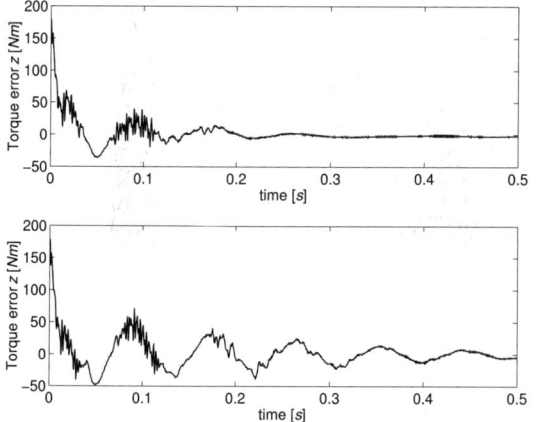

Fig. 4. Torque error z for both controllers: decoupling based controller (above), singular perturbation based controller (below)

VII. CONCLUSIONS

In this paper an impedance controller for flexible joint robots has been presented which is based on an exact decoupling of the torque dynamics. A stability analysis was given for the case of an impedance behaviour in which a desired stiffness and damping matrix can be chosen while the manipulator's inertial behaviour keeps unchanged. Finally, an experimental comparison of the proposed controller structure with a simpler singular perturbation based controller was given. The implementation of the proposed controller on the 7DOF DLR-light-weight-robot (see Fig. 5) is topic of our current research activity.

Fig. 5. 7DOF DLR-light-weight-robot (third generation)

ACKNOWLEDGMENT

This work was supported by the German Federal Ministry of Education and Research "BMBF-Leitprojekt MORPHA" number ITL 0902E.

VIII. REFERENCES

[1] A. Albu-Schäffer, C. Ott, U. Frese, and G. Hirzinger. Cartesian Impedance control of redundant robots: Recent results with the dlr-light-weight-arms. *IEEE Int. Conf. on Robotics and Automation*, 2003.

[2] G. Hirzinger, A. Albu-Schäffer, M. Hähnle, I. Schaefer, and N. Sporer. On a new generation of torque controlled light-weight robots. In *IEEE International Conference of Robotics and Automation*, pages 3356–3363, 2001.

[3] N. Hogan. Impedance control: An approach to manipulation, part I - theory, part II - implementation, part III - applications. *Journ. of Dyn. Systems, Measurement and Control*, 107:1–24, 1985.

[4] H. K. Khalil. *Nonlinear Systems*. Prentice Hall, second edition, 1996.

[5] O. Khatib. A unified approach for motion and force control of robot manipulators: The operational space formulation. *IEEE Journal of Robotics and Automation*, 3(1):1115–1120, February 1987.

[6] T. Lin and A. Goldenberg. Robust adaptive control of flexible joint robots with joint torque feedback. *IEEE Int. Conf. on Robotics and Automation*, pages 1229–1234, 1995.

[7] T. Lin and A. Goldenberg. A unified approach to motion and force control of flexible joint robots. *IEEE Int. Conf. on Robotics and Automation*, pages 1115–1120, 1996.

[8] A. Loria. Cascaded nonlinear time-varying systems: analysis and design. *Minicourse at the Congreso Internacional de Computacion, Cd. Mexico*, 2001.

[9] C. Ott, A. Albu-Schäffer, and G. Hirzinger. Comparison of adaptive and nonadaptive tracking control laws for a flexible joint manipulator. *IEEE Int. Conf. on Intelligent Robots and Systems*, 2002.

[10] V. Santibanez and R. Kelly. Strict lyapunov functions for control of robot manipulators. *Automatica*, 33(4):675–682, 1997.

[11] P. Seibert and R. Suarez. Global stabilization of nonlinear cascade systems. *Systems and Control Letters 14*, pages 347–352, 1990.

[12] M. Spong. Modeling and control of elastic joint robots. *IEEE Journal of Robotics and Automation*, RA-3:291–300, 1987.

[13] S. Stramigioli and H. Bruyninckx. Geometry and screw theory for constrained and unconstrained robot. *Tutorial at ICRA*, 2001.

Design and Simulation of Robust Composite Controllers for Flexible Joint Robots

H.D. Taghirad[†] **and M.A. Khosravi**
Advanced **R**obotics and **A**utomated **S**ystems (ARAS),
Department of Electrical Engineering, K.N. Toosi U. of Technology,
P.O. Box 16315–1355, Tehran, Iran. [†]E-mail: taghirad@saba.kntu.ac.ir

Abstract

In this paper the control of flexible joint manipulators is studied in detail. A composite control algorithm is proposed for the flexible joint robots, which consists of two main parts. Fast control, u_f, which guarantees that the fast dynamics remains asymptotically stable, and the corresponding integral manifold remains invariant. Slow control, u_s, itself consists of a robust PID designed based on the rigid model, and a corrective term designed based on the reduced flexible model. The stability of the overall closed loop system is proved to be UUB stable, by Lyapunov stability analysis. Finally, the effectiveness of the proposed control law is verified through simulations. It is shown that the proposed control law ensure the robust stability and performance, despite the modeling uncertainties.

I. Introduction

After the inception of harmonic drive, multiple–axis flexible robot manipulators are widely used in industrial and space applications. In early eighties researchers showed that the use of control algorithms developed based on rigid robot dynamics on real non–rigid robots is very limited and may even cause instability [15]. To avoid this problem, many researchers have proposed control algorithms based on slow and fast dynamics of the system. Among them, in adaptive methods many algorithms are developed for FJR's, in most of which a term due to the fast subsystem is added to the adaptive algorithm based on rigid models [3, 4]. In robust methods by considering model uncertainties the stability of the fast subsystem is first analyzed and by the use of robust control synthesis, a robust controller is designed for the slow subsystem [1, 7]. Hence, most of the research on FJR's are concentrated on nonlinear control schemes. In this paper we propose a new method based on the simple form of PID, and analyze the robust stability of the uncertain closed–loop system in the presence of structured and unstructured uncertainties. In this analysis we introduce an integral manifold plus a composite control law in order to restrain the integral manifold invariant and to satisfy asymptotic stability requirement. The control effort consists of three elements, the first element is designed for the fast subsystem, the second term is a robust PID control designed for the rigid subsystem and the third term is a corrective term designed based on the first order approximation of the reduced flexible system. Based on the Lyapunov stability theory the complete closed–loop system is proven to be UUB stable. In order to verify the effectiveness of the proposed design method, and to compare its results to that presented in the literature, simulation of single and two link flexible joint manipulators

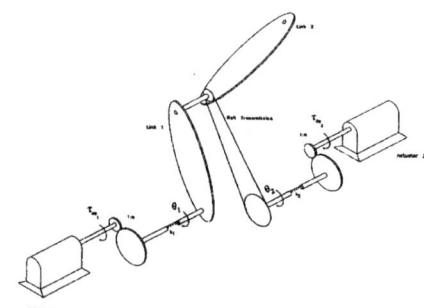

Fig. 1. Two-link Flexible Joint Manipulator.

are examined. It is shown in this study that the proposed control law ensure the robust stability and performance, despite the modeling uncertainties.

II. Flexible Joint Robot Modeling

Spong [13], has derived a nonlinear dynamical model for FJR using singular perturbation, in which the slow states are the position and velocities of the joints and the fast states are the forces and their derivatives. In order to model an N–axis robot manipulator with n revolute joints assume that: $\hat{q}_i : i = 1, 2, ..., n$ denote the position of i'th link and $\hat{q}_i : i = n+1, n+2, ..., 2n$ denote the position of the i'th actuator scaled by the actuator gear ratio. If the joint is rigid $\hat{q}_i = \hat{q}_{n+i} \forall i$. For flexible joint, if the flexibility is modeled with a linear torsional spring with constant k_i, the elastic force z_i is derived from:

$$z_i = k_i(\hat{q}_i - \hat{q}_{n+i}) \quad (1)$$

The spring constants k_i's are relatively large and rigidity is modeled by the limit $k_i \to \infty$. Let u_i denotes the generalized force applied by the i'th actuator and use the notation:

$$q = (\hat{q}_1, ..., \hat{q}_n, \hat{q}_{n+1}, ..., \hat{q}_{2n})^T = (q_1^T | q_2^T)^T \quad (2)$$

The equation of motion of the system can be written in the following form using Euler–Lagrange formulation.

$$\begin{cases} M(q_1)\ddot{q}_1 + N(q_1, \dot{q}_1) = K(q_2 - q_1) \\ J\ddot{q}_2 = K(q_1 - q_2) - D\dot{q}_2 + T_F + u \end{cases} \quad (3)$$

in which,

$$N(q_1, \dot{q}_1) = V_m(q_1\dot{q}_1)\dot{q}_1 + G(q_1) + F_d\dot{q}_1 + F_s(\dot{q}_1) + T_d \quad (4)$$

and K is the joint stiffness matrix, $M(q_1)$ is the mass matrix, $V_m(q_1\dot{q}_1)$ is the matrix of Coriolis and centrifugal terms, $G(q_1)$ is the vector of gravity terms, F_d is the viscous friction matrix, $F_s(\dot{q}_1)$ is the Coulomb friction vector, T_d is the vector of the joint bounded unmodeled dynamics, J is the actuator moments of inertia matrix, D is the actuator viscous friction matrix, and T_F is the actuator bounded unmodeled dynamics. For all revolute manipulators, it is shown in [2, 10], that

$$m_1 I \leq M(q_1) \leq m_2 I \quad ; \quad \|V_m(q_1\,\dot{q}_1)\| \leq \zeta_c \|\dot{q}_1\| \quad (5)$$

$$\|G(q_1)\| \leq \zeta_g \quad ; \quad \|F_d\dot{q}_1 + F_s(\dot{q}_1)\| = \zeta_{f0} + \zeta_{f1}\|q_1\| \quad (6)$$

$$j_1 I \leq J \leq j_2 I \quad ; \quad d_1 I \leq D \leq d_2 I \quad (7)$$

Moreover, if the perturbations are bounded:

$$\|T_d\| \leq \zeta_e \quad ; \quad \|T_F\| \leq \zeta_{f2} \quad (8)$$

in which $\zeta_{f2}, \zeta_e, d_2, d_1, j_2, j_1, \zeta_{f1}, \zeta_{f0}, \zeta_g, \zeta_c, m_2, m_1$ are positive real constants. If the joints are all rigid:

$$M_t(q)\ddot{q} + N_t(q,\dot{q}) = u_0 \quad (9)$$

in which $q = q_1$ and M_t is a positive definite matrix. This model is the model of FJR where $k \to \infty$ verifying that the FJR model is a singularly perturbed model of rigid system. Assume that all spring constants are equal the elastic forces of the springs can be calculated by:

$$z = k(q_1 - q_2), \quad K = kI \quad (10)$$

in order to use a small quantity for singular perturbation define $\epsilon = \frac{1}{k}$ by which for rigid system ($k \to \infty$) in this form we have $\epsilon \to 0$. Multiplying M^{-1} to the both side of 3 and taking $z = k(q_1-q_2)$, $q = q_1$, and using $\dot{q}_2 = \dot{q}_1 - \epsilon\dot{z}$:

$$\begin{cases} \ddot{q} = a_1(q,\dot{q}) + A_1(q)z \\ \epsilon\ddot{z} = a_2(q,\dot{q},\epsilon\dot{z}) + A_2(q)z + B_2 u \end{cases} \quad (11)$$

in which,

$$A_1 = -M^{-1}(q) \quad ; \quad a_1 = -M^{-1}(q)N(q,\dot{q}) \quad (12)$$

$$a_2 = -\epsilon J^{-1}D\dot{z} + J^{-1}D\dot{q} - J^{-1}T_F - M^{-1}(q)N(q,\dot{q}) \quad (13)$$

$$A_2 = -(M^{-1}(q) + J^{-1}) \;,\; B_2 = -J^{-1} \quad (14)$$

Equation 11 represents FJR as a nonlinear and coupled system. This representation includes both rigid and flexible subsystems in form of a singular perturbation model.

III. Reduced Flexible Model

The singular perturbation model of the FJR is given in Equation 11, This model represents the flexibility in the joints, however, the reduced order model is the model of rigid system, which can be easily derived from Equation 11 by setting $\epsilon = 0$. With some matrix manipulation it can be shown that:

$$(M+J)\ddot{q} + N - T_F + D\dot{q} = u_0$$

Rewrite this equation in this form:

$$M_t(q)\ddot{q} + N_t(q,\dot{q}) = u_0 \quad (15)$$

in which

$$M_t(q) = M(q) + J \quad (16)$$
$$N_t(q,\dot{q}) = N(q,\dot{q}) - T_F + D\dot{q} =$$
$$V_m(q,\dot{q})\dot{q} + G(q) + (F_d + D)\dot{q} + F_s(\dot{q}) + T_d - T_F \quad (17)$$

This representation introduces a 2n dimension manifold, M_o, which is called the rigid manifold. If $\epsilon \neq 0$ the produced manifold M_ϵ, which is a function of ϵ represents the flexible system. To define flexible manifold M_ϵ assume:

$$z = H(q,\dot{q},u,\epsilon) \qquad q\epsilon R^n, u\epsilon R^n, z\epsilon R^n \quad (18)$$
$$\dot{z} = \dot{H}(q,\dot{q},u,\epsilon) \qquad q\epsilon R^n, u\epsilon R^n, z\epsilon R^n \quad (19)$$

M_ϵ is an integral manifold for the flexible system if for each initial condition

$$\begin{cases} z(t) = \Delta \\ \dot{z}(t) = \Delta' \end{cases} \text{ and } \begin{cases} q(t) = \zeta \\ \dot{q}(t) = \zeta' \end{cases}$$

in M_ϵ all trajectories of $q(t)$ and $z(t)$ for $t > t_o$ remain in the manifold M_ϵ. In other words $\forall t > t_o$:

$$z(t) = H(q(t),\dot{q}(t),u(t),\epsilon) \quad (20)$$
$$\dot{z}(t) = \dot{H}(q(t),\dot{q}(t),u(t),\epsilon) \quad (21)$$

Now, the reduced flexible model can be derived by replacing $z, \dot{z}$ with $H, \dot{H}$ in Equation 11.

$$\ddot{q} = a_1(q,\dot{q}) + A_1(q)H(q,\dot{q},u,\epsilon) \quad (22)$$

The order of this equation is equal to the rigid system, however, this model includes the effects of flexibility in form of an invariant integral manifold embedded in itself. Hence, this reduced order model is not an approximation of the FJR model, but it represents its projection on the integral manifold.

IV. Composite Control

In order to have a valid reduced flexible model for the system, it is essential that the M_ϵ be an invariant manifold, or the fast dynamics be asymptotically stable. This can be satisfied using a composite control scheme [6]. In this framework the control effort u consists of two main parts, u_s the control effort for slow subsystem, and u_f the control effort for fast subsystem, as:

$$u = u_s(q,\dot{q},\epsilon) + u_f(\eta,\dot{\eta}) \quad (23)$$

in which $u_f(\eta,\dot{\eta})$ is designed such that the fast dynamics becomes asymptotically stable. η denotes the deviations of fast state variables from the integral manifold.

$$\eta = z - H(q,\dot{q},u_s,\epsilon) \quad (24)$$
$$\dot{\eta} = \dot{z} - \dot{H}(q,\dot{q},u_s,\epsilon) \quad (25)$$

The slow component of the control effort, $u_s(q,\dot{q},\epsilon)$, is also designed based on the reduced flexible model. We describe the design technique for u_f and u_s in the next subsections, respectively.

A. Fast Subsystem Dynamics and Control

Recall Equation 24; hence,

$$\epsilon\ddot{\eta} = [a_2(q,\dot{q},\epsilon\dot{z}) - a_2(q,\dot{q},\epsilon\dot{H})] + A_2(q)\eta + B_2 u_f \quad (26)$$

Substitute the value of a_2 and use fast time scale $\tau = \frac{t}{\sqrt{\epsilon}}$ with some manipulations we reach to [5]:

$$\epsilon\ddot{\eta} = A_2(q)\eta + B_2 u_f \quad (27)$$

and in state space form:

$$\epsilon \begin{bmatrix} \dot{\eta} \\ \ddot{\eta} \end{bmatrix} = \begin{bmatrix} \emptyset & \epsilon I \\ A_2(q) & \emptyset \end{bmatrix} \begin{bmatrix} \eta \\ \dot{\eta} \end{bmatrix} + \begin{bmatrix} \emptyset \\ B_2 \end{bmatrix} u_f \quad (28)$$

The flexible modes are not stable since the eigenvalues are on the imaginary axis. Hence, u_f must be designed such that the eigenvalues are shifted to the open left half plane in order to guarantee stability.

Theorem 1: The diagonal and positive definite matrices K_{pf} and K_{vf} exist such that the closed loop system including the subsystem 27 with the control effort $u_f = K_{pf}\eta + K_{vf}\dot{\eta}$ becomes globally asymptotically stable. (Proof in [17])

B. Control of Reduced Flexible Model

The reduced flexible model represents the effect of flexibility in the form of the flexible integral manifold. In this section a robust control algorithm is proposed for the system based on this model. In order to accurately derive a robust control law $u_s(q,\dot{q},\epsilon)$ for the system, manipulation of partial differential equation is necessary. To avoid complex manipulations, we propose deriving the robust control law $u_s(q,\dot{q},\epsilon)$ to any order of ϵ from the series expansion of the integral manifold to the same order of ϵ.

$$H(q,\dot{q},u_s,\epsilon) = H_0(q,\dot{q},u_s) + \epsilon H_1(q,\dot{q},u_s) + \cdots \quad (29)$$

and implement the controller $u_s(q,\dot{q},\epsilon)$ in the same form as:

$$u_s(q,\dot{q},\epsilon) = u_0(q,\dot{q}) + \epsilon u_1(q,\dot{q}) + \cdots \quad (30)$$

in which the functions $H_i(q,\dot{q},u_s), u_i(q,\dot{q}), i=0,1,\cdots$ are calculated iteratively without need to solve the partial differential equations. It is important to note that as $\epsilon \to 0$, u_s tends to rigid control, and H tends to rigid integral manifold. u_0 is designed using a robust design technique based on the rigid reduced order model ($\epsilon = 0$), and H_0 is calculated from:

$$H_0 = -A_2^{-1}(a_{20} + B_2 u_0) \quad (31)$$

in which:

$$a_{20} = a_2(q,\dot{q},0) = J^{-1}D\dot{q} - J^{-1}T_F(q,\dot{q}) - M^{-1}(q)N(q,\dot{q}) \quad (32)$$

Let:

$$a_2(q,\dot{q},\epsilon\dot{H}) = a_{20} + \epsilon\Delta a_2 + O(\epsilon^2)$$

in which a_{20} is given in Equation 32, and compare to Equation 13 we reach to:

$$\begin{cases} \Delta a_2 = -J^{-1}D\dot{H} \\ \Delta a_{20} = -J^{-1}D\dot{H}_0 \end{cases}$$

Hence,

$$\epsilon\ddot{H}_0 = a_{20} + A_2 H_0 + B_2 u_0 + \epsilon(\Delta a_{20} + A_2 H_1 + B_2 u_1) + O(\epsilon^2) \quad (33)$$

and,

$$\ddot{H}_0 = \Delta a_{20} + A_2 H_1 + B_2 u_1 \quad (34)$$

Therefore,

$$H_1 = A_2^{-1}(\ddot{H}_0 - \Delta a_{20} - B_2 u_1) \quad (35)$$

To calculate u_1 refer to reduced flexible model 22 and approximate it to the first power of ϵ:

$$\ddot{q} = a_1(q,\dot{q}) + A_1(q)H_0 + \epsilon A_1(q)A_2^{-1}(\ddot{H}_0 - \Delta a_{20} - B_2 u_1)$$

By factoring the equal powers of ϵ we reach to:

$$u_1 = B_2^{-1}(\ddot{H}_0 - \Delta a_{20}) \quad (36)$$

The only condition on robust control design is that u_0 must be at least twice differentiable. Finally, the control law for slow subsystem has the form:

$$u_s = u_0 + \epsilon u_1 \quad (37)$$

In which u_1 is called the corrective term which is derived through this subsection and u_0 is the robust control based on the rigid model elaborated in the next section.

C. Robust PID Control for Rigid model

In this section we first propose a robust PID controller based on the rigid model of the system and then prove its robust stability with respect to the model uncertainties. Recall the rigid model of the system from Equation 15, choose a PID controller for u_0:

$$u_0 = K_V \dot{e} + K_P e + K_I \int_0^t e(s)ds = Kx \quad (38)$$

in which

$$\begin{cases} e = q_d - q \\ K = [K_I \quad K_P \quad K_V] \\ x = [\int_0^t e^T(s)ds \quad e^T \quad \dot{e}^T]^T \end{cases}$$

Similar to [2, 11] and [12], assume:

$$\underline{m}_t I \leq M_t(q) \leq \overline{m}_t I \quad (39)$$

and put some limits on:

$$\|N_t\| \leq \beta_0 + \beta_1\|L\| + \beta_2\|L\|^2 \quad ; \quad \|V_m\| \leq \beta_3 + \beta_4\|L\| \quad (40)$$

in which $\|.\|$ is the Euclidean norm and $L = [e^T \quad \dot{e}^T]$. Implement the control law u_0 in 15 to get:

$$\dot{x} = Ax + B\Delta A \quad (41)$$

where

$$A = \begin{bmatrix} \emptyset & I_n & \emptyset \\ \emptyset & \emptyset & I_n \\ -M_t^{-1}K_I & -M_t^{-1}K_P & -M_t^{-1}K_V \end{bmatrix} \quad B = \begin{bmatrix} \emptyset \\ \emptyset \\ M_t^{-1} \end{bmatrix}$$

$$\Delta A = N_t + M_t \ddot{q}_d \quad (42)$$

To analyze the system robust stability consider the following Lyapunov function:

$$V(x) = x^T P x = \frac{1}{2}[\alpha_2 \int_0^t e(s)ds + \alpha_1 e + \dot{e}]^T . M_t .$$
$$[\alpha_2 \int_0^t e(s)ds + \alpha_1 e + \dot{e}] + w^T P_1 w \quad (43)$$

in which

$$w = \begin{bmatrix} \int_0^t e(s)ds \\ e \end{bmatrix} \quad P_1 = \frac{1}{2}\begin{bmatrix} \alpha_2 K_P + \alpha_1 K_I & \alpha_2 K_V + K_I \\ \alpha_2 K_V + K_I & \alpha_1 K_V + K_P \end{bmatrix}$$

Hence,

$$P = \frac{1}{2}\begin{bmatrix} \alpha_2 K_P + \alpha_1 K_I + \alpha_2^2 M_t & \alpha_2 K_V + K_I + \alpha_1\alpha_2 M_t & \alpha_2 M_t \\ \alpha_2 K_V + K_I + \alpha_1\alpha_2 M_t & \alpha_1 K_V + K_P + \alpha_1^2 M_t & \alpha_1 M_t \\ \alpha_2 M_t & \alpha_1 M_t & M_t \end{bmatrix}$$

Since M_t is a positive definite matrix, P is positive definite, if and only if, P_1 is positive definite.

Lemma 1: Assume the following inequalities hold:
$$\alpha_1 > 0 \quad \alpha_2 > 0 \quad \alpha_1 + \alpha_2 < 1$$
$$s_1 = \alpha_2(k_P - k_V) - (1 - \alpha_1)k_I - \alpha_2(1 + \alpha_1 - \alpha_2)\overline{m}_t > 0$$
$$s_2 = k_P + (\alpha_1 - \alpha_2)k_V - k_I - \alpha_1(1 + \alpha_2 - \alpha_1)\overline{m}_t > 0$$

Then P is positive definite and satisfies the following inequality (Rayleigh-Ritz)[9]:
$$\underline{\lambda}(P)\|x\|^2 \le V(x) \le \overline{\lambda}(P)\|x\|^2 \tag{44}$$

in which,
$$\underline{\lambda}(P) = min\{\frac{1 - \alpha_1 - \alpha_2}{2}\underline{m}_t, \frac{s_1}{2}, \frac{s_2}{2}\}$$
$$\overline{\lambda}(P) = max\{\frac{1 + \alpha_1 + \alpha_2}{2}\overline{m}_t, \frac{s_3}{2}, \frac{s_4}{2}\}$$

and
$$s_3 = \alpha_2(k_P + k_v) + (1 + \alpha_1)k_I + (1 + \alpha_1 + \alpha_2)\alpha_2\overline{m}_t$$
$$s_4 = \alpha_1\overline{m}_t(1 + \alpha_1 + \alpha_2) + (\alpha_1 + \alpha_2)k_V + k_P + k_I$$

Proof is based on Gershgorin theorem and is similar to that in [11]. with some manipulations we can show [5]:
$$\gamma = min\{\alpha_2 k_I, \alpha_1 k_P - \alpha_2 k_V - k_I, k_V\}$$

Now considering Equations 40, 42 and $\|L\| \le \|x\|$ then,
$$\begin{aligned}\xi_0 &= \alpha_2^{-1}\lambda_1\beta_0 + \alpha_2^{-1}\lambda_1\lambda_3\overline{m}_t \\ \xi_1 &= \gamma - \lambda_1\beta_3 - \lambda_2\overline{m}_t - \alpha_2^{-1}\lambda_1\beta_1 \\ \xi_2 &= \lambda_1\beta_4 + \alpha_2^{-1}\lambda_1\beta_2\end{aligned}$$

in which
$$\begin{cases}\lambda_1 = \lambda_{max}(R_1) \\ \lambda_2 = \lambda_{max}(R_2) \\ \lambda_3 = sup\|\ddot{q}_d\|\end{cases}$$

and $\lambda_{Min}, \lambda_{Max}$ are the least and largest eigenvalues, respectively, and
$$R_1 = \begin{bmatrix} \alpha_2^2 I & \alpha_1\alpha_2 I & \alpha_2 I \\ \alpha_1\alpha_2 I & \alpha_1^2 I & \alpha_1 I \\ \alpha_2 I & \alpha_1 I & I \end{bmatrix}$$
$$R_2 = \frac{1}{2}\begin{bmatrix} \emptyset & \alpha_2^2 I & \alpha_1\alpha_2 I \\ \alpha_2 I & 2\alpha_1\alpha_2 I & (\alpha_1^2 + \alpha_2)I \\ \alpha_1\alpha_2 I & (\alpha_1^2 + \alpha_2)I & \alpha_1 I \end{bmatrix}$$

According to the result obtained so far, we can prove the stability of the error system based on the following theorem.

Theorem 2: The error system 41 is stable of the form of UUB, if ξ_1 is chosen large enough.
The conditions are:
$$\xi_1 > 2\sqrt{\xi_0\xi_2}$$
$$\xi_1^2 + \xi_1\sqrt{\xi_1^2 - 4\xi_0\xi_2} > 2\xi_0\xi_2(1 + \sqrt{\frac{\overline{\lambda}(P)}{\underline{\lambda}(P)}})$$
$$\xi_1 + \sqrt{\xi_1^2 - 4\xi_0\xi_2} > 2\xi_2\|x_0\|\sqrt{\frac{\overline{\lambda}(P)}{\underline{\lambda}(P)}}$$

These conditions can be simply met by making ξ_1 large enough by choosing large enough control gains K_P, K_v, and K_I. (Proof in [17])

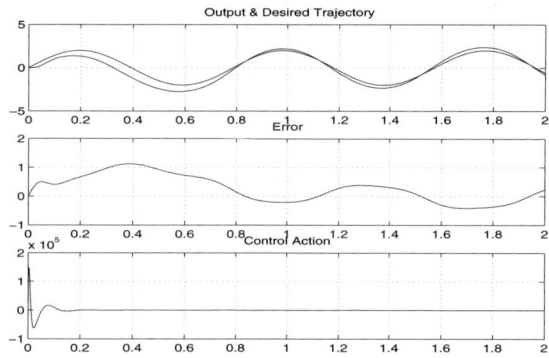

Fig. 2. Poor tracking performance of the closed loop system for perturbed model; Spong algorithm

V. Stability Analysis of the Complete Closed-loop System

The stability of the fast, and slow subsystems are separately analyzed in previous sections. However, the stability of the complete closed-loop system may not be guaranteed through these separate analysis [8]. In this section the stability of the complete system is analyzed. Recall the dynamic equations of the FJR Equation 11. The integral manifold and the control effort are chosen as:
$$\eta = z - H$$
$$H = H_0 + \epsilon H_1$$
$$u = u_s + u_f = u_0 + \epsilon u_1 + u_f$$

Combine these equations to Equation 11, 35, 31 and 38, and consider, $x = \begin{bmatrix}\int_0^t e(s)^T ds & e^T & \dot{e}^T\end{bmatrix}^T$; $y = \begin{bmatrix}\eta^T & \dot{\eta}^T\end{bmatrix}^T$ then,
$$\dot{x} = Ax + B\Delta A + C\begin{bmatrix}I & \emptyset\end{bmatrix}y \tag{45}$$
$$\epsilon\dot{y} = \tilde{A}y \tag{46}$$

in which,
$$A = \begin{bmatrix}\emptyset & I & \emptyset \\ \emptyset & \emptyset & I \\ -M_t^{-1}K_I & -M_t^{-1}K_P & -M_t^{-1}K_V\end{bmatrix}; B = \begin{bmatrix}\emptyset \\ \emptyset \\ M_t^{-1}\end{bmatrix}$$
$$\Delta A = N_t + M_t\ddot{q}_d$$
$$C = \begin{bmatrix}\emptyset \\ \emptyset \\ -A_1\end{bmatrix}; \tilde{A} = \begin{bmatrix}\emptyset & \epsilon I \\ A_2 + B_2 K_{pf} & -\epsilon J^{-1}D + B_2 K_{vf}\end{bmatrix}$$

Theorem 3: There exist diagonal and positive definite matrices K_{pf} and K_{vf} such that the closed loop system 46 becomes globally asymptotically stable. (Proof in [17])

Theorem 4: The closed-loop system of Equations 45 and 46 is UUB stable if K_{pf}, K_{vf}, and ξ_1 are chosen large enough. (Proof in [17])

The detail conditions on the PID controller parameter bounds to preseve the closed–loop stability, are given in [17]. However, the stability conditions met if the controller gains are selected high enough.

VI. Simulations

In order to verify the effectiveness of the algorithm a simulation study has been forwarded next. In the following simulation study, the results of the closed loop performance of a single, [14], and a two link flexible joint manipulator, [1], examined in the literature is compared to that of the proposed control algorithm.

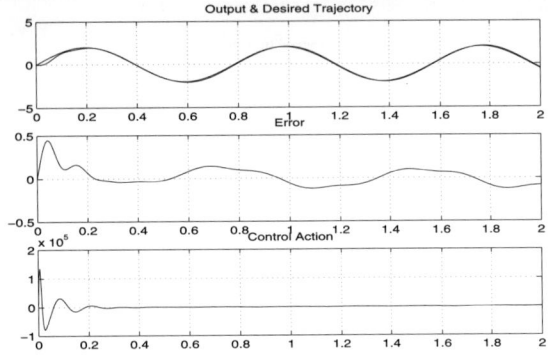

Fig. 3. Suitable tracking performance of the closed loop system for perturbed model; Proposed algorithm

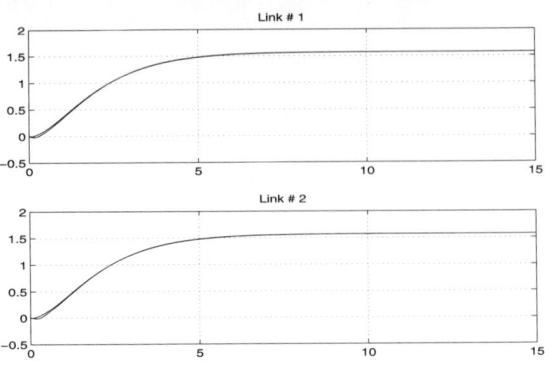

Fig. 4. Tracking performance of the closed loop system and perturbed model; Proposed algorithm

A. Single Link Flexible Joint Manipulator

Consider the single link flexible joint manipulator introduced in [14]. The dynamic equation of motion of this system is as following:

$$\begin{aligned} \dot{x}_1 &= x_2 \\ \dot{x}_2 &= \frac{-MgL}{I}\sin(x_1) - \frac{K}{I}(x_1 - x_3) \\ \dot{x}_3 &= x_4 \\ \dot{x}_4 &= \frac{K}{J}(x_1 - x_3) + \frac{1}{J}u \end{aligned} \qquad (47)$$

in which $x_1 = q_1$ and $x_2 = q_2$. By choosing $q_1 = q$ and $z = K(q_1 - q_2)$ as the elastic force, the model of the system can be rewritten in a singular perturbation form:

$$\begin{aligned} \ddot{q} &= \frac{-MgL}{I}\sin(q) - \frac{1}{I}z \\ \epsilon\ddot{z} &= \frac{-MgL}{I}\sin(q) - (\frac{1}{I} + \frac{1}{J})z - \frac{1}{J}u \end{aligned} \qquad (48)$$

in which $\epsilon = \frac{1}{K}$.

Spong has proposed a composite control law for this system, in which there exists two control components corresponding to the fast and slow dynamics. As it is illustrated in [14], the closed loop system became unstable, provided that only the corresponding rigid control effort u_0 is applied on the system. Moreover, the system becomes stable and the desired trajectory $q_d = \sin(8t)$ is well tracked, implementing the proposed composite control on the nominal model of the system. However, this algorithm is not robust to the model parameter variations. As illustrated in Figure 2 the tracking performance is getting quite poor for the maximum perturbation values for the parameters $I, J, M,$ and L. For the sake of comparison, the proposed robust PID controller may be now applied on the same system. The proposed control law is composed of three terms, in which the rigid control law is a PID controller whose coefficients satisfies the robust stability conditions elaborated in Theorem (4) as following:

$$u_o = 200\dot{e} + 500e + 100 \int_0^t e(s)ds.$$

The integral manifold would be:

$$H_o = -4.9\sin(q) - \frac{1}{2}u_o$$

and the corrective term corresponds to

$$u_1 = \ddot{H}_o.$$

The fast control law is a simple PD controller satisfying the robust stability conditions such as:

$$u_f = 5\eta + 5\dot{\eta}$$

in which η indicates the variation of z from the integral manifold H.

It is observed that by implementing the proposed control law, not only the system is well tracking the desired trajectory for the nominal parameters of the model [5], but also the robust stability and tracking performance of the system with maximum variation in its model parameters are preserved (Figure 3).

B. Multiple Link Flexible Joint Manipulator

Consider the two link Flexible Joint manipulator illustrated in Figure 1. In this manipulator Joint flexibility is modeled with a linear torsional spring with stiffness k. The equation of motion of this system and its parameters is given in [1]. Our proposed algorithm is applied to the system for comparison of the results. Hence, the reduced order first order model is evaluated as following:

$$\begin{bmatrix} 2.25 & 0.5\cos(\epsilon H_1^o) \\ 0.5\cos(\epsilon H_1^o) & 1.25 \end{bmatrix} \begin{bmatrix} \ddot{\theta}_1 \\ \ddot{\theta}_2 \end{bmatrix} + \begin{bmatrix} H_1^o + \epsilon H_1^1 \\ H_2^o + \epsilon H_2^1 \end{bmatrix}$$

$$\begin{bmatrix} 14.7\cos(\theta_1) + 0.5\dot{\theta}_2^2 \sin(\epsilon H_1^o) \\ 4.9\cos(\theta_2) - 0.5\dot{\theta}_1^2 \sin(\epsilon H_1^o) \end{bmatrix} = \begin{bmatrix} 0 \\ 0 \end{bmatrix}$$

In order to evaluate the fast dynamics caused by the joint flexibility, the normalized time variable $\tau = \frac{t}{\sqrt{\epsilon}}$ is used. Hence,

$$H_1^o = \ddot{\theta}_1 - u_1^o \ ; \ H_2^o = \ddot{\theta}_2 - u_2^o \qquad (49)$$

$$H_1^1 = -\ddot{H}_1^o - u_1^1 \ ; \ H_2^1 = -\ddot{H}_2^o - u_2^1 \qquad (50)$$

With expanding Equation 49 to the first order of ϵ we have:

$$H_1^1 = -0.5\dot{\theta}_2^2 \ H_1^o \ ; \ H_2^1 = -0.5\dot{\theta}_1^2 \ H_1^o \qquad (51)$$

And from Equation 50 we get:

$$u_1^1 = -0.5\dot{\theta}_2^2 \ H_1^o - \ddot{H}_1^o \ ; \ u_2^1 = -0.5\dot{\theta}_1^2 \ H_1^o - \ddot{H}_2^o \qquad (52)$$

Finally, the slow part of the control law will be calculated from:

$$u_{1s} = u_1^o + \epsilon u_1^1 \ ; \ u_{2s} = u_2^o + \epsilon u_2^1 \qquad (53)$$

The u_1^o, u_2^o are the rigid part of the control law and as elaborated before is robustly designed as a PID controller.

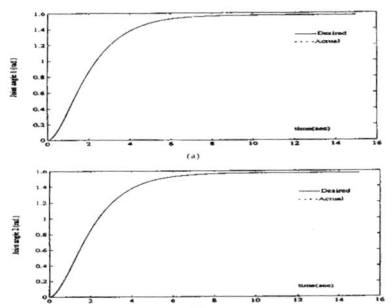

Fig. 5. Tracking performance of the closed loop system for nominal model; Al-Ashoor et al algorithm

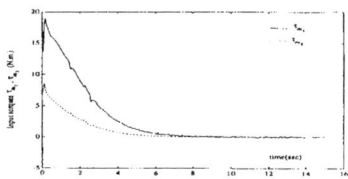

Fig. 6. Control effort for the closed loop system and nominal model; Al-Ashoor et al algorithm

In here we design the PID gains as following which satisfies the robust conditions:

$$u_1^o = 500e + 50\dot{e} + 50 \int_0^t e(s)ds \quad (54)$$

$$u_2^o = 200e + 50\dot{e} + 50 \int_0^t e(s)ds$$

The fast control law is also designed as a PD controller as:

$$u_{1f} = \dot{\eta}_1 + \eta_1 \;\; ; \;\; u_{2f} = \dot{\eta}_2 + \eta_2 \quad (55)$$

Finally the control law is composed from the east and slow parts:

$$u_1 = u_{1s} + u_{1f} \;\; ; \;\; u_2 = u_{2s} + u_{2f} \quad (56)$$

To have simulation results compared to [1], the reference signal is considered as:

$$\theta_i = 1.57 + 7.8539e^{-t} - 9.428e^{-t/1.2} \quad i = 1,2 \quad (57)$$

in which the joint angles reach to a final value of $\theta_i = \frac{\pi}{2}$ from zero initial state. Figure 4 illustrates the response of the perturbed system to our proposed composite control law. The system becomes stable, and the tracking performance is quite desirable, despite the 50% variation in model parameters. The control is limited to a maximum allowable bounds by adding a saturation block in the simulation. Al-Ashoor et al have used a robust–adaptive control law in addition to the composite law we introduced in this paper. Figure 5 illustrates the results obtained for the reference signal introduced in Equation 57 in [1]. This figure illustrates the tracking performance despite the bounded control effort illustrated in Figure 6. Comparing it to our result (Figure 4), similar performances are obtained. The only limitation exists in our proposed law compared to that in [1], is the amplitude of the control law in the initial time of the simulation. The adaptive law have smaller control effort in the beginning of the simulation, due to the adaptive nature of the algorithm, and using the information of the identified model of the system in the control law. This issue is under current investigation, and promising results are obtained by a $\mathcal{H}_\infty$ –based composite controller, in which the control effort can be limited to desirable bounds, [16].

VII. Conclusions

In this paper the control of flexible joint manipulators is examined in detail. In order to achieve the required performance a composite control algorithm is proposed, consisting of corresponding control law for fast and slow subsystems. A simple PD control is proposed for the fast subsystem, and it is proven that the fast subsystem becomes asymptotically stable. The slow subsystem itself is controlled through a robust PID controller designed based on the rigid model, and a correction term designed based on the reduced flexible model. The stability of the complete closed–loop system is analyzed and it is shown that the proposed controller is capable of robustly stabilizing the uncertain flexible joint manipulator. Finally, the effectiveness of the proposed control law is verified through simulations. are compared to that given in the literature, and the effectiveness of preserving the robust stability, and performance of the system is verified and compared relative to them.

References

[1] R.A. Al-Ashoor, R.V. Patel, and K. Khorasani. Robust adaptive controller design and stability analysis for flexible-joint manipulators. *IEEE Transactions on Systems, Man and Cybernetics*, 23(2):589–602, Mar-Apr 1993.

[2] J. J. Craig. *Adaptive Control of Mechanical Manipulators*. Addison-Wesely, 1988.

[3] F. Ghorbel and M. W. Spong. Stability Analysis of Adaptively Controlled Flexible Joint Manipulators. In *Proceedings of the 29th Conference on Decision and Control*, pages 2538–44, 1990.

[4] F. Ghorbel and M.W. Spong. Adaptive integral manifold control of flexible joint robot manipulators. *Proceeding of IEEE International Conference on Robotics and Automation*, 1:707–714, 1992.

[5] Mohammad. A. Khosravi. *Modeling and Robust Control of Flexible Joint Robots*. M.Eng. Thesis, Department of Electrical Engineering, K. N. Toosi University of Technology, Tehran, 2000.

[6] P. V. Kokotovic and H. K. Khalil. *Singular Perturbations in Systems and Control*. IEEE Press., 1986.

[7] Y. H. Chen M. C. Han. Robust Control Design For Uncertain Flexible-Joint Manipulators: A Singular Perturbation Approach. In *Proceedings of the 32th Conference on Decision and Control*, pages 611–16, 1993.

[8] J. O'Reilly P. V. Kokotovic, H. Khalil. *Singular Perturbation Methods In Control: Analysis and Design*. Academic Press., 1986.

[9] Z. Qu. *Nonlinear Robust Control of Uncertain systems*. John Wiley & Sons., 1998.

[10] Z. Qu and D. M. Dawson. *Robust Tracking Control of Robot Manipulators*. IEEE Press., 1996.

[11] Z. Qu and J. Dorsey. Robust PID Control of Robots. *International Journal of Robotics and Automation*, 6(4):228–35, 1991.

[12] Z. Qu and J. Dorsey. Robust Tracking Control of Robots by a Linear Feedback Law. *IEEE Transaction on Automatic Control*, 36(9):1081–4, September 1991.

[13] M.W. Spong. Modeling and control of elastic joint robots. *Journal of Dynamic Systems, Measurement and Control*, 109:310–319, 1987.

[14] M.W. Spong, J.Y. Hung, S. Bortoff, and F. Ghorbel. Comparison of feedback linearization and singular perturbation techniques for the control of flexible joint robots. *Proceedings of the American Control Conference*, 1:25–30, 1989.

[15] M.W. Spong, K. Khorasani, and P.V. Kokotovic. Integral manifold approach to the feedback control of flexible joint robots. *JRA*, RA-3(4):291–300, Aug 1987.

[16] H.D. Taghirad and M. Bakhshi. Composite H-infinity controller synthesis for flexible joint robotss. In *Proceedings of Int. Conf. IROS, 2002*.

[17] H.D. Taghirad and M. A. Khosravi. Stability analysis, and robust PID design for flexible joint robots. *Proceeding of the International Symposium on Robotics*, 1:144-149, 2000.

Design and experimental evaluation of a single robust position/force controller for a single flexible link (SFL) manipulator in collision

Kamyar Ziaei*, David W. L. Wang†

University of Waterloo
Dept of Electrical and Computer Engineering
Waterloo, Ontario, Canada

Abstract

This paper presents a design methodology for designing position/force controllers for a single flexible link (SFL) manipulator. The objective is to design a single controller that stabilizes the system during both constrained and unconstrained motion, avoiding problems that arise when using switching controllers. A coherent system identification based robust control design methodology is presented. As a robust controller, we consider the quantitative feedback theory (QFT) based design. The proposed method is very suitable since linear model of constrained SFL manipulator is a gross simplification and includes significant uncertainties that can be handled well in the QFT frame work.

1 Introduction

To perform many advanced tasks, the robot must come into contact with the environment. Examples of such tasks include manufacturing tasks such as grinding, deburring and polishing, and space applications such as the repair and maintenance of satellites. For these tasks, it is necessary to control both the position of the end-effector and the constrained force between the end-effector and the environment. There are generally three distinct stages in a robotic operation: free-space or unconstrained motion, constrained motion, and the transition phase. The control objective during the unconstrained segment of a general robotic task is to control the position of the end-point. The problem during the constrained segment is generally called hybrid position/force control and is more difficult, because it involves simultaneous control of the position of the end-point in a direction tangent to the surface of the environment and of the force exerted by the manipulator normal to surface of the environment. In the transition phase, the effect of the impact force generated during the transition from unconstrained to constrained segment of the task must be addressed. Position and force control of rigid link manipulators has been extensively studied separately in the past twenty years. However, the research on the contact transition control flexible link manipulators is not as mature. The mathematical modeling of a SFL manipulator in free motion is straight forward and has been well documented, see eg. [1]. In position control, it's common to use the clamped free or pinned free boundary conditions to solve the governing partial differential equations. In the constrained case, however the beam is in contact with an object and does not have a free end. In [2] and [3], Matsuno *et al.* show that the boundary conditions for the flexible beam in contact with the environment become non-homogeneous since they depend on the contact force, the input torque, as well as the contact position. This makes the flexible beam equations very difficult to solve and simplifications must be made. To overcome these modeling difficulties, we performed system identification on the manipulator to determine an accurate transfer function for free and constrained motions.

Some drawbacks of the identification of single flexible link manipulators using ARMA type models have been previously reported [4, 5]. In [4], it is shown that the standard ARMA model in unmodified form cannot satisfactorily describe the resonant modes of the system. To improve the identification results, the authors made several modifications to the model including pre-filtering and fixing some parameters based on spectral analysis. In [5], as an alternative to ARMA models, a frequency domain technique has been used to parameterize the transfer function of flexible link manipulators. Since only the magnitude response is used, the frequency domain identification method in [5] is only

---

*kziaei@kingcong.uwaterloo.ca
†dwang@kingcong.uwaterloo.ca

suitable for identifying minimum-phase transfer functions with slightly damped zeros such as the transfer function from the shaft velocity to tip acceleration. In the next section we present a newly developed system identification based on orthogonal basis functions.

2 System identification

2.1 Orthogonal basis functions

Consider the single-input single-output LTI system

$$y(k) = G(q)u(k) + v(k), \quad G(q) = \sum_{k=1}^{\infty} g_k q^{-k} \quad (1)$$

where $y(k)$ and $u(k)$ are the output and input signals and q denotes the shift operator: $q^{-1}u(k) \triangleq u(k-1)$. The noise process $\{v(k)\}$ is assumed to be stationary process with zero mean values. The objective of system identification is to find an estimate of $G(q)$ from observations of the input and output measurements. One popular method to identify $G(q)$ is to approximate it with a finite impulse response model (FIR), which is obtained by truncating the infinite sum in (1).

The approximation of stable dynamical systems using orthonormal functions has attracted a wide interest over the last decade [6, 7]. This is because models based on parametric orthonormal bases offer distinct advantages over existing model structures, such as FIR, AR and ARX models. An FIR model, expands the system transfer function using the delay operator to obtain a predictor model which is linear in parameter. However, such models are usually of very high order. By using *a priori* information of the approximate plant pole locations, a more appropriate, low order expansion model can be obtained. In this paper, we are using the so-called generalized orthogonal basis functions (GOBF). The basis functions are termed generalized orthonormal bases, because they encompass other orthonormal bases – such as *Laguerre* and *Kautz* bases – as restrictive special cases.

The advantage of generalized orthonormal bases as compared to other orthonormal basis functions such as the Laguerre and Kautz functions is the ability to include a priori estimates of real and complex pole locations of the model.

Using a set of basis functions, $\Lambda_i(q)$, a model of order n is given by

$$y(k) = \left(\sum_{i=1}^{n} \theta_i \Lambda_i(q)\right) u(k) \quad (2)$$

Note that, similar to the FIR model, the model in (2) is linear in estimation parameters, θ_i. In addition, if the set of basis functions, $\Lambda_i(q)$, correspond to orthonormal rational transfer functions, many nice properties of the FIR models including the numerical ones are preserved [7]. Also, the theoretical results of the FIR model including evaluation of the estimates variances can be extended to the more general model with orthonormal basis function in a straight-forward manner as shown in [6]. The set of basis functions, $\{\Lambda_i \in \mathcal{L}_2, i = 1, 2, \cdots\}$, forms a set of orthonormal functions if the inner product

$$\langle \Lambda_m, \Lambda_n \rangle = \frac{1}{2\pi} \int_{-\pi}^{\pi} \Lambda_m(e^{j\omega}) \overline{\Lambda}_n(e^{j\omega}) d\omega$$
$$= \begin{cases} 1 & \text{if } m = n \\ 0 & \text{otherwise} \end{cases} \quad (3)$$

The orthonormal basis functions proposed in [6] are given by

$$\Lambda_m(q) = \left(\frac{\sqrt{1-|\xi_m|^2}}{q-\xi_m}\right) \prod_{i=1}^{m-1} \left(\frac{1-\overline{\xi}_i q}{q-\xi_i}\right) \quad (4)$$
$$m = 1, \ldots, n$$

where ξ_i are the poles of the basis transfer functions, $\Lambda_i(q)$. They are chosen based on *a priori* knowledge of the system dynamics. If the system to be modeled includes resonant behavior, one should include complex conjugate poles in the basis functions. However, a complex conjugate pole pair, ξ_i and $\xi_{i+1} = \overline{\xi}_i$, results in complex impulse responses for filters $\Lambda_i(q)$ and $\Lambda_{i+1}(q)$ which are inappropriate in a system identification setting. Instead, by using a linear combination of $\Lambda_i(q)$ and $\Lambda_{i+1}(q)$ will produce real-valued impulse responses, $\tilde{\Lambda}_i(q)$ and $\tilde{\Lambda}_{i+1}(q)$, without compromising the orthonormality (see [6]). The basis functions are given by [8]:

$$\tilde{\Lambda}_i = \sqrt{\frac{1-|\xi_i|^2}{1-|\xi_{i-1}|^2}} \frac{(1-\overline{\xi}_{i-1}q)(\mu_1 q + \mu_2)}{q^2 - (\xi_i + \overline{\xi}_i)q + |\xi_i|^2} \Lambda_{i-1}(q)$$
(5a)

$$\tilde{\Lambda}_{i+1} = \sqrt{\frac{1-|\xi_i|^2}{1-|\xi_{i-1}|^2}} \frac{(1-\overline{\xi}_{i-1}q)(\mu_3 q + \mu_4)}{q^2 - (\xi_i + \xi_i)q + |\xi_i|^2} \Lambda_{i-1}(q)$$
(5b)

with the following being one possible set of solution for $\mu_i's$:

$$\mu_1 = \frac{|1-\xi_i^2|}{\sqrt{1+|\xi_i|^2}}, \qquad \mu_2 = 0$$

$$\mu_2 = -\frac{\xi_i + \overline{\xi}_i}{\sqrt{1+|\xi_i|^2}}, \qquad \mu_4 = \sqrt{1+|\xi_i|^2}.$$

2.2 Parameter estimation

Suppose, in our model, we fix n poles $\{\xi_1,\ldots,\xi_n\}$, which may be repeated poles for the sake of increasing the modeling accuracy. Then the basis functions are given by (4) or (5a) and (5b). As mentioned earlier, prior knowledge of the system can be used in selecting the pole locations. In [8], the authors suggest a global optimization technique which maybe used for finding or fine-tunning the pole locations. Given the orthonormal basis functions, the parameters, θ_i, in (2) can be estimated using the standard least-squares parameter estimates

$$\hat{\theta} = \left[\sum_{k=1}^{N} \phi(k)\phi(k)^T\right]^{-1} \sum_{k=1}^{N} \phi(k)y(k). \quad (6)$$

where $\phi(k)^T = [\Lambda_1(q)u(k),\ldots,\Lambda_n(q)u(k)]$ and $\hat{\theta} = [\hat{\theta}_1,\cdots,\hat{\theta}_n]^T$.

The system identification using orthogonal bases is applied to experimental data taken from a single flexible link manipulator in free motion and in contact. For free motion case, the object is to find the transfer function from the motor torque to tip position of the manipulator, and in constrained case, we want to find the transfer function from motor torque to the force exerted by the manipulator to the environment. The flexible link has a set of strain gauges to measure the tip deflection which can also be used as force sensor in constrained case [9]. The data were collected at a sampling frequency of 200 Hz. The input data used for the identification was a sine sweep (0.5Hz – 12Hz) for free motion. For the case when the manipulator was in contact with the environment, we used a pseudo random binary signal with a dc offset to ensure that the manipulator remains in contact during the data collection. The signal were filtered by a notch filter to reduce the frequency content of the signals around the first mode of vibration.

In free motion, we found that four poles are adequate to obtain a good model of the system: two real poles which describes the rigid body dynamics and a pair of complex conjugate poles describing first mode of vibration. For constrained case, we observed that higher order modes are excited so that we had to include three pairs of complex conjugate poles. Figures 1(a)-(b) show the comparison of the identified model and measured output. As can be seen the agreement is quite remarkable. In Fig 1(a), the deviation at low frequency is believed to be due to friction. We will used the identified models for robust controller design. In the next section, we briefly describe the stochastic embedded approach which we use for the estimation

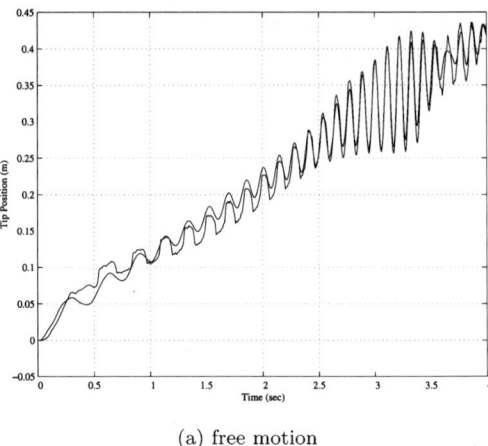

(a) free motion

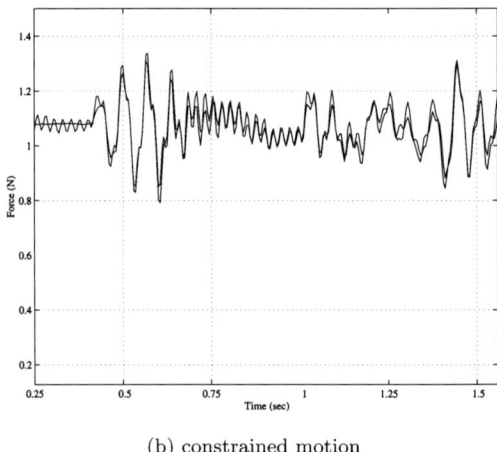

(b) constrained motion

Figure 1: Comparison of measured (solid) and identified model (dashed)

of model uncertainty that can be incorporated in our controller design.

3 Estimation of Modeling error using Stochastic Embedded Approach

Our identified models are merely low order approximations of the flexible link manipulator– a infinite dimensional system in nature. Thus, we need to include some quantification of the implied model imperfections. Frequency domain uncertainty estimates have gained considerable interest in the control community [10]. Unlike the classical identification methods, these new techniques enable the inclusion of both structural model errors (bias) and noise (variance) in the estimated uncertainty bound. However, in order to accomplish this, some prior knowledge of the model error must be available. One method that can be nicely combined with our QFT controller design framework is the stochastic embedding approach which we will briefly discuss. For a detailed presentation, readers are referred to [11] and the references therein.

The modeling error is caused by structural model error due to undermodeling (bias error) and the model error due to noise (variance error). Bias error is considered a deterministic quantity while variance error is of a stochastic nature. Consequently, the two components of the total error are fundamentally different quantities and require fundamentally different treatments. However, in order to make coherent estimation procedures of the total model error, most of the reported results in the literature have relied upon treating the undermodeling and the noise as similar quantities. The stochastic embedding approach assumes that the undermodeling is a realization of a stochastic process with known distribution.

Let the identified model be written as

$$G(q, \hat{\theta}) = \sum_{i=1}^{n} \hat{\theta}_i \Lambda_i(q). \quad (7)$$

There exist a unique value θ_0 such that the estimates $\hat{\theta} \to \hat{\theta}_0$ as the number of data $N \to \infty$.

In stochastic embedding approach, the bias error is treated as a random variable. Therefore, it makes the assumption that the true transfer function $G_T(j\omega T_s)$ is a stochastic process with $\mathrm{E}\{G_T(e^{j\omega T_s})\} = G(e^{j\omega T_s}, \theta_0)$ which can be decomposed as

$$G_T(e^{j\omega T_s}) = G(e^{j\omega T_s}, \theta_0) + G_\Delta(e^{j\omega T_s}). \quad (8)$$

where $G_\Delta(e^{j\omega T_s})$ represents the undermodeling with $\mathrm{E}\{G_\Delta(e^{j\omega T_s})\}=0$. The expectation $E\{\cdot\}$ means averaging over different realizations of the undermodeling [11]. We use an FIR model of order $L \leq N$ for $G_\Delta(e^{j\omega T_s})$:

$$G_\Delta(e^{j\omega T_s}) = \sum_{k=1}^{L} \eta(k) e^{-j\omega k T_s} = \Pi(e^{j\omega T_s})\eta \quad (9)$$

with

$$\Pi(e^{j\omega T_s}) = \begin{bmatrix} e^{-j\omega T_s}, e^{-2j\omega T_s}, \ldots, e^{-Lj\omega T_s} \end{bmatrix}$$
$$\eta = \begin{bmatrix} \eta(1), \eta(2), \ldots, \eta(L) \end{bmatrix}^T$$

Here $\eta(k)$ denotes the undermodeling impulse response. Note that in stochastic embedded approach $\eta(k)$ is treated as random variable. Rewriting (9) in vector form

$$G(e^{j\omega T_s}, \theta) = \Lambda(e^{j\omega T_s})\theta \quad (10)$$

with

$$\Lambda(e^{j\omega T_s}) = \begin{bmatrix} \Lambda_1(e^{j\omega T_s}), \Lambda_2(e^{j\omega T_s}), \ldots, \Lambda_n(e^{j\omega T_s}) \end{bmatrix}$$

From Eqns. (9), (10) and (8) we have:

$$G_T(e^{j\omega T_s}) - G(e^{j\omega T_s}, \hat{\theta}_N) = \Lambda(e^{j\omega T_s})(\theta_0 - \hat{\theta}_N) + \Pi\eta \quad (11)$$

Now, for estimation of the modeling error in frequency domain, we need to compute the variance of the undermodeling, i.e. $\left\{ |G_t(e^{j\omega T_s}) - G(e^{j\omega T_s}, \hat{\theta})|^2 \right\}$. In [11], Goodwin et al. showed that the undermodeling can be exactly quantified provided we obtain the covariances of noise and the undermodeling. This can be done under any assumption on the probability distribution for the undermodeling and the noise using maximum likelihood method. Furthermore, we can obtain confidence ellipses in the complex plane for our frequency response estimate $G(e^{j\omega T_s}, \hat{\theta})$. For the identified models of Section 2.2, we can calculate the uncertainties at each frequency and incorporate these in our controller design. For example, Fig. 2 displays (on a Nichols plot) the results of uncertainty estimation for the identified model of the flexible link manipulator in contact. The uncertainty is largest at higher frequency because of unmodeled high frequency dynamics.

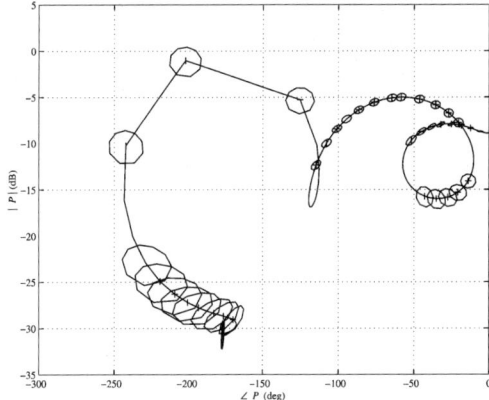

Figure 2: Nichols plot of the identified model of a flexible link manipulator contact including 90% confidence ellipses

4 Controller Design

In this Section we design the single non-switching robust controller for single flexible link manipulator for free and constrained motion.

For all the family of the transfer functions that we identified in Section 2.2, our objective is to design a strictly proper controller, $C(s)$, such that the following conditions are satisfied.

- Closed-loop robust stability: The condition for robust stability is given by

$$\left|\frac{C(j\omega)P(j\omega)}{1+C(j\omega)P(j\omega)}\right| \leq 1.4, \quad \forall \omega \in [0,\infty) \quad (12)$$

which implies ≈ 5.3dB gain margin and $\approx 50^o$ phase margin.

- plant output disturbance attenuation (*sensitivity*): Plant output is bounded by

$$\left|\frac{1}{1+C(j\omega)P(j\omega)}\right| \leq 0.9, \quad \forall \omega \leq 140) \quad (13)$$

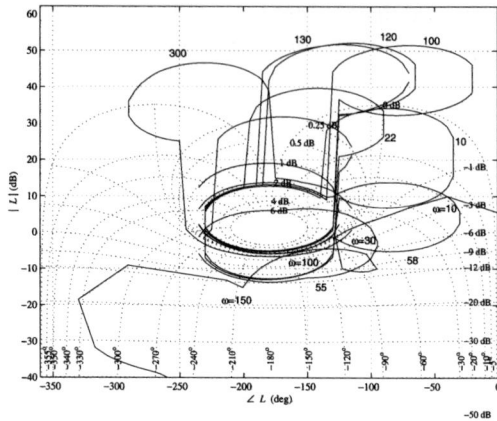

Figure 3: Nominal loop and QFT bounds on Nichols chart

Note that the allowable nominal loop will be limited by the above constraints. Further, we have not attempted to define tracking bounds, because we found that it was very difficult to define some reasonable tracking bounds that is satisfied in both free and constrained cases. The above design specifications can be translated into constraints on the nominal open-loop transfer function, $L_0(j\omega) = P_0(j\omega)C(j\omega)$ where $P_0(j\omega)$ is the nominal plant frequency response. These constraints are called QFT bounds and are usually shown on the Nichols chart [12]. We used the two identified models including the uncertainty estimates to compute the QFT bounds for the constraints (12) and (13). They are shown in Fig. 3 for selected frequencies. We use the identified model in constrained case as the nominal plant model, P_0, for the loop-shaping. The final loop-shaping of the system without violating the bounds is also shown in Fig. 3, where the following simple controller is used.

$$C(s) = \frac{3.3(\frac{s}{1.73}+1)}{\frac{s^2}{7.3^2}+\frac{1.6}{7.3}+1} \quad (14)$$

5 Experimental results

For the experimental results of this section, the robot was commanded to follow a trajectory in free space. An object is placed on this path so that the tip of the flexible link robot will contact the object. The collision occurs at approximately 2 seconds into the trajectory and the impact velocity is ≈ 0.3 m/sec. We do not use any prior knowledge about the object's position. This means that the manipulator does not make any attempt to slow down before hitting the object. In free motion, the feedback signal is the the tip position which is defined by: $y(t) = \theta(t) + w(x,t)$, where $\theta(t)$ is the motor angle and $w(x,t)$ is the deflection of the beam from the rest. Both arm deflection and contact force are measured using strain gauges without any additional force sensor. Once, the manipulator hits the environment, the collision is detected from a sudden change of strain gauges signal, and the feedback signal is switched form the measured position to the measured force. Also the desired trajectory is changed to maintain a contact of 1N force. However, the controller remains the same. Fig. 4 shows the position trajectory and the manipulators tip position during the experiment. The force response is shown in Fig. 5. Note that changes in force reading before the collision ($t < 2$ sec) is due to vibration induced changes in strain gauge signal.

6 Conclusions

In this paper, we proposed a coherent robust identification and controller design for a flexible link manipulator in collision. It is shown that system identification using orthonormal basis function is a powerful method for obtaining a good system model for controller design purposes. We used stochastic embedded approach to obtain uncertainty estimates for our iden-

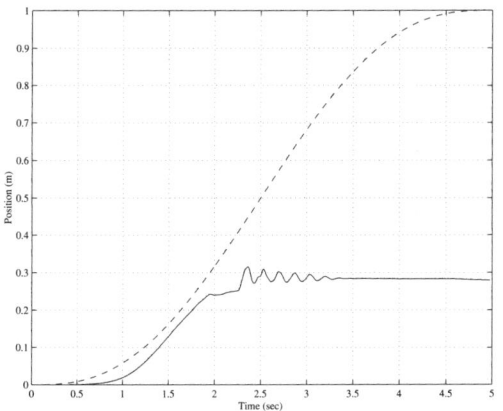

Figure 4: Desired (dashed) and actual (solid) position trajectory

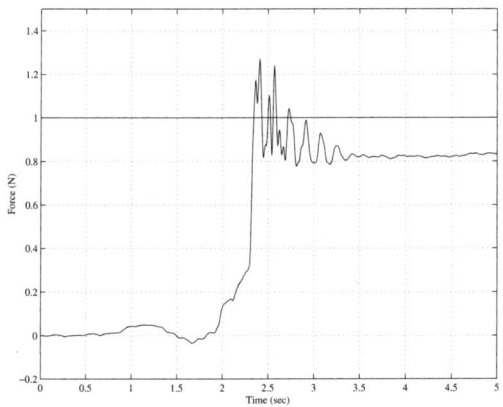

Figure 5: System Force response

tified models which can be nicely incorporated into our controller design. The QFT approach was then applied to design a robust controller that simultaneously stabilizes both position and force control of the single flexible link manipulator. The experimental results show that the proposed controller was able to stabilize the system in free motion, constrained motion and during the transition. The design methodology presented in this paper should be applicable to other control systems.

References

[1] D. Wang and M. Vidyasagar, "Transfer functions for single flexible link", *International Journal of Robotics Research*, vol. 10, no. 5, pp. 540–549, 1991.

[2] F. Matsuno, R. Asano, and Y. Sakawa, "Modeling and quasi-static hybrid position/force control of constrained planar two-link flexible manipulators", *IEEE Transactions on Robotics and Automation*, vol. 10, no. 3, pp. 287–297, 1994.

[3] F. Matsuno and S. Kasai, "Modeling and robust force control of constrained one-link flexible arms", *Journal of Robotic Systems*, vol. 15, no. 8, pp. 447–464, 1998.

[4] D. M. Rovner and R. H. Cannon, "Experiments towards on-line identification and control of a very flexible one-link manipulator", *The International Journal of Robotics Research*, vol. 6, no. 4, pp. 3–14, 1987.

[5] A. P. Tzes and S. Yurkovich, "Application and comparison of on-line identification methods for flexible manipulator control", *The International Journal of Robotics Research*, vol. 10, no. 5, pp. 515–527, 1991.

[6] B. Ninness and F. Gustafsson, "A unifying construction of orthonormal bases for system identification", *IEEE Transaction on Automatic Control*, vol. 42, no. 4, pp. 515–521, 1997.

[7] B. Wahlberg and P. M. Makila, "On approximation of stable linear dynamical systems using Laguerre and Kautz functions", *Automatica*, vol. 32, no. 5, pp. 693–708, 1996.

[8] K. Ziaei and D. W. L. Wang, "Application of orthonormal basis functions for system identification of a flexible-link manipulator", *Control Engineering Practice*, 2001.

[9] Moorehead S.J. and D.W.L. Wang, "Collision detection using a flexible link manipulator: A feasibility study", in *Proc. IEEE international Conference on Robotics and Automation*, 1996, pp. 804–809.

[10] J. Baillieul, "Editor", *IEEE Trans. Automatic Control, Special Issue on System Identification for Robust Control*, vol. 37, no. 7, 1992.

[11] G. C. Goodwin, M. Gevers, and B. Ninness, "Quantifying the error in estimated transfer functions with application to model order selection", *IEEE Trans. Automatic Control*, vol. 37, no. 7, pp. 953–963, 1992.

[12] O. Yaniv, *Quantitative feedback design of linear and nonlinear control systems*, Kluwer Academic Publishers, 1999, TJ 216 Y36 1999.

Estimation of the Flexural States of a Macro-Micro Manipulator Using Acceleration Data

K. Parsa, J. Angeles, and A. K. Misra
Department of Mechanical Engineering &
Centre for Intelligent Machines
McGill University, Montreal, Quebec, Canada
kourosh.parsa@mail.mcgill.ca, angeles@cim.mcgill.ca, arun.misra@mcgill.ca

Abstract—The subject of this paper is a state-estimation algorithm which uses the twist data—velocity and angular velocity—of the base of a micro-manipulator, placed on the end-effector of its macro counterpart, to estimate the flexural states of the flexible links of the macro-manipulator. The twist data are inferred from the acceleration signals delivered by an accelerometer array—a kinematically redundant array of tri-axial accelerometers. The array signals can also be utilized to calculate the translational and angular accelerations of the micro-manipulator base, which are in turn used to obtain a set of dynamics equations for the macro-manipulator, thus reducing the order of the dynamics model. Next, the dynamics equations of the macro-manipulator and the state-output relations are linearized, the latter in closed form so as to lower the computational cost in a control loop. The relations thus obtained are then used in an extended Kalman filter to estimate the flexural states of the system.

Keywords— Flexible manipulator, flexural coordinates, linearized kinematics, macro-micro manipulator, state estimation, twist vector.

I. Introduction

The development of long-boom, flexible manipulators carrying a smaller, rigid-link manipulator has motivated the ongoing research work reported here. A paradigm of these systems, often referred to as macro-micro manipulators, is Canadarm2, mounted on the International Space Station. Canadarm2 is poised to carry, at the end of its fully extended length of 17.5 m, a dual-arm manipulator with a total of 15 shorter links, called the Special-Purpose Dextrous Manipulator. The paper focuses on the development of state-estimators for this class of systems.

In this paper *flexural generalized coordinates*, or *flexural coordinates* in short, are understood as a set of real variables that can describe the deformed shape of a structurally flexible link or system in a *discretized* sense. These coordinates can be, for example, the nodal displacements in a finite-element mesh, the end-point displacements of the flexible links, or the generalized coordinates used in the assumed-modes method. These generalized coordinates along with their time-rates of change, *flexural generalized velocities*, constitute the *flexural states*.

Since the very early experiments on flexible manipulators, the problem of determining the flexural states has attracted many researchers' attention. In [1], an optical sensor was used to determine the end-link displacement of a single-link flexible manipulator, which was then used in a state-feedback control loop. The optimal control of a two-link, three-degree-of-freedom flexible manipulator was reported in [2], where the authors suggested the use of three-axis force and torque sensors at the proximal end of each link to estimate the flexural states using an observer. Other researchers have reported the use of strain gauges for the same purpose, among many others [3, 4, 5].

Another type of the sensor used for the estimation of flexural states is the accelerometer. In [6], the authors discuss a sliding observer for a system kinematically similar to that considered in [2]. The instrumentation system included two single-axis accelerometers attached at the end of each link. It appears, however, that the authors neglected the orientation change of the accelerometers, which can cause errors in the estimation results.

Now, let us assume that an *accelerometer array*—a kinematically redundant array of tri-axial accelerometers, a prototype of which is shown in Fig. 1—is attached to the base of the micro-manipulator of a macro-micro structure, and that the array data are processed to obtain the twist—velocity and angular velocity—of the base. In this paper, we intend to report on an algorithm which utilizes the accelerometer readouts and the twist data of the base to estimate the flexural states of the macro manipulator. The observer that we will use to estimate the flexural states from the twist data is an extended Kalman filter (EKF).

The structure of the observer proposed here is explained in Section II. Then, in Section III, we review some basic kinematics relations. The dynamics modeling of the system is the subject of Section IV. The discrete state-space dynamics equations are obtained in Section V. Next, the linear relations are derived in Section VI. The EKF relations are recalled in Section VII. Some simulation results are reported in Section VIII. The observability issue is discussed in Section IX, the paper then concluding in Section X.

All the vectors and cross-product matrices used here are expressed in their corresponding local frames, *overdots* representing *element-wise* time-derivatives.

II. STATE-ESTIMATOR STRUCTURE

State estimation using an observer requires both a set of dynamics equations, expressing the *modeled dynamics* of the state variations, and a set of measurement equations, otherwise known as state-output relations. The latter are algebraic relations in the states, the former, in continuous-time domain, comprising—usually—ordinary differential equations. Since the dynamics equations of a multilink manipulator are quite complicated, it would seem promising if the manipulator kinematic relations could be used as the modeled dynamics used in the observer, for the kinematics relations are far less involved algebraically. However, a kinematics-based observer has a major drawback: The kinematic relations, which relate the evolution of the states to the pose and twist of one or more bodies in the kinematic chain, are of a pure-integrator nature, thus exhibiting an inherently unstable behavior.

The dynamics equations of the entire macro-micro system, when used as the modeled dynamics for the observer of a macro manipulator, are to be integrated online at each sampling time. This is especially expensive computationally because of the usually twice-as-large dimension of the system of differential equations for the entire macro-micro system. For this very reason, we suggest using the macro-manipulator dynamics equations alone as the modeled dynamics. However, to be able to estimate the macro states, one needs to have the reaction wrench—force and moment—applied on the macro-manipulator end-effector by the base of the micro. This wrench can be determined if the dynamics equations of the micro-manipulator, as a floating system, and the acceleration of its base are available. This approach is explained in full detail in Section IV.

III. KINEMATIC RELATIONS

A. The Twist Vector

The twist vector of a body is understood here as a set of scalar variables that comprise the necessary and sufficient amount of information to determine the velocity field in the body. For a rigid body, the three components of the body angular velocity and those of the velocity vector of a landmark point of the body can be these variables.

We define below two different twist vectors for a rigid body: (*i*) Vector $\mathbf{t}$, briefly called the *twist*; and (*ii*) vector $\bar{\mathbf{t}}$, termed the *Cartesian twist*.

$$\mathbf{t} \triangleq \begin{bmatrix} \dot{\mathbf{p}} \\ \boldsymbol{\omega} \end{bmatrix}, \quad \bar{\mathbf{t}} \triangleq \begin{bmatrix} \mathbf{v} \\ \boldsymbol{\omega} \end{bmatrix}, \quad (1)$$

where $\mathbf{p}$, $\mathbf{v}$, and $\boldsymbol{\omega}$ are the position vector of the local-frame origin, its absolute velocity, and the frame angular velocity, respectively, all expressed in the local frame. Furthermore, according to the convention set forth in Section I, $\dot{\mathbf{p}}$ is the

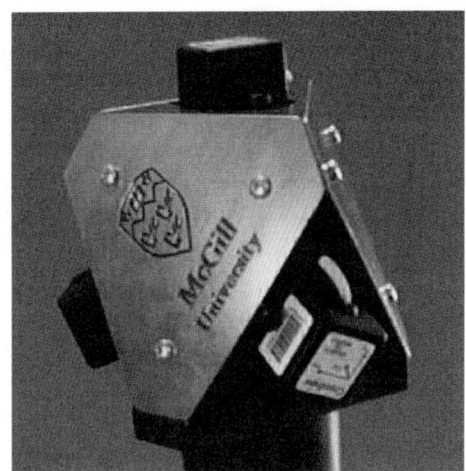

Fig. 1. An accelerometer array

element-wise time derivative of vector $\mathbf{p}$, and hence, it is, in general, different from $\mathbf{v}$.

From rigid-body kinematics, the relation between the twist and the Cartesian twist is

$$\bar{\mathbf{t}} = \begin{bmatrix} \dot{\mathbf{p}} + \boldsymbol{\omega} \times \mathbf{p} \\ \boldsymbol{\omega} \end{bmatrix} = \begin{bmatrix} \mathbf{1} & -\mathbf{P} \\ \mathbf{O} & \mathbf{1} \end{bmatrix} \mathbf{t} \quad (2)$$

in which $\mathbf{P} \triangleq \mathrm{CPM}(\mathbf{p})$ is the *cross-product matrix*[1], of vector $\mathbf{P}$, while $\mathbf{1}$ and $\mathbf{O}$ are the 3×3 identity and zero matrices, respectively.

Hence, for the micro-manipulator base, for example, the twist takes on the form

$$\mathbf{t}_{m0} \triangleq \begin{bmatrix} \dot{\mathbf{p}}_{m0}^T & \boldsymbol{\omega}_{m0}^T \end{bmatrix}^T \quad (3)$$

where $\mathbf{p}_{m0}$ and $\boldsymbol{\omega}_{m0}$ are the micro-manipulator base position and angular-velocity vectors, respectively. Consequently, we have

$$\bar{\mathbf{t}}_{m0} = \mathbf{R}\,\mathbf{t}_{m0}, \quad \mathbf{R} \triangleq \begin{bmatrix} \mathbf{1} & -\mathbf{P}_{m0} \\ \mathbf{O} & \mathbf{1} \end{bmatrix}, \quad (4)$$

in which $\bar{\mathbf{t}}_{m0}$ represents the Cartesian twist of the micro-manipulator base, and $\mathbf{P}_{m0} = \mathrm{CPM}(\mathbf{p}_{m0})$.

B. The Jacobian Matrix

In robot kinematics, the Jacobian $\mathbf{J}$ can be defined as the partial derivative of the twist vector with respect to the vector of generalized velocities, i.e.,

$$\mathbf{J} \triangleq \frac{\partial \mathbf{t}}{\partial \dot{\mathbf{q}}} \quad (5)$$

---

[1] The cross-product matrix $\mathbf{V}$ of a vector $\mathbf{v}$, not dependent upon $\mathbf{x}$, is the skew-symmetric matrix given by
$$\mathbf{V} \equiv \mathrm{CPM}(\mathbf{v}) \triangleq \frac{\partial (\mathbf{v} \times \mathbf{x})}{\partial \mathbf{x}}, \quad \forall \mathbf{x} \in \mathbb{R}^3 \Longrightarrow \mathbf{V}\mathbf{x} \equiv \mathbf{v} \times \mathbf{x}.$$

where $\mathbf{q}$ represents the vector of generalized coordinates, $\dot{\mathbf{q}}$ being the vector of generalized velocities. Obviously, since the relation between $\mathbf{t}$ and $\dot{\mathbf{q}}$ is linear, the Jacobian is independent of the generalized velocities.

Noticing the definition of the twist vector, one can partition the Jacobian matrix into two blocks, corresponding to the two parts of the twist vector, as

$$\mathbf{J} = \begin{bmatrix} \mathbf{J}_{\dot{p}} \\ \mathbf{J}_{\omega} \end{bmatrix}; \quad \mathbf{J}_{\dot{p}} \triangleq \frac{\partial \dot{\mathbf{p}}}{\partial \dot{\mathbf{q}}}, \quad \mathbf{J}_{\omega} \triangleq \frac{\partial \boldsymbol{\omega}}{\partial \dot{\mathbf{q}}}. \quad (6)$$

C. The Partial Derivatives of the Twist

We start by recalling a basic result.

Theorem 1: The partial derivative of the angular velocity $\boldsymbol{\omega}$ of a rigid body in a serial kinematic chain with respect to the chain generalized-coordinate vector $\mathbf{q}$ can be expressed in terms of the Jacobian $\mathbf{J}_{\omega}$, its time-rate, and the cross-product matrix $\boldsymbol{\Omega}$ of the body angular velocity as

$$\frac{\partial \boldsymbol{\omega}}{\partial \mathbf{q}} = \dot{\mathbf{J}}_{\omega} + \boldsymbol{\Omega}\,\mathbf{J}_{\omega}. \quad (7)$$

The above theorem is proven in [7], where the theorem is then used to derive closed-form relations for the partial derivatives of the Cartesian twist with respect to the generalized-coordinate and the generalized-velocity vectors.

Based on the above-mentioned closed-form relations, and denoting the vector of generalized coordinates of the macro-manipulator by $\boldsymbol{\psi}_{\mathrm{M}}$, we have

$$\frac{\partial \bar{\mathbf{t}}_{\mathrm{m}0}}{\partial \dot{\boldsymbol{\psi}}_{\mathrm{M}}} = \mathbf{R}\,\mathbf{J}_{\mathrm{M}}, \quad \frac{\partial \bar{\mathbf{t}}_{\mathrm{m}0}}{\partial \boldsymbol{\psi}_{\mathrm{M}}} = \mathbf{R}\,\dot{\mathbf{J}}_{\mathrm{M}} + \overline{\mathbf{R}}\,\mathbf{J}_{\mathrm{M}}, \quad (8)$$

$$\mathbf{J}_{\mathrm{M}} \triangleq \frac{\partial \mathbf{t}_{\mathrm{m}0}}{\partial \dot{\boldsymbol{\psi}}_{\mathrm{M}}}, \quad \overline{\mathbf{R}} \triangleq \begin{bmatrix} \boldsymbol{\Omega}_{\mathrm{m}0} & -\mathbf{P}_{\mathrm{m}0}\,\boldsymbol{\Omega}_{\mathrm{m}0} \\ \mathbf{O} & \boldsymbol{\Omega}_{\mathrm{m}0} \end{bmatrix}, \quad (9)$$

and $\boldsymbol{\Omega}_{\mathrm{m}0}$ is defined as the cross-product matrix of $\boldsymbol{\omega}_{\mathrm{m}0}$.

IV. DYNAMICS EQUATIONS

It has been shown [8] that the macro-manipulator dynamics model can be derived as

$$\mathbf{I}_{\mathrm{M}}\ddot{\boldsymbol{\psi}}_{\mathrm{M}} = \mathbf{w}_{\mathrm{M}}^{\mathrm{s}} + \mathbf{w}_{\mathrm{M}}^{\mathrm{ac}} - \left(\frac{\partial \mathbf{t}_{\mathrm{m}0}}{\partial \dot{\boldsymbol{\psi}}_{\mathrm{M}}}\right)^{T} \mathbf{w}_{\mathrm{m}0}^{\mathrm{ex}} \quad (10)$$

in which $\mathbf{I}_{\mathrm{M}}$ is the macro-manipulator mass matrix, and $\mathbf{w}_{\mathrm{M}}^{\mathrm{s}}$ represents the *system wrench*, containing all the Coriolis, gravity, centripetal, and flexural generalized forces; $\mathbf{w}_{\mathrm{M}}^{\mathrm{ac}}$ is the vector of actuation generalized forces; and $\mathbf{w}_{\mathrm{m}0}^{\mathrm{ex}}$ is the reaction wrench applied on the micro-manipulator base by the end-link of the macro-manipulator. Furthermore, if $\boldsymbol{\psi}_{\mathrm{M}}$ is chosen to be

$$\boldsymbol{\psi}_{\mathrm{M}} \triangleq \begin{bmatrix} \boldsymbol{\theta}_{\mathrm{M}}^{T} & \boldsymbol{\zeta}^{T} \end{bmatrix}^{T} \quad (11)$$

where $\boldsymbol{\theta}_{\mathrm{M}}$ and $\boldsymbol{\zeta}$ are the vector of joint variables and flexural generalized coordinates of the macro-manipulator, then

$$\mathbf{w}_{\mathrm{M}}^{\mathrm{ac}} = \begin{bmatrix} \boldsymbol{\tau}_{\mathrm{M}}^{T} & \mathbf{0}_{f}^{T} \end{bmatrix}^{T}, \quad (12)$$

$\boldsymbol{\tau}_{\mathrm{M}}$ and $\mathbf{0}_f$ being the vector of joint-actuation torques and the zero vector of dimension f, the number of flexural generalized coordinates, respectively.

As mentioned in Section II, the reaction wrench $\mathbf{w}_{\mathrm{m}0}^{\mathrm{ex}}$ is needed if the dynamics model (10) is to be used as the modeled dynamics. To calculate this wrench, at each instant, the acceleration measurements delivered by the accelerometer array, which is installed on the micro-manipulator base, is collected. Next, these data are processed, as explained in [9], to obtain the twist, the angular acceleration of the base, and the translational acceleration of the origin of its local frame. The information thus obtained is substituted into the dynamics model of the micro-manipulator to obtain the reaction wrench sought. This approach is formulated below.

First, we notice that the micro-manipulator dynamics model can be derived as [8]

$$\mathbf{I}_{\mathrm{m}}\dot{\mathbf{v}}_{\mathrm{m}} = \mathbf{w}_{\mathrm{m}}^{\mathrm{s}} + \mathbf{w}_{\mathrm{m}}^{\mathrm{ac}} + \bar{\mathbf{w}}_{\mathrm{m}0}^{\mathrm{ex}} + \mathbf{w}_{\mathrm{m}}^{\mathrm{ex}} \quad (13)$$

in which $\mathbf{I}_{\mathrm{m}}$ is the mass matrix, and $\mathbf{w}_{\mathrm{m}}^{\mathrm{s}}$, $\mathbf{w}_{\mathrm{m}}^{\mathrm{ac}}$, and $\mathbf{w}_{\mathrm{m}}^{\mathrm{ex}}$ represent the system, actuation, and end-effector external wrenches of the micro-manipulator, respectively. Moreover, the vectors $\mathbf{v}_{\mathrm{m}}$, $\mathbf{w}_{\mathrm{m}}^{\mathrm{ac}}$, and $\bar{\mathbf{w}}_{\mathrm{m}0}^{\mathrm{ex}}$ are given by

$$\mathbf{v}_{\mathrm{m}} = \begin{bmatrix} \mathbf{t}_{\mathrm{m}0} \\ \dot{\boldsymbol{\theta}}_{\mathrm{m}} \end{bmatrix}, \quad \mathbf{w}_{\mathrm{m}}^{\mathrm{ac}} = \begin{bmatrix} \mathbf{0}_6 \\ \boldsymbol{\tau}_{\mathrm{m}} \end{bmatrix}, \quad \bar{\mathbf{w}}_{\mathrm{m}0}^{\mathrm{ex}} = \begin{bmatrix} \mathbf{w}_{\mathrm{m}0}^{\mathrm{ex}} \\ \mathbf{0}_{n_{\mathrm{m}}} \end{bmatrix}, \quad (14)$$

where $\boldsymbol{\theta}_{\mathrm{m}}$ and $\boldsymbol{\tau}_{\mathrm{m}}$ are the vectors of the micro-manipulator joint variables and joint-actuation torques.

Then, partitioning the mass matrix and all the vectors of the dynamics model (13) corresponding to the two parts of $\mathbf{v}_{\mathrm{m}}$, we can rewrite the dynamics model as

$$\mathbf{M}_{\mathrm{bb}}^{\mathrm{m}}\dot{\mathbf{t}}_{\mathrm{m}0} + \mathbf{M}_{\mathrm{br}}^{\mathrm{m}}\ddot{\boldsymbol{\theta}}_{\mathrm{m}} = \mathbf{w}_{\mathrm{mb}}^{\mathrm{s}} + \mathbf{w}_{\mathrm{mb}}^{\mathrm{ex}} + \mathbf{w}_{\mathrm{m}0}^{\mathrm{ex}}, \quad (15a)$$

$$(\mathbf{M}_{\mathrm{br}}^{\mathrm{m}})^T\dot{\mathbf{t}}_{\mathrm{m}0} + \mathbf{M}_{\mathrm{rr}}^{\mathrm{m}}\ddot{\boldsymbol{\theta}}_{\mathrm{m}} = \mathbf{w}_{\mathrm{mr}}^{\mathrm{s}} + \mathbf{w}_{\mathrm{mr}}^{\mathrm{ex}} + \boldsymbol{\tau}_{\mathrm{m}}, \quad (15b)$$

in which $\mathbf{M}_{\mathrm{bb}}^{\mathrm{m}}$, $\mathbf{M}_{\mathrm{br}}^{\mathrm{m}}$, and $\mathbf{M}_{\mathrm{rr}}^{\mathrm{m}}$ are the blocks of the mass matrix $\mathbf{I}_{\mathrm{m}}$; $\mathbf{w}_{\mathrm{mb}}^{\mathrm{s}}$ and $\mathbf{w}_{\mathrm{br}}^{\mathrm{s}}$ are the sub-arrays of its system wrench; and $\mathbf{w}_{\mathrm{mb}}^{\mathrm{ex}}$ and $\mathbf{w}_{\mathrm{br}}^{\mathrm{ex}}$ are the sub-arrays of the end-effector external wrench, i.e.,

$$\mathbf{I}_{\mathrm{m}} \equiv \begin{bmatrix} \mathbf{M}_{\mathrm{bb}}^{\mathrm{m}} & \mathbf{M}_{\mathrm{br}}^{\mathrm{m}} \\ (\mathbf{M}_{\mathrm{br}}^{\mathrm{m}})^T & \mathbf{M}_{\mathrm{rr}}^{\mathrm{m}} \end{bmatrix}, \quad \mathbf{w}_{\mathrm{m}}^{\mathrm{s}} \equiv \begin{bmatrix} \mathbf{w}_{\mathrm{mb}}^{\mathrm{s}} \\ \mathbf{w}_{\mathrm{mr}}^{\mathrm{s}} \end{bmatrix}, \quad \mathbf{w}_{\mathrm{m}}^{\mathrm{ex}} \equiv \begin{bmatrix} \mathbf{w}_{\mathrm{mb}}^{\mathrm{ex}} \\ \mathbf{w}_{\mathrm{mr}}^{\mathrm{ex}} \end{bmatrix}.$$

Subscripts "b" and "r" above refer to the base and the joint degrees of freedom of the micro-manipulator, respectively.

Next, one can solve (15b) and (15a) for $\mathbf{M}_{\mathrm{rr}}^{\mathrm{m}}\ddot{\boldsymbol{\theta}}_{\mathrm{m}}$ and $\mathbf{w}_{\mathrm{m}0}^{\mathrm{ex}}$, respectively, to obtain

$$\mathbf{M}_{\mathrm{rr}}^{\mathrm{m}}\ddot{\boldsymbol{\theta}}_{\mathrm{m}} = -(\mathbf{M}_{\mathrm{br}}^{\mathrm{m}})^T\dot{\mathbf{t}}_{\mathrm{m}0} + \mathbf{w}_{\mathrm{mr}}^{\mathrm{s}} + \mathbf{w}_{\mathrm{mr}}^{\mathrm{ex}} + \boldsymbol{\tau}_{\mathrm{m}}, \quad (16a)$$

$$\mathbf{w}_{\mathrm{m}0}^{\mathrm{ex}} = \mathbf{M}_{\mathrm{bb}}^{\mathrm{m}}\dot{\mathbf{t}}_{\mathrm{m}0} + \mathbf{M}_{\mathrm{br}}^{\mathrm{m}}\ddot{\boldsymbol{\theta}}_{\mathrm{m}} - \mathbf{w}_{\mathrm{mb}}^{\mathrm{s}} - \mathbf{w}_{\mathrm{mb}}^{\mathrm{ex}}. \quad (16b)$$

Finally, if the translational and angular accelerations of the micro-manipulator base as well as its twist are determined using the accelerometer array, one can readily obtain $\dot{\mathbf{t}}_{\mathrm{m}0}$ and the other information needed for the right-hand sides of (16). Thereafter, solving the above linear equations sequentially is quite simple.

V. State-Space Dynamics Model

Let us denote the state vector of the macro-manipulator by $\mathbf{x}$, which is given by

$$\mathbf{x} \triangleq \begin{bmatrix} \mathbf{x}^1 \\ \mathbf{x}^2 \end{bmatrix}, \quad \mathbf{x}^1 \triangleq \boldsymbol{\psi}_M \equiv \begin{bmatrix} \boldsymbol{\theta}_M \\ \boldsymbol{\zeta} \end{bmatrix}, \quad \mathbf{x}^2 \triangleq \dot{\boldsymbol{\psi}}_M \equiv \begin{bmatrix} \dot{\boldsymbol{\theta}}_M \\ \dot{\boldsymbol{\zeta}} \end{bmatrix}. \quad (17)$$

Using the above definitions of the states, one can write the mathematical model of the macro-manipulator, (10), in state-space form:

$$\dot{\mathbf{x}}^1 = \mathbf{x}^2, \quad (18a)$$
$$\dot{\mathbf{x}}^2 = \mathbf{f}(\mathbf{x}) + \mathbf{B}(\mathbf{x})\,\mathbf{u}(t), \quad (18b)$$

with $\mathbf{f}(\mathbf{x})$, $\mathbf{B}(\mathbf{x})$, and $\mathbf{u}(t)$ defined below:

$$\mathbf{f}(\mathbf{x}) \triangleq \mathbf{H}(\mathbf{x}^1)\,\mathbf{w}_M^s(\mathbf{x}), \quad (19)$$
$$\mathbf{B}(\mathbf{x}) \triangleq \begin{bmatrix} \mathbf{H}_1 & -\mathbf{H}\mathbf{J}_M^T \end{bmatrix}, \quad (20)$$
$$\mathbf{u}(t) \triangleq \begin{bmatrix} \boldsymbol{\tau}_M \\ \mathbf{w}_{m0}^{ex} \end{bmatrix}, \quad (21)$$

in which matrix $\mathbf{H}$, defined as the inverse of the mass matrix $\mathbf{I}_M$, is divided into two blocks:

$$\mathbf{H} \equiv \begin{bmatrix} \mathbf{H}_1 & \mathbf{H}_2 \end{bmatrix} \triangleq \mathbf{I}_M^{-1}. \quad (22)$$

The first block, $\mathbf{H}_1$, is an $N_M \times n_M$ matrix, with N_M and n_M being the degree of freedom of the macro-manipulator and its number of links, respectively.

Hence, the discrete-time linearized form of (18), obtained by applying Euler's scheme on the linearized equations, can be written as

$$\mathbf{x}_{k+1}^1 = \mathbf{x}_k^1 + h\mathbf{x}_k^2 + \mathbf{m}_k^1, \quad (23a)$$
$$\mathbf{x}_{k+1}^2 = \mathbf{x}_k^2 + h\left.\frac{\partial \mathbf{f}}{\partial \mathbf{x}}\right|_{\hat{\mathbf{x}}_k}\mathbf{x}_k + \bar{\mathbf{u}}_k + \mathbf{m}_k^2, \quad (23b)$$

in which h is the sampling period, $\mathbf{m}_k^1$ and $\mathbf{m}_k^2$ are the sub-arrays of the vector $\mathbf{m}_k$ of the uncorrelated white-noise processes representing the unmodeled dynamics, and $\bar{\mathbf{u}}_k$ is a redefined input function given by

$$\bar{\mathbf{u}}_k \triangleq h\left[\mathbf{f}(\hat{\mathbf{x}}_k) - \left.\frac{\partial \mathbf{f}}{\partial \mathbf{x}}\right|_{\hat{\mathbf{x}}_k}\hat{\mathbf{x}}_k + \mathbf{B}(\hat{\mathbf{x}}_k)\,\mathbf{u}_k\right]. \quad (24)$$

Thus, in standard form, we have

$$\mathbf{x}_{k+1} = \mathbf{A}_k\,\mathbf{x}_k + \tilde{\mathbf{u}}_k + \mathbf{m}_k, \quad (25)$$
$$\mathbf{A}_k \triangleq \begin{bmatrix} \mathbf{1} & h\mathbf{1} \\ h(\partial \mathbf{f}/\partial \mathbf{x}^1) & \mathbf{1} + h(\partial \mathbf{f}/\partial \mathbf{x}^2) \end{bmatrix}_{\mathbf{x}=\hat{\mathbf{x}}_k}, \quad (26)$$
$$\tilde{\mathbf{u}}_k \triangleq \begin{bmatrix} \mathbf{0}_{N_M}^T & \bar{\mathbf{u}}_k^T \end{bmatrix}^T. \quad (27)$$

To complete the derivation of the state-space dynamics model, we obtain the partial derivatives of $\mathbf{f}$ with respect to $\mathbf{x}^1$ and $\mathbf{x}^2$, needed in (24 & 26), as

$$\frac{\partial \mathbf{f}}{\partial \mathbf{x}^1} = \mathbf{H}\left(\frac{\partial \mathbf{w}_M^s}{\partial \boldsymbol{\psi}_M} - \left.\frac{\partial (\mathbf{I}_M \mathbf{c})}{\partial \boldsymbol{\psi}_M}\right|_{\mathbf{c}=\mathbf{H}\mathbf{w}_M^s}\right), \quad (28a)$$

$$\frac{\partial \mathbf{f}}{\partial \mathbf{x}^2} = \mathbf{H}\frac{\partial \mathbf{w}_M^s}{\partial \dot{\boldsymbol{\psi}}_M}, \quad (28b)$$

where vector $\mathbf{c}$ is a dummy variable. To derive the partial derivative of $\mathbf{f}$ with respect to $\mathbf{x}^1$, we have used the well-known formula for the partial derivative of the inverse of a matrix.

VI. State-Output Relations

To obtain the state-output relations for the system, we denote the output vector by $\mathbf{y}$ and define it as

$$\mathbf{y} \triangleq \begin{bmatrix} \mathbf{y}^1 \\ \mathbf{y}^2 \end{bmatrix}, \quad \mathbf{y}^1 \triangleq \begin{bmatrix} \boldsymbol{\theta}_M \\ \dot{\boldsymbol{\theta}}_M \end{bmatrix}, \quad \mathbf{y}^2 \triangleq \bar{\mathbf{t}}_{m0} \quad (29)$$

where $\bar{\mathbf{t}}_{m0}$, as mentioned before, is the Cartesian twist of the micro-manipulator base. While the relation of $\mathbf{y}^1$ with the states of the system is evidently linear and readily known, for the entries of $\mathbf{y}^1$ are among the system states themselves, the state-output relations pertaining to $\mathbf{y}^2$ are indeed nonlinear and thus should be linearized. To this end, $\mathbf{y}^2$ is expanded up to first-order terms, namely,

$$\mathbf{y}^2(\mathbf{x}) \approx \mathbf{y}^2(\hat{\mathbf{x}}) + \left.\frac{\partial \mathbf{y}^2}{\partial \mathbf{x}}\right|_{\hat{\mathbf{x}}}(\mathbf{x} - \hat{\mathbf{x}}) \quad (30)$$

where $\hat{\mathbf{x}}$ represents the estimated state vector, and, apparently,

$$\frac{\partial \mathbf{y}^2}{\partial \mathbf{x}} \equiv \begin{bmatrix} \frac{\partial \mathbf{y}^2}{\partial \mathbf{x}^1} & \frac{\partial \mathbf{y}^2}{\partial \mathbf{x}^2} \end{bmatrix} = \begin{bmatrix} \frac{\partial \bar{\mathbf{t}}_{m0}}{\partial \boldsymbol{\psi}_M} & \frac{\partial \bar{\mathbf{t}}_{m0}}{\partial \dot{\boldsymbol{\psi}}_M} \end{bmatrix}. \quad (31)$$

The partial derivatives needed in (31) are available in closed form from (8).

Consequently, upon redefining the output vector, the state-output relation can be rewritten as

$$\bar{\mathbf{y}} = \mathbf{C}\,\mathbf{x} + \mathbf{n}(t), \quad (32)$$
$$\mathbf{C} \equiv \mathbf{C}(\hat{\mathbf{x}}) \triangleq \begin{bmatrix} \mathbf{E}_{n_M \times N_M} & \mathbf{O}_{n_M \times N_M} \\ \mathbf{R}\dot{\mathbf{J}}_M + \bar{\mathbf{R}}\mathbf{J}_M & \mathbf{R}\mathbf{J}_M \end{bmatrix}, \quad (33)$$

in which $\mathbf{n}(t)$ is the vector of the uncorrelated, white measurement-noise processes, and $\mathbf{E}_{n_M \times N_M}$ is a rectangular array of ones and zeros in which the largest left-hand-side block is the $n_M \times n_M$ identity matrix. The new output vector, $\bar{\mathbf{y}}$, is defined as

$$\bar{\mathbf{y}} \triangleq \mathbf{y}(\mathbf{x}) - \mathbf{y}(\hat{\mathbf{x}}) + \left.\frac{\partial \mathbf{y}}{\partial \mathbf{x}}\right|_{\hat{\mathbf{x}}}\hat{\mathbf{x}}. \quad (34)$$

Hence, the time-discretized form of the state-output relation, (32), can readily be written as

$$\bar{\mathbf{y}}_k = \mathbf{C}_k\,\mathbf{x}_k + \mathbf{n}_k. \quad (35)$$

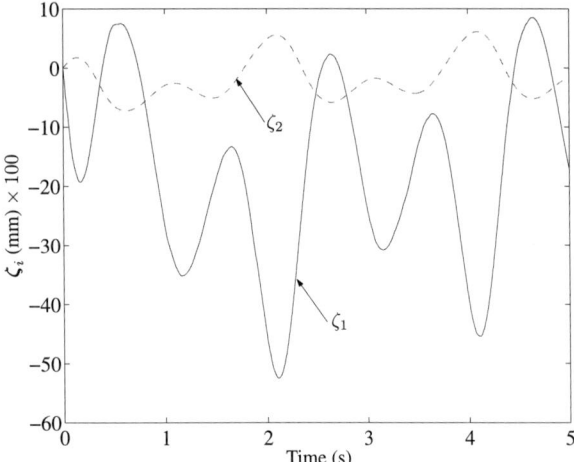

Fig. 2. The actual values of the flexural generalized coordinates

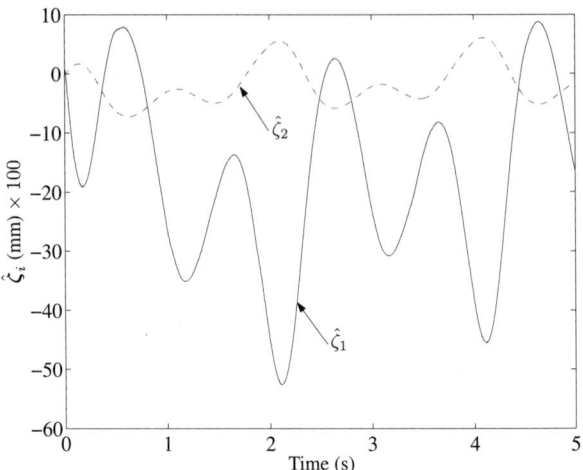

Fig. 3. The estimated values of the flexural generalized coordinates

VII. THE EXTENDED KALMAN FILTER

With the linearized governing equations available, one can use the EKF relations to obtain the state estimates. To this end, we resort to the relations given below, as derived in [10].

$$\mathbf{P}_{0,0} = \mathrm{Var}(\mathbf{x}_0), \qquad \hat{\mathbf{x}}_0 = \mathrm{E}(\mathbf{x}_0), \tag{36a}$$

$$\mathbf{P}_{k,k-1} = \mathbf{A}_{k-1}\mathbf{P}_{k-1,k-1}\mathbf{A}_{k-1}^T + \mathbf{\Gamma}_{k-1}\mathbf{Q}_{k-1}\mathbf{\Gamma}_{k-1}^T, \tag{36b}$$

$$\hat{\mathbf{x}}_{k|k-1}^1 = \hat{\mathbf{x}}_{k-1}^1 + h\,\hat{\mathbf{x}}_{k-1}^2, \tag{36c}$$

$$\hat{\mathbf{x}}_{k|k-1}^2 = \hat{\mathbf{x}}_{k-1}^2 + h\left[\mathbf{f}(\hat{\mathbf{x}}_{k-1}) + \mathbf{B}(\hat{\mathbf{x}}_{k-1})\,\mathbf{u}_{k-1}\right], \tag{36d}$$

$$\mathbf{G}_k = \mathbf{P}_{k,k-1}\mathbf{C}_k^T(\mathbf{C}_k\mathbf{P}_{k,k-1}\mathbf{C}_k^T + \mathbf{R}_k)^{-1}, \tag{36e}$$

$$\mathbf{P}_{k,k} = (\mathbf{1} - \mathbf{G}_k\mathbf{C}_k)\mathbf{P}_{k,k-1}, \tag{36f}$$

$$\hat{\mathbf{x}}_k = \hat{\mathbf{x}}_{k|k-1} + \mathbf{G}_k(\mathbf{y}_k - \hat{\mathbf{y}}_{k|k-1}), \tag{36g}$$

where $\mathbf{Q}_{k-1}$ and $\mathbf{R}_k$ are the covariance matrices of the uncorrelated white-noise processes $\mathbf{m}_{k-1}$ and $\mathbf{n}_k$, respectively.

VIII. SIMULATION RESULTS

The ideas explained here regarding the estimation of the flexural states using an accelerometer array are demonstrated by simulating the flexural motion of a planar $RRRR$ manipulator. The first two links of this manipulator—constituting the macro-manipulator—are assumed flexible, while the other two—making up the micro-manipulator—are assumed rigid. Moreover, we discretize the flexible links using the assumed-modes method, taking the clamped-free eigenfunctions as the shape-functions; the bending of each of the macro-manipulator links is described by one flexural generalized coordinate. The specifications of the links are given in Table I.

To excite flexural motion in the macro-manipulator, we assume that the joints 1, 2, and 4 are locked at $\theta_1 = 0$,

	#	Mass (kg)	Length (m)	EI (Nm^2)
Macro	1	1.0	1.0	93.266
	2	0.50	0.50	93.266
Micro	1	0.20	0.25	–
	2	0.20	0.25	–

TABLE I

THE LINK SPECIFICATIONS OF THE $RRRR$ MANIPULATOR

$\theta_2 = \pi/12$ (rad), and $\theta_4 = \pi/2$ (rad), while the motion of the third link is given by

$$\theta_3 = \left[\vartheta_1 \cos(\omega t) + \vartheta_2 \cos(2\omega t)\right] u(t) \tag{37}$$

where $\vartheta_1 = \pi/8$ (rad), $\vartheta_2 = \pi/12$ (rad), $\omega = \pi$ (rad/s), and $u(t)$ is the unit step function.

Applying our, we can estimate both the flexural states and the joint-rates, but we mainly focus here on the former. Figures 2 and 4 show the actual values of the flexural generalized coordinates and generalized velocities, respectively; the estimation results are displayed in Figs. 3 and 5. The states with a subscript 1 pertain to the first link, while those having a subscript 2 pertain to the second link. As seen from the graphs, the estimator has been able to estimate the flexural states with a high accuracy.

IX. OBSERVABILITY OF THE FLEXURAL MOTION

A complete observability analysis for the type of systems treated in this paper seems quite elusive, due to the nonlinear, highly complex nature of both dynamics equations and state-output relations. Nonetheless, to help us visualize how many flexural generalized coordinates one can estimate using one accelerometer array, consider a cantilever flexible link. If an accelerometer array is installed at the tip of such a link, the three components of the translational acceleration and those of the angular acceleration of the array, totalling six, can be used to estimate six flexural

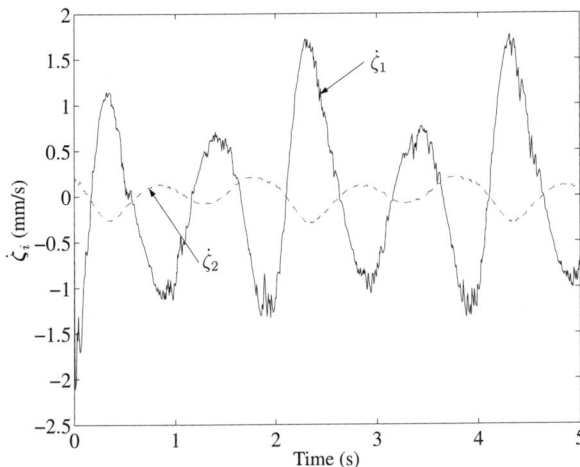

Fig. 4. The actual values of the flexural generalized velocities

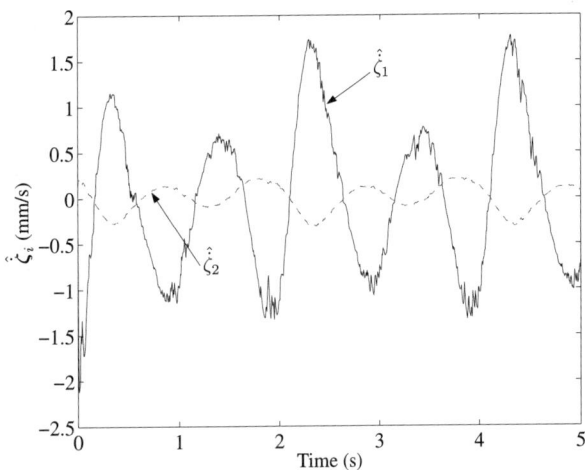

Fig. 5. The estimated values of the flexural generalized velocities

coordinates: Four flexural generalized coordinates describing bendings in two planes plus two flexural coordinates in each plane; one of the latter represents the torsional twist of the link, the other the elongation of the link axis. Consequently, if, in a particular case, more than two flexural coordinates are needed to effectively describe the bending of a link in one plane, then additional accelerometer arrays must be installed on the link along its length.

X. Conclusions

The estimation of the flexural states, the flexural generalized coordinates and generalized velocities, of the flexible links of a macro-micro manipulator was the subject of this paper. It was proposed that the data collected by an accelerometer array fixed to the micro-manipulator base be used to infer the translational and angular accelerations of the base as well as its translational and angular velocities. The acceleration data were used to separate the dynamics of the macro-manipulator from that of its micro counterpart. The velocity data, on the other hand, were taken as the outputs, and their relations to the flexural states were written and linearized in closed form. Then, having all the state-space equations in linearized form, we used them in an extended Kalman filter (EKF) to estimate the flexural states. Simulation results reported in this paper demonstrate that the proposed estimation algorithm can indeed produce reliable estimates of the states. Although we have made use of an EKF to estimate the states, the state-space equations derived here can be used in any other type of observer as well, especially since most of the observers devised for nonlinear systems are based upon linear state-output relations.

Acknowledgements

This research work was supported by the Natural Sciences and Engineering Research Council of Canada, under Research Grants OGP0004532 and OGP0000967.

References

[1] R. H. J. Cannon and E. Schmitz, "Initial experiments on the end-point control of a flexible one-link robot," *The Int. J. of Robotics Research*, vol. 3, pp. 62–75, Fall 1984.

[2] T. Fukuda and A. Arakawa, "Optimal control and sensitivity analysis of two links flexible arm with three degrees of freedom," in *Proc. 28th IEEE Conf. Decision and Control*, (Tampa, FL), pp. 2101–2106, December 1989.

[3] J. Carusone, K. S. Buchan, and G. M. T. D'Eluterio, "Experiments in end-effector tracking control for structurally flexible space manipulators," *IEEE Trans. Robotics and Automation*, vol. 9, pp. 553–560, October 1993.

[4] M. Moallem, R. V. Patel, and K. Khorasani, "An observer-based inverse dynamics control strategy for flexible multi-link manipulators," in *Proc. IEEE Conf. Decision and Control*, (Kobe, Japan), pp. 4112–4117, December 1996.

[5] A. Konno, M. Uchiyama, Y. Kito, and M. Murakami, "Configuration-dependent vibration controllability of flexible-link manipulators," *The Int. J. Robotics Research*, vol. 16, pp. 567–576, August 1997.

[6] A. S. Zaki and W. H. ElMaraghy, "A robust observer for flexible-link manipulators control," in *Proc. the American Control Conf.*, pp. 3334–3338, June 1995.

[7] K. Parsa, J. Angeles, and A. K. Misra, "Linearized kinematics for state estimation in robotics," in *Advances in Robot Kinematics: Theory and Applications* (J. Lenarčič and F. Thomas, eds.), pp. 39–48, Kluwer Academic Publishers, June 2002.

[8] K. Parsa, *Dynamics, State Estimation, and Control of Manipulators with Rigid and Flexible Sub-systems*. PhD thesis, McGill University, Montreal, Quebec, Canada, To be submitted in September 2002.

[9] K. Parsa, J. Angeles, and A. K. Misra, "Attitude calibration of an accelerometer array," in *Proc. IEEE Int. Conf. Robotics and Automation on CD Rom*, (Washington D.C., USA), May 2002.

[10] C. K. Chui and G. Chen, *Kalman Filtering with Real-Time Applications*. Springer Series in Information Sciences, Springer Verlag, New York, 2nd ed., 1991.

A New Impedance Control Concept for Elastic Joint Robots
- A case of a 1 DOF Robot with Programmable Linear Passive Impedance -

R. Ozawa[*1*2] and H. Kobayashi[*1]

[*1]School of Science and Technology, Meiji University
1-1-1 Higashi-mita, Tama, Kawasaki, Kangawa 214-8571 Japan
[*2]The Japan Society for the Promotion of Science
e-mail: {ryuta, kobayasi}@isc.meiji.ac.jp

Abstract - The purpose of this paper is to propose a new impedance control concept for elastic joint robots with programmable passive impedance devices in the transmission. The concept allows us to use the same index both for free motion and for contact task. We apply it to a one-DOF elastic joint robot and derive an adjustment law for the robot. The numerical simulations show that implementation of the concept can be realized without any dynamic models.

1 Introduction

Impedance control method [6] as well as Hybrid position/force control method [14] is one of the most famous force control methods. Impedance control method is usually executed with stiff robots by computational method. This is very useful in a well-structured environment, but is not efficient in a real environment, since there are the following problems:

Problem 1 Almost all impedance control methods require dynamics of both a controlled robot and an environment. But it is usually too difficult to identify the dynamics precisely.

Problem 2 The desired impedance is not clear. Furthermore, we must prepare different impedance parameters for free motion and for contact task.

Problem 3 Although impedance control is a unified approach to control both free motion task and compliant motion task, there is no unified way to select impedance parameters for both tasks.

For Problem 1, Tsuji *et al.* [17] have identified the impedance of an environment by neural network and Cheah and Wang [4] have done it by iterative learning. But these methods do not tell us how to decide desired impedance parameters.

Problem 2 is deeply related to design of impedance parameters. The design can be classified as follows;

(a) experimental design and

(b) analytical design with a theoretical index, e.g., the maximum power transfer index.

Most researchers have adopted design of (a). For example, Siciliano and Villani [15, p.41] have selected the parameters by taking account of estimated environment stiffness and desired force. Tsuji *et al.* [18] have measured data of human arm impedance to control a prosthetic forearm. But in these case they assume that we can find some optimal parameters independently of robot dynamics.

As examples of (b), Hogan [6] has used two indexes for a 1 DOF robot, i.e., the maximum power transfer index and an index to minimize deviations of the force and the position. Arimoto *et al.* [3] have proposed the maximum power transfer index for robotic fingers with elastic skin and used an adaptive technique. But these methods require precise dynamic models of both a robot and an environment.

Problem 3 is related to Problem 2, because we have to re-select the parameters from one task to another. In addition, we cannot use an index like the maximum power transfer for free motion because we have no load, i. e., no environment which the power is transferred to. As far as the authors know, there is no method to solve the above three problems at the same time.

On the other hand, Kovitz *et al.* [10] have proposed a concept of "*Programmable Passive Impedance (PPI)*" to realize robust and variable impedance by inserting PPI devices into transmission of robots. As a result, the joint elasticity become programmable. Hereafter we call robots with PPI devices in the transmission Programmable Elastic Joint Robot (PEJR). After their work, many researchers have developed similar devices for force control applications of robot arms [7, 9, 11, 12]. They have investigated mechanical properties of their devices in detail, but the usage of these devices in robotics has not matured yet, because of difficulty of designing the control systems. For example, a simple extension of computational techniques

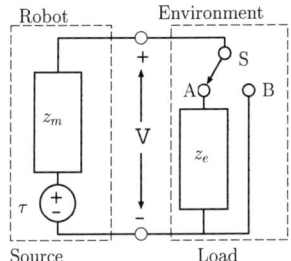

Figure 1: Circuit expression of conventional robots

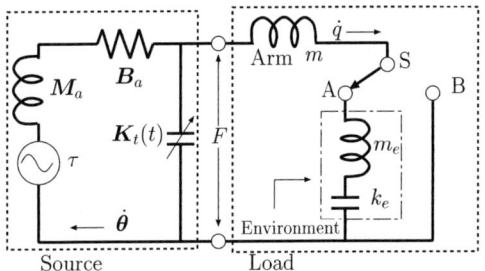

Figure 2: Circuit expression of PEJR

like Computed Torque method requires measurement of the joint acceleration and jerk. In addition, it diminishes innate effects of these PPI devices [13]. In this paper we propose a new impedance control method for a PEJR. We show that the concept allows us to solve the above problems at the same time. We apply this concept to a one DOF PEJR and derive an impedance adjustment law for the robot moving along a sinusoidal trajectory.

Section 2 shows a new circuit expression of a PEJR and define a problem studied in this paper. Section 3 formulates a case study of a 1 DOF PEJR, derives an impedance adjustment law of the compliance devices, and shows some simulations. We conclude in Sec. 4 and describe about future works in Sec. 5.

2 Impedance control for a PEJR

The impedance matching problem in an electric circuit is basically regarded as a decision problem of load impedance for source impedance. The typical solution is that load impedance must be the complex conjugate of source impedance [16, p.36].

Now we consider a circuit expression of conventional stiff robots. The simplified one is shown in Fig. 1. Here Z_m is the source dynamics and Z_e is the load dynamics. Conventional impedance control methods treat robot dynamics as a source and environment one as a load. In this case, the load dynamics is given and we can control only the source dynamics. Anderson and Spong [1] have asserted that manipulators should be controlled to respond as the duality* of the environment. These methods are useful to control contact force. But in the case of position control, the load is disappeared (S is switched from A to B in Fig. 1). So despite of the advantage of impedance control that can use both free motion and contact task, we must prepare a different criteria for each control mode.

Next we show a circuit expression for a PEJR in Fig. 2, where M_a is the motor inertia, B_a the motor viscosity, K_t the variable elasticity of the transmission,

*If the environment behaves as inertial, then the robot should behave as capacitive, and *vice versa*.

and θ the motor position. The important difference between Figs. 1 and 2 is in the environments; Figure 2 treats the robot's drive system and the transmission elasticity as a source, and the robot linkages and the environment as a load (also see Fig. 3). Thus the load remains even if the robot does not interact with the environment (S is switched from A to B in Fig. 2). Next we will consider about an impedance adjustment problem.

The basic strategy to control the system behaviors is to match the source impedance with the load one appropriately by adjusting the transmission stiffness. So we investigate the following problem.

Problem Statement We choose the drive system and the elastic transmission as a source, and both the linkages and the environment as a load, as shown in Fig. 2. Then find an adjustment law of the transmission stiffness such that the whole system behaves just like as a pure resistance.

This is similar to the standard impedance matching problem but a little different from it, because the load resistance is not necessarily equal to the source one. This problem requires to adjust impedance devices in order that the power factor [16, p.140] is unity.

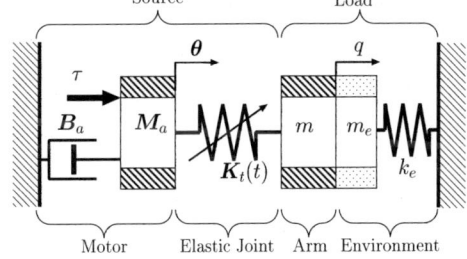

Figure 3: PEJR model

3 A case study of a 1DOF PEJR

We will show a solution of an impedance control problem for a 1 DOF PEJR, namely an adjustment law of PPI devices and position feedback gains assure the convergence to the anti-resonant point.

3.1 Linear PPI

In this section we explain linear PPI devices, which are linear springs and can be controlled the spring coefficient in real time. Figure 4 shows an image of performance of the device. The elastic force $f(x)$ increases linearly as x (the deflection of the transmission) increases. And we can control the incline within the hatched area. Examples of such devices are a Mechanical Impedance Adjuster developed by Morita et al. [11] for PEJRs and a non-linear spring developed by Hayashibara for tendon-driven robots [5]. We will deal with them hereafter and derive an adjustment law of the incline in section 3.3.

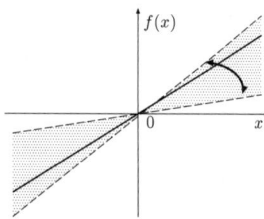

Figure 4: A performance of a linear PPI device; we can control the slop in real time.

3.2 Antiresonance of a 1 DOF PEJR

Figure 5 shows the two examples of 1 DOF PEJRs. The left one is a conventional PEJR arm and it has a PPI device in the joint axis, and the right one is so called a tendon-driven (or a wire-driven) robot arm [8] and each tendon has a PPI device.

The equations of motion of the robots are given by

$$m\ddot{q} + \bm{J}_j^T \bm{K}_t \bm{l} = -m_e \ddot{q} - k_e q, \quad (1)$$
$$\bm{M}_a \ddot{\bm{\theta}} + \bm{B}_a \dot{\bm{\theta}} + \bm{R}_a \bm{K}_t \bm{l} = \bm{\tau}. \quad (2)$$

Here $m, \bm{M}_a$ and m_e are the mass of the link, the motors and the environment, respectively, $\bm{B}_a$ is the motor's viscosity, $\bm{K}_t = \mathrm{diag}.\{k_1, k_2, \cdots, k_M\}$ is the spring coefficient matrix of the transmission, $q, \bm{\theta}$ and $\bm{l} = (\ell_1, \ell_2, \cdots, \ell_M)^T$ are the positions of the link and motor and the deflection in the PPI devices, respectively, $\bm{\tau}$ is the input torque, k_e is the stiffness of the environment. In the case of the PEJR shown in Fig. 5, $\bm{J}_j = -1$ and $\bm{R}_a$ means the gear ratio. On the other hand, in the tendon-driven one, $\bm{J}_j = (r_0, -r_0)^T$ and $\bm{R}_a = \mathrm{diag}. \{r_a, r_a\}$ where r_0 and r_a are the pulley radii on the joint and the motors, respectively. In both cases, $\bm{l}$ is given as follows;

$$\bm{l} = \bm{J}_j q + \bm{R}_a \bm{\theta}. \quad (3)$$

The motor input $\bm{\tau}$ is described by

$$\bm{\tau} = -\bm{R}_a \bm{A}_1 (\bm{R}_a \bm{\theta} + \bm{J}_j \alpha) - \bm{A}_2 \dot{\bm{\theta}}, \quad (4)$$

where α is a desired trajectory. Here we consider only the periodic function $\alpha = c e^{i\omega t}$. When we set that $q = c_1 e^{i\omega t}$ and $\bm{\theta} = \bm{c}_2 e^{i\omega t}$ in the steady state, then we can get the following equation.

$$\begin{pmatrix} x_{11} & \bm{J}_j^T \bm{K}_t \bm{R}_a \\ \bm{R}_a \bm{K}_t \bm{J}_j & \bm{X}_{22} \end{pmatrix} \begin{pmatrix} c_1 \\ \bm{c}_2 \end{pmatrix} = \begin{pmatrix} 0 \\ \bm{R}_a \bm{A}_1 \bm{J}_j \end{pmatrix} c, \quad (5)$$

where $x_{11}(i\omega) = -\bar{m}\omega^2 + (k_e + \bm{J}_j^T \bm{K}_t \bm{J}_j)$,

$$\bm{X}_{22}(i\omega) = -\omega^2 \bm{M}_a + \bm{R}_a(\bm{K}_t + \bm{A}_1)\bm{R}_a + i\omega \tilde{\bm{B}}_a,$$

$\tilde{\bm{B}}_a = \bm{A}_2 + \bm{B}_a$ and $\bar{m} = m + m_e$. So if we would like to make system behave as a pure resistor, the imaginary part of c_1 must be zero, that is,

$$\bm{J}_j^T \bm{K}_t \bm{J}_j = \bar{m}\omega^2 - k_e. \quad (6)$$

Equation (6) gives the antiresonant condition of this system. Then, c_1 is given by

$$c_1 = -(\bm{J}_j^T \bm{K}_t \bar{\bm{X}}_{22} \bm{K}_t \bm{J}_j)^{-1} \bm{J}_j^T \bm{K}_t \bar{\bm{X}}_{22} \bm{A}_1 \bm{J}_j c, \quad (7)$$

where $\bar{\bm{X}}_{22} = \bm{R}_a \bm{X}_{22} \bm{R}_a$. If $\bm{A}_1 = \bm{K}_t$, then the amplitude of the joint trajectory coincides with that of the desired one, but we have to note that the phase difference between two trajectories becomes π [rad].

3.3 An impedance adjustment law

We must adjust linear PPI devices in the PEJR in order to satsfy the property of Eq. (6). Here we propose an adjustment law of the impedance devices. First we define a regressor expression of the elasticity of the PPI devices as follows:

$$\bm{Y}(\bm{l})\bm{\phi}(t) \triangleq \bm{K}_t(t)\bm{l}, \quad (8)$$

where $\bm{Y}(\bm{l}) = \mathrm{diag}.(\ell_1, \ell_2, \ldots, \ell_M)$ and $\bm{\phi}(t) = (k_1(t), k_2(t), \cdots, k_M(t))^T$. $\bm{\phi}(t)$ is the elasticity of the

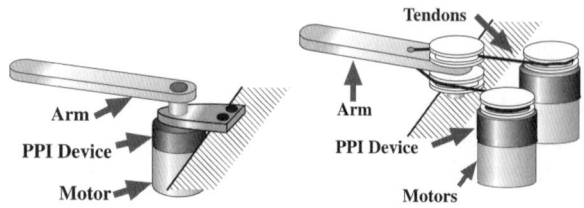

Figure 5: Example of a 1 DOF PEJRs; (left) a elastic joint robot, (right) a tendon-driven robot

PPI devices. Then the ajustment law of ϕ is given by

$$\phi(t) = \phi_0 + \beta \int_0^t \left[\mathbf{Y}^T \{\mathbf{R}_a \theta(\tau)\} \mathbf{R}_a \dot{\theta}(\tau) \right. \\
\left. + \mathbf{Y}^T \{\mathbf{J}_j \alpha(\tau) + \mathbf{l}(\tau)\} \{\mathbf{J}_j \dot{\alpha}(\tau) + \dot{\mathbf{l}}(\tau)\} \right. \\
\left. - \mathbf{Y}^T \{\mathbf{J}_j \alpha(\tau)\} \mathbf{J}_j \{\dot{q}(\tau) + \dot{\alpha}(\tau)\} \right] d\tau, \quad (9)$$

where ϕ_0 is an inital elasticity of the PPI devices and β is a positive number. We also adjust the diagonal elements of the position feedback gain matrix $\mathbf{A}_1$ in Eq. (4) using the same law as Eq. (9). Next we consider the stability of the system. We select the following candidate of a Lyapunov function

$$V = \frac{1}{2} \{ \bar{m}(\dot{q} + \dot{\alpha})^2 + \dot{\theta}^T \mathbf{M}_a \dot{\theta} + k_e(q + \alpha)^2 + \beta^{-1} \tilde{\phi}^T \tilde{\phi} \\
+ (\mathbf{J}_j \alpha + \mathbf{l})^T \hat{\mathbf{K}}_t (\mathbf{J}_j \alpha + \mathbf{l}) + \theta^T \mathbf{R}_a \hat{\mathbf{K}}_t \mathbf{R}_a \theta \}$$

where $\bar{m} = m + m_e$ and $\hat{\mathbf{K}}_t$ satisfys that

$$\mathbf{J}_j^T \hat{\mathbf{K}}_t \mathbf{J}_j = \bar{m}\omega^2 - k_e. \quad (10)$$

Here we define $\hat{\phi}$ as $\mathbf{Y}(\mathbf{l})\hat{\phi} = \hat{\mathbf{K}}_t \mathbf{l}$ in regressor expression and $\tilde{\phi} = \phi - \hat{\phi}$. Then the time derivative of V is given like this.

$$\frac{d}{dt} V = - \dot{\theta}^T \tilde{\mathbf{B}}_a \dot{\theta} + (\dot{q} + \dot{\alpha}) \left\{ \bar{m}\ddot{\alpha} + (\mathbf{J}_j^T \hat{\mathbf{K}}_t \mathbf{J}_j + k_e)\alpha \right\}.$$

Substituting Eq. (10) and $\ddot{\alpha} = -\omega^2 \alpha$ into the above equation, the time derivative of V is reduced to:

$$\frac{d}{dt} V = -\dot{\theta}^T \tilde{\mathbf{B}}_a \dot{\theta}. \quad (11)$$

This means that the motor angle θ approaches some constant angle $\bar{c}$, and eventually we have

$$\theta = \bar{c} \text{ and } \dot{\theta} = 0. \quad (12)$$

So we can easily see that the system is stable. The steady state analysis of Eq. (9) shows the convergence of ϕ to a constant value. Using Eqs. (1), (2) and (12) and LaSalle's invariance theorem [2, p.252], the maximum invariant set is given by

$$\lim_{t \to \infty} \left(q(t), \dot{q}(t), \theta(t), \dot{\theta}(t), \phi(t) \right) = \\ (-\alpha(t), -\dot{\alpha}(t), \mathbf{c}_3, 0, \mathbf{c}_4), \quad (13)$$

where $\mathbf{c}_3$ and $\mathbf{c}_4$ are constant vectors. Note that the desired trajectory α satisfies $\ddot{\alpha} = -\omega^2 \alpha$.

3.4 Simulation results

A 1 DOF robot driven with two tendons, shown in the right of Fig. 5, is used in the simulation. We

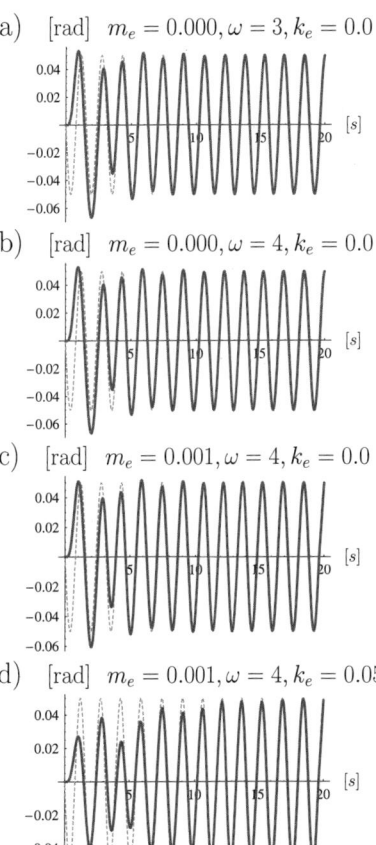

Figure 6: Joint trajectories. The Green dashed lines show the desired angle $-\alpha$ and the red lines show the joint angle q. Cases (a), (b) and (c) are free motion and case (d) is a contact task.

use $\alpha = 0.05 \sin \omega t$ [rad] as the desired trajectory. We set $\mathbf{J}_j = (-0.01, 0.01)^T$, $\mathbf{R}_a = 0.01 \mathbf{I}_2$, $\mathbf{M}_a = 0.001 \mathbf{I}_2$, $\mathbf{B}_a = \mathbf{A}_2 = 0.006 \mathbf{I}_2$, $m = 0.001$ and $\beta = 7 \times 10^8$, the initial value of $\mathbf{A}_1 = \mathbf{K}_t = 500 \mathbf{I}_2$. And we show the behaviors under the different sets of the frequency ω, the mass of the environment m_e and the stiffness of the environment k_e.

Figure 6 shows the joint trajectories. Cases (a), (b) and (c) are position control and case (d) is force control. The green dashed lines are the desired trajectories multiplied by -1, and the red solid lines are the joint trajectories. The joint trajectories converge to the desired ones in spite of the difference between the parameter sets. Figure 7 shows the trajectories of the joint stiffness. The green dashed lines are the joint stiffness given by Eq. (10) and the red solid lines are the transition of the joint stiffness. We can easily see the joint stiffness values also converge to the optimal ones by the adjustment law (9).

Figure 8 shows that the energy consumption disappears, once the systems converge to the antiresonant

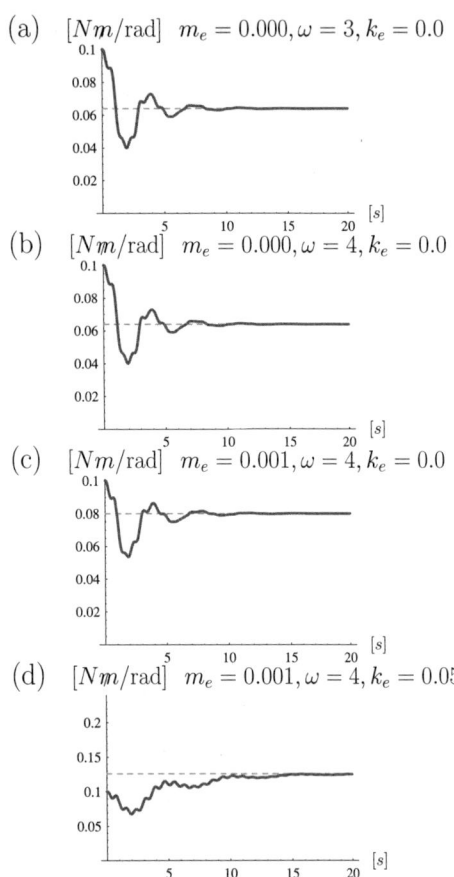

Figure 7: Joint stiffness. The green dashed lines show the antiresonant point described in Eq.(6), the red lines show the joint elasticity.

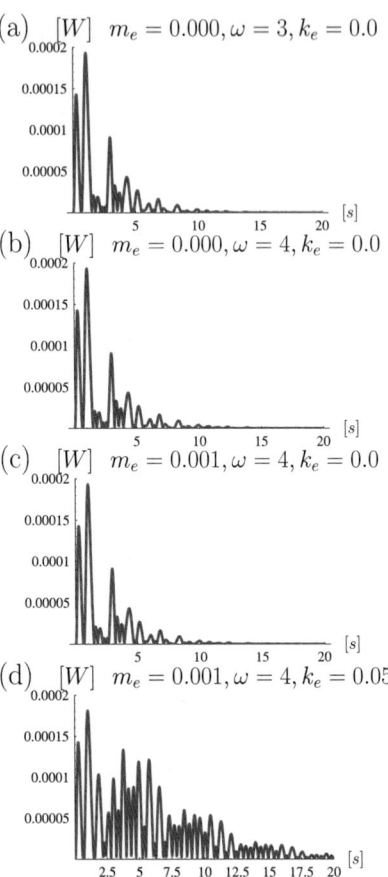

Figure 8: Supplied energy from the source to the load.

points, i. e., the steady states are selected in the optimal way that the power factors are unity.

4 Conclusion

In this paper we proposed a new impedance control concept for robots with PPIs at joint. We analised the one DOF PEJR and derived an impedance adjustment law, and showed the stability of the system under the law. We also confirmed the efficiency of the law by the simulations. The features are given as follows:

- Our new interpretation regards the drive system and the joint elasticity as a source and the manipulator and the environment as a load, therefore the load always exists in the circuit expression.

- When we treat the desired trajectory α as input and the joint velocity $\dot{q}$ as output, the system behaves like a pure resistance. This means that the power factor of the system is equal to unity.

- Almost all impedance control methods have executed by describing the impedance parameters explicitly. On the other hand, our method does not describe them but adjust the response optimally. This means that the optimal impedance is selected automatically. In addition, the control (4) and the adjustment law (9) contain only the internal variables and the desired path. Therefore they do not require the dynamics of the robot and the environment.

- Our method is effective for both free motion (cases (a), (b) and (c)) and contact task (case (d)).

- If we try to achieve the impedance matching by computational method, then we have to spend some energy to realize the spring-like motion. But our method does it by adjusting the impedance devices without any additional energy.

- While computational methods can move the robot along any continuous trajectories, our method limits the motion to the periodic ones.

So our method can be solved Problems 1, 2 and 3 in Introduction at the same time under some conditions.

5 Future Work

While it has been demonstrated that the new impedance control concept is valid, substantial works remain. The example we showed in this paper is extremely elementary. Therefore we have to extend it to non-linear and multi-DOF cases, and investigate the performance in more detail.

It is interesting to know that this interpretation is similar to one in ecological psychology [19], which asserts that people cannot recognize the inertia ellipsoids of their arms and the grasped objects separately. So this interpretation may also play an important role in human motion analysis.

Acknowledgment

We would like to thank Prof. C. K. Tse in The Hong Kong Polytechnic University for his variable advice about a basic circuit theory, and to thank the reviewers for their comments. This research has been supported by Grant-in-Aid for Scientific Research.

References

[1] R. J. Anderson and M. W. Spong. Hybrid impedance control of robotic manipulators. *IEEE Trans. on Robotics and Automation*, Vol. 4, No. 5, pp. 549–556, Oct. 1988.

[2] S. Arimoto. *Control theory of non-linear mechanical systems*. Oxford University Press, 1996.

[3] S. Arimoto, H. Y. Han, C. C. Cheah, and S. Kawamura. Extension of impedance matching to nonlinear dynamics of robotic tasks. *Sysmtems and Control Letters*, Vol. 36, pp. 109–119, 1999.

[4] C. C. Cheah and D. Wang. Learning impedance control for robotic manipulators. *IEEE Trans. on Robotics and Automation*, Vol. 14, No. 3, pp. 452–465, 1998.

[5] H. Hayashibara. Design of nonlinear spring for variable stiffness mechanism. In *Proc. of the Ammual Conf. of the Robotics Society of Japan*, 1M23, 2002, in Japanese.

[6] N. Hogan. Impedance control: An approach to manipulation. *ASME J. of Dynamic Systems, Measurement, and Control*, Vol. 107, pp. 1–24, 1985.

[7] K. Hyodo and H. Kobayashi. A study on tendon controlled wrist mechanism with nonlinear spring tensioner. *J. of the Robitics Society of Japan*, Vol. 11, No. 8, pp. 1244–1251, 1993, in Japanese.

[8] H. Kobayashi, K. Hyodo, and D. Ogane. On tendon-driven robotic mechanisms with redundant tendons. *The Inter. J. of Robotics Research*, Vol. 17, No. 5, pp. 561–571, May 1998.

[9] K. Koganezawa, M. Yamazaki, and N. Ishikawa. Mechanical stiffness control of tendon driven joints. *J. of the Robitics Society of Japan*, Vol. 18, No. 7, pp. 1003–1010, 2000, in Japanese.

[10] K. F. Laurin-Kovitz, J. E. Colgate, and S. D. R. Carnes. Design of components for programmable passive impedance. In *Proc. of IEEE Inter. Conf. on Robotics and Automation*, pp. 1476–1481, 1991.

[11] T. Morita and S. Sugano. New control method for robot joint by mechanical impedance adjuster *J. of the Robitics Society of Japan*, Vol. 14, No. 1, pp. 131–136, 1996, in Japanese.

[12] M. Okada, Y. Nakamura, and S. Ban. Design of programmable passive compliance shoulder mechanism. In *Proc. of the 2001 IEEE Inter. Conf. on Robotics & Automation*, pp. 348–353, 2001.

[13] R. Ozawa and H. Kobayashi. Response characteristics of elastic joint robots driven by various types of controllers against external disturbances. In *Proc. of The 6th Inter. Conf. on MOVIC*, pp. 420–425, 2002.

[14] M. H. Raibert and J. J. Craig. Hybrid position/force control of manipulators. *ASME J. of Dynamic Systems, Measurement, and Control*, Vol. 103, No. 2, pp. 126–133, 1981.

[15] B. Siciliano and L. Villani. *Robot Force Control*. Kluwer Academic Publishers, Boston, 1999.

[16] C. K. Tse. *Linear Circuit Analysis*. Addison-Wesley, 1998.

[17] T. Tsuji, H. Akamatsu, K. Harada, and M. Kaneko. On-line learning of robot arm impedance using neural networks. *J. of the Robitics Society of Japan*, Vol. 17, No. 2, pp. 234–241, 1999. in Japanese.

[18] T. Tsuji, H. Shigeyoshi, O. Fukuda, and O. Kaneko. Bio-mimetic control of an externally powered prosthetic forearm based on EMG signals. *Trans. of The Japan Society of Mechanical Engineers (C)*, Vol. 66, No. 648, pp. 2764–2771, 2000. in Japanese.

[19] M.T. Turvey. Dynamic touch. In *American Psycologist*, Vol. 51, pp. 1134–1152, 1996.

From Visuo-Motor Self Learning to Early Imitation – A Neural Architecture for Humanoid Learning

Yasuo Kuniyoshi　　Yasuaki Yorozu　　Masayuki Inaba　　Hirochika Inoue

Department of Mechano-Informatics, School of Information Science and Technology,
The University of Tokyo
7-3-1 Hongo, Bunkyo-ku, Tokyo 113-8656 Japan.
kuniyosh@isi.imi.i.u-tokyo.ac.jp

Abstract—Behavior imitation ability will be a key technology for future human friendly robots. In order to understand the principles and mechanisms of imitation, we take a synthetic cognitive developmental approach, starting with minimum components and create a system that can learn to imitate others. We developed a visuo-motor neural learning system which consists of orientation selective visual movement representation, distributed arm movement representation, and a high-dimensional temporal sequence learning mechanism. The vision and the movement representations model the findings in primate brain, i.e. macaque area MT(or human area V5) and the primary motor area. The learning mechanism is insipired by the finding that there are excessive connections in neonate brain. As our robot explores the visuo-motor self movement patterns, it learns coherent patterns as high-dimensional trajectory attractors. After the learning, a human comes in front of the robot showing arm movements which are similar to the ones in self learning. Although the robot has never seen or programmed to interpret human arm movement, and the detail of visual stimuli are very different, the robot identifies some of the patterns as similar to those in self learning, and responded by generating the previously learned arm movement. In other words, the robot exhibits early imitation ability based on self exploratory learning.

I. Introduction

Behavior imitation plays a crucial role in cognitive and social development of human infants.

In the real world, the model and the imitator never share exactly the same task situation nor the body characteristics. Therefore, an exact copying of bodily movement is neither possible nor meaningful. Imitation is establishing a common meaning between the behaviors of the model and the imitator which are different and meaningful to each individual. In other words, imitation requires extraction of meaningful features from the model's behavior, and re-establishing them by the imitator's own behavior[8], [9]. This means understanding the concept of similarity among real world events and in social contexts, also called as categorization, which serves as the foundation of symbols, reasoning, and social intelligence.

Among the animals of similar genotypes with humans, Chimpanzees, whose genotype is the closest to humans, are the only non-human primates which exhibit imitation abilities, with much less coherency and reliability compared to human infants. This also implies that imitation is a crucial phenomenon in understanding human intelligence.

In the robotics context, the realization of humanoid robots has introduced a strong demand for human friendly task intelligence. Human behavior and task requirements change drastically depending on the given situation. Therefore, fixed response patterns are not adequate. A crucial functionality would be to interpret and respond to an open-ended variety of human actions.

Understanding by creating imitation capabilities with robots, in other words, a synthetic study of imitation, will contribute to both understanding the foundation of human intelligence and to realizing human-friendly humanoid robots.

A. Related Work

There has been a growing interest in robotic imitation over the last several years. The early attempts dealt with specific tasks, e.g. block manipulation[8], [9], mobile robot navigation[6], [4], dynamic arm motion in a "kendama" play[12], and head motion[3], [5]). These systems have fixed specific mechanisms for each task.

Recently, psychophysical experiments are carried out to clarify what the imitator is watching when observing the model's arm gesture[10]. And an integrated model of imitation is proposed in the context of brain science[1]. The systematic understanding of effective features for extraction and re-instantiation is very important along with establishing an integrated model of the system behavior of many cognitive brain functions contributing to imitation.

These previous work provided important elements and perspectives to understanding and modeling imitation ability. However, they do not fully account for other important aspects of imitation. Imitation ability is adaptive by its nature and coupled tightly with cognitive development. Hence, we propose a new, developmental approach to deeper understanding of imitation ability.

B. Developmental Approach to Robotic Imitation

Fig. 1. Levels of Imitation

in order to understand the principles of human cognitive functions, it is very important to take a developmental perspective[2]. Imitation capability is no exception[7].

Piaget[14] established a theory of development of imitation in infants starting from reflexes and sensory-motor learning leading to purposive and symbolic level. Although his rigid stage transition hypothesis was problematic and experimentally refuted, the core idea that an interaction patterns of an autonomous system bootstraps itself by acquiring novel functionality and exploiting them to acquire further is still thought to capture an important aspect of human development.

We take a synthetic developmental approach where we build robotic systems that exhibit capabilities for imitation and its boot-strap learning process. In such approach, design and evaluation of the systems will benefit much from an appropriate definition of taxonomy or levels of apparent performance. It should reflect the differences in subtending mechanisms or interaction modes.

Figure 1 shows our taxonomy of imitation in a family task example. The family [1] is engaged in a garden cleaning task, sweeping and collecting fallen leaves. The leftmost girl is making her best effort watching the mother and copying her posture and movement as precisely as possible. However, the tool in the girl's hands is a mop. So the movement does not have much effect on the leaves. The boy in the middle grabbed a correct tool and manages to actually sweep the ground. However, he does not care about the overall goal of the task. He continues to sweep the ground with passion, and the leaves get scattered around. The boy on the right completely understands the purpose of the task. He failed to grab the same tool as his mother, but he found appropriate replacements – a different kind of sweeper and a dust tray. To effectively use the new tools he takes a very different posture from others. Nevertheless, he effectively contributes to the goal of the task which is shared with his mother.

Our claim is that the above example clarifies three important "modes" of imitation defined in terms of what is shared between the model and the imitator.

1) Appearance level: The posture and the movement of the body (and its extension, i.e. the manipulated tool), which are directly observable, are shared.
2) Action level: The unit causal relationship including the posture and movement, and its effect on the target objects is shared. Further incidents caused by the actions or underlying purpose is not considered.
3) Purposive task level: The overall goal of the task and its means-ends structure is understood. The overall goal is shared, and the rest of the task structure is flexibly adapted to a given situation.

It is important to note that we do not claim that human imitation ability develops through these three stages in a rigid order. Rather these are three different modes of cognitive processes representing different sets of mechanisms and different integration of information. All these modes can be active at the same time and have an integrated effect.

C. Acquisition of Neonatal Imitation

Our ultimate goal is to build a system which begins from level 0 (below Appearance level) and autonomously acquire all three levels through learning.

This paper focuses on the very early part where the system begins with no capability of imitation and acquires the appearance level capability

Meltzoff and Moore[11] discovered that very early neonates exhibit facial and manual imitation. They pose two alternative explanations; (1) Existence of an innate mechanism which represents the gestural postures of body parts in supra-modal terms, i.e. representations integrating visual, somatosensory, and motor domains. (2) Possibility of creating such representation through self exploratory "body babbling" during the fetus period. Meltzoff and Moore submit to the first hypothesis, proposing a mechanism called AIM.

In this paper, we support the second hypothesis above. Our aim is to show that assuming primitive configuration of fetuses and neonate's brain, the result

---

[1]The authors thank Kuniyoshi's family for their cooperation to this research.

of self exploratory sensory-motor learning can be immediately re-used in imitating a first-seen other human's gestural motion.

We attempt to achieve this by creating a robot system which exhibits such behavior. Although it is a minimalist model, it works on a real humanoid robot and responds to a real human gesture in the real world. It should serve as an initial model for a synthetic developmental study of imitation, which will clarify necessary modules to be added/acquired for more complex types of imitation.

II. Overview of the System

Imitation abilities of neonates discovered by Meltzoff and Moore implies the existence of a mechanism in newborns for matching visual patterns of other persons' body movements to self movements. However, recent findings about rich exploratory movements of fetuses and extensive neural development during this period suggest a possibility of self exploratory learning during the fetus period and its effect on newborn imitation abilities.

As a synthetic approach to this hypothesis, we developed a visuo-motor neural learning system which consists of orientation selective visual movement representation, distributed arm movement representation, and a high-dimensional temporal sequence learning mechanism (Fig.2). As our robot explores the visuo-motor self movement patterns, it learns coherent patterns as high-dimensional trajectory attractors. After the learning, a human comes in front of the robot showing arm movements which are similar to the ones in self learning. Although the robot has never seen or programmed to interpret human arm movement, and the detail of visual stimuli are very different, the robot identifies some of the patterns as similar to those in self learning, and responded by generating the previously learned arm movement. In other words, the robot exhibits early imitation ability based on self exploratory learning.

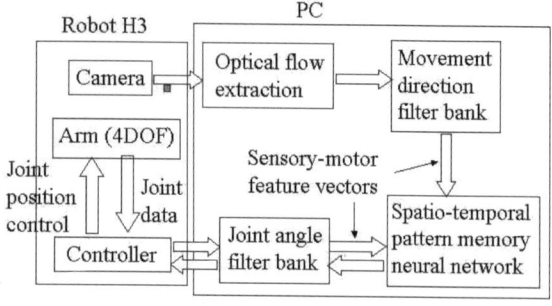

Fig. 2. System Configuration.

A. The Robot

Figure3 shows the robot "H3" used in our experiment. It has dual 4 DOF arms with 1 DOF grippers and a CCD camera on the 3DOF neck. This upper body is placed on a mobile base. This robot system was developed in our lab. The base mobility was not used in our experiments.

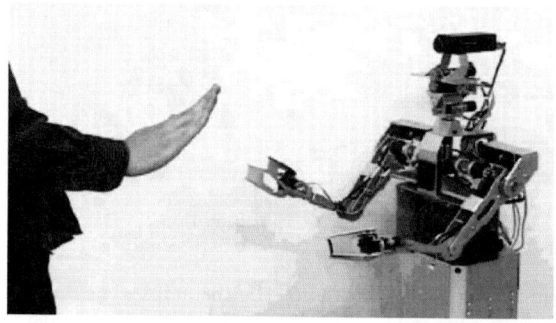

Fig. 3. The Robot H3 Observing Human Hand Movement.

III. Sensor Motor Interface

In neural learning, appropriate encoding of sensory-motor data is very important. Our neural network relies on similarity and continuity of incoming features to extract consistent patterns and to recognize them. So the encoded data should maintain some topology of the real data. On the other hand, regarding the purpose of this research, task specific encoding, such as identifying a chunk of temporal data as one of known primitives, should be avoided. In the following sections, we describe our encoding scheme which is very simple and general.

A. Visual features

As the most basic feature representing visual movement, we chose optical flow. In primates' brain, optical flow plays a central role in movement recognition. Our system extracts optical flow vectors at 400 points from a 256 by 256 image at frame rate.

The flow pattern should be encoded into a low-resolution scalar vector corresponding to an array of input neurons. We adopted a model of movement direction selective cells found in macaque brain area MT and in the corresponding human brain area V5. The optical flow array is covered by overlapping circular receptive fields of the direction selective cells. Within each receptive field, the flow vectors are weighted by a Gaussian function to emphasize the contribution of central flow vectors. Then a flow histogram is computed for 12 directions. This yields 12 scalar values for each receptive field(Fig.4). Let $\mathbf{p_i}$ be the flow vector at i-th point, and $\mathbf{u_j}(1 \geq j \geq 12)$ the unit vectors in 12 directions, then j-th directional component of i-th flow f_{ij} is:

$$f_{ij} = \begin{cases} \mathbf{p_i} \cdot \mathbf{u_j} & (\text{if } \mathbf{p_i} \cdot \mathbf{u_j} \geq 0) \\ 0 & (\text{if } \mathbf{p_i} \cdot \mathbf{u_j} < 0) \end{cases}$$

The directional selective cell response is obtained by the Gaussian weighted sum of the above components within the receptive field. Let r_i be the distance of i-th point from the center of the receptive field, then for i ranging within the receptive field;

$$F_{kj} = \sum_i f_{ij} e^{-ar_i^2}$$

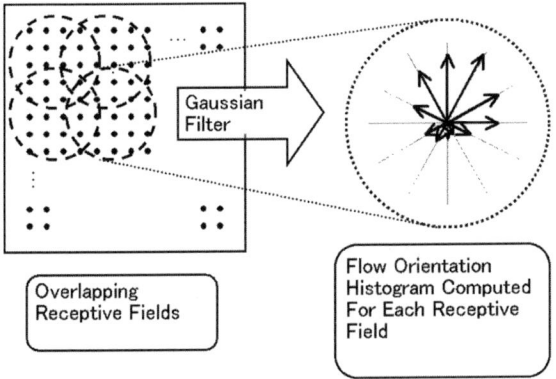

Fig. 4. Encoding Visual Motion Features with Movement Direction Selective Cells.

B. Joint Angle Features

The joint angle data of the robot arm is obtained from and fed to the arm controller as a four dimensional scalar vector. There are two conditions to be met in deciding the encoding scheme: (1) Small noise in encoded data should not result in a drastic change in the decoded (real) joint angles. (2) The encoding scheme must be reversible, i.e. it must be possible to decode the neural net state into real joint angles in order to generate arm movement.

We adopt the following distributed coding scheme. For each joint, we assign a set of neurons each representing a specific angle. The input real joint angle a_j generates a Gaussian potential centered at a_j over the neurons. Thus, the activation g_i of the i-th neuron representing angle v_i is:

$$g_i = c e^{-b(v_i - a_j)^2}$$

where c and b are constants.

In our current experiment, the neurons exhibit binary activation which acts as a threshold function for g_i. As a result, a constant number of neurons around a_j get activated(Fig.5).

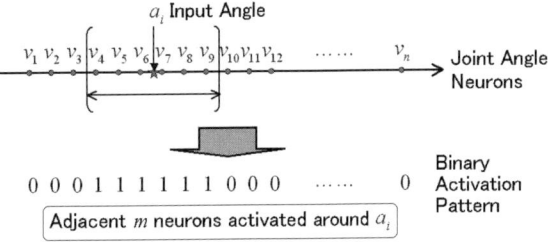

Fig. 5. Encoding Arm Joint Angle.

IV. Spatio-Temporal Pattern Learning

Based on the findings about excessive random connections in neonates brain, we hypothesize that they treat vision and motor data as "melted together" and extracts information structure in an amodal mannar. As a first step of modeling such a system, we combine the incoming vision and motor feature vectors into a single high-dimensional vector and feed to a uniformly connected large neural network which extracts spatio-temporal patterns from the sensory-motor-time space.

During learning, the vision and motor data are encoded as described above and fed to the neural network as a temporal sequence of high-dimensional scalar vectors. In our experiment, a 36 dimensional visual feature vector and a 150 dimensional joint angle vector are generated every 33 [msec]. Even for a brief gesture lasting a few seconds, a sequence of about 100 vector pairs is fed to the neural net. This is quite a challenging learning task for conventional sequence learning networks.

We adopt a new neuron model called "non-monotonic neural net" proposed by Morita et al.[13]. It is basically a dynamic continuous-valued Hopfield network consisting of neurons with a special non-monotonic output function(Fig.6):

$$f(u) = \frac{1 - e^{-cu}}{1 + e^{-cu}} \cdot \frac{1 + \kappa e^{c'(|u|-h)}}{1 + e^{c'(|u|-h)}}$$

Each neuron's behavior is described by the following equation:

$$\tau \frac{du_i}{dt} = -u_i + \sum_{j=1}^{n} w_{ij} y_j + z_i$$

where u_i is internal potential of i-th neuron, w_{ij} is the connection weight from j-th neuron, z_i is the external input signal, and τ is time constant. Our network has 1000 neurons, each connecting to 499 other neurons.

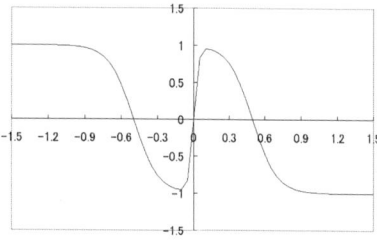

Fig. 6. Non monotonic output function.

The network adopts covariance rule for learning. Let r_i be the learning data, τ' be learning time constant with $\tau' \gg \tau$, the rule is described as the following:

$$\tau' \frac{dw_{ij}}{dt} = -w_{ij} + \alpha r_i y_j$$

A. Spatio-temporal pattern as trajectory attractor

In a standard Hopfield network, a static data is memorized as a point attractor. Morita's network learns a temporal sequence of high-dimensional vectors as a "trajectory attractor" over its high-dimensional state space. Figure7 illustrates the situation. As the learning proceeds for a static data, a point attractor is formed. Due to the non-monotonic output function, the attractor shape does not become too steep, avoiding fake memories and facilitating smooth transition of the network state. As the learning data vector continuously change over time, a trench like attractor is formed to reflect the temporal sequence of the data.

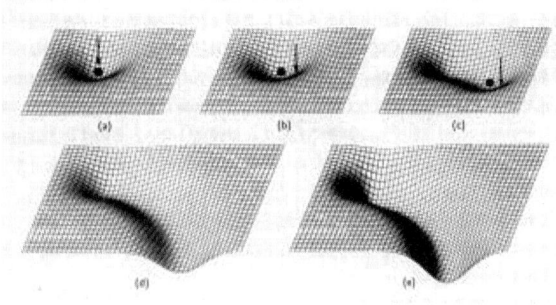

Fig. 7. Trajectory attractor (adopted from Morita96).

After learning, when the system observes a similar visual motion, it is encoded as a temporal sequence of visual feature vectors. It drives the network state close to the previously learned trajectory attractor. Then the state is trapped into the attractor and moves along it. Once the network state moves along the attractor, the projection of the state onto the motor dimensions is immediately output and decoded into arm motor commands.

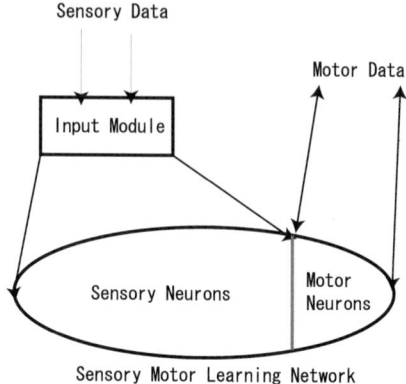

Fig. 8. Structure of the Neural Network

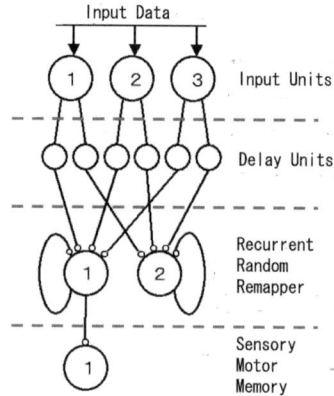

Fig. 9. Input Module

B. Handling Real Data

The robustness of Morita's network is quite high. In our preparatory experiments, we tested a network of 516 neurons, each connected to 150 other neurons. The degree of connection was deliberately lowered (from full connection) to evaluate the effect of the reduced degree of connection.

The sensor data vectors of 150 dimension and motor data vectors of 66 dimension were fed to the network in a mannar that each component of the vector is connected to one neuron. In the learning phase, 4 different temporal sequences of 135 sensory-motor vectors were learned by the network. After the learning, one of the 4 temporal sequences was selected and only the sensory components of the vectors were fed to the network, and the motor neuron outputs were examined.

Experiments with simulated data showed that the network correctly associates the sensory sequence with up to 30% noise with learned motor sequence. This shows that the network is robust enough to handle real data with substantial amount of noise.

However, we found two problems with the network: (1) Although the network tolerates random noise as long as the overall trajectory in sensory-motor space is smooth, it is quite fragile when the data changes abruptly, or when the data stays still. (2) In order to deal with long complex sequences, the network requires enough number of neurons. However in the original design of the network, the number cannot exceed the dimension of the sensory motor vector.

In order to deal with the above problems, we added an input module as described in figure 9. Each component of the input vector is fed to a set of delay units which are FIR filters with different decay parameters. The output of the delay units are connected to a large number of recurrent random remapper neurons. The connections between the two layers are random and fixed. The output of each remapper neuron is connected to one neuron in the main sensory motor learning network. The delay units has smoothes out the abrupt changes in the input sequences, and the remapper layer increases the dimensionality of the input vector which provide input signals to much larger number of neurons (in the learning network) than the original dimensionality of sensory data. Also, the recurrent connections of the remapper neurons add inertia to the temporal change in input sequences, reducing the problems with static part in the sequences.

By introducing the above module, the system became robust enough to learn and recognize real visuo-motor data. Our current learning system has the sensory vector dimension of 36, the motor vector dmiension of 150, 850 remapper neurons and the same number of sensory learning neurons. Thus the main sensory-motor learning network has 1000 neurons, and each of them is connected to 499 other neurons.

V. Experiments

The experiment consists of the following two phases:
1) The robot generates three kinds of arm movement patterns in front of its own camera. The patterns are (1) swinging the hand sideways, (2) swinging the hand vertically, (3) circular movement. The robot observes its own motion and learns the visuo-motor temporal patterns. The three patterns are learned by the same network.
2) After learning, a human stands in front of the robot, showing either kind of hand movement from the above repertoire. The robot observes the human hand movement, feeds the data into the network and drives the arm by the motor dimensions of the network state.

The scenes from the self learning and the imitation experiment are shown in Fig.10

It is important to note that the system has no built-in knowledge or algorithm for recognizing human hands, and its appearance and movement is substantially different from the learning data. Nevertheless, the robot did imitate a first-seen human arm movement. The robot is actually recognizing the pattern as similar to the one in its past experience with its own arm movement. Moreover, the same network responds selectively to three different patterns, generating correct arm movement patterns corresponding to each kind of human gesture.

Figure 11 shows some of the joint angle profiles of the arm movement during self learning phase. Figure 12 shows some of the profiles of the generated arm movement in response to given human gestures. The figure shows three trials for each type of arm movement patterns.

In 7 out of 9 trials shown in Fig.12, the system successfully recognized first-seen human arm movements as known patterns and generated correct arm movements. However, in Pattern1-Trial 2 and Pattern3-Trial2, the system failed to recognize the given patterns in the middle of the sequences. We made 20 trials and the overall success rate of correct imitation was 81%.

VI. Discussions

We constructed a visuo-motor learning system consisting of optical flow extraction, movement direction selective filter bank, distributed representation of arm joint angles, and a 1000-neuron non-monotonic dynamic neural network. It successfully learned long temporal sequences of high dimensional visuo-motor feature vectors as trajectory attractors in its state space.

Our robot first learned the visuo-motor correspondence of its own arm movement patterns. Of course the robot has no initial knowledge about its arm kinematics. After learning, a human comes in front of the robot, showing an arm movement which is similar to one of the patterns known to the robot. Although the appearance of the human hand and its detailed movement are different from any of the scenes which the robot's has ever seen, the robot selectively responded to the stimuli by properly imitating the movement patterns. The property of trajectory attractor and the global movement feature detector modeled after primate brain worked together to extract a global structure of the movement patterns and assimilating them to known patterns. As the system treats vision and motor modalities as a unity, recognizing a pattern immediately leads to generating the corresponding

Fig. 10. Scenes from self learning (left), robot's view during learning (middle), imitating human gesture after learning (right).

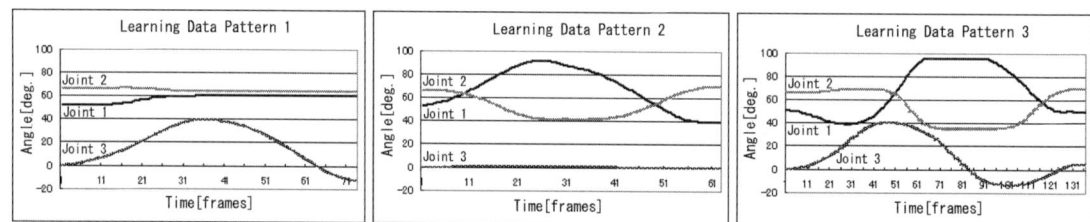

Fig. 11. The joint angle profile of the three arm movement patterns during self learning. Vertical axis indicate the angles [deg]. Horizontal axis indicate time [video frames]. Each line corresponds to a joint angle trajectory. Pattern 1: Horizontal swing (left), Pattern 2: Vertical swing (middle), Pattern 3: Circular movement (right).

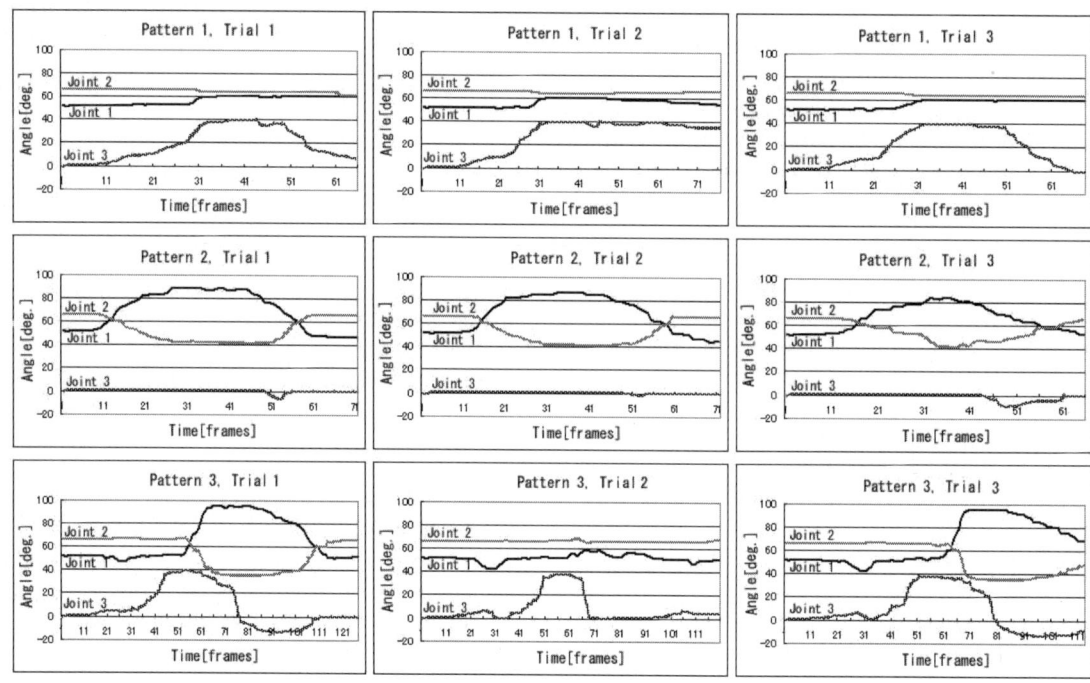

Fig. 12. The joint angle profile of the generated arm movement in response to the three arm movement patters given by a human demonstrator. Vertical axis indicate the angles [deg]. Horizontal axis indicate time [video frames]. Each line corresponds to a joint angle trajectory. Pattern 1: Horizontal swing (top row), Pattern 2: Vertical swing (middle row), Pattern 3: Circular movement (bottom row). Trial 1 (left column), Trial 2 (middle column), Trial 3 (right column).

arm movement. In other words, the robot "cannot help" imitating the target once it recognizes the pattern.

The above system does not assume any built-in

motor primitives or visual pattern primitives. It is a minimalist system and yet powerful enough to deal with real sensory motor data. Thus, it is useful as a testbed for exploring more complex cognitive functionalities subtending sophisticated imitation such as body schema, spatial representation and inferences, acquisition of novel motor primitives, visual attention, handling multiple events, and dealing with goal directed actions.

VII. REFERENCES

[1] M. A. Arbib, A. Billard, M. Iacoboni, and E. Oztop. Synthetic brain imaging: Grasping, mirror neurons and imitation. Neural Networks, 13:975–997, 2000.

[2] M. Asada, K. F. MacDorman, H. Ishiguro, and Y. Kuniyoshi. Cognitive developmental robotics as a new paradigm for the design of humanoid robots. Robotics and Autonomous Systems, 37(2-3):185–193, 2001.

[3] L. Berthouze, P. Bakker, and Y. Kuniyoshi. Development of oculo-motor control through robotic imitation. IEEE/RSJ International Conference on Robotics and Intelligent Systems, Osaka, Japan, pages 376–381, 1996.

[4] K. Dautenhahn. Getting to know each other – artificial social intelligence for autonomous robots. Robotics and Autonomous Systems, 16:333–356, 1995.

[5] J. Demiris, S. Rougeaux, G. M. Hayes, L. Berthouze, and Y. Kuniyoshi. Deferred imitation of human head movements by an active stereo vision head. In Proc. IEEE Int'l Workshop on Robot and Human Communication (ROMAN), pages 88–93, 1997.

[6] G.M. Hayes and J. Demiris. A robot controller using learning by imitation. In A. Borkowski and J. L. Crowley, editors, Proceedings of the 2nd International Symposium on Intelligent Robotic Systems, pages 198–204, Grenoble, France, 1994. LIFTA-IMAG.

[7] Y. Kuniyoshi. The science of imitation — towards physically and socially grounded intelligence —. Technical Report TR-94001, RWC, 1994.

[8] Y. Kuniyoshi, M. Inaba, and H. Inoue. Learning by watching: Extracting reusable task knowledge from visual observation of human performance. IEEE Trans. Robotics and Automation, 10(5), 1994.

[9] Y. Kuniyoshi and H. Inoue. Qualitative recognition of ongoing human action sequences. In Proc. IJCAI93, pages 1600–1609, 1993.

[10] M. J. Mataric and M. Pomplun. Fixation behavior in observation and imitation of human movement. Cognitive Brain Research, 7(2):191–202, 1998.

[11] A. N. Meltzoff and M. K. Moore. Imitation of facial and manual gestures by human neonates. Science, 198:75–78, 1977.

[12] H. Miyamoto, S. Schaal, F. Gandolfo, H. Gomi, Y. Koike, R. Osu, E. Nakano, Y. Wada, and M. Kawato. A kendama learning robot based on bi-directional theory. Neural Networks, 9(8):1281–1302, 1996.

[13] M. Morita. Memory and learning of sequential patterns by nonmonotone neural networks. Neural Networks, 9:1477–1489, 1996.

[14] J. Piaget. Play, Dreams and Imitation in Childhood. New York: W. W. Norton, 1962.

Learning About Objects Through Action - Initial Steps Towards Artificial Cognition

Paul Fitzpatrick*; Giorgio Metta*†; Lorenzo Natale†; Sajit Rao†; Giulio Sandini†
* AI-Lab, MIT, Cambridge, MA, U.S.A.
†LIRA-Lab, DIST, Univ of Genova, Viale Francesco Causa 13, Genova 16145, Italy
paulfitz@ai.mit.edu; {pasa, nat, sajit, sandini}@dist.unige.it

Abstract—Within the field of Neuro Robotics we are driven primarily by the desire to understand how humans and animals live and grow and solve every day's problems. To this aim we adopted a "learn by doing" approach by building artificial systems, e.g. robots that not only look like human beings but also represent a model of some brain process. They should, ideally, behave and interact like human beings (being situated). The main emphasis in robotics has been on systems that act as a reaction to an external stimulus (e.g. tracking, reaching), rather than as a result of an internal drive to explore or "understand" the environment. We think it is now appropriate to try to move from acting, in the sense explained above, to "understanding". As a starting point we addressed the problem of learning about the effects and consequences of self-generated actions. How does the robot learn how to pull an object toward itself or to push it away? How does the robot learn that spherical objects roll while a cube only slides if pushed? Interacting with objects is important because it implicitly explores object representation, event understanding, and can provide definition of objecthood that could not be grasped with a mere passive observation of the world. Further, learning to understand what one's own body can do is an essential step toward learning by imitation. In this view two actions are similar not only if their kinematics and dynamics are similar but rather if the effects on the external world are the same. Along this line of research we discuss some recent experiments performed at the AI-Lab at MIT and at the LIRA-Lab at the University of Genova on COG and Babybot respectively. We show how the humanoid robots can learn how to poke and prod objects to obtain a consistently repeatable effect (e.g. sliding in a given direction), to help visual segmentation, and to interpret a poking action performed by a human manipulator.

I. INTRODUCTION: LEARNING TO ACT ON OBJECTS

In order to explore the role of sensory information and motor skills in cognition, we are pursuing a developmental approach. The basic idea is that assembling something as complex as a cognitive system from scratch as a collection of modules is virtually impossible. Rather, as in humans, cognitive abilities develop over time layering over previous stages of development. Development may therefore not be just an artifact of biological systems, but a necessary way to manage complexity [1]. To this end we see three broad stages in the development of our humanoid robot platform: The first stage involves learning a *body-image* or body-schema [2] [3]. Having learned to distinguish its body from the rest of the world the robot can move on to the second stage, which is interaction with external objects. The third and final stage involves interpreting object-object interactions. An essential feature of this developmental program is that each stage strictly *requires* and *layers upon* the previous stages. In this paper we present results for the second and third stages in this developmental schema.

Results from neuroscience suggest that action and manipulation are fundamental for acquiring knowledge about objects [4] [5] [6]. These results point out that action and perception are possibly much more intertwined than what was once believed. It would be difficult to draw a separation for where perception ends and action starts. Even the distinction between motor and sensory areas tends to be very blurred.

Drawing more from the neural science literature, we now know that areas active when reaching and/or grasping present a mixed structure containing action and sensory related neurons. Arbib and colleagues [6] interpreted these responses as the neural analogue of the affordances of Gibson [7]. In Gibson's theory an affordance is a visual characteristic of an object which can elicit an action without necessarily involving an object recognition stage. It seems that areas AIP (parietal) and F5 (premotor/frontal) are active in such a way to provide the individual with a mechanism to detect affordances. F5 projects to F1 (primary motor cortex) and can therefore control behavior. In particular area AIP contains neurons that respond both when generating a grasping movement and when observing the object being grasped [5] [8]. Responses are congruent: e.g. a precision grip responsive neuron would also fire during the observation of a small object (for which the precision grip is a likely action). Similar neurons with slightly different temporal characteristics have been observed in F5.

Rizzolatti and coworkers [5] extensively probed area F5. They found another class of grasping neurons that also responded during observation of somebody else's

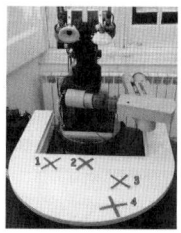 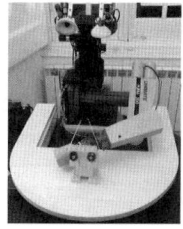 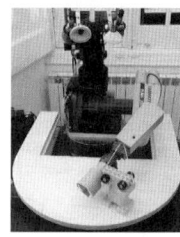

(a) initial arm-positions for target approach

(b) At the beginning of a pushing movement

(c) At the end of the movement

Fig. 1. The experimental setup

action. They called this newly discovered cells mirror neurons. This activation of F5 is coherent with the idea that the brain internally reproduces/simulates the observed actions. This can possibly form the basis for recognition of complicated biological motion and for mimicry of observed behaviors.

Taken together, these results suggest that the ability to visually interpret the motor-goal or behavioral/purpose of the action may be helped by the monkey's ability to perform that action (and achieve a similar motor goal) itself. Therefore, active exploration of the environment may be critical not only to subsequent performance of an action, but also for the interpretation of the actions of others. We argued at the beginning of this section that this "probing/exploring activity" is the second stage of a developmental sequence. It is at this stage when the ability to interact with object is formed, and includes the detection of affordances and the manipulation of objects. In the following sections we present an implementation of this stage. We will show how a humanoid robot

- Learns the effect of pushing/pulling actions on objects and uses this to drive goal-directed behavior.
- Acquires a particular affordance and behaviorally demonstrates this form of "understanding" about objects.
- Uses the knowledge of affordances, gained through exploration, to interpret human action and mimic the last observed action.

II. LEARNING TO PUSH/PULL/POKE OBJECTS

The goal of this experiment is to learn the effect of a set of simple pushing/pulling actions from different directions on a toy object, and then use the learned knowledge to move a new object in a desired goal-direction.

Figure 1 A Shows the experimental platform "Babybot", an upper torso humanoid robot at the LIRA Lab, Univ of Genova. Babybot has a 5 DOF head, and a 6 DOF arm, and 2 cameras whose Cartesian images are mapped to a log-polar format [9]. The robot also has a force sensitive wrist, and a metal stub for a hand. The target is placed directly in front of the robot on the play-table. Babybot starts from any of four different initial positions (shown in the figure) at the beginning of a trial run.

In a typical trial run the robot continuously tracks the target while reaching for it. The target (even if it is moving) is thus ideally always centered on the fovea, while the moving hand is tracked in peripheral vision. Figure 1 (B) shows the arm at one of its initial positions and (C) shows the end of the trial with the target having been pushed to one side.

During each such trial run, the time evolution of two event variables are continuously monitored: the initial proprioceptive hand-position, and then at the moment of contact (when the hand first touches the object) the direction of the *retinal* displacement of the target.

A. The Target Representation and Results of Learning

The purpose of the training phase is to learn a mapping from initial hand position to direction of target movement. Therefore, associated with each initial hand position is a direction map (a circular histogram) that summarizes the directions that the target moved in when approached from that position. After each trial the appropriate direction map is updated with the target motion for that particular trial.

Approximately 70 trials, distributed evenly across the four initial starting positions, were conducted. Figure 2 shows the four direction maps learned, one for each initial arm position considered. The maps plot the frequency with which the target moved in a particular direction at the moment of impact. Therefore longer radial lines in the plot point towards the most common direction of movement. As we can see, the maps are sharply tuned towards a dominant direction.

B. Testing the Learned Maps

The learned maps are used to drive motor planning in a straightforward manner as shown in Figure 3. The robot is presented with the usual target as before, but this time also with another toy nearby (Figure 3 a). The goal is to push the target towards the new toy. The robot first extracts the desired displacement vector from the scene, finds the direction map which is most tuned in that direction. The initial-hand position corresponding to that map, is used and the dynamics takes care of the rest, resulting in the motion of the target towards the desired direction, (Figure 3b,c).

A quantitative measure of the performance before and after learning is to look at the error angle between the

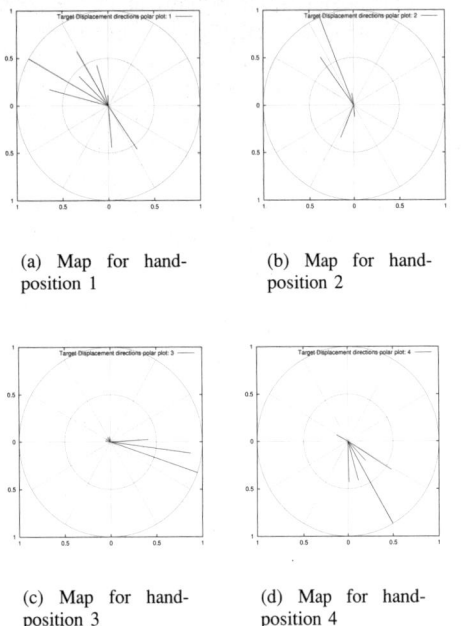

(a) Map for hand-position 1

(b) Map for hand-position 2

(c) Map for hand-position 3

(d) Map for hand-position 4

Fig. 2. The learned target-motion direction maps, one for each initial hand-position

  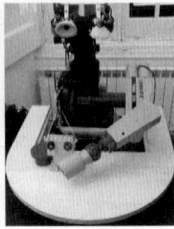

(a) The round toy is the new desired target position

(b) The learned maps are used to re-position the arm in preparation for pushing

(c) At the end of the movement

Fig. 3. The learned direction maps are used to drive goal-directed action

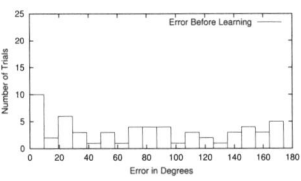 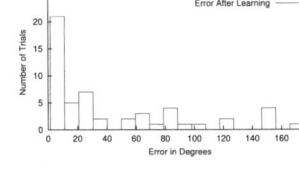

(a) Error distribution before learning

(b) Error distribution after learning

Fig. 4. Improvement in performance: Plots of the distribution of the angle between the desired direction and actual direction, before and after learning. Zero degrees indicates no error, while 180 degrees indicates maximum error.

desired direction of motion (of the target towards the goal) and the actual direction that the target moved in when pushed. First, as a baseline control case, 54 trials were run with the goal position (round toy) being varied randomly and the initial hand-position being chosen randomly among the four positions (i.e learned maps are not used to pick the appropriate hand position). Figure 4(a) shows the error plot. The distribution of errors is not completely flat as one would expect because the initial hand-positions are not uniformly distributed around the circle. Nevertheless, the histogram is not far from uniform. Doing the same experiment, but using the learned map to position the hand, yields the error plot shown in 4(b). As we can see the histogram is significantly skewed towards an error of 0, as a result of picking the correct initial-hand position from the learned map.

III. LEARNING OBJECT AFFORDANCES

In the previous section, we ignored both the identity of the object, and the gross shape (elongation) of the object, and learned possibly the simplest property of the behavior of the object; namely the instantaneous direction of motion as a result of a push/pull action.

In this section we take both identity, and some shape properties of the object into account and thereby have a more detailed characterization of the behavior. The uniqueness of the motion-signature of the object can used to identity the object itself, and can be associated with the visual appearance of the object. We describe an experiment where a robot acts on a small set of objects using a small motor repertoire consisting of four actions indicated for convenience as pull in, side tap, push away, and back slap.

During a training/exploration stage the robot performs several trials for each ⟨object, action⟩ pair and learns the motion signature of the behavior of the object for that action. We show how, in a more goal-directed mode, the robot uses it's knowledge of the object affordances to choose the appropriate action to make a given object roll. Finally we show how the robot is able to interpret the effect of a human action on an object, and respond with it's own action that produces a similar effect, i.e. mimicry.

The robot used for this experiment was "Cog", an upper torso humanoid robot [10] at the MIT A.I. Lab. Figure 5 shows the robot, Cog has two arms, each of which has six degrees of freedom. The joints are driven by series elastic

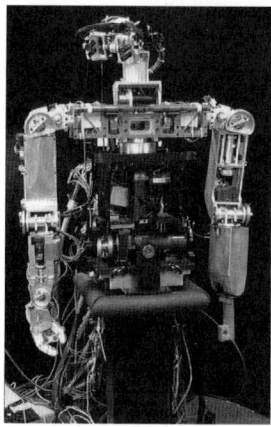 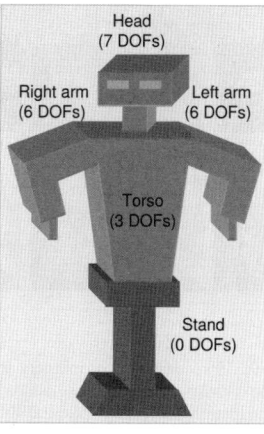

Fig. 5. Degrees of freedom of the robot Cog. The arms terminate in a primitive flipper. The head, torso, and arms together contain 22 degrees of freedom.

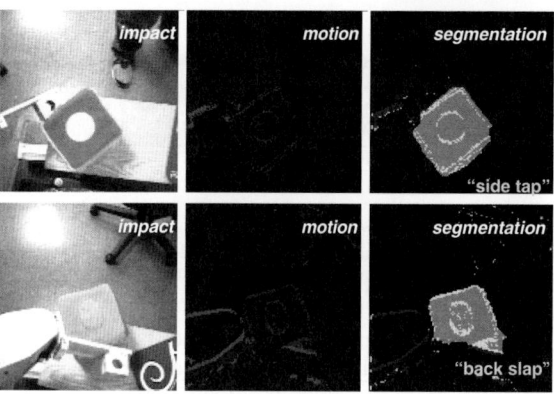

Fig. 6. Cog poking a cube. The top row shows the flipper poking the object from the side, turning it slightly. The second row shows Cog batting an object away. The images in the first column are frames at the moment of impact. The second column shows the the motion signal at the point of contact. The bright regions in the images in the final column show the segmentation produced for the object.

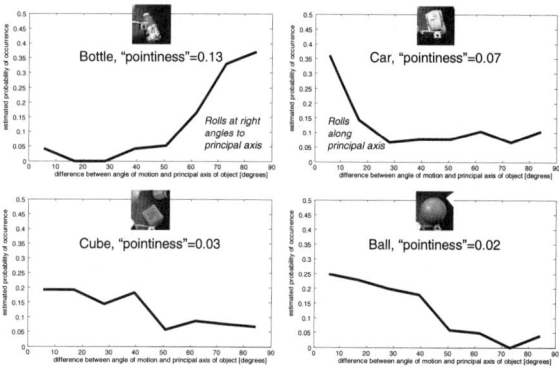

Fig. 7. Probability of observing a roll along a particular direction for the set of four objects used in Cog's experiments. Abscissas represent the difference between the principal axis of the object and the observed direction of movement. Ordinates are the estimated probability.

actuators - essentially a motor connected to its load via a spring. Cog's head has seven degrees of freedom and mounts four cameras and a gyroscope simulating part of the human vestibular system.

A. Characterizing Object Behavior

During the training phase, each of the four objects in this experiment (an orange juice bottle, a toy car, a cube, and a colored ball) is "poked" about a 100 times, i.e. roughly 25 repetitions of each action for each object. The visual feature of the resulting object behavior that is extracted is the instantaneous direction of motion of the object (as in the previous section) but this time *the angle is relative to the principle axis of inertia of the object*. The direction of motion of the object and principle axis of inertia are extracted as follows.

During a single poking operation, the arm is identified as the first object to move in the scene. Once the arm is identified, a sudden spread of motion in the image (spatially correlated with the end-effector) identifies a contact. Whatever was moving before this instant is considered background (the arm, other disturbances). Thereafter, the newly moving "blob" can be identified as the object. A further refinement is needed to fill in the gaps since motion detection is in general sparse and depends on the visual appearance/texture of the object. Two examples of poking and segmentations are shown in Figure 6.

Having segmented the object the principle axis of inertia is easily extracted, and its instantaneous direction of motion is gathered from the optical flow information of the pixels belonging to the object. Then the difference between these two angles is used to update a "motion-signature probability map" of the kind shown in Figure 7.

B. Results and Demonstration of Learning

For each ⟨object, action⟩ pair the representation of the object affordances or movement signature is in terms of a histogram of probabilities for each relative angle of motion. In other words the probabilities represent the likelihood of observing each of the objects rolling along a particular direction with respect to their principal axis. Figure 7 shows one set of such estimated probability maps learned from the exploration/training stage.

To provide a behavioral demonstration of the learned affordances, one of the known objects was placed in front of Cog, with Cog's task being to choose the action that would most likely make that particular object roll.

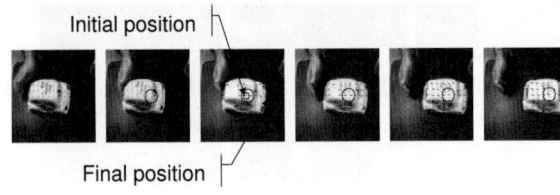

Fig. 8. An example of observed sequence. Frames around the instant of impact are shown. Initial and final position after 12 frames is indicated.

The object is recognized, localized and its orientation estimated (principal axis). Recognition and localization are based on the same color information collected during learning. Cog then uses its understanding of the affordance of the object (Figure 7) and of the geometry of poking to choose the action that is most likely to make the object roll. The object localization procedure has an error between 10° and 25° which proved to be tolerable for our experiment. We performed a simple qualitative test of the overall performance of the robot. Out of 100 trials the robot made 15 mistakes. Twelve of them were due to imprecise control: e.g. the end point touched the object earlier than expected moving the car outside the field of view. The remainders (3) were genuine mistakes due to misinterpretation of the object position/orientation.

This experiment represents an analogue to the response of F5/AIP as explained in Arbib's model [6] in that a specific structure of the robot detects the affordance of the object and links it to the generation of behavior. This is also the first stage of the development of more complex behaviors which relies on the understanding of objects as physical entities with specific properties.

C. Understanding the actions of others

An interesting question then is whether the system could extract useful information from seeing an object manipulated by someone else.

In fact, the same visual processing used for analyzing an active poking has been used to detect a contact and segment the object from the manipulator. The first obvious thing the robot can do is identify the action just observed with respect to its motor vocabulary. It is easily done by comparing the displacement of the object with the four possible actions and by choosing the action whose effects are closer to the observed displacement. This procedure is orders of magnitude simpler than trying to completely characterize the action in terms of the observed kinematics of the movement. Here the complexity of the data we need to obtain is somewhat proportional to the complexity of the goal rather than that of the structure/skill of the foreign manipulator.

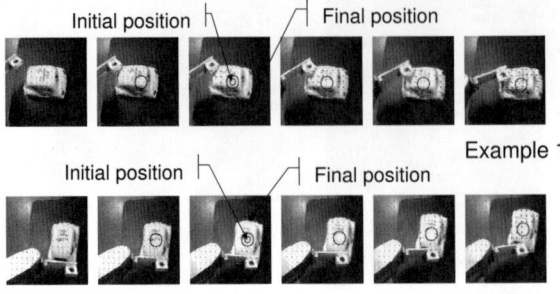

Fig. 9. Two examples of mimicry following the observation of Figure 8. Cog mimics the goal of the action (poking along the principal axis) rather than the trajectory followed by the toy car.

The robot can also mimic the observed behavior if it happens to see the same object again. This requires another bit of information. The angle between the affordance of the object (preferred direction of motion) and the observed displacement is measured. During mimicry the object is localized as in section III-A and the action which is more likely to produce the same observed angle (relative to the object) is generated. If, for example, the car was poked at right angle with respect to its principal axis Cog would mimic the action by poking the car at right angle. In spite of the fact that the car preferred behavior is to move along its principal axis. Examples of observation of poking and generation of mimicry actions are shown in Figure 8 and 9.

IV. CONCLUSION

All biological systems are embodied systems, and an important way they have for recognizing and differentiating between objects in the environment is by simply acting on them. Only repeated interactions (play!) with objects can reveal how they behave when acted upon (e.g. sliding vs rolling when poked). We have shown two experiments where, in a discovery mode, the visual system learns about the consequences of motor acts in terms of such features, and in a goal-directed mode the mapping may be inverted to select the motor act that causes a particular visual change. These two modes of learning; the consequences of a motor act, and selecting a motor act to achieve a certain result, are obviously intertwined, and together are what we mean by "learning to act". Furthermore, the same information can also be used to interpret the effect of a human-action on an object (as seen in mirror neurons), and thereafter select an appropriate action to mimic the effect on the object. The experiments together underline the central theme that learning to act on objects is very important, not only to get better at interacting with future objects, but also to interpret the actions of others.

V. ACKNOWLEDGMENTS

Work on Cog was funded by DARPA as part of the "Natural Tasking of Robots Based on Human Interaction Cues" project under contract number DABT 63-00-C-10102, and by the Nippon Telegraph and Telephone Corporation as part of the NTT/MIT Collaboration Agreement.

Work on BabyBot was funded by European Commission Information Society Technologies branch, as part of the "Cognitive Vision Systems" project under contract number IST-2000-29375, and also as part of the MIRROR project under contract number IST-2000-28159.

VI. REFERENCES

[1] G. Metta, G. Sandini, L. Natale, R. Manzotti, and F. Panerai, "Development in artificial systems," in *Proc. EDEC Symposium at the International Conference on Cognitive Science*, (Beijing, China), Aug. 2001.

[2] J. Lackner and P. DiZio, "Aspects of body self-calibration," *Trends in Cognitive Sciences*, vol. 4, July 2000.

[3] A. Sirigu, J. Grafman, K. Bressler, and T. Sunderland, "Multiple representations contribute to body knowledge processing. evidence from a case of autopagnosia," *Brain*, vol. 114, pp. 629–642, Feb 1991.

[4] M. Jeannerod, *The Cognitive Neuroscience of Action*. Cambridge Massachusetts and Oxford UK: Blackwell Publishers Inc., 1997.

[5] V. Gallese, L. Fadiga, L. Fogassi, and G. Rizzolatti, "Action recognition in the premotor cortex," *Brain*, vol. 119, pp. 593–609, 1996.

[6] A. Fagg and M. Arbib, "Modeling parietal-premotor interaction in primate control of grasping," *Neural Networks*, vol. 11, no. 7–8, pp. 1277–1303, 1998.

[7] J. Gibson, "The theory of affordances," in *Perceiving, acting and knowing: toward an ecological psychology* (R. Shaw and J. Bransford, eds.), pp. 67–82, Hillsdale NJ: Lawrence Erlbaum Associates Publishers, 1977.

[8] M. Jeannerod, M. Arbib, G. Rizzolatti, and H. Sakata, "Grasping objects: the cortical mechanisms of visuomotor tranformation," *Trends in Neurosciences*, vol. 18, no. 7, pp. 314–320, 1995.

[9] G. Sandini and V. Tagliasco, "An anthropomorphic retina-like structure for scene analysis," *Computer Vision, Graphics and Image Processing*, vol. 14, pp. 365–372, 1980.

[10] R. Brooks, C. Breazeal, M. Marjanovic, and B. Scassellati, "The Cog project: Building a humanoid robot," *Lecture Notes in Computer Science*, vol. 1562, pp. 52–87, 1999.

Motor Learning Model using Reinforcement Learning with Neural Internal Model

Jun Izawa, Toshiyuki Kondo, and Koji Ito
Department of Computational Intelligence and Systems
Science, Tokyo Institute of Technology(TIT)
4259, Nagatuta Midori, Yokohama Kanagawa 226-8502,Japan.
e-mail)izawa@ito.dis.titech.ac.jp

Abstract— The present paper proposes a learning control method for the musculoskeletal system of arm based on reinforcement learning. An optimization for the hand trajectory and muscle's force distribution is needed to acquire the reaching motion. The proposed architecture can acquire an optimized motion through learning the task. However, the biological control system composed of musculoskeletal system is not able to sense the state without time delay. The time delay causes instability of learning. The proposed scheme consists of the reinforcement learning part and neural internal model. Neural internal model is employed to compensate for the time delay by estimating the state of muscloskeletay system. Then, there must be a modeling error if some noise is included. Thus we introduce the minimum modeling error criterion for reinforcement learning, which gives not only the reduction of total muscle level but also the smoothness of the hand trajectory. The effectiveness and the biological plausibility of the present model is demonstrated by several simulations.

I. INTRODUCTION

Recently, reinforcement learning attracts attention as a learning method including a planning of movements[5][6][1]. J.C.Houk et al.[3] have proposed the hypotheses that the basal ganglia might involves actor-critic learning which is a kind of reinforcement learning. Moreover, some researchers on human movement analysis believe that reinforcement learning is crucial for understanding human movements[4]. It is difficult, however, to apply reinforcement learning to the system with any hidden state of the environment, that is, with non-Markov property. Biological control systems include the long time delay (at least 100ms) associated with neural transmission and neural computation, which means non-Markov property.

Now we propose on interacting motor learning method based on reinforcement learning, which introduces a neural internal model to compensate for the time delay. The internal model predicts the state of the environment, with learning the environmental dynamics.

II. MUSCULOSKELETAL SYSTEM

Now, we assume that the muscle force $T \in R^n$ is modeled as

$$T(l,\dot{l},u) = K(u)(l - l_r(u)) + B(u)\dot{l}, \qquad (1)$$

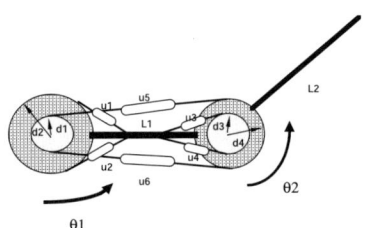

Fig. 2. Musculoskeletal arm

$$K(u) = diag(k_0 + k_i u_i), \qquad (2)$$
$$B(u) = diag(b_0 + b_i u_i), \qquad (3)$$
$$l_r(u) = [l_{r1}, l_{r2}, l_{r3}, \cdots, l_{rm}]^T, \qquad (4)$$
$$l_{ri} = l_0 + r u_i, \qquad (5)$$

where K, B are the coefficient matrices of elasticity and viscosity, respectively u is the muscle activation vector, m is the dimension of the muscle motor command vector, l is the muscle length vector, and l_r is the equilibrium length vector of the muscle. The relation of the infinitesimal displacements between the muscle length $l = (l_1, l_2, \cdots, l_m)^T$ and the joint angle $\theta = (\theta_1, \theta_2, \cdots, \theta_j)^T$ is given by

$$dl = G(\theta)d\theta, \qquad (6)$$

where G is the Jacobian matrix.

Then, from the principle of virtual work, the relation of the joint torque τ and the muscle tension T is obtained as follows.

$$\tau = -G^T T. \qquad (7)$$

We adopt , as an example for application, the musculoskeletal arm which consists of a two link arm with two joints and six muscles, where mono and bi-articular muscles are embedded as shown in Fig.2. The model is simple, but has the essential feature of a musculoskeletal system. Assuming that the moment arm on the adhesion point of the muscle is constant, we can obtain Jacobian G as follows.

$$G = \begin{bmatrix} -d_1 & d_2 & 0 & 0 & -d_5 & d_6 \\ 0 & 0 & -d_3 & d_4 & -d_7 & d_8 \end{bmatrix}^T. \qquad (8)$$

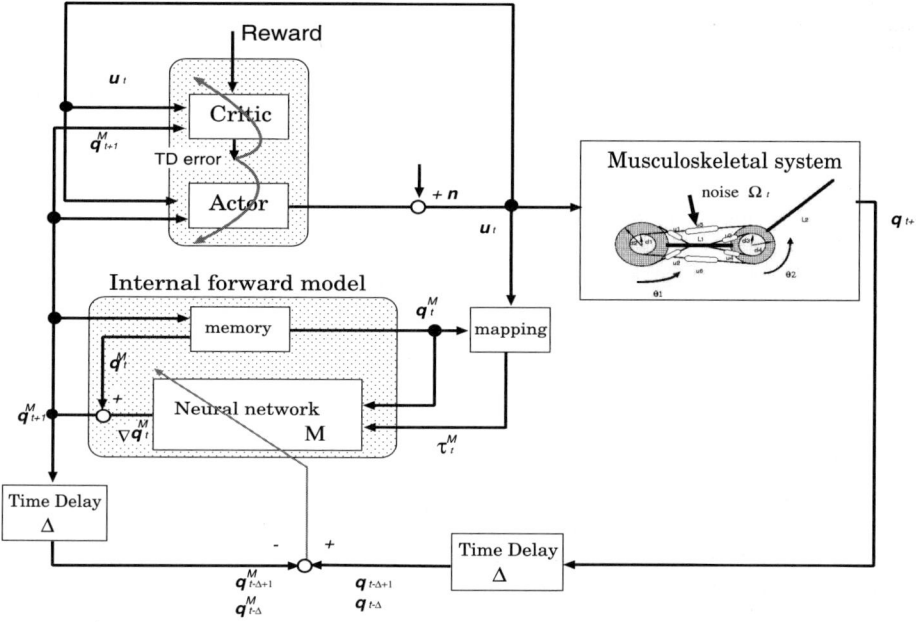

Fig. 1. Block diagram of proposed learning system. The diagram consists of the actor-critic network, musculoskeletal system of arm and internal models are embedded.

From (6) and (7), we obtain

$$\tau = G^T K_m G[-r(G^-)u - \theta] - G^T B_m G\dot{\theta} \quad (9)$$
$$= R(u)[\theta_v(u) - \theta] - D(u)\dot{\theta}, \quad (10)$$

where R and D are the coefficient matrices of elasticity and viscosity in the joint spaces, and θ_v is called the virtual equilibrium point.

A moment arm is defined as follows. Muscle 1 and 2 is 4.0cm. Muscle 3 and 4 is 2.5cm. Muscle 5 and 6 is 3.5cm. A parameter of muscle model is defined so that k, k_0, b, b_0, r are 100N/m, 2021N/m, 100Ns/m, 200Ns/m and 0.2 respectively.

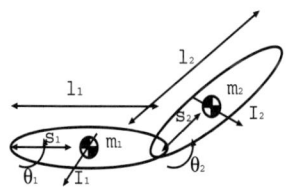

Fig. 3. Two joint arm

Dynamic equation of 2 link arm (Fig.3) is

$$\tau_1 = M_{11}\ddot{\theta}_1 + M_{12}\ddot{\theta}_2 + h_{122}\dot{\theta}_2 + 2h_{112}\dot{\theta}_1\dot{\theta}_2$$
$$\tau_2 = M_{21}\ddot{\theta}_1 + M22\ddot{\theta}_2 + h_{211}\dot{\theta}_1, \quad (11)$$

where,

$$M_{11} = m_1 s_1^2 + I_1 + m_2(l_1^2 + s_2^2 + 2l_1 s_2 \cos\theta_2) + I_2$$
$$M_{12} = M_{21} = m_2(s_2^2 + l_1 s_2 \cos\theta_2) + I_2$$
$$M_{22} = m_2 s_2^2 + I_2$$
$$h_{122} = h_{112} = -h_{211} = -m_2 l_1 s_2 \sin\theta_2. \quad (12)$$

Each parameter is shown in Tab.I

TABLE I
PARAMETERS OF TWO-LINK ARM

	Link1	Link2
m_i[kg]	1.59	1.44
l_i[m]	0.35	0.35
s_i[m]	0.18	0.21
I_i[kg·m^2]	1.63×10^{-2}	1.64×10^{-2}

III. SYSTEM ARCHITECTURE

Fig.1 shows an overview of the proposed architecture. The schematic diagram consists of Actor-Critic network[2], musculoskeletal system, and internal forward model.

The musculoskeletal system receives a motor command $u_t \in R^n$ and outputs the joint angles and angular velocities $q_{t+1} = (\theta_{t+1}, \dot{\theta}_{t+1}) \in R^m$ as the state values of the system. The motor command u_t consists of the output of the Actor-network and the search noise. Another noise Ω_t derived from a fluctuation in motor neurons is added to the motor command. The internal forward model predicts the state of the system q_{t+1}^M with receiving the motor command u_t. The state in the internal forward model q_t^M is an estimated value of q_t. The output of the neural network ∇q_t^M is a variation of the state value at each step. Thus, the state of the next step q_{t+1}^M is calculated by adding the variation ∇q_t^M to the current state q_t^M. The state of the musculoskeletal system is observed with a time delay Δ. The internal forward model consists of the neural network, the memory for current states and the mapping from the motor command to the joint torque.

The mapping from the motor command u_t to the joint torque τ_t^M is calculated based on Eq.(10). The weight parameter of the Actor-Critic network is updated so as to decrease the TD-error.

The state value received by the Actor-Critic is defined as s and distinct from the state value of the musculoskeletal system q because, when $n > m$, the system dose not obtain an optimal value through reinforcement learning based on only q. Thus, in the present paper, reinforcement learning is executed receiving $s_t = (q_t^M, u_{t-1})$ as a state value.

IV. REINFORCEMENT LEARNING WITH THE INTERNAL FORWARD MODEL

We applied the reinforcement learning with the internal forward model to the musculoskeletal system. Fig.2 shows an overview of the algorithm. The musculoskeletal system consists of two-link rigid body and six muscle-like actuators. The muscle model has the spring-like property and generates a muscle force by moving a rest length of the spring. It is, here, assumed that the insertion point of the muscle is constant irrespective of the change of the joint angle.

The initial position of the hand is adjusted such that the end point is at the start position defined in Fig.5. The initial value of u is preset to 0. The internal forward model includes the neural network(NN) whose units consists of sigmoidal function. The number of unit in middle layer of the neural network is twelve. The updating ratio of the network is 0.1. The Actor-Critic network is made of Gauss-Sigmoid NN[7]. The number of units in input layer of sigmoidal neural network embedded in Gauss-Sigmoid NN is 86. The number of units in meddle layer is twelve in Critic-network and is eighteen in Actor-network.

The initial weight of neural network is preset randomly except the weight between the middle layer and output layer which is preset to 0 except for bias unit. The weight related to bias unit is preset to 1.

The discount rate of the value function in reinforcement learning is 0.99. The search noise within the motor command n is made of low-pass filtered white noise with the variance σ. The low-pass filter is $\tau\dot{n} = -n + n_{in}$, where the time constant $\tau = 0.1$. The deviation of the noise is

---

Initialize the critic network and the actor network
Repeat(for each trial):
1) Initialize $s_0^M = (q_0^M, u_0)$
2) Repeat(for each step)
   a) Observe $s_t^M = (q_t^M, u_{t-1})$
   b) $u_t \leftarrow A(s_t^M, w_a) + n_t$
   c) Calculate the state of the system and the forward model,
   $$q_{t+1} \leftarrow P(q_t, u_t)$$
   $$q_{t+1}^M \leftarrow q_t^M + M(q_t^M, \tau_t^M, w_M)$$
   d) Update the Forward-model-network, M
      input : $q_{t-\Delta}^M, \tau_{t-\Delta}^M$
      target : $q_{t-\Delta+1} - q_{t-\Delta}$
   e) If $t \geq \Delta$, initialize $s_0^{M\prime} = (q_0^{M\prime}, u_0^\prime)$, start the internal learning process.
      i) Observe $s_{t-\Delta}^{M\prime} = (q_{t-\Delta}^{M\prime}, u_{t-1-\Delta}^\prime)$
      ii) $u_{t-\Delta}^\prime \leftarrow A(s_{t-\Delta}^{M\prime}, w_a) + n_{t-\Delta}$
      iii) Calculate the state of the forward model,
      $$q_{t-\Delta+1}^{M\prime} \leftarrow q_{t-\Delta}^{M\prime} + M(q_{t-\Delta}^{M\prime}, \tau_{t-\Delta}^\prime, w_M)$$
      iv) Calculate the TD-error,
      $$\delta_{t-\Delta}^\prime \leftarrow r_{t-\Delta+1}^\prime + \gamma \hat{V}(s_{t-\Delta+1}^{M\prime}, w_c) - \hat{V}(s_{t-\Delta}^{M\prime}, w_c)$$
      v) Train the Critic-network
         input : $s_{t-\Delta}^{M\prime}$
         target : $\hat{V}(s_{t-\Delta}^{M\prime}, w_c) + \alpha \delta_{t-\Delta}^\prime$
      vi) Train the Actor-network
         input : $s_{t-\Delta}^{M\prime}$
         target : $A(s_{t-\Delta}^{M\prime}, w_a) + \beta \delta_{t-\Delta}^\prime$.
$n_{t-\Delta}$

Fig. 4. Learning algorithm

20% of the output and adjusts to be 1% as the increase of $\hat{V}$.

The reaching motion is selected as an example showing the effectiveness of the architecture. The agent has to move the hand position from the start point to the goal area.

Now, it is assumed that a biological noise is included in the muscle activation. And it is assumed that the variance of the noise is proportional to the squared value of the motor command.

These causes a modeling error, which gives a bad effect on learning. From this, we introduce *minimum modeling error criterion* for the reaching motion. Then, the cost can be expressed by the reward definition as follows.

$$r_E = (x_t - x_t^M)^T(x_t - x_t^M), \qquad (13)$$

where, x^M is the estimated state of hand position. As a result, a reward definition in the present paper can be expressed as follows.

$$r = \begin{cases} 1 - c \cdot r_E, & \text{for } x_H \in S \cap x_H \in G \\ 0, & \text{for } x_H \in S \cap x_H \notin G \\ -1, & \text{for } x_H \notin S \end{cases} \quad (14)$$

where, x_H, S, G and c is the hand position, the working area, the goal area, weighting coefficient respectively.

V. SIMULATION

A. Learning based on the state prediction

The following results show that the state prediction with the internal forward model is effective for reinforcement learning. Learning of Actor-Critic network was executed in parallel with updating the weight parameter of internal forward model. The reward was given when the end point position reached in goal area (radius is 2cm). In this simulation, the variation in muscle activity and the minimum modeling error criterion is not included. Note that, $c = 0$.

First, the result for the system without time delay is shown in Fig.5. Although the movement of arm seems exploratory in the early stage of learning, the smooth hand path is acquired in the final stage. The learning is exactly converged.

Next, Fig.6 shows the hand paths of the system with time delay. Note that the system has no internal forward model and did not involve the state prediction. In each condition, (a)(b)(c), the system has time time delay in the state observation which is 50ms, 100ms or 200ms respectively. The longer time delay is, the more remarkable the vibration of the end point is. Moreover, learning a reaching motion failed in the case of 200ms.

On the other hand, as shown in Fig.7, the hand path is smooth when the system has the internal forward model to compensate for the time delay. The hand trajectory is smooth without vibration. If the radius of the goal area is smaller, the amplitude of vibration will be reduced, because it seems to be damped into goal area as seen in Fig.7. In addition, Fig.8 shows that the system with the state prediction can get more rewards than the system without the state prediction. Thus, the performance of the learning system can be improved by the state prediction when the system involves time delay in the state observation.

B. Minimum modeling error criterion

Next, we introduced the minimum modeling error criterion into the reinforcement learning with internal forward model. The noise was added to the motor command during leaning. The internal forward model was learned in parallel with reinforcement learning from the initial stage of learning. c in Eqn.14 is preset to 5. Then c is adjusted to be 10 as progress of learning because, if c is preset high in the initial stage, the system become difficult in acquiring enough reward.

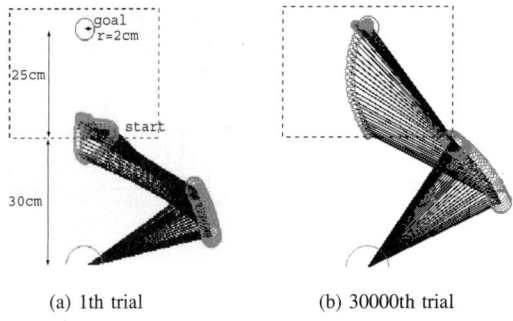

(a) 1th trial   (b) 30000th trial

Fig. 5. Stick picture of acquired reaching motions:(a) The picture of the 1st trial. At the initial state of the reaching motion, the distance between the shoulder joint and the hand position is 30cm, and the distance between the hand position and the center of target goal is 25cm. The radius of goal area is 2cm.(b)The picture of the 30000th trial. The reaching motion is acquired with roughly straight hand path.

Fig.9 shows the hand path and the tangential velocity after 30000 trials, in which the time delay is 100ms. The hand path is straight and the tangential velocity seems to be bell-shaped. Comparing with Fig.7, the peak of velocity decreases and the velocity profile becomes smooth. That is to say, the minimum modeling error criterion has much effect on the smoothness of hand trajectory.

Fig.10 shows the resultant muscle forces. Although we preset the motor command in the initial stage so that each muscle force can be $250N$ or so, the acquired muscle forces is between $30N$ and $100N$. These indicate that the total muscle force decreased as a result of leaning.

Fig.11 shows the total muscle force of six muscles at each trial. We simulated not only when the variance of the noise in muscle activation was proportional to the squared motor command but also when the variance of the noise was constant. The total muscles force decreases as a trial number when the variance correlates with the motor command. On the other hand, the total muscle force dose not decrease when the variance is constant, which means the system can not acquire the motion keeping the total muscle force low.

Accordingly, the reward given with the minimum modeling error criterion makes the hand path straight and makes tangential velocity smooth with decreasing the total muscle force.

Minimizing the modeling error at the goal area is identical to minimizing the variance of the hand position. Eq.(13) is related to the variance of hand position when the variance of noise added to the muscle activation is proportional to the squared motor command. Minimizing the modeling error is identical to minimizing the weighted squared summation of the motor command through time. This is why the total muscles force decreases. That is to

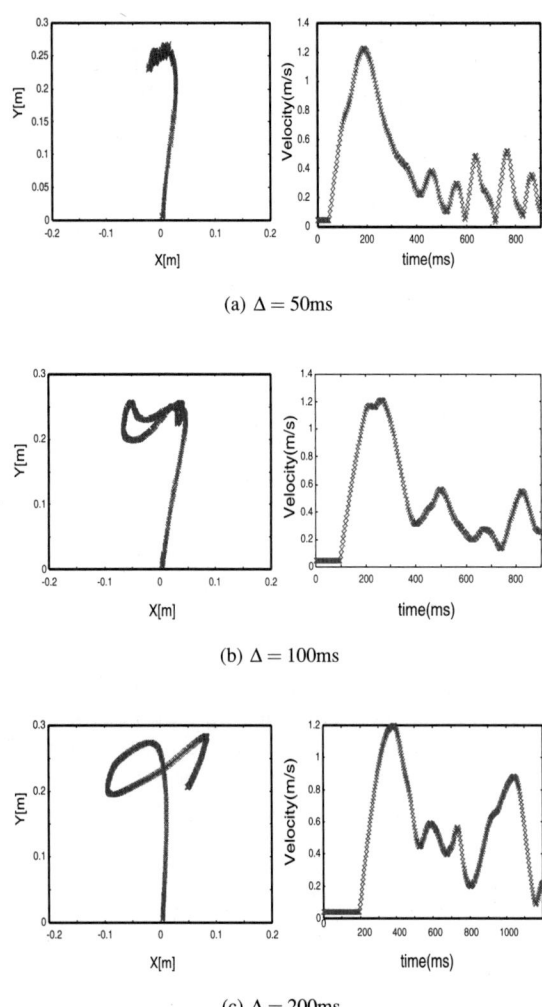

Fig. 6. Hand trajectory and tangential velocity after 15000 trials without forward model regardless of time delay

say, because the minimum modeling error criterion keeps the total muscle force low, the peak of the hand velocity decreases and the hand path becomes straight.

VI. CONCLUSION

The proposed method is essentially composed of the internal model and Actor-Critic network. The internal model is made up through the supervised learning in parallel with reinforcement learning. The difficulty caused by the time delay included in the neural transmission is solved by the state prediction with neural internal model. In addition, we proposed the minimum modeling error criterion as a reward definition. This enables the system obtain smooth movements with the low total muscle force which is reasonable for metabolic cost. Until now, some motor learning models with the forward model and

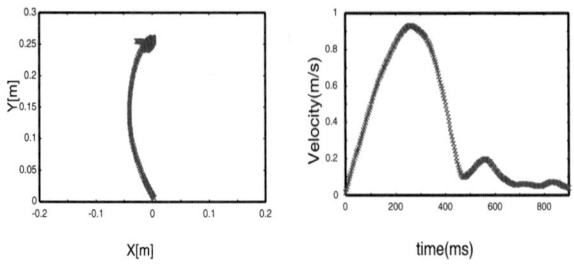

Fig. 7. Hand trajectory and tangential velocity after 30000 trials with the forward model to compensate for the time delay(200ms)

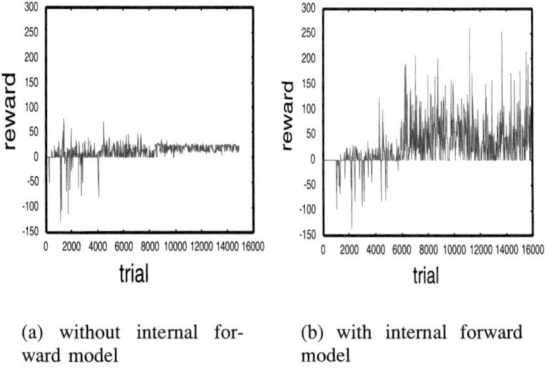

(a) without internal forward model

(b) with internal forward model

Fig. 8. Cumulative reward against trial number. If the reaching motion is realized very well, the cumulative reward per one trial is high level.

reinforcement learning has been proposed. However, those models involve no muscle systems and no time delay in neural transmission from the muscle skeletal systems to the central nervous systems. In the present paper, it is assumed that the time delay is known in advance. In the case of real biological motor learning, it is not probable that the delay time is given. Our future work is to propose a learning model with the unknown time delay.

ACKNOWLEDGMENT

A part of this research was supported by Takahashi Industrial and Economic Research Foundation, Grant-in-Aid(14350277,14750362) for scientific research,JSPS, and Mitutoyo Association for Science and Technology.

VII. REFERENCES

[1] K. Althoefer, B. Krekelberg, D. Husmeire, and L. Seneviratne. Reinforcement learning in a rule-based navigator for robotic manipulators. *Neurocomputing*, 37:51–70, 2001.

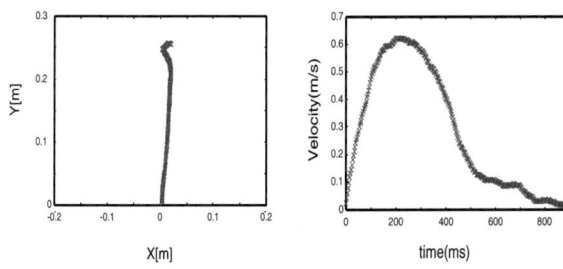

Fig. 9. Hand trajectory and tangential velocity after 30000 trials with the forward model to compensate for the time delay(100ms) and the minimum model prediction error criterion.

[2] A. F. Barto, R. S. Sutton, and C. W. Abderson. Neuronlike adaptive elements that can solve difficult learning control problem. *IEEE Transactions on Systems, Man, and Cybernetics*, 13(5):834–846, 1983.

[3] J. C. Houk, J. L. Adams, and A. G. Barto. A model of how the basal ganglia generate and use neural signals that predict reinforcement. In *Models of Information Processing in the Basal Ganglia*. MIT Press, Canbrigde, MA, USA, 1994.

[4] M.I. Jordan and D.M. Wolpert. Computational motor control. In M.Gazzaniga, editor, *The Cognitive Neurosciences*, pages 601–620. MIT Press, 1999.

[5] P. Martin and J.R. Millan. Learning reaching strategies through reinforcement for a sensor-based manipulator. *Neural Network*, 11:359–376, 1998.

[6] P. Martin and J.R. Millan. Robot arm reaching through neural inversions and reinforcement learning. *Robotics and Autonomous Systems*, 31:227–246, 2000.

[7] Katsunari Shibata, Masanori Sugisaka, and Koji Ito. Hand reaching movement acquired through reinforcement learning. In *Proc. of 2000 KACC (Korea Automatic Control Conf.)*, volume 90rd (CD-ROM), 2000.

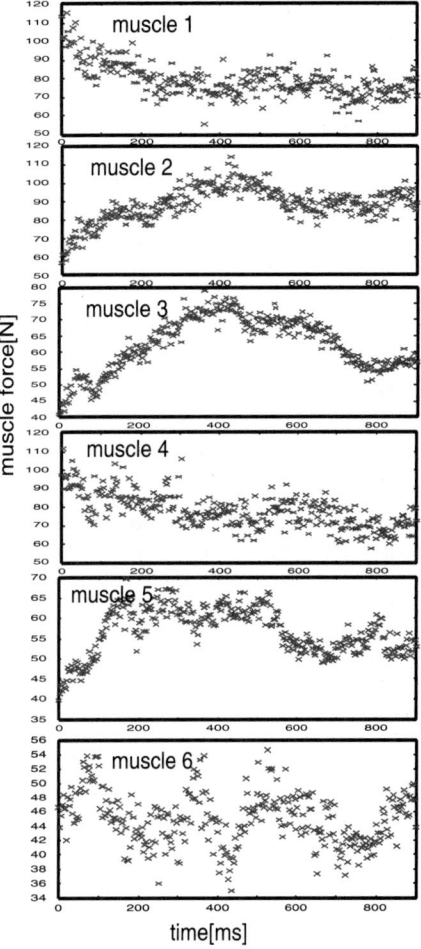

Fig. 10. Muscle force (30000 trials). The number of muscle corresponds to the number in Fig.1. The noise is added to the motor command launched from Actor-network. This gives a variance in muscle force.

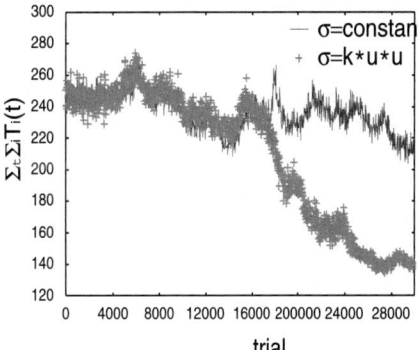

Fig. 11. Total muscle force summation against trials. The line plotted with "-" indicates the total muscle force when the variance of noise is independent from activation level. The line plotted with "+" indicates the total muscle force when the variance of noise depends on activation level.

Statistical Analysis and Comparison of Questionnaire Results of Subjective Evaluations of Seal Robot in Japan and U.K.

Takanori Shibata [1,2], Kazuyoshi Wada [1,3], and Kazuo Tanie [1,3]

1 National Institute of Advanced Industrial Science and Technology
1-1-1 Umezono, Tsukuba 305-8568, Japan
E-mail: shibata-takanori@aist.go.jp
2 PERESTO, JST
3 Univ. of Tsukuba

Abstract: This paper describes research on mental commit robot that seeks a different direction from industrial robot, and that is not so rigidly dependent on objective measures such as accuracy and speed. The main goal of this research is to explore a new area in robotics, with an emphasis on human-robot interaction. In the previous research, we categorized robots into four categories in terms of appearance. Then, we introduced a cat robot and a seal robot, and evaluated them by interviewing many people in Japan. The results showed that physical interaction improved subjective evaluation. Moreover, a priori knowledge of a subject has much influence into subjective interpretation and evaluation of mental commit robot. In this paper, 440 subjects evaluated the seal robot, Paro by questionnaires in an exhibition at Science Museum in London, U.K. This paper reports the results of statistical analysis of evaluation data, and compares them with those obtained in Japan.

Fig. 1 Seal Robot: Paro

Fig. 2 Demonstration at Science Museum in London

1. Introduction

Robots that coexist symbiotically with people are machines to affect human hearts and lay stress on people's subjective evaluation, although they sometimes do physical work [1]. Such robots should be designed as open systems considering unknown environment and the minds of users, not as closed systems and predetermined environments as with industrial robots. Because these are complicated requiring consideration to people and sensitivity, they cannot be designed by conventional scientific and technological concepts alone [2, 3].

We studied and developed artificial emotional creatures as the examples of robots that coexist with people [1, 3-9]. The artificial emotional creatures exist as subsidiaries in everyday life similar to pets, hold equal relationships with people, move by orders from people and act autonomously, and contact with people [10, 11], giving people pleasure and relaxation. They prevent mental illness by enriching and healing the humans' heart, give pleasure to daily leisure, and commit themselves to the humans' mental part. They are referred to as "mental commit robots".

Seal type mental commit robot was applied to therapy of children at pediatric hospital and assisting activity of elderly people at elderly institutions [12-14]. The results showed that there were psychological merit, physiological merit, and social merit by interaction with the seal robot.

A study was conducted by questionnaires in "Exhibition of Dream Technology" held for about 3 weeks from July to August 2000, on how people subjectively evaluate robots in human interaction, a characteristic of mental commit robots [15]. Seal robots were used for evaluation. Respondents were exhibition visitors, and after demonstrators explained the purpose and functions of the robot, they were requested to interact with the robot. Multiple questions about the robot were asked by questionnaires. The tabulation results of subjective evaluation provided high evaluation on the whole. In addition, important factors to be key points in robot evaluation were extracted from the results of principal component analysis, thus confirming relationship between them and factors such as shape and the feel emphasized in design. Respondents were grouped to find differences in evaluation among groups. These results revealed that the evaluation factors of "favorable impression by contacting" and "favorable impression by appearance" are important as those pertaining to evaluation in the design and production of mental commit robots to be highly appreciated in subjective evaluation.

Another study was conducted by questionnaires in "Japan: Gateway to the Future, Digital Technology Exhibition" held for about 7 weeks from February to March 2002 at the Science Museum in London, U.K., on how people subjectively evaluate robots in human interaction, a characteristic of mental commit robots. The way of study was the same as the previous one in Japan. This paper reports results of statistical analysis of evaluation data obtained in U.K., and compares them with those in Japan.

The developmental purpose and process of the mental commit robot are stated in Chapter 2. The method of subjective evaluation is explained in Chapter 3, statistical analysis of results of evaluation data are discussed in Chapter 4, and they are summarized in Chapter 5.

2. Seal Robot: Paro

2.1. Mental Commit Robot
Mental commit robots are intended to offer people not physical work or service but mental effects such as pleasure and relaxation as personal robots. Robots act independently with purposes and motives while receiving stimulation from the environment as with living organisms. Actions manifested by interaction with people are interpreted by people as if robots have hearts and feelings.

2.2. Previous Process
A basic psychological experiment was conducted on the subjective interpretation and evaluation of robot behavior by people in the interaction with the robot. This showed the importance of appropriately stimulating human senses and extracting associations. Sensor systems such as visual, hearing, and tactile senses for robots were studied and developed. In the tactile sense, a plane tactile sensor using an air bag was developed to cover the robot for bodily contact between people and robots. This can detect position and power when people contact the robot, and at the same time, it allows people to feel softness. Dog, cat, and seal robots were developed using sensors.

In the subjective evaluation of mental commit robots in this paper, a mental commit seal robot "Paro" version 6 was used.

2.3. Seal Robot "Paro"
The robot and major functions are shown in Fig.1. The appearance was designed with a harp seal baby as a model, and the surface is covered with pure white fur. A newly developed plane tactile sensor is inserted between the hard inside skeleton and fur to express the soft natural feel and to permit the measurement of human contact with the robot. The robot has the 4 senses of sight (light sensor), audition (determination of sound source direction and speech recognition), and equilibrium in addition to the above-stated tactile sense. Mobile parts are as follows: vertical and horizontal neck movements; front and rear paddle movements; and each eyelid movement important as facial expression. The robot acts by using the 3 elements of the internal states of the robot, sense information from sensors, and daily rhythm (morning, daytime, and night) to manifest activities through interaction with people.

3. Purposes and Method of Subjective Evaluation

3.1. Purpose of Subjective Evaluation
Subjective evaluation was done with cat robots for 88 respondents and seal robots for 785 respondents in the previous studies. The respondents were all from Japan. The objectives of this paper were that questionnaires from hundreds of persons in London, U.K., were conducted on a large scale to collect subjective evaluation data about interaction with robots and analyze them statistically, thus analyzing and classifying the tendency of robot evaluation and elements as evaluation points. It was also intended to collect requests, opinions, and impressions to the development of mental commit robots that are used for future R&D.

3.2 Method of Subjective Evaluation
In "Japan: Gateway to the Future, Digital Technology Exhibition" (visitors: 109,713) held at Science Museum in London, U.K. for 46 days from February 1st to March 16th, 2002, seal robots Paro were displayed. Demonstration was to be done through the whole period, so 2 Paros were prepared before experiments.

In the display booth, demonstrators conducted

Table 1 Question about Subjects

Personal questionnaire
1 Sex ?
2 Age ?
3 Nationality ?
4 Occupation ?
5 Do you like animals?
6 Are you owning any pets or do you want to own pets?
7 Why do you not have any pets?
8 Do you know about real baby seals?
9 How much free time do you have in daily life?
10 How much extra money do you have in daily life?

Table 2 Question about Subjective Evaluation of Paro

Evaluation questionnaire
1 Cute
2 Want to pet it
3 Want to talk to it
4 Has vitality
5 Easy to get friendly with
6 Has real expressions
7 Natural
8 Feels good to the touch
9 Fun to play with
10 Comfortable play with
11 Relaxing
12 Like
13 Needed in this world
14 Want it for myself
15 Would give as a present

four to five demonstrations a day. After the developmental process, purpose, and the functions of mental commit robots were explained for about 5 minutes, Paro was placed on the table to allow the audiences who visited the display booth to touch the robot freely. A questionnaire was given to visitors who after interaction, and accepted this filled out the form. Some 440 questionnaires were obtained during the first 11 days.

3.2.1. Questionnaire

The contents of the questionnaire are classified largely into questions about the respondent on the front of the questionnaire form (Table 1); questions of 5-grade evaluation for subjective evaluation on the front (Table 2); and questions about impressions and requests of Paro on the back. The answer time was different from person to person, but 5-10 minutes were necessary for each of front and back questions, so those on the back were asked only visitors with free time. Many children were included, so parents were requested to help them in answering questionnaires.

3.2.2. Condition of Demonstration (Interaction)

Paro was placed on the table to allow visitors to the display booth to interact freely with it. The front of the display booth was sometimes quite crowded and yet every cycle of demonstration took about 30 minutes to permit as many people as possible to experience interaction by turn. Interaction mainly involved the actions of contacting and stroking, and women hugged it in some cases. In other cases, visitors called its name, or brought their faces close to it.

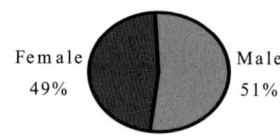

(a) Distribution by Gender

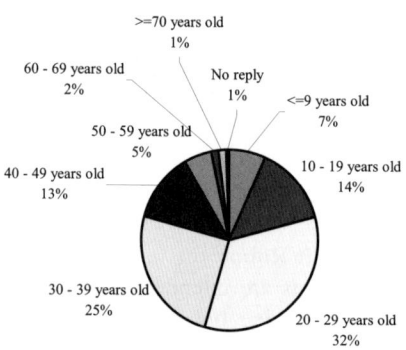

(b) Distribution by Age

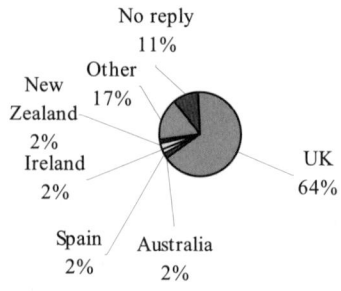

(c) Distribution by Nationality

Fig. 3 Proportion of Gender, Age, and Nationality of Respondents

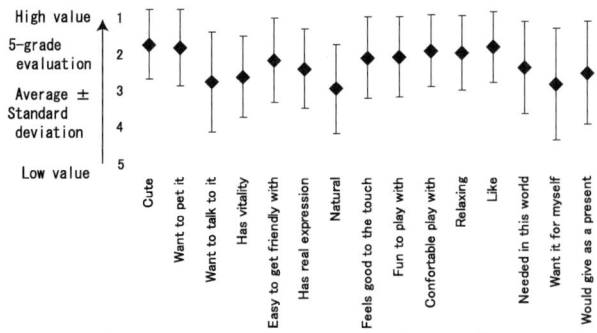

(a) 320 sheets at Science Museum in London, U.K.

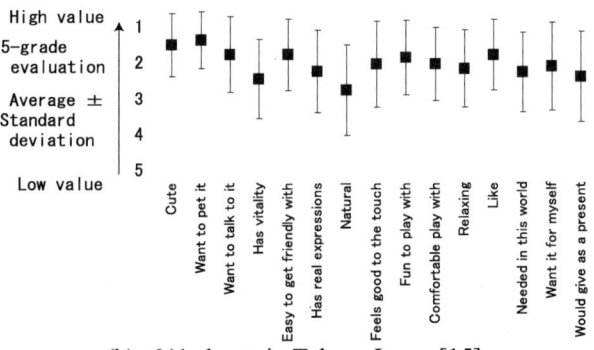

(b) 641 sheets in Tokyo, Japan [15]

Fig. 4 Average and Standard Deviation of Answers to 15 Questions about Subjective Evaluation

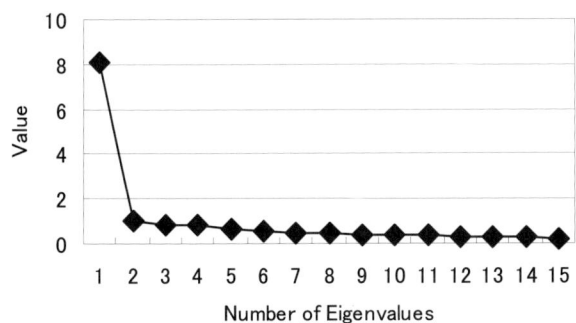

Fig. 5 Eiganvalues in Principal Component Analysis

4. Results and Discussion

4.1. Answer Results of Questionnaires

Fig.3 (a), (b), and (c) show the proportion of gender, age, and nation of 320 respondents who filled out complete answers to 15 questions. Fig.4 (a) gives the values of average and standard deviation of answers to the 15 questions of subjective evaluation at Science Museum in London, U.K. The results of subjective evaluation indicate that favorable answers were obtained for the most part, and the variables of "Cute," "Want to pet it," and "Like" are highly evaluated. The results were very similar to those in the previous study in Tokyo, Japan as shown in Fig. 4 (b).

4.2. Principal Component Analysis of Subjective Evaluation

Multivariate analysis was conducted to evaluate the answers of subjective evaluation comprehensively. In analysis, there is no external criterion and the contents of all 15 questions are explanatory variables for Paro evaluation; the multivariate analysis techniques of such variables include principal component analysis, factor analysis, and latent structure analysis.

Principal component analysis was used to do comprehensive evaluation by 15 explanatory variables, and because cumulative contribution becomes higher than the case of extracting the same number of factors by factor analysis. This analysis was conducted with 320 sheets consisting of complete answers to 15 questions.

4.3. Results and Discussion of Principal Component Analysis

The principal component analysis was done with 320 sheets of complete answers to 15 questions of subjective evaluation to extract one factor with eigen-value of 1 or higher (Fig.5). Kaiser normalization was done to enhance factor interpretation, and then one factor characterized by factor loading shown in Table 3 was obtained.

Table 3 One Factor Characterized Factor Loading

variable	One Factor
	Factor 1
Cute	0.740
Want to pet it	0.736
Want to talk to it	0.691
Has vitality	0.705
Easy to get friendly with	0.789
Has real expressions	0.682
Natural	0.770
Feels good to the touch	0.701
Fun to play with	0.807
Comfortable play with	0.784
Relaxing	0.745
Like	0.837
Needed in this world	0.616
Want it for myself	0.717
Would give as a present	0.640
%total Variance	53.732

4.3.1. Interpretation of Factors

The factor is characterized by the variables of "Easy to get friendly with," "Fun to play with," "Comfortable play with," and "Like." As shown in Fig. 4, average score of these factors are high (about 2).

4.3.2. Comparison of Factor Scores

In the next step, when factor was calculated to obtain the results of actual evaluation, the distribution of factor score for the whole respondents (320) became as in Fig.6.

To analyze factor scores in detail, respondents were grouped and average factor scores were compared between groups. Grouping was done with gender, age, likes and dislikes of animals, and nationality, and the factor score of each group was subjected to Wilcoxon's test, a nonparametric test as a test for mean difference, because the normality or equal variance of each group cannot be expected. The comparison of factor score averages between proportions of respondent number of each group is shown in Fig.7.

In grouping by gender, a significant difference was seen in the score, and women tended to give it high evaluation. In grouping by age, the factor was significant, and the group of less than 20 years old and more than 50 years old tended to evaluate it highly. As for likes and dislikes of animals, the scores of both factors were significant, and the group to dislike animals tended to give factor lower evaluation than the group that liked animals. In grouping by nationality, there was no significant difference.

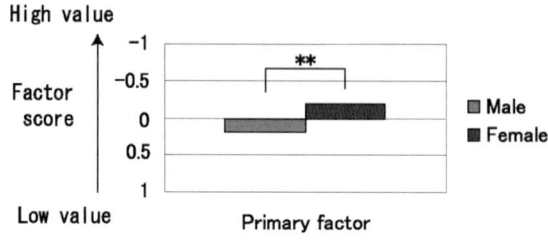

(a) Comparison by Gender

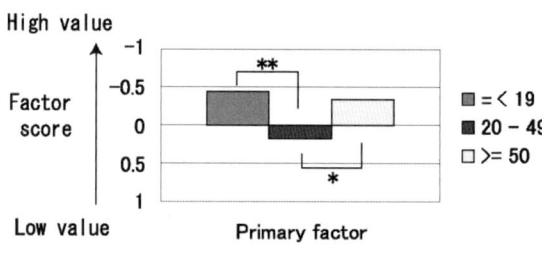

(b) Comparison by Age

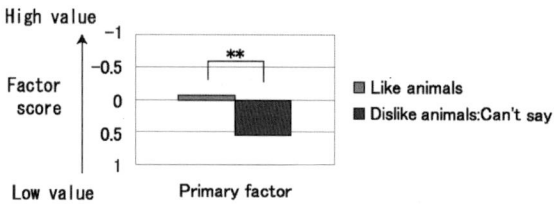

(c) Comparison by Likes and Dislikes Animals

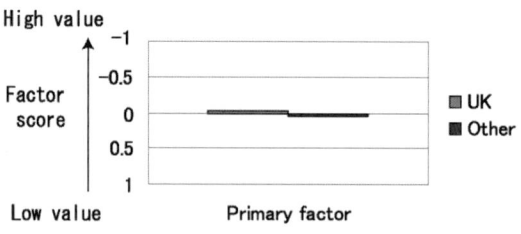

(d) Comparison by Nationality

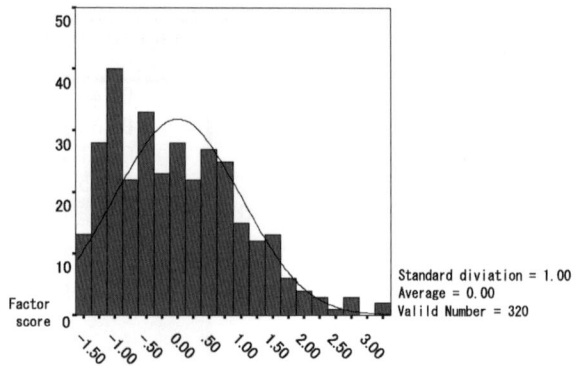

Fig. 6 Factor Score Distribution

Fig. 7 Proportion of Number of Each Group Member and Comparison of Factor Scores Average

5. Conclusions

A seal robot, "Paro," was introduced to many visitors in "Japan: Gateway to the Future, Digital Technology Exhibition" at Science Museum in London, U.K., and interaction with Paro was done to ask questionnaires about the subjective evaluation of Paro.

The tabulation results of subjective evaluation in U.K. provided high evaluation on the whole, and they were similar to those obtained in Japan. In addition, important factors to be key points in robot evaluation were extracted from the results of principal component analysis. Respondents were grouped to find differences in evaluation among groups. These results revealed that the evaluation factors of "favorable impression by contacting" and "favorable impression by appearance" are important as those pertaining to evaluation in the design and production of mental commit robots to be highly appreciated in subjective evaluation. As for Paro, there was not significant difference among nationality.

These results as requests and impressions obtained in questionnaires will be used for the improvement of Paro.

References:

[1] T. Shibata, et al., Emotional Robot for Intelligent System - Artificial Emotional Creature Project, Proc. of 5th IEEE Int'l Workshop on ROMAN, pp.466-471, 1996

[2] H. Petroski, Invention by Design, Harvard University Press, 1996.

[3] T. Shibata and R. Irie, Artificial Emotional Creature for Human-Robot Interaction - A New Direction for Intelligent System, Proc. of the IEEE/ASME Int'l Conf. on AIM'97, paper number 47 and 6 pages in CD-ROM Proc., 1997

[4] T. Shibata, et al., Artificial Emotional Creature for Human-Machine Interaction, Proc. of the IEEE Int'l Conf. on SMC, pp.2269-2274, 1997

[5] T. Tashima, S. Saito, M. Osumi, T. Kudo and T. Shibata, Interactive Pet Robot with Emotion Model, Proc. of the 16th Annual Conf. of the RSJ, Vol. 1, pp.11-12, 1998

[6] T. Shibata, T. Tashima, and K. Tanie, Emergence of Emotional Behavior through Physical Interaction between Human and Robot, Procs. of the 1999 IEEE Int'l Conf. on Robotics and Automation, 1999

[7] T. Shibata, T. Tashima, K. Tanie, Subjective Interpretation of Emotional Behavior through Physical Interaction between Human and Robot, Procs. of Systems, Man, and Cybernetics, pp.1024-1029, 1999

[8] T. Shibata, K. Tanie, Influence of A-Priori Knowledge in Subjective Interpretation and Evaluation by Short-Term Interaction with Mental Commit Robot, Proc. of the IEEE Int'l Conf. on Intelligent Robot and Systems, 2000

[9] T. Shibata, et al., Mental Commit Robot and its Application to Therapy of Children, Proc. of the IEEE/ASME Int'l Conf. on AIM'01, paper number 182 and 6 pages in CD-ROM Proc., 2001

[10] M. M. Baum, N. Bergstrom, N. F. Langston, L. Thoma, Physiological Effects of Human/Companion Animal Bonding, Nursing Research, Vol. 33. No. 3, pp.126-129, 1984

[11] J. Gammonley, J. Yates, Pet Projects Animal Assisted Therapy in Nursing Homes, Journal of Gerontological Nursing, Vol.17, No.1, pp.12-15, 1991

[12] T. Shibata, et al., Mental Commit Robot and its Application to Therapy of Children, Proc. of the IEEE/ASME Int'l Conf. on AIM'01 paper number 182 and 6 pages in CD-ROM Proc., 2001

[13] K. Wada, T. Shibata, T. Saito, K. Tanie, Robot Assisted Activity for Elderly People and Nurses at a Day Service Center, Proc. of the IEEE Int'l Conf. on Robotics and Automation, 2002

[14] T. Saito, T. Shibata, K. Wada, K. Tanie, Examination of Change of Stress Reaction by Urinary Tests of Elderly before and after Introduction of Mental Commit Robot to an Elderly Institution, Proc. of AROB, 2002

[15] T. Shibata, T. Mitsui, K. Wada, and K. Tanie, Subjective Evaluation of Seal Robot: Paro -Tabulation and Analysis of Questionnaire Results, Jour. of Robotics and Mechatronics, Vol. 14, No. 1, pp. 13-19, 2002

A Biologically Inspired Homeostatic Motion Controller for Autonomous Mobile Robots

Do-Young Yoon
Dept. of Electrical Eng., Korea Univ., & ISCRC, KIST. 39-1,Haweolgok -dong, Seongbuk-gu, Seoul, Korea.

Sang-Rok Oh
Intelligent System Control Research Center, Korea Institute of Science and Technology (KIST), 39-1,Haweolgok -dong, Seongbuk-gu, Seoul, Korea.

Gwi-Tae Park
Dept. of Electrical Engineering, Korea Univ., 1, 5-ga, Anam-dong, Seongbuk-gu, Seoul, Korea.

Bum-Jae You
Intelligent System Control Research Center, Korea Institute of Science and Technology (KIST), 39-1,Haweolgok -dong, Seongbuk-gu, Seoul, Korea.

Abstract - This paper proposes an autonomous motion controller for mobile robots. It combines a biologically inspired motion planner and a nonlinear feedback controller in order to make a simple and practical motion controller in natural unknown environments. The motion planner part is designed in the frame of behavior-based scheme and adopts the biological concept "homeostasis." It plays the role of determining desired temporal target posture and modifying controller's gain parameters. The other nonlinear controller part drives the robot to the given target posture from the planner. As a result, the stable nonlinear feedback controller in ideal obstacle free space becomes practical one in natural environments constrained by obstacles. The proposed controller is simple mainly due to its behavior-based characteristics and stable due to the feedback controller considering robot's physical constraints. Simulated results and some experimental feature for autonomous navigation of the differential drive mobile robot, named "MARI", are shown for the present.

I. INTRODUCTION

In order for a robot to operate fully autonomously, it must have ability of decision-making and execution of the decision. The former topic can be referred to as designing a high level planner[1] usually and is studied mainly in the field of artificial intelligence or robot architecture. The latter one can be referred to as designing a low level controller[1] and is studied in automatic control area. However, those two of planner and controller should be combined into a whole robot controller[1] with elaborate manner. In usual robot controller, a planner mainly views environmental information while controller focuses on robot's own action. So, a planner with the assumption that the controller executes its all commands ideally in every situation may not be practical since the robot has constraint in its own action. On the contrary, a controller assuming the ideal free environment is also impractical. This paper proposes an autonomous motion controller for differential drive mobile robot. It is composed of a planner and a controller, which are fused into a whole motion controller. For the mission of navigation in unknown environment, the proposed motion controller deals with the environmental constraint of obstacles and the robot's own constraint of nonholonomic maneuvering [1]. A Behavior-based [2] planner is adequate for the unknown environment since it less relies on prior global knowledge and plans temporal path on-line. The main feature of the proposed behavior-based planner is inspired by biological property of homeostatic strategy. Its main idea is explained in the subsequent section and more details are shown in [3]. It generates possible safe way from collision locally and modifies robot's velocity in harmony with the complexity of environment. A nonlinear feedback controller executes maneuvering to the posture given by the planner. The controller's gain parameters are modified according to the environmental constraints through the planner.

II. THE MOTION CONTROLLER

The proposed autonomous motion control system is composed as Fig. 1 below.

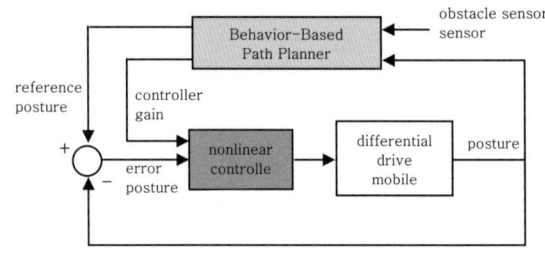

Fig. 1 Overall motion control system

The planner is assumed to have global goal posture $q_g = (x_g, y_g, \theta_g)^T$, and receives consecutively sensor data for detecting obstacles. With these two information, it generates temporal goal posture and then gives it to the controller as the reference posture $q_r = (x_r, y_r, \theta_r)^T$. The controller actuates the robot in order to stabilize to this given reference posture. While the controller tries to stabilize to the reference posture, the robot may collide with obstacles since the controller itself is developed for free space without considering obstacles. The planner knows

---
[1] In this paper, we use these terms to imply the meanings as follows: planner (decision maker), controller (executer), robot controller (whole, planner+controller), motion controller (whole, path planner+controller) in spite that somewhat different usage is possible in general.

information for obstacles and hence it modifies the controller's gains so that the controller stabilizes to the reference posture within a safe spatial and velocity boundary. In the meanwhile, the controller uses nonlinear feedback of current posture $q_c = (x_c, y_c, \theta_c)^T$.

III. THE BEHAVIOR-BASED PATH PLANNER

The planner does not have any prior knowledge such as a map, obstacle positions, etc. except for the goal posture. So, it can decide only next temporal goal posture q_r so that the whole linkage of consecutive temporal goal postures results in the path to the global goal posture. This job can be referred to as on-line step-by-step path planning. The next temporal posture is decided within the obstacle sensor range because the planner cannot know environmental situation beyond the sensor range and hence planning for that area may not be valid.

In the meanwhile, the planner developed by the frame of behavior-based scheme, where the whole needed job is decomposed as the "behaviors" and they are implemented as the form of sensing-action pair. On running time, the robot's resultant action reference is determined by coordination of the predefined behaviors reactively. Hence there is a need of effective strategies for these behavior decomposition and coordination. For the sake of this need, the "homeostatic" behavior-based planner has been proposed [3].

A. Mission versus Homeostasis of a Robot

Most of the behavior-based schemes treat all behaviors as a same kind of external (sub)mission for the robot to accomplish but the proposed scheme views the robot's behaviors in different sight. It views the robot's behaviors as one of two different kind of action. One is external mission behavior as usually viewed and the other is so called "homeostatic behaviors". The term homeostasis, quoted from biology [4], was used in somewhat extended meaning from its original strict meaning in order to imply the tendency or driving force to maintain desired state of the robot itself, where the desired states include mainly vitality of the robot such as safety from crash, battery level maintenance and etc. In other words, the homeostatic behaviors are those for fundamental necessary conditions for all another kind of missions to be executable. Let us apply and hence explain this concept through an example of autonomous navigation of a mobile robot as follows: A behavior set

$$\mathbf{B} = \{Move\text{-}To\text{-}Goal, Avoid\text{-}Obstacle, Stay\text{-}On\text{-}Path, Move\text{-}Ahead\}$$

is assumed to be given for navigation [5]. Here, *Move-To-Goal, Stay-On-Path* and *Move-Ahead* may be considered as mission behaviors while *Avoid-Obstacle* had better be viewed as homeostatic behavior rather than mission because robot's aim is primarily just to arrive goal but *Avoid-Obstacle* is not robot's explicit goal but it can be considered as a kind of necessary condition to achieve the goal. Expanding this example, robot behaviors can be categorized as one of homeostatic behaviors or mission behaviors. Now, the relation of these two kinds of behaviors will be used to coordinate behaviors in order to make the robot's resultant behavior, that is, action reference.

B. Behavior Decomposition and Coordination Algorithm

1) Behavior Decomposition: Design of Necessary Behaviors: Robot's necessary behaviors are decomposed and classified by two factors of homeostasis and "behavior sector". The behavior sector is introduced additionally for parallel execution of behaviors, which don't share same control objective so that they don't need to be coordinated. That is, behaviors, which share same control objective, are grouped together into one behavior sector. From these, a behavior set **B** can be expressed as follows:

$$\mathbf{B} = \bigcup_i \mathbf{B}_i,$$
$$\mathbf{B}_i = \mathbf{B}_{iH} \cup \mathbf{B}_{iM}, \mathbf{B}_{iH} \cap \mathbf{B}_{iM} = \mathbf{0},$$

where, $i = 1, 2, \cdots, n$,

n: number of behavior sector,

B : whole behavior set, (1)

$\mathbf{B}_i$: set of i th behavior sector,

$\mathbf{B}_{iH}$: homeostatic subset
 of i th behavior sector,

$\mathbf{B}_{iM}$: mission subset
 of i th behavior sector,

0 : null set.

Using frame of (1), behaviors can be decomposed into different set according to behavior designer. In this research, navigation is accomplished by only four simple behaviors: $\mathbf{B} = \{InSafety, ToGoal, InBound, InResolution,\}$, and the behavior set is decomposed as Fig. 2.

homeostatic subsets	mission subsets	behavior sectors	objectives
$\mathbf{B}_{1H}$ = {InSafety}	$\mathbf{B}_{1M}$ = {ToGoal}	$\mathbf{B}_1$	direction
$\mathbf{B}_{2H}$ = {InBound}	$\mathbf{B}_{2M}$ = **0**	$\mathbf{B}_2$	sweeping boundary
$\mathbf{B}_{3H}$ = {InResolution}	$\mathbf{B}_{3M}$ = **0**	$\mathbf{B}_3$	maximum velocity

Fig. 2 Decomposition of behaviors for behavior-based navigation.

In Fig. 2, the behavior *InSafety* plays a role of collision avoidance by inhibiting the robot from going to dangerous direction, *ToGoal* lets the robot move to goal direction, *InBound* modifies the controller's gains for the trajectory of the robot position to be within the safe spatial boundary during stabilizing to each temporal reference posture, and *InResolution* decides the robot's maximum velocity in harmony with clutter of the environment.

2) Behavior Coordination: Deciding Resultant Action Reference: Each decomposed behavior has two elements. One is its output value corresponding to robot's action and the other is the parameter, here named "satisfaction level", which is introduced and implemented in each behavior in order to compare significance of each behavior. It indicates degree of achievement of each behavior's own desire so that the behavior having lowest satisfaction level gets the right to be executed. In this paper, satisfaction levels made as real number in the range of minimum 0.0 to maximum 1.0. Example of a designed satisfaction level is shown in Fig. 3 for easy explanation.

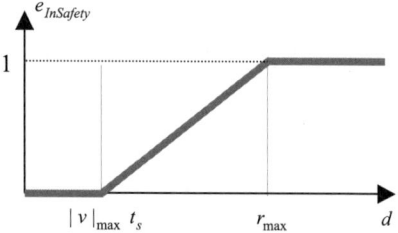

where, $e_{InSafety}$: satisfaction level
d : sensed distance
$|v|_{max}$: maximum possible velocity
t_s : sampling period
r_{max} : maximum possible range value of ranging sensor

Fig. 3 An example of the satisfaction level

Real behavior coordinations occur only inside of each behavior sectors while behavior coordinations among different sectors are not necessary, but they only have to be executed in independently. Behaviors in a same behavior sector are coordinated by following scheme. Referring to (1) and Fig. 2, let the name of each behavior be denoted as behavior itself and its satisfaction level at the same time for simplicity. Then coordination, which includes step for evaluation of satisfaction levels in order to explain overall step, is proposed for the sector *i* as follows:

1. Evaluate satisfaction levels of each behavior.
2. Using the value, select the most unsatisfied homeostatic and mission behavior respectively, that is, choose homeostatic candidate behavior b_{iH}^* and mission candidate b_{iM}^* such that

$$b_{iH}^* = \min_p(b_{iHp}), \ b_{iHp} \in \mathbf{B}_{iH},$$
$$b_{iM}^* = \min_q(b_{iMq}), \ b_{iMq} \in \mathbf{B}_{iM}.$$

3. Change (regulate) the "physical" states of the robot itself by executing b_{iH}^* selected at step 2.
4. Execute behavior b_{iM}^* as the final resultant action of sector *i* under the regulated state of the robot.

It is noted here that behaviors included multiple sectors are executed when it becomes the winner by the above manner in all sectors it belongs to.

In step 3, changing or regulating (term originated from it's original definition in biology) physical states of action means that the robot refits its parameters related to action such as strength of action, direction of action, etc. to the current environmental states. In Fig. 2, the behavior *InSafety* is the only one homeostatic behavior in sector 1 and hence it is the most unsatisfied homeostatic behavior. When the robot moves from free space into front of obstacles, *InSafety* changes robot's available moving direction from whole 360 degree into the range excluding the portion covered by obstacles. By this "homeostatic regulation process", the robot's physical body state itself becomes unmovable into obstacle-covered range virtually. This is illustrated in Fig. 4, the behavior *ToGoal* may forces the robot move direction "A" but, by homeostatic regulation, robot already has become non-movable to "A". Hence output of *ToGoal* in our scheme should reside in movable range and, at the same time, it is better for the output to be as closer to the goal direction as possible. So output of *ToGoal* is forced change into the direction "B".

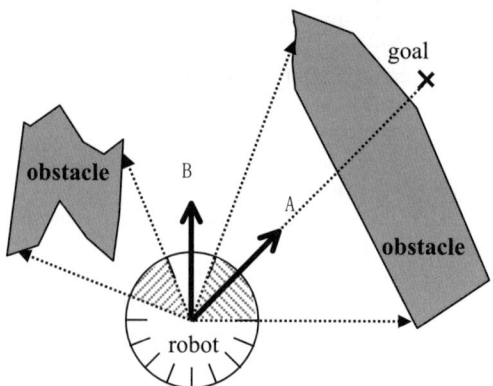

/// non-movable direction range
A moving direction without regulation
B moving direction regulated by *InSafety*

Fig. 4 Navigation with homeostatic regulation

On the other hand, maximum velocity is modified independently in behavior sector 3 regardless of other two sectors. The behavior *InResolution* in sector 3 modulates robot's maximum velocity to fit to current complexity of the environment in such a way as fine control, that is, low velocity when obstacles are nearby while coarse control, that is, fast velocity when obstacles are far away. The behavior *InBound* modifies controller gains in order to prevent possible collision during approaching to the given reference posture.

VI. THE CONTROLLER

Differential drive robots have kinematic constraint as (2) in its maneuvering,

$$\dot{q} = \begin{pmatrix} \dot{x} \\ \dot{y} \\ \dot{\theta} \end{pmatrix} = \begin{pmatrix} \cos\theta & 0 \\ \sin\theta & 0 \\ 0 & \omega \end{pmatrix} \begin{pmatrix} \upsilon \\ \omega \end{pmatrix}, \quad (2)$$

where $q, (x, y), \theta, \upsilon, \omega$ represent posture, position, heading angle, linear velocity, and angular velocity of the robot respectively. All angular values are represented within $[-\pi, \pi]$, which is same in entire this paper. (2) means a nonholonomic constraint as $\dot{y}\cos\theta - \dot{x}\sin\theta = 0$ and it restricts the robot's moving into the direction normal to the axis of the driving wheels. The control problem in this research is to stabilize the error between the virtual reference robot's posture q_r given by the planner and current posture q_c as shown in Fig. 5. Due to the nonholonomic constraint, the robot cannot reach the reference posture through ideal straight trajectory unless it uses policy of "stop-turn-go" in every temporal planning step. The controller uses posture feedback. For this, we choose the frame on the robot, that is, X_C-Y_C as the basis frame for all derived variables. This has proven to be effective during this research for development of the controller. The two error models are derived for two mode of forward and backward driving to the reference posture.

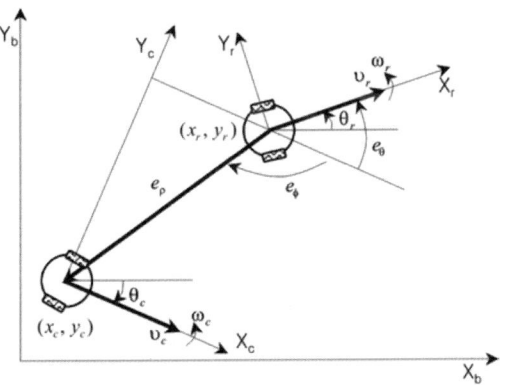

Fig. 5 Posture stabilization problem.

A. The Error Model for Forward Driving:

The posture error with respect to the robot frame is obtained as

$$e_q = \begin{pmatrix} e_x \\ e_y \\ e_\theta \end{pmatrix} = \begin{pmatrix} \cos\theta_c & \sin\theta_c & 0 \\ -\sin\theta_c & \cos\theta_c & 0 \\ 0 & 0 & 1 \end{pmatrix} \begin{pmatrix} x_c - x_r \\ y_c - y_r \\ \theta_r - \theta_c \end{pmatrix}, \quad (3)$$

and its dynamics follows as (4) from (2) and (3).

$$\dot{e}_q = \begin{pmatrix} \dot{e}_x \\ \dot{e}_y \\ \dot{e}_\theta \end{pmatrix} = \begin{pmatrix} 0 & \omega_c & 0 \\ -\omega_c & 0 & 0 \\ 0 & 0 & 0 \end{pmatrix} \begin{pmatrix} e_x \\ e_y \\ e_\theta \end{pmatrix} - \begin{pmatrix} \cos e_\theta \\ \sin e_\theta \\ 0 \end{pmatrix} \upsilon_r + \begin{pmatrix} 0 \\ 0 \\ 1 \end{pmatrix} \omega_r$$
$$+ \begin{pmatrix} 1 \\ 0 \\ 0 \end{pmatrix} \upsilon_c - \begin{pmatrix} 0 \\ 0 \\ 1 \end{pmatrix} \omega_c. \quad (4)$$

Then (4) is translated into polar coordinate with respect to the robot frame as (6) with,

$$\begin{pmatrix} e_\rho \\ e_\phi \\ e_\theta \end{pmatrix} = \begin{pmatrix} \sqrt{e_x^2 + e_y^2} \\ \tan^{-1}(e_y/e_x) \\ e_\theta \end{pmatrix}, \quad (5)$$

where the quadrant of e_ϕ is determined by signs of e_x and e_y, and the e_θ is same in both cartesian and polar coordinate.

$$\begin{pmatrix} \dot{e}_\rho \\ \dot{e}_\phi \\ \dot{e}_\theta \end{pmatrix} = \begin{pmatrix} \cos e_\phi & \sin e_\phi & 0 \\ -\dfrac{1}{e_\rho}\sin e_\phi & \dfrac{1}{e_\rho}\cos e_\phi & 0 \\ 0 & 0 & 1 \end{pmatrix} \begin{pmatrix} \dot{e}_x \\ \dot{e}_y \\ \dot{e}_\theta \end{pmatrix}. \quad (6)$$

In our navigation scheme, the reference posture is fixed in one planning step so that the robot's two velocity elements υ_c and ω_c are zero with respect to the robot frame itself. Using this, that is, (4) with $\upsilon_c = \omega_c = 0$ and (6) give us the final error dynamics in polar coordinate as (7)

$$\begin{pmatrix} \dot{e}_\rho \\ \dot{e}_\phi \\ \dot{e}_\theta \end{pmatrix} = \begin{pmatrix} -\upsilon_r \cos(e_\phi - e_\theta) \\ \dfrac{\upsilon_r}{e_\rho} \sin(e_\phi - e_\theta) \\ \omega_r \end{pmatrix}. \quad (7)$$

B. The Error Model for Backward Driving:

Differential drive robots have ability to move backward as well as forward and, backward driving is more preferable to forward for the half cases of the whole possible situation, and the adequate choice of forward or backward driving like (8) makes smooth trajectory without inversion of driving direction during approach to the target posture. The error model for the backward case is obtained in same manner through (3)~(7) with θ_r and θ_c are replaced with $\theta_r \pm \pi$ and $\theta_c \pm \pi$ respectively. The two error model are then merged into:

$$e = \begin{pmatrix} e_\rho \\ e_\phi + \xi \\ e_\theta \end{pmatrix}, \quad \begin{pmatrix} \dot{e}_\rho \\ \dot{e}_\phi \\ \dot{e}_\theta \end{pmatrix} = \begin{pmatrix} -\upsilon_r \cos(e_\phi - e_\theta) \\ \dfrac{\upsilon_r}{e_\rho} \sin(e_\phi - e_\theta) \\ \omega_r \end{pmatrix}, \quad (8)$$

where $\xi = \text{sgn}(e_\phi - e_\theta)(\zeta \pi / 2 - \pi / 2)$, and $\zeta = \text{sgn}(\pi / 2 - |e_\phi - e_\theta|)$, that is, ζ is equal to ± 1 (+1 is assigned for zero), and ξ is equal to $\pm \pi$ or zero. The Brockett's necessary condition for global asymptotical stability [6] cannot be applied, and hence a certain continuous time-invariant smooth feedback controller [7] for reference robot's velocity υ_r and ω_r can be obtained from (8), and the final controller for robot velocity υ_c and ω_c from υ_r and ω_r can be obtained by considering relative motion for coincidence between the robot frame and the reference frame as follow:

$$\upsilon_c = \upsilon_r, \quad \omega_c = \omega_r. \quad (9)$$

We developed the controller for the two different usages, one is for temporal reference posture on the way of navigation and the other is for final one step of docking into the global goal posture.

C. The controller for position stabilization

In the on-line step by step path planning scheme, the robot cannot know what heading angle is favorable for further step because the information beyond the sensor range cannot be obtained. In other words, the controller cannot get the full reference posture from the planner. It is a natural way to set the reference posture such that $e_\theta = e_\phi + \xi$ for the case. The controller is derived as:

$$\upsilon_c = \zeta k_1 e_\rho \cos(e_\phi + \xi - e_\theta),$$
$$\omega_c = k_1 \frac{1}{2} \sin 2(e_\phi + \xi - e_\theta) + k_2 \frac{(e_\phi + \xi - e_\theta)^2}{\sin(e_\phi + \xi - e_\theta)} \quad (10)$$
$$+ k_3 \sin(e_\phi + \xi - e_\theta),$$
$$k_1, k_2 > 0, \quad k_3 \geq 0,$$

In (10) ω_c is bounded since $|e_\phi + \xi - e_\theta| \leq \pi/2$ always from the definition of ξ in (8), and the controller stabilizes the posture error to $e_\rho = 0, e_\phi + \xi = e_\theta$ globally and asymptotically. Proof of this global and asymptotical stability shows that the controller can bring the robot to whatever reference position the planner generates while ignoring the nonholonomic constraint of the robot. The stability can be proved with the Lyapunov function $V(e_q) = \frac{1}{2} e_\rho^2 + 1 - \cos(e_\phi + \xi - e_\theta)$ [8].

D. The controller for posture stabilization

This controller is used for stabilizing a full posture, in other words, for docking that means approaching some position with a specific heading angle. This docking job can be needed for driving final goal posture such as battery charger dock. For this, the following controller is derived,

$$\upsilon_c = \zeta k_1 e_\rho \cos(e_\phi + \xi - e_\theta),$$
$$\omega_c = k_1 \{\frac{1}{2} \sin 2(e_\phi + \xi - e_\theta) + \cos(e_\phi + \xi - e_\theta)(e_\phi + \xi)\}$$
$$+ k_2 \frac{(e_\phi + \xi - e_\theta)^2}{\sin(e_\phi + \xi - e_\theta)} + k_3 \sin(e_\phi + \xi - e_\theta), \quad (11)$$
$$k_1 > 0, k_2, k_3 \geq 0, k_2 + k_3 \neq 0.$$

This controller is bounded and makes the zero posture error $[e_\rho, e_\phi, e_\theta]^T = [0, -\xi, 0]^T$ globally and asymptotically stable. Proof of the global and asymptotical stability shows that the controller can dock the robot into whatever reference posture the on-line planner generates while ignoring the nonholonomic constraint of the robot. The stability can be proved with the Lyapunov function $V = \frac{1}{2} e_\rho^2 + \frac{1}{2}(e_\phi + \xi)^2 + 1 - \cos(e_\phi + \xi - e_\theta)$ [8].

E. Modification of the Controller Gains

The planner's behavior *InBound* modifies parameter k of the controller preventing the sweeping trajectory from touching obstacles. Maximum sweeping width w_s from the straight line between current position and the reference position is hard to solve since it should integrate $\dot{x}$, and $\dot{y}$ in advance. So, we developed a practical method as below.

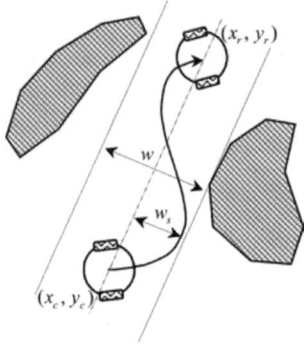

Fig. 6 Avoiding obstacle by modifying controller parameters

In (10) or (11), we can see that relative magnitudes between k_1 and (k_2, k_3) mainly determines the sweeping width during stabilization. First, determine the path width by closest distance between obstacles and the straight line between current position and the reference position. k is then determined by predefined law so that the sweeping width is less than the path width. The k-determining law is made from off-line experiments and simulation with various combinations of k. We made the law by obtaining w_s as

(k_2, k_3) changes within a certain practical boundary while k_1 is fixed. The other geometrical factors between current and reference postures related to sweeping width is fixed to worst case. As the result we obtained safe (k_2, k_3) as a function of w for the controller (10) and (11) respectively:

$k_2 = k_3 = -35.811w^3 + 42.758w^2 - 17.621w + 2.6911,$
$k_2 = k_3 = -11.175w^3 + 19.134w^2 - 11.375w + 2.5279,$
$k_1 = 0.3.$

V. SIMULATION AND EXPERIMENT

Simulations under perfect posture and obstacle information were performed and the experiment is under progress with our robot named "MARI" equipped with odometer with optical encoder and sonar sensor. Fig. 7 shows the robot. Fig. 8 shows successful autonomous determining ability of temporal reference points resulting in the overall path to global goal. Fig. 9 shows a docking with obstacle avoidance by modification of controller gains.

Fig. 7 The robot, "MARI," in the experimental environment

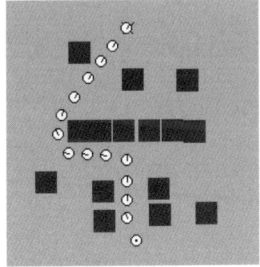

Fig. 8 Autonomous determining ability of temporal reference points

VI. CONCLUSIONS AND FUTURE WORKS

This paper proposed an autonomous robot control scheme covering high-level planner and low-level controller. The reasonable biologically inspired homeostatic strategy in planner and the stable controller under real constraint of obstacles are the main contributions. So, this research seems helpful for designing practical fully autonomous robotic controller, especially for a mobile robot. Real world experiments are on going and the fuzzy rule determining the amount of change of gains considering all factors are under development.

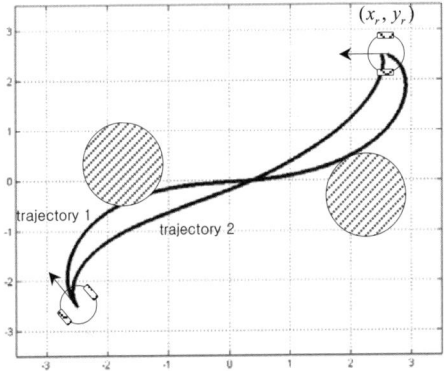

Fig 9 An example of docking with obstacle avoidance

VII. ACKNOWLEDGEMENT

This work is sponsored partially by National Research Lab Project by Ministry of Science and Technology, Korea. The authors hope to send sincere thanks to Myung Hwang-Bo and Sung-On Lee for helping in making the robot.

VIII. REFERENCES

[1] I. Kolmanovsky, N.H. McClamroch, "Developments in Nonholonomic Control Problems," *IEEE Control Systems Magazine*, Vol. 15, Issue: 6, Dec. 1995, pp. 20 -36.
[2] R.C. Arkin, *Behavior-based robotics*, The MIT Press, Cambridge; 1998, pp. 175-204.
[3] D.Y. Yoon, S.R. Oh, G.T. Park, B.J. You, "A Behaviour-Based approach to Reactive Navigation for autonomous robot," *in Proceedings of the 15th IFAC (International Federation of Automatic Control) World Congress*, 21st-26th July, 2002, Barcelona, SPAIN.
[4] R.E. Ricklefs, *Ecology*, Chiron Press, Portland, Oregon; 1973; pp. 137-156.
[5] R.C. Arkin, *Towards cosmopolitan robots: Intelligent navigation in extended man-made environments*. Ph.D. Dissertation, Umass, Amherst, MA; 1987.pp. 46-60.
[6] R.W. Brockett, "Asymptotic stability and feedback stabilization,' in R.W. Brockett, R.S. Millman, and H.J. Sussmann, Ed., *differential Geometric Control Theory*, Birkhauser, Boston;1983, pp. 181-191.
[7] C.C. de Wit, B. Siciliano, G.Bastin, Eds., *Theory of Robot Control*, Springer-Verlag, London; 1996, pp. 331-361.
[8] J. J. Slotine and W. Li, *Applied Nonlinear Control*, Prentice-Hall International, Inc, Englewood Cliffs, NJ; 1991, pp. 40-126.

The HIT/DLR Dexterous Hand: Work in Progress

X.H. Gao, M.H. Jin, L. Jiang, Z.W. Xie, P. He, L. Yang, Y.W. Liu, R. Wei, H.G. Cai,

H. Liu[a], J. Butterfass[a], M. Grebenstein[a], N. Seitz[a], G. Hirzinger[a]

Robot Research Institute, Harbin Institute of Technology (HIT), Harbin 150001, P.R. China
[a] Institute of Robotics and Mechatronics, German Aerospace Center(DLR), 82230 Wessling, Germany
Email: ljiang_hit@yahoo.com.cn, gxh1113@yahoo.com.cn, Hong.Liu@dlr.de

Abstract

This paper presents the current work progress of HIT/DLR Dexterous Hand. Based on the technology of DLR Hand II, HIT and DLR are jointly developing a smaller and easier manufactured robot hand. The prototype of one finger has been successfully built. The finger has three DOF and four joints, the last two joints are mechanically coupled by a rigid linkage. All the actuators are commercial brushless DC motors with integrated analog Hall sensors. DSP based control system is implemented in PCI bus architecture and the serial communication between the hand and DSP needs only 6 lines(4 lines power supply and 2 lines communication interface). The fingertip force can reach 10N.

Keywords: dexterous hand, multisensory, serial communcication

1 Introduction

The development of dexterous robot hand is a very challenging endeavor, which has been pursued by many researchers. Many dexterous robot hands have been built over the past three decades[1][2][3][4][5]. These devices make it possible for the robot to grasp and manipulate objects.

Since 1997 DLR has developed two generation of multisensory dexterous robot hands: DLR Hand I[4] and II[6]. Both hands are highly integrated multisensory mechtronic hands. Based on the experience of DLR Hand I, the DLR Hand II was designed toward stronger and more reliable. The number of cables between the hand and main microprocessor has been greatly reduced from more than 400 to only 8. The optimal combination of BLDC motors, harmonic drives, belt transmission and differential bevel gear transmission makes the fingertip force up to 30N. The extra degree of freedom of thumb enables the hand not only for power grasping but also for fine manipulation. It is well recognized that the DLR Hand II is one of the

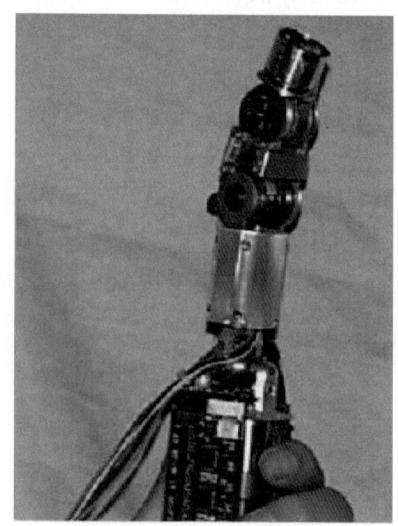

Fig.1: One finger of the HIT/DLR Hand

best robotic hand in the world. On the other side, however, because of its high integration it is not easy to manufacture such hand, especially for the actuator system where all the motors are specially designed and the analog hall sensors must be adhesived and calibrated carefully. Since 2001, based on the DLR's experience, DLR and HIT have been developing a smaller human-like dexterous robot hand. The goal will be to build a robot hand with less difficulty to reproduce it. To achieve this goal, several issues have to be considered, such as the fingertip output force, actuators and sensors. If the fingertip force must be greater than 30N, then it would be very difficult for other commercial motors to realize it in the same space occupation. For many cases one third of 30N fingertip force will be enough for many fine manipulation. Based on this basic understanding DLR and HIT are jointly engaged in developing a low-cost and easy manufacturing robot hand.

This paper presents to date progress on this project. The paper is organized as follows: Section 2 gives an overall description about the system. Section 3 describes the multisensory system. Integrated electronics and hand controller hardware are introduced in section 4 and section 5, respectively. Finally, section 6 addresses conclusion and future work.

2 Overall system description

To achieve a high degree of modularity, all four fingers are identical. A prototype of one finger of the HIT/DLR hand is shown in Fig.1. All actuators are integrated in the finger's base and the finger's body directly, the electronics and communication controllers are fully integrated in the finger's base in order to minimize weight and the amount of cables needed for a hand.

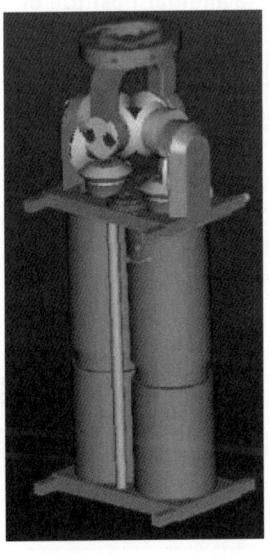

Fig.2: Base joint unit and its actuation system

The effectiveness of bevel gear differential transmission has been successfully demonstrated in the DLR Hand II. In order to save the work for special motors assembly with adhesive and calibration of analog Hall sensors, the first important issue is to select appropriate commercial motors with the power for 10N fingertip force. Through strict comparison the brushless DC motor (1628 BLDCM) with planetary gears transmission from Faulhaber Co. have been chosen for the base joint's actuators. And the motor itself has been applied in the finger's body. The motor measures 16mm in diameter and 28mm in length. There are 8 cables including three drive signals, three analog Hall sensors and corresponding power and ground.

Fig.2 shows the 3D structure of one base joint unit. The base joint with two degrees of freedom is of differential bevel gear type, the planetary drive gears are directly coupled to the BLDCM, and the bevel gears are connected to planetary gears via additional gear reduction of 2:1. For curling/extension motion the motors apply a synchronous motion to the bevel gears using the torque of both motors. For abduction/adduction the motors turn in contrary directions. This causes a curling motion on the fingertip using the torque of both motors and means we can use small motor and reducer while reaching bigger output force on the fingertip.

The middle link is actively actuated by a BLDCM in combination with a tiny harmonic drive gear. The harmonic drive measures 20mm in diameter and 13.4mm in length and the reduction

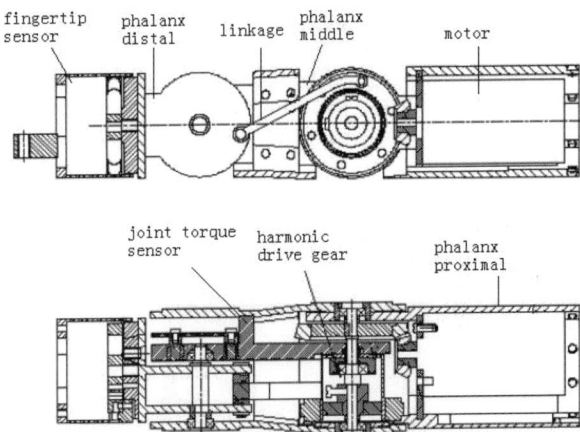

Fig.3: Finger unit with one integrated actuator and two coupled joints.

ratio is 1:80. The motions of middle phalanx and distal phalanx are not individually controllable, they are connected by means of the linkage, whose structure and parameters are optimized by simulation. The finger unit is shown in Fig.3.

Kinematic design of multifingered hand shows that the motion of the thumb and the fourth finger is absolutely necessary to improve the grasping performance in case of precision and power grasp. Therefore the hand will be designed with an additional degree of freedom in order to realize motion of the thumb relative to the palm. This enables to use the hand in different configurations.

3 Multisensory System

A dexterous robot hand needs as a minimum a set of force and position sensors to enable control schemes like position control and impedance control in autonomous operation and teleoperation. Compared to DLR Hand II, there is some improvement in the sensor system. Instead of contact potentiometer in each joint a contactless Hall sensor based joint position sensor has been developed. Also, base joint torques have been designed in flat form so that the whole height could be reduced. Sensor equipment of the HIT/DLR hand is shown in Table1.

Table1 Sensor equipment of one finger

Sensor type	Count/finger
Joint torque	3
Joint position	3
Motor position	3
Force/torque	1
Temperature	2

3.1 Joint torque sensor

Each joint is equipped with strain gauge based joint torque sensor. To reduce the length of a finger a new type base joint torque sensor with two degrees of freedom is designed (Fig.4). The torque sensor located in middle joint is integrated into the connecting part and can precisely measure the external torque.

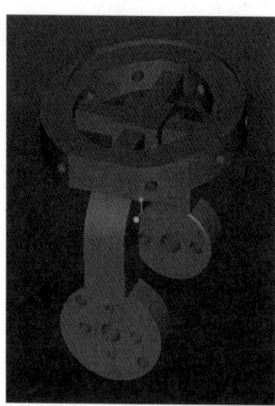

Fig.4: Base joint torque sensor

3.2 Joint position sensor

Since we can calculate the joint position from the motor Hall sensors, the joint position sensor would not be absolutely necessary. However, in the presence of the elasticity and hysteresis of the transmission system, joint position sensor can provide with a more accurate information of joint position, and it can eliminate the necessity of referencing the fingers after power up.

A new integrated 2-axis hall effect sensors, replacing conventional contact potentiometer and contactless incremental encoder, are integrated in every active joint of the HIT/DLR hand in order to measure joint position and meet the requirement. The heart of the sensor is a cross-shape Vertical Hall Sensor(VHS) featuring a total of five supply contacts and four signal contacts. The active zone of the device measures about 0.2 to 0.3mm in size. A permanent magnet mounted on the axis of a rotating shaft generates the magnetic field required for the measurement, working principle of the measuring system is shown in Fig.5. The magnetic field created by the permanent magnet is parallel to the sensor surface. With a rotation of the axis, the angle of the magnetic field rotates in the sensor plane. After offset compensation and pre-amplification, the two output signals are processed to yield absolute angular position or velocity.

Since base joint is of differential type, we have to calculate the joint position of the base joint from the hall element output values. Since there was no way to measure the joint position directly due to space restriction, we have to add a special measuring module.

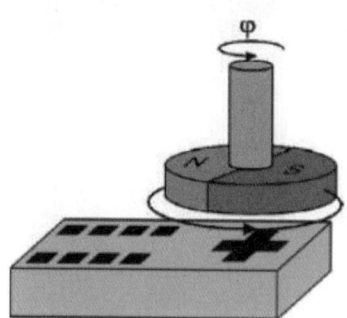

Fig.5: Principle of measuring system by using integrated hall sensor

3.3 Motor position sensor

Each BLDCM is equipped with three linear Hall effect sensors which are used for commutation of the motors. These sensors supply three sinusoidal signals with a constant phase shift of 120°. The position within the magnetic cycle of the motor can be calculated from arbitrary two sinusoidal signals. By additionally counting the cycles the relative position of the motor can be calculated, while the speed can be calculated by differentiation of the position signal.

3.4 Force/torque fingertip sensor

Besides the torque sensors in each joint we have integrated a miniaturized six dimensional strain gauge-based force/torque sensor with fully digital output into the fingertip. Signals from foil strain gauge bridges are amplified and converted to digital representations of the force and torque applied to the sensor. Signal processing circuit and A/D converter are mounted within the sensor body in order to achieve the integration of the sensor and improve the behavior of the signals. It needs only 6 wires including power supply.

Fig. 6, full digital 6dof fingertip force sensor

The force and torque measure ranges are 30N for F_x, F_y and F_z, 200Nmm for M_x, M_y and M_z, respectively. A 200% mechanical overload protection is provided in the structure. Fig. 6 shows an integrated fingertip sensor with electronics.

3.5 Temperature sensor

Several precision integrated-circuit temperature sensors are integrated into the finger. These sensors have small size and low quiescent circuit. Information from these temperature sensors are used to compensate for outputs of the above sensors and protect the system.

4 Integrated electronics

One major goal of the design of the HIT/DLR Hand was to achieve full integration of the electronics needed in the finger and the palm for minimizing weight and the amount of wires and enhancing the system reliability.

In finger unit we achieved the integration of electronics by using flexible printed circuit board(PCB) with appropriate bending space and shape(Fig.7). One end of the board is connected to the fingertip force/torque sensor, another is connected to the communication controller in the finger. Signal processing circuits of joint position sensor, joint torque sensor and base joint torque sensor and A/D converter are included in this board.

Fig.7 Flexible printed circuit board in the finger

In accordance with mechanical structure of the base joint, the base joint position sensor circuit board is compactly designed, which is connected with the communication controller in the finger and the power converter via board-to-board interfaces composed of headers and board stackers. The power converters for driving three motors in a finger are located directly beside the motors and they are galvanically decoupled from the sensor electronics in order to minimize the noise induced by the running motors.

A serial ADC with 8 analog channels and 12 bit resolution is used to convert the sensor signals as near as possible to the sensor conditioning circuitry into digital data. There are in total 23 sensor signals in a finger, in consideration of both the mechanical structure design and the fact that two signals from a joint position sensors and two signals from a motor must be collected at the same time for the best measuring accuracy, a 5 serial ADC are used. These distribution is as follows: one converter is located in the fingertip sensor, two in the base joint torque sensor circuitry and two in the power converter board.

5 Hand controller hardware

The design of control system has been greatly conditioned by the large amount of sensory information which is needed to be acquired and elaborated at run-time, therefore a DSP(Digital Signal Processor)-based real-time control system has been built. Electronics and communication architecture in a finger are shown in Fig.8. The kernel of the system is a commercially available processor board with a TMS320C6711. Some characteristics of this board are: 150MHz clock, floating point arithmetic unit, 2Mbyte of synchronous burst memory and 512K bytes flash program memory. This board is plugged in a PCI slot of a industrial computer via PCI bridge controller. The DSP board and industrial computer communicate via the Dual Port RAM (DPRAM).

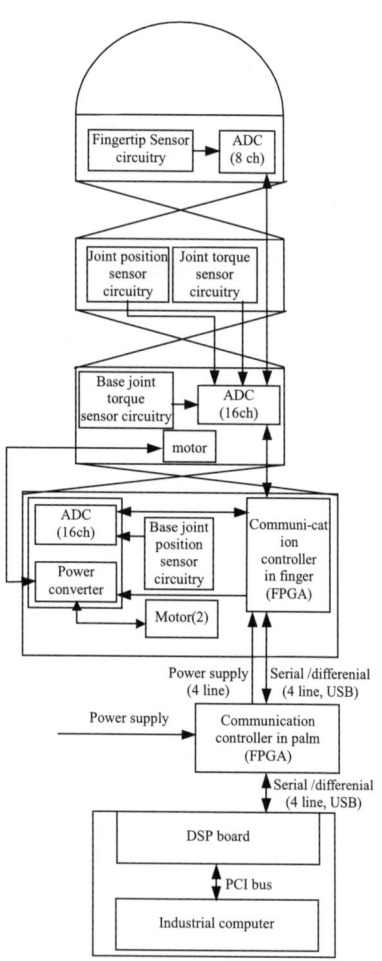

Fig.8: Electronics and communication architecture

In order to use the hand freely on different robot manipulators, reduce cables and the possibility of noise in the sensor signals, a fully integrated serial communication system, based on FPGA(Field Programmable Gate Array) has been developed.

Each finger holds one FPGA-based communication controller, which is responsible for the collection of all sensory information in the finger, the distribution of the data from the control scheme to three actuators for finger control, and the communication with the communication controller in the hand base. Furthermore this controller performs some signal processing task like digital filtering and calculating the motor speed. The communication controller in the hand base links the serial data stream of each finger to the data stream of the DSP board. Besides a four power supply the number of external cables of HIT/DLR hand is an four line communication interface since the data is transmitted via LVDS(Low Voltage Differential Signal).

The whole control architecture has been freshly tested. From a PC manual operator can command the finger's movement and also read the sensor data simultaneously.

6 Conclusion and Future work

A prototype of HIT/DLR Hand's finger and it's control architecture have been presented. The initial experiment proves the fingertip output force up to 10N. Future work will be concentrated on the redesign of the prototype and control algorithm.

Reference

[1] S.C.Jacobsen,J.E.Wood,D.F.Knutti,K.B.Biggers. The UTAH/MIT Dextrous Hand: Work in Progress. The International Journal of Robotics Research. 1984,3(4):21-50

[2] J.K.Salisbury,J.J.Craig. Articulated Hands: Force Control and Kinematic Issues. International Journal of Robotics Research. 1982,1(1):4-17

[3] J.Butterfass,G.Hirzinger,S.Knoch,H.Liu. DLR's Multisensory Articulated Hand, Part I: Hard-and Software Architecture. Proceedings of the IEEE International Conference on Robotics andAutomation. Leuven, Belgium. 1998:2081-2086

[4] H. Liu, J. Butterfass, S.Knoch, P.Meusel, G. Hirzinger. A New Control Strategy for DLR's Multisensory Articulated Hand. IEEE Control Systems. 1999,19(2): 47-54

[5] C.S.Lovchik,M.A.Difler. The Robonaut Hand: A Dextrous Robotic Hand for Space. Proceedings of the IEEE International Conference on Robotics and Automation. Detroit, Michigan. 1999: 907-912

[6] J. Butterfass, M. Grebenstein, H. Liu, G. Hirzinger: DLR-Hand II: Next Generation of a Dextrous Robot Hand, Proceedings of the IEEE ICRA, Seoul, Korea, 2001

Development of a Soft-fingertip and Its Modeling Based on Force Distribution

Kwi-Ho Park°, Byoung-Ho Kim†, and Shinichi Hirai°

° : Dept. of Robotics, Ritsumeikan Univ., Kusatsu, Shiga, Japan
† : JSPS Post-doc Fellow, Dept. of Robotics, Ritsumeikan Univ., Kusatsu, Shiga, Japan
(E-mail: hirai@se.ritsumei.ac.jp)

Abstract

In this paper, a hemisphere-shaped soft fingertip for soft fingers is developed and its modeling based on force distribution is presented. We first analyze the geometrical relation of the soft fingertip when it is deformed. Secondly, the force distribution of the soft fingertip is investigated by using a compressional strain mechanism. And then, we propose a nonlinear model of the soft fingertip which enables us to obtain the total contact force at the contact surface of each finger in manipulating tasks. Finally, the proposed force function is verified by experiment, where a tactile sensor is used to measure the contact force distribution in the contact surface and its total force. The developed soft fingertip and its force function can be usefully applied to soft-fingered manipulations.

Keywords : *development and modeling of a soft fingertip, force distribution, tactile sensor*

1 Introduction

Many grasping and manipulation algorithms have been presented for manipulating of an object by multi-fingered hands [1]-[3]. Tactile sensor-based manipulation algorithms [4, 5] were developed to consider the reliable information about the grasp geometry including contact positions. By the way, the deformation effect of objects or soft fingertips is a common issue in general robotic applications. Recently, many researchers have been interested about the deformation of an object and/or a soft fingertip in multi-fingered manipulating tasks and thus, soft manipulations by multi-fingered hands have been considered as an active research field [6]-[9]. Also, it is well-known that the field of developing a soft finger and/or a soft fingertip is a fundamental area in soft manipulations.

Related to the modeling of contact, compliant materials for robot fingers were tested [10]. Xydas et al. [11] presented a modeling of contact mechanics for soft fingers. Hiromitsu et al. [12] showed various type of contact shapes at the contact surface of a finger in a human grasping. From their observations of human grasping and contact mechanics, it is confirmed that when an object is being manipulated by a multi-fingered hand with soft fingertips, each fingertip force can be determined by considering the force distribution according to the deformation of soft fingertips. A strain-based silicone gel model was investigated [13]. For practical applications, however, those approaches are rather complex because their methodologies are based on a numerical algorithm. On the other hand, there are not so much researches related to develop a proper soft fingertip.

The objective of this paper is to present a simple model of a soft fingertip for practical soft-fingered applications. We first develop a soft fingertip like human skin for soft-fingered manipulations and also present a simple model of the developed soft fingertip based on the force distribution at the contact surface. Through experimental works, we verify the proposed model of the developed soft fingertip, where a tactile sensor and a tactile sensor signal processing system are used to measure the contact force distribution in the contact surface and its total force.

2 Development of a Soft Fingertip

The object manipulating system by a human hand shown in Fig. 1 is one of typical soft-fingered manipulations. Usually, a human hand has many soft fingertips and thus it can be applied to various manipulating tasks. There are various type of contact shapes at the contact surface of a finger in a human grasping [12]. Generally, we can experience that the contact between the object and each fingertip of the object-hand system is made by a surface contact and also, soft fingertips are appropriately deformed during the manipulation process. In order to implement those soft-fingered manipulations, a proper soft fingertip is necessary.

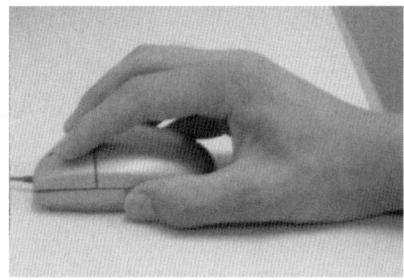

Fig. 1. An object manipulating system by a human hand.

For soft-fingered applications, we developed a hemisphere-shaped soft fingertip with radius of 0.01[m] as shown in Fig. 2, where a molding device is used to make our fingertip. The developed soft fingertip was made by using a Pringel compound of Exseal Co.(http://www.exseal.co.jp). The feature of our soft fingertip is similar to the skin of a human and thus, it can be applied to make a finger with soft fingertip.

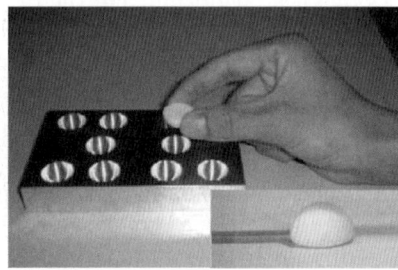

Fig. 2. A developed soft fingertip.

3 Geometrical Analysis of Soft Fingertips

When a soft fingertip is being contact into a rigid object, there exists some deformation. This section presents geometrical analysis of the deformation of soft fingertips. Particularly, we consider a typical one-dimensional contact model of common manipulating tasks as shown in Fig. 3.

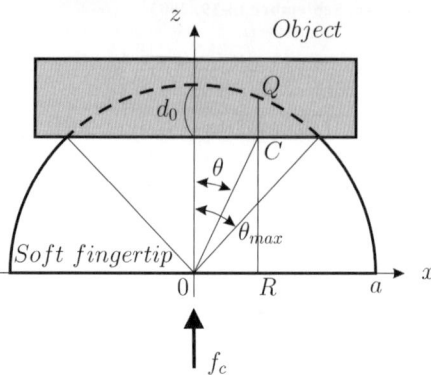

Fig. 3. A simple one-dimensional contact and deformation.

Fig. 3 shows a contact status of a soft fingertip which is contacting into an object to the normal direction. Specially, let us consider a hemisphere-shaped soft fingertip with radius of a. Then, the following relation of a circle is satisfactory on the surface;

$$x^2 + z^2 = a^2. \quad (1)$$

The x-position of a point, Q, on the surface of the fingertip in the two-dimensional space is given by

$$Q_x = (a - d_0)tan\theta \quad (2)$$

where d_0 denotes the deformation length of the soft fingertip and θ implies a touching angle shown in Fig. 3. By combining (1) and (2), the z-position of the point Q can be represented as

$$Q_z = \sqrt{a^2 - (a - d_0)^2 tan^2\theta}. \quad (3)$$

Then, the z-directional length of $\overline{QC}$ is given by

$$\overline{QC} = Q_z - (a - d_0). \quad (4)$$

Physically, the length of $\overline{QC}$ implies the deformation at a point on the contact surface of the soft fingertip.

Also, the maximum touching angle, θ_{max}, can be determined by

$$\theta_{max} = tan^{-1}\sqrt{\frac{a^2 - (a - d_0)^2}{(a - d_0)^2}}. \quad (5)$$

Usually, the contacting area of a soft fingertip is dependently increased when the fingertip is being deformed. Specifically, when it is deformed as d_0 as shown in Fig. 4, the radius of the contact surface observed at the contact point of C can be represented by

$$r = (a - d_0)tan\theta. \quad (6)$$

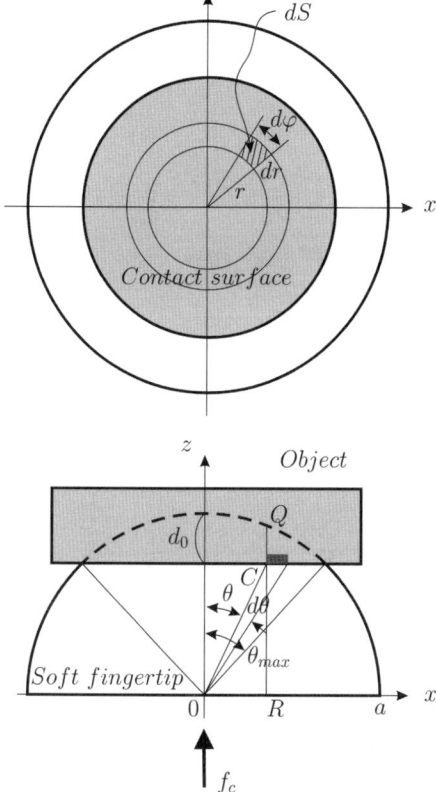

Fig. 4. The contacting area of a soft fingertip.

Also, the small change of the radius, dr, is given by

$$dr = \frac{(a-d_0)}{\cos^2\theta}d\theta, \quad (7)$$

Thus, the resultant area of the small increased contact surface due to the deformation, dS, can be obtained by

$$\begin{aligned} dS &= rdrd\varphi \\ &= \frac{(a-d_0)^2 \sin\theta}{\cos^3\theta}d\theta d\varphi. \end{aligned} \quad (8)$$

4 Modeling of Soft Fingertip

For soft-fingered object manipulations, this section presents a one-dimensional modeling of hemispherical-shaped soft fingertips as shown in Fig. 3. In this model, it is assumed that the shape of contact formed at the fingertip is symmetric to the normal axis of the fingertip.

In order to obtain a proper model of a soft fingertip with hemispherical shape, we first analyze the force relation of a cylindrical particle of the soft fingertip. The stress, σ, pressed on the soft particle shown in Fig. 5 is defined as

$$\sigma \equiv \frac{F}{S} \quad (9)$$

where F is the magnitude of the contact force that acts per contact area, S, of the particle.

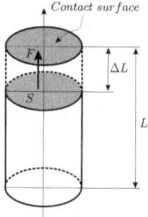

Fig. 5. A cylindrical particle of a soft fingertip.

When a small compressive contact force is applied to the soft particle, the length of the particle decreases, but the contact surface increases. From this physical phenomenon, the compressional strain, e, is defined as the fractional change in the length of the soft particle as follows:

$$e \equiv \frac{\Delta L}{L} \quad (10)$$

where e is dimensionless and it is negative if the stress is due to compression such as many robotic grasping tasks($\Delta L < 0$). Then, by using Young's modulus($Y = \sigma/e$) [14], the force relation acting on the soft particle is given by

$$F = Y\frac{\Delta L}{L}S. \quad (11)$$

On the other hand, the pressure acting on a particle is defined as

$$P \equiv \frac{F}{S}. \quad (12)$$

By combining (11) and (12), the resultant force relation acting on a particle can be expressed as

$$F = PS = Y\frac{\Delta L}{L}S. \quad (13)$$

Note that the contact force at the surface of the soft particle depends on the pressure distribution of the soft particle, the length and its change of the soft particle, the area of contact surface, and the Young's modulus of the soft particle.

Then, we will model the soft fingertip of Fig. 3 as a set of infinite soft particles addressed in advance.

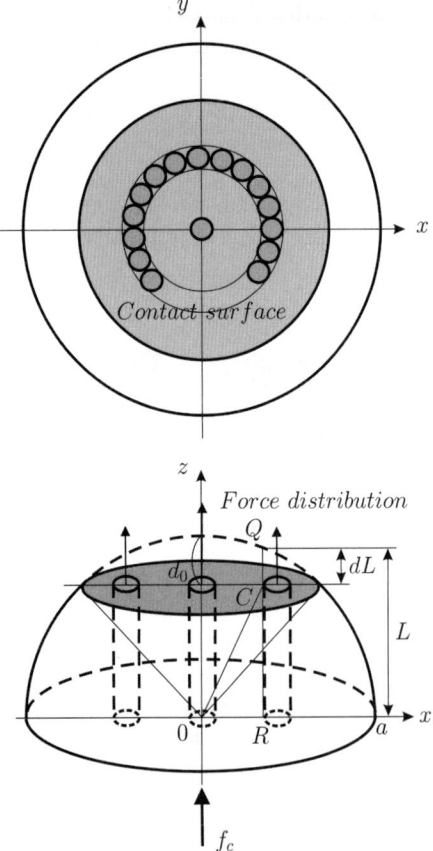

Fig. 6. A model of a soft fingertip based on the force distribution.

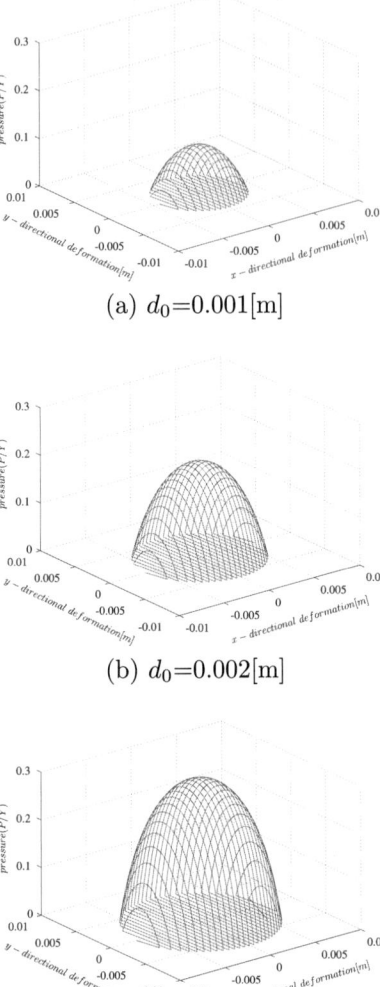

(a) $d_0=0.001$[m]

(b) $d_0=0.002$[m]

(c) $d_0=0.003$[m]

Fig. 7. Pressure distribution patterns at each deformation of the soft fingertip.

In this sense, the deformation of the fingertip can be modeled as a deformation set formed by the force distribution as shown in Fig. 6. Since the force of the fingertip is formed by the pressure, the pressure distribution for all particles will gives an idea to determine the total contacting force of the fingertip. Specifically, Figs. 7(a), (b), and (c) show theoretical pressure distribution patterns($P = Y\frac{\Delta L}{L}$) according to the deformation of 0.001[m], 0.002[m], and 0.003[m], respectively.

As a result, by integrating the force distribution of the soft fingertip, the total contact force at the contact surface can be expressed as follows:

$$\begin{aligned} f_c &= \int_0^{2\pi} \int_0^{\theta_{max}} P dS \\ &= \int_0^{2\pi} \int_0^{\theta_{max}} Y\frac{dL}{L} dS \\ &= Y\pi d_0^2 \end{aligned} \quad (14)$$

where Y denotes the Young's modulus of the fingertip and it can be estimated by experiment. And length parameters of the soft fingertip according to the deformation are given by

$$L = \sqrt{a^2 - (a-d_0)^2 tan^2\theta} \quad (15)$$

and

$$dL = \sqrt{a^2 - (a-d_0)^2 tan^2\theta} - (a-d_0). \quad (16)$$

From (14), it is pointed out that the fingertip force is proportional to the square of the deformation of the soft fingertip.

Next, in order to verify the proposed force relation given by (14), we performed experimental works by using the experimental setup shown in Fig. 8. The height gauge in Fig. 8 is to measure of the deformation of the fingertip and the tactile sensor system shown in Fig. 9 enables us to measure the contact force distribution in the contact surface and its total force. In this experiment, we used a tactile sensor(I-SCAN10×10) and a tactile sensor system of Nitta Ltd.(http://www.nitta.co.jp).

Fig. 8. Experimental setup.

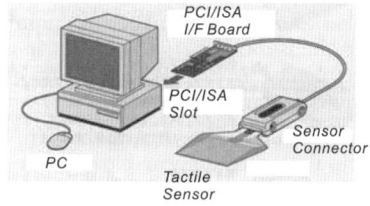

Fig. 9. A tactile sensor system.

By measuring the contact force distribution according to the deformation of the fingertip, we evaluated the elasticity parameter(Y) of the developed soft fingertip. The contact force distribution is obtained by integrating the pressure distribution in the contact surface. Through the experiment, we have obtained pressure distribution patterns of using the tactile sensor. Figs. 10(a), (b), and (c) show experimental pressure distribution patterns according to the deformation of $0.001[m]$, $0.002[m]$, and $0.003[m]$, respectively.

By using (14) and those pressure distributions, we can finally obtain the total contact force as the deformation. Fig. 11 shows both experimental force profile of the soft fingertip and theoretical force profile according to the deformation. From the experimental evaluation, the elastic modulus, Y, is estimated as $0.25[MN/m^2]$. Physically, if the value of Y is more smaller, the compliance of a soft fingertip is more larger. Thus, the elasticity of our soft fingertip is more soft than a common rubber of $7[MN/m^2]$ [14].

In many soft-fingered practical applications, the control performance of a given task may be dependently determined as this value.

Recently, we have developed a two-fingered hand with soft fingertips as shown in Fig. 12. In future, the developed soft fingertip and its modeling will be applied to soft-fingered manipulations using the developed hand.

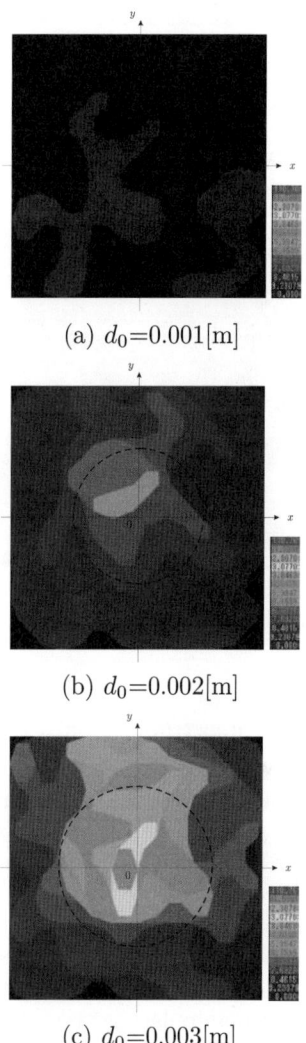

(a) $d_0=0.001[m]$

(b) $d_0=0.002[m]$

(c) $d_0=0.003[m]$

Fig. 10. Pressure distribution patterns(unit: Kpa) of the tactile sensor for each deformation, where the dotted circle implies the pressed contact surface.

5 Concluding Remarks

In this paper, a hemisphere-shaped soft fingertip for soft fingers was developed and a nonlinear force

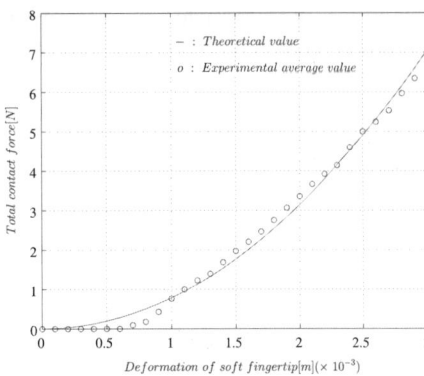

Fig. 11. Force profiles of the developed soft fingertip: experimental and theoretical results.

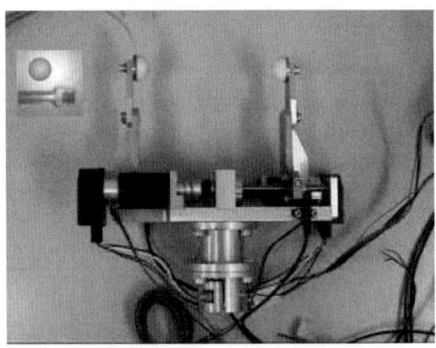

Fig. 12. A developed two-fingered hand with soft fingertips.

function of a soft fingertip according to the deformation was modeled by considering the force distribution in the contact surface. Through experimental evaluations, the proposed force function was verified, where a tactile sensor and a tactile sensor signal processing system were used to measure the contact force distribution in the contact surface and its total force. The proposed force function for soft fingertips enables us to obtain the total contact force at the contact surface of a finger in object manipulating tasks. Specifically, our model was considered in the one-dimensional contact of a finger. Additional work is to extend our approach to the two- and three-dimensional modelings of soft fingertips and also apply the developed soft fingertip to real soft-fingered applications.

Acknowledgments

This work was supported by the research fund of Japan Society for the Promotion of Science.

References

[1] A. B. A. Cole, J. E. Hauser, and S. S. Sastry, "Kinematics and control of multifingered hands with rolling contact," *IEEE Trans. on Automatic Control*, Vol. 34, No. 4, pp. 398-404, 1989.

[2] H. Maekawa, K. Tanie, and K. Komoria, "Dynamic grasping force control using tactile feedback for grasp of multifingered hand," *Proc. of IEEE Int. Conf. on Robotics and Automation*, pp. 2462-2469, 1996.

[3] X. -Z. Zheng, R. Nakashima, and T. Yoshikawa, "On dynamic control of finger sliding and object motion in manipulation with multifingered hands," *IEEE Trans. on Robotics and Automation*, Vol. 16, No. 5, pp. 469-481, 2000.

[4] H. Maekawa, K. Tanie, K. Komoria, and M. Kaneko, "Development of a finger-shaped tactile sensor and its evaluation by active touch," *Proc. of IEEE Int. Conf. on Robotics and Automation*, pp. 1327-1334, 1992.

[5] G. Kinoshita, Y. Kurimoto, H. Osumi, and K. Umeda, "Dynamic contact sensing of soft planar fingers with tactile sensors," *Proc. of IEEE Int. Conf. on Robotics and Automation*, pp. 565-570, 2001.

[6] S. Arimoto, P. A. N. Nguyen, H. -Y. Han, and Z. Doulgeri, "Dynamics and control of a set of dual fingers with soft tips," *Robotica*, Vol. 18, pp. 71-80, 2000.

[7] T. Yoshikawa and T. Watanabe, "Dynamic control of soft-finger hands for povoting an object in contact with the envoronment," *IEEE/RSJ Int. Conf. on Intelligent Robots and Systems*, pp. 324-329, 2000.

[8] T. Wada, S. Hirai, S. Kawamura, and N. Namiji, "Robust manipulation of deformable objects by a simple PID feedback," *Proc. of IEEE Int. Conf. on Robotics and Automation*, pp. 85-90, 2001.

[9] O. B. Bayazit, J. -M. Lien, and N. M. Amato, "Probabilistic roadmap motion planning for deformable objects," *Proc. of IEEE Int. Conf. on Robotics and Automation*, pp. 2126-2133, 2002.

[10] M. R. Cutkosky, J. M. Jordan, and P. K. Wright, "Skin materials for robotic fingers," *Proc. of IEEE Int. Conf. on Robotics and Automation*, pp. 1649-1654, 1987.

[11] N. Xydas and I. Kao, "Modeling of contact mechanics and friction limit surfaces for soft fingers in robotics, with experimental results," *Int. Jour. of Robotics Research*, Vol. 18, No. 8, pp. 941-950, 1999.

[12] S. Hiromitsu and T. Maeno, "Stick/slip distribution on the fingerpad and response of tactile receptors when human grasp an object," *Jour. of the Robotics Society of Japan*, Vol. 68-c, No. 667, pp. 202-207, 2002.

[13] Y. Satoh, F. Wakui, K. Sogabe, and N. Shimizu, "Steady-state response of silicon β gel and a mass system," *Dynamics and Design Conf. of Japan*, pp. 647-652, 2000.

[14] P. M. Fishbane, S. Gasiorowicz, and S. T. Thornton, *Physics for scientist and engineers*, Prentice Hall, Englewood Cliffs, New Jersey, 1993.

From Nominal To Robust Planning: The Plate-Ball Manipulation System

Giuseppe Oriolo*   Marilena Vendittelli*   Alessia Marigo°   Antonio Bicchi[†]

* Dipartimento di Informatica e Sistemistica, Università di Roma "La Sapienza"
Via Eudossiana 18, 00184 Roma, Italy, {oriolo,vendittelli}@dis.uniroma1.it

° IAC-CNR, Viale del Policlinico 137, 00161 Roma, Italy, marigo@iac.rm.cnr.it

[†] Interdept. Research Center "Enrico Piaggio", Università di Pisa
Via Diotisalvi 2, 56100 Pisa, Italy, bicchi@ing.unipi.it

Abstract

Robotic manipulation by rolling contacts is an appealing method for achieving dexterity with relatively simple hardware. While there exist techniques for planning motions of rigid bodies in rolling contact under nominal conditions, an inescapable challenge is the design of robust controllers of provable performance in the presence of model perturbations. As a preliminary step in this direction, we present in this paper an iterative robust planner of arbitrary accuracy for the plate-ball manipulation system subject to perturbations on the sphere radius. The basic tool is an exact geometric planner for the nominal system, whose repeated application guarantees the desired robustness property on the basis of the Iterative Steering paradigm. Simulation results under perturbed conditions show the effectiveness of the method.

1 Introduction

Rolling manipulation has recently attracted the interest of robotic researchers as a convenient way to design dextrous hands with simplified hardware (see [1, 2, 3] and the references therein). Here, dexterity indicates the capability to relocate and reorient a manipulated object by maintaining a firm grasp on it. A first prototype of a hand purposefully implementing rolling manipulation was presented in [2]. The nonholonomic nature of rolling contacts between rigid bodies guarantees the generic controllability of rolling pairs, i.e., that any two surfaces (with the only exception of surfaces that are mirror images of each other) can be arbitrarily reoriented and relocated by rolling. While such result is limited to smooth surfaces, the case of a polyhedral object to be manipulated was considered in [4].

The archetypal example of rolling manipulation is the plate-ball system [5, 6, 7, 8]: the ball (the manipulated object) can be brought to any contact configuration by maneuvering the upper plate (the first finger), while the lower plate (the second finger) is fixed. Despite its mechanical simplicity, the planning and control problems for this device already raise challenging theoretical issues. In fact, in addition to the well-known limitations due to nonholonomy (essentially, the lack of smooth stabilizability), the plate-ball system is neither flat nor nilpotentizable; therefore the classical techniques (e.g., see [9]) for planning and stabilization of nonholonomic systems cannot be applied.

To this date, only the planning problem has been attacked with some success; e.g., see the algorithms in [1, 6]. Like for any planner based on open-loop control, however, the successful execution of maneuvers is not preserved in the presence of perturbations — some sort of feedback is necessary to induce a degree of robustness. This advancement appears to be mandatory in order to fulfill the promise of rolling manipulation of providing a reliable technological solution.

The final objective of our research is to move from planning to robust stabilization by exploiting the mechanism of iteration as proposed in [10], i.e., sampling the system state and repeatedly applying the same planner at discrete instants. In addition to the simplicity of design, this general stabilization approach (IS, or Iterative Steering) has the advantage of driving the system along the predictable trajectories typical of the planner. Such feature is particularly useful in the presence of configuration space constraints, e.g., due to workspace obstacles.

A first step in the above direction was presented in [11], where we considered the problem of rolling a ball whose radius was only known up to some measurement error, and designed a robust controller for this system by iterating an approximate planner based on a nilpotent approximation of the dynamics. While simulations showed the effectiveness of such method in rejecting the radius perturbation, only local stability was guaranteed. Moreover, the formal proof that the controller satisfied the requirements of the IS paradigm required an additional condition on the contraction rate which eventually affected the convergence speed.

Here, we retain the general strategy (IS) but we change the basic tool, i.e., the planner. In particular, by adopting an exact (for the nominal system) planner based on geometric arguments — of interest in itself — we are able to derive a scheme that drives the perturbed plate-ball system from any configuration to the desired goal with arbitrary precision. However, as will be made clear in the paper, due to the specific nature of the nominal planning algorithm, an iterative robust planner is obtained rather than a stabilization method.

The paper is organized as follows. In Sect. 2, the model of the plate-ball system is briefly described. In Sect. 3, a spinning maneuver which achieves rotation of the ball around its vertical axis is proposed in two versions: a open-loop and a closed-loop version. The exact planner for the nominal system using the latter is devised in Sect. 4, while the iterative planner generated by the IS strategy is described in Sect. 5, where we also report simulation results confirming the achieved robustness. A short discussion on the perspectives of this work concludes the paper.

2 The model

The plate-ball system of Fig. 1 is a special case of rolling contact between regular surfaces (see[3] for a complete treatment). Its kinematic equations describe the evolution of the (local) coordinates of the contact points on the plate, $\alpha_p = (x, y) \in \mathbb{R}^2$, and on the ball, $\alpha_b = (u, v) \in \mathbb{R}^2$, as well as of the sphere orientation ψ with respect to the plane, given by the holonomy angle between the two Gauss frames associated to α_p and α_b.

Denoting by ρ its radius, the ball can be parameterized as

$$f(u,v) = \begin{pmatrix} \rho \cos v \cos u \\ \rho \cos v \sin u \\ \rho \sin v \end{pmatrix}$$

with $\{(u,v)| -\pi < u < \pi, -\pi/2 < v < \pi/2\}$. Following the derivation of Montana [5], one obtains the following kinematic equations

$$\begin{pmatrix} \dot{\alpha}_p \\ \dot{\alpha}_b \\ \dot{\psi} \end{pmatrix} = \begin{pmatrix} 0 & 1 \\ 1 & 0 \\ \frac{\cos \psi}{\rho \cos v} & -\frac{\sin \psi}{\rho \cos v} \\ -\frac{\sin \psi}{\rho} & -\frac{\cos \psi}{\rho} \\ \frac{\tan v \cos \psi}{\rho} & -\frac{\tan v \sin \psi}{\rho} \end{pmatrix} w \quad (1)$$

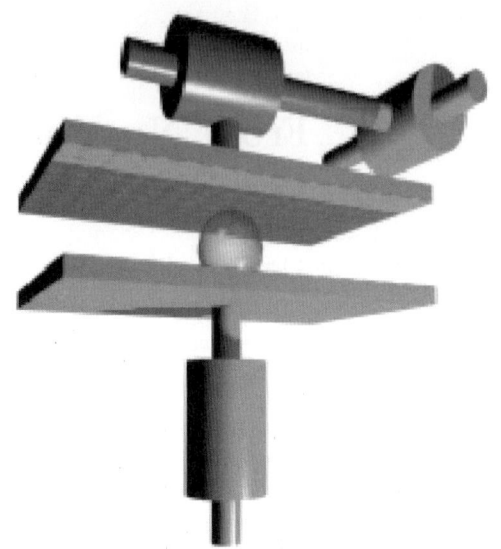

Figure 1: The plate-ball manipulation system

where w is the cartesian velocity $(\dot{y}, \dot{x})$ of the contact point on the plane, which we assume to be the control input.

3 The spinning maneuver

The planner to be presented in Sect. 4 requires the capability of 'spinning' the ball, i.e., changing ψ without altering the values of the other system coordinates. The perfect rolling assumption prevents a pure rotation of the ball around an axis which is perpendicular to the finger surface at the contact point, for this would violate the underlying nonholonomic constraint. However, due to the controllability of the system, for any angle $\Delta \psi$ there exists a control function $w(t)$ steering the system from the configuration $(\alpha_p, \alpha_b, \psi)$ to the configuration $(\alpha_p, \alpha_b, \psi + \Delta \psi)$ in finite time. In this section, we present two ways to compute such a control function: the first in open-loop, and the second — a slight modification of the first — in closed-loop. The utility of the closed-loop version in setting up our planner will be clarified in Sect. 4.

3.1 The open-loop maneuver

The spinning maneuver is obtained by a sequence of three control functions. Up to a change of coordinates, we can assume that at the beginning of the maneuver the contact between the ball and the lower finger occurs at the south pole (i.e., $v = -\frac{\pi}{2}$).

The first control function forces the the ball to roll along a geodesic ($u = \text{constant} = u_0$) so that the contact point is steered from the south pole to the particular parallel corresponding to $v = \bar{v}$, whose determination is discussed later. This is simply obtained by the following steering control:

$$w(t) = \begin{pmatrix} \sin\psi_0 \\ \cos\psi_0 \end{pmatrix}, \qquad (2)$$

where ψ_0 is the initial orientation of the ball. In order to reach $\bar{v}$, this steering control must be applied over a time interval $[0, \rho(\bar{v} + \pi/2)]$.

The second control drives the contact point along the same parallel until the contact point on the plane completes a circle. When this happens, u has reached a value $u_0 + \Delta u$, with Δu determined as follows.

Assuming without loss of generality $\bar{v} \in (-\pi/2, 0)$, the radius of the circle on the plane is $\rho \tan(\bar{v} + \frac{\pi}{2})$, so that its length is $2\pi\rho \tan(\bar{v} + \frac{\pi}{2})$. On the other hand, the parallel traced by the contact point on the sphere is a circle of radius $\rho \sin(\bar{v} + \frac{\pi}{2}) = \rho \cos\bar{v}$. Being the length of the path traced by the contact point on the sphere equal to the length of the path traced by the contact point on the plane, we have

$$2\pi\rho \tan(\bar{v} + \frac{\pi}{2}) = \Delta u \rho \cos\bar{v},$$

from which

$$\Delta u = -\frac{2\pi}{\sin\bar{v}}.$$

We can determine the net change $\Delta\psi$ that the sphere orientation undergoes at the end of the spinning maneuver as the integral of the gaussian curvature over the region bounded by the closed path traced by the contact point on the sphere (total curvature):

$$\begin{aligned}\Delta\psi &= \int_{u_0}^{u_0+\Delta u} \int_{-\frac{\pi}{2}}^{\bar{v}} K \|f_u \times f_v\| dv du \\ &= \int_{u_0}^{u_0+\Delta u} \int_{-\frac{\pi}{2}}^{\bar{v}} \cos v \, dv du = (1 + \sin\bar{v})\Delta u\end{aligned}$$

where K is the gaussian curvature of the sphere.

Plugging the expression of Δu in the latter equation we get

$$\Delta\psi = -2\pi \frac{1 + \sin\bar{v}}{\sin\bar{v}}$$

so that the parallel to be traveled in order to spin the ball by $\Delta\psi$ is identified by

$$\sin\bar{v} = -\frac{2\pi}{2\pi + \Delta\psi}.$$

It is easy to verify that a steering control realizing the rotation on the parallel is

$$w(t) = \begin{pmatrix} -\cos(\psi_0 + \frac{\tan\bar{v}}{\rho}t) \\ \sin(\psi_0 + \frac{\tan\bar{v}}{\rho}t) \end{pmatrix}, \qquad (3)$$

to be applied over a time interval $[0, 2\pi\rho/\tan\bar{v}]$.

The spinning maneuver is completed by a third control action (the opposite of function (2)) that simply brings the sphere back to the south pole along the geodesic $u = \text{constant} = u_0 + \Delta u$. At this point, while the contact point on the plane and on the sphere are back to the starting configuration, the orientation has been changed as desired.

3.2 The closed-loop maneuver

When the radius ρ of the sphere is not exactly known, the spinning maneuver so far described cannot be executed. In fact, the steering controls (2) and (3) require the value of ρ in their expression and/or duration. If the nominal value of ρ used to compute the controls is different from the actual value, there will be three consequences: *(i)* the value of v reached with control (2) is different from the desired $\bar{v}$, *(ii)* the sphere under control (3) does not roll along a parallel, and *(iii)* the path traced by the contact point on the plane is not a *closed* circle. In terms of the final configuration of the ball, this means not only that the desired orientation is not reached, but also that u, v, x and y do not go back to their initial values due in particular to *(ii)* and *(iii)*.

While the error in ψ is acceptable and will be recovered by the iterative version of the planner to be presented in Sect. 4, the non-cyclicity in the other variables would destroy the convergence of the planner. We therefore devise a modified version of the spinning maneuver that uses a closed-loop control to roll the ball along a parallel without knowing its radius.

Given the rolling equations (1), it is straightforward to verify that the closed-loop steering control

$$w(t) = \begin{pmatrix} -\cos\psi(t) \\ \sin\psi(t) \end{pmatrix}$$

yields $\dot{v} \equiv 0$ (i.e., $v(t) \equiv \bar{v}$), driving the contact point on the sphere along the parallel reached after the application of the first control (2). The other system equations can be integrated in closed form; in particular, the other coordinates of the ball are obtained as

$$\begin{aligned} u(t) &= u_0 - \frac{1}{\rho\cos\bar{v}} t \\ \psi(t) &= \psi_0 - \frac{\tan\bar{v}}{\rho} t, \end{aligned}$$

while the contact point on the plane will describe the circle

$$\begin{aligned} x(t) &= x_c + \frac{\rho}{\tan\bar{v}} \sin\left(\psi_0 + \frac{\tan\bar{v}}{\rho} t\right) \\ y(t) &= y_c + \frac{\rho}{\tan\bar{v}} \cos\left(\psi_0 + \frac{\tan\bar{v}}{\rho} t\right) \end{aligned} \qquad (4)$$

with radius $\rho/\tan\bar{v}$ and center in

$$x_c = x_0 - \frac{\rho}{\tan\bar{v}}\sin\psi_0$$
$$y_c = y_0 - \frac{\rho}{\tan\bar{v}}\cos\psi_0.$$

Equation (4) indicates that the circle is completed at time $\bar{t} = |\frac{2\pi\rho}{\tan\bar{v}}|$. Once again, the control duration time would depend on the ball radius and, hence, the 'parallel roll' is not robust yet w.r.t. perturbation on ρ. The desired robustness can be achieved by modifying the open-loop control (3) as follows

$$w(t) = \begin{pmatrix} -\cos\psi(t)s(\psi) \\ \sin\psi(t)s(\psi) \end{pmatrix}, \quad (5)$$

with

$$s(\psi) = 1 - \delta_{-1}(\psi - \psi_0 - 2\pi)$$

where δ_{-1} is the Heaviside step function.

4 The exact planner

Denote by $\mathcal{M}$ the plate-ball configuration space, which is locally diffeomorphic to $\mathbb{R}^2 \times \mathbb{R}^2 \times S^1$. Let $p_0 = (\alpha_p^0, \alpha_b^0, \psi^0)$ and $p_g = (\alpha_p^g, \alpha_b^g, \psi^g)$ be two points in $\mathcal{M}$. The algorithm steers the system from p_0 to p_g through the following intermediate configurations:

$$p_0 \stackrel{\text{Step1}}{\mapsto} p_1 = (\alpha_p^g, \alpha_b^1, \psi^1)$$
$$p_1 \stackrel{\text{Step2}}{\mapsto} p_2 = (\alpha_p^g, \alpha_b^g, \psi^2)$$
$$p_2 \stackrel{\text{Step3}}{\mapsto} p_g = (\alpha_p^g, \alpha_b^g, \psi^g)$$

Step 1 is simply executed by applying a constant control such that the velocity of α_p (the contact point on the plane) is a vector with the same direction of $(\alpha_p^g - \alpha_p^0)$ and unit norm:

$$w(t) = \frac{(\alpha_p^g - \alpha_p^0)}{\|\alpha_p^g - \alpha_p^0\|}, \quad t \in [0, T_1], \quad T_1 = \|\alpha_p^g - \alpha_p^0\|.$$

Step 2 is performed by alternating two maneuvers. Up to a change of coordinates, assume again that the contact point on the sphere reached at the end of Step 1 is the south pole. The first maneuver, realized by a constant control of the form

$$w(t) = w_u, \ t \in [0, T_2], \ T_2 = \frac{1}{4}\left|v^g + \frac{\pi}{2}\right|\rho, \ \|w_u\| = 1,$$

rolls the ball along an arc of the geodesic corresponding to u_g, which joins the initial and the desired contact point on the sphere. The second is a closed-loop spinning maneuver that rotates the ball of $\frac{\pi}{2}$ around the axis perpendicular to the finger through the point of contact as explained in Sect. 3.2. By repeating four times the two maneuvers in sequence, the contact point on the ball is steered to α_b^g while the contact point on the plane has come back to the initial point α_p^g, having traced a square of edge T_2.

Step 3 brings the last variable ψ to its desired value by using the closed-loop spinning maneuver to achieve a rotation $\Delta\psi = \psi^g - \psi^2$.

5 Robust planning by IS

As mentioned in the introduction, our idea is to robustify the planner by using the iteration mechanism. The theoretical framework of IS [10] indicates that a robust stabilizer can be obtained by iterating a planner with suitable properties, the most relevant of which is (Hölder-)continuity at the origin with respect to the desired reconfiguration. In practice, this property means that the configuration space path generated by the planner 'shrinks' and eventually vanishes when the desired reconfiguration goes to zero.

Without going into technical details, it is clear that the planner of the previous section does not possess this property[1], due to the repeated spinning maneuvers in Step 2, each of which adds $\pi/2$ to the current value of ψ by driving x, y along the same path on the plane (see Fig. 4). Therefore, the simple iteration of the whole planner does not yield a robust stabilizer.

However, an arbitrarily accurate robust planner can be obtained by iterating *separately* Steps 2 and 3 of the planner (note that Step 1 is insensitive to perturbations on the sphere radius) until the state error is below a given tolerance. The proof of convergence of the error with the iterations is lengthy and therefore omitted, but basically relies on a simple property of perturbed discrete-time systems [10, Lemma 1]. The same proof guarantees that the steering error converges exponentially to zero starting from any configuration.

We note that the adoption of the closed-loop version of the spinning maneuver in the planner is essential for guaranteeing robustness. In fact, if the sphere radius is not exactly known, the non-cyclicity in u, v, x and y of the open-loop spinning maneuver pointed out at the beginning of Sect. 3.2 would result in a *persistent* perturbation on the sampled error dynamics, which would destroy the convergence. All the other induced perturbations are instead non-persistent, and therefore rejected by the iteration mechanism itself.

To show the effectiveness of the iterative planner, we report the results of the simulated execution of a

---
[1] Actually, the same is true for any other *exact* planner in the literature; instead, the approximate planner used in [11] satisfies the continuity property.

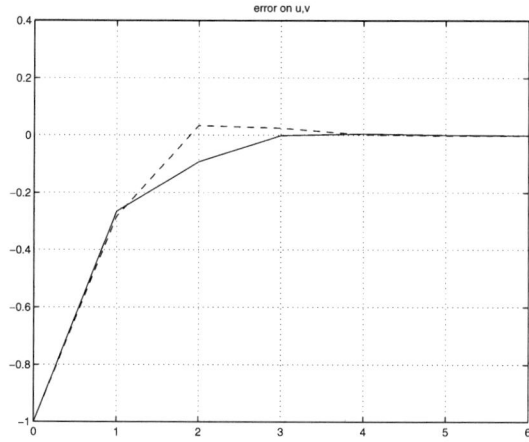

Figure 2: Error on u (solid) and v (dashed) at the end of each iteration of Step 2

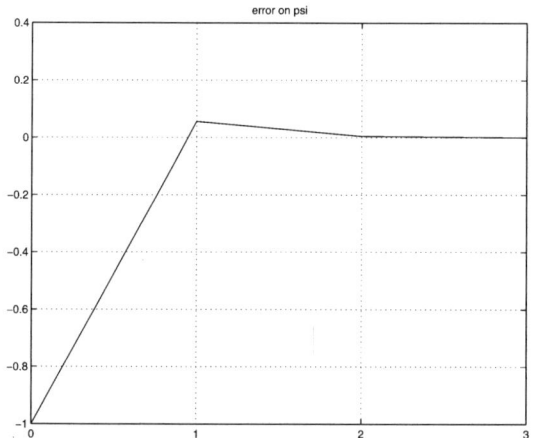

Figure 3: Error on ψ at the end of each iteration of Step 3

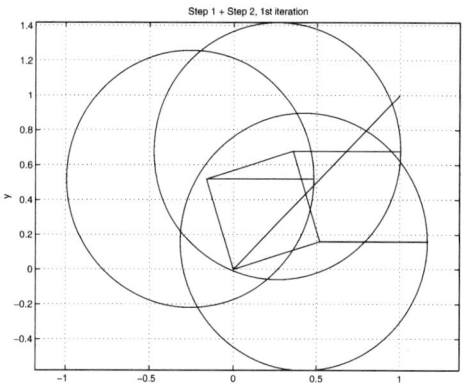

Figure 4: Path of the contact point on the plane during Step 1 and the first iteration of Step 2

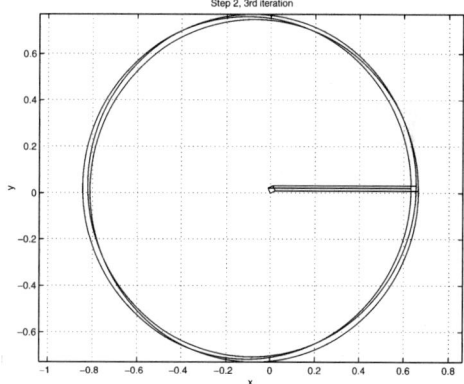

Figure 5: Path of the contact point on the plane during the third iteration of Step 2

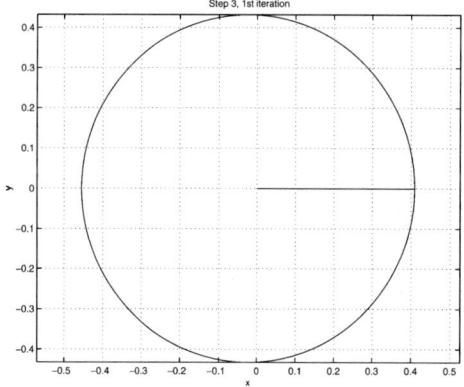

Figure 6: Path of the contact point on the plane during the first iteration of Step 3

steering task. The system has to reach the origin of the configuration space starting from the configuration $(1, 1, 1, 1, 1)$ (m,m,rad,rad,rad), despite a 5% perturbation on the nominal unit radius. The iteration of Steps 2 and 3 is interrupted as soon as the norm of the corresponding error (on u, v and ψ, respectively) is below 10^{-3}. Figure 2 reports the values of the error on u, v, during the iterations of Step 2, while Fig. 3 refers to error on ψ during the iterations of Step 3. Figures 4–7 show four paths of the contact point on the plane, corresponding respectively to Step 1 + the first iteration of Step 2, the third iteration of Step 2, the first iteration of Step 3, and the third iteration of Step 3. Note that the path on the plane contracts during the repeated execution of Step 3.

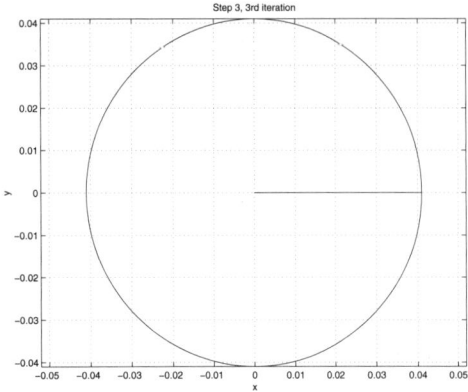

Figure 7: Path of the contact point on the plane during the third iteration of Step 3

6 Conclusions

As an intermediate result of our investigation aimed at deriving robust controllers for rolling manipulation mechanisms, we have presented a planner that can drive the plate-ball system to a desired configuration with arbitrary precision in spite of perturbations on the sphere radius. The planner relies on the repeated application of a steering algorithm that is exact for the nominal system and has been designed to guarantee its effectiveness within the iterative framework.

Future work on this planner includes the computation of explicit bounds on the admissible perturbation as well as its combination with a motion planning scheme to ensure the successful execution of maneuvers in the presence of obstacle. At a more general level, however, the achievement of our long-term objective (robust *stabilization*) will pass through the synthesis of nominal planners with the analytic properties required by the IS paradigm (Hölder-continuity of the steering control law with respect to the desired reconfiguration). In this respect, the so-called *local-local* property of [12] and *topological property* of [13] (of which the former is a relaxation) are of interest because they represent the topological counterpart of the Hölder-continuity condition. Therefore, it will be necessary to bridge steering controllers and geometrical planners by transferring algorithms and properties between them.

Acknowledgments

Work supported by the following contracts: MURST MISTRAL, IST-2001-37170 (RECSYS), IST-2001-38040 (TOUCH-HAPSYS), ASI I/R/124/02 (TEMA).

References

[1] A. Marigo and A. Bicchi, "Rolling bodies with regular surface: controllability theory and applications," *IEEE Trans. on Automatic Control*, vol. 45, pp. 1586–1599, 2000.

[2] A. Bicchi and R. Sorrentino, "Dexterous manipulation through rolling," *1995 IEEE Int. Conf. on Robotics and Automation*, pp. 452–457, 1995.

[3] R. M. Murray, Z. Li, and S. S. Sastry, *A Mathematical Introduction to Robotic Manipulation*, CRC Press, 1994.

[4] A. Marigo, Y. Chitour, and A. Bicchi, "Manipulation of polyhedral parts by rolling," *IEEE Int. Conf. on Robotics and Automation*, pp. 2992–2997, 1997.

[5] D. J. Montana, "The kinematics of contact and grasp," *Int. J. of Robotics Research*, vol. 7, no. 3, pp. 17–32, 1988.

[6] Z. Li and J. Canny, "Motion of two rigid bodies with rolling constraint," *IEEE Trans. on Robotics and Automation*, vol. 6, pp. 62–72, 1990.

[7] V. Jurdjevic, "The geometry of the plate-ball problem," *Arch. for Rational Mechanics and Analysis*, vol. 124, pp. 305–328, 1993.

[8] R. W. Brockett and L. Dai, "Non-holonomic kinematics and the role of elliptic functions in constructive controllability," in *Nonholonomic motion planning*, Z. Li and J. F. Canny, Eds., pp. 1–21. Kluwer Academic Publishers, 1993.

[9] J.-P. Laumond (Ed.), *Robot Motion Planning and Control*, Springer-Verlag, 1998.

[10] P. Lucibello and G. Oriolo, "Robust stabilization via iterative state steering with an application to chained-form systems," *Automatica*, vol. 37, pp. 71–79, 2001.

[11] G. Oriolo and M. Vendittelli, "Robust stabilization of the plate-ball manipulation system," *2001 IEEE Int. Conf. on Robotics and Automation*, pp. 91–96, 2001.

[12] A. Marigo and A. Bicchi, "A local-local planning algorithm for rolling objects," *2002 IEEE Int. Conf. on Robotics and Automation*, pp. 1759–1764, 2002.

[13] S. Sekhavat and J. P. Laumond, "Topological properties for collision free nonholonomic motion planning: the case of sinusoidal inputs for chained form systems," *IEEE Trans. on Robotics and Automation*, vol. 14, no. 5, pp. 671–680, 1998.

Motion Sensing for Robot Hands Using MIDS

Alan H. F. Lam[1,2], Raymond H. W. Lam[1,2], Wen J. Li[1,2,*], Martin Y. Y. Leung[2] and Yunhui Liu[2]

[1]*Centre for Micro and Nano Systems*
[2]*Dept. of Automation and Computer-Aided Engineering*
The Chinese University of Hong Kong, Hong Kong SAR

*Contacting Author: wen@acae.cuhk.edu.hk

Abstract – A novel computer input system - the Micro Input Devices System (MIDS) – is under development by merging MEMS sensors and existing wireless technologies. This system could potentially replace the functions of the mouse, pen, and keyboard as input devices to the computer. The system could also be used as a general wireless 3D motion sensing device. In this paper, we will present our work on using MIDS for motion sensing application of robot hands. MIDS is used to evaluate the performance of PD adaptive control and Impedance control schemes in manipulating a five-fingered robot hand and in manipulating this hand to grasp a ball. Experimental results indicate that MIDS is capable of obtaining real-time 3D acceleration/vibration data wirelessly for the robotic hand, hence eliminating the need to perform the time-consuming integration of the position sensor data to obtain acceleration. Moreover, our initial results also indicate that further exploration of this technology could eventually produce a new control-input device for robotic grasping manipulators. These results are presented in this paper.

I. INTRODUCTION

Nowadays, fingered robots are becoming more ubiquitous, e.g., in the manufacturing industry, robot hands are now used to handle many duplicated tasks such as grasping in product assembly line. Human-like robots such as Honda's humanoid robot: ASIMO [1] and Sony's SDR-4X [2] are commercially available in the market. These robots can be classified into two different types: controlled by the CPU automatically and manipulated by operators. For automated robot hands, motion sensing devices are essential to measuring the motion of the robot hands for quality control and system evaluation. For manual control robot hands, a user friendly input device can be useful to the operator in controlling manipulators such as grasping robotic hands. In this case, a multi-functional device is needed to handle both the motion sensing function and the control-input function. In this paper, we will show that our current research in developing the Micro Input Devices system (MIDS) will enable many new capabilities in terms of robotic sensing and control, including multi-functional device for sensing and controlling robotic hands.

We believe that by combining the advent in MEMS sensing and wireless technologies, it is possible to develop a novel computer input system that could enable multi-functional input tasks and allow the overall shrinkage in size of the graphical user interface (GUI) and text-based user interface (TUI) input devices. Experimental results from our prototype input system [3] and [4] (a similar input system was also proposed by K. Pister's group at Berkeley [5]) indicate that both GUI and TUI functions could be performed using existing MEMS-based motion detection sensors. In terms of mobile computing, we envision the MIDS to serve the functions of the present day mouse, light pen and keyboard such that it will allow users to input text, draw graphical image, move curser, and control drag and drop motion. Moreover, this system can be used to capture the motion of the hand, e.g., the motion of the robot hands and the fingers can be detected. This means that MIDS can potentially function as an external motion sensing device to give extra information to engineers to design or modify the controller of a robotic-hand manipulation system. Furthermore, we believe that this system can further be explored to ultimately replace the robotic gloves that are used today to interface with virtual and real robotic hands. This paper will describe the components of our system and present our encouraging results for robotic hand motion sensing experiments.

In robotics area, research in grasping using robot hands has been extensive in the past decade [6]-[9]. In order to study the dynamics of the robot hands and develop effective control algorithms for grasping tasks, K. Nagata et al. have developed a master hand for grasping information capturing [6]. Their master hand can be used to measure the motion of the human. The working principle is to measure the change of the kinetics contact between human fingers and the object so that it could give some reference information for development of grasping algorithm of robot hands. This is also one of the objectives of our project but different operation. Our proposed system, the Micro Input Devices System (MIDS), functions as a motion sensing device and control-input device for grasping robot hands. MIDS is not used to develop the control algorithm for grasping only; rather, it acts as a robot-human-interface that allows users to control the robot hands by their body motion. The details of MIDS are presented in the following sections.

II. MIDS: MICRO INPUT DEVICES SYSTEM

In terms of motion sensing, MIDS is a portable, cheap and small size motion detector for robot hands. It can give the motion information to the controller such that these data can be used to design or modify the control or the input data for neural network training scheme. On the other hand, MIDS is a wearable and multi-functional input device. Potentially, a MIDS (a system made of one or more MIDS components and other peripheral subsystems such as Bluetooth wireless transmitters, power-storage units, …etc) is able to measure acceleration, velocity, and position of the robot fingertips, and thus, allowing users to control the robot hand to do tasks such as grasping motion of real or virtual robotic hands.

MEMS sensors play a major role in our endeavour to develop a functional MIDS due to their low-cost and miniaturized size. We propose to use MEMS sensors to measure multi-dimensional force (acceleration) of each finger and hand, and wirelessly transmit these motion data to the computer for input information process. In this paper, the prototype of a MIDS suitable for motion sensing and control-input functions for robot hand is presented. The key subsystems of this prototype are described below.

A. System Description

Our prototype MIDS consists of 4 main subsystems: 1) the MIDS rings with MEMS multi-axes acceleration sensors, 2) the MIDS controller, 3) the wireless transmission interface board connected to a PC, and 4) the display interface program. The MIDS rings are positioned on the human/robot fingers (depending on the purpose: as a motion sensing device or a control-input device) and are electrically connected with a wireless transmission controller that acts as a communication link between the sensors and the PC. Potentially, wireless links could also be established between the MIDS rings and the MIDS controller. Inside the controller, a microprocessor is used to analyze and encode the sensor signals for wireless transmission. Another microprocessor is placed in the wireless transmission interface board (which is connected to the PC) to decode the data received from the wireless controller. For the motion sensing MIDS, it passes the received data to low-pass filter before sending to the PC. For control-input device, the received data is passed to a control algorithm and then converted the command signals to the PC. A display interface program is used to plot the sensor data and the control-input commands.

B. MEMS Sensors for Multi-axes Force Sensing

The most important subsystem of the MIDS is the MIDS ring. An illustration of the components of the MIDS ring is shown in Fig. 1.

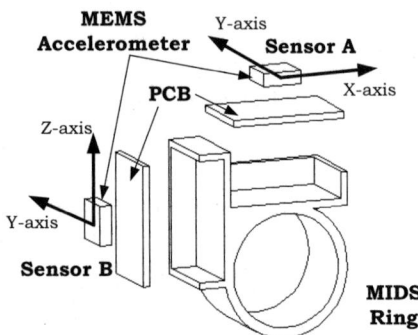

Fig. 1 Schematic diagram of a MIDS ring

Two dual-axis MEMS accelerometers (manufactured by Analog Devices Inc.) are mounted as shown in Fig. 1. Sensor A is placed at the top of the ring horizontally to measure fingertip accelerations in the x and y directions. Sensor B is placed at the side vertically to detect accelerations in the y and z directions. Therefore, sensor A can detect the plane motion of the fingertip and sensor B can detect the fingertip angle (relative to rotation about the mid-joint of a finger) and the vertical movement. The sensors employ the principle of relating the capacitance variation between the polysilicon comb-drives to acceleration to detect motion (as illustrated in Fig. 2)

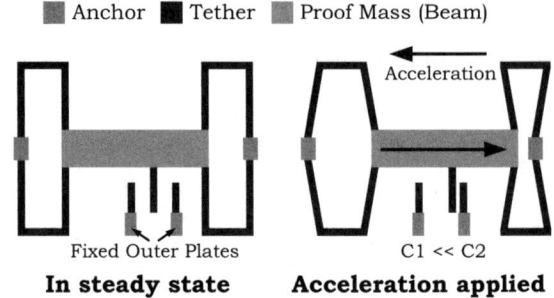

Fig. 2 Illustration of sensor operation

When movement is applied to a sensor, the proof mass is moved such that the capacitances between the two fixed outer plates (C1 and C2) are changed. The acceleration can then be determined by the ratio of the capacitances.

C. Wearable Wireless MIDS

The prototype of our MIDS is shown in Fig. 3 and Fig. 4. The ring-shape housing is made by a rapid prototyping machine (FDM1600 by StrataSys Inc.). The two MEMS accelerometers on a ring are connected to a signal conditioning circuits and are powered by the battery cell from the MIDS controller subsystem.

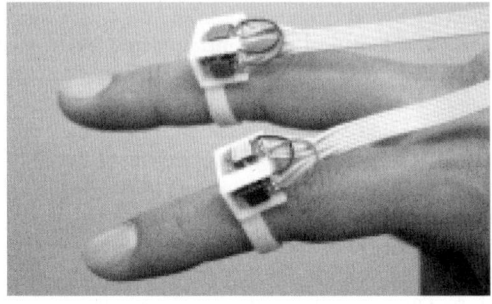

Fig. 3 Prototype of MIDS rings.

The entire wearable MIDS prototype is shown in Fig. 3. A microprocessor (AT90S8515) is used to count the duty cycles of the sensing signals and convert the signals to acceleration information. Then, a Radiometrix TX2 transmitter is used to transmit the packed signal sequentially. One the signal receiving end, the RX2 receiver passes the received data to another microprocessor, which unpacks the data and passes suitable commands to the PC from serial port. The data acquisition schematic is shown in Fig. 5.

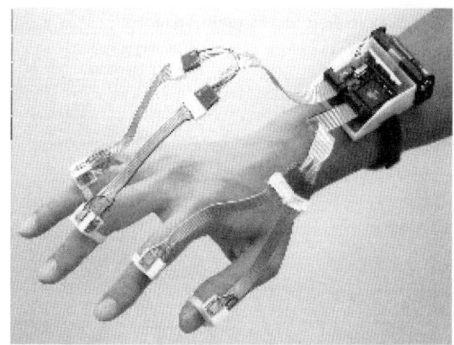

Fig. 4 Wearable wireless MIDS prototype

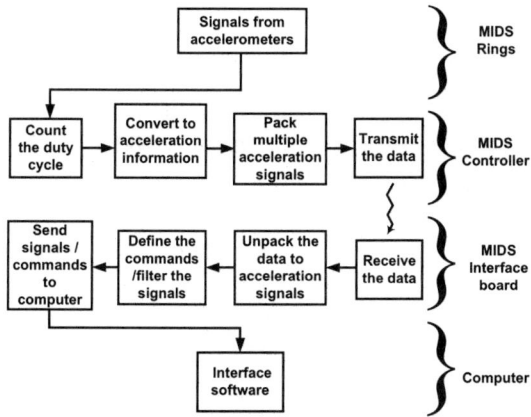

Fig. 5 Data acquisition schematic for the MIDS

The technical data of the sensors, transmitter, receiver and microprocessor is summarized in TABLE I. The resolution and the operating voltage of the overall system (MIDS) are also shown in TABLE I. The sampling frequency of MIDS is about 30 data sets per second (30Hz per data set), which is larger than the reaction rate of human (less than 20Hz), and means that MIDS is suitable for human-controlled computer input operation.

TABLE I
TECHNICAL DATA OF THE MIDS

Sensor	
Acceleration range (1g = 9.81ms^{-2})	$\pm$2g
Resolution	5mg
Bandwidth	5000 Hz
Temp. range	0 – 70 °C
Supply current	0.6mA
Transmitter and receiver	
Transmission rate	40kbps
Microprocessor	
Speed grade	4 MHz
Power consumption at 3V, 25 °C	Active: 3mA Idle mode: 1mA
MIDS	
Resolution (1 data set include 8 acceleration signals)	~30 data sets /sec
Operating voltage	3V – 6V (4.5V)
Operating power (for 4 sensors)	~ 0.025 W

III. EXPERIMENTS

Experiments were performed to demonstrate the motion detection in 3D space of our MIDS as it was used on a five-fingered robot hand. Two experiments have been performed: 1) grasping motions of two controlled systems – PD adaptive system and Impedance system and 2) grasping a ball using PD adaptive system. The results are shown below.

A. Experimental Setup

The manipulator used in the experiments is a five-fingered robot hand system at the Robot Control Laboratory of the Chinese University of Hong Kong. The robotic finger made by Yaskawa has three revolute joints driven by AC motors through a harmonic drive of 80:1 reduction ratio. The robot hand is controlled by a distributed DSP C40's system. The MIDS rings were positioned on the end-effectors of two robotic fingers. The installation and the configuration of the MIDS for the five-fingered robot hand are shown in Fig. 6 and Fig. 7. The x-y plane is parallel to the horizontal plane of the world coordinate frame and the z-axis is parallel to the vertical axis, which is consistent to Fig. 1.

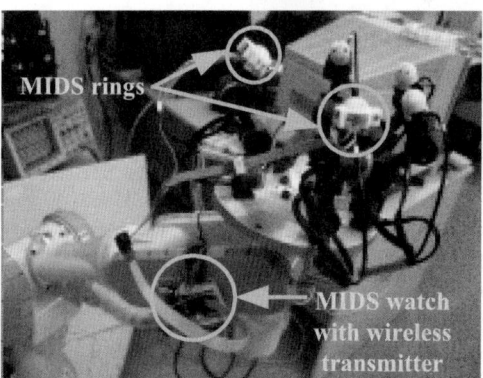

Fig. 6 Installation of MIDS for fingered robot hand

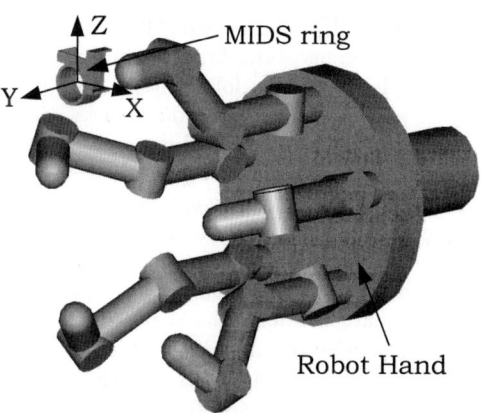

Fig. 7 MIDS configuration for robot hand motion detection

The experimental setup for the motion sensing test is shown in Fig. 8. The grasping motion is first detected by the MIDS rings. The acceleration signal then is passed to the microprocessor of the MIDS controller for signal encoding. After that, the packed signal is transmitted to the receiver through the wireless transmitter inside the MIDS controller. Once the wireless receiver receives the signal, another microprocessor inside the interface board will decode the received signal and then pass the acceleration signals to the PC through the serial port. We have also developed an interface program called MIDS Interface, which is able to display real-time signals from MIDS.

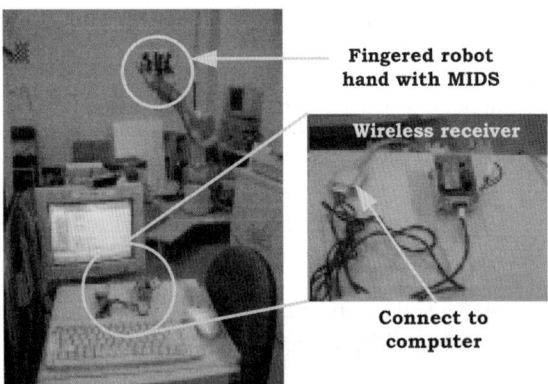

Fig. 8 Experimental setup for motion sensing of fingered robot hand

B. Experimental Results

The first experiment was to evaluate the performance of two control methods (PD adaptive control and Impedance control) for the grasping robot hand. The responses of the fingertip motion of the robot fingers are shown in Fig. 9. The responses are in 3D space (xyz directions as defined in Fig. 7). The results in Fig. 9 show that vibration occurred during the grasping motion. At the initial state, the acceleration values of x-direction and y-direction were $0ms^{-2}$ and the acceleration value of z-direction was $-10ms^{-2}$ (-g) due to gravity. At t = 1sec, the robot fingers rotated along x-axis from 0° to 90° and then still stayed for about 1sec, hence the acceleration values of y-direction and z-direction changed to $10ms^{-2}$ and $0ms^{-2}$, respectively. After that, the robot fingers returned to the initial position following the same path and kept moving periodically. In the whole process, vibrations in xyz-directions (especially in x-direction) for both PD adaptive system and Impedance control system were observed. However, the frequency and the amplitude of the PD adaptive system were higher than the Impedance control system. According to these results, the Impedance control system is slightly better than PD adaptive system. From this experiment, we have shown that the MIDS could be used as a wireless motion sensing device for a robotic grasping system. The experimental data can be used for control algorithm design or neural network training. In the short future, the control-input function of MIDS will be demonstrated for controlling the grasping robot manipulators.

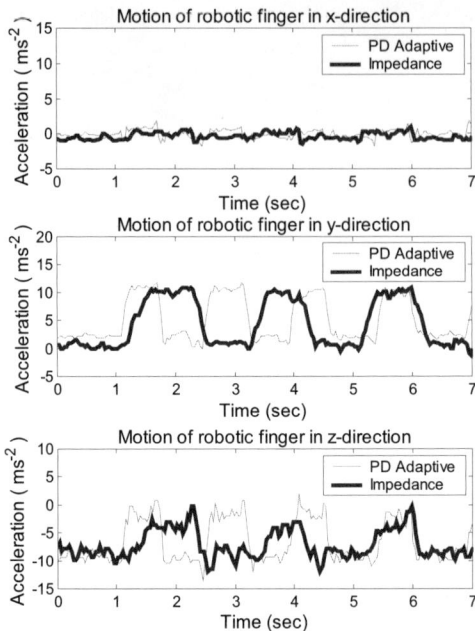

Fig. 9 Experimental results for the grasping motion of robot hand measured by MIDS

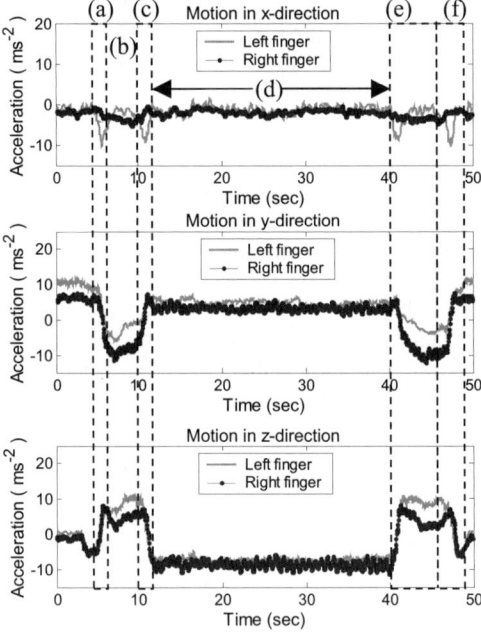

Fig. 10 Results of "Grasping a ball" experiment

The second experiment was to measure the response of the robot hand grasping a ball. The process of the experiment for robot hand grasping a ball is shown in Fig. 11. The robot hand first moved to the top of the ball and then grasped the ball. After holding the ball, the robot hand moved to the original position and then rotated the ball. After that, the robot hand moved down again to release the ball to the original position. The process of grasping motion started from (a) and ended at (f) in Fig. 11. The corresponding signals of the process from (a) to (f) can be seen in the Fig.

10. The responses of two robotic fingers in xyz direction are shown in Fig. 10. As shown, the acceleration variations during the time period from 4sec to 12sec and from 40sec to 48sec were very obvious. This is because the robot hand had moved from top to down to grasp the ball and then released the ball. The reason of the signals in the x-direction keeping nearly constant is that the motion of robot moving up and down was rotated by x-axis. For the left finger, there was an offset angle along z-direction so that the MIDS could measure a fraction of movement for the duration of the move period. During the grasping period (from 4sec to 6sec), a pulse in x and z directions was measured because the finger was rotated along the y-axis to grasp the ball. For the same reason, during the release of the ball, pulses in both x and z directions were measured (from 46sec to 48sec). From 12sec to 40sec, the ball was rotated. The vibration of the fingers during this rotation motion could also be measured.

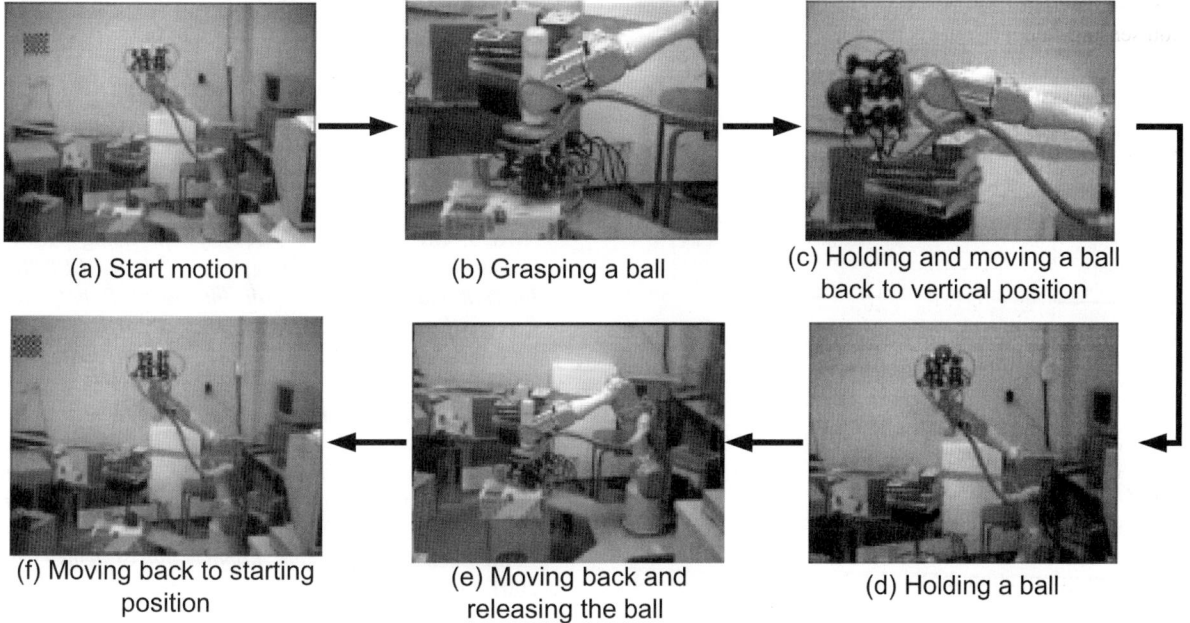

Fig. 11: Process for the second experiment: grasping a ball

IV. POTENTIAL APPLICATION

Many control methods, such as the PD adaptive control and the Impedance control, focus on the position and force control without considering vibration. According to the results of previous section, vibration problems still transpire in these position and force control methods. For instance, using a robot hand with traditional control methods, such as position and force control, to carry and transport a glass of water from point to point exhibits a high possibility of spilling due to vibration.

Currently, to obtain vibration or force information for the type of robot hand system tested above, we can use the encoders in the finger joints to measure the position and then calculate the second derivatives to get the acceleration values. These acceleration information can then be used to improve the control algorithm. However, the error can be significant due to differentiation of the collected data.

Another method is to install some accelerometers to measure the vibration. And then the measured signals can be fed back to a new controller that considers the acceleration. However, it is quite costly to install extra sensors to existing systems, and some systems may not be suitable to install extra sensors.

Our portable MIDS can be helpful for this purpose. It is easy to install the MIDS in the existing robotic system by using suitable design of rings. Then the MIDS can directly measure the acceleration of each finger. However, it is still costly to install MIDS to robotic systems with multiple machines.

Based on the needs of market, we proposed a method which allows the collection of acceleration information for many robotic systems while maintaining the cost at a reasonable level. Normally, most of the industrial robots, especially those serve in the factory assembly line, are trained to perform the same tasks in cycle, resulting in the anticipated repeated acceleration patterns. And also, there are many robotic systems in which joint angle sensors have been installed inside the robot hand for measuring the finger angles. Then the positions of the fingertips can be calculated by the joint angles. However, calculating second derivatives of the positions to get the accelerations faces the drawback of significance degeneracy for highly non-linear systems. In this case, we proposed a control algorithm using neural network to tackle this problem. Neural network training method can be used to get the mapping from the sensors' readings to the accelerations. The accelerations can be measured by our portable MIDS. Once the neural network model is built, it is not necessary to measure the acceleration anymore. After training, the MIDS can be removed from the machine because the acceleration values can be obtained directly by the neural network model. The neural network model can find the corresponding acceleration patterns based on the positions measured by joint angle sensors. Therefore, only one MIDS is required to provide the acceleration information to many robotic systems during the neural network training period. An illustration of a sample control algorithm using the above method is shown in Fig. 12. Hence, Our MIDS is a flexible motion sensing device for many robotic systems.

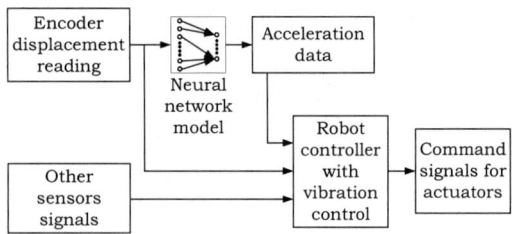

Fig. 12 A proposed Neural Network control system for robot hand fingertip with vibration control.

Since high accuracy joint angle sensors have been installed in the robot hands of most robotic systems, positions of the fingertips can be obtained by the joint angles. So, it is not necessary to do the integration for the acceleration signals from MIDS to get the velocity and displacement information for this kind of application. In the future, the acceleration integration will be discussed for other applications and using MIDS to control the robotic hand system will also be demonstrated.

V. CONCLUSION

A multi-functional portable micro input devices system has been successfully demonstrated to sense various motions of a grasping robot hand. The MIDS uses MEMS sensor to detect robot fingertip motions. Experimental results for testing two different control algorithms of a grasping robot hand and the grasping motion of a ball have been presented. The results indicate that the MIDS is applicable of measuring motion for robot hand systems. In the future, we will demonstrate the control-input function of MIDS for controlling a robotic grasping hand.

IV. ACKNOWLEDGMENTS

This work was funded by The Chinese University of Hong Kong with partial support from the Research Grants Council of the Hong Kong Special Administrative Region (Project no. CUHK4206/00E).

V. REFERENCES

[1] http://world.honda.com/robot/
[2] http://www.watch.impress.co.jp/pc/docs/2002/0320/sony.htm
[3] Alan H. F. Lam and Wen J. Li, "MIDS: GUI and TUI in mid-air using MEMS sensors", *in Proceedings of the International Conference on Control and Automation*, June 2002, pp. 1218-1222.
[4] Alan H. F. Lam, Wen J. Li, Yunhui Liu, and Ning Xi, "MIDS: Micro Input Devices System Using MEMS Sensors", *in Proceedings of the IEEE/RSJ International Conference on Intelligent Robots and Systems*, Switzerland October 2002, pp. 1184 - 1189.
[5] J. Perng, B. Fisher, S. Hollar, K. S. J. Pister, "Acceleration sensing glove," *in Proceedings of ISWC International Symposium on Wearable Computers*, San Francisco, Oct. 1999.
[6] K. Nagata, F. Saito, and T. Suehiro, "Development of the master hand for grasping information capturing", *in Proceedings of 2001 IEEE/RSJ International Conference on Intelligent Robots and Systems*, Oct. 2001, vol.3, pp. 1757 –1762.
[7] Dan Ding, Yun-Hui Liu, J. Zhang, and A. Knoll, "Computation of fingertip positions for a form-closure grasp", *in Proceedings of 2001 IEEE International Conference on Robotics and Automation*, May 2001, vol. 3, 2001pp. 2217 –2222.
[8] Y. Hasegawa, J. Matsuno, and T. Fukuda, "Regrasping behavior generation for rectangular solid object", *in Proceedings of 2000 IEEE International Conference on Robotics and Automation*, April 2000, vol.4, 2000 pp. 3567 –3572.
[9] C.C. Cheah, H.Y. Han, S. Kawamura, and S. Arimoto, "Grasping and position control for multi-fingered robot hands with uncertain Jacobian matrices", *in Proceedings of 1998 IEEE International Conference on Robotics and Automation*, May 1998 vol.3, pp. 2403 –2408.

Mechatronic Design of Innovative Fingers for Anthropomorphic Robot Hands

L. Biagiotti, F. Lotti, C. Melchiorri, G. Vassura

DEIS - DIEM, University of Bologna
Via Risorgimento 2,
40136 Bologna, Italy
{lbiagiotti, cmelchiorri}@deis.unibo.it, {fabrizio.lotti, gabriele.vassura}@mail.ing.unibo.it

Abstract—In this paper, a novel design approach for the development of robot hands is presented. This approach, that can be considered alternative to the "classical" one, takes into consideration compliant structures instead of rigid ones. Compliance effects, which were considered in the past as a "defect" to be mechanically eliminated, can be viceversa regarded as desired features and can be properly controlled in order to achieve desired properties from the robotic device. In particular, this is true for robot hands, where the mechanical complexity of "classical" design solutions has always originated complicated structures, often with low reliability and high costs. In this paper, an alternative solution to the design of dexterous robot hand is illustrated, considering a "mechatronic approach" for the integration of the mechanical structure, the sensory and electronic system, the control and the actuation part. Moreover, the preliminary experimental activity on a first prototype is reported and discussed. The results obtained so far, considering also reliability, costs and development time, are very encouraging, and allows to foresee a wider diffusion of dextrous hands for robotic applications.

Keywords: Articulated Hands, Compliance, Manipulation, Control, Linear Motors

I. INTRODUCTION

Thanks to the availability of advanced control structures, improved actuators, miniaturized electronics and sensors, compliance in robotic structures can be no longer considered as a relevant defect, to be eliminated by the designer by means of stiffness-oriented solutions, but begins to be regarded as a controllable property and, in some cases, a resource for improving the overall design and the robot dexterity. As a matter of fact, the design of advanced robotic systems is moving from the "classical" concept of precise and stiff structures, often heavy and very complex, to that of light and flexible ones, with the perspectives of increased performances, high mechanical simplification, and consistent cost reduction.

As a matter of fact, the accuracy on trajectories tracking can now be obtained by adopting smart control procedures based on distributed sensorial feedback, such as vision, position and force sensing. A wide branch of robotic research is explicitly oriented along this development line (see [1] for basic references) and has achieved so far important results.

This novel tendency applies also to the design and control of robotic hands, where reflex-based control procedures can be applied in order to compensate for kinematic inaccuracies due to structural and local compliance, similarly to what happens to human hands.

On the basis of this assumptions, the design of dexterous hands can be currently based on new criteria and needs:

- a mechatronic design approach becomes necessary, i.e. mechanical structures cannot be designed independently on the sensory and control systems, and *vice-versa*;
- mechanical compliance can be accepted, or purposely introduced, into the system design in order to increase the manipulation dexterity (this is the primary goal e.g. of compliant pads distributed on the contact surfaces) or to simplify the overall structure of the hand (it may be the case of design solutions that avoid a great number of parts to be assembled, like e.g. compliant articulations and so on).

This paper presents an activity currently under development at the University of Bologna for the design and implementation of a new dextrous robot hand based on these concepts. In particular, in Sec. II the general guidelines and main features of the new prototype are discussed. In Sec. III and in Sec. IV some experimental results and the control strategy are presented and commented, while in Sec. VI final remarks and plans for future activity are reported.

II. DEVELOPMENT OF A COMPLIANT FINGER ARCHITECTURE

A prototype of an articulated finger is currently under development at the laboratories of DEIS and DIEM at the University of Bologna. The main specifications of the

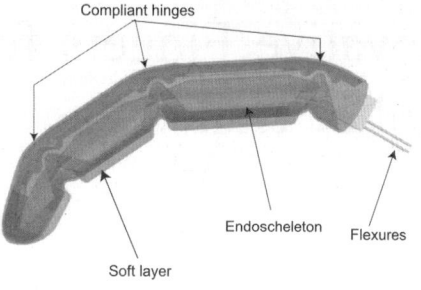

Fig. 1. CAD representation of the finger under development.

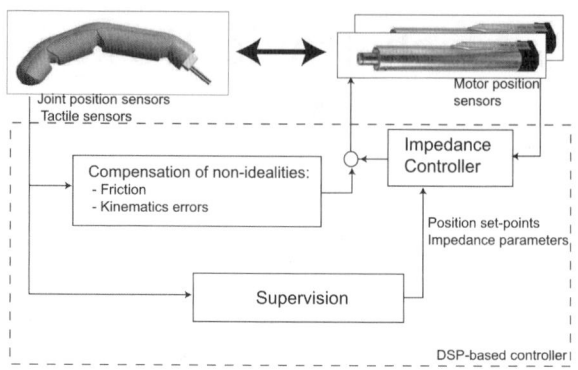

Fig. 2. Whole scheme of the finger.

proposed design are:

- the finger will be part of an anthropomorphic robotic hand conceived according to a concept of structural and functional integration with the carrying robot arm; this assumes that part of the functional components of the hand (actuators, ...) can be hosted inside the robot arm;
- the finger explicitly addresses an endoskeletal structure concept, so that it can host external compliant layers, like in the biological model of the human hand, in order to increase contact adaptability and grasp robustness and stability (see Fig. 1);
- actuation is provided by remote linear actuators (currently linear synchronous motors, in the future hopefully contractile artificial muscles or similar) with motion transmission obtained with flexible elements routed with low-friction linear guides, without any pulley or other non-biomorphic devices;
- the finger must integrate distributed sensory equipment in order to allow application of control procedure taking into account local and structural compliance;
- the mechanical structure must be compatible with the endoskeletal design and possibly must be simple, easy to be manufactured and assembled, cheap and reliable;
- the kinematic architecture of the finger must be compatible with manipulation tasks, that means to avoid under-actuation and to provide at least 3 controlled degrees of freedom per finger (2 parallel rotary joints for bending + 1 for adduction-abduction).

A scheme of the general architecture is shown in Fig. 2. The endoskeletal structure is purposely undefined, as well as the number of tendons and actuators, because different design solutions can be compatible with this scheme. In the literature, similar attempts to develop anthropomorphic endoskeletal hands have recently presented, see [2], [3], [4].

A. Mechanical framework of the finger

In the biological model, the endoskeletal articulated frame is obtained by separate bones connected by ligaments. Instead, in the design concept currently under evaluation, compliant mechanisms are considered, i.e. chains of rigid links connected through elastic hinges allowing relative motion between them. The use of compliant mechanisms and the investigation on their properties and design criteria have been rapidly growing in the last years, with significant applications in many fields, including MEMS (micro electro mechanical systems) and robotics [1]. In particular, application of compliant mechanism concepts to robotic end effectors has been so far limited to small-scale manipulation grippers [5]. The main interest of such a structure concerns the following aspects:

- the whole articulated structure can be composed by a single piece (e.g. a moulded plastic item) composed of rigid and elastic parts (hinges); this structure greatly simplifies manufacturing and assembly operations, reduces costs, enhances reliability;
- the structure can easily host both tendon guides and sensors with their wires;
- this endoskeleton is fully compatible with the integration of an external layer of soft material reproducing the tissues of the human hand.

Several morphological solutions have been defined and evaluated, according to a systematic design approach, [6], [7]. The structural scheme adopted in the present prototype implementation is shown in Fig. 3, while in Fig. 4 a view of the finger structure prototype is presented.

At present, only three parallel joints have been implemented: the the adduction-abduction joint is not present. The proximal and the medial joints are independently actuated, while the distal joint is coupled to the movement of the medial joint. Joint actuation is provided by remote motors, and transmission is obtained with guided flexures, that in the present implementation are integrated with the finger structure in PTFE, but can be obtained

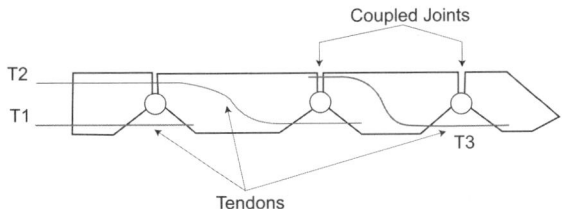

Fig. 3. Structural scheme of the endoskeleton.

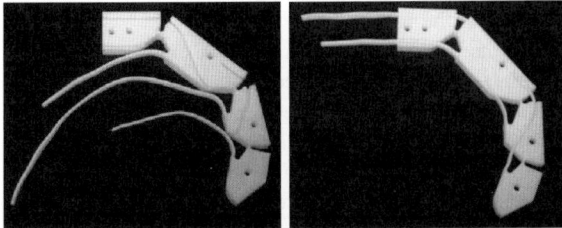

Fig. 4. Prototype of the finger.

in different material, e.g. high-strength steel. The way the two flexures act during the free motion (tension or compression) depends on the kinematic design and on the structural design of hinges, the stiffness of which must be overcome by flexures. By adopting a relaxed configuration of the finger frame with angles of $45°$ between each pair of links (like in the prototype under evaluation), it is necessary to apply compression loads only when the finger is opening, while only tension loads occur during closure and, most important, when a contact on the internal surfaces (normally used for grasping and manipulation) happens. In this way, problems of flexure buckling can be greatly reduced. Optimization of hinges, as to topological definition, material choice, correct sizing in order to obtain acceptable lifetime, is one of the major problems of structural design. In particular, low bending stiffness is desirable but, at the same time, torque and compression loads should not determine excessive strain. The present design is the evolution of early sizing attempts, but could be changed by the structural optimization analysis still in progress.

III. EXPERIMENTAL TESTS

In order to validate the design of the finger and draw useful observations about its properties, a laboratory activity has been developed. The experimental tests has concerned the different components of the system, in particular the endoskeleton and the material chosen for the pads.

A laboratory setup has been built in order to actuate the finger and verify possible drawbacks that such a structure may present. By means of this test bed, shown in Fig. 5.a, it is possible to impose linear displacements on flexures and control the position of the finger.

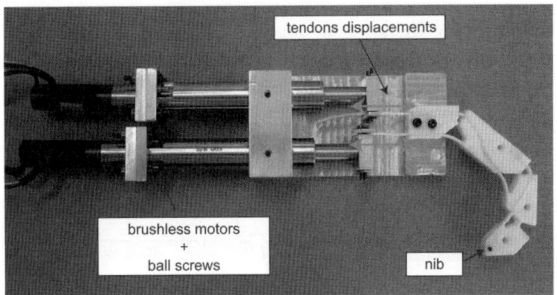

(a)

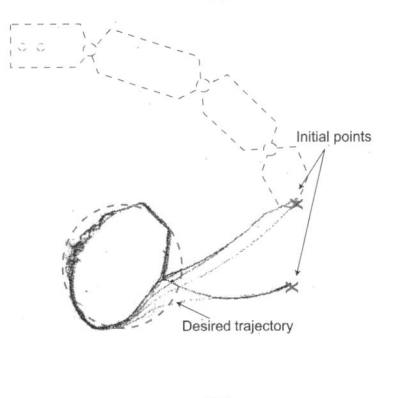

(b)

Fig. 5. Experimental setup for the endoskeleton (a) and trace of the desired motion (b).

Moreover, through a nib placed on the fingertip, the trajectory in the workspace can be recorded. This may be particularly useful in order to test the kinematic properties of the finger, comparing the desired movements with the real one.

Particular attention has been given to the hinges. As an analytical study has shown, their behavior can be roughly modelled with a rotational spring. The main difference between theory and experimental results involves the forward kinematic function. In fact, due to their peculiar structure, the hinges do not generate exactly rotative motions, being subject to undesired deformations and other non-idealities. Anyhow, the following kinematic relation between tendons displacement (l_T) and joint variables (θ) has been adopted to plan desired motions:

$$l_T = d\,\theta\,\frac{\sin(\frac{\theta_0-\theta}{2})}{\sin(\frac{\theta}{2})}$$

being θ_0 the joint angle corresponding to the reference configuration (upright finger), d a mechanical parameter shown in Fig. 6. This equation is computed in the ideal case of revolute pairs (see Fig. 6) and does not take into account nonlinear effects, as shown in Fig. 5.b where a repetitive circular motion of the fingertip is reported. Several motions are shown, with different initial positions:

although the tracking errors are not negligible (due to the non considered non-linearities), note that the behavior of the finger is perfectly repetitive.

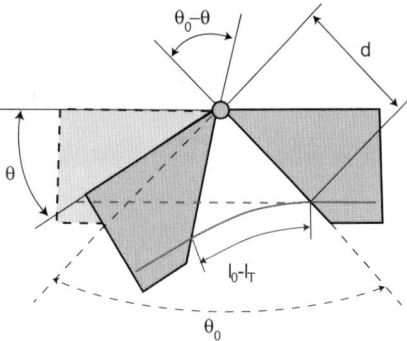

Fig. 6. Ideal model of the hinges.

Currently, the finger is tested without the outer layers of compliant material. Preliminary tests with an external compliant shell made of polyurethane gel (Technogel by Bayer) have shown an increase of the joint stiffness (depending on the actual thickness over each joint) of about 30%, effect that must be considered in actuation and control.

IV. FROM THE CONTROL TO THE ACTUATION SYSTEM

In the literature, several works have addressed the stabilization of robots with elastic joints (see e.g. [8], [9]), mainly focusing on tracking problems. Instead, the main tasks of a robot hand involve continuous interactions with environment. In this case, it is more natural to adopt a force control scheme, although force controllers [10] seem inadequate during motions in free space or when the interaction starts. Therefore, strategies that consider explicitly the control of interaction seem to be more adequate. In particular, an *impedance controller* appears to be a very suitable solution for this case. This is particularly true considering the mechanical structure we are dealing with. In fact, the compliance of the endoskeleton and of the soft pads may introduce instability phenomena [11], moreover, the kinematics is affected by not-negligible errors and therefore the controller must be robust with respect to these "side effects" of the proposed design.

As known, goal of an impedance controller is not to minimize the errors between a reference command input and the output (e.g. position or force), but rather its effect on a physical system is to modify its behavior and impose a desired dynamic relationship between forces and velocities. In particular, as shown in [12], if the behavior of a *simple impedance* is imposed on a manipulator, then stability is achieved when it grasps objects of arbitrary dynamic complexity, provided the initial stability of the objects. A simple way to impose a target impedance on a robot is with the following well-known control law [13]:

$$\mathbf{F_a} = -\mathbf{J(q)}^\mathbf{T}[\mathbf{K(X(q) - X_0)} - \mathbf{J(q)}^\mathbf{T}\mathbf{B(J(q)\dot{q})}]$$

where $\mathbf{F_a}$ is the vector of forces exerted by the actuators, $\mathbf{q}$ are the generalized coordinates of the system (in the specific case the length of the tendons), $\mathbf{X}(\cdot)$ and $\mathbf{J}(\cdot)$ are the kinematic transformation equations and their Jacobian, $\mathbf{K}(\cdot)$ and $\mathbf{B}(\cdot)$ represent the force/displacement and force/velocity relations, $\mathbf{X_0}$ is the vector of equilibrium positions of the manipulator.

Besides its properties of robustness during the interaction with the environment, this controller is very interesting because of its insensitiveness with respect to certain class of "disturbances" [12] characterizing the mechanical structure of the finger, shown in the previous sections. In particular, this algorithm is robust to large errors in manipulation kinematic equations and unmodeled interface dynamics (e.g. the soft pads on the finger's phalanges). Finally, this controller is based only on position/velocity feedbacks, according to the same requirements of simplicity, reliability and low-cost at the base of the whole project.

In order to implement this controller, the mechanical system must have some fundamental features. First the motion transmission must be back-drivable and with a friction as small as possible. Moreover, the sensors must be *co-located* [11], that is physically located in the same points of actuators. In this way the relationship between forces exerted by the motors $\mathbf{F_a}$ and velocities $\dot{\mathbf{q}}$, expressed by the impedance controller, are power consistent. In particular, considering a low stiffness of the transmission (e.g. with tendons, like in the present case) the placement of position sensors far from motors implies the presence of unstable modes in the control loop.

Normally, the transformation of a rotational motion into a linear one produces high friction effects and sometimes complicated and not-back-drivable mechanical solutions.

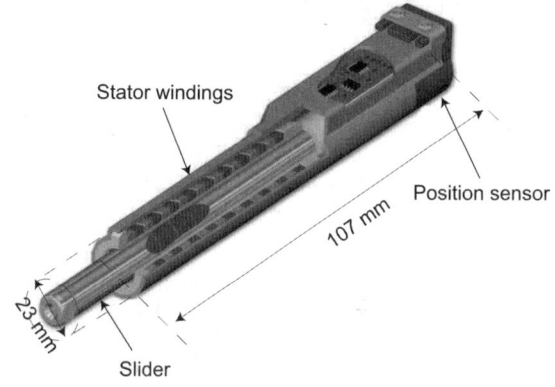

Fig. 7. Sketch of the linear motors used to actuate the finger.

For these reasons, linear motors have been chosen for the actuation system.

This kind of technology, not yet extensively adopted in robotics, allows a simple and direct connection with the flexures. In this way, it is possible to have back-drivable transmission chain and with limited frictions. Moreover, in the motors used for our application (produced by LinMot® [14]) the position sensor is integrated in the structure, that looks like a simple tubular stator with a moveable slider, see Fig. 7. Despite the quite small dimensions, and therefore suitable for an integration into the forearm of a complete hand/arm system, the performances are appropriate to actuate the finger, as shown in table I.

Peak force	33 N
Continuous force	9 N
Max accel.	280 m/s^2
Max velocity	2.4 m/s

TABLE I
PERFORMANCES OF THE LINEAR MOTORS.

V. SENSORY SYSTEM

Despite the fact that an impedance controller could perform interaction tasks and manipulation of objects without force information, based only on motor position sensors, in order to enhance the capability of the finger the adoption of a suitable set of sensors seems necessary. In particular, additional sensors must be used to eliminate position errors, mainly due to bending of tendons and undesired deformations of hinges, to compensate for inevitable friction phenomena and to allow fine manipulation tasks, which involve small forces. Therefore, as shown in

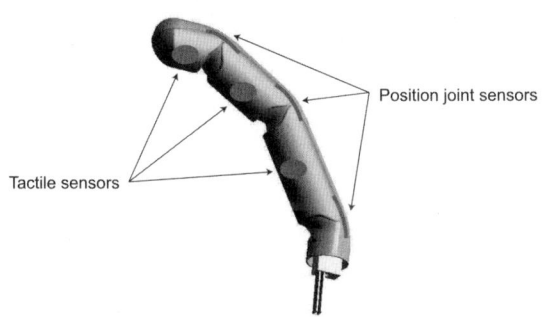

Fig. 8. Complete sensory equipment of the finger.

Fig. 8, the sensory system must include position sensors on the joints and distributed tactile sensors [15].

A. Position sensors

In order to know the relative positions between the links of the finger, a measure based only on tendon lengths is not sufficient. On the other hand, the peculiar structure of the joints (without a fixed rotation center) makes it difficult to find a suitable position sensor. In fact, the traditional robotic technologies (e.g. potentiometers or hall-effect based sensors) require well defined path.

A possible solution could be a special purpose sensor (see Fig. 9.a), built directly on the hinges of the endoskeleton. This solution is based on strain gauges, glued on the deformable structure. Nevertheless, in this manner it is possible only a partial compensation of kinematic errors, such as those produced by tendons bending, but not of errors directly imputable to the hinges.

An alternative solution, currently tested, is based on flex-sensors, Fig. 9.b, usually used in data gloves in order to measure the bending of human fingers. These sensors are based on piezoresistive effect, and provide a variation of resistance proportional to the bending angle. In this case, the position information can compensate for all the kinematic errors, because the sensor is physically separated from the finger structure.

Both solutions appear suitable according the the criteria of simplicity and reliability that characterize the project.

B. Force sensors

In order to exert small forces on the environment, implicit force control based on motor current(/force) control appears not to be a proper solution. In fact, because of the friction of the mechanical transmission and some limitations of this strategy, it is impossible to approach the resolution on force control of the human hand. Therefore, additional force/tactile sensors are necessary. Possible solutions can be:

- distributed tactile sensors under the soft "skin";
- intrinsic tactile sensors integrated in the endo-skeletal structure.

According to the former choice, several sensors must be placed in finger pads in order to measure when and where the contact occurs and the magnitude of the normal force. The latter solution is based on a measurement, by means of strain gauges, of the deformations of the finger structure (purposely designed) produced by external forces. In the robotic community, the last solution is normally preferred, because of its simplicity and reliability; nevertheless a

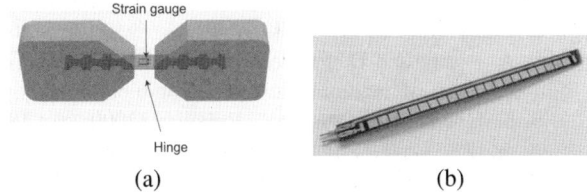

Fig. 9. Joint position sensors: special purpose sensor (a), commercial flex sensor (b).

deeper analysis seems necessary, in order to test the interactions between the soft layer and the sensors.

VI. Conclusions

In this paper we have presented the preliminary stage of an activity aiming at producing an innovative robot hand. The basic idea is that the *compliance* is not a side-effect of the mechanical design, but it could be a desirable feature of a device in order to obtain reliable and fine manipulation capabilities. Following this concept, the structure of a finger, based on so called "compliant mechanisms", has been design and the first prototype built and tested. The structure shows attractive characteristics, such as a small size, which allows the adoption of a visco-elastic cover (similar to the human hands) and an easy integration with the sensory system, and a simplicity, also in the production process, that could lead to a wider diffusion of robot hands. On the other side, the first experimental activity has shown some drawbacks of the prototype. In particular, because of a non-ideal behavior of the hinges and the flexures, the kinematics of the finger can not be easily modelled, and is affected by errors. Therefore, the control must be adequate, in order to guarantee the stability of the system in presence of uncertainties and unmodelled dynamics. At the same time, a suitable sensory system must enhance the performances of the finger. Mechanical structure, control algorithms and sensory equipment are designed together, according to a mechatronic approach, whose target is a simple, reliable and low-cost system.

We have proposed an architecture for the whole system (based on linear motors, joint bending sensors and tactile sensors) shown in Fig. 2, and the current work deals with its experimental implementation and validation.

VII. REFERENCES

[1] L.L. Howell, "Compliant mechanisms", John Wiley & Sons, 2001.

[2] Y.K. Lee and I. Shimoyama , "A skeletal Framework Artificial Hand Actuated by Pneumatic Artificial Muscles", in Proc. IEEE Int. Conf. on Robotics and Automation '99, ICRA 99, Michigan, May 1999.

[3] S. Schultz, C. Pylatiuk and G. Bretthauer , "A New Ultralight Robotic Hand", in Proc. IEEE Int. Conf. on Robotics and Automation '01, ICRA 01, Seoul, Korea, May 21-26, 2001.

[4] A. Bicchi and D. Prattichizzo , "Analysis and optimisation of tendinuos actuation for biomorphically designed robotic systems", Robotica (2000), vol.18, pp. 23-31.

[5] J. M. Goldfarb and N. Celanovic , "A flexure-based gripper for small-scale manipulation", Robotica (1999), Volume 17, pp.181-187.

[6] F. Lotti and G. Vassura , "A Novel Approach to Mechanical Design of Articulated Fingers for Robotic Hands", in Proc. IEEE/RSJ Int. Conf. on Intelligent Robots and Systems '02, IROS 02, Lausanne, Switzerland, 2002.

[7] F. Lotti and G. Vassura , "Sviluppo di Soluzioni Innovative per la Struttura Meccanica di Dita Articolate per Mani Robotiche", in Associazione Italiana per l'Analisi delle Sollecitazioni (AIAS) XXXI Convegno Nazionale, Parma, September 2002 (in italian).

[8] A. De Luca, "Dynamic Control of Robot with Joint Elasticity", in *Analisys and Control of Nonlinear Systems*, C.I. Byrnes, C.F. Martin, R.E. Saeks (Eds.), pp 61-70, North-Holland, Amsterdam, NL, 1988.

[9] P. Tomei, "A Simple PD Controller fo Robots with Elastic Joints", in IEEE Transactin on Automatic Control, Vol. 36, No. 10, October 1991.

[10] B. Siciliano and L. Villani , "Robot Force Control", Kluwer Academic Publishers, 1999.

[11] S.D. Eppinger and W.P. Seering, "Three Dynamic Problems in Robot Force Control", in IEEE Transactions on Robotics and Automation , Vol. 8, No. 6, December 1992.

[12] N. Hogan, "On the Stability of Manipulators Performing Contact Tasks", in IEEE Journal of Robotics and Automation, Vol. 4, No. 6, December 1988.

[13] N. Hogan, "Impedance Control: An approach to manipulation, Parts I-III", in ASME Journal of Dynamic Systems, Measurement and Control, Vol. 107, pp. 1-24, 1985.

[14] URL: http://www.linmot.com

[15] C. Melchiorri, "Tactile Sensing for Robotic Manipulation", in *Articulated and Mobile Robotics for SErvices and TEchnologies (RAMSETE)*, A. Bicchi, S. Nicosia, B. Siciliano, P. Valigi (Eds.), Springer Verlag, 2001.

Microassembly of 3-D MEMS Structures Utilizing a MEMS Microgripper with a Robotic Manipulator

Nikolai Dechev, William L. Cleghorn, James K. Mills

Department of Mechanical and Industrial Engineering, University of Toronto,
5 King's College Road, Toronto, Ontario, Canada, M5S 3G8
dechev@mie.utoronto.ca, cleghrn@mie.utoronto.ca, mills@mie.utoronto.ca

Abstract

This paper describes the process of bonding a MEMS (Micro-ElectroMechanical System) microgripper to the distal end of a robotic manipulator arm using a molten solder bonding technique. This task is part of ongoing work which involves the development of a general microassembly workstation. The goal of this workstation is to construct 3-D microstructures from MEMS sub-components. The microgrippers bonded using the method described here are 1.5 mm by 0.6 mm in size. The methodology behind the solder bonding approach is presented, along with the design of a custom soldering device referred to as the contact head. The contact head is used as the interface between the robotic manipulator and the microgripper. Experimental results are given in a qualitative discussion. An explanation of the bonding procedure using automated calibration is described, along with pictures from the associated microscopy system, and some scanning electron microscope images.

1. Introduction

This paper describes the process of bonding a MEMS (Micro-ElectroMechanical System) microgripper to the distal end of a robotic manipulator arm. The microgripper is mechanically and electrically bonded to the manipulator arm by solder bonding using tin-lead solder. The task of integrating the MEMS microgripper with the robotic manipulator is part of ongoing work described in [1], which involves the development of a general microassembly workstation. The goal of the workstation, shown in Figure 1, is to remove a micro-component from a MEMS chip, reorient the micro-component in space, move it to a secondary location, and join it to another micro-component. Micro-components are always fabricated parallel to a MEMS chip. However, after the workstation removes them from the MEMS chip, the workstation is able to rotate these micro-components in the α and β axes. In this way, complex 3-D microstructures can be assembled from a set of initially planar MEMS sub-components.

MEMS design has received considerable attention in recent years. This is due to the hopes of the industrial and research communities that MEMS designs can be used to mass-produce useful new products, similar to the way in which microelectronics research lead to the mass-production of IC (Integrated Circuit) chips in the past. However there is a fundamental difference between MEMS devices and IC devices. IC devices require no mechanically moving parts to function, while MEMS devices rely on micro-mechanical moving parts to provide their unique functionality.

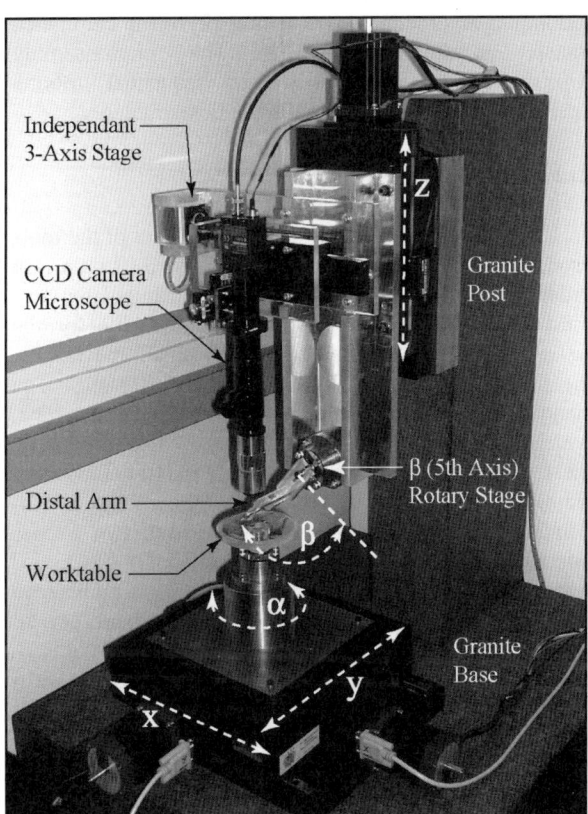

Figure 1. MJMP Microassembly Workstation.

Many of the MEMS devices being developed are fabricated using surface micromachining, which is adapted from IC fabrication technology. Surface micromachining, like IC fabrication, is based on the successive application and etching of thin films of material such as silicon nitride, silicon oxide, polysilicon and gold. Due to the use of thin film fabrication techniques, the MEMS parts that are created are relatively thin, in comparison with their length and width. For example, a typical MEMS part made of one layer of material could be up to 3000 microns long or wide, but only a maximum of 2 microns thick. Surface micromachining allows only 3 to 5 layers of material to be used as structural layers for the construction of micro-mechanical mechanisms. Also, surface micromaching fabricates micro-components in their working locations and orientations. Although some of these micro-components are able to translate slightly or to rotate, they remain in approximately the same region in which they were fabricated. These three limitations of surface micromaching place design restrictions on the type of devices that can be created. Another result of these limitations is that many MEMS devices operate along or near the plane of fabrication, which limits the diversity of device designs. A general process which can create fully 3-D MEMS would allow for more complex and useful devices. Through post-fabrication techniques, it is possible to create 3-D MEMS devices. Some of these techniques involve pop-up structures [2] or flip chip batch transfer microassembly [3]. The microassembly workstation described here uses sequential robotic operations to assemble 3-D MEMS microstructures.

2. Background

This paper will focus on the development of the tools used to grasp micro-components on a MEMS chip. In order to grasp micro-components, the manipulator arm must be equipped with an endeffector suitable for the task. Since MEMS parts can be very small and fragile, the design of the endeffector is very important. The standard tool for micromanipulation is the tungsten probe, which is 50 mm long or larger. Some research groups use these probes as endeffectors to move micro-components from one place to another, or to re-orient the micro-components [4]. Other groups have used two probes operating together, to manipulate small micro-blocks [5] with a kind of dexterous grasp. More advanced approaches include the use of HexSil or single crystal silicon fabricated milligrippers [6] to grasp the micro-components. The smallest of the HexSil grippers are 5 mm in length. All of these systems use endeffectors that can be categorized as macro-scale objects. That is, the human hand can be used to mount the endeffector to the manipulator system.

The work presented here uses surface micromachined endeffectors (microgrippers), which are 1.5 mm by 0.6 mm in size, as shown in Figure 2. The microgrippers are fabricated by the MUMPs (Multi-User MEMS Processes) surface micro-machining process [7]. Microgrippers of this size allow for some unique advantages [1] for grasping, over that of the macro-sized endeffectors. The micro-grasp tip of the microgripper is electrically driven to open and close, thereby allowing the microgripper to grasp other micro-components. Thermal resistive actuators [8] are used to convert electrical energy into mechanical energy that opens and closes the micro-grasp tip. The actuators are powered by applying +5 VDC to the central attachment pad and grounding the two outer attachment pads. Since these microgrippers are too small to be mounted by hand to the distal end of the manipulator (contact head), a solder bonding system has been developed, to bond the contact head to the three attachment pads.

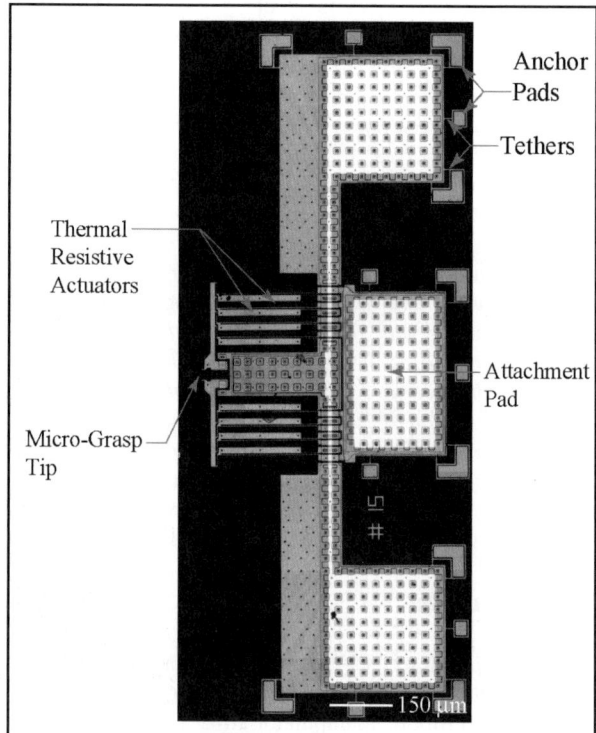

Figure 2. Composite Image of Actual Microgripper

3. The Manipulator System

A five axis robotic manipulator, named the MJMP (Manipulator and Joiner of Micro Parts) is the basis of the microassembly workstation, shown in Figure 1. The axes of the MJMP are split into two groups with the x, y and α axes as one group on the granite base, and the z and β axes as the second group mounted on the granite post.

The x, y and z axes are comprised of Danaher Precision Systems Ltd. crossed-roller bearing stages, driven by ball-bearing lead screws with a 2 mm lead. Vexta five-phase stepper motors are used for driving the three translation stages and are set to 20,000 steps per revolution, which provides a linear step distance of 0.1

μm. Linear encoders with a 0.1 μm resolution provide feedback to the operator. The translation stages have an open loop repeatability of +/- 0.2 μm. The α and β rotational axes are custom designed to have radial and axial runouts of less than 2 μm. This is achieved through the use of NSK-RHP P2 precision ball bearings. The motors driving the rotational axes have a resolution of 0.36°. In addition, there is an independent 3-axis manual translation stage on which the microscope system is mounted. This allows the microscope to be moved independently of the MJMP.

The three translation and two rotation axes are commanded by a 5 axis Galil card, interfaced to a personal computer. The control strategy for positioning the axes of the MJMP varies, depending on the task. Work involving experimental or unproven microassembly procedures is carried out by manual operator control using a joystick. The operator relies on the microscopy system for visual feedback, and a readout display showing the linear encoder positions. For proven microassembly procedures, such as the solder bonding technique to be discussed later, the MJMP uses automatic control and relies on a program, and data capture through the Galil card. The goal of the microassembly workstation is to automate all microassembly operations, once the procedures have been established.

This split configuration for the MJMP was chosen since it eliminates interference problems [1] between the macro-sized elements of the MJMP and the micro-sized components of the MEMS chip. This concept is illustrated in Figure 3. Note the contact head, which is used as the interface between the MJMP and the microgrippers. The contact head in Figure 3(a), is oriented 45 degrees below the horizontal. In this orientation the metal tips of the contact head are the lowest point of the manipulator attached to the granite post. This orientation allows the metal tips to probe the surface of the MEMS chip, and at the same time allows the metal tips to be in direct view of the microscope system.

The contact head is a custom designed soldering iron that solders itself to the MEMS microgrippers, thereby electrically and mechanically joining them the distal end of the manipulator arm. Figure 3(a) illustrates how the contact head is attached to the MJMP. The distal arm is attached to the β axis rotary stage. The head clamp is bolted to the distal arm, and is used to hold the contact head, which is securely clamped within it using two bolts. The metal tips protruding from the bottom of the contact head are made of copper and are used for the solder bond to the microgripper.

In order to bond a microgripper to the MJMP, the following procedure is used. With the distal arm in the orientation of Figure 3(a), the metal tips of the contact head are aligned in the x and y directions with the attachment pads of the microgripper. The alignment is done with the aid of the microscopy system. Next, a localization procedure, described later, is carried out to locate the height of the metal tips, above the MEMS chip. Next, the soldering procedure, described later, is carried out to solder the MEMS microgripper to the bottom side of the metal tips. After soldering, the z axis is commanded up, thereby removing the microgripper from the MEMS chip. The microgripper is fabricated such that it is only attached to the MEMS chip by tethers, as shown in Figure 2. The tethers are attached to anchor pads which are permanently attached to the MEMS chip. The tethers are strong enough to hold the microgripper onto the MEMS chip during shipping but are designed to breakaway during the solder bonding procedure, thereby freeing the microgripper from the MEMS chip.

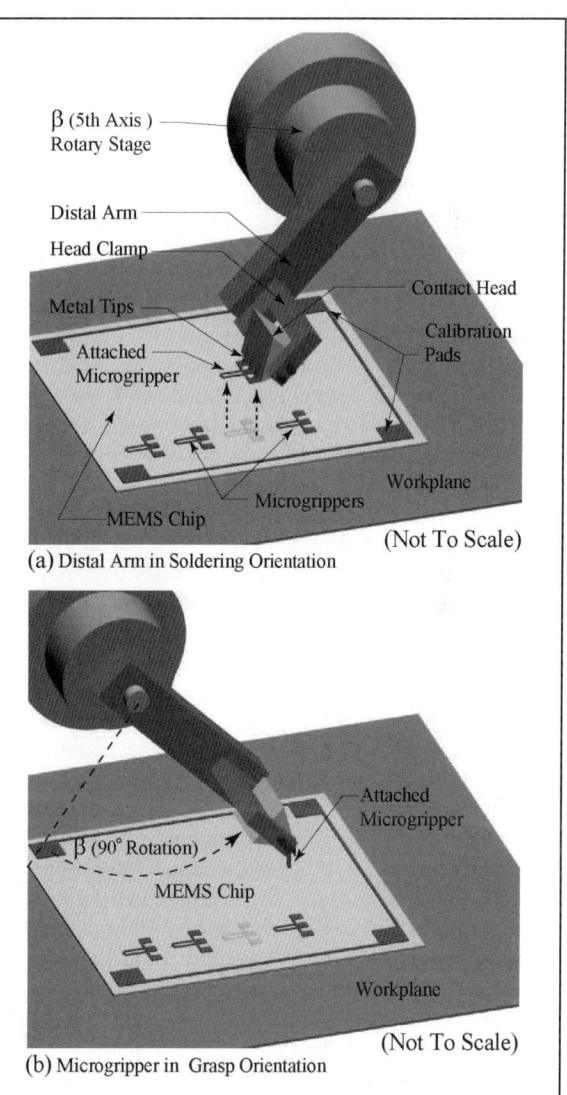

Figure 3. Illustration of β Axis and Contact Head

After a microgripper is soldered onto the metal tips of the contact head, as shown in Figure 3(a), the microgripper becomes the lowest point of the distal arm and is able to probe the MEMS chip, while in direct view of the microscope. Note that when the distal arm is rotated 90 degrees counter clockwise about the β axis, as shown in Figure 3(b), the contact head is again 45 degrees below the horizontal, although the microgripper is now perpendicular to the MEMS chip, however, it is again the lowest point on the distal arm. In this orientation, it becomes the endeffector of the MJMP, and is able to grasp other micro-components on the MEMS chip. Since the microscope is mounted on its own three axis stage, it is manually moved to ensure that the microgripper is always in direct view of the microscopy system.

4. Solder Bonding System

4.1 Design of the Contact Head

The goal of the bonding procedure is to mechanically and electrically fasten the microgripper to the contact head of the MJMP. Three separate electrical connections must be made with the microgripper, one for +5 VDC and the other two for ground. These three connections provide power to the thermal resistive actuators on the microgripper, allowing it to open and close the micro-grasp tip. Although glue could be used for bonding to the contact head, it does not conduct electricity well, and is not easily removed when a microgripper must be changed.

There are six design requirements for the contact head, which are: 1) rigidly hold three metal connectors that are electrically isolated from one another, 2) provide a uniform heat source to the metal connectors that is sufficient to melt tin-lead solder, 3) include a temperature sensor to control the heating of the contact head 4) ensure the material used for the contact head body does not melt, does not conduct current, and is a good conductor of heat 5) ensure the contact head body is mechanically strong enough to be clamped into the head clamp and lastly 6) ensure the entire contact head occupies a space/geometry small enough such that it does not interfere with the optics of the microscope system. The resulting contact head design is illustrated in Figure 4.

The first step in creating the contact head is the construction of the three metal strip cluster. This consists of three strips of gold plated copper, which are 200 microns thick. The three strips are placed in a plastic mold and separated from each other by 400 micron gaps. Castable ceramic material is used to form the structure around the three metal tips and the body of the contact head. A ceramic powder is mixed with water to form a liquid-ceramic fluid that is poured into a mold and flows to fill all the gaps. The ceramic is then allowed to set for 24 hours, is removed from the mold, and is cured in an oven at 250 °C. The resulting metal strip cluster is a solid ceramic cylinder with three embedded copper strips protruding from each side, all electrically isolated from each other, as shown in Figure 4(a). There are many different formulations for castable ceramics. The ceramic used for the contact head is based on Al_2O_3, can be heated up to 1600 °C, is an electrical insulator and is also a good thermal conductor, compared to ordinary ceramics.

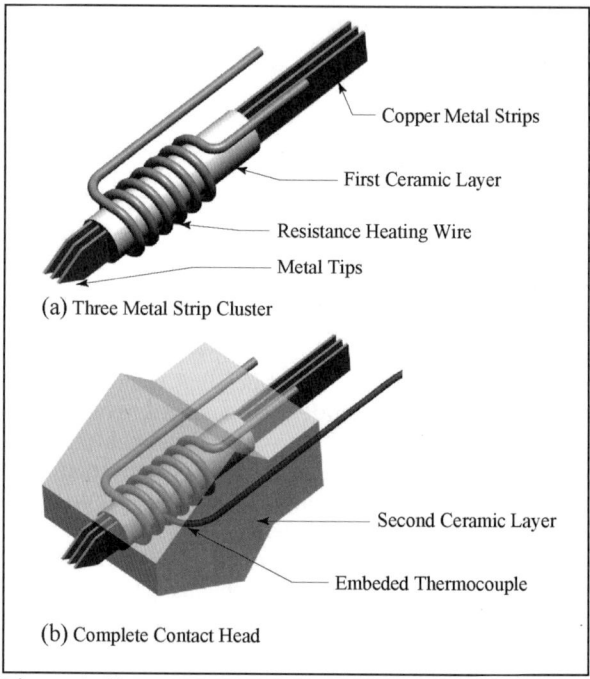

Figure 4. Illustration of Contact Head

After the metal strip cluster is complete, nickel-chromium electrical-resistance heating wire is wrapped around the cylinder to form a coil. There are a minimum number of loops required for the coil, in order to achieve sufficient heat. Square pin connectors are crimped onto the ends to the resistance wire. Finally, the metal strip cluster with the encircling resistance heating wire coil, and a J-Type thermocouple are set into a second plastic mold. Castable ceramic liquid is poured into the second mold, is allowed to set for 24 hours, removed from the mold and is cured in an oven, at 450 °C. The resulting structure is illustrated in Figure 4(b). The contact head is then tested to ensure there are no electrical shorts between the thermocouple, the nickel-chromium wire, and the three copper strips. Finally the contact head is connected a Partlow MIC 2000 temperature controller, and tested for heating. The contact head used in these experiments uses a 57 watt heating coil and can reach steady state temperatures of up to 500 °C. It is controlled by time proportional control, and keeps the temperature steady to within 2 °C. A picture of the actual contact head, mounted to the distal arm, is shown in Figure 5.

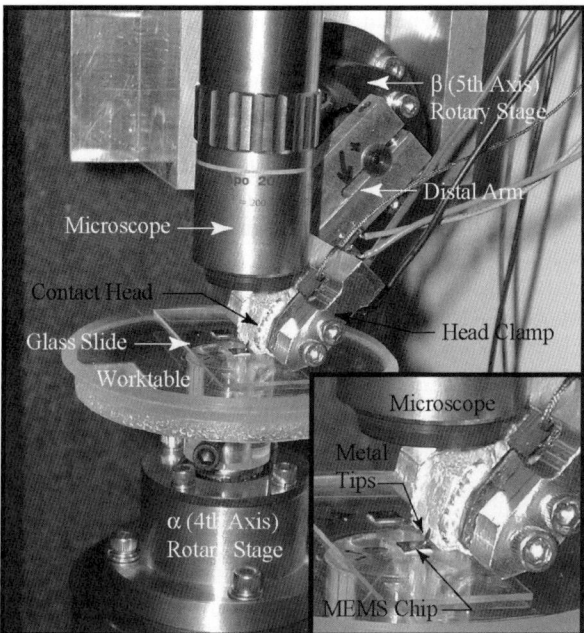

Figure 5. Contact Head within Distal Arm.

4.2 Integration of Contact Head with Distal Arm

Figure 3(a) illustrates the contact head held within the head clamp located on the distal arm of the manipulator. Under ordinary soldering conditions, the contact head is heated to 300 °C, and must be thermally isolated from the distal arm. Several layers of woven fiberglass fabric are placed on either side of the contact head, within the head clamp. These layers significantly reduce the flow of heat from the contact head to the distal arm, such that the distal arm is warm to the touch while the contact head is at 300 °C. Fiberglass fabric was chosen since it can be heated up to 700 °C, is not a hazardous material, and remains somewhat compliant allowing the contact head to thermally expand without damage to itself.

Before using a newly installed contact head, two additional steps are performed. Firstly, the contact head is subjected to a high temperature heating cycle to stabilize its position within the head clamp. Secondly, the entire distal arm assembly, with the clamped and thermally stabilized contact head, undergoes a precision grinding operation. The three metal tips are precision ground such that a line that intersects each of the three tips is parallel to the 5th axis. In this way, the three metal tips will touch down onto the surface of a MEMS chip at almost the same time.

4.3 Localization of MEMS Chip with the MJMP

The distance in the z-direction, between the microgrippers on the surface of the MEMS chip and the metal tips of the contact head, must be calibrated. This distance is critical in forming a solder bond. A glass slide, onto which the MEMS chip is bonded, is mounted onto the worktable of the α axis. The MEMS chips arrive from the fabrication facility bonded to glass slides, however, the gluing process results in an orientation of the MEMS chip such that the surface plane of the MEMS chip is not parallel to the plane of the glass slide. Therefore, the surface of the MEMS chip is not perpendicular to the α axis, when it is mounted on the worktable. Since the MEMS chip itself is single crystal silicon, it is flat. Therefore, by determining the z position of at least three different points on the chip, the z positions of all points of the chip are known. Four calibration pads, as shown in Figure 3(a), are designed into the perimeter of the MEMS chip. Each pad is comprised of a gold layer 300 by 300 microns, elevated 2.5 microns above the substrate, and all pads are electrically connected to each other.

To perform localization of the MEMS chip with respect to the MJMP, the vertical (z-axis) distance, from a calibration pad to the metal tips of the contact head, is measured. The measurement is made by the translation of the z axis of the MJMP, such that contact is made by one of the three metal tips of the contact head with a calibration pad. Contact is detected with the closure of a circuit loop created by the calibration pad and the metal tips. Upon contact, the encoder position of the x, y and z axes is recorded. Three such measurements are carried out to locate the plane of the MEMS chip with respect to the contact head, and a fourth measurement is used to provide redundancy so as to check for error.

4.4 Preparation of Contact Head for Soldering

The solder used for bonding is Sn63/Pb37, and has a melting temperature of 183 °C. This solder is used since it has the lowest melting temperature of any solder, and therefore minimizes the amount of heat necessary to perform the solder bonding operation. To apply the solder to the contact head, it is rotated to the position shown in Figure 3(b) and is heated to 300 °C. The solder is applied by hand directly to each of the three metal tips of the contact head. The correct amount of solder to apply is important. Too little solder will not create a solder bead large enough to allow a joint to be formed. Too much solder will cause a short circuit between two or three of the metal tips. Oxidation of the solder is also a problem. The solder bond must be performed quickly after the solder is applied, otherwise an oxide layer forms on the molten solder and can prevent a bond.

5. Experimental Results

Results showing a microgripper that has been successfully joined to the contact head of the MJMP and removed from the MEMS chip is shown in Figure 6. A majority of soldering attempts between the metal tips and the microgrippers are successful, using a technique described here as 'hot calibration'.

5.1 Manual Soldering of Microgrippers

Initial soldering attempts were unsuccessful due to the thermal expansion of the distal arm, and the unexpectedly large thermal expansion of the worktable, the glass slide and the MEMS chip. During the manual solder bonding process, the temperature of the various components is dynamic throughout the procedure.

In these experiments, the cold (unheated) contact head is lined up with the microgripper attachment pads. The contact head heater coil is turned on, until the temperature reaches 300 °C, at which point the solder is presumed to have melted and bonded to the microgripper, and the heater coil is turned off. Upon heating of the contact head, thermal expansion begins in the distal arm due to conduction, and in the worktable, glass slide and MEMS chip due to radiation and conduction. Because the temperature never achieves a steady state, the thermal expansion is difficult to predict. Due to the uncertainty of the thermal expansion, the exact tip position of the contact head is uncertain. Therefore, during a bonding experiment, if the contact head is closer to the MEMS chip than anticipated, it would crush the microgripper and break it. If the contact head is farther from the MEMS chip than anticipated, no bond would be made.

5.2 Automated Hot Calibration Soldering

The 'hot calibration' solder bonding approach is successful at joining the microgripper to the contact head. In this approach, the contact head temperature is held constant at 300 °C, using the temperature controller, while the calibration procedure of Section 4.3 is carried out. The thermal expansion of the contact head and distal arm reaches steady state, and is accounted for during the calibration procedure. However, thermal expansion of the worktable, the glass slide and the MEMS chip will be dynamic during the calibration. In order to account for the dynamic thermal expansion, the speed and motion of the calibration procedure is such that it matches the speed and motion of the solder bonding procedure. In this way, as the contact head approaches the MEMS chip, and makes contact, the rate of heating of the chip due to radiation, and the subsequent thermal expansion will be equal in both cases. Therefore the thermal expansion of system is accounted for by the calibration procedure. The matching of speed and motion is not possible without automatic control, and therefore the 'hot calibration procedure' is automated. It is assumed that the radiation heating of MEMS chip is even regardless which calibration pad is contacted, since the contact head is much larger than the chip. Since the position of all microgrippers is accurately known with respect to the calibration pads in the x and y directions, and the z position is calibrated, no manual alignment using the microscope is necessary. The coordinates of the microgripper to be joined are entered into the controller and the MJMP automatically aligns with the microgripper and moves down in the z-direction. The metal tips stop 5 microns above the calibrated height of the microgripper, since the height of the solder bead is not accounted for by the calibration procedure. The actual height of the solder bead is usually 30 to 50 microns in the z-direction, which ensures that the molten solder contacts the microgripper. An SEM (Scanning Electron Microscope) picture of a successfully bonded microgripper is shown in Figure 6.

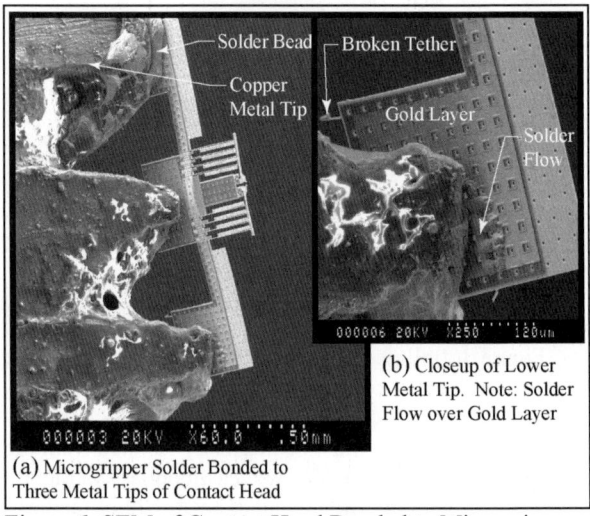

Figure 6. SEM of Contact Head Bonded to Microgripper

A voltage of +5VDC, at a frequency of 1 Hz was applied across the central metal tip and the outer two metal tips, of the contact head shown in Figure 6. The thermal actuators operate well, however, the desired gripper tip motion is not always achieved. The microgripper design of Figure 6 is sensitive to the relative thermal contraction between the three metal tips, which causes a slight bending in the microgripper. This bending is often sufficient to misalign the thermal actuators, leading to poor tip deflection.

Figure 7 is a series of images taken by the optical microscope (shown in Figure 5), of the microgripper. Note that the microgripper is positioned several microns above a micro-component on the surface of the MEMS chip. The depth of focus of the microscope image is 1.5 microns, therefore all objects closer or beyond this range appear out of focus. Figure 7(b) is the same image, but with the micro-component in focus. Figure 7(c) shows the microgripper tips located above a grid, used to locate the tips with respect to the MEMS chip, after the solder bonding operation. Figure 7(d) shows the central attachment pad of the microgripper. The dark blur on the lower right is the central metal tip of the contact head that is bonded to the pad. The field of view of all the images in Figure 7 is 320 by 240 microns. Figure 8 shows a series of SEM pictures of a successful microassembly operation. Figure 8(a) shows the initial fabrication position of a micro-part. The microgripper used for the operations of Figure 8(b) to (d) was solder bonded to the contact head using the same procedure described in this paper, however the operation of this microgripper is not described here. Figures 8(c) and (d) show completed assembly operations.

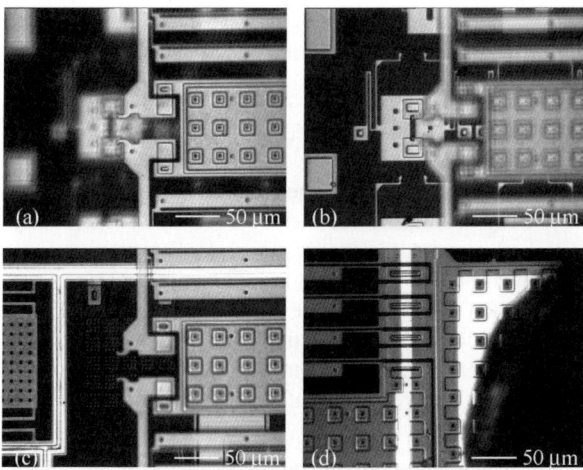

Figure 7. Microgripper Soldered to Metal Tip Above Chip

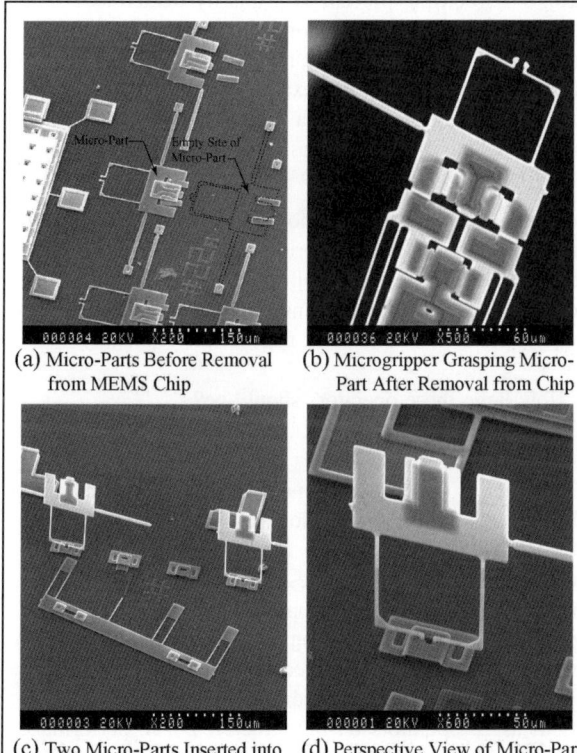

Figure 8. SEM Images of Microassembly Operation

6. Conclusion

The process of joining a MEMS microgripper to the distal end of a robotic manipulator arm using a hot calibration, molten solder bonding technique, has been described. The methodology and equipment used for the soldering approach were presented, along with the design of the contact head. The contact head acts as an interface between the microgripper and the MJMP, provides three electrical paths to the microgripper, and has a built in heater coil to solder the metal tips to the microgripper. The experimental results show an SEM image of a typical successful solder bond. Using this bonding technique various types of microgrippers can be joined to a robotic manipulator, to perform microassembly operations using MEMS micro-components.

Acknowledgements

This material is based upon work supported under a Natural Sciences and Engineering Research Council of Canada (NSERC) equipment grant. In addition, the authors wish to thank the Canadian Microelectronics Corporation (CMC) for allocating chip area.

References

[1] Nikolai Dechev, William L. Cleghorn, James K. Mills, "Micro-assembly of microelectromechanical components into 3-D MEMS", Canadian Journal of Electrical and Computer Eng., Vol. 27, No. 1, January 2002, pp. 7-15.
[2] Hui, Elliot E., Howe, Roger T., Rodgers, Steven M., "Single-Step Assembly of Complex 3-D Microstructures", *Proceedings IEEE Thirteenth Annual International Conference on Micro Electro Mechanical Systems*, Miyazaki, Japan, (23-27 Jan. 2000.) Piscataway, NJ, USA: IEEE, pp. 602-607, 2000.
[3] Singh, A., Horsely, D., Cohn, M., Pisano, A., Howe, R., "Batch Transfer of Microstructures Using Flip-Chip Solder Bonding", *IEEE Journal of Microelectromech. Systems*, Vol 8, No. 1, March 1999.
[4] Top Down Group, "MEMS at Zyvex", [online], [cited Sept 5, 2002], Available from World Wide Web: <http://www.zyvex.com/Research/MEMS.html>
[5] E. Shimada, J. A. Thompson, J. Yan, R. Wood, R. S. Fearing, "Prototyping MilliRobots Using Dextrous Microassembly and Folding", *Proc. ASME IMECE / DSCD*, Orlando, Florida, November 5-10, 2000., pp 1-8.
[6] Chris G. Keller, "MEMS Precision Instruments", [online], [cited Sept 5, 2002], Available from World Wide Web: <http://www.memspi.com/index.html>
[7] JDS Uniphase, MEMS Business Unit (Cronos), 3026 Cornwallis Road, Research Triangle Park, N.C. 27709
[8] J. Robert Reid, Victor M. Bright, John H. Comtois, "Force measurements of polysilicon thermal micro-actuators", *Proceedings SPIE*, Vol. 2882, Austin, Texas, pp. 296-306, Oct 14-15, 1996.

Micromanipulation Contact Transition Control by Selective Focusing and Microforce Control

Ge Yang
Department of Mechanical Engineering
University of Minnesota, Twin Cities
Minneapolis, MN 55455, USA
Email: gyang@me.umn.edu

Bradley J. Nelson
Institute of Robotics and Intelligent Systems
Swiss Federal Institute of Technology, Zurich
Zurich, Switzerland
Email: bradley.nelson@iris.mavt.ethz.ch

Abstract— A fundamental requirement of micromanipulation is to control the impact force and subsequently the contact force in the transition of the micromanipulator end-effector from noncontact to contact state. This is especially important in protecting fragile microstructures and preventing undesirable motion. This paper proposes a method of using the integration of selective focusing and microforce control to achieve fast transition control while minimizing impact force. The method is applied to contact transition in microassembly pick-and-place operations. The initial long-range approach motion of the end-effector towards its target is controlled based on focus measures computed from images captured through a microscope. When the end-effector comes into focus near the target, the system switches to microforce control to minimize impact force and to regulate the contact force. An optics model for microscope focusing is proposed to characterize the dynamic behavior of the end-effector images during the approach motion. The connection between this model and the scale-space theory of computer vision is emphasized. Three different focus measures are tested and compared in performance. The proposed method has been experimentally verified to be able to achieve fast transition control with minimal impact force.

I. INTRODUCTION

An elementary operation in micromanipulation is to move the micromanipulator end-effector, often a microgripper, into contact with its target. During this transition, it is important to control the impact force and subsequently the contact force in order to protect fragile microstructures such as MEMS devices and to prevent undesirable motion. Contact transition control at macroscale has been studied extensively [16]. Its major performance requirements include ensuring stability, minimizing impact force, regulating the resultant contact force, and maximizing efficiency. These requirements are contradictory. Basically, the methods used to satisfy these requirements can be classified into the following three categories:

- Changing the mechanical properties of the contacting objects, for example [1][23].
- Reducing impact energy by lowering the approach speed. Proximity sensors are used to measure the relative distance between the end-effector and its target, for example [11][15].
- Changing the dynamics of the transition process, for example [8][21][22].

Proximity sensing provides a natural and effective way to maximize the efficency of micromanipulation operations involving contact transition. This is because such efficiency depends on both the approach motion speed before the contact and the transition response speed in establishing the contact. With proximity sensing, a large approach speed can be used before the contact transition without increasing the final impact energy. In fact, humans routinely use vision for proximity sensing both at macroscale and microscale. For micromanipulation, humans usually use focus as a proximity measure since high resolution microscopes have limited depth-of-field. Focus measures provide depth information with the advantage of not requiring correspondence. Automatic micromanipulation and microassembly require the automation of this process for both efficiency and robustness.

Autofocusing for microscopy has been studied in [6] among others. Autofocusing for microassembly has been studied in [2][20]. The focusing process is primarily treated as a computer vision task in the past. Contact transition control is not addressed explicitly. In addition, relatively simple geometry or configuration is assumed. For example, the end-effector used is typically a long probe with a pointed tip. However, many microgrippers have 3D structures. The micromanipulation objects can also have complex 3D structures.

Focus measure selection is another important issue in using focus for proximity sensing. Various focus measure operators have be studied, for example [9][14][17][18]. However, a few major questions remain to be answered in order to make the focusing operation automatic at microscale, especially

- How large these operators should be?
- Where to place these operators?

This paper proposes a micromanipulation transition control method based on the integration of focus proximity sensing and microforce control. In order to address the above-mentioned limitations and questions, an optics model for microscope focusing is first proposed. This model shows the connection between the focusing process and the scale-space theory of computer vison. Based on this model, a selective focusing technique is proposed to guide the pacement of focus measure operators on 3D structures. The performance of several widely used focus measure operators are also tested and compared. The effectiveness of the integration of selective focusing and microforce control is experimentally verified. Emphasis is placed on testing algorithms under application-based experiment settings and on using simple control algorithms that do not require accurate dynamics models.

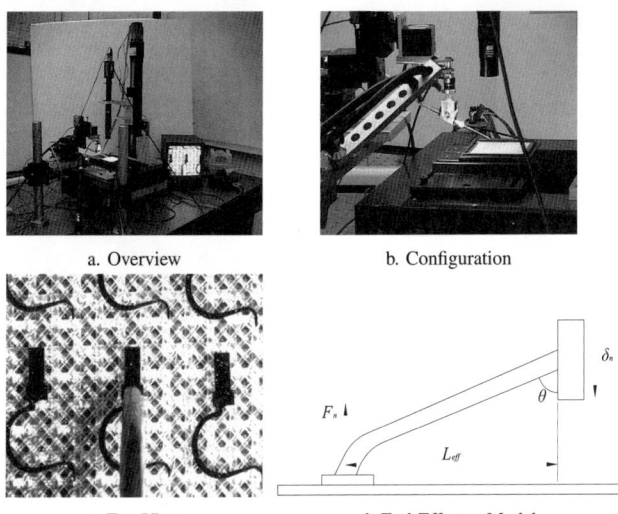

Fig. 1. Experiment Configuration

II. EXPERIMENT CONFIGURATION

One DOF micromanipulation pick-and-place operations are the targeted application. The experiment setup is shown in Figure 1. The objective is to control the contact force in picking up the metal parts from the vacuum release tray. Each metal part is approximately 500μm in width and 100μm in thickness. A steel hypodermic needle is machined to be used as the end-effector for vacuum pickup. (Figure 1.c shows the needle tip placed at the pickup position of a metal part.) The micromanipulator used is an adapted Sutter MP285. It is controlled by a Sensoray 626 I/O board. Microscopic vision feedback is provided by an Edmund Scientific VZM450i microscope with a SONY XC-75 CCD camera. The images are captured using a Matrox Corona framegrabber. The microscope is mounted on a vertical linear slide. The position of the microscope can be controlled with a repeatability of approximately 5μm. Microforce feedback is provided by an ATI Nano17 6DOF force/torque sensor. The entire system is controlled using a personal computer (2.2GHz Intel P4, 256M PC800 RDRAM).

The following assumptions are made:

- The geometry of the end-effector and the target are known. The image of the end-effector is a connected set.
- The aperture and magnification of the microscope remain constant. The illumination remains constant.
- The displacement of the end-effector is parallel to the microscope optical axis. That is, one DOF case is addressed.
- Parts are mesoscale[1]. The magnitude of the interactive force is in the range of milli-newton to micro-newton (referred to as microforce).

Since the parts are fixed on the membrane of the vacuum release tray in the experiment, other sticking forces can be neglected.

Other possible proximity sensing configurations include using a horizontal view to observe the distance between the end-effector and its target [15]. Other physical sensors can also

[1]We refer to 1 μm to 100 μm as 'microscale' and 100 μm to 1 mm as 'mesoscale'.

be used to measure the distance [11]. However, these schemes increase the complexity of the system and the end-effector. In addition, the vertical view is still necessary for alignment.

In practice, in order to prevent collision, the microscope is not focused on the manipulation target directly because the resolution of proximity provided by focusing is limited. Instead, after the microscope is focused on the micromanipulation target, it is moved a predetermined distance (shown in Section III) away from its in-focus position such that the end-effector comes into focus before it contacts the target. Initially the end-effector moves towards its target at a large yet constant speed. When the end-effector comes into focus, the system converts to microforce control. The microforce control algorithm is designed such that the approach speed of the micromanipulator is reduced to a small contant value to minimize impact force. This speed is maintained until the end-effector contacts the target surface. The microforce control algorithm will then regulate the contact force to a required constant value. A major advantage of this strategy is that it does not require prior depth information or the explicit computation of depth information.

III. SELECTIVE FOCUSING

For objects whose depth is larger than the depth-of-field of the microscope used, the definition of "in-focus" or "out-of-focus" is really dependent on what features are being observed. Therefore, for these objects, the focusing process is inherently *selective*. For example, in autofocusing cameras, such features are selected by the photographer through pointing the small AF window(s) on the region of interest. In order to make autofocusing possible, the end-effector design and/or experiment design must ensure the existence of features that can be chosen to determine focus without ambiguity. After the features are selected, autofocusing is primarily a search for the maximum of focus measures computed by applying the corresponding operators on the images [9]. In this process, three questions must be answered, namely: what operators to use, where to apply these operators, and how large these operators should be.

A. Selection of Focus Measure Operators

A variety of focusing measures have been proposed [9][14][17]. Three focus measures, namely image variance M_{IV}(Equation 1), sum of gradient magnitude M_{GS}(Equation 2), and sum-of-modified-Laplacian M_{SML}(Equation 3) have been reported to be among the most effective based on extensive experiments. Notice that M_{IV} and M_{GS} are linear operators while M_{SML} is a nonlinear operator [17]. The performance of M_{IV} and M_{GS} have been compared in [9][17] on planar objects. In this paper, these operators are tested and compared for their stability and resolution in 3D settings. The motivation will become more clear after the discussion of operator placement. In addition, to the best of our knowledge there is no reported performance comparison of M_{IV} and M_{GS} against M_{SML}.

$$M_{IV} = \frac{1}{|E|} \iint_E (I(x,y) - \mu)^2 dx dy \qquad (1)$$

$$M_{GS} = \frac{1}{|E|} \iint_E \|\nabla I(x,y)\|^2 dx dy \qquad (2)$$

No.1　　No.2　　No.3　　No.4　　No.5

Fig. 2. Sample Image Sequence at Increments of 3mm

$$M_{SML} = \frac{1}{|E|} \iint_E \left(\left|\frac{\partial^2 I}{\partial x^2}\right| + \left|\frac{\partial^2 I}{\partial y^2}\right| \right) dx dy \quad (3)$$

where E is the region of interest in the image $i(x,y)$. The mean image intensity in E is denoted by μ.

B. Placement of Focus Measure Operators

1) Introduction to the Problem: The fundamental requirement on any focus measure operator is that it is unimodal, meaning that it has only one peak value when the object of interest moves towards its in-focus position. This depends not only on the focus measure itself, but also on where to place it. From a physics' perspective, the operator must be placed on features with a unique in-focus position. A pointed probe tip satisfies this requirement.

Autofocusing for micromanipulation requires automatic placement of focus measure operators. As pointed out in [9], this requires active attention. For example, if the micromanipulation task requires the contact of the microgripper with surface $A_1B_1C_1D_1$ in Figure 3, the autofocusing algorithm must be able to selectively place the operators on the image of this surface rather than on that of $A_3B_3C_3D_3$. Krotkov [9] proposed using a linefinder as a preprocessing interest operator. However, when the object of attention is still far from its in-focus position, it is difficult to use method of this type since the edges are very weak. Such difficulty becomes even more remarkable at microscale, as shown by Figure 2. Five images are chosen from a squence. They correspond to the positions of the end-effector approaching its target at increments of 3mm. Notice that only light blobs and virtually no edges of the end-effector are observable in Figure 2.1 and Figure 2.2. (In order to make the problem clear, the background is set to be in focus.) It should be emphasized that such phenomenon happens not just when the object of interest moves a large distance. It happens whenever the microscope's depth-of-field is significantly smaller than the depth of the target or the end-effector displacement along the microscope optical axis. However, such cases are common because high resolution microscopes, required by high precision micromanipulation, must have large numerical aperture and thus small depth-of-field (Equation 4).

Figure 2 shows that initially the image of the end-effector undergoes significant changes in both intensity and shape. During this period, it is difficult to rely on features like edges to decide where to apply focus measure operators. In order to address this problem, a microscope optics model is first proposed to characterize the dynamic behavior of the images as the end-effector moves toward its in-focus position.

2) A Microscope Focusing Model: Taking the Sparrow limit for two-point sources as its definition and assuming incoherent illumination, the resolution D of a microscope is given by

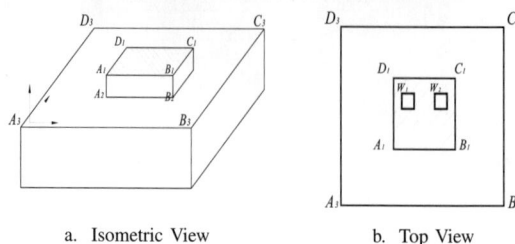

a. Isometric View　　b. Top View

Fig. 3. Illustration of Selective Focusing

$$D = \frac{0.47\lambda}{NA} \quad (4)$$

where λ is the wavelength. NA is the numerical aperture. The depth-of-field of a microscopes consists of two parts: one from wave (physical) optics and the other from geometrical optics [10].

$$d_{total} = d_{wave} + d_{geom} = \frac{n\lambda}{NA^2} + \frac{ne}{M \cdot NA} \quad (5)$$

where n is the refractive index, e is the smallest distance that can be resolved by the image detector, and M is the magnification. Since high resolution microscopes have large numerical aperture, they also have small depth-of-field. At high NA, the depth of field is determined primarily by wave optics, while at low NA, the depth of field is determined primarily by geometrical optics [10].

For a well-corrected optical system, the blurring effect due to wave optics can be characterized by the following point spread function F_{wave}, often referred to as an Airy disk function [4][7].

$$F_{wave}(r,L) = \frac{k_1}{L^2} \left[\frac{J_1(k_2 r/L)}{k_2 r/L} \right]^2 \quad (6)$$

where

$$k_1 = \frac{4\pi r_0^2}{\lambda^2} \quad k_2 = \frac{2\pi r_0}{\lambda} \quad (7)$$

Here $J_1(\cdot)$ is the first order Bessel function of the first kind. The distance on the image plane from the position of interest to the image center is denoted r. The aperture stop radius r_0 is kept constant in our experiments. L is the distance from the light source to the image point. An increase in L will lead to an increase in the scale of the function and a quadratic decrease in its magnitude. The larger the L, the stronger the smoothing effect caused by defocusing. As an object comes gradually into focus, the distance L decreases. Thus the scale is reduced.

As shown in Figure 4, F_{wave} can be well approximated by the following 2D Gauss function [7].

$$G_{wave}(r, \sigma_w) = \frac{k}{\sigma_w^2} e^{-\frac{1}{2}\left[\frac{r}{\sigma_w}\right]^2} \quad (8)$$

In particular, the approximation error in terms of $\|F_{wave}(r,L) - G_{wave}(r,\sigma_w)\|_\infty$ decreases monotonically as the scale L and σ_w increase.

The geometrical blurring function is given by [17]

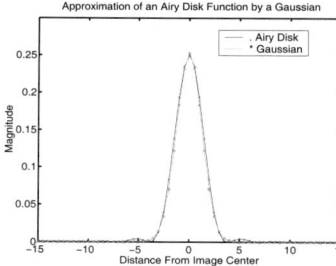

Fig. 4. Approximation of the Airy Disk Function by a Gaussian

$$G_{geom}(r, \sigma_g) = \frac{1}{2\pi\sigma_g^2} e^{-\frac{1}{2}\left[\frac{r}{\sigma_g}\right]^2} \quad (9)$$

The total point spread function is given by the following convolution [5]

$$G_{total}(r, \sigma) = G_{geom}(r, \sigma_g) \otimes G_{wave}(r, \sigma_w) \quad (10)$$

Because of the following convolution property of Gauss functions,

$$G(x, \sigma_1) \otimes G(x, \sigma_2) = G(x, \sqrt{\sigma_1^2 + \sigma_2^2}) \quad (11)$$

we know that $G_{total}(r, \sigma)$ is still a Gauss function with scale $\sqrt{\sigma_g^2 + \sigma_w^2}$.

3) Relation Between σ_g & σ_w and Object Displacement: The relation between the end-effector displacement and changes in σ_g and σ_w must be established. Assume that the end-effector displacement is parallel to the optical axis of the microscope. Also, assume thin lens model. Let f, u and v denote the focus length, the distance between the object and the principal plane, and the distance between the principal plane and the object image respectively. We have

$$\frac{1}{f} = \frac{1}{u} + \frac{1}{v} \quad (12)$$

Therefore, we have

$$v = \frac{f}{1 - f/u} \quad (13)$$

$$dv/dt = -\frac{1}{(u/f - 1)^2} du/dt \quad (14)$$

When the object is far from its in-focus position, v can be considered to be constant. So is the normal distance from the object image to the image detector plane. This distance is approximately equal to L when r is small. For the VZM450i microscope under fixed magnification, the distance between the image detector and the lens is fixed. From the proportional relation between σ_w and L, we have

$$d\sigma_w/dt \propto \frac{1}{(u/f - 1)^2} du/dt \quad (15)$$

Therefore, when the object is far from its in-focus position, σ_w is also approximately constant.

Let R denote the radius of the blur circle[17]; the following relation holds for G_{geom} [17],

$$\sigma_g \propto R \propto \left(\frac{1}{f} - \frac{1}{u} - \frac{1}{s}\right) \quad (16)$$

$$d\sigma_g/dt \propto \frac{1}{u^2} du/dt \quad (17)$$

where s is the distance between the principle plane and the image plane. Again, as the object is far from its in-focus position, σ_g tends to be constant.

To summarize, as the object is far away from it in-focus position, both σ_g and σ_w tend to be constant. As the object moves towards to its in-focus position, the scales σ_g and σ_w will both decrease. Their decrease rate will increase as the the object moves towards to its in-focus position. These conclusions are supported by experiment observations. The algorithm based on this model for automatic operator placement will be described in the following Subsection E.

4) Connection with the Scale-Space Theory: Based on the microscope focusing model proposed above, the change of the end-effector images can be treated as being smoothed by a series of Gauss filters with continuously varying kernels. Such dynamic behavior has been studied extensively using the scale-space theory of computer vision [13]. The original intent of the scale-space theory is to use the scale property of images to develop preprocessing techniques for early vision. However, from a mathematical point of view, the treatment for focusing is equivalent. In particular, this coincides with the common observation that when an object is out-of-focus only its large-scale features are observable.

This connection with the scale space theory can also be used to explain other phenomena commonly observed under microscope optics, such as rounding off of sharp edges, merging and expansion of blob features, etc. Some mathematical treatment can be found in [3]. This connection also guides the design of focus measure operators. In particular, it will give a lower bound of the operator window size, as discussed below.

C. Scale of Focus Measure Operators

The displacement of a corner feature is less than 2σ under a change of σ in scale [3]. Therefore, the minimum operator window size can not be smaller that 4σ. For a specific microscope configuration, σ can be directly computed. In practice, window size must be sufficiently large to accommodate drifting and other possible noise in images. However, in order to ensure unimodel property and high proximity resolution, the window size should not be too large to cover features of significantly different depth.

D. Focus Measure Performance Comparison

This subsection presents experimental evidence showing that the performance of focus measure operators is sensitive to their positions. Performance comparison is also presented. Two groups of 30×30 windows are manually placed on the same sequence of 41 images taken consecutively at increments of 100 μm as the vacuum gripper approaches the metal part (Figure 5). Alghouth image 26 and image 27 are both determined to be satisfactorily sharp in focus by human observations, image 26 seems to be slightly better. The unnormalized focus measures for the two groups of windows are compared in Figure 6 and 7. It is clear that incorrectly placed operators

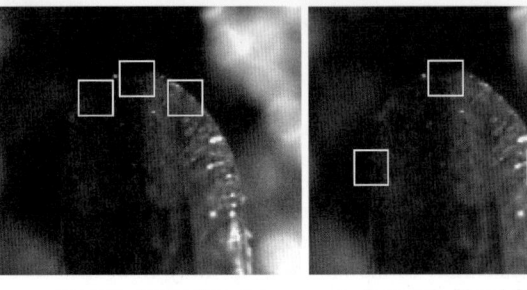

a. Correctly Placed Operators b. Incorrectly Placed Operators

Fig. 5. Placement of Focus Measure Operators

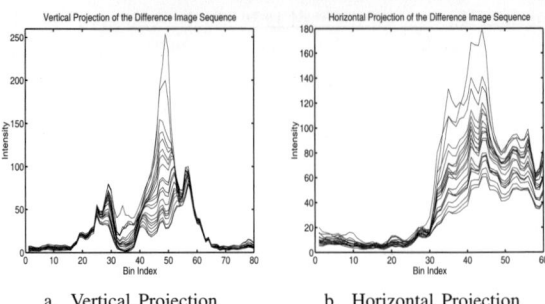

a. Vertical Projection b. Horizontal Projection

Fig. 8. Segmentation by Projection

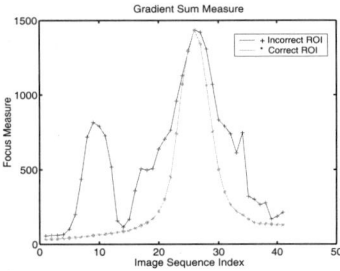

Fig. 6. Performance of M_{GS}

do not guarantee that there is only one peak in focus measure functions. In terms of performance, the M_{GS} operator provides the best focus resolution. The above phenomenon can be explained from the geometry of the needle. In Figure 5.b, the two lower windows are placed on regions that contains part of the vertical sidewall of the needle. Since the sidewall is roughly parallel to the microscope optical axis, each window contains more than one in-focus features as the needle moves. Notice that in Figure 5 the background is blurred. This is due to the displacement of the microscope, as explained in Section I. Theoretically, this distance can be computed using Equation 5. In practice it is difficult to get accurate results due to several factors. For example, the wavelength λ can only be assumed to be within a certain range. Therefore, a practical solution is to determine this distance by experiments. The microscope displacement d_{mic} is determined using the following equation

$$d_{mic} = c \cdot d_{exp} \quad (18)$$

where d_{exp} is the distance corresponding to 5% decrease from the peak value. c is a safety coefficient.

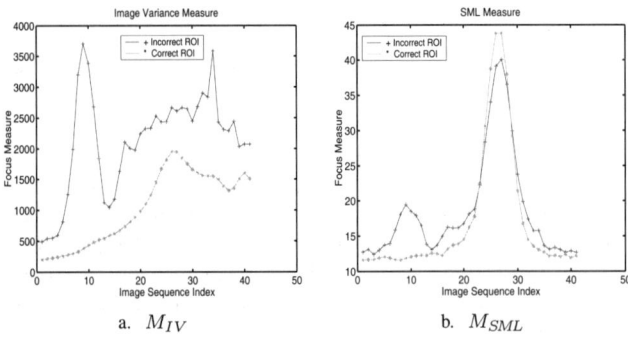

a. M_{IV} b. M_{SML}

Fig. 7. Performance of M_{IV} and M_{SML}

E. The Algorithm for Automatic Operator Placement

In preprocessing, a segment of the in-focus image of the needle tip is used to generate the template (Figure 9.b). The operator windows are placed on the template based on the knowledge of the geometry of the needle. Their positions relative to the template are fixed.

The algorithm consists of two major steps: preliminary segmentation and template matching. The initial segmentation is used to coarsely determine the region that contains the image of the end-effector. Then the focus measure operators are automatically placed by matching the template with the end-effector image. Details are given below.

1) Preliminary Segmentation: The horizontal axis and vertical axis are divided into M and N bins respectively. In our experiments, the bin size is 8 pixels. The objective is to accommodate small image drifting and random additive noise.

- Step 1. Capture and save the first image I_0.
- Step 2. Capture a new image I_k. Compute the difference image $D_k = |I_k - I_0|$. Compute the vertical projection of D_k along each column and the horizontal projections of D_k along each row. Sum the projection in each vertical and horizontal bin. Normalize each sum by the image row number and column number respectively. These projections are denoted by $d_{x,i}^k, i = 1..M$ and $d_{y,j}^k, j = 1..N$.
- Step 3. Identify those bins satisfying $d_{x,i}^k > \sigma$ and $|d_{x,i}^k - d_{x,i}^1| > \sigma$, and $d_{y,j}^k > \sigma$ and $|d_{y,i}^k - d_{y,i}^1| > \sigma$, where σ is the noise threshold.
- Step 4. Conservatively remove isolated bins, if any.
- Step 5. If current segmentation does not differ from the previous one, go to template matching. Otherwise, go to Step 2.

Figure 8 shows the change of the projections of a sequence of 20 images taken at increments of 400 μm towards the target. The threshold σ is measured by computing the intensity variance of a sequence of static images of the background. It can also be acquired from the difference image. Experiments show that it is not effective to use gradient image or image variance to perform this segmentation. This is due to the strong smoothing effect when the part is far from its in-focus position. Most of the high frequency components are filtered out. This makes the edges weak compared to those in the background.

2) Template Matching: The result of the preliminary segmentation is shown in Figure 9.a. The segmented rectangular region of interest is then divided into a regular grid of 5×5 rectangular cells. The purpose is to reduce the requirement on memory. The size of the cell is not significant as long as it is sufficiently small. For each newly captured image, the gradient sum operator M_{GS} is applied on each cell. The result is saved for the final retracing.

Without losing generality, the position of the $w \times h$ template $T(x,y)$ is identified as the coordinate of its upper left corner. The similarity measure $P(x,y)$ is a SSD (sum-of-squared-differences) type function. The matching processing is basically a search for the (x,y) that minimizes $P(x,y)$.

$$P(x,y) = \sum_{u=0}^{w} \sum_{v=0}^{h} |I(x+u, y+v) - T(u,v)| \quad (19)$$

- Step 1. Capture a new image I_k. Compute the difference image D_k. Then compute $d_{x,i}^k, i = 1..M$ and $d_{y,j}^k, j = 1..N$.
- Step 2. Compute the gradient image $G(x,y) = \|\nabla i(x,y)\|^2$. Compute the M_{GS} measure for each of the cell. Save the result.
- Step 3. If first time, start the template matching from a position based on the segmentation result. Otherwise start from previous position. Perform template matching.
- Step 4. Check if the template matching is consistent with the segmentation from the difference image projection. If template matching fails (templates contradicts the result of the segmentaion), use the symmetry axis of the segmented area to reinitiate the process. The resulting template position is (x_k, y_k).
- Step 5. If the distance $\sqrt{(x_k - x_{k-1})^2 + (y_k - y_{k-1})^2}$ between two consecutive template matching positions is less than a threshold σ_d, consider the operators placed.
- Step 6. Retrace the performance measure for the placed operator windows from saved gradient images for the cells. Analyze the change of focus measure. Stop after 5% decrease from peak is detected. Otherwise go back to Step 1.

The displacement threshold σ_d is chosen to be 3 pixels. This algorithm is based on the focusing model. When the part is close to its in-focus position, the scale will be significantly reduced. So will be the displacement of the end-effector image. The computation of the projection of the difference image is used to check the template matching position. Based on the scale-space theory, the axis of symmetry of the segmented area should roughly corresponds to the axis of symmetry of the needle image. Therefore, the symmetry axis of the segmented area can be used as a starting position to reinitiate the template matching process when necessary. When the focus measure decreases 5% from its peak value, the system converts to microforce control.

IV. MICROMANIPULATION CONTACT TRANSITION CONTROL

A. Modeling

Although the metal parts are placed on the membrane of the vacuum release tray, the mechanical support comes from the stiff supporting grid. Therefore, each part can be treated as a stiff surface. The machined hypodermic needle can be modeled using beam theory (Figure 1d). The model shows that the normal contact force F_n is linearly related to the vertical

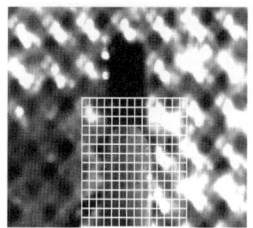
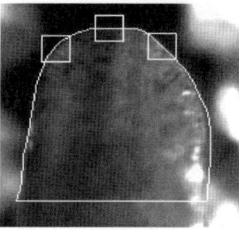

a. Preliminary Segmentation   b. Template Matching

Fig. 9. Segmentation and Template Matching

displacement δx of the micromanipulator, as described in the following equation.

$$F = \frac{3EI}{L^3 sin^2\theta} \delta_n = K_n \delta_n \quad (20)$$

where L is the length of the beam, and E and I are the Young's modulus and moment of inertia of the beam respectively. The coefficient K_n is calibrated to be 0.25mN/μm.

B. Microforce Measurement

The nominal force and torque resolution of ATI Nano17 are 1mN and 4mN·mm respectively. However, experiments show that the actual force resolution is significantly worse due to noise and perturbation. On the other hand, the end-effector consists of a long cantilever beam, which is common for micromanipulation to avoid collision and occlusion due to the limited working distance of microscope objectives. Because this configuration mechanically amplifies contact torques, contact force can be estimated from torque measurement. The estimated contact force is given by

$$\hat{f}_n = \frac{M}{L_{eff}} = \frac{\sqrt{M_x^2 + M_y^2}}{L_{eff}} = M_x \frac{\sqrt{1 + tan^2\alpha}}{L_{eff}} \quad (21)$$

When the contact force is small, the tangential force is generally negligible. If there is no rotation in pick-and-place operations, The contact force can be computed from M_x (or M_y) while $tan\alpha$ is determined through calibration using linear regression. Experiments show that the linear relation between the moment and the normal force is maintained to the force magnitude of at least 100mN. The effective distance L_{eff} in equation 21 is calibrated to be 52mm, leading to an equivalent force resolution of about 80 mN. Experiments show that the actual force resolution using this scheme is about 0.3 mN.

C. Contact Transition Controller Design and Experiments

A PI controller is used to control the MP285 micromanipulator through its joystick port. No separate controller is used for the transition phase. Performance comparison of this controller with other contact transition controllers is discussed in [22]. Since the magnitude of the control signal is proportional to the frequency of microstep motor actuation impulses, a further bilinear transform (Equation 23) is necessary to generate the actual control signal $\tilde{V}_c$ for a short transition time. The initial approach speed is 1000 μm/sec, while the final speed in transition control is reduced to 200 μm/sec. The target contact force is 20mN. Experiment data is shown in Figure 10.a. This is compared with the transition

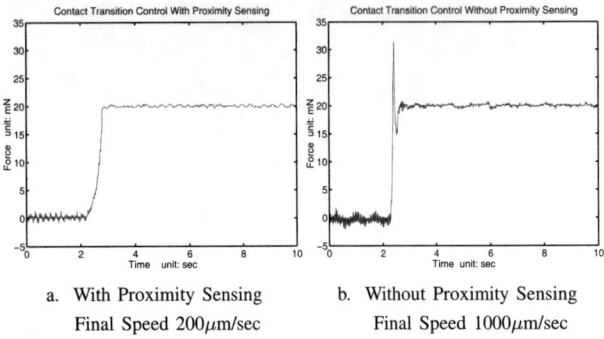

a. With Proximity Sensing Final Speed 200μm/sec
b. Without Proximity Sensing Final Speed 1000μm/sec

Fig. 10. Comparison of Contact Transition Control Performance

control response (Figure 10.b) of using approach speed 1000 μm/sec without proximity sensing. The PI controller has been retuned in this case for optimized performance. Notice that increased approach speed also increases the initial noise level. A large impact force spike is generated due to the large approach speed. By using proximity sensing, this large spike is eliminated without sacrificing efficiency since the large approach speed 1000 μm/sec can still be used.

$$V_c = K_p(f_{ref} - \hat{f}_n) + K_i \int (f_{ref} - \hat{f}_n) \quad (22)$$

$$\widetilde{V}_c = \begin{cases} \frac{\beta V_c}{V_c + \gamma} + V_{off}^+ & \text{if } V_c \geq 0 \\ \frac{\beta V_c}{V_c + \gamma} + V_{off}^- & \text{if } V_c < 0 \end{cases} \quad (23)$$

β equals the voltage that actuates the micromanipulator at its approach speed. $\gamma \ll 1$.

V. CONCLUSION

Microscopic vision and microforce feedback are natural complements to each other in many aspects. Their integration is fundamental to the automation of high precision micromanipulation and microassembly. A method based on this sensory information integration has been proposed for micromanipulation contact transition control. An optics model has been proposed for the focusing process. The model considers both geometric and wave optics. It shows that the dynamic behavior of the images can be described by using the framework of the scale-space theory. This relation plays a elementary role in determining the size and location of the focus operators. Mechanical amplification is used to partially overcome the constraints in implementing microforce feedback. Experimental results on microassembly pick-and-place operations confirm the effectiveness of this method.

In the immediate future, the work presented here will be extended to the case that the displacement is not parallel with the microscope optical axis. Another extension is to develop a more general framework for the segmentation and track under small depth-of-field. Although the template matching provides the possibility of extending the method to 3D cases, whether it is indeed the method of choice needs to be further studied. Model research work is also necessary on microscope modelling for other microscopy settings such as confocal microscopy and DIC (differential-interference-contrast) microscopy.

ACKNOWLEDGEMENT

This work was supported in part by the National Science Foundation through Grant Number 9996061.

REFERENCES

[1] P. N. Akella and M. R. Cutkosky, "Contact transition control with semiactive soft fingertips," IEEE Trans. on Robot. & Autom., vol.11, no.6, pp.859-867, 1995.

[2] S. Allegro, C. Chanel, and J. Jacot, "Autofocus for automated microassembly under a microscope," Proc. 1996 IEEE Int. Conf. on Image Processing, vol.2, pp.677-680.

[3] F. Bergholm, "Edge Focusing," IEEE Trans. Pattern Analysis & Machine Intelligence, vol.9, no.6, pp.726-741, 1987.

[4] M. Born and E. Wolf, *Principles of Optics*, 7th ed., Cambridge University Press, 1999.

[5] J. T. Feddema and R. W. Simon, "CAD-driven microassembly and visual servoing," Proc. 1998 IEEE Int. Conf. on Robot. & Autom., pp.1212-1219.

[6] J.-M. Geusebroek, F. Cornelissen, A. W. M. Smeulders and Hogo Geerts, "Robust autofocusing in microscopy," Cytometry, vol.36, no.1, pp.1-9, 2000.

[7] E. Hecht, *Optics*, 3rd ed., Addison Wesley Longman, 1998.

[8] J. M. Hyde and M. R. Cutkosky, "Controlling contact transition," IEEE Contr. Sys. Mag., vol.14, no.1, pp. 25-30, 1994.

[9] E. Krotkov, "Focusing," Int. J. Computer Vision, vol.1, no.3, pp.223-237, 1987.

[10] S. Inoue and K. R. Spring, *Video Microscopy*, Plenum Press, New York, 1997.

[11] Y.F. Li, "A sensor-based robot transition control strategy," Int. J. Robot. Res., vol. 15, no.2, pp.128-136, Apr. 1996.

[12] T. Lindeberg, "Detecting salient blob-like image structures and their scales with a scale-space primal sketch: A method for focus-of-attention", Int. J. Computer Vision, vol. 11, pp. 283–318, Dec. 1993.

[13] T. Lindeberg, *Scale-Space Theory in Computer Vision*, Kluwer Academic Publishers, 1994.

[14] S. K. Nayar and Y. Nakagawa, "Shape from Focus,", IEEE Trans. Pattern Analysis & Machine Intelligence, vol. 16, no. 8, pp. 824-831, August 1994.

[15] B. J. Nelson and P. Khosla, "Force and vision resolvability for assimilating disparate sensor feedback,", IEEE Trans. on Robot. & Autom., vol. 12, no. 5, pp. 714-731, October 1996.

[16] N. Sarkar, X. Yun, and R. Ellis, "Live-constraint-based control for contact transitions," IEEE Trans. on Robotic. & Autom., vol.14, no.5, pp.743-754, 1998.

[17] M. Subbarao, T. S. Choi, and A. Nikzad, "Focusing Techniques," J. Optical Engineering, vol.32, no.11, pp. 2824-2836, nov. 1993.

[18] M. Subbarao and J. K. Tyan, "Selecting the optimal focus measure for autofocusing and depth-from-focus," IEEE Trans. Pattern Analysis & Machine Intelligence, vol. 20, no. 8, pp. 864-870, August 1998.

[19] A. Sulzmann, P. Boillat and J. Jacot, "New developments in 3D computer vision for microassembly, " Microrobotics and Micromanipulation, A. Sulzmann and B. J. Nelson Eds., SPIE Proceedings, Vol. 3519, pp.36-47.

[20] T. Tanikawa, T. Arai, and Y. Hashimoto, "Development of vision system for two-fingered micromanipulation," Proc. 1997 IEEE/RSJ Int. Conf. Int. Robots & Systems, pp. 1051-1056.

[21] T. J. Tarn, Y. Wu, N. Xi, and A. Ishidori, "Force regulation and contact transition control," IEEE Contr. Sys. Mag., vol. 16, no.1, pp.32-40, 1996.

[22] R. Volpe and P. Khosla, " A theoretical and experimental investigation of impact control for manipulators," Int. J. Robot. Res., vol.12, no.4, pp.351-365, 1993.

[23] I. D. Walker, "The use of kinematic redundancy in reducing impact and contact effects in manipulation," Proc. 1990 IEEE Int. Conf. Robot. & Autom., pp.434-446.

[24] J. Z. Wang, J. Li, R. M. Gray, and G. Wiederhold, "Unsupervised multiresolution segmentation for images with low depth of field," IEEE Trans. Pattern Analysis & Machince Intelligence, vol. 23, no. 1, pp. 85-90, January 2001.

Two-Dimensional Signal Transmission Technology for Robotics

Hiroyuki Shinoda, Naoya Asamura, Mitsuhiro Hakozaki, and Xinyu Wang

Department of Information Physics and Computing
The University of Tokyo
7-3-1 Hongo, Bunkyo-ku, Tokyo 113-8656 Japan
{shino, asamura, hakozaki, wang}@alab.t.u-tokyo.ac.jp

Abstract

The forms of communication available now are categorized into the one or three dimensional. One dimensional communication includes metal wires and optical fibers in which the electro-magnetic field is confined in one dimensional medium. Wireless communication based on RF or optical connection emits electro-magnetic field in 3-D space. Now what if we have "two-dimensional communication" in which signals travels from one point to another point freely in elastic two-dimensional space using electromagnetic field confined in 2-D space? In this paper, we describe such a new technology of 2-D communication brings new paradigm to robotics. The methodologies of machine-design, system-integration, sensing, and computing will be drastically changed. We show architecture of the 2-D signal transmission based on relaying packets between communication chips on a thin sheet, the physical structure of the 2-D signal transmission, the protocols of the signal relay, and the results of the basic experiments.

Keywords; two-dimensional communication, diffusive signal transmission, packet communication, sensor network, robot skin, micromachine, ubiquitous, distributed computing

1. Introduction

The forms of communication available now are categorized into one or three dimensional devices. One dimensional communication includes metal wires and optical fibers in which the electro-magnetic field is confined in one dimensional medium. One dimensional communication causes a problem when a large number of elements are to be connected with each other. The wiring comes to need a major effort in system design including circuit design, connecting distributed sensors and actuators, powering micromachines, and distributed computation.

Wireless communication based on RF or optical connection can be classified into three-dimensional communication, as the electromagnetic field is released in three-dimensional space. Wireless communication had been considered to be a hope to solve the wiring problem [1,2,3,4,5]. The problem of it is that the electromagnetic field propagates beyond the target, which makes energy transmission difficult, and degrades the communication capacity when multiple elements communicate simultaneously.

Now what if we obtain a "two-dimensional communication" device in which signals from one point to another point travel freely in flexible two-dimensional space? In this paper, we show a method to realize such a 2-D signal transmission device. We describe how it changes the robotics in machine design, system integration, sensing, actuating and computing.

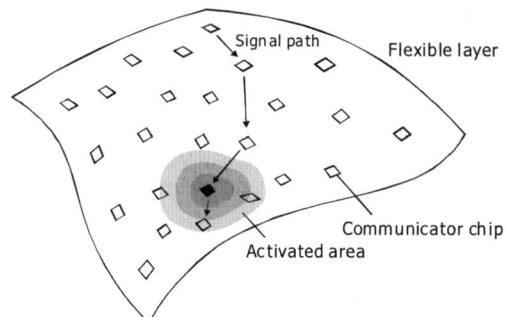

Fig. 1: Schematic diagram of two-dimensional signal transmission. Signals are transmitted in an isotropic 2-D space.

2. Diffusive signal transmission in 2-D layer

The two-dimensional communication we call here is the one in which signals are transmitted in two-dimensionally isotropic structure using electromagnetic field confined in 2-D space.

A structure of two-dimensional signal transmission proposed here is shown in **Fig. 1** and **Fig. 2**. Multiple LSI chips are distributed between flexible conductive layers of rubber or cloth. The chip can be a sensor or a connector to another functional device with a communication interface. The signal reaches a finite distance physically, and the packet is transmitted to an arbitrary point by being relayed (hopping) among the communicators as **Fig. 1** shows. Every element connected to the layers can communicate with each other without individual wires. The layer can be realized with elastic material allowing extension because the material of the 2D layer does not have to be conductive as much as a usual wire. A signal is transmitted from a chip to another neighboring chip by diffusive coupling. **Fig. 2** describes the cross-section of the communication layers. First we explain the principle of low-frequency signal transmission based on the structure type I in **Fig. 2** (a).

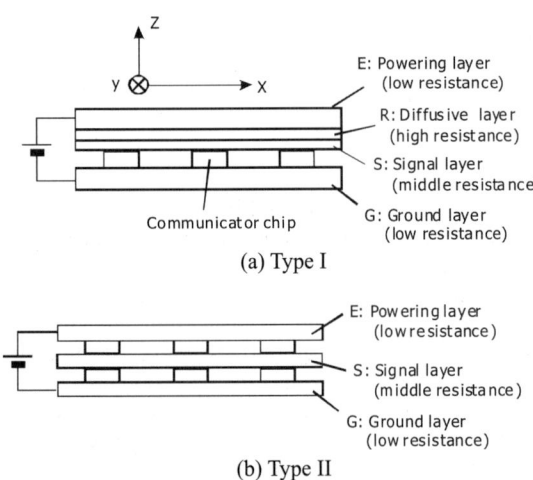

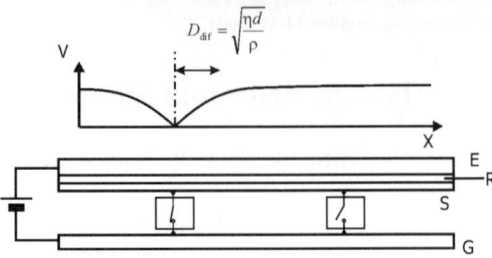

Fig. 3: Illustration of diffusive signal transmission in type I. A signal sent from a chip is detected only by chips inside the D_{dif} circle.

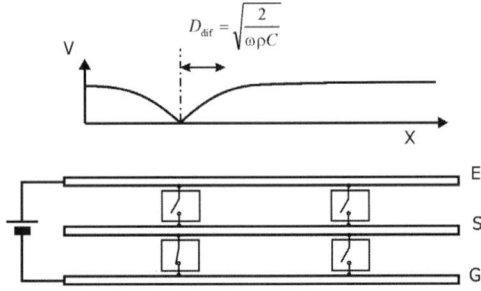

Fig. 2: Cross section of the 2-D signal transmission device. Each chip contacts with two layers in type I and three layers in type II.

Fig. 4: Illustration of diffusive signal transmission in type II.

[Diffusion mode I]

Suppose the lateral resistance of G and E layers is negligible. The sheet resistance of R layer in **Fig. 2** (a) is sufficiently larger than that of S layer. When the thickness of S layer is negligible, the alternative component of voltage distribution $V(x,y)$ on S layer satisfies the following diffusion equation

$$C\frac{\partial}{\partial t}V + \frac{1}{\eta d}V = \left(j\omega C + \frac{1}{\eta d}\right)V = \frac{1}{\rho}\left(\frac{\partial^2}{\partial x^2} + \frac{\partial^2}{\partial y^2}\right)V \quad (1)$$

where C [F/m²], η [Ωm], d [m], ρ [Ω], and ω [rad/s] are respectively capacitance between S layer and G-and-E layers, volume resistivity of R layer, the thickness of R layer, sheet resistivity of S layer, and angular frequency of the signal. The $V/\eta d$ implies the current density from the S layer to E layer through R layer.

At a low frequency where $j\omega C \ll 1/\eta d$, one dimensional solution $V(x)$ of Eq. (1) is written as

$$V = V_0 \exp\left(\pm \frac{x}{D_{\text{dif}}}\right)\exp(j\omega t) \quad (2)$$

where

$$D_{\text{dif}} = \sqrt{\frac{\eta d}{\rho}} \quad (3)$$

As this equation shows, electrical potential changes occurring at a point on S layer does not go beyond the diffusion length D_{dif}. Then a signal sent from a chip is detected only by chips inside a circle with the radius D_{dif}. This is illustrated in **Fig. 3**.

[Diffusion mode II]

A structure shown in **Fig. 2** (b) reduces power consumption. In this case the solution of Eq. (1) is written as

$$V = V_0 \exp\left(\pm \frac{1+j}{\sqrt{2}}\sqrt{\omega\rho C}\, x\right)\exp(j\omega t) \quad (4)$$

Then

$$D_{\text{dif}} = \sqrt{\frac{2}{\omega\rho C}} \quad (5)$$

gives the diffusion length. When $\omega = 10^7$ [rad/s], $\rho = 500$ [Ω], $C = 4\times 10^{-8}$ [F/m²], the diffusion length D_{dif} is given as 10 [cm].

[Two dimensional point current source]

The two-dimensional solution with cylindrical symmetry is given as follows. When an electrode with the radius r_0 (See **Fig. 5**) located at the origin has a potential $V_0 \exp(j\omega t)$, the potential at a distance r from the origin is written as

$$V(r) = V_0 \frac{J_0\left((1-j)\frac{r}{D_{\text{dif}}}\right) - jN_0\left((1-j)\frac{r}{D_{\text{dif}}}\right)}{J_0\left((1-j)\frac{r_0}{D_{\text{dif}}}\right) - jN_0\left((1-j)\frac{r_0}{D_{\text{dif}}}\right)} \exp(j\omega t) \quad (6)$$

using Bessel functions in diffusion mode II. **Fig. 5** (b) shows a function $|H_0(x-jx)| = |J_0(x-jx) - jN_0(x-jx)|$ to evaluate the equation. For diffusive mode I, we can obtain $V(r)$ by substituting D_{dif} for

$$D_{\text{dif}} = \frac{1+j}{\sqrt{\rho/(\eta d) + j\omega\rho C}} \quad (7)$$

When $r, r_0 \ll |D_{\text{dif}}|$, the $V(r)$ is approximately written as

$$V(r) \approx V_0 \log r / \log r_0. \tag{8}$$

If $r \gg D_{\text{dif}}$, the $V(r)$ decreases exponentially as the one dimensional case. Therefore, the diffusion length in the two-dimensional case is comparable to that in the one-dimensional case.

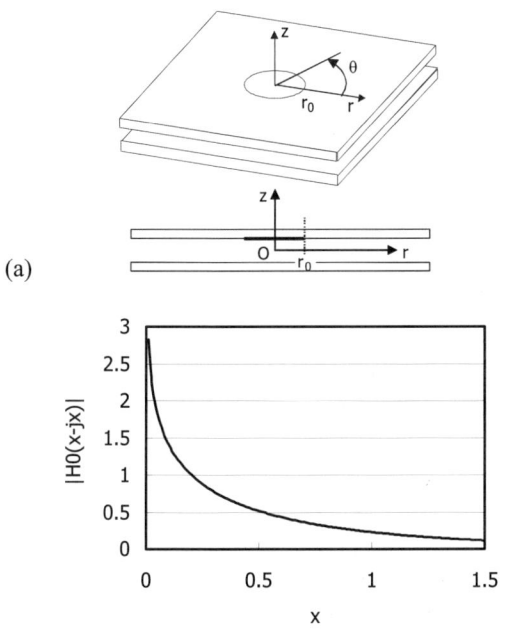

(a)

(b)

Fig. 5: (a): Definitions of r_0 and r. (b): Plots of a function $|H_0(x-jx)| = |J_0(x-jx) - jN_0(x-jx)|$ to evaluate Eq. (6).

[Electromagnetic wave radiation in the gap space]

When the switch in **Fig. 4** operates at a very high frequency so that the electromagnetic wavelength is smaller than the diffusion length, the gap between S and G (E) layer becomes a waveguide that allows TEM mode [6] to occur in which the electric field is perpendicular to the layer and magnetic field circles around the radiation point. The electromagnetic energy is localized in the gap between the two conductive layers. The current runs around the surface with the skin depth $\sqrt{2/\omega\mu\sigma}$ where μ and σ are the permeability and the conductivity of the conductive layers ($=1/\eta$), respectively. The magnetic field $\mathbf{B} = (0, B(r,z,t), 0)$ in the cylindrical coordinate is given as

$$B(r,z,t) = B_0 H_1^{(2)}(kr) \exp(j\omega t) \tag{9}$$

in the gap where

$$\text{Re}[k] \sim \omega/c \quad (c\text{: light velocity}), \tag{10}$$

$$\text{Im}[k] \sim \frac{1}{d}\sqrt{\frac{\varepsilon_0 \omega}{8\sigma}} \quad \text{(attenuation constant)}, \tag{11}$$

and

$$H_1^{(2)}(z) \equiv J_1(z) - jN_1(z). \tag{12}$$

The signal transmission by electromagnetic wave realizes short-delay and large-throughput communication. In the following sections, however, we discuss only diffusive signal transmission.

3. Two-dimensional signal transmission by distributed LSI chips

The proposed system transmits signals by relaying packets among neighboring communication chips. Comparing with other imaginable methods in which electromagnetic field travels in a long range passively in the 2-D medium, we notice the following advantages in relaying packets (hopping).

1) The excited domain on a signal path forms a one-dimensional chain. Then required energy for signal transmission is not wasted in unrelated area.
2) Multiple pairs of chips can communicate simultaneously without interference because the excited domains are localized into a one-dimensional chain.
3) If some area of the device damaged, a new path can be produced dynamically.
4) It is free from multi-path and reflection problems.

Next we summarize the relationship between the density of the chip distribution and the specification of communication.

Throughput and delay

Until a packet reaches a place as far as L, the packet is relayed

$$N \approx L/D_{\text{dif}} \tag{13}$$

times. As Eq. (5) shows, the diffusion length D_{dif} is proportional to $1/\sqrt{\omega}$ at a high frequency in type II. Therefore if we keep the chip spacing comparable to the diffusion length, signal delay of K bit packet is given as

$$T_D \approx KT_0 N \approx KL\sqrt{\frac{\rho C}{\omega}}, \tag{14}$$

where T_0 is the period for one bit supposed to be comparable to $1/\omega$. Then the signal delay decreases as the signal frequency is heighten though it results in increase of the number of chips and relaying.

On the other hand, the throughput is simply determined by the signal frequency, and is proportional to the signal frequency.

Energy consumption

Suppose the minimal change of electro-magnetic energy density detectable by a communication chip in the 2-D medium, is written as e [J/m²]. Then the energy consumed for one-bit transmission by N relaying is given as

$$esN \approx eD_{\text{dif}}^2 \frac{L}{D_{\text{dif}}} = eL\sqrt{\frac{1}{\rho C \omega}} \tag{15}$$

where s indicates the diffusion area comparable to D_{dif}^2. Then the energy consumption in the communication layer decreases as the signal frequency increases. For example, the energy consumed in the communication layer is given as 1 [nJ/m/bit] when $D_{\text{dif}} = 10$ [cm], $d = 1$ [mm], and the voltage amplitude in signals is 1 [V].

On the other hand, the total power consumption inside communication chips increases as the chip number and operation frequency increase. The tradeoff is a factor to determine the optimal chip density and signal frequency. In our first design of 10 MHz operation, the energy consumed by one chip is estimated at 1 mW. When we locate the chips with a density 10^4 [$1/m^2$], the total power consumption inside the chips amounts to 10 [W/m^2]. This total power consumption can be reduced with the technique of gated clock. In table 1, we summarize specification of the communication chip in type II that we are developing.

Table 1: Specification of the projected communication chip (Type II)

Signal throughput between neighboring chips	10 Mbps
Power consumption in signal layer	1 nJ/m/bit
Chip operation	10 MHz
Chip size	3 mm square
Power consumption in the chip	1 mW/chip
Diffusion length	10 cm

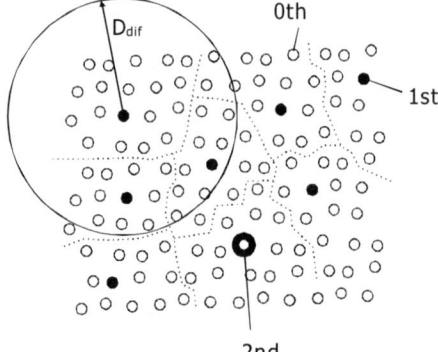

Fig. 6: Distribution of communication chips. The figure shows a system with a single 2nd-order chip.

4. Algorithm of signal relay

The system is based on packet communication [7,8]. The signal path between an arbitrary point and another point is established dynamically. The condition that density of communicators is very high and the diffusion length is limited is similar to that of wireless sensor network [3-5]. However signal power reduction is less important than that in wireless sensor network and topology changes of signal paths happen less frequently than that in ad hoc network. Here we show the outline of a hierarchical architecture in which all chips have global IDs and connection between arbitrary two chips is assured. The proposed method satisfies the following conditions.

1) Since vast amount of chips are used, the produced chip should have a uniform or a small variation of design. The ID numbers of the chips are also desired to be assigned dynamically using random number generators.
2) The signal paths (selection of chips in relaying) are established for all the chips in sufficiently short time.
3) The signal paths are established without a priori knowledge of the chip positions.

Table 2: Composition of packet

1	ID of the next receiver of this packet.
2	Command (forward, path_search, echo_requirement, echo_back, id_assign)
3	Length of header
4	Length of data
5	ID of the destination
6	ID of the sender of this packet
7	ID of the signal origin
8	Series of the IDs of the *n*-th order chips to the destination ⋮ Series of the IDs of the 1st order chips to the next 2nd order chip or the destination
9	Data

In our system, we prepare from 0th to *n*th order communication chips. In **Fig. 6**, the white circles indicate 0th communicators. The 0th chips are the terminals (sensors or connectors to outer devices) of this system, and they can communicate only with the 1st-order "parent" chip that supervises the 0th chips. One 1st-order chip can have M 0th-order children within the diffusion area in which the signal reach directly from the 1st-order chip. The spacing of 1st order communicators is comparable to or smaller than the diffusion length D_{dif}. The 2nd order chip can also have M 1st-order children. In our system, the M is set at 256. A *n*th-order chip ($n > 0$) possesses the following functions to relay signals.

1) It knows all the IDs of the (n-1)th order children of itself. It also knows the relay path (a series of IDs of (n-1)th order chips) to every children chip of it. It also knows the next chip toward its parent.
2) The *n*th order chip has all functions and information that it should have if it was a 0, 1, $\cdots, n-2$, or (*n*-1)th order chip. Therefore a *n*-th order chip also has a table of $n-2, n-3, \cdots, 0$ th order children as a $n-1, n-2, \cdots, 1$st order parent. Then a *n*th-order chip has an ID table of nM chips as the maximum.

A packet is composed of the contents shown in Table 2. A *n*th-order chip that received a packet behaves as follows.

1) If the ID of item 1 in Table 2 does not coincide with the receiver's ID, the receiver ignores the packet.
2) If the destination is included in the children table, it creates a packet adding the path data of item 8 in Table 2 and sends it. The "path data" includes ID series from 1st order to (*n*-1)th order. The "ID series of *k*th order" ($0 < k < n$) is the series of the chip IDs of *k*th order to the next (*k*+1)th order chip (if it exists).

3) If the destination (the goal of the transmission) designated in the packet is not included in the table of the children, it creates the path data to the nth-order chip indicated by the path data of item 8, and it sends the packet to the next chip. If the next nth-order chip is not indicated in item 8, it transmits the packet toward its parent.

Based on this algorithm, each chip can send a packet to an arbitrary chip in total of M^n chips. (Notice that packets are always relayed by the 1st-order chips physically.) The commands for establishing IDs and signal paths are summarized in item 2 of Table 2.

For establishing IDs of 1st-order chips, for example, a 2nd-order chip sends "echo_requirement" command to the 1st-order chips in the diffusion area as **Fig. 7** (a) shows. Next as **Fig. 7** (b) shows, the chips that received the echo_requirement command responds with "echo_back" after a certain waiting time following a random number. The 2nd-order chip gives an ID to the sender of the echo-back by "id-assign" command. Then as **Fig. 7** (c) shows, the 2nd-order chip sends "path_search" to the chip that obtained ID in the previous process. The chip that received the path-search creates echo-requirement, assigns the IDs around them, and tells the path from the 2nd-order parent. The newly assigned 1st-oder chip repeats this process until the 2nd-order chip obtains a certain number of 1st-order children.

Drawbacks of this method are that the packet has a relatively large overhead including path data, and that signals gather around high order chips. One practical solution is that we assign the x-y coordinate as the IDs to the first order chips in advance. Then the first order chips can decide the next receiver from the coordinate without second or higher order chips.

5. Prototype and basic experiment

Up to now, we have already designed the 0th and 1st order communication chips whose physical architecture is shown in **Fig. 8**. The floor plan shown in **Fig. 9** was obtained by Avant! (Synopsys) Apollo, based on CMOS 0.35 μm rule. All functions required for the 1st order chip with a 128 byte buffer were realized in a 3 mm by 2 mm area of digital circuit and 1.5 mm square area of input-output analogue circuits. The chip is fabricated by ROHM Ltd., Japan. The operation of the LSI of the 0th order chip was tested and the LSI was molded as **Fig. 10** shows.

Experimental results of diffusive attenuation of signals are shown in **Fig. 11**. A signal layer of sheet resistance 1 kΩ, distant from the ground layer by 0.2 mm was tested for 1, 5, and 10 MHz sinusoid. Detected voltage amplitudes at a distance r from the signal source are plotted in **Fig. 11**. The plots are the ratios of measured amplitude to the measured amplitude where r = 1 cm. Experimental results of attenuation coincided with the theory of Eq. (6).

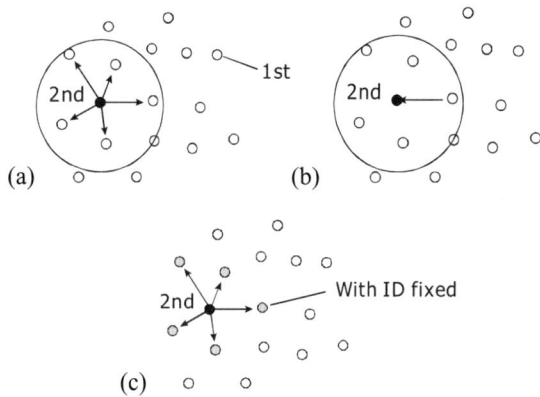

Fig. 7: Procedures of ID fixation. The 2nd-order parent sends "echo-requirement" first to neighboring 1st-order chips. Next the parent gives IDs to the neighbors based on the order of response. After fixing the IDs, the parents send "path-search" to the neighboring children to establish the IDs around them.

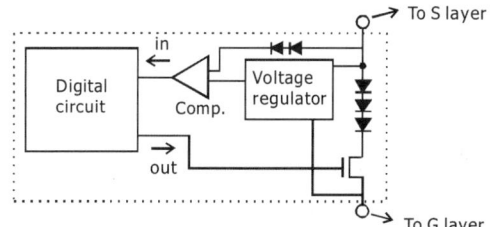

Fig. 8: The structure of the communication chip for Type I architecture.

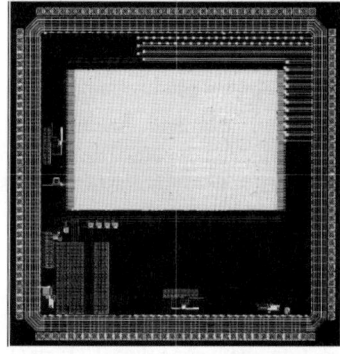

Fig. 9: The mask pattern we designed for a 1st-order communication chip. The chip size is 5 mm by 5mm with digital circuit of 3 mm by 2mm.

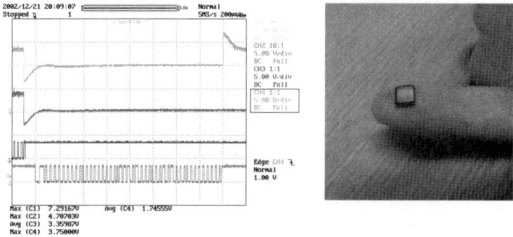

Fig. 10: The operation of the LSI of the 0th order chip of type I was tested (left) and the LSI chip was molded (right).

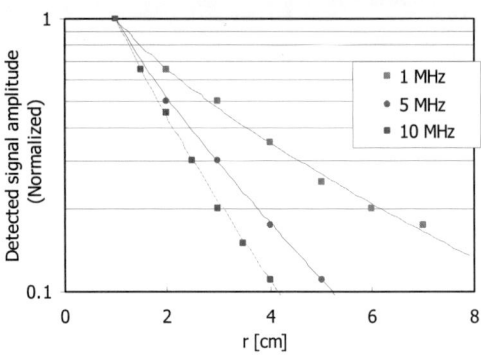

Fig. 11: Experimental results of diffusive signal attenuation for a signal layer with sheet resistance 1 kΩ distant from the ground layer by 0.2 mm. The ratios of measured voltage amplitudes to the measured amplitude where $r = 1$ cm are shown (r: distance from the voltage source).

6. Applications

A flexible 2-dimensional communication device in which signals can be transmitted with large throughput is useful in various aspects and scenes of robotics as follows.

1) Excluding complicated wires from robot system

If we cover a robot with this 2-D communication layer, elements of the robot such as computers, motors, and sensors can be connected to the device at arbitrary positions to communicate each other. In addition to the information exchange, they obtain energy by simply connecting the terminals to the E layer. The complicated wires combining those elements are removed.

2) A vast number of elements can be connected

A large number of elements including sensors, displays, and other functional parts can be connected and communicate at fast speed. An elastic tactile sensing device with more than 10^6 elements is realized.

3) Activating micromachines and small tags

Micromachines on the 2-D device can communicate each other and obtain energy through non-contact proximity connection between the 2-D device and micromachines using light or inductive coupling [9]. Objects with small communication interface chips without battery can communicate ubiquitously through a floor and a wall equipped with the 2-D device.

4) Wearable computing

Robot can wear a computing device formed into clothes. Signals with a frequency more than 1 MHz can be transmitted in the flexible clothes. The human can also wear it, and it can assist the communication between the human and the robot.

5) Wireless connection through a foot

Usual robots always have contact with floor by their feet. Then if they communicate through the feet with a floor equipped with the 2-D device, they can communicate with other robot as freely as the traditional wireless communication. The communication through the feet is free from the electromagnetic radiation. The communication capacity offered by the 2-D communication floor in which the signal path is a one-dimensional chain, will be far larger than that of 3-D communication sharing a single 3-D space for all communication elements.

6) Enabling a high-speed distributed computation

The elements connected to the communication layer can always communicate with each other if the elements follow the protocols of the 2-D communication. Computing element including memories, processors, and resistors in the processor can be connected through 2-D layers potentially. A flexible device can be a computer that also contains sensors, displays and other interfaces in it.

7. Summary

In this paper, we proposed a new concept of communication, "2-D" signal transmission. In a thin, flexible medium allowing extension, signals can travel with a large throughput. Basic concept and a method of realizing a device with 10 MHz-operation chips were described. A communication chip with 3 mm by 2 mm of digital circuit based on CMOS 0.35 μm rule was shown to be possible. Various applications of the device were described.

References

[1] M. Hakozaki, H. Oasa and H. Shinoda, "Telemetric Robot Skin," Proc. 1999 IEEE Int. Conf. on Robotics and Automation, pp. 957-961, 1999.

[2] C. Bisdikian, "An Overview of the Bluetooth Wireless Technology," IEEE Communications Magazine, December, pp. 86-94, 2001.

[3] J. Heidemann, F. Silva, C. Intanagonwiwat, R. Govindan, D. Estrin, and D. Ganesan, "Building Efficient Wireless Sensor Networks with Low-Level Naming," Proc. Symposium on Operating Systems Principles, pp. 146-159, 2001.

[4] E. Shih, S. H. Cho, N. Ickes, R. Min, A. Sinha, A. Wang, A. Chandrakasan "Physical layer driven protocol and algorithm design for energy-efficient wireless sensor networks" Proc. MOBICOM'01, pp. 272-287, 2001.

[5] V. Hsu, J. M. Kahn, and K. S. J. Pister, "Wireless Communications for Smart Dust," Electronics Research Laboratory Technical Memorandum Number M98/2, February, 1998.

[6] D. M. Pozar, Microwave Engineering, 2nd ed. John Wiley and Sons, 1997.

[7] A. S. Tanenbaum, Computer Networks, Third Edition, Prentice Hall, 1996.

[8] D. Comer, Internetworking with TCP/IP, Vol 1: Principles, Protocols and Architecture, Prentice Hall, 4th Ed, 2000.

[9] J. A. V. Arx and K. Najafi, "On-Chip Coils with Integrated Cores for Remote Inductive Powering of Integrated Microsystems," Proc. TRANSDUCERS '97, pp. 999-1002, 1997.

Micropeg Manipulation with a Compliant Microgripper

Woo Ho Lee[†], Byoung Hun Kang[†], Young Seok Oh[†], Harry Stephanou[†],
Arthur C. Sanderson[†], George Skidmore[‡], and Matthew Ellis[‡]

lee,kangb,ohys,hes,acs@cat.rpi.edu, gskidmore,mellis@zyvex.com

[†] Center for Automation Technologies,
Rensselaer Polytechnic Institute, Troy, New York, 12180
[‡] Zyvex Corporation, Richardson, Texas, 75081

Abstract

This paper presents analytical, simulation and experimental results from a study of compliant insertion tasks in microassembly. Gripper compliance is desirable to compensate for positional errors and to prevent the breakage of a gripper during assembly tasks. An analytical model is derived to study the motion and force profiles during compliant insertion. Thermal bimorph microgrippers with a compliant tip are designed and fabricated using a silicon DRIE process, and are mounted on a precision motion stage. A series of micropeg manipulation tasks such as pick up, rotation, and insertion are successfully performed. Finally, a comb structure is integrated in the gripper to calculate insertion force by measuring the deflection of a gripper, which is essential for automated microassembly.

1 Introduction

Recent progress in MEMS enables the mass production of microparts. In most cases, heterogeneous microparts produced from various fabrication processes need to be assembled and packaged in order to achieve a desired functionality. In microassembly, the most challenging issue is to understand scaling effects and to design a microassembly system that can handle microparts. Toward this goal, significant efforts are focused on the design of microgrippers and micromanipulators.

Researchers have investigated an adhesion force at the microscale to manipulate microparts [1,6,11]. A thermal bimorph actuation microgripper with force sensor is developed in [4]. Shimada et al. [8] presented dexterous tweezers and demonstrated micromanipulation tasks such as grasping and rolling. For a parallel assembly, Bohringer et al. [2] proposed programmable force fields to manipulate microparts with an array of microactuators. Skidmore et al. [9] proposed a method for exponential and massively parallel assembly. For force-guided assembly, it is necessary to measure the force at the micro or nano scale. Recently, Greminger and Nelson [5] detected the deflection of micro scale cantilever beam to measure nanonewtons force using visual information.

This paper presents modeling, analysis, gripper design, force sensing, and micromanipulation experiments for compliant insertion tasks in microassembly. An analytical model is derived to study the motion and force profiles during compliant insertion. A thermal bimorph method originally developed by Comtois and Bright [3] is adopted to actuate the gripper tip. Advantages of thermal bimorph actuation include larger forces and wider displacements. Thermal bimorph microgrippers with two types of compliant tips are designed and fabricated using the silicon DRIE (Deep Reactive Ion Etching) process, and mounted on a precision motion stage. A series of micropeg manipulation tasks such as pick up, rotation, and insertion are successfully performed. Finally, a comb structure integrated in the gripper is used to calculate the insertion force by measuring the deflection of the gripper.

2 Analytic Modeling of Compliant Insertion

A static force equation is derived to analyze the compliant insertion model in the XZ plane. A microgripper that holds a micropeg is represented by linear springs, k_x^g in the x-direction and k_z^g in the z-direction, and by a torsional spring k_θ^g in the y-direction. Fig. 1 shows an equivalent compliant insertion model that represents the gripper holding a flexible peg during contact with a hole. Initial work on insertion modeling and compliance design can be found in [7,10]. It is assumed that the micropeg has a linear stiffness k_x^p in the x-direction only. The assumption is reasonable because the peg is long in the z-direction, and it has a large deformation along the x-direction compared to the z-direction.

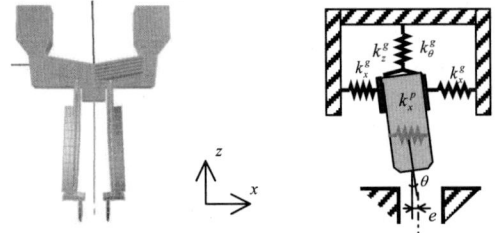

Figure 1: A compliant microgripper and its simplified model with a micropeg.

Fig. 2 illustrates a simplified free body diagram that is used to derive force equations for compliant insertion. Contact occurs as the gripper moves downward toward a hole, and reaction forces ($N, \mu N$) applied to the peg cause the deflection of the peg and the gripper.

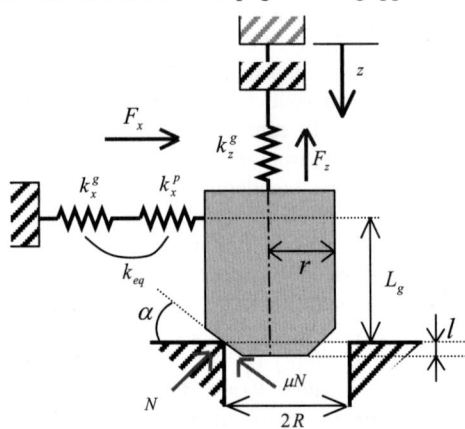

Figure 2: Simplified insertion model during contact.

Here, we set an equivalent stiffness, k_{eq}, and define an eccentricity ratio, c, as expressed in eq. (1).

$$k_{eq} = \frac{k_x^g k_x^p}{k_x^g + k_x^p}, \quad c = \frac{R-r}{R} \qquad -(1)$$

2.1 Insertion Modeling

Depending on contact force discontinuity, an insertion sequence can be divided into three different states [10]: chamfer crossing, one point contact, and two points contact as shown in Fig. 3. The motion and force equations are derived separately at each state by considering the different constraints.

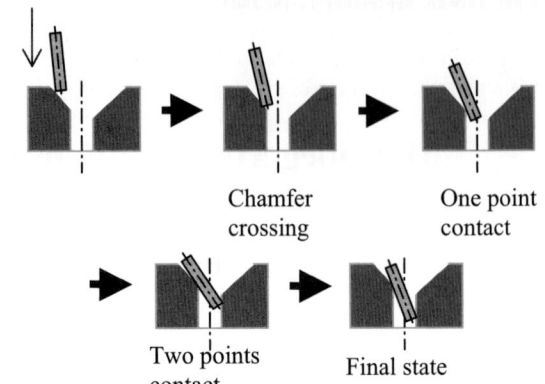

Figure 3: Insertion sequence of peg-in-hole.

In the chamfer crossing case, the rotation angle θ in the y-direction, the displacement in the x-direction, X, and the reaction forces in the x and z-directions (F_x, F_z) can be derived as eq. (2):

$$\theta = \theta_0 + \frac{k_{eq}\left(\frac{z}{\tan\alpha}\right)\left(L_g\begin{pmatrix}\sin\alpha\\-\mu\cos\alpha\end{pmatrix} - r\begin{pmatrix}\cos\alpha\\+\mu\sin\alpha\end{pmatrix}\right)}{\left(k_{eq}L_g^2 + k_\theta\right)\begin{pmatrix}\sin\alpha\\-\mu\cos\alpha\end{pmatrix} - k_{eq}L_g r\begin{pmatrix}\cos\alpha\\+\mu\sin\alpha\end{pmatrix}}$$

$$X = -\left(X_0 - \frac{k_\theta\left(\frac{z}{\tan\alpha}\right)(\sin\alpha - \mu\cos\alpha)}{\left(k_{eq}L_g^2 + k_\theta\right)\begin{pmatrix}\sin\alpha\\-\mu\cos\alpha\end{pmatrix} - k_{eq}L_g r\begin{pmatrix}\cos\alpha\\+\mu\sin\alpha\end{pmatrix}}\right)$$

$$F_x = k_x\left(\frac{k_\theta\left(\frac{z}{\tan\alpha}\right)(\sin\alpha - \mu\cos\alpha)}{\left(k_{eq}L_g^2 + k_\theta\right)\begin{pmatrix}\sin\alpha\\-\mu\cos\alpha\end{pmatrix} - k_{eq}L_g r\begin{pmatrix}\cos\alpha\\+\mu\sin\alpha\end{pmatrix}}\right)$$

$$F_z = \frac{k_{eq}k_\theta(\cos\alpha + \mu\sin\alpha)\left(\frac{z}{\tan\alpha}\right)}{\left(k_{eq}L_g^2 + k_\theta\right)(\sin\alpha - \mu\cos\alpha) - k_{eq}L_g r(\cos\alpha + \mu\sin\alpha)}$$

$$-(2)$$

where, $(\cdot)_0$ is an initial value and α is a chamfer angle. In the one-point contact case, we derive eq. (3):

$$\theta = \frac{k_{eq}(L_g - l - \mu r)\left((\varepsilon_0 - cR) + L_g\theta_0\right) + k_\theta\theta_0}{k_{eq}(L_g - l - \mu r)(L_g - l) + k_\theta}$$

$$X = -(cR + L_g\theta - l\theta)$$

$$F_x = k_{eq}(X_0 + X) \qquad -(3)$$

$$F_z = \frac{\mu k_{eq} k_\theta\left((\varepsilon_0 - cR) + l\theta_0\right)}{k_{eq}(L_g - l - \mu r)(L_g - l) + k_\theta}$$

In the two point contact case, we derive eq. (4). We assumed that two-point contact condition does not return to a one-point contact condition during insertion. Based on this assumption, the rotation angle in the y-direction and the displacement in the x-direction are calculated from the geometric constraint of the peg and the hole.

$$\theta = \frac{2cR}{l}$$

$$X = R - r\cos\theta - L_g \sin\theta$$

$$F_x = k_{eq} L_g \left(\theta_0 - \frac{2cR}{l}\right) + k_{eq}(\varepsilon_0 + cR)$$

$$F_z = \begin{cases} \left(\frac{2\mu}{l}\right)\left((k_{eq}L_g^2 + k_\theta)\left(\theta_0 - \frac{2cR}{l}\right) + k_{eq}L_g(\varepsilon_0 + cR)\right) \\ -\mu\left(1 + \frac{\mu d}{l}\right)\left(k_{eq}L_g\left(\theta_0 - \frac{2cR}{l}\right) + \frac{k_{eq}L_g(\varepsilon_0 + cR)}{L_g}\right) \end{cases}$$

-(4)

2.2 Slip between a Gripper and a Peg

A slip between a gripper and a peg may occur when the gripping force is not sufficient during an insertion. The slip condition is governed by several parameters such as the friction coefficient, gripping force, and reaction force. When we move a gripper in the z-direction to insert a peg into a hole, the gripping force is calculated using eq. (5), where, F_{go} is an initial gripping force, F_v is an additional gripping force generated by the applied input voltage, and F_z is a reaction force in the z-direction.

$$F_g = \mu(F_{go} + k_x^g X + F_v)$$ -(5)

A slip is generated between a gripper and a peg when $F_z - F_g > 0$ is satisfied during an insertion task.

3 Simulation

Simulation was performed to investigate the behavior of a peg and forces during compliant insertion. Using the displacement and force equation derived from the previous section, we calculate the displacement and force profile along the z-direction. Because the displacement and force profiles are piecewise continuous functions, there are discontinuities at the transition of each state through the entire insertion task.

Initial values of simulation parameters are as follows: friction coefficient of silicon ($\mu = 0.3$), the length of a peg ($L_g = 400\mu m$), the radius of a peg ($r = 55\mu m$), the radius of a hole ($R = 70\mu m$), Young's modulus of silicon ($E = 160GPa$), an initial rotation error ($\theta_0 = 10°$), and an initial position error ($\varepsilon_0 = 2R - r\cos\theta_0$). The stiffness of a gripper and a peg is computed using FEA simulation and those values are $k_x^g = 55.2 N/m$, $k_z^g = 8695.7 N/m$, $k_x^p = 81.9 N/m$, and $k_\theta^g = k_x^g L_g^2$.

3.1 Deflection Analysis

Fig. 4 shows the rotation and the x-directional displacement of a peg during insertion. Two cases of simulation are performed to compare results between a rigid peg model (graphs with a subscript c) and a flexible peg model (graphs with a subscript p). A rigid peg model considers only the compliance of the gripper. With an initial rotation error ($\theta_0 = 10°$), the rotation angle of a peg becomes larger during the chamfer crossing (state I) and the one-point contact (state II) states. During the one-point contact stage, the increasing rate of a rotation angle is lower than during the chamfer crossing state as the peg begins to slip into the hole. Differences between the two models during the chamfer crossing and the one-point contact states are significantly larger than in the two-point contact state. During the two-point contact state (state III), the rotation angle rapidly decreases and approaches zero due to the constraints of the peg and the hole.

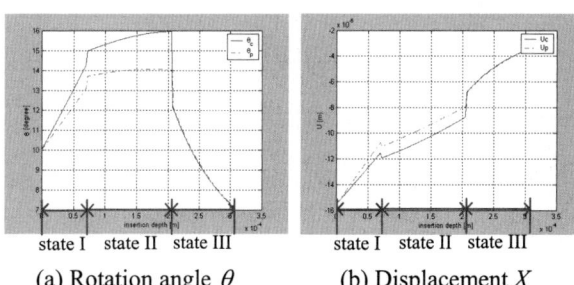

(a) Rotation angle θ (b) Displacement X

Figure 4: Deformation profile of a peg during insertion.

The x-directional displacement of the peg X moves linearly toward the hole during states I and II.

3.2 Force Analysis

The x-directional force profile shown in Fig. 5(a) has the same tendency as the displacement profile because the gripper and the peg are modeled as linearly connected springs. During the one-point contact state (state II), the z-directional force F_z is decreased as the peg begins to slip into the hole as shown in Fig. 5(b). In the transition phase between states II and III, there is a "force jump" due to an additional contact force from the inside of a hole. After the transition, F_z decreases and approaches zero as the rotation angle of the peg approaches zero. The result implies that a larger gripping force is required to successfully complete an insertion task during the two-point contact state.

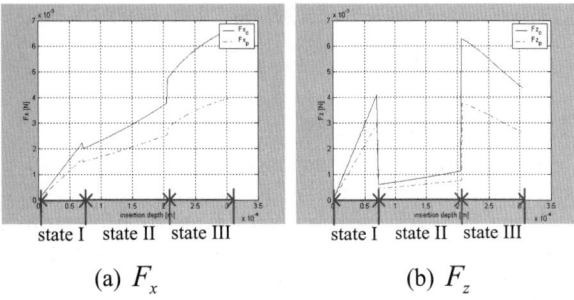

(a) F_x　　　　　(b) F_z

Figure 5: Force profiles during insertion.

Slip between the gripper and the peg can occur when F_z is greater than the gripping force F_g as shown in eq. (5). Fig. 6 illustrates slip and non-slip regions with an initial gripping force F_{go} = 9mN. Slip occurs at the end period of chamfer crossing (state I) and during the entire period of two-point contact (state III). In order to avoid slippage, a gripper should be designed to either have sufficient gripping force or an anti-slip mechanism. The result can also be used to predict whether slip occurs during insertion with some given initial position and rotation errors.

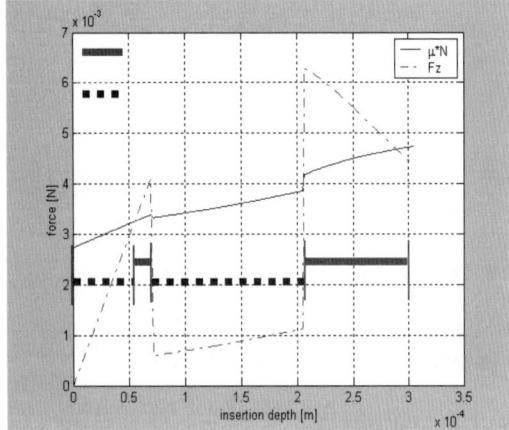

Figure 6: Slip force between gripper and peg.

4 DRIE Silicon Die Fabrication

A SOI (Silicon On Insulator) die that contains MEMS devices was fabricated using DRIE (Deep Reactive Ion Etching) process at Honeywell's MEMSPlus facility. Fig. 7 shows microgrippers, micropegs, and an array of holes on a silicon die whose size is 10×3mm². The thickness of the structural layer is 50μm, the sizes of the microgripper and the micropeg are 1200x900μm² and 300×100μm², respectively.

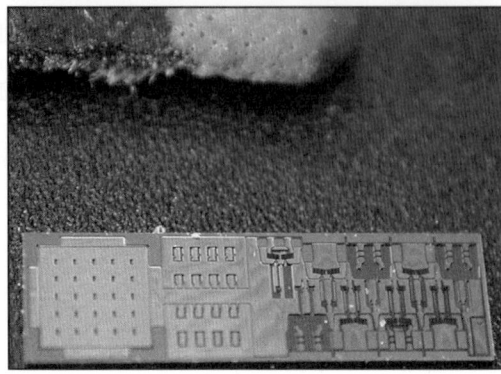

Figure 7: A DRIE fabricated silicon die with grippers, pegs and an array of hole.

Initially, microgrippers and micropegs are connected on the sidewall of the die by tether structures. A micro probe system is used to detach the grippers and the pegs out of the die by breaking tethers. The main advantage of the DRIE process is that we can build thicker structures. In practice, it is difficult to make a hole with a chamfer because the DRIE process can only make a vertical wall structure. Instead, a chamfer is made in the micropeg.

4.1 Thermal Bimorph Gripper

Thermal bimorph actuators are used in MEMS because of larger forces and deflections with lower input voltages. We obtained 20μm opening at the tip of the gripper with a 5-volt input.

4.2 Gripper Tip Design

Two different types of tips, spring and flexure, were designed to provide passive compliance to the tip of the gripper as shown in Fig. 8. In the case of the flexure structure, the center of rotation is fixed, so that it can be treated as a one-DOF revolute joint. This type of structure is stiffer than a spring structure, and hence generates less deformation, but is appropriate for precise manipulation.

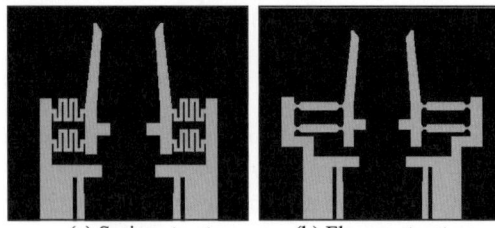

(a) Spring structure　　　(b) Flexure structure

Figure 8: Gripper tips with compliance structures.

Spring structures offer larger working ranges. The center of rotation depends on the amount of deformation and there is minor axial elongation of the spring structure.

Accordingly, it can be treated as a two-DOF joint with combined revolute and prismatic joints.

4.3 Gripper Integrated with Force Sensor

To actively compensate for misalignments between a peg and a hole during insertion, force information is needed in addition to the passive compliance of a gripper. The z-directional force is measured to monitor the insertion state. For this purpose, a comb structure is attached on the base part of a gripper as shown in Fig. 9. It measures capacitance changes caused by the deflection of the gripper. In addition, the capacitance can also be used to measure the opening of the gripper as the lower part of the gripper deflects during the opening.

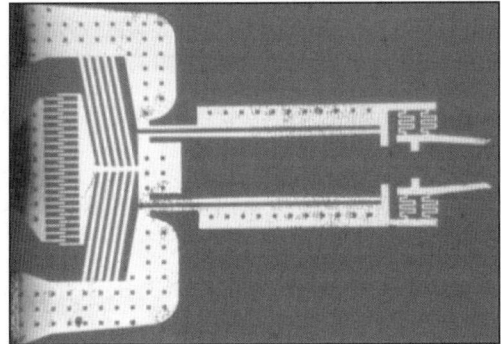

Figure 9: Gripper with a comb structure for force measurement.

5 Experiments

5.1 System Setup

Two CCD cameras, a fine motion stage, and a gripper driving circuit were set up on a vibration isolation table as shown in Fig. 10. For visualizing the peg-in-hole insertion experiment, two CCD zoom cameras are used to get top and side views. Both have 90mm working distance. One is located in front of a gripper to detect the location of a hole. The other is located on top of the gripper to detect misalignments between a gripper and a hole.

To manipulate the gripper and align it with the hole, two stages are used: 3 DOF xyz-translation stage and 3 DOF αβγ-rotation stage. The gripper is attached to a copper block that is mounted on the xyz-translation stage. A DRIE magazine chip is placed on the αβγ-rotation stage. The translation stage is used for positioning the gripper with the peg and the rotation stage is used for the rotation of the magazine chip. To operate the microgripper, an input voltage is supplied through the copper block. The copper block is designed for supplying input voltage as well as dissipating heat generated by the input voltage.

The two pads of the gripper are bonded on the copper block using conductive epoxy. The maximum operating temperature of the epoxy is around 120°C.

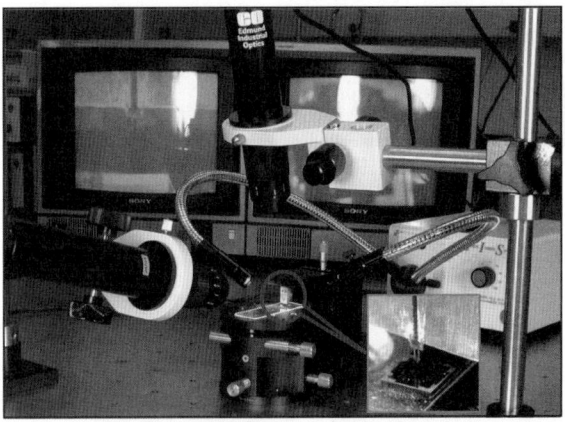

Figure 10: Insertion experimental setup

5.2 Manipulation: Pick up, Rotation, and Insertion Task

The initial position of the gripper can be within a few millimeters above the peg as shown in Fig. 11(a). Alignment of the gripper and the peg is performed using the xyz-translation and the αβγ-rotation stages.

The high temperature of the gripper (possibly around 500°C) may melt the epoxy or burn the gripper itself when a high voltage (8 volts) is supplied to open the gripper widely. Fig. 12 shows a melted silicon gripper that is covered with melted epoxy. An improved gripper design that can efficiently dissipate excessive heat is needed to avoid the problem. Due to the voltage limitation, the opening between the gripper tips was insufficient for a wide peg. It was however possible to pick up the peg because the gripper tips are designed for passive opening due to the compliance mechanism shown in Fig. 11(b).

To insert the peg in the vertical direction, it is necessary to rotate it. Since there is no rotational mechanism on the gripper, this was done using the friction force on the vertical wall of the silicon as shown in Fig. 11(c). Using xyz-translation, the gripper and peg are aligned above the hole of the DRIE magazine as shown in Fig. 11(d). Then, the rotational misalignment is adjusted by the xyz-rotation stages. Finally, the peg is inserted into the hole using the passive compliance of gripper as shown in Fig. 11(e).

Initial experiments were performed to explore a micropeg manipulation task, and further research is continuing to compare the experimental results to the analytic model.

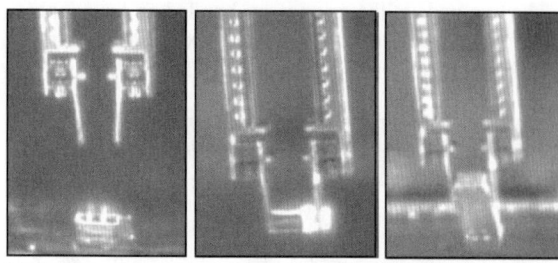

(a) Approach to a peg (b) Pick up a peg (c) Rotate a peg

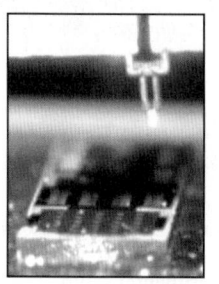 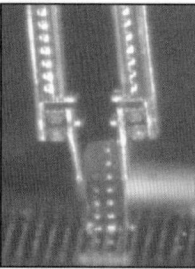

(d) Approach to hole   (e) Insert into hole

Figure 11: Peg-in-hole insertion experiment.

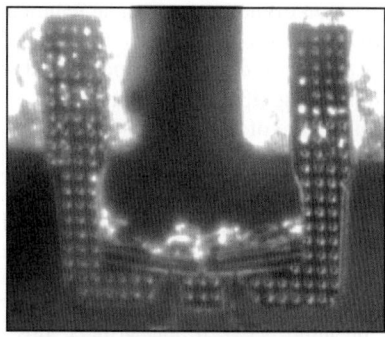

Figure 12: Gripper with melted silicon structure and epoxy caused by excessive heat at high input voltage.

6 Discussion

A microgripper with compliant tip was designed and used to demonstrate successful micropeg manipulation and insertion tasks. Benefits of a compliant gripper include misalignment compensation during micropeg manipulation and insertion, easy gripping force control due to lower stiffness, and early collision detection between the gripper and a part. Further work is needed to include scaling issues such as stiction, electrostatic force, and Van der Waals force in the analytical compliant insertion model. Additional work is also planned to optimize microgripper compliance, and develop task planning algorithms for force guided compliant insertion.

Acknowledgements

This paper is based upon work supported in part by ATP award number # 70NANB1H3021 from NIST. The authors are pleased to acknowledge the fabrication of a silicon die from Honeywell's MEMSplus.

References

[1] F. Arai and T. Fukuda, "Adhesion-type Micro Endeffector for Micromanipulation," in *Proc. of IEEE International Conference on Robotics and Automation*, pp.1472-1477, 1997.

[2] K. F. Bohringer, B. R. Donald, R. Mihalovich, and N. C. MacDonald, "Sensorless Manipulation using Massively Parallel Microfabricated Actuator Arrays," in *Proc. of IEEE International Conference on Robotics and Automation*, pp.826-833, 1998.

[3] J. H. Comtois, V. M. Bright, and M. W. Phipps, "Thermal Microactuators for Surface Micromachining Processes," in *Proc. of SPIE*, vol.2642, pp.10-21, 1995.

[4] G. Greitmann and R. A. Buser, "Tactile Microgripper for Automated Handling of Microparts," *Sensors and Actuators*, A.53, pp.410-415, 1996.

[5] M. A. Greminger, G. Yang, and B. J. Nelson, "Sensing Nanonewton Level Forces by Visually Tracking Structural Deformation," in *Proc. of International Conference on Robotics and Automation*, pp.1943-1948, 2002.

[6] D. S. Haliyo, Y. Rollot, and S. Regnier, "Manipulation of Micro-object using Adhesion Forces and Dynamical Effects," in *Proc. of International Conference on Robotics and Automation*, pp.1949-1954, 2002.

[7] M. Peshkin, "Programmed Compliance for Error Corrective Assembly," *IEEE Trans. on Robotics and Automation*, vol.6, no.4, pp.473-482, 1990.

[8] M. Shimada, J. A. Tompson, J. Yan, R. J. Wood, and R. S. Fearing, "Prototyping Millirobots using Dextrous Microassembly and Folding," in *Proc. ASME IMECE/DSCD*, vol.69-2, pp.933-940, 2000.

[9] G. D. Skidmore, M. Ellis, E. Parker, N. Sarkar, and R. Merkle, "Micro Assembly for Top Down Nanotechnology," in *Int. Symp. on Micromecha. and Human Science*, pp.3-9, 2000.

[10] D. E. Whitney, "Quasi-Static Assembly of Compliantly Supported Rigid Parts," *Journal of Dynamic Systems, Measurement, and Control*, vol.104, pp.65-77, 1982.

[11] Y. Zhou and B. Nelson, "Adhesion Force Modeling and Measurement for Micromanipulation," in *Proc. of SPIE*, vol.3519, pp.169-190, 1998.

Levitated micro-nano force sensor using diamagnetic materials

Mehdi Boukallel, Joël Abadie, Emmanuel Piat
Laboratoire d'Automatique de Besançon
UMR CNRS 6596
25, rue Alain Savary
25000 Besançon, France
E-mail : epiat@ens2m.fr, Web : www.lab.cnrs.fr

Abstract

Under suitable conditions, diamagnetic materials allow to achieve stable levitation of permanent magnets in entirely passive configuration. Using NdFeB magnets and diamagnetic materials such as graphite in a particular configuration, we build a passive levitated force sensor with a variable stiffness and linear output. The suspended part is used as the sensing device and two directions of force measurement are possible. The absence of friction makes the sensor highly sensitive and forces around nN can be measured. The established model of both magnetic and diamagnetic forces allows to calculate the applied force on the end point of the levitating device after measuring the position of the levitating part. This paper presents the description of the levitated sensor, force calculation and experimental results.

1 Introduction

Magnetic levitation is a suitable solution to design a new kind of microdevices (micromachines, micromotors, micromanipulators,...). Since microdevices are able to achieve submicron precision, mechanical friction induced when a displacement or rotation is made can severally reduce their performance. Thus, contact-free levitation in the parts where the friction is important seems to be an interesting solution.
The two basic possibilities to achieve magnetic levitation are either passive or active. The term active is used for systems with a feed back control loop used to stabilize the levitation, in opposition to passively levitated systems which do not require any control loop. In the active magnetic levitation, the use of sensors to measure the position of the suspended object is necessary (in general, one sensor per controlled DOF [1]).
It is possible to achieve passive levitation using diamagnetic materials. There are two classes of such materials : diamagnetic materials for which the relative magnetic permeability μ_r is slightly less than unity at room temperature, and materials in the superconductive state with $\mu_r=0$ at low temperature. Since scaling laws favour the use of microdevices at room temperature [2] and the necessity of maintaining a sufficiently low temperature can be troublesome, we chose to use the first class of diamagnetic materials to design the levitated force sensor. The diamagnetic materials can be used as the levitated part [3] or to stabilize levitation of a permanent magnets in the magnetic field of another one [3]. Because the first configuration needs a high magnetic field in the levitating area [3], we chose to use the second configuration, more easier to achieve, to build the suspension mechanism of the levitated force sensor.

2 Passive levitation using diamagnetic materials

According to Earnshaw theorem [4], stable free suspension of a permanent magnet in the magnetic field of another magnet is not possible. Earnshaw proved that a configuration consisting of bodies which attract or repel one another with a force proportional to the inverse square of the distance between them is unstable. The most complete theory of the possibility and the conditions for free levitation are given by Braunbeck [3]. He proved that free levitation in constant magnetic field is possible only with the use of materials with relative permeability $\mu_r < 1$ such like diamagnetic materials. This is principally due to the behaviour of diamagnetic materials in a magnetic field. When an external magnetic field is applied to a diamagnetic material, the latter becomes magnetized in the opposite direction of the applied magnetic field. For this reason, a force is produced which causes the diamagnetic material to be expelled from the magnetic field.

There are three basic configurations to achieve stable free levitation of a permanent magnets. Each of

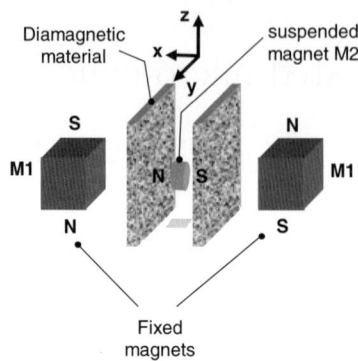

Figure 2: The suspension mechanism of the levitating sensor.

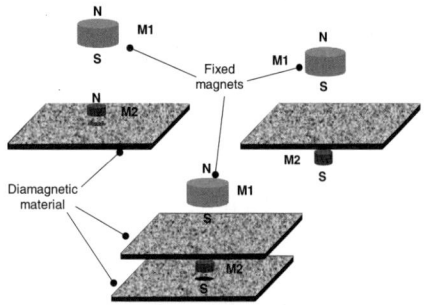

Figure 1: Basic configurations to achieve stable levitation of a permanent magnets.

them is represented in the figure 1. In the first configuration (on the left of the figure), it is possible to stabilize the equilibrium state of small magnet M_2 by placing a diamagnetic material closely below it. The latter exercises an upward force of repulsion upon M_2 which increase if $M2$ comes close to the diamagnetic material. Under some suitable conditions, any slight lowering of M_2 from the equilibrium state results in an increase in the repulsion exercised by the diamagnetic body which compensate the drop of the attraction force between M_1 and M_2. If a slight upward displacement of M_2 from the equilibrium state is made, the attractive force are less important than the weight of M_2. In the two remaining configurations, the diamagnetic levitation works by similar way as described in the upper part.

3 Description of the sensor

The suspension mechanism, called $L1$, of the sensor uses a combination of the three configurations quoted before. This new configuration is presented in the figure 2. We use two similar (material, geometry,...) magnets $M1$ placed such that their north and south poles are in the opposite direction. If a third magnet $M2$ is placed on the generated magnetic field between the two magnets $M1$ such that the attractive forces are compensated, then it appears clearly that the equilibrium state is unsta-

ble[1]. If a diamagnetic material is added to this configuration, the equilibrium state can be stabilized[2]. The suspension mechanism gives two degrees of freedom : displacement in the y and z direction.

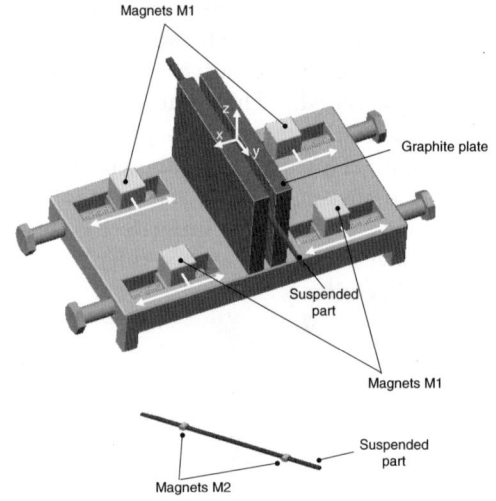

Figure 3: Diagram of the levitated sensor.

The entire configuration of the sensor is presented in figure 3. We use two suspension mechanisms $L1$ and $L2$ spaced in order to reduce the influence of each one through the other. The position of the four magnets $M1$ is independently adjustable along $\vec{x}$. Thus, it is possible to vary the stiffness of the sensor and also correct the possible

---

[1] A slight displacement of $M2$ on the left or on the right increase the magnetic forces between $M1$ and $M2$.

[2] The diamagnetic forces is already opposed to the displacement of $M2$ and then to the attractive forces produced by $M2$.

drift of the magnet property.

The magnets $M1$ are made of NdFeB with a remanent magnetic field of 1.3 T for a volume of 1 cm^3. The suspended cylindrical magnets $M2$ are made of NdFeB with a remanent magnetic field of 0.95 T. Magnetized through the thickness (cf. figure 2), the two suspended magnets (for $L1$ and $L2$) have a radius of 1.63 mm and a thickness of 2.34 mm. The mass of the suspended magnet is 34 mg. The sensing part is made with epoxy resin which is fixed to the two cylindrical magnets $M2$ (cf. figure 5). It is 10 cm in length with a variable section for a weight of 90 mg (the weight of both magnets $M2$ is included). The section of the end point of the sensing part is 0.5 mm by 0.5 mm. The diamagnetic material used is graphite with a diamagnetic susceptibility χ_m of -12 e^{-5} ($\mu_r = 1 + \chi_m$). The air gap between the two graphite plates is 3 mm. Figure 4 shows the builded sensor. We add a laser sensor with 4 mm range and 1 μm resolution behind the force sensor to measure the displacement of the levitating part in the y direction.

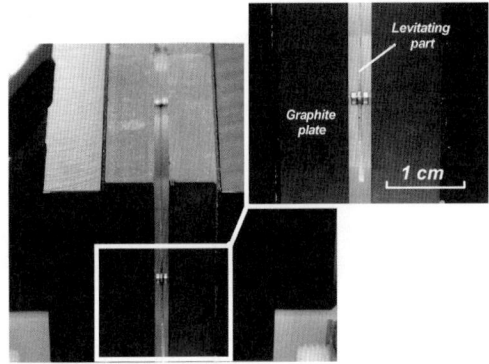

Figure 5: End point of the levitating sensor.

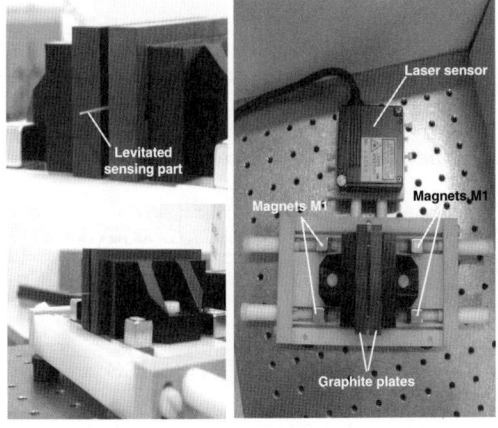

Figure 4: Levitating force sensor using diamagnetic materials.

4 Magnetic and diamagnetic forces

For the module $L1$, at the equilibrium state position five forces are applied on the suspended magnet $M2$ (cf. figure 6). The right square magnet $M1$ applies an attractive force, called $\overrightarrow{F_{att}}$, on the suspended magnet $M2$ along $\overrightarrow{x}$. Due to the symmetry of the builded prototype, the left square magnet $M1$ applies also an attractive force, called $\overrightarrow{F'_{att}}$ on the suspended magnet $M2$ but in the opposite direction. We note $\overrightarrow{F_T}$ the vectorial sum of $\overrightarrow{F_{att}}$ and $\overrightarrow{F'_{att}}$. The suspended magnet $M2$ is placed between two plates of graphite, the right and the left graphite plates apply respectively a repulsive force on $M2$, called $\overrightarrow{F_{dia}}$ and $\overrightarrow{F'_{dia}}$, in opposite direction along $\overrightarrow{x}$. These respective forces increase if $M2$ comes close to the right or the left graphite plate. The fifth force $\overrightarrow{P}$, acting along $\overrightarrow{z}$, is the half of the total weight of the suspended part. The same force description cited before is also valid for the module $L2$.

In order to evaluate both magnetic and diamagnetic forces, we define their formulation for the module $L1$.

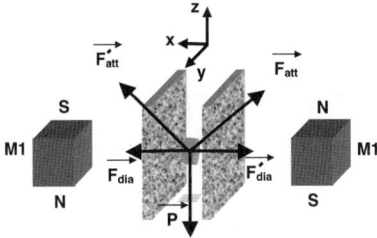

Figure 6: Forces applied at the equilibrium position.

4.1 Magnetic forces

We consider a small particle dv of the magnet $M2$ which has a magnetization $\overrightarrow{m}$. $\overrightarrow{B_1}$ denotes the magnetic field induction creating by the magnets $M1$. So, the force $\overrightarrow{df_T}$ applied by $M1$ in the volume element dv is given by :

$$\overrightarrow{df_T} = \overrightarrow{m} \cdot (\overrightarrow{\nabla} \overrightarrow{B_1}) dv \qquad (1)$$

with :

$$\overrightarrow{\nabla}\overrightarrow{B_1} = \begin{bmatrix} \frac{\partial B_{1,x}}{\partial x} & \frac{\partial B_{1,x}}{\partial y} & \frac{\partial B_{1,x}}{\partial z} \\ \frac{\partial B_{1,y}}{\partial x} & \frac{\partial B_{1,y}}{\partial y} & \frac{\partial B_{1,y}}{\partial z} \\ \frac{\partial B_{1,z}}{\partial x} & \frac{\partial B_{1,z}}{\partial y} & \frac{\partial B_{1,z}}{\partial z} \end{bmatrix} \quad (2)$$

The three components of the magnetic force are :

$$df_{Tx} = \left[m_x \frac{\partial B_{1,x}}{\partial x} + m_y \frac{\partial B_{1,y}}{\partial x} + m_z \frac{\partial B_{1,z}}{\partial x} \right] dv$$

$$df_{Ty} = \left[m_x \frac{\partial B_{1,x}}{\partial y} + m_y \frac{\partial B_{1,y}}{\partial y} + m_z \frac{\partial B_{1,z}}{\partial y} \right] dv$$

$$df_{Tz} = \left[m_x \frac{\partial B_{1,x}}{\partial z} + m_y \frac{\partial B_{1,y}}{\partial z} + m_z \frac{\partial B_{1,z}}{\partial z} \right] dv$$

(3)

Due to symmetry of the two modules $L1$ and $L2$, the suspended magnets $M2$ have no torque. Thus, their magnetization $\overrightarrow{m}$ have only a m_x component ($\overrightarrow{m} = m_x \overrightarrow{x}$). Equation (3) becomes :

$$df_{Tx} = \left[m_x \frac{\partial B_{1,x}}{\partial x} \right] dv$$

$$df_{Ty} = \left[m_x \frac{\partial B_{1,x}}{\partial y} \right] dv \quad (4)$$

$$df_{Tz} = \left[m_x \frac{\partial B_{1,x}}{\partial z} \right] dv$$

The magnetization $\overrightarrow{m}$ is constant in the entire volume V of the magnet $M2$. From (4), the force acting on the suspended magnet $M2$ can be written as :

$$F_{Tx} = m_x \iiint_V \frac{\partial B_{1,x}}{\partial x} dv$$

$$F_{Ty} = m_x \iiint_V \frac{\partial B_{1,x}}{\partial y} dv \quad (5)$$

$$F_{Tz} = m_x \iiint_V \frac{\partial B_{1,x}}{\partial z} dv$$

with,

$$\overrightarrow{F_T} = F_{Tx} \overrightarrow{x} + F_{Ty} \overrightarrow{y} + F_{Tz} \overrightarrow{z}$$

4.2 Diamagnetic forces

We consider a small particle dv' of the diamagnetic material which has an induced magnetization $\overrightarrow{m^d}$.

$\overrightarrow{B_T}$ denotes the magnetic field induction creating by both $M1$ and $M2$. So, the force $\overrightarrow{df_{dia}}$ applied by the volume element dv on $M2$ is given by :

$$\overrightarrow{df_{dia}} = \overrightarrow{m^d} \cdot (\overrightarrow{\nabla}\overrightarrow{B_2}) dv' \quad (6)$$

Due to the symmetry of the magnetic field and the *large* dimension of the diamagnetic plate, all the components df_y^d, df_z^d are compensated in the entire volume of the diamagnetic material. Therefore, only the df_x^d component acts on $M2$. Then :

$$df_x^d = \left[m_x^d \frac{\partial B_{2,x}}{\partial x} + m_y^d \frac{\partial B_{2,y}}{\partial x} + m_z^d \frac{\partial B_{2,z}}{\partial x} \right] dv'$$

(7)

The induced magnetization is $\overrightarrow{m^d} = \frac{\chi_m}{\mu_0}(\overrightarrow{B_1} + \overrightarrow{B_2})$. Equation (7) becomes :

$$df_x^d = \frac{\chi_m}{\mu_0} \left[B_{T,x} \frac{\partial B_{2,x}}{\partial x} + B_{T,y} \frac{\partial B_{2,y}}{\partial x} + B_{T,z} \frac{\partial B_{2,z}}{\partial x} \right] dv'$$

(8)

Because $\left\|\overrightarrow{B_1}\right\| \ll \left\|\overrightarrow{B_2}\right\|$, it comes :

$$df_x^d = \frac{\chi_m}{2\mu_0} \left[\frac{\partial B_{2,x}^2}{\partial x} + \frac{\partial B_{2,y}^2}{\partial x} + \frac{\partial B_{2,z}^2}{\partial x} \right] dv' \quad (9)$$

Finally, the total acting force is :

$$\overrightarrow{F_{dia}} = \left(\frac{\chi_m}{2\mu_0} \iiint_{V'} \left[\frac{\partial \left\|\overrightarrow{B_2}\right\|^2}{\partial x} \right] dv' \right) \cdot \overrightarrow{x} \quad (10)$$

5 Experimental results

All the magnetostatic simulations were made with flux 3D software. Figure 7 presents both simulated and experimental data of the magnetic flux induction for magnet $M1$ in the vertical direction ($\overrightarrow{z}$). Experimental data of the magnetic flux density are measured each 50 μm with a Tesla meter with 10 mT resolution.

Two directions of force measurement along $\overrightarrow{y}$ and $\overrightarrow{z}$ are possible with the force sensor. However, we have only studied in this paper the performance of the sensor along $\overrightarrow{y}$ and also for a fixed configuration where the spacing between magnets $M1$ in the two modules $L1$ and $L2$ is equal to 8 cm.

Figure 8 shows how it is possible to achieve force measurement along $\overrightarrow{y}$. If a force $F_{Py}.\overrightarrow{y}$ is applied on the end point of the suspended part, the new equilibrium state is obtained when $2F_{Ty} = -F_{Py}$. The calculation of F_{Ty} according to Δy allows to

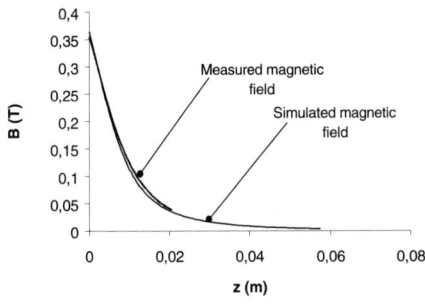

Figure 7: Simulated and experimental data of magnetic flux density.

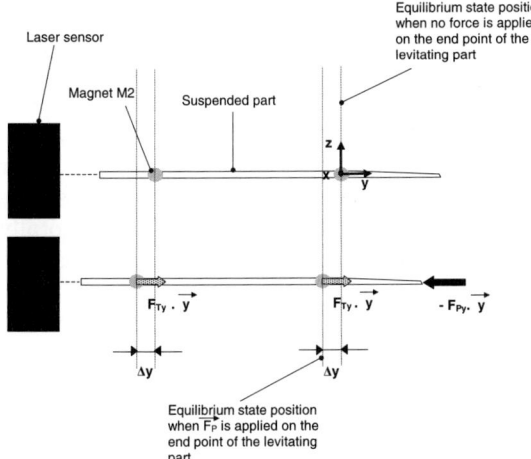

Figure 8: Force measurement along $\vec{y}$.

know the value of the applied force $F_{Py}.\vec{y}$.
Since the magnetization $\vec{m}$ is constant on the entire volume of the magnet $M2$, equations (5) show that components F_{Tx}, F_{Ty} and F_{Tz} of $\vec{F_T}$ acting on the levitating part are respectively proportional to $\frac{\partial B_{1,x}}{\partial x}, \frac{\partial B_{1,x}}{\partial y}$ and $\frac{\partial B_{1,x}}{\partial z}$.

Figure 9 shows evolution of the $B_{1,x}$ component, for $L1$, at $x = 0$ and y varying from -10 mm to 10 mm and z from -20 mm to 5 mm. In the area of levitation, $B_{1,x}$ is a parabolic curve along $\vec{y}$. Thus, its derivative form is linear along $\vec{y}$ which gives a linear output force variation F_{Ty} illustrated in figure 12. In order to obtain a good resolution, along $\vec{y}$, it is necessary to have the smallest F_{Ty} for a given displacement along $\vec{y}$. Thus (cf. equation (5)), it is necessary to have the smallest derivative form of $B_{1,x}$ along $\vec{y}$ which is the case because the magnetic field in the levitated area is quasi uniform (cf. figure 10).

Figure 10 presents the distribution of the F_{Ty}

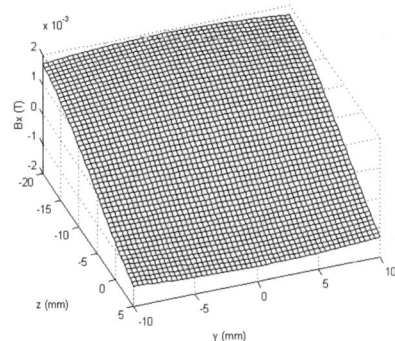

Figure 9: $B_{1,x}$ component of the magnetic flux density according to y and z for $L1$.

and F_{Tz} components of $\vec{F_T}$ for the module $L1$. The square form and the circular form on the figure represent respectively the position of the magnet $M1$ and the suspended magnet $M2$ at the equilibrium state when $\vec{F_P} = \vec{0}$. We can see that the position where the equilibrium was established is obtained for the conditions $F_{Ty} = 0$ and [3] $F_{Tz} = mg$. Since the magnetic field at the equilibrium state is quasi uniform, the variation of the F_{Ty} component of $\vec{F_T}$ is very small. The simulated data shows that the height of levitation of the suspended part is $z=2$ mm, experimental measurement made with a micrometer device lead to $z=2.05$ mm.

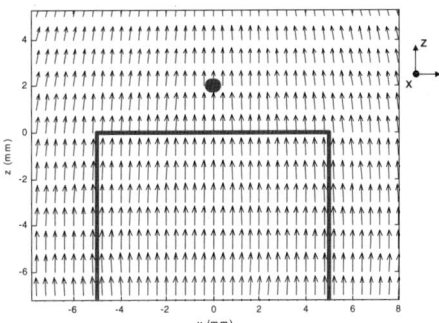

Figure 10: F_{Ty}, F_{Tz} distribution for $L1$.

In order to evaluate the force output given by the levitating sensor, figures 12 and 13 present respectively the evolution of the F_{Ty} and F_{Tz} components at the height of the levitation (z=2 mm) for $L1$.

In figure 12 we also plot the result of the linear interpolation made with Matlab software. The

---
[3] m is the half mass of the suspended part.

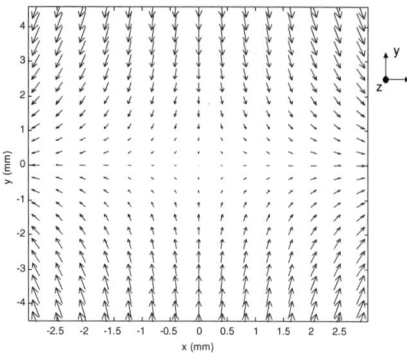

Figure 11: F_{Tx}, F_{Ty} distribution for $L1$.

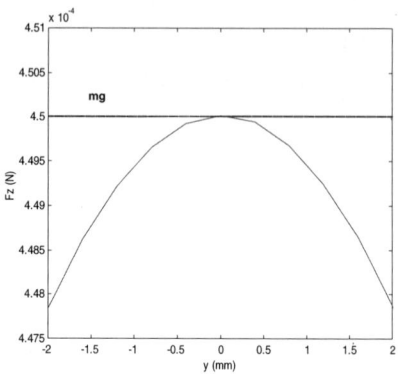

Figure 13: Evolution of F_{Tz} according to y at the height of levitation for $L1$.

measurement. The sensor uses a totally passive magnetic levitation mechanism. This is possible as the mechanism configuration uses diamagnetic materials and strong magnets in a particular configuration. Since the levitating part is used as the sensing part, the friction problem is totally suppressed which makes the sensor very sensitive. Force about 19 nN can be measured for a 1 μm displacement. Due to the configuration of the suspension mechanism a quasi uniform magnetic field is obtained in the area of the levitating part. Consequently, a linear output is observed in the range of the laser sensor.

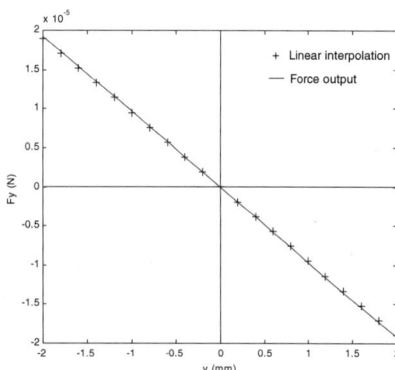

Figure 12: Evolution of F_{Ty} according to y at the height of levitation for $L1$.

stiffness of the sensor for that configuration [4] and for displacement along $\vec{y}$ is 0.019 N/m [5] in the 4 mm range of the laser sensor. In other words, if a displacement of 1 μm [6] along $\vec{y}$ is made, the simulated force is 19 nN. Figure 13 shows evolution of the F_{Tz} component, for $L1$, at the height of levitation compared to the weight of the suspended magnet. Only a small deviation of F_z is observed, the maximum deviation is 2 μN which corresponding to a variation less than 1 μm of the height of levitation.

Conclusion

In this paper we have presented a prototype of a micro-nano force sensor with two directions of

---

[4]The spacing between magnets $M1$ in the two modules $L1$ and $L2$ equal to 8 cm.

[5]This value is similar to the stiffness of Atomic Force Microscopes.

[6]This value is the actual resolution of the laser sensor used.

Acknowledgments

This research was supported by the LAB (Laboratoire d'Automatique de Besançon : LAB UMR CNRS 6596 - France - http://www.lab.cnrs.fr).

References

[1] Roland Moser, Jan Sandtner, and Hannes Bleuler. Diamagnetic suspension system for small rotors. *Journal of Micromechatronics*, 1(2):131–137, 2001.

[2] H. Bleuler. Active micro levitation. In *3rd Int. Symp. On Micromachine and Human Sci.*, pages 129–136, Nagoya - Japan, Oct. 1992.

[3] A.H.Boerdijik. Technical aspects of levitation. *Philips Research Reports*, 11:45–46, Dec. 1956.

[4] S. Earnshaw. On the nature of the molecular forces. *Trans. Cambridge phil.Soc.*, 7:97–112, 1842.

Design and Modeling of Classes of Spatial Reactionless Manipulators

Abbas Fattah, Ph.D., Research Scholar
Sunil K. Agrawal, Ph.D., Professor

Mechanical Systems Laboratory, Department of Mechanical Engineering
University of Delaware, Newark, DE 19716, U.S.A.
fattah, agrawal@me.udel.edu

Abstract – For conventional designs of robots, manipulator motions result in forces and moments on the base. These forces and moments may cause undesirable translation and rotation of the base. The objective of this paper is to systematically analyze the fundamentals of reactionless robots. Based on this analysis, designs of two distinct classes of spatial robots are proposed. The designs are achieved through appropriate choices of geometric and inertial parameters. Due to the underlying conservation laws, the trajectory must satisfy additional constraints. We illustrate the reactionless feature of these robots through computer simulations. Currently, we are fabricating reactionless robots to illustrate the underlying concepts.

1 Introduction

In the literature, a number of methods have been proposed for static balancing of machines through passive means using counterweights and springs ([6], [1]). These methods have been applied to serial and parallel mechanisms. Center of mass is an important property of a machine. At the turn of last century, the noted biomechanician Fischer investigated the use of auxiliary linkages to study the motion of center of mass of a human body [3]. In recent years, Agrawal and coworkers revised this concept and they fabricated a design where the center of mass of the system was located appropriately through the addition of parallelogram linkages [1].

Research has been reported on dynamic balancing of linkages, especially with four-bars ([4], [2]). Counterweights or idler-loops are used to minimize forces and moments transmitted to the base. Ricard and Gosselin [5] have recently applied dynamically balanced four-bar linkages as the legs of a 3-DOF planar parallel manipulator. In most research, dynamic balancing of the mechanisms is attained by proper choices of geometric and inertial parameters. The focus in this paper is to achieve the reactionless behavior by using passive joint connection between the manipulator and the base in addition to proper choices of geometric and inertial parameters.

The main contributions of this paper are: (i) systematic study of the necessary and sufficient conditions for reactionless machines; (ii) design and simulation study of classes of spatial manipulators using counterweights and auxiliary parallelograms to minimize the forces and moments transmitted to the base; (iii) attainment of zero moment transmission to the base along specified axes through appropriate choice of passive joints.

The organization of the paper is as follows: Section 2 reviews the dynamic behavior of coupled bodies in open-chain systems. Section 3 uses these dynamic equations to find the necessary and sufficient conditions for design of reactionless machines. Based on the underlying mathematics, a spatial open chain robot with counterweights is described in Section 4. A second design using auxiliary parallelograms is outlined in Section 5. Detailed mathematical models and simulation are presented for these two classes in Section 6.

2 Theoretical Background

First, we study robots in an open chain. The bodies are numbered from 1 through n as shown in Fig. 1. We consider the body i in the chain to be connected to body $i-1$ at O_i and to $i+1$ at O_{i+1}. The center of mass of the body i is labeled as C_i. At the interconnection point O_i, the force and moment applied on body i from $i-1$ are respectively labeled as $\mathbf{F}_i$ and $\mathbf{M}_i$.

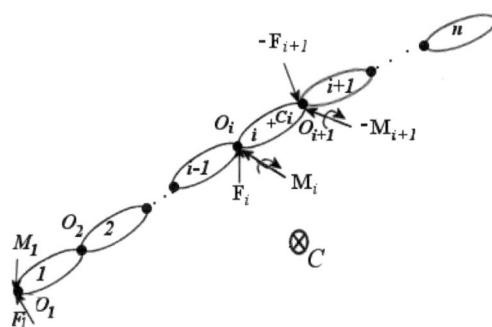

Figure 1: An open chain of bodies.

A free-body diagram of i reveals that $-\mathbf{F}_{i+1}$ and $-\mathbf{M}_{i+1}$ are respectively the force and moment applied from the body $i+1$ on i. The Newton-Euler equations of motion for body i are

$$\mathbf{F}_i - \mathbf{F}_{i+1} + m_i g \mathbf{n} = m_i \mathbf{a}_{C_i} \quad i=1,..,n, \quad (1)$$

$$\frac{d}{dt}(\mathbf{I}_{C_i} \cdot \omega_i) = \mathbf{M}_i - \mathbf{M}_{i+1} + \mathbf{r}_{C_i O_i} \times \mathbf{F}_i - \mathbf{r}_{C_i O_{i+1}} \times \mathbf{F}_{i+1} \quad i=1,..,n, \quad (2)$$

where $\mathbf{r}_{C_i O_i}$ is a vector from C_i to O_i, $\mathbf{r}_{C_i O_{i+1}}$ is a vector from C_i to O_{i+1}, m_i is the mass of body i, $\mathbf{a}_{C_i}$ is the acceleration of the center of mass C_i, g is the gravity constant, $\mathbf{n}$ is the gravity direction, $\mathbf{I}_{C_i}$ is the inertia dyadic of body i about its center of mass C_i, and ω_i is the angular velocity of body i. In this analysis, we assume that body 1 is connected to body 0 which applies force $\mathbf{F}_1$ and moment $\mathbf{M}_1$, respectively. Also, body n is the terminal body in the chain, and has no other external force and moment except for the force $\mathbf{F}_n$ and moment $\mathbf{M}_n$ applied from body $n-1$. Now, the following results are satisfied by the system:

I) The motion of the overall system is characterized by the following two equations:

$$\mathbf{F}_1 + Mg\mathbf{n} = M\mathbf{a}_C, \qquad (3)$$

$$\frac{d}{dt}\mathbf{H}_C = \mathbf{M}_1 + \mathbf{r}_{CO_1} \times \mathbf{F}_1, \qquad (4)$$

where $\mathbf{H}_C$ is the net angular momentum vector around the system center of mass C, M is the total mass of the system defined as $\sum_{i=1}^{n} m_i$ and $\mathbf{r}_{CO_1}$ is a vector from C to O_1. The expression for $\mathbf{H}_C$ is

$$\mathbf{H}_C = \sum_{i=1}^{n}[\mathbf{I}_{C_i} \cdot \omega_i + \mathbf{r}_{CC_i} \times m_i \mathbf{v}_{C_i}], \qquad (5)$$

where $\mathbf{v}_{C_i}$ is the velocity of the center of mass of C_i.

II) If O is an inertially fixed point and $\mathbf{H}_O = \sum_{i=1}^{n}[\mathbf{I}_{C_i} \cdot \omega_i + \mathbf{r}_{OC_i} \times m_i \mathbf{v}_{C_i}]$, it can be shown that $\frac{d}{dt}\mathbf{H}_O$ is given by the following expression:

$$\frac{d}{dt}\mathbf{H}_O = \mathbf{M}_1 + \mathbf{r}_{OC} \times Mg\mathbf{n} + \mathbf{r}_{OO_1} \times \mathbf{F}_1. \qquad (6)$$

3 Reactionless Machines

Eqs. (3)-(6) characterize the overall behavior of a machine connected to a base body. These equations can now be studied for the design of reactionless machines. In this section, we consider single contact systems.

For a single open-chain system, Eqs. (3)-(6) are the governing equations. In this section, we assume that the contact between the system and the ground happens either at C or O_1. These two points may be the same in some cases. One can now make the following observations for a single contact system:

1. If $\mathbf{a}_C = 0$, $\mathbf{v}_{CM}$ is a constant. Since a machine at some initial time is at rest, this constant must be zero. This condition also implies that the center of mass of the machine is inertially fixed.

Through counterbalancing, it is possible to design a machine such that its center of mass is at O_1 where it is connected to the ground. Counter balancing of the chain must start out from the last body. The center of mass of body n must be placed on the joint axis connecting n and $n-1$. On carrying out this procedure successively to other bodies in the chain, the center of mass of the whole system will get placed on the joint axis between bodies 0 and 1. As a result, the center of mass becomes inertially fixed during entire motion. An example of a spatial manipulator with this property is described in Section 4. A second design is when the center of mass is at a point different from O_1 and the ground connection of this machine is at this center of mass. Section 5 outlines an example of this class of spatial manipulators which uses auxiliary parallelograms to locate the center of mass. In both designs, $\mathbf{F}_1 = -Mg\mathbf{n}$ from Eq. (3) and the force transmitted to the ground is a non-zero constant equal to $Mg\mathbf{n}$.

2. If the center of mass of the machine is inertially fixed, the contact point $O_1 = C$. From Eq. (4),

$$\mathbf{M}_1 = \frac{d}{dt}\mathbf{H}_C \qquad (7)$$

Now, two distinct cases are possible:

(a) Through proper choice of geometric and inertial parameters, the machine is designed so that $\mathbf{H}_C = 0$ [5]. In this case, $\mathbf{M}_1 = 0$;

(b) If the connection between body 1 and O is through passive single or multiple degree-of-freedom joints, appropriate transmitted components of the moment between the system and the ground are zero. For example, if the connection at the center of mass between the system and the ground is through ball and socket joint, $\mathbf{M}_1 = 0$, i.e., there is no moment transfer between them. In this case, $\frac{d}{dt}\mathbf{H}_C = 0$. This will result in three scalar constraints to be satisfied by motion of the system. *In this case, the ground reaction is a constant and the ground moment is zero.*

4 Design of Spatial Open-Chain Manipulators

In this section, we illustrate the design procedure for a class of spatial open-chain manipulators through the example of a four link open-chain mechanism, as shown in Fig. 2(a). Joints 1 and 2 are located at point O_1 where the mechanism is connected to the ground as depicted in Fig. 2(b). Axis of joint i is along axis $\hat{\mathbf{z}}_{i-1}$. Joint 2 is passive and all other joints are actuated. Link i is denoted by $O_{i-1}O_i$ for $i = 2, 3, 4$. Therefore, the moment transmitted to the ground along this axis vanishes and this results in an equation which needs to be satisfied through proper motion planning.

For modeling purposes, an inertial frame $\mathcal{F}_0$ is located at point O_1 and consists of the unit vectors $\hat{\mathbf{x}}_0, \hat{\mathbf{y}}_0, \hat{\mathbf{z}}_0$. Similarly, a coordinate frame $\mathcal{F}_i$ is attached to link i at the point O_i with unit vectors $\hat{\mathbf{x}}_i, \hat{\mathbf{y}}_i, \hat{\mathbf{z}}_i$. The design steps are as follows:

1. Design the manipulator such that the center of mass of the whole system is located at the first joint, i.e., remains inertially fixed.

2. The angular momentum of the system is used to derive the moment transmitted to the ground.

3. Through appropriate choice of geometric and inertial parameters, certain components of angular momentum are set to zero.

4. The other remaining nonzero components of angular momentum are made to be zero through motion planning.

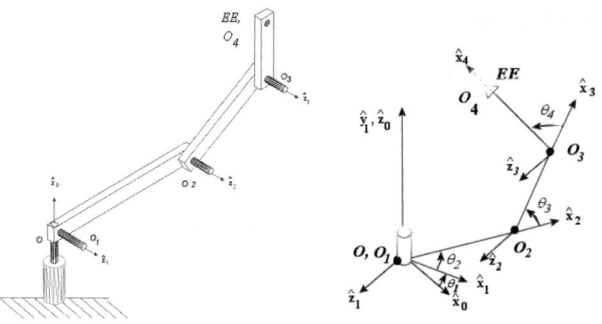

Figure 2: A four link open-chain mechanism (a) An AutoCad drawing; (b) A schematic modeling

---

4.1 Counter balancing of an Open-Chain

First, we outline a procedure to locate the center of mass of an open-chain at O_1. We start from the outermost link and add a counterweight to bring the center of mass of this link on the immediately preceeding joint axis. We repeat this process sequentially from n to 1 until the center of mass of the whole chain is at O_1. The mass of the counterweight for a link i is given by the relation

$$m_{cwi} = \sum_{j=i}^{n} m'_j + \sum_{j=i+1}^{n} m_{cwj}, \quad i=1,...,n \quad (8)$$

where m'_j is the mass of link j before adding counterweight and m_{cwj} is the counterweight for link j. The location of counterweight of link i and the center of mass of the link can be derived, respectively, as

$$l'_i = \frac{m'_i l_{ci} + \sum_{j=i+1}^{n} m_j l_i}{m_{cwi}}, \quad i=1,...,n \quad (9)$$

$$l_{*i} = \frac{-l'_i m_{cwi} + m'_i l_{ci}}{m_i} \quad i=1,..,n, \quad (10)$$

where l_{ci} is the center of mass of link i before adding counterweight to it and m_i is the mass of link i after adding counterweight, namely $m_i = m_{cwi} + m'_i$.

4.2 Angular Momentum of an Open-Chain

For the four-link manipulator, the angular momentum about the system center of mass, i.e., point O_1, can be written using Eq. (5) as

$$\begin{aligned}\mathbf{H}_{O_1} =\ & \dot{\theta}_1\{\mathbf{I}_1\hat{\mathbf{z}}_0 + \mathbf{I}_2\hat{\mathbf{z}}_0 + \mathbf{I}_3\hat{\mathbf{z}}_0 + \mathbf{I}_4\hat{\mathbf{z}}_0 + \overline{O_1C_1}\times(\hat{\mathbf{z}}_0\times\\ & m_1\overline{O_1C_1}) + \overline{O_1C_2}\times(\hat{\mathbf{z}}_0\times m_2\overline{O_1C_2}) + \overline{O_1C_3}\times\\ & (\hat{\mathbf{z}}_0\times m_3\overline{O_1C_3}) + \overline{O_1C_4}\times(\hat{\mathbf{z}}_0\times m_4\overline{O_1C_4})\} +\\ & \dot{\theta}_2\{\mathbf{I}_2\hat{\mathbf{z}}_1 + \mathbf{I}_3\hat{\mathbf{z}}_1 + \mathbf{I}_4\hat{\mathbf{z}}_1 + \overline{O_1C_2}\times(\hat{\mathbf{z}}_1\times\\ & m_2\overline{O_1C_2}) + \overline{O_1C_3}\times(\hat{\mathbf{z}}_1\times m_3\overline{O_1C_3}) + \overline{O_1C_4}\times\\ & (\hat{\mathbf{z}}_1\times m_4\overline{O_1C_4})\} + \dot{\theta}_3\{\mathbf{I}_3\hat{\mathbf{z}}_2 + \mathbf{I}_4\hat{\mathbf{z}}_2 + \overline{O_1C_3}\times\\ & (\hat{\mathbf{z}}_2\times m_3\overline{O_2C_3}) + \overline{O_1C_4}\times(\hat{\mathbf{z}}_2\times m_4\overline{O_2C_4})\} +\\ & \dot{\theta}_4\{\mathbf{I}_4\hat{\mathbf{z}}_3 + \overline{O_1C_4}\times(\hat{\mathbf{z}}_3\times m_4\overline{O_3C_4})\}\end{aligned} \quad (11)$$

where $\mathbf{I}_i$ is the inertia tensor of link i, $\overline{O_jC_i}$ is a position vector from O_j to the center of mass C_i of link i, and m_i is the mass of link i. We express this angular momentum in frame $\mathcal{F}_1$. Note that the inertia tensor $^{\mathcal{F}_i}\mathbf{I}_i$ is taken to be diagonal in the frame $\mathcal{F}_i$ with the expression $^{\mathcal{F}_i}\mathbf{I}_i = diag\begin{bmatrix}I_{i1} & I_{i2} & I_{i3}\end{bmatrix}$, where I_{i1}, I_{i2} and I_{i3} are the mass moment of inertias of link i along $\hat{\mathbf{x}}_i$, $\hat{\mathbf{y}}_i$ and $\hat{\mathbf{z}}_i$. This inertia tensor must be converted to $\mathcal{F}_1$.

Using Eq. (11), the angular momentum of the system about system center of mass $\mathbf{H}_{O_1} = \mathbf{H}_O = \mathbf{H}_C$ can be written as

$$\mathbf{H}_O = H^O_{x_1}\hat{\mathbf{x}}_1 + H^O_{y_1}\hat{\mathbf{y}}_1 + H^O_{z_1}\hat{\mathbf{z}}_1, \quad (12)$$

where $H^O_{x_1}$, $H^O_{y_1}$ and $H^O_{z_1}$ are

$$\begin{aligned}H^O_{x_1} =\ & \dot{\theta}_1[(I_{21}-I_{22}-m_2l_{*2}^2-m_3l_2^2-m_4l_2^2)s_2c_2 +\\ & (I_{31}-I_{32}-m_3l_{*3}^2-m_4l_3^2)s_{23}c_{23} + (I_{41}-I_{42})\\ & s_{234}c_{234} + (-m_3l_2l_{*3}-m_4l_2l_3)s_{223}]\end{aligned} \quad (13)$$

$$\begin{aligned}H^O_{y_1} =\ & \dot{\theta}_1[I_{12} + I_{21}s_2^2 + I_{22}c_2^2 + I_{31}s_{23}^2 + I_{32}c_{23}^2 +\\ & I_{41}s_{234}^2 + I_{42}c_{234}^2 + m_2l_{*2}^2c_2^2 + m_3(l_2c_2 +\\ & l_{*3}c_{23})^2 + m_4(l_2c_2 + l_3c_{23})^2]\end{aligned} \quad (14)$$

$$\begin{aligned}H^O_{z_1} =\ & \dot{\theta}_2[I_{23} + I_{33} + I_{43} + m_2l_{*2}^2 + m_3l_2^2 + m_3l_{*3}^2 +\\ & 2m_3l_2l_{*3}c_3 + m_4l_2^2 + m_4l_3^2 + 2m_4l_2l_3c_3] +\\ & \dot{\theta}_3[I_{33} + I_{43} + m_3l_2l_{*3}c_3 + m_3l_{*3}^2 +\\ & m_4l_3^2 + m_4l_2l_3c_3] + \dot{\theta}_4I_{43}.\end{aligned} \quad (15)$$

Here, s_{ijk} and c_{ijk} stand for $\sin(\theta_i+\theta_j+\theta_k)$ and $\cos(\theta_i+\theta_j+\theta_k)$, respectively. Also, the links are considered to be slender such that the center of mass of a link i is located at a distance l_{*i} from its joint axis, $i=2,3,4$, as defined in Eq. (10).

Using Eq. (7), the total moment transmitted to the base expressed in frame $\mathcal{F}_1$ is

$$\mathbf{M}_1 = \frac{^{\mathcal{F}_0}d}{dt}\mathbf{H}_{O_1} = M_{x_1}\hat{\mathbf{x}}_1 + M_{y_1}\hat{\mathbf{y}}_1 + M_{z_1}\hat{\mathbf{z}}_1 \quad (16)$$

where $M_{x_1} = (\frac{d}{dt}H^O_{x_1} + H^O_{z_1}\dot{\theta}_1)$, $M_{y_1} = \frac{d}{dt}H^O_{y_1}$, $M_{z_1} = (\frac{d}{dt}H^O_{z_1} - H^O_{x_1}\dot{\theta}_1)$.

The moment M_{z_1} is zero because joint 2 is passive and revolute. If $H^O_{x_1}$ and $H^O_{z_1}$ can be made to be zero, both M_{x_1} and M_{z_1} will vanish. However, it is impractical to make these terms identically zero. Hence, we adopt a two step procedure: (i) Select the geometric and inertial parameters such that $H^O_{x_1} = 0$, (ii) Choose the motion of the joints such that $H^O_{z_1} = 0$. Therefore, the only moment transmitted to the ground is M_{y_1} exerted by the actuator at joint 1. From Eq.(13), the following conditions are required to vanish $H^O_{x_1}$:

$$I_{21} - I_{22} - m_2l_{*2}^2 - m_3l_2^2 - m_4l_2^2 = 0 \quad (17)$$

$$I_{31} - I_{32} - m_3l_{*3}^2 - m_4l_3^2 = 0 \quad (18)$$

$$I_{41} - I_{42} = 0 \quad (19)$$

$$-m_3l_2l_{*3} - m_4l_2l_3 = 0 \quad (20)$$

The first three equations can be satisfied by using proper design of the links. To satisfy the equations, one could use offset length for counterweight on each link. Eq. (20) is satisfied through the procedure in Section 4.1.

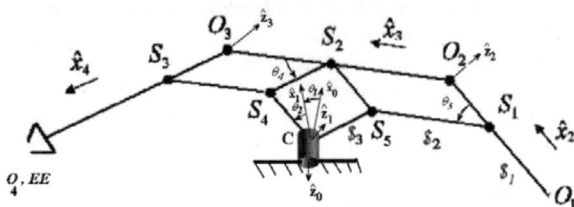

Figure 3: A spatial manipulator with auxiliary parallelograms

4.3 Planning of Open-Chain Manipulator

This section explains how planning is used to vanish $H_{z_1}^O$. Upon substitution of Eqs. (17)-(20) in Eq. (15), it simplifies to

$$A\dot\theta_2 + D\dot\theta_3 + F\dot\theta_4 = 0 \quad (21)$$

where A, D and F are

$$\begin{aligned} A &= I_{23} + I_{33} + I_{43} + m_2 l_{*2}^2 + m_3 l_2^2 + \\ &\quad m_3 l_{*3}^2 + m_4 l_2^2 + m_4 l_3^2 \quad (22)\\ D &= I_{33} + I_{43} + m_3 l_2 l_{*3} c_3 + m_3 l_{*3}^2 + m_4 l_3^2 \quad (23)\\ F &= I_{43} \quad (24) \end{aligned}$$

Eq.(21) is an integrable rate constraint on the motion of the last three joints of the system. If joints 3 and 4 are actively driven, the motion of joint 2 is

$$\dot\theta_2 = -\frac{D}{A}\dot\theta_3 - \frac{F}{A}\dot\theta_4 \quad \text{or} \quad \theta_2 = -\frac{D}{A}\theta_3 - \frac{F}{A}\theta_4 + C \quad (25)$$

where C is a constant of integration. Motion planning to satisfy this constraint is outlined in Section 6. In summary, with these choices of parameters and trajectory, the moment transmitted to the base has only one component, i.e., $M_{y1} = M_{z0}$. It is possible to extend this method for spatial manipulators with more than four links.

5 Manipulator with Auxiliary Parallelograms

In this section, we illustrate the design with a class of spatial manipulators with auxiliary parallelograms that locate the center of mass, as shown in Fig. 3 [1]. The center of mass is then used as the connection point for the system to the inertial frame by a revolute joint. The center of mass of the manipulator is located using a set of auxiliary parallelograms, as shown in Fig. 3. A set of scaled lengths $\$_i$ are computed from the description of the mechanism to construct the parallelograms and locate the center of mass of the whole system. Hence, the center of mass C becomes the inertially fixed point.

In this design, we consider three links with three auxiliary parallelograms which are connected to joint 1 at C as shown in Fig. 3. Joints 1 and 2 are located at C. The axes of joints 2, 3 and 4 are parallel to each other and are along $\hat{\mathbf{z}}_1$. The axis of joint 1 is along $\hat{\mathbf{z}}_0$ and is perpendicular to the axes of all three joints. An inertial frame $\mathcal{F}_0$ is located at C consisting of unit vectors $\hat{\mathbf{x}}_0$, $\hat{\mathbf{y}}_0$, $\hat{\mathbf{z}}_0$. A coordinate frame $\mathcal{F}_i$ is at point O_i with unit vectors $\hat{\mathbf{x}}_i$, $\hat{\mathbf{y}}_i$ and $\hat{\mathbf{z}}_i$. Link i is denoted by $O_i O_{i+1}$ for $i = 1, 2, 3$. Auxiliary links are defined as: $\bar{l}_{1a} = 1\bar{a} = S_5 S_2 = C S_4$, $l_{2a} = 2a = S_1 S_5$, $\bar{l}_{2a} = 2\bar{a} = S_4 S_3$ and $l_{3a} = 3a = S_5 C = S_2 S_4$. The joint 2 at C is passive and all other joints are actuated. Therefore, the moment transmitted to the ground along axis z_1 vanishes, i.e., $M_{z1}^C = 0$, similar to the observation in Section 3. The angular momentum of the whole system is derived about C in order to obtain the moment transmitted to the ground. The inertia and geometric parameters as well as trajectories for this manipulator are designed using procedures similar to Section 4.

5.1 Angular Momentum of Manipulator

Upon writing the total angular momentum for the system about C and simplifying the equation thus obtained, it can be shown that the system angular momentum about the system center of mass $\mathbf{H}_C$ expressed in frame $\mathcal{F}_1$ is

$$\mathbf{H}_C = H_{x_1}^C \hat{\mathbf{x}}_1 + H_{y_1}^C \hat{\mathbf{y}}_1 + H_{z_1}^C \hat{\mathbf{z}}_1 \quad (26)$$

where $H_{x_1}^C$, $H_{y_1}^C$, $H_{z_1}^C$ are

$$H_{x_1}^C = \dot\theta_1(A_1 s_2 c_2 + A_2 s_{23} c_{23} + A_3 s_{234} c_{234} - A_4 s_{223} - A_5 s_{2234} - A_6 s_{23234}) \quad (27)$$

$$\begin{aligned} H_{y_1}^C &= \dot\theta_1[(I_{11} + 2\bar{I}_{1a,1})s_2^2 + (I_{12} + 2\bar{I}_{1a,2})c_2^2 + \\ &\quad (I_{21} + I_{2a,1} + \bar{I}_{2a,1})s_{23}^2 + (I_{22} + I_{2a,2} + \bar{I}_{2a,2})c_{23}^2 + \\ &\quad (I_{31} + 2I_{3a,1})s_{234}^2 + (I_{32} + 2I_{3a,2})c_{234}^2 + m_1(-l_{*1}c_2 + \\ &\quad \$_2 c_2 + \$_2 c_{23} + \$_3 c_{234})^2 + m_2(-l_1 c_2 + \$_1 c_2 - l_{*2} c_{23} \\ &\quad + \$_2 c_{23} + \$_3 c_{234})^2 + m_3(l_1 c_2 - \$_1 c_2 + l_2 c_{23} - \\ &\quad \$_2 c_{23} + 2l_{*3} c_{234} - \$_3 c_{234})^2 + \overline{m}_{1a} \bar{l}_{1*a}^2 c_2^2 + \\ &\quad \overline{m}_{1a} \$_3^2 c_{234}^2 - 2\overline{m}_{1a} \bar{l}_{1*a} \$_3 c_2 c_{234} + m_{2a}(l_{2*a} c_{23} - \\ &\quad \$_2 c_{23} - \$_3 c_{234})^2 + \overline{m}_{2a}(l_1 c_2 - s_1 c_2 + \bar{l}_{2*a} c_{23})^2 \\ &\quad + m_{3a}(l_1 c_2 - \$_1 c_2 + l_{3*a} c_{234} - \$_3 c_{234})^2 + \\ &\quad m_{3a}(l_{3*a} - \$_3)^2 c_{234}^2] \quad (28) \end{aligned}$$

$$\begin{aligned} H_{z_1}^C &= (A_7 + 2A_4 c_3 + 2A_5 c_{34} + 2A_6 c_4)\dot\theta_2 + \\ &\quad (A_9 + A_4 c_3 + A_5 c_{34} + 2A_6 c_4)\dot\theta_3 \\ &\quad + (A_{10} + A_6 c_4 + A_5 c_{34})\dot\theta_4 \quad (29) \end{aligned}$$

where a symbol such as $I_{2a,1}$ represents the moment of inertia of auxiliary link $2a$ along $\hat{\mathbf{x}}_3$, $\bar{I}_{1a,2}$ is the moment of inertia of auxiliary link $1\bar{a}$ along $\hat{\mathbf{y}}_2$, symbols shown in Fig. 3.

Also, the links are considered to be slender such that the center of mass of a link i is located at a distance l_{*i} from its joint axis, $i = 1, 2, 3$. Moreover, the center of masses of auxiliary links $i\bar{a}$ is located at $\bar{l}_{i*a}$, $i = 1, 2$, and the center of mass of auxiliary link ia, $i = 2, 3$, is located at l_{i*a}. All coefficients in the above equations are functions of geometric and inertial parameters of the links.

Using Eq. (7), one can derive the moment transmitted to the base given by

$$\mathbf{M}_1^C = \frac{\mathcal{F}_0 d}{dt}\mathbf{H}_C = M_{x_1}^C \hat{\mathbf{x}}_1 + M_{y_1}^C \hat{\mathbf{y}}_1 + M_{z_1}^C \hat{\mathbf{z}}_1 \quad (30)$$

where $M^C_{x_1} = (\frac{d}{dt}H^C_{x_1} + H^C_{z_1}\dot{\theta}_1)$, $M^C_{y_1} = \frac{d}{dt}H^C_{y_1}$, $M^C_{z_1} = (\frac{d}{dt}H^C_{z_1} - H^C_{x_1}\dot{\theta}_1)$.

The moment $M^C_{z_1}$ is zero because of passive revolute joint at C. If $H^C_{x_1}$ and $H^C_{z_1}$ can be made to be zero, both $M^C_{x_1}$ and $M^C_{z_1}$ will vanish. However, it is impractical to make these terms identically zero. Hence, we adopt a two step procedure similar to the open-chain case: (i) Select the geometric and inertial parameters such that $H^C_{x_1} = 0$, (ii) Choose the motion of the joints such that $H^C_{z_1} = 0$. Therefore, the only moment transmitted to the ground is $M^C_{y_1}$ exerted by the actuator at joint 1.

5.2 Design and Planning of Manipulator

It can be readily determined from Eq.(27) that the conditions required to vanish $H^C_{x_1}$ are

$$A_1 = A_2 = A_3 = A_4 = A_5 = A_6 = 0 \quad (31)$$

We choose the geometric and inertial parameters of the links such that Eqs. (31) are satisfied. Since there are many unknowns and six equations to be set to zero, one uses approximate solution by using these geometric and inertial parameters. Upon solving these equations, it turns out that all equations are not exactly equal to zero and there are some errors in final results. However, we may decrease these errors by changing the geometric and inertia parameters.

Next, upon substitution of Eq. (31) in Eq. (29), one obtains the following equation for vanishing $H^C_{z_1}$

$$A_7 \dot{\theta}_2 + A_9 \dot{\theta}_3 + A_{10} \dot{\theta}_4 = 0 \quad (32)$$

Eq.(32) is a holonomic rate constraint equation on the motion of the last three joints of the parallelogram system. If joints 3 and 4 are actively driven, the motion of joint 2 is given as

$$\dot{\theta}_2 = -\frac{A_9}{A_7}\dot{\theta}_3 - \frac{A_{10}}{A_7}\dot{\theta}_4 \quad \text{or} \quad \theta_2 = -\frac{A_9}{A_7}\theta_3 - \frac{A_{10}}{A_7}\theta_4 + C_1, \quad (33)$$

where C_1 is a constant of integration. In summary, the moment transmitted to the base has only one component $M^C_{y_1}$.

6 Numerical Examples

This section describes numerical examples for models with open and closed chain. The task is to drive the manipulators from given positions and orientations of the end-effector to a desired final one. Note that only those end-effector orientations are allowed which can be obtained by rotation of the plane. Even though any motion can be chosen for the end-effector, we choose cycloidal motion for the joints. The moment transmitted to the base are computed using the procedure explained in Sections 4 and 5.

The two designs of spatial 4-link manipulators satisfy the holonomic Eqs. (25) and (33) respectively. Hence, only three out of four joints can be actively driven and the second joint is considered passive. The motion of first joint is independent of the other three joints. The actuated joints are assumed to have the following cycloidal motion

$$\theta_i = \theta_{i0} + (\theta_{if} - \theta_{i0}) * s \quad i = 1, 3, 4 \quad (34)$$

where θ_{i0} and θ_{if} are the initial and final values for θ_i and s is

$$s = \frac{1}{2\pi}[\frac{2\pi}{t_f}t - \sin(\frac{2\pi t}{t_f})]. \quad (35)$$

t_f is the final time for the motion. Using Eqs. (25) and (33), θ_2 is written for two models in terms of θ_3 and θ_4. The boundary conditions are chosen such that the constants C and C_1 in Eqs.(25) and (33) vanish. The motion of end-effector with respect to inertial frame can be written for open-chain manipulator as

$$\begin{aligned} x_e &= c_1(l_2c_2 + l_3c_{23} + l_4c_{234}) \\ y_e &= s_1(l_2c_2 + l_3c_{23} + l_4c_{234}) \\ z_e &= l_2s_2 + l_3s_{23} + l_4s_{234} \end{aligned} \quad (36)$$

and for the manipulator with auxiliary parallelograms as

$$\begin{aligned} x_e &= c_1(\bar{l}_{1a}c_2 + \bar{l}_{2a}c_{23} + \bar{l}_{3a}c_{234}) \\ y_e &= s_1(\bar{l}_{1a}c_2 + \bar{l}_{2a}c_{23} + \bar{l}_{3a}c_{234}) \\ z_e &= \bar{l}_{1a}s_2 + \bar{l}_{2a}s_{23} + \bar{l}_{3a}s_{234} \end{aligned} \quad (37)$$

Using the above motion, one can also compute the time history of moments transmitted to the base using Eqs. (16) and (30) respectively.

As an example, the initial and final joint angles in Eqs. (34) are chosen as: $\theta_{10} = -\pi/2, \theta_{1f} = \pi/2, \theta_{30} = 0, \theta_{3f} = 2\pi/3, \theta_{40} = 0, \theta_{4f} = \pi/2$ and the time period is $t_f = 5s$.

6.1 Open chain manipulator

The lengths and masses of the links are chosen as $l_2 = 0.2628$, $l_3 = 0.3329$, $l_4 = 0.1402$ $m'_2 = 0.540 l_2$, $m'_3 = 0.540 l_3$ and $m'_4 = 0.540 l_4$, respectively. All lengths and masses are expressed in MKS units. Using the counterbalanced design as mentioned in Section 4.1, one can obtain the detailed design of the manipulator as: m_2= .9461880, m_3=.5109480, m_4=.1514160, m_{cw2}=.8042760, m_{cw3}=.3311820, m_{cw4}=.0757080, l'_2=.2396148785, l'_3=.2425507337, l'_4=.07010000000, l''_2=.3440277073, l''_3=.3599308832, l''_4=.1070795187. Upon substitution of the joint trajectories from Eqs. (36) in Eq. (16), the time history of moments transmitted to base are shown in Figs. 4. The time history of end-effector motion with respect to inertial frame is computed using Eqs. (36) and are shown in Fig. 5.

6.2 Manipulator with Auxiliary Parallelograms

The details of the geometric and inertial parameters for the manipulator with augmented parallelograms are: $l_1 = 0.2628$, $l_2 = 0.3329$, $l_3 = 0.1402$, $m_1 = 0.247023$, $m_2 = 0.53396$, $m_3 = 0.512472$, $m_{2a} = 0.113893$, $m_{3a} = 0.0892$, $\overline{m}_{1a} = 0.0369$, $\overline{m}_{2a} = 0.359926$, $\$_1 = .2445594804$, $\$_2 = .2087011776$ and $\$_3 = .02342699352$. All lengths and masses are expressed in MKS units. The time history of moments transmitted to the base are derived by inserting Eqs. (37) into Eq. (30). The results are shown in Fig. 6. As shown, the moments transmitted to base in $\hat{\mathbf{x}}_1$ and $\hat{\mathbf{z}}_1$ directions are very small but not equal to zero because of the approximate solution that we have

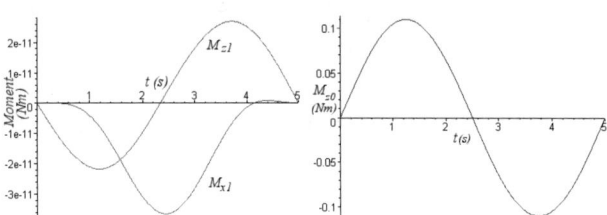

Figure 4: Moments M_{x1}, M_{z1} and M_{z0} for the designed open chain manipulator

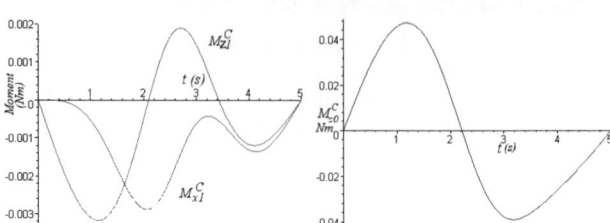

Figure 6: Moments M_{x1}^C, M_{z1}^C and M_{z0}^C for manipulator with auxiliary parallelograms

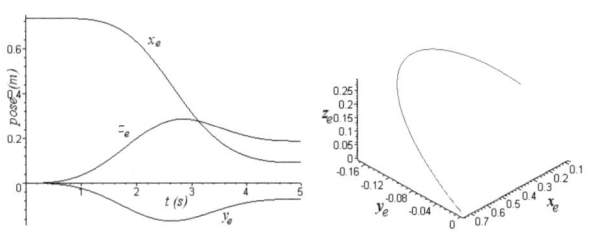

Figure 5: End-effector trajectory in the inertial frame for open chain manipulator

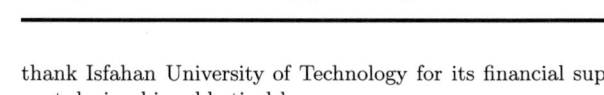

Figure 7: End-effector trajectory in the inertial frame for manipulator with auxiliary parallelograms

used to satisfy Eq. (31) in Section 5.2. it may be noted that it is possible to obtain better result using different geometric and inertia parameters. Fig. 7 shows the time history of end-effector motion in the inertial frame using Eqs. (37).

7 Conclusion

The paper provided a systematic study of necessary and sufficient conditions for design of classes of reactionless spatial robots. Detailed design and simulation were presented for spatial robots with open chains using counterweights and auxiliary parallelograms. It was shown that these designs do not transmit any extra force to the base besides the gravity force. Also, through appropriate choice of geometric and inertial parameters and motion planning, the transmitted moment to the base has only one component in the direction of the actuated joint while the other two components of the moment identically vanish. In conventional designs of robots, manipulator motions result in forces and moments on the base, while the proposed methodology yields designs which will not cause undesirable reaction forces or moments to the base. We believe that these robots will be specially valuable in space robotics since manipulator motions will not create disturbances on the base thereby keeping the position and attitude of the base constant during motion.

Acknowledgments: The authors thank National Science Foundation for support of this work through 'Presidential Faculty Fellows' program. The first author would also like to thank Isfahan University of Technology for its financial support during his sabbatical leave.

References

[1] Agrawal, S. K., Gardner, G., and Pledgie, S., "Design and Fabrication of a Gravity Balanced Planar Mechanism Using Auxiliary Parallelograms", *Journal of Mechanical Design, Transactions of the ASME*, Vol. 123, No. 4, 2001, 525-528.

[2] Arakelian, V. H. and Smith, M. R., "Complete Shaking Force and Shaking Moment Balancing of Linkages", *Mechanisms and Machine Theory*, Vol. 34, 1999, 1141-1153.

[3] Auerbach, F. and Hort, W., *Handbuch der Physikalischen und Technischen Mekanik*, J. A. Barth, Leipzig, 1928, 380-381 (reference to Fischer's work).

[4] Berkof, R. S., Lowen, G. G., "A New Method for Completely Force Balancing Simple Linkages", *Journal of Engineering for Industry*, Vol. 91, 1969, No. 1, pp.21-26.

[5] Ricard, R., Gosselin, C., M., "On the Development of Reactionless Parallel Manipulators", In Proceedings, ASME Design Engineering Technical Conferences, MECH-14098, 2000.

[6] Wang, J. and Gosselin, C., "Static Balancing of Spatial three degrees-of-freedom Parallel Mechanisms", *Mechanisms and Machine Theory*, Vol. 35, 1999, 437-452.

A Comparison of the Oxford and Manus Intelligent Hand prostheses

P.J.Kyberd[1], J.L.Pons[2]

Cybernetics Department[1],
Reading University,
Reading RG6 6AX (U.K.)

Instituto de Automática Industrial[2],
Consejo Superior de Investigaciones Científicas,
Arganda del Rey, Madrid 28500 (Spain)

Abstract

Historically, commercial hand prosthesis have adopted a low level of innovation mainly due to the strict conditions such a system must undergo. The difficult feedback to the prosthesis user has limited the functional range of commercial systems. Nevertheless, the use of advanced sensors in combination with performing hand mechanisms and microcontrollers could lead to more natural and functional prototypes. The Oxford and Manus intelligent hand prostheses are examples of innovative approaches. This paper compares and contrasts the technological solutions implemented in both systems to address the design conditions.

Keywords

Dextrous manipulation, artificial hands, Prosthetics, EMG control

1 Introduction

The design of dextrous hands is an inter-disciplinary field, in which, aspects related to hand kinematics and design, sensor integration, power and efficient actuator design, hand control and human-machine interfaces apply. Robotic dexterity has attracted a great deal of interest in recent years. Amongst the various diverse applications upper limb prosthetics is an especially challenging area. Several millions of years of evolution witness the progress of our hands to their current status. In addition, it is recognised, [1], that the processing power devoted to control the hand, as measured by the amount of motor cortex devoted to it, is as much as the legs and trunk combined. Thus it is essential when developing technical aids to study the human hand and its actions.

One of the first works in which a description of the taxonomy of human hands is given is due to *Schlesinger*, [2]. In this study, human grasping capabilitiy was classified according to grasped object size and shape into cylindrical grasp, precision grasp, hook prehension, tip grasp, spherical grasp and lateral hip,

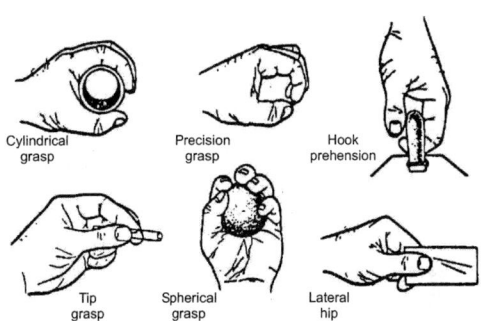

Figure 1: Basic grasp primitives according to Schlesinger: cylindrical grasp, precision grasp, hook prehension, tip grasp, shpherical grasp and lateral grasp

see figure 1. It has been stated that 90% of all common dextrous actions are covered with these six basic grasp primitives. Later, *Napier*, [3], introduced a functional classification of grasp dividing the categories according to the general action it was associated with, thus precision and power, and *Iberall*, [4], chose a classification that reflected the hand's posture which corresponds to the precision and cylindrical grasps in figure 1. The distinctive feature of human hand as compared to other primates' hands is the ability of the thumb to easily oppose the other four fingers. The thumb in opposition enables, for instance, precision and power grasps, while in non-opposition, lateral and hook grasps are available.

All these considerations have lead to commercial non-anthropomorphic prostheses implementing basic hand pinch trying to mimic the precision and power grasps. However, for anthropomorphic prostheses, the finger kinematics are kept simple in commercial prosthesis, i.e. no interphalangeal joints are provided and thus unnatural hand closing and opening patterns are achieved. Since grasp cannot adapt to object shape and size, i.e. hand is not compliant, very large tip forces are required to reach stable grasps. Research interest in hand prosthetics has focussed on mechanical design and advanced sensor technologies leaving the human control problems untouched. One particular area of investigation is the control of grasp compliance. Slip sensors have been developed to adapt and optimize grasp force as a function of slip detection. This approach tends to mimic human hand reflexes that allows dextrous grasps through different types of sensors. As an example, *Kyberd et al.*, [5], introduced a Hall effect based force sensor for automatic manipulation with slip detection capabilities.

As far as actuator technologies are concerned, different novel alternatives have been addressed. It was already in 1982, in the framework of a workshop held at MIT, [6], where it was pointed out the current actuator technologies are the most serious, long term impediment to the development of artificial hands. During these two decades Shape Memory Alloys have been studied as a possible actuator technology, [7, 8, 9], but efficiency, reliability and response time are claimed as the most limiting factors. On the other hand, high torque ultrasonic motors were also proposed as alternative technologies, [10]. Again, efficiency and high voltages required for operation are the main drawbacks even when the choice is more balanced as compared to traditional EM motors. Eventually, the design and analysis of any hand prothesis must include reference to control methods.

Commercial prosthesis are either body or electrically powered. Usually, electrically controlled prostheses make use of EMG electrodes to establish the human-machine interface, HMI. For practical usage the number of EMG channels is limited to two, but the implementation of pattern recognition approaches can potentially lead to a much higher number of control commands, [11, 12, 13].

As a summary, it could be said that hand prostheses should provide aesthetical solutions (light, silent, natural looking) to human like hand closing patterns. Ideally, the manipulation scheme should involve the user by means of biofeedback whilst allowing autonomous grasping reflexes (anti-slippage reflex). In addition a rugged construction should be attempted for it to be used under uncontrolled environments. This paper presents the technological comparison of the so-called Oxford and Manus intelligent hand prostheses. In a first section, the mechanical structure and design are discussed and compared. In the next section, system electronics and sensor integration will be presented and eventually in section 4 we will introduce the manipulation scheme. Finally, some conclusions will be drawn and some lines of future work outlined.

2 Hand mechanism

The hand mechanism should provide the mechanical support to allow as many grasping modes as possible.

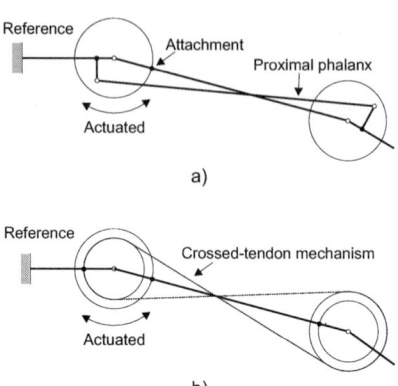

Figure 2: Detail of the interphalangeal coupling: a) rigid linkages used in the Oxford hand, b) cross-tendon mechanism implemented in the Manus hand

Aesthetics is a great concern of prothesis users, and as a consequence, the hand must be light, it should be as compact as possible, with all the driving parts within the envelope of the hand and silent.

The kinematic structure of both the Oxford and the Manus hands is based on two coupled active fingers (index and middle) plus an active thumb. In both cases, index and middle fingers are driven by a single motor and have coupled interphalangeal joints. However, whilst in the Oxford hand, the thumb motion is restricted to a single plane, i.e. flexion-extension, in the Manus approach an intermitent mechanism is used to provide thumb flexion-extension in two different planes, i.e. opposition and non-opposition.

The coupling of the interphalangeal joints is, in the case of the Oxford hand, provided by a four-bar mechanism as depicted in figure 2a. In practice, this is implemented by a rigid link from the outer face of the most proximal phalanx to the inner face of the most distal phalanx. The optimization of the four-bar mechanism dimensions provides a progressive flexion of the fingers resembling the natural hand motion.

Alternatively, the Manus mechanism is based on using a cross-tendon mechanism as schematically depicted in figure 2b. Stainless steel cables are used through pulleys to implement this mechanism. The relatively low stiffness of these cables results in a more compliant flexion motion of the fingers, however frequent maintenance is required and a tensioner mechanism was provided.

A very distinctive feature of the Oxford hand is that the index and middle fingers are linked through a whiffle-tree mechanism. It allows the hand a multipoint grip since the motion of each finger is not stopped by any contact that occurs in the other finger. The main advantage of this approach is that the load between hand and grasped object is spread on the surface

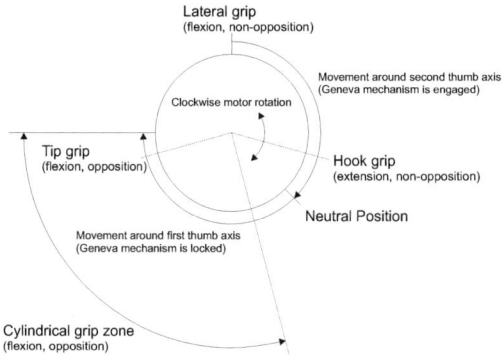

Figure 3: Schematic representation of the Geneva based thumb kinematics: grasping cycle

	Oxford	Manus
Hand Mechanism		
Precision	OK	OK
Power	OK	OK
Lateral	−	OK
Hook	−	OK
Compliant grasp	OK	−
Electronics & Manipulation scheme		
Microcontroller	LonWorks	Host: PIC 17C756 Local: PIC 16C76
Force sensors	FSR	Hall effect
Position sensors	Position servo	Hall effect
Slip sensors	Acoustic	
Hand control	1 Channel EMG	2 Channel EMG
State transition trigger	Contact & EMG	18-EMG words (Pattern recognised)
Anti-slippage reflex	OK	−

Table 1: Comparison of hand mechanism features

of the hand and, as a consequence, a lower grasping force could be used. In fact the maximum pinch force is for Oxford hand 45 N, while in the Manus hand, which is not compliant, 70 N is the pinch force. This, in turn, results in power savings and a reduced size and weight of driving mechanisms.

The anthropomorphic appearance is created by means of dummy fourth and fifth fingers. In the case of the Oxford hand these are fixed, however, in the Manus hand, these two fingers are made of a martensitic alloy that retains their shape when manually deformed. This allows for the manual adjustment by the user, without the penalty of fatigue of the finger's metal cores, which is a common failing of conventional commercial passive fingers.

An interesting feature of the Manus device is the use of a Geneva wheel based mechanism to drive the thumb in two different planes out of one single motor shaft. In this approach, the movements are transformed into a cycle. In this cycle two movement planes are defined. In the first plane, a cylindrical and tip grip can be realized, i.e. thumb is flexed in an opposition pattern. In the second plane, hook and lateral grips are possible, i.e. thumb is flexed in a non-opposition pattern. The key position in this cycle is the neutral position, at which the movement can change from one plane to the other. Figure 3a shows a scheme of the cycle. The cycle starts at the lateral grip, with the motor rotating clockwise. While the grip is opening, the thumb passes the hook grip position. At any point during this motion, the direction of rotation of the motor can reverse, so the grip closes again. If the motor continues rotating in the same direction, finally the thumb reaches the neutral position. Upon further rotation, the movement plane changes and the thumb reaches a cylindrical grip. Depending on the position of the fingers a tip grip can be reached or the cylindrical grip closes completely.

In both prostheses a natural appearance is achieved even when the Manus prototype is slightly overdimensioned, i.e. the size is approximately 20% larger than the average human hand which results in an overall weight of 1,200 grams including the socket.

3 Electronics and Sensors

Both the Oxford and the Manus hand prostheses control is based on compact microcontrollers, (see schematic in figures 4a and 4b), respectively. Oxford hand control is implemented around the LonWorks microcontroller. This can also link into a modular LonWorks bus based structure, when the hand is incorporated into a full arm [14]. Each joint has its own controller and the prosthesis also has an input processor in a manner more similar to MANUS. The Manus device uses a hierarchical structure comprising a host controller, the Microchip PIC 17C756, and two local controllers, both of them Microchip PIC 16C76. In the case of the Manus controller, the electronic hierarchical structure follows the hierarchical manipulation scheme that will be introduced in the next section.

3.1 Control electronics

The Oxford hand controller includes A/D conversion, digital input and output, processing and memory. It performs the signal acquisition from EMG analogue amplifiers, from force sensors at the finger tips and the measure of the digit flexion from the position servo. In addition, the high speed input capabilities of the chip allow to capture the slip signal made up of a stream of pulses.

The Oxford controller implements an RS2323 serial connection. This link is used both for diagnosis and for training purposes through a palmtop PC and the appropriate software [15]. The main concern addressed when designing the Oxford electronic controller was to minimize the current drain so that the hand could be operated for at least 12 hours on a single charge. The

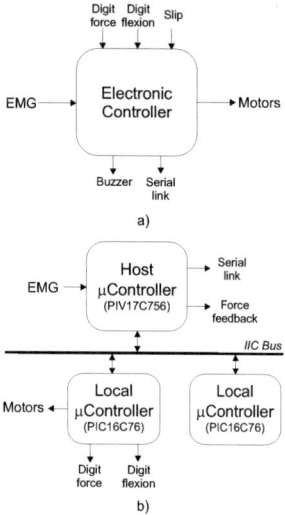

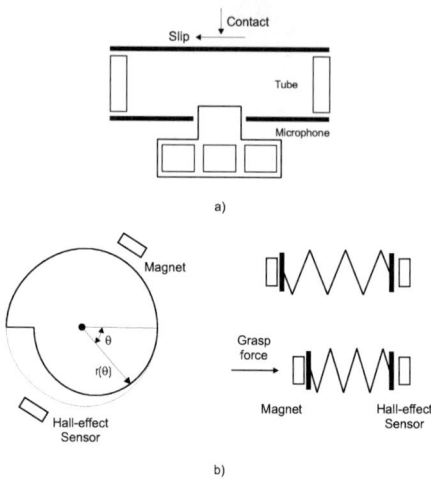

Figure 4: Schematic electronics: a) combination of a series of sensory inputs on a single controller in the Oxford hand, b) hierarchical approach implemented in the Manus hand

Figure 5: a) Schematic of the slip sensor, vibrations excited during slip are detected by the microphone, b) Schematic of the position and force sensors, change in magnetic field due to digit flexion and fingertip force are detected by Hall-effect pick-ups

PARK mode implemented in the manipulation scheme (see next section) allows the electronics to draw just 25 mA.

The Manus propotype is in a less advanced stage of development. It implements the hierarchical electronic structure depicted in figure 4b. The host controller is in charge of implementing EMG analogue acquisition, pattern recognition of EMG commands and overall control of the system. Local controllers perform low level control of each active hand axis. This involves signal acquisition from force sensors at the finger tips and from digit flexion sensors. They also implement the stiffness grasp control introduced in next section. The communication between host and local controllers is implemented via a IIC bus protocol by Philips.

The Manus electronics has built-in redundancy with greater processing power than immediately required. This gives the system additional flexibility, which was an important factor during development. In both approaches, the Oxford and the Manus hand, the motors are driven via the PWM outputs from the microcontroller and the local controllers respectively. In addition, a specific electronic driver was included in the Manus system to drive the ultrasonic motor that powers the wrist pronation-supination motion.

3.2 Sensors

The low level control of hand reflexes is implemented in the Oxford hand approach via force, position and slip sensors. The Manus approach limits the sensory information to force and position.

Several sensor technologies have been tested in both prostheses. The current Oxford hand uses Force Sensing Resistors, FSR, as force pick-ups at the fingertips and the position information is obtained from the position servo. A very interesting feature of the Oxford hand is the slip input. In this approach the slip is detected as an acoustic signal. The sliding condition between two surfaces can be detected as a low frequency vibration by a conventional microphone. The system is implemented around a rubber tube in contact with the external hand glove, see figure 5a. The vibration arising from sliding contact is mechanically connected to the air inside the tube while the mechanical impedance for external noise sources is much less matched. The resulting analogue signal is converted into a stream of pulses of 2.2 μm in width. The output from several sensors can then be summed up and the result is counted. The corresponding grasp force is made proportional to this value.

In the Manus approach, both force and position information is obtained from sensors based on Hall effect. Hall-effect pick-ups are placed opposite to small magnets. In the force sensors, the magnets are spring mounted so that a linear relationship between force and sensor output is obtained. In the position counterparts, a ferromagnetic cam is mounted on the active shaft between magnet and Hall effect pick-up. The iron to air ratio is modified according to shaft rotation, and eventually the reluctance of the magnetic circuit altered. This, in turn, results in a modified magnetic field at the pick-up location, see figure 5b.

4 Manipulation scheme

Both the Oxford and the Manus hand prosthesis are examples of EMG controlled prosthetic devices. Even when the manipulation scheme in both devices is radically different, they are both based on the same premise, namely, since it is difficult to include the user in the control loop through biofeedback the hand controller must perform autonomous coordination and control of the hand.

The control system utilized in the Oxford hand uses the Southampton Adaptive Manipulation Scheme (SAMS), [16]. In this approach, a three-level hierarchical scheme is implemented. The top level performs a supervisory control, in which simple commands are considered. The low level implements hand reflexes and individual joint control. In between, a coordination level is in charge of minimising grip force to save power consumption. This coordination is based on the compliance capabilities of the hand and looks for maximum contact area between hand and grasped object.

This control approach is based both on EMG information and contact information provided by force sensors. This way, if contact is first reached at the finger tips, a precision grip is adopted, while if contact is first established at the palm, a power grip is implemented. Two-channel EMG information obtained from opposing muscles at the forearm is used to drive the hand, a hand control state diagram can be seen in figure 6a.

Proportional POSITION control of the hand opening and closing is performed based on the tension detected at an extensor muscle. The flexor is used to switch to the TOUCH, HOLD or SQUEEZE modes. While in TOUCH mode, the controller automatically adapts grasp force to any change in orientation of the hand. Once a comfortable task orientation is reached, the flexor contraction is used to change to HOLD mode where sensors detect any possible slipping between hand and object and the force is accordingly adapted to counteract slipping. If slip detection is not desired, a second flexor contraction brings the controller to SQUEEZE mode where force is made proportional to muscle contraction. A PARK state exists in which the user can switch the batteries into a power saving status. Transitions between control states can be a consequence of EMG input or from sensors in the hand, for a detailed description of the control approach the reader is referred to [17, 16].

The state diagram for the Manus system can be seen in figure 6b. In this case, state transition is always associated to EMG input. The Manus manipulation scheme makes use of one single EMG channel, generally associated to flexors or extensors at the forearm. Each transition is made based on a EMG com-

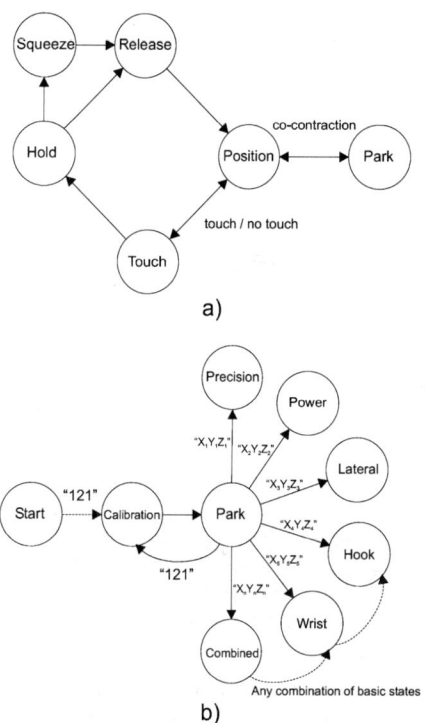

Figure 6: Control state diagram: a) Oxford approach transitions based on EMG or sensory information, b) Manus approach transitions based on coded EMG information

mand comprising three consecutive EMG bursts (muscle contractions). Two different EMG thresholds are used to establish digital levels for each EMG burst. Eventually, a command is generated with the following structure XYZ, where X, Y and Z can be 0, 1 or 2. Following this scheme, up to 18 different EMG commands can be used to triger control state transitions (all commands with $X = 0$ are rejected, since it would be not possible discerning $0YZ$ patterns from $YZ0$). A two-level hierarchical manipulation scheme is also implemented in Manus. The top level is in charge of the supervisory control of hand operation and of EMG decoding. The low level is in charge of the control of individual joints. In the Manus prosthesis a stiffness control scheme is implemented. In this control scheme, each finger is modeled as a spring whose neutral position is full flexion, the more extended the finger is the higher the exerted force on the object. Provided enough contact points are present, the resulting grasp is stable and it tends to counteract force unbalance due to orientation changes. The transition between stiffness control of grasp and hold status is trigered by a time delay since grasp command.

5 Conclusions

The constraints on the designer of a prosthesis are far more stringent than those of a robot manipulator. Not

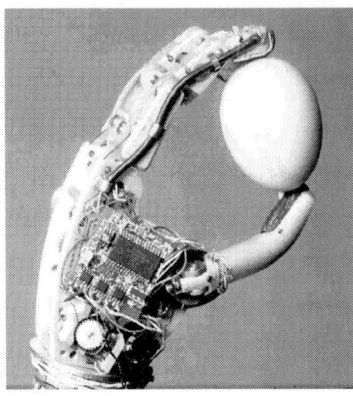

a)

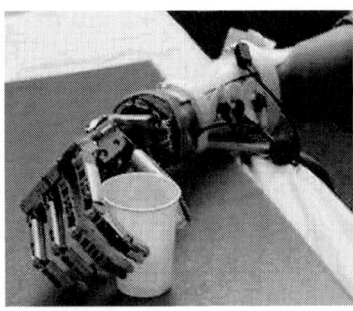

b)

Figure 7: Hands in operation: a) Oxford hand performing a precision grasp, b) Manus hand in a power grasp

merely the practical aspects of size, weight and price to make a practical system, but the environment that the hand must operate in is that which was designed around the natural solution, thus there is a tendancy for the solutions to be superficially similar.

The aim of this paper is showing current trends in the research on upper limb prosthesis through the comparison of two state-of-the-art prototypes. Therefore aspects related to mechanical design, sensors and electronics and manipulation schemes are addressed.

The concept of autonomous operation is implemented in both approaches. This highlights the inherent lack of biofeedback capabilities of EMG based manipulation schemes and points out the direction for further research.

Both systems have been subjected to clinical trials. Experimental evaluation has shown that the manipulation schemes are reasonably well accepted by the amputee. This paper discussed two devices created to fulfil a similar requirement, but have been achieved through different methods.

Acknowledgments

The authors would like to thank The Royal Society and Consejo Superior de Investigaciones Científicas that supported this research through their bilateral agreement.

References

[1] W. Penfield, G. Rasmussen, **The Cerebral Cortex of Man**, Macmillan, 1950.

[2] G. Schlesinger, **Der mechanische aufbau der knstlichen glieder**, *Borchrdt et al.* (eds.), Ersatzglieder und Arbeitshilfen fur Kriegsbeschadigte und Unfallverletzte, Springer, pp. 361-699, 1919.

[3] J.R. Napier, **The prehensile movements of the human hand**, *Journal of Bone and Joint Surgery*, 38B(4), pp. 902-913, 1956.

[4] T. Iberall, **Human prehension and dextrous robot hands**, *International Journal of Robotics Research*, 16(3), pp. 285-299, 1997.

[5] P.J. Kyberd, P.H. Chappel, **A force sensor for automatic manipulation based on the Hall effect**, *Meas. Sci. Technol.*, Vol. 4, pp. 281-287, 1993.

[6] J.M. Hollerbach, **Workshop on the design and control of dextrous hands**, *MIT-AI Memo No. 661*, 1982.

[7] J.L. Pons, H. Rodríguez, R. Ceres, W. Van Moorleghem, D. Reynaerts, **Study of SMA Actuation to Develop a Modular, User-adaptable Hand Prosthesis**, *Actuator'98*, pp. 95-104, Bremen, Germany, 1998.

[8] J-I Throndsen, **Characterisation of Shape Memory Alloys for use as an Actuator**, *Masters thesis*, Oxford University, 1996.

[9] Alciamr Soares, **Shape Memory Alloy Actuators for upper limb prosthesis**, *PhD thesis*, Edinburgh University, 1997.

[10] J.L. Pons, H. Rodríguez, A. Duarte, I. Luyckx, D. Reynaerts, R. Ceres, H. Van Brussel, **High torque ultrasonic motors for hand prosthetics: current status and trends**, *Actuator'2000*, pp. 285-288, Bremen, Germany, 2000.

[11] P. Herberts, C. Almstrm, R. Kadefors and P.D. Lawrence, **Hand Prosthesis Control via Myoelectric Patterns**, *Acta Orthopaedica scandinavia*, Vol. 44, pp. 389-409, 1973

[12] C. Almström, P. Herberts, **Clinical Applications of a Multifunctional Hand Prosthesis**, *Proceedings of the 5th International Symposium on External Control of Human Extremities*, ETAN, Dubrovnik, Yugoslavia, 1975

[13] J.L. Pons, R. Ceres, S. Levin, I. Markovitz, B. Saro, D. Reynaerts, D. Van Moorleghem, **Virtual reality training and EMG control of the MANUS hand prosthesis**, *to be published*, 2002

[14] A. Poulton, P.J. Kyberd and D. Gow, **Progress of a modular prosthetic arm**, In: *Universal Access and Assistive Technology* S. Keates, P. Langdon Eds, Springer, ISBN 1852335955, 2002.

[15] P.J.Kyberd, S.te Winkel, A.Poulton, **A Wireless Telemetry System for User Training of Artificial Arms**, *Prosthetics and Orthotics International*, Vol. 26, pp. 78-81, 2002.

[16] P.J. Kyberd, P.H. Chappel, **The Southampton hand: An intelligent myoelectric prosthesis**, *Journal of Rehabilitation Research and Development*, Vol. 31(4), pp. 326-334, 1994.

[17] P.J. Kyberd, M. Evans, S. te Winkel, **An intelligent anthropomorphic hand, with automatic grasp**, *Robotica*, Vol. 16, pp. 531-536, 1998.

Mechatronics Design and Kinematic Modelling of a Singularityless Omni-Directional Wheeled Mobile Robot

W.K. Loh   K.H. Low

School of Mechanical and Production Engineering
Nanyang Technological University
50 Nanyang Avenue, Singapore 639798
kenloh@pmail.ntu.edu.sg, mkhlow@ntu.edu.sg

Y.P. Leow

Singapore Institute of Manufacturing Technology
71 Nanyang Drive
Singapore 638075
ypleow@SIMTech.a-star.edu.sg

Abstract

This paper focuses on the kinematic modelling, mobility analysis and mechatronics design of a singularityless omni-directional wheeled mobile robot (OWMR). To achieve a singularityless motion, firstly, the kinematic model of a wheeled mobile robot (WMR), a platform equipped with three omni-directional wheels (ODW), is formulated. Based on this model, a smooth trajectory and control command for the 3-wheel OWMR can be obtained. The analysis on the mobility of a WMR is carried out using the functional matrix. It is shown that a WMR with three ODWs has a mobility of three. A WMR with three ODWs was designed and prototyped. Experimental results have been employed to verify the theoretical conclusions.

1 Introduction

Omni-directional drive is one of the hottest features wanted by most mobile robot builders. Conventional WMRs are restricted in their motion because they cannot move sideways without a preliminary manoeuvring [1]. Large variety of mechanisms have been developed in order to improve the manoeuvrability and steering accuracy of WMR. A design that uses three steerable drive centred wheels [2, 3, 4, 5, 6] has been documented. The wheels are controlled by two actuators, one for locomotion and one for steering, which is capable of continuously varying its orientation through 360°; as such the design may be termed omni-directional [7]. However, the limitation of this design is that the WMR has to keep changing wheels orientation, which cause time delay, while tracking a trajectory with discontinuous heading. Furthermore, the steering and driving of the wheels must be coordinated to prevent the wheels "fighting" against each other and to avoid possible singular configurations [3, 6].

WMRs with driven offset wheels [2, 8, 9, 10, 11, 12] are reported to be omni-directional; however, they entail one serious drawback – it is impossible to reorient the planes of the wheels, with the rolling disabled, without producing a drift of the platform [2]. Moreover, they need redundant actuators to avoid singularities [8, 13].

Another means to achieve perfect mobility is to use ODW, also known as *Mecanum* wheel, which allow independent control of WMR motion in longitudinal, lateral, rotation directions without the singular characteristics of conventional wheel. The ODWs are not steered, as the wheel is directly coupling with the actuated mounted on platform. This capability reduces the space and time required for WMR motion, which is a significant advantage in all applications, especially in space constrained, and obstacle intensive environments. An ODW consists of a central hub with a series of passive elliptical rollers, which are mounted at an angle to the wheel plane of rotation. A comprehensive discussion on the variants of ODWs can be found in the reference [13, 14].

Robots constructed with these wheels normally possess either three driven omni-directional wheels arranged in a △ or Y manner [15], or four driven omni-directional wheels arranged in a rectangular manner [16]. Such robots are omni-directional; however, their principal drawbacks are their low load-carrying capacity and the surmountable bump height, which depends on the roller diameter, rather than on the wheel diameter. Furthermore, they are sensitive to vertical vibration resulting from the successive shocks occurring each time a new roller contacts the ground. However, what these robot lacks in load carrying capacity and surmountable bump height, they make up with fluent manoeuvrability and no singularities.

2 Kinematic Modelling

Actual mechanical systems consist of many components or sub-systems. Likewise, the mathematical model of a WMR can be decomposed into many sub-systems, which can then be modelled and designed independently in parallel, thereby shortening the overall *design cycle* and *time-to-market*. In its simplest form, a WMR consists of a *platform* sub-system that is supported by a number w of wheel sub-systems. These sub-systems may be of the same type or hybrid. Thus, a WMR can be decomposed into the platform sub-system and several wheel sub-systems.

The concept of using the Jacobian matrix to relate the motion of each wheel to the motion of the robot and, subsequently combining the individual wheel equations to obtain the composite robot equation of motion was introduced in [16]. Along the same lines, a modular approach for modelling kinematically constrained multibody systems using Lagrange's equations was developed in [17]. The modelling of the WMRs in both cases is modular. The latter is, however, simpler due to the conciseness of the notation used.

In this section, the kinematic models of a WMR is derived. The WMR model consists of a platform sub-system and three homogeneous wheel sub-systems. The various sub-system models are derived separately and, subsequently, assembled to form the overall WMR model.

A comprehensive treatment on the different types of omni-directional wheels can be found in [13]. In our work here, we focus on omni-directional wheels in which the axis of the roller is orthogonal to the axis of the wheel. The main reason for doing so is the availability of such wheels from suppliers.

We carry out the kinematic analysis for a WMR whose detailed architecture, coordinate systems, geometry and variables used in the analysis are shown in Fig. 1. Since the plane of the wheel does not vary,

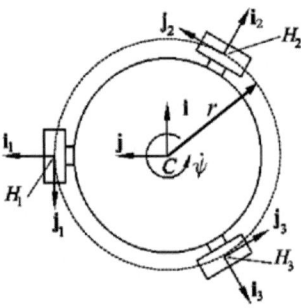

Figure 1: Notation for omni-directional platform.

it is more convenient to express the unit vectors of the wheel coordinates $\mathbf{j}_i$ in the body-frame such that

$$[\mathbf{j}_1]_\mathcal{B} = \begin{bmatrix} -\sin(90°) \\ \cos(90°) \end{bmatrix} = \begin{bmatrix} -1 \\ 0 \end{bmatrix} \quad (1)$$

$$[\mathbf{j}_2]_\mathcal{B} = \begin{bmatrix} -\sin(330°) \\ \cos(330°) \end{bmatrix} = \begin{bmatrix} \frac{1}{2} \\ \frac{\sqrt{3}}{2} \end{bmatrix} \quad (2)$$

$$[\mathbf{j}_3]_\mathcal{B} = \begin{bmatrix} -\sin(210°) \\ \cos(210°) \end{bmatrix} = \begin{bmatrix} \frac{1}{2} \\ -\frac{\sqrt{3}}{2} \end{bmatrix} \quad (3)$$

The velocity of H_i is

$$[\mathbf{j}_i \quad \mathbf{1}] \begin{bmatrix} r\dot{\psi} \\ \dot{\mathbf{c}} \end{bmatrix} = [\dot{\mathbf{h}}_i] \quad (4)$$

where r refers to the radii of the WMR, and $\dot{\mathbf{c}}$ is the velocity vector of point C.

The wheel kinematics can be derived with reference to the proposed wheel architecture, as shown in Fig. 2. The rolling angle of the ith wheel is denoted by θ_i and the radius of the wheel is denoted by r_w, which is the distance from the point H_i to the point of contact

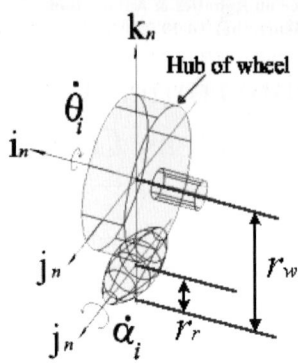

Figure 2: Notation for omni-directional wheel.

between the roller and the ground. Furthermore, the rolling angle and radius of the active roller are denoted by α_i and $\mathbf{r}_r$, respectively. The velocity of point H_i can be determined from the spinning of the wheels and rollers, such that

$$\dot{\mathbf{h}}_i = -r_w \dot{\theta}_i \mathbf{j}_i + r_r \dot{\alpha}_i \mathbf{i}_i \quad (5)$$

Equating the left-hand side of eq.(4) with that of eq.(5) we obtained the relationship between the velocity of the platform with the velocity of the wheel, which is

$$\dot{\mathbf{c}} + r\dot{\psi}\mathbf{j}_i = -r_w \dot{\theta}_i \mathbf{j}_i + r_r \dot{\alpha}_i \mathbf{i}_i \quad (6)$$

Only $\dot{\theta}_i$ can be controlled through the actuator of the ith wheel, while $\dot{\alpha}_i$ accommodates itself to allow for pure rolling of the roller with respect to the ground. Since all these rollers are unactuated, their velocities $\dot{\alpha}_i$ can be eliminated from eq.(6) by dot-multiplying throughout by $\mathbf{j}_i$, thereby deriving

$$\mathbf{j}_i^T \dot{\mathbf{c}} + r\dot{\psi} = -r_w \dot{\theta}_i \quad (7)$$

For $i = 1$ to 3, eq.(7) takes the form

$$\mathbf{J}_1 \mathbf{t} = \mathbf{J}_2 \dot{\boldsymbol{\theta}} \quad (8)$$

The two Jacobian matrices $\mathbf{J}_1$ and $\mathbf{J}_2$ of the WMR, the planar twist $\mathbf{t}$ of the platform and the actuated-joint-rate vector $\dot{\boldsymbol{\theta}}$ are:

$$\mathbf{J}_1 = \begin{bmatrix} 1 & \mathbf{j}_1^T \\ 1 & \mathbf{j}_2^T \\ 1 & \mathbf{j}_3^T \end{bmatrix}, \qquad \mathbf{J}_2 = -r_w \begin{bmatrix} 1 & 0 & 0 \\ 0 & 1 & 0 \\ 0 & 0 & 1 \end{bmatrix},$$

$$\mathbf{t} = \begin{bmatrix} r\dot{\psi} \\ \dot{x} \\ \dot{y} \end{bmatrix}, \text{ and } \dot{\boldsymbol{\theta}} = \begin{bmatrix} \dot{\theta}_1 \\ \dot{\theta}_2 \\ \dot{\theta}_3 \end{bmatrix},$$

where $\mathbf{J}_1$ and $\mathbf{J}_2$ are 3×3 matrices, while $\mathbf{t}$ and $\dot{\boldsymbol{\theta}}$ are 3×1 vectors.

3 Mobility Analysis

The pose of the platform on the plane consists of the triplet $[\psi \ \mathbf{c}^T]^T$. Therefore, this system has three degrees of freedom. If further constraints act on the platform, the number of degrees of freedom reduces accordingly. A minimum of three actuators will be required to generate motion. Moreover, a minimum of three wheels is sufficient to support and ensure stability of the platform. We analyse the mobility of a WMR with three omni-directional wheels using the functional matrix [18]. Equation (8) can be rearranged as

$$[\mathbf{J}_1 \ -\mathbf{J}_2] \begin{bmatrix} \mathbf{t} \\ \dot{\boldsymbol{\theta}} \end{bmatrix} = \mathbf{0} \qquad (9)$$

where $\mathbf{0}$ denotes the 6×1 zero vector. These relations are expressed in compact form as

$$\mathbf{F}\dot{\mathbf{q}} = \mathbf{0} \qquad (10)$$

The 3×6 functional matrix $\mathbf{F}$ is

$$\mathbf{F} = \begin{bmatrix} 1 & \mathbf{j}_1^T & r_w & 0 & 0 \\ 1 & \mathbf{j}_2^T & 0 & r_w & 0 \\ 1 & \mathbf{j}_3^T & 0 & 0 & r_w \end{bmatrix} \qquad (11)$$

while $\dot{\mathbf{q}}$ is

$$\dot{\mathbf{q}} = \begin{bmatrix} r\dot{\psi} & \dot{x} & \dot{y} & \dot{\theta}_1 & \dot{\theta}_2 & \dot{\theta}_3 \end{bmatrix}^T \qquad (12)$$

In this case, the functional matrix $\mathbf{F}$ is of full rank, that is, three, and its nullity is three [20]. Thus, a WMR with three omni-directional wheels has three degrees of freedom. Furthermore, since $\mathbf{J}_1$ is of full rank, therefore, this WMR does not experience any singularity. Note that the WMR modelled by Yi and Kim [9] does encounter singularity problem even though the WMR is able to perform omni-direction motion.

4 Design and Testing

4.1 Mechatronics Design

A WMR with three omni-directional wheels has been designed and constructed. As an example of typical robot, it involves an integrated design approach involving mechanical, control system design and software programming. Because the 3-wheel WMR was in fact a prototype, certain elements of the design were conceived, for this initial model, with flexibility in dimension, functionality, and availability. Our main focus was a good manoeuvrability such that it could track any path in configuration space (x, y, θ) in singularityfree manner.

4.2 Experimental Setup

The experimental setup and the prototype of the WMR with three omni-directional wheels [19], is shown in Fig. 3. The host terminal is a 300 MHz Pentium II. It runs the control user interface, ODMCS (Omni-Directional Motion Control System), which is written in Visual Basic. ODMCS comprises basic controls (translation and rotation) and a number of demonstration applications.

The 3-wheel WMR feedback control system illustrated in this work consists of a DC-power supply, 3 PIC-servo motion controller cards, a Z232-485 converter board, and a PC. The DC motors with incremental encoders and PIC-servo motor control boards are connected by 10-pin flat ribbon cable.

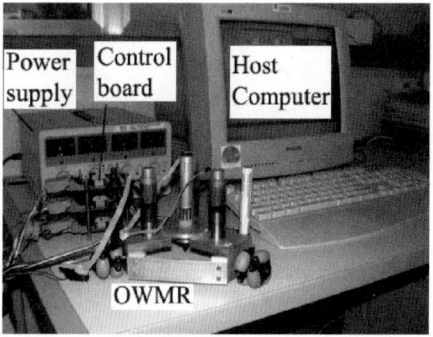

Figure 3: Experimental setup for the WMR with three omni-directional wheels.

The WMR was tested for its ability to execute four different types of motions. They are, namely, *rectilinear translation*; *curvilinear translation*; *rotation*; and *dual-path*.

In order to visualise the actual path travelled by the WMR, we attached two marker pens to the WMR. The first pen, denoted as "Pen C", is attached to the centre of the platform, while a second pen, denoted as "Pen P", is attached to one of the three plates connecting the wheels. The actual locations of the two pens with respect to the platform, is shown in Fig. 4. The derivation of kinematic equations required for path generation are shown in Appendix A. For simplicity, only *dual-path* and *curvilinear translation* are discussed.

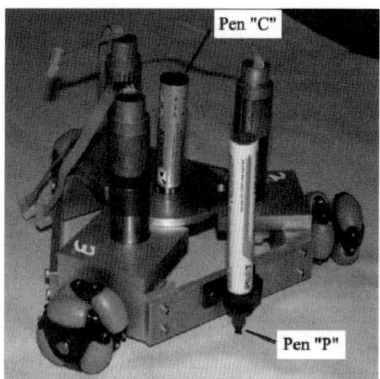

Figure 4: Locations of pens for tracing the path taken by the WMR.

4.3 WMR in Rectilinear Translation

In the first test, the WMR is tasked to move in a diamond shaped path without any change in orientation of the WMR. This type of translation, referred to as rectilinear translation, is shown in Fig. 5.

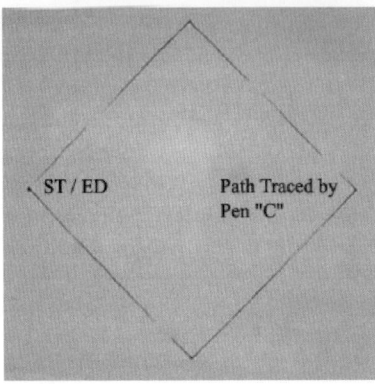

Figure 5: Traced path of WMR in rectilinear translation. (ST: start, ED: end)

4.4 WMR in Curvilinear Translation

In the second test, the WMR is tasked to move in a "S" shaped path and again without any change in orientation of the WMR, as shown in Fig. 6. This type of translation, referred to as curvilinear translation, in which all points on the WMR move along congruent curves.

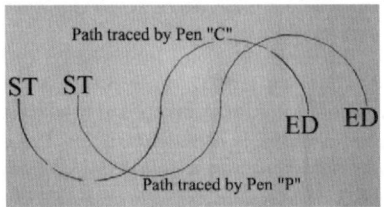

Figure 6: Traced path of WMR in curvilinear translation.

4.5 WMR in Rotation

In the third test, the WMR is tasked to move in a circular path. This time, the orientation of the WMR changes with respect to the fixed reference frame. This type of motion, referred to as rotation, in which all points on the WMR move along concentric curves, is shown in Fig. 7.

4.6 WMR in Dual-path

In the fourth test, the WMR is tasked to move in a straight path and at the same time rotate about its centre. This type of motion, we termed it as dual-path. The actual result is shown in Fig. 8. The path traced by "Pen C", which is attached to the centre of the WMR, is straight. On the other hand, the path traced by "Pen P", which is attached to the side of the WMR, is cycloid in shape. Figure 8 shows that the platform actually rotates by 180 degrees.

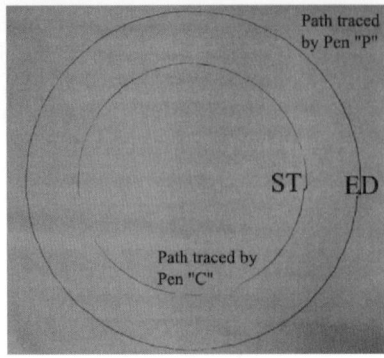

Figure 7: Traced path of WMR in rotation.

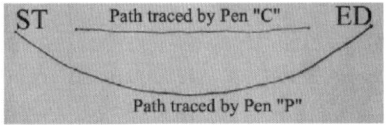

Figure 8: Traced path of WMR in dual-path.

4.7 Remarks

We have demonstrated the ability of a WMR with three omni-directional wheels, to execute four different types of motion, namely, rectilinear translation, curvilinear translation, rotation and combined translation and rotation. The results show that this class of WMR has a very fluent maneuverability, and is therefore, suitable for applications that requires quick changes to the WMR path of travel in any direction, without the risk of running into a singular configuration. It is also suitable for applications whereby the orientation of the WMR needs to be changed, while the WMR is still travelling in the given direction, without having to stop for re-orientation.

5 Conclusions

The work presented in this paper revolves around three main topics, namely, kinematic modelling, mobility analysis and mechatronics design of WMRs. In order to better understand the characteristics of the locomotion of WMRs, we derived the kinematic models of 3-wheel OWMR. We also studied the mobility of WMRs with three active omni-directional wheels. The analysis shows that WMRs with three omni-directional wheels have a mobility of three and are singular-free, as no steering angles are involved.

A WMR with three omni-directional wheels is designed and built. The prototype showed its omni-directional mobility by drawing various sharp turnings (rectilinear translation), smooth circle (curvilinear translation); as well as dual-path motion. The WMR of three degrees of freedom is fully demonstrated by its ability to execute a motion involving both translation and rotation simultaneously. Testings show that the constructed WMR is highly manoeuvrable; able to response to quick change in direction; and allows control to its orientation without running into a singular configuration. However, the ODW is more complex and difficult to fabricate compared to conventional wheels. It explained why ODW is not widely used in industrial application.

Acknowledgements

The authors would like to express their appreciation to Prof. Jorge Angeles, a visiting professor from the McGill University of Canada, for his contribution to the mobility analysis in this work. Thanks are also due to the technicians at the Robotic Research Center for their helps during the testing phase of the prototype.

References

[1] Killough, S.M. and Pin, F.G. 1992. Design of an omni-directional and holonomic wheeled platform prototype. *Proc. IEEE Int Conf on Robotics and Automation*, 84–90, Nice, France.

[2] Leow, Y.P., Angeles, J. and Low, K.H. 2000. A comparative mobility study of three-wheeled mobile robots. *Proc. The Sixth International Conference on Control, Automation, Robotics and Vision*, CD-ROM, Singapore.

[3] Leow, Y.P. and Low, K.H. 2001. Mobile platform with centred wheels. *Proc. Second Asian Symposium on Industrial Automation and Robotics*, 21–28, Bangkok, Thailand.

[4] Alexander, J.C. and Maddocks, J.H. 1989. On the kinematics of wheeled mobile robots. *The Int Journal of Robotics Research*, 8(5):15–27.

[5] Dong, S.K., Hyun, C.L. and Wook, H.K. 2000. Geometric kinematics modeling of omni-directional autonomous mobile robots and its applications. *Proc. IEEE Int Conf on Robotics and Automation*, 2033–2038, San Francisco, California.

[6] Leow, Y.P., Low, K.H. and Lim, S.Y. 2001. Kinematic modeling and analysis of mobile robots with centred wheels. *Proc. Sixth International Conference on Mechatronics Technology*, CD-ROM, Singapore.

[7] West, M. and Asada, H. 1994. Design of ball wheel vehicles with full mobility, invariant kinematics and dynamics and anti-slip control. *Proc. The 1994 ASME Design Technical Conference - 23rd Biennial Mechanisms Conference*, 72:377–384, Minneapolis, Minnesota.

[8] Campion, G., Bastin, G. and D'Andréa-Novel, B. 1993. Structural properties and classification of kinematic and dynamic models of wheeled mobile robots. *Proc. IEEE Int Conf on Robotics and Automation*, 462–469, Atlanta, Georgia.

[9] Yi, B.-J. and Kim, W. K. 2000. The kinematics for redundantly actuated omni-directional mobile robots. *Proc. of the IEEE Int Conf on Robotics and Automation*, 2485–2492, San Francisco, California.

[10] Ferrière, L., Raucent, B. and Fournier, A. 1996. Design of a mobile robot equipped with off-centred orientable wheels. *Proc. Research Workshop of ERNET*, 127–136, Darmstadt, Germany.

[11] Holmberg, R. and Khatib, O. 1999. Development of a holonomic mobile robot for mobile manipulation tasks. *Proc. Int Conf on Field and Service Robotics*, 268–273, Pittsburgh, Pennsylvania.

[12] Wada, M. and Mori, S. 1996. Modeling and control of a new type of omni-directional holonomic vehicle. *Proc. IEEE Int Conf on Robotics and Automation*, 265–270, Minneapolis, Minnesota.

[13] Fisette, P., Ferrière, L., Raucent, B. and Vaneghem, B. 2000. A multibody approach for modelling universal wheels of mobile robots. *Mechanism and Machine Theory*, 35:329–351.

[14] Byun, K.S., Kim, S.J. and Song, J.B. 2001. Design of continuous alternate wheels for omni-directional mobile robots. *Proc. IEEE Int Conf on Robotics and Automation*, 767–772, Seoul, South Korea.

[15] Saha, S.K., Angeles, J. and Darcovich, J. 1995. The design of kinematically isotropic rolling robots with omni-directional wheels. *Mechanism and Machine Theory*, 30(8):1127–1137.

[16] Muir, P.F. and Neuman, C.P. 1987. Kinematic modeling of wheeled mobile robots. *J. Robotic Systems*, 4(2):281–340.

[17] Ostrovskaya, S. 2001. *Dyanamics of Quasiholonomic and Nonholonomic Reconfigurable Rolling Robots*. Ph.D thesis, McGill University, Montreal, Canada.

[18] Freudenstein, F. 1962. On the variety of motions generated by mechanisms. *ASME J. of Engineering for Industry*, 84:156–160.

[19] Loh, W.K. 2002. Development of an Omni-directional Wheeled Mobile Robot. *Final Year Project Report*, School of Mechanical and Production Engineering, Nanyang Technological University, Singapore.

[20] Leow, Y.P., Low, K.H. and Loh, W.K. 2002. Kinematic Modelling and Analysis of Mobile Robots with Omni-Directional Wheels *Accepted for presentation at the Seventh International Conference*

on Control, Automation, Robotics and Vision, Singapore.

Appendix A: Path Generation

A.1: Dual-path

Let **C** be the position vector of the center of the disk point C, and the WMR has rotated counter-clockwise through an angle ψ. The no-slip condition requires that the horizontal displacement, d, equals to a circular arc as follows:

$$d = r\psi$$

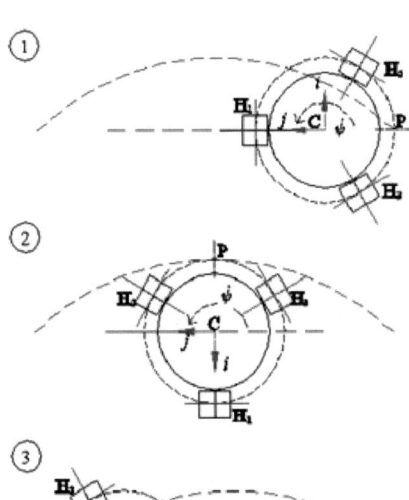

Figure 9: Dual-path motion.

Therefore, we have,
1. Position vector of point C

$$\mathbf{r}_{C_1} = r\psi \mathbf{j} \qquad (13)$$

2. Velocity vector of point C

$$\dot{\mathbf{c}} = \mathbf{v}_{C_1} = r\dot{\psi}\mathbf{j} \qquad (14)$$

3. Angular velocity of point P (constant rate)

$$\dot{\psi} = \frac{\psi}{t} \qquad (15)$$

The actuated joint-rate $\dot{\theta}_i$ can be determined based on $\dot{\mathbf{c}}$, and $\dot{\psi}$ given in eqs. (16) and (17), which are,

$$\dot{\theta}_i = \frac{r\dot{\psi}}{r_w t}\begin{bmatrix} 1 & \mathbf{j}_i^T \end{bmatrix}\begin{bmatrix} -1 \\ 0 \\ -1 \end{bmatrix} \qquad (16)$$

A.2: Curvilinear Translation

A body attached translating reference frame, Cij, is attached to the center of the WMR. The frame is always keeping the orientation of its frame same as the fixed frame OXY. The R refers to the radii of sine path. And, the angles, α and β, measured from the negative x-direction and positive in counter-clockwise direction. Since no rotation is needed, the $l\dot{\psi}$ term in eq.(7) should be eliminated. The new kinematic equation is,

$$\mathbf{j}_i^T \dot{\mathbf{c}} = -r\dot{\theta}_i \qquad (17)$$

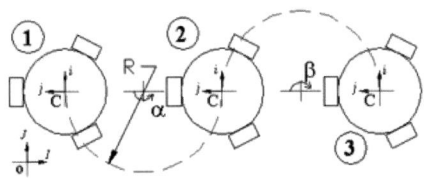

Figure 10: Curvilinear translation motion.

To map the coordinate system that set up in Fig. 1 and Fig. 10, we obtain the following relationships:

$$\psi = \alpha + 180° = 180° - \beta \qquad (18)$$
$$I = -\mathbf{j} \qquad (19)$$
$$J = \mathbf{i} \qquad (20)$$

From position 1 to 2, we have,
1. Position vector of point C

$$\mathbf{r}_{C_1} = -R\sin\alpha\,\mathbf{i} + R\cos\alpha\,\mathbf{j} \qquad (21)$$

2. Velocity vector of point C

$$\mathbf{v}_{C_1} = -R\dot{\alpha}\cos\alpha\,\mathbf{i} - R\dot{\alpha}\sin\alpha\,\mathbf{j} \qquad (22)$$

or,

$$\mathbf{v}_{C_1} = R\dot{\psi}\cos(\psi)\mathbf{i} - R\dot{\psi}\sin(\psi)\mathbf{j} \qquad (23)$$

From position 2 to 3, we have,
3. Position vector of point C

$$\mathbf{r}_{C_1} = -R\sin\beta\,\mathbf{i} - R\cos\beta\,\mathbf{j} \qquad (24)$$

4. Velocity vector of point C

$$\dot{\mathbf{c}} = \mathbf{v}_{C_1} = -R\dot{\beta}\cos\beta\,\mathbf{i} + R\dot{\beta}\sin\beta\,\mathbf{j} \qquad (25)$$

or,

$$\dot{\mathbf{c}} = \mathbf{v}_{C_1} = R\dot{\psi}\cos(\psi)\mathbf{i} + R\dot{\psi}\sin(\psi)\mathbf{j} \qquad (26)$$

The corresponding actuated joint-rate $\dot{\theta}$ can be determined based on $\dot{\mathbf{c}}$, given in eqs. (27) and (28), which are,

$$\dot{\theta}_i = -\frac{R\dot{\psi}}{r_w}\begin{bmatrix} 1 & \mathbf{j}_i^T \end{bmatrix}\begin{bmatrix} 0 \\ \cos\psi \\ k\sin\psi \end{bmatrix} \qquad (27)$$

where $k=-1$ (position 1 to position 2), and $k=1$ (position 2 to position 3).

Design of a 6 DOF Haptic Master for Teleoperation of a Mobile Manipulator

[†,††]Dongseok Ryu, [†,††]Changhyun Cho, [†]Munsang Kim, [††]Jae-Bok Song

[†]Advanced Robotics Research Center
Korea Institute of Science and Technology
39-1 Hawolgok-dong, Sungbuk-ku, Seoul 136-791, Korea
{sayryu, chcho, munsang}@kist.re.kr

[††]Dept. of Mechanical Eng., Korea University,
Anam-dong, Sungbuk-gu, Seoul 136-701, Korea
jbsong@korea.ac.kr

Abstract – This paper presents design of a 6 DOF haptic master which is based on a combination of a planar 3 DOF parallel mechanism and a spatial 3 DOF parallel mechanism. Since low inertia is vital to backdrivability and transparency of the haptic device, all actuators are placed on the base and thus some forces for haptic feedback are transmitted by tendon-driven mechanism. This device was intended to teleoperate a mobile manipulator, which requires planar 3 DOF motion for navigation of the vehicle and full 6 DOF motion for manipulation. Therefore, the proposed device is designed to have two modes of a planar task and a 3D task which are switchable without any change in hardware. This feature provides efficient actuation and reduces computational burden since only 3 actuators are involved in the case of a planar task. The paper deals with a detailed description of the proposed haptic master and Jacobian analysis. Some applications are given to verify the validity of this device.

I. INTRODUCTION

Haptic devices not only serve as a device for virtual reality and teleoperation but also provide a new paradigm as a human computer interface. For the past decades, various haptic devices have been developed. In the beginning, pointing devices capable of Cartesian 3 DOF motion (i.e., *x*, *y*, *z* translation) were mainly developed. Then substantial efforts have been directed toward the development of haptic devices which have 6 DOF force reflection capability and dexterity.

Haptic devices are similar to general manipulators in that they can generate force and torque at the end-effector against the external force acting on it. In addition to this force generating capability, however, haptic devices should be *backdrivable* so that the actuators should be able to follow the user's hand motion rapidly [1]. Since the user feels resisting forces caused by the actuator inertia and gears and transmissions incorporated in the interface, backdrivability requires minimal gear ratio and low actuator inertia. Another important feature of a haptic device is *transparency* [2]. This means that the haptic interface should transmit the interaction with the virtual environment to the user as faithfully as possible. Hence, transparency requires low inertia and high bandwidth of the device [3].

In designing a haptic device based on serial mechanisms, it is difficult to achieve precision and rigidity, because an error is accumulated from the base to the end-effector. Since an actuator to actuate each joint is usually placed near the corresponding joint, an increase in the number of DOFs in the serial configuration leads to an increase in inertia of the device, thus resulting in degradation of backdrivability and transparency. Due to these problems, it is desirable to limit serial configuration to devices with less than 3 DOF motion.

Recently, a great deal of work has been done in multi DOF haptic devices based on parallel mechanism [4-6]. Since these devices employ parallel mechanisms as a basic configuration, high stiffness and accuracy can be achieved with relative ease. Moreover, it is easy to make all actuators installed near the fixed base, and thus the mass and moment of inertia of the moving parts are greatly reduced, thus leading to enhanced backdrivability and transparency. However, parallel haptic devices have relatively small workspace compared to serial ones.

Some haptic applications require motion and force reflection in the whole 3D workspace (i.e., 6 DOF motion). Even for these applications, however, the device seldom needs the whole 6 DOF motion at all times during operation. In most 6 DOF haptic devices, all 6 actuators are activated to create force feedback even when only simple motion is desired, since the Cartesian space and joint space are closely coupled in the device. It is desirable, therefore, that a haptic device be designed so that only necessary DOFs are activated while other DOFs remain unactivated depending on the situations. This strategy has several advantages. First, mass and moment of inertia of the moving parts are reduced, which leads to improved backdrivability and transparency. Second, computational burden needed to solve kinematic and static equations involving all DOFs for determination of the end-effector posture and the required force reflection are greatly reduced. Third, it is energy efficient because only necessary actuators are activated.

In development of haptic devices, we need to consider not only their features such as the number of DOFs and their performance but also convenience and fitness in the light of applications. Although the proposed haptic master can be used for a variety of tasks, it was intended to teleoperate a mobile manipulator in the first place. So design concept was focused on which type of haptic device is most appropriate to teleoperation of a mobile manipulator. It was observed that most haptic devices tend to divide 3D motion into 3 DOFs in position (i.e., x, y, z translation) and 3 DOFs in orientation (e.g., roll, pitch, yaw). When 6 DOF motion is required, therefore, a 3 DOF mechanism capable of providing rotational motion is added at the end-effector of the 3 DOF positioning device.

A close observation of teleoperation of a mobile manipulator reveals that it requires planar 3 DOF motion for navigation of the vehicle and full 6 DOF motion for manipulation task. Therefore, the proposed device is designed to have two modes of a planar task and a 3D task which are switchable without any change in hardware. This research focuses on development of such haptic device. The proposed haptic master consists of two independently working mechanisms as shown in Fig. 1; one is a planar 3 DOF parallel manipulator and the other is a spatial 3 DOF parallel manipulator.

Fig. 1 Overview of the haptic master

The remainder of this paper is organized as follows. In Section 2, design of a proposed haptic device is introduced. Section 3 deals with Jacobain analysis of the device. Section 4 compares the proposed device with other devices and introduces simple applications. Section 5 concludes the research results.

II. DESIGN CONCEPT AND STRUCTURE

The proposed haptic master is composed of upper and lower mechanisms as shown in Fig. 1. Both mechanisms are constructed in the form of a parallel mechanism to achieve high stiffness and accuracy. The upper mechanism is attached serially to the end-effector of the lower one to form a 6 DOF haptic master. This structure allows actuators to be placed on the base, thus leading to good backdrivability because the actuators need not move during operation of the device. In what follows, the upper and lower mechanisms are discussed in detail.

The lower mechanism is designed to be a planar 3 DOF parallel manipulator. When selection of planar 3 DOF parallel manipulators is limited to three identical limb structures, seven planar 3 DOF parallel manipulators (RRR, RRP, RPR, PRR, RPP, PRP, and PPR) are feasible [7, 8] In the design of parallel manipulators such as a Stewart-Gough platform, prismatic joints (e.g., a linear ball screw) are employed for large force transmission to the moving platform. Since back drivability of a prismatic joint based on such a linear ball screw is not as good as that of a revolute joint, a RRR limb structure comprising 2 links and 3 revolute joints is adopted in this design as shown in Fig. 2. In Fig. 2, joints A, B, and C are active since they are actuated by electric motors, while other joints are passive. It is seen that actuation of active joints enables the platform to perform 3 DOF motion of (x, y, θ).

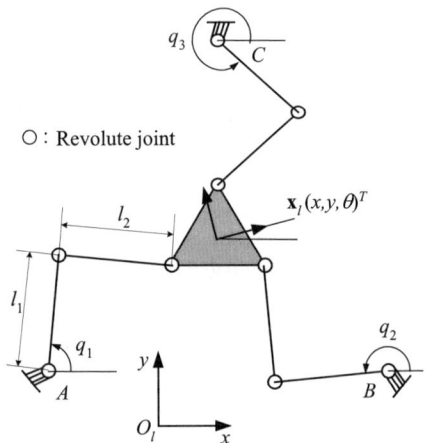

Fig. 2 Schematic of the lower mechanism.

The upper mechanism is designed as a spatial 3 DOF parallel manipulator. Among many spatial parallel devices [9-11], 3 RRS spatial parallel mechanism was employed as shown in Fig.3. In Fig. 3, joints D, E, and F are active joints actuated by electric motors, while other joints are passive. Actuation of active joints enables the triangular end-effector to perform 3 DOF motion of (z, ϕ, ψ).

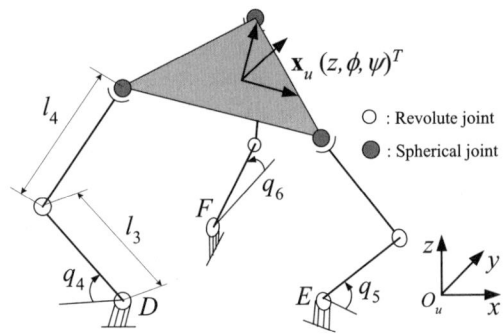

Fig. 3 Schematic of the upper mechanism.

Finally, a new 6 DOF haptic master is constructed by putting both mechanisms together serially as shown in Fig.4.

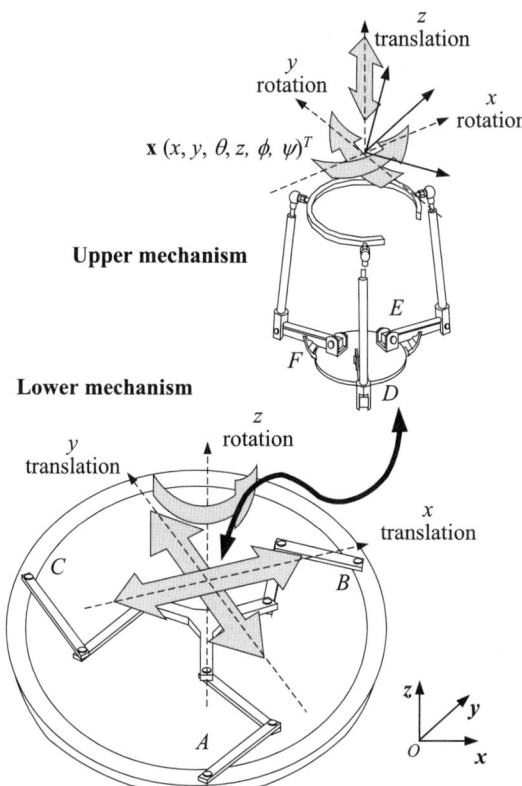

Fig. 4 Structure of the proposed haptic master.

It is noted that both upper and lower mechanisms are independent of each other in 3D motions. For instance, if only planar motion is required, 3 DOFs of the lower mechanism are sufficient to provide this motion. This is particularly important feature as for force reflection.

In most 6 DOF haptic devices, all 6 actuators are activated to create force reflection even when only simple motion is involved, since the Cartesian space and joint space are closely coupled. In the proposed haptic device, however, the upper and lower mechanisms are decoupled to some extent, so force reflection problem becomes simpler to implement.

If 3 actuators to drive the upper mechanism are placed on the end-effector of the lower mechanism, which corresponds to the base of the upper mechanism, they are forced to move with motion of the lower mechanism. In this case a user needs to move more mass due to these actuators, and thus back drivability of the device is significantly reduced. One solution to this problem is to adopt tendon-driven mechanism in which actuators can be placed on the base which is fixed during operation. As shown in Fig. 5, pulleys for tendon are placed at the revolute joints of the limbs of the lower mechanism. Thus q_4 for the upper mechanism is inevitably related to the motion of the lower mechanism.

$$q_i = (q_i' + \theta), \quad i = 4,5,6 \qquad (1)$$

where q_i' is the angle created by the motor for the lower mechanism and θ is a function of q_1, q_2 and q_3. It follows that control of the upper mechanism needs control of the lower mechanism as well as control of q_4', q_5' and q_6'.

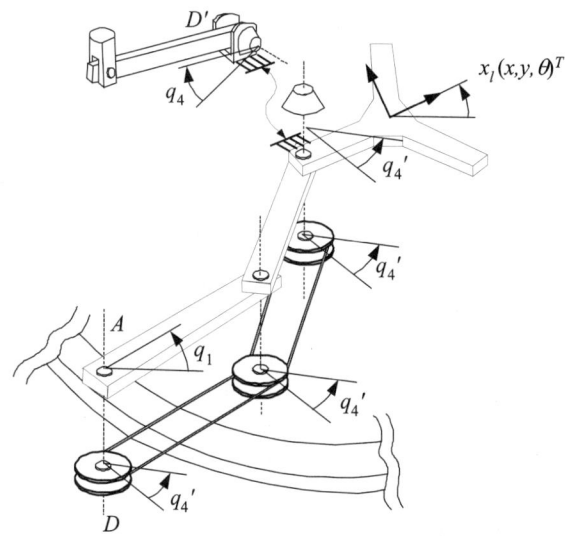

Fig. 5 Schematic of tendon-driven mechanism for force transmission.

Fig. 6 Photo of tendon-driven mechanism.

III. JACOBIAN ANALYSIS

In order to operate the haptic device, we need to find the kinematic equations which relate joint variables to Cartesian variables. Although the whole system is of 6 DOFs, the device can be decomposed into the upper and lower mechanisms, which makes derivation of kinematic equations easier. The resulting kinematic equations, however, are so lengthy that they are omitted here owing to limited space. The kinematic equations for each mechanism can be found in the literature, so only coupling between the upper and lower mechanisms will be discussed in this section.

Let the pose (i.e., position and orientation) of the end-effector be described by a vector $\mathbf{x}$ and the actuated joint variables by a vector $\mathbf{q}$ as follows:

$$\mathbf{x}_l = \{x \quad y \quad \theta\}^T, \quad \mathbf{x}_u = \{z \quad \phi \quad \psi\}^T$$
$$\mathbf{x} = \begin{Bmatrix} \mathbf{x}_l \\ \mathbf{x}_u \end{Bmatrix} \tag{2}$$

$$\mathbf{q}_l = \{q_1 \quad q_2 \quad q_3\}^T, \quad \mathbf{q}_u = \{q'_4 \quad q'_5 \quad q'_6\}^T$$
$$\mathbf{q} = \begin{Bmatrix} \mathbf{q}_l \\ \mathbf{q}_u \end{Bmatrix} \tag{3}$$

where the subscripts l and u denote the lower and upper mechanisms, respectively. Note that all the variables were shown in Fig. 3 and 4. The differential relations between the joint and the Cartesian vectors for each mechanism can be described by

$$\mathbf{J}_{ql} \cdot d\mathbf{q}_l = \mathbf{J}_{xl} \cdot d\mathbf{x}_l \tag{4}$$

$$\mathbf{J}_{qu} \cdot d\mathbf{q}_u = \mathbf{J}_{xu} \cdot d\mathbf{x}_u \tag{5}$$

where $\mathbf{J}_{ql}$ and $\mathbf{J}_{xl}$ are the Jacobians of the lower mechanism, and $\mathbf{J}_{qu}$ and $\mathbf{J}_{xu}$ are the Jacobians of the upper mechanism, respectively. Using the definitions of Eqs. (2) and (3), Eqs. (4) and (5) can be integrated into a single equation

$$\mathbf{J}_q \cdot d\mathbf{q} = \mathbf{J}_x \cdot d\mathbf{x} \tag{6}$$

where

$$\mathbf{J}_q = \begin{bmatrix} \mathbf{J}_{ql} & \mathbf{0} \\ \mathbf{0} & \mathbf{J}_{qu} \end{bmatrix} = \mathrm{diag}[J_{q1} \, J_{q2} \, J_{q3} \, J_{q4} \, J_{q5} \, J_{q6}] \tag{7}$$

$$\mathbf{J}_x = \begin{bmatrix} \mathbf{J}_{xl} & \mathbf{J}_{xc1} \\ \mathbf{J}_{xc2} & \mathbf{J}_{xu} \end{bmatrix} \tag{8}$$

where $\mathbf{J}_q$ and $\mathbf{J}_x$ are the Jacobians of the haptic master, and $\mathbf{J}_{xc1}$ and $\mathbf{J}_{xc2}$ are the off-diagonal elements associated with coupling of the upper and lower mechanisms.

Since Eq. (6) is related to the kinematic constraint imposed by the mechanisms, we need to investigate the nature of coupling between the upper and lower mechanisms. First, the lower mechanism is completely independent of the upper one. That is, motion of the upper mechanism alone does not induce any motion in the lower one. Therefore, Eq. (6) together with Eqs. (7) and (8) is converted into

$$\mathbf{J}_{ql} \, d\mathbf{q}_l = \mathbf{J}_{xl} \, d\mathbf{x}_l + \mathbf{J}_{xc1} \, d\mathbf{x}_u = \mathbf{J}_{xl} \, d\mathbf{x}_l \tag{9}$$

which means that the matrix $\mathbf{J}_{xc1} = \mathbf{0}$ since $d\mathbf{x}_u$ does not affect $d\mathbf{q}_l$. Second, the upper mechanism is partially dependent on the lower one. That is, z rotation (i,e., θ) of the lower mechanism affects motion of the upper mechanism due to coupling by tendon-driven nature, while x and y translations do not. In this case Eq. (6) becomes

$$\mathbf{J}_{qu} \, d\mathbf{q}_u = \mathbf{J}_{xu} \, d\mathbf{x}_u + \mathbf{J}_{xc2} \, d\mathbf{x}_l \tag{10}$$

where the third column of $\mathbf{J}_{xc2}$ is nonzero whereas the other two columns are zero, since only $d\theta$ affects $d\mathbf{q}_u$. Consequently, the matrix $\mathbf{J}_x$ is a lower triangular matrix.

From Eq. (6), the overall Jacobian matrix $\mathbf{J}$ can be defined as

$$d\mathbf{q} = \mathbf{J} \, d\mathbf{x} \tag{11}$$

where $\quad \mathbf{J} = \mathbf{J}_q^{-1} \cdot \mathbf{J}_x \tag{12}$

For force feedback, the relation between the joint torque vector τ and the force/moment vector at the end-effector $\mathbf{F}$ needs to be found. By the principle of virtual work, the following relation holds.

$$\mathbf{F}^T \cdot d\mathbf{x} = \tau^T \cdot d\mathbf{q} \tag{13}$$

where

$$\tau = \{\tau_1 \quad \tau_2 \quad \tau_3 \quad \tau_4 \quad \tau_5 \quad \tau_6\}^T \tag{14}$$

$$\mathbf{F} = \{F_x \quad F_y \quad F_\theta \quad F_z \quad F_\phi \quad F_\psi\}^T \tag{15}$$

Substitution of Eq. (12) into (13) yields

$$\mathbf{F} = \mathbf{J}^T \tau \tag{16}$$

or equivalently,

$$\tau = \mathbf{J}^{-T} \mathbf{F} \tag{17}$$

Singularity analysis is important when design of a robotic device is conducted. Since two Jacobians are involved in Eq. (5), these two matrices need to be analyzed. First, the matrix $\mathbf{J}_q$ becomes singular when $\det(\mathbf{J}_q) = 0$. Since $\mathbf{J}_q$ is a diagonal matrix, singularities occur when one of the diagonal elements vanishes, which arises whenever the device reaches its workspace boundary. If this type of singularity occurs, the device can resist forces or moments in some directions for zero joint torques.

Second, the matrix $\mathbf{J}_x$ becomes singular when $\det(\mathbf{J}_x) = 0$. Since $\mathbf{J}_x$ is a lower triangular matrix, its determinant becomes

$$\det(\mathbf{J}_x) = \det(\mathbf{J}_{xl}) \det(\mathbf{J}_{xu}) = 0 \tag{18}$$

Therefore, $\mathbf{J}_x$ becomes singular when either $\mathbf{J}_{xl}$ or $\mathbf{J}_{xu}$ or both are singular. If this type of singularity occurs, the device cannot resist forces or moments in some directions even for large joint torques.

Another important feature of the proposd design is the simplified computation of the inverse Jacobian matrix. In Eq. (17), $\mathbf{J}^{-T}$ needs to be obtained to compute the joint toqrues required for the desired end-effector force $\mathbf{F}$. From Eq. (12), $\mathbf{J}^{-T}$ becomes

$$\mathbf{J}^{-T} = \mathbf{J}_q \cdot \mathbf{J}_x^{-T} \tag{19}$$

From Eq. (8),

$$\mathbf{J}_x^T = \left[\begin{array}{c|c} \mathbf{J}_{xl} & \mathbf{J}_{xc2} \\ \hline \mathbf{0} & \mathbf{J}_{xu} \end{array}\right] \tag{20}$$

Since it is a upper triangular matrix, its inverse can be computed by the matrix inversion lemma as follows:

$$\mathbf{J}_x^{-T} = \left[\begin{array}{c|c} \mathbf{J}_{xl}^{-1} & \mathbf{J}_{xl}^{-1} \cdot \mathbf{J}_{xc2} \cdot \mathbf{J}_{xu}^{-1} \\ \hline \mathbf{0} & \mathbf{J}_{xu}^{-1} \end{array}\right] \tag{21}$$

It is noted that the inversions involved in Eq. (21) are done for 3 x 3 matrices, which are much simpler than those for 6 x 6 matrices. Hence, computational load of the inverse Jacobian is significantly reduced in comparison with 6 DOF devices of similar type.

IV. CONSTRUCTION AND APPLICATIONS

It is difficult to compare the proposed haptic master with other haptic devices, because each device has different structure and features. To evaluate the proposed device, comparison has been made with PHANToM, a well-known commercial haptic device [12] and is listed in Table 1.

Table 1 Comparison of the haptic master with PHANToM.

	Haptic master	PHANToM
DOF	6 (6 inputs / 6 outputs)	6 / 3 (6 inputs / 3-axis force display)
Weight	11kg (actuator 4kg)	
Workspace	Cylinder (110mm diameter, 100mm high)	Cube (80 x170 x 250 mm)
Force reflection	Force: continuous 20N, peak 30N force Torque: continuous 2Nm, peak 3Nm Torque	Force: continuous 1.5N, peak 10N force

The schematic diagram of teleoperation of a mobile manipulator is shown in Fig. 7. As mentioned before, the haptic master was devised for use in teleoperating a mobile manipulator. The operation can be classified into two modes: a planar mode and a 3D mode. The planar mode is selected when the mobile platform navigates in the environment. Since this mode requires only 3 DOF planar motion which can be provided by the lower mechanism alone, only the lower mechanism is utilized. When the operator grips the handle of a haptic master and moves it in the direction in which he intends to move, the joint angle is sensed by encoders and then fed to the DSP controller where the handle pose (i.e., position and orientation) is computed from the joint information. This pose information is transmitted to the host PC which generates the command to operate the mobile manipulator. Finally, the local controller at the mobile manipulator receives this command and enables it to conduct the desired operation which the operator intends to do.

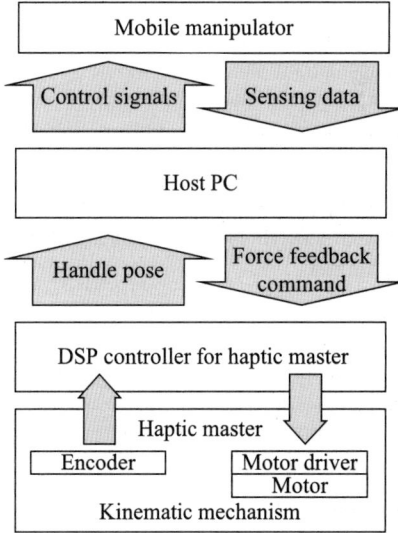

Fig. 7 Signal flow of teleoperation of a mobile manipulator with haptic master.

Once the mobile manipulator reaches the goal, it performs a given task using the manipulator, which requires full 6 DOF motion and possibly force reflection. For transmission of force and torque which the manipulator experiences in the environment, a force/torque sensor needs to be installed at the end-effector of the manipulator. The operator can perceive the environment by feeling forces transmitted through the handle of a haptic master.

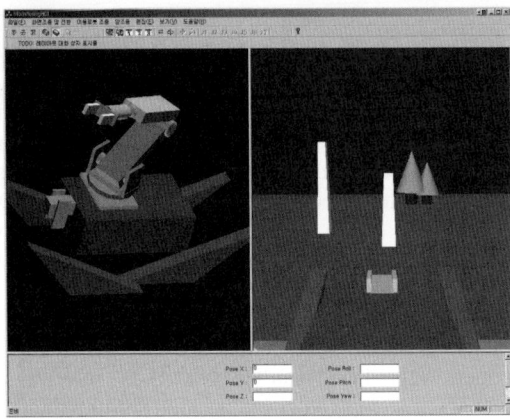

Fig. 8 Virtual environment for teleoperation of a mobile manipulator.

V. CONCLUSION

This paper presents design of a 6 DOF haptic master which is based on a combination of a planar 3 DOF parallel mechanism and a spatial 3 DOF parallel mechanism. From the analysis and actual construction of the haptic master, the following conclusions are drawn.

1. By placing all actuators on the fixed base and employing tendon-driven mechanism, backdrivability and transparency of the device have improved due to reduced inertia.

2. The haptic master was designed to have two independently operating mechanisms for partial use of total DOFs available. By this design efficient actuation and reduced computational burden are achieved since only 3 actuators are involved for the case of a planar task.

Currently research on improvement of force reflection capability is under way.

VI. REFERENCES

[1] G. Burdea, Force and touch feedback for virtual reality, A Wiley-Interscience Publication, 1996.

[2] D.A. Lawrence, and J.D. Chapel, "Performance trade-offs for hand controller design," Proc. of IEEE Int. Conf. on Robotics and Automation, pp. 3211-3216, 1994.

[3] J. E. Colgate, and J. M. Brown, "Factors affecting the Z-width of a haptic display," Proc. of IEEE Int. Conf. on Robotics and Automation, pp. 3205-3210, 1994.

[4] G.L. Long, and C.L. Collins, "A pantograph linkage parallel platform master hand controller for force-reflection," Proc. of IEEE Int. Conf. on Robotics and Automation, pp. 390-395, 1992.

[5] H. Nomo, and H. Iwata, "Presentation of multiple dimensional data by 6 DOF force display," Proc. of IEEE Int. Conf. on Intelligent Robots and System, pp. 1495-1500, 1993.

[6] Y. Tsumaki, H. Naruse, D. N. Nenchev, and M. Uchiyama, "Design of a compact 6-DOF haptic interface," Proc. of IEEE Int. Conf. on Robotics and Automation, pp. 2580-2585, 1998.

[7] L.-W. Tsai, Robot analysis: The mechanics of serial and parallel manipulators, A Wileys-Interscience Publication, 1999.

[8] C., Gosselin, and J., Angeles, "The optimum kinematic design of a planar three degree of freedom parallel manipulator," ASME Journal of Mechanisms, Transmissions, and Automation in Design, Vol.110, pp. 35-41, 1988.

[9] J.A. Carretero, M. Nahon, and R.P. Podhorodeski, "Workspace analysis of a 3-dof parallel mechanism," Proc. of IEEE Int. Conf. on Intelligent Robots and System, pp. 1021-26, 1998.

[10] L.-W. Tsai and S. Joshi, "Comparison study of architectures of four 3 degree-of-freedom translational parallel manipulator", Proc. of IEEE Int. Conf. on Robotics and Automation, pp. 1283-1288, 2001.

[11] E. Ottaviano, C. Gosselin, and M. Ceccarelli, "Singularity analysis of CaPaMan: A three-degree of freedom spatial parallel manipulator," Proc. of IEEE Int. Conf. on Robotics and Automation, pp. 1295-1300, 2001.

[12] T., Massie, and K. Salisbury, "The PHANToM haptic interface: a device for probing virtual objects" ASME Winter Annual Meeting, DSC-Vol.55-1, pp.295-300, 1994

A Passive Robot System for Measuring Spacesuit Joint Damping Parameters

H. Wang, X.H. Gao, M.H. Jin, L.B. Du,
J.D. Zhao, H.Y. Hu, H.G. Cai

Robot Research Institute

Harbin Institute of Technology

Harbin, 150001

P.R. China

T.Q. Li

Institute of Space
Medico-Engineering

Beijing, 100094

P.R. China

H. Liu

Institute of Robotics and
Mechatronics

German Aerospace Center, DLR

82230 Wessling

Germany

Abstract – This paper presents a novel passive robot system with 6 DOF force/torque sensor for measuring spacesuit joint damping parameters. Based on its special mechanical structure, a 3 DOF model of flexible IVA (intra vehicular activity) spacesuit joint has been built. Experimental results prove the effectiveness of the measuring principle. Potential application of the measuring system is discussed.

I Introduction

Besides guaranteeing to the safety of astronaut, another main demand of spacesuit is to provide astronaut joint mobility when they work in space. With the development of the space flight project, astronauts' working range is expanding constantly and their operations are to be more and more complex. Therefore enhancing joint mobility plays an important role in the design of spacesuit.

Spacesuit joint's mobility is the main manifestation of the mobility of spacesuit. Main components of IVA and EVA (extra vehicular activity) spacesuit are the joints of the shoulder, elbow and wrist. Joint mobility of spacesuit is determined by mechanical characteristics of these joints, which should have the lowest damping torque and widest moving range. However, when a suited astronaut moves, there is always an inevitably damping torque resulted from surplus pressure (pressure difference between inside and outside of the spacesuit), which is produced by spacesuit vacuum protection. In order to verify joint mobility and provide exact evaluation index for the joint structure design and manufacturing, the damping parameters need to be exact measured.

Traditionally, there are two kinds of methods for measuring the damping parameters of spacesuit [1]. One is putting the measuring device outside the spacesuit and another is putting inside. The former is the method connecting the spacesuit to a measuring device equipped with torque sensor and joint angle measuring apparatus, which can only measure a single joint of the spacesuit. This method is inconvenient for operation and can't provide coupling effect of different joints. The later is the method putting the measuring device and the drive system into the spacesuit, which directly measures the damping torque and angle of the spacesuit joint by driving the device. Such device can only match a specific dimension spacesuit. Moreover, if there is a malfunction, it will damage the very expensive spacesuit.

This paper presents a novel passive robot measuring system, which can measure all the joints simultaneously, and has no damage to the spacesuit. Also, with a little adjustment, it can fit different size of spacesuit.

Through two years hard working, Robot Research Institute (RRI) of HIT cooperating with Institute of Space Medico-Engineering (ISME) has developed a measuring system based on a passive robot. Paper [2] has described the robot design. Paper [3] has described the measuring principle of the system and the theoretical analysis of modeling the flexible joint of spacesuit. This paper will particularly emphasize on analyzing the experimental results on the basis of these two papers and discuss the problems in the measuring experiments. The paper is arranged as follows: Section II presents the measuring principle of the system, section III and section IV describe the model of the passive robot and spacesuit arm respectively, section V presents the method of system verification, section VI analyzes the experimental results and section VII discusses the potential application of the system.

II Measuring Principle

Figure 1 is the schematic diagram of measuring principle of the system [2,3]. Between the ends of spacesuit arm and passive robot there is a 6 DOF force/torque sensor, which is used to measure the interaction force and torque between the robot and spacesuit. When operator moves the handle at the end of the robot at a constant speed, spacesuit arm will move together with the robot. Because the position and orientation of the end point of spacesuit are related to ones of the passive robot, the position and orientation can be calculated by forward kinematics through robot joint angle encoders. 6 DOF force/torque sensor output is related to the spacesuit joint damping torques and depends on the configuration of the spacesuit.

Fig 1. Measuring principle of passive robot

The calculation process has four steps. Firstly, utilizing joint angles measured by the optical encoders in the passive robot joints, the position and orientation of the robot and the spacesuit's arm in robot coordinates system can be worked out according to forward kinematics. Secondly, the position and orientation of the spacesuit's arm in its own coordinates can be calculated by coordinate transformation from robot base coordinate system. Thirdly, the rotation angle of every spacesuit joint can be calculated by inverse kinematics. At last, damping torque of every spacesuit joint can be worked out through Jacobian matrix from the 6 DOF force/torque value of arm end [4]:

$$\tau = J^T \cdot F \qquad (1)$$

where J is the Jacobian matrix of spacesuit and F is the 6 DOF force/torque value.

III Modeling of the Passive Robot

The passive robot has six degrees of freedom with rotation joints. Figure 2 shows the coordinates of robot based on robot kinematics. The Denavit-Hartenberg parameters of the robot are shown in Table 1.

Transformation matrix between robot joints can be calculated from $A_{i-1}^i (i=1\cdots 6)$, the last transformation matrix A_6^7 represents translation transformation from coordinate O_6 to coordinate O_7. ($s\theta_i$ represents $\sin\theta_i$, $c\theta_i$ represents $\cos\theta_i$)

$$A_{i-1}^i = \begin{bmatrix} c\theta_i & -s\theta_i & 0 & a_{i-1} \\ s\theta_i c\alpha_{i-1} & c\theta_i c\alpha_{i-1} & -s\alpha_{i-1} & -s\alpha_{i-1}d_i \\ s\theta_i s\alpha_{i-1} & c\theta_i s\alpha_{i-1} & c\alpha_{i-1} & c\alpha_{i-1}d_i \\ 0 & 0 & 0 & 1 \end{bmatrix}$$

$$A_6^7 = \begin{bmatrix} 1 & 0 & 0 & 0 \\ 0 & 1 & 0 & l_6 \\ 0 & 0 & 1 & l_7 \\ 0 & 0 & 0 & 1 \end{bmatrix}$$

Transformation matrix of the robot is:

$$T = A_0^1 A_1^2 A_2^3 \cdots A_6^7 \qquad (2)$$

Table 1. D-H parameters of the passive robot

i	α_{i-1}	a_{i-1}	d_i	θ_i
1	$0°$	0	l_1	θ_1
2	$90°$	0	0	θ_2
3	$-90°$	0	l_2	θ_3
4	$90°$	0	l_3	θ_4
5	$-90°$	0	l_4	θ_5
6	$90°$	0	l_5	θ_6

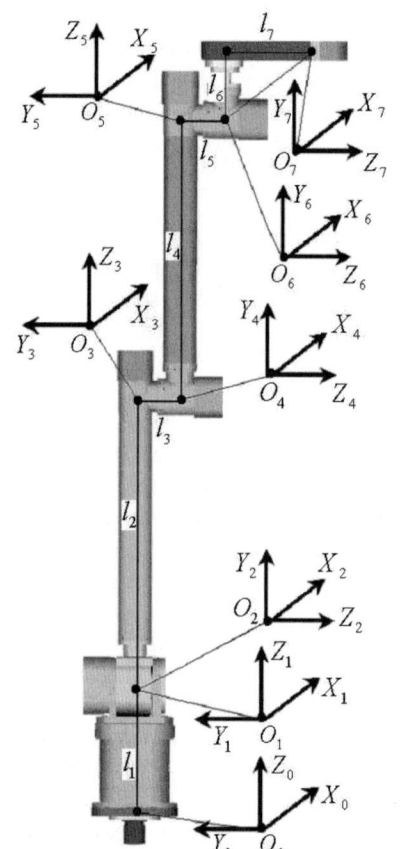

Fig 2. Coordinates of the passive robot

IV Modeling of the IVA Spacesuit Arm

IVA spacesuit arm has three degrees of freedom, which includes the abducent/adducent joint of shoulder, the extendible/bendy joint of elbow and the rotation joint of wrist.

Spacesuit arm is different from general robot whose components are rigid bodies. When the extendible/bendy joint angle changes, the rotation center of joint will change. In the measuring process, spacesuit will be charged to 0.04Mpa (the standard pressure when astronaut works). Although the motion of spacesuit arm is similar to the motion of rigid body in this situation, the change of the flexible joint's rotation center (extendible/bendy joint of elbow) can't be neglected. Consequently, the model of spacesuit arm must be built according to the mechanical structure of the flexible joint.

The coordinates of IVA spacesuit's left arm are shown in figure 3, in which Ls is the width of shoulder, La and Lb are the lengths of upper arm and forearm respectively. To make the model more accurate and the coordinate transformation more convenient, we set up five coordinates instead of three, among them two joint angles are constant (joint 1 and joint 3), which are related to the manufacturing of the spacesuit.

Figure 4 is the schematic diagram of the mechanical structure of the flexible elbow joint, in which Lr is the length of flexible joint when it is straight, La_0 and Lb_0 are the lengths of upper arm and forearm that are outside of the joint respectively. In the figure, ab is the fixed length, its center is O. When the joint is straight, this point is just the center of joint. When the joint bents, ab will turn into arc (length is constant). When benting to θ_4, O will move to O', while La and Lb will increase:

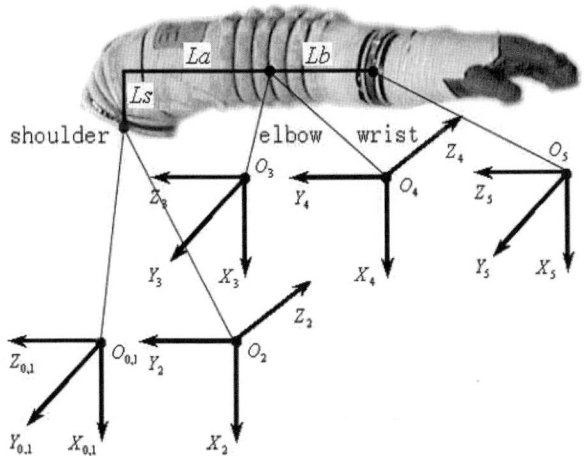

Fig 3. Coordinates of the IVA spacesuit's left arm

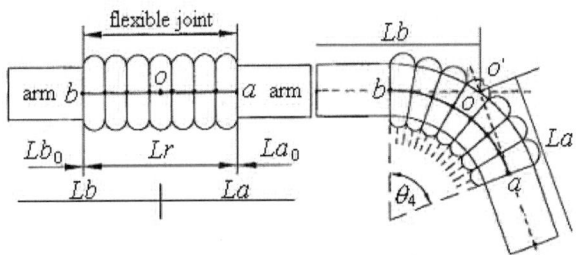

Fig 4. Schematic diagram of flexible elbow joint

$$aO' = bO' = \frac{Lr}{\theta_4} tg \frac{\theta_4}{2} \quad (3)$$

So
$$La = La_0 + \frac{Lr}{\theta_4} tg \frac{\theta_4}{2} \quad (4)$$

$$Lb = Lb_0 + \frac{Lr}{\theta_4} tg \frac{\theta_4}{2} \quad (5)$$

D-H parameters of the spacesuit's left arm are shown in Table 2.

Table 2. D-H parameters of spacesuit's left arm

i	α_{i-1}	a_{i-1}	d_i	θ_i
1	$0°$	0	0	θ_1 (constant)
2	$90°$	0	0	θ_2
3	$-90°$	$-Ls$	$-La$	θ_3 (constant)
4	$90°$	0	0	θ_4
5	$-90°$	0	$-Lb$	θ_5

V System Verification

To verify validity of the system, a 1-DOF measuring device has been constructed to measure the extendible/bendy joint of elbow. When measuring, only the elbow joint can move while other joints are fixed. An optical encoder is equipped on the axis of the device to measure the angle of elbow joint. At the conjunction of the device and the spacesuit, there is a torque sensor to measure torque when the joint moves. Principle of this device is shown in figure 5.

In the figure, F represents the force vertical to the spacesuit forearm, which is measured by the torque

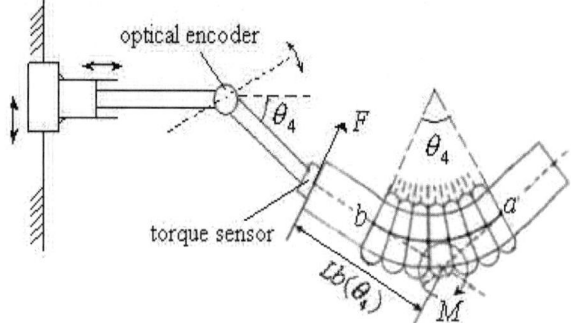

Fig 5. Schematic diagram of 1-DOF measuring device

sensor. θ_4 represents joint angle. $Lb(\theta_4)$ is the length of spacesuit forearm, which includes the constant length of straight arm and the length of the flexible joint(given above). Damping torque of the elbow joint is:

$$M = F \cdot Lb(\theta_4) \qquad (6)$$

Compared with the results measured by this device, the measuring system can be evaluated.

VI Experiments

The following are the experimental results of a charged empty spacesuit (no man inside).

(1) Measuring the shoulder joint

Figure 6 and figure 7 are original curves and B-spline curve of the damping torque of the left shoulder joint respectively. In the figure, the curve of damping torque versus its joint angle exhibits a significant degree of hysteresis. At the same joint angle, damping torque of the joint in abducing is greater than that in adducting.

(2) Measuring the elbow joint

Figure 8 plots the B-spline curve of the damping torque of the left elbow joint, which also has evident hysteresis feature. At the same joint angle, damping torque in bending is greater than that in extending.

These results reveal that the damping torque of the spacesuit's arm joint depends uniquely on the particular history and direction of joint rotation. [5] Hysteresis is the effect of elastic distortion stress when joint moves, which results from the special material and structure of the spacesuit joint.

(3) Measuring the wrist joint

Figure 9 plots the B-spline curve of the damping torque of the left wrist joint. It reveals that the damping torque of this joint is little, which is coincident with the fact.

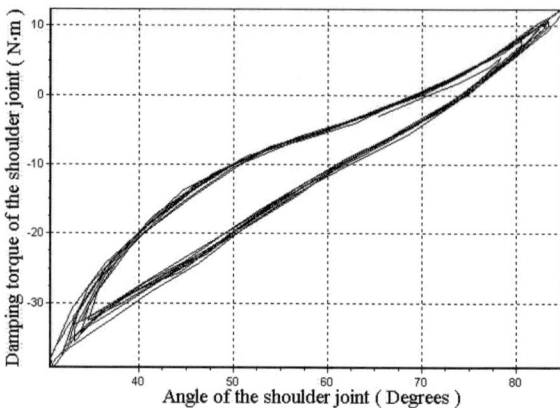

Fig 6. Original curves of the damping torque of the left shoulder joint

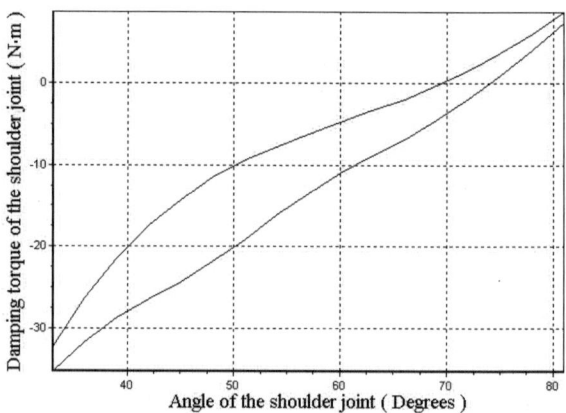

Fig 7. B-spline curve of the damping torque of the left shoulder joint

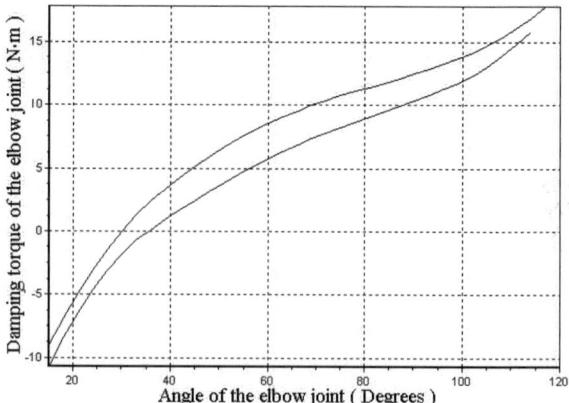

Fig 8. B-spline curve of the damping torque of the left elbow joint

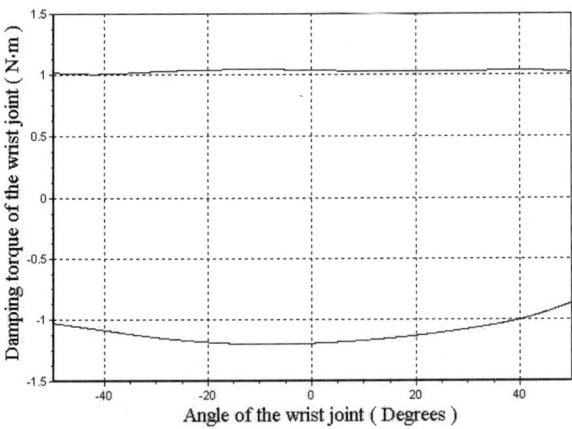

Fig 9. B-spline curve of the damping torque of the left wrist joint

(4) Comparative results

Figure 10 shows the comparative results of damping torque of the left elbow joint that are measured by the robot and by the 1-DOF measuring device.

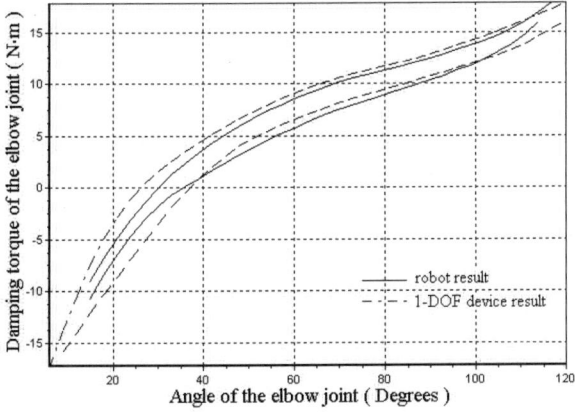

Fig.10 Comparative results of the damping torque of the left elbow joint that are measured by the robot and by the 1-DOF measuring device

In the figure, the solid line is the result measured by the robot, and the broken line is the result measured by the 1-DOF device. Although there is little difference between these two results, their trends are similar, and the curve measured by the robot is mostly inside the hysteresis area of the 1-DOF measuring device.

Experimental results verify the correctness and reliability of the idea. However, differences between the results of robot and 1-DOF device shows that the precision of this measuring system still need to be improved, there are some works to do as followed:
(1) The model of spacesuit arm is a key factor to the precision of the measuring results. Making the model perfect is our proceeding work. More accurate model of flexible joint of the spacesuit is essential for the measuring accuracy.
(2) The gravity torque of the components of spacesuit arm brings also errors to the results. Gravity compensation according to robot position and orientation should be taken into consideration.
(3) The inverse calculation of the spacesuit joints' angle is a difficult problem. A more precise and efficient algorithm is a very important issue.

VII Potential Application

Besides measuring spacesuit joint damping parameters, this measuring system can also be used in some other areas.
(1) Analyzing the reach envelope available to suited astronaut by measuring the position and orientation of the end of spacesuit arm. Thus it provides the work efficiency of the spacesuit.
(2) Following the principle of measuring spacesuit, the torque of human body's joint muscle can be measured by modeling human body as multi-stuffs.

VIII Conclusion

The passive robot system is efficient in measuring spacesuit joint damping parameters. The model of IVA spacesuit's arm realistically reflects its mechanical structure by mathematic expressions, and satisfies the demands of measuring and calculation. Experimental results prove the effectiveness of the measuring principle and the reliability of the measuring system.

IX References

[1]. H.C.Vykukal, B.Webbon. Pressure Suit Joint Analyzer. NASA Case. 1980,7:2~5.
[2]. Du Li-bin, Gao Xiao-hui, Liu Long, Li Tan-qiu. "The Measuring System of Joint Torque of Spacesuit Based on Robot." To be published on Space Medicine & Medical Engineering 2003(2).
[3]. Zhao Jing-dong, Jin Ming-he, Cai He-gao, Hu Hai-ying. "Study on The Measuring Method of Flexible Joint Torque of 3-DOF Spacesuit Based on Robot". Robot, 2002,24(7):753-755.
[4]. Cai Zi-xing. (2000). Robotics. Tsinghua university publishing company, Beijing, China.
[5]. D.J.Newman, P.B.Schmidt, D.Rahn, N.Badler, and D.Metaxas. "Modeling the Extravehicular Mobility Unit (EMU) Spacesuit: Physiological Implications for Extravehicular Activity (EVA)." 30th International Conference on Environmental Systems, Toulouse, France, July 10-13, 2000

Parameters Identification and Vibration Control for Modular Manipulators

Yangmin Li and Yugang Liu
Dept. of Electromechanical Engineering
University of Macau
Av. Padre Tomás Pereira S.J., Taipa
Macao S.A.R., P. R. China

Xiaoping Liu and Zhaoyang Peng
Automation School
Beijing University of Posts and Telecommunications
Beijing 100876
P. R. China

Abstract – The joint parameters of redundant manipulators are prerequisite data for effective dynamics control. An identification method via fuzzy theory and Genetic Algorithm has been presented to study modular redundant robots. The Genetic Algorithm is used in the fuzzy optimization expecting to obtain global optimal solutions. Experimental modal analysis and Finite Element Method have been exploited in dynamics modeling. The joint parameters of a 9-DOF modular redundant robot have been identified. Active vibration control has been approached to a simplified 4-DOF modular manipulator by DOF reduction to the 9-DOF modular manipulator.

1. INTRODUCTION

Extensive research papers can be found in the structure design, kinematics analysis, dynamic behaviour and effective control method study for modular manipulators[1][2], few papers focus on the parameters identification and vibration control of the modular robot[3][4]. The modular robot is composed of different modules via joints. The dynamic characteristics of the flexible joints have important influence on the dynamic characteristics of the whole robot. In order to get the dynamic model of the modular robot with high precision, it is necessary to identify the joint parameters: equivalent stiffness and damping coefficients. The identification of the physical parameters of the robot is much more complicated than the general mechanical structures whose joints are mostly composed of the connecting bolts or welding structure[5]-[8]. In this paper, a method of combining the Finite Element Method(FEM) with the experimental modal analysis is proposed. Because both FEM and experimental modal analysis existing errors, it is difficult to build accurate mathematical models, so the fuzzy theory is applied in the identification of the joint parameters of a 9-DOF modular redundant robot. Genetic Algorithm is used in the fuzzy optimization, which can make the identified results much more reasonable. In addition to parameter identification, active vibration control based on neural network and optimal control method has been carried out.

II. EXPERIMENTAL MODAL ANALYSIS

A. Experiment Procedures

The experiment is made with a 9-DOF modular redundant robot. The structure is excited in different points sequentially and the response is measured in a fixed point. The total number of tested points on the robot is 286. The hammer excitation method is used in this experiment. The excitation is carried on X, Y and Z direction respectively and the acceleration responses of the fixed point on the upper side of the robot are measured in X and Y direction with the accelerometers at the same time. The photo of the experiment is shown in Fig.1.

Fig. 1 Photo of the experiment

B. Results Analysis

The transfer functions between the excitation point and different response points can be obtained via the experiment, only one figure is selected as shown in Fig.2 because of page limitations.

Several lower orders of modes of the robot under the determined gesture are obtained by fitting the transfer function curves with the modal analysis software. The natural frequencies and damping ratios of the most important four orders of modes are shown in Table 1. The main deformation of the modular robot is bending vibration.

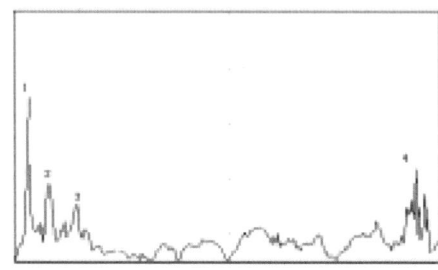

Fig.2 Transfer function for the 149th point in Z direction

Table 1 Natural frequencies and damping ratios

Mode No.	Natural Frequency(Hz)	Damping Ratio
1	28.258	0.03985
2	65.835	0.41387
3	102.993	0.07485
4	536.765	0.00000

C. Dynamic Modeling of the Redundant Robot

The dynamic model of the whole robot can be considered as the models of all modules and the joint surfaces using substructure synthesis method. The joints can be substituted with the equivalent elements of stiffness and the damping in the direction of x, y and z between the finite link points in

the joints. In the modeling of the whole structure with the shell elements, *1835* elements are classified. The measured points in the experiment should be corresponding to the nodes of the FEM. The matrices of stiffness and mass of every module can be obtained by the finite element model. The matrices of the whole robot can be derived by stacking the matrix of every module with the joint elements.
The equation of motion of the whole robot is

$$[M_0]\{\ddot{x}\}+[C_j]\{\dot{x}\}+([K_0]+[K_j])\{x\}=\{f\} \quad (1)$$

where: $[M_0]$ and $[K_0]$ is the mass and stiffness matrix composed by each module respectively in $[n \times n]$; $[C_j]$ and $[K_j]$ is joint damping and stiffness matrix composed by each joint respectively in $[n \times n]$; $\{\ddot{x}\}, \{\dot{x}\}$ and $\{x\}$ is the generalized acceleration, velocity and displacement vectors respectively in $[n \times 1]$; $\{f\}$ is the generalized force vector in $[n \times 1]$; n is the number of generalized coordinates. From equation (1), we can get the eigenvalues and eigenvectors of the robot.

III. IDENTIFICATION METHOD
A. Parameters Identification Model

The *ith* modal equation can be derived by the whole robot system dynamics equation (1) as follows

$$s_i^2\begin{bmatrix}M^{(1)} & & & \\ & M^{(2)} & & \\ & & \ddots & \\ & & & M^{(10)}\end{bmatrix}\begin{Bmatrix}\varphi^{(1)}\\ \varphi^{(2)}\\ \vdots \\ \varphi^{(10)}\end{Bmatrix}_i + s_i\begin{bmatrix}c_{j1} & -c_{j1} & & \\ -c_{j1} & c_{j1}+c_{j2} & -c_{j2} & \\ & -c_{j2} & \ddots & \\ & & & c_{j9}\end{bmatrix}\begin{Bmatrix}\varphi^{(1)}\\ \varphi^{(2)}\\ \vdots \\ \varphi^{(10)}\end{Bmatrix}_i +$$

$$\begin{bmatrix}K^{(1)}+k_{j1} & -k_{j1} & & \\ -k_{j1} & k_{j1}+K^{(2)}+k_{j2} & -k_{j2} & \\ & -k_{j2} & \ddots & \\ & & & k_{j9}+K^{(10)}\end{bmatrix}\begin{Bmatrix}\varphi^{(1)}\\ \varphi^{(2)}\\ \vdots \\ \varphi^{(10)}\end{Bmatrix}_i = 0 \quad (2)$$

where: $M^{(n)}$ and $K^{(n)}$ is the *nth* modular mass and stiffness matrix; c_{jn} and k_{jn} is the joint *n* damping matrix and stiffness matrix respectively; s_i is the measured complex frequency; $\{\varphi^{(1)}, \varphi^{(2)}, \cdots, \varphi^{(10)}\}_i^T$ is the complex mode shapes in the space of the *ith* mode.

In the following equations, subscript *i* represents the *ith* order mode, superscript *(n)* represents the module *n*. In the identification of the joint parameters, the sequence of DOF of the nodes of the model is adjusted firstly. Considering that only parts of the response of the nodes of the FEM are measured in the experiment, the nodes of the FEM can be divided into the measured nodes that do not exist in the joints, unmeasured nodes that do not exist in the joints and the measured joint nodes.

The equation of the module *n* can be written as

$$s_i^2 diag\{M_{j(n-1)jn}^{(n)} \quad M_{uu}^{(n)} \quad M_{mm}^{(n)} \quad M_{jnj(n+1)}^{(n)}\}\begin{bmatrix}\varphi_{j(n-1)}^{(n)} & \varphi_u^{(n)} & \varphi_m^{(n)} & \varphi_{jn}^{(n)}\end{bmatrix}^T +$$

$$\left\{\begin{bmatrix}-k_{j(n-1)}^{(n)} & k_{j(n-1)}^{(n)}+K_{j(n-1)j(n-1)}^{(n)} & K_{j(n-1)u}^{(n)} & K_{j(n-1)m}^{(n)} & K_{j(n-1)jn}^{(n)} & 0 \\ 0 & K_{uj(n-1)}^{(n)} & K_{uu}^{(n)} & K_{um}^{(n)} & K_{ujn}^{(n)} & 0 \\ 0 & K_{mj(n-1)}^{(n)} & K_{mu}^{(n)} & K_{mm}^{(n)} & K_{mjn}^{(n)} & 0 \\ 0 & K_{jnj(n-1)}^{(n)} & K_{jnu}^{(n)} & K_{jnm}^{(n)} & K_{jnjn}^{(n)}+k_{jn}^{(n)} & -k_{jn}^{(n)}\end{bmatrix} +\right.$$

$$\left. s_i\begin{bmatrix}-c_{j(n-1)}^{(n)} & c_{j(n-1)}^{(n)} & & & & \\ & & 0 & & & \\ & & & 0 & & \\ & & & & c_{jn}^{(n)} & -c_{jn}^{(n)}\end{bmatrix}\right\}\begin{Bmatrix}\varphi_{j(n-1)}^{(n)}\\ \varphi_{j(n-1)}^{(n)}\\ \varphi_u^{(n)}\\ \varphi_m^{(n)}\\ \varphi_{jn}^{(n)}\\ \varphi_{jn}^{(n+1)}\end{Bmatrix}_i = 0 \quad (3)$$

where the subscript *j(n-1)*, *jn* and *j(n+1)* represents the measured DOF of the nodes of the joint *n-1*, *n* and *n+1* respectively; subscript *m* and *u* represents the measured and unmeasured DOF of the nodes of the module respectively.

From the second row of the equation (3), we can get

$$(s_i^2 M_{uu}^{(n)} + K_{uu}^{(n)})\varphi_u^{(n)} = -K_{um}^{(n)}\varphi_m^{(n)} - K_{uj}^{(n)}\varphi_{jn}^{(n)} \quad (4)$$

Therefore $\varphi_u^{(n)}$ can be obtained from the equation (4).
From the last row of the equation (3), we can get

$$(s_i c_{jn}^{(n)} + k_{jn}^{(n)})\{\varphi_{jn}^{(n)} - \varphi_{jn}^{(n+1)}\} = -s_i^2 M_{jnj(n+1)}^{(n)}\varphi_{jn}^{(n)} - K_{jn}^{(n)}\varphi^{(n)} \quad (5)$$

where $K_{jn}^{(n)} = \begin{bmatrix}K_{jnj(n-1)}^{(n)} & K_{jnu}^{(n)} & K_{jnm}^{(n)} & K_{jnjn}^{(n)}\end{bmatrix}$.

So $c_{jn}^{(n)}$ and $k_{jn}^{(n)}$ can be derived from the equation (5). The new equation can be composed of the last row of the equation of every module.

$$s_i^2 diag\{M_{j1j2}^{(1)} \cdots M_{j9j10}^{(9)}\}\begin{bmatrix}\varphi_{j1}^{(1)} \cdots \varphi_{j9}^{(9)}\end{bmatrix}^T + diag\{K_{j1}^{(1)} \cdots K_{j9}^{(9)}\}\begin{bmatrix}\varphi_{j1}^{(1)} \cdots \varphi_{j9}^{(9)}\end{bmatrix}^T +$$

$$(s_i diag\{c_{j1}^{(1)} \cdots c_{j9}^{(9)}\} + diag\{k_{j1}^{(1)} \cdots k_{j9}^{(9)}\})\begin{bmatrix}\varphi_{j1}^{(1)} - \varphi_{j1}^{(2)} \cdots \varphi_{j9}^{(9)} - \varphi_{j9}^{(10)}\end{bmatrix}^T = 0 \quad (6)$$

if A_i and B_i are introduced, then the equation (6) can be written as

$$A_i + B_i\{KC_j\}_i = 0 \quad (7)$$

Where $\{KC_j\}_i = \{s_i c_{j1}^{(1)} + k_{j1}^{(1)}, s_i c_{j2}^{(2)} + k_{j2}^{(2)}, \cdots, s_i c_{j9}^{(9)} + k_{j9}^{(9)}\}_i^T$

$$\{KC_j\}_i = B_i^{-1}(-A_i) \quad (8)$$

if $B_i^{-1}(-A_i)$ is defined as

$$B_i^{-1}(-A_i) = \{a_1^i, a_2^i, \cdots a_9^i\}^T \quad (9)$$

The formula of the weighted factor for the joint *n* in the *ith* order of mode is

$$w_n^i = \frac{\left|\varphi_{jn}^{(n)} - \varphi_{jn}^{(n+1)}\right|_i}{\sum_{i=1}^{t}\left|\varphi_{jn}^{(n)} - \varphi_{jn}^{(n+1)}\right|_i} \quad (10)$$

where *t* represents the order of mode; w_n^i represents the weighted factor of the joint *n* in the *ith* mode; $\varphi_{jn}^{(n)}$ and $\varphi_{jn}^{(n+1)}$ represents the mode shapes of the joint *n* in the module *n* and module *n+1* respectively.

The joint parameters identified are represented as

$$\{KC_j\} = \left\{\sum_{j=1}^{t} w_1^i a_1^i, \sum_{j=1}^{t} w_2^i a_2^i, \cdots \sum_{j=1}^{t} w_9^i a_9^i\right\}^T \quad (11)$$

B. Identify Joint Parameters via Fuzzy Theory

Considering that there exist uncertain errors in the FEM and the experiment, the reasonable results can be identified using fuzzy equation. The equation (11) can be written as

$$f(\{KC_j\}) = \{KC_j\} - \left\{\sum_{j=1}^{t} w_1^i a_1^i, \sum_{j=1}^{t} w_2^i a_2^i, \cdots \sum_{j=1}^{t} w_9^i a_9^i\right\}^T \quad (12)$$

The equation(12) can be expressed by the following fuzzy form

$$\|f(\{KC_j\})\| \geq 0 \quad (13)$$

Its membership function is

$$\mu(x) = \begin{cases} 1 - x/\varepsilon & 0 \leq x \leq \varepsilon \\ 0 & x > \varepsilon \end{cases} \quad (14)$$

where $x = \|f(\{KC_j\})\|$.

C. Optimization of Joint Parameters

The objective function is the errors between natural frequencies and mode shapes obtained from the analytical model and the corresponding natural frequencies and shapes obtained from the experiment. The expression of the objective function is

$$\min u = \alpha \sum_{i=1}^{t}(f_i - f_{ie})^2 + \beta \sum_{i=1}^{t}\|\varphi_i - \varphi_{ie}\| \quad (15)$$

where f_i and f_{ie} is the *ith* order of natural frequency obtained from the analytical model and experiment respectively; φ_i and φ_{ie} is the *ith* order of mode shape obtained from the analytical model and experiment respectively; α and β is the weight coefficient respectively.

Equation (13) is the constraint condition for identifying the joint parameters with the fuzzy optimization method. When select $\mu(x) = \lambda$, we can get $x = (1-\lambda)\varepsilon$, where ε is defined to guarantee the solution in the range of the constraint.

Then the constraint condition becomes

$$\|f(\{KC_j\})\| = (1-\lambda)\varepsilon \quad (16)$$

When λ is determined within the range of 0 to 1 according to the step selected as 0.01 or smaller, in every determined λ, we can get a group of optimization solution $\{KC_j\}$, from this series of solutions, we can select the optimized and reasonable solution.

IV. IDENTIFICATION RESULTS ANALYSIS

Since the frequencies and mode shapes identified through the experiment are real values in this paper, only the joint stiffness results can be obtained via above method.

The analytical model of the joint 9 between the module 9 and the module 10 is shown in Fig.3. The other joints are similar to the joint 9, and are omitted here. In Fig.3, there are four pairs of nodes in the four corners, every pair are connected by three springs along the three axis X, Y and Z respectively. The stiffness coefficients of these springs are equivalent to the practical stiffness of the joint. The number in the nodes represents the number order of nodes of the finite element model. In Fig.3, the spring in the direction Z is shown, and those in the other directions are omitted. The stiffness in the direction Z represents the tension stiffness and those in the directions X and Y represent the shear stiffness. There are eight pairs of nodes in the joint 1, six pairs of nodes in the joint 2, four pairs of nodes in the other joints. The total number of pairs of nodes is forty-two. Supposed that the stiffness value in the three directions X, Y and Z between every pair of nodes in a joint be equal each other, then there are one hundred and twenty-six parameters to be identified.

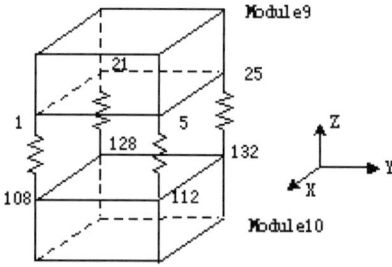

Fig. 3 Sketch map of module of the joint 9

The constraint range of the stiffness parameters is defined as 1.0e+3 to 1.0e+9 (KN/m). when $\varepsilon = 1.0e+3$, the value of the membership is 0.6. By using the fuzzy optimization method, the identified joint stiffness coefficients of the joint 9 is shown in the Table II, the identified parameters of the rest of joints are omitted. Several low order natural frequencies and mode shapes are shown in Table III.

Table II Results of the joint 9

Nodes of Joint		Joint stiffness($\times 10^7 N/m$)		
Module 9	Module 10	X	Y	Z
1	108	4.9517	5.3386	5.2973
5	112	4.8551	4.8319	4.8038
21	128	5.4201	5.0569	4.7360
25	132	5.4008	5.2117	5.1266

The errors of the natural frequencies between the analytical values and the experimental results are small, the largest error is 5.45% as shown in Table III. It demonstrates that the analytical model can substitute the real robot to some extent in the dynamic characteristics.

Table III Comparison of frequencies

Mode No.	Experimental values(HZ)	Analytical values(HZ)	Relative errors
1	28.258	29.800	5.45%
2	65.835	63.386	-3.72%
3	102.993	100.662	-2.26%
4	536.765	534.516	-0.42%

V. DYNAMIC MODELING OF THE MODULAR ROBOT

In order to simplify the structural dynamic model of the robot, the reduction of the number of DOF of every module is carried out. Considering that the flexibility of the robot comes from the joint mainly, the DOF called the master DOF that exist in the corners and the joints of the robot are kept, and the other DOF called the slave DOF are neglected. The vibration of the robot in the direction of x is the main vibration of this robot, the reduction to the model of the robot can be carried out via retain the joint DOF and the DOF of the corners in the direction of x.

The motion of equation of the simplified 4-DOF modular robot is

$$[M]\{\ddot{x}\}+[K]\{x\}=\{f\} \qquad (17)$$

where $[M]=[T]^T[\tilde{M}][T]$, $[K]=[T]^T[\tilde{K}][T]$ and $\{f\}=[T]^T[\tilde{f}][T]$ represents the mass matrix, stiffness matrix and the force vector of the reduced model respectively; $\tilde{M}$, $\theta_o^{(i)}$ and $\tilde{f}$ represents the original mass matrix, stiffness matrix and the force vector of the 9-DOF modular robot respectively; $[T]$ represents the transfer matrix.

VI. Active Vibration Control Method

In this paper, genetic algorithm is used to initiate the weights of the BP network so that the weights of the BP network can be closed to the global optimization point at the beginning. The BP network is used to adjust the error online.

A. State Equation of the Simplified Model of the Robot

The vibration control of the simplified model of the robot in the vertical gesture is studied by using the BP neural network algorithm. Let $x_1=\{x^{(1)} \ x^{(2)} \ \cdots \ x^{(n)}\}^T$ and $x_2=\{\dot{x}^{(1)} \ \dot{x}^{(2)} \ \cdots \ \dot{x}^{(n)}\}^T$, then Equation (17) can be written as

$$\begin{bmatrix}I & 0\\0 & M\end{bmatrix}\begin{Bmatrix}\dot{x}_1\\\dot{x}_2\end{Bmatrix}=\begin{bmatrix}0 & I\\-K & 0\end{bmatrix}\begin{Bmatrix}x_1\\x_2\end{Bmatrix}+\begin{bmatrix}0\\I\end{bmatrix}\{f\} \qquad (18)$$

Furthermore

$$\begin{Bmatrix}\dot{x}_1\\\dot{x}_2\end{Bmatrix}=\begin{bmatrix}0 & I\\-M^{-1}K & 0\end{bmatrix}\begin{Bmatrix}x_1\\x_2\end{Bmatrix}+\begin{bmatrix}0\\M^{-1}\end{bmatrix}\{f\} \qquad (19)$$

Define $s=\{x_1 \ x_2\}^T$ as the state vector, the state equation of the simplified robot model can be obtained

$$\dot{s} = As + Bu \qquad (20)$$
$$y = Cs \qquad (21)$$

where $A=\begin{bmatrix}0 & I\\-M^{-1}K & 0\end{bmatrix}$, $B=\begin{bmatrix}0\\M^{-1}\end{bmatrix}$ and $C = diag\{0\cdots1\cdots1\cdots0\}$ are the coefficient matrices of the state equation and the output equation, and in the matrix C, the number 1 denotes the positions of the sensors in the control system. $u=\{f\}=\{f_e\}+\{f_c\}$ is a vector of force composed by the external force vector and the control force vector.

B. Control System Structure

Control system block diagram is shown in Fig.4.

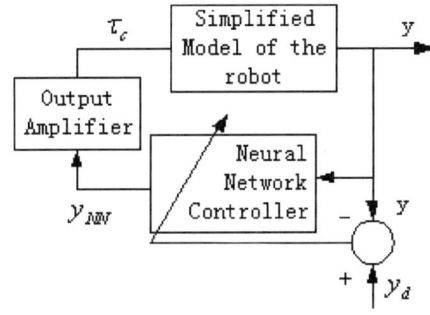

Fig.4 Control system diagram

The state values measured by the sensors are used as the input of the neural network controller, at the same time the weights of the neural network are adjusted via the difference between the practical state and the desired response. $\tau_c=\{\tau_{c1} \ \tau_{c2} \ \tau_{c3} \ \tau_{c4}\}^T$ are the vector of controlled moment; k is the coefficient of output amplifier.

C. BP Neural Network Controller Design

The BP network with four input layers, four output layers and ten hidden layer is designed as the active vibration controller. The outputs of the BP neural network are added on to the joints of the model in the form of moments.

The relationship between the input and output in every layer of the neural network is defined as

$$\tau_c^{(i)} = K \times y_{NN}^{(i)}, \ y_{NN}^{(i)} = a\left[\frac{1-\exp(-bv_o^{(i)})}{1+\exp(-bv_o^{(i)})}\right],$$

$$v_o^{(i)} = \sum_{j=1}^{j\leq 10} y_h^{(j)} w_{ho}^{(ij)} - \theta_o^{(i)}, \ y_h^{(j)} = a\left[\frac{1-\exp(-bv_h^{(j)})}{1+\exp(-bv_h^{(j)})}\right]$$

$$v_h^{(j)} = \sum_{m=1}^{4} y_{in}^{(m)} w_{ih}^{(jm)} - \theta_h^{(j)}, \ y_{in}^{(m)} = \begin{cases} y_i & (m=1)\\ y_j & (m=2)\\ y_{n+i} & (m=3)\\ y_{n+j} & (m=4)\end{cases} \qquad (22)$$

Where $v_h^{(j)}$ and $v_o^{(i)}$ represents the input of neurons of the hidden layer and the output layer respectively.

The state discrete equations corresponding to the equations (20) and (21) are

$$s(k+1) = G \cdot s(k) + H \cdot u(k) \qquad (23)$$
$$y(k) = C \cdot s(k) \qquad (24)$$

where $G=e^{A\cdot\delta t}$, $H=\int_0^{\delta t} e^{A\cdot t}dt \cdot B = A^{-1}(e^{A\cdot\delta t}-E)B$; A is a non-singular matrix; A and B matrices are time-invariant

3257

matrices, E is the unit matrix, δ_t is the sampling time; A, B and C are all matrices of constant coefficients.

Because the joints of the robot are all revolute joints, the forces have to be changed to moments to be added on to the model. Changing the force in equation (23) to moment we can obtain

$$s(k+1) = G \cdot s(k) + H \cdot D \cdot \{\tau_o(k) + \tau_c(k)\} \quad (25)$$

Where $s(k)$ and $s(k+1)$ are the state values in the *kth* and *(k+1)th* sampling instants respectively; $\tau_c(k)$ and $\tau_o(k)$ are the control moments and external forces acted on the robot after the *(k-1)th* and *kth* sampling; D is a coefficient matrix.

Since the aim of the vibration control is to reduce the vibration, the anticipated sampling result should be zero. Thus the difference between the sampling result and the anticipated result in the *kth* sampling is

$$\varepsilon(k) = y(k) - y_d = y(k) \quad (26)$$

In order to reduce the vibration validly, at the same time do not need to use much bigger moment, combining with the sub-optimal control, the object function for the neural network controller can be defined as

$$J = \sum_{k=0}^{N} \frac{1}{2} y^T(k) \cdot y(k) + \sum_{k=0}^{N-1} \frac{1}{2} \tau_c^T(k) \cdot \tau_c(k) \quad (27)$$

Since
$$x(1) = G \cdot x(0) + H \cdot D \cdot \tau_c(0)$$
$$x(2) = G \cdot x(1) + H \cdot D \cdot \tau_c(1)$$
$$= G^2 \cdot x(0) + G \cdot H \cdot D \cdot \tau_c(0) + H \cdot D \cdot \tau_c(1)$$
$$\ldots\ldots$$
$$x(N) = G^N \cdot x(0) + G^{N-1} \cdot H \cdot D \cdot \tau_c(0)$$
$$+ G^{N-2} \cdot H \cdot D \cdot \tau_c(1) + \cdots + H \cdot D \cdot \tau_c(N-1)$$

Thus
$$\frac{\partial x(N)}{\partial \tau_c(k)} = G^{N-1-k} \cdot H \cdot D \quad (k=0,1,2\ldots N-1)$$

Similarly, we can obtain,

$$\frac{\partial x(i)}{\partial \tau_c(k)} = G^{i-1-k} \cdot H \cdot D \quad (i=1, 2, \ldots N, k=0, 1\ldots i-1) \quad (28)$$

Therefore

$$\frac{\partial y(i)}{\partial \tau_c(k)} = C \cdot G^{i-1-k} \cdot H \cdot D \quad (29)$$

Comprehensive consideration of the equations (24), (27) and (29), we can obtain

$$\frac{\partial J}{\partial \tau_c(N-1)} = y^T(N) \cdot C \cdot H \cdot D + \tau_c(N-1)$$

$$\frac{\partial J}{\partial \tau_c(N-2)} = y^T(N) \cdot C \cdot G \cdot H \cdot D$$
$$+ y^T(N-1) \cdot C \cdot H \cdot D + \tau_c(N-2)$$

$$\frac{\partial J}{\partial \tau_c(0)} = y^T(N) \cdot C \cdot G^{N-1} \cdot H \cdot D + y^T(N-1) \cdot C \cdot G^{N-2} \cdot H \cdot D \quad (30)$$
$$+ \cdots + y^T(1) \cdot C \cdot H \cdot D + \tau_c(0)$$

The adjustable parameters of the neural network controller are $w_{ho}^{(ij)}$, $w_{ih}^{(jm)}$, $\theta_o^{(i)}$ and $\theta_h^{(j)}$; the gradients that the output $\tau_c^{(i)}$ with respect to these parameters are

$$\nabla_{w_{ho}^{(ij)}} J = \frac{\partial J}{\partial \tau_c^{(i)}(k)} \cdot \frac{\partial \tau_c^{(i)}(k)}{\partial w_{ho}^{(ij)}}$$
$$= T^{(i)} \cdot K \cdot \frac{b}{2a} \cdot (a + y_{NN}^{(i)}) \cdot (a - y_{NN}^{(i)}) \cdot y_h^{(j)}$$

$$\nabla_{\theta_o^{(i)}} J = \frac{\partial J}{\partial \tau_c^{(i)}(k)} \cdot \frac{\partial \tau_c^{(i)}(k)}{\partial \theta_o^{(i)}}$$
$$= T^{(i)} \cdot K \cdot \frac{b}{2a} \cdot (a + y_{NN}^{(i)}) \cdot (a - y_{NN}^{(i)}) \cdot (-1)$$

$$\nabla_{w_{ih}^{(jm)}} J = \sum_{i=1}^{i \leq 4} \left[\frac{\partial J}{\partial \tau_c^{(i)}(k)} \cdot \frac{\partial \tau_c^{(i)}(k)}{\partial w_{ih}^{(jm)}} \right]$$
$$= \sum_{i=1}^{i \leq 4} \left[T^{(i)} \cdot K \cdot \frac{b^2}{4a^2} \cdot (a + y_{NN}^{(i)}) \cdot (a - y_{NN}^{(i)}) \right.$$
$$\left. \cdot w_{ho}^{(ij)} \cdot (a + y_h^{(j)}) \cdot (a - y_h^{(j)}) \cdot y_{in}^{(m)} \right]$$

$$\nabla_{\theta_h^{(j)}} J = \sum_{i=1}^{i \leq 4} \left[\frac{\partial J}{\partial \tau_c^{(i)}(k)} \cdot \frac{\partial \tau_c^{(i)}(k)}{\partial \theta_h^{(j)}} \right]$$
$$= \sum_{i=1}^{i \leq 4} \left[T^{(i)} \cdot K \cdot \frac{b^2}{4a^2} \cdot (a + y_{NN}^{(i)}) \cdot (a - y_{NN}^{(i)}) \right.$$
$$\left. \cdot w_{ho}^{(ij)} \cdot (a + y_h^{(j)}) \cdot (a - y_h^{(j)}) \cdot (-1) \right]$$
$$(31)$$

Where $T^{(i)} = \frac{\partial J}{\partial \tau_c^{(i)}(k)}$ can be obtained from equation (30); $a=1.716$, $b=2/3$ are the parameters used in the hyperbolic tangent function of the activation function of the hidden layer.

The adjustment formulae for the neural network parameters can be obtained as follows

$$w_{ho}^{(ij)}(k) = w_{ho}^{(ij)}(k-1) - \eta \nabla_{w_{ho}^{(ij)}} J$$
$$\theta_o^{(i)}(k) = \theta_o^{(i)}(k-1) - \eta \nabla_{\theta_o^{(i)}} J$$
$$w_{ih}^{(jm)}(k) = w_{ih}^{(jm)}(k-1) - \eta \nabla_{w_{ih}^{(jm)}} J$$
$$\theta_h^{(j)}(k) = \theta_h^{(j)}(k-1) - \eta \nabla_{\theta_h^{(j)}} J \quad (32)$$

Where learning rate is $\eta = 1 \times 10^{-3}$.

D. Initialization of Weights and Thresholds of Neural Network

D.1 Encoding

The floating-point array vector is taken as decision-making variable whose values are the weights and thresholds of BP network.

D.2 Optimization Model

In this paper, group size is selected as 200 and the objective function is as followed

$$J = \frac{1}{2} y^T(N) \cdot y(N) + \sum_{k=0}^{N-1} \frac{1}{2} \left[y^T(k) \cdot y(k) + \tau_c^T(k) \cdot \tau_c(k) \right] \quad (33)$$

In order to keep consistent with survival of the fittest theory, the fitness function value is taken as: $fitness = \frac{1}{J}$.

D.3 Selection Operator

In this paper, proportional selection operator is used to decide which individual will be passed down to the next generation. Moreover, Elitist Model is used in this study.

D.4 Arithmetic Crossover Operator

Assuming that Y_A^t and Y_B^t are two old individuals, then two new individuals created by arithmetic crossover operator will be as followed

$$\begin{cases} Y_A^{t+1} = \alpha Y_B^t + (1-\alpha) Y_A^t \\ Y_B^{t+1} = \alpha Y_A^t + (1-\alpha) Y_B^t \end{cases} \quad (34)$$

Where $\alpha = 0.25$, crossover rate is $p_m = 0.6$.

D.5 Uniform mutation Operator

Assuming that the gene value on a mutation point be in the zone of $[u_{min}, u_{max}]$, mutation operator is replacing the old gene value with a random value in this zone. The parameters are selected as followed: $u_{min} = -1$, $u_{max} = 1$ and $p_m = 0.001$.

VII. SIMULATION RESULTS

Controlling simulation of the simplified modular robot is carried out. The state feedback point is the place where the sensors are put. Wilson-θ method is used to calculate the response in time domain. Sampling interval is *100μs* and the number of sampling points is 200. The output of BP network is changed into torques and they are put onto the joints of the simplified modules. Ten groups of simulation analysis are carried out using the GAs-based BP network controller and the classical BP network controller. The GAs-based BP network controller is more steady and easier to converge as shown in Fig.5 (a) and (b). The result shows that the GAs-based BP network controller overcomes the problem of local optimization effectively, and it is more appropriate for vibration control than the classical BP network.

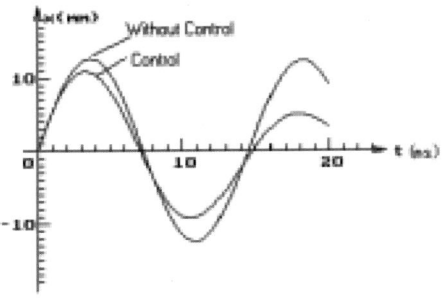

(a) Over 10 times of training

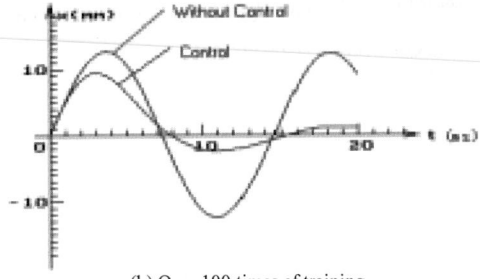

(b) Over 100 times of training

Fig.5 Controlling results using the BP neural network and genetic algorithm sub-optimal controller

VIII. CONCLUSIONS

The experimental modal analysis for a 9-DOF modular robot is carried out, and the dynamic behaviors of the robot are investigated. The joint stiffness parameters of the robot are identified by means of the method presented. The modeling of the 9-DOF modular redundant robot with the finite element method is carried out, which uses the joint parameters of stiffness and damping coefficients identified by the model of every module and the experimental modal values. An active vibration controller is designed for a simplified 4-DOF modular robot via combining the genetic algorithm, neural network with the sub-optimal control. It is shown that the neural network and genetic algorithm based sub-optimal controller is more steady and easier to converge than the classical neural network controller.

IX. ACKNOWLEDGMENTS

This project was funded by the Research Committee of University of Macau under grant no. RG 025/00-01S/LYM/FST and RG037/01-02S/LYM/FST.

X. REFERENCES

[1] D. Schmitz, P. Khosla, and T. Kanade, "The CMU Reconfigurable Modular Manipulator System" *Technical Report CMU-RI-TR-88-07*, Robotics Institute, Carnegie Mellon University, May, 1988.

[2] Y. Fei, X. Zhao and L. Song, "A method for modular robots generating dynamics automatically", *Robotica*, Vol.19, 2001, pp.59-66.

[3] X. Liu and B. Yang, "Structure vibration control by tuned distributed vibration damper", in 1999 *ASME Design Engineering Technique Conferences*, Las Vegas, Nevada, USA, Sept. 12-15, 1999, pp.1-9.

[4] X. Lu, Y. Tao and J. Zhou, "Neural network based active vibration control of flexible structures", *Int. J. Robot. Res.*, Vol.13, No.1, 2000, pp.107-111.

[5] F. Behi and D. Tesar, "Parametric identification for industrial manipulation using experimental modal analysis", *IEEE Transaction on Robotic and Automation* Vol.7, No.5, 1991, pp.642-652.

[6] Y. Li, C. Xie, Y. Liu, "Parameter identification on Puma 760 robot dynamics", *in The Second Asia Conference on Robotics and Its Application*, Beijing, P. R. China, Oct., 1994, pp.155-158.

[7] J. Wang and P. Sas, "A method for identifying parameters of mechanical joints", *ASME Journal of the Mechanical Design*, Vol.57, 1990, pp.31-39.

[8] M. Hashimoto and Y. Kiyosawa, "The effects of joint flexibility on robot motion control using joint torque sensors", *Inl.J. Robot. Autom.*, Vol.13, No.2, 1998, pp.43-47.

Experimental Identification and Evaluation of Performance of a 2DOF Haptic Display

Antonio Frisoli   Massimo Bergamasco

PERCRO - Scuola Superiore S. Anna - Pisa, Italy

e-mail: antony@sssup.it, bergamasco@sssup.it

Abstract

This paper presents a methodology for the evaluation of the performance of a given Haptic Display. The procedure can be carried out through the own proper sensors/actuators which equip the haptic display and allows to characterize the device from a static/dynamic point of view.

1 Introduction

Haptic Interfaces are robotic devices designed for conveying to the user a realistic sensation of touching an object in the virtual world. The characterization of performance for an Haptic Interface relies on the evaluation of several parameters which are inherent to the mechanical properties of the device, such as dynamic and force bandwidth, transparency and perceived mass, maximum stiffness [6, 13, 12].

When dynamic properties of a given Haptic Display are known, several analyses can be carried out, such as about the absolute stability of the device, when displaying virtual impedances/admittance to a human operator [1]. This allows to improve the design of the controller, which can be adapted to overcome the limitations of the mechanics. From the points presented above, it can be argued how relevant it is to experimentally characterize a given haptic display.

Most of techniques for measuring the performance of a robotic system are based on identification schemes, which usually require additional instrumentation, according to the signals to be measured. In robotics literature, the identification of unknown parameters is often carried out by sensorizing the system either with accelerometers [3, 8] or with force sensors [11]. In fact, by measuring the input-output gain between forces and accelerations, a more accurate estimate of the transfer function can be obtained.

When only co-located force/position measurements are available, such as commonly it is the case of industrial manipulators -seldom equipped with velocity sensors-, available techniques are based on exact-linearization and band-pass filtering [14]. In fact in practice it is quite difficult to measure indirectly velocity as well as higher order signals from position measurements, as they are corrupted by the noise. For instance the quantization error introduced by encoder measurements considerably worsens the quality of velocity signal [7] and renders unsuitable the derivation of the acceleration signal from position measurements only.

Another possible approach is the formulation of the dynamic equations in terms of an integral model, alternatively to the classical differential formulation, which does not require the measure of joint velocities [10]. However when the identification is restricted to a single joint, without nonlinearities due to gravity compensation and with consistent decoupling of the dynamics of different DOFs, the identification analysis can be conducted on each isolated dof, also in the case of parallel actuation, by assuming a LTI relation between force and positions, with good results in terms of estimate of the system model. Under the assumption of linearity, spectral analysis represents a powerful mean [16] for determining unknown parameters of the transfer function, since it allows to exclude the influence of random noise present on measured signals.

This paper presents a practical approach for measuring some of the most relevant dynamic properties of Haptic Displays, particular suitable for the characterization of mechanisms actuated through tendon drives. The method does not require a special sensorization of the device and can be carried out by only position measurements. Tendon drives are commonly adopted in haptics for their smooth behavior, since allow to ground the actuators with low friction and attaining high back-drivability. As main disadvantage, the dynamic response of the system is greatly influenced by the elasticity of tendons. As a case study, an experimental identification scheme has been applied to the PERCRO Isotropic Force Display, in order to assess its dynamic performance.

2 The Isotropic Force Display

The Isotropic Force Display is a 2-dof planar haptic device, with high kinematic isotropy over the worskpace, which has been realized at PERCRO [15]. It is composed of two rotary actuators, driving a closed 5-bar linkage by two pairs of opposed tendons realized through steel cables. The actuators are located apart from the linkages of the mechanism.

The starting terminal of a single tendon connects to the pulley mounted on the motor shaft, whilst its end terminal is attached to the grounded base link (the reader is referred to Figure 1). Guide pulleys of different radii route orderly the tendons clockwise or counter-clockwise over circular primitives centered at the joint axes. In the developed design all guide-pulleys are mounted on ball bearings, but the last one is bolted to the base link. The routing is then completed for each motor by a second tendon that is routed in a opposite way on the joint pulleys, in order to realize a pre-tensioned bi-directional tendon drive.

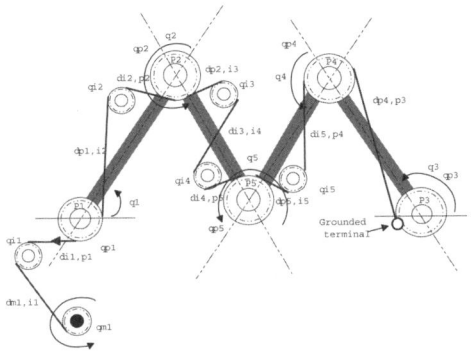

Figure 1: Kinematic representation of the system with relative notation

The tendon drive couples linearly the angular displacements of all the joints to the motor displacements. By actuating one motor, while the other one is locked, the mechanism motion is kinematically constrained by the tendon drive, as represented in Figure 4. The particular nature of the tendon drive allows to shape the force response of the system to achieve high value of kinematic isotropy over the workspace. Two high resolution encoders are placed on the motor joints for sensing the position of the end-effector in the workspace. The Isotropic Force Display is controlled through a DSpace 1103 real-time control board under the RTI Matlab environment.

Figure 2: The Isotropic Force Display

2.1 2-port linear representation

The dynamics of the system when it is constrained to move only along a given direction, e.g. by locking one actuator while the other is left free, can be described in terms of a 2-port linear system, as commonly done in other works, e.g. [1]. By reducing the number of degrees of freedom of the system, it is possible to analyze through standard tools for linear networks, properties such as transparency and unconditional stability of the system, when coupled to an operator.

If we consider the scheme given in Figure 1, each actuated degree of freedom can be modeled as in Figure 3. M_m and M_l represents respectively the equivalent motor and link mass. The tendon transmission is represented through a damping factor b_t (usually negligible) and a stiffness constant k_t, while b_p and b_m are respectively the intrinsic structural and the motor damping factors [1].

Such an hypothesis is acceptable when the dynamics of the idle pulleys is negligible with respect to the actuator dynamics and if the non-linear terms due to the Coriolis-centrifugal effects are considerably lower than the main inertial terms, when evaluated at the reference velocities.

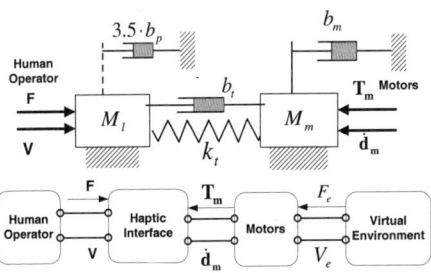

Figure 3: The 2-port linear model of the Haptic Interface system

The haptic interface acts as a two-ports networks which exchanges energy with the operator through the applied force F and velocity V and with the actuator through the tendon tension T and tendon displacement $\dot{d}_m$, measured at the motor location [1]. A general two-port system captures the relationship between *efforts* and *flows* at the two accessible terminal pairs. As the two-port system is a linear 4-terminal network, it can be represented through Y-parameters (Short Circuit Admittance Parameters) or H-parameters (Hybrid Parameters) [2], which are frequency dependent functions:

$$\begin{bmatrix} V \\ \dot{d}_m \end{bmatrix} = \begin{pmatrix} Y_{11} & Y_{12} \\ Y_{21} & Y_{22} \end{pmatrix} \begin{bmatrix} F \\ T_m \end{bmatrix} \quad Y-parameters \quad (1)$$

$$\begin{bmatrix} F \\ \dot{d}_m \end{bmatrix} = \begin{pmatrix} H_{11} & H_{12} \\ H_{21} & H_{22} \end{pmatrix} \begin{bmatrix} V \\ T_m \end{bmatrix} \quad H-parameters \quad (2)$$

The two representations are related by the following expression:

$$H = \begin{pmatrix} Y_{11}^{-1} & -Y_{11}^{-1} Y_{12} \\ Y_{11}^{-1} Y_{21} & Y_{22} - Y_{21} Y_{11}^{-1} Y_{12} \end{pmatrix} \quad (3)$$

---

[1]The coefficient 3.5 multiplying b_p stems from the solution of the dynamics for the general 2 DOF case

In particular Y_{11} represents the input admittance seen by the operator when $T_m = 0$, while Y_{22} represents the input admittance seen by the motor when $F = 0$. H_{12} is the transfer function between velocity at motor $\dot{d}_m$ and velocity at the end-effector V when the motor is free $T_m = 0$.

Note that the sign of effort and flow variables has been chosen such that the effort is forcing the flow inside the system. Moreover once the effort variable has been chosen, the positive sign of corresponding flow variable leads to an increase of system internal energy (see figure 3 for the sign convention).

After having solved the dynamic equations for the general case of the 2DOF mechanism, and linearized it in a point of the workspace (see [4, 5] for details), we found that the parameters for the hybrid formulation can be expressed for each actuator as:

$$H_{11} = \left.\frac{F}{V}\right|_{T_m=0} = \qquad (4)$$

$$\left(\frac{1}{\tau}\right)^2 \frac{s^3 M_l M_m + a_1 s^2 + a_2 s + (b_m + 3.5 b_p) k_t}{M_m s^2 + (2.5 b_p + b_m + b_t) s + k_t} \qquad (5)$$

$$H_{12} = \left.\frac{\dot{d}_m}{V}\right|_{T_m=0} = \frac{1}{\tau} \frac{b_t s + k_t}{M_m s^2 + (2.5 b_p + b_m + b_t) + k_t} \qquad (6)$$

$$H_{21} = \left.\frac{F}{T_m}\right|_{V=0} = -H_{12} \qquad (7)$$

with

$$a_1 = (M_l b_m + b_t M_m + 2.5 b_p M_l + b_t M_l + b_p M_m) \qquad (8)$$

$$a_2 = (b_t b_m + k_t M_m + k_t M_l + b_p b_m + 3.5 b_t b_p + 2.5 b_p^2) \qquad (9)$$

The constant τ represents the force/velocity transmission ratio between end-effector and motor displacements.

For the Y-parameters we have:

$$Y_{12} = \left.\frac{V}{T_m}\right|_{F=0} = -\tau \frac{b_t s + k_t}{s^3 M_l M_m + a_1 s^2 + a_2 s + (b_m + 3.5 b_p) k_t} \qquad (10)$$

$$Y_{22} = \left.\frac{\dot{d}_m}{T_m}\right|_{F=0} = \frac{s^2 M_l + (b_t + b_p) s + k_t}{s^3 M_l M_m + a_1 s^2 + a_2 s + (b_m + 3.5 b_p) k_t} \qquad (11)$$

3 The identification procedure

In order to test the linear model given by the above equations, the dynamic response of the system was experimentally investigated. The measures were made by using the available co-located actuators and position sensors. The identification procedure was applied separately to the two actuated dof, in the center position of the workspace. According to the scheme depicted in figure 4, one motor at a time was kept still in zero by a stiff proportional position control (simulating a virtual bi-directional slide), while the second one was excited with a given torque command (input signal). The resultant motor position (output signal) was read through the motor encoders and stored.

The excitation input signal for identification was a multisine composed with frequencies ranging from $1, 2, \ldots, 120\,Hz$. The number of points composing a period of the excitation signal was at most of 1000 points. The command torque was estimated through the current delivered by the driver to the motor. A sampling frequency of $f_s = 500\,Hz$ was used for running the simulation. Figure 5 shows a time domain representation of input data, with a typical output response. The amplitude of the excitation signal was set to $30\,Nmm$, which is quite above the threshold of static friction ($4 - 6\,Nmm$).

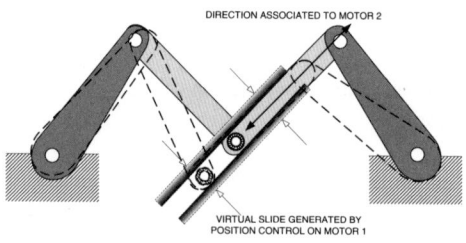

Figure 4: The identification procedure

Then following a standard approach in spectral analysis [16, 9], the magnitude of the Bode plot associated to the transfer function $G(j\omega)$ has be drawn on the basis of the output and input spectral density $\Phi_{yy}(\omega)$ and $\Phi_{xx}(\omega)$, and of autocorrelation functions (ACF) of input and outputs signals $\phi_{yy}(\omega)$ and $\phi_{xx}(\omega)$, defined as

$$\Phi_{xx}(\omega) = \int_{\infty}^{\infty} \phi_{xx} e^{-j\omega\tau} d\tau \qquad (12)$$

$$\phi_{xx}(\tau) = \lim_{T \to \infty} \int_{-T}^{+T} x(t) x(t + \tau) dt \qquad (13)$$

$$\Phi_{yy}(\omega) = \int_{\infty}^{\infty} \phi_{yy} e^{-j\omega\tau} d\tau \qquad (14)$$

$$\phi_{yy}(\tau) = \lim_{T \to \infty} \int_{-T}^{+T} y(t) y(t + \tau) dt \qquad (15)$$

The relations between the magnitude and phase of transfer function $G(j\omega)$, corresponding to Y_{22}/s (see (11)), are as follows:

$$\Phi_{yy}(\omega) = |G(j\omega)|^2 \Phi_{xx}(\omega) \implies M(\omega) = \sqrt{\frac{\Phi_{yy}(\omega)}{\Phi_{xx}(\omega)}} \qquad (16)$$

$$\Phi_{yx}(\omega) = G(j\omega) \Phi_{xx}(\omega) \implies \phi(\omega) = \angle\left(\frac{\Phi_{yx}(\omega)}{\Phi_{xx}(\omega)}\right) \qquad (17)$$

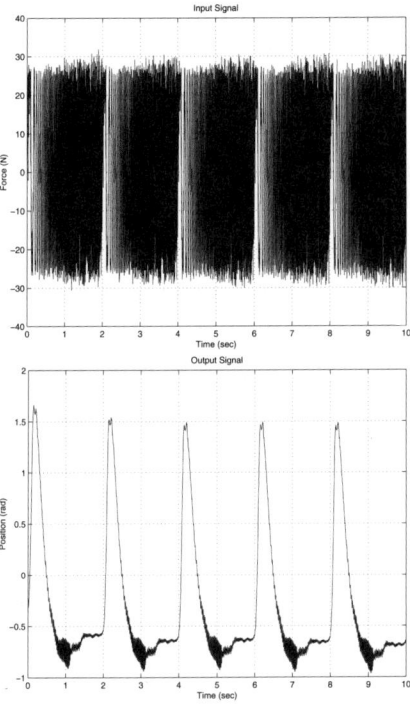

Figure 5: Input and output signals in time domain

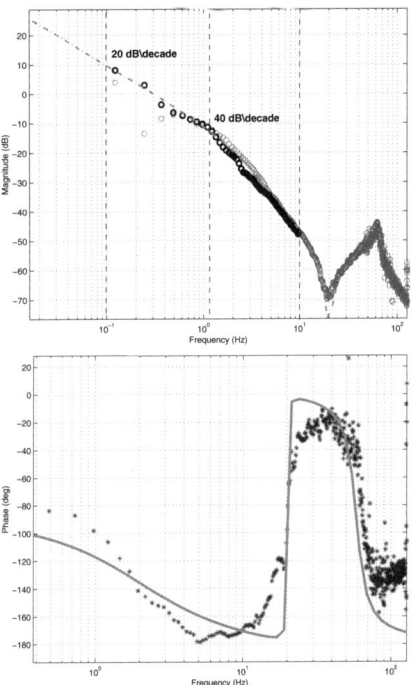

Figure 6: Experimental Bode plot of Y_{22}/s (see eq. (11))

The resultant experimental bode plot is shown in figure 6 for the case of one actuator.

It is evident the presence of a resonant mode at about $62\,Hz$, and the effect of the double zero of the transfer function, according to the expression given in (11).

An expression, called the *coherence function* is commonly evaluated to estimate the degree of distortion of the clean output caused by the presence of noise. The coherence function is defined as [16]:

$$\gamma_{xy}^2 = \frac{|\Phi_{xy}(\omega)^2|}{\Phi_{xx}(\omega)\Phi_{yy}(\omega)} \qquad (18)$$

If the coherence function value is approximately equal to 1.0 (generally ≥ 0.8), then experimental data are considered reliable, in the sense that they have overcome the noise disturbance [8]. Small values of coherence function imply lack of reliability of measured data.

As the plot of figure 7 reveals, the experimental diagram are reliable in the range of frequencies which have been used for exciting the system.

The results of scanning the system response over a wider range of frequencies (see figure 8), up to the limit value of the Nyquist frequency of $250\,Hz$ has confirmed that the system does not present unmodeled dynamic modes at frequency above the resonance frequency and within the range of $250\,Hz$. On the basis of the above experimental results, the frequency response based on the model of the Haptic Interface system has been assessed and tuned up. From the analysis of Bode plots of Figure 6, it can be argued the presence of a pole around $1\,Hz$, with the change in the slope of the magnitude plot. The superimposed modeled response, plotted in the same figure with continuous line, has been obtained by using a value of $b_p = 0.2 \cdot b_m$, which corresponds to a location of a pole at $1.9851\,Hz$. The value of b_m has been assumed known from motor datasheet. It can be seen from the Bode plot, that the mechanical resonant frequency ω_m and the anti-resonant frequency ω_l are respectively $\sim 62\,Hz$ and $20\,Hz$. If we calculate the resonance

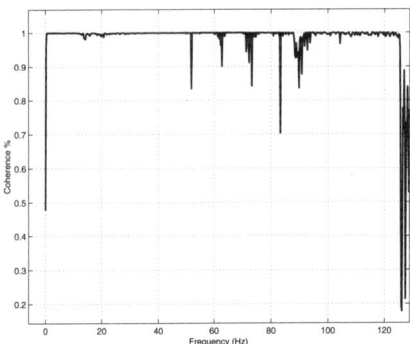

Figure 7: Coherence function for diagram of figure 6

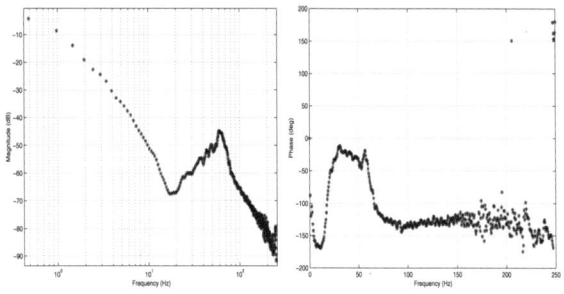

Figure 8: Bode plot of Figure 6 obtained for a scan up to the Nyquist frequency of 250 Hz

ratio R of the two frequencies [17], we can verify that the experimental results agree with CAD estimates. In fact from the value of motor and link equivalent inertias M_m and M_l, it holds:

$$\frac{1}{R} = \frac{\omega_l}{\omega_m} = 3.1 = \simeq \sqrt{\frac{\frac{1}{M_m} + \frac{1}{M_l}}{\frac{1}{M_l}}} = 2.9 \quad (19)$$

The stiffness constant k_t can be estimated from the value of the anti-resonant frequency ω_l [17] as:

$$k_t = (\omega_l)^2 M_l = 9.1622 \frac{N}{mm} \quad (20)$$

If r_m is the radius of the motor pulley, this value measured for linear motion of the tendon terminal, can be reduced to the equivalent angular value $k_m = k/r_m^2 = 0.5864 \text{Nm/rad}$. In order to verify the previous estimate of mechanical stiffness, an experimental static measure was done, by blocking the link against a rigid mechanical stop, and applying a linearly increasing torque to the motor. The displacement measured at motor was due to both the elasticity of tendon and structural compliance of the system. The coefficients of a

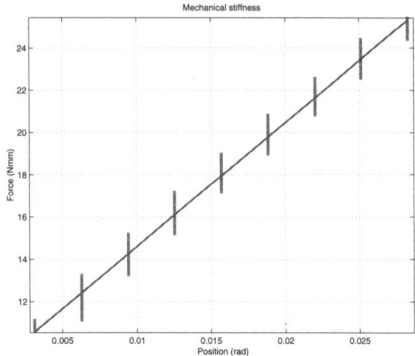

Figure 9: Mechanical stiffness at the motor side

straight line which fits the data $x(i)$ to $y(i)$, in a least squares sense, are given by $y = 0.586 \cdot x + 8.7371 \cdot 10^{-3}$ (see Figure 9). The empiric value $k = 0.586 \frac{Nm}{rad}$ agrees with previous results. The constant term 8.7371 Nmm is an estimate of the breakaway force at motor joint, due to static friction.

On the basis of the above conclusions, the hybrid parameters H_{ij} have been numerically estimated. From plots of figure 6 it is possible to observe that the theoretical model fits with the experimental with a good correlation. However it must be pointed out that the estimation of the damping factor is not precise, since non-linearities due to friction forces, present at low frequencies, determine a greater estimate with respect to the real value.

4 Transparency and Structural response

On the basis of the identification results, two common measures of performance for an haptic interface [3], the force and the structural dynamic response of the system, can be estimated. Figure 10(a) shows the frequency gain between the input torque T_m applied at the motor side and the transmitted force F at the end-effector, under the hypothesis that the end-point is held stationary in a given position, i.e. $V = 0$, corresponding to H_{12}, eq. (6). From this plot it is possible to have an estimate of the dynamic force bandwidth of the system [13]. For the Isotropic Force Display spectral components of the input force with frequencies greater than 70 Hz are damped by the system. Below 10 Hz forces and velocities are transmitted according to the law of statics/differential kinematics. Above this characteristic frequency, the jacobian transformation computed in static conditions approximates with some error the real mapping of forces/velocities between the two spaces. Figure 10(b) shows the estimated structural dynamics which is the ratio between the acceleration reached at the end-effector under a given exciting sinusoidal command torque T_m, while $F = 0$, so corresponding to sY_{12}, eq. (10). The structural response presents a resonant peak at ω_m, and is coherent with similar profiles obtained through real acceleration measures [3].

Figure 10(c) represents the impedance response of the system to a position disturbance at the end-point, corresponding to H_{11}, (4). Together with Figure 10(a) it provides a measure of the transparency of the device. An ideal system would produce a frequency response that mimics a point mass. As expected, the effect of viscous and Coulomb friction is that the impedance typically takes on a minimum value [13]. The impedance response of the system based on measurements on only one dof provides a correct estimation in the center of the workspace, where a good decoupling of dynamic effects of the two actuators is achieved.

5 Conclusions and future work

The model which has been tuned up revealed to be quite accurate and able to predict instability conditions, due to the discrete time implementation of constraint in Virtual Environments. The arise of instability was notices at frequencies close to the resonant values. Next step will be the study and design of a novel kind of discrete time-implemented impedance, which can preserve the stability for higher values of simulated stiffness. This work has been partially funded by EU under the 5FP within the IST-2000-29580 "PURE-FORM" project.

References

[1] R.J. Adams and B. Hannaford. Stable haptic interaction with virtual environments. *IEEE Transactions on Robotics and Automation*, 15(3):465–474, 1999.

[2] P.M. Chirlian. *Basic Network Theory*. Mc-Graw Hill Company, 1969.

[3] R. E. Ellis, O. M. Ismaeil, and M.G. Lipsett. Design and evaluation of a high-performance haptic interface. *Robotica*, 14:321–327, 1996.

[4] A. Frisoli. *Design and Modeling of Haptic Interfaces: an Integrated Approach*. PhD thesis, Scuola Superiore S. Anna - Pisa, Italy, 2002.

[5] A. Frisoli and M. Bergamasco. *Proceedings of RoManSy- The Theory and Practice of Robot Manipulators*, chapter Hamiltonian formulation of the constrained dynamics of a tendon driven parallel mechanism. Springer, 2002.

[6] V. Hayward and O.R. Astley. Performance measures for haptic interfaces. In *Robotics Research: The 7th International Symposium*, 1996.

[7] V. Hayward, J. Choksi, G. Lanvin, and C. Ramstein. *Advances in Robot Kinematics*, chapter Design and multi-objective optimization of a linkage for a haptic interface, pages 352–359. Kluwer Academics, 1994.

[8] S.J. Huang and C.Y. Chen. Measurement and analysis of structural dynamic properties of robotic joint transmission system. *Journal of Robotic Systems*, 10(1):103–22, 1993.

[9] I. Kollar. *Frequency Domain System Identification Toolbox*. The Mathworks, 1999.

[10] K. Kozlowski. *Modeling and Identification in Robotics*. Springer, 1991.

[11] C.D. Lee, D.A. Lawrence, and L.Y. Pao. A high-bandwidth force-controlled haptic interface. In *Proc. ASME IMECE 2000, Int. Mech. Eng. Conf. Exhib.- Symposium on Haptic Interfaces for Teleoperation and Virtual-Reality*, 2000.

[12] M. Moreyra and B. Hannaford. A practical measure of dynamic response of haptic devices. In *Proc. IEEE Intl. Conf. on Robotics and Automation*, 1998.

[13] J.B. Morrell and J.K. Salisbury. Performance measurements for robotic actuators. In *Proceedings of the ASME Dynamic Systems and Control Division*, 1996.

[14] M.T. Pham, M. Gautier, and P. Poignet. Identification of joint stiffness with bandpass filtering. In *Proceedings of ICRA*, 2001.

[15] G.M. Prisco, A. Frisoli, F. Salsedo, and M. Bergamasco. A novel tendon driven 5-bar linkage with a large isotropic workspace. In *Proc. ASME IMECE '99, Int. Mech. Eng. Conf. Exhib.*, 1999.

[16] J. Schwarzenbach and K.F. Gill. *System Modelling and Control*. Edward Arnold, 1978.

[17] K. Sugiura and Y. Hori. Vibration suppression in 2- and 3-mass system based on the feedback of imperfect derivative of the estimated torsional torque. *IEEE Trans. Industr. Electronics*, 43(1):56–64, 1996.

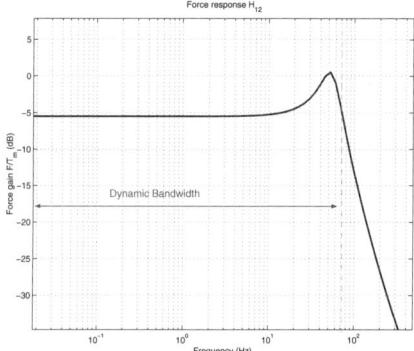

(a) Frequency-gain plot of the estimated force response H_{12}, eq. (6)

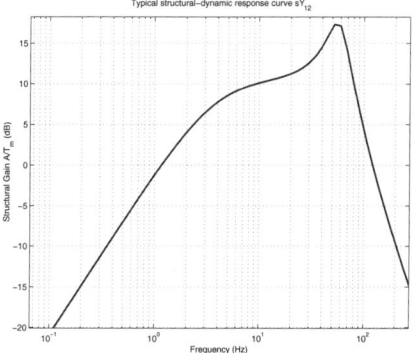

(b) Estimated structural-dynamic response sY_{12}, eq. (10)

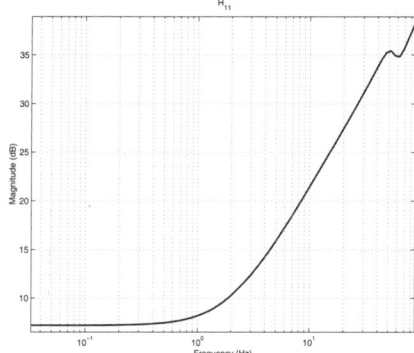

(c) Impedance response at the end effector H_{11}, eq. (4)

Figure 10: Estimate of some performance measure of Haptic Display

A Robust Friction Control Scheme of Robot Manipulators[*]

Jeng-Shi Chen and Jyh-Ching Juang[†]

Department of Electrical Engineering
National Cheng Kung University
Tainan, Taiwan, R.O.C.
Phone: 886-6-2757575 ext. 62313
Fax: 886-6-2345482
e-mail: juang@mail.ncku.edu.tw

ABSTRACT

In this paper, the tracking problem is considered for robot manipulators in the presence of static friction, bounded disturbance, and modeling uncertainties. A composite tracking control strategy is proposed in which an adaptive friction estimation is used to estimate the extent of friction and a robust controller is then designed to enhance the overall stability and robustness. Performance issues of robust adaptive friction control are illustrated in simulation made for a two degree of freedom planar robot manipulator.

1 Introduction

Tracking control of robot manipulators is the most common task in robotic applications. However, there are model uncertainties, such as unknown work load, unknown friction coefficient and inexact knowledge of link lengths and masses, which present a challenging control problem. In addition, the presence of frictions imposes a severe performance limitation for robot manipulators, leading to nonlinear friction phenomenon such as tracking lags, position errors, sluggish response, stick-slip motions, and in some cases limit cycles [1], [2] [3]. Therefore, an explicit account of these effects and their consequences is expected to result in improved performance.

Compensation of model uncertainties and frictions for robot tracking control has received a considerable amount of attentions in the literature. The control of uncertain system is usually accomplished using either an adaptive control or a robust control philosophy. However, the use of an adaptive control algorithm causes more complexity, especially for higher order nonlinear time-varying system such as robot manipulator system. On the other hand, the robust controllers allow for unstructured time-varying disturbances but they are unable to guarantee asymptotic tracking. Li [4] developed a design method of nonlinear H^∞ controller, by choosing a set of suitable penalty variables so that the HJI inequality is replaced by algebraic matrix Riccati inequality. In [5], a global H^∞ control design for nonlinear system with time-varying disturbances is introduced where a Lyapunov function is used to solve the H^∞ control problem instead of the HJI equation. As far as the dynamically friction model is not concerned, they all assumed the friction as a exogenous disturbance. Therefore, the good performance is determined by a small γ parameter and a large penalty output weighting matrix. This will result in a large feedback gain and excessive control input. Moreover, a more accurate representation of friction should allow the control gain to be decreased, since the friction disturbance terms could be compensated by the feedforward component of the control algorithm. In [1], a nonlinear estimator with respect to Coulomb friction coefficient was proposed. The method was then further analyzed in [6].

The paper attempts to extend the Lyapunov function based algorithm as established in [1] to nonlinear systems and address the associated robustness issues. In particular, nonlinear H^∞ control design approach is developed to address the control design problem for systems that are subject to parametric uncertainties, Coulomb frictions and Stribeck effects. More significantly, it is shown that the two control signals can be designed separately or sequentially. That is, there seems to exist a separation principle in friction observer and robust state feedback control design.

The paper is organized as follows. The main result concerning the design of a robot manipulator tracking composite control scheme that combines the merits of friction estimation and feedback control is presented in Section 2. The approach is extended to account for uncertainties in the system dynamics in Section 3. In section 4, performance issue of the controller are illustrated in a simulation study made for a two degree of freedom planar robot manipulator. The conclusions are given in Section 5.

2 Adaptive Friction Control Scheme

The classical model of friction is a regarding force that is a nonlinear, memoryless, odd function of velocity. In its simplest form, the Coulomb friction force that opposing the motion is the product of the friction coefficient and sign of the velocity. In this section, the tracking problem of the manipulator will be considered

---

[*]This work was supported by the National Science Council, Republic of China, under Contract NSC-91-2213-E-006-049.
[†]Corresponding author

to account for this type of friction effect.

Consider the dynamic equations of an n-link friction mechanical manipulator whose dynamical equation is given by

$$M(q)\ddot{q} + V(\dot{q}, q) + G(q) + F_c(\dot{q}) = \tau \quad (1)$$

where $q \in \mathbf{R}^n$ is the vector of generalized coordinates, $\tau \in \mathbf{R}^n$ is the control torque input vector, $M(q)$ is the inertia matrix which is symmetric and positive definite for all $q \in \mathbf{R}^n$, $V(q,\dot{q})\dot{q}$ is the centripetal/Coriolis torques, $G(q)$ is the gravitational torques, and the components $F_{ci}(\dot{q}_i)$, $i = 1, \cdots n$ of $F_c(\dot{q})$ are friction forces acting independently on each joint. A vital property of the inertia matrix M is that there exists positive scalars λ_1 and λ_2 such that

$$\lambda_1 I \leq M(q) \leq \lambda_2 I \quad (2)$$

Each friction force component $F_{ci}(\dot{q}_i)$ is assumed to have the following form

$$F_{ci}(\dot{q}_i) = d_i \, \text{sgn}(\dot{q}_i)$$

with d_i is an unknown constant coefficient and sgn is the signum function defined by

$$\text{sgn}(x) = \begin{cases} +1 & x > 0 \\ 0 & x = 0 \\ -1 & x < 0 \end{cases}$$

Then the dynamic friction can be expressed in the matrix form
$$F_c(\dot{q}) = \Sigma_1 d,$$

with $d = \{d_i\}$ the coefficient vector of the Coulomb friction and $\Sigma_1 = \text{diag}\{\text{sgn}(\dot{q}_i)\}$, a diagonal matrix with diagonal elements $\text{sgn}(\dot{q}_i)$'s.

The controller designed in this section is a modification of the computed-torque controller to account for the Coulomb friction. The closed-loop control configuration is illustrated in Fig. 1, and the design goal is to find the control law of τ or u such that q and $\dot{q}$ track the desired trajectory q_d and $\dot{q}_d$. By inspecting (1), it can be seen that if τ is chosen according to the form:

$$\tau = V(\dot{q}, q) + G(q) + M(q)u \quad (3)$$

then τ can be considered as an inner loop control variable as shown in Fig. 1. And, since the inertia matrix M is invertible, the combined system (1) and (3) reduces to:

$$\ddot{q} = u - M(q)^{-1}\Sigma_1 d \quad (4)$$

In terms of state space representation, the system (4) becomes
$$\dot{x} = Ax + Bu - BM^{-1}\Sigma_1 d \quad (5)$$

where

$$A = \begin{bmatrix} 0 & 0 \\ I & 0 \end{bmatrix}, \; B = \begin{bmatrix} I \\ 0 \end{bmatrix}, \; x = \begin{bmatrix} x_1 \\ x_2 \end{bmatrix} = \begin{bmatrix} \dot{q} \\ q \end{bmatrix}$$

Suppose that a desired trajectory $t \to q^d(t)$ is given in the joint space. We form the error vectors as

$$e_1 = x_1 - x_{1d} = \dot{q} - \dot{q}^d, \quad e_2 = x_2 - x_{2d} = q - q^d.$$

Define $e = x - x_d = \begin{bmatrix} x_1 \\ x_2 \end{bmatrix} - \begin{bmatrix} x_{1d} \\ x_{2d} \end{bmatrix} = \begin{bmatrix} e_1 \\ e_2 \end{bmatrix}$, the error dynamic equation is

$$\dot{e} = Ae + Bu - BM^{-1}\Sigma_1 d - B\ddot{q}^d. \quad (6)$$

In the following, it is assumed that the pair (A, B) is controllable. As the state variables that are related to the generation of friction forces are assumed to be the first n entries in the state, define the $n \times 2n$ matrix E as $E = \begin{bmatrix} I & 0 \end{bmatrix}$ where I is the $n \times n$ identity matrix. Note that the $n \times n$ matrix EB is of full rank.

To track the error state system, a composite control law is developed in which the control is assumed to contain two parts: one depends linearly on the state for the nominal stability and the other attempts to compensate for the nonlinear friction based on the estimator. Let $\hat{d}$ be the estimate of the friction coefficient vector d, the the control signal is represented as

$$u = \ddot{q}^d + Fe + M^{-1}\Sigma_1 \hat{d} \quad (7)$$

for some state feedback gain matrix F. With this control scheme, the closed-loop state equation can be written as

$$\dot{e} = (A + BF)e - BM^{-1}\Sigma_1 \tilde{d} \quad (8)$$

where $\tilde{d}$ is the friction estimation error which is defined as $\tilde{d} = d - \hat{d}$. The friction estimate $\hat{d} = \{\hat{d}_k\}$ is assumed to be of the following form [1] [2],

$$\hat{d}_k = \varsigma_k - g_k(|x_k|), \quad k = 1, \cdots, n$$

for some auxiliary variable ς_k and differential function $g_k(|x_k|)$. In vector form, the friction estimation scheme is

$$\hat{d} = \varsigma - \{g_k(|x_k|)\} \quad (9)$$

with $\varsigma = \{\varsigma_k\}$. Taking the derivative of the friction estimate leads to

$$\dot{\hat{d}}_k = \dot{\varsigma}_k - g'_k(|x_k|) \, \dot{x}_k \, \text{sgn}(x_k), \quad k = 1, \cdots, n.$$

Here, $g'_k(|x_k|)$ stands for the derivative of $g_k(|x_k|)$ with respect to its argument $|x_k|$. In terms of vector representation, the derivative of the friction estimate follows

$$\dot{\hat{d}} = \dot{\varsigma} - G_x \Sigma_1 E((A + BF)e + \dot{x}_d - BM^{-1}\Sigma_x \tilde{d}) \quad (10)$$

where $G_x = \text{diag}\left[g'_k(|x_k|)\right]$. Note that the two matrices G_x and Σ_1 commute.

The following theorem provides a criterion for closed-loop stability.

Theorem 1: *With respect to the system that is subject to unknown Coulomb friction, there exists a control law that leads to global asymptotic stability of the* closed-loop system provided that G_x in the friction estimation scheme (9) is designed such that

$$H_x \equiv -G_x EBM^{-1} - M^{-1}B^T E^T G_x < 0 \quad (11)$$

and there exists a positive definite matrix P such that

$$PA + A^T P + PBM^{-1}H_x^{-1}M^{-1}B^T P < 0 \quad (12)$$

for all x.

Proof: Consider the Lyapunov function candidate

$$V(e, \tilde{d}) = \frac{1}{2}\left[e^T Pe + \tilde{d}^T \tilde{d}\right]$$

for some symmetric, positive definite matrix P. From (10), suppose that the auxiliary variable ς is designed such that

$$\dot{\varsigma} = G_x \Sigma_1 E((A + BF)e + \dot{x}_d) \quad (13)$$

then the friction estimation error dynamic can be written as

$$\dot{\tilde{d}} = -\dot{\hat{d}} = -G_x \Sigma_1 EBM^{-1}\Sigma_1 \tilde{d}.$$

This, together with the closed-loop dynamics in (8), implies that the derivative of $V(e,\tilde{d})$ along the state trajectory is

$$\dot{V}(e,\tilde{d}) = \frac{1}{2}\left[\dot{e}^T Pe + e^T P\dot{e} + \dot{\tilde{d}}^T \tilde{d} + \tilde{d}^T \dot{\tilde{d}}\right]$$

Applying the Schur complement technique, it is equivalent to find a positive definite matrix P and a feedback gain F such that (11) is satisfied and

$$P(A+BF) + (A+BF)^T P - PBM^{-1}H_x^{-1}M^{-1}B^T P < 0 \quad (14)$$

By setting the feedback gain as

$$F = -M^{-1}(G_x EBM^{-1} + M^{-1}B^T E^T G_x)^{-1}M^{-1}B^T P.$$

The criterion (14) becomes (12). This completes the proof. ♠

3 Robust Adaptive Friction Control Scheme

It is known that model uncertainty due to unknown load or unknown parameters will degrade the system performance. In this section, we will investigate the design of controller of robot manipulators to account for friction and uncertainty at the same time. Suppose that the parameter matrices in the robot model is decomposed as

$$\begin{aligned} M(q) &= M_0(q) + \Delta M(q) \\ V(\dot{q},q) &= V_0(\dot{q},q) + \Delta V(\dot{q},q) \\ G(q) &= G_0(q) + \Delta G(q) \end{aligned}$$

where $M_0(q)$, $V_0(\dot{q},q)$ and $G_0(q)$ are the nominal parameter matrices and are assumed to be known, whereas the parameter perturbations $\Delta M(q)$, $\Delta V(\dot{q},q)$, and $\Delta G(q)$ are unknown. Here,

ΔM : the uncertainty of the $M(q)$ due to the changes of load;
ΔV : the perturbation of term $V(\dot{q},q)$ due to the change of load and viscus friction
ΔG : the perturbation of the gravitational force due to the configuration and changes of the total mass of the robotic manipulator

Therefore, the dynamical equation for the robotic system should take the following form

$$(M_0 + \Delta M)\ddot{q} + V_0 + \Delta V + G_0 + \Delta G + F_c = \tau. \quad (15)$$

Fig. 2 depicts the closed-loop system. The inner loop nonlinear control law based on the nominal model becomes:

$$\tau = M_0 u + V_0 + G_0 \quad (16)$$

where $M_0(q)$, $V_0(\dot{q},q)$ and $G_0(q)$ are the nominal parameter matrices. Substituting (16) into (15) yields:

$$\ddot{q} = u - M_0^{-1}F_c + w \quad (17)$$

where

$$w = M_0^{-1}(-\Delta M \ddot{q} - \Delta V - \Delta G) \quad (18)$$

is treated as the uncertainty. It can then be shown that the tracking error satisfies the equation

$$\dot{e} = Ae + Bu + Bw - BM_0^{-1}\Sigma_1 d - B\ddot{q}^d. \quad (19)$$

where e, A, and B are as in (6). The objective of the robust tracking design is to determine a control law such that the closed-loop system is tracked in the presence of unknown friction $\Sigma_1 d$ and uncertain w. The technique being employed for robust tracking is to construct a friction estimation scheme to estimate the extent of friction and then a robust tracking feedback scheme to achieve stability and robustness. It will be shown that the two design steps can be carried out separately or sequentially.

Let $z = C_1 e$ be a fictitious output, the robust tracking control design problem can be reformulated as the finding of a control u for the system (19) such that the L_2-norm gain from w to z is less than γ for any permissible friction $\Sigma_1 d$. The robust tracking controller assumes the same controller structure in (7) and friction estimator in (9) except that the relevant parameters are designed to meet the additional design requirements. With respect to the friction estimation scheme, let the auxiliary variable be

$$\dot{\varsigma} = G_x \Sigma_1 E((A+BF)e + \dot{x}_d) + Ne \quad (20)$$

for some matrix N to be determined. The closed-loop system becomes

$$\frac{d}{dt}\begin{bmatrix} e \\ \tilde{d} \end{bmatrix} = \underbrace{\begin{bmatrix} A+BF & -BM_0^{-1}\Sigma_1 \\ -N & -G_x \Sigma_1 EBM_0^{-1}\Sigma_1 \end{bmatrix}\begin{bmatrix} e \\ \tilde{d} \end{bmatrix}}_{f(e,\tilde{d})}$$

$$+ \underbrace{\begin{bmatrix} B \\ G_x \Sigma_1 EB \end{bmatrix}}_{g(e,\tilde{d})} w$$

$$z = \underbrace{\begin{bmatrix} C_1 & 0 \end{bmatrix}}_{h(e,\tilde{d})} \begin{bmatrix} e \\ \tilde{d} \end{bmatrix} \quad (21)$$

The above closed-loop system has an L_2-norm gain from w to z less than γ if and only if there exists a scalar C^1 function $V : \mathbf{R}^{2n+n} \to \mathbf{R}^+$ with $V(0) = 0$ such that the following Hamilton-Jacobi-Bellman inequality is satisfied [7]:

$$V_{e,\tilde{d}}f(e,\tilde{d}) + \tfrac{1}{2\gamma^2} V_{e,\tilde{d}} g(e,\tilde{d}) g^T(e,\tilde{d}) V_{e,\tilde{d}}^T \\ + \tfrac{1}{2} h^T(e,\tilde{d}) h(e,\tilde{d}) < 0. \quad (22)$$

where $V_{e,\tilde{d}}$ is the partial derivative of V with respect to the state vector $\begin{bmatrix} e \\ \tilde{d} \end{bmatrix}$. Consider the following Hamilton-Jacobi-Bellman function candidate

$$V(e,\tilde{d}) = \frac{1}{2}\left[e^T P e + \tilde{d}^T \Pi^{-1} \tilde{d} \right] \quad (23)$$

for a positive definite matrix P and a positive diagonal matrix Π. Substituting this candidate function into (22), it can be shown that a sufficient condition for robust stability is that the following matrix inequality is satisfied.

$$e^T \left(P(A+BF) + \tfrac{1}{2\gamma^2} g_{11} + \tfrac{1}{2} C_1^T C_1 \right) e \\ + \tilde{d}^T \left(-\Pi^{-1} N + \tfrac{1}{2\gamma^2} g_{21} \right) e \\ + e^T \left(-P(BM_0^{-1}\Sigma_1) + \tfrac{1}{2\gamma^2} g_{12} \right) \tilde{d} \\ + \tilde{d}^T \left(-\Pi^{-1} G_x \Sigma_1 E B M_0^{-1}\Sigma_1 + \tfrac{1}{2\gamma^2} g_{22} \right) \tilde{d} < 0$$

where

$$g_{11} = P(BB^T)P \\ g_{12} = P\left(B(G_x \Sigma_1 EB)^T\right)\Pi^{-1} \\ g_{21} = \Pi^{-1}(G_x \Sigma_1 EBB^T)P \\ g_{22} = \Pi^{-1}\left((G_x \Sigma_1 EB)(G_x \Sigma_1 EB)^T\right)\Pi^{-1}$$

The design is greatly simplified by setting the matrix N as

$$N = -\Pi \Sigma_1 M_0^{-1} B^T P + \gamma^{-2} \Sigma_1 G_x EBB^T P \quad (24)$$

In this case, the conditions for robust stability can be stated as

$$P(A+BF) + (A+BF)^T P \\ + \gamma^{-2} P (BB^T) P + C_1^T C_1 < 0 \quad (25)$$

and

$$-G_x EBM_0^{-1}\Pi - \Pi M_0^{-1} B^T E^T G_x \\ + \gamma^{-2} G_x EBB^T E^T G_x < 0 \quad (26)$$

Theorem 2: *The perturbed system under unknown Coulomb frictions is robustly tracked provided that there exist G_x and a positive diagonal Π such that for all x*

$$\begin{bmatrix} -G_x EBM_0^{-1}\Pi - \Pi M_0^{-1} B^T E^T G_x & G_x EB \\ B^T E^T G_x & -\gamma^2 I \end{bmatrix} < 0 \quad (27)$$

and there exists a positive definite matrix Q such that the following linear matrix inequality is satisfied

$$\begin{bmatrix} B^{\perp T} & 0 & 0 \\ 0 & I & 0 \\ 0 & 0 & I \end{bmatrix} \begin{bmatrix} AQ + QA^T & B & QC_1^T \\ B^T & -\gamma^2 I & 0 \\ C_1 Q & 0 & -I \end{bmatrix} \begin{bmatrix} B^\perp & 0 & 0 \\ 0 & I & 0 \\ 0 & 0 & I \end{bmatrix} < 0 \quad (28)$$

where $B^\perp$ is the orthogonal complement of B.

Proof: The condition (27) is essentially the same as (26). The design inequality (25) involves P and F. By applying standard linear matrix inequality techniques [8] to reduce the dependency on F and substituting $Q = P^{-1}$, (25) can be reformulated as (28). ♠

The resulting design linear matrix inequalities in (27) depends on the matrix M_0 and also on G_x. A sufficient condition of (27) is to find a G_x such that

$$0 < G_x < \frac{2\gamma^2}{\lambda_2}\Pi \quad (29)$$

for all x.

Note that in the absence of the friction terms, the above theorem suggests that the robust tracking design boils down to be the linear matrix inequality (28) which has appeared in LMI literature [9]. More significantly, through the auxiliary variable ς and the design of N, it appears that there exists a separation principle in the robust tracking of structure uncertainty and friction. The design steps are as follows. Firstly, the robust tracking with void friction terms is considered. By solving the linear matrix inequality (28), the state feedback gain F is determined. Secondly, the friction estimation scheme is designed so that the inequality (29) is satisfied. The two design steps are essentially conducted separately. The resulting estimator/controller is then synthesized by constructing the matrix N in (24) for the estimator.

4 Illustrative Examples

Consider a two-link planar manipulator shown in Fig. 3, where it is assumed that the link masses are concentrated at the ends of the links, with system parameters: link masses $m_1, m_2 (kg)$, lengths $l_1, l_2 (m)$, angular positions $q_1, q_2 (rad)$, applied torques $\tau_1, \tau_2 (Nm)$. The dynamic equation of a two-link rigid manipulator is given by

$$\tau = M(q)\ddot{q} + V(\dot{q},q) + G(q) + F_c(\dot{q})$$

where

$$M(q) = \begin{bmatrix} m_2 l_2^2 + 2m_2 l_1 l_2 c_2 + (m_1 + m_2) l_1^2 & m_2 l_2^2 + m_2 l_1 l_2 c_2 \\ m_2 l_2^2 + m_2 l_1 l_2 c_2 & m_2 l_2^2 \end{bmatrix}$$

$$V(\dot{q}, q) = m_2 l_1 l_2 s_2 \begin{bmatrix} -2\dot{q}_1 \dot{q}_2 - \dot{q}_2^2 \\ \dot{q}_1^2 \end{bmatrix}$$

$$G(q) = \begin{bmatrix} m_2 g l_2 c_{12} + (m_1 + m_2) g l_1 c_1 \\ m_2 g l_2 c_{12} \end{bmatrix}$$

$$F_c(\dot{q}) = \begin{bmatrix} d_1 \mathrm{sgn}(\dot{q}_1) \\ d_2 \mathrm{sgn}(\dot{q}_2) \end{bmatrix}$$

In the above, c_i, s_i, c_{ij}, and s_{ij} represents $\cos(q_i)$, $\sin(q_i)$, $\cos(q_i + q_j)$, and $\sin(q_i + q_j)$, respectively. The following parameters $m_1 = 1kg$, $m_2 = 1kg$, $l_1 = 1m$, $l_2 = 1m$, $d_1 = 3$, $d_2 = 2$, and $g = 9.8$ are given. Then the working area is $-\frac{2}{3}\pi(rad) \leq q_1 \leq \frac{2}{3}\pi(rad)$ and $-\frac{2}{3}\pi(rad) \leq q_2 \leq \frac{2}{3}\pi(rad)$. It may then be shown that $\lambda_1 = 1.0$ and $\lambda_2 = 6.0$. Let the desired trajectory used in all examples be described to travel from an initial location to a final location and then moves backward to the original location. The desired joint velocity profile is shown in Fig. 4. In the simulation examples, the initial values of the internal state are set as $\varsigma_1 = 4$ and $\varsigma_2 = 3$.

The objective of the design is to achieve stability, track the desired trajectory, and bounded L_2-norm gain from w to z. According to (29), with $\Pi = 20$

$$g(|x_1|) = \mu - \mu e^{-|x_1|} + \eta |x_1|$$

the design parameter μ and η must satisfies

$$\mu e^{-|\dot{q}|} + \eta < \frac{2\gamma^2}{\lambda_2}\Pi \quad \text{and} \quad \mu e^{-|\dot{q}|} + \eta > 0$$

for all $\dot{q}$. It suffices to set the output penalty weighting matrix $C_1 = \begin{bmatrix} 1 & 0 & 5 & 0 \\ 0 & 1 & 0 & 5 \end{bmatrix}$, $\gamma = 1$, $\mu = 5$ and $\eta = 0.5$. The design equation (28) is solved to give Q and then the robust state feedback gain F is determined. By using the LMI control toolbox [10], the following matrices are obtained:

$$Q = \begin{bmatrix} 2.5260 & 0 & -0.3436 & 0 \\ 0 & 2.5260 & 0 & -0.3436 \\ -0.3436 & 0 & 0.0943 & 0 \\ 0 & -0.3436 & 0 & 0.0943 \end{bmatrix},$$

$$F = \begin{bmatrix} -31.4116 & 0 & -114.5023 & 0 \\ 0 & -31.4116 & 0 & -114.5023 \end{bmatrix}.$$

In order to demonstrate the robustness of the robust friction compensation scheme, the following two cases are considered.

Case 1: Coulomb Friction + Stribeck Friction

Mathematically [3], the steady-state behavior of each friction force that includes Coulomb friction and Stribeck friction can be modeled as

$$f_k = f_{1k} + f_{2k} = d_k \,\mathrm{sgn}(x_k) + \beta_k e^{-(\frac{x_k}{\nu_k})^2} \mathrm{sgn}(x_k) \quad (30)$$

for some d_k, β_k and ν_k. Note that the coefficient ν_k is positive to reflect the decay phenomenon. The coefficient β_k is positive for a typical Stribeck friction modeling. In simulation, the friction model coefficients are set as $\nu_k = \frac{1.25}{180}\pi$ for $k = 1, 2$, $\beta_1 = 0.6$, and $\beta_2 = 0.4$. The tracking error response and the associated friction estimates are depicted in the Fig. 5. Satisfactory convergence response are observed for the robust controller when the system is subject to Coulomb and Stribeck frictions.

Case 2: Coulomb Friction+ Stribeck Friction + Parametric Uncertainties

Consider the same simulation setup as in the Case 1 except that the mass m_2 is perturbed to $m_2 + \Delta m_2$. Thus, the inertia matrix, centrifugal/Coriolis force, and gravity term are, respectively, perturbed by

$$\Delta M(q) = \Delta m_2 \begin{bmatrix} l_2^2 + 2l_1 l_2 c_2 + l_1^2 & l_2^2 + l_1 l_2 c_2 \\ l_2^2 + l_1 l_2 c_2 & l_2^2 \end{bmatrix}$$

$$\Delta V(\dot{q}, q) = \Delta m_2 l_1 l_2 s_2 \begin{bmatrix} -2\dot{q}_1 \dot{q}_2 - \dot{q}_2^2 \\ \dot{q}_1^2 \end{bmatrix}$$

$$\Delta G(q) = \Delta m_2 \begin{bmatrix} gl_2 c_{12} + gl_1 c_1 \\ gl_2 c_{12} \end{bmatrix}$$

The uncertain parameter is set to 10% of the nominal value. The error, as shown in Fig. 6, is within the specified bound and indeed can be reduced when the penalty weighting function or performance level γ is adjusted.

5 Conclusion

In this paper, a robust friction control algorithm is developed to achieve robust tracking of the manipulator that is subject to both unknown friction and parameter perturbation. The resulting dynamic controller combines the advantages of nonlinear H_∞ control and adaptive estimation. More importantly, it is shown that the design of the friction estimator and that of the robust control gain can be conducted separately. Effectiveness of the design procedure has been supported by the simulations made for two degree of freedom robot manipulator.

References

[1] A. Yazdizadeh, K. Khorasani, Adaptive Friction Compensation Based on the Lyapunov Scheme. Proceeding of the 1996 IEEE Conference on Control Application., pp. 1060-1065, September 1996.

[2] B. Friedland and Y.-J. Park, On adaptive Friction Compensation. IEEE Transactions on Automatic Control, Vol 37, No 10, pp. 1609-1612, 1992.

[3] A.-H. Brain, E. D. Pierre, C. Canudas de Wit., A Survey of Models, Analysis Tools and Compensation Method for Control of Machines with Friction. Automatica., Vol. 30, No. 7, pp. 1083-1138, 1994.

[4] S. Li, Nonlinear H_∞ Controller Design for a Class of Nonlinear Control System. 23rd International Conference on Industrial Electronic, Control, and Instrumentation, pp. 291-294, 1999.

[5] L. Acho, Y. Orlov and L. Aguilar, Global H_∞ Control Design for Tracking Control of Robot Manipulators. Proceedings of the American Control Conference, pp. 3986-3990, 2002.

[6] T.-L. Liao and T.-I Chien, An Exponentially Stable Adaptive Friction Compensator. IEEE Transactions on Automatic Control, Vol. 45, N0. 5, May 2000.

[7] A. Isidori and A. Astolfi, Disturbance attenuation and H_∞ control via measurement feedback in nonlinear systems. IEEE Transactions on Automatic Control, Vol 37, pp. 1283-1293, 1992.

[8] S. Boyd, L. El Ghaoui, E. Feron, and V. Balakrishnan, Linear Matrix Inequalities in System and Control Theory. SIAM, 1994.

[9] P. Gahinet and P. Apkarian, A Linear Matrix Inequality Approach to H_∞ Control. International Journal of Robust and Nonlinear Control, Pages 421-448, 1994.

[10] P. Gahinet, A Nemirovski, A. J. Laub, and M. Chilali, LMI Control Toolbox. Mathworks Inc., 1995.

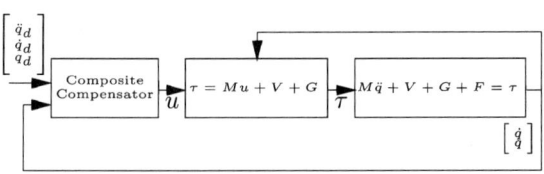

Figure 1: Control Configuration.

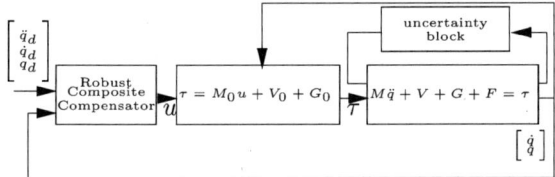

Figure 2: Robust Control Configuration.

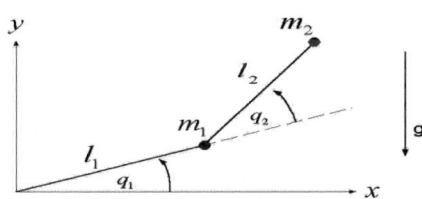

Figure 3: Two Link Planar Manipulator.

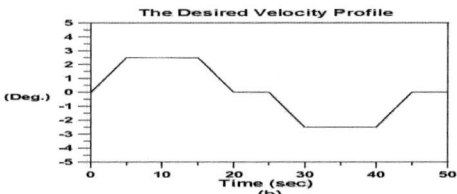

Figure 4: Desired position trajectory and velocity profile.

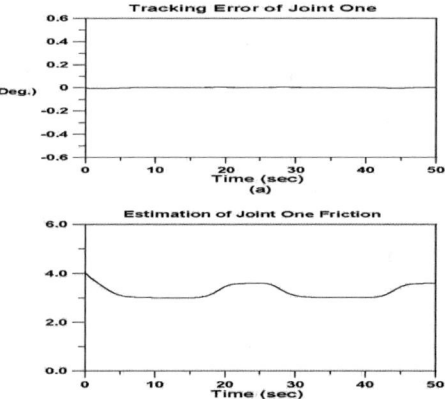

Figure 5: Responses of the robust controller for the system that is subject to Coulomb friction and Stribeck friction.

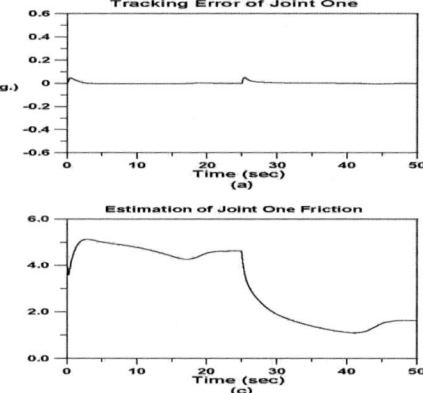

Figure 6: Responses of the robust controller for the system that is subject to Coulomb friction, Stribeck friction and parametric uncertainty.

Identification of the Dynamic Parameters of the Orthoglide

Sylvain GUEGAN, Wisama KHALIL, Philippe LEMOINE

Ecole Centrale de Nantes
IRCCyN U.M.R. C.N.R.S. 6597
1 Rue de la Noë, BP 92101, 44321 Nantes Cedex 03, FRANCE
wisama.khalil@irccyn.ec-nantes.fr

Abstract

This paper presents the experimental identification of the dynamic parameters of the Orthoglide [1], a 3-DOF parallel. The dynamic identification model is based on the inverse dynamic model, which is linear in the parameters. The model is computed in a closed form in terms of the Cartesian dynamic model elements of the legs and of the Newton-Euler equation of the platform. The base inertial parameters of the robot, which constitute the identifiable parameters, are given.

1 Introduction

The inverse dynamic model is important for high performance control algorithms, and the forward dynamic model is required for their simulation. For these two applications the numerical values of the dynamic parameters (inertial and friction) must be known. The determination of the base inertial parameters, which represent the only identifiable parameters [2], is treated in this paper by a numerical method [3]. This method is based on the QR decomposition of the observation matrix of the dynamic identification model of the robot. The experimental identification of the dynamic parameters is based on the use of a dynamic model linear in the parameters. This model permits to use the least squares solution to solve the estimation problem [4].

2 Kinematic modeling of the Orthoglide

The Orthoglide has three PRPaR identical legs (where P, R and Pa stand for Prismatic, Revolute and Parallelogram joint, respectively). Each leg is composed of six passive revolute joints and 1 active prismatic joint, (fig. 1). We define frame F_0 fixed with the base and frame F_P fixed with the mobile platform (fig. 2). Their origins are A_1 and P respectively. Their axes (x_0, y_0, z_0) and (x_P, y_P, z_P) are parallel. The base frames of the legs are defined by the frames F_{A1}, F_{A2} and F_{A3} (fig. 2), whose origins are A_1, A_2 and A_3 respectively. The z_{Ai} axes are along the prismatic joint axes. The Khalil and Kleinfinger notations [5], are used to describe the geometry of the system (fig. 3).

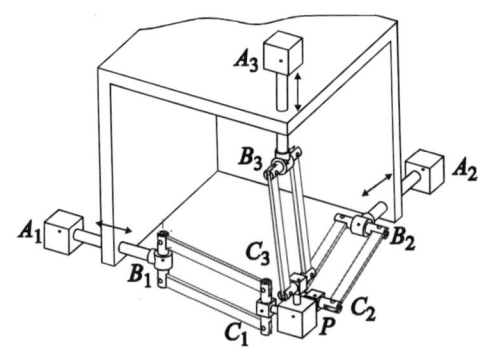

Fig. 1: Orthoglide kinematic architecture.

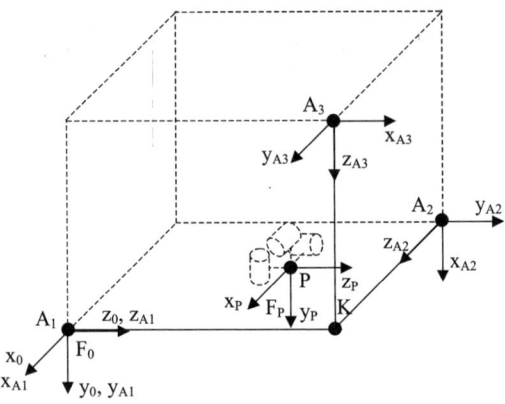

Fig. 2: Base frame, platform frame and leg frames.

The following notations are used:

L (3×1) vector of the motorized joint variables:

$$\mathbf{L} = \begin{bmatrix} q_{11} & q_{12} & q_{13} \end{bmatrix}^T;$$

$^0\mathbf{V}_p$ (3×1) vector of the linear velocity of the origin of the platform.

The derivative of **L** and $^0\mathbf{V}_p$ with respect to the time are denoted $\dot{\mathbf{L}}$ and $^0\dot{\mathbf{V}}_p$ respectively.

This work has been supported by the project MAX of the program ROBEA of the department STIC of the French CNRS.

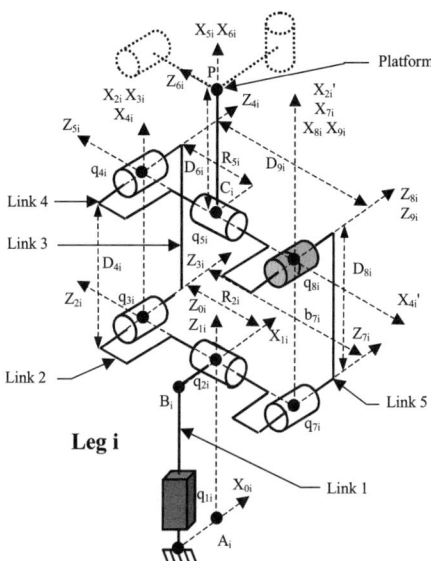

Fig. 3: Link frames of one leg.

The following kinematic models are presented in [6]:
i) The inverse kinematic model of the robot:
$$\dot{L} = {}^0J_p^{-1} \, {}^0V_p \tag{1}$$
Where ${}^0J_p^{-1}$ is the inverse Jacobian matrix of the Orthoglide, which is always regular in the working space.
ii) The inverse kinematic model of a leg i:
$$\dot{q}_i = {}^0J_i^{-1} \, {}^0V_p \tag{2}$$
$$\dot{q}_i = [\dot{q}_{1i} \quad \dot{q}_{2i} \quad \dot{q}_{3i}]^T \tag{3}$$
Where ${}^0J_i^{-1}$ is the inverse Jacobian matrix of the leg i.
The velocities of the other joints of each leg can be obtained in terms of $\dot{q}_i$ (see appendix).
iii) The second order inverse kinematic model of the leg:
$$\ddot{q}_i = {}^0J_i^{-1} \left({}^0\dot{V}_p - {}^0\dot{J}_i \dot{q}_i \right) \tag{4}$$

3 Inverse dynamic model

The inverse dynamic model gives the motorized forces, Γ_{robot}, in terms of the desired trajectory of the mobile platform ${}^0P_p, {}^0V_p, {}^0\dot{V}_p$. The dynamic model is computed in two steps. First we calculate, the reaction forces of the platform on the legs at point P, which is denoted by f_i, then the Newton-Euler equation of the platform is applied to obtain the motor forces [6].

The general form of the inverse dynamic model of a leg i, is written as (see appendix):
$$\Gamma_i = H_i(q_i, \dot{q}_i, \ddot{q}_i) + {}^0J_i^T \, {}^0f_i \tag{5}$$
Where:
H_i is the inverse dynamic model of leg i, when its terminal point is free.

Γ_i is composed of the independent torques/forces of the joints of the leg i, where Γ_{1i} and Γ_{2i} are zero:
$$\Gamma_i = [\Gamma_{1i} \quad \Gamma_{2i} \quad \Gamma_{3i}]^T = [\Gamma_{1i} \quad 0 \quad 0]^T \tag{6}$$
Using equation (5) the forces f_i can be written as:
$${}^0f_i = -H_{xi}(q_i, \dot{q}_i, \ddot{q}_i) + {}^0J_i^{-T} \Gamma_i \tag{7}$$
Where:
$$H_{xi}(q_i, \dot{q}_i, \ddot{q}_i) = {}^0J_i^{-T} H_i(q_i, \dot{q}_i, \ddot{q}_i) \tag{8}$$
H_{xi} is the inverse dynamic model with respect to the position Cartesian space at point P (fig.3) [7][8]. We show that [6]:
$${}^0f_i = -H_{xi}(q_i, \dot{q}_i, \ddot{q}_i) + {}^0J_{p[:,i]}^{-T} \Gamma_i \tag{9}$$
Where ${}^0J_{p[:,i]}^{-T}$ represents the ith column of the inverse transpose Jacobian matrix of the robot.

The Newton-Euler equation of the platform is written as (no rotation):
$${}^0F_p = {}^0\dot{V}_p M_p - M_p \, {}^0g \tag{10}$$
With:
${}^0g$ Acceleration of gravity, referred to frame F_0:
${}^0g = [0 \quad g \quad 0]^T$, $g = 9.81 \, m.s^{-2}$
M_p Mass of the platform;
${}^0F_p$ Total external forces on the platform.
From equations (9) and (10), the dynamic model is given by:
$$\Gamma_{robot} = {}^0J_p^T \left({}^0F_p + \sum_{i=1}^{3} \left[H_{xi}(q_i, \dot{q}_i, \ddot{q}_i) \right] \right) \tag{11}$$
Different methods can be used to calculate $H_i(q_i, \dot{q}_i, \ddot{q}_i)$ [9][10][11]. To reduce the computational cost, the customized Newton-Euler method, which is linear in the dynamic parameters is used [12].

4 Dynamic identification model

The dynamic model of each leg i can be represented as a linear function of the inertial and friction parameters of the leg K_i. Thus the equation (11) can be written as:
$$\Gamma_{robot} = D_{robot} K_{robot} \tag{12}$$
K_{robot} is the vector of the standard dynamic parameters of the Orthoglide:
$$K_{robot} = [M_P \quad K_1^T \quad K_2^T \quad K_3^T]^T \tag{13}$$
$$D_{robot} = [D_P \quad D_1 \quad D_2 \quad D_3] \tag{14}$$
K_i is the vector of the standard dynamic parameters of the leg i, such that:
$$K_i = [Ma_{1i} \quad Fs_{1i} \quad Fv_{1i} \quad \chi_i^T]^T \tag{15}$$
$$\chi_i = [\chi_{1i}^T \quad \cdots \quad \chi_{5i}^T]^T \tag{16}$$
Where:

- Ma_{1i} is the inertia of the rotor of motor i referred to the joint side;
- Fv_{1i} is the viscous friction parameter;
- Fs_{1i} is the coulomb friction parameter;
- χ_i is the vector of the inertial parameters of link i.

The standard inertial parameters of the link j (j = 1 to 5, fig. 3) of the leg i are collected in the (10×1) vector:

$$\chi_{ji} = \begin{bmatrix} XX_{ji} & XY_{ji} & XZ_{ji} & YY_{ji} & YZ_{ji} & ZZ_{ji} & MX_{ji} & MY_{ji} & MZ_{ji} & M_{ji} \end{bmatrix}^T \quad (17)$$

Where:
- $XX_{ji}, ..., ZZ_{ji}$ are the elements of the inertia matrix;
- $MX_{ji}, MY_{ji}, MZ_{ji}$ define the first moments of link ji;
- M_{ji} is the mass of link ji.

Thus, $\mathbf{K}_{robot}$ is a (160×1) vector and $\mathbf{D}_{robot}$ is a (3×160) matrix.

4.1 Base dynamic parameters of the robot

The base dynamic parameters represent the minimum number of parameters from which the dynamic model can be calculated. The dynamic model complexity is reduced when computed by the base dynamic parameters. Besides, they constitute the only identifiable parameters [3]. They can be obtained from the standard parameters, by eliminating the dynamic parameters that have no effect on the dynamic model and by grouping some others.

To determine them, we use a numerical method, which is based on the QR decomposition [4]. First we determine the base parameters of each leg, then we determine the effect of connecting the platform.

There are 14 base parameters for legs 1 and 2. They are given by (i = 1, 2): $Ma_{1Ri}, Fv_{1i}, Fs_{1i}, ZZ_{2Ri}, MX_{2i}, MY_{2Ri}, XX_{3Ri}, XY_{3Ri}, XZ_{3Ri}, YZ_{3Ri}, ZZ_{3Ri}, MX_{3Ri}, MY_{3Ri}, MX_{4i}$. Since, the prismatic joint of leg 3 is along gravity, there are 15 base parameters for leg 3, the grouped inertia Ma_{1R3} does not eliminate M_{1R3} (Whose effect on the force of motor 3 will be constant and equal to $-g.M_{1R3}$). The grouped relations are (the index R indicates that some parameters are grouped with that one):

$$Ma_{1Ri} = Ma_{1i} + M_{1i} + M_{2i} + M_{3i} + M_{4i} + M_{7i}$$
$$ZZ_{2Ri} = ZZ_{2i} + YY_{3i} + YY_{4i} + D_{4i}^2 M_{4i} + YY_{7i}$$
$$MY_{2Ri} = MY_{2i} + MZ_{3i} + MZ_{4i} + MZ_{7i}$$
$$XX_{3Ri} = XX_{3i} - YY_{3i} + XX_{7i} - YY_{7i} - D_{4i}^2 M_{4i}$$
$$XY_{3Ri} = XY_{3i} + XY_{7i}$$
$$XZ_{3Ri} = XZ_{3i} - D_{4i} MZ_{4i} + XZ_{7i} \quad (18)$$
$$YZ_{3Ri} = YZ_{3i} + YZ_{7i}$$
$$ZZ_{3Ri} = ZZ_{3i} + ZZ_{7i} + D_{4i}^2 M_{4i}$$
$$MX_{3Ri} = MX_{3i} + MX_{7i} + D_{4i} M_{4i}$$
$$MY_{3Ri} = MY_{3i} + MY_{7i}$$
$$M_{1R3} = M_{13} + M_{23} + M_{33} + M_{43} + M_{73}$$

D_{4i} is the distance between the axes of q_{3i} and q_{4i} (fig.3).

Taking into account the three legs and the platform, we obtain that the parameters MX_{3Ri} (i = 1, 2, 3) are grouped with the mass of the platform and with the parameters: $M_{1Ri3}, ZZ_{2Ri}, XX_{3Ri}$ and ZZ_{3Ri}:

$$M_{RP} = M_P + \frac{1}{D_{41}} MX_{3R1} + \frac{1}{D_{42}} MX_{3R2} + \frac{1}{D_{43}} MX_{3R3}$$

$$Ma_{1Ri} = Ma_{1Ri} - \frac{1}{D_{4i}} MX_{3Ri}$$

$$ZZ_{2Ri} = ZZ_{2Ri} - D_{4i} MX_{3Ri} \quad (19)$$

$$XX_{3Ri} = XX_{3Ri} + D_{4i} MX_{3Ri}$$

$$ZZ_{3Ri} = ZZ_{3Ri} - D_{4i} MX_{3Ri}$$

$$M_{1R3} = M_{1R3} - \frac{1}{D_{43}} MX_{3R3}$$

To understand the physical meaning of these grouped parameters, let us consider that the center of mass of links 3_i and 7_i is in the middle of $O_{3i}O_{4i}$ and $O_{7i}O_{8i}$ respectively. Thus:

$$MX_{3i} = \frac{D_{4i}}{2} M_{3i}, \quad MX_{7i} = \frac{D_{4i}}{2} M_{7i} \quad (20)$$

Using equations (18) and (20) into (19), we obtain:

$$M_{RP} = M_P + \sum_{i=1}^{3} \left[M_{4i} + \frac{1}{2} M_{3i} + \frac{1}{2} M_{7i} \right]$$

$$Ma_{1Ri} = Ma_{1i} + M_{1i} + M_{2i} + \frac{1}{2} M_{3i} + \frac{1}{2} M_{7i} \quad (21)$$

$$M_{1R3} = M_{13} + M_{23} + \frac{1}{2} M_{33} + \frac{1}{2} M_{73}$$

From equation (21), we show that:
- The masses M_{4i} are grouped entirely with the platform;
- The masses M_{3i} and M_{7i} are divided by two: one half is grouped with the platform and the other with Ma_{1Ri} and also with M_{1R3} when i = 3.

The base dynamic parameters of the Orthoglide are given in table 1, on which we added an offset on the motor forces. The masse M_{1R3} will be grouped with Off_3. Thus the total number of parameters is 43.

Table 1: Base inertial parameters of the Orthoglide

M_{RP}						
Ma_{1Ri}	Off_i	Fv_{1i}	Fs_{1i}	ZZ_{2Ri}	MX_{2i}	MY_{2Ri}
XX_{3Ri}	XY_{3Ri}	XZ_{3Ri}	YZ_{3Ri}	ZZ_{3Ri}	MY_{3Ri}	MX_{4i}

So, $\mathbf{K}_{Brob}$, which contains the base dynamic parameters of the robot, is written as:

$$\mathbf{K}_{Brob} = \begin{bmatrix} M_{RP} & \mathbf{K}_{B1}^T & \mathbf{K}_{B2}^T & \mathbf{K}_{B3}^T \end{bmatrix}^T \quad (22)$$

Where $\mathbf{K}_{Bi}$ is the vector of the base dynamic parameters of the leg i. The corresponding $\mathbf{D}_{Brob}$ matrix can be written as:

$$\mathbf{D}_{Brob} = \begin{bmatrix} \mathbf{D}_P & \mathbf{D}_{B1} & \mathbf{D}_{B2} & \mathbf{D}_{B3} \end{bmatrix} \quad (23)$$

4.2 Computation of D_{Brob}

The vector H_i of the leg i can be written as:

$$H_i(q_i, \dot{q}_i, \ddot{q}_i) = \begin{bmatrix} D_{mi} \\ 0 \quad D_{Li} \\ 0 \end{bmatrix} K_{Bi} \quad (24)$$

Where the elements of D_{mi} correspond to the base parameters corresponding to the motorized joint (Ma_{1Ri}, Fv_{1i}, Fs_{1i} and Off_i for i = 1 to 3):

$$D_{mi} = \begin{bmatrix} D_{Ma_{1Ri}} & D_{Fv_{1i}} & D_{Fs_{1i}} & D_{Off_i} \end{bmatrix} \quad (25)$$

With:
$D_{Ma_{1Ri}} = \ddot{q}_{1i}$, $D_{Fv_{1i}} = \dot{q}_{1i}$, $D_{Fs_{1i}} = \text{sign}(\dot{q}_{1i})$ and $D_{Off_i} = 1$.

The columns of D_{Li} correspond to the other base parameters.

Then, from (11), and (24), we deduce that:

$$D_P = {}^0J_P^T \left({}^0\dot{V}_P - {}^0g\right) \quad (26)$$

$$D_{B1} = \begin{bmatrix} D_{m1} \\ 0 & {}^0J_P^T \; {}^0J_1^{-T} \; D_{L1} \\ 0 \end{bmatrix} \quad (27)$$

$$D_{B2} = \begin{bmatrix} 0 \\ D_{m2} & {}^0J_P^T \; {}^0J_2^{-T} \; D_{L2} \\ 0 \end{bmatrix} \quad (28)$$

$$D_{B3} = \begin{bmatrix} 0 \\ 0 & {}^0J_P^T \; {}^0J_3^{-T} \; D_{L3} \\ D_{m3} \end{bmatrix} \quad (29)$$

4.3 Exploitation of the similarity of the legs

The complexity of the dynamic identification model could be reduced by making use of the similarity of the legs. Thus, the base parameters ZZ_{2Ri}, MX_{2i}, MY_{2Ri}, XX_{3Ri}, XY_{3Ri}, XZ_{3Ri}, YZ_{3Ri}, ZZ_{3Ri}, MY_{3Ri}, MX_{4i} of the three legs could be grouped together. Thus the matrix D_{Brob} becomes:

$$D_{sym} = \begin{bmatrix} D_P & \begin{matrix} D_{m1} & 0 & 0 \\ 0 & D_{m2} & 0 \\ 0 & 0 & D_{m3} \end{matrix} & D_{LS} \end{bmatrix} \quad (30)$$

$$D_{LS} = {}^0J_P^T \sum_{i=1}^{3}\left[{}^0J_i^{-T} D_{Li}\right]$$

K_{sym} is a (23×1) matrix containing the base inertial parameters of the Orthoglide:

$$K_{sym} = \quad (31)$$

$\begin{bmatrix} M_{RP} & Ma_{1R1}Fv_{11} & Fs_{11} & Off_1 \cdots Ma_{1R3} & Fv_{13} & Fs_{13} & Off_3 & ZZ_{2R} \cdots MX_4 \end{bmatrix}^T$

With:
$ZZ_{2R} = ZZ_{2R1} = ZZ_{2R2} = ZZ_{2R3}, \cdots, MX_4 = MX_{41} = MX_{42} = MX_{43}$

4.4 Identification of the base dynamic parameters

The identification has been carried out using least squares techniques on the dynamic model as described by Gautier in [13],[14]. To identify the base dynamic parameters some trajectories are sampled at different times. The matrix D_{sym} is calculated for each sample and all of them are collected in the matrix W, to obtain the following overdetermined linear system of equations:

$$Y = W K_{sym} + \rho \quad (32)$$

Where:
ρ is the modeling error;
W is the (3r×c) observation matrix and Y is the (3r×1) matrix corresponding to the joint forces for all the samples, with c = 23 and r represents the number of the samples.

The solution of the linear system of equations (32) gives the estimation of the base dynamic parameters.

5 Experimental results

The identification method has been experimentally carried out on the Orthoglide prototype of the IRCCyN. The motors are AC servomotors. The control system is based on a DSPACE 1103 digital signal-processing card. The sampling period is 2.5 ms.

5.1 Planning of the identification trajectory

To identify the base dynamic parameters the choice of the robot trajectory is very important in order to excite the different parameters [15]. The condition number of W has been used to select the best trajectory. This number measures the sensitivity of the solution with respect to the noise in the data. The Orthoglide motorized joints could be derived independently. So trajectories have been generated between random joint positions. By simulation we selected 10 random trajectories giving a good condition number. The trajectories are then executed on the real system and sampled with a period which is equal to 2.5 ms. The matrix D_{sym} is calculated for each sample and all of them are collected in the observation matrix W.

5.2 Estimation of the observation matrix

The computation of the D_{sym} matrix needs the estimation of the joint positions, velocities and accelerations. The joint positions are measured thanks to the digital encoder.
The joint positions have been filtered with a 4th-order low-pass Butterworth filter in both the forward and reverse directions to avoid phase distortion. The corresponding cut-off frequency is 100 Hz. The joint velocities and accelerations are calculated using numerical derivation based on a central difference

algorithm. In fact, low-pass filtering associated with a difference algorithm provides a pass-band filter to estimate derivatives at low frequency, which avoid derivating high frequency noise [13].

In order to eliminate high frequency torque noises and ripples from **Y**, the columns of **W** and the vector **Y** are filtered in a process called parallel filtering, using the function "decimate" of order 5 from Matlab [13].

5.3 Estimation of the dynamic parameters

The least squares solution has been applied to relation (32) in order to estimate the dynamic parameters:

$$\hat{\mathbf{K}}_{sym} = \mathbf{W}^+ \mathbf{Y} \quad (33)$$

Where $\mathbf{W}^+$ is the pseudo-inverse of $\mathbf{W}$.

The standard deviations are estimated considering the matrix **W** to be deterministic one, and ρ to be a zero mean additive independent noise, with standard deviation σ_ρ. The variance-covariance matrix of the estimation error and standard deviations can be calculated by [13]:

$$\mathbf{C}_{\hat{\mathbf{K}}} = \sigma_\rho^2 \left(\mathbf{W}^T \mathbf{W}\right)^{-1} \quad (34)$$

$$\sigma_{\hat{K}_i} = \sqrt{C_{\hat{K}_i}} \quad (35)$$

Where σ_ρ is obtained by the expression:

$$\hat{\sigma}_\rho^2 = \frac{\|\mathbf{Y} - \mathbf{W}\hat{\mathbf{K}}\|^2}{3r - c} \quad (36)$$

The relative standard deviation is given by:

$$\sigma_{\hat{K}ri} = 100 \frac{\sigma_{\hat{K}i}}{\hat{K}_i} \quad (37)$$

Table 2 gives the estimated base dynamic parameters and their relative standard deviations:

Table 2: Estimation of the base dynamic parameters

	$\hat{K}$	$\sigma_{\hat{K}ri}$		$\hat{K}$	$\sigma_{\hat{K}ri}$
M_{RP}	3.2555	3.4208	Off_3	-2.6104	-23.524
Ma_{1R1}	8.9738	0.4971	ZZ_{2R}	-2.8560	-31.5797
Fv_{11}	84.7633	0.5559	MX_2	-4.6706	-31.0705
Fs_{11}	54.35	0.4534	MY_{2R}	-0.0200	-29.4402
Ma_{1R2}	8.843	0.4904	XX_{3R}	1.5458	30.7643
Fv_{12}	67.1047	0.7118	XY_{3R}	0.0127	23.2455
Fs_{12}	80.5591	0.3105	XZ_{3R}	0.0378	14.5685
Ma_{13}	8.7612	0.5125	YZ_{3R}	-0.0088	-22.8285
Fv_{13}	83.7167	0.5669	ZZ_{3R}	-0.1356	-10.103
Fs_{13}	54.9855	0.4426	MY_{3R}	0.0120	43.8424
Off_1	-3.0074	-7.8658	MX_{4R}	4.5554	31.827
Off_2	-1.8386	-11.989			

Size of **W**: 65814 x 23
Condition number of **W**: 4805.57

A parameter with $\sigma_{\hat{K}ri} \geq 15$ is considered to be not good identified [15], it may have a little effect on the model and cannot be identified with acceptable accuracy on the actual trajectory. Taking into account that the parallelograms links are light and symmetric, and considering the grouped relations, we find that ZZ_{2R}, MX_2, ..., MX_4 are close to zero and can be neglected. Thus, we repeat the identification process with the following essential parameters (table 3):

Table 3: Estimation of the essential base dynamic parameters

	$\hat{K}$	$\sigma_{\hat{K}ri}$		$\hat{K}$	$\sigma_{\hat{K}ri}$
M_{RP}	1.5993	1.2601	Ma_{13}	8.6790	0.3487
Ma_{1R1}	8.7002	0.3128	Fv_{13}	83.7047	0.5700
Fv_{11}	84.6695	0.5594	Fs_{13}	54.9723	0.4451
Fs_{11}	54.4006	0.4554	Off_1	-2.5644	-4.8819
Ma_{1R2}	8.4992	0.3210	Off_2	-0.4616	-27.1756
Fv_{12}	66.888	0.7177	Off_3	-10.1975	2.3957
Fs_{12}	80.7283	0.3114			

Size of **W**: 65814 x 13
Condition number of **W**: 77.69

We note that the identified values are in accordance with respect to our knowledge of the system. The parameters M_{RP} is near the a priori value, which is 1.5 Kg. The estimation errors of each parameter $\sigma_{\hat{K}ri}$ are acceptable (except Off_2, which could be neglected). The condition number of the observation matrix is very good. We note that the effect of the rotor inertia, viscous friction and coulomb friction of the motorized joints are important compared to the effect of the base parameters of the parallelogram links. The dynamic model corresponding to the parameters of table 3 is easy to compute on line for control purpose.

5.4 Validation of the results

Two main validation procedures have been carried out:
i) The comparison of the estimated torques with respect to the measured torques on the trajectory that have been used in the identification, and with some other trajectories that have been not used in the identification.
ii) The addition of a payload on the platform to observe the evolution of the mass parameter of the platform M_{RP}.

All these tests show very good results.

6 Conclusion

This paper presents the identification of the dynamic parameters of the Orthoglide, a 3-DOF parallel robot that moves in the Cartesian space with fixed orientation. The dynamic identification model is based on the inverse dynamic model, which is linear in the parameters. The model is computed in terms of the Cartesian dynamic model elements of the legs and of the Newton-Euler equation of the platform. The base inertial parameters of the robot, which constitute the minimum number of inertial parameters, are determined using a numerical method using the QR decomposition. We proposed to make use of the similarity of the legs in order to reduce the number of parameters and to improve the condition number of the observation matrix. Experimental results are presented, and the validation is very good. Future work will concern the use of the identified model to control the robot using a dynamic control law.

References

[1] Wenger, P., and Chablat D., "Kinematic analysis of a new parallel machine tool : the Orthoglide", *Advances in Robot Kinematic, Kluwer Academic Publishers, 2000, pp.305-314.*

[2] Mayeda H., Yoshida K., Osuka K., "Base parameters of manipulator dynamic models", *IEEE Trans. on Robotics and Automation, Vol. RA-6(3), 1990, pp. 312-321.*

[3] Gautier, M., "Numerical calculation of the base inertial parameters", *IEEE Int. Conf. on Robotics and Automation, Cincinnati, U.S., 1990, pp.1020-1025*

[4] Gautier, M., and Khalil, W., "Direct calculation of minimum set of inertial parameters of serial robots", *IEEE Trans. on Robotics and Automation, Vol6(3), 1990, pp.368-373.*

[5] Khalil, W., and Kleinfinger, J.-F., "A new geometric notation for open and closed-loop robots", *Proc. IEEE Conf. on Robotics and Automation, San Francisco, US, 1986, pp.1174-1180.*

[6] Guegan, S., and Khalil, W., "Dynamic modeling of the Orthoglide", Advances in Robot Kinematic, Kluwer Academic Publishers, Caldes de Malavella, Spain, 2002, pp.387-396.

[7] Khatib, O., "A unified approach for motion and force control of robot manipulators: the operational space formulation", *IEEE J. of Robotics and Automation, Vol. RA-3(1), 1987, pp.43-53.*

[8] Lilly, K.W., and Orin D.E., "Efficient O(N) computation of the operational space inertia matrix", *Proc. IEEE Int. Conf. on Robotics and Automation, Cincinnati, US, 1990, pp.1014-1019.*

[9] Khalil, W., and Dombre, E., "Modeling, identification and control of robots", *Hermes Penton, 2002, London-Paris.*

[10] Luh, J.Y.S, Walker, M.W., and Paul, R.C.P., "On-line computational scheme for mechanical manipulators", *Proc. 2nd IFAC/IFIP Symp. on Information Control Problems in Manufacturing Technology, Stuttgart, 1979, pp.165-172.*

[11] Craig J.J., "Introduction to robotics: mechanics and control", *Addison Wesley Publishing Company, Reading, 1986.*

[12] Khalil, W., and Kleinfinger J.-F., "Minimum operations and minimum parameters of the dynamic model of tree structure robots", *IEEE J. of Robotics and Automation, Vol. RA-3(6), 1987, pp.517-526.*

[13] Gautier M., "A comparison of filtered models for dynamic identification of robots", *Proc. IEEE 35th Conf. on Decision and Control, Kobe, Japon, déc. 1996, pp. 875-880.*

[14] Gautier M., Khalil W., Restrepo P.P., "Identification of the dynamic parameters of a closed loop robot", *Proc. IEEE Int. Conf. on Robotics and Automation, Nagoya, mai 1995, pp. 3045-3050.*

[15] Gautier M., Khalil W., "Exciting trajectories for inertial parameters identification", *The Int. J. of Robotics Research, Vol. 11(4), 1992, pp. 362-375.*

Appendix : Dynamic model of a leg

Each leg has a planar parallelogram closed loop. The inverse dynamic model of the equivalent tree structure is obtained by cutting the revolute joint q_{8i} (i = 1, 2, 3), figure 3:

$$\mathbf{\Gamma}_{tr_i} = \mathbf{H}_{tr_i}(\mathbf{q}_i, \dot{\mathbf{q}}_i, \ddot{\mathbf{q}}_i) \qquad (38)$$

Leg i is isolated from the platform, so we can consider the variables q_{1i}, q_{2i} and q_{3i} to be independent. In the complete model of the robot, only q_{1i} is active and the torques Γ_{2i} and Γ_{3i} are zero.

Let the vector $\mathbf{q}_{a_i}$ be composed of the independent joints and the vector $\mathbf{q}_{p_i}$ be composed of the passive joints of leg i:

$$\mathbf{q}_{a_i} = \begin{bmatrix} q_{1i} & q_{2i} & q_{3i} \end{bmatrix}^T \qquad \mathbf{q}_{p_i} = \begin{bmatrix} q_{4i} & q_{5i} & q_{7i} \end{bmatrix}^T \quad (39)$$

The constraint equations of the loop are:

$$q_{4i} = -q_{3i}, \quad q_{5i} = -q_{2i} - \frac{\pi}{2}, \quad q_{7i} = q_{3i}, \quad q_{8i} = -q_{3i} \quad (40)$$

The dynamic model of the leg is obtained from Γ_{tr_i} and the constraint equations by [9]:

$$\begin{aligned}\mathbf{\Gamma}_i &= \mathbf{H}_i(\mathbf{q}_i, \dot{\mathbf{q}}_i, \ddot{\mathbf{q}}_i) \\ \mathbf{\Gamma}_i &= \begin{bmatrix} \Gamma_{1i} & \Gamma_{2i} & \Gamma_{3i} \end{bmatrix}^T = \mathbf{G}_i^T \mathbf{\Gamma}_{tr_i}\end{aligned} \qquad (41)$$

With: $\mathbf{G}_i = \dfrac{\partial \mathbf{q}_{tr_i}}{\partial \mathbf{q}_{a_i}}$ and $\mathbf{q}_{tr_i} = \begin{bmatrix} \mathbf{q}_{a_i}^T & \mathbf{q}_{p_i}^T \end{bmatrix}^T$, thus:

$$\mathbf{G}_i^T = \begin{bmatrix} 1 & 0 & 0 & 0 & 0 & 0 \\ 0 & 1 & 0 & 0 & -1 & 0 \\ 0 & 0 & 1 & -1 & 0 & 1 \end{bmatrix} \qquad (42)$$

The dynamic model of the tree structure is obtained by recursive symbolic Newton-Euler method [12].

Since all the parameters of the 5th and 7th links are grouped with the other links. Thus the matrix **G** can be reduced to the first 4 columns:

$$\mathbf{G}_i^T = \begin{bmatrix} 1 & 0 & 0 & 0 \\ 0 & 1 & 0 & 0 \\ 0 & 0 & 1 & -1 \end{bmatrix} \qquad (43)$$

And:

$$\mathbf{\Gamma}_{tr_i} = \begin{bmatrix} \Gamma_{tr_{1i}} & \cdots & \Gamma_{tr_{4i}} \end{bmatrix}^T \qquad (44)$$

Experimental dynamic identification of a fully parallel robot

Andrès Vivas* Philippe Poignet* Frédéric Marquet* François Pierrot* Maxime Gautier§

*Laboratoire d'Informatique, Robotique et Microélectronique de Montpellier - UMR CNRS 5506, 161 Rue Ada, 34392
Montpellier cedex 5, France

§Institut de Recherche en Communication et Cybernétique de Nantes, UMR CNRS 6597, 1 rue de la Noë, 44321
Nantes cedex 03, France

{vivas, poignet}@lirmm.fr

Abstract - This paper deals with the experimental identification of the dynamic parameters of parallel machines. The dynamic parameters are estimated by using the weighted least squares solution of an over determined linear system obtained from the sampling of the dynamic model along a closed loop exciting trajectory. Experimental results are exhibited for the H4 robot, a fully parallel structure providing 3 degrees of freedom (dof) in translation and 1 dof in rotation. A comparative study is performed depending on the available measurements i.e. different sensor locations (motor, end effector).

1. INTRODUCTION

After the works in parallel mechanisms introduced by Gough [1] or Steward [2], Clavel [3] proposed the Delta structure, a parallel robot dedicated to high-speed applications. In the same way, the "hexapod" [4], [5] has been used intensively in industry. This is due to the exceptional simplicity of the Delta 3-dof solution and the enormous research effort dedicated to the "hexapod". Many alternate designs have been proposed like the HexaM [6], which is an evolution of the Hexa robot [7]. For most pick-and-place applications, at least four dof are required (3 translations and 1 rotation to arrange the carried object in its final location). For the Delta robot, this is achieved thanks to an additional link between the base and the gripper, but it seems not to be as efficient as a parallel arrangement. On the other hand, 6-dof fully-parallel machines currently used in machining suffer from their complexity (they need at least 6 motors while the cutting process requires only 5 controlled axis plus the spindle rotation) and from their limited tilting angle. As an intermediate solution to these drawbacks, a 4-dof parallel mechanism – the H4 robot - have been proposed [8], [9]. Fig. 1 shows a photography of the H4 parallel robot. This machine is based on 4 independent active chains between the base and the nacelle; each chain is actuated by a brushless direct drive motor fixed on the base and equipped with an incremental position encoder. Thanks to its design, the mechanism is able to provide high performances. In order to achieve high speed and acceleration for pick-and-place applications or precise motion in machining tasks, an accurate dynamic modeling is required to increase the quality of their simulation in order to improve their design and to compute advanced model based robust controllers such as moving horizon control schemes. However the first difficulty is to estimate the physical parameters (mass, inertia and frictions), especially when the only available measurements are given by the incremental sensors located on the actuators.

Therefore, in this paper, we focus on the estimation of the dynamic parameters of the rigid multi body model. The parameters are estimated by a classical technique of weighted least squares [11], [14]. We mainly discuss two identification results depending on the available measurements. We compare the influence of the sensor locations on the estimation results of physical parameters: i) first, the nacelle acceleration is estimated through the computation of the kinematic model and its derivative ii) secondly, additional sensors (rotation and 3-axis acceleration sensors) located on the end effector provide further measurements.

The paper is organized as follows : Section 2 is dedicated to the geometric, kinematic and dynamic modelling. Section 3 recalls the basis of the identification method. Section 4 exhibits and discuss experimental identification results of the H4 robot. Finally, conclusions are given in section 5.

2. MODELING

A. Geometric and kinematic modelling

The Jacobian matrix and the forward geometric model are needed to compute the dynamic model (see section 2.2). Therefore we briefly present the way of computing the different relationship necessary to obtain these model and matrix. The design parameters of the robot are described on Fig. 2 where the following parameters have been chosen:

Fig. 1 H4 robot

$\alpha_1 = 0$; $\alpha_2 = \pi$; $\alpha_3 = 3\pi/2$; $\alpha_4 = 3\pi/2$
$u_1 = u_y$; $u_2 = -u_y$; $u_3 = u_x$; $u_4 = u_x$

The angles α_i describe the position of the four motors, L is the length of arms, l is the length of the forearms, θ the nacelle's angle, and d and h are the half lengths of the "H" forming the nacelle. O is the origin of the base frame and D is the origin of the nacelle frame. R gives the motor's position. The A_iB_i segments represent the arms of the robot and P_iB_i the forearms segments. The joint positions are represented by q_i.

To obtain the geometric model, it is necessary to express the different points of the mechanical system with respect to the origin O. The origin is fixed in the middle of the nacelle with the coordinates (x, y, z). In the Cartesian space, the end effector position is given by (x, y, z, θ).

$$OD = [x \ y \ z]^T \quad (1)$$

The vector that joins the absolute origin O and all of the forearms to the nacelle is:

$$OA_i = OD + DA_i = \begin{bmatrix} x \\ y \\ z \end{bmatrix} + DA_i \quad (2)$$

The DA_i segments can be expressed as:

$$DA_1 = \begin{bmatrix} h\cos\theta \\ h\sin\theta + d \\ 0 \end{bmatrix} ; DA_2 = \begin{bmatrix} -h\cos\theta \\ -h\sin\theta + d \\ 0 \end{bmatrix} \quad (3)$$

$$DA_3 = \begin{bmatrix} -h\cos\theta \\ -h\sin\theta - d \\ 0 \end{bmatrix} ; DA_4 = \begin{bmatrix} h\cos\theta \\ h\sin\theta - d \\ 0 \end{bmatrix} \quad (4)$$

Moreover, the vector that links the absolute origin and all of the arms to the forearms is:

$$OB_i = OP_i + P_iB_i \quad (5)$$

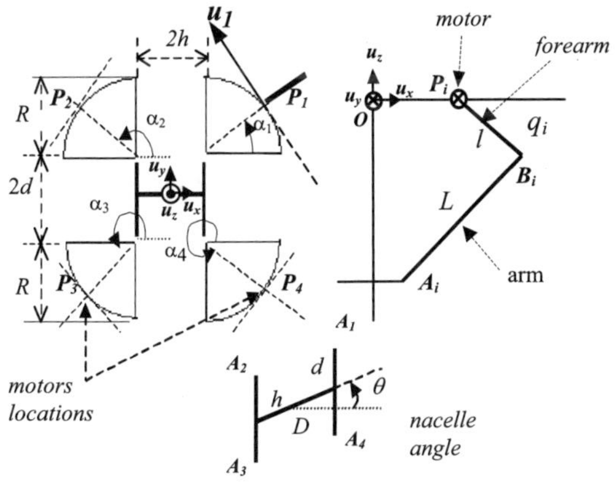

Fig 2. Design parameters

with:

$$P_iB_i = \begin{bmatrix} l\cos q_i \cos\alpha_i \\ l\cos q_i \sin\alpha_i \\ -l\sin q_i \end{bmatrix} \quad (6)$$

and actuator locations are:

$$OP_1 = \begin{bmatrix} h + R\cos\alpha_1 \\ d + R\sin\alpha_1 \\ 0 \end{bmatrix} ; OP_2 = \begin{bmatrix} -h + R\cos\alpha_2 \\ d + R\sin\alpha_2 \\ 0 \end{bmatrix}$$

$$OP_3 = \begin{bmatrix} -h + R\cos\alpha_3 \\ -d + R\sin\alpha_3 \\ 0 \end{bmatrix} ; OP_4 = \begin{bmatrix} h + R\cos\alpha_4 \\ -d + R\sin\alpha_4 \\ 0 \end{bmatrix} \quad (7)$$

Finally, arms coordinates are given by:

$$A_iB_i = A_iO + OB_i \quad (8)$$

The analytical forward position relationship is difficult to compute. Up to now, the simplest model we've got is a 8th degree polynomial equation. The forward model is then computed iteratively using the classical formula:

$$x_{n+1} = x_n + J(x_n, q_n)[q - q_n] \quad (9)$$

Where q is the convergence point and J is the robot Jacobian matrix. If the mechanism is not in a singular configuration, this expression is derived as follows [8], [9]:

$$J = J_x^{-1} J_q \quad (10)$$

Where:

$$J_x = \begin{bmatrix} A_1B_{1x} & A_1B_{1y} & A_1B_{1z} & (DC_1 \times A_1B_1)_z \\ A_2B_{2x} & A_2B_{2y} & A_2B_{2z} & (DC_2 \times A_2B_2)_z \\ A_3B_{3x} & A_3B_{3y} & A_3B_{3z} & (DC_3 \times A_3B_3)_z \\ A_4B_{4x} & A_4B_{4y} & A_4B_{4z} & (DC_4 \times A_4B_4)_z \end{bmatrix} \quad (11)$$

$$J_q = diag((P_iB_i \times A_iB_i) \cdot u_{mi}), \ i = 1,...4 \quad (12)$$

DC_i is the distance between the center of the nacelle and the center of the half lengths of the "H" that forms the nacelle.

B. Dynamic modelling

In first approximation, the dynamic model is computed by considering physical dynamics. Indeed, drive torques are mainly used to move the motor inertia, the fore-arms and the arms and the nacelle equipped with a machining tool. Because of the design, the fore-arm inertia can be considered as a part of the motor inertia and the arm (manufactured in carbon materials) effects are neglected [8], [9].

If Γ_{mot} is the (4x1) actuator torque vector, the basic equation of dynamics can be written as:

$$\Gamma_{mot} = I_{mot}\ddot{q} + J^T M(\ddot{x} - G) + F_v\dot{q} + F_c sign(\dot{q}) \quad (13)$$

where I_{mot} represents the motor's inertia matrix including the forearm's inertia, M a matrix containing the mass of the nacelle and its inertia, $\dot{q}$ is the (4x1) joint velocity vector, $\ddot{q}$ is (4x1) the joint acceleration vector, $\ddot{x}$ is the vector of cartesian accelerations, and G the gravity constant. Thanks to the design, the forearm's inertia is taken into account in the motor's inertia. F_v are the viscous friction coefficients and F_c are the Coulomb friction.

With:

$$I_{mot} = \begin{bmatrix} I_{mot1} & 0 & 0 & 0 \\ 0 & I_{mot2} & 0 & 0 \\ 0 & 0 & I_{mot3} & 0 \\ 0 & 0 & 0 & I_{mot4} \end{bmatrix} \quad (14)$$

$$M = \begin{bmatrix} M_{nac} & 0 & 0 & 0 \\ 0 & M_{nac} & 0 & 0 \\ 0 & 0 & M_{nac} & 0 \\ 0 & 0 & 0 & I_{bc} \end{bmatrix} \quad (15)$$

It is first assumed that the nacelle acceleration $\ddot{x} = [\ddot{x} \ \ddot{y} \ \ddot{z} \ \ddot{\theta}]^T$ and the motor position $q = [q_1 \ q_2 \ q_3 \ q_4]^T$ are directly measured. The dynamic equation can be rewritten in a relation linear to the dynamic parameters. By introducing $J^T = [J_{43}^T \ J_{41}^T]$, it follows:

$$\Gamma_{mot} = \begin{bmatrix} \ddot{q} & J_{43}^T \begin{bmatrix} \ddot{x} \\ \ddot{y} \\ \ddot{z} - G \end{bmatrix} & J_{41}^T \ddot{\theta} & \dot{q} & sign(\dot{q}) \end{bmatrix} X \quad (16)$$

where X is the vector of parameters:

$$X = [I_{mot1} \ I_{mot2} \ I_{mot3} \ I_{mot4} \ M_{nac} \ I_{bc} \\ F_{v1} \ F_{v2} \ F_{v3} \ F_{v4} \ F_{c1} \ F_{c2} \ F_{c3} \ F_{c4}]^T \quad (17)$$

If acceleration measurement $\ddot{x}$ is not available, $\ddot{x}$ can be evaluated by:

$$\ddot{x} = J\ddot{q} + \dot{J}\dot{q} = [\ddot{x}_q \ \ddot{y}_q \ \ddot{z}_q \ \ddot{\theta}_q]^T \quad (18)$$

where J depends on x and q, $\dot{J}$ is computed using a central difference algorithm.

Then, the second identification dynamic model is given by:

$$\Gamma_{mot} = \begin{bmatrix} \ddot{q} & J_{43}^T \begin{bmatrix} \ddot{x}_q \\ \ddot{y}_q \\ \ddot{z}_q - G \end{bmatrix} & J_{41}^T \ddot{\theta}_q & \dot{q} & sign(\dot{q}) \end{bmatrix} X \quad (19)$$

Both models are expressed in a general form:

$$Y = DX \quad (20)$$

where Y is the torque measurement vector, D is called the regressor and X is the vector of unknown parameters.

III. IDENTIFICATION METHOD

The identification technique, classically developed for robot manipulators, is applied for the parallel robot. Usually X is estimated as the least squares (LS) solution of an over-determined linear system obtained by sampling and filtering the dynamic model (20) along a trajectory $(q, \dot{q}, \ddot{q})$, considering that ρ is a zero mean additive independent noise, with a standard deviation σ_ρ such that:

$$C_{\rho\rho} = E(\rho^T \rho) = \sigma_\rho^2 I_r \quad (21)$$

where E is the expectation operator. I_r is the (rxr) identity matrix. The over-determined system is written as follows:

$$Y = WX + \rho \quad (22)$$

where Y is the (rx1) measurement vector, W is the (rxN) observation matrix, N is the number of parameters to identify. In fact Y is obtained by concatenation of n measurements vectors Y^j of the n motor torques with different error standard deviations. A better solution is to calculate the WLS solution of the global system (22). The r^j rows, corresponding to joint j equation, are weighted by the coefficient of the diagonal matrix of the error covariance matrix defined as follows:

$$C_{\rho\rho} = (G^T G)^{-1} \qquad G = diag(S) \quad (23)$$

G is a (rxr) diagonal matrix composed of the elements of S.

$$S = [S^1 ... S^n], S^j = \left[\frac{1}{\hat{\sigma}_\rho^j} ... \frac{1}{\hat{\sigma}_\rho^j}\right] \quad (24)$$

S^j is a ($1 \times r^j$) row matrix. An unbiased estimation $\hat{\sigma}_\rho^j$ is used from the regression on each joint j subsystem:

$$\hat{\sigma}_\rho^{j^2} = \frac{\|Y^j - \Phi^j \hat{\Theta}^j\|^2}{(r^j - N_p^j)} \quad (25)$$

$Y^j, \Phi^j, \Theta^j, r^j, Np^j$ are the measurement vector, the observation matrix, the number of equations and the number of minimum parameters for each joint j subsystem respectively.

The WLS vector solution $\hat{X}_w$ minimizes the Euclidean norm of the vector of weighted errors ρ:

$$\hat{X}_w = \underset{X}{Arg.min}\left[\rho^T G^T G \rho\right] \quad (26)$$

$\hat{X}_w$ and the corresponding standard deviations $\sigma_{\hat{X}wi}$ are calculated as the LS solution of (22) weighted by G:

$$Y_w = W_w X + \rho_w$$
$$Y_w = GY, \; W_w = GW, \; \rho_w = G\rho \quad (27)$$

Complete details concerning the WLS identification technique and its practical implementation can be found in [10], [11], [12].

IV. EXPERIMENTAL RESULTS

A. Determination of the amplifier gain

For industrial robots, torques are usually estimated using a linear relation between torque and voltage applied to the amplifier:

$$\Gamma_m = G_T V_T \quad (28)$$

where V_T is the current reference of the amplifier current loop and G_T the gain of the joint drive chain. A good estimation of G_T is important to obtain a good estimation of the physical parameters.

A force sensor, located on the nacelle, is used to measure directly the force produced at the arm end. Applying different input tensions to the amplifier, the resulting torques are measured and the gain is estimated (Table 1). In [10], other techniques to estimate the G_T values are given.

B. Dynamic parameters estimation

In this work, we are focused in the estimation of the following dynamic parameters:

$$X = [I_{mot1} \; I_{mot2} \; I_{mot3} \; I_{mot4} \; M_{nac} \; I_{bc}$$
$$F_{v1} \; F_{v2} \; F_{v3} \; F_{v4} \; F_{c1} \; F_{c2} \; F_{c3} \; F_{c4}] \quad (29)$$

TABLE 1
AMPLIFIERS GAINS

	Motor 1	Motor 2	Motor 3	Motor 4
G_T (N.m/V)	2.85	2.65	2.70	2.87

For computing the regressor, joint velocities and accelerations are estimated by a band pass filtering of the position. The band pass filtering is obtained by the product of a low pass filter in both the forward and the reverse direction (Butterworth) and a derivative filter obtained by a central difference algorithm, without phase shift. A parallel filtering is implemented to reject the high frequency ripples of the measured motor torques. Practical aspects of the derivative estimation and data filtering are completely detailed in [13] and [14].

In order to get good identification results, exciting trajectories containing slow motions (in such a case, friction will be preponderant) and high dynamic motions (inertia phenomena become preponderant) are generated. Finally, concatenation of these trajectories is used. Examples of generated trajectories are presented in Fig. 3.

1) Identification without additional sensor: Initially, the Cartesian accelerations are computed thanks (18), where the actuator velocities and accelerations have been computed from the actuators positions. Table 2 shows the obtained parameters of the dynamics model (19). The relative standard deviation ($\%\sigma_{\hat{x}r}$) is given.

The physical parameters are quite well estimated in comparison to the prior values of nacelle mass and the motor inertia (0.975 Kg and 0.012 $N.m^2$ respectively). However additional measurements are provided to improve estimation accuracy.

2) Identification using additional sensors: An accelerometer located on the nacelle gives the Cartesian accelerations on the

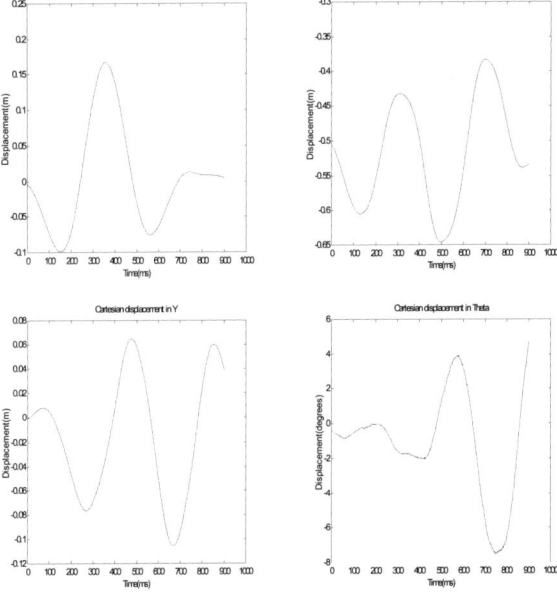

Fig. 3 Typical trajectories in the Cartesian space

TABLE 2
H4 DYNAMIC PARAMETERS WITHOUT ADDITIONAL SENSORS

Parameter	Estimated values	Units	%$\sigma_{\hat{x}r}$
I_{mot1}	0.0141	N.m²	2.6286
I_{mot2}	0.0120	N.m²	3.0444
I_{mot3}	0.0153	N.m²	1.6939
I_{mot4}	0.0213	N.m²	1.1933
M_{nac}	1.0492	Kg	0.4236
I_{bc}	0.0030	N.m²	3.5049
F_{v1}	0.1636	N.m.s/rad	5.6781
F_{v2}	0.0560	N.m.s/rad	15.5674
F_{v3}	0.0930	N.m.s/rad	6.5734
F_{v4}	0.0917	N.m.s/rad	6.4301
F_{c1}	1.1453	N.m	2.0450
F_{c2}	1.0950	N.m	2.0563
F_{c3}	0.7222	N.m	2.8366
F_{c4}	0.9932	N.m	2.0451

three axes (*x*, *y* and *z*) and a rotation sensor gives the central bar rotation. We use them to perform the identification straight with the model of (19). Fig. 4 shows the comparison of measured data obtained from the accelerometer and rotation sensor and those provided by (16). Rotation acceleration is numerically computed by a finite central difference.

Table 3 shows the fourteen estimated parameters of the dynamic model (16). In case of the rigid multi body model identification, additional sensors located on the end effector mainly improve the estimation of the viscous friction coefficients and their relative standard deviations.

C. Experimental validation

The validation of the identification results consists in comparing the measured torques with those obtained by computing the inverse dynamic model with the estimated parameters. Fig. 5 exhibits cross validations with new trajectories that have not been used previously for the identification. Simulation and measurements are very close. The dynamic parameters are quite well estimated.

These experiments show good results of the dynamic identification. The use of the accelerometer and rotation sensor are not very necessary for the rigid model.

V. CONCLUSION

In this paper experimental results related to the identification of physical dynamic parameters of a fully parallel robot are presented. Estimated values depending on the available measurements at the end effector are exhibited. The cross validation shows good identification results.

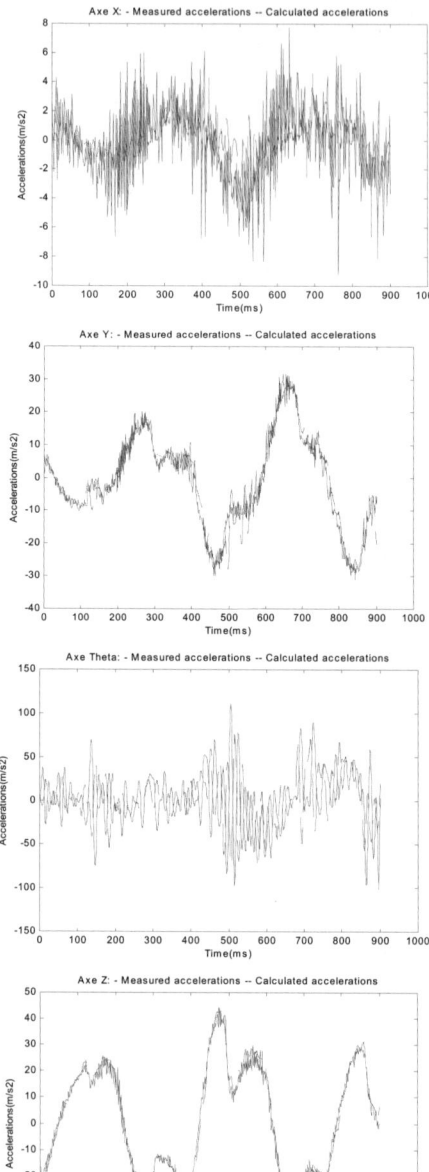

Fig. 4 Calculated and measured accelerations

However this structure working with high speed and acceleration presents important flexibility and slight differences during validation may be due to flexibility. Therefore, we are currently working on the introduction of lumped elasticities in the dynamic model. In the future, we will compare the results with other estimation methods and this complete dynamic model will be used in model based control scheme with moving horizon for machining tasks.

TABLE 3
ESTIMATED PARAMETERS USING ADDITIONAL SENSORS

Parameter	Estimated values	Units	$\%\sigma_{\hat{x}r}$
I_{mot1}	0.0167	$N.m^2$	2.3695
I_{mot2}	0.0164	$N.m^2$	2.3590
I_{mot3}	0.0176	$N.m^2$	1.5776
I_{mot4}	0.0234	$N.m^2$	1.1579
M_{nac}	0.984	Kg	0.4666
I_{bc}	0.0029	$N.m^2$	3.7311
F_{v1}	0.2112	N.m.s/rad	4.7212
F_{v2}	0.1236	N.m.s/rad	7.5670
F_{v3}	0.1266	N.m.s/rad	5.2000
F_{v4}	0.1133	N.m.s/rad	5.6255
F_{c1}	1.2186	N.m	2.0756
F_{c2}	1.0252	N.m	2.3623
F_{c3}	0.7902	N.m	2.7986
F_{c4}	1.0394	N.m	2.1046

VI. REFERENCES

[1] V.E. Gough, "Contribution to discussion of papers on research in automotive stability, control and tyre performance", *Proc. Auto Div., Inst. Mechanical Engineers*, 1956-1957.

[2] D. Stewart, "A plataform with 6 degrees of freedom", *Proc. of the Ins. of Mech. Engineers*, 180 (Part 1, 15), pp. 371-386, 1965.

[3] R. Clavel, "Une nouvelle structure de manipulateur parallèle pour la robotique légère", *APII*, 23 (6), pp. 501-519, 1989.

[4] J.-P. Merlet, *Les robots parallèles*, 2nd Edition, Hermes, 1997.

[5] H. K. Tönshoff, "A systematic comparison of parallel kinematics", *Keynote in proceedings of the first forum on parallel kinematic machines*, Milan, Italy, August 31 – September 1, 1998.

[6] F. Pierrot, T. Shibukawa, "From Hexa to HexaM", *Proc. IPK'98: Internationales Parallelkinematik-Kolloquium*, Zürich, June 4, pp. 75-84, 1998.

[7] F. Pierrot, P. Dauchez and A. Fournier, "Fast parallel robots", *Journal of robotics systems*, 8 (6), pp. 829-840, 1991.

[8] O. Company, F. Pierrot, "A new 3T-1R parallel robot", *ICAR'99*, Tokyo, Japan, October 25-27, pp. 557-562, 1999.

[9] F. Pierrot, F. Marquet, O. Company, T. Gil, "H4 Parallel Robot,: Modeling, Design and Preliminary Experiments", *Proc. of the 2001 IEEE International Conference one Robotics & Automation*, Seoul, Korea, May 21-26, 2001.

[10] P. Restrepo, "Contribution à la modélisation, identification et commande des robots à structures fermées: application au robot ACMA SR400", *Ph.D Thesis*, Ecole Centrale de Nantes, France, 1996.

[11] Ph. Poignet, M. Gautier, "Extended kalman filtering and weighted least squares dynamic identification of robots", *Control Engineering Practice*, 2001, vol. 9/12, pp. 1361-

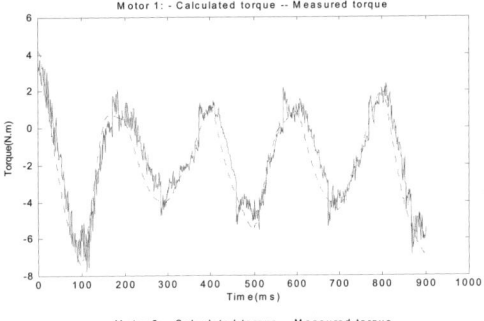
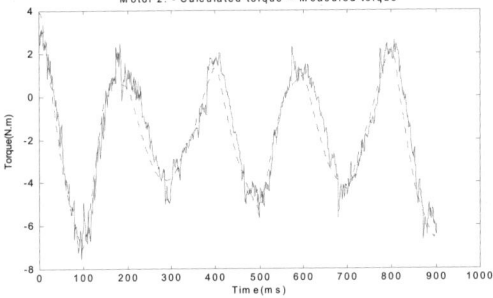
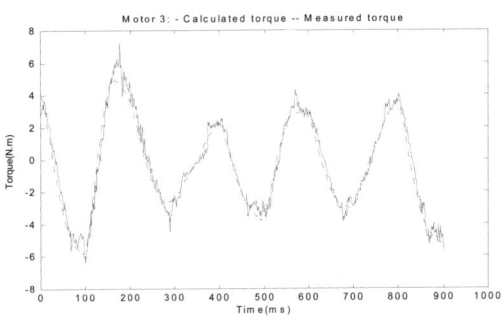
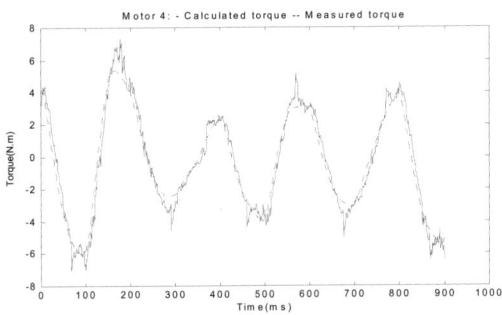

Fig. 5 Calculated and measured torques

1372.

[12] C. Canudas de Wit, B. Siciliano, G. Bastin, *Theory of robot control*, Springer, 1996.

[13] W. Khalil, E. Dombre, *Modeling, Identification and Control of Robots*, Hermes Penton Science, London, 2002.

[14] M.T. Pham, M. Gautier, Ph. Poignet, "Dynamic identification of high speed machine tools", *Proc. of the II International Seminar on Improving Machine Tools Performance*, La Baule, France, 2000.

Pattern-based Architecture for Building Mobile Robotics Remote Laboratories

A. Khamis   D.M. Rivero   F. Rodríguez   M. Salichs

Department of System Engineering and Automation
Carlos III University of Madrid
Avda. Universidad, 30 – 28911 Leganés, Madrid – Spain
{akhamis, mrivero, urbano, salichs}@ing.uc3m.es

Abstract – The building of remote laboratories for laboratory experiments in mobile robots requires expertise in a number of different disciplines, such as Internet programming, telematic and mechatronic systems, etc. Remote laboratories offer students access to complementary experiments, not available at their own university, as support to lectures. An intuitive user interface is required for inexperienced people to control the robot remotely. This paper describes a design pattern-based architecture to build remote laboratories for mobile robotics. The proposed remote laboratory is currently used to provide remote experiments on indoor mobile robotics, addressing different approaches to solve the main problems of mobile robotics, such as sensing, motion control, localization, world modeling, planning, etc. These experiments are being used in several mobile robotics and autonomous systems courses, at the undergraduate and graduate levels.

1. INTRODUCTION

The Internet has become a major global tool for communication and information sharing. It provides a global, integrated communication infrastructure that enables an easy implementation of distributed systems. For this reason, a great deal of attention is now being paid to the World Wide Web as a tool for building remote laboratories for tele-education on mechatronics. Robotics education provides an ideal field for tele-education systems because of its flexibility. Unlike traditional fields, robotics is still an emerging area. Relatively few programs exist at the graduate level, and even fewer at the undergraduate level. The courses in existence are still new and are open to rapid change and new approaches.

Remote laboratories can be used to provide a superior educational experience to a purely in-residence laboratory. These labs are not restricted to synchronized attendance by instructors and students; they have the potential to provide constant access whenever needed by student [1]. Very few remote laboratories for mobile robotics have been commercialized or moved outside the research laboratories to public access. The main problems of remote control robotics have been addressed in [2-3]. The XAVIER system, built from a set of standard components for communication, planning and behavior integration, has been in almost daily use since December 1995 [4]. The RHINO system developed at Bonn University is similar to the XAVIER system and has been used as a tour guide in museums [5]. Schilling has presented a model design for remote mobile robots [6]. These systems have been developed to cover certain subjects by providing online experiments or online classes without providing any kind of generic tools for the remote users by which they will be able to customize the experiment according to their needs. An integrated architecture for a tele-education system in the field of mobile robotics has been proposed in [7], which relies on three main processes: tutoring, authoring and administration. This system has proposed the use of assessment tools and the collaboration mechanisms.

II. SYSTEM DESCRIPTION

A. Remote Laboratory Overview

A remote laboratory can be defined as a network-based laboratory where user and real laboratory equipment are geographically separated and where the telecommunication technologies are used to give users access to laboratory equipment. By networking many remote laboratories, we can obtain a framework called virtual laboratory. The project IECAT at which we are participating is an example of such frameworks in the field of autonomous and teleoperated mechatronic systems [8]. These laboratories represent a coordinated set of experiments for students with hardware facilities physically spread over different locations, but accessible by students via the Internet. In the designing of these distance laboratories for robotic systems, a number of challenges must be addressed, particularly the telematics infrastructure which gives access to the experiments, as well as the user interface which provides the necessary interactivity with the remote hardware supporting the learning process of students through appropriate feedback [9]. To implement a remote laboratory, an Internet-based open-loop teleoperation model is commonly used as shown in Fig.1.

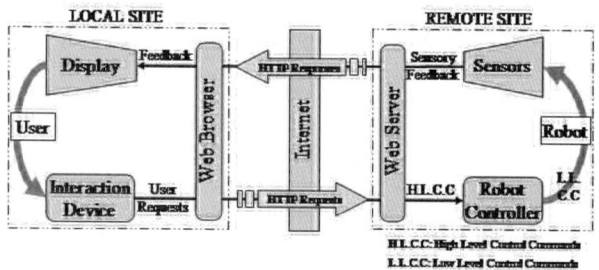

Fig.1 Internet-based Open-Loop Teleoperation Model

This model is based on the simple protocol commonly used in distributed computation "The Request/Response Protocol". The client interacts with the system using any Web browser to make the request. Client requests are translated to HTTP requests, which are satisfied by the Web server. These requests are converted to high-level control requests that are received by the robot controller which transmits them as low level control requests to execute the required task. Sensory feedback is required to give the user information about the remote robot's environment and the consequences of his/her commands. By using the concepts derived from this simple model, a remote laboratory can be developed to provide live performance experimentation.

B. Mobile Robot Description

The robot used is the indoor mobile robot B21 from Real World Interface [10] with mobility software. The B21 hardware consists of two main sections, the base and the enclosure. The base contains the batteries and the motors (4 high torque, 48 VDC servo motors with 4-wheel synchronous drive) which translate (90 cm/s with resolution 1 mm) and rotate (167 °/s with resolution 0.35°) the robot, as well as dedicated microprocessors which handle very low level functions such as dead reckoning. The enclosure contains two main computers, a power distribution system and a camera is mounted on the top of the enclosure. In addition, the B21 has many types of sensors such sonar, laser, tactile, infrared sensors.

III. HARDWARE ARCHITECTURE

Remote laboratories offer students access to complementary experiments, not available at their own university, as support to lectures. Thus they can experiment with remote hardware during a time slot which they can select according to their schedule. The implementation of the remote laboratory system should make efficient use of the resources in order to account for bandwidth and time delay restrictions. On the other hand, such systems should be implemented in an extendable way, which guarantees that the different software modules need not to be rewritten when new hardware is installed to be accessed through the experiments.

To obtain a maximum level of portability, an important design decision was that all interactions with the remote laboratory being developed could be accomplished with only a Web-browser as an interface; no additional software or plug-ins should be needed for the use of the laboratory. With Java, we can provide a cross-platform user interface for configuration, testing and visualization of our robot software system in action. Fig.2 shows the remote laboratory hardware architecture, which is multi-layered to facilitate quick start-up and an unprecedented level of code reuse and transportability.

A. User Layer

During the implementation phase, event-based control approach has been used to guarantee the system stability in the presence of network time delay. In this approach, non-time based motion reference is used. This reference is usually related directly to real time sensor measurements and the task and thus time delay will not have effect on the stability [11]. The role of the user in the control loop is just to activate of deactivate robotic skill and therefore the control loop is not sensitive to communication delay. The idea -of overcoming communication constraints by communicating at more abstract level and increasing the robot's autonomy- is fundamental to remote control via constrained communications.

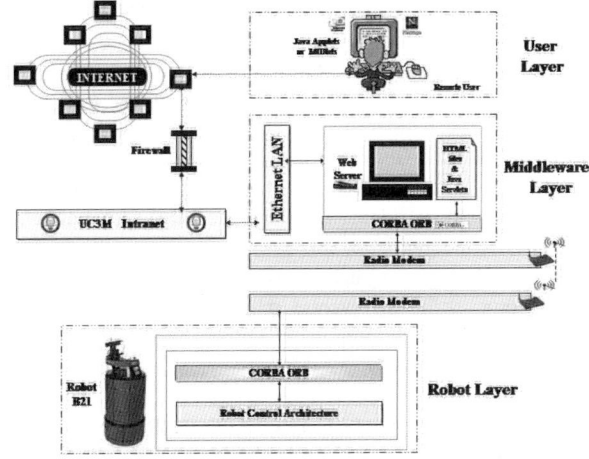

Fig.2 Hardware Architecture

Users can access the experiment using any Web-browser (Netscape or Internet Explorer). Users request are received by Java applets or MIDlets and events are sent to the corresponding Java Servlet in the Web server to activate/deactivate a robotic skill or to invoke sensory data.

B. Middleware Layer

In this layer, there is a PC linked internally to an Ethernet LAN and externally to the university's Intranet. It runs Apache Web server, which hosts the HTML files and two agents implemented based on Java servlets:

The *robot–on agent*, which deals with the real robot, when the robot is actually running to provide the required information to process certain ability selected by the user or required in an experiment's step. This agent contains two groups of java servlets, control servlets that send control commands to the robot and sensor servlets to invoke the sensory data. The other agent, which forms the middleware layer, is the *robot-off* agent. This agent represents systematic knowledge of mobile robotics by accessing to a database using Java Data Base Connectivity (JDBC). It contains also many servlets such as evaluation servlets, login servlets, etc...

C. Robot Layer

The robot layer has the robot skills servers, which have been implemented based on a two level architecture called AD (Deliberative and Automatic levels [12]. A skill represents the robot's ability to perform a particular task. They are all built-in robot action and perception capacities [13]. In the deliberative level there are skills capable of carrying out high level tasks, while at the automatic level there are skills in charge of interacting with the environment. The path planner, the environment modeller and the task supervisor are some of the skills included in the deliberative level. The sensorimotor and the sensorial skills are found in the automatic level. The first are in charge of the robot motion. The second ones detect the events needed to produce the sequencer transitions which manage the task performed by the robot.

In the AD architecture, skills are client-server modules. Each skill is implemented as a distributed object and it is activated by the deliberative level sequencers [14]. During the period of time in which the skills remain active, they can connect to data objects of other skills or to sensorial servers (Odometry, sonar, laser and camera). As a result, the skills can generate motor actions over the robot actuators (through a drive server), when considering sensorimotor skill, or events, when sensorial skills are considered. The skill outputs – actions and events – are stored in its data objects.

IV. SOFTWARE ARCHITECTURE

An intuitive user interface is required for inexperienced people to control the robot remotely. The proposed architecture is visual proxy pattern-based architecture. A design pattern is a formal description of a problem and its solution. Each pattern describes a problem which occurs over and over again in our environment, and then describes the core of the solution to that problem, in such a way that you can use this solution a million times over, without ever doing it the same way twice. By using design patterns, high reuse rate, i.e. the ratio of reused code to the total code, can be obtained.

The visual-proxy pattern is in some ways a specialization of the Presentation/Abstraction/Control (PAC) architecture [15], which can be used to build user interfaces for object-oriented systems and can guarantee high degree of extensibility and reusability of the software components [16]. The objective of PAC architecture is to separate the generation of the user interface entirely from the abstraction layer object to provide the reusability and extensibility facilitates. The PAC control object is passive with respect to message flow. The messages go directly from the visual proxy (presentation layer) to the abstraction-layer object that manufactured the proxy. In the visual proxy architecture, the encapsulation is still intact in the sense that the implementation of the abstraction-layer object can change without the outside world knowing about it.

Fig. 3 is a static model for the proposed architecture using unified modeling language (UML). This architecture consists of three tiers:

A. Client Tier

In this tier, the *ExperimentTool* class implements *User_interface* so it can produce visual proxies when asked.

It asks the other *CommandHandler* and the *DynamicRepository* classes for visual proxies as simple *JComponents*. This class contains a constructor and implementation of all methods required by the interface but it isn't a God class, where object-oriented systems tend to be networks of cooperating objects with no central God class that controls everything from above. It positions the asked proxies within the panel but does absolutely nothing else with them. This class is simply a passive vehicle for holding visual proxies.

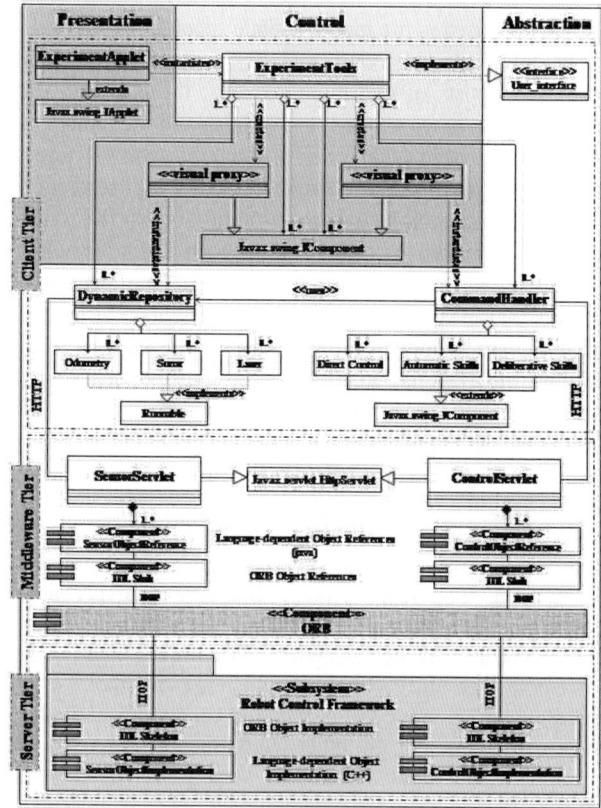

Fig. 3 Software Architecture

The proxies communicate directly with the abstraction-level objects that creates them and these abstraction layer objects can communicate with each other.

If the state of an abstraction-level class changes as a result of some user input, it sends a message to another of the abstraction-level classes, which may or may not choose to update its own user interface (its proxy) as a sequence.

The aggregation relationship between the *ExperimentTools* class and user interface components indicates that the *ExperimentTools* class may or may not have one or more interface component and it is aware of component it has, but the component is not aware for what experiment interface it is for. Here every component can be associated with one or more experiment interface.

ExperimentApplet instantiates the *ExperimentTools* to form the experiment interface seen by the user via any Web browser. *CommandHandler* class can be direct control, automatic skill, deliberative skill or a combination of different types of control classes in the same panel. A dependency relationship has been established between *CommandHandler* and *DynamicRepository* which specifies that a change in *DynamicRepository* may affect *CommandHandler* that use it but not necessarily the inverse *CommandHandler* may use the *DynamicRepository* class to invoke sensory data, which may be necessary to complete the control task, such as in the case of obstacle avoidance supervised control skill by using sensory data. *DynamicRepository* class provides information about sensor's status and can be used to remotely acquire the sensory data. This class is implemented as a thread by implementing the *Runnable* interface to provide updated sensory data in real time. This data may be odometry, which indicates the actual robot position and its translational and rotational velocities or ultrasonic sensor data or laser sensor data. Both *CommandHandler* and *DynamicRepository* classes interact with the middleware using http protocol to send the user requests to the robot server.

B. Middleware Tier

The communication between the servlets and the remote robot servers is done via the Object Request Broker (ORB) of the Common Object Request Broker Architecture (CORBA) where the Java servlet acts as a client to the robot skill. The ORB provides the communication via the unified interface language Interface Definition Language (IDL) and based on the Internet Inter-ORB protocol (IIOP) [17].

The decision to use CORBA as the distributed object architecture of the remote laboratory is based on a qualitative and quantitative comparison between the two most commonly used architectures, CORBA and RMI [18]. This study concluded that CORBA is suitable for large scale or partially Web-enabled applications where legacy support is needed and good performance under heavy client load is expected. Moreover, CORBA servers can be located at any Internet site. RMI, on the other hand, is suitable for small scale fully Web-enabled applications where legacy support can be managed with custom build or pre-built bridges, where ease of learning and ease of use is more critical than performance.

C. Server Tier

Server tier consists of subsystems based on the robot control framework which can be the commercial Mobility framework [10] or our own AD Architecture-based framework. By using these subsystems the ORB objects are implemented by an encapsulated and modular manner. The object to be implemented may be a sensor object to invoke the sensor data or a control object to send control commands to the robot actuators. IDL skeletons provide a language-independent object implementation and then communicate with the IDL stubs via ORB and based on IIOP protocol.

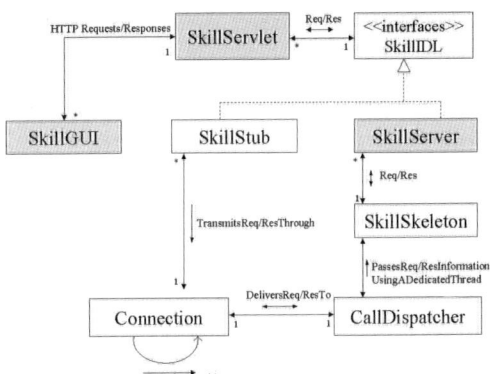

Fig. 4 Client/Middleware/Server Interaction

D. Client/Middleware/Server Interaction

The client/middleware interaction is based on Http Requests/Responses and the middleware/server interaction is implemented using the broker pattern (ORB). As shown in Fig. 4 to activate or deactivate a robotic skill, the user has to use the SkillGUI to send Http-based commands to the SkillServer through SkillServlet, which servers as an intermediate server between client and server. SkillServlet doesn't call the SkillServer directly but through an IDL interface (SkillIDL). This class calls methods of a proxy object that implements the SkillIDL interface. Because it calls methods through an interface, it doesn't need to be aware of the fact that it is calling the methods of a stub object that is a proxy for the SkillServer object rather than the SkillServer object itself.

The stub object encapsulates the details of how calls to the SkillServer object are made and its location. These details are transparent to SkillServlet objects. Also stub objects assemble information identifying the SkillServer object, the method being called and the values of the arguments into a message. On the other end of the connection, part of the message is interpreted by a CallDispatcher object and the rest by a skeleton object. The connection classes are responsible for the transport of message between the environment of a remote SkillServlet (Web Server) and the environment of the SkillServlet (Robot Server).

The CallDispatcher receives message through the connection object from a remote stub object and then passes each message to an instance of an appropriate skeleton class, which are created by the CallDispatcher. The CallDispatcher object are responsible for identifying the SkillServer object whose method will be called by the skeleton object.

The skeleton classes are responsible for calling the methods of SkillServer objects on behalf of remote SkillServlet objects. The skeleton object extracts the argument values from the message passed to it by the CallDispatcher object and then calls the indicated method passing it the given argument values. If the called message returns a response, then the skeleton object is responsible for creating a message that contains the return value and sending it back to the stub so that the stub object return it. If the called message throws an exception, the skeleton object is responsible for creating an appropriate message.

The SkillServer classes (implemented in C++) implement SkillIDL interface. Instances of this class can be called locally through the SkillIDL interface, or remotely through a stub object that also implements the same skill interface.

E. Error Detection & Recovery

The error detection and recovering is perhaps the most significant challenge to time-delayed telerobotics. In the remote laboratory, errors can be handled by using a three stage process: autonomous detection, shared diagnosis, and manual recovery. Errors are detected by using the visualization means such as streaming video, graphical models, sensory data or connection status panels. The diagnosis task is shared by the user and the system. In recovering from the error, the user can telecollaborate with a human at the remote site or can ask the necessary privileges to be able to telnet the remote servers to reboot them.

V. REMOTE LABORATORY IMPLEMENTATION

An online laboratory in a field such as mobile robotics must have live performance characteristic, not just virtual reality or simulation programs. The multi-layered architecture, described in the previous section, has been implemented in order to reach this goal. During the academic year 2001-2002, the developed laboratory has been used to update a postgraduate course about intelligent autonomous robots. Student feedback was gathered using an online questionnaire. The student responses were uniformly positive as to use of the different proposed teaching activities specially the use of the remote laboratory. Most of the student felt that the online experiments helped them to achieve a deeper and longer understanding of the subject material.

During the laboratory session, the student can communicate and teleoperate the experiment using a Java applet or MIDlet as shown in Fig.5. The task is then executed by a local control system and the results displayed on the operator's browser.

This client-level interface includes many operations that can be made on the robot such as position control, obtaining sensor data, drawing world maps and evaluate the errors. The following subsections describe some implemented experiments, which are being used in a postgraduate course on intelligent autonomous robots.

A. Direct Motion Control

This experiment aims at familiarizing the user with the mobile robot motion control and positioning by using different interaction elements such as PC or any Mobile Information Device Profile (MIDP)-enabled device as handheld PDA or cellular phones. The remote user can send direct control commands to move the robot forward, backward or to turn it left or right. Using a 2D model for the lab and the robot and the odometry data (actual location with respect to the initial point, translational and rotational robot velocities), the remote user will be able to view the effect of the sent commands.

A study has been done to measure the response time during the PC-based direct control experiment, which is the time collapsed between sending the motion commands and the motion start. Table I shows the response time when the experiment was ran from local site (Carlos III University of Madrid) and when it was run from remote sites (University of Applied Science - Germany and University of Reading - England).

TABLE I
RESPONSE TIME VARIATIONS

	Min. (ms)	Max. (ms)	Av. (ms)
Spain	121	137.76	127.2
Germany	273	354	293.51
England	394	2444	1157.45

The latency and the throughput of the Internet are highly unpredictable and inevitable. There are old qualitative studies [19] that show people seem to be able to compensate for (learn) small added delays, but cannot learn large ones (>100 msec) therefore the delay will be noticed by the user but it can be accepted for such type of educative application.

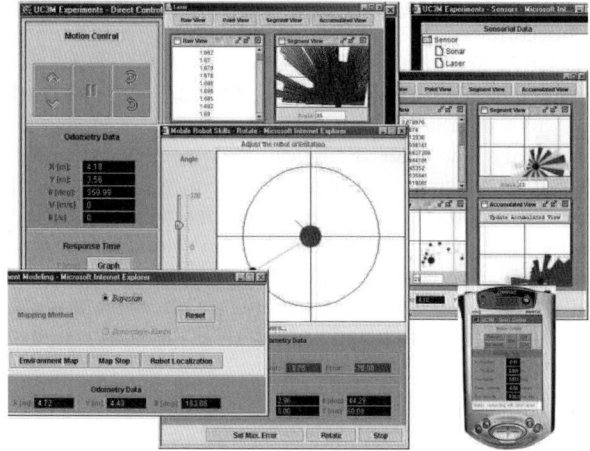

Fig. 5 Screenshot of Experiment Interfaces

B. Remote Activation for Motion Skills

The skills can be activated by execution orders produced by other skills or by a sequencer. They return data and events to the skills or sequencers which have activated them. A skill can send a report about its state while it is active or when it is deactivated. For example, the skill called *gotogoal* can provide information as to whether the robot has achieved the goal or not. When this skill is deactivated it might supply information about the error between the current robot position and the goal [14]. By using the proposed architecture, many interfaces have been implemented to remotely activate or deactivate motion skills such as go to goal, orientation control, wall following, obstacle avoidance. Skill sequencer can be used to generate complex skill by combing simple skills. It is responsible for deciding what skills have to be activated at each moment to avoid data inconsistence problems.

C. Sensorial Data Acquisition

The objective of this experiment is the environment perception using multi-sensor data (sonar and laser). The experiment is divided into two parts: without robot motion and with robot motion. The objective of the first part is to understand the operation of sonar and laser sensors and to be familiar with these readings. The second part aims at recognizing the real environment using the sensory data.

D. Environment Modelling using Sensorial Data

Building environment maps from sensory data is an important aspect of mobile robot navigation, particularly for those applications in which robots must function in unstructured environments. Ultrasonic range sensors are, superficially, an attractive sensor modality to use in building such maps, due mainly to their low cost, high speed and simple output. Elfes's algorithm [20] is used to generate environment maps from sonar data. In this algorithm range measurements from multiple viewpoints are combined in a two-dimensional 'occupancy grid'. Each cell in the grid is assigned a value indicating the probability that the cell is occupied. A self-localization algorithm [21] is used to estimate the robot's position by computing sets of poses which provide a maximal-quality match between a set of current sensor data and the constructed map. The remote user will be able to compare the result of the localization algorithm with the odometry data to determine the error.

E. Free Tour

An unguided tour has been implemented, which does not determine any order and tasks at all. This tour provides generic tools to the user and let him/her to customize the experiment according to his/her needs. These generic tools include 2D model for the robot and the lab, odometry data panel, sonar data panel, laser data panel, motion controller and low-level tele-programming editor.

VI. CONCLUSION

The paper describes a three-tiers architecture to build mobile robotics remote laboratories. The visual proxy pattern used to build the user interfaces of the remote laboratory, provides flexible user interfaces with minimal coupling relationships between subsystems. The generation of the user interface is entirely separated from the abstraction layer object to provide the reusability and extensibility facilitates.

Based on the described architecture, many interfaces have been implemented for simple automatic movement skills such as direct control (using PC or PDA or Mobile Phone), Go to point, orientation control and wall following skills. A sequencer has been used to combine simple skills to obtain complex skills such as go to point with obstacle avoidance.

VII. ACKNOWLEDGMENTS

The authors gratefully acknowledge the funds provided by the Spanish Government through the DIP2002-188 project of MCyT (Ministerio de Ciencia y Tecnología).

VIII. REFERENCES

[1] K. Forinash and R. Wisman, "The Viability of Distance Education Science Laboratories". *T.H.E. Journal*, vol. 29, 2001, No.2 September.
[2] C., Sayers. *Remote Control Robotics*. 1st Edition, Springer Verlag, 1999.
[3] D. Barney, and T. Ken, "Distributed Robotics over the Internet," *IEEE Robotics and Automation Magazine*, vol.7, no.2, pp.22-27, 2000.
[4] R. Simmons, J. Fernandez, R. Goodwin, S. Koeing, J. O'Sullivan, "Lessons Learned from Xavier". *IEEE Robotics & automation Magazine*, Vol.7, No.2, pp. 33-39, June 2000
[5] D. Schulz, W. Burgard, D. Fox, S. Thrun, A. Cremers, "Web Interfaces for Mobile Robots in Public Places". *IEEE Robotics & automation Magazine*, Vol.7, No.1, pp. 48-56, March 2000.
[6] K. Schilling, "Model Design for Remote Mobile Robots", *Proceedings of Tele-Education in Mechatronics Based on Virtual Laboratories*, July 18th - 21st, 2001, Weingarten, Germany.
[7] F. Rodríguez, A. Khamis and M. Salichs, "Design of a Remote Laboratory on Mobile Robots", *Internet-based Control Education, ibce01*, 12-14 Dec. 2001, UNED, Madrid, Spain.
[8] Innovative Educational Concepts for Autonomous and Teleoperated System (IECAT) http://www.ars.fh-weingarten.de/iecat/index.html
[9] A. Khamis, M. Pérez Vernet, K. Schilling, "A Remote Experiment on Motor Control of Mobile Robots", *the 10th Mediterranean Conference on Control and Automation, MED 2002*, Lisbon, Portugal, July 9-12, 2002.
[10] Real World Interface, http://www.irobot.com/rwi/
[11] N. Fung, W. Lo, and Y. Liu, "Improving Efficiency of Internet-based Teleoperation using Network QoS". *Proceedings of the 2002 IEEE International Conference on Robotics & Automation*, pp. 2707-2712, Washington, DC, May 2002.
[12] R. Barber, M.A. Salichs, "A new human based architecture for intelligent autonomous robots". *The Fourth IFAC Symposium on Intelligent Autonomous Vehicles*, p85-90. Sapporo, Japan. September 2001.
[13] R. Alami, R. Chatila, S. Fleury, M. Ghallab, and F. Ingrand. "An Architecture for Autonomy". *The International Journal of Robotics Research*, 17 (4): 315-337, 1998.
[14] M.A. Salichs, M.J. López, R. Barber, "Visual Approach Skill for a Mobile Robot using Learning and Fusion of Simple Skills", *Robotics and Autonomous Systems*, vol.38, 2002, pp.157-170.
[15] H. Rohbert, P. Sommerlad and M. Stal. *Pattern Oriented Software Architecture: A System of Patterns*. Jogn Wiley & Sons, 1996.
[16] http://www.javaworld.com/javaworld/jw-07-1999/jw-07-toolbox.html
[17] Object Management Group, http://www.omg.org/
[18] M. Juric, I. Rozman, and M. Hericko," Performance Comparison of CORBA and RMI", *Information and Software Technology*, No. 42, pp. 915-933, 2000.
[19] R. Marphy and E. Rogers, "Human-Robot Interaction", *Final Report for DARPA/NSF Study on Human-Robot Interaction*, http://www.csc.calpoly.edu/~erogers/HRI/HRI-report-final.html.
[20] A. Elfes, "Occupancy grids: A stochastic spatial-representation for active robot perception", in *Proceedings of the Sixth International Conference on uncertainty in AI*.
[21] R. Brown and B. Donald, "Mobile Robot self-Localization without Explicit Landmarks". *In Algorithmica*, vol. 26, pp. 515-559, 2000.

Digital Passive Geometric Telemanipulation

C. Secchi[1], S. Stramigioli[2], C. Fantuzzi[1]

[1]DISMI, Univ. of Modena and Reggio Emilia
Viale Allegri 13, 42100 Reggio Emilia, Italy
secchi.cristian@unimore.it
fantuzzi.cesare@unimore.it

[2]Drebbel Institute, Univ. of Twente
P.O.Box 217, 7500 AE Enschede, NL
S.Stramigioli@ieee.org

Abstract— In this paper we present an intrinsically passive telemanipulation scheme over a digital transmission line Internet-like. We present an analysis of the energetic behavior of the communication line both in case of loss of packages and in case of variable delay. The sample data nature of the passive controller is explicitly taken into account following the approach outlined in [9] and [10].

I. INTRODUCTION

In the design of a control system for telemanipulation there are two main problems to be addressed in order to avoid instability and poor performance: the communication channel and the interaction with an unknown environment. The first problem has been approached in several ways for fixed time delays in continuous time. In particular, a very elegant approach has been proposed in [1] in which the scattering theory is used to obtain a passive, and, therefore, not destabilizing, communication channel. In [6] the wave reflection problem arising with the scattering approach is solved and some adaptive techniques are used to improve the performances of the telemanipulation scheme.
For an overview and a comparison of the various techniques proposed in the literature, see, for instance, [2] and [5].
A very suitable tool to solve the second problem is passivity theory. A stable interaction can be obtained controlling the energy exchange between the interacting systems, which can be achieved using intrinsically passive controllers (IPC, [8]) for master and slave devices.
In [11] a general framework for the telemanipulation of port-Hamiltonian systems in a continuous domain is proposed: IPC is used to obtain a stable interaction and scattering theory to implement a passive communication channel. In [7] some improvements and extensions of the scheme to consider tele-grasping are proposed. The aim of this work is twofold. First, we want to explicitly consider the discrete nature of controllers, which are always implemented by means of a digital system. The passivity of the overall telemanipulation system has still to be granted and we will use the concepts presented in [10] and [9] to build a suitable passive scheme. Second, we want to study the behavior of the system in case of master and slave exchange data using an Internet-like communication channel. There are two problems to take into account: the strongly variable time delay and the possible unreliability due to the loss of packages associated with the channel.

The paper is organized as follows: in Sec.II we will give some background on both continuous and discrete port-Hamiltonian systems and on the energetical consistent way for connecting them, in Sec.III we will introduce a digital version of the scatterized communication channel and we will prove its passivity both in case of variable delay and in case of loss of packages. In Sec.IV we will present the overall sample data passive telemanipulation scheme and and in Sec.V we will present some simulations in order to verify our results. Finally, in Sec.VI some conclusions and future work will be addressed.

II. PORT-HAMILTONIAN SYSTEMS

A. Continuous port-Hamiltonian systems

We will now try to give an intuitive description of port-Hamiltonian systems using coordinates in order to concentrate on the prime contribution of the paper. More formal descriptions can be found in [12]. We can consider a port-Hamiltonian system as composed of a state manifold $\mathcal{X}$, an energy function $H : \mathcal{X} \to \Re$ corresponding to the internal energy, a network structure $D(x) = -D(x)^T$ whose graph has the mathematical structure of a Dirac structure([3]), which is in general a state dependent power continuous interconnection structure, and an interconnection port represented by an effort-flow pair $(e, f) \in V^* \times V$ which is geometrically characterized by dual vector elements. This port is used to interact energetically with the system. The power supplied through a port is equal to $e(f)$ or using coordinates to $e^T f$. We can furthermore split the interaction port in more sub-ports, each of which can be used to model different power flows. We will indicate with the subscript I the power ports by means of which the system interacts with the rest of the world, with the subscript C the power ports associated with the storage of energy and with the subscript R the power ports relative to the dissipative part. Summarizing, we have:

$$\begin{pmatrix} e_I \\ f_C \\ e_R \end{pmatrix} = D(x) \begin{pmatrix} f_I \\ e_C \\ f_R \end{pmatrix}$$

where $D(x)$ is a skew symmetric matrix representing the Dirac structure.

Due to the skew-symmetry of $D(x)$, we have, using coordinates:

$$P_I + P_C + P_R := e_I^T f_I + e_C^T f_C + e_R^T f_R = 0 \quad (1)$$

which is a power balance meaning that the total power coming out of the network structure should be always equal to zero.

A dissipating element of the system can be modeled using as characteristic equations $e_R = R(x) f_R$ with $R(x)$ a symmetric and positive semi-definite tensor.

If we furthermore set $\dot{x} = f_C$ and $e_C = \dfrac{\partial H}{\partial x}$, due to the previous power balance we obtain:

$$\dot{H} + f_R^T R(x) f_R = -e_I^T f_I$$

which clearly says that the supplied power $-e_I^T f_I$ equals the increase of internal energy plus the dissipated one.

B. Discrete port-Hamiltonian systems

For a lot of applications it is meaningful to find a discrete time representation of a physical system which can be used either as a virtual environment in haptics or as an IPC [8] in interacting tasks or in telemanipulation. Hereafter we will briefly review how to discretise a port-Hamiltonian system preserving its passivity. More details can be found in [10].

We can describe a discrete time port-Hamiltonian system as a continuous time port-Hamiltonian system in which the port variables are frozen for a sample interval T. In what follows we indicate with $v(k)$ the value of the discrete variable $v(t)$ corresponding to the interval $t \in [kT, (k+1)T]$.

If we rewrite Eq.(1) for the discrete case, we have:

$$e_I^T(k) f_I(k) + e_C^T(k) f_C(k) + e_R^T(k) f_R(k) = 0 \quad (2)$$

Furthermore, during the interval k, we have to consider a constant state $x(k)$ corresponding to the continuous time state $x(t)$. This implies that during the interval k, the dissipated energy will be equal to $T f_R^T(k) R(x(k)) f_R(k)$ and the supplied energy will be equal to $-T e_I^T(k) f_I(k)$. In order to be consistent with the energy flows, and as a consequence conserve passivity, we need therefore a jump in internal energy $\Delta H(k)$ from instant kT to instant $(k+1)T$ such that:

$$\Delta H(k) = -T f_R^T(k) R(x(k)) f_R(k) - T e_I^T(k) f_I(k)$$

This implies that the new discrete state $x(k+1)$ should belong to an energetical level such that:

$$H(x(k+1)) = H(x(k)) + \Delta H(k)$$

We can indicate the set of possible energetically consistent states, which can be found solving the previous equation in $x(k+1)$, as

$$I_{k+1} := \{ x \in \mathcal{X} \quad s.t. \quad H(x) = H(x(k)) + \Delta H(k) \}.$$

The set I_{k+1} can be either empty or have more solutions. The situation in which I_{k+1} is not empty is the most common. In this case we have to choose a state among the elements of I_{k+1}; we should choose a state which is "close", in some sense, to the current state $x(k)$. The distance between states can be characterized defining an affine connection on the state manifold $\mathcal{X}$. In this paper we used Euclidean coordinates and the Euclidean connection. In this case, the next state $x(k+1)$ is chosen as the intersection of I_{k+1} with the straight line passing from $x(k)$ and directed along $f_C(k)$.

If $I_{k+1} = \emptyset$, the proposed algorithm doesn't lead to any solutions. The correct dynamical behavior of the system and its passivity can be recovered with the *energy leap* strategy outlined in [10].

As a summary of the procedure just outlined, we hereafter algorithmically explain the way the discrete system can be integrated

1) Given an initial state $x(k)$, we set $e_C(k) = \dfrac{\partial H}{\partial x}(x_k)$.

2) Using the value of the system input $f_I(k)$ and the previously calculated $e_C(k)$, we can calculate $e_I(k)$, the output of the interaction port, and $f_C(k)$.

3) $f_C(k)$ is then used to calculate the next state $x(k+1)$ using the procedure explained at the beginning of this subsection.

C. Passive interconnection

Consider the port interconnection of a continuous time Hamiltonian system H_C and a discrete Hamiltonian system H_D (but the result of this subsection is independent of the nature of the energetically interconnected systems) through a sampler and zero-order hold as shown in Fig.1. If the sample&hold is not properly designed, it can happen that the process generates extra energy and that the passivity of the whole system is not assured even if the two interconnected systems are passive.

Suppose that H_C has an admittance causality (effort in/flow out) and therefore H_D has an impedance causality (flow in/effort out). We will have that:

$$e(t) = e_d(k) \qquad t \in [kT, (k+1)T]$$

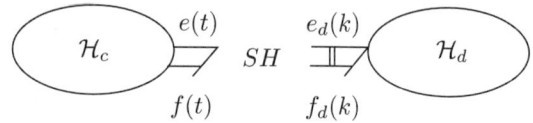

Fig. 1. The interconnection of discrete and continuous port-Hamiltonian systems

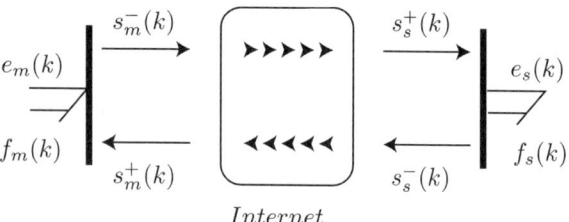

Fig. 2. The scatterized communication channel

The following theorem can be proved ([10])

Theorem 1 (Sample Data passivity): If we define for the interconnection port of H_D

$$f_d(k) := \frac{x(kT) - x((k+1)T)}{T}, \quad (3)$$

where $x()$ represents the integral of the continuous flow, we obtain an equivalence between the continuous time and discrete time energy flow in the sense that for each n:

$$\sum_{i=0}^{n-1} e_d^T(i) f_d(i) = -\int_0^{nT} e^T(s) f(s) ds \quad (4)$$

From the previous considerations, it is possible to understand that at each sampling time, we have an EXACT matching between the physical energy going into the continuous time system and the virtual energy coming from the discrete time port independently of the sample time T and of eventual intersample dynamics of the continuous system. It is remarkable that the choice reported in Eq.(3) which is very simple and at the same time attractable due to the fact that it corresponds to position measurements, in practice gives such a powerful and at the same trivial result. The only energy leakage is due to the fact that the discrete time system has no way what so ever to predict the value of the continuous time system at the interconnection port and this implies that only at the end of the sample period will have an exact measure of the energy it supplied to the continuous time system. But the amount of the eventually produced extra energy is exactly known and it can be compensated with a damping circuit or by a clever book-keeping strategy.

III. THE COMMUNICATION CHANNEL

Since we want to consider an Internet-like communication line, we have to consider a discrete communication channel and we will implement it by using scattering theory. It is possible to define a discrete time scattering. Each discrete time ports (either the master or the slave one) of the discrete communication channel illustrated in Fig.2 is characterized by an effort $e(k)$ and by a flow $f(k)$. The energy flowing into the system in one sample period is equal to:

$$E = T\langle e(k), f(k) \rangle$$

We can always make the following decomposition of the power flow into an incoming power wave and an outgoing power wave in such a way that:

$$\langle e, f \rangle = \frac{1}{2}\|s_Z^+\|^2 - \frac{1}{2}\|s_Z^-\|^2$$

By discrete integration the energy flow during one sample period is:

$$E(k) = T\langle e, f \rangle = \frac{T}{2}\|s_Z^+\|^2 - \frac{T}{2}\|s_Z^-\|^2$$

where T is the sample period. Thus we can interpret $\frac{T}{2}\|s_Z^+\|^2$ and $\frac{T}{2}\|s_Z^-\|^2$ as incoming and an outgoing energy packages respectively.

At each sample time the system will acquire the incoming energy quantum $\frac{T}{2}s_Z^+(k)$ and the discrete effort $e(k)$ and will calculate the discrete flow $f(k)$ and the discrete energy quantum $\frac{T}{2}s_Z^-(k)$ to transmit through the communication channel.

In [11] the mappings which allow to compute $s_Z^-(k)$ and $f(k)$ from $s_Z^+(k)$ and $e(k)$ are reported.

We will use the following notation for the discrete derivative and the discrete integral:

$$dg(k) = \frac{g(k+1) - g(k)}{T} \quad I_h^k g = \sum_{i=h}^{k-1} g(i) T$$

where g is a generic sequence.

In the following considerations we will assume that the protocol (i.e. the Internet protocol) used to implement the communication line doesn't change the order of the packages, such as the Internet TCP/IP protocol. The following result can be proven:

Proposition 1: In case of fixed transmission delays and no loss of packages, the discrete communication channel is lossless in a discrete sense.

Proof: The power flow into the communication channel is:

$$P_L(k) = \tfrac{1}{2}\|s_m^-(k)\|^2 + \tfrac{1}{2}\|s_s^-(k)\|^2 - \tfrac{1}{2}\|s_m^+(k)\|^2 -$$
$$-\tfrac{1}{2}\|s_s^+(k)\|^2$$

but

$$s_s^+(k) = s_m^-(k - d_m) \quad s_m^+(k) = s_s^-(k - d_s)$$

where d_m and d_s are respectively the delays associated to the communication between master and slave and slave and master respectively. We can therefore write.

$$P_L(k) = \tfrac{1}{2}\|s_m^-(k)\|^2 - \tfrac{1}{2}\|s_m^-(k-d_m)\|^2 + \tfrac{1}{2}\|s_s^-(k)\|^2 - \tfrac{1}{2}\|s_s^-(k-d_s)\|^2$$

By trivial computations it can be shown that, in general:

$$dI_{k-h}^k g = g(k) - g(k-h)$$

where g is a generic sequence. We can, therefore, write:

$$P_L(k) = d[I_{k-d_m}^k(\tfrac{1}{2}\|(s_m^-)\|^2) + I_{k-d_s}^k(\tfrac{1}{2}\|(s_s^-)\|^2)]$$

Defining as energy function of the communication channel

$$E_L(k) = [I_{k-d_m}^k(\tfrac{1}{2}\|s_m^-\|^2) + I_{k-d_s}^k(\tfrac{1}{2}\|s_s^-\|^2)]$$

we have that:

$$P_L(k) = dE_L(k)$$

which means that all the power flowing through the communication channel is stored. The communication channel is, therefore, lossless in a discrete sense. ∎

The unreliability source of an Internet-like communication channel follows from the fact that some packages could be lost during transmission because of some traffic problems or some troubles in the servers each package has to cross.

Let us investigate the energetic behavior of the communication channel in the case one package is lost in the transmission between master and slave. We will have that a 0 signal is received at the slave side instead of the correct value, namely $\tfrac{1}{2}s_m^-(k-d_m)$ and that:

$$P_L(k) = \tfrac{1}{2}\|s_m^-(k)\|^2 + \tfrac{1}{2}\|s_s^-(k)\|^2 - \tfrac{1}{2}\|s_m^+(k)\|^2$$

We can write:

$$P_L(k) = \tfrac{1}{2}\|s_m^-(k)\|^2 + \tfrac{1}{2}\|s_s^-(k)\|^2 - \tfrac{1}{2}\|s_m^+(k)\|^2 =$$

$$= \tfrac{1}{2}\|s_m^-(k)\|^2 + \tfrac{1}{2}\|s_s^-(k)\|^2 - \tfrac{1}{2}\|s_s^-(k-d_s)\|^2 -$$

$$- \tfrac{1}{2}\|s_m^-(k-d_m)\|^2 + \tfrac{1}{2}\|s_m^-(k-d_m)\|^2 =$$

$$= dE_L(k) + \tfrac{1}{2}\|s_m^-(k-d_m)\|^2$$

We have an additional term which is positive definite and which represents a dissipated power. When a package is lost in the transmission the channel dissipates an energy quantum corresponding to the value that it should have received from the master side. The behavior of the system, nonetheless, keep on being passive.

Let, now, $\mathcal{F}_{ms}$ and $\mathcal{F}_{sm}$ the set of time instants in which a package in the communication between master and slave is lost and the set of time instants in which a package in the communication between slave and master is lost respectively. The energetic behavior of the channel is:

$$P_L(k) = dE_L(k) + \alpha(\tfrac{1}{2}\|s_m^-(k-d_m)\|^2) + \beta(\tfrac{1}{2}\|s_s^-(k-d_s)\|^2)$$

where

$$\alpha = \begin{cases} 0 & k \notin \mathcal{F}_{ms} \\ 1 & k \in \mathcal{F}_{ms} \end{cases} \quad \beta = \begin{cases} 0 & k \notin \mathcal{F}_{sm} \\ 1 & k \in \mathcal{F}_{sm} \end{cases}$$

where α and β are coefficients that activate a power dissipation of the channel when a package gets lost.

Let us now consider the case in which the delay between master and slave is variable. Suppose that h is the minimum delay of the transmission between master and slave. Because of the variable delay, it can happen that some packages are so much delayed that the receiving queue is empty for some sample periods. In these cases we can always write:

$$P_L(k) = \tfrac{1}{2}\|s_m^-(k)\|^2 + \tfrac{1}{2}\|s_s^-(k)\|^2 - \tfrac{1}{2}\|s_m^+(k)\|^2 =$$

$$= \tfrac{1}{2}\|s_m^-(k)\|^2 + \tfrac{1}{2}\|s_s^-(k)\|^2 - \tfrac{1}{2}\|s_s^-(k-d_s)\|^2 -$$

$$- \tfrac{1}{2}\|s_m^-(k-h)\|^2 + \tfrac{1}{2}\|s_m^-(k-h)\|^2 =$$

$$= dE_L(k) + \tfrac{1}{2}\|s_m^-(k-h)\|^2 = dE_L(k) + P_d(k)$$

where now we define:

$$E_L(k) = [I_{k-h}^k(\tfrac{1}{2}(s_m^-)^2) + I_{k-d_s}^k(\tfrac{1}{2}(s_s^-)^2)]$$

as the energy function. We dissipate the energy quantum corresponding to the package we were expecting. At a time $k+j$ the system receives the package whose corresponding energy quantum was dissipated at time k. In this case it will be:

$$P_L(k+j) = dE_L(k+j) - \tfrac{1}{2}\|s_m^-(k-h)\|^2 + \tfrac{1}{2}\|s_m^-(k+j-h)\|^2 = dE_L(k+j) + P_p(k+j) + P_d(k+j)$$

At sampling period $k+j$, the delayed incoming package injects an extra energy $TP_p(k+j)$, on the other hand the missed package $\tfrac{1}{2}\|s_m^-(k+j-h)\|^2$, since we are assuming that only one package can be transmitted per each sample period, causes the dissipation of $TP_d(k+j)$ energy.

We can see that the delay of a package first implies a dissipation of an energy quantum and then an energy injection of the same energy quantum. It is, then, clear that

$$\sum_{n=0}^{\infty} P_p(n) + P_d(n) = 0$$

and, therefore, there is no global production of extra energy.

Both in case of fixed and variable delay the scatterized communication channel is lossless; the main difference is that when we have a constant delay energy is neither produced nor dissipated but simply stored, while when we have a variable delay the energy quanta associated to the delayed packages are first dissipated and then injected back to the systemSince the variable delay introduces a *finite* extra-delay on the packages, passivity is preserved. One could expect that the energy injection leads to some non passive (and therefore potentially unstable) behavior but this is not the case since before being injected the extra energy has already been dissipated.

In case of some packages are lost the behavior of the communication channel is dissipative since we do not have any energy injection to recover the dissipated quanta.

It is now straightforward to state the following:

Proposition 2: The scatterized communication channel is passive even in case of variable time delay and of loss of packages.

IV. THE TELEMANIPULATION SCHEME

We now have all the components to build a passive sampled data telemanipulation system over the Internet as illustrated in Fig.3 using bond-graph notation.

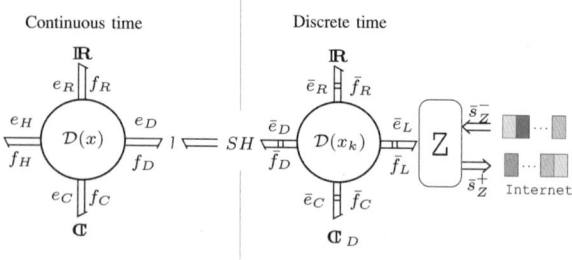

Fig. 3. The Passive Sample Data Telemanipulation Scheme.

The subscript H indicates the power port by means of which the system interacts with the rest of the world, the subscript R represents the dissipative port while the subscript C the port associated to a storage of energy. The subscript L indicates the power port associated to the communication channel and the subscript D the power port of interaction between the system and the controller. The barred power variables are discrete power variables. We have a continuous port-Hamiltonian system (the robot) connected in an energetic consistent way to a passive discrete controller, obtained by discretising the continuous IPC. The discrete communication channel is implemented by means of the scattering variables.

The energy of the telemanipulation system is composed of the physical energies stored into the physical systems,

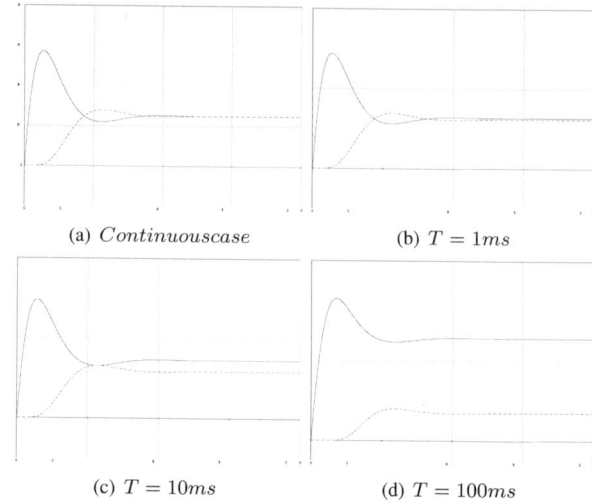

(a) $Continuous case$ (b) $T = 1ms$

(c) $T = 10ms$ (d) $T = 100ms$

Fig. 4. Free motion with different sample times. Position of the master (continuous) and of the slave (dashed)

the virtual energy stored into the discrete controllers and the virtual energy stored into the communication channel.

V. SIMULATIONS

In this section we will provide some simulations to state the validity of the proposed telemanipulation scheme.

Each robot (master and slave) is a 1 DOF system, a simple mass and is controlled by a discretized IPC, connected in a power consistent way to the continuous robot. The communication channel is implemented by means of the digital scattering strategy.

In the first simulation we applied an impulsive force to the master and we plotted the positions of master and slave; the delay is constant and it si equal to 0.5 seconds. The results of this simulation are shown in Fig.4 We can see that the smaller the sample time the closer is the behavior of the digital scheme to the continuous one, meaning that the discretisation algorithm is well posed. Moreover, we can notice that the bigger is the sample time the worse are the performances. This is because the information we are transmitting gets worse; nevertheless the behavior of the overall system is passive independently of the value of the sample period.

In the next simulation we are considering a communication channel which represents the Internet. The delay in the communication between master and slave and between slave and master is variable (with a mean of 0.5 sec.) and in the communication between master and slave every 2 seconds 10 packages are lost while in the communication between slave and master each 3 seconds 20 packages are lost. Once again we applied an impulsive force to the master and we plotted the positions of master and slave. The results are shown in figure Fig.5. The

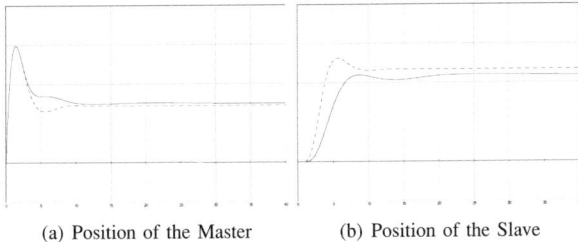

(a) Position of the Master   (b) Position of the Slave

Fig. 5. Variable delay and loss of packages

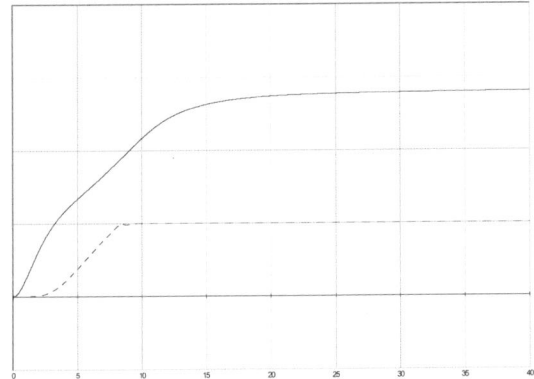

Fig. 6. Position of the master (continuous) ad of the slave (dashed)

dashed line represents the position of master and slave in case of constant delay (0.5 seconds) and of no loss of packages. The sample time is 10 ms. We can notice that the performances decrease but the stability is maintained. In the next simulation we implemented an interaction task. The master is pushed with a constant force and the slave interacts with a wall (implemented with a spring-damper system) posed at position $x = 0.1$. There is variable delay and loss of packages in the communication channel in both senses. The position of master and slave are shown in figure Fig.6. We can see that when the slave stops when it touches the wall. The force of interaction is reflected back to the master side and compensates the force applied to the master. In fact we can see that the position of the master is constant even if the operator is keeping on applying a force.

VI. CONCLUSIONS AND FUTURE WORK

In this work we extended the passive telemanipulation scheme proposed in [11] in order to take into account the sampled data nature of controllers. Furthermore we considered a digital unreliable communication channelas the Internet and we showed that, using scattering theory, it is possible to preserve passivity both in case of loss of packages and of variable transmission delay. Future research will deal with the definition of a meaningful affine connection to be used to integrate the discrete port-Hamiltonian, as partially explained in [4]. From a more practical point of view, we would like to implement in a real setup the proposed scheme.

Acknowledgments

This work has been done in the context of the European sponsored project GeoPlex with reference code IST-2001-34166. Further information are available at www.geoplex.cc.

VII. REFERENCES

[1] R. Anderson and M. Spong. Bilateral control of teleoperators with time delay. *IEEE Transactions on Automatic Control*, 34(5):494–501, 1989.

[2] P. Arcara and C. Melchiorri. A comparison of control schemes for teleoperation with time delay. In *Proceedings of IFAC TA 2001, Conf. on Telematics Applications*, 2001.

[3] T.J. Courant. Dirac manifolds. *Trans. American Math. Soc. 319*, pages 631–661, 1990.

[4] O. Gonzalez. Time integration and discrete hamiltonian systems. *Journal of Nonlinear Science*, 1996.

[5] C. Melchiorri and A. Eusebi. *Int. Summer School on Modeling and Control of Mechanisms and Robot*, chapter Telemanipulation: System Aspect and Control Issues. World Scientific, 1996.

[6] G. Niemeyer and J. Slotine. Stable adaptive teleoperation. *IEEE Journal of Oceanic Engineering*, 16(1):152–162, 1991.

[7] C. Secchi, S. Stramigioli, and C. Melchiorri. Geometric grasping and telemanipulation. In *Proceedings to IEEE Conference on Intelligent Robotic Systems*, Maui, Hawaii, USA, October 2001.

[8] S. Stramigioli. *Modeling and IPC Control of Interactive Mechanical Systems: a coordinate free approach*. Springer, London, 2001.

[9] S. Stramigioli. About the use of port concepts for passive geometric telemanipulation with varying time delays. In *Proceedings to Mechatronics Conference*, Enschede, The Netherlands, June 2002.

[10] S. Stramigioli, C. Secchi, A.J. van der Schaft, and C. Fantuzzi. A novel theory for sample data systems passivity. In *Proceedings to IEEE Conference on Intelligent Robotic Systems*, 2002.

[11] S. Stramigioli, A. van der Schaft, B. Maschke, and C. Melchiorri. Geometric scattering in robotic telemanipulation. *IEEE Transactions on Robotics and Automation*, 18(4), 2002.

[12] A.J. van der Schaft. L_2-*Gain and Passivity Techniques in Nonlinear Control*. Communication and Control Engineering. Springer Verlag, 2000.

Impedance Reflecting Rate Mode Teleoperation

F. Mobasser, K. Hashtrudi-Zaad*and S.E. Salcudean [†]

Department of Electrical and Computer Engineering,
Queen's University, Kingston, ON, Canada [khz @ ece.queensu.ca]
[†] Department of Electrical and Computer Engineering,
University of British Columbia, Vancouver, B.C., Canada [tims @ ece.ubc.ca]

Abstract

Transparent teleoperation under rate mode has proven to be difficult in terms of stability, performance and implementation. This is mainly due to the need for the exchange of derivatives and integral of measured positions and forces. This paper proposes a new control architecture designed based on the environment impedance reflection concept. The performance of this controller is compared to that of a conventional controller, under different operational conditions using both analytical methods and numerical simulations.

1 Introduction

After stability, the transparency performance of teleoperators is a vital factor in the design of bilateral teleoperation controllers. A system is known to be transparent, if there is a correspondence between the master and slave positions and forces [1]. In many applications due to safety issues in dealing with fast manipulators or due to the limitation of the master workspace, the system is designed to operate under rate (velocity) mode [2]. In these systems, the master position is interpreted as a velocity command for the slave. Shuttle arm [2] and hydraulic excavating machines [3] are good examples of this kind of systems. In rate mode operation, the kinematic correspondence between the master and slave is not limited to unity [4] and transparency is achieved if the impedance perceived by the operator matches the environment impedance [5].

As it is shown by Parker et al. [6], in two-channel force-rate architectures, an un-natural feel of the environment is felt by the operator when the kinematic correspondence between the master and slave is lost. This can be solved by sending the environment force derivative to the master. However this may cause a serious implementation issue as the measured environment force is often noisy. As a remedy, a uniform use of high-pass filters in all four-channel feedback paths has been proposed in [4]. The performance of this controller, as will be shown later, is prone significantly to communication-channel delays. In this paper, a control architecture for increased resistance in performance to noise and delay is proposed. This controller is designed based on the use of environment impedance reflection [7]. A variation of on-line recursive least squares identification technique will be employed for remote estimation of the environment impedance parameters in the presence of force measurement noise and abrupt changes in the impedance parameters [8]. The performance of the proposed controller under different operational conditions is compared to that of the controller in [4].

This paper is organized as follows: in Section 2, the dynamics of a general four-channel bilateral controller is presented. Furthermore, transparency conditions under rate mode and some implementation issues are discussed. Section 3 introduces new controller in addition to the controller proposed in [4]. It further introduces a recursive method for online estimation of the environment dynamic parameters. The impedance matching performance provided by the above controllers under different environment and communication-channel conditions is examined in Section 4. Moreover, the time-domain performance of these rate-mode controllers under the above conditions and in the presence of force measurement noise are simulated and compared in Section 5. Conclusions are drawn is Section 6.

2 System Representation and Transprency-Optimized Controller

Consider the block diagram of a general four-channel master-slave teleoperation control architecture as shown in Figure 1, in which T_d denotes the

---
*This work was supported by the National Science and Engineering Research of Canada (NSERC).

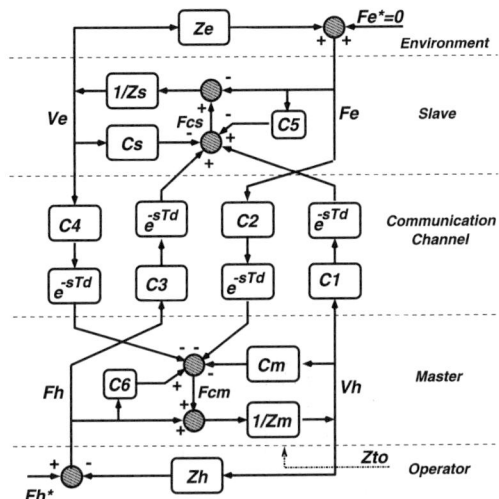

Figure 1: Block diagram of a four-channel controller.

communication-channel delay [5, 9]. Assuming being in contact with the operator and the environment, respectively, the master and slave dynamics are modelled by

$$Z_m V_h = F_h + F_{cm} \quad , \quad Z_s V_e = -F_e + F_{cs} \quad (1)$$

where $Z_m := M_m s$ and $Z_s := M_s s$ represent the linear-time-invariant (LTI) mass models of the force actuated master and slave manipulators, F_{cm} and F_{cs} are the controller commands, and V_h, V_e, F_h and F_e are the Laplace transforms of the master and slave velocities, the operator force on the master and the slave force exerted on the remote environment. The operator and environment are also modelled by LTI impedances Z_h and Z_e according to

$$F_h = F_h^\star - Z_h V_h \quad , \quad F_e = F_e^\star + Z_e V_e \quad (2)$$

where $F_h^\star$ and $F_e^\star(=0)$ are the Laplace transforms of the operator and environment exogenous force inputs, respectively. Here, $C_m := B_m + \frac{K_m}{s}$ and $C_s := B_s + \frac{K_s}{s}$ denote the local position controllers, C_5 and C_6 are the local force feedback controllers, and $C_1, \cdots, C_4$ denote the remote compensators. As seen from Figure 1, the master and slave control commands F_{cm} and F_{cs} are composed of local and remote signals according to

$$F_{cm} = -C_m V_h + C_6 F_h - C_4 e^{-sT_d} V_e - C_2 e^{-sT_d} F_e \quad (3)$$
$$F_{cs} = -C_s V_e - C_5 F_e + C_1 e^{-sT_d} V_h + C_3 e^{-sT_d} F_h \quad (4)$$

The master-slave closed-loop dynamic equations can be expressed as

$$Z_{cm} V_h + C_4 e^{-sT_d} V_e = (1+C_6) F_h - C_2 e^{-sT_d} F_e \quad (5)$$
$$C_1 e^{-sT_d} V_h - Z_{cs} V_e = -C_3 e^{-sT_d} F_h + (1+C_5) F_e \quad (6)$$

where $Z_{cm} := Z_m + C_m$ and $Z_{cs} := Z_s + C_s$. In case of negligible time-delay ($T_d \approx 0$), using (5)-(6), the transmitted impedance to the operator defined as $Z_{to} := \frac{F_h}{V_h}|_{F_e^\star=0}$, is derived as

$$Z_{to} = \frac{[Z_{cm} Z_{cs} + C_1 C_4] + [(1+C_5) Z_{cm} + C_1 C_2] Z_e}{[(1+C_6) Z_{cs} - C_3 C_4] + [(1+C_5)(1+C_6) - C_2 C_3] Z_e} \quad (7)$$

Transparency performance can be described quantitatively as a match between the environment impedance and the impedance transmitted to the operator, i.e. $Z_{to} := \frac{F_h}{V_h}|_{F_e^\star=0} = \frac{F_e}{V_e}|_{F_e^\star=0} := Z_e$ [5], pending on the kinematic correspondence between the master and slave.

2.1 Transparency under Rate Control

In rate control mode, master position is interpreted as a velocity command by slave. In practice, master position is integrated and sent as a position command to slave, that is $C_1 := \frac{1}{s} C_1'$. In this case, kinematic correspondence between the master and slave is defined as $X_h \equiv V_e$. If $C_1, \cdots, C_4$ are not functions of Z_e and $T_d \approx 0$, transparency, i.e. $Z_{to} = Z_e$ in (7), is achieved if and only if the *transparency-optimized control law*

$$\begin{cases} C_1 := \frac{1}{s} C_1' = \frac{1}{s} Z_{cs} \\ C_2 := s C_2' = s(1+C_6) \quad \text{Perfect} \\ C_3 := \frac{1}{s} C_3' = \frac{1}{s}(1+C_5) \quad \text{Transparency} \\ C_4 := s C_4' = -s Z_{cm} \quad \text{Condition-set} \end{cases} \quad (8)$$

and $(C_2', C_3') \neq (0,0)$ holds. This is the transparency-optimized control law under position mode proposed in [5, 9], with the exception that the signals from the master to the slave are integrated while the signals from the slave to the master have to be differentiated. As a result, there are the following issues with regard to the implementation of the transmitted optimized control law (8):

• If communication delays are significant, transparency is compromised as the transmitted impedance becomes

$$Z_{to} = \frac{[Z_{cm} Z_{cs} + C_1' C_4' e^{-2sT_d}] + [C_3' Z_{cm} + C_1' C_2' e^{-2sT_d}] Z_e}{[C_2' Z_{cs} - C_3' C_4' e^{-2sT_d}] + [C_2' C_3'(1 - e^{-2sT_d})] Z_e} \quad (9)$$

$$= \frac{[Z_{cm} Z_{cs}(1 - e^{-2sT_d})] + [C_3' Z_{cm} + C_2' Z_{cs} e^{-2sT_d}] Z_e}{[C_2' Z_{cs} + C_3' Z_{cm} e^{-2sT_d}] + [C_2' C_3'(1 - e^{-2sT_d})] Z_e} \quad (10)$$

• Although only position and velocity are needed to implement the remote position feedback $C_1 V_h = \frac{Z_{cs}}{s} V_h = (M_s + \frac{B_s}{s} + \frac{K_s}{s^2}) V_h$, the slave acceleration and jerk are required to realize $C_4 V_e = -(M_m s^2 + B_m s + K_m) V_e$. In practice, the velocity derived by differentiating measured position is noisy. Therefore, the acceleration and jerk become too contaminated and unreliable to be computed and used. In this case, the

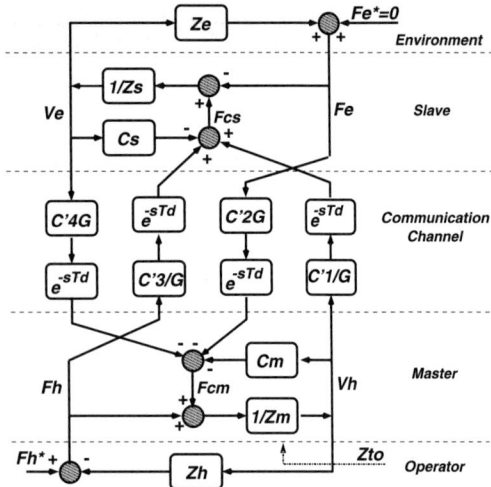

Figure 2: Block diagram of the USF controller.

practical Transparency-Optimized Controller (**TOC**) would only include $C_4 V_e = -K_m V_e$.

- The remote force feedback term $C_2 F_e = C_2' s F_e$ requires the numerical computation of the derivative of the environment force. Due to the noise and abrupt changes in the measured force, $\dot{F}_e$ can be extremely noisy, and therefore it is not recommended for use for stable transparent performance[1]. This point will be observed in Section 5 simulations, where the environment measured force is contaminated with white noise.

To solve the above performance and implementation issues, two control strategies, one reported in [4] and the other proposed in this paper will be introduced next. The performance of these controllers will be compared analytically and by simulations.

3 Alternative Rate Controllers

These bilateral controllers are in two types based on the use of uniform scaling/filtering and environment impedance reflection.

3.1 Controller Based on Uniform Scaling/Filtering (USF)

In this controller, as proposed by Zhu *et al.* in [2] and shown in Figure 2, $C_5 = C_6 = 0$ (no local force feedback) and the control parameters $C_1, \cdots, C_4$ are ideally chosen as $C_1 = \frac{Z_{cs}}{G}, C_2 = G, C_3 = \frac{1}{G}, C_4 = -GZ_{cm}$ to satisfy a general kinematic correspondence relation $V_h = GV_e$. Choosing G as a high-pass filter $G(s) := \frac{\alpha s}{s+\alpha}$ with large value for α, practically differentiates the input at low-frequencies, that is $G(s) \approx s, \forall \omega < \alpha$. Since the acceleration-based terms $Z_s V_h = M_s s V_h$ and $Z_m V_e = M_m s V_e$ in $C_1 V_h$ and $C_4 V_e$

---
[1]To avoid extra notation, some abuse of notation is done in showing time domain signals using their laplace transforms.

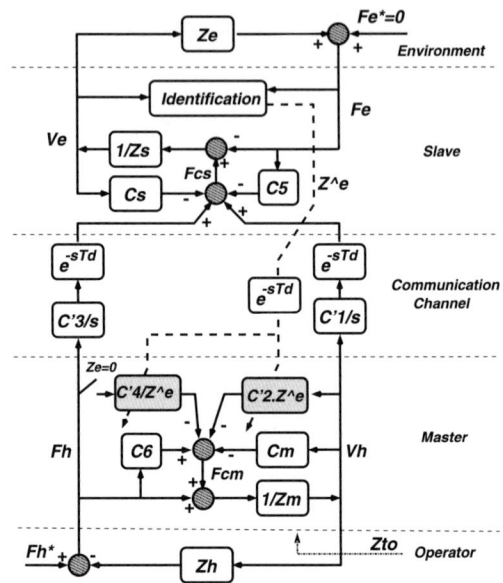

Figure 3: Block diagram of the EIR controller.

are not available, $C_1 = \frac{C_s}{G}$ and $C_4 = -GC_m$ are used in practice. Note that, due to the use of the high-pass filter $G(s)$, the term $B_m V_e$ is included in the controller. In addition, only F_e and not $\dot{F}_e$ needs to be used in the force feedback channel $C_2 F_e = C_2' F_e G$. The down side of this controller is the resulting kinematic correspondence relation ($X_h \equiv G X_e$) that may deviate significantly from unity ($X_h \equiv V_e$) if α is not chosen large enough. Using (9), the transmitted impedance is

$$Z_{to} = \frac{[Z_{cm}Z_{cs} - C_m C_s e^{-2sT_d}] + [Z_{cm} + C_s e^{-2sT_d}]Z_e}{[Z_{cs} + C_m e^{-2sT_d}] + [(1 - e^{-2sT_d})]Z_e} \quad (11)$$

In case of negligible delay and similar master and slave, $Z_{to} = Z_m + Z_e$, implying the operator feeling the master dynamics. This is due to the elimination of inertial terms in the coordinating force feedback terms $C_1 V_h$ and $C_4 V_e$.

3.2 Controller Based on Environment Impedance Reflection (EIR)

In this controller, which is shown in Figure 3, the environment impedance Z_e is used to replicate $C_2'(sF_e)$ and $C_m(sV_e)$. If the environment impedance is known, the force derivative sF_e is replaced with $Z_e V_h$ since $sF_e = s(Z_e V_e) = Z_e(sV_e) \approx Z_e V_h$. Due to practical purposes, only $(B_e + \frac{K_e}{s})V_h$ is utilized. Approximating the slave acceleration/velocity sV_e with the master velocity/position V_h is intuitively correct as in the rate control mode, the master position is interpreted as velocity command at the slave side and kinematic correspondence demands $V_h \equiv sV_e$. In a dual manner, sV_e can be approximated by $\frac{F_h}{Z_e}$, since $sV_e = -s\frac{F_e}{Z_e} = \frac{sF_e}{Z_e} \approx \frac{F_h}{Z_e}$. In this case, the remote compensators are realized according to

$$C_2 F_e \approx C_2'(B_e + \frac{K_e}{s})V_h \quad (12)$$

$$C_4 V_e \approx \begin{cases} 0 & , \text{ free motion} \\ -\frac{Z_{cm}}{Z_e}F_h & , \text{ otherwise} \end{cases} \quad (13)$$

In free motion operation where $Z_e = 0$, the slave position feedback has to be cut, as $\frac{Z_{cm}}{Z_e}F_h$ becomes ill-defined. This is shown by a switch in Figure 3. The elimination of $C_4 V_e$ may cause a significant position/velocity tracking error $X_h - V_e$ as $C_4 V_e = -Z_{cm}V_e$ has non-zero value and can not be replaced with zero in (13). In this case, the system is practically operating in open loop as both feedback signals (12) and (13) are nulled. This tracking error problem can be alleviated by using tighter local position controller at the slave (i.e. higher K_s). Using (5)-(6), the transmitted impedance to the operator can be found to be

$$Z_{to} = \begin{cases} \frac{Z_{cm}}{C_2'} & , \text{ free motion} \\ \left(\frac{Z_{cm}+C_2'(B_e+\frac{K_e}{s})}{Z_{cm}+C_2'Z_e}\right)Z_e & , \text{ otherwise} \end{cases} \quad (14)$$

As it is seen, Z_{to} is independent of the communication channel delay. This is true only if Z_e is constant and is known a priori such as in operations on structured homogenous environments or in cases where the homogenous environment impedance has been identified off-line. In case the environment impedance varies, Z_e has to be identified and transmitted to the master side on-line. This adds a delay of T_d to the environment impedance estimate $\hat{Z}_e$. In this case, (14) becomes

$$Z_{to} = \begin{cases} \frac{Z_{cm}}{C_2'} & , \text{ free motion} \\ \left(\frac{Z_{cm}+C_2'(B_e+\frac{K_e}{s})e^{-sT_d}}{Z_{cm}e^{-sT_d}+C_2'\hat{Z}_e}\right)\hat{Z}_e & , \text{ otherwise} \end{cases} \quad (15)$$

The effect of delay on Z_{to} is not significant since Z_{to} for soft and hard environments are approximated by $Z_e e^{sT_d}$ and $Z_e e^{-sT_d}$ respectively. The superior performance of the EIR controller in the presence of delay will be shown in the simulation results of Sections 4 and 5.

Note that after utilizing the controller (12)-(13), the only flow of information from the slave to the master is through the estimated environment impedance. This prevents the contact force noise to be transferred to the master. On the other hand, this may limit the system transparency unless Z_e is known precisely; otherwise, the environment impedance needs to be estimated truly at a high convergence rate. To directly include the slave contact information, more accurate approximations

$$C_2 F_e \approx C_2'[(M_e s + B_e)V_h + K_e V_e]$$
$$\approx C_2'(B_e V_h + K_e V_e) \quad (16)$$

$$C_4 V_e \approx -(M_m s(sV_e) + B_m s(V_e) + K_m V_e)$$
$$\approx \begin{cases} -K_m V_e & , \text{ free motion} \\ -(\frac{M_m s}{Z_e}F_h + \frac{B_m s}{Z_e}F_e + K_m V_e) \end{cases} \quad (17)$$

can be used. In free motion, part of $C_4 V_e$, i.e. $-K_m V_e$, is still fed back to the master resulting in reduced position/velocity error $X_h - V_e$. One should note that the above approximations are valid once the environment exogenous force input in (9) is zero so that $F_e = Z_e V_e$ holds.

On-line Identification Method: In the teleoperation simulation results reported in Section 5, Self-Perturbing Recursive Least Squares (SPRLS) online identification method is employed to identify environment impedance. SPRLS is a variation of the recursive least squares method. According to [8], in comparison to the other existing RLS algorithms, this method has superior performance in the presence of noise and disturbance. To apply SPRLS, the environment dynamic is discretized and parameterized to

$$y(k) = \boldsymbol{\phi}^T(k)\boldsymbol{\theta} \quad (18)$$

where $\boldsymbol{\phi}(k) = [\dot{V}_e(k) \ V_e(k) \ X_e(k)]^T$ is the regressor input, $y(k) = F_e(k)$ is the measured output and $\boldsymbol{\theta} = [M_e \ B_e \ K_e]^T$ is the parameter vector to be identified. To avoid the use of $\dot{V}_e(k)$, the two sides of (18) can be low-pass filtered as:

$$y_f(k) = \boldsymbol{\phi}_f^T(k)\boldsymbol{\theta} \quad (19)$$

where $y_f = \frac{y}{1+Ts}$ and $\boldsymbol{\phi}_f(k) = [\frac{1}{T}(V_e - \frac{V_e}{1+Ts}) \ \frac{V_e}{1+Ts} \ \frac{X_e}{1+Ts}]^T$. Considering the above assumptions, the SPRLS method is formulated as

$$\hat{\boldsymbol{\theta}}(k) = \hat{\boldsymbol{\theta}}(k-1) + \mathbf{K}(k)\varepsilon(k) \quad (20)$$

$$\varepsilon(k) = y_f(k) - \boldsymbol{\phi}_f^T(k)\boldsymbol{\theta}(k-1) \quad (21)$$

$$\mathbf{K}(k) = \frac{\mathbf{P}(k-1)\boldsymbol{\phi}_f(k)}{1 + \boldsymbol{\phi}_f^T(k)\mathbf{P}(k-1)\boldsymbol{\phi}_f(k)} \quad (22)$$

$$\mathbf{P}(k) = \mathbf{P}(k-1) - \mathbf{K}(k)\boldsymbol{\phi}_f^T(k)\mathbf{P}(k-1) \quad (23)$$
$$+ round[\gamma\varepsilon^2(k-1)]\mathbf{I} \quad (24)$$

with the initial condition
$$\mathbf{P}(0) = M\mathbf{I}$$
$$\hat{\boldsymbol{\theta}}(0) = [0\ 0\ 0]^T \quad (25)$$

where ε is the estimated output error, $\mathbf{K}$ is the adaptation gain vector, $\mathbf{P}$ is the covariance matrix, and M

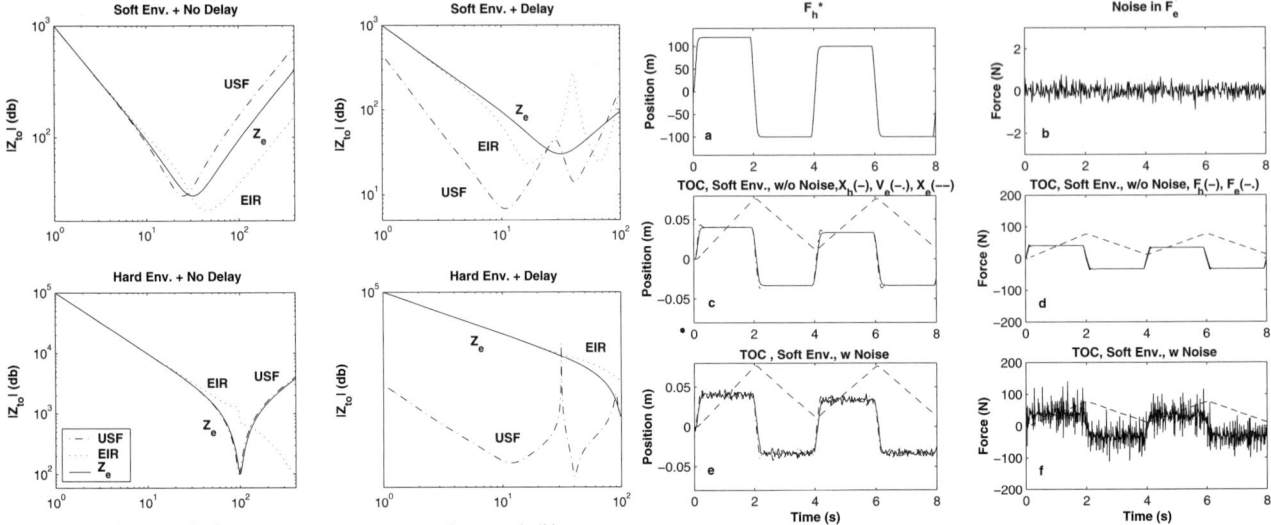

Figure 4: Magnitude Bode plot of Z_e and Z_{to} for the USF and EIR control architectures.

Figure 5: $F_h^\star$ and measurement noise in F_e (a,b). Position and force responses for TOC operating on a soft environment with/without the effect of noise (c-f).

is a very large number. γ determines the sensitivity of the algorithm to noise and is directly related to the standard deviation of the measurement noise. For fast track of the abrupt changes in environment parameters, the covariance matrix $\mathbf{P}$ is reset to $M\mathbf{I}_{3\times 3}$ when sudden change in the environment impedance is detected. This modification improves the performance and the convergence time significantly, specially in intermittent contact applications.

4 Impedance Matching Performance

Figure 4 depicts the magnitude Bode plots of the transmitted impedances for the above control architectures for a soft environment $Z_e = s + 30 + \frac{1000}{s}$ (top figures) and a hard environment $Z_e = 10s + 1000 + \frac{10^6}{s}$ (bottom figures). The right column figures show the case when a 100 (ms) of delay, modelled by a 5th-order Padé approximation, is introduced to the communication-channel.

In the case of negligible delay, both architectures perform the same at low frequencies below 10 (rad/s). The USF controller performs better at higher frequencies, especially on hard environments. The reason is that for USF, $Z_{to} = Z_m + Z_e$. Therefore, Z_e becomes dominant as it increases. From (14), the EIR controller is almost transparent for very soft environments. This can be observed in Figure 4 for low to moderate frequencies where acceleration term $M_e s$ is not significant. This is not the case in free motion operations as the operator feels the position controlled master dynamics $\frac{Z_{cm}}{C_2'}$ in (14).

As expected and seen from Figure 4 (right column), the EIR controller seems to be resistant to time delay. This is especially more visible at lower frequencies with the hard environment. This can mainly be traced back to the lack of direct environment force feedback. From the above discussion, the following recommendations can be made; i) When delay is negligible, any architecture can be used for operations at low frequencies (for higher frequencies, the USF controller is suggested), ii) When delay is substantial, the EIR controller is recommended.

5 Dynamic Simulations

In this section, the results of simulations conducted on a telerobotic system using different control architectures are reported and compared. These results simulate operations on the soft environment, characterized in Section 4. The simulation conditions include the absence or the presence of significant delay in the communication-channel ($T_d = 100$ (ms)) and noise in the environment force measurement F_e (as shown in Figure 5-b). The teleoperation system parameters are assumed to be $Z_h = 0.5s + 70 + \frac{2000}{s}, Z_m = Z_s = 0.7s, C_m = 6.2 + \frac{15.5}{s}, C_s = 20 + \frac{100}{s}$ and $\alpha = 50$. The periodic profile of the operator exogenous force, as shown in Figure 5-a, has been chosen to push the end-effector enough in the environment in the first cycle to avoid any disconnection in contact simulations. A set of simulations has also been conducted for operations involving intermittent contact.

A practical problem with implementation of the EIR controller is that just after entering the environ-

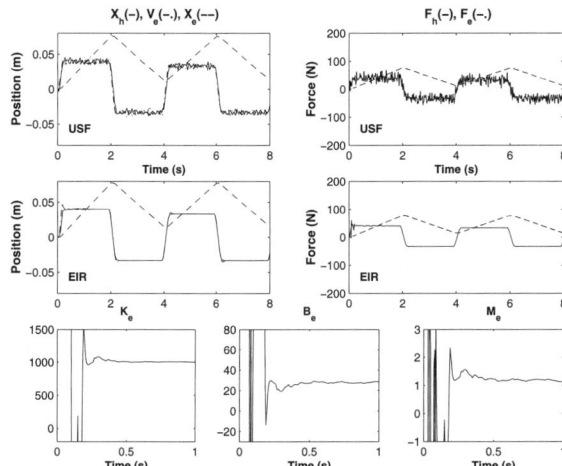

Figure 6: Position/velocity/force responses for contact mode with noise.

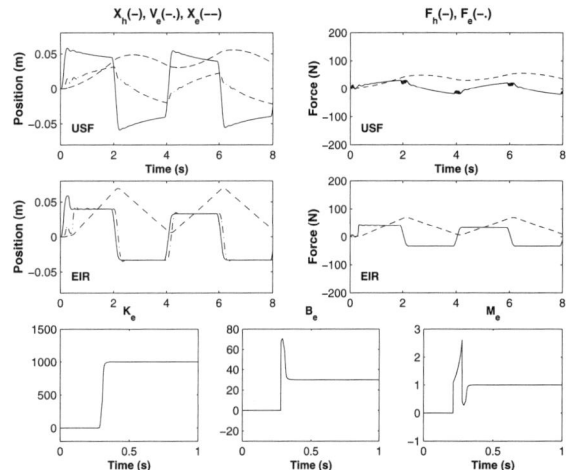

Figure 7: Position/velocity/force responses for contact mode with delay.

ment, $\hat{Z}_e$ in (13) is very small. This causes the operator to experience a substantial repelling force, resulting in the end effector pulling out of the environment at each trial. This issue can be resolved by limiting $H := \frac{Z_{cm}F_h}{\hat{Z}_e}$ to $-\beta$ and β according to

$$C_4 V_e \approx \begin{cases} 0 & \text{, free motion} \\ -sgn(H)min(|H|,\beta) & \text{, otherwise} \end{cases} \quad (26)$$

The value of β is determined by $\beta = \frac{K_m}{\underline{K}_e}\bar{F}_h$ where $\underline{K}_e$ is the lower bound on the environment stiffness and $\bar{F}_h$ is the maximum amplitude of F_h which is related to F_h^*.

One other suggestion to solve the above problem is to use a second identification routine with high initial conditions for $\hat{\theta}$ for updating $\hat{Z}_e$ in (13). In this case, right after contact, $\frac{Z_{cm}}{\hat{Z}_e}F_h$ is a very small value and as time grows, it is expected to reach its actual value. However in practice $\hat{Z}_e$ does not converge to Z_e monotonically and it shows fluctuations that pushes $\frac{Z_c m}{\hat{Z}_e}F_h$ to high values. This again necessitates the use of hard limits of (26).

Simulation Results: Figures 5-c,d show the position/velocity and force responses of the system controlled by the TOC law (8). In the absence of noise in F_e the matching between X_h and V_e is quite satisfactory. Note that $\int F_h$ should match F_e. The same performance have been derived for the USF and EIR controllers but is not reported in this paper.

Contact Simulations: Figures 5-e,f and 6 show the simulation results for TOC and other control architectures when noise is added to F_e. As for the TOC (Figures 5-e,f), the master variables X_h and especially F_h are extremely noisy and distorted ($|F_h| < 150$ N).

This undesirable behavior has been observed in simulations on hard contacts as well. Therefore, the TOC performance will not be pursued any further. From Figure 6, the USF controller is sensitive to noise; however, the effect is considerably lower than that of the TOC in Figures 5-e,f. This is mainly due to the fact that USF avoids using $\dot{F}_e$. The EIR is the least sensitive and the impedance parameters are truly identified. If local slave force feedback is used in the EIR controller, i.e. $C_5 = C_3 - 1 < 0$, then V_e and as a result X_h become only *slightly* contaminated. As shown in Figure 7, when the time delay of 100(ms) is added to the communication-channel (including $\hat{Z}_e$), the performance of USF controller degrades significantly. On the other hand, the environment impedance parameters are identified with fast convergence rate of 30(ms) and the EIR architecture shows remarkable resistance to time delay.

Intermittent Contact Simulations: Figure 8 simulates the results of intermittent contact with a soft environment located at 0.04 cm. The USF controller produces a better position tracking. The reason is the kinematic mismatch in the EIR controller in free motion. Figure 9 shows the results of the simulations with measurement noise in F_e. The effect of noise in F_h for USF controller is quite significant while EIR with no force feedback shows a better performance. Using the covariance resetting technique described in Section 3, the convergence time of the identification method decreases remarkably to 20(ms) with negligible measurement noise and 150(ms) in the presence of significant measurement noise.

The hard contact simulations conducted with the EIR controller did not produce a satisfactory perfor-

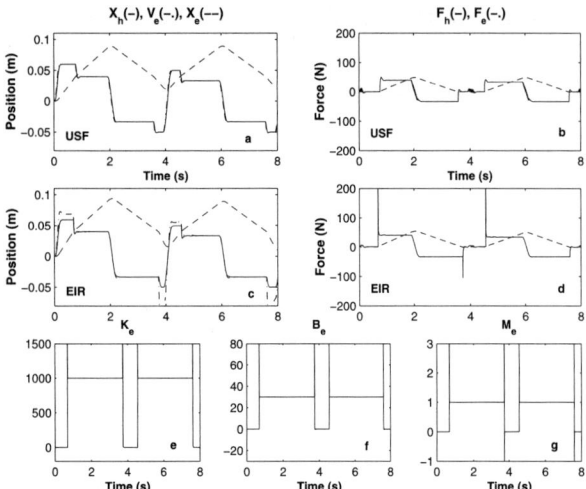

Figure 8: Position/velocity/force responses for intermittent contact.

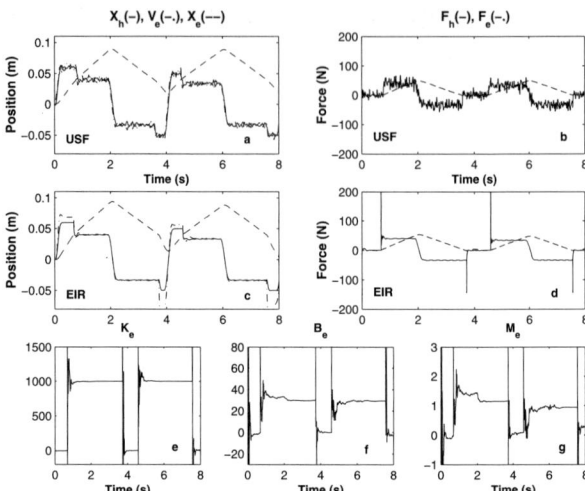

Figure 9: Position/velocity/force responses for intermittent contact in the presence of measurement noise.

mance. The reason traces back to the slow convergence rate of the available on-line identification methods. In case of negligible noise in hard contact applications, fast iterative batch identification method can be used to achieve the same performance as reported previously for the soft contact case.

6 Conclusions

Transparent teleoperation under rate mode requires the transmission of slave acceleration and contact force derivative to the master. This poses severe degradation of performance in the presence of delay and noise. As a remedy, in this paper, a control architecture based on the use of Environment Impedance Reflection (EIR) has been proposed. The performance of this controller has been compared to that of a controller designed based on uniform scaling/filtering (USF). The results of the analysis and simulations have confirmed each other; however, they did not point to a unique control architecture for all different operational conditions. In the case of negligible delays, both the USF and EIR architectures can be used for slow operations. For higher frequencies, the USF controller is suggested. When noise and delay are substantial, the EIR controller is recommended.

Future work should aim towards a search for fast identification algorithms for hard environments, that are robust against measurement noise. Furthermore, the above two controllers will be examined and evaluated on a hybrid master-slave experimental test-bed consisting of a PHANTOM haptic device.

References

[1] Y. Yokokohji, and T. Yoshikawa, "Bilateral control of master-slave manipulators for ideal kinesthetic coupling," *IEEE Trans. Rob. & Auto.*, Vol. 10, pp. 605-620, 1994.

[2] W.S. Kim, F. Tendick, S.R. Ellis, and L.W. Stark, "A comparison of position and rate control for telemanipulations with consideration of manipulator system dynamics," *IEEE J. Rob. & Auto.*, Vol. 3, No. 5, pp. 426-436, 1987.

[3] P.D. Lawrence, S.E. Salcudean, N. Sepehri, D. Chan, S. Bachmann, N. Parker, M. Zhu and R. Frenette, "Coordinated and force feedback control of hydraulic excavators," *Proc. Int. Symp. Exp. Rob. (ISER)*, Stanford, CA, 1995, pp. 114-121.

[4] S.E. Salcudean, M. Zhu, W.-H. Zhu and K. Hashtrudi-Zaad, "Transparent bilateral teleoperation under position and rate control," *Int. J. Rob. Res.*, Vol. 19, pp. 1185-1202, 2000.

[5] D.A. Lawrence, "Stability and transparency in bilateral teleoperation," *IEEE Trans. Rob. & Auto.*, Vol. 9, pp. 624-637, 1993.

[6] N.R. Parker, S.E. Salcudean and P.D. Lawrence, "Application of force feedback to heavy-duty hydraulic machines," *Proc. IEEE Int. Conf. Rob. & Auto.*, Atlanta, GA, 1993, pp. 375-381.

[7] B. Hannaford, "A design framework for teleoperators with kinesthetic feedback," *IEEE Trans. Rob. & Auto.*, Vol. 5, pp. 426-434, 1989.

[8] T.H. Hunt, "A comparative study of system identification techniques in the presence of parameter variation, noise, and data anomalies," *Proc. IEEE 39th Midwest symposium on Circuits and Systems*, pp. 593-596, 1996.

[9] K. Hashtrudi-Zaad and S.E. Salcudean, "Transparency in time-delayed systems and the effect of local force feedback for transparent teleoperation," *IEEE Trans. Rob. & Auto.*, Vol. 18, No. 1, pp. 108-114, 2002.

Laboratory Tools for Robotics and Automation Education

Claudio Cosma   Mirko Confente   Debora Botturi   Paolo Fiorini

{cosma,confente,debora,fiorni}@metropolis.sci.univr.it

Dipartimento di Informatica

Universitá di Verona

37134 Verona – Italy

Abstract— This paper describes our efforts and plans to develop a Virtual Laboratory for the education in Robotics and Automation. These efforts are characterized by the need of blending R&A subjects into a traditional Computer Science curriculum, thus forcing a specific selection of development topics. In this context, the Robotics Laboratory must provide basic as well as advanced experiments, to address the needs of students at different education levels. In this paper, we present the development of three main applications, to support Control Systems and Robotics classes, as well as thesis and dissertation research. Of particular interest is the effort in the area of teleoperation, preliminary to the opening, next year, of a new curriculum on Medical Informatics, in which Computer Assisted Surgery will play an important role.

I. INTRODUCTION AND PAST WORK

Robotic education has traditionally relied on standard classroom lectures combined with laboratory experiments to make the students understand the principles and the practice of this discipline. However the availability of fast computer connections and of powerful Internet servers makes it now possible to think of a hybrid approach in which students experiment the theory in virtual environments before starting the laboratory experiments. This approach has several advantages, including safety, unlimited laboratory availability, better curriculum integration and increased flexibility. The disadvantage of this approach is that it is not supported by standard equipment and commercial products, and that each laboratory must invest a significant amount of time to implement a useful set-up. During setup however, students in the laboratory acquire a valuable expertise in a very important field, i.e. remote control of complex electromechanical systems. Furthermore, the development of a virtual laboratory for robotics education offers significant benefits to Universities that do not have a large variety of engineering courses and laboratories, since the same laboratory can provide experimental support to a range of topics and classes requiring different expertise and schedule. In particular, our laboratory is in the process of developing a series of virtual experiments, connected to real laboratory experiments, which cover a few aspects of basic control theory, robotic analysis and programming, software architecture, teleoperation, and robot motion planning.

The design and development of virtual laboratories and experiments is made possible by the availability of technologies for the remote command and control of Internet-based devices. The basic technology elements have been gradually developed in the past few years and only now they are being put together in a consistent, integrated form. In particular, Internet-based communication is the key to allow users to interact with the laboratory experiments and exchange textual, audio and video information. As proposed in [1], teleoperation systems would form the base of several applications such as health care [2], tele-medicine, collaborative design [3], entertainment and teleconferencing. However, basis theoretical issues prevent a direct use of some of the experiments, since the communication time delay of the system may introduce dangerous instabilities. In fact, this is also an important area of research to which students can contribute, implementing and experimenting with the algorithms proposed in [4], [5], [6], for the compensation of time delay.

The specific applications that interface robots and the Web are very recent. One of the earliest examples of generalized use of a robot via Internet is the Mercury project [7] and the PumaPaint project [8]. These project did not have educational goals and explored various forms of interactions of people with remote robots. They were characterized by a Web interface allowing a limited set of commands.

In [9], [10], [11], [12], [13], [14] various examples are proposed of interfaces to remote robotic arms or vehicles via Internet. These applications experiment with the execution of robotic tasks, such as pick-and-place, tourguide in museums and in general interaction with people. A web site for motion planning of nonholonomic vehicles motions is described in [15]. In [16] the development is described of a laboratory similar to the one described in this paper is presented. Recently, a workshop was conducted on the subject of Education in Robotics [17], addressing the different areas of teaching using robots, learning with mobile robots and developing suitable teaching curricula.

Another essential element for the development of remote robotics experiments is the availability of a skilled and dedicated workforce. In our Department we could count on the support of the recently established (July

Fig. 1. Teleoperation bench

2001) IEEE Student Branch of Verona, that now counts about 30 student members. Further development will be to establish a Robotics Student Chapter, which will be the firs established in a Italian University. Although student members are interested in Vision, Electronics, and Information Theory more than 50% of members are also members of Robotics and Automation and Control System Societies, thus providing the necessary interest and motivation to support the development of remote laboratory experiments. During this year, for example, IEEE Student Members aided in the organization of a graduate course in robotics attended by many students coming from various member Countries of the European Community.

The efforts described in the paper take advantage of the available resources, both in terms of human power, equipment and technology, to build an infrastructure capable of offering remote laboratory experiments to University students. The paper is organized following the Automation and Robotics curriculum that the laboratory supports. In particular, Section II describes the experiments to introduce basic concepts of control system theory. Section III describes the development of graphical simulators for the learning of kinematic and dynamic aspects of robotics. Section IV describes the tools available for graduating students to experiment with advanced topics, such as software control architectures, operator interfaces, teleoperation systems and motion planning. Of course, many of the ideas described in the following Sections are still in the development phase and therefore Section V besides summarizing the paper presents our plans for future development and expansion of the virtual laboratory.

II. CONTROL THEORY EXPERIMENTS

To share efficiently applications and instrumentation, all laboratory experiments share the common theme of robotics and teleoperation. However, since control systems understanding is essential, we decided to limit experiment complexity to their vary basic structure, so that hardware and software architectures would not confuse and distract the students. To achieve this result we are developing the single-axis force reflecting teleoperation system shown in Figure 1. This system is characterized by two parts, each with an independent motor, controlled by a single computer running Windows NT and a commercial servo board. The first part, called the master on the left side of Figure 1, is a motor with an encoder and an appendix on the shaft, used as handle of the teleoperation master. The second part, the slave of the teleoperation, consists of a three joint manipulator, whose first joint is shown on the right side of Figure 1. The research purpose of this set up is to compare the performance of teleoperation architectures in a minimalist environment, where the single axis system, and the very simple software do not shield the true characteristics of the architecture. The experiment is well instrumented and allows extensive data collection.

Because of the simple structure of this setup, it is possible without much effort to use the two parts as the basis of several Control System experiments. By controlling the motor of each part separately, one can use the same set up to demonstrate the performance of basic motion control algorithms. The handle of the Master can be used to simulate a a fixed load. A variable load can be simulated by changing the configuration of the second and third joints of the Slave part of the system, which can also be done while the first joint is moving, thus creating interesting identification and control experiments. The two motors are controlled through a Web interface, which is used by students to identify motor parameters and then design simple controllers for position and velocity tracking. Currently we are testing the first joint of the Slave, to find its dynamical parameters, before adding the remaining joints.

It is interesting to note that this simple setup, developed by students as a class project, has shown unexpected difficulties and challenges in terms of identifying the right motor control board, adapting the software drivers supplied by the manufacturers, identifying the motors, whose parameters are obviously unknown, and so forth. Thus, the educational aspects of the project lies not only in the experiments that are carried out with the system, but also in developing and maintaining a robust and efficient setup.

III. ROBOTICS EXPERIMENTS

Robotics can be taught in a single course or in a variety of courses, thus education has to be tailored to a specific curriculum, even though many important subjects can be left out. In our Department, only one course in robotics is available to the students, and therefore it was decided

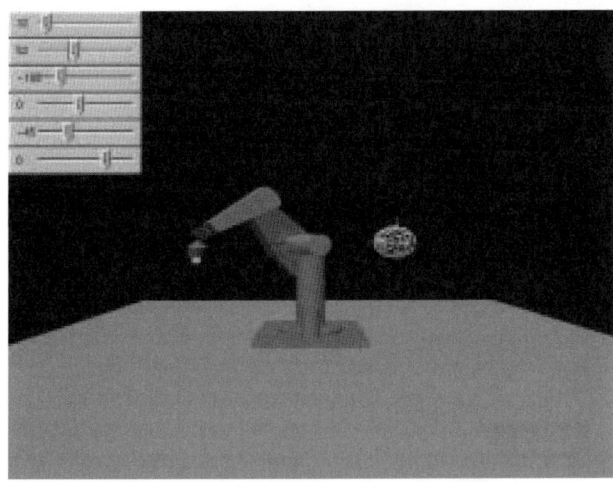

Fig. 2. Graphical simulation of the Puma robot

that the laboratory would cover only the main classical concepts, leaving the learning of more advanced topics to the preparation of the undergraduate, or graduate, thesis. We are fortunate that the laboratory is equipped with a variety of robots, acquired through loans, donations, and acquisitions. At the moment, course support is divided into two phases. The first, carried out in the computer laboratory, consists of developing algorithms to be tested on graphical simulations of a PUMA 560 manipulator and a Nomad200 mobile robot. Students have the opportunity to develop kinematic algorithms for the fixed robot and navigation and collision avoidance algorithms for the mobile robot. At a second time, working algorithms are tested on the real robots under very safe conditions. Figure 2 show the graphical interface used to test the algorithms for the fixed robot. The graphical interface can be connected to user-developed algorithms, both for direct and inverse kinematics, thus allowing the students to experiment with their own approach. Figure 3 shows the graphical interface used to develop and test motion planning algorithms. The environment is fixed, and represents a portion of the laboratory. Students can interface their algorithms to the graphical environment and test the quality of their code or of their implementation of known algorithms. Figure 3 shows an example of a program implementing the Tychonievich algorithm for trajectory planning in dynamic environments. At the moment only the first phase of the course support has been implemented. However, during the current year, class projects will be organized to extend this virtual approach and to let students become more familiar with the real equipment in the laboratory.

IV. SUPPORT FOR THESIS AND DISSERTATION RESEARCH

Since robotics includes a significant experimental part, most thesis work is developed in the laboratory, by de-veloping and testing new ideas within the scope of the research areas in the laboratory. Currently research is organized into three main areas: teleoperation, mobile robotics, and exploration robotics. These areas have emerged as those capable of attracting funds from various sponsors and therefore students learn very early on in their professional life, the fine art of balancing individual research interests with the hard deliverables of sponsored research. The main applications of teleoperation addressed in the laboratory refer to the needs of space exploration and surgical research. We have developed several algorithms for the compensation of time delay teleoperation under the sponsorship of the Italian Space Agency (ASI) in cooperation with the Robotics Laboratory of ENEA in Rome. However, the area in which we are investing the most, in terms of human and economic resources is the development of new procedures for Computer Assisted Surgery (CAS). Our laboratory is currently equipped with two joysticks capable of force reflection on all six axes, one of which is shown in Figure 4. These devices give Master and Doctoral students working in the laboratory a great opportunity, since force reflection is demanded by surgeons, and so far no commercial product is delivering it. Students are then in the position of cooperating with personnel of the nearby Medical School to evaluate the quality of their research with realistic experiments carried

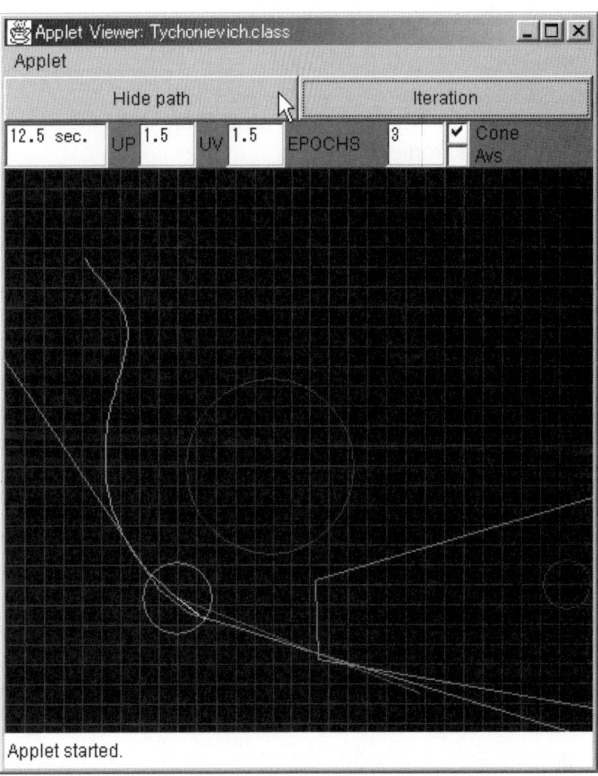

Fig. 3. Graphical simulation of motion planning algorithms

Fig. 4. One of the NASA-JPL force reflecting joysticks used in the laboratory.

type of automatic transport, i.e. the goods whose automatic transport would result in savings to the company, and thus justify the investment in the robotic transport system. If this project is successful, i.e. if tests will demonstrate the validity of the approach taken and that the operation selected can be successfully carried out by a robot, the Company is planning to extend this approach to several other plants and to introduce robots in other departments. The ultimate goal is to provide automatic monitoring and service to the production lines, achieving around-the-clock unassisted production and providing human workers with advanced robotic devices.

Finally, exploration robotics covers the broad area of design and algorithms for robots in extreme situations, such as Antarctica, space, and natural and man-made disasters. The main line of robot used in this research is represented by the family of hopping robots shown in Figure 6.

These devices, developed at NASA Jet Propulsion Laboratory in 1999 [18], are characterized by a pause between jumps to select the next hop direction and recharge the propulsion mechanism. They are a compromise between functionality and system's electromechanical complexity [19], [20]. The prototypes demonstrated that it's possible to develop a robot with mobility and sensing capabilities with only few actuators, provided that every operation is executed in a sequential manner. More precisely, the operational cycle of hopping robots is based on the orderly execution of self-righting, panning the camera to acquire images, recharge the thrusting mechanism in preparation

out either in virtual environments, or with the fixed manipulator in the laboratory.

Mobile robotics research addresses problems related to service robotics and, in particular, to the development of mobile robots for the transportation of small objects. Automated logistics is an area receiving a lot of attention from the academic world as well as the industrial world, and we are addressing this emerging research area by equipping our Nomad200 mobile robot with a gripping device suitable for picking up and holding small parcels. At the moment, we are using the gripping device of the Nomad, shown in Figure 5. This area of research is extremely rich, since it includes the development of task as well as motion planning algorithms, the development of gripping devices for parcels of unknown size, material and weight (within the limits set by the project requirements), and of grasping algorithms, to ensure that grasping is robust with respect to parcel position and orientation. A project currently under development addresses the problems of light logistics in the warehouse of a manufacturing company. One of the main challenges consists of developing a system that impacts as little as possible the existing plant. Students have been working with company personnel to identify the best

Fig. 5. The Nomad200 used for logistics development.

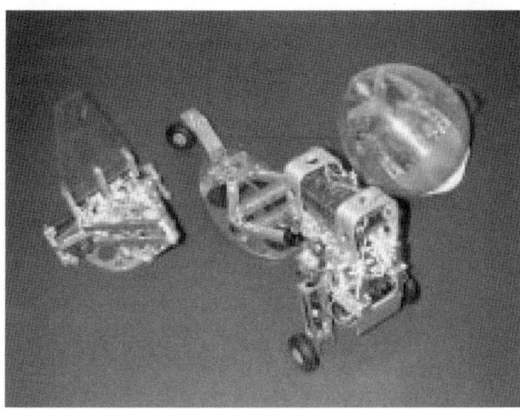

Fig. 6. Three generation of hopping robots

for the next jump, orient the robot in the desired direction, and execute the jump.

For these devices, students carrying out their Master thesis have analyzed and simulated a self-localization algorithm and a new stereo vision system, based on omnidirectional cameras. The objective of the localization algorithm is to be able to identify the landing position of a jumping robot, after landing, without the aid of sensors like radar and laser-scanner positioned on an external platform, and also without on-board vision, since

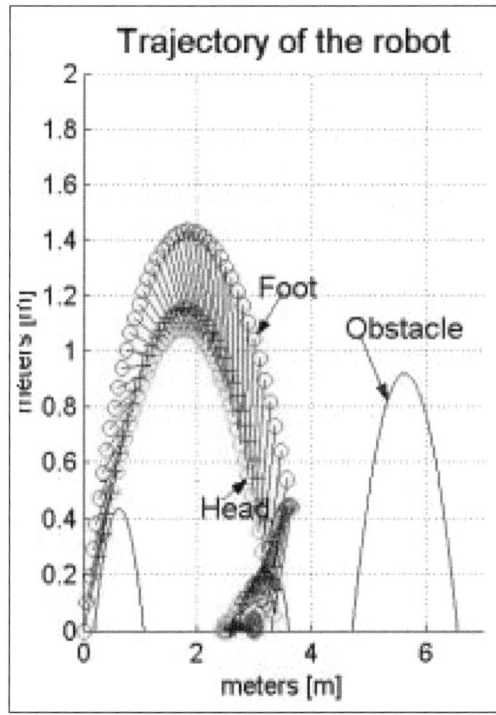

Fig. 7. Simulation of the hopper jump.

machine vision will be integrated at a later time. Currently, students have developed a simulation of robot flight and landing, assuming only the availability of accelerometers, a gyroscope and touch sensors, to detect the point of impact on the robot body. This keeps the sensor suite compatible with the robot characteristics and suitable for real hardware implementation. Since we are interested in outdoor localization and the proposed sensor suite lacks devices that can detect objects in the environment and extract features, we cannot use feature-based maps and triangulation techniques, commonly used in mobile robotics. Furthermore, the discontinuous locomotion modality of the robot prevents the use of simple, although inaccurate, odometric localization, typical of wheeled locomotion. Therefore, without visual references and odometry, the most immediate approach to localize the robot after a jump is an "a posteriori" reconstruction of its trajectory using data provided by the on board sensors before, during, and after the fly. Figure 7 show a simulation of a jump of the robot, used to estimate the landing position afer a jump. However, the estimation error in this probabilistic method can grow unbounded if it is not checked periodically. To overcome this problem, in a MAster's Thesis we have developed the geometrical analysis of a vision system that can be used in the development of a vision-based localization algorithm to reset the position estimate of the robot by using environment landmarks. The two methods combined will, eventually, provide a robust localization method for hopping robots characterized by discontinuous motion. The vision system for a hopping robot must take into consideration the specific nature of the robot motion, be robust to physical impacts, and provide enough information for precise navigation. For these reasons, we have selected a special type of omnidirectional vision system, the Panoramic Annular Lens (PAL), which has all the desired characteristics. Figure 8 shows the optical analysis of the stereo vision system proposed as the visual sensor for the hopping robots. We are hoping that in a near future this work could be continued under support of the European Space Agency (ESA).

Thus, also advanced reseach can be carried out using the set up in the laboratory, with satisfaction from the students who are able to produce innovative results.

V. CONCLUSIONS

The paper presents an overview of the experiments available in the robotics laboratory of the University of Verona (Italy) in support of robotics education. Since our University is not equipped with independent laboratories for all the disciplines contributing to robotics, this laboratory is used as a multipurpose facility, supporting education in basic Control System courses, Robotics courses, and the development of thesis for Master and Doctoral studies. By considering course support, the laboratory is

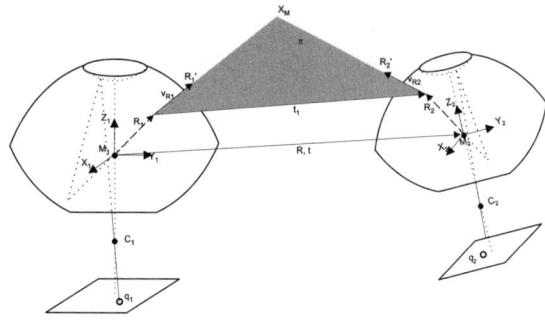

Fig. 8. The stereo omnidirectional vision system for hoppers.

clearly divided into three functional areas, control system experiments, robotics experiments, and research development. Since the research emphasis is on teleoperation, mobile robotics and exploration robotics, experiments are devised using elements of more complex setup. For example, control system experiments are carried out using a portion of a teleoperation system used also for quantitative analysis of teleoperation algorithms.

We think that in a few more months, we will be able to use the laboratory at its full potential, by completing some of the connections between the computer laboratory, whre students can develop applicaiton programs using graphical simulations, and the physical laboratory where the applications are tested on the real hardware. Safety is of paramount concern and therefore, before allowing the direct download of applicaitons to the robot servers and their unsupervised test by remote students, we must ensure that no safety requirements is violated.

VI. REFERENCES

[1] P. Fiorini and R. Oboe. Internet-based telerobotics: Problems and approaches. In *International Conference of Advanced Robotics (Icar'97)*, Monterey, CA, July 7-9 1997.

[2] P. Dario, E. Guglielmelli, and B. Allotta. Mobile robotic aids. In *IEEE International Conference on Robotics and Automation*, pages 17–24, Nagoya, Japan, May 27 1995. Workshop on Robots for Disabled and Elderly People.

[3] P. Buttolo, R. Oboe, and B. Hannaford. Architectures for shared haptic virtual environments. *Computer and Graphics*, 1997. accepted for publication.

[4] R.J. Anderson and M.W. Spong. Bilateral control of teleoperators with time delay. *IEEE Transaction on Automatic Control*, 34(5):494–501, May 1989.

[5] G. Niemeyer and J.E. Slotine. Stable adaptive teleoperation. *IEEE Journal of Oceanic Engineering*, 16(1):152–162, January 1991.

[6] A. Eusebi and C. Melchiorri. Stability analysis of bilateral teleoperation robotic systems. In *3rd European Control Conference (ECC'95)*, Rome, Italy, 1995.

[7] K. Goldberg, M. Mascha, S. Gettner, and N. Rothenberg. Desktop teleoperation via the world wide web. In *IEEE Int. Conf. Robotics and Automation*, 1995.

[8] M.R. Stein. Painting on the world wide web: The pumapaint project. Technical report, Carnegie Mellon University, http://yugo.mme.wilkes.edu/villanov/, 1996.

[9] B. Dalton and K. Taylor. A framework for internet robotics. In *Int. Conf. Intelligent Robots and systems (IROS): Workshop on web robots*, 1998.

[10] E. Paulos and Canny J. Delivering real reality to the world wide web via telerobotics. In *IEEE Int. Conf. Robotics and Automation*, 1996.

[11] R. Simmons. Xavier: an autonomous mobile robot on the web. In *Int. Conf. Intelligent Robots and systems (IROS): Workshop on web robots*, 1998.

[12] P. Saucy and F. Mondada. Khep-on-the-web: One year of acces to a mobile robot through the internet. In *Int. Conf. Intelligent Robots and systems (IROS): Workshop on web robots*, 1998.

[13] R. Siegward, C. Wannaz, P. Garcia, and F. Blank. Guiding mobile robots through the web. In *Int. Conf. Intelligent Robots and systems (IROS): Workshop on web robots*, 1998.

[14] Burgard W. et al. The interactive museum tourguide robot. In *15th Nat. Conf. Artificial Intelligence*, 1998.

[15] S. Piccinocchi, M. Ceccarelli, F. Piloni, and A. Bicchi. Interactive benchmark for planning algorithms on the web. In *IEEE Int. Conf. Robotics and Automation*, 1997.

[16] A. Bicchi et al. Breaking the lab's walls: telelaboratories at the yuniversity of pisa. In *IEEE Int. Conf. Robotics and Automation*, pages 1903–1909, May 2001.

[17] Casals. A and Grau A., editors. *1st Workshop on Robotics education and training*. Euron, Weingarten, Germany, 21 July 2001.

[18] J. Burdick and P. Fiorini. Minimalist jumping robots for celestial exploration. *International Journal of Robotics Research (accepted)*.

[19] E. Hale, N. Shara, J. Burdick, and P. Fiorini. A minimally actuated hopping rover for exploration of celestial bodies. volume 1, pages 420–427. IEEE Inr. Conf. on Robotics and Automation, May 2000.

[20] P. Fiorini, S. Hayati, M. Heverly, and J. Gensler. A hopping robot for planetary exploration. Proc. of IEEE Aerospace Conference, March 1999.

Passivity Analysis of Sampled-Data Interactive Systems

Ravi Hebbar and Wyatt S. Newman
Electrical Engineering and Computer Science Department
Case Western Reserve University
Cleveland, Ohio 44106
(rxh20@po.cwru.edu   wsn@po.cwru.edu)

Abstract: A method for passivity analysis of a class of nonlinear systems is described. For linear plants controlled by sampled-data controllers, the resulting interaction admittance is nonlinear. It is shown that satisfying the linear requirements for passivity is insufficient to guarantee passivity of the nonlinear system. To analyze such plants, we first show that if a waveform constituting a counter-example of passivity exists, then a counter-example may also be found within a limited class of functions. By restricting attention to this class of functions, a numerical passivity evaluation is feasible. Examples using the method are presented.
keywords: passivity, sampled-data, aliasing.

1. Introduction

For dynamically-interacting systems, such as a robot contacting an environment during an assembly operation or multiple robots collaborating in a manipulation task, issues of stability can be difficult to analyze. One very valuable tool is passivity analysis. If each dynamic subsystem (linear or nonlinear) satisfies the property of passivity with respect to its interaction port(s), then any arbitrarily-complex interaction among these subsystems is guaranteed to be stable [1]. This sweeping implication makes it attractive to design robots (and other dynamically-interacting machines) such that each subsystem satisfies passivity. This approach yields building blocks that can be assembled safely into more sophisticated systems.

For linear systems, passivity analysis is relatively straightforward as shown in [2]. However, for nonlinear systems, passivity is defined [3], but there are no general tools for evaluating passivity. This problem becomes apparent even in the seemingly benign case of a linear, time-invariant (LTI) system under computer control (sampled-data control). Use of a sample-and-hold operator makes the resulting system admittance (or impedance) nonlinear.

In this paper, we introduce a method for evaluating the passivity of non-linear systems of this form. It is shown that the nonlinearity is due to aliasing effects. We define sets of periodic functions, called *alias sets*, comprised of *alias functions*, and prove that a system of the stated form is non-passive if and only if there exists an alias function of the form described that provides a counter-example of passivity. Excitation of the system at its interaction port with such an alias function results in extraction of average power from the system, thus proving non-passivity. Thus, to prove passivity for this class of nonlinear systems, it is only necessary to evaluate the sets of alias functions. An approximate numerical method is introduced to perform such an evaluation, constituting a tool for passivity analysis of LTI systems with sampled-data controllers.

2. An illustrative example

Consider the system of Fig 1, which is the same case studied by Colgate and Schenkel in [4] in the context of passivity of sampled-data controlled systems. A pure mass is acted on by an actuator and by forces from the environment. The intent of the control law is to exert pure damping via the actuator, i.e. $F_m(t) = -B_v v(t)$. If this control law were executed ideally without sampling effects, then the controlled plant theoretically would be passive for arbitrarily large (positive) values of control damping, B_v.

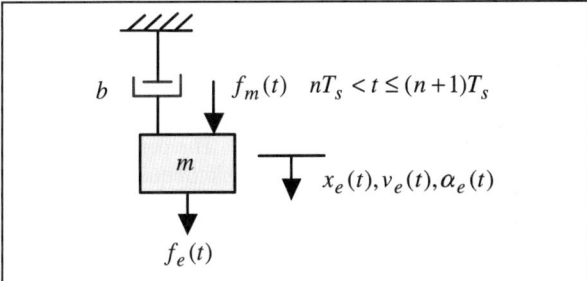

Figure 1: Mass-damper system subject to environment forces, $f_e(t)$, and motor forces, $f_m(t)$ under sampled-data control.

Consider, in contrast, the following discrete version of damping control:
$$F_m(t) = -B_v \left(x(nT_s) - x((n-1)T_s) \right)/T_s \, , nT_s < t \leq (n+1)T_s$$
This system was analyzed by Colgate and Schenkel [4], who proved that the controlled plant is only passive for values of B_v constrained by $|B_v| < b$.

What is particularly troublesome about the nonlinear sampling effects is that, it is hard to find analytic solutions for all but the simplest cases, and experimentally the waveforms that prove non-passivity can be difficult to find. Consider the example of Fig 1 using the proposed sampled-data control law with the following parameters:

$m = 1\,\text{kg}, \quad b = 100\,\text{N/m/s}, \quad B_v = 120\,\text{N/m/s}, \quad T_s = 0.1\,\text{s}$

Note that this choice of gain B_v exceeds the passivity bound identified by Colgate and Schenkel. The closed-loop poles of the system in the z-domain for the above-mentioned values are within the unit circle, and the plant is stable. Further, if we were to test this system with a shaker table and impedance head, a sweep of sinusoidal velocity excitations would produce the apparent impedance function (equivalent to a describing function [5]) shown in Fig 2. Both by analysis and computation, one would find there is no sinusoidal excitation that would extract net energy. According to linear passivity conditions, one would conclude that the system is passive.

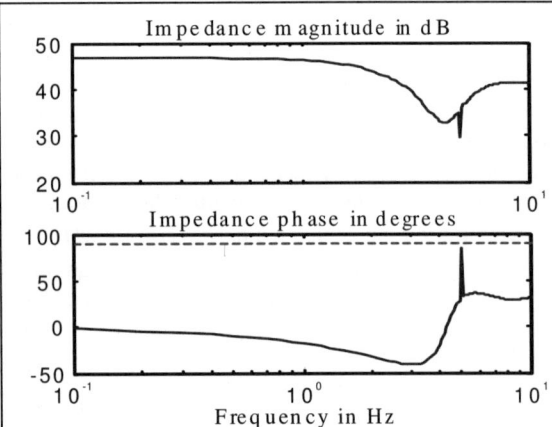

Figure 2: The impedance describing function of the plant of Fig 1 subject to sampled-data control. For each frequency, the environment port is driven to follow a pure sinusoidal velocity. The corresponding environment force contains multiple frequencies. The describing-function impedance is the magnitude and phase of the force response relative to the velocity magnitude and phase, considering only that component of the force response at the same frequency as the excitation velocity.

However, if the plant is excited by the (non-sinusoidal but periodic) velocity waveform shown in Fig 3, the system will respond with the control effort shown (updated at discrete times) and the periodic environment force waveform shown (comprised of many frequency components). The instantaneous power transfer (defined as positive *into* the system *from* the environment) is the product of environment force times velocity at the interaction port. This quantity is plotted along with the force and velocity profiles in Fig 3. The integral of the power over one period of excitation is the net energy delivered to (or extracted from) the system. We can define the average power as the energy per cycle divided by the period of excitation. Although the instantaneous power transfer in this example goes both positive and negative, the average power transfer in response to this excitation is negative (corresponding to net power extracted from the system). For this example, the instantaneous peak power is 45 W and the average rate of power extraction is 4.5 W. By repeating this excitation cyclically, arbitrary amounts of energy could be extracted from the system over time. By this counter-example, it is shown that the system can behave as an inexhaustible energy source, and thus the system is non-passive.

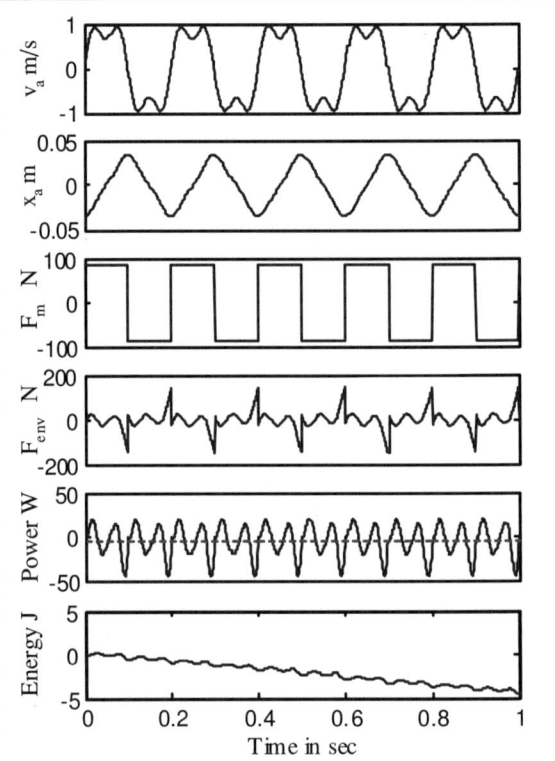

Figure 3: Response of the controlled system of Fig 1 to a periodic environment velocity profile for which net average power is extracted from the system. The example velocity waveform is the sum of two sinusoids.

This example illustrates the difficulty of evaluating passivity for computer-controlled systems. An exhaustive evaluation of responses to sinusoidal excitations is insufficient, and an exhaustive evaluation of all possible waveforms is infeasible. While Colgate and Schenkel were successful in deriving analytic passivity conditions for simple systems, extension of this approach to more complex systems seems intractable. We present an alternative method applicable to arbitrary LTI systems under sampled-data control, limited at present to consideration of a 1-DOF interaction port.

3. Passivity and periodic waveforms

We can show that, for our class of systems, if there exists a waveform of excitation that constitutes a counter-example of passivity, then there is also a *periodic* waveform that constitutes a counter-example. We will subsequently show that there must also be a counter-example from a restrictive class of periodic waveforms, which we will call *alias functions*. To prove this, we must rely on some specific properties of the class of systems considered. We assume the following:

1) The open-loop plant is a linear, time-invariant system with state function $\mathbf{x}(t)$ and state dynamics given by: $\dot{\mathbf{x}}(t) = \mathbf{A}\mathbf{x}(t) + \mathbf{b_1} f_e(t) + \mathbf{b_2} f_m(t)$, $f_e(t)$ is the force exerted by the environment on the system at the interaction port. The corresponding power variable is the velocity at the interaction port, $v_e(t)$.

2) The controller input is $f_m(t)$ which is assumed to be of the form encompassing generic sampled-data control systems based on system state and "U" delayed samples of the system state:

$$f_m(nT_s) = \sum_{u=0}^{U} \mathbf{K}_u \mathbf{x}((n-u)T_s)$$

where $\mathbf{K}_u = [K_{u1} \quad K_{u2} \quad \ldots\ldots \quad K_{uL}]$, $K_{ul} \in \Re$

3) It is assumed that it is possible to exert forces $f_e(t)$ to hold the environment port at the condition $x_e(t) = 0$ (i.e., it is possible to "clamp" the output).

4) It is assumed that the controller is stable in the sense that if $x_e(t)$ is forced to be held at 0 (by $f_e(t)$), then the system state will converge to an equilibrium state $\mathbf{x}_{ref}$.

5) It is assumed that there is a control policy that can be exerted by the environment force, $f_e(t)$, that can bring the closed-loop system from an arbitrary (bounded) initial condition to the equilibrium state $\mathbf{x}_{ref}$ in finite time with finite energy from the environment. Similarly, we assume a control policy $f_e(t)$ exists that can bring the system from $\mathbf{x}_{ref}$ to a reachable bounded state, $\mathbf{x}_1$, in finite time with finite energy.

If the plant is non-passive, then there exists some means of excitation that enables unlimited extraction of energy. Let us say that for such excitation, if it exists, the state at time t_1 is $\mathbf{x}_1$ and the state at time t_2 is $\mathbf{x}_2$. The energy required to achieve state $\mathbf{x}_1$ from $\mathbf{x}_{ref}$ is E_1. The energy required to bring the system from state $\mathbf{x}_2$ to $\mathbf{x}_{ref}$ is E_2. The energy transfer from time t_1 to time t_2 is E_{out} (E_{out}<0 for energy extracted from the system). Since the hypothesized excitation is able to extract unlimited energy, and since E_1 and E_2 are bounded, we can always choose some time t_2 such that $E_{out} < -|E_1 + E_2|$. We may then construct a waveform which: starts from rest at $\mathbf{x}_{ref}$; approaches state $\mathbf{x}_1$ in finite time (approach phase); follows the energy-extraction excitation waveform to bring the system state to $\mathbf{x}_2$ (energy extraction phase); and finally brings the system back to $\mathbf{x}_{ref}$ in finite time (return phase). In this sequence, net energy is extracted from the system, $E_{net} = E_{out} + E_1 + E_2 < 0$. Since the initial state is recovered, the sequence may be repeated cyclically to extract unlimited energy. Thus, if there exists *some* waveform of excitation that disproves passivity, then there also exists a *periodic* excitation that disproves passivity, as shown here by construction.

Note that it is also possible to hold the system at $\mathbf{x}_{ref}$ until a sampling instant and repeat the cycle at that time, thus assuring that the periodic waveform aligns with sample instants. Therefore, we can construct a periodic waveform as above with period $T_p = MT_s$, where T_s is the sample period of our controller and M is (for the convenience of notation) an even integer. For periodic waveforms of this type (starting and ending at $\mathbf{x}_{ref}$ at sample instants and resulting in net energy extraction each cycle), let us say the shortest such period (aligned with sampling instants) corresponds to $T_p = M_{min}T_s$. It is then also true that there is a periodic waveform with period $T_p = MT_s$, for all $M > M_{min}$, that extracts energy each cycle. To construct such alternative-period waveforms, it is sufficient to hold the example waveform at $\mathbf{x} = \mathbf{x}_{ref}$ for additional sample periods before repeating the pattern. Therefore, if there is some waveform that extracts energy from the system, then there is also a family of periodic waveforms with period MT_s, $M > M_{min}$, each of which extracts energy from the system.

4. Periodic excitations and alias sets

We have shown that the search for an excitation waveform that constitutes a counter-example of passivity can be restricted to periodic waveforms of period MT_s, where $M > M_{min}$. We can further reduce the search space by recognizing orthogonality of what we will define as *alias functions*. First, consider an arbitrary periodic waveform $v(t)$ of period ω_0, which we can describe in terms of a Fourier series:

$$v(t) = \sum_{k=0}^{\infty} A_k \cos(k\omega_0 t + \phi_k)$$

When $\omega_0 = 1/MT_s$, we may regroup the Fourier terms into $N+1$ groups (where M is an even integer and $N = M/2$), constituting $N+1$ *alias functions*, $g_i(t)$, with which we can alternatively describe our periodic function: $v(t) = \sum_{i=0}^{N} g_i(t)$

The ith alias function, $g_i(t)$, is a member of an *alias set*, defined in terms of the sample period, T_s, the number of samples per waveform period, M, and a harmonic number, i. The alias set $\mathfrak{A}_{i,M,T_s}(t)$ is comprised of all functions of the form:

$$\mathfrak{A}_{i,M,T_s}(t) = \sum_{j=-\infty}^{\infty} A_j \cos((jM+i)\omega_0 t + \phi_j)$$

Alias sets are defined with respect to the sample rate, T_s, of a controller. For a given system, this is a constant, and we will simply specify an alias set indexed by i and M, $\mathfrak{A}_{i,M}(t)$, implicitly assuming T_s.

Each term of the Fourier series belongs to one of the alias sets, 0 through i. The jth term of the ith alias function is identical to term $|jM+i|$ of the Fourier series (with a minus-sign correction on the phase for terms $j<0$). Thus, by construction, any Fourier series can be re-written uniquely as the sum of $N+1$ alias functions. The alias functions are so named because, when sampled at period T_s, the contributions of all terms of $\mathfrak{A}_{i,M}(t)$ for $j \neq 0$ are indistinguishable from the equivalent sampling of a sine wave at frequency $i\omega_0$. That is, all the higher frequencies within an alias function alias down to the lowest frequency of that function. This lowest frequency corresponds to a harmonic within the Nyquist range, of which there are N frequencies (plus 0).

The value to re-expressing a periodic function in terms of a sum of alias functions is that alias functions combine linearly with respect to average power (which is our test for passivity). We may choose to evaluate the power function with respect to velocity waveforms, $v(t)$. (Actually, the power function is a *functional*—a mapping from each function within a certain class onto a unique real number [6]). If we specify a periodic $v_e(t)$ (with periodicity aligned with sample instants) then there will follow a corresponding consistent periodic environment force, $f_e(t)$. Thus, we can define $P(v_e(t)) = \frac{1}{MT_s} \int_0^{MT_s} f_e(t) v_e(t) dt$ to be a mapping from the space of periodic functions $v_e(t)$ onto scalar values $P(v_e(t))$. For alias functions, the following properties are true:

(a) If $v_e(t)$ is a member of $\mathfrak{A}_{i,M}(t)$, then the consistent, steady-state $f_e(t)$ is also a member of $\mathfrak{A}_{i,M}(t)$.

(b) Given $v_e(t) = \sum_i v_i(t)$, where $v_i(t) \in \mathfrak{A}_{i,M}(t)$, then $P(v_e(t)) = \sum_i P(v_i(t))$.

(c) If there exists a periodic function $v(t)$ for which $P(v(t)) < 0$ (thus proving the system is non-passive), then there is also an alias function $v_i(t) \in \mathfrak{A}_{i,M}(t)$ for which $P(v_i(t)) < 0$.

With these properties of alias functions, we can confine our search for counter-example excitation waveforms to individual alias sets. This decomposition permits a numerical search procedure that approaches an exhaustive test for waveforms that can extract net energy from the system. If there are no alias functions that evaluate to $P(v_i(t)) < 0$, then the system is necessarily passive.

An outline of the proof of (a) is as follows. For $v_i(t) \in \mathfrak{A}_{i,M}(t)$, it can be shown that, in the steady state, the state vector of the linear system will be comprised of periodic functions that are also elements of $\mathfrak{A}_{i,M}(t)$. A state-based controller will sample this state at times nT_s and exert a sample-and-hold output for $f_m(t)$. The sample-and-hold function operating on a waveform from $\mathfrak{A}_{i,M}(t)$ will produce a controller waveform that is also an element of $\mathfrak{A}_{i,M}(t)$. For $v_i(t) \in \mathfrak{A}_{i,M}(t)$ and $f_m(t) \in \mathfrak{A}_{i,M}(t)$, it follows that the consistent $f_e(t) \in \mathfrak{A}_{i,M}(t)$. (Details of this derivation can be found in the appendix).

Property (b) is provable as follows. If $g_i(t) \in \mathfrak{A}_{i,M}(t)$ and $g_j(t) \in \mathfrak{A}_{j,M}(t)$, $i \neq j$, then integral $\int_0^{MT_s} g_i(t) g_j(t) dt = 0$. This follows from the fact that all frequencies in $g_i(t)$ are distinct from all frequencies in $g_j(t)$, and thus these functions are orthogonal with respect to integration over a common period. We have that $v_e(t) = \sum_i v_i(t)$, $v_i(t) \in \mathfrak{A}_{i,M}(t)$ and

$f_e(t) = \sum_i f_i(t)$, $f_i(t) \in \mathfrak{A}_{i,M}(t)$. Then, $\int_0^{MT_s} f_e(t) v_e(t) dt$

$= \int_0^{MT_s} \sum_i f_i(t) \sum_j v_j(t) dt = \sum_i \int_0^{MT_s} f_i(t) v_i(t) dt$. Thus

$P(v_e(t)) = \sum_i P(v_i(t))$, which proves assertion (b).

The proof of (c) follows. Assume there exists a periodic $v(t)$ with period MT_s and $P(v(t)) < 0$. We can re-write $v(t)$ as a sum of alias functions, $v(t) = \sum_i v_i(t)$, $v_i(t) \in \mathfrak{A}_{i,M}(t)$. Then $P(v(t)) = \sum_i P(v_i(t))$. To satisfy $P(v(t)) < 0$, it must be true that at least one of the alias functions $v_i(t)$ also satisfies $P(v_i(t)) < 0$.

Through this sequence of arguments, we have arrived at the following. For the nonlinear systems considered, if there is *any* waveform of environmental excitation that violates passivity, then there is also a periodic waveform $v_i(t) \in \mathfrak{A}_{i,M}(t)$ for which the average power function is negative, $P(v_i(t)) < 0$.

5. Application: a numerical search of alias sets

Through the definition of alias sets, we have reduced the search space for passivity counter-examples from arbitrary waveforms to more restrictive sets of waveforms (alias functions). However, there is still an infinity of waveforms to examine. In application, we approximate each candidate alias function as a truncated series of terms:

$$\mathfrak{A}_{i,M}(t) = \sum_{j=-\infty}^{\infty} A_j \cos\bigl((jM+i)\omega_0 t + \phi_j\bigr)$$

$$\approx \sum_{j=-V}^{V} A_j \cos\bigl((jM+i)\omega_0 t + \phi_j\bigr)$$

The set of approximate alias functions is spanned by the $2V+1$ coefficients, A_j, and the $2V+1$ phase shifts, ϕ_j.

If the interaction-port acceleration is bandwidth-limited, then the series expression for the velocity converges rapidly, and a truncated summation is a good approximation to a feasible waveform. For our example of Fig 3, the waveform we used to illustrate violation of passivity was an alias function consisting of only two terms.

In addition to the space of $4V+2$ parameters that describe our approximate set of alias functions within an alias set, there are $(M/2)+1$ alias sets to examine for each value of assumed periodicity, $\omega_0 = MT_s$. We do not know *a priori* what value to assume for M. However, we are assured that there is some index M_{min} such that our system is passive if and only if none of the alias sets $\mathfrak{A}_{i,M_{min}}(t)$ contains a counter-example. For all other choices of $M > M_{min}$, it is also true that the system is passive if and only if none of the alias sets in $\mathfrak{A}_{i,M_{min}}(t)$ contains a counter-example. Thus, if we pick M to be large enough, it is sufficient to examine the $(M/2)+1$ alias sets defined for period MT_s.

To execute our approximate numerical evaluation of passivity, we proceed as follows:

Initialization: Choose a precision of alias-function approximation (number of terms in the truncated summation). Choose "M" as large as practical (limited by computational capacity). Initialize harmonic number, i, to $i=0$.

Iteration:

- For harmonic i, pick a seed alias function from set $\mathfrak{A}_{i,M}(t)$ by assigning each frequency component an amplitude and phase; call this $v_e(t)$.
- Evaluate $f_e(t)$, the periodic environment force consistent with the controlled plant and $v_e(t)$.
- Evaluate $P_{avg}(v_e(t)) = \dfrac{1}{MT_s} \int_0^{MT_s} f_e(t) v_e(t) dt$ and

$$P_{rms}(v_e(t)) = \left[\frac{1}{MT_s} \int_0^{MT_s} (f_e(t) v_e(t))^2 dt \right]^{1/2}.$$

- Iterate on coefficients and phase shifts of the alias function to minimize $P_{avg}(v_e(t))/P_{rms}(v_e(t))$ within alias group $\mathfrak{A}_{i,M}(t)$. The waveform $v_e(t) = v_{ext}(t)$ that minimizes $P_{avg}(v_e(t))/P_{rms}(v_e(t))$ is the *extremal waveform* within alias set $\mathfrak{A}_{i,M}(t)$.
- If $P_{avg}(v_e(t))/P_{rms}(v_e(t)) < 0$, then stop; the plant is non-passive.
- Else, increment harmonic index, i; if $i = M/2$, then stop; the plant is passive (there are no counter examples in $\mathfrak{A}_{i,M}(t)$).
- Else, repeat for next harmonic number (repeat from "*Iteration*").

6. Example numerical evaluation of passivity

The above algorithm was invoked on the system of Fig 1. Per the analytic result of Colgate and Schenkel, the limit of the gain B_v for passivity is $-b < B_v < b$. Using our numerical search with truncated alias functions consisting of only three terms and $M=100$, the computed gain limit for passivity was $-b < B_v < 1.11b$. With 11 terms in the approximate alias functions and the same M, counter-examples of passivity were found when the following limits were violated $-b < B_v < 1.035b$. As the number of terms in the truncated series representation of alias functions increases, the numerical solution converges rapidly on the analytic solution.

As an extension, consider the same system, but with the actuator bandwidth limited. The additional dynamics is presumed to be a linear, first-order low-pass filter with a corner frequency of 0.2 Hz. There is no known analytic means to evaluate the control-gain bounds with respect to passivity for this simple extension. Using our numerical search, we were able to compute that the limiting gain B_v is approximately $-b < B_v < 10.87b$, which is considerably larger than the original example. For a first-order low-pass filter with a corner frequency of 0.5 Hz, the limiting B_v is approximately $-b < B_v < 5.34b$.

The numerical evaluation procedure is extensible to more complex systems. We are currently exploring extensions to explicit force control and impedance control algorithms.

7. Conclusions

In this paper we have developed a framework to analyze passivity of a class of nonlinear systems, consisting of linear systems under sampled-data control. It is shown that satisfying linear frequency-domain conditions is insufficient to prove passivity. We introduce an alternative method for evaluation of passivity of such systems. Specifically, it is shown that a system is non-passive if and only if there is a periodic excitation of a certain form (defined here as an *alias function*) that can extract energy from the system cyclically. By examining this restricted set of functions, we can evaluate if a system is passive or non-passive. The restricted search makes a numerical implementation feasible. Results of numerical evaluation are shown to converge to the known analytic solution for a simple case. A more complex example is also evaluated, illustrating extensibility of our method beyond the analytically solvable cases.

References:

1. J. E. Colgate, *The Control of Dynamically Interacting Systems*, Ph. D. thesis, Massachussettes Institute of Technology, Cambridge, MA, Aug. 1988.
2. M. Dohring, and W. Newman, "Admittance Enhancement in Force Feedback of Dynamic Systems", *Proc. of 2002 IEEE International Conference on Robotics and Automation*, Vol. 1, pp. 638-643, 2002.
3. R. J. Anderson, "Passive Computed Torque Algorithms for Robots", *Proc. of the 28th IEEE Conf. on Decision and Control*, Vol. 2, pp. 1638-1644, 1989.
4. J. E. Colgate, and G. Schenkel, "Passivity of a Class of Sampled-Data Systems: Application to Haptic Interfaces", *American Control Conference*, Vol. 3, pp. 3236-3240, 1994.
5. G. F. Franklin, J. D. Powell, and M. Workman, *Digital Control of Dynamic Systems*, 3rd edition, Addison-Wesley Publishing Company, p. 559, 1988.
6. D. E. Kirk, *Optimal Control Theory An Introduction*, Prentice Hall, p. 109, 1970.
7. B. C. Kuo, *Automatic Control Systems*, 6th edition, Prentice Hall, p. 102, 1991.

Acknowledgments

This work was supported by the National Science Foundation under NSF grant ECS 0200388. This support is gratefully acknowledged.

Appendix

Let $\mathbf{x}(t) = [x_1(t) \quad x_2(t) \quad \ldots \quad x_L(t)]^T$

$$f_m(nT_s) = \sum_{u=0}^{U} \mathbf{K}_u \mathbf{x}((n-u)T_s), \mathbf{K}_u = [K_{u1} \ldots K_{uL}]$$

If $x_l(t) \in \mathfrak{A}_{i,M}(t) \quad \forall l \leq L$, then $x_l((n-u)T_s)$

$$= \sum_{j=-\infty}^{\infty} A_j \cos(i\omega_0(n-u)T_s + \phi_j), \because M\omega_0 T_s = \omega_s T_s = 2\pi$$

$x_l((n-u)T_s) \in \mathfrak{A}_{i,M}(t)$ as seen from p.102 of [7]

Thus $f_m(nT_s) = \sum_{u=0}^{U} \mathbf{K}_u \mathbf{x}((n-u)T_s) \in \mathfrak{A}_{i,M}(t)$

$\dot{\mathbf{x}}(t) = \mathbf{A}\mathbf{x}(t) + \mathbf{b_1} f_e(t) + \mathbf{b_2} f_m(t)$

Since $(x_l(t) \in \mathbf{x}(t)) \in \mathfrak{A}_{i,M}(t)$, $(\dot{x}_l(t) \in \dot{\mathbf{x}}(t)) \in \mathfrak{A}_{i,M}(t)$, and $f_m(t) \in \mathfrak{A}_{i,M}(t)$, we can conclude that $f_e(t) \in \mathfrak{A}_{i,M}(t)$

For a stable system any $v_i(t) \in \mathfrak{A}_{i,M}(t)$, at the environment port results only in $f_m(t), f_e(t) \in \mathfrak{A}_{i,M}(t)$

$x_e = 0 \Rightarrow \mathbf{x} = \mathbf{x}_{ref}$, by stability

If for some $v_i(t) \in \mathfrak{A}_{i,M}(t)$, $\mathbf{x}(t) = \mathbf{x}_i(t) + \mathbf{x}_j(t)$, then,

$f_m(t) = g_i(t) + g_j(t)$ and $f_e(t) = h_i(t) + h_j(t)$

$$\Rightarrow \begin{cases} \dot{\mathbf{x}}_i(t) = \mathbf{A}\mathbf{x}_i(t) + \mathbf{b_1} h_i(t) + \mathbf{b_2} g_i(t) \\ \dot{\mathbf{x}}_j(t) = \mathbf{A}\mathbf{x}_j(t) + \mathbf{b_1} h_j(t) + \mathbf{b_2} g_j(t) \end{cases}$$

For $v_i(t) = 0$, the dynamics at the jth alias set still exists, which contradicts the assumption of stability. Thus if $v_i(t) \in \mathfrak{A}_{i,M}(t)$, then $f_m(t), f_e(t) \in \mathfrak{A}_{i,M}(t)$

Uncalibrated Visual Servoing Technique Using Large Residual

G. W. Kim	B. H. Lee	M. S. Kim
School of Electrical Engineering	School of Electrical Engineering	Human Robotics Research Center
Seoul National University	Seoul National University	Korea Institute of Science and
Seoul, Korea	Seoul, Korea	Technology(KIST), Seoul, Korea
E-mail : kgw0510@snu.ac.kr	E-mail : bhlee@asri.snu.ac.kr	E-mail : munsang@kist.re.kr

Abstract - This paper addresses a moving target tracking system with estimation of the image Jacobian without knowledge of camera configuration or robot kinematics. We propose an efficient algorithm to track a moving target using a numerical method and apply this algorithm to a robot system. The robot system is controlled using the nonlinear least squares optimization technique. The Full Newton's method and the secant approximation method are used to calculate joint angles. In this paper, large residuals of joint values are calculated using the secant approximation method. The image Jacobian is then estimated using the recursive least squares (RLS) algorithm. In addition, the velocity of the target influences the performance of the system because of the delay to process an image. Thus, we present a motion prediction algorithm for a moving target. The target position on the image plane is predicted using the autoregressive model (ARM). We then compare the performance of the small residual case with the performance of the large residual case using simulation. The experimental results show the improved performance using this prediction algorithm.

I. INTRODUCTION

Image-based visual servo control is compatible for dynamic environment because the system is little affected by the camera model or the kinematic model of a robot. It is necessary to calibrate the camera in order to calculate the image Jacobian. In industrial practice, it is difficult to calibrate the camera precisely. In order to minimize the effect of the system parameters, many researches for effective algorithms to estimate the image Jacobian has been carried out [5][6][9][12]. The method to estimate the image Jacobian using Broyden's method was originally developed by Hosoda [5]. In [6], [9], and [12], the authors demonstrated the model-independent target tracking using the extended methods of estimation of the image Jacobian.

Because image processing is a time-consuming work, there is a limitation in tracking a moving target and estimating the image Jacobian effectively. So the algorithm to predict a motion of a moving target has been focused mainly on improving the overall performance of the system [6][7][13][15].

In this paper, joint variables for robot tracking are induced as an iterative form by the Full Newton's method. The residual term is also calculated using the secant approximation method. The image Jacobian is estimated using the Broyden's method so as to reduce the burden of calibration process [6][9][12]. But image Jacobian estimated by its method can diverge or oscillate. The recursive least squares (RLS) algorithm, therefore, is applied to estimate the image Jacobian for enhancement of the system stability [9]. We also propose the algorithm to predict a motion of a moving target using the autoregressive model (ARM). And we analyze the performance of the zero (or small) residual case and the large residual case in simulation. Finally, experimental results show the improvement of the performance when the proposed algorithm is applied to a practical robot visual servoing problem.

II. IMAGED-BASED VISUAL SERVOING

A. Definition

According to the feedback input, Sanderson and Weiss [1] classified the visual servo system as image-based or position-based. In the image-based visual servo system, it is controlled by an image error defined by the difference between the reference feature and the target feature. Because this system does not estimate the 3D pose of the end-effector in the hand-eye system, it has low sensitivity for a camera calibration error. Therefore, the image error converges to zero if the closed-loop system is asymptotically stable under weak calibration.

In the image-based visual servo system, the change of feature parameters is connected to the change of parameters related to the robot end-effector by the image Jacobian.

B. Image Jacobian

The image Jacobian defines the relationship between the velocity of a robot end-effector and the change of an image feature. In the image-based visual servo control, control inputs are generated by an error defined by image features. The image Jacobian plays a role of a connection between the change of image feature parameters and the change of robot motion. Assume a robot with m-degrees of freedom. A point in the robot configuration space is represented by m-dimensional vector, $q = [q_1, \cdots q_m]^T$. Also a position of the end-effector in the task space is represented by p-dimensional vector, $r = [r_1, \cdots r_p]^T$. The relation between the change of robot joints ($\dot{q}$) and the change of the end-effector position in the robot task space ($\dot{r}$), is defined by the robot Jacobian, $J_R(q) \in R^{p \times m}$ as

$$\dot{r} = J_R(q)\dot{q} \tag{1}$$

The change of image feature parameters results from the change of robot position, thus the relation between them is needed. A point in the image feature space is represented by n-dimensional vector, $f = [f_1, \cdots f_n]^T$. The relation between the change of image features ($\dot{f}$) and the change of robot end-effector position ($\dot{r}$) is defined by the image Jacobian, $J(r) \in R^{n \times p}$ as

$$\dot{f} = J(r)\dot{r} \tag{2}$$

According to the combination of (1) and (2), it is possible to relate the change of image features to the change of joint angles directly. The composition of the robot and the image Jacobian produces a different form of the image Jacobian, $J_q(q) \in R^{n \times m}$ as

$$\dot{f} = J(r)\dot{r} = J(r)J_R(q)\dot{q} \equiv J_q(q)\dot{q} \tag{3}$$

III. UNCONSTRAINED OPTIMIZATION PROBLEM

A. Nonlinear Least Squares Optimization Problem

In the image plane, we can define a target position and an end-effector position as f^* and $f(q)$, respectively. Then the image error between two positions can be expressed as $e(q) = f(q) - f^*$. In order to minimize the error, the objective function that is a function of squared error is defined as

$$E(q) = \frac{1}{2}e^T(q)e(q) \tag{4}$$

This problem to find the optimal solution of the joint values to minimize the objective function can be defined as an optimization problem.

B. Full Newton's Method for Joint Control

If we assume the linear model in the neighborhood at a given joint vector q_k, the objective function can be approximated as [11][14]

$$\hat{E}(q) = E(q_k) + \nabla E(q_k)(q - q_k) + \cdots \tag{5}$$

The objective function, $E(q)$, is a convex function which is minimized at the robot configuration q^*. Assume that the objective function is second order differentiable for q. And the dimension of $e(q)$ is $m \times 1$. The first order differential term is shown as

$$\nabla E(q) = \sum_{i=1}^{m} e_i(q) \cdot \nabla e_i(q) = J(q)^T e(q) \tag{6}$$

And the second order differential term is shown as

$$\nabla^2 E(q) = \sum_{i=1}^{m} \left(\nabla e_i(q) \cdot \nabla e_i(q)^T + e_i(q) \cdot \nabla^2 e_i(q) \right) \tag{7}$$
$$= J(q)^T J(q) + R(q)$$

where $R(q) \equiv \sum_{i=1}^{m} e_i(q) \cdot \nabla^2 e_i(q)$

The objective function in (4) has the minimum value when the differential value of the objective function is equal to zero because the objective function is a convex function as mentioned above. The first order differential equation of $\hat{E}(q)$ is shown as

$$\frac{\partial \hat{E}(q)}{\partial q} = \nabla E(q_k) + \nabla^2 E(q_k)\Delta q \tag{8}$$

The joint value to minimize the objective function is shown in an iterative form as

$$q_{k+1} = q_k - (J(q_k)^T J(q_k) + R(q_k))^{-1} J(q_k)^T e(q_k) \tag{9}$$

C. Secant Approximation Method for Large Residual

If we assume small residual in (9), $R(q_k)$ can be ignored. If it has, however, a large value, the performance of the system will become worse. So we have to calculate a residual term for more precise model. When large residual is assumed, the residual, $R(q_k)$, is estimated using a secant approximation.

When the second order differential matrix called Hessian matrix is defined as $H(q) = \nabla^2 e(q)$, it is expressed by a generalized matrix, H_k, using secant approximation.

$$H_k(q_k - q_{k-1}) \approx \nabla e(q_k) - \nabla e(q_{k-1}) \quad (10)$$

Hessian matrix is able to be induced for an iterative form as it is mentioned above. H_k represents the kth step of Hessian matrix.

Therefore, if secant approximation is applied to residual, $R(q) \equiv \sum_{i=1}^{m} e_i(q) \cdot \nabla^2 e_i(q)$, it is shown as

$$\nabla^2 e_{k,i}(q_k) \approx R_{k,i} \quad (11)$$

$$R(q_k) \approx R_k = \sum_{i=1}^{m} e_{k,i}(q_k) \cdot R_{k,i} \quad (12)$$

Then each $R_{k,i}$ of above equations should be satisfied with (10).

$$R_{k,i}(q_k - q_{k-1}) = \nabla e_{k,i}(q_k) - \nabla e_{k-1,i}(q_{k-1}) \quad (13)$$

Multiply each side in (13) by $q_k - q_{k-1}$.

$$\begin{aligned}
R_k(q_k - q_{k-1}) &= \sum_{i=1}^{m} e_{k,i}(q_k) \cdot R_{k,i}(q_k - q_{k-1}) \\
&= \sum_{i=1}^{m} e_{k,i}(q_k) \cdot \left(\nabla e_{k,i}(q_k) - \nabla e_{k-1,i}(q_{k-1})\right) \\
&= (J(q_k) - J(q_{k-1}))^T e(q_k)
\end{aligned} \quad (14)$$

Using R_k given in (14), (9) can be rewritten as

$$\begin{aligned}
q_{k+1} &= q_k - \left(J(q_k)^T J(q_k) + R_k\right)^{-1} \cdot J(q_k)^T e(q_k) \\
R_k &= \frac{(J(q_k) - J(q_{k-1}))^T e(q_k)(q_k - q_{k-1})^T}{(q_k - q_{k-1})^T (q_k - q_{k-1})}
\end{aligned} \quad (15)$$

IV. ESTIMATE IMAGE JACOBIAN

A. Broyden's Method for Jacobian Estimation

The image Jacobian can be calculated through a process of calibration, but system parameters must be known exactly to calibrate. It is, however, nearly impossible to know precise parameters and sensitive to change system parameters under a dynamic environment. To overcome these disadvantages, many estimation algorithms have been proposed. We applied Broyden's method to estimate the image Jacobian [9][12].

In (15), if $\hat{J}(q_k)$ is defined as the kth image Jacobian to be estimated, it can be shown as

$$\frac{\partial e(q_k)}{\partial q} = \nabla e(q_k) = \hat{J}(q_k) \quad (16)$$

The Taylor series expansion of the error function in the neighborhood at q_k, $e(q) = f(q) - f^*$ is

$$e(q) = e(q_k) + \nabla e(q_k)(q - q_k) + \cdots \quad (17)$$

Approximate (17) by ignoring the high-order term and define the affine model as

$$M_k(q) = e(q_k) + \nabla e(q_k)(q - q_k) \quad (18)$$

The affine model is an approximated model of the error function so (18) is expressed as

$$(\hat{J}(q_k) - \hat{J}(q_{k-1}))\Delta q = \Delta e - \hat{J}(q_{k-1})\Delta q \quad (19)$$

In (19), the image Jacobian is estimated to minimize the norm of $\hat{J}(q_k) - \hat{J}(q_{k-1})$. It is minimized using the minimum norm solution [4]. That is

$$\hat{J}(q_k) = \hat{J}(q_{k-1}) + \frac{\left(\Delta e - \hat{J}(q_{k-1})\Delta q\right)\Delta q^T}{\Delta q^T \Delta q} \quad (20)$$

B. Recursive Least Squares Algorithm

In (20), the estimated image Jacobian can diverge and it has bad influence on the performance of the system. If we use the several past steps to estimate the image Jacobian, the performance of the system could be improved. For this reason, RLS algorithm is applied, and then the system can be controlled stably [1][9][12].

The cost function for a change of the affine model is

$$C(k) = \sum_{i=1}^{n} \lambda^{k-i} \|M_k(q_{i-1}) - M_{i-1}(q_{i-1})\|^2 \quad (21)$$

Calculate $\hat{J}(q_k)$ to minimize (21). That is

$$\begin{aligned}
\hat{J}(q_k) &= \hat{J}(q_{k-1}) + \frac{\left(\Delta e - \hat{J}(q_{k-1})\Delta q\right)\Delta q^T P_k}{\lambda + \Delta q^T P_k \Delta q} \\
P_k &= \frac{1}{\lambda}\left(P_{k-1} - \frac{P_{k-1}\Delta q \Delta q^T P_{k-1}}{\lambda + \Delta q^T P_{k-1} \Delta q}\right) \quad k = 1, 2, \ldots
\end{aligned} \quad (22)$$

The forgetting factor, $\lambda \in (0,1)$, is a rate of dependency for past data. This factor has influence on the performance of the system importantly.

V. TRACKING ALGORITHM

A. Definition of Autoregressive Model (ARM)

For a prediction of a moving target, the prediction model has to be defined. Using this model, we can predict the motion of a target at each image sample. Therefore, the position of a target can be defined in discrete time, and then the prediction model can be written as a form of a difference equation [8][10].

$$f(n) = \sum_{i=1}^{p} \phi_{p,i} f(n-i) + e(n) \quad (23)$$

$e(n)$ is a stationary white Gaussian noise and $\phi_{p,i}, i=1,\ldots,p$ is autoregressive parameters.

Assume a constant acceleration of a target. Then we can define the acceleration ARM for each direction which is x or y. Considering x direction, ARM is

$$\ddot{f}_x(n) = \alpha_{1,1} \ddot{f}_x(n-1) + e(n) \quad (24)$$

That is

$$f_x(n) = (2+\alpha_{1,1})f_x(n-1) + (-1-2\alpha_{1,1})f_x(n-2) \\ + \alpha_{1,1}f_x(n-3) + e(n) \quad (25)$$

Finally, the third-order ARM is obtained and the autoregressive parameters are

$$[\phi_{3,1} \ \phi_{3,2} \ \phi_{3,3}] = [2+\alpha_{1,1} \ -1-2\alpha_{1,1} \ \alpha_{1,1}] \quad (26)$$

B. Estimation of Autoregressive Parameters

In (25), we can predict a motion if the autoregressive parameters are known. These parameters are estimated using maximum likelihood estimator. Estimating $\alpha_{1,1}$ in (24) and simplifying

$$\hat{\alpha}_{1,1} = \frac{\sum_{i=4}^{N} \ddot{f}(i)\ddot{f}(i-1)}{\sum_{i=4}^{N} \ddot{f}^2(i-1)} \quad (27)$$

If $\hat{\alpha}_{1,1}$ is applied to (25) for each direction, a motion of a moving target could be predicted.

VI. SIMULATION

In this section, we show the improvement of the system performance of the large residual case. The simulation is performed under the same condition. The initial position of the robot end-effector is (250,50) and the motion of a target is $(130+100\sin(\omega t)+20\cos(\omega t), 100+50\cos(\omega t)+10\sin(\omega t))$ with the initial position at (150,150). The purpose of this simulation is to compare the performance of the large residual case with the performance of the zero residual case. Fig.1 represents respectively the pixel error and the trajectory for the zero residual case. And Fig.2 shows respectively the pixel error and the trajectory for the large residual case.

The results of this simulation are shown in Table I. We can easily compare the large residual case with the zero residual case. The plot of image errors in Table I is shown in Fig.3. Using large residual, we can reduce the error about 17%. This result shows that the model of the objective function using large residual is more accurate than the model using zero or small residual.

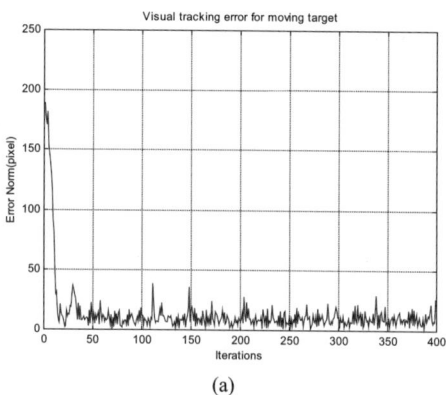

(a)

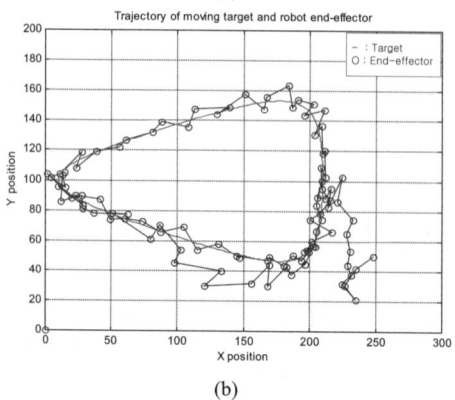

(b)

Fig.1 (a) Pixel error for zero-residual at $\omega = 2\pi \times 0.25$ rad/s (b) Trajectory of target and end-effector for zero residual

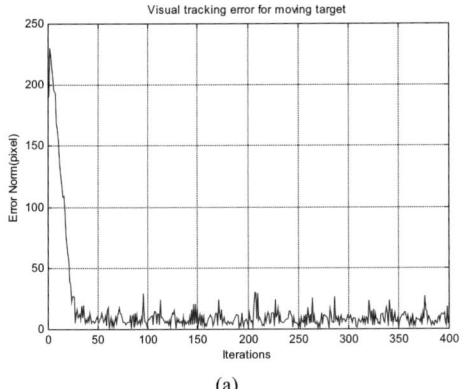

(a)

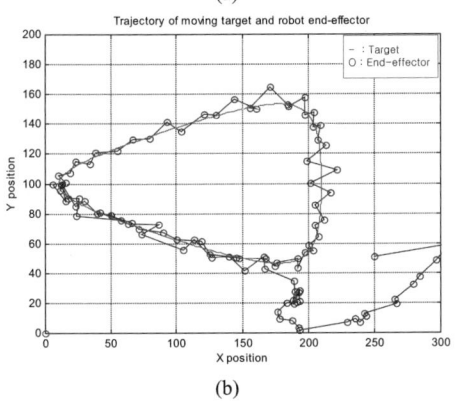

(b)

Fig.2 (a) Pixel error for large residual at $\omega = 2\pi \times 0.25$ rad/s (b) Trajectory of target and end-effector for large residual

TABLE I
AVERAGE ERROR NORM AT THE STEADY-STATE CONDITION[1]

Vel. of target (ω)	Zero residual	Large residual
$2\pi \times 0.1$	5.8699	4.2305
$2\pi \times 0.15$	7.0957	5.8786
$2\pi \times 0.2$	8.7039	7.0510
$2\pi \times 0.3$	11.4139	10.0392
$2\pi \times 0.4$	14.3233	12.9719
$2\pi \times 0.5$	21.4134	18.2784

The settling time of the zero residual case is faster than that of the large residual case shown in Fig.4. The reason is that the model of the objective function changes slowly in large residual.

VII. EXPERIMENTS

In these experiments, we use the 3DOF SNU-ERC DD robot and CCD camera. The Meteor II frame grabber is

---

[1] System state after settling time that is the first time when error norm is below 5

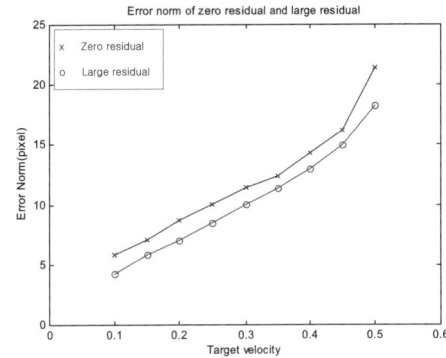

Fig.3 Results of error norm of zero residual case and large residual case

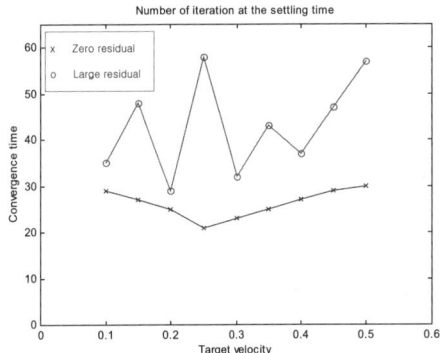

Fig.4 Settling time of zero residual case and large residual case

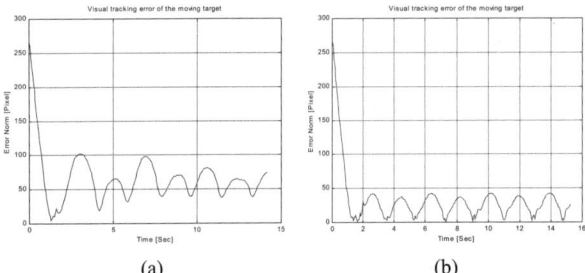

Fig.5 Pixel error at $\omega = 2\pi \times 0.2$ rad/s (a) the case not applied the motion prediction algorithm (b) the case applied the motion prediction algorithm

used to acquire the image. The resolution of the camera is 640x480 and 1 pixel ≈ 0.5*mm*. The frame grabber is installed in a Pentium II 800MHz PC. And the sampling period is 80*ms*. We analyze the performance when the tracking algorithm is applied or not. The trajectory of a motion of a target is $(320 + 120\sin(\omega t), 240 + 80\cos(\omega t))$. Fig.5 shows respectively the performance of the system applied the prediction algorithm and the performance of the system not applied the prediction algorithm. In Fig.6, the overall performance of two cases is shown.

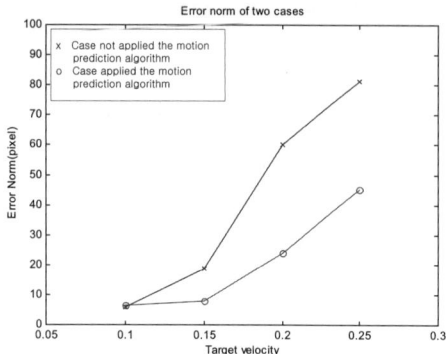

Fig.6 Results of error norm of the case applied ARM or not

The average pixel errors in Fig.5 are 60.2348 and 24.0851 respectively. And then the performance of the case applied the prediction algorithm (ARM) is better. Using this algorithm, we can reduce the effect of the time delay occurred by the image processing.

VIII. CONCLUSIONS

In this paper, the algorithm to estimate the image Jacobian without calibration is proposed.

Nonlinear least squares optimization problem is defined for the calibration-free visual servo system, and then the control system is proposed to minimize an image error. The recursive least squares (RLS) algorithm is used for estimating the image Jacobian in order to improve the stability of the system. We assume large residual for the model of the objective function.

In addition, autoregressive model (ARM) is defined so as to reduce an error raised by time delay produced by an image processing.

Finally, the simulation results show that the performance of large residual is better than that of zero or small residual. We also verify the improvement of the performance when the prediction algorithm is applied in experiments.

IX. ACKNOWLEDGMENTS

This work was supported in part by the HWRS-ERC at KAIST, the ASRI, and the BK21 Information Technology at Seoul National University.

X. REFERENCES

[1] D.Hager and P.I.Corke, "A Tutorial on Visual Servo Control," *IEEE Transactions on Robotics and Automation*, Vol. 12, No. 5, pp651-670, 1996

[2] A.C Sanderson, L.E.Weiss, and C.P.Neuman, "Dynamic sensor-based control of robots with visual feedback," *IEEE Transactions on Robotics and Automation*, vol. RA-3, pp404-417, Oct. 1987

[3] W.L.Brogan, *Modern Control Theory*, Prentice Hall, Englewood Cliffs, N.J., 1985

[4] H.Sutanto, R.Sharma, and V.Varma, "Image based Autodocking without Calibration," *Proceedings of the 1997 IEEE International Conference on Robotics and Automation*, pp974-979, 1997

[5] K.Hosoda and M.Asada, "Versatile Visual Servoing without Knowledge of True Jacobian," *IEEE/RSJ/GI International Conference on Intelligent Robots and Systems*, pp186-193, 1994

[6] J.A.Piepmeier, G.V.McMurray, and H.Lipkin, "Tracking a Moving Target with Model Independent Visual Servoing : a Predictive Estimation Approach," *Proceedings of the 1998 IEEE International Conference on Robotics & Automation*, pp2652-2657, 1998

[7] M.Asada, T.Tanaka, and K.Hosoda, "Adaptive Binocular Visual Servoing for Independently Moving Target Tracking," *Proceedings of the 2000 IEEE International Conference on Robotics and Automation*, pp2076-2081, 1997

[8] K.Shanmugan and A.Breipohl, *Random Signals : Detection, Estimation, and Data Analysis*, New York : Wiley, 1988

[9] J.A.Piepmeier, G.V.McMurray, and H.Lipkin, "A Dynamic Jacobian Estimation Method for Uncalibrated Visual Servoing," *Proceedings of the 1999 IEEE/ASME International Conference on Advanced Intelligent Mechatronics*, pp944-949, 1999

[10] A.Elnagar and K.Gupta, "Motion Prediction of Moving Objects Based on Autoregressive Model," *IEEE Transactions on Systems, Man, and Cybernetics*, Vol. 28, No. 6, pp803-810, 1998

[11] M.Avriel, *Nonlinear Programming-Analysis and Methods*, Prentice-Hall, 1976

[12] J.A.Piepmeier, G.V.McMurray, and H.Lipkin, "A Dynamic Quasi-Newton Method for Uncalibrated Visual Servoing," *Proceedings of the 1999 IEEE International Conference on Robotics and Automation*, pp1595-1600, 1999

[13] R.Kelly, F.Reyes, J.Moreno, and S.Hutchinson, "A Two Loops Direct Visual Control of Direct-Drive Planar Robots with Moving Target," *Proceedings of the 1999 IEEE International Conference on Robotics & Automation*, pp599-604, 1999

[14] P.G.Ciarlet and J.L.Lions, *Handbook of Numerical Analysis Vol. I*, North-Holland, New York, 1990

[15] S.Higashi, S.Komada, M.Ishida and T.Hori, "Obstacle Avoidance of Redundant Manipulators on Visual Servo System Using Estimated Image Features," *Proceedings of the 1998 5th International Workshop on Advanced Motion Control*, pp165-170, 1998

STOMP: A Software Architecture for the Design and Simulation of UAV-based Sensor Networks

Erik D. Jones[†]
edjones@ucdavis.edu

Randy S. Roberts[‡]
roberts38@llnl.gov

T. C. Steve Hsia[†]
hsia@ece.ucdavis.edu

[†]Dept. of Electrical & Computer Engineering, U.C. Davis, Davis, CA 95616 USA
[‡]Lawrence Livermore National Laboratory, Livermore CA 94550 USA

Abstract— This paper presents the Simulation, Tactical Operations and Mission Planning (STOMP) software architecture and framework for simulating, controlling and communicating with unmanned air vehicles (UAVs) servicing large distributed sensor networks. STOMP provides hardware-in-the-loop capability enabling real UAVs and sensors to feedback state information, route data and receive command and control requests while interacting with other real or virtual objects thereby enhancing support for simulation of dynamic and complex events.

I. INTRODUCTION

The use of unmanned air vehicles (UAVs) in sensing applications is becoming increasingly important. As the cost of UAVs decrease, it becomes practicable to use several UAVs to service large distributed networks to maintain robust communications, and decrease latency. As this trend continues it is critical to have tools that allow designers to simulate, control and prototype networks that use multiple UAVs. Such tools would allow designers to rapidly test and develop complex behaviors, conduct trade-off studies, and test new algorithms and hardware. To address these design and simulation issues, the Simulation, Tactical Operations and Mission Planning (STOMP) architecture was developed. STOMP is an application and framework designed to study and operate sensor networks where UAVs are fundamental to the collection of data.

As in [1] and [2], STOMP uses a visual editor to allow designers to assemble and configure each simulation. Designers may specify the state of every object in the system individually, in groups or globally and, using the graphical event configuration system, define events to trigger state changes at specified times in order to test complex behaviors and response to exceptions.

A simulated environment, however, is not sufficient to completely test and model all aspects of UAV control, response and cooperation[3]. It is simply not possible to simulate all of the dynamics of a full UAV and distributed network system. However, through a communication subsystem, STOMP connects the virtual environment to real-world hardware systems. Command, control and state information is exchanged with real UAVs and sensors, creating a hardware-in-the-loop feedback to both test new algorithms within the virtual simulation as well as hardware and software implementation in real UAVs and sensors.

Where previous hardware-in-the-loop (HIL) simulators have focused on modelling the behavior of a single entity in a dynamic environment, such as in [2] and [4], STOMP is designed to simulate and control large numbers of objects. Unlike [2], STOMP is implemented in C++ rather than Matlab, which has allowed STOMP to support larger networks with many more objects. Several cooperation and adaptation algorithms were modelled using Matlab[5] and even in a less demanding non-HIL mode, Matlab was unable to adequately perform when many objects existed in the model.

The framework for STOMP is designed around an object oriented architecture thereby allowing designers to adapt behaviors and algorithms of existing objects or assemble new objects rapidly and easily. This flexibility has enabled rapid prototyping and testing of new cooperation and path planning algorithms for large UAV networks such as those presented in [5].

In section II, a high-level functional overview of STOMP is presented. Predefined UAV and sensor object behaviors and features are described. The communication, event and display controllers are also covered. In section III, the software framework, class structure, key algorithms and data structures are described. Future expansion, development and use of STOMP is described in section IV.

II. FUNCTIONAL ARCHITECTURE

STOMP is designed to simulate UAVs, sensors and their interactions in a distributed sensor network under a variety of conditions. It is designed to easily implement control and cooperation architectures, and displays the reaction of the architecture to events that the designer can script into the simulation through a graphical interface. Through an internal communication controller, STOMP can feedback information from real UAVs and sensors using wireless Ethernet for data acquisition from sensors and a wireless serial interface for command, control and state information thereby providing hardware-in-the-loop simulations.

A functional block diagram of STOMP is illustrated in Figure 1. As shown in the figure, STOMP consists of the following main blocks: 1) Sensor and UAV objects (where sensor objects are depicted as circles, and UAV objects are depicted as hexagons); 2) a Communication Controller; 3) an Event Controller; and 4) a Display Controller. Details of the functions of these blocks are given in the proceeding subsections.

TABLE II

Static Properties	UAV	Sensor
IP Address	●	●
Battery Capacity	●	●
Battery Drain Rate	●	●
Memory Capacity	●	●
Fuel Capacity	●	–
Fuel Drain Rate	●	–
Mechanical Characteristics	●	–

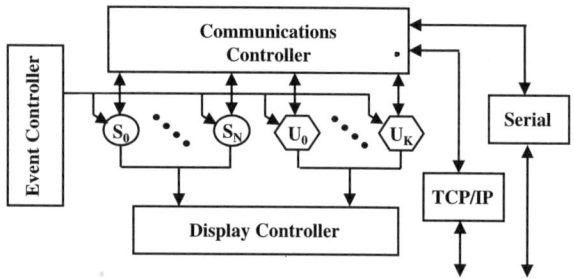

Fig. 1. Functional Block Diagram of STOMP

The UAV and sensor objects contain state information and algorithms relevant to the simulation and operation of UAVs and sensors in the network as given in Table I. Waypoints are the most fundamental building block of all STOMP simulation objects. The path planning controller and flight controllers use waypoints as means to control and calculate heading, position and plan routes. Waypoints contain only a single state variable, position. Position is stored in terms of decimal degrees of longitude and latitude. Depending on the configuration of each object,

TABLE I

State	UAV	Sensor	Waypoint
Position	●	●	●
Altitude	●	●	–
Heading	●	●	–
Speed	●	●	–
Battery Life	●	●	–
Memory Usage	●	●	–
Status	●	●	–
Fuel	●	–	–
Vertical Velocity	●	–	–

state updates may be obtained from the event controller (purely virtual), communication controller (purely real) or a combination (partially virtual). This feedback mechanism increases the richness of the dynamics that STOMP can simulate through the real and virtual interaction. Table II lists the static properties within the UAV and sensor objects. Static properties are set by the designer prior to running a simulation and define rates and characteristics used for communication and update of state information for pure or partially virtual UAVs and sensors.

A. UAV Objects

UAV objects are equipped with a flight controller, communication controller and path planning controller as shown in Figure 2. If an external hardware device is not connected to that particular UAV, the flight controller simulates the dynamics of flight, providing position updates, changes in heading, ground speed, vertical velocity, fuel status and drain rate, and battery life and drain rate. If the flight controller is to be connected to a real UAV, the designer will set the appropriate flag when setting the initial states and parameters for the UAV using a graphical interface. This UAV is then connected via a wireless serial interface to a MP2000 autopilot engineered by Micropilot[6]. Using a communication wrapper, the MP2000 receives waypoint information from STOMP in order to control the direction of flight to service either virtual or real sensors. STOMP polls the MP2000 in order to update state information within the simulation asynchronously while the simulation is stepping through time.

The path planning controller implements the cooperation architecture described in [5] and path planning algorithm described in [7]. The path planning controller receives necessary information about the state of the system from other UAVs and sensors via the communication controller. This controller provides waypoint information to the flight controller in order to direct the UAV to the next sensor or waypoint.

B. Sensor Objects

Sensors contain their specific data acquisition equipment and a communication controller. Although STOMP has been designed to send and receive any kind of data, the display controller was designed to display and allow the user to access image data. The wireless Ethernet network in real sensors and UAVs is managed using Mobile Mesh[8] routing software and uses TCP sockets. STOMP operates at the application layer, while Mobile Mesh operates at layer 3 of the OSI model, thus STOMP

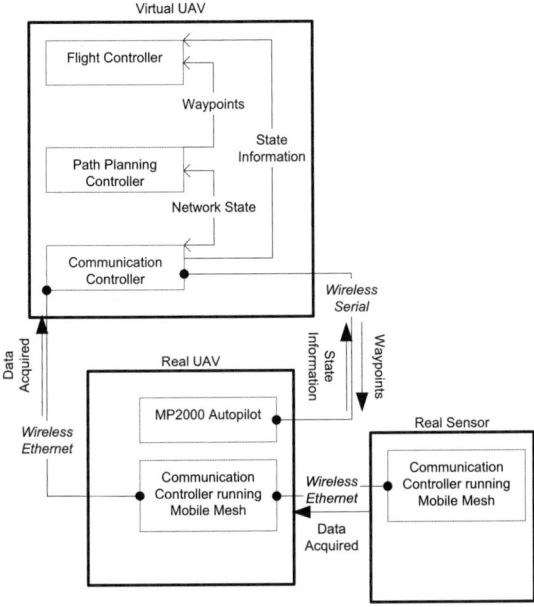

Fig. 2. Virtual UAV and Real UAV Communication

remains independent of the specific communication and routing methods used in the real world environment.

Each sensor also contains a communication controller that implements a subset of the features included in STOMP. Since sensors only need to unload their data to UAVs and subsequently into STOMP, the data transfer protocol and ICMP functionality are the only features that are implemented in both simulated and real sensor elements.

C. Communication Controller

The communication controller is central to STOMP. It coordinates all communication between the main event controller, and real and virtual UAVs and sensors. Since this controller is the gateway to external hardware, it also routes command and control requests to the MP2000 autopilot via a wireless serial interface. The MP2000 is polled periodically and state changes are updated for the associated UAV object. The state information is then visualized within the simulator, providing the operator a graphical overview of the state of the system. As the simulation progresses, the communication controller will upload new waypoints to the MP2000 as determined by the path planning controller contained within the partially virtual UAV as shown in Figure 2.

As either virtual or real UAVs fly close to their target sensor node, the communication controller will attempt to contact the sensor using ICMP over a wireless Ethernet interface. If initial contact can be established, the UAV's communication controller will attempt to communicate with the sensor's communication controller in order to exchange data. The proprietary protocol uses a connected TCP socket to send and receive the appropriate data. The protocol can resume broken file transfers and provides full error checking in order to guarantee rapid and reliable transmission over possibly unreliable connections. If communication is lost, the communication controller attempts to contact the sensor again using ICMP and resume the file transfer process. If there was no new data available at the sensor or the data was successfully transferred, the communication controller notifies the path planning object and the UAV is sent to the next waypoint.

D. Event Controller

The event controller initializes the simulation, and provides several facilities for scripting simulation scenarios using a graphical interface. Scripting scenarios allows designers to study the reaction of the cooperation and path planning algorithms to various events. An event consists of a change in state (position, velocity, heading, etc.), disabling or enabling a sensor or UAV and the time at which the event occurs. STOMP tracks time in terms of steps which do not necessarily correlate with physical times. Events that occur externally are received by the communication controller and processed asynchronously to the event controller.

The event controller also provides facilities for post-simulation analysis of network communications. As the simulation progresses, all state information of every element in the simulation (sensors and UAVs) is recorded, along with line-of-sight data computed from the Digital Terrain Elevation Data (DTED) data. This state information can be exported to a file and is suitable for postprocessing by other analysis tools such as Matlab, OMNet++[9] or OPNET^{TM}. In this manner, the packet-level behavior of the network can be studied.

E. Display Controller

STOMP is divided into two different display modes, the designer view and the simulator view. When a new simulation is being created, the designer view is used to place objects within the coordinate space of the DTED space represented as a color coded topographical relief map. Labels may be added at various positions, enabling the identification of useful landmarks on the map. The user may either input the coordinates directly or use the mouse and graphical interface to place objects. While in the designer view, object properties and initial states may also be set.

When the simulation is started, the simulator view appears in front of the designer view. In this mode, display controller provides the designer with visual feedback of the state of the network as it progresses. It also displays data collected by operational UAVs from deployed sensors. The positions of the UAVs and sensors, along with the current UAV paths, are displayed on top of shaded terrain maps loaded from the DTED. When new data is received from a sensor, the sensor icon changes color, and the image is displayed by clicking on the indicator.

III. SOFTWARE ARCHITECTURE

STOMP was written in C++ using a highly object oriented design methodology. The modularity of the design allows for new objects to be designed and integrated into the framework rapidly. Objects are derived from base classes and provide updated functionality in order to support object-specific features. An overview of the class hierarchy is shown in Figure 3. Programmers may use inheritance to create more feature rich objects tailored to a particular application. All STOMP objects provide an initialization function and a Serialize() function for saving and restoring their state on disk.

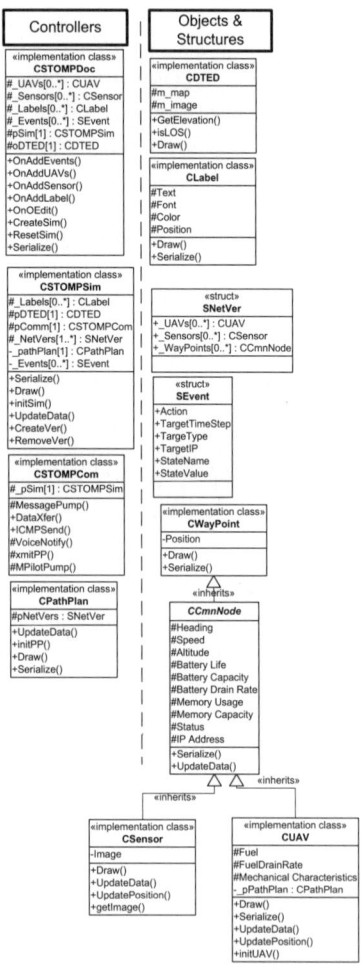

Fig. 3. UML Diagram of STOMP Framework

A. CSTOMPDoc - Scenario Class

The CSTOMPDoc class contains all necessary information to start a simulation. UAVs, sensors, labels, waypoints, events and display parameters are stored in this class. CSTOMPDoc provides the mechanism for storing and retrieving state information for the scenario at time index 0. When a simulation is created, objects are copied out of this class into the simulation/event controller class for processing beyond time index 0.

When a simulation is ready for execution, CSTOMPDoc creates a new thread, opens a new view window and initializes the simulation and event controller. Control is then passed to the new view allowing the designer to begin testing the scenario.

B. CSTOMPSim - Event and Simulation Class

CSTOMPSim is the main event and simulation controller for the STOMP framework. After a scenario is designed, all of the object information defined by the user is loaded into this class. CSTOMPSim then initializes all of the UAV and simulation objects, preparing them for synchronous state updates using the simulation clock which is controlled through the graphical interface.

CreateVer() is used to create the initial snapshot of the network for processing by the path planning controller in the leader UAV. As exceptions occur to the network, CreateVer() is used to create a new snapshot of the network for the leader UAV before transmitting its changes to the remaining UAVs that are controlled by CSTOMPSim. When the no more UAV objects are connected to a version, RemoveVer() is called to conserve memory and remove the stale version from the linked list.

CSTOMPSim creates a thread for the communication controller so that asynchronous updates to state information received from the MP2000 connected UAV or data received from external sensors can be processed independently of the simulation clock. This allows for the seamless integration of real and virtual devices without concerning the designer with timing issues. While a simulation can progress synchronously, a real UAV or sensor need not wait for STOMP's simulation clock pulse to continue with its next operation.

C. CSTOMPCom - Communication Controller Class

CSTOMPCom contains all of the logic to process communication requests received from the event controller, state updates or command and control to the MP2000 autopilot, and data transfer between UAVs and sensors, both virtual and real. The communication controller runs in an independent thread created by CSTOMPSim and can receive messages sent to its thread process, interrupts received from the MP2000 connected via a wireless serial interface or interrupts received via a TCP connected socket.

When an event occurs that requires the communication controller to take a particular action, such as to transmit new path planning information to all UAVs in the network, a message is sent to the process. Messages are processed in the MessagePump() function and action is taken to process the message and notify the event controller or target objects depending on the type of request received. CSTOMPSim, the UAV class (CUAV), and the sensor class (CSensor) provide mechanisms to lock internal data structures as needed, such as when processing asynchronous state updates received from external hardware. Data structure locking is the mechanism that prevents the asynchronous updates to interfere with the clocked events within CSTOMPSim.

Several functions within CSTOMPCom are called to take a particular action once a message has been properly decoded by MessagePump(). The DataXfer() function is called by a UAV to receive image data from real sensors after initial contact is verified using ICMPSend(). If new data has been received, CSTOMPCom will call VoiceNotify() to notify the operator using a synthesized voice that new data was received and identifies the sensor it was received from. If the leader UAV wishes to rebroadcast a new path plan, the UAV object will send a message which will trigger a call to xmitPP(). This function sends out new path plan and waypoint data to all UAVs in the network.

D. CUAV, CSensor - Simulated Objects

Simulated objects such as the UAV and sensors are contained in the classes CUAV and CSensor respectively. Each is derived from a common base class that contains behavior and information related to both objects and necessary for integration into STOMP. For example, every STOMP object displayed on the screen must supply a Draw() function and similarly, every object that is clocked within the simulator must provide an UpdateData() function. The interfaces for these functions is provided in the common base class. Simulated objects must also provide a Serialize() function for storing and retrieving its state information.

When the simulation clock triggers a call to UpdateData(), virtual UAVs and sensors update various state variables based upon static properties and rates set by the designer and dynamic models programmed into the function. UpdateData() calls UpdatePosition() which calculates the new position of the object based on its heading and speed. If it is a UAV object, the UpdatePosition() function checks to see if the UAV has arrived at a sensor and calls appropriate communication routines in CSTOMPCom to contact the sensor. Once CSTOMPCom has notified the UAV that its ready to proceed (after either a successful or unsuccessful attempt to communicate with the sensor), the UAV calculates a new heading to the next waypoint and proceeds.

Periodically, CSTOMPCom will call MPilotPump() to poll the MP2000 for new state information or to send new waypoint information if a MP2000 is associated with a UAV in the simulation. The virtual UAV that is connected to the MP2000 may trigger a call to MPilotPump() by sending the appropriate message to the CSTOMPCom thread if, for example, the network has been reorganized or the UAV needs to fly to another waypoint for some other reason.

E. CPathPlan - Path Planning Class

CSTOMPSim and every UAV object contains its own independent CPathPlan object. CPathPlan is the class that contains all of the cooperation and global path planning logic. During initialization, the path planning object is given a snap shot of the state of the network, including all UAV and sensor states. As needed, the path planning object will signal CSTOMPCom to transmit or retrieve information from real or virtual UAVs and sensors.

When exceptions in the network occur, such as the loss of a UAV or sensor, the communication object is first notified. It then creates a new version or snapshot of the network in CSTOMPSim. The leader[5] UAV is notified and given a pointer to the new state of the network. This new version is used by the leader to perform path planning and broadcast new commands and information to other UAVs. Like all STOMP simulation objects, main processing occurs in the clocked UpdateData() function.

IV. CONCLUSION AND FUTURE WORK

In this paper we have presented an overview of the STOMP application and framework. As UAVs become less expensive to build and implement, these types of devices become easier to deploy in order to service large distributed networks. In order to increase robustness or to decrease latency, cooperation between UAVs becomes necessary. STOMP is a feature rich environment through which a designer may test new algorithms involving cooperation, communication, command or control of these networks. The graphical interface provides a mechanism to quickly setup purely virtual, partially virtual or purely real environments quickly and easily. When configured for purely real objects, STOMP essentially acts as a ground station, providing visual feedback to the operator of the state of the network and access to remote data.

Since it is impossible at this time to model all of the dynamics of a real system in a lab environment, the STOMP framework provides the ability to interact with external devices in real world deployments. The feedback from external UAVs and sensors has provided an extremely useful mechanism to more accurately test the behavior of large networks with very little hardware.

Thus far STOMP has been used as a command and control unit for the real UAVs by providing waypoint information to the MP2000 autopilot and processing the cooperative behaviors within the simulator (partially virtual). In the next phase of our research, the cooperative behaviors that have been modelled thus far in STOMP as described in [5] and [7] will be implemented in real UAVs. STOMP will be used to fully test the behavior of real UAVs and sensors by providing simulated communication and behavior making the UAV and sensors believe they are communicating with a much larger network.

Advancing the modularity and increasing the scripting features available to STOMP designers has potential benefits for a wide range of applications beyond simple UAV/sensor communication networks. In many areas of mobile robotics where large scale deployments or testing is necessary, STOMP can reduce the time and cost of designing these networks, testing complex behaviors and implementing those features through command and control or autonomy by providing a flexible framework that can interact with real devices and simulate larger environments.

V. Acknowledgements

This work was performed under the auspices of the U. S. Department of Energy by the University of California, Lawrence Livermore National Laboratory under Contract No. W-7405-Eng-48 and under IUT-B522313 from the Laboratory Directed Research and Development program. The authors would like to thank Dave McCallen, Director of the Center for Complex and Distributed Systems, for his support of the project. The authors would also like to acknowledge the support provided by Alan Smith, Chad Goerzen, Matt Myrick, and Noah Reddell.

VI. REFERENCES

[1] D. MacKenzie, R. Arkin, and J. Cameron, "Multiagent mission specification and execution," *Autonomous Robots*, pp. 29–52, 1997.

[2] J. Lemaire and F. Devie, "A flexible hardware in the loop simulator for a long range autonomous underwater vehicle," *IEEE Oceanic Engineering Society Conference Proceedings*, vol. 3, pp. 1359–1363, 1998.

[3] H. Hagras, V. Callaghan, and M. Colley, "Outdoor mobile robot learning and adaptation," *IEEE Robotics and Automation Magazine*, vol. 8, pp. 53–69, September 2001.

[4] D. Carrijo, A. Oliva, and W. de Castro Leite Filho, "Hardware-in-loop simulation development," *International Journal of Modelling & Simulation*, vol. 22, pp. 167–175, 2002.

[5] C. Kent and R. Roberts, "Cooperation and path planning for unmanned air vehicles," *Lawrence Livermore National Laboratory. Technical Report UCRL-JC-149915*, September 2002.

[6] Micropilot Corporation, *MP2000 Autopilot*. [ONLINE] Available: http://www.micropilot.com.

[7] C. Cunningham and R. Roberts, "An adaptive path planning algorithm for cooperating unmanned air vehicles," *Proceedings of the 2001 IEEE International Conference on Robotics and Automation*, pp. 3981–3986, 2001.

[8] MITRE Research Group, *Providing Solutions For Mobile Adhoc Networking*. [ONLINE] Available: http://www.mitre.org/tech_transfer/mobilemesh.

[9] A. Varga, *OMNet++ Discrete Event Simulation Program*. [ONLINE] Available: http://www.hit.bme.hu/phd/vargaa/omnetpp.

Uncalibrated Robotic 3-D Hand-Eye Coordination Based on the Extended State Observer

Hongyu Ma   Jianbo Su
(mhy1@sina.com, jbsu@sjtu.edu.cn)
(Institute of Automation & Research Center of Intelligent Robotics,
Shanghai Jiao Tong University, Shanghai, 200030, P.R. China)

Abstracts - This paper studies a novel method to deal with the unknown image Jacobian matrix model for the uncalibrated robotic hand-eye coordination. An extended state observer is designed for online estimation of the image Jacobian matrix that is regarded as the system's unmodeled dynamics. A nonlinear feedback control law is then adopted for the coordination controller. With this scheme, a universal calibration-free controller is developed, which is independent from specific tasks and system configuration, for uncalibrated hand-eye coordination. Simulations and experiments of 3-D robotic visual positioning and tracking tasks show the validity and feasibility of the proposed control scheme.

I. Introduction

The Image Jacobian Matrix model has been proved to be an effective tool to approach the robotic visual servoing problem theoretically and practically. It directly bridges the visual sensing and the robot motion with linear relations [2], without knowing the calibration model [1] of the visual sensor(s) and the robot actuators. Thus it has been extensively studied in uncalibrated robotic hand-eye coordination. Yoshimi et al [3] demonstrated a 2D alignment of an eye-on-hand manipulator, which first showed the potential of the image Jacobian matrix model to solve calibration-free visual servoing problem. Since a linear matrix is used to approximate the nonlinear mapping between the visual feedback and the robot control temporarily and spatially, the Jacobian matrix is changing and needs online real-time estimation during the whole process of task fulfillment. Performance of the online estimation of the image Jacobian matrix is the key issue for the quality of the uncalibrated hand-eye coordination system [4]. However, the current estimation methods have problems such as estimation-lag, singularity, convergence and convergence speed [5]. Especially in dynamic circumstances, these problems become more serious. There are other efforts to deal with the online estimation of the image Jacobian matrix and the uncalibrated hand-eye coordination control. Among those, the artificial neural network is a main tool [6], [7]. Unfortunately, since the off-line training of the ANN needs to take as many training samples as possible from the whole robotic workspace, which has resulted in a large workload and a doubtful feasibility, this approach is restricted in practice.

To deal with the system uncertainties, the state observer has been proved to be an effective way in control theory. Thus if the image Jacobian matrix to be estimated online is considered as the system's unmodeled dynamics, the observer design theory can be employed to estimate the unknown image Jacobian matrix model as well as the unknown external disturbances. In [12], the state observer method is first introduced to estimate the image Jacobian matrix and a 2-D positioning task is addressed to show the feasibility and the primary effects of this control scheme.

This paper is to study the 3-D tracking problem of the uncalibrated hand-eye coordination system based on stereo vision feedback. In [12], the design of the coordination controller neglected the inherent coupled relations in the different parts of the system and separated the whole plant into 2 independent subsystems. Thus it only has limited performance when applied to the dynamic tracking problem. This problem is extensively considered in this paper. Simulation and experiment results demonstrate the feasibility and effectiveness of this control scheme.

II. Overall System Description

For a robotic hand-eye coordination system that contains n global visual feedbacks formed by n global cameras, the task is to design a robotic control to make the positions of the target and the gripper to coincide with each other in each image plane. Suppose that there is a point A in the motion space. Its position is $W = (x_w, y_w, z_w)^T$ in the robotic coordinate system and $P_i = (p_x^i, p_y^i)^T$ in the image coordinate system of the i-th camera. The relationship between the robot workspace and the image plane can be expressed as

$$\dot{P} = J \cdot U \quad (1)$$

where $U = \dot{W} = (u_x, u_y, u_z)$ is the velocity vector in the robotic coordinate system, which is also the system control input if point A identifies the gripper. If $\dot{P}_i = (\dot{p}_x^i, \dot{p}_y^i)^T$ is the velocity vector in the i-th image coordinate system of the point A, $\dot{P} = (\dot{P}_1^T, \cdots, \dot{P}_n^T)^T$ is the velocity vector formed by all the n image coordinates of A and its dimension is $2n \times 1$. $J_{2n \times 3} = (J_1^T, \cdots, J_n^T)^T$ is the image Jacobian matrix that describes the nonlinear differential mapping from robot motion space to the n image coordinates systems. As an example, the image Jacobian matrix formed by a stereo vision system can be expressed as:

$$J_{4\times 3} = \begin{bmatrix} J^1 \\ J^2 \end{bmatrix} = \begin{bmatrix} J^1_{11} & J^1_{12} & J^1_{13} \\ J^1_{21} & J^1_{22} & J^1_{23} \\ J^2_{11} & J^2_{12} & J^2_{13} \\ J^2_{21} & J^2_{22} & J^2_{23} \end{bmatrix} \quad (2)$$

where $J^i, (i=1,2)$ is the image Jacobian matrix for the i-th camera. And the image velocity vector $\dot{P}$ can correspondingly be described as:

$$\dot{P} = (\dot{p}_x^1, \dot{p}_y^1, \dot{p}_x^2, \dot{p}_y^2)^T \quad (3)$$

We take the unknown image Jacobian matrix as the system's unmodeled dynamics and take advantage of the extended state observer [10] to estimate it online. The design of the extended state observer does not rely on the system configuration and prescribed tasks of the system. Thus a general control scheme may result by composing of a state observer for image Jacobian matrix estimation and a specific strategy for coordination control. In this paper, we adopt a nonlinear state error feedback for coordination control introduced in [11].

III. Design of Coupled Controller

Suppose $\hat{J}$ is the initial estimation of system's image Jacobian matrix, (1) can be rewritten as:

$$\dot{P} = \hat{J} \cdot U + \alpha(t) \quad (4)$$

where

$$\alpha(t) = (J - \hat{J}) \cdot U + \xi(t) \quad (5)$$

is the sum of the model uncertainties and the unknown external disturbance of the system $\xi(t)$. Assume h is the sampling step, the discrete form can be described as

$$P(k+1) - P(k) = h \cdot (\hat{J} \cdot U(k) + \alpha(k)) \quad (6)$$

For a visual tracking system based on stereo vision, we first design a tracking differentiator (TD) to arrange the practical transient process for tracking the input signal. The form of TD can be depicted as follow:

$$\begin{cases} \varepsilon_1(k) = Z_1(k) - P^*(k) \\ Z_1(k+1) = Z_1(k) - h \cdot R \bullet fal(\varepsilon_1(k), \alpha_1, \delta_1) \end{cases} \quad (7)$$

where

$$fal(\varepsilon_i, \alpha_i, \delta_i) = \begin{cases} |\varepsilon_i|^{\alpha_i} sign(\varepsilon_i) & |\varepsilon_i| > \delta_i \\ \varepsilon_i / \delta_i^{1-\alpha_i} & |\varepsilon_i| \leq \delta_i \end{cases} \quad (8)$$

$\varepsilon_i, \alpha_i, \delta_i (i=1,\cdots,4)$ are all vectors whose dimensions are 4×1 and respectively corresponds to the x direction and y direction of the 2 image planes obtained by the 2 cameras. The dimensions of other parameter vectors appeared in (7) and (8) are the same. ' $\bullet$ ' means the multiplication between the corresponding elements of each vector and the result is also a vector.

$P^* = \begin{bmatrix} P^*_{1x} & P^*_{1y} & P^*_{2x} & P^*_{2y} \end{bmatrix}^T$ is the system's input as well as the input of TD. Z_1 is the output vector of TD and used to track the input signal P^*.

Design an extended state observer (ESO) to estimate the influence of $\alpha(t)$ led by system model uncertainties and unknown external disturbance. The form of ESO can be depicted as

$$\begin{cases} \varepsilon_2(k) = Z_2(k) - P(k) \\ Z_2(k+1) = Z_2(k) + h \cdot (Z_3(k) \\ \qquad - B_1 \bullet fal(\varepsilon_2(k), \alpha_2, \delta_2) + \hat{J} \cdot U(k)) \\ Z_3(k+1) = Z_3(k) - h \cdot B_2 \bullet fal(\varepsilon_2, \alpha_3, \delta_3) \end{cases} \quad (9)$$

where vector Z_2 is used to track the system output vector P and vector Z_3 is used to approximately estimated this influence of $\alpha(t)$ at every transient time, which is the key to fulfill the tasks with the approach of coupled coordination controller effectively.

The coordination controller is designed based on nonlinear state error feedback (NLSEF), which has the following form:

$$\begin{cases} \varepsilon_3(k) = Z_1(k) - Z_2(k) \\ U_0(k) = h \cdot K \bullet fal(\varepsilon_3(k), \alpha_4, \delta_4) \\ U(k) = (\hat{J}^T \cdot \hat{J})^{-1} \cdot \hat{J}^T (U_0(k) - Z_3(k)) \end{cases} \quad (10)$$

After compensation for the affect of $\alpha(t)$ which has been estimated by ESO, the system control input U, which is also the velocity vector of the gripper in robot motion space, can be obtained. The overall system control diagram is given in Fig.1. From Fig.1, we can see that the robot is controlled to converge the system errors in both x and y directions in image planes of the two cameras. The system control input U is used for all ESOs to estimate system's unmodeled dynamics, including image Jacobian matrix as well as the system external disturbances. Thus the inherent coupled relation existing in coordination system is not neglected, but embodied perfectly.

Suppose $P^*(k)$ is the position where the robot hand is desired to be at the time instant k in the image planes, namely, the system input. For the visual tracking task, we provide a first-order prediction for position of the moving object to eliminate the steady error. Then $P^*(k)$ can be provided by:

$$\begin{aligned} P^*(k) &= P^o(k) + (P^o(k) - P^o(k-1)) \\ &= 2P^o(k) - P^o(k-1) \end{aligned} \quad (11)$$

where $P^o(k)$ and $P^o(k-1)$ are the positions of the object in the image planes respectively at the sampling moments k and $k-1$.

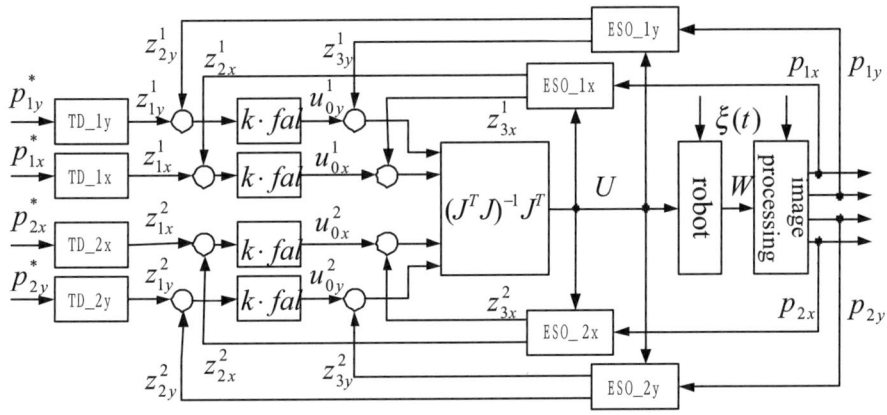

Fig 1: Structure of uncalibrated robotic hand-eye coordination based on extended state observer

IV. Simulations and Experiments

1 Simulations

In our simulations, a stereovision system of two cameras is used for the visual feedback of the robot working space. The poses of the two cameras with respect to the robotic base coordinate system are defined in Euler angles: $\psi_1 = 0.474$, $\theta_1 = -0.128$, $\varphi_1 = 0.242$ for the first camera and $\psi_2 = -0.486$, $\theta_2 = -0.065$, $\varphi_2 = -0.116$ for the second camera. The translational vectors are $T_1 = [-408\ 69\ 1065]^T$ mm for the first camera and $T_2 = [-396\ -134\ 974]^T$ mm for the second camera. The internal parameters of the two cameras are adopted from two true cameras in our lab with the focal lengths $f_1 = 4.72mm$, $f_2 = 5.13mm$ and pixel resolution $N_x = 4.9/582 mm/pixel$ and $N_y = 3.7/512 mm/pixel$ for the two cameras. In simulations, all the parameters of TD, ESO and NLSEF for the tracking control in x and y directions of the two cameras are selected to be the same, which is shown in Tab. 1. The initial estimation of the image Jacobian matrix is given in Tab.2.

Tab 1: Selection of controller parameters

TD			ESO						NLSEF		
α_{1x}^1	δ_{1x}^1	r_x^1	α_{2x}^1	δ_{2x}^1	b_{1x}^1	α_{3x}^1	δ_{3x}^1	b_{2x}^1	α_{4x}^1	δ_{4x}^1	k_x^1
0.5	5	20	0.5	3	10	0.5	5	20	0.5	5	12

Tab 2: Selection of the initial estimation of image Jacobian matrix

Cam.1						Cam.2					
$\hat{J}_{111}$	$\hat{J}_{112}$	$\hat{J}_{113}$	$\hat{J}_{121}$	$\hat{J}_{122}$	$\hat{J}_{123}$	$\hat{J}_{211}$	$\hat{J}_{212}$	$\hat{J}_{213}$	$\hat{J}_{221}$	$\hat{J}_{222}$	$\hat{J}_{223}$
0.7	-0.2	0	0.2	0.9	-0.5	0.7	0.1	0	-0.1	0.9	0.5

Suppose that the target moves upward spirally in the robot motion space which is unknown to the robotic controller. A normal distributed random noise with a maximum magnitude of ± 2 pixels and a zero average are added as the external disturbances in both x and y directions of each image plane during the image processing. The system response is given in Fig.2 and Fig.3. Fig.2 shows the tracking errors in the 2 image planes, which are both less than 5 pixels when tracking is stable. Fig.3 shows the variations of the target and the gripper's x, y, z coordinates in robot motion space, from which the effectiveness of the proposed control scheme is demonstrated.

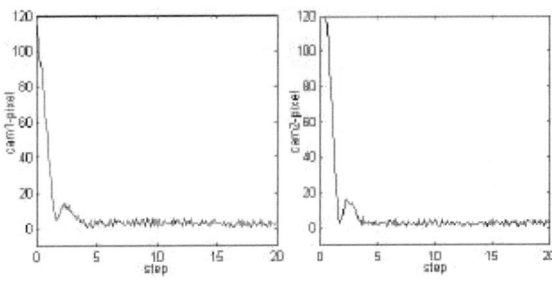

Fig 2a Tracking error from cam1   Fig 2b Tracking error from cam2

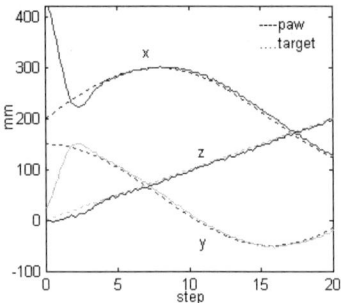

Fig 3: Tracking 3-D moving object

2 Experiments

The configuration of our experiment system is shown in Fig.4. A 4-DOF Adept 604s robot is adopted. The last link for rotation is locked so that the gripper mounted at the end of the last link can have three Degree-of-Freedoms. The stereo vision feedback is formed by 2 globally fixed cameras. The image size adopted in image processing is 320 × 320 pixels. To simplify image processing and object recognition, the gripper and the target are respectively identified by a red and a green color block. The positions of the target and the gripper are all defined as the centers of the color blocks in each image plane. The positions of the target and the gripper are all defined as the centers of the color blocks in each image plane. If the two blocks are close to each other and even overlap each other, the task is finished, based on this criterion we define a threshold, if the error between the gripper and the target is less than the threshold in each image, we think we have fulfilled the task. The sampling period is selected as 0.5s in our experiments.

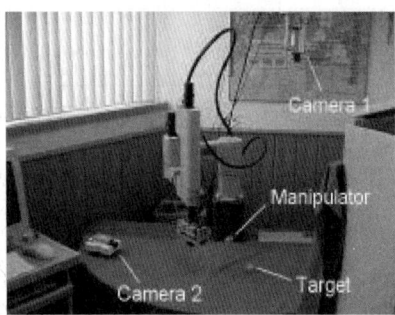

Fig 4: Setup of the experimental system

1 Object visual position

Figs.5-6 show the gripper moving trajectory in the 2 image planes observed by the 2 cameras. The position of the target is identified by '+' or 'x' and the gripper by 'o' or ' ' at every sample time. We can see from the figures that at time of 4.5s, the gripper arrives at the target with the edge of the red block touching the edge of the green block. Thus the error between the position of the gripper and that of the target in each image plane is less than the distance between the centers of the two color blocks defining the gripper and the target respectively. This error is less than 15 pixels when the task is completed.

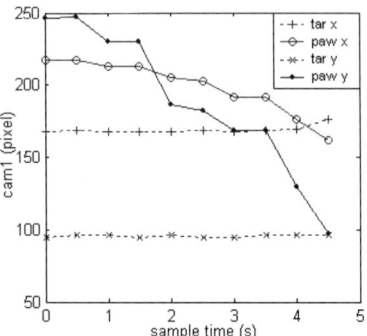

Fig 5: Visual position captured by Cam.1

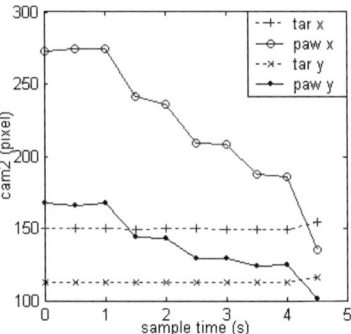

Fig 6: Visual position captured by Cam.2

2 Object visual tracking

Figs.7-12 demonstrate the result of object visual tracking experiment. Figs.7-10 are the scenes captured at the time of task beginning and task finishing obtained by the 2 cameras. Figs.11-12 shows the motion trajectory of both the target and the gripper in the 2 image planes. The target is making a nearly linear movement with the velocity of about 20 mm/s. The task is done at the time of 8.5s with the error between the positions of the target and the gripper in each image plane of about 28 pixels due to the size of the color blocks. In fact the gripper and the target has overlapped each other in each image, which we can see from fig.8 and fig.10, and also for the reason of overlapping, the distance between the center of the two blocks is larger than real one when completed. From the results of the experiments we can confirm the feasibility and effectiveness of this state observer-based approach for uncalibrated robotic hand-eye coordination system.

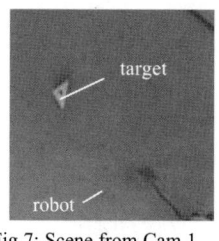

Fig 7: Scene from Cam.1
in the beginning of tracking

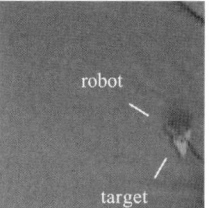

Fig 8: Scene from Cam.1
at the end of tracking

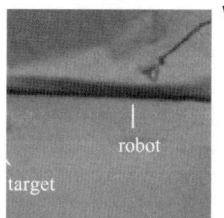

Fig 9: Scene from Cam.2
in the beginning of tracking

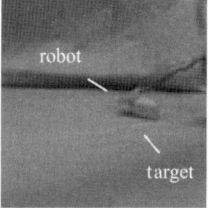

Fig 10: Scene from Cam.2
at the end of tracking

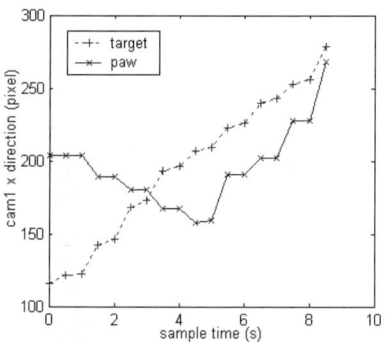

Fig 11a: Visual tracking trajectory by Cam.1 x direction

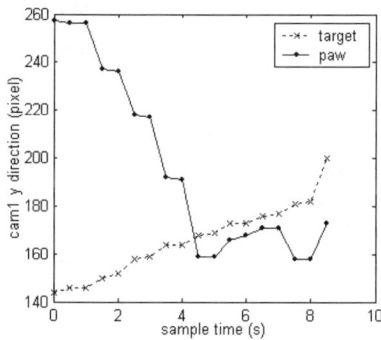

Fig 11b: Visual tracking trajectory by Cam.1 y direction

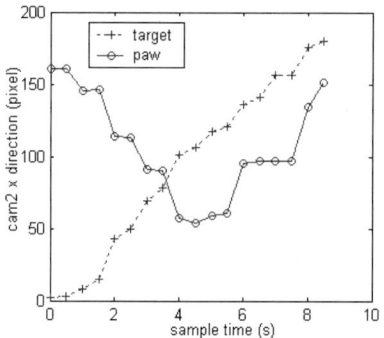

Fig 12a: Visual tracking trajectory by Cam.2 x direction

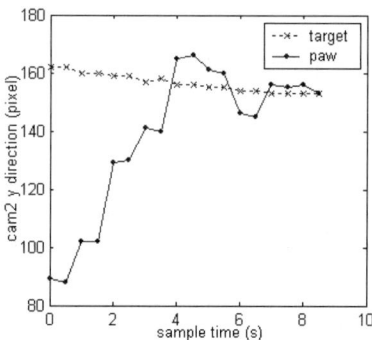

Fig 12b: Visual tracking trajectory by Cam.2 y direction

V. Conclusions

A novel approach for estimating image Jacobian matrix for uncalibrated robotic hand-eye coordination is presented in this paper. This approach, which is state observer based, is different from others in that the estimation procedure is independent from system configuration and prescribed tasks. Based on the new estimation method, a new control scheme for the uncalibrated robotic hand-eye coordination system is thus developed by a nonlinear state error feedback law, considering the coupled relationship inherently existing in the hand-eye coordination system. Extensive simulations and experiments demonstrate the effectiveness and the applicability of the proposed control scheme.

References

[1] Tsai, R.Y. An efficient and accurate camera calibration technique for 3D machine vision. In Proc. IEEE Computer Vision and Pattern Recognition, pp.364-374, 1986

[2] A.C. Sanderson, L.E. Weiss, and C.P. Neuman. Dynamic sensor-based control of robots with visual feedback. IEEE Trans. Robot. Automat., Vol. RA-3, No.5, pp.31-47, 1987

[3] Yoshimi B H and Allen P K. Alignment using an uncalibrated camera system. IEEE Trans. Robot.

Automat., Vol.11, No.4, pp.516~521, 1995

[4] Jiang Qian and Jianbo Su. Online Estimation of Image Jacobian Matrix by Kalman-Bucy Filter for Uncalibrated Stereo Vision Feedback. Proc. of IEEE Inter. Conf. on Robot. & Automation, 562-567, USA, 2002

[5] L. Hsu and P.L.S. Aquino. Adaptive visual tracking with uncertain manipulator dynamics and uncalibrated camera. Proc. of the 38th Conference on Decision & Control, pp.1248-1253, 1999

[6] H. Hashimoto, T. Kubota, M. Sato and F. Hurashima. Visual control of robotic manipulator based on neural networks. IEEE Trans. On Industrial Electronics, 39(6): 490-496, 1992

[7] Qielu Pan, Jianbo Su and Yugeng Xi. Uncalibrated 3D robotic visual tracking based on stereo vision. Robot (in Chinese), 22(4): 293-299, 2000

[8] Jingqing Han. Auto disturbance rejection controller and its applications. Control and Decision (in Chinese), 13(1): 19-23, 1998

[9] Jingqing Han and Wei Wang. Nonlinear tracking defferentiator. Journal of Sys. Sci. & Math. Sci. (in Chinese), 14(2): 177-183, 1994

[10] Jingqing Han. The extended state observer of a class of uncertain systems. Control and Decision (in Chinese), 10(1): 85-88, 1995

[11] Jingqing Han. Nonlinear state error feedback control law-NLSEF. Control and Decision (in Chinese), 10(3): 221-225, 1995

[12] Jianbo Su and Yugeng Xi. Uncalibrated Hand/Eye Coordination based on Auto Disturbance Rejection Controller Proc. Of IEEE International Conference on Decision and Control, 923-924, Las Vegas, Dec., 2002

Robust Visual Tracking Using a Fixed Multi-camera System

Vincenzo Lippiello, Bruno Siciliano, and Luigi Villani
PRISMA Lab – Dipartimento di Informatica e Sistemistica
Università degli Studi di Napoli Federico II
Via Claudio 21, 80125 Napoli, Italy
{vincenzo.lippiello,bruno.siciliano,luigi.villani}@unina.it

Abstract—The problem of tracking the position and orientation of a moving object using a stereo camera system is considered in this paper. A robust algorithm based on the extended Kalman filter is adopted, combined with an efficient selection technique of the object image features, based on Binary Space Partitioning tree geometric models. An experimental study is carried out using a vision system of two fixed cameras.

I. INTRODUCTION

A common problem in machine vision is that of tracking the 3D-pose (position and orientation) of moving objects in a wide workspace. Visual tracking applications require the adoption of motion estimation algorithms ensuring robustness in the presence of noise and stable tracking also in unstructured environments. Moreover, the information extracted from visual measurements must be available in real time.

A widely adopted approach to robust motion estimation from visual measurements is based on the extended Kalman filter, which represents a good trade-off between computational load and estimation accuracy [1], [2], [3]. When the object is moving in unstructured environments, good performance can be ensured using a multi-camera system, which guarantees information redundancy.

In this paper the extended Kalman filter is used to set up a visual tracking algorithm for a stereo camera system. The filter equations are derived according to the systematic procedure presented in [4], for the case of n video cameras fixed in the workspace. The algorithm requires the recognition of the object corners (feature points) from which the object pose can be computed using a simple point CAD model.

The estimation process is improved by adopting an efficient selection method of the feature points based on Binary Space Partitioning (BSP) trees [5]. The selection algorithm allows managing in real time a large amount of information provided by the stereo cameras system. Thus the information redundancy can be effectively exploited to improve accuracy and robustness of the visual tracking task.

Experimental tests are presented to show the computational feasibility and the effectiveness of the proposed approach.

II. EXTENDED KALMAN FILTER

Consider a system of n video cameras fixed in the workspace and a moving object. The geometry of the system is described in Fig. 1.

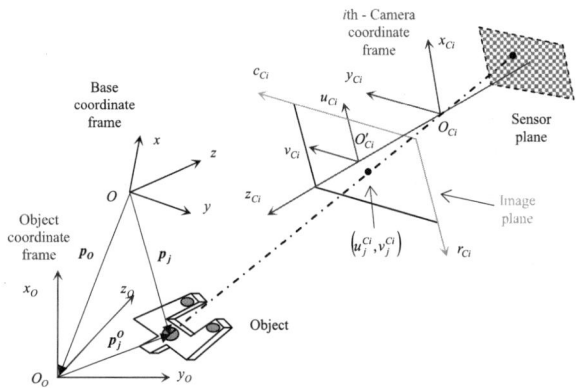

Fig. 1. Reference frames for the i-th camera and the object using the pinhole model.

A frame O_{ci}–$x_{ci}y_{ci}z_{ci}$ attached to the i-th camera (camera frame), with the z_{ci}-axis aligned to the optical axis and the origin in the optical center, is considered for each camera. The sensor plane is parallel to the $x_{ci}y_{ci}$-plane at a distance $-f_e^{ci}$ along the z_{ci}-axis, where f_e^{ci} is the effective focal length of the camera lens. The image plane is parallel to the $x_{ci}y_{ci}$-plane at a distance f_e^{ci} along the z_{ci}-axis. The intersection of the optical axis with the image plane defines the principal optic point O'_{ci}, which is the origin of the image frame O'_{ci}–$u_{ci}v_{ci}$ whose axes u_{ci} and v_{ci} are taken parallel to the axes x_{ci} and y_{ci} respectively.

The position of the origin and the rotation matrix of the i-th camera frame with respect to the base frame are denoted by o_{ci} and R_{ci} respectively. These quantities are constant because the cameras are assumed to be fixed to the workspace, and can be computed through a suitable calibration procedure [6].

The position and orientation of the object with respect to the base frame can be specified by defining a frame O_o–$x_oy_oz_o$ attached to the object and considering the coordinate vector of the origin $o_o = [x_o\ y_o\ z_o]^T$ and the rotation matrix $R_o(\varphi_o)$, where $\varphi_o = [\phi_o\ \alpha_o\ \psi_o]^T$ is the vector of the Roll, Pitch and Yaw angles. The components of the vectors o_o and φ_o are the six quantities to be estimated.

Consider m *feature points* on the object. It can be shown (see [4]) that the position of the j-th feature point in the i-th camera frame can be computed as

$$p_j^{ci} = R_{ci}^T(o_o - o_{ci} + R_o(\varphi_o)p_j^o), \quad (1)$$

where p_j^o is the position vector of the j-th feature point with respect to the object frame. This vector is constant and is assumed to be known form the object CAD model.

By folding the m equations (1) into the perspective transformation of the n cameras, a system of $2mn$ nonlinear equations is achieved. These equations depend on the measurements of the m feature points in the image planes of the n cameras, while the six components of the vectors o_o and ϕ_o are the unknown variables.

To solve these equation in real-time, the extended Kalman filter is adopted, which provides a recursive solution.

In order to write the Kalman filter equations, a discrete-time dynamic model of the object motion has to be considered. Assuming that the object velocity is constant over one sample period T, the model can be written in the form

$$w_k = Aw_{k-1} + \gamma_k \quad (2)$$

where $w = [x_o\ \dot{x}_o\ y_o\ \dot{y}_o\ z_o\ \dot{z}_o\ \phi_o\ \dot{\phi}_o\ \alpha_o\ \dot{\alpha}_o\ \psi_o\ \dot{\psi}_o]^T$ is the state vector, γ is the dynamic modeling error, and A is a (12×12) block diagonal matrix of the form

$$A = \text{diag}\left\{\begin{bmatrix}1 & T\\0 & 1\end{bmatrix}, \cdots, \begin{bmatrix}1 & T\\0 & 1\end{bmatrix}\right\}.$$

The output equation of Kalman filter is chosen as

$$\zeta_k = g(w_k) + \nu_k \quad (3)$$

with

$$g(w_k) = \left[\frac{x_1^{c1}}{z_1^{c1}}\ \frac{y_1^{c1}}{z_1^{c1}}\ \cdots\ \frac{x_1^{cn}}{z_1^{cn}}\ \frac{y_1^{cn}}{z_1^{cn}}\right.$$
$$\left.\cdots\ \frac{x_m^{c1}}{z_m^{c1}}\ \frac{y_m^{c1}}{z_m^{c1}}\ \cdots\ \frac{x_m^{cn}}{z_m^{cn}}\ \frac{y_m^{cn}}{z_m^{cn}}\right]_k^T,$$

where the coordinates of the feature points p_j^{ci} are computed from the state vector w_k via (1). In the above equation, ν_k is the measurement noise and the ζ_k is the vector of the *normalized* coordinates of the m feature points in the image plane of the n cameras.

Since the output model is nonlinear in the system state, it is required to linearize the output equations about the current state estimate at each sample time, considering the so-called extended Kalman filter. The recursive form of the extended Kalman filter equations is reported in [4].

III. PRE-SELECTION ALGORITHM

The accuracy of the estimate provided by the Kalman filter depends on the number of the available feature points. Inclusion of extra points will improve the estimation accuracy but will increase the computational cost. It has been shown that a number of five or six feature points, if properly chosen, may represent a good trade-off [2]. Automatic selection algorithms have been developed to find the optimal feature points [7]. In order to increase the efficiency of the selection algorithms, it is advisable to perform a pre-selection of the points that are visible to the camera at a given sample time. The pre-selection technique proposed in this paper is based on Binary Space Partitioning (BSP) trees.

A BSP tree is a data structure representing a recursive and hierarchical partition of a n-dimensional space into convex subspaces. It can be effectively adopted to represent the 3D CAD geometry of an object [8].

In order to build the tree, each object has to be modelled as a set of planar *polygons*; this means that the curved surfaces have to be approximated. Each polygon is characterized by a set of *feature points* (the vertices of the polygon) and by the vector normal to the plane leaving from the object. For each node of the tree, a *partition plane*, characterized by its normal vector and by a point, is chosen according to a specific criterion; the node is defined as the set containing the partition plane and all the polygons lying on it.

Each node is the root of two subtrees: the *front* subtree corresponding to the subset of all the polygons lying entirely on the front side of the partition plane (i.e., the side corresponding to the half-space containing the normal vector), and the *back* subtree corresponding to the subset of all the polygons lying entirely on the back side of the partition plane.

The construction procedure can be applied recursively to the two subsets by choosing, for each node, a new partition plane among those corresponding to the polygons contained in that subtree. The construction ends when all the polygons are placed in a node of the tree.

Further details on BSP trees and an example of construction can be found in [9].

Once a BSP tree representation of an object is available, it is possible to select the feature points of the object visible from a given camera position and orientation, by implementing a suitable visit algorithm of the tree. The algorithm can be applied recursively to all the nodes of the tree, starting from the root node as showed in Fig. 2.

When the algorithm processes a node, the current set of projections of the visible feature points on the image plane is updated by adding all the projections of the feature points of the polygons of the current node and eliminating all the projections of the feature points that are hidden by the projections of the polygons of the current node.

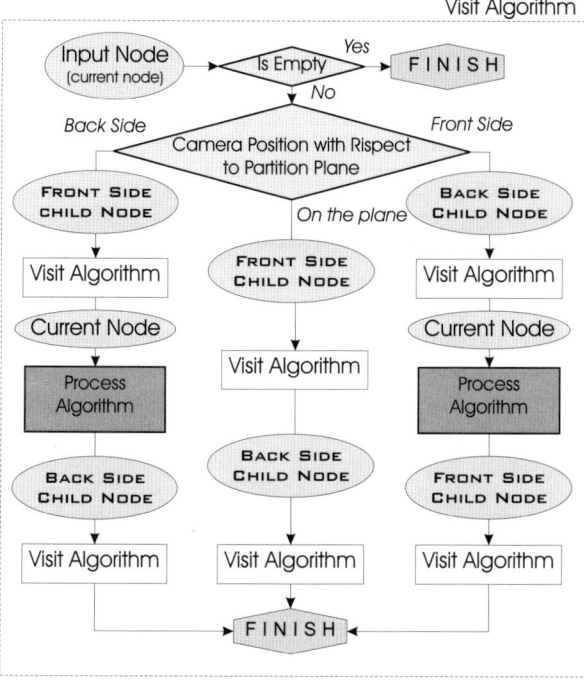
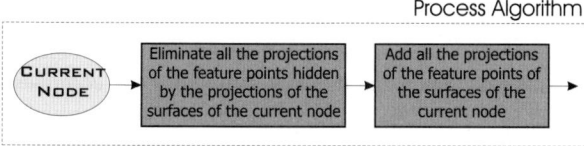

Fig. 2. Recursive visit algorithm of the BSP tree for the selection of visible feature points.

A windowing test is adopted to select the projections of the feature points that can be well localized. In particular, only the points that can be centered into suitable rectangular windows of the image plane are considered for the next step of selection, while the points that are out of the field of view of the camera, and the points that are too close each other or to the boundaries of the image plane, are discarded.

The number of remaining feature points after pre-selection and windowing test is typically too high with respect to the minimum number sufficient to achieve the best Kalman filter precision. It has been demonstrated that an optimal set of five or six feature points guarantees about the same precision as that of the case when an higher number of feature points is considered [2], [3].

The optimality of a given set of feature points is valued through the composition of suitably selected quality indexes into an optimal cost function. The quality indexes must be able to provide accuracy, robustness and to minimize the oscillations in the pose estimation variables. To achieve this goal it is necessary to ensure an optimal spatial distribution of the projections of the feature points on the image plan and to avoid chattering events between different optimal subsets of feature points chosen during the object motion. Moreover, in order to exploit the potentialities of a multi-camera system, it is important to achieve an optimal distribution of the feature points among the different cameras.

Without loss of generality, the case of two identical cameras is considered.

A first quality index is the measure of spatial distribution of the predicted projections on the image planes of a subset of q_i selected points for the i-th camera, $i = 1, 2$:

$$Q_{si} = \frac{1}{q_i} \sum_{k=1}^{q_i} \min_{\substack{j \in \{1, \ldots, q_i\} \\ j \neq k}} \|\boldsymbol{p}_j - \boldsymbol{p}_k\|.$$

Notice that $q = q_1 + q_2$ is chosen between 6 and 8 to prevent fault cases.

A second quality index is the measure of angular distribution of the predicted projections on the image planes of a subset of q_i selected points for the i-th camera, $i = 1, 2$:

$$Q_{ai} = 1 - \sum_{k=1}^{q_i} \left| \frac{\alpha_k}{2\pi} - \frac{1}{q_i} \right|$$

where α_k is the angle between the vector $\boldsymbol{p}_{k+1} - \boldsymbol{p}_{Ci}$ and the vector $\boldsymbol{p}_k - \boldsymbol{p}_{Ci}$, being $\boldsymbol{p}_{Ci}$ the central gravity point of the whole subset of feature points, and the q_i points of the subset are considered in a counter-clockwise ordered sequence with respect to $\boldsymbol{p}_{Ci}$, with $\boldsymbol{p}_{q_i+1} = \boldsymbol{p}_1$.

In order to avoid chattering phenomena, the following quality index, which introduces hysteresis effects on the change of the optimal combination of points, is considered

If a polygon is hidden from the camera (i.e., the angle between the normal vector to the polygon and the camera z-axis is not in the interval $]-\pi/2, \pi/2[$ or the polygon is behind the camera), the corresponding feature points are not added to the set.

At the end of the visit, the current set will contain all the projections of the feature points visible from the camera, while all the hidden feature points will be discarded. Notice that the visit algorithm updates the set by ordering the polygons with respect to the camera from the background to the foreground.

IV. SELECTION ALGORITHM

The pre-selection technique recognizes all the feature points that are visible from a camera view point. However, this does not ensure that all the visible points are "well" localizable, i.e., their positions can be effectively measured with a given accuracy. Moreover, the number of the well localizable feature points may be larger than the *optimal* number of points ensuring the best pose estimation accuracy.

for the i-th camera, $i = 1, 2$:

$$Q_{hi} = \begin{cases} 1 + \epsilon & \text{if actual combination = previous one} \\ 1 & \text{otherwise} \end{cases}$$

where ϵ is a positive constant.

In order to distribute the points among the two cameras, the following indexes are considered:

$$Q_e = 1 + \frac{2}{q}\left(\frac{2}{q} - 1\right)\left|q_1 - \frac{q}{2}\right|$$

$$Q_d = \frac{q_1/d_1 + q_2/d_2}{q/\min\{d_1, d_2\}}$$

where q_i is the number of points assigned to the i-th camera, and d_i is the distance of the i-th camera form the object, $i = 1, 2$. The first index ensures an equal distribution of points among the cameras, while the second index takes into account the distance of the cameras from the object.

The proposed quality indexes represent only some of the possible choices, but guarantee satisfactory performance when used with the pre-selection method and the windowing test presented above, for the case of two fixed cameras. Other examples of quality indexes are presented, e.g., in [7], and some of them can be added to the indexes adopted in this paper.

The cost function is chosen as

$$\boldsymbol{Q} = \frac{Q_e Q_d}{q}\left(q_1 Q_{s1} Q_{a1} Q_{h1} + q_2 Q_{s2} Q_{a2} Q_{h2}\right)$$

and must be evaluated for all the possible combinations of the visible points on q positions. In order to determine the optimal set at each sample time, the initial optimal combination of points is first evaluated off-line. Then, only the combinations that modify at most one point for camera with respect to the current optimal combination are tested on-line, thus achieving a considerable reduction of processing time.

V. ESTIMATION PROCEDURE

A functional chart of the estimation procedure is reported in Fig. 3. It is assumed that a BSP tree representation of the object is built off-line from the CAD model. A Kalman filter is used to estimate the corresponding pose with respect to the base frame at the next sample time. The feature points selection and windows placing operation can be detailed as follows. For each camera:

- **Step 1:** The visit algorithm described in the previous section is applied to the BSP tree of the object to find the set of all the feature points that are visible from the camera.
- **Step 2:** The resulting set of visible points is input to the algorithm for the selection of the optimal feature points.

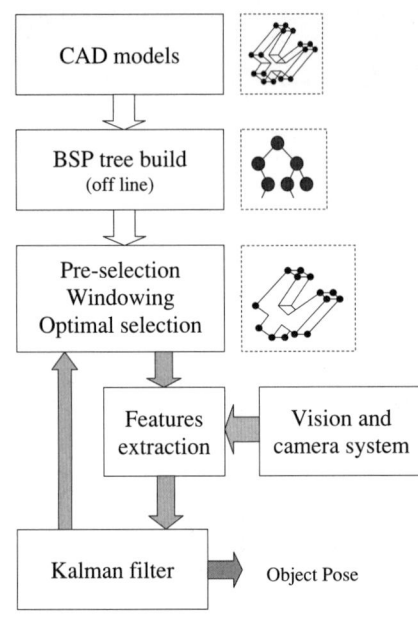

Fig. 3. Functional chart of the estimation procedure.

- **Step 3:** The location of the optimal feature points in the image plane at the next sample time is computed on the basis of the object pose estimation provided by the Kalman filter.
- **Step 4:** A dynamic windowing algorithm is executed to select the parts of the image plane to be input to the feature extraction algorithm.

At this point, all the image windows of the optimal selected points are elaborated using a feature extraction algorithm. The computed coordinates of the points in the image plane are input to the Kalman filter which provides the estimate of the actual object pose and the predicted pose at the next sample time used by the pre-selection algorithm.

Notice that the procedure described above can be extended to the case of multiple objects moving among obstacles of known geometry [9]; if the obstacles are moving with respect to the base frame, the corresponding motion variables can be estimated using Kalman filters.

VI. EXPERIMENTS

The experimental set-up is composed by a PC with Pentium IV 1.7GHz processor equipped with two MATROX Genesis boards, two SONY 8500CE B/W cameras, and a COMAU Smart3-S robot. The MATROX boards are used as frame grabbers and for a partial image processing (e.g., windows extraction from the image). The PC host is also used to realize the whole BSP structures management, the pre-selection algorithm, windows processing, the selection algorithm and the Kalman filtering. Some steps of image

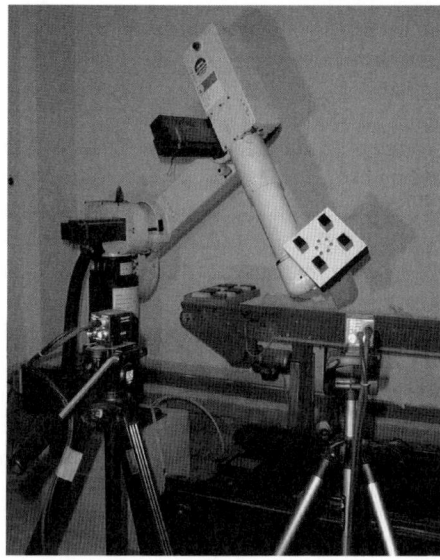

Fig. 4. COMAU Smart3-S robot and SONY 8500CE cameras.

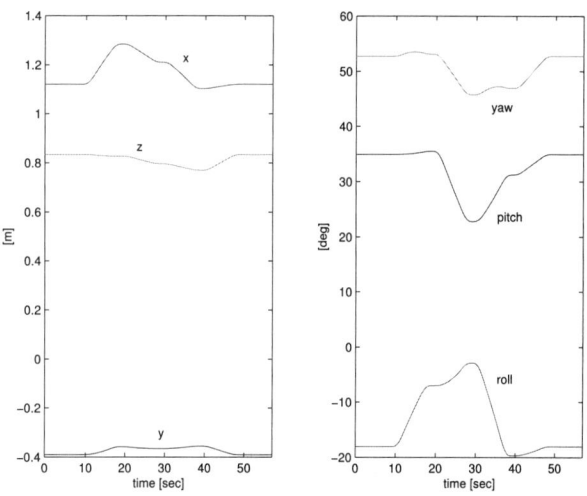

Fig. 5. Object trajectory. Left: Position trajectory. Right: Orientation trajectory

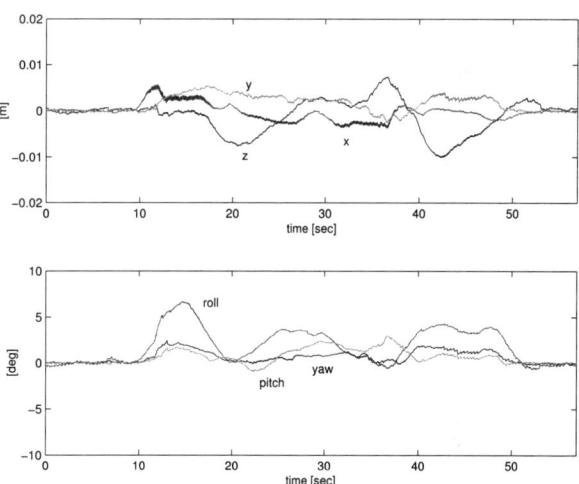

Fig. 6. Time history of the estimation errors. Top: Position errors. Bottom: Orientation errors

processing have been parallelized on the MATROX boards and on the PC, so as to reduce computational time. The robot is used to move an object in the visual space of the cameras; thus the object position and orientation with respect to the base frame of the robot can be computed from joint position measurements via the direct kinematic equation. In order to test the accuracy of the estimation provided by the Kalman filter, the cameras were calibrated with respect to the base frame of the robot using the calibration procedure presented in [6], where the robot is exploited to place a calibration pattern in some known pose of the visible space of the camera. The cameras resolution is 576×763 pixels and the nominal focal length of the lenses is 16 mm. The two cameras are disposed at a distance of about 150 cm from the object with an angle of about $\pi/2$ between the z_c axes. The sampling time used for estimation is limited by the camera frame rate, which is about 26 fps. No particular illumination equipment has been used, in order to test the robustness of the setup in the case of noisy visual measurements.

All the algorithms for BSP structure management, image processing and pose estimation have been implemented in ANSI C. The image features are the corners of the object, which can be extracted with high robustness in various environmental conditions. The feature extraction algorithm is based on Canny's method for edge detection and on a simple custom implementation of a corner detector [10]. In particular, to locate the position of a corner in a small window, all the straight segments are searched first, using an LSQ interpolator algorithm; then all the intersection points of these segments into the window are evaluated. The intersection points closer than a given threshold are considered as a unique average corner, due to the image noise. All the corners that are at a distance from the center of the window (which corresponds to the position of the corner so as predicted by the Kalman filter) greater than a maximum distance, are considered as fault measurements and are discarded. The maximum distance corresponds to the variance of the distance between the measured corner positions and those predicted by the Kalman filter.

The object used in the experiment has 40 vertices, which are all used as feature points. Fig. 4 shows the stereo vision system and the robot carrying the object.

The time history of the trajectory used for the experiment is represented in Fig. 5. The maximum linear

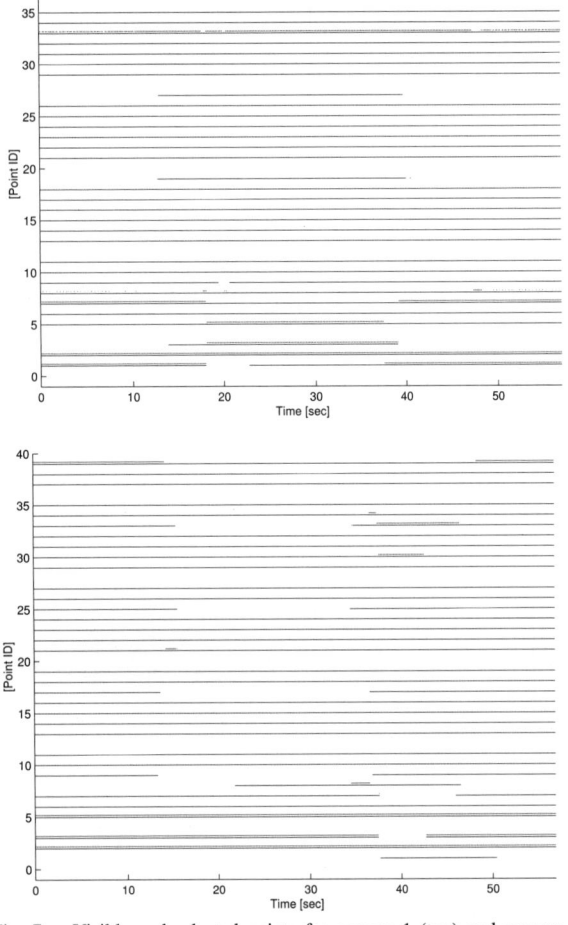

Fig. 7. Visible and selected points for camera 1 (top) and camera 2 (bottom). For each point: the bottom line indicates when it is visible; the top line indicates when it is selected for feature extraction.

velocity and angular velocity are about 3 cm/s and 3 deg/s respectively. The time history of the estimation errors is shown in Fig. 6. Noticeably, the accuracy of the system reaches the limit allowed by cameras calibration, for all the components of the motion. As it was expected, the errors for the motion components are of the same order of magnitude, thanks to the use of a stereo camera system.

In Fig. 7 the output of the whole selection algorithm, for the two cameras, is reported. For each of the 40 feature points, two horizontal lines are considered: a point of the bottom line indicates that the feature point was classified as visible by the pre-selection algorithm at a particular sample time; a point of the top line indicates that the visible feature point was chosen by the selection algorithm. Notice that 8 feature points are selected at each sample time in order to guarantee at least five or six measurements in the case of fault of the extraction algorithm for some of the points. Remarkably, 4 feature points for camera are chosen at each sampling time,

coherently with the symmetric disposition of the cameras with respect to the object.

VII. CONCLUSION

The problem of estimating the pose (position and orientation) of a moving object from visual measurements has been considered in this paper. The extended Kalman filter has been used to recursively compute an estimate of the motion variables from the measurements of the position of suitable feature points of the object. The efficiency of the algorithm has been improved by adopting a technique of selection of the visible feature points at each sample time based on a Binary Space Partition tree representation of the object geometry. The experiments on a stereo camera system have shown the effectiveness of the algorithm and have confirmed its practical feasibility.

Acknowledgments

This work was supported by MIUR and ASI.

VIII. REFERENCES

[1] S. Lee and Y. Kay, "An accurate estimation of 3-D position and orientation of a moving object for robot stereo vision: Kalman filter approach," *1990 IEEE Int. Conf. on Robotics and Automation*, pp. 414–419, 1990.

[2] J. Wang and J.W. Wilson, "3D relative position and orientation estimation using Kalman filter for robot control," *1992 IEEE Int. Conf. on Robotics and Automation*, pp. 2638–2645, 1992.

[3] J.W. Wilson, "Relative end-effector control using cartesian position based visual servoing," *IEEE Trans. on Robotics and Automation*, vol. 12, pp. 684–696, 1996.

[4] V. Lippiello, B. Siciliano, and L. Villani, "Position and orientation estimation based on Kalman filtering of stereo images," *2001 IEEE Int. Conf. on Control Applications*, pp. 702–707, 2001.

[5] V. Lippiello, B. Siciliano, and L. Villani, "Objects motion estimation via BSP tree modeling and Kalman filtering of stereo images," *2002 IEEE Int. Conf. on Robotics and Automation*, pp. 2969–2973, 2002.

[6] J. Weng, P. Cohen, and M. Herniou, "Camera calibration with distortion models ad accuracy evaluation," *IEEE Trans. on Pattern Analysis and Machine Intelligence*, vol. 14, pp. 965–980, 1992.

[7] F. Janabi-Sharifi and J.W. Wilson, "Automatic selection of image features for visual servoing," *IEEE Trans. on Robotics and Automation*, vol. 13, pp. 890-903, 1997.

[8] M. Paterson and F. Yao, "Efficient binary space partitions for hidden-surface removal and solid modeling," *Discrete and Computational Geometry*, vol. 5, pp. 485-503, 1990.

[9] V. Lippiello, B. Siciliano, and L. Villani, "A new method of image features pre-selection for real-time pose estimation based on Kalman filter," *2002 IEEE/RSJ Int. Conf. on Intelligent Robots and Systems*, pp. 372–377, 2002.

[10] J. Canny, "A Computational Approach to Edge Detection", *IEEE Trans. on Pattern Analysis and Machine Intelligence*, vol. 8, pp. 679–698, 1986.

Development of Piezoelectric Bending Actuators with Embedded Piezoelectric Sensors for Micromechanical Flapping Mechanisms*

Domenico Campolo, Ranjana Sahai, Ronald S. Fearing
{minmo, rsahai, ronf}@eecs.berkeley.edu
Department of EECS, University of California, Berkeley, CA 94720. USA

Abstract

This paper presents the fabrication and the testing of piezoelectric unimorph actuators with embedded piezoelectric sensors which are meant to be used for the actuation of the Micromechanical Flying Insect (MFI). First the fabrication process of a piezoelectric bending actuator comprising a standard unimorph and a rigid extension is described together with the advantages of adding such an extension. Then the convenience of obtaining an embedded piezoelectric sensor by a simple and inexpensive variation of the fabrication process is pointed out. A model for the sensor embedded into a unimorph actuator with rigid extension is derived together with its flat response band limits. Calibration steps are also outlined which allow, despite residual parasitic actuator-sensor coupling, the use of the actuator with the embedded sensor for measuring position and inertial forces when external mechanical structures are driven. An experiment is carried out which validates the model for the actuator/sensor device under desired operating conditions. Preliminary application of the fabricated device to the MFI is also presented where the mechanical power fed into the wing is estimated.

1 Introduction

Piezoelectric actuators are widely used in smart structure applications due to their high bandwidth, high output force, compact size, and high power density properties. For such reasons they are very appealing for mobile microrobotic applications such as the Micromechanical Flying Insect (MFI) [1] where, because of strict size/weight constraints, smart structures capable of both actuating and sensing are preferred. Since the technology needed to fabricate PZT based bending actuators was already available [4], the

---

*This work was funded by ONR MURI N00014-98-1-0671, DARPA.

possibility of integrating sensorial capabilities into the actuators themselves was investigated. Many works exist where piezoelectric thin patches are bonded to structures in order to sense the deformation of a specific area [2] but, in most of the cases, the process for fabricating unimorph actuators needs to be heavily modified. The idea of extending the capabilities of a standard unimorph actuator with a lateral sensor came from [6] although no hint was given for the fabrication involved.

The possibility of having the sensing section and the actuating section coexisting on the same piezoelectric layer, differentiated by simply patterning the electrode instead of aligning and bonding two different piezoelectric layers next to one another, was investigated. The sensor obtained in this way was affected by a strong electromechanical coupling between the actuating area and the sensing area, mainly a parasitic capacitance between the two electrodes. In order to shield such a capacitance, a third grounded electrode was then introduced in between the two sections.

2 Fabrication

Since the publication of [4], several improvements have been made to the fabrication of PZT based unimorph actuators. The only parameter unchanged is the ratio of thicknesses of the two layers constituting the unimorph since its choice was based upon the output energy optimization criteria [4].

The basic unimorph is obtained by bonding together stainless steel and PZT (PZT-5H, T105-H4E-602 ceramic single sheet, Piezo Systems, Inc.) respectively with thicknesses $t_s = 76.2 \mu m$ and $t_p = 127 \mu m$. Fabrication details can be found in [4].

A different commercial piezoelectric material (PZN-PT from TRS Ceramics, Inc.) will be eventually used for the actuation of the MFI because of its superior properties [4]. PZN-PT is a single crystal piezoelec-

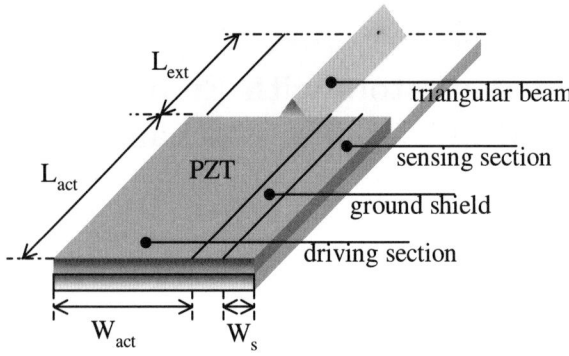

Figure 1: Piezoelectric actuator diagram comprising basic unimorph actuator and rigid extension.

tric material which is available in fixed dimensions. Difficulties arising in reshaping its size, without damaging its single crystal properties, impose the use of its original dimensions. On the contrary, PZT is a ceramic material, relatively inexpensive, which can be simply reshaped by laser cutting without perceptible loss of its properties. Although much larger, for testing purposes, an equivalent unimorph can be made out of PZT which provides, from the mechanical output, the same stiffness and blocking force theoretically provided by a PZN-PT based unimorph. Planar dimensions for the PZT based unimorph have been determined to be (width) $W_{act} = 6mm$ and (length) $L_{act} = 12mm$.

2.1 Rigid Extension

MFI [4] is a biomimetic project and a major design constraint is the wing beat resonance, determined to be at about $150Hz$. The stiffness of the actuator is therefore designed [4] to resonate, together with wing inertia reflected through an amplifying mechanism (4-bar mechanism), at this frequency. With the given fixed dimensions for the PZN-PT based unimorph, which translate into W_{act}, L_{act}, t_s and t_p for the PZT based equivalent one, the resulting stiffness would be too large. A rigid extension can be designed so that, by acting as a lever, it would provide larger free displacement at its tip together with lower blocking force, thus, leading to lower stiffness. In order to obtain the required stiffness, a rigid extension of length $L_{ext} = 8mm$ is needed. Rigidity of such an extension is a necessity. A flexible one would bend, storing part of the actuation energy instead of transmitting it to the wing. A first idea would be extending the very stainless steel layer which constitutes the uni-

morph but it can be easily seen that it would be even more compliant than the unimorph itself. A stiffer structure can be bonded on top of the extended stainless steel such as a hollow triangular beam as shown in Fig. 1. Such a folded structure is much stiffer than a planar structure. It can in fact be made out of thinner, therefore lighter, stainless steel ($12.5\mu m$) and still be considered rigid. The lighter the extension the more ideal it can be considered. As shown later, the inertia of the reflected wing is about $190mg$ while the inertia of the extension is less than $10mg$. The dynamics, i.e. the inertial terms, of the extension can then be neglected.

Advantages of adding a light and hollow rigid extension instead of having a longer unimorph with similar output mechanical behavior, i.e. stiffness and blocking force at the tip, can be summarized as:

- the final actuator will be lighter since a hollow beam replaces a comparably long section of the unimorph itself.

- higher first mode resonant frequency since also the equivalent mass of the actuator, which together with its stiffness defines the first resonance, is lighter.

- behaving as a lever, the rigid extension represents a first stage where the actuator tip displacement is pre-amplified. A pre-amplifying stage helps reducing nonlinearities deriving from a single high ratio amplifying stage (often needed in piezoelectric actuators based applications).

- higher energy density can be achieved since the extension converts a force load at the tip into a combination of the same force plus a torque (see Fig. 5) at the unimorph tip. Such a combination produces more bending throughout the unimorph than only the force would.

2.2 Side Sensor

As qualitatively described in [6], considering an actuator of width W_{act} and length L_{act} constituted by a piezoelectric layer bonded onto an elastic layer, another narrow piezoelectric strip of width $W_s \ll W_{act}$ and same length as the actuator can be bonded onto the elastic layer, side by side with the piezoelectric layer constituting the unimorph actuator itself. This way it will be subject to the same deflection as the actuator and via the piezoelectric effect a measure of the mean curvature of the actuator can be derived, as shown later.

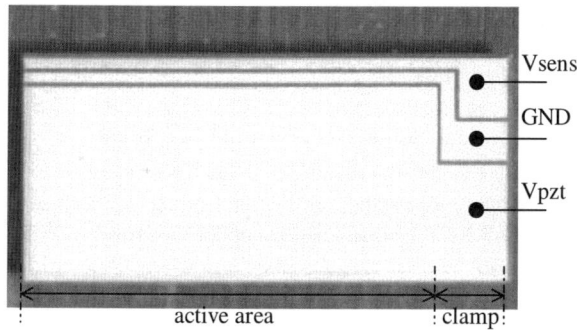

Figure 2: A photo the piezoelectric layer after laser cutting and top electrode patterning.

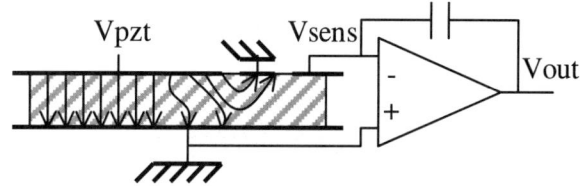

Figure 3: Front view (zy plane) of the piezoelectric layer and its electrical connection.

The sensor should be much narrower than the actuator in order not to affect the deflection. Ideally there should be no coupling between the piezoelectric layer of the unimorph and the sensing strip.

PZT ceramics can be easily shaped by laser cutting. A $532nm$ (green) laser beam can be generated by the QuikLaze micro-machining system (New Wave Research Inc.) which, at its maximum output power, is capable of cutting through PZT quickly enough not to overheat the sample. This allows precisely cutting out the piezoelectric pieces that will be lately bonded onto the stainless steel layer. By reducing the power of the laser beam, it is also possible to pattern the very thin layer of nickel that constitutes the electroding layer of the piezoelectric sheets. Since laser cutting the PZT pieces is a necessary step in the fabrication of the actuator, adding electrode patterning to the current process is a very inexpensive operation since no alignment has to be done and it takes relatively little time compared with the laser cutting itself.

Fig. 2 shows a photo of the piezoelectric layer whose top side electrode has been patterned with low power laser beam and then cut through with high power laser beam. Its whole length is divided into an active area which will be free to bend together with the elastic layer and the remaining area which, together with the elastic layer, will be clamped and kept from bending. The clamping area is where the wires will contact the electrodes, V_{pzt} for the actuator and V_{sens} for the sensor. The presence of a third electrode (GND) is explained below.

Differently from previous work [2], in order to keep the fabrication process simple and inexpensive, the actuator and the sensor coexist on the same piece of piezoelectric material. They only differ from one another because of the discontinuity of the electrodes, leading thus to a strong electromechanical coupling. As shown later, the sensor is meant to reveal a structural bending by producing polarization charge within its volume, proportional to the mean curvature of the unimorph actuator. An electromechanical coupling with the actuator section leads to extra induced charge proportional to the driving voltage and always in phase with it (at least at the frequencies of interest) while the structural bending would not always be in phase with the driving voltage because of the structural modes. The parasitic coupling will therefore result into an offset signal, also called feed-through, always in phase with the driving voltage and that can be taken into account during the calibration. Before calibrating, however, an effective way of shielding the electrical coupling is possible by introducing an electrically grounded strip between actuator and sensing section (referred to as GND in Fig. 2) which can act as a shield for the electric field line coming from the actuator section.

Fig. 3 shows electrical connections of the piezoelectric device with driving voltage V_{pzt} and the charge amplifier while the bottom electroded and the top shielding strip are electrically grounded. Because of the operational amplifier, the sensing electroded is also at virtual ground. The electric field in the piezoelectric layer of a standard unimorph actuator with a single top electrode can be considered vertical (z direction) but when a patterned electrode is considered, as in the case of interest, the field fringing can become significant. Fig. 3 schematically sketches how the most of the electric field lines are captured by the grounded strip, thus shielding the sensing area.

3 Model of a Unimorph Actuator Plus Rigid Extension

A working model for the unimorph actuator can be derived from [7]. Following [7], a coordinate system is

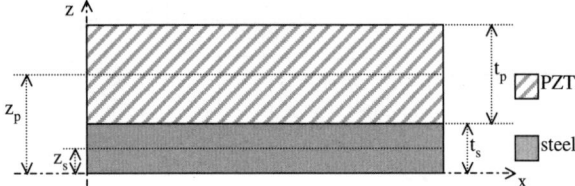

Figure 4: Coordinate system: the unimorph actuator is a long, narrow and uniform beam along the x axis. Non uniformity, i.e. different materials, is only assumed along the z direction.

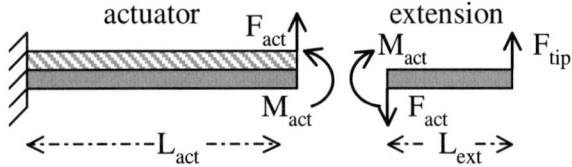

Figure 5: Free body diagram for the unimorph actuator (left) plus rigid extension (right).

adopted as in Fig. 4 such that $z = 0$ coincides with the base of the beam. Let z_p and z_s be respectively the coordinate of the center of piezoelectric and stainless steel layer, t_p and t_s their thicknesses, Y_p and Y_s be their Young's modulus. The neutral axis can be defined as (equation (11) in [7]):

$$z_n = \frac{z_s Y_s t_s + z_p Y_p t_p}{Y_s t_s + Y_p t_p} \quad (1)$$

while the moment of inertia is:

$$\begin{aligned} EI &= W_{act} Y_s \left((z_s - z_n)^2 t_s + \frac{t_s^3}{12} \right) \\ &+ W_{act} Y_p \left((z_p - z_n)^2 t_p + \frac{t_p^3}{12} \right) \end{aligned} \quad (2)$$

When bending occurs, z_n will actually depend upon x (the coordinate system is in fact defined upon the rest configuration of the unimorph) and also z_s and z_p but not $(z_s - z_n)$ nor $(z_p - z_n)$ which will be constant. Referring to the left side of Fig. 5, equation (26) in [7] can be reduced to the case where δ_{act} and α_{act}, respectively deflection and slope at the tip of the actuator, are to be determined as linear functions of F_{act} and M_{act}. Moreover, the effect of an applied voltage V_{pzt} can be seen as due to an equivalent mechanical moment $M_{pzt} = M_V V_{pzt}$ applied at the tip of the actuator, just like M_{act}, where M_V is the equivalent mechanical torque per unit voltage at the electrodes [7]. The following equation holds:

$$[\delta_{act}\ \alpha_{act}]^T = \mathbf{S_{act}} \left([F_{act}\ M_{act}]^T + [0\ M_{pzt}]^T \right) \quad (3)$$

where:

$$\mathbf{S_{act}} = \frac{L_{act}}{EI} \begin{bmatrix} \frac{L_{act}^2}{3} & \frac{L_{act}}{2} \\ \frac{L_{act}}{2} & 1 \end{bmatrix} \quad (4)$$

is the compliance matrix of the unimorph actuator.

Equation (3) simply assumes that the actuator operates at quasi-static conditions. Such an assumption is valid for frequencies much below the first resonant mode which, in the case under consideration, occurs after $500Hz$ while the working frequencies are at about $150Hz$.

As for the rigid extension in the right side of Fig. 5, the dynamical equations of a rigid body simply transform into a lever equation:

$$[F_{act}\ M_{act}]^T = [1\ L_{ext}]^T F_{tip} \quad (5)$$

if inertial terms are negligible. Equation (5) combined with (3) leads to:

$$[\delta_{act}\ \alpha_{act}]^T = \mathbf{S_{act}} \left([1\ L_{ext}]^T F_{tip} + [0\ M_{pzt}]^T \right) \quad (6)$$

As previously mentioned $M_{pzt} = M_V V_{pzt}$, i.e. after calibration, M_{pzt} can be directly derived by measuring the input voltage. The sensor, as shown later, will provide a measurement of α_{act}. Therefore, after algebraic manipulation, (6) can be rewritten as:

$$[\delta_{act}\ F_{tip}]^T = \mathbf{G_{act}}\ [\alpha_{act}\ V_{pzt}]^T \quad (7)$$

moreover, δ_{tip}, i.e. the displacement at the tip of the extension, is easily evaluated as:

$$\delta_{tip} = [1\ L_{ext}] \cdot [\delta_{act}\ \alpha_{act}]^T = \delta_{act} + \alpha_{act} L_{ext} \quad (8)$$

4 Model of Piezoelectric Sensor

The model for the charge across the grounded electrodes of a narrow strip of piezoelectric material bonded to a cantilevered beam subjected to a given nonuniform curvature is derived here similarly to [7]. Considering only the piezoelectric material, let S and T be respectively the x-axis components of the mechanical strain and stress and E and D be respectively the z-axis components of the electric field and the dielectric displacement. The other components are not considered, as in [7]. The constitutive piezoelectric equations are:

$$\begin{cases} S &= (Y_p^E)^{-1} T + d_{31} E \\ D &= d_{31} T + \epsilon^T E \end{cases} \quad (9)$$

where Y_p^E is the Young's modulus at constant electric field, d_{31} is piezoelectric coefficient and ϵ^T is the dielectric permittivity at constant stress. The first equation of (9) can be substituted into the second leading to:

$$D = d_{31}Y_p^E S + (\epsilon^T - Y_p^E d_{31}^2)E \quad (10)$$

As in [5] and [7], the strain is directly related to the neutral axis $z_n(x)$ by:

$$S = -(z - z_n)\frac{\partial^2 z_n(x)}{\partial x^2} \quad (11)$$

Integrating (10) along z from one electrode located at $z_1 = z_p - t_p/2$ to the other located at $z_2 = z_p + t_p/2$ and considering that $\int E\,dz = 0$ when extended between the two grounded electrodes z_1 and z_2:

$$\begin{aligned}\int_{z_1}^{z_2} D\,dz &= d_{31}Y_p^E \int_{z_1}^{z_2} S\,dz \\ &= -\frac{\partial^2 z_n(x)}{\partial x^2} d_{31}Y_p^E \int_{z_1}^{z_2}(z - z_n)\,dz\end{aligned} \quad (12)$$

Since divergence of the dielectric displacement must be null and only the z-axis component D is here considered, $D = constant$. Both integrals in (12) can now be evaluated, leading to:

$$D = -d_{31}Y_p^E(z_p - z_n)\frac{\partial^2 z_n(x)}{\partial x^2} \quad (13)$$

At each point x, the dielectric displacement D is proportional to the curvature $\frac{\partial^2 z_n(x)}{\partial x^2}$ which is the only term in (13) which varies along x. As in [5], in order to derive the dielectric charge built in the piezoelectric material, the flux of D across the area of one of the electrodes (i.e. $\Sigma = [0\ W_s] \times [0\ L_{act}]$) can be evaluated as:

$$\begin{aligned}Q &= \int\int_\Sigma D\,dxdy \\ &= -W_s d_{31} Y_p^E (z_p - z_n) \left.\frac{\partial z_n(x)}{\partial x}\right|_0^{L_{act}} \\ &= -W_s d_{31} Y_p^E (z_p - z_n) \alpha_{act}\end{aligned} \quad (14)$$

since the sensor is subjected to the same clamping conditions of the actuator, i.e. cantilevered, and therefore $\frac{\partial z_n(0)}{\partial x} = 0$ and $\frac{\partial z_n(L_{act})}{\partial x} = \alpha_{act}$.

5 Model Limits

Equation (14) relates the polarization charge built in the sensor directly to the slope α_{tip} at the end of unimorph actuator and before the extension. A simple way to measure the polarization charge is by reading the voltage that develops across the sensor electrodes when these are left disconnected. In reality piezoelectric materials are affected by several kind of losses, among these the dielectric ones. Considering the piezoelectric layer as a capacitor, if a static charge is initially placed across its electrodes, the capacitor will eventually discharge because of ohmic losses of the dielectric material constituting the capacitor itself. A piezoelectric layer can then be thought of as a pure capacitor C_0 with a resistance R_0 in parallel which takes into account the dielectric losses. Elementary circuit theory shows that such a configuration adds a zero in the origin and a pole, i.e. $s/(R_0 C_0 s + 1)$, to the final frequency response of the sensor. $R_0 C_0$ represents the discharging time constant of the lossy piezoelectric capacitor.

The pole at $(R_0 C_0)^{-1}$ is the lower limit for the frequency range at which the sensor provides a flat response. It is not possible to eliminate the zero in the origin, i.e. the sensor cannot be used at DC or in quasi-static conditions, but it is possible to keep the pole at low frequencies by using a charge amplifier, as shown in Fig. 3. This well known circuit pumps an external charge into the sensing electrode in order to neutralize the polarization charge and to keep the voltage across the piezoelectric layer at zero (virtual ground). In this way, since the voltage is virtually zero, there is no effect due to the dielectric losses. In fact, a charge amplifier as the one in Fig. 3 beside the capacitor C has to include a resistor R (not shown in the picture) in parallel. Thus instead of being limited by the $(R_0 C_0)^{-1}$ the band is limited by $(RC)^{-1}$ which is controllable and can be set to be small enough not to interfere with the frequency range of interest.

5.1 Band upper limit

The flat-response band upper limit of the actuator/sensor system is solely given by the first mode of the unimorph actuator. Such a frequency depends upon the geometrical dimensions of the actuator itself which should be designed so that the first resonant mode occurs much after the working bandwidth ($100 - 200Hz$ for the MFI). Having a flat response simply means that, at any frequency within the flat range, a static model for the actuator is sufficient, i.e. matrices $\mathbf{S_{act}}$ and $\mathbf{G_{act}}$ are frequency independent. After the first resonance, a non-flat response of the unimorph actuator could be theoretically considered and δ_{tip} and F_{tip} could still be derived by (7) but $\mathbf{G_{act}}$ would now be frequency dependent.

6 Calibration

Calibration is a necessary step in order to be able to practically use the actuator/sensor device. Both ac-

F_{tip}	V_{pzt}	δ_{tip}
0	150V	450μm
31.3mN	0	200μm

Table 1: DC measurement for characterizing the stiffness and the free displacement of the actuator.

tuator and sensor calibrations need to be performed but while the former can be done at DC, i.e. with a static procedure, the latter has to be done at AC, i.e. dynamically, since the sensor provides no response at DC.

As also described in [4], by means of an optical microscope, a digital video camera and a TV screen, it is possible to measure linear deflections ranging in the order of $10 - 10^3 \mu m$. For the actuator, described by (6), only EI and M_V ($M_{pzt} = M_V V_{pzt}$) need to be determined since L_{act} and L_{ext} are known. A bending actuator can be characterized at DC by its stiffness and its free displacement. For the former, zero voltage is applied at the electrodes while for the latter no force is applied at the tip. Table 1 reports typical values where $F_{tip} = 31.3mN$ is applied by hanging a $3g$ mass at the tip of the actuator. Measurements in Table 1 are sufficient to calibrate EI and M_V from (6) and (8).

Calibration for the sensor has to be performed at AC, i.e. dynamically, within the flat response band. With reference to Fig. 3, given an (AC) input voltage of amplitude $\tilde{V}_{pzt}$, a signal of amplitude $\tilde{V}_{out}$ can be detected from the charge amplifier and, via optical microscope, the corresponding displacement amplitudes $\tilde{\delta}_{tip}$ and $\tilde{\delta}_{act}$ can be measured. The last two measurements are necessary to infer, via (8), $\tilde{\alpha}_{act} = \frac{\tilde{\delta}_{tip} - \tilde{\delta}_{act}}{L_{ext}}$. Frequency response $H(j\omega) = \frac{V_{out}(j\omega)}{V_{pzt}(j\omega)}$ of the unloaded actuator can be measured by means of a Dynamic Signal Analyzer (HP3562A). For a given general input $V_{pzt}(j\omega)$, in the frequency domain, $\alpha_{act}(j\omega)$ can be evaluated as:

$$\begin{aligned}\alpha_{act}(j\omega) &= -\frac{\tilde{\alpha}_{act}}{\tilde{V}_{out}} V_{out}(j\omega) \\ &= -\frac{\tilde{\alpha}_{act}}{\tilde{V}_{out}} H(j\omega) V_{pzt}(j\omega)\end{aligned} \quad (15)$$

Equation (15) is justified by the fact that α_{act} is related to the polarization charge Q (and therefore to the voltage V_{out}) by a linear relationship as in (14). Such a multiplying factor is derived by the calibration constants $\tilde{\alpha}_{act}$ and $\tilde{V}_{out}$. The minus sign takes into account the inverting functionality of the charge amplifier in Fig. 3.

Once V_{pzt} is known and α_{act} is derived by (15), δ_{act} and F_{tip} can be determined from (7) while δ_{tip} is derived from (8).

As previously mentioned, the ground shield in Fig. 1 is designed to limit the direct coupling between actuator and sensor. Although greatly attenuated (actuators without such ground shield are affected by a very strong feed-through), it is not completely zeroed, resulting in a small offset signal always in phase[1] with V_{pzt}. Although not perceptible for quantities such as α_{act}, δ_{act} and δ_{tip}, it is evident for F_{tip} since, when no loading effect is applied at the tip, a non zero force $F_{tip} = F_0$ is measured from (7).

In order to counterbalance such an offset, (7) can be rewritten as:

$$[\delta_{act}\ F_{tip}]^T = \mathbf{G_{act}}\ [\alpha_{act}\ (1+\Delta_{off})V_{pzt}]^T \quad (16)$$

where $\mathbf{G_{act}}$ is now calibrated (since EI and M_V are calibrated) and the constant value Δ_{off} is used to introduce a counterbalancing offset always in phase with V_{pzt}. Δ_{off} is chosen in order to zero (or better minimize) F_{tip} at low frequencies (but still within the flat response frequency range), i.e. canceling out the force offset F_0.

7 Testing and Application

An actuator with an embedded sensor was fabricated as previously described. Its purpose is to be used for driving and sensing, via a 4-bar mechanism (basically a lever), a wing attached at the end of the 4-bar itself as in Fig. 7.

Since nonlinear behaviors are expected when dealing with large angular displacements of the wing (due to the 4-bar mechanism), the fabricated actuator/sensor device was first tested by means of simple point wise mass $m_{tip} = 190mg$ attached at the tip of the extension. A point wise mass was chosen in order to match the inertia of the wing as seen before the 4-bar structure, i.e. mimicking the real structure without introducing nonlinearities. The frequency response of such a system was obtained by sweeping, over the frequency range of interest for the MFI, a driving signal V_{pzt} of constant amplitude $10V$ peak-to-peak and by

---
[1] Since the circuit in Fig. 3 is meant to measure a charge, any parasitic capacitance C_{par} between the actuating electrode and the sensing one will generate a charge $Q_{par} = C_{par} \Delta V$, where ΔV is the voltage between the two electrodes. Because of the operational amplifier, the sensing electrode is kept at virtual ground, i.e. $\Delta V = V_{pzt}$. Q_{par} is then proportional to V_{pzt} and therefore always in phase with it.

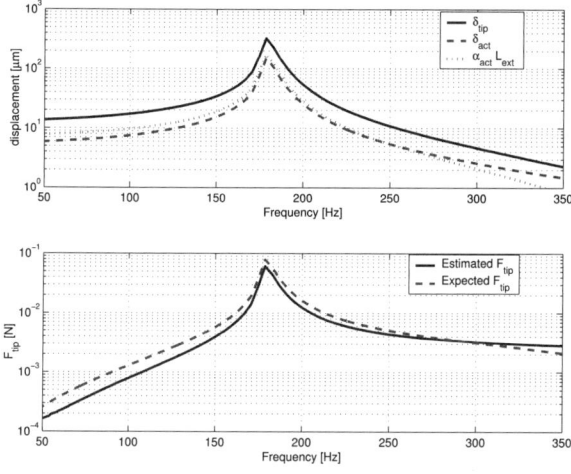

Figure 6: Amplitude versus frequency plots of (above) α_{act}, δ_{act}, δ_{tip} and (below) F_{tip} estimated and expected.

reading the output signal V_{out} out of a charge amplifier as the one in Fig. 3 (a $10nF$ capacitance was used with a $5M\Omega$ in parallel, leading to an expected lower band limit of about $3Hz$). By using (15) and (16), i.e. after calibration, δ_{act} and F_{tip} can be estimated and therefore δ_{tip} is also derived from (8). $\alpha_{act}L_{ext}$, δ_{act} and their sum δ_{tip} are shown in the upper plot in Fig. 6. In the lower plot the estimated F_{tip} and its expected value are compared, showing fairly good matching. The expected value corresponds to the inertial force due to m_{tip}, i.e. the mass itself times the second time derivative of δ_{tip} (acceleration of m_{tip}), which in the frequency domain can be expressed as $(j\omega)^2 \delta_{tip} m_{tip}$. For δ_{tip} the estimated value was used since it agreed well with what was measured via the optical microscope at various frequencies.

The test just described shows that the sensor behaves accordingly to the model when the actuator is subjected to operating conditions of interest for the MFI. It can now be tested with the MFI mechanical structure.

7.1 Application to the MFI

A simple diagram of the thorax of the MFI is shown in Fig. 7. A 4-bar mechanism is used as a mechanical transformer to convert the force and linear displacement of the actuator into the torque and the angular displacement needed to drive the wing. Large forces are transformed into small wing torques which are difficult to directly measure without wires

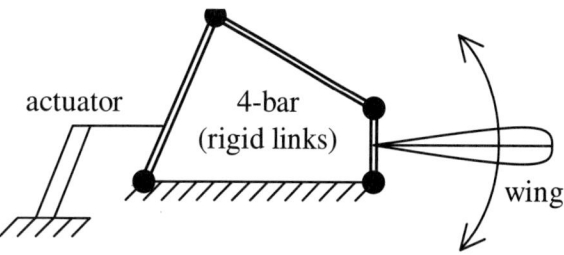

Figure 7: Simplified diagram of the MFI thorax comprising actuator, 4-bar (lever) mechanism and flapping wing.

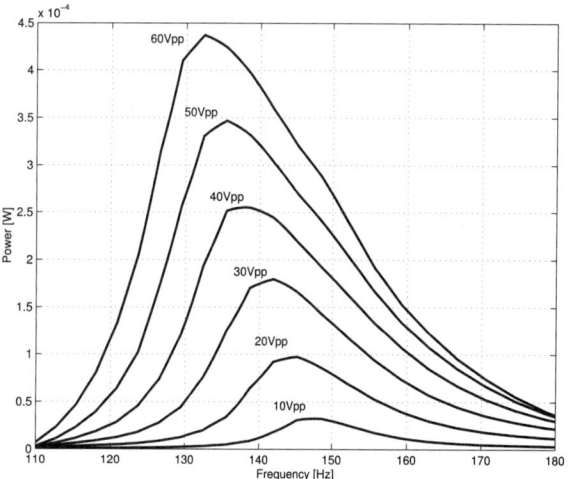

Figure 8: Power dissipated by the wing for incremental driving voltage.

on wings. Sensors are more conveniently placed on the actuating section, and the wing angle is inferred by knowing the 4-bar amplification ratio (although nonlinear). Fig. 8 shows, for increasing input voltage, the mechanical output power dissipated by the wing $\frac{1}{2} Re\left[v_{tip}(j\omega) F_{tip}^*(j\omega)\right]$ provided by the actuator, where $v_{tip}(j\omega) = j\omega\, \delta_{tip}(j\omega)$ is the velocity of the actuator tip (in the frequency domain) and the superscript * denotes the complex conjugate. Both $v_{tip}(j\omega)$ and $F_{tip}(j\omega)$ have been estimated from the input voltage V_{pzt} and the sensor output.

Fig. 8 also shows a typical nonlinear behavior, i.e. the resonant frequency shifts towards lower frequencies when the driving voltage amplitude is increased. The actuator, which is linear at the operating conditions of the experiment, is transformed into a nonlinear one by the saturation of the 4-bar transmission ratio. In particular, a softening of the transformed actuator occurs, i.e. for larger displacement the actuator becomes more compliant. It can be shown, [3], that a damped mass-spring system with a softening spring leads to frequency shift as the ones shown in Fig. 8.

8 Summary and Future Work

By patterning the electrodes of a piezoelectric layer by means of a low power laser beam, sensing capabilities were embedded into a standard piezoelectric unimorph actuator. In order to reduce the electromechanical coupling between actuating and sensing section an additional electrode was patterned and grounded providing a shielding effect that significantly reduced the coupling. A model was derived for the sensor which, together with the static model of the actuator, provided a simple means of estimating external force and displacement at the tip of the actuator itself. The calibration steps were outlined in order to practically use a fabricated device. A point wise mass was attached at the tip of the actuator and its weight was chosen in order to reproduce the operating conditions arising when an wing is driven, through a coupling mechanism, by the actuator. The force measured by the sensor matched the expected ones which were computed by perfectly knowing the nature of the load, i.e. the point wise mass. The actuator/sensor device was then used to drive a MFI thorax and the sensor signal was processed to derive the mechanical power fed into the wing, an important information for wing flapping mechanisms. As future work, the actuator/sensor device will be used to induce the wing to track the desired kinematics.

Acknowledgments

The authors would like to thank Srinath Avadhanula, Gabe Moy, Metin Sitti, Joe Yan and Robert J. Wood for useful discussions and assistance.

References

[1] R.S. Fearing, K.H. Chiang, M. Dickinson, D.L. Pick, M. Sitti, and J. Yan, "Wing Transmission for a Micromechanical Flying Insect," *Proc. of the IEEE Robotics and Automation Conf.*, pp. 1509-1515, San Francisco, CA, USA, April 2000.

[2] C. K. Lee, "Piezoelectric Laminates: Theory and Experiments for Distributed Sensors and Actuators," *Intelligent Structural Systems,* Kluwer Academic Publishers, pp. 75-167, December 1992.

[3] S. Sastry, *Nonlinear Systems: Analysis, Stability, and Control*, Springer-Verlag, New York, 1999.

[4] M. Sitti, D. Campolo, J. Yan, R.S. Fearing, T. Su, D. Taylor, and T. Sands, "Development of PZT and PZN-PT Based Unimorph Actuators for Micromechanical Flapping Mechanisms, " *IEEE Int. Conf. Robotics and Automation*, pp. 3839-3846, Seoul Korea, May 21-26, 2001.

[5] J. G. Smits and A. Ballato, "Dynamics Admittance Matrix of Piezoelectric Cantilever Bimorphs," *Journal of Microelectromechanical Systems*, pp. 105-112, vol. 3, 1994.

[6] Kenji Uchino, *Piezoelectric Actuators and Ultrasonic Motors*, Kluwer Academic Publishers, Boston, 1997.

[7] M. Weinberg, "Working equations for piezoelectric actuators and sensors," *Journal of Microelectromechanical Systems*, vol. 8, pp. 529-533, Dec. 1999.

Online Motion Planning using Laplace Potential Fields

Diego Alvarez   Juan C. Alvarez   Rafael C. González
{dalvarez, juan, corsino}@isa.uniovi.es
Electrical & Computer Engineering Dept.
University of Oviedo
33204 Gijón, Asturias, Spain

Abstract— Robot Navigation is an especially challenging problem when only online sensor information is available. The main problem is to guarantee global properties, such as algorithm convergence or trajectory optimality, based on local information. In this paper we present a new non-heuristic sensor-based planning algorithm, characterized by: 1) it is based in potential functions, allowing to introduce optimality criteria, 2) it is computed incrementally to introduce last sensor readings, and 3) it accounts for robot dynamics. The result is a method suitable for real-time navigation, it is intuitive and easy to understand, and produces smooth and safe trajectories.

I. INTRODUCTION

The basic robot motion planning problem consists of finding a collision-free path joining an initial and a final position on workspace. In an unknown terrain, the robot obtains local terrain information by continuously employing a sensor system [10].

For the incomplete information problem, there are several methods which have been proved to work well in practical situations, but do not guarantee convergence (heuristic algorithms).

As alternative, some researchers have proposed exact sensor based planning algorithms, see the seminal work of Lumelsky [9]. But these algorithms are not designed to optimize parameters such as distance or time.

This paper presents a new non-heuristic sensor-based planning method. It is based in potential functions computed incrementally, allowing to introduce an optimality criteria. The robot control commands are continuously being updated using sensor information while the robot travels to the goal. The method takes into account the robot dynamic motion restrictions, it is intuitive and easy to understand, and produces smooth and safe trajectories.

The paper is organized as follows. Section II reviews the literature concerned with non-heuristic motion planning with incomplete information, including Laplace's potential field approach. In Section III it is discussed issues related to the inclusion of robot dynamics in Laplace's potential methods. Section IV presents the proposed online planning algorithm, and Section V resumes the results and conclusions.

II. NON-HEURISTIC SENSOR-BASED MOTION PLANNING

Exact Sensor-Based Motion Planning algorithms compute collision-free paths while simultaneously they get new information about the environment.

Most of the real-time robot motion algorithms used in practice were developed to solve the *local obstacle avoidance* problem. In order to achieve global convergence, a hierarchical solution based on "planning + local avoidance" was used.

The first sensor based method with convergence guarantee was proposed by Lumelsky (*bug* algorithms) [9]. It computes the route to the goal using only local sensor information. Further extensions considered: robot dynamics [13], better sensing capabilities [8], and other experimental issues [2][7].

Another approach was to extend the local avoidance methods to provide global properties. Examples of these methods are extensions to the potential fields proposed by Khatib [6], such as the Global Dynamic Window [4]. They are based on the selection of different local fields using a set of heuristic rules. It makes difficult further convergence analysis.

For a complete information framework, potential fields can be calculated that guarantee convergence while producing optimized trajectories [11][3]. It would be interesting to see if any of these methods can be computed incrementally preserving their global properties.

Laplace Potential Fields guarantee convergence when complete information about the robot workspace is available [5]. This paper explores a method to incrementally compute such a field [12][14], while taking into account the dynamic constraints of the robot.

A. Motion planning based on Laplace's potential field

If a potential field verifies Laplace's equations:

$$\sum_{i=1}^{n} \frac{\delta^2 \phi}{\delta x_i^2} = 0 \qquad (1)$$

it is assured that it has not local minimum points [5]. By using Taylor's series expansion to solve this equation we

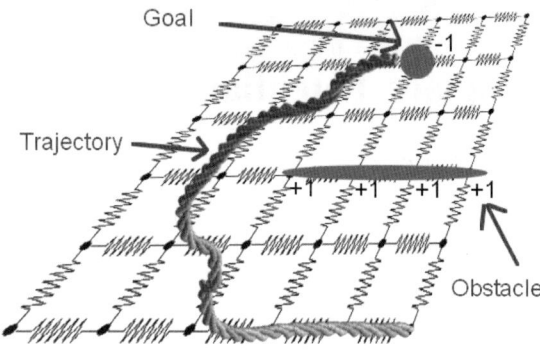

Fig. 1. Electric analogy representing Laplace's potential field. Trajectory traveled by a robot which moves toward the Goal while avoiding the Obstacle, following the minimum resistance path. Contour conditions are expressed as voltage sources applied to the Obstacle and Goal.

obtain the expression:

$$\phi(x_1,\ldots,x_n) = \phi(x_1-\delta,\ldots,x_n) + \ldots$$
$$\ldots + \phi(x_1+\delta,\ldots,x_n) + \phi(x_1,\ldots,x_n-\delta) + \ldots$$
$$\ldots + \phi(x_1,\ldots,x_n+\delta) \quad (2)$$

According to this solution, the computed potential field at each point is the mean value of the surrounding points of the grid. To compute the value of the field we only have to select the set of contour conditions which represent our problem and solve equation (1) by using a relaxation numerical algorithm. This numerical method sets an initial value for every point, and then evaluates equation (2) in all the points of the field, iterating this last step until achieving the convergence to the solution.

Contour conditions must be selected in a way that assure that there will not be collisions with the obstacles, and that the minimum of the field is placed at the goal. All the points of the grid which are inside an obstacle or within a forbidden region, are set to a *high* value (i.e. +1), while the point of the grid nearest to the goal is set to a *low* value (i.e. -1). The initial value for all the points of the grid which are not contour conditions must be set to an *intermediate* value (i.e. 0).

A way to understand Laplace's potential field is by using an electric analogy. Figure 1 shows how Laplace's field can be related to the voltage of resistors grid. The voltage at each grid node is the mean value of the nodes surrounding it, the same condition which appears when solving Laplace's equation (2). Contour conditions can be interpreted as voltage sources applied to the obstacles and to the goal.

Although the gradient lines of the potential field go along the correct direction to the goal, the magnitude of the gradient cannot be used as a motion command, because it has not a smooth variation. Due to this limitation,

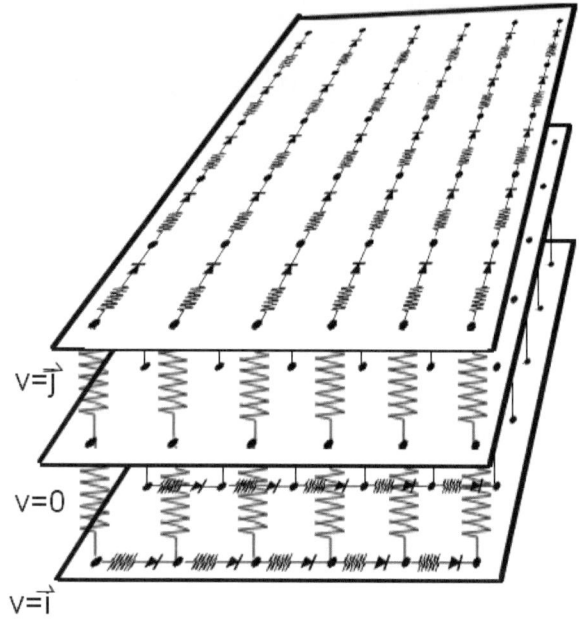

Fig. 2. Modified electric grid to include motion constraints. The three planes in the figure are related to three of the feasible velocities pair: $[v_x=0, v_y=1]$, $[v_x=0, v_y=0]$ and $[v_x=1, v_x=0]$. I.e. in the subplane $[v_x=0, v_y=1]$, the points of the grid are not connected along the x axis, and the movement is only allowed to a point of the grid with a bigger y coordinate.

Laplace's Potential field is only used to compute the motion direction, while the velocity values are selected using another criteria.

III. USING LAPLACE POTENTIAL FIELDS IN A ROBOT WITH DYNAMIC CONSTRAINTS

This potential field can be used by a free motion point But a real robot, with non-holonomic or dynamic constraints, needs a motion planning algorithm which computes a feasible trajectory, so that the motion constraints must be included in the potential field. To take this into account we must study:

1) How to create a potential field in the state-space of the robot including these constraints (Section III-A).
2) How to generate the motion commands using this potential field (Section III-B).
3) How to optimize the potential field to achieve the goal in a shorter time (Section III-C).

A. Potential Field Construction in State Space

To take into account nonholonomic or dynamic constraints, we can extend the potential field from the working space to the robot state space. If our robot is a free motion point with mass M, traveling in a bidimensional space $\Omega = (x,y)$, and commanded by a pair of forces (F_x, F_y)

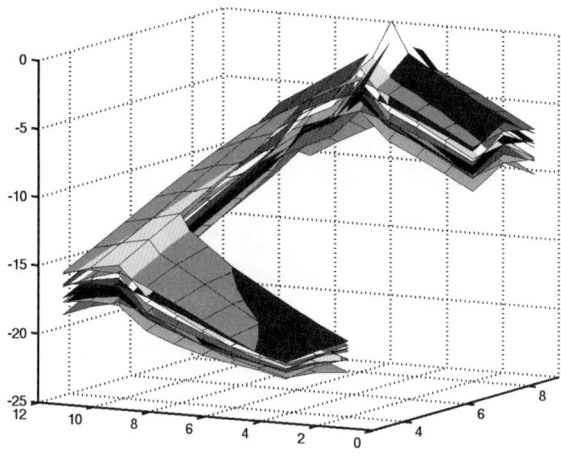

Fig. 3. In each point $[x_0, y_0]$ of the plane, the value of the potential field also depends on the velocities (v_x, v_y). For each pair $[v_{x0}, v_{y0}]$ of velocities, we have a potential surface on the (x, y) plane. In order to command the robot, the best choice is the pair of velocities $[v_{x0}, v_{y0}]$ which minimize the value of the potential field in the point $[x_0, y_0]$.

the associated state-space is:

$$E_1 = (x, y, v_x, v_y) \qquad (3)$$

The classic solution to Laplace's equation does not work in this new state-space. The problem can be seen in the next example: a robot must travel between an initial state $q_i = [x_1, y_1, 0, 0]$ and a final state $q_g = [x_2, y_2, 0, 0]$ in an environment without obstacles. Using equation (2), the solution will be a straight line between points q_i and q_g. But this line is placed in the subspace $v_x = v_y = 0$, so the robot must travel with zero velocity.

Therefore we must find a new set of contour conditions that allows us to solve the problem while taking into account the dynamic constraints. In the state-space of our example (3), coordinates v_x and v_y constrain the motions of the robot in the x and y components. These constraints must be included in the potential field, allowing the spread of the potential only in the directions where the movement is feasible. Using the electric analogy, the new Laplace's potential field would be related with a grid of resistors and diodes, and these diodes would only allow the movement of a charge in the unrestricted directions. Figure 2 illustrates a partial example of this grid.

This is only a variation in the contour conditions, so that it has no effect in the convergence properties of the method. The value of the field in each point is computed as the mean value of the reachable surrounding points.

B. Generation of Motion Commands

The next question to solve is the selection of the motion commands using the information of the field. If we select

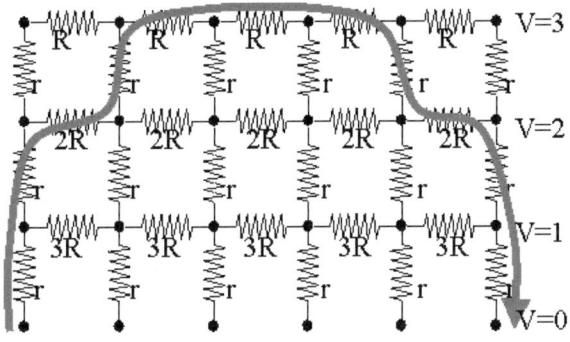

Fig. 4. The velocity variation along the trajectory, can be controlled adjusting the values of the resistors. If all the resistors of this grid are equal, then the trajectory would go along the shortest route of the grid. The values of the resistors for these hyperplanes are: R, $2R$, and $3R$ for $V = 3$, $V = 2$ and $V = 1$ respectively.

these commands using the gradient of the field, we would obtain a command vector in the space (x, y, v_x, v_y), which is the next desired state. However, we are not allowed to select the next state freely (we only have control over the forces F_x and F_y). The applied forces modify the velocities of the robot (v_x and v_y coordinates), while the position of the robot is modified according to these velocities. Therefore, the next command motion must be computed considering only the subspace:

$$E_{grad} = (x_0, y_0, v_x, v_y) \qquad (4)$$

where $[x_0, y_0]$ is the current position.

There are two options to select the force commands:
1) we can select the velocities $[v_{x0}, v_{y0}]$, which minimize the potential E_{grad} and select a force in order to achieve that velocity, or
2) we can calculate the gradient of the field E_{grad} in the (v_x, v_y) coordinates and use it as the force command $[F_{x0}, F_{y0}]$.

Figure 3 shows a potential field computed in the state space $E = (x, y, v_x, v_y)$. In this example, the allowed velocities are only $\{-1, 0, +1\}$ in each velocity component. The Figure shows nine surfaces corresponding to all the possible velocity combinations. We obtain the command motion to use by selecting at each point the pair of velocities which minimize the potential field, .

C. Optimization of the Potential field

To achieve a better control of the robot, there must be more velocity options than in the previous example $\{-1, 0, +1\}$. If we do not take this into account, the method will select a feasible trajectory, but using the lowest possible velocity. This problem is showed in Figure 4.

In that example we have a robot traveling in an unidimensional space x, with allowed velocities $v_x = \{0,1,2,3\}$. The selected trajectory will be the route with the lowest resistance value of the grid, and if all the resistors have the same value it will be the route with the minimum velocity. In order to achieve higher velocities, the value of the resistors must be inversely related to the velocity module. The values of the resistor connecting the different velocity hyperplanes (r in Figure 4), and the resistors placed in each velocity hyperplane $\{V=1, V=2, V=3\}$, can be adjusted to achieve an optimum velocity curve (related to the robot maximum accelerations).

IV. ONLINE APPLICATION IN A ROBOT WITH DYNAMIC CONSTRAINTS

In order to implement the previous method in an incremental way, better suited for sensor based motion, two things have to be settled: the model of the robot and the update mechanism.

A. Model of the robot

The robot model defines the potential field structure. For a circular mobile robot modeled as a disk, with mass M and inertia constant I and controlled by a force in the movement direction F, and a torque T, the associated state-space is:

$$E_2 = (x, y, \theta, v, \omega) \quad (5)$$

being its configuration space $\Omega_2 = (x, y, \theta)$.

In this subsection, we will repeat the points of Section III taken into account this new model of the robot.

1) Building the field: When building the potential field, we must consider that the component θ is a periodic one, and it can only take values in the interval $[0, 2\pi]$. It has two consequences:

- The grid used to compute the field must be divided in an integer number of sections (i.e. $\{\pi, \frac{3\pi}{4}, \frac{\pi}{2}, \frac{\pi}{4}, 0, -\frac{\pi}{4}, -\frac{\pi}{2}, -\frac{3\pi}{4}\}$).
- We must also remember that this is a periodic component when connecting the nodes of the grid. The nodes with minimum and maximum values of θ are neighbors. Due to this fact, when we compute the potential in the points with value $\theta = +\pi$ we must account for the values of the field in $\theta = +\frac{3\pi}{4}$ and $\theta = -\frac{3\pi}{4}$.

2) Command Generation: When we compute the motion commands, we must use the subspace

$$E_{grad2} = (x_0, y_0, \theta_0, v, \omega) \quad (6)$$

The commands can be the desired velocities $[v_0, \omega_0]$ or the pair $[F, T]$.

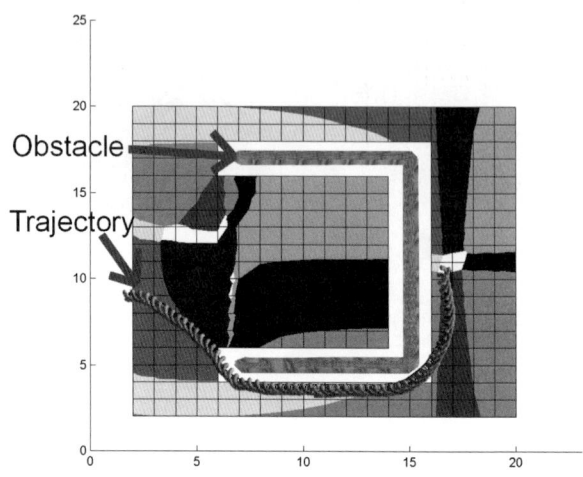

Fig. 5. Trajectory done by a mobile robot while avoiding a convex obstacle in its environment.

3) Optimization: When optimizing the potential field, the relations between the resistors are more complex than in the example of Section III. When computing the value of these resistors we must consider that:

- resistors connecting points with different (x,y) value must be inversely related to v,
- resistors connecting points with different θ value must be inversely related to ω, and
- the resistors connecting points with different v or ω values are chosen to achieve maximum linear and angular accelerations respectively.

In Figure 5, we can see a simulation of the trajectory traversed by a mobile robot controlled by Laplace's potential field as described.

B. Incremental update of the potential field

When the robot is traveling in a incomplete information framework, it must use the sensors aboard to detect all the relevant obstacles. The motion planning algorithm must include this information as soon as it is available, in order to avoid the collisions and to assure the convergence to the goal.

The method selected to solve the potential field equation (1) in the state-space of the previous robot model (5) is an iterative relaxation algorithm. In each iteration step the value of the potential field is recalculated until it converges to the solution. If complete information about the environment is available, the method will solve it, with a number of iterations proportional to the size of the environment and to the length of the shortest route to the goal.

Due to the properties of the relaxation method applied, we can include new sensorial data into the contour con-

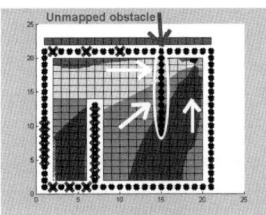
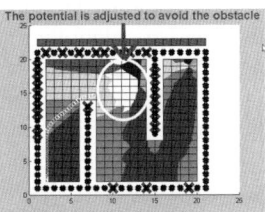

(a) Before seeing the second obstacle the robot has an initial *solution* to achieve the goal

(b) The robot sees the second obstacle, the potential field begins to change and the robot stops

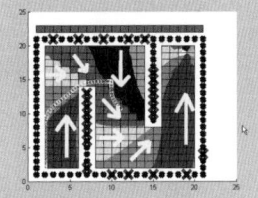
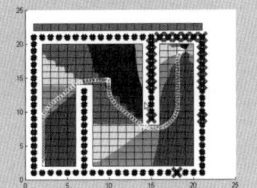

(c) The potential field is updated, and the robot begin to travel along the new solution

(d) Final trajectory to the goal

Fig. 6. The potential field changes when new obstacles are detected. When the robot sees the second wall, the potential field begins to be modified. Initially a stop region is created, avoiding the collision, and after a few iterations the potential field achieves the new solution.

ditions after each iteration. It naturally leads to the field update Algorithm 1.

Notice that the method was derived from an exact algorithm based on complete information, which does not depend of the initial values. After the appearance of each new obstacle the field is modified, but its convergence does not depend on initial conditions either. Therefore, the new motion commands do guarantee reaching the goal if no other obstacles are detected.

Finally, sensing capabilities must be taken into account to assure non collision with obstacles. When the sensors aboard detect a new obstacle, robot must have time enough to stop or to avoid it in spite of its dynamics and the time needed to upgrade the field [1].

V. RESULTS AND CONCLUSIONS

This paper presents a new non-heuristic sensor-based planning method based on potential functions computed incrementally. The method takes into account the robot dynamic motion restrictions, it is intuitive and easy to understand.

The method uses the values of the resistors in the grid to adjust the behavior of the robot to its nonholonomic and dynamic constraints. It is the most straightforward solution

Algorithm 1 Incremental Upgrade of the Laplace Field

1) Gather all the previous workspace information in the contour initial conditions
2) Solve the potential field in these conditions
3) Compute the motion commands (see Section III-B)
4) Apply motion commands
5) Gather information about the environment
6) If the robot has arrived to the Goal, send a stop command and finish the algorithm
7) Else, use the new information to update the field contour conditions
8) Update the potential field using Equation (2) in all the nodes of the grid
9) Return to Step 3

when the motion restrictions are related to the change of each coordinate. Other types of restrictions would require to adapt the method consequently.

A high computational load when working in large environments is the principal drawback of the proposed method. When working with very large maps, the potential field cannot be totally updated in only one control cycle.

The environment data is updated with each new sensor measure, upgrading the potential field at each iteration. The resulting trajectories are also continuously updated, as showed by the example in Figure 6. The potential field changes when new obstacles are detected. It creates a stop region avoiding the collision, and after a few iterations a new solution is reached. The final result is a smooth and a sure route to the goal.

VI. REFERENCES

[1] Juan C. Alvarez, Rafael C. González, Diego Alvarez, Andrei Shkel, and V. Lumelsky. Sensor management for local obstacle detection in mobile robots. In *IEEE/RSJ/GI Int. Conf. on Intelligent Robots and Systems*, Oct. 2002.

[2] Juan C. Alvarez, A. Shkel, and V. Lumelsky. Accounting for mobile robot dynamics in sensor–based motion planning. *Int. Journal of Robotics & Automation*, 16(3):132–141, 2001.

[3] Jérôme Barraquand, Bruno Langlacis, and Jean-Claude Latombe. Numerical potential field techniques for robot path planning. *IEEE Trans. Systems Man and Cybernetics*, march/april 1992.

[4] Oliver Brook and Oussama Khatib. High-speed navigation using the global dynamic window approach. In *IEEE Int. Conf. on Robotics and Automation*, 1999.

[5] Christopher I. Connolly and Roderic A. Grupen. On the applications of harmonic functions to robotics. *Journal of Robotic Systems*, 10:931–946, 1993.

[6] Oussama Khatib. Real-time obstacle avoidance for manipulators and mobile robots. *Int. J. Robotics Research*, 5(1):90–98, 1986.

[7] S. L. Laubach and J. W. Burdick. An autonomous sensor-based path-planner for planetary microrovers. In *IEEE Int. Conf. on Robotics and Automation*, pages 347–354, Detroit, MI, USA, 10-15 may 1999.

[8] V. Lumelsky and T. Skewis. Incorporating range sensing in the robot navigation function. *IEEE Trans. Robotics and Automation*, 20(5):1059–1069, sep 1990.

[9] V. J. Lumelsky and A. A. Stepanov. Dynamic path planning for a mobile automaton with limited information on the environment. *IEEE Trans. Autom. Control*, 31(11), nov 1986.

[10] Nageswara S. V. Rao, Srikumar Kareti, Weimin Shi, and S. Sitharama Iyengar. Robot navigation in unknown terrains: Introductory survey of non–heuristic algorithm. Technical Report ORNL/TM-12410, Oak Ridge National Lab., July 1993.

[11] Elon Rimon and Daniel E. Koditschek. Exact robot navigation using artificial potential functions. *IEEE Trans. Robotics and Automation*, 8(5):501–518, oct 1992.

[12] Keisuke Sato. Dead-lock free motion planning using the laplace potential field. Technical report, Robotics Labs. University of Tokyo, 1997.

[13] A. Shkel and V. Lumelsky. The jogger's problem: Control of dynamics in real-time motion planning. *Automatica*, 33(7):1219–1233, jul 1997.

[14] John S. Zelek and Martin D. Levine. Local-global concurrent path planning and execution. *IEEE Trans. Systems Man and Cybernetics*, 30:865–870, november 2000.

Potential-Based Path Planning for Robot Manipulators in 3-D Workspace

Chien-Chou Lin   Jen-Hui Chuang

Dept. of Computer and Information Science,
National Chiao Tung University,
Hsinchu, 30010, Taiwan, ROC

Abstract— A novel collision avoidance algorithm is proposed to solve the path-planning problem of a high DoF robot manipulators in 3-D workspace. The algorithm is based on a generalized potential field model of 3-D workspace. The approach computes, similar to that done in electrostatics, repulsive force and torque between manipulator and obstacles using the workspace information directly. Using these force and torque, a collision-free path of a manipulator can be obtained by locally adjusting the manipulator configuration for minimum potential. The proposed approach is efficient since these potential gradients are analytically tractable. Simulation results show that the proposed algorithm works well, in terms of computation time and collision avoidance, for manipulators up to 6 links.

I. INTRODUCTION

Path planning of a manipulator is to determine a collision-free trajectory from its original location and orientation (called starting configuration) to goal configuration [1]. Some planners adopt the configuration space (c-space) based approaches [2] [3] [4]. A point in the c-space indicates a configuration of manipulator which is usually encoded by a set of manipulator's parameters, e.g., joint angles between manipulator links. The forbidden regions in the c-space correspond to manipulator configurations which intersect obstacles. Thus, path planning is reduced to the problem of planning a path from the start to goal point in free space of the c-space.

Unlike c-space based approaches, workspace-based algorithms directly use spatial occupancy information of the workspace to solve path planning problem [5] [6] [7] [8] [9] [10] [11]. In addition to collision avoidance, some approaches try to find paths with minimum risk of collision. To that end, repulsive potential fields between manipulator and obstacles are used in [5] [7] [8] [9] and [11] to find the best match of their shapes in the path planning.(For a survey of related works please see also [7] and [12].)

In general, the potential function used to model the workspace is a scalar function of the distances between boundary points of the robot and those of obstacles. The gradient of such a scalar function, i.e., the repulsive force between the robot and obstacles, can be used to move the former away from latter making potential-based methods simple. In [5], an algorithm is developed to compute a safe and smooth object path by minimizing the potential function locally for obstacle avoidance, while the gross robot movement is subject to the constraints derived from the topology of the path given a priori.

In this paper, the potential field model presented in [6], as reviewed in the next section, is adopted to model the workspace for the path planning of robot manipulators. The approach computes, similar to that done in electrostatics, repulsive force and torque between objects in the workspace. A collision-free path of a manipulator can then be obtained by locally adjusting the manipulator configuration to search for minimum potential configurations using these force and torque. The proposed approach uses one or more guide planes (GPs) among obstacles in the free space as final or intermediate goals in the workspace for the manipulator to reach. These GPs provide the manipulator a general direction to move forward and also establish certain motion constraints for adjusting manipulator configuration during path planning, as discussed in Section 3. In Section 4, simulation results are presented for path planning of manipulators in different 3-D environments. Section 5 gives a summary of this work.

II. GENERALIZED POTENTIAL FIELDS IN THE 3-D SPACE

In [6], it is shown that the Newtonian potential, being harmonic in the 3-D space, can not prevent a charged point object from running into another object whose surface is uniformly charged. This is because the value of such a potential function is finite at the continuously charged surface. Subsequently, generalized potential models are proposed to assure collision avoidance between 3-D objects. The generalized potential function is inversely proportional to the distance between two point charges to the power of an integer and, as reviewed next, the potential and thus its gradient due to polyhedral surfaces can be calculated analytically. The path planner of manipulators proposed in this paper will use these results to evaluate the repulsion between manipulators and obstacles.

Consider a planar surface S in the 3-D space, the direction of its boundary, ΔS, is determined with respect to its

---

This work is supported by National Science Council of Taiwan under grant no. NSC89-2213-E-009-207 surface normal, $\hat{\mathbf{n}}$, by the right-hand rule, $\hat{\mathbf{u}} \times \hat{\mathbf{l}} = \hat{\mathbf{n}}$, where $\hat{\mathbf{u}}$ and $\hat{\mathbf{l}}$ are along the (outward) normal and tangential directions of ΔS, respectively. For the generalized potential function, the potential value at $\mathbf{r}$ is defined as

$$\int_S \frac{dS}{R^m}, \quad m \geq 2 \tag{1}$$

where $R = |\mathbf{r}' - \mathbf{r}|$, $\mathbf{r}' \in S$, and integer m is the *order* of the potential function. The basic procedure to evaluate the potential at $\mathbf{r}$ is similar to that outlined in [13] for the evaluation of the Newtonian potential ($m=1$) and can be summarized as follows:

(i) Write the integrand of the potential integral over S as surface divergence of some vector function.
(ii) Transform the integral into the one over ΔS based on the surface divergence theorem.
(iii) Evaluate the integral as the sum of line integrals over edges of ΔS.

Without the loss of generality, it is assumed that

$$d \stackrel{\Delta}{=} \hat{\mathbf{n}} \cdot (\mathbf{r} - \mathbf{r}') > 0 \tag{2}$$

which is equal to the distance from $\mathbf{r}$ to Q.

For (i), we have (see [6])

$$\frac{1}{R^m} = \nabla_S \cdot (f_m(R) \mathbf{P}) \tag{3}$$

where $\mathbf{P}$ is the position vector of $\mathbf{r}'$ with respect to the projection of $\mathbf{r}$ on Q, $\mathbf{r}_Q$, and

$$f_m(R) = \begin{cases} \dfrac{\log R}{R^2 - d^2}, & m = 2 \\ \dfrac{-1}{(m-2) R^{m-2}(R^2 - d^2)}, & m \neq 2 \end{cases} \tag{4}$$

Note that if $\mathbf{r}_Q$ is inside S, $f_m(R)$ will becomes singular for some $\mathbf{r}'' = \mathbf{r}_Q$ (and $R = d$). Let S_ε denote the intersection of S and a small circular region on Q of radius ε and centered at $\mathbf{r}_Q$, the potential due to S can be evaluated as

$$\begin{aligned}\int_S \frac{dS}{R^m} &= \lim_{\varepsilon \to 0} \left[\int_{S - S_\varepsilon} \nabla_S \cdot (f_m(R) \mathbf{P}) dS + \int_{S_\varepsilon} \frac{dS}{R^m} \right] \\ &= \sum_i \mathbf{P}_\mathbf{i}^0 \cdot \hat{\mathbf{u}}_\mathbf{i} \int_{C_i} f_{m,i}(l_i) dl_i + g_m(\alpha),\end{aligned} \tag{5}$$

where

$$f_{m,i}(l_i) = f_m(R = \sqrt{l_i^2 + d^2 + (P_i^0)^2}), \tag{6}$$

$$g_m(\alpha) = \begin{cases} \alpha \log d, & m = 2 \\ \dfrac{\alpha}{(m-2) d^{m-2}}, & m > 2 \end{cases} \tag{7}$$

P_i^0 is the distance between $\mathbf{r}_Q$ and edge segment C_i, l_i is measured from the projection of $\mathbf{r}$ on C_i along the direction of $\hat{\mathbf{l}}_i$, and α is the angular extent of the circumference of S_ε lying inside S as $\varepsilon \to 0$.

In this paper, only $m=3$ is considered and we have

$$\begin{aligned}\int_{C_i} f_{3,i}(l_i) dl &= \int_{l_i^-}^{l_i^+} f_{3,i}(l_i) dl \\ &= \frac{1}{P_i^0 d} \left[\tan^{-1} \frac{l_i^- d}{P_i^0 R^-} - \tan^{-1} \frac{l_i^+ d}{P_i^0 R^+} \right].\end{aligned} \tag{8}$$

with R^- and R^+ equal to the distances from $\mathbf{r}$ to the two end points of C_i, respectively. Thus, the repulsive force exerted on a point charge due to S can be found analytically by evaluating the gradient of

$$\Phi(x, y, z) = \frac{1}{z} \tan^{-1} \frac{xz}{y \sqrt{x^2 + y^2 + z^2}} \tag{9}$$

at some (x, y, z)'s.

For the potential-based path planning of manipulators, the evaluation of the repulsion between manipulators and obstacles involves the calculation of the repulsion between pairs of polygons; each pair has a polygon from the manipulator and the other from obstacles. For continuously charged object surfaces, it is obvious that the direct calculation of the potential between two polygons requires a quadruple integral. To simplify the mathematics, links of manipulator are approximately represented by a set of point samples on their surfaces in this paper. Usually, the sampling points are located on the vertexes and edges of links and their distribution should be as uniform as possible. The repulsion between manipulators and obstacles, in forms of repulsive force and torque, can then be estimated in closed form through superposition using the above analytical expressions.

III. PATH PLANNING

For a rough description of manipulator path, the proposed approach uses one or more guide planes (GPs) as final or intermediate goals in the 3-D workspace. The GPs are polygons among obstacles in the free space, providing the manipulator a general direction to move forward (see Fig. 1). A collision-free traversal of a given sequence of GPs by the end-effector is regarded as a global solution of the path planning problem of a manipulator.[1] In the planning procedure, intermediate GPs will also be added along the path each time when a given GP is not directly reachable.

A. Basic Procedure of Path Planning

Assuming that a guide plane GP_1 is given as an intermediate goal, the basic path planning procedure for moving the end-effector p' of a manipulator onto GP_1 include:

---

[1]In this paper, it is assumed the sequence of GPs are given in advance. Usually, the GPs can be obtained as a subset of the cross-sections associated with the *generalized cylinder* [14] representation of the passage.

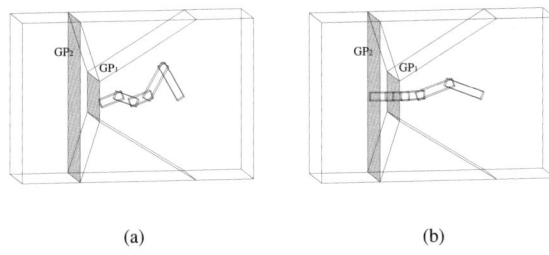

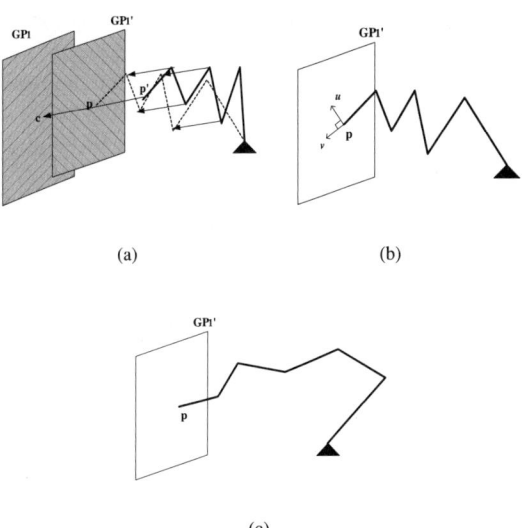

(a)   (b)

Fig. 1. A manipulator is moved toward the goal (not shown) by sequentially traversing a sequence of GPs.

(i) Translate the distal links[2] of the manipulator to move its end-effector p' toward the GP_1. If p' can not reach GP_1 directly, e.g., due to collision, a virtual intermediate plane GP_1' is inserted. (Fig. 2(a))

(ii) Find the minimum potential configuration of the manipulator for $p \in GP_1'$ by repeatedly executing :

  (a) Search for the minimum potential configuration of the manipulator with the distal link fixed in orientation. The minimization is performed by sliding p along two orthogonal directions on GP_1', e.g., $\vec{u}$ and $\vec{v}$. (Fig. 2(b))

  (b) Search for the minimum potential configuration of the manipulator with the end-effector p fixed in position. The minimization is performed by changing the orientation of the distal link with p as its rotation center. (Fig. 2(c))

(iii) Repeat (i) and (ii) until p reaches GP_1.

In general, there are different ways to change the manipulator configuration to move p toward GP_1. A simple translation of distal links is adopted as a preliminary implementation of (i). As shown in Fig. 2(a), the translation of the distal links is carried out to move the end-effector from p' to c, where c is the centroid of GP_1. If collision occurs or if the connectivity of manipulator can not be maintained, the distance of the translation is reduced until the translation is collision-free while the manipulator remain connected. Accordingly, a new GP is inserted, e.g., GP_1'. No configuration improvement to reduce the repulsive potential is considered at this stage.

As for the search for the minimum potential configuration of the manipulator in (ii), links of the manipulator are adjusted from the distal link to the base link using the repulsion experienced by the manipulator. The distal link has five DoFs, i.e., two for its location for $p \in GP_1'$ and three for its orientation. While each of other distal links

Fig. 2. Basic path planning procedure for a given GP (see text).

has three DoFs for its orientation, [3] the two base links, together, have at most three DoFs with their connecting joint being constrained to lie on a circle.

In (ii), the associated constrained optimization problem is divided into two iterative univariant optimization procedures, as in (ii-a) and (ii-b). In (ii-a), the distal link is fixed in its orientation (see Fig. 2(b)) as p slides on GP_1' to search for the minimal potential configuration and other distal links are sequentially adjusted in orientation, staring from the link connected to the distal link. In (ii-b), the distal link is adjusted in orientation while fixed in position (see Fig. 2(c)) and the procedure for adjusting the rest links is similar to that in (ii-a). For a particular GP, say GP_1' in Fig. 2, (ii-a) and (ii-b) are repeatedly performed until negligible changes in the manipulator configuration is obtained. Then, another intermediate GP which is closer to GP_1 is obtained with (i) and the process repeats. The path planning algorithm, as summarized below, ends as the end-effector reaches the given GP_1 or exist abnormally for an infeasible problem.

Algorithm End_Effector_to_GP

Step 0   Initialize $\delta = \overline{p'c_i}$ and $GP \leftarrow GP_i$.

Step 1   Translate the distal links of the manipulator with distance δ to move p' to p along the direction of $\overrightarrow{p'c_i}$. Find the smallest $n \geq 0$ such that $\delta \leftarrow \delta/2^n$ corresponds to a feasible and collision-free translation. If $\delta < \delta_{min}$, go to Step 6; otherwise, construct GP_i' with $GP_i' \parallel GP_i$ and $p \in GP_i'$, and let $GP \leftarrow GP_i'$.

---

[2] In step (i), an intermediate simple solution of the inverse kinematics problem is obtained by translating of all manipulator links except for the two base links, the base link and the link connected to it. In step (ii), the problem is solved by finding a sequence of sub-optimal solutions with monotonically decreasing potential. Finally, the minimal potential solution is found in (iii).

[3] In this paper, the spherical joint is adopted to connect links of a manipulator since its high DoF can take full advantage of the proposed potential minimization algorithm.

Step 2 Translate the distal link by sliding p on GP to minimize the potential.
Step 3 Adjust joint angle of the manipulator for the minimum potential configuration with p fixed in position.
Step 4 Go to Step 2 if the translation in Step 2 or the joint angle adjustment in Step 3 is not negligible.
Step 5 If p reaches GP_i, the planning is completed. Otherwise, $p' \leftarrow p$ and go to Step 1 with $\delta = p'c_i$.
Step 6 Exit and report that GP_i is unreachable.

For path planning involving multiple GPs, the above algorithm will be executed for each of them sequentially. It is assumed that the planning for a GP starts as the planning of the previous GP is completed. The path planning ends as the end-effector reach the goal, which is usually a (goal) GP in path planning problems considered in this paper.

B. Implementation Details

1) Step 1 of Algorithm End_effector_to_GP: In Step 1, the configuration of the two base links is determined simply by considering a planar motion on the plane formed by J_0 (the base joint), the old position of J_1, and the new position of J_2 such that an additional rotation will move J_2 to its new location. On the other hand, for simplicity, the end-effector is translated from p' to p by distance $\delta = max(p'c_i/2^n), n \geq 0$, along the direction of $\overrightarrow{p'c_i}$. A threshold, i.e., $\delta_{min} = 1\%$ of workspace size, is established to set a lower bound of the magnitude of allowed translation. A translation which requires a smaller movement indicates an infeasible path planning problem.

While no configuration improvement to reduce the repulsive potential is considered in the above translation procedure, Steps 2 and 3 minimize the potential through constrained optimization procedures. To that end, an intermediate GP, GP'_i, is added along the path to serve as a geometric constraint for successive adjustments of manipulator configuration. Again, as long as $p \in GP'_i$, the choice of the orientation of GP'_i is not unique. For simplicity, $GP'_i \parallel GP_i$ is adopted in our simulation.

2) Adjusting end-effector position in Step 2: Consider the forces exerted on the distal link lnk_n, as shown in more detail in Fig. 3. Let f_1 be the repulsive force exerted on lnk_n due to the repulsion between lnk_n and the obstacles, and f_2 be the force exerted on J_{n-1} due to the repulsive torque, denoted as τ_0, between other manipulator links (lnk_1 through lnk_{n-1}) and obstacles. For a univariant minimization approach, only one variable is adjusted at a time. To determine the minimum potential location of lnk_n as p sliding on GP, all of the joint angles of the manipulator, except the base joint J_0, are assumed to be fixed. Therefore, τ_0 can be calculated by considering a single rigid composite link formed by lnk_1 through lnk_{n-1} with respect to J_0. Thus, we have

$$f_2 = \frac{\tau_0}{l_0} \quad (10)$$

where l_0 is the length of $\overline{J_0 J_{n-1}}$.

To determine the direction in which p should slide on GP, and thus lnk_n should translate, to reduce the repulsive potential, the projection of the resultant force exerted on lnk_n along an arbitrary $\overrightarrow{u} \parallel GP$

$$f_u = f_{1u} + f_{2u} \quad (11)$$

is calculated, where subscript u denotes the projection along $\overrightarrow{u}$. A gradient-based binary search for the minimum potential location of p along $\overrightarrow{u}$ can then be performed using (11). The initial step of sliding is arbitrarily chosen as 10% of the workspace size. If the movement of lnk_n along $\overrightarrow{u}$ or f_u is negligible, e.g., the movement is smaller than 1% of workspace size, another minimizing scheme along $\overrightarrow{v} \in GP$, which is orthogonal to $\overrightarrow{u}$, is performed to minimize f_v. Step 2 ends when two consecutive binary searches along the two orthogonal directions result in negligible movement of lnk_n.

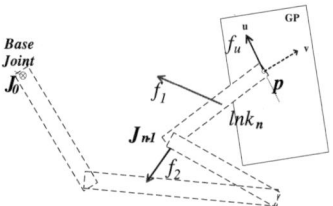

Fig. 3. Sliding p on GP, by translating lnk_n to reduce the repulsive potential.

In the above search processes, each time the position of p is changed, the orientation of rest links, i.e., joint angles at J_0 through J_{n-1}, need to be adjusted for connectivity and for minimum potential of the manipulator. Such a procedure is very similar to the joint angle adjustment performed in Step 3 of the algorithm, as discussed next.

3) Adjusting joint angle in Step 3: The direction in which the distal link should rotate is determined by the repulsive torque experienced by the distal link with respect to p. Let τ_n be the repulsive torque experienced by lnk_n with respect to p due to the repulsion between lnk_n and let f_2, as described in the previous subsection, be the force exerted on J_{n-1} due to the repulsive torque between other manipulator links and obstacles. The resultant torque experienced by lnk_n with respect to p is equal to

$$\tau_n^* = \tau_n + f_{2\perp} \cdot l_n \quad (12)$$

where l_n is the length of lnk_n and $f_{2\perp}$ is the projection of f_2 onto a plane perpendicular to lnk_n.

To find the minimum potential orientation of lnk_n for p fixed in position, gradient-based binary searches are performed using the projection of τ_n^* along three orthogonal axes of rotation, one axis at a time. For each binary search,

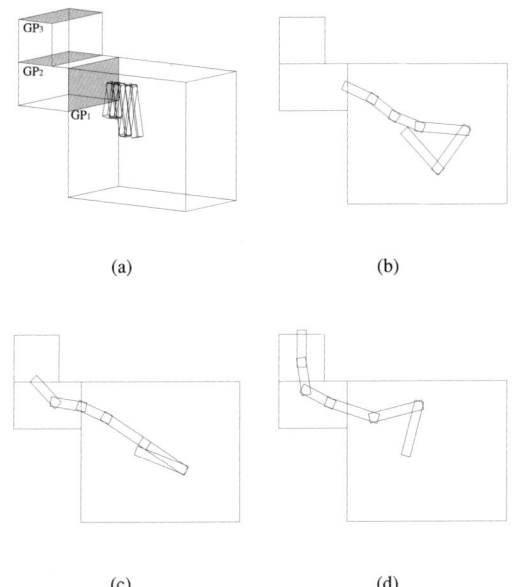

(a)   (b)

(c)   (d)

Fig. 4. A path planning example for a 6-link manipulator in a 3-GP workspace. (a) The initial configuration. (b)-(d) The intermediate configurations.

the initial rotating angle is arbitrarily chosen as 5°, while the accuracy in identifying the 1-D potential is chosen as 0.5^o. Each time the orientation of lnk_n is changed, the orientation of the rest links are adjusted recursively for connectivity and for minimum potential using τ_{n-1}^*, τ_{n-2}^*...., etc. Step 3 ends if a binary search using τ_n^* results in a negligible change in the orientation of lnk_n, i.e., 0.5^o.

IV. SIMULATION RESULTS

In this section, simulation results are presented for path planning performed on Pentium III (1GHz) for manipulators in 3D environment. Figs. 4(a) show the initial configuration for a 6-link manipulator in a 3-GP workspace wherein the GPs are shown as gray polygons. The simulation takes a total of 2.924 seconds to plan an 8-configuration collision-free path. Figs. 4(b)-(d) show side views of intermediate configurations of the manipulator, as its end-effector reaches each of the three GPs. Links of the manipulator basically lies in the middle of the workspace due to the repulsive potential model.

Figs. 5(a)(b) show the initial and final configurations, respectively, of the path planning for another 6-link manipulator stretching into a tapered passage with 4 GPs. The passage makes up, down and right turns at GP_1, GP_2 and GP_3, respectively. Figs. 5(c) shows the complete manipulator trajectory. In order to show the manipulator trajectory more clearly, configurations obtained by $End\_Effector\_to\_GP$ are shown in different gray levels,

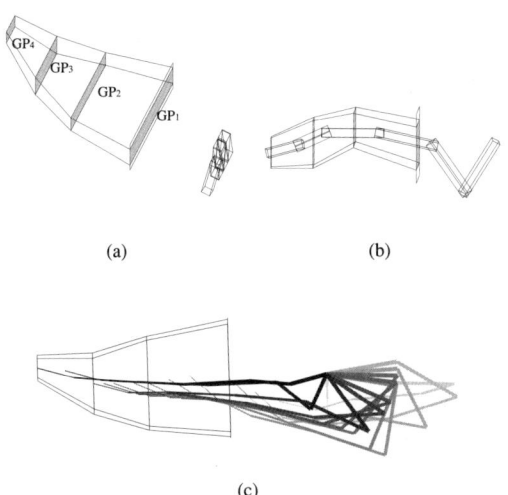

(a)   (b)

(c)

Fig. 5. A path planning example for a 6-link manipulator in a 4-GP workspace. (a) The initial configuration. (b) The final configuration. (c)The complete trajectory

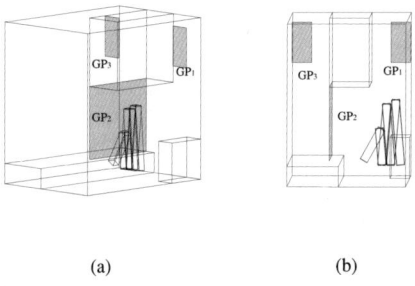

(a)   (b)

Fig. 6. Two different views of the initial configuration of a path planning example for a 6-link manipulator in a blocked workspace with 3 GPs.

i.e., the initial configuration is shown in white while final configuration is shown in black. It is readily observable that these trajectories are safe and smooth. The simulation takes a total of 26.648 seconds to plan the 12-configuration collision-free path.

Figs. 6(a)(b) show the initial configurations from two different views, respectively, of the path obtained for a 6-link manipulator working in a blocked workspace. Figs. 7-9 show the intermediate configurations and the partial trajectory as its end-effector reaches each of the three GPs, respectively. The simulation takes a total of 34.199 seconds to plan a 11-configuration collision-free path. Because the workspace is more complex, the path planning is more time-consuming than other examples.

V. SUMMARY

In this paper, a potential-based algorithm is proposed to solve path planning problems for robot manipulators

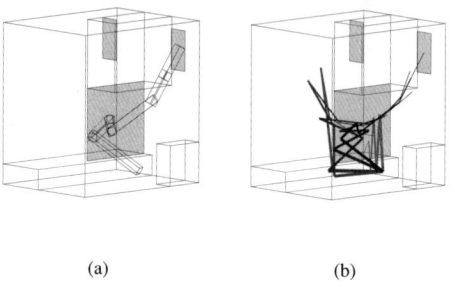

Fig. 7. (a) The intermediate configurations as manipulator end-effector reaches GP_1 in Fig. 6. (b) The partial trajectory between initial configuration and Fig. 7(a).

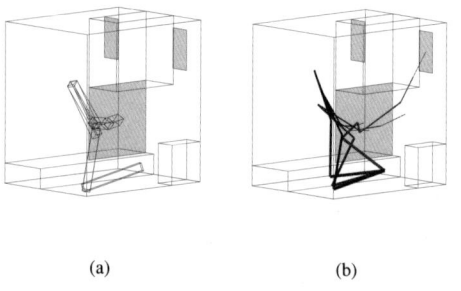

Fig. 8. (a) The intermediate configurations as its end-effector reaches GP_2 in Fig. 6. (b) The partial trajectory between Fig. 7(a) and Fig. 8(a).

in the 3-D workspace. The proposed algorithm uses an artificial potential field to model the workspace wherein obstacles surfaces are assumed to be charged uniformly and the manipulator is represented as a set of charged sampling points. The repulsive force and torque between manipulator and obstacles thus modeled are analytically tractable, which makes the algorithm efficient. To give a general direction of the path, a sequence of GPs to be reached by the manipulator are assumed to be given in advance in the workspace. As a GP is an intermediate or final goal for the end-effector of a manipulator to reach, it also helps to establish certain motion constraints for adjusting manipulator configuration during path planning. According to such constraints, the proposed approach

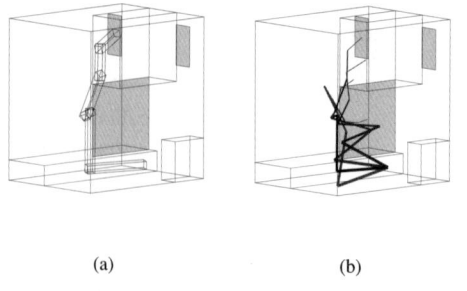

Fig. 9. (a) The intermediate configurations as its end-effector reaches GP_3 in Fig. 6. (b) The partial trajectory between Fig. 13(a) and Fig. 9(a).

derives the path for a manipulator by adjusting its configurations at different locations along the path to minimize the potential using the above force and torque. Simulations show that a path thus derived is always safe and spatially smooth.

VI. REFERENCES

[1] J. C. Latombe, "Motion planning: a journey of robots, molecules, digital actors, and other artifacts," *International Journal of Robotics Research*, vol. 18, no. 11, pp. 1119–1128, Nov. 1999.

[2] T. Lozano-Perez, "Spatial planning: a configuration space approach," *IEEE Trans. on Computers*, vol. C-32, no. 2, pp. 108–120, Feb. 1983.

[3] R. Brooks and T. Lozano-Preez, "A subdivision algorithm in configuration space for findpath with rotation," *IEEE Trans. on Sys., Man, Cybern.*, vol. 15, no. 2, pp. 224–233, March-April 1985.

[4] L. E. Kavraki, P. Svestka, J. C. Latombe, and M. H. Overmars, "Probabilistic roadmaps for path planning in high-dimensional configuration spaces," *IEEE Trans. on Robotics and Automation*, vol. 12, no. 4, pp. 566–580, Aug. 1996.

[5] J-H. Chuang and N. Ahuja, "An analytically tractable potential field model of free space and its application in obstacle avoidance," *IEEE Trans. on Sys., Man, Cybern., Part B: Cybernetics*, vol. 28, no. 5, pp. 729–736, 1998.

[6] J-H. Chuang, "Potential-based modeling of three-dimensional workspace of the obstacle avoidance," *IEEE Trans. on Robotics and Automation*, vol. 14, no. 5, pp. 778–785, 1998.

[7] Y. K. Hwang and N. Ahuja, "Gross motion planning a survey," *ACM Computer Survey*, vol. 24, no. 3, pp. 219–291, 1992.

[8] P. Khosla and R. Volpe, "Superquadric artificial potentials for obstacle avoidance and approach," *Proc. IEEE Intl. Conf. Robot. and Auto.*, pp. 1778–1784, 1988.

[9] J-H. Chuang, C-H. Tsai, W-H. Tsai, and C-Y. Yang, "Potential-based modeling of 2-D regions using nonuniform source distribution," *IEEE Trans. on Sys., Man, Cybern., Part A: Systems and Humans*, vol. 3, no. 2, pp. 197–202, 2000.

[10] K. P. Valavanis, T. Herbert, and N. C. Tsourveloudis, "Mobile robot navigation in 2-D dynamic environments using electrostatic potential fields," *IEEE Trans. on Sys., Man, Cybern., Part A: Systems and Humans*, vol. 30, no. 2, pp. 187–196, 2001.

[11] C-H. Tsai, J-S. Lee, and J-H. Chuang, "Path planning of 3-D objects using a new workspace model," *IEEE Trans. on Sys., Man, Cybern., Part B: Cybernetics*, vol. 31, no. 3, pp. 405–410, Aug. 2001.

[12] J. C. Latome, *Robot Motion Planning*, MA: Kluwer, Boston, 1991.

[13] D. R. Wilton, A. W. Glisson S. M. Rao, D. H. Schaubert, O. M. Al-Bundak, and C. M. Butler, "Potential integrals for uniform and linear source distributions on polygonal and polyhedral domains," *IEEE Trans. Antennas and Propagation*, vol. AP-32, no. 3, pp. 276–281, 1984.

[14] T. O. Binford, "Visual perception by computer," *Proc. Intl. Conf. on System and Controls*, December 1971, Miami.

Dual Dijkstra Search for Paths with Different Topologies

Yusuke Fujita[1], Yoshihiko Nakamura[2], Zvi Shiller[3]

e-mail: {[1]fujita,[2]nakamura}@ynl.t.u-tokyo.ac.jp, [3]shiller@barak-online.net

[1,2]Department of Mechano-Informatics, University of Tokyo

7-3-1 Hongo, Bunkyo-ku, Tokyo 113-0033 Japan

[3]The College of Judea and Samaria, Ariel 44837 Israel

Abstract

This paper describes a new search algorithm, Dual Dijkstra Search. From a given initial and final configuration, Dual Dijkstra Search finds various paths which have different topologies simultaneously. This alogrithm allows you to enumerate not only the optimal one but variety of meaningful candidates among local minimum paths. It is based on the algorithm of Dijkstra[10], which is popularly used to find an optimal solution. The method consists of two procedures: First computes local minima and ranks the paths in order of optimality. Then classify them with their topological properties and take out only the optimal paths in each groups. Computed examples include generating collision-free motion along 2D space and motion planning of 3DOF robot. We also proposed the idea of motion compression, which simplifies the high dimensional motion planning problem. Together with this idea, we applied Dual Dijkstra Search to 7DOF arm manipulation problem and succeeded in obtaining variety of motion candidates.

Key Words — Motion Planning, Path Planning, Topology, Local Minimum

1 Introduction

Early research on motion planning has focused on computing a single obstacle-free path in cluttered environments, with a focus on completeness and complexity [1]. The desire to produce algorithms that are complete (actually resolution complete), i.e. that find a path if one exists, resulted in algorithms that despite their relatively low (polynomial) complexity were computationally impractical when applied to high dimensional problems. The need to address practical motion planning problems in complex industrial environments [2] and for humanoids lead to the development of probabilistic and Monte Carlo search algorithms, typically known as the random roadmap approach[3]~[7]. While the probabilistic algorithms effectively trade-off completeness for computational efficiency, they have difficulties finding valid paths through narrow corridors in the free space, where the probability of sampling configurations is low.

The notion of a single path is impractical for redundant high degree-of-freedom systems, such as humanoid robots and computer graphics (CG) human figures, for which a single optimal solution may not represent the full span of their diverse valid motions for a given task. To enable diverse motion generation of such systems, it is desirable to compute several motion candidates among which the proper one can be selected for each circumstance.

To compute several motion candidates, we developed a new search algorithm, that we call the Dual Dijkstra Search (DDS). Unlike randomized algorithms, it searches over the entire configuration space to simultaneously generate the global optimal path, in addition to several distinct local minima. The first step of this approach runs the Dijkstra search twice to produce a list of ranked paths that span the entire graph in order of optimality. The number of paths in the list is the number of nodes in the graph. The second step prunes the initial set of paths to retain the local optimal path for each homotopy class, similar to the approach used in [9]. The computation of typical algorithms for K-best paths [10][11] is linear in the number of paths (K), which may be large if the paths are to span the entire graph. In contrast, the Dual Dijkstra Search computes paths that span the entire graph at a computational cost that is independent of the number of paths generated. The DDS requires two dijkstra searches (similar to $K = 2$) plus pruning. In this paper, we demonstrate the DDS for a 7 DOF arm.

2 Searching for paths of different homotopy classes

Most motion planning algorithms search in the configuration space for paths that connect the initial and final configurations. Paths of the same topology, or of the same homotopy class, are those that can be transformed from one to the other by continuous deformation, such as bending, stretching, twisting, and the like, without passing through the obstacles. The paths shown in Fig.1 cannot be transformed into the same path continuously and are hence topologically inequivalent.

Fig.1: Path with different topology

Typical algorithms for computing the "K shortest paths" [10][11] first search for the optimal path, then search for the second best, the third, and so on..., as shown in Fig.2. This approach has several disadvantages:

1. The computational cost grows linearly with K.

2. Paths are found in a decreasing order of optimality, generating many "similar" paths.

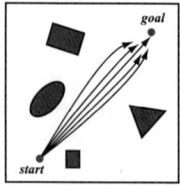   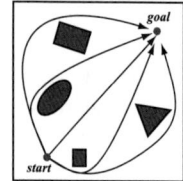

Fig.2: K Shortest Path Algorithm   Fig.3: Distinct Paths

Because of these disadvantages, generating paths of different topologies, like in Fig.3, is quite costly.

3 Dual Dijkstra Search

3.1 Dijkstra Search[8]

The Dijkstra search computes the shortest paths from the start node to all other nodes in the graph. For every node, it computes and stores, minimum cost to the start node and the previous node on the optimal path. By tracing back the previous node, like in Fig.4, it is possible to generate the optimal path. This algorithm is useful for generating only the optimal path between any two nodes, and cannot generate other candidates paths between the two nodes.

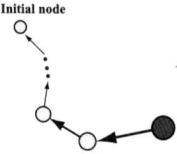

Fig.4: Pointer to previous node

3.2 Dual Dijkstra Search

The Dual Dijkstra Search algorithm consists of two steps. In the first step, for each node in the graph, we compute the shortest path between the start an goal nodes that passes through that node. In the second step, we select the shortest path in each homotopy class [9]. These steps are now described in some details.

3.2.1 Step 1

Step 1 is based on running the dijkstra search twice, as follows:

1. Run dijkstra search from the initial node as shown in Fig.5.

2. Run dijkstra search from the final node as shown in Fig.6.

3. For each node, add the costs computed by the two dijkstra searches.

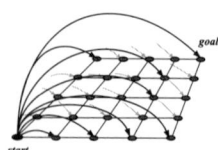   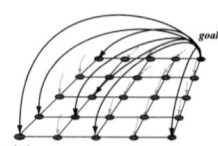

Fig.5: Path from start to all nodes   Fig.6: Path from goal to all nodes

After Step 1, each node stores:

1. The sum of the minimum costs to the start node and to the goal node, which is essentially the minimum cost for the path between the start and goal that passes through that node, as shown in Fig.7.

2. Pointer to the previous node on the path leading to the start node. In Fig.8, this is shown as pointer (a).

3. Pointer to the previous node on the path leading to the goal node. In Fig.8, this is shown as pointer (b).

Tracing the paths to the start and to the goal nodes produces the path from the start node, to passes through the via point, and reaches the goal node. Since each node represents a distinct path, the total number of paths that are generated in step 1 equals to the number of nodes on the graph.

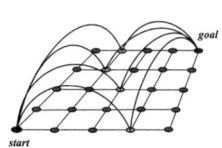

Fig.7: Path through a via point

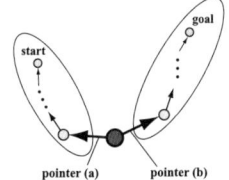

Fig.8: Pointers toward start and goal

3.2.2 Step 2

This step prunes the paths computed in Step 1 to retain only one path in each homotopy class. This is done by selecting the shortest path in a given set of paths and identifying *adjoint* paths, which have nodes that either coincide with, or are neighbors to at least three consecutive nodes of the best path. An *adjoint* path belongs to the same homotopy class as the best path since there is no "hole" (a node that does not belong to neither path) between the two paths. All paths in the set are checked successively in the order of their costs for adjoint nodes with the previously tested paths. If some node of the successive path does not adjoin any of the previous paths, that path becomes the shortest path in its own homotopy class:

1. Arrange an ordered list of all nodes according to the costs.

2. Select the minimum cost node from the list, and generate the shortest path that passes trough that node as shown in Fig.9-(1). Mark this path as the optimal path.

3. Attach flags to all nodes along the selected path and remove them from the list as shown in Fig.9-(2). Note that all the nodes along the path have the same cost.

4. Select the minimum cost node from the remaining list, and generate the shortest path through that node. And then,

   - If all the nodes along the path adjoin at least one flagged node, then discard of that path. The paths drawn in thin lines in Fig.9-(3) were discard.

   - If there is a node which does not adjoin any flagged node, mark that path. This path has different topology from the paths marked previously. The path generated through node F in Fig.9-(4) represents a newly marked path, having a different topology than the previous paths.

5. Repeat procedure 3 and 4 until all the nodes are flagged. Fig.9-(5) to Fig.9-(6) show schematically the remaining of the process until all nodes are either marked or discarded.

4 Implementation and Experiments

4.1 Implementation details

We implemented the DDS algorithm in a path planner to search for distinct paths in a given configuration space. The planner consists of the following three steps:

1. Construct a lattice graph in the configuration space.

2. Compute step 1 of Dual Dijkstra Search.

3. Compute step 2 of Dual Dijkstra Search.

To construct a lattice graph in the first step, we lay over the configuration space a uniform grid and assign cost to all edges. To move the path away from obstacles, we add a potential function to increase the cost near the obstacles.

4.2 Experimental results

We implemented a path planner in C++ for any given multi-dimensional space. First we randomly set obstacles in a two dimensional space as shown in Fig.10, and applied the planner. It produced the distinct paths shown in Fig.11. Every path in Fig.11 has a different topology and a local minimum of the corresponding homotopy class. The color in Fig.11 shows the cost distribution throughout the space. Nodes of

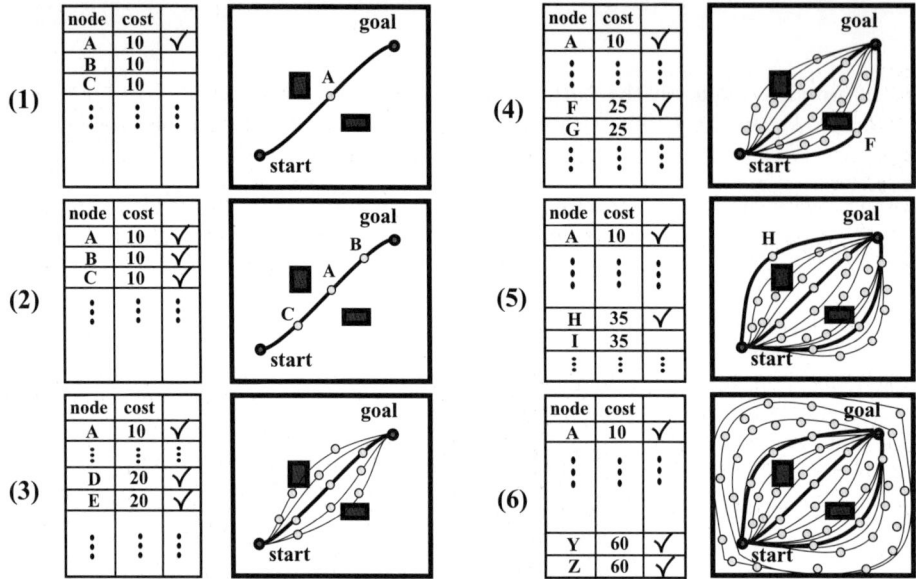

Fig.9: Procedure for step2

darker color generate the longest paths. The local optimal paths therefore pass through the bright areas in each topological regions.

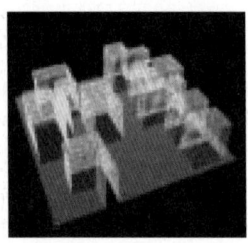

Fig.10: Environment with obstacles

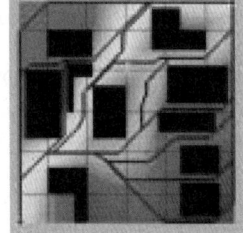

Fig.11: Path computed by Dual Dijkstra Search

In the following, we analyze the performance of the algorithm for 2 to 4 dimensional spaces. We randomly set some obstacles and applied the planner with several grid resolutions. All measurements were taken on Pentium 4, 2 Ghz machine with 256 MB of RAM. Table 1, Table 2, and Table 1 shows the computation time required in 2, 3, and 4-dimensional spaces respectively. P1, P2, and P3 in each graph corresponds to the three processes mentioned above.

As see from the tables, the computation time increases exponentially with the dimension of the space, and 3-dimensional configuration space seems to be the highest dimension in practical use.

4.3 Motion planning for a 3 DOF robot

We used this planner for a 3 DOF articulated robot. The obstacles were mapped to the configuration space

Table 1: Computation time for 2 dimensional space

resolution	P1(sec)	P2(sec)	P3(sec)
10	0	0	0
20	0.016	0	0
30	0.016	0	0
40	0.015	0	0
60	0.047	0	0.031

Table 2: Computation time for 3 dimensional space

resolution	P1(sec)	P2(sec)	P3(sec)
10	0.047	0	0
20	0.422	0.187	0.047
30	1.765	0.766	0.344
40	4.687	2.219	1.156
60	19.203	9.938	6.187

Table 3: Computation time for 4 dimensional space

resolution	P1(sec)	P2(sec)	P3(sec)
5	0.109	0	0
10	3.125	0.328	0.062
15	22.764	5.625	0.454
20	93.657	33.219	2.141

using the free collision detection software Coldet[12]. In this experiment, a floating ball served as an obstacle. The planner produced two different motions as shown bellow, each avoiding the obstacle differently.

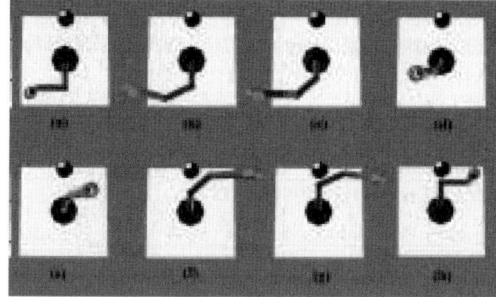

Fig.12: Motion 1

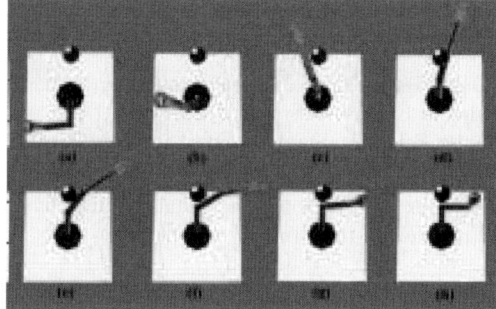

Fig.13: Motion 2

5 Motion Compression of High DOF Robots

The planner is impractical when we think of motion planning of high degree of freedom robots, because of the computational cost. If we can simplify the motion of such robots by representing their motion with a smaller number of parameter, we might find solution whit smaller computational cost.

5.1 Idea of motion compression

In order to simplify the motion of high DOF robots, we will represent their motion by the combination of "motion elements". We chose typical type of motion such as "raise hand", "extend hand to the left", and the like, as the motion elements. These motion elements correspond to the paths in configuration space,
and we call them principal trajectory. Preselect and store some principal trajectory, and by selecting and combining them, we can represent motion with a small number of parameter. It is important how to select the principal trajectory to combine, but let's suppose that we have selected appropriate three principal trajectory and continue the story.

Let p_1, p_2, and p_3 be the principal trajectory. We now normalize p_i with respect to time, and divide them into equal parts between the initial and final points along the trajectory. Each points on the trajectory signifies the posture at each moment. The main idea of simplifying robot's motion is to represent robot's posture as the interior point of the three points which corresponds to a same moment. What we are doing here is the approximation of the configuration space by set of planes which are made by connecting points on p_1, p_2, and p_3 as you can see in Fig.14. In Fig.14, a_i is the vector which represents the origin of each plane, and base vectors for each plane are b_i, and c_i. Fig.15 shows the i-th plane. j and k determines the ratio of combining p_1, p_2, and p_3. And now we can approximate the high dimensional configuration space with the subspace which can be represented by only three parameters, $(i\ j\ k)$, and can search for path in $(i\ j\ k)$ space.

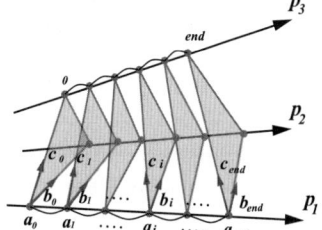 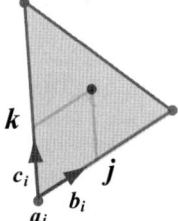

Fig.14: Search space   Fig.15: i-th surface

5.2 Experiments on motion compression with Dual Dijkstra Search

We used the idea of motion compression in order to apply Dual Dijkstra Search to motion planning of 7 DOF arm. We used three motions such as "extend arm to the left", "extend arm to the right", and "extend arm downward", as the principal trajectory, and we represented the 7 dimensional configuration space with only three parameters. we applied Dual Dijkstra Search in the space of those three parameters, and we obtained two different motions shown in Fig.16 and Fig.17. The way to avoid an obstacle differs in two figures, and these motions can be used as motion candidates.

Fig.16: Motion 1

Fig.17: Motion 2

6 Conclusion

In this paper, a new search algorithm, the Dual Dijkstra Search, was proposed to generate the local minimum path in noncrossing homotopy class. The planner generates paths that span the entire graph by applying the Dijkstra algorithm only twice, once from each end point. This is far cheaper than any typical K-best search algorithms, which may require a quite large K to generate paths that span the entire graph.

By representing robot motion with the combination of "motion elements", we simplified the high DOF problem into 3-dimensional motion planning problem. Using this idea of motion compression together with Dual Dijkstra Search, we obtained various motion candidates for a 7 DOF arm.

Acknowledgments

This research was supported by the CREST program, the Japan Science and Technology Corporation, and the Information-Technology Promotion Agency (IPA) Japan.

References

[1] J.-C.Latombe: *Robot Motion Planning*, Kluwer Academic Publishers, Boston, 1991.

[2] C. Van Geem, T. Simeon, J.-P. Laumond. "Mobility Analysis for feasibility Studies in CAD models of industrial environments", *IEEE Int. Conf. Robot. Autom.*, pp.1770-1775, Detroit, 1999.

[3] Lydia Kavraki and J.-C.Latombe, "Randomized Preprocessing of Configuration Space for Fast Path Plannin", *Proc. IEEE Intl. Conf. Robot. Autom.*, pp. 2138-2139, San Diego, 1994.

[4] Nancy M. Amato and Yan Wu, "A Randomized Roadmap Method for Path and Manipulation Planning", *Proc. IEEE Intl. Conf. Robot. Autom.*, pp. 113-120, 1996.

[5] Oliber Brock and Lydia E. Kavraki, "Decompositon-based Motion Planning: A Framework for Real-time Motion Planning in High-dimensional Configuration Spaces", *Proc. IEEE Intl. Conf. Robot. Autom.*, pp.1469-1474, 2001.

[6] David Hsu, "Randomized Single-Query Motion Planning in Expansive Spaces. PhD thesis, Dept. of Computer Science", Stanford Univ, Stanford, CA, 2000.

[7] J.J.Kuffner and S.M.LaValle: "RRT-Connect: An Efficient Approach to Single-Query Path Planning", *Proc. IEEE. Intl. Conf. Robot. Autom.*, 2000.

[8] E.W. Dijkstra: A note on two problems in connexion with graphs, Numerische Mathematik, 1:269–271, 1959.

[9] Z. Shiller and S. Dubowsky: "On Computing the Global Time Optimal Motions of Robotic Manipulators in the Presence of Obstacles", *IEEE trans. on Robotics and Automation*, 7(6):785-797, 1991.

[10] E.Q.V.Martions and J.L.E.Santos: "A new shortest paths ranking algorithm",
(http://www.mat.uc.pt/ẽqvm/cientificos/ investigacao/Artigos/K.ps.gz)

[11] E.L. Lawler: *Combinatorial Optimization*, Holt, Rinehart and Winston, New York, 98-100, 1976.

[12] ColDet: Free 3D Collision Detection Library, http://photoneffect.com/coldet/.

A Novel Potential-Based Path Planning of 3-D Articulated Robots with Moving Bases

Chien-Chou Lin   Chi-Chun Pan   Jen-Hui Chuang

Dept. of Computer and Information Science,
National Chiao Tung University,
Hsinchu, 30010, Taiwan, ROC

Abstract— This paper proposes a novel path planning algorithm of 3-D articulated robots with moving bases based on a generalized potential field model. The approach computes, similar to that done in electrostatics, repulsive force and torque between charged objects. A collision-free path can be obtained by locally adjusting the robot configuration to search for minimum potential configurations using these force and torque. The proposed approach is efficient since these potential gradients are analytically tractable. In order to speedup the computation, a sequential planning strategy is adopted. Simulation results show that the proposed algorithm works well, in terms of collision avoidance and computation efficiency.

I. INTRODUCTION

Path planning of a robot is to determine a collision-free trajectory from its original location and orientation (called starting configuration) to goal configuration [1]. Some planners adopt the configuration space (c-space) based approaches [2] [3] [4] [5]. A point in the c-space indicates a configuration of robot which is usually encoded by a set of a robot's parameters, e.g., joint angles between robot links. The forbidden regions in the c-space correspond to robot configurations which intersect obstacles. Thus, path planning of a robot is reduced to the problem of planning a path of a point in free space of the c-space.

Unlike c-space based approaches, geometric algorithms directly use spatial occupancy information of the workspace to solve path planning problem [6] [7] [8] [9] [10] [11] [12] [13]. Workspace-based algorithms usually extract relevant information about the free space and use them together with the robot geometry to find a path. In addition to collision avoidance, some approaches try to find paths with minimum risk of collision. To minimize such a risk, repulsive potential fields between robots and obstacles are used in [7] [8] [9] [10] [12] and [13] to match their shapes in the path planning.(For a survey of related works please see also [8] and [14].)

In general, the potential function used to model the workspace can be a scalar function of the distances between boundary points of the robot and those of obstacles. The gradient of such a scalar function, i.e., the repulsive force between the robot and obstacles, can be used to move the former away from latter making potential-based methods simple. In [9], an algorithm is developed to compute a safe and smooth object path by minimizing the potential function locally for obstacle avoidance, while the gross robot movement is subject to the constraints derived from the topology of the path given a priori. Since the potential is minimized for the obstacle avoidance purpose only, its local minima do not cause a problem in the path planning. Similar approach is generalized to the 3-D space for the path planning of rigid objects in [13].

In this paper, the generalized potential field model presented in [10], which is used in [13] and is reviewed in the next section, is adopted to model the workspace for the path planning of articulated robots. The proposed approach uses one or more guide planes (GPs) among obstacles in the free space as final or intermediate goals in the workspace for the articulated object to reach. These GPs provide the articulated object a general direction to move forward and also help to establish certain motion constraints for adjusting robot configuration for minimum potential during path planning, as discussed in Section 3. An implementation of the proposed approach based on a sequential planning strategy is also considered in this section. In Section 4, simulation results are presented for path planning performed for robots in different 3-D environments. Section 5 gives conclusions of this work.

II. GENERALIZED POTENTIAL FIELDS IN THE 3-D SPACE

In [10], it is shown that the Newtonian potential, being harmonic in the 3-D space, can not prevent a charged point object from running into another object whose surface is uniformly charged. This is because the value of such a potential function is finite at the continuously charged surface. Subsequently, generalized potential models are proposed to assure collision avoidance between 3-D objects. The generalized potential function is inversely proportional to the distance between two point charges to the power of an integer and, as reviewed next, the potential and thus its gradient due to polyhedral surfaces can be calculated analytically. The path planner of manipulators proposed in this paper will use these results to evaluate the repulsion between manipulators and obstacles.

This work is supported by National Science Council of Taiwan under grant no. NSC89-2213-E-009-151

Consider a planar surface S in the 3-D space, the direction of its boundary, ΔS, is determined with respect to its surface normal, $\hat{\mathbf{n}}$, by the right-hand rule, $\hat{\mathbf{u}} \times \hat{\mathbf{l}} = \hat{\mathbf{n}}$, where $\hat{\mathbf{u}}$ and $\hat{\mathbf{l}}$ are along the (outward) normal and tangential directions of ΔS, respectively. For the generalized potential function, the potential value at $\mathbf{r}$ is defined as

$$\int_S \frac{dS}{R^m}, \qquad m \geq 2 \tag{1}$$

where $R = |\mathbf{r}' - \mathbf{r}|$, $\mathbf{r}' \in S$, and integer m is the *order* of the potential function. The basic procedure to evaluate the potential at $\mathbf{r}$ is similar to that outlined in [15] for the evaluation of the Newtonian potential ($m=1$) and can be summarized as follows:

(i) Write the integrand of the potential integral over S as surface divergence of some vector function.

(ii) Transform the integral into the one over ΔS based on the surface divergence theorem.

(iii) Evaluate the integral as the sum of line integrals over edges of ΔS.

Without the loss of generality, it is assumed that

$$d \stackrel{\Delta}{=} \hat{\mathbf{n}} \cdot (\mathbf{r} - \mathbf{r}') > 0 \tag{2}$$

which is equal to the distance from $\mathbf{r}$ to Q.

For (i), we have (see [10])

$$\frac{1}{R^m} = \nabla_S \cdot (f_m(R)\mathbf{P}) \tag{3}$$

where $\mathbf{P}$ is the position vector of $\mathbf{r}'$ with respect to the projection of $\mathbf{r}$ on Q, $\mathbf{r}_Q$, and

$$f_m(R) = \begin{cases} \dfrac{\log R}{R^2 - d^2}, & m = 2 \\ \dfrac{-1}{(m-2)R^{m-2}(R^2 - d^2)}, & m \neq 2 \end{cases} \tag{4}$$

Note that if $\mathbf{r}_Q$ is inside S, $f_m(R)$ will becomes singular for some $\mathbf{r}'' = \mathbf{r}_Q$ (and $R = d$). Let S_ε denote the intersection of S and a small circular region on Q of radius ε and centered at $\mathbf{r}_Q$, the potential due to S can be evaluated as

$$\int_S \frac{dS}{R^m} = \lim_{\varepsilon \to 0} \left[\int_{S-S_\varepsilon} \nabla_S \cdot (f_m(R)\mathbf{P}) dS + \int_{S_\varepsilon} \frac{dS}{R^m} \right] \tag{5}$$

$$= \sum_i \mathbf{P}_i^0 \cdot \hat{\mathbf{u}}_i \int_{C_i} f_{m,i}(l_i) dl_i + g_m(\alpha),$$

where

$$f_{m,i}(l_i) = f_m(R = \sqrt{l_i^2 + d^2 + (P_i^0)^2}), \tag{6}$$

$$g_m(\alpha) = \begin{cases} \alpha \log d, & m = 2 \\ \dfrac{\alpha}{(m-2)d^{m-2}}, & m > 2 \end{cases} \tag{7}$$

P_i^0 is the distance between $\mathbf{r}_Q$ and edge segment C_i, l_i is measured from the projection of $\mathbf{r}$ on C_i along the direction

of $\hat{\mathbf{l}}_i$, and α is the angular extent of the circumference of S_ε lying inside S as $\varepsilon \to 0$.

In this paper, only $m=3$ is considered and we have

$$\int_{C_i} f_{3,i}(l_i) dl = \int_{l_i^-}^{l_i^+} f_{3,i}(l_i) dl \tag{8}$$

$$= \frac{1}{P_i^0 d} \left[\tan^{-1} \frac{l_i^- d}{P_i^0 R^-} - \tan^{-1} \frac{l_i^+ d}{P_i^0 R^+} \right].$$

with R^- and R^+ equal to the distances from $\mathbf{r}$ to the two end points of C_i, respectively. Thus, the repulsive force exerted on a point charge due to S can be found analytically by evaluating the gradient of the following function

$$\Phi(x, y, z) = \frac{1}{z} \tan^{-1} \frac{xz}{y\sqrt{x^2 + y^2 + z^2}} \tag{9}$$

at some (x, y, z)'s.

For the potential-based path planning of manipulators, the evaluation of the repulsion between manipulators and obstacles involves the calculation of the repulsion between pairs of polygons; each pair has a polygon from the manipulator and the other from obstacles. For continuously charged object surfaces, it is obvious that the direct calculation of the potential between two polygons requires a quadruple integral. To simplify the mathematics, links of manipulator are approximately represented by a set of point samples on their surfaces in this paper. Usually, the sampling points are located on the vertexes and edges of links and their distribution should be as uniform as possible. The repulsion between manipulators and obstacles, in forms of repulsive force and torque, can then be estimated in closed form through superposition using the above analytical expressions.

III. PATH PLANNING

To use the above results to solve the path planning problem considered in this paper, the proposed approach computes repulsive force and torque experienced by each rigid component, e.g., a link, of a 3-D articulated robot. A collision-free path of the articulated robot can then be obtained by locally adjusting its configuration along the path for minimum potential using these force and torque.[1]

For a rough description of the path of an articulated robot, the proposed approach identifies one or more guide planes (GPs) as its final or intermediate goals in the 3-D workspace. The GPs are polygons among obstacles in the free space, providing the articulated robot a general direction to move forward[2] (see Fig. 1). As for a descrip-

---

[1]In this paper, it is assume that links of the articulated robot are connected with spherical joints since the high DoF of such joints can take full advantage of the proposed potential minimization algorithm.

[2]In this paper, it is assumed the sequence of GPs are given in advance. Usually, for an elongated free space passage, like the one shown in Fig. 1, the GPs can be obtained as a subset of the cross-sections associated with the *generalized cylinder* [16] representation of the passage.

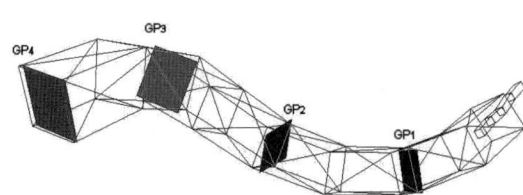

Fig. 1. A 3-D articulated robot is to move toward the goal by sequentially traversing a sequence of GPs.

tion of the configuration of the articulated robot, an n-link articulated robot is represented by its leading tip and the $n-1$ joints, as shown in Fig. 2. A sequential and collision-free traversal of a given sequence of GPs by these control points (and the links) is regarded as a global solution of the path planning problem of the articulated robot. In the planning procedure discussed next, intermediate GPs will also be added along the path each time when a given GP is not directly reachable.

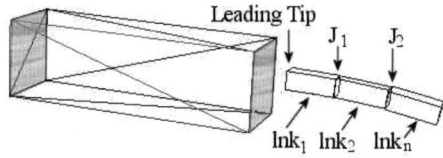

Fig. 2. The leading tip, joints and links of a 3-D articulated robot with a moving base moved toward the goal.

A. Basic Procedure of Path Planning

Assuming that a guide plane GP_1 is given as an intermediate goal, the basic path planning procedure for moving the leading tip, p', of an articulated robot onto GP_1 include (see Fig. 3):

(i) Translate links of the articulated robot to move p' toward GP_1. If p' can not reach GP_1 directly, e.g., due to collision, insert an intermediate plane GP'_1. (Fig. 3(a))

(ii) Search for the minimum potential configuration of the articulated robot by repeatedly executing:

  (a) Search for the minimum potential configuration of the robot with lnk_1 fixed in orientation by sliding p on GP'_1 (see Fig. 3(b)).

  (b) Search for the minimum potential configuration of the robot with p fixed in position (see Fig. 3(c)).

(iii) Repeat (i) and (ii) until p reaches GP_1.

In general, there are different ways to change the configuration of the articulated robot to move p' toward

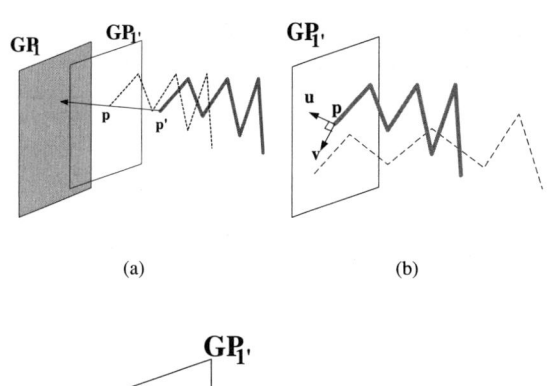

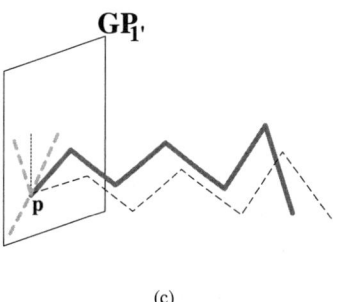

Fig. 3. Basic path planning procedure for a given GP (see text).

GP_1. A simple translation of all links is adopted in (i) as a preliminary implementation of the algorithm, as shown in Fig. 3(a). The direction of the translation is determined such that $\overrightarrow{p'p}$ is in the direction of the attractive force $\overrightarrow{F_{p'}}$[3] experience by p' due to GP_1. A collision check is performed for such a translation. If collision occurs, the distance of the translation is halved until the translation is collision-free. Accordingly, a new GP is inserted, e.g., GP'_1.

As for the search for the minimum potential configuration of the articulated robot in (ii), since lnk_1 has five DoFs for $p \in GP'_1$, the associated constrained optimization problem is divided into two iterative univariant optimization procedures. In (ii-a), lnk_1 is fixed in its orientation as p slides on GP'_1 to search for the minimal potential configuration and other links are sequentially adjusted in orientation, staring from lnk_2. In (ii-b), lnk_1 is adjusted in orientation while fixed in position and the procedure for adjusting the rest links is similar to that in (ii-a). For a particular GP, say GP'_1 in Fig. 3, (ii-a) and (ii-b) are repeatedly performed until negligible changes in the configuration of the articulated robot is obtained. Then, another intermediate GP which is closer to GP_1 is obtained with (i) and the process repeats. The path planning algorithm, as summarized below, ends as the leading tip reaches the given GP_1 or exist abnormally for an infeasible problem.

---

[3]The calculation of the direction of the attraction is similar to the calculation of the repulsion, except the direction is reversed.

Algorithm Articulated_Robot_to_GP

Step 0 Initialize $\delta = \delta_0$ and GP $\leftarrow GP_i$.

Step 1 Translate all links of the robot with distance δ to move p' to p along $\overrightarrow{F_{p'}}$. Find the smallest $n \geq 0$ such that $\delta \leftarrow \delta/2^n$ corresponds to a collision-free translation. If $\delta < \delta_{min}$, go to Step 6; otherwise, construct GP'_i with $GP'_i \perp \overrightarrow{F_p}$ and $p \in GP'_i$, and let GP $\leftarrow GP'_i$.

Step 2 Translate lnk_1, by sliding p on GP, and adjust the orientations of the rest links to minimize the potential.

Step 3 With p fixed in position, adjust the orientations of all links to minimize the potential.

Step 4 Go to Step 2 if the translation in Step 2 or the joint angle adjustment in Step 3 is not negligible.

Step 5 If p reaches GP_i, the planning is completed. Otherwise, $p' \leftarrow p$ and go to Step 1 with $\delta = \delta_0$.

Step 6 Exit and report that GP_i is unreachable.

For path planning involving multiple GPs, the above algorithm will be executed for each of them sequentially. It is assumed that the planning for a GP starts as the planning of the previous GP is accomplished. The path planning ends as the leading tip of the articulated robot reaches the goal, which is usually a (goal) GP in the path planning problems considered in this paper.

B. Implementation Details of Articulated_Robot_to_GP

1) Initialization and Step 1: In the initialization of *Articulated_Robot_to_GP*, the initial step size, δ_0, is arbitrarily chosen as 10% of workspace size. Generally, the number of robot configurations, and thus the computation time, depend on δ_0. For larger δ_0, less configurations will need to be calculated along the path if there is no collision. On the other hand, if δ_0 is too large, the computation time may actually increase because collisions may occur frequently. A threshold, i.e., $\delta_{min} = 1\%$ of workspace size, is established to set a lower bound of the magnitude of allowed translation. A translation which requires a smaller movement, $\delta < \delta_{min}$, indicates an infeasible path planning problem.

In Step 1 of *Articulated_Robot_to_GP*, the articulated robot is translated, as a rigid object, to move the leading tip from p' to p by distance δ along the direction of attraction, $\overrightarrow{F_{p'}}$, as shown in Fig. 4(a). In the far field, $\overrightarrow{F_{p'}}$ has a near spherical symmetry and p' is attracted toward GP_i as if GP_i is a point attractor. In the near field, $\overrightarrow{F_{p'}}$ will lead p' toward GP_i approximately in the normal direction of GP_i. In practice, since δ_0 is finite, instead of infinitely small, p' may reach GP_i at different locations from various directions.

While no configuration improvement to reduce the repulsive potential is considered in the above translation procedure, Steps 2 and 3 minimize the potential through constrained optimization procedures. To that end, an intermediate GP, GP'_i, is added along the path to serve as a geometric constraint for successive adjustments of the configuration of lnk_1. As long as $p \in GP'_i$, the choice of the orientation of GP'_i is not unique. A reasonable approach is adopted in our simulation wherein GP'_i is chosen to be perpendicular to $\overrightarrow{F_p}$, as shown in Fig. 4(b). Thus, the direction of translation to minimize the potential in Step 2, as discussed next, is perpendicular to that of the translation in the next execution, if necessary, of Step 1.

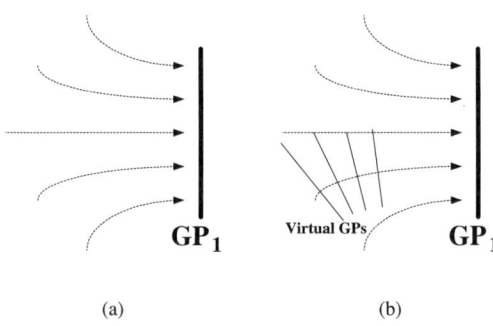

Fig. 4. The additional GPs and the moving direction of articulated robots (see text).

2) Adjusting the leading tip position in Step 2: In order to speed up the computation, a sequential planning strategy [17] which plan the motion of links sequentially is adopted in Step 2. Accordingly, the procedures of Step 2 first minimize the potential of lnk_1 whose translation is constrained in two dimensions by GP. As for the motion of lnk_i ($i > 1$), it is assumed that all links up to lnk_{i-1} has been planned, i.e., one end of lnk_i is fixed in position. Therefore, only the orientation of lnk_i, which has three degrees of freedom, need to be adjusted for minimum potential.

Consider the forces exerted on lnk_1, see Fig. 5 for example, since the minimization is constrained by $p \in GP$, only the projection of the resultant force experienced by lnk_1 on GP is taken into account. Let f_1 be the repulsive force exerted on lnk_1 due to the repulsion between lnk_1 and the obstacles. To determine the direction in which lnk_1 should translate to slide p on GP to reduce the repulsive potential, the projection of the resultant force exerted on lnk_1 along an arbitrary $\overrightarrow{u} \parallel GP$, $f_{1,u}$ is calculated.[4] A gradient-based binary search for the minimum potential location of p along $\overrightarrow{u}$ can then be performed. The initial step of sliding is arbitrarily chosen as 10% of the workspace size. If the movement of lnk_1 along $\overrightarrow{u}$ or $f_{1,u}$ is negligible, e.g., the movement is smaller than 1% of workspace size, another minimizing scheme along $\overrightarrow{v} \parallel GP$, which is orthogonal to $\overrightarrow{u}$, is performed to minimize $f_{1,v}$. Step 2 ends when two consecutive binary searches along the two orthogonal

[4]To move $link_1$ such that the potential will have maximum rate of reduction, $\overrightarrow{u}$ can be chosen as the projection of $\overrightarrow{f_1}$ on GP.

directions result in negligible movement of lnk_1. Once the optimal configuration of lnk_1 is determined, the procedure for adjusting the rest links is similar to that in Step 3, as discussed next.

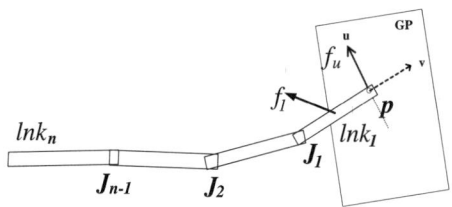

Fig. 5. Translating lnk_1 to slide p on GP'_1 to reduce the repulsiv potential.

3) Adjusting joint angle in Step 3: Once the minimum potential position of lnk_n is determined with Step 2 of *Articulated_Robot_to_GP*, another univariant procedure, which allows lnk_1 to adjust its orientation with p fixed in position is performed to reduce the potential further, as shown in Fig. 3(c). Under such a constraint, lnk_1 can rotate with respect to p to reduce the repulsive potential. The direction in which lnk_1 should rotate is determined by the repulsive torque experienced by lnk_1 with respect to p.

Let τ_1 be the repulsive torque experienced by lnk_1 with respect to p due to the repulsion between lnk_1 and obstacles. To find the minimum potential orientation of lnk_1 for p fixed in position, gradient-based binary searches are performed repeatedly using the projection of τ_1 along three orthogonal axes of rotation, e.g., $\tau_{1,u}$, $\tau_{1,v}$ and $\tau_{1,w}$, respectively. For each binary search, the initial rotating angle is arbitrarily chosen as 5°, while the accuracy in identifying the 1-D potential minimum is chosen as 0.5^o. Step 3 ends if a binary search results in a negligible change in the orientation of lnk_1, i.e., less than 0.5^o.

IV. SIMULATION RESULTS

In this section, simulation results are presented for path planning performed on Pentium III (500MHz) personal computer for articulated robots in 3-D environments. Fig. 6(a) shows the initial configuration of a 3-link articulated robot in a 3-GP workspace wherein the GPs are shown as black polygons. Fig. 6(b) shows the complete trajectory of the articulated robot which reaches the final GP safely. Since collisions occur frequently near the 90° turn of the passage for the δ_0 chosen, more configurations of the articulated robot are planned near the turn than those near start and goal. Due to the repulsive potential model, the trajectory of the articulated robot is smooth and lies near the middle of the workspace. The simulation takes a total of 18.266 seconds to plan the 12-configuration collision-free path.

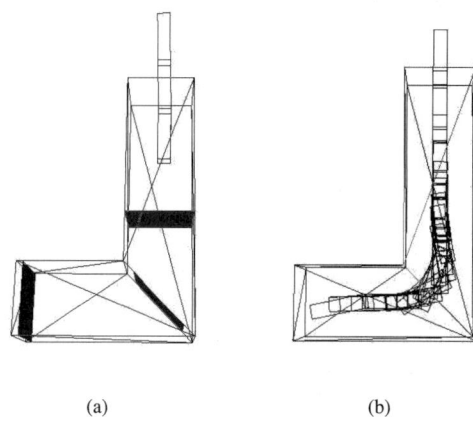

Fig. 6. A path planning example for a 3-link articulated robot in a 3-GP workspace. (a) The initial configuration. (b) The trajectory.

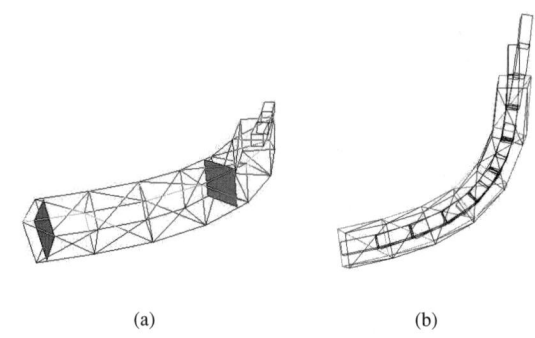

Fig. 7. A path planning example for a 3-link articulated robot in a 2-GP workspace. (a) The initial configuration. (b) The trajectory.

Figs. 7(a)(b) show the initial configuration and the final trajectory, respectively, of the path obtained for another 3-link articulated robot in a 2-GP workspace. The simulation takes 26.638 seconds to planning the 8-configuration path within a workspace with 48 triangles of obstacles. Since the repulsive force and torque is computed for each triangular obstacle face, the computation time for the path planning is roughly proportional to the number of triangles of obstacles. For example, the path planning within a similar workspace but with 248-triangle obstacles takes about 130 seconds.

Figs. 8(a)(b) show the initial configuration and final trajectory for the path planning of a 4-link articulated robot moving into a winding passage with 4 GPs. While the robot motions in Figs. 6 and 7 are essentially two-dimensional, the robot trajectory planned in 8 requires three-dimensional maneuvering. The simulation takes a total of 68.999 seconds to plan the 11-configuration collision-free path.

3369

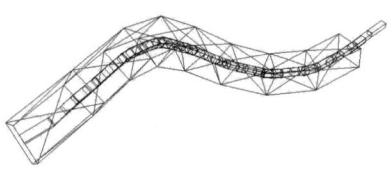

Fig. 8. The trajectory of the 4-link articulated robot shown in Fig. 1.

Figs. 9(a)(b) show the initial configuration and final trajectory of the path obtained for another 4-link articulated robot moving in a turnaround passage. The simulation takes a total of 151.518 seconds to plan the 24-configuration collision-free path. Since the passage of this example is more crooked, more intermediate GPs are added into the path, which in turn increases the computation time.

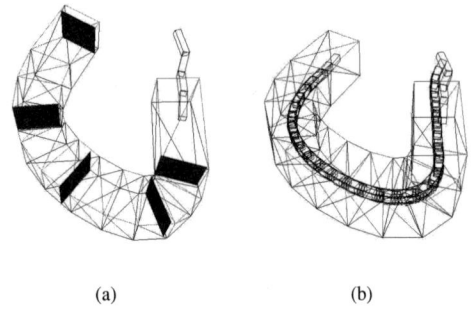

(a)   (b)

Fig. 9. A path planning example for a 4-link articulated robot in a 5-GP workspace. (a) The initial configuration. (b) The trajectory.

V. CONCLUSION

In this paper, a potential-based path planning of articulated robots with moving bases is proposed. According to the adopted potential model, surfaces of workspace obstacles are assumed to be uniformly charged and the links of the articulated robot are represented as a set of charged sampling points. The repulsive force and torque between links and obstacles thus modelled are analytically tractable. It is assumed that a sequence of GPs to be traversed by an articulated robot are given in advance in the workspace providing the general direction of the path. According to the motion constraints established by the GPs, the proposed approach derives the path for the articulated robot by adjusting its configurations at different locations along the path to minimize the potential using the above force and torque. Simulation results show that a path thus derived always stays away from obstacles and is spatially smooth.

VI. REFERENCES

[1] J. C. Latombe, "Motion planning: a journey of robots, molecules, digital actors, and other artifacts," *The Intl. Journal of Robotics Research*, vol. 18, no. 11, pp. 1119–1128, Nov. 1999.

[2] T. Lozano-Perez, "Spatial planning: a configuration space approach," *IEEE Trans. on Computers*, vol. C-32, no. 2, pp. 108–120, Feb. 1983.

[3] R. Brooks and T. Lozano-Preez, "A subdivision algorithm in configuration space for findpath with rotation," *IEEE Trans. on Sys., Man, Cybern.*, vol. 15, no. 2, pp. 224–233, March-April 1985.

[4] L. Kavraki and J. C. Latombe, "Randomized preprocessing of configuration space for fast path planning: Articulated robots," *Proc. IEEE/RSJ Intl. Conf. on Intelligent Robots and System*, pp. 1764–1772, 1994.

[5] N. C. Tsourveloudis, K. P. Valavanis, and T. Herbert, "Autonomous vehicle navigation utilizing electrostatic potential fields and fuzzy logic," *IEEE Trans. on Robot. and Auto.*, vol. 17, no. 4, pp. 490–497, 2001.

[6] O. Khatib, "Real-time obstacle avoidance for manipulators and mobile robots," *The Intl. Journal of Robotics Research*, vol. 5, no. 1, pp. 90–98, Spring 1986.

[7] P. Khosla and R. Volpe, "Superquadric artificial potentials for obstacle avoidance and approach," *Proc. IEEE Intl. Conf. on Robot. and Auto.*, pp. 1778–1784, 1988.

[8] Y. K. Hwang and N. Ahuja, "Gross motion planning a survey," *ACM Computation Survey*, vol. 24, no. 3, pp. 219–291, 1992.

[9] J-H. Chuang and N. Ahuja, "An analytically tractable potential field model of free space and its application in obstacle avoidance," *IEEE Trans. on Sys., Man, Cybern., Part B: Cybernetics*, vol. 28, no. 5, pp. 729–736, 1998.

[10] J-H. Chuang, "Potential-based modeling of three-dimensional workspace of the obstacle avoidance," *IEEE Trans. on Robot. and Auto.*, vol. 14, no. 5, pp. 778–785, 1998.

[11] A. Thanailakis, P. G. Tzionas, and P. G. Tsalides, "Collision-free path planning for a diamond-shaped robot using two-dimensional cellular automata," *IEEE Trans. on Robot. and Auto.*, vol. 13, pp. 237–250, 1997.

[12] J-H. Chuang, C-H. Tsai, W-H. Tsai, and C-Y. Yang, "Potential-based modeling of 2-D regions using nonuniform source distribution," *IEEE Trans. on Sys., Man, Cybern., Part A: Systems and Humans*, vol. 3, no. 2, pp. 197–202, 2000.

[13] C-H. Tsai, J-S. Lee, and J-H. Chuang, "Path planning of 3-D objects using a new workspace model," *IEEE Trans. on Sys., Man, Cybern., Part B: Cybernetics*, vol. 31, no. 3, pp. 405–410, Aug. 2001.

[14] J. C. Latome, *Robot Motion Planning*, MA: Kluwer, Boston, 1991.

[15] D. R. Wilton, A. W. Glisson S. M. Rao, D. H. Schaubert, O. M. Al-Bundak, and C. M. Butler, "Potential integrals for uniform and linear source distributions on polygonal and polyhedral domains," *IEEE Trans. Antennas and Propagation*, vol. AP-32, no. 3, pp. 276–281, 1984.

[16] T. O. Binford, "Visual perception by computer," *Proc. Intl. Conf. on System and Controls*, December 1971, Miami.

[17] K. K. Gupta, "Fast collision avoidance for manipulator arms: A sequential search strategy," *IEEE Trans. on Robot. and Auto.*, vol. 6, no. 5, pp. 522–532, Oct. 1990.

Improved Analysis of D*

Craig Tovey[1] Sam Greenberg[2] Sven Koenig[1]

[1] College of Computing
Georgia Institute of Technology
Atlanta, GA 30332-0280
{ctovey, skoenig}@cc.gatech.edu

[2] School of Mathematics
Georgia Institute of Technology
Atlanta, GA 30332-0160
SamIAm@math.gatech.edu

Abstract—D* is a planning method that always routes a robot in initially unknown terrain from its current location to a given goal location along a shortest presumed unblocked path. The robot moves along the path until it discovers new obstacles and then repeats the procedure. D* has been used on a large number of robots. It is therefore important to analyze the resulting travel distance. Previously, there has been only one analysis of D*, and it has two shortcomings. First, to prove the lower bound, it uses a physically unrealistic example graph which has distances that do not correspond to distances on a real map. We show that the lower bound is not smaller for grids, the kind of map-based graph on which D* is usually used. Second, there is a large gap between the upper and lower bounds on the travel distance. We considerably reduce this gap by decreasing the upper bound on arbitrary graphs, including grids. To summarize, we provide new, substantially tighter bounds on the travel distance of D* on grids, thus providing a realistic analysis for the way D* is actually used.

I. INTRODUCTION

Robot navigation problems in unknown terrain have been studied in both theoretical robotics and theoretical computer science. A good overview is given in [9]. However, empirical robotics researchers have often developed their own planning methods. These methods have been demonstrated on mobile robots that solve complex real-world tasks and perform well in practice. Consider, for example, a robot that has to move to a given goal location in unknown terrain. The sensors on-board a robot can typically sense the terrain only near its current location, and thus the robot has to interleave planning with movement to sense new parts of the terrain. This is called sensor-based planning [2]. Dynamic A* [11], [12], [13] (short: D*) is a sensor-based planning method that always moves the robot from its current location on a shortest presumed unblocked path to the goal location. A presumed unblocked path is a path that does not contain known obstacles. D* traverses the path until it discovers new obstacles. It then repeats the procedure, taking into account all the obstacles that it has learned about. If the robot reaches the goal location, it stops and reports success. If at any point it fails to find a presumed unblocked path from its current location to the goal location, it stops and reports that the goal location cannot be reached from the start location. Figure 1 gives

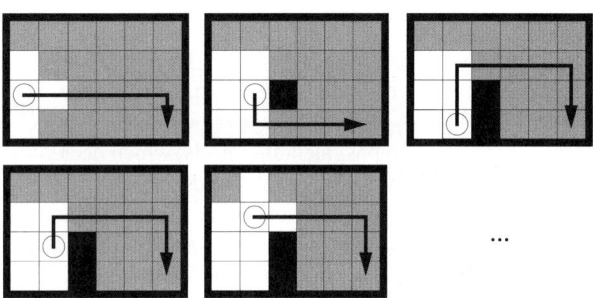

Fig. 1. D* Example

an example of the behavior of D*. White locations are unblocked, black locations are blocked, and grey locations have unknown status. The example assumes that the robot senses the status of its four neighboring locations and can then move to an unblocked neighboring location. The circle shows the current location of the robot. The goal location is in the lower right corner.

D* has been used outdoors on an autonomous high-mobility multi-wheeled vehicle that navigated 1,410 meters to the goal location in an unknown area of flat terrain with sparse mounds of slag as well as trees, bushes, rocks, and debris [15]. As a result of this demonstration, D* is now widely used in the DARPA Unmanned Ground Vehicle (UGV) program, for example, on the UGV Demo II vehicles. D* is also being integrated into Mars Rover prototypes, tactical mobile robot prototypes and other military robot prototypes for urban reconnaissance [3], [6], [16]. Furthermore, it has been used indoors on Nomad 150 mobile robots in robot-programming classes to reach a goal location in unknown mazes that were built with three-foot high, forty inch long cardboard walls [8], [7]. D* has also been used as the key method of various robot-navigation software [1], [14].

D* is so popular because its repeated replanning can be implemented efficiently [11], [12], [13] and easily [4]. Hence, evaluating the plan-execution times (that is, the travel distances of the robots) is crucial to determining the overall performance of the robots. In practice, the plan-execution times seem to be reasonably small. It is thus important to understand whether this is due to properties of the test terrains or whether the plan-execution times are indeed guaranteed to be good in any kind of terrain.

So far, there exists one analysis of D* [10]. It considers the worst-case travel distance of D* on arbitrary graphs $G = (V, E)$ and proves a lower bound of $\Omega(\frac{\log |V|}{\log \log |V|} |V|)$ steps and an upper bound of $O(|V|^2)$ steps. However, the analysis has two shortcomings. First, while D* can be used on arbitrary graphs (such as Voronoi graphs), it is typically used on grids. However, the example graph used to prove the lower bound was not a grid, and moreover used unrealistic edge lengths not related to physical distances. Thus, the lower bound could potentially be smaller for grids or other physically realizable graphs. We show that this is not the case. Second, there is a large gap between the upper and lower bounds on the travel distance. We reduce this gap by proving an upper bound of $O(|V|^{3/2})$ steps on arbitrary graphs, including grids, which decreases the gap considerably. The proof builds on our analysis of the upper bound on the worst-case travel distance of greedy mapping [5]. We thus provide new, substantially tighter bounds on the worst-case travel distance of D* on grids and a realistic analysis for the way D* is actually used.

II. ASSUMPTIONS

We study the travel distance of D* analytically. We assume that the robot is omni-directional, point-sized, equipped with only a radial short-distance sensor, and capable of error-free motion and sensing. The sensors on-board the robot uniquely identify its location and the neighboring locations and determine whether the neighboring locations contain obstacles. We model the terrain as graph. Vertices in the graph represent locations in the terrain. Traversing an edge in the graph corresponds to traveling from one location to an adjacent location in the terrain. We are interested in the quality of the plans determined by D* as a function of the number of vertices of the graph. We use the worst-case travel distance of the robot (in edge traversals or, synonymously, steps) to measure the plan quality because a small worst-case travel distance guarantees that the robot performs well in all terrains.

With these assumptions, we can formalize the behavior of D* as follows. We call a graph $G = (V, E)$ vertex-blocked by $B \subset V$ if B is the set of blocked vertices (that is, vertices that cannot be visited). On a finite (undirected) graph $G = (V, E)$ vertex-blocked by B, a robot has to reach a designated goal vertex. D* always moves the robot from its current vertex along a shortest presumed unblocked path to the goal vertex. A presumed unblocked path is one that contains no vertices which are known to be blocked. D* terminates when the robot reaches the goal vertex or there are no presumed unblocked paths to the goal vertex,

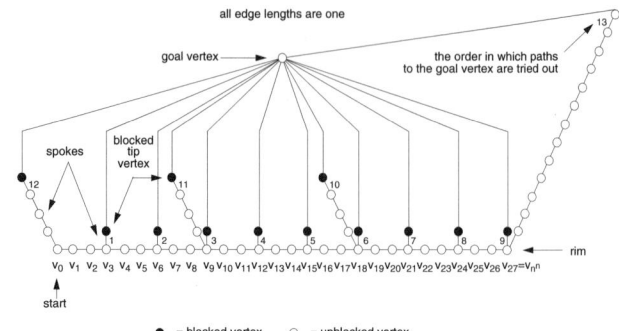

Fig. 2. Previous Example Graph for Lower Bound

in which case the goal vertex is unreachable from the start vertex since the graph is undirected. D* must terminate since each path it tries either reaches the goal or augments the set of known blocked vertices.

III. D*: LOWER BOUND ON GRIDS

We now prove a lower bound on the worst-case travel distance of D* on vertex-blocked grids. We first review the previous analysis, which employs the key idea of tricking the robot into traversing the same long path back and forth many times. Second, we give a visual and conceptual overview of how to transform that example into a grid without losing the key idea. Third, we explain exactly how the grid is constructed. Finally, we analyze the worst-case travel distance of D* on our grid graph, proving the same lower bound.

A. Previous Result

The previous analysis proved that the worst-case travel distance of D* is $\Omega(\frac{\log |V|}{\log \log |V|} |V|)$ steps on vertex-blocked graphs $G = (V, E)$ [10]. This lower bound is achieved with graphs of the structure shown in Figure 2. We now sketch the main idea of its construction, but with our own rim-and-spoke terminology, in order to introduce our much more complex grid construction.

The graph of Figure 2 consists of a long horizontal path of length d^d (where d is a integer parameter), which we call the "rim", and a set of "spokes" of varying lengths attached to the rim at various vertices. The uppermost "tip" vertex of each spoke is blocked and connected to the goal vertex by an edge. Note that the edges from the tips to the goal are physically unrealistic edges, because in the graph the robot can move to the goal from any tip in one step. The possible spoke lengths are $\sum_{i=0}^{h} d^i$ for the nonnegative integers h. We refer to a spoke of length $\sum_{i=0}^{h} d^i$ as a "class h spoke". Longer spokes are spaced farther apart from each other than short spokes. In particular, the vertices where class h spokes attach to the rim have distance d^{h+1} from each other. Hence, if the robot is at a vertex where a class h spoke attaches to the rim, then it is shorter to go to the goal along the rim to the next class h spoke, than it is to go via any class $h+1$ spoke.

The robot starts at vertex v_0. Initially, it traverses the rim from left to right, checking the class 0 spokes for a path to the goal vertex; then it returns along the rim from right to left, checking class 1 spokes for a path to the goal vertex, and so on. Each class forces the robot to traverse the rim once. If there are d classes, then the robot traverses the rim d times for a total travel distance of d^{d+1}. Since a computation shows that there are $|V| = O(d^d)$ vertices in the graph, the total travel distance is $\Omega(\frac{\log |V|}{\log \log |V|}|V|)$.

B. Conceptual Overview

The lower bound from the previous analysis could potentially be smaller on more realistic graphs such as the grid graphs on which D* is often used. We now prove that this is not the case with a much more complex construction of a grid that uses the key idea from the previous analysis, to fool the robot into traversing the lengthy rim many times. The robot moves back and forth, each time checking spokes of the next class for a path to the goal vertex because the end of each spoke is connected to the goal vertex. Each time, the robot finds that the tip vertex of the spoke is blocked and thus needs to traverse a different spoke. The spokes are spaced at particular distances from each other on the rim to trick the robot into visiting all the class h spokes before visiting any class $h+1$ spokes. However, the graph topology of the previous analysis cannot directly be embedded into a grid because the goal vertex must be simultaneously adjacent to the ends of many spokes of greatly different lengths, which moreover are placed at great distances from each other. In a grid, on the other hand, each cell is adjacent to at most four other cells, and the distance between two adjacent cells is always one. We use several ideas to modify the graph topology of the previous construction to be able to embed it into a grid. A rough conceptual sketch of these ideas is shown in Figure 3.

1) Attach each spoke at a separate vertex to the rim (Figure 3a). This eliminates the problem of a vertex on the rim being adjacent to too many other vertices. As long as longer class spokes are spaced far enough apart, the robot is still fooled into repeatedly traversing the rim.
2) Remove the very short spokes (Figure 3b). We have to place the goal vertex at some distance D from the rim, and we thus cannot construct spokes of length less than D. This is one of the reasons to use classes $0.8d$ to $0.9d$ instead of using classes 0 to d. As long as there are $\Omega(d)$ classes, the lower bound is the same order.
3) Move the spokes physically closer together, but maintain their distances from each other along the rim. We do this by "squeezing" the rim into an

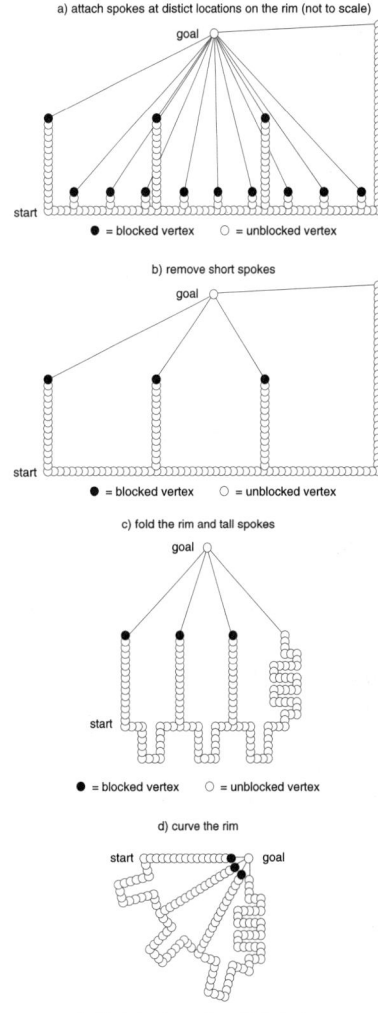

Fig. 3. Steps of the Transformation

accordion shape (Figure 3c). In particular, the sections of the rim between spokes get bent into long loops, which we call "columns."

4) Redesign the spokes so that they all have the same physical height, while maintaining their original lengths (Figure 3c). In particular, build a pair of blocked walls of the same height, with some space between them. Put a twisty path of the appropriate length in between the walls.
5) Once the spokes are fairly close and of equal height, bend the rim into part of a circular arc, pushing the tips of the spokes together towards the goal vertex (Figure 3d). It is not possible to squeeze too many distinct vertices into a small area on a grid, but this problem is solved by blocking the paths to the goal vertex a bit before the goal vertex.

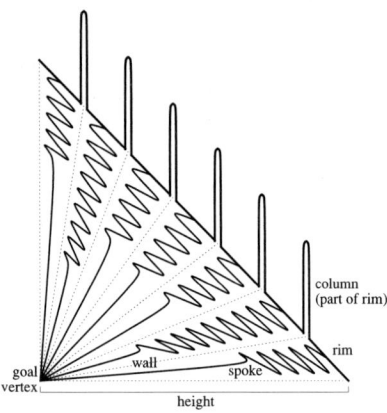

Fig. 4. Conceptual Figure with Two Spoke Classes

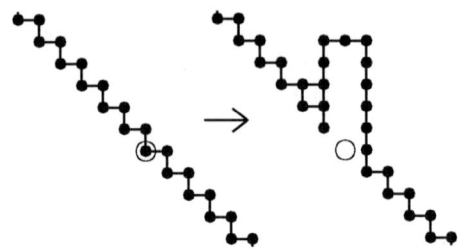

Fig. 5. A column of height 3.

C. Structural Overview

Figure 4 illustrates the structure of the actual construction. Place the goal vertex g at $(0,0)$. Place the rim on a diagonal from $(0,z^{0.7z})$ to about $(z^{0.7z},0)$, for some sufficiently large integer z. Note that the rim is two concentric quarter circles in the "taxicab" metric $\mathscr{L}_1$, so each point is within 1 of $z^{0.7z}$ from g. Along the rim, there will be "spoke-base points" and "column-base points", alternating. (Note: To avoid notational clutter in this paragraph and in the foregoing, we omit the "floor" operation notation. Here, for example, $z^{0.7z}$ really means $\lfloor z^{0.7z} \rfloor$.)

From each spoke-base, construct a windy path towards g. Each path has length $2z^j$ for some $j \in \{0.8z, 0.8z+1, \ldots, 0.9z\}$. We call such a path of length $2z^j$ a "spoke of class j". The graph will contain z^{z-j-1} spokes of class j, for each j.

At each column-base point, replace the point with a construction to add distance, increasing the steps needed to pass that point on the rim. This ensures that the distance between successive spokes of class j is z^{j+1}, as well as setting the total rim length to z^z. Put a blocked cell near the end of each spoke except one of class $0.9z$. The robot will be tempted by each of the spokes of class $0.8z$ in turn, following the rim for about z^z steps. The robot then will turn around and be tempted only by the spokes of the next class $0.8z+1$, again traversing the rim for approximately z^z steps, and so on.

D. Construction

Define the outer rim R such that

$$R := \{(x,y) \in \mathbb{Z}_{\geq 0} \times \mathbb{Z}_{\geq 0} : z^{0.7z} \leq x+y \leq z^{0.7z}+1\}.$$

For each $i \in \{0.8z, \ldots, 0.9z\}$, we will create z^{z-i-1} spokes of class i. Let $S := \frac{z^{0.2z} - z^{0.1z-1}}{z-1}$ be the number of total spokes. For $i \in \{1, \ldots, S\}$, define the i_{th} spoke-base, b_i, such that

$$b_i := \left(\frac{z^{0.7z}}{S}i + \frac{z^{0.7z}}{2S}, z^{0.7z} - \frac{z^{0.7z}}{S}i - \frac{z^{0.7z}}{2S}\right).$$

Therefore $b_i \in R$.

The taxicab distance between two adjacent spoke bases will be $2z^{0.7z}/S$, but to make the construction's key idea work, these distances must be longer for the robot as it moves in the graph. We increase the graph distances by inserting detour loops into the rim. Lemma 1 makes this idea precise.

Lemma 1: Given a function $t: \{1, \ldots, S\} \to \mathbb{Z}_{\geq 0}$ such that $t(i) - t(i-1) \geq \frac{2z^{0.7z}}{S+1} \forall i$, it is possible to modify R using only cells above R in the plane, such that $\forall j > i$, traveling along R from b_i to b_j takes between $(t(j) - t(i) - 4)$ and $(t(j) - t(i) + 4)$ steps.

Proof: For $i \in \{0, \ldots, S\}$, define the i_{th} column-base, c_i such that

$$c_i := \left(\frac{z^{0.7z}}{S}i, z^{0.7z} - \frac{z^{0.7z}}{S}i\right).$$

Therefore $c_i \in R$. At each of the column-bases, we remove the point itself and the point above it from the rim and add two paths traveling upwards, connected at the top. So, if the column-base is at (x,y), remove (x,y) and $(x,y+1)$ from R and add one path from $(x-1,y+2)$ to $(x-1,y+3+h)$ and another path from $(x+1,y)$ to $(x+1,y+3+h)$, with a connecting point at $(x,y+3+h)$. This increases the steps needed to cross this point in the rim by $2h$. We call such a construction a "column of height h". A column of height 3 is illustrated in Figure 5.

We now define an iterative algorithm for building the columns. For a fixed i, assume the previous columns have been built and let D be the current distance from b_1 to b_i. Let $h := \lfloor \frac{t(i) - t(1) - D}{2} \rfloor$ and build a column of height h at c_{i-1}. This ensures that the distance from b_1 to b_i is now $t(i) - t(1)$, up to the round-off error. Repeat the process for all later i. (Note that the recalculation from $t(1)$ here prevents accumulation of round-off error.) ∎

The fourth idea was to build each spoke as a twisty path of the appropriate length. Each spoke will consist of a wedge of sufficient area. The spokes will not overlap,

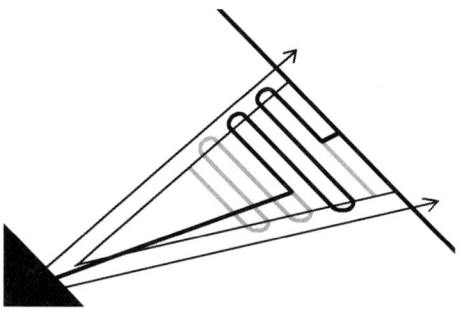

Fig. 6. The hypothetical path H_i, along with the shortening necessary to set the path length to $l(i)$.

except in a small unblocked triangular region near the goal vertex, within which all paths are direct.

Lemma 2: Given a function $l : \{1,\ldots,S\} \to \mathbb{Z}_{\geq 0}$ such that $z^{0.7z} \leq l(i) \leq z^{z-1}$, $\forall i$, it is possible to construct spokes from b_i such that the distance from b_i to t is between $(l(i)-4)$ and $(l(i)+4)$.

Proof: We would like to connect each of the spokes to g. However, the max degree of a grid prevents this. Therefore we add a triangle T such that

$$T := \{(x,y) : x,y \in \mathbb{Z}_{\geq 0} \bigwedge x+y \leq z^{0.5z}\}.$$

We may then simply connect the spokes to T. For each $i \in \{1,\ldots,S\}$, define the i_{th} "tip" vertex t_i such that

$$t_i := \left(\frac{z^{0.5z}}{S}i + \frac{z^{0.5z}}{2S}, z^{0.5z} - \frac{z^{0.5z}}{S}i - \frac{z^{0.5z}}{2S} \right).$$

Therefore $t_i \in T$. This will be the point that the i_{th} spoke connects to.

To construct the paths of length $l(i)$ for each i, construct a hypothetical path H_i from b_i of excessive length. Then, when building the actual graph, simply include as much of H_i as necessary before the graph takes a direct path to t_i. The point where the graph ignores H_i and instead switches to a direct path to t_i depends on $l(i)$. Block the cell on the path just before it reaches t_i.

Building H_i takes a bit of construction. Use Euclidean rays from g to partition the space between R and g into areas A_i and then create many path segments running parallel to R called "levels". H_i runs up and down these levels, traveling back and forth to increase length. Define a space C_i to give room to connect one level to the next without coming close to the rays. This is all illustrated in Figure 6. The triangle in the the lower left corner represents a region of unblocked cells, bordered by the t_i. Since all of the twisty paths have become direct paths by the time they reach their t_i, and the blockages occur prior to reaching t_i, it is OK for the spokes to overlap within this unblocked region.

For each $i \in \{0,\ldots,S\}$, define the i_{th} "ray" r_i to be the Euclidean line from $\left(\frac{z^{0.5z}}{S}i, z^{0.5z} - \frac{z^{0.5z}}{S}i\right)$ to c_i. Hence r_i goes from T to R. For each $i \in \{1,\ldots,S\}$, define the i_{th} area A_i to be the integer points between r_{i-1} and r_i. Define the i_{th} cushion C_i such that

$$C_i := \{(x,y) : x,y \in \mathbb{Z}_{\geq 0} \bigwedge d[(x,y),r_i] \leq 8\}.$$

For each $i \in \{1,\ldots,z^{0.3z}-2\}$ and each $j \in \{1\ldots,0.1z^{0.7z}\}$, define the level $l_{i,j}$ such that

$$l_{i,j} := \{(x,y) : z^{0.7z} - 6j \leq x+y \leq z^{0.7z} - 6j + 1\}$$
$$\cap \; (A_i - C_i - C_{i+1}).$$

Use levels $\{l_{i,0},\ldots,l_{i,0.1z^{0.7z}}\}$ to make H_i, using C_i and C_{i+1} to connect the levels. C_i is large enough to let H_i avoid crossing the ray.

The distance between c_i and c_{i+1} is $\frac{2z^{0.7z}}{S} \geq 2z^{0.3z}$. The levels are parallel and contained in a Euclidean triangle. The smallest is only one tenth of the way to the point, so each level is longer than $z^{0.3z}$. There are $0.1z^{0.7z}$ levels, so the total length of H_i is at least $0.1z^z$.

To build the actual spoke, define a function $s(p)$ for all points $p \in H_i$, such that $s(p)$ is the distance from b_i to t if we were to shorten H_i at p and take a direct path to t_i from p. (Note that this definition involves the actual distance in the graph, avoiding accumulated round-off error.) Let S_i be the point in H_i that minimizes $|s(S_i) - l(i)|$. For any two points p, p' in the same level, if $d[p,p'] = 2$, then $|s(p)-s(p')| = 2$, since $d[p,t] = d[p',t]$. Therefore, if S_i is contained in one of the levels, shortening H_i at S_i gives a spoke within 2 of $l(i)$. If S_i is contained in one of the cushions, we may be able to create an even more precise spoke. For any adjacent $c,c' \in C_i \cap H_i$, $|s(c)-s(c')| \leq 1$, as the path from c to t_i likely passes through c'. Hence, regardless of whether S_i appears in a level or in a cushion, we exceed the precision required by the Lemma. ∎

To finally build our graph, define a position function $p[i,j]$ such that

$$p[i,j] := z^{i+1}j + \frac{z^{0.8z+1}}{0.1z+1}(i - 0.8z).$$

This will be the "position" of the j_{th} spoke of class i. Order the spokes by this position function, so the first spoke is the one with lowest position, the second is the one with second lowest position, and so forth.

Define the length function l such that $l[k] := 2z^i$, where i is the class of the k_{th} spoke. Use this l with Lemma 1.

Define t such that $t[k] = p[i,j]$, where i and j are the coordinates for the k_{th} spoke. Use this t with Lemma 2.

E. Analysis

We are now ready to prove the following theorem.

Theorem 1: The worst-case travel distance of D* on vertex-blocked grids $G = (V,E)$ is $\Omega(\frac{\log|V|}{\log\log|V|}|V|)$ steps.

Proof: The distance the robot must travel to find a spoke of class i and then travel to g is at most $z^{i+1} + 2z^i$ steps. For any $j > i$, simply traveling a spoke of class j will take at least $2z^j \geq 2z^{i+1}$ steps. Hence the robot will walk to the smallest class spoke available, find a blocked cell, and go to the next of that class, traversing the rim.

By the placement of the spokes, notice that $\forall h, h' \in \{0.8z, \ldots, 0.9z\}$, if $h < h'$, then the rightmost h-class spoke is to the right of the rightmost h'-class spoke. Also the leftmost h-class spoke is to the left of the leftmost h'-class spoke. Hence, after visiting every spoke of class i, the robot turns around and finds the first spoke of class $i+1$. Each time the robot traverses the rim, it goes from the leftmost spoke of class i to the rightmost spoke of class i. This distance is more than $z^z - 2z^{i-1} = \Omega(z^z)$. The total travel distance (just on the rim) is therefore at least $z^{0.1z} \Omega(z^z) = \Omega(z^{z+1})$.

On the other hand, there are $\theta(z^z)$ vertices in the rim (including the columns). There are $O(z^{0.5z})$ vertices in T. In class i there are fewer than $\frac{z^z}{z^{i-1}}$ spokes, each with z^i vertices, so each class contains $O(z^{z-1})$ vertices. There are $0.1z$ classes, so there are $O(z^{z-.9})$ vertices in the spokes. Therefore n, the total number of vertices in the graph, is $\theta(z^z)$. If $n = \theta(z^z)$, then the total distance is $\Omega(z^{z+1}) = \Omega(nz) = \Omega(\frac{n \log(n)}{\log \log n})$. ∎

IV. D*: UPPER BOUND

We now prove an upper bound on the worst-case travel distance of D* on arbitrary vertex-blocked graphs, including grids. We first review the previous analysis that proved an upper bound of $O(|V|^2)$ steps on arbitrary vertex-blocked graphs $G = (V, E)$. We then explain our analysis, that builds on ideas of both the previous analysis of D* and our previous analysis of greedy mapping [5] but proves an upper bound of $O(|V|^{3/2})$ steps. Notice that the worst-case travel distance of any planning method that moves a robot from its current vertex to the goal vertex is $\Omega(|V|)$ because the robot has to visit every vertex in the worst case. Chronological backtracking is a planning method that achieves this bound. Thus, the best upper bound on the worst-case travel distance of any planning method is $O(|V|)$. Our upper bound of $O(|V|^{3/2})$ thus shows that D* is not very badly suboptimal.

A. Previous Result

The previous analysis showed that the worst-case travel distance of D* is $O(|V|^2)$ steps on arbitrary vertex-blocked graphs $G = (V, E)$ [10]. We now sketch the main idea of its proof in order to introduce our more complex proof of a smaller upper bound. We use the term *blockage* to mean a blocked vertex. A *blockage detection* occurs when the robot observes a blockage it did not know about before. When a blockage detection occurs, the robot detects all blockages adjacent to its current vertex. The robot always follows a shortest presumed unblocked path to the goal vertex until it either detects a blockage or stops. Each such path has a length of at most $|V| - 1$ steps. A blockage detection can never occur twice at the same vertex, so there can be no more than $|V|$ blockage detections. Thus, the total travel distance can be no longer than $|V|$ paths of length $|V| - 1$, each preceding a blockage detection, plus one more path of length $|V|$ leading to the goal vertex. Thus, the worst-case travel distance of D* on vertex-blocked graphs is at most $|V|^2$ steps.

B. New Analysis

The quadratic upper bound from the previous analysis leaves a large gap with the lower bound of the previous section. We now narrow the gap by proving that the worst-case travel distance of D* is $O(|V|^{3/2})$ steps. The proof of this tighter upper bound uses the same ideas of blockage detections and of conceptually separating the route of the robot into paths between successive blockage detections that we just used to justify the quadratic upper bound.

Theorem 2: The worst-case travel distance of D* on vertex-blocked graphs $G = (V, E)$ is (at most) $O(|V|^{3/2})$ steps.

Proof: We call a blockage detection at which the robot discovers that it cannot reach the goal vertex a *terminating* blockage detection. All other blockage detections are *nonterminating*. Let $b \leq |V|$ denote the number of nonterminating blockage detections that occur. Let c^i ($0 \leq i \leq b$) denote the vertex at which the robot is located during the ith blockage detection. (The zeroth blockage detection is considered to occur at the start vertex, just prior to the first edge traversal.) Let L_i ($1 \leq i \leq b$) denote the number of steps between the $(i-1)$st and the ith blockage detection.

The robot cannot take more than $|V|$ steps after the bth blockage detection, because it travels on a shortest presumed unblocked path from c^b to the goal vertex and stops either at the goal vertex or earlier at the terminating blockage detection. Then, the total number of steps taken by the robot is at most

$$|V| + \sum_{i=1}^{b} L_i. \tag{1}$$

For every vertex $v \in V$, let d_v^i ($0 \leq i \leq b$) denote the length of a shortest presumed unblocked path from vertex v to the goal vertex directly after the ith blockage detection. That is, d_v^i is the *presumed distance* from v to the goal vertex, just after the ith blockage detection occurs. We define the value d_v^i as unchanged from the value d_v^{i-1} if D* detects that it cannot reach the goal vertex from vertex v. If the presumed distance from a vertex v to the goal vertex is infinite right from the start, we set $d_v^0 = 0$. Note that $0 \leq d_v^i \leq |V| - 1$ and that d_v^i is nondecreasing in i.

We define $\phi^i = \sum_{v \in V} d_v^i \geq 0$. Note that ϕ^i is nondecreasing in i and $\phi^b \leq |V|^2$.

The main intuition behind the proof is that the robot "thinks" it is getting closer to the goal vertex with each step it takes along a presumed unblocked path, but gets "disappointed" by a blockage detection which increases the presumed distances d_v to the goal. Intuitively, if there are many large L_i the robot should get many big disappointments, which should greatly increase ϕ, and there is a limit on how large ϕ can get. This should force $\sum_i L_i$ and hence the total travel distance to be small. This intuition is similar to that in a bound proved for greedy mapping in [5]. However, there is a direct relationship between an analogous ϕ function and L_i for greedy mapping, but there is no such direct relationship for D*. Our proof is therefore considerably more complicated. We use a proxy Δ^i instead of counting the number of steps L_i directly.

We define $\Delta^i = d_{c^i}^i - d_{c^i}^{i-1} \geq 0$ ($1 \leq i \leq b$), that is, Δ^i is the amount by which the presumed distance from the robot to the goal vertex increases when the robot is at vertex c^i and detects the ith blockage.

Let D denote the presumed distance from the current vertex of the robot to the goal vertex (the present time will be clear from the context). Consider the motion of the robot from the $(i-1)$st to the ith blockage detection. The robot makes L_i steps, and D decreases by L_i during those steps. Then, when the robot is at vertex c^i, D increases by Δ^i. Therefore, D increases by a total of $\sum_{i=1}^{b}(\Delta^i - L_i)$ during the execution of D* from the start to directly after the bth blockage detection. On the other hand, $D = d_{c^0}^0$ at the start and $D = d_{c^b}^b$ directly after the bth blockage detection. Therefore, D increases by a total of $d_{c^b}^b - d_{c^0}^0$ during the execution of D* from the start to directly after the bth blockage detection. Thus, it holds that $d_{c^b}^b - d_{c^0}^0 = \sum_{i=1}^{b}(\Delta^i - L_i)$. Since $-|V| \leq d_{c^b}^b - d_{c^0}^0 \leq |V|$, it holds that $-|V| \leq \sum_{i=1}^{b}(\Delta^i - L_i) \leq |V|$ and $\sum_{i=1}^{b} L_i \leq |V| + \sum_{i=1}^{b} \Delta^i$. From the bound (1), it then follows that the total number of steps taken by the robot is at most

$$|V| + \sum_{i=1}^{b} L_i \leq 2|V| + \sum_{i=1}^{b} \Delta^i. \quad (2)$$

This bound is crucial since it relates Δ^i to the actual quantity of interest.

Let x_z denote any vertex at a presumed distance of z edges or less from c^i ($1 \leq i \leq b$), where the distance is the length of a shortest presumed unblocked path directly after the ith blockage detection. Then $d_{c^i}^i \leq z + d_{x_z}^i$ since, according to the triangle inequality, the presumed distance directly after the ith blockage detection from c^i to the goal vertex cannot be larger than the presumed distance from c^i via x_z to the goal vertex. Since blockage detections can only increase the presumed distances, x_z must also have been at a presumed distance of z edges or less from c^i directly after the $(i-1)$st blockage detection. Also, since the graph is undirected, the presumed distance from c^i to x_z equals the presumed distance from x_z to c^i. Then $d_{x_z}^{i-1} \leq z + d_{c^i}^{i-1}$ since, according to the triangle inequality, the presumed distance directly after the $(i-1)$st blockage detection from x_z to the goal vertex cannot be larger than the presumed distance from x_z via c^i to the goal vertex. Putting the two inequalities together, we get $d_{x_z}^i - d_{x_z}^{i-1} \geq (d_{c^i}^i - z) - (d_{c^i}^{i-1} + z) = (d_{c^i}^i - d_{c^i}^{i-1}) - 2z = \Delta^i - 2z$. Since the presumed distances $d_{x_z}^i$ are nondecreasing in i, it also holds that $d_{x_z}^i - d_{x_z}^{i-1} \geq 0$ and thus $d_{x_z}^i - d_{x_z}^{i-1} \geq \max(\Delta^i - 2z, 0)$.

For all $0 \leq z \leq d_{c^i}^i$ ($1 \leq i \leq b$), there must be at least $z+1$ vertices at a presumed distance of z edges or less from vertex c^i directly after the ith blockage detection, since there exists a presumed unblocked path from vertex c^i to the goal vertex. Let x_z ($0 \leq z \leq d_{c^i}^i$) denote a vertex at a presumed distance of z edges or less from vertex c^i directly after the ith blockage detection, with $x_z \neq x_{z'}$ for $z \neq z'$. Note that $x_0 = c^i$. Since $\Delta^i = d_{c^i}^i - d_{c^i}^{i-1} \leq d_{c^i}^i$, it holds that

$$\begin{aligned}
\phi^i - \phi^{i-1} &\geq \sum_{z=0}^{d_{c^i}^i}(d_{x_z}^i - d_{x_z}^{i-1}) \\
&= \Delta^i + \sum_{z=1}^{d_{c^i}^i}(d_{x_z}^i - d_{x_z}^{i-1}) \\
&\geq \Delta^i + \sum_{z=1}^{d_{c^i}^i}\max(\Delta^i - 2z, 0) \\
&\geq \Delta^i + \sum_{z=1}^{\Delta^i}\max(\Delta^i - 2z, 0) \\
&\geq \frac{(\Delta^i)^2}{4}.
\end{aligned}$$

Summing over i results in

$$\sum_{i=1}^{b} \frac{(\Delta^i)^2}{4} \leq \phi^b - \phi^0 \leq |V|^2.$$

We can now bound $\sum_{i=1}^{b} \Delta^i$. The values Δ^i ($1 \leq i \leq b$) are constrained by $\Delta^i \geq 0$ and $\sum_{i=1}^{b}(\Delta^i)^2 \leq 4|V|^2$. Calculus shows that maximizing $\sum_{i=1}^{b} \Delta^i$ subject to these constraints is achieved by $\Delta^i = 2|V|/\sqrt{b}$ for all $1 \leq i \leq b$. Thus, $\sum_{i=1}^{b} \Delta^i \leq 2b|V|/\sqrt{b} \leq 2|V|^{\frac{3}{2}}$ because $b \leq |V|$. Finally, the upper bound from Formula (2) implies that the total number of steps taken by the robot is at most $2|V| + 2|V|^{\frac{3}{2}}$. ∎

V. CONCLUSION

In this paper, we analyzed the worst-case travel distance of D*, a planning method that always moves a robot in initially unknown terrain from its current vertex on a shortest presumed unblocked path to a given goal vertex. We improved on the only existing analysis by proving a lower bound of $\Omega(\frac{\log|V|}{\log\log|V|}|V|)$ steps on vertex-blocked grids and an upper bound of $O(|V|^{3/2})$ steps on arbitrary vertex-blocked graphs, including grids. This provides new, substantially tighter bounds on the worst-case travel distance of D* on grids and thus a realistic analysis for the way D* is actually used.

ACKNOWLEDGMENTS

This research is partly supported by NSF awards under contracts IIS-9984827, IIS-0098807, and ITR/AP-0113881. The views and conclusions contained in this document are those of the authors and should not be interpreted as representing the official policies, either expressed or implied, of the sponsoring organizations and agencies or the U.S. government.

VI. REFERENCES

[1] B. Brumitt and A. Stentz. GRAMMPS: a generalized mission planner for multiple mobile robots. In *Proceedings of the International Conference on Robotics and Automation*, 1998.

[2] H. Choset and J. Burdick. Sensor based planning and nonsmooth analysis. In *Proceedings of the International Conference on Robotics and Automation*, pages 3034–3041, 1994.

[3] M. Hebert, R. McLachlan, and P. Chang. Experiments with driving modes for urban robots. In *Proceedings of the SPIE Mobile Robots*, 1999.

[4] S. Koenig and M. Likhachev. Improved fast replanning for robot navigation in unknown terrain. In *Proceedings of the International Conference on Robotics and Automation*, 2002.

[5] S. Koenig, C. Tovey, and W. Halliburton. Greedy mapping of terrain. In *Proceedings of the International Conference on Robotics and Automation*, pages 3594–3599, 2001.

[6] L. Matthies, Y. Xiong, R. Hogg, D. Zhu, A. Rankin, B. Kennedy, M. Hebert, R. Maclachlan, C. Won, T. Frost, G. Sukhatme, M. McHenry, and S. Goldberg. A portable, autonomous, urban reconnaissance robot. In *Proceedings of the International Conference on Intelligent Autonomous Systems*, 2000.

[7] I. Nourbakhsh. *Interleaving Planning and Execution for Autonomous Robots*. Kluwer Academic Publishers, 1997.

[8] I. Nourbakhsh and M. Genesereth. Assumptive planning and execution: a simple, working robot architecture. *Autonomous Robots Journal*, 3(1):49–67, 1996.

[9] N. Rao, S. Hareti, W. Shi, and S. Iyengar. Robot navigation in unknown terrains: Introductory survey of non-heuristic algorithms. Technical Report ORNL/TM–12410, Oak Ridge National Laboratory, Oak Ridge (Tennessee), 1993.

[10] Y. Smirnov. *Hybrid Algorithms for On-Line Search and Combinatorial Optimization Problems*. PhD thesis, School of Computer Science, Carnegie Mellon University, Pittsburgh (Pennsylvania), 1997. Available as Technical Report CMU-CS-97-171.

[11] A. Stentz. Optimal and efficient path planning for partially-known environments. In *Proceedings of the International Conference on Robotics and Automation*, pages 3310–3317, 1994.

[12] A. Stentz. The focussed D* algorithm for real-time replanning. In *Proceedings of the International Joint Conference on Artificial Intelligence*, pages 1652–1659, 1995.

[13] A. Stentz. Optimal and efficient path planning for unknown and dynamic environments. *International Journal of Robotics and Automation*, 10(3):89–100, 1995.

[14] A. Stentz. CD*: A real-time resolution optimal re-planner for globally constrained problems. In *Proceedings of the National Conference on Artificial Intelligence*, pages 605–612, 2002.

[15] A. Stentz and M. Hebert. A complete navigation system for goal acquisition in unknown environments. *Autonomous Robots*, 2(2):127–145, 1995.

[16] S. Thayer, B. Digney, M. Diaz, A. Stentz, B. Nabbe, and M. Hebert. Distributed robotic mapping of extreme environments. In *Proceedings of the SPIE: Mobile Robots XV and Telemanipulator and Telepresence Technologies VII*, volume 4195, 2000.

On the Nonlinear Controllability of a Quasiholonomic Mobile Robot

Alessio Salerno Jorge Angeles

Department of Mechanical Engineering & Centre for Intelligent Machines
McGill University, Montreal, Canada
salerno|angeles@cim.mcgill.ca

Abstract—We report on the controllability of a novel mobile robot which comprises two driving wheels and an intermediate body carrying the payload. By virtue of quasiholonomy, a concept introduced elsewhere, the robot is underactuated by design. One challenge here is the control of the motion of the intermediate body, which will tend to rotate about the wheel axis as the wheels are actuated. We prove that it is possible to completely control the robot using only the wheel motors, while tracking a desired trajectory, with apparent advantages in terms of cost, weight and efficiency. To do this, we show that every linearization of the robot dynamics model around an equilibrium point verifies the Kalman rank condition for controllability. Moreover, using modern results from nonlinear control theory, we prove that the robot is locally accessible and small-time locally controllable.

I. Introduction

Controllability is a key issue in the design and control of a robotic mechanical system, since it provides the minimum number of actuators needed to completely control the system. Particularly interesting is the controllability of quasiholonomic wheeled mobile robots (WMRs), a class of nonholonomic systems with mathematical models formally identical to their holonomic counterparts [1], [2]. This property makes these systems easier to control [2]; because of their non-integrable constraint equations, moreover these systems can be designed to be underactuated. Underactuation leads to a less expensive, lighter and more efficient robot.

We report on the controllability of *Quasimoro*, a novel quasiholonomic underactuated WMR which comprises two driving wheels and an intermediate body (IB) carrying the payload, without any caster wheel. To achieve quasiholonomy, the mass center of the central body is placed on the vertical passing through the midpoint of the line joining the wheel centers. Moreover, in order to cope with instability, the mass center of the IB is placed below the abovementioned line. The robot control system main tasks are: i) positioning and orienting the payload, supported by the IB, on a flat surface (primary task); ii) suppressing the oscillations of the IB (secondary task.) Now, while apparently it is possible to accomplish the primary task using only the wheel motors, accomplishing the secondary one needs further investigation.

We show that relying only on the kinematic model of *Quasimoro* it is not possible to completely control the robot. Hence it is necessary to resort to the dynamics model, which is nonlinear.

To analyze the controllability of a nonlinear system, we can proceed in two non-equivalent ways: i) linearize the mathematical model around an equilibrium point and then apply the Kalman rank condition for controllability (KRCC) [3]; or ii) use modern controllability criteria for nonlinear systems [4]. Following the first approach, it is shown that every linearization around points of equilibrium satisfies the KRCC. The application of the second approach requires some tools from differential geometry that will be briefly recalled. Applying the Lie algebra rank condition (LARC) and the Sussmann Theorem on small-time local controllability (STLC), we show that *Quasimoro* is locally accessible from any point of the state space and small-time locally controllable from any equilibrium point.

A system closely related to ours is the mobile inverted pendulum *JOE* [5]. However, a few remarkable differences between the two systems are to be highlighted:

a) to sense the orientation of the central body with respect to the vertical, *Quasimoro* uses a cost-effective inclinometer, while *JOE* uses a rate gyro;
b) the mass center of *Quasimoro* is placed below the line joining the wheel centers; *JOE*'s is placed above;
c) the integration of the readout signal of *JOE*'s rate gyro is sensitive to drift problems, which makes the robot slowly move forward (or backward) in order to catch the imaginary "fall" [5], a problem avoided in *Quasimoro* using an inclinometer;
d) the linear controllers used by *JOE* neglect the nonlinear forces associated with the motion of the robot; the controller accuracy thus quickly degrades as the speed increases, because many of the dynamic forces involved, such as Coriolis and centrifugal, vary as the square of the speed [6]. In the case of *Quasimoro* such a problem will be overcome using nonlinear feedback control.

Moreover, the controllability tests reported here are more general, as compared to the one used in [5]. In fact we derive here theoretical results that are generally applicable; prototype manufacturing will be reported in due course.

II. Fundamentals of Nonlinear Controllability

For details on the material recalled in this section, the reader is referred to [7], [8], and [9].

A multivariable, continuous, time-invariant nonlinear control system with m inputs $u_1, \ldots, u_m$ and p outputs $y_1, \ldots, y_p$ is described, in state-space form, by means of a model of the type

$$\dot{\mathbf{x}}(t) = \mathbf{k}(\mathbf{x}(t), \mathbf{u}(t)) \tag{1}$$
$$\mathbf{y}(t) = \mathbf{l}(\mathbf{x}(t), \mathbf{u}(t)) \tag{2}$$

where the state vector $\mathbf{x} \equiv [x_1 \cdots x_n]^T$ belongs to $\mathbb{R}^n$, while the output vector $\mathbf{y} \equiv [y_1 \cdots y_p]^T$ to $\mathbb{R}^p$ and each component of the input vector $\mathbf{u} \equiv [u_1 \cdots u_m]^T \in \mathbb{R}^m$ takes values in the class of piecewise-constant functions $\mathcal{U}$ over time.

A particular class of continuous-time nonlinear systems is known as *affine systems*, which takes the form [8]

$$\dot{\mathbf{x}}(t) = \mathbf{f}(\mathbf{x}(t)) + \sum_{i=1}^{m} \mathbf{g}_i(\mathbf{x}(t)) u_i(t) \tag{3}$$
$$y_i = h_i(\mathbf{x}(t)) \quad 1 \leq i \leq p, \tag{4}$$

where $\{h_i\}_1^p$ are real-valued functions defined on $\mathbb{R}^n$ and $\mathbf{f}, \mathbf{g}_1, \ldots, \mathbf{g}_m$ are smooth vector fields on $\mathbb{R}^n$; in particular, $\mathbf{f}$ is called the drift vector field, while $\{\mathbf{g}_i\}_1^m$ are called the input vector fields. Henceforth, we omit the time argument, when not essential, to simplify the notation.

Denoting the unique solution of eq. (3) at time $t \geq 0$ for a particular input $\mathbf{u}(\cdot)$ and initial condition $\mathbf{x}(0) = \mathbf{x}_0$ as $\mathbf{x}(T, 0, \mathbf{x}_0, \mathbf{u})$, we can state [7]

Definition II.1 (Controllability) The system (3), (4) is called controllable if, for any choice of $\mathbf{x}_1, \mathbf{x}_2 \in \mathbb{R}^n$, there exists a finite time T and an input $\mathbf{u} : [0, T] \to \mathcal{U}$ such that $\mathbf{x}(T, 0, \mathbf{x}_1, \mathbf{u}) = \mathbf{x}_2$.

To date, however, there are very few theoretical results concerning questions on the natural form of controllability defined above; we thus try two other structural characterizations of the system (3), (4), namely, local accessibility (LA) and STLC, that are related to the previous definition [4].

Definition II.2: Given a neighborhood $\mathcal{V}$ of $\mathbf{x}_0$, we define the set of states reachable at time τ from $\mathbf{x}_0$ with trajectories contained in $\mathcal{V}$ as $\mathcal{R}^{\mathcal{V}}(\mathbf{x}_0, \tau) \equiv \{\boldsymbol{\xi} \in \mathbb{R}^n \mid \exists \mathbf{u} : [0, T] \to \mathcal{U} \mid \mathbf{x}(\tau, 0, \mathbf{x}_0, \mathbf{u}) = \boldsymbol{\xi}$ and $\mathbf{x}(t, 0, \mathbf{x}_0, \mathbf{u}) \in \mathcal{V}, t \leq \tau\}$. The set of states reachable within time T from $\mathbf{x}_0$ with trajectories contained in $\mathcal{V}$ is defined, in turn, as

$$\mathcal{R}_T^{\mathcal{V}}(\mathbf{x}_0) \equiv \bigcup_{\tau \leq T} \mathcal{R}^{\mathcal{V}}(\mathbf{x}_0, \tau).$$

This definition leads naturally to the one below, which can be found in [4] and [7], respectively.

Definition II.3 (Local accessibility) The system (3), (4) is locally accessible from $\mathbf{x}_0$ if $\mathcal{R}_T^{\mathcal{V}}(\mathbf{x}_0)$ contains a non-empty open set of $\mathbb{R}^n$ for all neighborhoods $\mathcal{V}$ of $\mathbf{x}_0$ and all $T > 0$. If this holds for any $\mathbf{x}_0 \in \mathbb{R}^n$, then the system is called locally accessible.

Definition II.4 (Small-time local controllability) The system (3), (4) is small-time locally controllable from $\mathbf{x}_0$ if, for all neighborhoods $\mathcal{V}$ of $\mathbf{x}_0$ and all $T > 0$, $\mathcal{R}_T^{\mathcal{V}}(\mathbf{x}_0)$ contains a non-empty neighborhood of $\mathbf{x}_0$. If this holds for any $\mathbf{x}_0 \in \mathbb{R}^n$, then the system is called small-time locally controllable.

Now we discuss two conditions for LA and for STLC; beforehand we recall some tools from differential geometry.

A. Differential Geometry

All the definitions and propositions of this subsection can be found in [8].

Definition II.5 (Lie bracket) Given two smooth vector fields $\mathbf{g}_1$ and $\mathbf{g}_2$ on $\mathbb{R}^n$, the Lie bracket of $\mathbf{g}_1$ and $\mathbf{g}_2$ is the new vector field defined as

$$[\mathbf{g}_1, \mathbf{g}_2] \equiv \frac{\partial \mathbf{g}_2}{\partial \mathbf{x}} \mathbf{g}_1 - \frac{\partial \mathbf{g}_1}{\partial \mathbf{x}} \mathbf{g}_2.$$

Proposition II.1 (Properties of the Lie brackets) The properties of the Lie bracket of vector fields are listed below:

i) Bilinearity over $\mathbb{R}$, i.e., if $\mathbf{f}_1, \mathbf{f}_2, \mathbf{g}_1, \mathbf{g}_2$ are vector fields and r_1, r_2 real numbers, then

$$[r_1\mathbf{f}_1 + r_2\mathbf{f}_2, \mathbf{g}_1] = r_1[\mathbf{f}_1, \mathbf{g}_1] + r_2[\mathbf{f}_2, \mathbf{g}_1],$$
$$[\mathbf{f}_1, r_1\mathbf{g}_1 + r_2\mathbf{g}_2] = r_1[\mathbf{f}_1, \mathbf{g}_1] + r_2[\mathbf{f}_1, \mathbf{g}_2];$$

ii) Skew-commutativity, i.e. if $\mathbf{f}$ and $\mathbf{g}$ are vector fields, then $[\mathbf{f}, \mathbf{g}] = -[\mathbf{g}, \mathbf{f}]$;

iii) The Lie bracket satisfies the Jacobi identity, i.e. if $\mathbf{f}, \mathbf{g}, \mathbf{p}$ are vector fields, then $[\mathbf{f}, [\mathbf{g}, \mathbf{p}]] + [\mathbf{g}, [\mathbf{p}, \mathbf{f}]] + [\mathbf{p}, [\mathbf{f}, \mathbf{g}]] = \mathbf{0}$.

Definition II.6 (Lie algebra) A vector space $\mathcal{V}$ over $\mathbb{R}$ is a Lie algebra if, in addition to its vector space structure, it is possible to define a binary operation $\mathcal{V} \times \mathcal{V} \to \mathcal{V}$, called a bracket and written $[\cdot, \cdot]$, satisfying the three properties of Proposition II.1.

Definition II.7 (Distribution) Given m [smooth] vector fields $\mathbf{g}_1, \ldots, \mathbf{g}_m$, all defined on $\mathbb{R}^n$, the [smooth] distribution associated with $\mathbf{g}_1, \ldots, \mathbf{g}_m$ is the assignment to each $\mathbf{x} \in \mathbb{R}^n$ of the vector space spanned by vectors $\mathbf{g}_1(\mathbf{x}), \ldots, \mathbf{g}_m(\mathbf{x})$ and indicated by $\Delta \equiv \text{span}\{\mathbf{g}_1, \ldots, \mathbf{g}_m\}$. The "value" of Δ at $\mathbf{x}$ is denoted by $\Delta(\mathbf{x}) \equiv \text{span}\{\mathbf{g}_1(\mathbf{x}), \ldots, \mathbf{g}_m(\mathbf{x})\}$.

Definition II.8: A distribution Δ_1 contains a distribution Δ_2, which is expressed as $\Delta_1 \supset \Delta_2$, if $\Delta_1(\mathbf{x}) \supset \Delta_2(\mathbf{x})$ for all $\mathbf{x}$. A vector field $\mathbf{f}$ belongs to a distribution Δ, which is expressed as $\mathbf{f} \in \Delta$, if $\mathbf{f}(\mathbf{x}) \in \Delta(\mathbf{x})$ for all $\mathbf{x}$. The dimension of a distribution at $\mathbf{x}$ is the dimension of the space $\Delta(\mathbf{x})$. A distribution is involutive if the Lie bracket $[\boldsymbol{\tau}_1, \boldsymbol{\tau}_2]$ of any pair of vector fields $\boldsymbol{\tau}_1$ and $\boldsymbol{\tau}_2$ belonging to Δ is a vector field which belongs to Δ. The smallest involutive distribution containing Δ is called the involutive closure of Δ, denoted by $\overline{\Delta}$.

B. Controllability Tests

Over the past 30 years, the controllability of nonlinear systems has been investigated intensively; as a result, sufficient conditions for STLC have been derived. In the present work we resort to a general theorem for STLC by Sussmann, which contains all the above sufficient conditions as particular cases [10]. Before introducing this theorem we recall some preliminary definitions and a well-known theorem for LA.

B.1 Local Accessibility

Definition II.9 (Accessibility distribution) The accessibility distribution $\overline{\Delta}_{\mathcal{A}}$ of the nonlinear system (3), (4), is the involutive closure of the distribution $\Delta_{\mathcal{A}} \equiv \text{span}\{\mathbf{f}, \mathbf{g}_1, \ldots, \mathbf{g}_m\}$.

The computation of $\overline{\Delta}_{\mathcal{A}}$ may be achieved via an iterative procedure [4]: $\overline{\Delta}_{\mathcal{A}} = \text{span}\{\mathbf{v} \mid \mathbf{v} \in \Delta_i, \forall i \geq 1\}$ with $\Delta_1 \equiv \Delta_{\mathcal{A}}$, $\Delta_i \equiv \Delta_{i-1} + [\Delta_1, \Delta_{i-1}]$, $i \geq 2$, where $[\Delta_1, \Delta_{i-1}] \equiv \text{span}\{[\mathbf{g}, \mathbf{v}] \mid \mathbf{g} \in \Delta_1, \mathbf{v} \in \Delta_{i-1}\}$. The above procedure stops after k steps, where k is the smallest integer such that $\Delta_{k+1} = \Delta_k = \overline{\Delta}_{\mathcal{A}}$. Since $\dim \overline{\Delta}_{\mathcal{A}} \leq n$, it follows that one stops after at most $n - m$ steps [4].

An important result for testing the LA of nonlinear systems is the LARC (Lie Algebra Rank Condition) Theorem, also known as the Chow Theorem [11]

Theorem II.1 (LARC) If the LARC $\dim \overline{\Delta}_{\mathcal{A}}(\mathbf{x}_0) = n$ holds, then the system (3), (4) is locally accessible from $\mathbf{x}_0$. If LARC holds for all $\mathbf{x} \in \mathbb{R}^n$, then the system is locally accessible.

B.2 Small-Time Local Controllability

Unfortunately, for nonlinear systems with non-trivial drift the Chow Theorem provides only information about the LA and not the STLC of the system.

To date the strongest statement on STLC for nonlinear control systems with drift is due to Sussmann [10]. In order to state Sussmann's Theorem, we need a few concepts:

Let $\mathcal{X} \equiv \{\mathbf{f}, \mathbf{g}_1, \ldots, \mathbf{g}_m\}$ be a finite set of vector fields; we denote with $\text{Br}(\mathcal{X})$ the set of all possible iterated Lie brackets involving $\mathbf{f}, \mathbf{g}_1, \ldots, \mathbf{g}_m$ [12]; methods on generating $\text{Br}(\mathcal{X})$ are given in [13]. The problem in the generation of $\text{Br}(\mathcal{X})$ is that not all the Lie brackets are linearly independent because of the skew-commutativity property and the Jacobi identity [11]. A particular basis of iterated Lie brackets which takes into account the above properties is the Philip Hall basis; systematic procedures to construct a Philip Hall basis are included in [11], [14].

Definition II.10: Let $\mathcal{X} \equiv \{\mathbf{g}_0, \mathbf{g}_1, \ldots, \mathbf{g}_m\}$ be a finite set of vector fields; the degree of the Lie bracket $\mathbf{v} \in \text{Br}(\mathcal{X})$ is defined as

$$\delta(\mathbf{v}) \equiv \sum_{i=0}^{m} \delta^i(\mathbf{v})$$

where $\delta^i(\mathbf{v})$ is the number of occurrences of $\mathbf{g}_i$ in $\mathbf{v}$.

Now we have the theorem due to Sussmann[1].

Theorem II.2 (General theorem on STLC) Given a nonlinear system (3), (4) and $\mathbf{x}_0$ an equilibrium state, assume that $\mathcal{X} \equiv \{\mathbf{f}, \mathbf{g}_1, \ldots, \mathbf{g}_m\}$ satisfies the LARC at $\mathbf{x}_0$. Further assume that whenever $\mathbf{v} \in \text{Br}(\mathcal{X})$ is a bracket for which $\delta^0(\mathbf{v})$ is odd and $\delta^1(\mathbf{v}), \ldots, \delta^m(\mathbf{v})$ are all even, then $\mathbf{v}$ may be written as a linear combination of brackets of lower degree, the system (3), (4) being small-time locally controllable from $\mathbf{x}_0$.

With this theorem in mind, we define a "bad" bracket to be that bracket for which the drift vector field appears an odd number of times and for which the input vector fields appear each an even number of times (including zero times) [12].

III. KINEMATIC MODEL

In deriving the mathematical model of the robot we assume that it undergoes motion on a horizontal planar surface, which we will call $\mathcal{B}$. Moreover, the robot wheels are assumed to be always in contact with $\mathcal{B}$.

A simplified model of the robot is shown in Fig. 1. We have defined $\mathcal{A}$ and $\mathcal{A}'$ as the axis passing through the centers of the wheels and that parallel to the latter and passing through the mass center C_3 of the IB, respectively. The chassis of the IB is represented by a cylinder with axis of symmetry $\mathcal{D}$ which is normal to $\mathcal{A}$.

Two orthonormal triads of vectors, namely, $\{\mathbf{i}_0, \mathbf{j}_0, \mathbf{k}_0\}$ and $\{\mathbf{e}, \mathbf{m}, \mathbf{n}\}$ are defined. The triad $\{\mathbf{i}_0, \mathbf{j}_0, \mathbf{k}_0\}$ defines an inertial frame attached to the ground with origin O and with $\mathbf{k}_0$ normal to $\mathcal{B}$. The frame defined by $\{\mathbf{e}, \mathbf{m}, \mathbf{n}\}$ is attached to the IB and centered at the midpoint C of the line linking the mass centers of the two wheels, while $\mathbf{n}$ lies on $\mathcal{D}$, and $\mathbf{e}$ lies on $\mathcal{A}$.

We indicate with θ_1 and θ_2 the angular displacements of the two wheels, while $\mathbf{c}$ is the two-dimensional position vector of point C. We define θ_3 as the angle of rotation of the IB about $\mathcal{A}$.

Defining l as the distance between the centers of the wheels and r as the wheel radius, and assuming that the wheels roll without slipping on $\mathcal{B}$, the kinematic model of the robot can be written in the affine state-space form of eq. (3), where $\mathbf{u} \equiv [\ \dot{\theta}_1\ \ \dot{\theta}_2\ ]^T$,

$$\mathbf{x} \equiv [\ \mathbf{c}^T\ \ \psi\ \ \theta_3\ ]^T, \quad \mathbf{f} \equiv [\ 0\ \ 0\ \ 0\ \ \dot{\theta}_3\ ]^T,$$
$$\mathbf{g}_1 \equiv [\ -(r/2)\sin\psi\ \ (r/2)\cos\psi\ \ -\rho\ \ 0\ ]^T,$$
$$\mathbf{g}_2 \equiv [\ -(r/2)\sin\psi\ \ (r/2)\cos\psi\ \ \rho\ \ 0\ ]^T,$$

$\rho \equiv r/l$, and ψ is the orientation angle, i.e., the angle between vectors $\mathbf{i}_0$ and $\mathbf{e}$.

---

[1] The version we have written here is greatly simplified, but retains the content of the original theorem — more details can be found in [10].

J_1, J_2 [kg m^2]	m, m_3 [kg]	g [m/s^2]	d, r, l [m]
0.396, 0.693	3, 19	9.81	0.1, 0.3, 0.7

TABLE I

GEOMETRIC AND INERTIAL PARAMETERS OF *Quasimoro*

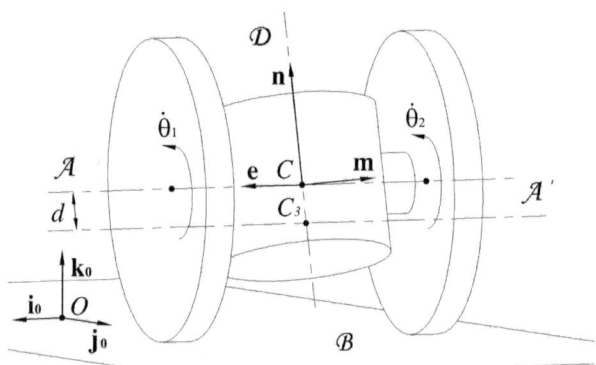

Fig. 1. Simplified model of *Quasimoro*

It is apparent, without resorting to any of the tests recalled in the previous section, that the state variable θ_3 cannot be controlled by using only the above kinematic model. Therefore, in order to completely control the robot, without increasing the number of actuators, we will have to test the controllability of the dynamics model.

IV. DYNAMICS MODEL

A. Controllability of the Linearized System

The dynamics model of *Quasimoro* can be written in the state-space form of eq. (1), with

$$\mathbf{u} \equiv [\begin{array}{cc} \tau_1 & \tau_2 \end{array}]^T, \tag{5}$$

$$\mathbf{x} \equiv [\begin{array}{cccccc} \theta_1 & \theta_2 & \theta_3 & \dot{\theta}_1 & \dot{\theta}_2 & \dot{\theta}_3 \end{array}]^T, \tag{6}$$

$$\mathbf{k}(\mathbf{x}, \mathbf{u}) \equiv \begin{bmatrix} x_4 \\ x_5 \\ x_6 \\ [C - (u_2 A - u_1 B)J_2 - T(u_1 - u_2)]/Q \\ [C - (u_1 A - u_2 B)J_2 - T(u_2 - u_1)]/Q \\ [K \cos x_3 u_1 + K \cos x_3 u_2 + D]/Q \end{bmatrix},$$

where

$$A \equiv \frac{1}{2}[\frac{m_3 r^2}{2} - (mr^2 + 2J_1)\rho^2], \quad K \equiv \frac{1}{2}m_3 r d,$$

$$B \equiv \frac{3}{2}mr^2 + \frac{1}{2}m_3 r^2 - A,$$

$$C \equiv (B - A)(x_6^2 J_2 + \cos x_3 m_3 g d) K \sin x_3,$$

$$D \equiv [2K^2 \cos x_3 \sin x_3 x_6^2 + (A + B)m_3 g d \sin x_3],$$

$$Q \equiv (B - A)(2T - J_2 B - J_2 A), \quad T \equiv K^2 \cos x_3^2,$$

while $\{x_i\}_1^6$ and $\{u_i\}_1^2$ are the i-th components of $\mathbf{x}$ and of $\mathbf{u}$, respectively; J_2 is the moment of inertia of the IB about $\mathcal{A}$, $\{\tau_i\}_1^2$ are the motor torques, m is the mass of each augmented wheel (i.e. the wheel along with the shaft which actuates it), m_3 is the augmented mass of the IB (i.e. taking into account also the payload), d is the distance between C and C_3, g is the gravity acceleration and J_1 is the moment of inertia of the IB about its axis of symmetry.

In order to derive the set of equilibrium states of the dynamical system of the robot at hand we set $\mathbf{k}(\mathbf{x}, \mathbf{u}) = \mathbf{0}_6$, with $\mathbf{u} = \mathbf{0}_2$ [15]. Hence, the set of equilibrium points is given by

$$\mathcal{X}_e = \{\mathbf{x} \equiv [\begin{array}{ccccc} x_1 & x_2 & x_3 & x_5 & x_6 \end{array}]^T \in \mathbb{R}^n | \\ x_i = 0 \;\; \forall i \geq 3\}. \tag{7}$$

From the structure of $\mathcal{X}_e$ it is apparent that every equilibrium state is equivalent to the others for controllability test purposes.

Linearizing around one of the states of equilibrium $\mathbf{x}_0 \equiv \mathbf{0}_6$ and considering the values indicated in Table I we obtain $\dot{\mathbf{x}} = \mathbf{A}\mathbf{x} + \mathbf{B}\mathbf{u}$, with

$$\mathbf{A} \equiv \left.\frac{\partial \mathbf{k}}{\partial \mathbf{x}}\right|_{\mathbf{x}=\mathbf{x}_0} = \begin{bmatrix} \mathbf{O}_{3\times 2} & \mathbf{0}_3 & \mathbf{1}_3 \\ \mathbf{O}_{3\times 2} & \mathbf{a} & \mathbf{O}_{3\times 3} \end{bmatrix},$$

$$\mathbf{B} \equiv \left.\frac{\partial \mathbf{k}}{\partial \mathbf{u}}\right|_{\mathbf{x}=\mathbf{x}_0} = \begin{bmatrix} \mathbf{0}_3 & \mathbf{0}_3 \\ \mathbf{b}_1 & \mathbf{b}_2 \end{bmatrix},$$

$$\mathbf{a} \equiv [\begin{array}{ccc} 7.478 & 7.478 & -33.061 \end{array}]^T,$$

$$\mathbf{b}_1 \equiv [\begin{array}{ccc} 1.321 & -0.346 & -0.401 \end{array}]^T,$$

$$\mathbf{b}_2 \equiv [\begin{array}{ccc} -0.346 & 1.321 & -0.401 \end{array}]^T.$$

where $\mathbf{0}_k$ is the k-dimensional zero vector while $\mathbf{O}_{i\times j}$ and $\mathbf{1}_k$ are the $i \times j$ zero matrix is the $k \times k$ identity matrix, respectively.

Computing the rank of the controllability matrix $\mathbf{C} \equiv [\begin{array}{cccccc} \mathbf{B} & \mathbf{AB} & \mathbf{A}^2\mathbf{B} & \mathbf{A}^3\mathbf{B} & \mathbf{A}^4\mathbf{B} & \mathbf{A}^5\mathbf{B} \end{array}]$, we have that rank($\mathbf{C}$) = 6, and hence, the linearized system is controllable according the KRCC. A well-known theorem in control theory states that if the linearization of a nonlinear system at an equilibrium state is controllable, then the nonlinear system is small-time locally controllable from the equilibrium state [10]. In view of this result we obtain

Result IV.1: Resorting to its dynamics model, *Quasimoro* is small-time locally controllable from any equilibrium state.

B. Controllability of the Nonlinear System

The verification of the KRCC for the linearized system is not an exhaustive analysis, because:

i) the controllability of the system has not been investigated for every point of the state space; and

ii) we have referred to a linear approximation of the dynamics model, while the system might show a highly nonlinear behavior.

Therefore, the LA and STLC should be tested to complete the analysis.

B.1 Local Accessibility Analysis

The dynamics model of the robot can be rewritten in the affine state-space form (3), with **x** and **u** defined as in eqs. (5)-(6) and

$$\mathbf{f} \equiv \begin{bmatrix} x_4 \\ x_5 \\ x_6 \\ -(x_6^2 J_2 + m_3 d \cos x_3 g) m_3 d \sin x_3 / rS \\ -(x_6^2 J_2 + m_3 d \cos x_3 g) m_3 d \sin x_3 / rS \\ \sin x_3 [g(m_3 + m) + m_3 d \cos x_3 x_6^2]/S \end{bmatrix}, \quad (8)$$

$$\mathbf{g}_1 \equiv \begin{bmatrix} 0 \\ 0 \\ 0 \\ (m_3^2 d^2 \cos x_3^2 l^2 - E)/(m_3 d r^2 FS) \\ P/(m_3 d r^2 FS) \\ \cos x_3 / r^2 S \end{bmatrix}, \quad (9)$$

$$\mathbf{g}_2 \equiv \begin{bmatrix} 0 \\ 0 \\ 0 \\ P/(m_3 d r^2 FS) \\ (m_3^2 d^2 \cos x_3^2 l^2 - E)/(m_3 d r^2 FS) \\ \cos x_3 / r^2 S \end{bmatrix}, \quad (10)$$

where $E \equiv (6J_2 ml^2 + J_2 m_3 l^2 + 2J_2 mr^2 + 4J_2 J_1)$, $F \equiv (2mr^2 + 3ml^2 + 4J_1)$, $P \equiv (J_2 m_3 l^2 - 2J_2 mr^2 - 4J_2 J_1 - m_3^2 d^2 \cos x_3 l^2)$, and $S \equiv [m_3^2 d^2 \cos x_3^2 - J_2(m_3 + 3m)/(m_3 d)]$.

The distribution defined by the vector fields $\mathbf{f}, \mathbf{g}_1, \mathbf{g}_2$, is $\Delta_\mathcal{A} \equiv \text{span}\{\mathbf{f}, \mathbf{g}_1, \mathbf{g}_2\}$. Now we follow the procedure described in Subsection II.B.1 in order to compute the involutive closure of $\Delta_\mathcal{A}$, that will be indicated with $\overline{\Delta}_\mathcal{A}$. The first step is to compute the dimension of the first distribution Δ_1, defined as

$$\Delta_1 \equiv \Delta_\mathcal{A} \equiv \text{span}\{\mathbf{f}, \mathbf{g}_1, \mathbf{g}_2\}. \quad (11)$$

By looking at eqs. (8)–(10) we infer that $\dim \Delta_1 = 3$ $\forall \mathbf{x} \notin \mathcal{X}_e$, where $\mathcal{X}_e$ is the set of equilibrium states defined in eq. (7). Hence, for the time being we will investigate the LA of the dynamical system at every point of the state-space, but those of equilibrium.

The next step consists in computing Δ_2, namely

$$\Delta_2 \equiv \Delta_1 + [\Delta_1, \Delta_1] = \text{span}\{\mathbf{f}, \mathbf{g}_1, \mathbf{g}_2\}$$
$$+ \text{span}\{[\mathbf{f}, \mathbf{g}_1], [\mathbf{f}, \mathbf{g}_2], [\mathbf{g}_1, \mathbf{f}], [\mathbf{g}_1, \mathbf{g}_2], [\mathbf{g}_2, \mathbf{f}], [\mathbf{g}_2, \mathbf{g}_1]\}$$
$$= \text{span}\{\mathbf{f}, \mathbf{g}_1, \mathbf{g}_2, [\mathbf{f}, \mathbf{g}_1], [\mathbf{f}, \mathbf{g}_2], [\mathbf{g}_1, \mathbf{g}_2]\},$$

where, in the last simplification, we have applied the skew-commutativity property and the Jacobi identity. Computing the Lie brackets, we obtain

$$[\mathbf{f}, \mathbf{g}_1] = \begin{bmatrix} -(m_3^2 d^2 \cos x_3^2 l^2 - E)/(m_3 d r^2 FS) \\ P/(m_3 d r^2 FS) \\ -m_3 d \cos x_3 /(m_3 d r^2 S) \\ 0 \\ 0 \\ -\sin x_3 x_6 / rS \end{bmatrix},$$

$$[\mathbf{f}, \mathbf{g}_2] = \begin{bmatrix} P/(m_3 d r^2 FS) \\ -(m_3^2 d^2 \cos x_3^2 l^2 - E)/(m_3 d r^2 FS) \\ m_3 d \cos x_3 /(m_3 d r^2 S) 0 \\ 0 \\ -\sin x_3 x_6 / rS \end{bmatrix},$$

$$[\mathbf{g}_1, \mathbf{g}_2] = \mathbf{0}_6,$$

whence, $\Delta_2 = \text{span}\{\mathbf{f}, \mathbf{g}_1, \mathbf{g}_2, [\mathbf{f}, \mathbf{g}_1], [\mathbf{f}, \mathbf{g}_2]\}$, $\dim \Delta_2 = 5$ $\forall \mathbf{x} \notin \mathcal{X}_e$. The distribution $\Delta_3 \equiv \Delta_2 + [\Delta_1, \Delta_2]$, reduces to $\Delta_3 = \text{span}\{\mathbf{f}, \mathbf{g}_1, \mathbf{g}_2, [\mathbf{f}, \mathbf{g}_1], [\mathbf{f}, \mathbf{g}_2], [\mathbf{g}_1, [\mathbf{f}, \mathbf{g}_1]]\}$, by applying the skew-commutativity property, the Jacobi identity, and knowing that $[\mathbf{g}_1, [\mathbf{f}, \mathbf{g}_1]] = [\mathbf{g}_2, [\mathbf{f}, \mathbf{g}_2]] = [\mathbf{g}_1, [\mathbf{f}, \mathbf{g}_2]] = [\mathbf{g}_2, [\mathbf{f}, \mathbf{g}_1]]$, and that

$$[\mathbf{f}, [\mathbf{f}, \mathbf{g}_1]] = [\mathbf{f}, [\mathbf{f}, \mathbf{g}_2]] \in \text{span}\{\mathbf{f}, \mathbf{g}_1, \mathbf{g}_2, [\mathbf{f}, \mathbf{g}_1], [\mathbf{f}, \mathbf{g}_2], [\mathbf{g}_1, [\mathbf{f}, \mathbf{g}_1]]\}.$$

We thus obtain $\dim \Delta_3 = 6$ $\forall \mathbf{x} \notin \mathcal{X}_e$. At this step the procedure ends, since there is no need to compute Δ_4 in order to verify that Δ_3 is involutive; this is so because the dimension of Δ_3 is equal to the dimension of the state vector, which is the maximum dimension that $\overline{\Delta}_\mathcal{A}$ can attain (see Subsection II.B.1). Thus, we can conclude that the accessibility distribution is given by

$$\overline{\Delta}_\mathcal{A} = \Delta_3 = \text{span}\{\mathbf{f}, \mathbf{g}_1, \mathbf{g}_2, [\mathbf{f}, \mathbf{g}_1], [\mathbf{f}, \mathbf{g}_2], [\mathbf{g}_1, [\mathbf{f}, \mathbf{g}_1]]\},$$

and that

Result IV.2: Resorting to the dynamics model, Quasimoro is locally accessible from any state different from those of equilibrium.

Now, in order to analyze the LA at the states of equilibrium we have to set up a new procedure. At equilibrium, the vector fields defined in eqs. (8)–(10) take the form

$$\mathbf{f} = \mathbf{0}_6, \quad (12)$$

$$\mathbf{g}_1 = \begin{bmatrix} 0 \\ 0 \\ 0 \\ (E - m_3^2 d^2 l^2)/r^2 FU \\ (2J_2 mr^2 + 4J_2 J_1 - J_2' m_3 l^2)/r^2 FU \\ -m_3 d / rU \end{bmatrix}, (13)$$

$$\mathbf{g}_2 = \begin{bmatrix} 0 \\ 0 \\ 0 \\ (2J_2 mr^2 + 4J_2 J_1 - J_2' m_3 l^2)/r^2 FU \\ (E - m_3^2 d^2 l^2)/r^2 FU \\ -m_3 d / rU \end{bmatrix}, (14)$$

with $U \equiv (-m_3^2 d^2 + J_2 m_3 + 3J_2 m)$. Defining $\Delta_{\mathcal{A}}$ and Δ_1 as in eq. (11) and looking at eqs. (12)–(14), we obtain that $\dim \Delta_1 = 2 \quad \forall \mathbf{x} \in \mathcal{X}_e$, while $\Delta_2 = \text{span}\{\mathbf{g}_1, \mathbf{g}_2, [\mathbf{f}, \mathbf{g}_1], [\mathbf{f}, \mathbf{g}_2]\}, \dim \Delta_2 = 4 \quad \forall \mathbf{x} \in \mathcal{X}_e$. The distribution Δ_3 and its dimension are given by

$$\Delta_3 = \text{span}\{\mathbf{g}_1, \mathbf{g}_2, [\mathbf{f}, \mathbf{g}_1], [\mathbf{f}, \mathbf{g}_2], [\mathbf{f}, [\mathbf{f}, \mathbf{g}_1]]\},$$
$$\dim \Delta_3 = 5 \quad \forall \mathbf{x} \in \mathcal{X}_e.$$

Computing the distribution Δ_4 we obtain

$$\Delta_4 = \text{span}\{\mathbf{g}_1, \mathbf{g}_2, [\mathbf{f}, \mathbf{g}_1], [\mathbf{f}, \mathbf{g}_2], [\mathbf{f}, [\mathbf{f}, \mathbf{g}_1]],$$
$$[\mathbf{f}, [\mathbf{f}, [\mathbf{f}, \mathbf{g}_1]]]\}, \dim \Delta_4 = 6 \; \forall \mathbf{x} \in \mathcal{X}_e.$$

The procedure stops, since the distributions used to build $\overline{\Delta}_{\mathcal{A}}$ have reached a dimension equal to the dimension of the state vector, hence proving that

Result IV.3: Resorting to its dynamics model, *Quasimoro* is locally accessible from any equilibrium state, according to the accessibility distribution

$$\overline{\Delta}_{\mathcal{A}} = \Delta_4 = \text{span}\{\mathbf{g}_1, \mathbf{g}_2, [\mathbf{f}, \mathbf{g}_1], [\mathbf{f}, \mathbf{g}_2], [\mathbf{f}, [\mathbf{f}, \mathbf{g}_1]],$$
$$[\mathbf{f}, [\mathbf{f}, [\mathbf{f}, \mathbf{g}_1]]]\}. \quad (15)$$

Therefore, merging the results IV.2 and IV.3, we can infer that the dynamical system of the WMR at hand is locally accessible from any point of the state space, i.e. *Quasimoro is locally accessible.*

B.2 Small-Time Local Controllability Analysis

As we saw from the accessibility distribution of eq. (15), the maximum order of bracketting reached is 4; hence, all the possible Lie brackets that can be obtained from $\mathbf{f}, \mathbf{g}_1$, and $\mathbf{g}_2$ of eqs. (12)–(14) up until order 4 have to be derived. The foregoing derivation has to be conducted in order to search for "bad" Lie brackets. To this end, we need a Philip Hall basis for a nilpotent algebra[2] of order 4 generated by three vector fields $\mathbf{f}$, $\mathbf{g}_1$ and $\mathbf{g}_2$. Using a software package for Lie algebraic operations we can compute the Philip Hall basis [16], the "bad" brackets of the foregoing basis being given by $\mathbf{f}$, $[\mathbf{g}_1, [\mathbf{f}, \mathbf{g}_1]]$, and $[\mathbf{g}_2, [\mathbf{f}, \mathbf{g}_2]]$. Using a symbolic package, Maple^{TM} for example, it can be shown that the three foregoing vector fields vanish; hence, all the "bad" brackets can be expressed in terms of lower-order Lie brackets.

Therefore, the sufficient condition of the Sussmann Theorem is satisfied and we can conclude that resorting to its dynamics model, *Quasimoro* is small-time locally controllable from any of the equilibrium states, thus validating the result IV.1.

V. Conclusions

The controllability analysis of an underactuated quasiholonomic WMR was reported here. It was shown that the kinematic model is not sufficient to completely control the system. Furthermore, the dynamics model of the robot linearized around an equilibrium point turns out to be controllable according the KRCC.

Moreover, an in-depth analysis of the dynamics model of the WMR shows that the system is locally accessible from every state and small-time locally controllable from any equilibrium state, thus proving that it is possible to control the robot by using only the wheel motors.

Finally, the results discussed here can be readily extended to the controllability analysis of a two-wheeled mobile inverted pendulum system.

VI. Acknowledgments

This work was made possible by NSERC, Canada's Natural Sciences and Engineering Research Council, under RG A4532. The Winter 2002 differential fee waiver granted to Alessio Salerno is acknowledged too.

References

[1] S. Ostrovskaya and J. Angeles, "Nonholonomic systems revisited within the framework of analytical mechanics," *Applied Mechanics Reviews*, vol. 51, no. 7, pp. 415–433, 1998.

[2] A. Salerno, S. Ostrovskaya, and J. Angeles, "The development of quasiholonomic wheeled robots," in *Proc. IEEE Int. Conf. Robotics & Automation*, Washington D.C., 2002.

[3] C. Chen, *Linear System Theory and Design*. New York: Oxford University Press, 1999.

[4] A. D. Luca and G. Oriolo, "Modelling and control of nonholonomic mechanical systems," in *Kinematics and Dynamics of Multibody Systems.* J. Angeles and A. Kecskemethy, (Eds.), springer-Verlag, Vienna, 1995.

[5] F. Grasser, A. D'Arrigo, S. Colombi, and A. Rufer, "Joe: A mobile, inverted pendulum," *IEEE Trans. Industrial Electronics*, vol. 49, no. 1, pp. 107–114, 2002.

[6] J. Slotine and W. Li, *Applied Nonlinear Control*. Englewood Cliffs, NJ: Prentice Hall, 1991.

[7] H. Nijmeijer and A. van der Schaft, *Nonlinear Dynamical Control Systems.* Berlin: Springer-Verlag, 1990.

[8] A. Isidori, *Nonlinear Control Systems*. Berlin: Springer-Verlag, 1989.

[9] S. Sastry, *Nonlinear Systems: Analysis, Stability, and Control*. New York: Springer, 1999.

[10] H. Sussmann, "General theorem on local controllability," *SIAM J. Control and Optimization*, vol. 25, no. 1, pp. 158–194, January 1987.

[11] R. Murray, Z. Li, and S. Sastry, *A Mathematical Introduction to Robotic Manipulation*. Boca Raton, FL: CRC Press, 1994.

[12] J. Ostrowski and J. Burdick, "Controllability tests for mechanical systems with symmetries and constraints," *J. Appl. Math. and Computer Science*, vol. 7, no. 2, pp. 305–331, 1997.

[13] A. Lewis, "Aspects of geometric mechanics and control of mechanical systems," Ph.D. dissertation, Caltech, Pasadena, CA, 1995.

[14] J. Serre, *Lie Algebra and Lie Groups*. New York: Springer-Verlag, 1992.

[15] H. Khalil, *Nonlinear Systems*. Upper Saddle River, NJ: Prentice Hall, 1996.

[16] M. Torres-Torriti, *Lie Tools Package — User's Guide, Version* 1.00, 2002. [Online]. Available: http://www.cim.mcgill.ca/~miguelu/

---

[2] A Lie algebra is defined to be *nilpotent* of order λ if there exists an integer λ such that all Lie brackets of length greater than λ are zero [11].

Exploiting Redundancy to Implement Multi-Objective Behavior

Yuandong Yang   Oliver Brock   Roderic A. Grupen

Laboratory for Perceptual Robotics
Department of Computer Science
University of Massachusetts Amherst
Email: {yuandong, oli, grupen}@cs.umass.edu

Abstract

Teams of robots can be redundant with respect to a given task. This redundancy can be exploited to pursue additional objectives during the execution of the task. In this paper, we describe a control-based method to exploit such redundancy for the execution of additional behavior, leading to the improvement of overall performance. The control-based method provides a suitable mechanism for combining controllers with different objectives. The mechanism ensures that the subordinate controllers do not interfere with the superior controllers. Thus it allows to build controllers exhibiting complex behavior from simple primitives, while maintaining their provable performance characteristics. The effectiveness of the framework is demonstrated by experiments with a multi-robot exploration task.

1 Introduction

Distributed control methods for multi-robot teams have been an active area of research. Individual members of a team are controlled independently in such a way that a desired behavior of the group emerges. Robots observe the world, abstract, and act independently and concurrently. The main objective and difficulty in designing distributed controllers is the coordination of team members.

Various methods have been introduced for distributed control tasks. We distinguish two approaches: top-down and bottom-up methods. In top-down methods, the controller is recursively divided into less complex units. The division continues until all units can be physically implemented. The resulting controller is generally well suited to solve the given task, but might fail to generalize easily to a different task.

Bottom-up methods, on the other hand, build small modules first. Higher-level behavior is achieved by combining lower-level modules. Once the lower-level modules are built, they can easily be combined into different controllers addressing varying task requirements. Consequently, bottom-up methods can be considered more versatile and flexible compared to top-down methods.

Within the framework of bottom-up methods, the construction process for controllers requires two steps: the design of versatile modules and their combination into controllers. The resulting modules should be robust, while methods to combine them should maintain their desirable properties. In this paper, we introduce a control-based framework to address these issues.

Our framework allows to specify a "universe" of controllers by combining *objective functions* representing a desired behavior, *state estimators* to perceive information about the state of the world, and *effectors* to modify the state of the world. Each combination of instantiations from these categories can be viewed as a control option. Desired behavior is achieved and its performance can be guaranteed by employing closed-loop disturbance rejection.

The control-based framework presented here also specifies how controllers are combined. The controllers are ranked based on their importance for the overall behavior. Actions of subordinate controllers are projected into the nullspace of superior controllers. This mechanism ensures that the performance of the superior controller remains unaffected, while performing additional behavior whenever possible.

We demonstrate the effectiveness of this control-based method for multi-objective robot control by applying it to the task of exploring an unknown environment with a team of robots. The experiments presented in Section 4 demonstrate how the redundancy of the team can be exploited to add secondary objectives, which significantly improve the overall performance of the team.

2 Related Work

We will restrict our discussion of previous work to those methods classified as bottom-up, as described above. The behavior-based control method [1][6] is one of the successful ones. It replaces behaviors couched in expensive representations with local control decisions and sensor information. The subsumption architecture arbitrates actions using inhibition and suppression mechanism between behavioral modules.

Several people have contributed to the area of behavior-based methods. Maes [5] developed a method by which robots learn how to combine pre-designed basic behaviors. Hoff [3] designed algorithms to learn both the basic behaviors with sub-goals and how to combine them to achieve more complex behaviors. Mataric [7] gave an impressive demonstration of behavior-based control on real robots. Two behaviors can be combined into new higher-level behaviors by behavior combination operators. Reinforcement learning is applied to let robots learn how to construct higher level behaviors automatically.

Behavior-based methods have their advantages, but we want to argue that they have some problems. The simple combination of two or more basic behaviors cannot ensure the execution of either behavior. Reinforcement learning can be employed to identify functioning combinations, but the controllers it chooses are unlikely to be generalizable and will probably fail if the environment or other conditions vary.

Huber and Grupen [4] introduced the idea of control-based methods. They can be still classified as behavior-based methods, but it is more formalized. Huber successfully applied the method to learning to control legged-robots [4]. The robots learn how to walk without falling, which is guaranteed by the virtue of discrete event specifications.

In this paper, we extend this technique following the work of Sweeney [9] that treats multiple mobile robots with multiple concurrent objectives; namely searching a multi-room floor plan while maintaining in a connected communication network.

3 Exploiting Redundancy using Control-Based Methods

3.1 Specifying Controllers

A set of controllers C can be described using a vocabulary of objectives Φ, sensors S, and effectors E:

$$C = \Phi \times S \times E$$

A specific controller c then is given by the tuple (ϕ, s, e) with $\phi \in \Phi$, $s \in S$, and $e \in E$. The objective function ϕ is measured using the sensor s and continuous closed-loop control is employed to optimize ϕ using the effector e. Closed-loop control ensures robust performance by continuous disturbance rejections. Note that this framework is very general and that computational resources, sensors, and effectors do not have to reside in physical proximity. Our notation for such a controller, also called control primitive, is given by

$$\phi|_e^s, \text{ with } \phi \in \Phi, s \in S, e \in E.$$

Applying this formalism to the domain of teams of mobile robots we will use the notation $\phi|_i^s$ throughout the remainder of this paper. The objective function ϕ is represented by an artificial potential field. Robot i uses a sensor s to descend the gradient of ϕ.

3.2 Combining controllers

Control primitives as described above perform single objectives, represented by ϕ in a robust manner. Our goal is to combine many such primitives into a complex controller without sacrificing robustness. We determine a ranking of control primitives in such a way that the behavior of a subordinate controller should never interfere with a superior controller. This can be accomplished by projecting the actions resulting of a subordinate controller into the nullspace of the superior controller.

If the task of a robot is specified by the vector $\dot{x}$ and the robot is redundant with respect to that task, the robot can perform a set of actions not affecting task performance. These actions are said to be in the nullspace of the Jacobian matrix J, associated with the task.

The control of the robot is composed from two components: one representing task behavior, $J^{\#}\dot{x}$, and a second one representing additional, subordinate behavior K. The subordinate behavior is projected into the nullspace $(I - JJ^{\#})$ of J to not affect the execution of the superior task:

$$\dot{\theta} = J^{\#}\dot{x} + (I - JJ^{\#})K, \quad (1)$$

where $J^{\#}$ designates the Moore-Penrose inverse of J, I is the identity matrix, $K = \frac{\delta p}{\delta x}$ represents the gradient of the potential function p of a subordinate controller, and $(I - JJ^{\#})$ is the nullspace projection of J.

We use the notation $\phi_1 \triangleleft \phi_2$, or ϕ_1 "subject to" ϕ_1, to express the nullspace relationship between two primitives ϕ_1 and ϕ_2. Primitive ϕ_1 is subordinate and must be performed in the nullspace of superior primitive ϕ_2. Consequently, actions resulting from ϕ_1 will only be performed if they do not affect the execution to ϕ_2.

The approach of combining control primitives permits to safely pursue multiple objectives as long as subordinate controllers do not affect the behavior of superior ones. The control primitives are constructed from sets of objective functions, sensors, and effectors; by employing closed-loop disturbance rejection the robustness of the primitive as well as the robustness of the combined controller can be guaranteed. We can continue this process of combining controllers, until the resulting nullspace does not suffice any more to perform a certain desired behavior.

3.3 Exploiting Redundancy

We will now apply the framework presented in the previous section to teams of mobile robots. In such teams each member can be treated as a sensor-effector pair – it has the ability to sense and change the environment. If the team resources exceed the requirement of the task, we say that the team exhibits *redundancy* with respect to that task. For example, in a single-objective exploration task, the robot always follows the steepest descent of the potential function representing the task in order to minimize the exploration time. A team of robots, however, will be redundant with respect to the exploration task.

In the remainder of the paper we will demonstrate how we use the framework introduced in this section to exploit the redundancy encountered in robot teams. We present three different concurrent controllers consisting of multiple control primitives. Each controller solves the exploration task. The controllers differ in additional, subordinate primitives, relying on redundancy to improve the overall performance during the exploration task. Exploration is performed in a multi-room floor plan represented by a grid, measuring 32 by 32 cells. The task of the robots is to explore all the cells with infrared sensors while maintaining a line-of-sight constraint. This constraint requires the robots to maintain an unobstructed line of sight and to exceed a given distance from each other. This constraint is motivated by communication requirements between robots using radio signals with limited range.

The robots need to plan its path to all unexplored regions while avoiding hitting any obstacles. We use harmonic functions[2, 8] to generate the motion of the robots. Harmonic functions have desirable properties in dynamic environments. All the obstacle cells are assigned a potential value of 1 and the goal cell is assigned a potential value of 0. We use dynamic programming to compute the potentials.

The following three controllers were used to demonstrate that redundancy can be exploited to increase performance. Each of them combines at least two behaviors for an exploration task performed by a team of two robots.

(A) Line-of-sight exploration controller: The line-of-sight exploration controller is implemented in Sweeney's work [9]. The exploration of the robots is frontier-based [10]: the robots are always heading to the biggest boundary between free space and unexplored space using artificial potentials and gradient descent. Equation 2 describes the line-of-sight exploration controller in the framework introduced above:

$$\Phi|^s_{\{L,F\}} = \{\phi|^s_L \triangleleft \phi|^{LOS_L}_F\}, \qquad (2)$$

where s refers to the infrared sensors of the robots, LOS stands for line of sight, and L, F refer to the leader and follower, respectively. The subordinate exploration primitive $\phi|^s_L$ of the leader is performed in the nullspace of (or "subject to") the primary line-of-sight control primitive $\phi|^{LOS_L}_F$. Any motions resulting from the exploration primitive will be projected into the nullspace of the line-of-sight primitive. Thus, the line-of-sight constraint will be always maintained.

The additional line-of-sight constraint might increase exploration time with respect to a single-robot exploration, because the leader has to ensure that the follower is able to maintain visibility. This controller will serve as our reference controller. In what follows we propose modifications to this controller with the goal of speeding up the exploration process. Improvement will be measured relative to the line-of-sight exploration controller, or reference controller.

(B) Line-of-sight exploration with collaborative search behavior: In this modification of the reference controller, the goal of the follower to perform its own exploration by maintaining the line-of-sight constraint to the leader. The line-of-sight constraint bounds the distance between the leader and the follower from above. The exploration primitive for the follower also imposes a lower bound. The follower is moving towards an unexplored region within those bounds, allowing it to assist the leader in exploring the maze while still maintaining the line-of-sight constraint. This also forces the motion of the follower to be mapped into the nullspace of the line-of-sight controller.

Equation 3 describes the line-of-sight controller with collaborative search behavior using the notation introduced earlier. A third control primitive $\phi|^s_F$ is added (see Equation 2). The collaborative search primitive is subject to the follower performing the line-of-sight control primitive. The motion of the follower will be projected into the nullspace of the line-of-sight primitive.

$$\Phi|_{\{L,F\}}^{s} = \{\phi|_{L}^{s} \triangleleft \phi|_{F}^{LOS_L} \triangleright \phi|_{F}^{s}\} \quad (3)$$

The controller in the equation 3 can be rewritten as separate controllers for the leader and follower: $\phi|_{L}^{s} \triangleleft \phi|_{F}^{LOS_L}$ and $\phi|_{F}^{LOS_L} \triangleright \phi|_{F}^{s}$. The first part can be interpreted as the leader exploring the environment, while "pulling" the follower along. The second part represents the follower "pushing" the leader ahead by proceeding into unexplored regions in which the line-of-sight constraint is maintained. The follower's explorative motion is mapped into the nullspace of $\phi|_{F}^{LOS_L}$, the line-of-sight controller, preventing any motions violating the primary LOS constraint.

In Figure 1 the traces of the two robots performing exploration using this controller are shown. In the middle of the figure, we can see that the follower moves to the right of the leader, assisting in the exploration of an unexplored region. Such behavior reduces exploration time.

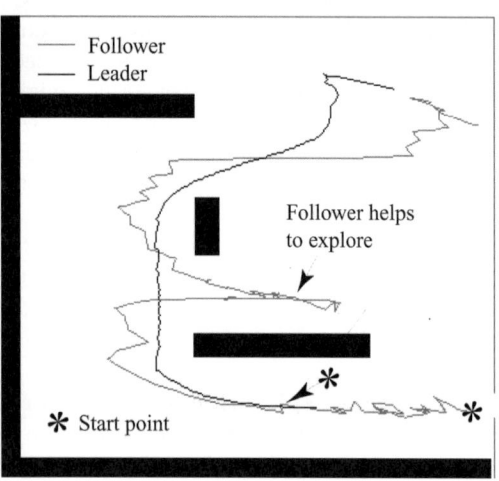

Figure 1: The traces of leader/follower with collaborative search

(C) **Line-of-sight exploration with collaborative search and role-change behavior:** Based on controller B, we try to further improve the performance by adding an additional behavior. We employ the ability of the leader and the follower to change roles during the exploration process. A role change can save exploration time when a change in direction occurs. Consider the example of exploring a room. At first, the follower follows the leader into that room. After the leader has completed the exploration, both robots need to exit the room. The leader's motion will force the follower out of the room. If the robots switch roles instead, however, the leader does not have to travel the distance to pass the follower and the exploration can resume faster. A role-change is always in the nullspace of the line-of-sight controller because it does not change the distance between the leader and follower.

The decision to change roles is based on the harmonic potential. We notice that the paths of the robots are mostly determined by obstacles in proximity. Once these have been detected, the corresponding potential surface for the leader and the follower will looks similar. As a result, the paths chosen by the two robots will be almost identical. A monitor is added to the line-of-sight exploration controller, monitoring if the leader and the follower are on similar paths and if the follower is ahead of the leader with respect to the direction of motion. If this situation occurs for longer than a predetermined interval, the leader and follower change roles. Note that the role-change always is in the nullspace of the line-of-sight behavior.

4 Experimental Results

We applied the three controllers introduced in the previous section to exploration tasks in 1920 randomly generated mazes. The result is shown in Table 1. The exploration time is measured in terms control cycles of the simulator required to complete the exploration. In each control cycle, the robots communicate with each other and execute an action. Figure 2 shows the traces of the robots controlled by the three controllers in an identical maze. For the line-of-sight controller, an average 870 of control cycles were needed to complete the exploration. Because the follower is fast enough to follow the leader, the exploration time with line-of-sight constraint is only a marginally longer than the exploration with a single robot.

We evaluate the three controllers presented here by comparison relative to the reference controller. The comparison is shown in Table 1.

The line-of-sight controller with collaborative search behavior exhibits an average improvement of 30% compared to the reference controller. This improvement is a result of the follower providing additional information about obstacles, effectively increasing the sensing range of the leader, and thus assisting with the exploration within line-of-sight distance. The standard deviation in the improvement can be explained by the fact that the environment initially is completely unknown. A short-term gain does not necessarily brings a long-term gain. Exploration decisions made locally might not be optimal globally. But

	LOS	LOS with CS	LOS with both CS and RC
average exploration control cycles	857.2	578.2	565.9
standard deviation	209.2	115.7	112.8
average improvement(percent)	N/A	30.0	31.5
standard deviation(percent)	N/A	17.9	17.4

Table 1: Comparison of three controllers in 1920 simulated exploration tasks. LOS stands for line-of-sight exploration; RC stands for role-change; CS stands for collaborative search.

	$\frac{1}{1}$ sensor range	$\frac{3}{4}$ sensor range	$\frac{1}{2}$ sensor range
infrared range in grid cells	8	6	4
average exploration control cycles	578.2	764.9	1204.9
standard deviation	115.7	120.0	187.1
average improvement(percent)	30.0	22.1	11.8
standard deviation(percent)	17.9	15.6	16.6

Table 2: Comparison of three line-of-sight controllers with collaborative search behavior, using differant infrared sensor ranges.

even within the range of the standard deviation there is considerable improvement compared to the reference controller.

The controller with collaborative search *and* role-change behavior resulted in an average improvement of 31.5%. This improvement is not significant relative to the controller with collaborative behavior. The small magnitude of this improvement can be explained as follows: Each time leader and follower change roles, they only save the exploration distance between the leader and the follower. These distances are small compared to the total required travel for exploration. Consequently, the exploration time gained by role-switching is marginal compared to the overall exploration time.

The degree to which the proposed controllers are able to reduce the overall exploration time of a given environment is dependent upon the sensing capabilities of the robots. To demonstrate this we performed experiments with different sensing ranges for the infrared sensors. Table 2 compares the performance of the line-of-sight exploration controller with collaborative for various sensor ranges. If the range is shortened to $\frac{3}{4}$ of the range for the experiments reported in Table 1, we only observe an average improvement of 22% with respect to the reference controller with the same sensor range. Further reducing the range to $\frac{1}{2}$ of the original one, reduces the average improvement to only 11%. This is due to the fact that in order to maintain the line-of-sight constraint, leader and follower have to be able to perceive each other. Reducing the range of the infrared sensors also reduces the distance they can be apart without violating the line-of-sight constraint.

This limits the capacity of the follower to perform independent exploration, thus deteriorating the overall performance.

These results presented in this section demonstrate that by exploiting the redundancy of teams of robot with respect to a given task using a control-based nullspace scheme, we can significantly improve the overall performance. Note that while the experiments described above are performed with only two collaborating robots, the general framework extends to an arbitrary number of additional robots. These could either have to maintain a line-of-sight constraint with the leader or adopt other followers as their leaders.

5 Conclusion

In this paper, we presented a formal framework to combine control-based methods to exploit the redundancy of a team of robots with respect to a given task. The framework allows the incremental integration of control primitives. By performing each additional behavior in the nullspace of the original controller, we can guarantee that the added behavior will not affect the performance of existing controllers. As a consequence, added controllers can be used to implement more complex behavior and to improve the overall performance in a robust manner.

We presented experiments, demonstrating the effectiveness of the proposed approach. The experiments apply the framework to the exemplary application of team-based exploration of unknown environments. Augmenting a reference controller for this task with additional behaviors, we were able to show significant improvements in exploration speed. The experiments conclusively show that our framework is capable of combining various behaviors in a prioritized manner to result in more sophisticated and advantageous overall behavior of the team.

Acknowledgments

The authors would like to thank John Sweeney for their helpful insights and discussion in preparing this paper. Work on this paper has been supported in part by MARS/SDR, NSF CDA-9703217, DARPA/ITO DABT63-99-1-0022, and DABT63-99-1-0004.

References

[1] Rodney A. Brooks. Intelligence without representation. *Artificial Intelligence Journal*, 47:139–159, 1991.

[2] C. Connolly and R. Grupen. Harmonic control. In *Proceedings of the IEEE International Symposium on Intelligent Control*, pages 503–506, Glasgow, Scotland, Aug. 1992.

[3] J. Hoff and G. Bekey. An architecture for behavior coordination learning. In *IEEE International Conference on Neural Networks*, pages 2375–2380, Perth, Australia, Nov 1995.

[4] M. Huber and R. A. Grupen. A control structure for learning locomotion gaits. In *Proceedings of Seventh International Symposium on Robotic and Applications*, Anchorage, AK, May 1998. TSI Press.

[5] P. Maes and R. A. Brooks. Learning to coordinate behaviors. In *National Conference on Artificial Intelligence*, pages 796–802, 1990.

[6] M. Mataric. Behavior-based control: Main properties and implications. In *Proceedings of the IEEE International Conference on Robotics and Autonomation, Workshop on Architectures for Intelligent Control Systems*, pages 46–54, 1992.

[7] M. J. Mataric. Interaction and intelligent behavior. Technical Report AITR-1495, MIT, 1994.

[8] E. Prestes, E. Silva, P. M. Engel, M. Trevisan, and M.A.P. Idiart. Exploration method using harmonic functions. *Journal of Neurophysiology*, 40:25–42, July 2002.

[9] J. Sweeney, TJ Brunette, Y. Yang, and R. Grupen. Coordinated teams of reactive mobile platforms. In *Proceedings of the IEEE International Conference on Robotics and Automation*, pages 299–304, 2002.

[10] B. Yamauchi. A frontier based approach for autonomous exploration. In *Proceedings of the IEEE International Symposium on Computational Intelligence in Robotics and Automation*, pages 146–151, Monterey, CA, July 1997.

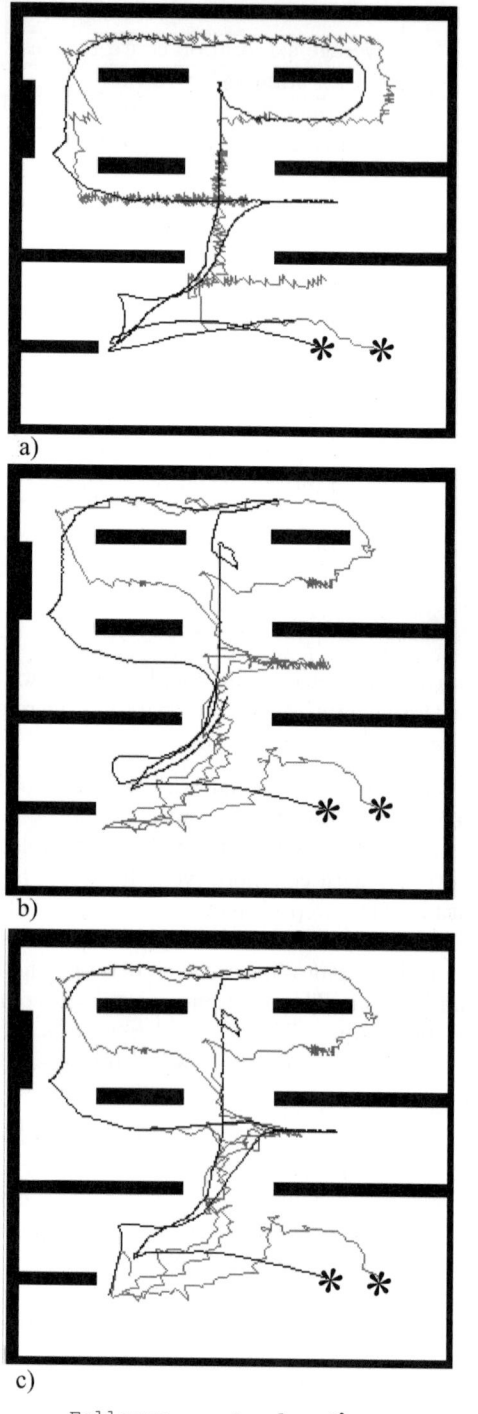

Figure 2: The traces of leader and follower controlled by the three controllers in the same maze: a) line-of-sight exploration, 720 control cycles; b) line-of-sight exploration with collaborative search, 531 control cycles; c) line-of-sight exploration with both collaborative search and role-change, 492 control cycles and 3 role-changes.

Bilateral time-scaling for control of task freedoms of a constrained nonholonomic system

Siddhartha S. Srinivasa   Michael A. Erdmann   Matthew T. Mason
siddh@cs.cmu.edu   me@cs.cmu.edu   mason@cs.cmu.edu
The Robotics Institute, Carnegie Mellon University, Pittsburgh, PA - 15213

Abstract—We explore the control of a nonholonomic robot subject to additional constraints on the state variables. In our problem, the user specifies the *path* of a subset of the state variables (the *task freedoms* $\mathbf{x}_P$), i.e. a curve $\mathbf{x}_P(s)$ where $s \in [0,1]$ is a parametrization that the user chooses. We control the *trajectory* of the task freedoms by specifying a *bilateral time-scaling* $s(t)$ which assigns a point on the path for each time t. The time-scaling is termed bilateral because there is no restriction on $\dot{s}(t)$, the task freedoms are allowed to move backwards along the path. We design a controller that satisfies the user directive and controls the remaining state variables (the *shape freedoms* $\mathbf{x}_R$) to satisfy the constraints. Furthermore, we attempt to reduce the number of control switchings, as these result in relatively large errors in our system state. If a constraint is close to being violated (at a *switching point*), we back up $\mathbf{x}_P$ along the path for a small time interval and move $\mathbf{x}_S$ to an open region. We show that there are a finite number of switching points for arbitrary task freedom paths. We implement our control scheme on the *Mobipulator* and discuss a generalization to arbitrary systems satisfying similar properties.

I. INTRODUCTION

The goal of the Mobipulator project is to build a desktop assistant - a robot that manipulates commonplace desktop items like paper and pencil. In [1], we described the hardware and software architecture of a robot with four independantly controlled wheels, none of them steered, called the Mobipulator.

In [1], we implemented a configuration space planner that moved the paper from a start location to a goal. When the robot has two of its wheels on paper (called the *hands*) and the other two on the desktop (called the *feet*), the robot is said to be in *dual-differential drive* mode (Fig. 1). In this mode, the motion of the paper is unconstrained - the wheel velocities span the space of the paper velocities. The configuration space planner treated the paper as a trailer hitched to the center of the hands and found paths for the paper while steering the robot away from obstacles. If the dual-differential drive mode was violated, the robot moved across the paper and continued the plan.

In this paper, we explore a more dynamic task - a user controls the path of the paper remotely. We devise a controller that executes the user's directive and maintains the robot in dual-differential drive mode.

We define the state variables that the user controls as

Fig. 1
DUAL-DIFFERENTIAL DRIVE MODE[1]

the *task freedoms* and the remaining state variables the *shape freedoms*. In our problem, the task freedoms are the pose of the paper $\mathbf{x}_P = (x_P \; y_P \; \theta_P)^T$ and the shape freedoms are the pose of the robot relative to the paper $\mathbf{x}_R = (x_R \; y_R \; \theta_R)^T$.

The decomposition of state variables into task and shape variables was used by Nakamura[2] to study robot arms. He termed the pose of the end effector as the *manipulation variable* and studied the control of joint angles to obtain desired paths.

In our problem, the user specifies the *path* of the paper, i.e. a curve $\mathbf{x}_P(s)$ in the task space where $s \in [0,1]$ is a parametrization that the user chooses. Note that there is no notion of time in the path specification. We control the *trajectory* of the task freedoms by specifying a *time-scaling* $s(t)$ which assigns a point on the path for each $t \in [0,T]$, where T is the time to completion of the path.

The concept of time scaling was used by Bowbrow et al.[3] to find time-optimal trajectories of a fully-actuated manipulator along a specified path, subject to limitations in actuator torque. They used a time-scaling $s(t)$ with a unilateral constraint $\dot{s} > 0$, i.e. the end-effector was not allowed to move backwards along the specified path.

We will show in §4 that for our problem we cannot impose the unilateral constraint on the time-scaling function and require bilateral time-scaling with $\dot{s}$ unconstrained.

For example, in Fig. 2, the user specifies the path ($ac-cb-bd$), we specify the time-scaling $s(t)$ and the resulting trajectory is ($ac-cb-bc-cb-bd$).

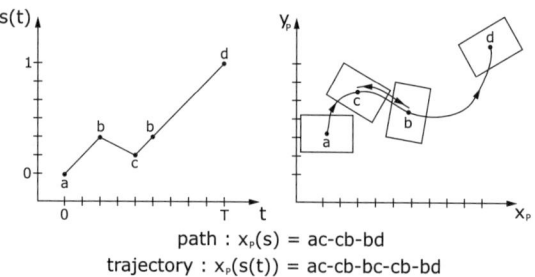

path : $x_P(s)$ = ac-cb-bd
trajectory : $x_P(s(t))$ = ac-cb-bc-cb-bd

Fig. 2
TIME-SCALING OF USER'S PATH

As the user is interested only in the motion of the paper, we have an extra degree of freedom - four wheels control the three paper degrees of freedom. The configuration space planner used the extra freedom to hitch the paper to the robot. We use it to maintain the shape freedoms in dual-differential drive. Intuitively, this is similar to a moving hitch placed optimally at each instant.

The idea of using redundant degrees of freedom to optimize performance was used by Baillieul et.al.[4] in the control of redundant manipulators. They used the *extended Jacobian* technique to move the end effector along a prescribed path *and* locally optimize an objective function.

In §4, we decompose the system into the task and the shape subsystems. We rewrite the shape system as a function of the task freedoms. The user controlled task freedoms appear as a drift term in the shape system. We use bilateral time-scaling to control this drift, and the extra degree of freedom to move the shape freedoms to satisfy dual-differential drive. In §5, we provide a control policy that reduces the number of control switches required. We explain the motivation for this policy in §2. §6 describes an implementation of the control law on the real robot and some of the problems faced. §7 explores a generalization for arbitrary nonholonomic systems.

II. BACKGROUND OF NONHOLONOMIC SYSTEMS

A general form of a nonholonomic system is given by:

$$\Sigma : \quad \dot{\mathbf{x}} = u_1 \mathbf{f_1}(\mathbf{x}) + \cdots + u_m \mathbf{f_m}(\mathbf{x}) \quad (1)$$

where $2 \leq m < n$, $\mathbf{x} = (x_1 \cdots x_n)^T$ is the state vector defined in an open subset S of $\mathbb{R}^n$, $u_i \in \mathbb{R}$ are the control inputs, and $\mathbf{f_1}, \cdots, \mathbf{f_m}$ are vector fields on S. Σ is said to be *completely nonholonomic* if the rank of the controllability Lie algebra generated by $u_1 \cdots u_m$ is n. A completely nonholonomic system is completely controllable (*Chow's theorem*[5]).

A. Motion planning

Motion planning for nonholonomic systems is complicated by the fact that not all motions are feasible, only those which satisfy the instantaneous nonholonomic constraints. Nevertheless, the completely nonholonomic assumption guarantees that feasible motions do exist which steer an arbitrary initial state to a final state. We refer the reader to Kolmanovsky et al.[6] for a detailed review of motion planning for nonholonomic systems.

The motion planning problem for nonholonomic systems can be defined as : for every pair of points (**p**,**q**) $\in S$, generate an open-loop control $\mathbf{u}(t) = (u_1 \cdots u_m)^T$ that steers **p** to **q**.

One approach is to consider the extended system :

$$\Sigma_e : \quad \dot{\mathbf{x}} = v_1 \mathbf{f_1}(\mathbf{x}) + \cdots + v_m \mathbf{f_m}(\mathbf{x}) + v_{m+1} \mathbf{f_{m+1}}(\mathbf{x}) + \cdots + v_r \mathbf{f_r}(\mathbf{x}) \quad (2)$$

where $\mathbf{f_{m+1}}, \cdots, \mathbf{f_r}$ are higher-order Lie brackets of the $\mathbf{f_i}$ chosen so that $\mathbf{f_1}(\mathbf{x}), \cdots, \mathbf{f_r}(\mathbf{x})$ span $\mathbb{R}^n$ for all $\mathbf{x} \in S$. (2) can be solved for the control $\mathbf{v}(t)$ that steers **p** to **q**.

The hard part is to generate Lie brackets from the control inputs. Fast switchings of piecewise constant or polynomial inputs[7] and high-frequency high-amplitude periodic control inputs[8] are some of the techniques used to generate motions in the directions of the Lie brackets.

B. Motions of the Mobipulator

One motion along a higher-order Lie bracket for the Mobipulator is akin to parallel parking the robot relative to the paper. This is achieved by spinning the wheels repeatedly forwards and backwards for small time intervals. We use rubber O-rings on the wheels to help better grip the paper. This also increases friction between the feet and the desktop and thus the wheels require a minimum torque before they start spinning. As a result, fast switchings of wheel torques cause a nonsmooth sideways motion and result in the wheels slipping. This produces errors in both the pose of the robot and the paper. Hence, we would like to avoid control switches for accurate motion.

III. PROBLEM STATEMENT

The kinematics for the Mobipulator in dual-diff drive mode can be described as :

$$M : \quad \begin{pmatrix} \dot{\mathbf{x}}_\mathbf{P} \\ \dot{\mathbf{x}}_\mathbf{R} \end{pmatrix} = \mathbf{A}(\mathbf{x_R}) \begin{pmatrix} \omega_1 \\ \omega_2 \\ \omega_3 \\ \omega_4 \end{pmatrix} = \mathbf{A}\,\omega \quad (3)$$

where $\mathbf{x_P} = (x_P\ y_P\ \theta_P)^T$ is the pose of the paper in the world frame, $\mathbf{x_R} = (x_R\ y_R\ \theta_R)^T$ is the pose of the robot *relative* to the paper, and the ω_i are the angular velocities of the wheels.

M requires the robot to be in dual-differential drive mode. The shaded region in Fig. 3 describes the allowable

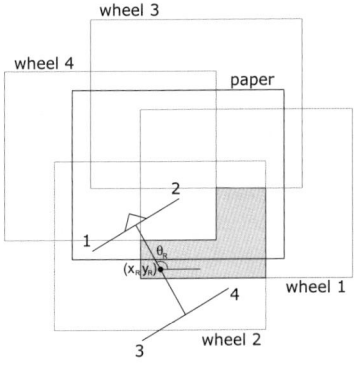

Fig. 3
ALLOWABLE $(x_R \; y_R)$ AT $\theta_R = 2\pi/3$

pose of the robot relative to the paper for a given θ_R. Each light rectangle represents the locus of the center of the robot with one of the wheels touching the edge of the paper. If the center lies in the interior of a light rectangle, the corresponding wheel lies in the interior of the paper. The shaded polygon is the locus of points with the manipulating wheels (1 and 2) on the paper and the locomoting wheels (3 and 4) on the desktop.

The allowable region depends only on the relative pose θ_R and the dimensions of the robot and the paper. The shape freedom constraints can thus be written as :

$$H: \quad \mathbf{h_1}(\theta_R) \leq \begin{pmatrix} x_R \\ y_R \end{pmatrix} \leq \mathbf{h_2}(\theta_R) \quad (4)$$

The problem can be stated as:

Given any user-specified paper path $\mathbf{x_P}(s)$, is it possible to construct wheel angular velocities ω_i that attain the specified $\mathbf{x_P}$ and satisfy the constraint H? Furthermore, is it possible to construct a control policy that reduces the number of control switches required?

IV. CONTROL OF SHAPE FREEDOMS

In this section we will prove that we can control the Mobipulator system to follow a user-specified $\mathbf{x_P}(s)$ and satisfy H. We will first consider the case where the user has control of the velocity $\mathbf{\dot{x}_P}(t)$ and show that the resulting system is *not* small time locally controllable (STLC). This negative result will provide us with insight to modify the user's control to a path $\mathbf{x_P}(s)$. We will show that the resulting system is STLC only if there are no constraints on $\dot{s}(t)$.

A. The shape and task subsystems

We decompose the Mobipulator system M into two subsystems - the task system M_t and the shape system M_s. M_t comprises of the task freedoms the user controls (the $\mathbf{x_P}$) and M_s comprises of the shape freedoms (the $\mathbf{x_R}$). We will analyze each of these subsystems separately.

$$M \quad : \quad \begin{pmatrix} \mathbf{\dot{x}_P} \\ \mathbf{\dot{x}_R} \end{pmatrix} = \mathbf{A}\,\omega = \begin{bmatrix} \mathbf{B} \\ \mathbf{C} \end{bmatrix} \omega$$

$$M_t \quad : \quad \mathbf{\dot{x}_P} = \mathbf{B}\,\omega$$

$$M_s \quad : \quad \mathbf{\dot{x}_R} = \mathbf{C}\,\omega$$

B. The task subsystem

We first observe that $\mathbf{B}$ is of rank 3. This is beacause the motion of $\mathbf{x_P}$ is unconstrained in M_t, i.e. if H is satisfied, any desired $\mathbf{\dot{x}_P}$ can be attained by a correct selection of the wheel velocities. We create an augmented rank 4 system by adding the nullspace vector of $\mathbf{B}$, $\mathbf{n_B}$, and an additional scalar α which we can control without affecting the task freedoms.

$$M_{ta}: \quad \begin{pmatrix} \mathbf{\dot{x}_P} \\ \alpha \end{pmatrix} = \begin{bmatrix} \mathbf{B} \\ \mathbf{n_B} \end{bmatrix} \omega \quad (5)$$

We can now invert the augmented system. Intuitively, α represents the one degree of freedom that we have in choosing ω. We will use this freedom to satisfy H.

$$\omega = \begin{bmatrix} \mathbf{B} \\ \mathbf{n_B} \end{bmatrix}^{-1} \begin{pmatrix} \mathbf{\dot{x}_P} \\ \alpha \end{pmatrix} \quad (6)$$

C. The shape subsystem

The shape freedoms can be rewritten in terms of the task freedoms using Eqn. 6.

$$\mathbf{\dot{x}_R} = \mathbf{C} \begin{bmatrix} \mathbf{B} \\ \mathbf{n_B} \end{bmatrix}^{-1} \begin{pmatrix} \mathbf{\dot{x}_P} \\ \alpha \end{pmatrix} \quad (7)$$

This can be arranged in a more intuitive form :

$$M_{sa}: \quad \mathbf{\dot{x}_R} = \mathbf{F}\,\mathbf{\dot{x}_P} + \mathbf{g}\,\alpha \quad (8)$$

From Eqn. 8 we can see that the motion of the task freedoms appears as a drift term $\mathbf{F}\,\mathbf{\dot{x}_P}$ in the shape system and the degree of freedom α appears as the scalar control.

We can gain further insight on M_{sa} by studying $\mathbf{g}$:

$$\mathbf{g} = \frac{c}{4} \begin{pmatrix} \cos(\theta_R) \\ \sin(\theta_R) \\ 0 \end{pmatrix} \quad (9)$$

where c is the radius of each wheel.

This field lets the robot move forwards and backwards relative to the paper. For example, if there was no drift (i.e. if $\mathbf{\dot{x}_P}$ were $\mathbf{0}$), then the only robot motion that does not manipulate the paper is this forwards-backwards motion.

D. A negative result

For H to be satisfied, we would like to be able to control the shape freedoms (described by M_{sa}) immaterial of the drift generated by the user-controlled task freedoms. If we allow the user to control $\mathbf{\dot{x}_P}(t)$, this requires M_{sa} to be STLC for all $\mathbf{\dot{x}_P}$ and $\mathbf{x_R}$. We use the following theorem to prove that this is *not* true.

Theorem 1 *The system M_{sa} is small-time locally controllable at $\mathbf{x_R}$ if and only if there exists some α^0 such that for all $\dot{\mathbf{x}}_\mathbf{P}$*

$$0 = \mathbf{F}\,\dot{\mathbf{x}}_\mathbf{P} + \mathbf{g}\,\alpha^0 \tag{10}$$

Proof : Refer Sussman[9]

Corollary 1 *The system M_{sa} is not STLC*

Proof : (10) is an overconstrained system with three equations and one variable. Any non-zero choice of $\dot{\mathbf{x}}_\mathbf{P}$ will yield no soulution for α^0.

Intuitively this means that the drift field will always dominate and can force the system towards a constraint boundary, and then violate it. Thus, if the user has control of $\dot{\mathbf{x}}_\mathbf{P}(t)$, it is not always possible to satisfy H.

E. Time scaling

We overcome this problem by allowing the user to control the path $\mathbf{x_P}(s)$. We control the trajectory with a time-scaling $s(t)$ that assigns a point on the path for each $t \in [0, T]$, where T is the time to completion of the path. The velocity at any time t is given by :

$$\dot{\mathbf{x}}_\mathbf{P}(\mathbf{s(t)}) = \mathbf{x}'_\mathbf{P}(\mathbf{s(t)})\,\dot{s}(t) \tag{11}$$

We control the the time-scaling rate $\dot{s}(t)$. This in turn provides us with control of the the velocity of the task freedoms along the path.

We first restrict $\dot{s}(t) > 0$, i.e. we cannot reverse the motion of the paper along the path. Denoting $\dot{s}(t)$ by β, we write the time-scaled shape system as :

$$M_{su}: \begin{pmatrix} \dot{\mathbf{x}}_\mathbf{R} \\ \dot{s} \end{pmatrix} = \begin{pmatrix} \mathbf{Fx'_P} \\ 1 \end{pmatrix} \beta + \begin{pmatrix} \mathbf{g} \\ 0 \end{pmatrix} \alpha \tag{12}$$

M_{su} is a control-affine system with one unilateral control $\beta > 0$. We can test STLC for this system :

Theorem 2 *The system M_{su} is small-time locally controllable at $\mathbf{x_R}$ if and only if there exists some α^0 and some $\beta^0 \neq 0$ such that for all $\mathbf{x_P}$*

$$0 = \begin{pmatrix} \mathbf{Fx'_P} \\ 1 \end{pmatrix} \beta^0 + \begin{pmatrix} \mathbf{g} \\ 0 \end{pmatrix} \alpha^0 \tag{13}$$

Proof : Refer Goodwine and Burdick[10]

Corollary 2 *The system M_{su} is not STLC*

Proof : The last row of (13) can be 0 for any arbitrary $\mathbf{x_P}$ only if $\beta^0 = 0$. Hence the system M_{su} is *not* STLC.

We now remove the restriction on $\dot{s}$. This results in a bilateral control β. The untestricted system M_{sb} looks like (12) but with β unrestricted. We rewrite it as :

$$M_{sb}: \begin{pmatrix} \dot{\mathbf{x}}_\mathbf{R} \\ \dot{s} \end{pmatrix} = \mathbf{f}_\beta\,\beta + \mathbf{f}_\alpha\,\alpha \tag{14}$$

We prove this system is STLC with the following theorem:

Theorem 3 *The system M_{sb} is small-time locally controllable at $\mathbf{x_R}$ if the controllability Lie algebra generated iterated Lie brackets of $\mathbf{f}_\beta$ and $\mathbf{f}_\alpha$ spans the state space.*

Proof : Refer Chow[5]

Corollary 3 *The system M_{sb} is STLC*

Proof : The brackets $[\mathbf{f}_\beta]$, $[\mathbf{f}_\alpha]$, $[\mathbf{f}_\beta, \mathbf{f}_\alpha]$ and $[\mathbf{f}_\beta, [\mathbf{f}_\beta, \mathbf{f}_\alpha]]$ span the 4-dimensional state space for all $\mathbf{x_P} \neq \mathbf{0}$.

F. Discussion

We have proved that we can follow any task freedom path and satisfy the constraints. To show STLC, we have had to use higher order brackets of $\mathbf{f}_\beta$ and $\mathbf{f}_\alpha$. An application of these brackets will cause control switchings and the resulting motion will not be smooth. Hence we must choose our control policy in such a way that we minimize the use of the brackets. We will describe one such policy in §5. Another point of concern is that since $\dot{s}$ is unrestricted, we can move forwards and backwards along our path. We will prove in §5 that any arbitrary path can be completed in a finite amount of time.

V. Control of the *Mobipulator*

Here we will propose control policies α and β, for the Mobipulator. We will show that control switchings can be restricted to a finite number of *switching points*.

A. Control policy

Recall that the control vector field $\mathbf{g}$ (Eqn. 9) allows the motion of the robot forwards and backwards relative to the paper. This is shown as the line LL' in Fig.4. We can use α to servo to any point on this line. LL' intersects the constraint boundaries at L_{max} where the robot is farthest from paper, and L_{min} where the robot is closest to the paper. Our policy servos the robot to the midpoint L_{mid}. This maximizes the minimum distance between the wheels and the edges of the paper and is a safe policy that is less severe to motion errors.

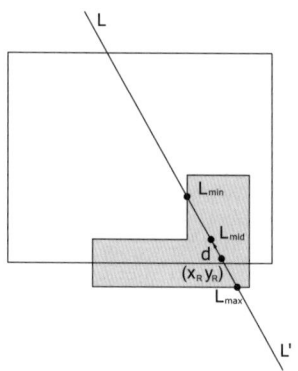

Fig. 4

Control policy for α

For β, we use the simple policy of $\beta = 1$ as long as H is satisfied. The next sub-section describes the policies chosen when H is violated.

B. Switching points

Note that with $\beta = 1$ we are deliberately allowing the paper to drift. We do this to avoid motions that require control switchings until imperative. H is violated when $L_{min} \to L_{max}$ and no further motion is possible along the control $\mathbf{g}$. This occurs either when a manipulating wheel and a locomoting wheel touch edges of the paper (Q and R in Fig.5) or when a manipulating wheel touches a corner of the paper (P in Fig.5). We term these points *switching points* as they mandate a switch in the control policy.

C. Control policy at switching points

At switching points, we move the paper backwards along the path for a small time interval δt by setting $\beta = -1$. Choice of α depends on whether we want net motion forwards or backwards. To escape from P, we need to move along x_R, while to escape from Q, we need to move along $-y_R$. Consider P. The motion of the robot along x_R is given by :

$$\dot{x}_R = \mathbf{F}(1)\mathbf{x}'_\mathbf{P}\beta + \mathbf{g}(1)\alpha \tag{15}$$

$\mathbf{F}(1)$ and $\mathbf{g}(1)$ are the first row of $\mathbf{F}$ and $\mathbf{g}$ respectively.

For small δt, we can approximate $\mathbf{F}(1)\mathbf{x}'_\mathbf{P}$ and $\mathbf{g}(1)$ as constants k_F and k_g respectively. At switching points we first set $\beta = -1$ and α to α_1. After δt, we set $\beta = 1$ and α to α_2. The net motion along x_R is given by the sum :

$$\begin{aligned}
\delta x_R \approx \dot{x}_R \delta t &= k_F \beta \delta t + k_g \alpha \delta t \tag{16}\\
\delta x_{R1} &\approx -k_F \delta t + k_g \alpha_1 \delta t \\
\delta x_{R2} &\approx k_F \delta t + k_g \alpha_2 \delta t \\
\delta x_{R1} + \delta x_{R2} &\approx k_g(\alpha_1 + \alpha_1)\delta t \tag{17}
\end{aligned}$$

By servoing to L_{max}, we can obtain the largest positive α and get the largest motion along x_R. Conversely, by servoing to L_{min}, we can obtain the largest negative α and get the largest motion along $-x_R$. Note that the net motion obtained is immaterial of the motion of the paper as k_F is cancelled during motion.

We continue the motions until the wheel is sufficiently clear of the constraining edge or vertex (defined by a threshold w shown in Fig.5). Since each of the abovementioned motions produces the same net motion along the desired direction, w will be attained in finite time.

VI. IMPLEMENTATION

The Mobipulator system has a camera that tracks colored fiducials marked on the robot and the paper, and provides pose information at 10 Hz. Our test path is a hexagon whose width is 8 times the width of the robot. The paper is rotated by $60°$ about its center at each vertex. Note that, with the chosen path and the starting pose of the robot, no switching points are encountered. We will describe the effect of switching points separately.

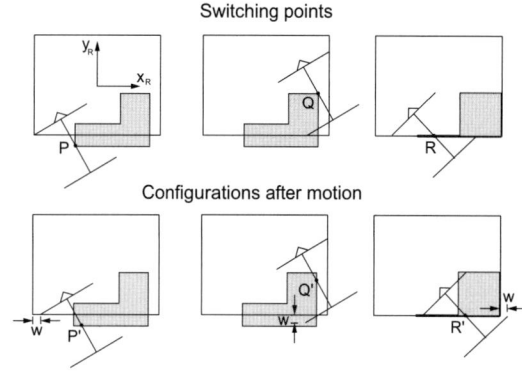

Fig. 5
SWITCHING POINTS

A. Open loop execution

We have written a simulator that inputs a user-specified path, executes our control policies and outputs wheel angular velocities. We fed the output to the robot. The intended path (the solid hexagon in Fig.6(b)) and two runs of the open-loop execution are shown. The path has large error because slip between the robot and the paper causes a loss of dual-differential drive. The error builds up as there is no feedback. The maximum error between the start and the goal was 16cm. and the angular error was $27°$. The robot took about 3 minutes to complete the path.

B. Error correction

Since the task system M_t is holonomic, errors in the motion of the paper can be corrected by proportional control. We used the pose information from the camera to servo the paper to its intended path. At every camera update, we computed the paper velocity required to servo to the path. This velocity was fed to the simulator which provided the necessary wheel angular velocities. We applied these velocities until the camera updated the pose again. Fig.6(c) shows four runs of the robot. The maximum deviation from the path was 4.6cm. and the maximum angular error was $4°$. The time to completion was comparable to the open-loop implementation.

C. Switching points

We ran separate trials to test the accuracy of motion at switching points. We ran 15 tests open-loop for the robot in switching point R. None of the tests were successful. This was because this motion required the wheels to be placed vey close to the edge of the paper. The accumulated slip caused a loss of dual-differential drive mode and the motion failed. When implemented with visual-servoing, 10 out of 15 tests were successful. However, the robot required 18 repetitions of the motion described in §5.3

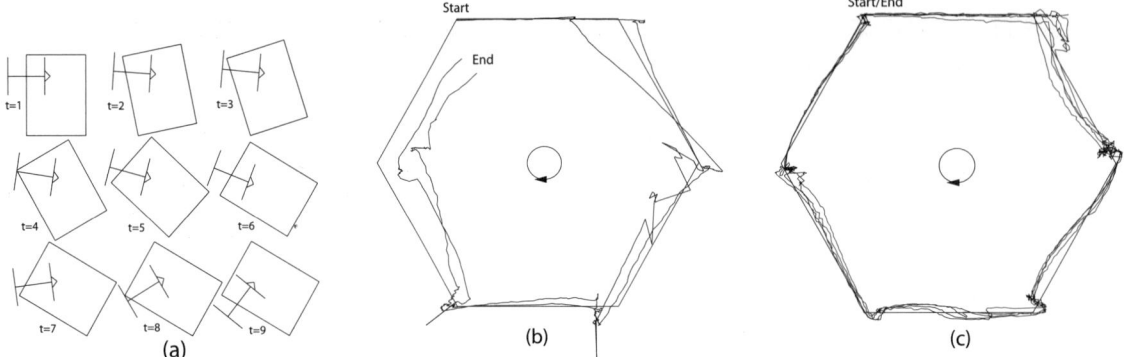

Figure 6: Implementation of the control policy to move the paper in a hexagon whos width is 8 times the width of the robot. The paper is rotated by $60°$ at each vertex. (a) shows the motion of the robot during one such rotation of the paper. (b) shows the paper pose for open-loop execution. (c) shows the paper pose when the control policy is run with visual feedback.

which took about 1 minute to complete. The resulting error in the location of the paper was 1.4cm.

VII. Discussion

Our method decomposes the system state into task freedoms and shape freedoms and analyzes each subsystem separately. By doing so, we reduce a nonholonomic control problem to two subproblems - one of following a holonomic path in task space and another of controlling the nonholonomic shape freedoms to satisfy constraints. We thus isolate the nonholonomy to a smaller set of state variables. By designing a suitable control policy in shape space we can also ensure that the task space path is followed smoothly with reduced control switchings. Furthermore, since the task system is holonomic, error correction is easy.

The method outlined in §4 can be applied to any nonholonomic system. The key lies in the fact that the matrix B is full rank. Given a nonholonomic system that looks like M, we can perform row operations on the A and generate a full rank B. The corresponding state variables can be treated as the task freedoms and the remaining state variables can be treated as the shape freedoms. We can then apply the same theorems outlined in §4 to prove STLC and generate control laws.

VIII. Future Work

We will work on policies and paths that minimize control switchings. Currently, we use the safe policy of using α to servo to L_{mid}. One can imagine a mixed policy that servoes the robot to L_{max} when the robot is close to a corner to increase the clearance between the robot and the paper. It is unclear as to whether there exists a single optimal policy for all possible paths. Another interesting problem arises when the controller has the freedom to choose the path - given a fixed policy, what is the path that minimizes the number of switching points.

IX. Acknowledgements

We thank Rashmi Patel for help with the implementation. This work was supported in part by NSF grants IIS-9820180, IIS-9900322, IIS-0082339 and IIS-0222875.

X. References

[1] Srinivasa, Baker, Sacks, Reshko, Mason, and Erdmann, "Experiments with nonholonomic manipulation," in *IEEE Int. Conf. on Rob. and Auto.*, 2001.

[2] Y. Nakamura, "Advanced robotics-redundancy and optimization," in *Addison-Wesley*, 1991.

[3] J.Bowbrow, S.Dubowsky, and J.Gibson, "Time-optimal control of robot manipulators along specified paths," *Int. J. of Rob. Research*, vol. 4, 1985.

[4] J.Baillieul, "Kinematic programming alternatives for redundant manipulators," in *IEEE Int. Conf. on Robotics and Automation*, 1985, pp. 722–728.

[5] W.Chow, "Uber systeme von linearen partiallen differentialgleichungen erster ordnung," *Math Ann.*, vol. 117, pp. 98–105, 1939.

[6] I.Kolmanovsky and N.McClamroch, "Developments in nonholonomic control problems," in *IEEE Control systems magazine*, 1995, pp. 20–36.

[7] G. Lafferriere and H. Sussmann, "A differential geometric approach to motion planning," in *Nonholonomic Motion Planning*, 1993, pp. 235–270.

[8] H. Sussmann and W.Liu, "Limits of highly oscillatory controls and approximation of general paths by admissible trajectories," in *IEEE International conference on decision and control*, 1991.

[9] H.Sussman, "Lie brackets and local controllability : A sufficient condition for local scalar-input systems," *SIAM J. on control and opt.*, vol. 21(5), 1978.

[10] B. Goodwine and J. Burdick, "Controllability with unilateral control inputs," in *IEEE Conference on Decision and Control*, 1996.

Development of a Mobile Robot for Visually Guided Handling of Material

T. I. James Tsay, M. S. Hsu, and R. X. Lin

Department of Mechanical Engineering,
National Cheng Kung University
Tainan, Taiwan 701, R.O.C.
tijtsay@mail.ncku.edu.tw

Abstract

Mobile robots frequently replace humans in handling and transporting wafer carriers in semiconductor production lines. A mobile robot is constructed in this paper. The developed mobile robot is primarily composed of a mobile base, a robot manipulator, and a vision system. Since the guidance control system of the mobile base inevitably causes positioning errors of the mobile base, this study employs the eye-in-hand vision system to provide visual information for controlling the manipulator of the mobile robot to grasp accurately stationary material during pick-and-place operations between a predefined station and the mobile robot. This work further proposes a position-based look-and-move task encoding control strategy for eye-in-hand vision architecture, that maintains all target features in the camera's field of view throughout the visual guiding. Moreover, the manipulator can quickly approach the material and precisely position the end-effector in the desired pose. Numerous techniques are required for implementing such a task, including image enhancement, edge detection, corner and centroid detection, camera model calibration method, robotic hand/eye calibration method, using a camera with controlled zoom and focus, and task encoding scheme. Finally, the theoretical results for the proposed control strategy are experimentally verified on the constructed mobile robot. Specific experimental demonstrations include grasping the target object with different locations on the station and grasping the target object tilted by different angles to the station.

1. Introduction

For production lines with a small production volume and period, and which must be quickly adapted to changes in the factory layout, a new material transfer unit, which is an AGV equipped with a robot manipulator, is used. Such units are becoming popular for use in material transfer systems. This new material transfer unit, which is also called a mobile robot, is sufficiently mobile that it can perform tasks in the production line very flexibly.

The semiconductor industry highly prioritizes maximizing productivity per square foot of clean room space. However, wafer diameters have changed from 200mm to 300mm and the weight of a cassette of 25 wafers is estimated to have increased from 3.4kg to 7.8kg. Transferring wafer carriers, boxes, cassettes, or FOUPs (front open unit pods), in the production lines is becoming more difficult and requiring more human effort. Moreover, transferring wafer carriers using material transfer units could be safer than using human operators. In a semiconductor factory, production lines typically have a small production volume and period, leading to frequent factory re-layouts. Thus, applying mobile robots in the production lines of a semiconductor factory is favorable. Mobile robots can not only increase productivity and save human resources, but also perform tasks safely, reliably, and flexibly, and support efficient handling and transportation.

Whenever the mobile base, following a predefined path, drives a manipulator to a specified station using a guidance control system, the guidance control system inevitably causes position and orientation errors of the mobile base. Uncertainties in either the location of the mobile base or the location of the object lead to a position mismatch, which usually results in failure of the following pick-and-place operation. A more direct way of dealing with this problem is to use a CCD camera to provide visual feedback. The camera is typically mounted on the end-effector of the manipulator. This configuration has attracted considerable interest in the field of vision-based manipulator control.

Hager and Corke [4] classified the visual feedback control systems into four main categories. Position-based control aims to eliminate the errors determined by the pose of the target with respect to the camera; the features extracted from the image planes are used to estimate the pose in space. The control law then sends the command to the joint-level controllers to drive the servomechanism. This control architecture is the so-called hierarchical control and is referred to as the dynamic position-based look-and-move control structure. The control architecture is referred to as the position-based visual servo (PBVS) control structure if the servomechanism is directly controlled by the control law mentioned above rather than by the joint-level controller. The errors to the control law in image-based control, are directly governed by the extracted features of the image planes. The control architecture is hierarchical and referred to as the dynamic image-based look-and-move control structure if the errors are then sent to the joint-level controller to drive the servomechanism. The visual servo controller eliminates the joint-level controller and directly controls the servomechanism. Such control architecture is termed an image-based visual servo (IBVS) control structure.

This study adopts an eye-in-hand vision system to provide visual information to a manipulator on a mobile robot, to compensate for the uncertainties in location due to the mobile base or the object. This study focuses mainly on developing a novel position-based look-and-move control strategy to guide the manipulator to approach the object and precisely position its end-effector in the desired pose. Several schemes are required to implement such a control strategy, including image enhancement, edge detection, corner detection, camera model and robotic hand/eye calibrations, and a task encoding scheme. Using such a control strategy, the mobile robot is expected to be able to travel between two stations without in-between stops and perform take-and-place operations on the non-planar ground.

The rest of this paper is organized as follows. Section 2 introduces the hardware architecture of the constructed mobile robot. Section 3 proposes a scheme for detecting corners and a centroid of the target. Section 4 explicates position-based task encoding scheme. Section 5 proposes the strategy for controlling a fast approaching to the target object and precisely positioning the end-effector in the desired pose. Section 6 presents experimental results. Finally, section 7 draws some conclusions.

2. Constructed Mobile Robot

A mobile robot was constructed in the Robotic Systems and Control Lab., National Cheng Kung University in 2001. It consists mainly of a mobile base, a robot manipulator, and a vision system.

The designed mobile base is balanced by four passive caster wheels at its four corners. The base is driven and steered by two powered wheels. Steering is accomplished by differential velocity control of the powered wheels. The mobile base employs three types of sensory devices - the optical encoder, the inductive proximity sensor, and the optical proximity sensor. Two optical encoders are attached to the wheels' motors and are used to extract the positions of the two powered wheels, for mobile base navigation. The inductive proximity sensor detects the position of the stations and forks in a predefined path. Ten optical proximity sensors are mounted facing down on the bottom of the mobile base, approximately 45 mm above the ground. Five are mounted at the front and the other five are mounted at the rear. These optical proximity sensors detect the path predefined by a long strip of reflective tape, to enable the guidance control system to navigate the mobile base. In the low level control loop, each drive wheel is controlled by one independent PI controller in the velocity control loop of the attached motor. The high level control loop determines the required velocities of the drive wheels using heuristic decision rules derived from human experience, so that steering is performed by the differential velocity control of the drive wheels.

The robot manipulator used here is the Mitsubishi RV-3AL industrial robot made in Japan, equipped with a CR2-532 controller. The robot manipulator is a vertical articulated manipulator with six degree-of-freedom (D.O.F.). Commands written in MELFA-BASIC IV robot language can be sent from a personal computer (PC) to the controller via an RS-232C connection or an Ethernet interface at 10Mbps. In this work, two Ethernet cards, one installed in each of the PC and the controller, are connected to a hub, using two straight Ethernet cables, to allow the PC to communicate with the controller over the Internet.

A vision system is installed in the mobile robot to guide the robot manipulator to perform pick-and-place operations. The vision system includes one CCD (Charge Coupled Device) camera and one frame grabber. A Sony FCB-IX10 color camera, which can be remotely operated via a RS232C linkage, is attached to the end-effector of the robot manipulator. Camera settings and control can be made from a computer via an RS-232C. In this work, only zoom and focus adjustments are performed on line. A Matrox Meteor-II/Standard frame grabber converts the images captured by the CCD camera into digital signals. The frame grabber features a 32-bit PCI bus master/slave host interface that has a peak transfer rate of 132 Mbytes/sec.

The designed mobile base, an industrial robot, and a vision system can be suitably integrated into a mobile robot, as depicted in Fig. 1. A three-deck rack is assembled on the top cover of the mobile base. The robot manipulator is mounted on the top of the rack. One CCD camera is attached to the end-effector of the robot manipulator. The first PC is placed on the third deck of the rack, while the second PC and one on-line uninterruptible power system (UPS) are placed on the second deck; the controller of the robot manipulator is placed on the first deck.

Apart from the communication between the two computers in the rack via the RS232 connection, the first computer adjusts the functions of the CCD camera through the RS-232C, processes the images captured by the installed frame grabber, conducts the robot manipulator according to the implemented visual servoing algorithm, and communicates with the host computer. Meanwhile, the second computer navigates the mobile base, according to the guidance control algorithm and controls the plug-in device. Two wireless modems, one connected to the first PC and the other to the host computer, communicate with the computers via RS-232 connections. These two modems are set to full-duplex transmission mode and their baud rate is set to 9600 bps. Information can be exchanged between the mobile robot and the host computer through these two modems. In applications, the mobile robot can behave autonomously, according to the assignment of tasks by the host computer, or can be remotely controlled by the operator, who presses some pre-specified buttons on the keyboard of the host computer on-line.

The electric power system of the mobile robot consists of one UPS and two sealed lead acid batteries which are connected in series. Two motor drivers, two optical encoders, ten optical proximity sensors, and one inductive proximity sensor are connected to these two batteries, and supplied with a power of 26 Ah each. Two PCs, one manipulator controller, one linear actuator, one CCD camera, and one wireless modem are connected to the UPS via independent electrical transformers. While the mobile base is moving, the UPS supplies battery backup power to all the connected equipment. The electric power should come from outside the mobile robot when the robot manipulator moves because the motion of the robot manipulator, carried by the mobile base, consumes much power. Accordingly, an automatic plug-in device is installed in the mobile base, while a socket with a funnel-shaped holder is assembled on the bottom of the station and is connected to the wall outlet, as shown in Fig. 2. In the plug-in device of the mobile base, a mechanism with an electrode that serves as a plug, is attached to the end of a linear actuator. After the robot manipulator is driven to a predetermined station by the mobile base, the robot manipulator enters the stage of pick-and-place operations. The specially designed plug can be driven out of the left side of the base by the linear actuator and inserted into the socket in the station, so that the on-line UPS supplies continuous, controlled power to all the connected equipment, particularly the power-consuming robot manipulator.

3. Corner and Centroid Detection

A lighting agent is designed, according to the backlighting technique, and mounted on the top of the station, as illustrated in Fig. 2, to intensify the top surface of the part and remove its shadow, thereby reducing the computation time spent in image processing. Three fluorescent tubes are installed inside the lighting agent, using tinfoil as a reflector. The lighting agent is covered with a translucent acrylic board, and the part is placed to achieve a high contrast. The high contrast, provided by backlighting using the lighting agent, makes identifying the edges of parts simpler. This section proposes an approach to detecting accurately the corners and centroid of a silhouette captured by a vision system. The detected corners and centroid serve as the target point-feature set in visually guided grasping, presented below.

Figure 3 presents the flowchart for locating the centroid and four corners of the target area. The captured image is first preprocessed using the thresholding technique and opening and closing operations to yield a clear binary image of the target area, from which the centroid can be obtained. The verticality mask is used to detect the verticality of the edges. After the verticality mask is applied to the entire image, the edges of the target area may be divided into many blobs, depending on the alignment of the edges. When the edges of the target area lie approximately $45°$ or $135°$ to the horizontal image axis, the number of blobs does not equal two since no vertical edges can be intensified, so the corners cannot be quickly located by the verticality mask. Accordingly, the Laplace operator is applied to estimate quickly the approximate locations of the corners. Except in the case considered above, the corners' approximate locations can be quickly estimated using the verticality mask. Once the approximate corner locations are determined, four square windows can be obtained from the four approximate locations to specify that all the edge points encircled by each window are on the same edge. The edge points of each edge yield a line equation. The precise location of each corner can then be obtained from the point of intersection of two adjacent lines.

4. Position-Based Task Encoding for the Robotic System with an Eye-in-Hand Configuration

Position-based task encoding architecture [2] is motivated by the following idea. When f_e equals g_e, then m_f equals m_g; f_e is the current target point-feature estimated in the workspace; g_e is the desired target point-feature estimated in the workspace; m_f is the currently observed target point-feature in the image space, and m_g is the desired target point-feature in the image space. This idea emphasizes "estimation" in the encoding. Then f_e and g_e are normally defined following calibration to estimate the camera model C_e.

$$f_e = C_e^{-1}(m_f)$$
$$g_e = C_e^{-1}(m_g)$$

where C_e^{-1} is the inverse of C_e when C_e is injective.

If m_f equals m_g, then f_a equals g_a. f_a is the current target point-feature in the workspace, and g_a is the desired target point-feature in the workspace.

$$f_a = C_a^{-1}(m_f)$$
$$g_a = C_a^{-1}(m_g)$$

where C_a^{-1} is the inverse of C_a when C_a is injective.

In position-based task encoding scheme, in an ideal case of perfect reconstruction, f_e represents the locations of features in the workspace. In this work, C_e is injective since a pinhole camera model is used, and C_a is injective. Accordingly, f_a approaches g_a as f_e approaches g_e.

5. Control Strategy for Approaching the Target Object

This study applies the concept of task encoding to solve the problem caused by the calibration errors, and adjusts the zoom and focus of the camera according to the distance from the camera to the target object to increase the accuracy of the target point-feature set. The employed strategy for controlling a manipulator with a single camera mounted on the end-effector as it approaches the target object, is based on the proposed position-based look-and-move task encoding structure. Position-based control requires the explicit calculation of the relative pose, including position and orientation, between the camera and the target object. This relative pose can be determined from the image of the target point-feature set, which is obtained via the camera.

5.1 Four Off-Line Tasks

Four off-line tasks must be executed in advance. These are camera calibration, hand/eye calibration, determining the desired relative pose between the camera and the object to fixing the camera on the object, and determining the desired relative pose between the camera and the object to grasp the object. A technique [7] for camera calibration is employed in this research. In the hand/eye calibration, a model plane world frame is assumed to have been selected for the camera model calibration. The operator drives the probe attached to the last link of the manipulator to the three marked points on the origin and two axes of the model plane world frame through a teaching box to determine the positions of these three points in relation to the manipulator base frame. Then, the relative location between the camera and the end-effector can be achieved through the definition for the coordinate transformations. During both of the following off-line tasks, the camera is adjusted to the wide-angle of view.

When determining the desired relative pose between the camera and the object to fixing the camera on the object, the end-effector is driven by a teaching box to a location, or the object is brought by a human operator to a location such that the object is separated from the end-effector by a safe distance (in this case, 30 cm) and can be fixated by the camera. Accordingly, the desired relative pose between the object and the camera can be determined from the observed image of the target point-feature set. Using the image of the target point-feature set, the homography between the object frame and the image frame can be estimated using a maximum likelihood estimation algorithm. Given the known intrinsic parameters obtained from off-line camera calibration and the estimated homography, the desired relative pose between the camera and the object, $T_o^{g'}$, can be obtained. The aim of this off-line task is to lock the centroid of the target area in the optical center of the image plane after step 1 of the control strategy is executed. Otherwise, in the following step, the target object may vanish from the field of view when the camera is set to the tele-angle of view.

In the determination of the desired relative pose between the camera and the object to grasping the object, the end-effector is driven by a teaching box to a location or the object is brought by a human operator to a location so that the gripper can grasp the object. With this setup, the desired pose of the object in relation to the camera can be determined from the observed image of the target point-feature set. The procedure by which the desired relative pose between the object and the camera, T_o^g, is obtained from the observed image of the target point-feature set is the same as that for fixing the camera on the object.

5.2 Position-Based Look-and-Move Task Encoding Structure

As illustrated in Fig. 4, the control law uses the current relative pose between the target object and the end-effector as feedback information and the desired relative pose between the target object and the end-effector as the input. The output of the control law is a relative motion command for the manipulator to move. This command is sent to the position controller of the manipulator, so that the desired relative pose between the target object and the effector is determined. The calculations pertaining to the control structure are listed in order as follows.

(1) Extracting the target point-feature set

In the feedback loop, a target point-feature set is extracted from the image by the proposed image processing algorithms, and used to determine the relative pose T_o^c. An approach is presented to keep all of the target point-feature set in the camera's field of view. The idea is to move the end-effector according to the position of the centroid in the image plane. If the centroid is on the right side or the left side of the image plane, then the end-effector keeps moving, following a relative motion command that it moves slightly in the positive or negative y direction, until all of the target point-feature set is on the image plane. If the centroid is on the top or the bottom of the image plane, then the end-effector follows a relative motion command to keep moving in the positive or negative x direction until all of the target point-feature set is on the image plane.

(2) Estimating T_o^c and T_o^g

Given the desired target point-feature set and the object model, the homography (H) between the object frame and the desired image frame can be estimated using the maximum likelihood estimation algorithm [7]. With the known intrinsic parameters of the camera model (A) and H, the extrinsic parameters, R and t can be obtained [7]. Therefore, the estimated T_o^g is,

$$T_o^g = \begin{bmatrix} R & t \\ 0 & 1 \end{bmatrix} = \begin{bmatrix} r_1 & r_2 & r_3 & t \\ 0 & 0 & 0 & 1 \end{bmatrix} \quad (1)$$

The camera observes the current target point-feature set. T_o^c can be estimated in the same way as T_o^g. Notably, T_o^c is estimated from the camera's intrinsic matrices A, corresponding to different zooms and foci. However, T_o^g is estimated using only a single intrinsic matrix A, for the wide-angle of view.

(3) Computing T_o^e and $T_o^{eDesired}$

T_o^e and $T_o^{eDesired}$ can be calculated using the following equation that involves two known matrices:

$$T_o^e = T_c^e T_o^c \quad (2)$$

$$T_o^{eDesired} = T_g^{eDesired} T_o^g \quad (3)$$

where T_c^e changes with the zoom and focus. The next subsection details the implementation of T_c^e. $T_o^{eDesired}$ is always $T_{c_w}^e$ and T_o^g is changed to $T_o^{g'}$, in step 1 of the control strategy.

(4) Computing $T_{eDesired}^e$

The control law uses the feedback of the current pose T_o^e and the desired pose $T_o^{eDesired}$ to generate a relationship between the current end-effector frame and the desired end-effector frame, which can be written as,

$$T_{eDesired}^e = T_o^e (T_o^{eDesired})^{-1} \quad (4)$$

$$\Rightarrow T_{eDesired}^e = \begin{bmatrix} R_{eDesired}^e & t_{eDesired}^e \\ 0 & 1 \end{bmatrix} = \begin{bmatrix} n_x & t_x & b_x & X \\ n_y & t_y & b_y & Y \\ n_z & t_z & b_z & Z \\ 0 & 0 & 0 & 1 \end{bmatrix} \quad (5)$$

$T_{eDesired}^e$ is chosen as the control error. If the current end-effector is translated by $t_{eDesired}^e$ and rotated by an angle $\theta_{eDesired}^e$ around a principal axis $\bar{P}_{eDesired}^e$, then the end-effector frame will coincide with the desired end-effector frame.

(5) Computing the relative motion command T_{eNew}^e

The movement by $t_{eDesired}^e$ and $\theta_{eDesired}^e$ may lead to a collision between the end-effector and the target object because of inevitable errors of the calibration. Therefore, the end-effector moves to new end-effector frame in each step of the control strategy and precisely positions the end-effector in the desired final pose. T_{eNew}^e can be determined by choosing a scale factor k, which is less than one, to weight $T_{eDesired}^e$ in each step of the control strategy.

$T_{eDesired}^e$ can be represented as a six element vector,

$$T_{eDesired}^e = [X \ Y \ Z \ A \ B \ C]^T \quad (6)$$

where $[X \ Y \ Z]^T$ represents the translation of $T_{eDesired}^e$ and $[A \ B \ C]^T$ represents the rotation of $T_{eDesired}^e$; A is yaw angle; B is pitch angle, and C is roll angle. $[A \ B \ C]^T$ can be obtained from (7):

$$A = \tan^{-1}\left(\frac{t_z}{b_z}\right), B = \tan^{-1}\left(\frac{-n_z}{b_z \cos A + t_z \sin A}\right), C = \tan^{-1}\left(\frac{n_y}{n_x}\right) \quad (7)$$

Accordingly, $T_{eNew}^e = k[X \ Y \ Z \ A \ B \ C]^T$ and the relative motion command $k[X \ Y \ Z \ A \ B \ C]^T$ is sent to the servo system of the manipulator.

5.3 Control Strategy

During the pick-and-place operation, the approach to the target is further divided into four steps. Step 1: The end-effector is driven to a location to fix the camera on the object. Step 2: The end-effector approximately approaches the target object. Step 3: The end-effector accurately approaches the target object. Step 4: The end-effector is slightly regulated to grasp the target object.

Step 1: Driving the end-effector to a location to fix the camera on the object

The end-effector is moved above the target object so that the centroid of the target area is in the optical center of the image plane. Therefore, the desired output pose of the end-effector is,

$$T_o^{eDesired1} = T_{g'}^{eDesired1} T_o^{g'} \quad (8)$$

where $T_{g'}^{eDesired1}$ is $T_{c_w}^e$ and the relative pose $T_o^{g'}$ can be obtained off-line.

First, the end-effector is moved above the station and the camera zooms out to a wide-angle view to search for the target object. Then, the current pose T_o^e is estimated using a number of techniques, including assigning a coordinate frame to the target object, point-feature set extraction, using a camera model in the wide-angle of view, and determining the relative pose $T_{c_w}^e$ from the off-line hand/eye calibration in the wide-angle of view.

The coordinate transformation from the desired end-effector frame to the current end-effector frame, can be written as,

$$T_{eDesired1}^e = T_o^e (T_o^{eDesired1})^{-1} \quad (9)$$

$T_{eDesired1}^e$ is formed into a pose vector $[X_1 \ Y_1 \ Z_1 \ A_1 \ B_1 \ C_1]^T$ with six elements. Z, A, and B are set to zero since the main aim of this step is to fix the centroid of target area in the center of view. Therefore, this step sends a relative motion command $T_{eNew}^e = [X_1 \ Y_1 \ 0 \ 0 \ 0 \ C_1]^T$ to the servo system of the manipulator. Then, the camera zooms in to the tele-angle of view while the end-effector is driven to a new pose.

Step 2: The end-effector approximately approaches the target object.

When the end-effector is moved above the target object, the desired pose of the end-effector changes so that the desired relative pose between the target object and the end-effector for picking up the target object is as follows.

$$T_o^{eDesired} = T_g^{eDesired} T_o^g \quad (10)$$

where $T_g^{eDesired}$ is $T_{c_w}^e$ and T_o^g can be obtained off-line.

The current end-effector pose T_o^e in step 2 can be estimated in the same way as T_o^e in step 1. The camera zooms in to the tele-angle of view to estimate the current pose T_o^e

using a different camera model and a different relative hand/eye pose $T_{c_t}^e$.

The coordinate transformation from the desired end-effector frame to the current end-effector frame, can be written as,

$$T_{eDesired}^e = T_o^e \left(T_o^{eDesired}\right)^{-1} \quad (11)$$

$T_{eDesired}^e$ also can be formed as a vector $[X_2 \ Y_2 \ Z_2 \ A_2 \ B_2 \ C_2]^T$ with six elements. Scale factors of 0.9 are chosen to weight the pose vector to prevent a collision between the end-effector and the target object. Thus, this step sends a relative motion command $T_{eNew}^e = 0.9[X_2 \ Y_2 \ Z_2 \ A_2 \ B_2 \ C_2]^T$ to the servo system of the manipulator. The camera zooms out to the standard-angle of view as the end-effector is driven to a new pose.

Step 3: The end-effector accurately approaches the target object.

The desired relative pose between the target object and the end-effector, $T_o^{eDesired}$, in steps 2, 3 and 4 are the same. The current end-effector's pose T_o^e in step 3 can be estimated in the same way as T_o^e in step 2, but using a different camera model and a different relative hand/eye pose $T_{c_s}^e$ in the standard-angle view.

From (11), $T_{eDesired}^e$ is calculated accurately in this step and scale factor of one is selected to weight the pose vector. A relative motion command $T_{eNew}^e = [X_3 \ Y_3 \ Z_3 \ A_3 \ B_3 \ C_3]^T$ is sent to the servo system of the manipulator. The camera zooms out to the wide-angle of view while the end-effector is driven to a new pose.

Step 4: The end-effector is slightly regulated to grasp the target object.

The current end-effector's pose T_o^e in step 4 can be estimated in the same way as T_o^e in step 1. Equation (11) is used to calculate $T_{eDesired}^e$. In this step, a scale factor of 0.5 is chosen to weight the pose vector and then send the relative motion command $T_{eNew}^e = 0.5[X_4 \ Y_4 \ Z_4 \ A_4 \ B_4 \ C_4]^T$ to drive the end-effector to a new pose. Step 4 is executed until the position error and the orientation error are both less than predefined values (in this study, 3 mm and 3 degrees). The end-effector then grasps the target object and moves it to the mobile robot. To define the position and orientation errors, $T_{eDesired}^{e_j}$ is written as follows.

$$T_{eDesired}^{e_j} = T_o^{e_j} \times (T_o^{eDesired})^{-1} = \begin{bmatrix} n_x & t_x & b_x & X \\ n_y & t_y & b_y & Y \\ n_z & t_z & b_z & Z \\ 0 & 0 & 0 & 1 \end{bmatrix} \quad (12)$$

The position and orientation errors are then defined below.

$$Error_{position}^j = \sqrt{X^2 + Y^2 + Z^2} \quad (13)$$

$$\theta_{error}^j = \tan^{-1}\left(\frac{\sqrt{(t_z - b_y)^2 + (b_x - n_z)^2 + (n_y - t_x)^2}}{n_x + t_y + b_z - 1}\right) \quad (14)$$

Equation (13) gives the position error and (14) defines the orientation error as the angle of rotation between the two coordinate frames about the principle axis.

6. Experimentation

The mobile base is fixed on the ground and the target object is moved to some specified locations relative to the station to simulate the situation of the application stage, to evaluate the performance of the mobile robot during the application stage, as depicted in Fig. 5. The target object was placed in nine different positions, separated by 20 cm. In each position, the target object was pointed in three different directions. Finally, the target object was also tilted by 3° and 6° to the station. Table 1 lists these possible locations of the target object. Executing all the steps of the control strategy approximately takes a minimum of approximately 6 seconds and a maximum of well over 13 seconds, as shown in Table 1.

After all steps of the control strategy are executed, the end-effector almost reaches the target. The final location of the end-effector, T_o^{eFinal}, is then recorded. Similarly, in each test, (13) gives the position error and (14) defines the orientation error as the angle of rotation between the two coordinate frames about the principle axis.

7. Conclusion

During pick-and-place operations between a predefined station and a mobile robot, the guidance control system inevitably causes position and orientation errors of the mobile base. Therefore, this study proposes a position-based look-and-move task encoding control strategy for controlling the end-effector of the eye-in-hand manipulator to approach and grasp the target object. Numerous techniques are involved and include (image enhancement, edge detection, corner and centroid detection, camera model calibration method, the robotic hand/eye calibration method, a camera with controlled zoom and focus, the task-encoding approach.

In image enhancement, this study uses a lighting agent as a light source. It can intensify the top surface of an object and remove its shadow. Therefore, the corners and centroid of the top surface of the object are used as feature points. This work further proposes an approach to detecting corners and the centroid accurately. These feature points are used to estimate the pose. However, the accuracy of the estimation of the pose is based on the well-known camera model and the relative hand/eye pose; the error in the illuminated pattern must be small. These requirements are difficult to meet. Accordingly, this study applies the concept of task encoding to solve the problem caused by the calibration errors, and proposes a new and simple approach to performing hand/eye calibration. Moreover, an adjustable zoom and focus function is added to the camera in the control strategy, to prevent perturbation of the image. A control strategy based on the position-based task encoding scheme is proposed and divided into four steps. The advantages of this control strategy are that it maintains all target features in the camera's field of view and quickly guides the end-effector toward the target object, and precisely position it in the desired pose, even for a target object arbitrarily located in the camera's field of view initially.

References

[1] J. Baeten, H. Bruynickx, and J. D. Schutter, "Tool/camera configurations for eye-in-hand hybrid vision/force control", *in Proceedings of the 2002 IEEE ICRA*, May 2002, pp.1704-1709.

[2] Wen-Chung Chang, J. P. Hespanha, A. S. Morse, and G. D. Hager, "Task re-encoding in vision-based control systems", *in Proceedings of IEEE Conference on Decision and Control*, vol. 1, 1997, pp.48–53.

[3] P. I. Corke and M. C. Good, "Dynamic effects in visual closed-loop systems", *IEEE Transactions on Robotics and Automation*, vol. 12, no.5, Oct.1996, pp.671-683.

[4] S. Hutchinson, G. D. Hager, and P. I. Corke, "A tutorial on visual servo control", *IEEE Transactions on Robotics*

and Automation, vol. 12, no. 5, Oct. 1996, pp.651-670.

[5] S. Matthias, "Towards autonomous robotic serving: using an integrated hand-arm-eye system for manipulating unknown objects", *Robotics and Autonomous Systems*, Elsevier Science, Vol. 26, January 1999, pp. 23-42.

[6] H. N. Nair and C. V. Stewart, "Robust focus ranging", *CVOR92*, 1992, pp. 309-314.

[7] Z. Zhang, "A flexible new technique for camera calibration", *IEEE Transactions on Pattern Analysis and Machine Intelligence*, vol. 22, no. 11, Nov. 2000, pp.1330 –1334.

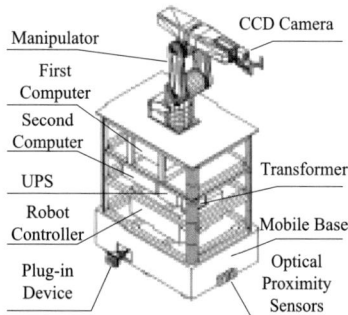

Fig. 1   Schematic of the mobile robot

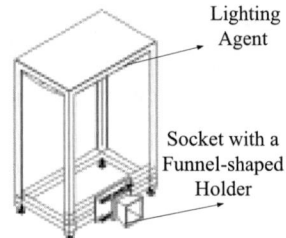

Fig. 2   Station assembled with a socket and a lighting agent

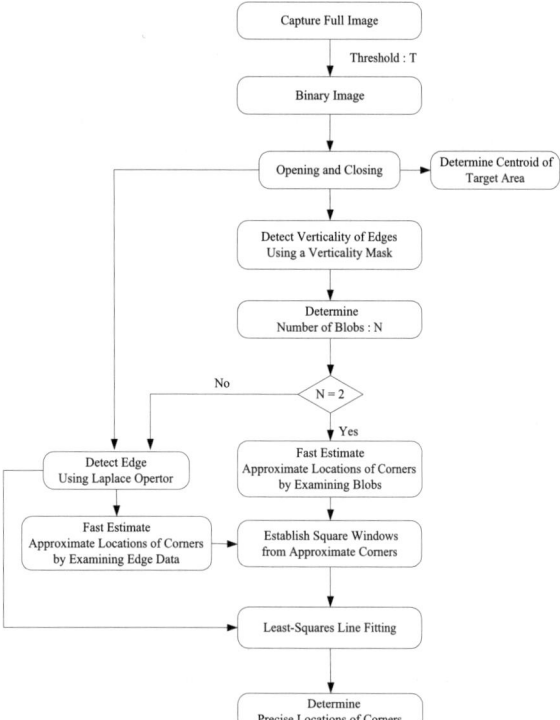

Fig. 3   Flowchart for detecting the centroid and four corners of the target area of an image

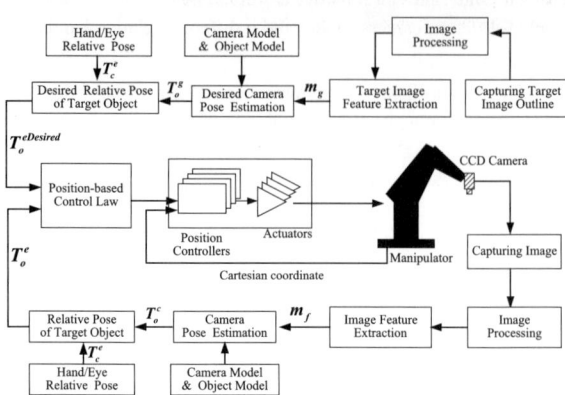

Fig. 4   Position-based look-and-move task encoding structure

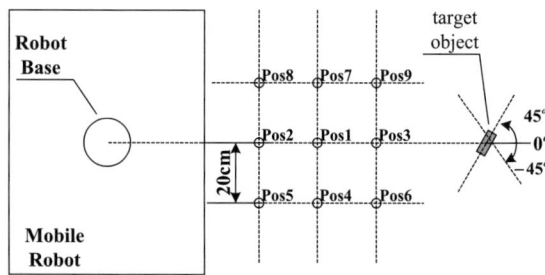

Fig. 5   Possible locations of the target object

Table 1   Pose error of the eye-in-hand manipulator

Target Object Locations	Position Error $Error_{position}$ (mm)	Orientation Error θ_{error} (degree)	Time Taken (ms)
Pos1 No Rotation	0.9881	0.20667	7311
Pos1 Rotation 45^0	1.08524	0.41285	8515
Pos1 Rotation -45^0	0.86094	0.84925	8305
Pos2 No Rotation	0.60153	0.86320	5816
Pos2 Rotation 45^0	1.02945	0.35308	8966
Pos2 Rotation -45^0	0.69213	0.16169	8069
Pos3 No Rotation	1.13784	0.54724	5954
Pos3 Rotation 45^0	0.95979	0.52956	7930
Pos3 Rotation -45^0	1.22811	0.56528	8783
Pos4 No Rotation	1.18799	0.72069	7111
Pos4 Rotation 45^0	1.20157	0.89720	6470
Pos4 Rotation -45^0	0.91066	0.96095	7486
Pos5 No Rotation	1.27777	0.79123	7174
Pos5 Rotation 45^0	0.92629	0.58872	6545
Pos5 Rotation -45^0	0.54059	0.40850	8023
Pos6 No Rotation	0.91731	0.28519	6524
Pos6 Rotation 45^0	1.03510	0.25414	6516
Pos6 Rotation -45^0	1.03734	0.76735	8113
Pos7 No Rotation	1.28678	0.47722	8055
Pos7 Rotation 45^0	0.95050	0.45956	6493
Pos7 Rotation -45^0	1.08605	0.55671	7108
Pos8 No Rotation	0.69097	0.37936	6390
Pos8 Rotation 45^0	0.85743	0.07677	8523
Pos8 Rotation -45^0	0.31304	0.38289	7027
Pos9 No Rotation	0.76156	0.54270	6508
Pos9 Rotation 45^0	1.10026	0.30462	7116
Pos9 Rotation -45^0	0.56873	0.46416	8139
Tilted 3°	0.90559	0.45732	11831
Tilted 6°	1.03510	0.25414	13032

Trajectory Planning of Mobile Manipulator with Stability Considerations

Seiji Furuno, Motoji Yamamoto and Akira Mohri

Department of Intelligent Machinery and Systems
Faculty of Engineering, Kyushu University
6-10-1 Hakozaki, Higashi-ku, Fukuoka 812-8581, Japan
E-mail : furuno@sc.mech.kyushu-u.ac.jp

Abstract

This paper presents methods of trajectory planning for a mobile manipulator with stability considerations. The proposed trajectory planning method is to generate a trajectory for the mobile manipulator from a given path of the end-effector considering stability. Then, we derive a dynamics model of the mobile manipulator considering it as the combined system of the manipulator and the mobile platform. ZMP criterion is used as an index for the system stability. The trajectory planning problem is formulated as an optimal control problem with some constraints. To solve the problem, we use a hierarchical gradient method which synthesizes the gradient function in a hierarchical manner based on the order of priority. The simulation results of the 2-link planar nonholonomic mobile manipulator are given to show the effectiveness of the proposed algorithm.

1 Introduction

A mobile manipulator composed of a manipulator and a mobile platform has a much larger workspace than a fixed-base manipulator due to the mobility provided by the mobile platform. But it is difficult to control, since it has high redundancy, nonholonomic constraints from the mobile platform, and dynamic interactions between the manipulator and the mobile platform.

Until now many papers treat the mobile manipulator's problems. Yamamoto et al.[1] have proposed a control method so that a end-effector tracks a desired trajectory compensating dynamic interactions. Kurisu et al.[2] have proposed trajectory planning and dynamic control methods when the end-effector's trajectory is given. Desai et al.[3] have proposed the trajectory planning method for two mobile manipulators handling an object in cooperation considering these dynamics.

However, these papers do not consider the problem of system stability. Since the mobile manipulator is a structurally unstable system, it has a possibility of overturn when moving at high speed or working by use of the manipulator. Therefore, it is very important to take the stability into considerations.

Some papers have studied the stability of the mobile manipulator. Dubowsky et al.[4] have proposed a method for the time optimal motion planning keeping the vehicle stationary. But the vehicle cannot move on the horizontal plane. Rey et al.[5] have proposed a algorithm for automatic tipover prediction and prevention, but the longitudinal plane model is used as the numerical model. Both studies do not consider the nonholonomic constraint on the vehicle. Huang et al.[6] have proposed stability control method based on ZMP (Zero Moment Point) criterion. In their paper, first the vehicle motion is derived considering its dynamics, manipulator workspace and system stability. Next the manipulator motion is derived considering its stability and configuration. However, since dynamic interaction arises between the manipulator and the vehicle, it is more desirable to consider the system combined rather than separated.

In this paper, we propose the method for planning the trajectory of the nonholonomic mobile manipulator from its end-effector's path is given considering the dynamic stability. Then ZMP criterion proposed by Huang is used as an index for the system stability. First we derive a dynamics model of the mobile manipulator considering it as the combined system of the manipulator and the mobile platform in section 2. In section 3, ZMP criterion is explained. The trajectory planning problem is formulated as an optimal control problem with some constraints in section 4. To solve the problem numerically, the hierarchical gradient method which was proposed by Iwamura et al.[7]

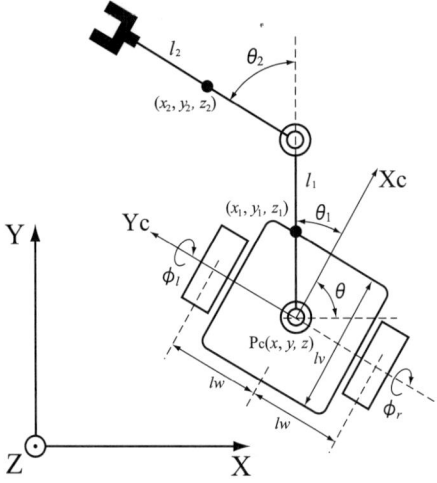

Fig.1 Mobile manipulator

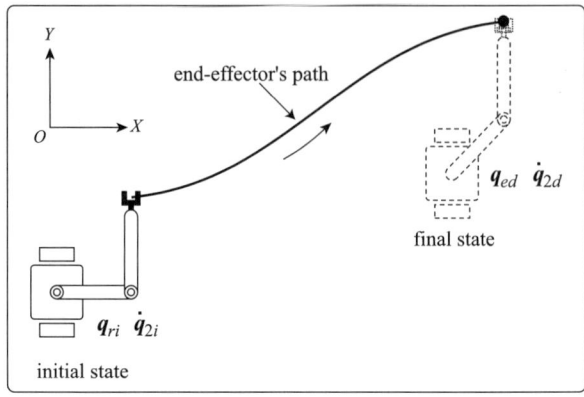

Fig.2 Trajectory planning problem

is used. Finally, Simulation results are given to show the effectiveness of the proposed method in section 6.

2 Modeling of Dynamic Equation

In this section, we derive the dynamic equation of the mobile manipulator consisting of the 2-link manipulator and wheel-type mobile platform shown by Fig.1. The dynamic equation is given by the following equation.

$$M(q)\ddot{q} + C(q,\dot{q}) = F \qquad (1)$$

where $M(q)$ is the inertia matrix and $C(q,\dot{q})$ is the centrifugal and Colioris vector. Their details are omitted. q and F are the following vectors.

$$q = (x, y, \theta, \phi_r, \phi_l, \theta_1, \theta_2)^T \qquad (2)$$
$$F = (0, 0, 0, \tau_{\phi_r}, \tau_{\phi_l}, \tau_{\theta_1}, \tau_{\theta_2})^T \qquad (3)$$

x, y are the coordinates of the center of gravity of the mobile platform, θ is the heading angle of the mobile platform, ϕ_r and ϕ_l are rotation angles of two driven wheels, θ_1 and θ_2 are joint angles of the manipulator, τ_{ϕ_r} and τ_{ϕ_l} are input torques of the platform, τ_{θ_1} and τ_{θ_2} are input torques of joints. Eq.(1) is derived using Lagrange equation. Details of derivation are omitted.

Next, the constraint equation to which the mobile platform is subjected is given by

$$A(q)\dot{q} = 0 \qquad (4)$$

Here, we derive the dynamic equation which satisfies the equation (1), (4) simultaneously using the matrix $B(q)$ which satisfies $A(q)B(q) = 0$.

$$\dot{q} = B(q)\dot{q}_2 \qquad (5)$$
$$\ddot{q}_2 = (B^T(q)M(q)B(q))^{-1}B^T(q)$$
$$\times (F - C(q,\dot{q}) - M(q)\dot{B}(q)\dot{q}_2) \qquad (6)$$

where $q_2 = (\phi_r, \phi_l, \theta_1, \theta_2)^T$. Since it is not necessary to take the rotation angles of wheels into considerations, Eq.(5), (6) can be rewritten as the following equations.

$$\dot{q}_r = B_r(q_r)\dot{q}_2 \qquad (7)$$
$$\ddot{q}_2 = (B^T(q_r)M(q_r)B(q_r))^{-1}B^T(q_r)$$
$$\times (F - C(q_r, \dot{q}_r) - M(q_r)\dot{B}(q_r)\dot{q}_2) \qquad (8)$$

where $q_r = (x, y, \theta, \theta_1, \theta_2)^T$.

3 Trajectory Planning Problem

In this paper, the method to generate the trajectory which satisfies system stability and executes manipulation task is discussed when an initial state (configuration and velocities), a final state (end-effector's position and velocities) and the end-effector's path are given as shown in Fig.2. In this section, the constraint conditions and performance index which are imposed on this trajectory planning problem are discussed.

3.1 Boundary condition

The boundary conditions in both ends of the trajectory are expressed as the following equations.

$$q_r(0) = q_{ri} \quad \dot{q}_2(0) = \dot{q}_{2i} \qquad (9)$$
$$q_e(t_f) = q_{ef} \quad \dot{q}_2(t_f) = \dot{q}_{2f} \qquad (10)$$

where q_{ri}, q_{2i} are initial states, q_{ef}, q_{2f} are final states and t_f is the final time. q_e is the end-effector's position and is expressed by following equations from Fig.1.

$$q_e = (x_e, \ y_e)^T \qquad (11)$$
$$x_e = x + l_1\cos(\theta + \theta_1) + l_2\cos(\theta + \theta_1 + \theta_2) \qquad (12)$$
$$y_e = y + l_1\sin(\theta + \theta_1) + l_2\sin(\theta + \theta_1 + \theta_2) \qquad (13)$$

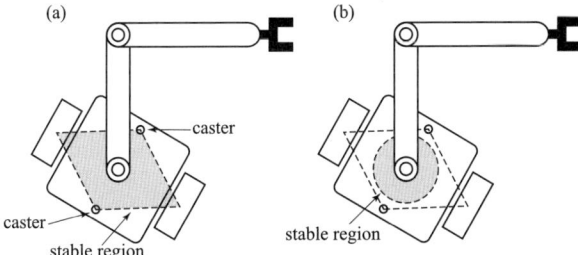

Fig.3 Stable region

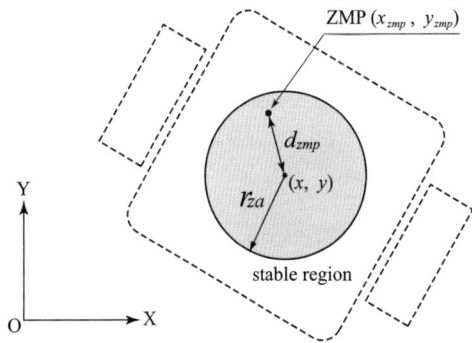

Fig.4 Stability degree

3.2 Path constraint

When the path to be tracked is given by $g_e(x, y) = 0$, the constraint condition for the end-effector to be on the path is expressed as a following equation.

$$g_e(x_e, y_e) = 0 \quad (14)$$

3.3 ZMP constraint

In this paper, in order to take the stability of the mobile manipulator into considerations, the concept of the stable region based on ZMP which Huang et al.[6] proposed is used. First, ZMP, the stable region and the stability degree are defined and ZMP constraint condition is given by using them.

ZMP. ZMP is the point on the floor where the resultant moments of the gravity, the inertial forces of the mobile manipulator and the external forces are zero. When a workspace is in a plane and there are no external forces and moments, ZMP (x_{zmp}, y_{zmp}) are given by following equations.

$$x_{zmp} = \frac{\sum_{i=0}^{2} m_i x_i g - \sum_{i=0}^{2} m_i \ddot{x}_i z_i}{\sum_{i=0}^{2} m_i g} \quad (15)$$

$$y_{zmp} = \frac{\sum_{i=0}^{2} m_i y_i g - \sum_{i=0}^{2} m_i \ddot{y}_i z_i}{\sum_{i=0}^{2} m_i g} \quad (16)$$

where m_i is the mass, x_i, y_i, z_i are the coordinates of the center of gravity, $\ddot{x}_i$ and $\ddot{y}_i$ are the acceleration of the center of gravity, g is the gravitational acceleration. Index i expresses each part, $i = 0$ is the mobile platform, $i = 1$ is first link and $i = 2$ is second link.

Stable Region. In order for the mobile manipulator to be always stable, ZMP must be within the stable region. The stable region is a quadrangle and it is expressed with the position relation between the wheels and the casters. For example, when two casters are attached to the mobile platform in front and rear, the stable region is expressed as a quadrangle which connects them (see Fig.3(a)). Here, for simplicity, the stable region is dealt with as a circle inscribed in the quadrangle (see Fig.3(b)).

Stability Degree. Next, the stability degree is defined using ZMP and the stable region. The stable region is defined to be the circle whose center is $P_c(x, y)$ and radius is r_{za}. The distance from $P_c(x, y)$ to $ZMP(x_{zmp}, y_{zmp})$ is defined as d_{zmp} (see Fig.4). d_{zmp} is expressed as follows.

$$d_{zmp} = \sqrt{(x_{zmp} - x)^2 + (y_{zmp} - y)^2} \quad (17)$$

Then, the stability degree h_z is defined as a following equation.

$$h_z = r_{za} - d_{zmp} \quad (18)$$

In order for the mobile manipulator to be stable, h_z must have a positive value. Therefore, the constraint condition to be stable is expressed as a following inequality.

$$h_z > 0 \quad (19)$$

3.4 Performance index

In order to make the wheel torque and the joint torque small, the performance index p is defined as follows.

$$P = \frac{1}{2} \int_0^{t_f} \boldsymbol{F}^T \boldsymbol{W} \boldsymbol{F} dt \quad (20)$$

where $\boldsymbol{W}$ is a weighting factor.

4 Optimal Control Problem

In this section, the trajectory planning problem defined in chapter 3 is formulated as an optimal control problem.

A state vector $\boldsymbol{x}$ and a control vector $\boldsymbol{u}$ are defined as follows.

$$\boldsymbol{x} = (x, y, \theta, \theta_1, \theta_2, \dot{\phi}_r, \dot{\phi}_l, \dot{\theta}_1, \dot{\theta}_2)^T \quad (21)$$

$$\boldsymbol{u} = (\tau_{\phi_r}, \tau_{\phi_l}, \tau_{\theta_1}, \tau_{\theta_2})^T \quad (22)$$

Then, the dynamic equation (7), (8) are rewritten as

$$\dot{x} = f(x, u); \quad x(0) = x_0 \quad (23)$$

where x_0 is an initial state. And the vector x_e is defined as follows.

$$x_e = (x_e, y_e, \dot{\phi}_r, \dot{\phi}_l, \dot{\theta}_1, \dot{\theta}_2)^T \quad (24)$$

Let x_{ed} be the desired final state. The terminal constraint is expressed as

$$r_t = x_e(t_f) - x_{ed} = 0 \quad (25)$$

Next, the path constraint condition (14) and the ZMP constraint condition (19) are rewritten as following equations using the state vector $x(t)$.

$$g_e(x(t)) = 0 \quad (26)$$
$$h_z(x(t)) \geq 0 \quad (27)$$

In order to satisfy Eq.(26) and Eq.(27) during the whole time ($0 \leq t \leq t_f$), following equations must be stisfied.

$$r_e = \int_0^{t_f} c_e(x(t))dt = 0 \quad (28)$$

$$r_z = \int_0^{t_f} c_z(x(t))dt = 0 \quad (29)$$

where $c_e(x(t)), c_z(x(t))$ are penalty functions which take positive values, when constraint conditions are not satisfied and they are expressed as following equations.

$$c_e(x(t)) = (g_e(x))^2 \quad (30)$$

$$c_z(x(t)) = \begin{cases} (h_z(x))^2 > 0 & \text{if } h_z(x) < 0 \\ 0 & \text{if } h_z(x) \geq 0 \end{cases} \quad (31)$$

The performance index is given by the following equation using the control vector u.

$$p = \int_0^{t_f} f_0(x(t), u(t))dt \quad (32)$$

$$= \frac{1}{2}\int_0^{t_f} u^T W_p u dt \quad (33)$$

where W_p is a weighting factor.

5 Hierarchical gradient method

In order to solve the optimal control problem given in chapter 4, the hierarchical gradient method is used. In this chapter, the method to synthesize the gradient function in a hierarchical manner considering the order of priority is presented. The number of constraint conditions is assumed to be three(terminal, path and ZMP constraints), they are set as r_1, r_2 and r_3 and the performance index is p.

If a input vector $u(t)$ is given, a trajectory $x(t)$ is obtained by solving (23). Then the constraint vector r_i ($i = 1, 2, 3$) can be calculated from (25), (28), (29). Therefore, the constraint vector r_i can be written as a fucntion of $u(t)$ as follows.

$$r_i = h_i(u(t)) \quad (i = 1, 2, 3) \quad (34)$$

From (34), the variation in r_i due to variation in the input vector u is expressed as follows (see Appendix).

$$\delta r_i = \int_0^{t_f} J_i \delta u dt \quad (i = 1, 2, 3) \quad (35)$$

The variation in p due to variation in the input vector u can be expressed by using the Hamiltonian H as follows (see Appendix).

$$\delta p = \int_0^{t_f} H_u \delta u dt \quad (36)$$

Where $H_u = \partial H/\partial u$. From (35) and (36), the gradient function is expressed as follows.[7]

$$\delta u = \alpha \Big[J_1 I_{11}^{-1} r_1 + K(r_2 - I_{21} I_{11}^{-1} r_1) $$
$$+ LR_3 + H_u - J_1^T I_{11}^{-1} \int_0^{t_f} J_1 H_u^T dt$$
$$- K \int_0^{t_f} \hat{J}_2 H_u^T dt$$
$$- L I_{33}^{-1} \int_0^{t_f} \hat{J}_3 H_u^T dt \Big] \quad (37)$$

$$K = \hat{J}_2^T I_{22}^{-1} - J_1^T I_{11}^{-1} I_{12} I_{22}^{-1} \quad (38)$$

$$L = \hat{J}_3^T - J_1^T I_{11}^{-1} I_{13} - K I_{23} \quad (39)$$

$$R_3 = r_3 - I_{31} I_{11}^{-1} r_1 - D(r_2 - I_{21} I_{11}^{-1} r_1) \quad (40)$$

$$D = I_{32} I_{22}^{-1} - I_{31} I_{11}^{-1} I_{12} I_{22}^{-1} \quad (41)$$

$$I_{i1} \equiv \int_0^{t_f} J_i J_1^T dt \quad (i = 1, 2, 3) \quad (42)$$

$$I_{1i} \equiv \int_0^{t_f} J_1 \hat{J}_i^T dt \quad (i = 2, 3) \quad (43)$$

$$I_{ij} \equiv \int_0^{t_f} \hat{J}_i \hat{J}_j^T dt \quad (i = 2, 3 \quad j = i, 3) \quad (44)$$

$$I_{32} \equiv \int_0^{t_f} J_3 \hat{J}_2^T dt \quad (45)$$

$$\hat{J}_2 \equiv J_2 - I_{21} I_{11}^{-1} J_1 \quad (46)$$

$$\hat{J}_3 \equiv J_3 - I_{31} I_{11}^{-1} J_1 - D \hat{J}_2 \quad (47)$$

Table 1 Mobile manipulator parameters

m_v	50.0 [kg]	I_{zv}	2.08 [kg·m²]	l_v	0.5 [m]
m_w	2.0 [kg]	I_{zw}	0.006 [kg·m²]	l_w	0.5 [m]
m_1	20.0 [kg]	I_{yw}	0.01 [kg·m²]	l_1	0.4 [m]
m_2	20.0 [kg]	I_{z1}	0.273 [kg·m²]	l_2	0.4 [m]
-	-	I_{z2}	0.273 [kg·m²]	-	-
z	0.5 [m]	z_1	1.0 [m]	z_2	1.0 [m]

6 Simulation Results

In this section, simulation results are shown. Parameters of the mobile manipulator are given in Table 1.

The end-effector's path is defined as the straight line expressed by a following equation.

$$g_x(x, y) = x = 0 \tag{48}$$

$\boldsymbol{x}_0$ and $\boldsymbol{x}_{ed}$ are assumed as follows.

$$\boldsymbol{x}_0 = (0, -0.566, \frac{\pi}{4}, 0, \frac{\pi}{2}, 0, 0, 0, 0)^T \tag{49}$$

$$\boldsymbol{x}_{ed} = (0, 2.8, 0, 0, 0, 0)^T \tag{50}$$

r_{za} is 0.16 [m], t_f is 6.0 [sec], $\boldsymbol{W}_p = \text{diag.}(1, 1, 1, 1)$.

6.1 Without ZMP constraint

First, we calculate the case where stability is not taken into considerations. The first-priority is given to the terminal constraint ($\boldsymbol{r}_1 = \boldsymbol{r}_t = 0$) and the second-priority is given to the path constraint ($\boldsymbol{r}_2 = \boldsymbol{r}_e = 0$). Fig.5(a) shows the trajectory, Fig.5(b) shows the d_{zmp} behavior and Fig.5(c) and (d) show angular velocities of wheels and joints. In this case, d_{zmp} is over the threshold value 0.16 [m] (a dotted line in Fig.5(b)). Therefore, it is difficult to execute the task on this trajectory stably.

6.2 With ZMP constraint

Next, we calculate the case where stability is taken into considerations. The first-priority is given to the terminal constraint ($\boldsymbol{r}_1 = \boldsymbol{r}_t = 0$), the second-priority is given to the path constraint ($\boldsymbol{r}_2 = \boldsymbol{r}_e = 0$) and the third-priority is given to ZMP constraint ($\boldsymbol{r}_3 = \boldsymbol{r}_z = 0$). Fig.6(a) shows the trajectory, Fig.6(b) shows the d_{zmp} behavior and Fig.6(c) and (d) show angular velocities of wheels and joints. In this case, ZMP is always within the stable region.

7 Conclusions

This paper discussed the trajectory planning problem of the mobile manipulator with stability considerations. First, we derive the dynamics of the mobile manipulator considering it as the combined system of the manipulator and the mobile platform. As an index for system stability, ZMP criterion is used. We formulate the trajectory planning problem as the optimal control problem with some constraints. To solve the problem numerically, we use the hierarchical gradient method based on the gradient function synthesized in hierarchical manner. Finally, simulation results were given to show the effectiveness of the proposed method.

Appendix

The derivation process of $\boldsymbol{J}_i$ and $\boldsymbol{H}_u$ is explained. The order of the priority is defined as $\boldsymbol{r}_1 = \boldsymbol{r}_t$, $\boldsymbol{r}_2 = \boldsymbol{r}_e$, $\boldsymbol{r}_3 = \boldsymbol{r}_z$. The performance index p is expressed by the following equation.

$$p = \int_0^{t_f} f_0(\boldsymbol{x}, \boldsymbol{u}) dt$$

First, define the error vector $\boldsymbol{e}_i(t)$ ($i = 1, 2, 3$) as follows.

$$\boldsymbol{e}_1(t) \equiv \boldsymbol{r}_1(t) = \boldsymbol{x}_e(t) - \boldsymbol{x}_{ed}$$
$$\boldsymbol{e}_2(t) \equiv \boldsymbol{r}_2(t) = \int_0^{t_f} c_e(\boldsymbol{x}(\tau)) d\tau$$
$$\boldsymbol{e}_3(t) \equiv \boldsymbol{r}_3(t) = \int_0^{t_f} c_z(\boldsymbol{x}(\tau)) d\tau$$

Then, $\delta \boldsymbol{r}_i$ is expressed as follows.

$$\delta \boldsymbol{r}_i = \int_0^{t_f} \boldsymbol{J}_i \delta \boldsymbol{u} dt, \quad (i = 1, 2, 3)$$
$$\boldsymbol{J}_i \equiv \boldsymbol{e}_{ix} \boldsymbol{f}_u$$

Where $\boldsymbol{f}_u = \partial \boldsymbol{f}/\partial \boldsymbol{u}$ and $\boldsymbol{e}_{ix}$ ($i = 1, 2, 3$) are given by following equations.

$$\dot{\boldsymbol{e}}_{1x} = -\boldsymbol{e}_{1x} \boldsymbol{f}_x; \quad \boldsymbol{e}_{1x}(t_f) = \boldsymbol{I}$$
$$\dot{\boldsymbol{e}}_{2x} = -\boldsymbol{e}_{2x} \boldsymbol{f}_x - \boldsymbol{c}_{ex}; \quad \boldsymbol{e}_{2x}(t_f) = 0$$
$$\dot{\boldsymbol{e}}_{3x} = -\boldsymbol{e}_{3x} \boldsymbol{f}_x - \boldsymbol{c}_{zx}; \quad \boldsymbol{e}_{3x}(t_f) = 0$$

where $\boldsymbol{f}_x = \partial \boldsymbol{f}/\partial \boldsymbol{x}$, $\boldsymbol{c}_{ex} = \partial c_e/\partial \boldsymbol{x}$, $\boldsymbol{c}_{zx} = \partial c_z/\partial \boldsymbol{x}$, $\boldsymbol{I}$ is $n \times n$ identity matrix (n is the dimension of the state vector $\boldsymbol{x}(t)$).

Next, we define the function H using the multiplier function ψ.

$$H(\boldsymbol{x}, \boldsymbol{u}, \psi) \equiv f_0(\boldsymbol{x}, \boldsymbol{u}) + \psi^T \boldsymbol{f}(\boldsymbol{x}, \boldsymbol{u})$$

And ψ satisfies the following equations.

$$\dot{\psi} = -\boldsymbol{H}_x; \quad \psi(t_f) = 0$$

where $\boldsymbol{H}_x = \partial \boldsymbol{H}/\partial \boldsymbol{x}$. Then δp is given by

$$\delta p = \int_0^{t_f} \boldsymbol{H}_u \delta \boldsymbol{u} dt$$

where $\boldsymbol{H}_u = \partial \boldsymbol{H}/\partial \boldsymbol{u}$.

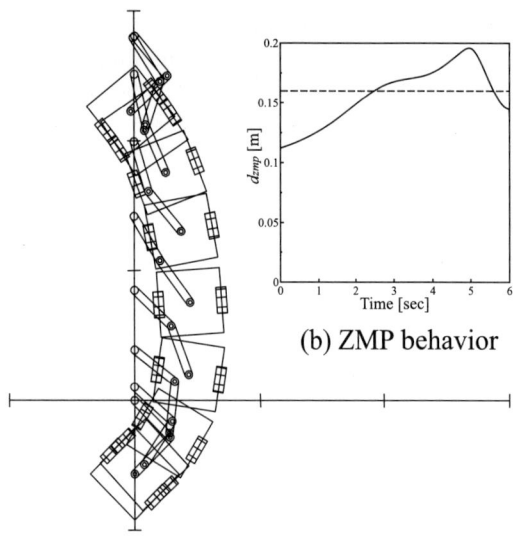

(b) ZMP behavior

(a) Trajectory of mobile manipulator

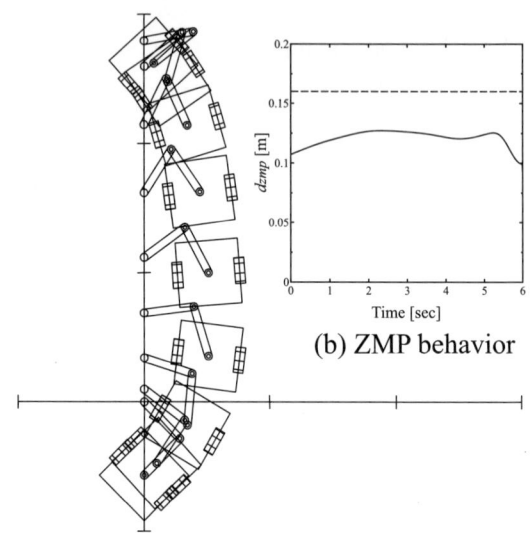

(b) ZMP behavior

(a) Trajectory of mobile manipulator

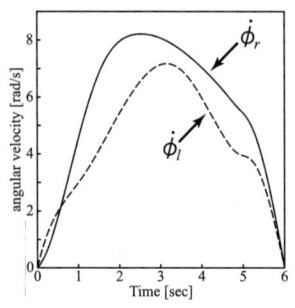

(c) Angular velocity of wheel

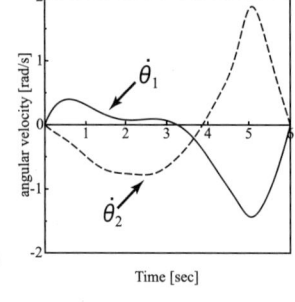

(d) Angular velocity of joint

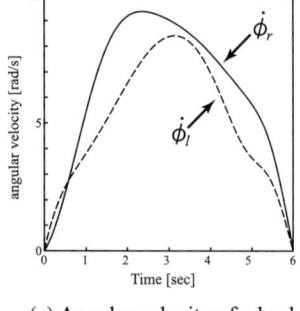

(c) Angular velocity of wheel

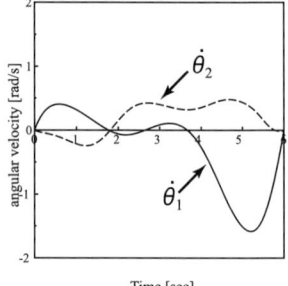

(d) Angular velocity of joint

Fig.5 Simulation result without ZMP constraint

Fig.6 Simulation result with ZMP constraint

References

[1] Y.Yamamoto and X.Yun, "Effect of the Dynamic Interaction on Coordinated Control of Mobile Manipulators", IEEE Trans. on Robotics and Automation, Vol12, No.5, pp.816-824 (1996)

[2] M. Kurisu and T. Yoshikawa, "Trajectory Planning and Dynamic control of a Mobile Manipulator", Trans. of the Japan Society of Mechanical Engineers, Vol.62, No.596-C, pp.242-248 (1996) (in Japanese).

[3] J.P.Desai and V.Kumar, "Nonholonomic motion planning for multiple mobile manipulators", Proc. IEEE Int. Conf. on Robotics and Automation, pp.3409-3441 (1997)

[4] S. Dubowsky and E. E. Vance "Planning Mobile Manipulator Motions Considering Vehicle Dynamic Stability Constraints", Proc. IEEE Int. Conf. on Robotics and Automation, pp.1271-1276 (1989)

[5] D. A. Rey and E. G. Papadopoulos "On-line Automatic Tipover for a Mobile Manipulator", Proc. IEEE/RSJ Int. Conf. on Intelligent Robots and Systems, pp.870-876 (1997)

[6] Q. Huang, S. Sugano and K. Tanie, "Motion Planning for a Mobile manipulator Considering Stability and Task Constraints", Proc. IEEE Int. Conf. on Robotics and Automation, pp.2192-2198 (1998)

[7] M. Iwamura, M. Yamamoto and A. Mohri, "A Gradient-Based Approach to Collision-Free Quasi-Optimal Trajectory Planning of Nonholonomic Systems", Proc. IEEE/RSJ Int. Conf. on Intelligent Robots and Systems, pp.1734-1740 (2000)

Motion Control for Vehicle with Unknown Operating Properties
- On-Line Data Acquisition and Motion Planning -

Kazuya Okawa[1], Shin'ichi Yuta[2]

[1] Chiba University, Chiba, Japan, okawa@faculty.chiba-u.jp
[2] University of Tsukuba, Tsukuba, Japan, yuta@roboken.esys.tsukuba.ac.jp

Abstract

*This paper describes a sequence of methods for controlling the vehicle with unknown operating properties, such as an air cushion vehicle, a submarine and an airship. We took up **Air Cushion Vehicle** (ACV) as an example of such a vehicle, and studied its motion planning based on acquired data. ACV in this paper has three simple unit motions (left-turn, right-turn and an advance), and can execute one of the unit motions optionally. After the execution, ACV measures its own position by itself and acquires data of how it moved in the real environment. Then it plans the motion that arrives at a target position using the acquired data. In this paper, we propose a sequence of these approaches and show the effectiveness of them by experiments.*

1. Introduction

By the wheel type robot, generally, an operator can get its motion from a robot's mechanism, the number of rotations of a wheel, and so on. Therefore, the operator can plan robot's motion based on the calculated motion model, in advance [1]. However, it is difficult to calculate and estimate a motion from rotations of a propeller of the vehicle, such as a helicopter [2], a submarine [3], an airship [4] and an air cushion vehicle [5].

From such a background, we focused on the vehicle with uncertain properties, like the vehicle that drives with propellers. And we did research in a method of motion control for those robots.

Many researchers already proposed the motion control methods for the vehicle that drives with propellers. However, almost all of them used a feedback control that was controlled for approaching to a target. In this case, although there is a merit of being applicable even if there is no model of a strict robot's motion.

We took up Air Cushion Vehicle (ACV) as an example of a vehicle with uncertain properties, and studied about a motion planning for the vehicle in order to reach the target position in the flat floor.

In this paper, we defined ACV's motions as three simple unit motions, such as left-turn, right-turn and an advance. At first, ACV executes each unit motion, then measures and acquires data of how it moved in the real environment. Secondly, it plans a sequence of motions that reach to the target position by switching the unit motions. Thirdly, ACV executes an appropriate motion based on the motion planning as it modifies its own position based on the measured position.

We propose a sequence of those approaches that are "Position Estimation by Omni-Directional Camera", "Data Acquisition of Motions" and "Motion-Planning based on Acquired Data". After that, we show an experimental demonstration in order to confirm the effectiveness of our approaches.

2. Experimental Environment and System

2.1 Experimental Environment

The experimental environment that we are using is shown in Fig. 1. There are two cylinders painted by the color of cyan and red, on the flat floor. ACV uses these cylinders as marks to measure its own position. In this paper, we defined that 1) the red mark is an origin of the coordinates; 2) the line from the origin to the cyan mark is a X-axis; 3) the axis, which intersects perpendicularly with an X-axis, is a Y-axis. ACV used it as the standard coordinates.

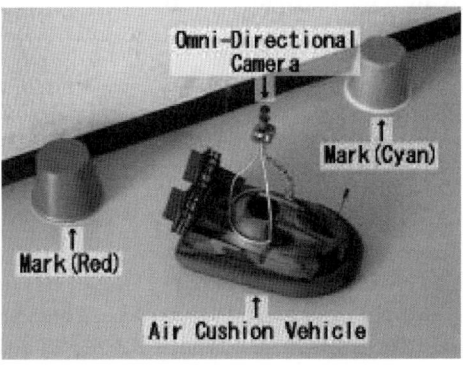

Fig.1: Air cushion vehicle and marks

2.2 Air Cushion Vehicle

Air Cushion Vehicle (ACV) that we are using is shown in Fig. 1. ACV has one propeller for hovering at the center, and two propellers for thrusting in the rear of it. Although the propeller for the hovering has stopped, it rotates whenever the propellers for the thrusting are driven. Therefore, ACV hovers and moves only when an operator inputs motion commands, but it will lose lift and will stop above the floor when there is no input.

ACV that we are using has an omni-directional camera that is a sensor for measuring its own position. The images acquired with the camera are transmitted to an external computer by radio. And the computer performs image processing, position calculation, a motion planning, and so on. After that, a motion command decided by the motion planning is broadcasted from the external computer to ACV by radio.

2.3 Motion Control

ACV has abilities to make various motions, if the rotation direction and rotation time of the propellers for thrusting are adjusted. In this paper, however, in order to simplify the problem to a motion planning, motion control is limited to the following three discrete motions, and these motions are defined as unit motions.

Motion 1: The left propeller for thrusting is clockwise rotated for 1 second.

Motion 2: The right propeller for thrusting is clockwise rotated for 1 second.

Motion 3: The both propellers for thrusting is clockwise rotated for 1 second at the same time.

After the previous motion is stopped completely, ACV can choose one of them optionally as a next motion.

3. Position Estimation by Omni-Directional Camera

3.1 Basic Idea for Position Estimation

ACV cannot calculate its own position by an odometer like a wheel type robot. So, we installed an omni-directional camera on it. The direction and distance from ACV to marks can be measured when ACV detects these marks in the environment by the camera. Therefore, ACV can estimate its own position by using the measured data (Fig.2).

In this approach, although a measurement error is included at every measuring, the error has an advantage of not accumulating. On the one hand, it has the fault about a

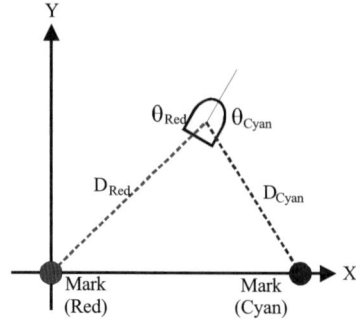

Fig. 2: Position estimation based on data measured by omni-directional camera

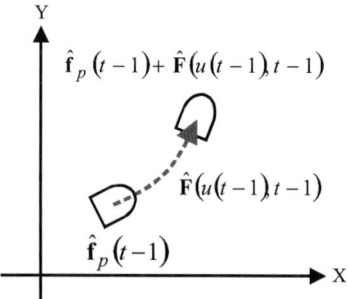

Fig. 3: Position estimation based on acquired motion data

measurement error that is included in the measurement position based on the omni-directional camera image. For example, the measurement error is included about 3cm in the measurement position when ACV is at a distance of 100cm from the marks.

In this paper, we propose the method for estimating the precise position in consideration of these features.

3.2 Position Estimation

We can estimate ACV's position of Time t from the measurement position $\mathbf{f}_m(t)$ of Time t, as shown in Fig. 2. On the other hand, we can estimate ACV's position of Time t by assuming that ACV moves the displacement $\hat{\mathbf{F}}(u(t-1),t-1)$ by unit motion $u(t-1)$ at the estimated position $\hat{\mathbf{f}}_p(t-1)$ at Time $t-1$, as shown in Fig. 3. We use information of these different positions in order to estimate accurate ACV's position $\hat{\mathbf{f}}_p(t)$ of Time t.

$$\hat{\mathbf{f}}_p(t) = \alpha \times \{\hat{\mathbf{f}}_p(t-1) + \hat{\mathbf{F}}(u(t-1),t-1)\} + (1-\alpha) \times \mathbf{f}_m(t) \quad (1)$$

$$\hat{\mathbf{F}}(w,t) = \begin{cases} \beta \times \hat{\mathbf{F}}(w,t-1) + (1-\beta) \times \{\hat{\mathbf{f}}_p(t) - \hat{\mathbf{f}}_p(t-1)\} \\ \qquad\qquad\qquad\qquad\qquad\qquad\text{if } w = u(t-1) \\ \hat{\mathbf{F}}(w,t-1) \qquad\qquad\qquad\quad\text{if } w \neq u(t-1) \end{cases} \quad (2)$$

where α and β are the learning parameters, and there is a limitation of values, $0 \leq \alpha, \beta \leq 1$. In this case, however, $\alpha = \beta = 0$ when Time $t = 0$ or 1.

3.3 Analysis of Estimation Error

In this section, we analyze an estimation error of the proposed position estimation method. In the beginning, we defined that an error $\mathbf{n}(t)$ including the measured position $\mathbf{f}_m(t)$ and an error $\mathbf{e}_p(t)$ including the estimated position $\hat{\mathbf{f}}_p(t)$ as the following;

$$\mathbf{n}(t) = \mathbf{f}_m(t) - \mathbf{f}_p(t) \tag{3}$$

$$\mathbf{e}_p(t) = \hat{\mathbf{f}}_p(t) - \mathbf{f}_p(t) \tag{4}$$

where $\mathbf{f}_p(t)$ is a pure position of Time t.

The measurement error $\mathbf{n}(t)$ has character of independence at every measurement. In other words, an average of the error $\overline{\mathbf{n}(t)}$ is 0; a variance of the error $\overline{\mathbf{n}(t) \times \mathbf{n}(t)}$ is σ and $\overline{\mathbf{n}(t) \times \mathbf{n}(t-1)}$ is 0.

$\mathbf{E}_p(w, t-1)$ is an error included in the estimated displacement $\hat{\mathbf{F}}(w, t-1)$ when motion w is executed at Time $t-1$. In this case, however, we assumed that ACV had chosen the same motion in every execution, in order to simplify a problem. The error is described by the following equation:

$$\begin{aligned}\mathbf{E}_p(w, t-1) &= \hat{\mathbf{F}}(w, t-1) - \mathbf{F}_p(w) \\ &= \frac{1}{\alpha} \times \mathbf{e}_p(t) - \mathbf{e}_p(t-1) - \frac{1-\alpha}{\alpha} \times \mathbf{n}(t)\end{aligned} \tag{5}$$

where $\mathbf{F}_p(w)$ is a pure of displacement when the motion w is executed.

Next, we calculated an error $\mathbf{e}_p(t)$ including in the estimated position $\hat{\mathbf{f}}_p(t)$ of Time t by using the above equations. The error is described by the following equation:

$$\begin{aligned}\mathbf{e}_p(t) &= (2\alpha + \beta - \alpha\beta) \times \mathbf{e}_p(t-1) - \alpha \times \mathbf{e}_p(t-2) \\ &\quad + (1-\alpha) \times \{\mathbf{n}(t) - \beta \times \mathbf{n}(t-1)\}\end{aligned} \tag{6}$$

where $\mathbf{e}_p(0) = \mathbf{n}(0)$ and $\mathbf{e}_p(1) = \mathbf{n}(1)$.

3.4 Confirmation of Position Estimation

In order to investigate the effectiveness of the proposed method, we apply the Monte Carlo method: because it is difficult to calculate a variance of estimated position error $\mathbf{e}_p(t)$ by recurrent formula. We assumed that the variance of measurement error is the random number of a regular gauss distribution defined as $\sigma = 1$. Also, we assumed that parameters of position (x, y, θ) are independence. In these conditions, we investigated the effectiveness from 100,000 times of states.

In advance, we investigated a relation between the learning parameters and the error variance because the error variance is influenced strongly to the learning parameters that are α and β. The result is shown in Fig. 4. When the learning parameters were investigated at interval of 0.01, we confirmed that there was a tendency for an error variance to become the smallest at $\alpha = 0.63$ and $\beta = 0.55$.

Then, we investigated a relation between trials and the error variance, when the learning parameters are $\alpha = 0.63$ and $\beta = 0.55$. The result is shown in Fig. 5. Although the error variance is over 1 when the number of trials is under 7 times, it is under 1 when the number of trials is over 7 times. It means that the estimated position error is less than the measured position error, because we assumed that the variance of measurement error was 1. The error variance was converged on 0.37 finally.

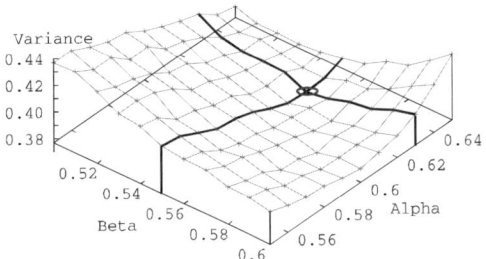

Fig. 4: Relation between learning parameters and error variance

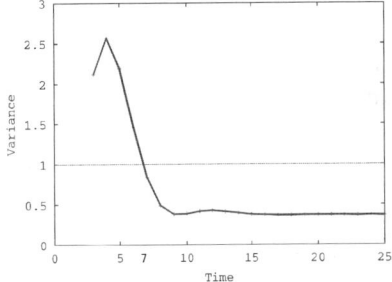

Fig. 5: Error variance in the estimated position

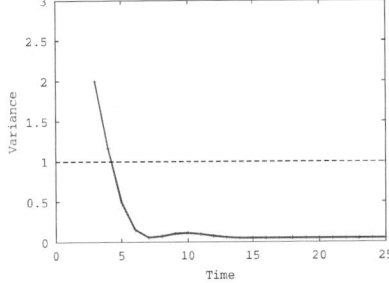

Fig. 6: Error variance in the estimated displacement

In these investigations, although we assumed that ACV executed the same motion at every execution, it will be applicable if each motion data is acquired of each.

In addition, we investigated similarly about the error variance included in displacement using the Monte Carlo method. The result is shown in Fig. 6. The error variance is less than 1 when the number of trials is over 4 times, and the error variance is finally converged on 0.22. It means that we can estimate accurately how ACV moves when it executes its motion, and therefore, it also means that we can use as the motion data in the motion planning.

4. Motion Planning based on Acquired Data

4.1 Basic Idea for Motion Planning

The purpose of this paper is to find motions in order to reach the target position by switching three unit motions. ACV that we are using in experiment moves on the flat floor, so its state can describe by three parameters (x, y, θ). In other words, its state can transfer to a point in the configuration space that is constructed by the three parameters (x, y, θ). As shown in Fig. 7, the left figure is the image of real environment, and the right figure is a configuration space.

The point in the configuration space moves when ACV moves and changes its state in the real environment. It means that the displacement by executing a motion in the real environment is described as a vector in the configuration space, as showing in Fig. 8. The vector corresponds $\mathbf{F}(w,t)$ to the value the equation (2).

The state that is estimated as the result of the executing motions is the end of point of vector sum in the configuration space, as shown in Fig. 9. Therefore, evaluation for the motion planning can be estimated as the distance in the configuration space.

4.2 Scaling of Configuration Space

Although the configuration space is constructed by three axes as position (x, y, θ), a unit of θ is different from units of position (x, y). Therefore, there is a problem in using the Euclid distance in the configuration space for evaluation then. In order to solve the problem, we adjusted scales of each axis by using tolerances that are judgments of whether ACV arrived at the target position.

When the tolerances of the position (x, y, θ) are defined $E_x[mm], E_y[mm], E_\theta[rad.]$, the scaling parameters are calculated as the following equation.

$$(X, Y, \theta) = \left(\frac{E_\theta}{E_x \times \pi}, \frac{E_\theta}{E_y \times \pi}, \frac{1}{\pi} \right) \quad (7)$$

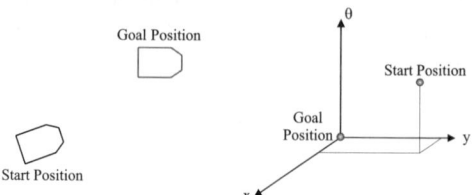

Fig. 7: Image of real environment and configuration space for motion planning

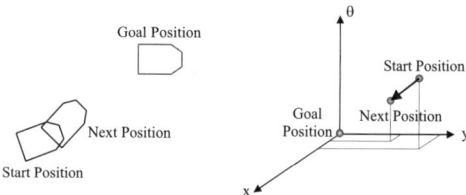

Fig. 8: Displacement and a vector in configuration space

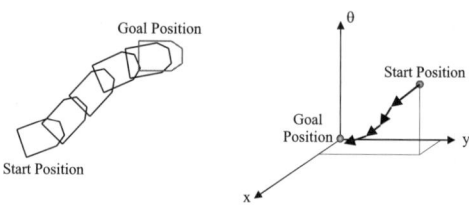

Fig. 9: Motion planning in configuration space

4.3 Motion Planning

The sequence of motions that arrives to the target position corresponds to finding a vector sum that reaches from the present point to the target point in the configuration space. We adapt a genetic algorithm (GA) as searching method of an appropriate sequence of motions.

5. Experiments and Results

We apply the proposed approaches to ACV, and confirm the effectiveness of them by experiments.

5.1 Confirmation of Position Estimation

In this section, we confirm the effectiveness of the position estimation method that was proposed in Chapter 3. We assumed that ACV executed the same motion continuously and estimated its own position by the proposed equations. We can confirm the effectiveness of the proposed equations by comparing the estimated position to the pure position. However, it is difficult to measure the pure position, so we compared the estimated position to the position measured from the omni-directional camera images. Although the position measured from the camera images includes errors, we considered that there is no problem because the error is not so big when ACV moves in neighborhood of marks.

5.1.1 Experimental Conditions

We put the mark of the color of red and cyan that was detached 50cm. ACV measured and calculated its own position based on the image from the omni-directional camera. After measuring the initial position $f_m(0)$ at time $t=0$ and performing Motion 1 $(=w)$, a position $f_m(1)$ is measured at time $t=1$. $\hat{F}(w,1)$ is computed by substituting these values based on the equation (2), and ACV's position is calculated from time $t=2$ by the equation (1).

5.1.2 Experimental Results

What displayed the position of ACV actually moved on the display is shown in Fig. 10. Circles of the color of red and cyan are marks placed in the environment, and the green line that connects them is equivalent to an x-axis. Blue and green rectangles mean the measured position and the estimated position, respectively.

We have confirmed in the beginning that the estimated position and the measured position were not corresponded. However we have confirmed that the measured position and the estimated position were corresponded gradually by correcting to the data of displacement of motion $1 (=w)$ and carrying out the estimated position using it.

As the result, we confirmed that the estimation position corresponded with the measurement position. It means that ACV could estimate its own position appropriately.

5.2 Confirmation of Motion Planning

In this section, we confirm the effectiveness of the proposed motion planning by actual experiments.

5.2.1 Experimental Conditions

We put the mark of the color of red and cyan that detached 50cm. And a target of position was assumed $(50[cm], 50[cm], 0[rad.])$ on the mark standard coordinate. An Initial position of ACV was put around at $(-30[cm], 20[cm], 0[rad.])$, but ACV measured and estimated its own position by using the omni-directional camera.

The displacement data of each unit motion were acquired in the previous experiments, as shown the arrows in Fig. 8.

5.2.2 Experimental Results

The result of experiment is shown in Fig. 11. The frame of the rectangle drawn on the floor means the target position, but ACV cannot detect it.

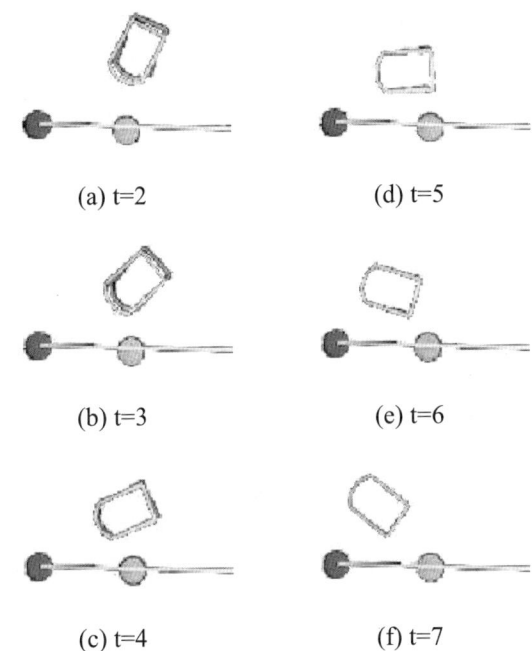

(a) t=2 (d) t=5
(b) t=3 (e) t=6
(c) t=4 (f) t=7

Fig. 10: Experimental Results for Position Estimation

At the time $t=1$, ACV could find a sequence of motions that reaches to the target positions in the configuration space, and then ACV executed a motion that is at a head of the sequence. As the result, ACV moved in the real environment. ACV calculated and estimated its own position based on the image by the omni-directional camera. In this experiment, the estimated position didn't correspond to the expected position. Therefore ACV executed the motion planning to find a sequence of motions that reaches to the target position, again. ACV repeated such sequence of processes. As the result, we confirmed that ACV could have reached to the target position at the time $t=7$.

Although we used a computer that has Celeron 466MHz, it used about 200ms for the motion planning of each step.

6. Summary

This paper described a sequence of approaches in order to control the vehicle with unknown operating properties. We took up air cushion vehicle as an example of it. Concretely, we proposed the position estimation by an omni-directional camera. And we proposed the motion planning based on the acquired motion data that was improved from the general genetic algorithm.

Moreover, we adapted the proposed approach to the actual ACV, and confirmed that ACV could have reached to the target position.

References

[1] J. Y. Wong, *Theory of Ground Vehicles*, A Wiley-Interscience Publication, ISBN 0-471-52496-4.

[2] K. Harbick, J. F. Montgomery and, G. S. Sukhatme, "Planar Spline Trajectory Following for an Autonomous Helicopter", IEEE Int'l Symposium on Computational Intelligence in Robotics an Automation '01, pp.408-413 (2001).

[3] E. W. Mc Gookin, D. J. Murray-Smith, and Y. Li, "Submarine Sliding Mode Controller Optimization using Genetic Algorithms", Int'l Conf. on Control '96, Exeter, UK, Vol.1, pp 424-429 (1996)

[4] K. Motoyama, K. Suzuki, M. Yamamoto, and A. Ohuchi, "Evolutionary Design of Learning State Space for Small Airship Control", From animals to animats: the sixth international conference on the simulation of adaptive behavior (SAB'00), J. A. Meyer et al. (Eds.), The MIT Press, pp. 307-316 (2000).

[5] I. Fantoni, R. Lozano, F. Mazenc, and K. Y. Pettersen, "Stabilization of a nonlinear underactuated hovercraft", International Journal of Robust and Nonlinear Control, Vol.10, No.8, pp.645-654 (2000).

[6] L. Davis, "Handbook of Genetic Algorithm", Van Nostrand (1990).

[t=3]

[t=4]

[t=5]

[t=1]

[t=6]

[t=2]

[t=7]

Fig. 11: Experiment in the real environment by ACV

Remote Control of a Mobile Robot Using Distance-Based Reflective Force

J. B. Park
School of Electrical Engineering
Seoul National University
Seoul, Korea
E-mail:pjb0922@snu.ac.kr

B. H. Lee
School of Electrical Engineering
Seoul National University
Seoul, Korea
E-mail:bhlee@asri.snu.ac.kr

M. S. Kim
Human Robotics Research Center
Korea Institute of Science and
Technology(KIST), Seoul, Korea
E-mail:munsang@kist.re.kr

Abstract - In this paper, a real-time obstacle avoidance method is discussed for a remote mobile robot controlled by a teleoperator. The method enables the remote user to drive the mobile robot without collisions. It consists of a real-time environment modeling algorithm using ultrasonic sensors, and a mobile robot control algorithm for obstacle avoidance. The former solves the limitations of the ultrasonic sensors such as noise sensitivity, poor directionality, and specular reflection for obtaining more accurate sensor data. Then the latter conducts wall or center following implemented by a fuzzy controller for obstacle avoidance using the environment model obtained by the previous algorithm. The command for obstacle avoidance is actually applied as a reflective force using a haptic device such as a force feedback joystick. The remote mobile robot, ROBHAZ-DT(RoBot in HAZardous environment - Double Track) developed by KIST(Korea Institute of Science and Technology), is employed as the mobile robot model. The normal and the maximum speed of the ROBHAZ-DT are 3.1km/h and 7.2km/h, respectively. The weight of the robot is 50kg. Simulations with the ROBHAZ-DT model are carried out to verify the performance of the proposed method.

I. INTRODUCTION

A teleoperation system with a mobile robot conventionally conducts operations in hazardous environments such as nuclear power plant, undersea, space and minefield instead of human beings. The ROBHAZ-DT developed by KIST is also a kind of robots to carry out teleoperations in hazardous environments, especially in the minefield. The teleoperator can drive the robot using a haptic device such as a master robot or a force feedback joystick apart from the teleoperation region. The haptic device can not only receive a user's input but also generate a reflective force and reflect it back to the user. To generate the proper reflective force, we must know some data about the remote environment. Ultrasonic sensors are usually used for obtaining the data of the environment in many mobile robot applications[1][4][5][7]. The sensors, however, have some limitations such as noise sensitivity, poor angular resolution, and specular reflection. The various environment representation methods have been studied by others such as certainty grid methods with ultrasonic sensors attached around the robot[5][7], where large memories and numerous sensors are needed. To solve these problems and to obtain the remote environment information for the ROBHAZ-DT with only six ultrasonic sensors, a real-time environment modeling algorithm is proposed. Then, wall[1][2][3] and center following methods are used for obstacle avoidance using the environment model previously obtained. Both of them are implemented by a fuzzy logic controller without considering dynamics of the robot. Finally the control command for obstacle avoidance is reflected back into the user as the force in a haptic device. Thus, the teleoperator feels the reflective force and maneuvers the robot safely among obstacles. Of course, the methods using reflective forces with a haptic deivce have already been studied by others[8][9]. They almost use the potential field theory[7] to generate reflective forces. The forces are simply generated by summation of the vectors whose magnitudes are proportional to the inverse of the square of the distance measured by ultrasonic sensors. Thus, although it enables the robot to avoid collisions, it cannot guide the user to drive the robot safely. But the proposed method in this paper enables the user to drive the robot safely following the contour of obstacles using wall or center following methods.

In Section II, we describe the kinematics of ROBHAZ-DT, the geometric model of the ultrasonic sensors attached to the robot, and the force feedback joystick model. In Section III, we describe the modeling algorithm of the remote environment and analyze the control states and the transition conditions of the robot. Then we use a fuzzy controller for wall and center following, and generate the reflective force for the force feedback joystick. In Section IV, simulation results are presented to show the effectiveness and validity of the proposed algorithm.

II. SYSTEM ARCHITECTURE

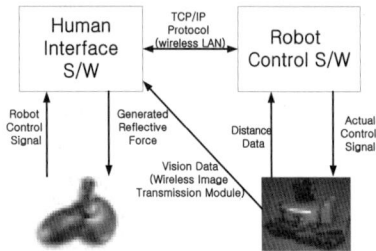

Fig.1 Configuration of the teleoperation system

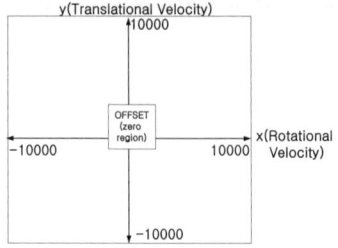

Fig.3 Axes of the joystick which match the translational and rotational velocities of the robot

The whole teleoperation system is divided into a human interface S/W(software) and a robot control S/W as shown in Fig.1.

The human interface S/W receives the user's inputs and generates reflective forces into the force feedback joystick. The mobile robot control S/W converts the user's inputs into actual control signals for the mobile robot using its kinematics and its state conditions. The translational and rotational velocities of the robot are independently controlled by the robot control S/W as shown in Fig.2. The human interface S/W communicates with the mobile robot control S/W by a wireless LAN using TCP/IP protocols with 100ms period considering the computation time of the robot control S/W. Thus the translational and rotational velocities of the robot can be expressed discretely with sampling period T_s (100ms) as follows.

If $t = k \cdot T_s (k = 0,1,2,3,\cdots)$ then,

$$v(t) \stackrel{\Delta}{=} v(k \cdot T_s) = v(k), \qquad (k \cdot T_s \leq t < (k+1) \cdot T_s) \quad (1)$$

$$w(t) \stackrel{\Delta}{=} w(k \cdot T_s) = w(k), \qquad (k \cdot T_s \leq t < (k+1) \cdot T_s) \quad (2)$$

$$\dot{v}(t) \stackrel{\Delta}{=} \dot{v}(k \cdot T_s) = \dot{v}(k) = 0, \quad (k \cdot T_s \leq t < (k+1) \cdot T_s) \quad (3)$$

$$\dot{w}(t) \stackrel{\Delta}{=} \dot{w}(k \cdot T_s) = \dot{w}(k) = 0, \quad (k \cdot T_s \leq t < (k+1)T_s) \quad (4)$$

Here, translational and rotational accelerations of the robot are defined as zero during sampling period because the translational and rotational velocities are constant during that period.

In Fig.1, vision data obtained by the CCD camera attached to the robot are directly transmitted by the wireless image transmission module.

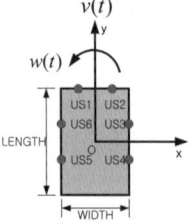

Fig.2 Axes of the translational and rotational velocities, and ultrasonic sensor arrangement attached to the robot

The size of the robot is defined by WIDTH and LENGTH, and six ultrasonic sensors attached to the ROBHAZ-DT perpendicularly as shown in Fig.2. Six circles expressed by US1,US2,…,US6 clockwise indicate ultrasonic sensors. Ultrasonic sensors have some limitations such as noise sensitivity, poor directionality and specular reflection that are well-defined in [6][7]. In order to obtain a proper environment model, we solve these problems caused by the limitations of ultrasonic sensors in Section III.

Wing Man Force Joystick that we utilize as a haptic device can receive a user's input and generate a reflective force. Y-axis and X-axis of the joystick match the translational and rotational velocities of the robot, respectively as shown in Fig.3. It is just the same with the reflective forces. We define OFFSET as safe region where the generated forces are not reflected to the user.

III. REMOTE MOBILE ROBOT CONTROL ALGORITHM

A. Environment Model Parameters

To maneuver the remote mobile robot, we must know information of the remote environment. First of all, the parameter of the front obstacles is defined by the shortest distance between the robot and the obstacles. To reduce the noise sensitivity and poor directionality of the ultrasonic sensors in front side, we use an approach similar to the error eliminating method developed by Johann Borenstein[6]. The random noise caused by noise sensitivity of the sensors can be rejected by comparing the difference between two consecutive readings of the same sensor as follows.

$$|D(k) - D(k-1)| < \varepsilon \quad (5)$$

where

$D(k)$ is a reading at time k.

ε is an acceptable constant for good reading, (not caused by external noise).

If (5) is not satisfied, the present reading is not good. Thus, the present reading is discarded. Then, the representative distance of the front obstacles is eventually determined by

$$|D_{US1}(k) - D_{US2}(k)| < \delta \qquad (6)$$

where

- $D_{US1}(k), D_{US2}(k)$ are distances between the robot and the front obstacles measured at time k by two front ultrasonic sensors, US1 and US2, respectively.
- δ is an acceptable constant for the same object measurement of the two front sensors, (two front sensors detect the same obstacle).

If the condition of (6) is satisfied, the average value of two front readings, $D_{us1}(k)$ and $D_{us2}(k)$ represents the distance $D_F(k)$ between the robot and the front obstacles as follows.

$$D_F(k) = \frac{D_{US1}(k) + D_{US2}(k)}{2} \qquad (7)$$

If the condition of (6) is not satisfied, the minimum value of two front readings represents the front distance $D_F(k)$ as follows.

$$D_F(k) = \min(D_{US1}(k), D_{US2}(k)) \qquad (8)$$

In the case of left or right obstacles, the noise sensitivity and poor directionality of ultrasonic sensors are reduced by the same way. The representative parameters of the left or right environment model consist of an angle between the robot heading direction and the surface of the obstacles, and a distance between the center of the robot and the surface of the obstacles. The tangent plane of the surface of the obstacles can be modeled as the wall. We explain only right wall case because the modeling algorithm for the left wall is similar to that for the right wall. First, we define three cases for measurement of the right wall. In Fig.4, case 1 illustrates that the robot moves straight forward without rotational motion. In this case, we can compute the representative parameters as follows.

From $\triangle ABC$, a right-angled triangle,

$$\Theta_R(k) = \tan^{-1}\left(\frac{D_R(k-1) - D_R(k)}{v(k-1) \cdot T_s}\right) \qquad (9)$$

$$D_{R\perp}(k) = \left(D_R(k) + \frac{WIDTH}{2}\right) \cdot \cos(\Theta_R(k)) \qquad (10)$$

where

- $D_R(k)$ is the representative distance measured by the right sensors.
- WIDTH is the width of the robot defined in Fig.2.

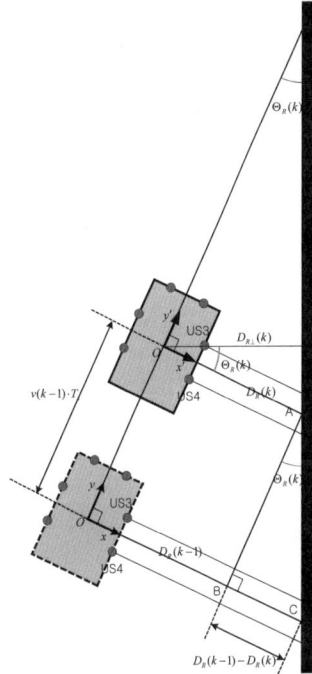

Fig.4 Geometry of the robot trace from time k-1 to time k for obtaining the environment model parameters in case 1, while the robot is moving straight forward without rotational motion. The solid and the dotted rectangles represent the robot at k and k-1, respectively.

Next, case 2 illustrates that the robot moves with both translational and rotational velocities. In this case, the angle $\Theta_R(k)$ between the robot heading direction and the right wall is estimated as follows.

$$\Theta_R(k) = \Theta_R(k-1) + w(k-1) \cdot T_s \qquad (11)$$

where

- $\Theta_R(k-1)$ is the angle between the robot's heading direction and the surface of the right obstacle at time k-1.
- $W(k-1)$ is the angular velocity of the robot at time k-1.

The distance between the robot's center and the right wall is also computed by (10) using the measured distance $D_R(k)$ at time k.

While the robot in case 1 and 2 can detect the right wall using the sensors in right-hand side at time k, the robot in case 3 cannot detect the wall because of the specular reflection of the ultrasonic sensors. Therefore we have to estimate both environment parameters, $\Theta_R(k)$ and $D_{R\perp}(k)$. First, $\Theta_R(k)$ can be estimated by (11) like case 2. Secondly, to solve the estimation problem of the distance $D_{R\perp}(k)$ we divide it into two problems according to the angular velocity $w(k-1)$ at time k-1.

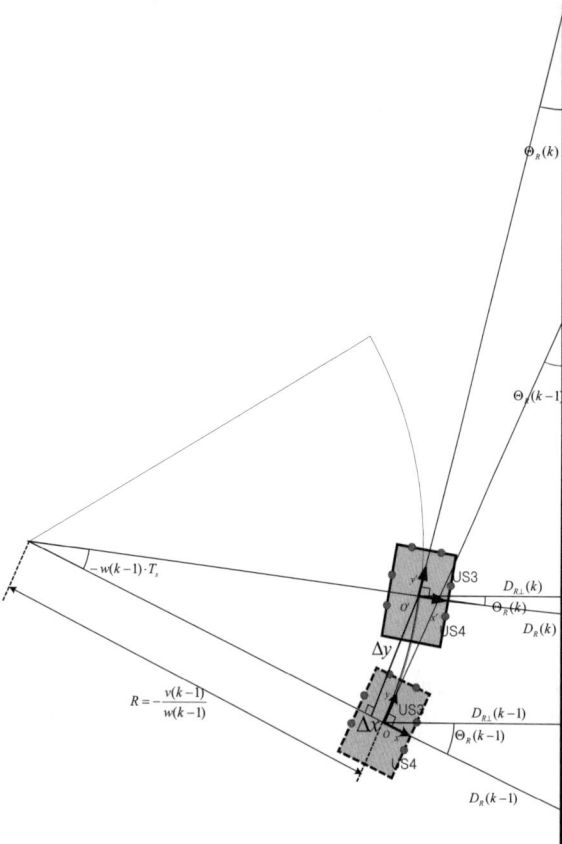

Fig.5 Geometry of the robot trace from time k-1 to time k for obtaining environment model parameters in case 2, while the robot is moving with both translational and rotational velocities. The solid and the dotted rectangles represent the robot at k and k-1, respectively.

In Fig.5, when the angular velocity of the robot is not zero, that is $w(k-1) \neq 0$, the distance $D_{R\perp}(k)$ is formulated as follows.

$$D_{R\perp}(k) = \frac{|-\Delta x \cdot \cos(\Theta_R(k-1)) + \Delta y \cdot \sin(\Theta_R(k-1)) - D_{R\perp}(k-1)|}{\sqrt{\cos^2(\Theta_R(k-1)) + \sin^2(\Theta_R(k-1))}} \quad (12)$$

$$= |-\Delta x \cdot \cos(\Theta_R(k-1)) + \Delta y \cdot \sin(\Theta_R(k-1)) - D_{R\perp}(k-1)|$$

where

$$\Delta x = -\frac{v(k-1)}{w(k-1)}(1 - \cos(-w(k-1) \cdot T_s)) \quad (13)$$

$$\Delta y = -\frac{v(k-1)}{w(k-1)} \cdot \sin(-w(k-1) \cdot T_s) \quad (14)$$

When the angular velocity of the robot is zero, that is $w(k-1) = 0$, $\Theta_R(k)$ can also be computed by (11). In this case, the distance $D_{R\perp}(k)$ can be easily computed as follows.

$$D_{R\perp}(k) = D_{R\perp}(k-1) - v(k) \cdot T_s \cdot \cos(90° - \Theta_R(k)) \quad (15)$$
$$= D_{R\perp}(k-1) - v(k) \cdot T_s \cdot \sin(\Theta_R(k))$$

TABLE I
FOUR STATES OF THE ROBOT ACCORDING TO THE ENVIRONMENT WHILE THE ROBOT IS MOVING

S1	*Free Moving State* - The user can freely maneuver the mobile robot.
S2	*Measure Initial Value State* - The desired angular velocity keeps zero until the reliable initial values are measured.
S3	*Wall Following State* - The desired angular velocity is determined by a fuzzy controller for wall following.
S4	*Center Following State* - The desired angular velocity is determined by a fuzzy controller for center following

B. Fuzzy Logic Controller and State Transition Diagram

To control the remote mobile robot actually, we generate desired velocities for obstacle avoidance using the environment model parameters obtained previously. In case of the front obstacles, the desired translational velocity is generated, which is proportional to the inverse of the front representative distance. Then, to generate the desired angular velocity, we define 4 control states such as S1(Free Moving State), S2(Measure Initial Value State), S3(Wall Following State), S4(Center Following State).

In S3 and S4, the desired angular velocity is determined by a fuzzy controller according to the robot states as shown in Table I. The fuzzy controller can make the robot follow the straight line defined as the reference line of wall or center following for obstacle avoidance. The inputs of the fuzzy controller are the distance D_{diff} between the robot's center and the reference line, and the angle Θ_R between the robot's heading angle and the wall. The output of the fuzzy controller is the desired angular velocity w_d. The fuzzy controller has the principles as follows.

1) If the distance between the robot's center and the reference line is big, the fuzzy controller modifies the angular velocity w_d for the robot to approach the reference line.
2) If the distance between the robot's center and the reference line is small, the fuzzy controller modifies the angular velocity w_d for the robot to run parallel with the reference line.

The crisp output of the proposed fuzzy controller is calculated by the center of the area method.

The state transition conditions of each control state in Fig.6 are defined in Table II.

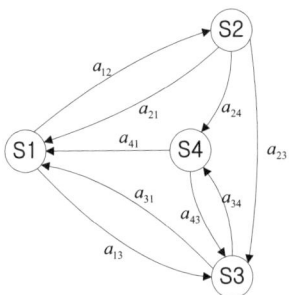

Fig.6 State transition diagram

TABLE II
STATE TRANSITION CONDITIONS FOR THE MOBILE ROBOT

a_{12}	The robot detects wall and prepares to measure initial values of the environment parameters.
a_{13}	The robot must avoid dangerous situation that the robot is near to the wall.
a_{21}	The wall disappears
a_{23}	The robot successfully measures the initial values or the robot must avoid dangerous situation that the robot is near to the wall.
a_{24}	A wall occurs on the opposite of the reference wall.
a_{31}	The wall disappears.
a_{34}	A wall occurs on the opposite of the reference wall.
a_{41}	Both walls disappear.
a_{43}	One of the walls disappears.

The desired velocities generated in the previous process are applied to the user as reflective forces. First of all, the translational velocity is applied as shown in Fig.7. If the robot speed is faster than the reference speed v_d the teleoperator feels the repulsive force with the magnitude F_{max} and the negative direction for reducing the robot speed. F_{max} can be set considering user's preference. In the figure, OFFSET means the safe region for neglecting the force generated by slight change of joystick position along y-axis.

In S2, S3 or S4, the teleoperator feels the repulsive force to keep the angular velocity as the reference angular velocity w_d as shown in Fig.8. Thus, the user feels the forces with the magnitude F_{max}, and positive or

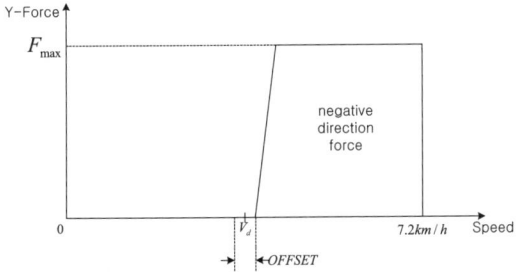

Fig.7 Reflective Force along y-axis for control speed

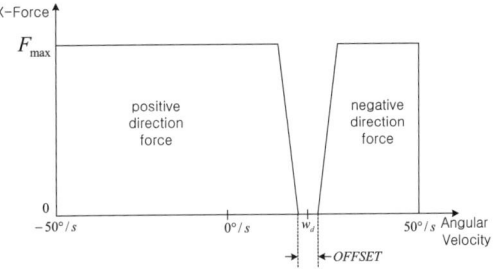

Fig.8 Reflective Force along x-axis for control angular velocity

negative direction for maintaining the angular velocity as w_d. OFFSET is defined for the same purpose of that of the y-axis force.

IV. SIMULATION RESULTS

To verify the effectiveness of the proposed algorithm, we perform a simulation. In the simulation, the robot is ordered to move from A(0sec) to H(18.0sec), where there are 2 walls and 1 rectangular type obstacle at the center as shown in Fig.9. The actual movement and the robot control states are shown as follows.

$$A(0\sec) \xrightarrow{S1} B(0.9\sec) \xrightarrow{S2} C(1.7\sec) \xrightarrow{S3}$$
$$D(6.8\sec) \xrightarrow{S1} E(9.8\sec) \xrightarrow{S3} F(12.1\sec) \xrightarrow{S4}$$
$$G(16.8\sec) \xrightarrow{S3} H(18.0\sec)$$

The operator can only see the display set-up remotely from the mobile robot. The reflective forces can be detected at the joystick when the robot approaches the obstacle or the side walls. In the simulation, the allowable maximum speed is set as 3.60km/h.

In Fig.10, the reference and the actual robot speed are described by a solid and a dotted line, respectively. From 6.3(sec) to 8.9(sec), the reference robot speed is reduced for avoiding collision with the front obstacle and the actual robot speed is kept under the reference speed.

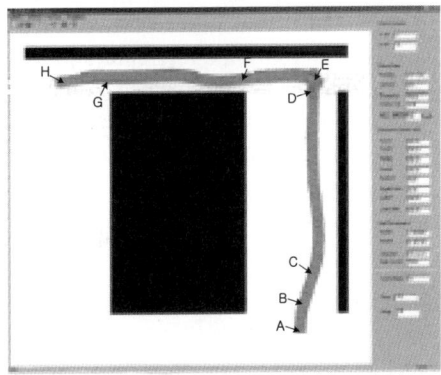

Fig.9 Operator display set-up

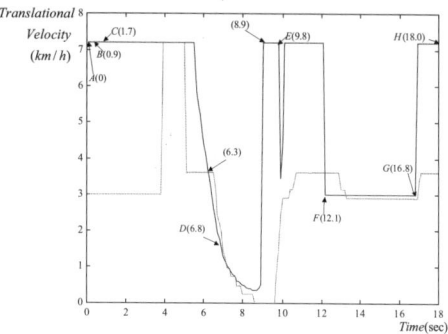

Fig.10 Ref.(solid) and Actual(dotted) velocities of the robot

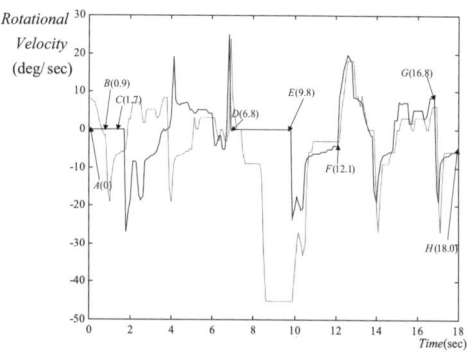

Fig.11 Ref.(solid) and Actual(dotted) angular velocities

From F(12.1sec) to G(16.8sec), the robot keeps the speed as 3.0km/h for moving with safety. Here, we define the initial value of the reference robot speed as 7.2km/h to remove the repulsive force effect. In Fig.11, the reference and the actual angular velocity are described by a solid and a dotted line, respectively. In S1, the actual angular velocity is not relative to the reference angular velocity since the reflective force is not generated. From B(0.9) to C(1.7), the actual velocity is going to following the reference velocity defined as zero for measuring initial values. At that time, the state transition condition a_{23} occurs. Thus, the robot begins to follow the wall and the difference between the reference and the actual velocities is reduced by degrees. From E(9.8sec) to H(18.0), the reference and the actual velocities have almost the same values. Although there are some error and delay, the robot moves well avoiding collisions with obstacles from the simulation.

V. CONCLUSIONS

The mobile robot control algorithm proposed in the paper is based on the distance between the ultrasonic sensors and the obstacles. For obstacle avoidance, the remote mobile robot control algorithms such as wall following and center following are implemented by a fuzzy logic controller without considering dynamics of the robot. The reference velocities determined by the proposed algorithms are applied to the user as the reflective forces. The simulation results show the performance and reliability of the proposed algorithms.

Further research includes the experiments using the actual mobile robot, ROBHAZ-DT. Sensor fusion techniques can be applied for acquiring high reliability of the mobile robot navigation.

VI. ACKNOWLEDGEMENTS

This work was supported in part by the HWRS-ERC at KAIST, the ASRI, and the BK21 Information Technology at Seoul National University.

VII. REFERENCES

[1] G. Bauzil, M. Briot, and P. Ribes, "A navigation sub system using ultrasonic sensors for the mobile robot Hilare," 1st Int. Conference on Robot Vision and Sensory Controls, Stratford-upon-Avon, UK, 1981, pp. 47-58 and pp. 681-698.

[2] G. Giralt, "Mobile robots," NATO ASI Series, vol. F11, Robotics and Artificial Intelligence. New York: Springer-Verlag, 1984, pp. 365-393.

[3] J. Iijima, S. Yuta, and Y. Kanayama, "Elementary functions of a self-contained robot 'YAMABICO 3.1'," Proc. 11th Int. Symp. Ind. Robots, Tokyo, 1983, pp. 211-218.

[4] A. Elfes, "A sonar-based mapping and navigation system," Technical Report, The Robotics Inst., Carnegie-Mellon Univ., pp. 25-30, 1985.

[5] H.P. Moravec, "Certainty grids for mobile robots," Technical Report Preprint, The Robotics Inst., Carnegie-Mellon Univ., 1986.

[6] Johann Borenstein and Yoram Koren, "Error Eliminating Rapid Ultrasonic Firing for Mobile Robot Obstacle Avoidance", IEEE Transactions on Robotics and Automation, Vol. 11, No. 1, February 1995.

[7] Johann Borenstein and Yorem Koren, "Real-Time Obstacle Avoidance for Fast Mobile Robots," IEEE Transactions on Systems, Man, and Cybernetics, Vol. 19, No. 5, September/October, 1989.

[8] Sung-Gi Hong, Ju-Jang Lee, and Seungho Kim, "Generating Artificial Force for Feedback Control of Teleoperated Mobile Robots," IEEE International Conference on Intelligent Robots and Systems.

[9] Jun-Pyo Hong, Oh-Sang Kwon, Eung-Hyuk Lee, Byung-Soo Kim, and Seung-Hong Hong, "Shared-Control and Force-Reflection Joystick Algorithm for the Door Passing of Mobile Robot or Powered Wheelchair," IEEE TENCON 1999.

Motion Planning for Humanoid Walking in a Layered Environment

Tsai-Yen Li
Computer Science Department
National Chengchi University,
Taipei, Taiwan, R.O.C.

Pei-Feng Chen
Computer Science Department
National Chengchi University,
Taipei, Taiwan, R.O.C.

Pei-Zhi Huang
Computer Science Department
National Chengchi University,
Taipei, Taiwan, R.O.C.

Abstract - Motion planning is one of the key capabilities for autonomous humanoid robots. Previous researches have focused on weight balancing, collision detection, and gait generation. Most planners either assume that the environment can be simplified to a 2D workspace or assume that the path is given. In this paper, we propose a motion planning system capable of generating both global and local motions for a humanoid robot in a layered or two and half dimensional environment. The planner can generate a gross motion that moves the humanoid vertically as well as horizontally to avoid obstacles in the environments. The gross motion is further realized by a local planner that determines the most efficient footsteps and locomotion over uneven terrain. If the local planner fails, the failure is feedback to the global planner to consider other alternative paths. The implemented humanoid planning system is an interactive tool that can compute collision-free motions for a humanoid robot in an on-line manner.

I. INTRODUCTION

The potential market of service and entertainment humanoid robots has attracted great research interests in the recent years. Several models of humanoid robots have been designed in research projects. Among the active research topics, enabling a humanoid robot to move autonomously with motion planning capability is one of the challenging problems that need to be addressed. An autonomous robot should be able to accept high-level human commands and walk in a real-life environment consisting of floors and stairs without colliding with environmental obstacles. A high-level command is something like "Move to location A on the second floor" while the robot is currently at some location B on the first floor, for example.

The motion for a humanoid robot to achieve a given goal is typically very complex because of the degrees of freedom involved and the contact constraint that needs to be maintained. Therefore, it is common to take a two-level planning approach to solve this problem. The first level only considers *global motion planning*, which is the motion planning of the whole body treated as a simple projected geometry. Given the gross motion from the first level, the second level only considers *local motion planning* that moves the legs of a humanoid robot to realize the corresponding gross motion in an efficient way.

In this paper, we propose a motion-planning system capable of generating efficient walking motions for a humanoid to reach a goal on a layered environment. We assume that the system is given an elevation and height description of the objects in the workspace and accept a goal-oriented command from a user. The system will generate a feasible global path and the associated locomotion that bring the humanoid to reach the goal as efficient as possible. At a first glace, the problem is similar to the general path-planning problem. However, since the definition of obstacles for this problem depends on the leg length of the humanoid and the local relative height, the problem definition deserves further clarification. A user may have personal preference on the paths if the goal can be reached via various paths of different heights. In addition, the motion plan proposed by a global motion planner may not always be feasible for locomotion arrangement. In this case, the interaction between the two levels of planning becomes an interesting problem.

The rest of the paper is organized as follows. After reviewing related work in motion planning and humanoid in the next section, we will describe in details the problem we consider in this paper. Then, we will then present our global and local motion planners in Section IV and V, respectively. In Section VI, we will present several examples from the simulation in our experiments. Finally, we will conclude our work with future directions in the last section.

II. RELATED WORK

The gross motion-planning problem was originally brought up in the context of robotics to generate collision-free path for mobile robots or manipulators. A survey of approaches to the problem can be found in [9]. Generally speaking, early research focuses on developing theoretical foundation and complete solutions for the problem [11]. Due to the curse of dimensionality, several researches in the last decade proposed practical solutions that can be applied to wider arrange of applications despite they usually lack completeness[1][2].

Many efficient planners have been proposed to solve the problem for objects with low degrees of freedom (DOF's) (typically less than or equal to four). Most of these planners are complete planners because they can always give a correct answer (success or failure) to the given problem. Among these planners, the potential-field based approach is the most popular one and is also the one used in our gross motion planner. An artificial potential field is typically used in the workspace as a heuristic to search the configuration space for a feasible path.[2]

The research of generating humanoid motions can be found in robotics and computer animation [3][10][14]. Although various aspects of motion generation have been studied, we will only concern the lower-body motion and the resulting body displacement. Early researches focus on generating a dynamically stable motion for a given path on a flat or uneven ground [4][5][7]. Although the locomotion for

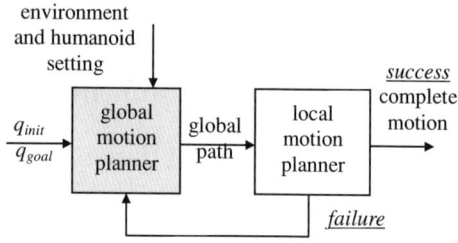

Fig. 1. Planning loop for a typical query of humanoid motion

regular walking can be computed kinematically, many approaches choose to use or modify motion-captured data due to the complexity of a human figure. Techniques such as motion warping [16] or dynamic filtering [15][17] are often used to ensure that the captured motions can be transformed into a dynamically feasible one. However, these techniques are not as flexible as kinematics-based methods in handling obstacles in an uneven terrain.

Not until recent years, the problem of gross motion planning for humanoid robots becomes one of the active research topics in robotics and computer animation [6][8][13]. In [6], a gross motion planner utilizing graphics hardware has been proposed to generate humanoid body motion on a flat ground in real time. Captured locomotion is used to move the humanoid along the generated global path. In [8], a biped robot can plan its footsteps amongst obstacles but cannot step onto them. In [13], a multi-layer grid is used to represent the configuration space for a humanoid with different locomotion such as walking and crawling. The humanoid may change its posture along a global path. In short, most gross motion planners for humanoid robots assume a flat ground and adopt canned motions for simplicity. However, the assumption is often over-restricted since a humanoid robot is more likely to work in a layered environment filled with objects of various shapes and heights as in the real life.

III. PROBLEM DESCRIPTION

According to motion granularity, the motion-planning problem usually can be classified into global (gross) motion planning and local (fine) motion planning. For the problem of walking on a layered environment for a humanoid, both types of planning needs to be considered in order to ensure that the desired task can be accomplished. Although the gross and fine motion planners can be designed separately and solved sequentially, we think they should be connected in a loop with feedbacks as shown in Fig. 1. The global path from the global motion planner is fed into the local motion planner to create corresponding footsteps and locomotion. However, the local motion planner may fail to generate locomotion for the given path. In this case, the planner should feedback the failure with reasons to the global planner to compute another global path. Taking this decoupled view can greatly reduce the complexity of such a planning problem. In the following subsections, we will describe the problems of global planning and local planning separately in

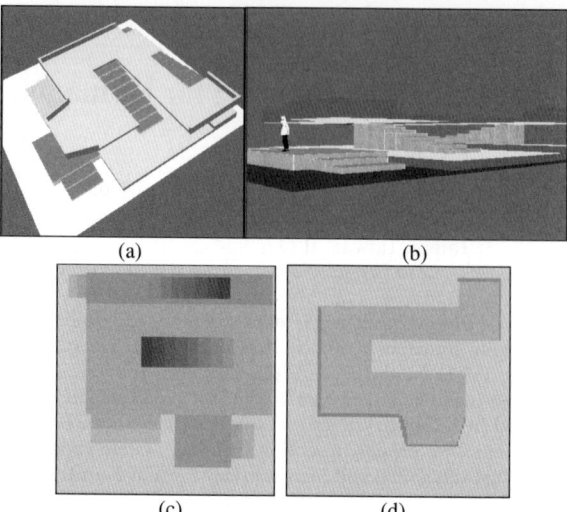

Fig. 2. (a) Top view of the workspace, (b) side view of the workspace, (c) and (d) are the height maps of first and second layers

more details.

A. Global Planning Problem

The global planner assumes that we are given a geometric description of the objects in the workspace as well as the geometric and kinematic description of a humanoid. The workspace contains multiple layers, and each layer is comprised of objects of various heights. Unlike the basic path-planning problem where the definition of obstacles is rather straightforward, the obstacles in our global planning problem are not explicitly given. Instead, an object is an obstacle to a humanoid only if there is no way for the humanoid to step onto or pass under the object due to the humanoid's height. In addition, a humanoid must stand on a large enough area in order to maintain a stable stance. If the ground of the workspace is described as a smooth surface, the slope of the surface cannot be too large to cause foot slippery. In summary, the planning problem is rather complex in real life, and we need to make reasonably assumptions to simplify the problem.

First, we assume a discrete workspace. The input to our planner could be a continuous function for the elevation of the ground and a polygonal description of the objects. However, we assume that we can convert these descriptions into several layers of elevation grids of some resolution. Each cell in a grid contains an elevation value for the whole cell in that layer. An example of workspace with the layered environment is shown in Fig. 2. The elevation for each cell in a layered grid (128x128) is represented by a gray-scale value in Fig. 2(c) and 2(d). We denote the height of a cell i at layer l by c_i^l, and the offset of layer l from some reference ground by d^l. Second, we assume that the resolution of the elevation grid is coarse enough for a humanoid's foot to step onto a cell. We also assume that the maximal height that a humanoid can step onto is denoted by h, which is a property of the given humanoid. Third, the height of the humanoid is H, and we assume that the humanoid does not bend its body to pass an obstacle for now. Fourth, we assume that a humanoid will not stay in the object boarder region for more than some designated units of time, m. This situation happens when the geometry of a humanoid intersects the boarder. This assumption is to make sure that the humanoid does not stay in the border region except for trespassing purpose. Fifth, we assume that the geometry of the humanoid can be simplified to an enclosing circle of radius r such that the orientation dimension can be ignored at planning time. We assume that a humanoid will always face forward and we can recover its orientation in a postprocessing step.

In summary, the objective of the global motion planner is to find a collision-free path for the body trunk of the humanoid to move from the initial configuration to the goal configuration in a two-and-half-dimensional space. The output of the planner is a global path that will be sent to the local planner for further processing.

B. Local Planning Problem

The local planner aims to find a feasible locomotion for the lower body of a humanoid with a given global path. We assume that the output path from the global planner is a 3D stepwise curve. This curve is a polyline comprised of a set of vertical or horizontal connected line segments. In other words, we temporarily ignore the orientation change of the path and stretch the path into a one-dimensional stair-like profile. According to the kinematic parameters of the humanoid, the local planner will generate a *feasible* and *efficient* plan for footstep placement and the corresponding locomotion for lower-body joints. A feasible motion plan must satisfy geometric and kinematic constraints. For example, the humanoid should be collision-free and all joints are within their joint limits. However, we do not use any explicit dynamics model for simplicity reasons, and we assume that this simplification does not cause dynamics feasibility problems in normal walking motions. By efficient plans, we mean that the path should be the most efficient in terms of energy consumption. An efficient motion usually also means a natural motion that a human normally takes.

The local planning problem described above is challenging because the number of possible arrangements (each arrangement consists of a set of footsteps) grows exponentially in the length of the global path (or number of footsteps) even if we restrict the possible footstep sizes to a limited number. However, according to our daily walking experience, we typically plan foot placement only for the next two or three steps instead of for the whole path. Therefore, it is reasonable to take an incremental approach where we call the local planner in every step to plan only for a few steps (two or three, typically) ahead. Another advantage of this approach is that we can allow the configuration of obstacles to change at run time without calling for global replanning immediately as long as the change does not prevent the local planner from generating feasible locomotion. Thus, we will redefine our local planning problem as finding a feasible locomotion for the next n steps with a given path profile. The planner should return failure and indicate the failure location along the path if it cannot find a feasible locomotion plan for the next n steps.

IV. GLOBAL MOTION PLANNING

We will now present our approach to solving the global motion-planning problem. In addition to being collision-free, all configurations along the path must be reachable according to the kinematic constraints, such as joint limits, of the humanoid robot. The path must also satisfy the stability constraint requiring that any continuous portion of the path cannot stay in the border region for longer than some period of time. In the following subsections, we will first compute a reachability map and then a collision map to represent the properties of the grids in the configuration space.

A. Reachability Map and Collision Map

Suppose that we are given the heights and offsets for a set of objects in the workspace. Offset is the base elevation where the object is placed. Different offsets mean different layers. For each layer, we compute a height map containing the elevation value of each cell above the layer in the workspace grid. As indicated in the previous section, a cell is considered an obstacle cell if and only if there are no ways to reach the cell from its neighbors under the height constraint. A cell could be unreachable because its height difference with its neighbors is too large for a humanoid to step onto or the clearance between this layer and the layer above is too small for a humanoid to fit in.

Given an initial configuration of the humanoid, we can compute a map, called *reachability map*, where obstacle regions are comprised of the unreachable cells. A reachability map consists of several slices with one slice for each layer. For example, Fig. 3(a) and 3(b) show the reachability map for the two layers of the example scene in Fig. 2. The black regions are the unreachable regions marked as obstacles. These slices could be "connected" if there exists an object whose height is large enough to bring the humanoid to step onto some neighboring cells of the above or below layer. We compute this map by a wave propagation algorithm similar to the one used to construct NF1 potential fields [9]. Suppose that the current cell under propagation is i. The algorithm advances to a neighboring cell i' at layer l or a neighboring layer l' only if the height difference between them is less than h ($\left|c_i^l - c_{i'}^l\right| < h$ or $\left|(d^l + c_i^l) - (d^{l'} + c_{i'}^{l'})\right| < h$) and the humanoid height H can fit into the clearance above cell i' at layer l or l' ($\left|d^{l+} - d^l - c_{i'}^l\right| > H$ or $\left|d^{l'+} - d^{l'} - c_{i'}^{l'}\right| > H$, where $l+$ denotes the layer above l). We can convert this map, built in

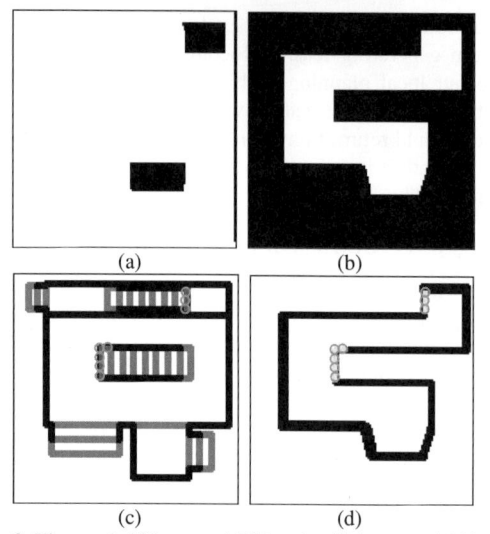

Fig. 3: The reachability map (a)(b) and collision map (c)(d) for layer one and two of the environment in Fig. 2.

```
STABLE_BFP()
1   install q_i in T;
2   INSERT(q_i, OPEN); mark q_i visited;
3   SUCCESS ← false;
4   while ¬ EMPTY(OPEN) and ¬ SUCCESS do
5     q ← FIRST(OPEN);
6     for every neighbor q' of q in the grid do
7       if LEGAL(q') then
8         if q' is unstable then
9           q'.cnt = q.cnt+1;
10        else
11          mark q' visited;
12        install q' in T with a pointer to q;
13        INSERT(q', OPEN);
14        if q' = q_g then SUCCESS ← true;
15  if SUCCESS then
16    return the backtracked feasible path
17  else return failure;
```

Fig. 4: The STABLE_BFP algorithm

the workspace, into its corresponding C-space by growing the obstacle regions with the humanoid's radius r. The resulting C-space map can then be used to build a potential field to guide the search in the planning process.

As mentioned in the previous subsection, we need to identify the regions where unstable situation might occur if a humanoid stay there for too long. A map describing the regions is called *collision map*. A cell in the collision map is defined as *unstable* if and only if the region covered by the enclosing circle of a humanoid contains cells with height difference less than h. The region comprised of unstable cells is called the *unstable region*. A cell is defined as *forbidden* if there exists height difference larger than h. The remaining cells in the collision map are the *free* cells. One can compute such a map by first identifying the cells with different heights in their neighborhood horizontally and vertically. We can then grow these cells by the radius r to form the final collision map. These two maps are used in the planning algorithm presented in the next subsection.

B. Path Planning Algorithm

The planning algorithm for computing the humanoid motion is shown Fig. 4. The STABLE_BFP algorithm is similar to the classical Best-First Planning (BFP) algorithm [9] for low-DOF problems. In an iteration of the search loop, we use the FIRST operation to select the most promising configuration q from the list of candidates (OPEN) for further exploration. We visit each neighbor q' of q and check their validity (via the LEGAL operation) for further consideration. A configuration is legal if it is collision-free (in the freespace of reachability map), marked unvisited, and temporarily stable. It is temporarily stable if and only if the humanoid has not entered the unstable region for a period longer than a given threshold determined by the user. This duration is kept as an instability counter for each cell in the unstable region when we propagate nodes in it. Note that the validity of a configuration in the unstable region depends on the instability counter of the parent configuration. If there are more than one possible parent configurations, we cannot exclude any of them. Therefore, in the STABLE_BFP algorithm, we do not mark a configuration visited if it is in the unstable region. A configuration in this region can be visited multiple times as long as its instability counter does not exceed the maximal value.

In the STABLE_BFP algorithm, we use FIRST operation to select the most promising configuration for further exploration. In the BFP planners, an artificial potential field is usually the only index for goodness. Planners with this approach can usually yield short paths. In our case, the height difference could be an important index as well since one may prefer climbing up or stepping down stairs to taking a longer path. Therefore, in the FIRST operation, we use a linear combination of both criteria (potential for horizontal measures and height difference from q_i for vertical measures) with weights specified by the user.

C. Postprocessing

If our planner succeeds in finding a feasible path for a humanoid, we need to perform two tasks in the postprocessing step. First, we convert the found path into a smooth one via a smoothing routine. The smoothing algorithm is very similar to the ones for typical path planners. One usually replaces a subpath in the original path continuously with a straight-line path segment. A major difference for smoothing a path in our new planner is on the metric for measuring distance. This metric is defined with the same criteria as in the FIRST procedure such that the user preference can be preserved. Finally, we have to recover the orientation parameter of the humanoid for each step in the path. If the path contains a sharp turn that is hard for a humanoid to follow, we can add additional steps into the path to slow down the orientation change.

V. LOCAL MOTION PLANNING

From the problem definition described in the previous section, we assume that the local planner is asked to plan an efficient motion only for the next few steps (two or three, in our settings) instead of for the whole global path. We will first describe how we compute a collision-free walk locomotion for a given footstep length and a path profile. Then we will describe how to plan the foot placement for future steps.

A. Kinematics-Based Locomotion Generation

In this subsection, we will describe how we generate a collision-free locomotion for humanoid walking with a given foot location for the next step. We use an inverse-kinematics approach to compute the locomotion. According to [3], walking motion for a human figure can be divided into two phases: *single support* and *double support*. In the double support phase, both legs touch the ground and mass center of the body shift gradually from the back leg to the front leg. In the single support phase, a stretched leg supports the body while the other leg swings forward. Keyframes are defined at the conjunction of the two phases or in the interior of the double support phase.

We use the following principles to compute the joint angles between two keyframes. In either phase, the stretched leg on the ground determines the pelvis location. In the case of double support, the joint angle of the other leg is determined by inverse kinematics between the foot and pelvis. In the single support phase, the foot trajectory for the swinging leg in the Cartesian space is determined first and the joint angles are then computed according to inverse kinematics.

Now, the remaining problem is how to compute a collision-free trajectory for the swinging foot. We use a Bezier curve with two fixed endpoints on the ground to represent the trajectory. The other two control points of a Bezier curve become the parameters that we can adjust to avoid collision with obstacles. We use a numerical approach that move the control points iteratively until a legal curve is found. However, the four-dimensional space spanned by the two control points is too large to search in an on-line manner. Therefore, we use the following rule to find the collision-free curve as quickly as possible and avoid an exhaustive search.

The locations of the control points start at some location slightly above the midpoint between two endpoints. When a collision between the curve and the environment is detected, the control points are moved upward and outward to lift the curve as shown in Fig. 5. The curve is divided into two parts (left and right) from the summit point. Each part computes the intersected points of the curve and moves the corresponding control point for some distance along the direction from the midpoint of two endpoints toward the intersection points. If one part does not cause intersection while the other does, its control point also moves upward for the same

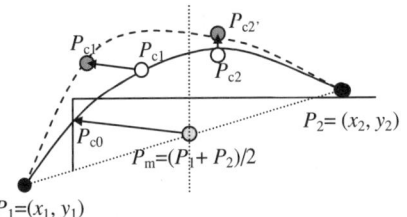

Fig. 5: Adjusting the Bezier curve for the swinging leg to avoid collisions

amount as the other control point moves. The search process stops when a legal curve is found or the locations of the control points have left a given legal region. In the second case, the planner returns failure.

B. Footstep Planning

We now consider the problem of finding foot placement. Assume that the local planner is called in every step of execution, and we will search for the most energy efficient locomotion for the next n steps, where n is set to, say, 3. We first discretize the range of all possible footstep sizes into m unit lengths for each step. Then we try to find a set of desirable footstep locations, called *footstep configuration*, denoted by $q_f=(s_1, s_2, s_3)$, where s_1, s_2, and s_3 are step lengths. The energy efficiency for q_f, denoted by $\mathit{Eff}(q_f)$, is defined by $(s_1+s_2+s_3)/E$, where E is the total energy consumption that will be defined later.

Depending on the available time for planning, one may choose to search for the global maximal or a local maximal configuration. In our on-line application, since the planner is called in every step of execution at run time, time efficiency is crucial. Therefore, we use a Best-First Search algorithm to find a collision-free locomotion plan for the next n steps with the maximal energy efficiency only locally. The algorithm starts from some neutral position initially and explores its neighbors for better configurations. We maintain a list of available configurations and choose the most promising configuration in each step for further exploration until a local maximal is reached. The program returns failure if no such a footstep configuration exists. In an iteration of the execution loop, we plan for the next few steps by taking the goal/current footstep configuration as the starting point for the next iteration. In most occasions, the previous footstep configuration usually provides a good initial guess that can quickly bring the search to a local maximum.

The energy E consumed in one step depends on several factors. For example, it depends on the walking speed, vertical and horizontal movements, and joint movements. Although the energy consumption for a given humanoid robot can be computed with a dynamic model, our real-time planning requirement discourages such an approach. Instead, we take a simpler approach of computing this energy from statistic data obtained from real human walking experiments. We assume that the energy consumption is composed of three parts: horizontal movement, vertical movement, and

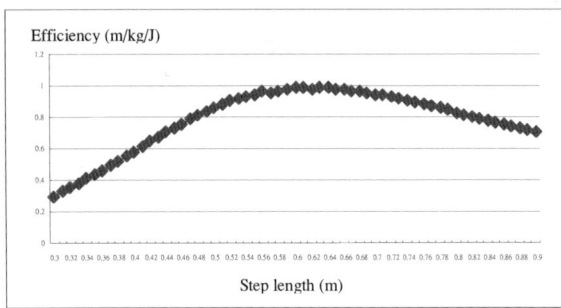

Fig. 6. Energy efficiency for different step lengths on a flat ground

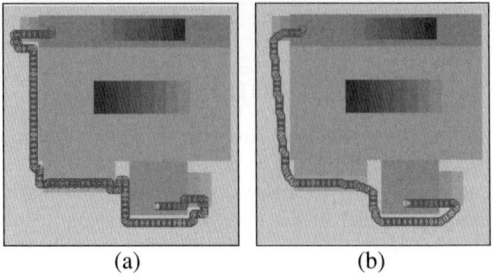

Fig. 7. An example of (a) unsmoothed and (b) smoothed global path in a layered scene

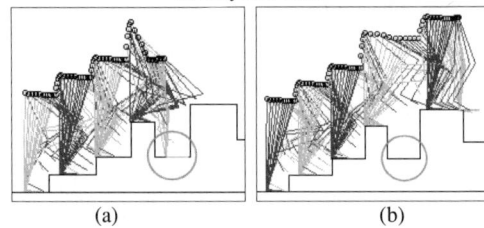

Fig. 8. Locomotion planning: (a) fails with one-step planning and (b) succeeds with two-step planning

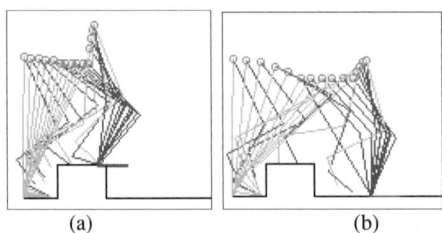

Fig. 9. Efficient locomotion generation: (a) stepping on is less energy efficient than (b) stride over

joint movement. That is, $E=\alpha*E_n+\beta*E_h+\gamma*E_r$, where E_n, E_h, and E_r are the energy for horizontal movement, vertical movement, and joint movement, respectively. We assume a normal walking speed so that we have a common ground for comparison. According to [12], the energy for normal walking on a flat ground can be computed with the following formulas: $E_n = E_w/n = (32+102*s^4)/n$, where s is the step size and n is the number of steps per minute. Experiments show that $s/n=0.007$ for adult male. Therefore, E_n can be simplified to be a function of s only. According to [12], the energy consumption for stepping upward is proportional to the stepping height. Third, for the same footstep size, a humanoid could raise legs for different amounts and therefore consume different amounts of energy. We take a weighted sum of changes on the joint angles to compute the extra energy consumption, E_r. The weights are mainly determined by the masses of the raised parts. With these formulas, we can compute the energy efficiency for different footstep lengths, as shown in Fig. 6. Since the statistic data is taken from experiments with real humans, the most efficient step length is around 60cm, which agrees with our daily experiences.

VI. EXPERIMENTS

We have implemented the global and local motion planners in Java and connected the planners to a VRML browser to display the final simulation results. All the planning times reported below are taken from experiments run on a regular 650MHz PC. The resolutions for the grid workspace and configuration space are all 128x128 in the global planner. The resolution for the locomotion is on the discretization of footstep length, which is divided into 13 units in the range of 30cm to 90cm. In the following subsections, we will use several examples to demonstrate the effectiveness of the planners.

A. An Example for Global Motion Planning

Fig. 7 shows an example of global motion generated for the scene of Fig. 2. Fig. 7(a) and 7(b) show the unsmoothed and smoothed global paths that take the humanoid from a stage to the other side of the platform.. Note that there are two ways to get to the goal location, and the planner generates the shorter path to reach the goal. The planning time for such an example is 220ms for preprocessing computation (such as building the reachability map, collision map, and potential field) and 20ms for path searching and smoothing.

B. An Example for Local Motion Planning

In Fig. 8, we show an example of local motion plan that takes the lower body of a humanoid to climb up a stair with a given ground profile. In the example of Fig. 8(a), the humanoid fails to make further moves after stepping down the stair because it only plans one step ahead. In Fig. 8(b), we plan two steps ahead and therefore avoid the problem by taking a larger step length. In Fig. 9, we compare the planning results with and without considering energy efficiency. For example, according to the energy model described in Section V, the motion in Fig. 9(b) is more efficient than the motion in Fig. 9(a) because it does not need to step onto the obstacle.

C. Integrated Example

We use the integrated example in Fig. 10 to demonstrate the interaction between the global and the local planners. The global motion generated initially upon a user request

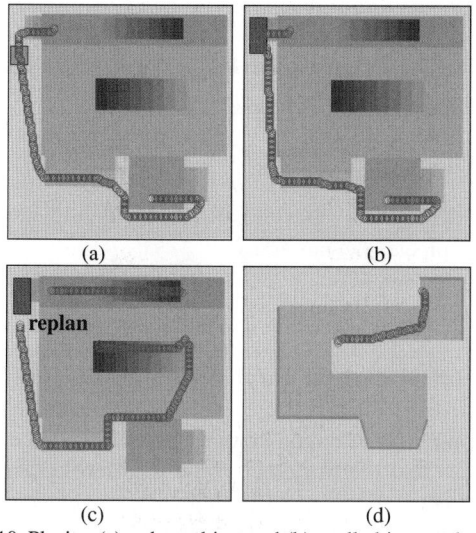

Fig. 10. Placing (a) a short object and (b) a tall object on the path at execution time. (c)(d) An alternative path is generated.

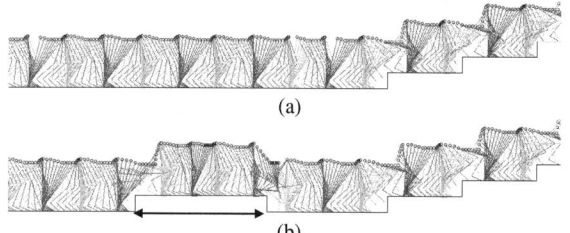

Fig. 11. Locomotion adapting to a dynamically placed object

may not succeed at execution time when it is sent to the local motion planner for further processing. For example, an object could be placed on the way of the global path at execution time as shown in Fig. 10(a) and 10(b). Although the new green square object in Fig. 10(a) is on the path, the local planner generates a locomotion plan that steps over the object, as shown in Fig. 11(b) (compared to the original plan in Fig. 11(a)). If the placed object is too tall (Fig. 10(b)), the local planner may fail at execution time. In this case, the local planner returns the failure point to the global planner in order to generate an alternative feasible path via the second floor, as shown in Fig. 10(c) and 10(d). The re-planning takes about 761ms including rebuilding all the maps.

VII. CONCLUSIONS AND FUTURE WORK

Building autonomous humanoid robots has been the goal of many applications in robotics and computer animation. Spatial reasoning is a key capability to enable a robot to accept high-level commands and move autonomously. In this paper, we have proposed a planning system composed of global and local motion planners that can generate feasible and efficient motion plans for a humanoid in a layered environment. Several simulation examples have been presented to demonstrate the capabilities of the planner.

In the current system, we assume that a humanoid robot can only perform normal walking motions. However, a real human usually can avoid obstacles with various kinds of locomotion and body motions. For example, a human can crawl or stoop to avoid upper-layer obstacles. We will consider this kind of situations as an extension of our work in the future.

VIII. ACKNOWLEDGMENT

This work was partially supported by National Science Council under contract NSC 91-2213-E-004-005.

IX. REFERENCES

[1] J. Barraquand, L. Kavraki, J.C. Latombe, T.Y. Li, and P. Raghavan, "A Random Sampling Scheme for Path Planning," *Intl. J. of Robotics Research*, 16(6), Dec. 1997.

[2] J. Barraquand and J. Latombe, "Robot Motion Planning: A Distributed Representation Approach," *Intl J. of Robotics Research*, 10:628-649, 1991.

[3] A. Bruderlin and T. W. Calvert, "Goal-Directed, Dynamic Animation of Human Walking," *Proc. of ACM SIGGRAPH*, 1989.

[4] Q. Huang, K. Kaneko, et al., "Balance Control of a Biped Robot Combining Off-line Pattern with Real-time Modification," *Proc. of IEEE Intl. Conf. on Robotics and Automation*, pp.3346-3352, April, 2000.

[5] H. Ko and N.I. Badler, "Animating Human Locomotion with Inverse Dynamics," *IEEE Transaction on Computer Graphics*, 16(2), pp.50-59. 1996.

[6] J. Kuffner, "Goal-Directed Navigation for Animated Characters Using Real-time Path Planning and Control" *Proc. of CAPTECH'98 Workshop on Modeling and Motion capture Techniques for Virtual Environments*, Springer-Verlag, 1998.

[7] J. Kuffner, et. al., "Motion Planning for Humanoid Robots under Obstacle and Dynamic Balance Constraints," *Proc. of IEEE Intl. Conf. on Robotics and Automation*, May 2001.

[8] J. Kuffner, et. al., "Footstep Planning Among Obstacles for Biped Robots," *Proc. of 2001 IEEE Intl. Conf. on Intelligent Robots and Systems (IROS 2001)*, 2001.

[9] J. Latombe, *Robot Motion Planning*, Kluwer, MA, 1991.

[10] N. Pollard, et. al., "Adapting Human Motion for the Control of a Humanoid Robot," *Proc. of 2002 IEEE Intl. Conf. on Robotics and Automation*, pp2265-2270, May 2002.

[11] J.H. Reif, "Complexity of the Mover's Problem and Generalizations," *Proc. of the 20th IEEE Symp. on Foundations of Computer Science*, pp. 421-427, 1979.

[12] J. Rose and J.G. Gamble, *Human Walking*, Williams and Wilkins, 1994.

[13] Z. Shiller, K. Yamane, Y. Nakamura, "Planning Motion Patterns of Human Figures Using a Multi-Layered Grid and the Dynamics Filter" *Proc. of IEEE Intl. Conf. on Robotics and Automation*, pp.1-8, May 2001.

[14] H. C. Sun and N. M. Dimitris, "Automating gait generation," *Proc. of ACM SIGGRAPH*, 2001.

[15] S. Tak, O. Song, and H.-S. Ko, "Motion Balance Filtering," *Proc. of the Eurographics Conf.*, 2000.

[16] A. Witkin and Z. Popovic, "Motion Warping," *Computer Graphics Proc.*, SIGGRAPH95, pp.105-108, 1995.

[17] K. Yamane and Y. Nakamura, "Dynamics Filter – Concept and Implementation of On-Line Motion Generator for Human Figures," *Proc. of IEEE Intl. Conf. on Robotics and Automation*, pp.688-695, April 2000.

Enhancing the Reactive Capabilities of Integrated Planning and Control with Cooperative Extended Kohonen Maps

Kian Hsiang Low[†], Wee Kheng Leow[†], and Marcelo H. Ang Jr.[‡]

[†]Dept. of Computer Science, and [‡]Dept. of Mechanical Engineering, National University of Singapore, Singapore

Email: [†]ieslkh, leowwk@comp.nus.edu.sg, [‡]mpeangh@nus.edu.sg

Abstract— Despite the many significant advances made in robot motion research, few works have focused on the tight integration of high-level deliberative planning with reactive control at the lowest level. In particular, the real-time performance of existing integrated planning and control architectures is still not optimal because the reactive control capabilities have not been fully realized. This paper aims to enhance the low-level reactive capabilities of integrated planning and control with Cooperative Extended Kohonen Maps for handling complex, unpredictable environments so that the workload of the high-level planner can be consequently eased. The enhancements include fine, smooth motion control, execution of more complex motion tasks such as overcoming unforeseen concave obstacles and traversing between closely spaced obstacles, and asynchronous execution of behaviors.

I. INTRODUCTION

Robot motion research has proceeded along two dichotomous streams: high-level deliberative planning and low-level reactive control. Deliberative planning uses a world model to generate an optimal sequence of collision-free actions that can achieve a globally specified goal in a complex static environment. However, in a dynamic environment, unforeseen obstacles may obstruct the action sequence, and replanning to react to these situations can be too computationally expensive. Reactive control directly couples sensed data to appropriate actions. It allows the robot to respond robustly and timely to unexpected obstacles and environmental changes but may be trapped by them. These two paradigms have their own strengths and weaknesses, which are rather complementary. Their union into one coherent integrated framework can potentially mitigate their respective drawbacks and yield the best of both approaches.

Nevertheless, developing such a unification methodology is non-trivial and recent proposals lack the capacity for real world use. In particular, the real-time performance of existing integrated planning and control architectures is still not optimal because the reactive control capabilities have not been fully realized. Often, the workload of the high-level planner far exceeds that of the low-level reactive controller ([6], [9]). The planner produces the exact motion path with detailed sequence of actions to be executed by the actuators. The reactive controller performs only a single task, i.e., simple local obstacle avoidance, by correcting the course of action.

Hence, the work presented in this paper aims to augment the low-level reactive capabilities of integrated planning and control for handling complex, unpredictable environments so that the workload of the high-level planner can be consequently eased. The following key enhancements distinguish our framework from existing architectures:

A. Perform fine, smooth and efficient motion control

A high degree of smoothness and precision in motion control is essential to the efficient execution of sophisticated tasks and the social interaction with humans. This can only be achieved with *continuous response encoding* (i.e., infinite set of responses) of very low-level velocity/torque control of motor/joint actuators. To do so, an alternative would be to encode all possible action combinations ([12], [14]) but it becomes computationally intractable with higher degrees of freedom. Our proposed architecture trains a self-organizing neural network to continuously sample the low-level configuration space. Other integrated architectures ([2], [9]) utilize potential fields [8] in their reactive controllers to encode continuous responses, which are subject to local minima problems [10]. In contrast, integrated architectures ([15], [16]) that employ *discrete response encoding* (i.e., finite, enumerated set of responses) encode high-level motion commands (e.g., {forward, left, right, ..., etc.}), which may not be physically realizable due to negligence of kinematic constraints (e.g., non-holonomy). Interpolation of these discrete commands may incur problems similar to potential fields.

B. Perform sophisticated motion tasks

Extensive simplification of the reactive mechanisms for selecting actions may unnecessarily deflate the reactive capabilities to achieving only simple tasks because information useful to selecting actions has not been fully exploited. For example, arbitration strategies [7] only allow one winning behavior among a group of competing ones to assume full control of the robot until the next selection cycle. This precludes the execution of several, possibly conflicting motion tasks (e.g., target reaching and obstacle avoidance) in parallel. Superposition techniques ([1], [4], [8]) perform a vector sum of action commands, each optimal to its respective behavior, to produce a combined output that may not guarantee the satisfaction of the overall motion task.

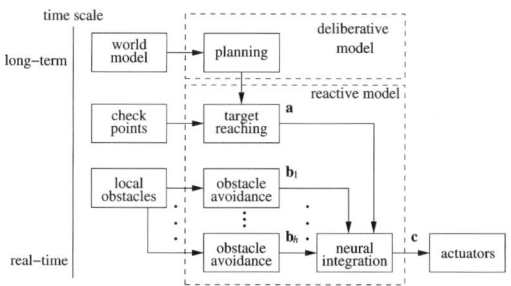

Fig. 1. A framework for the tight integration of planning, target reaching, and obstacle avoidance.

Problems of local minima and no passage between narrowly spaced obstacles may arise [10]. Orientation selection models ([3], [16]) face similar problems as distance information is not considered. This class of methods allow a robot with long range sensors to detect and avoid complex obstacles but the distant presence of a narrow doorway may be missed due to poor resolution at long range. It also cannot slow down while approaching obstacles. Our architecture presents a versatile action selection mechanism that utilizes both distance and angle information to execute complex motion tasks (e.g., avoiding complex obstacles) requiring a multitude of concurrently active behaviors.

C. Achieve asynchronous execution of behavioral modules

The asynchronous execution of behaviors, each at its fastest rate possible, is crucial to the preservation of reactive capabilities. Integrated architectures (e.g., [15], [16]) that coordinate their behaviors to operate directly on the action representation require them to be synchronized to produce a meaningful action. On the other hand, our architecture utilizes a time-independent action selection module to learn a neural map representation of the local workspace that can interface with asynchronized behaviors.

These enhancements to the reactive capabilities simplify the deliberative planner such that it is only required to produce a sequence of checkpoints in global workspace instead of a complete motion path in configuration space. The constraint of adhering strictly to a generated path is removed.

II. INTEGRATED PLANNING & CONTROL FRAMEWORK

A. Overview

Our integrated planning and control framework is illustrated in Fig. 1. At the highest level, the *planning* module produces a sequence of checkpoints from the start point to the goal using an approximate cell decomposition method in [13]. However, our algorithm operates in the robot's workspace instead of the configuration space. In essence, the free workspace is decomposed into much fewer cells than do other decomposition techniques to reduce the search time. Any two points in the cell can be traversed by reactive motion. This method is elaborated in a separate paper [11].

The reactive model consists of *Cooperative Extended Kohonen Maps* (EKMs), which are trained to produce a collision-free sequence of low-level (motor velocity) control commands that move the non-holonomic mobile robot from one checkpoint to the next. The *target reaching* module contains a *target localization EKM* that provides excitatory inputs to the *motor control EKM* in the *neural integration* module at and around locations of the sensed checkpoint. The *obstacle avoidance* modules contain *obstacle avoidance EKMs*, which provide inhibitory inputs to the motor control EKM at and around locations where obstacles are detected. The motor control EKM in the neural integration module combines these inputs from the EKMs and uses them to select appropriate actions that can negotiate unforeseen obstacles while reaching targets.

All the modules operate asynchronously at different rates. The planning module typically operates at the time scale of several seconds or minutes depending on task complexity. The target reaching module operates at about 256 ms between servo ticks while the obstacle avoidance module operates at intervals of 128 ms. The neural integration module is activated as and when neural activities are received. The asynchronous execution of modules is the key to preserving reactive capabilities while allowing improvement of performance by the deliberative planner. In fact, the planner can be removed and the resulting decapitated architecture degrades to a purely reactive system capable of less complex motion tasks.

B. Target Reaching

The target reaching module adopts an egocentric representation of the sensory input vector $\mathbf{u}_p = (\alpha, d)^T$ where α and d are the direction and the distance of a checkpoint relative to the robot's current location and heading. At the goal state at time T, $\mathbf{u}_p(T) = (\alpha, 0)^T$ for any α. It uses the target localization EKM to self-organize the sensory input space $\mathcal{U}$. Each neuron i in the EKM has a sensory weight vector $\mathbf{w}_i = (\alpha_i, d_i)^T$ that encodes a region in $\mathcal{U}$ centered at $\mathbf{w}_i$. Based on each incoming sensory input $\mathbf{u}_p$, the corresponding neuronal activities are sent to the motor control EKM in the neural integration module (Section II-D).

Target Localization

Given a sensory input vector $\mathbf{u}_p$ of a target location,

1) Determine the winning neuron s in the target localization EKM. Neuron s is the one whose sensory weight vector $\mathbf{w}_s = (\alpha_s, d_s)^T$ is nearest to the input $\mathbf{u}_p = (\alpha, d)^T$:

$$D(\mathbf{u}_p, \mathbf{w}_s) = \min_{i \in \mathcal{A}(\alpha)} D(\mathbf{u}_p, \mathbf{w}_i). \quad (1)$$

The difference $D(\mathbf{u}_p, \mathbf{w}_i)$ is a weighted difference

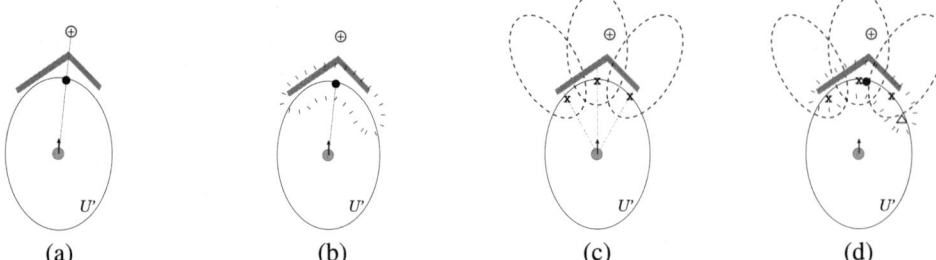

Fig. 2. Cooperative EKMs. (a) In response to the target ⊕, the nearest neuron (black dot) in the target localization EKM (ellipse) of the robot (gray circle) is activated. (b) The activated neuron produces a target field (dotted ellipse) in the motor control EKM. (c) Three of the robot's sensors detect obstacles and activate three neurons (crosses) in the obstacle localization EKMs, which produce the obstacle fields (dashed ellipses). (d) Subtraction of the obstacle fields from the target field results in the neuron at △ to become the winner in the motor control EKM, which moves the robot away from the obstacle.

between $\mathbf{u}_p$ and $\mathbf{w}_i$:

$$D(\mathbf{u}_p, \mathbf{w}_i) = \beta_\alpha (\alpha - \alpha_i)^2 + \beta_d (d - d_i)^2 \quad (2)$$

where β_α and β_d are constant parameters. The minimum in Eq. 1 is taken over the set $\mathcal{A}(\alpha)$ of neurons encoding very similar angles as α:

$$|\alpha - \alpha_i| \leq |\alpha - \alpha_j|, \quad \text{for each pair } i \in \mathcal{A}(\alpha), j \notin \mathcal{A}(\alpha) . \quad (3)$$

In other words, direction has priority over distance in the competition between EKM neurons. This method allows the robot to quickly orient itself to face the target while moving towards it. In the EKM, each neuron encodes a location $\mathbf{w}_i$ in the sensory input space $\mathcal{U}$. The region of $\mathcal{U}$ that encloses all the neurons is called the *local workspace* $\mathcal{U}'$. Even if the target falls outside $\mathcal{U}'$, the nearest neuron can still be activated (Fig. 2a).

2) Compute output activity a_i of neuron i in the target localization EKM.

$$a_i = G_a(\mathbf{w}_s, \mathbf{w}_i) \quad (4)$$

The function G_a is an elongated Gaussian:

$$G_a(\mathbf{w}_s, \mathbf{w}_i) = \exp(-(\frac{\alpha_s - \alpha_i}{\sigma_{a,\alpha}})^2 - (\frac{d_s - d_i}{\sigma_{a,d}})^2). \quad (5)$$

Parameter $\sigma_{a,d}$ is much smaller than $\sigma_{a,\alpha}$, making the Gaussian distance-sensitive and angle-insensitive. These parameter values elongate the Gaussian along the direction perpendicular to the target direction α_s (Fig. 2b). This elongated Gaussian is the *target field*, which plays an important role in avoiding local minima during obstacle avoidance.

C. Obstacle Avoidance

Each obstacle avoidance module contains an obstacle localization EKM that is self-organized in the same way as the target localization EKM (Section II-E); each neuron i in the obstacle localization EKMs has the same input weight vector $\mathbf{w}_i$ as the neuron i in the target localization EKM. The robot has h directed distance sensors around its body for detecting obstacles. Hence, each activated sensor encodes a fixed direction α_j and a variable distance d_j of the obstacle relative to the robot's heading and location. Each sensed input $\mathbf{u}_j = (\alpha_j, d_j)^T$ induces neuronal activities, which are sent to the motor control EKM in the neural integration module (Section II-D).

Obstacle Localization

For each sensed input $\mathbf{u}_j, j = 1, \ldots, h$,
1) Determine the winning neuron s' in the jth obstacle localization EKMs. Each sensor input $\mathbf{u}_j$ activates a winning neuron s' in the jth obstacle localization EKM, which is activated in the same manner as Step 1 of Target Localization (Section II-B).
2) Compute output activity $b_{i,j}$ of neuron i in the jth obstacle localization EKMs:

$$b_{i,j} = G_b(\mathbf{w}_{s'}, \mathbf{w}_i) \quad (6)$$

where

$$G_b(\mathbf{w}_{s'}, \mathbf{w}_i) = \exp(-(\frac{\alpha_{s'} - \alpha_i}{\sigma_{b,\alpha}})^2 - (\frac{d_{s'} - d_i}{\sigma_{b,d}(d_{s'}, d_i)})^2)$$

$$\sigma_{b,d}(d_{s'}, d_i) = \begin{cases} 3.5 & \text{if } d_i \geq d_{s'} \\ 0.035 & \text{otherwise.} \end{cases}$$
$$(7)$$

The function G_b is a Gaussian stretched along the obstacle direction $\alpha_{s'}$ so that motor control EKM neurons beyond the obstacle locations are also inhibited to indicate inaccessibility (Fig. 2c). If no obstacle is detected, $G_b = 0$. In the presence of an obstacle, the neurons in the obstacle localization EKMs at and near the obstacle locations will be activated to produce *obstacle fields* (Eq. 6). The neurons nearest to the obstacle locations have the strongest activities.

D. Neural Integration

The neural integration module contains a motor control EKM, which integrates the excitatory and inhibitory inputs from the neurons in the target and obstacle localization

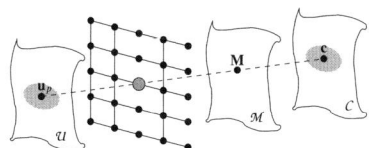

Fig. 3. Motor control EKM. The neurons map the sensory input space $\mathcal{U}$ indirectly to motor control space $\mathcal{C}$ through control parameter space $\mathcal{M}$.

EKMs respectively. It is trained to partition the sensory input space $\mathcal{U}$ into locally linear regions. Each neuron i in the motor control EKM is self-organized in the same way as the target and obstacle localization EKMs by encoding the same input weight vector $\mathbf{w}_i$ as the neuron i in those EKMs. It also has a set of output weights which encode the outputs produced by the neuron. However, unlike existing direct-mapping methods [17], the output weights $\mathbf{M}_i$ of neuron i of the motor control EKM represents control parameters in the parameter space $\mathcal{M}$ instead of the actual motor control vector (Fig. 3). The control parameter matrix $\mathbf{M}_i$ is mapped to the actual motor control vector $\mathbf{c}$ by a linear model (Eq. 10). Compared to direct-mapping EKM, indirect-mapping EKM can provide finer and smoother motion control. Detailed comparison and discussion have been reported in [11]. With indirect-mapping EKM, motor control is performed as follows:

Motor Control
1) Compute activity e_i of neuron i in the motor control EKM.
$$e_i = a_i - \sum_{j=1}^{h} b_{i,j} \quad (8)$$
where a_i is the excitatory input from the neuron in the target localization EKM (Section II-B) and $b_{i,j}$ is the inhibitory input from the neuron in the jth obstacle localization EKM (Section II-C).
2) Determine the winning neuron k in the motor control EKM. Neuron k is the one with the largest activity:
$$e_k = \max_i e_i . \quad (9)$$
3) Compute motor control vector $\mathbf{c}$ for target reaching:
$$\mathbf{c} = \begin{cases} \mathbf{M}_k \mathbf{u}_p & \text{if } -\mathbf{c}^* \leq \mathbf{M}_k \mathbf{u}_p \leq \mathbf{c}^* \text{ and } k = s \\ \mathbf{M}_k \mathbf{w}_k & \text{otherwise.} \end{cases} \quad (10)$$

The constant vector $\mathbf{c}^*$ denotes the upper limit of physically realizable motor control signal. For instance, for the Khepera robots, $\mathbf{c}$ consists of the motor speeds v_l and v_r of the robot's left and right wheels. In this case, we define $\mathbf{c} \leq \mathbf{c}^*$ if $v_l \leq v_l^*$ and $v_r \leq v_r^*$. Note that if $\mathbf{c}$ is beyond $\mathbf{c}^*$, simply saturating the wheel speeds does not work. For example, if the target is far away and not aligned with the robot's heading, then saturating both wheel speeds only moves the robot forward. Without correcting the robot's heading, the robot will not be able to reach the target. Hence, the winning neuron's input weights $\mathbf{w}_k$ are used to generate the physically realizable motor control output. This motor control would be the best substitution for the sensory input $\mathbf{u}_p$ because $\mathbf{w}_k$ is closest to $\mathbf{u}_p$ compared to other weights $\mathbf{w}_i, i \neq k$.

In activating the motor control EKM (Fig. 2d), the obstacle fields are subtracted from the target field (Eq. 8). If the target lies within the obstacle fields, the activation of the motor control EKM neurons close to the target location will be suppressed. Consequently, another neuron at a location that is not inhibited by the obstacle field becomes most highly activated (Fig. 2d). This neuron produces a control parameter that moves the robot away from the obstacle. While the robot moves around the obstacle, the target and obstacle localization EKMs are continuously updated with the current locations and directions of the target and obstacles. Their interactions with the motor control EKM produce fine, smooth, and accurate motion control of the robot to negotiate the obstacle and move towards the target until it reaches the goal state $\mathbf{u}_p(T)$ at time step T.

Recall that the various modules run asynchronously at different rates (Section II-A). In particular, the obstacle avoidance module runs at a faster rate than the target reaching and robot separation modules. During neural integration, the localization EKMs remain activated until they are updated asynchronously at the next sensing cycle. So, the motor control EKM can receive continuous inputs from the localization EKMs and is always able to produce a motor signal as and when new inputs are sensed.

E. Self-Organization of EKMs

In contrast to most existing methods, online training is adopted for the EKMs. Initially, the EKMs have not been trained and the motor control vectors $\mathbf{c}$ generated are inaccurate. Nevertheless, the EKMs self-organize, using these control vectors $\mathbf{c}$ and the corresponding robot displacements $\mathbf{v}$ produced by $\mathbf{c}$, to map $\mathbf{v}$ to $\mathbf{c}$ indirectly. As the robot moves around and learns the correct mapping, its sensorimotor control becomes more accurate. At this stage, the same online training mainly fine tunes the indirect mapping. The self-organized training algorithm (in obstacle-free environment) is as follows:

Self-Organized Training
Repeat
1) Get sensory input $\mathbf{u}_p$.
2) Execute target reaching procedure and move robot.
3) Get new sensory input $\mathbf{u}'_p$ and compute actual displacement $\mathbf{v}$ as a difference between $\mathbf{u}'_p$ and $\mathbf{u}_p$.
4) Use $\mathbf{v}$ as the training input to determine the winning neuron k (same as Step 1 of Target Reaching).
5) Adjust the weights $\mathbf{w}_i$ of neurons i in the neighborhood $\mathcal{N}_k$ of the winning neuron k towards $\mathbf{v}$:
$$\Delta \mathbf{w}_i = \eta G(k,i)(\mathbf{v} - \mathbf{w}_i) \quad (11)$$

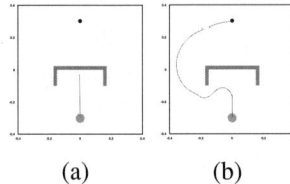

(a)   (b)

Fig. 4. Negotiating unforeseen concave obstacle. (a) The robot using Braitenberg obstacle avoidance and action superposition was trapped but (b) the one adopting the integrated architecture with cooperative EKMs successfully overcame the obstacle.

where $G(k,i)$ is a Gaussian function of the distance between the positions of neurons k and i in the EKM, and η is a constant learning rate.

6) Update the weights $\mathbf{M}_i$ of neurons i in the neighborhood $\mathcal{N}_k$ to minimize the error e:

$$e = \frac{1}{2}G(k,i)\|\mathbf{c} - \mathbf{M}_i\mathbf{v}\|^2 . \qquad (12)$$

That is, apply gradient descent to obtain

$$\Delta \mathbf{M}_i = -\eta \frac{\partial e}{\partial \mathbf{M}_i} = \eta\, G(k,i)(\mathbf{c} - \mathbf{M}_i\mathbf{v})\mathbf{v}^T . \qquad (13)$$

The target and obstacle localization EKMs self-organize in the same manner as the motor control EKM except that Step 6 is omitted. At each training cycle, the weights of the winning neuron k and its neighboring neurons i are modified. The amount of modification is proportional to the distance $G(k,i)$ between the neurons in the EKM. The input weights $\mathbf{w}_i$ are updated towards the actual displacement $\mathbf{v}$ and the control parameters $\mathbf{M}_i$ are updated so that they map the displacement $\mathbf{v}$ to the corresponding motor control $\mathbf{c}$. After self-organization has converged, the neurons will stabilize in a state such that $\mathbf{v} = \mathbf{w}_i$ and $\mathbf{c} = \mathbf{M}_i\mathbf{v} = \mathbf{M}_i\mathbf{w}_i$. For any winning neuron k, given the sensory input $\mathbf{u}_p = \mathbf{w}_k$, the neuron will produce a motor control output $\mathbf{c} = \mathbf{M}_k\mathbf{w}_k$ which yields a desired displacement of $\mathbf{v} = \mathbf{w}_k$. For a sensory input $\mathbf{u}_p \neq \mathbf{w}_k$ but close to $\mathbf{w}_k$, the motor control output $\mathbf{c} = \mathbf{M}_k\mathbf{u}_p$ produced by neuron k will still yield the correct displacement if linearity holds within the input region that activates neuron k. Thus, given enough neurons to produce an approximate linearization of the sensory input space $\mathcal{U}$, the indirect-mapping EKM can produce finer and smoother motion control than the direct-mapping EKM.

III. Experiments and Discussions

Quantitative evaluation of the performance of indirect-mapping EKM for motor control has already been presented in [11]. This section presents a qualitative evaluation of the enhanced reactive capabilities of the integrated architecture with cooperative EKMs. The experiments were performed using Webots, Khepera mobile robot simulator (http://www.cyberbotics.com), which incorporated

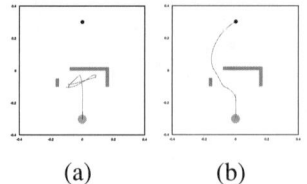

(a)   (b)

Fig. 5. Passing through unforeseen narrow doorway between closely spaced obstacles. (a) The robot using Braitenberg obstacle avoidance and action superposition was trapped but (b) the one adopting the integrated architecture with cooperative EKMs successfully passed through the narrow doorway to the goal.

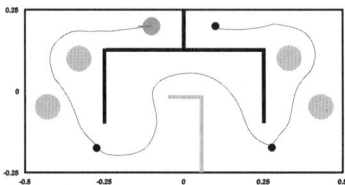

Fig. 6. Motion of robot (dark gray) in an environment with unforeseen static obstacles (light gray). The checkpoints (small black dots) were located at the doorways and the goal position. The robot can successfully navigate through the checkpoints to the goal by traversing between narrowly spaced convex obstacles in the first and the last room and overcoming a concave obstacle in the middle room.

10% noise in its sensors and actuators. 12 directed long-range sensors were also modelled around its body of radius 2.5cm. Each sensor had a range of 17.5cm, enabling the detection of obstacles at 20cm or nearer from robot center. To simulate noise, the sensors have a resolution of 0.5cm.

Two tests were performed to compare the integrated architecture using cooperative EKMs with that presented in [11]. The latter used Braitenberg's Type-3C vehicle [4] for obstacle avoidance and superposition (or vector sum) technique for action selection. For both architectures, the target reaching and obstacle avoidance modules ran at intervals of 256ms and 128 ms respectively. The robot's performance was assessed in an environment under two *unforeseen* conditions: (1) concave obstacle, and (2) narrow doorway between closely spaced obstacles.

In the first test (Fig. 4), the robot fitted with the Braitenberg scheme and superposition technique got trapped by the concave obstacle (Fig. 4a). The target reaching module tried to move the robot forward to reach the target while the obstacle avoidance module moved it backward to avoid the obstacle. The combined output cancelled each other's efforts. On the other hand, the robot that adopted the integrated architecture with cooperative EKMs successfully overcame the obstacle to reach the goal (Fig. 4b).

In the second test (Fig. 5), the robot endowed with the Braitenberg scheme and superposition technique could not pass through the narrow doorway between closely spaced obstacles (Fig. 5a). The repulsive forces from the walls counteracted the attractive force to the designated goal. In contrast, the robot that adopted the integrated architecture with cooperative EKMs successfully traversed through the narrow doorway to the goal (Fig. 5b).

These two tests demonstrate that for the architecture with

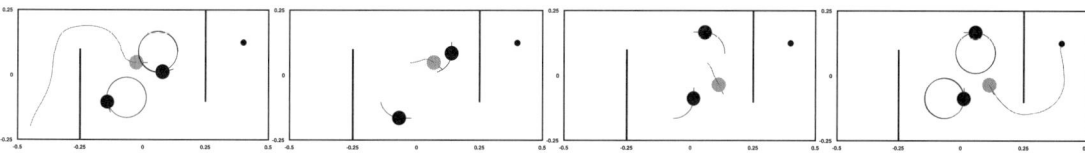
Fig. 7. Motion of robot (gray) in an environment with two unforeseen obstacles (black) moving in anticlockwise circular paths. The robot could successfully negotiate past the extended walls and the dynamic obstacles to reach the goal (small black dot).

action superposition mechanism, though each behavioral module proposes an action optimal to itself, the vector sum of these action commands produces a combined output that may not guarantee the satisfaction of any motion task. The neural integration method used in the integrated architecture with cooperative EKMs, however, considers the suboptimal preferences of each behavioral module via excitatory or inhibitory inputs and integrates them to determine an action that can satisfy each behavior to a certain degree. Such tightly coupled interaction between the behaviors and the action selection mechanism enable more complex motion tasks to be achieved.

The environment for the next test with unforeseen static obstacles consisted of three rooms connected by two doorways (Fig. 6). The robot began in the top corner of the leftmost room and was tasked to move into the narrow corner of the right-most room via three checkpoints plotted by the planner. It was regarded to have reached a checkpoint if it was less than 5mm from the checkpoint and was required to stop at the goal. The robot was able to navigate through the checkpoints to the goal by traversing between narrowly spaced convex obstacles in the first and the last room, and overcoming a concave obstacle in the middle room. The results of this test demonstrate the efficacy of the integrated planning and control architecture with cooperative EKMs in handling complex, unpredictable environments.

The environment for the last test also consisted of three rooms connected by two doorways (Fig. 7). The middle room housed two Khepera robots moving in anticlockwise circular paths. The robot began in the left-most room and was tasked to move to the right-most room without the help of the planner. Nevertheless, the robot was able to negotiate past the extended walls and the dynamic obstacles to reach the goal. This experiment further verified the enhanced reactive capabilities of the integrated architecture with cooperative EKMs in performing more complex motion tasks.

IV. CONCLUSION

By tightly integrating the target reaching, obstacle avoidance and neural integration modules with cooperative EKMs, the reactive capabilities of an integrated planning and control mobile robot architecture can be enhanced. This is extremely useful in complex, unpredictable environments where a robot controlled by this method can perform more complex motion tasks like negotiating unexpected concave and extended obstacles, and traversing between narrowly spaced obstacles. These characteristics distinguish our architecture from those integrated frameworks (e.g., [2], [5], [11]) that employ action superposition mechanisms ([1], [4], [8]), which can be easily trapped by unforeseen concave obstacles and narrowly spaced obstacles unless global replanning is executed. Our continuing research goal is to apply the integrated framework to the planning and control of robot manipulators.

ACKNOWLEDGMENTS

This research is supported by NUS R-252-000-018-112.

REFERENCES

[1] R. C. Arkin. Motor schema based mobile robot navigation: An approach to programming by behavior. In *Proc. ICRA*, pages 264–271, 1987.
[2] R. C. Arkin and T. Balch. AuRA: Principles and practices in review. *J. Experimental and Theoretical AI*, 9(2-3):175–189, 1997.
[3] J. Borenstein and Y. Koren. The vector field histogram: Fast obstacle avoidance for mobile robots. *IEEE Trans. Robot. Automat.*, 7(3):278–288, 1991.
[4] V. Braitenberg. *Vehicles: Experiments in Synthetic Psychology*. MIT Press, Cambridge, MA, 1984.
[5] O. Brock and O. Khatib. Executing motion plans for robots with many degrees of freedom in dynamic environments. In *Proc. ICRA*, volume 1, pages 1–6, 1998.
[6] O. Brock and O. Khatib. High-speed navigation using the global dynamic window approach. In *Proc. ICRA*, volume 1, pages 341–346, 1999.
[7] R. Brooks. A robust layered control system for a mobile robot. *IEEE J. Robot. Automat.*, 2(1):14–23, 1986.
[8] O. Khatib. Real-time obstacle avoidance for manipulators and mobile robots. In *Proc. ICRA*, pages 500–505, 1985.
[9] O. Khatib, S. Quinlan, and D. Williams. Robot planning and control. *Robotics and Autonomous Systems*, 21:249–261, 1997.
[10] Y. Koren and J. Borenstein. Potential field methods and their inherent limitations for mobile robot navigation. In *Proc. ICRA*, pages 1394–1404, 1991.
[11] K. H. Low, W. K. Leow, and M. H. Ang Jr. Integrated planning and control of mobile robot with self-organizing neural network. In *Proc. ICRA*, volume 4, pages 3870–3875, 2002.
[12] P. Pirjanian. Multiple objective behavior-based control. *Robotics and Autonomous Systems*, 31(1-2):53–60, 2000.
[13] S. Quinlan and O. Khatib. Towards real-time execution of motion tasks. In R. Chatila and G. Hirzinger, editors, *Experimental Robotics 2*. Springer-Verlag, 1993.
[14] J. Riekki and J. Röning. Reactive task execution by combining action maps. In *Proc. IROS*, volume 1, pages 224–230, 1997.
[15] J. K. Rosenblatt. DAMN: A distributed architecture for mobile navigation. *J. Experimental and Theoretical AI*, 9(2-3):339–360, 1997.
[16] A. Saffiotti, K. Konolige, and E. Ruspini. A multi-valued logic approach to integrating planning and control. *Artificial Intelligence*, 76(1-2):481–526, 1995.
[17] C. Touzet. Neural reinforcement learning for behavior synthesis. *Robotics and Autonomous Systems*, 22(3-4):251–281, 1997.

RoboDaemon - A device independent, network-oriented, modular mobile robot controller

Gregory Dudek and Robert Sim

Centre for Intelligent Machines, McGill University
3480 University St, Montréal, Québec, Canada H3A 2A7
{dudek,simra}@cim.mcgill.ca

Abstract—We discuss a software environment for multi-robot, multi-platform mobile robot control and simulation. Like others, we have observed that mobile robotics research is greatly facilitated by the availability of a suitable simulator for both vehicle kinematics as well as sensing, and have created an environment that permits this while allowing a large measure of device independence. By using a multi-processor internet-based architecture, our platform permits multiple users to use a variety of programming interfaces (visual, script-based or various application programming interfaces (API's)) to rapidly prototype methods to control multiple heterogeneous robots both in simulation and in real-world settings. We present an overview of our architecture and discuss its future directions.

I. INTRODUCTION

In this paper, we describe a software system and associated architecture for controlling groups of mobile robots. Robotics, and particularly mobile robotics, is a problem domain that entails an exceptional requirement for the integration of multiple software and hardware technologies. In order to construct a truly effective mobile robot, several extant problems in computational perception, artificial intelligence and discrete control must be solved simultaneously. As the mechanical systems for mobile robotics research have matured it has become increasingly feasible to implement robotics sensing and navigation techniques in a device-independent manner. Further, the use of a standardized interface between robot control software and, concomitantly, higher-level modules greatly facilitates research on the integration of and interactions between different components that constitute a mobile robot system. We will describe a mobile robot control architecture that facilitates the use and combination of different software subsystems and different hardware platforms, providing a uniform and simple, but versatile interface to the user.

In this paper we focus on a particular software package called *RoboDaemon* that serves as a nexus for both robots and other software subsystems. By virtue of its role as an interface between components, this software package plays a critical role in defining an entire architecture (the

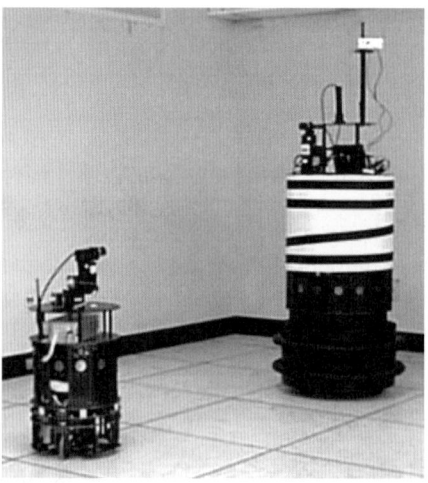

Fig. 1. A core problem: how to control multiple, heterogeneous robots through a uniform interface.

McGill Mobile Robotics Architecture, or MMRA) for mobile robot control. In this paper, however, we confine our attention to the role of this single package itself since it provides somewhat better focus. It should also be noted that the role of connecting different devices and subsystems has also been approached recently in the context of robot operating systems and special purpose programming languages. While there are common themes to our own in such work, this work is particularly well suited to experimentation and rapid prototyping as opposed to the development of high performance applications.

One of our objectives has been to construct an infrastructure that would allow researchers to combine software and hardware modules for different mobile robotic sub-problems. This has been necessary if only to permit the investigation of specific research problems that are our primary objective. In many ways, this mirrors the issues being faced by the community at large: to develop systems that integrate results not only from multiple individual researchers, but from multiple teams and institutions, each

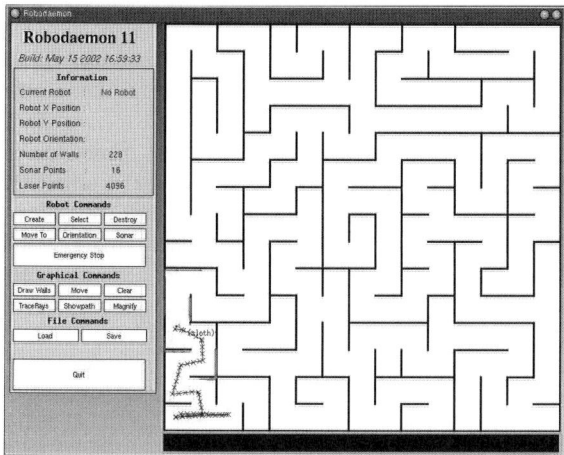

Fig. 2. The *RoboDaemon* graphical user interface

of whom may employ different robotic hardware or who may favour different programming environments. Such an objective involves motivating different individuals and groups to adopt components of one another's systems in the first place; that task motivates this work, but it is outside the scope of this discussion. Equally important, it presupposes that a mechanism and infrastructure exists that permits subsystems to be combined with limited effort, assuming there exists an underlying motive. (In fact, adding functionality to an existing subsystem is usually attractive to researchers if the infrastructure facilitates it.)

RoboDaemon, a core component of our infrastructure and control system, provides a variety of such services, namely the abstraction of robotic and sensing programming interfaces away from the low-level devices, the ability to develop robotic applications in a multi-user setting, and the ability to operate in full simulation when hardware resources are scarce, slow or expensive to operate. The objectives of our effort have many commonalities with other general-purpose robot control systems such as the *Task Control Architecture* [1], [2] (TCA) and the related systems TCX and IPC, the *Reactive Action Package* [3], [4] (RAP), the RAVE architecture [5], or even the subsumption architecture [6], [7].

A. Objectives

The objectives of our archtecture are to provide robotics researchers with an abstract programming layer that provides access to a variety of robotic and sensing platforms. Included among the goals of the project are:

1) **Hardware independence.** We have implemented robot drivers for a variety of vendors (including Nomadics, RWI and Cyberworks), providing users with an abtraction layer, allowing for vendor-independent application development (Figure 1).

2) **Multi-sensor platform.** Our architecture facilitates multiple sensing capabilities on each instantiated robot, including sonar, laser, infra-red sensing and video.

3) **Multi-user platform.** Our architecture employs a client-server model to allow multiple users to collaborate in the control of a single, or multiple, robots, working alone or together in the real world, or in one or more simulated worlds.

4) **Multi-interface platform.** We provide both graphical and command-line interfaces for user commands and feedback (Figure 2). Furthermore, we have implemented network-based clients for a variety of popular programming languages.

5) **Fully simulated environment.** When hardware resources are scarce or cumbersome to operate, or where time is of the essence, the user may opt to use simulators for any of the supported hardware platforms. Simulation includes modelling of robot kinematics (including noise models), sensor properties and the environment in which they operate. The user has the ability to instantiate simulators in parallel and real-time with real robots, or to operate in instantaneous mode, enabling the rapid evaluation of the outcome of user algorithms.

6) **Extensible environment.** Our architecture provides interfaces for extending functionality and behavior. A plug-in interface provides access to low-level functionality, enabling, for example, the addition of new graphical interfaces, collision-avoidance modes, or random map-generation features, while a module interface provides the ability to access external computational services, such as out-sourcing map or image data processing.

B. Outline

In the remainder of this paper we discuss related work, and the problem of promoting the integration of robotics solutions. We continue with an exposition of the architecture, provide an illustrative example, and discuss some of the realized benefits and challenges. Finally, we close with a discussion of open problems, shortcomings and findings from our work.

II. RELATED WORK

Several research groups have considered the issues of building systems to control mobile robots. While our work is directed primarily at rapid prototyping and the support of diverse hardware and software systems and interfaces, related work on architectures for mobile robots has served as an important precursor.

Several authors have considered task planning as it applies to robotic systems. While the range of contributions is too long to enumerate, systems that combine planning with exception management are particularly noteworthy. PRS (Procedural Reasoning System) [8] is a schema-based system that is particularly suited to goal and sub-goal management and dealing with the sequencing and interaction between subgoals. PRS interacts with a "lower level" egocentric representational framework called LPS - Local Perceptual Space which is core to the Saphira robot control package. The Saphira system [9], [8] uses LPS and shares several objectives and features with *RoboDaemon*. Saphira is also based on a client-server model of robot control whereby the robot provides a set of basic functions that can be used to interact with it. Like the system described here, it allows for extensibility on the part of end-users, although as far as we know the process is substantially more intensive than either the "plug-in" or scripting facilities we seek to provide.

The execution support language deals with task and goal management and provides rich mechanisms for dealing with failures and exceptional events [10], [11]. IPEM and similar systems also show their key strength in the domain of planning, plan monitoring and exception handling [12]. In our work we place very little emphasis on planning *per se* and assume it is handled by either a higher level supervisory system or (in the case of reactive planning) by systems below the level of abstraction of what we describe here.

ROGUE [13] is an architecture built on a real robot which provides algorithms for the integration of high-level planning, low-level robotic execution, and learning. Somewhat closer to this work is the Task Description Langauge (TDL) and its sibling Task Control Architecture (TCA) [14], [15]; architectures based on language extensions to C++. These provide mechanisms to facilitate planning, exception handling and inter-task communication and synchronization via TCA. Like the present work, TCA uses a centralised process control mechanism. Again, the emphasis in that work is more specifically on real time control and on planning than in the present work where a premium is put on ease of use, user interface and flexibility.

Several powerful commercial systems have also been developed, such as ControlShell by Real-Time Innovations. However, as observed in [16] their closed nature makes them difficult to integrate into an research environment except in special circumstances.

III. Promoting Integration

In many cases integrating the results of disparate researchers depends critically on having this objective in mind from the outset of a project life cycle. While it is possible to retrofit existing subsystems to work together, it is usually infeasible due to limits on available time and energy.

In addition, several projects have, understandably, focussed on the examination of what can be accomplished using a uniform consistent formalism which makes all the components of the systems adhere to a single methodology and communications mechanism. Systems based on the subsumption architecture, for example, were originally composed *in principle* of components that can *all* be described as simple finite-state automata [6], [7] and which communicate with one another using a specific class of abstraction mechanism[1]. While such an approach serves to exemplify the potential of a given approach, and even to simplify design, it can frustrate progress[2]. Further, since researchers are governed by their own peculiar tastes, it discourages widespread adoption.

In this work, we focus on factors that make the system readily acceptable to potential researchers working on particular subsystems. Three most important factors that have contributed to this acceptability are:

1) making basic functionality easy to learn and accomplish,
2) providing intuitive and informative, yet unobtrusive, feedback on system behaviour, and
3) providing readily available facilities that would be difficult to replicate otherwise.

a) Ease of acquisition: Persuading researchers to adopt a framework for system integration involves an undeniable component of salesmanship. While being easy to learn is not, in fact, a critical attribute for long-term users or system functionality, it has proven very important in attracting users initially (especially since they are often on tight schedules, even with their initial implementation efforts).

b) Informative feedback: A common difficulty in working with abstracted interfaces is the lack of informative diagnostic information when failures occur "under the hood". Users are also often frustrated by black-box interfaces that fail to produce an expected result. Finally, textual output can be useful to diagnose some failures but, in a robotics context and particularly in simulation, visual output can be far more compelling and informative.

c) Facilities: The third aspect that encourages adoption is the provision of facilities and features that would be cumbersome to replicate otherwise. In the case of robot controllers, essential features include a high level of control abstraction (so that the developer can ignore the underlying actuation problems in moving the robot), robust simulation and sensor modelling, and rich visualization tools for observing sensor output and obtaining

---

[1]Several highly capable systems have been constructed using this architecture and it is possible to relax the standard design constraints if necessary.

[2]For example, the widespread successful adoption of the C programming language is partly attributed to its frighteningly lax enforcement of stylistic and syntactic rules, for example with respect to type checking.

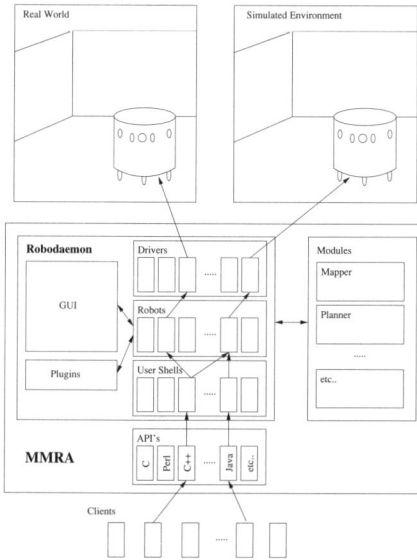

Fig. 3. *RoboDaemon* and the MMRA architecture.

Fig. 4. Simulated camera viewpoint.

feedback on the robot's state. It should be noted that this goal is often in competition with the ease of use criteria and it is important to keep this tension in mind when adding new features.

In addition to the attributes noted above, there are several additional features that we believe are important for a widely acceptable common infrastructure.

IV. SYSTEM ARCHITECTURE

Figure 3 provides an overview of the *RoboDaemon* package in the context of the MMRA architecture. *RoboDaemon*'s core role is to receive client connections and manage a set of instantiated robots. The robot instances may interface to real robots through device drivers and wireless connections, or may instead communicate with simulated counterparts. *RoboDaemon* also provides visualization tools and a set of plugins that define high-level robot behaviours. Finally, *RoboDaemon* provides a module interface that facilitates inter-process communication with subsystems that perform CPU-intensive processing tasks. In the remainder of this section we will discuss the various components of the architecture.

A. Simulation

Hardware simulation has become a major factor in many domains. Despite this, many robotic systems have only limited simulation abilities and only a few can *realistically* simulate the performance of their sensors. In many ways, a poor simulation is worse than none at all as it leads to false expectations. The advantages of an effective simulation are threefold: demand on scarce or high-maintenance resources is reduced; extensive sets of experiments can be readily developed, conducted and exhaustively examined; and experiments can be repeated when anomalies are detected.

There are three aspects to simulation in our architecture: robot modelling, sensor modelling, and environment modelling.

- **Robot models:** We have endeavoured to produce accurate kinematic models of our robots. This includes not only addressing the motion properties of the robot, but also idiosyncracies such as quantization error (for example, the differential drive on the Nomadics Scout exhibits significant quantization error when rotations are induced), and appropriate noise models for odometry error.
- **Sensor models:** We employ the sonar simulator described in [17], [18] which takes into account attenuation, beam width, multi-path reflections and material reflectivity. We have also developed a laser simulator and modeled noise properties as a function of distance, as well as accurate models of blind spots and the pose of the real sensor on our real robots. Finally, through our plugin and module interfaces, we have developed camera simulators that render the robot's view point using OpenGL or PovRay (Figure 4).
- **Environment models:** Part and parcel with sensor modelling is the problem of representing the environment with sufficient accuracy. Furthermore, different problem domains require different levels of realism. Our core environment facility employs planar polygons to represent walls and other obstacles. The reflectivity of these polygons can be controlled individually, and texture maps can be applied for visual sensors. Furthermore, we have additional tools for converting large sets of real sonar and laser data into line segment models for producing realistic simulations of planar environments.

B. Run-time Modes

RoboDaemon has a variety of run-time modes to suit the needs of developers:

- **Real-time vs instantaneous events:** While the robot simulators are event driven, it is possible to turn off the clock, allowing events to be processed without delay. The primary benefit of instantaneous mode is that long simulations can be sped up, allowing for faster debugging and completion times.
- **Synchronous vs asynchronous control:** *RoboDaemon* allows the user to define the level of interaction they will have with the robots. For many applications, stop and shoot mode is sufficient, not to mention straightforward to debug, whereas more intensive applications call for continuous velocity and acceleration control. A variety of safeguards are taken to ensure that a robot in asynchronous mode avoids collisions. These modes also apply equally to real and simulated agents.

C. User Interface

As we have mentioned, *RoboDaemon* supports both command-line and graphical user interaction. The server also facilitates remote interaction using telnet or one of the client API's. The graphical interface supports robot creation, point and click navigation, map editing and user-defined buttons for higher-level commands. Visualization is a key component of the debugging process and the path of the robot (both real, if known, and odometry-based), sensor measurements, and user-defined notes can all be plotted visually. Furthermore, the user is not restricted to using the built-in X11 client, but can turn any command-line connection to the server, local or remote, into a GUI event stream for contructing custom GUI's.

The command-line interface provides a slightly richer set of primitives to work with than the GUI. In addition to common navigation commands, the user can specify local variables and macros, execute shell escapes, read a script of commands from file, and examine and set various internal parameters controlling sensing and individual robots. The overall design is object oriented, such that each robot can be configured individually, and each connected user can set their own shell preferences.

D. Network-Based Client Support

As mentioned, the system implements a client-server model, allowing multiple clients to connect, possibly from remote locations. Our current set of client API's include C, C++, Perl, and Java. It is also possible to simply connect and supply commands in ASCII format. The versatility of this approach is obvious– multiple users can gain access to the robotic hardware, in the language of their preference.

E. Robot Drivers

Our multi-vendor implementation currently supports Nomadics Scouts and Nomad 200 robots, RWI B-12, and Cyberworks robots. We have implemented simulators for all of these models, as well as a generic cylindrical robot simulator (on which the Nomadics and RWI simulators are based), and an ever-popular Point-Robot Simulator, which is convenient for studying theoretical problems without the burden of additional safety constraints. Among our current projects is the development of a driver and simulator for Sony Aibo robots.

F. Embedded Plugins and External Modules

One of the core goals of this project is the facilitation of subsystem integration. This goal is often circumvented by incompatible programming interfaces and data types. We provide two venues for users to extend the functionality of the architecture, often with minimal revision to the imported system.

a) Plugins: Plugins provide a method for extending core *RoboDaemon* functionality. The plugin interface is an API that gives the programmer access to robot, gui and shell internals. Examples of ideal plugins are renderers for camera simulation, new GUI extensions, and path planners. A key feature of the plugin interface is that new plugins can be created and loaded dynamically without recompilation of the core system.

b) Modules: Modules provide an extremely powerful method for system integration. The module interface allows users to import commands from other running processes by interfacing with that process' command line interface.

We illustrate the interface by example. We have a software package, called Rhys, that processes range images in order to fit geometric models[19]. Rhys operates through a command line interface. The module interface allows the *RoboDaemon* user to run an instance of Rhys and import all of the commands available in Rhys to *RoboDaemon*. For example, if Rhys has a command called `'fit'`, one can call the command `'rhys.fit'` in *RoboDaemon*. Map and robot data is shared with the module using simple file formats, and it is possible for a "*RoboDaemon*-aware" application to send explicit commands back to *RoboDaemon* for execution. The module interface requires only that the subsystem can operate through a command-line interface. The key point to be noted is that modules can be developed independently of the MMRA architecture, in whichever programming environment the developer prefers (in fact, Rhys was developed long before the current version of *RoboDaemon* was written).

V. VALIDATION

Our experience with *RoboDaemon* has suggested it meets several of its key design goals. In particular, we

Fig. 5. Cooperative localization setup. (Figure courtesy Yiannis Rekleitis)

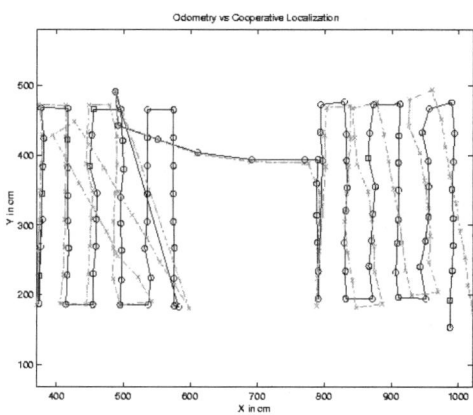

Fig. 6. Odometric (x) vs Tracker-corrected (o) trajectories of the robot.

Fig. 7. Robot viewpoint during exploration.

have observed that unfamiliar users can acquire a working knowledge and implement control algorithms rapidly.

Specifically, we have found that students in an introductory robotics course having no prior robot programming experience are able to learn the interface and design and test sensor-based robotics algorithms in limited time (for example wall following, obstance avoidance, etc.). Typically, students are able to produce working prototypes within a week (concurrent with other course work) assuming they already understand the algorithms to be used. The quality and utility of the simulation is further validated when the resulting algorithms were tested on the real robots in our laboratory with a roughly 85% success rate for algorithms that worked well in simulation (and failures due primarily to incorrectly set parameters).

RoboDaemon also meets the need for enabling subsystem integration. For the purposes of illustration, we present here an example of how the platform has been used to integrate research from various sources in our laboratory. The results of these experiments have been reported elsewhere [20].

Our recent work on contructing visual maps requires that a collection of training images be obtained with ground-truth position information as to where the images were collected[21]. Related research in our lab entails using multiple robots to perform "cooperative localization" whereby a laser range finder mounted on an observing robot is used to locate the pose of a second robot with a mounted target[20]. Furthermore, the cooperative localization method employs Rhys, the line segment fitting software described above.

The scenario we are faced with is one in which three processes (mapper, localizer, and line fitter) must interact to control two robots with distinctly different sensor configurations (one with a laser, the other with a camera) (Figure 5). *RoboDaemon* acts as the intermediary, servicing sensing requests and motion commands, enabling images to be added to the mapper and employing Rhys to outsource data processing. The localizer interfaces with *RoboDaemon* with the C API over a TCP/IP socket, while the mapper collects images through *RoboDaemon* and obtains the ground truth pose estimates with the Perl API. Rhys can operate in the background as a module, although in our example it was connected to directly by the localizer. Figure 6 illustrates the error-prone odometry-based trajectory of the robot versus the "ground truth" obtained from the laser setup. One of the images collected for the visual mapper is depicted in Figure 7.

VI. DISCUSSION

We have presented an architecture for robotic application development in a heterogeneous linguistic, vendor and user environment. Our architecture provides a rich set of facilities and features that encourage the integration of subsystems, and allows for both real-time control and maintenance of a collection of robots as well as full simulation of their kinematics and sensor properties. We believe that a core development and control environment such as the one presented here is essential to the succesful development of complete, functional robotic systems. Our future work involves the ongoing development of this platform, including increased integration with other robotic platforms and with other researchers' navigation, localization and map-building solutions.

VII. REFERENCES

[1] R. Simmons, "Structured control for autonomous robots," *IEEE Transactions on IEEE Transactions on Robotics and Automation*, vol. 10, February 1994.

[2] R. Simmons, R. Goodwin, K. Z. Haigh, S. Koenig, and J. O'Sullivan, "A layered architecture for office delivery robots," in *Proceedings of First International Conference on Autonomous Agents* (W. L. Johnson, ed.), (Marina del Rey, CA), pp. 245–252, ACM Press, New York, NY, February 1997.

[3] R. J. Firby, *Adaptive Execution in Complex Dynamic Domains*. New Haven, CT: Yale University Technical Report YALEU/CSD/RR 672, January 1989.

[4] R. J. Firby, "The rap lanauge manual," Animate Agent Project Working Note AAP-6, University of Chicago, March 1995.

[5] K. Dixon, J. Dolan, W. Huang, C. Paredis, and P. Khosla, "Rave: A real and virtual environment for multiple mobile robot systems," in *Proceedings of the IEEE/RSJ International Conference on Intelligent Robots and Systems (IROS'99)*, vol. 3, pp. 1360–1367, October 1999.

[6] R. A. Brooks, "Intelligence without representation," *Artificial Intelligence*, vol. 47, pp. 139–1159, 1991.

[7] R. Brooks, "A layered intelligent control system for a mobile robot," in *Robotics Research 3* (Faugeras and Giralt, eds.), pp. 365–372, MIT Press, 1986.

[8] K. Konolige and K. Myers, "The saphira architecture for autonomous mobile robots," in *AI and Mobile Robots* (D. K. R. P. B. R. Murphy, ed.), Cambridge, MA: MIT Press, 1998.

[9] K. Konolige, K. Myers, E. Ruspini, and A. Saffiotti, "The saphira architecture: A design for autonomy," *Journal of experimental & theoretical artificial intelligence*, vol. 9, pp. 215–235, 1997.

[10] B. Pell, E. Gat, R. Keesing, N. Muscettola, and B. Smith, "Plan execution for autonomous spacecraft," in *Proc. International Joint Conf on Artificial Intelligence*, AAAI Press, August 1997.

[11] R. Bonasso, J. Firby, E. Gat, D. Kortenkamp, D. Miller, and M. Slack, "Experiences with an architecture for intelligent, reactive agents," *Journal of Experimental and Theoretical Artificial Intelligence*, vol. 9, march 1997.

[12] J. A. Ambros-Ingerson and S. Steel, "Integrating planning, execution and monitoring," in *Proc. Seventh National Conference on Artificial Intelligence*, AAAI Press, 1988.

[13] K. Z. Haigh and M. M. Veloso, "Interleaving planning and robot execution for asynchronous user requests," *Autonomous Robots*, vol. 5, pp. 79–95, march 1998.

[14] R. Simmons, T. Smith, M. B. Dias, D. Goldberg, D. Hershberger, A. Stentz, and R. Zlot, "A layered architecture for coordination of mobile robots," *Multi-Robot Systems: From Swarms to Intelligent Automata*, p. Kluwer, 2002.

[15] E. Coste-Maniere and R. Simmons, "Architecture, the backbone of robotic systems," in *Proceedings of the IEEE Conference on Robotics and Automation*, (San Francisco, CA), IEEE press, April 2000.

[16] L. Petersson, D. Austin, and H. Christensen, "Dca: A distributed control architecture for robotics," in *Proceedings of IEEE/RSJ International Conference on Intelligent Robots and Systems*, pp. 2361–2368, IEEE Press, October 2001.

[17] D. Wilkes, G. Dudek, M. Jenkin, and E. Milios, "Modelling sonar range sensors," in *Advances in Machine Vision: Strategies and Applications* (C. Archibald and E. Petriu, eds.), Singapore: World Scientific Press, 1992.

[18] G. Dudek, "Shape classification and scale-space texture," *1992 Association for Research in Vision and Opthamology Program Book*, May 1992.

[19] P. MacKenzie and G. Dudek, "Precise positioning using model-based maps," in *Proceedings of the International Conference on Robotics and Automation*, (San Diego, CA), IEEE Press, 1994.

[20] I. Rekleitis, R. Sim, G. Dudek, and E. Milios, "Collaborative exploration for the construction of visual maps," in *IEEE/RSJ/ International Conference on Intelligent Robots and Systems*, vol. 3, (Maui, Hawaii, USA), pp. 1269–1274, IEEE/RSJ, IEEE,ISBN 0-7803-6614-X, October 2001. http://www.cim.mcgill.ca/~yiannis/Publications/iros01.pdf.

[21] R. Sim and G. Dudek, "Learning and evaluating visual features for pose estimation," in *Proceedings of the Seventh IEEE International Conference on Computer Vision(ICCV)*, (Kerkyra, Greece), IEEE Press, sept 1999.

A Vision-based Haptic Exploration

Hiromi T.Tanaka, Kiyotaka Kushihama and Naoki Ueda
Computer Vision Laboratory
Computer Science Dept.
Ritsumeikan University
{hiromi, kiyotaka, uedanaoki}@cv.cs.ritsumei.ac.jp

Shin-ichi Hirai
Robotics Department
Ritsumeikan University
hirai@se.ritsumei.ac.jp

Abstract

Real-world objects exhibit rich physical interaction behaviours on contact. Such behaviours depend on how heavy and hard it is when held, how its surface feels when touched, how it deforms on contact, etc. Recently, there are thus growing needs for haptic exploration to estimate and extract such physical object properties as mass, friction, elasticity, relational constraints etc.. In this paper, we propose a novel paradigm, we call haptic vision,which is a vision-based haptic exploration approach toward an automatic construction of reality-based virtual space simulator, by augmenting active vision with active touch. We apply this technique to mass, elational constraints estimation and elasticityr, and use these results to construct virtual object manipulation simulator. Experimental results show that feasibility and validity of the proposed approach.

1. Introduction

Recently, haptic interface has been intensively studied ([2]-[4]) for providing a sense of touch in virtual environments, which is an essential modality to explore, recognize and understand the real world. Real-world objects exhibit rich physical interaction behaviours on contact. Such behaviours depend on how heavy and hard it is when hold, how its surface feels when touched, how it deforms on contact, and how it moves when pushed, etc. These aspects of visual and haptic behaviour provide important interaction cues for manipulating and recognizing objects in virtual environments. There are thus growing needs for haptic exploration to estimate and extract physical object properties such as mass, friction, elasticity, relational constraints etc. Thus, we have proposed a novel paradigm, we call haptic vision, which is a vision-based haptic exploration approach toward an automatic construction of reality-based virtual space simulator. As Figure 1 shows, Haptic vision is an augmentation of active vision with active touch, which controls a contact to an object so that object's behaviours are caused most effectively, based on 3D shape and posture analysis by active vision. Physical object properties are then estimated through motion analysis on real-time range and color images observing object's behaviours against a known contact.

The work on physical object properties from interaction with robots had first introduced in the early 1980 [1]. While recent progress in haptic exploration with dextrous robot hands([2]-[6]), which requires complex robot control and grasping technique, we believe this is the first non-contact vision-based automatic approach for haptic exploration to model both geometrical and physical properties of real-world objects.

We apply this technique to mass, elasticity, and relational constraints ([7]) estimation, and use these results to construct virtual object manipulation simulator. Experimental results show that feasibility and validity of the proposed approach.

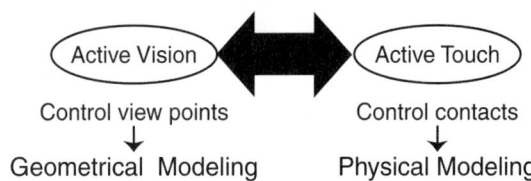

Figure 1. Haptic Vision

2. Haptic Vision

Haptic vision paradigm is motivated by recent development of a real-time handy laser range finder [9] which provides a function equivalent to human tactile sensing. That is, it acquires both static and dynamic geometrical information, i.e., it acquires 3D shape and deformation from real-time images, without contacting to the object and thus without dextrous robot hands.

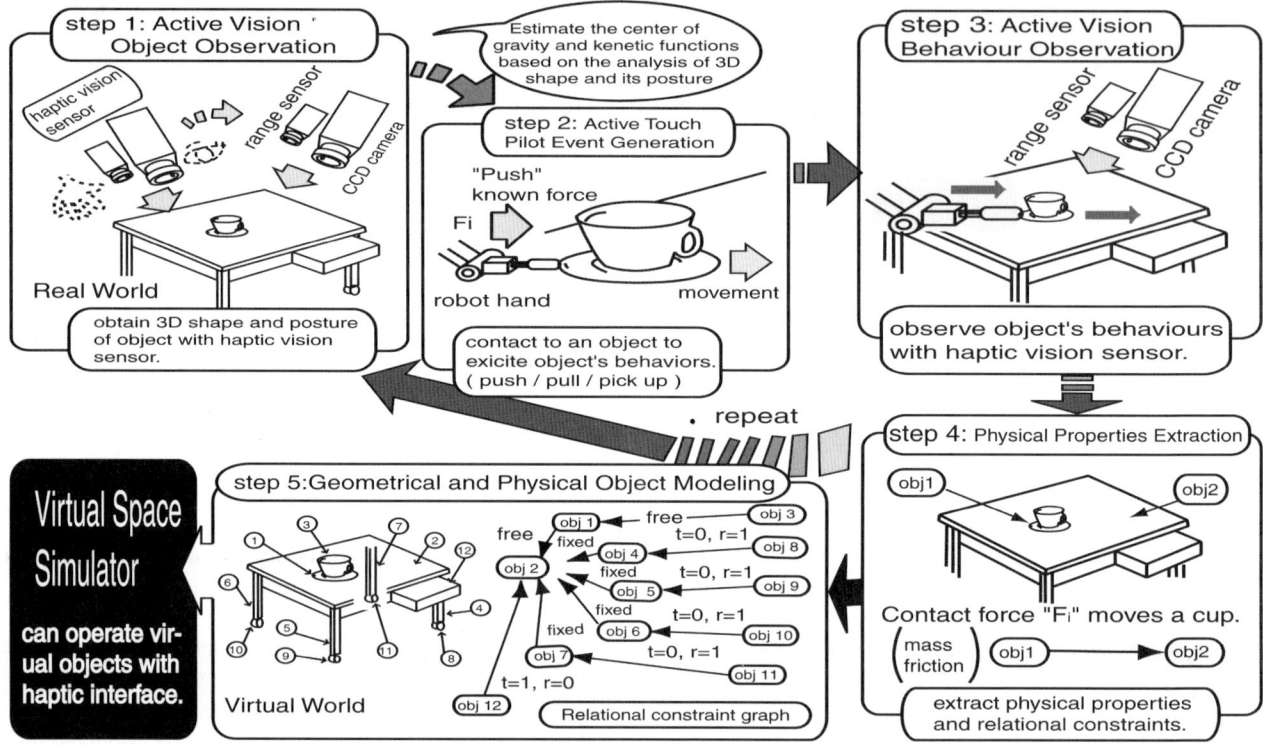

Figure 3. Haptic Vision Approach

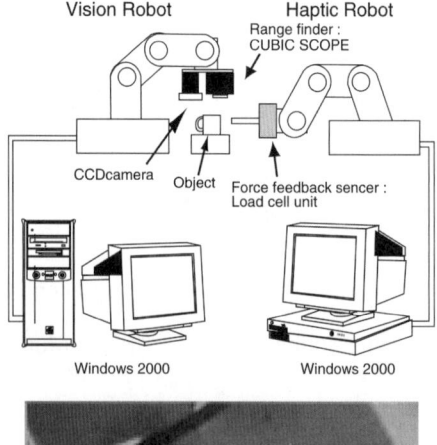

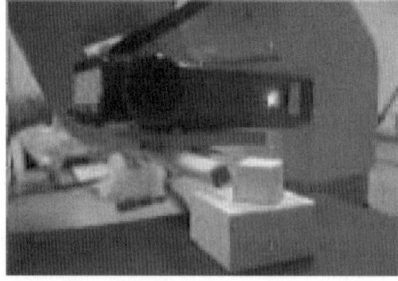

Figure 2. Haptic Vision System

Figure 2 shows our haptic vision system. Our haptic vision sensor, mounted on a robot hand, consists of a CCD camera and a real-time range finder where a CCD camera plays a roll of an "eye" to obtain a wide view of the scene, and a real-time range finder plays a roll of a "hand" to obtain 3D geometrical information by exploring surfaces of objects within a reach of a human hand. The other robot hand with a force-feed back sensor (a Load Cell Unit) makes a contact to cause object's behaviours.

Figure 3 shows our haptic vision approach.

In step1, we first observe an object by active vision to extract and model its geometrical properties such as 3D shape, surface texture and a posture using our haptic vision sensor.

In step 2, we make a contact to an object by active touch based on 3D shape and posture analysis by active vision. Such contact causes object's behaviours most effectively and stably. We call this dynamic scene of object's response as a pilot event where a prototypical behaviour due to the objective physical property is exhibited on response to a known contact force. We then estimate a next viewpoint from which the pilot event is observed most efficiently and stably, and move a haptic vision sensor to observe the pilot event by active vision, in step 3.

In step 4, physical object properties and relational constraints among objects are then estimated and extracted through motion analysis on real-time range and color images observing object's behaviours.

In step 5, we generate a scene representation as a relational constraint graph where each node represents an object with both geometrical and physical properties, and each arc represents adjacency relation with a degree of freedom in both rotation ($0 \leq r \leq 3$) and translation ($0 \leq t \leq 3$).

Above steps are repeated for each object in the scene until the scene representation for a reality-based virtual object manipulation will be completed.

3. Haptic Exploration

We first observe a man-made object in an indoor scene using our active shape inferring algorithm [10]. Our active vision system automatically acquires a set of principal views as shown in Fig. 4 based on the symmetry in stable postures, which are mostly orthographic and are efficiently used for 3D shape reconstruction.

3.1. Estimating Mass

Our approach to mass estimation is as follows. In the current stage, we assume that both static and dynamic friction coefficients, μ_s and μ_d, of an object are given.

In step 1, we first estimate a plane of symmetry S based on symmetry observe in both a top and a side views acquired by our active vision system, based on the 3D shape in a stable posture as shown in Fig. 4.

In step 2, we make a contact by "Push" operation by a robot hand, at a point P_c of the intersection of its surface and S, with the direction of a contact force F parallel to the horizontal plane and also included in the plane of symmetry, as shown in Fig. 5. Such contact force exerts on a center of friction and causes a pilot event for mass extraction where an object moves straight in the direction of F with no rotation and with no change in its posture.

In step 3, we measure transition of F during "Push" contact using a force-feedback sensor mounted on a robot hand. In general, a friction force F starts to increase at a contact point t_c, and rises up sharply until it reaches to the maximum friction $F\mu_s$ at t_s, at which the objects starts to move. Then, it drops a little, and goes into a steady state at $F\mu_d$, as shown in Fig. 6. We also track an object from a top view point during contact to confirm its straight movement, as shown in Fig. 7.

In step 4, we then estimate a mass M from $F\mu_s$ and $F\mu_d$ respectively as,

$$F\mu_s = \mu_s M g \quad (1)$$
$$F\mu_d = \mu_d M g \quad (2)$$

where g is a gravity force.

Fig. 8 and Table 1 shows mass estimation results of a ceramic coffee cup on a base surface of three material types, wood, rubber, and steel, respectively. Error rates show that mass estimation with μ_s are performed with reasonable accuracy except on a rubber surface of large friction coefficient.

Since we estimate a mass from eq.(1) and (2) assuming that friction coefficients μ_s and μ_d are known, the mass estimation accuracy thus depends on the stability of μ_s and μ_d in various environments where they are measured. We first estimate both μ_s and μ_d from eq.(3) and (4) by measuring an angle θ, a time T and a length L while an object is sliding, as shown in Fig. 9.

$$\mu_s = tan\theta \quad (3)$$
$$\mu_d = tan\theta - \frac{2L}{gt^2 cos\theta} \quad (4)$$

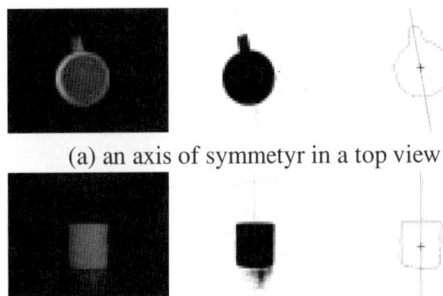

(a) an axis of symmetyr in a top view

(b) an axis of symmetyr in a side view

Figure 4. Symmetry Estimation from Principal Views and Silhouettes of a Cup Aquired by Our Active Vidion System

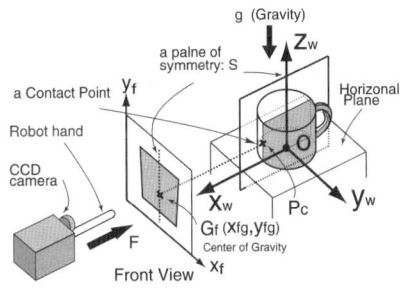

Figure 5. Contact for Mass Estimation

We then evaluate the estimated μ_s and μ_d to confirm their stabilities against changes in both temperature and humidity. Fig. 10 and 11 show Values the measurement results of μ_s and μ_d v.s. temperature and humidity changes, respectively. The results show that μ_s is more stable than μ_d.

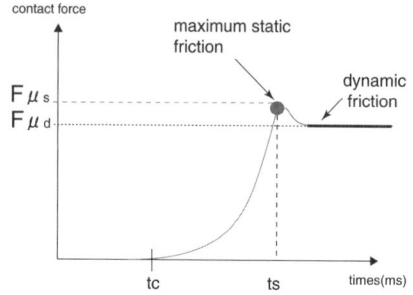

Figure 6. Contact Force Transition Graph

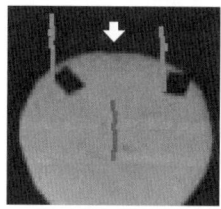

Figure 7. Trajectory of straight movement

Table 2 and Table 3 show the results of mass estimation on a ceramic cap, an aluminum block and a ceramic cup with craft tapes stuck on its bottom surface for static (μ_s) and dynamic (μ_d) friction coefficient. The stability evaluation of μ_s, and μ_d in Fig. 10 and in Fig. 11, and the mass estimation results in both Table 2 and 3 show that mass estimation with μ_s is more stable, with maximum error of 7.4010%.

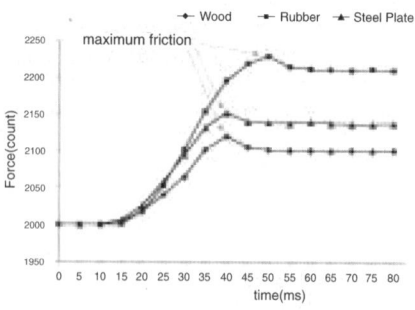

Figure 8. Contact Force Transition for a Cup on Wood, Rubber and Steel Plate Surfaces

3.2. Extracting Relational Constraints

Our approach to relational constraint extractions is as follows. We suppose that the scene consists of a set of 3D convex objects/parts.

In step 1, We first estimate boundaries between adjacent objects/parts along L-joint contours from a volume repre-

Table 1. Mass Estimation Results with μ_s

Base Surface	Contact Force (gw)	Static Friction Coefficient: μ_s	Estimated (g)	Real (g)	Error (%)
Wood	46.8	0.217	215.7	221	2.4
Rubber	94.9	0.533	178.0	221	19.5
Steel Plate	64.2	0.306	209.8	221	5.1

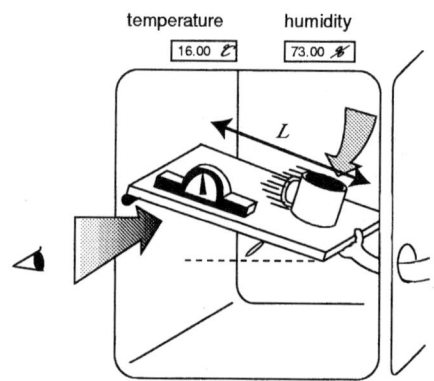

Figure 9. Measurement of friction coefficient

Table 2. Mass Estimation Results with μ_s

Object	Contact Force (gw)	Static Friction Coefficient: μ_s	Estimated (g)	Real (g)	Error (%)
Aluminum	174.9	0.2679	261.09	282	7.413
Ceramic	146.2	0.208	279.92	290	3.473
Craft tape	201.7	0.1943	403.59	378	6.77

Table 3. Mass Estimation Results with μ_d

Object	Contact Force: F (gw)	Dynamic Friction Coefficient: μ_d	Estimated M (g)	Real (g)	Error (%)
Aluminum	175.6	0.2674	262.52	282	6.905
Ceramic	110.5	0.203	212.47	290	26.73
Craft tape	197.2	0.19435	394.79	378	4.44

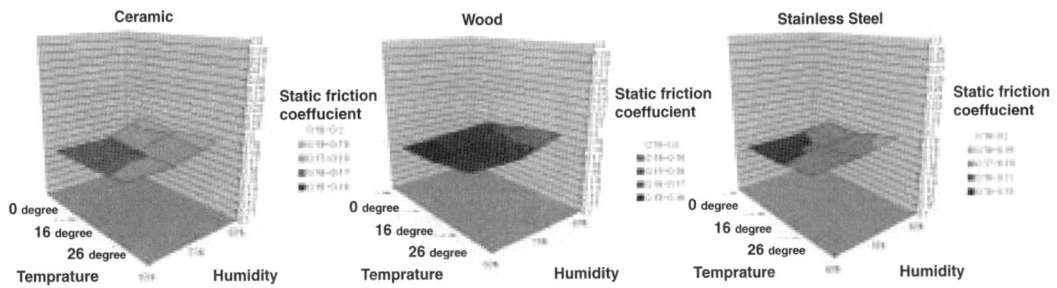

Figure 10. Static friction coefficients V.S. temperature & humidity change

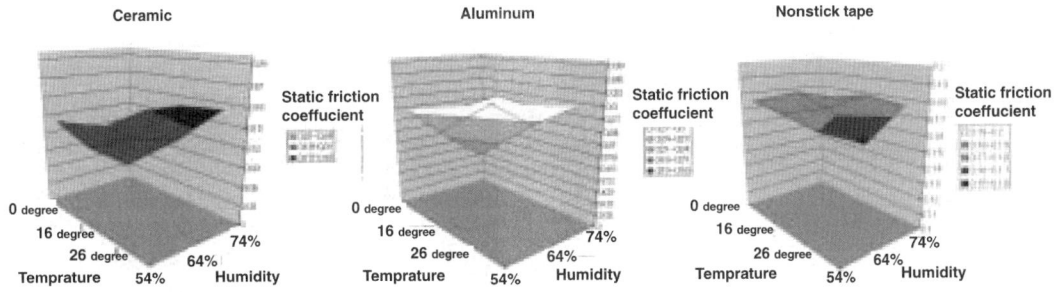

Figure 11. Dynamic friction coefficient V.S temperature & humidity change

sentation of the scene obtained from multi-view range images, as shown in Fig. 12. In current stage, we limit adjacency relations as support relations on the horizontal plane. Figure 13 shows various constraints for the set of DOFs. In the case of free support type, the DOF of an object will be 3 for the application of external force. If the contact is the type of rail such as drawer of a desk or window on a rail, the DOF of such object is 1. The contact of axis support type is a contact with rotational DOF of 1 only, that is the object can rotate but not allowed other movement. Cup on a saucer is an example of this type of contact. Note that we can hardly imagine a contact with translational DOF of 2 and rotational DOF of 0, or translational DOF of 1 and rotational DOF of 1. The priority shown in the table relates to how many Push operations are required to detect the relation or constraints between objects. In the cases of non-contacting support or axis support contact, we can set the contact point so that only one Push operation can detect the type of contact. This is the reason why the priority is the highest. However if you want to detect rail type contact, you have to try multiple Push operation. If there are multiple objects on a plane, you have to carefully select directions of multiple Push operation. Thus, the priority is lower for these cases.

In step 2, we make a contact at a corner along estimated L-joint boundaries, in order not to exert F on COG as shown in Fig. 14. Such contact excites a pilot event where one part is rotated and translated with maximum moment against the other part if they are separate and free each other, as shown in Fig. 15 (a), (b). Otherwise both parts move together if they consist a single object, as shown in Fig. 16 (a), (b).

In step3, we observe the pilot event in the horizontal plane from the top view point.

In step4, a relational constraint as a degree of freedom in both rotation and translation are estimated from its displacement, i.e., one degree of rotation freedom (r=1) and two degrees of translation freedom (t=2) between separate objects A and B, and no rotation and translation freedom (r=0, t=0) for a single glued object.

In step5, we add the above constraints into the scene representation as a adjacency relation of the arc between the parts, as shown in Fig. 15 (c) and Fig. 16 (c).

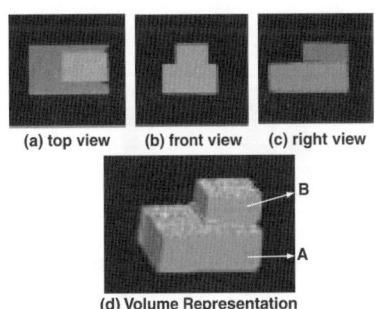

Figure 12. 3D Volume Representation

type	free	bounded	—	—	rail	axis
movement	free	partially free	—	—	straight	rotation
degree of freedom of translation	2	2	2	1	1	0
degree of freedom of rotation	1	1	0	1	0	1
priority	1	3	—	—	2	1
translation and rotation of "push"						

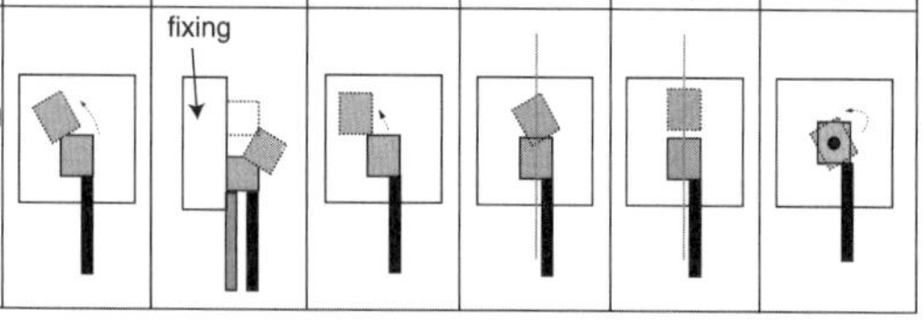

Figure 13. Classification of support-constraints

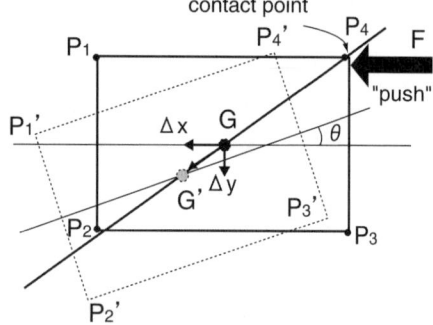

Figure 14. Contact for Relat. Const. Extraction

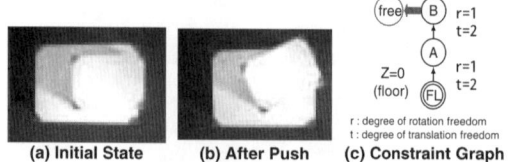

Figure 15. Push Effect on Separate Objects

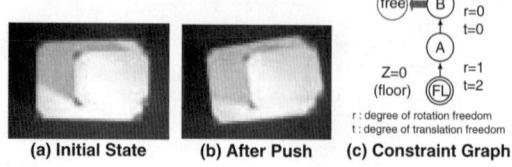

Figure 16. Push Effect on Single Object

3.3 Estimating Elasticity

Our approach to elasticity estimation by "Push" is as follows. When estimating elastic coefficient of an object using Push operation of the robot hand, we have to apply a contact force that deforms the it but does not make it move nor rotate since we have to know the deformation of elastic object. So, we have to apply a contact force perpendicular to the support plane that goes through the object's the center of gravity(COG) while keeping it on the support plane in the stable posture. If we limit the candidate support plane that is taken through the scene observation to level or horizontal one, we can define the contact force as perpendicular to the object and its the direction of a contact force goes through the 2D mass center of the top view contour image. Note that the x_t and y_t components of the 2D mass center of the top view contour image (x_t, y_t, z_t) should coincide with x_g and y_g components of the center of gravity (x_g, y_g, z_g). Thus, the elastic object to be observed should be plane symmetry and should have at least one plane of symmetry that is perpendicular to the support plane. We have adopted a cylindrical object for the elastic coefficient estimation, since cylinder is the shape that satisfies these conditions.

In step 1, Take a top view of the object by observing it from a perpendicular point. Then estimate the 2D mass center of the top view contour image and the plane of symmetry that includes the center of gravity.

In step 2, Apply an external force to the object by the Push operation of the robot hand. The speed of the Push operation is constant and the point of contact is at the intersection P_c of the object surface and a line perpendicular to the object that goes through the 2D mass center of the top view contour image G_{top} estimated from the top view shown in figure 17 (a). The direction of the force is downward, that is the action line of the force goes through the contact point P_c and the center of gravity that is perpendicular to the support plane.

In step 3, Apply an external force as shown in figure 18 while monitor its strength using a force feedback sensor attached onto the robot hand. Figure 18 is a graph that shows the transition of contact force when the object deforms by the external force F applied to the object. The period from t_c to t_s corresponds to the operation gradually applying downward force by pulling the force feedback sensor down. The period from t_s to t_e is the waiting time in which we are waiting until the elastic oscillation calms down while keeping the pressure constant. t_e is the time when the external force is removed.

In step 4, Observe the deformation using a camera and a range finder as shown in Figure 17 (b). We have observed the disposition of the point P_v, that is the intersection of object surface and a line parallel to the support plane which goes through the 2D mass center G_v of the 2D side view image, while placing the camera to the position (Pos. P_s) shown in Figure 17 (b). What we have to know is the disposition of the point P_v in the z-axis direction in the world coordinate system (X_w, Y_w, Z_w) shown in Figure 17 (b). The disposition (ε) can be obtained by subtracting H from H_{init}, where H_{init} is the original height of the object before applying force and H is the height observed in the image under pressure by converting the number of pixels that shows the extent of the object in Y_v direction in the side view image.

$$H = \frac{110}{480}h \quad (5)$$
$$\varepsilon = |H_{init} - H| \quad (6)$$

In step 5, Estimate the elastic coefficient E from the disposition (ε) and force F from the force feedback sensor using the following equation.

$$F = E\varepsilon \quad (7)$$

Figure 19 is a graph that shows the relation between the external force F and time t, and between the disposition (ε) and time t where t_c is the time when the external force F is started to be applied, the period from t_s to t_e is the period of constant pressure waiting the elastic oscillation calms down and t_e is the time when the external force is removed. Table 4 shows the result of experiment using two springs of different elastic coefficient. Through this experiment, we have confirmed that the elastic coefficient can be estimated stably within the error of 5 percent or so.

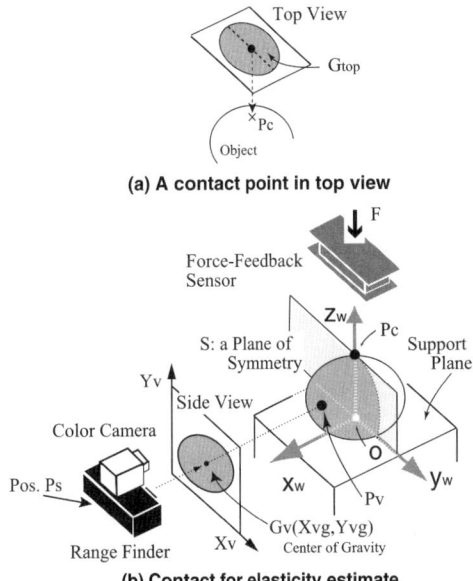

Figure 17. Contact for Elasticity Estimate

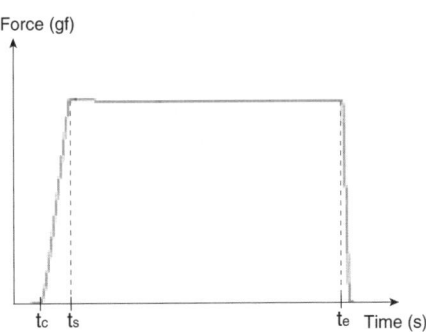

Figure 18. Contact Force Transition Graph for Elastic Body

Table 4. Elastic Coefficients Estimation Results

	Outside Diameter (mm)	Height (mm)	Estimated (kgf/mm)	Real (kgf/mm)	Error (%)
Spring 1	32.0	65.0	0.390	0.37	5.41
Spring 2	45.0	80.0	0.794	0.76	4.57

4. Reality-based Virtual Object Manipulation

We construct a virtual object manipulation simulator from a scene graph representation generated by haptic vision, with the number of objects, their 3D shape, mass and

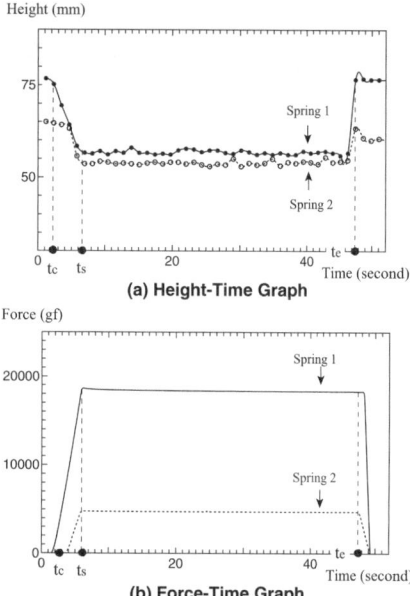

Figure 19. Contact Force and Height Transition for Spring

relational constraints. Figure 20 (a), (b) show a system configuration of our VR simulator and some scene of "Pick Up" manipulation with a Cyber Glove. Given the scene graph, our "Pick Up" simulation of the upper part causes no change in both rotation and translation of the lower part for two separate objects. Conversely the same "Pick Up" bring up the lower part together for a single (glued) object.

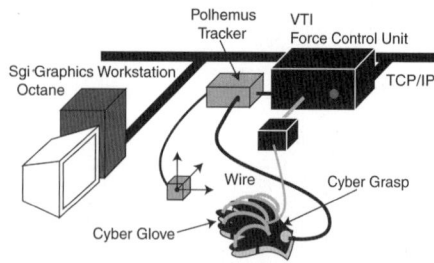

(a) system configuration

(b) "Pick Up" manipulating scene

Figure 20. Reality-Based Virtual Object Manipulation Simulator

5. Conclusion

We have proposed a haptic vision paradigm for a vision-based haptic exploration as an augmentation of active vision with active touch which causes object's prototypical behaviours to be observed by active vision. Preliminary experimental results, on mass estimation, relational constraints estimation, presented in companion paper [7], elasticity estimation, and automatic construction of virtual object manipulation simulator shows the feasibility and validity of the proposed approach.

References

[1] E. Krotkov, "Perception of Material Properties by Robotic Probing: Preliminary Investigations ," Proc. Intl. Joint Conf. Artifical Intelligence, Montreal, August, pp. 88-94, 1995.

[2] A.M. Okamura, M.L. Turner and M.R. Cutkosky, "Haptic Exploration of Objects with Rolling and Sliding," Proc. the 1997 IEEE ICRA, Vol. 3, pp. 2485-2490, 1997.

[3] A.M. Okamura, N. Smaby, and M.R. Cutkosky, "An Overview of Dexterous Manipulation," Proceedings of the 2000 IEEE International Conference on Robotics and Automation, Symposium on Dexterous Manipulation, Vol. 1, 2000, pp. 255-262, 2000.

[4] A. Bicchi, A. Marigo, and D. Prattichizzo, "Dexterity through rolling: Manipulation of unknown objects," Proceedings of the 1999 IEEE International Confe

[5] P.K. Allen and P. Michelman, "Acquisition and interpretation of 3-d sensor data from touch," IEEE Trans. on Robotics and Automation, 6(4), pp. 397-404, 1990.

[6] A.M. Okamura, M.A. Costa, M.L. Turner, C. Richard, and M.R. Cutkosky, "Haptic Surface Exploration," Experimental Robotics VI, Lecture Notes in Control and Information Sciences, Vol. 250, Springer-Verlag, pp. 423-432, 2000.

[7] Hiromi T. Tanaka and Kayoko Yamasaki, "Extracting Relational Constraints among Objects with Haptic Vision for Haptic-Interfaced Object Manipulation," Proc. the 12th ICAT, 2002,

[8] D.K. Pai, J. Lang, J.E. Lloyd, and R.J. Woodham, "Acme, a telerobotic active measurement facility," Experimental Robotics VI, Lecture Notes in Control and Information Sciences, Vol. 250, Springer-Verlag, 2000

[9] Yukio Sato and Masaki Otsuki, "Three-Dimensional Shape Reconstruction by Active Rangefinder," Proc. 1993 IEEE CVPR, pp. 142-147, 1993.

[10] Kengo Nishimura and Hiromi T. Tanaka, "Active Shape Inferring Based on the Symmetry in Stable Poses," Proc. the 1996 IEEE ICPR, pp. 136-139, 1996.

[11] Hiromi T. Tanaka and Kiyotaka Kushihama, "Haptic Vision - Vision-based Haptic Exploration," Proc. the 2002 IEEE ICPR, Vol. 2, pp. 852-855, 2002.

[12] Shinichi Tokumoto, Shinichi Hirai, and Hiromi Tanaka, "Constructing Virtual Rheological Objects," Proc. World Multiconference on Systemics, Cybernetics and Infomatics, pp.106-111, July, Auland, 2001

Vision Based Shape Estimation for Continuum Robots

Dr. Michael Hannan and Dr. Ian Walker

Dept. of Electrical and Computer Engineering, Clemson University, Clemson, 29634, USA

mhannan@ces.clemson.edu, ianw@ces.clemson.edu

Abstract— The investigation of continuum robots has become an area of considerable interest in the last several years. Unlike conventional robotic manipulators which bend in discrete locations, continuum robots bend over continuous sections. One of the main issues that is hampering research in this area is the determination of the robot's shape. In this paper we present a shape-determining scheme that is based on machine vision. The approach uses a high speed camera, an engineered environment, and image processing to determine the shape of our continuum robot called the Elephant's Trunk Manipulator. We present experimental results showing the effectiveness of the technique.

I. Introduction

Traditionally, robotic manipulators have been comprised of a small number of serially connected rigid links and actuated joints. Though these manipulators prove to be very effective for many tasks, they are not without their limitations. These limitations are due in the most part to their lack of maneuverability or total degrees of freedom. However, continuum style (i.e. continuous "backbone") robots, introduced in [6], exhibit a wide range of maneuverability, and can have a large number of degrees of freedom. Several different types of continuum manipulators can be seen in [1][2][4][5]. The motions exhibited by these continuum style robots are produced by bending the individual sections of the robot over their entire length; unlike traditional robots where the motions occur in discrete locations, i.e. joints. The motions displayed by continuum manipulators are often compared to those of biological manipulators such as trunks and tentacles. Continuum style robots can achieve postures that could only be obtainable by conventionally designed robots that contain many more degrees of freedom.

One of the biggest problems with continuum style robots is the sensing of the robot's configuration. Presently, there are no straightforward solutions for determining an accurate representation of these robot's shape. This is a potentially severe drawback which currently limits real-time implementation of these types of robots. The method that is used by the Elephant's Trunk at Clemson derives its shape based on its complex kinematic model and measurements of the change in length of its actuated cables [3]. Though this method proves to be reasonably accurate, a more precise measurement is desirable. A quite logical and intuitive approach to this shape determination problem is through the use of image processing. The concept is simple. First the robot is located in an image, and then the shape of the robot can be found. However, though the concept might be easy, the actual implementation of such a strategy is very difficult to achieve, especially if it is desired to accomplish it in real time. In this paper the concept of machine vision, with an emphasis on real time performance, is applied to help determine an accurate shape for the Elephant's Trunk Robotic Manipulator [3].

II. Experimental Setup

The experimental setup was composed of three main components: a Dalsa CCD high speed camera, a BitFlow Road Runner frame grabber, and an AMD 1300 MHz based PC. The Dalsa camera was selected due to its maximum

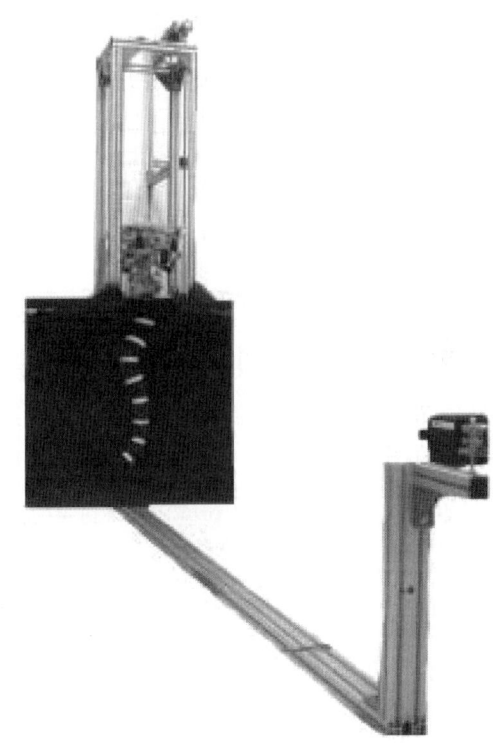

Fig. 1. Experimental Setup

frame rate of 955 frames per second. The camera was fixed at a distance of 3.05 m (10 ft) from the robot, see Figure 1, and thus giving a relatively wide view of the robot's workspace. To help facilitate real time performance the environment in the field of view of the camera was designed so that the desired regions of the robot could easily be located. This was done by making all non-important areas as dark as possible, and all the desired areas were colored as bright as possible, see Figure 1. This significantly reduces the amount of image processing that needs to be done, and thus decreases the overall time needed for image processing.

III. Image Processing

The Elephant's Trunk Manipulator is composed of four main sections. Each of these sections can bend over its

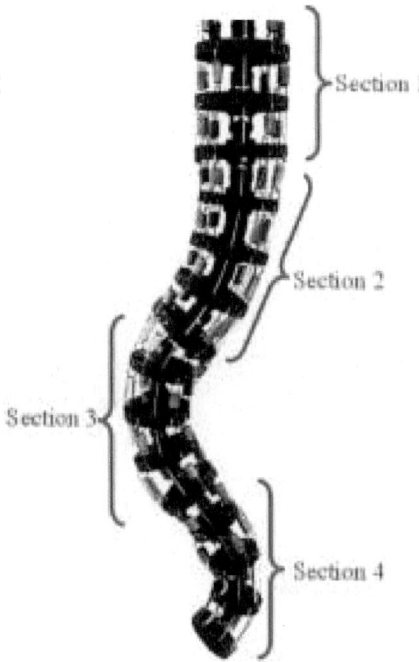

Fig. 2. Elephant's Trunk Robotic Manipulator

entire length as can be seen in figure 2. In this paper we only consider planar motion of the robot. The kinematics model the planar robot as a serial connection of four curves with constant curvature [3]. The objective is to determine a curvature for each section, and thus determine the shape of the robot. The fundamental principal exploited to accomplish this is that the curvature for a curve requires a minimum of three points [8]. In the case of a curve with constant curvature three points can completely define the entire curve. The most obvious example of a curve with constant curvature is that of a circle. Since the kinematic model of the Elephant's Trunk is based on constant curvature sections, each one of the sections can be expressed geometrically as part of a circle. Therefore, an appropriate strategy is to fit a circle to each of the sections. If three points can be determined from an image that represent the shape of one section, then a circle, and thus a curvature can be found for the section. The points that will yield the most accurate results for each section are the ones at each end of the section, and one point approximately in the center of the section, where all three points should be located on the center line of the section.

The algorithm used to find the three points is based on finding the center of each of three bands located on each section, see figure 3. One band is located at the beginning of a section, the next is located in the middle, and the third one is located at the end of the section. Standard image processing techniques are used to locate each band, and then find its center. This will give the three proper points need to define a circle, and thus the curvature for the section. This process is applied to all four sections of the robot, where the last band of one section is the first band of the next section. Since there are four sections, then a total of nine bands are needed to find all the points. However, the first band on the base section of the robot is fixed due to the design of the robot. Therefore, no band is needed for the first point of the first section since it can never move, and that is why in Figure 1 only eight bands are visible. In the next section the complete details involved in processing one frame of this strategy is presented.

A. Initial Processing

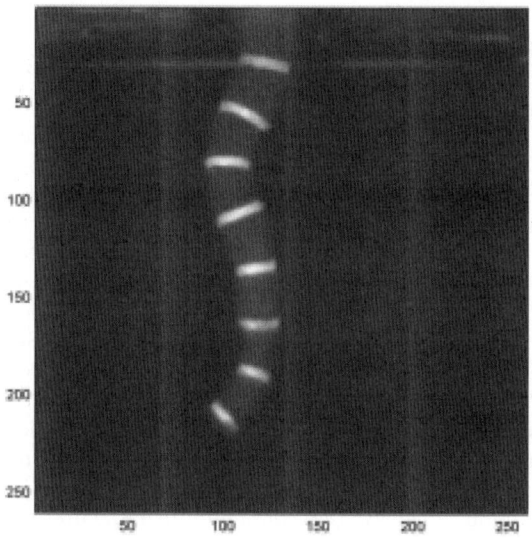

Fig. 3. Image Captured by the Camera

A typical image captured from the camera can be seen in Figure 3. Note, the image in Figure 3 was brightened by an intensity of 15 to make it viewable. Once the image is captured with the camera it needs to be processed such that the shape of the robot can be determined.

Initially it needs to be determined which pixels in the image are ones that are of value, and ones that are not. The first step is to use the process of thresholding. Thresholding an image will set all the pixels greater than a given intensity to one value of intensity, and any pixels with a value less than the given value to another intensity value. To help determine which value of intensity should be used in thresholding the histogram of the image is used. The histogram of the image captured by the camera is shown in Figure 4. It can be seen from the histogram that all of the pixel intensities are in the range from 5 to 25. Thus,

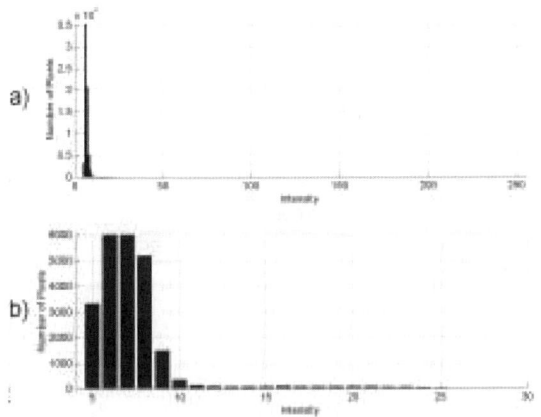

Fig. 4. Histogram of the Captured Image: a) Results for all 256 Levels of Intensity b) Closeup Examination of Important Pixel Intensities

the threshold value was initially picked in this range. After

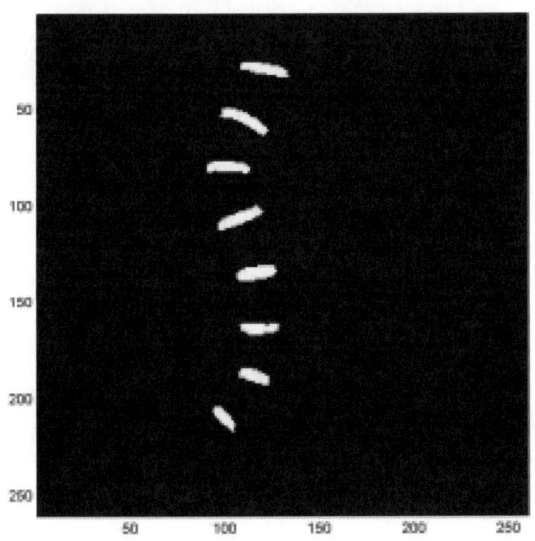

Fig. 5. Thresholded Image

some tuning, a threshold value of 12 was determined to yield the best results in separating useful pixels (bands) from the none useful ones (background). To threshold the image all the pixels with intensities above 12 had their intensity set to 256, and all the pixels with intensities less than 12 were set to 0. The resulting image after thresholding is shown in Figure 5.

Once the useful pixels have been found, the pixels must be sorted in such a way that they can be defined into logical regions. These regions can be defined using a process called segmentation. In segmenting the image an initial "seed" pixel is defined, and all the pixels that have connectivity with this seed pixel are defined as a region. The seed pixel simply starts out at the first pixel in the image, and after a region is "grown" around this seed pixel the location of the pixel is moved. Normally this seed pixel is simply shifted by one pixel from the previous location, but to save time the seed point was shifted by 12 pixels every time instead of just one. This can be done since the bands occupy a large number of pixels, and thus only one out of these many pixels needs to be found as a seed for the region to be properly determined. In segmenting the image in Figure 5, many different regions were found. Since only the regions that represent the eight bands are desired, another method was required to filter out the unwanted regions. This was accomplished by simply filtering out any regions that did not have an area between 50 and 175 pixels. The resulting segmented and filtered image is shown

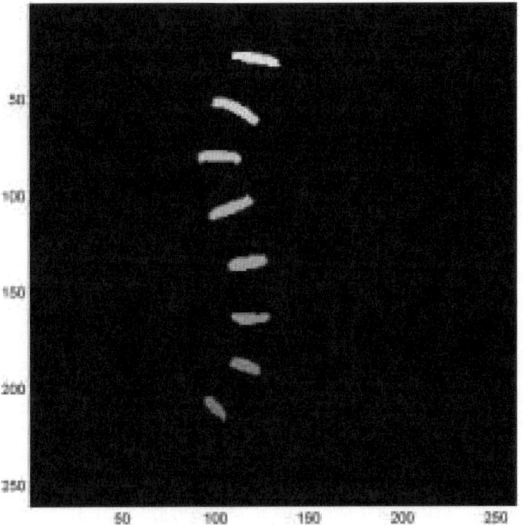

Fig. 6. Segmented Image

in Figure 6. Each region has its own pixel intensity, and this intensity is defined as the label for that region. It can be seen from Figure 6 that all eight desired regions were found, and each of which has its own label (color).

B. Determination of Curvature

Now that all eight desired regions have been located the process of determining the four curvatures for the robot can be started. To determine the curvature the center point of each section must first be determined. This was accomplished by determining the centroid (center of mass) of each of the region. As presented in [7] the centroid of each region given by its pixel coordinates of x_c and y_c can be calculated using the following equations

$$m_{pq} = \sum_{-\infty}^{\infty} \sum_{-\infty}^{\infty} i^p j^q f(i,j) \qquad (1)$$

$$x_c = \frac{m_{10}}{m_{00}}$$
$$y_c = \frac{m_{01}}{m_{00}}, \qquad (2)$$

where i and j are the pixel coordinates and $f(i,j)$ is the value of the pixel located at (i,j). The resulting centroids

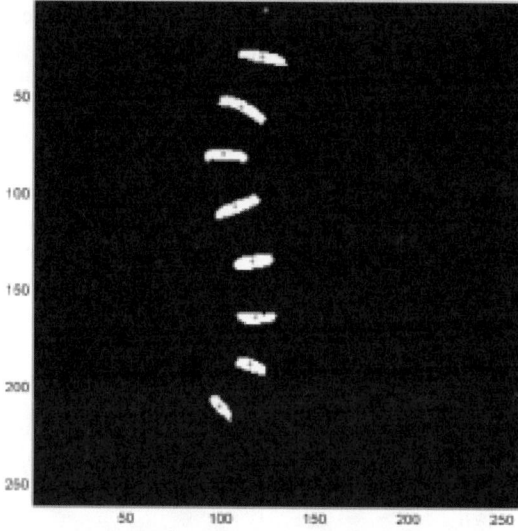

Fig. 7. Centroids of the Regions

for each band can be seen as single gray pixels in the regions shown in Figure 7.

Even though the centroids of each region have now been found, the curvature cannot yet be directly calculated for each section. The robot was designed such that each section could be defined as set of three regions. The first section is defined by the base location and regions one and two. The second section is defined by regions two, three, and four. The third sections is defined by regions four, five, and six. The fourth and final section is defined by regions six, seven, and eight. The problem is that so far the regions that have been determined from the image are in no specialized order. So, before the curvatures can be determined the regions must be ordered. Using the orientation of the principal axis for each region as a reference, the next section can be found by using a simple search algorithm. The orientation θ of the principal axis of a region is calculated as

$$m_{pq} = \sum_{-\infty}^{\infty}\sum_{-\infty}^{\infty} (i - x_c)^p (j - y_c)^q f(i,j) \quad (3)$$

$$\theta = \frac{1}{2}\arctan\left(\frac{2m_{11}}{m_{20} - m_{02}}\right). \quad (4)$$

Given the predefined shape of a region its principal axis will be perpendicular to the direction of the next region. The search algorithm uses the normal to the principal axis of the present region to determine where to look for the next region. The proper direction (±90 degrees) of the vector can be determined by comparing the normal vector's orientation to that of the orientation of the vector linking the present centroid's location to the previous centroid's location. Once the proper direction to search has been determined the algorithm then searches in an arc with a radius of 29 pixels over a range of 0.9 radians that is symmetric about the normal. The algorithm moves along the arc checking at each iteration what the pixel label is. If the label is that of the background, then the search continues. However, if the pixel is not labeled as background, then the label of the pixel is checked against the list of labels for the regions. The region the label belongs to is the next region in the order. This process is repeated until all of the regions are ordered. Figure 8 shows the arcs used to

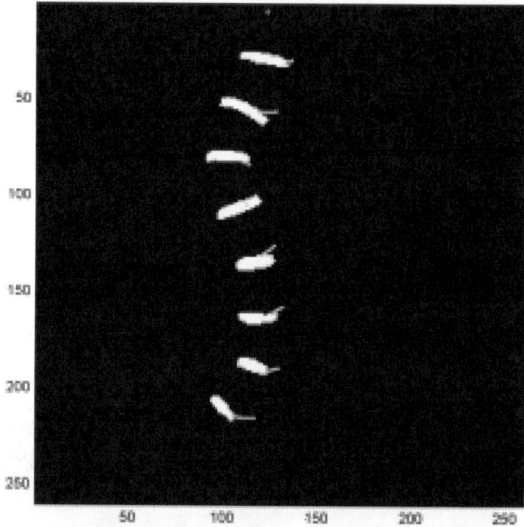

Fig. 8. Region Ordering Algorithm

find each region in the search algorithm. Note that the last ordered centroid corresponds to the end point of the manipulator, and thus provides an easy way to determine the error $\mathbf{x}_d - \mathbf{x}$ if needed.

Now that the order of the regions and their corresponding centroids have been determined, the curvature for the four sections of the robot can be found. Since a circle is defined to have a constant curvature, the curvature of each section can be found by fitting a circle through the three centroids that corresponds to each section. The general equation for a circle is

$$(x-a)^2 + (y-b)^2 = r^2, \quad (5)$$

where (a,b) is the location of the center of the circle in terms of (x,y) and r is the radius of the circle. Using the following substitution

$$\alpha = a^2 + b^2 - r^2. \quad (6)$$

Equation (5) can be rewritten as

$$2xa + 2yb - \alpha = x^2 + y^2. \quad (7)$$

Since Equation (7) is a linear equation it can be rewritten in terms of the three centroid locations as

$$\begin{bmatrix} 2x_1 & 2y_1 & -1 \\ 2x_2 & 2y_2 & -1 \\ 2x_3 & 2y_3 & -1 \end{bmatrix} \begin{bmatrix} a \\ b \\ \alpha \end{bmatrix} = \begin{bmatrix} x_1^2 + y_1^2 \\ x_2^2 + y_2^2 \\ x_3^2 + y_3^2 \end{bmatrix}, \quad (8)$$

where (x_1, y_1) is the location of the first centroid, (x_2, y_2) is the location of the second centroid, and (x_3, y_3) is the location of the third centroid. Since Equation (8) has three unknowns and three equations it can be solved for the parameters a, b, and α as

$$\begin{bmatrix} a \\ b \\ \alpha \end{bmatrix} = \begin{bmatrix} 2x_1 & 2y_1 & -1 \\ 2x_2 & 2y_2 & -1 \\ 2x_3 & 2y_3 & -1 \end{bmatrix}^{-1} \begin{bmatrix} x_1^2 + y_1^2 \\ x_2^2 + y_2^2 \\ x_3^2 + y_3^2 \end{bmatrix}. \quad (9)$$

Once α is found, r can be found by solving Equation (6). From [8], the curvature κ for the section is simply calculated as

$$\kappa = \frac{1}{r}. \quad (10)$$

This technique is then applied to all the sections to obtain their respective curvatures.

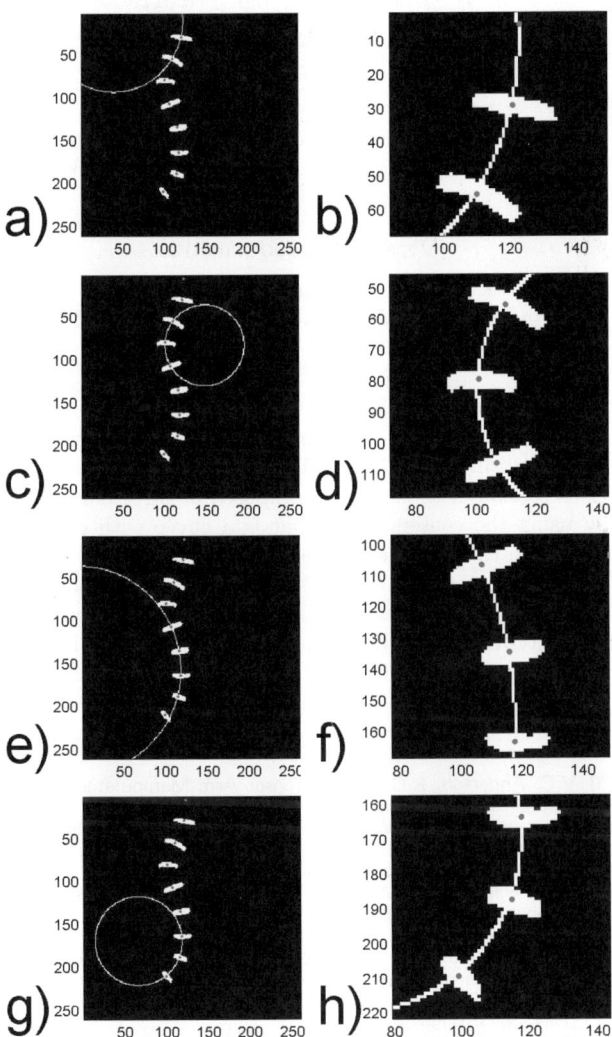

Fig. 9. Results of Circle Fitting Routine: a) Circle fit to Section 1 b) Closeup of Section 1 c) Circle fit to Section 2 d) Closeup of Section 2 e) Circle fit to Section 3 f) Closeup of Section 3 g) Circle fit to Section 4 e) Closeup of Section 4

To accomplish a constant update of curvatures in real time the image processing routine was run in parallel with the image capture routine. The two routines were synchronized by the capture routine signaling the processing routine when an image had just finished being captured. If the processing routine was not ready at this moment, it then waited until a new image was ready. This synchronization strategy was used to avoid data corruption. Thus, with the camera running at 955 fps, the only available frame rates for the image processing were $\frac{955}{n}$, where n is a positive integer. Due to optimization of both the environment and the image processing strategy the processing routine was able to run at 477.5 fps where it took about 1.67ms to process the image. Though this might not be considered real time, it is none the less very fast considering standard image processing routines typically run no faster than a few tens of frames per second. The frame rate is easily high enough to provide accurate and dependable information about the robot's shape to the controller. Though several optimizations were used to facilitate faster performance, this strategy proves that it is possible to use machine vision to determine the shape of a continuum robot.

C. Results

Figure 9 shows the circles fitted to each section. The curvature value for sections 1 to 4 are 0.012071, 0.021203, 0.008261, and 0.019422 where the units are $\frac{1}{pixel}$. The sign

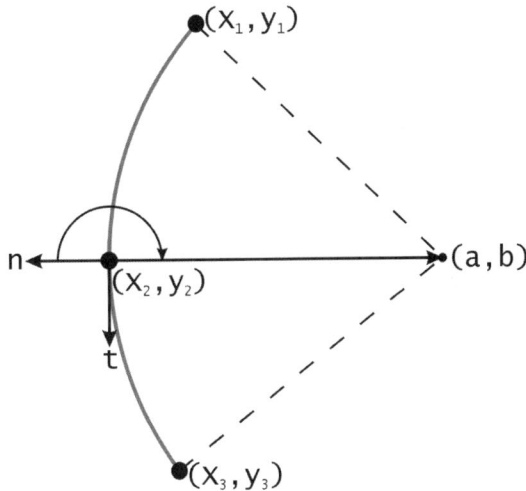

Fig. 10. Determination of the Sign of the Curvature

of the curvature can be calculated by comparing the direction of the principal normal of the Serret-Frenet frame at one of the centroids of one of the regions in the section to the direction of the vector from that centroid to the center of the calculated circle. The principal normal is simply the principal axis for the region, where the proper sign of the principal axis was determined during the centroid ordering algorithm. Referring to Figure 10, if the difference between the two vectors is 180 degrees than the curvature is positive, otherwise it is negative. Thus, the signed curvature values for sections 1 to 4 are +0.012071, -0.021203, +0.008261, and -0.019422.

To check the accuracy of the above results, they were compared to the curvature values determined by measuring the change in cable length. The values from the cable length strategy have the units of $\frac{1}{inch}$. Since the units between the two sets of curvatures are different, the curvatures from image processing can be converted to the units of $\frac{1}{inch}$. The conversion was experimentally verified to be 7.42 pixels per inch. All of the curvature values are listed in the following table.

Section	κ_v	κ_c	Δ	%
1	+0.05541	+0.08956	−0.03415	−38.1
2	−0.1085	−0.1573	−0.0488	+31.0
3	+0.05133	+0.06130	−0.00997	−16.3
4	+0.1311	+0.1441	−0.013	−9.02

where κ_v is the curvature from image processing $\left(\frac{1}{inch}\right)$, κ_c is the curvature derived from the change in cable length $\left(\frac{1}{inch}\right)$, $\Delta = \kappa_v - \kappa_c$, and % is the percent error between κ_v and κ_c, which is calculated as $\% = 100(\frac{\Delta}{\kappa_c})$. Two other configurations were measured, and their results are presented below.

Section	κ_v	κ_c	Δ	%
1	−0.0484	−0.0288	−0.0196	+68.1
2	−0.0448	−0.042	−0.0028	+6.67
3	+0.1068	+0.0963	+0.0105	+10.9
4	+0.1131	+0.0801	+0.033	+41.2

Section	κ_v	κ_c	Δ	%
1	−0.1201	−0.0724	−0.0477	+65.9
2	+0.0901	+0.0685	+0.216	+31.5
3	+0.1410	+0.1087	+0.0323	+29.7
4	−0.1285	−0.131	+0.0025	−1.91

The average magnitude of error for the three configurations is 57.4%, 23.1%, 19.0%, and 17.4% respectively for each section. The results show that there is a rather large difference between the curvature measurements of the cable strategy and the vision strategy. Thus, more experimentation needs to be done to help verify the accuracy of both the cable length based curvatures and the vision based curvatures. However, we are confident that most of the error arises from the cable measurement based method as discussed below.

The errors in the cable based curvatures may be attributed to the coupling of actuation between sections. As a section moves it is possible that previous sections can change shape without the encoder detecting any movement. This problem is based upon the actuation strategy only using one motor and a cable tensioning system to actuate one pair of cables. As a section bends only the tensioned side of the section is being actuated be the motor, and thus being measured by the encoder. If the a section further down the robot actuates in the same direction as the present section, the tensioning system can allow cable to be released on the tensioned side and retract cable on the untensioned side. Thus, the section's shape will change in a way that cannot be determined by the encoders.

Another problem that may be affecting the results is that when using the encoders for determining the curvature there is no easy way to determine what the initial robot's shape is. The simplest approach is to always start the robot with all four sections straight. This is the only position where the curvature for each section can be easily determined without any type of external measuring device. The problem is that even when the robot looks straight, it may not be exactly straight. Thus, there will be some error in the difference between the cable based curvatures and the vision based curvatures. The following table shows the difference between the two strategies when the robot looks to be in the straight position.

Section	κ_v	κ_c	Δ
1	−0.0115	0.0	−0.0115
2	0.0	0.0	0.0
3	0.0	0.0	0.0
4	0.0288	0.0	0.0288

Even in the straight position, there is an appreciable amount of error between the two different curvature measurements.

IV. Acknowledgements

This work was supported in part by NASA grant NAG5-9785, and in part by NSF/EPSCoR grant EPS-9630167.

V. Conclusion

We have presented a vision based method that can determine the shape of a continuum robot. The method uses image processing to find several distinct bands on the robot. Each section of the robot is composed of three bands, and by using the centers of the bands a curve with constant curvature can be fitted to the section. The serial combination of the curves describes the shape of the robot. The method demonstrates that it is possible to not only use image processing to determine a more accurate shape of a continuum style robot, but it can be done at very high speeds (real time).

References

[1] V.C. Anderson, R.C. Horn, "Tensor Arm Manipulator Design" *ASME paper 67-DE-57*.

[2] I. Gravagne, I.D. Walker, "On the Kinematics of Remotely-Actuated Continuum Robots," *IEEE Conf. on Robotics and Automation*, pp. 2544-2550, 2000.

[3] M.W. Hannan, I.D. Walker, "Analysis and Experiments with an Elephant's Trunk Robot" *International Journal of the Robotics Society of Japan*, Vol. 15, No. 8, pp. 847-858, 2001.

[4] H. Mochiyama, E. Shimemura, H. Kobayashi. "Shape Correspondence Between a Spatial Curve and a Manipulator with Hyper Degrees of Freedom". *IEEE Conf. on Intelligent Robots and Systems*, pp. 161-166, 1998.

[5] H. Ohno, S. Hirose, "Study on Slime Robot (Proposal of Slime Robot and Design of Slim Slime Robot)," *IEEE Conf. on Intelligent Robots and Systems*, pp. 2218-2223, 2000.

[6] G. Robinson, J.B.C. Davies, "Continuum Robots - A State of the Art," *IEEE Conf. on Robotics and Automation*, pp. 2849-2854, 1999.

[7] M. Sonka, V. Hlavac, and R. Boyle, *Image Processing, Analysis, and Machine Vision*, Brooks/Cole Publishing Company, 1999.

[8] D.J. Struik, *Lectures on Classical Differential Geometry*, Addison-Wesley Publishing Company, 1961.

Learning Human Control Strategy for Dynamically Stable Robots: Support Vector Machine Approach[*]

Yongsheng Ou and Yangsheng Xu
Department of Automation and Computer-Aided Engineering
The Chinese University of Hong Kong

Abstract

In this paper, we discuss the problem of how human control strategy can be represented as a parametric model using a Support Vector Machine (SVM), and how an SVM-based controller can be used to effectively control a dynamically stable system. We formulate the learning problem as a support vector regression and develop a new SVM learning structure to better implement human control strategy learning in control. The approach is fundamentally valuable in dealing with problems that normally dynamically stable robots experience, such as small sample data and local minima, and therefore is extremely useful in abstracting human controller for dynamic systems. The experimental study on the SVM approach with respect to other approaches clearly demonstrated the superiority of the SVM approach in terms of fedility, efficiency and effectiveness in implementation.

1 Introduction

Since system dynamics plays an integral role in the behavior of dynamically stable systems, an accurate analytic model is necessary in order to apply classical control techniques, yet such a model is often difficult to arrive at in practice. In general, these systems exhibit dynamics that are highly coupled, nonlinear, and vary substantially depending on precise configuration of the systems; moreover, friction and other difficult-to-model physical parameters often impact dynamically stable systems more severely than conventional quasi-static systems (e.g. four-wheel vehicles).

Humans have shown themselves to be extremely adept at mastering the complex and difficult control of the dynamically stable systems. Our group has performed significant research in the modelling of human control strategies through observation, or learning, of experimentally collected human training data [4]. Much of this prior research has, however, focussed on quasi-static systems, and therefore suffers from weaknesses that leave the current methods ill-equipped to cope with the challenges of autonomously controlling dynamically stable systems. This paper presents a new approach to the problem from a statistical angle. While Gyrover will serve as a principal platform for validating and testing the proposed work, we nevertheless anticipate straightforward transfer of the developed methods to other dynamically stabilized systems, such as biped walkers, bicycles, juggles, and others where human demonstration of desired behaviors is readily achievable.

Considerable research efforts have been directed toward learning control architecture using artificial neural networks (ANN) [5] or HMMs [1]. Despite success, human-based control methods based on previous methods face several problems. For example, ANNs and HMMs have problems related to local minima. This drawback leads to more oscillations in experimental results - compared with the training state data - which sometimes causes failure. Moreover, another problem is that ANNs often require large amounts of training data to make the training process to reach satisfactory levels of precision. However, it is inconvenient and sometimes impractical to obtain large sets of training samples. Furthermore, robot behavior is difficult to analyze and interpret mathematically, and is highly dependent on their fixed architectures. In this paper we discuss how a SVM can be used for abstracting human control strategy (HCS) problems.

2 Learning Human Control Through SVM

2.1 Support Vector Machine

In this paper, the term SVM will refer to both classification and regression methods, and the terms Support Vector Classification (SVC) and Support Vector Regression (SVR) will be used for specification. In SVR, the basic idea is to map the data X into a high-dimensional feature space $\mathcal{F}$ via a nonlinear mapping Φ, and to do linear regression in this space [7].

$$f(X) = (\omega \cdot \Phi(X)) + b \quad \text{with } \Phi : R^n \to \mathcal{F},\, \omega \in \mathcal{F}, \tag{1}$$

where b is a threshold. Thus, linear regression in a high dimensional (feature) space corresponds to nonlinear regression in the low dimensional input space

---

[*]This work is supported in part by Hong Kong Research Grant Council under the grants CUHK 4403/99E, CUHK 4228/01E and Hong Kong SAR Government under grand ITS/140/01.

R^n. Note that the dot product in Equation 1 between ω and $\Phi(X)$ would have to be computed in this high dimensional space (which is usually intractable), if we are not able to use the kernel that eventually leaves us with dot products that can be implicitly expressed in the low dimensional input space R^n. Since Φ is fixed, we determine ω from the data by minimizing the sum of the empirical risk $R_{emp}[f]$ and a complexity term $\|\omega\|^2$, which enforces flatness in feature space

$$\begin{aligned} R_{reg}[f] &= R_{emp}[f] + \lambda \|\omega\|^2 \\ &= \sum_{i=1}^{l} C(f(X_i) - y_i) + \lambda \|\omega\|^2, \end{aligned} \quad (2)$$

where l denotes the sample size $(x_1, ..., x_l)$, $C(.)$ is a loss function and λ is a regularization constant. For a large set of loss function, Equation 2 can be minimized by solving a quadratic programming problem, which is uniquely solvable [3]. It can be shown that the vector ω can be written in terms of the data points

$$\omega = \sum_{i=1}^{l} (\alpha_i - \alpha_i^*) \Phi(X_i), \quad (3)$$

with α_i, α_i^* being the solution of the aforementioned quadratic programming problem [7]. α_i and α_i^* have an intuitive interpretation as forces pushing and pulling the estimate $f(X_i)$ towards the measurements y_i [6]. Taking Equation 3 and Equation 1 into account, we are able to rewrite the whole problem in terms of dot products in the low dimensional input space

$$f(x) = \sum_{i=1}^{l} (\alpha_i - \alpha_i^*)(\Phi(X_i) \cdot \Phi(X)) + b. \quad (4)$$

For a more detailed reference on the theory and computation of SVM, readers are referred to [6].

3 SVM Learning Architecture

The idea of implicitly mapping data into a high-dimensional feature space has been a very fruitful one in the context of SVMs. Indeed, it is one of the major features which distinguish them from the general ANN algorithm.

3.1 Linear Relation in Feature Space

A polynomial kernel

$$k(Y, X) = (Y \cdot X + c)^d, \ c \geq 0 \quad (5)$$

can be shown to correspond to a map Φ into the space spanned by all products of order up to d (i.e., including those of an order smaller than d). For $d = 2$, $c = 0$ and $n = 2$, i.e., $X, Y \in R^2$, e.g., we have

$$(X \cdot Y)^2 = x_1^2 y_1^2 + x_2^2 y_2^2 + 2x_1 x_2 y_1 y_2 = (\Phi(X) \cdot \Phi(Y)),$$

define $\Phi(X) = (x_1^2, x_2^2, \sqrt{2} x_1 x_2)$.

For example, let us try to estimate a function

$$y = f(X) = x_1^2 + x_2^2 + 4x_1 x_2,$$

where $X \in R^2$ and $y \in R$. For a SVM learner, the input vector is (x_1, x_2). If we choose the above kernel with $d = 2$, $c = 0$ and $n = 2$, in the feature space the element vector is $(x_1^2, x_2^2, x_1 x_2)$. The input vector (x_1, x_2) has a nonlinear relation with the output y ($y = x_1^2 + x_2^2 + 4x_1 x_2$), whereas the feature space element vector has a linear relation with y as shown in Figure 1. For this relation, from Equation 4, it is faster to seek the optimal α_i and α_i^* from a smaller number of training samples than a Gaussian Radial Basis function (RBF) kernel SVM. This process decreases the time in seeking the optimization and the number of required support vectors.

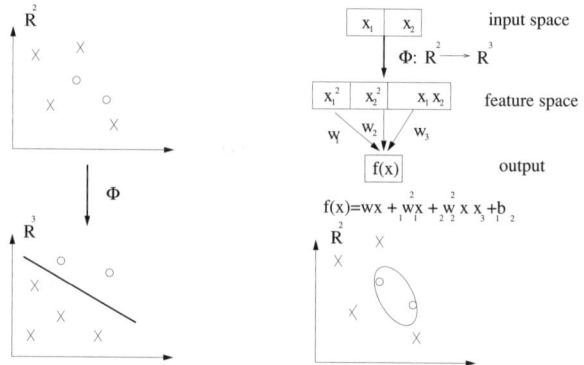

Figure 1: Linear relations in feature space.

3.2 Prior Knowledge in Robotic Systems

In many learning problems, such as pattern recognition and fault diagnosis, the relations between inputs and outputs are not simple and clear. In these cases, we have to choose relatively complex kernels such as an RBF kernel and use more training samples to train an SVM model.

However, and fortunately for robots and many other mechanical systems, we have extensive prior knowledge of dynamics and controller design. For example, a general dynamic equation of a robotic system is

$$M(q)\ddot{q} + N(q, \dot{q}) + G(q) = \tau,$$

where $q, N, G, \tau \in R^n$, $M \in R^{n \times n}$. Assuming that it is a position control, we can have a controller form as

$$\tau = M(q)^{-1}(-k_p \dot{q} - k_d(q - q_d) + N(q, \dot{q}) + G(q)) \quad (6)$$

where q_d is the control target vector. k_p and k_d can be chosen appropriately so that the characteristic roots of Equation 6 have negative real parts.

For example, let us consider the inverted pendulum system described in Section 5. Equation 10 is a typical example of this kind of controllers. An SVM learning controller can be produced from the sample data with the inputs $(\theta, \dot{\theta})$ and the output F. If we have no information about the controller form, we may choose an RBF kernel for training an SVM controller.

Based on the analysis above, we hope to build a linear relation on the feature space. Obviously, the SVM learning controller with an RBF kernel is not sufficient. If it was, then the feature space vector should have elements of

$$\{\dot{\theta}/C_\theta, \dot{\theta}C_\theta, \theta/C_\theta, \theta C_\theta, \dot{\theta}^2 S_\theta, S_\theta/C_\theta\}.$$

Present kernel functions can not transfer the input vector to this kind of feature space vector. To solve this problem, we have proposed a new extended space that is located between the input space and the feature space. For the above example, we set the new extended space as $(\theta, \dot{\theta}, S_\theta, C_\theta, 1/C_\theta)$. If q_i is an angle state variable, then $\sin(q)$ and $\cos(q)$ usually appear in the dynamic equations and controller. Moreover, in the example above, the $1/C_\theta$ term comes from $M(q)^{-1}$. The simulations in Section 5 demonstrate the advantages of this new method in comparison to the RBF kernel approach.

3.3 A New SVM Learning Architecture

We summarize our idea for a new SVM learning strategy for robotic state feedback systems in Figure 2. The formation of an extended space vector is mainly based on the control problems and the experience of users.

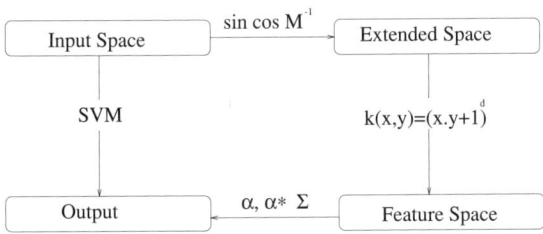

Figure 2: A new learning strategy for SVM implementation.

4 SVM-Based Learning

4.1 Training Example Collection

An SVM does not require a large amount of training samples as do most ANNs. Scholkopf [2] pointed out that the *actual risk* $R(w)$ of the learning machine is expressed as:

$$R(w) = \int \frac{1}{2} \|f_w(\mathbf{x}) - y\| dP(\mathbf{x}, y). \quad (7)$$

The problem is that $R(w)$ is unknown, since $P(\mathbf{x}, y)$ is unknown.

The straightforward approach to minimize the *empirical risk*,

$$R_{emp}(w) = \frac{1}{l} \sum_{i=1}^{l} \frac{1}{2} \|f_w(\mathbf{x})_i - y_i\|,$$

turns out not to guarantee a small actual risk $R(w)$, if the number l of training examples is limited. In other words: a small error on the training set does not necessarily imply a high *generalization* ability (i.e., a small error on an independent test set). This phenomenon is often referred to as *overfitting*. To solve the problem, novel statistical techniques have been developed during the last 30 years. For the learning problem, the *Structure Risk Minimization* (SRM) principle is based on the fact that for any $w \in \Lambda$ and $l > h$, with a probability of at least 1-η, the bound

$$R(w) \leq R_{emp}(w) + \Psi(\frac{h}{n}, \frac{log(\eta)}{n}) \quad (8)$$

holds, where the *confidence term* Ψ is defined as

$$\Psi(\frac{h}{n}, \frac{log(\eta)}{n}) = \sqrt{\frac{h(log\frac{2n}{h} + 1) - log(\eta/4)}{n}}.$$

The parameter h is called the VC (*Vapnik-Chervonenkis*) dimension of a set of functions, which describes the capacity of a set of functions. Usually, to decrease the $R_{emp}(w)$ to some bound, most ANNs with complex mathematical structure have a very high value of h. It is noted that when n/h is small (for example less than 20, the training sample is small in size), Ψ has a large value. When this occurs, performance poorly represents $R(w)$ with $R_{emp}(w)$. As a result, according to the SRM principle, a large training sample size is required to acquire a satisfactory learning machine. However, by mapping the inputs into a feature space using a relatively simple mathematical function, such as a polynomial kernel, a SVM has a small value for h, and at the same time maintains the $R_{emp}(w)$ in the same bound. To decrease $R(w)$ with Ψ, therefore, requires small n, which is enough also for small h.

4.2 Training Process

The first problem, before starting training, is to choose between SVR or SVC; the choice being dependent on the type of the control inputs. For example, our experimental system Gyrover has the control commands U_0 and U_1. Their values are scaled to the tile angle of the flywheel and the driving speed of Gyrover respectively. Thus, for this case, we will choose SVR.

The second problem is to select a kernel. There are several kinds of kernel appropriate for an SVM [6]. As mentioned in Section 3, the main function of kernels in an SVM is to map the input space to a high dimensional feature space, and at that space the feature elements have a linear relation with the learning output. Thus, it is better to test more kernels and choose the best one.

The third problem is about "scaling". Here, scaling refers to putting each column's data into a range between -1 and 1. It is important to void the outputs if they are seriously affected by some states, for the "unit" sake. Moreover, if some states are more important than others, we may enlarge their range to emphasize their effect.

4.3 Learning Precise Consideration

In practice applications, the control process of an SVM learner is much smoother than general ANN learners. The reason for this is that many ANNs and HMM methods exhibit the local minima problem, as shown in figure 3 whereas for an SVM the quadratic property guarantees the optimization functions are convex. Thus it must be the global minima. The only problem that needs to be considered concerns the linear relation on the feature space; if it is not linear the learning process may not converge.

Moreover, for the class of dynamically stable, but statically unstable robots such as Gyrover, high learning precision is required and vitally. This is because the learning controller will be controlling the robot and form a new dynamic system.

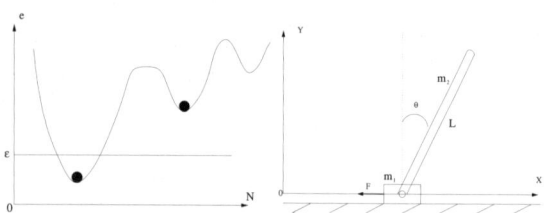

Figure 3: Local minima.   Figure 4: A pendular-robot system.

5 Simulation Study for a New SVM Learning Strategy

In these simulations, we compare the new support vector machine learning (SVM-NEW), addressed in Section 3 to an RBF kernel support vector machine (SVM-RBF) by learning from a model-based control for an inverted-pendulum system.

5.1 Inverted-pendular System

As shown in Figure 4, a pendulum can rotate around a block, which can slide along the X axis freely, i.e., there is no friction in the system.

Let $S_x := \sin(x)$, $C_x := \cos(x)$. The extended dynamic model of the system is

$$\begin{bmatrix} m_1 + m_2 & m_2 LC_\theta \\ m_2 LC_\theta & \frac{4}{3}m_2 L^2 \end{bmatrix} \begin{bmatrix} \ddot{x} \\ \ddot{\theta} \end{bmatrix} = \begin{bmatrix} m_2 LS_\theta \dot{\theta}^2 \\ m_2 LgS_\theta \end{bmatrix} + \begin{bmatrix} -F \\ 0 \end{bmatrix} \quad (9)$$

Our control target is to stabilize the pendulum to the vertical position. We do not consider the position of the block.

5.2 Model-based Control Strategies

Next, we addressed the control law as follows.

$$F = (-3H\dot{\theta} - 2H\theta + m_2 LS_\theta C_\theta \dot{\theta}^2 - (m_1 + m_2)gS_\theta)/C_\theta \quad (10)$$

where $H = \frac{4}{3}(m_1 + m_2)L - m_2 LC_\theta^2$.

Here, we omit the strict mathematical proof. The following simulations also verify it. For the purposes of simulation, the system parameters are as follows: $m_1 = 1.05kg$, $m_2 = 0.5kg$, $g = 9.8m/s^2$, $L = 1m$.

First, we use the model-based controller to manage the dynamic system. The initial parameters are as follows: $\dot{\theta}_0 = 1rad/s$, $\theta_0 = \pi/2.2rad/s$ and the control targets are $\dot{\theta}_d = 0$, $\theta_d = 0$. The time span of each control process is $T = 4.0s$, and the sampling time is $\delta T = 0.001s$. The results are shown in Figures ?? and 5 and the control input F is shown in Figure ??.

5.3 Comparison of SVM Learning Control Strategies

We conducted two simulations. In the first simulation, we chose an RBF kernel with the parameter $\gamma = 10$ and the kernel function $k(y, x) = e^{-\gamma|x-y|^2}$. θ, $\dot{\theta}$ are the inputs and the system control input F is the output of the SVM learning. We selected 400 training samples, or 1 in every 10 from among 4000 samples collected. We used about 21 minutes at the workstation to produce 187 support vectors from the training data.

We then changed the input $(\theta, \dot{\theta}, S_\theta, C_\theta, 1/C_\theta)$ and the kernel function $k(y, x) = (y \cdot x + 1)^5$. We selected the same training samples as in the RBF example. However, we used only about 6 minutes at the same workstation to produce 16 support vectors from the training data. Thus, the second simulation was much faster and produced less support vectors.

In these simulations, the SVM controllers sought to realize the same control targets that have been addressed above. The initial and constant parameters were not changed, The time span of the simulation was $T = 4.0s$, and the sampling time was $\delta T = 0.001s$.

The simulation results and comparisons with the above SVM controller are shown in Figure 5.

The figures clearly show that the new SVM has greater real-time capacity than the RBF one.

It is usual in learning human control that we have some prior knowledge about the learning objectives, especially for robotic systems. Even though we do

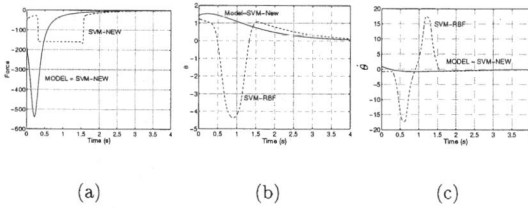

(a) (b) (c)

Figure 5: The comparisons of F, θ and $\dot{\theta}$.

not know what the equations are exactly, we often know the forms of the controllers. However, during the application of most ANNs, we seldom consider using this knowledge to adjust them to fit different systems because most ANNs have complex and relatively fixed mathematical structures.

Moreover, if we take the previous knowledge of robotic systems, we can modify the SVM learning architecture to make the learning process faster and more precise. These simulations clearly show the preponderance of this new SVM learning architecture.

6 Experiments

6.1 Experimental System – Gyrover

The single-wheel gyroscopically-stabilized robot, Gyrover, takes advantage of the dynamic stability of a single wheel. Figure 6 shows a photograph of the third Gyrover prototype.

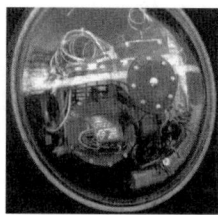 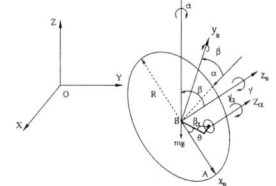

Figure 6: Gyrover: A single-wheel robot.

Figure 7: Definition of Gyrover's system parameters.

Gyrover is a sharp-edged wheel with an actuation mechanism fitted inside the rim. The actuation mechanism consists of three separate actuators: (1) a spin motor, which spins a suspended flywheel at a high rate and imparts dynamic stability to the robot; (2) a tilt motor, which steers Gyrover; and (3) a drive motor, which causes forward and/or backward acceleration by driving the single wheel directly.

Gyrover is a single-wheel mobile robot that is dynamically stable but statically unstable. It has both first-order and second-order nonholonomic constraints.

To represent the dynamics of Gyrover, we need to define the coordinate frames: three for position (X, Y, Z), and three for the single-wheel orientation (α, β, γ). The Euler angles (α, β, γ) represent the precession, lean and rolling angles of the wheel respectively. (β_a, γ_a) represent the lean and rolling angles of the flywheel respectively. They are illustrated in Figure 7.

6.2 Task and Experimental Description

The aim of this experiment is to compare the ability of an SVM and a general ANN in learning human control skills. We use a special control problem to illustrate and verify the previous analysis in modelling stochastic human control aspect.

The control problem consists of tracking Gyrover in a circle within a defined radius. We will build up the two learning controllers based on learning imparted from expert human demonstrations.

Achieving this goal requires two major control inputs: U_0 controlling the rolling speed of the single wheel $\dot{\gamma}$, and U_1 controlling the angular position of the flywheel β_a. During all experiments, we fix the value of U_0. The state variables β, $\dot{\beta}$, β_a, $\dot{\beta}_a$, $\dot{\alpha}$ are used during the training process. The state variable $\dot{\gamma}$ is removed because it was constant for most of the time and too noisy to produce good results. In order to construct a controller for tracking a circle, the trained model is adjusted in light of the above state variables, and its output is U_1.

A human expert controlled Gyrover to track a fixed 3-meter radius circular path two times and produced around 17000 and 18000 training samples respectively. Table 1 displays some raw sensor data from the human expert control process.

Input					Output
β	$\dot{\beta}$	β_a	$\dot{\beta}_a$	$\dot{\alpha}$	U_1
5.5034	0.9790	2.4023	0.8765	836.00	179.00
5.7185	1.2744	2.3657	1.4478	766.00	176.00
5.6012	-0.8374	2.1313	1.0767	566.00	170.00
5.1271	0.6641	2.1460	0.6030	554.00	170.00
5.9433	-0.4907	1.0425	1.3574	486.00	143.00

Table 1: Sample human control data.

6.3 Experimental Results and Discussions

We combined training sample data from both groups into a cascade neural network architecture with node-decoupled extended Kalman filtering learner (CNN) to obtain the weight matrix. Figure 8 is a section of a video tape portraying the CNN learning control and demonstrates the success of the controller and the oscillated performance of Gyrover. For an SVM

Figure 8: Video tape showing CNN learning control.

Figure 9: Video tape showing SVM learning control.

learning controller, we use only the first group of training data and select one in ten data sets. Vapnik's Polynomial kernel of order 2 is chosen and the input vector consists of current state variables (β_a, β, $\dot{\beta}_a$, $\dot{\beta}$, $\dot{\alpha}$). The output consists of control input U_1. Figure 9 is a section of video tape showing an SVM learning control. It demonstrates the success of the controller and the smoother performance of Gyrover than when learning control is by a CNN.

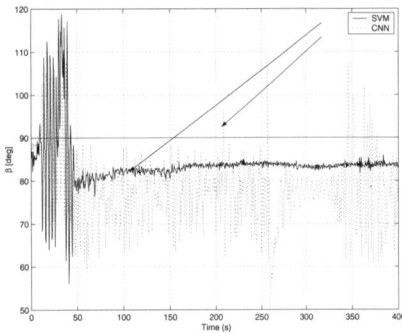

Figure 10: Lean angle β of SVM and CNN.

This phenomenon can also be seen from the experimental results of the lean angles β in CNN and SVM control in Figure 10. In comparing the CNN learner and SVM learner by testing the same set of human control training data, we know that the SVM learner produces smaller errors with regard to learning output and control input U_1, as shown in Figure 11 and Figure 12.

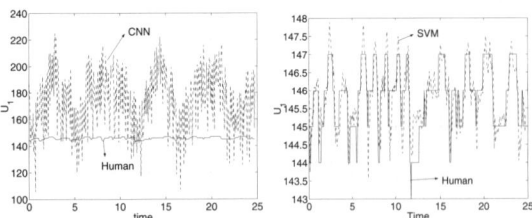

Figure 11: U_1 of Human control and CNN learner.   Figure 12: U_1 of Human control and SVM learner.

We have demonstrated that an SVM has better performance in human control strategy learning of dynamically stable systems than other ANNs. The reason for this is that many ANNs exhibit the local minima problem as shown in Figure 3. This causes the learning process to stop at the local minima before reaching the desired error bound. Thus, most ANN learning controllers have great difficulty in obtaining high precision, and experimental results are often much more oscillated. For an SVM, the quadratic optimization process guarantees the solution must be the global minima in feature space. Moreover, while the first SVM learning controller can work well, we require several training times to obtain a successful CNN learning controller.

7 Conclusion

In this paper, we employed the SVM approach based on statistical learning theory to overcome the limitations of the general ANN. The approach is fundamentally helpful in solving problems that normally a dynamically stable robot experiences, such as small sample data and local minima, and therefore is extremely useful in abstracting human controller for a dynamically stable system. The experimental comparison work verified our analysis and the superiority of the SVM method over the other methods.

Moreover, the study showed that based on the prior knowledge in robotic dynamics and control, our new SVM learning architecture for control strategy learning has better performance both in training process and in real-time implementation.

References

[1] J. Yang, Y. Xu and C. S. Chen, "Hidden Markov model approach to skill learning and its application to telerobotics," *IEEE Trans. on Robotic and Automation*, vol. 10, no. 5, pp. 621-631, 1994.

[2] S. Bernhard, C. J. C. Burges and A. Smola, "Advanced in kernel methods support vector learning," *Cambridge, MA, MIT Press*, 1998.

[3] A. Smola, "General cost function for support vector regression," *Proceedings of the Ninth Australian Conf. on Neural Networks*, pp. 79-83, 1998.

[4] M. C. Nechyba and Y. Xu, "Stochastic similarity for validating human control strategy models," *IEEE Trans. on Robotics and Automation*, vol. 14, no. 3, pp. 437-451, 1998.

[5] P. J. Antsaklis, Guest Editor, "Special issue on neural networks in control systems," *IEEE Contr. Syst. Mag.*, vol 12, no. 2, pp. 8-57, 1992.

[6] N. Cristianini and J. Shawe-Taylor, "A introduction to support vector machines and other kernel-based learning methods," *Cambridge University Press*, Cambridge, 2000.

[7] V. Vapnik, "The nature of statistical learning theory," *Springer-Verlag*, New York, 1995.

Perceptual Navigation Strategy: A Unified Approach to Interception of Ground Balls and Fly Balls

Keshav Mundhra, Thomas G. Sugar, Michael K. McBeath
Department of Mechanical and Aerospace Engineering, Psychology
Arizona State University
{kmundhra, tsugar, m.m}@asu.edu

Abstract

In previous work, we demonstrated the feasibility of perceptual navigation algorithms used to intercept fly balls. In this paper we expand the optical acceleration cancellation heuristic to also intercept ground balls. We used computer simulations and experiments with mobile robots with various ball trajectories and different initial positions of the fielder to test our new model. A new robotic system with improved vision and faster processing was developed for experimentally intercepting ground balls. The results support the generality of viewer based interception heuristics for both fly balls and ground balls.

1. Introduction

Human based algorithms to intercept fly balls have been research and studied. One theory suggests the path taken by a fielder is based on spatio-temporal relationships between the ball in space and the fielder on the ground. This strategy, called the optical acceleration cancellation (OAC) model, proposed that the fielder runs to maintain a constant increase in the tangent of the optical angle (gaze angle) while maintaining lateral alignment with the ball in order to intercept it. The same principle can be applied to intercept ground balls except that there is a constant decrease in the tangent of the optical angle measured from the horizon.

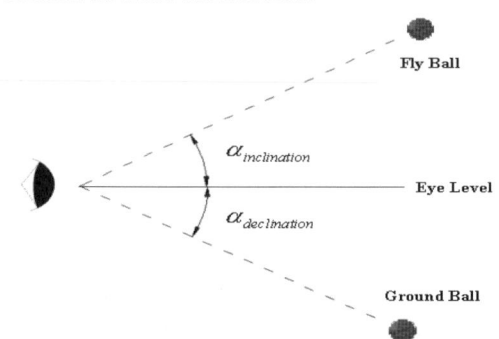

Experimentally, enhancements have been made to speed-up the image processing. Real time image processing requires tremendous processing power on the host computer in the form of clock speed of the processor and the memory available. An image processing program, Digital Video Robot (DVRobot) was developed to process images very quickly. The camera and the computer communicate using the IEEE-1394 protocol.

2. Literature Review

Chapman[1] in 1968 proposed that if α is the angle of gaze from a stationary fielder to a ball then the acceleration of $\tan(\alpha)$ is zero if the fielder is standing at the exact place where the ball will land. A fielder who starts from a place other than where the ball will land will eventually intercept the ball if he runs at a velocity such that the rate of change of $\tan \alpha$ is maintained to be a constant.

Oudejans and Michaels[2, 3] showed that the optical height (position of the ball on an imaginary plane located a fixed distance away from the fielder) increases at a fairly constant rate until just before the catch. This result was in agreement with Chapman's findings. McLeod and Diennes[4, 5] filmed skilled fielders as they moved forward or backward to catch balls projected at them. They found that $d^2/dt^2 (\tan \alpha)$ was maintained close to zero throughout the catching.

We propose to develop a unified approach to intercept ground balls and fly balls. Support for the hypothesis that the human visual system uses angular declination is provided by the experiments of Ooi [6]. They found that observers significantly underestimated distances when they were blindfolded and asked to walk up to a target they previewed with a pair of base-up prisms. Thus they confirmed that when angular declination was increased, the observers underestimated distance. After adapting to the base-up prisms however, the observers over estimated distances on prism removal. The conclusion that over-estimation of distance as an after effect of prism adaptation was due to lowered perceived eye level which reduced the objects angular declination below the horizon. Their work was based on the assumption that visual systems use the horizontal eye level as a reference for computing the angle of declination.

In robotics various algorithms have been proposed and tested to track objects in 3D space. One of the methods proposed by Schulz[7] uses a sample based data association filter to track a moving object with a robot. In one of their experiments the robot was able to track and generate trajectories of three people moving in a passage and it was able to follow one of these people. Their approach used probabilistic methods to deal with a varying number of objects.

Borgstadt [8], in another implementation of the OAC algorithm on mobile robots, collected acceleration data from the image, but this data was very noisy. In our previous work [9-12], we performed experiments with robots and computer simulations of perceptual navigation strategies using the OAC and the Linear Optical Trajectory (LOT) algorithms to intercept fly balls. We described active and passive models of the strategies based on the motion of the camera.

3. Mathematical Modeling and Simulation

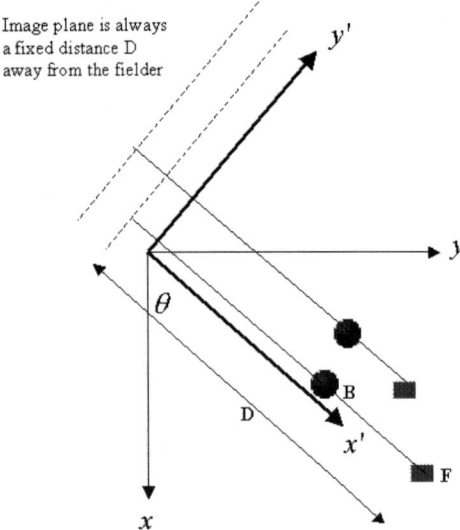

Figure 1. When the fielder first looks at the ball, the coordinate frame is rotated by an angle θ. This gives the new x', y' axis.

The original OAC strategy requires that the rate of change of the tangent of the optical angle to increase at a constant rate above the horizon. With ground balls, i.e. objects below the horizon ('eye level'), the same principle is applied but with α being measured from the eye level downward. Thus when α is zero degrees the fielder is looking straight ahead and when α is close to -90 degrees, the fielder is looking vertically down. Ooi [6] proposed that objects are at infinite distance when $\alpha = 0$ degrees and are extremely close when $\alpha = -90$ degrees.

For simulations in world coordinates and to simplify the calculations, the coordinate frame is rotated by an angle θ based on the initial position of the fielder relative to the ball. For the purpose of modeling, the robot is assumed to be an omni-directional robot. No skidding constraints are imposed. See Figure 1.

$$\theta = \tan^{-1}\left[\frac{y_{fi} - y_{bi}}{x_{fi} - x_{bi}}\right] \quad 3.1$$

All the subsequent calculations are done in this rotated coordinate frame. In the OAC algorithm

$$\frac{d}{dt}(\tan \alpha) = \text{Constant}$$

$$\tan \alpha = Ct + \frac{h_f}{D} \quad 3.2$$

D is the initial distance between the fielder and ball ($x'_{b0} - x'_{f0}$); t represents the time during the catching task; and h_f is the height of the fielder. From the geometry in Figure 10, it can be seen that

$$\tan \alpha = \frac{h_f}{x'_b - x'_f} \quad 3.3$$

Because the focal distance of the camera is fixed in our simulation, when an object is projected on the image plane the distance D from the fielder to the image will be fixed during the task.

$$\tan \alpha = \frac{z_{image}}{D}$$

The actual height of the target in the image plane is given by

$$z_{img\_act} = D \tan \alpha = \frac{Dh_f}{x'_b - x'_f} \quad 3.4$$

The desired height of the target as dictated by the OAC algorithm is given by

$$z_{img\_des} = D \tan \alpha = DCt + h_f \quad 3.5$$

The fielder will run at velocities to insure that z_{img_act} equals z_{img_des}. A proportional controller can be used to drive the error to zero.

$$\dot{x}'_f = K_{px}(z_{img\_des} - z_{img\_act})$$

$$\dot{x}'_f = K_{px}\left(DCt + h_f - \frac{Dh_f}{x'_b - x'_f}\right) \quad 3.6$$

For modeling purposes, a value for the constant C can be determined by differentiating equation 3.3.

$$\frac{h_f}{x'_b - x'_f} = Ct$$

Differentiating this expression and substituting the initial condition at time t_0 gives the following value for C.

$$\frac{-h_f(\dot{x}'_b - \dot{x}'_f)}{(x'_b - x'_f)^2} = C$$

At $t = t_0$, the fielder's velocity is zero ($\dot{x}_f = 0$) and the initial distance between the ball and the fielder is $D = x'_{f0} - x'_{b0}$ during the task. The value of C simplifies to

$$C = \frac{-h_f \dot{x}'_{b0}}{D^2} \quad 3.7$$

To maintain lateral alignment with the ball in the y direction another proportional controller is used.

$$\dot{y}'_f = K_{py}(y'_b - y'_f) \quad 3.8$$

The same control law is used for fly balls except the constant, h_f, is replaced with the actual height of the ball and since we neglect the height of the fielder in the fly ball case, the second term (h_f) in equation 3.6 is dropped. In the case of fly balls, the control law for the fielder is

$$\dot{x}'_f = K_{px}\left(DCt - \frac{Dz_b}{x'_b - x'_f}\right) \quad 3.9$$

3.1 Results from Simulation

The simulations were performed using Matlab™ and Simulink™. In the first set of simulations the OAC algorithm for ground balls is tested with a fixed ball path and the fielder approaches from four different initial positions ($\pm 75mm, \pm 75mm$). The ball starts from the origin and moves with a constant velocity of $0.002t$ m/s and $0.001t$ m/s in the x and y directions respectively. See Figure 2.

In the second set of simulations the fielder starts from the same initial position to intercept balls moving at constant speed in different directions. In all the simulations the ball starts from the origin (0,0) and the ball's trajectory is given by $x = \pm 0.002t$ m/s and $y = \pm 0.001t$ m/s. The controller gains, $K_{px} = 10$, $K_{py} = 0.1$, and the height of the fielder, $h_f = 2$ m, are constant. All simulations lasted for a period of 150 seconds. A high value of C enabled fast convergence of the fielder and ball trajectories. See Figure 3.

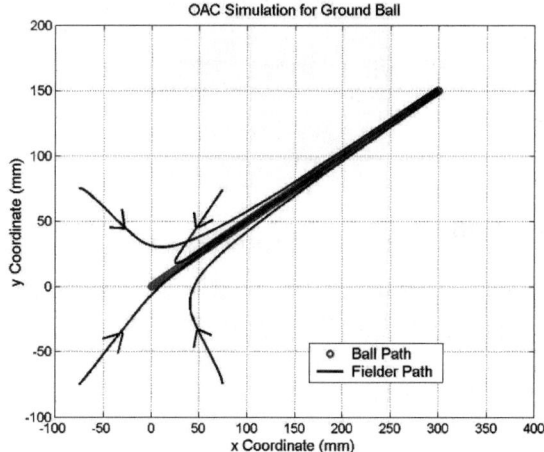

Figure 2. The fielder intercepting the ground ball from four different initial positions

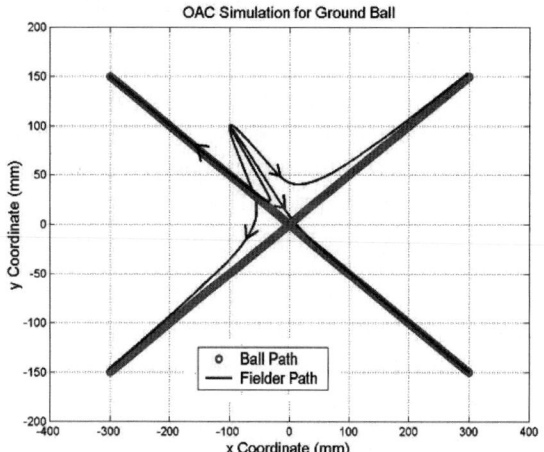

Figure 3. A fielder starts from a fixed initial position (-100,100) and the ball's motion starts at the origin and moves in different directions

4. Experiments

Figure 4. An extra computer added to the Nomad Scout robot for digital image processing.

In our experiments we used a Nomad Scout robot with an additional pan tilt mechanism from Directed Perception. The on-board computer mounted on the robot controls the drive wheels and the pan tilt mechanism. A SONY™ camcorder with a built in IEEE-1394 port sends the images to a laptop mounted on the robot. On the laptop, the DVRobot program tracks the target by matching the red, blue and green (RGB) intensities of pixels in the frame with a preset target RGB combination. A standard calculation is used to find the center of mass of all matching pixels. The laptop then transmits coordinate data to the on board computer running the navigation routine. See Figure 4.

In the implementation of the OAC algorithm, the initial motion of the ball is analyzed using the first few frames of the video. The rate at which the ball falls in the image plane is calculated. This gives an estimate of the constant C described in the modeling section. In the active OAC model, the camera is tilted downwards in accordance to equation 3.2. The velocity of the robot is proportional to the error in the target coordinate. The desired target coordinate is the center of the image. For example, the target is 180 and 120 pixels in the horizontal and vertical directions for an image of 360x240 pixels. In order to maintain alignment with the ball, the lateral velocity of the robot should be proportional to the error in the target coordinate in the horizontal axis of the image.

A fixed camera coordinate frame is used. As the robot rotates, the camera is panned in the opposite direction. In the analysis section, the coordinate frame in Figure 1 is not rotated.

The longitudinal and lateral velocities in the fixed camera coordinate system are transformed to the linear velocity and angular velocity (rotation) of the robot using a look-ahead controller.

$$\begin{bmatrix} v \\ \dot{\theta} \end{bmatrix} = \begin{bmatrix} \cos\theta & \sin\theta \\ -\sin\theta/L & \cos\theta/L \end{bmatrix} \begin{bmatrix} \dot{x}_f \\ \dot{y}_f \end{bmatrix}$$

In the experiments using a mobile robot, the ball was rolled in different directions in front of the robot and the path determined by the robot algorithm was recorded using wheel encoder data. The ball paths shown are approximations of the trajectory created by connecting the initial and final location of the ball with a straight line. The ball rolls at a fairly constant speed along that line.

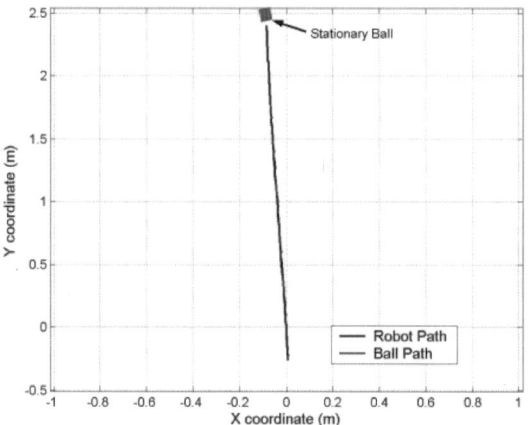

Figure 5. Trial with a stationary ball in front of the robot. The small sideways motion is due to lateral errors in the position of the ball in the image.

In all the trials the robot was able to intercept the ball. Note that there is an initial backwards motion of the robot in two of the trials. See Figures 8 and 9. In the start of these trials, the ball starts in the upper half of the image plane causing the robot to move backwards. This is analogous to a fielder being unsure whether the ball will rise as a fly ball or descend as a grounder. Ambiguity is largest at the start of a trial. As the camera tilts downward, the position of the ball rises in the image and the robot moves forward to intercept it.

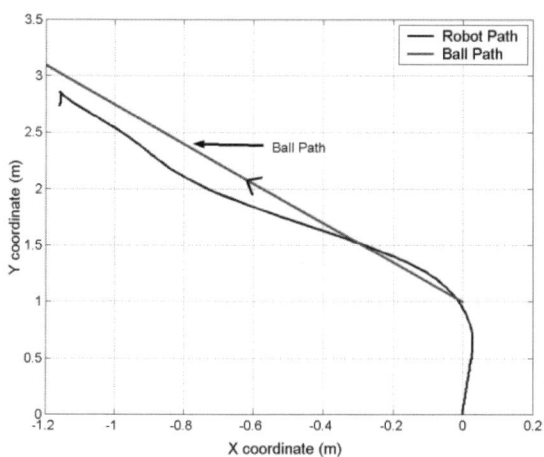

Figure 6. Ball moving away from the robot towards the left

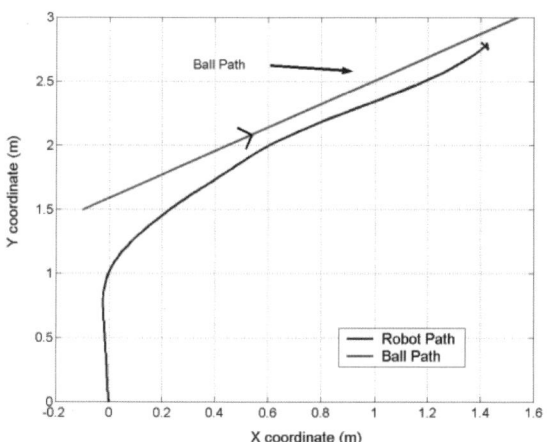

Figure 7. Ball moving away from the robot towards the right

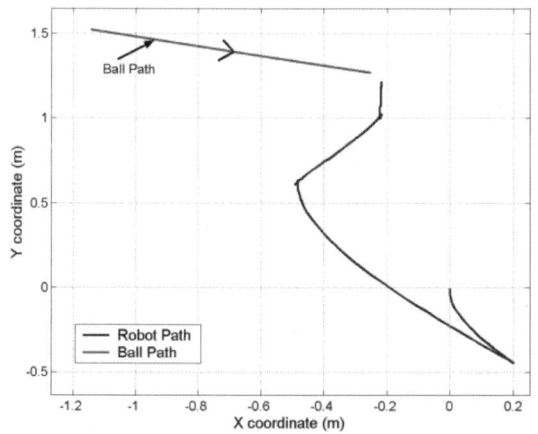

Figure 8. Ball moving towards the robot from the left

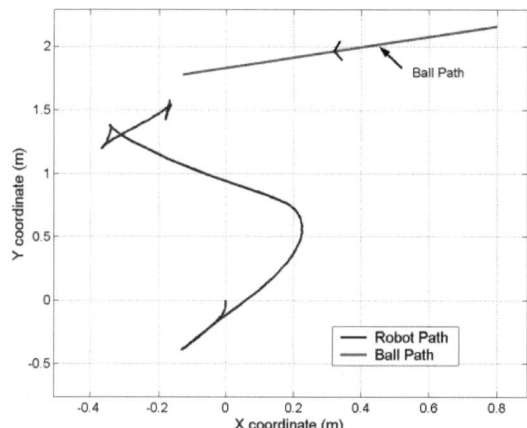

Figure 9. Ball moving towards the robot from the right

5. Conclusion

The simulations and results from experiments confirm the feasibility of the OAC strategy in intercepting ground balls. Thus, we have demonstrated a human based navigation strategy that can be used to intercept targets both above and below the horizon. With this principle of visual servoing, mobile robots can intercept projectiles with complicated and varying trajectories in three-dimensional space. The strategy uses optical geometry to compute the robot's path thus avoiding complex dynamic analysis.

We are currently working on a high-speed, real-time, robotic system that possesses both better processing techniques and higher navigation speed. Towards this goal we incorporated wireless-vision on a radio-operated, gas-powered car.

We believe that by understanding the way humans navigate to accomplish common tasks, mobile robotic systems can be created that navigate naturally in an unstructured environment. In the future, mobile robotic aids could use such navigational principles to help the elderly and the visually impaired. For example, a computer vision system mounted on a person will be able to detect whether an object is moving in a collision course and an avoidance routine will guide the person safely away from the danger.

Acknowledgements

Support from Arizona State University is gratefully acknowledged.

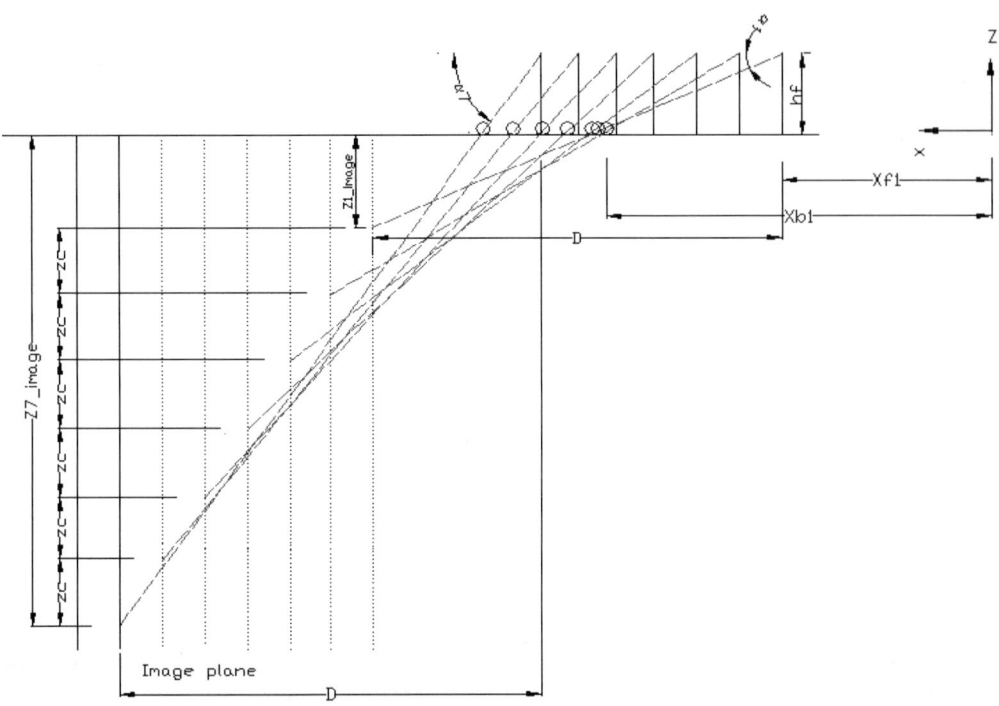

Figure 10-Pictorial representation of the optical angle cancellation algorithm for intercepting a ground ball. The projection of the ball in the image plane maintained a fixed distance, D, away from the fielder is falling at a constant rate.

References

[1] S. Chapman, "Catching a Baseball," *American Journal of Physics*, vol. 36, pp. 868-870, 1968.

[2] R. D. Oudejans, C. F. Michaels, F. C. Bakker, and M. A. Dolne, "The Relevance of Action in Perceiving Affordances: Perception of Catchableness of Fly Balls," *Journal of Experimental Psychology: Human Perception and Performance*, vol. 22, pp. 879-891, 1996.

[3] R. D. Oudejans, C. F. Michaels, and K. Davids, "Shedding Some Light on Catching in the Dark: Perceptual Mechanisms for Catching Fly Balls," *Journal of Experimental Psychology: Human Perception and Performance*, vol. 25, pp. 531, 1999.

[4] P. McLeod and Z. Dienes, "Do Fielders Know Where to Go to Catch the Base Ball or Only How to Get There?" *Journal of Experimental Psychology: Human Perception and Performance*, vol. 22, pp. 531-543, 1996.

[5] P. McLeod, N. Reed, and Z. Dienes, "Towards a Unified Fielder Theory: What We Do Not Know About How People Run to Catch a Ball," *Journal of Experimental Psychology: Human Perception and Performance*, vol. 27, pp. 1347, 2001.

[6] T. L. Ooi, B. Wu, and Z. J. He, "Distance Determined by the Angular Declination below the Horizon," *Nature*, vol. 414, pp. 197, November 2001.

[7] D. Schulz, W. Burgard, D. Fox, and A. B. Cremers, "Tracking Multiple Moving Objects with a Robot", *IEEE Computer Society Conference on Computer Vision and Pattern Recognition*, 2001

[8] J. A. Borgstadt and N. J. Ferrier, "Interception of a Projectile Using Human Vision Based Strategy," *International Conference on Robotics and Automation*, pp. 3189, April 2000.

[9] T. G. Sugar and M. K. McBeath, "Spatial Navigation Algorithm: Applications to Mobile Robotics," *Vision Interface Annual Conference*, 2001.

[10] A. Suluh, K. Mundhra, T. Sugar, and M. K. McBeath, "Spatial Interception for Mobile Robots," *International Conference on Robotics and Automation*, 2002.

[11] A. Suluh, T. Sugar, and M. K. McBeath, "Spatial Navigational Principles: Applications to Mobile Robotics," *International Conference on Robotics and Automation*, 2001.

[12] K. Mundhra, A. Suluh, T. G. Sugar, and M. K. McBeath, "Intercepting a Falling Object: Digital Video Robot," *International Conference on Robotics and Automation*, 2002.

Stereo Omnidirectional Vision for a Hopping Robot

Mirko Confente, Paolo Fiorini
Department of Computer Science
University of Verona
Strada Le Grazie 15, 37134 Verona, Italy

Giovanni Bianco
Information Services,
University of Verona,
Via S. Francesco, 37134 Verona, Italy

Abstract— This paper proposes a new geometrical structure for stereoscopic vision using omnidirectional cameras. The motivation of this work comes from the desire to equip a small hopping robot with an efficient and robust vision system to perform self localization during exploration missions. Because of size and weight constraints, we selected the Panoramic Annular Lens, for which no geometric model of stereo configuration was available. The paper describes the geometrical optical properties of the single lens and proposes a configuration for doing stereo vision with this lens. The analitical properties as well as the requirements of the complete system are discussed in the paper.

I. INTRODUCTION

With new missions to Mars to be launched in 2003, there is a renewed interest in planetary exploration with small robotic devices, as opposed to the current trend of large, multi-sensor exploration robots. So far, the most successfully mobility paradigm is the wheeled vehicle [1], that has been also proposed with inflatable wheels to increase exploration range [2], and made smaller in micro and nano robots [3]. Legged rovers have also been proposed to explore rough terrains. However, these approaches have not solved the conflicting requirements of designing small robots with high mobility. In fact, wheeled rovers can only overcome obstacles roughly the size of half a body length, thus the smaller the robot the higher the likelihood that it is blocked by a small rock. A solution to this problem was recently proposed by NASA-JPL and Caltech researchers who have developed the family of small hopping robots shown in Fig. 1, capable of coarse motion by jumping and fine motion by wheeling [4]. However, while the electro-mechanical design has progressed steadily, sensing and algorithm development have not been addressed yet. Because of the discontinuous motion of these robots and the unpredictable nature of the impact with the ground, traditional localization algorithms cannot be used. A possible approach to estimate the robot final position is to reconstruct the robot trajectory *a posteriori* based on appropriate sensor readings as discussed in [5].

This paper presents the geometrical analysis of the vision system that will be used in a complete localization algorithm to reset the position estimate of the hopping robot landing after a coarse estimation. The vision system for a hopping robot must take into consideration the specific nature of the robot motion, be robust to physical

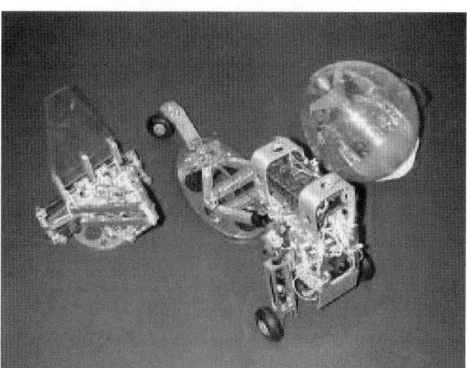

Fig. 1. Three generation of hopping robots

impacts, and provide enough information for precise navigation. We have selected a special type of omnidirectional vision system, which has all the desired characteristics. To introduce this system and the requirements imposed on its use by the hopping robot, in Section II the paper briefly describes the characteristics of the nominal configuration of a hopping robot. Then in Section III we present the biological background that motivates the approach selected. In Section IV we summarize the state of the art of omnidirectional vision research. In Section V we derive the geometrical model of the omnidirectional lens selected and in Section VI we derive the geometry of a motion-based stereo system. Finally in Section VII we summarize this research and describe our plans for future work.

II. THE HOPPING ROBOT REFERENCE CONFIGURATION

The research described in this paper has been developed using the hopping robot nicknamed "Frogbot" [2], as a reference model (Fig. 2). The main features of this family of hopping robot, developed at NASA Jet Propulsion Laboratory in 1999 [6], are summarized here for completeness. This robot is characterized by a pause between jumps to select the next hop direction and recharge the propulsion mechanism. This device was nicknamed Frogbot because of the similarity of its locomotory system with that of a frog. It is the result of a compromise between functionality and system's electromechanical complexity [2], [4], [6]. This prototype demonstrated that it is possible to develop a robot with mobility and sensing capabilities with only one actuator provided that every operation is executed in a sequential manner. More precisely, the Frogbot's

Fig. 2. The reference configuration of the hopping robot

operational cycle is based on the orderly execution of self-righting, panning the camera to acquire images, recharge the thrusting mechanism in preparation for the next jump, orient the robot in the desired direction, and execute the jump.

To consider a realistic use of the robot, the first problem that must be solved is clearly the availability of position information after a jump. In a companion paper [5] a solution to this problem is proposed by using a sensor suite consisting of accelerometers, gyro and impact sensors. However, the estimation error in this probabilistic method can grow unbounded if it is not checked periodically. In the following we present the motivations and the geometrical analysis of a vision system that will integrate and complement the other sensors hopefully providing a robust localization method for robots characterized by discontinuous motion.

III. BIOLOGICALLY INSPIRED LOCALIZATION METHOD

To identify a suitable localization method for a hopping robot, we have to satisfy very specific system requirements. In fact, for this type of robot, instruments have to be limited in number and weight, since we want to keep size and mass low, and the computational load must not to be too high, since on-board computing power is likely to be small. Therefore we propose a single camera system capable only to take a sequence of snapshots rather than a video, to limit the amount of memory needed. These minimalist requirements immediately point to biologically inspired algorithms, since they tend to be very efficient.

Many biology based robots are inspired by insects behavior. The reason is that insects have a limited capacity to elaborate input data, so their behavior can be more simply reproduced by a robot. In particular, in this research we take our inspiration from grasshoppers because their discontinuous travelling closely resembles the hopping robot motion. Similarly, we look at how grasshoppers localize themselves, to devise a localization method for the hopping robot.

Since we have decided to use a single camera in the robot system, we face two main problems: (1) the robot must keep a number of reference points in the field of view to compute its position; (2) the robot must compute its distance from the reference points. The first problem is solved by insects that have compound eyes providing a 360° vision. In this way, even when turning around their body center, they are able to keep track of the initial reference points, a task that is impossible for most mammals. The second problem can also find a solution in biology, since grasshoppers have their eyes so close to each other that they do not have stereo vision and cannot compute distance in this way. They compensate the lack of distance perception with a particular movement. When a grasshopper has to decide the next jump direction, or it is in front of an obstacle, it performs a motion called *peering* before jumping to the target [7]. In a typical peer, the tip of the abdoman remains stationary while the insect body moves from side to side in an arc. While the body is moving, the head counter-rotates against the body movement so that the net movement of the head is strictly translational, in a plane perpendicular to the line of sight. It was hypothesized in [7] that the insect is using motion parallax to judge distance to an object. Entomologists [8],

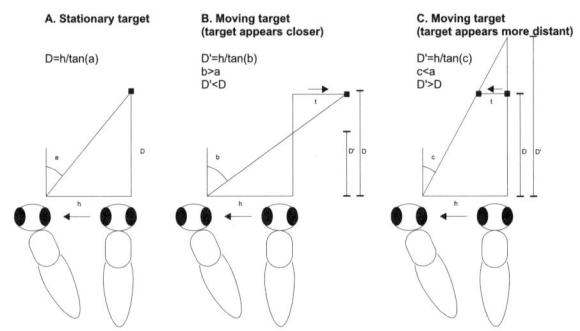

Fig. 3. Peering movement geometry.

[9] concluded that the principle of distance estimation in locusts is based on measurements of angle at the extent of the peers. By triangulation, the distance to the object can be estimated, as shown in Fig. 3.

Although many studies are available about hopping insects, none of them presents hypothesis on how these insects localize themselves to come back to their nest. We have chosen an omnidirectional visual sensor similar to insect vision and developed its geometrical model to find corresponding reference points and to calculate the robot displacement.

The base elements of our proposed localization method are the following: (1) use the lateral movement of the grasshopper to calculate distances from reference points; (2) use the epipolar geometry to estimate the rotation matrix and translation vector between 2 consecutive images;

(3) determine the position where the last snapshot was taken, utilizing the distance from known reference points; (4) assume a ballistic (parabolic) motion. In this paper we address only the geometrical analysis of the vision system and develop the epipolar geometry to compute stereo information.

IV. OMNIDIRECTIONAL VISION SYSTEMS

There are two good reasons to use panoramic imaging in motion estimation. First, the self motion (or egomotion) of a system can be estimated from displacements of images similarities. Using a standard perspective camera, correspondences can disappear because of the limited field of view of such a camera. Second, it is well known that motion estimation algorithms cannot, in some cases, distinguish a small translation of the camera from a small rotation. The confusion can be removed if a camera with a large field of view is used. Panoramic cameras can, in principle, obtain correspondences from everywhere in the image independently of the direction of motion. This intuitive reasoning has been corroborated by the result described in [10] showing that the motion estimation is almost never ambiguous if a spherical imaging surface is assumed. Two main families of panoramic cameras can be distinguished: panoramic vision systems based on moving cameras and motionless panoramic cameras. A pan-tilt mechanism is usually the main part of the first family. A standard perspective camera pans and tilts capturing a number of perspective shots of the scene. A panoramic image is then created as a composition of such images [11]. However, the pan-tilt camera requires a significant amount of time and it has to stop while capturing the shots that are used to compose the panoramic image, and this is not acceptable for an hopping robot. To the second family belong wide angle and mirror lenses. An application of special wide-angle lenses provides enhanced field of view too. This solution has the advantage of an easy assembly of lens and camera, however, the viewing angle is sometimes not sufficient, imaging model is difficult to define and the unique center of projection is preserved with difficulty.

A proper designed mirror assembled with a camera provides a panoramic view of the environment. Two main approaches appeared recently: (1) a mirror system constructed from several planar mirrors, a camera is assigned to each mirror plane; (2) one camera observes properly shaped convex or concave mirror [12]. Panoramic sensors with multiple cameras have satisfactory resolution but they are rather expensive and usually require complicated calibration and setup. Panoramic systems using an ordinary camera and a shaped mirror are simple, relatively cheap and easy to be calibrated and assembled. However they suffer from an inadequate resolution. Furthermore mirror cameras are not very robust and unsuitable for use in a hopping system. To satisfy the robustness requirement, we have chosen a particular lens, called Panoramic Annular Lens (PAL) [13] designed by Prof. Pal Greguss capable of giving the same viewing characteristics of a panoramic mirror camera, as described in next Section.

V. THE GEOMETRY OF PAL LENS

The geometry of the PAL imaging system is somewhat complex since there are two reflections and two refractions, as shown in Fig. 4. However, if we assume that the large circular mirror is an ellipsoid and that the small top mirror is a hyperboloid, and the setup of the two mirrors and a pinhole camera satisfies the fixed viewpoint constraint, we can obtain a rather elegant geometry of a single effective viewpoint under perspective projection. If that is the case, the real system can be modeled by the single-viewpoint geometry perfectly. The first refraction through

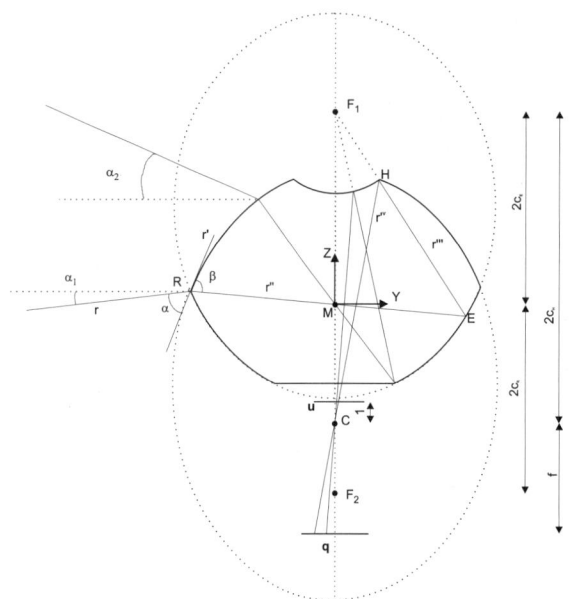

Fig. 4. Model of the PAL lens utilized.

the ellipsoidal surface changes the vertical viewing range from $[0, 90]$ to $[-\alpha_1, +\alpha_2]$, with $0 < \alpha_1 < \alpha_2 < 90$. The first reflection through the reflection ellipsoid moves the light ray to the hyperboloid and to the planar surface. The second refraction through the planar surface only moves the converging point of the ray collector lens, so we will ignore the second refraction.

We want to determine the relationship between the coordinates of a 3-D world point X_W in the space around the lens and its corresponding pixel q in the image plane as shown in Fig. 5. We express point X_W with respect to the lens reference system centered in M and call this point X_M. If X_M is in the visible space of the lens, the optical ray passing through X_M identifies one point R on the external

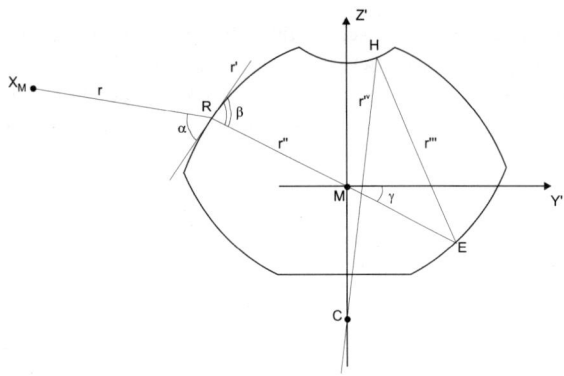

Fig. 5. Intersection of the lens with the plane $Y'Z'$. On this plane there are reflections and refractions.

surface of the lens. Inside the lens there are one refraction and two reflections on a plane S passing through point X_M, the origin M, and orthogonal to the plane $\{XY\}$ of the sensor, as shown in Fig. 5. All interesting points are positioned on plane S and so we can, without loss of generality, consider a 2-D reference system centered in M and coplanar with this plane. We obtain the space coordinates corresponding to a generic point $[y', z']^T$ on the plane S, using: $x = y'\cos\theta$, $y = y'\sin\theta$, $z = z'$, with $\theta = \arctan\left(\frac{y_M}{x_M}\right)$. To compute point R from point X_M (X'_M on plane S) we follow four steps. Knowing that lines r, r', r'' meet at point R and that these lines form angles α and β as shown in Fig. 5 related to each other by the refraction coefficient, we can find the coordinates of point R in the lens reference system in the following way:

- compute the intersection between line r'', passing through the origin and R, and the refraction ellipse;

$$\begin{cases} z' = m_1 y' & m_1 = \tan\gamma \\ \frac{(z'+c_r)^2}{a_r^2} + \frac{y'^2}{b_r^2} = 1 \end{cases}$$

- using line r, passing through points X_M, R and tangent line r' of the refraction ellipse in R, we compute α:

$$\alpha = \arctan\left(\frac{y'_R(y'_R - y'_M) + b_r^2(z'_R - z'_M)(z'_R + c_r)}{y'_R(z'_R - z'_M) - b_r^2(y'_R - y'_M)(z'_R + c_r)}\right)$$

- compute the angle β between the line r'', passing through the origin and point R and the tangent line r' of the refraction ellipse in R:

$$\beta = \arctan\left(\frac{y'_R + m_1 b_r^2(z'_R + c_r)}{m_1 y'_R - b_r^2(z'_R + c_r)}\right)$$

- compute the point R using the refraction coefficient of the external lens surface, and angles α and β.

Once R is computed, the point E can be found by intersecting the reflection ellipse and line r'' and therefore the point H on the reflection hyperboloid. We can find coordinates $\mathbf{u}$ on the normalized image plane using the fact that the vector from the camera center to the point H has a z coordinate equal to 1, since the normalized image plane is coplanar with the image plane and is located at unit distance from the camera center $\mathbf{u} = \frac{1}{z_C}R_C(H - t_C)$ with $z_C = R_{C_3}^T(H - t_C)$; where R_C and t_C represent the rotation matrix and translation vector between the lens and the camera reference system and $R_{C_3}^T$ is the third row of the matrix R_C. These coordinates can be expressed in the image plane coordinate system using intrinsic parameters combined into matrix K. We can determine pixel $\mathbf{q}$ corresponding to the spatial point X_M in the space outside the lens using $\mathbf{q} = K\mathbf{u}$.

The transformation from pixel $\mathbf{q}$ to the corresponding spatial point is carried out as follows. We first determine the optical ray:

- pixel coordinates are expressed in the normalized image plane reference frame using the calibration matrix K in the following manner: $\mathbf{u} = K^{-1}\mathbf{q}$
- point H is the intersection between hyperboloid and line r^{IV} passing through camera center C and pixel q is given by:

$$r^{IV} = \{\mathbf{C} + \lambda_H \mathbf{u} = \begin{bmatrix} 0 \\ 0 \\ (c_e - c_h) \end{bmatrix} + \lambda_H \begin{bmatrix} u \\ v \\ 1 \end{bmatrix}\}$$

$$H = C + \mathcal{H}(u)u \qquad \mathcal{H}(v) = \frac{b_h^2(c_h v_3 + a_h\|v\|)}{-a_h^2 v_1 - a_h^2 v_2 + b_h^2 v_3}$$

- point E on the ellipsoidal surface is given by the intersection between line r''', passing through F_1 and H, and the reflection ellipse:

$$E = F_1 + \mathcal{E}(H)H \qquad \mathcal{E}(v) = \frac{b_e^2(c_e v_3 + a_e\|v\|)}{a_e^2 v_1^2 + a_e^2 v_2^2 + b_e^2 v_3^2}$$

- point R on the refraction ellipsoid is computed using the intersection points of line r'' with the refraction ellipse:

$$R = \mathcal{R}(E)E \quad \text{with} \quad \mathcal{R}(v) = \frac{b_r^2(c_r v_3 + a_r\|v\|)}{a_r^2 v_1^2 + a_r^2 v_2^2 + a_r^2 v_3^2}$$

- from the value of R the plane orthogonal to the xMy plane and passing through R is computed. In this way we can determine the tangent line r' at the refraction ellipse in R and the angle β between r'', r' as:

$$\tan \widehat{r''r'} = \frac{m_{r'} - m_{r''}}{1 + m_{r'}m_{r''}} = \frac{-y'_R y'_E - b_r^2 z'_E(z'_R + c_r)}{y'_R z'_E + b_r^2 y'_E(z'_R + c_r)}$$

and in the same manner we can express the angle α.

- the equation of the line r, passing through point R on the lens is then given by $z' = my' + q$

$$\text{with} \quad m = \frac{y'_R + mb_r^2(z'_R + c_r)}{my'_R - b_r^2(z'_R + c_r)}, \quad q = z'_R - my'_R$$

This line identifies all space points X_M that are projected to pixel q.

VI. Epipolar Geometry for PAL Lens

Although researchers [14] have identified the usefulness of panoramic vision for robot navigation, only a few reference are available on the derivation of an appropriate geometry for stereo panoramic cameras. The epipolar geometry describes the relationship between a camera motion, a scene and the disparity in the panoramic images. Information about motion between camera positions can be estimated from points correspondences. In this section we show that the main concepts of the epipolar geometry can be successfully applied to the PAL lens. A similar work can be found in [15] for parabolic mirrors.

In the previous section we compute 3-D coordinates of a point R_1 on the refraction ellipsoid from a pixel q_1 on the image plane, where the index 1 refers to the first position of the lens. Now we project all points of the line passing through R_1 and X_M to the image plane of the second camera, as shown in Fig. 6. To do this we compute a plane passing through points R_1, its corresponding point R_{12} on the second lens and a point R'_1 on the line passing through R_1 and X_M. Coordinates of point R_1 can be expressed as a

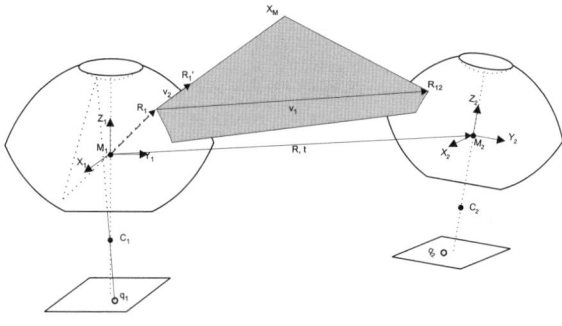

Fig. 6. Plane corresponding to pixel q_1 used to determine pixel q_2 on the second image plane.

function of pixel q_1. Using R_1 we are able to compute the equation of the line passing through R_1 and X_M and obtain a point R'_1 of this line. The third point R_{12} can be computed by projecting R_1 to the second lens and computing the intersection between the second lens and the ray passing through R_1. We can then express the coplanarity of these points in the coordinate system of the first lens:

$$a(x - x_{R_1}) + b(y - y_{R_1}) + c(z - z_{R_1}) = 0$$

with a, b, c components of vector $\vec{n}_1 = \vec{v}_1 \wedge \vec{v}_2$ normal to the plane and given by:

$$a = \begin{vmatrix} y_{v_1} & z_{v_1} \\ y_{v_2} & z_{v_2} \end{vmatrix} \quad b = \begin{vmatrix} z_{v_1} & x_{v_1} \\ z_{v_2} & x_{v_2} \end{vmatrix} \quad c = \begin{vmatrix} x_{v_1} & y_{v_1} \\ x_{v_2} & y_{v_2} \end{vmatrix}$$

Using the rotation matrix R between two lens reference systems we can express vector $\vec{n}_1$ and therefore the coplanarity condition in the second reference system. This plane intersects the second refraction ellipsoid and so we can find the points corresponding to the line passing through R_1 and X_M. Using the orthographic projection of a lens point to the xy plane of the second coordinate system centered at M_2 we can express these equations in matrix form: $r_2^T A_{r_2} r_2 = 0$. Using the relations derived in the previous Section we write the equation: $(E^T \mathscr{R}(E) N^T) A_{r_2} (N \mathscr{R}(E) E) = 0$ where N is a lower triangular matrix. This equation is a function of pixels $\mathbf{q}$ on the second image plane since $\mathscr{R}(E) E$ is a function of pixel corresponding to E, ellipsoid and hyperboloid parameters, and point R_1 given by pixel q_1. This equation defines the curve on which a point (in the second image plane) corresponding to pixel q_1 (in the first image plane) has to lie. In this manner from a pixel q_1 in the first lens image plane we are able to compute a set of pixels in the second image plane using the epipolar plane. This important practical relationship simplifies searching for correspondence of points to a limited area. We do not show a plot of the resulting epipolar curve since we do not have yet realistic lens parameters.

Usually, "egomotion estimation" algorithms recover the camera motion parameters R and t in an indirect manner [16]: they first estimates the essential matrix $E = RS$, with S antisymmetric matrix and then recover the motion parameters R, t with linear methods. This is possible because egomotion estimation algorithms use a plane passing through the center of the two reference systems, and therefore it is possible to tie R to t. In this manner every plane generated by pairs of points in the image planes includes the baseline and the translation vector on this line. In our case instead, the epipolar plane does not pass through the reference systems and so we have to estimate a common rotation matrix R and a translation vector $\vec{t}_i$ for every pairs of points R_1, R_2 on the lenses (Fig. 7). In the previous section we have seen the connection between a pixel q and its corresponding point R on the lens. Calling $\vec{t}_1$ the displacement of the first pair on the lenses, we can bind $\vec{t}_1$ to $\vec{t}$. As can be seen in Fig. 7, vectors $\vec{v}_{R_1}$, $\vec{v}_{R_2}$ and $\vec{t}_1$ are coplanar. We can

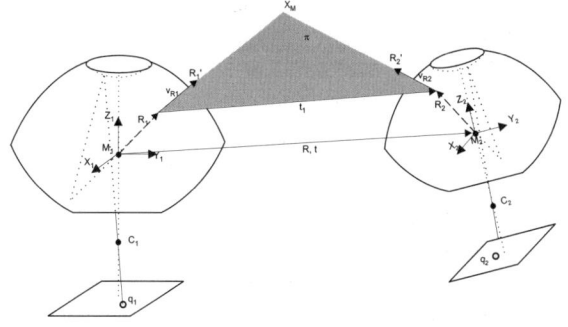

Fig. 7. Three coplanar vectors on the epipolar plane.

3471

express vector $\vec{t}_1$ as function of the unknown translation vector $\vec{t}$. The coplanarity condition of the plane passing through R_1 can be written as:

$$(\vec{t} + RR_2 - R_1) \cdot \vec{v}_{R_1} \wedge \vec{v}_{R_2} = 0$$

Using cross product and expressing the elements as a function of the unknown rotation matrix R and translation vector t we are able to derive the following equation for every pair of lens points R_1, R_2:

$$(t_x + r_{11}x_{R_2} + r_{12}y_{R_2} + r_{13}z_{R_2} - x_{R_1})n_x +$$
$$+(t_y + r_{21}x_{R_2} + r_{22}y_{R_2} + r_{23}z_{R_2} - y_{R_1})n_y +$$
$$+(t_z + r_{31}x_{R_2} + r_{32}y_{R_2} + r_{33}z_{R_2} - z_{R_1})n_z = 0$$

This is an equation with 12 unknown elements: 9 given by the rotation matrix R and 3 by the translation vector $\vec{t}$ ($\vec{n}$ is the normal vector of the epipolar plane π). We need at least eleven pairs of corresponding world points to obtain an egomotion estimation. Only eleven parameters are necessary since the translation vector t is given up to a scale factor ($\|t\| = 1$). In fact, we cannot recover the absolute scale of the scene without an additional external measurement, such knowing the distance between two points. Using the coplanarity condition we are able to compute the displacement of a point on the lens using a limited number of points in the image plane. This equation will be the basis of the localization algorithms for the hopping robot.

VII. CONCLUSION

In this paper we define some of the main constraints and geometrical relations of a visual sensing system for a hopping robot. The engineering decisions were suggested by similarities with the biological world, where some of the problems faced by a hopping robot are solved very efficiently. We propose a single camera vision system, able to collect stereo information from motion, similar to the peering movement of a grasshopper. By the same token, we adopt an omnidirectional vision system to simplify landmark tracking. The vision system is based on the Panoramic Annular Lens (PAL) for which we analyzed the optical geometry. We develop stereo relations for the PAL lens. The epipolar geometry for a pair of cameras with PAL is the main novel contribution of this work. The geometrical relations found constraint positions of corresponding points in panoramic images of the same scene and serve as the basis for a motion estimation algorithm that is the final goal of this research. With only 11 corresponding points in two images we are able to compute a motion estimation given by epipolar constraints. In the future we want to experiment with a prototype of the vision system and to integrate other types of sensors, such as giroscope and accelerometer, in the visual localization algorithm to achieve a better estimation of the robot position. In this manner we think that we will be able to develop a robust self localization method to compute the final position of a hopping robot.

VIII. REFERENCES

[1] http://mars.jpl.nasa.gov/tecnology/rovers/index.html.
[2] E. Hale, N. Shara, J. Burdick, and P. Fiorini. A minimally actuated hopping rover for exploration of celestial bodies. volume 1, pages 420–427. IEEE Int. Conf. on Robotics and Automation, May 2000.
[3] B. H. Wilcox and R. M. Jones. The MUSES-CN nanorover mission and related tecnology. volume 7, pages 287–295. Proc. of IEEE Aerospace Conf., 2000.
[4] P. Fiorini, S. Hayati, M. Heverly, and J. Gensler. A hopping robot for planetary exploration. Proc. of IEEE Aerospace Conference, March 1999.
[5] C. Cosma and P. Fiorini. Self localization of a hopping robot in unknown environment. Coimbra PT. Submitted to ICAR03.
[6] J. Burdick and P. Fiorini. Minimalist jumping robots for celestial exploration. *to appear Int. Journal of Robotics Research.*
[7] G. K. Wallace. Visual scanning in the desert locust schistocerca gregaria. volume 36 of *J. Exp. iol.*, pages 512–525, 1959.
[8] E. C. Sobel. The locust's use of motion parallax to measure distance. volume 167 of *J. Comp Physiol*, pages 579–588. University of Pennsylvania, USA, 1990.
[9] T. S. Collett. Peering: a locust behaviour pattern for obtaining motion parallax information. volume 76 of *J. Exp. Biol.*, pages 237–241. School of Biological Sciences, University of Sussex, UK, 1978.
[10] T. Brodsky, C. Fernmuller, and Y. Aloimonos. Directions of motion fields are hardly ever ambiguos. In *4th European Conference on Computer Vision*, volume 2, pages 119–128, Cambridge UK, 1996.
[11] H. Ishiguro, M. Yamamoto, and S. Tsuji. Omni-directional stereo. In *IEEE Transaction on Pattern Analysys and Machine Intelligence*, volume 14(2), pages 257–262, Feb. 1992.
[12] S. K. Nayar. Catadioptric omnidirectional camera. In *IEEE Int. Conference on Computer Vision and Pattern Recognition*, pages 482–488, Puerto Rico, USA, June 1997. IEEE Computer Society Press.
[13] http://www.manuf.bme.hu/~greguss/welcome.htm.
[14] Y. Yagi, Y. Nishizawa, and M. Yachida. Map-based navigation for a mobile robot with omnidirectional images sensor copis. In *IEEE Transaction on Robotics and Automation*, volume 11(5), pages 634–648, 1995.
[15] T. Svoboda. *Central Panoramic Cameras: Design, Geometry, Egomotion*. PhD thesis, Centre for Machine Perception, Czech Technical University, 1999.
[16] O. Faugeras. *3- DComputer Vision, A Geometric Viewpoint*. MIT Press, 1993.
[17] Z. Zhang, R. Deriche, O. Faugeras, and Q.T. Loung. A robust technique for matching two uncalibrated images through the recovery of the unknow epipolar geometry. Technical Report 2273, INRIA, 1995.
[18] Z. Zhu, E.M. Riseman, and A.R. Hanson. Geometrical modeling and real-time vision applications of a panoramic annular lens (PAL) camera system. Technical report, UMASS-Amherst, February 1999.

Real-time Tracking and Pose Estimation for Industrial Objects using Geometric Features

Youngrock Yoon, Guilherme N. DeSouza, Avinash C. Kak
Robot Vision Laboratory
Purdue University
West Lafayette, Indiana 47907
Email: {yoony,gdesouza,kak}@ecn.purdue.edu

Abstract—This paper presents a fast tracking algorithm capable of estimating the complete pose (6DOF) of an industrial object by using its circular-shape features. Since the algorithm is part of a real-time visual servoing system designed for assembly of automotive parts on-the-fly, the main constraints in the design of the algorithm were: speed and accuracy. That is: close to frame-rate performance, and error in pose estimation smaller than a few millimeters. The algorithm proposed uses only three model features, and yet it is very accurate and robust. For that reason both constraints were satisfied: the algorithm runs at 60 fps (30 fps for each stereo image) on a PIII-800MHz computer, and the pose of the object is calculated within an uncertainty of 2.4 mm in translation and 1.5 degree in rotation.

I. INTRODUCTION

There has been considerable interest in object tracking in the past few years. The applications of such systems vary enormously, ranging from: 1) tracking of different objects in video sequences[13]; 2) tracking of human bodies[10], human hands for sign language recognition[12], and faces for airport security[8]; 3) tracking of flying objects for military use; to finally 4) tracking of objects for automation of industrial processes[1]. However, when it comes to the last – automation of industrial processes – the task of designing a successful tracking algorithm becomes even more daunting. While most of the applications of object tracking can tolerate off-line processing and relatively large errors in pose estimation of the target object, the autonomous operation of an assembly line for manufacturing requires much higher levels of accuracy, robustness, and speed. Therefore, without an efficient tracking algorithm, it becomes virtually impossible, for example, to servo a robot to fasten bolts located on the cover of a car engine – as this engine continuously moves down the assembly line.

Using robots in moving assembly lines has obvious advantages: the improvements in productivity; the safety aspects of using robots in hazardous or repetitive tasks – which are not quite suitable for human workers – etc. Also, machine vision has been very useful in robotics, because it can be used to close the control loop around the pose of the robotic end-effector [7]. This visual feedback allows for more accurate retrieval and placement of parts during an assembly operation [4].

Unfortunately, visually guided robot control has been limited to applications with stationary calibrated workspaces, or to applications with the assembly line synchronized with the robot workspace – which creates the effect of the workspace being stationary with respect to the robot [2], [11]. The reason for this limitation in the applications is caused partially by a lack of fast tracking algorithms that can accurately guide the robot with respect to moving, dynamic workspaces.

In this paper, we describe such a fast and accurate visual tracking algorithm. While most tracking and pose estimation algorithms rely on CAD models or unstructured cloud of points to provide accuracy and redundancy[5], [9], our algorithm achieves very good accuracy using only three of the many similar geometric features of the object. By doing so, we also guarantee real-time performance which is indispensable for automation in dynamic workspaces. The target object used in our work is an engine cover as depicted in Fig. 1. This object has a metallic surface rich in circular shapes such as bolts, lug-holes, and cylindrical rods. That characteristic makes natural the choice of geometric features as visual cues for both tracking and 3D pose estimation of the object.

In Section 2, we describe the tracking algorithm, as well as the pose estimation algorithm. These two modules interact with our distributed visual servoing architecture, as described in detail in [1]. The results are presented in Section III followed by the conclusions and a discussion of future work presented in Section IV.

II. TRACKING SYSTEM

Our system is divided in two modules: the Stereo-vision Tracking module and the Pose Estimation module (Fig.3-a). The Stereo-vision Tracking module consists of a stereo model-based algorithm, which tracks three particular circular features of the target object and passes their stereo-corresponding pixel coordinates to the Pose Estimation module. An example of the processing performed by the Stereo-vision Tracking module is depicted in Fig.2, where the pixel coordinates of the features are indicated by cross-hairs. Based on these pixel coordinates, the Pose

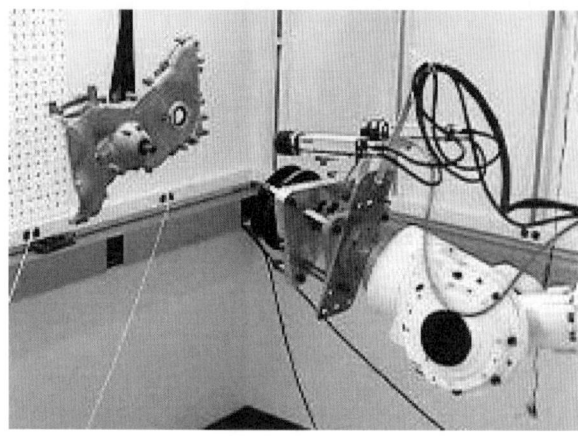

Fig. 1. The target object(engine block cover) and our stereo camera mounted on the robot end-effector

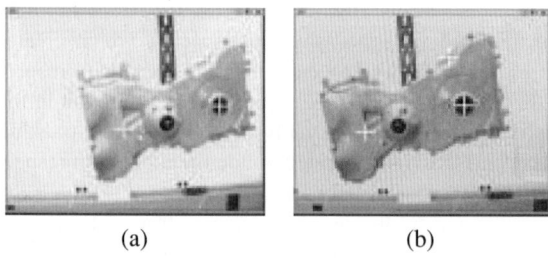

Fig. 2. Example of the Stereo-vision Tracking process - (a) left image, (b) right image

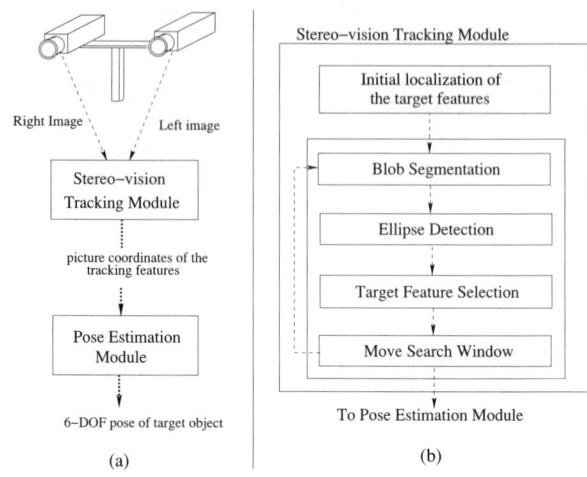

Fig. 3. System overview: (a) whole system (b) details of the Stereo-vision Tracking module

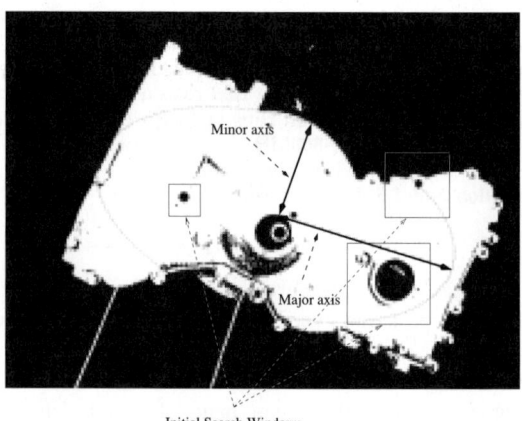

Fig. 4. Principal axes of the target blob and the initial positions of search windows for three target features.

Estimation module can calculate the 3D coordinates of the feature points using stereo reconstruction and estimate the complete pose (6-DOF) of the engine cover in the robot workspace.

Details of each module are presented in the next subsections.

A. Stereo-vision Tracking Module

In this section, we will present the tracking algorithm for only one of the stereo images (Fig.3-b.) This tracking algorithm performs in two distinct phases: initialization phase and tracking phase. During the initialization phase, the algorithm must roughly locate the center of mass of each of the three object features. Once these coordinates are determined, search windows around the three features can be defined. Those windows are used during the tracking phase of the algorithm to constrain the search for the features in the sequence of frames (Fig.4.)

In order to perform the initial localization of the features in the image space, the tracking module assumes that the whole engine cover can be seen[1]. Also, due to the elongated shape of the engine cover, two axes – the major and minor axes of the target object – can be easily obtained using principal components applied to the binarized image. These two axes are used to locate the features in the image space, since the pixel coordinates of the features can be easily expressed with respect to these two axes. Result of this initialization procedure is depicted on Fig.4. As we mentioned before, once the features are located in the image, search windows for each target feature can be defined. As one may infer, the initial position of the search window does not need to be exact. The only assumption during the initialization phase is that the initial position and size of the search window should be reasonably accurate for the search window to enclose the target feature. It is only in the tracking phase of the algorithm that the exact position of the target feature will be determined. Each search window has a size that equals

---

[1]This condition is satisfied by another module of our visual servoing system called Coarse Control[1].

twice the size of the target feature (blob) on top of which it is located.

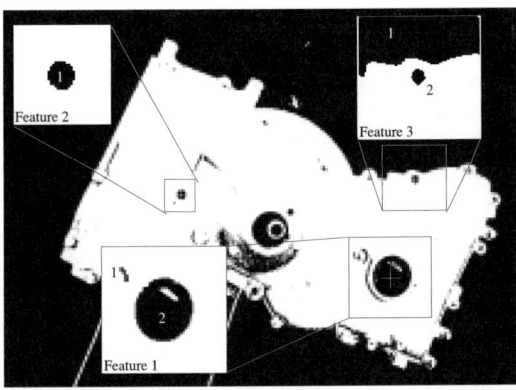

Fig. 5. Result of blob tracking. Numbered blobs shown in each enlarged search window are filtered by morphological filter.

During the tracking phase of the algorithm, many blobs may appear as candidate features inside the search windows. In order to decide which blobs represent the tracked features, a feature identification algorithm is applied to the pixels confined by the search windows. As the first step, this algorithm removes noise from the binarized image using a morphological filter (Fig. 5). Next, an ellipse detection procedure is employed to search for the circular features, which may be projected on the input image as ellipses. This procedure uses the fact that the Mahalanobis distance from a point on the border of an ideal ellipse to the center of the ellipse is always constant. The ellipse detection procedure is described in more detail by the following steps:
- Find the border pixels of each candidate blob using a border-following algorithm;
- Compute the center of mass of the border pixels and the covariance of their pixel coordinates. Let q_0 be the center of mass and M be the covariance matrix;
- For each pixel on the border, say q_i, calculate the Mahalanobis distance as defined by:

$$d_i = \sqrt{(q_i - q_0)^T M^{-1} (q_i - q_0)}$$

As we mentioned above, if the blob has an elliptical shape, this distance measure would be constant within a very small tolerance throughout all border pixels.
- Calculate the standard deviation of d_i, σ_d, and compare with an empirically obtained threshold. If this condition is satisfied, the candidate blob is accepted as an ellipse.

After all the blobs are tested, the blob with the minimum standard deviation is chosen as the target feature. To illustrate the results of this feature identification applied to Fig.5, we list in Table I the average distances and standard

TABLE I
MAHALANOBIS DISTANCE MEASURE FOR EACH OF BLOBS IN SEARCH WINDOW DISPLAYED IN FIGURE 5

Window number	Blob number	d Mean	σ_d
1	1	1.376526	0.499615
	2	1.413272	0.072097
2	1	1.417955	0.139090
3	1	1.382495	0.297839
	2	1.410550	0.101723

deviations for the blobs detected. In this example, blob 2 in window 1 is selected as feature 1, while blob 2 in window 3 is selected as feature 3. Since window 2 has only one candidate, if the blob passes the ellipse detection test it is automatically selected as feature 2.

Finally, during the tracking phase of the algorithm, the positions of the search windows in the next frame are updated based on the positions of the target features in the current frame. In this case, it is assumed that the movement of the target feature between frames will not exceed the size of the search window, which is determined using the actual motion speed of the target object in the assembly line in terms of the frame rate.

B. Pose Estimation

Assuming that the features can be tracked and their pixel coordinates can be calculated, it is the job of the Pose Estimation module to find the actual pose of the target object. Since the Stereo-vision Tracking module makes sure that the search window for a feature stays locked onto that feature, finding the stereo correspondence of the pixel coordinates becomes trivial. Therefore, obtaining the 3D coordinates of the features is easily accomplished using the stereo camera calibration [6] and simple stereo reconstruction methods [3].

Given the 3D coordinates of the features, the pose of the target object is defined by the homogeneous transformation matrix that relates the object reference frame and the robot end-effector reference frame – where the stereo cameras are mounted. This homogeneous transformation matrix can be decomposed into 6 parameters: x, y, z, yaw, pitch and roll (Euler 1). However, before we can calculate those six parameters, we need to find the object coordinate frame in end-effector coordinates. The object reference frame is defined with respect to the three model features (tracked features) as shown in Fig. 6 and is calculated by the Pose Estimation module as follows:
- Let the 3D coordinates of the three model features shown in Fig. 6 be P_1, P_2, P_3.
- The origin of object reference frame O is the point dividing the line $\overline{P_1 P_2}$ in half.
- The Y axis of the object reference frame is along the vector $\overrightarrow{OP_1}$.

- The X axis is along the vector defined by the cross product of the vectors $\overrightarrow{OP_1}$ and $\overrightarrow{OP_3}$.
- The Z axis lies on the plane of the three features and is calculated by the cross product of the X axis and the Y axis.

Since P_1, P_2, P_3 represent three vectors in the end-effector reference frame, the homogeneous transformation with respect to the object reference frame comes directly from the axes above.

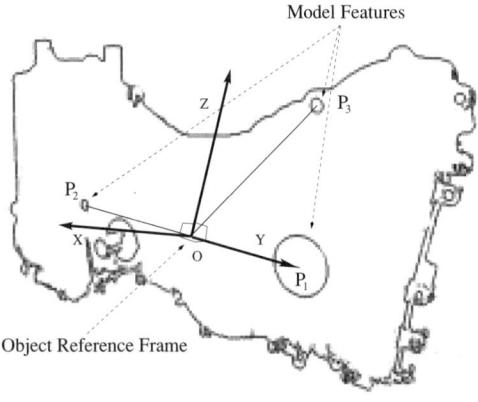

Fig. 6. Definition of object reference frame

III. EXPERIMENTAL RESULTS

Before we describe our error measurements, we must present the workspace on which the experiments were carried out. Our workspace consists of a target object placed in front of two stereo cameras mounted on a Kawasaki UX-120 robot end-effector as depicted in Fig. 1. Our system was implemented on a Linux environment running on a Pentium-III 800MHz with 512Mb of system memory. Two externally synchronized Pulnix TMC-7DSP cameras were connected to two Matrox Meteor image grabbers, which can grab pairs of stereo images at 30fps. The system can track all three object features at exactly 60fps (30fps per camera.)

In a system such as this one – for assembly of parts on the fly – the various errors in the tracking algorithm ultimately translate into how accurately the tracker positions the end-effector with respect to the target object. These errors originate mainly from: 1) the various calibration procedures such as hand-eye calibration, camera calibration, etc.; 2) the pose estimation using 3D reconstruction of the feature points; and 3) the ability of the system to detect and keep track of the target object. In [6], we have already presented a comprehensive calibration procedure that, for the same workspace, provided an accuracy of 1mm in 3D reconstruction of special objects (calibration patterns). Therefore, for this work we focused on the measurement of the errors specifically in 3D pose estimation and tracking.

The accuracy of the tracking algorithm was measured in two ways. First, we measured how the tracking algorithm performs the stereo reconstruction and pose estimation when the object is stationary. The second error measurement was regarding the pose estimation for a moving object.

1) Error Measurement for Stationary Target: The goal here was to measure the effects that algorithms such as binarizing, ellipse detection, etc. have in the pose estimation using 3D reconstruction as described in section II-B. Therefore, we ran the tracking algorithm and fixed both object and camera positions while we measured the uncertainties in pixel coordinates of the object features as well as the object pose in the world reference frame. These uncertainties reflect the errors of the algorithms above over a 5-minute sampling period, at 60fps.

In Table II we show the experimental results. For this experiment, the pose of the object was calculated at three different relative distances between object and cameras. For each of these distances, four positions of the camera were chosen with respect to the object: top, left, right, and bottom. Each position is about 20cm from the center position.

From the table above we notice that the uncertainty (σ) in the object pose is highly dependent on the uncertainty in the pixel coordinates of the features. Also from the table, the uncertainty in the position of the features seems to be unexpectedly independent of the distance between camera and target object, since at 70cm the error is smaller than at 60 or 80 cm. However, we attribute this behavior to other factors that may also affect the pose estimation, such as the proximity of the object to the vergence point of the cameras, the quality of the tracking for different sizes of the features as perceived in the image, and the illumination conditions (shade and reflectance for different angles/positions).

2) Error Measurement for Moving Target: For this experiment, measuring the error is more difficult than is the case for a stationary target. That is, it is not possible to calculate the ground truth for the object pose if the object is moving. Therefore, instead of measuring the error for a moving target, we fixed the target object and moved the camera (end-effector). The camera/end-effector moved along an arbitrary path, while the pose of the robot end-effector was monitored at each instant when an image was digitized. Fig. 7 depicts plots of the estimated and the actual values for each of the six components of the object pose – x, y, z, yaw, pitch, and roll – as the camera moves along the path.

Finally, in Table III, we summarize the statistics of the error in x, y, z, yaw, pitch, and roll. As one may

TABLE II
STATISTICS OF POSE ESTIMATION UNCERTAINTY FOR A STILL OBJECT; σ: STD, RANGE: MAX-MIN

Distance to object (mm)	Pose with respect to object	σ̄ among all six features (pixel)	Translational uncertainty (mm)		Rotational uncertainty (degree)					
					Yaw		Pitch		Roll	
			σ	range	σ	range	σ	range	σ	range
600	top	0.15	0.495	2.077	0.033	0.232	0.069	1.824	0.254	0.924
	left	0.29	0.566	2.918	0.035	0.270	0.448	3.005	0.264	1.389
	right	0.27	0.581	2.028	0.032	0.258	0.338	1.947	0.294	1.066
	bottom	0.24	0.280	2.450	0.077	0.414	0.314	3.360	0.109	1.047
700	top	0.22	0.099	0.301	0.077	0.255	0.390	2.629	0.019	0.138
	left	0.13	0.059	1.544	0.038	0.245	0.470	1.287	0.033	0.713
	right	0.04	0.076	1.368	0.008	0.370	0.080	1.294	0.035	0.559
	bottom	0.30	0.363	3.102	0.058	0.416	0.706	4.084	0.160	0.836
800	top	0.31	0.576	5.162	0.052	0.440	0.860	5.053	0.280	2.406
	left	0.17	0.160	1.993	0.026	0.314	0.827	5.414	0.077	0.904
	right	0.30	0.867	3.694	0.102	0.443	0.912	4.994	0.380	1.525
	bottom	0.15	0.539	3.948	0.076	0.708	0.206	3.586	0.253	2.414

observe, despite the occurrence of a few "off-the-curve" values (large *max abs error*'s in the table), the translational uncertainty is less than 2.5 mm, while the rotational uncertainty is less than 1.5 degree.

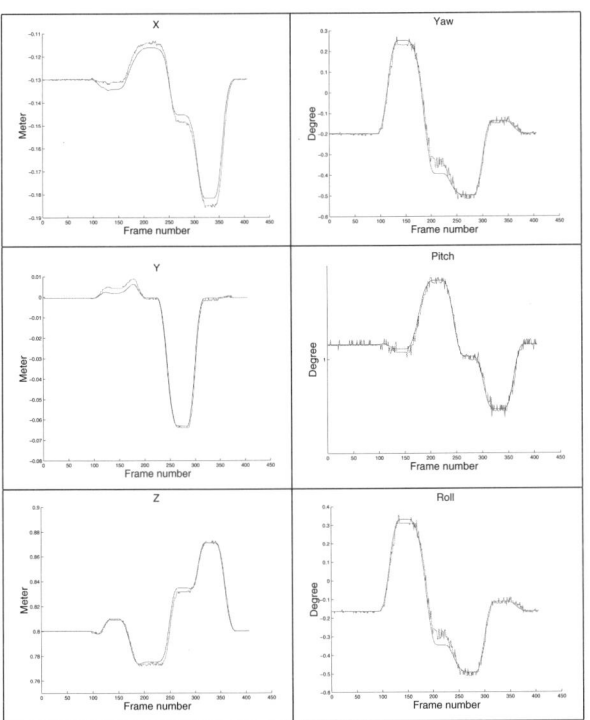

Fig. 7. Actual (blue plot) and estimated (red plot) values for the X, Y, Z, Yaw, Pitch, and Roll components of the object pose

TABLE III
MEAN AND STANDARD DEVIATION OF THE ERRORS BETWEEN ACTUAL AND ESTIMATED POSE.

	Mean	Std σ	Max abs error
X(mm)	-0.094	2.418	7.158
Y	0.190	1.630	5.396
Z	-0.703	2.210	9.298
Yaw(degree)	0.202	1.397	5.136
Pitch	-0.162	0.894	3.067
Roll	0.240	1.497	5.540

IV. CONCLUSION AND FUTURE WORKS

A real-time stereo tracking algorithm which can estimate the 6-DOF pose of an industrial object with high accuracy was presented. Integrated with the visual servoing system in [1], this tracking algorithm exposes a new frontier in automation of moving assembly lines.

In the future, we intend to study the effect of any changes in the vergence angle of the cameras on the target object's estimated pose. Also, because of the initialization phase, the application of this algorithm is currently constrained to a controlled environment with, for example, distinctive backgrounds, and consistent lighting conditions. While these constraints are not relevant to the tracking phase of the algorithm, these limitations need to be addressed in future work.

ACKNOWLEDGEMENT

The authors would like to thank Ford Motor Company for supporting this project.

V. REFERENCES

[1] G. N. DeSouza and A. C. Kak. A subsumptive, hierarchical, and distributed vision-based architecture for smart robotics. *IEEE Transactions on Robotics and Automation*, submitted, 2002.

[2] E. Ersu and S. Wienand. Vision system for robot guidance and quality measurement systems in automotive industry. *Industrial Robot*, 22(6):26–29, 1995.

[3] O. D. Faugeras. *Three-Dimensional Computer Vision*. MIT Press, 1993.

[4] J. T. Feddema and O. R. Mitchell. Vision-guided servoing with feature-based trajectory generation. *IEEE Transactions on Robotics and Automation*, 5(5):691–700, October 1989.

[5] C. Harris. Tracking with rigid models. In A. Blake and A. Yuille, editors, *Active vision*, pages 59–73, Chapter 4, 1992. MIT Press.

[6] R. Hirsh, G. N. DeSouza, and A. C. Kak. An iterative approach to the hand-eye and base-world calibration problem. In *Proceedings of 2001 IEEE International Conference on Robotics and Automation*, volume 1, pages 2171–2176, May 2001. Seoul, Korea.

[7] S. Hutchinson, G. D. Hager, and P. I. Corke. A tutorial on visual servo control. *IEEE Trans. on Robotics and Automation*, 12(5):651–670, Oct. 1996.

[8] I. Incorporated. Faceit(tm) face recognition technology. In *http://www.identix.com/products/pro_faceit.html http://www.cnn.com/2001/US/09/28/rec.airport.facial.screening*, 2002.

[9] E. Marchand, P. Bouthemy, F. Chaumette, and V. Moreau. Robust real-time visual tracking using a 2d-3d model-based approach. In *Proceedings of the 1999 IEEE International Conference on Computer Vision*, pages 262–268, Sept. 1999.

[10] T. B. Moeslunt and E. Granum. A survey of computer vision-based human motion capture. *Computer Vision and Image Understanding*, 81(3):231–268, 2001.

[11] T. Park and B. Lee. Dynamic tracking line: Feasible tracking region of a robot in conveyor systems. *IEEE Transactions on Systems, Man and Cybernetics*, 27(6):1022–1030, Dec. 1997.

[12] V. I. Pavlovic, R. Sharma, and T. S. Huang. Visual interpretation of hand gestures for human-computer interaction: a review. *IEEE Transactions on Pattern Analysis and Machine Intelligence*, 19(7):677–695, July 1997.

[13] Y. Rui, T. Huang, and S. Chang. Digital image/video library and mpeg-7: standardization and research issues. In *Proceedings of ICASSP 1998*, 1998.

Trajectory Generation for Constant Velocity Target Motion Estimation Using Monocular Vision

Eric W. Frew, Stephen M. Rock
Aerospace Robotics Laboratory
Department of Aeronautics and Astronautics
Stanford University
{ewf, rock}@sun-valley.stanford.edu

Abstract - The performance of monocular vision based target tracking is a strong function of camera motion. Without motion, the target estimation problem is unsolvable. By designing the camera path, the best possible estimator performance can be achieved. This paper describes a trajectory design method based on the predicted target state error covariance. This method uses a pyramid, breadth-first search algorithm to generate real-time paths that achieve a minimum uncertainty bound in fixed time or a desired uncertainty bound in minimum time.

1. INTRODUCTION

Monocular vision based target tracking is an important estimation method for autonomous field robots. Computer vision is an information-rich sensor that provides multiple capabilities including object detection, identification, and tracking. Single-camera solutions are useful for many vision applications. For example, micro air vehicles that provide aerial surveillance have size constraints that limit the available payload. The motion of a target object can be estimated from a single image track if the camera motion is known.

One important issue is the dependence of the target estimator performance on the camera motion. It is well known that for monocular vision based tracking the object motion is unobservable at any instant in time and that only certain camera paths yield solutions [1, 10]. Furthermore, favorable camera motions can be generated that achieve good estimator performance.

The trajectory design problem is not new and has been studied extensively for the related problem of passive sonar applications[2-8]. In the sonar literature, the estimation problem is referred to as bearings-only tracking or target motion analysis. The object is generally referred to as the target and the bearing sensor - the camera in this case - is referred to as the observer.

The concept of designing the observer trajectory in order to optimize estimator performance was first introduced by Hammel et al.[2]. They addressed the problem of trajectory design for a stationary target by using the Fisher Information Matrix (FIM) to describe the performance of an ideal estimator and by designing a numerical algorithm that minimized its determinant. They also derived lower bounds which could be used to generate observer paths analytically.

Oshman et al.[3] extended the stationary target problem by incorporating observer motion constraints. For the passive sonar application these constraints come mainly from hostility functions describing the operational environment.

Trajectory design for a moving target was studied by LeCadre and Jauffret [5]. They linearized the target motion analysis problem by transforming the measurement equations into a linear pseudomeasurement. They also used a discrete time formulation in order to reduce the analysis to multilinear algebra. They showed that the optimal path can be calculated for a target with arbitrary maneuvers under certain assumptions.

The problem addressed in this paper is the design of observer trajectories for monocular vision based tracking. Trajectories are desired that can be generated in real-time to minimize the estimate uncertainty in a given time or to minimize the time needed to achieve a specified error bound. Our approach is to extend the results of the sonar-based literature by addressing three separate issues.

First, the restricted camera field of view must be incorporated. In contrast to sonar systems that typically have a full 360 degree field of view, the computer vision system is constrained such that a measurement can only be taken if the camera is pointed towards the target. Some cameras cannot be moved independently from their heading and therefore cannot always remain pointed at the target (for example, a camera fixed to a ground rover). Hence, the trajectory design method must allow for motion that brings the target out of the camera field of view.

A second extension is needed in order to enable solutions of both the desired problems. Previous results have only minimized the uncertainty of the target estimate in given time. Here, the alternate problem is also included - minimizing the time taken to achieve a specified estimate uncertainty.

The final extension enables fast generation of the observer trajectories. The search space is too large to enable fully optimal solutions in reasonable time so a new, real-time strategy must be developed [8].

In order to accomplish these extensions a new optimization method is presented. A breadth-first search algorithm is used to generate a set of candidate trajectories that satisfy all constraints imposed on the system. A pyramid iteration scheme is used to calculate fast, sub-optimal paths. The optimization cost is defined using the predicted target error covariance matrix. Minimizing the determinant of this matrix is equivalent to maximizing the mutual information between the observer trajectory and the final target state [7].

The following sections present the new trajectory design method. The monocular vision based tracking estimator is first described in order to show its dependence on the observer motion. The details of the new trajectory design method are then presented, including the new cost function and new optimization method. A typical tracking scenario is simulated and the results show the successful estimation of target motion using the trajectory design method.

2. MONOCULAR VISION BASED MOTION ESTIMATION

Although the methods presented in this paper apply to the full three dimensional problem, only the 2-D planar case is presented. Hammel and Aidala[4] show that for the bearings-only tracking problem the 3-D results parallel the 2-D ones.

The geometry of the target motion estimation problem is shown in Figure 1. The target state is described by its position and velocity. The estimation state vector is

$$X = X_{target} = \begin{bmatrix} x & y & \dot{x} & \dot{y} \end{bmatrix}^T \quad (1)$$

For a constant velocity target, the state update equation is

$$X_{target}[k+1] = \Phi \cdot X_{target}[k] + \Gamma \cdot w[k] \quad (2)$$

$$\Phi = \begin{bmatrix} 1 & 0 & \Delta t & 0 \\ 0 & 1 & 0 & \Delta t \\ 0 & 0 & 1 & 0 \\ 0 & 0 & 0 & 1 \end{bmatrix} \quad \Gamma = \begin{bmatrix} 0.5\Delta t^2 & 0 \\ 0 & 0.5\Delta t^2 \\ \Delta t & 0 \\ 0 & \Delta t \end{bmatrix} \quad (3)$$

where $w[k]$ represents small accelerations that are treated as zero-mean Gaussian white process noise with covariance matrix $R_w = -E[ww^T] = \sigma_w \cdot I$.

This work assumes the motion of the observer is known. In this case, the motion of the target can be determined from a single point feature.

The computer vision measurement is the projection of the relative target position onto the image plane. The vision measurement is

$$z[k] = h(X[k]) = f \cdot (x_s/y_s) + v \quad (4)$$

$$X_s = \begin{bmatrix} T_{w2s} & 0 \\ 0 & T_{w2s} \end{bmatrix} \cdot (X_{target} - X_{observer})$$

$$= \begin{bmatrix} x_s & y_s & \dot{x}_s & \dot{y}_s \end{bmatrix}^T \quad (5)$$

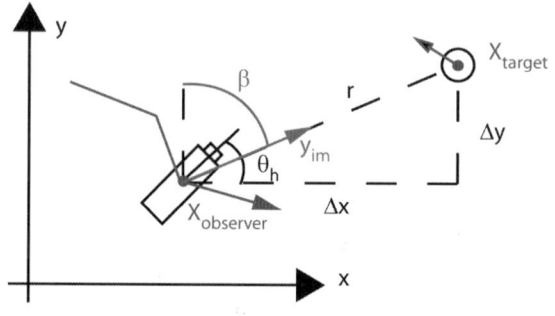

Figure 1. Problem geometry

$$T_{w2s} = \begin{bmatrix} \cos(\theta_h) & -\sin(\theta_h) \\ \sin(\theta_h) & \cos(\theta_h) \end{bmatrix} \quad (6)$$

where $v[k]$ is zero-mean Gaussian white measurement noise with covariance matrix $R_v = E[vv^T] = \sigma_v$. $X_{observer}$ is the known motion of the observer.

The linearized measurement matrix is

$$H(X) = \left.\frac{\partial h}{\partial X}\right|_{\tilde{X}} = f/y_s \cdot \begin{bmatrix} 1 & -x_s^2 \end{bmatrix} \cdot T_{w2s}. \quad (7)$$

The dependence of the target motion estimator on the observer motion appears in Equations (4) - (7). It is this dependence that is exploited by the trajectory design method in order to optimize the estimator performance.

Note, in practice the filter implementation is carried out in a transformed coordinate system. The results presented later in this paper use an Extended Kalman Filter (EKF) based on the modified polar coordinates proposed by Aidala and Hammel [9].

3. TRAJECTORY DESIGN

The trajectory design method is comprised of four steps. First, the permissible observer motion is defined in order to set the search space for the optimization. Many formulations are possible [5, 11] and the one presented here keeps the search space small. Second, the constraints that describe the limitations of the system are defined. Third, the cost functions are identified that specify the goals of the optimization. Finally, the optimization method finds the solution that satisfies the observer model, the constraints, and the cost function. The new method described here generates observer trajectories in real-time using an iterative search algorithm.

Observer Motion

The observer path is restricted to a finite number of maneuvers as shown in Figure 2. It has been shown that at least a single maneuver is needed in order to make the target motion observable [4]. The observer speed and maneuver duration are kept constant in order to reduce the complexity of the optimization. The time for each maneuver is distributed between a zero-radius turn

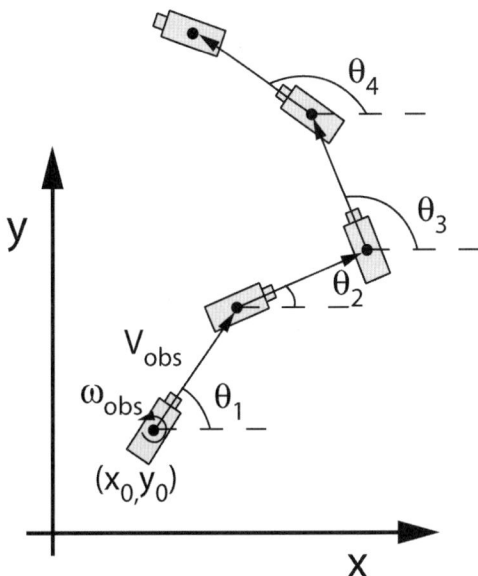

Figure 2. Trajectory parameterization

(T_{turn}) and a constant velocity traverse (T_{leg}) which follows - a longer turn leads to a shorter traverse. The size and number of maneuvers can be adjusted to make the observer path arbitrarily smooth. The free variable is the observer heading for each maneuver.

The equations describing the allowed observer motion for this work are as follows:

$$\begin{aligned} x_{obs}[k+1] &= x_{obs}[k] + VT_{leg}\cos(\theta_{k+1}) \\ y_{obs}[k+1] &= y_{obs}[k] + VT_{leg}\sin(\theta_{k+1}) \end{aligned} \quad (8)$$

$$\begin{aligned} T_{leg} &= T_{man} - T_{turn} \\ T_{man} &= T_{total}/n \\ T_{turn} &= (\theta_{k+1} - \theta_k)/\omega \end{aligned} \quad (9)$$

where θ is the observer heading, V is the observer speed, ω is the observer angular velocity, T_{man} is the duration of each maneuver, T_{total} is the total trajectory duration, and n is the number of maneuvers.

Optimization Problem

Two different optimization problems are formulated using the area of the predicted position uncertainty ellipse as defined by the position estimate error covariance matrix. This ellipse represents a region around the target estimate in which the true target state is located with some confidence. The area is used to describe the ellipse because it can easily be calculated from the determinant of the covariance matrix. Reducing this area is equivalent to reducing the uncertainty of the estimate.

The first optimization problem uses the area of the uncertainty ellipse as the cost function. It is minimized for a fixed number and duration of maneuvers and can be written:

$$\min_{\Theta}(A_{ellipse}(\Theta | T_{man}, n)). \quad (10)$$

For the second, the number of maneuvers is used as the cost. It is minimized for a fixed maneuver duration and specified final ellipse area:

$$\min_{\Theta}(n) \quad (11)$$

subject to $A_{ellipse}(\Theta, T_{man}, n) \leq A_{des}$.

The optimization method below will enable the use of either of these formulations.

The area of the position uncertainty ellipse is related to the target position estimate covariance by the expression

$$A_{ellipse} = \pi \cdot \sqrt{det(P_{cov})} \quad (12)$$

$$P_{cov} = \begin{bmatrix} P_{xx} & P_{xy} \\ P_{yx} & P_{yy} \end{bmatrix}$$
$$P_{xx} = E[(x-\tilde{x}) \cdot (x-\tilde{x})] \quad (13)$$
$$P_{yy} = E[(y-\tilde{y}) \cdot (y-\tilde{y})]$$
$$P_{xy} = P_{yx} = E[(x-\tilde{x}) \cdot (y-\tilde{y})]$$

The difficulty with the estimate error covariance is predicting its value beforehand. This is overcome by using the target estimate at the time the design method is invoked to predict the target motion. Using this motion, the estimator equations are propagated forward for each candidate trajectory. The resulting predicted error covariances are then used to calculate the costs for the trajectories.

Constraints

The vision field of view and the estimate uncertainty ellipse produce two limitations that must be incorporated into the trajectory design method. A field of view constraint is enforced when propagating the estimator equations, as described above. The estimate is only updated when the predicted target is in view of the candidate trajectory.

A second constraint prevents the observer from entering the uncertainty ellipsoid around the target estimate. Because the target could be anywhere within this ellipsoid with significant probability, it is necessary to include a constraint that keeps the observer out.

This constraint may be represented by

$$\begin{bmatrix} x_{est} - x_{obs} \\ y_{est} - y_{obs} \end{bmatrix}^T \cdot P_{cov} \cdot \begin{bmatrix} x_{est} - x_{obs} \\ y_{est} - y_{obs} \end{bmatrix} \geq c \quad (14)$$

where P_{cov} is the predicted position error covariance from Equation (13) and c is a scaling factor. The constraint is applied to every point along a candidate trajectory.

Optimization Method

The trajectory design problem is solved by performing a pyramid, breadth-first search. For either cost function, a breadth-first search tree is used to generate the list of candidate trajectories that satisfy all the given constraints and to calculate the predicted error covariances that describe those trajectories. The best candidate trajectory is identified and the process is repeated using this trajectory to narrow the search.

Although the ideal result of the optimization is the global minimum, this problem typically cannot be solved in a reasonable amount of time [8]. The search space has too many local minima for gradient-based methods and is too large for an exhaustive search. Instead, a sub-optimal result is obtained in real-time using an iterative pyramid scheme.

The free variables for the trajectory optimization are the headings of each maneuver. The heading space is discretized into a specified number of equal intervals. The number of intervals will determine the size of the search tree and the duration of the optimization process.

The search tree is expanded outward from the initial observer position in a breadth-first fashion. For each possible heading value, the observer motion and the estimator equations are propagated forward in time and the resulting position is checked against the constraints. If the new position satisfies the constraints it is added.

Each node in the search tree describes an observer state and the results of the propagated estimator up to that state. Because there is a unique path from the node back to the root of the tree, each node also represents a trajectory from the original observer state to the node state. Also, because the results of the estimator prediction to that point are stored at the node, the propagation of the estimator to the next node is only a single computation and the complete estimation does not need to be recalculated.

For the fixed-time minimum-uncertainty optimization the search tree is simply expanded to a depth that reaches the desired time. Once this level is reached the trajectory with the lowest cost is the solution.

For the fixed-uncertainty minimum-time optimization the search tree is expanded until the desired uncertainty is reached. Because the expansion is breadth first, the optimal trajectory will be the first one encountered that meets the desired uncertainty.

Once the optimal trajectory is found, the heading space is rediscretized about that result at a finer resolution. The breadth-first search is performed again and the new solution is found. This procedure is repeated a specified number of times until the final desired heading space resolution is achieved. In practice four to six iterations are sufficient to achieve a heading space resolution of less than one degree.

4. RESULTS

This section presents simulation results demonstrating the trajectory design method. The observer motion is based on a rover moving with a constant speed of 0.05 meters per second and a turning rate of 0.5 radians per second. The design method is invoked using the fixed-time minimum-uncertainty cost function. The observer is allowed five, 4.8 second maneuvers. The heading space is discretized into five intervals and the pyramid iteration is performed four times. For this simulation the computer vision system was modelled as a pinhole camera with a field of view of 20 degrees.

The observer begins at the location (-0.86, 0.11) meters and the target begins at (0.73, 0.62) and moves with a

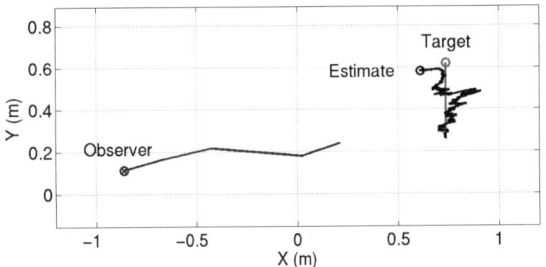

Figure 3. Observer, target, and estimate trajectories

constant velocity of (0.0, -0.015) meters per second The design method plans using these values for the target motion. The resulting observer path is shown in Figure 3 along with the true target path and the path of the target estimate over the course of the run The design method took 1.3 seconds to generate the observer trajectory.

The modified polar coordinate estimator [8] is used to calculate the position and velocity of the target. The trajectory design method is invoked after the estimator has run for several seconds. After this time the target estimate is at (0.61, 0.58) meters with a velocity of (0.04, 0.005) meters per second. Figure 3 shows the observer, target, and estimate curves from the time the design method is called.

There are three main features of the observer path. First, the observer moves toward the target throughout the path. As the range to the target decreases, the achievable bearing rate increases, allowing for wider triangulation and thus smaller error. Second, while moving toward the target, the observer also moves tangential to the line of sight to the target. It is this sideways motion that creates the bearing change needed to observe the target. Finally, the observer turns away from the target at least once. For nonholonomic vehicles, such as airplanes or rovers, this may cause the vision system to lose sight of the target. Procedures need to be implemented in order to reacquire the target should it get lost.

The results of the target motion estimator are shown in Figure 4 and Figure 5. The first figure shows the position error as a function of time. As expected the error in the y-dimension stays small throughout the run. In contrast, the error in the x-dimension, which corresponds closely to the target range, takes time to converge. However, by the end of the path the estimate error is small and the observer has successfully estimated the target motion. The second figure shows the velocity errors. Both components of the velocity converge throughout the run. The large spikes occur when the observer is stopped or turning and are an artifact of the estimator formulation.

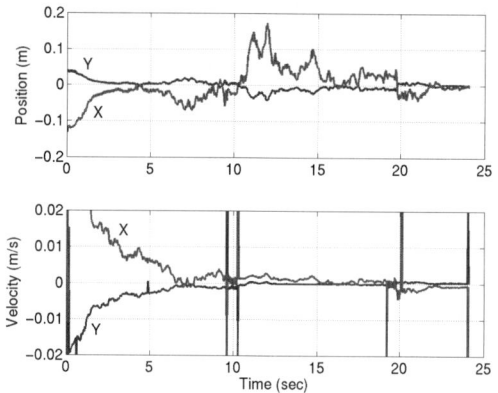

Figure 4. Target estimate errors

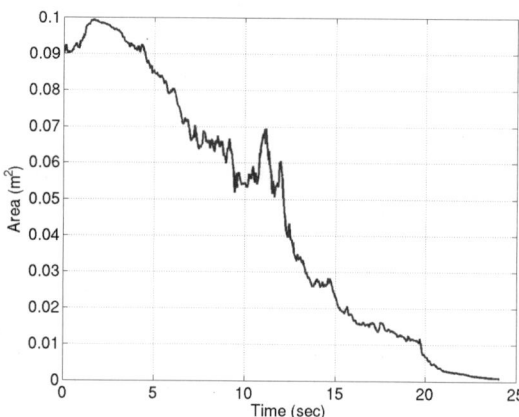

Figure 5. Area of the position uncertainty ellipse

Figure 5 shows a plot of the area of the uncertainty ellipse as a function of time. At the end of the observer trajectory the designer predicted an uncertainty area of 0.0082 square meters. The achieved error was 8.5e-04 square meters. This conservative result is due to the fact that the design method assumes measurements are only made at the end of each maneuver, while the estimator is actually run at 30 Hz.

The results of Figure 4 and Figure 5 show the successful estimation of the target motion. Furthermore, these results show that the predicted error covariance is a good description of the expected estimate performance and that basing the trajectory design method on this description is appropriate.

5. Conclusion

A method was presented for the generation of observer trajectories that improve monocular vision based estimation of a constant velocity target. Estimator performance is optimized by minimizing functions of the predicted target error covariance. In order to generate results in real-time, a sub-optimal iterative breadth-first search algorithm was described. Simulation results showed the success of the trajectory design method for a typical scenario.

6. References

[1] Ouyang Guanghui, Sun Jixiang, Li Hong, and Wang Wenhui. Estimating 3D motion and position of a point target. In *Proceedings of SPIE* 3137:386-394, 1997.

[2] S. E. Hammel, P. T. Liu, E. J. Hilliard, and K. F. Gong. Optimal observer motion for localization with bearing measurements. *Computers Math. Appl.*, 18(1-3):171-180, 1989.

[3] Yaakov Oshman and Pavel Davidson. Optimization of observer trajectories for bearings-only target localization. *IEEE Transactions on Aerospace and Electronic Systems*, 35(3):892-902, 1999.

[4] Sherry E. Hammel and Vincent J. Aidala. Observability requirements for three-dimensional tracking via angle measurements. *IEEE Transactions on Aerospace and Electronic Systems*, 21(2):200-207, 1985.

[5] J. P. Le Cadre and C. Jauffret. Discrete-time observability and estimability analysis for bearings-only target motion analysis. *IEEE Transactions on Aerospace and Electronic Systems*, 33(1):178-201, 1997.

[6] J. M. Passerieux and D.Van Cappel. Optimal observer maneuver for bearings-only tracking. *IEEE Transactions on Aerospace and Electronic Systems*, 34(3):777-788, 1998.

[7] Andrew Logothetis, Alf Isaksson, and Robin J. Evans. An information theoretic approach to observer design for bearings-only tracking. In *Proceedings of the 36th Conference on Decision and Control*, pages 3132-3137, San Diego CA, December 1997.

[8] Andrew Logothetis, Alf Isaksson, and Robin J. Evans. Comparison of suboptimal strategies for optimal own-ship maneuvers in bearings-only tracking. In *Proceedings of the American Control Conference*, pages 3334-3338, Philadelphia PA, June 1998.

[9] Vincent J. Aidala and Sherry E. Hammel. Utilization of modified polar coordinates for bearings-only tracking. *IEEE Transactions on Automatic Control*, 28(3):283-294, 1983.

[10] Eric W. Frew and Stephen Rock. Exploratory motion generation for monocular vision-based target localization. In *Proceedings of the 2002 IEEE Aerospace Conference*, CD-ROM only, Big Sky, MT, 2002.

[11] Andreas Huster, Eric W. Frew, and Stephen M. Rock. Relative position estimation for AUVs by fusing bearing and inertial rate measurements. In *Proceedings of the Oceans 2002 Conference*, pages 1857-1864, Biloxi, MS, October 2002.

Confluence of Parameters in Model Based Tracking

D. Kragic and H. I. Christensen
Centre for Autonomous Systems
Numerical Analysis and Computer Science
Royal Institute of Technology, Stockholm, Sweden
{danik,hic}@nada.kth.se [1]

Abstract

During the last decade, model based tracking of objects and its necessity in visual servoing and manipulation has been advocated in a number of systems [4], [7], [9], [12], [13], [14]. Most of these systems demonstrate robust performance for cases where either the background or the object are relatively uniform in color. In terms of manipulation, our basic interest is handling of everyday objects in domestic environments such as a home or an office.

In this paper, we consider a number of different parameters that effect the performance of a model–based tracking system. Parameters such as color channels, feature detection, validation gates, outliers rejection and feature selection are considered here and their affect to the overall system performance is discussed. Experimental evaluation shows how some of these parameters can successfully be evaluated (learned) on-line and consequently improve the performance of the system.

1 Introduction

For humans, everyday activities such as pointing, grasping, reaching, catching, various tool manipulation are strongly dependent on rich coordination between the eye and the hand. Each of these actions require attention to different attributes in the environment - while pointing requires only an approximate location of the object in the visual field, a reaching or grasping movement require more exact information about the object's pose. An extensive study of human visually guided grasps in [8] has shown that the human visuomotor system takes into account the three dimensional geometric features rather than the two dimensional projected image of the target objects to plan and control the required movements. Compared to most of the current robotic visual servoing systems, which are *image based* and based on 2D feature tracking, the information used by humans is much more complex and permits humans to operate in large range of environments. In terms of robotic manipulation, it is usually required to accurately estimate the pose of the object to, for example, allow the alignment of the robot arm with the object or to generate a feasible grasp and grasp the object. Using prior knowledge about the object properties such as size, texture or shape, a special representation can further increase the robustness of the tracking system. Realistic environments (tables, shelfs)

and natural objects (such as food packages, cups, etc.) offer us very little place for assumptions such as, for example, uniform color or simple texture attributes. In terms of pose estimation, a number of model–based tracking systems have been proposed [4], [12], [14], [13], [9], [7]. One common thing for all of them is the use of *wire–frame models* and *object features* to estimate current pose/velocity of the object. Most of them deal with tracking of particular targets (e.g. cars), usually uniform in color with a moderately varying backgrounds. Our paper provides an experimental appraisal of the parameter issue in a model based tracking system. The main objective is to present the different parameters that affect the performance of a model–based tracking system and integrate some of the ideas proposed in the above mentioned systems to achieve robustness in terms of tracking of textured objects in everyday environments. The system has successfully been used to: i) *estimate* the pose of an object to be grasped, and ii) *track* the pose of an object for cases of moving camera/moving robot tasks. The presented system uses a number of ideas proposed in other similar systems and integrates those to successfully cope with partial occlusions of the object and to mantain tracking of the object even in the case of significant rotational motion and background clutter. The ability to cope with occlusions and changes in the appearance of the object are two of the capabilities required for the design of a robust tracking and visual servoing system.

The paper is organized as follows. Section 2 outlines the different steps of our system. This is followed by a detailed description of different parameters that affect the performance in Section 3. An experimental evaluation is performed and it is further demonstrated how the parameters affect the overall system performance in Section 4. Finally, Section 5 summarizes the obtained results and presents avenues for our future research.

Figure 1: Example objects

---
[1] This research has been sponsored by the Swedish Foundation for Strategic Research through the Centre for Autonomous Systems.

2 System Overview

The tracking system presented in this work employs the classical *detect-match-update-predict* loop, [2]. The key problem to robust and precise object tracking are outliers caused by occlusions, cluttered background, specular reflections, shadows and texture. Approaches like condensation, [5], cope with outliers by taking a large number of sample hypotheses of the position of the tracked structure and a comparatively small number of edge measurements per sample. Our tracking system achieves robustness using a large number of measurements for a every pose hypothesis. The idea, similar to the ones proposed in [9] and [6], is to use an image motion model to account for motion between consecutive frames using normal displacements. Our approach relies on the estimation of normal flow for points (nodes) along lines and preserves the rigid structure of the object. As outlined in the introduction, the system is used for i) pose estimation, and ii) tracking. In our case, the basic differences between these two are:

i) Initial pose estimation retrieves the pose of the object at the beginning of the tracking (manipulation) sequence. Here, we assume that the initial guess is provided by an appearance based method or that a set of manual correspondences between the model and the object is available. The appearance based approach is briefly presented in Section 2.1. If, on the other hand, a set of correspondences is available, the iterative method proposed in [15] is used. This step is followed by an extension of [7] the nonlinear approach proposed in [16].

ii) Interframe pose change or pose tracking considers updating the pose of the object relative to the camera (or some other coordinate system) if there is a relative change in pose between these two. This is then used to *track* the pose of an object for cases where either (or both) camera and object are moving.

We have integrated both appearance based and geometrical models in our tracking system. After the object has been recognized and its position in the image is known, an appearance based method is employed to estimate its initial pose, [17].

2.1 Initialization

One of the problems to cope with during the initialization step is that the objects considered for manipulation are highly textured and therefore not suited for matching approaches based on, for example, line features [12], [14], [13]. The initialization step uses therefore the ideas proposed in [11]. During training, each image is projected as a point to the eigenspace and the corresponding pose of the object is stored with each point. Since the workspace of the robot is quite limited, a limited number of training images will suffice for most of the applications [17]. At run time, the pose parameters are found as the closest point on the pose manifold. Now, the wire-frame model of the object can be easily overlaid onto the image. Since a low number of images is used in the training process, pose parameters will not accurately correspond to the input image. Therefore, a local refinement method employed for tracking (Section 2.2) is used for the final fitting, Fig.2.

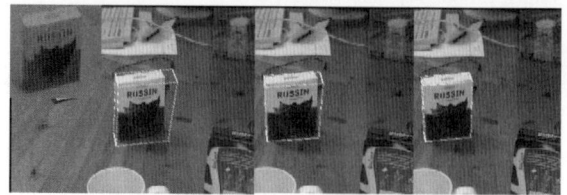

Figure 2: Fitting stage and change in pose: The image on the far left shows the nearest training image. Its pose is used as the starting value for the fitting process. The absolute change in pose parameters is: ΔX=4mm, ΔY=8mm, ΔZ=138mm, $\Delta\alpha$=23°, $\Delta\beta$=3°, $\Delta\gamma$=5°.

2.2 Pose Estimation

The system state vector is $\mathbf{x} = \left[X,Y,Z,\alpha,\beta,\gamma,\dot{X},\dot{Y},\dot{Z},\dot{\alpha},\dot{\beta},\dot{\gamma}\right]$ where α, β and γ represent roll, pitch and yaw angles. Using the ideas proposed in [9], normal flow along visible object features is used to find a geometric transformation of the object (relative change in pose) between two frames. Representing the pose of the object by a 4×4 homogeneous matrix $\mathbf{X}(\mathbf{R},\mathbf{t})$ and with an assumption of a small difference in pose between two adjacent frames, the relative change in pose is found as $\Delta\mathbf{X} = \sum \exp(g_i \mathbf{G}_i)$ where g_i represent the quantities of relative object motion observed in image and G_i are generators of a six-dimensional Lie group, each representing one of the six degrees of freedom of a rigid body.

2.3 Prediction and Update

The pose of the object is tracked over time using an $\alpha - \beta$ filter. Here, the pose of the target is used as measurement rather than image features as commonly used in the literature, [13]. This approach simplifies the structure of the filter which facilitates a computationally more efficient implementation.

2.4 Object Modeling

For most of the objects we want the robot to manipulate at this early stage, a simple polyhedral model will suffice. We have also integrated cones and cylinders in the system which allows us to deal with objects such as cups or plates. A model

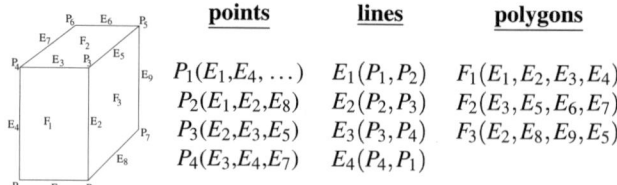

	points	**lines**	**polygons**
	$P_1(E_1,E_4,\ldots)$	$E_1(P_1,P_2)$	$F_1(E_1,E_2,E_3,E_4)$
	$P_2(E_1,E_2,E_8)$	$E_2(P_2,P_3)$	$F_2(E_3,E_5,E_6,E_7)$
	$P_3(E_2,E_3,E_5)$	$E_3(P_3,P_4)$	$F_3(E_2,E_8,E_9,E_5)$
	$P_4(E_3,E_4,E_7)$	$E_4(P_4,P_1)$	

Figure 3: An object is represented with points, lines and polygons. A similar schematic overview is presented in [12].

is constructed from a set of primitives, see Fig.3. In the simplest case, the primitives are the apparent object edges used to model the objects by points, lines and polygons defined both in the camera (3D) and image (2D) space. In addition, surface creases, markings, or regular texture patterns (circles, ellipses) can easily be integrated in the model. Given the current pose of the object, a hidden primitive removal is performed using *back face culling*, [18]. Assuming that the pose of the object changes a small fraction between frames, for optimization purposes, the visibility is not estimated in each frame.

3 Confluence of Parameters

We assume that the objects to be manipulated are placed on a table, shelf etc. In these situations, the background is fairly textured which, in the combination with textured objects, makes the process of feature detection relatively complicated. In terms of objects, surface patterns are commonly irregular in terms of shape (letters, flowers, etc.) which does not allows us to generate simple features (lines, ellipses) on the surfaces. Since an object is defined as a set of related primitives, which are related both in 2D and 3D, the robust improvements for the algorithm can be obtained at two levels: i) Obtaining measurements directly in the image and elimination of outlying measurements in the image, and ii) Pose update in 3D and elimination of the outliers based on pose parameters.

3.1 Outliers rejection in 2D

A number of systems use *tracking widows* for each feature of the model which are warped along the main feature extension, [13], [14], [12]. After this, edgels are extracted inside the window and used to fit the feature geometry to the data. Consequently, all pixels inside the window have to be processed to obtain feature candidates making this approach time consuming. In our system, using the predicted pose of the object, the visible edges of the object are projected onto the image. A number of control points (nodes) is generated along the edges. Assuming a small change in pose between two frames, for each control point correspondences are sought for to find the strongest image gradient in the vicinity of the control point. Because of the aperture problem, only the perpendicular distance along the edge is measurable. Therefore, it is sufficient to choose one of the eight cardinal directions closest to the direction of the line normal, allowing image search in one-dimension rather than two-dimensions (i.e. linear vs. quadratic complexity in the search range). Contrary to Kalman filter based methods, our method does not require in particular the introduction of a state model, noise variance of the state and measurement models, which are often crucial factors. However, both background and object texture properties will introduce a significant number of false positives. To improve the robustness with respect to the outliers, approaches such as RANSAC [4], factored sampling [5] and regularisation [7] have been proposed. We have decided to use the estimated normal displacements and fit a line through those using the least squares line fitting proposed in [1]. After that, the new normal displacements are estimated as the perpendicular distance between a node and the new line, see Fig.4.

3.2 Outliers rejection based on pose estimation

In combination with the outliers rejection considering normal displacements, we have also used the idea shown in Fig.5. Given the predicted pose of the object and the model, we can easily predict the parameters of all the features in the image. In

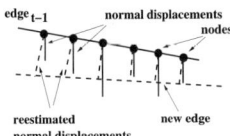

Figure 4: 2D outliers rejection: Using the old (predicted) position of the edge, normal displacements are estimated and new edge position found using least squares. The new line is used to reestimate the normal displacements.

this case, if there is a considerable difference in angle between the predicted and estimated edge, the edge is disregarded during in the pose estimation step. This allows us to successfully cope with shadows commonly occuring in the vicinity of edges. In addition, there are cases where two of the model edges may get matched to the same line in the image. In this case, the edge for which the predicted position is closest to the detected edge is used for pose estimation.

Figure 5: 2D fitting process where visible edges are shown in black and new, detected in white. Edge number 1 is disregarded in the pose estimation step, due to the large difference in angle compared to the previous frame.

3.3 Gradient threshold

Assuming that there is a considerable difference between the object and the background, one-dimensional search usually gives us satisfactory results. To cope with small differences in grey-level values between adjacent pixels, a threshold is commonly used. Inadequate threshold may results in false positives and an unsuccessful matching step. An empirical evaluation has shown that an adequate initial value for the threshold regarding all the features is 20. During tracking, this value is changed based on the average gradient estimated for all the generated nodes on a feature by feature basis.

3.4 Feature stability

In certain cases, some of the edges will be difficult to detect. This may happen if the background and the object are similar in color or if there is no enough difference between different facets of the object based on the lighting conditions, pose of the object, occlusion, etc. Consequently, the "goodness" of a feature will vary during tracking. We have introduced a confidence measure attached to each of the features depending on the frequency with which the feature is found during a tracking sequence. The confidence value is used to weight the feature's responses during pose estimation.

3.5 Number of nodes along features

Determining the number of nodes generated along the visible object contours is also one of the important parts of the system. In a model based tracking system there will always be a trade-off between the time required to estimate one cycle and the provided accuracy in terms of pose. There are two parameters that determine the number of nodes: i) current velocity of the object (or the change in pose) - if the change in pose between the frames is significant, the number of nodes is kept low, and ii) the confidence measure of a feature - for features with high confidence, a smaller number of nodes are generated.

3.6 Color space

Commonly, grey-level images are used during this type of tracking. It is widely known that in some cases, the blue channel contains lots of noise and is often disregarded. We have decided to investigate the idea of how much the information from different channels can help us during tracking. The basic idea is to base the evaluation on the sum of the gradients along the visible object features $\widehat{C} = \text{argmax}_C \left[\sum_{V_\mathbf{X}} |\nabla \mathbf{I}(\mathbf{p}) \, \mathbf{n}_e| \right]$ where $\nabla \mathbf{I}$ is the intensity (color value) gradient along the projected model edges, $\mathbf{n}_e$ is the edge normal, $\mathbf{p}$ is a point along the edge e and $V_\mathbf{X}$ are the visible edges of the model given the current pose $\mathbf{X}(\mathbf{R}, \mathbf{t})$. Considering the objects shown in Fig.1, we have estimated this value during a number of successful tracking sequences. During all the sequences the objects were placed on a table, see Fig. 8. The plots for each of the objects are shown in Fig.6. It can be seen that depending on the object properties, different channels gave different values for $\widehat{C}$. Consequently, the channel providing the highest value is chosen for tracking.

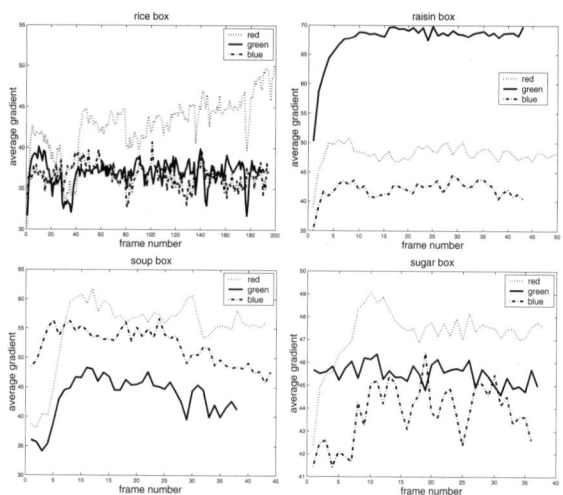

Figure 6: Average gradient plots for objects shown in Fig.1 using different color channels.

4 Experimental evaluation

The following sections show the performance of the system with the improvements proposed in the previous section. The system runs at frame rate on a standard Pentium PC.

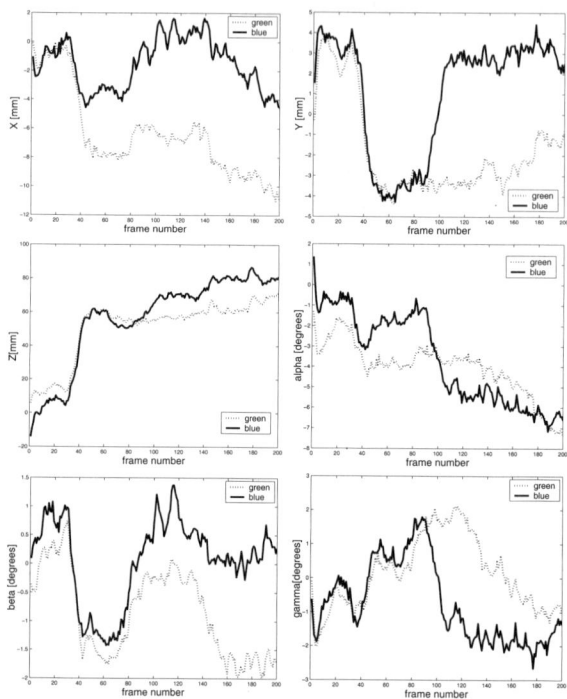

Figure 7: Error in pose when the object was tracked using different color channels. Here, the pose estimated from the red channel is used as the reference since it almost perfectly corresponds to the ground truth value.

4.1 Color space

The idea of using the channel giving the maximum average gradient was exploited in the following experiment where the system tracks a package of rice. Fig. 7 shows the difference in estimated pose parameters for each of the R,G,B channels. Here, the pose estimated from the R channel is used as the reference since it almost perfectly corresponds to the ground truth value. We will further investigate this basic idea for each of the features separately.

4.2 Tracking

Fig. 8 shows an example tracking sequence. The first two rows show a few images from the tracking sequence where the first row shows a unsuccessful and the second row an successful run. The estimated pose is overlaid in white. Here, the relative change in pose between the initial and the las frame is: $\Delta X=30$mm, $\Delta Y=10$mm, $\Delta Z=30$mm, $\Delta\alpha=8°$, $\Delta\beta=45°$, $\Delta\gamma=35°$. The goal here was to show how the system with the improvements proposed in the previous section copes with the significant changes in rotation. In the case of the first row, we have used the approach as proposed in [9] where normal displace-

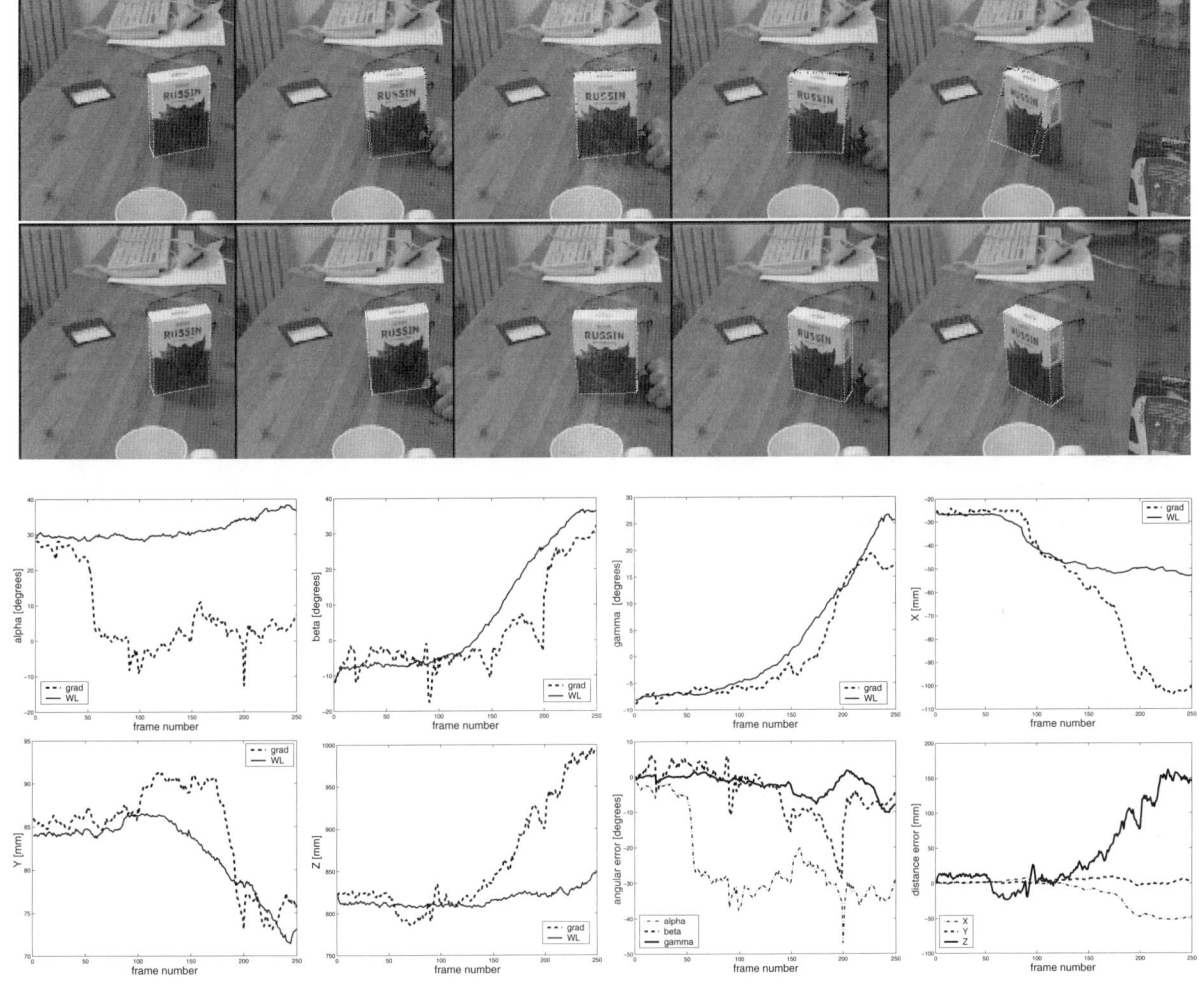

Figure 8: An example sequence where First row) the initial system without the improvements is used, and Second row) the system with the improvements proposed in Section3 is used. The last two rows show the estimated pose and their difference for both cases. Here, *grad* stands for the basic system, while *WL* denotes the improved system (see Section4.2 for detailed explanation).

ments are used for pose estimation. It can be seen that during the rotation, when one of the back faces comes to front, the tracker looses the object. The reason for this is that two nearby edges, belonging to that surface, get incorrectly matched. This does, however, not happen in the case of the improved system since because only the nearest edge is matched, and the other one is disregarded during pose estimation. The last two rows show the plots of each of the pose parameters as well as the error between them.

4.3 Local Fitting

Fig. 9 shows an example of the fitting stage. The two main differences between the examples are: i) the value of the minimum gradient required to estimate the normal flow as presented in Section 3.3, and ii) the ouliers rejection as presented in Section 3.1 and Section 3.2. The images in the upper row show an example of an unsuccessful fitting where only normal displacements are used in the pose estimation process. In addition, the threshold value for which a point is detected as edge was set to 10. This value was to low for this type of object and the background. In the second case (lower row), we have changed the gradient threshold to 20 and used outliers rejection as proposed in Section 3.1 and Section 3.2. The figure demonstrates a successful fitting step.

5 Summary and Conclusions

We have presented a model based tracking system and discussed its performance for cases of moderately textured objects in an everyday environment. The system relies on a simple geometrical model of the object to estimate its position and orientation in camera/robot coordinate system. Our approach integrates a number of ideas proposed in similar systems to achieve the robustness required for real–world applications. The main objective of the paper was the consideration of the

Figure 9: Two examples of fitting where the differences are i) the value of minimum gradient required to estimate normal displacements as presented in Section 3.3, and ii) the rejection of outliers using the ideas proposed in Section 3.1 and Section 3.2.

different parameters and their effect the system's performance. One of the key problems that has to be addressed in a tracking system are outliers. Textured background, shadows, occlusions, etc. will produce edges in the close proximity of the model edges. These are a particular problem for the traditional least-square fitting method used at this stage. Our future work will therefore consider the use of RANSAC that differs from the conventional least squares techniques as a small subset of data is used to estimate feature parameters. A problem that can occur for objects of simple geometry (boxes, cups) is that in some frames the number of detected features will give rise just to some of the pose (velocity) parameters. We will further investigate how, in this case, just the adequate pose parameters can be updated. Finally, our ultimate goal is to regain tracking after it has been lost. Our idea here is to integrate the appearance based method similar to the one used during the initialization step to allow continuous tracking and achieve a fault tolerant system.

References

[1] R.Deriche, R.Vaillant and O.Faugeras, "From noisy edge points to 3D reconstruction of a scene: A robust approach and its uncertainty analysis", *World Scientific*, vol. 2, pp. 225 - 232, Series in Machine Perception and Artificial Intelligence, 1992

[2] D. Kragic. "Visual Servoing for Manipulation: Robustness and Integration Issues", *PhD thesis*, Royal Institute of Technology, Stockholm, Sweden, 2001.

[3] S. Hutchinson, G.D. Hager and P. Corke "A tutorial on visual servo control", *IEEE TRA* **12**(5), pp. 651–670, 1996.

[4] M. Armstrong and A. Zisserman, "Robust object tracking" *Asian Conference on Computer Vision*, vol. I, pp 58–61,1995

[5] M. Isard and A. Blake, "Condensation – conditional density propagation for visual tracking", *IJCV*, 29:1, pp. 5–28, 1998.

[6] C. Harris and C. Stennett, "RAPID - a video rate object tracker", *Proc. British Machine Vision Conference*, pp. 73–77, 1990

[7] D.G. Lowe, "Robust model-based motion tracking through the integration of search and estimation", *IJCV*, 8:2, pp.113-122, 1992

[8] Y. Hu, R. Eagleson and M. Goodale, "Human visual servoing for reaching and grasping: The role of 3-D geometric features", *IEEE ICRA*, pp. 3209-3216, vol 3, 1999

[9] T.Drummond and R.Cipolla. Real-time tracking of multiple articulated structures in multiple views. *ECCV*, 2:20–36, 2000.

[10] D. Roobaert. "Pedagogical Support Vector Learning: A Pure Learning Appr oach to Object Recognition", PhD thesis, Royal Institute of Technology, Stockholm, Sweden, 2001

[11] S.Nayar, S.Nene, and H.Murase. Subspace methods for robot vision. *IEEE Trans. Robotics and Automation*, 12(5):750–758, 1996.

[12] M. Vincze, M. Ayromlou, and W. Kubinger. An integrating framework for robust real-time 3D object tracking. *ICVS'99*, 135–150, LNCS, H.I. Christensen (Ed.), vol. 1542, 1999

[13] P. Wunsch and G. Hirzinger. Real-time visual tracking of 3D objects with dynamic handling of occlusion. *ICRA'97*, 2:2868–2873

[14] D. Koller, K. Daniilidis, and H.H. Nagel. Model-based object tracking in monocular image sequences of road traffic scenes. *IJCV*, 10(3):257–281, 1993

[15] D. DeMenthon and L. Davis, "New exact and approximate solutions of the three–point perspective problem", *IEEE PAMI*, vol. 14, no. 11, pp.1100-1105, 1992

[16] H. Araujo, R. Canceroni and C. Brown, "A fully projective formulation for Lowe's tracking algorithm", *Technical report 641, The University of Rochester*, CS Department, Rochester, NY, 1996

[17] D. Kragic and H. Christensen, "Model Based Techniques for Robotic Servoing and Grasping" , *IEEE/RSJ IROS*, 2002

[18] J. Foley, A. van Dam, S. Feiner and J. Hughes, "Computer graphics - principles and practice", *Addison-Wesley*, 1990

# Visual Position Tracking using Dual Quaternions with Hand-Eye Motion Constraints

Tomas Olsson, Johan Bengtsson, Anders Robertsson, Rolf Johansson

Department of Automatic Control, Lund Institute of Technology, Lund University, SE-221 00 Lund Sweden;
E-mail: {Tomas.Olsson, Johan.Bengtsson, Anders.Robertsson, Rolf.Johansson}@control.lth.se

Abstract—In this paper a method for contour-based rigid body tracking with simultaneous camera calibration is developed. The method works for a single eye-in-hand camera with unknown hand-eye transformation, viewing a stationary object with unknown position. The method uses dual quaternions to express the relationship between the camera- and end-effector screws. It is shown how using the measured motion of the robot end-effector can improve the accuracy of the estimation, even if the relative position and orientation between sensor and actuator is completely unknown.

The method is evaluated in simulations on images from a real-time 3D rendering system. The system is shown to be able to track the pose of rigid objects and changes in intrinsic camera parameters, using only rough initial values for the parameters. The method is finally validated in an experiment using real images from a camera mounted on an industrial robot.

I. INTRODUCTION

A. Visual position tracking

Tracking and estimating the position of objects using measurements from one or several cameras has been an active research topic for many years. A special case is tracking of the position and orientation of rigid objects. Many methods for rigid body tracking work by minimizing some measure of the image space error as a function of the unknown position and orientation parameters. The minimization can for instance be performed using standard non-linear optimization methods such as Gauss-Newton or Levenberg-Marquardt. Another option is to use Kalman filtering techniques [1], [2].

The position and orientation can be parameterized in different ways, such as roll-pitch-yaw angles [1], quaternions or dual quaternions [2]. There are also various ways to measure the image space error, the most common measurements are the positions of point features [1] or line features [2], or point-to-contour errors [3], [4]. The point-to-contour method has a major advantage in that it does not require exact matching of features, only the error in the normal direction at a number of points on a contour. This only requires a one-dimensional search for features (edges).

In [3] it is shown that not only the position and orientation can be estimated, but also the intrinsic parameters of the camera (focal length, aspect ratio and principal point). In [4], on the other hand, it is pointed out that the problem of simultaneously tracking position and intrinsic parameters is ill-conditioned when the points of the object lie on a plane parallel to the image plane. This causes the Jacobian matrix, relating errors in position- and intrinsic parameters to image errors, to lose rank. Because of noise, this problem extends also to positions where the relative depth of the object points in the camera is small. A multi-camera tracking system is suggested as a possible solution to this problem.

B. Quaternions and dual quaternions

Unit quaternions [5], [6], [2] is a common representation of rotations. Similarly to real quaternions, dual quaternions are defined as $\check{\mathbf{q}} = (\check{q}^0, \check{\vec{\mathbf{q}}})$, where $\check{q}^0 = q^0 + \varepsilon q'^0$ is a dual number with $\varepsilon^2 = 0$, and where $\check{\vec{\mathbf{q}}} = \vec{\mathbf{q}} + \varepsilon \vec{\mathbf{q}}'$ is a dual vector. The dual quaternion operations are

$$\check{\mathbf{q}}_1 + \check{\mathbf{q}}_2 = (\check{q}_1^0 + \check{q}_2^0, \check{\vec{\mathbf{q}}}_1 + \check{\vec{\mathbf{q}}}_2) \tag{1}$$

$$k\check{\mathbf{q}} = (k\check{q}^0, k\check{\vec{\mathbf{q}}}) \tag{2}$$

$$\check{\mathbf{q}}_1 \check{\mathbf{q}}_2 = (\check{q}_1^0 \check{q}_2^0 - \check{\vec{\mathbf{q}}}_1^T \check{\vec{\mathbf{q}}}_2, \check{q}_1^0 \check{\vec{\mathbf{q}}}_2 + \check{q}_2^0 \check{\vec{\mathbf{q}}}_1 + \check{\vec{\mathbf{q}}}_1 \times \check{\vec{\mathbf{q}}}_2). \tag{3}$$

We will often write the dual quaternion as the sum of the real and dual parts $\mathbf{q} + \varepsilon \mathbf{q}'$. Its norm is given by $\|\check{\mathbf{q}}\|^2 = \check{\mathbf{q}}\bar{\check{\mathbf{q}}}$ with $\bar{\check{\mathbf{q}}} = \bar{\mathbf{q}} + \varepsilon \bar{\mathbf{q}}'$, and the unity conditions become

$$\mathbf{q}\bar{\mathbf{q}} = 1, \quad \bar{\mathbf{q}}\mathbf{q}' + \bar{\mathbf{q}}'\mathbf{q} = 0. \tag{4}$$

Unit dual quaternions can be used to represent general rigid transformations including translations, similarly to the way rotations can be represented by real quaternions. It can be shown, see [6], that the rigid transformation of a line through the point $\vec{\mathbf{p}}$, represented by its direction $\vec{\mathbf{n}}$ and moment $\vec{\mathbf{m}} = \vec{\mathbf{p}} \times \vec{\mathbf{n}}$, is given by $\check{\mathbf{q}}(\mathbf{n} + \varepsilon \mathbf{m})\bar{\check{\mathbf{q}}}$, where $\vec{\mathbf{n}}$ and $\vec{\mathbf{m}}$ are expressed as quaternions $\mathbf{n} = (0, \vec{\mathbf{n}})$ and $\mathbf{m} = (0, \vec{\mathbf{m}})$, respectively. The dual quaternion itself is $\mathbf{q} + \varepsilon \mathbf{q}'$, where $\mathbf{q}$ is the quaternion describing the rotation, and where $\mathbf{q}' = \mathbf{t}\mathbf{q}/2$ with $\mathbf{t} = (0, \vec{\mathbf{t}})$ being the translation.

C. Screws and robot motion constraints

Screws: According to Chasles' theorem [5], a general rigid transformation can be modeled as a rotation about

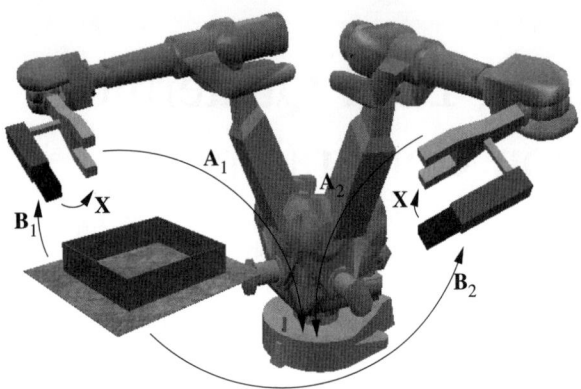

Fig. 1. Two different positions of the robot and relevant transformations.

an axis not through the origin and a translation along the rotation axis. The parameters of the screw are the direction $\vec{n}$ and the moment $\vec{m}$ of the screw axis line, the rotation angle θ, and the translation (pitch) d along $\vec{n}$. Together with the constraints $\vec{n}^T\vec{n} = 1$ and $\vec{n}^T\vec{m} = 0$ these parameters constitute the six degrees of freedom of a rigid transformation. It can be shown that the dual quaternion corresponding to the screw with parameters $\vec{n}$, $\vec{m}$, θ and d can be written as

$$\check{\mathbf{q}} = (\cos(\check{\theta}/2), \sin(\check{\theta}/2)\check{\mathbf{l}}), \quad (5)$$

where the dual angle is $\check{\theta} = \theta + \varepsilon d$, and the line is given by $\check{\mathbf{l}} = \vec{n} + \varepsilon \vec{m}$.

Robot motion constraints: The well known hand-eye equation

$$\mathbf{AX} = \mathbf{XB}, \quad (6)$$

with $\mathbf{A} = \mathbf{A}_2^{-1}\mathbf{A}_1$ and $\mathbf{B} = \mathbf{B}_2\mathbf{B}_1^{-1}$ from Fig. 1, can be written using dual quaternions as

$$\check{\mathbf{a}} = \check{\mathbf{q}}\check{\mathbf{b}}\bar{\check{\mathbf{q}}}. \quad (7)$$

In [6], it is shown that the scalar parts of $\check{\mathbf{a}}$ and $\check{\mathbf{b}}$ are equal, which can easily be shown as follows

$$\begin{aligned} Sc(\check{\mathbf{a}}) &= \frac{1}{2}(\check{\mathbf{a}} + \bar{\check{\mathbf{a}}}) = \frac{1}{2}(\check{\mathbf{q}}\check{\mathbf{b}}\bar{\check{\mathbf{q}}} + \check{\mathbf{q}}\bar{\check{\mathbf{b}}}\bar{\check{\mathbf{q}}}) = \\ &= \frac{1}{2}\check{\mathbf{q}}(\check{\mathbf{b}} + \bar{\check{\mathbf{b}}})\bar{\check{\mathbf{q}}} = Sc(\check{\mathbf{b}})\check{\mathbf{q}}\bar{\check{\mathbf{q}}} = Sc(\check{\mathbf{b}}). \end{aligned} \quad (8)$$

In terms of the screw parameters, Eq. (8) means that the angle and pitch of the camera screw and the robot end-effector screw must be equal [6]. This is known as the Screw Congruence Theorem, see [7].

D. Problem formulation

The purpose of this paper is to develop methods for real-time rigid body tracking with simultaneous calibration and tracking of intrinsic parameters. We intend to show that a dual quaternion parameterization of the object pose, together with measurements of the robot motion, can be used to formulate constraints on the estimated motion. The constraints can be expressed as linear equations in the states.

II. MODELING

Consider a manipulator with a single camera attached to its end-effector, viewing a stationary object. We assume that only rough initial values of the intrinsic camera parameters and the position/orientation of the object are known, but that a CAD model of the object is available. The motion of the robot end-effector is related to the motion of the camera through the hand-eye equation (6), where the relative sensor-actuator pose $\mathbf{X}$ is unknown. We assume that the camera can be modeled as a four parameter pinhole camera

$$\lambda \begin{pmatrix} u \\ v \\ 1 \end{pmatrix} = \begin{pmatrix} f & 0 & u_0 \\ 0 & \gamma f & v_0 \\ 0 & 0 & 1 \end{pmatrix} (\mathbf{R} \quad \mathbf{t}) \begin{pmatrix} X \\ Y \\ Z \\ 1 \end{pmatrix} \quad (9)$$

with λ corresponding to the depth of point (X, Y, Z) in the camera. The parameters to be estimated are f, γ, u_0, v_0, and some parameterizations of $\mathbf{R} \in SO(3)$ and $\mathbf{t} \in \mathbb{R}^3$.

A. Extended Kalman Filtering (EKF)

The motion of the system can be written as a non-linear discrete-time dynamic system

$$\mathbf{x}_{k+1} = \mathbf{f}(\mathbf{x}_k) \quad (10)$$
$$\mathbf{g}(\mathbf{p}_k, \mathbf{x}_k) = 0 \quad (11)$$

with $\mathbf{x}_k \in \mathbb{R}^n$ the state of the system, $\mathbf{p}_k \in \mathbb{R}^m$ a vector of measured outputs, $\mathbf{f}$ a known function describing the system dynamics, and $\mathbf{g}$ a known function relating the state to the output. The state vector is chosen as

$$\mathbf{x} = \begin{pmatrix} \mathbf{q}^T & \mathbf{q}'^T & f & \gamma & u_0 & v_0 & \omega^T & \mathbf{v}^T \end{pmatrix}^T \quad (12)$$

where $\mathbf{q}, \mathbf{q}' \in \mathbb{R}^4$ is the vector representation of the object-camera dual quaternion $\check{\mathbf{q}} = \mathbf{q} + \varepsilon \mathbf{q}'$, and $\mathbf{v}, \omega \in \mathbb{R}^3$ are the velocity and angular velocity.

Measurement model: In the function $\mathbf{g}(\mathbf{p}_k, \mathbf{x}_k)$ we have all the measurement equations, and the constraints on the state vector. For a point feature, denoted by index i, the measurement would be its image coordinates, denoted by $\mathbf{p}_{k,i} = (u_i, v_i)^T \stackrel{\text{def}}{=} \mathbf{h}_i(\mathbf{x})$, which is known from Eq. (9) to be related to the states by the equations

$$\begin{aligned} \hat{\mathbf{h}}_i(\mathbf{x}) &= \begin{pmatrix} \hat{u}_i \\ \hat{v}_i \\ \hat{w}_i \end{pmatrix} = \mathbf{K} \begin{pmatrix} \mathbf{r}_x^T(\mathbf{q}) & t_x(\check{\mathbf{q}}) \\ \mathbf{r}_y^T(\mathbf{q}) & t_y(\check{\mathbf{q}}) \\ \mathbf{r}_z^T(\mathbf{q}) & t_z(\check{\mathbf{q}}) \end{pmatrix} \begin{pmatrix} \mathbf{X}_i \\ 1 \end{pmatrix} = \\ &= \begin{pmatrix} f(\mathbf{r}_x^T(\mathbf{q})\mathbf{X}_i + t_x(\check{\mathbf{q}})) + u_0(\mathbf{r}_z^T(\mathbf{q})\mathbf{X}_i + t_z(\check{\mathbf{q}})) \\ \gamma f(\mathbf{r}_y^T(\mathbf{q})\mathbf{X}_i + t_y(\check{\mathbf{q}})) + v_0(\mathbf{r}_z^T(\mathbf{q})\mathbf{X}_i + t_z(\check{\mathbf{q}})) \\ \mathbf{r}_z^T(\mathbf{q})\mathbf{X}_i + t_z(\check{\mathbf{q}}) \end{pmatrix} \end{aligned} \quad (13)$$

where $\mathbf{K}$ is the matrix of intrinsic parameters in Eq. (9). The image space coordinates are given by

$$\mathbf{h}_i(\mathbf{x}) = \begin{pmatrix} u_i \\ v_i \end{pmatrix} = \begin{pmatrix} \hat{u}_i/\hat{w}_i \\ \hat{v}_i/\hat{w}_i \end{pmatrix}. \quad (14)$$

The rotation matrix can be calculated directly from the unit quaternion $\mathbf{q}$, and the translation can be obtained from $\mathbf{\check{q}}$ as

$$\mathbf{t} = 2\mathbf{q}'\bar{\mathbf{q}}. \quad (15)$$

If $\mathbf{p}_{k,i}$ are the measured image coordinates, we can write the measurement equation for this point as

$$\mathbf{g}_i(\mathbf{p}_{k,i}, \mathbf{x}_k) = \mathbf{p}_{k,i} - \mathbf{h}_i(\mathbf{x}_k) = 0 \quad (16)$$

This equation can be linearized around the predicted state $\mathbf{x}_k^{(p)}$, which gives the approximation

$$\mathbf{g}_i(\mathbf{p}_{k,i}, \mathbf{x}_k) \approx \mathbf{g}_i(\mathbf{p}_{k,i}, \mathbf{x}_k^{(p)}) + \frac{\partial \mathbf{g}_i}{\partial \mathbf{x}}(\mathbf{p}_{k,i}, \mathbf{x}_k^{(p)})(\mathbf{x}_k - \mathbf{x}_k^{(p)})$$
$$= \mathbf{p}_{k,i} - \mathbf{h}_i(\mathbf{x}_k^{(p)}) + \frac{\partial \mathbf{g}_i}{\partial \mathbf{x}}(\mathbf{p}_{k,i}, \mathbf{x}_k^{(p)})(\mathbf{x}_k - \mathbf{x}_k^{(p)}) \approx 0 \quad (17)$$

In our system however, the only image measurements available are the point-to-contour error in the predicted (local) normal direction of the contour, which can be approximated with the normal component of the error

$$\mathbf{g}_i^{(n)} = \mathbf{n}_i^T(\mathbf{p}_{k,i} - \mathbf{h}_i(\mathbf{x}_k^{(p)})) \quad (18)$$

Eq. (17) can then be rewritten as

$$0 = \mathbf{g}_i^{(n)}(\mathbf{p}_{k,i}, \mathbf{x}_k) \approx \mathbf{n}_i^T(\mathbf{p}_{k,i} - \mathbf{h}_i(\mathbf{x}_k^{(p)})) +$$
$$+ \mathbf{n}_i^T \frac{\partial \mathbf{g}_i}{\partial \mathbf{x}}(\mathbf{p}_{k,i}, \mathbf{x}_k^{(p)})(\mathbf{x}_k - \mathbf{x}_k^{(p)}) \quad (19)$$

which can be expressed on linear form as

$$\mathbf{y}_{k,i} = \mathbf{C}_{k,i}\mathbf{x}_k \quad (20)$$
$$\mathbf{C}_{k,i} = \mathbf{n}_i^T \frac{\partial \mathbf{g}_i}{\partial \mathbf{x}}(\mathbf{p}_{k,i}, \mathbf{x}_k^{(p)}) = -\mathbf{n}_i^T \frac{\partial \mathbf{h}_i}{\partial \mathbf{x}}(\mathbf{p}_{k,i}, \mathbf{x}_k^{(p)})$$
$$\mathbf{y}_{k,i} = -\mathbf{n}_i^T \frac{\partial \mathbf{h}_i}{\partial \mathbf{x}}(\mathbf{p}_{k,i}, \mathbf{x}_k^{(p)})\mathbf{x}_k^{(p)} - \mathbf{n}_i^T(\mathbf{p}_{k,i} - \mathbf{h}_i(\mathbf{x}_k^{(p)}))$$

The Jacobian of of $\mathbf{h}_i$ can be calculated by direct differentiation of Eq. (13) with respect to the elements of $\mathbf{x}$, combined with the equations

$$\begin{pmatrix} u_i' \\ v_i' \end{pmatrix} = \begin{pmatrix} \frac{\hat{u}_i'}{\hat{w}_i} - \frac{\hat{u}_i \hat{w}_i'}{\hat{w}_i^2} \\ \frac{\hat{v}_i'}{\hat{w}_i} - \frac{\hat{v}_i \hat{w}_i'}{\hat{w}_i^2} \end{pmatrix} \quad (21)$$

where $\hat{u}_i'$ denotes the differentiation of $\hat{u}_i$ with respect to the relevant quantity.

The constraints on the dual quaternion in Eq. (4) can be written on vector form

$$\mathbf{q}^T\mathbf{q} = 1, \quad \mathbf{q}^T\mathbf{q}' = 0 \quad (22)$$

and included by linearizing around the prediction $\mathbf{x}_k^{(p)}$, which gives us two linear equations in $\mathbf{q}$ and $\mathbf{q}'$

$$1 \approx \mathbf{q}^{(p)T}\mathbf{q}^{(p)} + 2\mathbf{q}^{(p)T}(\mathbf{q} - \mathbf{q}^{(p)}) \quad (23)$$
$$0 \approx -\mathbf{q}^{(p)T}\mathbf{q}'^{(p)} + \mathbf{q}'^{(p)T}\mathbf{q} + \mathbf{q}^{(p)T}\mathbf{q}', \quad (24)$$

which can be included among the output equations.

Including the robot motion constraints from Eq. (8) is also straightforward. Consider two different robot poses, represented by the the dual quaternions $\mathbf{\check{q}}_{B_1}$ and $\mathbf{\check{q}}_{B_2}$, and the corresponding relative object-camera poses $\mathbf{\check{q}}_{A_1}$ and $\mathbf{\check{q}}_{A_2}$. We know from Eq. (8) that the scalar parts of $\mathbf{\check{q}}_A = \mathbf{\bar{\check{q}}}_{A_2}\mathbf{\check{q}}_{A_1}$ and $\mathbf{\check{q}}_B = \mathbf{\check{q}}_{B_2}\mathbf{\bar{\check{q}}}_{B_1}$ must be equal. Define the scalar part of the relative robot pose $\mathbf{\check{q}}_B$ as $\check{q}_B^{(0)} = q_B^{(0)} + \varepsilon q_B'^{(0)}$, which can be calculated directly from the forward kinematics of the robot. The scalar part of $\mathbf{\check{q}}_A$ can be seen from Eq. (3) to be $\mathbf{q}_{A_1}^T\mathbf{q}_{A_2} + \varepsilon(\mathbf{q}_{A_1}'^T\mathbf{q}_{A_2} + \mathbf{q}_{A_1}^T\mathbf{q}_{A_2}')$, with the quaternions written on vector form. Setting the scalar parts equal gives us two more linear equations

$$\mathbf{q}_{A_1}^T\mathbf{q}_{A_2} = q_B^{(0)} \quad (25)$$
$$\mathbf{q}_{A_1}'^T\mathbf{q}_{A_2} + \mathbf{q}_{A_1}^T\mathbf{q}_{A_2}' = q_B'^{(0)}, \quad (26)$$

which can also be added to the system measurement equation, which can now be formulated as

$$\mathbf{y}_k = \mathbf{C}_k\mathbf{x}_k + \delta_k \quad (27)$$

where δ_k is a sequence of uncorrelated Gaussian noise, and the vector $\mathbf{y}_k$ and time-varying matrix $\mathbf{C}_k$ are obtained by stacking equations (19) for each edge search point, and adding constraints from Eq. (23)–(24) and (25)–(26). Any number of robot motion constraints can be added to the measurement equation. In general each position used gives us two independent constraints on the pose, meaning that three positions are sufficient to completely constrain the estimated pose. This can be compared to the problem of hand-eye calibration, where it is well known that three positions are necessary for the calculation of the hand-eye transformation [6].

State dynamics model: In the function $\mathbf{f}$ in the state update equation (11), we update the estimate of the dual quaternion using the equations

$$\dot{\mathbf{q}} = \frac{1}{2}\omega\mathbf{q} = \frac{1}{2}(0, \vec{\omega}_k)\mathbf{q} \quad (28)$$
$$\dot{\mathbf{q}}' = \frac{1}{2}\mathbf{t}\mathbf{q} + \frac{1}{2}\mathbf{t}\dot{\mathbf{q}} = \frac{1}{2}\mathbf{v}\mathbf{q} + \frac{1}{4}\mathbf{t}\omega\mathbf{q} \quad (29)$$

where the details can be found in [2]. Discretizing Eqs. (28) and (29) using sample time h, the noise-free state update equation becomes

$$\begin{pmatrix} \mathbf{q}_{k+1} \\ \mathbf{q}'_{k+1} \\ \mathbf{\check{K}}_{k+1} \\ \omega_{k+1} \\ \mathbf{v}_{k+1} \end{pmatrix} = \begin{pmatrix} \mathbf{I} & \mathbf{0} & \mathbf{0} & \frac{h}{2}\mathbf{Q}_k & \mathbf{0} \\ \mathbf{0} & \mathbf{I} & \mathbf{0} & \frac{h}{4}\mathbf{T}_k\mathbf{Q}_k & \frac{h}{2}\mathbf{Q}_k \\ \mathbf{0} & \mathbf{0} & \mathbf{I} & \mathbf{0} & \mathbf{0} \\ \mathbf{0} & \mathbf{0} & \mathbf{0} & \mathbf{I} & \mathbf{0} \\ \mathbf{0} & \mathbf{0} & \mathbf{0} & \mathbf{0} & \mathbf{I} \end{pmatrix} \begin{pmatrix} \mathbf{q}_k \\ \mathbf{q}'_k \\ \mathbf{\check{K}}_k \\ \omega_k \\ \mathbf{v}_k \end{pmatrix} \quad (30)$$

Fig. 2. Example of image with superimposed object model, where hidden features have been removed.

where $\tilde{\mathbf{K}} = (f, \gamma, u_0, v_0)^T$, and where the matrices $\mathbf{Q}_k$ and $\mathbf{T}_k$ correspond to the quaternion multiplications with $\mathbf{q}_k = (q_0, q_1, q_2, q_3)$ and $\mathbf{t}_k = (0, t_x, t_y, t_z) = 2\mathbf{q}'_k \bar{\mathbf{q}}_k$ in Eqs. (28) and (29). Adding uncorrelated Gaussian noise ε_k, Eq. (30) can be written as

$$\mathbf{x}_{k+1} = \mathbf{A}_k \mathbf{x}_k + \varepsilon_k. \quad (31)$$

The linearized equations in (30) and (27) can then be used to recursively update the state estimate using an Extended Kalman filter. The equations for the EKF can for instance be found in [8], [1], [2].

B. Object modeling and feature selection/localization

The object models consist of a number of planar surfaces connected at their edges, see Fig. 2. No assumptions are made about the shape of the planar surfaces, although in the experiments we use an object with only straight edges. At each step in the tracking visible object edges are selected, based on the predicted object pose and a pre-generated Binary Search Partitioning (BSP) tree description of the object, see [9]. The BSP tree recursively divides the surfaces in the object into "in front" and "behind", until we have a perfect front-to-back ordering. The surfaces are then processed front-to-back, each surface is clipped against all surfaces in front of it, and a number of search points are selected on each visible edge. The image position measurements are then obtained from a one-dimensional edge localization in the local edge normal direction at each point. The edges are found from the convolution with a differentiated Gauss kernel, at three different scales. To increase robustness only points where a clear single edge is detected are used by the tracker.

III. EXPERIMENTS

The algorithm is first evaluated in simulations using images generated using an image-generation program based on OpenGL, making it possible to simulate phenomena such as occlusion, specular reflections and noise from a cluttered background.

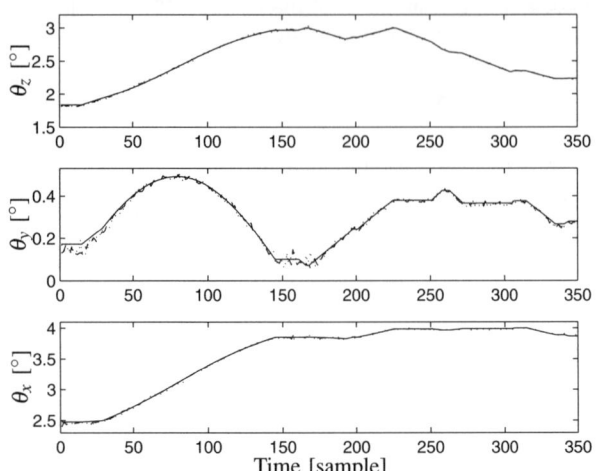

Fig. 3. Tracking of θ. The diagram shows the real orientation (*solid*), estimated orientation using four constraints (*dashed*), and estimated orientation using two constraints (*dotted*.)

The experiments are performed in two steps. The tracker is initialized with a poor initial guess for the intrinsic camera parameters, which is used to get a very rough estimate of the object-camera pose. We then run the tracker for a little over a second with the robot stationary to get a good initial estimate of the state. During the initialization phase the robot motion constraints are not used, since they would require a good initial guess for the object pose. When the state estimate has converged, the tracker is started, using the initial estimate of the state as $\check{\mathbf{q}}_{A_1}$ to constrain the position estimate according to Eq. (25)–(26).

IV. EXPERIMENTAL RESULTS

In this section the presented methods will be validated. There will be a comparison between when only using Eqs. (23) and (24) and when also using Eqs. (25) and (26). The first will be referred to as the *two constraint case* and the latter as the *four constraints case*. In the study we have looked at the mean of the absolute estimation error, which will be denoted with Δ. The number of edge search points varied between 100 and 250 during the motion.

A. Visual position tracking

Figs. 3 and 4 show the result of tracking the orientation θ and the translation $\mathbf{t}$. Both with four and with two constraints the tracking of the position is satisfactory. There are some differences in the tracking accuracy, see Table I. We see that with four constraints the mean error in the estimation of θ and $\mathbf{t}$ is reduced. Table I shows results using different conditions and number of constraints in the estimation of the object pose. For case 1 the only noise is from the image measurements. For case 2 extra noise $\in N(0,3)$ was added. The initial state covariance, $\mathbf{P}_0$, and

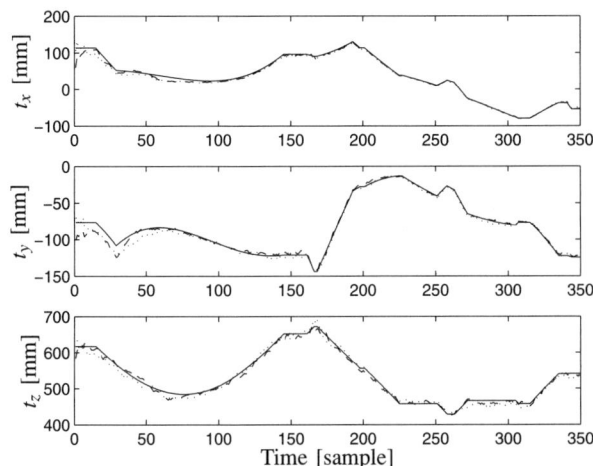

Fig. 4. Tracking of **t**. The diagram shows the real translation (*solid*), estimated translation using four constraints (*dashed*), and estimated translation using two constraints (*dotted*.)

state noise covariance, **Q**, for case 1 and 2 were set to

$$\mathbf{P}_0 = \text{diag}(0.1^2 \cdot \mathbf{1}_{1\times 8} \; 50^2 \; 0.1^2 \; 40^2 \; 40^2 \; \mathbf{0}_{1\times 6})$$
$$\mathbf{Q} = \text{diag}(\mathbf{0}_{1\times 8} \; 3^2 \; 0.01^2 \; 0.2^2 \; 0.2^2 \; 0.1^2 \cdot \mathbf{1}_{1\times 6}).$$

The output noise variance was set to $E(\delta_k^2) = 1$ in case 1 and 3 and to $E(\delta_k^2) = 3^2$ in case 2 and 4.

B. Varying focal length.

Fig. 5 shows results of when the focal length was varied between 300 and 600 pixels. Still the tracking of the focal length was successful, and the effect on the depth estimation was negligible.

C. Incorrect initial values

Figs. 6 and 7 show results of when the tracker starts with incorrect initial values, both for the intrinsic parameters and for the pose of the object. After approximately 30 samples the intrinsic parameters and the position have converged to their correct value. The intrinsic camera parameters in this experiment were $f = 400$, $\gamma = 1.0$, $u_0 = 320$, and $v_0 = 240$.

TABLE I
MEAN TRACKING ERRORS USING TWO AND FOUR CONSTRAINTS

Case	1		2	
# constr.	4	2	4	2
$\Delta\theta_z$ (°)	0.122	0.216	0.292	0.541
$\Delta\theta_y$ (°)	0.317	0.313	0.529	0.632
$\Delta\theta_x$ (°)	0.130	0.381	0.221	0.777
Δt_x (mm)	2.105	3.406	2.434	3.310
Δt_y (mm)	2.390	2.093	2.322	2.878
Δt_z (mm)	4.857	4.926	5.704	7.228
$\|\Delta\theta\|$ (°)	0.364	0.538	0.643	1.138
$\|\Delta t\|$ (mm)	5.808	6.345	6.622	8.455

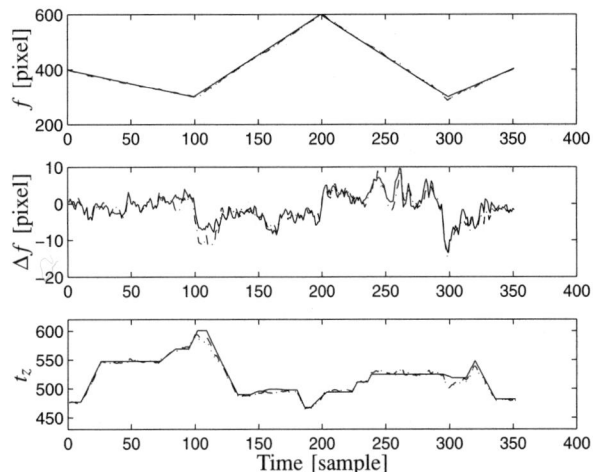

Fig. 5. Experiment where f is varying between 300 and 600 pixels. The diagram shows the real values (*solid*), estimation using four constraints (*dashed*), and estimation using two constraints (*dotted*.)

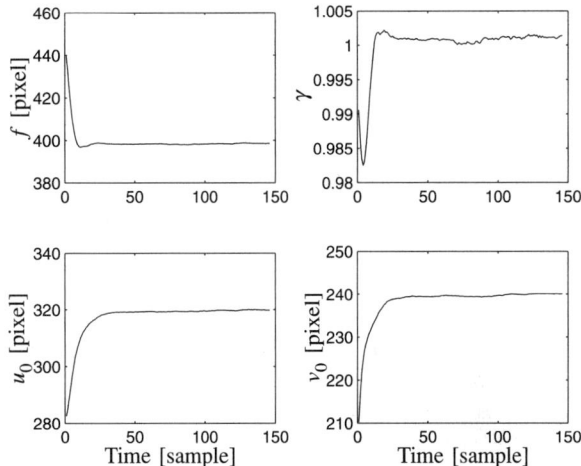

Fig. 6. Transient responses of estimates from incorrect initial values.

D. Real world experiments.

Fig. 8 show the results of an experiment using images from a Sony DFW-V300 640x480 pixels digital camera, see Fig. 2 for an example image. The camera was mounted on an ABB Irb2000 industrial robot. The top figures show the estimated focal length and principal point, which should be compared with the values $f = 1020$ pixels, $u_0 = 344$ pixels and $v_0 = 215$ pixels from an offline camera calibration. The lower figure shows the estimated position of the camera, where the lines indicate the direction of the camera z-axis.

V. DISCUSSION

The use of the robot motion constraints showed an improvement in the estimation of the parameters, even

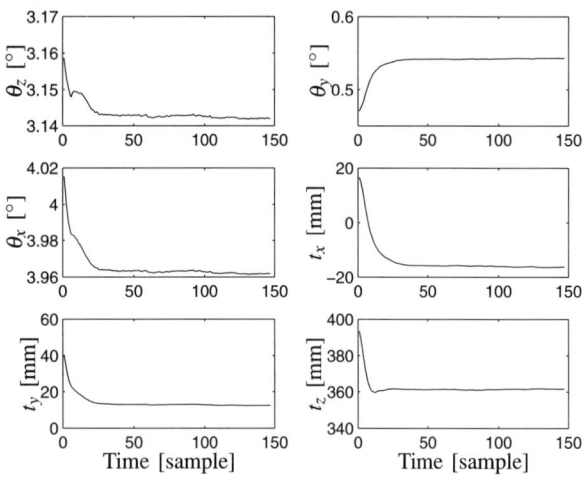

Fig. 7. Transient responses of estimates from incorrect initial values.

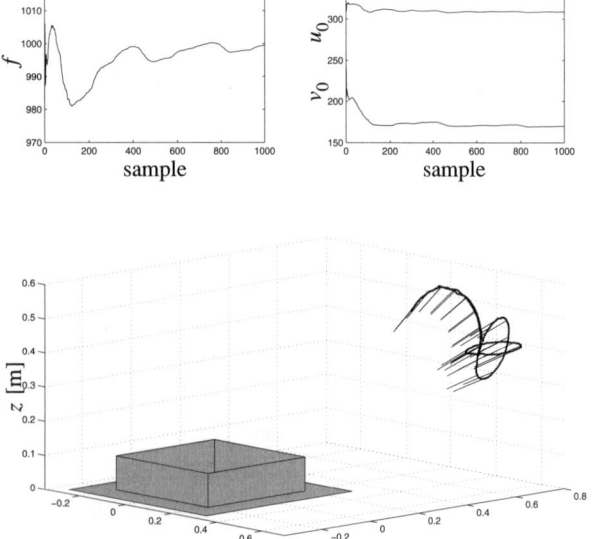

Fig. 8. Estimated focal length, principal point, and trajectory in the real world experiment.

though the hand-eye transformation was unknown. Additionally, it is also intuitively reasonable that the extra constraints should improve the robustness of the tracking against other error sources, such as errors due to the edge detector locking on to false edges. Experiments with multiple constraints show that three or more positions will constrain the position estimation completely, allowing for tracking of very fast motions.

The system is capable of performing a total calibration of all relevant parameters, based on only rough initial values. The method works well also on real world images with low intensity edges. The use of dual quaternions and robot motion constraints in this type of tracker has not previously been demonstrated. The fact that we are able to track even during changes in the intrinsic parameters is an advantage, for instance in vision-based control. This gives the practical advantage of allowing the system to dynamically change its field of view, allowing a wider range of motions.

One apparent drawback with our EKF-based algorithm is that the updating of the state covariance estimates is very time consuming when the number of outputs is large. Effectively, this limits the feasible number of edge search points to a value which is lower than for optimization based methods. The effective sampling interval varied between 66 ms and 91 ms during the simulation, where the variation can be explained by the varying number of search points. With some optimizations it is reasonably to assume that 30 frames/second is achievable for the real-world system, especially since there is no need for the relatively time consuming 3D rendering.

VI. CONCLUSION

We have developed methods for real-time rigid body tracking with simultaneous calibration and tracking of intrinsic parameters. We have shown that a dual quaternion parameterization can be used to formulate linear robot motion constraints on the estimated states, and that the extra constraints can help to reduce the tracking error. The method has been validated in simulations and experiments on an industrial robot.

VII. REFERENCES

[1] V. Lippiello, B. Siciliano, and L. Villani, "Objects motion estimation via BSP tree modeling and Kalman filtering of stereo images," in *IEEE Int. Conference on Robotics and Automation*, Washington D.C., 2002, pp. 2968–2973.

[2] J.S. Goddard, *Pose and motion estimation from vision using dual quaternion-based extended Kalman filtering*, Ph.D. thesis, University of Tennessee, Knoxville, TN, 1997.

[3] T. Drummond and R. Cipolla, "Real-time tracking of complex structures with on-line camera calibration," in *British Machine Vision Conference*, 1999, vol. 2, pp. 574–583.

[4] F. Martin and R. Horaud, "Multiple camera tracking of rigid objects," *International Journal of Robotics Research*, vol. 21, no. 2, pp. 97–113, February 2002.

[5] R.M. Murray, Z. Li, and S.S. Sastry, *A Mathematical Introduction to Robotic Manipulation*, CRC Press, Boca Raton, FL, 1994.

[6] K. Daniilidis, "Hand-eye calibration using dual quaternions," *International Journal of Robotics Research*, vol. 18, no. 3, pp. 286–298, 1999.

[7] H.H. Chen, "A screw motion approach to uniqueness analysis of head-eye geometry," in *IEEE Conference on Computer Vision and Pattern Recognition*, 1991, pp. 145–151.

[8] A. Gelb, Ed., *Applied Optimal Control*, The M.I.T. Press, Cambridge, Massachusetts, 1974.

[9] A. van Dam, L. Foley, S. Feiner, and J. Hughes, *Computer Graphics: Principles and Practice*, Addison-Wesley, Boston, MA, 2nd edition, 1991.

Fast 3D Tracking of Non-Rigid Objects

Nobuhiro Okada
Graduate School of Engineering,
Kyushu University,
Fukuoka, 812-8581, Japan

Martial Hebert
Robotics Institute,
Carnegie Mellon University,
Pittsburgh, PA 15213, USA

Abstract— A fast 3D tracking method of non-rigid objects is proposed. The method consists of fast ICP and modified RPM techniques for range images. By using the techniques, the authors aim real time 3D non-rigid object tracking. The authors developed an actual tracking system and show the effectiveness of the method by some experimental results.

I. INTRODUCTION

Tracking deformable surfaces in images has received a lot of attention in Computer Vision and Robotics. Although the problem is challenging if one uses 2-D video data, it becomes more manageable if one uses 3-D data. The latter is important in medical applications as well as in many inspection and industrial applications, in which sensors that provide direct measurement of 3-D surface shape are commonly used. In this paper, we investigate the implementation of tracking algorithms for recovering the motion - both rigid and non-rigid - of points on 3-D surfaces. We assume that consecutive observations of the surface are closely spaced in time, so that the motion is small and the search for correspondences between consecutive observations is limited to small areas. Also, we assume that the object is globally rigid and parts of the object are subjected to non-rigid deformations.

Non-rigid tracking is applicable to broad areas. Since it can create models of rigid or non-rigid objects, it offers a unified method for model creation. It can construct models that are useful for CG for synthesizing images of humans, for example. It provides ways to track and recognize human gestures and face expressions so that an easier man-machine interface can be built. Moreover, since each object can be measured while moving, a wide range of applications in the medical field can be considered.

For example, Okada et al. used extended ICP (Iterative Closest Point) algorithm in order to remove deformation contained in range images acquired while moving using a range finder set on a mobile robot [12]. They assumed, however, only uniform deformation.

DeCarlo et al. [7], Kakadiaris et al. [9], and Metaxas [10] used physics based modeling and deformable model for estimation of 3D shapes and motions, and applied them to recognition of gestures and face expressions. Haber et al. performed 3D reconstructions of heartbeat using multiple tagged 2D MR images and finite element model [8].

Well-established techniques exist for general registration. In this paper, we concentrate on two techniques, ICP for rigid registration, and RPM (Robust Point Matching) for non-rigid matching. Both ICP and RPM are classical algorithms and we do not re-define them here. Instead, the goal of the paper is to show how the algorithms can be used and modified for near real-time 3-D tracking and motion recovery in the non-rigid case. In particular, the RPM algorithm in its default version is not very amenable to real-time operation because it assumes no prior knowledge on the registration of the surfaces (i.e., predictions from motion at prior time instants). We propose a combination of fast ICP and modified RPM that is amenable to real-time tracking. The broader context of this work is the use of the next-generation range sensors which will have real-time acquisition rate. Our work is conducted in conjunction with the development of such a sensor [3], which will be able to acquire range data at hundreds of frame per second, thus making it possible to track complex motion in range images closely spaced in time. This paper is a step toward achieving non-rigid object tracking in real time for this type of sensor.

Our assumption can be summarized as follows: We adopt 1/30[sec] video rate as *real-time*, and aim to develop a system that can perform tracking between 2 successive range images which are acquired by a range finder that can measure at video rate. We also assume that, during a short interval between 3-D snapshots, the object is globally rigid and undergoes local, partial non-rigid deformation. For tracking, therefore, we first estimate the global rigid transformation, we then extract the non-rigidly deformed areas and estimate their deformation vectors. The three major techniques used to achieve 3D real-time non-rigid object tracking are: fast ICP, deformed areas extraction by a two-step ICP, and deformation vectors estimation by RPM. We describe the details of the three steps in the next three sections, emphasizing the optimizations and enhancements which we developed from the standard algorithms for this application.

II. FAST ICP

We use ICP algorithm as a first step toward non-rigid object tracking. The ICP algorithm was proposed originally by Chen et al.[4] and Besl et al.[1]. The algorithm chooses pairs of the nearest points from two range images and obtains the transformation to move one image so that the sum of the distances of the pairs is minimized. Many variants of ICP algorithm have been also developed in order for high-speed registration. Blais et al. used projection of points rather than searching closest points [2]. Neugebauer used a projection method to search corresponding points, adopted point-to-plane error, and performed simultaneous registration of multiple range images [11]. Finally, Rusinkiewicz et al. evaluated these techniques and developed high-speed ICP algorithms [15].

We chose the method developed by Neugebauer as the basis for the high-speed ICP algorithm [11]. He used point-to-plane error as the error of ICP, and used projection-based matching in order to find pairs of corresponding points. It is a high-speed ICP algorithm since it calculates the errors only for subsampled points. For details about Neugebauer's method, refer to [11].

We implemented the ICP algorithm based on this method, and performed registration of range images. The computer we used for all the experiments was a Linux machine with 1.2GHz Athlon processor and 256MB memory. We used 128×128 depth maps as range images. In the experiments, range images were created by mapping range data acquired with a Minolta Vivid700 onto 128×128 lattices. About $7,000 \sim 10,000$ points were included in the each image.

We used 500 as the number of points randomly chosen from an image in the ICP algorithm. The ICP algorithm converged in up to ten iterations for each pair of range images. The processing time to perform seven registrations per batch of eight range images was about 0.50[sec]. This time also includes reading the range image files and writing the registered and fused range images. The time to actually recover the registration transformations T's was about 0.2 [sec] for the seven registrations, which is close to our real-time target.

Eight of 128×128 range images taken by a range finder Minolta Vivid700 and the 3-D model created by fusing them are respectively shown in Fig.1 and Fig.2. The figures show that the ICP program could register the range images. We also show the result of ICP performed between Fig.1(1) and Fig.1(2) in Fig.3. The figure is a bottom view, light points indicate points on Fig.1(1), dark points indicate points on Fig.1(2), and lines indicate transfer vectors earned by ICP. In the figure, we can see that it could obtain the conversion vectors according to the rotation of the object.

III. EXTRACTION OF DEFORMED AREAS

In order to estimate deformation vectors, therefore, it is necessary to extract the non-rigid areas at first. We use the global transformation estimated by fast ICP for the extraction. Moreover, not only for extracting deformed

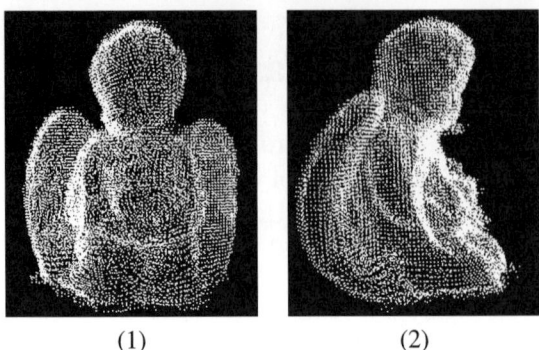

Fig. 2. Created model

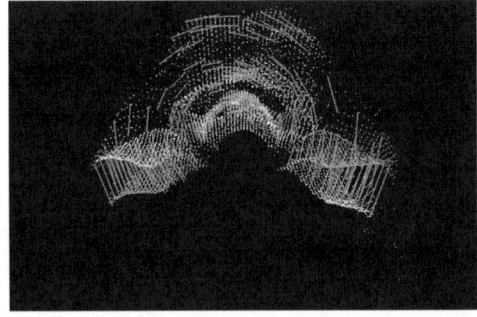

Fig. 3. Transform vectors between (1) and (2)

areas but also for reducing the ICP error caused by non-rigid deformation of the object, we developed a two-step ICP technique. As before, we assume that an object is rigid as a whole and contains non-rigid local deformations, and also that the deformations are small due to the small time interval between frames.

The flow of the two-step ICP is as follows (Fig.4): At first the algorithm estimates the global motion of the object using ICP under the assumption that the object is rigid. Since the object contains non-rigid areas, the estimated value may include errors. The algorithm, then, detects and removes areas that undergo additional non-rigid motion. Detection of deformed areas is performed as follows: Given a surface S_i at one time instant and the surface S_j observed at the next time instant after rigid motion and deformations are applied to S_i, the algorithm

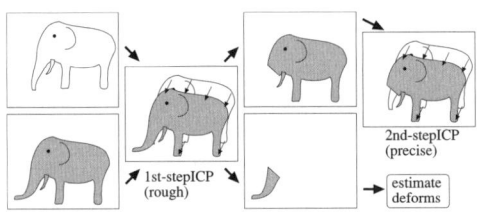

Fig. 4. Process of two-step ICP

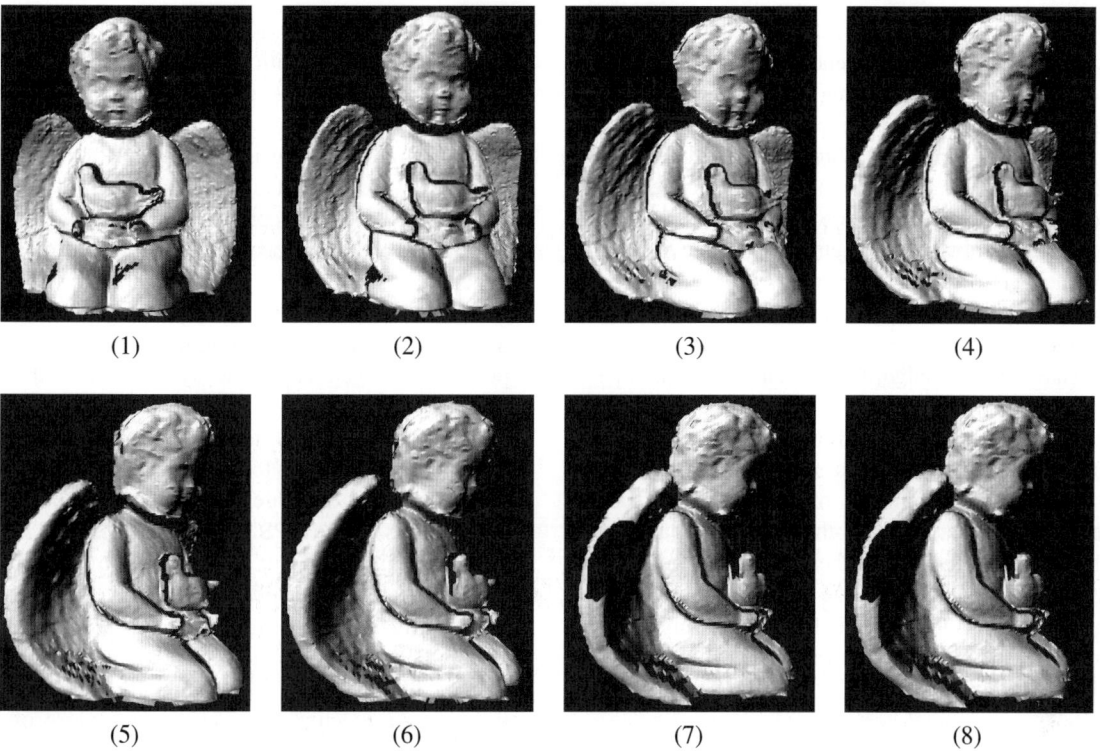

Fig. 1. Range images of an angel doll

inversely transforms S_j back into the coordinate system of S_i using the transformation estimated by the first step ICP, yielding a new surface S'_j, and projects each point on S'_j onto the lattice of S_i. For each point of S'_j, if there is no point on S_i around it, the algorithm decides that the point was subject to a non-rigid deformation. Finally, those points that are identified as part of the non-rigidly deformed part of the surface are eliminated, yielding a new surface S''_j.

The eliminated points are used for estimation of the deformation vectors. Here, although outliers caused by occlusions of the object or by noise may occur, the method used to estimate deformation vectors mentioned later is not significantly influenced by outliers. Finally, the rigid motion component between S_i and S''_j is estimated after removal of the non-rigidly deformed parts using ICP (the second step). It can be expected that the error of the estimated value is smaller than the error of the first step ICP since the deformed points are removed. It can, moreover, be expected that fewer iterations are needed in the second step by using the estimation of the first step as the initial value.

IV. ESTIMATION OF DEFORMATION

Given the global rigid transformation and the initial segmentation of the data into rigid and non-rigid parts estimated at the previous step, we can now estimate the deformation vectors in the non-rigid part of the data. We use the RPM (Robust Point Matching) algorithm developed by Rangarajan et al. and Chui et al. for the estimation of non-rigid deformation vectors [13] [14] [5] [6]. RPM is a robust algorithm that can match two sets of points under non-rigid deformation. Although RPM is an efficient algorithm, it is not designed for real-time operation. In particular, it is designed for general deformations between surfaces and does not take advantage of small incremental deformations. As a result, it is initialized with little or no knowledge of the correspondences between the two surfaces, leading to rather slow convergence. We propose below several extensions to RPM that make it suitable for fast registration. We discuss performance of the modified RPM at the end of this section.

A. Robust point matching

The RPM algorithm takes as input two sets of points X and V. X and V are $\{x_i, i = 1, 2, ..., I\}$ and $\{v_j, j = 1, 2, ..., J\}$, respectively, and the number of points may differ in each data set. These two data sets are matched by alternating two steps: first *soft* correspondences between points of X and points of V are established (a method commonly termed "softassign"), and, second, given the correspondences, the deformation between the two point sets is estimated using a thin plate spline (TPS) model using a least-squares method. The correspondences are loosely enforced initially and refined as the iterations proceed so that, upon convergence, each point on one surface has a single corresponding point on the other surface. The rate at which the correspondences are tightened is controlled by a simulated annealing schedule. Since softassign determines the correspondence between data sets, the exact correspondences are not needed in advance. It also allows deformations of an object since mapping between data sets is calculated using TPS. RPM has been shown to be robust and effective to register range images of a non-rigid object.

In the standard RPM algorithm, when softassign is performed the mapping is fixed, and when TPS is calculated the correspondence is fixed. Although it is usually very difficult to solve both problems simultaneously, solving one of them while keeping the other one fixed is easy. Even though the softassign and the TPS parts can reach optimal solutions individually, there is no guarantee that the combination will also be the optimal solution. The approach only guarantees that it finds local minima for the combined mapping and correspondence. For details about the algorithm, refer to [13] [14] [5] [6] [1].

B. Modified RPM algorithm

The standard RPM implementation is comparatively slow for range images that contain large data sets. The standard algorithm was modified in three ways to make it suitable for real-time operation:
1. Algorithmic improvements
2. Reduction of number of data points
3. Control of annealing step

Areas 2. and 3. were also addressed to some extent in Chui's original presentation. We apply and extend these techniques for further optimization of the algorithm below.

1) Algorithmic Improvements: The algorithmic improvements involve mainly the reduction of processing time by eliminating unnecessary calculations, by improving the sequence of computation, and by reducing the computation time through outlier removal. A key part of the RPM algorithm is the construction of a matrix of point pairs constructed from the data sets. The size of the matrix is n^2, where n is the number of data points. The entry (i, j) of the matrix is a function of the distance d between the two points i and j. In practice, a fast decaying function, such as $\exp(-d^2/T)$, is chosen so that points that are far away from each other contribute little to the overall objective function. For TPS recovery, the matrix is QR-decomposed. Two of the most time-consuming parts of RPM are the QR decomposition needed by the TPS fitting and the calculation of exp() in the update of the correspondence matrix.

In RPM, to calculate the deformation function using TPS, a QR decomposition of a matrix is performed. The standard QR decomposition is not necessary because only the portions of $n \times 4$ and 4×4 of the Q and R matrices are actually used, where n is the number of data points. To take advantage of this fact, we calculate only the necessary portions of the QR decomposition.

In the process of updating each element of correspondence matrix, there is a part that calculates $\exp(-d^2/T)$ using distances between corresponding points in the two data sets d^2 and the annealing temperature T. In practice, points that are already far away from each other, relative to the temperature, have no effect on the iterations, and calculating exp() is not necessary for them. If practice, the exponential term is set to zero when d^2/T becomes larger than a threshold. The threshold is chosen so that neglecting these points in the TPS does not affect the convergence of the algorithm. In our experiments, over 98 percent of the points are eliminated using this gating approach.

2) Reduction of number of data points: The processing time of RPM increases rapidly with the size of the data set. Specifically, the amount of computation is $O(n^3)$ because the program calculates products and inverse of matrices in order to update the TPS mapping.

Therefore, reducing the data set is critical in decreasing the computation time. However, an accurate result cannot be obtained in general by matching two decimated data sets and some care must be taken in how the data sets are decimated. The first step in reducing the data sets is to decimate one of the data sets, leaving the other one unchanged. This accelerates dramatically the cost of the matrix calculations in RPM since they depend on the number of the points contained in one of the two data sets.

However, since many points are still contained in the decimated data set, to match all the non-rigid points with another data set will still require substantial computation. To further accelerate processing, we divide the points included in the decimated data set into two or more data sets and individually perform RPM for each data set. In particular, only the set of points that is identified as undergoing a non-rigid transformation is processed, and that set of points is further divided into connected regions that are each processed separately. The processing time can be reduced further by dividing the decimated data set into individual areas. For example, supposing that a data set can be divided into m areas and that each area contains same number points, the calculation time of matrices for each area will become $1/m^3$ of the original calculation time and the whole becomes $1/m^2$. There remains, of course, the issue of ensuring consistency of motion between the subsets of the data. This issue is addressed below in Section IV-C.

3) Control of annealing step: The RPM algorithm estimates deformation vectors by performing iterative pro-

---

[1] A program which performs RPM written in MATLAB is distributed by Chui et al.
http://www.cise.ufl.edu/%7Eanand/students/chui/research.html

cessing, by lowering the annealing temperature in deterministic annealing. The number of iterations is determined by the annealing ratio, the initial temperature of annealing, and the terminal temperature, therefore the processing time is very strongly affected by these parameters. Adopting low initial temperature, high final temperature, and large annealing ratio is effective in order to shorten the processing time but increases the risk of converging to a poor local minimum.

Chui et al. used the square of the maximum distance between the points contained in the two range images as the initial annealing temperature. It was the value for allowing arbitrary matching in the absence of any prior information on the registration transformation between the surfaces. In our case, however, we already have a good estimate of the registration transformation, which brings the two surfaces in close proximity. Therefore, we can start the annealing using a lower initial temperature by using the square of the amount of expected deformation as the initial temperature. Chui et al. used the average of the square of distances between each point and its nearest point as the final annealing temperature in order to prevent overfitting. In practice, we observed that overfitting was still produced in our experiments and we used a slightly larger value for the final temperature. Chui et al. suggested that the value about $0.9 \sim 0.99$ was desirable as the annealing rate. Although a small annealing rate reduces the processing time, it may cause large errors by missing the global minimum of the objective function. Furthermore, in the original approach, update of the correspondence matrix and update of the transformation parameter are repeated five times for each annealing temperature. While the repetition time can be reduced in order to reduce processing time, this may produce large errors too by preventing proper convergence at each iteration. We estimated errors of RPM for annealing rate of 0.90, 0.93, 0.96 and 0.99, and the numbers of iterations of $1 \sim 5$. Then, we adopted 0.90 as the annealing rate, 2 times as the number of iterations.

C. Consistency constraints

Although RPM guarantees that the deformation vectors of points are continuous throughout each part of the data that has been processed separately, it is not sufficient in our case because we must also guarantee that the deformation vectors are continuous with motion vectors of points in rigid areas. Using the fact that the global motion is estimated by using the two-step ICP technique, we know that the algorithm only has to guarantee that the deformation vectors of non-rigid points vanish at the boundary with rigid areas after transformation of the range image using the transformation estimated by the two-step ICP.

We developed a mapping technique that enforces this constraint. In this approach we use not only all points $\{x\}$ contained in a range image S_i and the points $\{v\}$ in non-rigid areas in the other image S_j but also points $\{r\}$ that are at the boundary between the rigid and non-rigid areas of S_j. Then, the RPM process computes the mapping between data set $\{v,r\}$ and $\{x',r\}$ using the TPS mapping, where $\{x'\}$ is the data set obtained by matching data set $\{x\}$ with $\{v\}$ using the correspondence matrix produced by RPM. This mapping forces the points at the boundary between rigid and non-rigid regions to remain fixed. By construction, this mapping has vanishing motion vectors at the boundary points $\{r\}$, and, consequently, the deformation vectors are continuous at boundaries of rigid and non-rigid areas.

Even when we use this approach, RPM may output erroneous mapping. This error occurs because the method uses only points in non-rigid areas as data set $\{v\}$. Such an error will not occur if all the points of S_j, including rigid areas, are used as $\{v\}$. In principle, defining the constraining set $\{r\}$ as the set of boundary points is sufficient. For robustness reasons, however, it is best to spread out the set of constraining points over a larger area. In practice, we use not only the points at the boundary between rigid and non-rigid areas but also points uniformly sampled in rigid areas as data set $\{r\}$.

D. Examples

Some results of estimation of deformation vectors from range images are shown below. The results are obtained, as described above, after estimating the global rigid transformation by using the two-step ICP with fast ICP and running the modified RPM.

Range information acquired by a range finder is shown in Fig.5, which shows a test batch of 8 range images. We used the Minolta Vivid700 as a range finder. After changing the range information into 128×128 range images with grid size 1.0[mm], the two-step ICP program performed rigid registration and extraction of non-rigid areas, and made pairs of data sets used for RPM. One of the data sets had all the points included in the 128×128 range images, and the other had points in areas that were subject to non-rigid motion. The non-rigid data sets were also decimated using a 1/20 ratio. We also used points surrounding the deformed areas decimated using a 1/8 ratio and points in rigid areas decimated with 1/1000th of the points used as constraining points introduced in order to guarantee consistency of deformation vectors. We used 50 and 10 as the initial and final annealing temperatures, respectively, and 0.1 as λ in RPM. The processing time will greatly vary depending on the total number of points in the range images, the numbers of non-rigid areas, points in each area, and the constraining points. In our experiments, the number of points in each range image was about 10,000, the number of non-rigid areas in each image ranged from two to five regions,

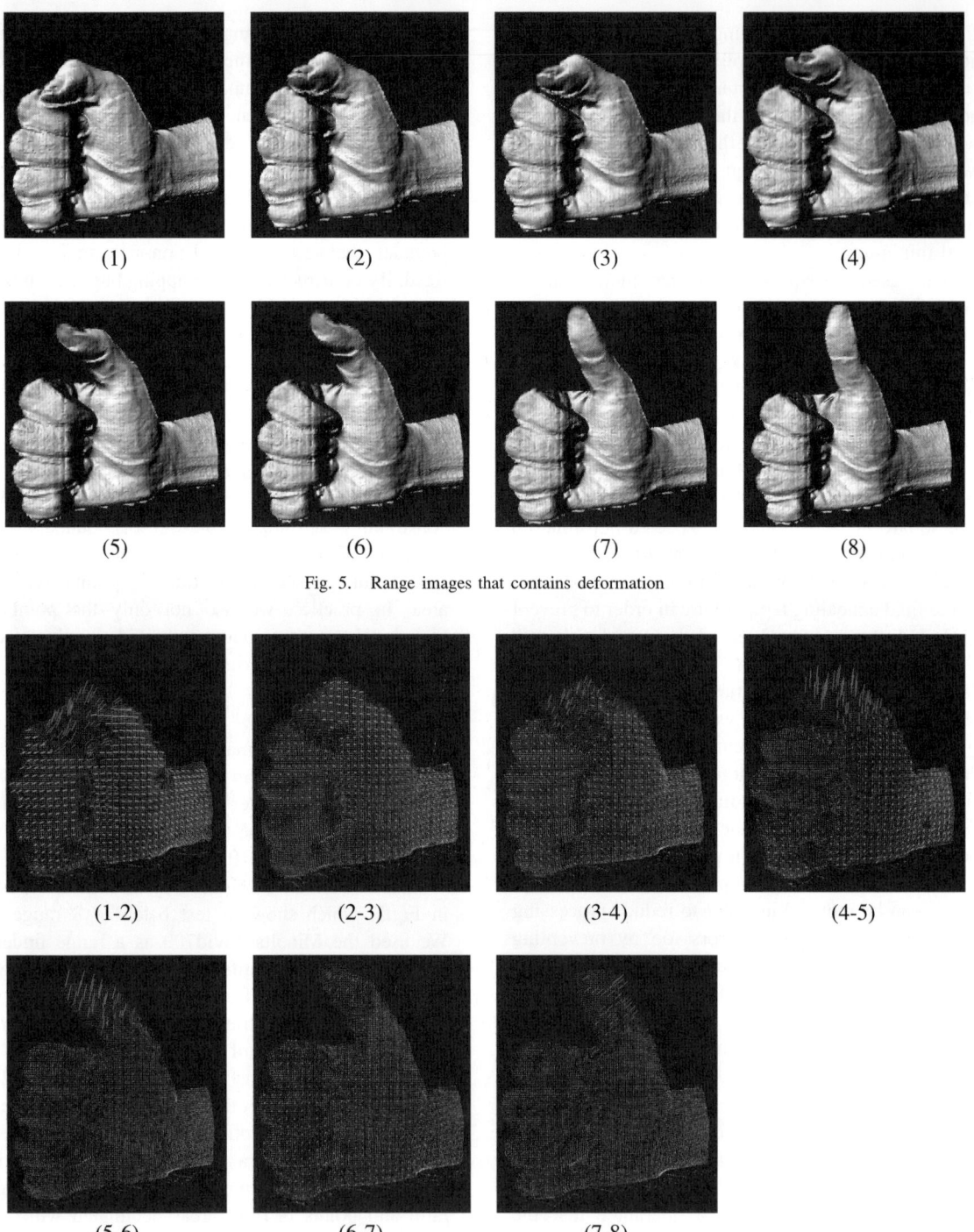

Fig. 5. Range images that contains deformation

Fig. 6. Deformation vectors estimated by RPM

and the number of points included in each non-rigid area ranged between 60 and 130. Under these conditions, our program took about 7.2[sec] to estimate deformation vectors for 8 range images. It is the actual processing time and does not include file reading and writing. Here, as in the experiments of the fast ICP, we used a Linux machine with a 1.2GHz Athlon and 256MB memory. The estimated deformation vectors are shown in Fig.6. The notation (i-j) means that it shows deformation vectors between (i) and (j) in Fig.5. In the figure, light points are points in (1) of Fig.6, slightly dark points are points in (2), and dark points are points that have been decided to be deformed. Dark vectors are deformation vectors estimated by RPM, and light vectors are transformation vectors of rigid areas estimated by ICP.

V. CONCLUSION

For the purpose of fast 3D tracking of deformable non-rigid objects, we proposed a combination of fast ICP, two-step ICP and estimation of deformation vectors using an accelerated version of the RPM algorithm. We used a method developed by Neugebauer as the fast ICP. In the technique, the process is especially accelerated by searching corresponding points using projection. We could register range image with 128×128 at about video rate ($1/30$ [sec]).

Next, in order to extract deformed areas in a range image and to estimate transformation of the object as a whole more correctly, we developed a two-step ICP technique. By using the result of the first ICP as the initial value of the second ICP, we could reduce the increase of the processing time caused by making ICP two steps. As output of this step, we separate the data into the part that follows the global rigid motion applied to the object from the part that follows a non-rigid motion. In order to estimate deformation vectors of points in areas where an object is deformed, we used robust point matching technique developed by Rangarajan et al. We showed that the deformation vector could be estimated using RPM. Moreover, we shortened the processing time significantly by improving the RPM algorithm and by using explicit constraints from the data and the motion.

As mentioned above, we could estimate the global transformation of the object and estimate the deformation vectors of points in the deformed areas from range images. Further refinement will focus on intelligent decimation of the data. More precisely, we will investigate a hierarchical approach in which the deformation vectors are estimated from reduced data set first and further refined using an expanded version of the data set until all the data is used in the estimation. The difficulty is the choice of an appropriate strategy for decimating the data set.

VI. REFERENCES

[1] P. J. Besl, and N. D. Mckay, "A Method for Registration of 3-D Shapes," *IEEE Trans. on PAMI*, vol.14, no.2, 1992, pp.239-256.

[2] G. Blais and M. D. Levine, "Registering Multiview Range Data to Create 3D Computer Objects," *IEEE Trans. on PAMI*, vol.17, no.8, 1995, pp. 820-824.

[3] V. Brajovic, K. Mori and N. Jankovic, "100 frames/s CMOS Range Image Sensor," *Digest of Technical Papers, IEEE Int. Solid-State Circuits Conf.*, 2001, pp. 256-257.

[4] Y. Chen, and G. Medioni, "Object Modeling by Registration of Multiple Range Images," *Proc. of the IEEE Int. Conf. on Robotics and Automation*, 1991, pp.2724-2729.

[5] H. Chui and A. Rangarajan, "A New Algorithm for Non-Rigid Point Matching," *Proc. of the IEEE Conf. on Computer Vision and Pattern Recognition (CVPR2000)*, vol.2, 2000, pp.44-51.

[6] H. Chui, "Non-rigid Point Matching: Algorithms, Extensions and Applications," PhD. Thesis, Yale University, 2001.

[7] D. DeCarlo, and D. Metaxas, "The Integration of Optical Flow and Deformable Models with Applications to Human Face Shape and Motion Estimation," *Proc. of the IEEE Computer Society on Computer Vision and Pattern Recognition*, 1996, pp.231-238.

[8] I. Haber, D. N. Metaxas, and L. Axel, "Using Tagged MRI to Reconstruct a 3D Heartbeat," *Computing in Medicine*, vol.2, issue5, 2000, pp.18-30.

[9] I. A. Kakadiaris, and D. Metaxas, "Model Based Estimation of 3D Human Motion with Occlusion Based on Active Multi-Viewpoint Selection," *Proc. of the IEEE Computer Society Conf. on Computer Vision and Pattern Recognition*, 1996, pp.81-87.

[10] D. Metaxas, "Deformable Model and HMM-Based Tracking, Analysis and Recognition of Gestures and Faces," *Proc. of Int. Workshop on Recognition, Analysis, and Tracking of Faces and Gestures in Real-Time Systems*, 1999, pp.136-140.

[11] P. J. Neugebauer, "Geometrical Cloning of 3D Objects via Simultaneous Registration of Multiple Range Images," *Proc. of the IEEE Int. Conf. on Shape Modeling and Applications (SMA '97)*, 1997, pp.130-139.

[12] N. Okada, E. Kondo, H. Zha, K. Morooka, and T. Nagata, "3-Dimensional Object Model Construction from Range Images Taken by a Range Finder on a Mobile Robot," *Proc. of the IEEE/RSJ Int. Conf. on Intelligent Robots and Systems (IROS98)*, vol.3, 1998, pp.1853-1858.

[13] A. Rangarajan, H. Chui, and F. L. Bookstein, "The Softassign Procrustes Matching Algorithm," *Information Processing in Medical Imaging, James Duncan and Gene Gindi, editors*, 1997, pp.29-42.

[14] A. Rangarajan, H. Chui, and E. Mjolsness, "A Relationship between Spline-based Deformable Models and Weighted Graphs in Non-rigid Matching," *Proc. of IEEE Computer Society Conf. on Computer Vision and Pattern Recognition (CVPR01)*, 2001, pp.I:897-904.

[15] S. Rusinkiewicz, and M. Levoy, "Efficient Variants of the ICP Algorithm," *Proc. of Third Int. Conf. on 3-D Digital Imaging and Modeling*, 2001, pp.145-152.

A General Framework for Automatic CAD-Guided Tool Planning for Surface Manufacturing

Heping Chen, Ning Xi, Weihua Sheng
Electrical and Computer Engineering Dept.
Michigan State University
East Lansing, MI

Yifan Chen, Allen Roche, Jeffrey Dahl
Scientific Research Lab
Ford Motor Company
Dearborn, MI

Abstract— Surface manufacturing is widely used in industry. Automatic CAD-guided tool planning has many applications in surface manufacturing, such as spray painting, spray forming, rapid tooling, cleaning and polishing. According to the material quantity requirements, these tasks can be categorized into two groups: the material uniformity problem and coverage problem. A general framework is developed to automatically generate trajectories of a free-form surface for these tasks. A given task is first transformed into one of the groups. Based on the CAD model of a free-form surface, constraints and tool model, a trajectory for a free-form surface is generated. Velocity optimization is discussed to optimize the material quantity. Simulations are performed to verify the developed framework. This framework can also be extended to other applications.

I. INTRODUCTION

Surface manufacturing is a process to add material to or remove material from surfaces of parts. Spray painting, spray forming, indirect rapid tooling, cleaning and polishing are typical examples in surface manufacturing. Tool planning for these applications is a challenging research topic. Typical teaching method requires the programmers to carry out extensive tests on a work cell and thus to improve the generated trajectories. This procedure is time consuming and tedious. Automatic off-line tool planning is desirable for these applications. Various methods have been developed to generate trajectories for these tasks. Although these tasks are different, they can be categorized into two groups: material uniformity problem and coverage problem. Material uniformity problem, such as spray painting, spray forming and rapid tooling, requires that the material quantity on a surface should satisfy some constraints. Coverage problem, such as cleaning and polishing, requires that every point on a surface has to be covered. Tool planning for spray painting is widely studied [1], [2], [3], [4]. Some tool planning methods can generate trajectories for free-form surfaces automatically, but can also satisfy the paint thickness requirements. Chen *et al.* [1] developed a simulation software to calculate paint thickness on a free-form surface. The tool planning for spray forming [5], [6] is quite similar to that of spray painting except satisfying area density requirements instead of thickness requirements. The area density requirements can be transferred into thickness requirements. Rapid tooling process is attracting more attention lately because the process can manufacture dies and tools rapidly. Chalmers [7] demonstrated an indirect rapid tooling process used in Ford Motor Company. Luo *et al.* [8] developed a path generation method for direct rapid tooling. There are different methods to generate trajectories for coverage problem. Sheng *et al.* [9] developed a method to automatically generate a trajectory such that every point on a surface can be covered. Huang [10] developed an optimal method to cover a surface by minimizing the turns. These methods can guarantee the coverage of a surface. Cleaning is an important process in cleaning electronic and mechanical components [11] or a surface [12]. All of these tasks need to generate trajectories such that the task constraints can be satisfied. Although tool planning methods are developed for some tasks, there is no general tool planning method. Also trajectory verification is not addressed in most of the papers.

Due to different tool models and constraints, there are some challenges to develop a framework for tool planning for these tasks. For uniformity problem, material quantity must satisfy certain constraints, but coverage problem does not have such a requirement. Some processes, such as rapid tooling, need to cover a surface many times. However, these tasks, both material uniformity problem and coverage problem, have some common points: the trajectories are generated using surface models, tool models and constraints; the tool trajectory is defined by a six dimensional vector which specifies the position and orientation of a tool at a time instant. From a practical point of view, it is desirable to develop a general framework for these tasks. In this paper, a general framework is developed to generate trajectories for different tasks. Verifications are performed to check if the generated trajectories satisfy the tasks constraints.

II. A GENERAL FRAMEWORK

A general framework for the automated CAD-guided tool planning system can be formulated as follows:
Given the CAD model of a free-form surface M, constraints Ω and tool model G, find a tool trajectory Ξ such

that the constraints are satisfied, i.e.

$$F(M, \Omega, G) = \Xi \quad (1)$$

Figure 1 shows the general framework to generate a trajectory for a free-form surface. Based on the CAD model of a free-form surface, tool model and constraints, tool trajectory planner will generate a trajectory. The trajectory is output to a simulation software to verify if the trajectory satisfies the constraints. The trajectory is also output to ROBCAD [13] to simulate the specified task.

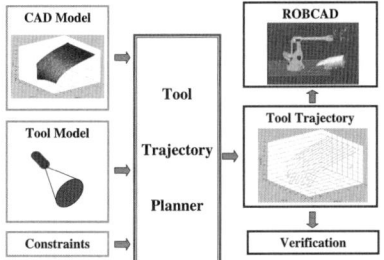

Fig. 1. The general framework for the CAD-guided tool planning system

The tool trajectory planner is the core of the general framework. Figure 2 shows how the tool trajectory planner works.

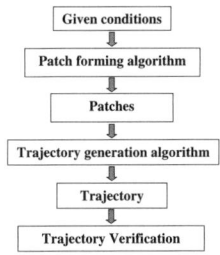

Fig. 2. The tool trajectory planner

For a given tasks, the CAD model of a free-form surface, constraints and tool model are given. Using the patch forming algorithm, patches are formed. The formed patches should satisfy the given constraints. Then using the tool planning algorithm, a trajectory is generated for each patch and the generated trajectories are integrated to form a trajectory for the free-form surface. Finally, the generated trajectory is verified to check if it satisfies the given constraints.

A. Task Conditions and Requirements

To obtain time-efficient tool trajectories and sufficiently utilize the workspace of the robot [1], a triangular approximation of a free-form surface is adopted in CAD modelling. The CAD model of a free-form surface can be formulated as:

$$M = \{T_i : i = 1, \cdots, N\} \quad (2)$$

where T_i is the ith triangle on the free-form surface; N is the number of triangles. Figure 3 shows the triangular approximation of a mold used in the algorithm implementation.

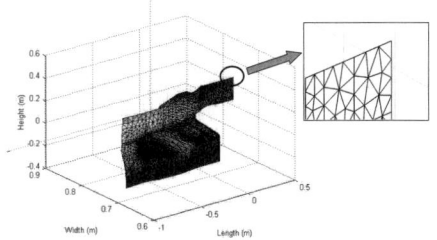

Fig. 3. The triangular approximation of a mold used in indirect rapid tooling

After a free-form surface is rendered into triangles, the normals of the triangles may not point to one side of the surface. We developed an algorithm [14] to adjust normals of the triangles such that all normals point to one side of a surface.

A general constraints can be expressed as follows:

$$\Omega = \{q_d, \Delta q_d, \omega\} \quad (3)$$

where q_d is the average material quantity constraint and Δq_d is the material quantity deviation from the average material quantity; ω is other constraints, such as material waste, cycle time, reachability, temperature and tool orientation constraints. For example, the temperature on a surface must be kept in a certain range during a rapid tooling process. Tool orientation constrain means that tool orientation should not change rapidly.

Spray painting, spray forming and indirect rapid tooling requires the generated trajectory to satisfy the material uniformity constraints, i.e. the material sprayed on a free-form surface must satisfy the average material quantity and its deviation constraints. Spray cleaning and polishing are coverage problems. For coverage problem, the material quantity sprayed on a point of a free-form surface is not important as long as the point is covered by the tool. Therefore, we can set the material quantity as a unit value if a point is covered, otherwise, it is 0 if it is not covered. Then the constraints can be expressed using a general formula, i.e.:

$$\begin{array}{ll} \Delta q_d \neq 0 & \text{Material uniformity problem} \\ q_d = 1, \Delta q_d = 0 & \text{Coverage problem} \end{array} \quad (4)$$

A typical tool model [1], [3], [15], [16] are adopted and shown in Figure 4(a). The typical profile of the material deposition rate can be roughly approximated by parabolic curves [3], [16] as shown in Figure 4(b). The material deposition rate on a flat surface can be modeled as:

$$G = f(r, \theta) \quad (5)$$

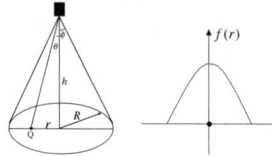

Fig. 4. (a) A tool model. (b) A tool profile

where r is the distance from a point to the tool center inside a cone. θ is the fan angle. R is the spray cone radius. Goodman [17] presented a method to measure the material deposition rate by covering a flat surface. Typically the tool standoff is kept constant [3], [4], [16], [18]. Therefore, the material deposition rate is only related to the distance r, i.e.

$$G = f(r) \quad (6)$$

The tool model is valid for coverage problem if the material deposition rate is set to be a constant. For some processes, such as polishing, the tool standoff is 0. The tool model is still valid.

B. Patch Forming Algorithm

Based on the given conditions, a subpart forming algorithm is developed to generate subparts. A feature filtering method is discussed to smooth the detail information of a free-form surface to form patches.

Chen *et al.* [14] developed a method to optimize the material quantity on a flat surface. This method is adopted here to calculate the path width d and tool velocity v. Suppose a flat surface S is sprayed, the material quantity of a point s on the flat surface is q_s. Then the mean square error of the material quantity has to be minimize to find the optimal d and v, i.e.:

$$\min_{d,v} E = \int (q_d - q_s)^2 dx \quad (7)$$

After minimization, the path width d and tool velocity v can be calculated.

After the material quantity on a flat surface is optimized, suppose the average, maximum and minimum material quantity are q_d, $\bar{q}_{max}$ and $\bar{q}_{min}$ respectively. Assume that the material quantity on the flat surface is projected to a free-form surface and the maximum deviation angle of the free-form surface relative to the normal of the flat surface is β_{th}. The maximum deviation angle is the maximum angle between the normals of the free-form surface and the flat surface. Then the material quantity q_s of each point s on the free-from surface can satisfy the following inequality without considering the tool standoff variation:

$$\bar{q}_{min} cos(\beta_{th}) \leq q_s \leq \bar{q}_{max} \quad (8)$$

If the material quantity of each point q_s in the free-form surface satisfies the constraints, i.e.,

$$|q_s - q_d| \leq \Delta q_d \quad (9)$$

then

$$\bar{q}_{max} - q_d \leq \Delta q_d \quad (10)$$
$$q_d - \bar{q}_{min} cos(\beta_{th}) \leq \Delta q_d \quad (11)$$

If equation (10) is always satisfied, the threshold angle β_{th} can be calculated using equation (11). This means, for any free-form surface, if the maximum deviation angle β satisfies $\beta \leq \beta_{th}$, the material thickness of any point on the free-form surface can satisfy the material quantity constraints.

For coverage problem, $\beta_{th} = 90^o$ when the tool orientation constraint is considered, i.e. the tool orientation can not be changed suddenly. The spray width and tool velocity can be arbitrarily chosen. Normally the spray width is chosen to be two thirds of the covering length. The tool velocity can be chosen to be half of the maximum velocity of a robot.

After the threshold angle β_{th} is obtained, subparts can be generated. A subpart is expressed as:

$$S_i = \{T_j \ | \ cos^{-1}(\vec{N}_j \bullet \vec{N}_k) < \beta_{th}, D(T_j, T_k) \leq R,$$
$$T_j \in M, T_k \in M\} \quad (12)$$

where, S_i is the ith subpart; $\vec{N}_j$ and $\vec{N}_k$ are the normals of the jth and kth triangles respectively; $D(T_j, T_k)$ is the distance between the centers of the jth and kth triangles. A seed triangle is arbitrarily chosen as the first triangle of a subpart. Surrounding triangles, whose distance to the seed triangle are less than the spray radius, are found. Then the angle between the normals of the seed triangle and each of the surrounding triangles is calculated. If the angle is less than the threshold angle β_{th}, the surrounding triangle is added to the subpart. After all of the surrounding triangles are checked, each of the newly added triangles is used as the seed triangle. The process continues until no more triangles can be added to the subpart. If there are remaining triangles, a seed is chosen from the remaining triangles and the subpart forming method is applied again to form a new subpart. The process for subpart forming continues until no triangles are left. Then the free-form surface is divided into one subpart or several subparts.

After finding subparts, the feature filtering algorithm is applied to generate patches. For a subpart, there are possibly some detail feature that will affect the generation of tool position and orientation. It is desirable to smooth these detail features before generating a trajectory. It is well-known that fourth order polynomial models can represent many useful 3D surfaces [19]. A 3L algorithm [19] is adopted here to fit a polynomial surface to the original surface data. After a polynomial surface is obtained, the original data set will be projected to the polynomial surface to get a new data set. The new data set will keep the original data structure without detail feature.

C. Tool Planning Algorithm

A tool trajectory includes a tool path and tool orientation. Here we will discuss how to generate a tool path first. After patches are found, a tool trajectory can be generated for each patch using the path width and the tool velocity. Sheng et al. [9] developed a bounding box method to generate a path for a patch. However, the method works well when a patch is regular. An improved bounding box method is proposed here to generate a tool path. Figure 5(a) shows a bounding box and a patch. Figure 5(b) shows a path generated using the improved bounding box method. The patch is cut using a series of

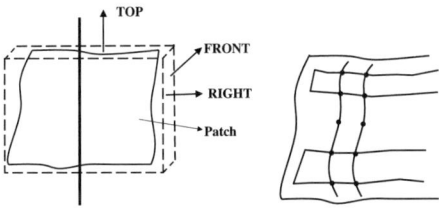

(a) (b)
Fig. 5. The improved bounding box method (a)A bounding box and a patch (b)the path of a patch

planes which are perpendicular to the RIGHT direction of the bounding box. A series of intersection lines will be generated on the patch. Each intersection line will be divide into segments using the path width. Then a series of points will be generated on each line. The points will be connected along the RIGHT direction as shown in Figure 5(b). Then a path is generated for the patch.

The tool orientation is determined based on the local geometry of a patch. Figure 6 shows part of a tool path and sample points. At each sample point, triangles whose

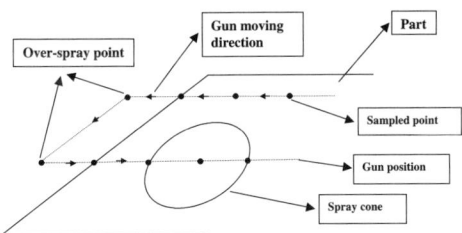

Fig. 6. Part of a tool path and a series sample points

distance to the sample point is less than the tool radius are found. The average normal of these triangles is defined as:

$$\vec{n}_a = \frac{\sum_{i=1}^{N} S_i \vec{n}_i}{||\sum_{i=1}^{n} S_i \vec{n}_i||}, \quad (13)$$

After finding the average normal, the tool orientation is the reverse direction of the average normal.

In the tool planning, the tool velocity is kept constant. The lower bound and upper bound of the material quantity is defined as:

$$\Delta q_{max} = q_{max} - q_d \quad (14)$$
$$\Delta q_{min} = q_d - q_{min}$$

where q_{max} and q_{min} are the maximum and minimum material quantity on a free-form surface respectively. According to equation (8), the upper bound of the material quantity is dependent on the maximum material quantity on a plane. However, the lower bound is dependent on both the minimum material quantity and the maximum deviation angle. The larger the maximum deviation angle is, the bigger the material quantity deviation is. This means the lower bound is larger than the upper bound. To minimize the material quantity deviation for the required material quantity, the lower bound has to be decreased. A method is developed here to decrease the lower bound by modifying the tool velocity. Figure 7 shows the material quantity projected from a plane to a free-form surface. Because the material on an area S on a plane is projected

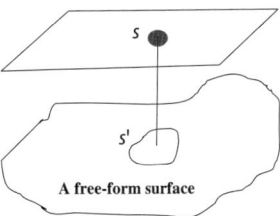

Fig. 7. The material quantity projected from a plane to a free-form surface

to an area S' on a free-form surface, the material quantity on the free-form surface is decreased. Therefore, the tool velocity has to be modified:

$$v' = v \frac{S}{S'} \quad (15)$$

where v' is the modified velocity.

An algorithm developed by Chen et al. [20] is adopted here to integrate the trajectories of the patches for a free-form surface. According to the algorithm, if the material quantity deviation from the required material quantity between any two patches is minimized, the material quantity deviation on a free-form surface with multiple patches is minimized. This means only the distance between any two paths has to be calculated such that the material quantity deviation is minimized. Using the distance, the patches can be integrated to form a trajectory for the free-form surface.

D. Trajectory Verification

Trajectory verification is an importance process because it will check if the task constraints are satisfied. An algorithm is developed to calculate the material quantity on a free-form surface. A typical model [3], [4] is adopted here to calculate the material quantity of a point on a free-form surface. Figure 8 shows a point on a free-form surface and a projected plane. The plane is generated using

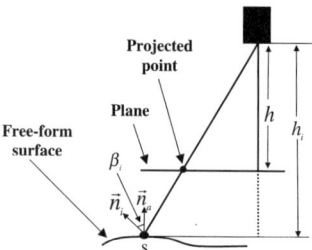

Fig. 8. A free-form surface and a projected plane: $\vec{n}_a$ is the tool direction; $\vec{n}_i$ is the normal of a triangle.

the tool direction and desired tool standoff. The actual material quantity q_s of the point s can be calculated as:

$$q_s = \bar{q}\left(\frac{h}{h_i}\right)^2 cos\beta_i \qquad (16)$$

where $\bar{q}$ is the material quantity of the projected point on the plane; h_i is the actual tool standoff and β_i is the deviation angle of the point. Using this model, the material quantity can be calculated.

III. IMPLEMENTATION AND TESTING

Since the coverage problem is a special case of the material uniformity problem, implementation for the coverage problem is not discussed here. Three material uniformity tasks, spray painting, spray forming and indirect rapid tooling, are implemented.

Assume that the required average material quantity is $q_d = 50$ and the material quantity deviation is 15. Units are not consider here since different tasks are presented. The spray radius $R = 50$. The material deposition rate is

$$f(r) = \frac{1}{10}(R^2 - r^2) \qquad (17)$$

After optimizing the material quantity deviation, the tool velocity, the overlap distance, the maximum and minimum material quantity are

$$v = 323.3, \quad d = 60.8$$
$$\bar{q}_{max} = 52.02, \quad \bar{q}_{min} = 48.05 \qquad (18)$$

From equation (11), the threshold angle is calculated: $\beta_{th} = 45.6°$.

The algorithm was implemented in C++ on a PC with Pentium III 860MHZ processor. Part of a car door, shown in Figure 9(a), was used to test the algorithm for spray painting. The triangular approximation was exported from GID (http://gid.cimne.upc.es/) with an error tolerance of 2 mm. The car door has 4853 triangles. Using the patch forming method, the car door form only one patch. The tool paths were generated using the bounding box method. The generated gun paths are shown in Figure 9(b). The gun direction was determined for each sample point.

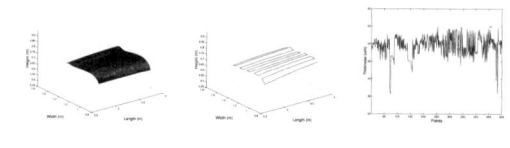

Fig. 9. Part of a car door: (a) The triangular approximation; (b) The generated path; (c) the simulated thickness

Simulations were performed to verify the generated trajectory. The paint thickness of randomly chosen points on the car door was computed and shown in Figure 9(c). The average, maximum, and minimum paint thickness are 49.4, 54.9, 35.6 respectively. The simulation results show that the trajectory generated using the developed algorithm can achieve required paint thickness.

Because glass fiber will be sprayed on a surface in spray forming process, area density is considered instead of thickness in spray painting process. The area density is defined as material weight on unit area. Part of a car frame is used to test the algorithm. Figure 10(a) shows the part; Figure 10(b) the generated path and Figure 10(c) the area density. The average, maximum and minimum area densities are 40.8, 44.5 and 36.0 respectively. The average area density is lower than the required density due to the surface curvature.

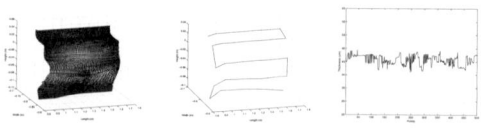

Fig. 10. Part of a car frame: (a) The triangular approximation; (b) the generated path; (c) the simulated thickness.

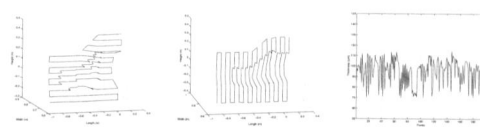

Fig. 11. (a), (b) The two perpendicular paths of a mold; (c) The simulated thickness.

In indirect rapid tooling, a part is sprayed many cycles. Typically two perpendicular trajectories are used and repeated to spray a mold. Figure 3 shows a mold rendered into triangles. Figure 11(a) and 11(b) show the two perpendicular paths.

Simulation was performed to calculate the metal thickness. Here only the two trajectories are used to perform the simulation. Figure 11(c) shows the metal deposited on the surface. The average, maximum, and minimum metal quantity were 93.4, 114.3 and 70.2.

The tool velocity is optimized using equation (15). Simulation is performed. Figure 12(a), 12(b) and 12(c) shows the simulation results for the spray painting, spray forming and indirect rapid tooling processes respectively. For spray painting, the results show that the lower bounds are decreased. The average, maximum and minimum thickness are 51.5, 59.4 and 40.6 respectively, The thickness deviation is decreased from 35% to 20%. For the spray forming process, the average, maximum and minimum area densities are 50.8, 55.2 and 41.5 respectively. Both the lower bound and average area density are much better. For the indirect rapid tooling process, the average, maximum and minimum metal thickness is 102.7, 80.1 and 120.4 respectively. The deviation of the metal quantity is decreased from 30% to 20%. Therefore, the simulation results show that the velocity optimization improves the material distribution on a surface.

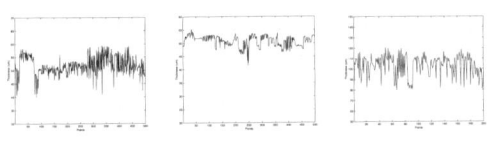

(a)   (b)   (c)

Fig. 12. The simulation results with velocity optimization: (a) spray painting; (b) spray forming (c) indirect rapid tooling.

IV. CONCLUSION

A frame work of the tool planning of different tasks is developed. Both material uniformity problem and coverage problem are considered in one constraint. Implementations are presented to generate different trajectories for different parts. Simulations are performed to calculate the material deposition on a free-form surface. Simulation results show that the generated trajectories satisfy the task constraints. Velocity optimization is discussed to minimize the lower bound. Simulation results are satisfactory. This framework can also be extended to other applications such as path planning for demining.

V. ACKNOWLEDGMENTS

Research is partially supported under NSF Grant IIS-9796300, IIS-9796287, EIA-9911077 and DMI 0115355. The authors would like to thank Tecnomatix Inc. for its providing us with the software ROBCAD.

VI. REFERENCES

[1] H. Chen, W. Sheng, N. Xi, M. Song, and Y. Chen. Automated robot trajectory planning for spray painting of free-form surfaces in automotive manufacturing. In *IEEE International Conference on Robotics and Automation*, volume 1, pages 450 –455, 2002.

[2] N. Asakawa and Y. Takeuchi. Teachingless spray-painting of sculptured surface by an industrial robot. In *IEEE International Conference on Robotics and Automation*, pages 1875 –1879, Albuquerque, New Mexico, April 1997.

[3] J. K. Antonio, R. Ramabhadran, and T. L. Ling. A framework for optimal trajectory planning for automated spray coating. *International Journal of Robotics and Automation*, 12(4):124–134, 1997.

[4] S. Suh, I. Woo, and S. Noh. Automatic trajectory planning system (atps) for spray painting robots. *Journal of Manufacturing Systems*, 10(5):396–406, 1991.

[5] L. F. Penin, C. Balaguer, J. M. Pastor, F. J. Rodriguez, A. Barrientos, and R. Aracil. Robotized spraying of prefabricated panels. *IEEE Robotics and Automation Magazine*, 5(3):18 –29, 1998.

[6] N. G. Chavka and J. Dahl. P4 preforming technology: Development of a high volume manufacturing method fo rfiber preforms. In *Society of Automotive Engineering*, 1998.

[7] R. E. Chalmers. Rapid tooling technology from ford country. *Manufacturing Engineering*, pages 36–38, Nov. 2001.

[8] R. C. Luo and J. H. Tzou. Investigation of a linear 2-d planar motor based rapid tooling. In *IEEE International Conference on Robotics and Automation*, pages 1471–1476, Washington, DC, USA, May 2002.

[9] W. Sheng, N. Xi, M. Song, Y. Chen, and P. MacNeille. Automated cad-guided robot path planning for spray painting of compound surfaces. In *IEEE/RSJ International Conference On Intelligent Robots And Systems*, pages 1918–1923, Takamutsa, Japan, October 2000.

[10] W. H. Huang. Optimal line-sweep-based decompositions for coverage algorithm. In *IEEE International Conference on Robotics and Automation*, pages 27–32, Seoul, Korea, May 2001.

[11] Y. K. Hwang, L. Meirans, and W. D. Drotning. Motion planning for robotic spray cleaning with environmentally safe solvents. In *IEEE/Tsukuba International Workshop on Advanced Robots*, pages 49–54, Tsukuba, Japan, Nov. 1993.

[12] H. Choset. Coverage of known spaces: The boustrophedon cellupdar decomposition. *Autonomos Robots*, 9:247–253, 2000.

[13] Tecnomatix. *ROBCAD/Paint Training*. Tecnomatix, Michigan, USA, 1999.

[14] H. Chen, W. Sheng, N. Xi, M. Song, and Y. Chen. Automated cad-guided robot trajectory planning for spray painting of free-form surfaces in automotive manufacturing. *Submitted to Journal of Robotic Systems*.

[15] E. Freund, D. Rokossa, and J. Rossmann. Process-orientated approach to an efficient off-line programming of industrial robots. In *Proceedings of the 24th Annual Conference of the IEEE Industrial Electronics Society, IECON '98.*, volume 1, pages 208 –213, 1998.

[16] W. Persoons and H. Van Brussel. Cad-based robotic coating with highly curved surfaces. In *International Symposium on Intelligent Robotics (ISIR'93)*, volume 14, pages 611–618, 1993.

[17] E. D. Goodman and L. T. W. Hoppensteradt. A method for accurate simulation of robotic spray application using empirical parameterization. In *IEEE International Conference on Robotics and Automation*, volume 2, pages 1357 –1368, Sacramento, California, April 1991.

[18] P. Hertling, L. Hog, R. Larsen, J. W. Perram, and H. G. Petersen. Task curve planning for painting robots. i. process modeling and calibration. *IEEE Transactions on Robotics and Automation*, 12(2):324 –330, April 1996.

[19] M. M. Blane, Z. Lei, H. Civi, and D. B. Cooper. The 3l algorithm for fitting implicit polynomial curves and surfaces to data. *IEEE Transaction on Pattern Analysis and Machine Intelligence*, 22(3):298–313, March 2000.

[20] H. Chen, N. Xi, Z. Wei, Y. Chen, and J. Dahl. Trajectory integration for a surface with multiple patches. In *Submitted to IEEE International Conference on Robotics and Automation*, 2003.

Conflict-Free Routing of AGVs on the Mesh Topology Based on a Discrete-Time Model

ZENG Jianyang
Center for Advanced Information Systems
School of Computer Engineering
Nanyang Technological University
Singapore 639798

HSU Wen-Jing
Center for Advanced Information Systems
School of Computer Engineering
Nanyang Technological University
Singapore 639798

Abstract — Automated Guided Vehicles (or AGVs for short) have become an important option in material handling. In many applications, such as container terminals, the service area is often arranged into rectangular blocks, which leads to a mesh-like path topology. Therefore, developing efficient algorithms for AGV routing on the mesh topology has become an important research topic. In this paper, we present a discrete time model, based on which a simple routing algorithm on the mesh topology is presented. The algorithm works by carefully choosing suitable parameters such that the vehicles using a same junction will arrive at different points in time, and hence no conflicts will occur during the routing; meanwhile, high routing performance can be achieved. Analyses of the task completion time and the requirements on timing control during the AGV routing are also presented.

1. INTRODUCTION

Automated Guided Vehicles (or AGVs for short) have become an important option in material handling [1-7, 9-11]. In many applications, such as container terminals[1, 9-11], the service area is often arranged into rectangular blocks, which leads to a mesh-like path topology. Therefore, developing efficient algorithms for AGV routing on this topology has become an important research topic.

There are many existing results about AGV [5]. However, relatively little has been known about routing on the mesh topology. [2-3] gave the analysis of time and space complexities for some basic AGV routing operations on 2D-mesh topology. The upper bounds of time and space complexities for AGV routing are $\Theta(n^2)$ and $\Theta(n^3)$ respectively, where n denotes the number of nodes in the path topology. However, the paper does not give the details of the routing algorithms and techniques to avoid congestion, conflicts, deadlocks, etc.

[6-7] presented different methods to schedule and route simultaneously in an $n \times n$ mesh-like path topology. The algorithms can schedule and route simultaneously up to $4n^2$ AGVs concurrently at one time. In these papers, the routing process is formulated as a sorting problem. Although there are no conflicts during the permutation, it requires $3n$ steps of well-defined physical moves, which requires AGVs to travel extra distance and consume extra energy to finish the tasks.

In this paper, we present a discrete time model on mesh topology for AGV routing. Based on this model, the routing algorithm is presented and time control requirement is analyzed. The key idea lies in making use of the regularity of the mesh, and hence the regularity of points of time when AGVs arrive at the intersections. By choosing a suitable speed for the AGVs along different directions, we can ensure that no conflicts among any AGVs will occur. We also analyze the algorithm in terms of bounds on the task completion time and requirements on the routing precision controls. By our design, the AGVs can advance directly to their destinations, unlike in [6-7], and hence high performance can also be ensured.

The remainder of the paper is organized as follows. Section 2 describes the mesh layout and the routing model. Section 3 gives the routing algorithm and the time control criteria to avoid conflicts. In Section 4, we analyze the performance of the routing algorithm. Section 5 gives an explicit method to derive the timing controls. Finally, Section 6 discusses possibilities of relaxing certain constraints and points out directions of future study.

II. DESCRIPTION OF MESH LAYOUT MODEL

In order to describe the marrow of our routing process clearly, we start with a simple but general model in which one yard block has only one station near an intersection of pathways (refer to Figure 1). In this mesh layout, there are in total $N \times N$ blocks, namely N blocks in each column and N blocks in each row. Each block has the same size. There are two paths with different directions between two adjacent blocks. Each Block has one Pick up-Drop off station(or P/D station for short), located at the upper right and upper top corner of the block. On the left-top side, there is a vehicle park where all AGVs are stationed initially and to which they will return upon completion of all tasks.

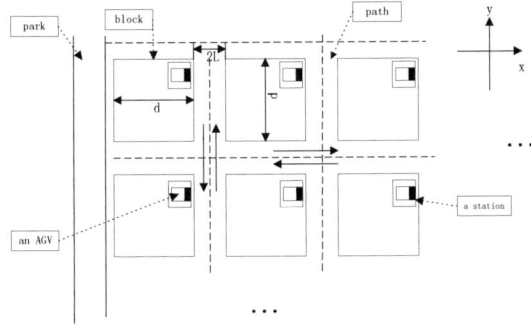

Figure.1 Realistic mesh layout

Although there are some important details for AGV routing, such as the size of the junction, the radius of $90°$ turn, the length of the AGV, etc[4-7], it is reasonable and realistic for us to simplify the mesh model for convenience of analysis and discussion. In the simplified mesh layout model, shown as Figure 2, there are N^2 junctions of pathways. A junction and the associated neighboring station are collectively regarded as a node. Each node is to assign it with the coordinates (x, y) as its address or ID, where x and y represent respectively the row and column IDs. This mesh layout is modeled by a graph $G=(V,E)$. The $N \times N$ vertices of the graph represent junction nodes, and the bi-directional edges represent two paths between two adjacent junction nodes, and the length of each edge is a constant.

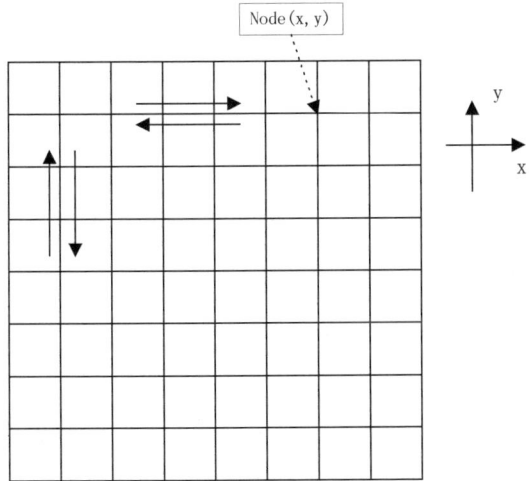

Figure.2 Simplified mesh routing model

We divide the AGV movements into three phases. In the first phase, let AGVs set out from the park to their pick up stations. In the second phase, let AGVs pick up loads and travel to their destinations and drop-off loads. In the third phase, let AGVs return to the park from their drop-off stations. Because it is easy for us to dispatch the AGV moving without any conflict in the first phase and the third phase, we can focus only on the second phase when the loaded AGVs move on the mesh layout.

We assume that the time can be divided into discrete units of time, and that each AGV always reaches every junction node at some discrete point on time. It is reasonable for us to make this assumption because the distance between two adjacent nodes is a constant, and we can adjust the speed of the AGVs to let them arrive at the junctions at multiples of the unit time.

We assume that when an AGV reaches its destination, it enters the buffer and leaves the mesh grid.

Based on the mesh layout model, we formally define the following.

Definition: $(x_1, y_1) = (x_2, y_2)$ if and only if $x_1 = x_2$ and $y_1 = y_2$.

Definition (Job): *A job is identified by an ordered pair $((PX,PY),(DX,DY))$, where (PX,PY) represenst the address of the pickup station, (DX,DY) represents the address of the drop-off station, and $(PX,PY) \neq (DX,DY)$.*

Assume that each job has a distinct origin and also a distinct(but different) destination, and an AGV is given only one job and any job is assigned to only one AGV.

Definition (Job Set): *A job set M denoting a set of k jobs, where $2 \leq k \leq \left\lfloor \dfrac{N^2}{2} \right\rfloor$, is defined as follows:*

$M=\{((PX_i, PY_i), (DX_i, DY_i))| 1 \leq PX_i, PY_i, DX_i, DY_i \leq N,$ for $i=1, 2, ..., k \}$.

According to the position of the origins and destinations of jobs, any given job set M can be divided into four subsets, denoted by M_x, M_y, M_{xy} respectively, such that

$M_x = \{((PX_i, PY_i),(DX_i, DY_i))| DX_i \neq PX_i, DY_i = PY_i$ for $i = 1,2,...,k \}$.
$M_y = \{((PX_i, PY_i),(DX_i, DY_i))| DY_i \neq PY_i, DX_i = PX_i,$ for $i = 1,2,...,k \}$.
$M_{xy} = \{((PX_i, PY_i),(DX_i, DY_i))| DY_i \neq PY_i, DX_i \neq PX_i,$ for $i = 1,2,...,k \}$.

We also divide M_{xy} into two subsets, denoted by M_{xy+} and M_{xy-}, such that

$M_{xy+} = \{((PX_i, PY_i),(DX_i, DY_i))| DY_i > PY_i, DX_i \neq PX_i,$ for $i = 1,2,...,k \}$.
$M_{xy-} = \{((PX_i, PY_i),(DX_i, DY_i))| DY_i < PY_i, DX_i \neq PX_i,$ for $i = 1,2,...,k \}$.

Accordingly, we have the following notations:

A_x: the set of AGVs that carry out jobs in M_x;
A_y: the set of AGVs that carry out jobs in M_y;
A_{xy+}: the set of AGVs that carry out jobs in M_{xy+};
A_{xy-}: the set of AGVs that carry out jobs in M_{xy-};

Definition (Direction of AGVs): $\vec{v}$ *is a unit vector which represents the direction of a given AGV, where* $\vec{v} \in \{+\vec{x}, -\vec{x}, +\vec{y}, -\vec{y}\}$. $\vec{v}_1 = \vec{v}_2$ *when* $\vec{v}_1$ *is in the same direction as* $\vec{v}_2$. $\vec{v}_1 = -\vec{v}_2$ *when* $\vec{v}_1$ *is in the opposite direction of* $\vec{v}_2$. $\vec{v}_1 \cdot \vec{v}_2 = 0$ *when* $\vec{v}_1$ *is in a vertical direction of* $\vec{v}_2$.

Definition (State of AGVs): $((x,y),t,\vec{v})$ *is the state of an AGV, where (x,y) represents the location in the mesh layout, and t represents the discrete time points of the AGV, and* $\vec{v} \in \{+\vec{x}, -\vec{x}, +\vec{y}, -\vec{y}\}$.

Definition (Collision): $((x_1,y_1),t_1,\vec{v}_1)$ *is the state of AGV1, and* $((x_2,y_2),t_2,\vec{v}_2)$ *is the status of AGV2. When* $t_1 = t_2$, $(x_1,y_1) = (x_2,y_2)$ *and* $\vec{v}_1 \in \{+\vec{x}, -\vec{x}, +\vec{y}, -\vec{y}\} - \{-\vec{v}_2\}$. *In this case, we say that AGV1 and AGV2 have a collision at* (x_1,y_1) *or* (x_2,y_2) *when* $t = t_1$ *on the mesh layout.*

III. CONFLICT-FREE ROUTING ALGORITHM

Based on the simplified mesh layout and the discrete time, the routing algorithm is presented as follows.

Let all AGVs set out from their pick up stations at the same time, when $t_0 = 0$.

Case a In the job set M_x. In this case, let the AGV travel along the row PY_i from (PX_i, PY_i) station to (DX_i, DY_i).

Case b In the job set M_y. In this case, let the AGV travel along the column PX_i from (PX_i, PY_i) station to (DX_i, DY_i).

Case c In the job set M_{xy+}. In this case, let the AGV firstly travel along the column PX_i from (PX_i, PY_i) station to (PX_i, DY_i) station. Then let it travel along the row DY_i from (PX_i, PY_i) station to (DX_i, DY_i) station.

Case d In the job set M_{xy-}. In this case, let the AGV firstly travel along the column PX_i from (PX_i, PY_i) station to (PX_i, DY_i). Then let it travel along the row DY_i from (PX_i, PY_i) station to (DX_i, DY_i) station.

The routing algorithm looks simple, and if we let AGVs travel in this rule at an arbitrary speed, it is very likely to have collisions on the mesh layout. However, as we will show shortly, if we control the time when each AGV reaches every junction node(we can do so by controlling the AGV's speed), AGVs can run on the mesh layout with the freedom of conflicts.

We let ΔT_{+x} denote the time required for an AGV to travel through one edge of the mesh along the $+\vec{x}$ direction. Let ΔT_{+x} ($\Delta T_{+y}, \Delta T_{-y}$) be defined similarly. We assume that AGVs travel at the speed v_{+x}, v_{-x}, v_{+y}, v_{-y} in these four cases respectively.

According to the preceding routing, we have the following conclusions.

Claim 3.1: *According to our routing algorithm, there is no conflict between any two AGVs belonging to the same set* A_x (*or* A_y).

[Proof]:
According to the definition of collision, and the assumption that each AGV has a distinct origin, it is quite clear that there is no conflict in the AGVs belonging to A_x (or A_y). □

Claim 3.2: *Based on the routing algorithm, any AGV will not run into conflict with other AGVs on the mesh layout, if the following relation is satisfied.*

(1) $$\frac{lcm(\Delta T_1, \Delta T_2)}{max(\Delta T_1, \Delta T_2)} \geq N \quad (3-1)$$

where ΔT_1 and ΔT_2 are any two permutations from $\{\Delta T_{+x}, \Delta T_{-x}, \Delta T_{+y}, \Delta T_{-y}\}$.

(2) $$\begin{cases} gcd(\Delta T_{+y}, \Delta T_{+x}) \nmid \Delta T_{-y} & (3-2) \\ gcd(\Delta T_{+y}, \Delta T_{-x}) \nmid \Delta T_{-y} & (3-3) \\ gcd(\Delta T_{-y}, \Delta T_{+x}) \nmid \Delta T_{+y} & (3-4) \\ gcd(\Delta T_{-y}, \Delta T_{-x}) \nmid \Delta T_{+y} & (3-5) \\ gcd(\Delta T_{\pm y}, \Delta T_{\pm x}) \geq N & (3-6) \end{cases}$$

here gcd the Greatest Common Divisor, and lcm the Least Common Multiple.(cf. Definition A.2 and Definition A.3).

[Proof]:
Firstly let us recall some definitions and theorems of number theory[8] which will be used in the discussions later.

Definition A.1 (Divisibility): *If a and b are integers, we say that a divides b if there is an integer c such that b=ac. If a divides b, we also say that a is a divisor or factor of b.*
Write $a|b$ if a divide b; otherwise, write $a \nmid b$ if a does not divide b.

Definition A.2 (The Greatest Common Divisor): *The greatest common divisor of two integers a and b, not both zero, is the largest positive integer that divides both a and b; it is denoted by gcd(a,b).*

Definition A.3 (The Least Common Multiple): *The least common multiple (gcd) of two integers a and b, is* the least positive integer divisible by both a and b; it is denoted by lcm(a,b).

Definition A.4 (Linear combination): *If a and b are integers, then a linear combination of a and b is a sum of the form ma+nb, where both m and n are integers.*

Definition A.5 (Linear diophantine equation): *A linear diophantine equation in two variables x and y is a diophantine equation of the form ax+by=c, where a, b and c are integers.*

Theorem A.1: *The greatest common divisor of the integers a and b, that are not both zero, is the least positive integer that is a linear combination of a and b.*

Theorem A.2: *Let a and b be positive integers. Then* $lcm(a,b) = \dfrac{ab}{gcd(a,b)}$.

Theorem A.3: *Let a and b be positive integers and $d=gcd(a,b)$. The equation ax+by=c has no integer solutions if $d \nmid c$. If $d|c$, then there are infinitely many integer solutions. Moreover, if $x=x_0$, $y=y_0$ is a particular solution of the equation, then all solutions are given by*
$$\begin{cases} x = x_0 + (b/d)n \\ y = y_0 - (a/d)n \end{cases}$$

Now continue with the proof of the claim 3.1. All cases of possible conflicts are shown in Figure 3. We omitted a few similar cases, which are symmetrical to some of the following cases).

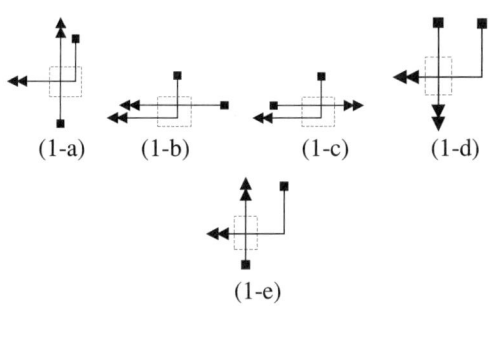

(1) $M_x - M_{xy}$ (or $M_y - M_{xy}$)

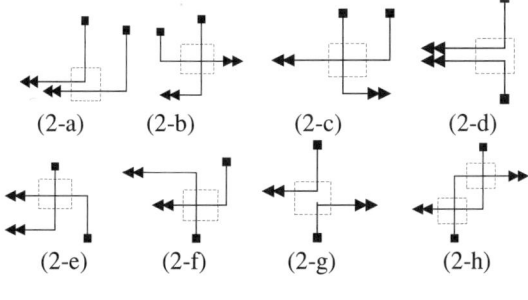

(2) $M_{xy} - M_{xy}$

Figure.3 All cases of possible conflicts

Assume the initial states of AGV1 and AGV2 respectively as follows.
AGV1: $((x_1,y_1),t_1=0,\vec{v}_1)$;
AGV2: $((x_2,y_2),t_2=0,\vec{v}_2)$.

When $(x_1,y_1)=(x_2,y_2)=(x',y')$, the states of AGV1 and AGV2 are respectively,
AGV1: $((x_1',y_1'),t_1',\vec{v}_1')$;
AGV2: $((x_2',y_2'),t_2',\vec{v}_2')$.

Now let us prove that $t_1' \neq t_2'$ in all cases of potential conflicts.

Case (1) This case covers (1-a), (1-b), (1-c), (2-d), (2-g). We have the following relations.
$$\begin{cases} t_1' = t_1 + i\Delta T_1 = i\Delta T_1, \\ t_2' = t_2 + j\Delta T_2 = j\Delta T_2, \end{cases}$$

where $0 \leq i, j \leq N-1$, and ΔT_1 and ΔT_2 are any two permutations from $\{\Delta T_{+x}, \Delta T_{-x}, \Delta T_{+y}, \Delta T_{-y}\}$.

According to the definition of lcm, we have
$$\begin{cases} i_{min} = \dfrac{lcm(\Delta T_1, \Delta T_2)}{\Delta T_1} \\ j_{min} = \dfrac{lcm(\Delta T_1, \Delta T_2)}{\Delta T_2} \end{cases}$$

According to Inequality (3-1), we know $i_{min}, j_{min} \geq N$, which conflict with the condition such that $0 \leq i, j \leq N-1$. Therefore in all these cases, $t_1' \neq t_2'$ for any i, j, where $0 \leq i, j \leq N-1$.

Case (2) This case covers (1-e), (2-a), (2-b), (2-c), (2-e), (2-f), (2-h). We have the following relations.
$$\begin{cases} t_1' = i\Delta T_{+y} + j\Delta T_{+x}, \\ t_2' = k\Delta T_{-y}, \end{cases}$$

or
$$\begin{cases} t_1' = i\Delta T_{+y} + j\Delta T_{-x}, \\ t_2' = k\Delta T_{-y}, \end{cases}$$

or
$$\begin{cases} t_1' = i\Delta T_{-y} + j\Delta T_{+x}, \\ t_2' = k\Delta T_{+y}, \end{cases}$$

or
$$\begin{cases} t_1' = i\Delta T_{-y} + j\Delta T_{-x}, \\ t_2' = k\Delta T_{+y}, \end{cases}$$

where $0 \leq i, j, k \leq N-1$.

These four relations are similar to each other, so we can focus on the first one.

Firstly we take a look at the following equation.

$$x\Delta T_{+y} + y\Delta T_{+x} = k\Delta T_{-y} \qquad (3\text{-}7)$$

where x and y are integers.

From Eq. (3-6) and Eq. (3-2), we have
$$gcd(\Delta T_{+y}, \Delta T_{+x}) \nmid k\Delta T_{-y}$$

According to Theorem A.3, we know that Eq. (3-7) has no integer solutions, so for any $i, j \in [0, N]$, $t_1' \neq t_2'$.

Case (3) This case covers (1-d). We have the following relation.

$$\begin{cases} t_1' = i\Delta T_{+y} + j\Delta T_{+x}, \\ t_2' = k\Delta T_{+y}, \end{cases}$$

In order to prove that $t_1' \neq t_2'$, we need to show that
$$i\Delta T_{+y} + j\Delta T_{+x} \neq k\Delta T_{+y},$$
namely, $j\Delta T_{+x} \neq |k-j|\Delta T_{+y}$.

We know that $0 \leq k - j \leq N - 1$, then this situation can be converted to the one in case (1) that we have proved.

Therefore, we conclude that in all cases, there is no any conflict using our routing algorithm and the time limit.

IV. TIME REQUIREMENT

Claim 4.1: *The time requirement T_r for all AGVs to transport all jobs is upper-bounded by*
$$2(N-1)\max\{\Delta T_{+x}, \Delta T_{-x}, \Delta T_{+y}, \Delta T_{-y}\}$$

[Proof]: Since all jobs are carried out in parallel, the time requirement for a job set M is determined by the most time-consuming job in the set. Formally, for any given job set, we have

$$T_r = \max\{T((PX_1, PY_1), (DX_1, DY_1)), T((PX_2, PY_2),$$
$$(DX_2, DY_2)), ..., T((PX_k, PY_k), (DX_k, DY_k))\},$$

where $T((PX_i, PY_i), (DX_i, DY_i))$ is the time requirement for AGVi to complete its job.

Assume that there exists job $((1,1), (N,N))$ which uses the most time and use $\max\{\Delta T_{+x}, \Delta T_{-x}, \Delta T_{+y}, \Delta T_{-y}\}$ time to go through one edge on the mesh layout, then we obtain the following relation.

$$T((1,1), (N,N)) = 2(N-1)\max\{\Delta T_{+x}, \Delta T_{-x}, \Delta T_{+y}, \Delta T_{-y}\}$$

Thus, the time requirement for a job set is upper-bounded by

$$T_r \leq T((1,1), (N,N))$$
$$= 2(N-1)\max\{\Delta T_{+x}, \Delta T_{-x}, \Delta T_{+y}, \Delta T_{-y}\} \qquad \square$$

Although our routing algorithm guarantees collision-freedom under some special criteria, the control system needs to know the time point when each AGV goes through every junction node. So it is necessary for us to consider the relation between different time points when each AGV goes through every junction node.

Definition (Time difference): *The time difference is the minus of two time points when two AGVs reach one special junction node. The minimum time difference is the minimum time difference for all AGVs at every junction node on the mesh layout.*

Claim 4.2: *The minimum time difference on the mesh layout is lower-bounded by.*

$$\min\{gcd(\Delta T_{+x}, \Delta T_{-x}), gcd(\Delta T_{+x}, \Delta T_{+y}), gcd(\Delta T_{+x}, \Delta T_{-y}),$$
$$gcd(\Delta T_{-x}, \Delta T_{+y}), gcd(\Delta T_{-x}, \Delta T_{-y}), gcd(\Delta T_{+y}, \Delta T_{-y})\}$$
namely,
$$\min_{\substack{x,y \in S \\ x \neq y}} \{gcd(x,y)\}$$
where $S = \{\Delta T_{+x}, \Delta T_{+y}, \Delta T_{-x}, \Delta T_{-y}\}$.

[Proof]:

To get the minimum time difference, we can find the minimum of the following equation:
$$i\Delta T_1 + j\Delta T_2$$
where i, j are integers, and $\Delta T_1, \Delta T_2$ are any two numbers from $\{\Delta T_{+x}, \Delta T_{-x}, \Delta T_{+y}, \Delta T_{-y}\}$.

According to Theorem A.1, the least positive integer of $i\Delta T_1 + j\Delta T_2$ is $gcd(\Delta T_1, \Delta T_2)$.

Thus for any ΔT_1, ΔT_2 from $\{\Delta T_{+x}, \Delta T_{-x}, \Delta T_{+y}, \Delta T_{-y}\}$, the minimum of the time difference is

$$\min\{gcd(\Delta T_{+x}, \Delta T_{-x}), gcd(\Delta T_{+x}, \Delta T_{+y}), gcd(\Delta T_{+x}, \Delta T_{-y}),$$
$$gcd(\Delta T_{-x}, \Delta T_{+y}), gcd(\Delta T_{-x}, \Delta T_{-y}), gcd(\Delta T_{+y}, \Delta T_{-y})\}$$
namely
$$\min_{\substack{x,y \in S \\ x \neq y}} \{gcd(x,y)\}$$
where $S = \{\Delta T_{+x}, \Delta T_{+y}, \Delta T_{-x}, \Delta T_{-y}\}$ $\square$

V. THE METHOD TO CONSTRUCT $\Delta T_{+x}, \Delta T_{+y}, \Delta T_{-x}, \Delta T_{-y}$

We introduce the following method to construct $\Delta T_{+x}, \Delta T_{+y}, \Delta T_{-x}, \Delta T_{-y}$, which satisfy the criteria of conflict-free routing.

We let $\Delta T_{+y} = P_1^{a+b} P_5^g$, $\Delta T_{-y} = P_2^{c+d} P_5^g$, $\Delta T_{-x} = P_1^a P_2^c P_3^e$ and $\Delta T_{+x} = P_1^a P_2^c P_4^f$

Where P_1, P_2, P_3 and P_4 are primes;

$a, b \geq \log_{P_1} N$;
$c, d \geq \log_{P_2} N$;
$e \geq \log_{P_3} N$;
$f \geq \log_{P_4} N$;
$g \geq \log_{P_5} N$.

Claim 5.2: *The values of $\Delta T_{+x}, \Delta T_{+y}, \Delta T_{-x}, \Delta T_{-y}$ constructed by this method satisfy the criteria of conflict-free routing.*

[Proof]:

According to Theorem A.2, we have

$$\frac{lcm(\Delta T_{+y}, \Delta T_{-y})}{\Delta T_{+y}} = \frac{\Delta T_{+y} \cdot \Delta T_{-y}}{gcd(\Delta T_{+y}, \Delta T_{-y}) \cdot \Delta T_{+y}}$$
$$= \frac{\Delta T_{-y}}{gcd(\Delta T_{+y}, \Delta T_{-y})} = \frac{P_2^{c+d} P_5^g}{P_5^g}$$
$$= P_2^{c+d} \geq N^2 \geq N$$

Similarly we can prove that for any two permutations ΔT_1 and ΔT_2 from $\{\Delta T_{+x}, \Delta T_{-x}, \Delta T_{+y}, \Delta T_{-y}\}$, the following relations are satisfied.

$$\frac{lcm(\Delta T_1, \Delta T_2)}{\Delta T_1} \geq N .$$

According to the construction method, we have

$$gcd(\Delta T_{+y}, \Delta T_{+x}) = gcd(P_1^{a+b} P_5^g, P_1^a P_2^c P_4^f) = P_1^a \geq N$$

Because $P_1^a \nmid P_2^{c+d} P_5^g$, we have

$$gcd(\Delta T_{+y}, \Delta T_{+x}) \nmid \Delta T_{-y} .$$

Similarly we can prove that the inequalities (3-3)-(3-6) are satisfied.

Therefore the values of $\Delta T_{+x}, \Delta T_{+y}, \Delta T_{-x}, \Delta T_{-y}$ constructed by this method satisfy all the criteria of conflict-free routing. □

The differences in $\Delta T_{+x}, \Delta T_{+y}, \Delta T_{-x}, \Delta T_{-y}$ have implications on the routing control. The smaller value of this difference generally means the more accurate timing when vehicles arrive at the junctions.

Claim 5.2: *The minimum time difference of $\Delta T_{+x}, \Delta T_{+y}, \Delta T_{-x}, \Delta T_{-y}$ constructed by this method $min\{P_1^a, P_2^c, P_5^g\}$, which is lower-bounded by $\Omega(N)$.*

[Proof]:

According to Claim 4.2, the minimum time difference on the mesh layout is lower-bounded by

$min\{gcd(\Delta T_{+x}, \Delta T_{-x}), gcd(\Delta T_{+x}, \Delta T_{+y}), gcd(\Delta T_{+x}, \Delta T_{-y}),$
$gcd(\Delta T_{-x}, \Delta T_{+y}), gcd(\Delta T_{-x}, \Delta T_{-y}), gcd(\Delta T_{+y}, \Delta T_{-y})\}$

Substituting into the values of $\Delta T_{+x}, \Delta T_{+y}, \Delta T_{-x}, \Delta T_{-y}$, we have

$min\{gcd(\Delta T_{+x}, \Delta T_{-x}), gcd(\Delta T_{+x}, \Delta T_{+y}), gcd(\Delta T_{+x}, \Delta T_{-y}),$
$gcd(\Delta T_{-x}, \Delta T_{+y}), gcd(\Delta T_{-x}, \Delta T_{-y}), gcd(\Delta T_{+y}, \Delta T_{-y})\}$
$= min\{P_1^a \cdot P_2^c, P_1^a, P_2^c, P_1^a, P_2^c, P_5^g\} = min\{P_1^a, P_2^c, P_5^g\}$

Therefore, the minimum time difference of $\Delta T_{+x}, \Delta T_{+y}, \Delta T_{-x}, \Delta T_{-y}$ constructed by this method is lower-bounded by $min\{P_1^a, P_2^c, P_5^g\}$. Because $P_1^a, P_2^c, P_5^g \geq N$, the minimum time difference is lower-bounded by $\Omega(N)$. □

From claim 4.1, the longest value of $\Delta T_{+x}, \Delta T_{+y}, \Delta T_{-x}, \Delta T_{-y}$ generally means a higher task completion time (for the given choice of time unit). The following result bounds this value.

Claim 5.3: *The values of $\Delta T_{+x}, \Delta T_{+y}, \Delta T_{-x}, \Delta T_{-y}$ are bounded by $\Theta(N^3)$.*

[Proof]:

According to the construction method, we have

$$\Delta T_{+y} = P_1^{a+b} P_5^g \geq N^2 \cdot N = N^3$$

Similarly, we can prove that $\Delta T_{-y} \geq N^3, \Delta T_{+x} \geq N^3, \Delta T_{-x} \geq N^3$.

If we keep the values of $\Delta T_{+x}, \Delta T_{+y}, \Delta T_{-x}, \Delta T_{-y}$ as small as possible, we can make them be bounded by $\Theta(N^3)$.

Therefore, the values of $\Delta T_{+x}, \Delta T_{+y}, \Delta T_{-x}, \Delta T_{-y}$ is bounded by $\Theta(N^3)$. □

We give a simple example to illustrate the construction. Let N=7, we can choose $\Delta T_{+y} = 2^{3+3} \cdot 13$, $\Delta T_{-y} = 3^{2+2} \cdot 13$, $\Delta T_{-x} = 2^3 \cdot 3^2 \cdot 7$, $\Delta T_{+x} = 2^3 \cdot 3^2 \cdot 11$. The minimum time difference of this case is $2^3 = 8$, and the ratio between the maximum and the minimum of $\Delta T_{+x}, \Delta T_{+y}, \Delta T_{-x}, \Delta T_{-y}$ is about 2.

VI. DISCUSSIONS AND CONCLUSIONS

We have presented a discrete time model for AGV routing on the mesh topology. In this model, each AGV is assumed to reach every junction node at discrete points in time. Based on this model, we proposed a routing algorithm which allows AGVs to travel at different multiples of the unit time along different directions, which guarantees the freedom of conflicts. The timing control requirement was analyzed, and the method to construct the multiples of the unit time was also introduced.

With our routing algorithm, all the AGVs can move directly towards their destinations without conflicts. Therefore, the overall routing performance is ensured. Moreover, since each AGV makes at most one turn during the entire routing process, the speed of each

AGVs is changed no more than once. Therefore, the energy requirement by our routing algorithm is also relatively low. In our routing model, each AGV on the mesh topology is assumed to be one point. However, there are some details that we must consider in actual implementations, such as the size of junction, the length of the AGV, etc. These considerations have a requirement of the minimum time difference, which can be adjusted by the control system. According to Claim 5.2, the minimum time difference is decided by $min\{P_1^a, P_2^c, P_5^g\}$. Thus we can choose the value of $\{P_1^a, P_2^c, P_5^g\}$ to increase the minimum time difference. Therefore, the discrete time model and the routing algorithm can be used in real mesh-like layout. Similarly, the task completion time can be controlled by choice of the units of time, the distance between intersections and/or the speeds of AGVs. For instance, by Claim 5.3, the maximum value of $\Delta T_{+x}, \Delta T_{+y}, \Delta T_{-x}, \Delta T_{-y}$ is bounded by $\Theta(N^3)$. As N increases, the time requirement to finish the jobs seems to increase quickly. However, we can choose a small unit of time to keep the actual time requirement low, as long as the minimum time difference for avoiding conflicts is satisfied.

We assumed that when an AGV reaches its destination, it enters the buffer and leaves the mesh grid. This assumption can be also relaxed. Usually, when an AGV enters the buffer of the P/D station, it takes some time for the vehicle to completely leave the mesh grid. The situation is similar when an AGV goes through the junction. However, as long as the time required for an AGV to enter the buffer of the station is less than the minimum time difference, there are still no conflicts during the AGV routing.

We assume that each AGV has distinct origin and also a distinct (but different) destination, namely, the pattern of AGV movement is permutation. This assumption can also be relaxed such that each AGV has multiple origins and multiple destinations. As long as we control the time points for AGVs to set out from the stations and let them enter the buffers at proper time point, we can still guarantee the freedom of conflicts.

There are many open issues for future research. Firstly, how to deal with the failure of AGVs? In our algorithm (as well as others), a single blockage will cause the failure of the whole system. Therefore, it is essential to consider fault-tolerant strategies. Secondly, if we allow each AGV to make more than one turn before it reaches its destination, we need more complicated scheme to avoid conflicts. Thirdly, we need to devise a method to decide the number of AGVs for the given jobs and to deal with idle AGVs.

VII. ACKNOWLEDGMENTS

We acknowledge the Maritime and Port Authority, A*STAR and Nanyang Technological University, all of Singapore, for their support of the research project.

VIII. REFERENCES

[1] Evers, J. J. M. and S. A. J. Koppers. Automatic guided vehicle traffic control at a container terminal. *Transportation Research Part A*, 30(1):21-34,1996.

[2] HSU, W.-J. and HUANG, S.-Y., 1994, Route planning of automated guided vehicles. Proceedings *of Intelligent Vehicles*, Paris, pp.479-485.

[3] Huang, S.-Y. and W.-J. Hsu. Routing automated guided vehicles on mesh like topologies. In *Proceedings of International Conference on Automation, Robotics and Computer Vision*, 1994.

[4] Qiu, L. and W.-J. Hsu, A bi-directional path layout for conflict-free routing of AGVs. *International Journal of Production Research*, 39(10): 2177-2195, 2001.

[5] Qiu, Ling, Wen-Jing Hsu, Shell-Ying Huang, and Han Wang, "Scheduling and Routing Algorithms for AGVs: a Survey". *International Journal of Production Research*, Vol. 40, No. 3, pp. 745-760, 2002.

[6] Qiu, Ling and Wen-Jing Hsu, "Routing AGVs on a Mesh-like Path Topology". In *Proceedings of the IEEE Intelligent Vehicles Symposium 2000 (IVS 2000)*, pp. 392-397, Dearborn, Michigan, USA, Oct. 3-5, 2000.

[7] Qiu, Ling and Wen-Jing Hsu, "Algorithms for Routing AGVs on a Mesh Topology". In *Proceedings of the 2000 European Conference on Parallel Computing (Euro-par 2000)*, pp. 595-599, Technical University of Munich, Munich, Germany, Aug. 29-Sep. 1, 2000.

[8] Thomsa Koshy, Elementary Number Theory with Applications. *Harcourt/Academic Press*, 2002.

[9] Ye, R., W.-J. Hsu, and V.-Y. Vee. Distributed routing and simulation of automated guided vehicles. In *Proceedings of TENCON 2000*, volume II, pages 315-320, Kuala Lumpur, Malaysia, September 24-27, 2000.

[10] Ye, R., V.-Y. Vee, W.-J. Hsu and S.N. Shah. Parallel simulation of AGVs in container port operations. In *Proceedings of 4th International Conference/Exhibition on High Performance Computing in Asia-pacific Region (HPC-ASIA 2000)*, volume I, pages 1058-1063, Beijing, China, May 14-17, 2000.

[11] Yu, X. and S.-Y. Huang, A Centralized Routing Algorithm for AGVS in Container Ports. In *Proceedings of the 4th International Conference on Computer Integrated Manufacturing*, Singapore, pages 589-600, 1997.

Development of an Automatic Mold Polishing System

Ming J. Tsai, Jau-Lung Chang, Jian-Feng Haung
Department of Mechanical Engineering
National Cheng-Kung University

Abstract—In this study, an automatic mold polishing system (AMPS), was developed. Software engineering technique was used to analyze the manual operation of polishing procedure to obtain the task specifications for the automatic system. Four main modules were designed in the system: geometric data processing, polishing process planning, polishing path planning, and the user interface modules.

The geometric data of molds can be input via three methods. The mold geometry and the polishing path can be drawn in 3D for verification. The AMPS can also generate process task list according to the mold polishing requirements. The task list includes all the procedures and optimal polishing parameters. The polishing tasks were then executed by a force controllable robot. The experiment shows that the resultant of surface roughness is closed to the prediction in the process planning.

Index Terms — Automatic Mold Polishing, Polishing Process Planning, Polishing Path Planning. If you have any suggestion, send e-mail to mjtsai@mail.ncku.edu.tw.

I. INTRODUCTION

Now a day, the mold industry should response to the human's need more quickly. However, in the manufacturing of molds, the most time consuming process is on the final stage – the polishing process. According to Lee [1] the polishing process of the mold not only affect the final appearance and the quality of the mold, it also occupies the total production time and cost up to 40%.

Almost all the polishing processes in the mold industry are conducted manually. The dust and noise produced by the polishing process are not good for human health. The monotonic nature of polishing process even lends itself to be rejected by workers. Besides, the polishing process needs exceptional skill that can only be taught by experienced master. For mass production of molds, automatic polishing of the mold is a necessary.

Weule and Timmermann [2] developed a mold polishing procedure and experimentally studied the optimal polishing parameters. Schmidt and Schauer [3] used the industrial robot for surface finishing of dies and molds. Lee, et. al. [1] developed of a polishing robot system. Lee, et. al. [4] used the CAM data for trajectory tracking of a polishing robot. Ahn, et. al. [5] developed a sensor integrated expert system for optimizing die polishing.

In our Lab., a serial of researches were conducted on automatic mold polishing. Juang [6] studied the general polishing process and the planar polishing path method for convex polygons. Lin [7] established an automatic robotic polishing system. Stone [8] used the Taguchi method to study the influence of the parameters and found optimal sets of polishing parameters. Chen [9] generated polishing path on free form surface. Wang [10] extended the process planning system into the plastic and cloth ring polishing process. Recently, Huang [11] studied the material remove rate under the influence of controllable polishing parameters. Two patents were obtained for the force/position control of the polishing tool [8][11].

However, those studies seem logically related but actually are piecemeal constructed. In this paper, system integration is conducted using the software engineering technique to construct the AMPS. The CAD data structure is designed for path and process planners. Fuzzy-neural is used to generate polishing schedule [8]. A dedicated 5 axes robot conducts the polishing tasks [11][12]. Experiments show that the polishing results satisfy the task requirements.

II. SYSTEM ANLYSIS

The main purpose of system analysis is to investigate the practical operation and find the specifications and the algorithm to achieve the goal for the software system.

A. Analysis of Polishing Process

The polishing process can be divided into four major steps: grinding, lapping, polish lapping, and polishing. The first step is considered as pre-process. ITRI [13] established a standard mold surface polishing process in Table 1. The basic arrangement of the polishing process along with different tool type and grin size is given. The polishing process planning in the AMPS is primarily based on this procedure. Several principles should be noticed during the arrangement of polishing process planning:

1) The polishing process should start with larger grain size and gradually go down to smaller grain size.
2) The polishing path should distribute evenly on the surface. No area should be over polished. Repeatedly path on the same surface will cause recess on the surface, as shown in Fig. 1.

3) The polishing path should pass through a point in as many directions as possible.
4) The polishing path should be free from collision with any object or to make undercut on the fillet surface, as shown in Fig. 2.

Table 1: The basic polishing process. [13]

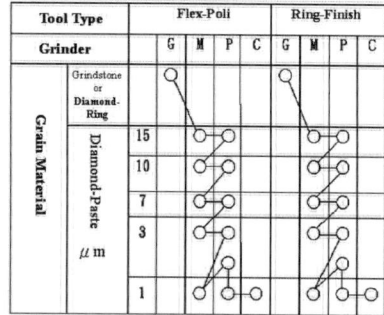

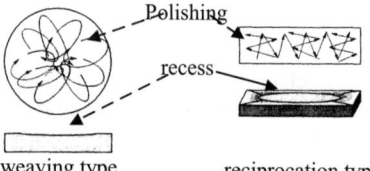

Fig 1. Wrong polishing path[14]

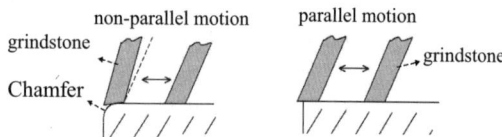

Fig 2. Wrong polishing action

B. System Main Functions

The system function should be well defined before the implementation be commenced. According to several years' of experiments and communications with the mold polishing industry, we design the AMPS to have four major functions. The four functions are developed into four main modules of the AMPS, As shown in Fig. 3.

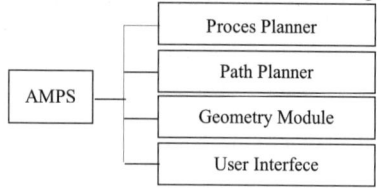

Fig. 3. System Main Functions

1) The Polishing Process Planner:

The AMPS should be capable to produce a process list according to the requirements of polishing task. The listed processes should be compliant to the experiments or the manual experience. So many parameters affect the polishing result. Users should simply input the goal of polishing and the information of the mold. AMPS should be able to generate a list of polishing procedure and output an "optimal" set of the controllable factors, such as the density of path, the grinder size of each step, etc. The "optimal" means the system can evaluate polishing efficiency to get a certain set of parameter. The process planner has six major components as shown in Fig. 4.

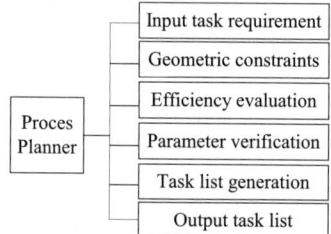

Fig. 4. Major components of the process planner

2) The Polishing Path Planner:

The path planner is to read and process the geometry data of molds. The path planner should be able to generate proper polish path according to the surface characteristic and polishing factors such as the tool type, the grinder size, etc. There are five major components for the path-planning module, as shown in Fig. 5.

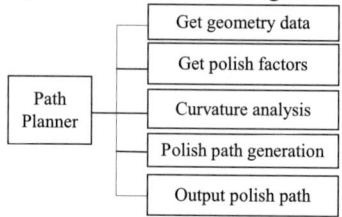

Fig. 5. Major components of the path planner

3) Geometry Data Input:

The robot should guide the polishing tool along the path while keeping normal to the surface. Geometry of the mold should be known for the system to generate the path. For convenient, three input methods are designed: file reading, keyboard input, and robot teaching, as shown in Fig. 6. The file input is implemented according to the IGES format. Some simple surfaces can be keyed in their respectively surfaces such as plane, cylinder, sphere, etc. can be taught via the robot. The module has 3D Display capability to verify the mold geometry and the path.

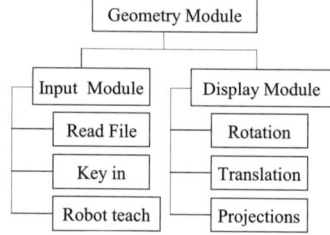

Fig. 6. Geometry input and display module

4) Human Interface

A user-friendly interface is a necessity for software systems. AMPS will provide interfaces for user to input the data, to display the mold geometry, to verify the generated path, and to monitor the polishing task being executed. Six categories of interface windows have been design for this module, as shown in Fig. 7.

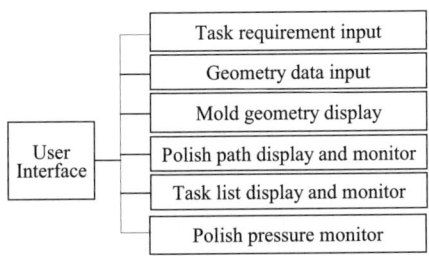

Fig 7. Human Interface module

C. Analysis of Data Flow

According to the required functions, the data flow in the system can be realized by the DFD (data flow diagram). DFD unveils the relationship of data among the software modules. DFD can be divided into different levels. The first level, as shown in Fig. 8, gives an overall picture of the system environment: the user and the polishing robot.

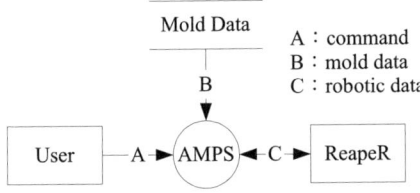

A: command, B: mold data, C: robotic path data
AMPS: Automatic Mold Polishing System
ReapeR: The 5 axes dedicated polishing robot

Fig. 8. Level_0 DFD

Level_1 and level_2 DFD expands the AMPS into internal process, the interface window, and the display window. The internal process is further divided into three: process planning, path planning, and the robotic execution modules. The display window includes geometry display and process list windows, as shown in Fig. 9.

III. SYSTEM STRUCTURE DESIGN

According to the required system function and data flow of AMPS, we can start to design the system structure.

A. Development environment

This system uses Visual C^{++} in PC for convenient access. Combined with MFC window architecture and the OpenGL graphic capability, we can design API (Window Application Interface) for displaying 3D mold geometry.

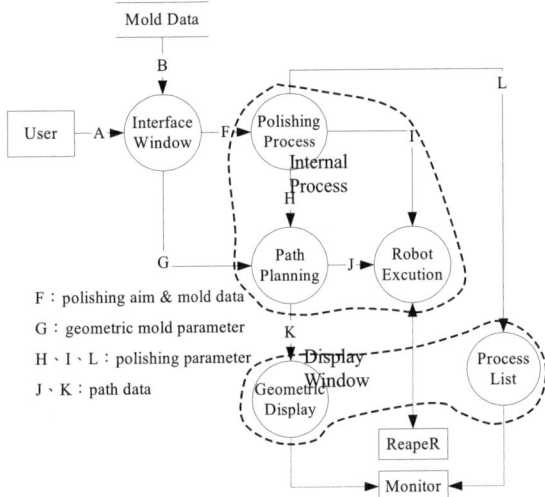

Fig 9. Level_2 DFD.

B. System Modules

The modular design has the advantages of easy integration, easy maintenance, and easy modification.

B.1 The Geometry Input and the Path Planner Modules

The geometric data can be input by three ways as described in previous section. The architecture of the module is shown in Fig. 10. The path planner considers the surface curvature and the grain size to decide the kind and the density of polishing path [12]. There are three kinds of polishing path. For planar surface, the cycloid path is used for greatest efficiency. For surfaces of revolution, fractal path is used, as shown in Fig. 11. For free form surface, scan line path is used.

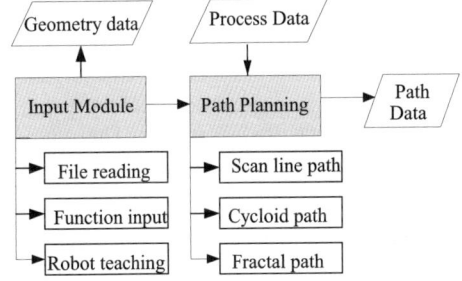

Fig. 10. The geometric input and path planner modules

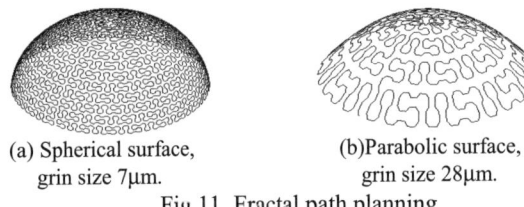

(a) Spherical surface, grin size 7μm.

(b) Parabolic surface, grin size 28μm.

Fig 11. Fractal path planning

B2. Process Planning Module

Process planner is the most important module of the AMPS. After input the task requirements, the module compares the material and initial surface roughness with the typical set of polishing curves. Each curve represents a set of polishing parameters. The polishing procedures are setup via finding the most efficient curve at a specific surface roughness. A task list is generated to produce an "optimal" set of polishing parameters for each procedure.

B.3 Display Module

As shown in Fig. 12, the module utilizes OpenGL to draw 3D mold geometry and the planned polishing path. The current robotic polishing path is simultaneously shown in real time for monitoring. The display screen also show the process list as well as the process data.

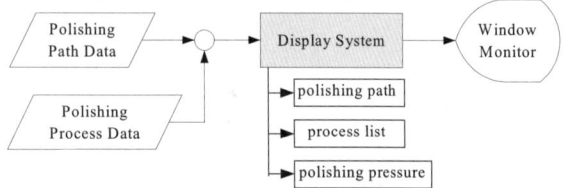

Fig. 12. Flowchart of display modules

IV. POLISHING PROCESS PLANNIGN

To build an automatic process planner, all the qualitative experience should be digitized into numbers. We should have an exact and clear measurement on the polishing parameters. The appraisal is surface roughness. The system uses the current surface roughness and the goal roughness to search for the optimal polishing curve of each procedure.

A. Determination of Polishing Parameters

The determination of polishing process is based on the previous studies that conducted by our laboratory from 1996 to 2001. The factors that affect the polishing results come from two sources. One is the kinematic parameters such as travel speed, feed rate, number of path, rpm of the tool, and tool weaving frequency, etc. The other is the setup of the polishing process, such as tool type, grinder type, grain size, and polishing pressure, etc.

B. Polishing Process Scheduling

Scheduling the polishing process is problematic because it is the most skillful item to be fulfilled by experienced worker. We studied the polishing process to find the procedure to switch from big grain size to smaller one. The relationship between the cycles of path in each grain size step and the surface roughness is called the R-n curve. The polishing efficiency is plotted vs. the cycle as the E-n curve. Then the R-E curves are computed using the above two curves. Given a start and a goal surface roughness, the process can be scheduled according the R-E curve by finding the most efficient curve for a surface roughness. Hence, a method that makes automatic process planning possible is developed.

B.1 Polishing Efficiency and Polishing Curves

The polishing efficiency is defined as: in a specific time and area, the amount of surface roughness reduced [8]:

$$P_E = -\frac{A\Delta R_a}{\Delta t} \quad (1)$$

Where P_E is the polishing efficiency, R_a is the average surface roughness, A is the area polished, and Δt is the time required to polish. Taguchi method is used to obtain optimal sets of polishing parameters. Experiments are then conduct using these sets of polishing parameters to generate the R-n curves. The R-n curve is fitted as:

$$R_a = ae^{bn} + c \quad (2)$$

Where n is the number of cycle executed, R_a is the surface roughness, and is the function of n. Coefficients a, b, and c are to be fitted. The efficiency to number of cycle (E-n) can be computed as:

$$P_E = -\frac{A\Delta R_a}{\Delta t} = -\frac{Aabe^{bn}}{T} = pe^{bn} \quad (3)$$

Let $p = -(Aab)/T$, A is the area polished (mm2), T is the total time required to a finish polish (sec). Then n is a function of P_E. The R-n and E-n curves of four typical polishing settings are given in Fig. 13. The relationship between the polishing efficiency and the surface roughness can be obtained as:

$$R_a = \frac{(-TP_E)}{Ab} + c = qP_E + c \quad (4)$$

Let $q = -T/(Ab)$, P_E is function of R_a, shown in Fig. 14.

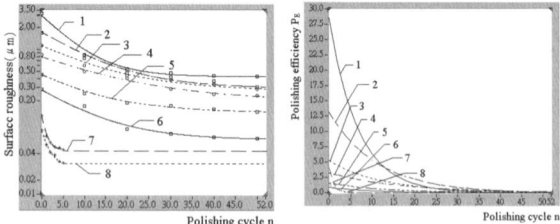

Fig. 13. The R-n curves and the E-n curves

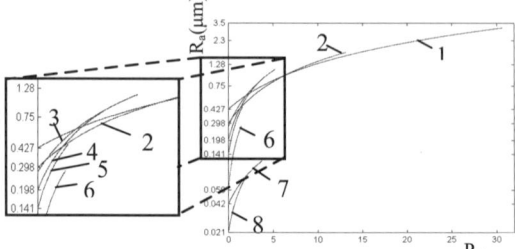

Fig. 14. The E-R curves

The three polishing curves are obtained from a mold that made of SKD-61 material. The polishing parameters are the typical "optimal" set according to the results from a serial of experiments using Taguchi Method. Some of the controllable polishing parameters are given in Table 2.

B.2 Automatic Process Planning

The polishing process should be changed while the surface roughness decreases. The change of process is according to the polishing efficiency. The E-R curve is very useful to schedule the process. For a surface roughness, there may be several parameter settings that can bring the roughness down to lower level, but only the right most one has the highest polishing efficiency. We just look down from a beginning surface roughness; find the right most curve; and this is the process setting for current process step. The next step is obtained when the roughness decreases to a certain level while another curve becomes the right most one. Then stop polishing with the current setting and change to the parameters of the new curve. Continue with this procedure until the desired surface roughness is reached. Then the whole polishing process can be scheduled.

Automatic polishing process planning can be fulfilled by solving the intersection of the curves or by searching the right most curves from the starting to the ending surface roughness. For example, if the starting surface roughness is 1.5μm, and 0.15μm is the desired, the process planning is shown in Fig. 15. In the figure, the black lines indicate the processes to be followed to get the goal roughness.

The problem left is how to decide the cycle needed for each polishing step. The cycles of each step required to execute may be computed using Equation (2). We just subtract the ending number n_e by the starting cycle number ns of each step and then round it to the integer. The formula can be obtained as:

$$n = n_e - n_s = \frac{\log(\frac{R_2 - c}{a})}{b} - \frac{\log(\frac{R_1 - c}{a})}{b} \quad (5)$$

Where R_1 is the starting roughness for each step, and R_2 is the ending roughness of that step.

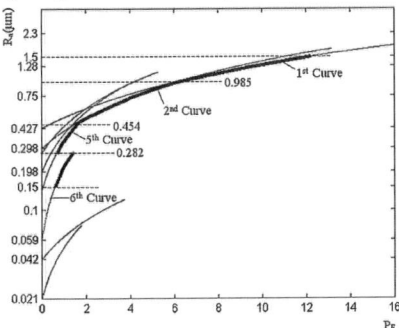

Fig. 15. The E-R curve for scheduling

V. THE USER INTERFACE DESIGN

The user interface design includes dialog windows, 2D, and 3D display window. Dialog windows give the user a convenient interface to input the task requirements. The 2D windows display the resultant task listing from the process planning. The listing displays all the polishing steps, the parameters for each step, and current polishing status. There are also two 2D windows to show the polishing pressure and force for monitoring in real time.

As shown in Figure 16, there are also two 3D windows to show the mold geometry and the generated polishing path. The object drawn in the 3D windows can be examined from different view angles and by zooming in and out. During the polishing process, the current robot path is superimposed on the mold geometry to compare with the polishing path.

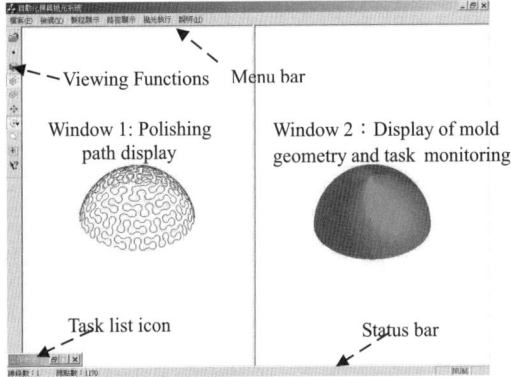

Fig. 16. 3D display windows.

VI. POLISHING EXPERIMENT

Experiments have been conducted on test mold to verify the system function. The mold has two surfaces, one is a plane and the other is a surface of revolution (parabolic). The material is SKD-61 with hardness HRC 45. The initial surface roughness is 0.457μm. The desired surface roughness is 0.040μm. A robot executes the polishing task with a force control mechanism, shown in Fig. 17. Fractal path on the revolving surface that produced from the path planner is shown in Fig. 11(b).

Figure 17: The polishing mechanism setup.

According to the task list, the polishing processes are conducted step by step. The stepwise polishing results are shown in Table 4. Three criteria for the system evaluation is defined and calculated. They are the polishing error ε, the stepwise fulfillment ∇_s, and the overall fulfillment ∇_a, as followings:

Polishing error: $\nabla_s = \dfrac{actual(R_1 - R_2)}{predicting(R_1 - R_2)} \times 100\%$ (6)

Stepwise fulfillment: $\varepsilon = \dfrac{actual(R_2) - predicting(R_2)}{predicting(R_1 - R_2)} \times 100\%$ (7)

Overall fulfillment: $\nabla_a = \dfrac{actual(R_{starting} - R_{ending})}{predicting(R_{starting} - R_{ending})} \times 100\%$ (8)

Since the resolution of the surface roughness instrument is 0.005μm Ra, the error appraisal is not a suitable judgment. The stepwise fulfillment and the overall fulfillment show how accuracy the AMPS can achieve. Fig. 18 shows the resultant mirror effect after automatic polishing.

Table 4. The stepwise polishing results

process	item	R_1 (μm)	Actual R_2 (μm)	Predicting R_2 (μm)	Δ (μm)	ε (%)	∇_s (%)	∇_a (%)
1	2nd set	0.457	0.453	0.454	-0.001	-33.3	133.3	133.3
2	5th set	0.454	0.287	0.282	0.005	2.91	97.09	97.14
3	6th set	0.282	0.135	0.120	0.015	9.26	90.74	95.55
4	7th set	0.120	0.078	0.076	0.002	4.55	95.45	99.48
5	8th set	0.076	0.041	0.040	0.001	2.78	97.22	99.76

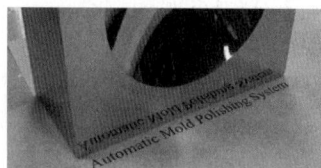

Fig. 18. The result of automatic polishing

VIII. DISCUSSION AND CONCLUSION

In this study, an automatic mold polishing system has been established. The main contributions of the AMPS to mold polishing industry are follows:

1) Using the software engineering technique, the computerization of the manual skill dependent polishing task is proven tangible.
2) The sequence of polishing process can be automatic scheduled by the computers.
3) The density of the polishing path is automatic adjusted based on the curvature and the grain size.
4) The software also provides display windows for user monitoring during the process execution.
5) Experiments show the fulfillment of polishing results can be achieved as high as 99.76%.

REFERENCE

[1] Lee, M. C., et. al., 1999, "Development of a user-friendly polishing robot system," Proc. 1999 IEEE/RSJ Int. Conf. on Intelligent Robots and System, v. 3, pp. 1914-1919。

[2] Weule, H., and Timmermann, S., 1990, "Automation of the Surface Finishing in the Manufacturing of Dies and Mold," Annals of the CIRP, v. 39, no. 1, pp. 299-303。

[3] Schmidt, J., Schauer, U., 1994, "Finishing of Dies and Moulds–an approach to quality-oriented automation with the help of industrial robots," Indu. Robot, (21)1, 28-31.

[4] Lee, M. C., et. al., 2001, "A robust trajectory tracking control of a polishing robot system based on CAM data," Rob. & Comp. Inte. Manuf., v. 17, pp. 177-183。

[5] Ahn, J.H., et. al., 2001, "Development of a sensor information integrated expert system for optimizing die polishing," Rob. & Comp. Int. Manuf., v17, pp. 269-276。

[6] Tsai, M. J., Juang, Jeng-Jei, Lin, Rung-Guei, 1996, "Process and Path Plannings for Automatic Mold Polishing System," Proc 13th Conf. Mech. Eng. CSME (Control), pp. 433-440.

[7] Tsai, M. J., Lin, Rung-Guei, Chen, Yuan-Jang, 1997, "Robotics Polishing Path Planning for Mold with Revolution Surface," Proc 14th National Conf. Mech. Eng. CSME (Manufacture & Production), pp. 39-45.

[8] Tsai, M. J., Stone, Duen-Jyh, Wang, Jiun-Chin, 1998, "Study on System Efficiency of Automatic Mold Polishing," Proc of the 15th Conf. Mech. Eng. CSME (Manufacture), pp. 307-314.

[9] Tsai, M. J., Chen, Yuan-Jang, 1999, "Automatic Polishing Path Planning for Mold with Free Form Surface," Proc of the 16th Conf. Mech. Eng. CSME (Control), pp. 211-218.

[10] Tsai, M. J., Wang, Jiun-Chin, Huang, Jian-Feng, 2000, "A Study on the Robotics Automatic Super Lapping Process for Mold with Revolution Surface," Proc of the 17th Conf. Mech. Eng. CSME (Control), pp. 305-313.

[11] Tsai, M. J., Huang, Hung-Lung, 2001, "Wearing Rate Consideration of Automatic Polishing Process on the Surface Model Reconstructed from NC Code," Proc of the 18th Conf. Mech. Eng. CSME (Control), pp.481-488.

[12] Jang, Jau-Lung, 2002, "Path and Process Plannings for an Automatic Mold Polishing System," Master Thesis of Dept of ME, National Cheng Kung University, Taiwan.

[13] Union Chemical Laboratory, ITRI, 1985, "Surface Finishing Technique of Molds," Industrial Development Bureau Ministry Economic Affairs

[14] Ou-Yang, Wei-Cheng, 1992, "Super finishing of Precision Molds," Chan-Hwa Sci. & Tech. Book Co., Taipei.

Locating and Checking of a BGA Pin's Position Using Gray Level

Chi-Wei Ruo   Ching-Long Shih

Department of Electrical Engineering
National Taiwan University of Science and Technology
43, Section 4, Keelung Road, Taipei, Taiwan, 106, R.O.C.
Email: shihcl@mail.ntust.edu.tw

Abstract

A machine vision system for SMT-mounting machine applications is usually a two-stage algorithm. It first measures the centroid and rotation angle of the SMD, and then checks each pin's area, position error, and grid coordinate location. In this paper, a set of complete procedures is proposed to locate and check the BGA image. During the locating procedures, we first calculate a threshold of the frame using an iterative threshold algorithm. If an object is found under this threshold, then the pin's area is calculated by the local threshold by using a momentum algorithm. After that, whether this object is a pin or not is decided by its neighbors' relative positions, the approximate rotation angle for finding the outer pins is calculated, and the centroid as well as the rotation angle of a BGA component is calculated by the rectangular least squares algorithm. The checking procedure also measures each pin's area using the momentum algorithm, then it calculates the radius of the moving sum using each pin's area, and finally measures the position error using a moving sum algorithm and judges each pin's type by gray level. The new method uses the gray level statistic information to solve the empty pad's problem and utilizes the symmetrical property of a circle to deal with the shape problem. Lastly, the CRC algorithm is used to check the correspondence between each pin and its pin type. This new method has a high accuracy and a low execution time. It can meet the crucial timing requirement of a high-speed SMT machine through experimental verification.

1. Introduction

The main function of an SMT (Surface Mounting Technology) mounting machine is to place the SMD (Surface Mounting Device) on the PCB through picking up and vision locating. Machine vision is used to measure the centroid and rotation angle on a variety of components. A machine vision program for an SMT-mounting machine application is a two-stage algorithm. It measures the centroid and rotation angle first, and then checks each pin's area, position error, and grid coordinate. The first stage must be robust enough to get the data when the picture is stained or corrupted, and the second stage must have both good accuracy and low execution time [1, 2].

There are already many research studies that focus on PCB checking or BGA PCB checking [3-7], but research that focuses on a mounting machine's vision problems are few. Barrtman [8] used a lead points' gray levels to calculate an SMD's centroid. Burel [9] applied neural networks to estimate an SMD's centroid. Hata used the variation of feature vectors to establish multiple dimension vectors and applied its difference to recognize an IC [10].

In this paper we propose a new algorithm for a high-speed mounting machine's BGA component locating and checking procedures. During the locating procedures, we first calculate a threshold of the frame using iterative threshold algorithm [14]. If an object is found under this threshold, then we calculate the pin's area by local threshold using a momentum algorithm [13]. After that, we decide whether this object is a pin or not by its neighbors' relative positions, and then we can calculate the approximate rotation angle for founding the outer pins, the centroid, and rotation angle of a BGA component which can be calculated by a least squares algorithm. When all pins are found, we use the new proposed least squares rectangle algorithm to calculate the BGA's centroid and rotation angle, which has a faster calculation speed and more robustness than using the 4 individual least squares lines method [3].

During the checking procedures, an iterative threshold is used to find some suspicious objects among a frame. If the distance from an object to its neighbors is one pitch and the lines that go through this object and its neighbors are perpendicular, then we can make sure that this object is a BGA pin. We propose a new method to measures the BGA pin's coordinate by a moving sum algorithm in the gray frame. The pin's type is also examined using its gray level statistic information.

All the binary algorithms in the locating and checking procedures are only used to judge the existence of a pin and to remove some fault objects, so that binary algorithms will not participate in the calculations which concern the numerical accuracy. Moreover, we put the geometrical information "BGA pin looks like a circle" into the algorithm to eliminate some noise from the BGA pin's image. Finally, each BGA pin type is verified by CRC algorithm[17], and each pin's position error is calculated. This proposed new BGA image locating and checking method has a high accuracy and a low execution time.

2. Measuring the center and rotation angle

The BGA locating algorithm first uses an iterative and momentum threshold algorithm to remove some fault noise, and then uses radial research to reduce the search time, and finally uses a least squares rectangle algorithm to measure the center and rotation angle of the BGA. We first present the key algorithm for obtaining the center and rotation angle of a BGA chip.

2.1 Least squares rectangle algorithm

We use outer rectangle pins here to calculate the position and then implement the least squares rectangle algorithm. This method first finds the pins on the BGA's outer rectangle and then classifies them into the groups U, D, R, and L. A rectangle can be described as four equations

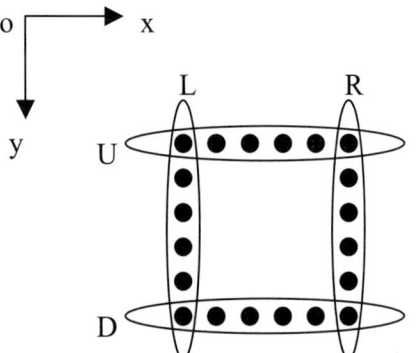

Figure 1. Least squares rectangle algorithm.

$$x + my + C_L = 0 \quad (1)$$
$$x + my + C_R = 0 \quad (2)$$
$$mx - y + C_u = 0 \quad (3)$$
$$mx - y + C_D = 0. \quad (4)$$

then

$$m = \frac{\sum_{p\in U} xy + \sum_{p\in D} xy - \sum_{p\in R} xy - \sum_{p\in L} xy - \frac{\sum_{p\in U}x \sum_{p\in U}y}{n_U} - \frac{\sum_{p\in D}x \sum_{p\in D}y}{n_D} + \frac{\sum_{p\in R}x \sum_{p\in R}y}{n_R} + \frac{\sum_{p\in L}x \sum_{p\in L}y}{n_L}}{\sum_{p\in U}x^2 + \sum_{p\in D}x^2 + \sum_{p\in R}y^2 + \sum_{p\in L}y^2 - \frac{(\sum_{p\in U}x)^2}{n_U} - \frac{(\sum_{p\in D}x)^2}{n_D} - \frac{(\sum_{p\in R}y)^2}{n_R} - \frac{(\sum_{p\in L}y)^2}{n_L}} \quad (5)$$

$$C_U = \frac{\sum_{P\in U}(y - mx)}{n_U} \quad (6)$$

$$C_D = \frac{\sum_{P\in D}(y - mx)}{n_D} \quad (7)$$

$$C_R = \frac{\sum_{P\in R}(-my - x)}{n_R} \quad (8)$$

$$C_L = \frac{\sum_{P\in L}(-my - x)}{n_L}, \quad (9)$$

where terms n_U, n_D, n_R, and n_L are the number of pins for each group U, D, R, and L, respectively. The centroid of the BGA is then located at

$$x_o = \frac{C_R + C_L + m(C_u + C_D)}{-2 - 2m^2} \quad (10)$$

$$y_o = \frac{-C_u - C_D + m(C_R + C_L)}{-2 - 2m^2}, \quad (11)$$

with a rotation angle of

$$\theta = \tan^{-1} m. \quad (12)$$

2.2. BGA locating flowchart

The program flowchart for BGA locating is shown as Fig. 2. First, we calculate a threshold of the frame using an iterative threshold algorithm [14]. Because the gray level's distribution range of a BGA pin is wider than the background's, the variance in the gray level of the BGA pins is bigger than the background gray level's variance. If the background is dark and the object is bright, then when we binarize this frame using the iterative threshold algorithm, the BGA pins' size will shrink. We find the outer bound by a radial search using this threshold.

If an object is found, then we calculate the pin's area by another local threshold using momentum algorithm [13] with the ROI being 1 pitch by 1 pitch. After that, we decide whether this object is a pin or not by the lengths and relative angles between its neighbors. Since a tin ball's diameter may change when the illumination changes, it is not reliable to judge an object by only using its area's size. A BGA pin must have at least 2 neighbors within its 4 adjacent neighbors and its neighbors must lie on the intersection of the grid lines.

We judge whether this object is a pin or not by the relative angles between 2 neighbors. When all pins are found, we calculate the approximated BGA rotation angle according to its neighbor's relative position. We then can get its outer rectangle pins using the approximate rotation angle. Finally, we calculate the rotation angle and position offset using the least squares rectangle method.

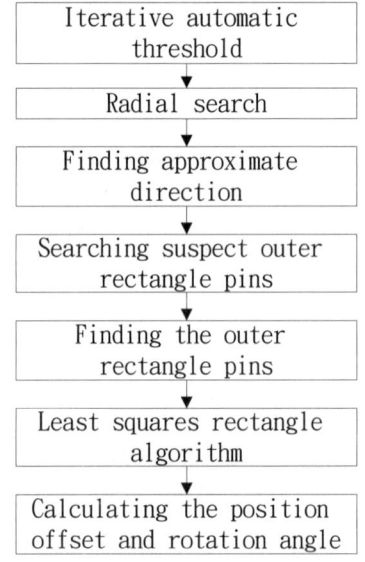

Figure 2. The program flowchart for BGA locating.

3. BGA image checking

In BGA image checking, the items checked for each pin are its area, center position error, and pin type on a grid coordinate position. The acceptable error distance square in this system is below $1\ Pixel^2$. The checking procedures first measure each pin's area using the momentum algorithm and labeling algorithm [16], and then decide whether this image is a null background, empty pad, or normal BGA pin by its gray level and area size (see Fig. 3). It calculates the radius of the moving sum under each pin's area and measures the pin's position error by the proposed sub-pixel moving sum algorithm (which will be discussed later). Finally it uses the CRC (cyclic redundancy check) algorithm [17] to check the correspondence between each pin and its pin type.

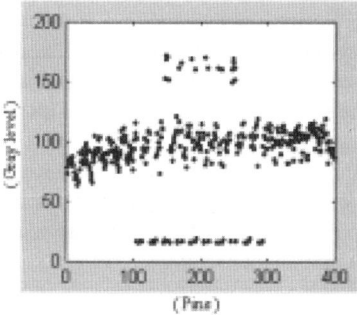

Figure 3. BGA pin's gray level: the brighter pins are empty pads, the lower parts are null but only background, and the normal pins' gray levels are between them.

3.1 Sub-pixel moving sum algorithm

Because a BGA pin looks like a circle, if a line passes through the circle, then we can calculate the moving sum along this line with the summation range of one BGA pin's diameter. The perpendicular line that goes through the midpoint of the maximum moving sum's interval must also go through the circle's origin. If we utilize this property to find out two independent perpendiculars, then the position of the BGA pin can be found at the intersection point.

The moving sum method achieves only pixel accuracy, but this accuracy is not good enough to satisfy the requirement and its accuracy must be improved to a sub-pixel's extent. The method to do that is using linear interpolation to get the gray levels between the two adjacent pixels and then calculate the maximum value of the sub-pixel moving sum. The moving sum $S(k)$ of n pixels along a line is defined as

$$S(k) = \sum_{i=k}^{n+k-1} G_i. \qquad (13)$$

As shown in Fig. 5, the solid circles represent the gray levels of the pixels, and the horizontal line's width represents a pixel's width. If the moving sum along a line has two or more maximums in different locations, then we choose the greatest k as its output. When the maximum of a moving sum is found at position x, then we can obtain

$$S(x) > S(x+1) \text{ and } S(x) \geq S(x-1);$$

and hence,

$$G_{x+n-1} > G_{x-1} \text{ and } G_x \geq G_{x+n}.$$

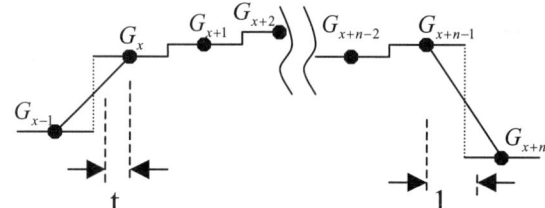

Figure 4. The moving sum along a line.

Because x is an integer, the fraction accuracy cannot be achieved. Therefore, we add t (where $0 \leq t < 1$) pixels' gray levels at the front of the moving sum and $(1-t)$ pixels' gray levels in the rear using a linear interpolation method as shown in Fig. 4, the fraction's moving sum $S_x(t)$ can be obtained as

$$S_x(t) = tG_x - \frac{t^2}{2}(G_x - G_{x-1}) + (1-t)G_{x+n-1}$$

$$+ \frac{(1-t)^2}{2}(G_{x+n} - G_{x+n-1}) + S(x) - \frac{G_x + G_{n+x-1}}{2} \qquad (14)$$

Because x is a constant, setting $S_x'(t) = 0$ yields

$$t_{max} = \frac{G_x - G_{x+n}}{G_x - G_{x-1} + G_{x+n-1} - G_{x+n}} \qquad (15)$$

and

$$0 < t_{max} \leq 1.$$

Therefore, the midpoint x_c of the maximum fraction moving sum's interval is

$$x_c = x + \frac{n}{2} - t_{max}. \qquad (16)$$

As a result, we have the following theorem.

Theorem 1

If n points integral moving sum $S(x)$ has a maximum, then n points fraction moving sum $S_x(t)$ must have a maximum at t_{max}, and $0 < t_{max} \leq 1$.

3.2 BGA pin's center position measurement

A single BGA pin's image is rather small and is often strained by noises. It is therefore necessary to calculate the moving sum from different directions to reduce the noise's influence. Choosing the 4 moving sum directions ($0°$, $90°$, $-45°$, and $45°$), and the directions of the perpendiculars are ($0°$, $90°$, $-45°$, and $45°$). Although 4 lines may generate 6 intersected points at most, we choose the point with the shortest distance among 3 lines as the output location. Selecting arbitrarily 3 lines among these perpendiculars, then an isosceles perpendicular triangle is generated. The shortest distance among the 3 lines is proportional to the triangle's shorter side length (Fig. 5).

Selecting 3 lines among 4 lines produces 4 situations. We calculate the short side length of the triangles for each situation and choose the shortest one, and then the circle origin can be found as shown in Table 1. The distances from this point to the 3 perpendiculars are all the same, but the distance to the fourth perpendicular is longer, and so the fourth perpendicular is treated as noise. A moving sum is sensitive to its accumulation's radius and each pin has a different area. Thus, we use its area to calculate its radius.

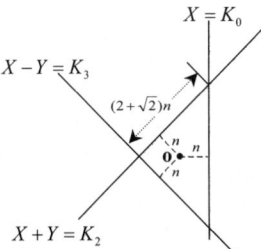

Figure 5. Isosceles perpendicular triangle, where the 4 perpendiculars at the midpoints be

$X = K_0$, $Y = K_1$, $X + Y = K_2$, and $X - Y = K_3$.

Table 1. The shortest side length and circle origin

Line removed	Triangles' shortest side length	Circle origin (X, Y)
0°	$\frac{\|2K_0 - K_2 - K_3\|}{\sqrt{2}}$	$\left(\frac{\sqrt{2}-1}{2}(K_2 + K_3) + (2-\sqrt{2})K_0, \frac{K_2 - K_3}{2}\right)$
90°	$\frac{\|2K_1 - K_2 + K_3\|}{\sqrt{2}}$	$\left(\frac{K_2 + K_3}{2}, \frac{\sqrt{2}-1}{2}(K_2 - K_3) + (2-\sqrt{2})K_1\right)$
45°	$\|K_0 - K_1 - K_3\|$	$\left(K_0 + \frac{-K_0 + K_1 + K_3}{2\sqrt{2}}, K_1 + \frac{K_0 - K_1 - K_3}{2\sqrt{2}}\right)$
−45°	$\|K_0 + K_1 - K_2\|$	$\left(K_0 + \frac{-K_0 - K_1 + K_2}{2\sqrt{2}}, K_1 + \frac{-K_0 - K_1 + K_2}{2\sqrt{2}}\right)$

3.3 BGA pin's arrangement check

Inside a BGA component, the pins are placed in a two-dimensional grid array, and it is necessary to check whether each pin is placed in the right grid coordinate. When we check a BGA pin using machine vision, the result is among one of 3 types: null, empty pad, and normal BGA pin. It is quite intrusive to save every pin's grid coordinate and type when checking, and this method causes two problems. First, it wastes memory space. Secondly, different components take different memory sizes. A database with a fixed size for each record is more reliable and faster, and the complexity of programming will become much easier.

The arrangement checking can be treated as an error-detecting problem. The most general algorithms for solving the error-detecting problems are parity check, checksum, and CRC. Among them, CRC has the highest accuracy. A 32-bit CRC is implemented here, and the fault acceptance rate is about $1/2^{32}$. Because the CRC's error-checking bytes have a fixed length, the programmer can use a database with a fixed size for each record. This method increases speed, reduces complexity, and the system's performance improves.

3.4 BGA image checking algorithm

The flowchart of checking a BGA image is shown in Fig. 6. The checking procedure first gets a single pin's image with the size of one pitch square, measures its area by a labeling algorithm, then decides whether this image is a null background, empty pad, or normal BGA pin by its gray level and area size, uses the moving sum algorithm to measure its center position, and then checks the pin's type on the pin grid coordinate position. After the calculation of each grid line's equation by the least squares error, we can obtain the intersection of these grid lines. We then calculate the pin's position error as the distance between measured pin's center position and one of the nearest grid line intersection (see Fig. 7). Finally, we decide whether it is a qualified component or not.

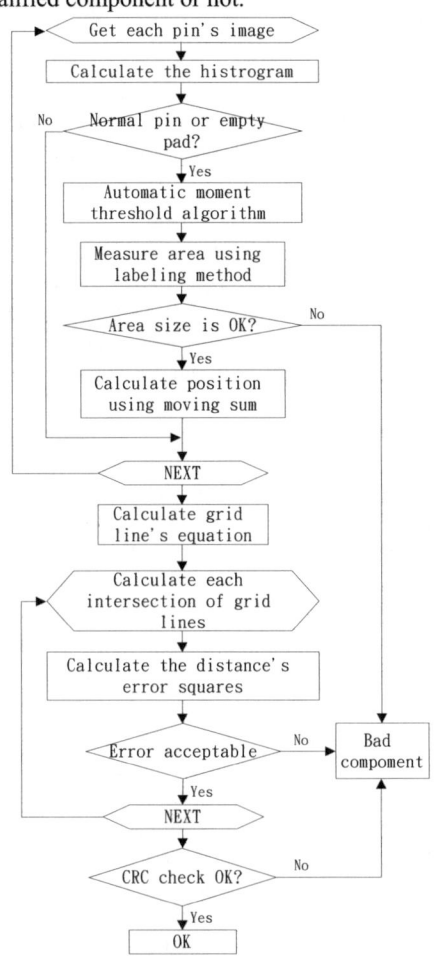

Figure 6. The program flowchart for BGA checking.

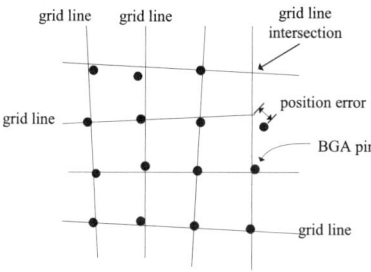

Figure 7. BGA pin position error.

4. Experimental results

The experimental image system's image capturing card is a Leutron PIC-PORT MONO H4 and the camera is a SONY XC55 progressive-scan camera, which has an extra shutter trigger input with a shutter speed range of 2 μsec to 250 msec. The Galil 1840 card is used for the system's motion control. To improve the efficiency, we capture the image while the SMT machine is still moving. We set the motion speed while capturing the image as 1m/sec and set the shutter speed as 10 μsec. Because the motion card has a sampling time of 250 μsec, we implement another motor-encoder capture card to latch the real time position. The motion card sends a capture signal to the image capturing card's extra shutter trigger input along with the motor-encoder capture card's latch input to capture the image and motor position simultaneously. This capture signal also starts a software interrupt to do the succeeding calculation works. The software is written in Visual C++6.0 and runs on the Windows NT 4.0. The execution time is measured by a PC K6III-400 computer and the total execution time is only 25.5 msec.

As shown in Fig. 8, when the moving sum algorithm is implemented, the maximum position error is reduced about 17~22%. Under a non-uniform illumination circumstance, the improvement is more significant than under a uniform illumination circumstance. The position error's accuracy using a fractional moving sum algorithm is about 0.1 $pixel^2$ in this system. If the position error is too small, then the improvement is saturated as shown in Table 2.

5 Conclusions

We have proposed herein a set of complete procedures for locating and checking the BGA image. In locating the BGA chip, the least squares rectangle algorithm has a faster calculation speed and better robustness than when using the traditional four individual least squares lines method. The moving sum algorithm in the gray frame is used to measure the positions of the BGA pins. This method has a lower sensitivity to image noise and has non-uniform illumination. Finally, the pin's arrangement check is largely simplified by using the CRC algorithm. The proposed methods are verified under the experimental test and their performances are both accurate and rapid. The BGA image system can filter out some noise; and it can also be used under a non-uniform illumination circumstance. The complete locating and checking procedures take only 25.5 milliseconds of execution time when we use a K6III-400 PC computer to place a 348-pin BGA component with an image size of 350x350.

Acknowledgement

This work was support by the National Science Council of Taiwan under Grant NSC 90-2212-E-011-045.

References

[1] G.T. Ayoub, "Machine vision in high-accuracy SMT autoplacers," *Circuits Manufacturing*, Vol. 30, No. 1, pp. 28-32, Jan. 1990.

[2] J. Woolstenhulme and E. Lubofsky, "Machine vision placement considerations," *Surface Mount Technology*, Vol. 14, No. 11, pp. 62-66, 2000.

[3] D.W.Capson and M.C.Tsang, "An experimental vision system for SMD component placement inspection," *IEEE Industrial Electronics Society 16th Annual Conference*, Vol. 1, pp. 815-820, 1990.

[4] C.H. Yeh and D.M. Tsai, "A rotation-invariant and non-referential approach for ball grid array (BGA) substrate conducting path inspection," *International Journal of Advanced Manufacturing Technology*, Vol. 17, No. 6, pp. 412-424, 2001.

[5] Y. Hara, N. Akiyama, and K. Karasaki, "Automatic inspection system for printed circuit boards," *IEEE Trans. on Pattern Analysis and Machine Intelligence*, Vol. 5, No. 6, pp. 623-630, 1983.

[6] G.A.W. West, "A system for the automatic visual inspection of bare-printed circuit boards," *IEEE Trans. on Systems, Man, and Cybernetics*, Vol. 14, No. 5, pp. 767-773, 1984.

[7] B. Benhabib, C.R. Charette, K.C. Smith, and A.M. Yip, "Automatic visual inspection of printed circuit boards: an experimental system," International Journal of Robotics and Automation, Vol. 5, No. 2, pp. 49-58, 1990.

[8] G. Burel, F. Bernard, and W.J. Venema, "Vision feedback for SMD placement using neural networks," *Proceedings of 1995 IEEE International Conference on Robotics and Automation,* Vol. 2, pp. 1491-1496, 1995. [9] J.P. Baartman, A.E. Brennemann, S.J. Buckley, and M.C. Moed. "Placing surface mount components using coarse/fine positioning and vision," *IEEE Trans. on Components Hybrids & Manufacturing Technology*, Vol. 13, No. 3, pp. 559-564, 1990.

[10] S. Hata, K. Hagimae, S. Hibi, and T. Gunji, "Assembled PCB visual inspection machine using image processor with DSP", *IECON'89 15th Annual Conference of IEEE Industrial Electronics*, Vol. 3, pp. 572-577, 1989.

[11] S. Belkasim, A. Ghazal, and O. Basir, "Edge enhanced optimun automatic thresholding," *Proceedings of 2000 ICS:Workshop on Image Processing and Pattern Recognition*, pp.78-85, 2000.

[12] J. Kittler and J. Illingworth, "An automatic threholding algorithm and its performance," *In Proc.*

seventh Int. Conference, Pattern recognition, Vol. 1, pp. 287-289, 1984.
[13] W.H. Tsai, "Moment-preserving thresholding : a new approach," *Computer Vision, Graphics & Image Processing*, Vol. 29, No. 3, pp. 377-393, 1985.
[14] T. W. Ridler, S. Calvard, "Picture thresholding using an iterative selection method," *IEEE Trans. on Systems, Man, and Cybernetics*, Vol. 8, No. 8, pp. 630-632, 1978.
[15] P. K. Shaoo, S.Soltani, and A.K.C. Wong, "A survey of thresholding techniques," *Computer Vision, Graphics, Image Processing*, Vol. 41, No. 2, pp. 233-260, 1988.
[16] R. Jain, R. Kasturi and B.G. Schunck, *Machine vision*, McGRAW-HILL,1995.
[17] J.E. Maze and B.R. Saltzberg, "Error-burst detection with randem CRC's," *IEEE Trans. on Communications*, Vol. 39, No. 8, pp.1175-1178, 1991.

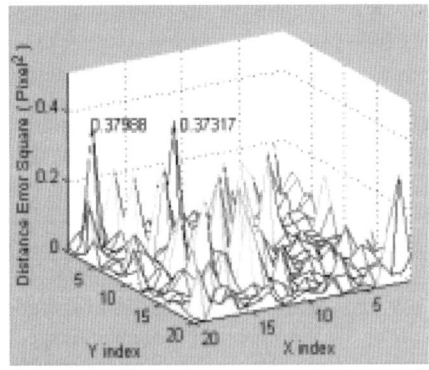
(a) uniform illumination with moving sum

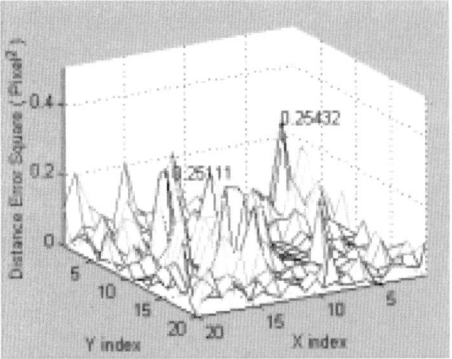
(b) non-uniform illumination with moving sum

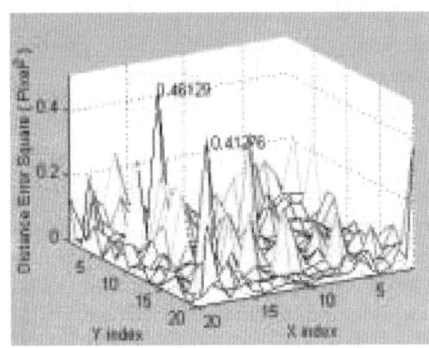
(c) uniform illumination without moving sum

(d) non-uniform illumination without moving sum

Figure 8. The experimental results of BGA pin position error.

Table 2. Error reduction percentage

Maximum pins	UNIFORM ILLUMINATION			NON-UNIFORM ILLUMINATION		
	Maximum pin error on average ($pixel^2$)		Error reduction	Maximum pin error on average ($pixel^2$)		Error reduction
	without moving sum	with moving sum		without moving sum	With moving sum	
1	0.461	0.380	17.6 %	0.328	0.254	22.5 %
2	0.437	0.377	13.8 %	0.327	0.253	22.7 %
5	0.380	0.325	14.4 %	0.291	0.246	15.6 %
10	0.312	0.294	5.6 %	0.256	0.236	7.9 %
20	0.254	0.248	2.6 %	0.214	0.210	2.0 %

High efficient robotic de-palletizing system for the non-flat ceramic industry

J. Norberto Pires
Mechanical Engineering Department
University of Coimbra
norberto@robotics.dem.uc.pt

Sérgio Paulo
Roca Cerâmica e Comércio, SA - PORTUGAL
sergio_paulo@roca.net

I. Abstract and Introduction

In this paper we describe a system designed to be used on a shop floor of a factory that produces non-flat ceramic products, where an extensive mixture of human and automatic labour is present. Today manufacturing setups rely increasingly on technology. It is common to have all sources of equipment on the shop floor, commanded by industrial PCs or PLCs connected by an industrial network to other factory resources. Also, the production systems are becoming more and more autonomous requiring less operator intervention in every day normal operation. That means using computers for controlling and supervision of the production systems, industrial networks and distributed software architectures [1]. It means also designing application software that is really distributed in the shop floor, taking advantage of the flexibility installed by using programmable equipment.

Non-flat ceramic products are commonly used in our homes and are mainly associated with personal care tasks. The industrial production of those ceramic products poses several problems to industrial automation, namely if robots are to be used. Basically, those problems arise from the characteristics of the ceramic pieces: non flat objects with high reflective surfaces, very difficult to grasp and handle due to the external configuration, very heavy and fragile, extensive surface sensitive to damage, high demand of quality on surface smoothness, etc. Also, the production setups for this type of products require very high quality and low cycle times, since this is a large scale industry that will only remain competitive if production rates are kept. Another restriction is related with the fact that this industry changes products very frequently, due to fashion tendencies in home decoration, etc. Also, there is an extensive mixture between automatic and human labour production, which is a difficult problem since Human-Machine Interfaces (HMI) are very demanding and a key issue in modern industrial automation systems.

The paper describes a prototype developed to de-palletize non-flat ceramic pieces in the final stage of production, just after they leave the high temperature oven, feeding the final human operated inspection lines. The prototype is installed at ROCA, a Spanish company operating in Portugal, and works with all their models of toilets and bidets.

The rest of the paper is organized as follows: section 2 briefly describes the problem and the defined objectives. Section 3, describes the solution adopted and briefly outlines the main characteristics of the software used for development. Section 4 explores the functionality of the obtained system, taking special attention to the HMI software. Finally conclusions are drawn in section 5.

II. Objectives

The main objective was to build a robotic system that could be used to feed the final inspection lines (Fig. 1). The system should be able to work with pallets composed by 4 levels of ceramic pieces, 8 pieces per level placed in a special order to keep pallet equilibrium, and with levels separated with pieces of hard paper. The rule used to arrange the pieces in the pallet is to place them alternatively one up – one down, starting from the ground level, then swap to one down – one up in the next level (Fig. 2) and keep the procedure in the proceeding levels. Levels are numerated from up to bottom, i.e., level 1 is the top level and level 4 is the bottom level.

Actually pallets are assembled manually by operators at the end of the high temperature oven. This means that the robotic system must be tolerant with possible medium-large palletizing errors, coming from misplaced pieces both in position and orientation, and showing also significant variations from level to level. Another important thing is that pallets are fed into the system by human operators using electro-mechanic pile drivers, which also introduces some variation in the pallet. Sometime in the future AGVs will be use to fulfill the task, reducing considerably the variations introduced and increasing efficiency of the system.

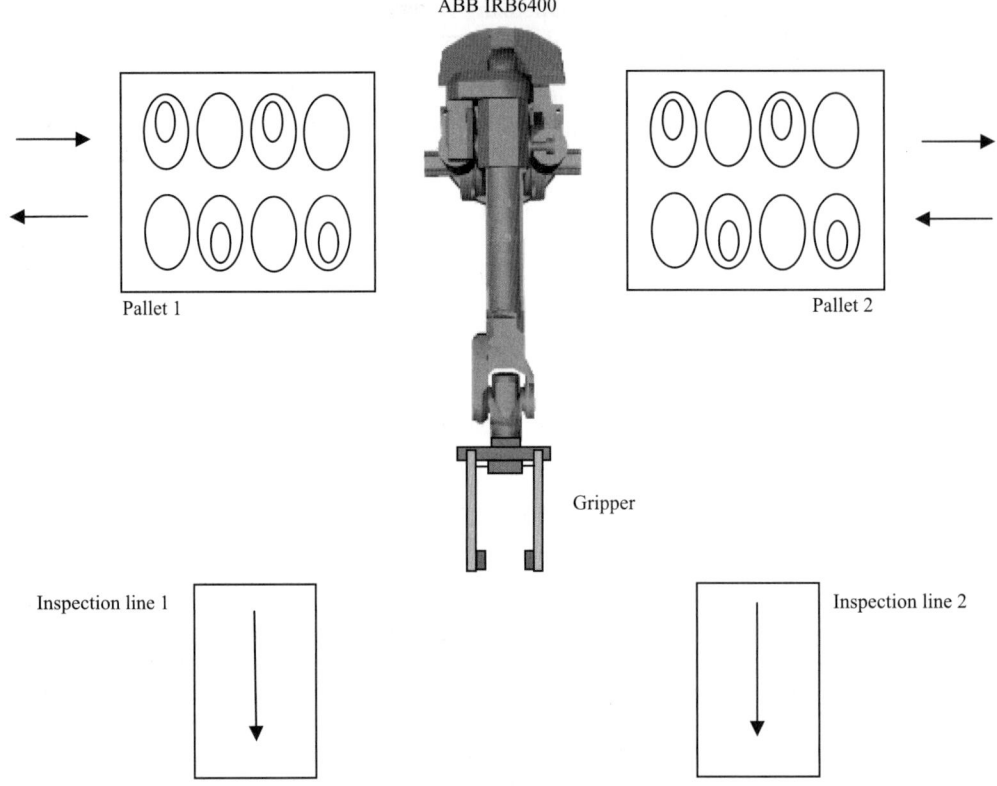

Figure 1 – Basic layout of the system.

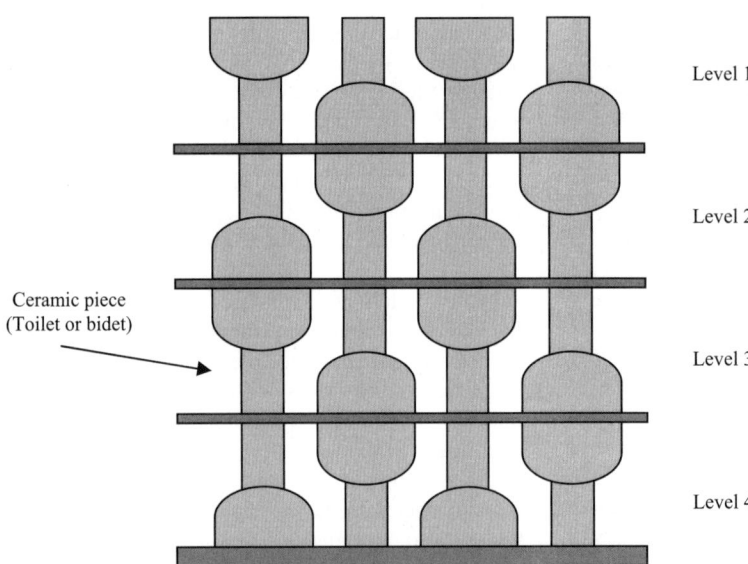

Figure 2 – Organization of ceramic pieces in the pallets.

Other important requirements include:

1. Possibility to easily introduce new product models;

2. Possibility to parameterize the operation on each model, so that best performance per model is achieved;

3. Possibility to change model under production without stopping production;

4. Possibility to monitor production using a graphical interface;

5. Obtain medium cycle times of about 12 seconds per piece, which means a new piece in each inspecting line every 24 seconds.

All these objects were addressed and a brief comment about each of them will be given in the next sections.

III. Software

Having these objectives in mind it was decided to operate the system using an external personal computer, using the teach pendant of the robot only for a few special routines not necessary in every day normal operations. Client-server software architecture was adopted. The robot controller software works as a server, exposing to the client a collection of services that constitute its basic functionality. A collection of services was designed to fulfill all the tasks required to the system, so that they could be called from the PC (Fig. 3).

The software architecture used in this work, was presented in detail elsewhere [2,3], and is distributed using a client-server model, based on software components developed to handle equipment functionality. Briefly, when we want to use some kind of equipment from a computer we need to write code and define data structures to handle all its functionality. We can then pack the software into libraries, which are not very easy to distribute being language dependant, or build a software control using one of the several standard architectures available (preferably ActiveX or JAVA). Using a software control means implementing methods and data structures that hide from the user all the tricky parts about how to have things done with some equipment, focusing only on using its functions in an easy way. Beside that, those components are easily integrated into new projects built with programming tools that can act as containers of that type of software controls, i.e., they can be added to new projects in a "visual" way.

IV. System Operation

The system is completely operated using a graphical panel running on the PC, built using the above mentioned ActiveX controls in Visual C++ (Fig. 4). When the system is started, the operator needs only to specify what product model will be used in each pallet, and if first pallets are fully assembled. Sometimes, due to production there are some non-fully assembled pallets on the shop floor, and there is the need to introduce those pallets in the system. To be able to do that, the software allows operator to specify the position and level of the first piece. That is however only possible on the first pallet, because the system resets definitions to the next pallets to avoid accidents, i.e., proceeding pallets are assumed to be fully assembled.

When the operator commands "automatic mode" the robot approaches the selected pallet in direction to the actual piece, searchers the piece border using laser sensors placed on the gripper, and fetches the ceramic piece. After that the robot places the piece in the first available inspection line, alternating inspection lines if they are both available, i.e., the robot tries to alternate between them, but if the selected one is not available then the other is used if available. If both inspection lines are occupied the robot waits for the first to became available. Figure 4 shows the interface used by the operator to command the system and monitor production. It shows the commands available, and the on-line production data that enables operators to follow production. All commands and events are logged into a log file, so that production managers can use it for production monitoring, planning, debugging, etc. The system uses also a database, organized in function of the model number, where all the data related to each model is stored. That data includes type of the piece, dimensions, height where the gripper should grab the piece, average position of the first piece of the pallet, height of the pallet, etc. Accessing and updating the database is done in "manual mode", selected in the PC interface. There is a "teaching" option that enables operator to introduce new models and parameterize the database for that model (Fig. 5). When that option is commanded, the robot pre-positions near the pallet and the operator can jog the robot using function keys to the desired position/orientation. Basically the de-palletizing operation is preformed step-by-step and the necessary parameters acquired in the process, inquiring the operator to correct and acknowledge when necessary. The operator is only asked to enter the "model number", and to teach the height and the width of the piece. The rest is automatic. After finishing this routine the model is introduce into the database, and the system can then work with that model number.

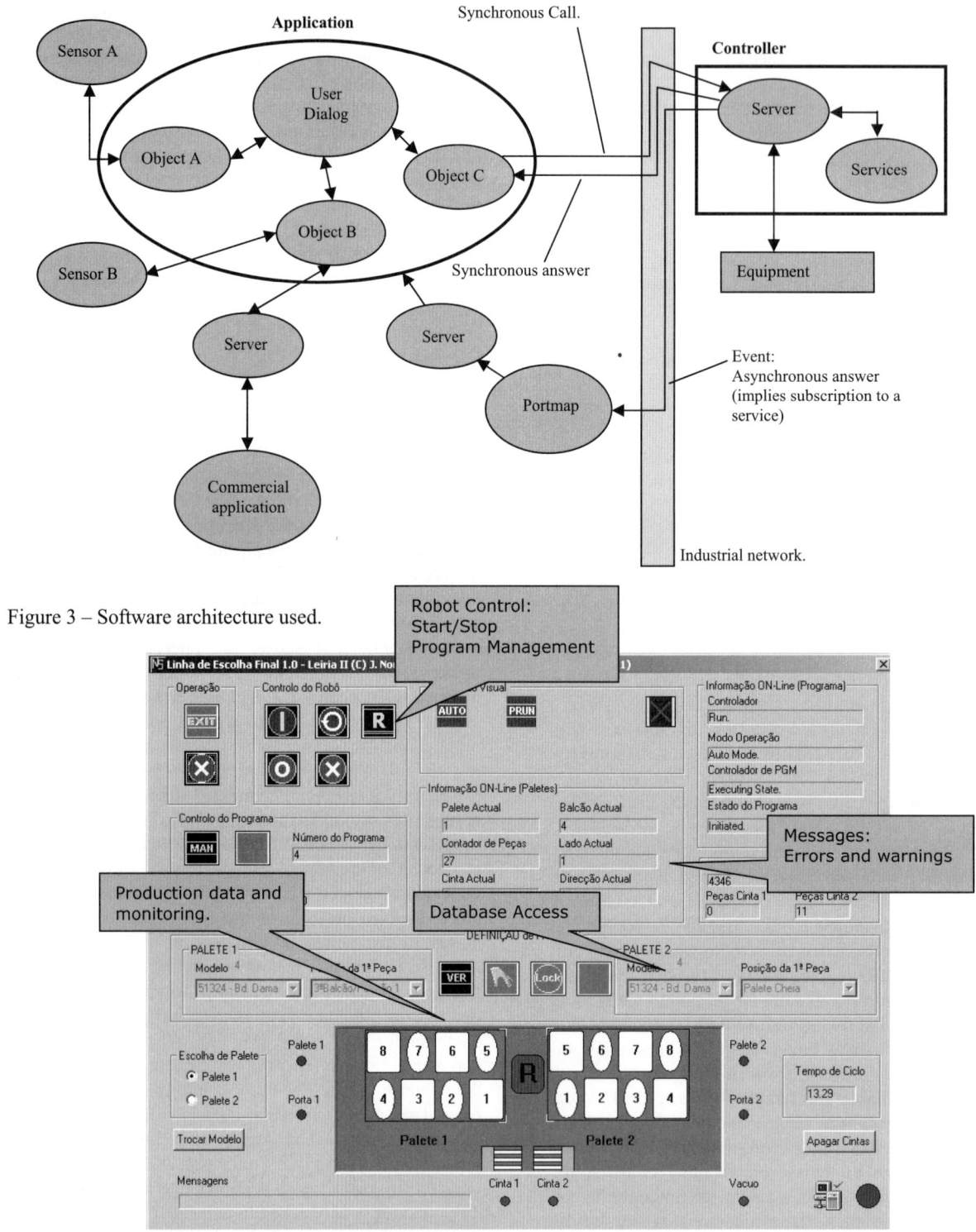

Figure 3 – Software architecture used.

Figure 4 – Interface used by the operator (labels in Portuguese).

V. Discussion

Figure 5 shows some aspects of the system in operation. The system showed to be very easy to operate, and the interface with operators showed to be efficient and easy to use. Operators adapted quickly and like to use it. One of the most important things about introducing robots into production systems, that share also a very high degree of human labour, is the interface between machines and humans. In some sense the interface should be easy to operate, but it should be designed in a way that there are no sudden moves, i.e., operation must start or resume slowly and take always paths well known by the operator. Predictability is an important thing. Safety is another important thing, since humans tend to adapt quickly and relax safety procedures. Because of that, intrusion into the cell must always act on the emergencies of the system as quickly as possible (using class 5 safety equipments), and should be logged for proper action. We adopted to add a special procedure for after emergency re-start that is time consuming, and requires operator login to resume. In that case, all emergencies are logged with the ID of the operator which prevents operators of being too much confident.

The average cycle time obtained is lower than 12 seconds, which complies with the requirement of one piece on each inspection line every 24 seconds.

Another important thing is the time to production, when a new model is introduced. Adding a model should be done easily by any operator, and should not take too long, so that normal production would not be affected. With the actual system, a new model is introduced and ready to be handle by the system in less than 5 minutes, including verification tests used to check if programming was well done.

The ideas presented were implemented on the system using an Ethernet network, connecting an industrial robot to a commanding PC and to the other resources of the shop-floor. The results show that using available technology and distributing functionality over the several components of the system, it is possible to explore the flexibility offered by actual automation equipment. Robots are a special case since they can be used to perform human like tasks and are programmable devices. Nevertheless, they are currently somehow limited, since the interfaces are non-standard and limited in functionality, difficult to use and program for advance remote applications. One way to easily distribute and integrate the code into new applications is by using software components, i.e., using ActiveX components, Java Beans, Corba components, etc. We usually use ActiveX do to the fact that they integrate well with Win32 environments. In fact we built several components for a variety of equipment (PLCs, CCD cameras, force/torque sensors, etc) using this language [2,3,7,10,11,12]. But other technologies could be used; the purpose here is on components and on integration with the environment chosen for operation, not in discussing the possibilities of each technology. Since we use win32, mainly Windows NT and 2000, which is an accepted standard in industry, ActiveX is somehow privileged because it was specially built for those environments and is based on DCOM like the operating system [8,9]. There is currently a discussion among researchers about open source software, as a way to give access to developers and allow implementation of their ideas. This is particularly necessary on research environments, where a good access to resources is needed. The problem is real and somehow urgent, since many features currently common on laboratories didn't reach industry yet due to system interface limitations (robot control systems are closed and with deficient software interfaces). Manufacturers should provide powerful APIs to enable user access to system resources, tailoring its behavior, from a remote computer. And that is a good business decision, because it enables third-party solutions for high demanding and non-traditional applications, which cannot be handled properly by general corporate products. That type of openness will be more important than open source, if it means support and full access to the robotic system (using currently available and accepted standards) by advanced users, which will contribute to the dissemination of the technology to non-traditional, and SME based applications. These types of applications are usually very tricky, very demanding on computer control and support, special sensors [11,12], etc., but not sufficiently important in the number of robots sold to motivate the interest of the robot manufacturers. That is the case of the example presented in this paper.

Figure 5 – Pictures of the system.

VI. Conclusions

In this paper an industrial robotic system designed for handling non-flat ceramic products was presented and briefly explained. In the process our approach to design software to this type of systems was introduced and some aspects discussed. Discussion about advantages of this type of approach was briefly outlined, along with recommendations that manufacturers of these types of automation equipments should follow in the near future.

The obtained system proved to be easy to use, efficient and integrated well in the existing production setup, which is fundamental in this type of environments with an extensive mixture between automatic and human labour.

VII. References

[1] Kusiak, A, "Computational Intelligence in Design and Manufacturing", John Wiley & Sons, 2000.

[2] Pires JN, Sá da Costa JMG, "Object Oriented and Distributed Approach for Programming Robotic Manufacturing Cells", IFAC Journal on Robotics and Computer Integrated Manufacturing, February 2000.

[3] Pires, JN, "Object-oriented and distributed programming of robotic and automation equipment", Industrial Robot, An International Journal, MCB University Press, July 2000.

[4] Bloomer J., "Power Programming with RPC", O'Reilly & Associates, Inc., 1992.

[5] Halsall F., "Data Communications, Computer Networks and Open Systems", Third Edition, Addison-Wesley, 1992.

[6] RAP, Service Protocol Definition, ABB Flexible Automation, 1996.

[7] Pires, JN, "EmailWare: A tool for e-manufacturing", Assembly Automation Journal, MCB University Press, Oxford, May 2001.

[8] Box D., "Essential COM", Addison-Wesley, 1998

[9] Rogerson D., "Inside COM", Microsoft Press, 1997.

[10] Pires JN, "Interfacing Robotic and Automation Equipment with Matlab", IEEE Robotics and Automation Magazine, September 2000.

[11] Pires, JN, "Force/torque sensing applied to industrial robotic deburring", Sensor Review Journal, MCB University Press, July 2002.

[12] Pires, JN, et al, "Object oriented and distributed software applied to industrial robotic welding", Industrial Robot, An International Journal, Volume 29, Number 2, 2002.

Multi-Objective Differential Evolution and Its Application to Enterprise Planning

Feng Xue
Arthur C. Sanderson
Robert J. Graves
Rensselaer's Electronics Agile Manufacturing Research Institute
Rensselaer Polytechnic Institute, Troy, NY 12180, USA
{xuef, sandea, graver}@rpi.edu

Abstract — Agility is important to modern enterprises. The effective coordination of large numbers of potential suppliers and manufacturers, demands a scientific methodology rather than just practical experience to make decisions on supply manufacturing planning problems. Particularly in cases where multiple decision objectives are important to process planning, empirical decisions are insufficient. This paper introduces formal methods to solve such multi-objective decision problems involved in general supply manufacturing planning, and specifically describes the extension of differential evolution methods to discrete problem domains. An enterprise planning problem with two objectives---cycle time and cost is used as a principal example. Such multi-objective optimization problems usually are very large and nonlinear. In this paper, the concept of differential evolution, which is well-known in single-objective continuous domain for its fast convergence and adaptive parameter setting, is extended to the discrete domain by introducing greedy probability, mutation probability, and crossover probability. Moreover, this concept is extended to discrete multi-objective optimization problem. The proposed discrete multi-objective differential evolution, or D-MODE algorithm is applied to obtain Pareto solutions of this general planning problem. A practical example in the electronics industry is used as an illustrative example to demonstrate the effectiveness of the proposed D-MODE.

1. Introduction

Advances in information technologies are driving fundamental changes in the processes and organizations of global enterprises. Innovations in software, networks, and database systems enable widely distributed organizations to integrate activities, share information, collaborate on decisions, and execute transactions. As a result, it is becoming increasingly uncommon for the creation of product and services in isolation, and they are being realized based on the creation of strategic and dynamic partnerships between suppliers, contract manufacturers, and customers. However, as the numbers of these distinct entities increase and they get more distributed, the complexity of forming efficient partnerships grows; it becomes more difficult to make ideal assignments with respect to multiple criteria including cost, time, and quality. Fundamental to this complexity is that each assignment has the potential to affect overall product cost, and product realization time, and therefore assignments cannot be considered independent of one another. Due to this complexity it is increasingly difficult to make these dynamic partnerships purely on the basis of prior experience, and it becomes necessary to develop efficient decision-making systems that can automate significant portions of the overall decision task.

As many other engineering applications, this supply manufacturing planning decision-making involves multiple criteria. The ideal solution is that one assignment can be identified which optimizes all criteria simultaneously. However, such ideal solutions can never be obtained in practical applications where outcome criteria may be fundamentally inconsistent. Optimal performance according to one objective, if such an optimum exists, often implies unacceptably low performance in one or more of the other objective dimensions, creating the need for compromise to be reached. In this paper, we consider the identification of multiple solutions that may be used to guide the final decision process.

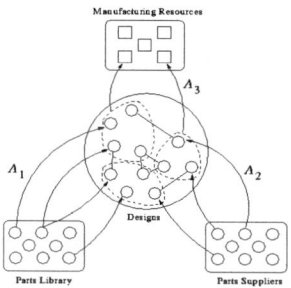

Figure 1: Structure of the design, supplier, manufacturing, planning decision problem. Lines with arrowheads indicate assignments. Dashed lines indicate aggregates. Identical parts in various designs have solid lines between them.

A model of the problem of integrated design, supplier and manufacturing planning for modular products where suppliers and manufacturing resources are network distributed is shown in Figure 1 and described in

[15][16][17]. This planning problem consists of three assignment problems (A1, A2, A3). The assignment problem A1 is the assignment of parts (from a parts library) to a design that satisfies a predetermined functional specification. Multiple designs that satisfy the functional specification are possible. The assignment problem A2 is the assignment of suppliers (from a list of available suppliers) who will supply the parts in a design, and the assignment problem A3 is the assignment of designs to available manufacturing resources. Each of these assignments contributes to overall product cost and product realization time, and has nonlinear (cannot be evaluated as weighted sums) effects on these measures. More detailed aggregation function and the related time and cost components can be found in [15]. A heuristic aggregation to combine product cost $C(x)$ and product realization time $T(x)$ for a complete design-supplier-manufacturing assignment x has been used for evaluation in this prior work

However, selection of an appropriate model of aggregating these two objectives and corresponding parameter to meet practical requirement is not possible in general. Each model would need to appropriately incorporate the nature of the problem itself and the preference structure of decision maker. Therefore, multi-objective optimization techniques must be developed.

In mathematical notation, a multi-objective optimization problem (MOOP) can be loosely posed as (without loss of any generality, minimization of all objectives is assumed):

$$\min Z(\mathbf{x}) = \begin{bmatrix} z_1(\mathbf{x}) \\ z_2(\mathbf{x}) \\ \vdots \\ z_k(\mathbf{x}) \end{bmatrix}, \quad \mathbf{x} \in \Omega$$

where $\Omega = \{\mathbf{x} \mid h(\mathbf{x}) = 0, g(\mathbf{x}) \leq 0\}$, and $\mathbf{x}$ is decision variable of dimension n. $Z: \Re^n \to \Re^k$, $h: \Re^n \to \Re^{m1}$, $g: \Re^n \to \Re^{m2}$, k is the number of objectives, $m1$ and $m2$ are the number of equality and inequality constraints, respectively.

In practical application, there is no solution that can minimize all the k objectives simultaneously. As a result, multi-objective optimization problems tend to be characterized by a family of alternatives that must be considered equivalent in the absence of information concerning the relevance of each objective relative to the others. These alternatives are referred to as Pareto optimal solutions, which have the same meaning with efficient, non-inferior, or non-dominated solutions. A Pareto optimal solution is defined as follows:

Definition: The vector $Z(\hat{\mathbf{x}})$ is said to dominate another vector $Z(\bar{\mathbf{x}})$, denoted by $Z(\hat{\mathbf{x}}) \prec Z(\bar{\mathbf{x}})$, if and only if $z_i(\mathbf{x}) \leq z_i(\mathbf{x})$ for all $i \in \{1, 2, \cdots, k\}$ and $z_j(\mathbf{x}) < z_j(\mathbf{x})$ for some $j \in \{1, 2, \cdots, k\}$. A solution $\mathbf{x}^* \in \Omega$ is said to be Pareto optimal solution for MOOP if and only if there does not exist $\mathbf{x} \in \Omega$ that satisfies $Z(\mathbf{x}) \prec Z(\mathbf{x}^*)$.

Evolutionary algorithms have gained a lot of interest on optimization (single objective) and learning area and have been applied to various practical problems. In recent 15 years, a new area of evolutionary multi-objective optimization has grown considerably. Evolutionary algorithms deal simultaneously with a set of possible solutions. This characteristic allows to find an entire set of Pareto optimal solutions in a single run of the algorithm, instead of have to perform a series of separate runs as in the case of the traditional mathematical programming techniques. Additionally, evolutionary algorithms are less susceptible to the shape or continuity of the Pareto front, whereas these two issues are concerns for mathematical programming techniques.

The Vector Evaluated Evolutionary Algorithm (VEEA) was the first practical approach for multi-criteria optimization using EAs, in which Schaffer extended Grefenstette's GENESIS program to include multiple criteria [11][12]. In this scheme, N/k sub-populations of equal size are assigned one to each criterion (where k is the number of criteria and N is the total population size). A modified selection operator performs proportional selection for each sub-population according to each objective function, while recombination and mutation operations are allowed to cross sub-population boundaries.

There also are other versions of evolutionary algorithms to attempt to promote the generation of multiple non-dominated solutions such as Fourman [5], Kursawe [9], Hajela and Lin [7]. However, none makes direct use of the actual definition of Pareto optimality. The concept of Pareto-based fitness assignment was first proposed by Goldberg [6], as a means of assigning equal probability of reproduction to all non-dominated individuals in the population. This method is consisted of assigning rank 1 to the non-dominated individuals and removing them from contention, then finding a new set of non-dominated individuals, ranked 2, and so forth.

Fonseca and Fleming [4] have proposed a slightly different scheme, whereby an individual's rank corresponds to the number of individuals in the current population by which it is dominated. Non-dominated individuals are, therefore, all assigned the same rank, while dominated ones are penalized according to the population density in the corresponding region of the trade-off surface. Srinivas and Deb [13] have implemented a similar sorting and fitness assignment procedure, called NSGA, but based on Goldberg's version of Pareto ranking. Horn et al. [8] proposed the tournament selection method based on Pareto dominance. The more recent algorithms include NSGA-II [3], and the SPEA algorithm [18].

2. Discrete Differential Evolution

Differential Evolution (DE) is a branch of evolutionary algorithm proposed by Storn and Price [14] for optimization problems over a continuous domain. DE is similar to (μ, λ) evolution strategy in which mutation plays the key role. For any selected individual, p_i, that undergoes mutation, the mutation operator is represented as follows:

$$p'_i = \gamma \cdot p_{best} + (1-\gamma)p_i + F \cdot \sum_{k=1}^{K}\left(p_{i_a^k} - p_{i_b^k}\right)$$

where p_{best} is the best individual in parent population, $\gamma \in [0,1]$ represents greediness of the operator, and K is the number of differentials used to generate the perturbation, F is the factor that scales the perturbation, $p_{i_a^k}$ and $p_{i_b^k}$ are randomly selected mutually distinct individuals in the parent population, and p'_i is the offspring. The basic idea of DE is to adapt the search step along the evolutionary process. At the beginning of evolution, the perturbation is big since parent individuals are far away to each other. As the evolutionary process proceeds to the final stage, the population converges to a small region and the perturbation becomes small. As a result, the adaptive search step allows the evolution algorithm to perform global search with a large search step at the beginning of evolutionary process and refine the population with a small search step at the end. The selection operator in DE selects the better of the parent and the offspring by comparing their fitness values:

$$p_i^{(t+1)} = \begin{cases} p_i'^{(t)} & \text{if } \Phi\left(p_i'^{(t)}\right) > \Phi\left(p_i^{(t)}\right) \\ p_i^{(t)} & \text{otherwise} \end{cases}.$$

In this paper, the DE concept is scaled to the discrete domain, and to the multi-objective optimization problem. In the basic differential evolution and its subsequent variants, DE allows mutation toward both the best individual and random perturbations. This is realized by forcing the individual to move in the direction of differential vector between the best individual and itself, and adding perturbation of differential vector among randomly selected individuals from the parent population. In a discrete domain, the decision variable is an n dimensional vector variable $\mathbf{x} = [x_1, x_2, \cdots x_n]$, $x_i \in \Omega_i$, where Ω_i's are a set of discrete vectors. In many situations, such as in various planning problem involved in supply manufacturing, the elements of the vector, Ω_i, are integer index, which have no physical meaning. In such situations, the differential vectors in traditional DE have no ordered physical interpretation. However, the main concept of DE, which is directing the individuals to current best solutions with adaptive perturbations, can be implemented in another way. In our discrete DE, this concept is realized by introducing three probabilities: *greedy probability* p_g, *mutation probability* p_m, and *crossover probability* p_c. The decision variable, $\mathbf{x}$, is represented in the evolutionary algorithm using a gene vector of length n that is the same as the real decision vector. The value in each allele position is as the corresponding value of the real decision vector. With this representation, the allele j of offspring of any individual p_i can be obtained in the reproduction operator as follows:

$$p'_{i_j} = \begin{cases} p_{best_j} & \text{if } p_g < rand() \\ \Omega_{j_{rand}} & \text{elseif } p_g + p_m < rand() \\ p_{a_j} & \text{elseif } p_g + p_m + p_c < rand() \\ p_{i_j} & \text{else} \end{cases}$$

where $rand()$ is a random number between 0 and 1, $\Omega_{j_{rand}}$ is a random selected value from Ω_j containing all possible values for allele j, p_a is a randomly selected individual from parent population that is distinct with p_i. For a single objective optimization problem, the p_{best} can be easily identified by choosing the individual with highest fitness value. In this way, the offspring reproduction operator introduced above captures the DE concept. The reproduction of each offspring is guided by the best individual by reproducing some of its gene information. The mutation part can be regarded as a constant small perturbation for the offspring generation. Reproducing gene information from other parent individuals is the adaptive perturbation that varies during the evolutionary process. At the beginning of the evolutionary process, this perturbation is large due to less similarity of the population, and small at the end due to more similarity of the population. This reproduction mechanism is as shown in Figure 2.

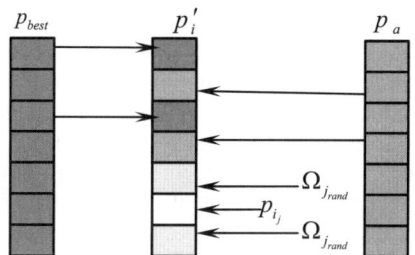

Figure 2: Discrete DE operator to produce offspring. The individual under operation gets some gene information from the "best" individual, and some from a selected individual from current parent population, and some from a random generation.

3. Multi-Objective Evolutionary Algorithm

Since the emergence of multi-objective evolutionary algorithms, there have been many variants. The representatives are multi-objective genetic algorithm (MOGA) due to Fonseca and Fleming [4], nondominated sorting genetic algorithm (NSGA) due to Srinivas and Deb [13], and niched Pareto genetic algorithm (NPGA) due to Horn et al. [8]. More recently, Zitzler and Thiele [18] proposed the strength Pareto evolutionary algorithm (SPEA), which has an external repository of global Pareto solutions with continuously update of this repository. This deterministic way itself is a complementary part of randomness of non-elitist evolutionary algorithms such as NSGA to keep the convergence to global Pareto solutions without having an effect on the stochastic properties of the evolutionary algorithm, though it is not the situation in SPEA.

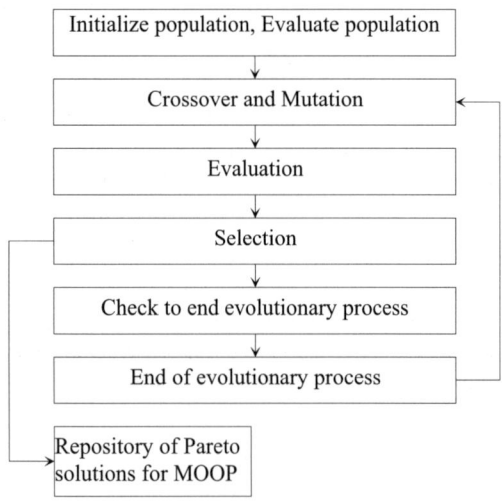

Figure 3: Flowchart of revised NSGA with Pareto solutions repository

The structure of a revised NSGA with this external repository population is shown in Figure 3. In NSGA, the non-dominated individuals are assigned rank 1 and removing them from contention, then a new set of non-dominated individuals are ranked 2, and so forth until all of the individuals are assigned a rank. A so called dummy fitness value is assigned each rank, with rank 1 having the highest fitness value. The same sharing techniques as used by Srinivas and Deb [13] is used here. After the selection of each generation, the new population is put into the repository and update the global Pareto solutions. The repository also crowds out solutions with close objective values in all dimensions.

4. Multi-Objective Differential Evolution

The traditional DE is extended to solve discrete problems by introducing the operator as described in section 2. It can also be scaled to discrete multi-objective optimization problem with careful design of selection of best individuals for production operator. Abbass el al [1] and Madavan [10] independently studied the extension of differential evolution to multi-objective optimization problem in continuous domain. As mentioned above, the best individual used in the production operator can easily be identified by choosing the individual with highest fitness value. However, in a multi-objective domain, the purpose of evolutionary algorithm is to identify a set of solutions, the so called Pareto optimal solutions. In this proposed discrete multi-objective differential evolution (D-MODE), a Pareto-based approach is introduced to implement the selection of the best individual for the production operator. At a certain generation of evolutionary algorithm, the population is sorted into several ranks. This is illustrated in the objective space for a bi-objective problem as shown in Figure 4. For any individual in the population, a set of non-dominated individuals, D_i, that dominates this individual can be identified. In the reproduction operator for a dominated solution in the parent population, the p_{best}, is chosen randomly from the set D_i. If the individual is already a non-dominated individual in the parent solution, the p_{best} will be itself. For a particular case as shown in Figure 4, all of the circled individuals would be the set of D_i, one of which would be the p_{best} for production operator of the bold solution x.

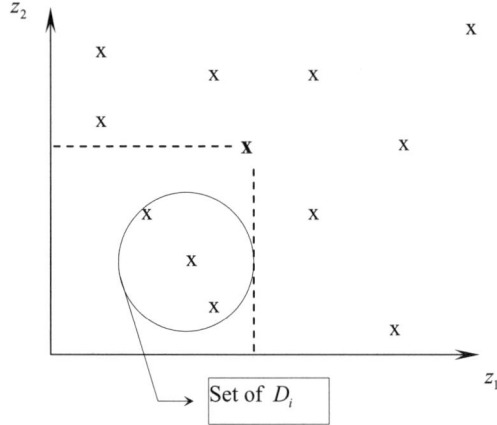

Figure 4: In order to realize the reproduction operator of a dominated individual in current generation, those individuals in first rank that dominate this individual are identified

In order to apply to multi-objective optimization, the Pareto-based fitness assignment and selection of NSGA-II introduced by Deb [3] is incorporated. The NSGA-II algorithm incorporates both an elite-preserving and an explicit diversity-preserving strategy. The population is sorted as in the NSGA. Instead of computing the niche count to add a penalty to the individuals crowded in a small region, the individuals within each non-dominated front that reside in the least crowded region in that front are assigned a higher rank. A crowding distance metric for a particular individual is obtained by calculating the summation of normalized distance along each objective between the individual and the surrounding individuals within the same non-dominated front. Such a crowding distance metric is used to estimate the density of solutions around such particular individual.

The original NSGA-II applies a $(\mu + \lambda)$ selection strategy. The individuals are first compared using Pareto rank, if the Pareto rank ties, the crowd distance metric is compared to fill the population of next generation. This strong elitism strategy, however, does not produce good results in our experiments. The authors [2] also point out the importance to keep diversity among non-dominated fronts to allow individuals in lower rank to enter the next generation. In the proposed D-MODE, there is another parameter σ_{crowd} to specify how close the solution is to its surrounding solutions in objective space in order to reduce its fitness. In fact, this parameter will prevent very similar individuals from entering next generation, which might lead to premature convergence.

5. Experimental Results

The proposed algorithm is applied to the design, supplier, manufacturing planning problem using design and supplier data from a real commercial electronic circuit board product, and data from three commercial manufacturing facilities.

In the experiment with seven modules, nine suppliers for each module, and four to six contract manufacturers, the total number of possible solution is $O(10^7)$. Both the revised NSGA and the proposed D-MODE are applied to find the Pareto set based on criteria of total product time and cost. Various experiments were conducted to simulate the different manufacturability, for instance, the available manufacturers might not have surface mount lines or mixture lines or through-hole lines, or they might have only surface mount lines or all three of them. Due to space limitation here, we only show the experimental results obtained from two of them, where there are only surface mount lines or no surface mount lines, denoted by Only-SM and No-SM, respectively. For a problem of this size, it is possible to identify the real Pareto set using exhaustive search, although it takes a long time to do so. In order to evaluate the performance, both revised NSGA and D-MODE were applied to this problem, and the real Pareto solutions for each possible situation are identified using exhaustive search for comparison. For such a bi-criteria problem the easiest way to compare the computed results with the real Pareto solution is to plot the real Pareto solutions and the Pareto solutions obtained by the evolutionary algorithm in the two dimensional objective plane.

For both of revised NSGA and D-MODE, the same population size of 200 is used and the same maximal generations of 200 are evolved. The real Pareto front and the computed Pareto front is plotted in the same plane as shown in Figure 5, 6, 7, 8, 9, 10. In all of these figures, a cross indicates a real Pareto solution, while a diamond indicates a computed solution. In Figure 5, computed results obtained using revised NSGA after 200 generations are plotted along with the real Pareto solution for Only-SM experiment; while the computed solutions for the same experiment using D-MODE after 100 and 200 generations are plotted in Figure 6 and Figure 7 respectively. In Figure 8, computed results obtained using revised NSGA after 200 generations are plotted along with the real Pareto solution for No-SM experiment; while the computed solutions for the same experiment using D-MODE after 100 and 200 generations are plotted in Figure 9 and Figure 10 respectively. For the Only-SM experiment, there are totally 21 Pareto solutions. The D-MODE finds 20 of them after 200 generations and 16 of them after 100 generations; while the revised NSGA finds only 14 of them after 200 generations. For the No-SM experiment, there are totally 18 Pareto solutions. The D-MODE finds 17 of them after 200 generations and 9 of them after 100 generations; while the revised NSGA only finds 10 of them after 200 generations. It can be seen that the results obtained using D-MODE are much better than those obtained using revised NSGA in terms of the number of Pareto solutions found and the convergence speed. The results obtained using D-MODE after 100 generations can even compete with the results obtained using revised NSGA after 200 generations.

It is interesting to note that the different performance of revised NSGA from No-SM to Only-SM experiment in terms of roughly fitting of the real Pareto front. It seems that this goodness of fitting is affected by the nature of the optimization problem. When the real Pareto front is roughly evenly distributed as in the No-SM experiment, the revised NSGA can identify an approximate Pareto front, though not real Pareto front, to roughly represent the trade-off nature among the multiple objectives of the optimization problem. In contrast, when the real Pareto front like the one of the Only-SM experiment does not possess roughly distributed solutions, the revised NSGA neglects a large part of solutions resulting in a bad representation of the trade-off nature of the optimization problem. Since the representation of trade-off nature is so important in decision making, this speculation poses an open question on choosing those continuous benchmark functions that always have smooth shape as test beds, which might have amenable properties for evolutionary algorithms to identify the Pareto front.

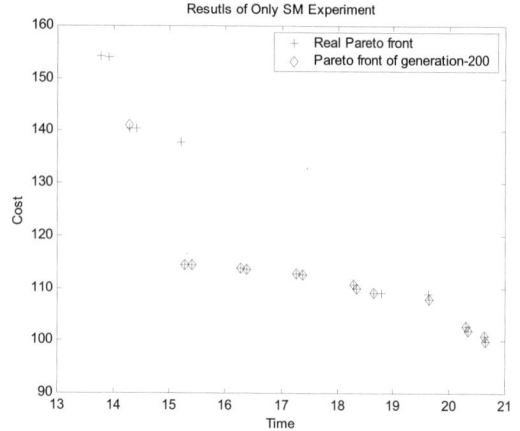

Figure 5: Pareto solutions obtained after 200 generations using revised NSGA and the real Pareto front for Only-SM

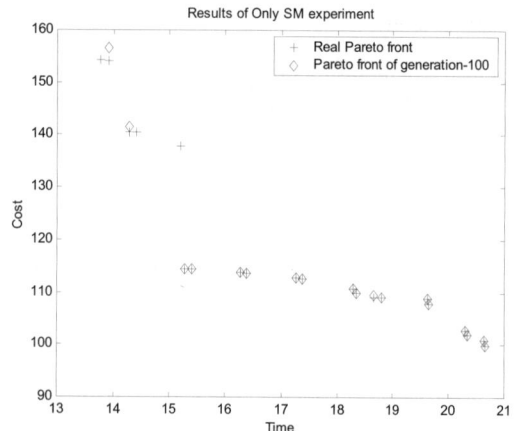

Figure 6: Pareto solutions obtained after 100 generations using D-MODE and the real Pareto front for Only-SM

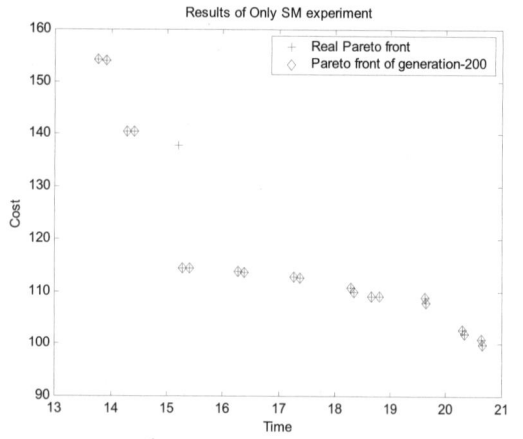

Figure 7: Pareto solutions obtained after 200 generations using D-MODE and the real Pareto front for Only-SM

6. Conclusions

Single objective evolutionary algorithms have been extensively applied to various practical problems. However, in many practical problems, there often are multiple objectives which cannot be optimized simultaneously. In this situation, it is essential to identify the trade-off solutions, i.e., the Pareto front, to facilitate the final decision. Multi-objective evolutionary algorithms have been developed to find such Pareto front. In this paper, the concept of differential evolution, which is well-known in single-objective continuous domain for its fast convergence and adaptive parameter setting, is extended to the discrete domain by introducing greedy probability, mutation probability, and crossover probability. Moreover, this concept is extended to the discrete multi-objective optimization problem. The preliminary testing of the proposed multi-objective differential evolution on an integrated design, supplier, manufacturing planning problem shows that this D-MODE is very effective in terms of convergence and the capability to identify Pareto solutions. The experimental results show that this D-MODE has much better performance compared with a revised NSGA, though further experiments need to be conducted to compare with more recent multi-objective evolutionary algorithms such as NSGA-II and SPGA. It is also noted that the experimental results in this paper pose an interesting question on choosing benchmark functions as test beds for multi-objective evolutionary algorithms.

Acknowledgements

This work has been conducted in the Electronics Agile Manufacturing Research Institute (EAMRI) at Rensselaer Polytechnic Institute. The EAMRI is partially funded by grant number DMI-#0121902 by National Science Foundation.

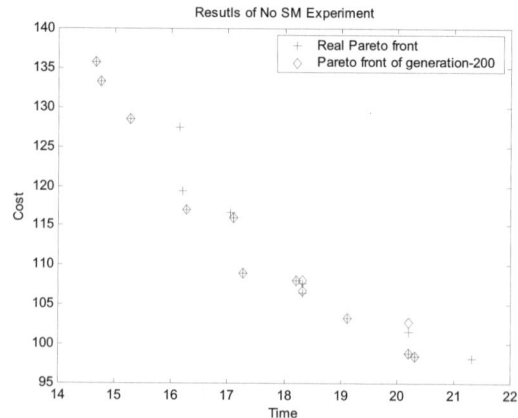

Figure 8: Pareto solutions obtained after 200 generations using revised NSGA and the real Pareto front for No-SM

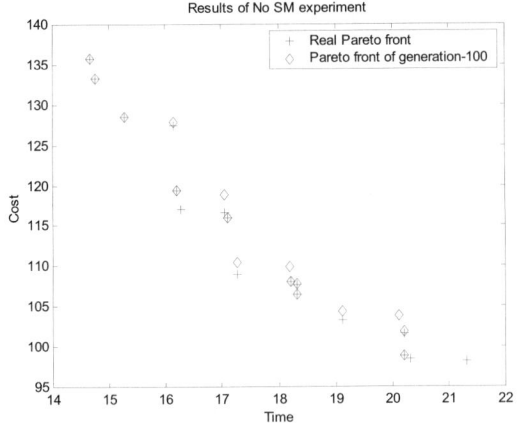

Figure 9: Pareto solutions obtained after 100 generations using D-MODE and the real Pareto front for No-SM

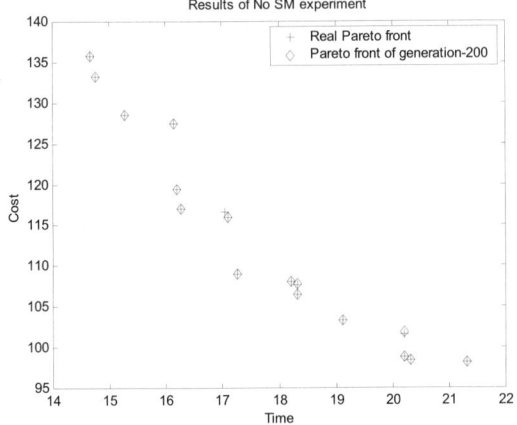

Figure10: Pareto solutions obtained after 200 generations using D-MODE and the real Pareto front for No-SM

Reference:

[1] Abbass, H A, Saker, R, and Newton C, "PDE: A Pareto-frontier differential evolution approach for multi-objective optimization problems", *Proc. of IEEE Congress on Evolutionary Computation*, pp971-978, 2001.

[2] Deb, K and Goel, T "Controlled elitist non-dominated sorting genetic algorithms for better convergence ," *Proc. of the First Int. Conf. on Evolutionary Multi-Criterion Optimization*, pp67-81, 2001

[3] Deb, K, Pratap, A, Agarwal, S, Meyarivan, T, "A fast and elitist multiobjective genetic algorithm: NSGA-II", *IEEE Trans. on Evolutionary Computation*, vol. 6, No. 2, pp182-197, 2002.

[4] Fonseca C M and Fleming, P J, "Genetic algorithms for multiobjective optimization: formulation, discussion, and generalization", *Genetic Algorithms: Proc. of the Fifth Int. Conf.*, Forrest S ed., pp416-423, 1993.

[5] Fourman, M P, "Compaction of symbolic layout using genetic algorithms" In Grefenstette JJ ed., *Proc. of the First Int. Conf. on Genetic Algorithms*, pp141-153, 1985.

[6] Goldberg, D E, *Genetic algorithms in search, optimization and machine learning*, Addison-Wesley, 1989.

[7] Hajela, P, and Lin, C-Y, "Genetic search strategies in multicriterion optimal design", *Structural Optimization*, 4, pp99-107, 1992.

[8] Horn J, Nafpiolitis N and Goldberg D, "A niched Pareto genetic algorithm for multiobjective optimization", *Proc. of the First IEEE Conf. on Evolutionary Computation*, pp82-87, 1994.

[9] Kursawe, F, "Avariant of evolution strategies for vector optimization, In Schwefel H-P and R. Manner eds., vol. 496 of *Lecture notes in computer science*, pp193-197, 1991.

[10] Madavan N K, "Multiobjective optimization using a Pareto differential evolution approach", *Proc. of IEEE Congress on Evolutionary Computation*, pp1145-1150, 2002.

[11] Schaffer J D, "Multiple objective optimization with vector evaluated genetic algorithm", In Grefenstette J J ed., *Genetic Algorithms and Their Applications: Proc. of the First Int. Conf. on Genetic Algorithms*, pp93-100, 1985.

[12] Schaffer J D and Grefenstette J J, Multi-objective learning via genetic algorithms, *Proc. of the Ninth Int. Joint Conf. on Artificial Intelligence*, pp593-595, 1985.

[13] Srinivas, N and Deb, K, "Multiobjective optimization using nondominated sorting in genetic algorithms", *Evolutionary Computation*, 2(3), pp221-248, 1995.

[14] Storn, R and Price, K, "Differential evolution: a simple and efficient adaptive scheme for global optimization over continuous spaces", Technical Report TR-95-012, International Computer Science Institute, Berkeley.

[15] Subbu, R, *Network-based distributed planning using coevolutionary algorithms*, Ph.D thesis, Rensselaer Polytechnic Institute, 2001.

[16] Subbu, R, Hocaoğlu, C, and Sanderson, A C, "A virtual design environment using evolutionary agents", In *Proc. of the IEEE Int. Conf. on Robotics and Automation*, 1998.

[17] Subbu, R, Sanderson, A C, Hocaoğlu, C, and Graves, R J, "Distributed virtual design environment using intelligent agent architecture", In *Proc. of the Industrial Engineering Research Conf.*, 1998.

[18] Zitzler, E and Thiele, L, "Multiobjective evolutionary algorithms: a comparative case study and the strength Pareto approach", *IEEE Trans. on Evolutionary Computation*, vol. 3, No. 4, pp257-271, 1999.

Task Planning with Transportation Constraints: Approximation Bounds, Implementation and Experiments

Ovidiu Daescu, Derek Soeder and R. N. Uma
Department of Computer Science, University of
Texas at Dallas
M/S EC 31, Richardson, TX 75083-0688
U.S.A.
{daescu,dsoeder,rnuma}@utdallas.edu

Abstract—In this paper we consider the problem of planning the execution of a set of tasks, where each task has associated some processing time and has to be transported to some destination. The problem arises in various applications in production planning and scheduling. We address several variants of the problem and discuss implementations of some of the proposed solutions. The objective is to compute a processing-and-delivery schedule so as to minimize the sum of delivery completion times. All the variants we consider are NP-hard. We experimentally evaluate the performance of some heuristics and show that the *shortest processing time* (SPT) heuristic consistently yields good schedules. We also briefly discuss results on scheduling with transportation in planar weighted subdivisions, that specifically address the computation of transportation times.

I. INTRODUCTION

Task scheduling research has mainly focused on efficiently assigning machine-time pairs to tasks, without taking into account transportation costs to specific destinations. Robot path planning research has addressed various "transportation" scenarios, assuming that robots are available at any given time. In many planning applications, however, the scheduling decisions are dependent on the cost of transporting the tasks from their originating source to the processing machine, the cost of transporting the semi-finished tasks from one machine to another and the cost of transporting the finished tasks to their respective destinations. In this paper, we address several variants of such problems, integrating the cost of transporting finished tasks to destinations into the planning decision. We discuss implementations, experimentally evaluate the performance of some heuristics and show that the *shortest processing time* (SPT) heuristic consistently yields good schedules.

A. Problem Definition and Notation

In this section we introduce the basic model and related notations. We assume that n tasks are initiated at a processing site, and task j has associated a pair (p_j, W_j), $j = 1, 2, \ldots, n$, where p_j denotes the processing time at the site and W_j is a destination to be delivered after processing. The tasks are to be processed and then transported to their respective destinations using v robots, each of which can carry at most c tasks at a time. The processing site is modeled as an identical parallel machine environment, that is, the machines are identical and that job j requires p_j units of processing time irrespective of the machine it is processed on. We assume we have an infinite storage space at the processing site to store processed tasks waiting to be transported. The goal is to schedule the processing and delivery of tasks so as to minimize the sum of delivery completion times, or equivalently, to minimize the average delivery completion time. The delivery completion time of a task is defined as the time at which the task is delivered to its destination. We assume the transportation times are known, as a result of some path planning preprocessing step (such as computing shortest path maps), and the time to have available a specific transportation time is negligible when compared to the actual transportation time. We denote by t_j^1 the time a robot takes to transport task j from the processing site to its destination and by t_j^2 the time a robot takes to return from task j's destination to the processing site. In general, t_j^1 need not necessarily be equal to t_j^2. Using the scheduling notation of [1] and extending the notation of [2], we denote the scheduling-with-transportation problems studied in this paper as $P \to W_j | v, c, t_j^1 \neq t_j^2 | \Sigma D_j$, where D_j is the *delivery completion time* of task j and W_j denotes the destination of task j. If all tasks have the same destination then $t_j^1 = t_1$ and $t_j^2 = t_2$, for all j.

B. Our Results

The results we present pertain to off-line problems that are known to be NP-hard. To the best of our knowledge, our results are the first constant-factor approximation algorithms, implementations and experiments for these problems with the objective of minimizing the sum of delivery completion times. In Section II, we discuss the worst-case performance of some simple heuristics designed in the context of pure scheduling problems. Specifically, using the *shortest processing time* (SPT) heuristic, we can compute a delivery schedule whose cost is guaranteed to be within a factor of 2 of optimal, for the case of one robot with capacity one.

In Section III we briefly consider scheduling with transportation in a planar weighted subdivision, where $t_j^1 = t_j^2$ and $t_i^1 \neq t_j^1$ for $i \neq j$. In Section IV, we discuss implementations and present an experimental study of a number of heuristics. We observe that the simple SPT heuristic yields near-optimal solutions on a variety of randomly generated data.

C. Related Work

From a worst-case perspective, several groups of researchers have studied "pure" scheduling problems [3], [4], [5], [6], [7], [8], where the transportation cost is trivial, and "pure" transportation problems [9], [10], [11], [12], where the scheduling cost is trivial. However, there is very little work in analyzing performance of algorithms for scheduling problems under transportation constraints. The work that we are aware of in this regard can be broadly categorized into two groups. The research in the first group has focused on flow shop models where the cost of transporting the semi-finished tasks from one machine to the next has been considered [13], [14], [15], [16]. Since the flow shop scheduling problems, even without tranportation constraints, are notoriously intractable, most of this research has focused on designing practical heuristics. The research in the second group ([17], [18]) has looked at scheduling problems where each task j also has a delivery time t_j associated with it. The delivery time is essentially the cost of transporting the finished task to its respective destination. All of this research has focused on the problem of minimizing the maximum delivery completion time. Hall and Shmoys [17] studied single machine models and gave constant-factor approximation algorithms and approximation schemes based on extended Jackson's rule [19]. Woeginger [18] studied a parallel machine model and gave constant-factor approximation algorithms based on variants of list scheduling. In both models, all tasks may be simultaneously delivered which in the context of our model is equivalent to having n robots each of capacity 1 to transport the n tasks. Recently, Lee and Chen [2] analysed a variety of scheduling problems under transportation constraints — both tranportation capacity and transportation time. They considered two kinds of transportation costs – (i) time to transport a semi-finished task from one machine to another as in flow shop models and (ii) time to transport a finished task to the customer/warehouse. They gave hardness results for several variants for both kinds of transportation costs. In this paper, we consider only the second kind of transportation costs.

II. SIMPLE HEURISTICS

In this section, we analyze the worst case performance of two simple well-known heuristics, namely, *Shortest Processing Time* (SPT) and *Longest Processing Time* (LPT) for some special cases of the above problem. We modify the SPT (LPT) heuristic as follows for the problem at hand. We schedule the tasks for processing on machines according to the SPT (LPT) rule and deliver the tasks in the order in which they complete. Clearly, these heuristics run in polynomial time.

We also discuss the *Shortest Remaining Processing Time* (SRPT) rule, which always processes the task with the shortest remaining processing time.

We will use the following notation: C_j^H denotes the completion time of task j under heuristic H and D_j^H denotes the delivery completion time of task j under heuristic H.

A. Performance of SPT

We first consider the special case of one robot of capacity c, transportation times $t_j^1 = t_1$ and $t_j^2 = t_2$ for $j = 1, \ldots, n$; that is, all tasks are destined to the same destination. This problem is denoted $P \to W | v = 1, c, t_1 \neq t_2 | \sum D_j$. Observe that without the transportation component, the problem reduces to $P || \sum C_j$ for which the SPT heuristic yields an optimal schedule [20].

B. Lower Bounds

We have two lower bounds. The first lower bound (denoted LB_1) is obtained under the relaxation that we have a pure scheduling problem. This is equivalent to having n robots, one per task. For this case, SPT yields an optimal schedule. Hence, $D_j^{SPT} = C_j^{SPT} + t_1$. Summing over all tasks, $LB_1 = \sum D_j^{SPT} = \sum C_j^{SPT} + nt_1$. The second lower bound (denoted LB_2) is obtained under the relaxation that we have a pure transportation problem. This is equivalent to having $p_j = 0$ for all tasks. It is easy to see that the optimal solution will transport c tasks in each of the first $B - 1$ batches and u tasks in the last batch, where $B = \lceil \frac{n}{c} \rceil$ and $u = n - (B-1)c \leq c$. We have, $D_1 = \cdots = D_c = t_1$, $D_{c+1} = \cdots = D_{2c} = 2t_1 + t_2$, $\ldots$, $D_{(B-2)c+1} = \cdots = D_{(B-1)c} = (B-1)t_1 + (B-2)t_2$ and $D_{(B-1)c+1} = \cdots = D_{(B-1)c+u} = Bt_1 + (B-1)t_2$. Summing over all tasks, $LB_2 = \sum D_j = c\frac{B(B-1)}{2}t_1 + c\frac{(B-1)(B-2)}{2}t_2 + uBt_1 + u(B-1)t_2 \geq c\frac{B(B-1)}{2}t_1 + c\frac{(B-1)(B-2)}{2}t_2 + u(B-1)(t_1 + t_2)$.

C. Analysis

As per SPT, the vehicle will transport $(B-1)$ full batches of c jobs each followed by the last (partial) batch containing u jobs. Let the jobs be numbered $1, \ldots, c, \ldots, ((B-1)c+1), \ldots, ((B-1)c+u)$. Let $wait_b^{max}$ be the maximum waiting time for jobs in batch b.

Theorem 2.1: SPT is a polynomial time $(c+1)$-approximation of $P \to W | v = 1, c, t_1 \neq t_2 | \sum D_j$.

Proof:

$$\begin{aligned}\sum D_j &= cD_c + cD_{2c} + cD_{3c} + \cdots + cD_{(B-1)c} + uD_{(B-1)c+u} \\ &= c\sum_{b=1}^{B-1} D_{bc} + uD_{(B-1)c+u} \\ &\leq c\sum_{b=1}^{B-1}[C_{bc}^{SPT} + \text{wait}_b^{max} + t_1] \\ &\quad + u[C_{(B-1)c+u}^{SPT} + \text{wait}_B^{max} + t_1] \\ &\leq [c\sum_{b=1}^{B-1} C_{bc}^{SPT} + c(\frac{(B-1)(B-2)}{2})(t_1+t_2) \\ &\quad + c(B-1)t_1] + [cC_{(B-1)c+u}^{SPT} + u(B-1)(t_1+t_2) + ct_1] \\ &\leq c[(\sum_{b=1}^{B-1} C_{bc}^{SPT} + C_{(B-1)c+u}^{SPT}) + Bt_1] \\ &\quad + [c(\frac{(B-1)(B-2)}{2})(t_1+t_2) + u(B-1)(t_1+t_2)] \\ &\leq cLB_1 + LB_2 \\ &\leq (c+1)OPT\end{aligned}$$

∎

The above analysis is fairly tight as is demonstrated by the following example. Consider $(c-1)$ jobs of size ε and one job of size p such that $p > t_1 + t_2$. We will refer to a job of size ε as a "small" job and a job of size p as a "big" job. Let the number of machines be m. The SPT algorithm processes all the c jobs and then transports them in one batch. The small jobs complete their processing by time $\frac{(c-1)\varepsilon}{m}$. The cost of the SPT algorithm is, therefore, $c[\frac{(c-1)\varepsilon}{m} + p + t_1]$. The cost of the optimal algorithm will be at most the cost of transporting the small jobs in one batch and the big job in a second batch. Hence, $OPT \leq (c-1)[\frac{(c-1)\varepsilon}{m} + t_1] + \frac{(c-1)\varepsilon}{m} + p + t_1$. Since $p > t_1 + t_2$, the robot has returned to the machine site before the completion time of the big job. $OPT \leq c[\frac{(c-1)\varepsilon}{m} + t_1] + p$. Taking the limit as ε goes to zero and p goes to infinity, the ratio SPT/OPT is at least c.

For the special case where the robots's capacity is 1, we have the following stronger result is possible.

Corollary 2.2: For $P \to W|v=1, c=1, t_1 \neq t_2|\sum D_j$, SPT is a polynomial time 2-approximation.

D. Performance of LPT

For the pure scheduling problem $P||\sum C_j$, it does not make sense to consider the LPT heuristic which schedules the longest processing time task first. However, for scheduling under transportation constraints, it is worthwhile to consider LPT, because it enables us to overlap the processing time of the longest task with the transportation time of smaller tasks. We only consider a weak bound for LPT. Following the same reasoning as in the previous section, LPT is a polynomial time $(c\frac{p_{max}}{p_{min}} + 1)$-approximation algorithm because $\sum C_j^{LPT} \leq \frac{p_{max}}{p_{min}}\sum C_j^{SPT}$.

E. Performance of SRPT

We consider trivial release dates and identical weights. This problem is denoted $P \to W_j|v=1, c=1, t_j^1 \neq t_j^2|\sum D_j$. Let us refer to this problem as $\mathscr{P}$. We decompose problem $\mathscr{P}$ into a sequence of two subproblems $\mathscr{U}$ and $\mathscr{V}$. The subproblem $\mathscr{U}$ corresponds to the machine scheduling phase $P||\sum C_j^{\mathscr{U}}$. The subproblem $\mathscr{V}$ is $1|r_j|\sum C_j^{\mathscr{V}}$ where $r_j^{\mathscr{V}} = C_j^{\mathscr{U}}$, $p_j^{\mathscr{V}} = t_j^1 + t_j^2$, and corresponds to the transportation phase. For the above subproblems, we have the following known results. Problem $\mathscr{U}$ is solved optimally using the SPT heuristic [20]. For problem $\mathscr{V}$, an algorithm of [3] converts a preemptive schedule to a non-preemptive schedule yielding a 2-approximation algorithm (the algorithm schedules tasks non-preemptively in the order of their preemptive completion time, ensuring that release date constraints are obeyed).

It is well known that the *Shortest Remaining Processing Time* (SRPT) rule, which always processes the task with the shortest remaining processing time, solves $1|r_j, pmtn|\sum C_j$ optimally in polynomial time.

We analyze the following algorithm: apply the SPT rule to subproblem $\mathscr{U}$. Apply the SRPT rule followed by CONVERT algorithm [3] to subproblem $\mathscr{V}$. Note that in the SRPT schedule, all tasks processed in the interval $[r_j, C_j^P]$ have processing times less than $p_j^{\mathscr{V}}$ and hence, in our algorithm, will be delivered before task j.

The delivery completion time of task j under problem $\mathscr{P}$ is given by $D_j^{\mathscr{P}} \leq C_j^P + \sum_{k:C_k^P \leq C_j^P} p_k + t_j^1$. But $C_j^P \leq C_j^{\mathscr{U}} + \sum_{k:C_k^P \leq C_j^P} p_k$ and $C_j^{\mathscr{U}} = C_j^{SPT}$. Hence $D_j^{\mathscr{P}} \leq (C_j^{SPT} + t_j^1) + 2\sum_{i:i\to j}(t_i^1 + t_i^2)$ where $i \to j$ denotes tasks i delivered before j. Using lower bounds analogous to those in Section II and summing over all tasks j, we have $\sum D_j \leq LB_1 + 2LB_2 \leq 3OPT$.

Theorem 2.3: There exists a polynomial time 3-approximation for $P \to W_j|v=1, c=1, t_j^1 \neq t_j^2|\sum D_j$.

Theorem 2.4: Under the notations above, given a ρ-approximation algorithm for $\mathscr{U}$, there exists a $(\rho+2)$-approximation algorithm for $\mathscr{P}$.

III. TRANSPORTATION TIME

Assume the processing site s is a point site in a planar subdivision $R = \{R_1, R_2, \ldots, R_n\}$, with n weighted convex regions. Once a task j is processed at s, it should be transported to some destination region R_j of R, along a straight line segment L joining s and R_j. We define the transportation time of task j from s to R_j along L as $T(j) = \sum_{L \cap R_i \neq \phi} w_i * d_i(L)$, where w_i is the (positive) weight of R_i and $d_i(L)$ is the Euclidean length of L within region R_i. The line segment L is such that $T(j)$ is minimized. The problem appears in emergency interventions and resource constrained applications where the cost of a turn (power consumption, etc.) maybe very high.

We find the optimal transportation time $T(j)$ for each region of R that is not adjacent to s. These times are stored in a table $\mathcal{T}$ to allow retrieval in constant time. An incoming task j at s is given by a pair (p_j, R_j) of processing time p_j and destination region R_j. From this pair, a new pair $(p_j, T(j))$ is formed in constant time, by retrieving $T(j)$ from $\mathcal{T}$. The latter pair becomes the input for the scheduling with transportation problem. Note that in this formulation the transportation times of various tasks are not the same. However, the time to go from the processing site to the destination is the same as the return time. We consider only the case $P \to W_j | v = 1, c = 1, t_j^1 = t_j^2 | \sum D_j$.

The lemma below givs an upper bound on computing transportation times.

Lemma 3.1: For a pair (j, R_j), the optimal $T(j)$ can be found in $O(n^2 \log^3 n)$ time.

IV. EXPERIMENTAL STUDY

In this section, we present the results of an experimental study of the algorithms discussed in the previous sections together with some additional heuristics. We focus on the problems $P \to W | v = 1, c, t_1 \neq t_2 | \sum D_j$ and $P \to W_j | v = 1, c = 1, t_j^1 \neq t_j^2 | \sum D_j$. Our goal in this study is to understand the quality of performance of the algorithms with worst-case guarantees together with other heuristics.

A. Heuristics and Experimental Design

In all of the following algorithms, if the robot has capacity $c > 1$, we dispatch the tasks so that all except the last batch are full and only the last batch may be partially full. For the problem $P \to W | v = 1, c, t_1 \neq t_2 | \sum D_j$, we consider the following heuristics: (*i*) SPT, LPT: these algorithms are as discussed in Section II; (*ii*) JR (*Johnson-like Rule*): this rule tries to capture the spirit of Johnson's algorithm for the two-machine flow shop problem $F2||C_{max}$ [21]. We essentially schedule the tasks in SPT order on γ fraction of the machines and in LPT order on the remaining $(1 - \gamma)$ fraction of the machines. In our experiments, we set γ to 0.5. It will be interesting to try out other values for γ; (*iii*) MWR (*Minimum Wait Rule*): The delivery time of a task can be bounded as $D_j \leq C_j +$ (time job j waits for a robot) $+ t_1$. In the variants we have studied so far, C_j can be computed optimally. Hence the inefficiency of the solution is largely due to the waiting time of the task. This heuristic was designed to minimize that component. MWR greedily selects a task for scheduling so that it minimizes the time a task has to wait for a robot or the time a robot has to wait for tasks. Essentially, the idea is, when the robot is busy, we want to process the longest job so as to maximize the overlap between the processing time of the job and the transportation time of the robot. That is, we choose a job that minimizes min{ job's waiting time after processing, robot's idle time waiting for a job to finish processing}. When the robot is idle, we want to process the shortest job so that the robot is kept waiting the least. For the problem $P \to W_j | v = 1, c = 1, t_j^1 \neq t_j^2 | \sum D_j$, we evaluate the SRPT heuristic from previous section.

Our experimental design follows that of [22]. Our parameters are set as follows — number of tasks $n \in \{20, 30, 50, 100\}$, distribution of $p_j \in \{\text{uniform, normal, bimodal}\}$, t_j^1 and t_j^2 are generated uniformly in the range $[1, p_{max}]$, number of robots $v = 1$ and capacity $c = \{1, 4\}$ and number of machines $m \in \{2, 3, 4, 5\}$. We generated 10 instances for each combination of parameters. We consider only trivial release dates and identical task weights.

B. Results

For the case of robot with capacity $c = 1$, we observe that out of a total of 120 instances, the MWR rule computes an optimal solution in all but 2 of them. This gives an indication of the strength of our lower bounds on these instances. From Table I it is seen that the algorithms in order of decreasing performance are MWR, JR, SPT and LPT. Moreover, JR and SPT are highly competitive to MWR.

For the case of robot with capacity $c = 4$, we observe that the performances of all heuristics deteriorate compared to their $c = 1$ counterparts. This is because, our delivery strategy has at most one partial batch. When we inspect the instance leading to the worst performance of SPT ($n = 20$ in Table II), we observe that delivering each task individually will drastically improve the quality of the schedule. Another reason for the degraded performance is because of our weak lower bounds (weaker than the case $c = 1$). From Table II it is seen that the algorithms in order of decreasing performance are SPT, JR, LPT and MWR.

We also implemented the SRPT heuristic to observe the quality of delivery schedules it generates. From Table III we observe that the SRPT heuristic on average yields a schedule within 4% of optimal.

C. Experimental Data

As in [22], we also observed that although we considered three different distributions for generating processing times, in the hope of generating different sorts of behavior, there were not significant qualitative differences in the performance on different distributions. Therefore, in presenting results, we group into one set all p_j distributions for a given n. Also, the most interesting results are obtained for smaller values of n. The number of machines did not impact the results much. Hence, we present results only for $n = 20, 30$ and $m = 2$.

V. CONCLUSIONS

We conclude that simple scheduling heuristics yield high quality, almost optimal delivery schedules on the data we have considered. It will be interesting to compare the performance of these simple heuristics against the more sophisticated LP-based heuristics and also the performance of these simple heuristics on harder data sets.

VI. ACKNOWLEDGMENTS

This research was partially supported by a Clark Foundation Research Initiation Grant from the University of Texas at Dallas. The authors would also like to thank Mansoor Mohsin and Joel Wein for helpful discussions.

VII. REFERENCES

[1] R. Graham, E. Lawler, J. Lenstra, and A. R. Kan, "Optimization and approximation in deterministic sequencing and scheduling: a survey," *Annals of Discrete Mathematics*, vol. 5, pp. 287–326, 1979.

[2] C.-Y. Lee and Z.-L. Chen, "Machine scheduling with transportation constraints," *Journal of Scheduling*, vol. 4, no. 1, pp. 3–24, 2001.

[3] C. Phillips, C. Stein, and J. Wein, "Minimizing average completion time in the presence of release dates," *Mathematical Programming B*, vol. 82, pp. 199–224, June 1998.

[4] L. A. Hall, A. S. Schulz, D. B. Shmoys, and J. Wein, "Scheduling to minimize average completion time: Off-line and on-line approximation algorithms," *Mathematics of Operations Research*, vol. 3, pp. 513–544, Aug. 1997.

[5] S. Chakrabarti, C. Phillips, A. S. Schulz, D. Shmoys, C. Stein, and J. Wein, "Improved approximation algorithms for minsum criteria," in *Proceedings of the 1996 International Colloquium on Automata, Languages and Programming, Lecture Notes in Computer Science 1099*. Berlin: Springer-Verlag, 1996, pp. 646–657.

[6] C. Chekuri, R. Motwani, B. Natarajan, and C. Stein, "Approximation techniques for average completion time scheduling," in *Proceedings of the 8th ACM-SIAM Symposium on Discrete Algorithms*, Jan. 1997, pp. 609–618.

[7] M. Goemans, "Improved approximation algorithms for scheduling with release dates," in *Proceedings of the 8th ACM-SIAM Symposium on Discrete Algorithms*, 1997, pp. 591–598.

[8] M. Goemans, M. Queyranne, A. Schulz, M. Skutella, and Y. Wang, "Single machine scheduling with release dates," 1997, Preprint.

[9] P. Chalasani, R. Motwani, and A. Rao, "Algorithms for robot grasp and delivery," in *2nd International workshop on algorithmic foundations of robotics*, 1996.

[10] S. Anily and J. Bramel, "Approximation algorithms for the capacitated traveling salesman problem with pick-ups and deliveries," 1997, manuscript.

[11] M. Charikar, S. Khuller, and B. Raghavachari, "Algorithms for capacitated vehicle routing," in *Proceedings of the 30th Annual ACM Symposium on Theory of Computing*, 1998, pp. 349–358.

[12] M. Charikar and B. Raghavachari, "The finite capacity dial-a-ride problem," in *Proceedings of the 39th Annual Symposium on Foundations of Computer Science*, 1998, pp. 458–467.

[13] P. Maggu, G. Das, and R. Kumar, "On equivalent job-for-job block in 2xn sequencing problem with transportation times," *Journal of the operations research society of Japan*, vol. 24, pp. 136–146, 1981.

[14] P. Maggu, M. Singhal, N. Mohammad, and S.K.Yadav, "On n-job, 2-machine flow shop scheduling problem with arbitrary time lags and transportation times of jobs," *Journal of the operations research society of Japan*, vol. 25, pp. 219–227, 1982.

[15] W. Yu, "The two-machine flow shop problem with delays and the one-machine total tardiness problem," Ph.D. dissertation, Eindhoven University of Technology, Eindhoven, The Netherlands, 1996.

[16] T. Ganesharajah, N. G. Hall, and C. Sriskandarajah, "Design and operational issues in AGV-served manufacturing systems," *Annals of Operations Research*, vol. 76, pp. 109–154, 1998.

[17] L. A. Hall and D. B. Shmoys, "Jackson's rule for single-machine scheduling: making a good heuristic better," *Mathematics of Operations Research*, vol. 17, no. 1, pp. 22–35, Feb. 1992.

[18] G. Woeginger, "Heuristics for parallel machine scheduling with delivery times," *Acta Informatica*, vol. 31, pp. 503–512, 1994.

[19] J. R. Jackson, "Scheduling a production line to minimize maximum tardiness," Management Science Research Project, UCLA, Research Report 43, 1955.

[20] R. Conway, W. Maxwell, and L. Miller, *Theory of Scheduling*. Addison-Wesley, 1967.

[21] S. M. Johnson, "Optimal two- and three-stage production schedules with setup times included," *Naval Research Logistics Quarterly*, pp. 61–68, 1954.

[22] M. W. P. Savelsbergh, R. N. Uma, and J. Wein, "An experimental study of LP-based approximation algorithms for scheduling problems," in *Proceedings of the 9th ACM-SIAM Symposium on Discrete Algorithms*, 1998, pp. 453–462.

$m=2$ and capacity $c=1$								
Heuristic	$n=20$				$n=30$			
	Mean	Std.Dev.	Min	Max	Mean	Std.Dev.	Min	Max
SPT	1.006	0.032	1.000	1.180	1.005	0.025	1.000	1.142
LPT	1.074	0.059	1.030	1.313	1.073	0.095	1.020	1.538
JR	1.000	0.003	1.000	1.014	1.002	0.011	1.000	1.059
MWR	1.000	0.000	1.000	1.000	1.000	0.000	1.000	1.000

TABLE I

PERFORMANCE OF HEURISTICS WITH RESPECT TO $\sum D_j$ OBJECTIVE AVERAGED OVER 30 INSTANCES. NUMBER OF JOBS $n=20,30$, NUMBER OF MACHINES $m=2$ AND ONE VEHICLE OF CAPACITY $c=1$.

$m=2$ and capacity $c=4$								
Heuristic	$n=20$				$n=30$			
	Mean	Std.Dev.	Min	Max	Mean	Std.Dev.	Min	Max
SPT	2.003	0.609	1.618	4.423	1.222	0.366	1.044	2.988
LPT	2.763	0.960	2.062	6.038	1.790	0.697	1.224	4.919
JR	2.191	0.775	1.691	5.192	1.324	0.452	1.077	3.404
MWR	3.915	1.009	2.764	6.065	2.710	0.599	1.630	4.280

TABLE II

PERFORMANCE OF HEURISTICS WITH RESPECT TO $\sum D_j$ OBJECTIVE AVERAGED OVER 30 INSTANCES. NUMBER OF JOBS $n=20,30$, NUMBER OF MACHINES $m=2$ AND ONE VEHICLE OF CAPACITY $c=4$.

SRPT and capacity $c=1$								
# machines	$n=20$				$n=30$			
	Mean	Std.Dev.	Min	Max	Mean	Std.Dev.	Min	Max
$m=2$	1.039	0.023	1.010	1.092	1.026	0.011	1.007	1.054
$m=3$	1.041	0.019	1.010	1.070	1.025	0.012	1.007	1.058
$m=4$	1.038	0.019	1.010	1.085	1.023	0.011	1.007	1.044
$m=5$	1.039	0.015	1.010	1.071	1.022	0.012	1.007	1.053

TABLE III

PERFORMANCE OF SRPT WITH RESPECT TO $\sum D_j$ OBJECTIVE AVERAGED OVER 30 INSTANCES. NUMBER OF JOBS $n=20,30$, NUMBER OF MACHINES $m=2,3,4,5$ AND ONE VEHICLE OF CAPACITY $c=1$.

Improvement of Product Sustainability

Meimei Gao, MengChu Zhou
Department of Electrical & Computer Engineering
New Jersey Institute of Technology
Newark, NJ 07102, USA
Email: mg2@njit.edu and zhou@njit.edu

Fei-Yue Wang
Institute of Automation, CAS, Beijing 100080, China and
Systems and Industrial Engineering Department
University of Arizona, Tucson, Arizona 85721, USA
Email: feiyue@sie.arizona.edu

Abstract – Increasing global population and consumption are causing declining natural and social systems. Sustainable development addresses these issues by integrating strategies for economic successes, environmental quality, and social equity. A Sustainability Target Method (STM) provides a practical sustainability target for individual businesses and products through determining the relative indicator resource productivity for environmental performance and the absolute indicator Eco-Efficiency. Based on the indicators in STM, this paper formulates the component selection problem such that a company can improve sustainability of its products most effectively through improving components' performance. The approach is illustrated through an example.

I. INTRODUCTION

Increasing public concern and statutory regulations about the environment and sustainable development are making environmental protection an important issue for industry. It is very important to develop and implement an unambiguous and quantitative measure for environmental impacts of products, processes and activities. There are various environmental impact assessment methods that are comprehensive in nature and generate a single numerical value reflecting the composite magnitude of global impact associated with a specific product. These include Eco-indicator 95 [4], Eco-indicator 99 [5], Ecological Footprint [7], and Sustainability Target Method (STM) [1,2].

Compared to other environmental impact assessment methods, the Sustainability Target Method (STM) shifts the focus from environmental impact to sustainability. STM explicitly considers economic or market value and is equally applicable to analyzing products/processes, services or facilities. By directly considering economic value, resource productivity measures can be generated, and comparison can be drawn between products or services that are not necessarily "functionally equivalent" but are instead economically equivalent. Thus, the STM approach provides a basis for decision making in different suppliers and manufacturers.

Based on the STM approach, this paper investigates the approaches to support a product's manufacturer to improve its product sustainability. A sensitivity analysis based decision making approach is proposed to provide a company with improvement suggestions on which component should be given the first priority consideration and requires most urgent improvement such that the product is most sustainable. The paper is organized as follows: Section II introduces the STM. Section III discusses the relationship between a product's STM indicators and its components' based on product tree structure. Section IV proposes a sensitivity analysis-based decision making approach to improve a product's sustainability. A case study and conclusions are given in Section V and VI, respectively.

II. SUSTAINABILITY TARGET METHOD

The STM approach utilizes the earth's carrying capacity estimates and economic information to provide a practical sustainability target for individual businesses and products. The principle of this approach is to establish a relationship between carrying capacity and both the environmental impact and economic value of products/processes. Initially formulated at Lucent Technologies Bell Laboratories, the STM is under continuing development through the collaboration of the Multi-lifecycle Engineering Research Center (MERC) at New Jersey Institute of Technology (NJIT) with Lucent Technologies and Agere Systems. The key and unique feature of this method is the link between carrying capacity, economic value, and environmental impact to provide an absolute or "target" criterion for sustainability that is practical for use by business [3].

This section gives a brief description about this approach. The basic STM concept and key parameter estimation/calculation methods are described in detail in [1,2].

STM involves interpreting different types of environmental impact based on a single dimensionless indicator, i.e., the environmental impact per unit production rate, as denoted by EI_{PR}. Its computation is accomplished via normalization by using impact reference levels that relate impact with economic value and sustainability. EI_{PR} provides the basis for calculating the relative indicator Resource Productivity (RP) for environmental performance and the absolute indicator Eco-Efficiency (EE) for sustainability.

A. Environmental Impact (EI)

The Environmental Impact (EI) resulting from an activity, such as manufacturing a product or providing a service, is quantified by normalizing each associated impact using an impact reference level that relates in a specific way to sustainability. EI is aggregated by adding the ratios to obtain a single dimensionless indicator of total impact.

$$EI = I_1/I_{R1} + I_2/I_{R2} + \cdots + I_N/I_{RN}$$

Each impact reference level I_R is the level at which the impact is environmentally sustainable at a rate of value generation V_R, i.e., the value reference level. Both the impact level and impact reference level are expressed per unit time (e.g., kg/year for emissions), and the same for V_R (e.g., \$/year).

For production rate P (product items per unit time), the environmental impact per unit production rate EI_{PR} is then:

$$EI_{PR} = EI/P = I_1^*/I_{R1} + I_2^*/I_{R2} + \cdots + I_N^*/I_{RN}$$

where each $I_i^* = I_i/P$ is the impact quantity per product item (e.g., kg CO_2 emissions per kWh of energy or kg of material).

B. Resource Productivity (RP)

Resource Productivity (RP) is the production rate achieved per unit environmental impact:

$$RP = P/EI = 1/EI_{PR}$$

RP serves as a relative indicator of environmental performance that allows for direct comparison, such as between products or alternative product designs. The larger its RP, the better a product. An increasing trend in the resource productivity indicator RP would be an indication that the firm is moving in the direction of using resources more efficiently and with less environmental impact.

C. Eco-Efficiency (EE)

Eco-Efficiency (EE) is defined as:

$$EE = \beta/EI$$

where $\beta = V/V_R$ (value ratio); V is value creation, e.g. revenue (\$/year). The value of the product is established by the market. V_R is the value reference level corresponding to the impact reference levels. Then,

$$EE = (V/V_R)/EI$$
$$= (V_{PR} \times P)/(V_R \times EI)$$
$$= V_{PR} \times (P/EI)/V_R$$
$$= V_{PR} \times RP/V_R$$

i.e. $EE = V_{PR} \times RP/V_R$

where V_{PR} is the revenue per product unit or value added per unit. EE is an absolute indicator of sustainability. It is the ratio of the actual rate of value generation ($V = V_{PR} \times P$) to the rate that is sustainable given the level of environmental impact ($V_R \times EI$). It can be interpreted as the benefit achieved compared to the minimum allowable benefit given the level of environmental impact incurred. Thus, the criterion for sustainability is $EE \geq 100\%$ ($EE > 100\%$ simply indicates even less impact than the sustainable level given the value being provided). EE serves as an absolute indicator of sustainability.

Definition: A product is sustainable if its $EE \geq 100\%$.

The STM provides a practical sustainability target for individual businesses and products through determining the relative indicator RP for environmental performance and the absolute indicator EE. We expect the values of RP and EE as high as possible. The higher these values, the higher production rate achieved per unit of environmental impact and more sustainable. The methods to compute these values can be found in [2,3] and [8], and is out of the scope of this paper.

III. PRODUCT TREE STRUCTURE

In general, a product can be viewed as the one that consists of its components. Thus, a company can improve its product's performances through improving

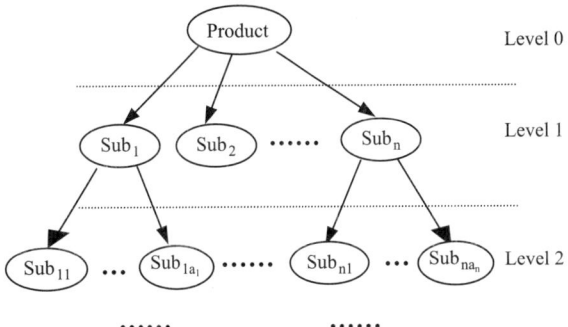

Fig.1 Tree structure of a product

the components' performances. One of the most important performances a company is facing to improve is sustainability, which integrates the strategies for economic success, environmental quality and social equity [6]. Given the STM information of the components, how a company is able to utilize the information to improve its product's sustainability is a decision-making issue we discuss in this paper. We first introduce a product tree structure because a product can be the component of another product and some products/components may be provided from their suppliers. The structure of a product can be represented as a tree shown in Fig. 1 [9]. The root of the tree is the product itself. The assembly at level k can be composed of several subassemblies/parts at level k+1, which are called immediate components of the assembly, and so forth. An assembly or component that cannot be separated is a leaf node. If it is made of a single material, it is called a part; otherwise, it is a final subassembly (FS) [10].

Theorem: Assuming that the value of each component in a product tree is fixed, the RP and EE of the product is the largest if the RP of each of its components is the largest.

Proof:
Assume that an assembly in a product tree has N immediate components. The quantity of Component i is n_i, and RP associated to Component i is RP_i, i=1, 2,..., N. Then, RP and EE of the assembly are

$$RP = \frac{1}{\frac{n_1}{RP_1} + \frac{n_2}{RP_2} + \cdots + \frac{n_N}{RP_N} + EI_{PR-Assembly}} \quad (1)$$

where $EI_{PR-Assembly}$ is the environmental impact produced in assembling the assembly from its immediate components.

$$EE = V_{PR} \times RP / V_R \quad (2)$$

where V_{PR} is the accumulated value per assembly.

According to Eq.(1), an assembly will have higher RP value if its immediate components have higher RP values. Since V_{PR} is fixed, V_{PR}/V_R is a constant. According to Eq.(2), for the assembly, a larger value of RP is corresponding to a larger EE value. Hence, an assembly will have the largest RP/EE value if its immediate components have the largest RP value. Applying this rule from leaf level to root, we can conclude that if all the components in a product tree have their largest RP values, RP and EE of the product are the largest. ♦

According to the above theorem, if a company attempts to make a product most sustainable, it should try to improve RPs of the components as much as possible. An issue is raised in this attempt. The problem is which component should be given the first priority consideration and requires most urgent improvement, such that the product is most sustainable. In the next section, we will utilize STM indicators and a sensitivity analysis approach to help a company to make such a decision.

IV. SENSITIVITY ANALYSIS BASED ON STM INDICATORS

Based on the STM approach, a sensitivity analysis approach can be used to determine the RP improvement of which component has the most significant effect on the product's RP and EE.

Without loss of the generality, this paper assumes that a company product is composed of N types of components – component i, i=1,...,N. The quality of Component i in it is n_i; the unit value of Component i is V_{PR_i}; and the RP and EE indicators associated to Component i are RP_i and EE_i respectively, i=1,...,N. Please note that these components are not necessary to be the product's immediate components.

A. The Product's RP Analysis

$$RP = \frac{1}{\frac{n_1}{RP_1} + \frac{n_2}{RP_2} + \cdots + \frac{n_N}{RP_N} + EI_{PR-Assembly}}$$

Where $EI_{PR-Assembly}$ is the environmental impact produced in assembling a product from its components. RP is the production rate achieved per unit environmental impact for the product.

For brevity and clarity, let $x = \frac{n_A}{RP_A} + \frac{n_B}{RP_B} + \cdots + \frac{n_N}{RP_N} + EI_{PR-Assembly}$, which is the environmental impact per unit production rate of the product.

The sensitivity of RP_i's effect on the product's RP is obtained by taking the first deviation of RP with respect to RP_i:

$$\frac{d(RP)}{d(RP_i)} = (-\frac{1}{x^2}) \cdot \frac{dx}{d(RP_i)}$$

$$= (-\frac{1}{x^2}) \cdot n_i \cdot (-\frac{1}{RP_i^2}) = \frac{1}{x^2} \cdot \frac{n_i}{RP_i^2},$$

TABLE I
MAIN COMPONENTS OF HANDY BOARD

ID	Component Name	n	RP	n/RP²	C	n/(RP²*C)
1	Nicad battery	1	6	0.028	0.5	0.056
2	AC/DC adaptor	1	26	0.0015	0.05	0.03
3	LED	15	355	0.00012	0.05	0.0024
4	32K static CMOS RAM	1	26	0.0015	0.2	0.0075
5	Motor driver	2	34	0.0017	0.05	0.034
6	Voltage regulator	3	112	0.00024	0.1	0.0024
7	RS232 converter	1	52	0.0004	0.05	0.008
8	6811 Microprocessor	1	13	0.0059	0.2	0.0295
9	16x2 LCD	1	15	0.0044	0.05	0.088
10	SPDT switch	2	23	0.0038	0.05	0.075
11	DIP socket	9	100	0.0009	0.05	0.018
12	Connector	1	65	0.00024	0.05	0.0048
13	Infrared demodulator	1	34	0.00087	0.05	0.017
14	PCB Base Board	1	20	0.0025	0.1	0.025

which is the change of RP of the product per unit change of RP of Component i.

A company may improve RP of different components with different efforts. Assuming C_i is the cost per unit change in RP_i. Considering this factor, the sensitivity of RP_i's effect on the change of RP of the product with unit cost is:

$$\frac{d(RP)}{d(RP_i) \cdot C_i} = \frac{1}{x^2} \cdot \frac{n_i}{RP_i^2 \cdot C_i}$$

B. The Product's EE Analysis

$EE = V_{PR} \times RP / V_R$

where V_{PR} is the accumulated value per unit product. Assuming that V_{PR} is fixed, V_{PR}/V_R is a constant. Thus, The sensitivity of RP_i's effect on the product's EE is

$$\frac{d(EE)}{d(RP_i)} = \frac{V_{PR}}{V_R} \cdot \frac{d(RP)}{d(RP_i)} = \frac{V_{PR}}{V_R} \cdot \frac{1}{x^2} \cdot \frac{n_i}{RP_i^2}$$

which is the change of product's EE per unit RP change of Component i.

Considering the cost per unit change in RP_i, the sensitivity of RP_i's effect on the change of EE of the product with unit cost is:

$$\frac{d(EE)}{d(RP_i) \cdot C_i} = \frac{V_{PR}}{V_R} \cdot \frac{1}{x^2} \cdot \frac{n_i}{RP_i^2 \cdot C_i}$$

The company should try to improve the RP of the component that can generate the largest improvement to its product's RP and EE. From RP and EE analysis, we achieve the following consistent conclusions:

(1) A company is producing a product, which is composed of N types of components. The number of type-i component is n_i, the RP associated to it is RP_i, and the cost per unit change in RP_i is C_i, i=1,..,N. In order to make the product more sustainable, the company should choose the component with the maximum value of

$\frac{n_i}{RP_i^2 \cdot C_i}$ first to improve the component's RP.

(2) In the case that the cost for resource productivity improvement is not an issue or unknown, a company should select the component with the maximum value of

$\frac{n_i}{RP_i^2}$ to improve first.

V. A CASE STUDY

Fig.2 The Handy Board

In this section, we use "Handy Board" as an example (Fig. 2) to illustrate the above analysis and verify the above conclusion. The Handy Board is a 68HC11-based controller board designed for an experimental mobile robot. The Handy Board is based on the 52-pin Motorola MC68HC11 processor, and includes 32K of battery-backed static RAM, four outputs for DC motors, a connector system that allows active sensors to be individually plugged into the board, an LCD screen, and an integrated, rechargeable battery pack. This design is ideal for experimental robotic project, but the Handy Board can serve other embedded control applications. The main components in the board are listed in Table I, where RP and C are the environmental impact and costs per unit change of RP respectively.

We assume that the environmental impact produced in assembling the product from its components $EI_{PR-assembly} = 0.07$; unit value of the product $V_{PR} = \$88$ per unit product; value reference level $V_R = \$100$ per unit environmental impact.

Initially, the RP and EE of the product:

$$RP = \frac{1}{\sum_{i=1}^{14} \frac{n_i}{RP_i} + EI_{PR-Assembly}}$$

$$= \frac{1}{0.83 + 0.07} = 1.1$$

$EE = V_{PR} \times RP / V_R = 88 \times 1.1 / 100 = 0.968$

In order to improve the product's sustainability, we can try to improve its components' RPs. From Table I and Conclusion (1) in Section IV, we can know that the company should focus on Component No. 9 -16x2 LCD first, because it has the maximum value of $n/(RP^2*C)$, which means that it has the most significant effect on the whole product with per unit cost. When the cost per unit change of a component's RP is not an issue or unknown, according to Table I and Conclusion (2) in Section IV, the company should focus on Nicad battery because it has the maximum value of n/RP^2.

Assume that the RPs of LCD and battery are increased by 10 per unit environmental impact. We analyze their effect on the product's sustainability respectively.

First consider that LCD's RP is increased by 10 per unit environmental impact, and RPs of other components' RPs keep unchanged, then the RP and EE of the product become:

$$RP = \frac{1}{\sum_{i=1}^{14} \frac{n_i}{RP_i} + EI_{PR-Assembly}}$$
$$= 1.15$$

$EE = V_{PR} \times RP / V_R = 88 \times 1.15 / 100 = 1.012$

The cost for improvement is:

$\text{cost} = C_A \cdot \Delta RP_A = 0.05 \times 10 = 0.5$

$\Delta RP = 1.15 - 1.1 = 0.05$

$\Delta EE = 1.012 - 0.968 = 0.044$

$\dfrac{\Delta RP}{\text{cost}} = \dfrac{0.05}{0.5} = 0.1$

$\dfrac{\Delta EE}{\text{cost}} = \dfrac{0.044}{0.5} = 0.088$

In the same way, if RP of battery is increased by 10 per unit environmental impact, and RPs of other components keep unchanged, then the RP and EE of the product become:

$RP = 1.26$

$EE = 1.109$

$\text{cost} = 5$

$\Delta RP = 1.26 - 1.1 = 0.16$

$\Delta EE = 1.109 - 0.968 = 0.141$

$\dfrac{\Delta RP}{\text{cost}} = 0.032$

$\dfrac{\Delta EE}{\text{cost}} = 0.023$

From the above calculation, we can see that the battery has more significant effect on product's sustainability by the same performance improvement as the LCD. However, if the cost for RP improvement is considered, the LCD has more significant effect on product's sustainability by the same performance improvement as the battery with per unit cost. The results are consistent with the conclusions we derived in Section IV.

VI. CONCLUSIONS

"Sustainable development" through eco-efficiency has become the rallying cry of a new breed of companies who see competitive advantages in conservation of resources and environmental stewardship. How to make its products most sustainable has become a company's one of the most important decisions to make. Based on STM, this paper generalizes the ideas and concepts presented in our previous work [11]. It for the first time investigates the approaches to support a company to improve its product's sustainability using the product's components' STM indicators, and adopts sensitivity analysis method to provide a company with improvement suggestions on which components should receive more attention and perform urgent improvement to make its products more sustainable. The future work includes investigating the application of STM in supply chain management and lifecycle assessment, such as supplier selection, analysis of service, use and recycling activities as well as the design and manufacturing processes. Decision-making approaches when uncertain information about components is involved are also among future research issues.

ACKNOWLEDGEMENT

This research is partially supported by the Multi-lifecycle Engineering Research Center (MERC) at New Jersey Institute of Technology (NJIT) and its industrial partners including Lucent Technologies, Agere Systems, Panasonic, and IBM. The second author is partially supported by the National Outstanding Young Scientist Research Award (Class B) from the National Natural Science Foundation of China. The third author is partially supported by the Oversea Outstanding Talent Program from the State Planning Committee and the Chinese Academy of Sciences, and the National Outstanding Young Scientist Research Award (Class A) from the National Natural Science Foundation of China.

REFERENCES

[1] Dickinson D. A., "A Proposed Universal Environmental Metric," Bell Laboratories Internal Memorandum, August 1999.

[2] Dickinson D. A, Mosovsky J. A., and Morabito J. M., "Sustainability: An evaluation & Target Method for Business – Summary & Reference Levels," Lucent Technologies Bell Laboratories Technical Memorandum, May 2001.

[3] Dickinson, D. A., Mosovsky, J. A., Caudill, R.J., Watts, D.J., and Morabito, J.M., "Application of the Sustainability Target Method: Supply Line Case Studies," *2002 IEEE International Symposium on Electronics and the Environment*, 2002, Page(s): 139 -143.

[4] Goedkoop M., "The Eco-indicator 95 Final Report," Pre Consultants, the Nertherlands, 1998

[5] Goedkoop M., and Sepriensma R., "The Eco-indicator 99, A Damage Oriented Method for Life Cycle Impact Assessment, Methodology Report," Second Edition, Pre Consultants, the Nertherlands, April 2000.

[6] Mosovsky J., Dickinson D. A., and Morabito J., "Creating competitive advantage through resource productivity, eco-efficiency, and sustainability in the supply chain", *Proceedings of the 2000 IEEE International Symposium on Electronics and the Environment*, 2000, pp230-237

[7] Wackernagal et al, "National Natural Capital Accounting with the Ecological Footprint Concept," Ecological Economics, Vol.29, pp375-390, 1999

[8] Yossapol C., Axe L., and Watts D. et al, "Carrying capacity estimates for assessing environmental performance and sustainability," *Proceeding of 2002 IEEE International Symposium on Electronics and the Environment*, 2002, pp. 32-37

[9] Zhou M. C, Caudill R. J., Sebastian D., and Zhang B., "Multi-lifecycle Product Recovery for Electronic Products," *Journal of Electronics Manufacturing*, Vol. 9, No.1, pp. 1-15, March 1999.

[10] Gao, M. Zhou M. C., and Caudill R. J., "Integration of Disassembly Leveling and Bin Assignment for Demanufacturing Automation," *IEEE Transactions on Robotics and Automation,* Vol.18, No.6, pp. 867-874, December 2002.

[11] Gao, M. Zhou M. C., D. A. Dickinson and Caudill R. J., "Product Sustainability Improvement Based on STM Indicators of Product Components," to appear in *Proceeding of 2002 IEEE International Symposium on Electronics and the Environment*, May 2003.

A Data Mining Based Clustering Approach to Group Technology

Mu-Chen Chen
Institute of Commerce Automation
and Management
National Taipei University of
Technology
Taipei, Taiwan, ROC

Hsiao-Pin Wu
Information System Department
Chinese Petroleum Corporation
Taipei, Taiwan, ROC

Chia-Ping Lin
Institute of Commerce Automation
and Management
National Taipei University of
Technology
Taipei, Taiwan, ROC

Abstract – Cellular manufacturing is an essential application of group technology (GT) in which families of parts are produced in manufacturing cells. This paper describes the development of a cell formation approach based on association rule mining and 0-1 integer programming. It is valuable to find the important associations among machines such that the occurrence of some machines in a machine cell will cause the occurrence of other machines in the same cell. A clustering model using the discovered association data is formulated to maximize the closeness measures among machines within each cell. From the results of three medium-sized problems, the proposed approach shows its ability to find quality solutions of cell formation problems.

Keywords: Cellular manufacturing, association rule, 0-1 integer programming.

I. INTRODUCTION

The primary application of group technology (GT) is in the cellular manufacturing design. The grouping of machine cells and their associated part families so as to minimize the cost of material handling is a major step in cellular manufacturing. Each manufacturing cell includes a group of various machines that are physically close together and can expectantly process a family of parts completely.

Hyer and Wemmerlov [12] divided cellular manufacturing problems into two categories: system structure problems and system operation problems. System structure problems include: (1) grouping parts into families; (2) grouping machines and processes into cells; (3) selection of tools, fixtures and pallets; (4) selection of material handling equipment; and (5) choice of equipment layout. The first two issues are commonly addressed together, and are also referred to as cell formation (CF). The operational problems include: (1) formation of maintenance and inspection policies; and (2) design of procedures for production planning, scheduling, and control.

The reduction in set-up time and material handling in cellular manufacturing can be realized by take advantage of the similarities of production process rather than by increasing the lot size. This leads to significant advantages in work-in-progress reduction, throughput time reduction, material handling cost reduction, scheduling simplification, product flow simplification, and quality improvement [11]. For further details of cellular manufacturing, readers are referred to Burbidge [4], Offodile *et al.* [20] and Brandon [3].

The cell formation problem is NP-complete and involves complicated combinatorial optimization [13]. Hence, most of the procedures are heuristic in nature. Many approaches have been applied to resolve the cell formation problem. They include array-based clustering methods, hierarchical and non-hierarchical clustering methods, graph theoretic methods and mathematical programming [10, 12, 13, 16, 17, 20, 21, 22, 25]. Some modern heuristics such as neural networks, simulated annealing, genetic algorithms and Tabu search [25] have been applied to the design of cellular manufacturing systems.

In modern data mining techniques, the association rule induction method can identify the closeness measures among items. When arranging machine cells in cellular manufacturing processes, data mining can be employed to discover beneficial information from the manufacturing process data. In data mining, association rules are descriptive patterns of the form $X \Rightarrow Y$, where X and Y are statements about the values of attributes of an instance in a database. The most famous example of an association rule was *market basket analysis*, in which the market basket consists of the set of items purchased by a customer on a single store visit. In cellular manufacturing, the association rule induction can be adopted to identify the closeness among machines.

The patterns of parts being processed on machines can be analyzed with association rules. With respect to these rules, the configuration of machine cells in manufacturing systems can be determined. Machines processing the same family of parts could have close associations, and these can be grouped jointly. The association function can be applied to the machine/component process data and performed to find relationships among machines. A closeness based clustering method is used to group machine cells by 0-1 integer programming. This paper aims to develop a cell formation approach based on association rule mining and clustering approach for designing the cellular manufacturing systems.

II. THE GT CELL FORMATION

Cell formation arranges the production around machine cells that are capable of manufacturing part families with similar process requirements. In the production flow analysis, cell formation techniques generally operate on the machine-part matrix (machine/component incidence matrix, m/c matrix), selection of product mix, and data collection on parts in the product mix [3, 5].

The binary version of the cell formation problem is

considered herein. It is usually illustrated by an m/c matrix $\mathbf{A}_{m \times n} = \{a_{ij}\}$, in which m is the number of machines, and n is number of components (parts). The m/c matrix represents the machining requirements of parts in the product mix. Each element in the m/c matrix can assume two values, '0' and '1'. A '1' entry indicates that the part of the corresponding column has an operation on the machine of the corresponding row. A '0' entry indicates the opposite. The m/c matrix with zero/one entries is the major input to cell formation algorithms, which form the machine-part groups. The information given by the m/c matrix shown in Figure I illustrate a case that produces 6 parts using 5 machines. Zeros are not displayed to avoid crowding the matrix. The value of $a_{2,3}$ is equal to '1'; thus, Part 3 needs an operation on Machine 2. In contrast, Part 3 does not need an operation on Machine 1 since $a_{1,3}$ is equal to '0'.

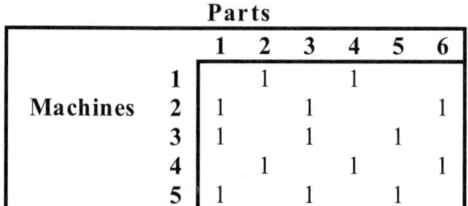

Figure I. Binary machine-component matrix.

The primary objective of a cell formation algorithm in the binary version of the problem is building completely independent cells, i.e., cells in which all parts included in a part family are only processed within the corresponding machine cell. However, this is a case rarely encountered in practice. The grouped cells are usually dependent, i.e., parts have inter-cell moves. The evaluation criterion to judge the goodness of solutions is grouping efficacy. The grouping efficacy measure assesses the quality of diagonalization. The grouping efficacy measure, Γ, is calculated using the following formula [14]:

$$\Gamma = 1 - \frac{e_v + e_x}{e + e_v} = \frac{e - e_x}{e + e_v} \qquad (1)$$

where e = total number of non-zero entries in the m/c matrix;

e_x = total number of non-zero entries outside the diagonal blocks (exceptional elements);

e_v = total number of zero entries inside the diagonal blocks (voids).

III. MINING ASSOCIATIONS AMONG MACHINES

Data mining techniques search through a database without a specific pre-determined hypothesis to generate implicit, previously unknown, and potentially useful information such as knowledge rules, constraints and regularities [8]. An association function is an operation against the set of records returning relationships among the collection of items.

When arranging machine cells by considering manufacturing processes, association rule mining can be employed to discover closeness measures (support values) among machines in the manufacturing process data. The cell formation problem in m/c matrix can be adequately transformed to take a format similar to that of the aforementioned basket analysis in Boolean transaction data. The machines processing the same family of parts could have close associations, and these machines may form a machine cell.

Since data mining creates many challenging research issues, it is necessary to perform dedicated studies to discover new data mining methods. Apriori algorithm is a relative new approach referred to as one of the solutions for association rule learning. The mission of Apriori can be affirmed as follows [8]: find all sets of items (*itemsets*) that have transaction support above minimum support. The support for an itemset is the number of transactions that contain the itemset. Itemsets with minimum support are called *large* itemsets.

In cellular manufacturing, the association rules can be used to appropriately group the machines with respect to their associations. Machines that frequently process a certain set of parts have close associations. The more associative the machines are, the closer are the machines. An efficient algorithm is required that restricts the search space and checks only a subset of all association rules, but, if possible, without missing important rules. Apriori algorithm is one such algorithm. It was developed by Agrawal *et al.* [1] and is implemented in a specific way in the cell formation addressed herein. The part of a corresponding column in the incidence matrix can be taken as a transaction identifier, and the machine of the corresponding row that processes the part can be taken as items purchased together in the basket.

IV. THE CLUSTERING MODEL

Machines that frequently process a certain set of parts have close associations. The machine grouping model stems from the general formulation of clustering problems [18]. The notations of the model are listed as follows:

s_{ij} = support value (closeness measure) between machines i and j;

$X_{ij} = \begin{cases} 1 \text{ if machine } i \text{ is assigned to the cell with} \\ \quad \text{machine } j \text{ as the cell medium;} \\ 0 \text{ otherwise;} \end{cases}$

$Y_j = \begin{cases} 1, \text{ if order } j \text{ is chosen as a cell median;} \\ 0, \text{ otherwise;} \end{cases}$

L_m = lower bound of machine cell size;

U_m = upper bound of machine cell size;
K = number of cells under consideration;
m = number of machines;
n = number of components;

The 0-1 integer model for grouping machines into cells takes the form as follows:

(P1) Maximize $\sum_{i=1}^{m}\sum_{j=1}^{m} s_{ij} X_{ij}$ (2)

Subject to: $\sum_{j=1}^{m} X_{ij} = 1, \quad i=1,2,...,m,$ (3)

$X_{ij} \leq Y_j, \quad i,j=1,2,...,m,$ (4)

$\sum_{j=1}^{m} Y_j = K;$ (5)

$Y_j L_m \leq \sum_{i=1}^{m} X_{ij} \leq Y_j U_m, \quad j=1,2,...,m,$ (6)

$X_{ij} = 0, 1, \quad i,j=1,2,...,m,$ (7)

$Y_j = 0, 1, \quad j=1,2,...,m.$ (8)

In the above model, the objective function (2) serves to select K machines as cell medians such that sum of association measures from all machines to their respective cell medians are maximized. The solution of this model results in the assignment of machines to cells maximizing association measures of the machines in the cells. Constraints (3) ensure that each machine belongs exactly to one machine cell. Constraints (4) and (5) are used to limit the number of machines cell as K. Constraints (6) ensures that at least L_m machines must be assigned to each machine cell, and at most U_m machines are assigned to each machine cell. In order to avoid the formation of a singleton machine cell, L_m is usually set to 2. Constraints (7) and (8) guarantee the binary solution for machine assignment. This formulation involves $(m+m^2)$ 0-1 variables and $(1+m+m^2)$ constraints.

In general, we can take $s_{ij} = s_{ji}$ for $i,j=1,2,...,m$ and $s_{ij} = 100$ if $i=j$. Therefore, Model P1 can be modified to a compact formulation (Liu 1999), and it is formulated as follows (Model P2):

(P2) Maximize $\sum_{i=1}^{m}\sum_{j=1}^{m} s_{ij} X_{ij}$ (9)

Subject to: $\sum_{j=1}^{m} X_{ij} = 1, \quad i=1,2,...,m,$ (10)

$X_{ij} \leq Y_j, \quad i=1,2,...,m, j=1,2,...,i,$ (11)

$\sum_{j=1}^{m} Y_j = K;$ (12)

$Y_j L_m \leq \sum_{i=1}^{m} X_{ij} \leq Y_j U_m, \quad j=1,2,...,m,$ (13)

$X_{ij} = 0, 1, \quad i=1,2,...,m, j=1,2,...,i,$ (14)

$Y_j = 0, 1, \quad j=1,2,...,m.$ (15)

As opposed to Model P1, the compact formulation involves $(m(m+1)/2+m)$ 0-1 variables and $(1+m+m(1+m)/2)$ constraints.

V. THE CELL FORMATION PROCEDURE

In this section, a cell formation procedure based on association rule mining and 0-1 integer programming is developed to identify machine groups and part families simultaneously with minimum manual judgment. The algorithm steps for cell formation are presented as follows.

Step 1: Organize machine-component data
(a). Read m/c matrix $\mathbf{A}_{m \times n}$.
(b). Take the part of the corresponding column in m/c matrix as a record identifier, and the machine of the corresponding row as an item.

Step 2: Proceed Apriori algorithm for obtaining association values among machines
(a). Discover the set of large 2-itemsets, $\mathbf{L}_2$.
(b). Generate support s_{ij} and confidence c_{ij} for each element in $\mathbf{L}_2$, i.e. each pair of Machines i and j (M_i and M_j), $i, j = 1, 2, \cdots, m$.
(c). Compute the minimum support s_{min} as the average value of s_{ij}, and the minimum confidence c_{min} as the average value of c_{ij}.
(d). Obtain the association rule set $\mathbf{R}$ by using the property of minimum support and minimum confidence.

Step 3: Solve the clustering model for grouping machine cells
(a). Set $K = \lfloor m/5 \rfloor$.
(b). Solve the clustering model with K machine cells, $\mathbf{F}_k = 1, 2, ..., K$.

Step 4. Group part families
(a). Maintain the machine cells obtained from Step 3.
(b). For each column (part), calculate the number of operations on each machine cell w_{kj}: $w_{kj} = \sum_{\forall M_i \in \mathbf{F}_k} a_{ij}$.
(c). For each part, assign it to Cell k ($\mathbf{F}_k$) with the highest value of w_{kj}. If any tie exists, it is broken in a random manner.

Step 5. Calculate the grouping efficacy
(a). Calculate the grouping efficacy of cell formation with

K machine cells and part families by using equation (1), and maintain the configuration with the highest grouping efficacy.

(b). If $K \leq \lceil m/3 \rceil$, set $K = K+1$ and return to Step 3. Otherwise, proceed to Step 6.

Step 6. Terminate

Return the cell formation with K machine cells and part families with the highest grouping efficacy.

The task-relevant data in the cell formation problem is the 0-1 incidence matrix (m/c matrix). It is input to the proposed algorithm in Step 1. The m/c matrix is the set of process cards. Each column in the m/c matrix represents a record in the association mining by Apriori algorithm. Each element in the row is a binary variable representing that a machine (item) is present.

In Step 2, the affinities of each pair of machines are calculated, so that only large 2-itemsets, $\mathbf{L}_2$, are generated. In cell formation, an association rule is an implication of the form $M_i \Rightarrow M_j$ [support = s%, confidence = c%]. Provided that support and confidence are higher than the minimum support and the minimum confidence, it implies that parts are frequently processed on Machine M_j, and also on Machine M_i. Therefore, these two machines are potentially required to be placed in the same machine cell. The minimum support and the minimum confidence are calculated as the average values of support and confidence.

Step 3 groups the machine cells by solving the association based clustering model so as to maximize the support values among the machines in a cell. The number of clusters is assigned between [$\lfloor m/5 \rfloor$, $\lceil m/3 \rceil$]. After the machine cells have been grouped, this algorithm proceeds to Step 4 to build part families. A part is assigned to a machine cell provided that most of the operations of this part are processed on this cell. In Step 5, the grouping efficacy is adopted to evaluate the performance of the diagonal block of m/c matrix. The algorithm terminates in Step 6, and outputs the cell formation with the highest grouping efficacy.

VI. NUMERICAL EXAMPLES

The approach has been implemented using the ILOG OPL 3.0 on an IBM compatible PC with a Pentium 4 processor. These three problems, referred to as Examples 1-3 [2, 6, 7], include different instances of the cell formation problem in terms of size and difficulty. In addition, results obtained from the existing methods are available in the cell formation literature.

The grouping efficacy (refer to equation (1)) is used as the evaluation criterion to test the proposed cell formation algorithm. The grouping efficacy has been frequently used in the literature and results of these three examples are available for comparison. Table I summarizes the computational results for the three data sets. Observing the results obtained by the proposed approach, they compare favorably with ZODIAC [7], GRAFICS [23], MST [19], MST-GRAFICS [24] and GA-TSP [9]. However, the proposed approach may not generate better solutions than GP-SLCA (Dimopoulos and Mort 2001) in terms of grouping efficacy.

GP-SLCA and MST allow the existence of singleton clusters (machine cells involving only a single machine). Since singleton clusters can result in a higher grouping efficacy value, GP-SLCA and MST generally have a higher degree of grouping efficacy in comparison with other algorithms. This occurs because grouping efficacy places more emphasis on minimizing voids in the cells than on preventing exceptional elements. In the real world cellular manufacturing systems, singleton clusters are only allowed in some particular situations. Therefore, GP-SLCA and MST may configure an impractical cellular layout. Another main limitation of GP-SLCA [11] is its high degree of computational complexity in relation to alternative cell formation algorithms since GP-SLCA is based on genetic programming. The information about the CPU time of GP-SLCA was not given in Dimopoulos and Mort [11]. However, the GP-based approach usually necessitates a vast amount of computational requirement to sufficiently converge to a solution. The CPU time required for the proposed algorithm ranges from 2.0 seconds for the smallest problem to 16.0 seconds for the largest problem on an IBM compatible PC with a Pentium 4 processor. From the computational results, the proposed approach shows its ability to find quality solutions for the cell formation problems.

Table I. Group efficacy of test examples.

No.	1	2	3
Size	16 X 30	24 X 40	40 X 100
Our method	0.708	0.842	1.0
ZODIAC	0.686	0.839	1.0
GRAFICS	0.675	0.839	1.0
MST-GRAFICS	0.644	N/A	N/A
MST	N/A	0.831	1.0
GA-TSP	N/A	0.84	1.0
GP-SLCA	0.7	0.84	1.0

VII. CONCLUSION

In this paper, an association clustering approach to cell formation problems by using association rule mining and 0-1 integer programming is developed to find the effective configurations for cellular manufacturing systems. By applying association rules to cell formation problems, certain sets of machines (machine groups) that frequently process some parts together can be inducted. A data mining technique referred to as association rule induction is utilized herein to find the association rules among machines from the process database. Machine cells are grouped by solving the association based clustering model so as to maximize the support values among the machines in a cell. Comparing the present results to those of several alternative approaches, the proposed methodology is both effective and efficient in solving cell formation problems.

VIII. REFERENCES

[1] R. Agrawal, T. Imielinski and A. Swami, "Mining association rules between sets of items in large databases," In *Proceedings of 1993 ACM-SIGMOD International Conference on Management of Data*, 1993, pp. 207-216.

[2] F.F. Boctor, "A linear formulation of the machine-part cell formation problem," *International Journal of Production Research*, vol. 29, no. 2, 1991, pp.343-356.

[3] J.A. Brandon, *Cellular Manufacturing: Integrating Technology and Management*, Wiley, NY, 1996.

[4] J.L. Burbidge, *Production Flow Analysis*, Oxford, Clarendon, 1989.

[5] J.L. Burbidge, "Production flow analysis for planning group technology," *Journal of Operations Management*, vol. 10, no. 1, 1992, pp. 5-27.

[6] M.P. Chandrasekharan and R. Rajagopalan, "ZODIAC-an algorithm for concurrent formation of part families and machine-cells," *International Journal of Production Research*, vol. 25, no. 6, 1987, pp. 835-850.

[7] M.P. Chandrasekharan, and R. Rajagopalan, "GROUPABILITY: an analysis of the properties of binary data matrices for group technology," *International Journal of Production Research*, vol. 27, no. 6, 1989, pp.1035-1052.

[8] M.-S. Chen, J. Han, and P.S. Yu, "Data Mining and: an overview from a database perspective," *IEEE Transactions on Knowledge and Data Engineering*, vol. 8, no. 6, 1996, pp. 866-883.

[9] C.H. Cheng, Y.P. Gupta, W.H. Lee and K.F. Wong, "A TSP-based heuristic for forming machine groups and part families," *International Journal of Production Research*, vol. 36,1998, pp. 1325-1337.

[10] C.H. Chu, "Clustering analysis in manufacturing cell formation," *OMEGA: International Journal of Management Science*, vol. 17, 1989, pp. 289-295.

[11] C. Dimopoulos and N. Mort, "A hierarchical clustering methodology based on genetic programming for the solution of simple cell-formation problems," *International Journal of Production Research*, vol. 39, no. 1, 2001, pp. 1-19.

[12] N.L. Hyer and U. Wemmerlov, "Group technology oriented coding systems: structures, application and implementation," *Production and Inventory Management*, vol. 26, no. 1, 1985, pp.125-147.

[13] J.R. King, and V. Nakornchai, "Machine- component group formation in group technology: review and extension," *International Journal of Production Research*, vol. 20, no. 1, 1982, pp.117-133.

[14] C.S. Kumar and M.P. Chandrasekharan, "Grouping efficacy: a quantitative criterion for goodness of block diagonal forms of binary matrices in group technology," *International Journal of Production Research*, vol. 28, no. 2, 1990, pp. 603-612.

[15] C.-M. Liu, "Clustering techniques for stock location and order-picking in a distribution center," *Computers and Operations Research*, vol. 26, 1999, pp. 989-1002.

[16] S.A., Mansouri, S.M.M. Husseini and S.T. Newman, "A review of the modern approaches to multi-criteria cell design," *International Journal of Production Research*, vol. 38, no. 5, 2000, pp. 1201-1218.

[17] C. Mosier and L. Taube, "The facets of group technology and their impacts on implementation-a state-of-the-art survey," *OMEGA: International Journal of Management Science*, vol. 13, 1985, pp. 381-389.

[18] J. M. Mulvey and H.P. Crowder, "Cluster analysis: an application of Lagrangian relaxation," *Management Science*, vol. 25, 1979, pp. 329-340.

[19] S.M. Ng, "Worse-case analysis of an algorithm for cellular manufacturing," *European Journal of Operational Research*, vol. 69, no. 3, 1993, pp. 384-398.

[20] O.F. Offodile, A. Mehrez and J. Grznar, "Cellular manufacturing: a taxonomic review framework," *Journal of Manufacturing Systems*, vol. 13, no. 2, 1994, 196-220.

[21] H.K. Seifoddini and M. Djassemi, "Determination of a flexibility range for quality index for formation of cellular manufacturing systems under product mixed variations," *International Journal of Production Research*, vol. 35, 1997, pp. 3349-3366.

[22] H.M. Selim, R.G. Askin and A.J. Vakharia, "Cell formation in group technology: review, evaluation and directions for future research," *Computers and Industrial Engineering* vol. 34, no. 1, 1998, pp. 3-20.

[23] G. Srinivasan and T.T. Narendran, "GRAFICS-a non-hierarchical clustering algorithm for group technology," *International Journal of Production Research*, vol. 29, no. 3, 1991, pp. 463-478.

[24] G. Srinivasan, A clustering algorithm for machine cell formation in group technology. *International Journal of Production Research*, vol. 32, 1994, pp. 2149-2158.

[25] V. Venugopal, "Soft-computing-based approaches to the group technology problem: a state-of-the-art-review," *International Journal of Production Research*, vol. 37, no. 14, 1999, pp. 3335-3357.

Modular Petri Net Based Modeling, Analysis and Synthesis of Dedicated Production Systems

G. J. Tsinarakis, K. P. Valavanis, N. C. Tsourveloudis
Technical University of Crete
Intelligent Systems and Robotics Laboratory
Dept. of Production Engineering and Management
Chania, Greece GR-73100

Abstract — Ordinary t-timed Petri Nets are used for modeling, analysis and synthesis of random topology production systems and networks. Each production system is first decomposed into production line (transfer chain), assembly, disassembly and parallel machines modules and then their corresponding modular Petri Net models are derived. The overall system PN model is obtained via synthesis of the generic modules satisfying system constraints. P- and T- invariants are calculated and given a random topology production system, the total number of the system PN model nodes (places, transitions) is calculated from the corresponding generic PN modules. Results show the applicability of the proposed methodology.

I. INTRODUCTION

Petri Nets and their modifications are widely used to study DEDS, production systems and networks. A Petri Net (PN) is defined as the five-tuple: $PN=\{P, T, I, O, m_0\}$, where $P=\{p_1, p_2 ... p_n\}$ is a finite set of places, $T=\{t_1, t_2... t_m\}$ is a finite set of transitions, $P \cup T=V$, where V is the set of vertices and $P \cap T = \emptyset$. $I: (P \times T) \rightarrow N$ is an input function and $O: (P \times T) \rightarrow N$ an output function with N a set of non-negative integers, and m_0 the PN initial marking. PN structural and behavioral properties capture precedence relations and structural interactions between system components [1]-[16].

In this paper, (which is the natural outgrowth of previous reported research [15], [17]) t-timed ordinary modular PNs are utilized for modeling, analysis, synthesis and simulation of random topology dedicated production systems, in which machines fail and are repaired randomly. Four generic PN modules, corresponding to the *production line* (or *transfer chain*), *assembly*, *disassembly* and *parallel machines* modules are derived. The overall system PN model is obtained via synthesis of the component models considering simultaneously overall system constraints, and the Martinez-Silva algorithm [19] is used to calculate P- and T- invariants. Given a random topology production system, the total number of the system PN model nodes (places, transitions) is calculated from the corresponding generic PN modules.

Paper contributions include: i) modular PN based approach is independent of the system architecture and structure, ii) the model construction method may be applied to any configuration DEDS, and iii) analysis and synthesis of any complex system is accomplished in terms of analysis and synthesis of the 4 basic PN modules.

Section 2 presents the generic modules; Section 3 derives their PN models; Section 4 presents the overall system PN model; Section 5 presents simulation results, while Section 6 concludes the paper.

II. PRODUCTION SYSTEM GENERIC MODULES

A production system may be viewed as a network of machines/workstations and buffers. Random machine breakdowns disturb the production process and starvation or blocking may occur affecting the downstream and upstream buffer levels. Events that may occur in a production network include changes in buffer states (full or empty) and changes in machine states (up or down). When a machine breaks down preceding machines remain operating until one of their downstream buffers is filled. Similarly, succeeding machines continue processing until their upstream buffers are empty.

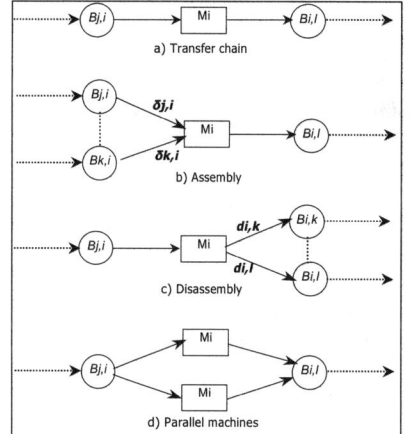

Fig.1 Fundamental, generic, modules

The production floor modeling approach introduced and explained in [15], [17] is extended so that every production system or network is decomposed into the line (chain), assembly, disassembly and parallel machines module, the simplest version of which is shown in Figure 1 (circles and rectangles represent buffers and machines; notation is straightforward). These modules, if connected to each other may represent manufacturing networks of various layouts.

Generalizations of the four generic modules are obvious; transfer chains may contain n machines ($n+1$ buffers), assembly (disassembly) modules may have n input (output) buffers and parallel machine module may have n machines.

III. PETRI NET MODELS OF GENERIC MODULES

The four basic PN models corresponding to the four generic modules of Figure 1 (called *generic PNs* from now on) are shown in Figures 2-5. Timed transitions are presented as white rectangles, while immediate transitions as black rectangles. All transition input and output arc weights are equal to 1. Table 1 explains the meaning of each place and transition. Places $p_0 - p_5$ and transitions $t_1 - t_4$ have the same meaning in all four generic PNs. Transitions correspond to system activities resulting in state changes, while places correspond to resource (machine, parts) availability or state (machine up, down, working, free). Table 2 shows PN module complexity for the general case of n machines in each module.

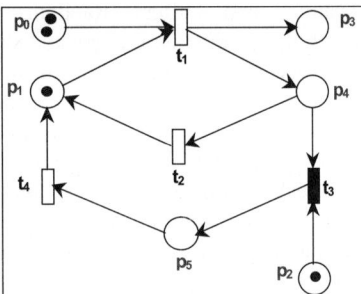

Fig.2 Production line (chain) PN module

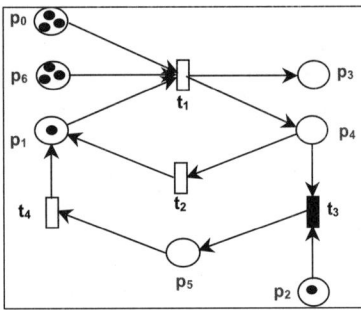

Fig.3 Assembly PN module

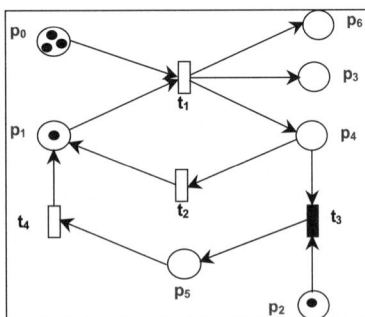

Fig. 4 Disassembly PN module

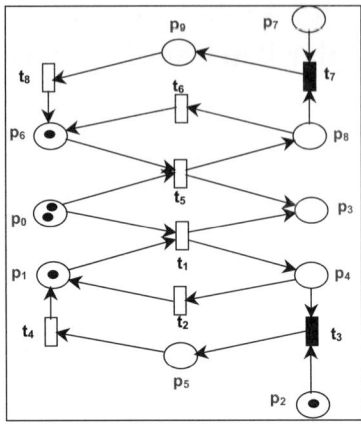

Fig. 5 Parallel machines PN module

The generic PN modules have been derived based on the following mostly realistic assumptions: i) buffers have finite capacities, ii) machines operate at a given speed that may change at specific moments according to events taking place in the system, iii) setup times and transportation times of pieces through the production system are negligible compared to production times, iv) machine breakdowns may happen infinitely often, but only after the completion of a production cycle. Tokens shown in the generic PN modules are for demonstration purposes only: the token in p_1 indicates that a machine is free and operates (next piece production may begin), but when one production cycle terminates, the next cycle starts either immediately (through t_2) or after the appearance of a machine breakdown and its repair (path p_2, t_3, p_5, t_4).

A. Discussion

Considering the four generic PN modules as shown in Figure 1 (not for simulation purposes) with any finite initial marking m_0, several observations are made: a) As long as there is part availability in the input buffer(s), all four generic PNs after the completion of one production cycle return to the state of starting a new cycle; b) the parts number in the initial buffer(s) defines the exact number of production cycles; c) all modules are *k-bounded*; d) modules are non-conservative (transition t_3 consumes two tokens and produces one, in assembly and disassembly module transition t_1 is also non conservative, in parallel machines module t_7 is non conservative); e) modules are non-persistent (firing of t_3 may disable t_2); f) modules are not repetitive and not consistent.

For the transfer chain, the upper limit for *k* is defined as *min* {*max* {C_0, C_3}, ($m_0(p_0)+m_0(p_3)$)}, where C_i is the maximum capacity of p_i. Maximum number of tokens in a place is the minimum of maximum capacity of two buffers and the sum of the initial tokens in these places.

For the assembly module, the upper limit for *k* is different since there are at two input buffers, defined as *min* {*max* {C_0, C_3, C_6}, ($m_0(p_3) + max$ { $m_0(p_0)$, $m_0(p_6)$})}.

For the disassembly assembly module, the upper limit for k is calculated considering two output buffers as $min\ \{max\ \{C_0,\ C_3,\ C_6\},\ (m_0(p_0)\ +\ max\{\ m_0(p_3),\ m_0(p_6)\})\}$.

B. Invariants

The Martinez-Silva algorithm [19] is used to calculate the minimal P- and T- invariants (all other invariants are linear combinations of minimal). P-invariants are nonzero nonnegative integer solutions X of the equation $X^T A = 0$ that also satisfy $X^T m = X^T m_0$ where X is an n_p-element vector, m_0 the initial marking of the net and m a marking of the reachability set of m_0, $R(m_0)$. T-invariants are the nonzero nonnegative integer solutions Y of the matrix equation $AY = 0$ where Y is n_t-element vector. There are $(n_p$-$r)$ basic P-invariants and $(n_t$-$r)$ T-invariants, where $r=rank(A)$. P-invariants express a notion of token conservation in sets of places for all reachable markings without enumeration of the reachability set. T-invariants describe a transition firing sequence s, such that $m_j \rightarrow m_j$.

Node	Model	Meaning
P_0	Common	Parts available in initial buffer
p_1	Common	Machine available to process part
p_2	Common	Machine breakdown
p_3	Common	Parts in final buffer
p_4	Common	Machine finished process of a part
p_5	Common	Machine out of order
p_6	Assembly	B type parts available in corresponding initial buffer
p_6	Disassembly	B type parts in the corresponding final buffer
p_6	Parallel machines	Second Machine (M_2) available to process part
p_7	Parallel machines	Machine M_2 breakdown
p_8	Parallel machines	Machine finished process of a part
p_9	Parallel machines	Machine M_2 out of order
t_1	Common	Machine processing (producing) part
t_2	Common	Empty machine's signal return
t_3	Common	Machine breaks down
t_4	Common	Machine has been repaired and is available to produce again
t_5	Parallel machines	Machine M_2 is processing (producing) part
t_6	Parallel machines	Empty machine's signal return for M_2
t_7	Parallel machines	Machine M_2 breaks down
t_8	Parallel machines	Machine M_2 has been repaired and is available to produce again

Table 1 Basic modules node (P and T) explanation

Model	Nodes type	Basic (generic) model	Generalized model (n components)
Transfer chain	P	6	$5 * n + 1$
	T	4	$4 * n$
Assembly	P	7	$n + 5$
	T	4	4
Disassembly	P	7	$n + 5$
	T	4	4
Parallel machines	P	10	$2 + 4 * n$
	T	8	$4 * n$

Table 2 Complexity of generalized PN modules for n machines

The transfer chain module has 2 P-invariants and no T-invariant. The P-invariants are $\{1\ 0\ 0\ 1\ 0\ 0\}$ and $\{0\ 1\ 0\ 0\ 1\ 1\}$ resulting in $m(p_0) + m(p_3) = n_1$ and $m(p_1) + m(p_4) + m(p_5) = 1$. The first P-invariant guarantees that the sum of parts in the initial and in the final buffer is constant and equal to the initial sum of parts in these buffers n_1, while the second shows 3 mutually exclusive machine states (machine ready to process part, empty or machine breakdown). The two above P-invariants are common for all four modules; however the other modules have one more. The third P-invariant of the assembly module is $m(p_6)+m(p_3)=n_2$ and refers to the sum of tokens of the second initial buffer and final buffer that is equal to the initial sum of parts in these two buffers. The third P-invariant of the disassembly module is $m(p_0)+m(p_6)=n_2$ and refers to the sum of tokens of the initial buffer and the second final buffer. The third P-invariant of the two parallel machines module is $m(p_6)+m(p_8)+m(p_9)=1$ and refers to the mutually exclusive states of the second machine M_2, same with the ones of the first machine.

IV. PN MODULE SYNTHESIS

The synthesis procedure of the simplest two transfer chains is shown in Figure 6. Generalizations are provided.

Observing Figure 6, it is obvious that places p_3 and p_6 are fused in place p_{3-6}. The total number of places is reduced by one, while transitions are equal to the total of each module transitions. The combined PN input places are reduced by one (p_{3-6} is an internal place, not input buffer any more). The maximum capacity of p_{3-6} may be defined as $C_{3-6} = min\{C_3,\ C_6\}$ or $C_{3-6} = max\{C_3,\ C_6\}$ or with any number in between (based on system constraints). Obviously, $m_0(p_{3-6})=m_0(p_3)+m_0(p_6)$.

The combined PN properties may be detected accordingly by simulation and use of appropriate tools. There exist three P-invariants (two are identical with the individual module P-invariants). Two refer to the mutually exclusive states of the combined PN given by equations $m(p_1)+m(p_4)+m(p_5)=1$ and $m(p_7)+m(p_{10})+m(p_{11})=1$. The third refers to the preservation of the total number of parts in the PN and is given by $m(p_0) + m(p_{3-6}) + m(p_9) = n_1$, where n_1 is the initial sum of parts (tokens) in the three places. Synthesis of other generic PN modules is obtained in a similar way but due to space limitations are omitted.

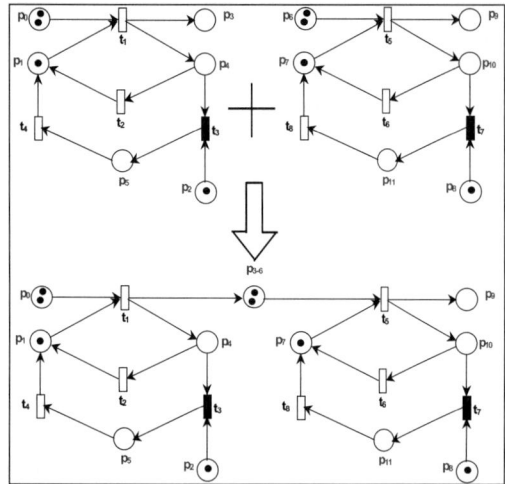

Fig. 6 Synthesis of two generic transfer chain modules

A. Generalizations

It is possible to calculate the number of nodes of a random topology production system PN model from the corresponding PN modules that compose it and from the number of external parts entering the system. The latter number is necessary to compute the number of fused places in the individual modules connection points.

Consider first that a production system combined PN model is derived in terms of the four generic Petri Net modules; that is n_1 modules of transfer chain, n_2 modules of two-piece assembly, n_3 modules of two-piece disassembly, n_4 modules of two-parallel machines, and that there are n_5 (external, non-fused) input places. The combined PN model consists of $4*(n_1+n_2+n_3)+8*n_4$ transitions. The total number of generic PN modules places when considered separately is $6*n_1+7*(n_2+n_3)+10*n_4$. Fusion of places at connection points reduces the number of places by $n_1+n_3+n_4+2*n_2-n_5$ (number of places that are not external inputs). Thus, the total number of the combined net places is $5*(n_1+n_2)+6*n_3+9*n_4+n_5$.

Considering next individual PN modules as shown in Table 2, the total number of transitions is given by

$$4*(n_2+n_3)+\sum_{i=1}^{n_1}4*l_i+\sum_{j=1}^{n_4}4*l_j,$$ where l_i and l_j are the transfer chain and parallel machine module components. The total number of places is calculated as

$$\sum_{i_1=1}^{n_1}(5*l_{i_1}+1)+\sum_{i_2=1}^{n_2}(l_{i_2}+5)+\sum_{i_3=1}^{n_3}(l_{i_3}+5)+$$
$$+\sum_{i_4=1}^{n_4}(4*l_{i_4}+2)-(n_1+n_3+n_4+\sum_{i_2=1}^{n_2}l_{i_2}-n_5)=$$
$$=\sum_{i_1=1}^{n_1}5*l_{i_1}+\sum_{i_3=1}^{n_3}l_{i_3}+\sum_{i_4=1}^{n_4}4*l_{i_4}+5*n_2+4*n_3+n_4+n_5,$$

where l_{i1}, l_{i2}, l_{i3}, l_{i4} show the number of machines (chain), number of input and output buffers in the assembly and disassembly modules, and the number of parallel machines in the parallel machine module, respectively.

V. A CASE STUDY

The production system of Figure 7 with its PN model shown in Figure 8 is used as a case study [15], [17]. The system is composed of 2 transfer chains, 2 assembly modules and 1 disassembly module. Parts enter the system through initial buffer (Module 1), while parts reaching the final buffer after machine M_5 are final parts ready to be removed from the system. PN consists of 28 places and 20 transitions. 5 transitions are immediate corresponding to potential machine breakdowns. There are 2 external input places p_0 and p_{25}. Parts reaching p_{29} are finished parts.

From Figure 9, it is obvious that internal buffer p_{9-19} is full of parts for a large percentage of the system function, potentially resulting in frequent machine (M_4) blockage, while other buffers (like p_{22-32}) do not even reach half of their capacity. Changing buffer capacities, for example by reducing arbitrarily the capacity of buffer p_{22-32} to 3 (from 8) and by repeating the simulation with the rest of the parameter values the same, the simulation is terminated after 547 steps with total duration 228 time units. Figure 10 shows the internal buffers levels during this simulation.

Next reduction of the buffer p_{16-26} capacity from 8 to 5 is tried. Simulation is terminated after 553 steps and the total time is 229 time units. The mean production time after these two changes is 6.94 time units. Figure 11 shows the results. This process may be repeated (trial and error) until all buffers work at their capacities.

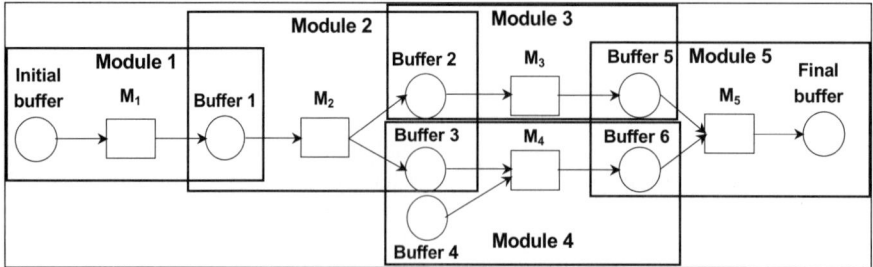

Fig. 7 A Production System and its module decomposition

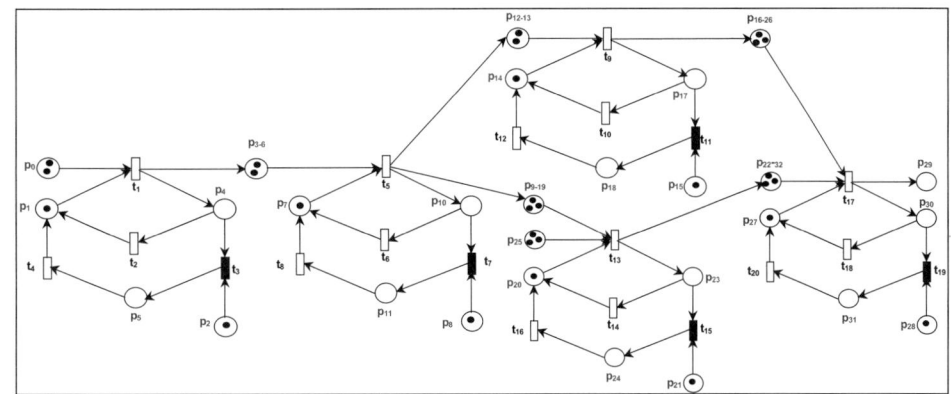

Fig.8 Overall system Petri Net model

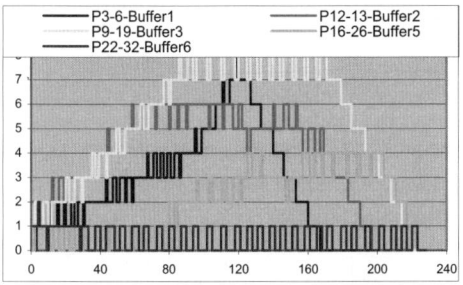

Fig. 9 Internal buffer levels during simulation

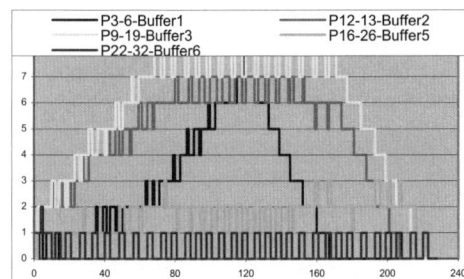

Fig. 10 New buffer levels-reduced capacity of p_{22-32}

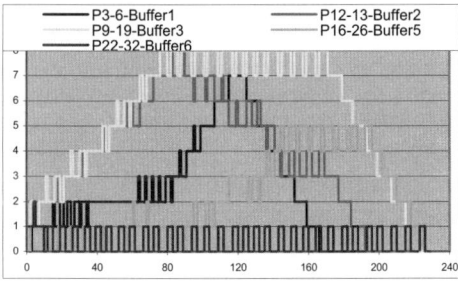

Fig. 11 New buffer levels-reduced capacity of p_{16-26}

System performance may also be improved by changing machine times, reducing idle periods, etc. By reducing the production time of machine M_4 by 1 time unit, the simulation is completed after 475 steps and the total time needed to produce the 33 pieces is 196 time units. The mean production time is 5.94 time units, 1 time unit less than before (15% performance improvement). The new buffer levels are shown in Figure 12.

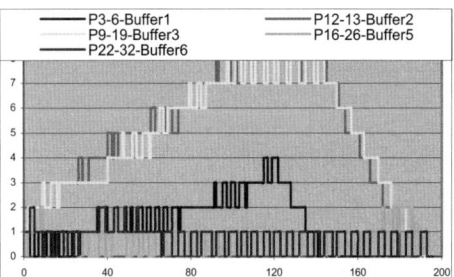

Fig. 12 Buffer levels after changing M_4 production rate.

Further system performance improvement may be obtained by studying machine breakdown behaviour. Reducing the frequency of machine breakdown appearance (with preventive machine maintenance) will increase system operational periods. Let's consider that the mean time of the exponential breakdown appearance for machine M_4 is 6 time units instead of 3. In this case the simulation is completed in 459 steps with total duration of 189 time units and mean production time of 5.73 time units, 0.21 time units less than before. Figure 13 shows the corresponding buffer levels during simulation.

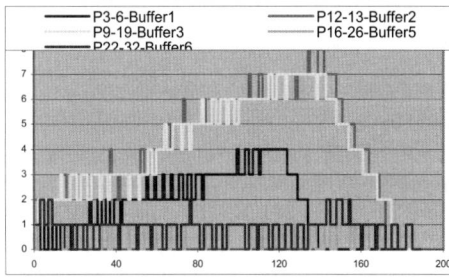

Fig. 13 New buffer levels after changing M_4 breakdowns mean time

Additionally, for this simulation, the corresponding machine operation rates are approximately 63.5% for M_1, 49.2% for M_2, 52.4% for M_3, 67.7% for M_4 and 70% for M_5. This shows that M_2 and M_3 are idle for longer time periods in comparison with the other machines (almost for half of the operational time of the system) and so they may be used for other activities as well. Figure 14 shows how production has changed as function of the changes made.

An interesting point that concerns production systems and is obvious also by simulations, is that the behaviour of the net is heavily determined by the slowest machine, as other machines of the net are obligated to follow its production rhythms through the appearance and spread of blockages and starvation phenomena.

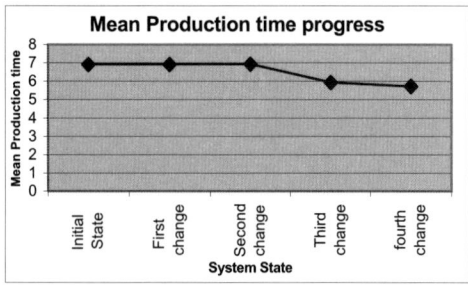

Fig. 14 Progress of the mean production cycle time in relation with the described system changes

VI. CONCLUSIONS

Modular PNs have been used for modelling, analysis and synthesis of random topology production networks. Four generic modules have been considered and their corresponding generic PN modules have been derived. Generalizations have been provided and expressions for the number of system nodes for random configuration systems have been calculated. P-invariants provide further insight to production systems study. Simulation results demonstrate the effectiveness of the proposed method.

VII. ACKNOWLEDGEMENTS

This work has been partially supported by a research grant from the GSRT/EU, EPAN M.4.3 Greece-Slovenia. The authors wish to thank Henryk Anschuetz for the provision of the full version of HPSIM Petri net simulator.

V. REFERENCES

[1] Balduzzi F., Giua A., Menga G., First – Order Hybrid Petri Nets: A model for Optimization and Control, *IEEE Transactions on Robotics and Automation*, Vol. 16, No. 4, August 2000.

[2] Domg M., Chen F. F., Process modelling and analysis of manufacturing supply chain networks using object-oriented Petri nets, *Robotics and Computer Integrated Manufacturing* 17, pp. 121 – 129, 2001.

[3] Frey G., Assembly line sequencing based on Petri-net T-invariants, *Control Engineering Practice* 8, pp. 63-69, 2000.

[4] Jehng Wern-Kueir, Petri net models applied to analyze automatic sequential pressing systems, *Journal of Materials Processing Technology* 120, pp. 115 – 125, 2002.

[5] Jeng M.D., Xie X., Modeling and Analysis of Semiconductor Manufacturing Systems With Degraded Behavior Using Petri Nets and Siphons, *IEEE Transactions on Robotics and Automation*, Vol. 17, No. 5, October 2001.

[6] Valavanis K., On the Hierarchical Modeling Analysis and Simulation of Flexible Manufacturing Systems with Extended Petri Nets, *IEEE Transactions on Systems, Man and Cybernetics*, Vol. 20, No. 1, February 1990.

[7] Desrochers A., Al – Jaar R., *Applications of Petri Nets in Manufacturing Systems –Modeling, Control and Performance Analysis*, IEEE Press, 1995.

[8] Moody J., Antsaklis P., *Supervisory Control of Discrete Event Systems Using Petri Nets*, Kluwer Academic Publications, 1998.

[9] David R., Alla H., *Petri Nets & Grafcet – Tools for modeling discrete event systems*, Prentice Hall 1992

[10] Proth J. M., Xiaolan X., *Petri Nets, A Tool for Design and Management of Manufacturing Systems*, John Wiley & Sons, 1996.

[11] Zhou M.C., DiCesare F., *Petri Net Synthesis for Discrete Event Control of Manufacturing Systems*, Kluwer Academic Publishers, 1993.

[12] Murata T., Petri Nets: Properties, Analysis and Applications, *Proceedings IEEE*, Vol. 77, No. 4, pp. 541 – 580, April 1989.

[13] Gu T., Bahri P., A survey of Petri net applications in batch processes, *Computers in Industry* 47, pp. 99 – 111, 2002.

[14] Valette R., Chezalviel – Pradin B., Girault F., *An Introduction to Petri net theory, Fuzziness in Petri Nets*, Eds. J. Cardoso, H. Carmago, *Studies in Fuzziness and Soft Computing*, Vol. 22, pp. 3 – 24, Physica Verlag 1999.

[15] Tsourveloudis N., Dretoulakis E., Ioannidis S., Fuzzy work-in-process inventory control of unreliable manufacturing systems, *Information Sciences*, Vol. 27, pp. 69 – 83, 2000.

[16] Christensen S., Petrucci L., Modular Analysis of Petri Nets, *The Computer Journal*, Vol. 43, No. 3, pp. 224 – 242, 2000.

[17] Ioannidis S., Tsourveloudis N., Valavanis K., Fuzzy Supervisory Control of Manufacturing Systems, *IEEE Transactions on Robotics and Automation* (Revised – Under review).

[18] Govil M., Fu M., Queueing Theory in Manufacturing: A Survey, *Journal of Manufacturing Systems*, Vol. 18, No. 3, 1999.

[19] Martinez J., Silva M., A Simple and fast algorithm to obtain all invariants of generalized Petri net, *Application and Theory of Petri Nets*, Vol. 52, pp. 301 – 311, New York: Springer-Verlag, 1982.

[20] Anschuetz Henryk, HPSIM, http://home.t-online.de/home/henryk.a/petrinet/e/hpsim_e.htm.

Robust Modeling of Dynamic Environment based on Robot Embodiment

Kuniaki NODA[1], Mototaka SUZUKI[1], Naofumi TSUCHIYA[1], Yuki SUGA[1], Tetsuya OGATA[1][2],
and Shigeki SUGANO[1]

1) Humanoid Robotics Institute, Waseda University
3-4-1 Ookubo Shinjuku Tokyo 169-8555, Japan
{march, mottaka, tsuchiya, ysuga, sugano}
@paradise.mech.waseda.ac.jp

2) RIKEN Brain Science Institute
Hirosawa, 2-1 Wako-shi, Saitama 351-0198, Japan
ogata@brain.riken.go.jp

Abstract - **Recent studies on embodied cognitive science have shown us the possibility of emergence of more complex and nontrivial behaviors with quite simple designs if the designer takes the dynamics of the system-environment interaction into account properly. In this paper, we report our tentative classification experiments of several objects using the human-like autonomous robot, "WAMOEBA-2Ri". As modeling the environment, we focus on not only static aspects of the environment but also dynamic aspects of it including that of the system own. The visualized result of this experiment shows the integration of multimodal sensor dataset acquired by the system-environment interaction ("grasping") enable robust categorization of several objects. Finally, in discussion, we demonstrate a possible application to making "invariance in motion" emerge consequently by extending this approach.**

1. Introduction

Our goal is to make robots which can move around and select their action to achieve a variety of purposes in the environment changing everymoment, and survive consequently. Particularly, the ability to behave suited to accomplishing many ends in our habitat and communicate with us smoothly is mostly required. On the other hand, almost all of the creatures on the earth have behaved adaptively to eat, and protect their own body from the predators and inherited their genes since about 550 million years ago. Therefore, when we design such a robot which can generate adaptive behaviors, bio-inspired approach, abstracting the principles from living creatures, is one of the most hopeful way to do that [1].

It is most important for such an adaptive robot to percept the change of their environment appropriately, and to behave rationally in response to the perception to adapt to the ecological niche. R. Brooks argued that evolution has concentrated its time to get the ability to move around in a dynamic environment, sensing the surroundings to a degree sufficient to achieve the necessary maintenance of life and reproduction - it is much harder [2]. In addition, recent studies in sensory-motor coordination show us how an agent reduces its sensor input space so that it can abstract the regularity from the sensor input. For the design of an agent which enables sensory-motor coordination, the evolutionary approach is also quite encouraging [3][4].

The essence of the concept of sensory-motor coordination is that active reaching out to the environment by an embodied agent makes the sensor inputs more structured and does learning it easier. It allows the cost and complexity of the successive computations to be reduced [5]. To generate an adaptive behavior against an object in the changing environment, an agent has to get meaningful information and couple them with its motor system. Besides, the agent has to have a model of the object, which is based on the multimodal sensor inputs through the sensors of the robot own, to generate an adaptive behavior against it. For example, P. Dario et al. have shown that the integration of visual and tactile information can be applied to a specific disassembly task successfully [6]. But, to have such models or representations is not the eventual purpose of an adaptive robot in the dynamic environment. Considering behaviors in the real world, it is not necessary nor desired to model the environment minutely either [2]. The purpose to have a model of the environment, which has the spatiotemporal correlation (i.e. spatial and temporal correlation) of multimodal sensory inputs, is to select an appropriate action suited to the situation in the dynamic environment.

Based on these ideas described above, we report our trial to model the dynamic environment and evaluate it using the human-like autonomous robot, WAMOEBA-2Ri (Fig. 1). Our basic idea as modeling is to shrink multi-dimensional aspects of the dynamically changing environment into the bodily data of the robot that is very easy to deal with.

Fig. 1. Human-like autonomous robot WAMOEBA-2Ri which has 7 degree-of-freedom arms.

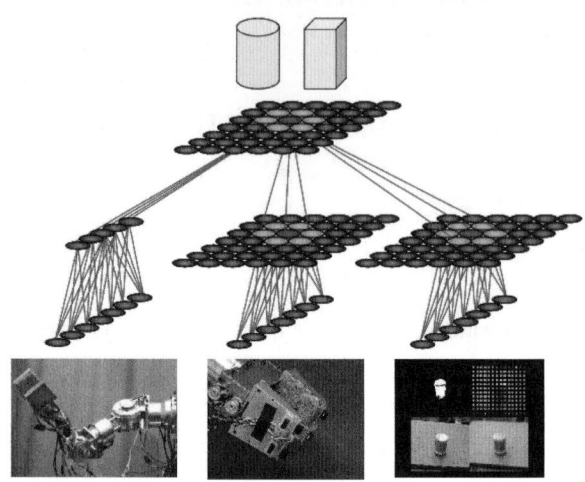

Fig. 2. Integration of Kohonen Maps.

2. Modeling of Dynamic Environment

2.1. Active Interaction: Grasping

Proper perception needs proper action. We should not discuss only the mechanism of perception as an independent function, but consider it as an aspect of the coupled system with behaviors of agents in the environment [6]. Well-suited sensory-motor coordination reduces the complexity of sensor input space and make them more structured. For human beings, grasping behavior is one of the most effective ones to percept an unknown object. When an infant finds something novel in the environment, he/she sometimes grasps it and brings it close to his/her face. This behavior enables the infant to observe the object more closely and normalize the scale of the object automatically [7]. Then, starts interacting with it using his/her own body (e. g. turn it, put it into his/her mouth, throw it away). This process is quite dynamic, active and flexible. In this course, he/she gets multimodal data about what the object is.

These physical interactions are also quite useful for autonomous robots when they learn or adapt to the dynamic environment. Interaction between an autonomous robot which has a proper sensory-motor system and the dynamic environment leads to getting a spatiotemporal correlation of its multimodal sensor data. Only through its multiple sensor inputs acquired by its own behavior, an autonomous robot is able to have a meaningful model of the changing environment for generating adaptive behaviors.

2.2. Integration of Multimodal Sensor Data

Physical interaction between an autonomous agent and the environment generates a coherence between multiple sensor data. This coherence plays an important role for the agent to behave to the object rationally because the coherence of the multiple sensor inputs which are coupled with an each actuator activate them temporally, and potentiate an adaptive coordinated behavior using its whole body consequently (Fig. 2).

3. Experiment

3.1. WAMOEBA-2Ri

This study has developed an autonomous robot named WAMOEBA-2Ri to investigate the robot-human emotional communication and autonomous intelligence of robot systems inspired by the mechanism of the endocrine system into both the hardware and the software [8][9]. WAMOEBA-2Ri is a wheel-type independent robot which consists of 20 degree-of-freedom in total.

WAMOEBA-2Ri has the sensors to acquire not only the external information by the CCD cameras, the microphones, and the torque sensors but also the internal information such as the voltage of the battery, the electrical current, the temperature of the motor et al. More detailed data of the hardware of WAMOEBA-2Ri are shown in Table 1. WAMOEBA-2Ri evaluates the information by using the self-preservation functions corresponding to these sensors. The function is a kind of fuzzy membership function which evaluates the durability of the robot hardware. Based on the result of the evaluation, WAMOEBA-2Ri can constantly control the sensor range, the motor speed, the cooling fan output, and the power switches of every motor and sensor. WAMOEBA-2Ri communicates with human beings in emotional ways by the expressions produced by these internal and external body changes.

	WAMOEBA-2R
Dimensions	1390(H) x 990(L) x 770(W) mm
Weight	Approx. 130 kg
Operating Time	Approx. 50 min
Max speed	3.5 km/h
Payload	2 kgf/hand
External DOF — Neck	2
External DOF — Vehicle	2
External DOF — Arm	4 x 2 = 8
External DOF — Hand	1 x 2 = 2
Internal DOF — Cooling Fan	10
Internal DOF — Power Switches	4
External Sensors — Image Input	CCD Cameras x 2
External Sensors — Audio Input	Microphones x 3
External Sensors — Audio Output	Speaker
External Sensors — Distance Detection	Ultrasonic Sensors x 4
External Sensors — Joint Torque	Torque Sensors x 6
External Sensors — Grip Detection	Photoelectric Sensors x 2 / Pressure Sensors x 2
External Sensors — Object Detection	Touch Sensors x 8
Internal Sensors — Temperature	Thermometric Sensors x 8
Internal Sensors — Battery Voltage	Voltage Sensor
Internal Sensors — Motor Current	Current Sensor
Material	Duralumin, Aluminum
CPU	Pentium III (500MHz) x 2
OS	RT-Linux

Table 1. Here showing specific data on the hardware of WAMOEBA-2Ri.

3.2. Grasping

From the point of view of evolution, the eventual goal of vision system is not only to recongnize objects, but to enable an autonomoous agent to emerge its behavior effectively in the real world [10]. In this experiment, we designed the grasping behavior as a part of sensory-motor coordination coupled with the sensor input. Actually as a reflective action caused by the color input of the robot vision. The vision system of this robot can search an area which color is more specific in comparison with the other and scale is suitable for grasping. Consequently, this robot can grasp the suitable object wherever it is on the table. As designing this vision-reflection system, our methodology is also based on the idea about evolution of color vision in primates. D. Osorio et al. claimed that the development of color vision enabled primates to discriminate fruits against the background [11]. Discriminating an unique colored area against the environment is also an important competency for an autonomous robot which moves around in the human habitat. Grasping behavior is shown in Fig. 3.

3.3. Experimental Task: Classification

Categorization is the underlying competency of intelligence. It is important for categorization by a robot to emerge a meaningful behavior which can exploit the mechanism of sensory-motor coordination. The essential

Fig. 3. A close view of WAMOEBA-2Ri grasping an object. The position of the object is movable on the table. Capturing the object by the two cameras activates the reaching behavior.

mechanism of categorization is the parallel sampling of an environment by the sensor maps through same or different sensor modalities. The task of this experiment is to classify three objects using the own sensors and actuators of the robot, WAMOEBA-2Ri. The appearance of the objects and experimental environment is shown in Fig. 4, and the characteristics of the three objects are summarized in Table 2. After grasping the object, the robot holds out its arm horizontally. This is because capturing object by two cameras activates the behavior. Holding out its arm horizontally makes the values of torque sensor at its shoulder joint maximum. That means that the torque sensor becomes most sensitive to the difference between the weights of the three objects at that moment. This is an easy example of sensory-motor coordination (we, human beings hardly do that when comparing the weights of two objects thanks to our dextrous, quite sensitive hands and arms without such an effort).

Then, the input data of the torque sensor maximized in this way, the color data i.e. YUV values of the object are calculated, values of the pressure sensors at the gripper, and width of the gripper (hereinafter, these pressure and width data are combined and called "hand data") are integrated using Kohonen maps in this experiment. Using Kohonen maps allow the robot to organize the models of the three objects based on its embodiment without the designer's intervention because of the self-organizing characteristic of the map. To investigate the robustness of this method, we added the every type of noise to the integrated dataset of the three objects and examined robustness of maps for categorization tasks under that various conditions. The noise is generated by random numbers whose maximum amplitude is 0.5 and added to the original normalized data from each sensor modality.

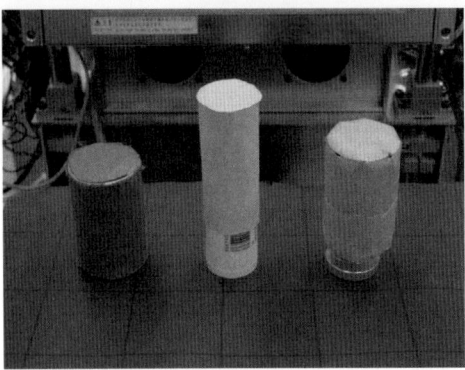

Fig. 4. Three objects for classification experiment. These objects have different characteristics which are summarized in Table 1. Hereinafter we call these objects a, b, and c from left to right.

object	color	stiffness	diameter	weight
a	red	hard	wide	heavy
b	yellow	hard	thin	light
c	green	soft	wide	light

Table 2. Characteristics of three objects. These characteristics are different apparently for us, but that is not necessarily for the other.

3.4. Result

Kohonen maps displaying distributions of firing rates of the neurons in the integration layer are shown in Fig. 5. Three maps at the top row clearly indicate that the multimodal sensor data have been integrated successfully and a slight glance at the three maps enables us to classify them obviously as well. The other maps shows the robust modeling of the three objects based on the robot embodiment even if these objects moved anywhere on the table. The robot could generate the embodied models of these objects using its own sensory-motor system in the dynamic environment. By grasping an object, the robot could organize a robust model that is composed by the data of color of the image, pressure at the hand, width of the gripper, and torque loaded with the shoulder of the object. Compared with the integrated data with no noise, the map of the integrated data of object 'b' with hand noise (in the middle of third row from the top) show the noise of hand data have much more influence on the integration data 'b', but it is clear that the integration of multimodal sensor dataset results in robust organized map that enable the classification task easily against any disturbance in this experiment.

3.5. Additional Experiment

To investigate the robustness of this multimodal integration that we examined, we compared the result of it with the one using an only single layered Kohonen map to integrate the multimodal sensor inputs. In this experiment, same type of random noise described at 3.3 was added to each sensor modalities. The integrated data with no noise were also classified clearly even in the single layered one, but the integrated data with hand noise were disturbed a lot particularly. Visual comparison between multi-layered Kohonen maps and single layered ones shows us the difference easily and clearly (data not shown). However, to investigate that difference quantitatively, we calculated Euclidean distances between the integrated data with no noise and the other ones with some kinds of noises in both cases. This calculation means to test of coincidences about the firing rate of whole neurons. The result of calculation is also shown in Fig. 6.

This graph shows the data with hand noise is much more disturbed in the case of single layered maps. Additionally, it shows that the noise of sensor input affects the disturbance on the Kohonen map in proportion to the dimension of sensor in the single layered ones. In the Kohonen algorithm, the influence on the organized map is determined by the ratio of the component data in the input vector. Therefore, in the multi-layered maps, abstracting sensor inputs from the lower maps and equalizing the dimensions of them results in much smaller influence of some kinds of noise on the organization of the maps. This result also means that it is not always necessary to use Kohonen map as a lower layer to equalize the dimension of sensor inputs. The other algorithm which is able to abstract sensor inputs can play the same role as Kohonen map in this experiment.

4. Discussion

In this experiment, the robot dealt with the static data of the objects and organized a model of the environment by integrating Kohonen maps which are assigned to each sensor modality. By grasping objects with its own sensory-motor system, the robot could reduce the dynamics of the objects and enabled to convert them into the easy, static, and bodily data from its own body. As a result, the multimodal integrated dataset led to robust categorization of the three.

By the way, the environment doesn't have only such a static aspect, but also the invariances (i.e. the unalterable characteristic emerged during a change) that is generated

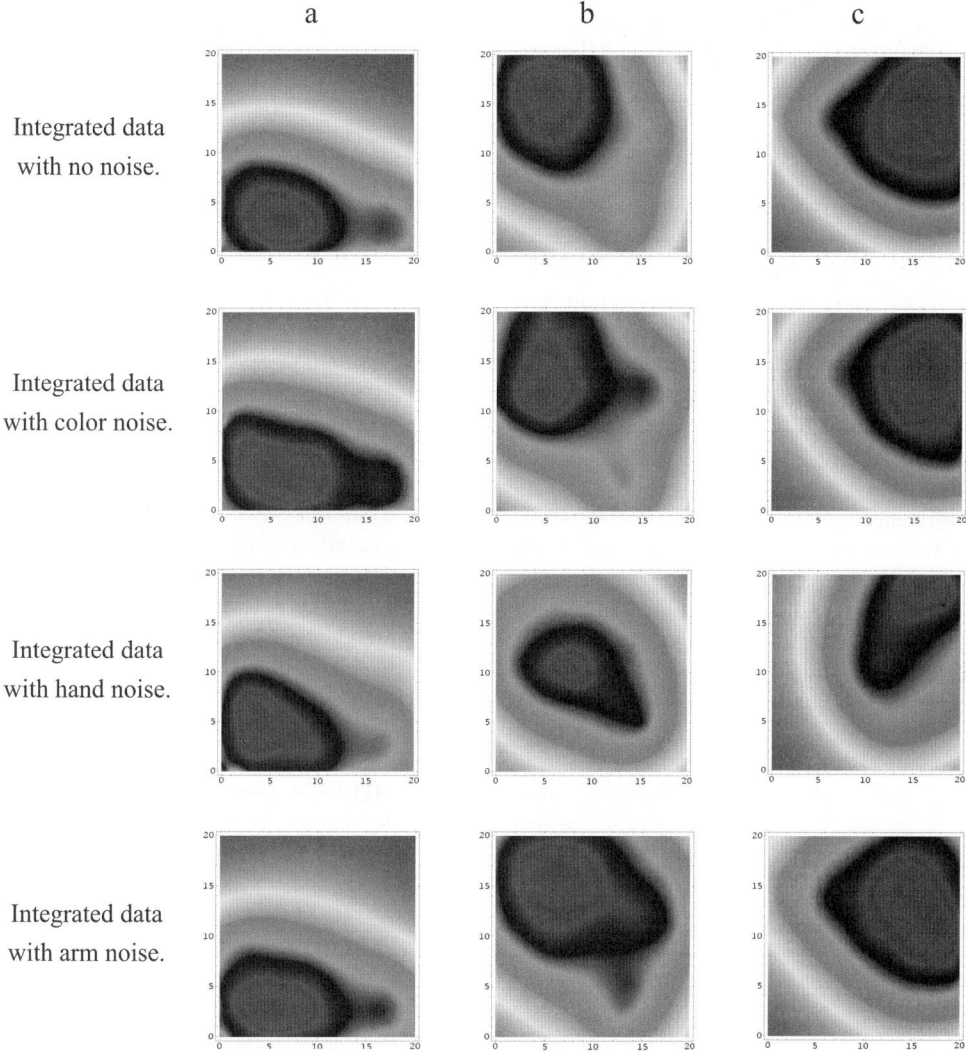

Fig. 5. Kohonen maps displaying the distribution of the firing rate in the integration layer.

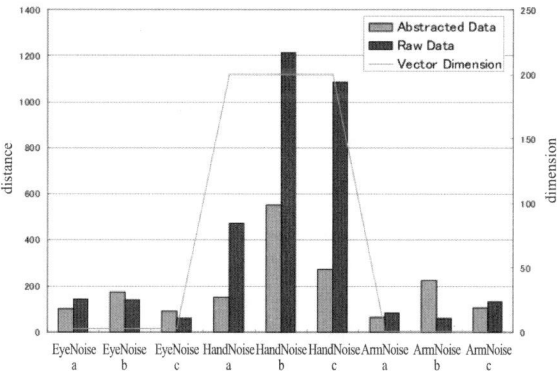

Fig. 6. Comparison between multi-layered Kohonen maps and single layered ones on robustness to classify the three objects.

in the system-environment acquaintance. In case of human beings, some invariances emerging when a dynamical system composed of our bodies and objects for perception generate appropriate behavior. In other words, we find "invariance in motion". For living creatures that have perception competency, it dominates their behaviors to percept the macro patterns of the environment. It is quite important for robots moving around in a dynamic environment to percept such a invariance directly because the invariance represent macro property of the environment.

Now we have to consider how an autonomous agent acquires such invariances in the dynamic environment using its competencies. As an easy example that we can try easily with the robot, we conceived an additional task, shaking the grasped object.

First, two objects (e.g. cylinders in this paper) with same weight and different height are prepared. Shaking behavior shows us a characteristic property of the object. The difference in height of these objects appears in the difference in time-series data of the torque loaded with the moving (shaking) joint. This is caused by the difference between two objects in moment of inertia of the dynamical system which consists of its arm and the object. Here the moment of inertia is an invariance which emerges through the behavior, "shaking the object". An agent never percept this property of the objects without moving dynamically.

In this experiment, the robot can percept the difference in moment of inertia between two objects directly by the shaking behavior based on its embodiment. This is an easy example of emergence of environmental properties caused by dynamics, that is invariances. As we described above, we have to consider perception for an autonomous robot as a competency to control its behavior suited to macro properties in the dynamic environment, and explore further emergence of such a dynamic sensory-motor coordination.

5. Conclusion

We proposed the robust way to model the dynamic changing environment through the physical interaction (grasping in this paper) followed by the active reaching coupled with the visual input and abstraction of multimodal sensor data based on robot embodiment. As an evaluation experiment, we reported the result of a classification experiment of objects which could move everywhere on the table using the human-like autonomous robot, WAMOEBA-2Ri. Finally, we discuss the possibility to develop this method into the further level that dynamic data generated by negotiating between the system (robot) and the environment are taken into account.

We described that it is one of the hopeful approaches to use an autonomous robot which has an appropriate sensory-motor system and the dynamics generated by a variety of active robot-environment interaction. It is not until we start from this stage that we can step forward to the hard problem on generating active and flexible behaviors in autonomous systems because the common essential of both two characteristics is being dynamic.

References

[1] B. Webb and T. R. Consi, *Biorobotics*. Cambridge, MA: MIT Press, 2001.

[2] R. A. Brooks, "Intelligence without Representation," *Artificial Intelligence*, 47, 139-160, 1991.

[3] S. Nolfi, "Adaptation as a more powerful tool than decomposition and integration," in: T. Fagarty and G. Venturini (Eds.), *Proceedings of the Workshop on Evolutionary Computing and Machine Learning, 13th International Conference on Machine Learning*, University of Bari, Italy, 1996.

[4] R. D. Beer, "Toward the evolution of dynamical neural networks for minimally cognitive behavior," in P. Maes, M. Mataric, J.-A. Meyer, J. Pollack, and S. W. Wilson (Eds.), From animals to animats: *Proceedings of the 4th International Conference on Simulation of Adaptive Behavior* (pp. 421-429). Cambridge, MA: MIT Press, 1996.

[5] R. Bajcsy, "Active Perception", *IEEE Proceedings*, 76, pp 996-1005, 1988.

[6] P. Dario, M. Rucci, C. Guadagnini, C. Laschi, "Integrating Visual and Tactile Information in Disassembly Tasks", *Proceedings of the '93 ICAR International Conference on Advanced Robotics*, pp. 191-196, 1993.

[7] J. J. Gibson, The ecological approach to visual perception. Boston: Houghton Mifflin, 1979.

[8] E. Bushnell and J. P. Boudreau, "Motor development in the mind: The potential role of motor abilities as a determinant of aspects of perceptual development", *Child Development*, 64, pp. 1005-1021, 1993.

[9] S. Sugano and T. Ogata, "Emergence of Mind in Robots for Human Interface -Research Methodology and Robot Model," in *Proceeding of IEEE International Conference on Robotics and Automation (ICRA'96)*, pp. 1191-1198, 1996.

[10] T. Ogata and S. Sugano, "Emotional Communication Between Humans and the Autonomous Robot which has the Emotion Model," in *Proceeding of IEEE International Conference on Robotics and Automation (ICRA'99)*, pp. 3177-3182, 1999.

[11] R. Pfeifer and C. Scheier, Understanding Intelligence. Cambridge, MA: MIT Press, 1999.

[12] D. Osorio and M. Vorobyev, "Colour vision as an adaptation to frugivory in primates," *Proceeding of the Royal Society of London*, B, 263:593-599, 1996.

Strategy Acquirement by Survival Robots in Outdoor Environment

Pitoyo Hartono[1], Keishiro Tabe[2], Kenji Suzuki[2], Shuji Hashimoto[2]

[1]Advance Research Institute for Science and Engineering Waseda University
[2]Department of Applied Physics, School of Science and Engineering Waseda University
Ohkubo 3-4-1 Shinjuku-ku Tokyo 169-8555
Japan
Email: {hartono,tab,kenji,shuji}@shalab.phys.waseda.ac.jp

Abstract – In this research we propose a model of autonomous robot that is able to survive in real world outdoor environment without human intervention. For an autonomous robot, energy is a very important factor, so it is only natural that the ability of the robot to accumulate energy should be used as evaluation of the robot's survival strategy.

The robot in this study is equipped with a solar panel, and a model of neural network ensemble. The solar panel enables the robot to autonomously accumulate energy in the outdoor environment, while the neural network ensemble enables the robot to acquire strategies that are needed in order to survive in the environment. The accumulated energy is considered as the feedback from the environment to measure the efficiency of the strategy exerted by the robot, and used as learning guidance for the robot to survive.

1. INTRODUCTION

In this research we construct a number of autonomous robots that acquire the strategies to survive in real world outdoor environments through a learning mechanism. The survival ability mentioned throughout this paper refers to the ability of the robots to autonomously maintain their energy amount to a certain level that ensures their ability for executing their strategies. The robots are equipped with solar panels, and a model of neural network ensemble that enables the robots to execute different strategies to cope with dynamic environments.

A number of learning algorithms to acquire strategies in certain environments have been proposed [1,2,3,4], but most of them are constructed on the assumption that the environments are stationary or a subclass of Markov Decision Processes (MPDs), which are not necessarily a good representation of real world environment. We also aware of a number of studies of using evolutionary process and learning process to train robots to perform some task [5,6,7,8,9], but most of the existing research deal with simulated environments or artificial environments. Our study differs from these studies, because in this study the task of the robot is to survive in the real world outdoor environment without human interventions.

For autonomous robots that have to survive in real world environments without human help, energy is a very important factor. Most of the studies of autonomous robots focus on the ability of the robots to make decisions without human help or the ability of the robot to generate good running strategy through their interaction with their living environments. We consider that the ability of energy accumulation without human helps is very critical for autonomous robots. Consequently, the ability of the robot for accumulating energy should be used as a parameter to evaluate the strategy exerted by the robots in order to survive in their living environment. A good strategy is a strategy that allows the robots to efficiently accumulate energy, which consequently ensures their survival, while a strategy that consistently decreases the energy level of the robot should be considered as a bad one.

We notice the studies of robots' behaviour acquirement that take the residual energy level of the robots into consideration [5,6]. Our proposed robots can be distinguished from these studies, because our robot autonomously learns to acquire the energy by referring to the difference between the accumulated and consumed energy that can be a good evaluation of the robot's strategy in the current environment, and a trigger for the switching of strategy in case of environmental change.

We implement a neural network ensemble model, consisting of a number of independent neural networks. The ensemble can automatically train each of its members to acquire a strategy (a function that maps sensory input into a certain action of the robot), and executes strategy switching to deal with unpredictable dynamic environment.

The hardware structure of the robot used in our study is explained in Section 2. The neural network ensemble that controls the behavior of the robot is explained in Section 3. Then the experimental result in the real outdoor environment is given in Section 4. The conclusions and discussions are presented in the final section.

II. ROBOT'S STRUCTURE

In this study, considering that the robot will have no access to energy sources provided by human, the robot is only equipped with a number of simple sensors to make the energy consumption minimum. The external and internal structures and the specifications of the robot are shown in Fig.1, Fig.2 and Table 1, respectively.

The robot has a solar panel attached to its top. The sensory inputs from the environment are obtained through the 4 CdS light sensors that are attached in the front, back, right site and left site of the robot. The robot is equipped with one chip microcontroller PIC, which measures the amount of accumulated and consumed energy, and executing the robot moving strategy by referring to the sensory input and the difference between energy accumulation and consumption. Two DC motors are set in the front site of the robot. The strategies acquired by the robot are expressed in the form of neural networks' parameters stored in the EEPROMs as shown in Fig. 2.

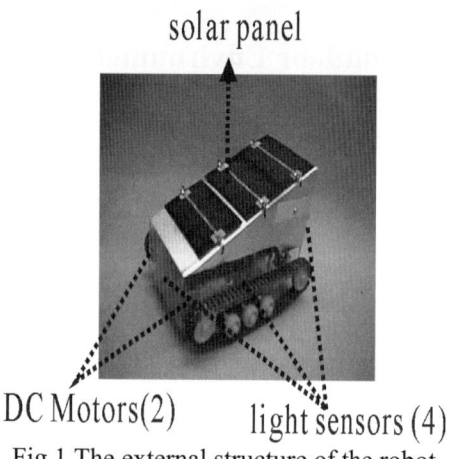

Fig.1 The external structure of the robot

TABLE 1.
ROBOT'S SPECIFICATIONS

Weight	0.7 [kg]
Dimension	21[cm] x 15[cm] x 15[cm]
Power consumption	Running: 2W, Stationary: 0.03W
Rechargeable battery	3500[mAh] at 3.6[V]
Solar panel	400[mAh] at 4.5 V
Sensors	4 CdS light sensors

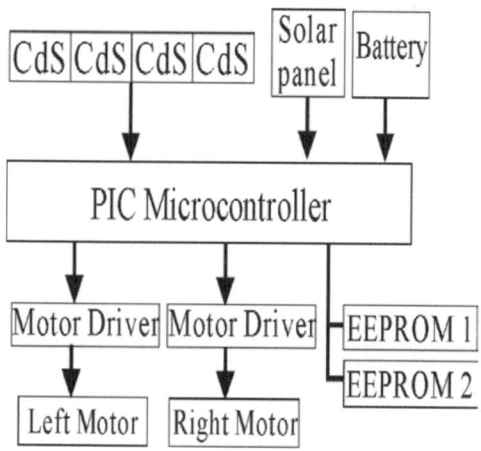

Fig.2 The internal structure of the robot

III. NEURAL NETWORK ENSEMBLE

The robot in this study is required to have a number of strategies to deal with dynamic environments and the disadvantages in its ability to collect information because of the limitation in the number and varieties of the sensors. It is clear that a single controller will be insufficient for the robot to survive in real world environment. For example, the robot may be required to execute two different strategies when running on two different surfaces, although the sensory inputs to the robot maybe the same. Because the robot cannot sense the difference in the surfaces, the switching of the strategies should be triggered by the feedback from the environment, which is the value of the energy accumulation.

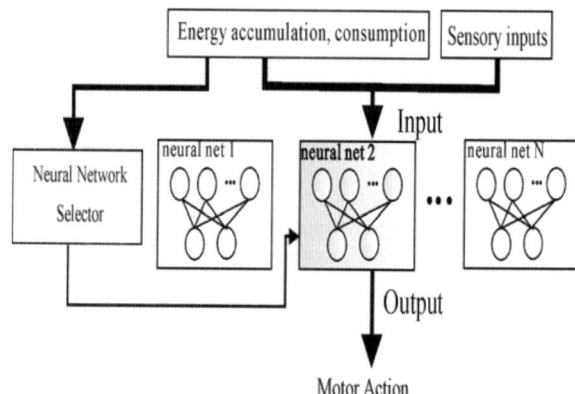

Fig.3 Neural Network Ensemble Controller

We propose a model of neural network ensemble consisting of a number of Perceptrons[12,13,14], each can be automatically trained to acquire one strategy.

A neural network ensemble with N members is illustrated in Fig. 3. The neural network ensemble is equipped with a neural network selector that selects one active member based on the evaluation of the chosen strategy. The strategy used by the robot is the one of the active member. The evaluation of the currently active strategy is the difference between the amounts of the energy accumulation and the energy consumption. As the consequence of a particular strategy taken by the robot, if the robot acquired a positive energy accumulation, then it can be taken as an indication that the robot is executing a correct strategy, so the strategy has to be reinforced by correcting the weights of the active member, while a negative accumulation indicates that the strategy is wrong, hence strategy switching should be triggered. Because the weight corrections are only executed for the currently active member, different members develop different strategies.

The input to the neural network is a 6 dimensional vector X of which elements are the sensory information from the four light sensors, the energy accumulation and consumption. . Each motor attached in the robot can take one of three possible actions, which are forward, stop and backward, so robot is able to generate 9 kinds of movements. Consequently the output of the neural network is 9 dimensional Y, where each of its components is associated with one action that could be generated by the robot. The structure of each neural network in the ensemble is shown in Fig. 4.

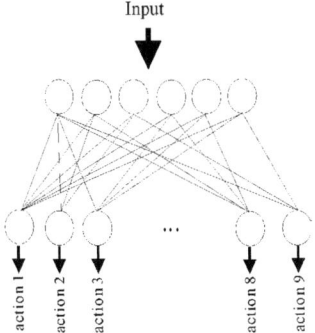

Fig. 4 Neural Network's structure

Each neural network has 6 input units that are connected to 9 output units.

The j-th element of the output of the s-th neural network (provided that the neural network is the active one) in the ensemble at time t, $y_j^s(t)$ is calculated as follows,

$$y_j^s(t) = \sum_{i=1}^{6} w_{ij}^s(t) x_i(t) \quad j \in \{1,2,\cdots,9\} \quad (1)$$

where w_{ij}^s is the weight between the i-th input unit and the j-th output unit in the s-th member, and x_i is the value of the i-th sensory input. The robot choses an action $act(t)$ whose associated output unit produces maximum value, as follows,

$$win(t) = \arg\max_j \{y_j^s(t)\} \quad (2)$$
$$act(t) = action\ win(t)$$

Provided that the s-th member is an active member at time t, its connection weights are updated based on the modified Hebbian learning method, as follows,

$$w_{ij}^s(t) = w_{ij}^s(t-1) + \eta\ x_i(t-1) E_\tau^s(t)$$
$$E_\tau^s(t) = E_{acc}(t) - E_{con}(t) \quad (3)$$

where η is the learning rate, $E_{acc}(t)$ and $E_{con}(t)$ are the accumulated and consumed energy at time t, respectively. τ is the length of the evaluation cycle.

The evaluation for a strategy taken by the robot at the n-th cycle is as follows,

$$E^s(n) = \sum_{t=(n-1)\tau}^{\tau} E_{(n-1)\tau}^s(t) \quad (4)$$

The weights of non-active members are not updated.

Positive value for E^s keeps the presently active member to further run and reinforce its strategy, while negative E^s triggers a switch of strategy. The switching of members is regulated by the following rule,

For $E^s < 0$

$$P_{continue} = 1 + \mu E^s$$
$$P_{switch} = -\frac{\mu}{N-1} E^s$$

For $E^s \geq 0$ $\quad (5)$

$$P_{continue} = 1$$
$$P_{switch} = 0$$

where μ is a normalization constant, $P_{continue}$ is the probability that the currently running strategy will continue to run and P_{switch} denotes the probability of strategy switching. When the switch occurs, the active member will be chosen randomly among the previously inactive $N-1$ members.

The life cycle of the robot is shown as follows,

Start
  Neural network ensemble initialisation
  while (survive){
    get sensory input
    get evaluation
    if(evaluation < 0){
      calculate switching probability
      if (switched){
        select a new strategy
        run the new strategy
      }
      else{
        continue running
      }
    }
    else{
      update weight
      continue running
    }
  }

IV. EXPERIMENT

Experiment with a robot is done in the outdoor environment inside the university campus, as shown in Fig. 5.

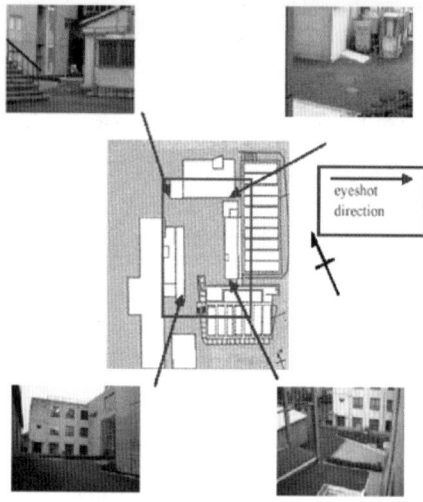

Fig. 5 Outdoor environment for the experiment

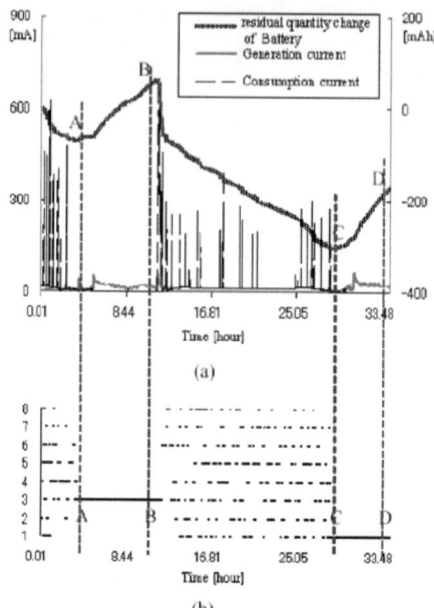

Fig. 6 Experiment Result

The experiment was conducted for 3 days during which the various weather conditions were observed.

We used a neural network ensemble with 8 members. The weights of the neural networks were initialised randomly. The values of normalization constant μ and learning rate η were empirically set to 0.06 and 0.04, respectively. The evaluation is given every 1 second.

Fig. 6 (a) shows the fluctuation of the robot's battery charge in the period of 33 hours around the second day of the experiment. High energy consumption is an indication of high activity of the robot. Fig. 6 (b) shows the active member at a given time. The duration A-B in Fig 6 is the daytime period. In this period the robot was able to intensely accumulate its energy. It is obvious that before the duration A-B (in the early morning), the energy consumption was high, because at that time the robot actively searching for sunspots. It is also clear that the 3rd neural network in the ensemble has the most efficient strategy for the robot to survive in this condition. In the duration B-C, where the robot was clearly unable to accumulate energy due to the shortage of sunray, the switching of members in the ensemble became frequent. We also noticed that the robots calmed down after a while, this is due to the activation of a strategy of energy saving. The robot acquired a sleep-like behavior. In C-D period, the robot was able to intensely recharge its energy, using the strategy generated by the 1st member of the ensemble. In this experiment, two different members (the 3rd member and the 1st one) were activated to run in the similar time periods in two different days. The possible reason for the behavior, is that although A-B and C-D are approximately the same time periods, the weather condition was not the same, or the robot was in the different positions, so these conditions were interpreted as different environments by the robot that consequently triggered different strategy.

V. CONCLUSION AND DISCUSSION

In this experiment we have constructed an autonomous robot that has the ability to obtain a family of strategies to survive in dynamic environments. The ability to acquire these strategies is based on the interaction between the robot and the environment, in which the environment gives feedback to the robot in the form of evaluation of the accumulated energy.

Strategy acquirement for survival without human intervention is one of the most important aspects for autonomous robot s that have to run in real world dynamic environments, especially when human have no prior knowledge, so that pre-programmed behavior is not possible. The ability to switch strategy in dealing with the environmental change is also a large contribution in increasing the robustness of the robots, because some physical or internal failures will be treated as environmental change for the robot and trigger the activation of different strategy that can deal efficiently with the conditions of the robots. In this case, the robot may not be running perfectly, but failure is not necessarily fatal for the robots. We also showed that the robot is able to acquire strategy for running in real world outdoor environment, and at the same time accumulate energy to ensure their survival.

From our preliminary experiments, we notice that the robot acquired a number of animal-like behavior, such as "sun-bathing" in the daylight, and sleep-like behavior in the night when there is no possibility for energy accumulation.

In the near future we plan to build a group of survival robots, each with slightly different external structures (for example the form of the wheel, the type of the sensors, the location of sensors, the efficiency of solar panels, etc), and internal structures (for example, different learning algorithms, different neural network's structures, learning parameters, etc). We also plan to implement a more sophisticated strategy switching mechanism [15,16,17]. These differences can be likened to diversity inside a particular biological species. It will be interesting to observe how different kind of individual robots acquired different kind of survival mechanisms. It is also of interest to observe the development of survival mechanisms of not only for a particular robot as an individual but also on a group of robots that have some kind of communication methods between them.

Although our main aim is to develop autonomous robots that can efficiently adapting to their living environment by self-acquiring survival strategies, we also expect to develop the robots, so that they can be used as a research platform to understand the process of strategy acquirement and the emergence of intelligence of living organism.

VI. REFERENCES

[1] K. Miyazaki and S. Kobayashi, "On the rationality of profit sharing in partially observable Markov Decision Processes, Proceedings of The 5th International Conference on Information System Analysis and Synthesis, 1999, pp. 190-197.

[2] C. Watkins, and P. Dayan, "Q-learning", Machine Learning Vol. 8, 1992, pp. 279-292.

[3] L. Chrisman, "Reinforcement learning with perceptual aliasing: the perceptual distinction approach", Proceedings of the 10th National Conference on Artificial Intelligence, 1995, pp. 387-395.

[4] R. McCallum, "Instance-based utile distinction for reinforcement learning with hidden state", Proceedings of the 12th International Conference on Machine Learning, 1995, pp. 387-395.

[5] K. Kondo and I. Ishikawa, "Hebbian Learning of a behavior network and its application to a collision avoidance by an autonomous agent", Systems, Control and Information, Vol. 15, No. 7, 2002, pp. 350-358. (in Japanese).

[6] D. Floreano and J. Urzelai, "Evolutionary robots with on-line self-organization and behavioural fitness", Neural Networks Vol. 13, 2000, pp. 431-443.

[7] L. Steels., "Emergent functionality in robotic agents through on-line evolution", In: Brooks, R. and Maes, P. (eds) Artificial Life IV. Proceedings of the Fourth International Workshop on the Synthesis and Simulation of Living Systems, 1994, pp. 8-14.

[8] S. Nolfi and D. Parisi, "Learning to adapt to changing environment in evolving neural networks", Adaptive Behavior Vol. 5, No. 1, 1997, pp. 75-98.

[9] S. Nolfi and D. Marocco, "Evolving robot able to integrate sensory-motor information over time", Theory in Biosciences, Vol. 120, 2001, pp. 287-310.

[10] L. Steels, "The artificial life roots of artificial intelligence, Artificial Life Journal, Vol. 1, Nos. 1-2, pp. 89-125, 1993.

[11] T. Ogata, and S. Sugano, "Emotional behavior adjustment system in robots", Proceedings of IEEE International Workshop on Robot and Human Communication, 1997, pp. 352-357.

[12] F. Rosenblatt, "Two theorem of statistical separability in the perceptron", Proceedings of a Symposium on the Mechanism of Thought Process, 1959, pp. 421-456.

[13] F. Rosenblat, Principal of Neurodynamics", Spartan Book, 1962.

[14] M. Minsky and S. Papert, Perceptron, MIT Press, 1969.

[15] P. Hartono and S. Hashimoto, "Temperature switching in neural network ensemble", Journal of Signal Processing, Vol. 4, No. 5, 2000, pp. 395-402.

[16] P. Hartono and S. Hashimoto, "Neural network ensemble with temperature control", Proceedings International Joint Conference on Neural Networks 1999, pp. II-4073-4078.

[17] P. Hartono and S. Hashimoto, "Extracting the Principal Behavior of a Probabilistic Supervisor Through Neural Network Ensemble, International Journal of Neural Systems, Vol. 12, Nos. 3 & 4, 2002, pp. 291-301.

A Bio-inspired Approach for Regulating Visco-elastic Properties of a Robot Arm

L. Zollo (1), B. Siciliano (2), E. Guglielmelli (1), P. Dario (1)

(1) Scuola Superiore Sant'Anna
ARTS Lab c/o Polo Sant'Anna Valdera
Viale Rinaldo Piaggio 34
56025 Pontedera (Pisa), Italy
Email: {l.zollo, e.guglielmelli, p.dario}@arts.sssup.it

(2) PRISMA Lab
Dipartimento di Informatica e Sistemistica
Università degli Studi di Napoli Federico II
Via Claudio 21, 80125 Napoli, Italy
Email: siciliano@unina.it

Abstract

Neurophysiological studies show that humans possess the capability of generating appropriate motor behaviors to different uncertain environmental conditions by combining a forward action, produced by the internal forward dynamic model, and a feedback control, realising the transformation from sensory information to motor commands.

To this regard, a control system based on the combination of a feedforward and a feedback control loop has been developed in order to provide a robot arm with human-like adaptation capabilities.

The work analyses the role of biological coactivation in the mechanism of adjustable visco-elastic arm properties and proposes a function for the evaluation of the robot arm coactivation based on the measure of the position error and the interaction force. The coactivation function is used to update the proportional and derivative parameters of the feedback controller and, consequently, the arm visco-elasticity in unpredictable environmental conditions.

Finally, experimental results on the evolution of the coactivation in the adaptation and de-adaptation phases are provided in the last section of the paper.

1 Introduction

Robotics-to-biology relation is quite intricate, since not only robotics can derive inspiration from biology, but it can even provide a useful means for better understanding human beings and biological behaviors in general (Fig. 1).

On one side, robotics can be considered as a means to test biological models on a platform and update the model through the validation mechanism. In this sense, robotics is a means of advancement in biological knowledge.

On the other side, biology can represent an inspiration source for the development of new control

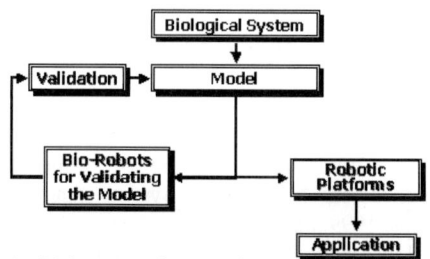

Figure 1: Biology-to-robotics scheme

models for robotic platforms and, in this case, biology is a means to increase robot performance.

The work presented in this paper has originated as an attempt of applying biological models to robots in order to overcome some limitations of the standard control strategies, but in a further step it would investigate the possibility to exploit the proposed controlled platform in a validation mechanism of biological models.

Particularly, the human motor control is taken into consideration in order to realise a robot arm motor control.

The study of neurophisiological works suggests a motor control model for the human arm mainly based on a combination of two basic types of transformations; the first one is the sensory-to-motor transformation, realised by an internal inverse dynamic model, while the second one is the motor-to-sensory transformation, mimed in an internal forward dynamic model [1].

The sensory-to-motor transformation provides the motor commands to reach the desired state, hence it is generally used as a feedback controller. In biological systems feedback control can cause instability in rapid movements due to the significant delay in the transmission of neural information [2] and, moreover, experimental evidence shows that the feedforward control (based on the internal forward model) plays a fundamental role in the biological motor control [3]–[5]. Hence, the human arm control seems to combine feedforward and feedback control in order to ensure stability in low as well as fast movements (by feedforward) [1] and to allow unskilled movements and interaction with the unknown environment by feedback [6], [7].

The work presented here proposes a control scheme for an anthropomorphic robot arm which, inspired by the above biological considerations, provides an easy solution to the adaptability of a robot arm to unpredictable interactions with the environment. In accordance with the biological motor control system, the proposed control scheme includes both a feedforward control loop (to improve stability) and a feedback control loop (aimed at providing the arm with adequate motor commands in the interaction as well as in the free space).

The developed feedback controller is based on a proportional-derivative plus gravity compensation control scheme [8]–[10] and introduces an adaptive law for updating the proportional-derivative parameters during the motion. The law for PD modulation is inspired by the adjustable mechanism of the visco-elastic property in the human arm, which allows regulating the muscular activity in accordance with the movement error [6].

The concept of variable visco-elastic property in the process of learning the arm internal model was exploited in [11], [12]. Particularly, an adjustable visco-elastic parameter modulation mechanism directly depending on the value of the control error was proposed in [6]. In such a way, when the learning process starts the visco-elastic property value is high but it decreases during learning jointly with the control error.

The PD parameter adaptive law takes into account the link between the visco-elastic property and the control error [6], but enriches it with the dependence on the value of interaction force in conditions of impact or constrained motion. In particular, the PD parameters are modulated through a unique parameter, called *coactivation*, which is updated on the base of the measures of the actual position error and the sensed interaction force. The coactivation allows modulating the arm visco-elastic property in accordance with the variability of the context of movements.

Evidence is provided in the experimental session which focuses on the role of coactivation in the regulation of the behavior of an 8-degree-of-freedom robot arm moving in a variable environment.

Finally, as regards the realization of the feedforward controller, the forward dynamic model is assumed to be known and the forward action is realised without learning.

2 Theoretical description of the control scheme

The study of important works on the biological motor control suggests a control scheme for an anthropomorphic robot arm which could take inspiration from the activation mechanisms of the human arm.

Particularly, the controller proposed in this section is inspired by the considerations on the cooperation between a feedforward control mechanism and a feedback control mechanism in the biological control and focuses on the role of the feedback loop in the regulation of the arm visco-elastic property.

Basically, the proposed arm controller consists of two torque contributions (Fig. 2): the first one (τ_{FF}) is produced by the feedforward loop through the forward dynamic model and the second one (τ_{FB}) is evaluated by the feedback controller through a proportional-derivative plus gravity compensation control law.

As regards the realization of the forward action, the arm dynamical model is assumed partially or completely known so that the corresponding torque vector could be calculated as

$$\tau_{FF} = B(q_d)\ddot{q}_d + C(q_d, \dot{q}_d)\dot{q}_d + F\dot{q}_d \quad (1)$$

where

- $B(q) \in R^{n \times n}$ is the joint inertia matrix (n is the joint space dimension);
- q_d is the $n \times 1$ desired joint vector;
- $\ddot{q}_d, \dot{q}_d$ are the joint acceleration and velocity vectors;
- $C(q, \dot{q})\dot{q} \in R^{n \times 1}$ is the vector of centrifugal and Coriolis torques;
- F is the diagonal, positive definite matrix of joint viscosity coefficients.

As regards the evaluation of the arm motor commands in the feedback loop, a proportional-derivative action in the joint space (2) is proposed, with more emphasis on the adjustable mechanism of K_P and K_D parameters:

$$\tau_{FB} = K_P(c)\tilde{q} - K_D(c)\dot{q} + g(q) \quad (2)$$

where K_P and K_D are respectively the proportional and derivative matrix gains which allow regulating the robot stiffness and dampness. The quantity $\tilde{q} = q_d - q$ is the joint position error and $g(q)$ is the estimate of the gravitational torques acting on the joints.

The proposed control law is addressed to improve the process of gain adjustment of the PD control in the joint space by mimicking the biological motor control acting on the human arm joints and agonist/antagonist muscles. Hence, the proportional and derivative gains are assumed to depend on a unique parameter c, capable to modulate both the stiffness and the damping factor of the robot arm. In this way, the choice of an adequate updating law for the c factor allows modulating the visco-elastic property of all the arm joints. The c factor is derived by the analysis of the human arm muscular activity and is called *coactivation*.

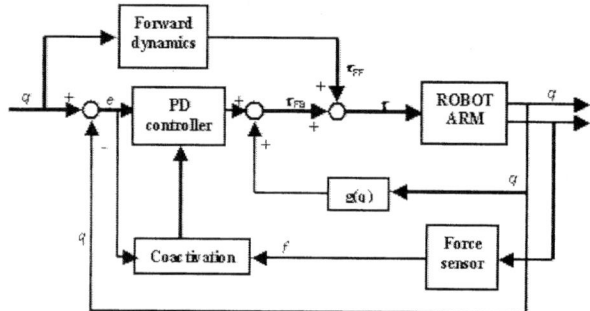

Figure 2: PD plus gravity compensation control scheme

3 Coactivation-based PD adaptive law

As explained in [13], [14], the spinal reflexes (expecially the stretch reflex) are strictly depending on the stiffness or, more specifically, on the level of coactivation between the agonist and antagonist muscles. In the human arm, the visco-elastic behavior is regulated by the activity of alfa and gamma motoneurons, coordinated by the coactivation mechanism. For instance, a low level of alfa activation causes also a low level of gamma activation and, consequenly, a low muscular activity.

Since gamma motoneurons are responsible for the sensitivity to variations in position or in velocity, the coordinated activity of alfa and gamma motoneurones allows the muscle contraction as well as the perception of the contraction. The proper activation of the two types of motoneurons is managed by the coactivation mechanism (Figs. 3, 4).

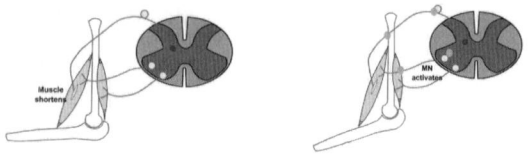

Figure 3: Muscle contraction *Figure 4: Muscle extension*

The concept of coactivation is often introduced in human arm modeling in order to describe the visco-elastic behavior of the flexor and extensor muscles [6], [14], [15] and develop a mechanism of neural motor learning based on the regulation of the visco-elastic parameters [6], [11].

Here instead, the concept of coactivation c is used to update the proportional and derivative gains of the control law both in the unconstrained and in constrained motion, by exploiting the variation of the arm visco-elastic properties with c. Thus, the idea is that the control system for the robot arm provides the joint actuation torques (through (2)) and modulates the arm stiffness and damping properties by c, as in the human arm the control system produces motor commands depending on c which regulates the muscle activity. An appropriate choice for the c value could improve the arm accuracy in the free space, by increasing the stiffness, and increase the arm compliance and elasticity when an external bound is sensed, by decreasing the stiffness.

In both cases, the gains of diagonal matrices K_P and K_D start from an initial value (empirically evaluated) and evolve by increasing or decreasing as a function of the coactivation factor.

In case of unconstrained motion, the zero force measured by the force sensors indicates that the arm can freely move in the execution of the task. Hence, the arm is required to be highly accurate and the i-th element of K_P has to increase from its minimum value as

$$k_P(c) = \begin{cases} k_{Pmin} & \text{if } c = 0 \\ k_{Pmin} + \bar{k}_p c & \text{if } c \neq 0, \ k_P < k_{Pmax} \\ k_{Pmax} & \text{if } c \neq 0, \ k_P \geq k_{Pmax} \end{cases} \quad (3)$$

The updating law for c is an increasing monotonic function of the sole position error

$$c = \alpha \sqrt{\tilde{q}^T \tilde{q}} \quad (4)$$

where $\bar{k}_p$ is the coefficient which allows regulating the level of coactivation for each joint. k_{Pmin} is the minimum gain allowing a quite accurate motion and k_{Pmax} is the maximum gain ensuring stability in the motion; α is a positive coefficient.

Also the i-th element of K_D matrix is characterised by an adaptable law similar to (3), but with a slower increasing rate, i.e.

$$k_D(c) = \begin{cases} k_{Dmin} & \text{if } c = 0 \\ k_{Dmin} + \bar{k}_d c & \text{if } c \neq 0, \ k_D < k_{Dmax} \\ k_{Dmax} & \text{if } c \neq 0, \ k_D \geq k_{Dmax} \end{cases} \quad (5)$$

When the sensors measure an interaction force acting on the arm, the k_P function has to decrease from its initial value in order to ensure more flexibility in the interaction direction and make the arm capable to adapt to the constraints, i.e.

$$k_P(c) = k_{Pin} \bar{h} c \quad (6)$$

$$c = c_{min} + \alpha \frac{1}{\sqrt{f^T f}} \quad \text{if } f^T f \neq 0 \quad (7)$$

where k_{Pin} is the initial value for the proportional parameter and $\bar{h}$ is a scalar coefficient.

The coactivation function (7) is always positive but inversely depending on the interaction force. When the force value f increases, the coactivation decays until a minimum value c_{min}, which has to be different from zero, in order to allow the arm to move, but sufficiently low to ensure a high level of compliance.

Experimental trials on stiffness evolution during the constrained motion pointed out the importance of setting a constant minimum value for c, c_{min}, to ensure stability of the robot arm even in the presence of very high interaction forces. As regards the superior limit, it is natural to choose it unitary, since in the interaction stiffness has to decrease from its initial value.

Like in neurophysiological studies, two evaluation parameters (adaptation and de-adaptation) are introduced to verify the validity of the experimental results. The term "adaptation" refers to "a process in which a system recovers previously learned skills after a change in the operating environment" [16]. In situations of interaction, de-adaptation is the phase in which the contact with the environment is finished, the measured interaction force is null and the robot visco-elastic properties are newly regulated by (4).

The de-adaptation to dynamic perturbations, like in humans, is quite faster than adaptation. In fact, in humans, adaptation requires more time and more movements to adapt than de-adaptation [17], [18]; in the robot arm, adaptation requires the time to reach an adequate level of coactivation in the interaction, while the de-adaptation is quite rapid.

4 Experimental setup and data analysis

The developed control scheme was tested on the Dexter robot arm [19], [20] made of 8 rotational joints and actuated through a steel cable mechanical transmission system.

The experimental setup includes the robot arm, with its hardware controller, and a 3-component force sensor mounted on the arm wrist (Fig. 5). The sensor is used to verify the presence of dynamic perturbations on the trajectory and switch the arm behavior from a stiff and accurate behavior to a more flexible and adaptive one and vice versa.

The main objective of the experimental session is to demonstrate how efficiently a unique parameter is able to regulate the PD gains in the joint space during the two phases of adaptation and de-adaptation, like in the biological motor control. Hence, the scenario chosen for the trials varies from the free space to the constrained space in an unpredictable way, in order to record the variation of the coactivation and of the corresponding visco-elastic properties during the motion.

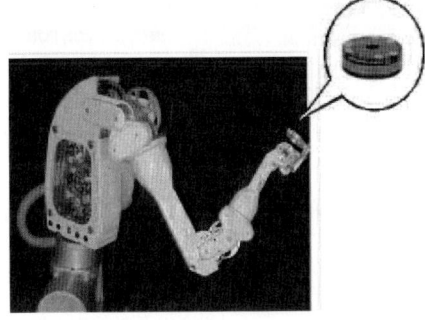

Figure 5: Experimental setup

The experimental session includes 4 different phases. The first one is addressed to evaluate the coactivation mechanism in the free space and to show how it influences the robot accuracy. Hence, the c graph is reported together with the position error (Figs. 6, 7) in order to evidentiate that, when the robot realises that it is free to move, accuracy increases during the motion together with the coactivation. In particular, the error decreases from 0.015 m to an average value of 0.0025 m. The corresponding K_P parameters for all the joints increase too from their initial value, depending on the position error (Fig. 8).

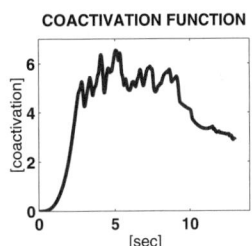 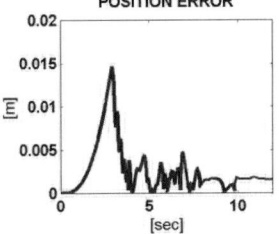

Figure 6: Coactivation function in the free space   *Figure 7: Position error in the free space*

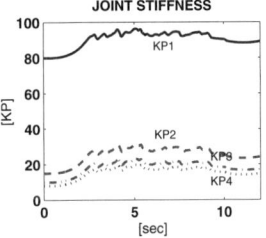 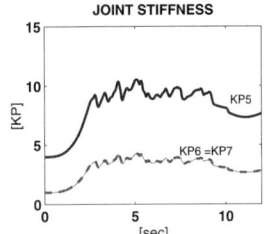

Figure 8: Joint stiffness in the free space

The second phase is aimed at monitoring the arm behavior when a contraint is sensed at the terminal point. The arm is bounded during the whole duration of the task and the coactivation, modulated by the sensed interaction forces, is recorded. The graphs show the norm of the measured interaction force which causes a decreasing evolution for c (Figs. 9 and 10) and the corresponding variations of K_P (Fig. 11). Compliance increases evidently with respect to the free space, since K_P gains notably decay from their initial values.

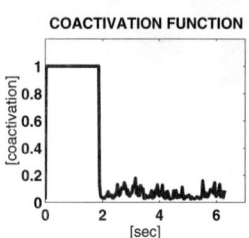
Figure 9: Coactivation function in the constrained motion

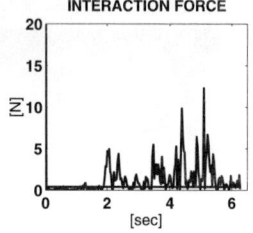
Figure 10: Interaction force in the constrained motion

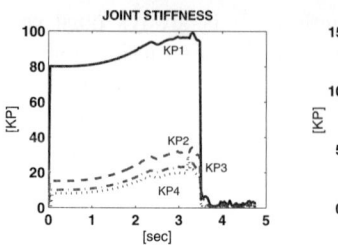
Figure 14: Joint stiffness in the adaptation phase

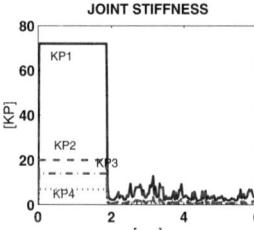

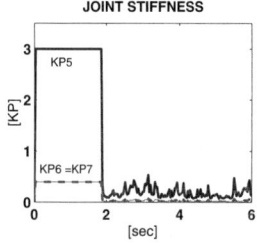

Figure 11: Joint stiffness in the constrained motion

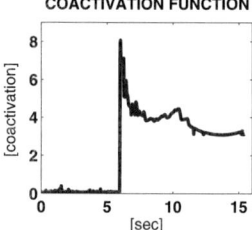
Figure 15: Coactivation function in the phase of de-adaptation

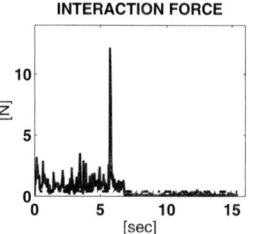
Figure 16: Interaction force in the phase of de-adaptation

As the last two phases, the adaptation and de-adaptation mechanisms are considered. This means analysing the transition from the accurate behavior in unconstrained environments to the adaptable behavior in unexpected constrained workspace (adaptation) and the reverse transition (de-adaptation). To this regard, the coactivation and the interaction forces are registered during both the adaptation and the de-adaptation and graphically compared.

As previously explained, in humans, the adaptation mechanism requires more time than the de-adaptation [16]. In the experimental trials, the robot arm coactivation increases and decreases rapidly during both two phases, but in the de-adaptation the variation is instantaneous (Figs. 15–17) while in the adaptation it is lightly slower (Figs. 12–14). Particularly, the high rate of decay in the adaptation is due to the high value of force measured in the impact. In this case, the role of c_{min} is of paramount importance to preserve the arm control and is empirically determined.

The transition from the free space to the constrained space is managed by the sensed interaction force. Due to the force sensor sensitivity, however, the force value which causes switching from the adaptation to the de-adaptation phase and vice versa is not 0 N but is empirically set to 0.4 N, as evident in the force graphs.

As a final consideration on the carried out experimental session, it can be concluded that the reported experimental results could be particularly meaningful if thoughts in tasks of learning by doing, where a teacher can intervene by modifying the robot actual trajectory and teach it the right way to execute a task in conditions of complete safety. Moreover, monitoring the human level of force and stiffness in the same tasks of interaction could be useful to set the right rate of variability of coactivation in the adaptation/de-adaptation phases.

5 Conclusions

The paper focuses on the importance of biology in the development of new control solutions which could overcome some limitations of the standard control strategies. To this regard, a control scheme for an 8-degree-of-freedom robot arm which mimics the human motor control model has been developed by including both a feedforward and a feedback control loop.

The work poses particular attention on the feedback control, realised by a PD action plus gravity compensation, by proposing a human-like mechanism (the coactivation) to regulate the visco-elastic arm properties. The benefit of using the coactivation consists of choosing a single function to modulate the proportional and derivative gains for all the arm joints.

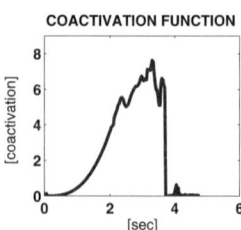
Figure 12: Coactivation function in the adaptation phase

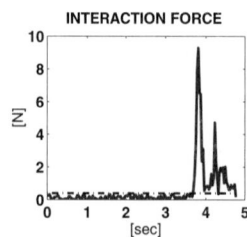
Figure 13: Interaction force in the adaptation phase

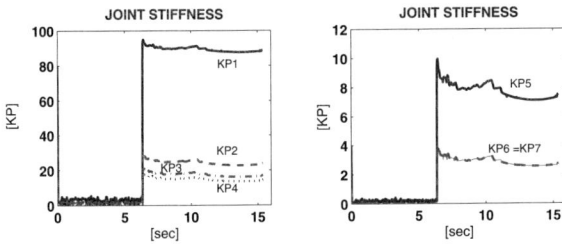

Figure 17: Joint stiffness in the de-adaptation phase

The experimental section emphasizes the role of the adjustable visco-elastic mechanism both in the free space (by increasing the robot accuracy) and in the hindered environment (by adapting to the constraints). Moreover the two phases of adaptation to a constraint and de-adaptation are analysed.

In view of the role that the developed robotic platform could play in a deeper biological knowledge, future studies will investigate if and how the human coactivation depends on the force information (as verified on the robot arm) and even if vision and other sensory information can contribute to the adjustable coactivation mechanism.

Moreover, the development of a feedforward control, based on a partial knowledge of the dynamic model or on an adaptive law, could provide information on the complexity of the internal model necessary for humans to realise a stable motor control even in fast movements.

References

[1] M. I. Jordan, D. M. Wolpert, "Computational motor control", *The Cognitive Neurosciences*, M. Gazzaniga (Ed.), 2nd Ed. Cambridge, MIT Press, 1999.

[2] R. C. Miall, D. J. Weir, D. M. Wolpert, J. F. Stein, "Is the cerebellum a Smith Predictor?", *Journal of Motor Behavior*, 1993, Vol. 25, pp. 203–216.

[3] A. Polit, E. Bizzi, "Characteristics of the motor programs underlying arm movements in monkeys", *Journal of Neurophysiology*, 1979, Vol. 42, pp. 183–194.

[4] C. Ghez, J. Gordon, M. F. Ghilardi, C. N. Christakos, S. E. Cooper, "Role of proprioceptive input in programming of arm trajectories", *Cold Spring Symposia on Quantitative Biology LV*, Cold Spring Harbor Laboratory Press, 1990, pp.837-847.

[5] E. Bizzi, N. Accornero, W. Chapple, N. Hogan, "Posture control and trajectory formation during arm movement", *Journal of Neuroscience*, 1984, Vol. 4, pp. 2738–2744.

[6] M. Katayama, S. Inoue, M. Kawato, "A strategy of motor learning using adjustable parameters for arm movement", *International Conference of the IEEE Engineering in Medicine and Biology Society*, 1998, Vol. 20, pp. 2370–2373.

[7] N. Schweighofer, M. A. Arbib, M. Kawato, "Role of the cerebellum in reaching movements in humans. I. Distributed inverse dynamics control", *European Journal of Neuroscience*, 1998, Vol. 10, pp. 86–94.

[8] J.K. Salisbury, "Active stiffness control of a manipulator in Cartesian coordinates", *19th IEEE Conference on Decision and Control*, Albuquerque, NM, 1980, Vol. 1, pp. 95–100.

[9] H. Kazerooni, P.K. Houpt and T.B. Sheridan, "Robust compliant motion for manipulators, Part 1: the fundamental concepts of compliant motion, Part 2: Design method", *IEEE Journal of Robotics and Automation*, 1986, Vol. 2, pp. 83–105.

[10] L. Sciavicco, B. Siciliano, *Modelling and Control of Robot Manipulators*, 2nd Ed., Springer-Verlag, London, UK 2000.

[11] T. D. Sanger, "Neural network learning control of robot manipulators using gradually increasing task difficulty", *IEEE Transactions on Robotics and Automation*, 1994, Vol. 10, pp. 323–333.

[12] M. Katayama, M. Kawato, "A neural control model that learns virtual trajectories for multi-joint arm movements", *Biological Cybernetics*, 1998.

[13] F. Baldissera, H. Hultborn, M. Illert, "Integration in spinal neuronal system", *Handbook of Physiology, Section 1: The nervous system*, J. M. Brookhart, V. B. Mountcastle, V. S. Brooks, S. R. Geiger (Eds.), 1981, Vol. 2, pp. 509–595.

[14] S. J. Serres, T. E. Milner, "Wrist muscle activation patterns and stiffness associated with stable and unstable mechanical loads", *Experimental Brain Research*, 1991, Vol. 86, pp. 451–458.

[15] M. Katayama, M. Kawato, "Virtual trajectory and stiffness ellipse during force trajectory control using a parallel-hierarchical neural network model", *Advanced Robotics*, 1991, pp. 1186–1194.

[16] M.A. Conditt, F. Gandolfo, F. Mussa-Ivaldi, "The motor system does not learn the dynamics of the arm by rote memorization of past experience", *Journal of Neurophysiology*, 1997, Vol. 78, pp. 554–560.

[17] R.B. Welch, "Adaptation of space perception", *Handbook of perception and human performance*, K. R. Boff, L. Kaufman, J. P. Thomas (Eds.), 1986.

[18] R. Shadmehr, F. Mussa-Ivaldi, "Adaptive representation of dynamics during learning of a motor task", *Journal of Neuroscience*, 1994, Vol. 15, pp. 3208–3224.

[19] L. Zollo, B. Siciliano, C. Laschi, G. Teti, P. Dario, "Compliant control for a cable-actuated anthropomorphic robot arm: an experimental validation of different solutions", *IEEE International Conference on Robotics and Automation*, Washington, DC, 2002, pp. 1836–1841.

[20] L. Zollo, B. Siciliano, C. Laschi, G. Teti, P. Dario, "An experimental study on compliance control for a redundant personal robot arm", to be published in *Robotics and Autonomous Systems*, 2002.

Realization of an Autonomous Search for Sound Blowing Parameters for an Anthropomorphic Flutist Robot

Shuzo Isoda*, Manabu Maeda*, Yuji Hiramatsu*, Yu Ogura*
Hideaki Takanobu**, Atsuo Takanishi***,****, Kunimitsu Wakamatsu*****

*Graduate School of Science and Engineering, Waseda University, Tokyo, Japan
**Department of Mechanical Systems Engineering, Kogakuin University, Tokyo, Japan
***Department of Mechanical Engineering, Waseda University, Tokyo, Japan
****Humanoid Robotics Institute, Waseda University, Tokyo, Japan
*****Professional Flutist
takanisi@waseda.jp,   http://www.takanishi.mech.waseda.ac.jp/

Abstract

In this paper, we describe an anthropomorphic flutist robot that it's expected to have human-like organs from the mechanical hardware and control functions. We developed a "General Position" searching algorithm, which is based on the relative distance between the robot's mouth and the flute mouthpiece. It is possible for the robot, using the General Position, to blow a whole tone. Thereafter, we developed an autonomous searching algorithm for the lip part parameters after setting the General Position, and an algorithm for evaluating the blowing sound. The robot could stably play a flute using these algorithms.

1. Introduction

We are researching and developing humanoid robots aiming at symbiosis between humans and robots. To achieve this, it is important for humans and robots to exchange emotional and sensitive information.

We developed a flute-playing robot for musical applications. When humans play the flute, they exert various expressions by cooperative movements of their body and face. We reproduced these flute-playing movements as closely as possible. Furthermore, from the Neuro-Robotics view of point, we aim to clarify the mechanisms of playing a flute and how the emotions and sensitive information can be exchanged between humans and robots.

Fig.1 shows the Anthropomorphic Flutist Robot WF-3RIX, which we began to develop in 1990. To date we have developed we developed the lung for breathing, mouth for shaping the air beam, the relative positioning mechanism between the embouchure hole and the mouth, the double tonguing mechanism, the vibrating mechanism, the fingers that can create a trill for all tones, and the musical performance system to correspond to MIDI accompaniment data[1]. In addition, to stabilize the flute playing, we developed the "General Position" searching algorithm, the algorithm for autonomous searching for the lip part parameters (tone searching), and the FFT tool for evaluating the blowing sound in real-time. The algorithms are described below.

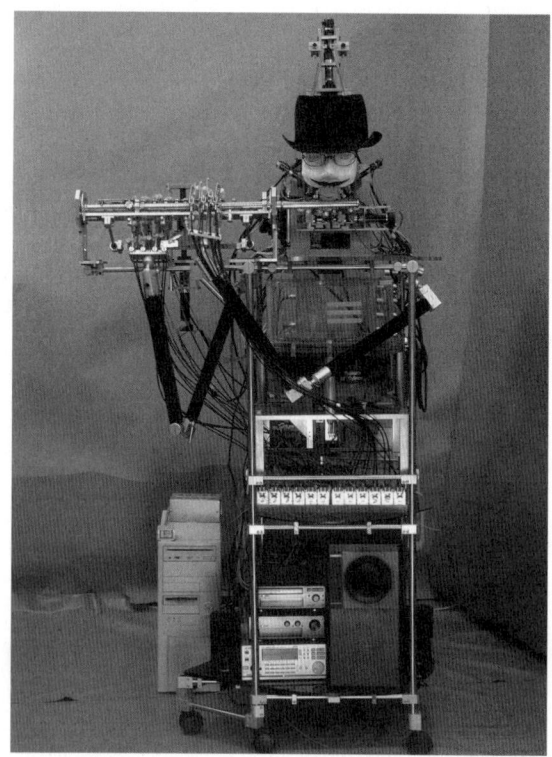

Fig. 1 Anthropomorphic Flutist Robot WF-3RIX

2. Robot's Parameters

The "Position Control Parameters" and the "Air Beam Control Parameters" are show in **Fig. 2**. The parameters X, Z and θ, called Position Control Parameters, determine the relative position between the embouchure hole and the mouth. The Air Beam Control Parameters are given by the air beam velocity (u), the mouth thickness control (α and β), the mouth width control (w) and the upper lip control (t).

3. Musical Performance System

This robot can perform music using the musical performance system (**Fig. 3**). The Sequencer PC gives MIDI output data and the MIDI Tone Generator module performs the accompaniment. The Robot Controller PC receives the timing clock from the MIDI data. The robot receives two kinds of information: "Robot Data" (the parameter for each tone) and "Music Data"(the score for this robot). The Robot Controller PC converts the "Music Data" into "Robot Data" parameters outputs, which are synchronized by the timing clock. The performance of this system, for playing music, is presented in [1].

4. General Position Searching Algorithm

4.1 Composition of a hearing system

To enable this robot to autonomously start playing a flute, we implemented the "General Position" algorithm. A professional flutist defined the "General Position" as the relative distance between the mouth and the flute mouthpiece, which makes it possible to blow a whole tone.

Because the robot needs to analyze the blowing sound by itself, we developed a hearing system and an algorithm for analyzing the blowing sound.

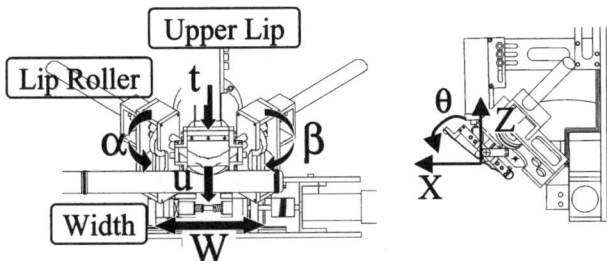

Fig. 2 Robot Data Parameters for Flute Playing

The hearing system was implemented using a dynamic microphone, with a frequency bandwidth from 40 to 15000[Hz], and a sound card that is used for digital recording. The tone-range for a flute is about 3 octaves and the frequency of the fundamental tones (**C4-C7**) is about from 263[Hz] to 2100[Hz]. It is sufficient to consider the seventh harmonic.

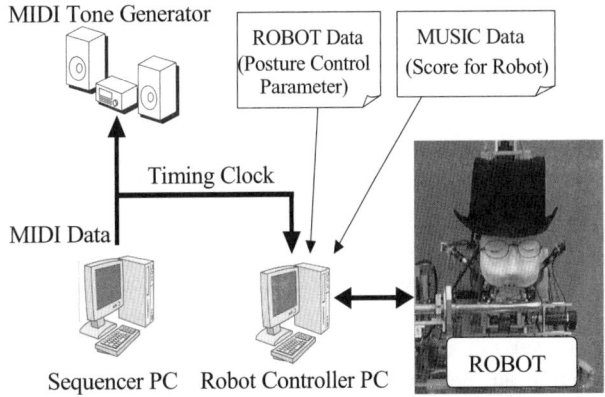

Fig. 3 Musical Performance System

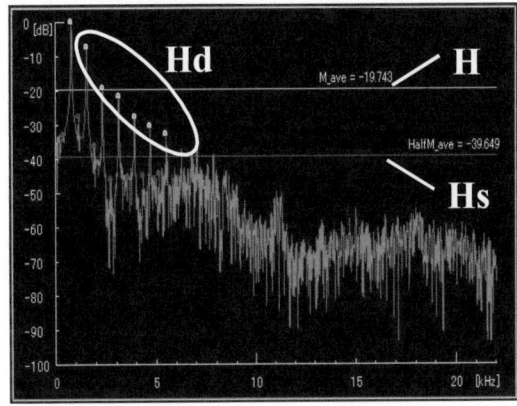

Fig. 4 Function Value

$$F(Note, Pitch) = V + w1_R \times (H - Hs) + w2_R \times Hd + \alpha_R$$

$$R[PitchRange] = \{Low, Middle, Hight\}$$

V : Volume [dB]
H : Harmonic Average [dB]
Hs: Semi-Harmonic Average [dB]
Hd: Even Harmonic Level - Odd Harmonic Level [dB]
α : Calibration Value
w1, w2 : Weight Value

Using this system, we analyze the blowing data by FFT. The evaluation function of the blowing sounds is shown in **Fig. 4**. This function is based on the volume level, and we added to this function two terms for evaluating the sound quality: "**H-Hs**" and "**Hd**."

The second term "**Hs**" is the average of harmonic in decibels based on the half fundamental frequency. We estimate the difference of the average harmonic in decibels: M[dB] (that is the amplitude of the fundamental tone that is assumed to be 0[dB]), and the over second harmonics that are transformed into decibel units, then the average of the value of the relative amplitude is calculated.

We evaluate how many harmonics appear in blowing sound, using the evaluation function.

"**Hd**" is the difference between the even harmonic level (**Le**[dB]) and odd harmonic level (**Lo**[dB]). The characteristic of this function is based on level volume. Moreover we can establish the same level evaluation function for the whole tone using the calibration value **a**, and using α for the decibel units.

Comparison of the tone areas (low, intermediate and high), revealed that the "**Hd**" value strongly affects the actual sound in low and intermediate areas. On the other hand, the "**H-Hs**" value strongly affects the sound in the high sound area. Therefore, we set the weight values as follows: **w1** = 0.5 and **w2** = 1.0 in the low and intermediate area, and in the high area, **w1** = 1.0 and **w2** = 0.5. This FFT tool corresponds to the robot's musical performance, while simultaneously it's evaluating the blowing sound [2]. This is possible because the FFT tool is synchronized with the timing clock from the MIDI.

4.2 Method of Searching for General Position

The "General Position" method is presented in **Fig. 5** to determine. The harmonic relations are shown below:

1) **C4**(263[Hz]) → **C5**(526[Hz]) → **G5**(788[Hz]) → **C6**(1051[Hz])
2) **D4**(295[Hz]) → **D5**(590[Hz]) → **A5**(884[Hz])
3) **F4**(351[Hz]) → **F5**(702[Hz]) → **C6**(1051[Hz])
4) **E4**(331[Hz]) → **E5** (662[Hz]) → **B5**(992[Hz])

At first, we choose one pitch from the above (from 1 to 4). For example, if we choose pitch 4, we have the following pattern:

1st Step: With the **E4** (fundamental tone) fingering, the robot is set at a position in which it can blow **E4** tone.

2nd Step: With the **E4** fingering, only air beam velocity (*u*) is increased. We search for the positions in which the robot can blow **E5** and **B5**.

3rd Step: With the **B5** fingering, the robot blows **B5**. The robot searches for the position in which there is no difference in the **B5** pitch between the **E4** fingering and the **B5** fingering.

Thus, we find the "General Position" using those processes.

4.3 Algorithm of Searching for General Position

We will explain the algorithm to search for the **E4**, **C4**, **D4** and **F4** based on the General Positions method described in 4.2 **Fig. 5** shows the General Position algorithm to search for **E4**. We search for three overlapping areas. First, we search the "Position Control Parameters" for each tone (**C4** to **C7**) using the orthogonal table. Then, the "Position Control Parameters" are fixed to the fundamental tone position (in this case **E4**). Then the robot searches for a suitable position for blowing **E4**, **E5** and **B5** with the **E4** fingering, as it is shown in **Fig. 6**.

The connecting line vector between **E4** and **E5** is evenly divided, and the position is moved step-by-step from vector **E4** toward **E5**. Ensuing, the robot checks whether the robot can blow both **E4** and **E5** in every moved position. The robot repeats this process until the robot

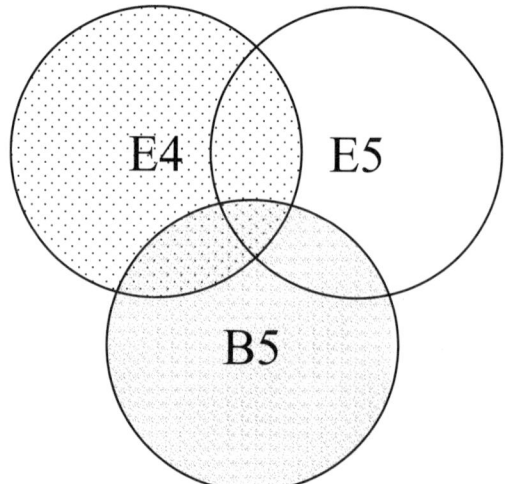

Fig.5 Image of Searching Algorithm (E4 Based)

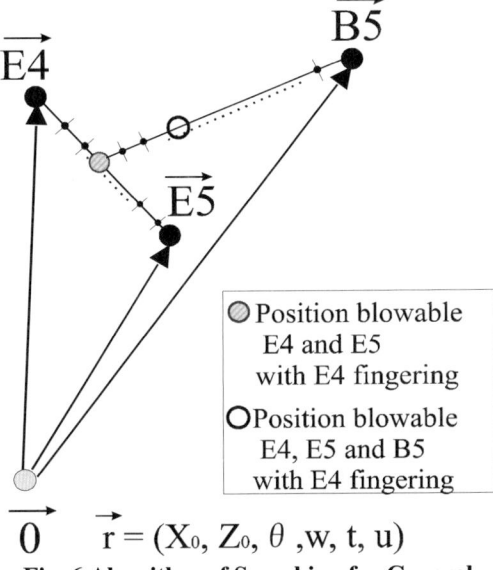

Fig. 6 Algorithm of Searching for General Position (E4 Based)

finds the position in which it robot can blow both **E4** and **E5**.

Then, the line from that position toward vector **B5** is evenly divided in the same way. By the same process, the robot searches for the position in which it can blow **E4**, **E5** and **B5**.

Finally, the robot confirms the **B5** pitch if there is any difference between that given with the **E4** fingering and the one given by the **B5** fingering.

In addition, to search for the actual "General Position" of **E4**, it is also necessary to search for other fundamental tones (**C4**, **D4** and **F4**) using the "General Position" method in the same way. It is necessary to narrow the solution from several fundamental tones to converge to one actual "General Position"; as shown in **Fig.7**.

We make the robot move to one of the **C4** based "General Position." From this position, the robot checks whether this position fulfills other fundamental tone attributes only by changing "Air Beam Velocity." If the searching fails, the robot calculates the slide and step size from the present position towards the "General Position" candidate that includes the failed tone and then it moves by one step. On the next position, the robot checks if the conditions for the all-fundamental tones are fulfilled. Again, the robot calculates the slide and step size from the present position toward the "General Position" and repeats the sequence. The robot repeats these processes until it finds the position where it can blow all the fundamental tones. Then, the robot saves that position as

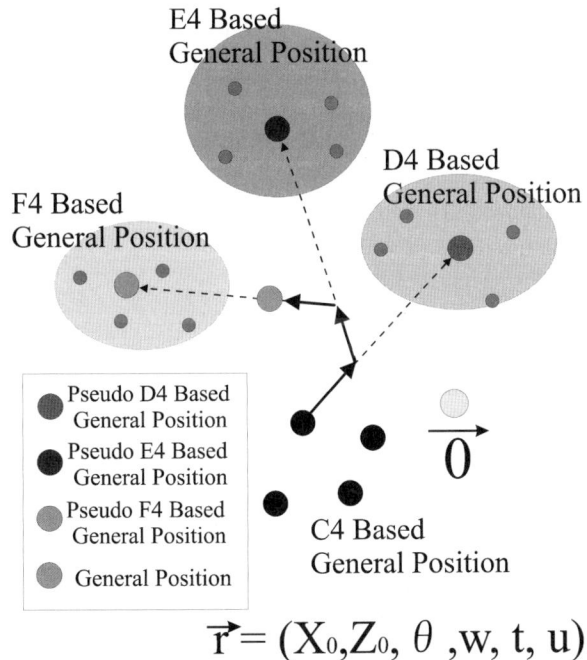

Fig. 7 Algorithm to Search for General Position

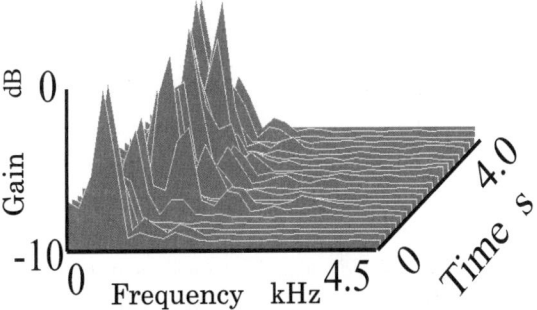

Fig. 8(a) Continuous Blowing Experiment (FFT) without General Position Searching Algorism

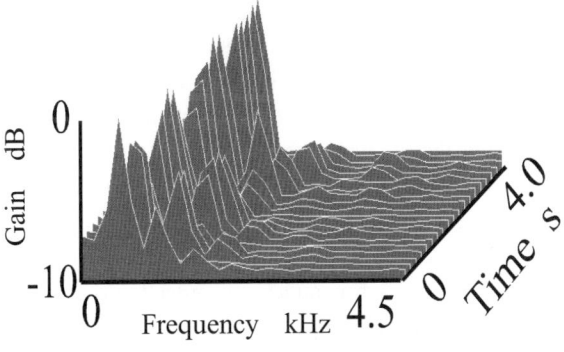

Fig. 8(b) Continuous Blowing Experiment (FFT) with General Position Searching Algorism

the "General Position".

4.4 "General Position" searching experiment

We made the robot search for the "General Position," using the searching method described before. The robot searched from four different starting positions, to verify results independently of the starting position. The robot started searching for the "General Position" from quite different positions within $\Delta X<1.65$[mm], $\Delta Z<1.06$[mm] and $\Delta\theta<1.77$[deg]. In addition, we compared the FFT analysis from the continuous blowing sounds (**C4** to **C7**) obtained from the previous "Position Control Parameters" and those obtained from the "General Position" searching algorithm (**Fig. 8 (a)**, **(b)**). Unlike the former, in the latter result, the spectrum of the fundamental tone frequency increases with time and each sound is on pitch.

5. Autonomous searching for "Air Beam Parameters"

5.1 Autonomous algorithm to search for the "Air Beam Parameters"

Using real-time the FFT tool and the evaluation function, we developed an autonomous algorithm to search for the "Air Beam Parameters" after setting the "General Position."

First, the robot loads the "Music Data" and picks up the tones that are needed for playing music and then it starts to search from the lowest tone. We set the allowable range of the evaluation function value. Using this algorithm, the robot searches the "Air Beam Parameters" (α, β, **u**, **t**, **w** in **Fig. 2**) that are within that range. Blowing sounds considered unstable, low sound pressure or low quality were considered outside that range.

After moving to the "General Position," the robot finds each movable range (where the evaluation function value is within the allowable range) by sequentially changing the mouth thickness and width, and the upper lip parameters. Then, after getting each movable ranges with the "Air Beam Parameters", they are divided. Finally, the robot changes each parameter in order to determine and evaluate the blowing sounds for a given period of time. The robot determines the final position where it can blow with a stable sound and a good attack (**Fig. 9**).

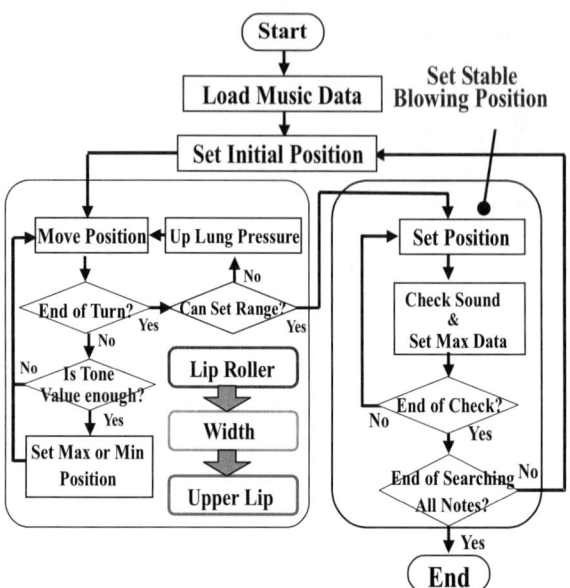

Fig. 9 Flowchart of autonomic searching air beam parameters

5.2 Searching Experiment and Results

We performed an experiment using the autonomous algorithm to search for the "Air Beam Parameters." We chose some musical compositions and the robot searched for the tones, from MIDI data, that were needed to play the tones. When searching of the "General Position," the robot blew for 2.0[s] and it calculated the blowing ratio (the stable blowing time divided by 2.0[s]). In this experiment, if the blowing ratio was below 80[%], the blowing sound was considered unstable. If the blowing ratio was more than 85[%], then the sound was recognized as a stable sound with a good attack.

5.3 Experimental Results

The robot performed the "Flute Quartet Kv.298" composed by Mozart using the developed algorithm. The results of these experiments are shown in **Fig. 10**. This is the graph of the sound ratio and pitch note needed for playing each note. After setting the "General Position" parameter, the robot searched for "Air Beam Parameters".

In most cases, the blowing ratios were over 85%. This result shows that the robot is good at finding the parameters for a stable blowing sound with a good attack.

6. Conclusions

1) We developed an algorithm to search for the "General Position."
2) We developed an autonomous algorithm to search for the "Air Beam Parameters" after setting the "General Position" using the real-time FFT tool
3) The robot played the "Flute Quartet Kv.298" composed by Mozart using the developed algorithms.
4) The robot control parameters were easy to determine than in previous versions [3]. Moreover, the blowing sound quality was improved and the musical performance was more stable.

Acknowledgment

A part of this research was done at the Humanoid Robotics Institute (HRI), Waseda University.

The authors would like to express thanks to Okino Industries LTD, OSADA ELECTRIC CO. LTD, SHARP CORPORATION, Sony Corporation, Tomy Company, LTD and ZMP Inc. for their financial support for HRI. Finally, we would like to express thanks to SolidWorks Corp., MURAMATSU Inc., CHUKOH CHEMICAL INDUSTRIES LTD. and Advanced Research Institute for Science and Engineering of Waseda University for their supports. This research was supported (in part) by a Gifu-in-Aid for the WABOT-HOUSE Project by Gifu Prefecture.

Reference

[1] Atsuo Tankanishi, et al: Development of an Anthropomorphic Flutist Robot WF-3RIV, International Computer Music Conference, pp328-331, 1998

[2] Ando Yukinori: Drive conditions of the flute and their influence upon harmonic structure of generated tone (An experimental study of the flute II), the journal of the Acoustical Society of Japan (in Japanese), Vol.26, No.7, pp297-305, 1970

[3] Atsuo Takanishi, et al: Development of an Anthropomorphic Robot – Autonomic searching method of General Position -, Proceedings of the 17th Annual Conference of the Robotics Society of Japan, pp1225-1226, 1999

[4] Atsuo Takanishi, et al: Research on Autonomous Musical Performance of an Anthropomorphic Flutist Robot, Proceedings of the 19th Annual Conference of the Robotics Society of Japan, pp, 2001

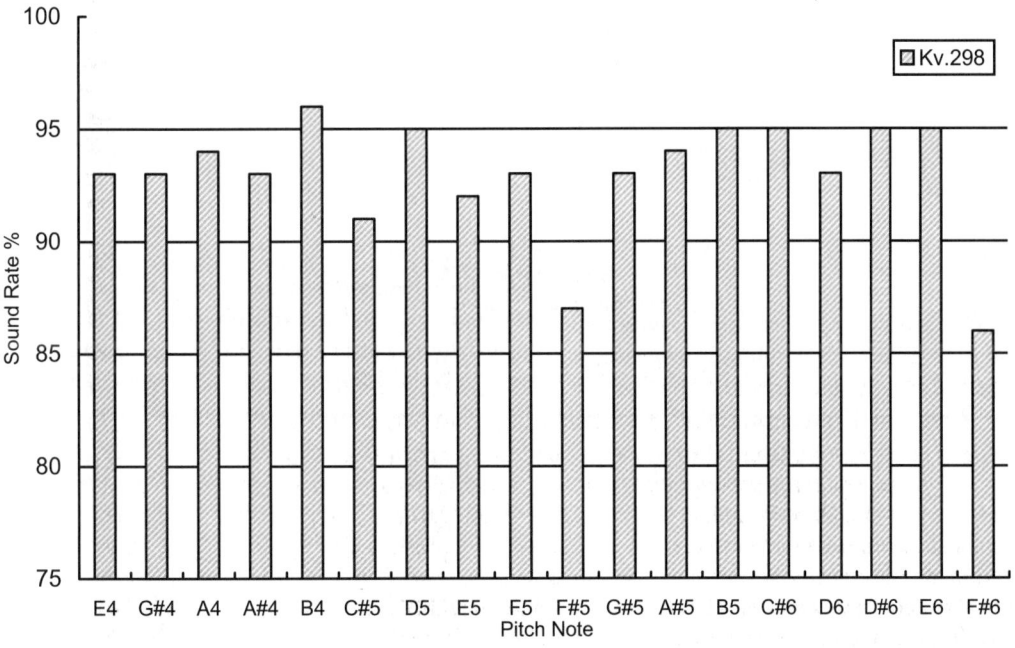

Fig. 10 Result of Searching Experiment

A New Mental Model for Humanoid Robots for Human Friendly Communication
- Introduction of Learning System, Mood Vector and Second Order Equations of Emotion -

Hiroyasu Miwa[*1], Tetsuya Okuchi[*1], Kazuko Itoh[*1], Hideaki Takanobu[*2, *4], Atsuo Takanishi[*3, *4]

*1 Graduate School of Science and Engineering, Waseda University
*2 Department of Mechanical Systems Engineering, Kogakuin University
*3 Department of Mechanical Engineering, Waseda University
*4 Humanoid Robotics Institute, Waseda University

#59-308, 3-4-1 Okubo, Shinjuku-ku, Tokyo, 169-8555 Japan
Tel: +81-3-5286-3257, Fax: +81-3-5273-2209
takanisi@waseda.jp, http://www.takanishi.mech.waseda.ac.jp/

Abstract

The authors have been developing human-like head robots in order to develop new head mechanisms and functions for a humanoid robot that has the ability to communicate naturally with a human by expressing human-like emotion. Furthermore, the interaction between humans and robots is one of the essential factors for the Neuro-Robotics. We believed that the Mental Dynamics caused by the stimuli from the internal and external environment is important in expressing emotion. Therefore, we newly developed a human-like head robot WE-4 (Waseda Eye No.4) in 2002. The "Learning System", the "Mood Vector" and the "Second Order Equations of Emotion" were introduced to the mental model of the robot. Moreover, an internal clock was introduced as an autonomic nerve system to express the activation component of the Mood Vector. And, we realized the expression of mental transition caused by the stimuli from the internal and external environment of the robot. In this paper, we describe the new mental model of a human-like head robot WE-4.

1. Introduction

Currently, the most practical robots are being used in the manufacturing industry as industrial robots. However, in the future, it is expected that personal robots will be widely used in home or town. And, communication between humans and robots will be more important than the current situation. Moreover, it is necessary for personal robots in the future to adapt to its partner and environment as well as to naturally communicate with humans. Therefore, we have been developing human-like head robots in order to develop new head mechanisms and functions for a humanoid robot that has the ability to express emotions and to communicate with humans in a human-like manner from the view points of NEURO-ROBOTICS.

The head robot is currently under active research in the field of robotics. Brooks developed a head robot, which expresses facial expressions using its eyes, eyelids, eyebrows and mouth. It can communicate with humans using visual information from CCD cameras [1]. Kobayashi developed a head robot that uses the fourteen Action Units of Ekman [2] [3]. It can express six basic facial expressions as quickly as a human using 24-DOF for facial expressions with air compressors for actuators. Also it can recognize human facial expressions using CCD cameras and reciprocate the same facial expressions back [4] [5].

The authors have been developing WE-3 (Waseda Eye No.3) series since 1995. And, coordinated head-eye motion with V.O.R. (Vestibular-Ocular Reflex) [6] [7], depth perception using the angle of convergence between the two eyes, adjustment to the brightness of an object with the eyelids, and human-like expressions using the eyebrows, lips and jaw were accomplished. We also produced auditory, cutaneous and olfactory sensations [8] [9]. In addition, we introduced the Equations of Emotion, the Robot Personality and a mental model which has three independent parameters to the robot [10].

In 2002, we developed a new human-like head robot WE-4 (Waseda Eye No.4). WE-4 has a more similar shape to a human than the previous robot, WE-3RV. We produced emotional expressions with not only the face but also the upper-half part of its body [11]. Moreover, we newly added the "Learning System", the "Mood Vector" and the "Second Order Equations of Emotion" to the robot's mental model. And, we realized that the expression of the mental transition caused by the stimuli from the internal and external environments of the robot. In this paper, we describe the new mental model of WE-4.

2. Mental Model

The Mental Dynamics, which is the mental transition caused by the stimuli from the internal and external environment of the robot, is extremely important in the emotional expression. But, the previous robot, WE-3RV,

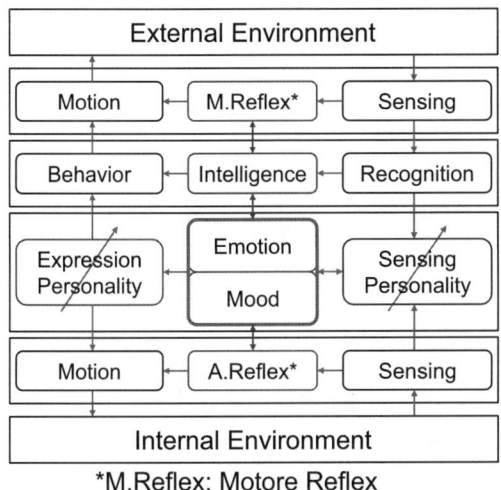

*M.Reflex: Motore Reflex
*A.Reflex: Autonomous Reflex

Fig. 1 Basic Information Processing Structure

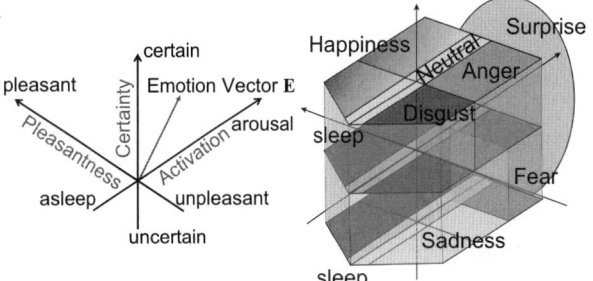

Fig. 2 3D Mental Space   Fig. 3 Region Mapping of Emotion

Table 1 Sensing Personality Table

Stimulus	Sensation		SPTa	SPTp	SPTc
Visual	Loose Sight of the Target		0	-	-
	Dazzling Light		+	-	0
	Target is Near		0	-	+
	Target Color	1	+	+	+
		2	+	-	+
		3	0	0	0
		4	0	0	0
Tactile	Pushed		+	-	0
	Stroked		+	+	+
	Hit		+	-	-
Auditory	Loud Sound		+	-	0
Temperature	Cold		0	+	-
	Hot		0	-	0
Olfactory	Alcohol		+	+	+
	Ammonia		+	-	0
	Smoke		+	-	+
	No Sense		+	-	0

was not capable of expressing the Mental Dynamics. Therefore, we newly added the "Learning System", the "Mood Vector" and the "Second Order Equations of Emotion" to a human-like head robot WE-4. And, we introduced an information flow into the robot shown in Fig. 1. We also adopted the 3D mental space, which consists of a pleasantness axis, an activation axis and a certainty axis, shown in Fig. 2. The vector **E** named the "Emotion Vector" expresses the mental state of WE-4. Furthermore, we mapped out seven different emotions in the 3D mental space as in Fig. 3 [10].

2.1 Learning System

In psychology, the human personality consists of two factors: the genetic factors and the environmental factors. The former are inborn, but the latter are obtained by experiences which are learned throughout a person's life [12]. Therefore, we introduced the "Learning System" in order for the robot to learn the experiences and to construct its personality based on its experiences dynamically.

There are the unconditioned stimuli and the conditioned stimuli as the external stimuli. The unconditioned stimuli are genetic factors. And, they affect to emotion constantly. But, the conditioned stimuli are environmental factors and their influences on emotion are variable. We introduced the SPT (Sensing Personality Table) to the robot mental model. The SPT determines how stimuli works in the robot's mental state [10]. We constructed the Learning System by autonomously changing the SPT for the conditioned stimuli based on its experiences. The SPT for the conditioned stimuli were defined in the equation (1). It was modeled on the Rescorla-Wagner Theory [13] which is a mathematical model of the respondent conditioning.

$$\Delta SPT_{Target} = \alpha(SPT_{US} - SPT_{Target}) \quad (1)$$

SPT_{US} : *SPT for Unconditioned Stimulus*

α : *Learning Rate*

SPT_{Target} : *SPT for Conditionned Stimulus*

The Equation (1) said that the SPT for the conditioned stimulus either dynamically increases or decreases based on the total of SPT for the unconditioned stimuli when a robot senses the conditioned stimulus with a strong emotion.

Table 1 shows an example of the SPT implemented to a human-like head robot WE-4. We defined two "Color" stimuli of visual stimuli as the conditioned stimuli and other stimuli as the unconditioned stimuli. And, we assigned the initial conditions of SPT for the conditioned stimuli as zero. By changing the SPT, it is possible to obtain a wide variety of Sensing Personalities.

2.2 Mood Vector

The human mental state is affected not only by the emotion but also by the mood. Here, the emotion is a strong change that is caused in a short time. The mood is a weak change that is caused in a long time. The emotion and the mood are closely related [14]. Therefore, in order to express the mood, we newly introduce the "Mood Vector" **M** that consists of a pleasantness axis and an activation axis. And, we redefined the emotional appraisal for the stimuli as the equation (3). The emotional appraisal for the external stimuli is determined by the SPT and the Mood Vector.

$$Mood\ Vector: M = (p_M, a_M, 0) \quad (2)$$
$$Emotional\ Appraisal = f(M, SPT) \quad (3)$$
$$= k_m \times M + SPT$$

k_m : Mood Influence Matrix

Furthermore, we considered that the current mental state influenced the pleasantness component of the Mood Vector. So, it is described by the integral of the pleasantness component of the Emotion Vector as the equation (4).

$$p_M = \int p_E dt$$
$$Mood\ Vector: M = (p_M, a_M, 0) \quad (4)$$
$$Emotion\ Vector: E = (p_E, a_E, c_E)$$

On the other hand, we considered that the activation component of the Mood Vector is similar to human biological rhythm. Therefore, we introduced the internal clock that is a kind of automatic nerve system in order to describe the activation component of the Mood Vector. We used the Van del Pol equation which is basic equation of the self-excited oscillation. And, we described the activation component of the Mood Vector using the equation (5). We realized that the expression of the internal clock like a human.

$$\ddot{a}_M + (1 - a_M^2)\dot{a}_M + a_M = 0 \quad (5)$$

2.3 Second Order Equations of Emotion

In dynamics, the motion of the objects is described by the equation of motion. We considered that the mental dynamics which is a transition of a human mental state might be expressed by similar equations to the equation of motion. Therefore, we expanded the equations of emotion into the second order differential equation which is modeled on the equation of motion shown as equation (6). We also introduced three Emotional Coefficient Matrixes, which are the Emotional Inertia, Emotional Viscosity and Emotional Elasticity Matrixes.

$$M\ddot{E} + \Gamma\dot{E} + KE = Emotional\ Appraisal \quad (6)$$

M : Emotional Inertia Matrix
Γ : Emotional Viscosity Matrix
K : Emotional Elasticity Matrix

The robot can express the transient state of the mental state after the robot senses the stimuli from the environment, compared with the previous equations of emotion [10] which could only express the stationary state.

Moreover, the robot can express different reactions to the same stimuli by changing the Emotional Coefficient Matrixes. We can easily assign various robot personalities. And, we can obtain the complex and various mental trajectories which were impossible to be expressed by the previous equations of emotion (e.g. the oscillated personality).

(a) Whole View   (b) Head Part
Fig. 4 WE-4

3. Robot Hardware

Fig. 4 presents the hardware overview of the human-like head robot WE-4 (Waseda Eye No.4). It has 29-DOF (Degrees of Freedom) as shown in Table 2 and has sensors, shown in Table 3, which serve as visual, auditory, olfactory and tactile sensations. WE-4 expresses its facial expression using its eyebrows, lips, jaw, facial color and voice shown in Fig. 5. Furthermore, it can use its neck, waist and lung motions to express emotions.

4. Experimental Evaluation

The authors implemented the new mental model into a human-like head robot WE-4, and, we evaluated the learning system, the mood vector and second order equations of emotion.

4.1 Experiment of Learning Emotional Appraisal

Experiments were conducted to evaluate the learning system. In the experiments, we showed a "Color" stimulus as a conditioned stimulus with "Stroke" that is a pleasant unconditioned stimulus or "Hit" that is an unpleasant unconditioned stimulus. And, WE-4 learned the SPT of the "Color". Then we obtained the learning curve of WE-4. We

(a) Neutral (b) Happiness (c) Anger

(d) Surprise (e) Disgust (f) Sadness (g) Fear

Fig. 5 Seven Basic Facial Expressions

Table 2 DOF Configuration

Part	DOF
Neck	4
Eyes	3
Eyelids	6
Eyebrows	8
Lips	4
Jaw	1
Lung	1
Waist	2
Total	29

Table 3 Sensors on WE-4

Sensation		Device	Quantity
Vision		CCD Camera	2
Auditory		Microphone	2
Cutaneous	Tactile	FSR	26
	Temperature	Thermistor	1
Olfactory		Semiconductor Gas Sensor	4

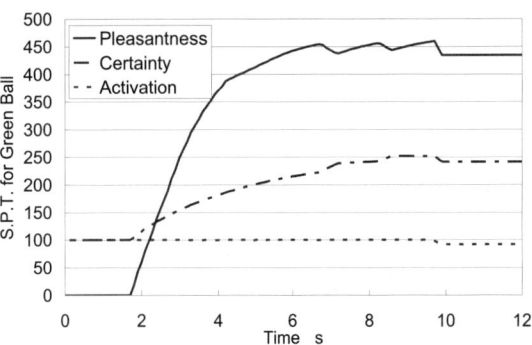

(a) Learning for Pleasant Stimulus

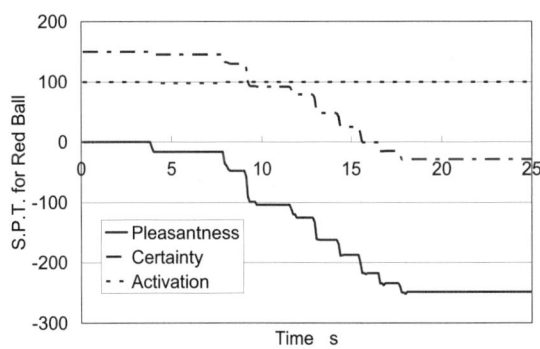

(b) Learning for Unpleasant Stimulus

Fig. 6 Result of Learning Experiment

used "Red" and "Green" balls as "Color" stimuli, and their initial state of SPT was zero. Fig. 6 shows the experimental results. We set the learning rate $\alpha = 0.01$.

As a result, the SPT for the conditioned stimulus continually increased in experiment (a) where the pleasant unconditioned stimuli were shown because the robot was continually stroked. However the SPT for the conditioned stimulus decreased in a staircase pattern in experiment (b) where the unpleasant unconditioned stimuli were shown because the hitting was intermittent stimulus. Therefore, we confirmed that WE-4 can learn its SPT for the conditioned stimuli based on the Learning System modeled on Rescorla-Wagner Theory by recognizing the conditioned

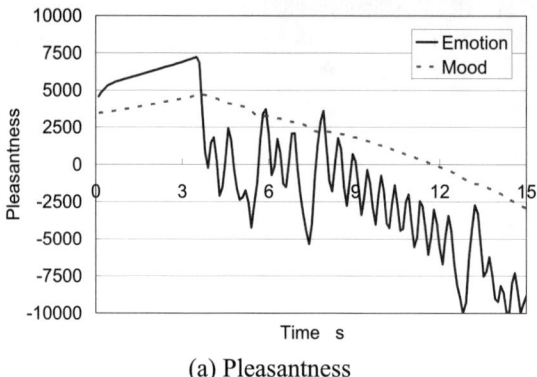

(a) Pleasantness

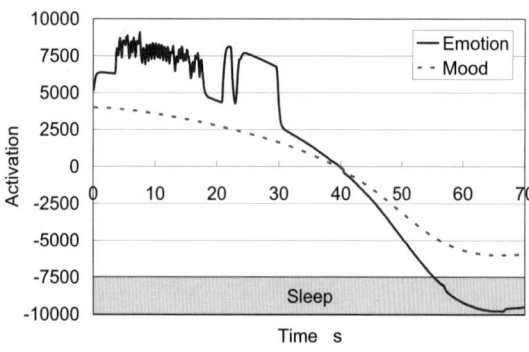

(b) Activation

Fig. 7 Result of Mood Experiment

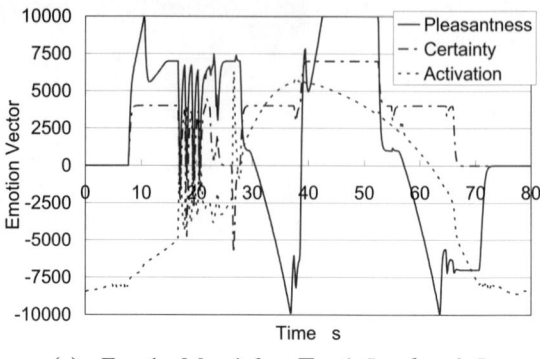

(a) $Exp.1: M = 1.0, \ \Gamma = 1.5, \ k = 0.5$

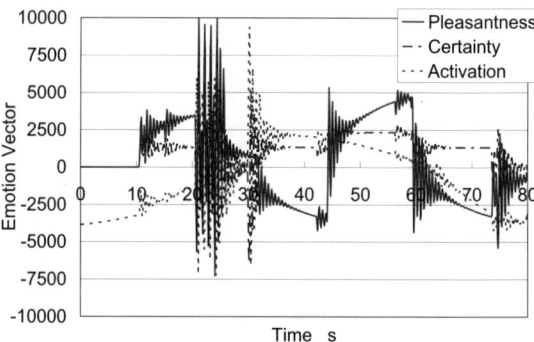

(b) $Exp.2: M = 1.0, \ \Gamma = 0.5, \ k = 1.5$

Fig. 8 Result of Emotional Experiment

stimuli with the unconditioned stimuli.

4.2 Experiment of Mood

We evaluated the interaction between the mood and the emotion. Before the experiment, we showed the pleasant stimuli to WE-4 in order to keep emotion and mood pleasant and arousal state. Then, we hit the WE-4. After that, we didn't show any stimulus to WE-4. Finally, we stopped the robot when the robot went to sleep. We obtained the Mood Vector and the Emotion Vector through the experiment. The experimental results are shown in Fig. 7.

As a result, concerning the pleasantness component, when WE-4 was hit, the mental state of WE-4 quickly became unpleasant. But, WE-4 became pleasant state again due to the Mood Vector remaining in the pleasant state. However, the Mood Vector slowly decreased by continuous hitting. We considered that because the integration time in equation (4) is from robot start until the present time, WE-4 kept pleasant mood although the Emotion Vector moved to unpleasant state by hitting.

On the other hand, concerning the activation component, the Mood Vector decreased according to the self-excited oscillation. The Emotion Vector also moved to low activation state according to the Mood Vector. And, WE-4 finally went to sleep.

Therefore, we confirmed that the robot can change its mental state by the internal stimuli, the external stimuli and the mood. And, we thought that the mood is effective to express long term mental reaction.

4.3 Experiment of Mental Model

We loaded the new mental model described in the second chapter into WE-4. And, we obtained the robot's mental states and emotional levels when the various stimuli were shown to WE-4. The following is the process of the experiments.

(1) Visual: "Green" (Pleasant, CS)
(2) Tactile: "Hit" (Unpleasant, US)
(3) Auditory: "Loud Sound" (Activation, US)
(4) Visual: "Red" (Unpleasant, CS)
(5) Olfactory: "Alcohol" (Pleasant, US)

Where,
  CS: Conditioned Stimulus
  US: Unconditioned Stimulus

WE-4 had learned the visual stimulus "Green" as a pleasant stimulus and "Red" as an unpleasant stimulus before the experiments. And, we prepared two personalities by changing the Emotional Coefficient Matrixes to WE-4. The following were the Emotional Coefficient Matrixes, and, the experimental results are shown in Fig. 8.

$$Exp.1: M = 1.0, \quad \Gamma = 1.5, \quad K = 0.5$$
$$Exp.2: M = 1.0, \quad \Gamma = 0.5, \quad K = 1.5$$

As a result, WE-4 had a calm personality in Exp. 1 because the mental trajectory was stable. However, the mental trajectory in Exp. 2 was oscillated and WE-4's personality became unstable personality. Moreover, throughout both experiments, WE-4 dynamically expressed its mental state affected by the visual, auditory, cutaneous and olfactory stimuli with facial expressions, facial color and neck motions. But, it was obviously confirmed from the robot motion that the mental state of WE-4 in Exp. 1 was more stable than the mental state of WE-4 in Exp. 2.

The obvious differences throughout the mental transition were caused by changing the Emotional Coefficient Matrixes. Therefore, we confirmed that the new mental model was effective to express the mental dynamics and to obtain a wide variety of the Robot Personalities using the new mental model.

5. Conclusions and Future Work

(1) We introduced the "Learning System" so that the robot learned its SPT for the conditioned stimuli according to the robot's experience.
(2) We introduced the "Mood Vector", and, the robot determined the emotional appraisal for the external and internal stimuli based on its SPT and Mood.
(3) We also introduced the internal clock to express the biological rhythm based on the Van del Pol Equation.
(4) We developed the "Second Order Equations of Emotion" which was modeled on the equation of motion. And, the mental dynamics were able to be expressed by the robot.

The authors introduced a new mental model for a humanoid robot. But, an evaluation system was not developed concerning the interaction between humans and the robot. In the future, we will develop the evaluation system to evaluate the robot mental model. Moreover, we will improve a human-like head robot WE-4 for more variable expressions.

Acknowledgment

A part of this research was done at the Humanoid Robotics Institute (HRI), Waseda University. The authors would like to express thanks to Okino Industries LTD, OSADA ELECTRIC CO.,LTD, SHARP CORPORATION, Sony Corporation, Tomy Company,LTD and ZMP INC. for their financial support for HRI. And, this research was supported by a Grant-in-Aid for the WABOT-HOUSE Project by Gifu Prefecture. We also would like to express thanks to NTT Docomo, SolidWorks Corp., Advanced Research Institute for Science and Engineering of Waseda University, Prof. Hiroshi Kimura, Waseda University for their supports to our robot.

References

[1] Cynthia Breazeal, Brian Scassellati: How to build robots that make friends and influence people, IROS99, pp.858-863, 1999
[2] Paul Ekman, Wallace V.Friesen: Facial Action Coding System, Consulting Psychologists Press Inc., 1978
[3] Tsutomu Kudo, P.Ekman, W.V. Friesen; Hyojo Bunseki Nyumon -Hyojo ni Kakusareta Imi wo Saguru - (Japanese), Seishin Shobo, 1987
[4] Hiroshi Kobayashi, et al.: Study on Face Robot for Active Human Interface - Mechanisms on Face Robot and Facial Expressions of 6 Basic Emotions -, the Journal of the Robotics Society of Japan Vol.12 No.1, pp.155-163, 1994
[5] Hiroshi Kobayashi, Fumio Hara: Real Time Dynamic Control of 6 Basic Facial Expressions on Face Robot, the Journal of the Robotics Society of Japan Vol.14 No.5, pp.677-685, 1996
[6] Kazutaka Mitobe, et al.: Consideration of Associated Movements of head and Eyes to optic and Acoustic Stimulation, The institute of electronics, information and communication engineers, Vol.91, pp.81-87, 1992
[7] Laurutis V.P., Robinson D.A.: The vestibulo-ocular reflex during human saccadic eye movements, J. Physiol., 373, pp.209-233, 1986
[8] Atsuo Takanishi, et al.: "Development of an Anthropomorphic Head-Eye Robot with Two Eyes", Proceedings of the IEEE/RSJ International Conference on Intelligent Robots and Systems, pp.799-804, 1997
[9] Hiroyasu Miwa, et al.: "Human-like Robot Head that has Olfactory Sensation and Facial Color Expression", Proceedings of the 2001 IEEE International Conference on Robotics and Automation, pp.459-464, 2001
[10] Hiroyasu Miwa, et al.: "Experimental Study on Robot Personality for Humanoid Head Robot", Proceedings of the 2001 IEEE/RSJ International Conference on Intelligent Robots and Systems, pp.1183-1188, 2001
[11] Hiroyasu Miwa, et al: "Development of a New Human-like Head Robot WE-4", 2002 IEEE/RSJ International Conference on Intelligent Robots and Systems, 2002 (to appear)
[12] G. W. Allport: "Pattern and Growth in Personality", Rinehart and Winston, 1961
[13] Hiroshi Imada: "Gakusyu no Shinrigaku (Japanese) ", pp.107-1224, Baifukan, 1996
[14] Kiyoshi Suzuki: "Shinrigaku: Keiken to Kodo no Kagaku (Japanese)", Nakanishiya Shuppan, 1991

Planning Velocities of Free Sliding Objects for Dynamic Manipulation

Qingguo Li and Shahram Payandeh
Experimental Robotics Laboratory, School of Engineering Science, Simon Fraser University
Burnaby, B.C., V5A 1S6, Canada

Abstract— In this paper, a novel numerical approach is proposed to solve the initial velocities of the free sliding object for given initial and final configurations. To find the desired initial velocities for free sliding objects is a key step for implementing dynamic manipulation. In order to plan the initial velocities, the motion of free sliding objects is modeled as a set of 6 first order differential equations, and the planning problem is formulated as a free boundary value problem(FBVP). Through a simple transformation, the FBVP is reduced to a standard Two-point boundary value(TPBV) problem. Quasi-Newton based optimization procedures are utilized to solve the planning problem. Unlike existing approaches, the proposed method does not require qualitative motion characteristics, thus it can be used for objects with general shape and arbitrary pressure distribution. Simulation results on polygonal objects with three to five vertices are used to demonstrate the planning method.

I. INTRODUCTION

In robot manipulation, one of basic tasks is to move object from initial configuration to goal configuration. One possible method is to grasp objects rigidly and then move them. The other is to move objects by nonprehensive techniques such as pushing, throwing, batting and striking[7]. Dynamic manipulation includes hopping, juggling, tapping, and batting, and impulse planar manipulation. Impulse planar manipulations were studied in [6][5][4]. The problem was decomposed into the impact problem and the inverse sliding problem. The inverse sliding problem is to determine the initial velocities required for the object to slide to the desired displacement(translation and rotation) based on the dynamics of sliding object. The dynamics of sliding motion of disks and rings has been studied in [9], some properties of the motion were proposed. In [3], concept of limit surface was introduced to study the relation between the motion of the slider and the frictional force. Owing to the complex dynamics of the free sliding object, for a given displacement of the object, it was founded that it is impossible to determine the desired velocities analytically.

The inverse sliding problem for the class of axisymmetric objects is addressed in [5]. Axisymmetric objects are those which have a pressure distribution that is a function of radius of the object only, and have the property that they always slide in a straight line. The planning is to find the initial linear velocity along the line, and the associated rotational velocity. Using the properties of monotonicity of displacement with respect to the initial velocities, a numerical approach is developed to find the desired initial velocities through subdividing the initial velocity space. However, the monotonicity properties do not hold for nonaxismmetric objects, which limits the applications of the impulse manipulation method. Recently the impulse manipulation has been extended to polygonal objects in [4]. A new set of qualitative dynamic characteristics of the motion are derived to relate the initial velocities and the displacement of the object. Heuristic rules were developed to search for the desired initial velocities in the 3-dimension initial velocity space. In above two approaches, first qualitative relationships between the initial velocities and displacement were derived, then the searching algorithm for the velocities are developed through bisection of the initial velocity space. All of above methods depends on some characteristics of the motion, they can only be applied to a specific class of objects, Besides, the algorithms used only qualitative heuristic information, the convergence is always slow.

Instead of using qualitative information, this paper proposed a new method to solve the free sliding problem using optimization techniques. Under Coulomb's assumptions on friction, a set of differential equations that govern the motion of the object can be derived for a given object. In the free sliding problem, the initial configuration and final goal configuration are known, and we know that at final goal configuration, the velocities of the object are zero, while the traveling time of the object is unspecified. Based on this observation, the free sliding problem can be formulated as a free boundary value problem. In order to use the well known existing techniques, the problem is reduced to a standard two-point boundary value problem and solved by using simple shooting methods[1]. The shooting method is implemented by integrating the initial value solver with optimization routines. Here we use the Quasi-Newton's method as the optimization routine. In general shooting methods, computing of Jacobian in optimization routine is always time consuming because several initial value problems have to be solved. In order to reduce the cost for computing Jacobian, we implement the Quasi-Newton's method with the Broyden update. In which the Jacobian is updated recursively without solving the initial value problems in each iteration.

The method proposed here is different from the previous methods[6][4]. In previous methods, first qualitative relationships between the initial velocities and displacement were derived based on the equations of free sliding objects, then the searching algorithm for the velocities are developed through bisection of the initial velocity space. These methods utilized same strategy for specific type objects or parts which satisfy the qualitative criteria obtained through analysis. For other kind of objects with different geometry or different pressure distribution, new criteria need to be derived to direct the search direction for the initial velocities. Just as the bisection method for nonlinear equation, the convergence speed of these methods is always slow. In this paper, instead of using qualitative properties of the motion, quantitative information is used to find desired initial velocities.

The remainder of the paper is organized as follows: model of the sliding motion is derived in section 2, in section 3 we formulate the velocity planning problem as free boundary value problem, and give the standard two-point boundary value(TPBV) formation. Section 4 presents the planning algorithm and implementation. Simulation results are carried out in section 5, and section 6 concludes the paper.

II. MODEL OF SLIDING MOTION

The motion of an free sliding object on a horizontal plane is governed by the friction force between the object and the plane. The friction force and torque can be calculated by integrating through each infinitesimal element of the object. Consider an object on the plane as shown in Fig. 1. XOY is the global coordinate, xoy is local coordinate associated with the object, and assign o at the center of mass. The configuration space of the object is defined as (X_o, Y_o, θ). Where (X_o, Y_o) gives the position of the local coordinate, while θ denotes the orientation of the object.

Assume the linear velocity of o is $v = (\dot{X}_o, \dot{Y}_o)$, we assume that on each infinitesimal element of the object, there acts a force of friction. Denote A as infinitesimal element of the object located at (x, y) in local coordinate, and r is the position vector from o to A, dm is the mass of element A calculated as $dm = \rho(x,y)dxdy$, $\rho(x,y)$ is the pressure distribution function over the area. $\omega = \dot{\theta}$ the angular velocity of object in the count-clockwise direction, μ is the friction coefficient, g is the acceleration of gravity. By integrating friction forces and torque over overall contact area, the frictional forces and torque acting on the object can be expressed as

$$F_X = -\mu \cdot g \cdot \rho \iint \frac{A_x}{\sqrt{B}} dxdy \tag{1}$$

$$F_Y = -\mu \cdot g \cdot \rho \iint \frac{A_y}{\sqrt{B}} dxdy \tag{2}$$

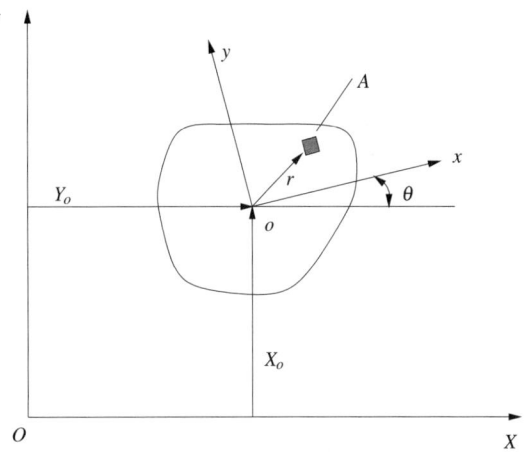

Fig. 1. Motion of planar object

$$T = -\mu \cdot g \cdot \rho \iint \frac{A_t}{\sqrt{B}} dxdy \tag{3}$$

Where

$$A_x = \dot{X}_o - \cos\theta \cdot \omega \cdot y - \sin\theta \cdot \omega \cdot x \tag{4}$$

$$A_y = \dot{Y}_o - \sin\theta \cdot \omega \cdot y + \cos\theta \cdot \omega \cdot x \tag{5}$$

$$\begin{aligned} A_t = &-(\dot{X}_o x + \dot{Y}_o y)\sin\theta \\ &-(\dot{X}_o y - \dot{Y}_o x)\cos\theta + \omega \cdot y^2 + \omega \cdot x^2 \end{aligned} \tag{6}$$

$$\begin{aligned} B = &(\dot{X}_o - \cos\theta \cdot \omega \cdot y - \sin\theta \cdot \omega \cdot x)^2 \\ &+ (\dot{Y}_o - \sin\theta \cdot \omega \cdot y + \cos\theta \cdot \omega \cdot x)^2 \end{aligned} \tag{7}$$

The motion of the sliding object subject to the friction can be written as

$$\begin{aligned} \ddot{X}_o &= F_X/m \\ \ddot{Y}_o &= F_Y/m \\ \ddot{\theta} &= T/I \end{aligned} \tag{8}$$

where m is the mass of the object, while I is the mass moment of inertia refer to the center of mass. Denote $\bar{x} = (x_1, x_2, x_3, x_4, x_5, x_6) = (\dot{X}_o, X_o, \dot{Y}_o, Y_o, \dot{\theta}, \theta)$ as the state variable, (8) can be rewritten as the state space form

$$\begin{bmatrix} \dot{x}_1 \\ \dot{x}_2 \\ \dot{x}_3 \\ \dot{x}_4 \\ \dot{x}_5 \\ \dot{x}_6 \end{bmatrix} = \begin{bmatrix} x_2 \\ \frac{F_X}{m} \\ x_4 \\ \frac{F_Y}{m} \\ x_6 \\ \frac{T}{I} \end{bmatrix} = \begin{bmatrix} f_1(\bar{x}) \\ f_2(\bar{x}) \\ f_3(\bar{x}) \\ f_4(\bar{x}) \\ f_5(\bar{x}) \\ f_6(\bar{x}) \end{bmatrix} \tag{9}$$

If we want to plan to desired initial velocities for free sliding object using the model (8), we must set up a small ε, and consider the object stopped when $v_0 = \sqrt{\dot{X}^2 + \dot{Y}^2 + \dot{\theta}^2} = \varepsilon$, this gives us the traveling time T_f.

III. FORMULATION OF THE PLANNING PROBLEM

The planning problem can be formulated as: An object with known geometry, mass distribution, and friction properties slides on a supporting surface. It starts from initial configuration (X_i, Y_i, θ_i), slowing down to rest at a final configuration (X_f, Y_f, θ_f). Given the initial and final configurations, determine the initial velocities.

The motion of the object on a plane is governed by (9), which is a system of 6 first-order differential equations. There exists no analytical solution for determining the initial velocities for a given displacement. However the initial velocities can only be found by using numerical procedures.

For a given initial configuration (X_i, Y_i, θ_i), and velocities $(\dot{X}_i, \dot{Y}_i, \dot{\theta}_i)$ at time $t = 0$, the trajectory of the motion can be uniquely determined, such that the object occupies the final configuration (X_f, Y_f, θ_f). The solution can be found by numerical integrating the system (9). This is known as the initial value problem. However for the planning problem, the initial configuration (X_i, Y_i, θ_i) at $t = 0$ and final goal configuration (X_f, Y_f, θ_f) are known, while the traveling time T_f of the object is unknown. If both the starting time and ending time for (9) are specified, the problem becomes a standard two-point boundary value (TPBV) problem in differential equation literature[1]. Since in the planning problem, only starting time $t = 0$ is specified, we need to use another boundary condition to determine the ending time T_f, the condition is that the velocities at the ending point T_f decay to zero. Here the system (9) has a solution satisfying 7 boundary conditions

$$\begin{cases} X_o(0) = X_i & X_o(T_f) = X_f \\ Y_o(0) = Y_i & Y_o(T_f) = Y_f \\ \theta_o(0) = \theta_i & \theta_o(T_f) = \theta_f \\ v_0(T_f) = 0 & \end{cases} \quad (10)$$

The statement of this problem is referred as *free boundary value problem*.

The free boundary value problem can be transformed to a TPBV problems by introducing new independent variables. Here in place of time t, we introduce a new independent variable τ, such that

$$t = \tau T_f \quad 0 \le \tau \le 1 \quad (11)$$

$$\dot{T}_f = \frac{dT_f}{d\tau} = 0 \quad (12)$$

In (11), we know that when τ varies from $\tau = 0$ to $\tau = 1$, the system will travel from $t = 0$ to $t = T_f$. And T_f is independent to τ.

After substituting (11) into (9), and augmenting (12) to (9), the motion of the object respect to τ can be written as

$$\begin{cases} \frac{dx_1}{d\tau} = T_f \cdot f_1(\bar{x}) \\ \frac{dx_2}{d\tau} = T_f \cdot f_2(\bar{x}) \\ \vdots \\ \frac{dx_6}{d\tau} = T_f \cdot f_6(\bar{x}) \\ \dot{T}_f = 0 \end{cases} \quad (13)$$

The system of 7 differential equations is now in standard TPBV form with τ varying between the known limits 0 and 1. And the boundary conditions are

$$\begin{cases} X_o(0) = X_i & X_o(1) = X_f \\ Y_o(0) = Y_i & Y_o(1) = Y_f \\ \theta_o(0) = \theta_i & \theta_o(1) = \theta_f \\ v_0(T_f) = 0 & \end{cases} \quad (14)$$

The planning problem is to determine initial velocities $(\dot{X}_o(0), \dot{Y}_o(0), \dot{\theta}(0))$ and the traveling time T_f, such that the solution of (13) satisfy boundary conditions (14).

The methods for solving TPBV problems fall into three categories: (1). the shooting method, (2) the difference method, (3). the variational method[8]. The shooting method is an extension of the initial value techniques. Its advantages are conceptual simplicity and it allows taking advantage of available initial value ordinary differential equations solvers.

IV. PLANNING THE VELOCITIES OF FREE SLIDING

A. Planning algorithm

In this section, the initial velocities are planned using shooting method to the TPBV formulation (13) and (14).

The associated initial value problem is defined as

$$\begin{cases} \frac{dx_1}{d\tau} = T_f \cdot f_1(\bar{x}) \\ \frac{dx_2}{d\tau} = T_f \cdot f_2(\bar{x}) \\ \vdots \\ \frac{dx_6}{d\tau} = T_f \cdot f_6(\bar{x}) \\ \frac{dT_f}{d\tau} = 0 \end{cases} \quad (15)$$

with the initial conditions at $\tau = 0$,

$$\begin{cases} X_o(0) = X_i & \dot{X}_o(0) = \dot{X}_i \\ Y_o(0) = Y_i & \dot{Y}_o(0) = \dot{Y}_i \\ \theta_o(0) = \theta_i & \dot{\theta}_o(0) = \dot{\theta}_i \\ T_f(0) = T_{fi} & \end{cases} \quad (16)$$

We know from the boundary condition (14) that $X_o(0) = X_i, Y_o(0) = Y_i, \theta_o(0) = \theta_i$ are known, and the configuration $X_o(1), Y_o(1), \theta_o(1)$ and the linear velocity $\dot{X}_o(1)$ at $\tau = 1$ are functions of $s = (\sigma_1, \sigma_2, \sigma_3, \sigma_4)^T = (\dot{X}_o(0), \dot{Y}_o(0), \dot{\theta}(0), T_f(0))$ at $\tau = 0$ in initial value problem. In order to solve the TPBV problem of (13), (14), we need to determine a starting velocities and traveling time $T_f(0)$ as $s = (\dot{X}_o(0), \dot{Y}_o(0), \dot{\theta}(0), T_f(0))$ for the initial value problem (15), (16), such that the solution obeys the boundary conditions (14) at the other end $\tau = 1$ as

$$X_o(1,s) = X_f \quad Y_o(1,s) = Y_f$$
$$\theta_o(1,s) = \theta_f \quad \dot{X}_o(1,s) = 0 \quad (17)$$

rewriting (17) as a vector function form

$$F(s) = \begin{bmatrix} X_o(1,s) - X_f & Y_o(1,s) - Y_f \\ \theta_o(1,s) - \theta_f & v_0(1,s) \end{bmatrix}^T \quad (18)$$

where $s = (\sigma_1, \sigma_2, \sigma_3, \sigma_4)^T$.

Solving the TPBV problem is equivalent to finding a solution of s as $\bar{s} = (\bar{\sigma}_1, \bar{\sigma}_2, \bar{\sigma}_3, \bar{\sigma}_4)$ such that

$$F(\bar{s}) = 0 \quad (19)$$

This nonlinear equations (18) can be solved by means of the general Newton's method

$$s^{(i+1)} = s^{(i)} - DF(s^{(i)})^{-1} \cdot F(s^{(i)}) \quad (20)$$

In each iteration step, one has to compute $F(s^{(i)})$, and the Jacobian matrix

$$DF(s^{(i)}) = \left[\frac{\partial F_j}{\partial \sigma_k} \right]_{s=s^{(i)}} \quad (21)$$

and the solution $d^{(i)} = s^{(i)} - s^{(i+1)}$ of the linear system of equation $DF(s^{(i)})d^{(i)} = F(s^{(i)})$. For the computation of $F(s^{(i)})$, one must solve the initial value problem (15),(16) for $s = s^{(i)} = (\sigma_1^{(i)}, \sigma_2^{(i)}, \sigma_3^{(i)}, \sigma_4^{(i)})$, the Jacobian $DF(s^{(i)})$ can not be calculated analytically, and it will be approximated by the matrix

$$\Delta F(s^{(i)}) = \begin{bmatrix} \Delta F_1(s^{(i)}) & \cdots & \Delta F_2(s^{(i)}) \end{bmatrix} \quad (22)$$

where

$$\Delta F_j(s^{(i)}) = \frac{1}{\Delta \sigma_j^{(i)}} (F_j(\sigma_1^{(i)}, \cdots, \sigma_j^{(i)} + \Delta \sigma_j^{(i)}, \cdots, \sigma_4^{(i)})$$
$$- F_j(\sigma_1^{(i)}, \cdots, \sigma_j^{(i)}, \cdots, \sigma_4^{(i)}))$$
$$\text{for} \quad j = 1, 2, 3, 4 \quad (23)$$

As the computation of $F(s^{(i)})$, the calculation of $F(\sigma_1^{(i)}, \cdots, \sigma_j^{(i)} + \Delta \sigma_j^{(i)}, \cdots, \sigma_4^{(i)})$ requires to solve the corresponding initial value problems (15), (16) with initial conditions $s = (\sigma_1^{(i)}, \cdots, \sigma_j^{(i)} + \Delta \sigma_j^{(i)}, \cdots, \sigma_4^{(i)})$. The approximate Newton's method is carried out as

$$s^{(i+1)} = s^{(i)} - \Delta F(s^{(i)})^{-1} \cdot F(s^{(i)}) \quad (24)$$

The planning algorithm is summarized as
1) Choose a starting vector $s^{(0)}$,
   For $i = 0, 1, 2, \cdots$, repeat steps 2-4,
2) Determine $X_o(1, s^{(i)}), Y_o(1, s^{(i)}), \theta_o(1, s^{(i)}), v_0(1, s^{(i)})$ by solving initial value problem (15), (16), then compute $F(s^{(i)})$ according to (18).
3) Choose $\Delta \sigma_j, j = 1, \cdots, 4$, and determine $X_o(1, s^{(i)} + \Delta \sigma_j e_j), Y_o(1, s^{(i)} + \Delta \sigma_j e_j), \theta_o(1, s^{(i)} + \Delta \sigma_j e_j), v_0(1, s^{(i)} + \Delta \sigma_j e_j)$ by solving 4 initial value problems (15), (16) for

$$s = s^{(i)} + \Delta \sigma_j e_j = \begin{bmatrix} \sigma_1^{(i)}, \cdots, \sigma_j^{(i)} + \Delta \sigma_j, \cdots, \sigma_4^{(i)} \end{bmatrix}^T \quad (25)$$

where e_j is a 4 dimensional vector of following form

$$e_j(i) = \begin{cases} 1 & i = j \\ 0 & i \neq j \end{cases} \quad (26)$$

4) Compute $\Delta F(s^{(i)})$ by means of (22),(23), and also the solution $d^{(i)}$ of the system of linear equations

$$\Delta F(s^{(i)}) d^{(i)} = -F(s^{(i)}) \quad (27)$$

and update

$$s^{(i+1)} = s^{(i)} + d^{(i)} \quad (28)$$

Algorithm 1: Shooting method without Broyden's update

In each step of the method, 5 initial value problems and a 4 dimensional system of linear equations needs to be solved.

B. Shooting with Broyden's Update

In each iteration of the algorithm, the Jacobian is computed numerically according to (22),(23), which needs solving 4 initial value problem. The computing of initial value problem is always time consuming. In order to reduce the computational cost, we consider to use the Broyden's method[2] to approximate the Jacobian matrix, the local convergence has been proven in [2]. In Broyden's method, The Jacobian matrix is updated in each iteration instead of computing numerically using (22) and (23). The planning algorithm with Broyden's update is:

1) Choose a starting vector $s^{(0)}$,
2) Determine $X_o(1, s^{(0)}), Y_o(1, s^{(0)}), \theta_o(1, s^{(0)}), v_0(1, s^{(0)})$ by solving initial value problem, then compute $F(s^{(0)})$ according to (18).
3) Choose $\Delta \sigma_j, j = 1, \cdots, 4$, and determine $X_o(1, s^{(0)} + \Delta \sigma_j e_j), Y_o(1, s^{(0)} + \Delta \sigma_j e_j), \theta_o(1, s^{(0)} + \Delta \sigma_j e_j), v_0(1, s^{(0)} + \Delta \sigma_j e_j)$ by solving 4 initial value problems for

$$s_j^{(0)} = s^{(0)} + \Delta \sigma_j e_j = \begin{bmatrix} \sigma_1^{(0)}, \cdots, \sigma_j^{(0)} + \Delta \sigma_j, \cdots, \sigma_4^{(0)} \end{bmatrix}^T \quad (29)$$

where e_j is a 4 dimensional vector of following form

$$e_j(i) = \begin{cases} 1 & i = j \\ 0 & i \neq j \end{cases} \quad (30)$$

4) Compute $\Delta F(s^{(0)})$ by means of (22), (23).
   For $i = 0, 1, 2, \cdots$, repeat steps 5-8,
5) Compute the solution $d^{(i)}$ of the system of linear equations

$$\Delta F(s^{(i)}) d^{(i)} = -F(s^{(i)}) \quad (31)$$

and update
$$s^{(i+1)} = s^{(i)} + d^{(i)} \qquad (32)$$

6) Determine $X_o(1, s^{(i+1)}), Y_o(1, s^{(i+1)}), \theta_o(1, s^{(i+1)})$, $v_0(1, s^{(i+1)})$ by solving initial value problem (15), (16), then compute $F(s^{(i+1)})$ according to (18).

7) Compute
$$y^{(i)} = F(s^{(i+1)}) - F(s^{(i)}) \qquad (33)$$

where $y^{(i)}$ is the difference of the function $F(.)$ during iteration $i+1$ and i, which provide the information for computing Jacobian.

8) Update Jacobian
$$\Delta F(s^{(i+1)}) = \Delta F(s^{(i)}) + \frac{(y^{(i)} - \Delta F(s^{(i)}) d^{(i)})(d^{(i)})^T}{(d^{(i)})^T d^{(i)}} \qquad (34)$$

Algorithm 2: Shooting method with Broyden's update

In algorithm 1, the computation of Jacobian is approximated by (22), which needs to solve 4 additional initial value problem in each iteration. Instead, the Jacobian is updated through (34) in algorithm 2. This algorithm only needs to solve one initial value problem in each iteration, which reduces the computational cost dramatically, which is 25 percent of algorithm 2.

The optimization procedures used in the planning algorithms belong to the class of local optimization methods, which exhibit local convergence. The choice of initial guess of velocities $\mathbf{s}^{(0)}$ will affect the convergence of the algorithm. And one reasonable choice is that we assume the object slides along the X and Y coordinate, and the initial linear velocities $\dot{X}_o(0), \dot{Y}_o(0)$ are chosen as

$$\begin{aligned} \dot{X}_o(0) &= \sqrt{2\mu g X_f} \\ \dot{Y}_o(0) &= \sqrt{2\mu g Y_f} \end{aligned} \qquad (35)$$

and the initial $\dot{\theta}(0)$ is chosen under the assumption that the object has pure rotation, that is,

$$\dot{\theta}_o(0) = \sqrt{2\mu g \theta_f \iint (\sqrt{x^2+y^2} \rho(x,y)) dx dy / I} \qquad (36)$$

V. SIMULATION RESULTS

Simulations are carried out with planning algorithm 2 on polygonal objects with 3-5 vertices, even and uneven pressure distribution are considered.

A. Planning for object with even pressure distribution

Consider a triangle object with edges $c = 0.2m, b = 0.17m$, angle $A = \pi/6$, pressure distribution $\rho = 200kg/m^2$, friction coefficient $\mu = 0.5$. The initial configuration is $[X_0 \ Y_0 \ \theta_0] = [0 \ 0 \ 0]$, the goal configuration is $[X_f \ Y_f \ \theta_f] = [1.5m \ 2m \ 1rad]$. Using the planning

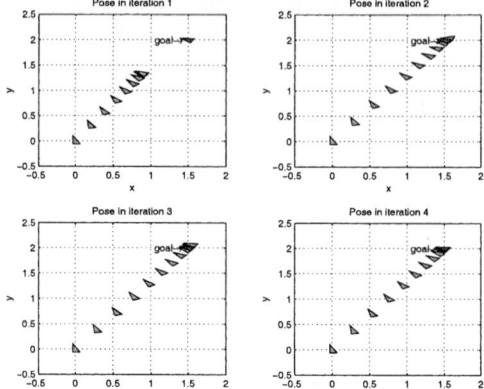

Fig. 2. Trajectory of object vs. iteration

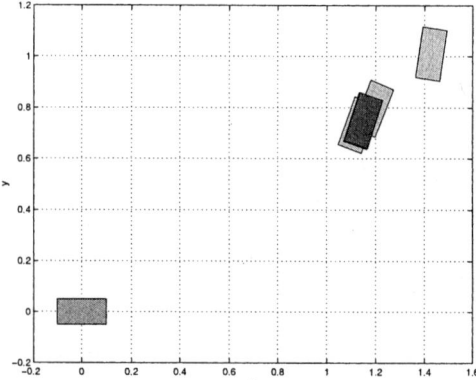

Fig. 3. Final configuration vs. Iteration

algorithm 2 proposed in this paper, the desired velocities were found in 4 iterations, the desired velocities are $[\dot{X} \ \dot{Y} \ \omega] = [2.9686m/s \ 3.9595m/s \ 1.7257m/s]$, and the trajectories of the object in each iteration are shown in Fig. 2.

Consider a rectangular object with even pressure distribution $\rho(x,y) = 200kg/m^2$, and dimension is $0.2m \times 0.1m$, The goal configuration is $[X_f \ Y_f \ \theta_f] = [1.5m \ 2m \ 1rad]$. Using the planning algorithm proposed in this paper, with initial guess of velocities as $s^{(0)} = [3.2m/s \ 2.3m/s \ 2.8rad/s]$. The configuration vs. iteration is shown in Fig. 3. It is clear that the planner find the desired velocities that lead the object to the goal configuration. The planning algorithm can obtain the desired initial velocities $[\dot{X} \ \dot{Y} \ \omega] = [2.88m/s \ 1.9m/s \ 2.77m/s]$ in 4 iterations.

Consider a pentagon with vertices located at
$(0,0), (0.2,0), (0.175,0.1), (0.15,0.2), (0.1,0.15)$ in global coordinate, and the pressure distribution function $\rho = 200kg/m^2$, The goal configuration is

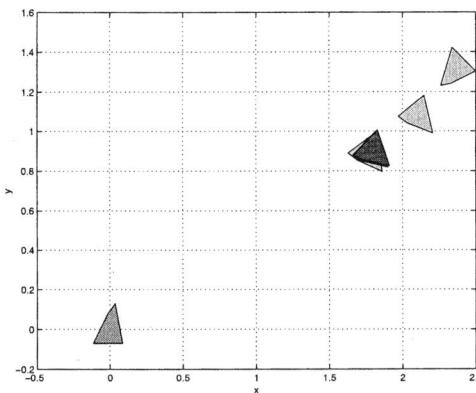

Fig. 4. Final configuration vs. Iteration

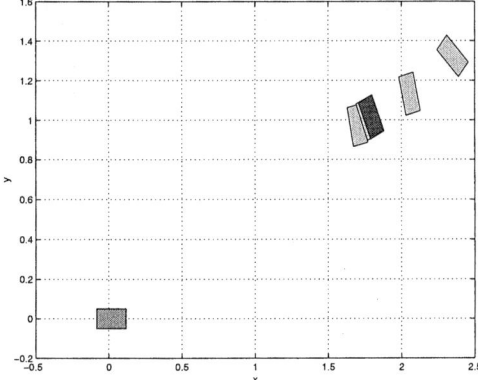

Fig. 5. Final configuration vs. Iteration

$[X_f \ Y_f \ \theta_f] = [1.8m \ 0.9m \ 2.0rad]$. Using the planning algorithm proposed in this paper, with initial guess of velocities as $s^{(0)} = [4.5m/s \ 2.5m/s \ 3.5rad/s]$. The configuration vs. iteration is shown in Fig. 4. The planning algorithm can find the desired initial velocities $[\dot{X} \ \dot{Y} \ \omega] = [3.97m/s \ 1.98m/s \ 3.18m/s]$.

B. Planning for object with uneven pressure distribution

Consider the rectangular object used in last section, and dimension is $0.2m \times 0.1m$, now consider the uneven pressure distribution $\rho(x,y) = 100kg/m^2$ for $x < 0.1$ and $rho(x,y) = 200kg/m^2$ for $x > 0.1$. The goal configuration is $[X_f \ Y_f \ \theta_f] = [1.8m \ 1m \ 2rad]$. Using the planning algorithm proposed in this paper, with initial guess of velocities as $s^{(0)} = [4.5m/s \ 2.5m/s \ 3.5rad/s]$. The final configuration vs. iteration is shown in Fig. 5. The planning algorithm can obtain the desired initial velocities $[\dot{X} \ \dot{Y} \ \omega] = [3.93m/s \ 2.19m/s \ 3.46m/s]$. It is clear that the planning algorithm works also for the objects with uneven pressure distribution.

VI. CONCLUSION

In this paper, a novel computational approach has been proposed to solve the initial velocities for the free sliding objects on a plane. The velocity planning problem is formulated as a free boundary value problem, and solved by using nonlinear optimization techniques. In the proposed method, quantitative information is used to search the desired velocities, no motion characteristics such as monotonicity of force and torque are needed, this method can be used for objects with general shapes and pressure distribution. The planning method is verified using numerical simulation on polygonal objects with 3-5 vertices under even and uneven pressure distribution. The proposed method gives fast convergence. The proposed planning method can be used in any impulse based manipulation or other manipulations which requires to solve initial velocities. As a part of future work, Experiment setup is being developed in order to show the practicality of the proposed method.

VII. REFERENCES

[1] U.M. Ascher and L.R. Petzold. *Computer methods for ordinary differential equations and differential-algebraic equations.* SIAM, Philadelphia, 1998.

[2] J.E. Dennis and JR.R.B. Schnabel. *Numerical methods for unconstrained optimization and nonlinear equations.* Prentice-Hall, Inc., Englewood Cliffs, N.J., 1983.

[3] S. Goyal, A. Ruina, and J.P. Papadopoulos. Planar sliding with dry friction. part 1. limit surface and moment function. *Wear*, 143:307–330, 1991.

[4] I. Han and S. Park. Impulsive motion planning for posing and orienting a polygonal part. *Intl. J. of Robotics Research*, 20(3):249–262, 2001.

[5] W.H. Huang. *Impulsive manipulation.* PhD thesis, Robotics Institute, Carnegie Mellon University, Pittsburgh, PA, 1997.

[6] W.H. Huang and M.T. Mason. Mechanics, planning, and control for tapping. *Intl. J. of Robotics Research*, 19(4):883–894, 2000.

[7] M.T. Mason. Progress in nonprehensile manipulation. *Intl. J. of Robotics Research*, 18(11):1129–1141, 1999.

[8] J. Stoer and R. Bulirsch. *Introduction to numerical analysis, Translated by R. Bartels, W. Gautschi, and C. Witzgall.* Springer, Berlin, 1993.

[9] K. Voyenli and E. Eriksen. On the motion of an ice hockey puck. *American Journal of Physics*, 53(12):1149–1153, 1985.

Experiments in Nonsmooth Control of Distributed Manipulation

T. D. Murphey, J. W. Burdick, J. Burgess, A. Homyk
Engineering and Applied Science, California Institute of Technology
Mail Code 104-44, Pasadena, CA 91125 USA
{murphey,jwb,burgess,homyk}@robotics.caltech.edu

Abstract— This paper describes an experimental modular distributed manipulation system upon which one can implement a variety of control schemes. We have shown elsewhere that when one includes the nonsmooth effects of friction into a model of distributed manipulation, nonsmooth feedback laws must generally be used to control distributed manipulators. We summarize results obtained with this experimental system that confirm the validity of control schemes proposed by the authors in recent papers (see [13, 14]). We describe the control algorithms in some detail and include specifics of the experimental set-up and experimental results.

I. Introduction

Distributed manipulators usually consist of an array of similar or identical actuators combined together with a control strategy to create net movement of an object or objects. The goal of many distributed manipulation systems is to allow precise positioning of planar objects from all possible starting configurations. Such "smart conveyors" can be used for separating and precisely positioning parts for the purpose of assembly. Distributed manipulator actuation methods ranges from air jets, rotating wheels, and electrostatics on the macroscale, to MEMS and flexible scilia at the microscale.

As reviewed below, a number of strategies have been proposed for controlling distributed manipulation systems. In this paper we describe an experimental test-bed that was designed to evaluate and validate such control systems. Our modular system can emulate a reasonably large class of distributed manipulators that generate motion through rolling and sliding frictional contact between the moving object and actuator surfaces. In such cases friction forces and intermittent contact play an important role in the overall system dynamics, leading to non-smooth dynamical system behavior. In previous work we presented non-smooth dynamical models for describing such systems [11], and non-smooth control laws [13, 14] that provably stabilize these systems. We have applied these techniques to our experimental system, and experimental results presented in this paper validite these methods.

Methods to design distributed manipulation control systems have been proposed in several works, including [3, 5, 6]. A common approach is based on the notion of *programmable vector fields* [2, 4]. In this methodology, one makes the possibly unrealistic assumption that the array's control capability can be idealized as a continuous distribution of forces across the array surface. In this abstraction, the manipulated object moves under the influence of these forces. The control design problem reduces

Fig. 1
THE FADM SYSTEM

to the selection of a continuous force field distribution that will locally transport the object to a prescribed position, and then stabilize it at that configuration. To implement the control strategy on the real array, one must adapt the continuous vector field control to the real (and discrete) actuator array.

The programmable vector field approach has been experimentally shown to work in MEMS-fabricated actuator arrays, where the array elements are "small" and "close" together relative to the size of the object being manipulated [2, 9]. This approach is additionally well suited to distributed air jets, where the aerodynamics effectively "smooth out" the resulting forces on the object. However, in cases where only a small number of actuators are in contact with the manipulated object (i.e., the continuous actuation approximation is poor) or the coefficient of friction μ is very high, the continuous approximation has been shown experimentally not to work as well [8]. In these cases, the continuous approximation does not adequately incorporate the physics of the actual array and the object/array interface.

These experimental observations previously led us to explicitly incorporate frictional and discontinuous contact effects into the analysis and control of distributed manipulation (and the related case of overconstrained wheeled vehicles [11]). We showed that under very simple and general assumptions on the friction model, the PFF approach leads to unstable systems when implemented on actual distributed manipulation arrays that have frictional contact [13]. The instability arises in the object's orientation at the equilibrium configuration. In Ref. [12] we presented a nonsmooth control law that locally stabilizes this instability in a provably correct way. In [14] we showed how to combine the programmable vector field approach and the local control law of [12] to obtain a globally exponentially controllable distributed manipulation control system.

In this paper, we briefly review these theories (Section II), present the design of an experimental distributed manipulation test-bed (Section III), develop explicity non-smooth control laws for this system (Section IV), and demonstrate these prior theoretical results on our prototype via detailed experiments (Section V).

II. Background

Many actual or proposed distributed manipulator implementations rely upon physical contact between the manipulated object and the driving elements. Examples include driving wheels, fingers, cilia, or flaps. To explicity investigate, incorporate, and control the complex frictional contact phenomena inherent in such systems, one needs to develop general modelling schemes that can capture these phenomena. One could resort to a general Lagrangian modelling approach that accounts for the contact effects through Lagrange multipliers. Instead, we sought to develop a general modelling scheme that captures the salient physical features, while also leading to equations that are amenable to control analysis.

To realize this goal, we use a "power dissipation model" (PDM) approach to model the governing dynamics of a distributed manipulation system involving a discrete number of frictional contacts. One can show that this method almost always produces unique models [11] that are relatively easy to compute, and to which one can apply control system analysis methods. Since the method is a quasi-static modeling method, it produces first-order governing equations, instead of second order equations that are associated with Lagrange's equations.

We assume that the moving body and actuator elements that contact the object can be modelled as rigid bodies making point contacts that are governed by the Coulomb friction law at each contact point. Let q denote the configuration of the array/object system, consisting of the object's planar location, and the variables that describe the state of each actuator element. Under these conditions, the relative motion of each contact between the object and an actuator array element can be modeled in the form $\omega(q)\dot{q}$. If $\omega(q)\dot{q} = 0$, the contact is not slipping, while if $\omega(q)\dot{q} \neq 0$, then $\omega(q)\dot{q}$ describes the slipping velocity.

In general, the moving object will be in contact with the actuator array at many points. From kinematic considerations, one or more of the contact points must be in a slipping state, thereby dissipating energy. The *power dissipation function* measures the object's total energy dissipation due to contact slippage.

Definition II.1. The *Dissipation* or *Friction Functional* for an n-contact state is defined to be

$$\mathcal{D} = \sum_{i=1}^{n} \mu_i N_i \mid \omega(q)\dot{q} \mid \qquad \text{(II.1)}$$

with μ_i and N_i being the Coulomb friction coefficient and normal force at the ith contact, which are assumed known.

Since there will generally not exist a motion where all of the contacts can be simultaneously slipless, we are lead to the following concept for finding the governing motions.

Power Dissipation Principle: With $\dot{q}$ small, an object's motion at any given instant is the one that minimizes $\mathcal{D}$. Assuming that the motion of the actuator array's variables are known, the *power dissipation method* assumes that the object's motion at each instant is the one that instantaneously minimizes power dissipation due to contact slippage. This method is adapted from the work of [1] on wheeled vehicles, and more details can be found in [11]. For a greater discussion of the formal characteristics of the PDM, and a discussion of the relationship between the PDM and Lagrangian approaches for such a system, see [11].

When one applies the PDM method, the governing equations that result take the following form.

Definition II.2. A system is a *multiple model driftless affine system (MMDA)* if it can be expressed in the form

$$\dot{x} = f_{\sigma_1}(x)u_1 + f_{\sigma_2}(x)u_2 + \cdots + f_{\sigma_n}(x)u_n \qquad \text{(II.2)}$$

where for any x and t, $f_{\sigma_i}(x) \in \{g_{\alpha_i}(x) | \alpha_i \in I_i\}$, with I_i an index set and f_i measurable in (x, t) and g_i analytic in (x, t) for all i. $\Diamond$

An MMDA is a driftless affine nonlinear control system where each control vector field may "switch" back and forth between different elements of a finite set. In our case, this switching corresponds to the switching between different contact states between the object and the array surface elements (i.e., different sets of slipping contacts) due to variations in contact geometry, surface friction properties, and normal loading. In [11] it was shown that the PDM generically leads to MMDA systems as in Definition II.2.

A. Equations of Motion

This section sketches the application of the power dissipation method to the example of a planar array of driven wheels. The i_{th} actuator is located at (x_i, y_i), has an orientation with respect to the origin of θ_i, and the velocity input at that actuator is u_i. I.e., the ith wheel is spinning at speed u_i. Moreover, let g_i be the transformation in $SE(2)$ (the special euclidean group) from the origin to a reference frame associated with the ith actuator. The relative velocity of each contact point between the wheel and moving object can be expressed as $\Omega(q)\dot{q}$, where q is the configuration of the object in $SE(2)$ and

$$\Omega_i(q) = \left[Ad^T_{g_i^{-1}} \begin{pmatrix} 1 \\ 0 \\ 0 \\ 0 \end{pmatrix} \begin{pmatrix} \cos\theta_i \\ \sin\theta_i \\ 0 \end{pmatrix} u_i \atop Ad^T_{g_i^{-1}} \begin{pmatrix} 0 \\ 1 \\ 0 \end{pmatrix} \quad 0 \right] \qquad \text{(II.3)}$$

and
$$g_i = \begin{bmatrix} R(\theta_i) & \begin{pmatrix} x_i \\ y_i \end{pmatrix} \\ 0 & 1 \end{bmatrix} \in SE(2)$$

is the homogeneous representation of the ith actuator node's configuration relative to a fixed reference frame, $Ad(\cdot)$ is the adjoint transformation which transforms velocities from one coordinate frame to another, and $R(\cdot) \in SO(2)$ (that is, R is a member of the special orthogonal group).

We now consider, through an intuitive discussion that can be backed up by analysis, the application of the PDM to this problem.

First note that the minimum of the power dissipation function only occurs when three of the contact constraints are satisfied (i.e. $\Omega_i(q)\dot{q} = 0$ for three choices of the index i). The constraints satisfied are precisely those which would otherwise dissipate the most energy if they were violated. The contact states that dissipate the most energy are those associated with the potential constraints having the largest three normal forces $\alpha_i = N_i \mu_i$. Based on these observations, if the center of mass determines the normal forces (based on assumptions about surface uniformity, etc.), and if $\mu(x,y)$ is uniform, then the object's motion satisfies whichever constraints are closest to its center of mass. That is, the particular region in which the center of mass lies determines the first two actively satisfied constraints. The third actively satisfied constraint is determined by the friction model. The system equations are found by solving for the annihilator of the constraint $\Omega(q)$. If the coefficient of friction for sideways slipping, μ_S, is less than the coefficient of friction for rolling slipping, μ_R, and if the nearest actuator to the center of mass is indexed by i and the second nearest is indexed by j, then the governing equations are:

$$\begin{bmatrix} \dot{x} \\ \dot{y} \\ \dot{\theta} \end{bmatrix} = \begin{bmatrix} \frac{u_i[s_j((x_i-x_j)c_i+y_is_i)+c_ic_jy_j]-u_jy_i}{(x_j-x_i)s_j+(y_i-y_j)c_j} \\ \frac{u_jx_i-u_i[c_ic_jx_i+s_i(x_js_j+(y_i-y_j)c_j)]}{(x_j-x_i)s_j+(y_i-y_j)c_j} \\ \frac{u_j-u_i\cos(\theta_i-\theta_j)}{(x_i-x_j)s_j+(y_j-y_i)c_j} \end{bmatrix} \quad (II.4)$$

where $c_i = \cos(\theta_i)$, $s_i = \sin(\theta_i)$, etc. It should be noted that here the index notation should be thought of as mapping (i,j) pairs to equations of motion in some neighborhood (not necessarily small) around the ith and j^{th} actuator. The transition between the equations of motion determined by actuators i and j to equations of motion determined by actuators k and l will in general be determined by the location of center of mass. This in turn leads to the state space being divided up by transition boundaries between different sets of equations of motion. To write this as an MMDA system, we may rewrite the above system as

$$\begin{bmatrix} \dot{x} \\ \dot{y} \\ \dot{\theta} \end{bmatrix} = f_1 u_i + f_2 u_j \quad (II.5)$$

where

$$f_1 \in \begin{bmatrix} \frac{-y_i}{(x_j-x_i)s_j+(y_i-y_j)c_j} \\ \frac{x_i}{(x_j-x_i)s_j+(y_i-y_j)c_j} \\ \frac{1}{(x_i-x_j)s_j+(y_j-y_i)c_j} \end{bmatrix} \quad f_2 \in \begin{bmatrix} \frac{s_j((x_i-x_j)c_i+y_is_i)+c_ic_jy_j}{(x_j-x_i)s_j+(y_i-y_j)c_j} \\ \frac{-c_ic_jx_i-s_i(x_js_j-(y_i-y_j)c_j)}{(x_j-x_i)s_j+(y_i-y_j)c_j} \\ \frac{-\cos(\theta_i-\theta_j)}{(x_i-x_j)s_j+(y_j-y_i)c_j} \end{bmatrix}$$

As the trajectory $q(t)$ crosses a boundary between one region where the equations of motion are determined by actuators i and j to a region where the equations of motion are determined by actuators k and l, we must allow f_1 and f_2 to be multivalued (hence the inclusion $\in$ instead of equality in Equation (II.5)).

B. Review of Relevant Theory

This section briefly describes three previous results that form the basis of the experiments in Section V. An *elliptic vector velocity field* is one of the form $\Psi(x,y) = (-ax, -by)$ (these are common structures in the open loop theory). The first result states that although elliptic vector fields cannot stabilize (II.5), with full state feedback the equations can be stabilized (see [13]).

Theorem II.1. *The system (II.5) is unstable using an elliptic vector velocity field* $\Psi(x,y) : \mathbb{R}^2 \to \mathbb{R}^2$, *but is locally stabilizable through a discontinuous feedback law.*

We should comment that the instability in Theorem II.1 is only true for this subset of open loop strategies, and is not necessarily true for all open loop strategies. It nevertheless motivates our work in closed loop control, both in this paper and elsewhere. This result only guarantees local stability, however, so a more global theory must be established for purposes of implementation. In this case, it is possible to use elements of the open loop theory of programmable vector fields to do so. In particular, the major goal of [14] was to combine the programmable vector field approach with our local feedback law implied by Theorem II.1. We use a programmable vector field

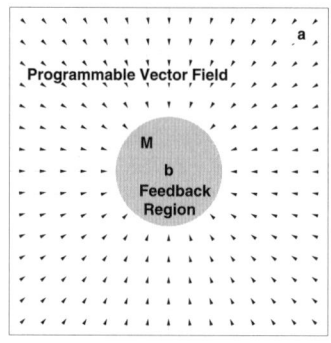

Fig. 2

A LASALLE INVARIANCE THEOREM

to govern the object's gross motions far away from the equilibrium point, and a locally stabilizing feedback law in the vicinity of the equilibrium configuration. Consider Fig. 2. The intuition behind this approach is that if we can move a package from one point $a \in \mathbb{R}^2$ to an equilibrium point $b \in \mathbb{R}^2$, and if we have feedback in a closed

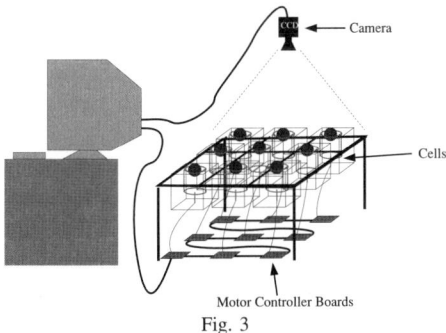

Fig. 3
CARTOON OF EXPERIMENT

neighborhood $\mathcal{M}$ of point b, we can wait to reorient the package until it has entered $\mathcal{M}$.

To effect this blending, we developed a variation of the classical LaSalle Invariance Principle which we then used as the basis for the following theorem (see [14]).

Theorem II.2. *Given a discrete planar array geometry, an elliptic vector velocity field $\Psi(x,y): \mathbb{R}^2 \to \mathbb{R}^2$ outside of $\mathcal{M} = B_\epsilon \times S_1$ (where $B_\epsilon \in \mathbb{R}^2$ is a ball of radius ϵ) for some $\epsilon > 0$, and a locally stabilizing feedback law (which we know exists by Theorem II.1) the solutions to the governing equations given by the PDM are globally stable.*

III. The Experimental Setup

An experimental prototype has been developed for testing the results in Section II and other proposed algorithms. A photograph of the system can be seen in Fig. 1, and a schematic of the system in Fig. 3. Our modular design is based on a cell concept. Each cell contains two actuators. One actuator drives a wheel that contacts the moving object, while the other actuator orients the wheel axis, (see Figure 4). Note that the orienting axis of each cell can be fixed so that the system can simulate simpler systems with a fixed driving wheel orientation. The driving wheel has a four-inch radius, and is made of soft foam rubber to accentuate the friction reaction force. These wheels satisfy the preferred friction distribution rule described in [13].

These cells can be easily repositioned into different configurations in the supporting structure so as to simulate different types of systems. As seen in Fig. 1, the Fully Actuated Distributed Manipulation (FADM) system is deployed with a total of nine cells. More cells can be added as needed.

Both actuators consist of Pittman brushless 12V motors, which are connected to JR-Kerr Pic-Servo-3PH motor controller boards. All 18 motor boards are connected through a daisy chain configuration to a central computer through one of its serial ports.

The position of the manipulated object is obtained and tracked visually. A Sony XC-73 monochrome CCD camera with a Cosmicar C60607 6 mm lens is used for the vision system. Images are captured by an Imagenation PXC-200 framegrabber card. For position acquisition and tracking, a rght triangle is affixed to the moving

Fig. 4
PHOTOGRAPH OF CELL. ACTUATED MOTIONS ARE MARKED IN BLUE.

object. Feature tracking software (written in C) developed at Caltech's Computational Vision Laboratory (see [7]) is used to find and track this triangle. Because of the communication delays required to send control signals to all motor controller boards in the daisy chain system, only six to seven iterations per second can be realized at present.

IV. Feedback Algorithms

This section presents in detail the algorithm used in the underactuated distributed manipulation experiment found in Section V. For a more rigorous treatment of the design of this control law, see [10]. The discontinous feedback law is based on designing control laws for each model in the governing multiple model system. Then, a supervisory controller switches between control law depending upon the current system state. The control laws and switching scheme are chosen to guarantee stability. This methodology allows the control design to be relatively simple, even for relatively complex systems.

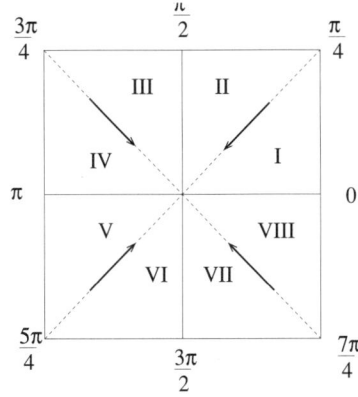

Fig. 5
ALGORITHM FOR NONSMOOTH FEEDBACK

Consider Figure 5, which might represent a portion

of a distributed manipulator near a desired equilibrium point. This region has four actuators (corresponding to the inputs $u_1, \ldots, u_4$ and represented in the figure by arrows) located at $(\pm 1, \pm 1)$, all pointed towards the origin. An analysis of this system using the PDM method shows that the region can be divided into 16 distinct regions. In the eight open regions, labeled **I** − **VIII**, one contact state holds. The other 8 boundary regions, labeled $0 - 2\pi$ in increments of $\frac{\pi}{4}$ denote the boundaries between contact states. Depending on which one of these 16 regions the center of mass of the object resides in, the algorithm chooses a different appropriate control law. In each one of these regions a control law calculated from the Lyapunov function $k(x^2+y^2+\theta^2)$ by solving $\dot{V}=-1$ for u_i, where k is some constant to be chosen during implementation. These control laws can be found in Table I. The "Region" Column shows the region of applicability, and the "Control Law" section shows the explicit control law for each actuator. For regions **I-VIII**, $u_3 = u_1$ and $u_4 = u_2$. For regions $0 - 7\pi/4$ whichever control satisfies $u_i = k\theta$, the control of the same parity (i.e. even or odd index) satisfies $u_j = -k\theta$, while the other two controls are equal.

The globally stabilizing control law in Theorem II.2 is actually quite simple. For our purposes, we chose a radius around the origin of .25 meters to be the "feedback" region, $\mathcal{M}$. Outside of $\mathcal{M}$ we defined the vector field ψ to be $\psi = (-x, -y)$ and inside of $\mathcal{M}$ we used feedback to stabilize the object to the desired position and orientation.

V. Experiments

Here we summarize experimental results that illustrate the theory reviewed in the previous sections [1]. The goal of the experiment was to stabilize an object from a random initial condition to the final configuration $(x_f, y_f, \theta_f) = (0 \text{ m}, 0 \text{ m}, 0 \text{ rad})$. For these experiments, the manipulated object is a piece of plexiglass. This choice allows us to view the actuator actions during manipulation while still ensuring reasonable amounts of friction. The plexiglass has a white piece of paper with a black triangle on it (which is identified by the tracking software) as can be seen in the movie snapshots in Figures 9 and 11. We should note that the plexiglass covers the majority of the viewable area. These snapshots also include outlines of the "goal" triangle position. The following paragraphs summarize each experimental effort and result.

A. Programmable Vector Field

This experiment uses our 9 cell experiment to implement an open loop elliptic vector velocity field. The main point to notice here is that the object did not reorient vey much despite the fact that we were trying to get it to reorient by approximately π radians. Nevertheless, the open loop method did successfully stabilize the x and y coordinates of the object's location. We should also make

[1]Movies of these and other experiments can be found at http://www.cds.caltech.edu/~murphey/experiment/.

Region		Control Law
I	u_1	$\frac{-u_4\left(\theta+x-y\right)+k\left(\theta^2+x^2+y^2\right)}{x+y}$
	u_2	$k\theta$
II	u_1	$\frac{u_2\left(\theta+x-y\right)+k\left(\theta^2+x^2+y^2\right)}{x+y}$
	u_2	$-k\theta$
III	u_1	$k\theta$
	u_2	$\frac{u_1\left(\theta+x+y\right)-k\left(\theta^2+x^2-y^2\right)}{x-y}$
IV	u_1	$-k\theta$
	u_2	$\frac{u_3\left(\theta+x+y\right)+k\left(\theta^2+x^2+y^2\right)}{x-y}$
V	u_1	$\frac{u_2\left(\theta-x+y\right)-k\left(\theta^2+x^2+y^2\right)}{x+y}$
	u_2	$k\theta$
VI	u_1	$-\frac{u_4\left(\theta-x+y\right)+k\left(\theta^2+x^2+y^2\right)}{x+y}$
	u_2	$-k\theta$
VII	u_1	$k\theta$
	u_2	$\frac{u_3\left(-\theta+x+y\right)+k\left(\theta^2+x^2+y^2\right)}{x-y}$
VIII	u_1	$-k\theta$
	u_2	$\frac{u_1\left(\theta-x-y\right)+k\left(\theta^2+x^2+y^2\right)}{x-y}$
0	u_1	$k\theta$
	u_2	$\frac{-u_1\left((-1+\delta)\theta+x+y\right)+k\left(\theta^2+x^2+y^2\right)}{\delta\theta+x-y}$
$\pi/4$	u_1	$\frac{-(-u_2+\delta(u_2+u_4))\left(\theta+x-y\right)+k\left(\theta^2+x^2+y^2\right)}{x+y}$
	u_2	$-k\theta$
$\pi/2$	u_1	$\frac{u_2\left(\delta\theta+x-y\right)+k\left(\theta^2+x^2+y^2\right)}{(1-\delta)\theta+x+y}$
	u_2	$-k\theta$
$3\pi/4$	u_1	$k\theta$
	u_2	$\frac{(-u_3+\delta(u_1+u_3))\left(\theta+x+y\right)-k\left(\theta^2+x^2+y^2\right)}{x-y}$
π	u_1	$k\theta$
	u_2	$-\frac{u_3\left(\delta\theta+x+y\right)+k\left(\theta^2+x^2+y^2\right)}{(-1+\delta)\theta+x-y}$
$5\pi/4$	u_1	$\frac{(-u_4+\delta(u_2+u_4))\left(\theta-x+y\right)-k\left(\theta^2+x^2+y^2\right)}{x+y}$
	u_2	$-k\theta$
$3\pi/2$	u_1	$-\frac{u_4\left(\delta\theta-x+y\right)+k\left(\theta^2+x^2+y^2\right)}{(-1+\delta)\theta+x+y}$
	u_2	$-k\theta$
$7\pi/4$	u_1	$k\theta$
	u_2	$\frac{-(-u_1+\delta(u_1+u_3))\left(\theta-x-y\right)+k\left(\theta^2+x^2+y^2\right)}{x-y}$

TABLE I

LIST OF CONTROL LAWS

clear that this experiment in no way proves that *other* open loop methods may not be successful, but is intended for purposes of comparison to the result that follow.

B. Local Nonsmooth Feedback

This experiment duplicates the geometry of Fig. 5. The goal is to stabilize the supported object to the origin under the strict conditions that the wheel orientations are fixed. The nonsmooth feedback law used in this experiment was precisely that described in the previous section. Something to notice in this experiment is how smoothly the θ variable is driven to 0. This is somewhat surprising considering that the θ variable is precisely the unstable mode in the open loop case (see [13]). However, this is precisely why the experimental results have rather large oscillations as the trajectory is stabilized. As the object gets very close to the origin of $\mathbb{R}^2$, the system becomes difficult to control, so if the trajectory gets too close to the origin of $\mathbb{R}^2$ before the

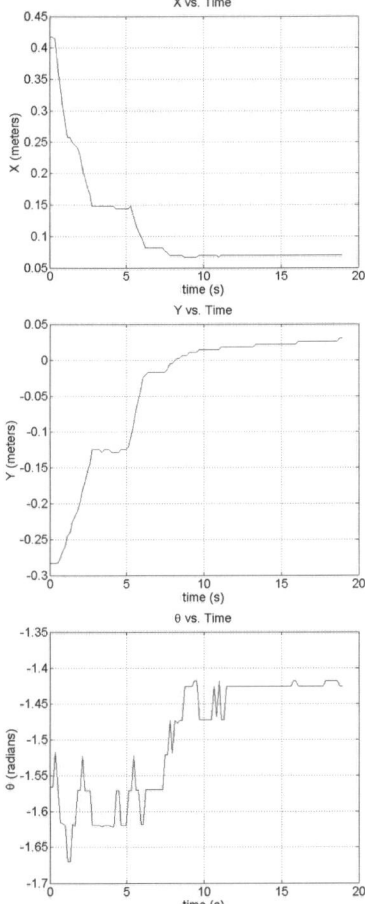

Fig. 6
OPEN LOOP CONTROL

Fig. 7
OPEN LOOP MOVIE SNAPSHOTS

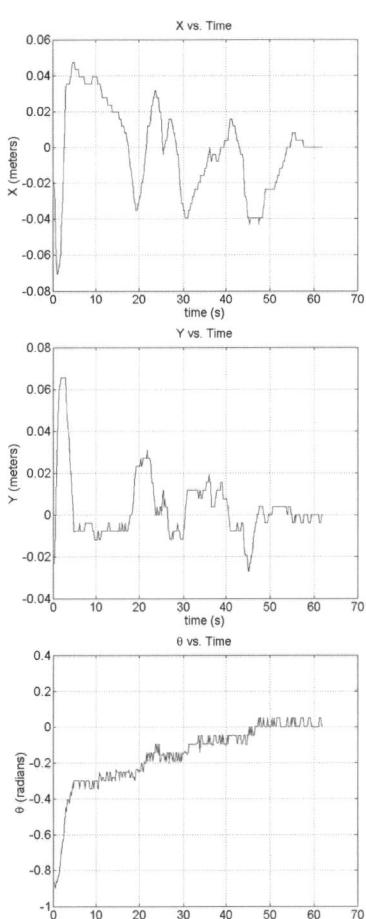

Fig. 8
UNDER-ACTUATED FEEDBACK CONTROL

Fig. 9
UNDER-ACTUATED MOVIE SNAPSHOTS

θ value is very near the desired one, the controls naturally "push" the system away from the origin of $\mathbb{R}^2$ until the orientation is close to the desired value.

C. Global Invariance

This experiment used nine FADM actuators arranged in a regular array (see Fig. 11). We defined a feedback region consisting of a 0.25 meter radius circle centered on the goal position. Outside of this region, object motions are governed by a programmable vector field. Inside of this region, we use a locally stabilizing feedback law that takes advantage of the fact that all of the wheels can be individually steered. This leads to exceptional performance in $\mathcal{M}$. See the companion paper [15] for details of this locally stabilizing control law.

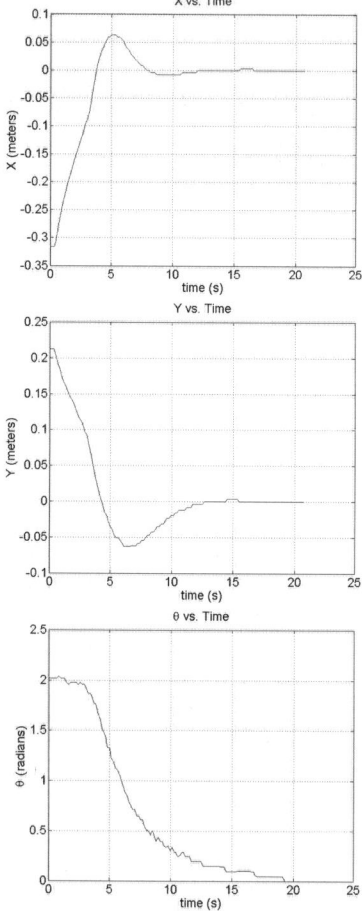

Fig. 10

COMBINING THE PROGRAMMABLE VECTOR FIELD WITH LOCAL FEEDBACK

Fig. 11

GLOBAL STABILIZATION MOVIE SNAPSHOTS

VI. Summary

The experimental results presented here confirm that the modeling and control methodologies in [13, 14] are valid. This indicates several important facts. First, it means that the effects of friction and intermittent contact are important to understanding many distributed manipulation problems. Second, the control algorithms relevant to this class of systems are typically nonsmooth, but still relatively simple. Basic Lyapunov techniques adapted to the nonsmooth setting proved to be more than sufficient for designing and implementing controllers for our distributed manipulation system. A next important step will be to implement the theory demonstrated here on the micro-scale, where friction and limited actuation are very important factors in control. In the future, these methods will be adapted to account for more realistic conditions at the MEMS level, making micro assembly and other micro tasks more feasible.

Acknowledgements: We would like to thank David Choi for his help in building the experiment. We are grateful to the National Science Foundation (grant NSF9402726) for helping to support this research through its Engineering Research Center (ERC) program.

References

[1] J.C. Alexander and J.H. Maddocks. On the kinematics of wheeled vehicles. *The International Journal of Robotics Research*, 8(5):15–27, October 1989.

[2] K.F. Böhringer, R. G. Brown, B. R. Donald, J.S. Jennings, and D. Rus. Sensorless manipulation using transverse vibrations of a plate. In *Proc. IEEE Int. Conf. on Robotics and Automation*, pages 1989–1996, Nagoya, Japan, 1995.

[3] K.F. Böhringer, B.R. Donald, L.E. Kavraki, and F. Lamiraux. A distributed, universal device for planar parts feeding: unique part orientation in programmable force fields. In *Distributed Manipulation*, pages 1–28. Kluwer, 2000.

[4] M. Coutinho and P. Will. The intelligent motion surface: a hardware/software tool for the assembly of meso-scale devices. In *IEEE Int. Conf. on Robotics and Automation (ICRA)*, 1997. Albuquerque, New Mexico.

[5] M.A. Erdmann and M.T. Mason. An exploration of sensorless manipulation. *IEEE Journal of Robotics and Automation*, 4(4), 1988.

[6] K.Y. Goldberg. Orienting polygonal parts without sensing. *Algorithmica*, 143(2/3/4):201–225, 1993.

[7] Luis Goncalves and Enrico Di Bernardo. Software for motion tracking. Developed at the Caltech Vision Lab, 2000.

[8] J. Luntz, W. Messner, and H. Choset. *Distributed Manipulation*, chapter Discreteness Issues in Actuator Arrays. Kluwer Academic Publishers, 2000.

[9] H. Fujita M. Ataka, A. Omodaka. A biomimetic micro motion system. In *Transducers - Digest International Conference on Solid State Sensors and Actuators*, pages 38–41, 1993. Pacifico, Yokohama, Japan.

[10] T. D. Murphey. *Control of Multiple Model Systems*. PhD thesis, California Institute of Technology, May 2002.

[11] T. D. Murphey and J. W. Burdick. Issues in controllability and motion planning for overconstrained wheeled vehicles. In *Proc. Int. Conf. Math. Theory of Networks and Systems (MTNS)*, Perpignan, France, 2000.

[12] T. D. Murphey and J. W. Burdick. Global stability for distributed systems with changing contact states. In *Proc. IEEE Int. Conf. on Intelligent Robots and Systems*, Hawaii, 2001.

[13] T. D. Murphey and J. W. Burdick. On the stability and design of distributed systems. In *Proc. IEEE Int. Conf. on Robotics and Automation*, Seoul, Korea, 2001.

[14] T. D. Murphey and J. W. Burdick. Global exponential stabilizability for distributed manipulation. In *Proc. IEEE Int. Conf. on Robotics and Automation*, Washington D.C., 2002.

[15] T. D. Murphey and J. W. Burdick. Smooth feedback control algorithms for distributed manipulation. In *Submitted to IEEE Int. Conf. on Robotics and Automation*, 2003.

Multi-agent Cooperative Manipulation with Uncertainty: A Neural Net-Based Game Theoretic Approach

Qingguo Li and Shahram Payandeh
Experimental Robotics Laboratory(ERL)
School of Engineering Science
Simon Fraser University, 8888 University Drive
Burnaby, B.C., Canada V5A 1S6
qlib@cs.sfu.ca shahram@cs.sfu.ca

Abstract— This paper proposes a novel planning method for multi-agent dynamic manipulation on a plane. The objective of planning is to find optimal forces exerted on the object by agents with which the object can follow a given trajectory. The main contributions of the proposed approach is: First, through integrating of noncooperative game and neural-net approximation, the planner can deal with unknown pressure distribution effectively. Second, by introducing cooperative game between agents, the forces exerted by agents distributed optimally. Based on the dynamic model of the pushed object, the planing problem is solved in two levels hierarchically. In the lower control level, generalized force inputs are designed by using minimax technique to achieve the tracking performance. In the coordination level, cooperative game is formulated between agents to distribute the generalized force, and the objective of the game is to minimize the worst case interaction force between agents and object. Simulations are carried out for the three-agent cooperative manipulation, results demonstrate the effectiveness of the proposed planning method.

I. INTRODUCTION

In robot manipulation, one of the basic task is to move object from an initial configuration to a goal configuration. One possible method is to grasp objects rigidly and then move them. The other is to move objects by using nonprehensile techniques[7]. Dynamic manipulation includes hopping, juggling, tapping, and batting, and impulse planar manipulation. Most work in dynamic manipulation were considering to use single agent to manipulate the objects, and the planning is carried out under the assumption of known friction and pressure distribution. Asides, owning to the industrial needs and desire to understand human cooperative behavior, there has been much concern on multi-agent manipulation. In [3], manipulation protocols for a small team of mobile robots are developed to push large boxes. In [1], distributed cooperative strategies are proposed for a group of cooperative behavior-based mobile robots to handle an object. The common assumption they made is the motion of the object under pushing is quasi-static, and none of them discussed the problem of optimality of cooperation between robotic agents. In order to address the dynamic cooperative behavior, in [4], a nonprehensile cooperative dynamic manipulation model is proposed, in this model, two robotic agents make point contact with the object, also relaxing the quasi-static assumption. The motion of the object under two agents pushing was modeled as a nonlinear system. Here we assume the finger agents are position/force controllable. The objective of cooperative planning is to find suitable interaction forces between agents and object, and the object will follow a given trajectory while it is pushed by agents with planned forces. The interest of the paper is to find an optimal solution such that the planned forces satisfy given criterion. One centralized planning approach has been proposed in [5], backstepping technique and quadratic programming are integrated to solve the planning problem, the assumption is that there is no uncertainty on pressure distribution. However, in practice the pressure distribution of the pushed object can not be known exactly as *a priori*, this will introduce uncertain frictional forces to the system. The main objective of this paper is to deal with the dynamic multi-agent manipulation planning problem under unknown pressure distribution. Neural networks possess a number of useful properties, such as the universal approximation capability, performance improvement through on- and offline learning, and distributed processing, which motivates its applications in control and signal processing. In order to deal with the effect of approximation error on the tracking error, several neural network-based H^∞ control schemes for nonlinear systems have been proposed [6][2]. Motivated by these factors, neural net-based H^∞ technique is utilized to cope with the uncertainty of pressure distribution in the cooperative planning problem.

In the dynamic pushing model, the forces exerted by agents need to obey some constraints. First, in order to avoid sliding between fingers and object, a constraint imposed on the force vector is that it must lie inside the friction cone of the contact surface. The other constraint is that in pushing action, the force vector can only direct inside the object. Owning to these constraints, it is difficult to solve the planning problem and find optimal forces simultaneously based on the pushing model. Instead, by introducing generalized forces(wrench) into the model, the overall planning process is divided hierarchically into lower control level and higher coordination level. In the lower level, we need to find generalized forces applied to the object, such that the object follows the given trajectory and obtains desired velocities, this is called a tracking problem in control literature. The design procedure is divided into three steps. At first, a linear nominal control design is obtained via full-state linearization with desired eigenvalue assignment. Next, neural network systems are constructed to approximate and eliminate the uncertainties related to pressure distribution. Finally, a minimax control scheme is specified to optimally attenuate the worst-case effect of the uncertainties of residual due to approximation to achieve optimal tracking performance.

In the coordination level, the task is to distribute the generalized forces among agents optimally. The formulation of coordination depends on the choice of performance index. Here each robot agent try to minimize the worst-case interaction force between itself and the object, each agent can make its decision and has its own objective, thus the coordination problem can be casted as a cooperative game be tween agents. The cooperative game can be formulated as a minimax problem, the goal is to minimize the maximal interaction forces.

II. Problem Formulation

In this paper, we consider the cooperative manipulation using multiple robotic agents pushing an object, and each robotic agent make frictional point contact with the object. Consider the planar motion of the object under three-agent pushing, XOY is the global coordinate, xoy is the local coordinate associated with the object, and assign o at the center of mass. The configuration space of the object is defined as (X_o, Y_o, θ). Where (X_o, Y_o) gives the position of the local coordinate, while θ denotes the orientation of the object. Denote the state of the system as $\bar{x} = (x_1, x_2, \cdots, x_6)^T = (X_o, \dot{X}_o, Y_o, \dot{Y}_o, \theta, \dot{\theta})^T$, we can derive following affine nonlinear system for three-agent dynamic manipulation

$$\begin{cases} \begin{bmatrix} \dot{x}_1 \\ \dot{x}_3 \\ \dot{x}_5 \end{bmatrix} = \begin{bmatrix} x_2 \\ x_4 \\ x_6 \end{bmatrix} \\ \begin{bmatrix} \dot{x}_2 \\ \dot{x}_4 \\ \dot{x}_6 \end{bmatrix} = \begin{bmatrix} \frac{F_X}{m} \\ \frac{F_Y}{m} \\ \frac{T}{I_0} \end{bmatrix} + Mt \begin{bmatrix} \cos x_5 & -\sin x_5 & 0 \\ \sin x_5 & \cos x_5 & 0 \\ 0 & 0 & 1 \end{bmatrix} \cdot \\ \begin{bmatrix} W_1 & W_2 & W_3 \end{bmatrix} \begin{bmatrix} F_{1n} \\ F_{1t} \\ F_{2n} \\ F_{2t} \\ F_{3n} \\ F_{3t} \end{bmatrix} \end{cases} \quad (1)$$

where

$$Mt = \begin{bmatrix} \frac{1}{m} & 0 & 0 \\ 0 & \frac{1}{m} & 0 \\ 0 & 0 & \frac{1}{I} \end{bmatrix}$$

$F_{in}, F_{it}, i=1,2,3$ are the normal and tangential components of the contact force generated by agents. W_i is the wrench matrix for the agent that transfers the pushing forces into global coordinate. It transforms the contact force into local coordinate. F_X, F_Y are the friction forces in global coordinate, and T is the torque respect to the center of mass generated by the friction force. Due to the uncertainty of pressure distribution, the friction forces and torque are always unknown. However, they are functions of the linear and rotational velocities. In this paper, we will use neural network as universal approximator to estimate the functions, and cancel the effect of the uncertainties. (please refer to [5] for the derivation of the model)

The dynamical cooperative manipulation planning problem can be described as follows: Given an object on a plane at a known initial configuration, find the forces exerted by each finger (or agent), such that the object move to a specified goal configuration with specific velocities. The inertial forces are not negligible compared to quasi-static manipulation, the planning and cooperation of the fingers must be developed based on the dynamic model of motion described by (1). For manipulation, there also exist constraints on the inputs, for example, the pushing direction must lay inside the friction cone, otherwise there will be sliding between the fingers and object; the agent can only push the object, without pulling. The cooperative manipulation planning problem is formulated as follows:

Given a task of a continously differentiable and uniformly bounded trajectory $\mathbf{q}_d = (X_{or}, Y_{or}, \theta_r)$ and corresponding velocities $\dot{\mathbf{q}}_d = (\dot{X}_{or}, \dot{Y}_{or}, \dot{\theta}_r)$, design the state feedback control input $\mathbf{u}(t) = (F_{1n}, F_{1t}, F_{2n}, F_{2t}, F_{3n}, F_{3t})^T$ such that the state of the system (1) follows the desired trajectory and minimize worst-case normal interaction forces at any instant time t. The objective function $M(\mathbf{u}(t))$ is chosen as

$$M(\mathbf{u}(t)) = \min_{u(t)} \max_i \{F_{in}\} \quad i=1,2,3 \quad (2)$$

III. Planning for Cooperative Manipulation

The object on a plane under pushing have three degrees of freedom, and it will need three independent control inputs to fully control the object to follow any given trajectory. Here we introduce F_{OX}, F_{OY} and T_O as three generalized control inputs to the system, and let $\bar{\mathbf{u}} = (\bar{u}_1, \bar{u}_2, \bar{u}_3)^T = (F_{OX}, F_{OY}, T_O)^T$, and rearrange the system (1) as

$$\begin{bmatrix} \dot{x}_1 \\ \dot{x}_3 \\ \dot{x}_5 \\ \dot{x}_2 \\ \dot{x}_4 \\ \dot{x}_6 \end{bmatrix} = \begin{bmatrix} x_2 \\ x_4 \\ x_6 \\ \frac{F_X}{m} \\ \frac{F_Y}{m} \\ \frac{T}{I_0} \end{bmatrix} + \begin{bmatrix} 0 & 0 & 0 \\ 0 & 0 & 0 \\ 0 & 0 & 0 \\ \frac{1}{m} & 0 & 0 \\ 0 & \frac{1}{m} & 0 \\ 0 & 0 & \frac{1}{T} \end{bmatrix} \bar{\mathbf{u}} \quad (3)$$

and with these control inputs, the system is fully actuated.

From (1), we know that the generalized control $\bar{\mathbf{u}}$ and original force inputs satisfy following relations

$$\bar{\mathbf{u}} = \begin{bmatrix} \cos x_5 & -\sin x_5 & 0 \\ \sin x_5 & \cos x_5 & 0 \\ 0 & 0 & 1 \end{bmatrix} \begin{bmatrix} \mathbf{W}_1 & \mathbf{W}_2 & \mathbf{W}_3 \end{bmatrix} \begin{bmatrix} F_{1n} \\ F_{1t} \\ F_{2n} \\ F_{2t} \\ F_{3n} \\ F_{3t} \end{bmatrix} \quad (4)$$

Thus the planning of the force inputs can be divided into two phases hierarchically. First, generalized force and torque controller $\bar{\mathbf{u}}$ for the uncertain system (3) will be developed using neural net-based H^∞ tracking control design in order to cope with the uncertainty of pressure distribution. Then the coordination will be carried out on (4) to distribute the generalized control into forces generated by each finger agent. By considering the constraints, a cooperative game will be formulated to solve the problem.

A. Neural Net-based H^∞ Design for The Generalized Controller

Denote the configuration of the object as $\mathbf{q}_1 = (X_o, Y_o, \theta)$, and $\mathbf{q}_2 = (\dot{X}_o, \dot{Y}_o, \dot{\theta})$. By introducing the state vector

$$\mathbf{q} = \begin{bmatrix} \mathbf{q}_1 \\ \mathbf{q}_2 \end{bmatrix} \quad (5)$$

By considering the fact that the friction forces and torque are only functions of the velocities $\mathbf{q}_2$, the equation (3) is transferred into the following standard form:

$$\dot{\mathbf{q}} = \begin{bmatrix} \mathbf{q}_2 \\ f(\mathbf{q}_2) \end{bmatrix} + \begin{bmatrix} 0 \\ Mt \end{bmatrix} \bar{\mathbf{u}} \quad (6)$$

where $f(\mathbf{q}_2) = (\frac{F_X(\mathbf{q}_2)}{m} \ \frac{F_Y(\mathbf{q}_2)}{m} \ \frac{T(\mathbf{q}_2)}{I_0})^T$.

Therefore we can define

$$\tilde{\mathbf{q}}_1 = \mathbf{q}_1 - \mathbf{q}_d \quad (7)$$

as the tracking error that we would like to drive to zero. Since the nonlinearities appear together with the control in equation (6), feedback linearization is easy to design to cancel the nonlinearities. However, the approach is possible only while all the nonlinearities are well known. In oder to design the controller under unknown uncertainties, we need to estimate the uncertainties, here neural networks are introduced to approximate the nonlinearities $f(\mathbf{q}_2)$ as $\hat{f}(\mathbf{q}_2, \Theta)$. Based on the approximation, the state feedback control law can be designed as

$$\bar{\mathbf{u}} = M^{-1}(\ddot{\mathbf{q}}_d - K_1\tilde{\mathbf{q}}_1 - K_2\dot{\tilde{\mathbf{q}}}_1 + u_0 - \hat{f}(\mathbf{q}_2, \Theta)) \quad (8)$$

where $\ddot{\mathbf{q}}_d$ denotes the second order derivative of $\mathbf{q}_d$, Θ is a vector containing the tunable network parameters, and K_1, K_2 are 3×3 matrix to be designed and u_0 is an auxiliary control signal yet to be specified.

Let

$$\hat{f}(\mathbf{q}_2, \Theta) = \begin{bmatrix} \hat{f}_1(\mathbf{q}_2, \theta_1) \\ \hat{f}_2(\mathbf{q}_2, \theta_2) \\ \hat{f}_3(\mathbf{q}_2, \theta_3) \end{bmatrix} \quad (9)$$

be the neural network to approximate the nonlinearities $f(\mathbf{q}_2)$, and the neural networks $\hat{f}_k(\mathbf{q}_2, \theta_k), (k=1,2,3)$ are composed of nonlinear neurons in hidden layers and linear neurons in the input and output layers, i.e,

$$\hat{f}_k(\mathbf{q}_2, \theta_k) = \sum_{i=1}^{p_k} \theta_{ki} H(\sum_{j=1}^{N} \omega_{ij}^k \mathbf{q}_{2j} + m_i^k), \quad k=1,2,3 \quad (10)$$

For simplicity, only consider θ_k as the adjustable weighting parameters. denote

$$\theta_k = \begin{bmatrix} \theta_{k1} \\ \vdots \\ \theta_{kp_k} \end{bmatrix} \quad \text{and} \quad \xi_k = \begin{bmatrix} H(\sum_{j=1}^{N} \omega_{1j}^k \mathbf{q}_{2j} + m_1^k) \\ \vdots \\ H(\sum_{j=1}^{N} \omega_{p_k j}^k \mathbf{q}_{2j} + m_{p_k}^k) \end{bmatrix} \quad (11)$$

Here we choose the following hyperbolic tangent function as nonlinear neurons functions

$$H(\mathbf{q}_2) = \frac{e^{\delta(\mathbf{q}_2)} - e^{-\delta(\mathbf{q}_2)}}{e^{\delta(\mathbf{q}_2)} + e^{-\delta(\mathbf{q}_2)}} \quad (12)$$

where $\delta(.)$ is a function of $\mathbf{q}_2$. The weights ω_{ij}^k and the biases m_i^k for $1 \leq i \leq p_k, 1 \leq j \leq 3$, and $1 \leq k \leq 3$, are specified beforehand in this application, where p_k is the number of nonlinear neurons in the neural network system $\hat{f}(\mathbf{q}_2, \theta_k)$. The neural-network system $\hat{f}(\mathbf{q}_2, \Theta)$ can be denoted as

$$\hat{f}(\mathbf{q}_2, \Theta) = \begin{bmatrix} \xi_1^T & 0 & 0 \\ 0 & \xi_2^T & 0 \\ 0 & 0 & \xi_3^T \end{bmatrix} \begin{bmatrix} \theta_1 \\ \theta_2 \\ \theta_3 \end{bmatrix} = \Xi\Theta \quad (13)$$

Now define the optimal parameter estimates Θ^* as follows

$$\Theta^* = arg \min_{\Theta \in \Omega_\theta} \max_{\mathbf{q}_2 \in \Omega} \|\hat{f}(\mathbf{q}_2, \Theta) - f(\mathbf{q}_2)\| \quad (14)$$

where $\|.\|$ denotes the Euclidean norm, Ω_θ and Ω denote the largest compact sets that Θ and $\mathbf{q}_2$ belong to, respectively.

Substituting (8) and (7) into (6), yields,

$$\begin{aligned} \dot{\tilde{\mathbf{q}}}_1 &= \mathbf{q}_2 \\ \dot{\mathbf{q}}_2 &= \mathbf{q}_d^{(2)} - K_1\tilde{\mathbf{q}}_1 - K_2\dot{\tilde{\mathbf{q}}}_1 \\ &\quad + u_0 + f(\mathbf{q}_2) - \hat{f}(\mathbf{q}_2) \\ &= \mathbf{q}_d^{(2)} - K_1\tilde{\mathbf{q}}_1 - K_2\dot{\tilde{\mathbf{q}}}_1 \\ &\quad + u_0 + \Xi\Theta^* - \Xi\Theta + (f(\mathbf{q}_2) - \Xi\Theta^*) \end{aligned} \quad (15)$$

Let us denote

$$\dot{e} = \begin{bmatrix} \tilde{\mathbf{q}}_1 \\ \dot{\tilde{\mathbf{q}}}_1 \end{bmatrix} = \begin{bmatrix} \mathbf{q}_1 - \mathbf{q}_d \\ \mathbf{q}_2 - \dot{\mathbf{q}}_d \end{bmatrix} \quad (16)$$

as the state error vector. And let

$$d(t) = f(\mathbf{q}_2) - \Xi\Theta^* \quad (17)$$

be the optimal approximation error, and $\tilde{\Theta}(t)$ is defined as

$$\tilde{\Theta}(t) = \Theta^* - \Theta(t) \tag{18}$$

From (7), (15) and (16), we obtain

$$\begin{bmatrix} \dot{\tilde{q}}_1 \\ \ddot{\tilde{q}}_1 \end{bmatrix} = \begin{bmatrix} \dot{\tilde{q}}_1 \\ -K_1\tilde{q}_1 - K_2\dot{\tilde{q}}_1 + u_0 + (\Xi\Theta^* - \Xi\Theta) + d \end{bmatrix} \tag{19}$$

the system in (19) represents the tracking error dynamics, it is a nominal linear uncertain system, the nonlinear uncertainties are modeled by d. A more convenient form of (19) could be

$$\dot{e} = Ae + Bu_0 + B(\Xi\Theta^* - \Xi\Theta) + Bd \tag{20}$$

where

$$A = \begin{bmatrix} 0 & I \\ -K_1 & -K_2 \end{bmatrix} \quad B = \begin{bmatrix} 0 \\ I \end{bmatrix} \tag{21}$$

The control parameters K_1, K_2 are selected such that A is Hurwitz and has desired eigenvalues such that the tracking dynamic (20) has a desired response if the system is free of uncertainty.

Theorem 1: For the system (3), if the control $\bar{u}(t)$ is chosen as

$$\bar{u} = Mt^{-1}(q_d^{(2)} - K_1\tilde{q}_1 - K_2\dot{\tilde{q}}_1 + u_0 - \hat{f}(q_2,\Theta)) \tag{22}$$

with

$$\dot{\Theta} = \eta \Xi^T B^T P e(t) \tag{23}$$

$$u_0 = -R^{-1}B^T P e(t) \tag{24}$$

where $R = R^T > 0$ is a weighting matrix and $P = P^T > 0$ is the solution of the following algebraic Riccati-like equation

$$PA + A^TP + Q - PB(R^{-1} - \frac{1}{\rho^2}I)B^TP = 0 \tag{25}$$

Then the minimax tracking performance is guaranteed for a prescribed ρ and the corresponding worst-case $d^*(t)$ is of the form

$$d^*(t) = \frac{1}{\rho^2}B^T P e(t) \tag{26}$$

Till now, we have designed the generalized controller for minimax tracking performance, our next task is to redistribute it to the agents.

B. Coordination as A Minimax Problem

For a generalized controller (wrench) (22) obtained from H^∞ design, how to distribute the wrench between agents is the objective of the higher level coordination. The coordination between agents can be considered as game between agents. Each agent has an objective, i.e., minimizing the normal interaction force between itself and object, and can make its own decision through choosing the interaction forces component F_{in} and F_{it} for agent i. If we consider the cooperation between agents, there will be a Pareto solution to the game. In order to obtain a single best compromise Pareto solution to the game, we consider the worst case interaction force between agents and object, the coordination problem can be modeled as a minimax optimization problem

$$\min_u \max_i \{F_{1n}, F_{2n}, F_{3n}\}$$
$$\text{Subject to} \quad \mu_i F_{in} - |F_{it}| \geq 0, \quad i = 1,2,3$$
$$F_{in} \geq 0, \quad i = 1,2,3$$

$$\begin{bmatrix} \cos x_5 & -\sin x_5 & 0 \\ \sin x_5 & \cos x_5 & 0 \\ 0 & 0 & 1 \end{bmatrix} \begin{bmatrix} W_1 & W_2 & W_3 \end{bmatrix} \begin{bmatrix} F_{1n} \\ F_{1t} \\ F_{2n} \\ F_{2t} \\ F_{3n} \\ F_{3t} \end{bmatrix}$$

$$= \begin{bmatrix} F_{OX} \\ F_{OY} \\ T_O \end{bmatrix} \tag{27}$$

where $\mathbf{u}(t) = (F_{1n}, F_{1t}, F_{2n}, F_{2t}, F_{3n}, F_{3t})^T$.

This problem can be solved by using standrad multi-objective optimization routines or solved by linear programming through introducing slack variable.

IV. SIMULATION RESULTS

In previous section, we proposed a new planning method for dynamic multi-agent manipulation. Here simulations will be carried out by using a triangular object under two-agent or three agent pushing, the object is shown in Fig. 1. where $a = 0.2m, b = 0.17m$, angle $A = \pi/6$, we assume the uniform distribution of mass, the mass density $\rho = 200kg/m^2$, and mass is $m = 1.7kg$, the mass moment of inertia is $I_0 = 0.0037$. The friction coefficient between the object and surface is $\mu = 0.5$, friction coefficient between the fingers and object is $\mu_i = 0.8$. The local coordinate is located at the center of mass and fixed on the object, and assign the global coordinate at the same location, but fixed on the supporting plane.

Consider agent 1 pushing at edge AC, agent 2 pushing at BC, agent 3 pushing at AB, assign contact coordinate to each finger, the associated contact normal and tangential vectors are written as

$$\begin{aligned} \mathbf{n}_1 &= (\tfrac{1}{2}, -\tfrac{\sqrt{3}}{2})^T & \mathbf{t}_1 &= (\tfrac{\sqrt{3}}{2}, \tfrac{1}{2})^T \\ \mathbf{n}_2 &= (-\tfrac{\sqrt{3}}{2}, -\tfrac{1}{2})^T & \mathbf{t}_2 &= (\tfrac{1}{2}, -\tfrac{\sqrt{3}}{2})^T \\ \mathbf{n}_3 &= (0, 1)^T & \mathbf{t}_3 &= (-1, 0)^T \end{aligned} \tag{28}$$

the contact points between fingers and object are given as

$$\begin{aligned} L_1 &= (-0.0167 \quad 0.0289) \\ L_2 &= (0.0500 \quad 0.00289) \\ L_3 &= (-0.0400 \quad -0.0283) \end{aligned} \tag{29}$$

computing the wrench matrices, we get

$$W_1 = \begin{bmatrix} 0.5000 & 0.8660 \\ -0.8660 & 0.5000 \\ 0.0476 & -0.0333 \end{bmatrix} \tag{30}$$

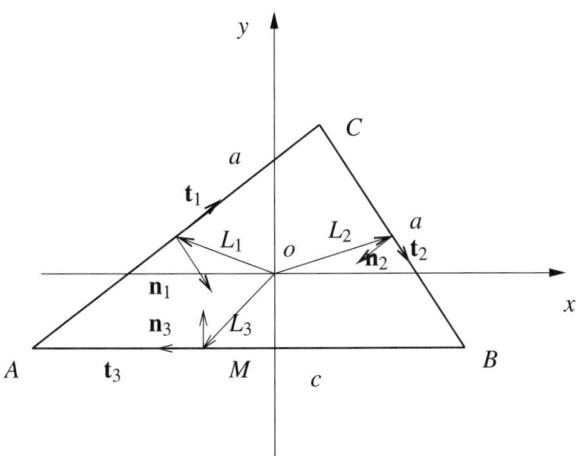

Fig. 1. Manipulation of triangular object

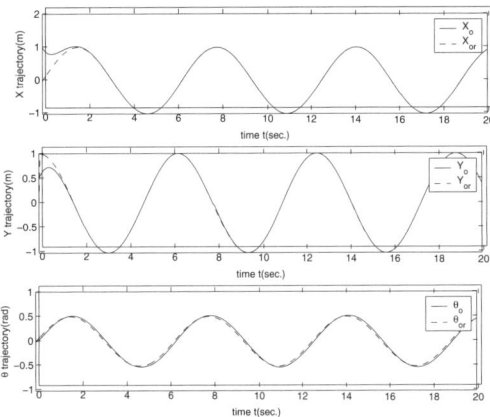

Fig. 2. H^∞ tracking results of positions ($\rho = 0.05$)

$$\mathbf{W}_2 = \begin{bmatrix} -0.8660 & 0.5000 \\ -0.5000 & -0.8660 \\ -0.0278 & 0.0481 \end{bmatrix} \quad (31)$$

$$\mathbf{W}_3 = \begin{bmatrix} 0 & -1.000 \\ 1.000 & 0 \\ -0.0400 & -0.0283 \end{bmatrix} \quad (32)$$

When we consider the translation and rotation of the object, three agents are needed to manipulate the object following predefined trajectory. The trajectories are given as

$$\begin{array}{lll} X_{or}(t) = \sin(t) & \dot{X}_{or}(t) = \cos(t) & \ddot{X}_{or}(t) = -\sin(t) \\ Y_{or}(t) = \cos(t) & \dot{Y}_{or}(t) = -\sin(t) & \ddot{Y}_{or}(t) = -\cos(t) \\ \theta_{or}(t) = \tfrac{1}{2}\sin(t) & \dot{\theta}_{or}(t) = \tfrac{1}{2}\cos(t) & \ddot{\theta}_{or}(t) = -\tfrac{1}{2}\sin(t) \end{array} \quad (33)$$

And we consider the uncertainties F_x, F_y, T associated with pressure distribution entering the system (1) randomly, and $F_x \sim [-0.5F_{xn}, 0.5F_{xn}]$, $F_y \sim [-0.5F_{yn}, 0.5F_{yn}]$, $T \sim [-0.5T_n, 0.5T_n]$, where F_{xn}, F_{yn}, T_n are the friction forces and torque generated under the assumption with known pressure distribution that is given above. And the initial condition $X_o(0) = 1, \dot{X}_o = 0, Y_o(0) = 0.5, \dot{Y}_o = 0, \theta_o(0) = 0.2, \dot{\theta}_o = 0$.

First, the H^∞ robust controller design is give by following steps.

1) Specify

$$K_1 = \begin{bmatrix} 3 & 0 & 0 \\ 0 & 2 & 0 \\ 0 & 0 & 8 \end{bmatrix} \quad K_2 = \begin{bmatrix} 9 & 0 & 0 \\ 0 & 7 & 0 \\ 0 & 0 & 5 \end{bmatrix}$$

2) select the attenuation level $\rho = 0,05$, respectively. And $Q = diag[100I_3, \quad 100I_3]$, and $R = \rho^2 I$. Solve the Riccati-like equation (25) using MATLAB function "are", we get P.

3) select neural network structure as (11). And for simplicity reason, we choose $\omega_{ij}^k = 1$ for all i, j, k.

i.e.,

$$\omega = \begin{bmatrix} 1 & 1 & 1 \end{bmatrix}$$

The neural network systems are selected to be $\xi_i = [\xi_{i,1}, \cdots, \xi_{i,17}]^T$ for $i = 1, 2, 3$, where $\xi_{i,j}$ is nonlinear neural function and has following form,

$$\xi_{i,j} = \frac{e^{\omega^T \mathbf{q}_2 - 4 + 0.5(j-1)} - e^{\omega^T \mathbf{q}_2 + 4 - 0.5(j-1)}}{e^{\omega^T \mathbf{q}_2 - 4 + 0.5(j-1)} + e^{\omega^T \mathbf{q}_2 + 4 - 0.5(j-1)}}$$

for $i = 1, 2, 3$, $j = 1, \cdots, 17$. And

$$\Theta = \begin{bmatrix} \theta_1 \\ \theta_2 \\ \theta_3 \end{bmatrix}, \quad \Xi = \begin{bmatrix} \xi_1^T & 0 & 0 \\ 0 & \xi_2^T & 0 \\ 0 & 0 & \xi_3^T \end{bmatrix}$$

where $\theta_i = [\theta_{i,1}, \cdots, \theta_{i,17}]^T$ for $i = 1, 2, 3$. The initial guess of $\Theta(0) = 0$.

4) Compute the H^∞ controller according (22) and (24). and update the weights of neural networks according to (23).

Fig. 2- 4 present the simulation results of neural network-based H^∞ tracking. It is clear that the proposed controller can control the uncertain system tracking a given trajectory. Fig. 4 shows the generalized forces. After obtaining the generalized forces, a minimax optimization problem can be set up as (27). The optimal force distribution is shown in Fig. 5. The resulting forces satisfy the performance index (2).

V. CONCLUSIONS AND FUTURE WORKS

This paper proposed a planning method dynamic cooperative multi-agent manipulation with unknown pressure distribution. The coordination between agents is performed hierarchically. In lower level, the generalized force inputs are designed by integrating linearization with neural network-based H^∞ technique. In higher coordination level, a cooperative game is solved to distribute the generalized force between agents, and obtain minimax performance.

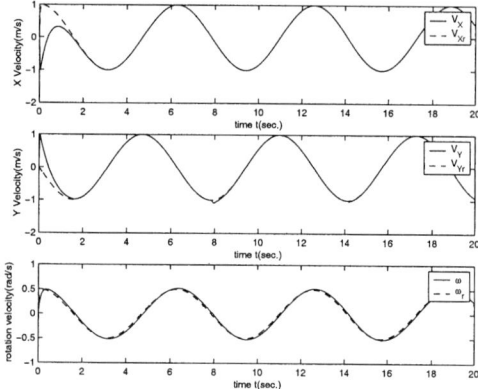

Fig. 3. H^∞ tracking results of velocities ($\rho = 0.05$)

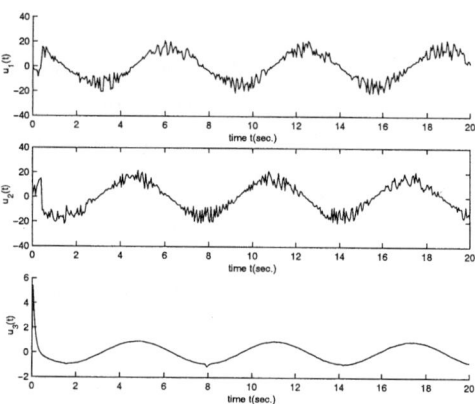

Fig. 4. Generalized forces $\bar{\mathbf{u}}(t)(\rho = 0.05)$

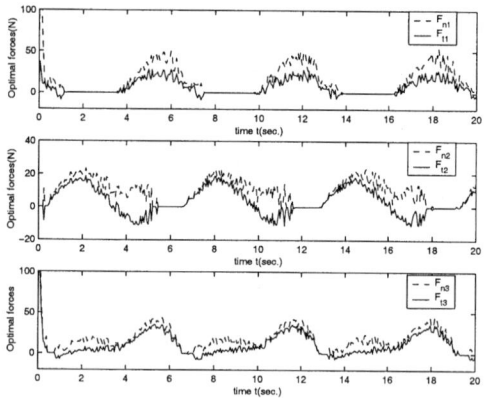

Fig. 5. Force coordination under three-agent manipulation ($\rho = 0.05$)

The coordination under three-agent manipulation is studied. The hierarchical structure for coordination is flexible, when more agents are added for manipulation, the redesign of lower level generalized force controller is not needed. Simulation results demonstrate the proposed coordination method. The proposed approach does not consider the dynamics of the agents, the integration of the dynamics of agents and object in planning will be addressed and how to plan contact points optimally between agent and object is also very interesting.

VI. REFERENCES

[1] M.N. Ahmadabadi and E Nakano. A constrain and move approach to distributed object manipulation. *IEEE Trans. on Robotics and Automation*, 17(2):157–172, 2001.

[2] Y. Chen and B. Chen. A nonlinear adaptive $\mathbf{H}^\infty$ tracking control design in robotic systems via neural networks. *IEEE Trans. on Control Systems Technology*, 5(1):13–29, 1997.

[3] D.R. Donald, J. Jennings, and D. Rus. Information invariant for distributed manipulation. *Intl. J. of Robotics Research*, 16(5):673–702, 1997.

[4] Q. Li and S. Payandeh. Modeling and analysis of dynamic planar multi-agent manipulation. In *IEEE International Symposium on Computational Intelligence in Robotics and Automation*, pages 200–205, July 2001.

[5] Q. Li and S. Payandeh. Centralized cooperative planning for dynamic multi-agent planar manipulation. In *Proc. 2002 IEEE Intl. Conf. on Decision and Control*, pages 2836–2841, Dec 2002.

[6] C. Lin and T. Lin. An $\mathbf{H}^\infty$ design approach for neural net-based control schemes. *IEEE Trans. on Automatic Control*, 46(10):1599–1605, 2001.

[7] M.T. Mason. Progress in nonprehensile manipulation. *Intl. J. of Robotics Research*, 18(11):1129–1141, 1999.

Cartop Manipulation

Wesley H. Huang   Kartik Babu   Jonathan A. Bandlow

Department of Computer Science
Rensselaer Polytechnic Institute, Troy, New York USA
`{whuang,babuk,bandlj}@cs.rpi.edu`

Abstract

This paper explores a novel mode of mobile manipulation using a car-like robot whose chassis does not extend above the height of its wheels. It transports an object that rests on top of the wheels, so as the robot moves forwards, the object moves forward relative to the robot. The robot must maneuver so that the object stays on top of the robot, or a second robot must be in position to receive the robot as it rolls off the first. We describe the mechanics of this system, describe its basic maneuvers and manipulation techniques, present a motion planner for generating plans to transport an object, and give results of our experimental implementation of this system.

1 Introduction

Manipulation is an important capability for mobile robots. Many tasks require objects to be transported, reoriented, or aligned. There has been a great deal of research on robotic manipulators and manipulation as well as on numerous aspects of mobile robotics. Researchers in the area of mobile manipulation have often combined the two by affixing a manipulator to a mobile robotic platform.

In these systems, the division between the "manipulation" and the "mobility" is very distinct. In human beings, however, the division is not so distinct — people sometimes locomote with their hands and manipulate with their feet! Human beings have a far greater repertoire of both manipulation and locomotion strategies in addition to simply having better "hardware."

We are interested in exploring a more natural integration between manipulation and mobility. In this line of inquiry, we have built small car-like robots which have a chassis that is shorter than the wheels. This allows us to place an object on top of the wheels. As a robot drives forwards, the object is pushed ahead. An illustration of our robots is shown in Figure 1.

In this paper we describe a variety of modes of manipulation that are possible with this mechanism: from basic maneuvers and dynamic manipulation to motion planning for transportation tasks.

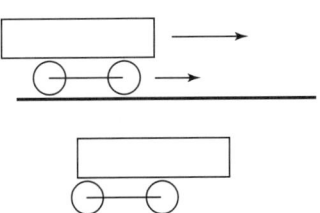

Figure 1: A cartop manipulator transporting an object.

1.1 Related work

There have been a variety of approaches to mobile manipulation explored in the robotics literature. (Due to space constraints, we omit a review of work using mobile robots to push objects and instead focus on other methods.)

One approach has been to incorporate an industrial manipulator into a mobile platform [4, 12, 8, 11]. This work has tended to focus on developing force or compliant control that can take advantage of the redundancies introduced by the mobile base.

Others have explored mechanically simple manipulators that require multiple robots to cooperate in order to transport objects. One line of research is that by Sugar and Kumar [10] who developed compliant manipulators that transport objects in a squeeze grasp between two robots. Huang and Holden [3] used mobile robots with two degree-of-freedom "palm" manipulators to pick up an object.

This work is most closely related to the "mobipulator" project of Mason and colleagues [7, 1, 9]. In this project, a "dual differential drive robot" is used to manipulate sheets of paper and other objects on a desktop. The robot consists of four wheels, each independently driven, none steered. With a sheet of paper under two of the wheels, the robot can manipulate the paper and drive around on the desk independently. Our work is similar in that the same hardware that is used for locomotion (i.e. the wheels) is also used for manipulation.

Another related work is the "Stickey" desktop robot developed by Rus et al. [6]. This robot features a "foot" with adhesive tape that can pick up a single sheet of paper and move it to another location. Coordination between this foot and the wheels of the robot allow it to "unstick" the foot from the paper.

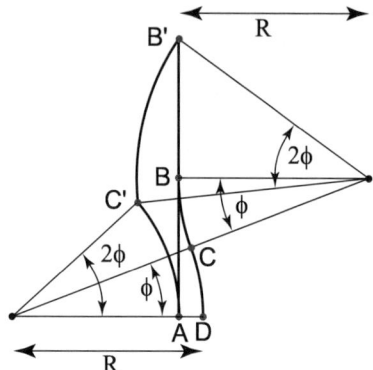

Figure 2: A parallel parking maneuver. The robot moves to the points A, B, C, and D; the object moves to the points A, B', C', and then back to A. For clarity, only the path of the reference point for the robot and the object are shown.

2 Basic motion & maneuvers

The basic design of a car-like robot involves a differential for the wheels that are driven (usually the rear wheels) and an Ackerman steering linkage for the front wheels. The Ackerman steering turns the front wheels unequally so that all wheels travel about a common center of rotation. The differential allows the driven wheels to roll at different velocities. The whole design is to prevent any of the wheels from slipping on the ground; however, it also serves us well in preventing the object from slipping relative to the robot.

As a wheel rolls forward it pushes the vehicle along. If an object is resting on top if the wheel, it is pushed ahead at the same rate relative to the vehicle. This means that the object will travel twice as far (and twice as fast) as the vehicle, relative to the world. This will hold whether the vehicle is going in a straight line or around a curve. In the latter case, since it travels twice as far along the arc (assume a constant turning radius), it will rotate twice as much as the robot.

One challenge for this "cartop manipulation" is that if the robot keeps going, the object will roll off the wheels. We will put this characteristic to good use in Section 3, but for now we will concentrate on limited excursion maneuvers that keep the object resting on a single robot.

This section analyzes the object behavior when the mobile manipulator executes the standard car-like robot maneuvers: parallel parking and the three-point turn. Perhaps most surprisingly, a parallel parking maneuver leaves the object in the same configuration while the robot moves sideways. In the three-point turn, both the robot and the object return to the same location, but the object will have rotated twice as much as the robot.

2.1 Parallel parking

We consider the parallel parking maneuver illustrated in Figure 2 that consists of a straight forwards segment of distance d followed by a backwards right turn and then a backwards left turn, both of angle ϕ about the respective center of rotation. The net effect for the robot is a displacement sideways of h. Given a value of d and the turning radius R, we have:

$$\sin \phi = \frac{d}{2R} \quad (1)$$

The robot displacement is then:

$$h = 2R(1 - \cos \phi) \quad (2)$$

We represent the position of the robot using the homogeneous coordinate vector $[\ kx\ \ ky\ \ k\ ]$ where the value of k is often taken to be 1. We assume the robot starts facing in the positive y direction with the center of its rear axle (the robot's reference point) at the origin.

Since the robot starts at the origin, so initially the robot and the object start at the point $\mathsf{A} = [\ 0\ \ 0\ \ 1\ ]^T$.

As the robot moves forwards as distance d to point $\mathsf{B} = [\ 0\ \ d\ \ 1\ ]^T$, the object moves forwards a distance of $2d$ (with respect to the world frame) to point B'. This transformation is represented by the matrix:

$$S = \begin{bmatrix} 1 & 0 & 0 \\ 0 & 1 & 2d \\ 0 & 0 & 1 \end{bmatrix} \quad (3)$$

so we now have $\mathsf{B}' = S\mathsf{A} = [\ 0\ \ 2d\ \ 1\ ]^T$.

Next, the backwards right turn rotates both the robot and the object about the point (R, d). While the robot undergoes a rotation of ϕ to point $\mathsf{C} = [\ R(1-\cos\phi)\ \ d/2\ \ 1\ ]^T$, the object undergoes a rotation of 2ϕ to point C'. This transformation is represented by the matrix:

$$T = \begin{bmatrix} \cos 2\phi & -\sin 2\phi & R \\ \sin 2\phi & \cos 2\phi & d \\ 0 & 0 & 1 \end{bmatrix} \begin{bmatrix} 1 & 0 & -R \\ 0 & 1 & -d \\ 0 & 0 & 1 \end{bmatrix} \quad (4)$$

so we can conclude that:

$$\mathsf{C}' = T\mathsf{B}' = \begin{bmatrix} -R\cos 2\phi - d\sin 2\phi + R \\ -R\sin 2\phi + d\cos 2\phi + d \\ 1 \end{bmatrix} \quad (5)$$

Finally, the backwards left turn rotates the robot an angle $-\phi$ about the point $(-R(2\cos\phi - 1), 0)$ to point $\mathsf{D} = [\ 2R(1-\cos\phi)\ \ 0\ \ 1\ ]^T$, and the object undergoes a rotation of -2ϕ to point D'. This is represented by the matrix:

$$U = \begin{bmatrix} \cos 2\phi & \sin 2\phi & -R(2\cos\phi - 1) \\ -\sin 2\phi & \cos 2\phi & 0 \\ 0 & 0 & 1 \end{bmatrix}$$
$$\times \begin{bmatrix} 1 & 0 & R(2\cos\phi - 1) \\ 0 & 1 & 0 \\ 0 & 0 & 1 \end{bmatrix} \quad (6)$$

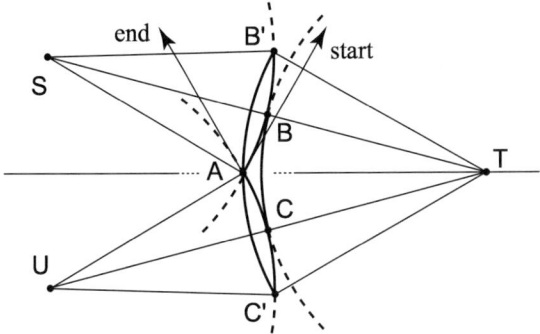

Figure 3: Construction for a three-point turn.

We can then compute the point $D' = UC'$ is then given by:

$$D' = \begin{bmatrix} d\sin 2\phi + 2R\cos\phi(\cos 2\phi - 1) \\ d(1 + \cos 2\phi) - 2R\cos\phi\sin 2\phi \\ 1 \end{bmatrix} = \begin{bmatrix} 0 \\ 0 \\ 1 \end{bmatrix} \quad (7)$$

Straightforward (if tedious) multiplication of the matrices and a few trigonometric identities leads to the intermediate result above. Application of a few more trigonometric identities and substitution using Equation 1 reveals that the object has in fact returned to its starting position!

2.2 Three-point turn

For the three point turn, it is simple to argue from the symmetry of the maneuver that the object undergoes twice as much rotation as the robot and, like the robot, returns to its starting position. We describe the construction of the three-point turn illustrated in Figure 3 in order to demonstrate its symmetry.

First, we draw the circles S and U through the starting point A and tangent to the starting and ending orientation, respectively. We must then draw the circle T tangent to circles S and U. All circles have a radius R, the turning radius of the robot. Note that the line TA is the perpendicular bisector of the line segment that could connect points S and U. It will serve as the line of symmetry for this construction.

The robot first travels forwards on the circle S to the point B which is the tangent point between circles S and T. The object also rotates about the point S, but twice as far to point B'. We then know that $\angle ASB' = 2\angle ASB$.

Next, the robot and object both rotate about point T: the robot to point C and the object to point C'. Because of symmetry about the line ST, $\angle BTA = \angle B'TB$ and therefore $\angle B'TA = 2\angle BTA$. Similarly, $\angle ATC' = 2\angle ATC$. Therefore we can conclude that $\angle B'TC' = 2\angle BTC$.

Finally, the robot and object both rotate about point U: the robot from point C and the object from C'. Due to symmetry about the line UT, we know that $\angle C'UC = \angle CUA$ and therefore $\angle C'UA = 2\angle CUA$.

At each step, the object undergoes a rotation twice that

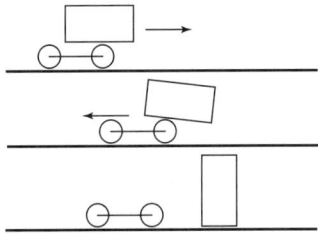

Figure 4: Dynamically standing up an object

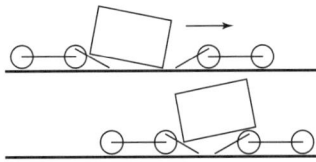

Figure 5: A "cow catcher" can be used to pick up objects off the ground.

of the robot, so when the robot and object have returned to their original starting position, the object will have rotated twice as far as the robot.

3 Dynamic & other motions

This mobile manipulator is capable of a number of types of dynamic motion. With sufficient speed, it can throw an object forwards that will then slide on the ground. Our hardware implementation was designed so that it could throw an object fast enough that it could slide for a half meter.

Another dynamic task is that of standing an object up. The basic approach is that the robot should drive forwards, the center of mass should pass the front wheels, and then the robot should quickly reverse. Friction from the wheels will impart a torque to the object and can stand it up. See Figure 4 for an illustration.

With the addition of some simple hardware, these mobile manipulators can pick up objects off the ground. Figure 5 illustrates a "cow catcher" on the front of one cartop manipulator and on the rear of another. This cow catcher is simply a ramp from the ground up to the wheel of the robot. Should the object slide up the cow catcher of the rear robot, the wheels should push it forward. When the object slides up the cowcatcher of the front robot, the wheel rotation should pull the object on top of the robot.

4 Motion planning

For these cartop manipulators to transport an object over any distance, they must be able to hand off the object from one robot to another. They should also work in an environment where they will have to avoid obstacles while transporting an object; this requires motion planning for both the object to be moved and for the individual robots.

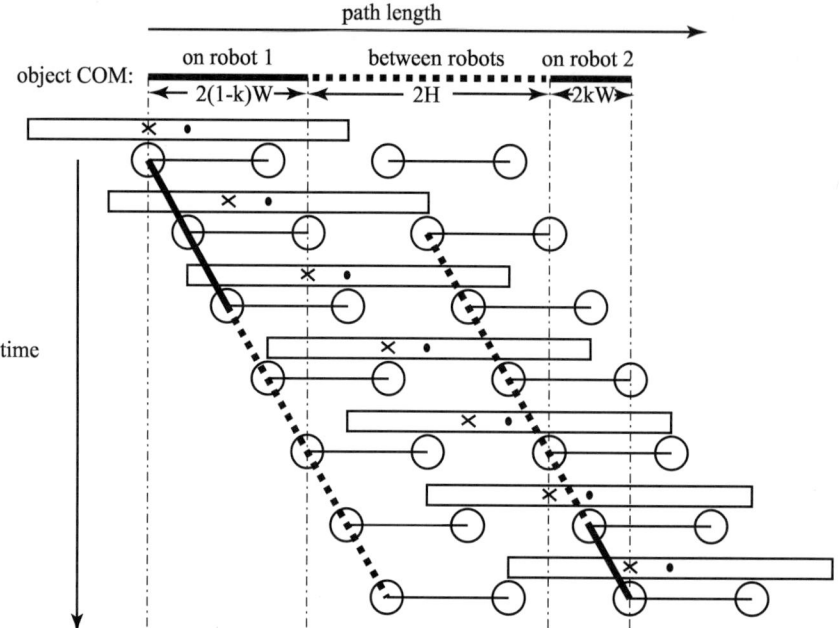

Figure 7: A timeline for a handoff cycle.

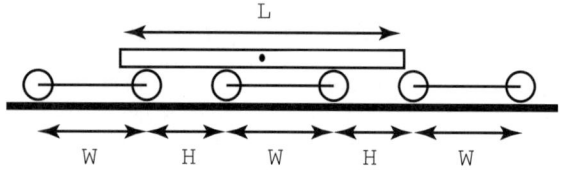

Figure 6: Handoff relationships

4.1 Handoffs

Suppose we are transporting an object of length L with a line of identical cartop manipulators with a wheelbase W and wheels of radius r separated by a "handoff distance" $H > 2r$, the distance from the centers of the front wheel of one robot to the rear wheel of the next. (See Figure 6 for an illustration.)

There are two basic constraints that ensure the object is supported by at least one wheel on either side of the center of mass (COM) which we assume is in the center of the object:

$$W \leq \frac{L}{2} \quad (8)$$

$$H \leq \frac{L}{2} \quad (9)$$

The limit on L given n robots is:

$$L < H + (n-1)(W+H) \quad (10)$$

We will require that the object COM start at some fraction k of the distance between the front and back wheels of a robot and that it will stop in this same relative position on the last robot at the end of a segment. Figure 7 shows an illustration of a single handoff, showing the location of each robot over time. Note that we use a reference point for the object that is not at its center of mass so that this reference point coincides with the robot reference point (the middle of the rear axle) at the start and the end.

Requiring the object to start and end each segment with the object and robot reference points coincident provides a certain decoupling between segments of path, e.g. between straight line segments and curved segments. In order for this to occur, each segment must have a length that is an integral multiple of $2(W+H)$. We may choose an H for each segment subject to the constraints above which, combined, give:

$$\frac{1}{n}L - \frac{n-1}{n}W < H \leq \frac{1}{2}L \quad (11)$$

However, one important factor affecting handoffs is that the curvature of the path may not change while two or more robots are supporting the object — otherwise, the robots would be traveling about different rotation centers and the object would slip on the wheels of one or both robots. In order to transition from, say, a straight line segment to a curved segment, some maneuvers, illustrated in Figure 8, would be required to remove all but one supporting robot. This robot can then change its steering direction, and finally, an additional maneuver is needed to "insert" a second robot under the object ahead of the first robot.

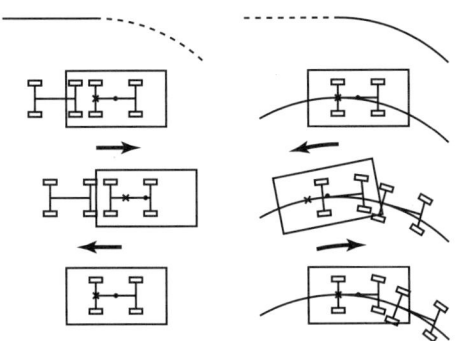
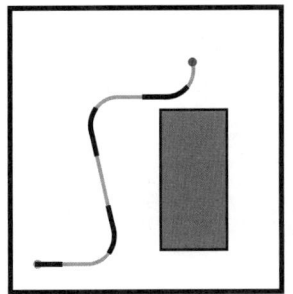
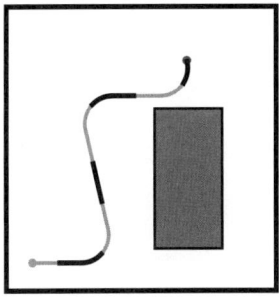

Figure 9: Path segments for two robots

Figure 8: Maneuvers that may be required at a transition from a straight to curved segment. On the left, the robots drive forward until the rear robot can drive away, and then the lead robot drives backward to the transition point. Similarly, on the right, the robot drives backwards along the same arc until the next robot can be put in place.

4.2 Nonholonomic motion planners

In order to plan robot motions for a transportation task, we first plan a path for the object, treating it as a car-like robot with the same turning radius as the robots. The object path is divided into 1-handoff length segments, and then paths for the individual robots are planned.

We have used Barraquand and Latombe's approach to nonholonomic motion planning [2] for planning the object path. This planner searches from the start configuration, and at each step, tries going forwards and backwards from a configuration at each steering angle (straight and maximum left and right). The distance of each attempted motion is set to be the length for a single handoff. The advantage of this planner is that we are guaranteed an integral number of handoffs, each of length $2(W + H)$. The only drawback is that a path is only guaranteed to reach the vicinity of the goal, not the goal itself.

For a $2(W + H)$ segment on the object path, the first robot carries the object a distance of $(1 - k)W + \frac{L}{2}$, and the second robot carries the object for $kW + \frac{L}{2}$. However the second robot of one segment is the first robot of the next segment. We have chosen $k = \frac{1}{2}$ to simplify the path decomposition if the path reverses direction.

The paths for each individual robot are planned using the approach described by Latombe [5]. This planner starts with a holonomic path and recursively refines it using Reeds & Shepp curves. We use a different planner here because the robots must move to the exact point to receive the object handoff and because there is no constraint on the length of path segments.

Figure 9 shows an object path and its decomposition into segments for each robot. The individual robot paths are shown in Figures 10 and 11. These paths are for robots with a 15.6 cm wheelbase and an object length of 40 cm. Only one robot supports the object any curvature change, maneuvers are not required at transition points.

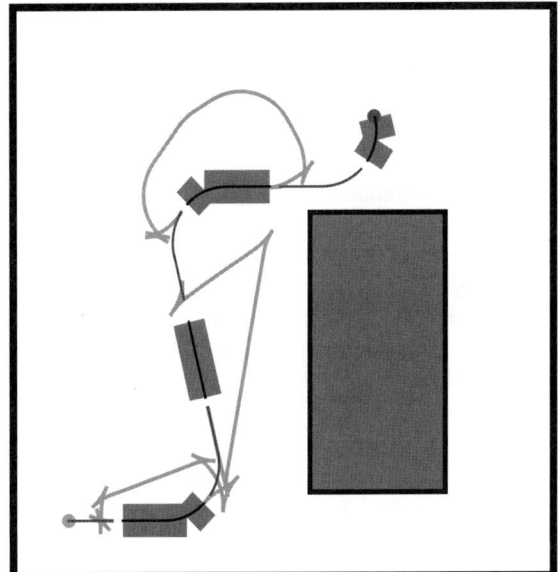

Figure 10: Complete path for the first robot

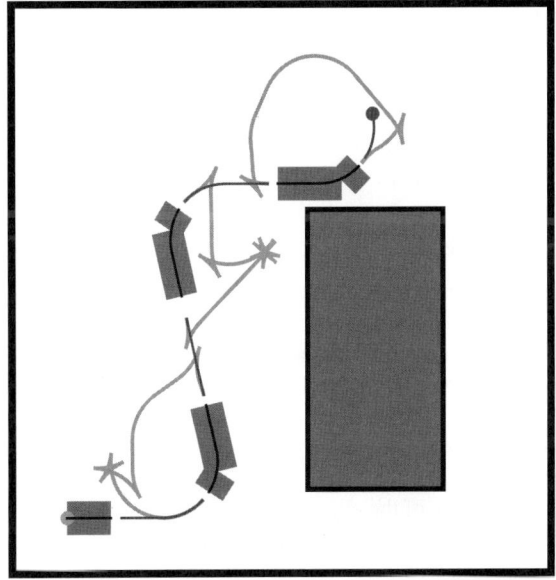

Figure 11: Complete path for the second robot

Figure 12: One of our "cartop manipulators"

5 Experimental hardware

We have built two cartop manipulators and have conducted a number of preliminary experiments.

The robots use a differential, steering blocks, wheel hubs, and several other components from model racing cars, but the chassis and wheels were custom made. A Maxon motor drives the robot while a hobby servo handles the steering. Optical quadrature encoders measure the rotation of the right rear wheel and the steering angle of the right front wheel. The wheels are made of Delrin with a #329 O-ring used as a tire. Figure 12 shows a picture of one of these robots.

A desktop computer controls the robots through a tether which also provides power to the robot. The drive motors are driven by AMC servo amplifiers, and the hobby servos are driven via a Pontech card. A ServoToGo card provides the D/A converters and hardware support for the encoders. While this arrangement has decreased development time, tether management may become an issue.

We have experimentally demonstrated that that these robots can dynamically stand up an object and have demonstrated the parallel parking and K-turn maneuvers. We expect to demonstrate the remaining manipulation skills described in this paper within the next few weeks.

6 Conclusions

This paper has described a novel method of mobile manipulation using a car-like mobile robot: the same wheels used for driving the robot are used to manipulate an object placed on top of the wheels. As the robot drives forwards, the object is pushed forwards relative to the robot. The basic maneuvers for a car-like robot have somewhat surprising results — the parallel parking maneuver results in no net motion of the object while the robot moves sideways underneath it, and the three-point turn results in object rotation twice that of the robot.

We have described a number of different types of manipulation that these robots are capable of, including dynamically standing up an object, picking an object up off the ground, and handing an object off to another robot. Our motion planner for this system first plans a path for the object, treating it as a car-like robot, and then plans paths for multiple robots that repeatedly hand the object off to each other. We have built two of these mobile manipulators and have experimentally demonstrated several types of manipulation.

Acknowledgements

Thanks to Drew Housten for help with the robots and to Kris Beevers for implementing the basic nonholonomic motion planners. Thanks also to Matt Mason & colleagues for sharing details of their "mobipulator" hardware. This work was supported by the National Science Foundation through award IIS–9983642.

References

[1] D. Balkcom and M. Mason. Progress in desktop robotics. In *The Eleventh Yale Workshop on Adaptive and Learning Systems*, 2001.

[2] J. Barraquand and J.-C. Latombe. Nonholonomic multibody mobile robots: Controllability and motion planning in the presence of obstacles. *Algorithmica*, 10:121–155, 1993.

[3] W. H. Huang and G. F. Holden. Nonprehensile palmar manipulation with a mobile robot. In *IEEE/RSJ IROS 2001*, volume 1, pages 114–119.

[4] O. Khatib. Mobile manipulation: the robotic assistant. *Robotics and Autonomous Systems*, 26(2–3):175–183, February 1999.

[5] J. C. Latombe. A fast path planner for a car-like indoor mobile robot. In *Ninth National Conference on Artificial Intelligence*, volume 2, pages 659–665, 1991.

[6] M. T. Mason, D. K. Pai, D. Rus, L. R. Taylor, and M. A. Erdmann. Experiments with desktop mobile manipulators. In *International Symposium on Experimental Robotics*, pages 37–46, 1999.

[7] M. T. Mason, D. K. Pai, D. Rus, L. R. Taylor, and M. A. Erdmann. A mobile manipulator. In *IEEE ICRA*, volume 3, pages 2322–2327, 1999.

[8] H. Seraji. A unified approach to motion control of mobile manipulators. *International Journal of Robotics Research*, 17(2):107–118, 1998.

[9] S. Srinivasa, C. Baker, E. Sacks, G. Reshko, M. Mason, and M. Erdmann. Experiments with nonholonomic manipulation. In *IEEE ICRA 2002*.

[10] T. G. Sugar and V. Kumar. Control of cooperating mobile manipulators. *IEEE Transactions on Robotics and Automation*, 18(1):94–103, February 2002.

[11] J. Tan and N. Xi. Integrated task planning and control for mobile manipulators. In *IEEE ICRA 2002*, volume 1, pages 382–387.

[12] Y. Yamamoto and X. Yun. Coordinating locomotion and manipulation of a mobile manipulator. *IEEE Transactions on Automatic Control*, 39(6):1326–1332, June 1994.

Smooth Feedback Control Algorithms for Distributed Manipulators

T.D. Murphey, J.W. Burdick

Engineering and Applied Science, California Institute of Technology
Mail Code 104-44, Pasadena, CA 91125 USA
{murphey,jwb}@robotics.caltech.edu

Abstract— This paper introduces a smooth control algorithm for controlling fully actuated distributed manipulation systems that operate by frictional contact. The control law scales linearly with the number of actuators and is simple to implement. Moreover, we prove that control law has desirable robustness properties in the presence of the nonsmooth mechanics inherent in distributed manipulation systems that rely upon frictional contact. This algorithm has been implemented on an experimental distributed manipulation test-bed, whose structure is briefly reviewed. The experimental results confirm the validity and performance of the algorithm.

I. INTRODUCTION

A distributed manipulation system typically consists of a large number of similar or identical actuators arranged in a planar array, combined with a control strategy to create net movement of an object or objects placed on the array. Many distributed manipulation systems are designed to allow precise positioning of planar objects from arbitrary initial configurations. With this capability, a distributed manipulator is a "smart conveyor" that can be used for separating parts and precisely positioning them for the purpose of assembly operations. Distributed manipulation systems offer potential for micro-assembly using MEMS technology [20]. Distributed manipulator actuation methods ranges from air jets and wheels on the macroscale, to microelectromechanical systems (MEMS) and flexible cilia at the microscale.

These machines are typically relatively easy to build, but the systematic control of such devices can be quite difficult, leading to a recent increase in the study of distributed manipulation control systems [2, 8, 9, 10, 13, 19, 20]. Distributed manipulation systems are in a sense massively *over-actuated*. They can potentially have thousands of inputs (though only a subset of the actuators typically influence the moving object at any given instant), while the output consists of the states of the manipulated object. For good general references on distributed manipulation, see [4, 6, 11].

This paper addresses the issue of control system design for a class of planar distributed manipulation systems whose physical operation involves rolling and sliding contacts between the moving object and the actuator surfaces. As reviewed in Section II, the governing equations of such systems are generally non-smooth [14]. In general, one needs to use discontinuous control laws to stabilize such systems. In prior work [16, 17, 15], we introduced non-smooth control laws for stabilizing this class of distributed manipulation systems. In this paper we show that while the governing equations can often be nonsmooth, in the special case of full actuation (whose definition is given shortly) there exists a *smooth* solution to the control design problem. Moreover, we prove that this solution is robust and stable in the presence of the non-smooth governing mechanics. We have implemented this algorithm on an experimental system that is described more fully in Section IV (and shown schematically in Figure 1. The excellent performance of the experimental implementation confirms our theoretical predictions.

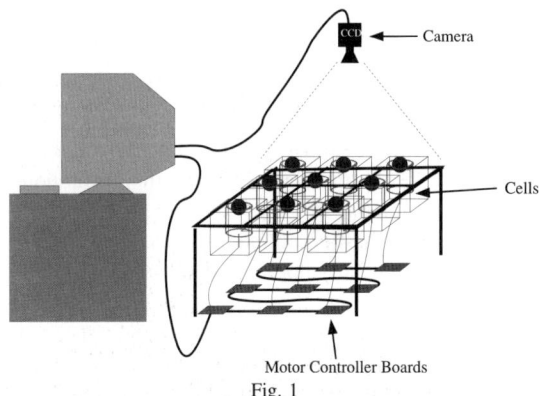

Fig. 1

A DISTRIBUTED MANIPULATION EXPERIMENT

This paper has the following structure. Section II reviews relevant background on distributed manipulation mechanics and control. Section III considers the case of "full actuation"- when all the actuators can be steered and driven. It gives an extremely simple, scale-able algorithm that is provably globally exponentially stabilizing, thus showing that it is highly desirable to have a fully actuated distributed manipulator.

Section IV summarizes experimental results obtained when this algorithm is implemented on the prototype system that is summarized in Section sec-our-experiment. A companion paper [18] describes this prototype test-bed in more detail, and summarizes experiments that validate our previously proposed non-smooth algorithms [15, 17] for the non-fully-actuated case.

II. BACKGROUND

The commonly used *programmable force field* (PFF) approach [3, 7] for distributed manipulator control is based on a continuous "force field" abstraction which

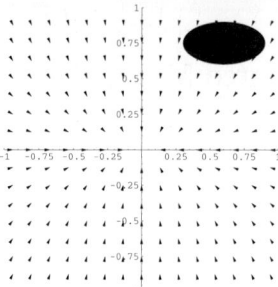

Fig. 2
PROGRAMMABLE FORCE FIELD

assumes that at each point on the manipulation surface one can specify the manipulation force at that point (see Figure II for an example of a continuous sink and an elliptical object). The dynamics of the moving object are obtained by integrating the continuous force field to get a total force on the part. To use the controls on an actual array, where the manipulation forces will be generated at discrete points, one must adapt the continuous approximation to the geometry of a given discrete array. For a good reference, see [5].

A. Nonsmooth modeling

When only a small number of actuators are in contact with the object being manipulated (i.e., the continuous actuation approximation is poor) or the coefficient of friction μ is very high, the continuous approximation has been shown experimentally not to work as well (see Luntz et al. [12]). In these cases, the continuous approximation does not adequately incorporate the physics of the actual array and the object/array interface.

These experimental observations previously led us to investigate the nonsmooth properties that arise due to frictional stick/slip phenomena at the interface between the actuation surfaces and the manipulated object [16]. We showed that under very simple and general assumptions on the friction model, the PFF approach leads to unstable systems when implemented on actual distributed manipulation arrays that have frictional contacts. In Ref. [15, 17] we presented local and global nonsmooth control algorithms that fix this instability problem. These controls are particularly relevant to the case where the distributed manipulation system is not fully actuated. In the case of Figure 1, this situation corresponds to not allowing the actuating wheels to be steered. The implementation of the result presented here is very simple by comparison, and moreover leads to very easy implementation, but is only appropriate to fully actuated systems (i.e, when the actuating wheels in the example of Fig. 1 are steerable).

We now give an overview of the modeling methodology we use. These results are needed as part of the proof of robustness and stability in Section III. The rationale behind this modeling technique is discussed in [14, 16]. Let q denote the configuration of the distributed manipulator system. The configuration q includes the object's position and orientation, as well as the variables that describe the state of each actuator. The relative motion of each contact between the part and an actuator array element can be modeled in the form $\omega(q)\dot{q}$. If $\omega(q)\dot{q} = 0$, the contact is not slipping (i.e., that contact is a nonholonomic constraint), while if $\omega(q)\dot{q} \neq 0$, then $\omega(q)\dot{q}$ describes the slipping velocity. In general, the moving part will be in contact with the actuator array at many points. From kinematic considerations, one or more of the contact points must be in a slipping state, thereby dissipating energy. The *power dissipation function* measures the part's total energy dissipation due to contact slippage.

Definition II.1. The *Dissipation* or *Friction Functional* for an n-contact state is defined to be

$$\mathcal{D} = \sum_{i=1}^{n} \mu_i N_i \mid \omega_i(q)\dot{q} \mid \qquad \text{(II.1)}$$

where μ_i and N_i are the Coulomb friction coefficient and normal force at the ith contact, which are assumed known.

Since there will generally not exist a motion where all of the contacts can be simultaneously slipless (although our current interest will be those cases when such a condition exists), we are lead to the following concept for finding the governing equations.

Power Dissipation Principle: With $\dot{q}$ small, a part's motion at any given instant is the one that minimizes $\mathcal{D}$.

The *power dissipation method* assumes that the part's motion at each instant is the one that instantaneously minimizes power dissipation due to contact slippage. This method is adapted from the work of [1] on wheeled vehicles. In [14], we showed that the power dissipation approach generically leads to multiple model systems defined next.

Definition II.2. A control system Σ is said to be a *multiple model driftless affine system (MMDA)* if it can be expressed in the form

$$\Sigma: \quad \dot{q} = f_1(q)u_1 + f_2(q)u_2 + \cdots + f_m(q)u_m \qquad \text{(II.2)}$$

where for any q and t, $f_i \in \{g_{\alpha_i} | \alpha_i \in I_i\}$, with I_i an index set, g_{α_i} analytic in (q,t) for all α_i, and the controls $u_i \in \mathbb{R}$ piecewise constant and bounded for all i. Moreover, letting σ_i denote the "switching signals" associated with f_i (which will be referred to as "MMDA maps"),

$$\sigma_i: \begin{array}{ccc} Q \times \mathbb{R} & \longrightarrow & \mathbb{N} \\ (q,t) & \longrightarrow & \alpha_i \end{array}$$

then the σ_i are measurable in (q,t).

An MMDA system is a driftless affine nonlinear control system where each control vector fields may "switch" back and forth between different elements of a finite set. The σ_i which regulate this switching may not be known, so we have no guarantees about the nature of the switching except that it is, by assumption, measurable. In the case of distributed manipulation, this switching corresponds to the switching among different contact states (i.e., different sets of slipping contacts) due to variations in contact geometry and surface friction properties. Practically, the switching in contact states can not be predicted in advance, and it is very difficult to measure in practice.

In the case of an n actuator array, we will have $2n$ potential constraints $\omega^i(q)$. The minimum of $\mathcal{D}$ is precisely the choice of $\dot{q}$ that annihilates the three constraints with the most dissipation. These, of course, will change over time because μ_i and N_i depend implicitly on the configuration, $q \in Q$. For our purposes, all we need to know is that an n actuator system will have $\binom{2n}{3}$ possible models. Note also that each of these models depends smoothly on the inputs u^k. This will be important in Theorem III.1 when we prove that the proposed control design is robust with respect to switching between models due to actuator uncertainties.

III. MAIN RESULT

This section describes in detail how a globally stabilizing *smooth* controller can be constructed. This approach requires that the distributed manipulator be fully actuated - i.e., at each actuator location the actuator can be oriented in any direction, and produce a velocity in that direction of arbitrary magnitude. The former is the important part while the latter can be made more realistic by using saturation functions.

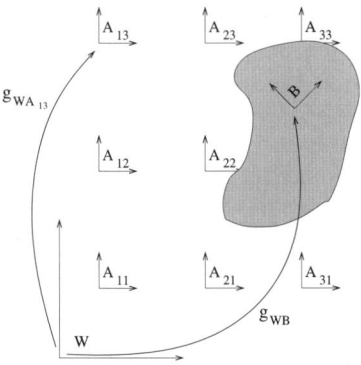

Fig. 3

RIGID BODY VELOCITIES

Consider Figure 3, which depicts an abstraction of a 9 cell experimental system (Section IV). Let W denote a fixed reference frame, let B denote a frame rigidly attached to the moving object, and let A_{ij} denote an "actuator frame" fixed at the point of contact between the actuator located at (x_i, y_j) and the object. This last frame has a fixed orientation with respect to W. Let the rigid body transformation from W to B be denoted by g_{WB} and the rigid body transformation form the W to A_{ij} be denoted by $g_{WA_{ij}}$. Recall that the g_{ab} are defined by

$$g_{ab} = \begin{bmatrix} R(\theta) & \begin{matrix} x_{ab} \\ y_{ab} \end{matrix} \\ 0 & 1 \end{bmatrix}$$

where $R(\theta) \in SO(2)$ describes the relative orientation of frame b with respect to frame a, and x_{ab} and y_{ab} are the translations going from frame a to frame b. The relative velocity $V_{body} = (\dot{x}_{body}, \dot{y}_{body}, \dot{\theta}_{body})$ of a point above actuator A_{ij} on the body is:

$$Ad_{g_{WA_{ij}}} \dot{q}$$

where in $SE(2)$ the Adjoint operator Ad_g is defined by

$$Ad_g = \begin{bmatrix} R(\theta) & \begin{matrix} y_{ab} \\ -x_{ab} \end{matrix} \\ 0 & 1 \end{bmatrix}.$$

We adopt the following control Lyapunov approach to the control design. Suppose we are given a Lyapunov function on $SE(2)$, denoted by $V(\cdot)$, and define target dynamics of the form:

$$\dot{q} = -\frac{\partial V(q)}{\partial q}.$$

This system is trivially exponentially stable. The velocity $\dot{q}$ is mapped to the actuators in order to obtain a feedback law.

We continue with a particular choice of Lyapunov function: $V(x, y, \theta) = k_1 x^2 + k_2 y^2 + k_3 \theta^2$ for $k_i > 0$. The Adjoint operator mapping velocities from W to the A_{ij} when the actuator frames are oriented parallel to the world frame is:

$$Ad_{g_{WA_{ij}}} = \begin{bmatrix} Id & \begin{matrix} y_j \\ -x_i \end{matrix} \\ 0 & 1 \end{bmatrix}$$

where Id is the identity. Transforming the velocity into the actuator frame yields $Ad_{g_{WA_{ij}}} \cdot (-\frac{\partial V(q)}{\partial q})$. Substituting in for $\frac{\partial V(q)}{\partial q}$, the actuator velocities should be

$$\begin{bmatrix} k_3 y_i(\theta - \theta_d) - k_1(x - x_d) \\ -k_3 x_i(\theta - \theta_d) - k_2(y - y_d) \\ -k_3 \theta \end{bmatrix}$$

where x_d, y_d, θ_d are the desired values and x, y, θ are the state feedback values. To transform this into wheel velocities and wheel orientations for the particular example found here, calculate the magnitude and direction of the (x, y) velocity. This gives for each actuator:

$$\theta_{ij} = \tan^{-1}\left(\frac{-k_3 x_i(\theta - \theta_d) - k_2(y - y_d)}{k_3 y_i(\theta - \theta_d) - k_1(x - x_d)}\right) \quad \text{(III.1)}$$

and

$$v_{ij} = \sqrt{\begin{array}{c}(-k_3 x_i(\theta - \theta_d) - k_2(y - y_d))^2 \\ +(k_3 y_i(\theta - \theta_d) - k_1(x - x_d))^2\end{array}}$$

(III.2)

where θ_{ij} is the orientation of the (i, j) actuator and v_{ij} is the wheel velocity of that actuator. So, given all the actuator locations, one computes Equations (III.1) and (III.2) for each actuator, and the feedback law is complete.

Now we consider the robustness of this feedback law with respect to the multiple model system that arises if the actuators are not all perfectly aligned. In particular, if we consider the controls obtained above to be the *desired* controls u_d^k and what we actually obtain are u^k, then we get slightly different dynamics. This brings us to the following theorem.

Theorem III.1. *There exists $\delta > 0$ such that if for $t > T$ we have $|u^k(t) - u_d^k(t)| < \delta \ \forall \ k$ for some T, then the solutions to the MMDA system predicted by the PDM are exponentially stable using the controls from Equations (III.1) and (III.2).*

Proof: First, we know that for the choice of controls u_d^k we have
$$\dot{V} = \frac{\partial V}{\partial q} \dot{q} < 0.$$
Therefore, in a sufficiently small neighborhood of $\dot{q} \in T_qQ$ (denoted by $B(\dot{q}, \epsilon) \subset T_qQ$) we have
$$\dot{V} = \frac{\partial V}{\partial q} \dot{q}_\epsilon < 0 \ \forall \ \dot{q}_\epsilon \in B(q, \epsilon).$$
(This is a simple consequence of the continuity of the expression $\dot{V}$ along a continuous path between $\dot{q}$ and any other element of T_qQ.)

Now, a sufficient condition for stability of a multiple model system is that all of the individual models not only be individually stable, but additionally all satisfy the same Lyapunov function (see [17]). We will use this fact here, and show that for sufficiently small δ all the multiple models will be in $B(\dot{q}, \epsilon)$, thereby ensuring overall stability of the nonsmooth system.

For a given set of inputs u^k we know that we have a corresponding set of kinematic constraints $\omega_i(q)$, and that the PDM implies that a subset of these satisfying $\omega_i(q)\dot{q} = 0$ will define the actual kinematics. In the case of $u^k = u_d^k$, we get precisely the desired dynamics. Because these kinematic constraints $\omega_i(q)$ depend continuously on the inputs, for any choice of ϵ' limiting how much we will allow the $\omega_i(q)$ to vary (and hence how much $\dot{q}$ can vary), we can always choose a δ such that $|u^k(t) - u_d^k(t)| < \delta$ accordingly. Therefore, we can always choose δ small enough such that $\dot{q} \in B(\dot{q}_d, \epsilon)$. This completes the proof. ∎

This theorem implies that even if the actuators start out in a kinematically incompatible state, as long as they converge to within some δ of the desired actuator state, the system will keep its stability properties. We should also note that this can easily be extended to exponential stability in a similar fashion.

Experimental results in Section IV illustrate that this method works extremely well. However, in the case where one does not have full actuation, one must ask if this control law has any analogs. In general the answer is no, see [17] for details.

IV. Experimental Results

An experimental apparatus has been developed for testing the results in this paper and in previous work by the authors. A photograph can be seen in Figure 4. The design is a modular one based on a basic cell design. Each

Fig. 4
Front of the FADM System

cell contains two actuators. One actuator orients the wheel axis, while the other actuator drives the wheel rotation. These cells can be repositioned easily into different configurations. The Fully Actuated Distributed Manipulation (FADM) system shown in Figure 4 is configured with a total of nine cells—though more can be easily added. For more details on the experimental setup, please refer to the companion paper [18].

We include an illustrative experimental[1] result (found in Figure 5). The goal of the experiment was point stabilization from a random initial condition to the origin (of $SE(2)$). We put the controlled object down in a random initial configuration (in this case $(x_0, y_0, \theta_0) = (0.4 \text{ m}, -0.4 \text{ m}, 2.6 \text{ rad})$) and initiated the experiment with the actuators all in the same initial conditions of $\theta = 0$ (relative to the world coordinate axes). The goal then was to stabilize the object to a final position of $(x_f, y_f, \theta_f) = (0 \text{ m}, 0 \text{ m}, 0 \text{ rad})$. The actually achieved final position was $(x_f, y_f, \theta_f) = (0.01 \text{ m}, 0.01 \text{ m}, 0.05 \text{ rad})$. The figure panels depict the x, y, and θ trajectories as functions of time and a plot of the (x, y) trajectory in the plane for a rectangular plexiglass object.

Notice that the translational stability of the origin is maintained, while the rotational dynamics are stabilized to $\theta = 0$ due to the feedback law. The important point to notice is the smoothness of the trajectory. This experiment indicates that when a distributed array is fully actuated, the feedback law in Equation (III.2) works extremely well. More importantly it is computationally very simple, and the number of computations scales linearly with the number of actuators. The feedback law has good disturbance rejection properties, as can be seen by the fact that the object is stabilized despite the fact that the actuator initial conditions are not compatible with the desired motion, hence verifying the result in Theorem III.1.

---

[1]Movies of these and other experiments can be found at *http://www.cds.caltech.edu/~murphey/experiment/*.

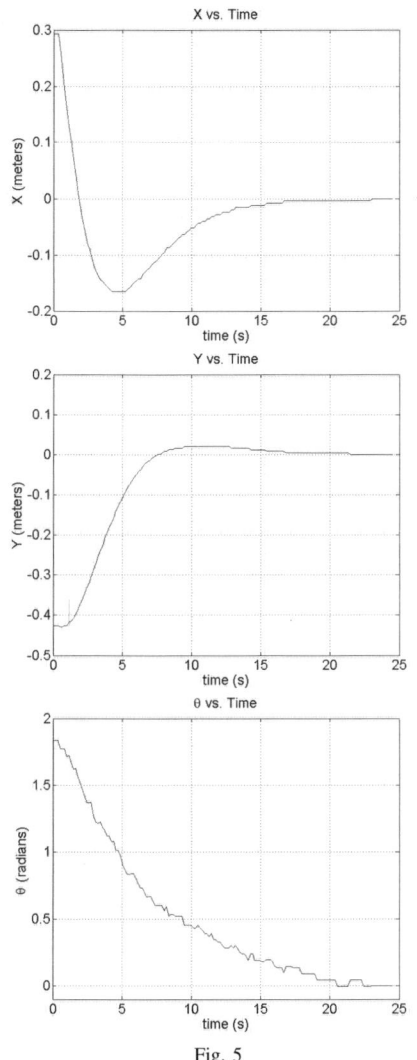

Fig. 5

GRAPHS OF THE x, y, AND θ DYNAMICS OF THE EXPERIMENTAL RESULTS

V. SUMMARY

The very simple algorithm presented in this paper provides a different closed-loop approach to the control of distributed manipulation systems. Prior work, particularly the programmable vector field approach, has assumed that all the distributed system's actuators move sufficiently fast that they slip all of the time. This causes significant stress on both the object being manipulated and the actuators themselves. In the case of fully actuated systems, it is possible to ensure that *all* of the relevant actuators contact the moving body without slip. This requires less energy for a given motion and moreover induces smaller forces on the object and actuators. This approach also raises many additional questions. For instance, for a given problem with nonsmooth mechanics, what is an analytical test to guarantee that a smooth solution (like the one presented here) exists? Is such a solution unique? Answers to these questions are part of ongoing research.

Acknowledgements: This work was partially supported by a grant from the National Science Foundation (grant NSF9402726) through its Engineering Research Center (ERC) program.

REFERENCES

[1] J.C. Alexander and J.H. Maddocks. On the kinematics of wheeled vehicles. *The International Journal of Robotics Research*, 8(5):15–27, October 1989.

[2] K. Bhattacharya and R.D. James. A theory of thin films of martensitic materials with applications to microactuators. *J. Mech. Phys. Solids*, 47:531–576, 1999.

[3] K.F. Böhringer, R. G. Brown, B. R. Donald, J.S. Jennings, and D. Rus. Sensorless manipulation using transverse vibrations of a plate. In *Proc. IEEE Int. Conf. on Robotics and Automation*, pages 1989–1996, Nagoya, Japan, 1995.

[4] K.F. Böhringer and H. Choset, editors. *Distributed Manipulation*. Kluwer, 2000.

[5] K.F. Böhringer, B.R. Donald, L.E. Kavraki, and F. Lamiraux. A distributed, universal device for planar parts feeding: unique part orientation in programmable force fields. In *Distributed Manipulation*, pages 1–28. Kluwer, 2000.

[6] D. Rus B.R. Donald, J. Jennings. *Algorithmic Foundations of Robotics (WAFR)*, chapter Information invariants for distributed manipulation, pages 431–459. A.K. Peters, Ltd, Wellesley, MA, 1995.

[7] M. Coutinho and P. Will. The intelligent motion surface: a hardware/software tool for the assembly of meso-scale devices. In *IEEE Int. Conf. on Robotics and Automation (ICRA)*, 1997. Albuquerque, New Mexico.

[8] H. Fujita. Group work of microactuators. In *International Advanced Robot Program Workshop on Micromachine Technologies and Systems*, pages 24–31, Tokyo, Japan, 1993.

[9] P. Krulevitch, A.P. Lee, P.B. Ramsey, J.Trevino J. Hamilton, and M.A. Northrup. Thin film shape memory alloy microactuators. *Journal of Micro Electro Mechanical Systems*, 5(270), 1996.

[10] A.P. Lee, C.F. McConaghy, P. Krulevitch, G.E. Sommargren, and J. Trevino. Electrostatic comb drive for vertical actuation. In *Proceedings of Micromachined Devices and Components, SPIE 1997 Symposium on Micromachining and Microfabrication*, volume 109, Austin, Texas, 1997.

[11] J. Luntz, W. Messner, and H Choset. Velocity field design for parcel manipulation on the modular distributed manipulation system. In *Proceedings of the IEEE International Conference on Robotics and Automation (ICRA)*, 1999.

[12] J. Luntz, W. Messner, and H. Choset. *Distributed Manipulation*, chapter Discreteness Issues in Actuator Arrays. Kluwer Academic Publishers, 2000.

[13] H. Fujita M. Ataka, A. Omodaka. A biomimetic micro motion system. In *Transducers - Digest International Conference on Solid State Sensors and Actuators*, pages 38–41, 1993. Pacifico, Yokohama, Japan.

[14] T. D. Murphey and J. W. Burdick. Issues in controllability and motion planning for overconstrained wheeled vehicles. In *Proc. Int. Conf. Math. Theory of Networks and Systems (MTNS)*, Perpignan, France, 2000.

[15] T. D. Murphey and J. W. Burdick. Global stability for distributed systems with changing contact states. In *Proc. IEEE Int. Conf. on Intelligent Robots and Systems*, Hawaii, 2001.

[16] T. D. Murphey and J. W. Burdick. On the stability and design of distributed systems. In *Proc. IEEE Int. Conf. on Robotics and Automation*, Seoul, Korea, 2001.

[17] T. D. Murphey and J. W. Burdick. Global exponential stabilizability for distributed manipulation. In *Proc. IEEE Int. Conf. on Robotics and Automation*, Washington D.C., 2002.

[18] T. D. Murphey and J. W. Burdick. Experiments in nonsmooth control of distributed manipulation. In *Submitted to IEEE Int. Conf. on Robotics and Automation*, 2003.

[19] K.P. Seward, P. Krulevitch, H.D. Ackler, and P. Ramsey. A new mechanical characterization method for microactuators applied to shape memory films. In *10th International Conference on Solid State Sensors and Actuators, Transducers '99*, Sendai, Japan, 1999.

[20] J. W. Suh, R. B. Darling, K. F. Böhringer, B. R. Donald, H. Baltes, and G. T. A. Kovacs. CMOS integrated organic ciliary array as a general-purpose micromanipulation tool for small objects. *Journal of Microelectromechanical Systems*, 8(4):483–496, December 1999.

Nanotube Devices Fabricated in a Nano Laboratory

Lixin DONG, Fumihito ARAI, Masahiro NAKAJIMA, Pou LIU, and Toshio FUKUDA

Department of Micro System Engineering, Nagoya University
Furo-cho 1, Chikusa-ku, Nagoya 464-8603, JAPAN
dong@robo.mein.nagoya-u.ac.jp, arai@mein.nagoya-u.ac.jp, nakajima@robo.mein.nagoya-u.ac.jp,
liupou@robo.mein.nagoya-u.ac.jp, fukuda@mein.nagoya-u.ac.jp

Abstract:
A nano laboratory—a prototype nano manufacturing system—is presented, which is composed with a nanorobotic manipulation system with 4 units and 16 degrees-of-freedom (DOFs), a nano fabrication system based on electron-beam-induced deposition (EBID) with an internal or external precursor evaporation reservoir equipped with a thermal field emission electron source or a nanotube cold cathode, and a real-time observation and measurement system based on a field emission scanning electron microscope (FESEM) equipped with 3-4 conventional atomic force microscope (AFM) cantilevers, piezoresistive levers or nanotube probes. Nanotube devices including a mass flow sensor, a linear bearing, and nanotube scissors are fabricated in the nano laboratory.

Key words: carbon nanotubes, nanodevices, nanorobotic manipulations, nano laboratory, electron-beam-induced deposition

1. INTRODUCTION

Over the past decade, carbon nanotubes (CNTs) [1] have been extensively explored. These explorations have revealed many exceptional properties of nanotubes (briefly summarized in Table 1) and proposed broad potential applications for them [2]. In bulk state, nanotubes can be used to synthesize conductive and high-strength composites, to fabricate field emission devices (flat display, lamp, gas discharge tube, x-ray source, microwave generator, etc.), to save and convert electrochemical energy (supercapacitor, battery cathode, electromechanical actuator, etc.), to store hydrogen, and so on. However, the most promising applications of nanotubes that have deepest implications for molecular nanotechnology need to maneuver the tubes individually to build complex nanodevices. Such devices include nanoelectronics and nano electromechanical systems (NEMS) (concisely listed in Table 2).

Almost all of such applications and many unlisted potential ones of nanotubes for nanoelectronics and NEMS involve charactering, placing, deforming, modifying, and/or connecting nanotubes. Although chemical synthesis may provide a way for patterned structure of nanotubes in large-scale [3], self-assembly may generate better regular structures, we still lack of capability to construct irregular complex nanotube devices. Nanomanipulation [4], especially nanorobotic manipulation[5-9], with its "bottom up" nature, is the most promising way for this purpose.

Table 1 Property of CNTs

Property	Item	Data
Geometrical	Layers	Single-walled nanotubes (SWNTs) or Multiwalled nanotubes (MWNTs)
	Aspect Ratio	10-1000
	Diameter	~0.4nm to >3nm (SWNTs) ~1.4 to >100nm (MWNTs)
	Length	Several μm (Rope up to cm)
Mechanical	Young's Modulus	~1 TPa (steel: 0.2TPa)
	Tensile Strength	45GPa (steel: 2GPa)
	Density	$1.33 \sim 1.4 g/cm^3$ (Al: $2.7 g/cm^3$)
Electronic	Conductivity	Metallic/Semi-conductivity
	Current Carrying Capacity	~$1TA/cm^3$ (Cu: $1GA/cm^3$)
	Field Emission	Activate Phosphorus at 1~3V
Thermal	Heat Transmission	>3kW/mK (Diamond: 2kW/mK)

Table 2 Applications of CNTs in Nanodevices

Device	Fabrication
Diode	Rectifying diode: a kink junction [10].
Transistor	Room-temperature (RT) field-effect transistors (FETs): a semiconducting SWNT placed between two electrodes (source and drain) with a submerged gate [11]. RT single electron transistor (SETs) [12]: a short (~20 nm) nanotube section.
ICs	Hybrid logic circuits: two nanotube transistors placed on lithographically fabricated electrodes: [13, 14]. Pure nanotube circuits [15 - 17]: interconnected nanotubes (intermolecular and intramolecular junctions [18-20]).
Switch	Pushing and releasing a suspended nanotube [21].
Memory	Electromechanical nonvolatile memory: suspended cross-junctions [22].
Probe	Manually assembly [23], chemical vapor deposition (CVD) [24], controlled assembly [25], and picking up a tube from vertically aligned SWNTs [26].
Tweezers	Assemble two CNT bundles on a glass fiber [27].
Scissors	Nanorobotic assembly and shape modification [28].
Sensor	Chemical sensor: a semiconducting SWNTs [29].
Bearing	Open a MWNT by electric pulses [30, 31].

2. NANO LABORATORY

2.1 General Description

A nano laboratory is presented as shown in Fig.1, which consists of a nanorobotic manipulation system (Fig. 2), an instrumentation system (a filed emission scanning electron microscope (FESEM) and a conventional atomic force microscope cantilever or a piezolever), and a nanofabrication system based on electron-beam-induced deposition (EBID) or manipulations. The specifications of the nano laboratory are listed in Table 3. The nano laboratory can be applied for manipulating nano materials—mainly but not limited to CNTs, fabricating nano building blocks, assembling nano devices, *in situ* analyzing the properties of such materials, building blocks and devices. The functions of it are summarized in Table 4, and many have been demonstrated elsewhere as shown in the references in the table.

Here we show metallic EBID for obtaining conductive deposits by applying tungsten hexacarbonyl ($W(CO)_6$) as precursors, and improved destructive fabrication for better length control of a MWNT by using surface van der Waals forces for interlayer clamping.

Table 3 Specifications of Nano Laboratory

Item	Specification
Nanorobotic Manipulation System	
DOFs	Total: 16 DOFs Unit1: 3 DOFs (x, y and β; coarse) Unit2: 1 DOF (z; coarse), 3-DOF (x, y and z; fine) Unit3: 6 DOFs (x, y, z, α, β, γ; ultrafine) Unit 4: 3 DOFs (z, α, β; fine)
Actuators	4 Picomotors™ (Units 1& 2) 9 PZTs (Units 2& 3) 7 Nanomotors™ (Units 2 & 4)
End-effectors	3 AFM cantilevers+1 substrate or 4 AFM cantilevers
Working space	18mmx18mmx12mmx360° (coarse, fine), 26μmx22μmx35μm (ultrafine)
Positioning resolution	30nm(coarse), 2mrad (coarse), 2nm(fine), sub-nm (ultrafine)
Nano Instrumentation System	
FESEM	Imaging resolution: 1.5 nm
AFM Cantilever	Stiffness constant: 0.03nN/nm
Piezolever	Stain gauge built-in
Nanofabrication System	
EBID	FESEM emitter: T-FE, CNT emitter, Gas introduction system

Table 4 Functions of Nano Laboratory

Function	Involved Manipulations
Property Characterization	Mechanical properties: buckling [8] or stretching
	Electric properties: placing between two probes (electrodes)
Nanofabrication	EBID with a CNT emitter and parallel EBID [32]
	Destructive fabrication: breaking [33]
	Shape modification: deforming by bending and buckling, and fixing with EBID [28]
Nanoassembly	Picking up by controlling intermolecular and surface forces, and forces [7]
	Connecting with van der Waals [8]
	Soldering with EBID [33]
	Bonding through mechanochemical systhesis [34]

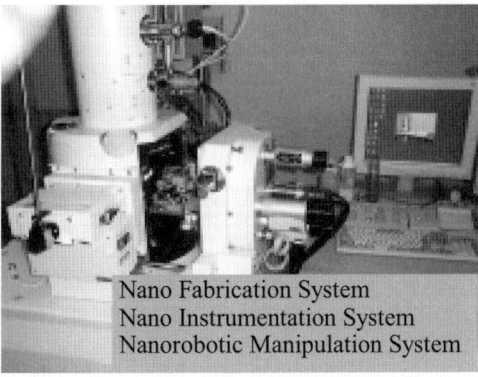

Fig.1 Nano Laboratory

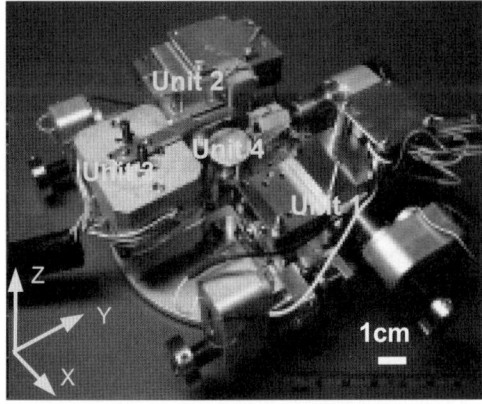

Fig.2 Nanorobotic Manipulators

2.2 EBID with $W(CO)_6$ as Precursor

We have demonstrated that EBID can be used for the fabrication of nanostructures, the connection of nanotubes [33], and shape fixing of an elastic deformed tube [28]. We also showed that nanotubes were ideal emitters for EBID [32], and the deposition efficiency could be improved much if CNT emitter arrays were available. However, these demonstrations adopted pump oil (contaminations) as precursors, and the deposits were amorphous carbon that has no conductivity. To confirm this, a nanowire is fabricated as shown in Fig.3. Bi-terminal measurement shows it is insulate. To apply EBID for the fabrications of electrodes or conductive spots, precursors with metallic components are needed. The deposit from EBID with $W(CO)_6$ have been shown conductive [35], so it is suitable to use it as precursors for metallic deposits. $W(CO)_6$ is white crystalline powder at room temperature and can sublime at this temperature. Fig.4 shows a nanowire fabricated by internal vaporizing $W(CO)_6$ in the vacuum chamber of the FESEM. Conductivity measurement shows that the resistance is 80.1kΩ, and the resistivity is 0.08Ωcm (average cross section: 100nm×100nm). It is also found that the deposition rate and the spot sizes depend on the beam current of FESEM emitter. Large current brings out high vertical growth rate but large spot size as shown in Fig.5.

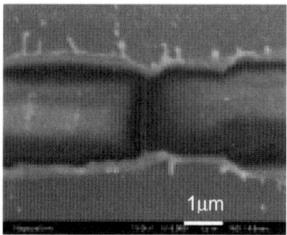 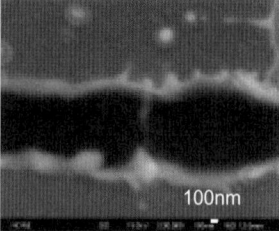

Fig.3 Insulate Nanowire   Fig.4 Conductive Nanowire

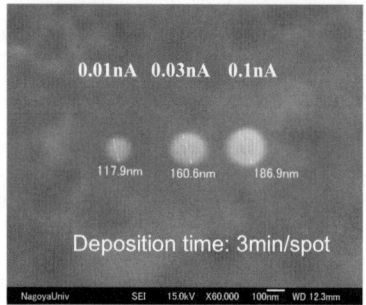

Fig.5 Deposition Rate at Different Beam Current

Better control for the conductivity is still a challenge because it involves beam current, gas/vapor flux and many other factors.

2.3 Improved Length Control of a MWNT through Destructive Fabrication by Surface Force Clamping

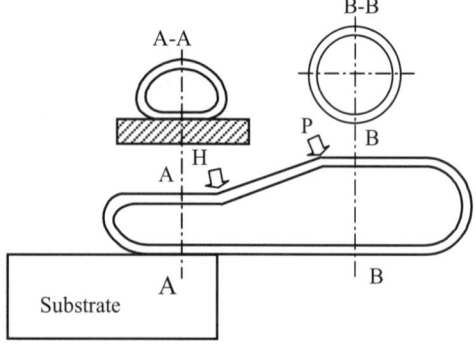

Fig. 6. Deformation of MWNTs by Surface van der Waals Forces

Destructive fabrication provides a method to get useful building blocks [33]. However, it is difficult to apply it for getting desired lengths of tubes. A perfect MWNT can be broken just at any place in its length. Although a tube with a defect would help us to determine where to break, it is unreasonable to desire that a defect always appear at the site at will. Plastic deformation can make defects on a tube, and hence can help us to determine the obtainable lengths.

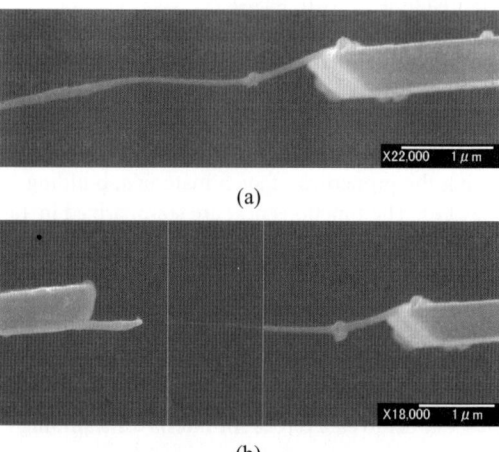

Fig.7 Improved Destructive Fabrication

Here we show a simpler method by placing a tube on a surface as shown in Fig.6. It has been demonstrated that surface van der Waals forces will deform a nanotube placed on a surface [36] as that shown in the section A-A in Fig.6. Deformation of the tube is determined by the number of layers of the tube and the surface property of the substrate that influence the Hamaker constant between the tube and the substrate. Anyway, the deformation will cause the tube to be stressed, and heptagons and pentagons may appear at points H and P when the stress is large enough. This phenomenon can help us to predict the breaking position, i.e., the most possible breaking sites will be in the section between points H and P. By adjusting the length placed on the substrate; we can determine the length left after breaking.

Fig.7 shows a result by using this technique. The nanotube is picked up and fixed on an AFM cantilever at its right end (Fig.7(a)). By placing the left end of it onto an Au coated silicon substrate, we can perform destructive fabrication by using the above-mentioned technique. Fig.7 (b) shows the result. We can find that the breaking occurred in a point resembling point H as shown in Fig.6.

3. NANOTUBE DEVICES

3.1 Nanotube Mass Flow Sensors

Measurement of ultra small flux of gases is an important and challenge problem. Silicon based microelectromechanical system (MEMS) can provide cantilevered thin wire as the transducer. But it is still difficult to fabricate vertical Silicon beam for measuring mass flow as small as several sccm. Like a nanotube pN force sensor, a cantilevered nanotube with very large aspect ratio is also possible to be used as a transducer for ultra-small gas flow.

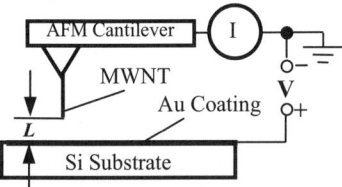

Fig.8 Design of a Mass Flow Sensor

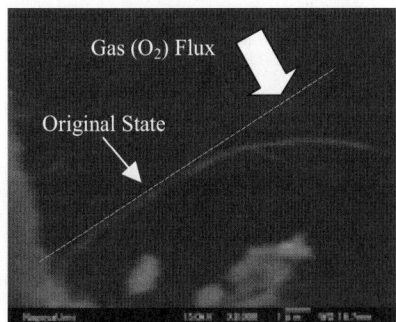

Fig.9 A Mass Flow Sensor

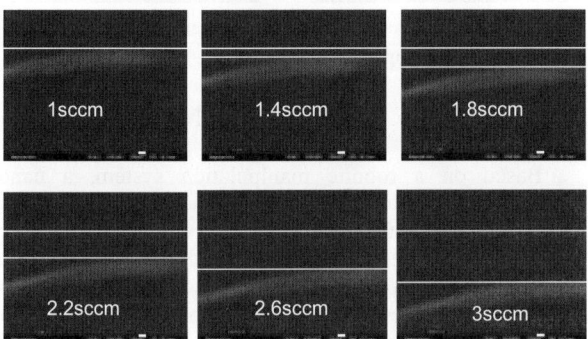

Fig.10 Deformation of Nanotube Sensor (Scale bars: 100nm)

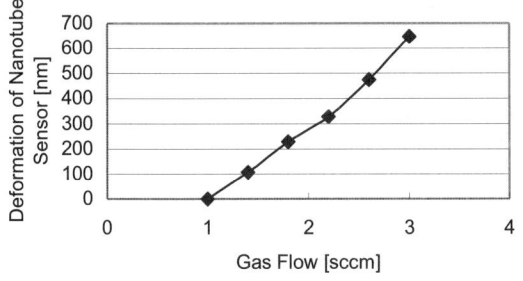

Fig.11 Calibration of Nanotube Mass Flow Sensor

Fig.8 shows a design. By measuring the field emission current or tunneling current (as the gap L<~1nm), it is possible to detect the deflection of the nanotube caused by gas flow. Fig.9 shows a cantilevered nanotube bundle (length is about 10μm). When a flow of O_2 gas comes on it, deflections as shown in Fig.10 appeared. The relations between the mass flow (in sccm) and the deflection of the nanotube is shown in Fig.11. It can be found that the nanotube mass flow sensor is quite sensitive. Resolution is 0.93×10^{-3} sccm/nm.

3.2 Nanotube Linear Bearings

Through destructive fabrications, typically, a layered structure and a sharpened structure can be obtained [33].

As shown in Fig.12(a), a MWNT is supported between a substrate (left end) and an AFM cantilever (right end), which are in turn fixed on the Units 1 and 2 of the manipulation system. Fig.12(b) shows a zoomed up image of the central blocked part of Fig.12(a), and the inset in it shows its structure schematically. It can be found that the nanotube has a thinner neck (part B in Fig. 12(b)) that was formed by destructive fabrication, i.e., by moving the cantilever to the right. To move it more in the same direction, a motion like a linear bearing is observed as shown in Fig.12(c), and it is shown schematically by the inset.

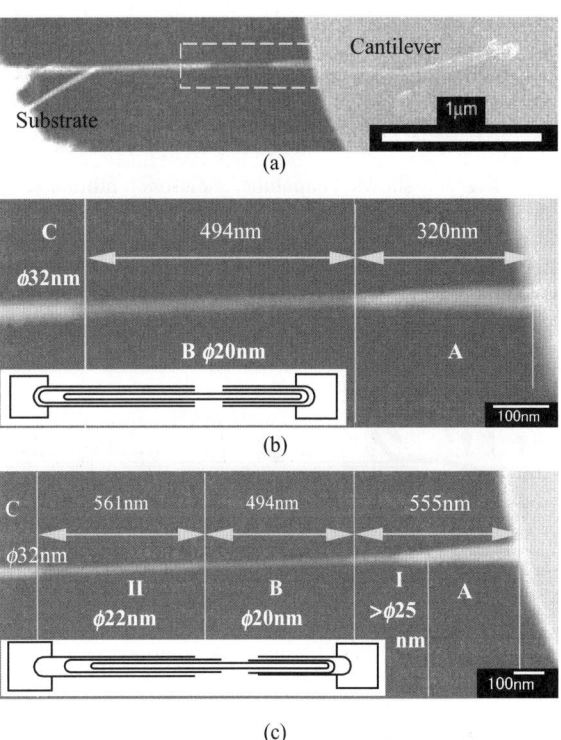

Fig.12 Nanotube Linear Bearing

By comparing Fig.12(b) and (c), we can find that part B remained unchanged in its length and diameter, while its two ends brought out two new parts I and II from parts A and B, respectively. Part II seems have uniform diameter (φ22nm), while part I is a tapered structure with the smallest diameter φ25nm. The interlayer friction has been predicted very small [30], but the direct measurement of the friction remains a challenging problem.

3.3 Nanotube Scissors

Nanotube scissors are designed for probing the conductivity or cutting a nanotube between the two arms with saturated current [28]. A new design by applying conductive EBID deposits is shown in Fig.13. To improve the cutting accuracy, it is necessary to modify the opening between the two arms (*g*). The key technique is the shape modification of nanotubes.

The schematic diagram of shape modification is shown in Fig.14. For a CNT, the maximum angular displacement will appear at the fixed left end under pure bending or at the middle point under pure buckling. A combination of these two kinds of loads (Fig.14(a)) will achieve a controllable position of kinked point and a desired kink angle *θ*. If the deformation is within the elastic limit of the nanotube, it will recover as the load released. To avoid this, EBID can be applied at the kinked point to fix the shape as shown in Fig.14 (b). Fig.15 shows a nanotube being bent and buckled under combined bending and buckling. Fig.16 shows an example for shape modifications. The MWNT is fixed on an AFM cantilever on its right end, and it is bent by attaching its left end to a substrate and moving the AFM cantilever downward, another EBID deposit fixed the shape of the kinked structure.

Fig.17 shows nanotube scissors fabricated by assembling two tubes on a commercially available AFM cantilever. Before shape modification, the opening between the two arms is larger than 1μm, whereas it is 85.6nm after modification.

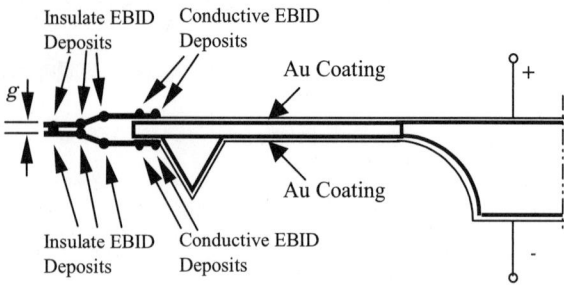

Fig.13 Nanotube Scissors

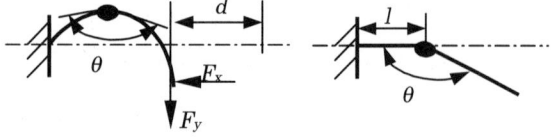

(a) Bending and buckling (b) Released state
Fig.14 Shape modification of CNT

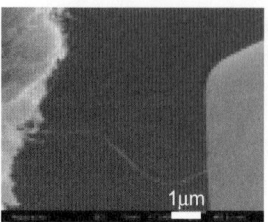 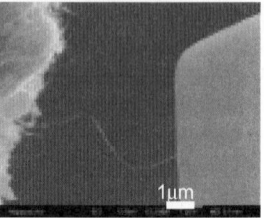

(a) Bending and buckling (b) 3-D bending and buckling
Fig.15 Deformations of CNTs under bending and buckling

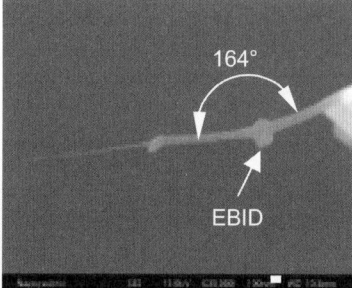

Fig.16 Shape Modification of a Nanotube (Scale bar: 100nm)

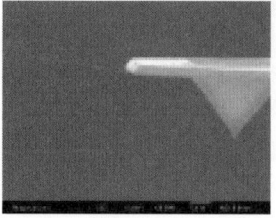 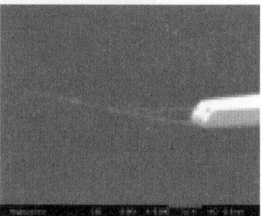

(a) Before Shape Modification (b) After Shape Modification
Fig.17 Nanotube Scissors (Scale bars: 1μm)

4. CONCLUSIONS

Based on a robotic manipulation system, a nano laboratory—a prototype nano manufacturing system has been presented, which is composed with a nanorobotic manipulation system with 4 units and 16 DOFs, a nano fabrication system based on EBID with an internal or external precursor evaporation reservoir equipped with a thermal field emission electron source or a nanotube cold cathode, and a real-time observation and measurement system based on FESEM equipped with 3-4 conventional AFM cantilevers, piezoresistive levers or nanotube probes. Nanotube devices including a mass flow sensor, a linear bearing, and nanotube scissors are fabricated in the nano laboratory. The nanotube mass flow sensor is a cantilevered nanotube, which is calibrated by observing the SEM images of the deflection of the cantilevered tube by introducing O_2 gas into the vacuum chamber of the FESEM. Result shows the nanotube mass flow sensor is sensitive for gas flow less than 3sccm (resolution: 0.93×10^{-3} sccm/nm). Independent sensor structure based on the measurement of tunneling current is presented. A linear nanotube bearing is prepared with destructive fabrication, and an improved technique is presented. It also provides a sample for the research on nanotribology. Nanotube scissors have been constructed by assembling two tubes onto a commercially available AFM cantilever to form a multiple functioned end-effector. Shape modification makes it possible to get an opening as small as several tens of nanometers. Conductivity of the scissors has been improved by EBID with $W(CO)_6$ precursor including metallic elements.

ACKNOWLEDGMENTS

This work was supported in part by the Scientific Research Fund of the Ministry of Education of Japan. We are grateful to Prof. Y. Saito at Mie University for providing us with MWNT samples.

REFERENCES

[1] S. Iijima, "Helical microtubules of graphitic carbon," *Nature*, vol.354, pp.56-58, 1991.

[2] R.H. Baughman, A.A. Zakhidov, W.A. de Heer, "Carbon nanotubes --the route toward applications", *Science*, vol.297, pp.787-792, 2002.

[3] N. R. Franklin, Y. M. Li, R. J. Chen, A. Javey, and H. J. Dai, "Patterned growth of single-walled carbon nanotubes on full 4-inch wafers," *Appl. Phys. Lett.*, vol.79, pp.4571-4573, 2001.

[4] D.M. Eigler and E.K. Schweizer, "Positioning Single Atoms with a Scanning Tunneling Microscope", *Nature*, vol.344, pp.524-526, 1990.

[5] M. F Yu, M. J. Dyer, G. D. Skidmore, H. W. Rohrs, X. K. Lu, K. D. Ausman, J. R. Von Ehr, and R. S. Ruoff, "Three-dimensional manipulation of carbon nanotubes under a scanning electron microscope", *Nanotechnology*, vol.10, pp.244-252, 1999.

[6] L.X. Dong, F. Arai and T. Fukuda, "3D nanorobotic manipulation of nano-order objects inside SEM", in *Proc. of the 2000 International Symposium on Micromechatronics and Human Science*, Nagoya, Japan, 2000, pp.151-156.

[7] L.X. Dong, F. Arai, and T. Fukuda, "3-D nanorobotic manipulations of nanometer scale objects", *J. of Robotics and Mechatronics*, vol.13, pp.146-153, 2001.

[8] L.X. Dong, F. Arai, and T. Fukuda, "3D nanorobotic manipulations of multi-walled carbon nanotubes," in *Proc. of the 2001 IEEE International Conf. on Robotics and Automation (ICRA2001)*, Seoul, Korea, 2001, pp.632-637.

[9] L.X. Dong, F. Arai, and T. Fukuda, "Three-dimensional nanorobotic manipulations of carbon nanotubes", *J. of Robotics and Mechatronics (JSME)*, vol.14, No.3, pp.245-252, 2002.

[10] Z. Yao, H.W.C. Postma, L. Balents and C. Dekker, "Carbon nanotube intramolecular junctions", *Nature*, vol.402, pp.273-276, 1999.

[11] S. J. Tans, A. R. M. Verchueren and C. Dekker, "Room-temperature transistor based on a single carbon nanotube," *Nature*, vol.393, pp.49-52, 1998.

[12] H. W. Ch. Postma, T. Teepen, Z. Yao, M. Grifoni, C. Dekker, "Carbon nanotube single-electron transistors at room temperature", *Science*, vol.293, pp.76-79, 2001.

[13] Y. Huang, X.F. Duan, Y. Cui, L. J. Lauhon, K.-H. Kim, and C. M. Lieber, "Logic gates and computation from assembled nanowire building blocks", *Science*, vol.294, pp. 1313-1317, 2001.

[14] A. Bachtold, P. Hadley, T. Nakanishi, and C. Dekker, "Logic circuits with carbon nanotube transistors", *Science*, vol.294, pp.1317-1320, 2001.

[15] R. Saito, G. Dresselhaus and M. S. Dresselhaus, "Tunneling conductance of connected carbon nanotubes," *Phys. Rev. B.*, vol.53, pp.2044-2050, 1996.

[16] L. Chico, V.H. Crespi, L.X. Benedict, S.G. Louie and M.L. Cohen, "Pure carbon nanoscale devices: nanotube heterojunctions", *Phys. Rev. Lett.*, vol.76, pp.971-974, 1996.

[17] M. Menon and D. Srivastava, "Carbon nanotube 'T junctions': nanoscale metal-semiconductor-metal contact devices," *Phys. Rev. Lett.*, vol.79, pp.4453-4456, 1997.

[18] M. S. Fuhrer, J. Nygård, L. Shih, M. Forero, Y.-G. Yoon, M. S. C. Mazzoni, H.J. Choi, J. Ihm, S.G. Louie, A. Zettl and P.L. McEuen, "Crossed nanotube junctions," *Science*, vol.288, pp.494-497, 2000.

[19] L.X. Dong, F. Arai, and T. Fukuda, "Inter-process measurement of MWNT rigidity and fabrication of MWNT junctions through nanorobotic manipulations", in *American Institute of Physics Conference Proceedings 590: Nanonetwork Materials: Fullerenes, Nanotubes, and Related Materials*, 2001, pp.71-74.

[20] T. Fukuda, F. Arai, and L.X. Dong, "Fabrication and property analysis of MWNT junctions through nanorobotic manipulations", *Int'l J. of Nonlinear Sciences and Numerical Simulation*, vol.3, pp.753-758, 2002.

[21] T.M. Tombler, C.W. Zhou, L. Alexseyev, J. Kong, H.J. Dai, L. Liu, C.S. Jayanthi, M.J. Tang and S.Y. Wu, "Reversible electromechanical characteristics of carbon nanotubes under local-probe manipulation", *Nature*, vol.405, pp.769-772, 2000.

[22] T. Rueckes, K. Kim, E. Joselevich, G.Y. Treng, C. L. Cheung and C.M. Lieber, "Carbon nanotube-based nonvolatile random access memory for molecular computing science", *Science*, vol.289, pp.94-97, 2000.

[23] H.J. Dai, J.H. Hafner, A.G. Rinzler, D.T. Colbert and R.E. Smalley, "Nanotubes as nanoprobes in scanning probe microscopy," *Nature*, vol.384, pp.147-150, 1996.

[24] J.H. Hafner, C.L. Cheung and C.M. Lieber, "Growth of nanotubes for probe microscopy tips", *Nature*, vol.398, pp.761-762, 1999.

[25] H. Nishijima, S. Kamo, S. Akita, Y. Nakayama, K. I. Hohmura, S. H. Yoshimura, and K. Takeyasu, "Carbon-nanotube tips for scanning probe microscopy: preparation by a controlled process and observation of deoxyribonucleic acid", *Appl. Phys. Lett.*, vol.74, pp.4061-4063, 1999.

[26] J. H. Hafner, C.-L. Cheung, T. H. Oosterkamp, and C. M. Lieber, "High-yield assembly of individual single-walled carbon nanotube tips for scanning probe microscopies", *J. Phys. Chem. B*, vol.105, pp.743-746, 2001.

[27] P. Kim, and C.M. Lieber, "Nanotube nanotweezers," *Science*, vol.286, pp.2148-2150, 1999.

[28] L.X. Dong, F. Arai, and T. Fukuda, "Shape modification of carbon nanotubes and its applications in nanotube scissors", in *Proc. of IEEE Int. Conf. on Nanotechnology (IEEE-NANO2002)*, Washington, U.S.A., Aug.26-28, 2002, pp.443-446.

[29] J. Kong, N.R. Franklin, C.W. Zhou, M.G. Chapline, S. Peng, K.J. Cho and H.J. Dai, "Nanotube molecular wires as chemical sensors", *Science*, vol.287, pp.622-625, 2000.

[30] J. Cumings and A. Zettl, "Low-friction nanoscale linear bearing realized from multiwall carbon nanotubes", *Science*, vol.289, pp.602-604, 2000.

[31] J. Cumings, P.G. Collins and A. Zettl, "Peeling and sharpening multiwall nanotubes", *Nature*, vol.406, p.586, 2000.

[32] L.X. Dong, F. Arai, and T. Fukuda, "Electron-beam-induced deposition with carbon nanotube emitters", *Appl. Phys. Lett.*, vol.81, pp.1919-1921, 2002.

[33] L.X. Dong, F. Arai, and T. Fukuda, "Three-dimensional nanoassembly of multi-walled carbon nanotubes through nanorobotic manipulations by using electron-beam-induced deposition," in *Proc. of the 1st IEEE Conf. of Nanotechnology (IEEE NANO2001)*, Maui, Hawaii, 2001, pp.93-98.

[34] L.X. Dong, F. Arai, and T. Fukuda, "3D nanoassembly of carbon nanotubes through nanorobotic manipulations", in *Proc. of the 2002 IEEE Int'l Conf. on Robotics & Automation (ICRA2002)*, Washington, U.S.A., May 11-15, 2002, pp.1477-1482.

[35] H.W. P. Koops, J. Kretz, M. Rudolph, M. Weber, G. Dahm and K.L. Lee, "Characterization and application of materials grown by electron-beam-induced deposition," *Jpn. J. Appl. Phys.*, vol.33-1(12B), pp.7099-7107, 1994.

[36] T. Hertel, R.E. Walkup and P. Avouris, "Deformation of carbon nanotubes by surface van der Waals forces", *Phys. Rev. B*, vol.58, pp.13870-13873, 1998.

Nano/Micro Technologies for Single Molecule Manipulation and Detection

Tza-Huei Wang[1,3] and Chih-Ming Ho[2]

[1]Mechanical Engineering Department and Biomedical Engineering Department
The Johns Hopkins University
[2]Mechanical and Aerospace Engineering Department
University of California, Los Angeles
[3]To whom correspondence should be addressed. E-mail: thwang@jhu.edu

Abstract -A sensitive, rapid, and efficient single-molecule detection method was developed by combining fluorescence correlation spectroscopy (FCS) and on-chip molecular manipulation techniques. The highly restricted measurement volume in the FCS system significantly reduces the intrinsic background noise and facilitates single-molecule sensitivity. A microchannel integrated with multiple 3-D electrodes was fabricated and used for molecular sensing and manipulation by which individual DNA molecules were fast transported to and sequentially focused within the tiny probing volume for measuring single-molecular information, thereby enhancing both the efficiency and rates of detection.

1. INTRODUCTION

Recent advance in optical imaging and fluorescence techniques has enabled the observation of dynamic behaviors of single molecules to measure the conformational information and molecular interactions at the level of individual molecules. These techniques offer an effective way to measure time trajectories of individual molecules and their conformational fluctuations so that otherwise hidden intermediate steps in the molecular interactions can be identified and heterogeneities among different molecules of a population can be understood. Consequently, new fundamental information can be obtained to facilitate exploration and study of important biological process such as protein folding, DNA/RNA binding, and gene expression. As only very limited signal intensity can be obtained from a molecule, downscaling of measurement volume for reducing intrinsic noise becomes a prerequisite for single-molecule detection so that the signal from a single molecule can surpass the background noise and be identified. Several detection methods including fluorescence correlation spectroscopy (FCS) [1], total internal reflection (TIR) microscopy [2], near-field scanning optical microscopy (NSOM) [3], and stimulated emission depletion (STED) [4] can provide an extremely small observation volume of about a femtoliter (1 fL=10^{-15}L) and have been used for single-molecule detection and measurement. Background radiation caused by Rayleigh and Raman scattering of the laser beam by the solvents is minimized due to the small number of solvent molecules present in such a small volume, so that the fluorescence bursts of single molecules that flow into the probing volume can be identified above the background.

Although the small probing volume in the aforementioned single-molecule detection methods largely reduces the background noise, it limits the detention efficiency (fraction of molecules in the analyzed sample that is actually detected) and the measurement rates. For example, the efficiency of single-molecule detection using the FCS technique to detect molecules in a conventional (50-100 μm i.d.) capillary columns is much less than 1%. The increasing application of single-molecule detection and its requirement of quick acquisition of molecular information have necessitated the development of highly efficient and rapid single-molecule measurement methods. In this paper, we demonstrated the development and implement of an electrokinetic molecular manipulation technique for single-molecule detection that significantly enhanced the detection efficiency and measurement rates.

Several techniques for cellular and molecular manipulation have been reported in the past few years, such as hydrodynamic stream focusing by creating a sheath flow [5], and molecular focusing by electric current [6]. The difficulties of using the hydrodynamic focusing and electric-current focusing techniques for single-molecule detection is the alignment of focused stream within the tiny probing region. Any slight difference in conductivity and flow rate between the two focusing channels may cause the stream to drift off center. Physical narrowing of microstructures [7] may also be applied to improve the detection efficiency of single-molecule detection. However, the rise of surface-to-volume ratio will enhance the adsorption of the analyte at the surfaces that interfere with the measurements, as well as cause molecules clogged in the limited flow passage.

2. METHODS

In this paper, we demonstrated an electrokinetic molecular focusing technique for FCS-based single-molecule detection by using 3-D electrodes [8]. In the FCS system, molecules are detected within a tiny focused illumination volume in the femtoliter range that is resulted from the pinhole rejection and light diffraction (Fig. 1).

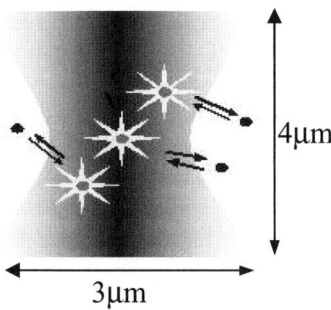

Figure 1 A focused laser illumination region of a FCS system. Molecules are excited and emit fluorescence within the femtoliter volume.

Fluorophore-labeled DNA molecules are excited by a focused laser beam. The subsequent shift of emission spectrum with respect to the excitation spectrum allows identification of presence of the target molecules. DNA molecules are loaded into a microchannel; they are then individually transported to and detected within a laser probing region downstream such that single-molecule information from a large population of molecules can be sequentially obtained. As shown in Fig 2, a pair of electrodes was fabricated in the inlet and outlet reservoirs respectively for transporting molecules using electrophoresis (EP). A set of three electrodes including two side electrodes and a middle electrode (Fig. 2) was fabricated in the middle of the microchannel to focus molecules to the laser probing region for measurements and to improve the detection efficiency. In the design, electric field is produced toward to probing region by applying proper potential between the two side electrodes and the middle electrode. Before DNA molecules passing through the laser probing region they are guided towards the middle electrode for detection by the applied electric field. Since the molecules are precisely focused to the downstream end of the middle electrode, which is designed as the focal region of the FCS system, more passing molecules can be individually sensed. This method overcomes the off-center problem encountered by the hydrodynamic and electric current focusing methods.

3. DESIGN AND FABRICATION

In order for preventing clogging and for minimizing the surface adsorption that interferes with measurements, the cross-sectional dimension of the microchannel used for the FCS-based single-molecule detection needs to be much larger than the laser probing volume (diameter of ~3 μm and depth of focus of ~4μm in the system used). The microchannel was designed as 120 μm-wide and 30 μm-deep. The laser probing region was focused close to the middle electrode. Consequently, the optimal design of molecular focusing is to make electrodes that can generate more uniform and stronger

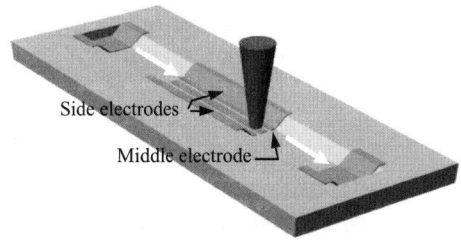

Figure 2 Conceptual schematics of a microchannel embedded with multifunction electrodes. A pair of electrodes in inlet and outlet electrodes are used for molecular transport. A set of three electrodes (two side electrodes and a middle electrode) are used for molecular focusing.

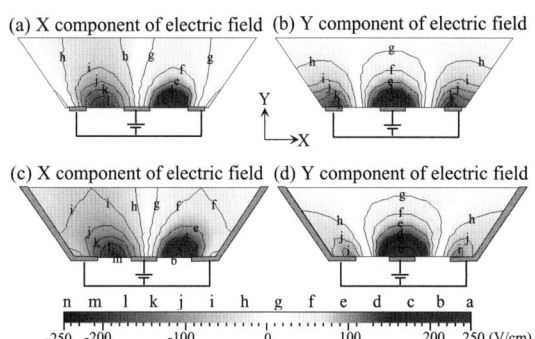

Figure 3 Simulation of electric fields generated from 2-D electrodes((a) and(b)) , and from 3-D electrodes((c) and (d)). The applied voltage set for simulation is 1V.

fields inside the microchannel and toward the tiny probing region to minimize the low-field regions, therefore maximize the number of molecules being detected.

We designed both 2-D electrodes and 3-D electrodes and compared their resulting electric field distributions. In the 2-D design (Fig. 3(a), (b)) whose electrodes are fabricated on bottom of a microchannel, the generated fields in the regions away from the middle electrode in the channel are too weak to efficiently transport molecules to the probing region (Fig. 3(a)). Negative charged DNA molecules near the side-walls and upper corners do not expose to higher enough electric fields to move them to the middle region, where upward electric fields are generated to push them further downward (Fig. 3(b)). As a result, most of them can not be transported toward the middle electrode for detection. However, in the 3-D design (Fig. 3(c), (d)) that contains two side electrodes covering the side-walls and a middle electrode on the bottom of the microchannel. Stronger electric fields are more uniformly and widely created from the middle to the side-electrodes, minimizing the ineffective regions in the design of the 2-D electrodes. As shown in Fig. 3 (c) and (d), for the 3-D design all DNA molecules (negative charged) except those in middle region experience outward fields and are electrophoretically moved inward to the middle region where upward fields are produced (Fig. 3(d)). Consequently, they are transported downward to the bottom and are focused to middle of the middle electrode.

Another two electrodes were made in the inlet and outlet reservoirs to move DNA molecules in the microchannel using EP forces (Fig. 2). Since molecules in upper regions experience lower electric fields (Fig. 2(a), 2(c)), it takes longer time for them to reach the probing region that is aligned on the downstream end of the middle electrode. Some of them may have already flown away before they are moved to the probing region, and are not able to be detected. Therefore, to have higher detection efficiency the focusing electrodes need to be extended forward enough to allow sufficient focusing time for the molecules to reach the probing region and be detected. The length of the focusing electrodes needs to be designed according the magnitude of applied focusing fields, the ratio between strengths of moving electric fields and focusing electric fields, and the geometry of the electrodes. As the distance between electrodes is small and ranges from only 20 to 100 µm, even the applied voltage is as low as 1 V the generated electric fields can be greater than 100 V/cm. Since low voltage is applied, the focusing can be implemented without worrying the bubble generation problems due to electrolysis.

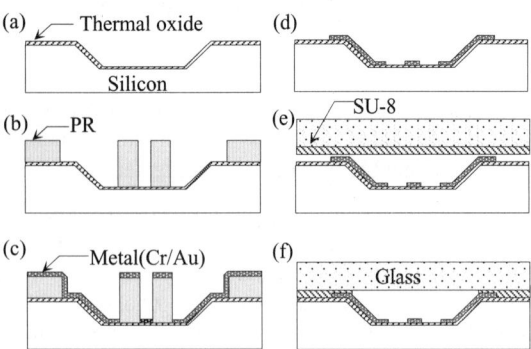

Figure 4 Process flow of a molecular focusing chip with 3-D electrodes.

Silicon wafers were chosen as substrate materials for fabricating microchannels because smooth and tapered side-walls can be formed and better metallic coverage can be achieved by the subsequent deposition of metallic thin films [9]. As shown in Fig. 4, a trapezoid microchannel was fabricated by anisotropic etching using KOH etcher. The channel width on top is 120 µm, the depth is 30 µm. After thermal silicon oxide was grown to a thickness of 5000Å for electric insulation, a 200 Å /2000Å Cr/Au layer was deposited by e-beam evaporation. To pattern 3-D Cr/Au electrodes on the top and bottom of the channel 10 µm thick PR, AZ4620, was over-exposed and developed for lift-off [9]. The width of the middle electrode was fabricated to be 20 µm with 20 µm space from the side electrodes. Pre-drilled Pyrex glass plates were spin-coated with SU-8 and cured at 85 ^{0}C for 3 minutes before bonding. The channel chips were then bonded with the glass plates, and the bonded chips were further cured at 95 ^{0}C for 10 minutes to get rid of excess solvent for UV exposure and patterning for clearing purposes. A picture of 3-channel array on a chip is shown in Fig. 5(a), and a SEM picture of a microchannel with a set of 3-D focusing electrodes is shown in Fig. 5(b).

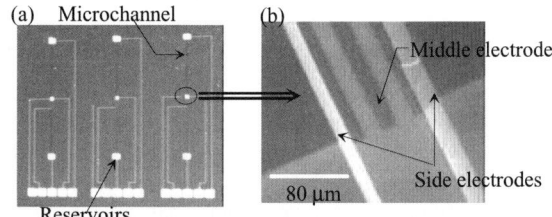

Figure 5 (a) A picture of the molecular focusing chip with three channels, (b) A SEM picture of the 3-D focusing electrodes.

4. EXPERIMENTAL SETUP AND SAMPLE PREPARATION

A fluorescence microscopic system is used to record the time trajectories of the analyte and to characterize the molecular focusing effects. A FCS system is then implemented to perform the single-molecule detection and to determine the improvement of detection efficiency.

In the fluorescence microscopic system, a fluorescence microscope (Olympus IX70) that has large pixel sizes was used for the focusing experiment. The microscope was equipped with a 100 W mercury lamp and a 100X oil immersion objective with N.A. 1.25 (Olympus) and was used to observe the trajectories of DNA molecules and fluorescence particles. The time-dependent images of the molecules and particles were recorded to video by an intensified CCD camera (Vedioscope ICCD 350F) and digitized with a video capture card (Truevision TARGA 1000).

In the FCS system (Fig. 6), a light beam (0.15 mW, 488nm) from an air-cooled Ar ion laser (Melles Griot, 35LAL415-220) passes into a beam expander (Melles Griot, 09LBZ010) and a band pass filter (Omega, XF1073). It then reflects from a dichroic beam splitter (Omega, Xf2037) to a 50X, 0.90 N.A. oil immersion objective (Olympus), which focuses the beam to a 1 mm spot within the channel. Fluorescence is collected by the same objective, passes through a dichroic beam splitter, filtered by a bandpass filter (Omega, XF3003), focused by a focusing lens (Newport, PAC052), focused through a 100 mm pinhole and finally collected by an avalanche photodiode (Perkin Elmer, SPCM-AQR). Amplified TTL level pulses were counted by a PC plug-in board (PC100D, Advanced Research Instruments) and stored. A program written in C was make to perform data acquisition and to set bin width (integration time per data point) of counting. Bin widths could be set to power of 2, eg. 64, 128, or 256 ms per bin. A correlation function is programmed and used for characterizing the kinetic information of molecules from the single-molecule measurements.

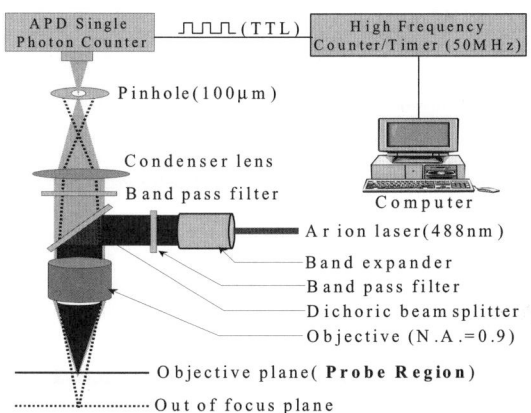

Figure 6 Schematics of a FCS setup used in the experiments

Both micro Carboxylate-modified Latex (CML) spheres and DNA molecules were used for determining the focusing capability of the developed microchannel with 3-D electrodes. The CML (Carboxylate-modified Latex) spheres with diameter 1 μm were purchased from IDT (Portland, Oregon). They were carboxylate-modified with negative surface charges and had been pre-loaded with a yellow-green fluorescence dye that allowed single-particle observation by fluorescence microscopy. In order to capture the trajectories of individual particles, the spheres were diluted with DI water to a final concentration of ~ 3×10^8 spheres/mL for the experiment. The DNA molecules used for the experiments include λ DNA digest (6 different DNA fragments: 3530, 4878, 5643, 5806, 7421, 21226 bp) and λ DNA (48 Kbp) and were purchased from Sigma. DNA sample was diluted in 1X Tris-borate buffer (TBE) to the desired concentrations and was stained with YOYO-1 iodide (Molecular Probes, Y3601). The YOYO-1 solution was diluted in DI water to the desired concentration before mixing with the DNA. The base pair : dye molecule ratio was kept as 5:1 for all the DNA solution to have better signal to noise ratio. The resulting solution was incubated for 30-60 min at room temperature in the dark and diluted in TBE to give a final concentration for the single molecule detection experiments.

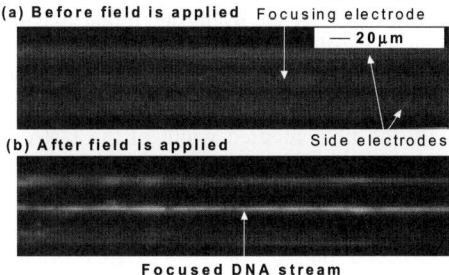
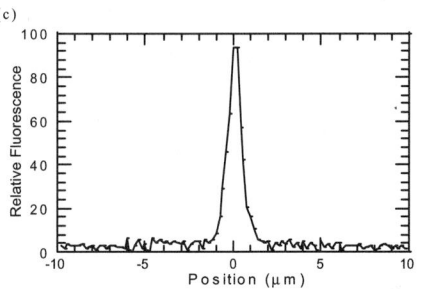

Figure 7 (a) and (b) CCD images of l DNA digest solution in a quasi-static condition, (c) Fluorescence intensity profile across the microchannel in the middle of the channel in Fig. 8(c)

5. RESULTS AND DISCUSSION

A microscopy coupled with a intensified CCD camera is first applied for characterizing DNA focusing in the microchannel embedded with 3-D electrodes before the tests of FCS single-molecule detection are performed.

Fig. 7 shows the result of applying potential to focus λ DNA digest (2×10^{-11} M) in a microchannel in a static flow condition. The upper width of the microchannel was 120 μm and the depth is 30 μm. Both the width of the focusing electrode and the distance between the focusing and side electrodes are 20 μm. In the fluorescence pictures (Fig. 7(a)), the three bright strips represent the aluminum electrodes because of their high fluorescence background. The upper and lower strips are the side electrodes while the middle strip represents the focusing electrode. Before applying the focusing electrical fields, the DNA molecules were uniformly distributed in the microchannel, resulting in a dim and wide DNA stream (Fig. 7(a)). After the potential was applied, the DNA molecules were confined inward and downward in the microchannel, forming a much brighter and narrower stream in the middle of the focusing electrode (Fig.7 (b)). Fig. 7(c) shows the fluorescence intensity profile across the microchannel in the longitudinal middle of the electrode-patterned region that was measured according to the image of Fig. 7(b). The stronger background fluorescence reflected from the Al electrodes was subtracted from the initial intensity to create a uniform background. The resulting profile is then normalized in order to compare the relative single levels. The fluorescence peak represents the focused DNA stream with the final focused width of the DNA stream around 2μm. To avoid polarization occurring on the electrodes, an AC voltage was superimposed on a DC potential during the electromolecular focusing. A resulting 10 KHz pulsed voltage V_{pp}=1.4V with 0.7 V offset was applied for the experiment. The time series pictures in Fig. 8 shows that the 1 μm CML particles (3×10^8/ml) were gradually transported to the focal plane and were confined and lined up in the middle of the focusing electrode in a quasi-steady condition after a 100 KHz pulsed focusing potential, V_{pp}= 1.0 V with 0.5 V offset, was applied. Due to the longer fluorescence bleaching time of CML particles, it is possible to record the trajectory of a single particle with consistent fluoresce intensity within a four-second focusing period.

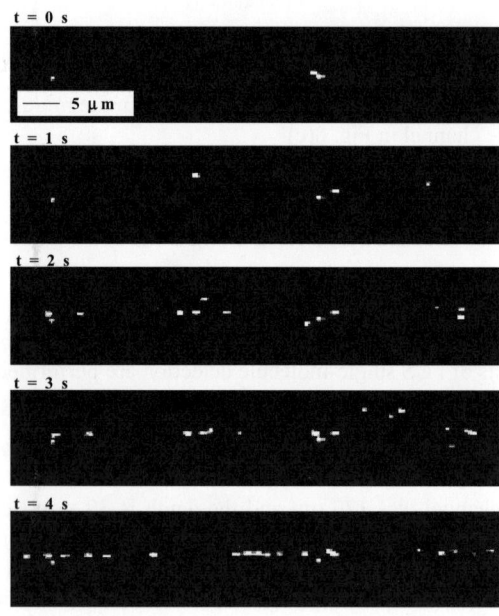

Figure 8 Trajectories of 1 mm CML particles after a focusing voltage is applied. The particles are lined up in the middle of the microchannel

To characterize the improvement of detection efficiency for single-molecule detection by using the same focusing chips, the FCS system equipped with an APD based single-photon counter was used. The e^{-2} probing volume in the system was approximated to be a cylinder 3.0 μm in diameter and 4 μm in height that gives a volume of 28 fL. A more diluted λ DNA solution, 10^{-14} M stained with YOYO-1 was flowed through the microchannel. When DNA solutions were introduced, discrete fluorescence bursts were seen due to the passage of individual DNA molecules through the focused laser beam. The probability of more than one DNA molecule simultaneously occupying the probing volume can be calculated using the concentration (10^{-14} M) and the probing volume of the implemented confocal LIF (28 fL).

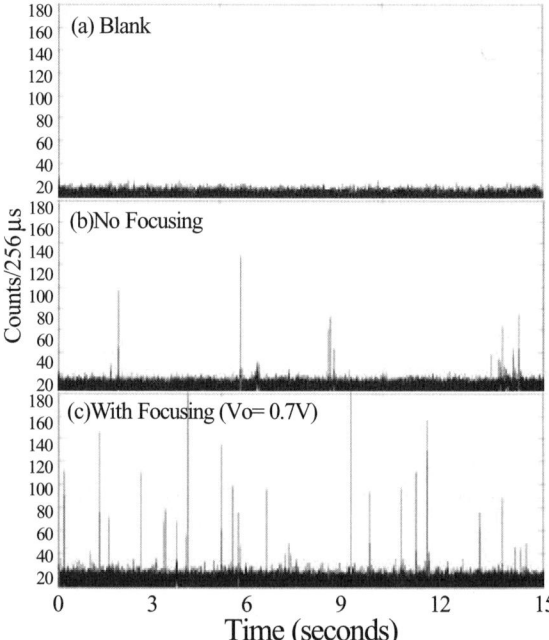

Figure 9 Fluorescence bursts in the detection of single λ DNA molecules.

Since the calculated probability is as low as 1.68×10^{-4}, the observed fluorescence bursts can be attributed to single molecules of DNA.

Fig. 9 presents a fifteen-second span of single molecule bursts events while DNA passing through the probing volume. The time bin for each counting was set as 256 μs. When the blank solution was flowed through the channel, no large fluorescence bursts were observed (Fig. 9(a)). When DNA solution was introduced, discrete burst events began to be seen. The single molecule bursts were rarely observed without proper molecular focusing (Fig. 9(b)). After the electric field was applied to concentrate the DNA molecules in the probing region, the frequency of the single molecule bursts was greatly increased (Fig. 9(c)). The autocorrelation functions were calculated to demonstrate the presence of non-Poissonian bursts due to single molecules (Fig. 10). The formula used to calculate the autocorrelation function was $G(\tau)=(1/N)\Sigma n(t)n(t+\tau)$, where G is the autocorrelation, N is the size of the data set, n is the value at time t, and τ is the offset. The autocorrelations in figure show that the magnitude of the autocorrelations increases when the electric focusing fields were applied. Based on the experimental data, the average improvement of detection efficiency can be more than 5-fold. Further improvements can be achieved by optimizing experimental factors such as the geometry of channel and electrodes, and electric conditions.

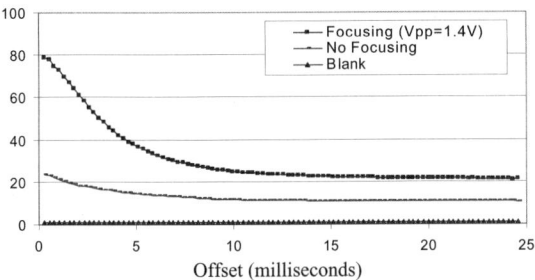

Figure 10 Autocorrelation functions calculated from the λ DNA solution.

6. CONCLUSION

Application of an electrokinetic molecular manipulation technique to FCS-based single-molecule detection was demonstrated to be an effective way to improve detection efficiency and to enhance rates of molecular information acquisition. Multiple electrodes were integrated in a microchannel both to rapid transport individual molecules for measurements and to focus molecules to the tiny measurement region for increasing the detection efficiency. Using a 3-D electrode design for the molecular focusing, it was demonstrated that flowing DNA fragments in a 120 µm-wide microchannel was focused to 2 µm-wide stream that is smaller than the probing size in the FCS system. In the single-molecule detection experiments, the frequency of detection (molecular events per second) was largely enhanced that shows both detection efficiency and measurement rates were improved.

6. REFERENCES

[1.] C. Zander, J. Enderlein, R.A. Keller, "Single Molecule Detection in Solution : Methods and Applications", Wiley-VCH, Berlin, 2002

[2.] T.E. Starr. N.L. Thompson, "Total internal reflection with fluorescence correlation spectroscopy: combined surface reaction and solution diffusion", Biophys J. 2001 Mar;80(3):1575-84.

[3.] F. de Lange et al., "Cell biology beyond the diffraction limit: near-field scanning optical microscopy", J Cell Sci. 2001 Dec;114(Pt 23):4153-60.

[4.] T. A. Klar, " Fluorescence microscopy with diffraction resolution barrier broken by stimulated emission", Proc Natl Acad Sci U S A. 2000 Jul 18;97(15):8206-10.

[5.] A. Castro and F.R. Fairfield, and E.B. Shera," Fluorescence Detection and Size Measurement of Single DNA Molecules, Anal. Chem. 69, 3915-3920 (1997)

[6.] S. C. Jacobson and J. M. Ramsey, "Electrokinetic Focusing in Microfabricated Channel Structures",Anal. Chem. 69, 3212-3217 (1997)

[7.] Y. H. Lee, R. G. Maus, B. W. Smith, J.D. Winefordner," Fluorescence Detection of a Single Molecule in a Capillary", Anal.Chem. 66, 4142-4149 (1994)

[8.] T. H. Wang, P. K. Wong, C. M. Ho, "Electrical Molecular Focusing for Laser Induced Fluorescence Based Single Molecule Detection", 15th IEEE International Conference on Micro Electro Mechanical System, MEMS 2002, 15-18 (2002)

[9.] T.H. Wang, S. Masset, and C. M. Ho, " A Zepto Mole DNA Microsensor", 14th IEEE International Conference on Micro Electro Mechanical System, MEMS 2001, 431-434 (2001)

Platform Technology for manipulation of Cells, Proteins and DNA

Gwo-Bin Lee, Long-Ming Fu
Department of Engineering Science
National Cheng Kung University, Tainan, Taiwan

Abstract — MEMS has been an enabling technology for biomedical applications recently. Not only does it provide an instrument to obtain information on molecular level, but it also allows us to manipulate bio-molecules efficiently. Micro devices and systems for manipulation of biological objects such as cells, proteins and DNA have been demonstrated. In this study, we report a technique using dielectrophoretic forces to manipulate micro particles. First the concept of several manipulation techniques will be reviewed. Then we will focus on novel manipulation modes. With the help of these biochips, we are able to perform manipulation of micro particles/cells.

I. INTRODUCTION

Recently, MEMS (Micro-electro-mechanical-systems) technologies and micromachining techniques have been popular for biomedical applications. Not only does it allow an access of information on molecular level, but it also provides new instrumentation for bio-analytical applications. In fact, it has made substantial impacts and attracted lots of interests in the field called "Bio-MEMS". Small bio-particles and cells could be collected, manipulated and separated using a micromachined chip [1]. Furthermore, micro devices dealing with several operation procedures such as sampling, injection, mixing, separation, detection could be integrated on a chip such that a micro total analysis system (μ-TAS) could be realized [2].

Using ac electrokinetic techniques to manipulate and separate bio-particles has been proved to an efficient way for bio-medical applications. For example, non-uniform alternating electric fields could be used to induce motion in polarizable particles/cells by generating so-called dielectrophoresis (DEP) forces [3]. Since most biological cells and macromolecules behave as dielectric particles in the external alternating electric fields, DEP has been found to have many useful biological applications such as particle separation, levitation, manipulation and characterization. Dielectrophoretic separation and trapping of sub-micrometer-scale bio-particles were also feasible [4].

In addition to simple electrode layout, a series of bar-shape electrodes could generate a so-called travelling-wave dielectrophoretic (twDEP) force. Masuda *et al.* [5] reported the principles of the travelling electric field and showed that it could be used to induce controllable motion of bio-particles. Red blood cells were successfully manipulated by applying a three-phase driving voltage at a low frequency (0.1-100 Hz) using a series of bar-shaped electrodes. Later on, Fuhr *et al.* [6] showed that a high-frequency traveling electric field was capable of generating a twDEP force and inducing a linear motion of pollen and cellulose particles. Similarly, Huang *et al.* [7] reported that the twDEP force could be used to manipulate yeast cells. A mixture of bacteria and cells could be separated successfully using the twDEP forces. Moreover, the twDEP devices could be integrated with optical detectors to form an integrated microsystem capable of particle separation and on-line detection. Thus, a so-called "lab-on-a-chip" could be realized [8].

The objective of this study is to design and fabricate a twDEP chip for the manipulation of micro-particles and cells. The optimum operation of the traveling electric field to control the trace of the cells was explored systematically. Using numerical simulations, it is possible to determine key parameters for controlling the motion of the bio-particles. Several crucial operating parameters such as electric potential distributions and dielectrophoretic forces were studied. Furthermore, the trace of the cells could be also investigated. A simple and reliable micromachining technique was utilized to fabricate the microchips to verify the performance of the twDEP chips. A new packaging method was also presented. At last, several innovative operation modes to manipulate micro-particles and yeast cells were demonstrated. Details regarding design, fabrication and operation of the twDEP chips will be discussed in the following sections.

II. MATERIALS AND CHIP FABRICATION

The twDEP electrodes were fabricated on commercially-available microscope glass slides with dimensions of 80×30×1 mm³ (Marienfeld, Germany). Prior to microfabrication of the glass chips, the slides were first annealed at 400℃ for 4 hours to relieve their internal residual stress. Photomasks to define bar-shaped electrodes were generated using a layout software (AutoCAD) and were printed on a transparent film using a high-resolution laser printer (10,000 dpi). All chemicals used in this study were regent grade (J D Baker, Philadelphia, USA) and process solutions were prepared using deionized (DI) water.

For micro-particle experiments, the suspension medium (NaCl solution in 280 mM mannitol) was adjusted to have a conductivity ranging from 0.2 mS/m to 1.6 S/m. Polystyrene latex beads used for the particle-manipulation experiments were purchased from Interfacial Dynamics Corporation (Portland, USA) and Duke Scientific Corporation (California, USA). The size of the beads ranges from 1.1 to 20 μm. Yeast cells of Candida albicans (Culture Collection and Research Center, Taiwan) were prepared for 48 hours at 30℃ in a culture medium. The size of the yeast cells sizes ranges from 1 to 6 μm.

A simplified fabrication process of the chip is shown in Fig. 1 and described as follows. Detail information could be found in Ref [9]. Starting with standard wafer cleaning process (Fig. 1a), glass substrates were immersed in a Piranha solution (H_2SO_4 (%) : H_2O_2 (%)= 3:1, 120℃) for 10 min, then rinsed in DI water and blown-dry with

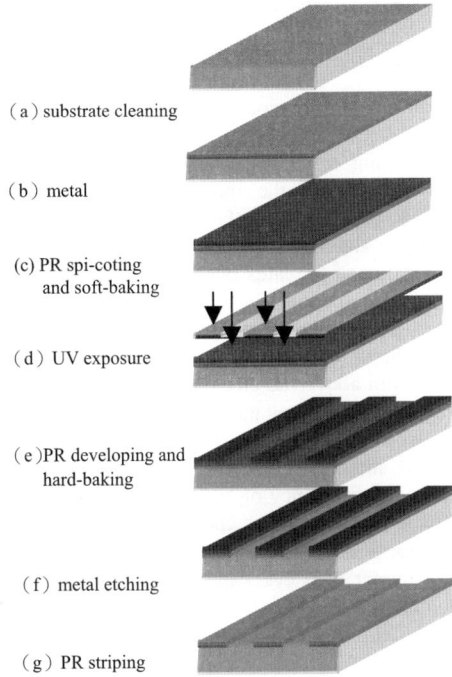

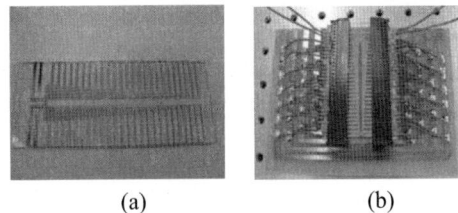

Fig.2 The photograph of (a) the travelling-wave dielectrophoresis chip and (b) a packaged micro cell chip system.

nitrogen gas. An adhesion layer of 0.02-μm chromium (Cr) was first deposited by E-beam evaporation onto the glass substrate, followed by deposition of a 0.2-μm layer of gold (Fig. 1b). Then the bar-shaped array microelectrodes were patterned using standard photolithography and metal etching process (Fig. 1(c)-(f)). At last, photoresist layer was striped by KOH solution at 50 ℃ (Fig. 1f). Electrode array with layout patterns were finally formed using the above-mentioned process. In this study, an upper PDMS plate with a micro channel (80 μm in width and 100 μm in depth) was used to cover the cell chips. Detail information of PDMS curing and de-molding process could be found in Ref. [10]. The PDMS plate was oxygen-plasma-treated and then bonded with the glass substrates.

Figure 2(a) shows a photograph of the micro twDEP chip. The detail information regarding electrode geometry could be found in Fig. 3(a). Basically, each electrode is 30 μm in length, 10 μm in width and 0.2 μm in thickness. The separation between the electrodes on the same side is 10 μm. The microelectrode arrays with different pitches and geometry were designed and fabricated to evaluate the

Fig.1 Fabrication process of the micro cell chips on glass substrates

performance of the chips. A simple and reliable method for packaging of the microchips using standard PCI (peripheral component interconnect) slots has been used. The bonding pads on the twDEP chips were patterned to have the same pitch of 0.7 mm as that on the PCE slots such that time-consuming wire bonding procedures could be eliminated (Fig. 2b).

III. THEORY

A. Potential field

Figure 3 depicts a portion of the electrode arrays showing the geometry and the applied electrical field. Note that only an area of 80 μm x 80 μm is shown in Fig. 3. Sinusoidal excitation signals with a peak magnitude of 10 V and sequential phase changes are also shown in this figure. Neighboring electrodes on the same side have a phase lag of 90° and electrodes opposite to each other have a phase lag of 180°. Therefore, the particles were observed to move along the center of the channel in a direction opposite to that of the traveling electric wave.

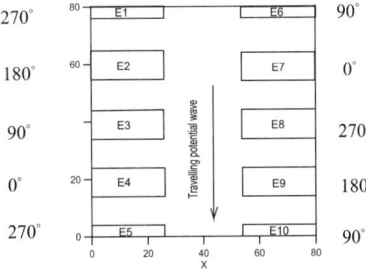

Fig. 3 The basic electrode geometry and phase sequences applied on the electrodes.

The instantaneous potential fields around the electrodes were obtained by solving the Laplace equation, which could be represented as follows.

$$\frac{\partial^2 \phi}{\partial x^2} + \frac{\partial^2 \phi}{\partial y^2} + \frac{\partial^2 \phi}{\partial z^2} = 0 \quad (1)$$

With known boundary conditions for ϕ, equation (1) was solved by using the Point Successive Over-Relaxation method (PSOR) iterative finite-difference technique [11]. Therefore, 3-dimesional distribution of the electric field in the DEP chip could be obtained.

B. Force field and Travelling-wave dielectrophoresis (twDEP)

For a cell, a dipole moment $m(t)$ can be induced by the external electric fields. Three spatial components can be defined as follows [12]:

$$m(t) = m_x(t)a_x + m_y(t)a_y + m_z(t)a_z \quad (2)$$

where a_x, a_y and a_z are the unit vectors in the x, y, and z directions, respectively. The $m_x(t)$, $m_y(t)$ and $m_z(t)$ are the magnitudes of the induced dipole moment in each corresponding direction. Taking $m_x(t)$ for example, the components of the dipole moment have the following form:

$$m_x(t) = 4\pi\varepsilon_m r^3 [\cos(\omega t + \varphi_x)\mathrm{Re}(f_{CM}) \\ - \sin(\omega t + \varphi_x)\mathrm{Im}(f_{CM})]E_{x0} \quad (3)$$

where ε_m is the absolute permittivity of the medium, r is the radius of the cell, ω is the angular frequency of the voltages applied on the electrodes. E_{x0} and φ_x are the magnitude and the phase angle of the x component of the electric field. The terms Re and Im are the real and imaginary components, respectively, of the Clausius-Mossotti factor f_{CM}, which is defined as

$$f_{CM} = \frac{\varepsilon_p^* - \varepsilon_m^*}{\varepsilon_p^* + 2\varepsilon_m^*} \quad (4)$$

where ε_p^* and ε_m^* are the complex permittivities of the particle and the suspending medium, respectively. The time-dependent dielectrophoretic force acting to the particle is given by:

$$F(t) = (m(t) \cdot \nabla)E(t) \quad (5)$$
$$= F_x(t)a_x + F_y(t)a_y + F_z(t)a_z$$

where

$$F_x(t) = m_x(t)\frac{\partial E_x(t)}{\partial x} + m_y(t)\frac{\partial E_x(t)}{\partial y} + m_z(t)\frac{\partial E_x(t)}{\partial z} \quad (6)$$

Following the same procedure, similar formulae for the force components $F_y(t)$ and $F_z(t)$ can be derived. Summing up all the force components and assuming that the phase factor on the electrodes array is a constant ($\nabla\varphi = 2\pi/\lambda$), the dielectrophoresis force acting on a particle can be represented as

$$F(t) = 2\pi\varepsilon r^3 [\operatorname{Re}(f_{CM})\nabla E^2 + (\frac{2\pi}{\lambda})\operatorname{Im}(f_{CM})E^2] \quad (7)$$

In equation (7), there are two different forces acting on a cell. For the force term ($\operatorname{Re}(f_{CM})$), it pulls or pushes the cell towards the strong or weak field regions, depending on whether $\operatorname{Re}(f_{CM})$ is positive or negative. For the force term ($\operatorname{Im}(f_{CM})$), it moves the cell along the channel, which is related to the DEP forces.

C. Velocity of a moving cell

A particle moving in a fluid experiences a drag force and the dielectrophoretic forces simultaneously under ac electrokinetic fields [13]. At equilibrium conditions, the traveling-wave dielectrophoresis force is balanced by the viscous drag, which could be represented by Stoke's equation. Therefore, considering travelling-wave dielectrophoresis in a medium of viscosity η, the velocity (u) of a cell travelling along the array of electrodes can be represented as follows

$$u = -\frac{2\pi\varepsilon_m r^2}{3\lambda\eta}\operatorname{Im}(f_{CM})E^2 = F(t)/6\pi\eta \quad (8)$$

It can be clearly seen that the velocity of the cell is a function of the cell radius, the strength of electric field, wavelength of the travelling field, the viscosity of the medium and the imaginary part of the Clausius-Mosotti factor. In this study, the velocity of a cell will be numerically simulated first. Then experiments using array electrodes with different driving voltages will be conducted to verify the capability of the cell chips.

IV. RESULTS AND DISCUSSION

As described in the previous section, the difference of the conductivity between the suspending medium and the particles and driving frequency of the applied travelling electric field are two of the most parameters for the manipulation of the particles. For the experiments of moving particles, values of 2.55 and 78 were used for the relative permitivities of the latex beads and the suspending medium (NaCl). The conductivity of the latex beads and suspending medium is 10 and 1 mS/m, respectively.

Figure 4 shows electric potential contours in X-Y plane, located 3 μm above the electrode surface, generated using a sinusoidal driving voltage as shown in Fig. 3. The separation between electrodes is 10 μm and an area of 80 μm x 80 μm is simulated. The driving voltage is 10 V peak-to peak with a frequency of 50 kHz. In this study, the average radius of the particles is chosen to be 3 μm. Note that each figure in Fig. 4 have a 30° phase (phase factor β) shift and one quarter of a cycle is shown. In Figure 4, a continuous animation of such plots through quarter a cycle reveals that the field minima are located at the central part of the channel, travelling in synchrony with the applied travelling potential wave, whereas the field maxima at the electrode edges travel in the opposite direction (against the travelling potential wave as shown in Fig. 3).

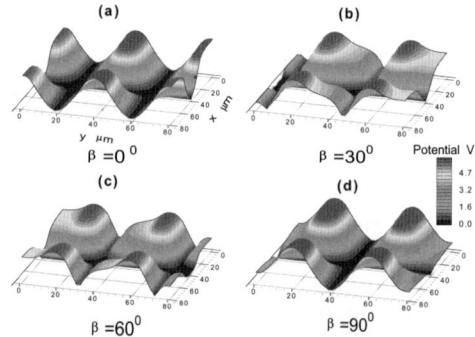

Fig.4 The electric potential contours in X-Y plane, located 3 μm above the electrode surface, which are generated as a sinusoidal driving voltage applied on the electrodes propagating with a 30° step for a quarter of one cycle.

The frequency window (*f*1) around 50 kHz with a medium conductivity of 15 mS/m shown in Fig. 5(a) indicates that latex beads can experience the combination of a negative dielectrophoretic force (see Fig. 5(a)) and a significant travelling-wave-dielectrophoresis force. The fact that the $\operatorname{Im}(f_{CM})$ of the induced dipole moment is positive also indicates, based on equation (8), that the latex beads will travel in a direction opposite to that of the applied electric travelling field. This founding is consistent with the experimental data shown in Fig. 6, obtained at a driving frequency of 50 kHz and suspending medium conductivity of 15 mS/m. Moreover, again in agreement with the experiment, when the conductivity of the suspending medium is reduced to 0.2 mS/m and a frequency range (*f*2) for travelling particles around 1~10 MHz is used (see Fig. 5(a)). Figure 5(b) shows distributions of twDEP forces with a driving frequency of 50 kHz. It is noted that the particles will travel against the direction of the travelling electric field along the central region of the channel between electrode arrays.

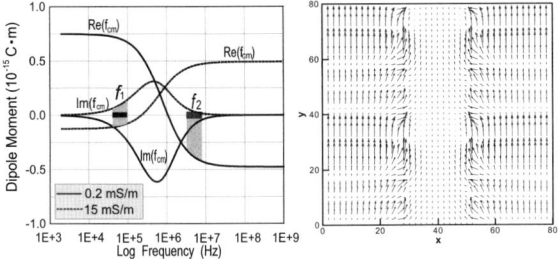

Fig.5 (a) Plots of the real and imagine parts of Clausius-Mossotti factors for latex beads in suspending medium with different conductivities and (b) travelling-wave dielectrophoresis (twDEP) force.

Then Crank-Nicolson method [14] was used to predict the trace of the micro particles from velocity fields (dx/dt=u). The velocity field can be obtained from equation (9) with a frequency of 50 kHz and a medium conductivity of 15 mS/m. There are 10 traces calculated while micro particles are injected at different locations at initial time t. At time $t + \Delta t$, the micro particle moves a distance $\Delta t \times u$ in the appropriate direction. The velocity u here is obtained by linear interpolation among the surrounding grid points. This is done at both time t and $t + \Delta t$ and then the two values are time-averaged to obtain the velocity of the micro particle over the time span Δt. Figure 6(a)~6(c) shows experimental data about the trace of micro particles injected at several locations. The tested particles are polystyrene beads with a diameter of 5 μm. The numerical simulation and experimental data are consistent as shown in Fig. 6(d).

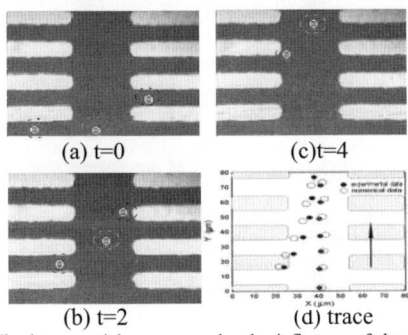

Fig.6 The latex particles move under the influence of the twDEP force along the central region of the channel between opposite electrode arrays, with a driving voltage of 10 V peak-to-peak and a frequency of 50 kHz The conductivity of the medium is 15 mS/m.

After the electric fields, the force distributions and the trace of the micro particles were numerically simulated, we then focused on new manipulation modes of the micro particles. In addition to pulling, pushing and traveling of the micro particle, two innovative manipulation modes have been proposed in this study. With the same electrode geometry shown in Fig. 3, some interesting manipulation modes could be achieved. The applied driving frequency is 50 kHz, the conductivity of the suspending medium is 20 mS/m and each electrode is applied 10 V peak-to-peak except the electrode E3 (E3 electrode is off, other electrodes are on) in this time span. At next time span Δt, the off electrode was switched to E4 and other electrodes were turned on, resulting in a negative dielectrophoretic force as shown in Fig. 7. Figure 8(a) shows electric potential distribution of this manipulation mode with the electrode E3 off. The electric potential shows a symmetric distribution near the "off" electrode. While one of four electrodes was switched off, the region around the electrode had the lowest electric potential field. Since the dielectrophoretic force is proportional to square of the strength of the electric filed, it implies that the other electrodes have strong negative DEP forces, repelling particles towards this "off" electrode. The force distribution of this manipulation mode could be found in Fig. 8(b). Using this concept and switching electrodes off in sequence, one can move particles step by step. The velocity of the moving particles could be then controlled by the driving frequency. Figure 9 shows experimental data of the manipulated micro particles. Note that micro particles on the same electrode move at the same time and stay on the same electrode.

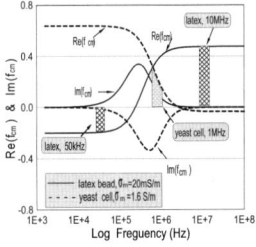

Fig.7 Plot of the real and imagine parts of Clausius-Mossotti factors for latex beads and yeast cells in suspending medium with different conductivities

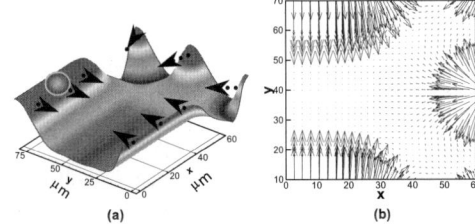

Fig.8 (a) The electric potential distributions, and (b) force vector for yeast cells travelling step by step.

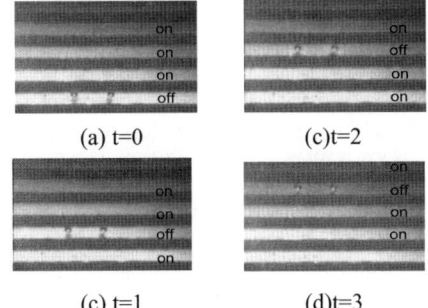

Fig.9 Experimental data for yeast cells travelling step by step

Another interesting mode to manipulate micro particles is to use the positive dielectrophoretic forces. The particles were manipulated at higher driving frequencies (>1 MHz, see Fig. 7) than ones for the negative dielectrophoretic forces. In this case, only one electrode was turned on, leaving the other three electrodes off. It was found that the particles were attracted from the lowest potential area to a higher potential area, where the electrode was turned on. When the electrode was switched on in sequence, the particles moved accordingly. For example, the electrodes E3, E6 and E10 (see Fig. 3) were switched on and other electrodes were switched off. The particles were attracted from the lowest potential area to the higher potential area where the electrodes were turned on. Figure 10 (a) represents electric potential contours of this manipulation mode with the electrodes E3, E6 and E10 on. In this case, the applied frequency is 10MHz, the conductivity of the suspending medium is 20 mS/m and each electrode is applied 10 V peak-to-peak. The potential field at three "on" electrodes is higher than that at "off" electrodes. Thus, the magnitudes of the positive dielectrophoretic forces are bigger at these "on" electrodes (Fig. 10(b)). Using this concept, the particles could move back and forth while the electrodes were switched on in sequence as shown in Fig. 11.

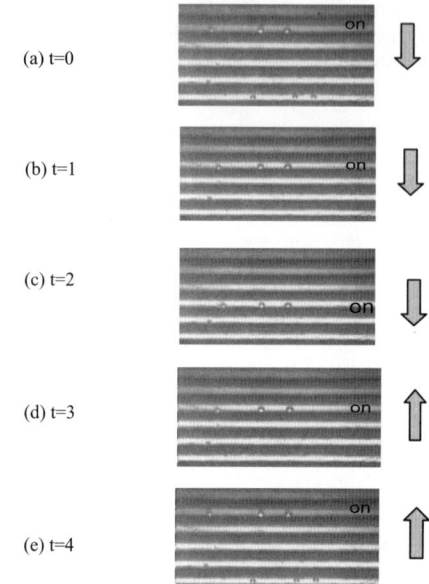

Fig.11 Experimental data for yeast cells travelling back and forth

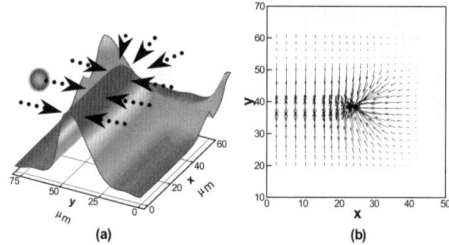

Fig.10 (a) The electric potential distributions, and (b) force vector for yeast cells travelling back and forth.

In addition to the latex beads, bio-objects, yeast cells, were also used to demonstrate the capability of the cell chips. For the manipulation of yeast cells, the positive DEP forces (see Fig. 7) could be used to collect and align yeast cells. Figure 15 shows that yeast cells could be collected on one electrode and aligned to a fine line while a driving voltage of 10 V peak-to-peak with a frequency of 1 MHz was applied. Relative permitivity of the yeast cell is 70 and conductivity of the suspending medium is 1.6 S/m. In this manipulation, before the power was switched on, the yeast cells were distributed randomly (see Fig. 12(a)). While the power was turned on and the frequency is smaller than 1.5 MHz, the yeast cells were centralized on the electrodes and linearly aligned at the center of the electrode (see Fig. 12(b)) by the positive DEP force. Under a high-frequency (lager than 2MHz, see Fig. 7) driving mode, all the yeast cells were attracted to the edges of the electrodes. Figure 13(b) shows the yeast cells of this manipulation mode to be attracted to the edges using a driving frequency of 10MHz.

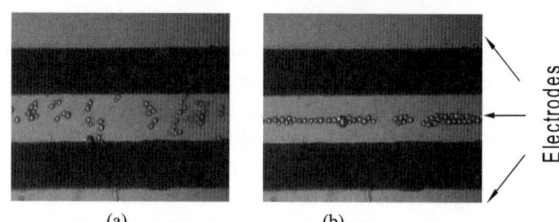

Fig.12 (a) Cells randomly distributed on the electrode. (b) Cells are collected and aligned under positive DEP forces.

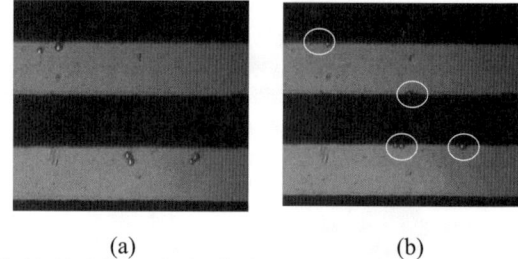

Fig.13 (a) Cells randomly distributed on the electrode. (b) Cells are attracted on the edge of the electrodes.

V. CONCLUSIONS

This paper presented a systematic study of various manipulation modes for micro particles using dielectrophoretic forces. The numerical models provided a useful tool for simulating the travellling potential wave. The formulae presented in this paper confirmed that the micro particles could be successfully controlled through the application of the travelling potential wave. The proposed numerical approach has been verified experimentally. The results indicated that the particles trace from the proposed numerical simulation was in good agreement with the experimental results. Furthermore, this study reported two novel manipulation modes, which controlled the electrodes power on/off for the DEP device system. The outcomes of this study could have a substantial impact on the development of bio-analytical systems using the dielectrophoretic forces.

VI. ACKNOWLEDGEMENTS

Authors would like to thank NCKU MEMS Center for access of equipments. Dr. Gwo-Bin Lee would like to thank partial financial supports from National Science Council (NSC 91-2323-B-006-005) and MOE Program for Promoting Academic Excellence of Universities under the grant number EX-91-E-FA09-5-4.

VII. REFERENCES

[1] N. G. Green and H. Morgan, "Dielectrophoresis of submicrometer latex spheres- experimental results," *J. Phys. Chem. B*, 103, 1999, pp. 41-50.

[2] D. J. Harrison and A Berg van den, "Micro Total Analysis Systems," Kluwer Academic Publishers. 1998.

[3] H. A. Pohl, *Dielectrophoresis,* Cambridge: Cambridge University Press, 1978.

[4] T. Schnelle, T. Müller, G. Gradl, S. G. Shirley and G. Fuhr, "Dielectrophoretic manipulation of suspended submicron particles," *Electrophoresis*, 21, 2000, pp. 66-73.

[5] S. Masuda, M. Washizu and I. Kawabata, "Movement of blood cells in liquid by non-uniform travelling field," *IEEE Trans. Ind. Appl.*, 24, 1988, pp. 217-222.

[6] G. Fuhr, R. Hagedorn, and T. Muller, "Linear motion of dielectric particles and living cells in microfabricated structures induced by traveling electric fields," *Proceeding of IEEE MEMS*, 1991, pp. 259-264.

[7] G. Fuhr, R. Hagedorn, and T. Muller, "Linear motion of dielectric particles and living cells in microfabricated structures induced by traveling electric fields," *Proceeding of IEEE MEMS*, 1991, pp. 259-264.

[8] L. Cui and H. Morgan, "Design and fabrication of t raveling wave dielectrophoresis structures," *J. Micromech. Microeng.*, 10, 2000, pp. 72-79.

[9] C. H. Lin, G. B. Lee, Y. H. Lin and G. L. Chang, "A fast prototyping process for fabrication of microfluidic systems on soda-lime glass," *J. Micromech. Microeng.*, 11, 2001, pp. 726-732.

[10] C. H. Chiou, G. B. Lee, H. T. Hsu, P. W. Chen and P. C. Liao, "Micro Devices Integrated with Microchannels and Electrospray Nozzles using PDMS Casting Techniques, " *Journal of Sensors and Actuators B: Chemical*. 2002, to be published.

[11] R. -J. Yang, L. -M. Fu and Y. -C. Lin, "Electroosmotic Flow in Microchannels," *J. of Colloid and Interface Science*, 239, 2001, pp. 98-105.

[12] H. Morgan, N. G. Green, M. P. Hughes, W. Monaghan and T. C. Tan, "Large-area travelling-wave dielectrophoresis particle separator," *J. Micromech. Microeng.*, 7, 1997, pp.65-70.

[13] Y. Huang, X-B. Wang, J. A. Tame and R. Pethig, "Electrokinetic behaviour of colloidal particles in travelling electric fields: studies using yeast cells," *J. Phys. D: Appl. Phys.*, 26, 1993, pp. 312-322.

[14] R. J. Yang and L. M. Fu, "Thermal and flow analysis of a heated electronic component," *Int. J. Heat Mass Transfer*, 44, 2001, pp.2264-2275.

3-D Nanomanipulation Using Atomic Force Microscopy

Guangyong Li, Ning Xi, Mengmeng Yu, Wai Keung Fung
Department of Electrical and Computer Engineering
Michigan State University
East Lansing, MI 48824, USA
E-mail: liguangy@msu.edu.

Abstract—The use of atomic force microscope (AFM) as a nanomanipulator has been evolving for various kinds of nanomanipulation tasks. Due to the bow effect of the piezo scanner of the AFM, the AFM space is different from the Cartesian space. In this paper, different 3-D nanomanipulation tasks using AFM such as nanolithography, pushing and cutting are discussed. 3-D path planning are performed directly in the AFM space and the 3-D paths are generated based on the 3-D topography information of the surface represented in the AFM space. This approach can avoid the mappings between the AFM space and Cartesian space in planning. By following the generated motion paths, the tip can either follow the topography of the surface or move across the surface by avoiding collision with bumps. Nanomanipulation using this method can be considered as the "true" 3-D operations since the cantilever tip can be controlled to follow any desired 3-D trajectory within the range of AFM space. The experimental study shows the effectiveness of the planning and control scheme.

I. INTRODUCTION

Since the invention of Atomic Force Microscope (AFM) [1], it has become the standard technique in imaging various sample surfaces down to the nanometer scale in ambient or fluid mediums. Besides its capability to characterize surfaces in nanometer scale, it has been demonstrated recently that AFM can be used to modify surfaces and manipulate nano-sized structures. By using AFM, thin oxide structures have been rearranged on the underlying surface by increasing applied load while scanning [2]. The sled-type motion of C_{60} islands during imaging has been studied in [3]. In [4], it was demonstrated that AFM can be used to deliberately move gold clusters on a smooth surface. Applications of AFM to manipulate and position nanometer-sized particles with nanometer precision were discussed in [5]. Using AFM to construct arbitrary patterns of gold nanoparticles was reported in [6]. The problem with these manipulation schemes is that they can only manipulate two-dimensional nanostructures on a smooth substrate surface. The surface tilt must be carefully removed before manipulation. The solution for this problem is to use techniques developed for three-dimensional nanomanipulation.

A promising method for 3-D nanomanipulation is to build a small nanomanipulator inside the vacuum capsule of a scanning electron microscope (SEM). Piezoelectric vacuum manipulators constructed inside the SEM have the ability to manipulate objects along the three linear degrees of freedom using the AFM tip as the end-effector [7]. Several kinds of manipulation of carbon nanotubes were performed using this kind of device [8], [9]. The obvious advantage of this method is that operation can be monitored in real-time. Another advantage of this method is that multi-end effectors can be built inside the SEM to achieve more degrees of freedom. The manipulation can be performed between the end-effectors without the need of a substrate. However, the manipulation accuracy of this method is not comparable to using the AFM since it has no feedback control during the manipulation. Since the samples are placed in vacuum and exposed to electron beam with high energy, this technique cannot be used to manipulate live biological samples. The expense of a SEM, ultrahigh vacuum condition, and space limitation inside the SEM vacuum capsule also impede the wide application of this method.

A scheme was proposed in [10] such that the nanoparticles were imaged in tapping mode and pushed on a surface in contact mode, while the normal feedback was switched on. This method can be considered as the beginning of the 3-D nanomanipulation in the sense that the cantilever tip follows the topography of the surface using internal feedback control. This method takes the risk of breaking the cantilever when switching on and off the vibration of the cantilever, changing the gains of the feedback loop, changing other parameters such as set-point, tip velocity, while the tip is touching the sample surface. Recently, some researchers are trying to combine the AFM with virtual reality interface and haptic devices to facilitate nanomanipulation. By introducing a virtual environment of the samples, nanomanipulation using the AFM become much easier. Besides 3-D synthetic visual feedback is provided for the operator, a 1-DOF haptic device had also been constructed in [11] for haptic display. However, this method suffers from false force due to the reflection from the sample surface. The false signal may mislead the operator to have an erroneous feedback. In [12], the AFM is also connected to a Phantom^{TM} stylus. In imaging mode, the topography data scanned from the AFM are sent to the Phantom^{TM} controller, and the operator can "feel" the topography of the sample. In manipulation mode, the Phantom^{TM} can be used to move the tip over the surface while keeping the internal normal force feedback on. However, it is still not clear whether the mapping from topography information to force information is helpful to the operation or not. These three methods can be considered as "semi" 3-D nanomanipulation since the cantilever tip can be controlled only to follow the topography of the surface either by the internal force feedback control loop or by the operator through the haptic devices.

Sometimes, it is not enough to let the cantilever tip follow the topography only. Arbitrary AFM tip paths may be desired, for example, moving the tip across a bumpy surface or a tilted surface. In this paper, by modeling the sample surface within the AFM space, the motion paths are generated by path planning. By following the generated motion paths, the tip can either follow the topography of the surface or move across the surface by avoiding collision with bumps. Using this approach, the AFM nanomanipulator can perform various nanomanipulation tasks such as lithography, pushing, cutting, and so on.

II. MODELING OF THE TASK SPACE IN THE AFM COORDINATES

Due to the scanning mechanism of the piezo tube, the scanning data that represent the surface topography are usually not consistent with the Cartesian coordinate system. The scanning trajectory of the piezo tube end is physically a concave bow in Cartesian space, while it is consider to be a straight line in AFM space. Therefore, topography represented in AFM space is always convoluted with a convex bow. A flat surface in a Cartesian space will map to a convex bow in an AFM space as shown in Fig. 1. The magnitude of the curvature depends on the scanning size. The bow effect can be negligible for small scanning size within several microns but it becomes significant for large scanning size. For scanning size of $90\mu m \times 90\mu m$, the uncorrected bow is about 60nm. The traditional examples of nanomanipulation based on AFM do not take this bow effect into consideration. Most of them work well because they are only 2-D operation. For 3-D manipulation, this bow effect has to be considered, else significant position errors may result in Z direction. However, if modeling and manipulation are performed under the same AFM coordinators, the error introduced by the bow effect can be avoided.

In order to determine the precise position of the features or objects on the surface, a coordinate frame

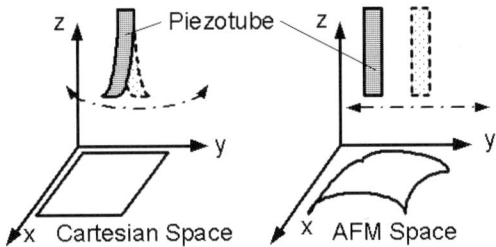

Fig. 1. The difference between Cartesian space and the AFM space: the scanning trajectory of the piezo tube is a concave curve viewed under the Cartesian space; the scanning trajectory of the piezo tube end is considered as a straight line if viewed from the AFM space. Therefore, a flat planes in Cartesian space maps to a convex plane in AFM space

must be set up on the sample surface. For conveniences, the center of the frame is usually assigned to the center point on the surface, which is the point first touched by the tip after tip engaged on the surface. The fast scan axis is defined as the X-axis and Y-axis is defined to be along the slow scanning direction. The Z direction is normal to the X-Y plane.

Modeling of the surface is simpler than the usual post-scanning image processing although they use some similar techniques to modify the scanning data such as removing noise. The usual post-scanning image processing attempts to generate a better view of the features on the surface by modifying the image such as flattening and de-noising the image, while the main purpose of modeling of the surface is to capture the precise position of the features within AFM space. The topography data must be left unchanged for the surface modeling. By turning off all software image correcting functions of the AFM, the topography of the surface is rendered by a matrix, M, with dimension of $N \times N$, where N depends on the sampling rate and usually equals to 128, 256, 512, or even 1024 for fine imaging. The relation between the 3-D representation and the matrix representation, which are shown in Fig. 2, are described by the following equations

$$\begin{cases} x &= (i - \text{int}(N/2))L/(N-1) \\ y &= -(j - \text{int}(N/2))L/(N-1) \\ z &= M(i,j) - M(\text{int}(N/2), \text{int}(N/2)) \end{cases} \quad (1)$$

where L is the scanning size, and int() is a rounding function.

III. TRAJECTORY TRACKING IN 3-D NANOMANIPULATION

The nanomanipulation using AFM is usually performed on a 2-D surface [4], [5], [6]. The tip motion in Z direction is blocked by turning off the normal force feedback, thus the Z value remain constant during manipulation. This approach requires a very

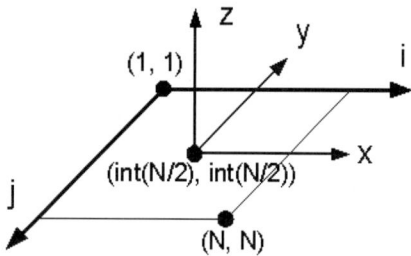

Fig. 2. The relation between the 3-D representation and the matrix representation

flat sample surface and the tilt of the surface must be removed. By keeping the normal force feedback on during manipulation [10], [12], or using the haptic devices with the aid of virtual reality [12], [11], the movement of the cantilever tip can be constrained to follow the surface topography in three dimensions. These methods have been proven very useful and successful in pushing and cutting nano-objects. Since the tip needs to be pressed into the surface during pushing as the dash line shown in Fig. 3, it is usually worn out quickly and contaminated easily. The lateral resolution of the imaging may decrease after each operation, because the same tip is used for manipulation and imaging.

When the attractive forces between the objects and the surface are large enough, the objects become unpushable because the tip tries to follow the topography as the dash line shown in Fig. 3. On the other hand, if the tip follows the dash-dot line in Fig. 3, it can either push the objects to the goal or cut the object, depending on the magnitude of the attractive forces between objects and the surface. In the following subsection, we discuss how to find the ideal paths for different cases.

A. Nanolithography in 3-D Workspace

The projection of the path for nanolithography to the X-Y plane is the designed pattern. The motion trajectory of the tip is approximated by a sequence of m line segments. The total $(m+1)$ points are composed of the joints of the segments plus the starting and ending points of the motion trajectory in X-Y plane. For example, If a letter "M" shown in Fig. 4(a) is a pattern that needs to be inscribed on a surface, the 2 dimensional path follows the sequence of linear segments from Point 1 to Point 7 or inversely as shown in Fig. 4 (b).

There are two different ways to implement the planning path in three dimensions. Method (1): keep the normal force feedback on but set a large constant force for the tip to push into the surface. When the

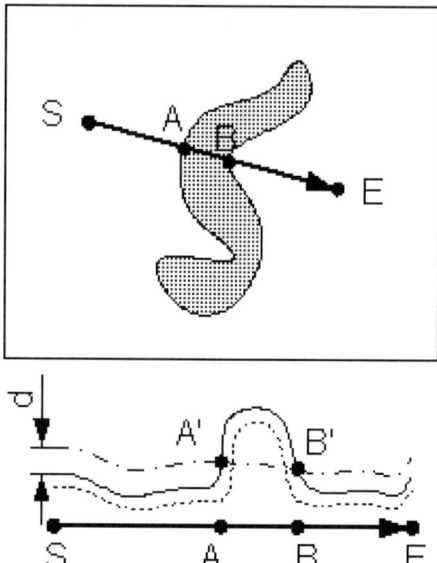

Fig. 3. The possible paths during pushing or cutting: The solid line is the surface topography convoluted with object; The dash line is the possible motion path of the tip; The dash-dot line is the ideal motion path of the tip

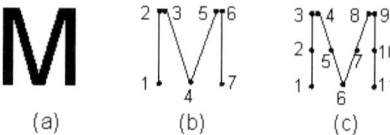

Fig. 4. The pattern "M" and its 2-D nanolithography path

tip is moving in the X and Y directions, it will automatically follow the topography of the surface while maintaining the constant contact force with the surface by the force feedback. This method is simple and successful for nano-inscribing, but is not realizable for other cases which need non-contact with the surface. Method (2): turn off the normal force feedback but generate a 3-D path based on the surface modeling data. Since the X, Y coordinates of the points defined on the path are already finely designed, we only need to obtain the Z coordinates of the path to perform 3-D motion. The Z coordinates can be found as

$$z = M(i,j) - M(\text{int}(N/2), \text{int}(N/2)) + d \quad (2)$$

where d is the depth, which is either positive or negative. $d > 0$ means that the tip moves above the surface; $d < 0$ means that the tip presses into the surface during moving. The index i, j are determined by

$$\begin{cases} i = \text{int}((N-1)x/L) + \text{int}(N/2) \\ j = \text{int}(N/2) - \text{int}((N-1)y/L) \end{cases} \quad (3)$$

Sometimes, in order to obtain a finer nanolithography result on a bumpy surface, longer segments may need to break into several shorter segments as shown in Fig. 4 (c).

B. Pushing and Cutting in 3-D Environments

Comparing with nanolithography, pushing and cutting are more difficult to implement at nanometer scale since friction at the nanometer scale differs significantly from that at the macro scale [13]. The experimental results in [14] show that friction at the nanometer scale is an intrinsic properties of a particular interface. An object undergoing lateral motion while in contact with a second object can either roll or slide. An in-plane rotation (sliding) of the carbon nanotube was observed on a mica surface by side-on pushes while both sliding and rolling were observed on a graphite surface [15]. There is no much difference between pushing and cutting for path planning problem, because whether the objects can be pushed or cut depends on the property of the interface between the objects and the substrate surface. If the friction is small, the objects can be pushed away to the goal either by sliding or rolling; if the friction is large enough, the objects may be dissected or unaltered. The paths are similarly designed both for pushing and cutting. How to design a scheme which can distinguish between pushing and cutting without changing the interface property is still an open problem.

The two different ways used in nanolithography can implement pushing/cutting paths in three dimensions. As shown in Fig. 3, if the tip is driven from point S directly to point E, the force feedback loop will guarantee the tip contacting the surface when Method (1) is used. However, since Method (1) wears out and contaminates the tip easily, Method (2) is preferred. In Method (2), the path is designed based on the three segments SA, AB and BE. The same path planning techniques as in nanolithography are used for SA and BE, while a straight line $A'B'$ as shown in Fig. 3 is designed for segment AB. Sometimes, a negative depth d is used to improve the performance without wearing out and contaminating the tip so quickly.

IV. EXPERIMENTS AND VERIFICATION

In order to verify the effectiveness of the proposed path planning method and control scheme for nanomanipulation using AFM, Various kinds of nanomanipulation tasks such as lithography, pushing, cutting are performed.

A. System Configuration for Nanomanipulation

A NanoScope IV atomic force microscope (Digital Instruments Inc., Santa Barbara, CA) was used for AFM imaging and manipulation. The AFM has a closed-loop scanning head with position correction by optical sensors in X-Y direction. The scanning head is equipped with a scanner (G type) with a maximum XY scan range of $90\mu m \times 90\mu m$ and a Z range of 5 μm. Cantilevers are standard microfabricated V-shaped, 200 μm long with 3μm tip height, silicon nitride Nanoprobes (Digital Instruments Inc.). In order to perform 3-D nanomanipulation, peripheral devices are installed with the AFM manipulation system. An optical microscope, a CCD camera, a haptic device (Phantom^{TM}) are included in the nanomanipulation system. The configuration of the system is shown in Fig. 5.

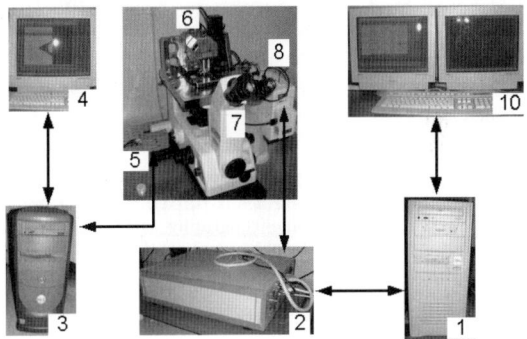

Fig. 5. The configuration of the nanomanipulation system: 1, main computer; 2, NanoScope IV controller; 3, second computer; 4, single monitor; 5, CCD camera; 6, AFM closed-loop head; 7, inverted optical microscope; 8, Bioscope controller; 9, twin-monitor

The AFM closed-loop head is controlled by the NanoScope IV controller through the Bioscope controller. The NanoScope IV controller is connected to the main computer which is responsible for running the control program of the system and also provides an interface for user to change control parameters and view the real-time data with the twin monitors. The purpose of the second computer is to decrease the burden on the main computer and guarantee real-time operation. The inverted optical microscope and the CCD camera help the operator to locate the tip and adjust the laser in the scanning head, and it can also search for the interesting area on a transparent sample by adjusting the stage which holds the sample.

B. Trajectory Tracking in 3-D Nanolithography

Using the system in Fig. 5, a 2-D trajectory tracking nanolithography is first carried out on a tilted surface. From the result, shown in Fig. 6(a), it can be seen that the inscribed depths of the three letters are different due to the surface tilt. The right letter "U", which is inscribe on the higher side, is inscribed deeper; while the left letter "M", which is on the lower side, is inscribed shallower. Fig. 6(b) shows the nanolithography result using 3-D trajectory tracking scheme. It is clear that the depth of the three letters are almost the same. Fig. 7 shows a more smooth nanolithography using 3-D trajectory tracking scheme. We can see that even the surface is tilted, the depth of the letters are unchanged.

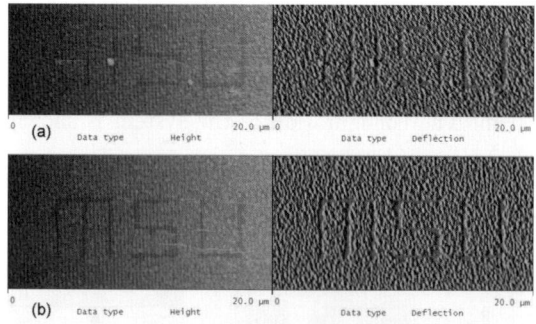

Fig. 6. (a) Nanolithography using 2-D trajectory control: left image is the height signal (surface is tilted), and right image is the deflection signal. (b) Nanolithography using 3-D trajectory control: left image is the height signal (surface is tilted), and right image is the deflection signal.

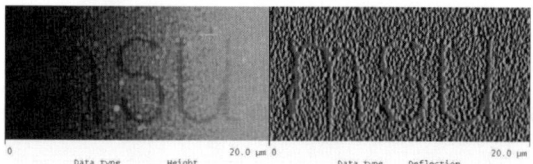

Fig. 7. Nanolithography using 3-D trajectory control: left image is the height signal (surface is tilted), and right image is the deflection signal.

C. Pushing and Cutting in 3-D Space

Using the system in Fig. 5, several pushing and cutting experiments have been carried out. By designing 3-D paths, a gold particle can be selectively pushed out from a cluster as shown in Fig. 8. Another pushing example is a multi-wall carbon nanotube, which is on a gold-coated surface. Whether the carbon nanotubes are pushed or cut depends strongly on the interface between the nanotubes and the surface. If the nanotube is pushed on its strong points, it will moving around as shown in Fig. 9. If the nanotube is pushed on its weak point, it will be broken as shown in Fig. 10. Therefore, it is also very important to determine where to push or cut the nanotube.

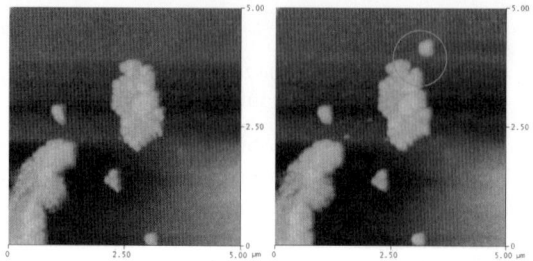

Fig. 8. Pushing a gold particle out of a cluster: left image shows original cluster of gold particles, and right image shows one of the particle is pushed away from the cluster.

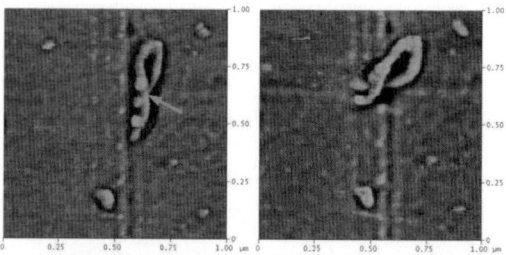

Fig. 9. Pushing a carbon nanotube using 3-D trajectory control: left image shows original orientation of the carbon nanotube, and right image shows the orientation of the carbon nanotube after pushing. The arrow indicates the pushing direction.

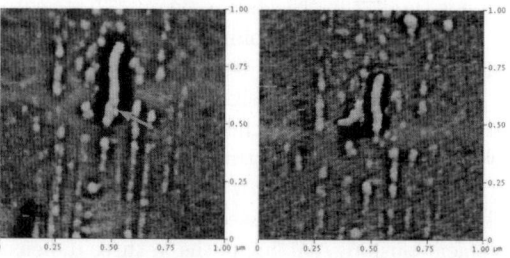

Fig. 10. Cutting a carbon nanotube using 3-D trajectory control: left image shows original shape of the carbon nanotube, and right image shows that the carbon nanotube has been cut into 2 pieces. The arrow indicates the cutting direction.

V. CONCLUSION

By considering the sample surface representation in the AFM space, a 3-D path planning method has been proposed. The motion paths are generated based on the modeling of the surface within the AFM space. By following the generated motion paths, the tip can either follow the topography of the surface or move across the surface by avoiding collision with bumps. Different 3-D nanomanipulation tasks using AFM such as nanolithography, pushing and cutting have been discussed. The 3-D nanomanipulation performed in the experiment verifies the effectiveness of the proposed path planning method and control scheme.

VI. ACKNOWLEDGMENTS

This research work is partially supported under NSF Grants IIS-9796300, IIS-9796287 and EIA-9911077.

VII. REFERENCES

[1] G. Binning, C. F. Quate, and C. Gerber. Atomic force microscope. Physical Review Letters, Vol. 56(9):930–933, 1986.

[2] Y. Kim and C. M. Lieber. Machining oxide thin films with an atomic force microscope: pattern and objective formation on the nanometer scale. Science, Vol. 257(5068):375–377, 1992.

[3] R. Luthi, E. Meyer, H. Haefke, L. Howald, W. Gutmannsbauer, and H.-J. Guntherodt. Sled-type motion on the nanometer scale: Determination of dissipation and cohesive energies of c_{60}. Science, Vol. 266(5193):1979–1981, 1994.

[4] D. M. Schaefer, R. Reifenberger, A. Patil, and R. P. Andres. Fabrication of two-dimensional arrays of nanometer-size clusters with the atomic force microscope. Applied Physics Letters, Vol. 66:1012–1014, February 1995.

[5] T. Junno, K. Deppert, L. Montelius, and L. Samuelson. Controlled manipulation of nanoparticles with an atomic force microscope. Applied Physics Letters, Vol. 66(26):3627–3629, June 1995.

[6] A. A. G. Requicha, C. Baur, A. Bugacov, B. C. Gazen, B. Koel, A. Madhukar, T. R. Ramachandran, R. Resch, and P. Will. Nanorobotic assembly of two-dimensional structures. In Proc. IEEE Int. Conf. Robotics and Automation, pages 3368–3374, Leuven, Belgium, May 1998.

[7] M.-F. Yu, M. J. Dyer, G. D. Skidmore, H. W. Rohrs, X.-K. Lu, K. D. Ausman, J. R. Von Ehr, and R. S. Ruoff. Three-dimensional manipulation of carbon nanotubes under a scanning electron microscope. Nanotechnology, Vol. 10:244–252, 1999.

[8] L. Dong, F. Arai, and T. Fukuda. 3d nanorobotic manipulations of multi-walled carbon nanotubes. In Proc. IEEE Int. Conf. Robotics and Automation, pages 632–637, Seoul, Korea, May 2001.

[9] L. Dong, F. Arai, and T. Fukuda. 3d nanoassembly carbon nanotubes through nanorobotic manipulation. In Proc. IEEE Int. Conf. Robotics and Automation, pages 1477–1482, Washington DC, USA, May 2002.

[10] L. T. Hansen, A. Kuhle, A. H. Sorensen, J. Bohr, and P. E. Lindelof. A technique for positioning nanoparticles using an atomic force microscope. Nanotechnology, Vol. 9:337–342, 1998.

[11] M. Sitti and H. Hashimoto. Tele-nanorobotics using atomic force microscope. In Proc. IEEE Int. Conf. Intelligent Robots and Systems, pages 1739–1746, Victoria, B. C., Canada, October 1998.

[12] M. Guthold, M. R. Falvo, W. G. Matthews, S. washburn S. Paulson, and D. A. Erie. Controlled manipulation of molecular samples with the nanomanipulator. IEEE/ASME Transactions on Mechatronics, Vol. 5(2):189–198, June 2000.

[13] G. V. Dedkov. Experimental and theoretical aspects of the modern nanotribology. Physical Status of Solid (a), Vol. 179:3–75, 2000.

[14] M. R. Falvo and R. Superfine. Mechanics and friction at the nanometer scale. Journal of Nanoparticles Research, Vol. 2:237–248, 2000.

[15] M. R. Falvo, R. M. Taylor, A. Helser, V. Chi, F. P. Brooks, S. Washburn, and R. Superfine. Nanometer-scale rolling and sliding of carbon nanotubes. Nature, Vol. 397:236–238, January 1999.

Bundled Carbon Nanotubes as Electronic Circuit and Sensing Elements

Victor T. S. Wong and Wen J. Li*

Centre for Micro and Nano Systems
The Chinese University of Hong Kong
Hong Kong SAR

*Contacting Author: wen@acae.cuhk.edu.hk

Abstract - Bundled multi-walled carbon nanotubes (MWNT) were successfully and repeatably manipulated by AC electrophoresis to form resistive elements between Au microelectrodes and were demonstrated to potentially serve as novel thermal and anemometrical sensor as well as simple electronic circuit elements. We have measured the temperature coefficient of resistance (TCR) of these MWNT bundles and also integrated them into constant current configuration for dynamic characterization. The I-V measurements of the resulting devices revealed that their power consumption were in μW range. Besides, the frequency response of the testing devices was generally over 100 kHz in constant current mode operation. Using the same technique, bundled MWNT was manipulated between three terminal microelectrodes to form simple potential dividing device. This device was capable of dividing the input potential into 2.7:1 ratio. Based on these experimental evidences, carbon nanotube is a promising material for fabricating ultra low power consumption devices for future sensing and electronic applications. Hence, we are currently developing fast and low cost MEMS fabrication processes to incorporate carbon nanotubes as sensing elements for various types of micro sensors.

1. INTRODUCTION

Power consumption is one of the most important engineering considerations in designing electrical circuits and systems. Hugh amount of efforts have been placed to minimize the power consumption of electrical systems, since high power consumption implies high heat dissipation which is undesirable in many applications. A typical example is the wall shear stress measurement in aerodynamic applications [1]. Excessive heat dissipation from a hot film anemometrical sensor will disturb the minute fluidic motion, crippling its ability to sense true fluidic parameters. With our preliminary experimental findings on bundled MWNT devices, we found that the devices can be operated at μW range, which is ultra low power consumption for applications like shear stress and thermal sensing (e.g., in the order of mW range for typically MEMS polysilicon devices [2]).

Carbon nanotubes (CNT), since discovered in 1991 by Sumio Iijima [3], have been extensively studied for their electrical [4] and mechanical properties [5]. In order to build a CNT based device, technique to manipulate the CNT has to be developed. Typical manipulation technique is by atomic force microscopy [6]. However, this pick-and-place technique is time consuming, though the technique has very high positioning accuracy. Past demonstrations by K. Yamamoto et al. showed that carbon nanotube can be manipulated by AC and DC electric field [7,8]. Besides, a recent report from L. A. Nagahara et al. demonstrated the individual single-walled carbon nanotube (SWNT) manipulation using nano-electrodes by AC bias voltage [9]. By using similar technique (i.e., AC electrophoresis), we have successfully manipulated bundled carbon nanotubes to form resistive elements between Au microelectrodes for sensing and electronic circuits efficiently. This paper reports the technique to form bundled MWNT resistive element between Au electrodes and our preliminary experimental findings on the electrical characterizations such as frequency response and I-V characteristics of the bundled MWNT devices. The results indicate that the carbon nanotube is promising to be used as high performance and low power consumption devices for future electronic and sensing applications.

2. FORMATION OF CNT ELEMENTS BY AC ELECTROPHORESIS

2.1 Fabrication of Microelectrodes

Array of Au microelectrodes with different geometrical shapes (see *Fig.1*) were fabricated on a 1.8 X 1.8 cm² glass substrate and silicon substrate for carbon nanotube manipulation. To start with, the substrate was first coated with Cr and Au inside e-beam evaporator and then spun with AZ5214 positive photoresist and was patterned to desired geometries. After the patterning procedure, the Au and Cr were then etched by Au etchant ($KI : I_2 : H_2O = 4:1:80$) and Cr etchant, respectively. The photoresist was then removed by acetone or oxygen plasma. Detailed parameters for the photolithography procedures can be found in [10].

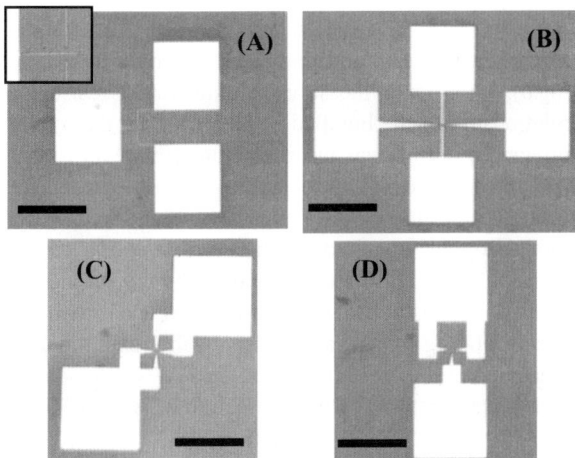

Fig.1. Different microelectrode geometries for MWNT manipulations, A) three-terminal microelectrodes to form three terminals potential dividing device (inset showing the gap (~ 5 μm) between the microelectrodes), B) four-terminal microelectrodes to form simple resistive bundled MWNT device, C) Cross microelectrodes to form parallel bundled MWNT geometry, D) T microelectrodes to form V-shape bundled MWNT geometry. (Scale Bar = 200 μm)

2.2 AC Electrophoretic Manipulation of CNT

2.2.1 Theoretical Background

AC electrophoresis (or dielectrophoresis) is a phenomenon where neutral particles undergoing mechanical motion inside a non-uniform AC electric field (see *Fig.2*). Detailed descriptions on AC electrophoresis can be found in [11]. The dielectrophoretic force imparted on the particles can be described by the following equation:

$$\vec{F}_{DEP} = \frac{1}{2}\vec{\alpha} V \nabla |\vec{E}|^2 \quad (1)$$

where $\vec{\alpha}$ is the polarizability of the particles, which is a frequency dependent term. V is the volume of the particles and $\nabla = \vec{i}\frac{\partial}{\partial x} + \vec{j}\frac{\partial}{\partial y} + \vec{k}\frac{\partial}{\partial z}$ is the gradient operator. $|\vec{E}|$ is the magnitude of the electric field strength. Equation (1) reveals that the generated force is dependent of the gradient of electric field rather than the direction of electric field. Besides, the polarizability function also determines whether the force generated is attractive (positive dielectrophoresis) or repulsive (negative dielectrophoresis).

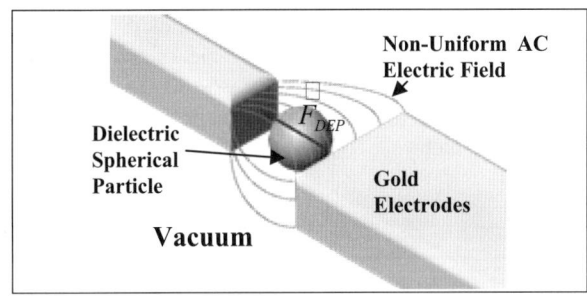

Fig.2. Under non-uniform AC electric field, dielectrophoretic force induced on the neutral particle cause the mechanical motion.

2.2.2 Experimental Details

The MWNT we used in the experiments was ordered commercially from [12] (prepared by chemical vapour deposition). The axial dimension and the diameter of the MWNT was 1 – 10 μm and 10 – 30 nm, respectively. Prior to the MWNT manipulation, 50 mg of the sample was ultrasonically dispersed in 500 mL ethanol solution and the resulting solution was diluted to 0.01 mg/mL for later usage.

After the Au microelectrodes were wire-bonded to the external circuits, approximately 10 μL of the MWNT/ethanol solution was transferred to the substrate by 6 mL gas syringe (see *Fig.3*). Then the Au microelectrodes were excited by AC voltage source (typically, 16 V peak-to-peak at 1 MHz). The ethanol was evaporated away leaving the MWNT to reside between the gap of the microelectrodes (see *Fig.4*). In *Fig.4*, room temperature resistance of 6.12 kΩ between microelectrodes was found upon two probe measurement which suggested that connection had been formed between two microelectrodes.

We experimentally found that the resistances of the MWNT bundles were sample dependent (i.e., different MWNT samples have different room temperature resistances) and the two probe room temperature resistances of the samples were typically ranging from several kΩ to several hundred kΩ. Since the conductivity of CNTs were dependent on their lattice geometries during their growth process, the conductivities of individual CNT cannot be well controlled, which results in the variation of conductivities in individual CNT. During the AC electrophoresis process to form MWNT bundles across microelectrodes, the MWNT was randomly connected between microelectrodes. Therefore, it is logically followed that different MWNT samples exhibited different conductivities.

Besides, the direction of the bundled MWNT alignments can be controlled by the microelectrode geometries. For example, in *Fig.5*, the MWNT bundles can be aligned into V-shape geometry and parallel geometry.

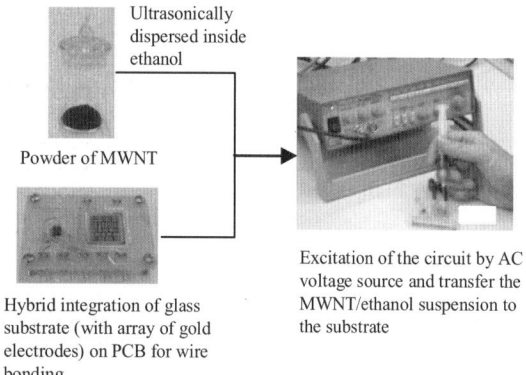

Fig.3. Experimental process flow showing the fabrication of MWNT based circuit elements.

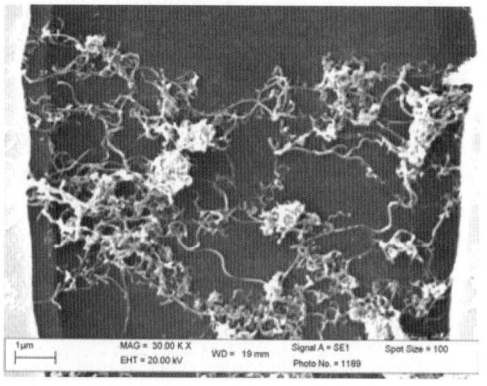

Fig.4. Scanning electron microscopic image (SEM) showing the MWNT connections between Au microelectrodes.

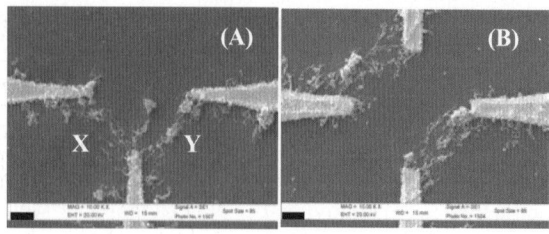

Fig.5. Special alignments of MWNT bundles. A) V-shape geometry. MWNT linkages X and Y were aligned in about 90°. B) MWNT linkages were aligned into parallel geometry. (Scale Bar = 3 μm).

3. CNT AS THERMAL SENSING ELEMENT

3.1 Thermal Sensitivity

Bundled MWNT was served as sensing element driven in constant current mode configuration (see *Fig.6*). In order to measure the temperature-resistance relationship of the bundled MWNT device, the hybrid integrated circuit was put inside an oven (Lab-Line® L-C Oven) and the temperature of the environment was kept monitored by the Fluke type K thermocouples attached on the surface of the circuit board. The TCR was then obtained by measuring the change of resistance of the bundled MWNT with the corresponding temperature. From the experimental measurements on a typical bundled MWNT device, its resistance dropped with temperature, which is in agreement with [13] (i.e. negative TCR). Interestingly, the TCR measurements of all of our testing devices did not converge but the ranges were generally within -0.1 to -0.2 %/°C (see *Fig.7*). Considerably drifting in the room temperature resistances of the device were observed in different measurements. We suspect the variations were contributed by the mismatch in thermal coefficient of expansion between the Au electrodes and the bundled MWNT, causing some of the MWNT linkages detach from the Au electrodes, or due to contaminations such as moisture during measurements. In order to form a more robust protection for MWNT bundles, we are currently developing a process to embed MWNT bundles between parylene layers to see its effectiveness (see *Fig.8*). Nevertheless, the temperature-resistance dependency of bundled MWNT implied its thermal sensing capability. Besides, from the I-V measurement of the bulk MWNT device, the current required to induce the self heating of the device was in μA range at several volts which suggested the power consumption of the device was in μW range (see *Fig.9*).

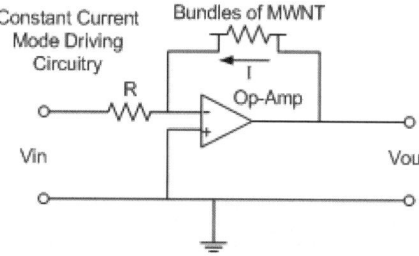

Fig.6. Schematic diagram showing the typical constant current mode configuration.

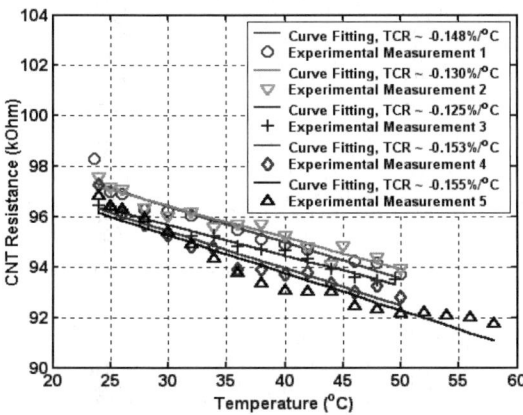

Fig.7. TCR for a typical bundled MWNT device in five different measurements.

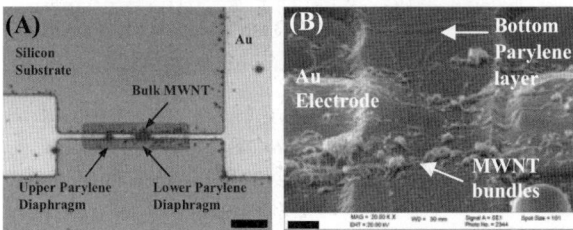

Fig.8. A) Optical microscopic image showing the prototyping parylene/MWNT/parylene device (Scale Bar = 20 μm). B) SEM image (tilted at 60°) showing MWNT bundles resting on parylene bottom layer in MWNT/parylene configurations (without top parylene layer coating for clear visualization for MWNT bundles) (Scale Bar = 1 μm).

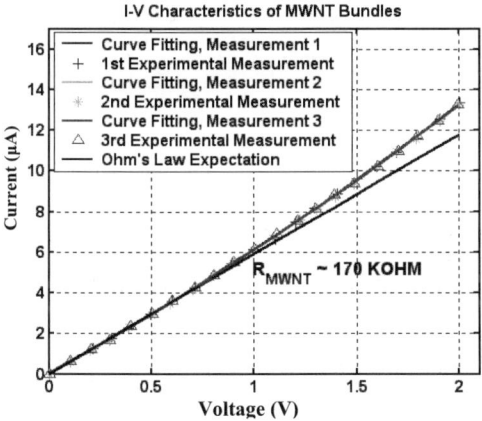

Fig.9. I-V characteristics of a typical bundled MWNT device. Three repeated measurements were performed to validate its repeatability. The straight line is the theoretical expectation using Ohm's Law and the room temperature resistance of bundled MWNT in our testing sample was about 170 kΩ.

3.2 Frequency Response

In order to pick up small variations of the sensing environment, sensors with fast frequency response are highly desired. To test the frequency response of the bundled MWNT device, input square wave of 2 V peak-to-peak at 10 kHz was fed into the negative input terminal of the circuit shown in Fig.6 and the output response was determined (see *Fig.10*.). From our experimental measurements, bundled MWNT device exhibited very fast frequency response. Using the approximation between the time constant and cutoff frequency [2],

$$f_c = 1/1.5 t_c \qquad (1)$$

where f_c is the cutoff frequency, t_c is the time constant of the response, and therefore the estimated cutoff frequency of the device was about 177 kHz (see *Fig.10*). As a comparison, typical frequency response of MEMS poly-silicon sensors in constant current mode configuration without frequency compensation is around several hundred Hz to several kHz [2, 14].

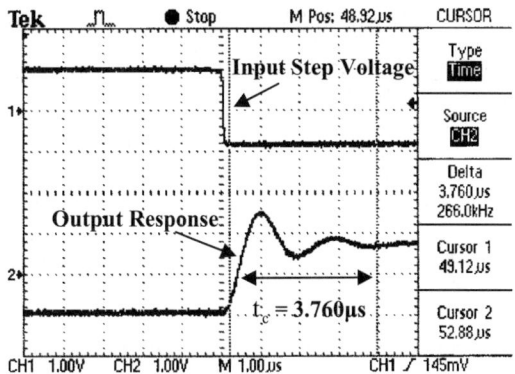

Fig.10. Frequency response of MWNT bundles in constant current mode configuration.

3.3 Bundled CNT as Sensing Element for Anemometry Sensing

A proof-of-concept experiment was performed to validate the fluid sensing ability of bundled MWNT device in μW operating power. The CNT sensor was placed perpendicular to a flow source with constant outlet velocity. The distance between the source and the sensor was then varied to induce different impinging velocities on the device (similar to Hiemenz flow) (see *Fig.11*). Although the flow environment was not well-controlled, results do clearly indicate the response of the CNT sensor to different impinging velocities (see *Fig.12*). Wind tunnel testing on the sensors will be performed in the later stage to determine their sensitivities.

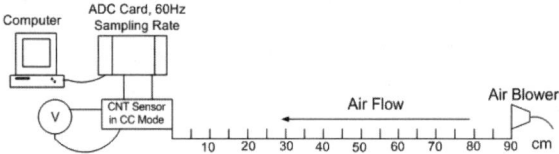

Fig.11. Schematic diagram showing the experimental setup for simple air blowing testing. The CNT sensor was placed in normal direction to the air flow.

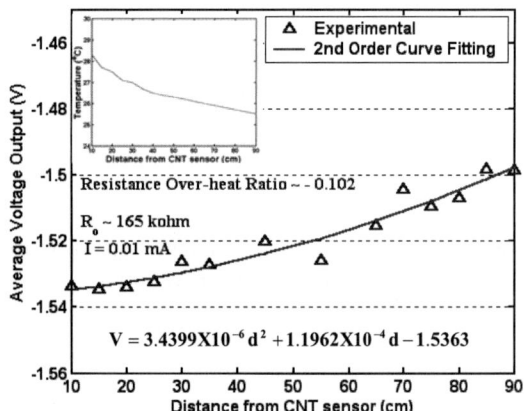

Fig.12. Output voltage variations with the air flow at different locations between the sensor and the air blower. Inset showing the temperature of the air flow at different locations and the range was generally within $2\,^{\circ}C$. The operating power of the bundled MWNT sensor was about 15 µm.

4. CNT AS POTENTIAL DIVIDING DEVICE

Using the technique reported in Section 2, bundled MWNT were manipulated by a three terminals Au microelectrodes (see *Fig.1*A) to form simple potential dividing circuit (see *Fig.13.*). Two probe room temperature resistivity measurements for the terminals 1-2, 1-3 and 3-2 were about 1141 kΩ, 307 kΩ and 833 kΩ respectively.

We have calculated the expected voltage output of the device using the potential dividing formula of a resistive circuit,

$$V_{13} = \frac{R_{13}}{R_{12}} \cdot V_{12}, \quad V_{32} = \frac{R_{32}}{R_{12}} \cdot V_{12} \qquad (2)$$

where R_{12}, R_{13} and R_{32} are the resistance across the terminal 1-2, terminal 1-3 and terminal 3-2 respectively. V_{12} is the voltage applied across terminal 1-2. The MWNT linkages incorporated with the three terminals microelectrodes can be used as a potential dividing circuit. The device was capable to switch the input potential into a ratio about 2.7:1 (see *Fig.14*). The I-V measurements on terminal 1-3 and terminal 3-2 revealed the power consumption of the bundled MWNT device was in µW range (see *Fig.15*).

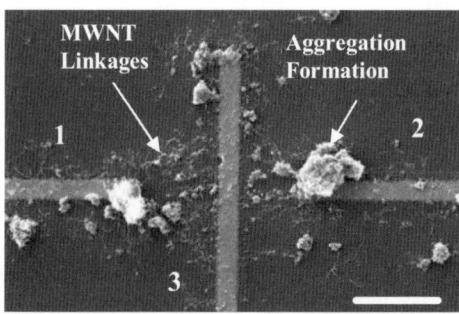

Fig.13. SEM image showing the formation of MWNT linkages with three terminals microelectrodes. Aggregation was observed and this was due to the instability of MWNT in ethanol medium. Terminals 1, 2 and 3 were indicated in the figure. (Scale Bar = 6 µm)

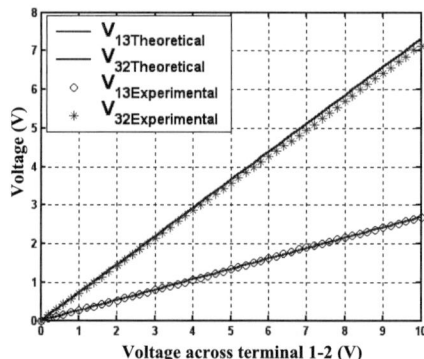

Fig.14. Comparison between the theoretical calculations and experimental results on the potential switching capability of a MWNT potential dividing circuit.

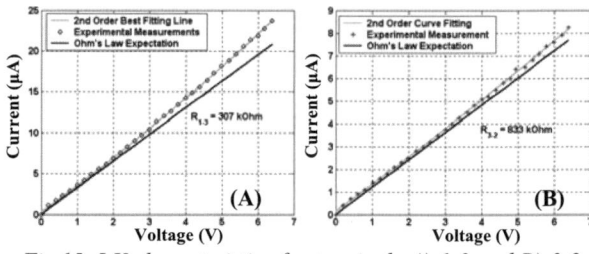

Fig.15. I-V characteristics for terminals A) 1-3 and B) 3-2. Due to the electrical conductivity dependency on different combinations of MWNT bundles, the non-linearity begun at different applied voltages in terminal 1-3 and 3-2 respectively.

5. CONCLUSION

A technique to form bundled MWNT resistive elements between Au electrodes was presented. The TCR measurements and the frequency response measurement of the bundled MWNT based device showed that bundled MWNT can be served as sensing element for high performance thermal sensing applications. Besides, the operating power of the resulting devices is in µW range which is ultra low power consumption for applications like shear stress sensing system. Apart from this, a bundled MWNT based potential dividing circuit was built with potential switching capability of 2.7:1 ratio and power consumption in µW range. From our demonstrations, MWNT can be used for ultra low power consumption sensing and electrical circuit applications.

6. ACKNOWLEDGEMENTS

The authors would like to thank the Chinese University of Hong Kong (CUHK) for providing funding for this project. Besides, the authors would like to sincerely thank Dr. W.Y. Cheung of Department of Electronic Engineering of CUHK, Ms. Catherine Yeung of Department of Physics of CUHK, Mr. H.Y. Chan, Ms. W.L. Zhou, and Mr. Johnny M.H. Lee of Centre for Micro and Nano Systems of CUHK for their help and discussion for the project.

7. REFERENCES

[1] J.B. Huang, C. Liu, F. Jiang, S. Tung, Y.C. Tai, C.M. Ho, "Fluidic Shear Stress Measurement Using Surface-Micromachined Sensors", Proceedings of IEEE Region 10 International Conference on Microelectronics and VLSI, (TENCON '95), pp. 16 – 19 (1995).

[2] C. Liu, J.B. Huang, Z. Zhu, F. Jiang, S. Tung, Y.C. Tai, C.M. Ho, "A Micromachined Flow Shear-Stress Sensor based on Thermal Transfer Principle", Journal of Microelectromechanical Systems, Vol. 8, No. 1, pp. 90 – 99 (1999).

[3] S. Iijima, "Helical Microtubules of Graphitic Carbon", Nature, Vol. 354, pp. 56 – 58 (1991).

[4] S. Frank, P. Poncharal, Z.L. Wang, W.A. de Heer, "Carbon Nanotube Quantum Resistors", Science, Vol. 280, pp. 1744 – 1746 (1998).

[5] E.W. Wong, P.E. Sheehan, C.M. Lieber, "Nanobeam Mechanics: Elasticity, Strength, and Toughness of Nanorods and Nanotubes", Science, Vol. 277, pp.1971 – 1975 (1997).

[6] T. Shiokawa, K. Tsukagoshi, K. Ishibashi, Y. Aoyagi, "Nanostructure Construction in Single-walled Carbon Nanotubes by AFM Manipulation", Proceedings of Microprocesses and Nanotechnology Conference 2001, pp. 164 – 165 (2001).

[7] K. Yamamoto, S. Akita, Y. Nakayama, "Orientation of Carbon Nanotubes Using Electrophoresis", Japanese Journal of Applied Physics, Vol.35, L917-L918 (1996).

[8] K. Yamamoto, S. Akita, Y. Nakayama, "Orientation and Purification of Carbon Nanotubes Using AC Electrophoresis", Journal of Physics D: Applied Physics, Vol. 31, L34-L36 (1998).

[9] L.A. Nagahara, I. Amlani, J. Lewenstein and R.K. Tsui, "Directed Placement of Suspended Carbon Nanotubes for Nanometers-scale Assembly", Applied Physics Letters, Vol. 80, No. 20, pp. 3826 – 3828 (2002).

[10] V.T.S. Wong, W.J. Li, "Dependence of AC Electrophoresis Carbon Nanotube Manipulation on Microelectrode Geometry", International Journal of Non-linear Sciences and Numerical Simulation, Vol. 3, Nos. 3 – 4, pp. 769 -774 (2002).

[11] H.A. Pohl, "Dielectrophoresis: The Behaviour of Neutral Matter in Nonuniform Electric Fields", Cambridge University Press (1978).

[12] Sun Nanotech Co Ltd, Beijing, P.R. China.

[13] T.W. Ebbesen, H.J. Lezec, H. Hiura, J.W. Bennett, H.F. Ghaemi, T. Thio, "Electrical Conductivity of Individual Carbon Nanotubes", Nature, Vol. 382, pp. 54 – 56 (1996).

[14] J.B. Huang, F.K. Jiang, Y.C. Tai, C.M. Ho, "MEMS-based Thermal Shear-stress Sensor with Self-frequency Compensation", Measurement Science and Technology, Vol. 10, No. 8, pp. 687 – 696 (1999).

Stiffness Analysis of the Humanoid Robot WABIAN-RIV: Modelling

Giuseppe Carbone[1,3] Hun-ok Lim[2,3] Atsuo Takanishi[3] Marco Ceccarelli[1]

[1]Laboratory of Robotics and Mechatronics
DiMSAT, University of Cassino
Via Di Biasio 43
03043 Cassino (Fr), Italy
carbone/ceccarelli@ing.unicas.it

[2]Department of System Design Engineering
Kanagawa Institute of Technology
1030 Shimoogino, Atugi
Kanagawa, 243-0292, Japan,
holim@sd.kanagawa-it.ac.jp

[3]Humanoid Robotics Institute
Waseda University
3-4-1 Ookubo, Shinjuku-ku
Tokyo, 169-8555, Japan
takanisi@waseda.jp

Abstract – In this paper a humanoid robot named as WABIAN-RIV (WAseda BIpedal humANoid Refined IV) is analyzed in terms of stiffness characteristics. This paper proposes basic models and a formulation in order to deduce the stiffness matrix as a function of the most important stiffness parameters of the WABIAN architecture. The proposed formulation is useful for numerical estimation of stiffness performances.

I. INTRODUCTION

Many research groups are working in the development of humanoid robots [1-7]. A humanoid robot is a complex mechatronic system requiring developments in many branches of science and engineering [8,9].

As regards the mechanical design, stiffness performances are a very important aspect that has to be considered [10-13]. In fact, in a robot if the stiffness of links and joints are inadequate, external forces and moments may cause large deflections in the links, which are undesirable from the viewpoint of both accuracy and payload performances. Therefore, a stiffness analysis is carried out on different types of robots in order to evaluate their stiffness performances [12-17]. Moreover, once a proper stiffness model and formulation have been defined they can be used also for design purposes in order to find an optimum compromise between weight of links and stiffness performances [18,19].

The above-mentioned considerations strongly suggest to perform stiffness analysis also on humanoid robots. This paper describes a stiffness analysis of the humanoid robot WABIAN-RIV by defining proper kinematic, stiffness and static models in order to deduce its stiffness matrix. The proposed formulation can be numerically implemented in order to obtain an estimation of the stiffness performances.

II. THE HUMANOID ROBOT WABIAN-RIV

WABIAN-RIV has been developed for human-robot cooperation work as shown in Fig.1. It has a total of forty-three mechanical dofs [7]; two six dof legs, two seven dof arms, two three dof hands, a four dof neck, two two dof eyes and a three dof trunk. The motion range of each link has been designed to be as human like as possible. The structure of WABIAN-R-IV has been built by using two different aluminum alloys: EDS (extra super duralluminum) and GIGAS. These alloys have been developed by YKK company in order to guarantee high stiffness and low weight [20].

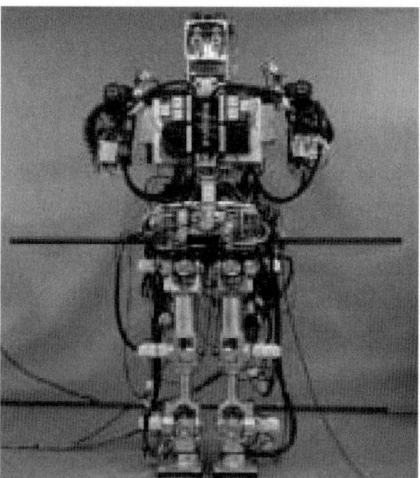

Fig.1 WABIAN-RIV.

By using WABIAN series, a variety of walking has been achieved such as dynamic forward and backward walking, marching in place, dancing, carrying a load, and emotional walking, [7,21,22].

III. A KINEMATICS MODEL FOR WABIAN-RIV

By examining the mechanical design of WABIAN-RIV one can deduce its overall kinematic model as shown in Fig.2.

The Denavit-Hartenberg convention has been used in order to define position and orientation of the link coordinate frames shown in Figs.3 to 6. In particular, Figs.3, 4, 5, and 6 show the detailed kinematic models of right leg, right arm, waist, trunk, neck, and shoulders, respectively. The kinematic models of left leg and left arm can be considered as the same of right leg and arm, respectively.

By using the above-mentioned parameters, the transformation matrix $\mathbf{T}^{i-1}_i$ that expresses the relation between frame i and i-1 can be written as

$$\mathbf{T}^{i-1}_i = \begin{bmatrix} c\theta_i & -s\theta_i & 0 & a_{i-1} \\ s\theta_i\, c\alpha_{i-1} & c\theta_i\, c\alpha_{i-1} & -s\alpha_{i-1} & -d_i\, s\alpha_{i-1} \\ s\theta_i\, s\alpha_{i-1} & c\theta_i\, s\alpha_{i-1} & c\alpha_{i-1} & d_i\, c\alpha_{i-1} \\ 0 & 0 & 0 & 1 \end{bmatrix} \quad (1)$$

where $c\alpha_i$, $s\alpha_i$, $c\theta_i$ and $s\theta_i$ are shorthand of $\cos\alpha_i$, $\sin\alpha_i$, $\cos\theta_i$ and $\sin\theta_i$, respectively. The transformation matrices for right leg, left leg, waist, trunk, right arm, left arm and shoulders can be derived by using the related D-H parameters into Eq. (1).

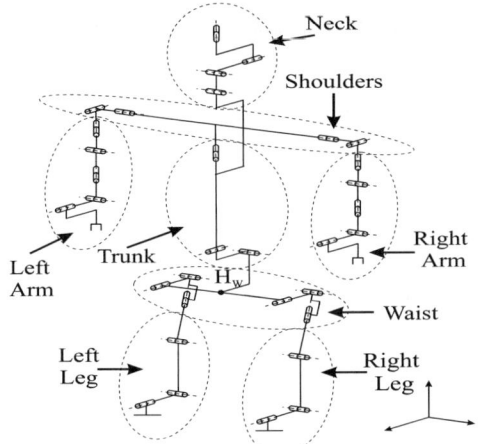

Fig.2 A Kinematic model for WABIAN-RIV.

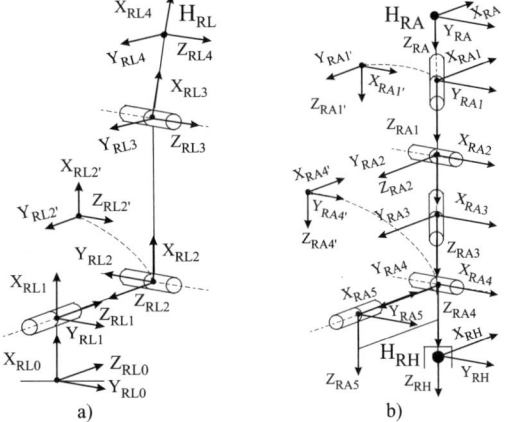

Fig.3 Kinematic model for WABIAN-RIV: a) right leg, b) right arm.

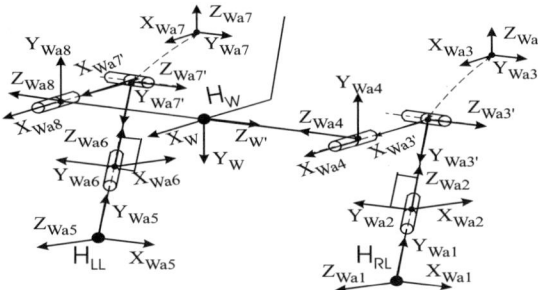

Fig.4 Kinematic model for WABIAN-RIV waist.

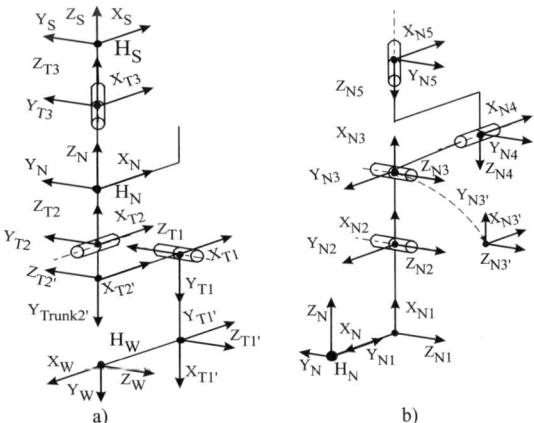

Fig.5 Kinematic model for WABIAN-RIV: a) trunk, b) neck.

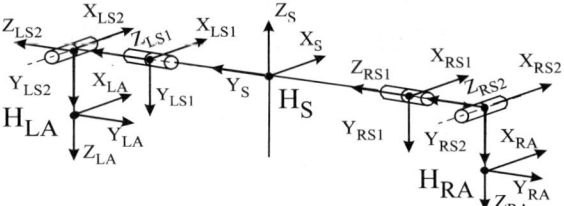

Fig.6 Kinematic model for WABIAN-RIV shoulders.

IV. A FORMULATION FOR STIFFNESS ANALYSIS OF WABIAN-RIV.

In order to make the analysis easier, WABIAN-RIV has been assumed to be composed of the following subcomponents: right leg, left leg, right arm, left arm, shoulders, trunk and neck as shown in Fig.2. According to this assumption, the compliance displacements of the overall system can be computed as the sum of the compliance displacements on each subcomponent by using the superposition principle.

In the model proposed in the following the displacements that are due to backlashes can be considered as negligible. Moreover, flectional and torsion displacement of links have been modeled as a rotation about the respective motor axis and considered as additional component of the motor lumped stiffness parameter itself.

In the analysis it has been also assumed that two external wrenches $\mathbf{W_R}$ and $\mathbf{W_L}$ act on the right and left hand of WABIAN-RIV, respectively. In fact, this can be considered as a general case since the WABIAN-RIV has been designed to carry loads by using one or both hands. Moreover, the stiffness analysis of the neck can be neglected, since the neck has not to carry any load.

By using the above-mentioned assumptions, the 6x1 compliance displacement vector of WABIAN-RIV at its right hand $\Delta \mathbf{S_{WR}}$ can be computed as

$$\Delta \mathbf{S_{WR}} = \Delta \mathbf{S_{RL}} + \Delta \mathbf{S_{LL}} + \Delta \mathbf{S_{Wa}} + \Delta \mathbf{S_T} + \Delta \mathbf{S_S} + \Delta \mathbf{S_{RA}} \quad (2)$$

where the compliance displacements on right leg, left leg, waist, trunk, shoulders and right arm can be computed as

$$\Delta \mathbf{S_{RL}} = \mathbf{K_{RL}^{-1}} \mathbf{M_{WRL}} \mathbf{W_R} \; ; \; \Delta \mathbf{S_{LL}} = \mathbf{K_{LL}^{-1}} \mathbf{M_{WLL}} \mathbf{W_R} ;$$
$$\Delta \mathbf{S_{Wa}} = \mathbf{K_{Wa}^{-1}} \mathbf{M_{WWa}} \mathbf{W_R} \; ; \; \Delta \mathbf{S_T} = \mathbf{K_T^{-1}} \mathbf{M_{WT}} \mathbf{W_R} \quad (3)$$
$$\Delta \mathbf{S_S} = \mathbf{K_S^{-1}} \mathbf{M_{WS}} \mathbf{W_R} \; ; \; \Delta \mathbf{S_{RA}} = \mathbf{K_{RA}^{-1}} \mathbf{W_R}$$

where $\mathbf{M_{WRL}}$, $\mathbf{M_{WLL}}$, $\mathbf{M_{WWa}}$, $\mathbf{M_{Wt}}$, and $\mathbf{M_{WS}}$ are 6x6 matrices, which can be obtained by means of static analysis, giving the forces and torques acting on the right leg, left leg, waist, trunk, shoulder and right arm, respectively, due to the external wrenches $\mathbf{W_R}$ and $\mathbf{W_L}$ and due to the masses of WABIAN-RIV. The matrices $\mathbf{K_{RL}}$, $\mathbf{K_{LL}}$, $\mathbf{K_{Wa}}$, $\mathbf{K_T}$, $\mathbf{K_S}$ and $\mathbf{K_{RA}}$ are 6x6 stiffness matrices for right leg, left leg, waist, trunk, shoulders and right arm, respectively, when they are described with respect to the reference frame attached to the right hand.

Thus, for the stiffness matrix for the right hand operation $\mathbf{K_{WR}}$ one can write

$$\mathbf{K}_{WR}^{-1} = \mathbf{K}_{RL}^{-1} \mathbf{M}_{WRL} + \mathbf{K}_{RL}^{-1} \mathbf{M}_{WLL} + \mathbf{K}_{Wa}^{-1} \mathbf{M}_{WWa}$$
$$+ \mathbf{K}_{T}^{-1} \mathbf{M}_{WT} + \mathbf{K}_{S}^{-1} \mathbf{M}_{WS} + \mathbf{K}_{RA}^{-1} \mathbf{M}_{WRA} \quad (4)$$

Similarly, for the stiffness matrix for the left hand operation $\mathbf{K}_{WL}$ one can write

$$\mathbf{K}_{WL}^{-1} = \mathbf{K}_{RL}^{-1} \mathbf{M}_{WRL} + \mathbf{K}_{RL}^{-1} \mathbf{M}_{WLL} + \mathbf{K}_{Wa}^{-1} \mathbf{M}_{WWa}$$
$$+ \mathbf{K}_{T}^{-1} \mathbf{M}_{WT} + \mathbf{K}_{S}^{-1} \mathbf{M}_{WS} + \mathbf{K}_{RA}^{-1} \mathbf{M}_{WLA} \quad (5)$$

It is worthy to note that the two 6x1 vectors of the wrenches $\mathbf{W}_R$ and $\mathbf{W}_L$ and the two 6x1 vectors of the compliance displacements $\Delta\mathbf{S}_{WR}$ and $\Delta\mathbf{S}_{WL}$ can be combined in 12x1 vectors. Therefore, by using Eqs. (4) and (5) a 12x12 stiffness matrix can be written as

$$\mathbf{K}_W = \begin{bmatrix} \mathbf{K}_{WR} & 0 \\ 0 & \mathbf{K}_{WL} \end{bmatrix} \quad (6)$$

A. Stiffness matrix for right leg

For the 6x6 stiffness matrix $\mathbf{K}_{RL}$ the following equation can be written

$$\mathbf{W}_{RL} = \mathbf{K}_{RL} \Delta\mathbf{L}_{RL} \quad (7)$$

where $\mathbf{W}_{RL}$ is a 6x1 vector of the static wrench acting on the point H_{RL}, Figs.3 and 4, of the right leg, and $\Delta\mathbf{L}_{RL} = (\Delta X_{RL4}, \Delta Y_{RL4}, \Delta Z_{RL4}, \Delta\gamma_{RL4}, \Delta\beta_{RL4}, \Delta\alpha_{RL4})^t$ is a 6x1 vector whose components are the linear and angular compliance displacements measured in the reference frame having the origin in H_{RL} itself. $\mathbf{W}_{RL}$ can be also computed as

$$\mathbf{W}_{RL} = \mathbf{W}_{TRL} \mathbf{W}_{IRL} \quad (8)$$

where $\mathbf{W}_{TRL}$ is a 6x6 matrix expressing the relation between $\mathbf{W}_{RL}$ and $\mathbf{W}_{IRL}$; $\mathbf{W}_{IRL}$ is a 6x1 vector whose components are the active forces and torques in each link of the right leg. This vector can be expressed as a function of the 6x1 vector $\Delta\mathbf{l}_{RL} = (\Delta x_{RL0}, \Delta x_{RL1}, \Delta x_{RL2}, \Delta\theta_{RL1}, \Delta\theta_{RL2}, \Delta\theta_{RL3})^t$, whose components are the compliance displacements and rotations in each link, in the form

$$\mathbf{W}_{IRL} = \mathbf{K}_{LRL} \Delta\mathbf{l}_{RL} \quad (9)$$

where $\mathbf{K}_{LRL}$ is a 6x6 diagonal matrix whose components are the lumped stiffness parameters defined in the model of Fig.7a).

The vector $\Delta\mathbf{l}_{RL}$ can be related to $\Delta\mathbf{L}_{RL}$ through

$$\Delta\mathbf{l}_{RL} = \mathbf{T}_{KRL} \Delta\mathbf{L}_{RL} \quad (10)$$

where $\mathbf{T}_{KRL}$ is a 6x6 matrix giving the relation between the compliance displacements occurring in each link and the compliance displacements in the point H_{RL}.

By using Eqs. (8,9,10) and Eq.(7) the expression of $\mathbf{K}_{RL}$ can be computed as

$$\mathbf{K}_{RL} = \mathbf{W}_{TRL} \mathbf{K}_{LRL} \mathbf{T}_{KRL} \quad (11)$$

Therefore, the 6x6 stiffness matrix $\mathbf{K}_{RL}$ can be computed once $\mathbf{W}_{TRL}$, $\mathbf{K}_{LRL}$, and $\mathbf{T}_{KRL}$ are known.

As regards the matrix $\mathbf{T}_{KRL}$, if it is invertible the Eq. (10) can give

$$\Delta\mathbf{L}_{RL} = \mathbf{T}_{KRL}^{-1} \Delta\mathbf{l}_{RL} \quad (12)$$

Nevertheless, the vector $\Delta\mathbf{L}_{RL}$ can be computed as a function of the vector $\Delta\mathbf{l}_{RL}$ by using the Kinematics of the right leg. In fact, one can write

$$\begin{bmatrix} \Delta X_{RL4} \\ \Delta Y_{RL4} \\ \Delta Z_{RL4} \end{bmatrix} = \mathbf{R}_{RL3}^{RL4} \begin{bmatrix} \Delta X_{RL3} \\ \Delta Y_{RL3} \\ \Delta Z_{RL3} \end{bmatrix}$$
$$\begin{bmatrix} \Delta X_{RL4} \\ \Delta Y_{RL4} \\ \Delta Z_{RL4} \end{bmatrix} = \mathbf{R}_{RL2}^{RL2'} \mathbf{R}_{RL2'}^{RL3} \mathbf{R}_{RL3}^{RL4} \begin{bmatrix} \Delta X_{RL2} \\ \Delta Y_{RL2} \\ \Delta Z_{RL2} \end{bmatrix} \quad (13)$$
$$\begin{bmatrix} \Delta X_{RL4} \\ \Delta Y_{RL4} \\ \Delta Z_{RL4} \end{bmatrix} = \mathbf{R}_{RL1}^{RL2} \mathbf{R}_{RL2}^{RL2'} \mathbf{R}_{RL2'}^{RL3} \mathbf{R}_{RL3}^{RL4} \begin{bmatrix} \Delta X_{RL1} \\ \Delta Y_{RL1} \\ \Delta Z_{RL1} \end{bmatrix}$$

where $\mathbf{R}_i^{i+1}$ is a 3x3 matrix that can be computed as the inverse of the $\mathbf{R}_{i+1}^i$ matrix given by extracting first three rows and columns from Eq. (1). Similarly, for the angular compliance displacements one can write

$$\begin{bmatrix} \Delta\gamma_{RL4} \\ \Delta\beta_{RL4} \\ \Delta\alpha_{RL4} \end{bmatrix} = \mathbf{R}_{RL3}^{RL4} \begin{bmatrix} \Delta\gamma_{RL3} \\ \Delta\beta_{RL3} \\ \Delta\theta_{RL3} \end{bmatrix}$$
$$\begin{bmatrix} \Delta\gamma_{RL4} \\ \Delta\beta_{RL4} \\ \Delta\alpha_{RL4} \end{bmatrix} = \mathbf{R}_{RL2}^{RL2'} \mathbf{R}_{RL2'}^{RL3} \mathbf{R}_{RL3}^{RL4} \begin{bmatrix} \Delta\gamma_{RL2} \\ \Delta\beta_{RL2} \\ \Delta\theta_{RL2} \end{bmatrix} \quad (14)$$
$$\begin{bmatrix} \Delta\gamma_{RL4} \\ \Delta\beta_{RL4} \\ \Delta\alpha_{RL4} \end{bmatrix} = \mathbf{R}_{RL0}^{RL1} \mathbf{R}_{RL1}^{RL2} \mathbf{R}_{RL2}^{RL2'} \mathbf{R}_{RL2'}^{RL3} \mathbf{R}_{RL3}^{RL4} \begin{bmatrix} \Delta\gamma_{RL0} \\ \Delta\beta_{RL0} \\ \Delta\theta_{RL0} \end{bmatrix}$$

Therefore, by extracting the suitable columns from Eqs. (13) and (14), and taking into account Eq. (12) the 6x6 matrix $\mathbf{T}^{-1}_{KRL}$ can be written in the form

$$\mathbf{T}_{KRL}^{-1} = \begin{bmatrix} s_{1RL} & c\theta_{RL3} & 1 & 0 & 0 & 0 \\ s_{2RL} & s\theta_{RL3} & 0 & 0 & 0 & 0 \\ -s\theta_{RL3} & 0 & 0 & 0 & 0 & 0 \\ 0 & 0 & 0 & -s\theta_{RL2} & 0 & 0 \\ 0 & 0 & 0 & -c\theta_{RL2} & 0 & 0 \\ 0 & 0 & 0 & 0 & 1 & 1 \end{bmatrix} \quad (15)$$

with

$$s_{1RL} = (c\theta_{RL1} c\theta_{RL2} - s\theta_{RL1} s\theta_{RL2}) c\theta_{RL3} \quad (16)$$
$$s_{2RL} = -(c\theta_{RL1} s\theta_{RL2} - s\theta_{RL1} c\theta_{RL2}) c\theta_{RL3}$$

As regards the matrix $\mathbf{W}_{TRL}$, if it is invertible the Eq.(11) can be written as

$$\mathbf{W}_{IRL} = \mathbf{W}_{TRL}^{-1} \mathbf{W}_{RL} \quad (17)$$

The matrix $\mathbf{W}^{-1}_{TRL}$ can be computed by carrying out a static analysis of the right leg through the simplified model of the static forces acting on the right leg shown in Fig.7b). By using this model and the kinematic model of Fig.4 the static equations can be written as

$$F_{RLi} = R_{RL(i+1)}^{RLi} F_{RL(i+1)} + P_{RLi} \qquad (18)$$
$$N_{RLi} = R_{RL(i+1)}^{RLi} N_{RLi} + L_{RL(i+1)}^{RLi} \times F_{RLi}$$

where R_{i+1}^{i} is a rotation matrix given by the first three rows and columns of the transformation matrices in the Eq. (1); L_{i+1}^{i} is the vector expressing the location of the higher numbered link to lower numbered link; i=0,1,2,3. It is worthy to note that F_{RL4} and N_{RL4} are 3x1 vectors, whose components are the forces and torques components of the 6x1 vector W_{RL}, respectively. The components of the 6x1 vector W_{IRL} are F_{RL0x}, F_{RL2x}, F_{RL3x}, N_{RL1z}, N_{RL2z}, N_{RL3z}. Therefore, W^{-1}_{TRL} matrix is given by computing these components from Eq. (18).

Since superposition principle is applicable for WABIAN-RIV, the effect of weights can be considered as an additive component of the wrench W_{RL} acting on H_{RL} in the form

$$W_{RL} = W'_{RL} + Weq_{RL} \qquad (19)$$

where Weq_{RL} is a wrench equivalent to the effect of weights and W'_{RL} is the wrench given through Eq. (18) when the effect of weights is neglected. In this way, W_{TRL}^{-1} can be written as

$$W_{TRL}^{-1} = \begin{bmatrix} b_{1RL} & b_{2RL} & s\theta_{RL1} & 0 & 0 & 0 \\ c\theta_{RL3} & -s\theta_{RL3} & 0 & 0 & 0 & 0 \\ 1 & 0 & 0 & 0 & 0 & 0 \\ 0 & 0 & b_{3RL} & b_{4RL} & b_{1RL}/c\theta_{RL1} & 0 \\ x_{RL2}s\theta_3 & b_{5RL} & 0 & 0 & 0 & 1 \\ 0 & x_{RL3} & 0 & 0 & 0 & 1 \end{bmatrix} \qquad (20)$$

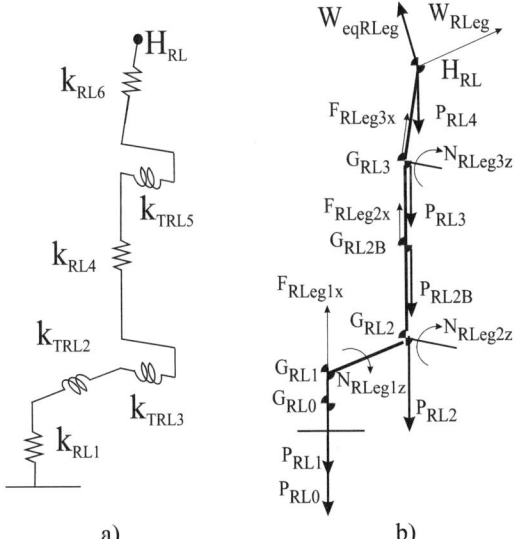

Fig.7 Simplified models for right leg of WABIAN-RIV: a) stiffness model, b) static model.

where

$$b_{1RL} = c\theta_{RL1}(c\theta_{RL2}c\theta_{RL3} - s\theta_{RL2}s\theta_{RL3})$$
$$b_{2RL} = -c\theta_{RL1}(c\theta_{RL2}s\theta_{RL3} - s\theta_{RL2}c\theta_{RL3})$$
$$b_{3RL} = -b_{1RL}x_{RL3} - c\theta_2 x_{RL2} \qquad (21)$$
$$b_{4RL} = s\theta_{RL2}c\theta_{RL3} + c\theta_{RL2}s\theta_{RL3}$$
$$b_{5RL} = x_{RL3} + x_{RL2}c\theta_3$$

In this case, the wrench Weq_{RL} can be computed by writing the equilibrium equations in X_{RL0}, Y_{RL0}, Z_{RL0} direction and along X_{RL0}-axis, Y_{RL0}-axis and Z_{RL0}-axis on point G_{RL0} for the model shown in Fig.7b) as

$$Weq_{RLFx} = P_{0RLx} + P_{1RLx} + P_{2RLx} + P_{2BRLx} + P_{3RLx} + P_{4RLx}$$
$$Weq_{RLFy} = 0 \; ; \quad Weq_{RLFz} = 0 \; ; \quad Weq_{RLTx} = 0$$
$$P_{RL2x} \times [G_{RLo}G_{RL2}]_{ZRL0} + P_{RL2Bx} \times [G_{RLo}G_{RL2B}]_{ZRL0} +$$
$$+ P_{RL3x} \times [G_{RLo}G_{RL3}]_{ZRL0} +$$
$$+ P_{RL4x} \times [G_{RLo}H_{RL}]_{ZRL0} = W_{eqLTy} + \qquad (22)$$
$$+ W_{eqLFx} \times [G_{RLo}H_{RL}]_{ZRL0}$$
$$P_{RL2Bx} \times [G_{RLo}G_{RL2B}]_{YRL0} + P_{RL3x} \times [G_{RLo}G_{RL3}]_{YRL0} +$$
$$+ P_{RL4x} \times [G_{RLo}H_{RL}]_{YRL0} =$$
$$= W_{eqLTz} + W_{eqLFx} \times [G_{RLo}H_{RL}]_{YRL0}$$

where P_{0RL}, P_{1RL}, P_{2RL}, P_{2BRL}, P_{3RL} and P_{4RL} are given by the weights of links and actuators lumped in G_{0RL}, G_{1RL}, G_{2RL}, G_{2BRL}, G_{3RL} and G_{4RL}, respectively; $[G_{RLi}G_{RLj}]_{kRL0}$ stands for the distance between points G_{RLi} and G_{RLj} in k_{RL0}-axis direction with k=X, Y, Z.

It is worthy to note that if the leg is in standing position Eq. (22) yields to $Weq_{RLFx}=0$ or if the leg is up it yields $Re_{RLx}=0$. However, the compliance displacements of the right leg when it is up does not contribute to the compliance displacements in the hands of WABIAN. Therefore, it can be assumed $Weq_{RLFx}=0$.

By using Eq. (22) and the above-mentioned assumptions the non zero components of the wrench Weq_{RL} in the {RL0} reference frame can be written as

$$Weq_{RLFx} = P_{0RLx} + P_{1RLx} + P_{2RLx} + P_{2BRLx} + P_{3RLx} + P_{4RLx}$$
$$Weq_{RLTy} = g[m_{RL2}d_{RL2} + m_{RL2B}(-\tfrac{1}{2} s\theta_{RL2}a_{RL2} + d_{RL2})$$
$$+ m_{RL3}(-s\theta_{RL2}a_{RL2} + d_{RL2}) - s\theta_{RL2}a_{RL2} + d_{RL2} \qquad (23)$$
$$Weq_{RLTz} = g[(\tfrac{1}{2} m_{RL2B} + m_{RL3})s\theta_{RL1}c\theta_{RL2}a_{RL2} +$$
$$+ m_4((s\theta_{RL1}c\theta_{RL2}c\theta_{RL3} - s\theta_{RL1}$$
$$(s\theta_{RL2}s\theta_{RL3})a_{RL3} + c\theta_{RL2}a_{RL2}))]$$

By using Eq. (23) and the inverse of the rotation matrices given by Eq. (1) the wrench Weq_{RL} in the {RL4} reference frame can be straightforward computed.

As regards the matrix K_{LRL}, it can be written as

$$\mathbf{K}_{LRL} = \begin{bmatrix} K_{RL1} & 0 & 0 & 0 & 0 & 0 \\ 0 & K_{RL4} & 0 & 0 & 0 & 0 \\ 0 & 0 & K_{RL6} & 0 & 0 & 0 \\ 0 & 0 & 0 & K_{TRL2} & 0 & 0 \\ 0 & 0 & 0 & 0 & K_{TRL3} & 0 \\ 0 & 0 & 0 & 0 & 0 & K_{TRL5} \end{bmatrix} \quad (24)$$

where K_{RL1}, K_{RL4}, K_{RL6}, K_{TRL2}, K_{TRL3} and K_{TRL5} are the lumped stiffness parameters shown in the model of Fig.7a).

B. Stiffness matrix for the other subcomponents

The stiffness matrix for each subcomponent can be computed as proposed in the previous section for the right leg. In particular, Eq. (11) can be written in the general form

$$\mathbf{K}_i = \mathbf{W}_{Ti} \mathbf{K}_{Li} \mathbf{T}_{Ki} \quad (25)$$

where the index i refers to one of the subcomponents that have been defined; $\mathbf{K}_{Li}$ is a 6x6 diagonal matrix whose components are the lumped stiffness parameters defined in the models reported in Figs.8a), 9a), 10a) and 11a) for trunk, right arm, waist and shoulders, respectively; $\mathbf{T}_{Ki}$ is a 6x6 matrix obtained by considering the kinematic models reported in Figs.3 to 6; $\mathbf{W}_{Ti}$ is a 6x6 matrix obtained through a static analysis for the related subcomponent that can be carried out by using the models reported in Figs.8b), 9b), 10b) and 11b) for trunk, right arm, waist and shoulders, respectively.

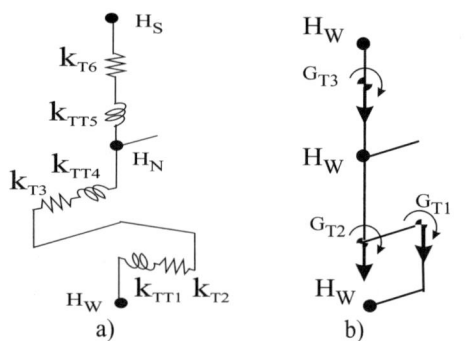

Fig. 8 Simplified models for trunk of WABIAN-RIV: a) stiffness model, b) static model.

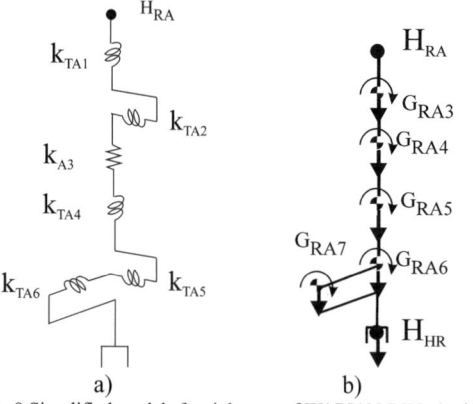

Fig.9 Simplified models for right arm of WABIAN-RIV: a) stiffness model, b) static model.

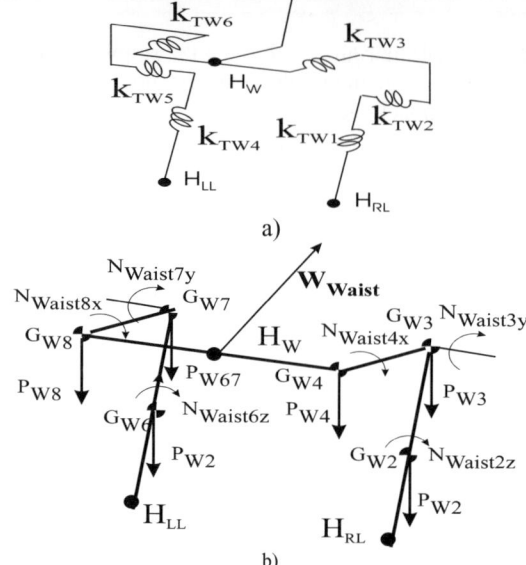

Fig.10 Simplified models for waist of WABIAN-RIV: a) stiffness model, b) static model.

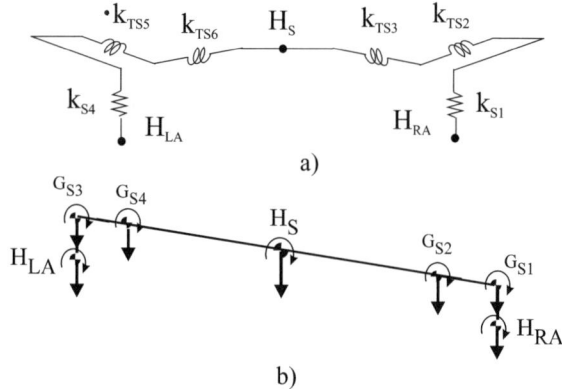

Fig.11 Simplified models for shoulders of WABIAN-RIV: a) stiffness model, b) static model.

V. A FORMULATION FOR STIFFNESS ANALYSIS OF WABIAN-RIV.

The stiffness matrices for each subcomponents can be computed through Eq. (25). By substituting the stiffness matrices for all the subcomponents in Eqs. (4) to (6) the stiffness matrix can be straightforward derived.

Once the stiffness matrix $\mathbf{K_W}$ is known, stiffness performances can be defined in different ways. For example, by using the determinant of the stiffness matrix as proposed in [16, 17] or by using stiffness indices as proposed for example in [19]. The stiffness performances can be also defined by means of the compliance displacements. In fact, if the stiffness matrix $\mathbf{K_W}$ is invertible one can write

$$\Delta \mathbf{S_W} = \mathbf{K_W}^{-1} \mathbf{W_g} \quad (26)$$

where $\mathbf{W_g}$ is a given 12x1 wrench. The compliance displacements computed through Eq. (26) give also an estimation of the accuracy of the manipulator in the case that the wrench $\mathbf{W_g}$ acts on it.

Eqs. (4) to (26) can be used even for design purposes in optimum design procedures. In fact, the optimum set of lumped stiffness parameters for a desired overall stiffness can be computed by inverting the proposed formulation. It is worth noting that formulations similar to the one proposed in this paper can be derived for most of the existing robotic system. Therefore, the above-mentioned consideration can be extent to other humanoid robots giving a general approach to improve their design and control.

VI. CONCLUSIONS.

In this paper a stiffness analysis of WABIAN-RIV (WAseda BIpedal humaNoid Refined IV) has been carried out based on proper kinematic, stiffness and static models of legs, waist, trunk, shoulders, neck and arms. By using these models a formulation for computing the stiffness matrix and compliance displacements is deduced. This formulation is suitable even for calculating the stiffness matrix in the ZMP (Zero Moment Point) during walking phases. The proposed formulation can be implemented in numerical algorithms in order to estimate stiffness performances of WABIAN-RIV. Similar models and formulation can be derived for other humanoid robots giving a general approach to improve their design and control.

IV. ACKNOWLEDGMENTS

The first author is thankful to the Italian Ministry MIUR for the research project RIME in the program PRIN01 and the Italian National Research Council CNR for the grant 203.21.04, which have permitted him to spend periods of study in the years 2002 and 2003 at the Humanoid Robotics Institute, Waseda University, Tokyo.

V. REFERENCES

[1] Bundeswehr University of Munich homepage, http://www.unibw-muenchen.de/hermes/

[2] Honda Motor Corporation homepage, http://www.honda.co.jp/ASIMO.

[3] KAIST homepage, http://mind.kaist.ac.kr/3_re/HumanRobot/HumanRobot.htm

[4] Kitano Symbiotic Systems Project homepage, http://www.symbio.jst.go.jp/PINO/index.html

[5] M.I.T. homepage, http://www.ai.mit.edu/projects/humanoid-robotics- group/index.html

[6] Sony Press Release, http://www.sony.co.jp/en/SonyInfo/News/Press/200011/00-057E2/

[7] Waseda University homepage, http://www.humanoid.waseda.ac.jp.

[8] Fukuda T., Michelini R., Potkonjak V., Tzafestas S., Valavanis K., Vukobratovic M., How far Away is Artificial Man?, *IEEE Robotics & Automation Magazine*, Vol.7, n.1, 2001, pp.66-73.

[9] Rosheim M.E., *Robot Evolution. The Development of Anthrobotics*, John Wiley & Sons, New York, 1994.

[10] Rivin E.I., *Mechanical Design of Robots*, McGraw-Hill, New York, 1988, pp.120-204.

[11] Tsai L-W., *Robot Analysis: the Mechanics of Serial and Parallel Manipulators*, John Wiley & Sons, New York, 1999.

[12] Duffy J., *Statics and Kinematics with Applications to Robotics*, Cambridge University Press, Cambridge, 1996, pp.153-169.

[13] Gosselin C., Stiffness Mapping for Parallel Manipulators, *IEEE Transactions on Robotics and Automation*, vol.6, n.3, 1990, pp.377-382.

[14] Kim H.Y., Streit D.A., Configuration Dependent Stiffness of the Puma 560 Manipulator: Analytical and Experimental Results, *Mechanism and Machine Theory*, vol.30, n.8, 1995, pp.1269-1277.

[15] Huang T., Mei M.P., Zhao X.Y., Zhou L.H., Zhang D.W., Zeng Z.P., Whitehouse D.J., Stiffness Estimation of a Tripod-Based Parallel Kinematic Machine; in *Proceedings of the IEEE International Conference on Robotics and Automation ICRA2001*, Seoul, vol.4, pp.3280-3285, 2001.

[16] Carbone G., Ceccarelli M., Teolis M., A Numerical Evaluation of the Stiffness of CaHyMan (Cassino Hybrid Manipulator), *Electronic Journal of Computation Kinematics EJCK*, http://www-sop.inria. fr/coprin/EJCK/EJCK.html, Vol. 1, n.1, 2002, Paper No.14.

[17] Ceccarelli M., Carbone G, A Stiffness Analysis for CaPaMan (Cassino Parallel Manipulator), *Mechanism and Machine Theory*, vol.37, n.5, 2002, pp.427-439.

[18] Chakarov D. Optimization Synthesis of Parallel Manipulators with Desired Stiffness, *Journal of Theoretical and Applied Mechanics*, Vol.28, n.4, 1998.

[19] Liu X.-J., Jin Z.-L., Gao F., Optimum Design of 3-Dof Spherical Parallel Manipulators with Respect to the Conditioning and Stiffness Indices, *Mechanism and Machine Theory*, Vol.35, n.9, 2000, pp.1257-1267.

[20] YKK Co. R. & D. Division Homepage, http://www.ykk.com/english/activities/research.html

[21] Lim H., Ishii A., Takanishi A., Emotion-based Walking of a Biped Humanoid Robot, in *Proceedings of the 13th CISM-IFToMM Symposium on Theory and Practice of Robots and Manipulators Ro.Man.Sy 2000*, Springer-Verlag, Wien, pp.295-306, 2000.

[22] Lim H-O., Setiawan S.A., Takanishi A., "Balance and Impedance Control for Biped Humanoid Robot Locomotion", in *Proceedings of the IEEE/RSJ International Conference on Intelligent Robots and Systems IROS 2001*, Maui, pp.494-499, 2001.

Auto-calibration for a Parallel Manipulator with Sensor Redundancy

Y.K. Yiu, J. Meng and Z.X. Li
Dept. of EEE, Hong Kong University of Science and Technology
(e-mail: eeyoyo@ust.hk, fax (852)2358-1485)

Abstract—In this paper, we propose two algorithms for the auto-calibration of the home position or the joint angle offsets for a parallel manipulator by utilizing the extra sensor(s) information (sensor redundancy), sampling over the workspace, and optimizing a suitably chosen cost function, without resorting to any other external equipment. Meanwhile, a measure or estimate of the precision of the machine is also obtained. It is very useful and convenient if the machine needs frequent re-calibration. Simulations and experiments are also performed to show the effectiveness of the algorithms.

I. INTRODUCTION

For any manipulator, to have an accurate and precise motion, there is always the need of calibration. For serial manipulators, it is normally done by precise external equipment such as CMM (Computer Measuring Machine) or laser interferometer. The equipment is expensive, and the calibration process is very time-consuming. For some sensors such as incremental encoders, there is no memory of the absolute position, and re-calibration is needed each time when the manipulator re-starts. Therefore, there is a great need of a fast, convenient and inexpensive method for calibration. Here, we propose a fast calibration method for parallel manipulators with sensor redundancy without resorting to any external equipment. Meanwhile we will also obtain a measure of the precision of the manipulator.

A good survey of the different kinematic analysis for both serial and parallel manipulators is given by Nielsen [8]. [1] reviewed the most notable advances in this field for the past 40 years. The focus of most of the papers in this respect is on solving the forward kinematics problems of a particular parallel manipulators, especially those of the Stewart platform and its variants, [3], [2], [9]. But there are also some general results for the planar parallel manipulator. The forward kinematics is the problem of determining the pose of the end-effector given the actuator coordinates. This is mathematically equivalent to solving a set of nonlinear equations. A number of numerical and algebraic techniques are proposed [4].

A mechanical system is said to be with sensor redundancy, if the number of sensors is more than the dof (degree-of-freedom) of the system, see, e.g., [10], [11] for details. In literature, sensor redundancy is used mostly to (1) eliminate multiple solutions of the forward kinematics problem, (2) speed up the computation of solutions for the forward kinematics problem, see, e.g., [6], [5]. [7] used the extra actuator(s) to share the load by optimizing some cost function with static force balance or dynamics considerations. In [11], the extra sensor(s) information is used to obtain a more accurate estimate of the end-effector coordinates in a purely kinematics consideration. However, here we want to explore the potential use of the extra sensor(s) information (sensor redundancy) in auto- or self- calibration of the manipulator without resorting to any other external equipment.

A. Outline of the Paper

In the next section, the calibration problem and the experimental platform will be described. Then, in Section III, the auto-calibration problem will be formulated and analyzed. In Section IV, two algorithms will be proposed to solve the auto-calibration problem. Finally in Section V, simulations and experiments results will be reported.

II. THE CALIBRATION PROBLEM

For the 2-dof parallel manipulator in our lab, its home position is roughly calibrated by manually moving the end-effector near a mark, each time when it is started. Thus, there may be up to several mm error for the home position. Furthermore, although the motors used are equipped with absolute encoders, the absolute position readings may be lost when there are some errors such as long power shortage, backup battery error, runaway error, and encoder reading error. Therefore, it is not convenient to calibrate the machine so often with expensive external equipment. We will show that with three encoders, we can calibrate the machine quickly by first taking a number of sample positions and then computing the home-position or the angle offsets by optimizing a cost function.

Fig. 1. A 2-DOF redundantly actuated parallel manipulator

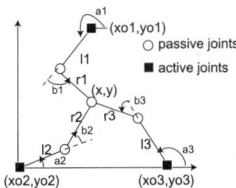

Fig. 2. Definitions for angles and lengths

A. Platform Description

The experimental platform is a 2-dof parallel manipulator with three actuators together with three internal encoders. It is built in the Machine Intelligence Laboratory in the Hong Kong University of Science and Technology. Its photo is shown in Figure 1. Definitions of lengths and angles are shown in Figure 2. We have identified the tree system or reduced system by cutting the common point for the multiple joints as shown in Figure 3. As a result, we have three 2-link robots, constrained in such a way that their end-effectors (the multiple joints) coincided with one another (A=B=C). The whole platform hardware system as depicted in Figure 4 consists of a PC, a motion control card GT400 from Googol Technology, three AC servo motors with drivers, and three 25:1 gear-boxes.

III. PROBLEM FORMULATION

Let us consider the 2-dof parallel manipulator in Figure 2 for an illustration. Let the end-effector coordinates be (x, y). Add the unknown angle offsets a_{oi}, which is to be determined after calibration, to the sensor joint angles $\tilde{a}_i$ to form the angles a_i. Let the passive joint angles with respect to the x-axis be apb_i (a_i plus b_i). Let (x_{oi}, y_{oi}) be the coordinates of the

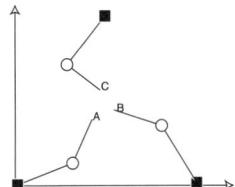

Fig. 3. The tree system

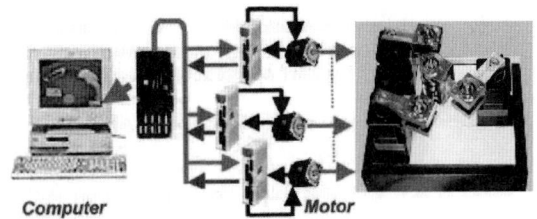

Fig. 4. The control hardware system for the 2-dof parallel manipulator

joint axis a_i. Then for a perfect parallel manipulator the following loop constraint equations are satisfied:

$$\begin{bmatrix} x \\ y \end{bmatrix} = \begin{bmatrix} x_{oi} + l_i cos(a_{oi} + \tilde{a}_i) + r_i cos(apb_i) \\ y_{oi} + l_i sin(a_{oi} + \tilde{a}_i) + r_i sin(apb_i) \end{bmatrix} \quad (1)$$

for $i = 1, 2$ if normally actuated, and for $i = 1, 2, 3$ if redundantly actuated with one additional actuator (or measured with one redundant sensor).

Here it can be easily seen why it is a 2-dof system, since the number of joint angles of the reduced tree system or of all the serial chains(a_i, b_i for $i = 1, 2, 3$) is six, while the number of effective constraint equations is four, their difference thus accounts for the two actual dof of the system.

Suppose now we take n sample points in the workspace, and let the corresponding angle readings be $\tilde{a}_i^j$ where $j = 1, 2, .., n$ are for different samples, $i = 1, 2$ or $i = 1, 2, 3$ for each serial link. Let the corresponding end-effector coordinates be (x^j, y^j), and the passive joint angles be apb_i^j, then we have:

$$\begin{bmatrix} x^j \\ y^j \end{bmatrix} = \begin{bmatrix} x_{oi} + l_i cos(a_{oi} + \tilde{a}_i^j) + r_i cos(apb_i^j) \\ y_{oi} + l_i sin(a_{oi} + \tilde{a}_i^j) + r_i sin(apb_i^j) \end{bmatrix}. \quad (2)$$

A. Problem Analysis

If there are only two sensors, i.e., the same as the dof of the system, then, the number of unknowns is $2n$(for x^j, y^j) $+ 2$(for a_{oi}) $+ 2n$(for apb_i^j), while, the number of equations is $4n$. Thus, it is impossible to determine the additional 3 unknown angle offsets a_{oi} by taking more samples.

If there are three sensors as in the case of our parallel manipulator, then, the number of unknowns is $2n$(for x^j, y^j) $+3$(for a_{oi}) $+3n$(for apb_i^j), while, the number of equations is $6n$.

Thus, in principle if we increase n, the number of samples, till the number of equations is equal to the number of unknowns ,i.e., when $n = 3$, then we can solve for all the unknowns and eventually calibrate the machines and obtain the angle offsets a_{oi}.

However, in practice, as there are always noise and other uncertainties, it is more feasible to find the angle offsets a_{oi} by optimizing some cost functions. One of the most popular cost functions is the sum of square:

$$J = \sum_{j=1}^{n}\sum_{i=1}^{3}\{(x^j - x_{oi} - l_i cos(a_{oi} + \tilde{a}_i^j))^2 - r_i cos(apb_i^j))^2 \\ + (y^j - y_{oi} - l_i sin(a_{oi} + \tilde{a}_i^j) - r_i sin(apb_i^j))^2\} \quad (3)$$

for $n \geq 3$.

Therefore, given the sensors readings $\tilde{a}_i^j$ for $i = 1, 2, 3$ and $j = 1, 2, 3, .., n$, if we regard $x^j, y^j, a_{oi}, apb_i^j$ as variables, then we can try to solve the following problem

$$\min_{x^j, y^j, a_{oi}, apb_i^j} J(x^j, y^j, a_{oi}, apb_i^j) \quad (4)$$

Let J_{min} be the value of this minimum J. J_{min} is actually a measure of the precision of the machine. We define the 'kinematics model root meant square error'(KMRMSE) which has the unit of length (mm) by the following equation:

$$\text{KMRMSE} = \sqrt{J/n} \quad (5)$$

where n is the number of sample points taken.

It should be noted that minimizing J is the same as minimizing the KMRMSE.

IV. ALGORITHMS

Two algorithms for auto-calibration (computing a_{oi}) based on the above analysis is described below:

A. Auto-calibration Algorithm 1

Step 1. Computing the Initial Values

1) The end-effector is manually moved to the approximate home position ,i.e., $(x^1, y^1) = (62, 92)$ in our case.
2) Read the corresponding sensors readings as our first sample $\tilde{a}_i^1$.
3) Solve the inverse kinematics problems for the home position to obtain an estimate for a_i^1 and compute the corresponding estimated angles offset $a_{oi} = a_i^1 - \tilde{a}_i^1$.
4) Take the other $n-1$ samples $\tilde{a}_i^j$ for $j = 2, 3, .., n$
5) For each of the n samples, with the initial estimate of a_{oi}, compute $a_i^j = \tilde{a}_i^j + a_{oi}$.
6) Using a forward kinematics map (see [10], [11] for a number of different forward kinematics maps for this platform) , compute the first estimate of x^j, y^j, apb_i^j
7) $(x^j, y^j, a_{oi}, apb_i^j)$ as computed above will be used as the initial value for the iterative optimization procedure in the Step 2.

Step 2. Iterative Optimization
Solve the below optimization problem with initial estimate computed in Step 1:

$$\min_{x^j, y^j, a_{oi}, apb_i^j} J(x^j, y^j, a_{oi}, apb_i^j) \quad (6)$$

To shorten the development time, we have used the MATLAB function 'fmincon' which is for solving general non-linear constrained optimization problems. A more specialized optimization procedure will be more efficient, but this needs further research.

The convergence rate of Algorithm 1 is very slow. However, this does not cause much problem as calibration is normally a once-for-all procedure and can be done off-line. Furthermore, it can be seen that the computation of the error does not depend on which forward kinematics map is actually used, as all x, y, apb_i as well as the angle offsets a_{oi} are regarded as variables for the optimization. If we consider only a_{oi} as the variables for optimization, and use one of the forward kinematics maps to compute the other variables x, y, apb_i(regarded as functions of a_{oi}), then we have Algorithm 2 which converges much faster and depends on the actual forward kinematics map in use.

B. Auto-calibration Algorithm 2(Depends on Forward Kinematics Map)

Step 1. Computing the Initial Values

1) The end-effector is manually moved to the approximate home position ,i.e., $(x^1, y^1) = (62, 92)$ in our case.
2) Read the corresponding sensors readings as our first sample $\tilde{a}_i^1$.
3) Solve the inverse kinematics problems for the home position to obtain an estimate for a_i^1 and compute the corresponding estimated angles offset $a_{oi} = a_i^1 - \tilde{a}_i^1$.
4) Take the other $n-1$ samples $\tilde{a}_i^j$ for $j = 2, 3, .., n$

Step 2. Iterative Optimization

1) Only a_{oi} is regarded as the variables for optimization.
2) Other variables x^j, y^j, apb_i^j are computed from the sensors readings $\tilde{a}_i^j$ and the optimization variable a_{oi} using a chosen forward kinematics map.
3) The following optimization problem is solved by iterative method with the initial values computed in Step 1.

$$\min_{a_{oi}} J(a_{oi}) \quad (7)$$

In principle, we may regard other kinematics parameters as variables such as l_i, r_i, x_{oi}, y_{oi} as well. In this way, we may even calibrate those kinematics parameters by taking more samples with redundant sensor(s) and performing optimization. However, the simulation result of this is not very accurate, and this requires further research.

V. SIMULATION AND EXPERIMENT RESULTS

We performed a number of simulations and experiments to verify the above algorithms.

(i) The following simulations are performed:
1) Assume the kinematics model is exact (l_i, r_i, x_{oi}, y_{oi} are exactly known), and preset a fixed angle offset a_{oi}, assume a mild home position error to compute the rough estimate of a_{oi}, use Algorithm 1 to recover the exact a_{oi}.
2) Assume the kinematics model is exact (l_i, r_i, x_{oi}, y_{oi} are exactly known), and preset a fixed angle offset a_{oi}, assume a mild home position error to compute the rough estimate of a_{oi}, use Algorithm 2 to recover the exact a_{oi}.
3) Suppose we only have a rough estimate of the kinematics parameters l_i, r_i, x_{oi}, y_{oi} as well as a_{oi}, use an Algorithm-2-like optimization procedure to recover them all.

(ii) The following experiments are performed:
1) Use Algorithm 1 to recover a_{oi} for number of samples $n = 17$ and $n = 54$.
2) Use Algorithm 2 to recover a_{oi} with the forward kinematics map f_{123} (see [10], [11] for its definition) for $n = 17$ and $n = 54$.

The detailed results are reported as follows:

A. Simulation Results for Algorithm 1

- Number of sample points: 10
- The true initial position (regarded as home position $(62, 92)$): $(65, 95)$ (3 mm error)
- Results are shown in Table I.
- Final estimate is almost the same as the true value.
- KMRMSE converges from 0.8399 mm to 0.000108187 mm in 150 iterations (see Figure 5). The convergence rate is much slower than that of Algorithm 2 as shown in Figure 5.

B. Simulation for Algorithm 2

- Number of sample points: 10
- The true initial position (regarded as home position $(62, 92)$): $(65, 95)$ (3 mm error)
- Results are shown in Table II.
- Final estimate is almost the same as the true value.

TABLE I
Simulation results for Algorithm 1

Variables	True value	Initial Est.	Final Est.
a_{o1} (degree)	11.11111	7.5980	11.1113
a_{o2} (degree)	-22.22222	-18.9263	-22.2226
a_{o3} (degree)	33.33333	36.0679	33.3335
x^0 (mm)	65	62	65.0006
y^0 (mm)	95	92	94.9997

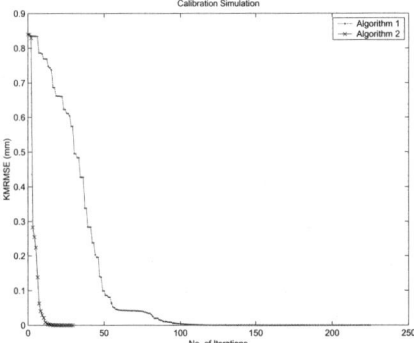

Fig. 5. Calibration simulation for Algorithm 1 and 2

- KMRMSE converges from 0.8399 mm to 0.000188832 mm in 21 iterations (see Figure 5). The convergence rate is much faster than that of Algorithm 1 as shown in Figure 5.

TABLE II
Simulation results for Algorithm 2

Variables	True value	Initial Est.	Final Est.
a_{o1} (degree)	11.11111	7.5980	11.1111
a_{o2} (degree)	-22.22222	-18.9263	-22.2223
a_{o3} (degree)	33.33333	36.0679	33.3333

C. Simulation for Calibrating the Kinematics Parameters

The algorithm used is a modified version of Algorithm 2. However, the kinematics parameters (l_i, r_i, x_{oi}, y_{oi}) are also regarded as variables for the optimization, hoping that this kinematics parameters as well as the angle offsets a_{oi} can be calibrated accurately. A total of 17 sample points are taken. The variables x_{o2}, y_{o2}, y_{o3} are fixed in order to remove the symmetry due to planar rotation and planar translation ($SE(2)$) of the coordinate transformations, thus, the dimension of the problem is reduced by 3. Therefore, the origin is always taken to be (x_{o2}, y_{o2}) of the joint axis a_2, and the x-axis always passes through the joint axis of a_3. The estimated results for the kinematics parameters are not very accurate, but those for the angle offsets a_{oi} are still acceptable. The reasons

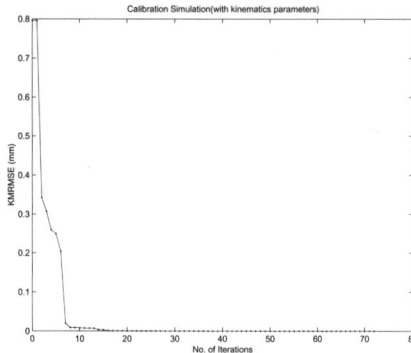

Fig. 6. Calibration simulation(with kinematics parameters)

why the estimates are not accurate for kinematics parameters are still not very well-understood.
- Number of sample points: 17
- Result is shown in Table III
- KMRMSE converges from 0.7964 mm to 0.000104815 mm in 26 iterations
- Final estimates of a_{oi} are acceptable, while those of the kinematics parameters $(l_i, r_i, x_{oi}, y_{oi})$ are not very good.

TABLE III

Simulation results for Algorithm 2 with kinematics parameters

Variables	True Value	Initial Est.	Final Est.
a_{o1} (degree)	11.11111	7.5905	11.1144
a_{o2} (degree)	-22.22222	-18.940	-22.2350
a_{o3} (degree)	33.33333	36.0307	33.3274
l_1 (mm)	70.0100	70.0000	70.0005
r_1 (mm)	69.9800	70.0000	70.0182
l_2 (mm)	70.0300	70.0000	69.9662
r_2 (mm)	69.9600	70.0000	70.0182
l_3 (mm)	70.0500	70.0000	69.9974
r_3 (mm)	69.9900	70.0000	70.0182
x_{o1} (mm)	61.9800	62.0000	61.9922
y_{o1} (mm)	184.0300	184.0000	184.0106
*x_{o2} (mm)	*0	*0	*0
*y_{o2} (mm)	*0	*0	*0
x_{o3} (mm)	123.9600	124.0000	123.9906
*y_{o3} (mm)	*0	*0	*0

* - Fixed to remove symmetry.

D. Experiment for Algorithm 1
- Numbers of samples are 16 and 54.
- Results are shown in Table IV for n=16 and in Table V for n=54.
- KMRMSE converges from 2.1576 mm to 0.1120 mm in 279 iterations for n=16 (see Figure 7).
- KMRMSE converges from 1.8935 mm to 0.1735 mm in 3503 iterations for n=54 (see Figure 8).
- Convergence rates of Algorithm 1 are much slower than those of Algorithm 2 (see Figure 7 and compare Figure 8 and Figure 9.)

TABLE IV

Experiment result for Algorithm 1 (n=16)

Variables	Initial Estimate	Final Estimate
a_{o1} (degree)	-105.7435	-106.6543
a_{o2} (degree)	8.9093	11.7025
a_{o3} (degree)	-60.1662	-63.8143
x^0 (mm)	62	57.0101
y^0 (mm)	92	96.3202

TABLE V

Experiment result for Algorithm 1 (n=54)

Variables	Initial Estimate	Final Estimate
a_{o1} (degree)	-105.4288	-105.4191
a_{o2} (degree)	8.8407	10.5865
a_{o3} (degree)	-60.8236	-64.4441
x^0 (mm)	62	58.3439
y^0 (mm)	92	96.3251

E. Experiment for Algorithm 2
- Numbers of samples are 16 and 54.
- Results are shown in Table VI for n=16 and in Table VII for n=54.
- For n=16, KMRMSE converges from 2.1576 mm to 0.1231 mm in 15 iterations ,i.e., a 175% improvement in precision. See Figure 7.
- For n=54, KMRMSE converges from 1.8935 mm to 0.0808477 mm in 18 iterations ,i.e., a 230% improvement in precision. See Figure 9.
- For n=16, the final KMRMSE is about $120\mu m$ which is reasonable as compared with the error (in the order of $131\mu m$ due to the 9-arc-minute backlash in the gearbox) in the simulation results of RMSE (see [11]) on the optimal forward kinematics map. This is approaching the machine precision limit that the manipulator structure allows.
- For n=54, the final KMRMSE is about $80\mu m$ which is reasonable as compared with the error (in the order of $131\mu m$ due to the 9-arc-minute backlash in the gearbox) in the simulation results of RMSE (see [11]) on the optimal forward kinematics map.
- The final KMRMSE is reduced from $120\mu m$ to $80\mu m$ when the number of samples increases from n=16 to n=54. However, there will be a limit for KMRMSE to decrease by increasing the number of samples. This limit is indeed a measure of the machine precision limited by the kinematics structure of the manipulator.

TABLE VI

Experiment result for Algorithm 2 (n=16)

Variables	Initial Estimate	Final Estimate
a_{o1} (degree)	-105.7435	-106.5857
a_{o2} (degree)	8.9093	11.6523
a_{o3} (degree)	-60.1662	-63.8658

TABLE VII

Experiment result for Algorithm 2 (n=54)

Variables	Initial Estimate	Final Estimate
a_{o1} (degree)	-105.4288	-106.5754
a_{o2} (degree)	8.8407	11.6962
a_{o3} (degree)	-60.8236	-63.8037

VI. CONCLUSION

We propose two different algorithms for auto- or self-calibration of parallel manipulator with sensor redundancy without resorting any other external equipment. The algorithms compute the home position or the joint angle offsets by utilizing the extra sensor information, sampling a number of points over the workspace, and optimizing the least square error cost function. Meanwhile the final value for the least square error is a useful indication of the precision of the machine. Various simulations and experiments are also performed to verify the effectiveness of the proposed algorithms.

VII. REFERENCES

[1] Arthur G. Erdman. Modern Kinematics - Developments in the Last Forty Years. John Wesley & Sons, Inc., 1993.

[2] M.L. Husty. An algorithm for solving the direct kinematics of general stewart-gough platforms. Mech. Mach. Theory, 31(4), 1996.

[3] R. Kamra, D. Kohli, and A.K. Dhingra. Displacement analysis of a six-dof parallel manipulator with 3r3p and 4r2p chains. In Proc. Of DETC 98, New York: American Society of Mechanical Engineers, 1998.

[4] T.Y Lee and J.K. Shim. Algebraic elimination-based real-time forward kinematics of the 6-6 stewart platform with planar base and platform. In Proceedings of IEEE International Conference on Robotics and Automation, pages 1301–1306, 2001.

[5] G.F. Liu, X.Z. Wu, and Z.X. Li. Inertial equivalent principle and adaptive control of redundant parallel manipulators. In Proceedings of IEEE International Conference on Robotics and Automation, 2002.

[6] G.F. Liu, Y.K. Yiu, and Z.X. Li. Inertial equivalent principle and adaptive control of redundant parallel manipulators. In American Control Conference, 2002.

[7] Y. Nakamura and M. Ghodoussi. Dynamics computation of closed-link robot mechanisms with nonredundant and redundant actuators. IEEE Transactions on Robotics and Automation, 5(3):294–302, 1989.

[8] J. Nielsen and B. Roth. On the kinematic analysis of robotic mechanisms. International Journal of Robotics Ressearch, 18(12), 1999.

[9] C.W. Wampler. Numerical continuation methods for solving polynomial systems arising in kinematics. J. of Mech. Design, 112(59), 1990.

[10] Y.K. Yiu. Geometry, Dynamics and Control of Parallel Manipulators. PhD thesis, EEE Dept., The Hong Kong University of Science and Technology, 2002.

[11] Y.K.Yiu and Z.X. Li. Optimal forward kinematics map for a parallel manipulator with sensor redundancy. In IEEE Int. Symposium on Computational Intelligence in Robotics and Automation (CIRA2003)(submitted), 2003.

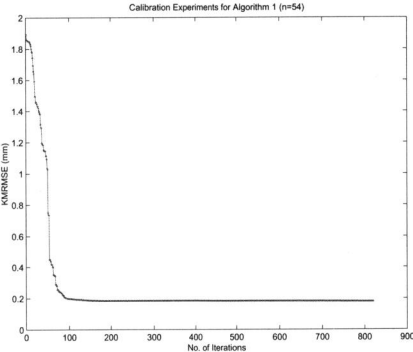

Fig. 8. Calibration graph for Algorithm 1 (n=54)

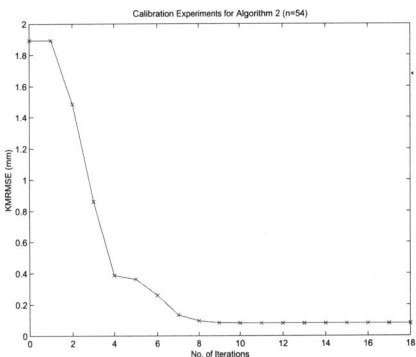

Fig. 9. Calibration graph for Algorithm 2 (n=54)

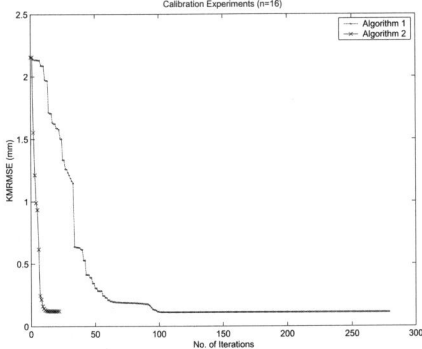

Fig. 7. Calibration graph for Algorithm 1 and 2 (n=16)

Optimum Force Balancing with Mass Distribution and a Single Elastic Element for a Five-bar Parallel Manipulator

Gürsel Alıcı and Bijan Shirinzadeh
Robotics & Mechatronics Research Laboratory
Department of Mechanical Engineering
Monash University
3800, VIC, Australia
E-mail: gursel.alici@eng.monash.edu.au

Abstract -- This paper deals with optimum force balancing of a planar parallel manipulator, articulated with revolute joints, by properly distributing link masses, and connecting only one spring between its two coupler links. After conducting the static force analysis of the mechanism, the force balancing is formulated as an optimisation problem such that a sum-squared values of bearing forces is minimized throughout an operation range of the manipulator, provided that a set of balancing constraints consisting of balancing conditions, the sizes of some inertial and geometric parameters are satisfied. Optimisation results indicate that the proposed optimisation approach is systematic, versatile and easy to implement for the optimum balancing of the parallel manipulator and other more general parallel manipulators. This work contributes to previously published work from the point of view of being a step towards optimum design of parallel manipulators, which is currently lacking in the literature.

1. INTRODUCTION

When the weights of the links of articulated mechanisms are not balanced, their performance is heavily impaired, and large actuator forces are demanded to move their links. The object of this study is, therefore, the optimum force balancing[1] of a revolute-jointed planar parallel mechanism by proper selection of mass distribution of the links, the size and attachment point of an elastic element- spring. Force balancing can be achieved by determining balancing parameters resulting in a constant total potential energy of the manipulator for all of its configurations. This follows that the manipulator is in equilibrium with zero actuator force, and less powerful and smaller actuators can be employed to move it. Of course, distributing or redistributing (in the form of adding counterweights) the mass of the links, and elastic elements increase the inertia forces as well as bearing and ground forces. We, therefore, employ a non-linear programming method in the selection of the mass distributions of the links, spring size and attachment points to ensure that the manipulator is optimum with respect to the bearing and ground forces. A mechanism having a poor geometry and mechanical advantage (transmission angles) is likely to have a poor global performance even though it is properly balanced. Therefore, the link dimensions obtained from another optimisation procedure [1] based on high kinematic accuracy, dexterity and singularity avoidance capabilities with a maximum of high mechanical advantage are used for force balancing.

Although there is a wealth of literature on balancing of single input mechanisms [2-6], planar and spatial parallel manipulators [7-10], and optimisation procedures based on the minimization of the force and moment transmitted to the ground [11-13], a little has been published on the optimum force balancing of planar parallel manipulators based on the minimization of all bearing and support (ground) forces. Yan et. all [3] have reported on a balancing method that combines kinematic synthesis, dynamic design, and input speed trajectory design for four-bar linkages. A wide account of studies on balancing methodologies for single degree of freedom mechanisms is provided in [4]. In a recent study [14], the mass distribution of a four-bar mechanism with small clearances at its three passive joints has been optimized without adding any counterweights to any links.

The work most relevant to this study includes that of Jean and Gosselin [7] who have reported on the static balancing of planar parallel manipulators based on the mass distribution of links, without paying attention to the minimization of the other forces in the mechanism. Later, Laliberte et. all [8] have studied static balancing of 3-DOF planar parallel manipulators and presented balancing conditions based on the mass distribution of the links and elastic elements. Static balancing of spatial parallel platform type mechanisms based on elastic elements only has been investigated by Ebert-Uhoff et.all [10], and least restrictive balancing conditions are determined. Wang and Gosselin [9] have described static balancing conditions for spatial 3-DOF parallel mechanisms by employing mass distribution of the links and springs. The mathematical framework for employing elastic elements to statically balance mechanisms has been derived by Streit and Gilmore [15]. It has been stressed that when elastic elements such as springs together with some counterweights are exploited for balancing, the total potential energy of the system consisting of elastic and gravitational potential energies is constant for all the configurations of the mechanism, and most importantly, a much smaller counterweight is required for balancing. When balancing with counterweights (or mass redistribution of mechanism

---

[1] Force balancing, gravity balancing and static balancing are used interchangeably in the mechanism literature.

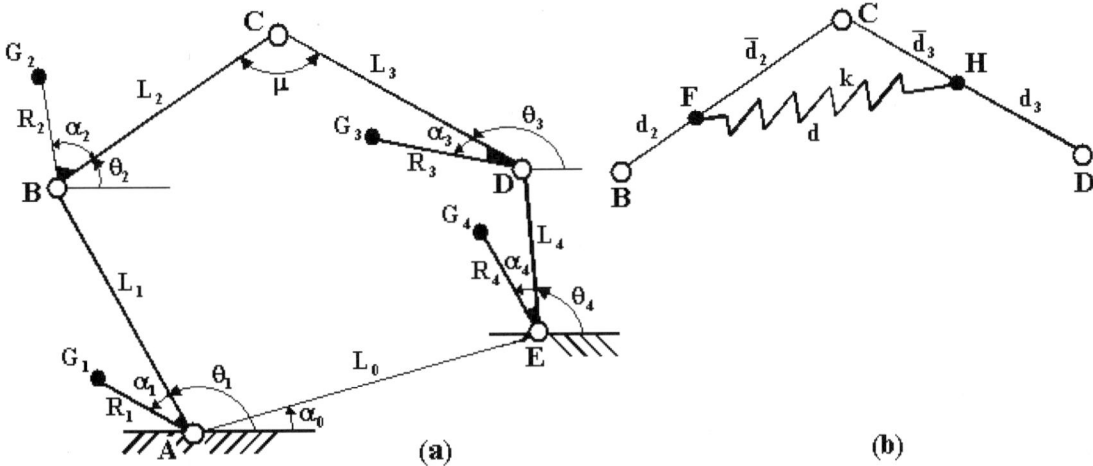

Fig. 1: Planar manipulator with force balancing parameters. The angle α_i is from the longitudinal axis of the ith link.

links) and elastic elements/springs are compared to each other, balancing with springs has the advantages of adapting to different payloads and allowing the manipulator to come to a new equilibrium configuration. When counterweights only are employed for balancing; the manipulator will collapse under a payload different from the design value. Further, counterweight balancing requires either a large mass or a large moment arm. While the heavy mass tends to increase the overall weight of the mechanism, the large moment arm restricts the usable workspace of the mechanism. Both options are not desirable. If it is not impossible, spring balancing should be preferred to counterweight balancing.

We will follow the mathematical formulation derived by Berkof and Lowen [5,16] to determine the necessary optimisation constraints for a constant potential energy of the system. Static force analysis of the manipulator is accomplished by using matrix method. The objective function employed is the sum-squared values of the bearing and ground forces calculated for a practical range of operation of the mechanism. The optimisation procedure is implemented in MATLAB using the *constr( )* function(In the new version of the MATLAB Optimisation Toolbox, it has been superseded by *fmincon*( )), which performs nonlinear optimisation based on a sequential quadratic programming method [17]. A set of optimisation trials were accomplished to prove that this optimisation methodology was an efficient, versatile and systematic procedure, and could be employed to determine the best possible balancing parameters consisting of mass distribution and elastic element. However, it must be kept in mind that the optimum solution depends on the initial values assigned to the balancing parameters and it does not necessarily represent a global minimum.

II. FIVE-BAR MANIPULATOR WITH REVOLUTE JOINTS

The five-bar planar manipulator considered in this study is shown in Fig. 1, where its two joints (A and E) connected to the ground are active and the others are passive joints. The input motions of the active joints can be independent from each other or be provided via a set of gears maintaining a specified phase angle between the two active [1]. Analytical expressions for the coordinates of the output point C, where the end effector is assembled to, are obtained for the provided joint inputs θ_1 and θ_4, and the specified link lengths L_0, L_1, L_2, L_3, L_4 and the angle α_0 [18]. It must be noted that for a parallel RRRRR manipulator $L_1 = L_4$ and $L_2 = L_3$. The angle α_i and the radial distance R_i describe the center of mass G_i of the ith link. The spring with a stiffness constant k of is located between the links 2 and 3. Its attachment point to the links is described by d_2, d_3.

III. STATIC FORCE ANALYSIS

The aim of static force analysis is to obtain analytical expressions for the forces acting on the bearings A, B, C, D, E while the manipulator is in equilibrium. The spring force is equal to

$$F_s = k(d - d_0) \tag{1}$$

where d_0 and d indicate the initial and final length of the spring. As explained in [15], it is reasonable for force balancing to assume that the initial length of the springs is zero. This can be physically achieved by placing the spring outside the line between the links via a wire and a pulley system [8,15].

For the sake of brevity, the free-body diagrams of the links are omitted here. It is assumed that the spring is massless and linear, and there is no friction in the system. The forces $F_{Ax}, F_{Ay}, F_{Bx}, F_{By}, F_{Cx}, F_{Cy}, F_{Dx}, F_{Dy}, F_{Ex}, F_{Ey}$ acting at the joints are obtained from the equilibrium equations written for each link of the manipulator. The resulting equations are put in a matrix-vector form;

$$[M][F] = [F_g] \quad (2)$$

where M, F, and F_g denote the square matrix of known mechanism dimensional parameters and joint angles, the vector of unknown forces, and the vector of gravitational and spring forces, respectively. As seen from the determinant of M given below, it is singular when $\theta_4 = \frac{\theta_2 + \theta_3}{2} + \frac{\pi}{2}(2p+1)$ for p =0,1.. or $\theta_1 = \theta_2$ or $\theta_2 = \theta_3$.

$$|M| = -2L_2^2 L_4^2 \cos\left(\theta_4 - \frac{\theta_2 + \theta_3}{2}\right) \sin(\theta_1 - \theta_2) \sin\left(\frac{\theta_2 - \theta_3}{2}\right) \quad (3)$$

Equation (2) is solved for the unknown forces by assuming that the matrix M is non-singular throughout the operation range of the manipulator.

IV. FORMULATION OF OPTIMUM FORCE BALANCING

The aim of a typical balancing procedure is to minimize the forces transmitted to the ground. In addition to this, the object of the optimisation procedure implemented in this study is to minimize all the reaction forces. As argued before, if it is not impossible, balancing with elastic elements should be preferred to counterweight balancing or mass distribution. With this in mind, the force-balancing conditions are determined for a combination of spring and mass distribution of the links.

In this case, a force-balanced mechanism requires that the total potential energy consisting of gravitational and elastic potential energies is constant for all configurations of the mechanism. The total potential energy is formulated as

$$V = M_T \, g R_g + \frac{1}{2} k \, d^2 \quad (4)$$

where M_T is the total mass of the moving links, R_g is the vertical component of the position vector $\vec{R}$, given by (8), describing the mass center of the mechanism. Using the notation given in Fig. 1, the position vector $\vec{R}$ for the mass center of the mechanism is

$$\vec{R} = \frac{1}{M_T} \sum_{i=0}^{4} m_i \vec{r}_i \quad (5)$$

where $\vec{r}_i$ is the position vector for the mass center G_i of the ith moving link with a mass of m_i with respect to the reference point A. The individual position vectors are expressed in complex numbers as

$$\vec{r}_1 = R_1 e^{j(\theta_1 + \alpha_1)}, \quad \vec{r}_2 = L_1 e^{j\theta_1} + R_2 e^{j(\theta_2 + \alpha_2)},$$
$$\vec{r}_3 = L_0 e^{j\alpha_0} + L_4 e^{j\theta_4} + R_3 e^{j(\theta_3 + \alpha_3)}, \quad (6)$$
$$\vec{r}_4 = L_0 e^{j\alpha_0} + R_4 e^{j(\theta_4 + \alpha_4)},$$

The exponential terms are related to each other by the loop closure equation

$$L_1 e^{j\theta_1} + L_2 e^{j\theta_2} = L_0 e^{j\alpha_0} + L_3 e^{j\theta_3} + L_4 e^{j\theta_4} \quad (7)$$

The exponential term $e^{j\theta_3}$ (or $e^{j\theta_2}$) is extracted from (7), and then substituted into (6). Now, the position vectors in (6) are put into (5) from which the position vector describing the overall mass center of the mechanism is obtained as

$$\vec{R} = \frac{1}{M_T} \begin{bmatrix} e^{j\theta_1}\left(m_1 R_1 e^{j\alpha_1} + \frac{m_3 R_3 L_1}{L_3} e^{j\alpha_3} + m_2 L_1\right) \\ + e^{j\theta_2}\left(m_2 R_2 e^{j\alpha_2} + \frac{m_3 R_3 L_2}{L_3} e^{j\alpha_3}\right) \\ + e^{j\theta_4}\left(m_4 R_4 e^{j\alpha_4} - \frac{m_3 R_3 L_4}{L_3} e^{j\alpha_3} + m_3 L_4\right) \\ + L_0 e^{j\alpha_0}\left(-\frac{m_3 R_3}{L_3} e^{j\alpha_3} + m_3 + m_4\right) \end{bmatrix} \quad (8)$$

As explained in Section 3, the elongation or retraction of the spring in the potential energy described by (4) is the spring length d expressed as

$$d^2 = \bar{d}_2^2 + \bar{d}_3^2 - 2\bar{d}_2 \bar{d}_3 \cos \mu \quad (9)$$

When (9) and the component of $\vec{R}$ associated with the gravitational acceleration are substituted into (4), the total potential energy of the system becomes

$$V = g \begin{bmatrix} \sin\theta_1\left(m_1 R_1 e^{j\alpha_1} + \frac{m_3 R_3 L_1}{L_3} e^{j\alpha_3} + m_2 L_1\right) \\ + \sin\theta_2\left(m_2 R_2 e^{j\alpha_2} + \frac{m_3 R_3 L_2}{L_3} e^{j\alpha_3}\right) \\ + \sin\theta_4\left(m_4 R_4 e^{j\alpha_4} - \frac{m_3 R_3 L_4}{L_3} e^{j\alpha_3} + m_3 L_4\right) \\ + L_0 \sin\alpha_0\left(-\frac{m_3 R_3}{L_3} e^{j\alpha_3} + m_3 + m_4\right) \end{bmatrix} + \frac{1}{2}k\left[\bar{d}_2^2 + \bar{d}_3^2\right] + \frac{1}{2}k\left[-2\bar{d}_2 \bar{d}_3 \cos \mu\right] \quad (10)$$

With reference to Fig. 1,

$$\cos \mu = \frac{L_2^2 + L_3^2 - Q^2}{2L_3 L_2} \quad (11)$$

where Q is the distance between point B and D, and is expressed as

$$Q^2 = L_0^2 + 2L_0 L_1 (\cos\theta_4 - \cos\theta_1) + 2L_1^2 [1 - \cos(\theta_1 - \theta_4)] \quad (12)$$

If the coefficients of $\sin\theta_z$ (for z =1,2,4) and the last term in (10) vanish, the potential energy of the system will be constant. If it is assumed that $\theta_1 - \theta_4$ is a constant, which is needed to make the last part of (12) constant, after necessary mathematical operations, the following force balancing conditions will be found;

$$g\left(m_1 R_1 \cos\alpha_1 + \frac{m_3 R_3 L_1}{L_3}\cos\alpha_3 + m_2 L_1\right) \\ - k\frac{\overline{d_2}\,\overline{d_3}}{L_3 L_2} L_0 L_1 = 0 \quad (13)$$

$$g\left(m_1 R_1 \sin\alpha_1 + \frac{m_3 R_3 L_1}{L_3}\sin\alpha_3\right) = 0 \quad (14)$$

$$g\left(m_2 R_2 \cos\alpha_2 + \frac{m_3 R_3 L_2}{L_3}\cos\alpha_3\right) = 0 \quad (15)$$

$$g\left(m_2 R_2 \sin\alpha_2 + \frac{m_3 R_3 L_2}{L_3}\sin\alpha_3\right) = 0 \quad (16)$$

$$g\left(m_4 R_4 \cos\alpha_4 - \frac{m_3 R_3 L_4}{L_3}\cos\alpha_3 + m_3 L_4\right) \\ + k\frac{\overline{d_2}\,\overline{d_3}}{L_3 L_2} L_0 L_1 = 0 \quad (17)$$

$$g\left(m_4 R_4 \sin\alpha_4 - \frac{m_3 R_3 L_4}{L_3}\sin\alpha_3\right) = 0 \quad (18)$$

From Eqs.15 and 16,

$$\alpha_2 = \alpha_3 \text{ and } R_3 = -\left(\frac{m_2}{m_3}\right)\left(\frac{L_3}{L_2}\right) R_2 \quad (19)$$

This is an important observation for the mass distribution of links 2 and 3, which have to obey (19) for force balancing. Depending on the mass and the link length ratios, R_3 is evaluated in terms of negative R_2, which implies that the locations of the mass centers for link 2 and link 3 are separated from each other by 180^0. This might restrict the usable workspace of the manipulator. But, it can be avoided by imposing tight constraints on the sizes of the radial distances R_2 and R_3. The remaining four conditions (Eqs.13,14,17,18) are taken as the constraints which must be satisfied by the balancing parameters while minimizing an optimisation function described in the next subsection.

A. Objective Function

Similar to the optimisation functions used before by others [3,11-12,14] for optimum balancing of single degree freedom mechanisms, a sum-squared discrete values of all reaction forces in the manipulator is adopted as the objective function (OF)

$$OF = \frac{1}{n}\sqrt{\sum_{i=1}^{n}\left(F_A^2 + F_B^2 + F_C^2 + F_D^2 + F_E^2\right)} \quad (20)$$

where n is the number of the discrete values of the operation range of manipulator divided into. The goal of the optimisation is to determine the numerical values of m_i, R_i, α_i, (for i= 1,2,3,4), k, d_2, d_3 of the manipulator minimizing the objective function and satisfying the constraints given in the next subsection.

B. Constraints

In order to limit the solution, the objective function is subjected to the following constraints, in addition to the constraints imposed by (13),(14),(17-19):

1. $-L_z \leq R_z \leq L_z$ for $z = 1\cdots 4$,
2. $1 \leq m_z \leq 5$ for $z = 1\cdots 4$,
3. $0 \leq \alpha_z \leq 180^0$ for $z = 1\cdots 4$,
4. $-L_z \leq d_z \leq L_z$ for $z = 2, 3$,
5. $100 \leq k \leq 200$

Hence, force balancing of the manipulator is formulated as a constrained nonlinear optimisation problem. A computer program based on a sequential quadratic programming method is prepared in MATLAB using the *constr( )* function to accomplish the constrained minimization of the OF as a function of the balancing parameters, starting with an initial value for each parameter.

V. NUMERICAL RESULTS AND DISCUSSION

As it is well known that a mechanism with a poor geometry and transmission angles, which is the angle μ between L_2 and L_3 of Fig. 1, will likely have a bad overall performance. With this in mind, the link lengths of the mechanism are obtained from another optimisation procedure based on the minimization of the overall deviation of the condition number of the manipulator Jacobian matrix from the ideal/isotropic condition number throughout the

TABLE 1
THE SOLUTIONS OBTAINED FROM THE CONSTRAINED OPTIMISATION OF FORCE BALANCING.

Solution No.	Initial Values $\begin{pmatrix} R_1, R_2, R_4, m_1, m_2, m_4, \\ \alpha_1, \alpha_2, \alpha_4, d_2, d_3, k \end{pmatrix}$	Optimised Values $\begin{pmatrix} R_1, R_2, R_4, m_1, m_2, m_4, \\ \alpha_1, \alpha_2, \alpha_4, d_2, d_3, k \end{pmatrix}$	Objective Function
1	$L_1/6, L_2/6, L_4/6, 3.0, 1.0, 3.0,$ $\pi/10, \pi/10, \pi/10, L_2/1.2,$ $L_3/1.2, 80$	$0.0, 21.9512, -7.2375,$ $1.0, 1.0, 2.6560, 0.0005, 0.0,$ $0.0, 23.5461, 23.6106, 102.5560$	10.9912
2	$L_1, L_2/10, L_4, 3.0, 1.0, 3.0,$ $\pi/10, \pi/10, \pi/10, L_2/1.2,$ $L_3/1.2, 120$	$-8.9068, -23.9729, -0.0703,$ $3.2478, 1.5165, 3.1703, 0.0, 2\pi,$ $0.0, 23.7566, 23.7863, 108.0785$	11.17026
3	$L_1/10, L_2/10, L_4/10, 2.0, 2.0, 2.0,$ $\pi/10, \pi/10, \pi/10, L_2/1.2,$ $L_3/1.2, 120$	$-1.7050, -23.9729, -7.3194,$ $2.1689, 2.0962, 5.0, 0.0, 0.0,$ $0.0, 21.1941, 21.9319, 126.2722$	19.5307

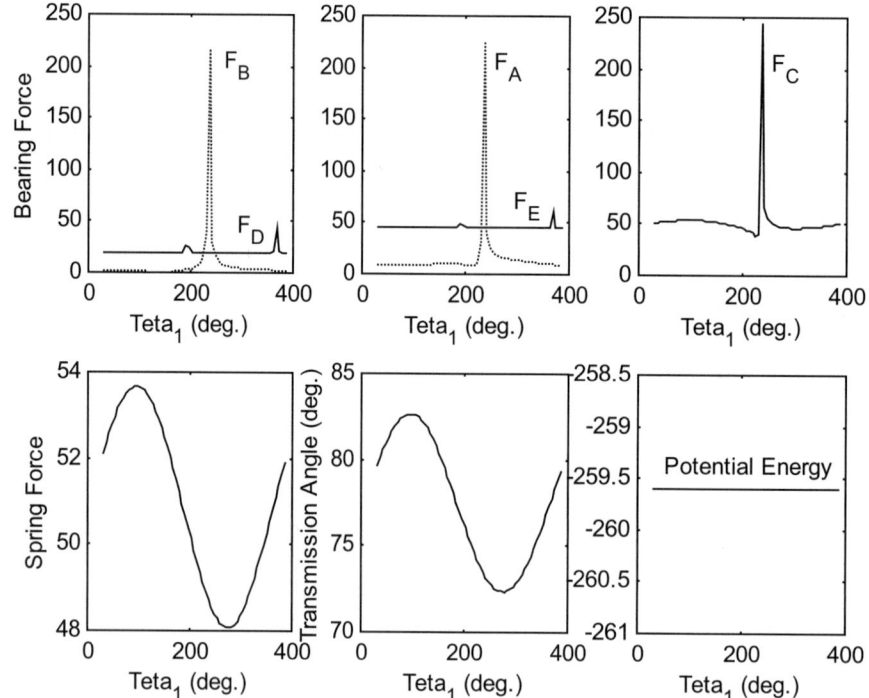

Fig. 2: The variation of reaction forces, spring force, transmission angle and the total potential energy of the balanced manipulator with θ_1 for the optimum balancing parameters given in the first row of Table 1.

workspace of the manipulator, [1] provided that the transmission angle μ is within the desirable range of $50^\circ \leq \mu \leq 130^\circ$. The planar parallel manipulator with $\alpha_0 = 0^\circ$ and the link lengths of $L_0 = 29.9748$, $L_1 = 9.6113$, and $L_2 = 23.9729$ has resulted in a transmission angle varying continuously between 72.3493° and 82.6156°. These link lengths are utilised in this study while selecting the optimum values of force balancing parameters. With reference to (19), $m_3 = m_2$ is selected such that $R_3 = -R_2$.

This has reduced the number of balancing parameters to twelve. For $30^\circ \leq \theta_1 \leq 390^\circ$, $\theta_4 = \theta_1 - 10^\circ$, with step sizes of 0.1 rad (i.e., n = 63), and $\alpha_0 = 0^\circ$, many optimisation trials with different initial values for the balancing parameters were conducted. Different initial values gave different solutions satisfying the objective function and the balancing constraints. Sets of optimised balancing parameters together with the initial values used are given in Table 1. Note that the units are arbitrary length, mass and spring constant units. But the units for the angles are radian. These values describe the mass distribution (m_z, R_z, α_z), spring stiffness constant, and attachment points for the balancing. The variation of reaction/bearing forces, spring force, transmission angle and the total potential energy of the balanced manipulator with θ_1 are shown in Fig.2.

In order to demonstrate the effectiveness of the optimisation procedure, the numerical values of the balancing parameters[2] given for a planar parallel manipulator (basically the same manipulator considered in this study) by Jean and Gosselin [7] are used to calculate the bearing forces. It is determined that although the parameters satisfy the necessary balancing conditions expressed by (13),(14),(17-19), the average values of the bearing forces is significantly higher than those obtained from the optimisation procedure presented above.

As given in (3), when $\theta_4 = \frac{\theta_2 + \theta_3}{2} + \frac{\pi}{2}(2p+1)$ for p =0,1.. or $\theta_1 = \theta_2$ or $\theta_2 = \theta_3$, the forces in the mechanism tend to go to infinity. This is an undesired situation that must be avoided by selecting proper link lengths during the design stage. When the mechanism is in these singular configurations ($\theta_1 = \theta_2$ or $\theta_2 = \theta_3$), the transmission angle requirement is also violated. This problem was considered during the optimisation procedure for determining link lengths [1].

The numerical values of the balancing parameters change with their initial values as well as the chosen link lengths, and therefore they do not necessarily represent a global minimum. However, by combining structural optimisation based on kinematic accuracy, dexterity, and singularity avoidance capabilities [1] and optimum dynamic balancing of the manipulator as a single optimisation problem with multi-objective functions, a step towards optimum design of parallel manipulators will be taken, which is currently lacking in the literature. We will report on this issue later in another study.

VI. CONCLUSIONS

We have presented an optimum balancing method based on the minimization of a sum-squared values of bearing forces throughout a practical operation range of the manipulator, provided that a set of balancing constraints consisting of balancing conditions, the sizes of inertial and geometric parameters are satisfied. Reducing the magnitude of the forces in a mechanism provides numerous advantages including an increase in the life of bearings and springs, less powerful and smaller actuators to move the mechanism, less shaking force and moment transmitted to the ground, and decreased wear in the mechanism components. A set of optimisation trials were performed to prove that the proposed optimisation approach was systematic, versatile and easy to implement for the optimum balancing of the parallel manipulator and more general parallel manipulators. This work contributes to previously published work from being a step towards optimum design of parallel manipulators.

VII. REFERENCES

[1] G. Alici and B. Shirinzadeh, "Constrained structural optimisation of a revolute-jointed planar parallel manipulator", *2003 IEEE/ASME International Conference on Advanced Intelligent Mechatronics*, July 2003, Japan.

[2] M. Skreiner, "Dynamic analysis used to complete the design of a mechanism", *J. of Mechanisms*, Vol. 5, pp.105 – 119, 1970.

[3] H. S. Yan, and R. C. Soong, "Kinematic and dynamic design of four-bar linkages by links counterweighing with variable input speed", *Mech. and Machine Theory*, Vol:36, pp. 1051 – 1071, 2001.

[4] G. G. Lowen, F. R. Tepper, and R. S. Berkof, "Balancing of linkages – An update", *Mech. and Machine Theory*, Vol:18, No. 3, pp. 213 – 220, 1983.

[5] R. S. Berkof, and G. G. Lowen, "A new method for completely force balancing simple linkages", *ASME J. of Engineering for Industry*, Vol.91, pp. 21 – 26, February 1969.

[6] C. Bagci, "Complete balancing of space mechanisms – shaking force balancing", *ASME J. of Mechanisms, Trans., and Auto. in Design*, Vol.110, No.12, pp. 609 – 616, December 1983.

[7] M. Jean, and C. M. Gosselin, "Static balancing of planar parallel manipulators", *Proceedings of the 1996 IEEE Int. Conference on Robotics and Automation*, pp. 3732 -- 3737, Minneapolis, Minnesota, April 1996.

[8] T. Laliberte, C. M. Gosselin, and M. Jean, "static balancing of 3-DOF planar parallel mechanisms", *IEEE / ASME Trans. on Mechatronics*, Vol: 4, No: 4, pp. 363 – 377, December 1999.

[9] J. Wang, and C. M. Gosselin, "Static balancing of spatial three-degree-of-freedom parallel mechanisms", *Mech. and Machine Theory*, Vol:34, No:3, pp. 437 – 452, 1999.

[10] I. Ebert-Uphoff, C. M. Gosselin, and T. Laliberte, "Static balancing of spatial platform mechanisms –revisited", *ASME J. of Mechanical Design*, Vol.122, pp. 43 –51, March 2000.

[11] B. Porter, and D. J. Sanger, "Synthesis of dynamically optimum four-bar linkages", *Procs. of Conference on Mechanisms, Paper C69/72*, pp. 24 – 28, Institution of Mechanical Engineers, 1972.

[12] T. W. Lee, and C. Cheng, "Optimum balancing of combined shaking force, shaking moment, and torque fluctuations in high-speed linkages", *ASME J. of Mechanisms, Trans., and Auto. in Design*,Vol. 106, pp. 242 – 251, June 1984.

[13] H. C. Yen, "Balancing of high speed machinery", *ASME J. of Engineering for Industry*,Vol.89, pp. 111 – 118, February 1967.

[14] B. Feng, N. Morita, and T. Torii, "A new optimisation method for dynamic design of planar linkage with clearances at joints – optimising the mass distribution of links to reduce the change of joint forces", *ASME J. of Mechanical Design*, Vol.124, pp. 68 –73, March 2002.

[15] D. A. Streit, and B.J. Gilmore, "Perfect spring equilibrators for rotatable bodies", *ASME J. of Mechanisms, Trans., and Autom. in Design*, Vol.111, No.4, pp. 451 – 458, 1989.

[16] G. G. Lowen, and R. S. Berkof, "Survey of investigations into the balancing of linkages", *J. of Mechanisms*, Vol.3, No.4, pp.221 – 231, 1968.

[17] A. Grace, *"Optimization Toolbox, for Use with MATLAB"*, The MATH WORKS Inc., 1992.

[18] G. Alici, "An Inverse Position Analysis of Five-bar Planar Parallel Manipulators", *Robotica*, 20:2, pp.195--201, 2002.

---

[2] Mass distribution parameters only

Kinematics and Dynamics of a Cable-like Hyper-flexible Manipulator

Hiromi Mochiyama[†] and Takahiro Suzuki[‡]

[†] National Defense Academy
Dept. of Mechanical Systems Engineering
Hashirimizu 1-10-20, Yokosuka, Kanagawa 239-8686, Japan
motiyama@nda.ac.jp

[‡] Institute of Industrial Science, University of Tokyo
Dept. of Information and Systems
Komaba 4-6-1, Meguro, Tokyo 153-8505, Japan
suzukitk@iis.u-tokyo.ac.jp

Abstract— A Hyper-Flexible Manipulator (HFM, for short) is a kind of continuum robots with a simple mechanical structure like a cable, rope and string, which are useful tools utilized everywhere in various forms. In this paper, in order to achieve dexterous and useful manipulation by this type of robot, we discuss kinematics and dynamics of an HFM. We rigorously derive a spatial, nonlinear and continuum dynamics model with an underactuated mechanism using special kinematics based on curve geometry and theory of robot manipulation.

I. INTRODUCTION

A. Motivation and Purpose

A Hyper-Flexible Manipulator (HFM, for short) is a kind of continuum robots with a simple mechanical structure like a cable, rope and string. This type of robotic manipulator has the following features as control systems:

- A HFM system is *infinite-dimensional* because it has infinite kinematic degrees of freedom.
- A HFM system is *nonlinear* because it allows very large deformation.
- A HFM system is *underactuated* because we cannot drive all the kinematic DOF independently.

There is no general theory to control such a mechanical system. It seems hopeless to achieve successful dynamic manipulation of HFMs. However, it is worth remarking the following two facts about HFMs:

1. A cable, rope, and string are very useful tools utilized everywhere in various forms.
2. In dexterous manipulation, we sometimes make our some joints free (or nearly free), which is equivalent to an underactuated mechanism.

For the first fact, examples include a cable of a crane truck in construction work, a line of fishing, a climbing rope hanging from a helicopter in rescue work, a hanging cable in blimp transportation, and a whipping cord for rotating a top. We have to notice that usefulness of this type of tool comes from not only the static property being lightweighted and tough but also the dynamics property such as unilateral nature in force transmission. If we used a long rigid bar instead of a crane cable, the system would be dangerous because much of impact force at the tip of the rigid bar transmits through it to a crane operator. Also, we can easily imagine that it is very difficult to apply moment to a rotating top by whipping with a rigid body.

As to the second fact above, it is necessary to notice that an HFM can be regarded as a mechanical system with infinite number of free rotational joints distributed continuously along its body. Osumi proposed a cooperative manipulation method where free joints are intentionally utilized to avoid undesirable internal forces [9]. Aiyama used free joints for the purpose of an environment-contacting task [1]. Remember that we make our wrist joints free positively when we throw a ball. The underactuated property may be key to achieving dexterous manipulation.

The purpose of this research is to provide a manipulation theory of a hyper-flexible manipulator for realizing dexterous and useful manipulation. To achieve this, we need to capture the essential properties in kinematics and dynamics of the robot. Thus in this paper, we discuss kinematics and dynamics modeling of an HFM.

B. Related Works

Dynamics modeling of an HFM like a cable can be categorized into the following three ways:

1. Regard it as a beam
2. Take a finite element method
3. Approximate it as a serial chain of rigid bodies with a large number of degrees of freedom

In the first modeling, an HFM is expressed by a partial differential equation, which means infinite dimensionality of an HFM is preserved. This model is a linear model based on the assumption that its deformation is very small, which may not be satisfied for a very flexible HFM. To avoid this drawback, Gravagne and Walker consider a continuous chain of planar beams as a model for their tentacle-like manipulator [3], but only a planar model was reported.

The second one, so called FEM modeling, is very effective method to simulate dynamical behavior of the system numerically. Suzumori et al. modeled the Flexible Microactuator by FEM and studied its dynamical behavior [10]. However, FEM models are not appropriate for analysis of control systems we want to do here.

The last model is easy to handle because its useful properties are clarified due to robotics research in the past. Ichikawa and Hashimoto regarded a cable as a planar serial rigid chain, and verified the effectiveness of their model by identification and position control experiments [4]. However, we cannot see infinite-dimensional property of an

HFM through this model, where there may be essentials of the system.

C. Our Approach

In this paper, we rigorously derive a spatial, nonlinear and continuum dynamics model with an underactuated mechanism. Our model of an HFM is on the extended line of a serial rigid chain model, that is, we regard an HFM as the limit of a serial rigid chain as the number of kinematic degrees of freedom goes to infinity. To derive the model along this line, we need special kinematics where curve geometry and theory of robot manipulation are combined. As for utilization of curve geometry for robot manipulation, see Chirikjian's work [2]. Wakamatsu and Wada also use it for manipulation of linear flexible objects [11]. To obtain a dynamics model, we first interpret the physical meaning of the serial rigid chain dynamics. Then, we apply the interpretation to the infinite dimensional case. We also provide simple planar case dynamics of an HFM for better understanding of the essential properties of the robot.

The organization of this paper is as follows. In Section 2, we discuss kinematics of a hyper-flexible manipulator based on curve geometry. In Section 3, using this kinematics, we derive an exact spatial, nonlinear and continuum dynamics model of an HFM. In Section 4, we analyze planar motion of the derived dynamics. In Section 5, the results of this paper are summarized.

II. KINEMATICS

A. Manipulator Geometry

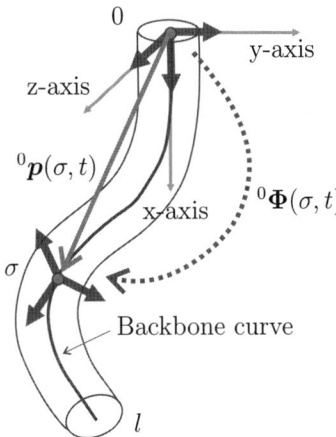

Fig. 1. Geometry of a hyper-flexible manipulator

Consider a hyper-flexible manipulator depicted in Fig. 1. The curve passing through the geometric centers of cross sections of the arm is called *the backbone curve*. We express the backbone curve by $\mathsf{p} : [0 \ l] \ni \sigma \mapsto \mathsf{p}(\sigma) \in E^3$, where constant l means the total length of a manipulator and E^3 denotes a Euclidean space. We attach the extended Frenet frame $\mathbf{\Phi}(\sigma)$ to each point on the curve $\mathsf{p}(\sigma)$. The extended Frenet frame is obtained by rotating the Frenet frame [5] along the arm shape [6].

Let $^0\mathbf{\Phi}(\sigma) \in SO(3)$ be the orientation matrix of the extended Frenet frame at σ, $\mathbf{\Phi}(\sigma)$, relative to the base frame $\mathbf{\Phi}(0)$. The change rate of this matrix along a backbone curve can be characterized by a three dimensional vector, $\boldsymbol{a}(\sigma, t) \in \Re^3$, as follows:

$$^0\mathbf{\Phi}_\sigma(\sigma, t) = {}^0\mathbf{\Phi}(\sigma, t) [\boldsymbol{a}(\sigma, t) \times] \quad (1)$$

where subscript σ denotes the partial spatial derivative along the curve, and $[\cdot \times]$ means the linear operator generating a skew symmetric matrix form a three-dimensional vector [1]. We call $\boldsymbol{a}(\sigma, t)$ the frame rate vector at σ along a backbone curve.

Let $^0\boldsymbol{p}(\sigma) \in \Re^3$ be the position vector of the point on the backbone curve at σ, $\mathsf{p}(\sigma)$, relative to the origin $\mathsf{p}(0)$ as viewed from the frame $\mathbf{\Phi}(0)$. This vector relates to the frame orientation matrices as

$$^0\boldsymbol{p}(\sigma, t) = \int_0^\sigma {}^0\mathbf{\Phi}(\eta, t) \boldsymbol{e}_\mathrm{x} d\eta, \quad (2)$$

where $\boldsymbol{e}_\mathrm{x} := [1 \ 0 \ 0]^T$. Define $^\xi\mathbf{\Phi}(\sigma) \in SO(3)$ and $^\xi\boldsymbol{p}(\sigma) \in \Re^3$ by $^\xi\mathbf{\Phi}(\sigma, t) := \{^0\mathbf{\Phi}(\xi, t)\}^T \{^0\mathbf{\Phi}(\sigma, t)\}, ^\xi\boldsymbol{p}(\sigma, t) := \{^0\mathbf{\Phi}(\xi, t)\}^T \{^0\boldsymbol{p}(\sigma, t)\}$, which are the orientation matrix of $\mathbf{\Phi}(\sigma)$ and the position vector of $\mathsf{p}(\sigma)$ from $\mathsf{p}(0)$ as viewed from $\mathbf{\Phi}(\xi)$ respectively. We abbreviate $^\sigma\mathbf{\Phi}(\sigma)$ and $^\sigma\boldsymbol{p}(\sigma)$ as $\mathbf{\Phi}(\sigma)$ and $\boldsymbol{p}(\sigma)$.

B. Axis Matrix

Let $\boldsymbol{\theta}(\sigma, t) \in \Re^n$ be the variable vector at σ which corresponds to the joint angle vector of a serial-chain manipulator, where n means the number of degrees of freedom of a manipulator at σ. Suppose that the relationship between this variable vector and the frame rate vector is expressed by mapping $\boldsymbol{f} : \Re^n \to \Re^3$ as

$$\boldsymbol{a}(\sigma, t) = \boldsymbol{f}(\boldsymbol{\theta}(\sigma, t)). \quad (3)$$

Here we define $\boldsymbol{A}(\sigma, t) \in \Re^{3 \times n}$ as the Jacobian matrix of $\boldsymbol{f}$, i.e.,

$$\boldsymbol{A}(\sigma, t) := \frac{\partial \boldsymbol{f}(\boldsymbol{\theta}(\sigma, t))}{\partial \boldsymbol{\theta}(\sigma, t)}. \quad (4)$$

We call $\boldsymbol{A}(\sigma, t)$ the rotational axis matrix at σ because each column vector of this matrix corresponds to the joint axis of a serial-chain manipulator.

Note that we can choose $\boldsymbol{\theta}$ freely because a cable has no joint actually. One of reasonable choices is to align (virtual) axes with the axes of the extended Frenet frame. By this choice, we can obtain a constant axis matrix as follows:

Case I: $n = 1$(Planar Case)

Choose $\theta \in \Re$ such that $\boldsymbol{a}(\sigma, t) = \boldsymbol{e}_\mathrm{z} \theta(\sigma, t)$, then we have a constant axis matrix as $\boldsymbol{A}(\sigma, t) = \boldsymbol{e}_\mathrm{z}$. In this case, $\theta(\sigma, t)$ denotes the curvature of the planar backbone curve.

---

[1] The symbole $[\cdot \times]$ is defined by $[\boldsymbol{i} \times] := \begin{bmatrix} 0 & -i_\mathrm{z} & i_\mathrm{y} \\ i_\mathrm{z} & 0 & -i_\mathrm{x} \\ -i_\mathrm{y} & i_\mathrm{x} & 0 \end{bmatrix}$ for vector $\boldsymbol{i} = [i_\mathrm{x} \ i_\mathrm{y} \ i_\mathrm{z}]^T \in \Re^3$. Note that $[\boldsymbol{i} \times] \boldsymbol{j} = \boldsymbol{i} \times \boldsymbol{j}$ for vector $\boldsymbol{i}, \boldsymbol{j} \in \Re^3$, where the latter '$\times$' means the outer product of vectors in a Euclidean space.

Case II: $n = 2$

We choose $\boldsymbol{\theta} \in \Re^2$ such that $\boldsymbol{a}(\sigma, t) = [\ \boldsymbol{e}_\mathrm{x}\ \ \boldsymbol{e}_\mathrm{z}\ ]\boldsymbol{\theta}(\sigma, t)$, which means $\boldsymbol{A}(\sigma, t) = [\ \boldsymbol{e}_\mathrm{x}\ \ \boldsymbol{e}_\mathrm{z}\ ]$. In this case, $\boldsymbol{\theta}$ consists of the curvature and torsion of the backbone curve.

Case III: $n = 3$

Take the choice $\boldsymbol{\theta}(\sigma, t) = \boldsymbol{a}(\sigma, t)$, then we have $\boldsymbol{A}(\sigma, t) = \boldsymbol{I}_3$, where $\boldsymbol{I}_3 \in \Re^{3 \times 3}$ denotes the identify matrix.

C. The Manipulator Jacobian Operator

We define the angular velocity vector at σ as viewed from $\Phi(0)$, $^0\boldsymbol{\omega}(\sigma, t) \in \Re^3$, by

$$^0\boldsymbol{\omega}(\sigma, t) = \int_0^\sigma {}^0\boldsymbol{a}_t(\eta, t)d\eta, \quad (5)$$

where subscript t denotes the partial time derivative.

Let $\boldsymbol{v}(\sigma, t) \in \Re^6$ be the vector of the translational and angular velocity of the hyper-flexible manipulator at σ as viewed from $\Phi(\sigma)$, This value can be described by

$$\boldsymbol{v}(\sigma, t) = \begin{bmatrix} \boldsymbol{p}_t(\sigma, t) \\ \boldsymbol{\omega}(\sigma, t) \end{bmatrix} = \mathcal{J}_{(\sigma, t)}\{\boldsymbol{\theta}_t\}, \quad (6)$$

where $\mathcal{J}_{(\sigma, t)}\{\bullet\}$ is called *the manipulator Jacobian operator*. The definition of this operator can be written as

$$\mathcal{J}_{(\sigma, t)}\{\bullet\} := \int_0^\sigma \mathrm{Ad}_{g(\sigma, \eta, t)} \bar{\boldsymbol{A}}(\eta, t)(\bullet)d\eta.$$

where $\bar{\boldsymbol{A}}(\eta, t) \in \Re^{6 \times n}$ is the extended axis matrix which includes the effect of the translational axes:

$$\bar{\boldsymbol{A}}(\eta, t) := \begin{bmatrix} \boldsymbol{0} \\ \boldsymbol{A}(\eta, t) \end{bmatrix}, \quad (7)$$

and $\mathrm{Ad}_{g(\sigma, \eta, t)} \in \Re^{6 \times 6}$ is the adjoint transformation matrix in terms of the rigid body transformation $g(\sigma, \eta, t) \in SE(3)$:

$$\mathrm{Ad}_{g(\sigma, \eta, t)} := \begin{bmatrix} {}^\sigma\boldsymbol{\Phi}(\eta, t) & [{}^\sigma\boldsymbol{p}(\eta, t) - {}^\sigma\boldsymbol{p}(\sigma, t)\times]{}^\sigma\boldsymbol{\Phi}(\eta, t) \\ \boldsymbol{0} & {}^\sigma\boldsymbol{\Phi}(\eta, t) \end{bmatrix}. \quad (8)$$

The manipulator Jacobian operator is the counterpart of a column vector of the manipulator Jacobian matrix of a serial chain manipulator.

The transposed operator [2] of the manipulator Jacobian operator, $\mathcal{J}_{(\sigma, t)}^T$, can be expressed by

$$\mathcal{J}_{(\sigma, t)}^T\{\bullet\} = \bar{\boldsymbol{A}}^T(\sigma, t)\int_\sigma^l \mathrm{Ad}_{g(\eta, \sigma, t)}^T(\bullet)d\eta \quad (11)$$

[2] Let $\langle \cdot, \cdot \rangle$ mean the inner product of two vector functions on $[0\ l]$ defined by

$$\langle \boldsymbol{y}, \boldsymbol{x} \rangle := \int_0^l \boldsymbol{y}(\sigma)^T \boldsymbol{x}(\sigma)d\sigma. \quad (9)$$

The operator $\boldsymbol{G}$ is said to be the transposed operator of $\boldsymbol{F}$ if the following relationship holds:

$$\langle \boldsymbol{y}, \boldsymbol{F}\boldsymbol{x} \rangle = \langle \boldsymbol{G}\boldsymbol{y}, \boldsymbol{x} \rangle \quad (10)$$

and we write $\boldsymbol{G} = \boldsymbol{F}^T$.

Let $\boldsymbol{\tau}(\sigma, t) \in \Re^3$ be the virtual joint torque vector corresponding to the virtual joint angle vector $\boldsymbol{\theta}(\sigma, t)$, where the word "virtual" is used for expressing that a cable-like HFM has no joint actually. Let $\Delta \boldsymbol{w}(\sigma, t)$ be the external wrench vector at σ. The principle of virtual work allows us to have the following relation:

$$\boldsymbol{\tau}(\sigma, t) = \mathcal{J}_{(\sigma, t)}^T\{\Delta \boldsymbol{w}\}. \quad (12)$$

III. Dynamics

We derive a dynamics model of an HFM by taking a limit of the serial rigid chain dynamics as the number of kinematic degrees of freedom goes to infinity. First we consider a slice of an HFM which is regarded as the limit of a rigid link.

A. Slices

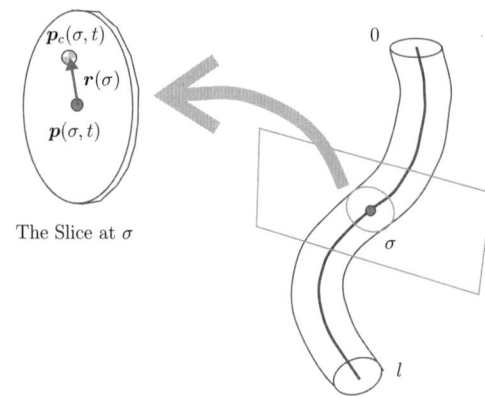

Fig. 2. Slice of a hyper-flexible manipulator

Consider the slice of a hyper-flexible manipulator with an infinitesimal width perpendicular to the backbone curve at σ as depicted in Fig. 2.

Let $m(\sigma) \in R_+$, $m(\sigma)\boldsymbol{r}(\sigma) \in \Re^3$, and $\boldsymbol{I}(\sigma) \in \Re^{3 \times 3}$ be the mass, the first moment of inertia, and the inertia tensor of the slice. The vector $\boldsymbol{r}(\sigma)$ is time-invariant and defined by $\boldsymbol{r}(\sigma) := \boldsymbol{p}_c(\sigma, t) - \boldsymbol{p}(\sigma, t)$ where $\boldsymbol{p}_c(\sigma, t)$ is the position vector of the center of mass of the slice.

The inertial wrench of slice σ, $\boldsymbol{\mu}(\sigma, t) \in \Re^6$, can be expressed by

$$\boldsymbol{\mu}(\sigma, t) = \begin{bmatrix} \boldsymbol{\mu}_\mathrm{f}(\sigma, t) \\ \boldsymbol{\mu}_\mathrm{n}(\sigma, t) \end{bmatrix}$$

$$:= \begin{bmatrix} m(\sigma)\boldsymbol{p}_{tt}(\sigma, t) \\ + \{[\boldsymbol{\omega}_t(\sigma, t)\times] + [\boldsymbol{\omega}(\sigma, t)\times]^2\}m(\sigma)\boldsymbol{r}(\sigma) \\ \boldsymbol{I}(\sigma)\boldsymbol{\omega}_t(\sigma, t) + \boldsymbol{\omega}(\sigma, t)\times \boldsymbol{I}(\sigma)\boldsymbol{\omega}(\sigma, t) \\ + m(\sigma)\boldsymbol{r}(\sigma)\times \boldsymbol{p}_{tt}(\sigma, t) \end{bmatrix}. \quad (13)$$

In the following sections, we will consider the dynamics of an HFM based on slices.

B. Dynamics of an HFM

Here we have a short review of serial rigid chain dynamics. The dynamics of a serial chain of N rigid bodies can be described by

$$^i\boldsymbol{w}_i = {}^i\boldsymbol{\mu}_i + \mathrm{Ad}_{i,i+1}^T {}^{i+1}\boldsymbol{w}_{i+1} \quad (14)$$
$$\boldsymbol{\tau}_i = \bar{\boldsymbol{A}}_i^T {}^i\boldsymbol{w}_i \quad (15)$$

where ${}^i\boldsymbol{w}_i \in \Re^6$ is the inertial wrench which transmits from link i to link $(i+1)$, ${}^i\boldsymbol{\mu}_i \in \Re^6$ is the inertial wrench of the i-th link, $\mathrm{Ad}_{i,i+1}^T \in \Re^{6\times 6}$ is the adjoint transformation matrix from i-th to $(i+1)$-th link, $\bar{\boldsymbol{A}}_i \in \Re^{3\times n}$ is the i-th joint axis matrix, and $\boldsymbol{\tau}_i \in \Re^n$ is the i-th joint torque. From the equations above, we have

$$\bar{\boldsymbol{A}}_i^T \sum_{j=i+1}^N \mathrm{Ad}_{i,j}^T \boldsymbol{\mu}_j = \boldsymbol{\tau}_i \quad (16)$$

where we use the boundary condition ${}^{N+1}\boldsymbol{w}_{N+1} = \boldsymbol{0}$, and the transition property of the adjoint transformation matrix $\mathrm{Ad}_{i,i+2} = \mathrm{Ad}_{i,i+1}\mathrm{Ad}_{i+1,i+2}$. The above equation expresses the balance that a joint axis torque is equal to one obtained by filtering the inertial wrench of all the links in the tip side from the joint by the transposed axis matrix. Note that the transposed adjoint transformation matrix plays a role to convert an inertial wrench effect at some link to another one.

By applying the above physical interpretation to an HFM, we have the following equation of motion at σ:

$$\bar{\boldsymbol{A}}^T(\sigma,t) \int_\sigma^l \mathrm{Ad}_{g(\eta,\sigma,t)}^T \boldsymbol{\mu}(\eta,t) d\eta = \boldsymbol{\tau}(\sigma,t), \quad (17)$$

which is equivalent to

$$\boldsymbol{A}^T(\sigma,t) \int_\sigma^l {}^\sigma\boldsymbol{\mu}_\mathrm{n}(\eta,t)$$
$$+ ({}^\sigma\boldsymbol{p}(\eta,t) - {}^\sigma\boldsymbol{p}(\sigma,t)) \times {}^\sigma\boldsymbol{\mu}_\mathrm{f}(\eta,t) d\eta = \boldsymbol{\tau}(\sigma,t). \quad (18)$$

Note that the obtained equations are infinite dimensional because these include the curve parameter $\sigma \in [0\ l]$.

In the case that $\boldsymbol{A}(\sigma,t) = \boldsymbol{I}_3$, differentiating the above equation by σ, we obtain the following simple expression:

$$\boldsymbol{\mu}_\mathrm{n}(\sigma,t) + \boldsymbol{e}_\mathrm{x} \times \int_\sigma^l {}^\sigma\boldsymbol{\mu}_\mathrm{f}(\eta,t) d\eta$$
$$= \boldsymbol{\tau}(\sigma,t) \times \boldsymbol{\theta}(\sigma,t) - \boldsymbol{\tau}_\sigma(\sigma,t). \quad (19)$$

C. Gravity Effect and Underactuated Control Structure

We do not expect that we can apply the axis torque, which means $\boldsymbol{\tau}(\sigma,t) \equiv \boldsymbol{0}$. Instead, assume that we can apply translational acceleration to the base as an input. Moreover, we take the gravity effect explicitly into account here. Let ${}^0\boldsymbol{u} \in \Re^3$ be the input translational base acceleration, and ${}^0\boldsymbol{g} := [\bar{g}\ 0\ 0]^T$, where $\bar{g}$ is the gravity acceleration constant. The external torque by the gravity and the input can be described by

$$\mathcal{J}_{(\sigma,t)}^T \left\{ m(\cdot) {}^0\boldsymbol{\Phi}^T(\cdot,t) \begin{bmatrix} {}^0\boldsymbol{g} + {}^0\boldsymbol{u} \\ \boldsymbol{0} \end{bmatrix} \right\}. \quad (20)$$

Taking the axis torque to be zero and adding instead the above term to the right-hand side for (18), we obtain

$$\boldsymbol{A}^T(\sigma,t) \int_\sigma^l {}^\sigma\boldsymbol{\mu}_\mathrm{n}(\eta,t) + ({}^\sigma\boldsymbol{p}(\eta,t) - {}^\sigma\boldsymbol{p}(\sigma,t))$$
$$\times \{{}^\sigma\boldsymbol{\mu}_\mathrm{f}(\eta,t) - m(\eta)({}^\sigma\boldsymbol{g} + {}^\sigma\boldsymbol{u})\} d\eta = \boldsymbol{0}. \quad (21)$$

Considering the reasonable choice that $\boldsymbol{A}(\sigma,t) = \boldsymbol{I}_3$ again, we have

$$\boldsymbol{\mu}_\mathrm{n}(\sigma,t) + \boldsymbol{e}_\mathrm{x} \times \left\{ \int_\sigma^l {}^\sigma\boldsymbol{\mu}_\mathrm{f}(\eta,t) d\eta - \tilde{m}(\sigma)({}^\sigma\boldsymbol{g} + {}^\sigma\boldsymbol{u}) \right\}$$
$$= \boldsymbol{0}, \quad (22)$$

where $\tilde{m}(\sigma) := \int_\sigma^l m(\eta) d\eta$, i.e., the weight of the part of an HFM from σ to the tip.

Note that the obtained systems are hyper-underactuated because the control input ${}^0\boldsymbol{u}$ is independent of σ. We have at most three inputs against the infinite number of kinematic degrees of freedom.

For better understanding of the obtained dynamics of an HFM, we see its planar motion in the next section.

IV. Motion in a Plane

In this section, we consider a planar HFM with uniform mass density.

A. Uniformity of Mass Density

Consider a hyper-flexible manipulator with the total weight m[kg], and the total length l[m]. Assume that the planar HFM is cylindrical in shape with radius r[m] when we stretch it, that is, its backbone curve is in a straight line. Also assume that it has a uniform mass density. The line mass density of slice σ, $m(\sigma)$[kg/m], can be given by

$$m(\sigma) = \frac{m}{l} =: \bar{\mu}. \quad (23)$$

Due to symmetry of cylinder and uniformity of mass density, the first moment of inertia of any slice becomes zero, i.e.,

$$\boldsymbol{r}(\sigma) = \boldsymbol{0}. \quad (24)$$

The inertia tensor of the slice is given by

$$\boldsymbol{I}(\sigma) = \bar{I}_\mathrm{zz} \begin{bmatrix} 2 & 0 & 0 \\ 0 & 1 & 0 \\ 0 & 0 & 1 \end{bmatrix} \text{ where } \bar{I}_\mathrm{zz} := \frac{\bar{\mu} r^2}{4}. \quad (25)$$

B. Planar Constraint

As mentioned in Section 2, in a planar case, $n=1$ and any rotational axis is z-directional, i.e.,

$$\boldsymbol{A}(\sigma,t) = \boldsymbol{e}_\mathrm{z} \text{ and } \boldsymbol{a}(\sigma,t) = \theta(\sigma,t)\boldsymbol{e}_\mathrm{z}, \quad (26)$$

where the variable $\theta(\sigma,t) \in \Re$ means the curvature of the backbone curve. In this case, the angler velocity is also always z-directional:

$$^0\boldsymbol{\omega}(\sigma,t) = \phi(\sigma,t)\boldsymbol{e}_\mathrm{z} \quad (27)$$

where $\phi(\sigma,t) \in \Re$ is the absolute angle defined by

$$\phi(\sigma,t) := \int_0^\sigma \theta(\eta,t)d\eta. \tag{28}$$

The frame orientation matrix can be expressed by

$$^\sigma\boldsymbol{\Phi}(\xi,t) = \begin{bmatrix} \cos{^\sigma\phi(\xi,t)} & -\sin{^\sigma\phi(\xi,t)} & 0 \\ \sin{^\sigma\phi(\xi,t)} & \cos{^\sigma\phi(\xi,t)} & 0 \\ 0 & 0 & 1 \end{bmatrix} \tag{29}$$

where $^\sigma\phi(\xi,t) := \phi(\xi,t) - \phi(\sigma,t)$. Note that

$$^\sigma\boldsymbol{\Phi}(\xi,t)\boldsymbol{e}_z = \boldsymbol{e}_z \tag{30}$$

for arbitrary σ and ξ, which makes calculations extremely simple.

We have expressions of the inertial force and moment of slice σ as follows:

$$\boldsymbol{\mu}_n(\sigma,t) = \bar{I}_{zz}\phi_{tt}(\sigma,t)\boldsymbol{e}_z, \tag{31}$$
$$\boldsymbol{\mu}_f(\sigma,t) = \bar{\mu}\boldsymbol{p}_{tt}(\sigma,t). \tag{32}$$

Suppose that we can apply only the horizontal acceleration to the base as an input, which can be expressed by

$$^0\boldsymbol{u} = \boldsymbol{e}_y u \tag{33}$$

where $u \in \Re$ is the control input.

C. Planar Dynamics w.r.t. the Absolute Angle ϕ

Using the above relations in this section for (21), we have the following planar motion dynamics:

$$\int_\sigma^l \bar{I}_{zz}\phi_{tt}(\eta,t)d\eta$$
$$+ \int_\sigma^l \boldsymbol{e}_z^T \left\{ (^0\boldsymbol{p}(\eta,t) - {}^0\boldsymbol{p}(\sigma,t)) \times \bar{\mu}{}^0\boldsymbol{p}_{tt}(\eta,t) \right\} d\eta$$
$$= \int_\sigma^l \boldsymbol{e}_z^T \left\{ (^0\boldsymbol{p}(\eta,t) - {}^0\boldsymbol{p}(\sigma,t)) \times \bar{\mu}\bar{g}\boldsymbol{e}_x \right\} d\eta$$
$$+ \int_\sigma^l \boldsymbol{e}_z^T \left\{ (^0\boldsymbol{p}(\eta,t) - {}^0\boldsymbol{p}(\sigma,t)) \times \bar{\mu}u\boldsymbol{e}_y \right\} d\eta. \tag{34}$$

In the left-hand side, the first term expresses the inertial effect of rotational motion of all the slices from σ to the tip, while the second one means the translational inertial effect. The first term is not dominant because $r \ll l$ in an HFM usually, which makes $\bar{I}_{zz}$ negligibly small. In the right-hand side, the former term describes the effect of the gravity, while the latter is generated by the base translational acceleration input. From this expression, we can see the geometric meaning of the planar dynamics.

The translational acceleration can be represented with respect to ϕ as

$$^0\boldsymbol{p}_{tt}(\eta,t) = [\boldsymbol{e}_z \times] \int_0^\eta \phi_{tt}(\xi,t){}^0\boldsymbol{\Phi}(\xi,t)\boldsymbol{e}_x d\xi$$
$$- \int_0^\eta \phi_t^2(\xi,t){}^0\boldsymbol{\Phi}(\xi,t)\boldsymbol{e}_x d\xi, \tag{35}$$

Then, by further calculation, we obtain the following planar dynamics expression:

$$\bar{I}_{zz}\int_\sigma^l \phi_{tt}(\eta,t)d\eta + \int_0^l \int_\sigma^l \bar{m}(\max(\eta,\xi))$$
$$\left\{ \cos{^\eta\phi(\xi,t)} \cdot \phi_{tt}(\xi,t) - \sin{^\eta\phi(\xi,t)} \cdot \phi_t(\xi,t)^2 \right\} d\eta d\xi$$
$$+ \int_\sigma^l \bar{m}(\eta)\sin\phi(\eta,t)d\eta \, \bar{g} = \int_\sigma^l \bar{m}(\eta)\cos\phi(\eta,t)d\eta \, u, \tag{36}$$

where $\bar{m}(\sigma) := \bar{\mu}(l-\sigma)$ is the uniform mass density version of $\tilde{m}(\sigma)$.

We will show that this planar dynamics equation reduces to the one-link pendulum equation under a certain restriction. Consider the constraint on the curvature as

$$\theta(\sigma,t) = \theta(0,t)\delta(\sigma), \tag{37}$$

where $\delta()$ is the delta function, which leads to

$$\phi(\sigma,t) = \int_0^\sigma \theta(\eta,t)d\eta = \theta(0,t), \tag{38}$$

i.e., all the slices are in the same direction. Under the constraint above, the planar dynamics (36) reduces to

$$\left(\frac{mr^2}{4} + \frac{ml^2}{3}\right)\theta_{tt}(0,t)$$
$$+ \frac{m\bar{g}l}{2}\sin\theta(0,t) = \frac{ml}{2}\cos\theta(0,t)u, \tag{39}$$

which is exactly the equation of motion of the planar pendulum with a cylindrical rod.

Differentiating (36) by $(-\sigma)$, we have a simpler expression

$$\bar{I}_{zz}\phi_{tt}(\sigma,t) + \int_0^l \bar{m}(\max(\sigma,\xi))$$
$$\left\{ \cos{^\sigma\phi(\xi,t)} \cdot \phi_{tt}(\xi,t) - \sin{^\sigma\phi(\xi,t)} \cdot \phi_t(\xi,t)^2 \right\} d\xi$$
$$+ \bar{m}(\sigma)\sin\phi(\sigma,t)\,\bar{g} = \bar{m}(\sigma)\cos\phi(\sigma,t)\,u \tag{40}$$

Note that the HFM dynamics (40) does not reduce to the pendulum equation (39) because (40) requires the smoothness of the curvature function.

D. Planar Dynamics w.r.t. the Curvature θ

The translational acceleration with respect to θ can be calculated as follows:

$$^0\boldsymbol{p}_{tt}(\eta,t) = \int_0^\eta [\boldsymbol{e}_z \times]\left\{ {}^0\boldsymbol{p}(\eta,t) - {}^0\boldsymbol{p}(\iota,t) \right\}\theta_{tt}(\iota,t)d\iota$$
$$- \int_0^\eta \int_0^\eta \left\{ {}^0\boldsymbol{p}(\eta,t) - {}^0\boldsymbol{p}(\max(\iota_1,\iota_2),t) \right\}$$
$$\theta_t(\iota_1,t)\theta_t(\iota_2,t)d\iota_1 d\iota_2 \tag{41}$$

Then we obtain the following planar dynamics with respect to θ, which is a familiar representation in robotics community:

$$\int_0^l \mathcal{M}(\sigma,\eta,t)\theta_{tt}(\eta,t)d\eta + \int_0^l \mathcal{C}(\sigma,\eta,t)\theta_t(\eta,t)d\eta$$
$$+ \mathcal{G}(\sigma,t) = \mathcal{U}(\sigma,t), \tag{42}$$

where $\mathcal{M}(\sigma, \eta, t) \in \Re$ and $\mathcal{C}(\sigma, \eta, t) \in \Re$ can be expressed by

$$\mathcal{M}(\sigma, \eta, t) = \bar{I}_{zz}\{l - \max(\sigma, \eta)\}$$
$$+ \int_{\max(\sigma,\eta)}^{l} \boldsymbol{e}_z^T \left[{}^0\boldsymbol{p}(\sigma,t) - {}^0\boldsymbol{p}(\xi,t)\times\right]$$
$$\bar{\mu}\left[\{{}^0\boldsymbol{p}(\xi,t) - {}^0\boldsymbol{p}(\eta,t)\}\times\right]\boldsymbol{e}_z d\xi, \quad (43)$$

$$\mathcal{C}(\sigma, \eta, t) := \int_0^l \Gamma(\sigma, \eta, \xi, t)\theta_t(\xi, t)d\xi \quad (44)$$

where $\Gamma(\sigma, \eta, \xi, t) \in \Re$ is represented by

$$\Gamma(\sigma, \eta, \xi, t) = \int_{\max(\sigma,\eta,\xi)}^{l} \boldsymbol{e}_z^T \left[{}^0\boldsymbol{p}(\sigma,t) - {}^0\boldsymbol{p}(\iota,t)\times\right]$$
$$\bar{\mu}\left\{{}^0\boldsymbol{p}(\iota,t) - {}^0\boldsymbol{p}(\max(\eta,\xi),t)\right\}d\iota. \quad (45)$$

The terms $\mathcal{G}(\sigma, t) \in \Re$ and $\mathcal{U}(\sigma, t) \in \Re$ have already appeared in (34) and (36), i.e., $\mathcal{G}(\sigma, t)$ corresponds to the first term of the right-hand side of (34) or the third term of the left-hand side of (36), while $\mathcal{U}(\sigma, t) \in \Re$ means the last term of (34) or (36). The values $\mathcal{M}(\sigma, \eta, t)$ and $\mathcal{C}(\sigma, \eta, t)$ are the correspondents of the component of the inertia matrix and the Coriolis matrix respectively. From (43)-(45), we can understand how the curvature acceleration and quadratic velocity contribute to the axis torque geometrically.

Further calculations allows us to have simpler expressions of these values:

$$\mathcal{M}(\sigma, \eta, t) = \bar{I}_{zz}\{l - \max(\sigma, \eta)\}$$
$$+ \int_\eta^l \int_\sigma^l \bar{m}(\max(\iota, \xi)) \cos{}^\iota\phi(\xi, t) d\iota d\xi \quad (46)$$

$$\Gamma(\sigma, \eta, \xi, t) =$$
$$- \int_{\max(\eta,\xi)}^{l} \int_\sigma^l \bar{m}(\max(\iota, \nu)) \sin{}^\iota\phi(\nu, t) d\iota d\nu \quad (47)$$

In the same manner as the previous subsection, differentiating (42) by $(-\sigma)$, we have

$$\int_0^l \mathcal{M}_{-\sigma}(\sigma, \eta, t)\theta_{tt}(\eta, t)d\eta + \int_0^l \mathcal{C}_{-\sigma}(\sigma, \eta, t)\theta_t(\eta, t)d\eta$$
$$+ \mathcal{G}_{-\sigma}(\sigma, t) = \mathcal{U}_{-\sigma}(\sigma, t), \quad (48)$$

where $\mathcal{M}_{-\sigma}(\eta, \xi, t)$ and $\mathcal{C}_{-\sigma}(\sigma, \xi, t)$ become simpler as

$$\mathcal{M}_{-\sigma}(\sigma, \eta, t) =$$
$$\bar{I}_{zz} + \int_\eta^l \bar{m}(\max(\sigma, \xi)) \cos{}^\sigma\phi(\xi, t) d\xi, \quad (49)$$

$$\mathcal{C}_{-\sigma}(\sigma, \eta, t) := \int_0^l \Gamma_{-\sigma}(\sigma, \eta, \xi, t)\theta_t(\xi, t) d\xi, \quad (50)$$

where

$$\Gamma_{-\sigma}(\sigma, \eta, \xi, t) =$$
$$- \int_{\max(\eta,\nu)}^{l} \bar{m}(\max(\sigma, \nu)) \sin{}^\sigma\phi(\nu, t) d\nu. \quad (51)$$

Note that simpler expressions are obtained by taking the partial derivative along the curve due to a continuum model.

V. CONCLUSION

In this paper, we derived an exact spatial dynamics model of a hyper-flexible manipulator like a cable using special kinematics based on curve geometry and theory of robot manipulation. The derived model is nonlinear and infinite-dimensional, but has preferable properties. The model is rich in physical and geometric meanings because it is expressed by spatial vectors and their operations. Moreover, we can use the partial derivative along the curve. This operation often makes mathematical expressions very simple. We also show the hyper-underactuated mechanism of an HFM.

For numerical simulation, we need to discretize the model spatially along the curve. It is reasonable to use a serial rigid chain model with considerably many degrees of freedom as a simulation model of an HFM. We have already done a numerical simulation using a planar serial rigid chain model with 50 degrees of freedom, and checked its string-like motion [7]. Of course, we have to check the validity of the derived model experimentally in the future.

The next step for establishing the manipulation theory for an HFM is to capture the essentials of the kinematics and dynamics. We are now studying the Hamilton structure of the system [7].

REFERENCES

[1] Aiyama, Y.: Environment-Contacting Task by Position-Controlled Manipulator using Free-Joint Structure, *Proc. of the 2001 IEEE/ASME International Conference on Advanced Intelligent Mechatronics*, 816/821, 2001.

[2] Chirikjian, G.S. and J.W. Burdick: A Modal Approach to Hyper-Redundant Manipulator Kinematics, *IEEE Trans. on Robotics and Automation*, Vol.10, No.3, 343/354, 1994.

[3] Gravagne, I.A., C.D. Rahn and I.D. Walker: Good Vibrations: A Vibration Damping Setpoint Controller for Continuum Robots, *Proc. of the 2001 IEEE International Conference on Robotics and Automation*, 3877/3884, 2001.

[4] Ichikawa, T. and M. Hashimoto: Dynamic Manipulation of a String by a Robot Manipulator, Proc. of the 19th Annual Conference of Robotics Society of Japan, 1243/1244, 2001. (in Japanese)

[5] Kobayashi, S.: *Differential Geometry of Curves and Surfaces*, Shokabo, 1995. (in Japanese)

[6] Mochiyama, H. and T. Suzuki: Kinematics of a Hyper-Flexible Manipulator, *Proc. of SICE Annual Conference on Control Systems*, 5/8, 2002. (in Japanese)

[7] Mochiyama, H. and T. Suzuki: Passivity-based Damping Control of a Hyper-Flexible Manipulator, *to appear in* Proc. of the 8th Robotics Symposia, 2003. (in Japanese)

[8] Murray, R.M., Z. Li and S.S. Sastry: *A Mathematical Introduction to Robotic Manipulation*, CRC Press, 1994.

[9] Osumi, H., M. Ono, M. Fujibayashi and M. Kagatani: Cooperative System for Multiple Position-controlled Robots with Free Joint Mechanisms, *Proc. of the 1997 IEEE International Conference on Robotics and Automation*, 1484/1489, 1997.

[10] Suzumori, K., T. Maeda, H. Watanabe and T. Hisada: Fiberless Flexible Microactuator Designed by Finite-Element Method, *IEEE/ASME Trans. on Mechatronics*, Vol. 2, No. 4, 281/286, 1997.

[11] Wakamatsu, H. and T. Wada: Modeling of String Objects for Their Manipulation, *Journal of the Robotics Society of Japan*, Vol. 16, No. 2, 145/148, 1998. (in Japanese)

Development of Parallel Manipulator "NINJA" with Ultra-High-Acceleration

Kiyoshi Nagai†   *Masaharu Matsumoto†   *Ken'ichiro Kimura†   Ban Masuhara†

†Ritsumeikan University, Department of Robotics,
Noji-higashi 1-1-1, Kusatsu, Shiga, 525-8577 JAPAN

Abstract

This paper describes the process of developing the parallel mechanism "NINJA", which is designed to achieve an acceleration rate over 100 [G] (G: gravitational acceleration). To design this mechanism, we introduce a two-step design method (an optimum design method). The technical part of this method consists of formulating the design problem of Step 1, where we assume and select the design parameters, which should satisfy the required specifications, and of Step 2, where we optimize the parameters under the constraint conditions based on the required specifications. To demonstrate the acceleration performance, we present experimental results using the prototype NINJA.

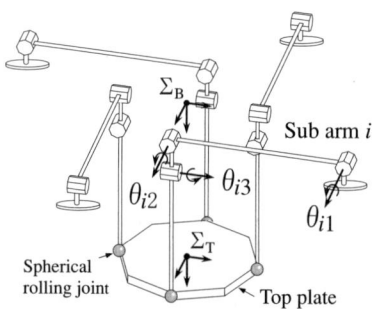

Fig. 1: 6 D.O.F. parallel mechanism NINJA

1 Introduction

We aimed at designing the first robot that achieves an acceleration performance over 100 [G]. High acceleration performances are required, for example, for bonding semiconductor-tip terminals with a lead line. This procedure demands high speed and accuracy in a narrow workspace. The acceleration and deceleration motion take up most part of the motion time and, therefore, achieving a high acceleration rate is important.

We adopted a parallel mechanism to achieve high acceleration and accuracy because this mechanism is designed to be lightweight by arranging actuators on a base. Mechanism endpoint deflections caused by the inertia force during acceleration are small due to the structure, which disperses force added at the mechanism endpoint and averages the position errors of each sub arm. However, the motion range is narrow because the sub arm endpoints are connected to each other.

Figure 1 shows our proposed 6 D.O.F. parallel mechanism NINJA. The four sub arms are connected to a common top plate.

In previous studies related to parallel mechanisms with a high speed motion, DELTA[1], HEXA[2],[3],

---

*Matsumoto and Kimura were graduate students of Mechanical Systems Course in Ritsumeikan University

FALCON[4] and RP-1AH[5] are developed, but these did not achieve performances over 100 [G].

In designing a highly efficient, defining the optimum design parameters through trial and error is difficult because many parameters relate mutually. Therefore, various design methods have been proposed[6]–[9]. We think that a systematic method is needed to design a robot with a high acceleration rate, including designing the dynamics of the robot. We have developed a two-step design method for NINJA. The framework of this method replaces the general optimum design method. In the two-step design method, after formulating the design problem, a desirable mechanism or system is developed using design parameters derived from an optimization method.

In this paper, we describe the design process of NINJA using a two-step design method and we evaluate experiments on a parallel mechanism prototype.

2 Mechanism Design Problem

2.1 Trade-off in mechanism design and the two-step design method

For example, if a constant length link is designed to be thin and light to enable a large acceleration, the link looses stiffness, and vice versa.

We think that to explicitly solve such a trade-off relationship, a design problem should be formulated and the design parameters should be determined by using an optimization method. Based on this thought, we developed a two-step design method and designed a prototype parallel mechanism.

The flow of the two-step design method is shown in **Fig. 2**. The technical part of this method consists of the formulation of a design problem, the assumption and selection (Step 1), and the optimization (Step 2). The whole two-step design method consists of a technical part and of procedures where designers confirm the propriety of obtained parameters and examine whether to accept performance realized by those parameters. This method takes advantage of replacing optimization of design parameters relied on designers with mathematical optimization methods.

In formulating a design problem, the designer must determine which design parameters to optimize. Potential parameters should be difficult to determine in trade-off and enhance the design's performance. To find the candidates, the designer may perform a sensitivity analysis of the design parameters [9] or a preliminary experiment. Designer's knowledge and experience should be adopted about the parameters which are difficult formulation, such as construction of a mechanism or selection of machine components.

The calculations in Step 2 for the optimization are difficult when assumed or selected mechanism models are described in detail. Since simple models are needed, we limit this design method to schematic designs as a basis for developing further ideas.

2.2 Formulating a design problem

In this design, the following prerequisites were given. First, the mechanism shown in **Fig. 1** is adopted. Second, the degree of freedom (D.O.F.) at the endpoint is six. Third, the workspace is large enough to move the trajectory shown in **Fig. 3** periodically between A and B, such as in Adept motion.

The requirements were to realize a) acceleration over 100 [G] and b) high accuracy. Since a) was the first requirement, we set the performance index as the maximum endpoint acceleration $\dot{v}_T(\alpha)$ where $\alpha(=[\alpha_1,\cdots,\alpha_n]^T)$ means the design parameters. In addition, since the endpoint velocity should not be overhead within the motion range, we determined as the constraint condition that the maximum endpoint velocity $v_{T\max}$ should be more than the peak of the endpoint velocity in the cyclic motion v_{Tp}. The $v_{T\max}$ is calculated from the actuator's maximum velocity, link length, and reduction ratio. The v_{Tp} is calculated from the peak value under the condition that a robot performs the cyclic motion with a velocity triangle profile in 100 [G]. However, instead of using $v_{T\max}$ as the constraint condition in the optimization, we checked whether the $v_{T\max}$ satisfied the constraint using the acquired design parameters following Step 2.

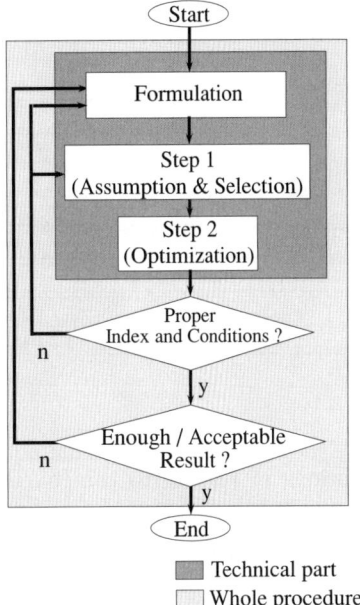

Fig. 2: Flow of the two step design method

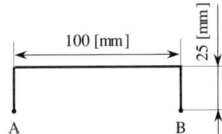

Fig. 3: Trajectory for the cyclic motion

On the other hand, the second requirement b) high accuracy does not only depend on the mechanism design. We determined as the constraint condition that the maximum deflection caused by inertia force $\delta_{T\max}$ should be within an allowable range of $\delta_a = 0.1 \times 10^{-3}$ [m]. This condition does not satisfy b) thoroughly, but depends on only mechanism design.

The design problem is formulated as follows:

$$\begin{aligned} &\text{maximize} \quad \dot{v}_T(\alpha) \\ &\text{subject to} \quad \delta_T(\alpha) - \delta_a \leq 0, \end{aligned}$$

where

$$\dot{v}_T(\alpha) = M^{-1}(J^T A^T \tau_A - h) \quad (1)$$
$$\delta_T(\alpha) = K^{-1}(M_T \dot{v}_T + h_T) \quad (2)$$

The details of eqs. 1 and 2 are shown in Appendix A. The requirement specifications and their quantitative

Table 1: Requirement specifications of NINJA

Prerequisite	
D.O.F.	6 D.O.F.
Workspace	$100(L) \times 100(W) \times 30(H)$[mm]
Perfomance index	
Acceleration	$\dot{v}_{Tx}, \dot{v}_{Ty}, \dot{v}_{Tz}$
Constraint condition	
Velocity	$v_{Tx\max} \geq v_{Txp}$, $v_{Ty\max} \geq v_{Typ}$, $v_{Tz\max} \geq v_{Tzp}$
Deflection	$\delta_{Tx\max}, \delta_{Ty\max}, \delta_{Tz\max} \leq 0.1$ [mm]

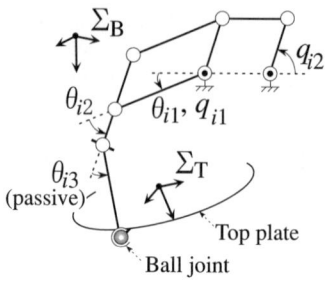

Fig. 4: Kinematic model of the sub arm

expressions are shown in **Table 1**. We did not consider these specifications in respect to components of the endpoint attitude in this design.

3   Design of Parallel Mechanism "NINJA"

This section describes the process of solving the formulated design problem of the parallel mechanism NINJA based on the two-step design method.

[Step 1]

Assumption of a mechanism We designed NINJA using the mechanism shown in **Fig. 1** as the prerequisite but if we find a more appropriate mechanism, we will adopt that. The parallel mechanism NINJA is composed of four sub arms and a top plate. This mechanism, which supports the top plate with the four arms, achieves a larger acceleration and maintains its stiffness, therefore, the endpoint deflection by inertia force is expected to decrease.

Assumption of a sub arm We assumed the sub arm model given in **Fig.4**. A sub arm has two actuated joints (● in the figure) and a passive joint, and it has three D.O.F. in total with respect to position (not including three D.O.F. of the ball joint). To arrange the actuators on the base, we adopted a parallel link structure to the sub arm itself because even though a wire or a belt has small inertia, the accuracy decreases when it is stretched.

Assumption of link shape We assumed the link parameters defined in **Fig. B-1** in Appendix B. The lengths of link1c, link2a, and link2b were each designed to satisfy the required workspace specifications. As the optimization parameters, the lengths of link1a, link1b, and link3 were each defined in relation to the acceleration and speed, and the width and depth of all links were defined in relation to the endpoint deflection.

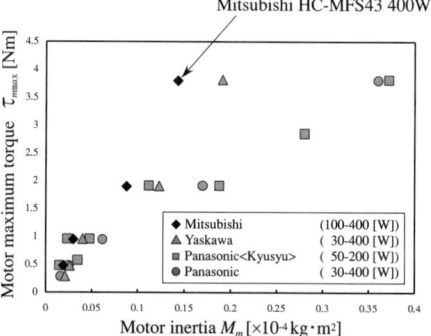

Fig. 5: Inertia-torque graph of each servo-motor

Selection of actuators We selected AC servo motors as the actuators based on their catalogue information. **Figure 5** shows the relationship between the motor inertia M_m and the maximum motor torque $\tau_{m\max}$. The graph implies that motors with a smaller M_m and a larger $\tau_{m\max}$ achieve a higher acceleration rate. Including our preliminary experiment [10], as a result, we adopted a HC-MFS43 AC servo motor (400 [W], MITSUBISHI ELECTRIC CO., LTD [11]) which has a large capacity.

Selection of gears First, we examined the different gear types and selected spiral gears because they have a larger feasible transmission force than spur gears of equal size. Second, we investigated the stage number corresponding to the reduction ratio. When a large reduction ratio and a small stage number are selected or a small reduction ratio and a large stage number, the gear inertia increases. Equation 3 represents the relationship between the reduction ratio ^jN_i (sub arm no.: $i = 1, \cdots, 4$, stage reduction: $j = 1, \cdots, 3$) and the converted moment of inertia $^jM_{Ai}$ around the output axis.

$$\begin{aligned} 1\text{ stage} &: {}^1M_{Ai} = M_{Aim}{}^1N_i^2 + {}^1m_{Ai} \\ 2\text{ stage} &: {}^2M_{Ai} = {}^1M_{Ai}{}^2N_i^2 + {}^2m_{Ai} \\ 3\text{ stage} &: {}^3M_{Ai} = {}^2M_{Ai}{}^3N_i^2 + {}^3m_{Ai} \end{aligned} \quad (3)$$

where M_{Aim} is the moment of inertia of the motor axis, M_{Ai} is the total moment of inertia consisting of M_{Aim}, shafts, and gears on the input side, and $^jm_{Ai}$ is the sum of moments of inertia of shafts and of gears on

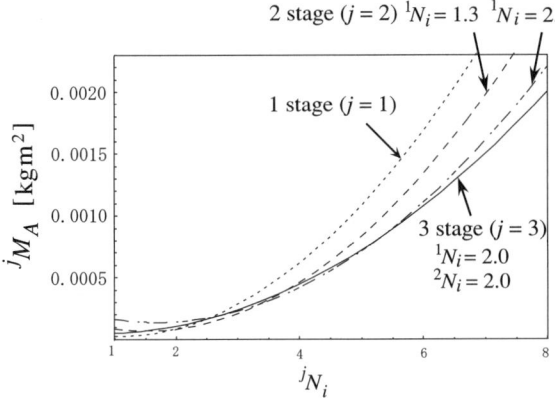

Fig. 6: $^jM_{Ai} - {}^jN_i$ graph of the sub arm i

Table 2: Design parameters defined in Step 1

Design parameters for optimization
Reduction ratio, lengths of links1a, 1b, and 3, link width and depth
Fixed design parameters in optimization
Max. motor torque, properties depending on the material including density or Young's modulus, mass and moment of inertia (ball bearings/ ball joints/ joint shafts), mass and moment of inertia of actuating system depending on reduction rate (gears/ motor's rotor/ shafts), length of links1c, 2a, 2b, and 2c, joint diameter

the output side. We calculated $^jM_{Ai}$ from eq. 3 by using the data of some actual gears. **Figure 6** shows the result fitted with a quadratic function. The result indicates that the appropriate numbers of stage reductions can be selected by corresponding to condition ^jN_i. We defined the reduction ratio ^jN_i in relation to the acceleration as the design parameter in the optimization.

Selection of materials The desirable material properties are small density and high stiffness to achieve a high acceleration rate and to simultaneously decrease the endpoint deflection. We chose a strong aluminium alloy as the material for links1a–1c and links2a–2c (**Fig.B-1** refers to each link name) and a magnesium alloy for link3 based on their acceleration performance in particular.

Selection of ball joints We selected SRJ006C (HEPHAIST SEIKO CO., LTD [12]), which has a large swing angle ±45 [deg] to obtain a large workspace.

Table 2 shows the design parameters we assumed or selected in Step 1.

[Step 2]

The optimization of the design parameters was applied to one sub arm instead of to the whole mechanism, in consideration of the sub arm's symmetrical arrangement. Choosing to optimize the sub arm implies that we neglect the interference terms caused by each sub arm motion, with the result that the acceleration and deflection estimated by the optimization method are not precise. However, if the main terms in a motion are considered, we think the acquired performance almost satisfies the requirements. We describe the optimization problems applied to the sub arm below.

Based on the formulation, we determined the design parameters to maximize the performance index $\dot{v}_T$ under the constraint of the endpoint deflection. In the calculation of $\dot{v}_T$, the maximum acceleration in horizontal direction $\dot{v}_{T x\max}$ (the direction based on the $\sum_{SBi}$ in **Fig.B-1**) and in vertical direction $\dot{v}_{Ty\max}$ were examined under the basis attitude of the sub arm (see **Fig.B-1**) and the endpoint speed $v_T = 0$. We neglected the interference terms of the inertia force, the centrifugal terms, and the Coriolis and the gravity terms. The $\dot{v}_{Tx\max}$ and $\dot{v}_{Ty\max}$ were derived from eqs. 4 and 5:

$$\dot{v}_{Tx\max} = \frac{{}^jN_i \tau_{m\max}(L_{2V}+L_{3V})}{M'_{Si11}+{}^jM_{Ai}} \quad (4)$$

$$\dot{v}_{Ty\max} = \frac{{}^jN_i \tau_{m\max}(L_1+L_{2H})}{M'_{Si22}+{}^jM_{Ai}} \quad (5)$$

where L_* is the link length shown in **Fig.B-1** and M'_{Sikk} is the moment of inertia (the actuated joint no. $k = 1, 2$) including the diagonal component M_{Sikk} of the inertia matrix of sub arm i $\boldsymbol{M}_{Si}$ and the moment of inertia around the actuated joint, which is converted by the inertia of the links and joints concerning the movement in the specified direction. M'_{Si11} consists of M_{Si11} and the inertia of half of the top plate, the two ball joints, and another link3. M'_{Si22} consists of M_{Si22} and the inertia of a quarter of the top plate and, one ball joint. $^jM_{Ai}$ is expressed independently from M'_{Sikk} because $^jM_{Ai}$ changes its value depending on ^jN_i. **Figure 7** shows $\dot{v}_{Tx\max}$ and $\dot{v}_{Ty\max}$ calculated by eqs. 4 and 5 by changing the link length and the reduction ratio.

In the calculation of the endpoint deflection, the deflection of every link was examined under the inertia force caused by an 100 [G] acceleration at the endpoint. We determined the link length, the width, and the height as the endpoint deflection, which is converted by the link deflection, to satisfy the allowable range 0.1×10^{-3} [m].

[Following Step 2]

We adjusted the design parameters to satisfy both requirement specifications, the endpoint acceleration and

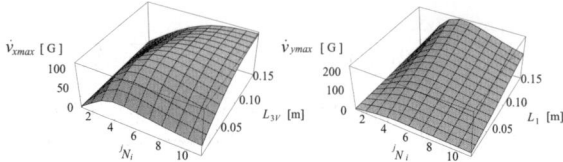

Fig. 7: Plotted acceleration graph by changing the link length and the reduction rate (left: $\dot{v}_{x\max}$, right: $\dot{v}_{y\max}$)

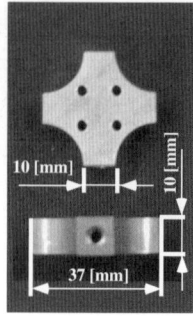

Fig. 8: Sub arm (left) and top plate (right)

Fig. 9: Prototype parallel mechanism, NINJA

the deflection, and as a result, in relation to the endpoint acceleration, $\dot{v}_{Tx\max}$ is 102.3 [G] and $\dot{v}_{Ty\max}$ is 119.7 [G] when the reduction ratio of the actuated joints 1 and 2 are $^2N_i = 4.0$ and $^3N_i = 6.8$, respectively. The link lengths L_1, L_{2H}, L_{2V}, and L_{3V} are 60, 20, 20, and 95 [mm], respectively, while in relation to the endpoint deflection, δ_{Tx} is 0.31×10^{-3} [m] and δ_{Ty} is 0.06×10^{-3} [m]. Though referring to **Table 2** δ_{Tx} does not satisfy the demands, we accept δ_{Tx}.

The 100[G] acceleration is not guaranteed in every direction because we only examined the acceleration performance in the specified direction. We expect the performance to be almost or less than 100 [G] near the basis attitude of the sub arm.

We confirmed whether the endpoint speed satisfies the demand specification. In the simple case where the maximum endpoint velocity $v_{T\max}$ in the basis attitude was calculated from the acquired link length, the reduction ratio, and the maximum instantaneous speed of the motor (5175 [rpm]≃542 [rad/s]), $v_{Tx\max}$ and $v_{Ty\max}$ are 9.2 and 10.8 [m/s], respectively, whereas the peak of the endpoint velocity v_{Txp} and v_{Typ} are 9.9 and 5.5 [m/s], respectively under the condition the sub arm performs the cyclic motion shown in **Fig. 3** with a velocity triangle profile in 100[G]. Though referring to **Table 2** $v_{Tx\max}$ does not satisfy the demand, we accept it.

Table B-1 in Appendix B shows the link parameters.

In this design, we ignored the loss of the friction due to the small reduction ratio but, essentially, we should consider it in the calculation of the acceleration and speed.

Figures 8 and **9** show the sub arm prototype and the NINJA parallel mechanism, respectively.

4 Experimental Evaluation

4.1 Position control of the sub arm

In the experiment, we investigated the performance of the prototype sub arm when the maximum desired acceleration 100 [G] ($\simeq 980$ [m/s^2]) was given to move a 20[mm] distance on the straight trajectory in the horizontal direction (X axis based on $\sum_{SBi}$ in **Fig. B-1**). The same experiment was also conducted in the vertical direction (Y axis).

The magnitude of the desired acceleration profile was given by a sine wave:

$$||\ddot{\boldsymbol{r}}_d|| = A \sin \omega t \qquad (6)$$

where $\ddot{\boldsymbol{r}}_d (= [\ddot{r}_{dx}\ \ddot{r}_{dy}]^T)$ is the desired acceleration vector, A is the amplitude of the acceleration, and ω is the angular frequency, which is determined from A and the moving distance. We set $A = 980$ [m/s^2] and $\omega = 555$ [rad/s]. The desired endpoint position $\boldsymbol{r}_d$ at time t is derived from eq. 6. To use the P-D control law in eq.7 with respect to the joint angle $\boldsymbol{q}_d$, $\boldsymbol{q}_d$ was calculated from $\boldsymbol{r}_d$ by using inverse kinematics.

$$\dot{q}_{adji} = K_{pi}(q_{di} - q_i) - K_{vi}\dot{q}_i \qquad (7)$$

where $\dot{q}_{adji}$ is the control input in the velocity control mode of the servo driver, K_{pi} and K_{vi} are the feedback gains and suffix $i(=1,2)$ is the actuated joint number.

These gains were determined by trial and error. The passive joint (J3) was fixed with an acrylic resin plate (17 [g]), as shown in **Fig. 8**. The top plate was not attached to the sub arm. We used a timer implemented

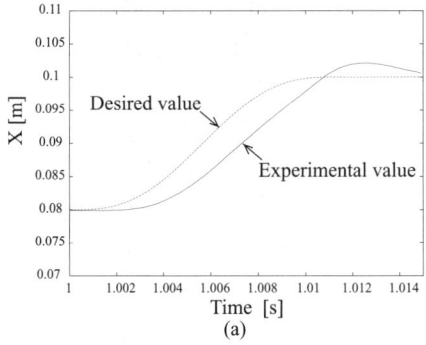

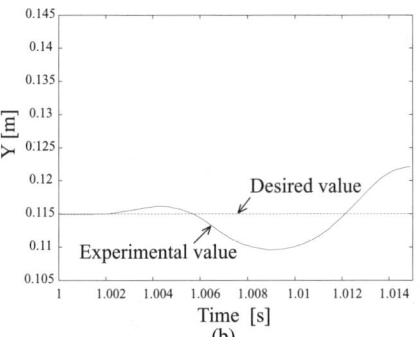

Fig. 10: Movement in X direction

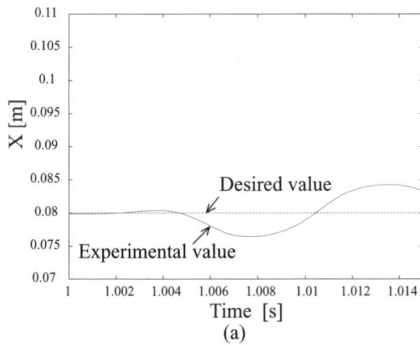

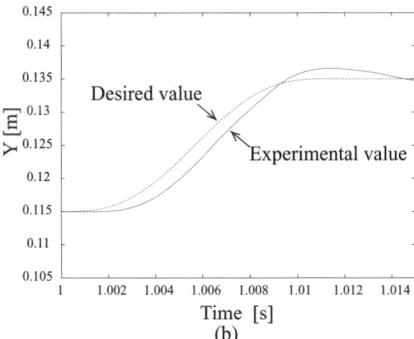

Fig. 11: Movement in Y direction

in an A/D converter to keep the sampling time constant (0.15 [ms]).

Figures 10 and **11** show the results of when the sub arm moved in X axis and Y axis direction. The time data include the time it took to move to the basis attitude. From the endpoint following trajectory, we confirmed that the sub arm achieved a high speed though about 2 [ms] delay occurred by the servo driver in the beginning of the motion. **Figures 10(b)** and **11(a)** show that the interference by the links influenced each other greatly. We need to develop an appropriate control method or system design to improve the performances.

4.2 Acceleration performance of Parallel Mechanism NINJA

We investigated the acceleration performance of parallel mechanism NINJA experimentally. In the experiments, Time and acceleration for the endpoint of NINJA to move 5.0 [mm] under the maximum velocity input of a motor driver lasting 5.0 [ms]. We computed the position of the endpoint of NINJA in the horizontal direction (X axis) and the vertical direction (Y axis) using joint angle data (the sampling time : 0.12 [ms]).

In the experimental results, the required time was 2.5 [ms] for the motion along the X axis (**Fig. 12(a)**) and 2.1 [ms] for the motion along the Y axis (**Fig. 13(a)**). The maximum velocities along X axis and Y axis were 2.3 [m/s] and 3.3 [m/s], respectively. Similarly, the maximum accelerations were 128.2 [G] (**Fig. 12(b)**) and 171.6 [G] (**Fig. 13(b)**), respectively. Here the position, velocity, and acceleration signals are computed using encoder signals and lowpass filters with a cut-off frequency of 400 [Hz].

The experimental results show the endpoint accelerations are over 100 [G], and these results means that NINJA has a high acceleration performance. We are now introducing an acceleration sensor to detect the exact acceleration.

5 Conclusion

This paper addressed the following topics:

1. We formulated the design problems of a prototype parallel mechanism NINJA to satisfy requirement specifications such as achieving an acceleration rate over 100[G].

2. To solve the design problems, we implemented a two-step design method to design the prototype NINJA. We determined the combination of design parameters that enable an endpoint acceleration over 100 [G] under the constraint within an allowable endpoint deflection. We produced the prototype NINJA based on those parameters.

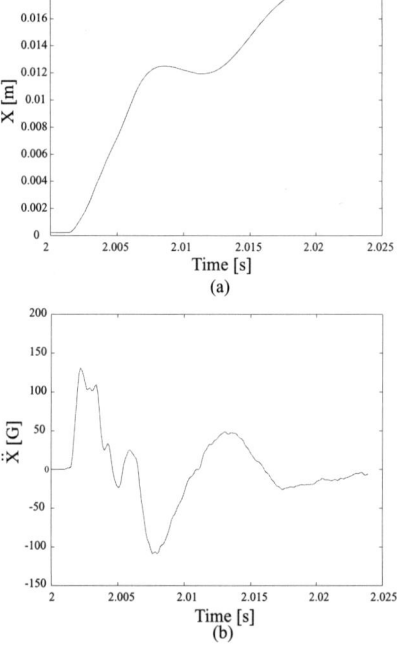

Fig. 12: Position and acceleration along the X axis

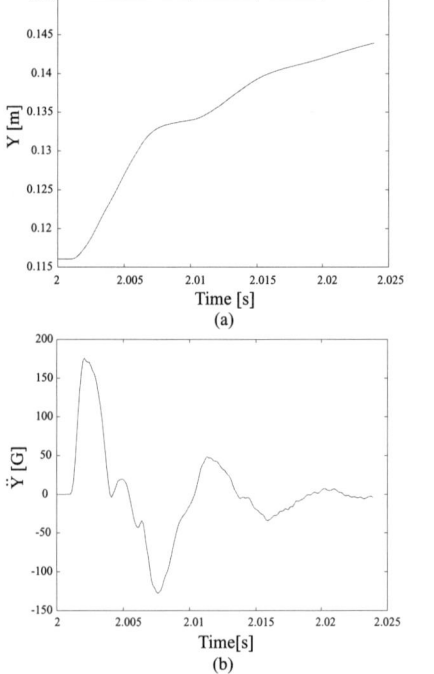

Fig. 13: Position and acceleration along the Y axis

3. We tested the performance of the sub arm and the prototype NINJA. The experimental results show the prototype NINJA had a high acceleration performance.

We only considered main components concerning motion in horizontal or vertical direction and we ignored influences of interfering terms caused by the link inertia force to simplify the calculation for the optimization. In our future work, we will examine the performance of the prototype NINJA in detail, and improve the mechanism by using an appropriate optimization method.

Acknowledgments

We thank Mr. Nakamura Fumihiko, Mr. Yamamoto Naruhito and all members of Nakamura Machining Works, JAPAN for the machining process of our parallel mechanism. We also thank Mr. Sugie Yutaka for deriving the basic equations and Mr. Oura Ryoichi for the experiments outlined in section 4.

References

[1] R. Clavel, "DELTA. A fast robot with parallel geometry", *Proc. Int. Symposium on Industrial Robots*, pp. 91-100, 1988.

[2] F. Pierrot, M. Uchiyama, et al., "A New Design of a Very Fast 6-DOF Parallel Robot", *J. of Robotics Mechatronics*, Vol.2, No.4, pp. 92-99, 1990.

[3] M. Uchiyama, et al., "Development of a 6-DOF High-Speed Parallel Robot HEXA", (In Japanese), *J. of Robotics Society of Japan*, Vol.12, No.3, pp. 451-458, 1994.

[4] S. Kawamura, et al., "Development of an Ultrahigh Speed Robot FALCON Using Parallel Wire Drive Systems", (In Japanese), *J.of Robotics Society of Japan*, Vol.15, No.1, pp. 82-89, 1997.

[5] J. Matsuyama, "High Speed and High Accuracy Robot with Closed-Loop Mechanism", *30th ISR*, Vol.1, pp. 95-100, 1999.

[6] K. H. Hunt, "Structural Kinematics of In-Parallel-Actuated Robot-Arms", *Trans. ASME*, J. of Mechanism, Transmissions, and Automation in Design, Vol.105, pp. 705-712, 1983.

[7] V. Potkonjak and M. Vukobratovic, "Computer-aided design of manipulation robots via multi-parameter optimization", *Mechanism and Machine Theory*, 18, 6 pp. 431-438, 1983.

[8] V. Potkonjak et al., "Interactive procedure for computer-aided design of industrial robots mechanisms", *Proc. of 13th ISIR*, pp. 16-85-94, 1983.

[9] K. Inoue, et al., "Study on Total Computer-aided Design System for Robot Manipulators", (In Japanese), *J.of Robotics Society of Japan*, Vol.14, No.5, pp. 710-719, 1996.

[10] K. Nagai and M. Matsumoto, "Design Method of a Driving Mechanism with Ultra High Acceleration", (In Japanese), *Proc. of the 17th Annual Conf. of the Robotics Society of Japan*, pp. 969-970, 1999.

[11] http://www.nagoya.melco.co.jp/

[12] http://www.hephaist.co.jp/

Appendix

A Basic Equations

A.1 Dynamics of mechanism

The dynamic equation of the sub arm and the top plate are represented as follows:

$$M_S \ddot{q}_S + h_S + J_S^T F_S = \tau_S \quad \text{(A-1)}$$
$$M_T \dot{v}_T + h_T = T_T \quad \text{(A-2)}$$
$$\tau_S = A^T \tau_A \quad \text{(A-3)}$$
$$T_T = J_T^T F_S \quad \text{(A-4)}$$

where $\ddot{q}_S \in \Re^{12}$ is the angular acceleration vector of all the sub arms including the passive joints, $\dot{v}_T \in \Re^6$ is the endpoint acceleration vector, $\tau_S \in \Re^{12}$ is the joint torque vector of all the sub arms including the passive joints, $\tau_A \in \Re^8$ is the active joint torque vector selected in τ_S, $F_S \in \Re^{12}$ is the force vector at all the sub arms endpoints, $T_T \in \Re^6$ is the force vector on the top plate, $M_S \in \Re^{12 \times 12}$ and $M_T \in \Re^{6 \times 6}$ are the inertia matrices, $h_S \in \Re^{12}$ and $h_T \in \Re^6$ contain the centrifugal, the Coriolis, and the gravity terms, and $J_S \in \Re^{12 \times 12}$ and $J_T \in \Re^{12 \times 6}$ are the Jacobian matrices (T means transpose). The suffixes S and T mean all the sub arms and the top plate, respectively. $A \in \Re^{8 \times 12}$ is the coefficient matrix composed of 0 and 1 between τ_S and τ_A.

Considering the velocity vector at the combination point between the sub arms and the top plate, we introduce the following notation:

$$J_S \dot{q}_S = J_T v_T \quad \text{(A-5)}$$

From eq. A-5, the Jacobian matrix $J \in \Re^{12 \times 6}$ defined as the whole parallel mechanism is shown in eq. A-6. The relationship between $\dot{q}_S$ and $\ddot{q}_S$ and v_T and $\dot{v}_T$ is shown in eqs. A-7 and A-8:

$$J = J_S^{-1} J_T \quad \text{(A-6)}$$
$$\dot{q}_S = J v_T \quad \text{(A-7)}$$
$$\ddot{q}_S = J \dot{v}_T + \dot{J} v_T \quad \text{(A-8)}$$

A.2 Endpoint Acceleration

From eqs. A-1 to A-4 and A-6, the dynamic equation of the whole parallel mechanism is given in eq. A-9:

$$M \dot{v}_T + h = J^T A^T \tau_A \quad \text{(A-9)}$$
$$h = h_T + J^T (M_S \dot{J} v_T + h_S) \quad \text{(A-10)}$$
$$M = M_T + J^T M_S J \quad \text{(A-11)}$$

where $M \in \Re^{6 \times 6}$ is the inertia matrix and $h \in \Re^6$ is a nonlinear term with respect to the whole mechanism. We obtain $\dot{v}_T$ from eq. A-9.

$$\dot{v}_T = M^{-1}(J^T A^T \tau_A - h) \quad \text{(A-12)}$$

A.3 Endpoint Deflection

The stiffness at endpoint $K \in \Re^{6 \times 6}$ consists of the sum of the joint stiffness matrices of the sub arm i $K_{qi} \in \Re^{12 \times 12}$, defined by a material property:

$$K = \sum_{i=1}^{4} J_i^T K_{qi} J_i \quad \text{(A-13)}$$

where $J_i = J_{Si}^{-1} J_{Ti}$. The endpoint deflection caused by the inertia force is expressed in eq. A-14.

$$\delta_T = K^{-1} T_T$$
$$= K^{-1}(M_T \dot{v}_T + h_T) \quad \text{(A-14)}$$

B Link data of NINJA

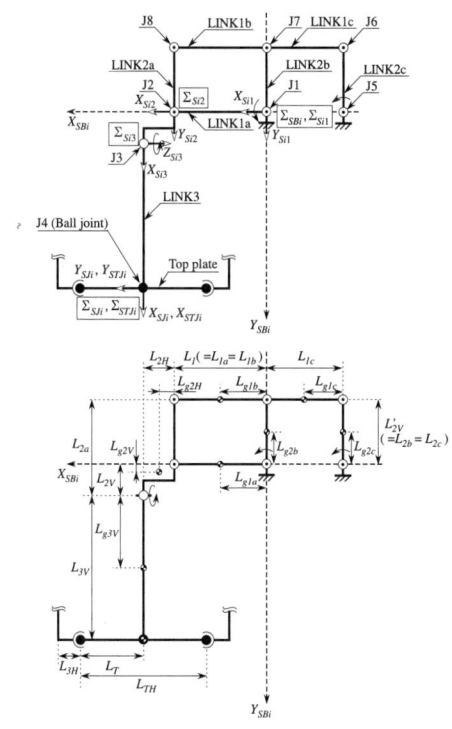

Fig. B-1: Definition of the link parameters and coordinates

Table B-1: Measurements of each link

Link	Length[mm]	Width[mm]	Depth[mm]	Weight[g]
1a	60	16	12	46
1b	60	10	10	24
1c	50	10	6	15
2a	40	17	7	43
2b*	40	18	7	20
2c	40	18	5.5*	33
3	95	18	9	58*

* Link2b consists of a couple of the same link, which holds links1a, 1b, and 1c. Link2c in depth direction is shaped like a "U". The link3 weight contains the ball joint weight 36 [g].

Robotic Force Control using Observer-based Strict Positive Real Impedance Control

Rolf Johansson and Anders Robertsson

Department of Automatic Control
Lund Institute of Technology, Lund University
PO Box 118, SE-221 00 Lund, Sweden
Phone +46 46 222 8791, Fax +46 46 138118
E-mail: Rolf.Johansson@control.lth.se, Anders.Robertsson@control.lth.se

Abstract—This paper presents theoretical and experimental results on observer-based impedance control for force control without velocity feedback. As the velocity may not available to measurement, which is often the case for industrial robots, an observer was designed to reconstruct velocity in such a way that it be useful for stabilizing feedback control and to modification of the damping in the impedance relationship. A good model of the robot joint used was obtained by system identification. Experiments were carried out on an ABB industrial robot 2000 to demonstrate results on observer-based SPR feedback applied in the design. Stability was shown via a modified Popov criterion.

Keywords: Impedance control, Observers, Stability, Popov criterion, Robotics.

I. INTRODUCTION

Today industrial robots are used in a wide range of applications. Many of these applications will naturally require the robot to come into contact with a physical environment—*e.g.*, welding, grinding, and drilling. This interaction with the environment will set constraints on the geometric paths that can be followed, a situation referred to as constrained motion.

To avoid excessive contact forces the robot trajectory must be planned with high accuracy. This is, however, often impossible because of geometric uncertainty and finite positioning accuracy. A way to solve this conflict is to let the robot manipulator be force controlled—*e.g.*, impedance control control and hybrid force/position control [21].

Among robot force control methods used, impedance control is aimed at control of the dynamic relation between position error and force error in interaction similar to Newton's second law of motion [9], [13], [2], [21]. The impedance relationship between force F and position x used in this paper is represented by the equation

$$F = K \cdot x + D \cdot \dot{x} \quad (1)$$

where the positive constants K and D represent design parameters for stiffness and damping, respectively. One way to achieve this relation is to control the following impedance variable to zero

$$\begin{aligned} z(t) &= Kx(t) + D\dot{x}(t) - F(t), \\ Z(s) &= (K+sD)X(s) - F(s) \end{aligned} \quad (2)$$

In its simplest form, such control can be accomplished using an ordinary PI-controller which involves feedback of z. A problem, however, is that the velocity is often not available for measurement and that differentiating feedback is error prone. Although stability and robustness may be improved using more sophisticated model-based control, poor force impact models and other aspects of insufficient world modeling still constitute a challenge to model-based force control [13], [2], [21].

This paper presents a method to improve impedance control by means observer-based feedback in one dimension [20], [8]. Our approach is based on strict positive real or feedback positive real (SPR/FPR) observer-based feedback design with a modified Popov criterion used for the stability analysis. In Sec. II, the control law and the design procedure are described and Sec. II is devoted to the stability analysis. Sections IV and V describe the experimental setup and the identification of a model for the system. Simulations and an experiment are presented in Sec. VI followed by a discussion and conclusions.

II. THE CONTROL LAW

Molander and Willems provided a design procedure for L—i.e., design for nonlinear state-feedback control—with specified gain margin [17]. They made a characterization of the conditions for stability of feedback systems with a high gain margin

$$\dot{x} = Ax + Bu, \quad z = Lx, \quad u = -f(z,t) \quad (3)$$

with $f(\cdot, \cdot)$ enclosed in a sector $[K_1, K_2]$. The following procedure was suggested to find a state-feedback vector L such that the closed-loop system will tolerate any

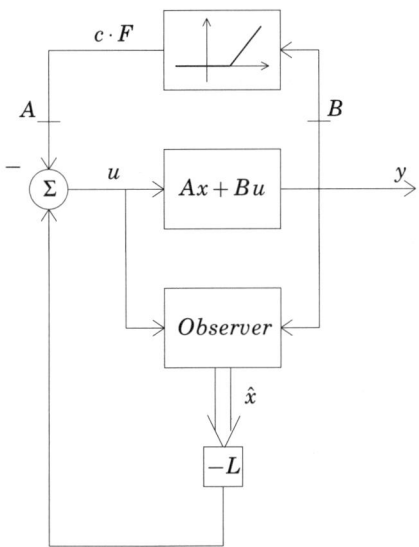

Fig. 1. Block diagram of the impedance control. The Popov plot of the transfer function from A to B is used to analyze stability for the contact force nonlinearity.

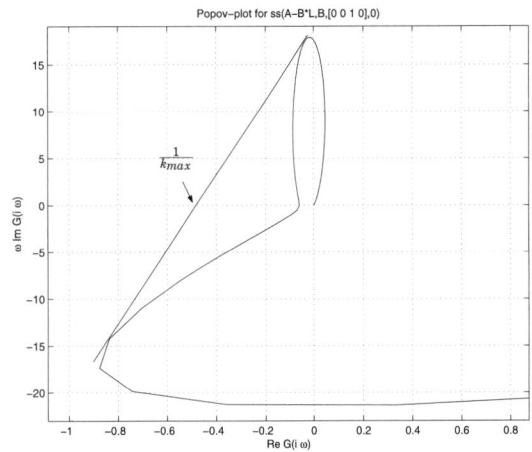

Fig. 2. Popov plot for the transfer function from A to B in Fig. 1 for the example. The maximum slope of the nonlinearity is given by k_{max}. For the quite compliant environment used in the experiments the stability margin is large.

$f(\cdot,\cdot)$ enclosed in a sector $[K_1, K_2]$. Synthesis of a state-feedback vector L with a robustness sector $[K_1, \infty)$ follows from

- Pick a matrix $Q = Q^T > 0$ such that (A,Q) is observable;
- Solve the Riccati equation $PA + A^T P - 2K_1 PBB^T P + Q = 0$ and take $L = B^T P$

which can be recognized as an FPR condition [15]—i.e., the stability condition will be that of an SPR condition on $L(sI - A + K_1 BL)^{-1} B$. The design procedure was based on a circle-criterion proof and involved a solution of a Riccati equation. In the context of observer-based state feedback control, the controllability condition presents a problem of application. In [12], such SPR/FPR design has been generalized to the case of observer-based state feedback control.

Strict Positive Real Design

Now consider the state feedback control law

$$u = -L\hat{x} - cF \qquad (4)$$

where F is the measured contact force and c is a constant. The state estimates, $\hat{x}$, are given by the full-order observer

$$\dot{\hat{x}}(t) = A\hat{x}(t) + Bu(t) + K_f(y(t) - C\hat{x}(t)) \qquad (5)$$

where the observer gain, K_f, is chosen using ordinary pole-placement of the eigenvalues of $A - K_f C$, with the system matrices (A,C) given by (15).

The control law (4) is equivalent to proportional control of the output variable (2), since the states in the state-space model of the robot joint are position and velocity.

The state feedback matrix, L, is chosen according to the results in [12], i.e.,

$$L = R^{-1} B^T P_c, \quad P_c = P_c^T > 0 \qquad (6)$$

where P_c is the solution to the Riccati equation

$$P_c(A - BL) + (A - BL)^T P_c = -Q_c - P_c BR^{-1} B^T P_c \qquad (7)$$

$Q_c > 0$ and $R > 0$ are design matrices, which represent penalties on the states and the control signal, respectively. The stiffness and damping in the impedance relation are modified indirectly by the choice of the state penalty matrix, Q_c, which is chosen as

$$Q_c = \begin{bmatrix} K_p & 0 & 0 & 0 \\ 0 & D_p & 0 & 0 \\ 0 & 0 & 0.01 & 0 \\ 0 & 0 & 0 & 0.01 \end{bmatrix} \qquad (8)$$

The penalty, K_p, on the position will affect the stiffness in the impedance relation, whereas D_p will affect the damping.

III. STABILITY ANALYSIS

Stability concerning the contact force nonlinearity is analyzed using the Popov criterion. A block diagram of the system under the impedance control (4) is shown in Fig. 1. Stability is analyzed by plotting the Popov curve for the transfer function from A to B. Then it is possible to determine a sector $[0, k_{max}]$ in which the contact force nonlinearity must be contained in order to guarantee stability [14]. The stability sector will be affected by the choice of the state feedback L.

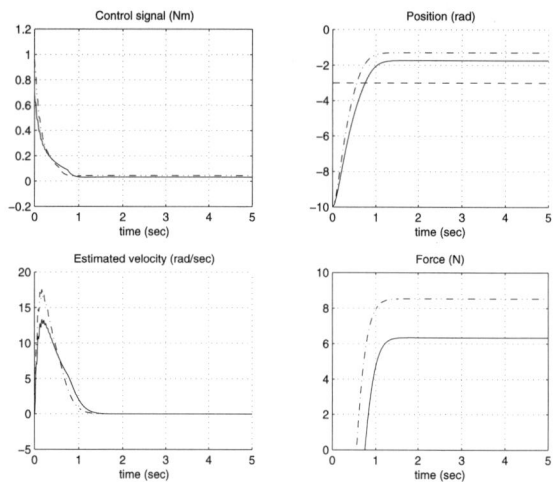

Fig. 3. Simulation of observer-based impedance control for two different choices of the penalty K_p. The solid curve corresponds to $K_p = 5$ and the dash-dotted to $K_p = 10$. The dashed line in the position plot marks the location of the screen. It is seen that a larger K_p gives higher stiffness, leading to a larger contact force. $D_p = 1$ in both simulations.

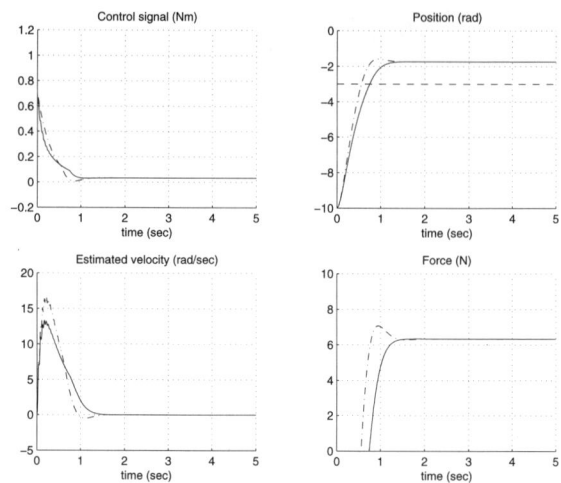

Fig. 4. Simulation of observer-based impedance control for two different choices of the penalty D_p. The solid curve corresponds to $D_p = 1$ and the dash-dotted to $D_p = 0.1$. The dashed line in the position plot marks the location of the screen. It is seen that a larger D_p increases the damping during the transient. $K_p = 5$ in both simulations.

Example In the design, the following matrices have been used

$$Q_c = \begin{bmatrix} 10 & 0 & 0 & 0 \\ 0 & 1 & 0 & 0 \\ 0 & 0 & 0.01 & 0 \\ 0 & 0 & 0 & 0.01 \end{bmatrix}, \quad R = 1000 \quad (9)$$

resulting in the state feedback gain

$$L = \begin{bmatrix} -0.0795, & 0.0191, & 0.1796, & 0.0054 \end{bmatrix} \quad (10)$$

The observer gain matrix, K_f, was chosen as

$$K_f = \begin{bmatrix} 270, & 16700, & 78, & -1190 \end{bmatrix}^T \quad (11)$$

Fig. 2 shows the Popov plot of the transfer function from A to B in Fig. 1 using the gain matrices above. To guarantee stability the slope of the nonlinearity, i.e., $c \cdot k$, must be less than $|k_{max}|$. The stiffness of the environment is $k = 5$ N/rad as mentioned in Sec. IV and $c = 0.01$. From Fig. 2, it can be concluded that the stability margin is large in this case.

IV. EXPERIMENTAL SET-UP

The experiments are performed in the Robotics Lab at the Department of Automatic Control in Lund using an ABB industrial robot Irb 2000 (Fig. 5). The controller is implemented in Matlab/Simulink, compiled and dynamically linked to the Open Robot Control System [18]. The forces are measured using a wrist-mounted, DSP-based force/torque sensor from JR3.

The physical constraint is represented by a vertical screen as seen in Fig. 5, and the impedance is controlled perpendicular to this screen using joint one, i.e., the base joint, of the robot. The situation can be modeled as in Fig. 6, where x_c is the location of the screen, x_∞ the stationary position, and x_r the desired position in the case of unconstrained motion. In the following x_r is zero. The stationary position will depend on the relation between the environmental stiffness and the robot stiffness as specified by the impedance relation. If the robot is made very stiff, then x_∞ will be close to x_r, whereas a stiff environment will lead to x_∞ being close to x_c.

The contact force was modeled as a regular spring, i.e.,

$$F = \begin{cases} 0 & x \leq x_c, \\ k \cdot (x - x_c) & x > x_c. \end{cases} \quad (12)$$

The stiffness, k, of the screen used in the experiments was 5 N/rad. The position measurement was given in radians on the motor side, and using the gear ratio of -71.44 the actual values of the robot arm were computed.

V. MODELING

A good model of the base joint is needed in order to design the observer. This model is obtained by system identification. The joint is modeled as two rotating masses connected by a spring-damper, reflecting the flexibility of the gear-box and the axis (Fig. 7). The angular position and velocity on the motor side are denoted φ_1 and ω_1, whereas φ_2 and ω_2 denote the corresponding quantities of the robot arm. The process input is the torque, τ, applied by the motor and the measured process output is the angular position on the motor side, φ_1. By introducing the state variables

$$x_1 = \varphi_1, \quad x_2 = \omega_1, \quad x_3 = \varphi_2, \quad x_4 = \omega_2 \quad (13)$$

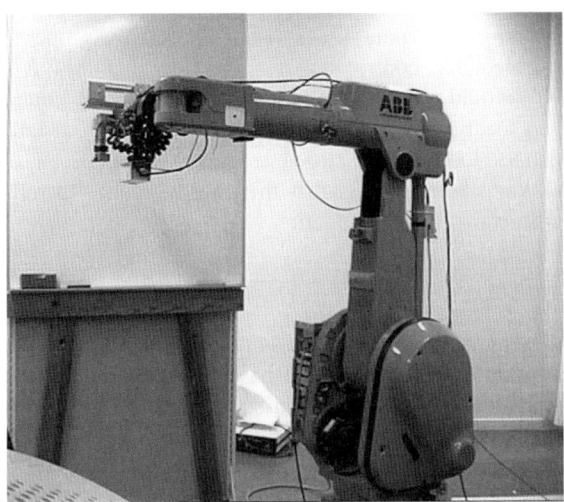

Fig. 5. The experimental setup. An ABB industrial robot Irb 2000 with an open control system is used, and the impedance control is performed perpendicular to the screen. The base joint (joint one) of the robot was used.

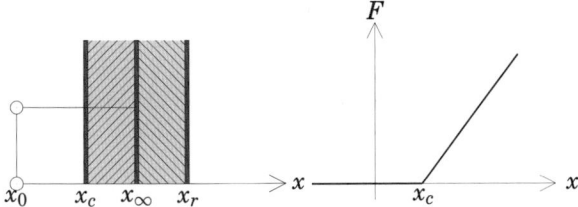

Fig. 6. A screen representing a physical constraint. x_c is the location of the screen, x_∞ the stationary position, and x_r the desired position in the case of unconstrained motion (*left*). The contact force nonlinearity (*right*): When the robot is in contact with the environment, the force can be modeled as a linear spring, the force being zero without contact.

and the input $u(t) = \tau(t)$, the system can be written on state-space form as

$$\dot{x} = \begin{bmatrix} 0 & 1 & 0 & 0 \\ -\frac{k_{12}}{J_1} & -\frac{D_1+d_{12}}{J_1} & \frac{k_{12}}{J_1} & \frac{d_{12}}{J_1} \\ 0 & 0 & 0 & 1 \\ \frac{k_{12}}{J_2} & \frac{d_{12}}{J_2} & -\frac{k_{12}}{J_2} & -\frac{D_2+d_{12}}{J_2} \end{bmatrix} x + \begin{bmatrix} 0 \\ \frac{1}{J_1} \\ 0 \\ 0 \end{bmatrix} u$$

$$y = \begin{bmatrix} 1 & 0 & 0 & 0 \end{bmatrix} x(t) \quad (14)$$

The numerical values of the coefficients in the state-space model above are estimated by *prediction error method* in Matlab [16]. The final identified model is given by

$$\dot{x} = \begin{bmatrix} 0 & 1 & 0 & 0 \\ -11476 & -8 & 11476 & 2 \\ 0 & 0 & 0 & 1 \\ 8445 & 2 & -8445 & -9 \end{bmatrix} x + \begin{bmatrix} 0 \\ 700 \\ 0 \\ 0 \end{bmatrix} u$$

$$y = \begin{bmatrix} 1 & 0 & 0 & 0 \end{bmatrix} x \quad (15)$$

Model validation is shown in Fig. 8 comparing true

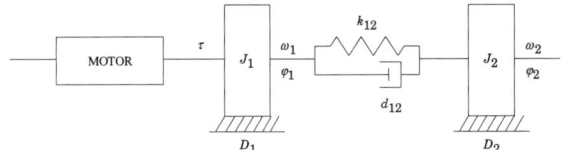

Fig. 7. Physical model of the robot joint. The angular position on the motor side, φ_1, is measurable. The input is the torque, τ, actuated by the motor.

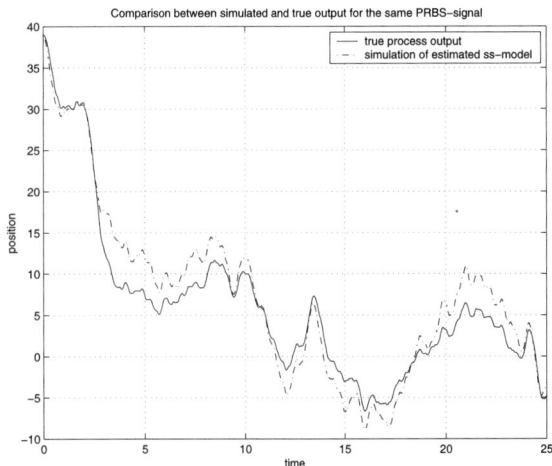

Fig. 8. Comparison of true process output with a simulation of the estimated state-space model. The same PRBS-signal has been used as input in both cases.

process output with a simulation of the state-space model (15).

VI. SIMULATIONS AND EXPERIMENT

The dash-dotted curve in Fig. 3 shows a simulation using the design in the example above. The dashed line in the position plot marks the location of the screen. By modifying the penalty K_p the stiffness of the impedance relation is changed, which is shown by the solid line. Fig. 4 shows a simulation examining the influence of the penalty D_p. It is seen that the damping during the transient is affected, but that the stationary force is the same in both cases. The simulations were done in Matlab/Simulink.

An experiment on the real robot is shown in Fig. 9. This experiment corresponds to the case analyzed in the example, and is the same as the dash-dotted curve in the simulation in Fig. 3. The initial force transient before contact is an inertia force. The correspondence between experiment and simulation is good. As predicted in the analysis, stability is preserved at contact.

VII. DISCUSSION

Recently, it has been shown that relaxation of the controllability and observability conditions imposed in the

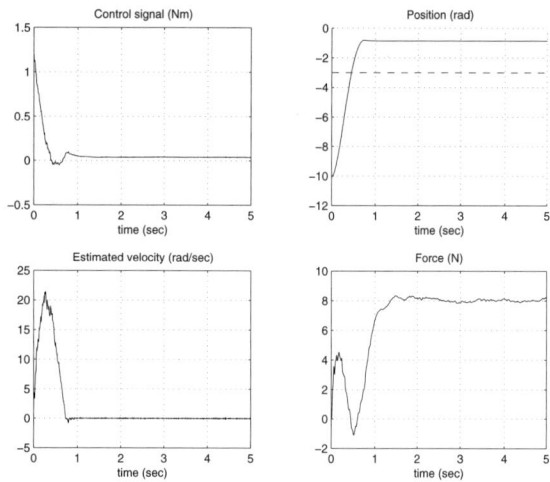

Fig. 9. Experiment of observer-based impedance control corresponding to the design analyzed in the example, i.e., $K_p = 10$ and $D_p = 1$. The initial force transient before contact is an inertia force. The dashed line in the position plot marks the location of the screen.

Yakubovich-Kalman-Popov (YKP) lemma can be made for a class of nonlinear systems described by a linear time-invariant system (LTI) with a feedback-connected cone-bounded nonlinear element [12]. This approach was used in order to achieve robustness in impedance control design.

An obvious extension would be to apply the results to full-dimensional wrenches. This would include use of the Jacobian to translate forces from the task space to the operational space, as well as full dynamics for the robot and gravity compensation. The simple linear model of the robot would not be sufficient and the coupling between the links would require nonlinear observer-based control [19]. Another possible extension would be to apply the results on other force control schemes such as parallel force/position control or hybrid force/position control [2], [5], [6], [21].

The approach to modification of the relative-degree and SPR properties is related to the 'parallel feedforward' as proposed in the context of adaptive control [1]. Another related idea is passification by means of shunting introduced by Fradkov [4]. All these approaches represent derivation of a loop-transfer function with SPR properties for a control object without SPR properties by means of dynamic extensions or observers. This idea that combines attractively with the observer-based SPR design used here. The Bar-Kana approach [1] starts with the following transfer functions

$$G_0(s), G_R(s), \Rightarrow G_{cl}(s) = \frac{G_0(s)}{1 + G_R(s)G_0(s)}$$

$$G_1(s) = G_0(s) + \frac{1}{G_R(s)} \quad (16)$$

Assuming that some $G_R(s)$—not necessarily proper or implementable control transfer function such as $(K + sD)$—would provide stabilizing SPR feedback control when feedback interconnected with $G_0(s)$, then a stable feedback loop can be closed around $G_1(s)$ of Eq. (16). The key observation is that the implementable 'parallel feedforward' transfer function

$$G_P(s) = \frac{1}{G_R(s)} \quad (17)$$

will achieve stable feedback without differentiation. Design of a mechanical device based on a related idea of impedance matching has been suggested by Dohring and Newman [3]. Combination of the 'parallel feedforward' [1] and the SPR design [12] may further increase robustness properties.

VIII. CONCLUSIONS

An approach to observer-based impedance control in one dimension has been presented. Impedance control is a robot force control technique aimed at control of the relation between position and force, rather than force control or position control separately. This technique has proved useful in dealing with geometric uncertainty, i.e., when the exact location of the environment is unknown. As industrial robots are often equipped with position sensors only, the velocity had to be reconstructed. A full-order observer was used to estimate the velocity, and the model of the robot joint needed for the observer dynamics was obtained by system identification.

The approach taken, observer-based impedance control, was based on results impedance control combined with observer-based SPR feedback [9], [12]. The design included the solution of a Riccati equation, and the stiffness and damping in the impedance relation were modified by the choice of weights in the design matrix. Stability was proven by means of a modified Popov criterion. The theoretical results were verified by simulations and experiments, the experiments being carried out on an ABB industrial robot Irb 2000.

IX. REFERENCES

[1] I. Bar-Kana. Parallel feedforward and simplified adaptive control. *Int J. Adaptive Control and Signal Processing*, 1:95–109, 1987.

[2] S. Chiaverini, B. Siciliano, and L. Villani. A survey of robot interaction control schemes with experimental comparison. *IEEE/ASME Trans. Mechatronics*, 4:273–285, 1999.

[3] M. Dohring and W. Newman. Admittance enhancement in force feedback of dynamic systems. In *Proc. 2002 IEEE Int. Conf. Robotics and Automation*, pages 638–643, Washington, DC, May 2002.

[4] A. Fradkov. Nonlinear feedback control: Regulation, tracking, oscillations. In *1st IFAC Workshop on New Trends in Control Systems Design*, pages 426–431, Smolenice, 1994.

[5] D.M. Gorinevsky, A.M. Formalsky, and A.Y. Schneider. *Force Control of Robotics Systems*. CRC Press, New York, 1997.

[6] A. Hemami and G. Zeng. An overview of robot force control. *Robotica*, 15:473–482, 1997.

[7] D. Henriksson. Observer-based impedance control in robotics. Master's thesis, Dept. Automatic Control, Lund University, 2000.

[8] D. Henriksson, R. Johansson, and A. Robertsson. Observer-based impedance control in robotics. In *Proc. 5th IFAC Symp. Nonlinear Control Systems (NOLCOS 2001)*, pages 360–365, St. Petersburg, Russia, July 4-6 2001.

[9] N. Hogan. Impedance control: An approach to manipulation, Parts I-III. *J. Dynamic Systems, Measurement, and Control, ASME*, 107:1–24, 1985.

[10] R. Johansson. *System Modeling and Identification*. Prentice Hall, Englewood Cliffs, NJ, 1993.

[11] R. Johansson and A. Robertsson. Observer-based SPR feedback control system design. 5th IFAC Symp. Nonlinear Control Systems (NOLCOS 2001), St. Petersburg, Russia, 2001.

[12] R. Johansson and A. Robertsson. Observer-based strict positive real (SPR) feedback control systems design. *Automatica*, 38:1557–1564, 2002.

[13] R. Johansson and M.W. Spong. Quadratic optimization of impedance control. In *Proc. IEEE Conf. Robotics and Automation*, pages 616–621, San Diego, CA, 1994.

[14] H.K. Khalil. *Nonlinear Systems*. Prentice-Hall, Upper Saddle River, NJ, 2nd edition, 1996.

[15] P. V. Kokotović and H. J. Sussman. A positive real condition for global stabilization of nonlinear systems. *Systems and Control Letters*, 13:125–133, 1989.

[16] L. Ljung. *System Identification Toolbox; User's Guide*. The MathWorks Inc., Natick, MA, 1991.

[17] P. Molander and J.C. Willems. Synthesis of state feedback control laws with a specified gain and phase margin. *IEEE Trans. Automatic Control*, AC-25:928–931, 1980.

[18] K. Nilsson and R. Johansson. Integrated architecture for industrial robot programming and control. *J. Robotics and Autonomous Systems*, 29:205–226, 1999.

[19] A. Robertsson. *On Observer-Based Control of Nonlinear Systems*. PhD thesis, Dept. Automatic Control, Lund Inst. Tech., 1999.

[20] B. Siciliano and L. Villani. Parallel force/position controller with observer for robot manipulators. In *Proc. 36th Conf. Decision and Control*, pages 1335–1340, 1997.

[21] B. Siciliano and L. Villani. *Robot Force Control*. Kluwer Academic Publishers, Dordrecht, NL, 1999.

Acknowledgement: The authors thank Dan Henriksson, M. Sc., for the contributions made during his master thesis project [7].

This research was sponsored by Nutek under the research program 'Complex Technical Systems'.

Impact When Robots Act Wisely

Eunjeong Lee (eunjeonglee@ieee.org)
Korea Advanced Institute Science and Technology

Juyi Park (juyi.park@vanderbilt.edu)
Vanderbilt University

Cheryl B. Schrader (schrader@ieee.org)
University of Texas at San Antonio

Pyung-Hun Chang (phchang@kaist.ac.kr)
Korea Advanced Institute Science and Technology

Abstract – For stabilization of a robot manipulator upon collision with a stiff environment, a nonlinear bang-bang impact controller is developed. Under this control, a robot can successfully achieve contact tasks without changing control algorithm or controller gains throughout all three modes: free space, transition and constrained motion. It uses a robust hybrid impedance/time-delay control algorithm to absorb impact forces and stabilize the system. This control input alternates with zero when no environment force is sensed due to loss of contact. This alternation of control action repeats until the impact transient subsides and steady state is established. After impact transient, the hybrid impedance/time-delay control algorithm is utilized. This bang-bang control method provides stable interaction between a robot with severe nonlinear joint friction and a stiff environment and achieves rapid response while minimizing force overshoots. During contact transition, we employ one simple control algorithm with same gains that switches only to zero, while other controllers use more than one control algorithms or different control gains. It is shown via experiments that overall controlled performance is superior or comparable to existing impact force control techniques.

1. INTRODUCTION

As technology develops, robots are expected to perform more sophisticated and diverse tasks. In most cases except surveillance, we want robots to manipulate the world/environments they encounter for our benefit, not just maneuver around in it avoiding obstacles. In general, such manipulations involve active interaction with an unknown or changing environment whether intentional or accidental, and this requires making and breaking contact with objects frequently. When robots move from free-space to constrained motion, impact generally occurs and the ability to achieve stable and smooth contact transition becomes crucial for successful manipulation. In contrast to its importance, however, relatively few research works have addressed the area of impact control [1-18].

In this paper, we propose an interaction controller that makes a robot establish stable contact and achieve the desired state/dynamics in a seamless fashion without changing a control algorithm or gains. We call it a nonlinear bang-bang impact force control (NBBIC), which uses a hybrid impedance/time-delay control algorithm (also called natural admittance/time-delay control (NAC/TDC)). The hybrid impedance/time-delay controller is chosen because it compensates for disturbances effectively, thus absorbing impact energy quickly. During contact transition, NBBIC can employ only one control algorithm with same gains that switches to zero control input, while other controllers use more than one control algorithms or different control gains. For example, while a rover robot moves quickly to explore the Mars surface, it may accidentally encounter a hard object. Upon collision, the NBBIC can quickly absorb impact energy without damaging the robot itself or the object. Then, the robot can perform contact tasks with the object, such as shaking hands or drilling.

Consider a task which involves establishing a desired dynamic interaction with a stiff environment. Such contact generates reaction forces. But the attempt to employ a reckless strategy of applying increased forces to reject them as disturbance may only aggravate the situation. Instead the approach should be to exert control action to accept/absorb impact forces while achieving the desired dynamic response. In our controller, therefore, we use a mitigated control action by on-off control instead of continually pushing the environment during contact transient. The off state of control input helps reduce further impact vibrations by letting the impact oscillations subside naturally.

The organization of this paper is as follows: Section II describes the integration of hybrid impedance/time-delay control with the proposed bang-bang impact control. Section III verifies the performances of NAC/TDC via experiments and compares the experimental results of NBBIC with other existing controllers. Section IV discusses conclusions and suggestions for future work.

II. CONTROL DESIGN

In this section, a brief description of natural admittance control and time delay control is presented along with their control laws. Additionally, this section develops a hybrid natural admittance/time delay (NAC/TDC) control.

A. Natural Admittance Control

Under natural admittance control (NAC), the target dynamics, Y_{des}, is chosen to be smaller than the maximum target admittance which does not violate the passivity constraint as follows [18-20]:

$$Y_{des} = \frac{1}{(K_{des}/s) + B_{des} + M_s s} < \frac{1}{M_s s} \quad (1)$$

where M_s, K_{des}, and B_{des} are the end-point mass and desired stiffness and damping, respectively. The simplest form of natural admittance control that achieves the target dynamics of (1) can be described as follows:

$$F = G_v(v_{cmd} - v) + (\frac{1}{s} K_{des} + B_{des})(v_{des} - v) \quad (2)$$

where

$$v_{cmd} = \frac{F_s + (K_{des}/s + B_{des})(v_{des} - v)}{M_s s}$$

and v_{des} and v represent the desired velocity and the velocity at the motor port, respectively. In (2), the first term corrects deviations of the actual response from the modeled response, v_{cmd}. The second term is the feedforward term which imposes desired dynamics implicitly. While a robot tries to achieve desired target dynamics of (1), the desired dynamics generate a virtual force on the end effector composed of virtual spring force and virtual damping force. Therefore, this virtual force should be also accounted for in the force feedback loop through v_{cmd}. In reality, however, this target admittance cannot be achieved due to undesirable dynamic effects such as friction. To mask these effects, the sensed environment force F_s is fed back as a velocity command.

However, like many other interaction controllers, NAC does not achieve the desired performance due to inherent nonlinear dynamics, modeling uncertainties and digital sampling. This problem can be solved by using a feedforward compensator, which does not affect stability but does affect the transient response and the equilibrium state [21]. However, this process is complicated and system dependent.

Better compensation may be achieved by a simpler estimation technique that evaluates a function representing the effect of uncertainties. Youcef-Toumi proposed a time delay control (TDC) for such a purpose [22]. Hsia also used a similar approach [23]. The technique uses a recent past observation of the system's response and its control input to directly estimate the unknown dynamics and unexpected disturbances at any given instant through time delay. In the next section, a hybrid impedance/time-delay control algorithm will be developed to enhance natural admittance control by combining it with time delay control.

B. Design of Hybrid Impedance/Time-Delay Control

Consider a single-input single-output single DOF robotic system that can be described by the nonlinear dynamic equation:

$$\ddot{x}(t) = f(x, \dot{x}, t) + h(x, \dot{x}, t) + b(x, \dot{x}, t)u(t) + d(t) \quad (3)$$

where x, $\dot{x}$ and $\ddot{x}$ are states, $u(t)$ is a control input, $b(x, \dot{x}, t)$, is a control distribution term, $d(t)$ represents unknown disturbances, and $f(x, \dot{x}, t)$ and $h(x, \dot{x}, t)$ represent known and unknown nonlinear dynamics of the system, respectively. The variable t represents time. The term $h(x, \dot{x}, t)$ includes actuator saturation and stiction as well as Coulomb friction and nonlinear spring characteristics in the transmission. The system output is the variable x and the reference model for x is defined by:

$$\ddot{x}_r(t) = c_r x_r(t) + a_r \dot{x}_r(t) + b_r r(t) \quad (4)$$

where $x_r(t)$, $\dot{x}_r(t)$ and $\ddot{x}_r(t)$ are model states, $r(t)$ is a reference input and c_r, a_r and b_r are known constants.

If we set b equal to a constant for the case where the control distribution term is assumed known as in [24], and assume that no information is available on the nonlinear dynamics, that is f is zero, the time delay control law can be derived as follows [25]:

$$u(t) = u(t-L) + (1/b)[-\ddot{x}(t-L) - K_p(x_r(t) - x(t)) \\ - K_v(\dot{x}_r(t) - \dot{x}(t)) + c_r x(t) + a_r \dot{x}(t) + b_r r(t)] \quad (5)$$

In this particular time delay control law, the delayed control action term and the delayed acceleration term attempt to compensate for the nonlinear dynamics and disturbances, while the desired reference model is followed by adjusting the error dynamics.

Now, let us examine the structure of the natural admittance control law in the time domain,

$$u(t) = K_{des}(x_{des}(t) - x(t)) + B_{des}(\dot{x}_{des}(t) - \dot{x}(t)) + G_v(\dot{x}_{cmd}(t) - \dot{x}(t)) \quad (6)$$

where

$$\dot{x}_{cmd}(t) = \int \frac{F_s(t) + K_{des}(x_{des}(t) - x(t)) + B_{des}(\dot{x}_{des}(t) - \dot{x}(t))}{M_s} dt \quad (7)$$

Notice that the term, $\dot{x}_{cmd}(t)$, can be regarded as a reference input, $r(t)$, in the time delay control law. The stiffness and damping terms in natural admittance control are essentially the position and velocity error terms in time delay control, respectively. Therefore, if we set:

$$\begin{aligned} a_r = b_r = G_v \\ r(t) = \dot{x}_{cmd}(t), \end{aligned} \quad (8)$$

we can nest the natural admittance control loop inside the time delay control loop. Thus, the hybrid impedance/time-delay control law becomes:

$$u(t) = u(t-L) + (1/b)[-\ddot{x}(t-L) + G_v(\dot{x}_{cmd}(t) - \dot{x}(t)) \\ + K_{des}(x_{des}(t) - x(t)) + B_{des}(\dot{x}_{des}(t) - \dot{x}(t)) + c_r x(t)] \quad (9)$$

where

$$\dot{x}_{cmd}(t) = \int \frac{F_s(t) + K_{des}(x_{des}(t) - x(t)) + B_{des}(\dot{x}_{des}(t) - \dot{x}(t))}{M_s} dt \quad (10)$$

Hybrid NAC/TDC control achieves optimal responsiveness and provides good trajectory following since the nonlinear effects and disturbances are attenuated by a direct estimation technique. In order to implement the control law only one system parameter, the mass, needs to be estimated.

C. Nonlinear Bang-Bang Impact Control

Based on the discussion in the previous section, a nonlinear bang-bang impact controller is proposed. During free-space motion, a natural-admittance/time-delay control (NAC/TDC) is used. During impact transient, NAC/TDC alternates with zero control input. When contact is broken due to bouncing, no control input is applied, and when contact is made, NAC/TDC is used. This bang-bang control approach repeats until steady state is attained. NAC/TDC is used again after contact is established. NAC/TDC is also used to bring a robot back to contact with an environment when it stops in free space due to zero control input during contact transient. The resulting control strategy for a single input system is described as follows.

1) In Unconstrained Motion : NAC/TDC
   If $F_s > F_{impact}$, then NAC/TDC. (11)

2) In Contact Transition : Bang-Bang Control
   If the manipulator and the environment are in contact,
   i.e., $F_s > F_{sw}$, then NAC/TDC. (12a)
   If the manipulator and the environment are out of contact,
   i.e., $F_s < F_{sw}$, then $u(t) = 0$. (12b)
   If $F_s < F_{sw}$ & $|v| < v_{threshold}$, then NAC/TDC. (12c)

3) After Impact Transient: NAC/TDC
   If $F_s > F_{sw}$, then NAC/TDC. (13)

F_{impact}, F_{sw} and $v_{threshold}$ are threshold values to detect impact, switching and zero velocity, respectively, and are dependent on the sensitivity of the torque and position sensors.

However, the zero control input of (12b) alone cannot achieve the desired dynamics if a robot stops while it is out of contact with the environment after bouncing off from the environment. In most cases, the nonlinear spring characteristics of robot transmissions and link can bring the robot back to the environment after impact due to its high restoring force. If the restoring spring force cannot overcome friction and inertia of the robot, however, it may stop in free space after it bounces off from the environment. In order to prevent it in free space, (12c) is added.

III. EXPERIMENTS

A. Experimental Setup

An experimental setup was constructed as shown in Figure 1 and Figure 2. The aluminum robot arm is attached to a Himmelstein model MCRT 28002T(5-2), 500*lb-in* (56.5 *Nm*) range, non-contact rotating torque transducer via a Tran-torque coupling. The torquemeter is used to measure the contact force at the tip of the aluminum link when it interacts with an environment. The other shaft of the torquemeter is coupled to the HD Systems harmonic drive CSF-20-1000-2A-GR (gear ratio of 100:1) via a Thomas miniature flexible disc coupling, which provides relatively high torsional stiffness along the shaft axis and compliance along all five remaining degrees of freedom. The harmonic drive is attached to a Maxon Precision Motors DC motor 148867, which is connected to a Hewlett Packard HEDM-5500 two channel incremental optical encoder with a resolution of 1,000 counts-per-revolution with quadrature output (0.0016 *rad/pulse*).

For data acquisition and control, a motion control interface card, Precision MicroDynamics model MFIO-4A Dual, is inserted into a PC Pentium III (500 *Mhz*). This I/O card has four digital-to-analog converters (DAC), four analog-to-digital converters (ADC) and four encoder interface channels used for input/output between the computer and external devices.

For controller implementation, control algorithms were written in C++ and implemented on a real time OS, QNX RTP, at a sampling rate of 1 *kHz*. The control command generated by the computer is sent to a PWM motor amplifier, Maxon Precision Motors ADS 50/10, through the DAC. The sensed motor position is imported into the computer through the encoder interface channel. The torquemeter is connected to a Himmelstein model 701 universal strain gauge amplifier and the measured torque transducer analog input is processed via the I/O board ADC.

The total inertia of the arm was estimated as 19.5 *kg* through a series of experiments and the total static friction force was measured to be 1.96 *Nm* by torquemeter. An aluminum plate and a Styrofoam block were used to make hard and soft environments, respectively.

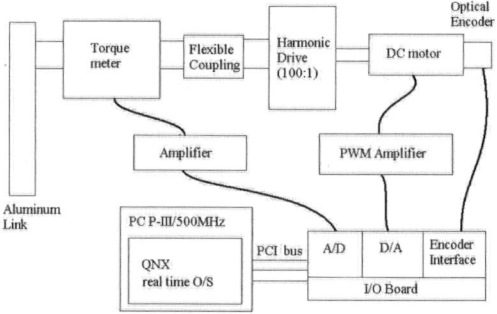

Figure 1. Schematic of experimental setup

Figure 2. Experimental setup

B. Experiments for Impact Control

In this section, the performance of the proposed bang-bang impact controller is compared with that of other existing impact force control techniques. A brief description of the compared impact control methods is provided including a mathematical formula of the control law tested. The control approaches are nonlinear bang-bang impact control, natural admittance control, explicit force control with negative force feedback, active impact damping control, event-driven switching control and integral gain explicit force control. The experiment has three stages of motion: free space motion, impact and constrained motion. The control performances are compared when the robot hits the environment with an impact velocity of 1.75 *m/sec*. The control parameters are adjusted to produce the best performance and still achieve stable force control.

Strategies for Impact Control:

Natural Admittance Control. The simplest form of natural admittance control of (9) and (10) is used. The control parameters are set to be $G_v = 1180 Ns/m$, $K_{des} = 5.9 \times 10^4 N/m$ and $B_{des} = 2900 Ns/m$.

Explicit Force Control with Negative Force Feedback. Explicit force control used in this experiment has the form:

$$u(t) = K_f (F_{des} - F_s) + K_v (\dot{x}_{des} - \dot{x}) \quad (14)$$

Force feedback gain, K_f, and velocity damping gain, K_v, are set to be 1.5 and 590 *Ns/m*, respectively. Reference force command, F_{des}, is set to 10 *N*.

Active Impact Damping Control. Active impact damping control [1] consists of two phases: an active damping control from free space motion until impact transient subsides (15) and a proportional-derivative force control after impact phase is over (16):

$$u(t) = K_v (\dot{x}_{des} - \dot{x}) \quad (15)$$
$$u(t) = F_{des} + K_f (F_{des} - F_s) - K_u \dot{x} \quad (16)$$

where F_{des} is a desired force command, K_f is a force feedback gain, and K_v is the velocity gain for the active damping employed. The control parameters are adjusted to be $F_{des} = 10 N$, $K_f = 0.2$ and $K_v = 290 Ns/m$.

Integral Gain Explicit Force Control. Youcef-Toumi and Gutz [9] demonstrated that integral force compensation with velocity feedback is effective for control of impacts. However, the idea of using integral action for impact control is not recommended because integrator wind-up is likely to occur due to the loss of contact. This can cause severe hopping on the constraint surface. The control law is

$$u(t) = K_{fi} \int_0^t (F_{des} - F_s) dt - K_v \dot{x}, \quad (17)$$

where K_{fi} is an integral force feedback gain and set to be $20 \, s^{-1}$. F_{des} and K_v are set to be 10 *N* and 1180 *Ns/m*, respectively.

Event-Driven Switching Control. Control rules for the event-driven switching control [6] are shown in (18) for free motion and in (19) for constrained motion

$$u(t) = -K_v \dot{x} + K_p (x_{sw} - x) \quad (18)$$
$$u(t) = -K_d \dot{x} + K_f (F_{des} - F_s) + K_I \int_{t_{sw}}^t (F_{des} - F_s) dt \quad (19)$$

where x_{sw} is link position stored when it began to contact the environment. The parameters are adjusted to be $K_v = 590 Ns/m$, $K_p = 5900 N/m$, $F_{des} = 10 N$, $K_d = 2900 Ns/m$, $K_f = 1.0$ and $K_I = 40 s^{-1}$ to have the shortest settling time.

Experimental Results for Impact Force Control:

In this section, the control performance of the nonlinear bang-bang impact control (NBBIC) is compared to other existing impact control strategies via experiments when the impact velocity is 1.75 *m/s*, which was the maximum velocity our setup can generate. We changed control parameters such as desired velocity of each controller so that all controllers can achieve a same impact velocity. Our objective was to investigate how well each controller performs upon hitting an environment at the same velocity. For the bang-bang control, the control parameters are adjusted to be $G_v = 1180 Ns/m$, $K_{des} = 5.9 \times 10^4 N/m$, $B_{des} = 2900 Ns/m$ and $1/b = 5.9 kg$ for NAC/TDC.

Figure 3 shows system responses when the link contacted the hard environment of an aluminum plate. For the same impact velocity of 1.75 *m/s*, all controllers experience an impact force of approximately 160 *N*. Figures 3 (a) and 3 (b) display that the settling time of NAC and NAC/TDC is very long (approximately 1 *sec*) and contact forces increased suddenly due to integral wind-up. In Figure 3 (c), however, we can see that our bang-bang control scheme effectively shortened the settling time down to 0.17 seconds.

The performance of NAC and NAC/TDC can be improved by applying an anti-windup scheme, which resets the control law integral term whenever the control input reaches its saturation limit. Figure 3 (d) shows experimental results for NAC/TDC with anti integral-windup. This result displays a shorter settling time (0.22 *secs*) than the result without anti-windup scheme, but still longer than that of nonlinear bang-bang control. NAC is observed to exhibit similar performance with anti integral-windup.

Figure 4 presents system responses for the control performance of other existing impact control strategies when the impact velocity is 1.75 *m/s*. It shows that all controllers achieve successful impact control against a very stiff environment.

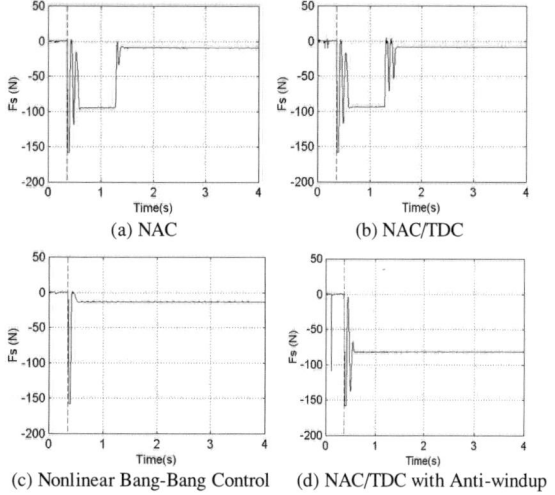

Figure 3. Experimental results for impact control (Dashed line indicates impact time.)

Figure 4 (a) illustrates the control performance of active impact damping control. The impact transient lasts about 0.5 *secs*. Figure 4 (b) illustrates the system response for explicit force control with negative force feedback. After impact, force oscillations subside rapidly within approximately 0.18 seconds. However, simulation results exhibit severe oscillations both in contact force and control input in the presence of large Coulomb friction in the joint [26]. Figure 4 (c) shows that event-driven switching controller makes the impact force vibration damp out quickly within 0.18 *secs*. The integral gain explicit force control (IGEFC) exhibits the longest settling time of 0.68 *secs* (Figure 4 (d)) and the largest contact force among the impact controllers compared. Note that there is a secondary jump in the contact force before it reaches steady state after impact due to the force error integral windup in IGEFC. The experimental results demonstrate that the proposed bang-bang impact control is better than or comparable to other existing impact control techniques for our system.

The NBBIC can perform a seamless contact manipulation that involves impact with only one control algorithm and the same gains in all three contact modes. An experiment was performed to demonstrate this advantage. First, the robot arm was commanded to travel down toward an aluminum plate on the table and hit the plate with impact velocity 1.75 *m/s*. Then, one end of the plate was moved up and down off the table while the robot link kept contact with the plate. For NAC/TDC, the control parameters are adjusted to be $G_v = 5900 Ns/m$, $K_{des} = 5.9 \times 10^4 N/m$, $B_{des} = 2900 Ns/m$ and $1/b = 5.9 kg$, and remained the same throughout the experiment. Our experiments show that the gains for a stable free motion also create a stable contact with the aluminum plate and the range of stable gains is wide. Figure 5 shows that the one-link arm quickly absorbs impact energy and then performs a stable contact task. For a soft environment, the control input exhibits more noise, but the control performance was the same as that for a hard environment. Thus, our experiments demonstrate that bang-bang impact control with constant gains is robust for soft and hard environments for a wide range of gains making it attractive in an unstructured environment.

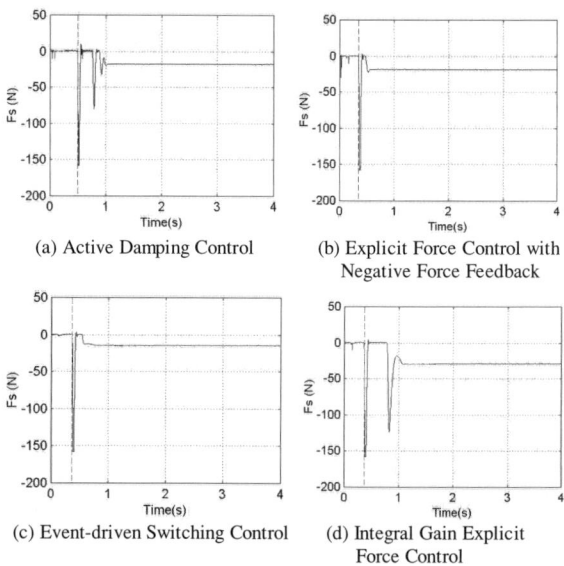

Figure 4. Experimental results for impact control (Dashed line indicates the impact time.)

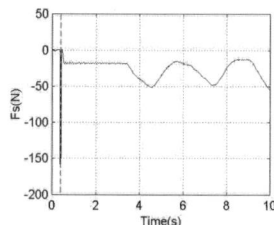

Figure 5. Emulation of desired dynamics after impact (Dashed line indicates the impact time.)

IV. CONCLUSIONS

In this paper, a hybrid impedance/time-delay control and a nonlinear bang-bang impact controller are developed and their performances compared to other impact control techniques via experiments. Experimental results show that the performance of the proposed nonlinear bang-bang impact control is superior or comparable to other existing impact control techniques.

It is shown via experiments that under NBBIC, a robot can successfully interact with an environment without changing control algorithm and control gains throughout all three modes: free space, transition and constrained motion. Our experiments demonstrate that the proposed nonlinear bang-bang impact controller can be effective in an unstructured environment because it has a wide range of gains that can make a stable contact both with soft and stiff environment without changing gains if the gains are set for stable free space motion. It is also very effective for robots with severe nonlinearities such as friction and joint flexibility. The proposed controller can be used for mobile robots and cooperative robot tasks as well as for robot interactions in space exploration.

Future work includes the experimental verification of the proposed control method for multi-degree of freedom robots as well as a more formal representation of stability analysis.

V. REFERENCES

[1] O. Khatib and J. Burdick, "Motion and Force Control of Robot Manipulators", *Proceedings of the 1986 IEEE International Conference on Robotics and Automation*, San Francisco, CA, pp. 1381-1386, 1986.

[2] N. Hogan, 1987, "Stable Execution of Contact Tasks Using Impedance Control", *Proceedings of the 1987 IEEE International Conference on Robotics and Automation*, Raleigh, NC, pp. 1047-1054, 1987.

[3] H.P. Qian and J. De Schutter, "Introducing Active Linear and Nonlinear Damping to Enable Stable High Gain Force Control in Case of Stiff Contact", *Proceedings of the 1992 IEEE International Conference on Robotics and Automation*, Nice, France, pp. 1374-1379, 1992.

[4] J.M. Hyde and M.R. Cutkosky, "Contact Transition Control: An Experimental Study", *Proceedings of the 1993 IEEE International Conference on Robotics and Automation*, Atlanta, GA, pp. 363-368, 1993.

[5] G.T. Marth, T.J. Tarn and A.K. Bejczy, "Stable Phase Transition Control for Robot Arm Motion", *Proceedings of the 1993 IEEE International Conference on Robotics and Automation*, Atlanta, GA, pp. 355-362, 1993.

[6] G.T. Marth, T.J. Tarn and A.K. Bejczy, "An Event Based Approach to Impact Control: Theory and Experiments", *Proceedings of the 1994 IEEE International Conference on Robotics and Automation*, San Diego, CA, pp. 918-923, 1994.

[7] R. Volpe and P. Khosla, "A Theoretical and Experimental Investigation of Impact Control for Manipulators", *The International Journal of Robotics Research*, Vol. 12, No. 4, pp. 351-365, August, 1993.

[8] R. Volpe and P. Khosla, "A Theoretical and Experimental Investigation of Explicit Force Control Strategies for Manipulators", *IEEE Transactions on Automatic Control*, Vol. 38, No. 11, pp. 1634-1650, November, 1993.

[9] K. Youcef-Toumi and D.A. Gutz, "Impact and Force Control: Modeling and Experiments", ASME *Journal of Dynamic Systems, Measurement, and Control*, Vol. 116, pp. 89-98, 1994.

[10] Y. Xu, J.M. Hollerbach and D. Ma, "Force and Contact Transition Control Using Nonlinear PD Control", *Proceedings of the 1994 IEEE International Conference on Robotics and Automation*, San Diego, CA, pp. 924-930, 1994.

[11] C. Carreras and I.D. Walker, "Sensitivity to Parametric Uncertainty in Robot Impact", *Proceedings of the 2001 IEEE/ASME International Conference on Advanced Intelligent Mechatronics*, Como, Italy, 2001.

[12] M. Shibata and T. Natori, "Impact Force Reduction for Biped Robot Based on Decoupling COG Control Scheme," *Proceedings of the International Workshop on Advanced Motion Control*, Nagoya, Japan, pp.612-617, 2000.

[13] L.B. Wee, and M.W. Walker, "On the Dynamics of Contact between Space Robots and Configuration Control for Impact Minimization," *IEEE Transactions on Robotics and Automation*, Vol. 9, No. 5, pp.581-591, Oct. 1993.

[14] M. Indri and A. Tornambe, "Impact Model and Control of Two Multi-DOF Cooperating Manipulators," *IEEE Transactions on Automatic Control*, Vol. 44, No. 6, pp.1297-1303, June 1999.

[15] B. Yu, "Modeling, Control Design and Mechatronic Implementation of Constrained Robots for Surface Finishing Applications," Ph.D. Dissertation, Oklahoma State University, 2000.

[16] K. Itabashi, K. Yamada, T. Suzuki, and S. Okuma, "Contact Force Control Based on Learning Operation without Switching the Servo Mode at the Instant of the Impact," *Transactions of the Japan Society of Mechanical Engineers*, Part C, Vol. 62, 596, pp.1473-1479, 1996.

[17] B. Brogliato, *Nonsmooth Mechanics*, Springer-Verlag, London, 1999.

[18] L. Menini and A. Tornambe, "Control of Mechanical Systems Subject to Non-Smooth Impacts," *Annual Reviews in Control*, Vol. 25, pp. 25-42, 2001.

[19] J.E. Colgate, "The Control of Dynamically Interacting Systems", *Ph.D. dissertation*, Dept. of Mechanical Engineering, M.I.T., Cambridge, MA, 1988.

[20] W.S. Newman, "Stability and Performance Limits of Interaction Controllers", ASME *Journal of Dynamic Systems, Measurement, and Control*, Vol. 114, pp. 563-570, 1992.

[21] W.S. Newman, "Stability and Performance Limits of Interaction Controllers", Technical Report TR-92-101, Center for Automation and Intelligent Systems Research, Case Western Reserve University, Cleveland, OH, 1992.

[22] K. Youcef-Toumi and O. Ito, "A Time Delay Controller for Systems With Unknown Dynamics", ASME *Journal of Dynamic Systems, Measurement, and Control*, Vol. 112, pp. 133-142, 1990.

[23] T.C. Hsia, T.A. Lasky, and Z.Y. Guo, "Robust Independent Robot Joint Control: Design and Experimentation," *Proceedings of the 1986 IEEE International Conference on Robotics and Automation*, Philadelphia, PA, pp. 1329-1334, 1988.

[24] P.H. Chang and J.W. Lee, "A Model Reference Observer for Time-Delay Control and Its Application to Robot Trajectory Control," IEEE Transactions on Control Systems Technology, Vol. 4, No. 1, pp. 2-10, Jan. 1996.

[25] E. Lee, "Force and Impact Control for Robot Manipulators with Unknown Dynamics and Disturbances," Ph.D. Dissertation, Department of Mechanical and Aerospace Engineering, Case Western Reserve University, Cleveland, Ohio, August, 1994.

[26] E. Lee, K.A. Loparo and R.D. Quinn, "A Nonlinear Bang-Bang Impact Force Control," *Proceedings of the 35th IEEE Conference on Decision and Control*, Kobe, Japan, December 1996.

Stiffness Control of a Three-link Redundant Planar Manipulator Using the Conservative Congruence Transformation (CCT)

Yanmei Li and Imin Kao
Department of Mechanical Engineering
SUNY at Stony Brook, Stony Brook, NY 11794-2300

Abstract—
In this paper, the Conservative Congruence Transformation (CCT), $\mathbf{K}_\theta - \mathbf{K}_g = \mathbf{J}_\theta^T \mathbf{K}_p \mathbf{J}_\theta$, is applied to a 3-link redundant planar manipulator. Since 3-link planar manipulator has one degree of redundancy, one constraint is allowed to be used to define the parameters of the system. Different constraints can be used to meet the specific requirements of manipulation. In this paper, two constraints are employed to obtain solution for redundant manipulation. One involves maintaining the most distal link along specific orientations; the other requires that the moment at the end-effector be always zero. In the latter case, the orientation of the distal link is decided in such a way that this constraint is satisfied. Numerical simulation and results are presented to illustrate that CCT is indeed a general and correct stiffness mapping in the analysis of redundant manipulators. The incorrect results of the conventional formulation, $\mathbf{K}_\theta = \mathbf{J}_\theta^T \mathbf{K}_p \mathbf{J}_\theta$, are also computed and compared with the results of the CCT theory.

I. INTRODUCTION

The theory of conservative congruence transformation (CCT) between the joint space and the $R^{3\times 3}$ Cartesian space for stiffness control was presented by Chen and Kao [1], [2], [3]. The CCT relates the mapping between the stiffness matrices of the joint and Cartesian spaces, and can be expressed by the following equation

$$\mathbf{J}_\theta^T \mathbf{K}_p \mathbf{J}_\theta = \mathbf{K}_\theta - \mathbf{K}_g \quad (1)$$

where $\mathbf{J}_\theta$ is the Jacobian matrix defined as $d\mathbf{x} = \mathbf{J}_\theta d\theta$, $\mathbf{K}_p$ and $\mathbf{K}_\theta$ are the Cartesian and joint stiffness matrices, respectively, and $\mathbf{K}_g = \left[\frac{\partial \mathbf{J}_\theta^T}{\partial \theta_n}\mathbf{f}\right] = \left[\left(\frac{\partial \mathbf{J}_\theta^T}{\partial \theta_1}\mathbf{f}\right) \left(\frac{\partial \mathbf{J}_\theta^T}{\partial \theta_2}\mathbf{f}\right) \ldots \left(\frac{\partial \mathbf{J}_\theta^T}{\partial \theta_n}\mathbf{f}\right)\right]$ is called the effective stiffness matrix due to the change in geometry of grasping and manipulation under the presence of external forces. The conventional formulation for the mapping between the Cartesian and joint stiffness matrices was given by the following equation [4]

$$\cdot \mathbf{J}_\theta^T \mathbf{K}_p \mathbf{J}_\theta = \mathbf{K}_\theta \quad (2)$$

In [1], [5], [6], the authors showed that the conventional formulation is in general an incorrect mapping that leads to inconsistency in fundamental conservative properties of stiffness control, although equation (2) has generally been accepted and widely used in robotics community. Chen and Kao [1] showed that equation (2) is only valid when robotic manipulators are at their unloaded equilibrium configuration, *i.e.*, without external force. Li and Kao [5] provided a theoretical proof to show that the CCT is the correct and general relationship for congruence mapping of stiffness control in robotics. Huang, Hung and Kao [6] studied the CCT for parallel manipulators with an emphasis on the Steward-Gough platform.

Redundant manipulators have more degrees of freedom than what is required by a task. Therefore, multiple sets of joint displacements can yield the same end-effector motion. These extra degrees of freedom, called the *redundancy*, of a robotic manipulator are the sources of their potential dexterity and also the cause of theoretical impediments limiting their practical utilization. Previously, the authors in [7] studied the distribution of singularity of redundant parallel manipulators. Optimal kinematic and dynamic control algorithms are designed and implemented, which made use of the redundancy of the parallel mechanism. The antagonistic stiffness with redundant actuation was studied in [8]. In [9], the authors studied the effects of actuator stiffness upon inverse position solutions for kinematically redundant manipulators (orientation of the end-effector is ignored). A kinematic control strategy for redundant serial manipulators was presented in [10], which yields repeatability in the joint space of a redundant serial manipulator whose end-effector undergoes some general cyclic type motion. In [11], a stiffness control method was developed for redundant manipulators, which effectively adjusts the task space stiffness and preserves the con-

---

Research Assistant; Email: yalli@ic.sunysb.edu
Associate Professor of Mechanical Engineering; Email: kao@mal.eng.sunysb.edu

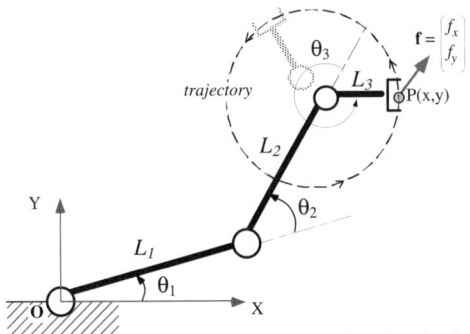

Fig. 1. A three-link redundant 3R planar manipulator for simulation.

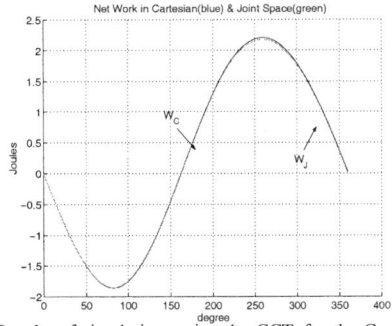

Fig. 2. Results of simulation, using the CCT, for the Cartesian-based stiffness control. Consistent results were obtained.

figuration stability.

In this paper, two different constraints were applied to a planar 3-link redundant manipulator to illustrate that the CCT is indeed a correct mapping for stiffness control in robotics, even for the redundant manipulators. Simulation results and comparison with the incorrect conventional stiffness control theory are also presented.

II. APPLYING CCT TO A THREE-LINK REDUNDANT PLANAR MANIPULATOR

In this section, we will apply the conservative congruence transformation (CCT) to a three-link redundant planar manipulator to illustrate the conservative property of the stiffness control. For a 3-link manipulator with force redundancy, *i.e.*, the dimension of the joint torque is three, $\tau = (\tau_1, \tau_2, \tau_3)^T$, while the dimension of the Cartesian bias force is two, $\mathbf{f} = (f_x, f_y)^T$, we can control the orientation of the end-effector of a three-link planar manipulator through the extra degree of freedom of the redundant joint to meet certain criterion.

In the following, we employ the CCT theory to the redundant manipulator with two different criteria: (i) maintain the distal link to be along the radial direction of a predefined circular path, and (ii) stipulate the moment at the end effector to be zero throughout the manipulation. Two control schemes are used: Cartesian-based control and joint-based control [5].

A. Criterion (i): radial distal link

The geometry of the 3R redundant manipulator is shown in Figure 1. The lengths of the links of the manipulator are $L_1 = 0.30\,m$, $L_2 = 0.23\,m$ and $L_3 = 0.05\,m$. We denote the position of the end-effector as $\mathbf{x} = [x, y]^T$ in the Cartesian space and $\theta = [\theta_1, \theta_2, , \theta_3]^T$ are the joint parameters. The end effector follows a closed path (circular trajectory) as shown in the figure.

From Figure 1, the Cartesian position of the end-effector can be easily obtained.

$$\begin{aligned} x &= L_1 c_1 + L_2 c_{12} + L_3 c_{123} \\ y &= L_1 s_1 + L_2 s_{12} + L_3 s_{123} \end{aligned} \quad (3)$$

with $c_1 = \cos\theta_1$, $s_1 = \sin\theta_1$, $c_{12} = \cos(\theta_1 + \theta_2)$, $\cdots$ etc. The Jacobian matrix becomes

$$\mathbf{J}_\theta = \begin{bmatrix} J_{11} & J_{12} & J_{13} \\ J_{21} & J_{22} & J_{23} \end{bmatrix} \quad (4)$$

where $J_{11} = -(L_1 s_1 + L_2 s_{12} + L_3 s_{123})$, $J_{12} = -(L_2 s_{12} + L_3 s_{123})$, $J_{13} = -L_3 s_{123}$, $J_{21} = L_1 c_1 + L_2 c_{12} + L_3 c_{123}$, $J_{22} = L_2 c_{12} + L_3 c_{123}$, and $J_{23} = L_3 c_{123}$. The Cartesian force, $\mathbf{f}$, and joint torque, τ, are related by

$$\tau = \mathbf{J}_\theta^T \mathbf{f} \quad (5)$$

In the following simulation, the initial orientation of the distal (or third) link is maintained in horizontal orientation. After that, it is specified to align with the radial direction of the circular trajectory for the upper half circle ($0 \to \pi$). When the end-effector moves to the lower half of the circle, the orientation of the distal link changes from $\pi \to 0$.

Cartesian-based stiffness control

In the numerical simulation under the Cartesian-based stiffness control, the trajectory of the end-effector is a circle with a radius of $R = 0.08m$, centered at $(x, y) = (0.37, 0.33)m$. The point P of the end-effector moves along the circular trajectory in counterclockwise sense, starting from the initial position at $(x_0, y_0) = (0.45, 0.33)m$ and finally returning to the same initial position. The prescribed conservative Cartesian stiffness matrix and the bias force are

$$\mathbf{K}_p = \begin{bmatrix} 300 & 0 \\ 0 & 250 \end{bmatrix} N/m \qquad \mathbf{f} = [20\ -25]^T\,N$$

In this case, we explicitly require the moment at the end effector to be zero by specifying the force and $\mathbf{K}_p$ as 2-space quantities. The results of numerical simulation are shown in Figures 2 and 3. From Figures 2 and 3, the following observations are in order:

- When the CCT is employed, the net works done in both the Cartesian and joint spaces of this redundant manipulator, W_C and W_J, are identical to each other.

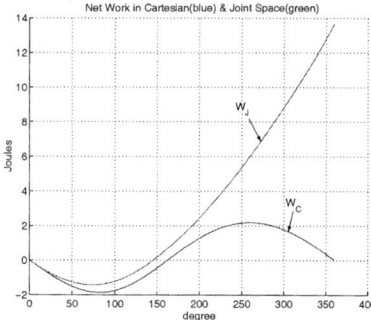

Fig. 3. Results of simulation, using the conventional mapping, for the Cartesian-based stiffness control.

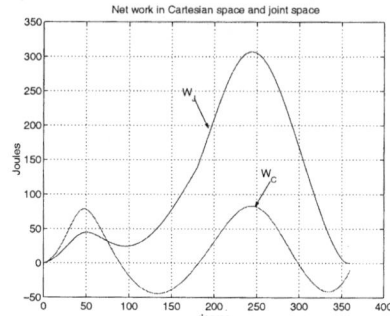

Fig. 4. Results of simulation, using the CCT, for the joint-based stiffness control.

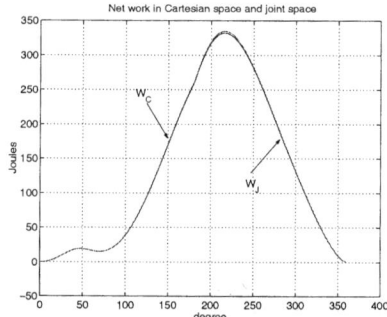

Fig. 5. Results of simulation, using the CCT, for the joint-based stiffness control, by considering the work done by the moment at the end effector.

In addition, since we are considering a conservative system ($\mathbf{K}_p$ is conservative), so we expect that the net work done in both Cartesian space and joint space are equal to zero when returned to the starting point after a full circle. From Figure 2, we can see that the results are correct.

- When the incorrect conventional formulation is used, the net work done in the joint space in Figure 3 is not zero, although the net work done in the Cartesian space remains zero. This indicates that energy is not conservative using the conventional formulation. Furthermore, the work done in Cartesian space and in joint space are not identical. The discrepancy reveals the inconsistency of the conventional mapping.
- It is important to note that equation (1) used in this simulation does not involve the generalized inverse of $\mathbf{J}_\theta$.

Joint-based stiffness control

In applying the joint-based control scheme, the center of the circular trajectory is at $(x, y) = (0.30, 0.30)m$, with a radius of $0.15m$. The initial position of the end-effector is at $(x_0, y_0) = (0.45, 0.30)m$. The joint stiffness matrix $\mathbf{K}_\theta$ and the joint torque τ are specified first as follows:

$$\mathbf{K}_\theta = \begin{bmatrix} 300 & 0 & 0 \\ 0 & 250 & 0 \\ 0 & 0 & 5 \end{bmatrix} N \cdot m \quad \tau = [10 \ -5 \ 5]^T N \cdot m$$

The initial simulation results are plotted in Figure 4 (using the CCT in equation 1). The results of the conventional formulation show large discrepancy and are not plotted here.

Unlike the Cartesian-based control scheme, the joint-based control strategy in such redundant manipulator involves the Penrose-Moore generalized inverse of the non-square Jacobian matrix in order to obtain the Cartesian stiffness matrix, i.e.,

$$\mathbf{K}_p = (\mathbf{J}^T)^* (\mathbf{K}_\theta - \mathbf{K}_g)(\mathbf{J}_\theta)^*$$

where '*' denotes the generalized inverse. Due to the generalized inverse of the non-square Jacobian matrix, part of the information for stiffness control is lost due to the consideration of partial space and application of the generalized inverse. Thus, we can see from Figure 4, when CCT is used, that there is certain degree of discrepancy between the work done in the Cartesian space and joint space.

In order to recover the lost information for stiffness control due to the application of generalized inverse of the Jacobian matrix, we add an additional row of $[1 \ 1 \ 1]^T$ to the Jacobian matrix in equation (4) in order to consider the work done by the moment at the end-effector. This is necessary because the non-zero moment applied at the end effector and the resulting work done can no longer be neglected. In this modified approach, the Cartesian coordinates are taken as $[x, y, \phi]$, and the Cartesian bias force are taken as $\mathbf{f} = [f_x, f_y, m]^T$. Figure 5 shows the rectified results. In this figure, the net work done in Cartesian space and joint space are exactly the same, by considering the work done by the moment of the end effector.

B. Criterion (ii): zero moment of end effector

In an alternate implementation of the redundant manipulator, we require that the moment of the end effector maintained at zero throughout the manipulation.

The figure of such setup is similar to that in Figure 1, except that the end effector will traverse an elliptical trajectory of semi major and minor axes of $0.06\,m$ and $0.02\,m$, respectively. The lengths of the links of the manipulator are the same, $L_1 = 0.30\,m$, $L_2 = 0.23\,m$ and $L_3 = 0.05\,m$.

From equation (5), we can deduce

$$\mathbf{f} = \mathbf{J}_\theta^{-T}\tau = \mathbf{B}\,\tau \quad (6)$$

In order for the moment of the end-effector to be zero, *i.e.*, $m = 0$, we have to consider equation (6) with the Jacobian matrix augmented by an additional row of $[1\ 1\ 1]$ to account for the moment and angular displacement of the end effector. This results in an additional equation corresponding to the the Jacobian matrix of

$$\phi = \theta_1 + \theta_2 + \theta_3 \quad (7)$$

We use the notation of $B_{ij} = \mathbf{B}(i,j)$ to denote the $(i,j)^{th}$ element of the matrix $\mathbf{J}_\theta^{-T}$. Hence, the joint torque must satisfy the following condition

$$m = B_{31}\tau_1 + B_{32}\tau_2 + B_{33}\tau_3 = 0 \quad (8)$$

Substituting $\tau = \tau_0 + \mathbf{K}_\theta d\theta$ into equations (8) and (6), we have

$$\epsilon = -(P_{31}d\theta_1 + P_{32}d\theta_2 + P_{33}d\theta_3) \quad (9)$$
$$= B_{31}\tau_{01} + B_{32}\tau_{02} + B_{33}\tau_{03} \quad (10)$$

with

$$P_{31} = B_{31}K_{q,11} + B_{32}K_{q,21} + B_{33}K_{q,31}$$
$$P_{32} = B_{31}K_{q,12} + B_{32}K_{q,22} + B_{33}K_{q,32}$$
$$P_{33} = B_{31}K_{q,13} + B_{32}K_{q,23} + B_{33}K_{q,33}$$

where $K_{q,ij}$ is the $(i,j)^{th}$ element of $\mathbf{K}_\theta$. Combining equation (8) with the definition of the Jacobian matrix, we have

$$\begin{bmatrix} dx \\ dy \\ \epsilon \end{bmatrix} = \mathbf{P}\begin{bmatrix} d\theta_1 \\ d\theta_2 \\ d\theta_3 \end{bmatrix} = \begin{bmatrix} J_{11} & J_{12} & J_{13} \\ J_{21} & J_{22} & J_{23} \\ P_{31} & P_{32} & P_{33} \end{bmatrix}\begin{bmatrix} d\theta_1 \\ d\theta_2 \\ d\theta_3 \end{bmatrix} \quad (11)$$

Thus, the displacement in the joint space, subject to the zero-moment constraint at the end effector, is obtained

$$\begin{bmatrix} d\theta_1 \\ d\theta_2 \\ d\theta_3 \end{bmatrix} = \mathbf{P}^{-1}\begin{bmatrix} dx \\ dy \\ \epsilon \end{bmatrix} \quad (12)$$

Cartesian-based control scheme

Here, the Cartesian stiffness matrix $\mathbf{K}_p$ and force $\mathbf{f}$ are specified, and equation (1) is used to determine the joint stiffness matrix $\mathbf{K}_\theta$. The displacement of the joint variable $d\theta$ is specified by equation (12), so that the criterion $m = 0$ is satisfied at each step.

Joint-based control scheme

The joint-based control solves for the Cartesian stiffness matrix $\mathbf{K}_p$ with specified joint stiffness matrix $\mathbf{K}_\theta$, via an inverted relationship given by equation (1). The joint displacement is also specified by equation (12) to satisfy $m = 0$.

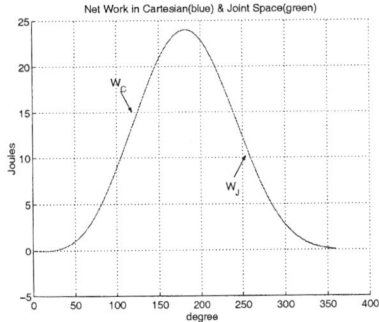

Fig. 6. Net work done over a closed loop using CCT under the Cartesian-based control scheme, by employing criterion (ii).

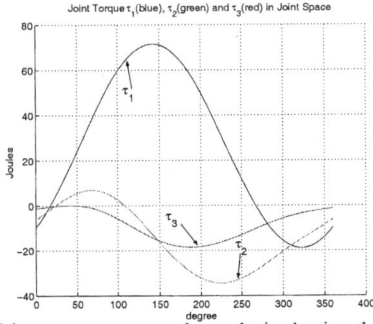

Fig. 7. Joint torques τ_1, τ_2 and τ_3, obtained using the CCT under Cartesian-based control scheme, by employing criterion (ii).

III. RESULTS OF NUMERICAL SIMULATION USING Criterion (ii)

A. Cartesian-based control scheme

1) Using CCT: In this simulation, the prescribed conservative Cartesian stiffness matrix and the bias force are

$$\mathbf{K}_p = \begin{bmatrix} 3000 & 0 & 0 \\ 0 & 2500 & 0 \\ 0 & 0 & 0 \end{bmatrix} N/m \quad \mathbf{f} = [-20N\ -25N\ 0]^T$$

Note that the moment and associated components are explicitly set as zero. The results of numerical simulation are shown in Figures 6 to 9.

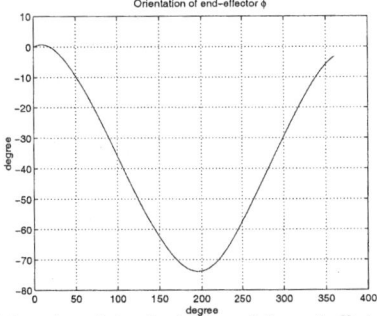

Fig. 8. Orientation of the distal link and the end-effector, ϕ, due to the imposition of $m = 0$ constraint with the CCT under Cartesian-based control scheme.

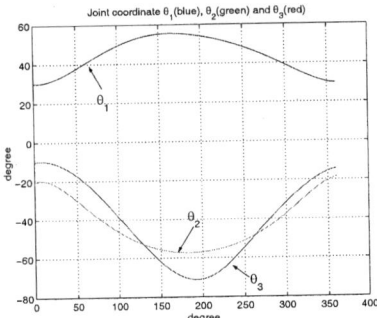

Fig. 9. Joint coordinates θ_1, θ_2 and θ_3, using CCT under Cartesian-based control scheme. Note that they do not return to the initial orientation due to the redundancy and the chosen constraint.

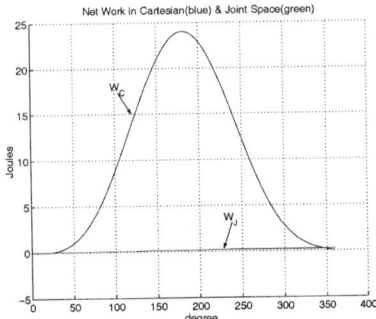

Fig. 10. Net work done over a closed loop using conventional formulation under the Cartesian-based control scheme.

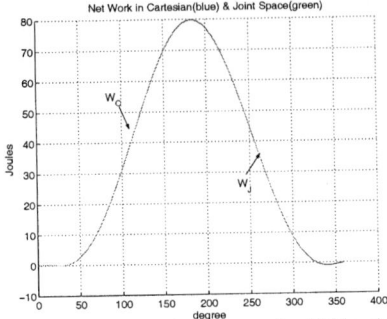

Fig. 11. Net work done over a closed loop using CCT under joint-based control scheme, $W_J = W_C$

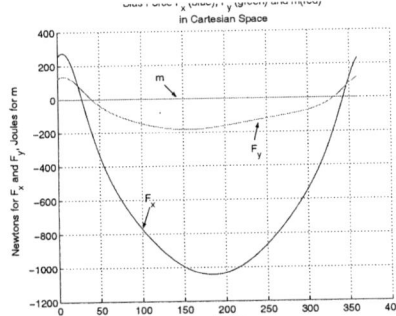

Fig. 12. The external force f_x, f_y and m applied at the end effector, using the CCT under joint-based control scheme

2) Using the conventional formulation: Using conventional formulation, the simulation results are shown in figures 10.

3) Discussions: From the above results, the following observations are in order:

- When the CCT is employed, the work done in both spaces are identical; that is, $W_J = W_C$. If the two work profiles integrated in different spaces are different, it means that work is either generated or destroyed by simply using difference reference coordinates. This is obviously a violation of very fundamental properties of physics.
- When the incorrect conventional formulation is used, the work done in Cartesian space and in joint space are very different. This huge discrepancy reveals the inconsistency of the conventional mapping.
- We notice that the three torques obtained using the CCT in Figure 7 all return to their original values. This is consistent with the behavior of a conservative system. On the other hand, such curves from the conventional formulation did not end with the same starting values.
- The results in Figures 8 and 9 show that the joint parameters do not necessarily return to their original values. This is due to the redundancy of the 3R planar manipulator and the constraint that we applied. By using different constraints, we will arrived at different end results. This is an expected result of redundant manipulators.

B. Joint-based control scheme

1) Using CCT: With the *joint-based control* scheme, the joint stiffness matrix $\mathbf{K}_\theta$ and the joint torque τ are specified first:

$$\mathbf{K}_\theta = \begin{bmatrix} 250 & 0 & 0 \\ 0 & 200 & 0 \\ 0 & 0 & 50 \end{bmatrix} N \cdot m \quad \tau = [20\ 25\ 6.35]^T\ N \cdot m$$

The simulation results are plotted in Figures 11 and 12.

2) Using conventional formulation: Using the conventional formulation, the simulation results are shown in Figures 13 and 14.

3) Discussions:

- As expected, the CCT renders consistent results in both work done, and the force and moment resulting from the manipulation (Figures 11 and 12).
- The work done in Cartesian space and joint space when the conventional formulation is used again shows large discrepancy, as found in Figure 13. In addition, the work obtained in the Cartesian space did not return to zero (the error is $-0.733\ Joules$ for the specific initial condition) when the manipulator returns to its starting position and orientation.
- The external force f_y from integration $\mathbf{f} = \int \mathbf{K}_p d\mathbf{x}$ is compared with the force obtained by $\mathbf{f} = \mathbf{J}_\theta^{-T} \tau$

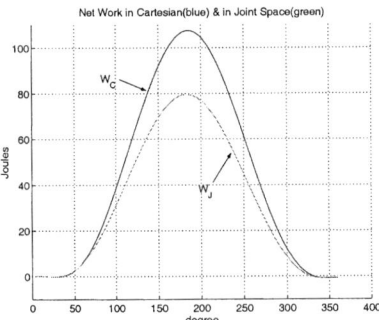

Fig. 13. Net work done over a closed loop using the conventional formulation under joint-based control scheme, $W_J \neq W_C$, and W_C did not return to zero as it is supposed to be

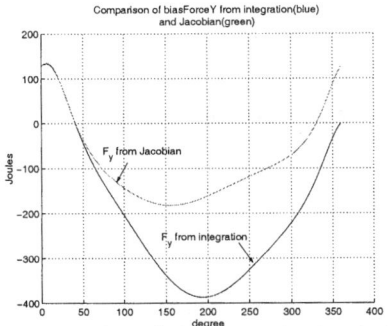

Fig. 14. The external force f_y from integration done in the Cartesian space, as compared with the force obtained by $\mathbf{f} = \mathbf{J}_\theta^{-T} \tau$ from Jacobian matrix, using conventional formulation under joint-based control scheme. A large discrepancy is observed also in other components of the force/moment, although only f_y is shown here.

(Figure 14), when the conventional formulation is employed. As a result of the incorrect conventional mapping, a very significant discrepancy is observed. The same is true for the other components of the force/moment, although only f_y is plotted here. On the other hand, the forces/moment match with each other perfectly when the CCT is employed (Figure 12).

IV. CONCLUSIONS

In this paper, we present the results of the CCT as applied to the redundant planar 3-link serial manipulator. The degree of redundancy was exploited to impose different constraints on the system. Two different constraints were applied to illustrate that the CCT is indeed a correct mapping for stiffness control in robotics, even for the redundant manipulators.

Furthermore, the application of generalized inverse, in conjunction with the physical insights such generalized inverse brings, should be carefully considered in redundant manipulators. Otherwise, incorrect results can be found if only partial algebra and vector space is considered with the generalized inverse. This, however, can be rectified by careful consideration of parameters involved in the process.

For example, as illustrated earlier the work done due to the moment at the end effector should be accounted for, in addition to the work done due to force in the Cartesian space when the generalized inverse is considered via the joint-based control.

It is important that appropriate constraints are chosen for manipulation of redundant manipulators. This paper reports that the CCT can be employed for redundant manipulators to yield consistent and correct results.

V. ACKNOWLEDGMENT

This research has been supported by the National Science Foundation under grant number IIS9906890.

REFERENCES

[1] S.-F. Chen and I. Kao, "Conservative congruence transformation for joint and cartesian stiffness matrices of robotic hands and fingers," *the International Journal of Robotics Research*, vol. 19, no. 9, pp. 835–847, September 2000.

[2] S.-F. Chen and I. Kao, "Simulation of conservative congruence transformation: Conservative properties in the joint and cartesian spaces," in *IEEE International Conference on Robotics and Automation*, April 2000, pp. 1283–1288.

[3] Shih-Feng Chen, *Fundamental Properties Of Stiffness Control In Grasping And Dextrous Manipulation In Robotics*, Ph.D. thesis, State University of New York at Stony Brook, May 2000.

[4] J. K. Salisbury, "Active stiffness control of a manipulator in cartesian coordinates," in *Proceedings 19th IEEE Conference on Decision and Control*, Albuquerque, NM, December 1980, pp. 87–97.

[5] Y. Li and I. Kao, "On the stiffness control and congruence transformation using the conservative congruence transformation (CCT)," in *Proc. of IEEE Robotics and Automation Conference*, Seoul, Korea, May 2001, pp. 3937–3942.

[6] C. T. Huang, W. H. Hung, and I. Kao, "New conservative stiffness mapping for the steward-gough platform," in *IEEE International Conference on Robotics and Automation*, May 2002, pp. 823–828.

[7] G. F. Liu, H. Cheng, Z. H. Xiong, X. Z. Wu, Y. L. Wu, and Z.X. Li, "Distribution of singularity and optimal control of redundant parallel manipuloators," in *the 2001 IEEE International Conference on Intelligent Robots and Systems*, Maui, HI, Oct 2001, pp. 177–182.

[8] B-J Yi, S-R Oh, and I. H. Suh, "Five-bar finger mechanism involving redundant actuators: Analysis and its applications," *IEEE Transactions on Robotics and Automation*, vol. 15, no. 6, pp. 1001–1010, 1999.

[9] J. A. Kuo, D. J. Sanger, and D. R. Kerr, "Effects of actuator stiffness upon inverse position solutions for kinematically redundant manipulators," in *ASME, Design Engineering Division*, 1994, pp. 419–425.

[10] Y. S. Chung, M. Griffis, and J. Duffy, "Unique joint displacement generation for redundant robotic systems," in *ASME, Design Engineering Division*, Scottsdale, AZ, USA, Sep 1992, pp. 637–641, ASME.

[11] H. R. Choi, W. K. Chung, and Y. Youm, "Stiffness analysis and control of redundant manipulators," in *Proc. of IEEE Robotics and Automation Conference*, San Diego, CA, USA, May 1994, pp. 689–695.

Cartesian Impedance Control of Redundant Robots: Recent Results with the DLR-Light-Weight-Arms

Alin Albu-Schäffer, Christian Ott, Udo Frese and Gerd Hirzinger

DLR - German Aerospace Center
Institute of Robotics and Mechatronics
alin.albu-schaeffer@dlr.de, christian.ott@dlr.de

Abstract— This paper addresses the problem of impedance control for flexible joint robots based on a singular perturbation approach. Some aspects of the impedance controller, which turned out to be of high practical relevance during applications are then addressed, such as the implementation of nullspace stiffness for redundant manipulators, the avoiding of mass matrix decoupling and the related design of the desired damping matrix. Finally, the proposed methods are validated through measurements on the DLR robot.

I. Introduction

The topic of impedance control is a traditional one in robotics. It provides a very suitable framework for controlling robots in contact with an unknown environment [5]. However, the focus on this field is renewed because of applications such as force feedback, or service and medical robotics, where safety of interaction is crucial. Those applications require also highly light-weight arms, which have minimal impact inertia and can be mounted on mobile systems. Because of the inherent flexibility of the light-weight structures, it becomes necessary to bring together the control field of flexible joint robots with that of compliant control. In [1], various Cartesian compliant control strategies (admittance, impedance and stiffness control) were compared and implemented on the DLR light-weight robots. Because of the rather slow Cartesian sampling rate (6ms), classical impedance control had the poorest performance. This changed significantly by increasing the sampling rate to 1ms, enabling the implementation of high quality impedance control. The recent results with the Cartesian impedance controller are the topics of this paper.

The singular perturbation theory is used to extend existing impedance control methods from the rigid body robot case to the flexible joint structure. This simple, but efficient method in case of robots with moderate elasticity uses a fast joint level torque control loop, which is receiving its desired torque from a classical Cartesian impedance controller. The DLR 7DOF light-weight arms (Fig. 1) with integrated joint torque sensors are very well suited for this kind of control algorithms [4]. The paper discusses several methods of providing a nullspace stiffness for the redundant manipulator in addition to a Cartesian stiffness. The difference between the methods is illustrated with

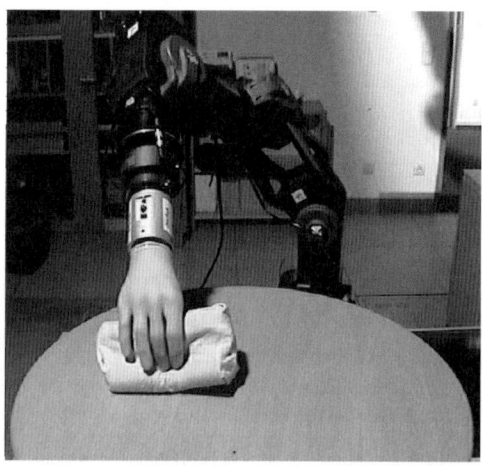

Fig. 1. DLR light-weight robot, generation II, using impedance control in a table wiping application.

measurements. Another practically important aspect was the design of the desired damping matrix in the case that the Cartesian mass matrix is not explicitly decoupled by the controller. Two different approaches to this problem are presented and were verified in experiments.

II. Control of the Flexible Joint Model

In this paper a model of a robot with n flexible joints is considered as proposed by Spong [11]:

$$M(q)\ddot{q} + C(q,\dot{q})\dot{q} + g(q) = K(\theta - q) + \tau_{ext}, \quad (1)$$
$$B\ddot{\theta} + K(\theta - q) = \tau_m. \quad (2)$$

Herein $q \in \Re^n$ and $\theta \in \Re^n$ are the vectors of link positions and the motor positions, respectively. $M(q)$ denotes the manipulator's mass matrix, $g(q)$ the gravity torques and $C(q,\dot{q})\dot{q}$ is the vector of Coriolis and centrifugal torques. K and B are diagonal matrices which contain the joint stiffness and the motor inertias. $\tau = K(\theta - q)$ is the vector of joint torques and τ_m the motor torque vector which is regarded as the control input. Finally τ_{ext} is a vector of external torques, which are exhibited by the manipulator's environment.

One common approach to the control of a flexible joint manipulator is the singular perturbation approach, in which the flexible joint model is virtually split up into a fast subsystem for the joint torques τ and a slow subsystem for the link positions q. Based on these two subsystems it is possible to design an inner loop controller for the joint torque τ, and an outer loop controller for the link position q separately, without the need to refer to the complete flexible model.

The application of the singular perturbation theory to flexible joint robots has been widely described in the literature and is not in the scope of this paper (see, e.g., [7], [8] for more details). Herein only the resulting controller structure, which will be used in the following sections as a basis for the implementation of different impedance control laws, shall be given briefly.

In [9] it has been shown that, given a desired torque vector τ_d, a state feedback controller of the form

$$\tau_m = \tau_d - K_T(\tau - \tau_d) - K_S\dot{\tau} , \quad (3)$$

with positiv definite contoller matrices K_T and K_S, can stabilize the torque dynamics around the equilibrium point $\tau = \tau_d$, and leads (under a singular perturbation consideration) to the following link dynamics:

$$\bar{M}(q)\ddot{q} + C(q,\dot{q})\dot{q} + g(q) = \tau_d + \tau_{ext} \quad (4)$$

with $\bar{M}(q) = (M(q) + (I + K_T)^{-1}B)$.

The desired torque vector τ_d in (4) can now be used for the realization of a Cartesian impedance controller as will be described in the next section.

III. CARTESIAN IMPEDANCE CONTROL

Based on the singular perturbation analysis, which was outlined in the last section, a rigid joint robot model of the form (4) may be considered for the design of the Cartesian impedance controller. A detailed study of appropriate Cartesian impedance control laws for nonredundant and redundant rigid robots is given, e.g., in [6].

In the following it is assumed that the manipulator's end-effector position and orientation can be described by a set of local coordinates $x \in \Re^m$, and the forward kinematics $x = f(q)$ is known. The relevant mappings between joint and Cartesian velocities and accelerations can be computed via the Jacobian $J(q) = \frac{\partial f(q)}{\partial q}$ as

$$\dot{x} = J(q)\dot{q} , \quad (5)$$
$$\ddot{x} = J(q)\ddot{q} + \dot{J}(q)\dot{q} . \quad (6)$$

Notice that in this paper only the nonsingular case is treated, thus it is assumed that the manipulator's Jacobian $J(q)$ has full row rank in the considered region of the workspace. A description of an appropriate singularity treatment can be found in [6].

The deviation of the actual Cartesian position from the desired equilibrium point $x_d(t)$ is denoted by $e_x = x - x_d(t)$. Then the goal for the impedance controller is to alter the system dynamics (4) such that, in presence of external forces and torques at the end-effector $F_{ext} \in \Re^m$, the closed loop behaviour is given by

$$\Lambda_d\ddot{e}_x + D_d\dot{e}_x + K_d e_x = F_{ext} , \quad (7)$$

with a desired mass Λ_d, a desired damping D_d and a desired stiffness matrix K_d.

In absence of further forces on the arm, the relationship between the external torques τ_{ext} and the forces and torques at the end-effector F_{ext} is given by:

$$\tau_{ext} = J(q)^T F_{ext} . \quad (8)$$

From (6) and (4) one can see that the relationship between the Cartesian accelerations $\ddot{x}$ and the joint torques τ is given by:

$$\ddot{x} - \dot{J}(q)\dot{q} + J(q)\bar{M}(q)^{-1}(C(q,\dot{q})\dot{q} + g(q)) = $$
$$J(q)\bar{M}(q)^{-1}(\tau_d + \tau_{ext}) . \quad (9)$$

If the desired torque vector τ_d is chosen as

$$\tau_d = J(q)^T F_\tau + C(q,\dot{q})\dot{q} + g(q) , \quad (10)$$

with $F_\tau \in \Re^m$ as a new control input vector, then the resulting Cartesian behaviour of the robot can be written as:

$$\Lambda(q)(\ddot{x} - \dot{J}(q)\dot{q}) = F_\tau + F_{ext} \quad (11)$$

with

$$\Lambda(q) = (J(q)\bar{M}(q)^{-1}J(q)^T)^{-1} \quad (12)$$

as an equivalent Cartesian mass matrix.

With (11) and (7) it follows that the feedback law

$$F_\tau = \Lambda(q)\ddot{x}_d - \Lambda(q)\Lambda_d^{-1}(D_d\dot{e}_x + K_d e_x) + $$
$$(\Lambda(q)\Lambda_d^{-1} - I)F_{ext} - \Lambda(q)\dot{J}(q)\dot{q} \quad (13)$$

leads to the desired closed loop behaviour (7).

In [6] it has been shown that, in principle, the same feedback law as described above may also be used in the case of a kinematically redundant manipulator ($m \neq n$). But it is well known that, in the redundant case, also those motions of the joints have to be considered which are embedded in the nullspace of the Jacobian $J(q)$ and which do not affect the end-effector position and orientation. Notice that these motions have been eliminated in (9) by the premultiplication with $J(q)$. For a formal analysis of this situation it is necessary to introduce additional nullspace-coordinates n which, together with the Cartesian coordinates x, admit a complete description of the robot's dynamics. An interesting set of such nullspace-coordinates with some advantageous properties was introduced by Park in [10].

In order to keep the computational complexity low, in the next section an appropriate extention of the Cartesian impedance controller (10) and (13) is treated, which can be used for the control of the manipulator's nullspace motion without a formal introduction of additional coordinates.

IV. Nullspace Stiffness

In this section it is assumed that the desired nullspace behaviour can be characterized in joint space by a desired positive definite stiffness matrix K_N and a positive definite damping matrix D_N as well as a desired equilibrium point q_N. From these a desired torque $\tau_{d,N} = -K_N(q - q_N) - D_N \dot{q}$ is computed according to a joint level PD-controller. In order to prevent interference with the Cartesian impedance behaviour, this desired torque has to be projected into the manipulator's nullspace by a properly chosen matrix $N(q)$. The desired torque is then composed as

$$\tau_d = \tau_{d,cart} + N(q)\tau_{d,N} \quad (14)$$

with $\tau_{d,cart}$ as the impedance controller torque from (10). In the following, three different nullspace projection matrices of different complexities are compared.

A. Static Nullspace Projection

Let $V(q)$ denote a full rank left annihilator of $J^T(q)$, i.e., $V(q)J^T(q) = 0$. Then a projection matrix of the form

$$N_1(q) = V(q)^T V(q) \quad (15)$$

may be used to project the desired torque into the nullspace of the manipulator's Jacobian via $\tau_0 = N_1(q)\tau_{d,N}$.
Notice that in practice the matrix $V(q)$ may be computed by a singular value decomposition of the Jacobian. Notice also that, in principle, $V(q)$ could also be used for the construction of nullspace coordinates n, which were mentioned in the last section, via $\dot{n} = V(q)\dot{q}$.

B. Dynamically Consistent Projection

It is well known that a static nullspace mapping is not sufficient in order to get a nullspace torque which does not affect the Cartesian behaviour. This can be easily seen by regarding the dynamical equations. A sufficient condition for a nullspace mapping to be consistent with the equations of motion is given by

$$J(q)\bar{M}(q)^{-1}N(q) = 0 \; , \quad (16)$$

as can be seen from (9). Obviously, this condition can be fulfilled for example with a mapping of the form

$$N_2(q) = \bar{M}(q)V(q)^T V(q) \quad (17)$$

in which the statical nullspace projection matrix is premultiplied, and thus scaled, by the manipulator's mass matrix. Another dynamically consistent nullspace mapping, which fits very well in the framework of operational space control, was proposed by Khatib [6]:

$$N_3(q) = (I - J^T(q)\Lambda(q)J(q)\bar{M}^{-1}(q)) \; . \quad (18)$$

This mapping has the conceptual advantage of being a projection matrix ($N_3 N_3 = N_3$), thus avoiding the above mentioned scaling from N_2. In order to implement this mapping, the Cartesian mass matrix is needed as well as the inverse of the joint mass matrix. But it is important to notice that these values have to be computed also for the implementation of the general impedance control law in (13).

V. Avoiding the Decoupling of the Inertial Behaviour

The decoupling of the inertial behaviour and the command of an arbitrary desired mass matrix in (7) using the controller (13) turns out to be very difficult to realize in practice. Looking at (13), it is clear that the decoupling can not be applied around the robot singularities, since the singularity of $J(q)$ implies through (12) that some of the elements of $\Lambda(q)$ will tend to infinity. This would in turn lead to infinite desired joint torques using (13), (10). But even in regions of the workspace where $J(q)$ is well-conditioned, the experimental success was limited in terms of the range of reachable values and of the decoupling accuracy. Possible reasons therefore are in our opinion:

- The influence of joint friction, which cannot be completely eliminated by the joint torque controller and a feed-forward friction compensation.
- The limited accuracy of the dynamical model.
- The additional need to measure the Cartesian force at the end-effector F_{ext}, which is in our setup only available with lower sampling rate of 6 ms.

As an alternative to (13), by focusing only on the implementation of Cartesian stiffness and damping, the following simpler control law was preferred:

$$\begin{aligned} F_\tau &= \Lambda(q)\ddot{x}_d - D_d\dot{e}_x - K_d e_x \\ &\quad - \bar{C}(q,\dot{q})\dot{e}_x - \Lambda(q)\dot{J}(q)\dot{q} \quad (19) \\ \tau_d &= J(q)^T F_\tau + C(q,\dot{q})\dot{q} + g(q) \; , \quad (20) \end{aligned}$$

where $\bar{C}(q,\dot{q})$ can be an arbitrary matrix, for which the skew symmetry of $\dot{\Lambda}(q) - 2\bar{C}(q,\dot{q})$ holds, e.i. $\bar{C}(q,\dot{q}) = 1/2\dot{\Lambda}(q)$. This leads to the following closed loop dynamics:

$$\Lambda(q)\ddot{e}_x + D_d\dot{e}_x + K_d e_x + \bar{C}(q,\dot{q})\dot{e}_x = F_{ext} \; . \quad (21)$$

Equation (21) represents a passive mapping from the external force F_{ext} to the velocity error $\dot{e}_x$, ensuring the stability of the system in free motion and in feedback interconnection with a passive environment.

VI. Damping Design

While imposing the closed loop dynamics (21) to the robot, the question raises up, how to design the desired damping matrix D_d depending on the desired stiffness K_d and the actual Cartesian mass matrix of the arm $\Lambda(q)$.

What the user may wish to specify for an application, is a well defined damping behaviour in every Cartesian direction (e.g., a critically damped one). It is then obvious that the damping matrix D_d cannot be constant, but has to be chosen as a function of $\Lambda(q)$. In the following, the variation of $\Lambda(q)$ will be considered slow, so that its derivative can be neglected, reducing (21) in the absence of external forces to

$$\Lambda(q)\ddot{e}_x + D_d \dot{e}_x + K_d e_x = 0. \quad (22)$$

All matrices which will be derived from $\Lambda(q)$ are assumed to be quasi-static as well.

A. Factorization Design

If the eigenvalues of the impedance dynamics should be all real, it can be easily seen that this can be achieved by a damping matrix of the form $D_d(q) = A(q)K_{d1} + K_{d1}A(q)$, where $A(q)$ and K_{d1} are defined as $A(q)A(q) = \Lambda(q)$ and $K_{d1}K_{d1} = K_d$ respectively. (22) can then be factorized as

$$A(q)\big(A(q)\ddot{e}_x + K_{d1}\dot{e}_x\big) + \quad (23)$$
$$+ K_{d1}\big(A(q)\dot{e}_x + K_{d1}e_x\big) = 0. \quad (24)$$

With the substitution $A(q)\dot{e}_x + K_{d1}e_x = w$, this leads to the system

$$A(q)\dot{e}_x + K_{d1}e_x = w \quad (25)$$
$$A(q)\dot{w} + K_{d1}w = 0, \quad (26)$$

which has n pairs of equal, real eigenvalues. An heuristic design approach may then be to choose a general damping design of the form

$$D_d(q) = A(q)D_\xi K_{d1} + K_{d1} D_\xi A(q), \quad (27)$$

where $D_\xi = \text{diag}\{\xi_i\}$ is a diagonal matrix and $0 \le \xi_i \le 1$ (0 for undamped behaviour and 1 for real eigenvalues). For a wide stiffness range this approach leads indeed to numerical eigenvalues with a damping very close to the desired one.

B. Double Diagonalization Design

An more elegant approach to the design of the damping matrix can be developed based on the generalized eigenvalue problem known from matrix algebra [3]:
Given a symmetric positive definite $n \times n$ matrix Λ and a symmetric $n \times n$ matrix K_d, a $n \times n$ nonsingular matrix Q can be found, such that $\Lambda = QQ^T$ and $K_d = QK_{d0}Q^T$ for some diagonal matrix K_{d0}.
By choosing the damping matrix as:

$$D_d(q) = 2Q(q)D_\xi K_{d0}^{1/2} Q^T(q), \quad (28)$$

an error dynamics of the form

$$Q(q)Q^T(q)\ddot{e}_x + 2Q(q)D_\xi d K_{d0}^{1/2} Q^T(q)\dot{e}_x + \quad (29)$$
$$Q(q)K_{d0}Q^T(q)e_x = 0.$$

can be obtained. This leads in the new coordinates $w = Q^T(q)e_x$ to a system of n decoupled equations with the requested damping behaviour:

$$\ddot{w} + 2D_\xi K_{d0}^{1/2}\dot{w} + K_{d0}w = 0. \quad (30)$$

As mentioned before, $Q(q)$ is here assumed to be quasi-static. Notice that the decoupling can be achieved only in some particular directions, which are given by $Q(q)$ and hence depend on $\Lambda(q)$ and K_d. This is not surprising, since the controller (19) does not provide a decoupling of the mass matrix. Consequently, if the system is excited, oscillations will occur decoupled and with the desired damping behaviour only in those special directions. This is the main limitation of the method.

VII. Experimental Results

A. Nullspace Behaviour

In this experiment the effects of the different nullspace projections from section IV on the Cartesian behaviour are treated. A desired Cartesian impedance behaviour was chosen, which is characterized by the stiffness values in table I and damping values of $\xi_i = 0.7$. In this experiment

TABLE I

Commanded values for the diagonal Cartesian stiffness matrix in the first experiment.

x	y	z	roll [1]	pitch	yaw
1000	1000	1000	300	300	300
$\frac{N}{m}$	$\frac{N}{m}$	$\frac{N}{m}$	$\frac{Nm}{rad}$	$\frac{Nm}{rad}$	$\frac{Nm}{rad}$

no external forces were present, and therefore, for an ideal nullspace projection, there should be no Cartesian end-effector deviation from the equilibrium point x_d independently of the manipulator's nullspace motion.
The desired equilibrium point for the nullspace motion was simply chosen as $q_N = 0$. The nullspace stiffness and damping matrices were chosen as diagonal matrices $K_N = k_N I$ and $D_N = d_N I$. The initial configuration of the arm can be seen in Fig. 2.

In order to generate an oscillating nullspace motion, a negative value was chosen for the damping factor together with a positive value for the stiffness k_N

$$d_N = -0.5 k_N^{1/2}, \quad (31)$$
$$k_N = 50 \ Nm/rad. \quad (32)$$

The resulting end-effector motion during the oscillating nullspace motion can then be used for an evaluation of the nullspace projections. In this comparison, only the translational motion of the end-effector shall be used. Then the considered Cartesian error $||e_x||_t$ denotes the

---

[1] The values for the rotational stiffness were deliberately chosen high, so that the particular representation of orientations has no significant effects on the results, as pointed out in [2], [12].

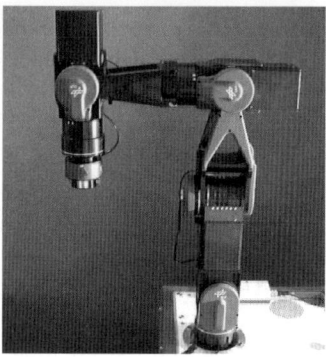

Fig. 2. Initial configuration of the arm for nullspace and damping design experiments.

Euclidean norm of the translational components of $x - x_d$ only. In Fig. 3 these Cartesian errors are shown for the different nullspace projection matrices. Herein the projection matrices were switched online between N_1, N_3, and N_2 during the nullspace oscillation. One can see that the Cartesian errors are considerably larger in case of the static nullspace projection via N_1, while the errors with the dynamically consistent projection matrices N_2 and N_3 are of similar magnitude.

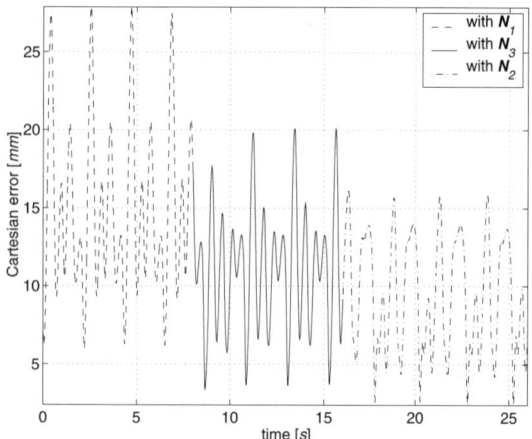

Fig. 3. Cartesian Error $||e_x||_t$ for the different nullspace projections.

In order to visualize also the nullspace motion, which was present during the generation of Fig. 3, the projection of the joint position deviation $q - q_N$ into the nullspace is given in Fig. 4. While the amplitudes of the oscillations in the nullspace for N_1 are the lowest (Fig. 4), they result in high Cartesian errors (Fig. 3). One can also see that the period of the oscillation is slightly different for the different nullspace projections. This results from the different weightings of the nullspace stiffness and the damping value by the projection matrices.

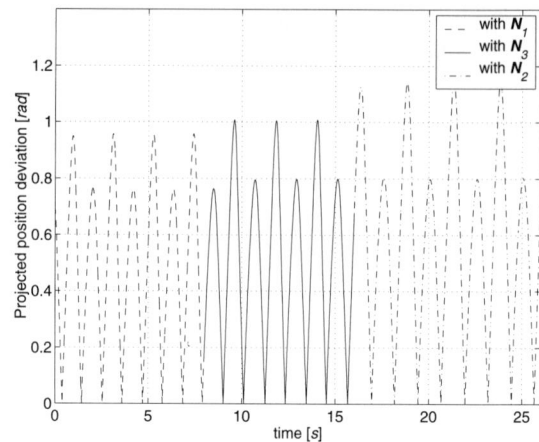

Fig. 4. Projected Joint Error $||V(q)(q - q_N)||_2$ for the different nullspace projections.

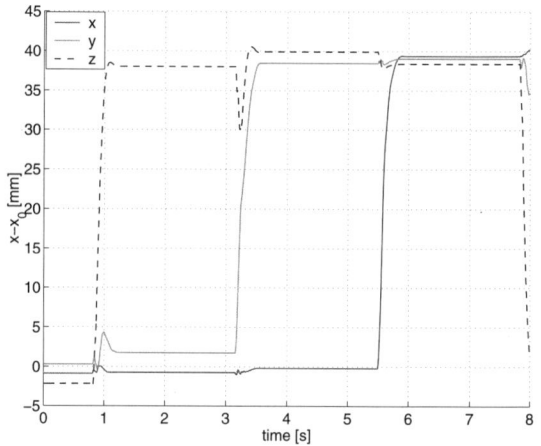

Fig. 5. Position step responses with the factorization damping design for $\xi_x = \xi_y = \xi_z = 0.7$.

B. Damping Design

The effectiveness of the two proposed damping design methods from Sect. VI is evaluated based on step responses which are executed successively in the three translational directions. During the experiment, the desired stiffness values from table II were used.

TABLE II

COMMANDED VALUES FOR THE DIAGONAL CARTESIAN STIFFNESS MATRIX AND THE NULL-SPACE STIFFNESS.

x	y	z	roll	pitch	yaw	null-space
4000 $\frac{N}{m}$	4000 $\frac{N}{m}$	4000 $\frac{N}{m}$	300 $\frac{Nm}{rad}$	300 $\frac{Nm}{rad}$	300 $\frac{Nm}{rad}$	0.0 $\frac{Nm}{rad}$

The damping values are set to $\xi_{\text{roll}} = \xi_{\text{pitch}} = \xi_{\text{yaw}} = 0.7$ for the rotations and to $\xi_n = 0.0$ for the null-space. In Fig. 5, all the step responses are critically damped using the factorization damping design. Since the Cartesian mass

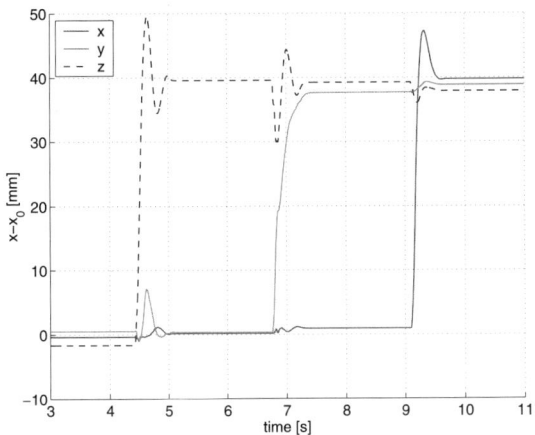

Fig. 6. Position step responses with a starting point x_0, with the factorization damping design for $\xi_x = 0.3$, $\xi_y = 1.0$, $\xi_z = 0.3$.

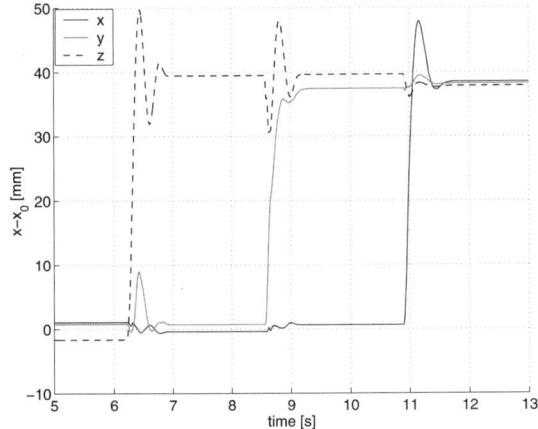

Fig. 7. Position step responses with a starting point x_0, with the double diagonalization damping design for $\xi_x = 0.3, \xi_y = 1.0, \xi_z = 0.3$.

matrix is not decoupled, it can be seen that a step in one coordinate is exciting also the other directions, but the disturbance is rapidly rejected. The steady-state errors are caused by the lack of perfect compensation of Coulomb friction. Figure 6 and 7 show measurements with the factorization design and the double diagonalization design, respectively, both with the damping set to $\xi_x = 0.3, \xi_y = 1.0, \xi_z = 0.3$. The performance of both methods is similar and corresponds well to the desired one.

VIII. SUMMARY

In this paper various practically relevant aspects of impedance control for redundant robots have been treated. The actual implementation of the impedance controllers on a flexible joint robot was based on a singular perturbation approach. Three different kinds of nullspace projections for the realization of a nullspace stiffness were described. This aspect has already been treated in detail in the literature and the focus herein lies on the experimental comparison. Another aspect, which we consider as quite important from a practical point of view, is the design of a suitable damping matrix. This design problem arises when a decoupling of the manipulator's mass behaviour is not possible or not desired. A rather heuristic method as well as a method based on double diagonalization of symmetric positive definite matrices have been presented and experimentally compared.

ACKNOWLEDGMENT

This work was supported by the German Federal Ministry of Education and Research "BMBF-Leitprojekt MORPHA" number ITL 0902E.

IX. REFERENCES

[1] A. Albu-Schäffer and G. Hirzinger. Cartesian impedance control techniques for torque controlled light-weight robots. *IEEE International Conference of Robotics and Automation*, pages 657–663, 2002.
[2] F. Caccavale, C. Natale, B. Siciliano, and L. Villani. Six-dof impedance control based on angle/axis representations. *IEEE Transactions on Robotics and Automation*, 15(2):289–299, 1999.
[3] D. Harville. *Matrix Algebra from a Statistician's Perspective*. Springer-Verlag, 1997.
[4] G. Hirzinger, A. Albu-Schäffer, M. Hähnle, I. Schaefer, and N. Sporer. On a new generation of torque controlled light-weight robots. *IEEE International Conference of Robotics and Automation*, pages 3356–3363, 2001.
[5] N. Hogan. Impedance control: An approach to manipulation, part I - theory, part II - implementation, part III - applications. *Journ. of Dyn. Systems, Measurement and Control*, 107:1–24, 1985.
[6] O. Khatib. A unified approach for motion and force control of robot manipulators: The operational space formulation. *IEEE Journal of Robotics and Automation*, 3(1):1115–1120, February 1987.
[7] P. V. Kokotovic, H. K. Khalil, and J. O'Reilly. *Singular Perturbation Methods in Control: Analysis and Design*. Academic Press Inc., 1986.
[8] A. D. Luca and P. Tomei. *Theory of Robot Control*, chapter Elastic Joints, pages 219–261. Springer-Verlag, Berlin, 1996.
[9] C. Ott, A. Albu-Schäffer, and G. Hirzinger. Comparison of adaptive and nonadaptive tracking control laws for a flexible joint manipulator. *IEEE Int. Conf. on Intelligent Robots and Systems*, 2002.
[10] J. Park, W. Chung, and Y. Youm. On dynamical decoupling of kinematically redundant manipulators. *IEEE Int. Conf. on Intelligent Robots and Systems*, 3(1):1495–1500, 1999.
[11] M. Spong. Modeling and control of elastic joint robots. *IEEE Journal of Robotics and Automation*, RA-3:291–300, 1987.
[12] S. Stramigioli and H. Bruyninckx. Geometry and screw theory for constrained and unconstrained robot. *Tutorial at ICRA*, 2001.

The Passivity of Natural Admittance Control Implementations

Mark Dohring (med5@po.cwru.edu), Wyatt Newman (wsn@po.cwru.edu)
Department of Electrical Engineering and Computer Science
Case Western Reserve University
Cleveland, OH 44106

Abstract— Natural Admittance Control (NAC) has been proposed as a means of implementing responsive force control that is both gentle and rapid while maintaining an end-effector admittance that is passive. Practical implementations, however, must deviate from the ideal NAC formulation, impacting the passivity of the manipulator. Issues arise particularly in the case where NAC is retrofitted to an existing industrial robot controller. Command input may be restricted to only position commands, and the servo algorithm may be fixed. Direct torque control is often not possible, precluding the use of torque feed-forward. Finally, the feedback sensors signals themselves often contain filters. Each of these three effects is examined and their effects on NAC implementations are identified by considering a simplified, continuous-time model of a typical NAC implementation.

Keywords— Passivity, NAC, analysis, parameter spaces.

I. Introduction

For robotic tasks involving extensive contact with the environment, such as force guided assembly, it is desirable that those interactions be both gentle and rapid. Impedance control was introduced by Hogan [1] as a useful viewpoint from which to design interaction controllers. As originally proposed, impedance control seemed to offer the ability to choose mass, damping, and compliant properties for a controlled system without limitations on the ranges of these properties. Experimentally, however, it was found that the ranges of these parameters for which contact stability was assured were surprisingly narrow. Colgate [2] analyzed impedance controllers with respect to passivity and recognized a fundamental limitation on the range of target masses for which the system would be stable in stiff contact. This restriction was attributable to non-collocation of sensors and actuators (e.g., resonances between the actuator and the force sensor). More recently, Colgate and Schenkel also showed that sampled-data effects further imposed limits on the ability to mimic passive physical dynamics [3].

Within the limitations exposed by Colgate, Newman proposed a control scheme, "Natural Admittance Control" (NAC), that used force-sensor feedback to suppress the effect of friction to make an interaction controller more responsive without violating passivity. Under certain idealized conditions, this control formulation was shown to satisfy passivity.

In this presentation, we illuminate three more effects limiting the performance of interaction controllers, analyzed specifically in the context of NAC. We will show that these effects limit the space of dynamic parameters that can be emulated while preserving passivity. The effects considered occur commonly in industrial robot controllers retrofitted with NAC. Specifically, such controllers commonly present an interface limited to position command inputs (with, internally, a proportional-plus-derivative position controller). The dynamics of this internal control imposes limits on passive NAC retrofits. Additionally, industrial robot controllers typically do not allow for torque feed-forward (direct torque control). This restriction imposes further limits on achievable NAC performance. The third effect discussed is the influence of a mis-match between velocity filtering and force filtering, either from inherent sensor dynamics, or deliberate filtering for noise reduction or anti-aliasing. It will be shown that, while such filtering is necessary, velocity and force low-pass filters should be matched as closely as possible to maximize achievable interaction performance.

II. Natural Admittance Control Formulation

A general design approach for NAC controllers was introduced in [4]. A modified version of NAC is presented here, which offers a more intuitive perspective.

In the following presentation, it is assumed that all parameters may be functions of complex frequency s unless otherwise specified. The system being controlled by NAC is a single degree of freedom system with possibly complex internal dynamics modeled by an admittance matrix. This system interacts with other systems through two ports— an actuator port, labeled port 1, and an environment port, labeled port 2. The system itself is composed entirely of passive elements; that is, the admittance matrix is the admittance of a passive system. All actuation occurs at port 1 and all interaction with the environment occurs at port 2. A controller uses feedback measurements of the port variables to derive an actuator input F_1 based on the values of F_2 and v_1 using general, frequency-dependent feedback compensators. An additional velocity signal v_a is also incorporated into the control to provide an input set-point to the closed-loop system.

A suitable compensator is designed for this signal to "attract" the system to the input trajectory defined by v_a, the attractor velocity.

The plant to be controlled is represented by a two-port admittance matrix relating actuator and environment forces to actuator and environment velocities.

$$\begin{bmatrix} v_1 \\ v_2 \end{bmatrix} = \begin{bmatrix} Y_{11} & Y_{12} \\ Y_{21} & Y_{22} \end{bmatrix} \begin{bmatrix} F_1 \\ F_2 \end{bmatrix} \quad (1)$$

where v_1 and v_2 are the actuator and environment velocities, respectively, and F_1 and F_2 are the actuator and environment forces, respectively.

The control law defines a new quantity v_I, the "ideal" velocity, as

$$v_I = G_{f_1} F_2 + G_{f_2} G_a (v_a - v_1) \quad (2)$$

which is used in the control law

$$F_1 = G_v (v_I - v_1) + G_a (v_a - v_1) \quad (3)$$

This formulation is representative of most current implementations of NAC. G_a is now a virtual force generator that creates the dynamics between the attractor and the environment port of the manipulator. G_{f_1} and G_{f_2} are usually appropriate integrating functions translating endpoint and virtual attractor forces into a velocity command. In most implementations, these two functions are equal. Note that v_1 is chosen as the feedback velocity for the virtual attractor force. This makes all of the velocity feedback signals collocated with the actuator and eliminates the need to measure v_2. This choice also eliminates the need to transform the direct feed-forward term in the control law with a transfer function from F_2 to F_1. The virtual force term expresses a virtual attractor that is dynamically at the actuator, but kinematically can be at the end effector through the use of an appropriate kinematic transform.

At this point, the compensators G_{f_1}, G_{f_2}, G_v, and G_a are arbitrary. However, by examining Equation 2, which defines the ideal velocity, some intelligent choices can be made. It is clear that G_a should have units of impedance, translating a velocity error into a force. Both G_{f_1} and G_{f_2} should have units of admittance, transforming forces into velocities. Ideally, one would choose G_v, G_{f_1}, and G_{f_2} to be large. If G_{f_1} and G_{f_2} are large, then the plant would emulate a large (i.e. responsive) admittance, responding rapidly to environment forces F_2 and to virtual forces due to the attractor and virtual impedance, G_a. Further, if G_v were very large, then the plant should follow the ideal velocity more closely, and thus the plant would emulate the target admittances (specified by G_{f_1} and G_{f_2}) closely. The design objective is then to maximize the G_f's (to mimic admittance responsiveness), to maximize G_v (to mimic the target admittance as closely as possible), and to choose G_a (attractor impedance) appropriate for a given task. However, these choices should be made subject to the restriction that the resulting actual plant dynamics satisfies passivity. In addition, when retrofitting NAC via a closed controller, the interface typically restricts implementation to specification of the attractor velocity and precludes the use of direct torque feed-forward.

We next consider a deliberately simplified dynamic system and examine the influence of controller retrofit restrictions on the achievable range of system performance subject to satisfying passivity.

III. ADMITTANCE DESIGN LIMITATIONS OF NATURAL ADMITTANCE CONTROL

In this analysis, we consider a one degree-of-freedom (DOF) system consisting of a single mass with a linear damper to ground. The mass is acted on by actuator forces, F_1, and environment forces, F_2. The actuator and controller are presumed to be ideal (linear and unlimited bandwidth). The plant model can be expressed as:

$$P(s) = \frac{1}{Ms + B} \quad (4)$$

For our NAC formulation, we choose $G_{f_1} = G_{f_2} = G_f$ corresponding to the desired admittance of:

$$G_{f_1} = G_{f_2} = G_f = \frac{1}{M_f s + B_f} \quad (5)$$

where M_f is the target mass and B_f is the target viscous friction. Nominally, these are estimates of the plant inertia, M, and damping, B, respectively. Changing M_f and B_f attempts to modify the apparent manipulator inertia and damping.

We choose the attractor impedance to be a simple combination of a spring and a damper, i.e.:

$$G_a = B_a + \frac{K_a}{s} \quad (6)$$

where K_a is the target stiffness and B_a is the target attractor damping. However, we presume we are not free to choose the form of the velocity controller, G_v, as it is implemented within a closed system, controllable only via a velocity set-point. We presume the velocity controller has the form:

$$G_v = B_v + \frac{K_v}{s} \quad (7)$$

This P-I velocity controller corresponds to an equivalent P-D position controller, which is commonly the default control algorithm, with K_v representing the integral velocity gain and B_v the proportional velocity gain.

Equations 1, 2, and 3 are combined to get an expression for the closed loop admittance at the environment port, with the plant model P substituting for all of the terms in the admittance matrix:

$$Y_2 = P \frac{1 + G_v G_f}{1 + P G_v + P(1 + G_v G_f) G_a} \quad (8)$$

This controller includes a direct feed-forward of the virtual forces generated by the attractor in the second term of Equation 3. In controllers lacking direct torque feed-forward, Equation 3 must be modified by dropping the second term:

$$F_1 = G_v(v_I - v_1) \quad (9)$$

In the non-feed-forward case, the closed loop environment port admittance becomes:

$$Y_2 = P \frac{1 + G_v G_f}{1 + PG_v(1 + G_a G_f)} \quad (10)$$

Plugging in the expressions for P, G_v, G_f, and G_a generates a rather complicated expression in either the feed-forward or non-feed-forward case. We first consider the case where the controller includes the feed-forward term in Equation 3. The admittance of Equation 8 becomes:

$$Y_2 = \frac{M_f s^3 + (B_f + B_v)s^2 + K_v}{a_4 s^4 + a_3 s^3 + a_2 s^2 + a_1 s + a_0} \quad (11)$$

where

$$a_4 = MM_f \quad (12)$$
$$a_3 = MB_f + M_f B + B_v M_f + B_a M_f \quad (13)$$
$$a_2 = BB_f + B_v B_f + K_v M_f + B_a B_f + K_a M_f$$
$$\quad + B_v B_a \quad (14)$$
$$a_1 = K_v B_f + K_a B_f + B_v K_a + B_a K_v \quad (15)$$
$$a_0 = K_v K_a \quad (16)$$

If force feed-forward is not available the control law in Equation 9 is used instead, producing the admittance:

$$Y_2 = \frac{M_f s^3 + (B_f + B_v)s^2 + K_v s}{a_4 s^4 + a_3 s^3 + a_2 s^2 + a_1 s + a_0} \quad (17)$$

where

$$a_4 = MM_f \quad (18)$$
$$a_3 = MB_f + M_f B + B_v M_f \quad (19)$$
$$a_2 = BB_f + B_v B_f + K_v M_f + B_v B_a \quad (20)$$
$$a_1 = K_v B_f + B_v K_a + B_a K_v \quad (21)$$
$$a_0 = K_v K_a \quad (22)$$

These expressions for the end-point admittance of the simple system are difficult to analyze in general, but for specific cases they can yield some interesting results. In order to be passive, the linear, rational function of the complex frequency s, $Y_2(s)$ must meet the requirements of a positive real function [5].
- It is analytic in the right-half s-plane.
- $\Re\{Y(s)\} \geq 0$, when $\Re\{s\} = 0$
- Any j-axis poles are simple and have positive real residues.

The condition requiring the poles and zeros to be in the left-half s-plane allows the quick elimination of a candidate function. If that condition is met, the remaining test, that $\Re\{Y_2(j\omega)\} \geq 0$, $\forall \omega$ must be performed. Given a rational function $Y(s) = \frac{Y_{\text{num}}}{Y_{\text{den}}}$, it is equivalent to showing that

$$\Re\{Y_{\text{num}}(j\omega)\} \Re\{Y_{\text{den}}(j\omega)\}$$
$$+ \Im\{Y_{\text{num}}(j\omega)\} \Im\{Y_{\text{den}}(j\omega)\} \geq 0, \forall \omega \quad (23)$$

Given numerical values of the parameters for the system and the controller, whether or not the system is passive can be evaluated numerically using a computer program, or the expression for $Y(s)$ can be tested using the root locus technique proposed by Colgate and Hogan in [6], combining the admittance with all possible pure spring and all possible pure mass environments, using the feedback gain K to represent the magnitude of the spring or the mass. The analytical expression can also be tackled by fixing certain parameters to simplify the expression sufficiently. In this way, analytical bounds for maintaining passivity can be calculated on the remaining parameters.

A. Parameter space search example

The non-feed-forward case was tested for a selection of parameters presented in Table I. Each set of parameters was fed into a Simulink™ model of the system and its controller. The admittance transfer function at the environment port was extracted using the Linear Analysis function, and the root locus test was used to determine whether the admittance was passive. For each parameter set, one of three results

Trial	M	B	M_f	B_f	K_v	B_v	K_a	B_a
1	1	1	1	1	1	1	1	1
2	1	1	1	1	6	10	1	1
3	3	5	3	0	10	1	3	3.5
4	3	5	3	5	10	1	3	3.5
5	3	5	3	4	10	1	3	3.5
6	3	5	3	3	10	1	3	3.5
7	3	5	3	2	10	1	3	3.5
8	3	5	3	1	10	1	3	3.5
9	3	5	3	0	9	1	3	3.5
10	3	5	3	0	8	1	3	3.5
11	3	5	3	0	3	1	3	3.5
12	3	5	3	0	0	1	3	3.5
13	3	5	3	0	100	10	1	1

TABLE I

Various candidate NAC implementations evaluated according to whether the resulting admittance was passive, non-passive but stable, or unstable. Trials 3, 8, and 13 were non-passive but stable. Trials 9, 10, and 11 were unstable in free motion. Trials 1, 2, 4, 5, 6, 7, and 12 were passive.

was possible. Either the system had a passive driving-point admittance, the system had a non-passive driving point admittance but was stable open loop (not in contact with an environment), or the system was unstable open loop. The primary result here shows a problem with setting the NAC model damping, B_f, to zero when the velocity controller integral gain, K_v, was non-zero. In other words, whether the closed-loop admittance is passive depends on the form of the velocity controller, G_v, whenever the target dynamics attempt to reduce the system's apparent viscous friction. In the case where an existing controller contains an integral velocity term, restrictions are placed on the reduction of apparent viscous friction.

B. Analysis of inertia reduction with NAC

To evaluate the ability of Natural Admittance Control to reduce the apparent inertia, the full, closed-loop admittance function was evaluated with no attractor force feed-forward with $K_v = 0$, $B_f = 0$, and $M_f = \mu M$. Under these conditions, the driving-point admittance function is

$$Y_2 = \frac{\mu M s^3 + B_v s^2}{\mu M^2 s^4 + \mu M (B + B_v) s^3 + B_v B_a s^2 + B_v K_a s} \quad (24)$$

Applying the condition in Equation 23 yields the following inequality that must be satisfied if the admittance is to be passive.

$$\mu M^2 (B + B_v)\omega^4 - \mu M B_v K_a \omega^2 + B_v^2 B_a \omega^2 + \mu M^2 B_v \omega^4 \geq 0, \ \forall \omega \quad (25)$$

This implies that

$$\mu \geq \frac{B_v}{B + B_v} \quad (26)$$

$$\mu \leq \frac{B_v B_a}{M K_a} \quad (27)$$

must be satisfied simultaneously to produce a passive admittance. The ratio of the target inertia, M_f to the actual inertia M, is constrained on both sides and must be neither too small, nor too large. Note also, that choosing $B_a = 0$, the target damping equal to zero, results in a non-passive system, regardless of the choice of μ since the second condition becomes $\mu < 0$. A too large value of K_a reduces the maximum allowable μ. Without virtual force feed-forward, there is a maximum allowable attractor stiffness beyond which the system becomes non-passive, no matter what target inertia is selected. This causes problems when NAC is used to emulate a position controller by using a high attractor stiffness.

This result may be compared to the case where attractor feed-forward is used. In this case, the admittance function becomes:

$$Y_2 = \frac{\mu M s^3 + B_v s^2}{Y_{\text{den}}} \quad (28)$$

where

$$\begin{aligned} Y_{\text{den}} &= \mu M^2 s^4 + \mu M (B + B_v + B_a) s^3 \\ &\quad + (\mu M K_a + B_v B_a) s^2 + B_v K_a s \end{aligned} \quad (29)$$

resulting in conditions to make the admittance passive

$$\mu \geq \frac{B_v}{B + B_v + B_a} \quad (30)$$

$$B_v^2 B_a \geq 0 \quad (31)$$

where B is the inherent system viscous damping, B_v is the velocity controller gain, and B_a is the virtual attractor damping. The second condition is not a condition on μ at all, rather it specifies that the target attractor virtual damping be positive. We see that with feed-forward the maximum μ is unbounded, and there are now no restrictions on the attractor stiffness, K_a.

However, in both the feed-forward and non-feed-forward cases, a minimum value is established for μ implying that the target inertia, M_f must be larger than some fraction of the actual inertia, determined by the system viscous damping and the controller velocity gains. B_v is likely to be large to enforce good tracking of the ideal velocity, implying that the minimum μ will be close to unity. The feed-forward case appears to offer an easing of this restriction by the presence of B_a in the denominator of Equation 30. However, this is not the case in practice since B_a (and indeed the entire controller) is often implemented digitally, introducing additional effects from sampling. Colgate and Schenkel, in [3] have identified upper limits on damping gains in a class of sampled-data systems.

Setting $M = 3$, $B = 5$, $B_v = 10$, $B_a = 1$, and $K_a = 1$, the non-feed-forward NAC case requires $\mu \geq \frac{2}{3}$ and $\mu \leq \frac{10}{3}$. In the case for NAC with feed-forward attractor force, the requirement is that $\mu \geq \frac{5}{8}$. As a practical matter, B_v will often be large compared to both B and B_a. In these cases, the minimum achievable μ becomes very close to unity.

C. Numerical searches in NAC parameter space

The NAC model under discussion has as parameters $M, B, M_f, B_f, K_v, B_v, K_a$, and B_a, and a binary condition of whether attractor force feed-forward is used, defining two, eight-dimensional, parameter sub-spaces. A full search in this space is impractical. However, searches through subspaces can provide insight into controller designs.

A search was conducted using the simple NAC model and the following fixed parameters: $M = 3.6$, $B = 3.0$, $B_v = 1400$, $K_a = 250$, $B_a = 33.33$, and no attractor virtual force feed-forward. These parameter settings were based on an implementation of NAC for the ParaDex manipulator ([7], [8], [9]). Search was conducted on the parameters K_v, M_f, and B_f, the velocity controller position gain, the target inertia, and the target model viscous friction, respectively. Search in the K_v direction was scaled logarithmically with the square root of the ratio of K_v to M, the undamped natural frequency of the inner velocity loop. The special case of $K_v = 0$ was added to the search list. Search in the M_f and B_f directions was scaled linearly as the ratios between the target model parameters and the actual plant parameters, i.e. $\frac{B_f}{B}$ and $\frac{M_f}{M}$.

At each search point in this 3-dimensional subspace, the closed loop admittance was generated analytically and tested to determine whether it was passive. In this automated search the tests for positive realness were applied directly by evaluating the pole and zero locations and using a calculation of the frequency response to check that the phase was bounded. Unstable plants were also identified, but in the results presented here, none were found.

Slices of data are shown in Figures 1, 2, and 3.

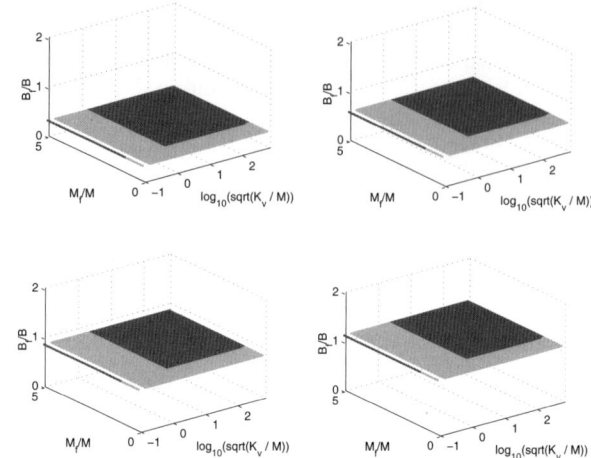

Fig. 2. Selected slices of 3-D parameter search. The darker, green dots indicate passive NAC implementations. The lighter, cyan dots indicate non-passive, but stable implementations. Note: The line of data in the plane where $\log_{10}(\sqrt{(K_v/M)}) = -1$ is actually data for $K_v = 0$.

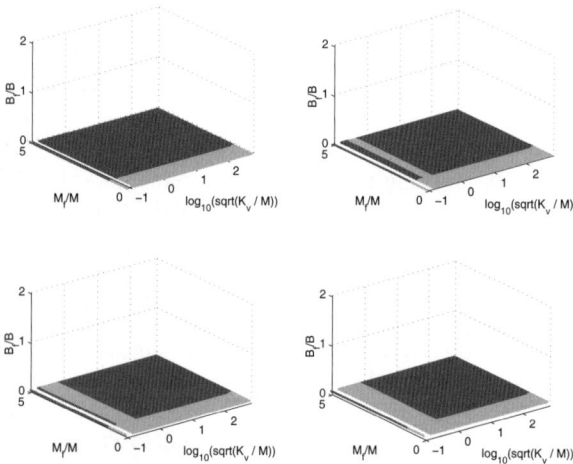

Fig. 1. Selected slices of 3-D parameter search. The darker dots indicate passive NAC implementations. The lighter dots indicate non-passive, but stable implementations. No unstable implementations are present in this search space. Note: The line of data in the plane where $\log_{10}(\sqrt{(K_v/M)}) = -1$ is actually data for $K_v = 0$. As B_f varies from 0 to 0.3, a band of non-passive systems develops for certain, low values of K_v

The figures illustrate the lower bound on the mass ratio, μ, generating non-passive admittances whenever μ becomes smaller than one. The upper bound on μ is not visible in the graphs as the parameters chosen would place that boundary near $\mu = 51.8$.

IV. Feedback sensor dynamics and NAC

The third effect considered was that of feedback sensor dynamics. If the feedback sensors used for environment force and actuator velocity have dynamics associated with them, the standard NAC formulation requires that they be matched to guarantee a passive admittance. Even low-pass, anti-aliasing filters must be taken into account. If there is a low-pass filter on the force, there must be a low-pass filter on the velocity signal. A complication is that the velocity signal

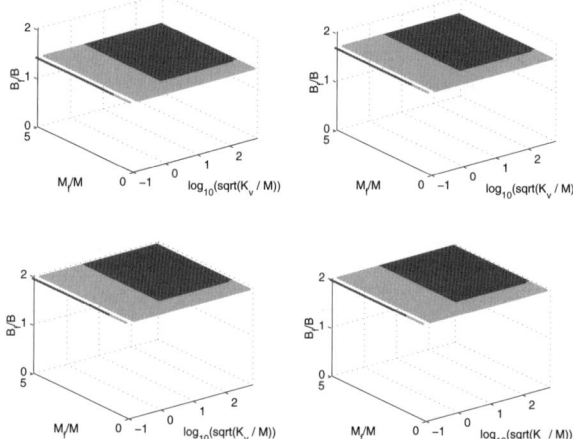

Fig. 3. Selected slices of 3-D parameter search. The darker dots indicate passive NAC implementations. The lighter dots indicate non-passive, but stable implementations. Note: The line of data in the plane where $\log_{10}(\sqrt{(K_v/M)}) = -1$ is actually data for $K_v = 0$.

is often derived from a numerical differentiation of a discrete position signal, such as from an encoder. The filter on the force signal is often analog. We will again restrict ourselves to the continuous-time case in our examination.

If a mismatch in filtering on the force and velocity feedback signals occurs, the resulting closed-loop, driving-point admittance may not be passive. Consider the simplified model used in previous example, except replace the feedback velocity, v_1 with a filtered version

$$v_{1f} = G_{vf}v_1 \qquad (32)$$

where G_{vf} represents the velocity filter dynamics, and

replace the feedback force, F_2 with a filtered version

$$F_{2f} = G_{ff} F_2 \qquad (33)$$

where G_{ff} represents the force filter dynamics. The driving-point admittance with feed-forward becomes

$$Y_2 = \frac{P(1 + G_v G_f G_{ff})}{1 + PG_v G_{vf} + P(1 + G_v G_f G_{vf})G_a} \qquad (34)$$

We choose the NAC target model dynamics to be equal to the plant dynamics to isolate the sensor filtering effect. With $G_f = P$, the admittance becomes:

$$Y_2 = \frac{P}{1 + PG_a} \frac{1 + G_v P G_{ff}}{1 + G_v P G_{vf}} \qquad (35)$$

Consider the case when $M = 1$, $B = 0$, $M_f = 1$, $B_f = 0$, $K_a = 1$, $B_a = 0$, $K_v = 0$, and $B_v = 600$. Let the filter on the force signal be a fourth-order Bessel filter with a cutoff frequency of 100 Hz. Figure 4 shows the results of using four different velocity filters: a unity-gain, zero-phase, all-pass filter (i.e. no

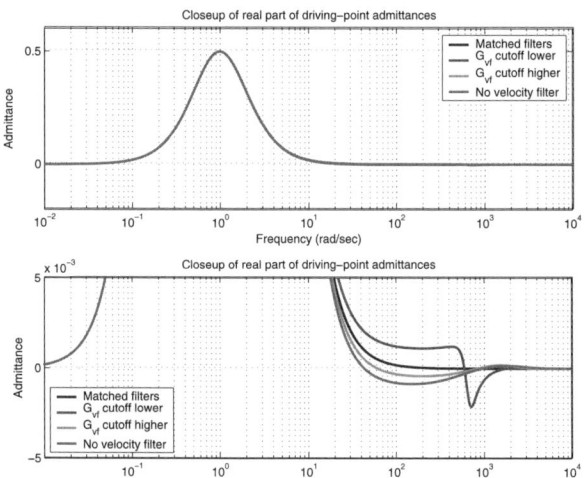

Fig. 4. Admittance real part for different velocity feedback signal low-pass filters. Included are the cases where the velocity filter was identically matched to the force feedback signal filter, where the velocity filter cutoff frequency was lower than the force filter cutoff frequency, were the velocity cutoff frequency was higher than the force cutoff frequency, and where the velocity filter was removed entirely.

filter) and 3 fourth-order Bessel low-pass filters with various cutoff frequencies. One was chosen to match the force feedback sensor cutoff frequency of 100 Hz. A second was set to a cutoff frequency of 50 Hz, and a third was set to a cutoff frequency of 200 Hz. The resulting admittances shown in the figure reveal non-passive behavior for each case where the velocity filter is not matched precisely with the force filter.

V. Conclusions

In this paper we have examined the passivity of Natural Admittance Control implementations. Even restricting ourselves to a plant model consisting only of a mass and a damper with no resonances, there are several effects that place restrictions on Natural Admittance Control's ability to maintain a passive endpoint admittance. Using a PI velocity controller (PD position) places restrictions on viscous friction masking. To effectively mask friction in the manipulator, the inner velocity loop should use a pure, proportional gain only. Feed-forward injection of the attractor dynamics into the control effort has an impact on NAC's ability to modulate the system inertia. NAC alone is not capable of significant inertia reduction either with or without feed-forward, but, in addition, the lack of feed-forward could make it impossible to find any target inertia for which NAC produces a passive admittance if the attractor stiffness is set too high. Finally, a mis-match in the dynamics of the force and velocity feedback sensor signals may also cause a non-passive admittance.

VI. Acknowledgments

This work was supported by the National Science Foundation under NSF grand ECS 0200388. This support is gratefully acknowledged.

References

[1] Neville Hogan, "Impedance control: An approach to manipulation: Parts I, II, and III–theory, implementation, and applications," *Journal of Dynamic Systems Measurement, and Control*, vol. 107, pp. 1–24, March 1985.
[2] J. E. Colgate, *The Control of Dynamically Interacting Systems*, Ph.D. thesis, Massachussettes Institute of Technology, Cambridge, MA, August 1988.
[3] J. E. Colgate and G. Schenkel, "Passivity of a class of sampled-data systems: Application to haptic interfaces," in *American Control Conference*, 1994, vol. 3, pp. 3236–3240.
[4] Wyatt S. Newman, "Stability and performance limits of interaction controllers," *Transactions of the ASME Journal of Dynamic Systems, Measurement, and Control*, 1992.
[5] Ernst A. Guilleman, *Synthesis of Passive Networks*, John Wiley & Sons, Inc., 1957.
[6] James E. Colgate and Neville Hogan, "On the stability of a manipulator interacting with its environment," in *Proceedings of the Twenty Fifth Annual Allerton Conference on Communication, Control, and Computing*, March 1988, pp. 821–828.
[7] Yuesong Wang, Wyatt S. Newman, and Robert S. Stoughton, "Workspace analysis of the paradex robot-a novel, closed-chain, kinematically-redundant manipulator," in *Proceedings of the 2000 IEEE International Conference on Robotics and Automation*, 2000, vol. 3, pp. 2392–2397.
[8] David P. Gravel and Wyatt S. Newman, "Flexible robotic assembly efforts at ford motor company," in *Proceedings of the IEEE International Symposium on Intelligent Control*, 2001, pp. 173–182.
[9] Daniel M. Morris, Ravi Hebbar, and Wyatt S. Newman, "Force guided assemblies using a novel parallel manipulator," in *Proceedings of the 2001 IEEE International Conference on Robotics and Automation*, 2001, vol. 1, pp. 325–330.

Interactive rendering of deformable objects based on a filling sphere modeling approach

François Conti
conti@robotics.stanford.edu

Oussama Khatib
ok@robotics.stanford.edu

Charles Baur
charles.baur@epfl.ch

Robotics Laboratory
Computer Science Department
Stanford University
Stanford, California 94305, USA

VRAI Group
IPR - LSRO
EPFL
CH - 1015 Lausanne, Switzerland

Abstract

Mass-spring systems have widely and effectively been used for modeling in real-time deformable objects. Easier to implement and faster than finite elements, these systems, on the other side, suffer from several drawbacks when coming to render physically believable behaviors. Neither isotropic or anisotropic materials can be controlled easily and the large number of springs and mass points composing the model makes it fastidious to define parameters to control elongation, flexion and torsion at a macroscopic level. Another weakness is that most of the materials found in nature maintain a constant or quasi-constant volume during deformations; unfortunately, mass-spring models do not have this property.

In this paper, we extend the current state-of-the-art in soft tissue simulation by introducing a six-degree of freedom macroscopic elastic sphere described by mass, inertia and volumetric properties. Spheres are placed along the medial axis transform of the object whose centers are connected by a skeleton composed of a set of three-dimensional elastic links. Spheres represent internal mass, volume and control the global deformation of the object. The surface is modeled by setting point masses on the mesh nodes and damped springs on the mesh edges. These nodes are connected to the skeleton by individual elastic links, which control volume conservation and transfer forces between the surface and volumetric model. Using this framework we also present an efficient method to approximate collision detection between multiple bodies in real-time.

1 Introduction

Rendering deformable objects in a realistic manner is a challenging task in computer simulation. While rigid bodies can be accurately described with a limited number of parameters, deformable objects, on the other side, due their internal structure, require a much larger set of numbers to express their state. During these last two decades, a wide variety of approaches have been presented in soft tissue simulation. Mass-spring system techniques have widely and effectively been used for modeling deformable objects where described by a set of mass points dispersed throughout the object and interconnected with each other through a network of springs. These systems are easy to model, to construct and have well understood physics. They are also well suited for parallel computation, making it possible to run complex environments in real-time for interactive simulations. (A surgery simulator for instance). On the other side, mass-spring systems have some drawbacks. Incompressible volumetric objects and high stiffness materials have poor stability requiring small time steps during the numerical integration process, which considerably slows down the simulation. A second category of techniques is finite elements methods, which offer a method with much higher accuracy to solve continuum models. In FEM, unlike mass-spring methods where the equilibrium equation is discretized and solved at each finite mass point, objects are divided into unitary surface (2D) or volumetric (3D) elements joined at discrete node points where a continuous equilibrium equation is approximated over each element. While finite element methods generate a more physically realistic behavior, at the same time they require much more numerical computation and therefore are difficult to use for real-time simulations.

Designing an interactive haptic simulator is a difficult challenge. If physical accuracy is a key factor towards rendering physically believable scenes, nevertheless, real-time control remains indispensable to produce interactive simulation; users will be disturbed if latencies or interrupts are introduced to the user input.

In this work, we extend the state-of-the-art in mass-spring concepts, by proposing an alternative method to model deformable objects. This method generates and connects together both a surface and volumetric representation based on the medial axis skeleton of a solid. This new algorithm is appealing because it decouples local and global deformation and renders them together with variable levels of resolution.

The paper is organized as follows. In paragraph 2, we introduce the different aspects and details of the algorithm. Collisions between multiple bodies and point-contact haptic interaction are discussed in paragraphs 3 and 4. Implementation of our system and experimental results are showed in paragraph 5 and 6 and finally a description of our future work and a conclusion are presented in paragraph 7.

2 Object Modelling

Given a closed surface, defined by a set of triangle, we construct the following attributes:

- A medial axis skeleton, obtained by using a valid MAT on the original surface of the object. (Figure 1a).

- A volumetric model, composed of filling spheres placed along the medial axis skeleton and inter-connected together via three-dimensional elastic links. (Figure 1b).

- A surface model, realized by setting point masses on the mesh nodes and damped springs on the mesh edges. Nodes are connected to the volumetric model (skeleton composed of spheres and links) through individual elastic links. (Figure 1c).

The following paragraphs (2.1, 2.2 and 2.3) describe in detail the algorithms developed for each stage. Figure 1 illustrates the procedure on a 2D object. For the remainder of the paper we will consider 3D objects.

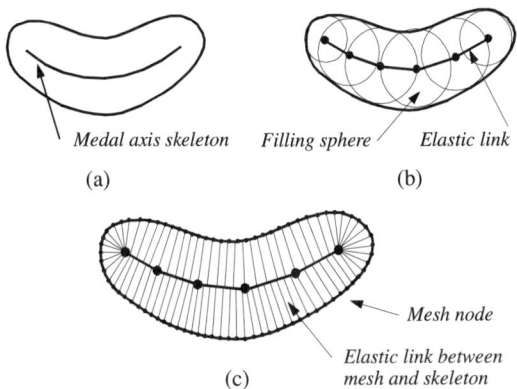

Figure 1 - Medial Axis Skeleton and its corresponding surface (a). Volumetric model composed of spheres and links centered along the skeleton (b). Connections between surfaces nodes and skeleton (c)

2.1 Medial axis transform

The notion of skeleton was introduced by H. Blum [2] as a result of the Medial Axis Transform (MAT) or Symmetry Axis Transform (SAT). The MAT determines the closest boundary points for each point in an object. An inner point belongs to the skeleton if it has at least two closest boundary points. A very illustrative definition of the skeleton is given by the prairie-fire analogy: the boundary of an object is set on fire and the skeleton is the loci where the fire fronts meet and quench each other.

To approximate a skeleton we implemented a method presented by Li and Al. [7] based on an edge-collapsing algorithm. At every round the triangle edge (u,v) with shortest Euclidean distance of the mesh and any associated faces collapse into a point at the average location of its endpoints u and v, and triangles incident to the edge are removed. During the process, whenever an edge (u,v) is not incident to any triangle, it is designated as a skeletal edge and vertices u and v maintain their positions until the end of the process. The process iterates until all triangles have been collapsed to edges or vertices. In other words, we are left with only skeletal edges, whose union is the skeleton of the mesh. This method was selected among other skeletonization techniques for its performance in speed and ease of implementation with surface based objects. In figure 2, we illustrate a clapsing edge example based on a 466 triangles mesh representing a gallbladder; the generated skeleton is composed of seven nodes and six links. By using a Pentium III - 1GHz computer, the complete skeletonization process took 1.5 seconds.

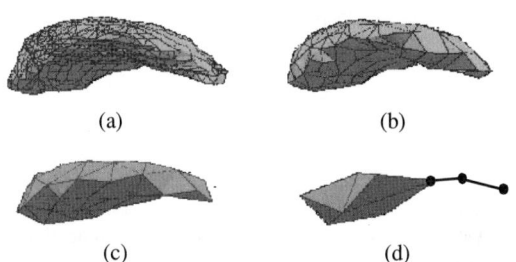

Figure 2 - Skeletonization of a triangle based mesh using a clapsing edge algorithm. The original mesh is composed of 466 triangles and collapses into a skeleton composed of 7 nodes and 6 links

2.2 Volume Modeling

A set of filling spheres compose the volumetric model. Spheres are placed inside the object, along the medial axis transform, and are linked together via elastic links. A sphere represents a portion of mass of an object and is described by a reference frame R, mass m and radius r. Spheres can freely translate and rotate in space unless constraints are applied; for instance, by constraining translation on certain spheres, it is possible to attach deformable bodies to ground points. Elastic links, connecting adjacent spheres together, are composed of six damped springs controlling *elongation, flexion* and *torsion*. They are described by a centerline λ going from *sphere 0* to *sphere 1*; two perpendicular unitary vectors $\vec{u}_i$ and $\vec{v}_i$ are attached to both spheres and lie in plane S_r; planes S_i are perpendicular to centerline λ and values α_t and β_t express the angles between centerline λ and vectors u_t and v_t. The different degrees of freedom of a link are described here bellow:

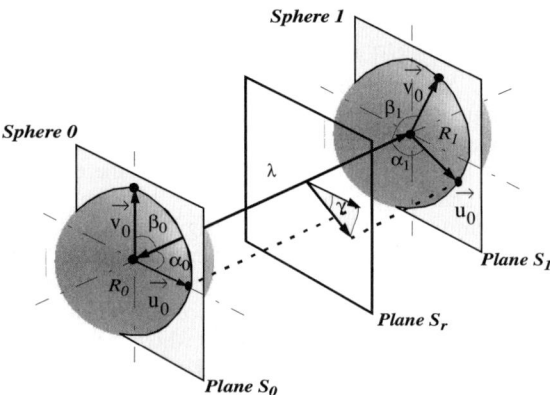

Figure 3 - Two spheres connected with an an elastic link.

Elongation is controlled by an axial spring connecting the centers of each sphere along centerline λ. Compression and elongation forces are applied on both spheres. See figure 4.

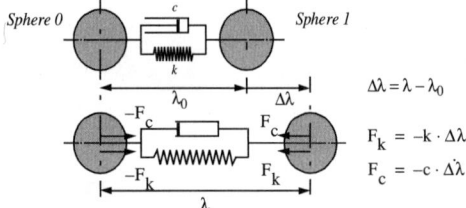

Figure 4 - Axial damped spring controlling elongation.

Flexion is expressed for each sphere by angles α_t and β_t around axis $\vec{v}_i$ and $\vec{u}_i$. Four angular springs control the orientation of each sphere in relation with the centerline λ. Forces and torques applied on each sphere are illustrates in figure 5.

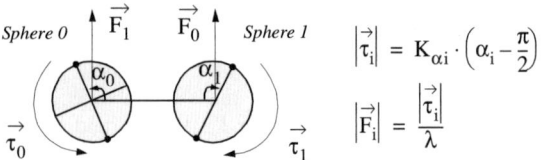

Figure 5 - Illustration of flexion arround axis $\vec{u}$

Finally, torsion is managed by an individual angular spring around centerline λ. Torsion angle γ is measured by projecting vectors $\vec{u}_0$ and $\vec{u}_1$ on plane Sr as illustrated in figure 3. Torques issued from torsion are applied on center of both spheres.

2.3 Surface Modeling

The surface model is created in two stages. First, point masses are set on the mesh nodes (vertices) and damped springs on the mesh edges, Secondly, all the nodes are connected to the skeleton by searching for the nearest link or nearest sphere. Two types of connections may occur:

- If a sphere center happens to be the nearest skeleton point, the surface node is directly attached via an elastic link to the reference frame R of the corresponding sphere. When the sphere rotates or translates, the surface mesh automatically follows the motion.

- If the nearest skeleton point happens to be located on a link, the surface node is projected and attached onto the centerline λ. During the deformation of a link, length λ is modified due to elongation or compression; nodes connected along the link are simply redistributed along the link. During torsion, the surface nodes rotate around the centerline depending of their position along λ a torsion angle γ.

Any forces applied on the surface triangles first generate a local deformation; afterwards, thanks to surface-skeleton links, forces propagate to the global model and affect the overall shape of the object.

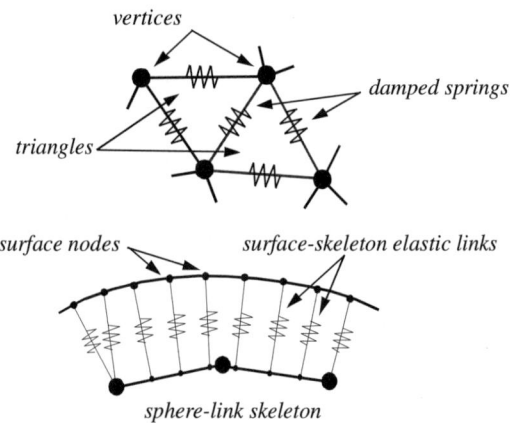

Figure 6 - The surface model is formed by a network of springs located at each triangle edge of the object. The surface nodes are connected to the skeleton via elastic links.

3 Multiple Body Collision

Fast and accurate collision detection between general polygonal models is a fundamental problem in computer-simulated environments. Most of the previous work in collision detection between polygonal models has focused on algorithms for convex polyhedra. However, they are not applicable in the case of deformable objects, since these structures are generally non-convex, and deform over time. Among the collision detection methods that are applicable to more general polygonal models, almost all of the optimizations rely on a pre-computed hierarchy of bounding volumes. The solutions range from axis-aligned box trees, sphere trees, to BSP trees. All these techniques, which perform very efficient rejection tests, may considerably slow down when objects are very close, causing interactive applications to become unresponsive and thus ineffective.

In our framework, we simplify collision detection between deformable objects by only considering impacts between filling spheres. This method requires much less computation compared to other techniques searching for collisions between triangles. Elastic collisions are computed between colliding spheres and directly influence the volumetric model. Since no collision detection is performed directly on the object's mesh, inaccuracies may be observed depending of the resolution of the skeleton. Figure 7 illustrates two objects in contact with each other; the contour of *object 1* slightly interpenetrates *object 2*. Even if inaccuracies are noticeable, this method does not degrade the overall performance of the simulation. This method will not perform correctly for skeletons that do not enclose the overall volume of the object (See Dolphin in figure 10); in these cases, objects may intersect via the empty gaps contained between adjacent spheres.

Complex scenes, containing several dozens of objects, may require computing hundreds or even thousands of collision checks between spheres, thus affecting the overall performance of the simulation. To reduce computational time during collision checking, we can consider for certain applications (i.e. a surgery simulator), that some objects or bodies only move in a limited range of space. Given this condition we can draw, for each object, a limited list of neighbors with whom potential collisions may occur and is worth checking.

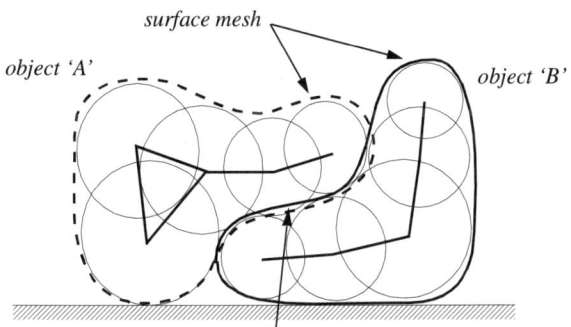

Figure 7 - Collision detection between two objects using the spherical representation.

4 Haptic Interaction

Early haptic rendering systems modeled surface contacts by generating a repulsive force proportional to the amount of penetration into an obstacle. While these penalty-based methods work well to model simple obstacles, such as planes or spheres, a number of difficulties are encountered when trying to extend these models to display more complex environments.

An alternative is not to look at the penetration of the user's finger into the object at all, but instead to constrain the motions of a substitute virtual object. In the method proposed by Ruspini et. Al [4], a representative object, a "proxy," substitutes in the virtual environment for the physical finger or probe. The "virtual proxy" can be viewed as if connected to the user's real finger by a stiff spring. As the user moves his/her finger in the workspace of the haptic device he/she may pass into or through one or more of the virtual obstacles. The proxy, however, is stopped by the obstacles and quickly moves to a position that minimizes its distance to the user's finger position. The haptic device is used to generate the forces of the virtual spring which appears to the user as the constraint forces caused by contact with a real environment. This approach is similar to the method for the "gob-object" first proposed by Zilles et. al [3] but does not require apriori knowledge of the surface topology. In our framework, the forces resulting from the finger-proxy model are directly applied to the mass-points composing triangles in contact with the finger. A weighted distribution based on the contact position of the finger on the triangle determines the amount of force applied to each vertex. An example of the finger-proxy interacting with a deformable object is illustrated in figure 8.

Because virtual objects are normally constructed with a large number of triangles, a naïve test based on checking if each primitive is in the path of the proxy would be prohibitively expensive. Instead a hierarchical bounding representation for the object is constructed to take advantages of the spatial coherence inherent in the environment. In our framework, we make use of the filling spheres model to generate a bounding box representation. Each sphere is attributed a second larger radius to contains all linked triangles. During collision check, we first search for any intersecting spheres. If one is selected we then check for eventual collision with every triangle inside the sphere. By directly using our spherical model for collision detection, we remove any extra computation necessary for updating a bounding box representation when an object changes shape.

A cache is maintained to avoid calling the low-level check multiple times for the same primitive during the same iteration. Some spatial coherence, such as a list of sphere or triangle neighbors, can be used to reduce the computation time between successive calls to the collision algorithm.

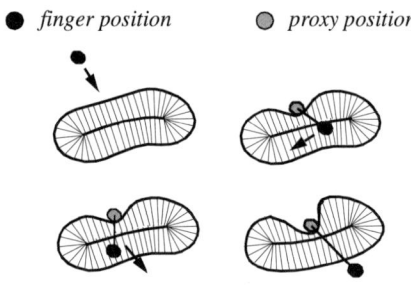

Figure 8 - finger-proxy example on a deformable object

5 Implementation

Our current system runs on a Dual 1-GHZ Pentium III Computer under Windows XP. Two main threads compose the backbone of our application: the *haptics engine* and the *dynamics simulator*. The separation of both processes was first proposed by Adachi et al. [13]. Decoupling the low-level force servo loop from the high-level control is important since the haptic servo loop must run at a very high rate, typically greater than 1000 Hz, to achieve a high quality force display. Our dynamics simulator typically runs at a much slower rate (~20-100Hz). The haptics engine uses the finger/proxy model, presented in paragraph 4, to compute forces between user's fingers and virtual objects; this information is transferred to the dynamics engine to generate physical interactions. Since the physical models are only updated at 20-50 Hz, the haptics engine performs a blending transition between the two last computed frames to avoid user to feel discrete steps between two iterations of the dynamics simulator. Even though a small delay is introduced (< 50 ms) between the user input and the simulation output, the effect is not perceived if time-steps remain small enough (< 100ms).

All physical models are generated offline for each object and loaded into memory during the startup process of the application. An XML script file describes the position, orientation and physical characteristics of each object in the scene.

6 Results

We developed several applications to demonstrate each component of the framework, some of which are available on our website.

In figure 10, we present a virtual Dolphin with its corresponding skeleton generated by placing filling spheres along the medial axis transform. Links interconnect neighbor spheres together. The mesh is composed of 1120 triangles and a connection between the mesh and the skeleton is performed for every vertex. A dynamic behavior was programmed for each Dolphin by controlling the leading sphere, located in the head of the Dolphin, along a sine wave function. The effects were propagated along the rest of the body and gave the illusion of a Dolphin swimming in an ocean. To demonstrate the low amount of computation required for this application, we performed a real-time simulation on a Pentium Pro 200 MHz computer equipped with an AccelGraphics video card for fast OpenGL rendering. A second application was performed on a much faster computer (Dual 1-GHz Pentium III), with a complete medical landscape containing a liver, digestive system, stomach and vesicle. 411 spheres composed the dynamic skeletons of the organs. Surface and global deformation were performed in real time, as well as haptic interaction between the user and the virtual environment. Figure 11 presents a close view of the user grabbing the surface of a gallbladder. We observe the local and global deformations being rendered.

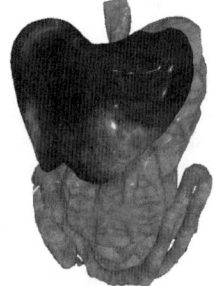 

Figure 9 - Laparoscopic surgery simulator. (Left) Overall view of the virtual organs. (Right) Workstation and force feedback haptic device

7 Conclusion

In this paper, we presented a technique to compute deformations of virtual objects by decoupling surface and volumetric representation. By introducing a new method based on six degrees of freedom filling spheres, we generated realistic global deformations on complex models with minimal computation. We also presented a methodology, based on the medial axis transform, to construct physical models at adjustable resolutions. Finally, we combined the global model with a surface spring model to handle local deformations.

While this approach has shown promising results for applications requiring real-time interactivity, nevertheless many problems remain when cutting procedures are realized. In these situations, updating the mesh and skeleton is a complex task and often computationally expensive.

Future work includes an optimized mechanism to update the model during cutting-procedures, volume conservation and comparison of this approach with finer methods such as FEMs.

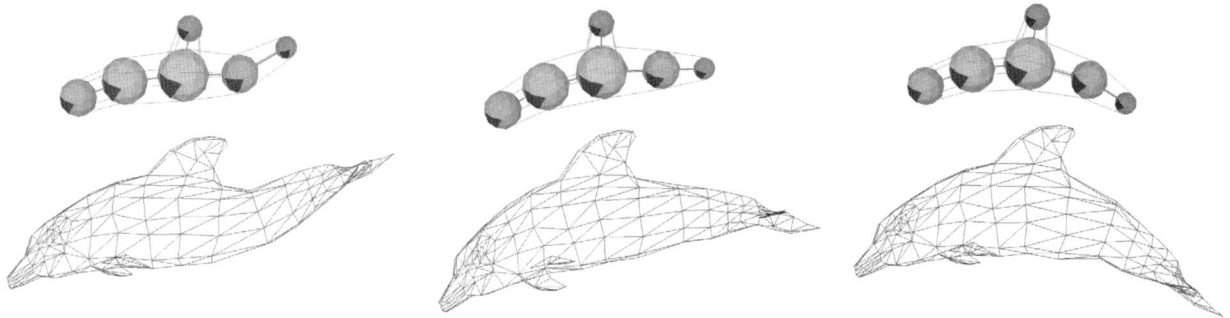

Figure 10 - Representation of a dolphin with a skeleton composed of 6 spheres and 5 links

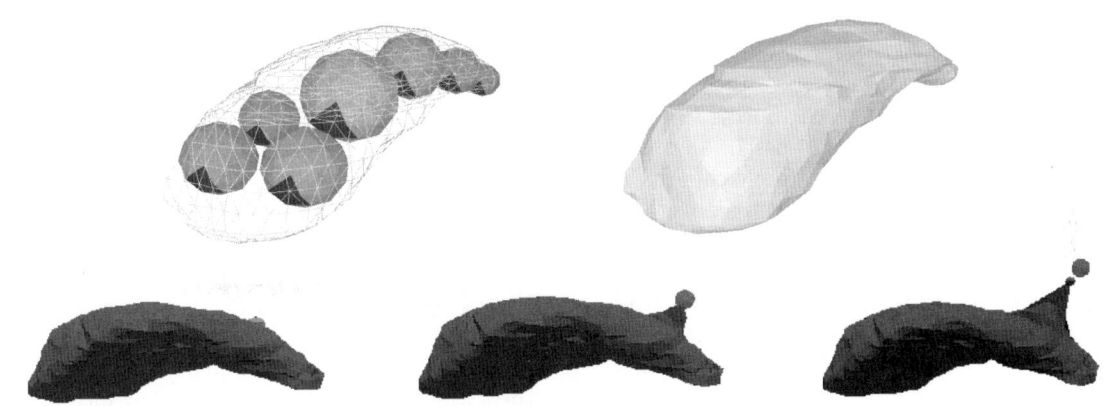

Figure 11 - Surface deformation of a virtual vesicule composed of 466 triangles. The skeleton is composed of 7 spheres and 7 links

References

[1] Francois Conti, *"Deformation of virtual objects"*, Master thesis, EPFL, Lausanne, Switzerland, 1999.

[2] H.Blum *"A transformation for extracting new descriptors of shape."*. In W. Wathen-Dunn, editor, Models for the perception of Speech and Visual Form, pages 362-380. M.I.T. Press, Cambridge, MA, 1967.

[3] C.Zilles, J.Salisbury. *"Constraint-based God-object Method for Haptic Display"*, ASME Haptic Interfaces for Virtual Environment and Teleoperator Systems. Dynamic Systems and Control, 1994, vol.1., pp 146,150.

[4] Ruspini, Diego, Krasimir Kolarov, Oussama Khatib, ``The Haptic Display of Complex Graphical Environments.'' SIGGRAPH 97 Proceedings, (August 1997), pp. 345-352.

[5] Nina Amenta, Sunghee Choi and Ravi Kolluri. *"The power Crust"*. ACM Symposium on Solid Modeling and Applications, 2001

[6] Hoppe H. *"Progressive Mesh"*, Proceeding of ACM SIGGRAPH, pp. 99-108, 1996

[7] Xuaeto Li, Tong Wing Woon, Tiow Seng Tan and Zhiyong Huang *"Decomposing Polygon Meshes for Interactive Applications"*

[8] David Bourguignon and Marie-Paul Cani. *"Controlling Anistropy in Mass-Spring Systems"*

[9] Sarah F. F. Gibson and Brin Mirtich. *"A Survey of Deformable Objects in Computer Graphics"*

[10] A. Van Gelder. *"Approximate Simulation of Elastic Me branes by Triangulated Spring Meshes."*. Journal of Graphics Tools, 3(2):21-41, 1998

[11] Kolja Kaehler, Joerg Haber and Hans-Peter Seidel. *"Dynamic Refinement of Deformable Triangle Meshes for Rendering"*

[12] Jaso J. Corso, Jatin Chhugani and Allison M Okamura. *"Interactive Haptic Rendering of Deformable Surfaces Based on the Medial Axis Transform"*. Eurohaptics, UK, July 8-10, 2002 Eurohaptics Proceedings, pp. 92-98

[13] Adachi, Y., Kumano, T., Ogino K., *"Intermediate Representation for Stiff Virtual Objects."* Proc. IEEE Virtual Reality

Passivity-Based High-Fidelity Haptic Rendering of Contact

Mohsen Mahvash and Vincent Hayward

Center For Intelligent Machines
McGill University, Montréal, Québec, H3A 2A7 Canada
mahvash|hayward@cim.mcgill.ca

Abstract

A method is described whereby the virtual haptic interaction with deformable elastic objects is created in terms of two processes: a slow process which carries out the simulation, and a fast process to render forces. Passivity theory is used to design an update strategy which reproduces exactly pre-computed responses between a tool and an object. This yields a design procedure for adjustable local models which guarantee the passivity of the interaction while preserving fidelity. Two examples of local models are given and some experimental results are reported.

1 Introduction

Creating a force reflecting virtual environment which simulates the details of actual contacts is a challenge. Moreover, the stability of an artificial haptic interaction is usually lost due to the discrete nature of the simulation, even when the update rate is in principle sufficiently high to represent the physics of a contact. We propose a method which provides stability as well as fidelity of the interaction in a multi-rate setting. A low rate process passes the contact responses to a fast control unit which provides passivity while preserving the details of the response. The computations done by the fast control unit are minimal, so this unit can be conveniently implemented using ordinary computing platforms or embedded processors.

The design of a passivity-based control strategy using model updates is first discussed in the general case. The technique is exemplified for the case of "poking" a surface and for the case of 3D interaction of a given tool with a virtual elastic object which can be inhomogeneous and/or anisotropic. The simulation also includes friction.

A useful approach to the stability of a haptic interaction is based on passivity because a system model of the operator is not required in the design. What is required instead is passivity of the artificial system which creates the virtual haptic interaction and that of the operator. Stability is achieved by providing passivity for all the components of the system. This is similar to what happens in the physical world when objects come into contact. Passivity and fidelity of the artificial system are sufficient to create realistic artificial contacts whether the interaction is stable or not.

2 Related Work

Anderson and Spong pioneered the use of passivity theory in teleoperation systems. They introduced the basic notions needed to analyze force feedback systems in terms of passivity [2]. A complete system was divided into four passive subsystems and a non-passive communication block. This made it possible to design a control such that the complete system remained passive. Niemeyer and Slotine adopted a similar framework and successfully used the notion of wave variables to provide stability as well as explicit design tradeoffs regarding fidelity [12]. This framework was also used by Yokokohji et al. to deal with fluctuating time delays [14].

Colgate and Shenkel used passivity to study the the effect of discrete time implementation of a virtual environment and derived a sufficient condition for stability [4]. Miller et al. later derived a sufficient condition for stability for a broad class of non-linear and time delayed virtual environments [11]. To provide stability, the approach is to introduce a virtual coupling to combat the effects of delay and discretization [1, 3, 11]. Recently, Hannaford and Ryu employed a different strategy involving a passivity observer coupled to a time-varying passivity controller designed to inject damping into the system [5]. In all these approaches, fidelity may be compromised to gain stabillity.

The excellent survey by Salisbury et al. describes

many important issues in haptic rendering, including the need for fidelity in some applications [13]. The authors suggest that calculating forces from a potential field can provide passivity for contact interaction with three dimensional virtual bodies, but the design of these fields was left as an open question.

The present paper adopts notions developed for teleoperation and uses them for haptics, considering that the remote environment, the slave robot, and the communication link are replaced by the computational simulation of a tool interacting with a body. Thus, the five basic subsystems of teleoperation are replaced by three in haptics as seen next.

3 Passivity in Haptic Interaction Systems

Definition: A system with input v, output f and initial energy $E(0)$, $v(t), f(t) \in \mathbb{R}^n$ is passive if [8]:

$$\int_0^t f(\tau)^\top v(\tau)\, d\tau + E(0) \geq 0 \quad (1)$$

for all functions v, f, and $t \geq 0$, see Figure 1.

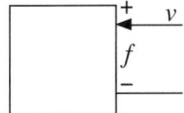

Figure 1: A block element.

A haptic interaction system is represented in block diagram format in the Figure 2.

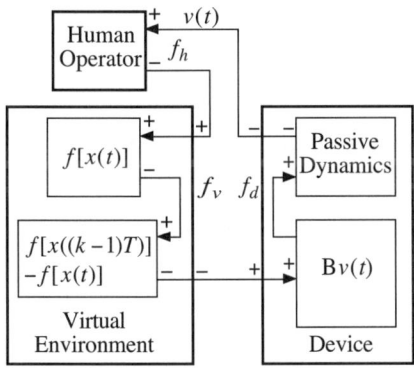

Figure 2: Haptic interaction system block diagram.

The complete system consists of three subsystems: the human operator, the virtual environment and the haptic device. No matter how well constructed, a device always has some residual damping. The device is thus represented by two sub-blocks: a damper $\mathbf{B}\,v$, where $\mathbf{B} > 0$, and a block which represents the other device dynamics which, excluding the force transducing element, are passive. The virtual environment is represented by two sub-blocks: $f[x(t)]$, the continuous time realization of a virtual environment and $f[x((k-1)T)] - f[x(t)]$, the difference between the discrete and the continuous time realizations of the same environment.

The damper sub-block and the block which represents the effect of delay and discretization can be lumped into one subsystem as in Figure 3. If the continuous realization of the virtual environment is passive, since the device dynamics are passive, the haptic interaction will be passive if the subsystem of Figure 3 is also passive.

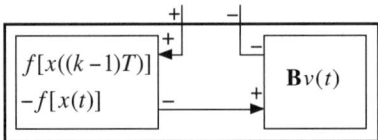

Figure 3: Subsystem of the complete system of Fig.2.

The subsystem of Figure 3 will be passive, if

$$2T\left[f(x_1) - f(x_2)\right]^\top (x_1 - x_2) < (x_1 - x_2)^\top \mathbf{B}\,(x_1 - x_2), \quad \forall x_1, x_2 \in \mathbb{R}^3. \quad (2)$$

Substituting x_1 by $x(\tau)$, x_2 by $x((k-1)T)$, and $(x_1 - x_2)$ by $[\tau - (k-1)T]v(\tau)$, for small T, Eq. (2) yields:

$$\frac{2T}{\tau - (k-1)T}\left[f(x(\tau)) - f(x((k-1)T))\right]^\top v(\tau) < v(\tau)^\top \mathbf{B}\,v(\tau). \quad (3)$$

Then, for $(k+1)T \geq \tau \geq kT$:

$$\left[f(x(\tau)) - f(x((k-1)T))\right]^\top v(\tau) < v(\tau)^\top \mathbf{B}\,v(\tau) \quad (4)$$

Integrating both sides of Eq. (4) gives:

$$\int_0^t \left[f(x(\tau)) - f(x((k-1)T))\right]^\top v(\tau)\, d\tau < \int_0^t v(\tau)^\top \mathbf{B}\,v(\tau)\, d\tau, \quad (5)$$

that is

$$\int_0^t \left[f(x((k-1)T)) - f(x(\tau)) + \mathbf{B}\,v(\tau)\right]^\top v(\tau)\, d\tau > 0. \quad (6)$$

Eq. 6 is the passivity condition for the subsystem of Fig. 3 for $E(0) = 0$. For x_1 close to x_2:

$$[f(x_1) - f(x_2)]^\top \approx \mathbf{J}_f(x_1 - x_2), \quad (7)$$

Cond. (2) can be simplified to:

$$\mathbf{B} > 2T\,\mathbf{J}_f(x) \quad (8)$$

Cond. (2) or Cond. (8) mean that the passivity of the haptic interaction, as seen from the end-effector of a device, for a passive virtual environment $f(x)$ can be achieved by a sufficiently high rate force update and without the need for virtual coupling.

4 Passive Model Update

We show that $f(x)$ can be passively re-created in terms of force updating models running at high-rate but updated at low rate by a slow process, see Fig. 4. Let i be the time slot used by the low rate process from time t_i to t_{i+1} and let $f_i(x)$ be the local model during that time slot:

$$f(x) = f_i(x), \quad t_i \le t < t_{i+1} \quad (9)$$

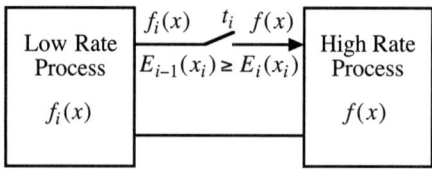

Figure 4: Multi rate haptic rendering.

The $f_i(x)$'s are conservative passive models which locally fits the contact simulation responses calculated by the low rate process.

The passivity condition for $f(x(t))$ is:

$$E(t) = \int_0^t f(x(\tau))^\top v(\tau)\, d\tau \ge 0 \quad (10)$$

$E(t)$ can be written in terms of f_i:

$$E(t) = \sum_{i=0}^{n-1}\left(\int_{t_i}^{t_{i+1}} f_i(x(\tau))^\top v(\tau)\, d\tau\right)$$
$$+ \int_{t_n}^{t} f_i(x(\tau))^\top v(\tau)\, d\tau. \quad (11)$$

For $f_i(x)$ conservative, $\nabla \times f_i = 0$,

$$\int_{t_i}^{t_{i+1}} f(x(\tau))^\top v(\tau)\, d\tau = \quad (12)$$

$$\int_{t_i}^{t_{i+1}} f(x(\tau))^\top \frac{dx(\tau)}{d\tau} d\tau = \int_C f(x)^\top dx \quad (13)$$

$$= \int_{x_i}^{x_{i+1}} f(y)^\top dy = E_i(x_{i+1}) - E_i(x_i), \quad (14)$$

where $x_i = x(t_i)$ and where $E_i(x)$ is defined by

$$E_i(x) = \int_{x_0}^{x} f_i(y)^\top dy. \quad (15)$$

Using Eq. (14), Eq (11) can be written:

$$E(t) = \sum_{i=0}^{n-1}[E_i(x_{i+1}) - E_i(x_i)] + E_n(x) - E_n(x_n)$$
$$= \sum_{i=1}^{n} E_{i-1}(x_i) - \sum_{i=0}^{n-1} E_i(x_i) - E_n(x_n) + E_n(x)$$
$$= \sum_{i=1}^{n}[E_{i-1}(x_i) - E_i(x_i)] + E_n(x). \quad (16)$$

Suppose now that f_i is adjusted at time t_i such that

$$E_{i-1}(x_i) \ge E_i(x_i), \quad (17)$$

then

$$E(t) \ge E_n(x) \ge 0, \quad (18)$$

therefore f is passive.

Several strategies can be used to insure that Cond. (17) is met. In the next Section, two examples are discussed.

5 Passive Rendering Examples

The first example shows how poking a virtual surface with a virtual tool can be rendered with high fidelity. The second shows how the theory can be applied to the case of interaction with deformable bodies. The two examples used different passive update strategies.

5.1 Local Linear Impedances And Adjusted Relaxed States

5.1.1 Strategy

The local models are selected (among other possibilities) to be linear impedances that locally fit single valued simulation responses. Such linear impedances are associated with an energy which is easily evaluated and which depends on three quantities: a state, the value of the state for which the energy is zero, i.e., the "relaxed state", and the impedance itself. The relaxed states are adjusted to preserve the energy condition Cond. (17).

For interactions along the poking dimension, linear local models are given by:

$$f_i(x) = \begin{cases} 0 & x \leq c_i, \\ k_i(x - c_i) & x > c_i, \end{cases} \quad (19)$$

where the c_i are the relaxed states and the k_i are the impedances locally fitting the output force of the low rate process. The value of k_i can be expressed in terms of the desired output force f_i at time t_i:

$$k_i = \frac{f_{i-1} - f_i}{x_{i-1} - x_i}. \quad (20)$$

$E_i(x)$ for f_i is given by

$$E_i(x) = \frac{1}{2} k_i (x - c_i)^2. \quad (21)$$

The equality of Cond. (17) is used to find an energy response equal to the desired response:

$$\frac{1}{2} k_{i-1}(x_i - c_{i-1})^2 = \frac{1}{2} k_i (x_i - c_i)^2. \quad (22)$$

Solving for c_i gives a value that guarantees a passive update:

$$c_i = (1 - \sqrt{\frac{k_{i-1}}{k_i}}) x_i + \sqrt{\frac{k_{i-1}}{k_i}} c_{i-1}. \quad (23)$$

When $k_i = 0$, i.e. moving in free space, k_i and c_i are updated by:

$$\begin{aligned} k_i = 0, \ c_i = x, & \quad \text{if } E_{i-1} = 0, \\ k_i = k_{i-1}, \ c_i = c_{i-1}, & \quad \text{if } E_{i-1} > 0. \end{aligned} \quad (24)$$

5.1.2 One Dimensional Interaction With Nonlinear Response

The strategy of last section was implemented to evaluate the gain in stability performance in comparison with other strategies.

The forces were generated by a PenCat/Pro™ haptic device (Immersion Canada Inc.). This device was powered by Lorenz flat-coil actuators driving the end-effector in a single stage via a planar linkage. These actuators were in turn powered by current amplifiers. As a result, the device had very little damping and friction, a best case for fidelity, but a worse case for inherent stability margin in the sense discussed in Section 3.

The simulation program consisted of two independent real-time threads running under RTLinux-3. One thread provided for rendering the forces and the other for computing the contact simulation. Here, the virtual environment is a nonlinear wall with k measured in N/m and x in m.

$$f(x) = \begin{cases} 0 & 0 \leq x \\ 100\, k\, x^2 & 0 < x \leq 0.005 \\ k(x - 0.000025) & 0.005 < x \end{cases} \quad (25)$$

The response parameters are chosen such that k alone defines an appropriate stability margin as defined by Cond. 8. To see that, consider that inside the quadratic portion of the response when $x \leq 0.005$, $\mathbf{J}_f$ reduces to $200\, k\, x$ and therefore is bounded by k. Thus, the same margin applies to the two portions of the response.

In order to create repeatable experimental conditions, the handle of the device was loaded by a rubber band stretched to create an equilibrium around 1 N. Then, k was increase until small disturbances determined the onset of limit-cycles. This was tested for three different methods:

1. A single process at various rates.
2. Two processes. The first one computed a response as in Eq. (25) at various rates as above, and a second rendered forces at a rate of 2,000 Hz using linear interpolation from one update to the next.
3. Two processes as described in the last section. The first one was the same as above, and the second one also ran at 2,000 Hz but implemented the passive updates strategy.

The results listed in Table 1 show the largest impedance values which could be rendered in a stable fashion for all conditions.

The results indicate that the linear interpolation method performs better than the single update method. The passive update method, however, gave a constant stability margin even when the update rate decreased from 2,000 to 100 Hz. The results also show that the passivity based approach strategy could realize larger impedances than the other methods.

Rate (Hz)	Single thread k (N/m)	Linear interpolation k (N/m)	Passive update k (N/m)
2000	170	170	212
1000	120	160	212
500	93	127	212
250	63	110	212
200	57	104	212
100	38	85	212

Table 1: Achievable impedances for three strategies.

5.2 Deformable Body With Nonlinear Response

The same strategy can be extended to the three dimensional haptic rendering of deformable bodies. It could be applied in settings which use FEM, BEM, or LEM methods for the simulation of deformation, provided that a low rate process provides local impedances to a high rate process (*e.g.* [7]). The development of a passive update technique for these cases is possible but is outside the scope of this paper. On the other hand, methods based on pre-calculated responses, as described in [9, 10], are simple enough to be treated in this section.

5.2.1 Pre-Calculated Response of a Deformable Body

A deformable body at rest is represented by a freeform surface meshed into triangular elements. The interaction of a tool with the body is encoded as a finite set of normal force-deflection responses stored for each node (penetration in a direction normal to the undeformed surface with a given tool). Each deflection maps to a force that has a magnitude and a direction. An assumption that we make is that the force direction is assumed to remain invariant with deflection (if needed, a dependency could be included), therefore, each of these responses requires the specification of a vector and of a single-valued function $f_i(\delta)$. This is depicted by Figure 5a. During interaction, a point of contact x_c is known on the flat surface of the patch. A vector u_n, termed the *response normal vector*, is defined by:

$$u_n = \sum_{j=1,2,3} n_i^j(x_c)\, u_i^j, \qquad (26)$$

where the u_i^j are the unit deflection normal vectors of the responses at node j of the element i, and where n_i^j is an interpolation function for element i for each node j which depends on x_c. A common choice in continuous mechanics that we use here is based on natural coordinates:

$$n_i^j(x) = \frac{A_j(x)}{A^i} \qquad (27)$$

where A^i is the area of the element i and A_j is the area of the triangle formed by the contact point and two nodes as defined in Figure 5b. A key property of the interpolation approach is to ensure continuity of the normal response vector over the surface of the body. This vector is almost always different from the geometrical normal vector defined by the patch.

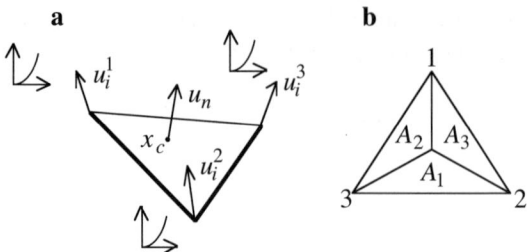

Figure 5: For each triangular element i, the vertices are associated to a normal $u_i^j, j = 1, 2, 3$ and to a response along this normal (**a**). A point inside a triangle defines three areas labeled as indicated (**b**).

5.2.2 Slow Simulating Process And Fast Rendering Process

The role of the low rate simulation process is to supply the vertices locations, the deflection normals at these vertices, and the force deflection curves of one element at a time. The active element, i.e. the element in contact with the tool, is determined using an interference detection method similar to the god object approach in haptics [15]. Since the elements are large relatively to a typical tool speed, this can be done at low rate; 100 Hz is sufficient (moving at 1 m/s over 10 mm patches).

The interaction force, however, is evaluated at high rate. The normal component is interpolated from three pre-calculated force-deflection responses, each representing the response of an *actual* contact (not an approximation) of a given body interacting with a given tool:

$$f_{in}(x) = \sum_{j=1,2,3} n_i^j(x_c)\, f_i^j(\delta_n(x_c)) \qquad (28)$$

An element switch is allowed when the contact point x_c crosses the edges of the active element. If available, a new element is accepted and the point x_c moved to it. This is accomplished with dual-threaded code communicating by a first-in-first-out (FIFO) queue. Two cases can arise: if x_c belongs to a neighboring patch found in the FIFO, a new element is accepted, otherwise the rendering process carries on with same patch. This gives a correct response at a wrong place, but there is no loss of continuity nor of passivity even if there is loss of synchronization between the movement of the tool and the simulation process.

We now add friction to the local model. Inside element i:

$$f_i(x) = f_{in}(x)u_n + f_{it}(x)u_t \qquad (29)$$

The contact point x_c is a function of the measured virtual tool position x set by the user. We first look at the case when there is no sliding, that is, when x_c is invariant. Referring to Figure 6, the deflection δ at the point of contact is given by:

$$\delta = x - x_c = \delta_n u_n + \delta_t u_t, \quad (30)$$

where δ_n and δ_t are the components of δ in a local frame. The penetration is defined by δ_n. The component δ_t may be viewed as presliding displacement in friction modeling [6].[1] When there is no sliding, $\delta_t < \delta_L$, where δ_L is the displacement under which there is no sliding.

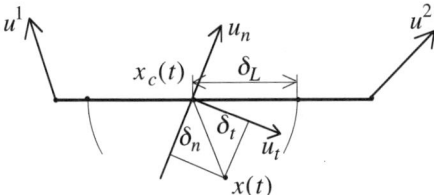

Figure 6: For clarity, only the two-dimensional case is illustrated. The extension to three dimensions is straightforward. u_n is defined in terms of u^1 and u^2. This defines a frame with origin at the point of contact x_c. When δ_t reaches δ_L, sliding occurs.

The pre-displacement δ_t is simply mapped onto the patch surface to determine the movements of x_c such that δ_t never exceeds δ_L [6]. The friction force at the sliding limit is $f_{it}(x) = \mu f_{in}(x)$ where μ is the limiting friction coefficient. Coulomb's law is invoked to assume that the friction force is independent from the area of contact.

5.2.3 Passivity

When poking without sliding, point x_c is created by projecting the x onto the surface of the active element, but only when x is outside the previous active element. From the time of creation of point x_C until the next creation, δ is replaced by $\delta - \delta_0$ where δ_0 is the initial value of δ. This adjustment of the relaxed state ensures that Cond (17) is satisfied.

During sliding, x_c moves within the element surface. If the energy lost due to sliding friction is greater than the change of $E_i(x)$ due to the sliding movement of contact point, the energy $E(t)$ of the interaction must be greater than $E_i(x)$, so passivity is preserved. During element switch, since the change in force is continuous, even with updates at low rate, Cond (17) is

---

[1] In [9], it is verified that δ_t is largely independent from δ_n for rubbery materials. It is probably also the case for many biomaterials.

satisfied which eliminates the need for further adjustment of the local model.

5.2.4 Implementation

The model was tested with a deformable cylindrical virtual body. The normal responses were:

$$f(\delta_n) = \begin{cases} 0 & 0 \leq \delta_n \\ 100\ k\delta_n^2 & 0 < \delta_n \leq 0.005 \\ k(\delta_n - 0.000025) & 0.005 < \delta_n \end{cases} \quad (31)$$

In this model, the system returns the nonlinear response at high rate and performs simulation at low rate. Figure 7 shows the user interface used in the test. The low rate process was running at only 100 Hz, yet the simulation was stable for $k_{max} = 170$ N/m for the fast process running at 2,000 Hz, as found in the experiment described in the previous section.

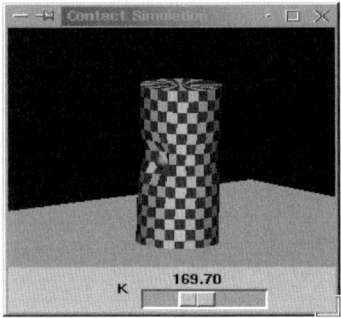

Figure 7: Tool contact simulation.

5.3 Conclusion

It was shown that low-rate yet stable haptic simulations involving tools interacting with deformable bodies could be achieved without compromising fidelity. In particular, various tool shapes and anisotropic and/or in-homogenous body can be haptically represented using a pre-calculated force-deflection response approach. This was achieved using a dual-threaded system having a slow thread executing the simulation itself and a high rate thread locally rendering the interaction. Passivity theory was used to constructively design two different local update strategies that guaranteed passivity under a wide range of circumstances.

Work is under way to extend the same framework to multiple contacts, and to tool movements with six degrees of freedom. It should also be extended to time varying dissipative virtual environments involving viscosity, plasticity, cutting and other forms of damage insofar as the application requires the simulation of these phenomena.

Acknowledgements

This research was funded by the project "Reality-based Modeling and Simulation of Physical Systems in Virtual Environments" of IRIS, the Institute for Robotics and Intelligent Systems (Canada's Network of Centers of Excellence). Additional funding is provided by NSERC, the Natural Sciences and Engineering Council of Canada, in the form of an operating grant for the second author.

References

[1] Adams, R. J., and Hannaford, B. 1999. Stable Haptic Interaction With Virtual Environments. *IEEE T. Robot. Automat.* 15:465–474.

[2] Anderson, R. J. and Spong., M. 1989. Bilateral Control of Teleoperators with Time Delay. *IEEE T. on Aut. Control.* 34(5):494–501.

[3] Brown, J. M, and Colgate, J. E. 1997. Passive Implementation of Multi-body Simulations for Haptic Display. Proc. *International Mechanical Engineering Congress and Exhibition,* Vol. 61, pp.85–92.

[4] Colgate, J. E., and Schenkel, G.G. 1997. Passivity of a Class of Sampled-Data Systems: Application to Haptic Interfaces. *J. of Robotic Systems.* 14(1):37-47.

[5] Hannaford, B., Ryu, J.-H. 2002. Stable Haptic Interaction with Virtual Environments. *IEEE T. Robot. and Automat.,* 18(1).

[6] Hayward, V., Armstrong, B., 2000. A New Computational Model of Friction Applied to Haptic Rendering. In *Experimental Robotics VI*, P. I. Corke and J. Trevelyan (Eds.), Lecture Notes in Control and Information Sciences, Vol. 250, Springer-Verlag, pp. 403-412.

[7] James, D. L., Pai D. K. 2001. A Unified Treatment of Elastostatic and Rigid Contact Simulation for Real Time Haptics. *Haptics-e*, 2(1).

[8] Lozano, R., Brogliato, B., Egeland, O., Mashke, B. 2000. *Dissipative Systems, Analysis and Control, Theory and Applications.* Springer Verlag: New York.

[9] Mahvash, M., Hayward, V., Lloyd, J. E. 2002. Haptic Rendering of Tool Contact. Proc. *Eurohaptics 2002.* pp. 110–115.

[10] Mahvash, M., Hayward, V. 2001. Haptic Rendering of Cutting, A Fracture Mechanics Approach. *Haptics-e.* 2(3).

[11] Miller, B., Colgate, J. E., Freeman, R. A. 1999. Guaranteed Stability of Haptic Systems with Nonlinear Virtual Environments. *IEEE T. on Robot. Automat.*

[12] Niemeyer, G., Slotine, J. J. 1991. Stable Adaptive Teleoperation. *IEEE J. Ocean. Eng.*, 16:152–162.

[13] Salisbury, K., Brock, D., Massie T., Swarup, N., Zilles, C. 1995. Haptic Rendering: Programming Touch Interaction with Virtual Objects. Proc. *Symposium on Interactive 3D Graphics*, ACM. pp. 123–130.

[14] Y. Yokokohji, Y., Imaida, T., Yoshikawa, T. 2000. Bilateral Control with Energy Balance Monitoring Under Time-varying Communication Delay, Proc. *IEEE Int. Conf. Robot. Automat.* pp. 2684–2689.

[15] Zilles, C. B., Salisbury, J. K. 1995. A Constraint-based God Object Method for Haptic Display. Proc. *IEEE Int. Conf. Intel. Rob. and Syst.,* Vol. 3, pp. 146–151.

On the Calibration of Deformation Model of Rheology Object by A Modified Randomized Algorithm

Hiroshi Noborio, Ryo Enoki, Shohei Nishimoto and Takumi Tanemura

Department of Engineering Informatics
Osaka Electro-Communication University
Hatsu-Cho 18-8, Neyagawa, Osaka 572-8530, Japan
nobori@noblab.osakac.ac.jp

Abstract

There are many kinds of rheology objects in our living life. If such rheology objects are individually and completely modeled, we can flexibly deal with such various objects in a factory by a robotic manipulator controlling six degrees-of-freedom forces and moments and also operate them at a 3-D graphics world in a house by a human via a wonderful haptic devise feeling the forces and moments. If such a system is developed after acquiring a rheology dynamic model, we can enjoy a clay work in a 3-D virtual environment.

For this purpose, we calibrate deformation model of a rheology object by a modified randomized algorithm based on experimental data. The data is measured from a bread material pushed by a robotic manipulator exactly. The model is a 3-D voxel/lattice structure with many elements, and each element consists of two dampers and one spring. By changing three coefficients of dampers and spring, we can describe various material properties concerning to viscosity and elasticity. In this paper, by minimizing shape difference between real and virtual rheology objects in the near-optimal algorithm, we find a better set of coefficients. Using the calibrated model, we can feel three degrees-of-freedom forces and three degrees-of-freedom moments attracted from a calibrated virtual object in a haptic devise and synchronously we can watch shape deformation of the object in a 3-D graphics animation.

1 Introduction

Dynamic animation is indispensable in robotics and virtual reality, which is aggressively used in teleoperation, humanoid, assembly, task planning, game, amusement and so on. The animation is quickly made in PC (e.g., Pentium 4. 2.26GHz) with a graphics accelerator (e.g., NVIDIA Quadro 2EX, 32MB for OpenGL). They are powerful and cheap for making a sequence of 30 full color images. Even though their motions are extremely complicated, the sequence dynamically describes object movements per the second. From this background, we can easily develop a software to make a graphics animation whose quality of image is high concerning to rendering such as lighting, shading, texture mapping and so on. As an example of this, we use a 3-D graphics software OpenGL in a personal computer (CPU: Pentium4 2.26GHz, Main memory: 1024MB) with a 3-D graphics acceleration board (NVIDIA Quadro 2EX, 32MB) in order to illustrate such a wonderful 3-D graphics animation quickly.

However, it is unfortunately difficult for us to generate exact trajectories of moving objects. The reason is that every object is always affected by complex physical properties in our living space. Some properties are known, but the others are approximated or unknown. For this reason, modeling and calibrating many kinds of physical properties are important in the dynamic animation. In general, the animation includes two behaviors: contact and non-contact behaviors. If two rigid bodies collide with each other, we can calculate a sequence of contact behaviors based on Coulomb and Hertz models. Concerning to this, researchers have proposed many kinds of models for making friction force or impulse and contact force or impulse [1],[2],[3],[4],[5],[6]. If many rigid bodies are connected each other, we can calculate a sequence of non-contact behaviors based on the Newton-Euler equations. To calibrate its dynamic parameters, we use two kinds of approaches using a few special motions and a huge number of general motions, respectively [7],[8],[9].

As contrasted with these works, modeling and calibrating a rheology object is quite backward. This work started before the decade [10],[11],[12], but has not been well developed. Several kinds of elastic objects have been modeled and animated in computer animation or virtual reality. However, some forget many kinds of viscosities, and the others could not be calibrated experimentally [13],[14],[15],[16]. As an exception, a 2-D pixel/lattice structure was calibrated under few experimental results, whose component consists of Voigt model (parallel damper and spring) and Maxwell model (serial damper and spring) [17], and then its similar 3-D voxel/lattice structure was calibrated under few experimental results, whose component consists of Voigt model (parallel damper and spring) and one adaptive damper [18]. The models are unfortunately calibrated under few data, and especially coefficients of dampers and springs are calibrated individually. In general, they are strongly depending on each other so as to produce many kinds of material properties. By changing a set of four or three coefficients flexibly, viscosity and elasticity of a rheology object totally increases or decreases. They synchronously appear by changing a set of four or three coefficients. For this reason, we adopt the similar 3-D voxel/lattice structure whose element consists of two dampers and one non-adaptive spring. Then, we calibrate three coefficients synchronously in our efficient randomized algorithm under many kinds of experimental data.

For this purpose, we firstly measure a sequence of deformed shapes several times while pushing a rheology object by a robotic manipulator and synchronously observing it by two stereo vision systems *Digiclops*. Secondly, we minimize a sequence of differences of deformed shapes between virtual and real rheology objects in the calibration. Finally, we watch deformation of the rheology object in a 3-D graphics animation by *OpenGL* and feel its three reactive forces and three reactive moments by six-degrees-of-freedom robot arm *Joyarm*. The system is useful in several practical areas. For example, modeling and calibrating an arbitrary real rheology object can be used in robot assembly [19],[20], the virtual system is frequently used at surgery simulations in medical engineering [21],[21],[22]. Also, they are the basic technique in many application areas [23],[24],[25],[26].

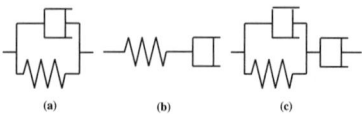

Figure 1 : Visco-elastic elements: (a) Voigt model. (b) Maxcell model. (c) An element with two dampers and one spring

In this paper, section 2 describes Voigt, Maxwell models, and then our element. In addition, we build a 3-D voxel/lattice structure for representing shape deformations and force propagations. Section 3 explains how to calibrate three coefficients of two dampers and one spring in our 3-D structure. The set of coefficients is used in dynamic equation represented as quadratic differential equation. This can be approximately calculated by the Runge-Kutta method. In section 4, we describe two experimental results for a same visco-elastic object by two kinds of pushing. First of all, we explain how to evaluate shape difference between real and virtual visco-elastic object. Secondly, by an efficient randomized algorithm using a sequence of shape differences, we calibrate three coefficients in order to construct a virtual rheology object accurately. Finally, we will give a few conclusions and future problems in section 5.

2 Model of Visco-Elastic Object

In this section, we will briefly explain a classic model, which products many material properties of some visco-elastic object. As shown in Fig.1(a),(b),(c), famous Voigt and Maxwell models and our component have been used for representing a visco-elastic object. Voigt model consists of spring and damper, which connects neighbor mass points in parallel. On the other hand, Maxwell model is a sequence of spring and damper between neighbor mass points. For example, the mixture is adopted for expressing a rheology object [17]. In this paper, we use an element which consists of Voigt model and one damper.

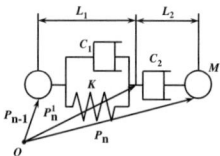

Figure 2 : Our element consists of Voigt model and a damper serially.

2.1 Our mass-spring-damper element

In Fig.2, we formulate some behaviors of our element. First of all, let O be the origin of coordinate system. Let P_{n-1} and P_n be coordinates of element endpoints. Spring and damper coefficients in the Voigt part are denoted as K and C_1, respectively. The other damper coefficient in the element is denoted as C_2. The natural length of Voigt part is given by L_1. Let M be mass at each endpoint. Let P_m be position of connecting point between Voigt and damper parts. Furthermore, $d_n = P_n - P_{n-1}$ is defined. Since these positions P_{n-1}, P_m and P_n exist on a straight line, P_m can be defined by a parameter k as follows: $P_m = kd_n + P_{n-1}$. Here, time varying direction vector is defined as $e_n = d_n/|d_n|$, and also time varying length coefficient is defined as $Z_n = k|d_n|$.

Let F_e be a force applied to a mass point P_n by the classic model. The force F_e equals to a force acting in the Voigt part. Thus, we have the following equation.

$$F_e = -C_1 \dot{Z}_n e_n - K(Z_n - L_1)e_n \qquad (1)$$

Also, the force F_e coincides the force acting to a damper part. In consequence, we obtain the following equation.

$$F_e = -C_2(\frac{d}{dt}(|d_n| - Z_n))e_n \qquad (2)$$

Here, a force applied to a mass point P_n is defined as F_a. Consequently, the dynamic equation of the mass point P_n is denoted as

$$M\ddot{P}_n = F_e + F_a \qquad (3)$$

From three equations (1), (2) and (3), we calculate the dynamic equation of three element model. First of all, by eliminating F_e in the equations (1) and (2), we directly obtain the parameter k. By substituting k into each of equations (1) and (2), we can obtain the value of vector F_e and consequently generate a motion of mass point P_n. The purpose of this research is to select a better set of three coefficients for at least two pushing a given visco-elastic object in a 3-D environment by two kinds of randomized algorithms.

In this paper, while changing values of three coefficients randomly in reasonable ranges by an efficient randomized algorithm, we are seeking for the best set. When the pushing is strong or weak, the best set of three coefficients a little bit differs from each other. However, as shown in our experimental results, the difference is not so large. For this reason, we use one best set for all pushing. Moreover, the classic model generates a wonderful sequence of object shapes, which is totally affected by values of three coefficients.

2.2 3-D voxel/lattice structure

A visco-elastic object deforms in a 3-D environment. In order to describe several kinds of deformations flexibly, we adopt a symmetric 3-D voxel/lattice structure. For this purpose, let us distribute mass points in a natural shape of a visco-elastic object at the same intervals along x, y, and z axes (Fig.3(a)). Let N be the number of mass points and M_{object} be total mass of the object. Therefore, each mass point is given by $M = M_{object}/N$.

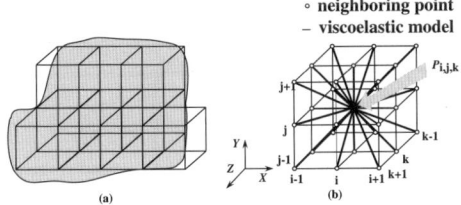

Figure 3 : (a) A 3-D voxel/lattice structure of a visco-elastic object (b) Neighboring lattice mass points.

Our elements are inserted between all neighboring mass points as illustrated in Fig.3(b). Namely, the elements are arranged, whose distances are 1, $\sqrt{2}$, and $\sqrt{3}$ (The unit is the distance between horizontal and vertical neighbor mass points). Visco-elastic deformation of an object can be represented by deformation of all the elements. Let $P_{i,j,k}$ be position vector corresponding to mass point (i,j,k). Let us derive motion equation of a mass point at $P_{i,j,k}$. Force acting on $P_{i,j,k}$ by the element between $P_{i,j,k}$ and its neighbor point $P_{i+\alpha,j+\beta,k+\gamma}$ is denoted by $F_{i,j,k}^{\alpha,\beta,\gamma}$. Then, total internal force acting on $P_{i,j,k}$ is given by the sum of $F_{i,j,k}^{\alpha,\beta,\gamma}$, that is,

$$F_{i,j,k}^e = \sum_{\substack{\alpha,\beta,\gamma \in \{-1,0,1\} \\ (\alpha,\beta,\gamma) \neq (0,0,0)}} F_{i,j,k}^{\alpha,\beta,\gamma} \qquad (4)$$

The force $F_{i,j,k}^{\alpha,\beta,\gamma}$ can be computed using a procedure explained in paragraph 2.1. Thus, force $F_{i,j,k}^e$ can be computed by summing all forces. Let $F_{i,j,k}^\alpha$ be a total external force acting on $P_{i,j,k}$. Thus, the equation of motion is described as follows:

$$M\ddot{P}_{i,j,k} = F_{i,j,k}^e + F_{i,j,k}^\alpha \qquad (5)$$

By solving a set of equations corresponding to all mass points consisting of the model, we can compute deformation of a visco-elastic object. By calculating successively forces among neighbor mass points, we can obtain forces on all mass points of the above 3-D voxel/lattice structure. Each force between neighbor mass points, position and velocity of each mass point are calculated by the quadratic differential equation. This can be done by the Runge-Kutta method. By these techniques, we can simulate deformation of a pushed visco-elastic object virtually in a 3-D graphics environment.

A 3-D voxel/lattice structure is as follows: The voxel structure consists of $6 \times 4 \times 6$ mass points (Fig.4). All elements are inserted between all neighboring mass points. Therefore, there are $5 \times 3 \times 5$ elements whose distances are 1, and there are $6 \times 4 \times 6 \times 2$ elements whose distances are $\sqrt{2}$. Moreover, we add the lattice structure into the voxel structure. Therefore, there are $5 \times 3 \times 5 \times 4$ elements whose distances are $\sqrt{3}$.

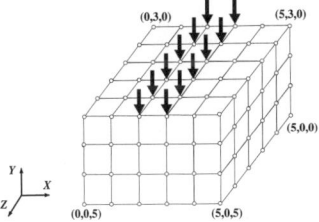

Figure 4 : 12 upper mass points of a rheology object are synchronously pushed by a rectangular object located on the tip of a robotic manipulator.

Let us compute deformation of the model when 12 upper mass points of a rheology object are synchronously pushed by a rectangular object located on the tip of a robotic manipulator (Fig.4). The sum of all mass points is really measured by $M = 6.0$ under $M_{object} = 840$ [g] and $N = 144$ [points].

Also, three coefficients $K[gf/cm^3]$, $C_1[gfs/cm^3]$ and $C_2[gfs/cm^3]$ are experimentally initialized as 400, 2000 and 2000, respectively, in two types of pushing. Finally, we should note that the bottom of the object is fixed to the space. This means 36 bottom mass points of a rheology object are received by the same repulsive forces of calculated attractive forces from its floor.

3 Two Kinds of Randomized Algorithms for Calibrating K, C1 and C2

In the last section, we construct a basic model for representing relations of forces, velocities and positions of many mass points in a rheology object. The model always needs two coefficients C_1 and C_2 of different damper parts and one coefficient K of a spring part. Changing the set of coefficients means changing material of the rheology object. Unfortunately, a rheology object has many aspects depending on absolute magnitude and time difference of a given outer force. This is an interesting property of the rheology object, which differs from rigid, plastic and elastic objects.

In this section, we calibrate K, C_1 and C_2 by minimizing shape differences between real and virtual rheology objects in two kinds of randomized algorithms. Finally, we should note that we did not use force differences acted from real and virtual rheology objects because force information is quite noisy (its magnitude and orientation include 10 and more percent errors).

3.1 How to calculate shape difference between real and virtual rheology objects

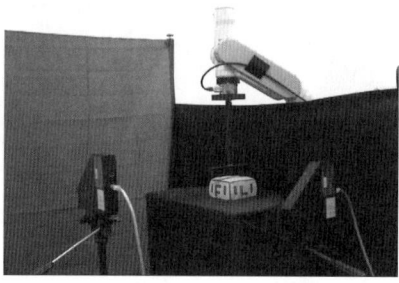

Figure 5 : An experiment system: A real rheology object is pushed by a rectangular object located at the tip of a robotic manipulator, and side deformations are simultaneously measured by two *Digicrops* cameras.

In this research, a rheology object is made by mixing wheat flour and water (Fig.6(a)). The scale of the object is denoted as 10cm × 6cm × 10 cm. The rheology object is pushed by a rectangular object located at the tip of a robotic manipulator (Fig.5). The deformation, that is, the sequence of shapes of

Figure 6 : (a) A real rheology object is built by mixing wheat flour, food red and water. (b) Its virtual rheology object is shown by a point-based computer graphics under about three or more thousand surface points measured by two *Digicrops* cameras.

the real rheology object is measured by the support of two stereo vision camera systems *Digiclops* and its SDK (Software Development Kit) *Triclops* (provided by Point Grey Research Inc, Canada). Each provides real-time 3-D digital image for capturing shapes of the object. A set of about three or more thousand points is captured as shape of the object (Fig.6(b)). Each image has 240 × 320 pixels with 24bit full color. The set is obtained three times per one second. An average error for capturing our rheology object is about 0.05cm if the distance from *Digiclops* to a rheology object is about 60cm. This error decreases experimentally by changing surface texture of the object and location of the camera and lighting without highlight and shadow. Especially, surface texture is artificially made by mixing food red into our rheology object. Then because of the latter reason, two sides of the rheology object are focused and measured by two stereo vision camera systems *Digiclops*.

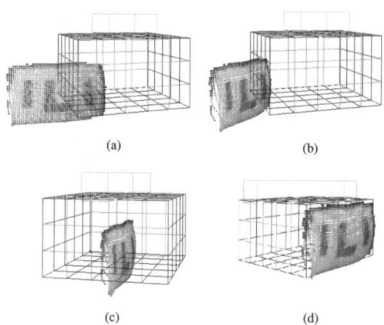

Figure 7 : The real and virtual coordinate systems are coincident with each other by matching their vertices and color landmarks.

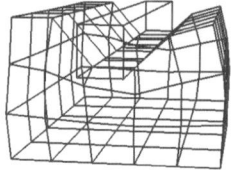

Figure 8 : A simulation system in PC: A virtual rheology object is pushed by a rectangular object, and deformations of five sides are calculated by a quadratic differential equation in our 3-D voxel/lattice structure.

Then, we calculate the sum of minimum distances from a captured point to the nearest surface of a virtual rheology object as follows: First of all, a virtual object (its coordinate system) is coincident with a real object (its coordinate system) by using their corners and landmarks as illustrated in Fig.7. A virtual object consists of $5 \times 3 \times 5$ hexahedrons individually deformed from cubes (Fig.8). Each hexahedron has six patches which are classified into real and virtual patches. A real patch is always outside a virtual rheology object, and a virtual patch is always inside it. We firstly determine whether a captured point is inside each hexahedron or not. On the assumption that each hexahedron is convex shape, if a captured point is always located in the opposite side of the normal vector of each patch of a hexahedron, the point is inside the hexahedron. In this case, the nearest point is always on one of real patches (is never on edges or vertexes) (Fig.9(a)). Therefore, we only calculate the minimum of shortest distances against real patches. On the other hand, if the point is outside all hexahedrons, we should calculate the minimum of $5 \times 3 \times 5$ shortest distances for all hexahedrons by Lin-Canny closest features algorithm [27] (Fig.9(b)).

Then, after neglecting minimum distances smaller than the average error $0.05cm$ of $Digiclops$, we summarize the other minimum distances from about three or more thousand captured points to their nearest surface of the rheology object. In our calibration, we use the sum S for four sets of captured points during and after each pushing. By minimizing the total S in an efficient randomized algorithm, we can obtain a better set of three coefficients K, C_1 and C_2.

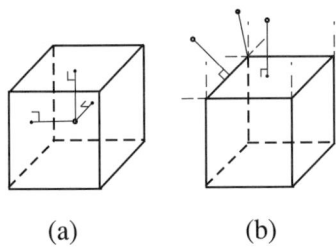

(a)   (b)

Figure 9: (a) If a captured point is inside a hexahedron, we calculate the minimum distance from the point to six patches around the hexahedron. (b) Otherwise, we calculate the minimum distance from a captured point to all hexahedrons.

In this section, we optimize K, C_1 and C_2 by using a steepest descendent method and two kinds of randomized algorithms. The global (former) randomized algorithm gets local minima from initial points uniformly selected in a given 3-D search space. On the other hand, the local (latter) randomized algorithm finally picks up a better local minimum around the best of all local minima selected in the former [28].

3.2 A Steepest Descendent Method

1. Two parameters T_{cal} and T_{ran} are given in advance.

2. Initialize coefficients in a 3-D search space.

3. We calculate shape difference S between real and virtual rheology objects.

4. In order to find all the possible neighbors, we decrease and increase K, C_1 and C_2 by Δ. In this 3-D case, we obtain eight possibilities, that is, ($K+\Delta$, $C_1+\Delta$, $C_2+\Delta$), ($K+\Delta$, $C_1+\Delta$, $C_2-\Delta$), ($K+\Delta$, $C_1-\Delta$, $C_2+\Delta$), ($K+\Delta$, $C_1-\Delta$, $C_2-\Delta$), ($K-\Delta$, $C_1+\Delta$, $C_2+\Delta$), ($K-\Delta$, $C_1+\Delta$, $C_2-\Delta$), ($K-\Delta$, $C_1-\Delta$, $C_2+\Delta$) and ($K-\Delta$, $C_1-\Delta$, $C_2-\Delta$). Then, using S for each neighbor, we select the minimum of sums at all the possible neighbors.

5. If the minimum is smaller than S obtained in step 3, we move to the neighbor with the minimum by decreasing or increasing K, C_1 and C_2 by Δ, and return to step 3. Otherwise, the algorithm finishes.

3.3 A Local Randomized Algorithm

1. By the steepest descendent method described above, we get one of the local minima. Then, if calculation time equals to or is smaller than T_{cal}, the algorithm ends with the smallest S, otherwise, move to step 2.

2. We randomly increase and decrease three coefficients K, C_1 and C_2 T_{ran} times by Δ. Then, return to step 1.

3.4 A Global Randomized Algorithm

1. Up to a time threshold T_{cal}, we randomly select initial points within a space whose ranges are $K^{min} \leq K \leq K^{max}$, $C_1^{min} \leq C_1 \leq C_1^{max}$ and $C_2^{min} \leq C_2 \leq C_2^{max}$ in 3-D search space. The density of the initial points is always uniform.

2. For all initial points, we get local or global minima by the steepest descendent method described above. Then, the algorithm ends with the smallest S of all the minima.

3.5 An Efficient Algorithm

1. By a global randomized algorithm, we globally find a better local minimum in a given 3-D search space.

2. By a local randomized algorithm from an initial point selected by step 1, we locally find a better local minimum near the initial point [28].

4 Comparative Results

In this section, we calibrate three coefficients of a virtual rheology object during deformations by two types of pushing a real rheology object, which are illustrated in Fig.10. The difference is only the velocity of pushing (direction and orientation are the same). In this research, our rheology object consists of wheat flour and water. During and after each pushing, deformation of the virtual object is visualized by a 3-D graphics software OpenGL in a personal computer (CPU: Pentium4 2.26GHz, Main memory: 1024MB) with a 3-D graphics acceleration board (NVIDIA Quadro 2EX, 32MB). Also in our virtual reality system, 3-D repulsive forces and 3-D repulsive moments of a rectangular object from a pushed virtual object can be felt by a *Joyarm* (Fig.11).

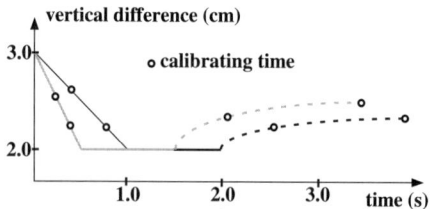

Figure 10 : Weak and strong pushing a rheology object. Weak (first) pushing is described by whole and dotted black line, and strong (second) pushing is shown by whole and dotted gray line.

Figure 11 : A human operator watches deformation of a rheology object by a 3-D graphics software *OpenGL* and feels its three forces and three moments by a six-degrees-of-freedom robotic arm *Joyarm*.

4.1 General Properties

In this paragraph, we describe a global aspect of a rheology object by changing three coefficients. As shown in Fig.12(a), as long as each of K, C_1 and C_2 increases, shape difference S decreases. However, if $C_1 + C_2$ increases extremely, a virtual rheology object cannot converge to a reasonable shape whose neighbor positions of masses are frequently reversed (Fig.12(b)). On the observation, we select search space $[K^{min}, K^{max}], [C_1^{min}, C_1^{max}], [C_2^{min}, C_2^{max}]$ limited by $[100,2000], [1000,8000], [5000,20000]$, respectively.

4.2 The First Pushing

In our efficient 3-D randomized algorithm, we set $\Delta = 1$, $T_{cal} = 30$ $[hour]$, $T_{ran} = 1000[number]$. After the calibration, we finally obtain coefficients $K[gf/cm^3]$, $C_1[gfs/cm^3]$ and $C_2[gfs/cm^3]$ as 1831, 4732 and 18524.

In the figures 13 and 14, we describe shape transformation of initial and calibrated virtual objects against the real object, respectively. A set of dark gray areas means shape differences are larger than 0.25cm, and another set of light gray areas means

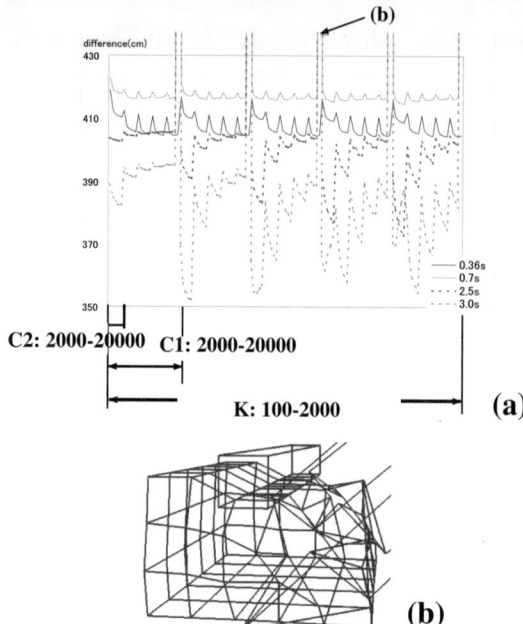

Figure 12 : (a) Shape difference S changes when K, C_1 and C_2 synchronously change within reasonable available ranges. (b) If $C_1 + C_2$ increases extremely, our rheology model does not converge to an unique shape.

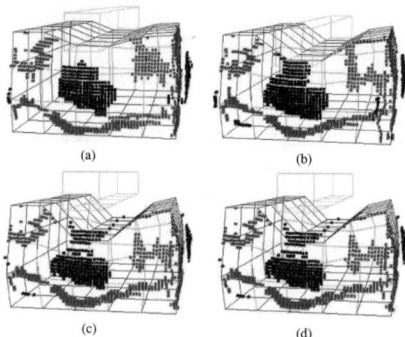

Figure 13 : (a),(b),(c),(d) Four shape differences between real and virtual objects by the first pushing, whose coefficients are initially given as reasonable values $K = 400$ and $C_1 = C_2 = 2000$. A set of dark gray areas means shape differences are larger than 0.25cm, and another set of light gray areas means the differences are less than 0.05cm.

the differences are less than 0.05cm. Therefore, we can see that differences between a calibrated object and its real object are smaller than differences between a non-calibrated object and the real object.

Also in the figure 15, we describe the distribution of differences of minimum distances from about three or more thousand captured points to initial and calibrated virtual objects. In the figure 15(a),(b),(c),(d), numbers of captured points are 34, 14, 90, 101, whose differences are less than 0.1cm, on the other hand, numbers of captured points are -47, -22, -33, 4, whose differences are more than 0.25cm, respectively. From these results, we can see that almost all numbers increase in relatively small differences, on the other hand, almost all numbers decrease in relatively large differences. This means a calibrated object is better than its non-calibrated object against the real object. As a result, the calibration leads a virtual rheology object that has high viscosities and elasticity. Moreover, by the comparison between calibrated and initial virtual rheology objects with the real rheology object, a calibrated rheology object looks like the real rheology object, and therefore our calibration is meaningful.

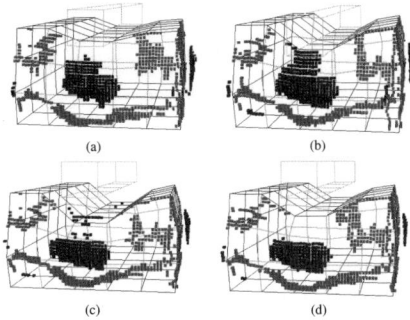

Figure 14 : (a),(b),(c),(d) Four shape differences between real and virtual objects by the first pushing, whose coefficients are completely calibrated as $K = 1831$, $C_1 = 4732$ and $C_2 = 18524$.

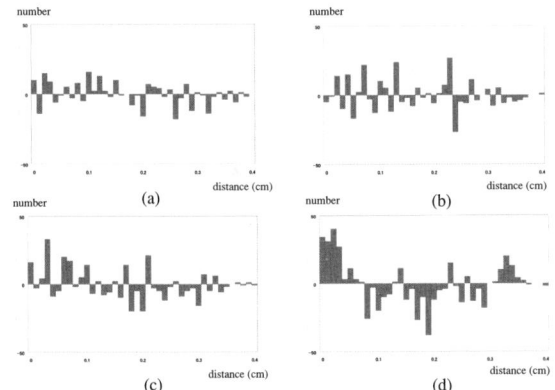

Figure 15 : (a),(b),(c),(d) (the number of captured points at each distance between a calibrated virtual object and its real object) - (the number of captured points at each distance between an initial virtual object and its real object) at four sampling times pushing the objects firstly. In each figure, plus numbers appear in relatively small distances (differences), on the other hand, minus numbers appear in relatively large distances (differences). This means a calibrated object is better than its non-calibrated object against the real object.

4.3 The Second Pushing

In the same algorithm, we use same parameters $\Delta = 1$, $T_{cal} = 30$ $[hour]$, $T_{ran} = 1000[number]$, and finally we obtain calibrated coefficients $K[gf/cm^3]$, $C_1[gfs/cm^3]$ and $C_2[gfs/cm^3]$ as 1398, 6345 and 18524.

In the figures 16 and 17, we compare shape transformation of initial and calibrated virtual objects against the real object, respectively. A set of dark gray areas means shape differences are larger than 0.25cm, and another set of light gray areas means the differences are less than 0.05cm. Therefore, we can see that differences between a calibrated object and its real object are smaller than differences between a non-calibrated object and the real object.

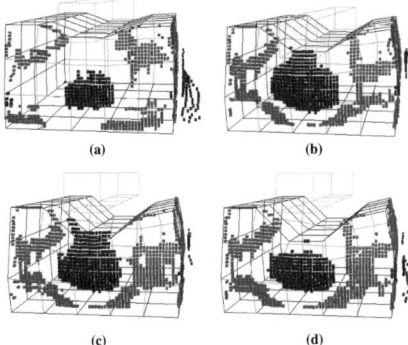

Figure 16 : (a),(b),(c),(d) Four shape differences between real and virtual objects by the second pushing, whose coefficients are initially given as reasonable values $K = 400$ and $C_1 = C_2 = 2000$.

Also in the figure 18, we compare the distribution of differences of minimum distances from about three or more thousand captured points to initial and calibrated virtual objects. In the figure 18(a),(b),(c),(d), numbers of captured points are 87, 40, 35, 107, whose differences are less than 0.1cm, on the other hand, numbers of captured points are -40, -47, 21, -58, whose differences are more than 0.25cm, respectively. From these results, we can see that almost all numbers increase in relatively small differences, on the other hand, almost all numbers decrease in relatively large differences. This means a calibrated object is better than its non-calibrated object against the real object. As a result, a calibrated rheology object is similar to the real rheology object, and therefore our calibration is meaningful.

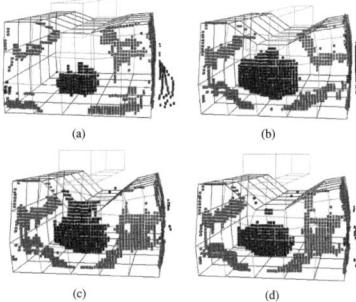

Figure 17 : (a),(b),(c),(d) Four shape differences between real and virtual objects by the second pushing, whose coefficients are completely calibrated as $K = 1398$, $C_1 = 6345$ and $C_2 = 18524$.

5 Conclusions

In this paper, we select a classic 3-D voxel/lattice structure with many mass-damper-spring components as a visco-elastic object, and than calibrate three coefficients of two damper and one spring in each component by an efficient randomized algorithm based on many experimental results. By these approaches, we can watch many deformations by a 3-D graphics animation ($OpenGL$) and feed many forces/moments by a haptics ($Joyarm$). In future, we try to investigate the deformation of rheology object by many kinds of pushing (i.e., translation movements along

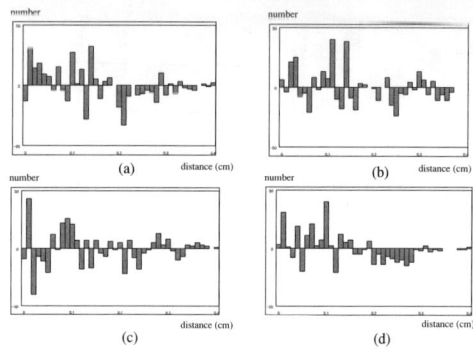

Figure 18 : (a),(b),(c),(d) (the number of captured points at each distance between a calibrated virtual object and its real object) - (the number of captured points at each distance between an initial virtual object and its real object) at four sampling times under pushing the objects secondly. In each figure, plus numbers appear in relatively small distances (differences), on the other hand, minus numbers appear in relatively large distances (differences). This means a calibrated object is better than its non-calibrated object against the real object.

X, Y and Z axes, and also rotation movements centered at X, Y and Z axes, and their combinations). Moreover, another structure (e.g., nested or non-nested tetrahedral meshes) and another component (e.g. mass-spring component, Voigt and Maxwell component) should be tested by our calibration. Finally, as the calibration algorithm, we should try to use another optimal algorithm such as GA (generic algorithm).

Acknowledgments

The authors thank Professor S.Hirai and H.Tanaka (Ritsumeikan University) for the fruitful discussion. This research is supported in part by 2002 Grants-in-aid for Scientific Research from the Ministry of Education, Science and Culture, Japan (No.14550247).

References

[1] D.Baraff, "Analytical Models for Dynamic Simulation of Non-penetrating Rigid Bodies," *Computer Graphics (Proc. SIGGRAPH)*, Vol.23, pp.223-232, 1989.

[2] D.Baraff, "Issues in Computing Contact Forces for Non-Penetrating Rigid Bodies," *Algorithmica*, Vol.10, pp.292-352, 1993.

[3] D.Baraff, "Fast Contact Force Computation Non-Penetrating Rigid Bodies," *Computer Graphics (Proc. SIGGRAPH)*, Vol.28, pp.23-34, 1994.

[4] B.V.Mirtich, "Impulse-Based Dynamic Simulation of Rigid Body Systems," *Ph.D Dissertation*, University of California at Berleley, 1996.

[5] R.Enoki and H.Noborio, "A Comparative Study of Many Randomized Algorithms to Calibrate Virtual Contact and Friction Force Models," *Proc. of the IEEE/RSJ Int. Conf. Intelligent Robots and Syatems*, pp.3042-3047, 2002.

[6] T.Iguchi, N.Katsuyama, H.Noborio and S.Hirai, "Computing and Calibrating Collision Impulses and its Application for Air Hockey Game," *Proc. of the IEEE/RSJ Int. Conf. Intelligent Robots and Syatems*, pp.2219-2226, 2002.

[7] H.Mayeda, K.Osuka and A.Kangawa, "A New Identification Method for Serial Manipulator Arms," *Proc. of the IFAC World Congress*, pp.74-79, 1984.

[8] H.Kawasaki and K.Nishimura, "Terminal-Link Parameter Estimation of Robotic Manipulators," *IEEE Journal of Robotics and Automation*, pp.485-490, 1988.

[9] T.Otsuki, T.Iguchi, Y.Murata and H.Noborio, "On the Identification of Robot Parameters by the Classic Calibration Algorithms and Error Absorbing Trees," *Proc. of the IEEE/RSJ Int. Conf. Intelligent Robots and Syatems*, pp.1916-1923, 2002.

[10] A.H.Barr, "Global and local deformations of solid primitives," *Computer Graphics (Proc. SIGGRAPH)*, Vol.18, pp.21-30, 1984.

[11] T.Sederberg and S.Parry, "Free-Form Deformation of Solid Gemetic Models," *Computer Graphics (Proc. SIGGRAPH)*, Vol.20, pp.151-160, 1986.

[12] D.Terzopoulos and K.Fleisher, "Modeling Inelastic Deformation: Viscoelasticity, Plasticity, Fracture," *Computer Graphics (Proc. SIGGRAPH)*, Vol.22, No.4, pp.269-278, 1988.

[13] A.Joukhader, A.Deguet and C.Laugie, "A Collision Model for Rigid and Deformable Bodies," *Proc. IEEE Int. Conf. on Robotics and Automation*, pp.982-988, 1998.

[14] Y.Chai and G.R.Luecke, "Virtual Clay Modeling Using the ISU Exoskeleton," *Proc. IEEE Virtual Reality Annual International Symposium*, pp.76-80, 1998.

[15] G.Debunne, M.Desbrun, M.P.Cani and A.Barr, "Adaptive Simulation of Soft Bodies in Real-Time," *Computer Animation 2000*, pp.133-144, 2000.

[16] G.Debunne, M.Desbrun, M.P.Cani and A.Barr, "Dynamic Real-Time Deformations using Space and Time Adaptive Sampling," *Computer Graphics (Proc. SIGGRAPH)*, pp.31-36, 2001.

[17] S.Tokumoto, Y.Fujita and S.Hirai, "Deformation Modeling of Viscoelastic Objects for Their Shape Control," *Proc. of the IEEE Int. Conf. on Robotics and Automation*, pp.1050-1057, 1999.

[18] S.Tokumoto, S.Hirai and H.Tanaka, "Constructing Virtual Rheological Objects," *Proc. World Multiconference on Systemics, Cybernetics and Infomatics*, pp.106-111, 2001.

[19] Y.F.Zheng, R.Pei and C.Chen, "Strategies for Automatic Assembly of Deformable Objects," *Proc. of the IEEE Int. Conf. on Robotics and Automation*, pp.2598-2603, 1991.

[20] T.Wada, S.Hirai and S.Kawamura, "Indirect Simultaneous Positioning Operations of Extensionally Deformable Objects," *Proc. IEEE/RSJ Int. Conf. on Intelligent Robots and Systems*, pp.1333-1338, 1998.

[21] S.Cotin, H.Delingette and N.Ayache, "Real-Time Elastic Deformations of Soft Tissues for Surgery Simulation," *IEEE Transactions on Visualization and Computer Graphics*, Vol.5, pp.62-73, 1999.

[22] S.P.DiMaio and S.E.Salcudean, "Needle Insertion Modelling and Simulation," *Proc. of the IEEE Int. Conf. on Robotics and Automation*, pp.2098-2105, 2002.

[23] S. Gibson "Using linked volumes to model object collision, deformation cutting, carving, and joining," *IEEE Visualization and Computer Graphics*, pp.169-177, 1999.

[24] E.Anshelevich, S.Owens, F.Lamiraux and L.Kavraki, "Deformable volumes in path planning applications," *Proc. of the IEEE Int. Conf. on Robotics and Automation*, pp.2290-2295, 2000.

[25] F.Lamiraux and L.Kavraki, "Planning paths for elastic objects under manipulation constraints," *Int. J. Robotics Research*, Vol.20, No.3, pp.188-208, 2001.

[26] O.B.Bayazit J.L.Nancy and M. Amato, "Probabilistic Roadmap Motion Planning for Deformable Objects," *Proc. of the IEEE Int. Conf. on Robotics and Automation*, pp.2126-2133, 2002.

[27] M.C.Lin and J.F.Canny, "A Fast Algorithm for Incremental Distance Calculation," *Proc. of the IEEE Int. Conf. on Robotics and Automation*, pp.1008-1014, 1991.

[28] J.-C.Latombe, "Robot motion planning", *Kluwer Academic Publishers.*, 1991.

Constructing Rheologically Deformable Virtual Objects

Masafumi KIMURA, Yuuta SUGIYAMA, Seiji TOMOKUNI, and Shinichi HIRAI

Dept. of Robotics, Ritsumeikan Univ., Kusatsu, Shiga 525-8577, Japan
E-mail: hirai@se.ritsumei.ac.jp
http://www.ritsumei.ac.jp/se/~hirai/

Abstract

A physical modeling of rheological objects is presented. Objects showing rheological nature involve foods and biological tissues yet no systematic approach to build their virtual objects is not established. In this article, we will construct 2D/3D virtual rheological objects.

Keywords: deformation, modeling, rheological objects

1 Introduction

Deformable objects can be categorized into 1) viscoelastic objects, 2) plastic objects, and 3) rheological objects. Construction of virtual viscoelastic objects and virtual plastic objects has been studied extensively in computer-aided surgery and computer graphics. On the other hand, construction of virtual rheological objects has not been studied yet though various objects such as type of food and biological tissues in the real world tend to deform rheologically.

In this article, we will develop a systematic and coherent method to construct virtual rheological objects. First, we will summarize the properties of rheological deformation. Secondly, we will select rheological elements appropriate for the construction of virtual rheological objects. Thirdly, we will apply the particle-based modeling to virtual rheological objects. We will then investigate the topology maintenance of virtual rheological objects. Finally, we will simulate the physical interaction among rheological objects.

Related works Solid mechanics has also been studied for a long time to formulate the deformation of solid bodies [1]. Solid mechanics basically focuses on the local deformation of solid bodies rather than the global deformation of objects. Rheology has been studied for past several decades and fruitful results have been obtained [2]. Rheology focuses on one-dimensional deformation rather than two-dimensional (2D) or three-dimensional (3D) deformation as well.

Modeling of global object deformation has been extensively studied in computer graphics and virtual reality. Elasticity theory has been applied to the modeling of deformable objects in physically based modeling [3, 4]. Introduction of finite element method (FEM) has extended these works; geometrically-nonlinear FEM and rotation-invariant nonlinear FEM are applied to the modeling of global deformation [5, 6]. Explicit FEM approach with Green strain is proposed to

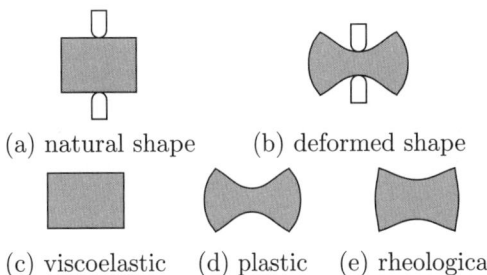

(a) natural shape   (b) deformed shape

(c) viscoelastic   (d) plastic   (e) rheological

Figure 1: Viscoelastic, plastic, and rheological deformations

perform realtime computation of global deformation [7] Boundary element method (BEM) has been introduced to the modeling of deformable objects [8, 9]. BEM approach is applicable to uniform objects alone but can reduce the computation time, resulting in realtime simulation of global object deformation. Nonlinear shell theory has been applied to the modeling of fabric deformation [10]. Particle-based model of a cloth has been proposed for the drape simulation of the cloth [11]. Implicit numerical integration has been introduced to the particle-based cloth model to reduce the computation time [12]. Modeling of plastic objects has been studied in computer graphics [13] and has been applied to computer crafts [14, 15].

This article contributes to the modeling of rheologically deformable objects. We will first formulate the dynamic behavior of rheological elements. Particle-based modeling [16, 17] will be applied to describe the 2D/3D object deformation. Then, topology maintenance of virtual rheological objects will be addressed. Finally, penalty method [18, 19] will be applied to simulate the physical interaction among rheological objects.

2 Rheological Deformation

Objects deform in response to forces applied to the objects. Objects can be categorized into three groups with respect to their deformation. Assume that a natural shape of an object is as given in Figure 1-(a). On applying external forces, the object deforms as in Figure 1-(b). Let us release the applied force and exam-

Table 1: Inferring rules of residual deformation

serial	residual	non-residual
residual	residual	residual
non-residual	residual	non-residual

(a) serial

parallel	residual	non-residual
residual	residual	non-residual
non-residual	non-residual	non-residual

(b) parallel

Table 2: Inferring rules of bouncing deformation

serial	bouncing	non-bouncing
bouncing	bouncing	bouncing
non-bouncing	bouncing	non-bouncing

(a) serial

parallel	bouncing	non-bouncing
bouncing	bouncing	bouncing
non-bouncing	bouncing	non-bouncing

(b) parallel

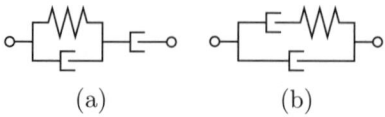

Figure 2: Minimal elements describing rheological deformation

ine the stable shape after the release. Deformation of *viscoelastic objects* is completely lost and their stable shape coincides with their natural shape, as illustrated in Figure 1-(c). Namely, viscoelastic objects have no *residual deformation*. Deformation of *plastic objects* completely remains and their stable shape coincides with their deformed shape under the applied forces, as shown in Figure 1-(d). Namely, plastic objects have no *bouncing deformation*. Objects with residual deformation and bouncing deformation are referred to as *rheological objects*. Deformation of rheological objects is partially lost after the applied forces are released, as illustrated in Figure 1-(e). Various objects including foods and tissues are categorized into rheological objects.

3 Modeling of Rheological Elements
3.1 Selection of Rheological Elements

Rheological objects deform according to forces applied to the objects. The relationship between the applied forces and the object deformation can be described in a physical model. Let us introduce an elastic element and a viscous element so that a physical model can describe the time-dependent deformation of a rheological object. Various deformation properties are then described by combinations of the two fundamental elements. These combinations are referred to as *rheological elements*. Next, we have to select rheological elements appropriate for virtual rheological objects.

As for the deformation properties of a rheological object, recall that 1) residual deformation is involved, 2) bouncing displacement is involved, and 3) vibrations decrease. Let us examine whether individual rheological elements satisfy the first condition. Let us investigate a rheological element consisting of two elements connected in series. If either of the elements has residual deformation, the connected element has residual deformation as well. If neither of the two has residual deformation, the connected element has no residual deformation. These inferences are summarized in Table 1-(a). Let us investigate a rheological element consisting of two elements connected in parallel. If both of the elements have residual deformation, the connected element has residual deformation. If either of the elements has no residual deformation, the connected element has no residual deformation. These inferences are summarized in Table 1-(b). Note that a viscous element has residual deformation while an elastic element has no residual deformation. Thus, we can determine whether a given rheological element has residual deformation or not using Table 1.

Similarly, let us examine whether individual rheological elements satisfy the second condition. Let us investigate a rheological element consisting of two elements connected in series. If either of the elements has bouncing deformation, the connected element also has bouncing deformation. If neither of the two has bouncing deformation, the connected element has no bouncing deformation. These inferences are summarized in Table 2-(a). Let us investigate a rheological element consisting of two elements connected in parallel. If either of the elements has bouncing deformation, the connected element has bouncing deformation as well. If neither of the two elements has bouncing deformation, the connected element has no bouncing deformation. These inferences are summarized in Table 2-(b). Note that an elastic element has bouncing deformation while a viscous element has no bouncing deformation. Thus, we can determine whether a given rheological element has bouncing deformation or not using Table 2.

Now let us examine whether individual rheological elements satisfy the third condition. Note that elements connected in parallel have the same displacement. Thus, a set of elements connected in parallel is referred to as a *part* of a rheological element. Each rheological element can be then regarded as a series of parts. If a part involves viscous elements, any vibration on the part converges to zero. On the other hand, vibration on a part without viscous elements oscillates and does not converge to zero. Thus, we find that all parts must involve viscous elements to satisfy the third condition.

Consequently, we find that rheological elements listed in Figure 2 satisfy the three conditions and consist of the minimal fundamental elements. In this arti-

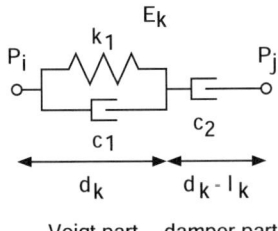

Figure 3: Three-element model

cle, we will use the rheological element shown in Figure 2-(a), which is referred to as a *three-element model*.

3.2 Formulation of the Rheological Element

We will apply the particle-based modeling [16, 17] to describe the deformation of rheological objects. We will describe a 2D/3D object shape by a combination of triangles/tetrahedra. Vertices of individual triangles/tetrahedra correspond to mass particles and their edges correspond to rheological elements. Then, the object model consists of a set of mass particles and a set of rheological elements among the particles. Let P_0 through P_{N-1} be mass particles, where N denotes the number of the particles. Let E_0 through E_{M-1} be rheological elements, where M specifies the number of the elements. Each rheological element connects two mass particles. One particle is referred to as the starting particle of the element while the other is referred to as its end particle.

Let us formulate the dynamic equation of a virtual rheological object. We will apply the three-element model, illustrated in Figure 2-(a), to rheological elements. Two mass particles P_i and P_j are connected by a rheological element E_k, as shown in Figure 3. This rheological element consists of two parts; the left part is referred to as a Voigt part and the right part is referred to as a damper part. Let P_i be the starting particle of the element while P_j be its end particle. Let x_i be the position of particle P_i, v_i be its velocity, and m_i be its mass. Let l_k be the length of element E_k and d_k be the length of its Voigt part. Then, the length of its damper part coincides with $l_k - d_k$. We find that state variables of the mechanical system shown in the figure are given by x_i, v_i, x_j, v_j, and d_k. Let L_k be the natural length of the Voigt part of element E_k. Since the extension of the Voigt part is given by $d_k - L_k$, the magnitude of a force generated by the Voigt part is described as follows:

$$f_k^{voigt} = -k_1\{d_k - L_k\} - c_1\dot{d}_k, \quad (1)$$

where k_1 represents the stiffness of the Voigt part and c_1 denotes its viscosity. Recalling that the length of the damper part is given by $l_k - d_k$, the magnitude of a force generated by the damper part is described as follows:

$$f_k = -c_2\{\dot{l}_k - \dot{d}_k\}, \quad (2)$$

where c_2 represents the viscosity of the damper part. Since the forces generated by the two parts coincide

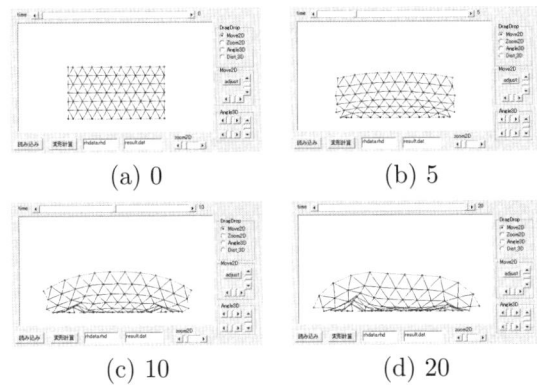

(a) 0   (b) 5

(c) 10   (d) 20

Figure 4: Simulation with linear damper

with each other, say, $f_k^{voigt} = f_k$, we have the following equation:

$$\dot{d}_k = \frac{-k_1\{d_k - L_k\} + c_2\dot{l}_k}{(c_1 + c_2)}. \quad (3)$$

Length l_k satisfies $l_k^2 = (\boldsymbol{x}_i - \boldsymbol{x}_j) \cdot (\boldsymbol{x}_i - \boldsymbol{x}_j)$. Differentiating this equation with respect to time yields

$$\dot{l}_k = \frac{(\boldsymbol{x}_i - \boldsymbol{x}_j) \cdot (\boldsymbol{v}_i - \boldsymbol{v}_j)}{l_k}. \quad (4)$$

Note that a force applied to particle P_i by element E_k is described as $f_k \boldsymbol{e}_k$ while a force applied to particle P_j is given by $-f_k \boldsymbol{e}_k$, where $\boldsymbol{e}_k$ is a unit vector from the starting particle to the end particle, which is given by $\boldsymbol{e}_k = (\boldsymbol{x}_j - \boldsymbol{x}_i)/l_k$.

Let R_i be a set of rheological elements with particle P_i as their starting particle and S_i be a set of rheological elements with particle P_i as their end particle. Then, any element involved in set R_i applies force $f_k \boldsymbol{e}_k$ to mass particle P_i while any element involved in set S_i applies force $-f_k \boldsymbol{e}_k$ to the mass particle. Consequently, we find that the dynamic equation of particle P_i is described as follows:

$$m_i \dot{\boldsymbol{v}}_i = \sum_{k \in R_i} f_k \boldsymbol{e}_k - \sum_{k \in S_i} f_k \boldsymbol{e}_k + \boldsymbol{F}_i^{ext}, \quad (5)$$

where $\boldsymbol{F}_i^{ext}$ be the resultant of external forces applied to particle P_i. On solving differential equations (3) and (5) numerically, we can compute the deformation of a rheological object.

3.3 Force-dependent Nonlinear Damper

Rheological elements involve damper parts to describe residual deformation. Conventional three-element model applies a linear damper to its damper part. Displacement of the linear damper continues increasing or decreasing as long as a force is exerted on the damper. This property is, however, inadequate for the introduction of gravity into virtual rheological objects. A virtual rheological object involving linear damper parts continues deforming as long as gravity forces are exerted on its mass particles. Thus, it

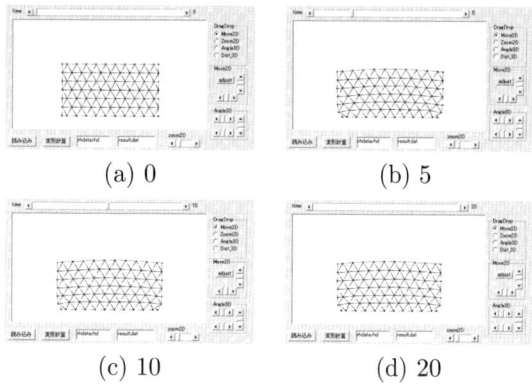

Figure 5: Simulation with force-dependent nonlinear damper

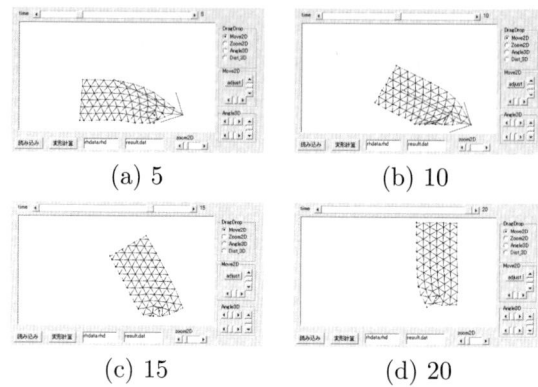

Figure 6: Simulation without gravity

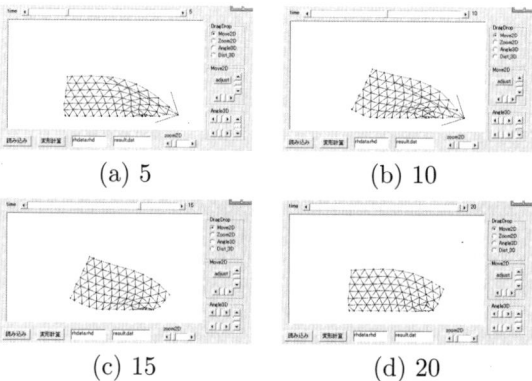

Figure 7: Simulation with gravity

turns out that the shape of a virtual rheological object collapses under gravity. Figure 4 simulates the deformation of a virtual rheological object involving linear damper parts under gravity. As shown in the figure, the object continues deforming and the shape of the object collapses finally. Thus, we will introduce a force-dependent nonlinear damper into the damper part of a three-element model. The viscosity of a force-dependent nonlinear damper changes according to the magnitude of a force applied to it. The viscosity corresponding to the gravity force applied to a particle must take a large value so that the shape of a virtual rheological object does not collapse under gravity. In this article, viscosity of a force-dependent nonlinear damper is given as follows:

$$c_2 = \begin{cases} c_{max} & (f \leq f_1) \\ Ae^{-Bf} & (f_1 \leq f \leq f_2) \\ c_{min} & (f \geq f_2) \end{cases},$$

where f denotes a force generated by the damper part, f_1, f_2, A, and B are constants, $c_{max} = Ae^{-Bf_1}$, and $c_{min} = Ae^{-Bf_2}$. Figure 5 simulates the deformation of a virtual rheological object involving force-dependent nonlinear dampers. We find that the shape of the object does not collapse under gravity.

On introducing force-dependent nonlinear dampers into a virtual rheological object, we can appropriately compute the deformation of the object regardless of body forces such as gravity and electromagnetic forces. Figures 6 and 7 show the deformation and the motion of a rheological object on a rigid table caused by an external force. Gravity does not work in Figure 6 while it works in Figure 7. Without gravity, only a reaction force from the table is applied to the object after the exerted external force is lost. Thus, the object moves upward, as shown in Figure 6-(d). Under gravity, the left bottom region of the object makes contact with the table again after the external force is lost because a reaction force and a gravitational force are applied to the object.

4 Particle-based Modeling of Rheological Objects

4.1 Particle-based Model

We have introduced particle-based approach [16, 17] to describe 2D/3D deformation of rheological objects. In this section, we will describe the topology of a virtual rheological object. The particle-based model involves a set of mass particles and a set of rheological elements among the particles. Each element has its starting and end particles. This implies that each element is directed from the starting particle to the end particle. Consequently, topology of a virtual rheological object is described by a directed graph. Nodes in the graph represent mass particles and the arcs describe directed rheological elements. Moreover, a 2D/3D shape is described by a connection of triangles/tetrahedra, which must be involved in object topology. Figure 8 shows a simple example of a 2D discrete element model. This model consists of 4 mass particles, 5 rheological elements, and 2 triangles. Table 3 provides the topology of this model. The starting and end particles are specified for each element. Each triangle consists of three arcs in the positive direction or in the negative direction. The symbol following each element denotes its direction in the triangle.

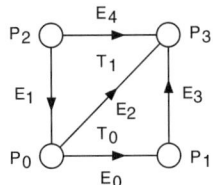

Figure 8: Example of 2D discrete element model

Table 3: Description of discrete element model

element	start point	end point
E_0	P_0	P_1
E_1	P_2	P_0
E_2	P_0	P_3
E_3	P_1	P_3
E_4	P_2	P_3

triangle	arcs					
T_0	E_0	+	E_3	−	E_2	−
T_1	E_2	+	E_4	−	E_1	+

4.2 Topology Maintenance

Topological connection among mass particles in a virtual rheological object must be consistent so that the deformation of the object can be computed appropriately. Since the dynamic equations given in eqs.(3) and (5) are solved numerically at discrete time points, the topological connection often collapses, resulting in failure of the computation.

Recall that a 2D virtual rheological object consists of triangles. In order to maintain the consistency in the topological connection, we have to distinguish a triangle from its reflection. Noting that edges of a triangle are directed, we find that a signed distance between a vertex of a triangle and its opposite edge can be defined, as illustrated in Figure 9. Let $\boldsymbol{n}_{ij}^k$ be a unit vector perpendicular to edge P_iP_j and directed to vertex P_k at the natural state of a virtual rheological object, as shown in Figure 9-(a). The signed distance between vertex P_k and edge P_iP_j is then formulated as follows:

$$d_{ij}^k = \overrightarrow{P_iP_k} \cdot \boldsymbol{n}_{ij}^k. \quad (6)$$

This signed distance must be larger than a small positive value so that the topological connection among particles P_i, P_j, and P_k is consistent. In other words, when the signed distance is shorter than the small positive value, particle P_k must be guided so that the signed distance increases. Thus, we will introduce the following artificial force generated by a virtual Voigt model between vertex P_k and edge P_iP_j:

$$\boldsymbol{f}_{ij}^k = \begin{cases} \boldsymbol{0} & (d_{ij}^k > \epsilon) \\ \{-K(d_{ij}^k - \epsilon) - C\dot{d}_{ij}^k\}\boldsymbol{n}_{ij}^k & (d_{ij}^k \leq \epsilon) \end{cases}, \quad (7)$$

where K and C denote elasticity and viscosity of the

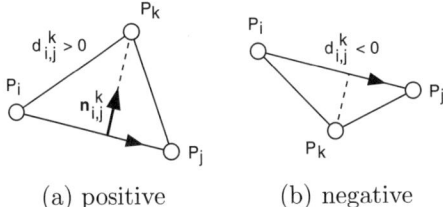

Figure 9: Signed distance between vertex and edge

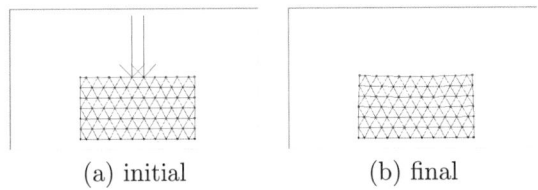

Figure 10: Modeling of viscoelastic object

virtual Voigt model, respectively, and ϵ represents a small positive threshold. When the signed distance d_{ij}^k is below threshold ϵ, the above artificial force is applied to particle P_k to increase the signed distance. The discussion can be extended to a 3D virtual rheological object by introducing a signed distance between a vertex of a tetrahedron and its opposite face.

The topological connection of a virtual rheological object may also collapse due to an inappropriate value of $a_k = d_k/l_k$, which specifies the ratio between the length of the Voigt part and that of the damper part. Note that ratio a_k must satisfy the condition $0 \leq a_k \leq 1$. Solving dynamic equations numerically at discrete time points often breaks this condition and the consistency in the topological connection is lost. Thus, we will define the possible minimum value a_{min} and the possible maximum value a_{max} of variable a_k. Namely, the following condition is imposed on ratio a_k:

$$a_{min} \leq a_k \leq a_{max}. \quad (8)$$

When the value of length d_k is below $a_{min}l_k$ during the computation process, the minimum value $a_{min}l_k$ is substituted into length d_k. When the value of length d_k exceeds $a_{max}l_k$ during the computation process, the maximum value $a_{max}l_k$ is substituted into length d_k.

5 Simulating Rheological Objects
5.1 Elasticity and plasticity

Our approach can describe not only rheological objects but viscoelastic objects and plastic objects in a systematic and coherent manner. Viscoelastic objects can be described by the Voigt part alone. Recall that the ratio of the Voigt part is described by a_k and the condition given in eq.(8) is imposed on the ratio. Thus, viscoelastic objects can be described simply by substituting 1 into both a_{min} and a_{max}. Figure 10 shows an example of the modeling of a viscoelastic object.

Plastic objects can be described by single dampers. On substituting 0 into elasticity k_1 of the Voigt part, a

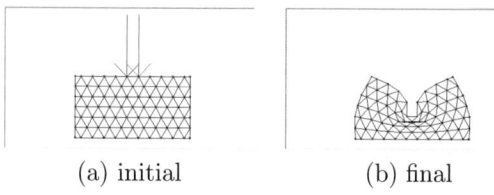

(a) initial (b) final

Figure 11: Modeling of plastic object

three-element model coincides with a serial connection of viscous elements. The viscosity of the Voigt part is given by constant c_1 and the viscosity of the damper part is given by c_2, which depends on the force applied to the part. The resultant viscosity of the serial connection is then described by $c_1 c_2/(c_1 + c_2)$. This resultant viscosity is mainly governed by the smaller viscosity. Namely, if $c_1 \ll c_2$, the resultant viscosity is almost equal to c_1, and vice versa. Since the viscosity corresponding to the gravity force applied to a particle must take a large value, the corresponding force-dependent nonlinear damper should be dominant, say, $c_1 \gg c_{max}$. Consequently, we find that plastic objects can be described by the following condition:

$$k_1 = 0, \quad c_1 \gg c_{max}.$$

The viscosity of an object is then specified by c_{min}. Figure 11 shows an example of the modeling of a plastic object.

5.2 Physical Interaction among Rheological Objects

Let us describe the physical interaction among rheological objects in contact. Collisions between two rheological objects and collisions between two regions of a rheological object cause reaction forces at the contacting regions. We will apply the penalty method [18, 19] to compute the reaction forces. Namely, reaction forces can be simulated by introducing artificial forces similar to eq.(7). In a 2D model, we will define an artificial force between a vertex and an edge on the surfaces. When the signed distance between a vertex and an edge is below a threshold and the foot of the perpendicular of the vertex is on the edge, a force generated by a virtual Voigt model is applied to the vertex and to the edge in the opposite directions. In a 3D model, we will define a artificial force between a vertex and a triangle on the surfaces.

Figure 12 shows the collision of two regions of a virtual rheological object. An external force is exerted on the object at the center of its top face and is lost after a while. Two regions of the object are in contact with each other, as shown in Figure 12-(b). The two regions suffer no interference and the object deformation can be computed well, as shown in Figure 12-(c) and (d). Figure 13 shows the deformation of a rheological object pressed by a rigid object. A rigid object is put on a rheological object and is removed after a while. The rigid object is modeled as a viscoelastic object with large elasticity. The deformation process can be computed successfully as shown in the figure.

Let us simulate kinetic friction forces caused by the collision between two regions of rheological objects.

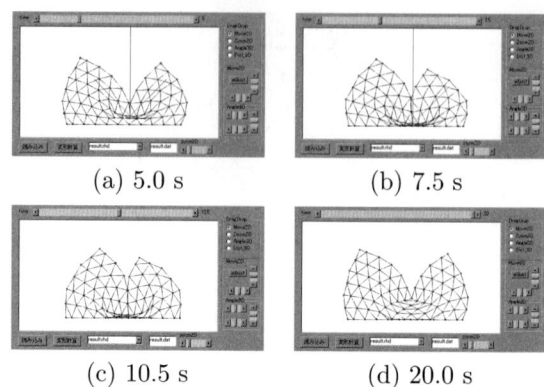

(a) 5.0 s (b) 7.5 s

(c) 10.5 s (d) 20.0 s

Figure 12: Self-collision of rheological object

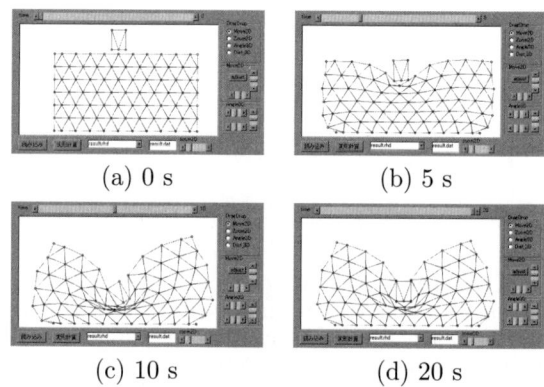

(a) 0 s (b) 5 s

(c) 10 s (d) 20 s

Figure 13: Deformation of rheological object pressed by rigid object

The magnitude of the kinetic friction can be computed by the Coulomb-Amonton law. The direction of the friction is determined by the relative velocity between a vertex and an edge in the 2D model or between a vertex and a triangle in the 3D model. Figure 14 demonstrates the effect of friction. In this demonstration, a small viscoelastic object is on a rheological object. Let us apply an external force to the right side of the rheological object. The small object then slides on the surface in the right direction. The coefficient of kinetic friction is equal to 0.0 in Figure 14-(a) while it is equal to 0.3 in Figure 14-(b). Since the surface is frictionless in Figure 14-(a) whereas it is frictional in Figure 14-(b), the small object slides more in Figure 14-(a) than it does in Figure 14-(b). As shown in the figure, we can simulate the kinetic friction forces between the two regions.

Figure 15 demonstrates the collision among six rheological objects under gravity. The proposed approach can simulate the physical interaction among multiple rheological objects, as shown in the figure.

6 Concluding Remarks

We have developed a systematic and coherent method to construct virtual rheological objects. First,

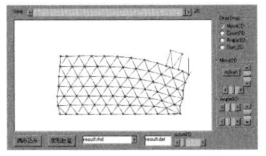 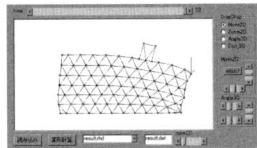

(a) frictionless surface   (b) frictional surface

Figure 14: Description of kinetic friction

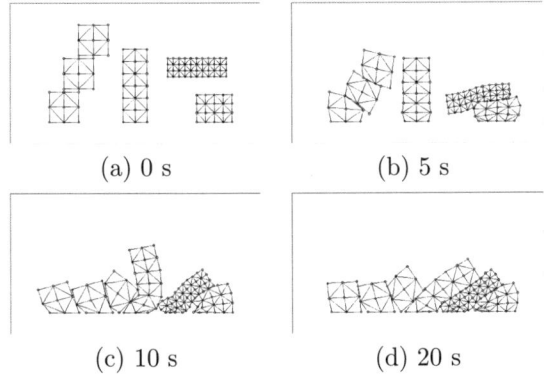

(a) 0 s   (b) 5 s

(c) 10 s   (d) 20 s

Figure 15: Collision among objects

we selected rheological elements appropriate for describing the deformation of rheological objects. It turned out that the three-element model with a force-dependent damper is appropriate. Secondly, we applied the discrete element approach to virtual rheological objects. We found that introducing a signed distance into a triangle/tetrahedron of a virtual rheological object and controlling the length ratio of the Voigt part are essential in order to keep the model topology consistent. Thirdly, we described the physical interaction among rheological objects. We have shown that our approach can describe viscoelastic, plastic, and rheological deformations in a systematic and coherent manner. It also turned out that the contact among rheological objects can be simulated appropriately.

Future issues include 1) identification of physical parameters in virtual rheological objects, 2) realtime computation of rheological deformation, and 3) dynamic simulation of food forming process and human mastication process.

References

[1] Fung, Y. C., *Foundations of Solid Mechanics*, Prentice-Hall, 1965.

[2] Barnes, H. A., Hutton, J. F., and Walters, K., *An Introduction to Rheology*, Elsevier Science Publishers, 1989.

[3] Terzopoulos, D., Platt, J., Barr, A., and Fleischer, K., *Elastically Deformable Models*, Computer Graphics, Vol.21, No.4, pp.205–214, 1987.

[4] Terzopoulos, D. and Witkin, A., *Deformable Models – Physically Based Models with Rigid and Deformable Components*, IEEE Computer Graphics and Applications, November, pp.41–51, 1988.

[5] Zhuang, Y. and Canny, J., *Haptic Interaction of Global Deformations*, Proc. IEEE Int. Conf. Robotics and Automation, pp.2428–2433, 2000.

[6] Picinbono, G., Delingette, H., and Ayache, N., *Non-linear and Anisotropic Elastic Soft Tissue Models for Medical Simulation*, Proc. IEEE Int. Conf. Robotics and Automation, pp.1371–1376, 2001.

[7] Debunne, G, Desbrun, M, Cani, M.-P., and Barr, A. H., *Dynamic Real-Time Deformations using Space & Time Adaptive Sampling*, Computer Graphics Proceedings (SIGGRAPH 2001), pp.31–36, 2001.

[8] James, D. L. and Pai, D. K., *ArtDefo – Accurate Real Time Deformable Objects*, Computer Graphics Proceedings (SIGGRAPH 1999), pp.65–72, 1999.

[9] James, D. L. and Pai, D. K., *Real Time Simulation of Multizone Elastokinematic Models*, Proc. IEEE Int. Conf. Robotics and Automation, pp.927–932, 2002.

[10] Eischen, J. W., Deng, S., and Clapp, T. G., *Finite-Element Modeling and Control of Flexible Fabric Parts*, IEEE Computer Graphics and Applications, pp.71–80, 1996.

[11] Eberhardt, B., Weber, A., and Strasswer, W., *A Fast, Flexible, Particle-System Model for Cloth Draping*, IEEE Computer Graphics and Applications, pp.52–59, 1996.

[12] Baraff, D. and Witkin, A., *Large Steps in Cloth Simulation*, Computer Graphics Proceedings (SIGGRAPH 1998), pp.43–54, 1998.

[13] Terzopoulos, D., and Fleisher, K., *Modeling Inelastic Deformation: Viscoelasticity, Plasticity, Fracture*, Computer Graphics, Vol.22, No.4, pp.269–278, Alberquerque, May, 1988.

[14] Galyean, T. A., and Hughes, J. F., *Sculpting: An Interactive Volumetric Modeling Technique*, Computer Graphics, Vol.25, No.4, pp.267–274, 1991.

[15] Chai, Y., and Luecke, G. R., *Virtual Clay Modeling Using the ISU Exoskeleton*, Proc. IEEE Virtual Reality Annual Int. Symp., pp.76–80, 1998.

[16] Witkin, A. and Welch, W., *Fast Animation and Control of Non-rigid Structures*, Computer Graphics Proceedings (SIGGRAPH 1990), pp.243–252, 1990.

[17] Frugoli, G., Galimberti, A., Rizzi, C., and Bordegoni, M., *Discrete Element Approach for Non-rigid Material Modeling*, Henrich, D. and Wörn, H. eds., *Robot Manipulation of Deformable Objects*, Springer–Verlag, Advanced Manufacturing Series, pp.29–41, 2000.

[18] Moore, M. and Wilhelms, J., *Collision Detection and Response for Computer Animation*, Computer Graphics, Vol. 22, No. 4, pp.289–298, 1988.

[19] Joukhader, A., Deguet, A., and Laugie, C., *A Collision Model for Rigid and Deformable Bodies*, Proc. IEEE Int. Conf. on Robotics and Automation, pp.982–988, Leuven, May, 1998.

Post-Stabilization for Rigid Body Simulation with Contact and Constraints

Michael B. Cline* and Dinesh K. Pai*†
* University of British Columbia, Vancouver, Canada.
† Rutgers, the State University of NJ, Piscataway, NJ
{cline|pai}@cs.ubc.ca

Abstract—Rigid body dynamics with contact constraints can be solved locally using linear complementarity techniques. However, these techniques do not impose the original constraints and need stabilization. In this paper we show how constraint stabilization can also be done in a complementarity framework. Our technique effectively eliminates the drift problem for both equality and inequality constraints and requires no parameter tweaking. We describe results from an implemented system, and compare the new technique to the well known Baumgarte stabilization.

I. Introduction

Dynamic simulation of rigid body contact is central to many aspects of robotics, including manipulation, design, and haptic interaction in virtual environments. Dynamics with contact constraints and friction lead to differential algebraic equations and inequalities, which are difficult to solve. Recently, these contact problems have been effectively formulated as linear complementarity problems (LCPs) [Lot82], [Bar94], [ST96], [AP97]. However, these techniques used "reduced index" formulations of the dynamics [AP98]. They only satisfy the constraints locally and can drift away from the constraint, and must be "stabilized" to continue to satisfy the constraint. In this paper we show how this stabilization can be performed in an LCP framework as well, and ensure that the stabilization steps also satisfy the inequalities due to contact and friction constraints. We provide examples from an implemented system which show that this technique is effective in stabilizing both equality and inequality constraints due to contact.

The remainder of the paper is organized as follows. In Sec. II we describe some related work. Sec. III provides a brief but complete description of LCP-based formulations of rigid body contact mechanics. Sec. IV describes constraint stabilization in general. Sec. V describes our new technique for post-stabilization in an LCP framework. Sec. VI describes the results, and compares the new approach to previous approaches.

II. Related Work

We draw from related work in two areas. The first area is the work on stabilization of ordinary differential

Supported in part by grants from IRIS NCE.

equations with invariants. The stabilization method that we will present is based on work by Ascher and Chin [Asc97] [ACR94]. These papers discuss stabilization methods in general and post-stabilization in particular as the favoured method.

The other area of related work is the literature on rigid body dynamics with contact, where the constrained dynamics equations and inequalities are formulated as a linear complementarity problem. The first paper to pose the contact problem as an LCP was published by Lötstedt [Lot82]. Baraff presented a method [Bar94] that used an LCP algorithm by Cottle and Dantzig [CD68] to solve contact and static friction forces at interactive rates.

One of the difficulties with earlier LCP methods was that there is no guarantee of the existence of a solution, in the presence of contact with Coulomb friction. Indeed, there are known configurations of rigid bodies for which no solution exists, such as the Painlevé paradox. Using a model which allows impulsive forces turned out to be the key to avoiding these problems. Stewart and Trinkle [ST96] introduced a time-stepping scheme that combines the acceleration-level LCP with a time step, to obtain an LCP where the variables are velocities and impulses. Anitescu and Potra [AP97], described a modification, which we describe in detail below, that guaranteed solvability regardless of the configuration or number of contacts.

We unify these two areas of research by showing how stabilization for simulations with contact can be done by formulating the post-stabilization problem as linear complementarity problem.

III. Background

A. Notation

We use a translation vector and a quaternion to represent the position and orientation of the rigid body. We append these together in a vector $\mathbf{p}$. The velocity $\mathbf{v}$ of the body is described by a vector containing the linear and angular velocity of the body. The state of the i'th body is described by the vectors $\mathbf{p}_i = (p_{ix} p_{iy} p_{iz} q_{ix} q_{iy} q_{iz} q_{iw})^T$ and $\mathbf{v}_i = (\omega_{ix} \omega_{iy} \omega_{iz} v_{ix} v_{iy} v_{iz})^T$.

We shall write the Newton-Euler equations of motion of body i as

$$\mathbf{f}_i = \mathbf{M}_i \mathbf{a}_i \qquad (1)$$

where $\mathbf{f}_i$ is the wrench (torque and force) acting on body i, which includes the Coriolis forces, $\mathbf{M}_i$ is the the spatial inertia matrix of the body, and $\mathbf{a}_i = \dot{\mathbf{v}}_i$. When dealing with a system of many bodies, we shall use the symbols $\mathbf{f}$, $\mathbf{a}$, and $\mathbf{M}$ to indicate vectors and matrices that contain information for many bodies.

B. Constrained Dynamics

Position constraints will be described by a *constraint function* $\mathbf{g}(\mathbf{p})$, which is a function which maps $\mathbf{p}$, the position vector of the rigid bodies, to a point in $\mathbb{R}^n$, where n is the number of degrees of freedom that the constraint removes from the system. If the constraint function returns a zero vector, then the position $\mathbf{p}$ satisfies the constraint.

Position constraints can be divided into *equality constraints* (e.g., joint constraints), where the constraint is $\mathbf{g}(\mathbf{p}) = 0$, and *inequality constraints* (e.g., contact constraints), where $\mathbf{g}(\mathbf{p}) \geq 0$. For the rest of this section, we will just be talking about equality constraints. We will talk about contact constraints separately in the next section.

Constraining the position of an object also constrains its velocity. Velocity constraints are of the form

$$\frac{d\mathbf{g}}{dt} = \mathbf{J}\mathbf{v} + \mathbf{c} = \begin{pmatrix} \mathbf{J}_1 & \mathbf{J}_2 & \cdots & \mathbf{J}_n \end{pmatrix} \begin{pmatrix} \mathbf{v}_1 \\ \mathbf{v}_2 \\ \vdots \\ \mathbf{v}_n \end{pmatrix} + \mathbf{c} = \mathbf{0}. \quad (2)$$

The matrix $\mathbf{J}$ is called the constraint's *Jacobian matrix*, which we refer to simply as the Jacobian. The Jacobian is a function of the current position of the bodies involved in the constraint. In practice, most constraints will only involve one or two bodies, so most of the matrices $\mathbf{J}_1 \ldots \mathbf{J}_n$ will be zero matrices.

A constraint on the acceleration can be found by taking the derivative again. Doing so results in an equation of similar form to Equation 2: $\mathbf{J}\dot{\mathbf{v}} + \mathbf{k} = \mathbf{0}$. The acceleration and velocity constraints use the same Jacobian matrix, but different terms $\mathbf{c}$ and $\mathbf{k}$.

When there are many constraints, we will often write all constraints simultaneously. For a system with n bodies, and m constraints, our velocity constraint equation looks like

$$\frac{d\mathbf{g}}{dt} = \begin{pmatrix} \mathbf{g}_1 \\ \mathbf{g}_2 \\ \vdots \\ \mathbf{g}_m \end{pmatrix} = \mathbf{J}\mathbf{v} + \mathbf{c} = \begin{pmatrix} \mathbf{J}_{11} & \mathbf{J}_{12} & \cdots & \mathbf{J}_{1n} \\ \mathbf{J}_{21} & \mathbf{J}_{22} & & \mathbf{J}_{2n} \\ \vdots & & \ddots & \vdots \\ \mathbf{J}_{m1} & \mathbf{J}_{m2} & \cdots & \mathbf{J}_{mn} \end{pmatrix} \begin{pmatrix} \mathbf{v}_1 \\ \mathbf{v}_2 \\ \vdots \\ \mathbf{v}_n \end{pmatrix} + \mathbf{c} = \mathbf{0}. \quad (3)$$

The constraint Jacobian matrix $\mathbf{J}$ has a dual use. In addition to relating the velocities to the rate of change of the constraint function $\mathbf{g}$, the rows of $\mathbf{J}$ act as basis vectors for constraint forces. Thus, when we solve for the constraint forces, we actually just need to solve for the coefficient vector λ (whose components are the *Lagrange multipliers*) that contains the magnitudes of the forces that correspond to each of these basis vectors.

The total force acting on the system is the sum of the external forces $\mathbf{F}_{ext}$ (in which we include Coriolis forces) and the constraint forces $\mathbf{J}^T \lambda$. Combining the Newton-Euler equations,

$$\mathbf{F} = \mathbf{J}^T \lambda + \mathbf{F}_{ext} = \mathbf{M}\dot{\mathbf{v}},$$

with the constraint equations $\mathbf{J}\dot{\mathbf{v}} + \mathbf{k} = 0$, we get the following system

$$\begin{pmatrix} \mathbf{M} & -\mathbf{J}^T \\ \mathbf{J} & 0 \end{pmatrix} \begin{pmatrix} \dot{\mathbf{v}} \\ \lambda \end{pmatrix} = \begin{pmatrix} \mathbf{F}_{ext} \\ -\mathbf{k} \end{pmatrix}, \quad (4)$$

which we can solve for the acceleration $\dot{\mathbf{v}}$ and the Lagrange multipliers λ. Baraff [Bar96] shows how to solve this system in linear time by exploiting the sparse structure of the matrix on the left hand side of Equation 4. This sparseness exploitation was previously done in the MEXX system [LNPE92].

1) Incorporating Contact Constraints: Contact constraints are different from joint constraints, and require special treatment. There are a few important features that set contact constraints apart from other constraints:

- Contact forces can push bodies apart, but cannot pull bodies towards each other. This leads to inequalities in the constraint equations, whereas other types of constraints are equalities.
- If there are many points of contact between a pair of bodies, there may be many solutions for the contact forces that are plausible.
- In the presence of Coulomb friction, a solution to the dynamics equations may not exist.

There are several methods for handling contact in rigid body simulations. Mirtich [Mir98] gives a summary of several different methods, and explains the advantages and disadvantages of each.

We focus on the *Linear Complementarity Problem* approach, first introduced in the context of constrained mechanical systems by Lötstedt [Lot82], and introduced to the computer graphics community by Baraff [Bar94].

To describe the contact model we use, we introduce the following terminology.

- A *contact* consists of a pair of *contact points*, one point attached to one rigid body, and the other point attached to another rigid body. The contact points are sufficiently close together for our collision detection algorithm to report a collision.
- A *contact normal* is a unit vector that is normal to one or both of the surfaces at the contact points.
- A *contact wrench* $\mathbf{J}^T \lambda$ is a wrench which prevents the two rigid bodies from interpenetrating.
- The *separation distance* of a contact is the normal component of the displacement between the two contact points. It is negative when the bodies are interpenetrating at the contact. The constraint function $\mathbf{g}(\mathbf{p})$ for a contact constraint returns the separation distance. The constraint is satisfied for $\mathbf{g} \geq 0$.
- The *relative normal acceleration a* of a contact is the second derivative of the separation distance with respect to time. The acceleration constraint is satisfied when $a = \mathbf{J}\dot{\mathbf{v}} + \mathbf{k} \geq 0$.

Let $\mathbf{a}$ be a vector containing the relative normal accelerations $\{a_1,...,a_n\}$ for all contacts, and λ be a vector of contact force multipliers $\{\lambda_1,...,\lambda_n\}$. The vectors $\mathbf{a}$ and λ are linearly related. Additionally, we have three constraints:

1) The relative normal accelerations must be positive: $\mathbf{a} = \mathbf{J}\dot{\mathbf{v}} + \mathbf{k} \geq \mathbf{0}$.
2) The contact force magnitudes must be positive (so as to push the bodies apart): $\lambda \geq \mathbf{0}$
3) For each contact i, only one of a_i, λ_i can be nonzero.

The problem of finding a solution to a linear equation given such constraints is called a *Linear Complementarity Problem* (LCP), which we discuss in Section III-B.2.

If we let $\mathbf{J}_e$ be the Jacobian of equality constraints, and $\mathbf{J}_c$ be the Jacobian of contact constraints, we can extend Equation 4 to include the contact constraints as follows [ST96]:

$$\begin{pmatrix}\mathbf{0}\\\mathbf{0}\\\mathbf{a}\end{pmatrix} - \begin{pmatrix}\mathbf{M} & -\mathbf{J}_e^T & -\mathbf{J}_c^T\\\mathbf{J}_e & 0 & 0\\\mathbf{J}_c & 0 & 0\end{pmatrix}\begin{pmatrix}\dot{\mathbf{v}}\\\lambda_e\\\lambda_c\end{pmatrix} = \begin{pmatrix}\mathbf{F}_{ext}\\-\mathbf{k}_e\\-\mathbf{k}_c\end{pmatrix}, \quad (5)$$
$$\mathbf{a} \geq \mathbf{0}, \lambda_c \geq \mathbf{0}, \mathbf{a}^T\lambda_c = \mathbf{0}.$$

In this form, we have a *mixed LCP*, meaning that only some of the rows have complementarity constraints on them. To obtain a pure LCP, we must eliminate $\dot{\mathbf{v}}$ and λ_e by solving for those in terms of λ_c.

2) Linear Complementarity Problems: A *Linear Complementarity Problem* (LCP) is, given a $n \times n$ matrix $\mathbf{M}$ and a n-vector $\mathbf{q}$, the problem of finding values for the variables $\mathbf{z} = \{z_1, z_2, .., z_n\}$ and $\mathbf{w} = \{w_1, w_2, .., w_n\}$ such that

$$\mathbf{w} = \mathbf{Mz} + \mathbf{q}, \quad (6)$$

and, for all i from 1 to n, $z_i \geq 0$, $w_i \geq 0$ and $z_i w_i = 0$. (These three constraints are called the *complementarity constraints*). We solve the linear complementarity problems using Lemke's algorithm [CPS92], [Mur88].

3) Solvable LCPs for Contact with Friction: It is well known that that when Coulomb friction is added, the acceleration-level dynamics equations (Equation 5) fail to have a solution in certain configurations, even for situations where there is only one contact involved.

Anitescu and Potra [AP97] present a time-stepping method which combines the acceleration-level LCP with an integration step for the velocities, arriving at a method which has velocities and impulses as unknowns, rather than accelerations and forces. Their method is guaranteed to have a solution, regardless of the configuration and number of contacts.

To discretize the system (5), Anitescu and Potra apply a forward Euler step on the velocities: $\dot{\mathbf{v}} \approx (\mathbf{v}_{next} - \mathbf{v}_{current})/h$, where $\mathbf{v}_{current}$ and $\mathbf{v}_{next}$ are the velocities at the beginning of the current time step, and the next time step, respectively, and h is the time step size. We then arrive at the mixed LCP

$$\begin{pmatrix}\mathbf{0}\\\mathbf{0}\\\mathbf{a}\end{pmatrix} - \begin{pmatrix}\mathbf{M} & -\mathbf{J}_e^T & -\mathbf{J}_c^T\\\mathbf{J}_e & 0 & 0\\\mathbf{J}_c & 0 & 0\end{pmatrix}\begin{pmatrix}\mathbf{v}_{next}\\\lambda_e\\\lambda_c\end{pmatrix} = \begin{pmatrix}\mathbf{Mv}_{current} + h\mathbf{F}_{ext}\\-\mathbf{k}_e\\-\mathbf{k}_c\end{pmatrix},$$
$$\mathbf{a} \geq \mathbf{0}, \lambda_c \geq \mathbf{0}, \mathbf{a}^T\lambda_c = \mathbf{0}. \quad (7)$$

Coulomb friction is introduced by adding some additional forces and constraints to the LCP (7).

Essentially, these complementarity conditions guarantee the following properties that we expect from Coulomb friction:

- If the contact impulse lies inside the friction cone (but not on its surface), then the bodies are not exhibiting relative tangential motion.
- If the bodies are in relative tangential motion, then the contact impulse lies on the surface of the friction cone, and exhibits negative work.
- As the approximation of the friction cone becomes closer to the ideal circular friction cone, the friction becomes closer to directly opposing the relative tangential motion.

Extending Equation 7 to include these friction constraints, we get the following linear complementarity problem

$$\begin{pmatrix}\mathbf{0}\\\mathbf{0}\\\mathbf{a}\\\sigma\\\zeta\end{pmatrix} - \begin{pmatrix}\mathbf{M} & -\mathbf{J}_e^T & -\mathbf{J}_c^T & -\mathbf{J}_f^T & 0\\\mathbf{J}_e & 0 & 0 & 0 & 0\\\mathbf{J}_c & 0 & 0 & 0 & 0\\\mathbf{J}_f & 0 & 0 & 0 & \mathbf{E}\\0 & 0 & \mu & -\mathbf{E}^T & 0\end{pmatrix}\begin{pmatrix}\mathbf{v}_{next}\\\lambda_e\\\lambda_c\\\lambda_f\\\gamma\end{pmatrix} = \begin{pmatrix}\mathbf{Mv}_{current}+h\mathbf{F}_{ext}\\-\mathbf{k}_e\\-\mathbf{k}_c\\0\\0\end{pmatrix},$$

$$\begin{pmatrix}\mathbf{a}\\\sigma\\\zeta\end{pmatrix} \geq \mathbf{0}, \begin{pmatrix}\lambda_c\\\lambda_f\\\gamma\end{pmatrix} \geq \mathbf{0}, \begin{pmatrix}\mathbf{a}\\\sigma\\\zeta\end{pmatrix}^T\begin{pmatrix}\lambda_c\\\lambda_f\\\gamma\end{pmatrix} = \mathbf{0}.$$
(8)

Due to lack of space, we refer the reader to [Ani97], [Cli02] for full details; here it is sufficient to observe that the general form of the LCP remains the same.

IV. CONSTRAINT STABILIZATION

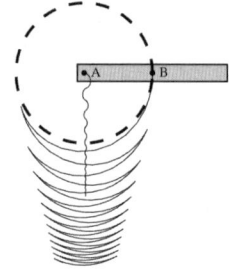

Fig. 1. An unstabilized simulation of a swinging pendulum.

Figure 1 is an example situation that motivates the need for stabilization. The figure shows a simple two-dimensional simulation of a swinging pendulum: one rigid body with one constraint. The only external force is that of gravity. The constraint is a hinge joint constraint that anchors point *A* to a fixed point. Over time, however, point *A* actually drifts, due to numerical integration errors. The centre of mass, rather than staying on the circular path that it is supposedly constrained to (shown with a dashed line), continues to fall further from that path as time progresses.

We can lessen this problem by using higher order, more accurate integration scheme (the ones used here are only first-order accurate), or by using smaller time steps. However, even using the best integration methods and small time steps, numerical drift can never be completely eliminated. Furthermore, it may not always be feasible to use the more complicated integrators or decrease the step size and still achieve acceptable performance. Even if we cannot achieve very high numerical accuracy, we still wish to have simulations that are plausible – ones where the constraints are always met. This section describes methods for counteracting constraint drift using stabilization.

Before continuing, we need to define our problem more clearly. *Stability* is a word that is used with a variety of meanings in different contexts in the differential equation literature. By the word *stabilization* here, we mean *stabilization of an ODE with respect to an invariant set*. We shall now explain what this means.

The position constraint equation coupled with the constrained Newton-Euler equations (see Equation 4) together are an example of a *differential-algebraic equation* (DAE). This is a term for a system of equations containing both differential equations and algebraic equations. DAEs are a relatively new area of research, and the methods for solving them directly are somewhat difficult. The usual approach is to convert the DAE into an equivalent ordinary differential equation (ODE) which can be solved by conventional techniques. In our case, we can do this by differentiating the position constraint equations twice, to arrive at a constraint in terms of accelerations, then substituting this acceleration constraint into the Newton-Euler equations. In doing this, we arrive at an ODE. The approach we use is actually slightly different, as we have described in Section III-B.3.[1]

The penalty of this approach is that we lose the constraints on the position and velocity. Upon discretization of the ODE, we introduce numerical errors, and we will experience *drift* away from the *constraint manifold*. The constraint manifold (also known as the *invariant set*) is the subset of the state space in which the constraints are satisfied. To counteract this, we must stabilize the ODE with respect to the invariant set. In other words, we must alter the ODE so that it has the same solutions as the original whenever $\mathbf{g}(\mathbf{p}) = 0$, but whenever $\mathbf{g}(\mathbf{p}) \neq 0$ the solutions is attracted towards the invariant set.

In this section we describe two methods of stabilization, in the context of rigid body simulation. First we will discuss *Baumgarte* stabilization, a method that is very popular because of its simplicity. We will then introduce *Post-Stabilization*, which is based on the work of Ascher et al. [Asc97] [ACR94]

A. Baumgarte Stabilization

Baumgarte's stabilization technique [Bau72] is one of the most familiar and commonly used methods, because of its simplicity. The idea here is to replace the acceleration constraint equation $\ddot{\mathbf{g}} = \mathbf{J}\dot{\mathbf{v}} + \mathbf{k} = 0$ with some a linear combination of the acceleration, velocity and position constraint equations:

$$0 = \ddot{\mathbf{g}} + \alpha \dot{\mathbf{g}} + \beta \mathbf{g}, \quad (9)$$

which creates a more stable ODE. If the velocity and position constraints are satisfied, the last two terms on the right hand side vanish, and we are left with the original acceleration constraint equation. A physical interpretation of this method is that we are adding additional correction forces, proportional to the error in the velocity and position constraints, to counteract drift.

Because our implementation uses only velocity-level constraints rather than acceleration constraints (see Section III-B.3), we use an even simpler version of Baumgarte stabilization where we replace the velocity constraint $\dot{\mathbf{g}} = \mathbf{J}\mathbf{v} + \mathbf{c} = 0$ with $\dot{\mathbf{g}} + \alpha \mathbf{g} = \mathbf{J}\mathbf{v} + (\mathbf{c} + \alpha \mathbf{g}) = 0$.

The main difficulty of using Baumgarte stabilization is that it is not always easy to find an appropriate value for the constants α and β.

B. Post-Stabilization

Another approach to stabilization is to follow each integration step with a stabilization step. The stabilization step takes the result of the integration step as input, and gives a correction so that the end result is closer to the constraint manifold. Post-stabilization methods are discussed in detail and compared to other stabilization methods by Ascher et al. [ACR94]. Here we give interpretation of post-stabilization that fits into Ascher's broader definition.

Let **p** be the position of a set of rigid bodies after the integration step. Let **g** be the constraint function (see Section III-B). In general, due to numerical drift, $\mathbf{g}(\mathbf{p}) \neq \mathbf{0}$. Let $\mathbf{G} = \frac{\partial \mathbf{g}}{\partial \mathbf{p}}$. In our stabilization step, we wish to find some **dp** such that $\mathbf{g}(\mathbf{p} + \mathbf{dp}) = 0$. Assuming **dp** will be small, we can make the approximation that

$$\mathbf{g}(\mathbf{p} + \mathbf{dp}) \approx \mathbf{g}(\mathbf{p}) + \mathbf{G}(\mathbf{p})\mathbf{dp}. \quad (10)$$

Rearranging this, we see that the stabilization term **dp** should satisfy

$$\mathbf{G}\mathbf{dp} = -\mathbf{g}(\mathbf{p}). \quad (11)$$

In general, **G** is not square, so $\mathbf{G}^{-1}$ does not exist. One way to solve for **dp** is to use the pseudoinverse of **G**:

$$\mathbf{dp} = -(\mathbf{G}^T(\mathbf{G}\mathbf{G}^T)^{-1})\mathbf{g}(\mathbf{p}). \quad (12)$$

This idea works fine as long as $\mathbf{G}\mathbf{G}^T$ is nonsingular, which is usually the case for a system that contains only equality constraints. However, we often run into singularities when contact constraints are involved. This

---

[1] We use *time-stepping* methods, where a numerical integration step is built into the system of equations we solve, and the equations are given in terms of velocities and impulses, rather than forces and accelerations.

is because the collision detector may find many contact points between a pair of objects, leading to constraints that are redundant. One approach to deal with singularities is to use a pseudoinverse formula based on singular value decomposition of $\mathbf{G}$. By truncating the small (nearly zero) singular values, we can find a pseudoinverse of $\mathbf{G}$ even when $\mathbf{GG}^T$ is singular.

Using a singular value decomposition in each time step, however, would be expensive, and it does not take the inequality constraints involved in contact into account. Instead, we find it natural to pose the post-stabilization problem as an LCP, just as we do with the dynamics equations. An additional benefit is that we get a more physically meaningful pseudoinverse. We explain this method in the next section.

V. Fitting Post-Stabilization into the Dynamics LCP Framework

As mentioned above, in the absence of contact constraints, we can find the stabilization term using the pseudoinverse of $\mathbf{G}(\mathbf{p})$ (Equation 12). Doing so is equivalent to solving the system

$$-\begin{pmatrix} \mathbf{I} & -\mathbf{G}(\mathbf{p})^T \\ \mathbf{G}(\mathbf{p}) & 0 \end{pmatrix} \begin{pmatrix} \mathbf{dp} \\ \lambda \end{pmatrix} = \begin{pmatrix} 0 \\ \mathbf{g}(\mathbf{p}) \end{pmatrix}. \quad (13)$$

In other words, we can express the problem of finding the post-stabilization step as a problem of finding Lagrange multipliers, just as we did with the dynamics equations (see Equation 4). We can make 13 look more like Equation 4 by doing two things:

1) Replace $\mathbf{G}(\mathbf{p})$ with $\mathbf{J}(\mathbf{p})$. These two matrices are both constraint Jacobians. The difference is that $\mathbf{G}$ multiplies with changes in position (7-vectors consisting of a translation, plus a quaternion), whereas $\mathbf{J}$ multiplies with twists, which are 6-vectors. As a result of using $\mathbf{J}$, the post-stabilization step $\mathbf{dp}$ would be a twist and needs to be converted back into our position representation. [2]
2) Replace the identity matrix with the mass matrix $\mathbf{M}$. By doing so, we are no longer using the pseudoinverse $\mathbf{G}^T(\mathbf{GG}^T)^{-1}$, but a weighted pseudoinverse, $\mathbf{M}^{-1}\mathbf{G}^T(\mathbf{GM}^{-1}\mathbf{G}^T)^{-1}$. This corresponds to favouring the position change that requires the least amount of energy. This way, for example, a rotation around an axis with low moment of inertia would be favoured over a rotation around an axis with high moment of inertia.

Making these two changes gives us

$$-\begin{pmatrix} \mathbf{M} & -\mathbf{J}(\mathbf{p})^T \\ \mathbf{J}(\mathbf{p}) & 0 \end{pmatrix} \begin{pmatrix} \mathbf{dp} \\ \lambda \end{pmatrix} = \begin{pmatrix} 0 \\ \mathbf{g}(\mathbf{p}) \end{pmatrix}. \quad (14)$$

Similar to our dynamics equations for systems with contact, it is natural to place complementarity constraints on the variables having to do with the contact constraints.

[2]Note that in two-dimensional rigid body simulations, $\mathbf{G}$ and $\mathbf{J}$ are identical, so this somewhat confusing distinction can be ignored.

- The post-step should never pull contacting bodies towards each other at the contact points, only push them apart: $\lambda_c \geq \mathbf{0}$
- If contact constraint functions were evaluated after adding the post step, the result must not be negative, but may be positive: $\mathbf{g}_c^+ = (\mathbf{g}_c^- + \mathbf{J}_c \mathbf{dp}) \geq 0$ (we use superscripts "-" and "+" to denote "before" and "after" the post-step. That is, $\mathbf{g}_c^- = \mathbf{g}(\mathbf{p})$, and $\mathbf{g}_c^+$ approximates $\mathbf{g}(\mathbf{p}+\mathbf{dp})$)
- $\lambda_c^T \mathbf{g}_c^+ = \mathbf{0}$: This constraint roughly means that for each contact c, either we are pushing the bodies apart at c or contact c's constraint will be satisfied in the absence of any push at c.

Adding the stabilization for contact constraints, we have the LCP

$$\begin{pmatrix} 0 \\ 0 \\ \mathbf{g}_c^+ \end{pmatrix} - \begin{pmatrix} \mathbf{M} & -\mathbf{J}_e^T & -\mathbf{J}_c^T \\ \mathbf{J}_e & 0 & 0 \\ \mathbf{J}_c & 0 & 0 \end{pmatrix} \begin{pmatrix} \mathbf{dp} \\ \lambda_e \\ \lambda_c \end{pmatrix} = \begin{pmatrix} 0 \\ \mathbf{g}_e^- \\ \mathbf{g}_c^- \end{pmatrix}, \quad (15)$$

$$\mathbf{g}_c^+ \geq \mathbf{0}, \lambda_c \geq \mathbf{0}, \lambda_c^T \mathbf{g}_c^+ = \mathbf{0}.$$

VI. Results and Discussion

A. 6-Link Chain

In the following test of our stabilization method, we simulate a 6 link chain falling freely under gravity (see figure 2). At time 0, the chain is completely horizontal. We observe the change in constraint error over time using one of three test conditions for comparison: no stabilization, Baumgarte stabilization, or post-stabilization. We evaluate the constraint function $\mathbf{g}(\mathbf{p})$ for each constraint, and use the maximum absolute value as a measure of constraint error. This number roughly the largest joint separation distance. For comparison, each link in the chain is 100mm long. The time step size is 0.001 seconds.

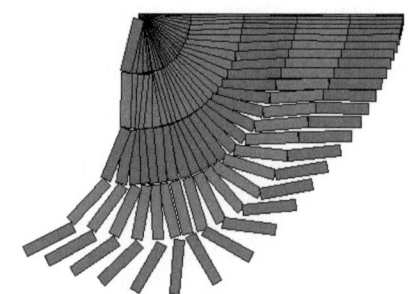

Fig. 2. A simulation of a 6-link chain falling freely under gravity. Screen clearing between frames is turned off to show motion over time. There is no constraint stabilization, so error in the constraints is evident as time progresses (note separation between the lowest two links of the chain).

Figure 3 shows us how the simulation behaves in the absence of any stabilization. The error grows over time as the joints of the chain separate. The graph has stair steps because of the periodic nature of the swinging chain, which behaves something like a pendulum. The error grows rapidly when the chain is moving more quickly.

By the end of 600 time steps (0.6 seconds), the error has climbed to around 20mm.

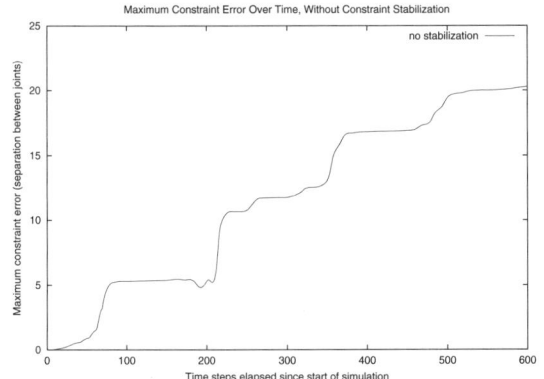

Fig. 3. Maximum constraint error over time, for a simulation of a swinging 6-link chain without any stabilization

In figures 4 and 5, we see the effect of Baumgarte stabilization on the error. The choice of constant used in Baumgarte stabilization has a large effect on how well the stabilization works. In figure 4, we show two choices of constant (50 and 200) that seemed to work best for this situation. Below 50, the error became much larger. With constant of 50, the maximum error is around 5mm, and when the constant is 200, we get slightly better results, with the error never exceeding 3mm.

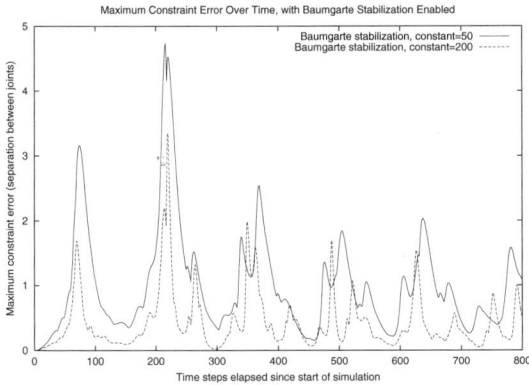

Fig. 4. Maximum constraint error over time, for a simulation of a swinging 6-link chain with Baumgarte stabilization, using a constant of 50 or 200

If the constant is raised higher, the error does not continue to go down. Instead, we start to notice another problem: the chain starts to jiggle unrealistically. This corresponds to the error correction term growing so large that it overshoots its goal in a single time step. Figure 5 shows an example of this, using Baumgarte stabilization with a constant of 2000. When one watches this simulation, one notices the chain bouncing around rapidly due to the large stabilization forces. Correspondingly, the error graph is quite jagged, although quantitatively the error is not much worse than when a constant of 50 is used.

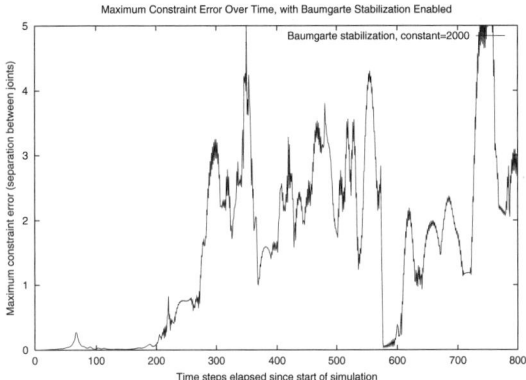

Fig. 5. Maximum constraint error over time, for a simulation of a swinging 6-link chain with Baumgarte stabilization, using a constant of 2000

We show the results of using post-stabilization in figures 6 and 7. Figure 6 shows two curves: one is the constraint error measured in each step *before* the postStabilization step is applied, and the other is the error *after* the postStabilization step. Note that the scale in figure 6 is only one tenth that of the scale in the Baumgarte stabilization graphs above. Even in this scale, the "after postStabilization" curve is barely perceptible. Figure 7 shows just this curve, on an even smaller scale. The constraint error in this curve never goes above 0.01mm.

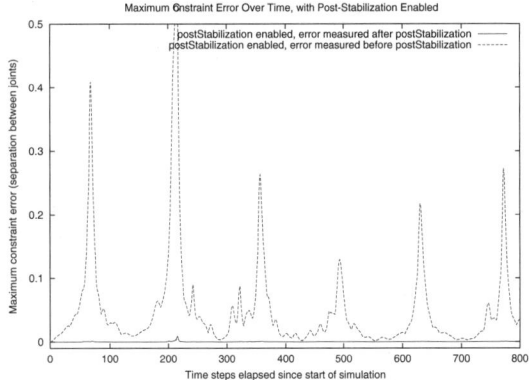

Fig. 6. Maximum constraint error over time, for a simulation of a swinging 6-link chain with post-stabilization enabled. Error is shown both before and after applying the post-stabilization step.

B. Stabilization of Contact Constraints

Contact constraints also have problems with constraint error. A contact constraint is said to have error if the separation distance at the contact is less than the *contact tolerance*. Figure 8 shows contact constraint error over a period of time when a rigid rectangle is pushed around on the screen by the user. Over this time period, the body experiences many different kinds of contact: impacts, resting contacts, sliding contacts and rolling contacts,

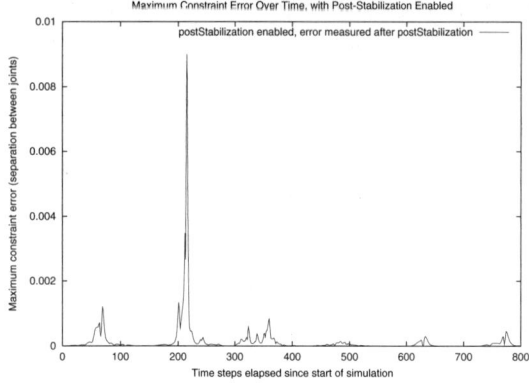

Fig. 7. Maximum constraint error over time, for a simulation of a swinging 6-link chain with post-stabilization enabled. This graph is a close-up of the smaller curve in figure 6

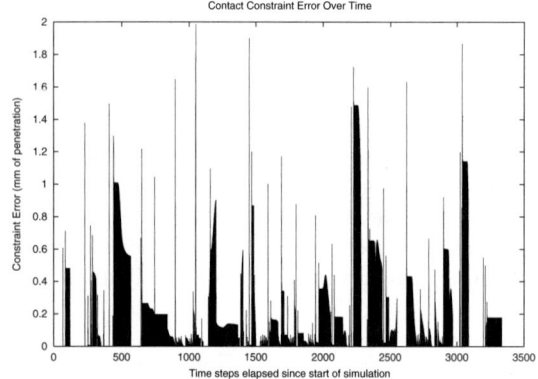

Fig. 8. Contact constraint error over time for a rigid rectangle undergoing user interaction. No stabilization is used.

sometimes with just one point of contact, sometimes with more than one (see figure 9 for some examples of typical motions).

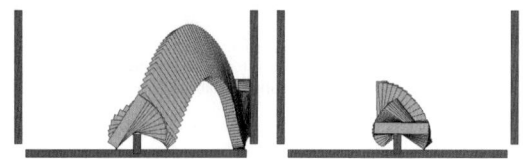

Fig. 9. Two screen captures from a contact simulation with a single moving rigid rectangle. User is pushing the rectangle interactively using repulsive forces at the mouse cursor.

When we run the same simulation with post-stabilization enabled, the constraint errors are eliminated completely. The post-stabilization scales well to simulations with many bodies and many contacts, such as the simulation shown in figure 10.

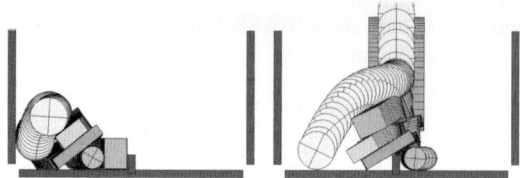

Fig. 10. Two screen captures from a contact simulation with five moving rigid bodies.

C. Advantages and Disadvantages of Post-Stabilization

Comparing Baumgarte stabilization with our post-stabilization method, here are some of the trade-offs:

- Baumgarte stabilization uses special constants that must be carefully set by hand. Post-stabilization does not require any tweaking.
- With post-stabilization, the constraint error usually is eliminated in the same time step that it is created. With Baumgarte stabilization, the constraint error cannot begin to affect the motion until one step later. Usually it takes several time steps for the error to dissipate completely. The amount of time it takes to absorb the constraint error depends on the values for the special constants. If the constants are too small, then the error will dissipate slowly. If the constants are too large, the stabilization will over-correct, causing jerky oscillations of the bodies.
- Baumgarte stabilization is easier to implement, and involves very little extra computation per time step. It only involves measuring the constraint error and adjusting the constant $\mathbf{k}$ in the constraint equation $\mathbf{Jv} + \mathbf{k} = 0$. Post-stabilization actually requires formulating and solving another mixed LCP in addition to the dynamics LCP. This is quite a large penalty because solving these LCPs is one of the main bottlenecks in simulation performance.[3]
- Post-stabilization can perform poorly when there are large errors in the constraints. This is because the linear approximation made in Equation 10 becomes a poor approximation. Usually this results in a disturbing crash of the simulation, with the bodies flying off wildly. It is also possible for the bodies to jump into interpenetration during the post-step in some circumstances. This can be helped some by restricting the size of the post-step, and taking multiple post-steps in the same time step, but with the penalties of having to solve more systems, and having to choose thresholds. In our experience, Baumgarte stabilization seems to perform better when there is a lot of constraint error, especially if the constants used are not too large.

---

[3]Recently, (after the present work was completed) Anitescu and Hart [AH02] have shown how the stabilization described in the paper can be done while solving a modification of the LCP (7). This saves the cost of solving a second LCP.

VII. CONCLUSION

We described a new technique for performing post-stabilization of contact constraints in a linear complementarity framework. Our approach requires no parameter tweaking and is effective in eliminating the drift problem. We presented examples that show the effectiveness of the technique, and discussed the tradeoffs in comparison to previous techniques. Thus constraint stabilization and dynamics simulation are unified in the same framework, which simplifies the design of contact simulation software.

VIII. REFERENCES

[ACR94] Uri M. Ascher, Hongsheng Chin, and Sebastian Reich. Stabilization of DAEs and invariant manifolds. *Numerische Mathematik*, 67(2):131–149, 1994.

[AH02] Mihai Anitescu and Gary D. Hart. A constraint-stabilized time-stepping approach for multi-body dynamics with contact and friction. *Reports on computational mathematics, ANL/MCS-P1002-1002, Mathematics and Computer Science division, Argonne National Laboratory,* December 2002.

[Ani97] Mihai Anitescu. *Modeling Rigid Multi Body Dynamics with Contact and Friction.* PhD thesis, University of Iowa, 1997.

[AP97] Mihai Anitescu and Florian Potra. Formulating dynamic multi-rigid-body contact problems with friction as solvable linear complementarity problems. *Nonlinear Dynamics*, 1997.

[AP98] Uri M. Ascher and Linda R. Petzold. *Computer Methods for Ordinary Differential Equations and Differential-Algebraic Equations.* Society for Industrial and Applied Mathematics, 1998.

[Asc97] Uri M. Ascher. Stabilization of invariants of discretized differential systems. *Numerical Algorithms*, 14(1–3):1–24, 1997.

[Bar94] David Baraff. Fast contact force computation for nonpenetrating rigid bodies. *Computer Graphics*, 28(Annual Conference Series):23–34, 1994.

[Bar96] David Baraff. Linear-time dynamics using Lagrange multipliers. *Computer Graphics*, 30(Annual Conference Series):137–146, 1996.

[Bau72] J. Baumgarte. Stabilization of constraints and integrals of motion in dynamical systems. *Computer Methods in Applied Mechanics and Engineering*, (1):1–16, 1972.

[CD68] R. W. Cottle and G. B. Dantzig. Complementary pivot theory of mathematical programming. *Linear Algebra and Appl.*, (1), 1968.

[Cli02] Michael B. Cline. Rigid body simulation with contact and constraints. Master's thesis, University of British Columbia, July 2002.

[CPS92] R.W. Cottle, J.S. Pang, and R.E. Stone. *The Linear Complementarity Problem.* Academic Press, Inc., 1992.

[LNPE92] C. Lubich, U. Nowak, U. Pohle, and Ch. Engstler. Mexx – numerical software for the integration of constrained mechanical multibody systems. Technical Report SC92-12, Konrad-Zuse-Zentrum fur Informationstechnik, 1992.

[Lot82] P. Lotstedt. Mechanical systems of rigid bodies subject to unilateral constraints. *SIAM Journal on Applied Mathematics*, 1982.

[Mir98] B. Mirtich. Rigid body contact: Collision detection to force computation, 1998.

[Mur88] Katta G. Murty. *Linear Complementarity, Linear and Nonlinear Programming*[4]. Heldermann Verlag, Berlin, 1988.

[ST96] D. E. Stewart and J. C. Trinkle. An implicit time-stepping scheme for rigid body dynamics with inelastic collisions and coulomb friction. *Internat. J. Numer. Methods Engineering*, 1996.

---

[4]This book is out of print, but is currently available online at http://ioe.engin.umich.edu/people/fac/books/murty/linear_complementarity_webbook/

Point-to-Point Paths Generation for Wheeled Mobile Robots

Diogo P. F. Pedrosa   Adelardo A. D. Medeiros   Pablo J. Alsina

UFRN – CT – DCA
59072-970, Natal, RN, Brazil
[diogo, adelardo, pablo]@dca.ufrn.br

Abstract

This paper proposes a point-to-point path generation method consistent with the non-holonomic constraints of a two-wheeled mobile robot. The generated path is described by continuous curves, instead of the traditional chaining of different curves. Parametric polynomials of third degree are used to calculate the robot configuration variables, $x(\lambda)$ and $y(\lambda)$. The orientation angle, $\theta(\lambda)$, is imposed to respect the non-holonomic constraint. The free polynomial coefficients are used to refine the path, avoiding maximal or minimal values of $x(\lambda)$ and $y(\lambda)$ which tends to generate shorter paths, avoiding unnecessary motions.

1 Introduction

Path generation is one of the main problems in mobile robot navigation. Frequently, this generation starts by planning what we call a *geometric path* (a path avoiding collisions with obstacles). The main techniques to find these geometric paths were compiled by Latombe [10]. They are based on well-founded algorithms and widely used in robotics systems [2, 3, 8, 12].

When a robot has kinematics constraints, the paths computed by these classic geometric planners are not directly executable by the robot. To solve this problem, it is necessary to adapt the geometric path, finding an *admissible path*. After that, the desired velocities are incorporated into the admissible path to generate the robot trajectory. Finally, the trajectory can be executed by the control level. An overview of a complete navigation robot system based on this approach is shown in fig. 2. This work proposes a technique to be used in the path adaptation module of such a system (level 2 in fig. 2): we do not deal neither with the obstacle avoidance problem (level 1) nor with the execution control of the robot trajectory (level 4).

One well-known method to compute admissible paths was proposed by Laumond *et al* [11]:

- we divide the geometric path in n segments, creating $n+1$ vertices;
- we compute *point-to-point paths* between all pairs of adjacent vertices, without considering collisions; the point-to-point paths must be directly executable by the robot;
- we check if these point-to-point paths introduce collisions with obstacles; in affirmative case, we increase the number n of segments and repeat the process; if not, this set of chained point-to-point paths is the definitive admissible path to the robot.

This approach requires calculating point-to-point paths between two distinct configurations of the robot. Many solutions were proposed [4, 5, 6, 7, 15, 17, 18]: most of them are direct or indirectly influenced by the seminal technique proposed by Reeds and Shepp [16]. They proved that the shortest path between two distinct configurations for a car which can move forward and backward is composed by arcs of circle of minimum radius and line segments. For a two wheeled mobile robot as in fig. 1, however, the minimum turning radius is zero, degenerating the arc of circle motion into a rotation about the robot's geometric center. In this case, the point-to-point paths calculated using the Reeds and Shepp's technique are composed of:

- a rotation about its geometric center to point to the next desired position;
- a straight line motion to this position;
- a rotation to reach the final orientation;

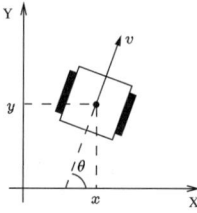

Figure 1: Wheeled mobile robot with null turning radius and configuration variables x, y and θ

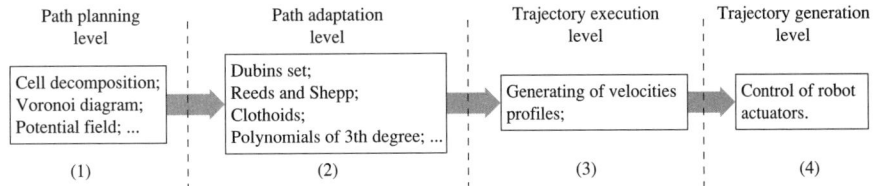

Figure 2: A robot navigation system based on adaptation of geometric paths

In some dynamics situations where the goal can quickly change its position, like soccer robots environments, the Reeds and Shepp's paths are not appropriate because the robot should stop its current line trajectory to adjust itself to a new orientation or can stand rotating for a long time trying to point to the moving goal. To avoid these inconveniences, this work proposes a alternative method to link two distinct configurations in the robot workspace.

The proposed technique is based on using parametric polynomials of third degree to calculate the configuration variables $x(\lambda)$ and $y(\lambda)$. The orientation angle $\theta(\lambda)$ is imposed to coincide with the path tangent because of the non-holonomic constraints. The λ parameter has values between (0, 1), where these extremal values indicate the initial and final configurations, respectively. We choose polynomials of third degree because they have the minimum degree where there are free parameters to refine the paths, as shown in the section 5. An interesting characteristic of the proposed method in real time application is the path determination by a closed formula of fast calculus, permiting the path determination at each sampling step, while Reeds and Shepp's method, for example, has 48 possible paths which must be analyzed in order to find the shortest path.

In section 2, the proposed point-to-point path generator is presented; in section 3, the mathematical singularities are analyzed; in section 4, it is shown how to refine the path in accord to the criterion presented in the section 5; some examples and conclusions are shown in sections 6 and 7, respectively.

2 Paths Generation

Point-to-point paths generation has to consider the non-holonomic constraints. In the case of robots with differential drive[1], the orientation angle must point to its linear velocity vector. We propose generating the configuration variables x and y by parametric polynomials of third degree with unknown coefficients (eqs. 1). Many others works has also used polynomials to generate feasible paths [1, 9, 13].

$$\begin{aligned} x(\lambda) &= a_0 + a_1\lambda + a_2\lambda^2 + a_3\lambda^3 \\ y(\lambda) &= b_0 + b_1\lambda + b_2\lambda^2 + b_3\lambda^3 \end{aligned} \quad (1)$$

The orientation angle θ is imposed to satisfy the non-holonomic constraints (eq. 2):

$$\theta(\lambda) = \tan^{-1}\left(\frac{dy/d\lambda}{dx/d\lambda}\right) = \tan^{-1}[d(\lambda)] \quad (2)$$

where $d(\lambda)$ is introduced to simplify the notation.

$$d(\lambda) = \tan[\theta(\lambda)] = \frac{b_1 + 2b_2\lambda + 3b_3\lambda^2}{a_1 + 2a_2\lambda + 3a_3\lambda^2} \quad (3)$$

In eqs. 1 and 3, λ varies in the interval (0, 1). When $\lambda = 0$, the robot is on its initial configuration (x_i, y_i, $d_i = \tan(\theta_i)$) and similarly when $\lambda = 1$, the robot is on its final configuration (x_f, y_f, $d_f = \tan(\theta_f)$). Applying these contour conditions to eqs. 1 and 3, a linear system with 6 equations and 8 variables is obtained. This system is shown in eq. 4.

$$\begin{cases} a_0 = x_i \\ b_0 = y_i \\ b_1 = d_i a_1 \\ a_0 + a_1 + a_2 + a_3 = x_f \\ b_0 + b_1 + b_2 + b_3 = y_f \\ b_1 + 2b_2 + 3b_3 = d_f(a_1 + 2a_2 + 3a_3) \end{cases} \quad (4)$$

We can arbitrate two of the eight coefficients to solve this linear system. Choosing a_1 and a_2 as the free variables, we can deduce the others coefficients without division operations, avoiding division-by-zero singularities. The solution is given by eq. 5, where $\Delta x = x_f - x_i$ and $\Delta y = y_f - y_i$.

$$\begin{cases} a_0 = x_i \\ a_3 = \Delta x - a_1 - a_2 \\ b_0 = y_i \\ b_1 = d_i a_1 \\ b_2 = 3(\Delta y - d_f \Delta x) + 2(d_f - d_i)a_1 + d_f a_2 \\ b_3 = 3d_f \Delta x - 2\Delta y - (2d_f - d_i)a_1 - d_f a_2 \end{cases} \quad (5)$$

---
[1]Two parallel and independently driven wheels.

Any values can be attributed to the coefficients a_1 and a_2, resulting in a path that respects the non-holonomic constraints and the contour conditions. However, the generated paths sometimes will not be adequate, as shown in section 4. This leads us to use some criterion to find good values to the free coeficients.

3 Mathematical Singularities

The result in eq. 5 is applicable to most situations, excepting when θ_i and/or θ_f are equals to $\pm\pi/2$, because d_i and d_f tend to $\pm\infty$. Thus, there are three situations where the eq. 3 has to be analyzed in order to redefine the system 4.

The first case occurs when both θ_i and θ_f are equals to $\pm\pi/2$. The new resulting system is given by eqs. 6, with b_1 and b_2 as free variables.

$$\begin{cases} a_0 = x_i \\ a_1 = 0 \\ a_2 = 3\Delta x \\ a_3 = -2\Delta x \\ b_0 = y_i \\ b_3 = \Delta y - b_1 - b_2 \end{cases} \quad (6)$$

The second case occurs if only $\theta_i = \pm\pi/2$. The free coefficients a_3 and b_3 are chosen, resulting in eqs. 7.

$$\begin{cases} a_0 = x_i \\ a_1 = 0 \\ a_2 = \Delta x - a_3 \\ b_0 = y_i \\ b_1 = 2(\Delta y - d_f \Delta x) - d_f a_3 + b_3 \\ b_2 = (2d_f \Delta x - \Delta y) + d_f a_3 - 2b_3 \end{cases} \quad (7)$$

Finally, the third case occurs when only θ_f is equal to $\pm\pi/2$. We can choose a_1 and b_2 as free variables and the result is shown in eqs. 8.

$$\begin{cases} a_0 = x_i \\ a_2 = 3\Delta x - 2a_1 \\ a_3 = a_1 - 2\Delta x \\ b_0 = y_i \\ b_1 = d_i a_1 \\ b_3 = \Delta y - d_i a_1 - b_2 \end{cases} \quad (8)$$

Similarly to the general case, any values to the free coefficients will result in paths respecting the non-holonomic constraints. However, these paths are not necessarily appropriate to the mobile robot.

4 Non-Adequate Paths

Although the paths generated by eqs. 5, 6, 7 and 8 satisfy the non-holonomic constraints and the contour conditions, sometimes these paths are not appropriate. An example is shown in the fig. 3, where the a_1 and a_2 values ($a_1 = -1.6863$ and $a_2 = 2.4863$) were specially chosen to generate a non-adequate path.

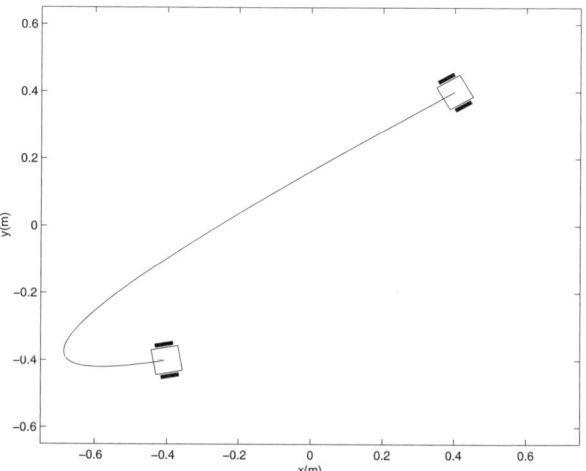

Figure 3: Example of non adequate path

This path is not "intelligent" because it contains a unnecessary backward movement with relation to the final configuration, that could be attained by a completely forward movement. Thus, it is important to use a criterion to calculate the free coefficients in order to improve the path generation.

5 Improvement Criterion

Several improvement criteria can be used to calculate the free coefficients, like minimal length, minimal or maximal rotation radius, etc. The criterion we decide to adopt in this article is to force the $x(\lambda)$ and $y(\lambda)$ polynomials (eqs. 1) to be monotonically increasing or decreasing between $0 < \lambda < 1$. As we do not allow the polynomials having maximum or minimum points in the interval, we eliminate unnecessary backward movements. Another interesting benefit is that the paths are bounded by a rectangle defined by the initial and final positions, if this criterion could be satisfied in both x and y directions. Thus, considering the following derivatives with relation to λ (eqs. 9):

$$\begin{aligned} dx/d\lambda &= a_1 + 2a_2\lambda + 3a_3\lambda^2 \\ dy/d\lambda &= b_1 + 2b_2\lambda + 3b_3\lambda^2 \end{aligned} \quad (9)$$

their roots must satisfy at least one of the following conditions:

- both roots are complex; or
- both roots are smaller or equals to 0; or
- both roots are greater or equals to 1; or
- one root is smaller or equal to 0 and the other is greater or equal to 1.

The first condition implies the discriminant (Δ) must be negative; the other three conditions can be mathematically calculated using the Routh-Hurwitz algorithm [14]. Applying them separately to eqs. 9 we can find two distinct systems of inequations which bound regions in the free coefficients' space which improve the path in the x and y directions. These free coefficients are different for the general and singular cases and a separated analyze will be realized.

5.1 Regions for the General Case

The region $a_1 \times a_2$ satisfying the improvement criterion for $x(\lambda)$ is bounded by the system of ineqs. 10. For $\Delta x > 0$, this region can be visualized in fig. 4.

$$\begin{cases} a_1 \geq 0 \\ a_2 \geq -a_1 \\ a_2 \leq 3\Delta x - 2a_1 \end{cases} \quad (10)$$

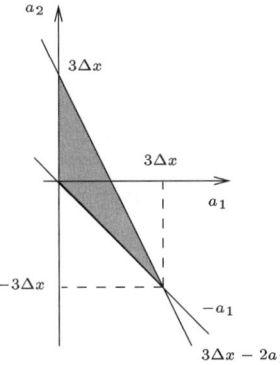

Figure 4: Region satisfying the x constraints

A similar calculus can be realized for the y direction, obtaining the region bounded by the system of ineqs. 11. Considering Δx, Δy, d_i and d_f positives this region can be shown in fig. 5.

$$\begin{cases} d_i a_1 \geq 0 \\ d_f a_2 \geq (d_i - 2d_f)a_1 + 3(d_f \Delta x - \Delta y) \\ d_f a_2 \leq 3d_f \Delta x - 2d_f a_1 \end{cases} \quad (11)$$

To improve the path in both directions, it is necessary to find a region derived from the intersection

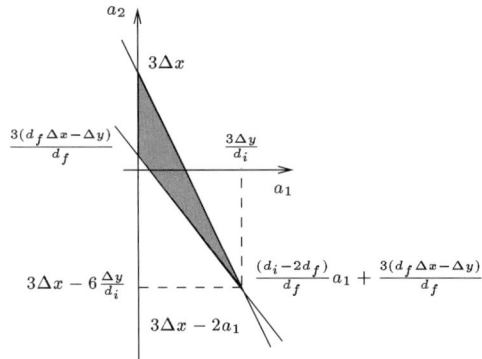

Figure 5: Region satisfying the y-constraints

between the regions for the x and y directions. In the examples shown in figs. 4 and 5 this intersection exists, but this is not always the case. Analyzing the systems 10 and 11 we can conclude that an intersection only exists when the initial and final orientations are pointing to inside of a rectangle defined by the initial and final positions. If this does not occur, the path can only be improved in one of the two directions. Figure 6 illustrates these two situations.

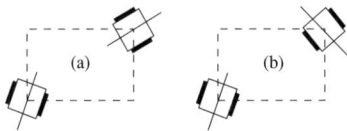

Figure 6: Situations where (a) it is and (b) it is not possible improving the path in both directions

5.2 Regions for the Singular Cases

The same improvement criterion imposed to the general case is applied to the cases where there are singularities, shown in section 3.

Considering the case where θ_i and θ_f are equals to $\pm\pi/2$ the region that improves the path in the y direction is defined by ineqs. 12. The criterion is always satisfied in the x direction.

$$\begin{cases} b_1 \geq 0 \\ b_2 \geq -b_1 \\ b_2 \leq -2b_1 + 3\Delta y \end{cases} \quad (12)$$

When only $\theta_i = \pm\pi/2$ the region that satisfies the criterion in the x direction is:

$$-2\Delta x \leq a_3 \leq \Delta x$$

while for the direction y is:

$$\begin{cases} a_3 \geq -2\Delta x \\ b_3 \leq \Delta y \\ b_3 \geq d_f a_3 - 2(\Delta y - d_f \Delta x) \end{cases}$$

Finally, when only $\theta_f = \pm \pi/2$ the result in the x direction is:

$$0 \leq a_1 \leq 3\Delta x$$

and to the direction y the region is:

$$\begin{cases} a_1 \geq 0 \\ b_2 \geq -d_i a_1 \\ b_2 \leq -2d_i a_1 + 3\Delta y \end{cases}$$

6 Examples

Figure 7 shows a path where the improvement criterion could be applied in both directions. The initial and final configurations are the same used in fig. 3. As we can see, this result exhibits a better performance.

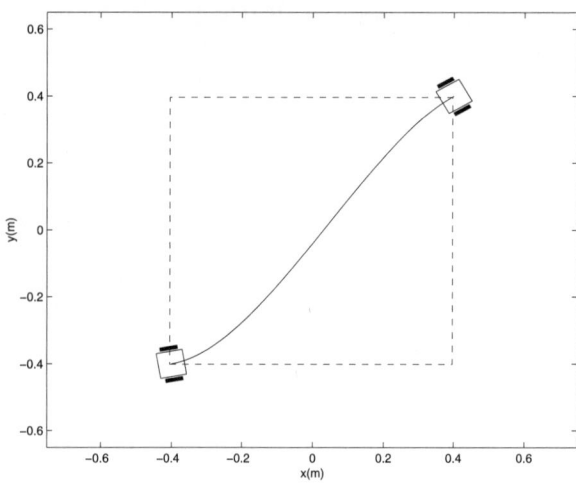

Figure 7: Improved path satisfying the criterion in both x and y directions

Figure 8 exemplifies a situation where it is not possible attending the improvement criterion in both x and y directions simultaneously. Thus, we show two paths where the first one is improved in the x direction and the second one, in the y direction.

7 Conclusions

The main contribution of this work is a point-to-point path generator which obeys the non-holonomic constraints for a wheeled mobile robot with differential

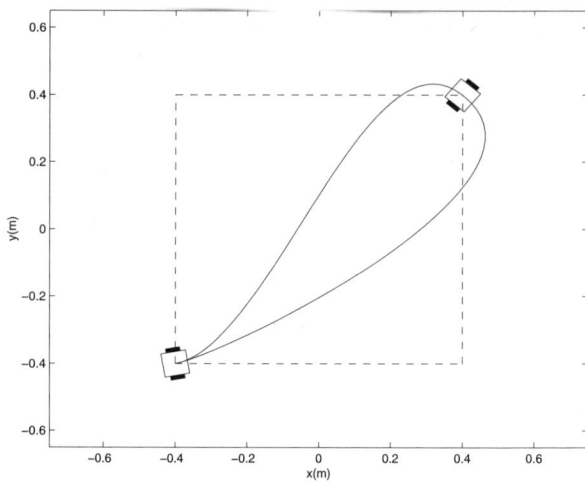

Figure 8: Improved paths satisfying the criterion in the x and y directions, separately

drive. The paths are described by continuous polynomials functions and not by a chaining of different segments. They are easily calculated, allowing a new path to be found at each sampling step. As there are no phases in the path execution, a new path can be easily adopted during the execution of a previous one.

The proposed method can be used to convert geometric paths in feasible paths in such a way they can be executed by non-holonomic robots. In environments without or with sparse obstacles (as a robot soccer environment), the geometric planner can be suppressed and the proposed method can be used as the unique path planner, combined with local avoiding collision techniques.

The adopted improvement criterion guarantees a monotonic reduction of the distance from the current robot position to the goal position, in the better situation. In the worst case the criterion guarantees the improvement in one of the x or y directions.

Possible extensions of this work include velocity profile generators for the point-to-point paths, comparative tests with others adapters and defining some other improvement criteria. Higher order polynomials can also be used to guarantee multi-criteria improvement.

8 Acknowledgments

This work was partially supported by the Brazilian agencies FINEP and CNPq.

References

[1] Akira Arakawa, Masayuki Hiyama, Takashi Emura, and Yoshiyuki Kagami. Trajectory generation for wheeled mobile robot based on landmarks. In *Proceedings of the 1995 IEEE International Conference on Systems, Man and Cybernetics*, Vancouver, Canada, October 1995.

[2] Howie Choset and Joel Burdick. Sensor based planning, part 01: The generalized voronoi graph. In *Proceedings of the IEEE International Conference of Robotics and Automation*, Nagoya, Japan, 1995.

[3] Howie Choset and Joel Burdick. Sensor based planning, part 02: Incremental construction of the generalized voronoi graph. In *Proceedings of the IEEE International Conference of Robotics and Automation*, Nagoya, Japan, 1995.

[4] Guy Desaulniers and François Soumis. An efficient algorithm to find a shortest path for a car-like robot. *IEEE Transactions on Robotics and Automation*, 11(6), December 1995.

[5] Adam W. Divelbiss and John T. Wen. A path space approach to nonholonomic motion planning in the presence of obstacles. *IEEE Transactions on Robotics and Automation*, 13(3), June 1997.

[6] Sara Fleury, Philippe Souères, Jean-Paul Laumond, and Raja Chatila. Primitives for smoothing mobile robot trajectories. *IEEE Transactions on Robotics and Automation*, 11(3), June 1995.

[7] Atsushi Fujimori, Peter N. Nikiforuk, and Madan M. Gupta. Adaptive navigation of mobile robots with obstacle avoidance. *IEEE Transactions on Robotics and Automation*, 13(4), August 1997.

[8] Yong K. Hwang and Narendra Ahuja. A potential field approach to path planning. *IEEE Transactions on Robotics and Automation*, 8(1), February 1992.

[9] Maher Khatib, Hazem Jaouni, Raja Chatila, and Jean-Paul Laumond. Dynamic path modification for car-like nonholonomic mobile robots. In *Proceedings of 1997 IEEE International Conference on Robotics and Automation*, Albuquerque, New Mexico, USA, April 1997.

[10] Jean-Claude Latombe. *Robot Motion Planning*. Kluwer Academic Press, 1991.

[11] Jean-Paul Laumond, Paul E. Jacobs, Michel Taïx, and Richard M. Murray. A motion planner for nonholonomic mobile robots. *IEEE Transactions on Robotics and Automation*, 10(5), October 1994.

[12] J. L. Díaz de León and J. H. Sossa. Automatic path planning for a mobile robot among obstacles of arbitrary shape. *IEEE Transactions on Systems, Man, and Cybernetics*, 28(3), June 1998. Part B: Cybernetics.

[13] Mark B. Milam, Kudah Mushambi, and Richard M. Murray. A new computational approach to real-time trajectory generation for constrained mechanical systems. In *Proceedings of the 39th IEEE Conference on Decision and Control*, Sydney, Australia, December 2000.

[14] Katsuhiko Ogata. *Modern Control Engeneering*. Pretence Hall, 1997. Third Edition.

[15] Leszek Podsędkowski, Jacek Nowakowski, Marec Idzikowski, and Istvan Vizvary. A new solution for path planning in partially known or unknown enviroment for nonholonomic mobile robots. *Robotics and Autonomous Systems*, 34, 2001.

[16] J. A. Reeds and L. A. Shepp. Optimal paths for a car that goes both forwards and backwards. *Pacific Journal of Mathematics*, 145(2), 1990.

[17] A. Scheuer and Th. Fraichard. Continuous-curvature path planning for car-like vehicles. In *Proceedings of the IEEE International Conference on Intelligent Robots and Systems*, Grenoble, France, September 1997.

[18] Andrei M. Shkel and Vladimir Lumelsky. Classification of the dubins set. *Robotics and Autonomous Systems*, 34, 2001.

Optimization-based Formation Reconfiguration Planning For Autonomous Vehicles

Shannon Zelinski
Department of EECS
University of California
Berkeley, CA 94720
shannonz@eecs.berkeley.edu

T. John Koo
Department of EECS
University of California
Berkeley, CA 94720
koo@eecs.berkeley.edu

Shankar Sastry
Department of EECS
University of California
Berkeley, CA 94720
sastry@eecs.berkeley.edu

Abstract

Given a group of autonomous vehicles, an initial configuration, a final configuration, a set of inter- and intra- vehicle constraints, and a time for reconfiguration, the Formation Reconfiguration Planning problem is focused on determining a nominal input trajectory for each vehicle such that the group can start from the initial configuration and reach its final configuration at the specified time while satisfying the set of inter- and intra-vehicle constraints. In this paper, we are interested in solving the Formation Reconfiguration Planning problem for a specific class of systems and a particular form of input signals so that the problem can be reformulated as an optimization problem which can be solved more efficiently, especially for a large group of vehicles.

1 Introduction

Advances in sensing, communication and computation are revolutionizing the development of advanced control technologies for distributed, multi-vehicle systems. These advances also enable the conduct of missions deemed impossible in the recent past. Autonomous formations have applications anywhere there is a task to be done requiring a group effort with minimal human supervision. Space applications benefit from formation control of satellites to perform distributed observations. In automated highway systems (AHS), cars organize themselves in platoons to increase highway throughput. Groups of unmanned aerial vehicles (UAVs) perform search and rescue collectively in restricted areas where human intervention is dangerous. To perform deep sea exploration, autonomous underwater vehicles (AUVs) must maintain tight formations due to limited bandwidth for communication and low visibility. In order to perform a set of predetermined missions, each vehicle is equipped with the necessary sensing, communication, and computation capabilities.

Recent years have seen the emergence of autonomous formation planning and control as a topic of great interest. In [1] and [2], by considering vehicles as linear systems, the problem of stabilizing a set of vehicles according to a given graph structure by utilizing only relative measurements is studied. The feasibility for keeping vehicles in such a given formation is studied in [1]. [4] also studies the feasibility for keeping vehicles in formation, while also considering a kinematical model of vehicle. Given a formation and vehicle dynamics, various control strategies have been developed based on information flow and group organization. In [2], a distributed control law for stabilizing the formation is derived for keeping a feasible formation around the group equilibrium point. A control design that preserves mesh stability of a group of vehicle is presented in [3] where a leader-follower organization is considered in the control design and only local information is used. In [5] a different control strategy based on virtual leaders and artificial potentials in order to keep a stable formation is considered. A method of formation reconfiguration planning and control for a group of vehicles in order to avoid obstacles is presented in [6]. The reconfiguration planning is based on enumeration of formation graphs to obtain next possible formation that can be used to navigate in the environment.

We are interested in the formation reconfiguration problem. In particular, we first consider if the problem is feasible assuming that all the information is accessible. Given a feasible problem, we will then be able to derive the necessary information flow and group organization for decentralized formation planning and control. Here, we are interested in solving the feasibility problem. The *Formation Reconfiguration Planning* (FRP) problem addressed in this paper is:

Problem 1.1 *Given a group of autonomous vehicles, an initial configuration, a final configuration, a set of inter- and intra- vehicle constraints, and a time for reconfiguration determine a nominal input trajectory for each vehicle such that the group can start from the initial configuration and reach its final configuration at the specified time while satisfying the set of inter- and intra-vehicle constraints.*

In this paper, we are interested in solving the FRP problem for a specific class of systems and a particular form of input signals so that we can represent the problem as an optimization problem that can be solved more efficiently especially for a large group of vehicles.

In particular, a point mass model is used to model dynamics of each autonomous vehicle. Although the simple dynamical model is used, the results can be naturally extended to systems that can be feedback linearized [7] such as S/VTOL aircraft [8], PVTOL aircraft [9] and helicopters[10].

The paper structure is as follows. We will begin by formulating the formation reconfiguration planning problem. Then we will discuss our approach to solving the problem. This will be followed by a design example using our solution and a presentation of some results gathered by simulating the design example. The paper will end with conclusions.

2 Problem Formulation

Consider a group of autonomous vehicles with the following dynamics

$$\dot{x}_i(t) = f_i(x_i(t), u_i(t)) \quad (1)$$

where the ith vehicle state $x_i \in \mathbb{R}^n$, the ith vehicle input $u_i \in \mathbb{R}^m$ and $i = 1, \cdots, N$. Each $f_i : \mathbb{R}^n \times \mathbb{R}^m \to \mathbb{R}^n$ is assumed to be as smooth as needed. The admissible input for each vehicle is specified by an input constraint $b_i(u_i(t)) \leq \alpha_i$ such as $\|u_i(t)\| \leq \alpha_i$. Denoting the group state as $x = [x_1^T, \cdots, x_N^T]^T$ and the group input as $u = [u_1^T, \cdots, u_N^T]^T$, the group dynamics can be rewritten as

$$\dot{x}(t) = F(x(t), u(t)) \quad (2)$$

where

$$F(x(t), u(t)) = \begin{bmatrix} f_1(x_1(t), u_1(t)) \\ \vdots \\ f_N(x_N(t), u_N(t)) \end{bmatrix}$$

with $x \in \mathbb{R}^{nN}$ and $u \in \mathbb{R}^{mN}$. Assume that all the inter- and intra-vehicle constraints are specified as a set of group state constraints $c_j(x(t)) \leq \beta_j$ for $j = 1, \cdots, M$. Especially, since we are interested in generating collision-free paths, a minimal separation requirement between vehicles is introduced such that each vehicle can keep a safe distance from any vehicle in the group. Thus, the minimal separation requirement for each pair of vehicles can be encoded as a group state constraint and hence there are $\frac{N(N-1)}{2}$ minimal separation constraints.

Define the group configuration at time t as $g(t) = [x^T(t), u^T(t)]^T$ which specifies the state and input conditions for all the vehicles in the group at time t.

In a mission, a cost function is given as a part of the mission specification and in general can be written as $J = \varphi(x(T), T) + \int_0^T L(x(t), u(t), t)dt$ where $\varphi(x(T), T)$ and $L(x(t), u(t), t)$ define the terminal cost and the running cost, respectively. Hence, if there are feasible solutions for the FRP problem as specified in Problem 1.1, it is desirable to find the optimal one with respect to the given cost function. Now, we restate our FRP problem as follows:

Problem 2.1 (FRP Problem) *Given a group dynamics, an initial group configuration g_s, a final group configuration g_f, a set of inter- and intra-vehicle constraints $b_i(u_i(t)) \leq \alpha_i$ for $i = 1, \ldots, N$ and $c_j(x(t)) \leq \beta_j$ for $j = 1, \ldots, M$, and the time for reconfiguration T, does there exit a group input $a(t)$ for $t \in [0, T]$ such that the group starting from $g(0) = g_s$ can reach $g(T) = g_f$ while satisfying the set of inter- and intra- vehicle constraints? If so, then select the group input $a(t)$ over $[0, T]$ which produces minimal value for a given cost function.*

The Formation Reconfiguration Planning (FRP) problem can be formulated as an optimal control problem[11, 12, 13] with dynamical and algebraic constraints as follows:

$$\min_{u(t)} J = \varphi(x(T), T) + \int_0^T L(x(t), u(t), t)dt \quad (3)$$

subject to

$$\dot{x}(t) = F(x(t), u(t)) \quad (4)$$
$$g(0) = g_s \quad (5)$$
$$g(T) = g_f \quad (6)$$
$$b_i(u_i(t)) \leq \alpha_i \ \forall t \in [0, T] \ \forall i \in \{1, \ldots, N\} \quad (7)$$
$$c_j(x(t)) \leq \beta_j \ \forall t \in [0, T] \ \forall j \in \{1, \ldots, M\}. \quad (8)$$

The optimal control problem in principle can be solved by applying standard techniques described in [11, 12, 13] based on calculus of variations or on Pontryagin's maximum principle. However, for a large group of vehicles these techniques become computationally inefficient since the performance of these techniques scales poorly not only with the number of states but also with the number of inter- and intra- constraints which increase rapidly with the number of vehicles. For example, the number of minimal separation constraints grows in the order of $O(N^2)$.

In this paper, we are interested in solving the FRP problem for a specific class of systems and a particular form of input signals so that the problem can be reformulated as an optimization problem that can be solved more efficiently especially for a large group of vehicles.

3 Approach

In order to reduce the problem complexity, we parameterize each input signal $u_i(t)$ over the interval $[0, T]$ by a set of parameters $\theta_{ki} \in \mathbb{R}^m$ for $k = 1, \cdots, K$. Hence, for $t \in [0, T]$ we have

$$u_i(t) = \sum_{k=0}^{K} \theta_{ki}\omega_k(t)$$

where $\omega_k : \mathbb{R} \to \mathbb{R}$ are basis functions for input $u_i(t)$. Many families of basis functions such as B-splines can be chosen and each would provide different advantages on representation and efficiency. This approach is proposed in [14] for solving many motion planning problems. In this paper, polynomials are used as basis functions and hence

$$u_i(t) = \sum_{k=0}^{K} \theta_{ki} t^k \quad (9)$$

where $\omega_k(t) = t^k$ for $k = 1, \ldots, K$. However, the selection of the order of polynomials K is problem dependent. In the next section, we will show how to pick the order of polynomials for an application. Therefore, the FRP problem becomes:

$$\min_{\theta_{01},\ldots,\theta_{KN}} J = \varphi(x(T), T) + \int_0^T L(x(t), u(t), t) dt \quad (10)$$

subject to

$$\dot{x}(t) = F(x(t), u(t)) \quad (11)$$
$$g(0) = g_s \quad (12)$$
$$g(T) = g_f \quad (13)$$
$$b_i(u_i(t)) \leq \alpha_i \; \forall t \in [0,T] \; \forall i \in \{1,\ldots,N\} \quad (14)$$
$$c_j(x(t)) \leq \beta_j \; \forall t \in [0,T] \; \forall j \in \{1,\ldots,M\}. \quad (15)$$

Depending on applications, various cost functions could be considered. However, the same set of constraints specified by (10)-(15) has to be satisfied regardless of which cost function is chosen. Once the feasible parameter range is obtained, then one can solve the FRP problem by searching for minimal value of the cost function over the range. Here, we are interested in the existence of solution for the FRP problem.

Problem 3.1 (Existence of Solution for FRP)
Given the FRP problem specified by (10)-(15), does there exit a set of parameters $\theta_{11}, \ldots, \theta_{KN}$ such that all the constraints (11)-(15) can be satisfied? If so, then determine the feasible range of the parameters.

For certain classes of systems, Problem 3.1 can be solved by using computational tools.

Theorem 3.2 *Given the FRP problem specified by (10)-(15) if $F(x(t), u(t)) = Ax(t) + Bu(t)$, where A is a $nN \times nN$ nilpotent matrix, B is a $nN \times mN$ matrix, and constraints specified by (12)-(15) are semi-algebraic constraints, there exits a computational procedure that decides whether there exits a set of parameters that satisfies (12)-(15).*

The system equation $\dot{x}(t) = Ax(t) + Bu(t)$ with nilpotent matrix A and polynomial input $u(t)$ belongs to a family of linear differential equations with decidable reachability problem [15]. Theorem 3.2 can be proved by posing the reachability computation as a quantifier elimination problem in the decidable theory of the reals. There are quantifier elimination tools that can perform symbolic computation and answer the existence problem. Since the problem is proved to be decidable for this class of systems, the computation is guaranteed to terminate in finite steps.

A clear illustration on how to apply the theory is provided in [15] for a single robot navigation problem. Furthermore, a feasible range of the parameters is also provided. The current algorithms for solving quantifier elimination are not able to handle problems with a large number of constraints or high order polynomials.

4 Design example

Now, we focus on the point-mass dynamics of N vehicles. The dynamics of each vehicle is then specified by a double integrator which is:

$$\begin{bmatrix} \dot{p}_i(t) \\ \dot{v}_i(t) \end{bmatrix} = \begin{bmatrix} v_i(t) \\ a_i(t) \end{bmatrix} \quad (16)$$

where $p_i, v_i, a_i \in \mathbb{R}^3$ and $i = 1, \cdots, N$. Define $x_i(t) = [p_i^T(t) \; v_i^T(t)]^T$, $u_i(t) = a_i(t)$. Hence, (16) can be written as $\dot{x}_i(t) = A_i x_i(t) + B_i u_i(t)$ with

$$A_i = \begin{bmatrix} 0_{3\times3} & I_{3\times3} \\ 0_{3\times3} & 0_{3\times3} \end{bmatrix}, \; B_i = \begin{bmatrix} 0_{3\times3} \\ I_{3\times3} \end{bmatrix}.$$

Thus, the group dynamics can be written as $\dot{x}(t) = Ax(t) + Bu(t)$ where $A \in \mathbb{R}^{nN \times nN}$ and $B \in \mathbb{R}^{nN \times nM}$ with $x = [x_1^T \cdots x_N^T]^T \in \mathbb{R}^{nN}$ and $u = [u_1^T \cdots u_N^T]^T \in \mathbb{R}^{nM}$.

Given the input signals, the state trajectories of each vehicle can be derived according to the vehicle dynamics. In particular, we have the following equations $v_i(t) = \int_0^t a_i(t)dt + v_i(0)$ and $p_i(t) = \int_0^t v_i(t)dt + p_i(0)$ Therefore, if the input trajectories and the initial conditions are provided, we can derive the state trajectories. The input signals are parameterized as polynomials of time. $a_i(t)$ is the acceleration vector of the ith vehicle and it can represented as:

$$a_i(t) = \sum_{k=0}^{K} a_{ki} t^k \quad (17)$$

where K is the order of the polynomial, and $a_{0i} \cdots a_{Ki}$ are the parameter vectors for the ith vehicle. As with $a_i(t)$, all of the a_{ki} parameter vectors are in $\mathbb{R}^3$.

Given the initial configuration g_s, the final configuration g_f, and the time for reconfiguration T, the state trajectories are constrained by the four vector equations $a_i(0) = a_{i0}$, $a_i(T) = \sum_{k=0}^{K} a_{ki}T^k$, $v_i(T) = \int_0^T a_i(t)dt + v_i(0)$, $p_i(T) = \int_0^T v_i(t)dt + p_i(0)$ for $i = 1, \ldots, N$. Therefore, in order to obtain feasible solutions for the FRP problem by considering only the dynamical and configuration constraints, $K \geq 4$. The necessary order of polynomials K thus depends on the remaining constraints.

Now, we are ready to formulate the FRP problem for autonomous vehicles. The objective is to determine the parameters for the input trajectories, minimizing the input energy, subject to dynamical, configuration, minimum vehicle proximity and maximum acceleration constraints. In general, other cost functions and constraints could be used, but we found energy, minimum proximity, and maximum acceleration to be most necessary to this problem.

$$\min_{a_{01} \cdots a_{KN}} J = \sum_{i=1}^{N} \frac{1}{T} \int_0^T a_i^T(t) W_i a_i(t) dt \quad (18)$$

subject to

$$\dot{x}(t) = Ax(t) + Bu(t) \quad (19)$$
$$g(0) = g_s \quad (20)$$
$$g(T) = g_f \quad (21)$$
$$\|p_i(t) - p_j(t)\| \geq \epsilon \ \forall t \in [0,T], \forall_{i \neq j} i,j \quad (22)$$
$$\|a_i(t)\| \leq \alpha, \ \forall t \in [0,T], \forall i \quad (23)$$

where $i, j = 1, \cdots, N$, ϵ is the minimum allowable distance between vehicles, and α is the maximum allowable acceleration input at any time. W_i is a diagonal matrix of weighting constants for the ith vehicle.

W_i helps shape the cost function to different scenarios. For example if one wants to restrict movement of a vehicle in the z plane, one would make the diagonal entries of W_i associated with acceleration in the z direction higher than the rest. A higher weight can be given to all entries of W for any vehicle with special energy limitations due to low fuel and make it move slower than the other vehicles. In the following section we will show several examples of how Wi can be used to shape the result.

It can been easily shown that the FRP problem for autonomous vehicles satisfies the conditions specified in Theorem 3.2. Thus, given an order of the polynomials K, one can apply Theorem 3.2 to determine whether the Kth polynomials would be sufficient for solving the FRP problem.

5 Optimization results

In order to simplify the problem further, we have chosen to use 4th-order polynomials for the parameterization of the acceleration trajectories. Hence, there is just one free vector for each vehicle. Without loss of generality, we choose this free parameter vector to be a_{1i}. Define the optimization vector q to be the composite of all free parameters a_{1i} given by

$$q = [a_{11}^T \cdots a_{1N}^T]^T. \quad (24)$$

Instead of writing the state and input trajectories as functions of q and t, we simply express them as $p_i(t), v_i(t)$, and $a_i(t)$ to avoid introducing new notations. We can now express the FRP related optimization problem as

$$\min_q J = \sum_{i=1}^{N} \frac{1}{T} \int_0^T a_i^T(t) W_i a_i(t) dt \quad (25)$$

subject to

$$\|p_i(t) - p_j(t)\| \geq \epsilon \ \forall t \in [0,T], \forall_{i \neq j} i,j \quad (26)$$
$$\|a_i(t)\| \leq \alpha, \ \forall t \in [0,T], \forall i \quad (27)$$

where $i, j = 1, \cdots, N$. The FRP related optimization problem was solved using a constrained optimization algorithm. We will discuss further in the final version of this paper. In order to illustrate the effectiveness of the weighting matrix W_i, we perform the same FRP related optimization problem in three different cases varying only W_i. The given formation configurations and constraint constants are as follows.

$$a_i(0) = a_i(T) = [0\ 0\ 0]^T\ \forall i \in \{1 \cdots 3\}$$
$$v_i(0) = v_i(T) = [0\ 0\ 0]^T\ \forall i \in \{1 \cdots 3\}$$
$$[p_1^T(0)\ p_2^T(0)\ p_3^T(0)]^T = [0\ 0\ 10\ 0\ 10\ 10\ 10\ 0\ 10]^T$$
$$[p_1^T(T)\ p_2^T(T)\ p_3^T(T)]^T = [0\ 0\ 10\ 10\ 0\ 10\ 0\ 10\ 10]^T$$
$$T = 30, \epsilon = 9.9, \alpha = 3$$

In case (a), all vehicles and directions are given equal weighting. $W_i = I_{3\times3}$ for all vehicles. Figure 1(a) shows a simulation of the position trajectory result for case (a). Since vehicle 1 was given the same desired final position as its initial position, and no other restrictions were made, it does not move. However, vehicles 2 and 3 curve their position trajectory to stay at least an ϵ distance away from vehicle 1. The q found for the solution in case (a) is $q = [-0.0012\ 0\ 0\ 0.0492\ 0.0065\ 0\ 0.0047\ 0.051\ 0]$.

In case (b), vehicle 2 and 3 are given higher weighting than vehicle one such that $W_1 = I_{3\times3}$ and $W_2 = W_3 = 5 \cdot I_{3\times3}$. Figure 1(b) shows a simulation of the position trajectory result for case (b). In Figure 1(b) it is more optimal for vehicles 2 and 3 to find a path that uses less energy than vehicle 1. Therefore, even though vehicle 1 ultimately ends up in the same position, it moves away from vehicle 2 and 3 during the formation reconfiguration in order to remain ϵ distance away. This illustrates the centralized nature of the optimization solution of the FRP problem by showing how one vehicle sacrifices for the good of the group. The q found for the solution in case (b) is $q = [-0.0343\ 0\ 0\ 0.0303\ 0.002\ 0\ -0.0132\ 0.0444\ 0]$.

In case (c), the x and y directions of all vehicles are given a higher weighting than the z direction. $W_i = diag(5\ 5\ 1)$ for all vehicles. Figure 1(c) shows a simulation of the position trajectory result for case (c). In Figure 1(c) once again, since vehicle 2 and 3 must accelerate in x and y but are inhibited in the x and y directions, it is optimal for them to find a path that uses less energy than vehicle 1. It is still up to vehicle 1 to move away from vehicle 2 and 3 during the formation reconfiguration. However this time, it moves away in the z direction because it is more costly to move in either the x or y direction. Once again, this illustrates the centralized nature of the optimization solution of the FRP problem by showing how one vehicle sacrifices for the good of the group. The q found for the solution in case (c) is $q = [0\ 0\ 0.0767\ 0.404\ 0.004\ 0\ 0.004\ 0.0404\ 0]$.

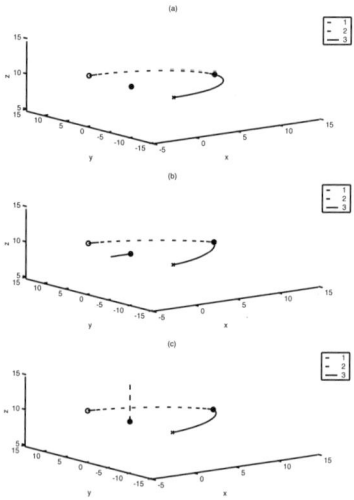

Figure 1: FRP solutions of the same problem using different W_i matrices. The circles and crosses represent the initial and final position of each vehicle respectively. (a). All vehicles and directions are given identical weighting. (b). Vehicles 2 and 3 are given higher weighting than vehicle 1. (c). The x and y directions are given higher weighting than the z direction for all vehicles.

The FRP optimization method can also be used to change formation in the presence of obstacles. All that is needed is extra constraints describing the obstacles. We assume that the space occupied in $\mathbb{R}^3$ by an obstacle can be described by an inequality as:

$$O_j = \{p_0 \in \mathbb{R}^3 : D_j(p_0) > \gamma_j\} \quad (28)$$

where $D_j : \mathbb{R}^3 \to \mathbb{R}$, $p_0 = [p_{x0} p_{y0} p_{z0}]^T \mathbb{R}^3$ and $\gamma_j \in \mathbb{R}$. Even though there is only one obstacle, the obstacle constraints apply to all vehicles. Therefore for each obstacle constraint, there are N additional constraints added to the FRP problem, i.e.

$$d_j(x_i) \leq \gamma_j \quad \text{for } i = 1, \ldots, N \quad (29)$$

where $d_j(x_i) = (D_j(\Pi x_i))$ with the projection matrix $\Pi = [I_{3\times 3}\ 0_{3\times 3}]$. Figure 2 shows how a formation of three vehicles can reconfigure around a circular obstacle in the x y plane. The two solutions have the same FRP

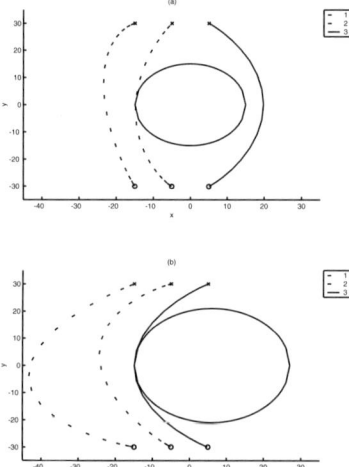

Figure 2: Two examples of a formation avoiding cylindrical obstacles. The circles and crosses represent the initial and final position of each vehicle respectively. (a). One vehicle breaks from the group to avoid the obstacle. (b) All three vehicles stay together when avoiding the obstacle.

specifications given by

$$a_i(0) = a_i(T) = v_i(0) = v_i(T) = [0\ 0\ 0]^T\ \forall i$$
$$[p_1^T(0)\ p_2^T(0)\ p_3^T(0)] =$$
$$[-15\ -30\ 10\ -5\ -30\ 10\ 5\ -30\ 10]$$
$$[p_1^T(T)\ p_2^T(T)\ p_3^T(T)] =$$
$$[-15\ 30\ 10\ -5\ 30\ 10\ 5\ 30\ 10]$$
$$T = 30,\ \epsilon = 9.9,\ \alpha = 3,\ W_i = I^{3\times 3},\ \forall i$$

Results shown in Figure 2(a) and 2(b) were given additional obstacle constraints. The obstacles considered in Figure 2(a) and (b) are given by

$$p_{x0}^2 + p_{y0}^2 < 15^2 \text{ and } (p_{x0} - 6)^2 + p_{y0}^2 < 21^2$$

respectively, each describing a circular obstacle of radius 15 and 21 respectively. In Figure 2(a) the obstacle is such that vehicle 3 breaks off from the formation in order to find a more optimal path around the obstacle. In 2(b) it is more optimal for vehicle 3 to stay with the group even though the trajectories of vehicle 1 and 2 increase in energy in order to accommodate the presence of vehicle 3. The FRP solution found for these two examples are (a)

$q = [-0.1199\ 0.2465\ 0\ -0.1402\ 0.104\ 0\ 0.2107\ 0.1341\ 0]$, and (b) $q = [-0.403\ 0.0772\ 0\ -0.2757\ 0.1912\ 0\ -0.2844\ 0.1315\ 0]$. Even though a formation may be capable of finding its way around a given obstacle, it may be more beneficial to avoid the obstacle in two FRP stages in order the use of high order of polynomials. Figure 3 shows a two stage example of how a formation can perform a sequence of reconfigurations in order to avoid a set of obstacles.

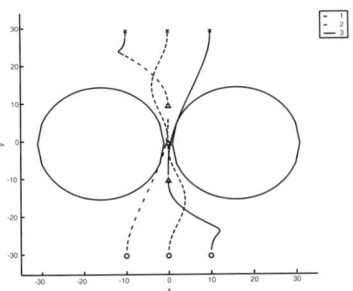

Figure 3: A formation performs a two stage FRP to move through a set of obstacles. The circles and crosses represent the initial and final configurations of the entire FRP problem respectively. The triangles represent the middle configuration as both the final configuration of the first stage and the initial configuration of the second stage of the FRP problem.

6 Conclusions

Optimization has proved to be a successful solution to the FRP problem. Our method of implementation is general and portable allowing for use in a wide range of applications for coordinated robots. For example, our method could easily be transported to two dimensions for ground robot coordination. This centralized control scheme has limitations in applications where formations are very large or communication is disrupted. For such applications, a decentralized control scheme is preferred. We are currently working on a decentralized approach to the FRP problem where each vehicle produces its own localized solution based on only local sensor information about its neighboring vehicles. As expected, this is proving to be a more complex problem. Therefore, centralized control is preferred in applications for smaller fully connected formations.

References

[1] J. A. Fax, R. M. Murray. "Graph Laplacians and Stabilization of Vehicle Formations", CDS Technical Report 01-007. In *Proceedings of IFAC World Congress*, Barcelona, Spain, July 2002.

[2] J. A. Fax, R. M. Murray. "Information Flow and Cooperative Control of Vehicle Formations". In *Proceedings of IFAC World Congress*, Barcelona, Spain, July 2002.

[3] A. Pant, P. Seiler, M. Broucke, T.J. Koo, J.K. Hedric. "Coordination and Control of Autonomous Vehicles". Submitted to *IEEE Transactions on Robotics and Automation*, March 2001.

[4] P. Tabuada, G. J. Pappas, P. Lima. "Feasible Formations of Multi-Agent Systems". In *Proceedings of the American Control Conference*, pp.56-61, Arlington, VA, June 2001.

[5] N. E. Leonard, E. Fiorelli, "Virtual Leaders, Artificial Potentials and Coordinated Control of Groups". In *Proceedings of the IEEE Conference on Decision and Control*, pp.2968-2973, Orlando, FL, December 2001.

[6] J. P. Desai, J. P. Ostrowski, V. Kumar. "Control of Changes in Formation for a Team of Mobile Robots". In *Proceedings of IEEE International Conference on Robotics and Automation*, pp. 1556-61, Detroit, MI, May 1999.

[7] A. Isidori. *Nonlinear Control Systems*. Springer-Verlag, 1995.

[8] G. Meyer, R. Su, L.R. Hunt. *Application of nonlinear transformations to automatic flight control*. Automatica, Vol. 20, No. 1, pp. 103-107, 1984.

[9] J. Hauser, S. Sastry, G. Meyer. *Nonlinear control design for slightly nonminimum phase system: Application to V/STOL aircraft*. Automatica, Vol. (28), No. 4, pp. 665-679, 1992.

[10] T. J. Koo, S. Sastry. Output Tracking Control Design of a Helicopter Model Based on Approximate Linearization. In *Proceedings of the 37th Conference on Decision and Control*, pp.3635-40, Tampa, Florida, December 1998.

[11] A. E. Bryson, Y.-C. Ho. *Applied Optimal Control: Optimization, Estimation, and Control*. Hemisphere Printing Co., 1975.

[12] G. Leitmann. *The Calculus of Variations and Optimal Control*. Plenum Press, New York, 1981.

[13] L.C. Young. *Optimal Control Theory*. Chelsea, second edition, 1980.

[14] C. Fernandes, L. Gurvits, Z. X. Li. A Variational Approach to Optimal Nonholonomic Motion Planning. In *Proceedings of the 1991 IEEE International Conference on Robotics and Automation*, pp.680-685, Sacramento, California, April 1991.

[15] G. Lafferriere, G.J. Pappas, and S. Yovine. Symbolic Reachability Computations for Families of Linear Vector Fields, Journal of Symbolic Computation, 32(3): 231-253, September 2001.

On the use of low-discrepancy sequences in non-holonomic motion planning

Abraham Sánchez
and René Zapata
LIRMM - University Montpellier II
161, rue Ada 34392, Montpellier - France

Claudio Lanzoni
LAR-DEIS
University of Bologna
Italy

Abstract— In this article, a recently developed approach for robot motion planning is extended and applied to non-holonomic mobile robots. This approach replace random sampling by deterministic one. We present several implementations of PRM-based planners. 1) Classical PRM with deterministic sampling and random sampling, 2) Deterministic and random Lazy-PRM, and 3) Lattice-based PRM. We have used several low-discrepancy sequences (Halton, Hammersley, Faure, and Sobol) and low-discrepancy lattices. Experimental results show that the deterministic variants of the PRM offer performance advantages in comparison to the original PRM.

I. INTRODUCTION

A recent trend in motion planning has been the development of randomized planners [2], [10], [11], [21], [1], [22], [17]. The main tradeoff for using randomization is that these planners are more efficient than their algebraic counterparts and can handle large degree of freedom problems, but at expense of completeness. Even though randomized planners are not complete[1], a notion of probabilistic or asymptotic completeness has been established for many of them (see for example [3], [9]).

A similar trend occurred many years ago in the area of the numerical integration, the development of *Quasi-Monte Carlo* methods. There are quasi-Monte Carlo methods not only for numerical integration, but also for various other numerical problems. In fact, for many Monte Carlo methods, it is possible to develop corresponding quasi-Monte Carlo methods as their deterministic versions. Invariably, the basic idea is to replace the random samples in the Monte Carlo method by deterministic points that are well suited for the problem at hand. Quasi-Monte Carlo methods have first been proposed in the 1950s, and their theory has since developed vigorously. However, for a rather long time, these methods remained the province of specialists. Deterministic sampling ideas have improved computational methods in many areas, including integration [20], [8], optimization [16], computer graphics [19], etc.

Branicky et al. [7], [14] propose the use of quasi-random sequences to question the value of randomization in the PRM context. Quasi-random variants of PRM-based planners have been analyzed, as classical PRM and lattice-based lazy planners. Bohlin presents a resolution complete planner based on an implicit grid [5]. Both of these papers approach the path planning problem by deterministically searching through a discrete set of points.

The main contribution of this paper is to extend the results presented in [7], [14] to the case of non-holonomic motion planning for car-like robots. This approach remarkably improves the PRM-based planners, by replacing random sampling. There are several sophisticated sampling strategies in the PRM framework [11], [1], [9], [6], [17], [22]. Many of these strategies require complex geometric operations that are difficult to implement in high-dimensional configuration spaces. Indeed, only the visibility sampling [17] and the work presented in [21] consider non-holonomic problems.

II. LOW-DISCREPANCY POINT SETS AND SEQUENCES

Quasi-Monte Carlo methods are one of the most successful applications of discrepancy theory, which is branch of pure number theory. While Monte Carlo methods use random numbers to provide probabilistic error bounds via the central limit theorem, quasi-Monte Carlo methods use *low-discrepancy sequences* (some authors speak of quasi-random points and quasi-random sequences) to allow deterministic bounds via the Koksma-Hlawka theorem. The idea is to use point sets, not randomly distributed, but very uniformly distributed. The extent to which the points are uniform has been mathematically defined as their *discrepancy*.

The historical origin of discrepancy theory is the theory of uniform distribution developed by H. Weyl and other mathematicians in the early days of the 20th century. While the latter deals with the uniformity of infinite sequences of points, the former deals with the uniformity of finite sequences. Finite sequences always have some irregularity from the ideal uniformity due to their finiteness. Discrepancy is a mathematical notion for measuring such irregularity. For this reason, discrepancy theory is sometimes called as the theory of irregularities of distribution. The formal definition of discrepancy is as follows:

$$D_N(P) = \sup_E \left| \frac{A(E;N)}{N} - \lambda(E) \right|, \quad (1)$$

where the supremum is taken over over all the subsets of $X = [0,1]^d$ of the form $E = [0, t_1) \times \cdots \times [0, t_d)$, $0 \leq t_j \leq 1$, $1 \leq j \leq d$, λ denotes the Lebesgue measure; and $(A;N)$

[1]A motion planner is *complete* if it always produces a path when one exists, and returns failure when one does not.

denotes the number of the x_j that are contained in E. Let P, a set of N d-dimensional sample points.

For many practical purposes it is preferable to work with infinite low-discrepancy point sets. In the PRM context, an infinite sequence is sometimes very convenient because it provides samples incrementally (which makes it easier to replace a random sequence). Such sequences include the Halton sequence (1960), Sobol' (1967), and Faure sequence (1981/82).

Similarly to the notion of discrepancy, it is possible to quantify the denseness of N points in the unit cube. The *dispersion* is defined by

$$d_N(P;X) = \sup_{x \in X} \min_{1 \leq n \leq N} d(x, x_N). \quad (2)$$

If $B(x;r)$ denotes the closed ball with center $x \in X$ and radius r, then $d_N(P;X)$ may also be described as the infimum of all radii $r \geq 0$ such that the balls $B(x_1;r), \ldots, B(x_N;r)$ cover X. Dispersion is useful in the analysis of the PRM because it indicates whether samples can be connected for a given connection radius parameter [14].

The following result provides an easy connection between the dispersion and the discrepancy [16]:

$$d_N(P) \leq D_N(P)^{\frac{1}{d}}. \quad (3)$$

Thus every low-discrepancy point set (or sequence) is a low-dispersion point set (or sequence), but not conversely [15]. In the multidimensional case, we can determine for every fixed N the minimum value of the dispersion $d_N(P)$ relative to the maximum metric. The following result of Sukharev contains the relevant information (see [16] for more details).

Theorem 1: *For any point set P of N points in $[0,1]^d$, we have*

$$d_N(P) \geq \frac{1}{2\lfloor N^{1/d} \rfloor}.$$

Furthermore, for every N and d, there exists a P for which equality holds.

Based on sampling theory, we observe that one must generate an exponential number of samples whether the PRM uses random or deterministic sampling. Next subsections present the construction of low-discrepancy sequences; they are easily implemented and fast to be computed in practice.

A. Halton sequence and Hammersley point set

1) Van der Corput sequence: Let $b \geq 2$ an arbitrary integer and $\mathbb{Z} = \{0, \ldots, b-1\}$ the residue system modulo b. The radical inverse function in base b is defined as

$$\phi_b : \mathbb{N} \to [0,1], \quad \phi_b(n) = \sum_{j=0}^{\infty} a_j(n) b^{-j-1},$$

where $n = \sum_{j=0}^{\infty} a_j(n) b^j$ is the b-adic expansion of the integer n. For every integer $b \geq 2$, the Van der Corput sequence in base b is the one-dimensional sequence $S_b = (x_n)_{n>0}$ defined as $x_n = \phi_b(n)$. The generalized Van der Corput sequence is obtained by permuting the digits $a_j(n)$ in the b-adic expansion.

2) Halton sequence: For a given dimension $d \geq 1$, let $b_1, \ldots, b_d$ be integers ≥ 2. Then, using the radical inverse function ϕ_b, we define the *Halton sequence* in the bases $b_1, \ldots, b_d$ as the sequence $x_0, x_1, \ldots$ with

$$x_n = (\phi_{b_1}(n), \ldots, \phi_{b_d}(n)) \text{ for all } n \geq 0.$$

For $d = 1$ this definition reduces to that of a Van der Corput sequence.

3) Hammersley point set: Hammersley point set is a closed-sequence variant of the Halton sequence.

For a dimension $d \geq 2$ and for integers $N \geq 1$ and $b_1, \ldots, b_{d-1} \geq 2$, the N-element *Hammersley point set* in the bases $b_1, \ldots, b_{d-1}$ is given by

$$x_n = \left(\frac{n}{N}, \phi_{b_1}(n), \ldots, \phi_{b_{d-1}}(n) \right) \text{ for } n = 0, 1, \ldots N-1.$$

B. Sobol and Faure sequences

In order to describe the construction, let us fix the dimension $d \geq 2$; we will construct points $x_0, x_1, x_2, \ldots$ in the unit cube I^d, and for every $k \in \mathbb{N}$ and $i \in \{1, \ldots, d\}$ let x_k^i denote the i-th coordinate of x_k, i.e., $x_k = (x_k^1, \ldots, x_k^d)$.

1) Sobol sequence: We first need d different irreducible polynomials $p_i \in \mathbb{F}_2[x]$, where $\mathbb{F}_2$ is the finite field with two elements (such polynomials can be found in lists). For convenience we set an a priori upper bound 2^m for the length of the computed part of the sequence (this bound is absolutely not necessary, but it makes the construction easier to be implemented). Now we construct for every $i \in \{1, \ldots, d\}$ the sequence $x_0^i, x_1^i, x_2^i, \ldots, x_k^i, \ldots$ for $k < 2^m$ in the same way:

Let be i fixed and $p_i(x) = x^r + a_1 x^{r-1} + \cdots + a_{r-1} x + 1$ the i-th of the given polynomials, then we choose arbitrary $v_1, \ldots, v_r$ with $1 \leq \frac{v_j}{2^{m-j}} \leq 2^j - 1$ and odd integer, e.g. $v_j = 2^{m-j} \cdot (2^j - 1)$. After this we compute for $j = r+1, \ldots, m$ (in this order)

$$v_j = a_1 v_{j-1} \oplus a_2 v_{j-2} \oplus \cdots a_{r-1} v_{j-r+1} \oplus v_{j-r} \oplus \frac{v_{j-r}}{2^r},$$

where $\oplus$ is the bitwise xor-operation (remark: for every j is $\frac{v_j}{2^{m-j}}$ an odd integer). Having once computed $v_1, \ldots, v_m$, we can compute x_k^i in two ways,

- either directly by $x_k^i = \frac{g_1 v_1 \oplus \cdots \oplus g_m v_m}{2^m}$, where $g_j = b_j \oplus b_{j+1}$ and $b_j \in \{0,1\}$ the coefficients in the binary representation of $k = \sum_{j \geq 1} b_j 2^{j-1}$,
- or, when x_{k-1}^i is given, by $x_k^i = \frac{(x_{k-1}^i \cdot 2^m) \oplus v_c}{2^m}$, where $c = min\{j | b_j = 1\}$ and $b_j \in \{0,1\}$ the coefficients in the binary representation of $k = \sum_{j \geq 1} b_j 2^{j-1}$.

2) Faure sequence: First let be p prime number with $p \geq d$. Then it would be convenient (but not necessary) to set an a priori upper bound p^m for the length of the computed part of the sequence and to precompute for all i,j with $0 \leq j \leq i \leq m$ the binary coefficients modulo p, i.e. $c_{i,j} = \binom{i}{j} \mod p$ (efficiently, one uses the recurrence $c_{i,j} = c_{i-1,j} + c_{i-1,j-1} \mod p$), $c_{i,j} = 0$ for $i < j$. For $k < p^m$: With $k \equiv \sum_{j=0}^{\bar{m}-1} b_j p^j$

the p-adic representation of k, $b_j \in \{0,...,p-1\}$ and $m \in \{0,...,m\}$, $b_{\bar{m}-1} \neq 0$, we set

$$x_k^1 = \sum_{j=0}^{\bar{m}-1} \frac{b_j}{p^{j+1}};$$

C. Lattice point sets

Quasi-Monte Carlo methods rely on sets of points that are evenly distributed over the domain of interest. One effective way of obtaining such sets is to use the lattice point method.

Let n be an integer ≥ 2 and $\mathbf{a} = (a_1,\ldots,a_d)$ be an integer vector modulo n. A set of the form

$$P_n = \{\{\mathbf{ak}/\mathbf{n}\} = (\{\mathbf{a_1 k}/\mathbf{n}\},\cdots,\{\mathbf{a_d k}/\mathbf{n}\})| k = 1,\ldots,n\}$$

is called a *lattice point set*, where $\{\mathbf{x}\}$ denotes the fractional part of x. The vector $\mathbf{a}$ is called a lattice point or generator of the set.

Sloan and Kachoyan generalized the above construction to obtain what they called integration lattices. Such lattices include rectangular grids as special cases. The monographs of Niederreiter [16] and Sloan and Joe [20] provide comprehensive expositions of the theory of integration lattices, and the review by Hickernell [8] gives some more recent results. As one can see, the formula for the lattice point set is simple to program. The difficulty lies in finding a good value of $\mathbf{a}$, such that the points in the set are evenly spread over the unit cube.

We refer to one particular lattice as the *Sukharev grid* [14], if it is constructed as follows for some N such that $k = N^{1/d}$ is an integer. Decompose X into N cubes of width $1/k$ so that a tiling of $k \times k \times \cdots \times k$ is constructed. Place a sample at the center of each cube. In comparison to a standard grid (starting at the origin), the sukharev grid does not waste points along the boundary of X.

III. THE BASIC PRM

PRM was developed by Kavraki, Svestka, Latombe and Overmars [11]. This approach solves the problem in two steps for robots with many degrees of freedom operating in static environment. The first step is a *learning phase* where a roadmap is randomly generated to infiltrate the free configuration space ($\mathcal{CS}_{free}$). The learning process relies on a fast local method to connect the individual configurations. After the learning phase, the second step is the *query phase* where the algorithm attempts to generate a path between two free configurations. The algorithm to construct a roadmap with N vertices is as follows:

```
1  n ← 0
2  while n < N
3     q ← SAMPLING_STRATEGY()
4     if q ∈ CS_free
5        U ← NEIGHBOR_STRATEGY.FIND_NEIGHBORS(q)
6        V ← V ∪ q
7        foreach u ∈ U
8           if LOCAL_PLANNER.VERIFY_PATH(q, u) = true
9              E ← E ∪ (q, u)
10       n ← n + 1
```

The algorithm is simple and straightforward. A 2D example is shown in Figure 1, for which $N = 500$. A simple glance with the Figure is enough to highlight the poor quality of the distribution of a typical random sequence. Indeed, by its nature, such a sequence presents many singularities: certain areas contain significant clusters of points, whereas broad zones remain entirely empty. The Figure also confirms the well-known fact that narrow passages in configuration space are notoriously difficult to find at random.

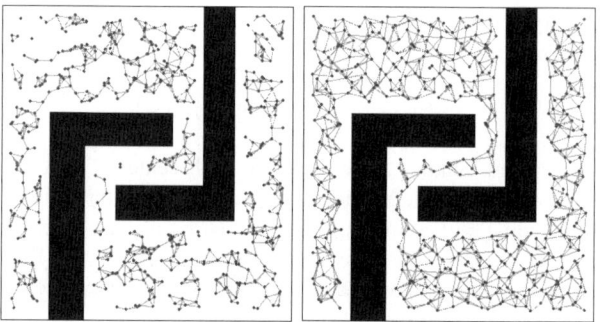

Fig. 1. Random Sampling vs. deterministic sampling. Each uses 500 samples and the same connection radius, 0.05. Deterministic sampling is based on Faure sequence.

A. Comparing PRM and DRM experimentally

The approach in the DRM context, like the authors in [7], [14] is to simply replace the pseudo-random sample generator that appears in Line 3 of the PRM construction algorithm with a low-discrepancy, deterministic sampling method. In actual practice, random numbers are generated by a deterministic algorithm that is implemented in the computer, and so we are really working with pseudo-random numbers.

Much research has been done on motion planning for non-holonomic car-like robots (see [12] for a review). Within PRM framework, Švetska and Overmars use RTR paths as local method [21]. An alternative to RTR local method is to use a local method that constructs the shortest car-like paths connecting its argument configurations [18]. Another randomized strategy that has been used for non-holonomic planning is the RRT approach [13].

Since a randomized algorithm rarely behaves identically on repeated trials, it is hard to compare a randomized algorithm with a deterministic one. This difficulty is met by comparing the performance of the deterministic PRM version with the "average" performance of k iterations of the classic PRM.

Another way to evaluate low-discrepancy point sets is the dispersion of a point set, Branicky et al. [7] suggest visualizing dispersion by placing a ball with radius equal to the dispersion at each sample point. If the dispersion is small, then a query can be easily connected to the sample points; also, the sample points are thought to cover the space densely, in some sense.

Table I shows the results of experiments performed on a difficult scene (Figure 2). The connection radius and the number of neighbors are given respectively in first and second

TABLE I

A COMPARISON BETWEEN PRM AND DRM FOR A NARROW CORRIDOR PROBLEM.

Rad	K	PRM	DRM-H/H	DRM-F	DRM-S	Factor
0.10	7	417	350	350	350	1.19
0.15	10	283	350	200	350	1.42
0.20	15	243	250	200	250	1.22
0.30	15	214	225	200	235	1.07
0.40	10	121	125	100	100	1.21

TABLE II

COMPARISONS OF THE NUMBER OF NODES FOR THE LAZY PRM VS. THE LAZY DRM.

Test	Min	Max	Avg	Lazy DRM
(a,b)	80	170	128	100
(c,d)	100	150	125	100
(e,f)	100	200	147	100

columns. The number of nodes required to find a path is shown for the different versions of DRM and 100 averaged trials of the PRM. In all experiments, we have used Halton/Hammersley(H/H), Sobol (S) and Faure (F) points for DRM; pseudo-random numbers were generated using the linear congruential generator. The final column indicates the improvement factor of deterministic sampling over random sampling, in terms of the number of nodes. In Figure 2, we show a path computed in this scene. We observed during the experimental tests that the election of the connection radius and the number of neighbors have an important influence in the roadmap construction step.

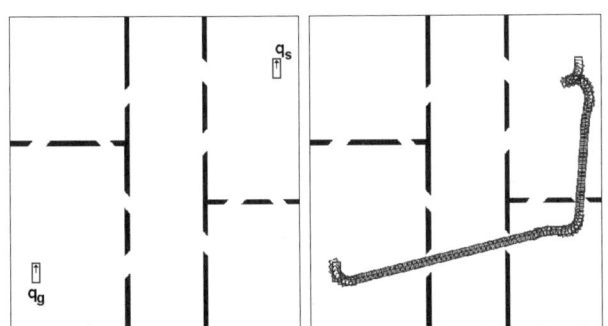

Fig. 2. This scene is difficult, the only way to get from the start configuration to the goal configuration, is by passing through some of the narrow passages. A path computed is shown.

IV. LAZY PRM

Computing paths using the local method is the most time-consuming step in the learning phase. We would like to avoid such computations as much as possible. A recent variant called lazy-PRM has been proposed [4], the idea is not to test whether the paths are collision-free unless it is really needed. The goal of this variant is to minimize the number of collision checks. The algorithm builds a roadmap, whose nodes are the user-defined start and goal configuration, initially it assumes that all nodes and edges in the roadmap are collision-free and searches the roadmap for a shortest path between these configurations. If a collision with the obstacles occurs, the corresponding nodes and edges are removed from the roadmap and a new shortest path search procedure is applied or new nodes and edges are added to roadmap. The process is repeated until a collision-free path is found. The resulting planner is sometimes very efficient in comparison to the original PRM. The rational behind this is that for most paths we only need to consider a small part of the graph before a solution is found.

A. Comparing lazy-PRM and lazy-QRM

Our approach in the Lazy-DRM context is to simply replace the pseudo-random sample generator with a low-discrepancy, deterministic sampling method.

The tests are performed in a "corridor-like" scene, consisting of 21 "rooms" and a "hallway" (see Figure 3). The main difficulty is the large number of narrow passages that connect the rooms to the hallway. The rectangular robot can only just pass through. Also, it is difficult for the robot to reverse its orientation when in the hallway. As a query test set, we take $\{(a,b),(c,d),(e,f)\}$. We have used Halton/Hammersley, Sobol and Faure points as inputs to both algorithms to solve this test set. Lazy-PRM uses the linear congruential method for generating pseudo-random points. For the lazy-PRM, we performed 150 trials on each test set.

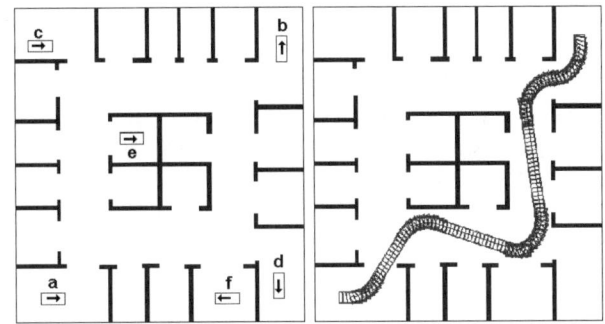

Fig. 3. A corridor-like scene, and its configuration test set. A path computed by the planner is shown.

Table II shows the minimum, maximum and average number of nodes for the lazy PRM in the first three columns. The final column shows the number used by the lazy DRM. In this case, the three low-discrepancy sequences used the same number of nodes. Table III compares computation times.

B. Lazy LRM

A difficulty when use the lazy PRM approach is that without prior knowledge about the particular problem, it is difficult to determine how many vertices should appear in the roadmap. Another difficulty is how many nodes should be added if the

TABLE III
COMPARISON OF THE RUNNING TIMES FOR THE LAZY PRM VS. THE LAZY DRM.

Test	Min	Max	Avg	PreCmp	PreCmp H/H	Query H/H	PreCmp S	Query S	PreCmp F	Query F
(a,b)	0.46	1.58	1.62	3.15	2.40	0.89	2.85	0.82	1.97	0.62
(c,d)	0.65	2.56	1.46	2.95	2.44	0.88	2.07	0.68	2.37	0.52
(e,f)	0.87	3.40	1.58	6.09	2.26	0.76	2.45	0.57	2.75	0.64

TABLE IV
COMPARISONS BETWEEN LAZY LRM, LAZY DRM AND 50 TRIALS OF LAZY PRM.

	lazy PRM	H/H	lazy DRM F	S	lazy LRM
PreCmp	10.14	5.18	4.59	11.63	0.65
Query	0.96	0.66	0.68	0.57	0.57
Nodes	210	140	160	250	100

query fails. These difficulties also motivated one of the lazy PRM authors to recently make a grid-based variant [5]. We know that lattices have a regular, well-defined neighborhood structure. This allows the initial roadmap to be implicitly defined with little or no pre-computation because all vertices, neighboring vertices and edges are defined implicitly by lattice rules [7]. Given that lattices have discrepancy bounds that are as good as the best bounds for non-lattice sample sets, we can make a lazy LRM (lattice-based roadmap) that simultaneously obtains the low-discrepancy benefits observed in the lazy DRM and the dramatic reduction in pre-computation time.

Figure 4 shows a very difficult scene to compare lazy LRM with the other planners (Lazy PRM and lazy DRM). It contains many narrow passages, it provides little freedom of movement for the robot, and any path solving the problem is very long. Table IV gives a comparison of running times for the different versions.

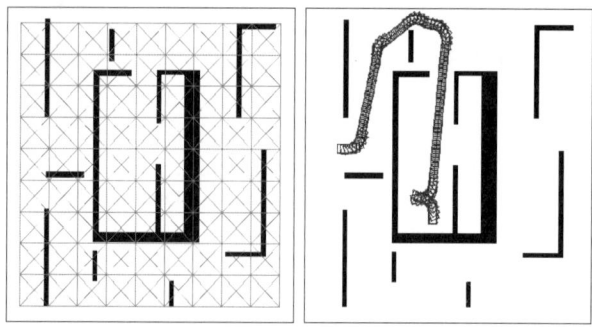

Fig. 4. Left: A lazy roadmap based on 100 low-discrepancy lattice points. Right: A feasible computed path.

V. DISCUSSION AND FUTURE WORK

We believe that randomization is useful in many contexts. Its value, nevertheless, depends greatly on the paradigm within it is used. Randomization does not appear to be advantageous in the PRM context, according to our experiments in non-holonomic motion planning and theoretical analysis presented by the authors in [7], [14]. Deterministic sampling enables the DRM, lazy DRM and lazy LRM to be *resolution complete* [7], [14]: if it possible to solve the query, they eventually solve it.

The lattice-based lazy PRM shows dramatic performance improvements, primarily because it exploits the neighborhood structure of the lattice to avoid the pre-computation required by the randomized lazy PRM.

The discrepancy and the dispersion are well-known measures for the irregularity of distribution of a sequence, both are relevant in the PRM context. Numerous low-discrepancy point sets and sequences have been proposed. They can be generated easily and quickly. For practical use, there are three different types of low-discrepancy sequences or point sets: Halton sequences, lattice rules, and (t,k)−sequences. The last one includes almost all important sequences such as Sobol' sequences, Faure sequences, Niederreiter-Xing sequences, etc. For all the three classes of sequences, randomization reportedly enhances the practical performances considerably. While randomization just guarantees good performances of such scrambled sequences on average, *derandomization* is considered as a process to look for those sequences which are theoretically guaranteed to be always good in terms of the convergence rate. If there is a "universal" sequence that always performs well for almost all problems, it would be ideal. Future work includes the study of these approaches.

Discrepancy is interesting in many respects, but its computation is known to be very difficult. The complexity of the well-known algorithms is exponential. One might have hoped that these difficulties are partially counterbalanced by the existence of upper bounds for some especial types of low-discrepancy sequences. Unfortunately, the minimum number of points for which these classical bounds become meaningful grows exponentially with the dimension.

We claim that deterministic sampling is suitable to capture the connectivity of configuration spaces with narrow passages. We will develop hybrid sampling methods (one can imagine The Gaussian sampling with deterministic sampling?).

Planner implementations were done in Java and Matlab

(partially using C) and the tests were performed on a Intel © Pentium III processor based PC running at 866 Mhz with 128 MB RAM.

PRM is a highly-flexible method. It is possible to apply it to some particular robot type, all that is needed is a local method that computes feasible paths for this robot type and some induced metric. Much research in PRM field has been done, not only in robotics, its application has extended to other fields as animation, computer games, virtual environments, and maintenance planning and training in industrial CAD systems.

In this work we have not tried to dismiss to all developed work in the PRM framework but rather we presented *some simple ideas* to obtain better results. The different improvements suggested are difficult to compare, since each author used his or her own implementation of PRM and used different test scenes, both in terms of environment and the robot type used.

Currently we are working on the application of the approach to mobile manipulators, and furthermore we are planning extensions of the method in various direction, aimed at solving more difficult motion planning problems (for instance, motion planning in scenes with multiple robots). Figure 5 presents preliminary results.

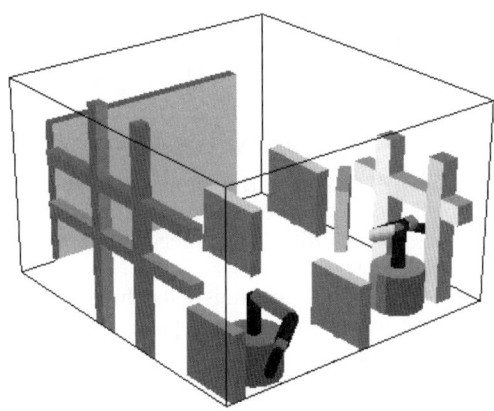

Fig. 5. A 8-dof nonholonomic mobile manipulator.

Acknowledgment

Abraham Sánchez is on leave from Computer Science Dept. University of Puebla-México, and is supported by PROMEP-BUAP-2000-19 (PROMEP/BUAP-447).

Claudio Lanzoni is supported by LAR-DEIS, University of Bologna, Italy. René Zapata is partially supported by Action de Collaboration France-México (ECOS-Nord, Action M01 M01). The authors are grateful to H. Niederreiter and F. J. Hickernell for providing suitable literature. Abraham Sánchez thanks Steven Lavalle for helpful discussions.

References

[1] N. M. Amato, O. Bayazit, L. K. Dale, C. Jones, and D. Vallejo. OBPRM: An obstacle-based PRM for 3D workspaces, *Proc. Workshop on the Algorithmic Foundations of Robotics*, pp. 155-168, 1998.

[2] J. Barraquand, and J-C. Latombe. Robot motion planning: A distributed representation approach, *The International Journal of Robotics Research*, Vol. 10, No. 6, pp. 628-649, 1991.

[3] J. Barraquand, L. Kavraki, J-C. Latombe, R. Motwani, Tsai-Yen Li, P. Raghavan. A Random sampling scheme for path planning, *The International Journal of Robotics Research*, Vol. 16, No. 6, pp. 759-774, December 1997.

[4] R. Bohlin and L. Kavraki. Path planning using lazy PRM, *Proc. of the IEEE Robotics and Automation Conference*, pp. 521-528, 2000.

[5] R. Bohlin. Path planning in practice: Lazy evaluation on a multi-resolution grid, *Proc. of the IEEE International Conference on Intelligent Robots and Systems*, pp. 49-54, 2001.

[6] V. Boor, M. H. Overmars, and A. Frank van der Stappen. The gaussian sampling strategy for probabilistic roadmap planners, *Proc. of the IEEE Robotics and Automation Conference*, pp. 1018-1023, 1999.

[7] M. S. Branicky, S. M. LaValle, K. Olson, L. Yang. Quasi-randomized path planning. *Proc. of the IEEE Robotics and Automation Conference*, pp. 1481-1487, 2001.

[8] F. J. Hickernell. Lattice rules: How well do they measure up?, *Random and quasi-random point sets*, P. Bickel (editor), Springer Verlag, pp. 109-166, 1998.

[9] D. Hsu, L. E. Kavraki, J-C. Latombe, R. Motwani, and S. Sorkin. On finding narrow passages with probabilistic roadmap planners, *Proc. Workshop on the Algorithmic Foundations of Robotics*, pp. 141-154, 1998.

[10] L. E. Kavraki. *Random networks in configuration space for fast path planning*, PhD Thesis, Stanford University, 1995.

[11] L. E. Kavraki, P. Švetska, J-C. Latombe, M. H. Overmars. Probabilistic roadmaps for path planning in high-dimensional configuration spaces, *IEEE Transactions on Robotics and Automation*, Vol. 12, No. 4, pp. 566-580, August 1996.

[12] J-P. Laumond (Ed.). Robot motion planning and control, *Lecture Notes in Control and Information Sciences*, Springer Verlag, 1998.

[13] S. M. LaValle and J. J. Kuffner. Randomized kinodynamic planning, *Proc. of the IEEE Robotics and Automation Conference*, pp. 473-479, 1999.

[14] S. M. LaValle and M. S. Branicky. On the relationship between classical grid search and probabilistic roadmaps, *Proc. Workshop on the Algorithmic Foundations of Robotics*, 2002.

[15] H. Niederreiter. Low-discrepancy and low-dispersion sequences, *Journal of Number Theory*, Vol. 30, No. 1, pp. 51-70, 1988.

[16] H. Niederreiter. *Random number generation and Quasi-Monte Carlo methods*. Society for Industrial and Applied Mathematics, 1992.

[17] C. Nissoux, T. Siméon, and J-P. Laumond. Visibility based probabilistic roadmaps, *Proc. of the IEEE International Conference on Intelligent Robots and Systems*, pp. 1316-1321, 1999.

[18] J. A. Reeds and R. A. Shepp. Optimal paths for a car that goes both forward and backwards, *Pacific Journal of Mathematics*, Vol. 145, No. 2, pp. 367-393, 1990.

[19] P. Shirley. Discrepancy as a quality measure for sample distributions, *Proceedings of Eurographics 91*, pp. 183-193, 1991.

[20] I. H. Sloan and S. Joe. *Lattice methods for multiple integration*. Oxford Science Publications, Englewood Cliffs, NJ, 1994.

[21] P. Švestka, and M. H. Overmars. Motion planning for car-like robots using a probabilistic learning approach, *The International Journal of Robotics Research*, Vol. 16, No. 2, pp. 119-143, April 1997.

[22] S. A. Wilmarth, N. M. Amato, and P. F. Stiller. MAPRM: A probabilistic roadmap planner with sampling on the medial axis of the free space, *Proc. of the IEEE Robotics and Automation Conference*, pp. 1024-1031, 1999.

Smooth Path Planning by Using Visibility Graph-like Method

Tomomi KITO*, Jun OTA*, Rie KATSUKI*, Takahisa MIZUTA*, Tamio ARAI*
Tsuyoshi UEYAMA**, and Tsuyoshi NISHIYAMA**

* Department of Precision Engineering, School of Engineering, The University of Tokyo
7-3-1 Hongo, Bunkyo-ku, Tokyo 113-8656, JAPAN
{kito, ota, rie, mizuta, arai}@prince.pe.u-tokyo.ac.jp
** DENSO WAVE Inc. 1-1 Showa-cho, Kariya-shi, Aichi 448-8661, JAPAN
{tsuyoshi.ueyama, tsuyoshi.nishiyama}@denso-wave.co.jp

Abstract

To achieve smooth motion of car-like robots, it is necessary to generate paths that satisfy the following conditions: maximum curvature, maximum curvature derivative, and curvature continuity. Another requirement is that human operators can manipulate the robots with ease.

In this paper, a path expression methodology consisting of line segments, circular arcs and clothoid arcs is presented. In addition, a method of global path generation with a visibility graph is proposed. To establish this method, the following steps are proposed: (a) the arrangement of sub-goals (middle points) and (b) the construction of the graph for path generation.

By using the proposed method, the paths were shortened 14% on average.

1. Introduction

Generating smooth paths is essential for mobile robot navigation. Since smooth motion may prevent skidding and improve the robot's dead reckoning ability, a robot can move for a long distance without receiving extra visual or range information. Thus, this problem is the focus in this paper. Specifically, we propose a path-generating method for a mobile robot to reach from a certain initial configuration to a target configuration as fast as possible under known environment without collision (Fig.1).

As demands for paths, a special emphasis is placed on the three conditions that follow:
(1) The mechanical constraints of a robot should be satisfied.
The robot in Fig.1 is subject to non-holonomic constraints, namely, it can only move in a tangent direction [1]. At this time, the first constraint of the path is the limitation of the steering angle. The minimum turning radius or the maximum curvature κ_{max} exists in paths. Second, since the angular velocity of the steering is upper-bounded, it is necessary to give a maximum to the curvature derivative σ_{max}.
(2) Paths should have a continuous curvature.
Paths that have a continuous curvature which track easier without skidding [2].
Some are only tasked with moving forward, for example, [8] while others are tasked with moving backwards and forward [10]. In this paper, the focus is on moving forward because the shift of two movements may cause positioning errors.
(3) The operational commands should be easy for users to manage.
An intuitive expression of paths is necessary. In this sense, the number of parameters for paths should be small.

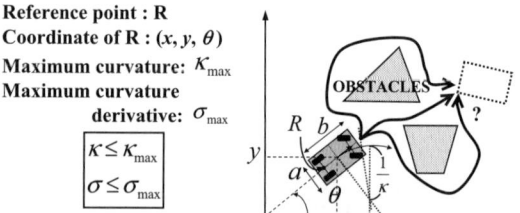

Reference point : R
Coordinate of R : (x, y, θ)
Maximum curvature: κ_{max}
Maximum curvature derivative: σ_{max}

$\kappa \leq \kappa_{max}$
$\sigma \leq \sigma_{max}$

Fig. 1: Setting of the environment and the car-like robot

The objective of this study is to propose a method for the generation of a collision-free path connecting two given configurations while fulfilling the three conditions reported above. The challenges are:
(a) How to express paths that satisfy the three conditions.
(b) How to derive the semi-optimal paths within the practical calculation cost.

2. The Problem Setting

2.1 Related Work

Numerous studies have been done to solve similar problems. In this section, the studies are classified according to (a).

- Circular arcs and line segments
Dubins proved that the shortest path is made up of circular arcs with minimum radius and line segments [3]. Jacobs et al. [4] have applied Dubins' curves to (b) by using a graph search algorithm in the configuration space. However, computing configuration space involves high calculation costs. Moreover, Dubins' curves have no continuity of curvature.
- Spline-curves
Splines are curves with continuous curvature and are dealt with in several path-generating methods [5][6]. The main defect of spline curves is that their curvature profiles are complicated. For this reason, the generated pattern of operation is difficult for users to understand,

i.e., the expression of paths with spline curves cannot satisfy condition (3).
- Clothoid curves
  Clothoid curves with a curvature that is linear to length satisfy condition (3). Kanayama et al. have generated continuous curvature paths by connecting two straight lines with a clothoid pair [7]. However, the path generated by this method could be much longer than needed because its curvature variation was more gradual than necessity.
- Circular arcs and line segments interpolated by clothoid curves
  Scheuer et al. have proposed a method of interpolating the knot of line segments and a circular arc of minimum radius by the clothoid arc of maximum curvature and maximum curvature derivative [8]. Using this method, paths became extensible from discontinuous paths to continuous paths. However, it is not always possible to connect two line segments with a clothoid arc, as doing so depends on the angle between the two segments. This is because only some specific values of curvature and curvature derivative are utilized for clothoid arc generation. In addition, the main drawback is that Probabilistic Road Map (PRM) method [9] was used to overcome the challenge expressed in (b). By using PRM method, there is no guarantee of which paths are acquired in complicated environment with a large number of obstacles. In addition, the calculation costs would be extremely high.

2.2 Conceptual Design

On the basis of the discussion in section 2.1, in this research, the following approaches are taken when dealing with the points raised in (a) and (b).
(a): Express paths using line segments, circular arcs, and clothoid arcs[8]. The generated paths are to satisfy the three conditions. The problem of connecting two line segments was solved by adequately tuning the values of the curvature and curvature derivative of the clothoid arcs.
(b): Derive the semi-optimal paths by using the visibility graph-like method in work space (not in the configuration space [4]). In this method, the start and goal configurations and the neighborhood of vertices of the obstacles must be set in relation to the nodes. All the nodes correspond to the robot's positions.

The problems are divided into two categories: (A) global path generation, and (B) local path generation. Concrete problems are specified for in each category:
(A-1) How to arrange the nodes considering the robot's shape and assignment of obstacles.
(A-2) How to search for semi-optimal paths.
(B-1) How to generate arcs connecting two nodes:
(B-2) How to link two arcs with a common node: generate the paths that keep curvature continuity between the two arcs.

In order to make the plot coherent, the topic (B) is discussed before (A). (B) is mentioned in Chapter 3, and (A) is dealt in Chapter 4.

3 Local Path Generation

3.1 Path Generation by Line Segments and Circular Arcs

In this section, the outline of the procedure, which derives the shortest path connecting two configurations (expressed with vectors), is described under the condition of a minimum radius. This is a preliminary step toward approach (a). The paths generated by this procedure are, namely, four candidates of Dubins' curves [3].

1. Prepare two tangent circles to two vectors. Each circle has a clockwise or counter-clockwise direction along the vector.
2. Draw the common tangents of the circles of the start vector and of the goal vector. In light of the circles' directions, there are at most four tangents. When a clockwise circle and a counter-clockwise circle overlap, there is no common tangent.
3. Generate at most four paths made up of the circular arcs and common tangent (Fig.2).

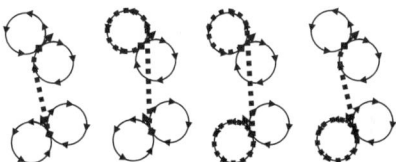

Fig.2 Four paths connecting two configurations

3.2 Path Interpolation Using Clothoid Arcs

In this section, the technique interpolating clothoid arcs into the paths mentioned in section 3.1 is stated under the condition of maximum curvature, curvature derivative, and curvature continuity (approach (a)).

3.2.1 The Outline of the Existent Technique [8]

The technique of [8] when the constraints of the maximum curvature κ_{max} and the maximum curvature derivative σ_{max} are given (Fig.3) is briefly described. The curvature profile of this path is shown in Fig.4.

1) Generate the path made up of line segments and circular arcs by the technique mentioned in section 3.1. The circles used here are to C_1 with radius $R > 1 / \kappa_{max}$ (defined by the constraints). Let the angle of two tangents be τ.
2) Prepare a circle C_2 which has the same center as C_1 with radius $1 / \kappa_{max}$ for each C_1.
3) Move the tangents t in parallel. t is decided from the constraints.
4) Consider the clothoid arc whose start point is q_0, the intersection of the line and C_1, and whose curvature changes 0 to κ_{max} with the curvature derivative σ_{max}. The length of this arc is $l = \kappa_{max} / \sigma_{max}$, the deflection is $\phi = \kappa_{max}^2 / \sigma_{max}$. The end point of it is set to q_1 on C_2.
5) Consider the arc along C_2 from q_1 of which the central angle is $\tau - 2\phi$ as a part of the path.
6) Connect the clothoid arc whose curvature derivative is $-\sigma_{max}$ to q_2, the end point of the circular arc. The end

point of the clothoid arc q_3 corresponds to the intersection of the line and C_1.

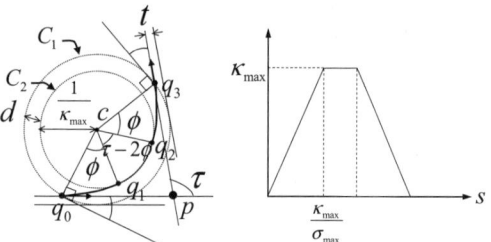

Fig.3 Interpolation by the clothoid arcs   **Fig.4 Curvature profile**

The difference of the radius of C_1 and C_2 is set as follows:

$$d = R - \frac{1}{\kappa_{max}} \quad (1)$$

3.2.2 Extension of the Existent Technique

The clothoid arcs cannot be introduced when $\tau - 2\phi$ by using the above-mentioned technique. We propose the technique by which the paths can be generated even when $\tau - 2\phi$. Although Fraichard et al. have solved this problem by introducing a backward motion in [10], it is not used here for the reason already explained.

The following two techniques can be considered:
(i) The maximum curvature is set to $\kappa < \kappa_{max}$ while the curvature derivative is σ_{max} (constant).
(ii) The maximum curvature is set to $\kappa < \kappa_{max}$ while the curvature derivative is $\sigma < \sigma_{max}$.

The curvature profiles of these two are shown in Fig.5. We should notice that there is condition that the two edges of the clothoid arc must be on q_0-p and on p-q_3 (Fig.3). Furthermore, the length of the clothoid arc should be as short as possible from the viewpoint of the ease of mobile robots' tracking. If (ii) is used, there are possibility of not fulfilling the above-mentioned condition because the arc length would be too long. Inevitably, (i) is adopted. The following formulas are realized at this time.

$$\frac{\kappa^2}{\sigma_{max}} = \frac{\tau}{2}$$

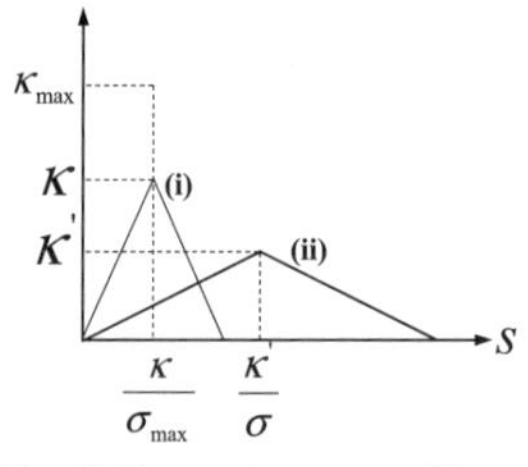

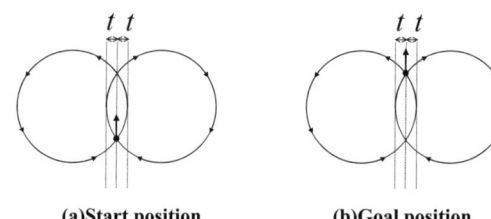

Fig.5 Curvature profiles 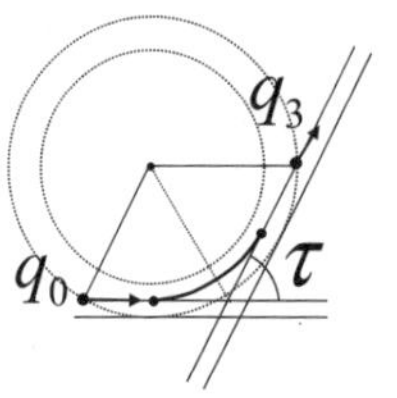   **Fig.6 Clothoid arcs ($\tau - 2\phi$)**

This clothoid-pair appears in Fig.6. By this technique, any clothoid arcs can be placed into the space between C_1 and C_2 (they could not go into C_2 as shown above).

4. Global Path Generation

In this chapter, we establish an algorithm that generates the global shortest path that avoids obstacles.

4.1 Node Arrangement for Start and Goal Positions

The following two circles are respectively arranged to the start position and the goal position as nodes (Fig.7). The circle in the figure is equivalent to C_1. By this arrangement, each point could be connected to clothoid arcs of both directions.

(a) Start position   (b) Goal position

Fig.7 Circles arranged to the start and the goal points

4.2 Arrangement of Middle Points

4.2.1 Decision of the Offset Considering a Robot's Configuration

As stated above, the path never goes into C_2. In a case in which the robot runs on the circumference of C_2, the robot goes $a/2$ inside C_2 (taking the robot's configuration into account). Thus, the distance into which the robot goes into C_1 is calculated as follows (by using (1)):

$$D = d + \frac{a}{2} \quad (2)$$

Consequently, there is assurance that the robot would never go into C_3, the circle of the same center as C_1, whose radius is D smaller than C_1 (Fig.8). The D is determined to the offset.

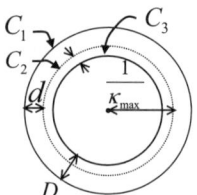

Fig.8 Threefold circle of the same center

4.2.2 Arrangement of Middle Points to the Vertices of Obstacles

In this section, the following procedure of arranging middle points as nodes to each vertex of obstacles is described. We take V_i for each vertex of obstacles (when the total number of vertices is n, i=1,2…,n).
(1) Add a circle with radius D to each vertex as an offset

(Fig.9-(a). Vertex V_i is taken as an example.).
(2) Form the tangents of these circles discretely. When each circle is discretized by δ, the number of tangents for each vertex is $2\pi/\delta$. We define the tangents of the circle added to V_i as T_{ij} ($j=1,2…, 2\pi/\delta$). T_{ij} is shown in Fig.9-(b) as an example.
(3) Arrange the threefold circle, C_1, C_2, C_3 so that C_1 touches each tangent (the inner side of C_1 should touch the vertex.). By this arrangement, each C_3 should pass through the vertex. An example of this arrangement is shown in Fig.9-(c). This threefold circle is to T_{ij}, which is one of the tangents (Fig.9-(b)).

All the threefold circles arranged to each vertex in such a procedure are considered to be middle points. Since, in precise terms, the arrangements show circles rather than points, they should henceforth be referred to as middle circles. These middle circles are all to be nodes.

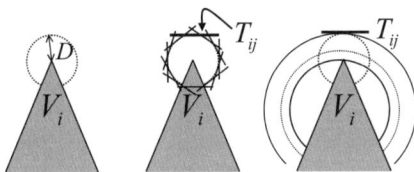

(a) Addition of offset (b) Tangent (c) Threefold circle

Fig.9 The arrangement procedure of middle circles

4.2.3 Node Arrangement by Translating the Start and Goal Nodes

As described in section 3.1, a common tangent does not exist when two circles of an opposite direction overlap ([3][4]). In order to generate the path between two such circles, the following technique is suggested.

First, we explain the procedure along the example shown in Fig.10. Now, M_{ij} (the tangent circle to T_{ij}) and one of the start circles overlap (Fig.10-(a)). The start position is moved forward until the two circles are in contact (Fig.10-(b)). The moved circles are defined as the middle circles. The original start position does not change. In the case of the goal circles, the procedure is the same as in the case of the start circles, except that the goal circle moves backward rather than forward. The reason that the start position is allowed to move forward and the goal position backward is that the path needs to pass the original start and goal positions.

Arranging these middle circles generates a new candidate for the path. The validity of designing these nodes is shown in the simulation of section 5.1.

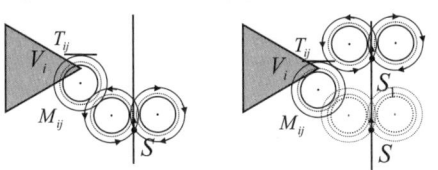

(a) Initial situation (b) Addition of the new node

Fig.10 Node-generation process

4.3 Path-Generating Algorithm
4.3.1 Search Procedure

In this section, the procedure for generating the shortest path is described by using the visibility graph-like method.
(1) The arrangement of obstacles is given in advance. Arrange two circles in the start position S and the goal position G (Fig.11-(a). Each circle is a node).
(2) Arrange middle circles (nodes) to each V_i in the procedure shown in 4.2.2 (Fig.11-(b)). It is discretized by $\pi/3$ unit for the sake of clarity in the figure.
(3) Arrange the middle circles to the translational positions by the procedure described in 4.2.3 (Fig.11-(c)).
(4) Construct the search graph by connecting nodes with arcs (Fig.11-(d)). In this regard, the arcs do not exist between the nodes made by movement of the same point (for example, between G_1 and G_2) or between the nodes on the same vertex (for example, between V_1-a and V_1-b). In the figure, all the arrows mean having stretched the arcs to all the components of the group. The components in the same group are not connected with arcs.
(5) Assign the path length as the cost to each arc (details about the assignment of the cost are given in the following section).
(6) Generate the shortest path based upon a graph search algorithm such as Dijkstra's.

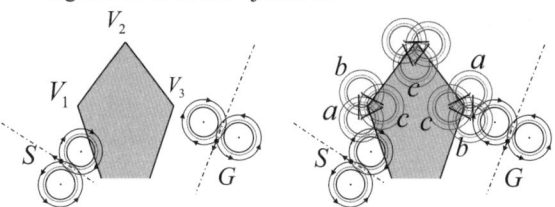

(a) Initial situation (b) Arrangement of middle circles

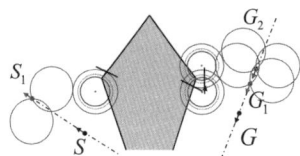

(c) Translation of the start and goal nodes

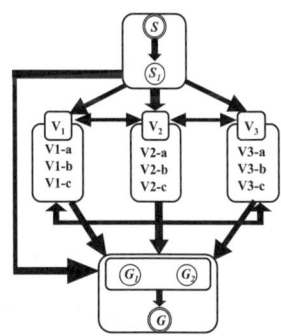

(d) Structure of the search graph

Fig.11 Visibility graph-like method

4.3.2 Assignment of Cost

A cost assigned to each arc can be defined by length of a sub-path, which means a path between two configurations. To begin with, a translational length is assigned as a cost to the arc if the arc connects the start or goal circle with the middle circle formed by translating that circle (for example in Fig.11, the cost of the arc that connects S and S_1 is the length from S to S_1). The other sub-paths contain clothoid curves. Since calculating the length of clothoid arcs involves high costs, the length of a clothoid arc is regarded as that of a circular arc by approximation.

The most important part of this argument is that both neighboring arcs influence a sub-path. In order to utilize Dijkstra's algorithm, only two nodes connected by the arc should decide the cost assigned to the arc. The other nodes should not influence it.

In order to cope with this problem, the method of cost assignment is proposed as follows. In order to make it easy to understand, this method is described with the use of a concrete example.

(1) The given environment is shown in Fig.12-(a). The definition of symbols is the same as used previously.
(2) Assign the length shown in Fig.12-(b) as the cost to the arc that connects S and M_1. As shown, the portion that is on the M_1 of this sub-path is from the intersection of the line segment and M_1 to the contact point with T_1.
(3) In the same way, the cost of the arc that connects M_{1i} and M_{2j} is the length from the contact point of M_{1i} and T_{1i}, through M_{1i} and the line segment, to the contact point of M_{2j} and T_{2j}.
(4) As can be seen, the intersection with the line segment is behind the contact point with T_{1i} on M_{1i}. In such a case, that is to say, when the arc from the contact point with T_{1i} to the intersection with the line segment meets an obstacle, this part of the path will have negative length(∗ in Fig.12-(d)).

By using this method, each arc can have its own independent cost, and the existent search algorithm is applicable to this graph.

The proposed algorithm assures sub-optimality because:
(a) The combination of Dubins' curve and the visibility graph makes it possible to derive nearly optimal paths.
(b) The length of clothoid arcs is not evaluated.
(c) Middle points at the obstacle edges are discretized.

4.4 Computational Complexity

Let the number of vertices of the obstacles in the environment be n, the number of nodes generated by translation of the start and goal nodes be a, b, and the number into which every vertex is discretized be t. The complexity of generating clothoid arcs is $O(tn)$. The total number of nodes in the search graph is $a+b+tn$. Hence, the complexity of the search is $O((a+b+tn)^2)$. In the proposed graph, especially, the complexity is considerably less than $O((a+b+tn)^2)$ because the actual number of arcs is small.

5. Simulation

5.1 Path Generation in Intricate Environment

A comparison between the paths generated by using the method proposed in Chapter 4 and the existent method [8] was made (Figs.13-15). The case in which the goal and middle circles which have an opposite direction overlap is shown in Fig.13-(a). The case in which the start and goal circles which have an opposite direction overlap is shown in Fig.13-(b). Figure 14 shows the results of the application of two methods to Fig.13-(a). By applying the method proposed in section 4.2.3, the generated path became approximately 64% shorter than by the former method. Figure 15 shows the result of the application of the proposed method to Fig.13-(b). In this case, the existent method did not generate a path.

Consequently, the necessity of the proposed method could be demonstrated.

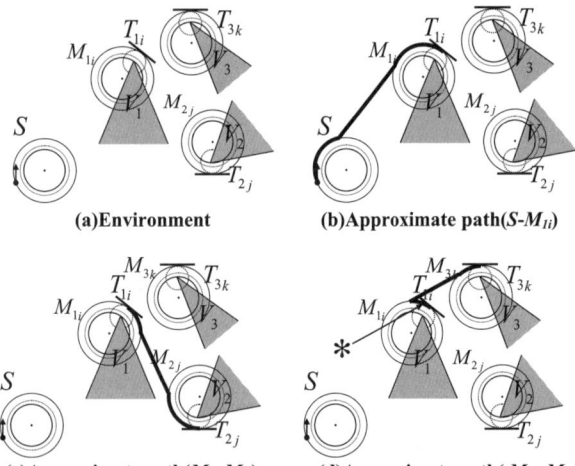

Fig.12 Approximate paths for the cost assignment

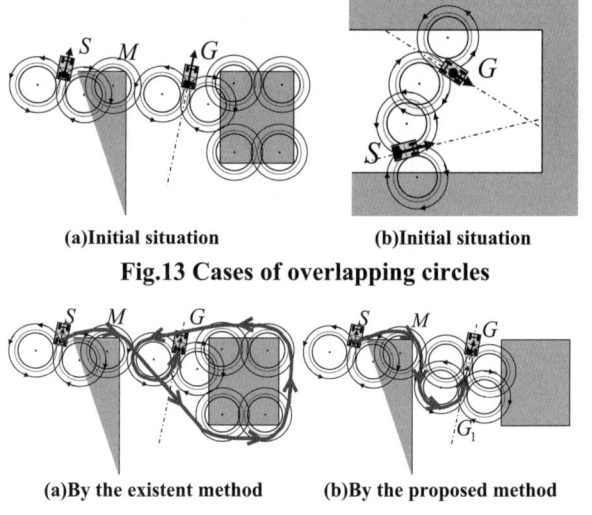

(a)Initial situation (b)Initial situation
Fig.13 Cases of overlapping circles

(a)By the existent method (b)By the proposed method
Fig.14 Results of the application to the case shown in Fig.13-(a)

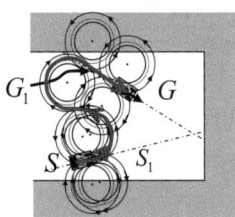

Fig.15 Result of the application to the case shown in Fig.13-(b)

5.2 Path Generation by using the visibility graph-like method in an environment with obstacles

The algorithm was applied to the same environment discussed in [8]. It is a 40m square workspace with 5 obstacles. The robot is 2.5m long and 1.5m wide. The maximum curvature and the maximum curvature derivative are defined as $0.2m^{-1}$ and $0.05m^{-2}$ respectively. In [8], four kinds of experimental results are shown. The method proposed here was applied to these four. As a result, the paths were shorter than those in the existent method, by approximately 10%, 13%, 14%, and 20%. The case in which the 20% shorter path was generated appears in Fig.16. The paths generated by the existent method are drawn by estimating the figure in [8].

The paths generated by the proposed method contained sub-paths in the condition τ-2ϕ which the existent method could not have dealt with. Furthermore, the paths that pass nearer the inner obstacles could be obtained.

 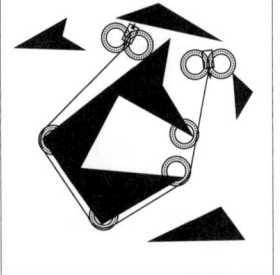

(a) By the former method[8]   (b) By the proposed method

Fig.16 Comparison between the generated paths

6. Conclusion

In this paper, a method of sub-optimal path generation has proposed in which the mechanical constraints of a robot have been considered: the maximum curvature, the maximum curvature derivative, and the curvature continuity. The generated paths are made of line segments, arcs, and clothoid arcs. By using this method, paths that could not have been created with the existent method can be generated. In the simulation, a path 20% shorter could be generated, while the average was 14%.

References

[1] J.-P. Laumond, P. E. Jacobs, M. Taïx, R. M. Murray: "A Motion Planner for Nonholonomic Mobile Robots," IEEE Trans. Robotics and Automation, Vol.10, No.5, pp.577-593, 1994.

[2] D. H. Shin, A. Ollero: "Mobile Robot Path Planning for Fine-Grained and Smooth Path Specifications," Journal of Robotic Systems, Vol. 12, No.7, pp. 491-503, 1995.

[3] L. E. Dubins: "On Curves of Minimal Length with a Constraint on Average Curvature, and with Prescribed Initial and Terminal Positions and Tangents," American Journal of Mathematics, Vol.79, pp. 497-516, 1957.

[4] P. Jacobs, J. Canny: "Planning Smooth Paths for Mobile Robots," Proc. of IEEE Int. Conf. on Robotics and Automation, AZ, pp. 2-7, 1989.

[5] K. Komoriya, K. Tanie: "Trajectory Design and Control of a Wheel-type Mobile Robot using B-spline Curve," Proc. of IEEE/RSJ Int. Workshop on Intelligent Robots and Systems, pp. 398-405, 1989.

[6] H. Dellingette, M. Hébert, K.Ikeuchi: "Trajectory generation with Curvature Constraint based on Energy Minimization," Proc. of IEEE/RSJ Int. Workshop on Intelligent Robots and Systems, Vol. 1, pp.206-211, 1991.

[7] Y. Kanayama, N. Miyake: "Trajectory Generation for Mobile Robots," Proc. of IEEE Int. Symp. Rob. Res., pp. 333-340, 1985.

[8] A.Scheuer, Th.Fraichard: "Continuous-Curvature Path Planning for Car-Like Vehicles," Proc. of IEEE/RSJ Int. Conf. on Intelligent Robots and Systems, Vol. 2, pp. 997-1003, 1997.

[9] L. E. Kavraki, P. Švestka, J. -C.Latombe, M. H. Overmars: "Probabilistic Roadmaps for Path Planning in High-Dimentional Configuration Spaces," Trans. of IEEE Int. Conf. on Robotics and Automation, Vol. 12, No.4, pp. 566-580, 1996.

[10] Th.Fraichard, J. -M. Ahuactzin: "Smooth Path Planning for Cars," Proc. of IEEE Int. Conf. on Robotics and Automation, pp. 3722-3727, 2001.

Implementation of Autonomous Fuzzy Garage-Parking Control by an FPGA-Based Car-Like Mobile Robot Using Infrared Sensors

Tzuu-Hseng S. Li, Shih-Jie Chang, and Yi-Xiang Chen

IC²S Laboratory, Department of Electrical Engineering,
National Cheng-Kung University, Tainan 70101, Taiwan, R.O.C.
E-mail: thsli@mail.ncku.edu.tw

Abstract

In this paper, the concepts of car maneuvers, fuzzy logic control (FLC), and sensor-based behavior are merged to implement the human-like driving skills in the garage-parking task by an autonomous car-like mobile robot (CLMR). We decompose the garage-parking control into four modes to synthesize the fuzzy garage parking control (FGPC). Computer simulation results illustrate the effectiveness of the proposed control schemes. The setup of the CLMR is provided, where the FGPC is implemented on a field-programmable gate array (FPGA) chip. Finally, the real-time experiments of the FPGA-based CLMR for the garage-parking task demonstrate the feasibility in a practical car maneuvering.

Keywords: Car-Like Mobile Robot, Fuzzy Logic Control, FPGA, Fuzzy Garage Parking Control, Real-Time Implementation.

I. Introduction

The parking problem of a CLMR has been investigated immensely for the last decades. Many researches [1-16] have proposed various control strategies on this topic. Each method has its own advantages and most of these researches have been applied to the real parking by a mobile robot. Numerous studies [1-5] adopt the FLC method to solve this problem. They usually derive fuzzy rules to model the parking experiences of a skilled driver to perform the parking task. Some researchers also have combined the FLC with other algorithms, for example, neural network [4] or genetic algorithm [5], to improve the feasibility of the autonomous parking control. [6-9] present some other maneuverable methods for parking problems. These methods consist of several stages or generate several maneuvers by using the specified paths (like circles, straight line segments, sinusoids, and *etc.*) from a starting location to a goal position.

Steering a car is confined with the conditions of the car's own capability of mechanism and the environment. Due to these reasons, we do not expect to design a continuously global controller for a car to perform all the maneuvering behaviors. Alternatively, we figure out some local controllers by a few experienced rules to maneuver in the specified region. Thus, to integrate all the local controllers, which work well in a respected region, as a whole controller is our objective.

Recently, using a reconfigurable FPGA to implement an algorithm is very attractive because FPGA offers a compromise between ASIC hardware and general-purpose processors. The FPGA chip is the preferred device for many researchers and engineers because it can be reconfigured virtually and instantaneously to produce a physical circuit, which can be evaluated in real time. Numerous literature [10-11] have implemented the fuzzy logic control on the FPGA or the VLSI chip for the mobile robots, where [11] adopts a low cost microcontroller and an FPGA to real-time guide the autonomous vehicle.

In this paper, we will adopt six infrared sensors to measure the relative distances between the CLMR and the circumstance. The FGPC is realized on the FPGA such that the CLMR can autonomously perform the garage-parking task.

II. Design of the FGPC

The motion of the CLMR is addressed firstly in the section. The design procedure of the FGPC is proposed. Finally, Computer simulation results are given to show the feasibility.

1. Kinematic model of a CLMR

Consider a kinematic model of the CLMR shown in Figure 1, where the rear wheels are fixed parallel to car body and allowed to roll or spin but not slip. All the corresponding parameters of the CLMR depicted in Figure 3 are defined as follows, (x_f, y_f) is the position of the front wheel center of CLMR, (x_r, y_r) is the position of the rear wheel center of CLMR, ϕ is the orientation of the steering wheels with respect to the frame of CLMR, θ is the angle between vehicle frame orientation and X-axis, and l is the wheelbase of CLMR.

The rear wheel kinematic equation of the CLMR is described by [7]

$$\begin{aligned}\dot{x}_r &= v\cos\theta\cos\phi \\ \dot{y}_r &= v\sin\theta\cos\phi \\ \dot{\theta} &= v\frac{\sin\phi}{l}\end{aligned} \quad (1)$$

where v is the speed of the front wheels. Equation (1) presents the backward movement of the CLMR. The front-wheel kinematic equation denoting forward motion of the CLMR is depicted as

$$\dot{x}_f = v\cos(\theta + \phi)$$
$$\dot{y}_f = v\sin(\theta + \phi) \qquad (2)$$
$$\dot{\theta} = v\frac{\sin\phi}{l}$$

2. Design procedure of the FGPC

For garage parking problem, a driver always passes the garage and then backs into it. In fact, backward-garage parking needs less narrow width of the garage than that in forward-garage parking. According to our driving experience, we will first pass the garage for a little distance to the starting position that the CLMR will begin to back into the garage. This action will make us easily back the car into garage. Then we make the steering wheel turn right and move backwards to the garage. When a car entirely backs into the garage, we will correct the car back and forth to an appropriate position. The process of the garage-parking behavior is shown in Figure 2. We decompose the process of the garage parking into following four modes.

Approach mode: The CLMR is moving on the lane towards to the garage. The CLMR will keep it on the centerline of the lane.

Pass mode: The CLMR is keeping on passing the garage until it reaches at the starting-parking (SP) position.

Parking mode: The CTMR is backing into the garage along the wall until the whole or most part enters the garage.

Correction mode: The CLMR examines its position to see whether it is properly staying in the garage via sensors. If the position of the robot does not meet the desired destination, it moves forwards or backwards in the parking lot to correct its parking position.

The design procedures of the FGPC for garage parking problem is separated into steering angle and speed controllers. The steering angle control and the speed control of the CLMR are described as follows.

1) Steering angle control

As an experienced driver drives a car on a straight road, he might be able to roughly describe his driving strategy on the straight road in terms of some linguistic rules. For example, if the car is a little far away from the centerline of the lane (CLL) and on the left-hand side of the CLL, one can steer slightly to the right. If the car is so far away from the CLL and on its left-hand side, one can steer largely to the right. On the contrary, if the car is a little (very) far away from the CLL and on its right-hand side, one can steer slightly (largely) to the left. For another conditions, if the car is moving towards to the CLL from its right-hand side or left-hand side, we will maintain this direction till the car is close to the CLL. Moreover, the closer the car is to the CLL, the smaller the steering angle is.

According these experiences, we can derive a two-input-single-output FLC scheme to command the steering angle of the front wheels. First, we introduce x_d and x_e as the input linguistic variables. As the robot is moving forwards, variable x_d is defined as $SR_1 - SL_1$ and variable x_e is defined as $SR_1 - SR_2$; as the robot is moving backwards, variable x_d is defined as $SR_2 - SL_2$ and variable x_e is defined as $SR_2 - SR_1$, where SR_1 is the distance between the front-right wheel of the CLMR and the wall, SR_2 is the distance between the rear-right wheel of the CLMR and the wall, SL_1 is the distance between the front-left wheel of the CLMR and the wall, and SL_2 is the distance between the rear-left wheel of the CLMR and the wall.

Variable x_d presents whether the CLMR is on the centerline between two walls or not. Variable x_e denotes the orientation of CLMR with respect to the wall, that is, the value of x_e represents the inclination situation of the CLMR between the walls. If the wall is not existent in the left-hand side of the CLMR, variable x_d is defined as $SR_1 - DS$ for moving forwards or x_d is defined as $SR_2 - DS$ for moving backwards, and where DS is the desired distance between the CLMR and the right-hand wall. The output of the FLC is the steering angle ϕ.

Input variables x_d and x_e are decomposed into five fuzzy partitions with triangular membership functions, and output variable ϕ is the fuzzy singleton-type membership function with five partitions. The partitions and the shapes of the membership functions are shown in Figure 3. According to the experienced driver described previously, fuzzy reasoning rules for the steering angle control can be summarized in Table 1. The defuzzication strategy is the weighted average method.

2) Speed control

For speed control, the driving skill can be achieved by the following situations. If the car is very far away from the front wall and the steering wheel command is ZE, we will speed up the car. If the car is also very far away from the front wall but the steering wheel command is a little turning to the right-hand side (or the left-hand side), we can just hold the car to a normal speed. If the car is just a little away from the front wall and the steering wheel command is ZE, we will also hold the car to a normal speed. If the car is gradually close to the front wall, we can slow down the car.

According these experiences, we can derive the

two-input-single-output FLC scheme to command the speed of the CLMR. Because the design procedure of the FLC for speed control is similar to that for steering angle control, the entirely design procedure for speed control process is omitted here for saving the length of the paper.

When the CLMR moves in the approach mode and the pass mode, the forward steering angle control will be utilized until the CLMR enters the parking mode. When the CLMR enters the parking mode, the CLMR makes the steering wheel turn right as possible as it can and moves backwards to the garage. When the CLMR is in the correction mode, we exploit the backward and forward steering angle control to correct the CLMR to an appropriate position. If the CLMR is very close to the rear of the garage, we let the CLMR go forwards by using the forward steering angle control to correct the position of the CLMR. Suppose the front wheels of the CLMR exceed the garage, we let the CLMR move backwards to persist correction. If the CLMR still cannot park in the appropriate position, we drive the CLMR back and forth until the CLMR is about the center of the garage. The speed control is used in all parking procedure.

3. Computer simulation results

The computer simulation results are given to demonstrate the effectiveness of the proposed FGPC. We exploit the kinematic equations (1) and (2) of the CLMR and calculate the relative distances (SF, SR, SR_1, SR_2, SL_1, and SL_2) among the CLMR and the wall. Figure 4 illustrates a typical example of the CLMR by applying the FGPC schemes, where the CLMR parks in the garage successfully.

III. Hardware Architecture of the CLMR

When the CLMR travels in the working environment, it will receive the data of the environment via six infrared sensors to recognize the information of the environment. The A/D converter transforms the data into digital signals. The controller reads the digital data once the conversion of A/D converter has been done and generates two control commands. One command is the PWM signal to the DC servomotor for controlling the steering angle of the CLMR. The other one will be sent to the DC motor for controlling the speed of the CLMR through the motor driver IC and D/A converter. The VHDL is programmed on the PC and then download to the FPGA chip via parallel port.

The hardware architecture of the CLMR consists of the following four parts, chassis mechanism, FPGA experiment board, motor driver, A/D and D/A converter, and sensor. The appearance of the real CLMR is shown in Figure 5. In order to simulate the behaviors of a real car, we adopt a 1/10th-scale four-wheeled vehicle with front-wheel drive and front steering wheels as the chassis mechanism of the CLMR. The chip of the FPGA experiment board is EPF6024ACT144-3 that is the Flex Series of the Altera chip and is manufactured by Galaxy Far East Corp. This chip can provide 24,000 gates and 117 I/O ports. The design of the control scheme is implemented on the FPGA chip.

For speed control of the CLMR, a driver IC is used to offer control signals in order to drive the DC motor. TA7291P DC motor driver IC manufactured by TOSHIBA Co., Ltd., is applied to drive DC motor and it also provides current-limiting and over-voltage protection circuit. For steering angle control of the CLMR, a DC servomotor is equipped to rotate the steering angle of front wheels. Basically, the position of a DC servomotor can be controlled by PWM signal. A three-wires DC servomotor is applied here, and one wire of this DC servomotor is the input of the PWM signal, which is generated by the FPGA chip.

Sensors are merely transducers that the FPGA chip can directly read data through the I/O port on the chip. Thus, A/D converter is used to transform the analog data into digital data that FGPA chip can process. The ADC0804 IC manufactured by NS Co. Ltd. is used in the experiment. Because we can alter voltage of the TA7291P motor driver IC for controlling the velocity of the CLMR, we need a D/A converter to transform digital signals from the FPGA chip into analog signal. The DAC0800 IC used here is also manufactured by NS Co. Ltd. But the output signal of the DAC0800 IC is analog current, we adopt the current-to-voltage converter LM1458 OP to give the voltage command.

The CLMR should be able to navigate in an unknown environment based on the information of sensors. UF 66 MG manufacturing by TELCO International Ltd. is the infrared sensor and gives the output voltage proportional to the reflection distance. There are six infrared sensors mounted on the CLMR. Four infrared sensors are mounted above the position of right wheels and left wheels of the CLMR, respectively. For avoid colliding with obstacles, we place one infrared sensor in front of the robot as well as one in the rear part. The purpose of this arrangement is that the infrared sensors can acquire sufficient data about the real circumstances so that the CLMR is able to correctly go forwards or backwards, stop, and make a left or right turn. By using six infrared sensors, the information of distances (SF, SR, SR_1, SR_2, SL_1, and SL_2) can be got.

IV. Circuit Design on the FPGA Chip

The hardware circuit architecture of the FGPC realized on the FPGA chip is shown in Figure 6. The entire I/O port for this chip includes 49 pins for input port and 12 pins for output port. In the input port, there are eight pins from each infrared sensor via A/D converter and one pin from the "CK" of the quartz oscillator. In the output port, there are two pins to the "IN$_1$" pin and the "IN$_2$" pin to the driver IC of the DC motor, one pin to the "WR" pin of the A/D converter, one pin to the DC servo motor, and 8 pins to the data input of the D/A converter. The FGPC contains three modules, including the PWM fitness module and two FLC modules. The modules of the FGPC are described as follows.

1. PWM fitness module

The PWM signal is provided to control the steering angle by controlling the DC servomotor. We utilize a up-counter whose output value is compared with a reference value. If the counted value is less than the reference value, then the output of the comparator exports '0'; otherwise, '1' is exported. We just need to set the reference value to choose a fit PWM signal. Owing to that the range of steering angle is -30 degree to +30 degree, we take 5 degrees as one unit to divide the range into 13 types of the PWM signals. When the PWM fitness circuit gets the crisp output from the FLC module, we use the multiplexer as the PWM selector to match the corresponding crisp output, and then the steering angel of the CLMR can be controlled.

2. Design of FLC module for steering angle control

The hardware architecture of the FLC module is also implemented on the FPGA chip. The hardware architecture includes fuzzification sub-module, DML sub-module, and defuzzification sub-module. We design one FLC module for steering angle control and one FLC module for speed control, respectively. Each sub-module of the FLC module for steering angle control will be described in the following paragraphs.

1) Fuzzification sub-module

Taking into account of different fuzzy input definitions described in Section II, when the CLMR goes forwards, we use the forward FWFC, on the contrary, we adopt the backward FWFC. We utilize the outside A/D converter to convert the analog voltage values measured by infrared sensors into digital data. The digital fuzzy input variables are appropriately chosen by the judgment circuit. Then the digital data stream are mapping to fit the universes of discourses.

2) DML sub-module

In this stage, the antecedent fitness block and inference rule-base block are utilized to implement the DML sub-module. The fuzzification sub-module will export the excited membership values and the excited fuzzy linguistic labels to the DML sub-module for fuzzy inference operations. Because a minimum operator is considered for realization of composition operation of the FLC in the antecedent fitness block, the "Min" circuit is implemented to compare the minimum value and obtain the antecedent values of the fuzzy rules. In the FLC module, the antecedent and the consequence of these IF-THEN rules are associated with fuzzy conditional statements. Design concept of inference rule base block is realized in the look-up table. In this DML block, it will output four control weighted values and four labels linguistic terms of the triggered rules to the next defuzzication sub-module.

3) Defuzzication sub-module

The defuzzification method is based on weighted average method mentioned in previous section. The design concept of the defuzzification sub-module is to figure out one crisp output based on excited four rules. Thus, the blocks of defuzzification sub-module implemented here include four multipliers, several adders, and one division divider. To speed up the calculation process of the multiplier, we adopt the parallel-type multiplier, which may take more volume. The multiplier includes several multiplexers, several registers, and one adder. The divider is the last circuit of the defuzzication sub-module. It composes of shifter, dividend register, divisor register, remainder register, quotient register, subtractor register, and logic circuit. The defuzzication sub-module will be implemented by integrating all circuits after realization of the divider.

4) Latch circuit

The pipeline concept is that we can cut a complex and large circuit to several sub-module circuits and use the latch circuit as the partition between short paths. The delay time between each data segment will be shorter and the operation clock frequency of the entire circuit can be speeded up. It is known that suppose the delay time and/or the sequence of transmission for each sub-module circuit is different, then unstable output signal and/or unexpected results may occur. The latch circuit is in fact set up at the ends of fuzzification, DML, and defuzzication sub-modules. Its objective is that the input signals of the present stage will not vary with the outputs changes of the former-stage sub-module. Finally, the implementation of the digital FLC module is shown in Figure 7, where all the sub-modules and the pipeline data latch architecture are included.

3. Design of FLC module for speed control

Because the hardware architecture of the FLC for speed control is the same as that for steering angle control, the entirely design procedure for speed control process are the same as those in previous sub-sections and is omitted here for saving the length of the paper.

V. Real-Time Implementation

Computer simulation results discussed in Section II show that the FGPC can successfully execute the garage-parking behavior. The hardware architecture of the CLMR and the digital circuit design on the FPGA chip have been respectively established in Section III and IV. In this section, we want to realize the FGPC on the CLMR in a real test ground. The dimensions of the investigated CLMR are length 380mm, width 240mm, and weight 4.2Kg, respectively. The actually experimental photographs of the FGPC on the CLMR are shown in Figure 8, where 12 sequential image stills captured by the handheld CCD camera are given. One can find from these pictures that the proposed FGPC can successfully complete the garage-parking mission.

VI. Conclusion

In this paper, an FGPC method has been presented to emulate human-like driving skills. This method, which is based on the car maneuvers, FLCs and infrared sensors, has been synthesized into the FGPC. We have also designed and implemented an autonomous CLMR, where chassis mechanism, motor driver, FPGA experiment board, A/D and D/A converter, and infrared sensor have been set up. The FGPC has been realized by the FPGA chip that is set up on the CLMR. All computer simulations and practically experimental results demonstrate that the propounded FGPC is indeed effective and feasible for practical application to real car maneuvers.

Acknowledgement

This work is supported by the National Science Council, Taiwan, Republic of China, under grants NSC90-2213-E006-052 and NSC91-2213-E006-026.

References

[1] M. Sugeno and K. Murakami, "An Experimental Study on Fuzzy Parking Control Using a Model Car," *Industrial Applications of Fuzzy Control* (M. Sugeno ed.), North-Holland, pp. 105-124, 1985.

[2] M. Ohkita, H. Mitita, M. Miura, and H. Kuono, "Traveling Experiment of an Autonomous Mobile Robot for a Flush Parking," *Proc. 2nd IEEE Conf. on Fuzzy Systems*, Francisco, Vol. 2, 1993, pp. 327-332.

[3] S. -J. Chang and T. -H. S. Li, "Design and implementation of fuzzy garage-parking control for a car-type mobile robot," *J. of Intelligent and Robotic Systems*, Vol. 34, 2002, pp. 175-194.

[4] W. A. Daxwanger and G. K. Schmidt, "Skill-Based Visual Parking Control Using Neural and Fuzzy Networks," *IEEE Int. Conf. on System, Man, and Cybernetics*, Vol. 2, 1995, pp. 1659-1664.

[5] D. Leitch and P. J. Probert, "New Techniques for Genetic Development of a Class of Fuzzy Controllers," *IEEE Trans. on System, Man, and Cybernetics*, Vol. 28, pp. 112-123, 1998.

[6] R. M. Murray and S. S. Sastry, "Nonholonomic Motion Planning: Steering Using Sinusoids," *IEEE Trans. on Automatic Control*, pp. 700-716, 1993.

[7] J. P. Laumond, P. E. Jacobs, M. Taix, and R.M. Murray, "A Motion Planner for Nonholonomic Mobile Robots," *IEEE Trans. on Robots and Automation*, No. 5, pp. 577-593, Vol. 10, 1994.

[8] I. E. Paromtchik and C. Laugire, "Motion Generation and Control for Parking an Autonomous Vehicle," *Proc. 1996 IEEE Conf. on Robotics and Automation*, Minneapolis, MN, Vol. 4, 1996, pp. 3117-3122.

[9] K. Jiang and L. D. Seneviratne, "A Sensor Guided Autonomous Parking System for Nonholonomic Mobile Robots," *Proc. 1999 IEEE Int. Conf. on Robotics and Automation*, Vol. 1, 1999, pp. 311-316.

[10] D. Kim, "An Implementation of Fuzzy Logic Controller on the Reconfigurable FPGA System," *IEEE Trans. on Industrial Electronic*, Vol. 47, No. 3 pp. 703-715, 2000.

[11] J. L. Arroyabe, G. Aranguren, "Autonomous Vehicle Guidance With Fuzzy Algorithm," *Proc. IECON 2000*, Vol. 3, 2000, pp. 1503-1508.

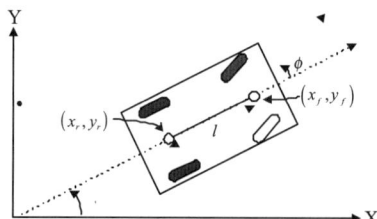

Figure 1. Kinematic model of a CLMR.

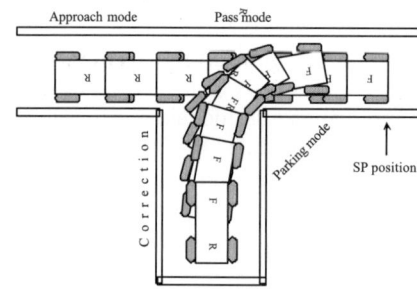

Figure 2. Process of the backward-garage parking.

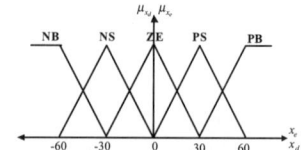

(a) Membership functions of x_d and x_e.

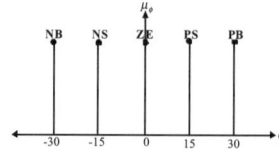

(b) Membership function of ϕ.

Figure 3. Fuzzy membership functions of the input-output variables for the steering angle control.

Table 1. Fuzzy rule table for the steering angle control.

x_e \ x_d ϕ	NB	NS	ZE	PS	PB
NB	NB	NB	NS	NS	ZE
NS	NB	NS	NS	ZE	PS
ZE	NS	NS	ZE	PS	PS
PS	NS	ZE	PS	PS	PB
PB	NE	PS	PS	PB	PB

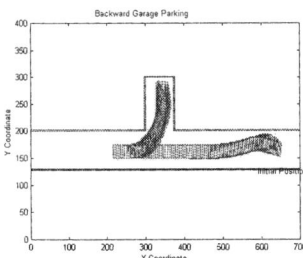

Figure 4. Simulation results with Initial posture $(x_f, y_f, \theta) = (650, 165, -10^0)$.

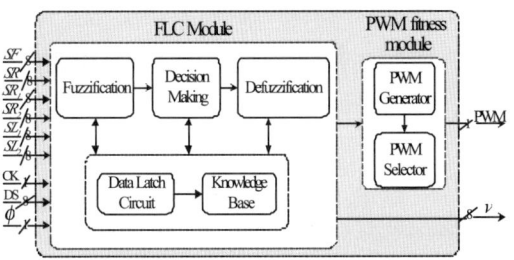

Figure 6. Hardware circuit architecture of the FBC

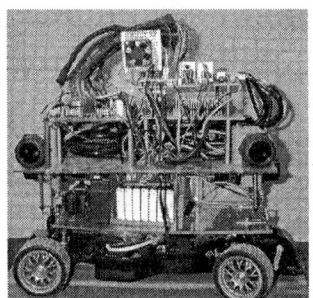

Figure 5. Actual appearance of the CLMR.

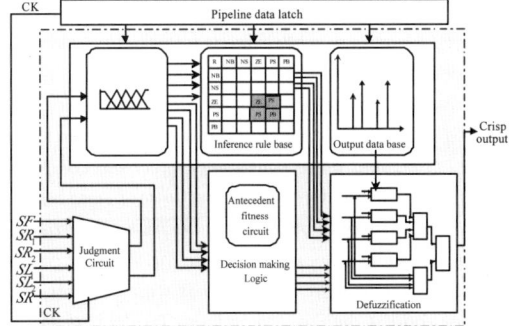

Figure 7. Implementation circuit of the FLC module.

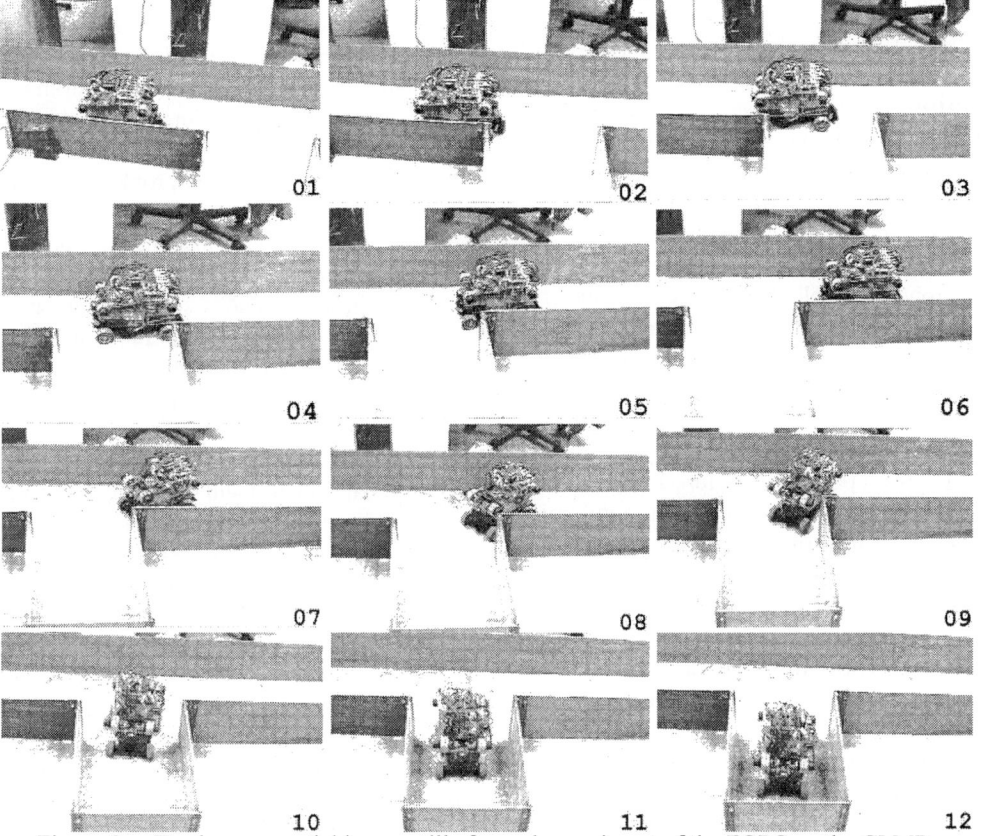

Figure 8. Twelve sequential image stills for real experiment of the FGPC on the CLMR.

On Energy-minimizing Paths on Terrains for a Mobile Robot

Zheng Sun and John Reif

Department of Computer Science
Duke University
Durham, NC 27708, USA
{sunz, reif}@cs.duke.edu

Abstract *In this paper we discuss the problem of computing optimal paths on terrains for a mobile robot. The cost of a path is defined to be the energy expended due to both friction and gravity. The model allows for ranges of impermissible traversal directions caused by overturn danger or power limitations. This model is interesting and challenging as it incorporates constraints found in realistic situations and these constraints affect the computation of optimal paths. We give some upper and lower bound results on the combinatorial size of energy-minimizing paths on terrains. We also present an efficient approximation algorithm that computes for two given points a path whose cost is within a user-defined relative error ratio. Compared to previous results with the same approach, this algorithm improves the time complexity by using (a) a discretization with reduced size, and (b) an improved discrete algorithm for finding optimal paths in the discretization. We present some preliminary experimental results to demonstrate the efficiency of our algorithm. We also provide a similar discretization for the same model but under less restricted assumptions.*

1 Introduction

With the growth of geographical information systems (GIS), now it is possible to find a terrain map (such as the one for Kaweah River basin shown in Figure 1) for virtually any location in the world. The availability of these high-resolution maps makes computing energy-minimizing paths for mobile robots possible. However, despite the obvious potential applications in both commercial and military areas, there has not been enough interest in the energy-minimizing path problem as compared to its practical significance in the real world. The extensively studied Euclidean shortest path problems fail to capture some of the characteristics of optimal path planning for a mobile robot, such as the variance of the friction coefficient in different areas, the limitations of the driving force of the robot, as well as the stability of the robot on a steep plane.

In this paper we study the problem of computing energy-minimizing paths (or optimal paths) on terrains for a mobile robot. Our work is based on the model introduced by Rowe and Ross [10, 9, 8], who defined the cost of a path to be the energy loss due to both friction and gravity. This model adds anisotropism to optimal path planning by taking into consideration impermissible traversal directions resulted from overturn danger or power limitations. This problem is a generalization of the weighted terrain optimal path problem, yet conceivably more difficult.

We provide a couple of complexity results on the combinatorial size of energy-minimizing paths on terrains. We

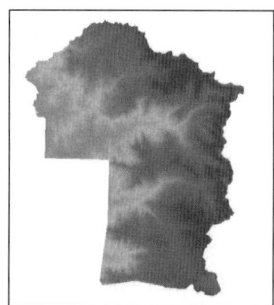

Figure 1. Terrain map of Kaweah River basin

show that any optimal path on a weighted terrain contains $O(n^2)$ segments. Here n is the number of faces in the terrain. With the introduction of anisotropism, however, we can construct a terrain with specified points s and t such that any optimal path connecting s and t contains exponential number of segments. These complexity results not only are of theoretical interest, but also have implications on the implementation of approximation algorithms.

To compute an approximate optimal path, we adopt the same discretization approach used by Lanthier, Maheshwari and Sack [3] for this problem. By placing discrete points (which we call *Steiner points*) on the boundaries of terrain faces and interconnecting these points by edges with appropriate weights, we reduce the original optimal path problem in a continuous space to computing an optimal discrete path (a path with the minimum total weight) in a discrete graph $\mathcal{G}$. The optimal discrete path found is then converted to a path in the original space as an approximate solution. This discretization approach is also used for the weighted region optimal path problem [1, 6, 2] as well as the optimal path problem in the presence of flows [7].

Lanthier, Maheshwari, and Sack [3] used Dijkstra's algorithm to compute an optimal discrete path in $\mathcal{G}$. If m Steiner points are placed on each boundary edge, their approximation algorithm has a time complexity of $O(nm^2 + nm \log(nm))$. We prove that the problem of computing an optimal discrete path in $\mathcal{G}$ can be solved in $O(nm \log(nm))$ time by using the BUSHWHACK algorithm [6, 12]. BUSHWHACK is a discrete search algorithm that can efficiently compute optimal discrete paths by exploiting the geometric properties of the discretization. We also show that the discretization used by [3] can be reduced while still guaranteeing the same asymptotic error bound as in [3]. By combining this discretization of reduced size with the BUSHWHACK algorithm, our approximation algorithm not only has an improved time complexity over the result in [3], but also is less dependent on various geometric parameters, such as the minimum angle between two adjacent boundary edges of a terrain face, and the maximum angle of a special range.are supported by the preliminary experimental results presented in Section 6.

We extend our work to steep terrains on which a robot can only move downhill. In this case the optimal path problem is even more challenging, as a terrain face not only can have different cost metric in various directions, but also may become an "anisotropic obstacle" that blocks any upward movement of a robot.

Although this model addresses some of the characteristics of optimal path planning for mobile robot that are not considered by Euclidean shortest path problems, it still does not take into consideration nonholonomic constraints found in real contexts. We refer readers to [4] for a review of works on nonholonomic motion planning.

2 Preliminaries
2.1 The Physical Model
We now describe the model first developed by Rowe and Ross [10]. Let r be a terrain face with a gradient of ϕ and let μ be the friction coefficient between the mobile robot and the surface of r. Following the notation of [3] we define $w = \mu \cdot cos\phi$ to be the "weight" of r. For a robot traveling on r with an inclination angle of φ (as shown in Figure 2), the energy cost is defined to be $mg(\mu \cos \phi + \sin \varphi) \cdot l = mg(w + \sin \varphi) \cdot l$, where mg is the weight of the robot, and l is the traveled distance. According to Rowe and Ross [10], "this formula was confirmed experimentally within 1% for wheeled vehicles on slopes of less than 20% in [11]."

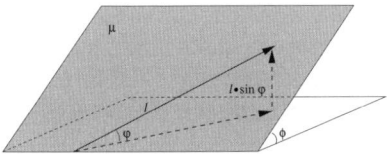

Figure 2. Energy cost

The problem is to find an energy-minimizing path from a given source point s to a given destination point t. The model assumes no acceleration during the entire course and no energy cost for making turns. We assume that each terrain face is a triangular region, and that source s and destination t are vertices of some terrain faces. We use n to denote the total number of all terrain faces.

There are three *impermissible ranges* defined for each terrain face, as shown in Figure 3. The *impermissible force range* indicates the range of uphill traversal directions that are too steep for a robot to climb. The other two are the *sideslope overturn ranges*, which include the forbidden directions that can cause overturn when the projection of the robot's center of gravity falls outside the convex hull of the support points. Each boundary angle of an impermissible range is called a *critical impermissibility angle*.

Another special case occurs when a robot is traveling downhill with such an inclination angle φ that $w + \sin \varphi < 0$. This will cause the robot to gain energy and accelerate. Therefore, the robot has to apply a braking force of $-mg(w + \sin \varphi)$ to avoid acceleration. The range in which the robot has to use a braking force is called *braking range*. The two boundary angles of the braking range, *critical braking angles*, can be computed by finding the solution φ_0 of the equation $w + \sin \varphi = 0$. The robot expends no energy when traveling in a direction that is inside the braking range, as the energy gained by going downhill is exactly offset by the energy expended for braking.

We call the impermissible force range, the sideslope overturn ranges, and the braking range *special ranges* of a terrain face. These special ranges are fixed for all points in that face. We define the "angle" of a range to be the angle between the two rays defining the boundary of the range. Let α_1, α_2 and α_3 be the angles of the impermissible force range, each sideslope overturn range and the braking range, respectively. There are four *regular ranges*, each of which is between any two adjacent special ranges. The energy cost formula $mg(w + \sin \varphi) \cdot l$ applies only when a robot is traveling in a direction that is inside a regular range.

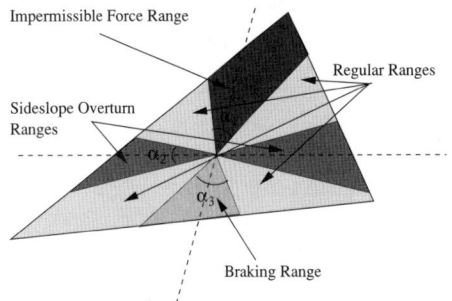

Figure 3. Impermissible, braking and regular ranges

To *effectively* move in a direction that is inside a impermissible range, a robot has to take a zigzag path with alternating directions that are inside regular ranges. It is therefore implicitly assumed that a robot can freely switch between two directions inside regular ranges, even if there is an impermissible range in between.

We say a range is *degenerate* if the size of the range is zero. A regular range can be degenerate if the two neighboring special ranges overlap with each other. A terrain face r is *regular* if all four special ranges are degenerate inside r; otherwise, r is *irregular*. If all the faces of a terrain are regular, we say it is a *regular terrain*; otherwise, it is an *irregular terrain*.

In this paper we consider the optimal path problem on irregular terrains. We are particularly interested in the case where the angle of one of the special ranges is close to π. As we shall see later in this paper, the closer the size of a special range is to π, the larger the difference in cost metrics can be between two directions inside that range.

An irregular terrain face r is *totally traversable* if it is always possible to travel between two points in r through a (straight or zigzag) path that lies entirely inside r. This can be guaranteed if the two upper regular ranges are not degenerate, which means $\alpha_1 + \alpha_2 < \pi$. Otherwise, the impermissible force range of r would overlap with the two sideslope overturn ranges, forming a combined impermissible range with an angle equal to $\pi + \alpha_2$. For any direction $\overrightarrow{vv'}$ inside this range, there is no feasible path connecting v and v' that lies entirely inside r, as $\overrightarrow{vv'}$ cannot be expressed as a nonnegative linear combination of permissible directions. However, as long as $\alpha_2 < \pi$, either the braking range or the two lower regular ranges are not degenerate, and therefore there will still be permissible traversal directions inside r. In this case, r is *partially traversable*. We say a terrain is a *totally traversable terrain* if each terrain face is totally traversable; otherwise, it is a *partially traversable terrain*.

2.2 Construction of Discrete Graph $\mathcal{G}$

In this section, we show that we can construct a directed graph $\mathcal{G}$ in such a manner that, for any discrete path p from s to t in $\mathcal{G}$, there exists a path p' from s to t in the original continuous space with the same cost. Furthermore, p' can be computed from p in linear time.

We discretize the original space by introducing Steiner points on boundary edges. For each terrain face r we add a number of Steiner points on each boundary edge of r. We construct a graph $\mathcal{G}$ that includes all Steiner points and vertices as nodes. For any two such nodes v, v' on the boundary of r, we add a directed edge $\overrightarrow{vv'}$ in $\mathcal{G}$. Furthermore, edge $\overrightarrow{vv'}$ is assigned a weight $d_r(v, v')$ equal to the cost of a *face-wise optimal* path connecting v and v'. A path is face-wise optimal if it is an energy-minimizing path, according to the cost metric of r, among all paths from v to v' that lie entirely inside r.

Now we consider how to compute the weight of edge $\overrightarrow{vv'}$. Let φ be the inclination angle of vector $\overrightarrow{vv'}$ and let l be the length of $\overrightarrow{vv'}$. Recall that, if φ is in a regular range, the energy cost formula is $mg(w + \sin \varphi) \cdot l$ for traveling from v to v' following a straight line. The first part $mg \cdot w \cdot l$ of the formula represents the energy expended due to the force of friction (friction cost), whereas the second part $mg \cdot \sin \varphi \cdot l$ represents the energy expended due to gravity (gravity cost). As the total gravity cost for traveling from s to t is always the altitude difference between the two points regardless of the path taken, we can extract the gravity cost from the cost formula. This leads to a simplified model, in which the cost of traveling for distance l with an inclination angle of φ is $mg \cdot w \cdot l$ (if φ is in a regular range). Similarly, if φ is in the braking range, the energy cost is zero for the robot. Therefore, in the simplified model, after removing the (negative) gravity cost of $mg \cdot \sin \varphi \cdot l$, the cost formula becomes $-mg \cdot \sin \varphi \cdot l$, which is the energy cost for braking.

In the following discussion, whenever we refer to the cost of a path, we mean the cost defined by the simplified model. This strategy is also used by [3]. With the simplified model, we have the following properties for face-wise optimal paths:

Claim 1 *Let v, v' be two points in a terrain face r. Let α be the angle of the range to which $\overrightarrow{vv'}$ belongs and let θ be the angle between $\overrightarrow{vv'}$ and the ray bisecting the range. Then if direction $\overrightarrow{vv'}$ is inside* **a)** *a regular range, the face-wise optimal is the straight line path $\overline{vv'}$, and the cost is $d_r(v, v') = mg \cdot w \cdot |\overline{vv'}|$;* **b)** *an impermissible range, any zigzag path with alternating directions of the two critical impermissibility angles of the range is face-wise optimal, and the cost is $d_r(v, v') = mg \cdot w \cdot |\overline{vv'}| \cdot \frac{\cos \theta}{\cos(\alpha/2)}$;* **c)** *the braking range, any path whose traversal direction remains in the braking range is face-wise optimal, and the cost is $d_r(v, v') = -mg \cdot \sin \varphi \cdot |\overline{vv'}| = -mg \cdot \sin \varphi_0 \cdot \frac{\sin \varphi}{\sin \varphi_0} \cdot |\overline{vv'}| = mg \cdot w \cdot |\overline{vv'}| \cdot \frac{\cos \theta}{\cos(\alpha/2)}$.*

The above formulae reveal two properties of face-wise optimal paths: (a) $d_r(v, v')$ can be stated in a uniform form for both impermissible ranges and the braking range, although α may represent different values; and (b) inside each special range $d_r(v, v')$ is proportional to $|\overline{vv'}| \cdot \cos \theta$, the Euclidean length of the projection of $\overrightarrow{vv'}$ on the ray bisecting the range.

3 Bound on Number of Segments of An Optimal Path

In Section 2 we showed how to construct a graph $\mathcal{G}$ from a terrain. $\mathcal{G}$ is totally dependent on the discretization, i.e., the placement of Steiner points on edges. An easy to implement discretization scheme is the uniform discretization, which places Steiner points with an equal distance on each edge. We can properly choose the distance between two adjacent Steiner points on edges of a terrain face r so that, for any crossing segment in r with cost C, there exists a neighboring *approximation segment* (a segment that connects two Steiner points) with cost $C + \epsilon$, for some user-specified ϵ. Therefore, for any optimal path with cost C_{opt}, there exists an approximate optimal path with cost no more than $C_{opt} + k \cdot \epsilon$, where k is the number of segments of the optimal path.

To construct a uniform discretization that has a constant (additive) error bound, we need to find an upper bound on the number of segments for all optimal paths. We first study the case of regular terrains. Recall that any face in a regular terrain has no impermissible or braking range. Therefore, a regular terrain is equivalent to a weighted terrain.

Mitchell and Papadimitriou [5] showed that an optimal path in a weighted planar subdivision has $O(n^2)$ segments. Using their general proof technique, we proved the same result for the weighted terrain case. The proofs of two important lemmas they used have to be modified significantly to fit into the weighted terrain case. Due to the limit of the space, we leave the proof of the theorem to the full version of the paper [13] and only state the result here:

Theorem 1 *Any optimal path on a regular terrain has $O(n^2)$ segments.*

The proof of the above theorem is not applicable to irregular terrains due to the anisotropism introduced by the special ranges. If the impermissible force range of a terrain face has an angle close to π, an optimal path p_{opt} may have to zigzag for arbitrary number of times before reaching the destination point. Therefore, the number of segments of an optimal path can only be bounded by the geometric parameters of the terrain faces.

However, since inside any terrain face the face-wise optimal path between two points v and v' can be computed directly, we can treat each subpath of p_{opt} that is face-wise optimal as one "virtual segment." The total error of an approximate optimal path of p_{opt} is not dependent on the number of segments of p_{opt}, but rather the number of virtual segments of p_{opt}, that is, the number of times p_{opt} switches from one region to another. In case there are multiple optimal paths from s to t, we are interested in only the optimal path with the least number of virtual segments. For example, on the surface of a pyramid with four identical faces, an optimal path connecting a point at the bottom of the pyramid to the apex can switch regions (faces) for infinite number of times while looping around the apex. However, there also exists another optimal path that stays in the same region while zigzagging towards the apex, and therefore has only one virtual segment.

For partially traversable terrains, we have:

Theorem 2 *If the input problem is specified with a total of N bits, for two points s and t on a partially traversable terrain, an optimal path with the least number of virtual segments, among all optimal paths that connect s and t, can contain $\Omega(2^{cN})$ virtual segments for some constant c.*

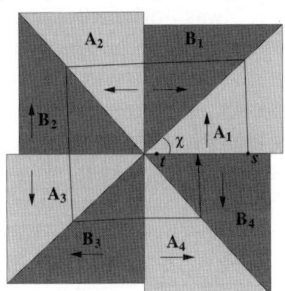

Figure 4. Problem instance

Proof Abstract: As shown in Figure 4, we construct an optimal path problem on a partially traversable terrain with eight faces. In the figure the arrow in each terrain face defines the upward direction in that terrain face. Each A_i, as shown in Figure 5, is totally traversable, and has a braking range with an angle of $2^{-c_2 N}$, no impermissible force range and a single regular range with an angle of $2^{-c_2 N}$ that combines the two upward regular ranges. Each B_i is partially traversable, with all four regular ranges degenerate, resulting in a combined impermissible range with an angle of $2\pi - 2^{-c_1 N}$ and a braking range with an angle of $2^{-c_1 N}$. An optimal path from s to t will contain alternating uphill and downhill segments. By carefully choosing the parameters of A_i and B_i, we can force any optimal path from s to t to move counterclockwise as shown in the figure. If we let $\tan \chi = 1 - 2^{c_3 N}$ for some constant c_3, each loop will take the robot $O(2^{-cN})$ closer to t for some constant c. Therefore, any optimal path from s to t will have to switch regions for $\Omega(2^{cN})$ times. ∎

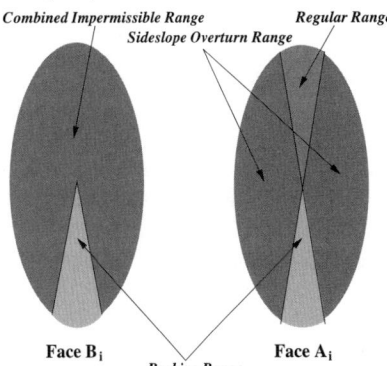

Figure 5. Ranges of A_i and B_i

In the above construction, we used four partially traversable terrain faces. It occurs to us that such an example cannot be constructed without using partially traversable terrain faces. We have the following conjecture:

Conjecture 1 *For any two points s and t on a totally traversable terrain, there exists an optimal path connecting s and t with $O(n^2)$ virtual segments.*

4 An Improved Approximation Algorithm

The uniform discretization does not guarantee a relative (multiplicative) error bound for the approximation. Lanthier, Maheshwari and Sack [3] developed a logarithmic discretization scheme that guarantees a $O(\epsilon')$-approximation for any optimal path. Here $\epsilon' = \frac{w_{max}\epsilon}{w_{min}\cos(\alpha_{max}/2)}$, where α_{max} is the maximum angle of all special ranges, and w_{max} (w_{min}) is the maximum (minimum, respectively) weight of all terrain faces. To achieve this error bound, the discretization needs to place $O(\log_{\delta_{min}}(L/R) + \log_{\mathcal{F}}(R/L))$ Steiner points on each edge. Here L is the length of the longest edge and $\delta_{min} = 1 + \sin \theta_{min}$, where θ_{min} is the minimum angle between any two adjacent edges of any terrain face. For each vertex v, let $f_1, f_2, \cdots, f_d$ be the faces incident to v and let h_v be the minimum distance between v and edges of f_i's not incident to v. We let $r_v = \epsilon \cdot h_v$ and R be the minimum of all r_v's. Furthermore, $\mathcal{F}$ is a parameter dependent on θ_{min} and some other geometric parameters.

The discretization scheme proposed by [3] adds Steiner points in three stages. In Stage 1, Steiner points are placed using the algorithm of [1] to ensure that the Euclidean distance between any two adjacent Steiner points on an edge is at most ϵ times the length of any face-crossing segment with one end between them. In Stages 2 and 3 some additional Steiner points are added for each of the braking and regular ranges.

In the following we show that the Steiner points added in Stage 1 alone can guarantee the same error bound of $O(\epsilon')$. In [3] the following assumption is implicitly used, and we will use the same assumption here.

Assumption 1 *Each terrain face r is totally traversable. Otherwise, the terrain face is considered to be non-traversable.*

We first briefly describe the placement of Steiner points (we refer readers to [1] for details). Let $e = \overline{vu}$ be a boundary edge of a terrain face. The Steiner points $v_1, v_2, \cdots, v_k$ are placed from v to u on e in such a way that $|\overline{vv_i}| = \epsilon h_v \cdot \delta^{i-1}$. Here δ is defined to be $1 + \epsilon \sin\theta_v$ if $\theta_v < \frac{\pi}{2}$, or $1 + \epsilon$ if otherwise, where θ_v is the smallest angle between any two incident edges of v. We add Steiner points $u_1, u_2, \cdots, u_{k'}$ from u to v on e in an analogous manner.

To prove that the discretization can guarantee an $O(\epsilon')$-approximation, we need to show that, for any optimal path p_{opt} in the original space, we can construct an approximation path in $\mathcal{G}$ with a cost at most $(1 + O(\epsilon')) \cdot \|p_{opt}\|$.

Our construction of approximation path is different from the one used in [3]. Suppose $\overline{vv'}$ is a segment of an optimal path ("optimal segment") in a terrain face r, where v is between two Steiner points u_1, u_2 on edge e and v' is between u'_1, u'_2 on edge e'. The construction scheme in [3] will choose either $\overline{u_1 u'_2}$ or $\overline{u_2 u'_1}$, whichever is in the same directional range as $\overrightarrow{vv'}$, as the approximation segment for $\overline{vv'}$. The addition of Steiner points in Stages 2 and 3 is to guarantee that at least one of $\overrightarrow{u_1 u'_2}$ and $\overrightarrow{u_2 u'_1}$ is in the same directional range as $\overrightarrow{vv'}$. Observe that, according to this method, the approximation segments corresponding to two consecutive optimal segments are not necessarily connected. In this case, an additional "joint" segment, which connects two adjacent Steiner points, is added to the approximation path.

Our construction scheme does not require that an approximation segment be in the same range as the corresponding optimal segment. We show that the cost of an approximation segment can be bounded regardless of its direction by using the following lemma:

Lemma 1 *For each $\overline{u_i u'_j}$, $i = 1, 2, j = 1, 2$, $d_r(u_i, u'_j) \leq (1 + \frac{3\epsilon}{\cos(\max\{\alpha_1/2, \alpha_2/2, \alpha_3/2\})}) \cdot d_r(v, v')$ for $\epsilon \leq \frac{1}{4}$.*

Here again α_1, α_2 and α_3 are the angles of the impermissible force range, each sideslope overturn range and the braking range, respectively. This lemma is equivalent to the corresponding lemma in [3], although we achieve so without using Steiner points that they add in Stages 2 and 3. Due to the limit of space, we leave the proof of this lemma to the full version of the paper [13].

One difference between our path construction and the one used in [3] is that each of segments $\overline{u_1 u'_1}$, $\overline{u_1 u'_2}$, $\overline{u_2 u'_1}$, and $\overline{u_2 u'_2}$ can be used as an approximation segment for $\overline{vv'}$. We will pick a segment so that it is connected to the approximation segment corresponding to the previous optimal segment of $\overline{vv'}$. Therefore, we can avoid adding the "joint" segments as the path construction in [3] does.

In the above, we assume that the optimal segment is a face-crossing segment with each of the two end points between two Steiner points. For other types of optimal segments, we pick approximation segments in the same way as in [3] and their costs can be bounded in an analogous manner. Therefore, Steiner points added in Stage 1 can guarantee an $O(\epsilon')$-approximation. (Recall that $\epsilon' = \frac{w_{max}\epsilon}{w_{min}\cos(\alpha_{max}/2)}$. The extra factor $\frac{w_{max}}{w_{min}}$ is introduced for bounding approximation segments corresponding to other types of optimal segments.) The number of Steiner points added in the first stage is $O\left(\log_{\delta_{min}}(L/R)\right)$ per edge.

After we construct the discretization, the next step is to use a discrete search algorithm to find an optimal discrete path in the resulting weighted graph $\mathcal{G}$. The approximation algorithm provided in [3] uses Dijkstra's algorithm, which takes $O(nm^2 + nm\log(nm))$ time, as $\mathcal{G}$ contains $O(nm)$ nodes and $O(nm^2)$ edges. Here n is the number of triangular regions and m is the number of Steiner points placed on each boundary edge of each region.

BUSHWHACK is also an algorithm for computing optimal paths in a weighted graph. Unlike Dijkstra's algorithm, which can be applied to an arbitrary weighted graph, BUSHWHACK is adept at finding optimal paths in a discretization of a space with certain geometric properties. By exploiting these geometric properties, BUSHWHACK is able to avoid accessing most of the edges in the discretization and thus reduce the time complexity to $O(nm\log m)$.

BUSHWHACK was originally designed for the weighted region optimal path problem [6]. Sun and Reif [12] later generalized BUSHWHACK to optimal path problems in a class of *piecewise pseudo-Euclidean spaces*. In [13] we show that BUSHWHACK is also applicable to the problem presented in this paper.

Combining the discretization mentioned above and the BUSHWHACK algorithm, we have the following theorem:

Theorem 3 *An $O(\epsilon')$-approximation of an energy-minimizing path can be computed in $O(nm'\log(nm'))$ time, where $m' = O\left(\log_{\delta_{min}}(L/R)\right)$.*

This is an improvement over the result presented in [3], which has a time complexity of $O(nm^2 + nm\log(nm))$ with $m = O(\log_{\delta_{min}}(L/R) + \log_{\mathcal{F}}(R/L))$. Our algorithm reduces not only the size of the discretization (from $O(\log_{\delta_{min}}(L/R) + \log_{\mathcal{F}}(R/L))$ to $O(\log_{\delta_{min}}(L/R))$ Steiner points per edge), but also the dependency of the time complexity on the size of the discretization. As the size of the discretization is decided by not only the user-specified ϵ but also a number of geometric parameters, it can be very large even for a moderate ϵ.

5 Partially Traversable Terrain Faces

As mentioned above, we assume as in [3] that each terrain face is totally traversable. Here we briefly discuss the case in which partially traversable terrain faces are allowed.

In this case, the path construction scheme described in the previous section may no longer be valid. Let v, v', u_1, u_2, u_1' and u_2' be defined as previously. Without loss of generality, we assume that the approximation segment for the previous optimal segment of $\overline{vv'}$ uses u_1 as one of its end points. It is possible that (as shown in Figure 6) although $\overrightarrow{vv'}$ is a permissible direction, both $\overrightarrow{u_1u_1'}$ and $\overrightarrow{u_1u_2'}$ are inside the combined impermissible range. Therefore, neither $\overline{u_1u_1'}$ nor $\overline{u_1u_2'}$ can serve as the approximation segment for $\overline{vv'}$, as it is impossible to travel from u_1 to u_1' (u_2').

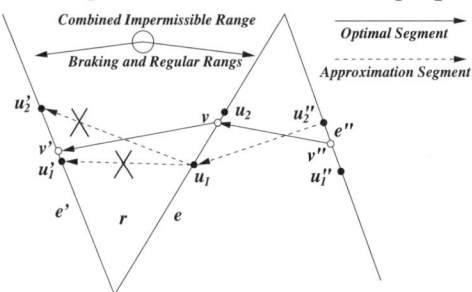

Figure 6. No valid approximation segment

To ensure that at least one approximation segment is available for each optimal segment, the discretization scheme needs to satisfy the following property:

Property 1 *Let e and e' be two boundary edges of a partially traversable terrain face r. Let u_1 and u_2 (respectively u_1' and u_2') be two adjacent Steiner points on e (respectively e'). If there exist two points $v \in \overline{u_1u_2}$ and $v' \in \overline{u_1'u_2'}$ so that direction $\overrightarrow{vv'}$ is not inside the combined impermissible range, then for $i=1,2$ at least one of $\overrightarrow{u_iu_1'}$ and $\overrightarrow{u_iu_2'}$ is not inside the impermissible range either.*

To construct such a discretization, we need another stage, Stage 1.b, to add a series of additional Steiner points for each Steiner point added in Stage 1. Let $\overrightarrow{V_1}$ and $\overrightarrow{V_2}$ be the two boundary angles of the combined impermissible range. For any Steiner point v on boundary edge e of terrain face r, if the ray from v with direction $-\overrightarrow{V_i}$, $i=1,2$, intersects another boundary edge e' of r, we add the intersection point as a Steiner point on e'. We apply this rule to all Steiner points recursively until no more Steiner points can be generated.

Observe that the Steiner points spawned by v on edge e (e') form a geometric series along e (e', respectively) with a ratio no less than $\mathcal{K} = \frac{1+\tan(\theta_{min}/2)\tan(\alpha_{2,min}/2)}{1-\tan(\theta_{min}/2)\tan(\alpha_{2,min}/2)}$, where $\alpha_{2,min}$ is the minimum α_2 among all partially traversable terrain faces, and θ_{min} is again the minimum angle between any two adjacent edges of any terrain face. When both θ_{min} and $\alpha_{2,min}$ are small, $\mathcal{K}$ is asymptotically $1 + \frac{\theta_{min} \cdot \alpha_{2,min}}{2}$. For any Steiner point v added in Stage 1, there are no more than $O(\log_\mathcal{K}(L/R))$ Steiner points spawned in Stage 1.b. This process is illustrated in Figure 7.

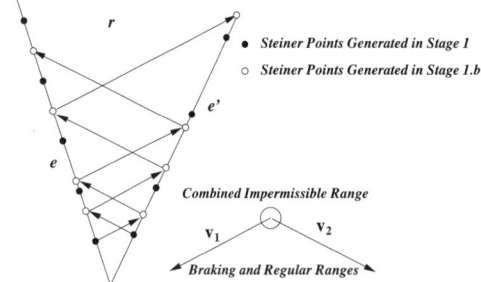

Figure 7. Additional Steiner points

It seems as though that the total number of Steiner points required for this discretization is $O(n \log_\mathcal{K}(L/R) \cdot \log_{\delta_{min}}(L/R))$. Recall that the Steiner points added on e in Stage 1 also form a geometric series with a ratio no less than δ_{min}. By adjusting the ratio properly, we can reduce the number of Steiner points added in Stage 1.b as most of the Steiner points spawned will coincide with existing Steiner points. Omitting the proof, we claim that the number of Steiner points can be bounded as the following:

Theorem 4 *To construct a discretization that guarantees an $O(\epsilon')$-approximation for the case in which partially traversable terrain faces are allowed, the total number of Steiner points required is $O(n \log_\mathcal{K}(L/R))$ if $\mathcal{K} \leq \delta_{min}$, or $O(n \log_{(\delta_{min}+1)/2}(L/R))$ otherwise.*

6 Preliminary Experimental Results

To compare the performance of our approximation algorithm described in Section 4 with that of the previous work, we implemented both algorithms using Java. The experimental results were acquired from a PC workstation with 1.8 Mhz Pentium processor and 1GB memory.

One dilemma we faced in presenting experimental results is the choice of experimental data. On one hand, we prefer real data as a randomly generated terrain may have unexpected impact on the performance of the algorithms. On the other hand, we want to avoid terrain maps modeled using triangular irregular networks (TINs). Recall that, for a given terrain of n faces and a user-defined ϵ, the size of the resulting discretization is dependent on not only ϵ and n, but also a number of other geometric characteristics. Therefore, in an experiment on a group of TINs, one TIN with some extremely skewed triangular faces may produce more Steiner points than all other TINs combined, and therefore the running time of an algorithm on this TIN will undesirably dominate the result of the entire experiment.

We chose to use triangular meshes generated from Digital Elevation Map (DEM). The advantage of using such triangular meshes is that each triangular face will not be too skewed, as its projection on the x-y plane is an isosceles right triangle. For our experiment, we used the map of Kaweah River basin in DEM ascii format, which is a 1424x1163 grid with 30m between two neighboring grid points. We took from the map 30 different 50x50 patches, and converted each of them into a triangular terrain by connecting two grid points diagonally for each grid cell.

We used the example of a car-like robot with a specific shape. The dimension of the robot is considered to be insignificant as compared to that of the terrain, but the ratio between the height and width, which is chosen to be 2:1, is used to compute the sideslope overturn range for any given terrain face. A friction coefficient is randomly picked for each terrain face, but only from a small range, again, to avoid producing skewed data. We also defined the maximum driving force of the robot. With the friction coefficient and the maximum driving force, we can compute the impermissible force range as well as the braking range for each terrain face.

Table 1. Speedup ratio for different ϵ.

ϵ	0.25	0.15	0.1	0.075	0.05
Speedup Ratio	2.03	1.19	0.81	0.54	0.35

For each generated terrain, we handpicked the points closest to the upper left and lower right corners as source and destination points respectively. For $\epsilon = 0.25, 0.15, 0.1, 0.075$ and 0.05, we timed the performance of the our algorithm, which uses BUSHWHACK along with the reduced discretization method, and the algorithm presented in [3] on finding an $\frac{w_{max}\epsilon}{w_{min}\cos(\alpha_{max}/2)}$-approximate optimal path. The results are shown in Table 1. Here we only provide the values of the speedup ratio, which is defined to be the ratio between average running time of our algorithm and that of the previous algorithm, for different ϵ. For example, for $\epsilon = 0.075$, the speedup ratio is 0.54, meaning that our algorithm in average takes 54% of the running time as compared to the previous algorithm. Note that the advantage of our algorithm becomes more significant as ϵ decreases, a finding consistent with the fact that BUSHWHACK is less dependent on ϵ than the standard Dijkstra-based algorithm is.

7 Conclusion and Future Work

In this paper we studied the energy-minimizing path problem. We provided some complexity results on the combinatorial size of energy-minimizing paths under various assumptions. We also presented an improved approximation algorithm using a smaller discretization as well as a more efficient discrete search algorithm. Experimental results show that our algorithm has a significant performance improvement over the previous algorithm when ϵ is small.

One remaining question is the complexity of the combinatorial size of energy-minimizing paths on totally traversable terrains. We believe that the upper bound should be $\Omega(n^2)$, the same as that of the weighted terrains, although proving so seems to be very hard.

We also plan to extend our experimentation work. In particular, we would like to study the impact of the improved discretization method on TINs. Imaginably, a TIN that contains skewed triangular faces may require significantly more Steiner points than does a triangular mesh generated from DEM for a given ϵ. The relative performance of our approximation algorithm may be improved due to the following reasons: a) the size of discretization of our algorithm is less dependent on various geometric parameters; and b) the complexity of our algorithm is less dependent on the number of Steiner points per boundary edge.

Acknowledgments This work is supported by DARPA/AFSOR Contract F30602-01-2-0561, NSF Grant EIA-0218376, NSF Grant EIA-0218359,NSF-11S-01-94604, NSF ITR Grant EIA-0086015.

References

[1] L. Aleksandrov, M. Lanthier, A. Maheshwari, and J.-R. Sack. An ϵ-approximation algorithm for weighted shortest paths on polyhedral surfaces. *Lecture Notes in Computer Science*, 1432:11–22, 1998.

[2] L. Aleksandrov, A. Maheshwari, and J.-R. Sack. Approximation algorithms for geometric shortest path problems. In *Proceedings of the 32nd Annual ACM Symposium on Theory of Computing*, pages 286–295, May 21–23 2000.

[3] M. Lanthier, A. Maheshwari, and J.-R. Sack. Shortest anisotropic paths on terrains. *Lecture Notes in Computer Science*, 1644:524–533, 1999.

[4] J. P. Laumond, editor. *Robot motion planning and control*. Number 229 in Lecture Notes in Control and Information Science. Springer, 1998.

[5] J. S. B. Mitchell and C. H. Papadimitriou. The weighted region problem: Finding shortest paths through a weighted planar subdivision. *Journal of the ACM*, 38(1):18–73, Jan. 1991.

[6] J. Reif and Z. Sun. An efficient approximation algorithm for weighted region shortest path problem. In *Proceedings of the 4th Workshop on Algorithmic Foundations of Robotics*, pages 191–203, Mar. 16–18 2000.

[7] J. Reif and Z. Sun. Movement planning in the presence of flows. *Lecture Notes in Computer Science*, 2125:450–461, 2001.

[8] N. C. Rowe. Obtaining optimal mobile-robot paths with non-smooth anisotropic cost functions using qualitative-state reasoning. *International Journal of Robotics Research*, 16(3):375–399, June 1997.

[9] N. C. Rowe and Y. Kanayama. Near-minimum-energy paths on a vertical-axis cone with anisotropic friction and gravity effects. *International Journal of Robotics Research*, 13(5):408–433, Oct. 1994.

[10] N. C. Rowe and R. S. Ross. Optimal grid-free path planning across arbitrarily contoured terrain with anisotropic friction and gravity effects. *IEEE Transactions on Robotics and Automation*, 6(5):540–553, Oct. 1990.

[11] A. A. Rula and C. C. Nuttall. An analysis of ground mobility models. Technical Report M-71-4, U.S. Army Engineer Waterways Experiment Station, 1971.

[12] Z. Sun and J. Reif. BUSHWHACK: An approximation algorithm for minimal paths through pseudo-Euclidean spaces. *Lecture Notes in Computer Science*, 2223:160–171, 2001.

[13] Z. Sun and J. Reif. On energy-minimizing paths on terrains for a mobile robot. Available at http://www.cs.duke.edu/~sunz/Papers/anisotropic.pdf, 2003.

Optimal Strategies to Track and Capture a Predictable Target

Alon Efrat   Héctor H. González-Baños[†]   Stephen G. Kobourov   Lingeshwaran Palaniappan

Department of Computer Science
University of Arizona, Tucson, AZ 85721
e-mail: {alon, kobourov, lingesh}@cs.arizona.edu

[†] Honda Research Institute USA, Inc.
Mountain View, California 94041
e-mail: hhg@honda-ri.com

Abstract—We present an $O(n \log^{1+\varepsilon} n)$-time algorithm for computing the optimal robot motion that maintains line-of-sight visibility between a target moving inside a polygon with n vertices which may contain holes. The motion is optimal for the tracking robot (the observer) in the sense that the target either remains visible for the longest possible time, or it is captured by the observer in the minimum time when feasible. Thus, the algorithm maximizes the minimum time-to-escape. Our algorithm assumes that the target moves along a known path. Thus, it is an off-line algorithm. Our theoretical results for the algorithm's runtime assume that the target is moving along a shortest path from its source to its destination. This assumption, however is not required to prove the optimality of the computed solution, hence the algorithm remains correct for the general case.

I. Introduction

In this paper, we consider the problem of keeping a target visible to a mobile observer for the longest possible time. Variations of this tracking problem arise in different applications such as visual servoing [1], [2], computer assisted surgery [3] and visibility-based planning of sensor control strategies [4].

A planning problem for maintaining line-of-sight visibility between two agents is considered in [5]. A dynamic programming approach generates the motions for a mobile robot (the observer) that tracks a moving target. If the target moves predictably, an optimal tracking motion can be computed off-line. If the target's motion is unpredictable, on-line strategies are used instead [6], [7]. Although on-line techniques are more desirable in a robotics context, off-line strategies remain useful in applications involving graphical simulations, supervision of industrial robots (which often follow known paths), and surveillance of vehicles constrained to move in road-maps.

Most off-line algorithms reported in the literature are not computationally efficient, requiring several seconds to compute even simple scenarios. In [8], a problem similar to the off-line tracking problem is modeled in the configuration-time space relative to the target. Their work takes practical considerations for on-line use into account and reduces planning time to a few seconds. But like [5], the work in [8] discretizes the space in order to approximate the optimal path. This process is costly, although heuristic techniques can speed the computation.

A related problem is sensor placement, previously addressed in [9] and [10] for visual tracking and exploration. Several works exists in regards to this problem. For example, the work in [11] defines a measure of motion observability based on the relationship between differential changes in the robot position and the corresponding differential changes in the observed visual features.

In this paper, we consider the problem of maintaining visibility inside a polygonal region P which may contain holes. The observer loses the target as soon as the line of sight between them is broken. The observer is modeled as a holonomic robot with a maximal speed of v_{obs}, fitted with an omnidirectional sensor of unlimited range. Target detection and robot localization are assumed to be perfect. The target is initially in view and moves along a known path until a time t_{stop}.

We present an $O(n \log^{1+\varepsilon} n)$ algorithm for finding the optimal path π^* given an initial observer position M_0. Here $\varepsilon > 0$ is an arbitrarily small constant, and n is the number of vertices of P. The path is optimal in the sense that it maximizes the time t_{esc} when the observer first loses the target. If the target is never lost, then π^* minimizes the distance between the observer and the target at every time $t < t_{stop}$. If this distance ever becomes 0 the target is said to be *captured*.

Although it is not required by our algorithm, if the target follows the shortest path to its destination the presentation becomes simpler and it is easier to obtain a bound on the running time. However, the algorithm computes an optimal path even when this assumption does not hold.

II. Problem Formulation

The 2-D target-tracking problem consists in computing the motion of an observer such that a target moving inside a planar workspace remains in view. We follow a similar notation to the one used in [6], which in turn follows from the standard notation used in motion planning [12].

The observer and the target move in a bounded Euclidean subspace $\mathcal{W} \subset \Re^2$ (the workspace). The free configuration spaces for the observer and the target are denoted by $\mathcal{C}^o$ and $\mathcal{C}^t$, respectively. The observer is assumed to be a point, therefore $\mathcal{C}^o = \mathcal{W}$. The configuration space of the target is also assumed to be two-dimensional, but the target is forbidden from moving arbitrarily close to an obstacle. In order to simplify our discussion, we represent the target as a small square with constant orientation.

Our work does not address system dynamics, therefore the state space $\mathcal{X}$ is equal to the Cartesian product $\mathcal{C}^o \times \mathcal{C}^t$. Define $q(t) \in \mathcal{C}^o$ as the observer's configuration at time t.

Let f be the transition equation for the observer: $\dot{q}(t) = f(q, u)$, where u is the vector of control inputs at t. We assume that $|\dot{q}(t)|$ is bounded by v_{obs}. For non-holonomic robots, it is useful to think of u as the pair (v, ω), where v is the speed of the observer and ω its change in bearing. Again, $|v|$ is bounded by v_{obs}. In this paper, we assume that steering is instantaneous. Although is sometimes possible to neglect steering delays, its inclusion in our framework is a topic for future work.

We represent the workspace as a polygon (possibly with holes). Therefore, from now on we use P instead of $\mathcal{W}$ to emphasize our choice of representation.

Visibility Model The observer is assumed to be fitted with an idealized omnidirectional sensor. Hence, a target is visible *iff* the line-of-sight between the target and the observer is un-obstructed.

Let $\mathcal{V}(q) \subseteq P$ be the set of locations from which the target is visible to an observer located at q. This set is the *visibility region* at q. An important concept in the formulation is that of *visibility sweeping line* ($\ell(t)$), defined as the chord (a segment fully contained in P connecting two points on ∂P) passing through the target and a reflex vertex of P. At any time t, the observer must be in one of the half-planes bounded by $\ell(t)$ in order to see the target; see Figure 1. Although there could be $\Theta(n)$ visibility lines, we later show that at any given time the observer's path is influenced by at most two of these.

Target Predictability and Optimal Paths The target is supposed to be *predictable*. That is, the configuration of the target $q^t(t) \in \mathcal{C}^t$ is known for all time $t < t_{stop}$, where t_{stop} is the target *stopping time*. Therefore, our algorithm computes an *off-line* strategy. It is important to note that although the target is predictable, it may be unavoidable for the observer to lose the target for a particular choice of initial conditions and target velocities. In this scenario, a useful notion is that of *escape time* (t_{esc}) —the time when the observer first loses the target.

Alternatively, the observer is sometimes able to capture the target —i.e., $q(t) = q^t(t)$ for some $t < t_{stop}$. The *capture time* (t_{cap}) is when the target is first captured.

In this paper, a path is optimal if it satisfies the following definition:

Definition 2.1: An observer path π^* is optimal if it is the shortest path among all possible paths that maximize t_{esc}. For the cases when t_{esc} does not exist (i.e., the observer never loses track of the target), then π^* is optimal if it minimizes t_{cap}. If capture before t_{stop} is not possible, then π^* minimizes the final separation between the target and the observer.

III. OVERVIEW OF THE ALGORITHM

Later in the paper, we will show that the optimal path π^* is composed of *straight-line segments* and *leaning* curves, as described below:

Straight-line segments: The observer moves towards a (carefully chosen) point M in a straight line without ever losing sight of the target.

Leaning curves: This is a curve defined by the connected

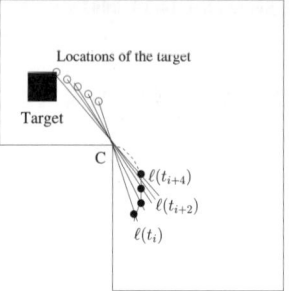

Fig. 1. The leaning curve: The observer is approaches the point C while remaining on $\ell(t)$ as it rotates about C.

part of the shortest path to a vertex C of P followed by the observer while moving at maximal speed, constrained to be on the sweeping line $\ell(t)$ rotating around C as the target moves along a straight line; see Figure 1.

Without any loss of generality, assume that the observer is at the origin at time $t = 0$. We now proceed to define the concept of a π-path.

Definition 3.1 (π-path): The path $\pi(q, t) \subset P$ is a π-*path* if it connects the origin to q and satisfies the following properties:

a. The target is always visible to the observer.
b. π is the shortest path that satisfies the above.
c. The observer traverses π at the top speed v_{obs}.
d. The observer arrives at q in time t.

Note that a path $\pi(q, t)$ does not necessarily exists for every combination of q and t. The following lemma establishes the uniqueness of a π-path:

Lemma 3.2: $\pi(q, t)$ is unique for any t and $q \in P$.

Proof: Assume that two distinct π-paths $\pi_1(q, t)$ and $\pi_2(q, t)$ exists, both reaching q at the arrival time t_f. Fix $0 < \alpha < 1$. Let $\pi_1(t)$ and $\pi_2(t)$ be the location of the observer at time t along the two paths, and define $\pi(t) = \alpha \pi_1(t) + (1 - \alpha) \pi_2(t)$. Note that the speed along π is always less or equal to v_{obs}, but there is an instant when the speed is strictly smaller since π_1 and π_2 both start at 0 (else the paths are identical). π is then shorter than π_1 or π_2, which are allegedly π-paths. Hence, π is infeasible in terms of target visibility and an obstacle exists between π_1 and π_2; see Figure 2.

Now, assume that at t_0 there is a value α for which $\alpha \pi_1(t) + (1 - \alpha) \pi_2(t)$ does not see the target, and t_0 is the earliest time when said α exists. Consider the triangle $T(t) = \Delta \pi_T(t) \pi_1(t) \pi_2(t)$ (where $\pi_T(t)$ is the target's position at t). Select δ such that $T(t)$ does not intersect any part of ∂P for any $0 < t \leq t_0 - \delta$ (see Figure 2). That is, $T(t_0 - \delta)$ is entirely below the obstacle. However, for a time arbitrarily close to the arrival time t_f, the triangle $T(t_f - \epsilon)$ is entirely above the obstacle. Therefore, the triangle $T(t)$ must cross the obstacle, and either the edge $(\pi_T(t), \pi_1(t))$ or the edge $(\pi_T(t), \pi_2(t))$ intersect the obstacle at some t. But this contradicts our assumption that both $\pi_1(t)$ and $\pi_2(t)$ see the target for all t. ∎

From here on we omit t in the notation of $\pi(q, t)$, and denote this path as $\pi(q)$. Let $\Gamma(t)$ be the subset of all points q such that the length of $\pi(q)$ is $v_{obs} t$. $\Gamma(t)$ contains the arrival point for the optimal path π^* at time t.

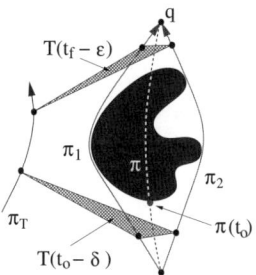

Fig. 2. π-paths are unique: $T(t)$ must cross the obstacle to go from configuration $T(t_0 - \delta)$ to $T(t_f - \epsilon)$, contradicting the assumption that both π_1 and π_2 see the target at all times.

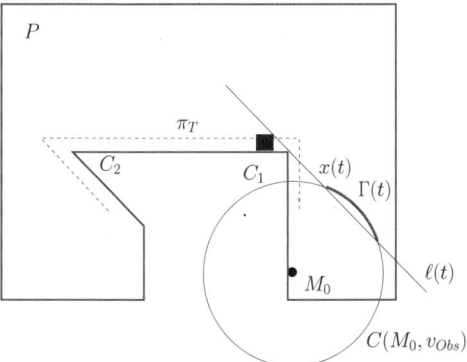

Fig. 3. A simple example: target follows path π_T. $\Gamma(t)$ is also shown.

A. Simple Example

Before we present the general algorithm for finding the optimal path π^*, we first describe a simple example, illustrated in Figure 3. This scenario was simulated and screen shots are shown in Figure 6(a)-(d). The target starts a distance d_0 from the observer, and moves up along the right wall of the polygon, until it reaches the vertex C_1. After rounding up this vertex the target proceeds to go left until it reaches vertex C_2. Afterwards, the target continues around C_2, going down until it reaches its destination. Let the initial position of the observer be M_0.

First Critical Point: M_1

Initially, for small values of t, the possible locations for the observer are all the points of P bounded by the circle $C(M_0, v_{\text{obs}}t)$ of radius $v_{\text{obs}}t$ and center M_0. In this simple case, $\Gamma(t)$ is a portion of this circle. Let $\ell(t)$ denote the visibility sweeping line to the target, once the target moves beyond the vertex C_1. That is, the target at time t is visible from all points above $\ell(t)$ and but not from those below $\ell(t)$; see Figure 3.

Let $\mathbf{x}(t)$ denote the higher intersection point of $C(M_0, v_{\text{obs}}t)$ with $\ell(t)$. As the target moves to the left the point $\mathbf{x}(t)$ moves to the right; see Figure 4.

Let $\theta(t)$ be the angle defined by the segment $\overline{M_0 \mathbf{x}(t)}$ and a vertical line. Initially, θ increases with t. After $\theta(t)$ reaches a maximum either the observer loses the target or $\theta(t)$ decreases again. Assume the latter. Let t_1 be the time at which $\theta(t)$ reaches its maximum, and let M_1 be

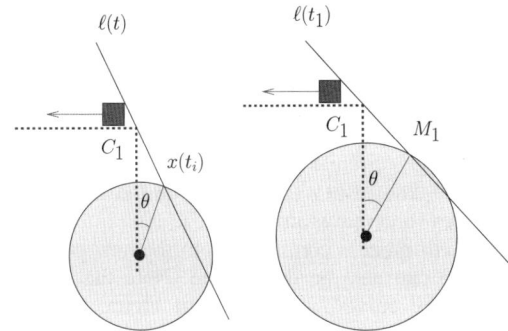

Fig. 4. The angle θ achieves its maximum at time t_1. We set M_1 to be the upper intersection point of $C(M_0, v_{\text{obs}}t_1) \cap \ell(t_1)$.

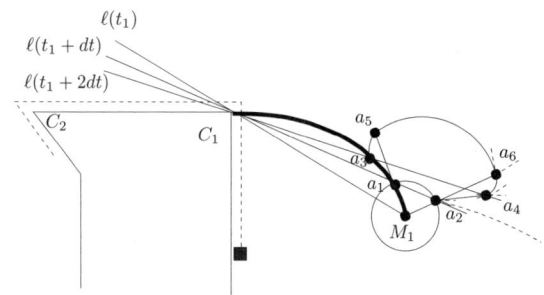

Fig. 5. Computing the leaning curve between M_1 and C_1.

the location of $\mathbf{x}(t_1)$. This point is significant because the observer's optimal path passes through M_1.

It's possible that $\ell(t)$ becomes tangent to $C(M_0, v_{\text{obs}}t)$ at time $t = t_1$. For $t'' > t_1$ we have $C(M_0, v_{\text{obs}}t'') \cap \ell(t'') = \emptyset$. That is, the circle $C(M_0, v_{\text{obs}}t'')$ and $\ell(t'')$ do not intersect after time t_1, and target visibility is broken. The observer is doomed to lose the target.

Thus, M_1 is either a tangent point or a local maximum. In either case is optimal to go directly to M_1 because doing so ensures that the target is seen for as long as possible. The difference is that π^* terminates at M_1 in the former case, but continues after M_1 in the latter.

After M_1: Leaning Curves

The optimal path after point M_1 becomes more complicated. In order to make the explanation simpler, we will assume that $\Gamma(t_1)$ contains only the point M_1. In general, M_1 may only be a local maximum, and $\Gamma(t_1)$ must include the circular arc above $\ell(t = t_1)$ (the complete algorithm acknowledges this). But for the simple example shown in Figure 3 we are confident that M_1 also happens to be the global maximum and that π^* passes through this point.

We proceed as follows: Let dt be an infinitesimal time interval. Draw a circle of radius $v_{\text{obs}}dt$ centered at M_1. This circle intersects $\ell(t_1 + dt)$ at points a_1 and a_2; see Figure 5. $\Gamma(t_1 + dt)$ becomes the arc $a_1 a_2$.

Next, we draw circles of radius $v_{\text{obs}}dt$ at every point on the edge $a_1 a_2$. Again, we consider the intersection of these circles with $\ell(t_1 + 2dt)$ and take all the points above $\ell(t_1 + 2dt)$. $\Gamma(t_1 + 2dt)$ is then composed of 3 edges:

The first one is an arc defined by the circle of radius $v_{obs}2dt$ centered at M_1, bounded by the segments $\overline{M_1a_1}$ and $\overline{M_1a_2}$. The second is the arc defined by the circle of radius $v_{obs}dt$ centered at point a_1. The third is the arc defined by the circle of $v_{obs}dt$ centered at point a_2. The points on these 3 circular arcs above $\ell(t_1+2dt)$ define the curve $\Gamma(t_1+2dt)$. Let a_3, a_5, a_6, a_4 be the intersections of these arcs (ordered as in Figure 5).

The growth process continues from the new points and $\Gamma(t_1+3dt)$ can also be determined. Note that the path formed by the curve $M_1, a_1, a_3...$ in Figure 5 forms a lower envelope, below which the target is not visible. This is exactly the *leaning curve*.

By definition, if the observer moves along the leaning curve the target will remain in sight. However, the observer may deviate from the leaning curve before reaching C_1. This will happen if the target makes a turn around a second vertex C_2.

Critical Point M_2

Following an analysis similar to that for M_1, we compute a second maximal point M_2 determined by C_2.

Figure 6(a)-(d) shows the sequence of tangents to the leaning curve as time increases. The observer may decide to depart the leaning curve along a tangent anywhere between M_1 and the vertex C_1. In fact, these tangential paths are π-paths. Therefore, the locus of points reachable by the observer at $t > t_1$ following such strategy is part of $\Gamma(t)$. We call this locus a ρ-edge of $\Gamma(t)$. Every point in ρ-edge is induced by a tangent to the leaning curve.

Next, we consider the intersections of the sweeping lines $\ell_2(t)$ defined by C_2 with the ρ-edges. At time t, the ρ-edge intersects $\ell_2(t)$ at two points. Consider the one closest to C_2. As t increases, the intersection achieves a maximum in the sense that it corresponds to the earliest tangent to the leaning curve; see Figure 6(d). This point is analogous to the maximal θ when computing M_1. We mark this point as M_2.

Similar to the computation of M_1, the point M_2 may be a tangent to the ρ-edge, in which case the observer is doomed to lose the target. Going to M_2 nevertheless keeps the target in view for as long as possible, thus π^* terminates at M_2. If M_2 is not tangent to the ρ-edge then this point becomes the new vertex of π^*.

So far the computed path contains the following edges: A segment connecting the observer's initial position M_0 to M_1; a leaning curve connecting M_1 to the departure point; and a segment connecting this point to M_2.

Once the observer reaches M_2 we grow a new leaning curve; Figure 6(d). The process continues until a future sweeping line fails to intersect $\Gamma(t)$, the target is caught, or the target stops. In the first case we lose the target, but π^* maximizes the escape time. In the latter cases, the observer either captures the target in the minimum time or minimizes the final distance to the target.

B. Approximating Leaning Curves

A leaning curve is defined by the sequence of intersection points between the visibility sweeping line $\ell(t)$

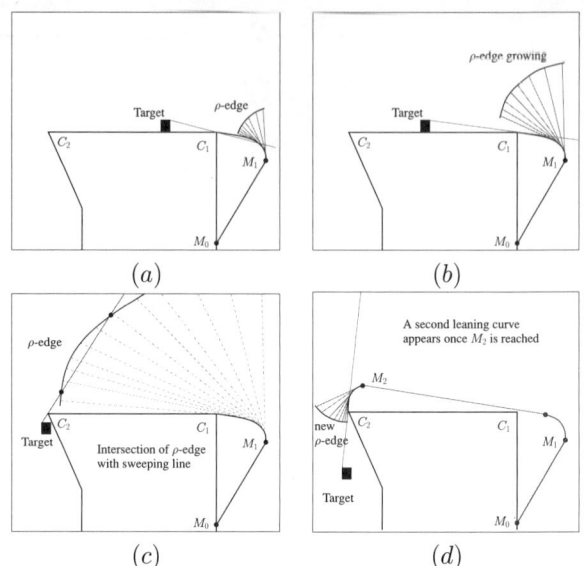

Fig. 6. Evolution of a leaning curve through time: (a) initial growth of a ρ-edge; (b) a fully-formed ρ-edge; (c) the intersection between $\ell(t)$ and the ρ-edge; (d) critical point M_2 and the resultant optimal path π^*.

and $\Gamma(t)$ as $\ell(t)$ rotates around a vertex of P. To our knowledge, this curve has no close-form solution.

The leaning curve and the ρ-edge can be approximated as follows. Select a small dt (ideally within 2% of the duration of the leaning curve). Calculate $\ell(t+dt)$ for the chosen value of dt (the line $\ell(t)$ is known because the target is predictable). Consider the following scenario:

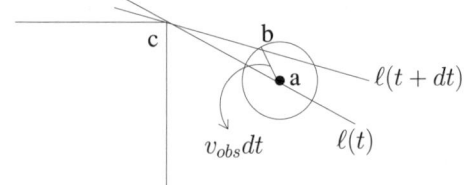

Now compute a circle of radius $v_{obs}dt$. Find the intersection of this circle with $\ell(t+dt)$. Take the point b which is closer to the vertex C and use it as the next starting point. The slope between the points a and b approximates the slope of the tangent at b.[1]

IV. OPTIMAL TARGET TRACKING ALGORITHM

Let M_0 and $\pi_T(0)$ be the initial positions of the observer and the target, respectively. The algorithm computes the evolution of the set $\Gamma(t)$ as the target moves at maximum speed v_T along π_T. At time t, $\Gamma(t)$ contains the candidate points through which π^* might pass. As shown below, $\Gamma(t)$ is a connected path in the plane composed of pieces we refer to as *edges*. Each edge has a constant description complexity.

---

[1]Alternatively, we could write the differential equation of the curve and use an ODE solver. This will be more efficient, but the geometric method illustrates better the mechanics of the leaning curve.

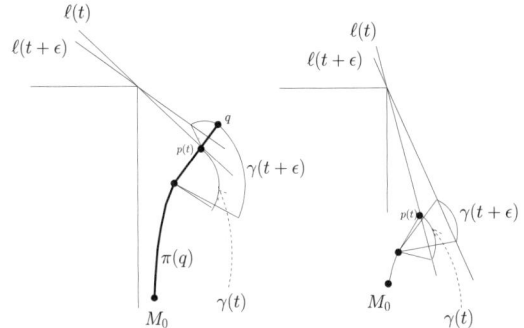

Fig. 7. Characterization of the path $\Gamma(t)$: (a) the curve $\gamma(t)$ is stretched by $p(t)$; (b) the curve $\gamma(t)$ shrinks and the point $p(t)$ does not contribute to $\gamma(t+\epsilon)$.

Informally, $\Gamma(t+\Delta)$ is computed from $\Gamma(t)$ by growing disks of radius $v_{obs}\Delta$ at every point of $\Gamma(t)$, taking the union of these disks, and defining $\Gamma(t + \Delta)$ as the set of points in the boundary of the union that see the target at $t+\Delta$. Matters are complicated by the edges and vertices of P, and the sweeping lines induced by the target's motion. If $\Gamma(t)$ becomes empty at any $t < t_{stop}$ the observer will lose the target. Otherwise, we have an optimal path that keeps the target in view until its final destination.

Calculating $\Gamma(t + \Delta)$ by discretizing t and growing circles $\forall\, q \in \Gamma(t)$ is indeed possible. But this procedure is highly inefficient. Instead, our algorithm maintains a set $\mathcal{O}$ of obstacles which implicitly define $\Gamma(t)$ in the following sense: Every π-path $\pi(q)$ can be described as the shortest path $\pi'(q)$ connecting M_0 to q such that $\pi'(q) \cap \mathcal{O} = \emptyset$. The algorithm finds all the critical events that affect the shape of $\Gamma(t)$ as the target moves, and stores these events in a priority queue Q. After each combinatorial change, we fetch a new event from Q, update $\Gamma(t)$ and $\mathcal{O}$, compute future events, and insert these into Q.

$\Gamma(t)$ is made up of two kinds of *edges*: circular arcs and ρ-edges. At any instant, one or two visibility sweeping lines define the endpoints of one or two edges. Let $\gamma(t)$ be one of these edges, and let $p(t)$ be its endpoint on the sweeping line $\ell(t)$. We say that $p(t)$ *stretches* $\gamma(t)$ if there is a point $q \in \Gamma(t+\varepsilon)$ such that $\pi(q)$ fails to go through $p(t)$ for some small $\varepsilon > 0$ (i.e., the last segment of $\pi(p(t+\varepsilon))$ does not intersect $\gamma(t)$). Otherwise, we say that $p(t)$ *shrinks* $\gamma(t)$. See Figure 7(a)-(b).

We say that a point $p(t_0) \in \gamma(t_0)$ is an M-point if $p(t)$ shrinks $\gamma(t)$ slightly before t_0, and $p(t)$ stretches $\gamma(t)$ slightly after t_0. It can be shown that every ρ-edge contains at most one M-point. We assume that given a description of a ρ-edge, the M-point for this edge can be computed in time $O(1)$. Note that some numerical instability may occur because ρ-edges are approximated, and the computation of a ρ-edge is affected by numerical errors in a previous ρ-edge. However, this is not a problem, since a good approximation scheme produces enough accuracy.

A. Types of Obstacles

(1) Edges of ∂P: A simple type of obstacle is an edge of the polygon that breaks the line of sight between the observer and the target.

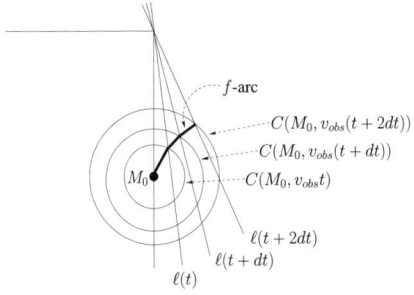

Fig. 8. An illustration of an f-arc.

(2) f-arcs: These are formed by the intersection point $p(t)$ of some edge $\gamma(t)$ of $\Gamma(t)$ with a visibility sweeping line $\ell(t)$ as time progresses while $p(t)$ is shrinking $\gamma(t)$; see Figure 8. The shortest path π^* never intersects an f-arc, except possibly at the endpoints of the f-arc. Thus we are concerned only with the endpoints of f-arcs. The f-arcs can be computed in $O(1)$ time given the description of $\gamma(t)$ and $\ell(t)$.

(3) δ-arc: These are formed by the trace of the shortest path that an observer can take on its way to a reflex vertex of P such that it remains on the sweeping line $\ell(t)$ as the target moves along a straight line. The leaning curves introduced earlier compose the δ-arcs. A δ-arc usually connects an M-point (defined above) to the reflex vertex of P that define $\ell(t)$; see Figure 5. A leaning curve is always a connected part of a δ-arc.

B. Types of Edges of $\Gamma(t)$

(1) Circular edges. A circular edge is denoted by its center, its angular span, and its radius at time t; see Figure 9(a). The centers for circular edges of $\Gamma(t)$ can be one of three different types:

  a) the starting point of the observer,
  b) an M-point,
  c) a reflex vertex of the polygon P.

(2) ρ-edges. Describing these edges is more involved. Each ρ-edge is induced by a δ-arc d. A point q is on the ρ-edge $\gamma(t)$ induced by d if the last segment e of $\pi(q)$ is a straight segment of length $v_{obs}(t - t')$, connecting q to a point q' of d, where $\Gamma(t)$ reaches q' at time t'; see Figure 10. Since $\pi(q)$ is a shortest path, the tangent to r at q' (if it exists) must be in the direction of e.

Now that we have defined the obstacles and the edges of $\Gamma(t)$ we can describe the data structure needed to maintain and update the event queue. At time t, we maintain the curve $\Gamma(t)$ as a list of the endpoints of the different edges of $\Gamma(t)$, together with the parameters for each edge. For circular edges these parameters are clear, while ρ-edges are stored as pointers to the parameters of the δ-arcs they emerged from.

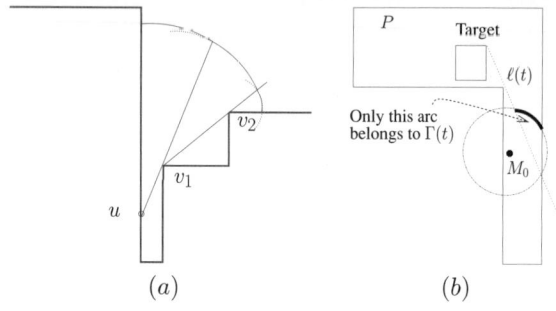

Fig. 9. (a): The circular edges which are parts of $\Gamma(t)$ are emerging from the vertices of P. (b): $\Gamma(t)$ slightly after an event at which a circular edge of $\Gamma(t)$ hits an edge of P. Only the thick edge closer to the target is stored in $\Gamma(t)$.

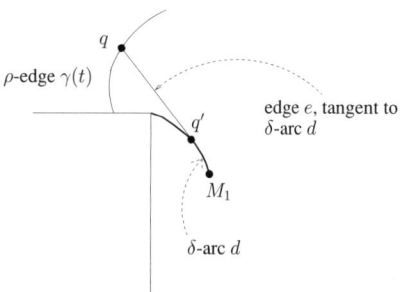

Fig. 10. Defining a ρ-edge. An observer following $\pi(q)$ leaves the δ-arc at time t', and the length of $\overline{qq'}$ is $(t-t')v_{\text{obs}}$.

C. Types of Events as t Increases

(1) $\Gamma(t)$ runs into an edge e of ∂P or into a sweeping line $\ell(t)$; see Figure 9(b). In this case, an edge of $\Gamma(t)$ is split into two. Since the observer tries to stay as close as possible to the target, only the edge closest to the target has to be maintained. Let $\gamma(t)$ denote this piece. (Note: the points on ∂P or on $\ell(t)$ are not on $\Gamma(t)$, since they are not endpoints of optimal paths.) We also compute and insert into Q the future time at which a combinatorial event defined by $\gamma(t)$ and e or $\ell(t)$ occurs, (e.g., $\ell(t)$ reaches the endpoint of $\gamma(t)$). In the case of an intersection with $\ell(t)$, we also insert into Q the M-point that occurs on $\gamma(t)$, if it exists.

(2) An edge of $\Gamma(t)$ runs into a reflex vertex C; see Figure 9(b). In this case we begin growing a circular edge at C, and insert it into $\Gamma(t)$.

(3) The first or last edge $\gamma(t)$ of $\Gamma(t)$ has degenerated to a single point. Let $\gamma'(t)$ be the adjacent edge of $\gamma(t)$ along $\Gamma(t)$. Upon reaching this event, $\ell(t)$ begins intersecting $\gamma'(t)$. We compute the future events for $\gamma'(t)$ as in the first case.

(4) The intersection point of a visibility sweeping line and an edge $\gamma(t)$ of $\Gamma(t)$ reaches a M-point q. We insert a new ρ-edge into $\Gamma(t)$ from this point, and if needed, a new circular edge. Note that $\gamma(t)$ must be the first or the last edge of $\Gamma(t)$. We also insert a new δ-arc into the list of objects $\mathcal{O}$. The other

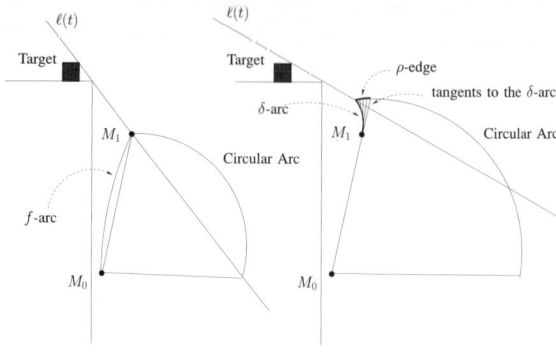

Fig. 11. The M point is reached and a ρ-edge is grown.

endpoints of these ρ-edges and δ-arcs are not known at this point, and will be revealed as time increases; see Figure 11.

(5) The intersection point of a visibility sweeping line $\ell(t)$ and the first or last edge $\gamma(t) \in \Gamma(t)$ intersects an edge of ∂P at a point q. In this case, the observer loses the target at point q, and $\pi(q)$ is the optimal path; see Lemma 4.1. A similar event occurs if $\ell(t)$ becomes tangent to $\gamma(t)$.

(6) A δ-arc d defined by the visibility sweeping line $\ell(t)$ reaches the vertex C of P around which $\ell(t)$ rotates. We finish inserting this arc into $\mathcal{O}$.

(7) The target has reached a bend-point on its path π_T (i.e., the target has changed its direction), or one of the visibility sweeping lines $\ell(t)$ hits a (new) vertex v of P. We re-compute the times of all future events in the queue in which $\ell(t)$ is involved. For each intersection point $p(t)$ of $\ell(t)$ and $\Gamma(t)$ we check whether it stretches or shrinks the edge $\gamma(t)$ of $\Gamma(t)$ that it intersects. In the former case, we insert a new ρ-edge into $\Gamma(t)$ and a new δ-arc into $\mathcal{O}$. In the latter case, we gradually shrink $\gamma(t)$ and insert a new f-arc into $\mathcal{O}$; see Figure 12.

Lemma 4.1: Let g be the f-arc determined by a visibility sweeping line $\ell(t)$, that is closer to the target. Assume that g intersects an edge of ∂P in the interior of g, and let q be the intersection point. Then $\pi(q)$ is the optimal path π^*, and the observer would lose the target at this point.

Proof: Assume that this intersection occurs at time t'. Clearly, if that observer follows $\pi(q)$ then it sees the target for all $t \leq t'$. On the other hand, it is not hard to verify that if t'' is slightly larger than t', then any point that the observer can reach in time t'' (not necessarily along a shortest path) is hidden from the target either by $\ell(t'')$ or by ∂P. ∎

Termination of the algorithm: The events of type 5 above takes care of the case where the observer loses the target. The other possibility is that the target reaches its destination at time t_M, and $\Gamma(t_M)$ is not empty. In this case, we take the point $q \in \Gamma(t_M)$ which is the closest to the final position of the target. Finally, the output of the algorithm is reconstructed by tracing back from q, through the objects of $\mathcal{O}$ around which $\pi(q)$ bends. For this, every edge of $\Gamma(t)$ and object of $\mathcal{O}$ "knows" from which obstacle of $\mathcal{O}$ it emerges.

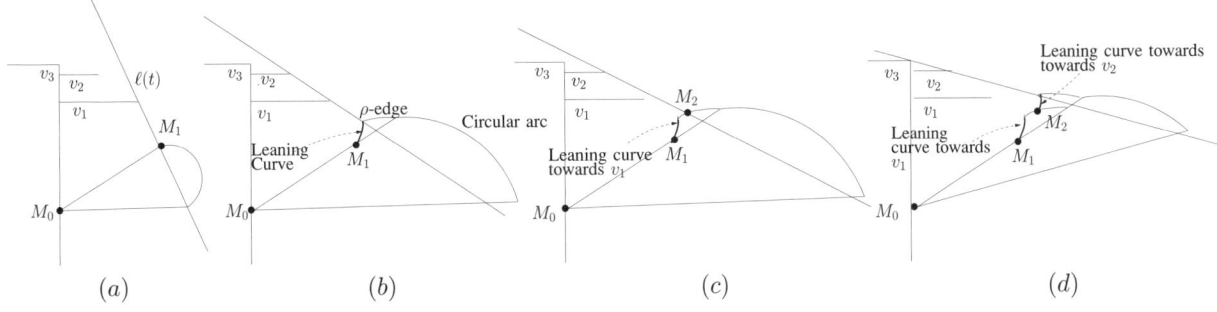

Fig. 12. (a) The first M-point, M_1; (b) a leaning curve is started towards v_1. Now the visibility sweeping line changes from v_1 to v_2 and; (c) a new M-point, M_2, is reached; (d) A leaning curve toward v_2 is started. and the visibility line changes from v_2 to v_3.

D. Structure of $\Gamma(t)$

Lemma 4.2: $\Gamma(t)$ is a connected curve.

Proof: $\Gamma(t)$ starts as a circular edge and it goes through combinatorial changes only at certain discrete events. The lemma holds for $t = 0$, as $\Gamma(t)$ is a single point. It is easy to show that if the lemma holds before each of the events listed above, then it holds afterwards as well. ∎

We define a *flight plan* Π to be a path of the observer in the plane, together with the speed of the target at each point along this path. A *legal flight plan* is a flight plan at which the observer moves at velocity $\leq v_{\text{obs}}$ starting from M_0, seeing the target at any point and time along its way.

Let $\mathcal{F}(t)$ denote the set of all points that the observer can reach at time t, following a legal flight plan. We say that $\Gamma(t)$ *hits* a segment e at time t' and point q if $\Gamma(t')$ intersects e at q, while there is no other point of $\mathcal{F}(t') \cap e$ in a small enough vicinity of q.

Lemma 4.3: For every segment $e \subseteq P$, the intersection $e \cap \mathcal{F}(t)$ consists of a single connected interval.

Proof: Consider two points $q_1, q_2 \in e \cap \mathcal{F}(t)$. Let Π_1 and Π_2 be two legal flight plans leading to q_1 and q_2, resp. Let point q be on e between q_1 and q_2, whose distance from q_1 is $\alpha \|q_1 - q_2\|$ for some $0 \leq \alpha \leq 1$.

Define $\Pi(t') = \alpha \Pi_1(t') + (1 - \alpha) \Pi_2(t')$. Then Π is a flight plan from M_0 to q, and the velocity of the observer along Π is $\leq v_{\text{obs}}$. Moreover, for every $0 \leq t' < t$, both $\Pi_1(t')$ and $\Pi_2(t')$ are on the sides of the sweeping lines from where the target is visible at time t'. Thus, $\Pi(t')$ also sees the target which implies that Π is a legal flight plan, and hence $q \in \mathcal{F}(t)$. ∎

We say that a connected curve γ is convex if the line segment connecting every two points on γ does not intersect γ in its interior.

Corollary 4.4: $\Gamma(t)$ is convex for every t.

Lemma 4.5: There are no two times $t_1 < t_2$ at which $\Gamma(t)$ hits e.

Proof: We show this by contradiction. Assume that $\Gamma(t_1)$ hits e at q_1 and $\Gamma(t_2)$ hits e at q_2. If at time t_2 there are other points of $\mathcal{F}(t_2)$ on e (except q_2) then by Lemma 4.3 there are also points of $\mathcal{F}(t_2)$ arbitrary close to q_2, which is a contradiction.

Let $E = \{q \in e \mid q \in \mathcal{F}(t), t < t_2\}$. If $q_2 \in E$ then the observer reaches q_2 at an earlier time, hence q_2 cannot be in $\Gamma(t_2)$. If $q_2 \notin E$ then let q' be the closest point of E to q_2, and let t' be the last time that q' sees the target. At this time a visibility sweeping line $\ell(t')$ passes through q', and $\ell(t')$ separates q' from q_2. Otherwise, an observer located at q' at time t' could have reached closer to q_2. But in this case, it is not hard to verify that q_2 could not see the target for any time $t > t'$. That is, it cannot be on $\Gamma(t_2)$, which is a contradiction. ∎

E. Implementation and Running Time

We construct a triangulation $\mathcal{T}$ of P. Once $\Gamma(t)$ intersects a triangle $\Delta \in \mathcal{T}$ we compute the future time at which $\Gamma(t)$ hits each of the remaining edges of Δ, or hits an edge or vertex of P adjacent to Δ. This takes $O(\log n)$ time per a triangle edge using Corollary 4.6 below. Here n is the number of edges in the polygon P.

Corollary 4.6: Assume that in the time interval $[t_1, t_2]$ neither $\Gamma(t)$ nor the visibility sweeping line $\ell(t)$ go through any combinatorial changes. Given a visibility sweeping line $\ell(t)$, we can find in time $O(\log n)$ the earliest time $t' \in [t_1, t_2]$ and point at which $\ell(t')$ intersects $\Gamma(t')$, if t' exists.

Proof: We perform a binary search on the order of the vertices of $\Gamma(t)$, as follows: we pick one such vertex $u(t)$ and we observe that $u(t)$ moves along a curved path in the plane as t changes. We can compute in time $O(1)$ the time t_u of the intersection point of $u(t)$ and $\ell(t)$, and by checking the directions in which $\ell(t_u)$ stabs $\Gamma(t_u)$, deduce on which side of $u(t)$ the next vertex should be picked. ∎

Lemma 4.7: Each vertex defines at most one circular edge of $\Gamma(t)$.

Proof: Assume that v is the center of two circular edges. These edges cannot exists at the same time and have the same radius, because of the way we maintain $\Gamma(t)$. On the other hand, if they are defined for two different times t_1, t_2, then v must be in both $\Gamma(t_1), \Gamma(t_2)$, contradicting Lemma 3.2. ∎

Lemma 4.8: The total number of events of type 7 that occur in the course of the algorithm is $O(n)$.

Proof: Let v be a vertex of P. Observe that in the course of the algorithm, there is at most one connected time interval $[t_1, t_2]$ in which v sees the target. In other words, once v has stopped seeing the target, it will never see it again. To verify this, consider $\mathcal{V}(v)$, the visibility polygon of v and note that since the target moves along a shortest path, the union of its locations inside $Vis(v)$ must be connected.

Now consider an event of type 7 as defined above. We need to bound the number of times that a visibility sweeping line moves from rotating about vertex v to another vertex, v'. Once this happens, v would not see the target anymore, hence this event can happen only $O(n)$ times. ∎

Lemma 4.9: The total number of edges of $\Gamma(t)$, that appear in the course of the algorithm is $O(n)$.

Proof: The bound on the number of circular edges, centered at vertices of P, is $O(n)$ (from Lemma 4.7). To bound the number of other ρ edges of $\Gamma(t)$, recall that each of their endpoints occurs when one of the sweeping lines goes through an event of type 7. Since the number of these events is $O(n)$ (from Lemma 4.8), we obtain the desired bound. Note that this bound also bounds the number of objects in $\mathcal{O}$. ∎

Lemma 4.10: The number of events of type 1 is $O(n)$.

Proof: The bound on the number of times that $\Gamma(t)$ hits a edge of P follows from Lemma 4.5. The bound on the number of times a visibility sweeping line can hit $\Gamma(t)$ follows from the fact that this event can happen only after some other type of combinatorial event (either the target changed its direction, or the sweeping line began rotating around another vertex of P, or the sweeping line began rotating around another vertex of the target). There are only $O(n)$ such events. ∎

We also need to efficiently handle when a sweeping line $\ell(t)$ rotating around v hits and proceeds to rotate around a new vertex v'. For this we maintain a list containing all vertices of P above $\ell(t)$, and a complementary list for those below $\ell(t)$. Each list is maintained using the data structure for dynamic convex hull of Chan [13]. As the target moves along a straight line, the next vertex intersected by $\ell(t)$ can be computed in time $O(\log n)$. Afterwards, the structure is updated in time $O(\log^{1+\varepsilon} n)$. Again, we use the fact that once a vertex of P fails to see the target it would never see it again. Thus, the total time required here is $O(n \log^{1+\varepsilon} n)$. Combining all of the above results yields the following theorem:

Theorem 4.11: We can find the optimal path π^* in time $O(n \log^{1+\varepsilon} n)$.

V. CONCLUSION AND FUTURE WORK

This paper introduced an $O(n \log^{1+\varepsilon} n)$-time algorithm for computing the optimal robot motion strategy that maintains line-of-sight visibility of a target moving inside a polygon with n vertices. The computed path maximizes t_{esc} (and minimizes t_{cap} when the target cannot escape). Our algorithm does not consider system dynamics. Nevertheless, the algorithm could prove useful in practice since t_{esc} for the kinematic case is sometimes an adequate approximation of t_{esc} for the dynamic case.

We ignored non-holonomic constraints in our analysis. However, this is not a serious shortcoming for non-holonomic robots capable of steering quickly because the paths computed by our algorithm tend to be smooth (i.e., few sharp turns). However, incorporating non-holonomic constraints is a topic for future research.

The inclusion of *range constraints* to the visibility model will be an important extension to our work to address a typical limitation in most sensors. A more difficult problem is the inclusion of *field-of-view constraints* and remove the assumption of omnidirectional vision. In this problem, the tangent of π^* must always be directed (roughly) towards the target in order to keep it in view.

Although the assumption that the target's path is known a priori is quite strong, we believe that our algorithm could be the basis of an on-line algorithm that maximizes the *true minimum time-to-escape*, defined as the shortest escape time among all possible target strategies in response to a sequence of observer actions. Worst-case tracking becomes a minimax problem for unpredictable targets. We believe that our algorithm may be a first step in designing a strategy for this general problem.

Acknowledgements This research was partially supported by NSF grant, ACR-0222920.

VI. REFERENCES

[1] S. Hutchinson, G. D. Gager, and P. I. Corke, "A tutorial on visual servo control," *IEEE Transactions on Robotics and Automation*, vol. 12, no. 5, pp. 651–670, 1996.

[2] N. P. Papanikolopoulos, P. K. Khosla, and T. Kanade, "Visual tracking of a moving target by a camera mounted on a robot: A combination of control and vision," *IEEE Transactions on Robotics and Automation*, 1993.

[3] S. M. Lavallée, J. Troccaz, L. Gaborit, A. L. Benabid P. Cinquin, and D. Hoffmann, "Image guided operating robot: A clinical application in stereotactic neurosurgery," in *Computer Integrated Surgery: Technology and Clinical Applications*, R.H.Taylor, S. Lavallée, G. Burdea, and R. Mösges, Eds. 1995, pp. 342–351, MIT Press.

[4] A. J. Briggs and B. R. Donald, "Visibility-based planning of sensor control strategies," *Algorithmica*, vol. 26, no. 3-4, pp. 364–388, 2000.

[5] S. M. LaValle, H. H. González-Baños, C. Becker, and J.-C. Latombe, "Motion strategies for maintaining visibility of a moving target," in *IEEE Conference on Robotics and Automation*, 1997.

[6] H. Gonzalez-Banos, C.-Y. Lee, and J.-C. Latombe, "Real-time combinatorial tracking of a target moving unpredictably among obstacles," in *IEEE Conference on Robotics and Automation*, 2002.

[7] P. Fabiani and J. C. Latombe, "Dealing with geometric constraints in game-theoretic planning," in *Proc. Int. Joint Conf. on Artif. Intell.*, 1999, pp. 942–947.

[8] T.-Y. Li and T.-H. Yu, "Planning object tracking motions," in *IEEE Conference on Robotics and Automation*, 1999.

[9] B. Nelson and P. K. Khosla, "Integrating sensor placement and visual tracking strategies," in *IEEE Conference on Robotics and Automation*, 1994.

[10] K. N. Kutulakos, C. R. Dyer, and V. J. Lumelsky, "Provable strategies for vision-guided exploration in three dimensions," in *IEEE Conference Robotics and Automation*, 1994.

[11] R. Sharma and S. Hutchinson, "On the observability of robot motion under active camera control," in *IEEE Conference on Robotics and Automation*, 1994.

[12] J.-C. Latombe, *Robot Motion Planning*, Kluwer Academic Publishers, Boston, MA, 1991.

[13] Chan, "Dynamic planar convex hull operations in near-logarithmaic amortized time," *JACM: Journal of the ACM*, vol. 48, 2001.

Planning Multi-Goal Tours for Robot Arms

Mitul Saha[1] Gildardo Sánchez-Ante[2] Jean-Claude Latombe[1]

(1) Computer Science Department, Stanford University, Stanford, CA, USA. Emails: {mitul, latombe}@cs.stanford.edu
(2) Computer Science Department, ITESM-Campus Guadalajara, Guadalajara, México. Email: gildardo@itesm.mx

Abstract

This paper considers the following multi-goal motion planning problem: a robot arm must reach several goal configurations in some sequence, but this sequence is not given. Instead, the robot's planner must compute an optimal or near-optimal path through the goals. This problem occurs, for instance, in spot-welding, inspection, and measurement tasks. It combines two computationally hard sub-problems: the shortest-path and traveling-salesman problems. This paper describes a greedy algorithm that operates under the assumption that the number of goals is relatively small (a few dozen at most) and the computational cost of finding a good path between two goals dominates that of finding a good tour in a graph with edges of given costs. Although the algorithm computes a quadratic number of goal-to-goal paths in the worst case, it is much faster in practice.

1 Introduction

A robot must reach several input goal configurations in some sequence, but this sequence is not given. Instead, a planner must determine the sequence that yields a path of minimal length through the goals (e.g., a path requiring minimal execution time). The robot stops at each goal, where its end-effector performs some operation. The time taken by this operation is independent of the goal ordering. This multi-goal planning (MGP) problem occurs frequently in practice, e.g., in spot-welding, inspection, and measurement tasks. Figure 1 shows a typical spot-welding workstation in an automotive body shop. The robot arm must bring the welding gun at successive positions where it performs welding operations to assemble body parts.

The MGP problem combines two computationally hard problems: the shortest-path and the traveling-salesman (TSP) problems. The former must be solved to determine optimal collision-free paths between pairs of goals and their lengths. The latter refers to the computation of an optimal tour in a graph whose edges are weighted by known goal-to-goal path lengths. Even

Figure 1: Spot-welding station

the basic Euclidean shortest-path problem among polyhedral obstacles is NP-hard [3]. Finding a shortest tour through points in the Euclidean plane is also NP-hard [10, 12]. However, for both problems, there exist efficient approximation and/or heuristic algorithms to compute sub-optimal, but still satisfactory solutions.

We assume that two functions are available:

- PATH – Given two robot configurations, this function returns a "good" collision-free path between them if they can be connected without collision. It indicates that no path exists otherwise.

- TOUR – Given a collection of points with given distances between all pairs, this function returns a "good" tour through these points that start and end at the same point.

A naive approach to compute a multi-goal path is to apply PATH to every pair of goals – hence, a quadratic number of times – before running TOUR once. However, in robotics problems, the number of goals typically ranges between 5 and 50, which is considered small for most approximate TSP algorithms. In contrast, robot paths must be computed in complex c-spaces, so that the running time of PATH is usually much greater than that of TOUR. Thus, we wish to use PATH sparingly, even if this requires running TOUR (or a portion of it) more frequently. We do this by setting the length of a path between two goals to a lower bound until PATH is applied to this pair of goals. Each call of TOUR applies to a graph of goals in which some edge lengths are exact (i.e., have been computed by PATH) and some edge lengths are lower bounds. After each run of TOUR, PATH is invoked to obtain the actual path length between selected pairs of goals. In the worst-case, PATH is applied to all pairs of goals. But, in practice, our algorithm avoids many calls to PATH and returns a tour in much less time than the naive approach. We show that when TOUR is based on the TSP algorithm given in [5, 12] that "doubles" the minimum spanning tree of the goals, our planning algorithm computes a tour whose length is within a given factor 2α, where $\alpha \geq 1$ is an input real number, of the tour computed by the naive approach.

2 Relation to Previous Work

(a) In [14] the MGP problem is considered for a coordinate measuring machine (CMM). Each goal defines a position of the CMM's probe. The proposed algorithm uses a multi-query PRM (Probabilistic Road-Map) path planner [9] to compute goal-to-goal paths. A roadmap is pre-computed over the probe's free space. Its nodes are the goal configurations, plus additional configurations picked at random until all the goals are in the same connected component. A search algorithm then extracts the shortest path (in the roadmap) between every two goals. This yields a reduced, complete graph whose nodes are exactly the goal configurations and in which each edge is labelled by the length of the path extracted between the two goals it connects. A TSP algorithm computes a near-optimal tour from this graph. As the lengths labelling the edges satisfy the triangular inequality, polynomial-time TSP algorithms are available to compute a tour within a constant ratio of the optimal tour [5, 11].

This method is based on a set of coherent choices. The CMM problems considered in [14] are formulated in quasi-planar c-space (where a configuration is the position of the tip of the measuring probe). In the two examples given in [14], the roadmaps only contain 100 nodes (including the goals). Thus, pre-computing them does not take much time. Moreover, once a roadmap is available, the search of a shortest path between every two goals is very fast. This justifies the fact that all goal-to-goal paths are computed before a tour is extracted.

Our multi-goal planner is based on different assumptions. In many applications (like spot welding), the cost of pre-computing a roadmap is high, due to both the high dimensionality of the c-space and the geometric complexity of the obstacles and the robot. It may then take a large number of goal-to-goal planning operations – often several hundreds – for a multi-query PRM planner to be more efficient in amortized time than a single-query PRM planner that does not pre-compute a roadmap [8, 13]. For this reason, the PATH function in our multi-goal planning algorithm is based on a single-query PRM planner (more details will be given in Section 5). But each evaluation of PATH is expedited by re-using the roadmap-construction work done at previous evaluations. Furthermore, in most problems, whether a multi- or single-query PRM planner is used, the average amortized computational cost per goal-to-goal planning operation is large relative to that of running a TSP algorithm. Hence, our algorithm strives to compute a tour without computing a path between all pairs of goals. This is consistent with the fact that our function PATH is based on a single-query PRM planner that does not pay the up-front cost of pre-computing a roadmap.

(b) The multi-goal planner described in [15] tries to avoid computing a collision-free path between every pair of goals. But, unlike ours, it neither uses lower bounds on path lengths, nor the information contained in intermediate tours computed by the TSP algorithm to guide the choice of the goal pairs between which paths are computed. Instead, this planner loops on the following operations: select pairs of goals using a technique like random selection or nearest-pair selection, compute a collision-free path between each pair, run the TSP algorithm. Once the TSP algorithm has returned a first complete tour, the planner can stop at any time, with each new computed tour being at least as good as the previous one. But, if the planner is stopped before all goal-to-goal paths have been generated, then there is no formal guarantee on the length of the last computed tour. In contrast, our algorithm provides such a guarantee (2α approximation factor). The planner in [15] uses a best-first search technique to generate goal-to-goal paths; this technique is inherently limited to searching low-dimensional c-spaces, which could restrict the range of problems it can handle.

(c) The multi-goal planner in [6] is aimed at generating inspection tours, for example finding cracks in large structures. The robot, which is modeled as a point in a polygonal free space F, must go to successive goal positions such that it can see every point in the boundary of F from at least one of these positions. This problem is also known as the watchman-route problem. Unlike in our problem, the goals are not given. In [6] they are computed using the approximate randomized "art-gallery" algorithm given in [7]. Next, the visibility graph of these goals and the vertices of F's boundary is computed and the shortest path between every two goals is computed [11]. An approximation TSP algorithm extracts a tour that is at most twice as long as the optimal tour. An extension to 3-D workspace is discussed in [6].

(d) Other variants of MGP problems, with no or few collision-avoidance constraints, have been addressed using general optimization techniques such as simulated annealing [4] or genetic algorithms [2].

3 Problem Formulation

We let C denote the c-space of a robot and $F \subseteq C$ its collision-free subset. We let ℓ stand for the selected measure that maps any given path τ in C to its length $\ell(\tau)$. Let $g_0, g_1, ..., g_r$ be $r+1$ input configurations in F called *goals*, with g_0 being also both the *start* and *end* configuration of the robot. These $r+1$ configurations are all distinct.

Any path in F joining two distinct goals g_i and g_j is called a *goal-to-goal* path. Any loop path in F starting and ending at g_0, and passing through every goal g_i, $i = 1$ to r, is called a *multi-goal path*. So, a multi-goal path τ is a sequence of $r+1$ goal-to-goal paths. The length of τ is assumed to be additive, i.e., it is the sum of the lengths (measured by ℓ) of the goal-to-goal paths it contains. This is reasonable since in most applications the robot stops at each goal; so, the goals divide the multi-goal path into pieces whose lengths can be independently computed. The *multi-goal planning problem* is to find the shortest multi-goal path, or an approximation of it.

The function PATH defined as follows is given. For any two goals g_i and g_j, PATH(g_i, g_j) returns a path in F joining g_i and g_j, if these configurations lie in the same component of F. For general robot arms, no efficient algorithm is available to compute a goal-to-goal path that is guaranteed to be within some approximation factor of the shortest path. So, PATH is by necessity a heuristic algorithm and we assume that the goal-to-goal paths it returns are "good enough." In our implementation, PATH is a single-query PRM planner with an optimization post-processing step (Section 5). Without loss of generality, we assume that the three paths returned by PATH between any three goals satisfy the triangular inequality. This assumption can always be trivially enforced, if needed. We also as-

---
Algorithm TOUR(G)
1. $T \leftarrow$ MIN-SPANNING-TREE(G)
2. Return PREORDER-WALK(T)

Figure 2: TOUR algorithm

sume that the planner finds a path between any two goals whenever one exists. Hence, if it fails to find a path between two goals, the MGP problem has no solution. (In fact, our implementation of PATH is only probabilistically complete, but this has never been a problem in our experiments.)

We define the *goal graph* to be the complete non-directed graph $G = (V, E, c)$ with weighted edges. $V = \{g_0, g_1, ..., g_r\}$ is the set of vertices and $E = \{\{g_i, g_j\} \mid i, j \in \{0, 1, ..., r\}\}$ the set of edges. Each edge $\{g_i, g_j\}$ is weighted by a positive real number $c(g_i, g_j)$, the edge's *cost*. A TOUR of G is any list $\pi = (g_0, g_{i_1}, ..., g_{i_r}, g_0)$ of all vertices starting and ending at g_0. Hence, $<i_1, ..., i_r>$ is a permutation of $\{1, ..., r\}$. The *cost* $c(\pi)$ of π is the sum of the costs of the edges along the tour, hence:

$$c(\pi) = \sum_{k=1}^{k=r+1} c(g_{i_{k-1}}, g_{i_k}) \text{ with } i_0 = i_{r+1} = 0.$$

Similarly, we define the cost $c(T)$ of a spanning tree T of G as the sum of the costs of the edges contained in T. A *minimum* tour (resp., minimum spanning tree) is one that has minimum cost over all possible tours (resp., spanning trees).

The function TOUR defined in Figure 2 computes a tour of G by "doubling" the minimum spanning tree of G [5, 12]. This TSP algorithm was also used in the multi-goal planners described in [6, 14]. In this function, MIN-SPANNING-TREE(G) returns a minimum spanning tree T of G and PREORDER-WALK(T) returns the list of vertices of G visited in a preorder walk of T with root g_0. (A preorder walk of a tree recursively visits every vertex in the tree, starting at the root, listing a vertex when it is first encountered.) The cost of T is no greater than that of the minimum tour. Moreover, when edge costs satisfy the triangular inequality, the cost of the tour PREORDER-WALK(T) is no greater than twice that of T, hence is within approximation factor 2 of the cost of the minimum tour [5]. Since G is complete, both MIN-SPANNING-TREE and TOUR run in time $\Theta(r^2)$.

Consider the NAIVE-MGP algorithm in Figure 3. To compute a multi-goal path, it first weighs every edge $\{g_i, g_j\}$ of the goal graph G by the length of the goal-

Algorithm NAIVE-MGP(G)
1. Set G to (V, E, c)
2. $V \leftarrow \{g_0, g_1, ..., g_r\}$
3. $E = \{\{g_i, g_j\} \mid i, j \in \{0, 1, ..., r\}\}$
4. For every $e = \{g_i, g_j\} \in E$ set $c(e)$ to
   $\ell(\text{PATH}(g_i, g_j))$
5. Return multi-goal path determined by TOUR(G)

Figure 3: Naive multi-goal planning algorithm

Algorithm GREEDY-MGP
1. Set G to (V, E, c)
2. $V \leftarrow \{g_0, g_1, ..., g_r\}$
3. $E = \{\{g_i, g_j\} \mid i, j \in \{0, 1, ..., r\}\}$
4. For every $e = \{g_i, g_j\} \in E$ set $c(e)$ to the length
   of the shortest path in C between g_i and g_j
5. Repeat
   a. $T \leftarrow$ MIN-SPANNING-TREE(G)
   b. $\kappa \leftarrow c(T)$
   c. Repeat while $c(T) \leq \alpha \times \kappa$
      i. If all edges in T are exact
         then return multi-goal path determined
         by PREORDER-WALK(T)
      ii. Pick a non-exact edge $e = \{g_i, g_j\}$ in T
      iii. Reset $c(e)$ to $\ell(\text{PATH}(g_i, g_j))$

Figure 4: Greedy multi-goal planning algorithm

to-goal path between g_i and g_j generated by PATH. (If PATH ever fails to find a path between two goals, then NAIVE-MGP returns *failure*, since no solution can possibly be found.) Then, it evaluates TOUR(G) once. The concatenation of the goal-to-goal paths between every two successive goals in the tour returned by TOUR(G) is the multi-goal path computed by NAIVE-MGP. Let $\mathcal{L}_N$ stand for the length of this path.

NAIVE-MGP always evaluates PATH $r(r+1)/2$ times. But we are interested in instances of the MGP problem where PATH is considerably more expensive to evaluate than TOUR. In robotics applications, this situation occurs frequently because r is usually rather small, while F is high-dimensional and geometrically complex. Hence, we wish to use PATH sparingly, even if this leads to evaluating TOUR (or just MIN-SPANNING-TREE) more often. Simultaneously, we would like the length of the generated multi-goal path to remain within a bounded factor of $\mathcal{L}_N$.

4 Greedy Planning Algorithm

To reduce the number of PATH evaluations, we create the goal graph $G = (V, E, c)$ with the cost $c(g_i, g_j)$ of every edge $\{g_i, g_j\}$ initialized to the length of the shortest path joining g_i and g_j in C. Since the shortest path in C may not be collision-free, this length is a lower-bound approximation of $\ell(\text{PATH}(g_i, g_j))$. Usually, this lower bound is very easy to calculate, as for most measures ℓ the shortest path in C between two configurations is the straight line segment joining them. Then, we successively adjust edge costs by evaluating PATH for edges picked in the current minimum spanning tree of G. Once the cost $c(g_i, g_j)$ of an edge $\{g_i, g_j\}$ has been updated to $\ell(\text{PATH}(g_i, g_j))$, this edge $\{g_i, g_j\}$ is said to be *exact*. This yields the algorithm GREEDY-MGP of Figure 4.

Steps 1-4 initialize the goal graph G. Hence, just after performing Step 4, all edges are non-exact. Step 5 first computes the minimum spanning tree T of G. Next, it evaluates PATH for non-exact edges of T, until either all edges are exact (in which case it exits with a solution), or the new cost of the tree is greater than α times its original cost. In the latter case, it loops by computing the new minimum spanning tree. (If PATH ever fails to find a path at Step 5.c.iii, then GREEDY-MGP returns *failure*.)

For a given MGP problem, let $\mathcal{L}_N$ and $\mathcal{L}_G$ denote the lengths of the solutions computed by NAIVE-MGP and GREEDY-MGP, respectively. Assume that, whenever these two algorithms evaluate PATH for the same two goals g_i and g_j, each evaluation yields the same goal-to-goal path. Then, we have:

Claim 1: *The lengths $\mathcal{L}_N$ and $\mathcal{L}_G$ of the paths computed by* NAIVE-MGP *and* GREEDY-MGP *verify*

$$\mathcal{L}_G \leq 2\alpha \times \mathcal{L}_N.$$

Proof: Let G_{ex} denote the goal graph in which all edges are exact, hence weighted by the lengths of the paths computed by PATH. Let T_{ex} be the minimum spanning tree of G_{ex}. Recall that the length $\mathcal{L}_N$ of the multi-goal path computed by NAIVE-MGP verifies $c(T_{ex}) \leq \mathcal{L}_N \leq 2c(T_{ex})$.

During the evaluation of GREEDY-MGP, the cost of every edge of G is always less than or equal to the cost of the same edge in G_{ex}. Thus, at Step 5, we have $\kappa \leq c(T_{ex})$. Step 5.c computes goal-to-goal paths for non-exact edges in T, until either the cost of T grows larger than $\alpha \times \kappa$, or all edges in T are exact. In this second case, we have $c(T) \leq \alpha \times \kappa \leq \alpha \times c(T_{ex})$. Since the length $\mathcal{L}_G$ of the multi-goal path determined by

PREORDER-WALK(T) at Step 5.c.i is guaranteed to be within factor 2 of $c(T)$, we have $\mathcal{L}_G \leq 2\alpha \times c(T_{ex})$, hence $\mathcal{L}_G \leq 2\alpha \times \mathcal{L}_N$. ∎

Note that the tour PREORDER-WALK(T) computed at Step 5.c.i may contain edges that are still not exact. For those edges, it is necessary to evaluate PATH in order to obtain a multi-goal path. Then, the goal graph may still contain many edges that are non-exact and the triangular inequality may not be satisfied by every triplet of nodes. But all what matters for the tour computed at Step 5.c.i to have cost within factor 2 of the cost of T is that the triangular inequality holds for the edges in T and its preorder walk.

In the worst case, GREEDY-MGP evaluates PATH $r(r+1)/2$ times. However, the experimental results of Section 6 show that in practice PATH is usually evaluated much fewer times.

5 Implementation of PATH

For robot arms, no algorithm is available to efficiently compute goal-to-goal paths that are within a guaranteed factor of shortest paths. Instead, we use a heuristic two-phase approach: first, compute a collision-free path; next, optimize this path. This approach was introduced in [1].

Our implementation of PATH uses the single-query PRM planner described in [13] to generate an initial collision-free path between two given goals g_i and g_j. This planner works by incrementally growing two trees of sampled configurations, called milestones, that are rooted at g_i and g_j, respectively. At every iteration, it picks a milestone m at random in one of the two trees and samples configurations in a neighborhood of m until one of them, m', tests collision-free. It installs m' as a child of m in its tree and creates a connection between m' and the closest milestone in the other tree, thus establishing a path of milestones between g_i and g_j. The planner then checks that the connection between every two successive milestones on this path is collision-free (each connection is a straight line segment in c-space). If all connections are collision-free, the planner returns the path, otherwise it keeps growing the trees of milestones. The planner exits with failure if it has not found a path after generating a given maximum number of milestones. The planner is probabilistically complete, with fast convergence rate [8, 13]. Failure to find a path, while one exists, has not been an issue in our experiments.

The milestone trees grown by the planner are not bi-

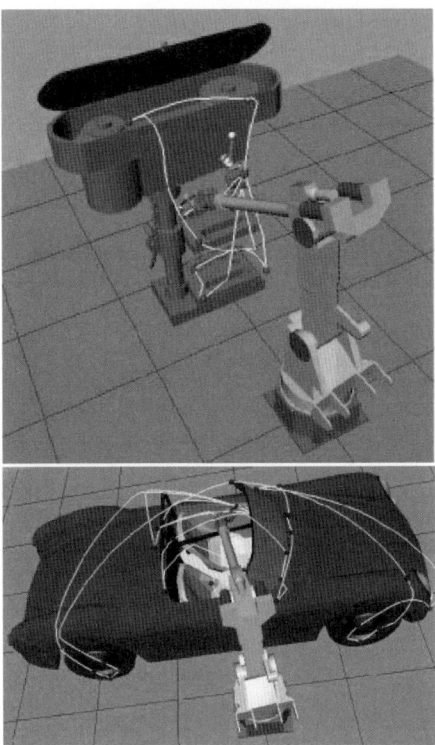

Figure 5: Examples 1 and 2 (10 and 30 goals)

ased in any particular direction. So, rather than discarding the two trees produced by a run of the planner, we store them. When PATH is invoked again, if any of the two goals in the new pair of goals has already been considered before, then the corresponding tree is reused by the planner, thus saving considerable amount of work. As more evaluations of PATH are performed, each evaluation, on average, takes less time.

Several optimizers can be used to improve a path generated by the PRM planner. For instance, the variational optimization technique used in [1] iteratively deforms the path to minimize the time needed to execute the path, given the robot's dynamic model and torque limits in the joints. Our implementation of PATH uses a simpler (and faster) optimizer. A path generated by the PRM planner is a polygonal line in c-space. We use the L_∞ metric weighted by the inverse of the maximum velocity of each joint to measure the length of each segment. Thus, the length of a path is the time to execute this path assuming that, along each segment, all joints move at constant velocity, with the joint needing the most time moving at maximum velocity. In practice, this length is only a lower-bound approximation of the actual time needed to execute the path. With this weighted L_∞ metric, the shortest path in c-space between any two configurations is the straight segment. The optimizer repeatedly replaces sub-paths by straight segments whenever these segments are collision-free. Though the outcome is at best locally optimal, it is usually quite satisfactory.

6 Experimental Results

We first compare GREEDY-MGP and NAIVE-MGP on two examples shown in Figure 5, with 10 and 30 goals, respectively. We proceed as follows. After every run of GREEDY-MGP, we run NAIVE-MGP with the same seed of the random-number generator. In addition, during the run of GREEDY-MGP, we note the sequence of pairs of goals on which PATH is invoked. Then NAIVE-MGP first computes goal-to-goal paths in this same order, before computing paths between all remaining pairs of goal (in any order). So, over the two runs, we are guaranteed that the same query PATH(g_i, g_j) made by GREEDY-MGP and NAIVE-MGP yields the same goal-to-goal path.

	NAIVE-MGP		GREEDY-MGP	
	Ex. 1	Ex. 2	Ex. 1	Ex.2
Total-time	37.89	310.50	12.49	43.51
Total-length	4.67	11.21	6.11	12.31
#PATH	55	465	21	55
#TOUR/MST	1	1	9	32
Time-PATH	37.6	306.25	8.92	13.27
Time-TOUR/MST	0.23	4.17	3.01	29.42

Table 1: Comparative performance results obtained with NAIVE-MGP and GREEDY-MGP ($\alpha = 1$)

Table 1 gives performance data obtained with NAIVE-MGP and GREEDY-MGP, when the parameter α is set to 1. The data are averages over 50 runs of each planner, with different seeds of the random-number generator (with every seed used to run both planners as indicated above). The rows indicate the total running times of a planner (in seconds), the length of the generated solution, the number of evaluations of PATH, the number of evaluations of MIN-SPANNING-TREE or TOUR, and the times spent in those evaluations. These results indicate that GREEDY-MGP is significantly faster than NAIVE-MGP, while producing paths that are, on average, only marginally longer. This gain derives from the fact that GREEDY-MGP calls PATH much more sparingly. Our experiments (which include many other examples) indicate that in general the relative speedup obtained with GREEDY-MGP grows when the number of goals increases.

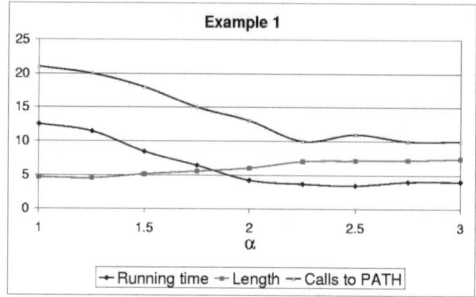

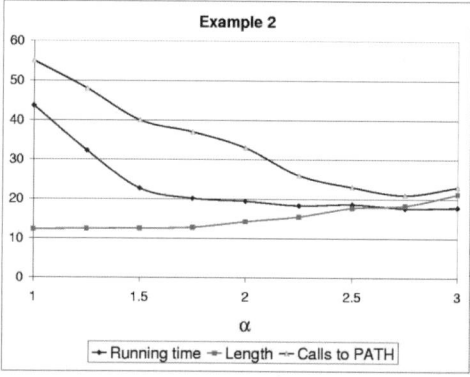

Figure 6: Influence of parameter α

We now analyze the impact of the parameter α on the performance of GREEDY-MGP. The two plots in Figure 6 were obtained for the previous two examples. Each contains three curves that respectively show the total running time of GREEDY-MGP, the length of the computed multi-goal path, and the number of evaluations of PATH when α takes the values 1, 1.5, 2, 2.5, and 3. Again, each value is an average over 50 runs. The plots indicate that the running time of GREEDY-MGP first decreases sharply, then levels out. On the other hand, the average lengths of the generated paths only increase moderately.

7 Conclusion

This paper describes a new multi-goal planning algorithm, GREEDY-MGP, that operates under the assumption that the cost of finding a good path between any two goals dominates that of finding a good tour in a graph with edges of given weights. Our algorithm uses the minimum spanning tree of the goal graph to decide which goal-to-goal paths to compute. The solution it returns is within 2α (where $\alpha \geq 1$) of the cost of the solution returned by the naive algorithm that first computes all goal-to-goal paths. Experiments show that in general our algorithm is much faster than the naive algorithm or the algorithm previously proposed in [14]. Its running time also drops sharply when α is set to values slightly greater than 1. The approximation factor 2α holds for the classic TSP algorithm that "doubles" the minimum spanning tree of the goal graph. Other TSP algorithms [10] might yield a more efficient GREEDY-MGP in practice, but we have not experimented with any of them. Most would not guarantee a 2α approximation factor.

There are several directions in which this work could be extended in the future. In some applications, a partial ordering is imposed on the input goals. How to efficiently incorporate this ordering into the multi-goal planner? In some cases, it is more convenient to specify the goals as configurations (positions and orientations) of the robot's end-effector. Since the inverse kinematics of a robot arm may have several solutions, each goal configuration of the end-effector may map into several goal configurations of the robot, but only one of these would have to be visited by the robot. Choosing among alternative robot goals would add to the complexity of the planner. Finally, it would be interesting to investigate the multi-robot case, where several robots sharing portions of their workspaces must reach goals without colliding with each other.

Acknowledgments: This research was partially funded by a gift from General Motors Research. This paper has benefited from discussions with Prof. Fritz Prinz, Dr. Fabian Schwarzer and Yu-Chi Chang.

References

[1] J.E. Bobrow. Optimal Robot Path Planning Using the Minimum-Time Criterion. *IEEE J. Robotics and Autom.*, 4(4):443-450, 1988.

[2] M. Bonert, L.H. Shu, and B. Benhabib. Motion Planning for Multi-Robot Assembly Systems. *Proc. ASME Design Engineering Technical Conf.*, Las Vegas, NV, Sept. 1999.

[3] J.F. Canny and J. Reif. New Lower Bound Techniques for Robot Motion Planning Problems. *Proc. IEEE Conf. on Foundations of Computer Science*, pp. 39-48, 1987.

[4] B. Cao, G.I. Dodds, and G.W. Irwin. A Practical Approach to Near Time-Optimal Inspection-Task-Sequence Planning for Two Cooperative Industrial Robot Arms. *Int. J. of Robotics Res.*, 17(8):858-867, 1998.

[5] T.H. Cormen, C.E. Leiserson, and R.L. Rivest. *Introduction to Algorithms*, The MIT Press, Cambridge, MA, 1990.

[6] T. Danner and L.E. Kavraki. Randomized Planning for Short Inspection Paths. *Proc. IEEE Int. Conf. on Robotics and Autom.*, San Francisco, CA, 2000.

[7] H.H. Gonzalez-Baños and J.C. Latombe. A Randomized Art-Gallery Algorithm for Sensor Placement. *Proc. ACM Symp. on Computational Geometry*, pp. 232-240, 2000.

[8] D. Hsu. *Randomized Single-Query Motion Planning in Expansive Space*. Ph.D. Thesis, Computer Sc. Dept., Stanford Univ., Stanford, CA, 2000.

[9] L.E. Kavraki, P. Svetska, J.C. Latombe, and M. Overmars. Probabilistic Roadmaps for Path Planning in High-Dimensional Configuration Spaces. *IEEE Tr. on Robotics and Autom.*, 12(4):566-580, 1996.

[10] D.S. Johnson and L.A. McGeoch. The Traveling Salesman Problem: A Case Study in Local Optimization. In *Local Search in Combinatorial Optimization*, E.H.L. Aarts and J.K. Lenstra (eds.), John Wiley and Sons, London, pp. 215-310, 1997.

[11] J.S.B. Mitchell. Shortest Paths and Networks. In *Discrete and Computational Geometry*, J.E. Goodman and J. O'Rourke (eds.), CRC Press, Boca Raton, FL, pp. 445-466, 1997.

[12] D.J. Rosenkrantz, R.E. Stearns, and P.M. Lewis II. An Analysis of Several Heuristics for the Traveling Salesman Problem. *SIAM J. of Computing*, 6(3):563-581, 1977.

[13] G. Sánchez and J.C. Latombe. On Delaying Collision Checking in PRM Planning – Application to Multi-Robot Coordination. *Int. J. of Robotics Res.*, 21(1):5-26, 2002.

[14] S.N. Spitz and A.A.G. Requicha. Multiple-Goals Path Planning for Coordinate Measuring Machines. *Proc. IEEE Int. Conf. Robotics and Autom.*, San Francisco, CA, 2000.

[15] C. Wurll, D. Henrich, and H. Wörn. Multi-Goal Path Planning for Industrial Robots. *Proc. Int. Conf. on Robotics and Applications (RA'99)*, Santa Barbara, CA, 1999.

Trajectory Generation for Vehicles Moving with Constraints on a Complex Terrain

Ken-Jui Tsao, Li-Sheng Wang, Po-Ting Kuo
Institute of Applied Mechanics
National Taiwan University
Taipei, Taiwan, ROC

Fan-Ren Chang
Department of Electrical Engineering
National Taiwan University
Taipei, Taiwan, ROC

Abstract – In this paper[1], the methodology of generating an optimal trajectory on a complex terrain for a specific vehicle is proposed. The possible paths are constrained by the limitations on the terrain and the capability of the vehicle. To deal with these constraints, the notions of forbidden point, forbidden direction, and forbidden path are introduced. After certain constants are specified, the method of dynamic programming is then invoked to find the optimal solution. If the target is beyond the maximal range of the vehicle, appropriate service stations are selected by using the auction algorithm. To speed up the computation process, the ideas of bi-spiral scheme and instant update are employed. With all the techniques at hand, numerical results show that the proposed method can generate the desired trajectory efficiently.

1. Introduction

For a vehicle moving on a complex terrain, its capability restricts the possible paths that it can follow. A jeep can climb a steeper slope than a passenger car, while a mobile robot may be able to climb even steeper slope. Different wear on the tire leads to different limit on the angle of turn. How to accommodate these constraints on the generation of optimal trajectory on a complex terrain is the main theme of this paper.

The constraints on the path may come from either the topology of the terrain or the vehicle itself. On the terrain, there may be some hazardous region or congested area that the vehicle cannot enter. There may be some service stations that the vehicle may need to pass. On the part of the vehicle, sharp turns may not be able to trace and it is preferred to move on a flat path rather than a slope. As a result, the classical algorithm of finding the shortest path needs to be modified. For the shortest-path problem, various techniques have been used to find the optimal solution. These include the Label correcting method [1], D'ijkstra's method [3], method of dynamic programming [6][7][8][9], and the auction algorithm [2]. If the dynamics of the vehicle is taken into consideration, various scheme such as the Hamilton-Jacobi's method can be adopted, as reviewed in [10]. However, if the complicated dynamical behavior is considered through certain constraints, for the corresponding path planning problems in which other characteristics, in addition to the length of the path, must be considered, the method of dynamic programming provides the most flexible platform based on the principle of optimality. The idea of dynamic programming has been used in [5] to find the optimal path of a vehicle on a terrain by including the consideration of forbidden region and the slope. In this paper, similar idea is adopted to find the optimal solution by considering more realistic limitations and the capability of the vehicle.

In particular, the notions of forbidden directions and forbidden paths are introduced in this paper. The limitations of rate of climb, minimal angle of turn, and maximal distance of travel are incorporated into the algorithm. Depending on the slope of the path, the scaled distance is used to perform the optimization process, so that flatter paths are preferred. If the length of the optimal path exceeds a certain bound, the vehicle needs to pass through some service stations. For this problem, we first transform it to a classical salesman problem and then use the auction algorithm to select the suitable service stations. Moreover, to enhance the computational speed, the notion of bi-spiral scheme is introduced and the technique of instant update is applied. From the numerical results shown is this paper, the methodology proposed here can indeed generate an optimal trajectory that meets all the above-mentioned requirements.

The rest of this paper is organized as follows. In Section 2, the underlying problem of path planning on a complex terrain is described in more detail. Section 3 presents the setup for realizing those constraints and the process of incorporating them into the method of dynamic programming. The algorithm for the selection of service stations and the bi-spiral scheme are also discussed. Numerical results are shown in Section 4, which shows the effectiveness of the proposed method. Finally, some concluding remarks are given in Section 5.

2. Problem Description

Consider a complex terrain, such as in Figure 1, consisting of geometric points with each position being specified by three components (x, y, z), with the corresponding axes being pointing toward the east, the north, and upward, respectively. The horizontal plane on which the terrain resides may be divided into small rectangles with a matrix of grid points (x_i, y_j), $i = 1,...,l; j = 1,...,n$, which are equally spaced with distance a, cf. Fig. 1. The point on the terrain corresponding to node (i, j) is represented by $P(i, j)$ and the height at that point is given by $z(i, j)$. There may be some regions on the terrain that the vehicle is prohibited to pass, such as a

---

[1] Corresponding Author: Li-Sheng Wang, email: wangli@gauss.iam.ntu.edu.tw

hazardous area, a lake or a building, which is identified as the set Ω. Moreover, it may be necessary for the vehicle to stop somewhere during the travel for service stations, such as gas stations. The locations of the service stations on the terrain are denoted by nodes $g_1,...,g_N$.

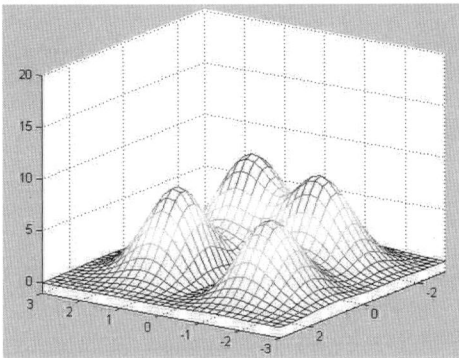

Fig. 1 complex terrain

The problem is to seek efficiently the optimal trajectory from a starting node $s=(i_s,j_s)$ to a target node $t=(i_t,j_t)$ without entering Ω such that the designated vehicle is able to follow. Different types of vehicles, such as a passenger car, a jeep, or a truck, have different capabilities, such that their motions are subject to various constrains. First, the initial direction of the path must be compatible with the heading of the vehicle at the node s. There may be some other nodes for which the direction of passing is limited. Secondly, the vehicle may not be able to move on a steep slope. This renders the limitation on the rate of climb of the trajectory. Thirdly, the vehicle may not be able to make sharp turns, and hence the radius of curvature of the path is constrained. Moreover, the maximum distance of travel may require the vehicle to stop at some service station if the destination is too far to reach with one gas tank. In addition, flat paths are preferred than the paths with steeper slope to save the energy. Other constraints may be imposed on the trajectory of motion. Those listed above shall be accommodated in the generation of the optimal path as discussed in the next section.

in which Ω is defined in the previous section. In order to describe the move from one node to its neighbors, the virtual move m is defined to be in the set of

$$M^+ = \{1,2,3,4,6,7,8,9\}$$

such that $m=1$ represents the move from (i,j) to $(i-1,j-1)$, cf. Fig. 2, in which the meanings of the other values of m are also given. Moreover, we use $m=5$ to denote the degenerate virtual move which refers to the move from one node to itself.

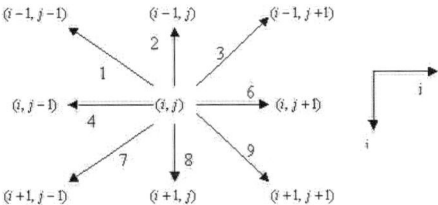

Fig. 2 Virtual Moves

For the virtual move m, the maps φ,ϕ are defined such that $(\varphi(i,m),\phi(j,m))$ is the destined node from (i,j) via m. The distance and the slope of the virtual move m from (i,j) are then computed from, respectively

$$dist(i,j,m) = [a^2(\varphi(i,m)-i)^2 + a^2(\phi(j,m)-j)^2 + (z(\varphi(i,m),\phi(j,m))-z(i,j))^2]^{1/2}$$

$$\alpha(i,j,m) = \tan^{-1}(\frac{z(\varphi(i,m),\phi(j,m))-z(i,j)}{a\sqrt{(\varphi(i,m)-i)^2+(\phi(j,m)-j)^2}})$$

Let the limitation on the rate of climb of the trajectory be represented by the maximal upward slope $\overline{\alpha}$ and maximal downward slope $\underline{\alpha}$, which are determined by the vehicle. To realize the forbidden region, the limitation on the rate of climb, and the preference of the flat paths, the scaled distance function for the virtual move m from (i,j) is defined as

3. The algorithm

To solve the problem described above, we adopt the idea of dynamic programming as discussed in [5]. First, we define the forbidden function F as

$$F(i,j) = \begin{cases} 0, & \text{if } (i,j) \in \Omega, \\ 1, & \text{otherwise,} \end{cases} \quad (1)$$

$$D(i,j,m) = \begin{cases} -1, & \text{if} \begin{cases} \varphi(i,m) = -1 \text{ or } > l \\ \text{or } \phi(j,m) = -1 \text{ or } > n \\ \text{or } F(i,j) = 0 \\ \text{or } F(\varphi(i,m), \phi(j,m)) = 0 \\ \text{or } \alpha(i,j,m) > \overline{\alpha} \\ \text{or } \alpha(i,j,m) < \underline{\alpha} \end{cases} \\ (w_0 |\alpha(i,j,m)| + 1) \cdot dist(i,j,m) \\ \quad , \text{otherwise} \end{cases} \quad (2)$$

in which the scale factor w_0 is specified according to the preference on the slope of the path. It is seen that if the slope of the virtual move m is too large, the distance of the move is set to -1, which indicates that the corresponding virtual move is prohibited. Moreover, by adjusting the scale factor, the preference of the flatter paths can be realized.

As discussed in the previous section, the directions of passing through some nodes are limited. The set of such nodes, denoted by $b \in B$, contains at least the starting node s. Since the direction of entering b and that of leaving b may have different constraints, we expand the set of virtual moves to include their reversals, denoted by

$$M^- = \{-1, -2, -3, -4, -6, -7, -8, -9\},$$

cf. Fig. 2. The maps φ and ϕ are defined accordingly for $m < 0$. The forbidden directions of passing through the node b are then given by a set $A(i_b, j_b) \subset M^+ \cup M^-$, in which the positive value refers to leaving and negative value refers to entering. If the direction is not allowed, the corresponding virtual move must be prohibited, and the scaled distance function is updated according to the following rule

$$\begin{aligned} &\text{for } m \in A(i_b, j_b), \\ &\text{if } m > 0 \text{ then } D(i_b, j_b, m) = -1 \\ &\text{if } m < 0 \quad \text{then} \\ &\quad D(\varphi(i_b, m), \phi(j_b, m), 10 + m) = -1 \end{aligned} \quad (3)$$

Note that for entering forbidden directions, the corresponding virtual moves from $(\varphi(i_b, m), \phi(j_b, m))$ is set to be prohibited.

Next, we characterize the constraint of minimal angle of turn for a designated vehicle. The turn is assumed to be composed of two consecutive line segments between nodes. Consider two consecutive virtual moves m, m' such that

$$(i,j) \xrightarrow{m} (\hat{i}, \hat{j}) \xrightarrow{m'} (\tilde{i}, \tilde{j}).$$

The angle of turn is then given by

$$\beta(i,j,m,m') = \angle \overrightarrow{P(i,j)P(\hat{i},\hat{j})}, \overrightarrow{P(\hat{i},\hat{j})P(\tilde{i},\tilde{j})}$$

i.e. the angle intersected by the segments $\overrightarrow{P(i,j)P(\hat{i},\hat{j})}$ and $\overrightarrow{P(\hat{i},\hat{j})P(\tilde{i},\tilde{j})}$. The constraint of turning may be then realized by requiring

$$\beta(i,j,m,m') > \beta_0$$

where β_0 is some constant determined by the type of the vehicle. The path is forbidden if the previous condition is violated.

With the above setup, we now ready to invoke the method of dynamic programming by a recursive process. Since the direction of move is essential in our consideration, we define the total cost starting from (i,j) in the direction of m within k moves by $C(k,i,j,m)$. At the step $k = 0$, the cost function is initialized to be 0 at the node t, and -1 at all other nodes for all virtual moves. As the algorithm proceeds, the value of the cost function is updated according to the following procedure. At step k, at the node (i,j) in the direction m such that $D(i,j,m) \neq -1$, we first construct the set of admissible second moves as

$$\begin{aligned} S(k,i,j,m) &= \{m' \in M^+ : \\ &\quad D(\varphi(i,m), \phi(j,m), m') \neq -1, \\ &\quad C(k, \varphi(i,m), \phi(j,m), m') \neq -1, \\ &\quad \beta(i,j,m,m') > \beta_0 \}. \end{aligned} \quad (4)$$

The cost function at step $k+1$ is then found as follows

$$\begin{aligned} &\text{if } S(k,i,j,m) \text{ is empty} \\ &\quad \text{set } C(k+1,i,j,m) = -1 \\ &\text{else} \\ &\quad C(k+1,i,j,m) \\ &\quad = \min_{m' \in S(k,i,j,m)} (D(i,j,m) + C(k, \varphi(i,m), \phi(i,m), m')) \\ &\quad \chi(k+1,i,j,m) \\ &\quad = \arg\min_{m' \in S(k,i,j,m)} (D(i,j,m) + C(k, \varphi(i,m), \phi(i,m), m')) \\ &\text{if } C(k,i,j,m) \leq C(k+1,i,j,m) \\ &\quad \text{then } C(k+1,i,j,m) = C(k,i,j,m), \\ &\quad \chi(k+1,i,j,m) = 5 \end{aligned} \quad (5)$$

It is noted that in the second part of the previous procedure, if the above minimization process yields a cost for step $k+1$ is greater than or equal to that for step k, further virtual move is not appropriate, and hence degenerate move occurs.

The above iteration proceeds until all the virtual moves become degenerate, which is equivalent to the condition,

$$C(\overline{k}+1,i,j,m) = C(\overline{k},i,j,m), \quad \forall i,j,m.$$

At the final stage, say step $\overline{k}$, the minimal cost of the optimal path starting from any node (i, j) to the target t in any direction m is obtained. We may then start with s and utilize the map γ to trace back the optimal path as follow:

$$\begin{aligned}
& m^* = \arg\min_{m \in M^*}\{C(\overline{k},i_s,j_s,m)\} \\
& i^* = i_s \\
& j^* = j_s \\
& \text{for } q=0 \text{ to } \overline{k}-1 \\
& \quad \text{if } \gamma(\overline{k}-q,i^*,j^*,m^*) \neq 5 \\
& \quad \text{then} \\
& \quad\quad i^* \leftarrow \phi(i^*,m^*) \\
& \quad\quad j^* \leftarrow \phi(j^*,m^*) \\
& \quad\quad m^* \leftarrow \gamma(\overline{k}-p,i^*,j^*,m^*)
\end{aligned} \quad (6)$$

The established path shall then be the optimal one satisfying the constraints discussed in Section 2 except the maximal distance of travel.

To meet the requirement of maximal range R_{max} for a specific vehicle, we first compute the length of the optimal path found in the above process, say R_{opt}. If $R_{opt} > R_{max}$, then we need to find suitable intermediate service stations. The strategy is to seek the optimal path for each pair of nodes in the set $\{s, g_1, ..., g_N, t\}$. Deleting those paths with length greater than R_{max}, the problem becomes the classical salesman problem as the one shown in Fig. 3. Many shortest-distance methods can be then used to obtain the optimal solution. Here we adopt the auction algorithm as discussed in [2] for its flexibility in dealing with the changes of distance functions.

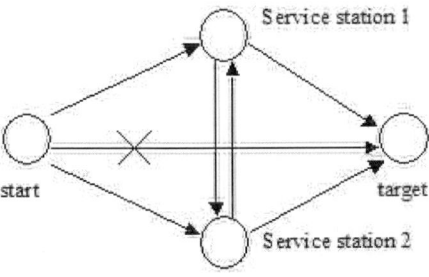

Fig. 3 Salesman Problem and Auction Algorithm

The above process may be very time-consuming if the terrain is large. In order to speed up the computation process, we adopt the bi-spiral scheme in the update process of the cost function. The sequence of nodes visited in each step k to apply the process (5) starts with a neighboring node of the target t, then proceeds spirally outward. Once all the nodes are visited, the update process reverses its direction to spirally inward return to the target. This sequence can significantly reduce the total number of steps, especially when the desired path is curved back and forth. Moreover, in the minimization process of (5), we could use the updated $C(k+1, \varphi(i,m), \phi(j,m), m')$, instead of $C(k, \varphi(i,m), \phi(j,m), m')$ if the node $(\varphi(i,m), \phi(j,m))$ has been visited at step $k+1$. This idea of instant update is similar to that of Gauss-Siedel method [4] in solving a set of algebraic equations, and can enhance the efficiency of the algorithm prominently.

4. Numerical Result

For the example of complex terrain shown in Fig. 1, we are now ready to apply the algorithm developed in Section 3 to find the optimal trajectory satisfying the constraints. First, we divide the horizontal plane into 31*31 grid nodes with grid length $a = 5$ (meter), and consider the case that $R_{max} = \infty$, i.e. the vehicle can move to any point on the terrain without service. Let the capability of the vehicle and the preference of slope impose the constraint on the possible path through the following constants:

$$\overline{\alpha} = 0.7, \underline{\alpha} = -0.8, \beta_0 = 120^0, w_0 = 4.0.$$

Incorporating these constants in the algorithm, the optimal trajectory from the node $s = (5, 21)$ to the target $t = (26, 18)$ is then obtained as shown in Fig. 4. The trajectory winds around the hills to reach the target.

If the constraint of maximal range is set as $R_{max} = 250$ (meter), the vehicle cannot reach the target without service. The strategy outlined in Section 4 using the auction algorithm is then followed to find suitable intermediate service stations. The result is shown in Fig. 5, in which the service stations are marked by big circles. For the same set of the starting node and the target node, another trajectory is chosen to pass by a service station. From these examples, it is seen that the methodology proposed in this paper can indeed give rise to an optimal trajectory satisfying all the constraints described before.

3807

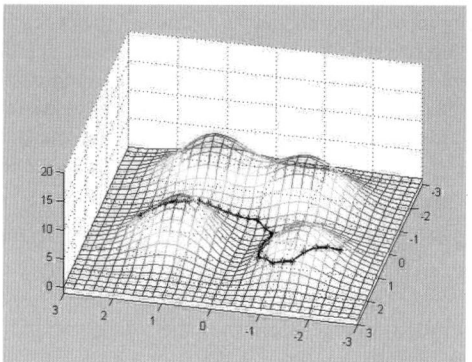

Fig. 4. Optimal Trajectory with Infinite Maximum Range

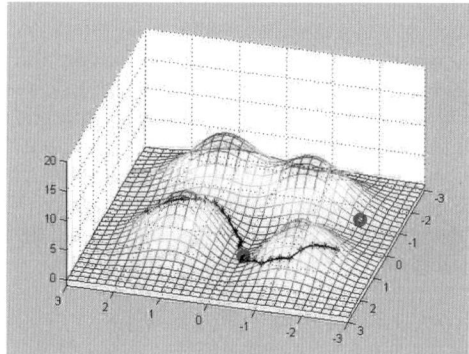

Fig. 5. Optimal Trajectory with Limited Maximum Range

5. Conclusion

This paper presents a methodology to generate an optimal path on a complex 3-D terrain that a specific vehicle is safe and capable to follow. Numerical results show that the method is effective and efficient. The algorithm can be used for a variety of vehicles, such as a passenger car, a truck, or a mobile robot, by choosing the appropriate constants in the algorithm. In addition to those constraints considered here, there may be many other limitations or preference to be included. The effect of the roughness of the terrain on the selection of path is an example. The incorporation of other constraints in the algorithm and the desire of reducing the computational load pave the way of future development of the methodology.

6. References

[1] Bertsekas, D.P., *Dynamic Programming and Optimal Control,* Athena Scientific, 1995.

[2] Bertsekas, D.P., "An Auction Algorithm for Shortest Paths," *SIAM J. on Optimization*, vol. 1, pp.425-447, 1991.

[3] Dijkstra, E.W., "A Note on Two Problems in Connection with Graphs," *Numer. Math.*, vol. 1, pp. 269-271, 1959.

[4] Gerald, C.F., Wheatley, P.O., *Applied Numerical Analysis*, Addison-Wesley Publishing Company, 1994.

[5] Kwok, K.S., Driessen, B.J., "Path Planning for Complex Terrain Navigation Via Dynamic Programming," *Proc. of American Control Conference, S.D., Cal.*, pp. 2941-2944, 1999.

[6] Lam, K.P., Tong, C.W., "Connectionist Network for Dynamic Programming Problems", *IEE Proc.-Comput. Digit. Tech.*, vol. 144, No. 3, pp. 163-167, 1997.

[7] Larson, R.E., Casti, J.L., *Principles of Dynamic Programming - Part II Advanced Theory and applications*, Marcel Dekker Inc., 1982.

[8] Lewis, F.L., Syrmos, V.L., *Optimal Control*, John Wiley & Sons, 1995.

[9] Ravindran, A., Phillips, D.T., Solberg, J.J., *Operations Research - Principles and Practice*, John Wiley & Sons, 1987.

[10] Tsitsiklis, J.N., "Efficient Algorithms for Globally Optimal Trajectories," *IEEE Trans. on Aumatic Control,* vol. 40, No. 9, 1995, pp.1528-1538.

Robot Motion Planning Using Adaptive Random Walks

Stefano Carpin[†]
School of Engineering and Science
International University of Bremen
Germany
s.carpin@iu-bremen.de

Gianluigi Pillonetto
Department of Information Engineering
The University of Padova
Italy
giapi@dei.unipd.it

Abstract—We propose a novel motion planning algorithm based on adaptive random walks. The proposed algorithm turns out to be easy to implement and the solution it produces can be easily and efficiently optimized. Furthermore the algorithm can incorporate adaptive components, so that the developer is not required to specify all the parameters of the random distributions involved, and the algorithm itself can adapt to the environment it is moving in. Proofs of the theoretical soundness of the algorithm are provided as well as implementation details. Numerical comparisons with well known algorithms illustrate its effectiveness.

I. INTRODUCTION

In the last years the problem of robot motion planning received further attention as a consequence of two possibly synergistic events. The massive introduction of randomized motion planners following the basilar work presented in [9] allowed to effectively face problems with a high number of degrees of freedom, not always solvable by formerly developed planners. Moreover, motion planners are being used for a number of applications that are beyond traditional robotics, like digital character motion generation, bioinformatics and many others (see [13]). The randomized approach appears to be the current main stream of research in motion planning and originated a wave of new algorithms which allows to study a wide variety of different problems. Indeed probabilistic planners are also being used to solve problems involving kinodynamic constraints ([15]). Recently there has also been an impulse to study algorithms not only probabilistic complete, but also resolution complete and to speculate on a possible turn back to efficient deterministic approaches ([4, 7]). The continuous flow of innovative tools for efficiently searching configuration state spaces confirms that as efficient planners are introduced, the frontier of applications is also being pushed further. Think for example at the problem of multi-robot motion planning ([6]), or protein folding and ligand binding, where problem instances with hundreds of degrees of freedom are common and are being tackled with randomized motion planners ([16]).

In this context we developed a new motion planner based on random walks. Preliminary results illustrate that, in spite of the widespread believe that random walks poorly perform in robot motion planning problems, by carefully choosing the random distributions involved it is possible to efficiently solve difficult problems with a good performance. In addition to that, the planner is well suited for the introduction of adaptive components, to let it adjust its random components in order to better address the environment where it moves. The paper is organized as follows. Section II illustrates the algorithm and section III provides details about simulations and numerical results. The theoretical soundness of the algorithm is sketched in IV, while conclusions are offered in section V.

II. THE RANDOM WALK ALGORITHM

For clarity we formalize the motion planning problem we wish to solve. We are given an n-dimensional space of configurations C and let C_{free} be the subset of free configurations of C. Let $x_{start} \in C_{free}$ and let $X_{goal} \subset C_{free}$. Our goal is to find a path connecting x_{start} with X_{goal}, i.e. a continuous function $f : [0,1] \to C_{free}$ such that $f(0) = x_{start}$ and $f(1) \in X_{goal}$.

Most of the randomized planners used so far uses uniform sampling over the entire configuration space (see however [3] for an example of planner using a Gaussian distribution). Instead, the algorithm we propose tries to find a solution by building a random walk growing from x_{start}. At every step a new sample is generated in the neighborhood of the last point in accordance with a Gaussian distribution[1] and if the segment connecting them lies entirely in C_{free}, the point becomes the last point in the walk, otherwise it is discarded. The basic version of the algorithm is illustrated in algorithm 1.

A crucial point of the algorithm consists of establishing the probability distribution used to generate samples. For this aim, the v_k used in line 4 indicates a n-dimensional Gaussian vector with zero−mean vector and covariance matrix Σ_k. As the search of the path is carried out in a possibly high dimensional configuration space whose topological shape is usually unknown, it can be non trivial to fix the values of Σ_k. For this reason it is possible to let the algorithm start with arbitrary values of Σ_k and let them evolve while it runs. As it will be shown

[†]with the Department of Information Engineering of the University of Padova while doing this work

[1]while using both uniform and Gaussian distribution yields effective planners, we will concentrate our discussion exclusively on Gaussian distributions. This because, in addition of exhibiting better performance and being more suited for adaptivity, it allows easier proofs of the probabilistic convergence of the algorithm (see section IV)

Algorithm 1 Basic Random Walk Based Motion Planner
1: $k \leftarrow 0$
2: $x_k \leftarrow x_{start}$
3: **while** NOT $x_k \in X_{goal}$ **do**
4:     Generate a new sample $s \leftarrow x_k + v_k$
5:     **if** the segment connecting x_k and s lies entirely in C_{free} **then**
6:         $k \leftarrow k+1$
7:         $x_k \leftarrow s$
8:     **else**
9:         discard the sample s
10:     **end if**
11:     Update the covariance matrix Σ
12: **end while**

in section IV the conditions required for the probabilistic convergence of the algorithm are very mild so that a wide variety of update rules can be used while setting the values. The adaptive rule we are using is the following (see [8]). Let

$$\overline{x}_k = \frac{1}{H} \sum_{i=k-H}^{k-1} x_i \quad (1)$$

be the average of the last H accepted samples. Given a square p-dimensional matrix M let m_{ij} be its generic element in position i,j. We define $diag(M)$ as follows

$$diag(M) = \begin{bmatrix} m_{11} & 0 & \cdots & \cdots & 0 \\ 0 & m_{22} & 0 & \cdots & 0 \\ \vdots & \cdots & \cdots & \cdots & \vdots \\ 0 & \cdots & 0 & m_{p-1,p-1} & 0 \\ 0 & \cdots & \cdots & 0 & m_{pp} \end{bmatrix} \quad (2)$$

i.e. the matrix obtained from M by setting to 0 all the elements outside the main diagonal. The update rule for Σ_k is then

$$\Sigma_k = \max\left(diag\left(\frac{1}{H}\left(\sum_{i=k-H}^{k-1} x_i x_i^T - H\overline{x}_k \overline{x}_k^T\right)\right), \Sigma_{MIN}\right) \quad (3)$$

where the function max returns a matrix whose generic element is the greatest of the corresponding elements in the two argument matrixes, and Σ_{MIN} is a diagonal constant matrix with strictly positive elements on the main diagonal. It then follows that Σ_k is diagonal too. In this preliminary stage, the choice of working with diagonal matrixes, i.e. to ignore correlations, has been done to get a simple implementation, but as illustrated in section IV the probabilistic completeness of the algorithm is guaranteed under less restrictive hypothesis. By dealing with diagonal matrix we get that the variance of Gaussian random vector used is the the variance of the last H accepted samples. Of course choosing a good value for H is important for getting a good algorithm performance and is definitely a point which needs more investigation.

One of the most important aspects of the random walk algorithm is that its performance is linear in the size of the number of samples generated, i.e. the time spent to generate a new sample is always the same and does not depend on the size of the previously generated samples set. This is different from most of the formerly developed algorithms, like probabilistic roadmaps (PRM) and rapidly exploring random trees (RRT), where the performance is usually quadratic in the number of generated samples (see however [1] for a version of the RRT algorithm where the time needed to generate a new sample is logarithmic in the number of already generated samples).

A natural technique to speed up the termination process is to let the algorithm perform a bidirectional search, with two random walks being expanded, one growing from the start point and the other from the goal region and to periodically verify if it is possible to join them. In order to avoid the examination of all the points, the connect trial is performed only between the last two generated points in the two walks. As already observed, the expedient of bidirectional search allows a substantial gain in the performance (see [11]). In addition to this, further opportunism can be introduced by letting the algorithm to periodically try to connect the last sample with the goal point (or the start point if in a bidirectional search we consider the walk growing towards the start point).

When the algorithm terminates, i.e. when the last generated point is in X_{goal}, the sequence of segments connecting x_i with x_{i+1} indeed makes up a path which solves the problem. However the quality of such path is extremely poor, because it includes a wide number of useless motions. Indeed the path obtained resembles a Brownian motion. For this reason a postprocessing stage is needed in order to smooth the generated trajectory. We use a *divide and conquer* algorithm similar to binary search (see algorithm 2). The *pushback* operations used therein append the given point to the end of the list.

Algorithm 2 Solution Smoothing
1: **SMOOTH**$(D, first, last, S)$
2: INPUT: D vector of points to smooth
3: INPUT: $first, last$ extremes of D to be optimized
4: INPUT/OUTPUT: S list with the smoothed sequence
5: **if** $first = last$ **then**
6:     $S.pushback(D[first])$
7: **else if** $first = last - 1$ **then**
8:     $S.pushback(D[first])$
9:     $S.pushback(D[last])$
10: **else if** the segment connecting $D[first]$ and $D[last]$ lies entirely C_{free} **then**
11:     $S.pushback(D[first])$
12:     $S.pushback(D[last])$
13: **else**
14:     SMOOTH$(D, first, (first+last)/2, S)$
15:     SMOOTH$(D, (first+last)/2+1, last, S)$
16: **end if**

The smoothing procedure is iterated until it is not able to further smooth the trajectory. It is worth noting that due to the good performance of the algorithm itself, the smoothing step turns out to be extremely fast and few iterative steps are needed. This will be clearly illustrated in section III.

III. SIMULATION DETAILS AND NUMERICAL RESULTS

The proposed algorithm has been developed and integrated into the MSL software developed at the university of Illinois ([14]), in order to compare it with state of the art motion planning algorithms over a set of standard problems. The MSL includes a wide range of variants of RRT based motion planners as well as the basic PRM motion planner. Many of the predefined problems included in the MSL involve complicated three dimensional objects moving in difficult environments (see figure 1 for an example). MSL performs collision detection using the PQP library developed at the University of North Carolina ([12]).

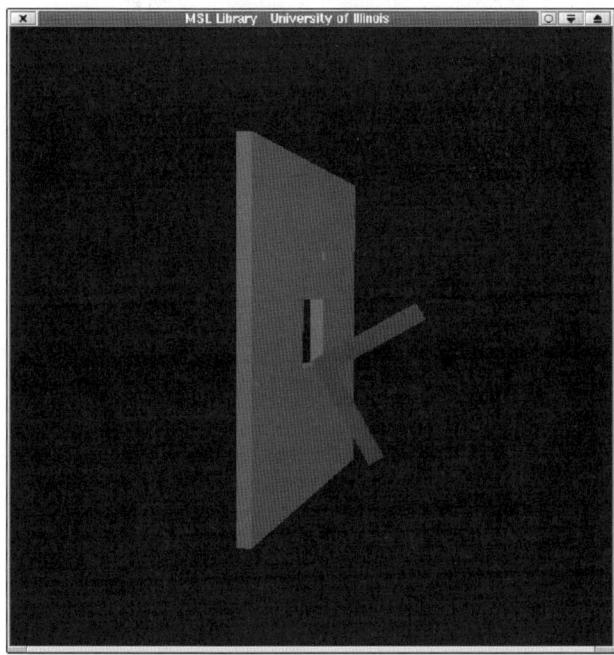

Fig. 1. Getting the L-shaped object from one side to the other side of the barrier is one of the many example problems provided with the MSL software. This example is called 3drigid3.

We first analyze the performance of the various steps of the proposed algorithm. All the numerical results concerning the random walk planner illustrated in this section refer to the bidirectional opportunistic planner previously described. The computer used is a Sun Ultra 10 workstation, working at 450 Mhz and with 256 Mbytes of RAM. History size for adaptivity, i.e. H, has been fixed to 10, while the square roots of diagonal values of Σ_{MIN} are set to one fifth of the difference between the maximal and minimal values which can be assumed by the corresponding degree of freedom.

Table I reports the data relative to 9 of the standard environments provided with the MSL. For lack of space we can not give an indepth description of every environment, but we use the same names given in the MSL, so the reader can refer to its documentation. We just state that the first 4 examples involve the model of a car with 3 degrees of freedom moving in a 2 dimensional environment (see figure 2), while the others refer to three dimensional objects moving in 3 dimensional environments (with 6 degrees of freedom). Column T_1 is the time spent to find the solution (seconds) and column N_1 is the number of samples in such solution, while column T_S is the time spent to smooth the solution and N_S is the number samples in the smoothed solution. T_{tot} is the sum of T_1 and T_s.

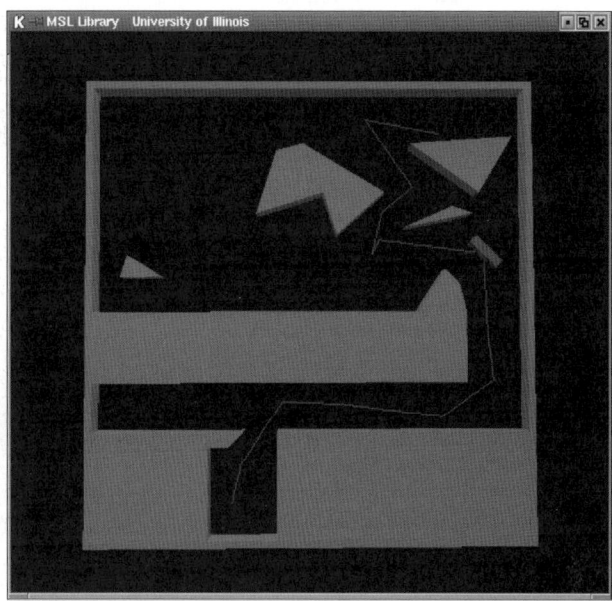

Fig. 2. Car2, one of the environments involving a car moving in a 3d environment

Environment	T_1	N_1	T_S	N_S	T_{tot}
Car 1	0.106	91.81	0.021	8.34	0.127
Car 2	1.419	311.92	0.154	16.14	1.566
Car 3	0.080	25.46	0.039	7.82	0.119
Car 4	0.013	22.24	0.006	7.76	0.019
Cage	26.54	512.84	2.22	14.32	28.76
Wrench	9.10	138.46	2.37	23.3	11.46
3drigid 1	0.236	896	0.133	5.94	0.369
3drigid 2	52.17	3940.7	1.36	15.27	52.53
3drigid 3	104.02	8247.2	0.453	7.15	104.47

TABLE I

TIME SPENT IN THE VARIOUS STEPS OF THE RANDOM WALK ALGORITHMS. DATA ARE AVERAGED OVER 100 TRIALS.

Two aspects are evident. First, the time spent by the algorithm is linear in the size of the generated random walk (compare columns T_1 and N_1). Second, even if the path produced by the random walk includes a great number of useless motions, significant improvements can be gained with the very fast post processing algorithm illustrated. We then compare the random walk algorithm with the basic PRM motion planner. Table II compares the overall time spent by both algorithms to find the solution. Again, time is expressed in seconds and data have been averaged over 100 trials.

It is well known that better versions of PRM exists (like lazy PRMs, see [2]), but it is however evident that adaptive random walks are indeed competitive. Finally we compare the perfor-

Environment	PRM	Random Walk
Car 1	0.369	0.127
Car 2	2.643	1.566
Car 3	0.858	0.119
Car 4	0.293	0.019
Cage	3442.76	28.76
Wrench	2526.26	11.46
3drigid 1	566.37	0.369
3drigid 2	871.96	52.53
3drigid 3	917.05	104.47

TABLE II

COMPARISON BETWEEN PRM AND RANDOM WALKS

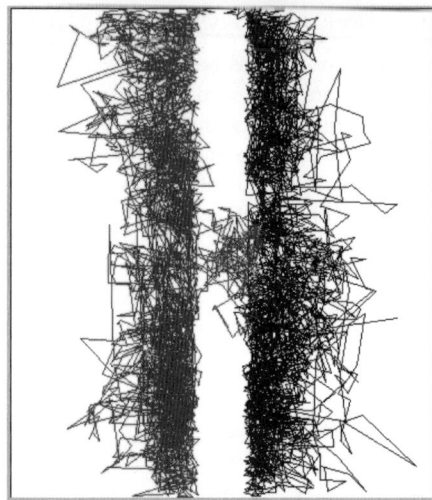

mance of the random walk planner with the RRTConCon algorithm provided in the MSL. RRTConCon is an RRT bidirectional greedy planner based on RRT ([11]). It has to be pointed out that in this case we compare the performance just for the last 5 problems which are holonomic problems. In fact, while RRTs are well suited for kinodynamic motion planning problems, the current version of the planner we are proposing does not handle kinodynamic constraints. Further investigation on this aspect is needed. So it is not fair to compare them over the examples involving the car, which is nonholonomic, as either our algorithm ignores the kinodynamic constraints or RRTs perform poorly when ignoring such constraints. Instead, while comparing random walks with PRM the comparison has been possible since they both ignore kinodynamic constraints.

Environment	RRT	Random Walk
Cage	6.29	28.76
Wrench	3.46	11.46
3drigid 1	0.25	0.369
3drigid 2	69.01	52.53
3drigid 3	89.55	104.47

TABLE III

COMPARISON BETWEEN RRTCONCON AND RANDOM WALKS

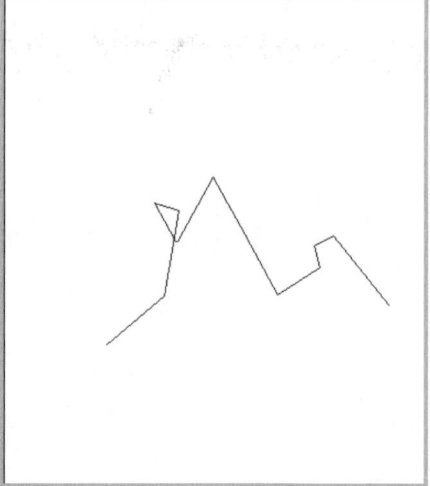

Fig. 3. Planned path before and after smoothing

In this case we can observe that in some cases RRTs are better, but there also exists environments where the opposite is true or where their performance is comparable. Of course no definitive conclusion can be drawn from a restricted set of examples. Moreover both algorithms have been run without trying to find out the optimal values for the many parameters involved.

We instead wish to outline that in spite of the wide skepticism against random walks, interesting performance emerges. Furthermore, by adding adaptive components it is possible to get a variable sampling resolution, which is indeed believed to be one of the most promising research direction. It should be outlined that in the case of single direction search adaptive random walks outperform RRT based motion planners. Finally, figure 3 illustrates an example of trajectory smoothing. It can be observed that even if the initial trajectory is extremely uneven, algorithm 2 succeeds in cutting out almost all useless motions.

IV. THEORETICAL FOUNDATIONS

In this section we provide the formalism and the proofs about the probabilistic convergence of the random walk algorithm introduced in the former sections. For lack of space not all the proofs are provided herein. The interested reader is referred to [5].

A. Preliminaries

We start defining a probability space as the triplet (Ω, Γ, η) where Ω is the sample space, whose generic element is denoted ω, Γ is a $\sigma-algebra$ on Ω and η a probability measure on Γ. Let C be $[0,1]^n$, whose generic element is denoted x and equipped with the $\sigma-algebra$ $B(C)$ consisting of all the Borel sets in $\mathbb{R}^n$ which are contained in C. We will also denote with $\mu(A)$ the Lebesque measure of a generic set A in $B(C)$. Moreover, we will denote $N_{x,\Sigma}(y)$ the Gaussian probability density as function of y, having mean x and covariance matrix Σ.

Definition 1: We define $B^+(C)$ as corresponding to the sets in $B(C)$ whose measure is strictly positive with respect to μ. Hereby, we will denote with C_{free} a fixed open set belonging to $B^+(C)$.

Definition 2: We call $\{x_1, x_2\}$ an admissible couple if $tx_1 + (1-t)x_2 \in C_{free}$ for $t \in [0,1]$.

We end this sub-section by introducing the function $g : C \times C \to C$ such that:

$$\begin{cases} g(x_1, x_2) = x_1 + x_2 & \text{if } \{x_1, x_1 + x_2\} \\ & \text{is an admissible couple} \\ g(x_1, x_2) = x_1 & \text{otherwise} \end{cases}$$

B. Adaptive Random Walk algorithm

We define the discrete time stochastic process $\{X_k\}_{k=0,1,2,...}$ on (Ω, Γ, η) and taking values on C, by the following recursive formulas:

$$\begin{cases} X_0(\omega) = x_{start} & \text{with } x_{start} \in C_{free} \\ X_k(\omega) = g(X_{k-1}(\omega), v_k(\omega)) & \text{for k=1,2,...} \end{cases} \quad (4)$$

where for every k the random vector $v_k(\omega)$ is normal, zero-mean with covariance matrix $\Sigma_k(\omega)$ (the dependence of a random variable on ω will be hereby implicit). We call the stochastic process in eq. 4, Adaptive Random Walk (ARW).

Definition 3: We define the random vector H_k as corresponding to $\{X_0, X_1, ..., X_k\}$. We also denote with $\sigma(H_k)$ the $\sigma - algebra$ generated by H_k ([10]).

Assumption 4: For every k, v_k is independent from $\sigma(H_{k-1})$ once Σ_k is known. Moreover, every entry of Σ_k is a random variable which is measurable with respect to $\sigma(H_{k-1})$

It comes from Assumption 4, that there exists for every k a function $L_k : C^k \to \mathbb{R}^{n \times n}$ such that $\Sigma_k = L_k(H_{k-1})$. We call L_k the learning rule of the random walk at instant k.

Assumption 5: For every k, $\varepsilon_1 I_n \leq \Sigma_k \leq \varepsilon_2 I_n$ where I_n is the $n \times n$ identity matrix and $\varepsilon_1, \varepsilon_2$ are strictly positive numbers.

Definition 6: Given H_{k-1} and assigned $x \in C$ and $A \in B(C)$, we denote with $P^m_{k,H(k-1)}(x, A)$ the m-step transition kernel of the chain, i.e. once known the history of the chain until instant k, $P^m_{k,H(k-1)}(x, A)$ provides the probability that ARW takes value in A at instant $k+m$.

Definition 7: Assigned a set q belonging to $B^+(C)$ and contained in C_{free}, we denote with $V(q)$ the maximal open set of points $x \in C_{free}$ such that for every y belonging to q, $\{x,y\}$ is an admissible couple.

We point out that the definition of $V(q)$ is similar to the one defined in [4].

C. Probabilistic completeness of the Adaptive Random Walk algorithm

Lemma 8: Let q a set belonging to $B^+(C)$ and contained in C_{free}. Then, given an arbitrary $D \in B^+(C)$ contained in $V(q)$, there exists a strictly positive m such that for every $x \in V(q)$, k and H_{k-1}:

$$P^2_{k,H_{k-1}}(x, D) \geq m$$

Proof: By definition of $V(q)$, we have that for every $x \in V(q)$, k and H_{k-1}:

$$P^1_{k,H_{k-1}}(x, q) \geq \int_q N_{x,\Sigma_k}(y) dy \geq l\mu(q) \doteq m_1$$

where l is a real number which uniformly bounds from below the function to be integrated on q (note that the existence of l comes from Assumption 5). Moreover, we also have that for every $x \in q$, k and H_{k-1}:

$$P^1_{k,H_{k-1}}(x, D) \geq \int_D N_{x,\Sigma_k}(y) dy \geq l\mu(D) \doteq m_2$$

Combining the two inequalities, we have that for every $x \in V(q)$, k and H_{k-1}:

$$P^2_{k,D}(x, D) \geq m_1 m_2$$

Remark 9: It is easy to note that when $V(q)$ is convex, we could replace $P^2_{k,D}(x, D) \geq m$ with $P^1_{k,D}(x, D) \geq m$ in the statement of Lemma 8.

Definition 10: Let T and U be subsets of C_{free}. We say that T and U are adjacent (and we write $Adj(T, U)$) when the following holds:

$$\overline{T} \cap \overline{U} \cap C_{free} \neq \emptyset$$

Lemma 11: Let A_i and A_j belong to $B^+(C)$ and to C_{free}, such that $Adj(V(A_i), V(A_j))$. Then, there exists a strictly positive m, such that for every $x \in V(A_i)$, k and H_{k-1}:

$$P^3_{k,H_{k-1}}(x, V(A_j)) \geq m$$

Proof: Since $Adj(V(A_i), V(A_j))$, there exists a point

$$x_0 \in \overline{A_i} \cap \overline{A_j} \cap C_{free}$$

Then, since C_{free} is open, there exists a ball $B_\varepsilon(x_0)$ with radius ε and center x_0 which is entirely in C_{free}. We define the set $D_1 = V(A_i) \cap B_\varepsilon(x_0)$. Clearly, we have $D_1 \in B^+(V(A_i))$. Then, by applying Lemma 8, there exists a strictly positive constant m_1 such that for every $x \in V(A_i)$, k and H_{k-1}:

$$P^2_{k,H_{k-1}}(x, D_1) \geq m_1$$

Let $D = V(A_j) \cap B_\varepsilon(x_0)$. Again, by applying Lemma 8 and by considering that $B_\varepsilon(x_0)$ is convex, there exists a strictly positive constant m_2, such that for every $x \in D_1$, k and H_{k-1}:

$$P^1_{k,H_{k-1}}(x, D) \geq m_2$$

Composing the two inequalities, we have that for every $x \in V(A_i)$, k and H_{k-1}:

$$P^3_{k,H_{k-1}}(x, D) \geq m_1 m_2$$

which completes the proof.

Definition 12: A triplet $\{C_{free}, x_{start}, x_{goal}\}$ is a solvable instance of the motion planning problem if the set

$$S(C_{free}, x_{start}, x_{goal}) = \{f : [0,1] \to C_{free}$$

$$\text{such that } f(0) = x_{start}, f(1) = x_{goal}, f \in C^0\}$$

is not empty.

Definition 13: Let $\{C_{free}, x_{start}, x_{goal}\}$ a solvable instance of the motion planning problem, let

$f \in S(C_{free}, x_{start}, x_{goal})$ and let A be a subset of C_{free}. We say that a solution f crosses A once if the set

$$t_a = \{t \in [0,1] \text{ such that } f(t) \in A\}$$

is an interval (either open, closed or half open).

Theorem 14: Let $\{C_{free}, x_{start}, x_{goal}\}$ be a solvable instance of the motion planning problem. Let us suppose that there exists a finite sequence $A_1, A_2, ..., A_k$, where every A_i belongs to $B^+(C)$ and is contained in C_{free}, such that

$$C_{free} = \bigcup_{i=1}^{k} V(A_i)$$

Then, there exists $f \in S(C_{free}, x_{start}, x_{goal})$ that crosses at most once $V(A_i)$ for $i = 1, 2, ..., k$.

Proof: Omitted.

Theorem 15: Let C_{free} be connected and such that there exists a finite sequence $A_1, A_2, ..., A_k$, where every A_i belongs to $B^+(C)$, is contained in C_{free} and

$$C_{free} = \bigcup_{i=1}^{k} V(A_i)$$

Then, for each $x_{start} \in C_{free}$ and X_{goal} belonging to $B^+(C)$ and to C_{free}, the algorithm ARW started in x_{start} will reach X_{goal} with probability 1.

Proof: Without loss of generality we suppose that X_{goal} is entirely included in one of the sets of the sequence $V(A_1), V(A_2), ..., V(A_k)$ denoted $V(A_l)$. In the light of Lemma 8 and 11, there exists a strictly positive m such that for every $V(A_i)$ and $V(A_k)$ with $Adj(V(A_i), V(A_k))$ and for every $x \in V(A_i)$, k and H_{k-1}:

$$P^3_{k,H_{k-1}}(x, V(A_k)) \geq m$$

and for every $x \in V(A_l)$, k and H_{k-1}:

$$P^2_{k,H_{k-1}}(x, X_{goal}) \geq m$$

This, in addition with Theorem 14, allows us to conclude that there exists constants h and s, independent from $x \in C_{free}$, k and H_{k-1} such that

$$\sum_{r=1}^{h} P^r_{k,H_{k-1}}(x, X_{goal}) \geq s > 0$$

It follows that the probability that ARW has never entered the set X_{goal} after Zh steps is less or equal to $(1-s)^Z$. Clearly, when Z diverges, this probability goes to zero.

V. Conclusions and Future Work

We introduced a random walk based adaptive motion planner. The algorithm builds a random walk according to a Gaussian sampling over the configuration space. As the produced path can be very uneven, an efficient post processing step is used, yielding a fast smoothing of the produced path. A simple but effective technique for letting the algorithm adapt the parameters of the random distribution used have been devised. Of course, it could also be possible to use more refined techniques, for example using correlation of the last samples, i.e. to use a non diagonal Σ_k. A sketch of the theoretical soundness of the algorithm has been provided. Numerical results illustrate that indeed for some problems the adaptive random walk planner appears to be competitive with previously developed algorithms. In the near future we plan to apply the algorithm to bioinformatics problems like protein folding. In that context, we expect the adaptivity to be better exploited, thanks to the continuous nature of the energetic levels concerning proteins instead of the boolean nature of the configuration space found in robot motion planning.

Acknowledgments

We thank professor Steve LaValle for making available the MSL software and Peng Cheng for useful hints.

References

[1] A. Atramentov and S.M. LaValle. Efficient nearest neighbor searching for motion planning. In *Proceedings of the IEEE Conference on Robotics and Automation*, pages 632–637, Washington, May 2002.

[2] R. Bohlin and L.E. Kavraki. Path planning using lazy prm. In *Proceedings of the IEEE International Conference on Robotics and Automation*, pages 1469–1474, Seoul, May 2001.

[3] V. Boor, M.H. Overmars, and A.F. van der Stappen. The gaussian sampling theory for probabilistic roadmap planners. In *Proceedings of the IEEE International Conference on Robotics and Automation*, pages 1018–1023, Detroit, May 1999.

[4] M. Branicky, S. M. LaValle, K. Olsen, and L. Yang. Deterministic vs. probabilistic roadmaps. Submitted to IEEE Transactions on Pattern Analysis and Machine Intelligence.

[5] S. Carpin. *Advanced techniques for randomized robot motion planning*. PhD thesis, The University of Padova, December 2002.

[6] S. Carpin and E. Pagello. Exploiting multi-robot geometry for efficient randomized motion planning. In Maria Gini et al., editor, *Intelligent Autonomous Systems 7*. IOS Press, 2002.

[7] P. Cheng and S.M. LaValle. Deterministic resolution complete rapidly-exploring random tree. In *Proceedings of the IEEE International Conference on Robotics and Automation*, Washington, May 2002.

[8] H. Haario, E. Saksman, and J. Tamminen. Adaptive proposal distribution for random walk metropolis algorithms. *Computational Statistics*, 14, 1998.

[9] L.E. Kavraki, P. Švestka, J.C. Latombe, and M.H. Overmars. Probabilistic roadmaps for path planning in high-dimensional configuration spaces. *IEEE Transactions on Robotics and Automation*, 12(4):566–580, 1996.

[10] B. Øksendal. *Stochastic Differential Equations*. Springer, 1998.

[11] J.J. Kuffner and S.M. LaValle. Rrt-connect: An efficient approach to single-query path planning. In *Proceedings of the IEEE Conference on Robotics and Automation*, pages 995–1001, San Francisco, April 2001.

[12] E. Larsen, S. Gottschalk, M.C. Lin, and D. Manocha. Fast proximity queries with swept sphere volumes. Technical Report TR99-018, Department of Computer Science, University of N. Carolina, Chapel Hill, 1999.

[13] J.C. Latombe. Motion planning: A journey of robots, molecules, digital actors, and other artifacts. *The International Journal of Robotics Research - Special Issue on Robotics at the Millennium*, 18(11):1119–1128, 1999.

[14] S.M. LaValle. Msl - the motion strategy library software. http://msl.cs.uiuc.edu.

[15] S.M. LaValle and J.J. Kuffner. Randomized kinodynamic planning. *International Journal of Robotics Research*, 20(5):378–400, 2001.

[16] G. Song and N.M. Amato. Using motion planning to study protein folding. In *Proceedings the 5th International Conference on Computational Molecular Biology (RECOMB)*, pages 287–296, 2001.

FSW (Feasible Solution of Wrench) for Multi-legged Robots

SAIDA Takao, Yasuyoshi YOKOKOHJI, Tsuneo YOSHIKAWA

*Department of Mechanical Engineering, Graduate School of Engineering,
Kyoto University, Kyoto, 606-8501 JAPAN.
stlab@ieee.org,{yokokoji,yoshi}@mech.kyoto-u.ac.jp*

Abstract

In this paper, we focuse on a problem how to confirm a feasible condition of applied force to a multi-legged robot on rough terrain. The problem often appeals as a subject of walking stability criterion for a legged robot. ZMP (Zero Moment Point) is one of the well-known criteria. It shows the condition as footprints of a legged robot, but it cannot be defined on the rough terrain. Therefore, we suggest a new criterion FSW (Feasible Solution of Wrench), which gives the feasible condition even on the rough terrain from the viewpoint of "wrench" – a special representation of force screw. And we present two short examples of FSW for a biped robot, how to analyse the validity of ZMP on stairs and how to design a force trajectory on rough terrain.

1. Introduction

Multi-legged robots are expected to find ways into various works and to collaborate with humans in the future. But, as yet so far, it is a difficult problem of them to accomplish their stabile mobilities in any environment. ZMP (Zero Moment Point) [1] is one of famous stability criteria for a walking robot [2–5]. Its physical feature is known as "CoP (center of pressure)" between the ground and the feet of the robot [6]. But it cannot be defined when the robot moves on plural contact planes, for example, going up stairs or opening a door. To avoid the disadvantage of CoP (ZMP) in such situation, Kogami [7] suggested the enhanced ZMP constraint and Sugihara [8] proposed the ZMP on the virtual horizontal plane. But their validities or physical meanings have not been confirmed.

On the other hand, to analyse singularities of parallel robot manipulators or to measure qualities of multi-fingered robot hands, another force criterion – wrench – is often adopted [9–13]. Its definition is not depending on geometric features.

From the veiwpoint of wrench, we propose a new criterion *FSW (Feasible Solution of Wrench)* to confirm feasible conditions of applied forces to multi-legged robots even on rough terrain as well as on a single plane. This paper is organized as follows. Sec.2 introduces some important features of ZMP and wrench, and then, Sec.3 defines FSW. Sec.4 describes some characteristics of FSW. Finally, Sec.5 shows two examples of FSW.

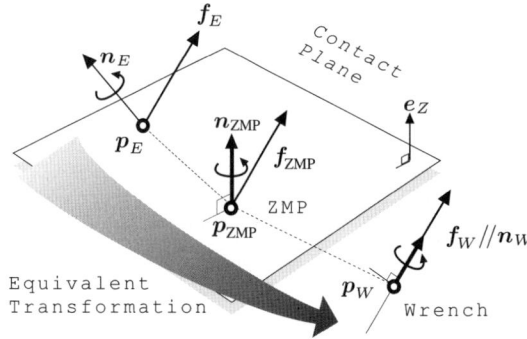

Fig. 1. Relationship among ZMP, wrench and a pair of force and moment on a contact plane. They are equivalent force screws.

2. ZMP and Wrench

In this section, we introduce two important indices, ZMP and wrench. To discuss following derivations simply, let p, f, n and e denote position, force, moment and direction cosine ($\mathcal{R}^{3\times 1}$), respectively.

2.1. ZMP (Zero Moment Point)

ZMP (Zero Moment Point) [1] is a special point where the applied moment is parallel to the normal of the contact plane (Fig.1). Now, suppose that force f_E and moment n_E are applied at point p_E on the contact plane. From this configuration, ZMP p_{ZMP} is specified as

$$p_{\text{ZMP}} = \frac{(e_Z \times n_E)}{(e_Z^T f_E)} + p_E \quad \in \mathcal{R}^{3\times 1} \quad (1)$$

$$n_{\text{ZMP}} = \frac{(e_E^T n_E)}{(e_E^T e_Z)} e_Z \quad \in \mathcal{R}^{3\times 1} \quad (2)$$

$$f_{\text{ZMP}} = f_E \quad (= f_E e_E) \quad \in \mathcal{R}^{3\times 1} \quad (3)$$

where f_{ZMP} and n_{ZMP} are the force and the moment observed at the ZMP. e_Z is the unit normal of the contact plane. $f_E (\in \mathcal{R}^1)$ and e_E are the magnitude and the direction cosine of f_E. In addition, the symbol "$\times$" and the upper-right subscript "T" mean the cross product and the transpose of vector, respectively.

ZMP is equivalent to "CoP (center of pressure)" on the contact plane [6]. It is only available within the minimum convex hull enclosing the contact region between the end-effector (foot) and the external object (ground). But its definition has following problems:

1) It cannot be clearly defined on rough terrain because e_Z cannot be identified in such condition.
2) It is not explicitly sensitive to any shearing force on the contact plane because the dominator of Eqn.(1) is insensitive against the shearing force.

2.2. Wrench

"Wrench" is a special representation of force screw [9, 10]. Its force and moment are parallel to each other (Fig.1). Its line of action is referred to as "wrench axis", p_W. From f_E, n_E and p_E, the axis p_W is specified as follows:

$$p_W = \frac{(e_E \times n_E)}{(e_E^T f_E)} + \eta\, e_E + p_E \quad \in \mathcal{R}^{3\times 1} \quad (4)$$

$$n_W = (e_E^T n_E)\, e_E \quad \in \mathcal{R}^{3\times 1} \quad (5)$$

$$f_W = f_E \quad (= f_E\, e_E) \quad \in \mathcal{R}^{3\times 1} \quad (6)$$

where f_W and n_W are the force and the moment observed on the wrench axis p_W. η is an arbitrary scalar parameter.

"Pitch", ρ_W, is another fundamental component of the wrench. It gives the ratio between f_W and n_W.

$$\rho_W = \frac{1}{f_E}(e_E^T n_E) \quad \in \mathcal{R}^1. \quad (7)$$

Substituting Eqn.(7) into (5), we get

$$n_W = \rho_W f_W. \quad (8)$$

Furthermore, wrench has following characteristics:
1) It can be defined even on plural contact planes.
2) The nearest point of its axis from the observation point can evaluate the effect of applied shearing forces on the contact plane.
3) Its axis passes through CoP (ZMP) if and only if e_E is parallel to e_Z or ρ_W is zero.
4) Its moment has minimum Euclidean norm in the equivalent representations of the force screw.

3. FSW(Feasible Solution of Wrench)

In this section, we suggest a new criterion *FSW (Feasible Solution of Wrench)*. FSW is derived from axis and pitch of wrench. It gives an insight for a multi-legged robot to keep its dynamic equilibrium condition on rough terrain.

3.1. Definitions of FSW

Suppose that f_E, n_E are the resultant force and moment at the observation point p_O, respectively (Fig.2). To discuss simply, let p_O locate at the origin of a coordinate system. Then, f_E and n_E are specified as

$$f_E = \sum_i f_{Ei} \quad (9)$$

$$n_E = \sum_i (n_{Ei} + p_{Ei} \times f_{Ei}) \quad (10)$$

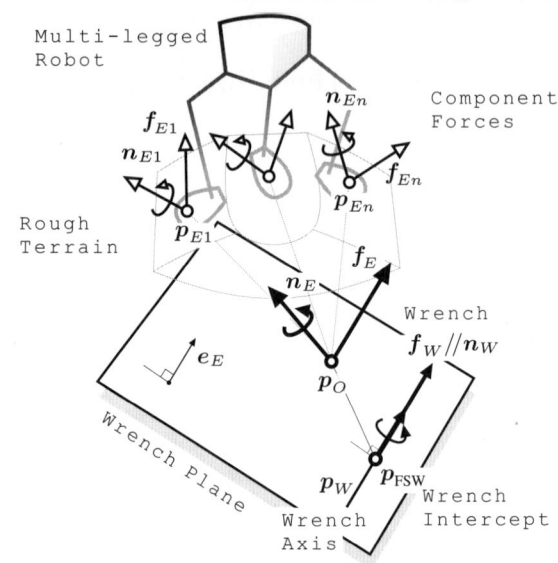

Fig. 2. Component forces, resultant force and moment at observation point, and wrench.

where f_{Ei} and n_{Ei} denote the applied force and moment to the multi-legged robot at the i-th application point p_{Ei}.

Substituting Eqn.(9),(10) into (4)~(6), we get the wrench for the resultant force screw. In Eqn.(4), let p_FSW be $p_W|_{\eta=0}$. The point p_FSW can be regarded as the intercept where a plane cuts out the wrench axis p_W. The plane has e_E as its normal, and it passes through p_E. So that, we can give two definitions about wrench.

Def. 1—Wrench Plane, Wrench Intercept: "Wrench Plane" is the plane which contains an observation point of resultant force and moment. Its normal equals to the wrench axis of the force screw. "Wrench Intercept" is the intercept where the plane cuts through the axis.

Substituting Eqn.(9) and (10) into p_FSW, we get

$$\begin{aligned} p_\text{FSW} &= \left(\sum_i \frac{f_{Ezi}}{f_E} p_{Exyi}\right) + \left(-\sum_i p_{Ezi}\frac{f_{Exyi}}{f_E}\right) \\ &\quad + \left(e_E \times \sum_i \frac{n_{Exyi}}{f_E}\right) + p_O \\ &= p_\text{n-FSW} + p_\text{t-FSW} + p_\text{m-FSW} + p_O \end{aligned} \quad (11)$$

where $p_\text{n-FSW}, p_\text{t-FSW}, p_\text{m-FSW}$ are the first, second and third element in the right side of p_FSW, respectively. In the same way, let ρ_FSW be the pitch at p_FSW. Substituting Eqn.(9) and (10) into (7), ρ_FSW is specified as

$$\rho_\text{FSW} = \left\{e_E^T\left(\sum_i p_{Exyi} \times \frac{f_{Exyi}}{f_E}\right)\right\} + \left(\sum_i \frac{n_{Ezi}}{f_E}\right). \quad (12)$$

Where e_E is also the direction cosine of the resultant force f_E in Eqn.(9). In addition, we define the bias filter matrix

$E_E = [I_{3\times3} - e_E e_E^T] \in \mathcal{R}^{3\times3}$. It gives

$$f_{Ezi} = e_E^T f_{Ei}, \qquad f_{Exyi} = E_E f_{Ei} \qquad (13)$$
$$p_{Ezi} = e_E^T p_{Ei}, \qquad p_{Exyi} = E_E p_{Ei} \qquad (14)$$
$$n_{Ezi} = e_E^T n_{Ei}, \qquad n_{Exyi} = E_E n_{Ei} \qquad (15)$$

where $I_{3\times3}$ is the $\mathcal{R}^{3\times3}$ identity matrix.

Then, we define FSW, the intercept p_{FSW}, its components $p_{n\text{-}FSW}, p_{t\text{-}FSW}, p_{m\text{-}FSW}$, and the pitch ρ_{FSW} as follows.

Def. 2—FSW: "FSW (Feasible Solution of Wrench)" is the set of the feasible reaction wrenches which can act as the external forces on a multi-legged robot. The wrench is specified from the component forces and moments acting between external objects and the end-effectors of the robot. It is not restricted from force-transmission mechanisms between the objects and the robot, for instance, point contact, flat contact, plural surface contact or magnetic field action. But of course, it must keep constraints between them, for example, torque limit or frictional condition. And FSW is composed of following "i-FSW" and "p-FSW".

Def. 3—i-FSW: "i-FSW (intercept FSW)" is the set of feasible wrench intercepts. Its element is denoted as p_{FSW}.

Def. 4—n-FSW, t-FSW, m-FSW: "n-FSW (normal force FSW)", "t-FSW (tangential force FSW)" and "m-FSW (moment FSW)" are components of i-FSW. They denote the effects of normal component forces, tangential component forces and component moments on the wrench plane. Their elements are denoted as $p_{n\text{-}FSW}, p_{t\text{-}FSW}, p_{m\text{-}FSW}$. The sum of them represents p_{FSW}.

Def. 5—p-FSW: "p-FSW (pitch FSW)" is the set of feasible wrench pitches. Its element is denoted as ρ_{FSW}.

4. Characteristics of FSW

4.1. CoP (ZMP) Domain Compatiblility

In Eqn.(11), $p_{n\text{-}FSW}$ in n-FSW is a kind of CoP (ZMP). Its equation is similar to the problem to solve the center of gravity in distributed mass systems.

If there is no $f_{Ezi}f_{Ezj} < 0 (\forall i, j)$, $p_{n\text{-}FSW}$ is the CoP on the wrench plane. And the point is in the minimum convex hull which encloses the footprints projected onto the wrench plane (Fig.3). But, however, if there exists $f_{Ezi}f_{Ezj} < 0 (i \neq j)$, $p_{n\text{-}FSW}$ is an external dividing point. Its domain is not closed. This case usually takes place in the grasping operation of multi-fingered robots.

Unless the resultant force f_E is canceled, n-FSW exists.

4.2. Effect of Shearing Forces and Steps

$p_{t\text{-}FSW}$ in t-FSW indicates the effect that shearing forces act at uneven heights on the feet. This kind of domain does not appear in the set of CoP (ZMP). As its special condition, if there exists p_{Ez} such that $p_{Ez} = p_{Ezi}(\forall i)$, $p_{t\text{-}FSW}$

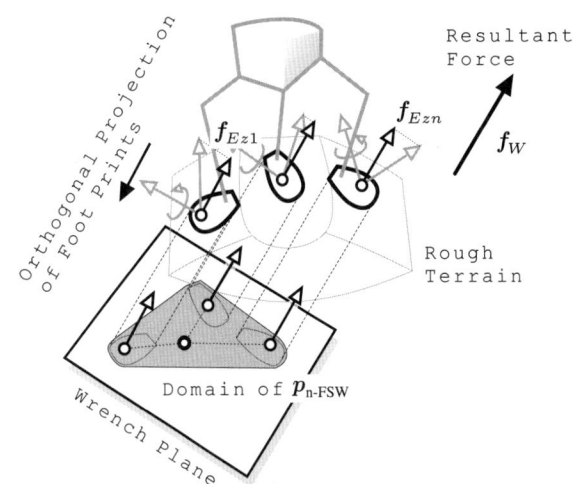

Fig. 3. n-FSW of a multi-legged robot is similar to the domain of ZMP if there is no $f_{Ezi}f_{Ezj} < 0(\forall i, j)$ in Eqn.(11).

becomes 0. Substitute Eqn.(9),(13) into (11), we get

$$\begin{aligned} p_{t\text{-}FSW} &= -\sum_i p_{Ezi}\frac{f_{Exyi}}{f_E} = -\frac{p_{Ez}}{f_E}E_E\sum_i f_{Ei} \\ &= -\frac{p_{Ez}}{f_E}E_E f_E e_E = 0. \end{aligned} \qquad (16)$$

We define Eqn.(16) is "zero t-FSW". It also represents a special condition between the resultant force and the component forces. This problem will be mentioned in Sec.5.2.

4.3. Normal Vector of Each Contact Plane

Now, we assume that each contact surface is flat between the external object and the end-effector of the multi-legged robot. And we also assume that each application point p_{Ei} is CoP (ZMP) on the surface. Let the normal vector of the each contact surface be e_{Ei}. Then, the applied component moment n_{Ei} can be simplified as

$$n_{Ei} = n_{Ei} e_{Ei}. \qquad (17)$$

Eqn.(17) and (11) simplify $p_{m\text{-}FSW}$ as

$$p_{m\text{-}FSW} = \sum_i \frac{n_{Ei}}{f_E} e_E \times e_{Ei}. \qquad (18)$$

This shows the relationship between the resultant force and the topographic feature of the external object. And $p_{m\text{-}FSW}$ can be independent of $p_{n\text{-}FSW}$ and $p_{t\text{-}FSW}$.

4.4. Observation Invariance

FSW is invariant for the observation point p_O because the wrench – the element of FSW – is invariant for the point. It is shortly proved as follows. In Eqn.(1) and (2), the dominator marks its maximum when the normal e_Z is parallel to the resultant force f_E. In Eqn.(4) and (5), the wrench always keeps such condition because the normal of the wrench plane is defined parallel to the force. So that, the domain of FSW keeps its shape even if p_O was changed.

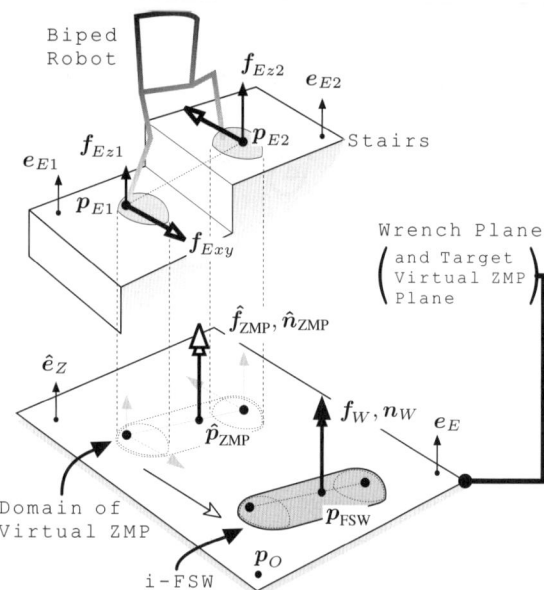

Fig. 4. Going up stairs with a couple of two reciprocal shearing forces.

4.5. Stability and Projected Footprints

Although p_{FSW} is out of the footprints projected onto the wrench plane, it does not mean any locomotion instability due to the remainder of t-FSW and m-FSW. But, however, even if p_{FSW} is within the i-FSW, the legged robots might be falling down when they break the restriction of the p-FSW. The detail of this subject is our future work.

5. Application of FSW

As mentioned in Sec.2.1, ZMP cannot be defined on rough terrain. Therefore, instead of ZMP, we will apply FSW and show two examples for a biped – going up stairs and walking on rough terrain.

5.1. Going up/down stairs

To go up stairs with a ZMP criterion, biped robots need some techniques. One of the convenient techniques is to make a "virtual ZMP" on a "virtual plane" [7, 8]. But the ZMP cannot estimate any effect of shearing force, and it is not invariant for the plane.

In this section, we show the difference between FSW and virtual ZMP on stairs in terms of tumbling error moment. In Fig.4, suppose following items on each foot:

1) Each foot has flat contact on each stair.
2) e_{Ei} is the normal vector of each stair.
3) Application point p_{Ei} is CoP (ZMP) and fixed.
4) Moment n_{Ei} equals to $\mathbf{0}$.
5) Shearing force f_{Exyi} is fixed.
6) Two shearings are canceled ($f_{Exy1} + f_{Exy2} = \mathbf{0}$).

To compute the two application points, i-FSW p_{FSW} and virutal ZMP $\hat{p}_{\mathrm{ZMP}}$, let $f_{Exy} = f_{Exy1} = -f_{Exy2}$. Then, substituting f_{Exy} into Eqn.(11) denotes p_{FSW} as

$$p_{\mathrm{FSW}} = \left(\sum_{i=1}^{2} \frac{f_{Ezi}}{f_E} p_{Exyi} \right) + p_O$$
$$+ \left((p_{Ez2} - p_{Ez1}) \frac{f_{Exy}}{f_E} \right). \quad (19)$$

To compute the virtual ZMP [7, 8] (let it be $\hat{p}_{\mathrm{ZMP}}$), neglect the height of the application points on the target virtual plane. In this application, it means to neglect p_{Ezi} from Eqn.(1). Then, $\hat{p}_{\mathrm{ZMP}}$ is specified as

$$\hat{p}_{\mathrm{ZMP}} = \left(\sum_{i=1}^{2} \frac{f_{Ezi}}{f_E} p_{Exyi} \right) + p_O \quad (20)$$

where we assumed that the direction of resultant force e_E, the normal of stair e_{Ei} and the normal of the target virtual plane $\hat{e}_Z$ lie in the same direction.

In Eqn.(19) and (20), there is a deviation between p_{FSW} and $\hat{p}_{\mathrm{ZMP}}$ on the target plane. But it is inconsistent on the force screws at the points. This problem is specified as

$$f_W = \hat{f}_{\mathrm{ZMP}} (= f_E), \qquad n_W = \hat{n}_{\mathrm{ZMP}} \quad (21)$$

in other words, the following tumbling moment Δn does not appear in the performance of $\hat{p}_{\mathrm{ZMP}}$,

$$\Delta n = (p_{\mathrm{FSW}} - \hat{p}_{\mathrm{ZMP}}) \times f_E$$
$$= (p_{Ez2} - p_{Ez1})(f_{Exy} \times e_E). \quad (22)$$

Note that even if the biped is under control with such virtual ZMP criterion, Δn must be compensated only just by local feedback control of each joint. But FSW provides a good criterion even if the robot walks over the undulations of the stairs.

5.2. Irregular Terrain

ZMP control scheme can effectively stabilize the locomotion of biped robots on single plane [2–4] but not on rough terrain. Mainly, this problem lies on the design of force trajectory with the robot. So that, in this section, we show an application of the force trajectory design as an inverse problem of p_{FSW} in f_E and n_E.

Now, we assume followings to solve the inverse problem on the rough terrain by FSW:

1) Each support foot lies flat on the rough terrain.
2) Application point p_{Ei} is CoP (ZMP) and fixed.
3) e_{Ei} is the normal vector of each contact plane and it is outward from the rough terrain.
4) Coulomb friction model is adopted between the feet and the terrain. And the friction model is static one.

And the schematic view of this subject is shown in Fig.5.

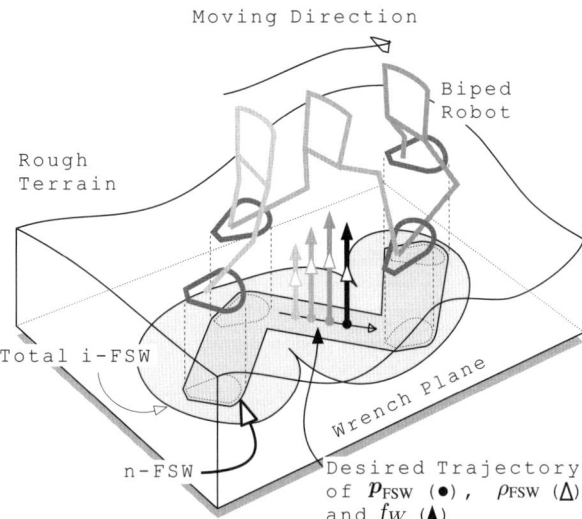

Fig. 5. Schematic view of walking robot on rough terrain with FSW.

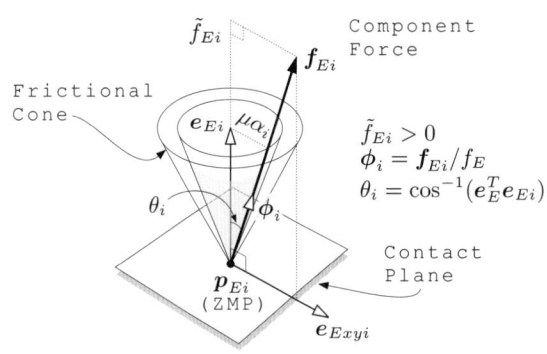

Fig. 6. Parameters of frictional cones on the contact plane.

Let $\phi_i \in \mathcal{R}^{3\times 1}$ be (f_{Ei}/f_E). Then, ϕ_i specifies e_E, $p_{\text{t-FSW}}$ and $p_{\text{n-FSW}}$ in Eqn.(11) as follows:

$$e_E = f_E/f_E = \sum_i f_{Ei}/f_E = \sum_i \phi_i \quad (23)$$

$$p_{\text{t-FSW}} = \sum_i -p_{Ezi} E_E \phi_i \quad (24)$$

$$p_{\text{n-FSW}} = \sum_i p_{Exyi} e_E^T \phi_i . \quad (25)$$

With respect to the biped robot, the inverse problem of Eqn.(24) with (23) for ϕ_i can be solved as

$$\begin{bmatrix} \phi_1 \\ \phi_2 \end{bmatrix} = \begin{bmatrix} \lambda e_E \\ (1-\lambda)e_E \end{bmatrix} + \frac{1}{(p_{Ez1}-p_{Ez2})} \begin{bmatrix} p_{\text{t-FSW}} \\ -p_{\text{t-FSW}} \end{bmatrix} \quad (26)$$

while λ is an arbitrary scalar parameter. The parameter can be specified from another parameter ζ:

$$\lambda = \frac{\zeta - p_{Ez2}}{p_{Ez1} - p_{Ez2}} \quad (27)$$

where ζ represents the virtual height of application point from the wrench plane. Therefore, λ shows the dividing ratio between the two application points p_{E1} and p_{E1}.

Note that even if $p_{Ez1} = p_{Ez2}$ (it means $p_{\text{t-FSW}} = 0$ as mentioned in Sec.4.2), ϕ_i can be solved in Eqn.(26) and λ still conserves the meaning of the ratio. And also, the next equation (28) is always satisfied.

Substituting Eqn.(26) into (25), $p_{\text{n-FSW}}$ is simplified as

$$p_{\text{n-FSW}} = \lambda p_{Exy1} + (1-\lambda) p_{Exy2} . \quad (28)$$

After all, Eqn.(28) means that, whatever $p_{\text{t-FSW}}$ is, only λ explicitly decides $p_{\text{n-FSW}}$.

As mentioned at the top of this section, we assume Coulomb friction model on each foot. The model restricts the applied forces on the feet. This condition is shown in Fig.6. Consider static translational friction at p_{Ei}, the component force f_{Ei} can be specified as

$$f_{Ei} = f_E \phi_i = \tilde{f}_{Ei}(e_{Ei} + \mu \alpha_i e_{Exyi}) \quad (29)$$

where e_{Exyi} is the unit orthogonal vector to e_{Ei}. $\tilde{f}_{Ei}$ is the pushing intensity. μ is the translational friction coefficient. And $\tilde{f}_{Ei}$ should be $\tilde{f}_{Ei} \geq 0$, α_i should be $0 \leq \alpha_i \leq 1$.

Substitute Eqn.(26) into (29), we get the frictional condition on the each foot as follows:

$$\alpha_i = \frac{1}{\mu} \frac{e_{Exyi}^T \{\lambda_i e_E + \Delta p_{Ezi} p_{\text{t-FSW}}\}}{e_{Ei}^T \{\lambda_i e_E + \Delta p_{Ezi} p_{\text{t-FSW}}\}} \quad (30)$$

where $\Delta p_{Ez1} = (p_{Ez1} - p_{Ez2})$, $\Delta p_{Ez2} = (p_{Ez2} - p_{Ez1})$, $\lambda_1 = \lambda$, $\lambda_2 = (1-\lambda)$.

Now, we assume another item to solve simply the inverse problem in this application.

- $p_{\text{t-FSW}} = 0$ (zero t-FSW).

It simplifies ϕ_i, f_{Ei} and α_i in Eqn.(26), (9) and (30) as:

$$\phi_i = \lambda_i e_E \quad (31)$$
$$f_{Ei} = \lambda_i f_E = \lambda_i f_E e_E \quad (32)$$
$$\alpha_i = \frac{1}{\mu} \frac{e_{Exyi}^T e_E}{e_{Ei}^T e_E} = \frac{1}{\mu} \tan \theta_i . \quad (33)$$

where θ_i is the angle of friction. It interprets the limit of α_i into $0 \leq \cos \theta_i \leq \cos(\tan^{-1} \mu)$.

To keep $\tilde{f}_{Ei} \geq 0$ in Eqn.(29), it requires that λ_i must be $0 \leq \lambda_i \leq 1$ in Eqn.(32). So that, ζ should be

$$\min_i p_{Ezi} \leq \zeta \leq \max_i p_{Ezi} . \quad (34)$$

This constraint and Eqn.(28) set $p_{\text{n-FSW}}$ within the minimum convex hull of the footprints of the biped robot. Moreover, Eqn.(32) requires that the resultant force f_E should be parallel with the each component force f_{Ei} acting on the foot of the robot (Fig.7).

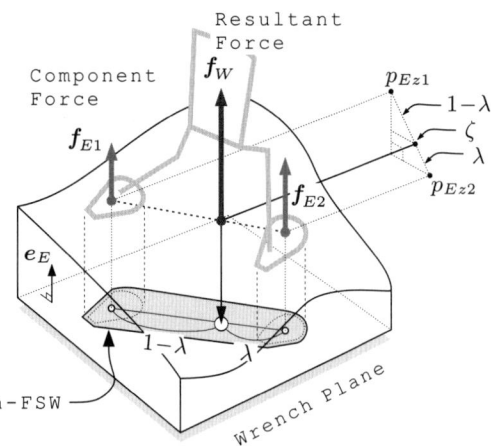

Fig. 7. Relationship between force distribution and parameter λ. The applied forces are parallel to each other.

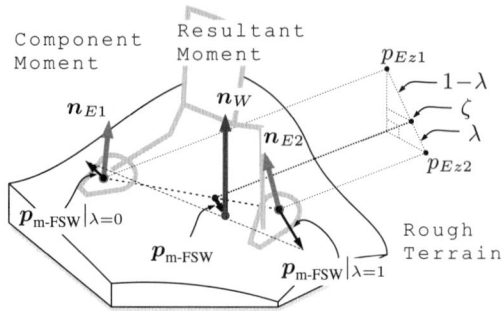

Fig. 8. Relationship between moment distribution and parameter λ.

Next, consider static rotational friction at p_{Ei}. The component moment n_{Ei} acting on the each foot can be specified as

$$n_{Ei} = \beta_i \nu \tilde{f}_{Ei} e_{Ei} \quad (35)$$

where, ν is the rotational friction coefficient. β_i should be $-1 \leq \beta_i \leq 1$. And remember that we assumed the point p_{Ei} as the CoP(ZMP) on the each foot.

Eqn.(32) gives n_{Ei} as

$$n_{Ei} = \lambda_i \beta_i f_E \left(\nu e_{Ei} e_{Ei}^T e_E \right) \quad (36)$$

where $\tilde{f}_{Ei} = \lambda_i e_{Ei}^T e_E$ is easily specified from Eqn.(29) and (32). Immediately, as to the resultant moment of the wrench, n_W in Eqn.(5), we get

$$n_W = \left\{ e_E^T \sum_i (n_{Ei} + p_{Ei} \times f_{Ei}) \right\} e_E$$
$$= \sum_i \lambda_i \beta_i f_E \left(\nu e_E (e_{Ei}^T e_E)^2 \right) . \quad (37)$$

In the same way, in Eqn.(11) and (12), $p_{\text{m-FSW}}$ and ρ_{FSW} are specified as follows:

$$p_{\text{m-FSW}} = \sum_i \lambda_i \beta_i \left(\nu (e_E \times e_{Ei}) (e_E^T e_{Ei}) \right) \quad (38)$$

$$\rho_{\text{FSW}} = \sum_i \lambda_i \beta_i \left(\nu (e_E^T e_{Ei})^2 \right) . \quad (39)$$

Fig.8 shows the schematic view of $p_{\text{m-FSW}}$ and n_W.

The results specify p_{FSW}, f_W, n_W. It means that we can treat a lot of cases on trajectory designs for biped robots on rough terrain. And for example, to employ the results, let λ be a function of time and then solve a locomotion of the robot on $\lambda(t)$. Its scheme is similar to the force distribution planning in 2D locomotion proposed by Pratt [5]. Whereas, a simulation or an experiment of this application will be our future work.

6. Conclusion

In this paper, we have suggested a new criterion *FSW (Feasible Solution of Wrench)* for multi-legged robots. The criterion shows the feasible condition of forces applied to the robot even on rough terrain, from the viewpoint of wrench. FSW is composed of some domains which have different physical meanings on the wrench plane: n-FSW (pressure), t-FSW (shearing and step), m-FSW (moment) and p-FSW (force-moment ratio).

We have shown the usefulness of FSW by explaining two typical applications for a biped robot on rough terrain – going up stairs and walking on the terrain. Our future research will be focused on how to generate and stabilize the motion of multi-legged robots with FSW over rough terrain.

References

[1] Miomir Vukobratović, et al. : "Contrbution to the synthesis of biped gait," IEEE Trans. on Bio-Med. Eng., vol.16, no.1 (1969).
[2] Kazuo Hirai, et al. : "The development of honda humanoid robot," IEEE Int. Conf. on Robotics and Automation (1998).
[3] Jin'ichi Yamaguchi, et al. : "Development of a bipedal humanoid robot - control method of whole body cooperative dynamics biped walking -," IEEE Int. Conf. on Robotics and Automation (1999).
[4] Koichi NISHIWAKI, et al. : "Online mixture and connection of basic motions for humanoid walking control by footprint specification," IEEE Int. Conf. on Robotics and Automation, pp.21–26 (2001).
[5] Jerry E. Pratt : Exploiting Inherent Robustness and Natural Dynamics in the Control of Bipedal Walking Robots. Ph.D.Thesis, Comp. Sci. Dept., MIT, Cambridge, Mass. (2000).
[6] Ambarish Goswami : "Postural stability of biped robots and the foot-rotation indicator (FRI) point," Int. J. Robot. Research, vol.18, no.6, pp.523–533 (1999).
[7] Satoshi Kogami, et al. : "A fast generation method of a dynamically stable humanoid robot trajectory with enhanced ZMP constraint," IEEE International Conference on Humanoid Robotics (2000).
[8] Tomomichi Sugihara, et al. : "Realtime humanoid motion gereration through ZMP manipulation," IEEE Int. Conf. on Robotics and Automation, pp.1404–1409 (2002).
[9] Ferdinand P. Beer, et al. : Vector Mechanics for Engineers, STATICS (2nd ed.). McGraw-Hill (1972).
[10] M.S. Ohwovoriole, et al. : "An extension of screw theory," Journal of Mechanical Design, vol.103, pp.725–735 (1981).
[11] M. Teichmann : "A grasp metric invariant under rigid motions," IEEE Int. Conf. on Robotics and Automation, pp.2143–2148 (1996).
[12] P. L. McAllister, et al. : "An eigenscrew analysis of mechanism compliance," IEEE Int. Conf. on Robotics and Automation, pp.3308–3313 (2000).
[13] Curtis L. Collins, et al. : "On the duality of twist wrench distributions in serial and parallel chain robot manipulators," IEEE Int. Conf. on Robotics and Automation, pp.526–531 (1995).

Intelligent Control of an Experimental Articulated Leg for a Galloping Machine

Luther R. Palmer[1], David E. Orin[1], Duane W. Marhefka[1],
James P. Schmiedeler[2], Kenneth J. Waldron[3]

[1] Department of Electrical Engineering, The Ohio State University (palmer.216 / orin.1 / marhefka.2@osu.edu)
[2] Department of Mechanical and Industrial Engineering, University of Iowa (jschmied@engineering.uiowa.edu)
[3] Department of Mechanical Engineering, Stanford University (waldron@cdr.stanford.edu)

Abstract

Intelligent controllers are being used with increasing effectiveness on complex systems. This work verifies the effectiveness of fuzzy control, an intelligent method, on a single, articulated-leg that was designed to be used on a high-speed galloping quadruped. Intelligent methods are compared to other control methods in simulation and on the OSU DASH (Dynamic Articulated Structure for High-performance) leg. It is shown that the intelligent controllers outperform non-learning methods. Using fuzzy control, the OSU DASH leg performs stable hopping on a treadmill moving at 2.0 m/s.

1 Introduction

Biological systems of varying sizes and travel speeds use legs to traverse all types of terrain. Conversely, man-made wheeled and tracked vehicles cannot operate on approximately half of the Earth's land surface [1]. Wheeled vehicles, however, presently hold a significant speed advantage over legged systems. If legged vehicles are to reach their full potential as seen in the cheetah, as the fastest land animal, and the mountain goat for its speed over rough terrain, the feasibility of high speeds must be investigated.

For energy reasons, most mid-sized, four-legged mammals use the gallop at high speeds. It is also known that mammals transition from a trot to a gallop at a speed directly related to their body mass [2]. Although running machines have been developed [3, 4, 5, 6], none surpassed its transition speed to benefit from galloping.

Real-time control becomes another non-trivial issue as legged machines increase in speed. Intelligent control algorithms may require more processing power than traditional controllers but can yield better results on increasingly complex legged machines. Fuzzy control, a type of intelligent control, incorporates the user's heuristic knowledge of the system, which reduces the need for accurate system identification and kinematic modeling of the system.

Raibert developed the fastest four-legged machine built to date, a trotting and bounding machine capable of moving at 2.9 m/s [3], but this machine did not surpass Hegland and Taylor's predicted speed to save energy by galloping [2]. Raibert's machine used prismatic legs with pneumatic and hydraulic actuation and relied only on foot placement at touchdown and leg thrust during stance to govern the movement of the body.

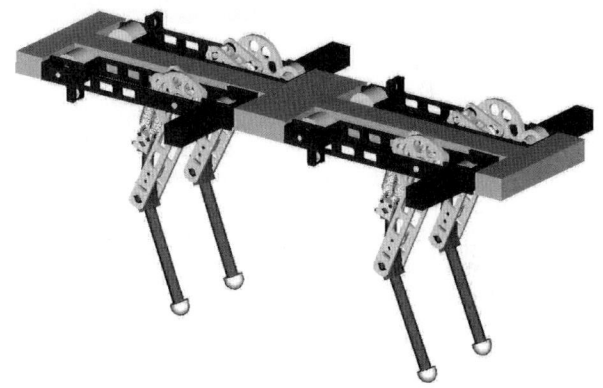

Figure 1: *Solid model of the quadruped design [8].*

Marhefka [7] simulated a planar quadruped that sustained speeds close to 7 m/s. This system used prismatic legs with electrical actuation and moved at speeds high enough to benefit from galloping. Marhefka also used a direct adaptive fuzzy control approach to stabilize the gallop. Schmiedeler [8] designed a quadruped (Fig. 1) with articulated legs that could also gallop at high speeds. Using articulated legs, the motion of the quadruped can be compared more closely to biological systems. The body is also designed with forward weight distribution to match the systems seen in nature. The leg contains springs to provide energy on recoil as described by Alexander [9]. Other motivations for the leg design were OLLIE [10] and the bow leg hopping robot at Carnegie Mellon University [11].

Marhefka simulated Schmiedeler's design in a single leg configuration that would eventually be used to test a prototype leg. The actual leg setup is shown in Fig. 2. Marhefka's simulated leg used a modified Raibert controller adapted for articulated legs and hopped at 5 m/s which satisfied the speed criterion. Marhefka noted several deficiencies with the Raibert controller and recommended the use of a fuzzy controller similar to the one used on his galloping quadruped.

The main objective of this work is to continue the efforts of Marhefka and Schmiedeler toward the assembly of a gal-

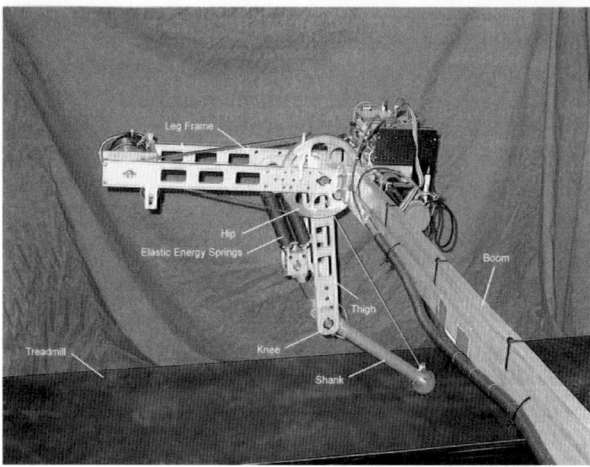

Figure 2: OSU DASH leg.

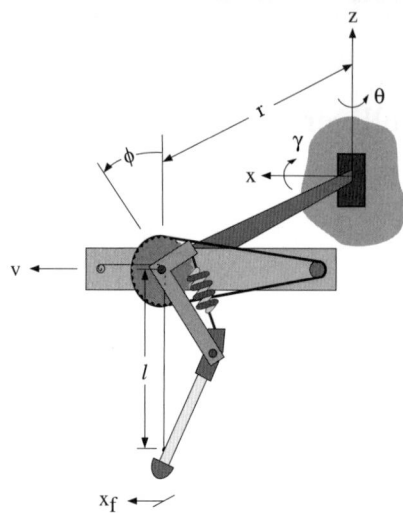

Figure 3: Kinematic diagram of the leg. A cable is actively controlled to retract the leg during flight, which extends the springs to add energy to the system. The energy is released when the foot is on the ground to offset the system losses.

loping quadruped. This includes 1) the development of an intelligent control system for the articulated leg in simulation and 2) the real-time control for an actual leg on a treadmill. This work also shows the usefulness of intelligent control on complex systems. A Levenberg-Marquardt (LM) learning algorithm is implemented with Marhefka's modified Raibert controller to improve the performance of the system. A fuzzy controller, which does not rely on any kinematic or dynamic analysis of the system, is also implemented.

The next section details the leg model used in this work and Section 3 develops the control methods used. Section 4 describes the system in simulation with the results shown for both controllers. The OSU DASH (Dynamic Articulated Structure for High-performance) leg is then described with the corresponding results.

2 Galloping Leg Model

The four primary leg assemblies are the thigh, shank, springs, and frame. Referring to Fig. 3, the L-shaped link which rotates about the hip is called the thigh. The angle between the lower part of the thigh and the vertical axis is described by ϕ. The link which rotates about the knee and is attached to the foot is called the shank. The angle between the upper and lower part of the shank is 25°. The thigh, shank, and leg extension springs form a closed kinematic loop. The leg has a nominal height of 64 cm and a mass of 13 kg. A cable, which effectively runs from the hip to the foot, is actively controlled to retract the leg during flight. This retraction extends the springs to store energy, which is used to overcome energy loss in the system. This energy is released when the foot is on the ground. The measurement of the cable length, l, from the hip to the foot connection can be used in kinematic equations to compute the length of the springs, which is directly related to the energy in the system. The thigh is connected to the frame by a revolute hip joint. For the one-legged system, the frame will be referred to as the body, which has a forward velocity, v.

A boom is rigidly connected to the frame to prevent lateral movement and pitch. The base of the boom is connected to a wall by a mount allowing rotation about a horizontal axis, measured as γ. The boom attachment to this mount allows rotation about a vertical axis, measured as θ. This rotation is referred to as the boom swing angle and gives the boom two degrees of freedom at the wall mount. Although the motion of the leg is circular around the radius of the boom, planar movement during one stride can be approximated when a long boom is used. The boom used in this project is 2.46 m. The forward foot position, x_f, is a measurement of the foot position relative to the boom coordinate system. Marhefka [7, pp. 148-150] provides a more detailed description of the system.

A 6-foot treadmill is positioned under the leg. The body velocity is measured with respect to the ground. The projected velocity of the body is computed as this body velocity plus the velocity of the treadmill.

3 Intelligent Control

The goal of this project is to achieve velocity control of the leg. Raibert [3] showed that the trajectory of the leg is determined by the horizontal placement of the foot relative to the hip and the energy in the springs at touchdown. These two parameters are dictated by the thigh angle and cable length, which are the two outputs of the controller. The system controller is called once per cycle at the top of body flight (TOF) and these outputs are used as setpoints for individual joint controllers between control cycles. Joint position and velocity are considered to be continuously controlled when the leg is in the air, but leg velocity and height are only controlled once per cycle. The inputs to the discrete controller are the present and desired body height and velocity.

Because the leg jumps on a treadmill, a simple proportional control algorithm is used to maintain the position of the leg on the treadmill: $v_d = K_p(\theta_d - \theta)$. The gain, K_p, is experimentally tuned to a desired correction rate for position error.

Next in this section, Raibert's original controller and its modifications for this leg will be presented. The fuzzy controller is then described. The resulting intelligent controllers use the same inputs and provide the same outputs for the system.

3.1 Modified Raibert Controller with LM Learning

Marhefka's [12] modified Raibert controller computes the forward touchdown position of the foot by

$$x_f = K_1 \frac{T_s}{2} v + K_2 (v - v_d) + x_{bias}, \quad (1)$$

where T_s is the time of the previous stance period, v is the TOF body velocity, and v_d is the desired TOF velocity. K_1 and K_2 are experimentally tuned gains. The first term of this equation estimates the foot placement required for running at constant speed and the second term corrects velocity errors. The x_{bias} term was added by Marhefka as an offset to maintain zero velocity. By observation, stance time does not vary much so $T_s/2$ can be included in one coefficient for v. The new equation, with a change of coefficient names, is

$$x_f = \alpha_1 v + \alpha_2 (v - v_d) + \alpha_3. \quad (2)$$

An integral style control on the height error is used to control the cable length output, which is directly related to the energy in the springs.

In complex systems, experimentally tuning multiple parameters can be tedious and lead to a sub-optimal solution. Intelligent algorithms can help tune these coefficients on line. Levenberg-Marquardt (LM) is a derivative of the Gauss-Newton learning method used to solve least squares problems [13]. It will be used to tune the α_1 and α_2 gains in Eq. 2.

The error signal, ϵ, to be minimized is

$$\epsilon = y - F(\mathbf{p}, \boldsymbol{\alpha}), \quad (3)$$

where y is the unknown best forward foot touchdown position for the present system states, $\mathbf{p}$. The function $F(\mathbf{p}, \boldsymbol{\alpha})$ represents Eq. 2 as the output of the modified Raibert controller dependent upon α_1, and α_2. The first step in Gauss-Newton is to linearize the error, $\epsilon(\boldsymbol{\alpha})$, around the current value of $\boldsymbol{\alpha}_j$. This is done using a truncated Taylor series expansion to produce $\hat{\epsilon}(\boldsymbol{\alpha}, \boldsymbol{\alpha}_j)$. The second step in Gauss-Newton is to compute the squared norm, $J_q(\boldsymbol{\alpha})$, of this linearized error

$$J_q(\boldsymbol{\alpha}) = \frac{1}{2} \hat{\epsilon}(\boldsymbol{\alpha}, \boldsymbol{\alpha}_j)^2, \quad (4)$$

and minimize it by determining $\boldsymbol{\alpha}_{j+1}$:

$$\begin{aligned} \boldsymbol{\alpha}_{j+1} &= \arg \min_{\boldsymbol{\alpha}} J_q(\boldsymbol{\alpha}) \\ &= \arg \min_{\boldsymbol{\alpha}} \frac{1}{2} \hat{\epsilon}(\boldsymbol{\alpha}, \boldsymbol{\alpha}_j)^2. \end{aligned} \quad (5)$$

Here, "$\arg \min_{\boldsymbol{\alpha}}$" is mathematical notation for the value of $\boldsymbol{\alpha}$ ("argument") that minimizes the norm. This is now a least squares problem which has the solution [13]

$$\begin{aligned} \boldsymbol{\alpha}_{j+1} = \boldsymbol{\alpha}_j - \\ \left(\nabla \epsilon(\boldsymbol{\alpha}_j) \nabla \epsilon(\boldsymbol{\alpha}_j)^T \right)^{-1} \nabla \epsilon(\boldsymbol{\alpha}_j) \epsilon(\boldsymbol{\alpha}_j). \end{aligned} \quad (6)$$

To avoid problems with computing the inverse in Eq. 6, the method is implemented as

$$\begin{aligned} \boldsymbol{\alpha}_{j+1} = \boldsymbol{\alpha}_j - \\ \left(\nabla \epsilon(\boldsymbol{\alpha}_j) \nabla \epsilon(\boldsymbol{\alpha}_j)^T + \boldsymbol{\Lambda}_j \right)^{-1} \nabla \epsilon(\boldsymbol{\alpha}_j) \epsilon(\boldsymbol{\alpha}_j) \end{aligned} \quad (7)$$

where $\boldsymbol{\Lambda}_j$ is a 2×2 matrix in our case, such that the matrix to be inverted is positive definite. In the LM method, $\boldsymbol{\Lambda}_j$ is a diagonal matrix whose elements, λ_1 and λ_2, can also be used to control the update step size.

Passino [13] suggests that if a system's input to output ratio gain is bounded, a direct adaptive approach should be capable of stabilizing the system. To assume a bounded ratio on this system is reasonable, because the geometry of the leg limits the amount of energy that can be input into the springs per cycle, which bounds the output height and velocity. The error, ϵ, used in Eq. 7 is not available for computing updates. Because of their monotonic relationship, the system error, $e = v_d - v$, is used instead of ϵ with good results. This will be verified later in simulation and on the prototype leg. The resulting update formula for our system is then

$$\alpha_{m_{j+1}} = \alpha_{m_j} + \frac{p_m}{p_m^2 + \lambda_m} e_j \quad m=1,2 \quad (8)$$

where
$\begin{aligned} e_j &= \text{system error,} \\ \lambda_m &= \text{step size control variables,} \\ p_1 &= v, \text{ and} \\ p_2 &= (v - v_d). \end{aligned}$

λ_1 and λ_2 ensure that the updates are bounded when p_1 or p_2 is small. The update is computed immediately before the controller is called at the beginning of the next cycle. The new coefficients, $\boldsymbol{\alpha}_{j+1}$, are then used in Eq. 2 to compute the setpoints for the following touchdown.

3.2 Fuzzy Controller

Another controller developed for the leg is a direct adaptive fuzzy controller. The structure of this control system is illustrated in Fig. 4. The fuzzy controller, the process, and the adaptation mechanism make up the three main parts of the system. The controller, as stated before, is only called once per jump cycle and outputs a setpoint for cable length and a setpoint for horizontal foot placement.

The control starts with fuzzification by mapping an input into one or more membership functions. The triangular input membership functions used to characterize body velocity are shown in Fig. 5. If the leg velocity is 0.15 rad/s, then $\mu_{0.0}^v = \mu_{0.3}^v = 0.5$, and all other membership functions for that input become zero.

The fuzzy rule-base is a table of controller outputs for every combination of input membership functions. The number of

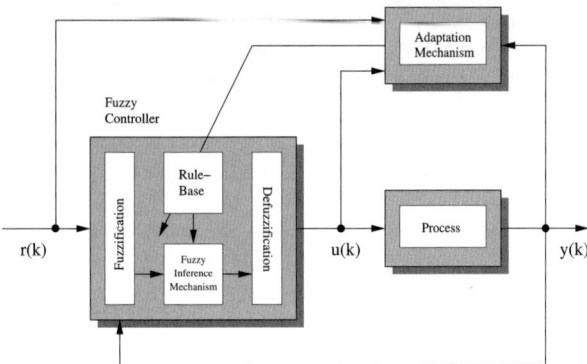

Figure 4: Structure of the Direct Adaptive Fuzzy Control System [7].

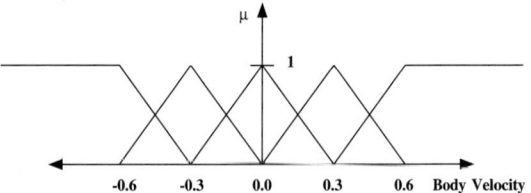

Figure 5: Example membership functions for the leg velocity.

rules is then equal to the product of the number of membership functions for each rule.

The inference mechanism is the next step in the fuzzy controller. This mechanism determines the applicability of each rule to the current inputs. The product is used to determine the certainty, μ_i, that the premise of rule i is currently applicable. The certainty of rule i whose premise is:

If velocity is '0.0 rad/s' and desired velocity is '0.5 rad/s' and height is '0.78 m' and desired height is '0.73 m/s'

would be:

$$\mu_i = \mu_{0.0}^v \times \mu_{0.5}^{v_d} \times \mu_{0.78}^h \times \mu_{0.73}^{h_d} \,. \qquad (9)$$

The last component of the fuzzy control is defuzzification. This process combines the recommendations of each rule in an output based upon rule certainties. Center average defuzzification is used and the output y is given by

$$y_k = \frac{\sum_i c_{i,k} \mu_i}{\sum_i \mu_i} \,, \qquad (10)$$

where μ_i is the premise certainty of rule i, and $c_{i,k}$ is the kth output of rule i. This equation shows a summation over all rules. Each rule output center is multiplied by its certainty, which weights the controller output toward the rule most applicable. Using triangular membership functions without center overlap limits the number of nonzero certainties in each input to two, making the maximum number of nonzero certainties that need to be included in Eq. 10 equal to 2^n, where n is the number of inputs. Adding membership functions to an input will not affect the amount of computation because only two membership functions are on in each input. Adding inputs, however, will increase the computation in Eqs. 9 and 10.

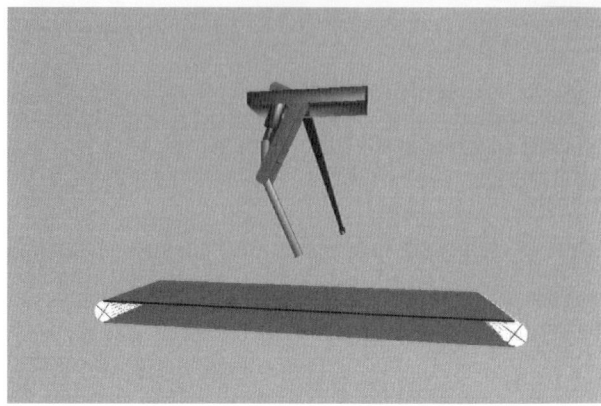

Figure 6: One-leg system in simulation.

The adaptation mechanism modifies the rule output centers to correct velocity errors. Immediately before the controller is called, the current system state is compared to the state desired at the previous cycle. The kth output for rule i, $c_{i,k}$ is updated as a factor of this error by

$$c_{i,k_{j+1}} = c_{i,k_j} + K_c \mu_{i_j} e_j \,, \qquad (11)$$

where K_c = adaptation gain,
μ_{i_j} = certainty of rule i, and
e_j = system error of cycle j.

K_c is tuned experimentally. Note that the certainty of rule i is used to scale the update size. This applies more change to the rule outputs that were more applicable. This certainty is nonzero for only 2^n rules meaning that only the rules applied to the previous controller outputs are updated by the present error.

This complex machine does not require complex algorithms for control. Both controllers are derived from a basic understanding of the machine. Aside from the learning mechanisms, the Raibert controller stores three parameters and the fuzzy controller stores an output for each rule, always totaling more than three for this leg. Raibert's method offers an advantage because only three parameters need to be stored. The computation per cycle is also less for this method. With more stored parameters and more computation, the fuzzy controller should perform better.

4 The Leg in Simulation

A screen capture of the simulation is shown in Fig. 6. The simulation algorithms used are from the DynaMechs [14] simulation library, a highly user-integrable package for general robotic systems with support for open and closed-loop kinematics. Unless otherwise stated, the treadmill is moving at 3 m/s under the leg.

Data for the leg controlled by the modified Raibert controller [7] with and without Levenberg-Marquardt (LM) learning is

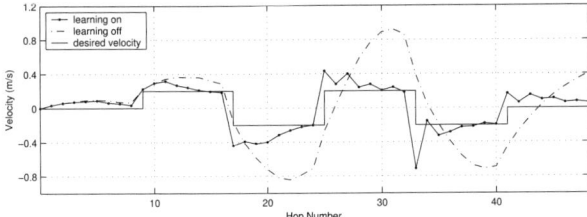

Figure 7: Modified Raibert controller with LM learning. Response of the system to desired velocity.

Input	Membership Function Centers	Units
v	-0.5, -0.25, 0.0, 0.25, 0.5	m/s
v_d	-0.4, -0.2, 0.0, 0.2, 0.4	m/s
h	76, 79, 82	cm
θ	-20, 0, 20	deg

Table 1: Fuzzy controller input membership function centers.

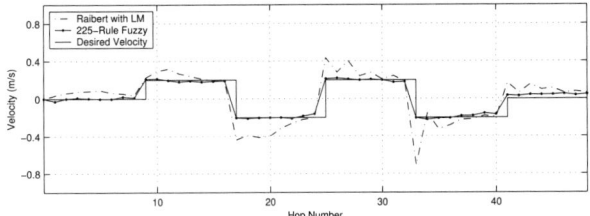

Figure 8: Modified Raibert controller with LM learning compared to a 225-rule fuzzy controller.

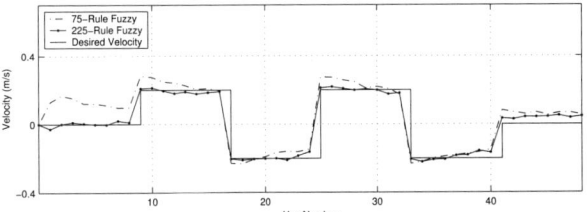

Figure 9: Simplified fuzzy controller. Response of the system to desired velocity using a 75-rule and a 225-rule fuzzy controller. The horizontal position of the leg on the treadmill was removed as an input to reduce the computation of the algorithm.

shown in Fig. 7. Without using online learning, the modified Raibert controller maintains stability but the leg does not quickly reach the desired velocity. The gains in Eq. 2, α_1, α_2, and α_3, were experimentally tuned to -0.05, -0.075, and 0.02 respectively. With online learning active, α_1 varied between 0.25 and 0.41, and α_2 varied between -0.157 and -0.137. The non-settling nature of α_1 and α_2 indicates that the position of the leg on the treadmill or the direction of leg motion affects the performance of the controller. Because of its simplicity, the modified Raibert controller cannot compensate for these effects. Additional terms can be added to the control structure but that would involve learning more coefficients online.

The fuzzy controller with inputs and membership function centers shown in Table 1 can compensate for these effects because appropriate inputs were added. The results in Fig. 8 show the modified Raibert controller with LM learning compared to a 225-rule fuzzy controller. The fuzzy controller outperforms the modified Raibert algorithm by showing a near deadbeat response to changes in desired velocity.

As stated in the discussion of fuzzy control, decreasing the number of inputs will significantly reduce the amount of computation to generate controller outputs. The prototype leg will have limited processing power so less computation by the fuzzy algorithm is preferred. The position of the leg on the treadmill is eliminated to reduce the algorithm computation. The number of rules falls from 225 to 75, and the maximum number of rules active at one time using the triangular membership functions drops from sixteen to eight. Only eight rule certainties need to be computed and only eight multiplications are needed in the defuzzification step.

The results of this controller simplification are presented in Fig. 9. The simplified 75-rule controller does not perform as well as the expanded controller but is still a significant improvement over the modified Raibert controller with LM learning.

The simulator was also used to verify the effectiveness of fuzzy control at high speeds. Figure 10 shows the data collected with the treadmill running at 5 m/s. At high speeds, training errors become amplified but the adaptation mechanism can be used to improve the results. The adaptation gain, K_c, is 0.125. This speed of 5 m/s is well above the predicted speed the quadruped needs to run to benefit from galloping.

5 Real-Time Control of the OSU DASH Leg

A brief description of the hardware and software is given, followed by the results on the OSU DASH leg.

5.1 Hardware

The Kameleon board from K-Team is the microcontroller used on the OSU DASH leg. A Motorola MC68376 performs the processing with a clock speed of 20 MHz. The accompanying robotics extension board (REB) provides convenient interfacing to the motors and encoders. The thigh axis is actuated by a 3-phase, brushless DC motor with a maximum torque of 5.68 Nm. The large pulley at the hip axis serves to increase the torque output of the motor by a 5.33:1 ratio. The cable axis is actuated by a similar brushless DC motor with a maximum torque of 11.8 Nm which is increased by a 10:1 gearbox. A 1 in diameter pulley is used to wrap the cable. Each motor has a 1000-count optical encoder to measure position. Optical encoders are also attached to the two boom degrees of freedom to monitor horizontal and vertical motion of the body. Quadrature encoding is used on the board to increase the resolution of all four encoders by a factor of four.

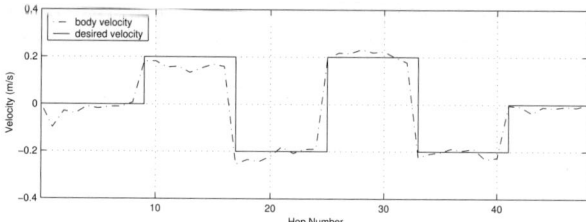

Figure 10: High-speed running using fuzzy control at a treadmill velocity of 5 m/s. The adaptation mechanism was enabled to improve the results.

Input	Membership Function Centers
$\dot{\theta}_b$	-0.6, -0.3, 0.0, 0.3, 0.6
$\dot{\theta}_d$	-0.6, -0.3, 0.0, 0.3, 0.6

Table 2: Fuzzy controller input membership function centers: $\dot{\theta}_b$ = body belocity, and $\dot{\theta}_d$ = desired velocity. Horizontal body position and height have been eliminated as inputs to reduce the processing required by the algorithm. Both velocities are measured in $counts/t_c$, which is the velocity measure of encoder counts per control time step.

The control system depends upon reliable and immediate detection of ground contact. A microswitch is used at the foot to detect touchdown and liftoff. The leg hops on a 6 ft long Jog-A-Dog treadmill. The velocity of the treadmill is manually operated with no sensing done by the leg. It has a maximum speed of approximately 5.0 m/s.

5.2 Software

The Kameleon and REB boards were shipped with an operating system and useful control subroutines downloaded to the board. Position control subroutines with trapezoidal velocity trajectories are called every 2.5 ms by the operating system. The code contains the supervisory leg phase controller which monitors the states and updates the phases of the leg. The code also contains the intelligent control routines which are called once per cycle and the joint servo routines.

The input membership function centers for the fuzzy controller used on the OSU DASH leg are shown in Table 2. The velocity is now measured in $counts/t_c$ which are counts per control time step. Converting these measurements into m/s adds unnecessary multiplications. The 25 rules for this controller are significantly less than the 225-rules used in simulation.

An elliptic filter is used to smooth the horizontal body velocity measurement. The success of both the modified Raibert controller [7] with Levenberg-Marquardt (LM) learning and the fuzzy controller depend on a good velocity measurement of the body. This filter is called every 10 ms to reduce the burden on the processor.

5.3 Results

The treadmill is turned off for all tests unless otherwise mentioned. The single output of the intelligent controller to the

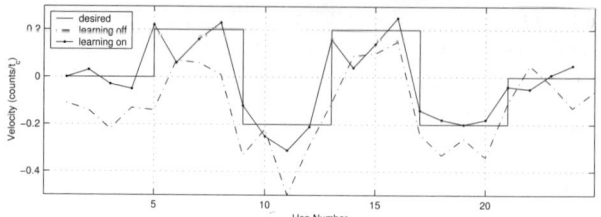

Figure 11: Modified Raibert controller with LM learning on the OSU DASH leg.

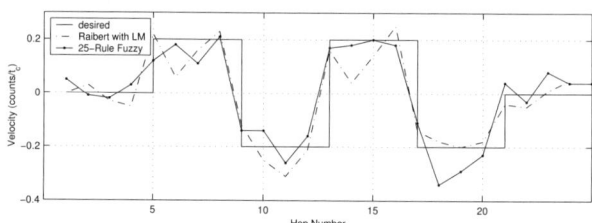

Figure 12: Modified Raibert controller with LM learning compared to a 25-rule fuzzy controller on the OSU DASH leg.

system is the desired thigh position. For these experiments, the same cable length is used for every hop. The thigh angle controller cannot actuate the thigh angle at touchdown to exactly match the desired thigh angle. The error is usually within $\pm 4.5°$ (± 50 counts) which is close enough to exhibit good control. Occasionally, the thigh control system does not perform well and the thigh error is closer to $9°$ (± 100 counts). Although the boom attachment restricts the frame from pitching forward or backward, flexure in the boom and play in the connections allow the frame to rock back and forth while jumping. This pitching, which cannot be measured, may differ slightly from one cycle to the next and cause small errors. These factors may result in poor leg performance from suitable controller outputs. For this reason, the trend of the data is a better assessment of the controller performance than an isolated hop. The results presented below were chosen as good representations of the leg performance under the specified controller.

Figure 11 displays the results of the modified Raibert controller with learning on and off. With learning off, the leg is stable enough to stay on the treadmill but does not approach the desired velocity consistently. With LM learning engaged, the leg velocity moves closer to the desired velocity.

The 25-rule fuzzy controller is compared to the modified Raibert controller with LM learning. The leg response to these controllers is presented in Fig. 12. The leg performs slightly better using the fuzzy controller. The mean of the squared error for the fuzzy controller is 0.0033, and the mean for the squared error using the modified Raibert controller with LM is 0.0038. This difference is small but consistent from test to test.

The 25-rule fuzzy controller was applied to the OSU DASH leg with the treadmill moving at 2.0 m/s underneath the body. The leg jumped stably on the treadmill with the addition of a control feature. As the foot approaches the ground, a constant voltage is applied to the thigh motor to force the linear foot velocity at touchdown to match the treadmill velocity. This phase has been observed in biological systems but the reasons why have not been fully investigated [15]. Without this retract phase, the OSU DASH leg was not able to jump on a treadmill moving at 2.0 m/s.

6 Summary

This work has provided a good foundation for further work towards the development and control of a galloping quadruped. This work verified the leg design and control algorithms by implementing them in real hardware. Issues were solved on this system and potential problems for the quadruped were formulated. The success of this leg also provided confidence in the feasibility of a quadruped. The OSU DASH leg will continue to be tested as much more can be learned from this system.

Acknowledgments

Support was provided by grant no. IIS-0208664 from the National Science Foundation to The Ohio State University.

References

[1] Anonymous, "Logistical Vehicle Off-Road Mobility," Tech. Rep., Project TCCO 62-5, U.S. Army Transportation Combat Developments Agency, Fort Eustis, VA., February 1967.

[2] N. C. Heglund and C. R. Taylor, "Speed, stride frequency and energy cost per stride: How do they change with body size and gait?," *Journal of Experimental Biology*, vol. 138, pp. 301–318, 1988.

[3] M. H. Raibert, *Legged Robots that Balance*, MIT Press, Cambridge, Massachusetts, 1986.

[4] J. Furusho, S. Akihito, S. Masamichi, and K. Eichi, "Realization of bounce gait in a quadruped robot with articular-joint-type-legs," in *Proceedings of the IEEE International Conference on Robotics and Automation*, Nagoya, Japan, 1995, pp. 697–702.

[5] M. Buehler, R. Battaglia, A. Cocosco, G. Hawker, J. Sarkis, and K. Yamazaki, "SCOUT: A simple quadruped that walks, climbs, and runs," in *Proceedings of the IEEE International Conference on Robotics and Automation*, Leuven, Belgium, 1998, pp. 1707–1712.

[6] H. Kimura, S. Akiyama, and K. Sakurama, "Realization of dynamic walking and running of the quadruped using neural oscillator," *Autonomous Robots*, vol. 7, pp. 247–258, 1999.

[7] D. W. Marhefka, *Fuzzy Control and Dynamic Simulation of a Quadruped Galloping Machine*, Ph.D. thesis, The Ohio State University, Columbus, Ohio, 2000.

[8] J. P. Schmiedeler, *The Mechanics of and Robotic Design for Quadrupedal Galloping*, Ph.D. thesis, The Ohio State University, Columbus, Ohio, 2001.

[9] R. McN. Alexander, "Three uses for springs in legged locomotion," *International Journal of Robotics Research*, vol. 9, pp. 53–61, 1990.

[10] H. DeMan, D. Lefeber, and J. Vermeulen, "Design and control of a robot with one articulated leg for locomotion on irregular terrain," in *Proceedings of the Twelfth CISM-IFToMM Symposium on Theory and Practice of Robots and Manipulators*, Vienna, 1998, Springer Verlag, pp. 417–424.

[11] H. B. Brown Jr. and G. Zeglin, "The bow leg hopping robot," in *Proceedings of the IEEE International Conference on Robotics and Automation*, Leuven, Belgium, 1998, pp. 781–786.

[12] D. W. Marhefka and D. E. Orin, "Fuzzy control of quadrupedal running," in *Proceedings of the IEEE International Conference on Robotics and Automation*, San Francisco, CA, 2000, pp. 3063–3069.

[13] K. M. Passino, "Biomimicry for optimization, control, and automation," unpublished manuscript, The Ohio State University, 2001.

[14] S. McMillan, D. E. Orin, and R. B. McGhee, "A computational framework for simulation of underwater robotic vehicle systems," *Autonomous Systems*, vol. 3, pp. 253–268, 1996.

[15] H. M. Herr and T. A. McMahon, "A trotting horse model," *International Journal of Robotics Research*, vol. 19, no. 6, pp. 566–581, 2000.

Implementing Configuration Dependent Gaits in a Self-Reconfigurable Robot

K. Støy[1],* W.-M. Shen[2], and P. Will[2]

kaspers@mip.sdu.dk, shen@isi.edu, and will@isi.edu

[1]The Adaptronics Group, The Maersk Institute, University of Southern Denmark
Campusvej 55, DK-5230 Odense M, Denmark

[2]USC Information Sciences Institute and Computer Science Department
4676 Admiralty way, Marina del Rey, CA 90292, USA

Abstract

In this paper we examine locomotion in the context of self-reconfigurable robots. Self-reconfigurable robots are robots built from many connected modules. A self-reconfigurable robot can change its shape and configuration by changing the way these modules are connected. The focus of this paper is to understand how several locomotion gaits can be represented in such a robot and how the robot can select one of these gaits depending on its configuration. We implement a control system based on role based control in a physical self-reconfigurable robot built from seven modules. In several experiments we successfully demonstrate that when the robot is manually reconfigured from a chain to a quadruped configuration the robot changes gait from a sidewinder snake gait to a quadruped walking gait. We conclude that role based control is a promising control method for controlling locomotion of self-reconfigurable robots.

1 Introduction

Self-reconfigurable robots are robots built from potentially many connected modules (see Figure 1 for a photo of such a module or refer to one of several physical realizations of self-reconfigurable robots [7, 8, 10, 11, 12, 13, 14, 15, 16, 23, 26]). Self-reconfigurable robots can autonomously change the way in which these modules are connected and through this self-reconfiguration process change their shape to fit the task-environment. Due to this capability self-reconfigurable robots are useful in task-environments where versatility is of importance [11]. A robot exploring the surface of a planet might for instance self-reconfigure into a snake to explore small caves and later into a rolling track to efficiently cover the distance back to the station.

*The work reported here was performed while visiting USC Information Sciences Institute

Figure 1: A CONRO module. The male child connectors are at the bottom right corner of the photo. The female connector is in the top left corner partly hidden from view. The two axes shows the modules two degrees of freedom.

Another desirable feature of self-reconfigurable robots is robustness [11]. The idea is that if one module fails another one can replace it. However in order to achieve this the system has to be built from identical modules. It is an open research question if the system should be built from just one kind of modules or a few, but for the sake of robustness it is important that there are modules to replace the ones which fail. In order to create a versatile and robust self-reconfigurable robot the control system also has to support these features. This implies that modules should run identical programs to make sure they can replace each other. Furthermore distributed control is mandatory to avoid a single point of failure. These constraints on the control system makes the control of self-reconfigurable robots a significant challenge.

Related work focus on how to reconfigure from one given shape to another. Centralized methods exist which plan how to reconfigure from one given shape to another [6, 5, 15] or distributed methods [11, 22]. Also methods exist where the shape emerges due to interaction between the modules and the environment [1]. In this work we focus on locomotion of self-reconfigurable robots. There are two broad classes of locomotion algorithms for self-reconfigurable robots: "water-flow" algorithms where the robot locomotes by having the modules climbing over each other [8, 2, 3]. However, for robots covering large distances this type of locomotion might not be sufficiently fast and energy efficient. Therefore it is important to develop systems which are able to locomote in a fixed configuration. This area is fairly well understood and solutions exist based on gait control tables [24, 25], hormones [16, 17], and roles [18, 19].

The pitfall of studying the control of self-reconfigurable robots in a fixed configuration is that in a fixed configuration it is not important to design a versatile and robust control system in order to make the system work. Therefore there is a risk of producing a control system that does not fit in the context of self-reconfigurable robots. In order to avoid this pitfall we investigate how several locomotion gaits can be represented in one self-reconfigurable system and how the system can change between these gaits during reconfiguration. In particular we extend our previous work by showing how role based control can be used in one system to represent a gait similar to that of a sidewinder snake and a quadruped walker. In role based control each module in the system plays a role. A role controls the motion of the module and takes care of the coordination with connected modules. Which role to play is decided based on both local configuration and the roles being played by neighboring connected modules. Specific gaits are not represented globally, but emerge due to the interaction between role playing modules. This means that as the robot is manually reconfigured the locomotion pattern changes gradually. In experiments with a real self-reconfigurable robot built from seven modules we demonstrate that the robot is able to make a transition from a sidewinder gait to a walker gait when it is manually reconfigured from a chain configuration to a quadruped configuration.

2 Role Based Control

Role based control consists of two algorithms. The role playing algorithm controls the motion of a module and the coordination and communication with connected modules. The role selection algorithm selects which role a module plays based on the local configuration and the roles being played by connected modules. We will first now describe these two algorithms.

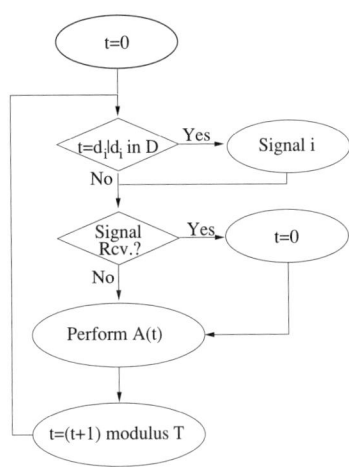

Figure 2: A flow diagram of the role playing algorithm. Refer to Section 2.1 for a description.

2.1 Role Playing

It is assumed that one of the connectors of a module is defined as a parent connector $p \in S$ where S is the set consisting of the connectors of a module. The remaining connectors are defined as child connectors $C = S\backslash\{p\}$. Modules can only be connected by connecting the parent connector to a child connector of another module. Note that the CONRO modules (see Figure 1) cannot physically be connected in any way that breaks these assumptions. The reason being that the female connector can be defined as the parent connector and the three male connectors as child connectors. Furthermore it is assumed that there are no loops in the system. These assumptions essentially limits the allowable configurations to tree configurations with well-defined parent-child relationships.

A role consists of three components. The first component is a function $A(t)$ that specifies the joint angles of a module given an integer $t \in [0:T]$, where T is the period of the motion and the second component that needs to be specified. The third component is a set of delays D. A delay $d_i \in D$ specifies the delay between the child connected to connector $i \in C$ and the parent. That is, if the parent is at step t_{parent} the child is at $t_{child} = (t_{parent} - d_i) \; modulus \; T$.

Given these assumptions a module plays a role by following the role playing algorithm outlined in Figure 2. Initially t is set to zero. If it is time to send a signal because $t = d_i$ a signal is sent through connector i. If a signal is received from the parent, through connector p, t is reset. This insures that the module is delayed as described above compared to the parent. Finally the positions of the servos are updated and t is incremented after which the next iteration is initiated.

In earlier work we have shown that this basic role playing algorithm is sufficient to produce locomotion patterns similar to that of a caterpillar and a sidewinder snake [18].

2.2 Role Selection

In more complex locomotion patterns it cannot be assumed that all modules play the same role. This implies that modules need to be able to play different roles and be able to select between them. This introduces the need for a role selection algorithm. The selection is based on the local configuration and which roles connected modules are playing. Before we introduce the role selection algorithm we will introduce some functions. The function $C : S \rightarrow Boolean$ returns true if a module is connected to connector $i \in S$ and false otherwise. The function $NC : S \rightarrow S$ returns the connector of the neighbor to which the connector $i \in S$ of this module is connected. Given that R is the set of roles that a module can play the function $NR : S \rightarrow R$ returns the role being played by the module connected to connector $i \in S$.

The local configuration is specified by: 1) The subset of connectors to which modules are connected $\{i \in S | C(i)\}$ 2) The connector of the parent to which the module is connected $NC(p)$ if $C(p)$. The local configuration is sufficient to select roles in simple configurations.

In configurations where the role cannot only be selected based on the local configuration the role is selected based on the set of roles being played by connected modules $\{r \in R | i \in S \land C(i) \land R(i) = r\}$. For instance five modules might be connected in a chain to form an arm. In this configuration the modules might have to play the roles of a shoulder, upper arm, elbow, lower arm, and wrist. The three modules in the middle cannot decided which role to play based on the local configuration, because the local configuration is identical for all of them. However if a module finds out that the parent is playing the shoulder role it can select the upper arm role and so on. In a tree configuration the root can determine it is the root based on the local configuration. The root can then instruct its children that they are children of the root and where they are connected. By induction we can see that it is possible to use this mechanism to decide a modules position in the global configuration if needed.

By combining these two selection mechanisms it can be decided as locally as possible what role a module should play which is important if the system is to scale.

3 The CONRO Robot

In the experiments reported in Section 5 we use the CONRO self-reconfigurable robot. The CONRO robot has been developed at USC's Information Sciences Institute [4, 9] (see Figure 1). The modules are roughly shaped as rectangular boxes measuring 10cm x 4.5cm x 4.5cm and weigh 100grams. The modules have a female connector located at one end by definition facing south s and three male connectors located at the other end facing east e, west w, and north n making $S_{conro} = \{s, e, w, n\}$. The parent connector p is defined to be the female connector s. Each connector has an infra-red transmitter and receiver used for local communication and sensing. The modules have two controllable degrees of freedom: pitch (up and down) and yaw (side to side). Processing is taken care of by an onboard Basic Stamp 2 processor. The modules have onboard batteries, but these do not supply enough power for the experiments reported here and therefore the modules are powered through cables.

4 Implementation

We will now show how to combine the sidewinder gait and the walking gait previously reported in [18, 19]. In order to perform these gaits four different roles are needed: sidewinder (sw), spine (sp), east leg (eleg), and west leg (wleg). That is $R = \{sw, sp, eleg, wleg\}$. These roles are listed below. The focus of this paper is to understand how these roles are combined and not how to design roles so therefore if you want more details about these role definitions please refer to [21].

$$A(sw, t) = \begin{cases} pitch(t) &= 20° \cos(\frac{2\pi}{T} t) \\ yaw(t) &= 50° \sin(\frac{2\pi}{T} t) \end{cases}$$
$$d_{north} = \frac{T}{5}$$
$$T = 180$$

(1)

$$A(sp, t) = \begin{cases} pitch(t) &= 0° \\ yaw(t) &= 25° \cos(\frac{2\pi}{T} t + \pi) \end{cases}$$
$$d_{east} = \frac{T}{4}$$
$$d_{south} = \frac{2T}{4}$$
$$d_{west} = \frac{3T}{4}$$
$$T = 180$$

The east leg role (eleg) is shown below. The west leg role (wleg) is identical except that t is replaced by $2\pi - t$.

$$A(eleg, t) = \begin{cases} pitch(t) &= 35° \cos(\frac{2\pi}{T} t) - 55° \\ yaw(t) &= 40° \sin(\frac{2\pi}{T} t) \end{cases}$$
$$T = 180$$

sw → sp,   if $C(e) \lor C(w)$
    → wleg, if $C(p) \land NC(p) = w$
    → eleg, if $C(p) \land NC(p) = e$

sp → sw,   if $\neg(C(e) \lor C(w))$
    → wleg, if $C(p) \land NC(p) = w$
    → eleg, if $C(p) \land NC(p) = e$

leg → sw,   if $(C(p) \land R(p) = sw) \lor C(n)$
    → sp,   if $C(p) \land R(p) = sp$

Figure 3: This figure shows the rules used to decide when to change between roles. The roles R are sidewinder (sw), spine (sp), east leg ($eleg$), and west leg($wleg$). The transitions for the west leg and the east leg are the same and are therefore both represented by leg. The CONRO module have the connectors $S = \{e, w, n, s\}$ where the parent connector $p = s$. The function $C : S \rightarrow Boolean$ returns a boolean value that indicates if a modules is connected to the specified connector. The function $NC : S \rightarrow R$ returns the connector of the parent module to which this module is connected. The function $R : S \rightarrow R$ returns the role being played by the module connected to the specified connector.

While playing these roles the local configuration and the roles being played by connected modules are monitored. This information is used to decide when to change role. The rules are as described below and shown in Figure 3.

A module playing the sidewinder role changes to the spine role if a module is connected to either the east or the west connector. If it is connected to the east or west side of a module it plays the role of the corresponding leg.

The rules for the spine role are similar to those of the sidewinder role except that if both modules to the sides are detached it changes role to a sidewinder module.

The legs decide to change role to sidewinder or spine if they are connected to the north connector of a module playing one of these roles. Additionally if a leg has a module connected on its north connector it changes role to sidewinder. This last transition is made to make sure that a leg placed at the root of the configuration tree discovers that and plays an appropriate role.

5 Experiments

In order to evaluate this system we conducted a series of experiments with a self-reconfigurable robot made from seven CONRO modules. The robot is initially configured into a long chain. The root module is started using an infrared signal. The root module signals its child after $T/5$ and so on. After $6T/5$ corresponding to approximately three

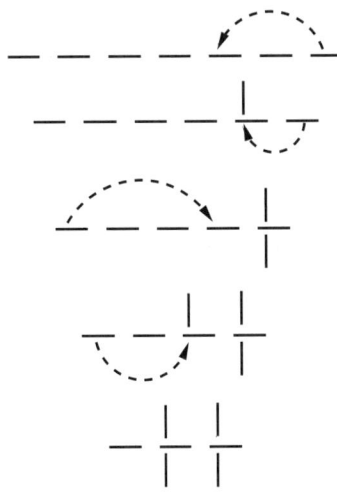

Figure 4: The reconfiguration process of experiment 1. The solid lines represent modules. The robot starts in a chain configuration as shown at the top. The modules are then manually reconfigured as indicated by the dashed arrows until the quadruped configuration shown at the bottom is obtained.

seconds all modules are synchronized and take part in the sidewinder gait. The synchronization time is increased by one period per synchronization signal missed. However in the experiments reported this was not observed. It was then recorded how long it takes to move 63cm using the sidewinder gait. The robot is then picked up and by-hand reconfigured into a four legged walker. In each experiment the reconfiguration sequence is different. The reconfiguration sequence for experiment 1 can be seen in Figure 4. When the system is synchronized again the robot is allowed to walk 87cm and this time is also measured. Note that throughout each experiment the modules are not reset. Three snapshots from an experiment is shown in Figure 5.

We repeated the experiment four times. It took on average 12.3±1.0sec to cover 63cm corresponding to 5.1cm/sec as a sidewinder. It took 36.5±2.4sec to walk 87cm corresponding to 2.4cm/sec. Detailed data can be found in Figure 6. Videos of these experiments can be found on the web page: http://www.isi.edu/conro.

6 Discussion

In the implementation section we saw that it is easy to extend role based control to be able to represent two gaits and a mechanism to select between them. The implementation is minimal which is also indicated by the fact that the main loop is implemented using only 350 lines of Basic Stamp 2 code including code for communication and

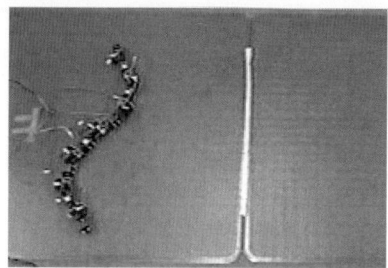 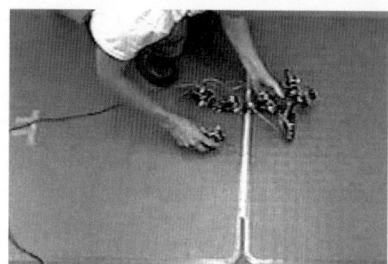 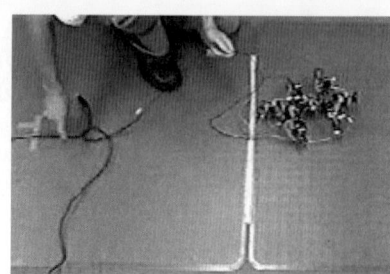

Figure 5: The robot first covers a distance of 63cm using a sidewinder gait (left). The robot is then manually reconfigured into a quadruped walker (middle). Finally the robot walks 87cm (right). Note that the cables only provide power.

motor control. The algorithm is also efficient. It manages to keep the modules synchronized only using constant time per period meaning that the locomotion speed is independent of the number of modules as shown in [19]. The time required to synchronize the system after a role change is proportional to $O(h)$ where h is the height of the configuration tree.

In role based control it is possible for a module to change role in response to any configuration change even if it does not occur in the immediate neighborhood of the module. This feature makes it possible to design a system that always produces an optimal locomotion pattern given a configuration. However a problem in realizing this is the increased complexity of the system, because of the potentially large number of roles and role selection conditions.

In role based control it is also possible for a module to change role based on a role change of another module. This feature facilitates turn taking where for instance the left leg explicitly waits for the right leg to be on the ground before it starts to move. In the implementation presented here this feature is not needed, because the period is constant for all roles and therefore the synchronization of the legs is automatically handled by role based control. However this might not the case in situations where the period depends on sensor input.

The experiments demonstrate that the system is robust to communication errors and reconfiguration which are important features of algorithms designed for self-reconfigurable robots. It might be worth to try to understand how this extreme robustness comes about. In Shen et al. [17] hormones are created by one module and the actions of the rest of the modules are executed and synchronized as this hormone is propagated through the configuration tree and back. The next hormone is generated when the previous hormone is propagated back to the creator. This means that critical state information is only represented one place in the system - in the hormone. The consequence is that if a hormone is lost the system stops (note that this event is likely to happen during reconfiguration). In role based control this is avoided, because each module essentially acts as the master of its subtree and signals are resent every period. In Yim [24, 25] modules have IDs and pick actions based on this ID which means each module has to be placed in a specific position in the configuration tree in order for the system to work. Furthermore the system is synchronized by a centralized master which means that if the master fails the system fails. This leads us to conclude that role based control gains its robustness from the distributed representation of the state of the gait and the fact that the modules are not assumed to have IDs.

In the work presented here roles changes occur in response to changes in the local configuration or role changes of connected modules. However a role change can also take place in response to elapsed time or sensor input as investigated in [20]. This might be a possible way to proceed in order to implement more complex behaviors.

In the current implementation the robot is reconfigured by hand. This is of course not satisfactory. We are currently working on using role based control to control the self-reconfiguration process itself. This would give a simple uniform algorithm able to control both self-reconfiguration and locomotion.

experiment	sidewinder	walker
1	13	25
2	11	30
3	13	28
4	12	30
avg	12.3	36.5
std.dev.	1.0	2.4

Figure 6: This figure shows the experimental data for four experiments. The second column indicated the time (seconds) it took for the robot to synchronize and move 63cm using the sidewinder gait. The third column indicates the time (seconds) it to took to locomote 87cm using the walker gait. The last two rows indicate respectively the average and standard deviation of these four experiments.

7 Summary

In this paper we have investigated how multiple gaits can be implemented in the same system using role based control. We have demonstrated that a self-reconfigurable robot made from seven CONRO modules is able to change from a sidewinder gait to a walker gait when the robot is manually reconfigured from a chain to a quadruped walker. The reconfiguration is done by hand, but without resetting or reprogramming the modules. Finally we have discussed that the key aspects of the system which make it possible to change between gaits include a distributed representation of the state and a system that does not rely on IDs.

Acknowledgments

This research is supported by the DARPA contract DAAN02-98-C-4032, the AFOSR contract F49620-01-1-0020, the EU contract IST-20001-33060, and the Danish Technical Research Council contract 26-01-0088.

References

[1] H. Bojinov, A. Casal, and T. Hogg. Emergent structures in modular self-reconfigurable robots. In *Proceedings, IEEE Int. Conf. on Robotics & Automation (ICRA'00)*, volume 2, pages 1734–1741, San Francisco, California, USA, 2000.

[2] Z. Butler, K. Kotay, D. Rus, and K. Tomita. Cellular automata for decentralized control of self-reconfigurable robots. In *Proceedings, IEEE Int. Conf. on Robotics and Automation (ICRA'01), workshop on Modular Self-Reconfigurable Robots*, Seoul, Korea, 2001.

[3] Z. Butler, K. Kotay, D. Rus, and K. Tomita. Generic decentralized control for a class of self-reconfigurable robots. In *Proceedings, IEEE Int. Conf. on Robotics and Automation (ICRA'02)*, pages 809–815, Washington, DC, USA, 2002.

[4] A. Castano, R. Chokkalingam, and P. Will. Autonomous and self-sufficient conro modules for reconfigurable robots. In *Proceedings, 5th Int. Symposium on Distributed Autonomous Robotic Systems (DARS'00)*, pages 155–164, Knoxville, Texas, USA, 2000.

[5] C.-J. Chiang and G.S. Chirikjian. Modular robot motion planning using similarity metrics. *Autonomous Robots*, 10(1):91–106, 2001.

[6] G. Chirikjian, A. Pamecha, and I. Ebert-Uphoff. Evaluating efficiency of self-reconfiguration in a class of modular robots. *Robotics Systems*, 13:317–338, 1996.

[7] G.S. Chirikjian. Metamorphic hyper-redundant manipulators. In *Proceedings, 1993 JSME Int. Conf. on Advanced Mechatronics*, pages 467–472, Tokyo, Japan, 1993.

[8] K. Hosokawa, T. Tsujimori, T. Fujii, H. Kaetsu, H. Asama, Y. Kuroda, and I. Endo. Self-organizing collective robots with morphogenesis in a vertical plane. In *Proceedings, IEEE Int. Conf. on Robotics & Automation (ICRA'98)*, pages 2858–2863, Leuven, Belgium, 1998.

[9] B. Khoshnevis, B. Kovac, W.-M. Shen, and P. Will. Reconnectable joints for self-reconfigurable robots. In *Proceedings, IEEE/RSJ Int. Conf. on Intelligent Robots and Systems (IROS'01)*, pages 584–589, Maui, Hawaii, USA, 2001.

[10] K. Kotay, D. Rus, M. Vona, and C. McGray. The self-reconfiguring robotic molecule. In *Proceedings, IEEE Int. Conf. on Robotics & Automation (ICRA'98)*, pages 424–431, Leuven, Belgium, 1998.

[11] S. Murata, H. Kurokawa, and S. Kokaji. Self-assembling machine. In *Proceedings, IEEE Int. Conf. on Robotics & Automation (ICRA'94)*, pages 441–448, San Diego, USA, 1994.

[12] S. Murata, H. Kurokawa, E. Yoshida, K. Tomita, and S. Kokaji. A 3-d self-reconfigurable structure. In *Proceedings, IEEE Int. Conf. on Robotics & Automation (ICRA'98)*, pages 432–439, Leuven, Belgium, 1998.

[13] S. Murata, E. Yoshida, K. Tomita, H. Kurokawa, A. Kamimura, and S. Kokaji. Hardware design of modular robotic system. In *Proceedings, IEEE/RSJ Int. Conf. on Intelligent Robots and Systems (IROS'00)*, pages 2210–2217, Takamatsu, Japan, 2000.

[14] A. Pamecha, C. Chiang, D. Stein, and G.S. Chirikjian. Design and implementation of metamorphic robots. In *Proceedings, ASME Design Engineering Technical Conf. and Computers in Engineering Conf.*, pages 1–10, Irvine, USA, 1996.

[15] D. Rus and M. Vona. Crystalline robots: Self-reconfiguration with compressible unit modules. *Autonomous Robots*, 10(1):107–124, 2001.

[16] W.-M. Shen, B. Salemi, and P. Will. Hormone-based control for self-reconfigurable robots. In *Proceedings, Int. Conf. on Autonomous Agents*, pages 1–8, Barcelona, Spain, 2000.

[17] W.-M. Shen, B. Salemi, and P. Will. Hormones for self-reconfigurable robots. In *Proceedings, Int. Conf. on Intelligent Autonomous Systems (IAS-6)*, pages 918–925, Venice, Italy, 2000.

[18] K. Støy, W.-M. Shen, and P. Will. Global locomotion from local interaction in self-reconfigurable robots. In *Proceedings, 7th Int. Conf. on Intelligent Autonomous Systems (IAS-7)*, pages 309–316, Marina del Rey, California, USA, 2002.

[19] K. Støy, W.-M. Shen, and P. Will. How to make a self-reconfigurable robot run. In *Proceedings, First Int. Joint Conf. on Autonomous Agents & Multiagent Systems (AAMAS'02)*, pages 813–820, Bologna, Italy, 2002.

[20] K. Støy, W.-M. Shen, and P. Will. On the use of sensors in self-reconfigurable robots. In *Proceedings, Seventh Int. Conf. on The Simulation of Adaptive behavior (SAB'02)*, pages 48–57, Edinburgh, UK, 2002.

[21] K. Støy, W.-M. Shen, and P. Will. Using role based control to produce locomotion in chain-type self-reconfigurable robot. *IEEE Transactions on Mechatronics, special issue on self-reconfigurable robots*, 7(4):410–417, Dec 2002.

[22] K. Tomita, S. Murata, H. Kurokawa, E. Yoshida, and S. Kokaji. A self-assembly and self-repair method for a distributed mechanical system. *IEEE Transactions on Robotics and Automation*, 15(6):1035–1045, Dec 1999.

[23] C. Ünsal and P.K. Khosla. Mechatronic design of a modular self-reconfiguring robotic system. In *Proceedings, IEEE Int. Conf. on Robotics & Automation (ICRA'00)*, pages 1742–1747, San Francisco, USA, 2000.

[24] M. Yim. *Locomotion with a unit-modular reconfigurable robot*. PhD thesis, Department of Mechanical Engineering, Stanford University, 1994.

[25] M. Yim. New locomotion gaits. In *Proceedings, Int. Conf. on Robotics & Automation (ICRA'94)*, pages 2508–2514, San Diego, California, USA, 1994.

[26] M. Yim, D.G. Duff, and K.D. Roufas. Polybot: A modular reconfigurable robot. In *Proceedings, IEEE Int. Conf. on Robotics & Automation (ICRA'00)*, pages 514–520, San Francisco, USA, 2000.

Controlling a Marionette with Human Motion Capture Data

Katsu Yamane, Jessica K. Hodgins, and H. Benjamin Brown
The Robotics Institute, Carnegie Mellon University
E-mail: {kyamane|jkh|hbb}@cs.cmu.edu

Abstract

In this paper, we present a method for controlling a motorized, string-driven marionette using motion capture data from human actors. The motion data must be adapted for the marionette because its kinematic and dynamic properties differ from those of the human actor in degrees of freedom, limb length, workspace, mass distribution, sensors, and actuators. This adaptation is accomplished via an inverse kinematics algorithm that takes into account marker positions, joint motion ranges, string constraints, and potential energy. We also apply a feedforward controller to prevent extraneous swings of the hands. Experimental results show that our approach enables the marionette to perform motions that are qualitatively similar to the original human motion capture data.

1 Introduction

Entertainment is one of the more immediately practical applications of humanoid robots and several robots have recently been developed for this purpose [1, 2, 3]. In this paper, we explore the use of an inexpensive entertainment robot controlled by motion capture data with the goal of making such robots readily available and easily programmable. The robot is a marionette where the length and pivot points of the strings are controlled by eight servo motors that bring the hands and the feet of the marionette to the desired positions (Figure 1).

Standard marionettes are puppets with strings operated by a human performer's hands and fingers. Creating appealing motion with such a puppet is difficult and requires extensive practice. Although for our marionette the servo motors move the strings, programming a robotic version of such a device by hand to produce expressive gestures would also be difficult. We solve this problem by using full-body human motion data to drive the motion. The human motion data is recorded as the positions of markers in three dimensions while an actor tells a story with his or her hands. After adaptation, the data are used to drive the motion of the marionette by taking into account the mar-

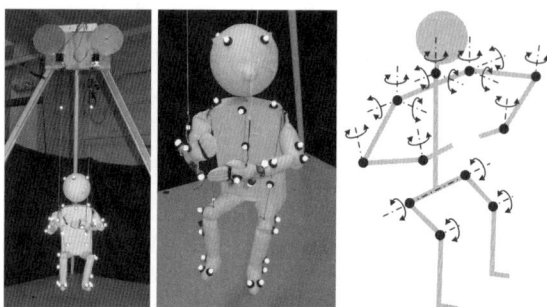

Figure 1: The motor-driven marionette and its model. The marionette is about 60cm tall. The shoulder and elbow joints have cloth stops to prevent unrealistic joint angles.

ionette's swing dynamics. These adaptations and dynamics compensation are necessary because the marionette has many fewer degrees of freedom and much smaller size than the human actor, strings rather than actuators at the joints, and no sensors except for the rotation of the motors.

The method described in this paper consists of four steps: (1) identify the swing dynamics of the hands and design a feedforward controller to prevent swinging and obtain a desired response, (2) obtain the translation, orientation, and scaling parameters that map the measured marker positions for the human motion into the marionette's workspace, (3) apply the controller to modify the mapped marker positions to prevent swing, and (4) compute the motor commands to bring the (virtual) markers attached to the marionette to the revised positions computed in step (3).

The relationship between the four steps is illustrated in Figure 2. In steps (2) and (4), we solve the inverse kinematics problem with many different constraints including marker positions, joint motion ranges, strings, and gravity. This algorithm is an extension of the first author's previous work [4]. In step (1), we model the dynamics of swing by capturing the response to a step input of each desired marker position and then use that response to design a feedforward controller to compensate for the swing motion.

As a demonstration of the algorithm, we include exper-

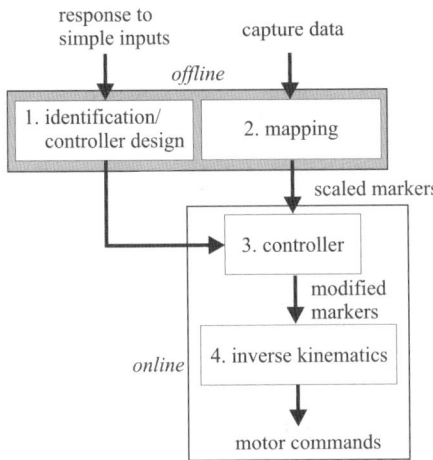

Figure 2: Overview of the marionette control system. The blocks in the top gray box are processed offline: identification/controller design for each marionette and mapping for each motion sequence. Those in the bottom white box is processed in realtime during a performance.

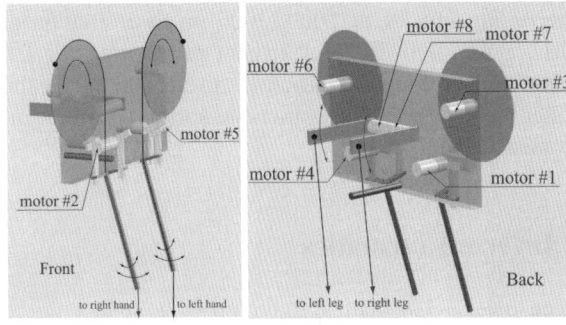

Figure 3: Closeup of the motors and pulleys; front of marionette (left), back (right).

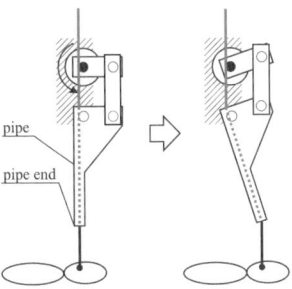

Figure 4: The mechanism for moving the hand in the horizontal direction.

imental results to compare the marionette motion to that of human actors. The motions of the marionette and the human actor are similar enough to distinguish different styles for the same story.

2 Related Work

Hoffmann [5] developed a human-scale marionette and controlled it to perform dancing motions using human data. The size and controllable degrees of freedom of the marionette are much closer to those of human than ours. The research is therefore focused more on image processing for measuring human motion than on mapping between human and marionette motions.

Mapping motion data is a common problem in applying motion capture data to a real robot or to a virtual character. The factors considered in previous work include joint angle and velocity limits [6], kinematic constraints [7], and physical consistency [8, 9]. However, the original and target human figures are usually more similar in degrees of freedom, dimensions, and actuators than the marionette is to a human actor.

The mechanism and dynamics of a string-driven marionette are quite similar to those of wired structures such as a crane. A number of researchers have worked on controlling a crane to bring an object to a desired position without significant oscillations [10]. This work assumes that the position of the object is known through the direction of the wire. Although we measure the position of the hand and feet for recording the swing dynamics, we do not have this information during a performance. Our accuracy requirements are much less because the marionette is gesturing in free space rather than precisely positioning an object.

3 Experimental Setup

The marionette is modeled as a 17DOF kinematic chain (Figure 1 right). Closeups of the motors and pulleys are shown in Figure 3. The marionette has eight servo motors (Airtronics Servo 94102); six control the arms and two control the legs. The motors are controlled by position commands sent from a serial port of a PC via an 8-channel controller board (Pontech SV203).

Motors 3 and 6 change the length of the strings connecting the hands. Motors 7 and 8 move the knee up and down. Motors 1, 2, 4 and 5 move the hands in horizontal directions by rotating the "pipes" and moving the pipe ends via four independent planar linkages (Figure 4).

We used a commercially available motion capture system from Vicon for capturing the actor's performance and the marionette's motion for identification of the swing dynamics. The system has nine cameras, each capable of recording images of 1000×1000 pixels resolution at 120Hz. We used different camera placements for the human subject and the marionette to accommodate the smaller workspace of the marionette and to ensure accurate measurements.

4 Inverse Kinematics

The inverse kinematics algorithm is used to enforce constraints to bring the markers representing the desired motion into the workspace of the marionette and to determine the motor angles that satisfy the desired marker positions and the physical constraints, including the desired marker positions, joint motion ranges, length and orientation of the strings, and potential energy. The potential energy constraint is introduced to model the effect of gravity. The inverse kinematics algorithm computes the joint angles and the motor commands that locally optimize the constraints. Because the algorithm was described in a previous paper [4], we present a short outline here.

Often, all the constraints cannot be satisfied due to the singularity of the configuration or to inconsistencies between the constraints. Therefore, the user is asked to divide the constraints into two groups: those that must be satisfied and those where some error is acceptable. The algorithm applies singularity-robust (SR) inverse [11] (also known as damped pseudo inverse [12]) to the lower-priority constraints. As described below, the SR-inverse distributes the error among the lower-priority constraints according to the given weights so that the resulting joint velocity does not become too large even if there are singularities or inconsistencies in the constraints.

We design a feedback controller for each constraint to ensure that the lower-priority constraints are satisfied as much as possible and to eliminate integration errors in both higher- and lower-priority constraints. The controller computes the required velocity when constraints are violated. For example, the feedback controller to bring a link to its reference position p^{ref} is $\dot{p}^{des} = k_p(p^{ref} - p)$ where k_p is a positive gain, p is the current position, and $\dot{p}^{des}$ is the desired velocity. Note that this velocity is not always realized for lower-priority constraints due to the nature of the SR-inverse algorithm.

With n_1 higher-priority constraints and n_2 lower-priority constraints, we have the following equations in generalized velocity $\dot{\theta}$:

$$J_1\dot{\theta} = v_1^{des} \quad (1)$$
$$J_2\dot{\theta} = v_2^{des} \quad (2)$$

where v_1^{des} and v_2^{des} are the desired velocities corresponding to higher- and lower-priority constraints respectively,

and J_1 and J_2 are the Jacobian matrices of the constraints with respect to θ.

We solve this equation for the generalized velocity as follows. First, we compute the set of exact solutions of Eq.(1) by

$$\dot{\theta} = J_1^\sharp v_1^{des} + (I - J_1^\sharp J_1)y \quad (3)$$

where $J_1^\sharp$ is the pseudo inverse of J_1, I is the identity matrix of the appropriate size, and y is an arbitrary vector. We can rewrite this equation as $\dot{\theta} = \dot{\theta}_1 + Wy$ where $\dot{\theta}_1 \triangleq J_1^\sharp v_1^{des}$, and $W \triangleq I - J_1^\sharp J_1$. Next, we compute the y with which $\dot{\theta}$ would satisfy Eq.(2) as closely as possible by

$$y = (J_2W)^*(v_2^{des} - J_2\dot{\theta}_1) \quad (4)$$

where $(J_2W)^*$ is the SR-inverse of J_2W. Finally, the generalized velocity $\dot{\theta}$ is computed by substituting y into Eq.(3), which is then integrated to compute the generalized coordinates in the next step.

In order to add a new constraint, we must design a feedback controller to compute the desired velocity and derive the corresponding Jacobian matrix. We describe the string and potential energy constraints in detail because the other constraints were described in the earlier paper [4].

4.1 String Constraints

Each string has a start point, an end point, and some number of intermediate points (Figure 5 left). The string can slide back and forth at the intermediate points. The current length of a string, l, must always be equal to or smaller than its nominal length l_0. l is computed by summing the length of all segments:

$$\begin{aligned} l &= \sum_{i=0}^{N-1} l_i = \sum_{i=0}^{N-1} |p_{i+1} - p_i| \\ &= \sum_{i=0}^{N-1} \sqrt{(p_{i+1} - p_i)^T(p_{i+1} - p_i)} \end{aligned} \quad (5)$$

where N is the number of segments, l_i ($0 \le i \le N-1$) is the length of segment i, p_i ($0 \le i \le N$) is the position of the i-th point. The Jacobian matrix of l with respect to the generalized coordinates θ is computed by

$$\begin{aligned} J_{str} &= \frac{\partial l}{\partial \theta} = \sum \frac{\partial l_i}{\partial \theta} \\ &= \sum \frac{1}{l_i}(p_{i+1} - p_i)^T \left(\frac{\partial p_{i+1}}{\partial \theta} - \frac{\partial p_i}{\partial \theta} \right) \\ &= \sum \frac{1}{l_i}(p_{i+1} - p_i)^T (J_{i+1} - J_i). \end{aligned} \quad (6)$$

Note that the Jacobian matrix is not defined for segments with $l_i = 0$, although we never encounter such situations in

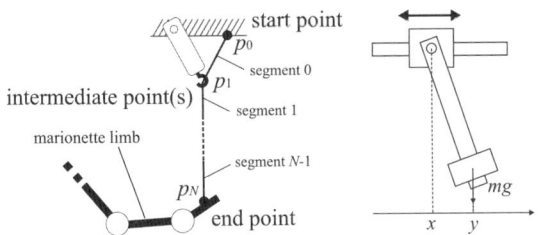

Figure 5: String models for inverse kinematics (left) and swing controller (right).

physical mechanisms. The feedback law for a string constraint is $v_{str}^{des} = k_{str}(l_0 - l)$.

In addition to the length, we also constrain the $(N-1)$th segment of the string to be vertical due to gravity. The two points p_{N-1} and p_N should then be vertical:

$$h_1 \cdot d_{N-1,N} = 0, \quad h_2 \cdot d_{N-1,N} = 0 \quad (7)$$

where h_1 and h_2 are independent unit vectors in the horizontal plane (e.g. $h_1 = (1\ 0\ 0)^T, h_2 = (0\ 1\ 0)^T$ if the gravity is in z direction) and $d_{N-1,N}$ is the unit vector from p_{N-1} to p_N, namely $d_{N-1,N} = (p_N - p_{N-1})/l_{N-1}$. The Jacobian matrix for this constraint is

$$J_v = \frac{1}{l_{N-1}} \begin{pmatrix} h_1^T \\ h_2^T \end{pmatrix} (J_N - J_{N-1}). \quad (8)$$

The desired velocity for this constraint is

$$v_v^{des} = -k_v \begin{pmatrix} h_1^T \\ h_2^T \end{pmatrix} d_{N-1,N} \quad (9)$$

where k_v is a positive gain.

4.2 Potential Energy

Because the joints that do not have strings directly attached to them will bend downward due to gravity, we also constrain the potential energy to be as small as possible by constraining the center of mass of the whole body to be as low as possible. The Jacobian matrix for this constraint is computed as $J_{pe} = d_g^T J_{COM}$ where $d_g = (0\ 0\ 1)^T$ is the unit vector in the direction of the gravity and J_{COM} is the Jacobian matrix of the center of mass with respect to the generalized coordinates. A method for computing J_{COM} can be found in [13]. The desired velocity for this constraint is a negative constant $-k_{pe}$.

5 Mapping

Before applying the measured marker positions to a marionette, we need to map them into new positions not only to adapt to the size of the marionette but also to comply with such physical constraints as the strings. Our marionette, for example, does not have a mechanism to move the pelvis. Therefore, if the captured motion contains translation or rotation of the pelvis, the motion should be translated or rotated so that the pelvis motion is eliminated.

In this section, we describe an algorithm to compute seven parameters for translation, rotation, and scaling that map the measured marker positions into new positions that satisfy the constraints of the inverse kinematics model described in Section 4. We compute the mapping parameters independently for each frame rather than using the same parameters for all frames. Although it might seem natural to fix a parameter such as scaling for a particular human actor, we have found that, because of the marionette's limited range of motion, using the best possible mapping for each frame is preferable to using a fixed mapping bounded by the most difficult posture in the motion clip.

Suppose we use N markers in frame k as reference and denote the positions of the markers attached to the marionette by $p_{k,i}^M$, those of the captured markers by $p_{k,i}^C$ and those of the mapped markers by $p_{k,i}^S$ ($i = 1 \ldots N$). We represent the translational, rotational, and scaling parameters of frame k by a position vector t_k, a 3-by-3 rotation matrix R_k, and a scalar s_k, respectively. Using these parameters, we compute the mapped position $p_{k,i}^S$ of marker i from its original captured position $p_{k,i}^C$ as $p_{k,i}^S = s_k R_k p_{k,i}^C + t_k$.

The system first computes the scaling, translation, and orientation parameters that minimize the total square distance between the measured markers and the virtual markers on the marionette in a fixed configuration. We can then use the inverse kinematics algorithm to compute the joint angles and string lengths that provide the best match between the two sets of markers. These two steps are repeated a number of times to refine the result.

The system finds the translation, rotation, and scaling parameters t_k, R_k, and s_k that minimize the evaluation function

$$J_k = \frac{1}{2} \sum_{i=1}^{N} |p_{k,i}^S - p_{k,i}^M|^2. \quad (10)$$

$p_{k,i}^M$ are constant because the configuration of the marionette is fixed during this frame by frame computation.

We combine the unknowns into one variable $q_k \in \mathbf{R}^7$, where the rotation matrix is represented by three independent variables whose time derivatives correspond to the angular velocity. We then use the common gradient method [14] to compute the optimum q_k incrementally as

$$q_k = q_k + \Delta q_k, \quad \Delta q_k = -k \frac{\partial J_k}{\partial q_k}. \quad (11)$$

The partial derivative of the mapped position $p_{k,i}^S$ with respect to q_k is computed as

$$H_{i,k} \triangleq \frac{\partial p_{k,i}^S}{\partial q_k} = \begin{pmatrix} I_3 & | & [r\times] & | & R_k p_{k,i}^C \end{pmatrix} \quad (12)$$

where I_3 is the 3-by-3 identity matrix, $r \triangleq s_k R_k p_{k,i}^C$, and $[r\times]$ is the cross product matrix of r. Using $H_{k,i}$, the partial derivative of J_k is computed by

$$\frac{\partial J_k}{\partial q_k} = (p_{k,i}^S - p_{k,i}^M)^T H_{k,i}. \quad (13)$$

We apply this process to each frame independently by starting from the same initial guess. Using the result of previous frame as the initial guess would reduce the computation time, but the algorithm might not recover from a failure to obtain good mapping parameters in one frame due to, for example, missing markers. Regardless of the initial guess, the resulting mapping parameters may not be continuous because the algorithm is finding only a local minima. Small discontinuities are not a problem, however, because the marker positions are "filtered" by the feedback controller and by the SR-inverse used in the inverse kinematics computation.

6 Controlling Swing

If the mapped motion is applied directly to the marionette, the hands of the marionette will swing and the motion will not be a good match to that of the human actor. We solve this problem by building a simple linear model for the swing dynamics and experimentally identifying its parameters. An alternative approach would be to model the full dynamics of the marionette, but this tactic is not practical because of uncertainty in the model parameters and the limitations of the motors and sensors. Because marionette is made of wood and cloth, it is difficult to precisely determine the mass, inertia, and friction parameters of the joints. The joints are cleverly designed to prevent unrealistic joint angles (Figure 1), but this design also makes modeling of the system more difficult. The motors are inexpensive hobby servos and do not provide precise control. Furthermore, we do not have sensors that measure the current state of the marionette during a performance.

For the simple model of the dynamics, we make three assumptions. (1) Swinging of the hands occurs in the horizontal plane. Pulling the hands or legs up or down does not create a swinging motion. (2) The motion of a hand along the x axis (left/right) and the y axis (forward/back) are independently controlled by one motor each. (3) There is no coupling between the swinging of the left and right hands. These simplifying assumptions allow us to model swing as four independent systems, two for each hand.

Some situations occur in which the second and third assumptions do not hold. The hand marker sometimes moves along a circular trajectory rather than a straight line. The markers with fixed inputs inevitably move slightly when other markers are moved, violating the last assumption. Both problems are most likely to occur when the hand is relatively close to the body because the stiffness of the elbow and shoulder joints forces the hand away from the body.

6.1 Modeling of Swing Dynamics

In this section we describe the swing dynamics model that, when combined with the feedback controller of the inverse kinematics algorithm in Section 4, predicts the swing motion.

The inverse kinematics algorithm included a proportional controller, where the velocity of the pipe end $\dot{x}$ is computed from the current position of the pipe end x and the desired position u as $\dot{x} = k(u - x)$ where k is a constant gain. Therefore the transfer function from the marker trajectory to the motion of the pipe end takes the form

$$x = \frac{1}{a_{ik}s + 1} u \quad (14)$$

where u is the input (marker trajectory), x is the output (motion of the pipe end), s is the Laplace transformation operator, and a_{ik} is the parameter that determines the amount of delay.

The motion of the hand for a given trajectory of the pipe end can be modeled as a pendulum with a moving base (Figure 5 right). Using the length of the pendulum l and the damping term d, the equation of motion of the pendulum under gravity g is linearized around $x = y$ as $\ddot{y} = l/g(x - y) + d(\dot{x} - \dot{y})$. In general, therefore, the transfer function from the motion of the pipe end to the marker motion is written as

$$y = \frac{b_s s + 1}{a_s s^2 + b_s s + 1} x \quad (15)$$

where y is the output (actual marker trajectory) and a_s and b_s are the parameters that determine the frequency and damping respectively.

Combining Eqs.(14) and (15), the system computing the desired marker trajectory from the actual trajectory will be a 3rd-order system. We estimate the three parameters from motion capture data.

The gains of both systems were assumed to be 1, which turned out to be not true, probably because the joint motion ranges of the pipe prevented the pipe end from reaching the desired position or because the stiffness of the arm joints did not allow the string to be perpendicular. We decided not to consider these model errors because the desired marker position is not achievable if it violates the joint range constraint and the stiffness strongly depends on the configuration of the arm making the system too complicated.

6.2 Feedforward Controller

The feedforward controller K is formed by connecting the desired response G_D and the inverse of the estimated model P_m in series, that is, $K = G_D P_m^{-1}$. In order for the controller to be proper (the order of the denominator of the transfer function is larger than that of the numerator), the order of G_D must be larger than 2. We selected a 3rd-order G_D so that the output of the controller is continuous. We can also improve the response of the total system by selecting G_D with a smaller delay. In practice, however, we cannot use an arbitrarily fast G_D because as the gain of the controller increases, it becomes sensitive to modeling errors.

The parameters of the string dynamics model, a_s and b_s, depend on the length of the strings; therefore, we repeat the identification process for several different heights for each hand and design a controller for each model. We then apply the weighted sum of the outputs of the three controllers, where the weights are determined according to the actual height during a performance.

7 Results

The inverse kinematics computation to obtain the motor commands was repeated four times for each frame to ensure convergence. The total computation time was about 36ms per frame on a laptop PC with a Mobile PentiumIII 1GHz processor. Motor commands were sent every 50ms.

Based on the inverse kinematics computation, we developed an online control interface for the marionette. The model consists of nine string length constraints, joint motion range constraints for eight joints, two string direction constraints, and the potential energy constraint. The user can select a marker and drag it to any position. The inverse kinematics algorithm then computes the motor commands and the joint angles to move the marker to the specified position. Figure 6 shows several snapshots of the marionette model and the corresponding postures of the actual marionette.

The swing controller was designed for three different heights (-0.59m, -0.44m, and -0.29m, measured from the center of the panel where the motors and pulleys are attached). We had a total of twelve controllers for the x and y directions of both hands. Table 1 lists the parameter sets for the right hand in the x direction. The parameters were tuned manually, although it should also be possible to apply standard system identification techniques [15].

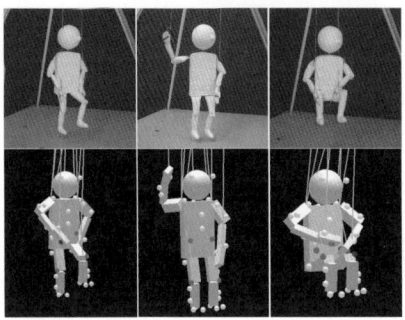

Figure 6: Postures generated by the interactive interface. Above: marionette, below: simulation.

Table 1: Parameters of the string dynamics models for the x direction of the right hand.

height [m]	-0.59	-0.44	-0.29
a_{ik}		0.8	
a_s	0.063	0.061	0.059
b_s	0.07	0.06	0.02

Figures 7 and 8 show the results of identification, controller design, and verification processes at the height of -0.29m. We used the motion capture system to measure the motion of the pipe end and right hand when a step input in x direction (left to right) was given as the desired marker trajectory (Figure 7). Then we designed a swing controller with the desired response $G_D = 1/(0.2s+1)^3$.

Finally, the designed controller was applied to the same desired marker trajectory used for the identification and the response was measured (Figure 8). The swing controller reduced the width of the first vibration by 40%. The trajectory of the hand without the swing controller is different from that used for parameter identification (Figure 7), although we used the same reference trajectory. This discrepancy probably explains why the controller could not remove the vibration completely, thereby illustrating that a small difference in the configuration results in a relatively large difference in the swing dynamics due to the stiffness of the arms.

To test the motion of the marionette on a longer performance, we recorded the motions of two actors for two stories: "Who Killed Cockrobin?" and "Alaska." Figure 9 compares the motions based on "Alaska" performed by actor 1. We used 32 reference markers and the two steps for mapping (computing approximate parameters and computing exact parameters) were repeated up to 500 times at each frame. The iteration was suspended if the total error of the marker positions were larger than the previous iteration. The computation time was approximately 5 seconds

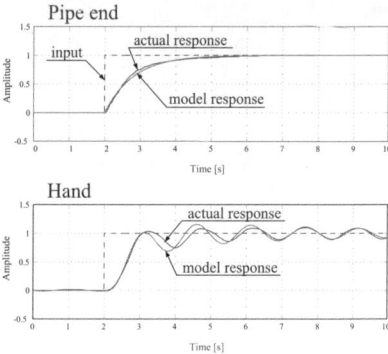

Figure 7: Actual and model responses to a step input. The amplitude of each motion is normalized. The hand of the marionette comes close to its head at this height.

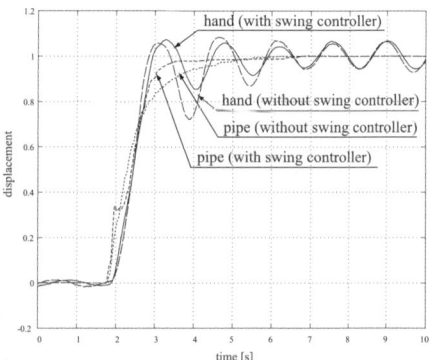

Figure 8: Response of the pipe and the hand to a step input.

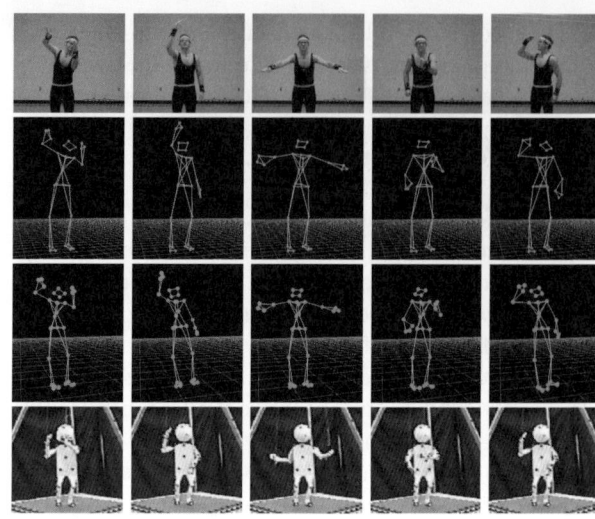

Figure 9: From the top: performance of actor 1 for "Alaska," the motion capture data, mapped marker positions, and the marionette's motion.

Figure 10: Marionette's motion for "Alaska" performed by actor 2.

per frame.

Figures 9 and 10 illustrate the same story performed by two different actors. The gestures are taken from approximately the same point in the story. The motion in Figure 11 is based on a different story performed by actor 1. The video clips are available online at http://humanoids.cs.cmu.edu/projects/marionette/.html, which also includes comparisons between the motions with and without the swing controller. The marionette's feet touch the floor as in a real performance.

8 Discussion

The motions of the actor and the marionette showed good correspondence, and we were able to distinguish two different styles for the same story (Figures 9 and 10). However, significant differences between the actor's and the marionette's postures were sometimes visible because of the limited range of motion of the pipes (for example the middle column of Figure 9). The marionette also had difficulty with fast motions because of the latency in the feedback controller of the inverse kinematics computation. This problem could be solved with a faster computer that could execute more iterations per step of the inverse kinematics computation, thereby increasing the stability of the computation and allowing larger gains.

Although the swing controller had a significant effect in isolated experiments, its effect during longer performances was quite small. We believe this discrepancy occurred because the stiffness of the arms is highly dependent on the configuration and this effect was not taken into account in the swing model. We could include this effect by testing the response of the system for both pipe position and string length.

The examples in this paper were limited to motions where the actor was told to stand in place during the perfor-

Figure 11: Marionette's motion for "Who Killed Cockrobin?" performed by actor 1.

mance. We could extend the range of feasible motions by adding more controllable strings and degrees of freedom. For example, a motor to control the string connecting the back would allow the marionette to bow. We could also add a pair of strings and motors to control the elbows independently or to move the entire marionette as a human operator would do for walking. In the construction of marionettes for human-operated performances extra strings are often added to enable a particular set of behaviors for that marionette's character.

We did not consider self collisions between the puppet and the strings or interaction with the environment. In the motions shown here we did not encounter situations where self collisions caused significant change of motion, but this issue is a serious concern in the design of performance marionettes with clothing that may catch on the strings. We kept the feet in contact with the floor to reduce the swing of the pelvis but did not explicitly consider contact with the environment in the control system. If the marionette had the additional degrees of freedom for such whole-body motions as walking, modeling of the interaction with the environment would be essential.

We explored two interfaces for driving the marionette: direct input of marker positions for realtime control and offline processing of human motion data. A third alternative would have been to capture a human-operated marionette performance to take advantage of the talent of a professional operator. The control scheme for this interface would presumably be significantly less complex because the motions would already be appropriate to the dynamics of the marionette. Such a system, however, could not easily be operated by an untrained user. In contrast, the control scheme described in this paper enables a naive user to program a motorized marionette to create entertaining performances simply by performing the gestures in a motion capture system.

Acknowledgments The authors would like to thank Rory Macey and Justin Macey for their assistance in capturing the human and marionette motions. This research was supported in part by NSF 0196089 and 0196217.

References

[1] "The Honda Humanoid Robot ASIMO," http://world.honda.com/ASIMO/.

[2] Y. Kuroki, T. Ishida, J. Yamaguchi, M. Fujita, and T. Doi1, "A Small Biped Entertainment Robot," in *Proceedings of Humanoids 2001*, Tokyo, Japan, November 2001.

[3] "Sarcos High Performance Robots," http://www.sarcos.com/entspec_highperfrobot.html.

[4] K. Yamane and Y. Nakamura, "Synergetic CG Choreography through Constraining and Deconstraining at Will," in *Proceedings of the IEEE International Conference on Robotics and Automation*, Washington DC, May 2002, pp. 855–862.

[5] G. Hoffmann, "Teach-In of a Robot by Showing the Motion," in *IEEE International Conference on Image Processing*, 1996, pp. 529–532.

[6] N. Pollard, J. Hodgins, M. Riley, and C. Atkeson, "Adapting Human Motion for the Control of a Humanoid Robot," in *Proceedings of the IEEE International Conference on Robotics and Automation*, Washington DC, May 2002, pp. 1390–1397.

[7] M. Gleicher, "Retargetting Motion to New Characters," in *Proceedings of SIGGRAPH '98*, 1998, pp. 33–42.

[8] S. Tak, O. Song, and H. Ko, "Motion balance filtering," *Eurographics 2000, Computer Graphics Forum*, vol. 19, no. 3, pp. 437–446, 2000.

[9] K. Yamane and Y. Nakamura, "Dynamics Filter—Concept and Implementation of On-Line Motion Generator for Human Figures," in *Proceedings of the IEEE International Conference on Robotics and Automation*, vol. 1, San Francisco, CA, April 2000, pp. 688–695.

[10] C. Rahn, F. Zhang, S. Joshi, and D. Dawson, "Asymptotically stabilizing angle feedback for a flexible cable gantry crane," *ASME Journal of Dynamic Systems, Measurement, and Control*, vol. 121, pp. 563–566, September 1999.

[11] Y. Nakamura and H. Hanafusa, "Inverse Kinematics Solutions with Singularity Robustness for Robot Manipulator Control," *Journal of Dynamic Systems, Measurement, and Control*, vol. 108, pp. 163–171, 1986.

[12] A. Maciejewski, "Dealing with the Ill-conditioned Equations of Motion for Articulated Figures," *IEEE Computer Graphics and Applications*, vol. 10, no. 3, pp. 63–71, May 1990.

[13] T. Sugihara, Y. Nakamura, and H. Inoue, "Realtime Humanoid Motion Generation through ZMP Manipulation based on Inverted Pendulum Control," in *Proceedings of the IEEE International Conference on Robotics and Automation*, Washington DC, May 2002, pp. 1404–1409.

[14] W. Press, S. Teukolsky, W. Vetterling, and B. Flannery, *Numerical Recipes in C Second Edition*. Cambridge, UK: Cambridge University Press, 1999.

[15] L. Ljung, *System Identification – Theory for the User*. Prentice – Hall, 1987.

Achieving Periodic Leg Trajectories to Evolve a Quadruped Gallop

Darren P. Krasny and David E. Orin

Department of Electrical Engineering, The Ohio State University, Columbus, OH, 43210
(e-mail: krasny.1@osu.edu, orin.1@osu.edu)

Abstract—For most large quadrupedal mammals, galloping is the preferred gait for high-speed locomotion. In this paper we evolve a gallop gait in a simulated quadruped robot at speeds from 3.0 to 10.0 m/s. To do so, we must generate periodic trajectories for the body and legs. An evolutionary algorithm known as set-based stochastic optimization (SBSO) is used to find the body trajectory while alternative methods are used to find periodic leg trajectories. The focus of this paper will be to evaluate three different methods for generating periodic leg trajectories. The combined solutions for the body and legs yield biological characteristics that are emergent properties of the underlying high-speed dynamic running gait.

1 Introduction

Because of the efficiency of dynamic running gaits like the gallop, there is considerable interest in developing robots capable of this type of locomotion [1, 2]. Perhaps the most notable early example of a quadruped robot capable of dynamic running was developed by Marc Raibert in the 1980's [3]. Raibert's machine had prismatic legs with pneumatic and hydraulic actuators and was capable of trotting, pacing, and bounding at speeds up to 2.9 m/s. To date, Raibert holds the record for the fastest speed for dynamic running in a quadruped robot [4]. While Raibert's work has laid the foundation for studies in dynamic running, there remains the task of developing even faster, more biomimetic quadruped robots capable of true high-speed gaits like the gallop.

Current research conducted at The Ohio State University and Stanford University is focused on developing just such a machine [5]. As part of that research, we are currently investigating the dynamics of high-speed quadrupedal galloping using a simulated robot with compliant, articulated legs and an asymmetric mass distribution similar to that of a goat [4]. An evolutionary approach is used to generate periodic trajectories for the body over a single stride using a planar dynamic simulation [6]. Because finding periodic leg trajectories involves increasing the dimensionality of the search space, extending the evolutionary search to find periodic trajectories for the body *and* the legs was found to be very difficult. While others have successfully searched large-dimensional spaces to generate stable quadruped gaits, the gaits under investigation were low-speed, statically stable, and were not required to closely resemble their biological analogs [7]. Here we wish to generate high-speed, dynamic running gaits that are *biomimetic* in character – a challenging task for the evolutionary algorithm. The particular goal of this paper is to explore two alternative leg methods, including a recursive least-squares neural network approach (hereafter, the "RLS/neural approach") and a simplified approach, and compare the performance of these methods with the evolutionary approach. Each method must find the proper parameters for each leg such that the beginning and ending leg angle and length are approximately equal.

The next section explains the evolutionary algorithm, followed by a description of the optimization problem. The dynamic model is explained after that, followed by descriptions of each of the periodic leg methods, including a comparative analysis of their performance. Next, the overall results for the gallop are presented using both the RLS/neural and simplified methods in parallel. Finally, the results of this study are summarized.

2 The SBSO Algorithm

The SBSO algorithm is a direct search evolutionary strategy that explores the parameter space in a parallel fashion using a gaussian cloud of individuals distributed about the current generation's elite individual. Figure 1 depicts the general process for several generations. The SBSO algorithm is similar to the standard genetic algorithm (GA), although it can be implemented without the "overhead" associated with imitating biological evolution.

The development for set-based stochastic optimization provided here is based on that given in [6, 8]. Similar to the genetic algorithm, SBSO defines the concepts of a "population" and an "individual," although in a simpler fashion. For example, the population of S members at generation k is defined in the SBSO algorithm as follows:

$$P(k) = \{\phi^j(k) : j = 1, 2, \ldots, S\}, \quad (1)$$

where each individual of the population has a set of parameters

$$\phi^j(k) = \left[\phi_1^j(k), \phi_2^j(k), \ldots, \phi_p^j(k)\right]^T, \quad (2)$$

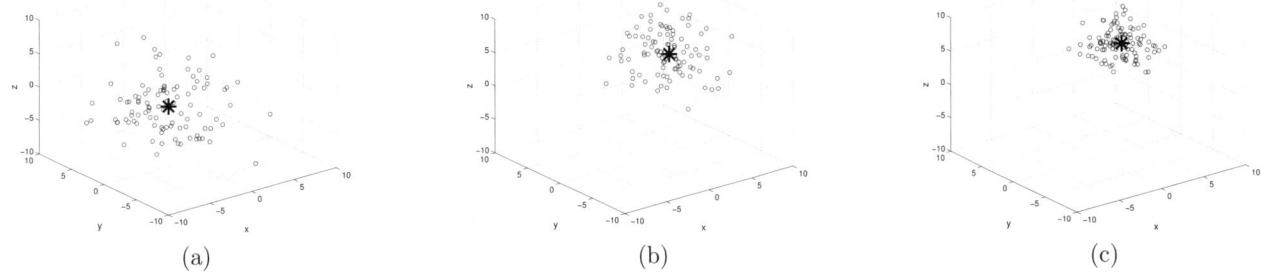

Figure 1: Illustration of the SBSO algorithm: (a) at generation $k = 0$, (b) $k = 1$, and (c) $k = 2$. The "*" in each plot represents the elite individual about which the search cloud is generated. The size of the search cloud decreases according to a pre-specified decay rate.

where $\phi_i^j(k) \in \Omega_i$ and Ω_i is the domain for parameter i. The population is initialized randomly and each individual is evaluated using the cost function $J(\phi)$. The best individual $\phi^{j^*}(k)$ with respect to $J(\phi)$ for the current generation k is then passed down to the next generation as follows:

$$\phi^{j^*}(k+1) = \phi^{j^*}(k) \ . \tag{3}$$

The remainder of the population is then generated using the following:

$$\phi^j(k+1) = \phi^{j^*}(k+1) + \mathbf{\Gamma} r^j \ , \tag{4}$$

for $j \neq j^*$ and $j = 1, \ldots, S - M$, where r^j is a $p \times 1$ vector containing random numbers with normal distribution, zero mean, and unit variance, $\mathbf{\Gamma}$ is a $p \times p$ diagonal matrix containing the scaling factors for each parameter, and M is the number of mutants in the population. The scaling matrix $\mathbf{\Gamma}$ is applied to the random number vector r^j to produce the desired "cloud" size around the fittest individual. *Projection* is used to keep ϕ^j within a convex-bounded domain Ω, where Ω contains the domains Ω_i for $i = 1, \ldots, p$.

The last step in the algorithm is mutation, where the last M individuals ϕ^j for $j \neq j^*$, $j = S - M + 1, \ldots, S$ are assigned parameter values from a uniform distribution on the entire parameter space. Again, projection is used to keep the individuals within Ω.

Once the new generation is created, it is evaluated again using $J(\phi)$, and a new elite individual is found. This process continues until a suitable solution is found or a certain number of generations N have been produced.

In order to improve the performance of the SBSO algorithm, several heuristic modifications were added. First, an adaptive $\mathbf{\Gamma}$ parameter was chosen so that the search cloud gradually decreases with each new generation, allowing the SBSO algorithm to narrow in on a solution. The scaling matrix $\mathbf{\Gamma}$ decreases with a specified decay rate as follows:

$$\mathbf{\Gamma}(k+1) = (1 - \delta_\Gamma)\mathbf{\Gamma}(k) \tag{5}$$

where δ_Γ is the scaling factor decay rate. If $\mathbf{\Gamma}$ becomes less than a minimum value $\mathbf{\Gamma}_m$, $\mathbf{\Gamma} = \mathbf{\Gamma}_m$. $\mathbf{\Gamma}$ is initialized with $\mathbf{\Gamma}_0$, where the values in $\mathbf{\Gamma}_0$ are chosen to be very large with respect to the domain Ω to generate an approximately uniform distribution over the parameter space. Parameter values can be found in [6].

Because the simulation can often fail to *bootstrap* (i.e., find an initial solution), $\mathbf{\Gamma}$ is decreased from its initial value only after the algorithm has made some improvement to ϕ^{j^*}. If no improvement in ϕ^{j^*} has been made over the past five generations, $\mathbf{\Gamma}$ is reset to its initial value $\mathbf{\Gamma}_0$ to enlarge the search space.

Finally, to facilitate the evolutionary search for each new running speed (or "rule"), the solution for the last rule is used to initialize the current rule. This initialization method was found to reduce the search time and increase the average convergence rate. Table 1 summarizes the SBSO algorithm with the modifications described above.

3 The Optimization Problem

The overall goal of this work is to analyze the dynamic characteristics of the gallop by finding optimal control parameters that produce a periodic stride with respect to body and leg trajectories. An example of a typical gallop is shown in Fig. 2. Following a long flight phase, called the "gathered flight phase" (because the legs are gathered more closely under the body), the two hind feet come down in succession, followed by the two front feet in succession[1].

The SBSO algorithm described above is used to find the appropriate control parameters and initial conditions to produce a periodic solution with respect to the *body's* trajectory for each galloping speed from 3.0 to 10.0 m/s (using 0.5 m/s increments). A fitness function is used to penalize the error between the actual and desired ending states of the body in order to drive the algorithm to convergence. Additional terms are included in the fitness function to minimize the total amount of energy expended

---

[1] At faster speeds, there may also be an additional flight phase between the rear and fore footfalls. In addition, the footfall sequence may be transverse (e.g., RR-LR-RF-LF) or rotary (e.g., RR-LR-LF-RF). For a planar simulation, both gait styles are equivalent. A transverse gait was arbitrarily chosen for the simulation.

Table 1: Summary of the modified SBSO algorithm.

1. Choose population size S, number of generations N, number of mutants M, and number of rules R. Choose parameter domain $\mathbf{\Omega}$, initial scaling matrix $\mathbf{\Gamma}_0$, minimum scaling matrix $\mathbf{\Gamma}_m$, and scaling factor decay rate δ_Γ.
2. For $m = 1, \ldots, R$:
   (a) Initialize rule: if $m > 1$, copy $\boldsymbol{\phi}^{j^*}$ from last rule to $\boldsymbol{\phi}^{j^*}(1)$; otherwise, choose $\boldsymbol{\phi}^{j^*}(1)$ from a uniform random distribution. Set $\mathbf{\Gamma} = \mathbf{\Gamma}_0$.
   (b) For $k = 1, \ldots, N$:
      (1) Create new generation: $\boldsymbol{\phi}^j(k) = \boldsymbol{\phi}^{j^*}(k) + \mathbf{\Gamma} \boldsymbol{r}^j$, for $j = 1, \ldots, S - M$, $j \neq j^*$, where $\boldsymbol{r}^j$ is a $p \times 1$ vector of random numbers with normal distribution, zero mean, and unit variance.
      (2) Project $\boldsymbol{\phi}^j(k)$ onto $\mathbf{\Omega}$ for $j = 1, \ldots, S - M$, $j \neq j^*$.
      (3) (Mutation) Select M individuals $\boldsymbol{\phi}^j$ for $j = S - M + 1, \ldots, S$, $j \neq j^*$, and assign parameter values from a uniform distribution on the parameter space. Project the M individuals onto $\mathbf{\Omega}$.
      (4) Evaluate $J(\boldsymbol{\phi}^j)$ for $j = 1, \ldots, S$.
      (5) Find $\boldsymbol{\phi}^{j^*}$ such that $J(\boldsymbol{\phi}^{j^*}) = \min_{\boldsymbol{\phi}^j} J(\boldsymbol{\phi}^j)$ [a].
      (6) If convergence requirements met, exit k-loop. Otherwise, continue.
      (7) (Selection) Create next generation's elite individual: $\boldsymbol{\phi}^{j^*}(k+1) = \boldsymbol{\phi}^{j^*}(k)$.
      (8) If progress has been made with respect to $J(\boldsymbol{\phi}^{j^*})$, update $\mathbf{\Gamma}$: $\mathbf{\Gamma}(k+1) = (1 - \delta_\Gamma)\mathbf{\Gamma}(k)$. If $\mathbf{\Gamma}$ becomes less than a minimum value $\mathbf{\Gamma}_m$, $\mathbf{\Gamma} = \mathbf{\Gamma}_m$. If no progress has been made, $\mathbf{\Gamma}(k+1) = \mathbf{\Gamma}_0$.

[a] Note that the best fitness in this problem corresponds to the minimum of $J(\boldsymbol{\phi})$.

Figure 2: The gallop is the preferred gait for high-speed locomotion for most large quadrupedal mammals [9]. (Illustration from [10].)

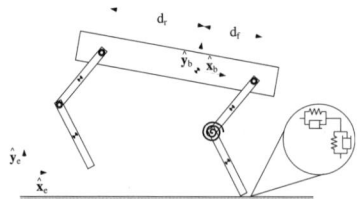

Figure 3: Dynamic model of the planar quadruped showing two of the four legs simulated. A compliant leg is simulated when the leg contacts the ground.

per stride (hereafter, "cost of transport," or COT) and the peak forces experienced in the legs. The complete specifications for the optimization problem are given in [6]. Due to space limitations, we will assume here that the optimization problem for the body trajectory has been solved, leaving the task of finding periodic leg trajectories, described below. Before discussing this, however, we will first present the dynamic model for the simulation.

4 The Dynamic Model

Figure 3 shows the dynamic model of the planar quadruped. Parameters can be found in [6] and are based in large part on [4]. The quadruped has a total mass of $61.3\,kg$, of which $46.3\,kg$ is comprised of the body. The center of mass is situated towards the front of the machine, similar to quadrupedal mammals, such that $d_f = 0.25\,m$ and $d_r = 0.57\,m$. A planar model was selected because the velocity component in the direction perpendicular to the plane of progression is typically small for quadrupedal running gaits [11]. As shown in the figure, the model is simulated differently depending on whether it is in flight or contacting the ground[2]. During flight the legs are simulated as simple rotational links with mass and inertial characteristics. During contact, however, a torsional spring is added at the knee joint to produce a compliant leg similar to that found in biological systems [13, 14, 15]. The springs at these joints allow for the storage of energy similar to the way biological systems store energy in muscles, tendons, and ligaments [14]. During flight it is assumed that there is no storage of energy; therefore, the spring is not included. To compensate for impact losses at the ground, energy is injected into the knee springs during stance to provide an upward thrust and restore the galloping height. The total energy per

[2] A nonlinear damping model based on [12] is used to model the ground contacts.

stride, E, includes actuator energy for each joint plus the energy injected into the knee springs during stance [6].

5 The Periodic-Leg Methods

While the SBSO algorithm is used to find periodic solutions for the trajectory of the body, a separate, more efficient method is sought to find periodic solutions for the legs. For a periodic solution with respect to the legs, the initial and final lengths and angles must be within a certain tolerance, as described in Table 2. Once the evolutionary algorithm has found a solution which satisfies both the body convergence requirements in [6] *and* the leg convergence requirements for all legs, the current rule will converge.

Finding periodic leg solutions requires initializing each leg somewhere along its transfer spline[3]. This requires specifying the estimated lift-off angle of the leg and the proportion of the transfer spline completed by the top-of-flight (TOF), which adds eight additional control parameters to the evolutionary parameter set. Three different methods were explored to find these control parameters, as explained below.

5.1 The Evolutionary Leg Approach

The evolutionary approach is simply an augmentation of the SBSO algorithm described above. This approach requires additional convergence requirements for the legs, additional cost components in the fitness function, and additional control parameters.

Two additional terms are included in the fitness function to minimize the discrepancy between initial and final leg lengths and angles. The fitness function is as follows:

$$\bar{J}(\phi^j) = J + \sum_{i=1}^{4} w_l |l_{f_i} - l_{0_i}| + \sum_{i=1}^{4} w_a |\theta_{v_{f_i}} - \theta_{v_{0_i}}|, \quad (6)$$

where J is the fitness function for the body and can be found in [6], $w_l = 62.5$, $w_a = 20.0$, l_{0_i} and l_{f_i} are the initial and final leg lengths (measured from hip to foot), and $\theta_{v_{0_i}}$ and $\theta_{v_{f_i}}$ are the initial and final leg angles (measured between the hip and the foot).

To generate periodic solutions for the legs, two control parameters per leg must be added to the set of evolutionary parameters for the body. The additional evolutionary leg parameters and their associated domains are listed in Table 3.

The eight additional evolutionary search parameters for the evolutionary leg approach yields an 18-dimensional search space. Finding "good," biomimetic solutions in such a high-dimensional space is quite challenging.

5.2 The RLS/Neural Approach

The RLS/neural approach uses on-line function approximation to find the estimated lift-off angles, θ_{LO_i}, and the proportion of the leg transfer splines completed at the next TOF, t_i^*, for each of the four legs. Each function has a set of five inputs, the first four of which are the initial conditions of the body, and the last is the pre-touchdown leg angle (θ_{l_i}). Each of the eight functions are approximated using its own neural network.

For simplicity, one structure was chosen for all of the neural networks. It was found through trial and error that a two-layer network with two neurons in the hidden layer and one neuron in the output layer performed satisfactorily in predicting θ_{LO} and t^*. Experimentation with additional neurons in the hidden layer yielded no noticeable performance gains. Log-sigmoid activation functions were employed in the hidden layer and linear functions were used in the output layer.

Figure 4 shows the structure of the neural network used for each of the eight leg functions. In the first (or hidden) layer, there are a total of ten weight values and two bias values used to compute the layer outputs a_1^1 and a_2^1 shown in Fig. 4. For log-sigmoid activation functions in the hidden layer, the output a_i^1 is calculated as follows:

$$a_i^1 = \frac{1}{1 + \exp^{-(\boldsymbol{w}_i^1 \boldsymbol{x} + b_i^1)}} \quad \text{for} \quad i = 1, 2, \quad (7)$$

where $\boldsymbol{w}_i^1$ is a 1×5 vector of weight values for the i-th neuron, $\boldsymbol{x}$ is the 5×1 input vector, and b_i^1 is the bias value for the i-th neuron. The input vector $\boldsymbol{x}$ consists of the initial conditions of the body and the nominal pre-touchdown leg angle for the current leg:

$$\boldsymbol{x} = [h_0 \quad v_0 \quad \alpha_0 \quad \omega_0 \quad \theta_l]^T. \quad (8)$$

In abbreviated notation, the vector output of the first

---

[3]The transfer spline is a cubic spline with a 0.25 sec duration that smoothly transfers the leg from its lift-off position back to the pre-touchdown angle. The leg is shortened and re-lengthened during leg transfer to avoid ground interference.

Table 2: Leg convergence requirements.

Initial Condition	Tolerance
Leg length (l_0)	$\pm 0.05\,m$
Leg angle (θ_{v_0})	$\pm 5.0\,\deg$

Table 3: Evolutionary parameters for leg i, $i = 1, \ldots, 4$, for the evolutionary leg strategy.

Param.	Description	Range
t_i^*	Transfer spline time	$[0.0, 1.0]$
θ_{LO_i}	lift-off angles	$[-2.5, -1.0]$ (rad)

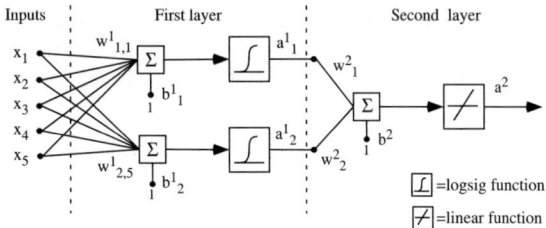

Figure 4: Neural net structure for one leg function.

layer is often written $\boldsymbol{a}^1$, where $\boldsymbol{a}^1 = [a_1^1 \ a_2^1]^T$.

In the second (or output) layer, the output a^2 is calculated using the outputs of the hidden layer as the arguments to the linear activation function:

$$a^2 = \boldsymbol{w}^2 \boldsymbol{a}^1 + b^2 \quad , \qquad (9)$$

where $\boldsymbol{w}^2 = [w_1^2 \ w_2^2]$ and b^2 is the bias value. The output a^2 thus represents the estimate for the function being approximated.

To facilitate the function approximation method, the weights and bias values in the hidden layer of the neural network were selected somewhat arbitrarily so that the unknown weight and bias values in the output layer would enter the neural network in a linear fashion. This is a useful arrangement because there exist good methods to tune "linear-in-parameters" structures [8]. To establish a linear-in-parameters network, the weight values $w_{i,j}^1 \ \forall i, j$ were set to 1.0, and the bias values were set to $b_1^1 = -5.0$ and $b_2^1 = 10.0$. The bias values were chosen to cover the range of expected inputs. Since the output layer consists of a linear activation function, the output of the entire neural network can now be represented as a linear function:

$$a^2 = \boldsymbol{\xi}^T \boldsymbol{\zeta} \quad , \qquad (10)$$

where $\boldsymbol{\xi} = [w_1^2 \ w_2^2 \ b_2]^T$, $\boldsymbol{\zeta} = [(\boldsymbol{a}^1)^T \ 1]^T$, and $\boldsymbol{a}^1$ is a known function of the inputs $\boldsymbol{x}$, the components of which are calculated by using Eq. (7). Thus by specifying the hidden layer weights and biases, the problem of tuning the neural network has been reduced to finding the three output layer parameters. For eight leg functions, this means that 24 parameters must be found.

The recursive least squares method can be used to update all parameters that enter the neural network linearly (i.e., in $\boldsymbol{\xi}$) by minimizing the error between the function estimate a^2 and the actual value y. While there exist other methods for computing the best $\boldsymbol{\xi}$, the recursive least squares approach offers the advantage that updates can be computed in real-time as each new *training pair* (i.e., $(\boldsymbol{x}, y)$) is received (as opposed to batch approaches, where the data is gathered first, then processed off-line). The RLS method used here also employs a weighting factor λ so that more recent data has more of an effect on the update. The update formulas for the weighted RLS method are given in [6].

In summary, the RLS/neural method is a powerful technique that also offers much potential for performance improvement because of the flexibility in adjusting the structure or type of the neural network. The next section describes an alternative approach to finding periodic solutions for the legs without using on-line function approximation.

5.3 The Simplified Approach

The simplified approach relies on the assumption that small changes in the initial leg trajectories do not greatly affect the body dynamics given the same set of initial conditions for the body and the same control parameters. For example, assume that the current stride is initialized with a certain set of leg states $\boldsymbol{\theta}_{s_0}$, where $\boldsymbol{\theta}_{s_0}$ contains the eight control parameters θ_{LO_i} and t_i^*, for $i = 1, \ldots, 4$, which are required to initialize the legs somewhere along their respective transfer splines. Then, at the end of the stride (at TOF), the final leg states actually achieved during the stride are recorded in $\boldsymbol{\theta}_{s_f}$. The simplified method assumes that if $\boldsymbol{\theta}_{s_f}$ is "fed back" to the evolutionary algorithm and the stride is repeated with exactly the same body initial conditions and control parameters, the result will be a new solution with approximately periodic leg trajectories and a body trajectory similar to the original stride.

The advantage of the simplified method is that virtually no additional computation is required to implement the method. On the other hand, the limitations are quite obvious. The assumption that one set of leg states will provide a good initialization for all individuals in a generation is obviously not very accurate unless the final leg states are approximately the same across the individuals in the population. This occurs only when the search cloud is relatively small (see Fig. 1). Furthermore, the assumption that changes in the initial leg trajectories have little impact on body dynamics is not entirely accurate. In fact, changing the initial leg trajectories seems to have the most noticeable effect on the body's angular velocity during flight.

In spite of its limitations, the simplified method can perform quite well and is used in parallel with the RLS/neural approach to generate periodic leg and body solutions for each gait. Section 6 examines the performance of the simplified method against the other two methods in finding periodic leg and body solutions for the gallop.

6 Results for the Periodic-Leg Methods

In this section each of the three periodic-leg methods described in Section 5 is evaluated by running 100 trials using the gallop. The maximum average peak force level in the legs was constrained to be $1800.0 \, N$. The leg convergence requirements are listed in Table 2, and the additional evolutionary parameters for the legs are listed in

Table 3. In evaluating the methods, three criteria were used, ranked in order of importance: quality of solutions (with respect to COT), convergence rate, and simulation time.

The COT graph and convergence rate for the three methods are shown in Figure 5. Both plots show that the evolutionary method was the worst performer across the board, with generally the highest COT values, worst average convergence rate (9.7%), and the longest simulation time (46 min per trial). The simplified method had the highest average convergence rate (35.3%), while the RLS/neural method had a slightly lower rate (24.8%). Both the RLS/neural and simplified methods generally performed the same with respect to quality of solutions and simulation time (around 28 min per trial for each). For this reason, both the RLS/neural and simplfed methods were used in parallel for the remainder of the testing, the results of which are presented next.

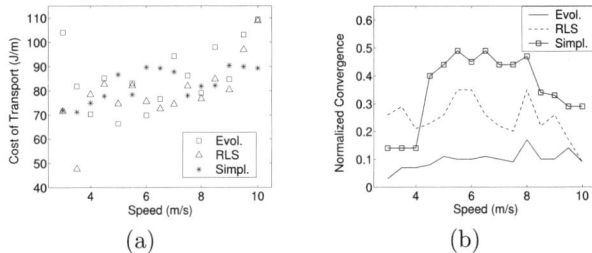

Figure 5: Comparison of (a) COT, and (b) convergence rate for the periodic-leg methods. For (b), the evolutionary method had an average convergence rate of 0.097, RLS/neural, 0.248, and simplified, 0.353.

7 Results for the Gallop

In this section the results of the evolutionary search for the gallop are presented. A total of 200 trials were run for the gallop, half using the RLS/neural approach and the other half using the simplified approach, as stated in Section 5. Optimal solutions with respect to COT were selected by processing the data using a MATLAB script.

Figure 6 shows the optimal pre-touchdown leg angles chosen for each speed. The plots show clear trends in leg angle as speed increases. The front legs (1 and 2) tend to be placed increasingly forward with respect to the hip joint as running speed increases. The rear legs, on the other hand, have more of a parabolic relation with speed. This would suggest that each pair of legs has a different function for galloping.

The front leg angles increase with speed to transform kinetic energy into elastic potential energy. As the forward body speed increases, the front feet hit the ground with a larger amount of horizontal kinetic energy. In order to capture this energy in elastic potential, the leg angle must be swept forward as speed increases to generate maximum leg compression. In essence, the quadruped "pole-vaults" over the front legs to push it forward and upward during the beginning of gathered flight. This characteristic has been well-documented in biological quadrupeds [11, 16]. The pole-vaulting behavior allows kinetic energy to be transformed to elastic potential, which is then re-converted to kinetic and gravitational potential energy.

The stride period and the stride length for the gallop are shown in Fig. 7. The first plot shows that the stride period is scattered about a relatively flat trend line with respect to running speed. On the other hand, the second plot shows that stride length increases almost linearly with speed. This important result suggests that

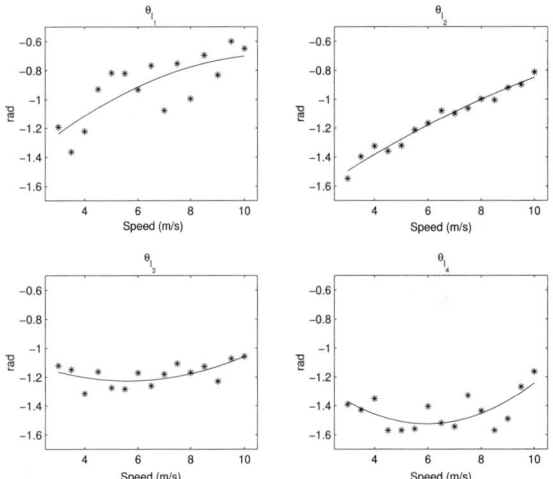

Figure 6: Optimal pre-touchdown leg angles θ_{l_i}, $i = 1, \ldots 4$, for the gallop.

the basic mechanism for increasing speed for the gallop is to increase stride length instead of decreasing stride period. This is one of the key characteristics of the gait which has been observed across a wide variety of animal species [17]. The similarity of this feature between the simulated results and biological observations is, perhaps, the most important indicator that the gait solutions found by the evolutionary algorithm are essentially biological in character. Fig. 8 shows several screen shots from the gallop simulation demonstrating both the realism of the gait and its periodicity.

8 Conclusion

In generating periodic solutions for the gallop, it was necessary to use additional methods to ensure periodicity of the leg trajectories. Out of this grew two successful approaches, the RLS/neural and simplified approaches, both of which could generate good solutions with reasonable convergence rates. While the simplified approach had a higher convergence rate overall, the RLS/neural approach performed comparably with respect to the quality of solutions found for the gallop and seems to offer the most

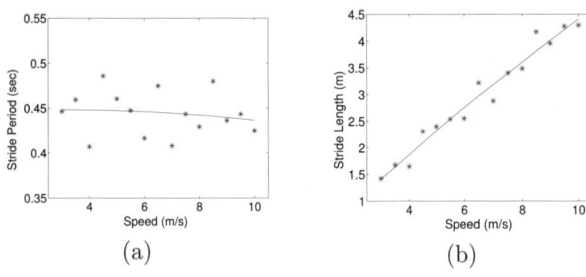

Figure 7: (a) Optimal stride period and (b) stride length for the gallop.

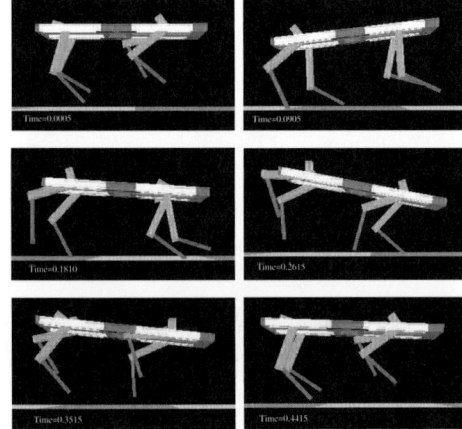

Figure 8: Screen shots of the gallop.

potential in performance improvements because of its flexible neural network structure.

The results presented in this work demonstrate the power of evolutionary search, especially when coupled with online learning techniques, to find solutions to very complex, high-dimensional optimization problems. The gait pattern that emerged from the search was realistic not only in appearance, but also in its characteristic trends. The data revealed from this testing provides an important foundation for the future implementation of high-speed galloping in an actual quadrupedal robot.

Acknowledgments

Support was provided by grant no. IIS-9907121 from the National Science Foundation to The Ohio State University and through the National Science and Engineering Graduate Fellowship program.

References

[1] D. W. Marhefka, D. E. Orin, J. P. Schmiedeler, and K. J. Waldron, "Intelligent control of quadruped gallops," *Accepted for publication in IEEE/ASME Transactions on Mechatronics*, 2003.

[2] J. P. Schmiedeler and K. J. Waldron, "Mechanical design of a quadrupedal galloping machine," in *Proceedings of the Tenth World Congress on the Theory of Machines and Mechanisms*, pp. 1985–1990, Oulu University Press, 1999.

[3] M. H. Raibert, "Trotting, pacing, and bounding by a quadruped robot," *Journal of Biomechanics*, vol. 23, pp. 79–98, 1990.

[4] J. P. Schmiedeler, *The Mechanics of and Robotic Design for Quadrupedal Galloping*. PhD thesis, The Ohio State University, Columbus, Ohio, 2001.

[5] J. P. Schmiedeler, D. W. Marhefka, D. E. Orin, and K. J. Waldron, "A study of quadruped gallops," in *Proceedings of 2001 NSF Design, Service and Manufacturing Grantees and Research Conference*, (Tampa, Florida), 2001.

[6] D. P. Krasny, "An analysis of high-speed running gaits in a quadruped robot using an evolutionary optimization strategy," Master's thesis, The Ohio State University, Columbus, Ohio, 2002.

[7] G. S. Hornby, S. Takamura, J. Yokono, O. Hanagata, T. Yamamoto, and M. Fujita, "Evolving robust gaits with AIBO," in *Proceedings of IEEE International Conference on Robotics and Automation*, pp. 3040–3045, 2000.

[8] K. M. Passino, "Biomimicry for optimization, control, and automation." unpublished manuscript, The Ohio State University, 2001.

[9] D. F. Hoyt and C. R. Taylor, "Gait and the energetics of locomotion in horses," *Nature*, vol. 292, pp. 239–240, 1981.

[10] P. P. Gambaryan, *How Mammals Run: Anatomical Adaptations*. New York: John Wiley & Sons, 1974.

[11] M. Pandy, V. Kumar, N. Berme, and K. Waldron, "The dynamics of quadrupedal locomotion," *ASME Journal of Biomechanical Engineering*, vol. 110, pp. 230–237, 1988.

[12] D. W. Marhefka and D. E. Orin, "A compliant contact model with nonlinear damping for simulation of robotic systems," *IEEE Transactions on Systems, Man, and Cybernetics*, vol. 29, pp. 566–572, November 1999.

[13] T. A. McMahon, "The role of compliance in mammalian running gaits," *Journal of Experimental Biology*, vol. 115, pp. 263–282, 1985.

[14] C. T. Farley, J. Glasheen, and T. A. McMahon, "Running springs: Speed and animal size," *Journal of Experimental Biology*, vol. 185, pp. 71–86, 1993.

[15] R. Blickhan and R. J. Full, "Similarity in multilegged locomotion: Bouncing like a monopode," *Journal of Comparative Physiology A*, vol. 173, pp. 509–517, 1993.

[16] G. A. Cavagna, N. C. Heglund, and C. R. Taylor, "Mechanical work in terrestrial locomotion: two basic mechanisms for minimizing energy expenditure," *American Journal of Physiology*, vol. 233, pp. R243–R261, 1977.

[17] N. C. Heglund and C. R. Taylor, "Speed, stride frequency and energy cost per stride: How do they change with body size and gait?," *Journal of Experimental Biology*, vol. 138, pp. 301–318, 1988.

DEVELOPMENT OF OMNI-DIRECTIONAL VEHICLE WITH STEP-CLIMBING ABILITY

Daisuke Chugo*, Kuniaki Kawabata[†], Hayato Kaetsu[†],
Hajime Asama[††/†] and Taketoshi Mishima*

*Saitama University, 255, Shimo-Ookubo, Saitama-shi, Saitama 338-8570, Japan
[†]RIKEN (The Institute of Physical and Chemical Research), 2-1 Hirosawa, Wako-shi, Saitama 351-0198, Japan
[††]The University of Tokyo, 4-6-1 Komaba, Meguro-ku, Tokyo 153-8904, Japan
Phone: +81-48-467-4274, Fax: +81-48-467-5480, E-mail: chugo@cel.riken.go.jp

Abstract: In this study, we propose a new holonomic mobile mechanism which is capable of step-climbing. This mechanism realizes to move in any direction on the flat floor and pass over steps and slopes in one (fore and hind) direction. The vehicle equips with seven omni-directional wheels with cylindrical free rollers and a passive body axis that provides to change the shape of the body on the rough terrain. There is no need for additional actuators or sensors for the passive body axis. In our previous work, we constructed a prototype mobile mechanism and analyzed its kinematic characteristics. In this paper, we design a new control scheme and improve the mechanism for passing over the steps more stably. The performance of the prototype system is verified through experiments and computer simulations.

Keywords: Omni-Directional Mobile Robot, Holonomic Robot, Irregular Terrain, Passive Suspension

1 INTRODUCTION

Omni-directional moving ability is required for mobile robots, because it might become more important for mobile robots to accomplish various tasks in the restricted space in short time. The holonomic omni-directional mobile robot can move in all directions at any time. This paper addresses an omni-directional mobile system with step-climbing ability.

The motivation of this work originates in demand in the nuclear power plants and large-scale institutions. There are a lot of narrow spaces, so high mobility and easy planning of operation [1] are necessary. Furthermore, the ability to move on uneven floor like the steps and the slopes is needed, since there are a lot of the steps and slopes in such a plant and the exact position of them is unknown. However, it is difficult for almost all of conventional omni-directional mobile robots to pass over an irregular terrain.

In previous research, various types of omni-directional mobile robots have been proposed: legged robots, ball-shaped wheel robots, crawler robots, etc.

The legged robots [2] can move in all directions and pass over the steps. However, the legged robots have a complex mechanism which is difficult to control. The robot must know an exact position of the step when the robot passes over it. The maximum speed of the legged robots is usually much slower than that of the wheel type mobile robots.

The robot with ball-shaped wheels can run in all directions. [3] However, the positioning is not accurate and it is not suitable for running on the rough grounds. The special crawler mechanism [4] for the omni-directional mobile robot can move successfully on the rough terrain, but it cannot climb over the large steps.

Thus, we proposed an omni-directional mobile vehicle system capable of climbing. [5] This system has omni-directional wheels and the rocker-bogie suspension system. [6] The vehicle can move in any direction on the flat floor and pass over steps or slopes in one direction.

The key idea of this vehicle system is a combination of an omni-directional wheel and a rocker-bogie suspension system. The rocker-bogie suspension system has the passive body axis and it enables the vehicle to pass over the rough terrain easily.

The main advantage of our vehicle system is the easy operation. Our vehicle can climb the step with a simple control. There is no need for additional actuators and sensors. The vehicle is not required the environmental model information that includes an exact position of steps or slopes.

In this paper, we describe two topics in our research. One point, we design control scheme for the vehicle system. The control scheme enable the vehicle to realizes holonomic running and step-climbing using its mechanism efficiently. The other point, we improve the mechanism of the prototype vehicle system for step-climbing more stably.

This paper is organized as follows: the mechanical design and the kinematic model on the vehicle are discussed in section 2; the design of control software is considered in section 3; the improvement of its mechanism and simulation results are presented in section 4; the implementation and experimental results are shown in section 5; in section 6, we conclude this paper.

2 SYSTEM CONFIGURATION

The ability to move on the rough terrain and pass over the steps is required for mobile robots in order to operate under hazardous environment such as the nuclear power plants and large-scale institutions.

We propose a new mechanism capable of holonomic running and climbing though it is simple and reasonable.

2.1 Mechanical Design

The high-speed motion and easy control and maintenance are required for the mobile vehicle to be used practically in nuclear power plants. Thus, it is necessary that the mechanism and the control system of the vehicle are simple and the vehicle can move at high speed in all directions.

The vehicle is equipped with the special wheels, as it is shown in Figure 1. This wheel is formed by combining twelve small and large cylindrical free rollers and generates omni-directional motion. [7][8]

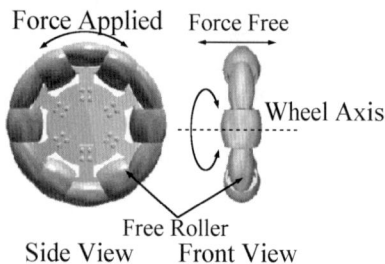

Figure 1. The special wheel with free rollers

A passive suspension system called Rocker-Bogie suspension system is used. The vehicle has free rotational joints, so the shape of it can be easily transformed when a force is added to its body or the wheels as shown in Figure 2. The vehicle must climb the step head-on by restrictions of its mechanism. The appropriate orientation can be easily obtained because the vehicle can move in all direction.

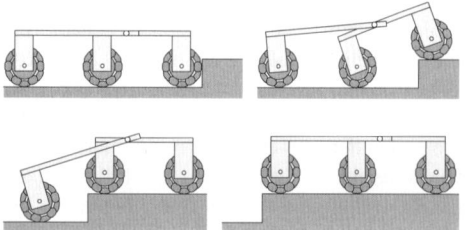

Figure 2. Rocker-Bogie suspension system

In case of the planetary rovers, at least six wheels are necessary to pass over large steps. [9]

The omni-directional wheels must be arranged in parallel and crossed in order to generate omni-directional motion. Therefore, we propose the robot that has seven universal wheels as shown in Figure 3.

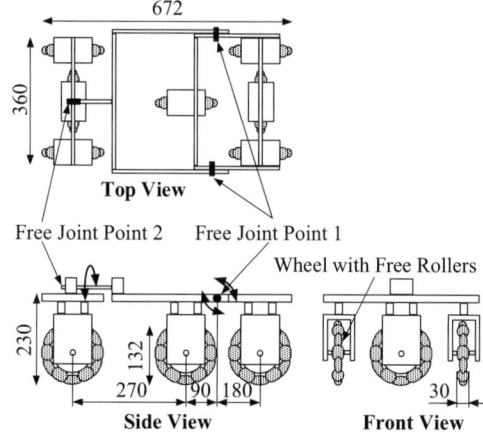

Figure 3. The mobile vehicle with seven omni-directional wheels

2.2 Kinematics

A description of the vehicle kinematics is shown in Figure 4. The coordinates, the length of each links, and the rotation speed of each wheel when the vehicle is on the flat floor are given as follows.

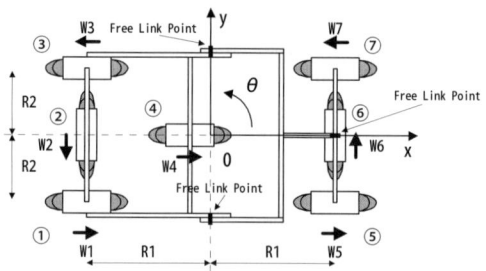

Figure 4. Coordinates of the robot

The following notations are used:
r : radius of the wheels [mm]
ω_i : rotation speed of the wheel i [rad/s]
V_i : rotation speed of the actuator i [rad/s]
k : gear ratio between the actuator and the wheel

The relations between the speed of each wheel and the speed of each actuator are:

$$\omega_i = kV_i \ (i = 1...7) \tag{1}$$

$\dot{X} = \begin{bmatrix} \dot{x} & \dot{y} & \dot{\theta} \end{bmatrix}^T$ is vehicle travel speed vector in the basic coordinates shown in Figure 4, and $V = \begin{bmatrix} V_1 & \sim & V_7 \end{bmatrix}^T$ is the vector for the rotation speed of each wheel. The relationship of $\dot{X}$ and V is as follows:

$$\dot{X} = J \cdot V, \qquad (2)$$

where

$$J = kr \cdot \begin{bmatrix} \frac{1}{5} & 0 & -\frac{1}{5} & \frac{1}{5} & \frac{1}{5} & 0 & -\frac{1}{5} \\ 0 & -\frac{1}{2} & 0 & 0 & 0 & \frac{1}{2} & 0 \\ \frac{1}{6R_2} & \frac{1}{6R_1} & \frac{1}{6R_2} & 0 & \frac{1}{6R_2} & \frac{1}{6R_1} & \frac{1}{6R_2} \end{bmatrix}$$

J is the Jacobi matrix with respect to $\dot{X}$.

The wheels No.1 and 5, and the wheels No.3 and 7 are on the same straight lines, respectively. Thus, this kinematic constrains determine the following equation:

$$\omega_1 = \omega_5 \qquad (3)$$

$$\omega_3 = \omega_7 \qquad (4)$$

Above, we can control the velocity of the vehicle by controlling the rotating speed of each wheel:

$$V = J^+ \cdot \dot{X}, \qquad (5)$$

J^+ is pseudoinverse, where

$$J^+ = (J^T J)^{-1} J^T = \frac{1}{kr} \cdot \begin{bmatrix} 1 & 0 & R_2 \\ 0 & -1 & R_1 \\ -1 & 0 & R_2 \\ 1 & 0 & 0 \\ 1 & 0 & R_2 \\ 0 & 1 & R_1 \\ -1 & 0 & R_2 \end{bmatrix}$$

A running test was done to verify the performance of the prototype vehicle. As the result of test, the prototype vehicle can move in omni-direction and climb the step of 80mm, with the maximum.

2.3 A prototype vehicle

The overview of the developed holonomic omni-directional vehicle is shown in Figure 5. The size of the vehicle is 750[mm](L) x 540[mm](W) x 520[mm](H) and the weight is about 22[kg]. (The weight of four batteries is included.) The block diagram of the electronics is shown in Figure 6.

Figure 5. The improved prototype robot

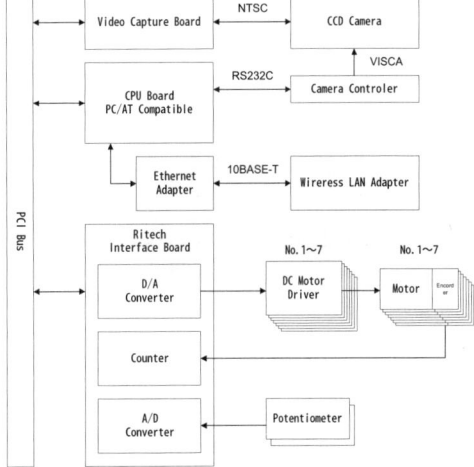

Figure 6. The block diagram of the control system that is mounted on the vehicle

3 CONTROL SYSTEM

In order to use the vehicle in the nuclear power plant efficently, we designed the control system with the following plans.

- The vehicle can be controlled by simple parameters (for example, only motion velocity of x, y and θ)
- The vehicle can move and pass over the step without information related to exact step position.

In general, it is difficult to estimate enviromental information preciously and more complicated control is required to sense such information. Therefore, the robot which require such information (e.g., the legged type) is disadvantageous for quick motion. The developed robot in such a plant must move intricately and quickly, so the vehicle should be able to move with simple control method.

Thus, the control of the vehicle needs to realize the following conditions.

- The vehicle can move with sufficient accuracy on the flat foor for step-climbing.

The vehicle must climb the step head on, so it needs to change direction head on before pass over it. Furthermore, the vehicle must move more than 1 meter with sufficient accuracy by dead reckoning, because the vehicle requires a run-up about 1[m] when it passes over the step. By preliminary tests, when its height is 6[cm], it is allowed to 30[degrees] and when its height is more than 8[cm], it is allowed to less than 5[degrees].

- The vehicle can climb the step stably, not it does not come to be shown in Figure 7.

Figure 7. The view of unstable situation

In the plant, there are a lot of steps. Their height are about 5[cm], so the vehicle must pass over it or more high. When the vehicle climbs the step in preliminary tests, the middle wheel float and balance is broken down as shown in Figure 7. So, we design that the vehicle can climb 100[mm] height stably.

3.1 PID Based Control System

The vehicle has seven wheels with DC motors and the motors are driven by velocity command. However, it is difficult to take the synchronization among the wheels when the load concentrates on the part of them.

In order to fulfill the two above-mentioned conditions, we propose new control scheme shown in Figure 8. PID control is utilized and added two improvements.

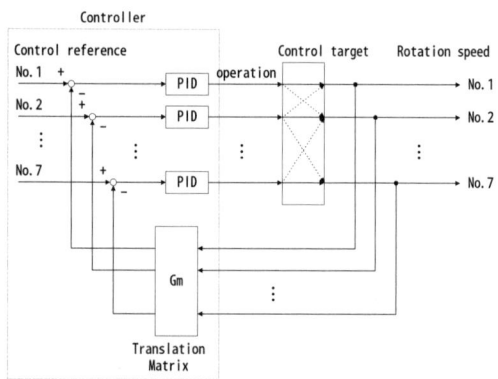

Figure 8. Contorl Diagram

In PID control, the integral term is effective for an offset dissolution and the derivative term is effective for disturbance. Both of terms are contrary. When the vehicle runs on the flat floor, the integral term is more important. On the other hand, when the vehicle pass over the step, the derivative term is more important, too. Two conditions have the relation of a trade-off and it is very difficult to balance two parameters in PID control.

Thus, we propose a new control law. Our approach is not only one control mode, but also two or more parameter sets and changes them according to a situation. Here, we prepare two parameter sets.

- For climbing the step: First priority matter is to stable running, the derivative term is taken large. Furthermore, we improved new synchronous control for each wheels and adopt this rule describing in the following chapter.
- For moving on the flat floor: First priority matter is to control accurately, so the integral term is taken large.

Using two parameter sets, we design control law so that two conditions may be fulfilled. The vehicle detects the step by the following equation:

$$E_i = V_i - v_i \qquad (6)$$

where the he following notations are used:
V_i: control reference of the actuator i (rad/s)
v_i: rotation speed of the actuator i (rad/s)

The controller judges by whether this value exceeds the fixed value. When the vehicle climbs the step, this value may be increase by a heavy load. So, if this value is larger than the fix value, the vehicle detects the step.

3.2 Control for Step-Climbing

When the vehicle climbs steps, the load applied to each actuator is too heavy, so the synchronous control among the wheels is very difficult. If synchronous control of each wheel is not perfect, the robot will become unstable. For instance, the vehicle bends in the direction that the locker-bogie mechanism does not expect as shown in Figure 7.

In order to cope with this problem, we improved the control system which synchronizes among the wheels. We installed the conversion matrix (Gm) shown in Figure 8. Gm is expressed by the following equation:

$$G_m = \begin{bmatrix} 1-4k/5 & 0 & k/5 & k/5 & k/5 & 0 & k/5 \\ 0 & 1 & 0 & 0 & 0 & 0 & 0 \\ k/5 & 0 & 1-4k/5 & k/5 & k/5 & 0 & k/5 \\ k/5 & 0 & k/5 & 1-4k/5 & k/5 & 0 & k/5 \\ k/5 & 0 & k/5 & k/5 & 1-4k/5 & 0 & k/5 \\ 0 & 0 & 0 & 0 & 0 & 1 & 0 \\ k/5 & 0 & k/5 & k/5 & k/5 & 0 & 1-4k/5 \end{bmatrix} \quad (7)$$

where *k* is Arbitrary constant (0<k<1).

When the vehicle detects the step, it uses equation 7. On the other hand, when it runs on the flat floor, it uses equation 8.

$$G_m = \begin{bmatrix} 1 & 0 & 0 & 0 & 0 & 0 & 0 \\ 0 & 1 & 0 & 0 & 0 & 0 & 0 \\ 0 & 0 & 1 & 0 & 0 & 0 & 0 \\ 0 & 0 & 0 & 1 & 0 & 0 & 0 \\ 0 & 0 & 0 & 0 & 1 & 0 & 0 \\ 0 & 0 & 0 & 0 & 0 & 1 & 0 \\ 0 & 0 & 0 & 0 & 0 & 0 & 1 \end{bmatrix} \quad (8)$$

This applies to only wheels that are attached in the advance direction. (No.1, 3, 4, 5 and 7, as it is shown in Figure 3.) By using this scheme, the load is distributed to all wheels and the synchronous control among the wheels becomes easy when the vehicle climb the step.

4 MECHANICAL IMPROVEMENT

We designed the vehicle to climb the step stably. The previous chapter, we approach from the control scheme. However, there is a problem in its mechanism, so the prototype vehicle cannot climb the step smoothly. We improve the prototype mechanism and realize step climbing stably using both the software scheme and the mechanical improvement for this vehicle system.

The rocker-bogie suspension system is very important for step climbing. But this mechanism on the vehicle, the position of a free joint is unsuitable, so the mechanism has not demonstrated sufficient performance. We improved the position of a free joint and the effect was verified by computer simulation.

4.1 Improvement of hardware

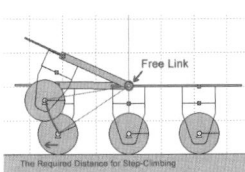 

(a) Simulation Model   (b) Prototype Robot
Figure 9. Prototype Robot (Previous Type)

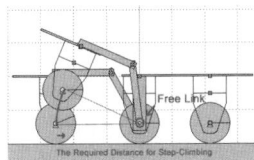 

(a) Simulation Model   (b) Prototype Robot
Figure 10. Improved Prototype Robot

The prototype system which was developed in our previous work, had a free joint in upper part of the body shown in Figure 9. However, the posture of the vehicle becomes unstable shown in Figure 7 when it climbs the step with this mechanism. Because a backslash occurs shown in Figure 9.

So, we change the position of a free joint from the upper part of the body to the axis of a middle wheel shown as Figure 10. The backslash and the downward rotation moment are canceled by this improvement.

4.2 Simulation result

By this simulation, we verified the improved mechanism works well. Simulation parameters are shown in Table 1. The result of the simulation, the improved model can climb the step with 150[mm] height, but the previous one can climb only 80[mm]. Furthermore, the previous model break the balance as shown in Figure 11, but the improve one doesn't break. As mentioned above, the improved vehicle realizes more stable running for step climbing.

The output of each wheel is shown in Figure 12. Comparing with the improved model, as for the previous one, the output inclines toward the rear wheel, and the load of step climbing is not distributed. On the other hand, in the improvement model, load is distributed, thus the improved model can climb the step more easily.

Table 1. Simulation Parameters

Length	Front 195[mm] Rear 400[mm]
Weight	Front 7.8[kg] Rear 13.8[kg]
Wheel diameter	132[mm]
Distance between wheels	Front 255[mm] (Previous one = 215[mm]) Rear 215[mm]
Motor control mode	Speed control mode
Running speed	0.25[m/sec]

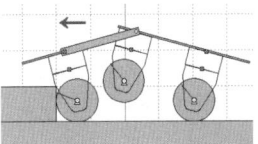

Figure 11. Simulation result of the previous model

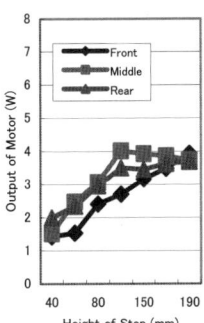 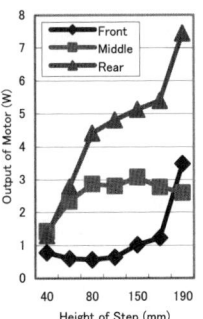

(a) New Model   (b) Previous Model
Figure 12. Motor output of Each Wheel

5 EXPERIMENT

We conducted the following two experiments.

- Moving in all directions on the flat floor.

In this experiment, we test the running ability of the vehicle on the flat foor. The result of it, the vehicle can move in omni-direction. We measured the move error of the vehicle shown in Table 2. The vehicle fulfills the required motion accuracy to climb the step.

Table 2. Motion error when the vehicle moves 1.0[m] forward. [mm]

	X Direction	Y Direction	θ Rotation
Advance	4.9	5.7	1.0
Sternway	5.6	13.0	0.7
Transverse	13.5	16.5	1.1
Revolution	1.0	3.6	3.3
Slanting	15.0	5.0	3.7

- Step-climbing function.

In this experiment, the vehicle pass over the step as shown in Figure 13. The vehicle runs at 0.25m/sec and a run-up is about 1 meter.

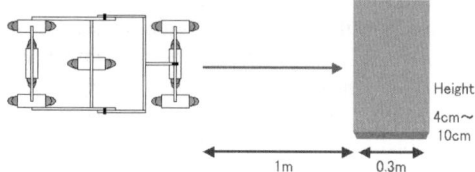

Figure 13. The step-climbing experiment

As the result of the experiment, the vehicle can climb up and down the step more smoothly than previous model due to the new control law and the improvement mechanism. The vehicle can pass over the 100[mm] height maximum and doesn't break the balance shown in Figure 5.

Figure 14 shows the rotation speed of each wheel when the vehicle climbs 50[mm] height. Each wheel is synchronized and the control mode is changed in suitable timing. Thus, the prototype system is efficiency for step climbing.

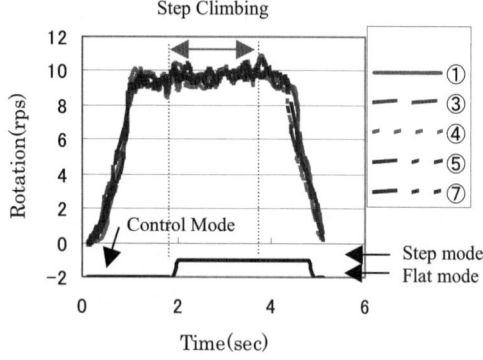

Figure 14. The rotation of each wheel

6 CONCLUSIONS

The omni-directional mobile vehicle system which is capable of climbing was discussed. The mechanism and system kinematics were described. New control scheme, which realizes holonomic moving and step climbing, was designed. Improvement of its mechanism and its simulation results were discussed. The implementation of the system and our experimental results were described. For future works, we will add the external sensors on the vehicle to detect the step. And a motion planning technique based on the environment information will be discussed.

REFERENCES

[1] G. Campion, G. Bastin and B.D. Andrea-Novel. Structual Properties and Classification of Kinematic and Dynamic Models of Wheeled Mobile Robots. In: IEEE Transactions on Robotics and Automation, Vol.12, No.1, pp.47-62, 1996.

[2] G. Endo and S. Hirose. Study on Roller-Walker: System Integration and Basic Experiments, In: Proc. ICRA, pp.2032-2037, 1999.

[3] M. Wada and H. Asada. Design and Control of a Variable Footpoint Mechanism for Holonomic Omnidirectional Vehicles and its Application to Wheelchairs. In: IEEE Transactions on Robotics and Automation, Vol.15, No.6, pp.978-989, 1999.

[4] S. Hirose and S. Amano. The VUTON: High Payload, High Efficiency Holonomic Omni-Directional Vehicle. In: Proc. of the 6th Symposium on Robotics Research, pp.253-260, 1993.

[5] A. Yamazaki, A.S. Development of a step-climbing omni-directional mobile robot. In: Proc. of the 2001 Int. Conf. on Field and Service Robotics, pp.327-332, 2001.

[6] H.W.Stone. Mars Pathfinder Microrover -a Small, Low-Cost, Low Power Space craft. In: Proc.of AIAA Forum on Advanced Developments in Space Robotics. 1996.

[7] B. Carisle. An Omni-Directional Mobile Robot. Developments in Robotics 1983, In: IFS Publications Ltd., pp.79-87, 1983.

[8] H. Asama, A.S. Development of an Omni-Directional Mobile Robot with 3 DOF Decoupling Drive Mechanism. In: Proc. of the 1995 IEEE Int. Conf. on Robotics and Automation, pp.1925-1930, 1995.

[9] Y. Uchida, K. Furuichi and S. Hirose. Fundamental Performance of 6 Wheeled off-road Vehicle HELIOS-V. In: Proc. of the 1999 IEEE Int. Conf. on Robotics and Automation, pp.2336-2341, 1999.

A Distributed Route Planning Method for Multiple Mobile Robots using Lagrangian Decomposition Technique

Tatsushi Nishi, Masakazu Ando, Masami Konishi and Jun Imai
Department of Electrical and Electronic Engineering, Okayama University.
E-mail: nishi@cntr.elec.okayama-u.ac.jp

Abstract— For the transportation in semiconductor fabricating bay, route planning of multiple AGVs(Automated Guided Vehicles) is expected to minimize the total transportation time without collision and deadlock among AGVs. In this paper, we propose a distributed route planning method for multiple mobile robots using Lagrangian decomposition technique. The proposed method has a characteristic that each mobile robot individually creates a near optimal route through the repetitive data exchange among the AGVs and the local optimization of its route using Dijkstra's algorithm. The proposed method is successively applied to transportation route planning problem in semiconductor fabricating bay. The optimality of the solution generated by the proposed method is evaluated by using the duality gap derived by using Lagrangian relaxation method. A near optimal solution within 5% of duality gap for a large scale transportation system consisting of 143 nodes and 15 AGVs can be obtained only within five seconds of computation time. The proposed method is implemented on 3 AGVs system and the route plan is derived taking the size of AGV into account. It is experimentally shown that the proposed method can be found to be effective for various types of problems despite the fact that each route for AGV is created without considering the entire objective function.

Keywords—Distributed autonomous robotics systems, multiple mobile robots, route planning, semiconductor fabricating bay

I. INTRODUCTION

In many semiconductor fabricating bay, multiple AGVs(Auto-Guided Vehicles) are widely used in transportation. The requests for transportation are given in every few seconds. It is required to derive a conflict-free route plan which minimizes the total transportation time in a few seconds. However, in recent years, the layout for transportation in semiconductor fabrication is growing rapidly with the expansion of the size of silicon wafer. The route planning problem for multiple mobile robots becomes increasingly difficult with respect to the increase of the number of AGVs. The route planning algorithm for semiconductor fabricating bay has been extensively studied and a number of algorithm have been proposed for multiple robot motion planning algorithms[1][2]. These algorithms are often characterized by centralized or decentralized approach. The centralized approach determines the routes for all of the robots by single decision maker[3]. On the other hand, the decentralized approach autonomously determines the route dissolving the conflicts and collecting information from other robots[4][5][6][7]. The main approach for decentralized motion planning is priority assignment by using the information of the environment, which [4] uses a hybrid control architecture to solve the coordination problem prioritizing the robots. [5] decomposes the problem into path planning and velocity planning for dynamic environment. However, most of these decentralized algorithms are based on the use of heuristics such as Market Economy [6], Ant system[7] for solving the coordination and dissolving the conflicts among AGVs. Our approach uses the Lagrangian decomposition method for coordinating the motion of the multiple mobile robots.

Lagrangian decomposition method removes the complicating constraints from the constrained set and replaces with a penalty term in the objective function, which can be decomposed into several tractable sub-problems. Recently, scheduling method based on the use of Lagrangian decomposition method has widely been used to improve the computation efficiency with near-optimal solution[8]. In this method, the improvement of the Lagrangian multiplier value and generation of a solution of each sub-problem is iteratively repeated. The Lagrangian decomposition method is successfully applied to a machine-oriented decentralized scheduling method[9].

This paper proposes a distributed route planning method for multiple mobile robots using Lagrangian decomposition technique. The main feature of the algorithm is that data exchanges among the robots and the re-route planning for each robot are iteratively repeated until a feasible route plan for the entire AGV is derived. The contribution of this paper is that the distributed route planning algorithm is systematically constructed by using the idea of Lagrangian decomposition technique and the optimality of the results obtained by the distributed route planning method is evaluated by using the duality gap derived by solving the dual problem using Lagrangian relaxation. The proposed method is implemented on experimental system consisting of 3 AGVs. The computation time is considerably shorter than that of the conventional method. Therefore it may be capable to apply to real time route planning for multiple mobile robots.

This paper is organized by the following sections. In Section II, the route planning problem in transportation is defined and the problem is formulated as a mixed integer linear problem. In Section III, the distributed route planning algorithm using Lagrangian decomposition technique is proposed. Numerical results and the optimality of the solution of the proposed method are described in Section IV. The implementation of experimental robots considering the geometrical size of the robots is described in Section V, which includes the simulation and experimental results.

II. Formulation of Route Planning Problem as Mixed Integer Problem

A. Route Planning Problem in Transportation

The layout model for two dimensional railed transportation system treated in this paper is shown in Fig. 1. The model of the transportation system consists of several nodes and arcs. Each node represents a place in which each AGV can stop or turn. Each arc represents the rail for AGVs. The following conditions are assumed in this problem.

- A request for transportation is given to each AGV a priori. A loading node and an unloading node is given for each AGV. Each AGV knows its starting and ending position.
- Each AGV knows the map for the pre-defined layout model. Each rail is two way traffic and two AGVs cannot travel on a rail at the same time.
- Each AGV can stop or turn only at the node.
- The velocity of the AGV is constant and the turning time can be included in the traveling time.
- The objective in transportation is to minimize the sum of the transportation time.

The route planning problem can be defined as: Given the loading place and unloading place for each AGV, the initial place for each AGV, to determine the route for AGVs so as to minimize the total transportation time satisfying the deadlock and collision avoidance constraints.

B. Problem Formulation

In this section, the route planning problem is formulated as a mixed integer linear programming problem. A route $X^k \in X$ of a AGV $k \in V$ can be described by a

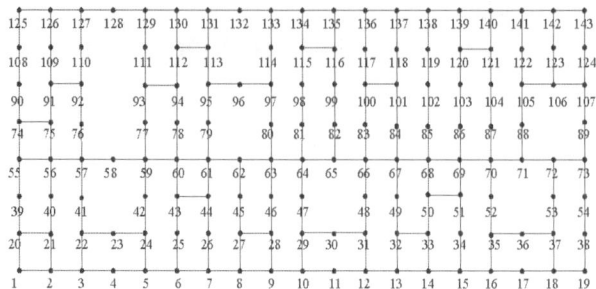

Fig. 1. Example of transportation system

set of variables as $X^k = \{x_{i,j,t}^k : i,j \in Q, k \in V, t \in T\}$ where Q is the set of nodes and V is the set of AGVs in the transportation system. A binary variable $x_{i,j,t}^k \in \{0,1\}$ denotes whether AGV k travels from node i to node j in time period t. N_i is the set of nodes connected to node i.

The constraints are shown in the following equations.

$$\sum_{j \notin N_i} x_{i,j,t}^k = 0 \qquad (\forall k, \forall i, \forall t) \qquad (1)$$

$$\sum_{j \in N_i} x_{i,j,t}^k \leq 1 \qquad (\forall k, \forall i, \forall t) \qquad (2)$$

$$\sum_{j \in N_i} x_{j,i,t}^k = \sum_{n \in N_i} x_{i,n,t+1}^k \quad (\forall k, \forall i, \forall t) \qquad (3)$$

$$\sum_{j \in N_{S_k}} x_{S_k,j,0}^k = 1 \qquad (\forall k) \qquad (4)$$

Here, S_k denotes the starting node for AGV k. Eq. (1) indicates that each AGV can travel on the arc to the connected node. Eq. (2) indicates that more than an AGV can not travel into a neighbor node at the same time. Eq. (3) indicates the time continuity constraints of the movement of AGVs which satisfies whether AGV k travels from node j to node i in time period t or not. Eq. (4) indicates the initial condition of the place of AGVs. The collision and interference avoidance constraints can be written as the following equations.

$$\sum_{k \in V} \sum_{j \in N_i} x_{j,i,t}^k \leq 1 \qquad (\forall i, \forall t) \qquad (5)$$

$$\sum_{k \in V} (x_{i,j,t}^k + x_{j,i,t}^k) \leq 1 \qquad (\forall i, \forall j, \forall t) \qquad (6)$$

Eq. (5) is the capacity constraints of a node indicating that more than one AGV can not travel into a node at the same time. Eq. (6) is the capacity constraints of a arc indicating that two AGV can not travel on a arc at a time.

The objective function is the sum of the transportation time for each AGV. To make the mathematical representation easier, we assume that the traveling time is equal to the distance between the neighbor two nodes in this formulation. However, this assumption need not to be made in the experimental system in section V. $\delta_{k,t} \in \{1,0\}$ becomes 1 if the AGV k has not arrived at the ending node in time period t, and becomes zero if the AGV k has arrived at the ending node. Thus, the variable $\delta_{k,t}$ satisfies the following Eqs. (7)-(9).

$$\sum_{i \in N_{G_k}} x^k_{i,G_k,t} \leq M(1-\delta_{k,t}) \qquad (\forall k, \forall t) \quad (7)$$

$$\sum_{i \in N_{G_k}} x^k_{i,G_k,t} \geq 1 - \delta_{k,t} \qquad (\forall k, \forall t) \quad (8)$$

$$-\delta_{k,t} + \delta_{k,t+1} \leq 0 \qquad (\forall k, \forall t) \quad (9)$$

M is the maximum value of left-side in Eq. (7) [5 in this problem]. Eq. (9) indicates $\delta_{k,t}$ becomes zero after the AGV k has arrived at the ending node. By using the variable $\delta_{k,t}$, the total transportation time can be written as $\sum_k \sum_t \delta_{k,t}$ (For example, the transportation time 3 for AGV 1 can be expressed by $\delta_{1,1} + \delta_{1,2} + \delta_{1,3}$ and $\delta_{1,1} = \delta_{1,2} = \delta_{1,3} = 1$).

The route planning problem is formulated as:

$$\min_{\{x^k_{i,j,t}\}} \sum_k \sum_t \delta_{k,t} \quad (10)$$

subject to (1)-(9) and $\delta_{k,t} = \{0,1\} \; (\forall \; k, t)$

III. Distributed Route Planning method using Lagrangian Decomposition Technique

A. AGV-based Decomposition of Route Planning Problem using Lagrangian Decomposition

The capacity constraints (5) and (6) are relaxed by using non-negative Lagrangian multipliers $\{\lambda_{i,k}\}$, $\{\phi_{i,j,t}\}$. The relaxed problem ($RP1$) is formulated as:

$$(RP1): \min_{\{x^k_{i,j,t}\}} L \quad (11)$$

$$L = \sum_k \sum_t \delta_{k,t} + \sum_{i \in Q} \sum_t \lambda_{i,t} \left(\sum_k \sum_{j \in N_i} x^k_{j,i,t} - 1 \right)$$
$$+ \sum_{i \in Q} \sum_{j \in Q} \sum_t \phi_{i,j,t} \left[\sum_k (x^k_{i,j,t} + x^k_{j,i,t}) - 1 \right] \quad (12)$$

Eq. (12) can be rewritten as:

$$L = \sum_k L_k - \sum_t \sum_{i \in Q} \lambda_{i,t} - \sum_t \sum_{i \in Q} \sum_{j \in Q} \phi_{i,j,t} \quad (13)$$

where, $\quad L_k = \sum_t \delta_{k,t} + \sum_t \lambda_{P_{(k,t)},t}$
$$+ \sum_t (\phi_{P_{(k,t-1)},P_{(k,t)},t} + \phi_{P_{(k,t)},P_{(k,t-1)},t}) \quad (14)$$

$P_{(k,t)}$ represents the node in which AGV k arrives at the end of time period t. This formulation indicates that the dual problem can be decomposable for sub-problem for each AGV k. The dual problem (DP) is formulated as:

$$(DP) \max_{\{\lambda_{i,t}, \phi_{i,j,t}\}} q(\lambda_{i,t}, \phi_{i,j,t}) \quad (15)$$

$$q(\lambda_{i,t}, \phi_{i,j,t}) = \left(\min_{\{x^k_{i,j,t}\}} \sum_k L_k \right) - \sum_t \sum_{i \in N} \lambda_{i,t}$$
$$- \sum_t \sum_{i \in N} \sum_{j \in N} \phi_{i,j,t} \quad (16)$$

From the above decomposition scheme, the route planning algorithm with the Lagrangian decomposition method has been developed. In this algorithm, the step of solving the sub-problems for each AGV for fixed Lagrangian multipliers and the step of solving the dual problem by updating the Lagrange multipliers are iteratively repeated until a near optimal solution is derived.

The objective function L_k for AGV-subproblem consists of the transportation time for each AGV, and Lagrangian multipliers indicating the costs for the using the node and the arc for AGV k. The Lagrangian multipliers $\{\lambda_{i,t}\}, \{\phi_{i,j,t}\}$ are updated after solving the sub-problem for each AGV. For the original Lagrangian decomposition technique, however, the dual solution may not always be a feasible solution and it is not guaranteed to obtain a feasible solution by using the Lagrangian decomposition method. Therefore, we embedded an penalty function ensuring the generation of feasible solution for each AGV-subproblem and the penalty coefficient ρ is increased after AGV-subproblem is solved.

$$L_k = \sum_t \delta_{k,t} + \sum_t \lambda_{P_{(k,t)},t}$$
$$+ \sum_t (\phi_{P_{(k,t-1)},P_{(k,t)},t} + \phi_{P_{(k,t)},P_{(k,t-1)},t})$$
$$+ \sum_{l \in V | l \neq k} \alpha^n_{k,l} (C^1_{k,l} + C^2_{k,l}) \quad (17)$$

$$C^1_{k,l} = \sum_t \sum_{i \in N_{P_{(k,t)}}} \bar{x}^l_{i,P_{(k,t)},t} \quad (18)$$

$$C^2_{k,l} = \sum_t \bar{x}^l_{P_{(k,t)},P_{(k,t-1)},t} \quad (19)$$

where, $\{\bar{x}^l_{i,j,t}\}$ represents the set of nodes included in the route of AGV l obtained by exchanging the data with other AGVs. $\alpha_{k,l}^n$ represents the penalty coefficient for violating the constraints between AGV k and AGV l. By increasing the penalty function in each iteration, the penalty function becomes zero when a feasible route plan for the entire AGV is derived.

The Lagrangian multipliers are updated by the following equations in the direction of sub-gradient of the dual function $q(\lambda_{i,t}, \phi_{i,j,t})$.

$$\lambda_{i,t}^{n+1} = \max\left\{0, \lambda_{i,t}^n + \Delta\lambda\left(\sum_k \sum_{j\in N_i} x_{j,i,t}^k - 1\right)\right\} \quad (20)$$

$$\phi_{i,j,t}^{n+1} = \max\left\{0, \phi_{i,j,t}^n + \Delta\phi(\sum_k (x_{i,j,t}^k + x_{j,i,t}^k) - 1\right\} \quad (21)$$
$$(\forall\ i,j,t)$$

For the calculation of sub-gradient: $\sum_k \sum_{j\in N_i} x_{j,i,t}^k - 1$ and $\sum_k (x_{i,j,t}^k + x_{j,i,t}^k) - 1$, the solutions derived at the other AGV l are necessary. Therefore, each AGV communicates with other AGVs and updates own Lagrangian multiplier values.

B. Distributed Route Planning Algorithm for Multiple Mobile Robots

A distributed optimization algorithm is explained in detail in this section.

The overall algorithm of the distributed route planning method consists of the following steps.

Step 1: Initialization of parameters ρ, and Lagrangian multipliers $\{\lambda_{i,t}\}$, $\{\phi_{i,j,t}\}$.

Step 2: Initial route planning for each AGV
Each AGV generates an initial route without taking other AGVs into account.

Step 3: Data exchange among the AGVs
Each AGV is capable to communicate with other AGVs and exchanging the data by using wireless communication. Each AGV exchanges the data of the position of other AGVs $\{x_{i,j,t}^l\}$.

Step 4: Convergence check
The derived route is feasible and the derived route has not been updated from a previous solution, each AGV stops the algorithm and the derived solution is regarded as a final solution. The derived route plan is sent to each AGV.

Step 5: Updating the value of Lagrangian multipliers
To solve the dual problem for each AGV in the distributed environment, each AGV updates the value of Lagrangian multipliers using Eqs. (20) and (21). The sub-gradients can be calculated by the solutions of all the other AGVs obtained in Step 3.

Step 6: Updating the value of penalty coefficient
If the route obtained at the previous iteration is infeasible, the penalty coefficients for each AGV which violates the constraints are increased according to the following equation. n represents the number of iteration.

$$\alpha_{k,l}^{n+1} = \alpha_{k,l}^n + \Delta\alpha \sum_{l\in V | l\neq k} (C_{k,l}^1 + C_{k,l}^2) \quad (22)$$

Step 7: Decision on whether the Step 8 is skipped or not
To make the convergence easier, the re-route planning of an AGV is skipped randomly. The detail of this step is explained in Section III-C.

Step 8: Re-route planning for each AGV
By using the position data of other AGVs obtained by Step 3, each AGV optimizes its own objective function and then return to Step 3.

C. Skipping of re-route planning

The step of skipping of re-route planning is explained in this section. Fig. 2 shows a simple route planning problem consisting of two AGVs. An initial position for each AGV represents the starting node and the arrowed node represents the ending node.

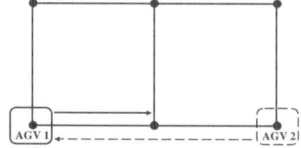

Fig. 2. Transportation demand and its initial route for each AGV

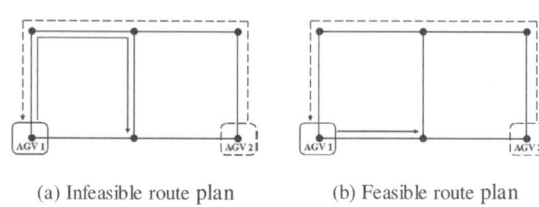

(a) Infeasible route plan (b) Feasible route plan

Fig. 3. An example of intermediate result obtained by the proposed method

If each AGV creates the route without considering other AGV, the route obtained by the initial route planning becomes infeasible as shown in Fig. 2. At the next step, the position of the AGV is exchanged between the AGVs and each AGV executes re-route planning individually. However, if the re-route planning is executed concurrently, each AGV selects route shown in Fig. 3(a) so as to minimize the penalty of violating the capacity constraints. Furthermore, at the next step, each AGV re-selects the route shown in Fig. 2. This oscillation often occurs and makes the convergence of the algorithm difficult. Therefore, each AGV skips the re-route planning randomly with a predetermined probability (20-40%), which is determined by several numerical experiments. By using this procedure, we confirmed by numerical examples that the algorithm can easily make convergence and feasible route plan shown in Fig. 3(b) can be obtained in every case.

D. Re-route planning algorithm

The route planning problem for each AGV is a motion planning problem for single mobile robot which has been intensively studied. Several optimization algorithms such as A* search, generic algorithm, or simulated annealing can be adopted for solving single AGV sub-problem. In our approach, Dijkstra's algorithm is used for single AGV sub-problem.

IV. NUMERICAL EXAMPLES

A. Example problem

The route planning problem for 7 mobile robots in a transportation system with 143 nodes is treated as an example problem. In order to evaluate the optimality of the solution generated by the proposed method, a static problem is solved. In the static problem, the transportation request is given to each AGV a priori as shown in Table. I. S node and G node represents the starting node and the ending node respectively.

TABLE I
TRANSPORTATION REQUEST FOR EACH AGV

AGV No.	1	2	3	4	5	6	7
S node	1	125	96	105	52	104	26
G node	99	38	49	92	23	109	123

The parameters used for the proposed method are $\Delta\alpha$, the skipping ratio, $\Delta\lambda$ and $\Delta\phi$. $\Delta\alpha = 0.8$, the skipping ratio is set to 25% and $\Delta\lambda, \Delta\phi = 0$ for reducing the computation time in the example problem. The results for the case of $\Delta\lambda, \Delta\phi > 0$ are not shown in this paper for space limitation. An example of the result of route planning is shown in Fig. 4. It demonstrates that the route plan without collision and deadlock is obtained by the proposed algorithm. The computation time for deriving a solution for this problem is 1 s (Pentium III 1GHz, 256MB memory is used).

B. Optimality of the solution obtained by the proposed method

The optimal solution cannot be obtained for the example problem by using branch and bound method

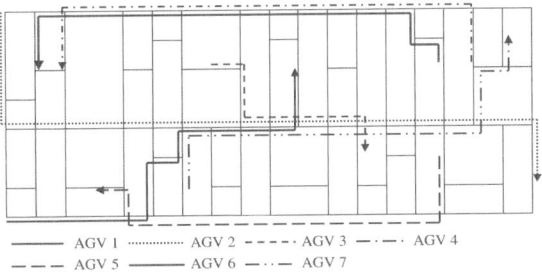

Fig. 4. The results of route planning for the example problem

when the number of AGV and the number of time horizon are increased. We have been confirmed that mixed integer programming solver can obtain an optimal solution for the problem within 29 nodes and 7 AGVs. The optimal solution may not be found even for the large scale problems. Therefore, in order to evaluate the quality of solutions derived by the proposed method, we obtained a lower bound of the problem by using Lagrangian relaxation explained in Section II and calculated the duality gap expressed by Eq. (23). The sub-gradient search algorithm is used to solve the dual problem shown in Eq. (15).

$$D = \frac{J - L^*}{L^*} \times 100 \ [\%] \quad (23)$$

J is the value of total transportation time obtained by the proposed method. L^* is the lower bound of the problem obtained by using Lagrangian relaxation. The duality gap D becomes zero if the solution obtained by the proposed method is the optimal solution.

In order to examine the duality gap for various types of problems, 15 problems are created. The number of AGV is changed from 1 to 15 for each problem. For each problem, 10 types of requests are generated and solved by the proposed method. Fig. 5 shows the average value of ten times of calculation by the proposed method. As shown in Fig. 5, the duality gap becomes larger when the number of AGV is increased. This is because when the number of AGV is increased, the deadlock and interference among AGVs frequently occurred and then it becomes difficult to obtain the optimal solution. However, the duality gap is within 5% for all the cases for the proposed method. Therefore, the proposed method is shown to be able to generate a near optimal solution for the example problem.

C. Comparison with the conventional method considering the entire AGV

The computation time of the proposed method for deriving a near optimal solution is compared with that of the conventional method in this section. The algorithm of the conventional method is that feasible

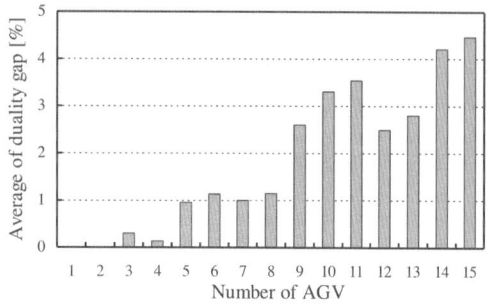

Fig. 5. Relationship between the number of AGV and average of duality gap

route for the entire AGV is always generated taking the entire AGV into account. The simulated annealing method is used in order not to be trapped into a bad local optimum. The candidate of route is generated randomly. But the probability of selecting the nodes is changed according to the position of the other AGV so as not to generate meaningless solutions.

Fig. 6 shows the computation time for both methods. The number of AGV is changed from 1 to 15. For the conventional method, the computation time for deriving a near optimal solution is much larger than that of the proposed method. This is because it takes much computation time to generate a feasible route when the number of AGV is increased. For the proposed method, the computation time is within 5 s for all the cases. The proposed method has also shown to be effective for deriving a near optimal solution compared with the conventional method.

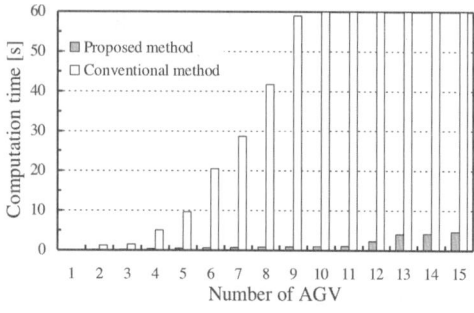

Fig. 6. Comparison of computation time

V. IMPREMENTATION OF DISTRIBUTED ROUTE PLANNING METHOD TAKING GEOMETRICAL SIZE OF MOBILE ROBOTS INTO ACCOUNT

A. Transportation bay in laboratory experiment

In the previous sections, the geometrical size of AGV is regarded as sufficiently small compared with the transportation bay. However, in an actual transportation system such as semiconductor fabricating bay, the size of the layout of transportation system is limited. Therefore, the size of the AGV has to be taken into account to avoid the collision and deadlock among AGVs. The size of experimental transportation bay and the size of experimental AGV are shown in Fig. 7.

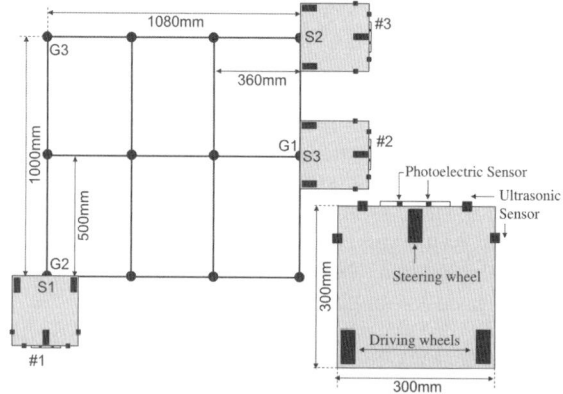

Fig. 7. Geometrical size of experimental system

B. Penalty function for each AGV reflecting geometrical size

The proposed method can individually generate the route for each robot. Therefore, the proposed algorithm can easily implemented taking the size of each AGV into account. The penalty term for violating the collision avoidance constraint is transformed from the node capacity constraint into the geometrical region constraint around the position of the robot. For the distributed route planning method, it becomes easy to modify the system and easy to deal with the addition of the AGV by installing the same software.

C. Results of simulation

The proposed algorithm is applied to the route planning problem with 3 experimental AGVs. The simulations are carried out for the experimental transportation system with 12 nodes shown in Fig. 7. The horizontal distance between nodes is 1080mm in total and the vertical distance is 1000mm in total. S and G in Fig. 7 and Fig. 8 represents the starting and ending node respectively. The traveling time required to perform each action is set to 30 unit for 90 degree turn and 60 unit for 180 degree turn based on the experimental results (1 unit is the time the robot moves 10mm). The total computation time for deriving a route plan for the proposed method is 1 second even though the size of AGV is considered in the optimization at each AGV. Fig. 8 shows the results of the movement of AGVs in the simulation.

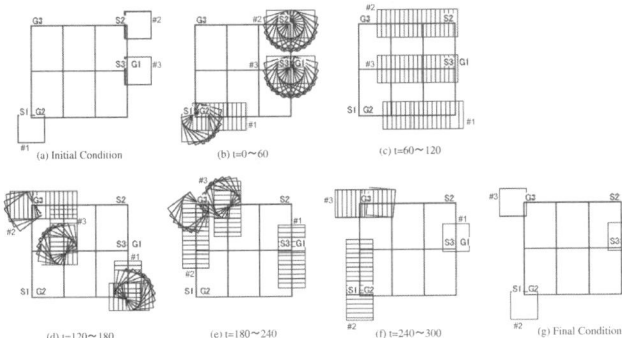

Fig. 8. The sequece of motion of 3 AGVs obtained in the simulation

D. Results of experiment

The route obtained by the proposed algorithm is sent to the experimental system. Each AGV has encoders for localization. There are white lines between the nodes in the experimental system. Each AGV has photoelectric sensor is used to track the trajectory of white lines. Fig. 9 shows the snapshot of the results obtained by the experiments. It has been demonstrated that three AGVs successively moved to each ending point without collision and deadlock among AGVs.

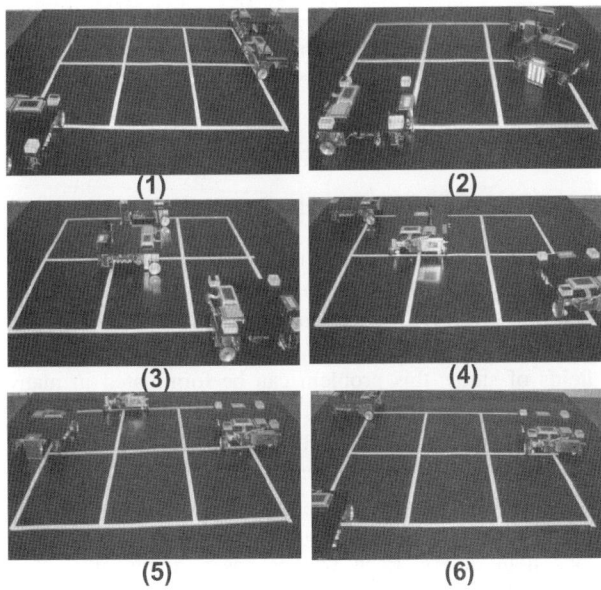

Fig. 9. Snapshot of the experimental results

VI. Conclusion

We have been proposed a distributed route planning method for multiple mobile robots using Lagrangian decomposition method. The proposed method can derive a feasible route which minimizes the total transportation time repeating the local optimization at each AGV and data exchange among the AGVs. The results have shown that the solution with the 5 % of duality gap can be obtained by the proposed method within five seconds of computation time. The proposed method has been implemented on the experimental AGVs system. It has been demonstrated that the AGVs can successively travel from the starting position to the ending position without collision and deadlock. The promising feature of the proposed method is the short computation time for deriving the solution for the large scale problems. By adopting the parallel processor, the proposed method has a possibility for reducing the computation time. Therefore, the proposed method can easily applicable to the real time route planning problems for large scale transportation system in semiconductor fabricating bay. The application of the proposed method to the dynamic route planning problem is one of our future works.

Acknowledgments

The authors gratefully acknowledge the funding provided by Electric Technology Research Foundation of Chugoku.

References

[1] J. A. Fernandez, J. Gonzalez, L. Mandow, J. L. Perez-de-la-Cruz, "Mobile Robot Path Planning: A Multicriteria Approach" Engineering Application of Artificial Intelligence, Vol 12, 543-554, 1999.
[2] M. Watanabe, M. Furukawa and Y. Kakakzu, "Intelligent AGV Driving Toward an Autonomous Decentralized Manufacturing System", Robotics and Computer Integrated Manufacturing, 17, 57-64, 2001.
[3] S. Akella and S. Hutchinson, "Coordinating the Motions of Multiple Robots with Specified Trajectories", Proceedings of the 2002 IEEE International Conference on Robotics and Automation, 624-631, 2002.
[4] K. Azarm and G. Schmidt, "Conflict-Free Motion of Multiple Mobile Robots Based on Decentralized Motion Planning and Negotiation", Proceedings of the 1997 IEEE International Conference on Robotics and Automation, 3526-3533, 1997.
[5] Y. Guo and L. E. Parker, "A Distributed and Optimal Motion Planning Approach for Multiple Mobile Robots", Proceedings of the 2002 IEEE International Conference on Robotics and Automation, 2612-2618, 2002.
[6] R. Zlot, A. Strenz, M.B. Dias, S. Thayer, "Multi-Robot Exploration Controlled by a Market Economy", Proceedings of the 2002 IEEE International Conference on Robotics and Automation, 3016-3023, 2002.
[7] Dorigo, M.,Maniezzo, V.,Colorni, A., "Ant System: Optimization by a Colony of Cooperating Agents", IEEE Transactions on System, Man, and Cybernetics-Part B, Vol. 26, No. 1, 1-13, 1996.
[8] Hoitomt, D.J.,Luh, P.B.,Pattipati, K.R., "A Practical Approach to Job-Shop Scheduling Problems", IEEE Transactions on Robotics and Automation, Vol. 9, No. 1, pp. 1-13 (1993).
[9] Nishi, T. et al., "Machine Oriented Decentralized Scheduling Method using Lagrangian Decomposition and Coordination Technique", Proceedings of the IEEE International Conference on Robotics and Automation,pp. 4173-4178 (2002).

Multi-Robot Task Allocation: Analyzing the Complexity and Optimality of Key Architectures

Brian P. Gerkey   Maja J Matarić
Computer Science Department
University of Southern California
Los Angeles, CA 90089-0781, USA
{bgerkey|mataric}@cs.usc.edu

Abstract— Important theoretical aspects of multi-robot coordination mechanisms have, to date, been largely ignored. To address part of this negligence, we focus on the problem of multi-robot task allocation. We give a formal, domain-independent, statement of the problem and show it to be an instance of another, well-studied, optimization problem. In this light, we analyze several recently proposed approaches to multi-robot task allocation, describing their fundamental characteristics in such a way that they can be objectively studied, compared, and evaluated.

I. INTRODUCTION

Since the early 1990s, the problem of task allocation in multi-robot systems has received significant and increasing interest in the research community. As researchers design, build, and use cooperative multi-robot systems, they invariably encounter the question: "which robot should execute which task?" This question must be answered, even for relatively simple multi-robot systems, and the importance of task allocation grows with the complexity, in size and capability, of the system under study. Even in the simplest case of homogeneous robots with fixed, identical roles, intelligent allocation of tasks is required for good system performance, if only to minimize physical interference.

Of course, task allocation need not be explicit; it may instead emerge from the interactions of the robots (physical and otherwise), as is the case with coordination methods put forward by proponents of *swarm robotics* [9]. Such an emergent system, when constructed skillfully, can be extremely effective and may provide the simplest and most elegant solution to a problem. However, it is a solution to a *specific* problem, and if robots are to be generally useful, we believe that they must be capable of solving a variety of problems.

Over the years, a significant body of work has been done on *explicit* multi-robot task allocation (MRTA), generally involving task-oriented inter-robot communication. A variety of such architectures have been proposed; while they are sometimes evaluated experimentally, they are rarely subject to formal analysis. It is coordination architectures of this type on which we focus our analysis in this paper. Our aim is to address two key shortcomings of the MRTA work to date[1]:

- computation and communication requirements are generally unknown
- aside from experimental validation in specific domains, there is no characterization of the *solution quality* that can be expected

The rest of this paper is organized as follows. In the next section we give a formal statement of the general problem of multi-robot task allocation, by way of reduction to an instance of a well-known optimization problem from Operations Research. In Section III, we analyze some significant MRTA architectures that have been proposed to date by considering them to be algorithms for (approximately) solving the underlying optimization problem. By doing so we will gain a deeper understanding of how these approaches function and what cost/benefit tradeoffs they introduce. We conclude in Section IV with a summary and a brief discussion of extensions and future directions for our work.

II. PROBLEM STATEMENT

We claim that multi-robot task allocation can be reduced to an instance of the *Optimal Assignment Problem* (OAP) [11], a well-known problem from Operations Research. A recurring special case of particular interest in several fields of study, this problem can be formulated in many ways. Given our application domain, it is fitting to describe the problem in terms of jobs and workers. There are n workers, each looking for one job, and m available jobs, each requiring one worker. The jobs can be of different priorities, meaning that it is more important to fill some jobs than others. Each worker has a nonnegative skill rating estimating his/her performance for each potential job (if a worker is incapable of undertaking a job, then the worker is assigned a rating of zero for that job). The problem is to assign workers to jobs in order to maximize the overall expected performance, taking into account the priorities of the jobs and the skill ratings of the workers.

[1]In a recent important paper [21], Pynadath & Tambe perform a similar critique and analysis of strategies for multi-agent teamwork.

Our multi-robot task allocation problem can be posed as an assignment problem in the following way: given n robots, m prioritized (i.e., weighted) single-robot tasks, and estimates of how well each robot can be expected to perform each task, assign robots to tasks so as maximize overall expected performance. However, because the problem of task allocation is a dynamic decision problem that varies in time with phenomena including environmental changes, we cannot be content with this static assignment problem. Thus we complete our reduction by iteratively solving the static assignment problem over time.

Of course, the cost of running the assignment algorithm must be taken into account. At one extreme, a costless algorithm can be executed arbitrarily fast, ensuring an optimal assignment over time. At the other extreme, an expensive algorithm that can only be executed once will produce a static assignment that is only initially optimal and will degrade over time. Finally there is the question of how many tasks are considered for (re)assignment at each iteration. In order to create and maintain an optimal allocation, the assignment algorithm must consider (and potentially reassign) every task in the system. Such an inclusive approach can be computationally expensive and, indeed, some implemented approaches to MRTA use heuristics to determine a subset of tasks that will be considered in a particular iteration.

Together, the cost of the static algorithm, the frequency with which it is executed, and the manner in which tasks are considered for (re)assignment will determine the overall computational and communication overhead of the system, as well as the solution quality. Thus it is these characteristics of MRTA architectures that we examine in Section III. Before continuing with a formal statement of our problem, we undertake a necessary aside regarding *utility*.

A. Utility

Utility is a unifying, if sometimes implicit, concept in economics [10], game theory [17], and operations research [2], as well as multi-robot coordination (see Section III). The idea is that each individual can somehow internally estimate the value (or the cost) of executing an action. It is variously called fitness, valuation, and cost. Since the exact formulation varies from system to system, we now give an instructive and generic, yet practical, definition of utility for multi-robot systems.

We assume that each robot is capable of estimating its fitness for every task of which it is capable. This estimation includes two factors, both task- and robot-dependent:

- expected quality of task execution, given the method and equipment to be used (e.g., the accuracy of the map that will be produced using a laser range-finder)
- expected resource cost, given the spatio-temporal requirements of the task (e.g., the power that will be required to drive the motors and laser range-finder in order to map the building)

Given a robot R and a task T, if R is capable of executing T, then we can define, on some standardized scale, Q_{RT} and C_{RT} as the quality and cost, respectively, expected to result from the execution of T by R. We can now define a combined, nonnegative utility measure[2]:

$$U_{RT} = \begin{cases} Q_{RT} - C_{RT} & \text{if } R \text{ is capable of executing } T \text{ and } Q_{RT} > C_{RT} \\ 0 & \text{otherwise} \end{cases}$$

The robots' utility estimates will be inexact for a number of reasons, including sensor noise, general uncertainty, and environmental change. These unavoidable characteristics of the multi-robot domain will necessarily limit the efficiency with which coordination can be achieved. We treat this limit as exogenous, on the assumption that lower-level robot control has already been made as reliable, robust, and precise as possible and thus that we are incapable of improving it. When we later discuss "optimal" allocation solutions; we mean "optimal" in the sense that, given the union of all information available in the system (with the concomitant noise, uncertainty, and inaccuracy), it is impossible to construct a solution with higher overall utility.

Note that although we have not discussed planning or learning, our definition of utility permits their introduction. The addition of a predictive model, for example, will (presumably) improve the accuracy of utility estimates by considering expected future events. To achieve the best system performance, such techniques should be used whenever possible.

B. Formalism

We are now ready to state our MRTA problem as an instance of the OAP. Formally, we are given:

- the set of n robots, denoted $I_1, \ldots, I_n$
- the set of m prioritized tasks, denoted $J_1, \ldots, J_m$ and their relative weights[3] $w_1, \ldots, w_m$
- U_{ij}, the nonnegative utility of robot I_i for task J_j, $1 \leq i \leq n, 1 \leq j \leq m$

We assume:

- Each robot I_i is capable of executing at most one task at any given time.
- Each task J_j requires exactly one robot to execute it.

These assumptions, though somewhat restrictive, are necessary in order to reduce MRTA to the classical OAP, which is given in terms of single-worker jobs and single-job workers. We are currently working on a more general

---

[2] We thank Michael Wellman for suggesting this formulation.

[3] If the tasks are of equal priority (as is often the case), then the weights are all equal to 1.

formulation that will allow us to relax these assumptions. Regardless, in most existing MRTA work (including the architectures that we study in Section III), these same assumptions are made, though often implicitly.

The problem is to find an optimal allocation of robots to tasks. An allocation is a set of robot-task pairs:

$$(i_1, j_1) \ldots (i_k, j_k), \ 1 <= k <= \min(m, n)$$

Given our assumptions, for an allocation to be *feasible* the robots $i_1 \ldots i_k$ and the tasks $j_1 \ldots j_k$ must be unique. The benefit (i.e., expected performance) of an allocation is the weighted utility sum:

$$U = \sum_{m=1}^{k} U_{i_m j_m} w_{j_m}$$

We can now cast our problem as an integral linear program [11]: find n^2 nonnegative integers α_{ij} that maximize

$$\sum_{i,j} \alpha_{ij} U_{ij} w_j \qquad (1)$$

subject to

$$\sum_i \alpha_{ij} = 1, \quad \forall j$$
$$\sum_j \alpha_{ij} = 1, \quad \forall i \qquad (2)$$

The sum (1) is just the overall system utility, while (2) enforces the constraint that we are working with single-robot tasks and single-task robots (note that since α_{ij} are integers they must all be either 0 or 1). Given an optimal solution to this problem (i.e., a set of integers α_{ij} that maximizes (1) subject to (2)), we construct an optimal task allocation by assigning robot i to task j only when $\alpha_{ij} = 1$.

By creating a *linear* program, we restrict the space of task allocation problems that we can model in one way: the function to be maximized (1) must be linear. Importantly, there is no such restriction on the manner in which the components of that function are derived. That is, individual utilities can be computed in any arbitrary way, but they must be combined linearly.

C. Scheduling formulation

The OAP, and thus MRTA, can also be seen from a scheduling perspective. Phrased in Brucker's terminology [4], our problem is in the class of scheduling problems described by:

$$R \,||\, \sum w_i C_i \qquad (3)$$

That is, the system is composed of heterogeneous parallel machines and overall performance is computed as a weighted sum of the utility values for the individual tasks (Brucker uses C_i as the utility estimate of the machine assigned to task i and w_i as the scalar weight for task i).

The problem class (3) is a superset of the simpler case of identical parallel machines:

$$P \,||\, \sum w_i C_i \qquad (4)$$

Problems in the class (4) are known to be $\mathcal{NP}$-hard [4], and thus so are problems in the class (3).

Fortunately, we can simplify our problem by making two domain-specific observations. First, we recognize that the MRTA problem is a degenerate scheduling problem. Whereas in scheduling one must assign tasks to machines over time, in MRTA we consider only a single time-slot at each iteration. Second, we can incorporate the task weights directly into the utility estimates if we make the reasonable assumption that the task weights are known to the robots and can be used in utility estimation. Given a utility estimate U_{ij} for robot i and task j and a scalar task weight w_j, we can define a new weighted utility estimate:

$$U'_{ij} = w_j U_{ij}$$

For a single time-slot, we can trivially make this change of variables for each of the mn utilities U_{ij} in $O(mn)$ time. Thus, our particular problem reduces to an unweighted scheduling problem, becoming an instance of the class:

$$R \,||\, \sum C_i \qquad (5)$$

Problems in the class (5) are known to be polynomially solvable, for example by Bruno et al.'s job scheduling algorithm [5], which runs in $O(mn^3)$ time. There also exist specialized algorithms that solve the OAP even faster, such as Kuhn's Hungarian method [13], which runs in $O(mn^2)$ time.

This result is important, because it suggests that we can develop practical, efficient mechanisms for making *optimal* allocations of tasks in multi-robot systems. If instead our problem were $\mathcal{NP}$-hard then we could only realistically expect to employ heuristic, potentially sub-optimal algorithms of the sort used to date for MRTA, and which we analyze in the next section.

Parker has shown [19] that a variant of MRTA, which she calls the ALLIANCE Efficiency Problem (AEP), is $\mathcal{NP}$-hard by restriction to the the $\mathcal{NP}$-complete problem PARTITION. This result does not contradict our preceding analysis of MRTA, for the AEP is in fact harder than the instantaneous task allocation problem that we consider in this paper. In the AEP, given is a set of tasks making up a mission, and the objective is to allocate a subset of these tasks to each robot so as to minimize the maximum time taken by a robot to serially execute its allocated tasks. Thus in order to solve the AEP, one must construct a time-extended schedule of tasks for each robot. This problem is clearly an instance of the scheduling problem:

$$R \,||\, C_{max} \qquad (6)$$

which is known to be $\mathcal{NP}$-hard [4]. Thus we have corroborated Parker's conclusion with a scheduling analysis.

III. ANALYSIS

Having given a formal statement of the MRTA problem, we are now in a position to analyze some of the key task allocation architectures from the literature. In this section we examine six approaches to MRTA, focusing on three characteristics:

- computation requirements [7]
- communication requirements [14]
- task consideration

In part because of trends in the research community that stress the importance of experimental validation with physical robots, such theoretical aspects of multi-robot coordination mechanisms have been largely ignored. However, they are vitally important to the study, comparison, and objective evaluation of the mechanisms. The large-scale and long-term behavior of the system will be strongly determined by the fundamental characteristics of the underlying algorithm(s). Thus we endeavor to derive and explicate those characteristics here. Before we continue, however, it will be necessary to explain the methodology that we use in our analysis.

A. Methodology

As we stated earlier, the key to effective task allocation for multi-robot systems is to iterate the assignment, in order to deal with changes in the tasks, the robots, and the environment. The architectures under study achieve this iteration in different ways, along two dimensions. First, while some approaches allow assignment and reassignment of all tasks at each iteration, some never reassign tasks (or at least only reassign them because of robot failure). Second, some approaches periodically consider all tasks simultaneously, while others consider single tasks sequentially as they are offered for (re)assignment. Thus, when we discuss complexities, we state them in terms of iterations, though the details of an "iteration" may vary across architectures.

We determine computation requirements, or running time, in the usual way, as the number of times that some dominant operation is repeated. For our domain that operation is usually either a calculation or comparison of utility, and running time is stated as a function of n and m, the numbers of robots and tasks, respectively. Since modern robots have significant processing capabilities onboard and can easily work in parallel, we assume that the computational load is evenly distributed over the robots, and state the running time as it is *for each robot*. For example, if we need to find for each robot the task with the highest utility, then the running time is $O(m)$, because each robot performs m comparisons, in parallel.

We determine communication requirements as the total number of inter-robot messages sent over the network. We do not consider message sizes, on the assumption that they are generally small (e.g., single scalar utility values) and approximately the same for different algorithms. We also assume that a perfect shared broadcast communication medium is in use and that messages are always broadcast, rather than unicast. So if, for example, each robot must tell every other robot its own highest utility value then the overhead is $O(n)$, because each robot makes a single broadcast.

B. The architectures

We have chosen for study six MRTA architectures that have been validated on either physical or simulated robots. Our choices are somewhat subjective, for there are a great many more architectures in the literature. However, we believe that we have gathered a set of approaches that is fairly representative of the work to date. In the following sections, we analyze these architectures; our results are summarized in Table I.

1) ALLIANCE and BLE: We begin our analysis with the behavior-based [15] ALLIANCE architecture [20], one of the earliest and best-known approaches to MRTA. At each iteration, all tasks are considered for (re)assignment, based on the robots' utility estimates. In this case, utilities are distributed among measures of *acquiescence* and *impatience*. For example, when a robot is currently executing a task, its utility for that task is decreased over time by its own increasing acquiescence and by the increasing impatience of the other robots. Similarly, a robot's utility for a task that is being executed by another robot is increased over time by its own impatience and by the other robot's acquiescence.

By splitting the utility estimation in this way, the ALLIANCE architecture decreases communication overhead. Since each robot is effectively modeling internally the progress of the others, the robots need not broadcast their utilities for each task (as is the case with many of the approaches described below). Specifically, assuming that all available tasks are currently underway, each engaged robot broadcasts only a heartbeat message each iteration, yielding a communication overhead of $O(m)$ per iteration. A significant drawback to this approach is that a variety of parameters that govern the robots' update rules for impatience and acquiescence must be carefully tuned. This problem was addressed in ALLIANCE in part by introducing parameter learning.

With regard to computation, each robot executes a greedy task-selection algorithm: for each available task, compare its own utility to that of every other robot and select the shortest task for which it is most capable (thus tasks are implicitly prioritized by length). This algorithm can be executed in $O(mn)$ time per iteration.

Name	Computational Requirements / iteration	Communication Requirements / iteration	Task Consideration
ALLIANCE [20]	$O(mn)$	$O(m)$	simultaneous, reassignment
BLE [22]	$O(mn)$	$O(mn)$	simultaneous, reassignment
M+ [3]	$O(mn)$	$O(mn)$	simultaneous, no reassignment
MURDOCH [12]	$O(1)$ / bidder $O(n)$ / auctioneer	$O(n)$	sequential, no reassignment
First-price auctions [23]	$O(1)$ / bidder $O(n)$ / auctioneer	$O(n)$	sequential, reassignment
Dynamic role assignment [6]	$O(1)$ / bidder $O(n)$ / auctioneer	$O(n)$	sequential, reassignment

TABLE I

Summary of selected MRTA architectures. Shown here are the computational and communication requirements for six key architectures. Note that "iteration" has a different meaning depending on whether tasks are considered simultaneously or sequentially.

Broadcast of Local Eligibility (BLE) [22] is another behavior-based approach to MRTA, with fixed-priority tasks. For each task, each robot has a corresponding behavior that is capable of executing the task, as well as estimating the robot's utility for the task. Utilities are computed in a task-specific manner as a function of relevant sensor data. These utilities are periodically broadcast to the other robots, with all tasks simultaneously considered for (re)assignment.

Since each robot must broadcast its utility for each task, we have communication overhead of $O(mn)$ per iteration. Upon receipt of the other robots' utilities, each robot executes a simple greedy algorithm: find the highest-priority task for which it is most fit. This algorithm requires each robot to compare, for each task, its own utility to that of every other robot, resulting in running time of $O(mn)$ per iteration.

The BLE algorithm has also been used to coordinate the actions of the Azzurra Robot Team (ART) [1] in a soccer domain. We note that, if task priorities are incorporated into utility estimates (see Section II-C) and all utility estimates have been gathered in a central table, both the ALLIANCE and BLE task algorithms can equivalently be stated in the following way:

1) Find the robot-task pair (i, j) with the highest utility.
2) Assign robot i to task j and remove them from consideration.
3) Go to step 1.

Exactly this greedy algorithm, operating on a global blackboard, has also been used in a recent study of the impact of communication and coordination on MRTA [18].

2) Auction-based approaches: The M+ system [3] achieves task allocation by use of a variant of the well-known Contract Net Protocol (CNP) [8]. The basic idea of the CNP is that when a task is available, it is put up for auction, and candidate robots make "bids" that are their task-specific utility estimates. The highest bidder (i.e., the best-fit robot) wins a contract for the task and proceeds to execute it.

In the M+ system, each robot considers, at each iteration, all currently available tasks. For each task, each robot uses a planner to compute its utility and announces the resulting value to the other robots. With each robot broadcasting its utility for each task, we have communication overhead of $O(mn)$ per iteration.

Upon receipt of the other robots' utilities, each robot executes essentially the same greedy task-selection algorithm that is used in ALLIANCE: find those tasks for which its utility is highest among all robots and pick from that set the highest-utility task. This algorithm can be executed in $O(mn)$ time per iteration.

Similar to M+, the MURDOCH task allocation mechanism [12] also employs a variant of CNP. Tasks are allocated by first-price auction [16] sequentially as they are stochastically introduced to the system; reassignment is not allowed. Utility is computed in a task-specific manner, as a function of relevant sensor inputs.

For each task auction, each available robot broadcasts its bid (i.e., utility), yielding communication overhead of $O(n)$ per iteration. Because of the asymmetric nature of MURDOCH's auctions, the running time varies between the bidders and the auctioneer. Each bidder need only compute its utility, while the auctioneer must find the highest utility among the bidders. Thus computational overhead per iteration is $O(1)$ for bidders and $O(n)$ for the auctioneer.

Another CNP-based approach to MRTA, this one applied to multi-robot exploration, is described in [23]. Tasks are allocated by first-price auction sequentially, but reassignment is allowed. Utilities are computed in a task-specific manner, in this case as a function of the estimated time required to travel to a target location. As with MURDOCH, each task auction requires each robot to broadcast its utility, resulting in communication overhead of $O(n)$ per iteration. Similarly, the computational overhead is asymmetric: $O(1)$ for bidders and $O(n)$ for the auctioneer (per iteration).

Finally, the dynamic role assignment architecture described in [6] is another CNP-based approach to MRTA. The complexities are the same as for MURDOCH.

That such auction-based allocation methods work in practice is not surprising, for it is well known that synthetic economic systems can be used to solve a variety of optimization problems. In fact, an appropriately constructed price-based market system (which the previously described architectures approximate to varying degrees) can optimally solve assignment problems. At equilibrium, such a market optimizes costs in the so-called *dual* of the original OAP, resulting in an optimal allocation [11], [2].

C. Solution quality

We have yet to characterize the results that can be expected from the architectures that we have analyzed. Unfortunately, such a characterization is difficult, if not impossible, to make. The crux of this difficulty is that all of the architectures execute some kind of *greedy* algorithm for task allocation. The solution quality of greedy optimization algorithms can be difficult to define, because it can depend strongly on the nature of the input. In the MRTA domain, the input (as described in Section II-B), is the set of robots, the set of tasks, and the environment that governs their evolution.

Although each of the approaches discussed in the previous section may in some cases produce allocations that are close to (or even equal to) an optimal allocation, it is trivial to construct pathological inputs that elicit arbitrarily sub-optimal allocations[4]. For example, consider the following matrix, giving the utilities (already weighted by task priority) of robots A and B for tasks x and y:

	x	y
A	100	99
B	99	1

(7)

Using either ALLIANCE or BLE, task x would be assigned to robot A, leaving task y for robot B. This solution

[4]In the parlance of algorithmic analysis, we are showing by counterexample that the MRTA problem, as it is approached by the architectures under study, lacks the *greedy-choice property*, which is a prerequisite for a greedy algorithm to produce an optimal solution [7].

yields the maximally sub-optimal value of 101, far less than the obvious optimal solution.

Similarly, CNP-based approaches are highly susceptible to the time ordering of tasks. The problem with such sequential architectures is really a time-extended analogue of the problem with simultaneous architectures such as ALLIANCE and BLE. Still working with the same utility matrix (7), imagine that the tasks x and y are introduced, in that order, to MURDOCH. Task x would be auctioned off to the highest bidder, robot A, and task y would be left for robot B, resulting in the same poor allocation.

Thus we go no further than to classify the algorithms analyzed in Section III-B as greedy, and assert that they should be expected to produce nearly identical solutions, subject to the following observations:

- Simultaneous consideration of tasks will produce allocations that are at least as good as those produced with sequential consideration.
- Allowing reassignment of previously assigned tasks will produce allocations that are at least good as those produced without reassignment.

We are currently working to establish more informative performance bounds for these algorithms.

IV. CONCLUSION

With the goal of bringing some objective grounding to an emerging area of research that has, to date, been largely experimental, we have presented a formal study of the problem of multi-robot task allocation (MRTA). We have given a domain-independent statement of the problem and shown that it can be understood as an instance of the well-known optimal assignment problem (OAP). By this reduction, as well as a related scheduling analysis, we have shown that it is possible to *optimally* solve the original MRTA problem in polynomial time. By interpreting them as algorithms for solving the underlying OAP, we have analyzed the computation and communication requirements and (sub-)optimality of several robot task-allocation architectures from the literature.

There are many ways in which to exploit this information. For example, when building a multi-robot system, one can use the results presented in Table I in order to select an appropriate task-allocation architecture (if necessary, the analysis in Section III-B can be easily extended to include other architectures). Similarly, when designing a new method for MRTA, our definition of the problem and our exposition on previous approaches may prove useful.

For our own research, we plan to pursue the opportunities provided by the substantial body of work regarding the OAP that is available in other fields, including operations research, economics, and game theory. We are currently investigating the applicability to the robot domain of a wide variety of efficient, optimal assignment algorithms, both distributed and centralized. Based in part on their communication and computation requirements (which are generally well-known), we plan to adapt, implement, and experiment with some of these algorithms in the domain of multi-robot task allocation.

ACKNOWLEDGMENTS

The research reported here was conducted at the Interaction Lab, part of the Center for Robotics and Embedded Systems (CRES) at USC. The work is supported in part by the Intel Foundation, DARPA Grant DABT63-99-1-0015 (MARS), and DARPA Grant F30602-00-2-0573 (TASK). We thank Michael Wellman, Herbert Dawid, and Gaurav Sukhatme for their insightful comments.

V. REFERENCES

[1] Giovanni Adorni, Andrea Bonarini, Giorgio Clemente, Daniele Nardi, Enrico Pagello, and Maurizio Piaggio. ART'00 - Azzurra robot team for the year 2000. In Peter Stone, Tucker Balch, and Gerhard Kraetzschmar, editors, *RoboCup 2000: Robot Soccer World Cup IV, LNCS 2019*, pages 559–562. Springer-Verlag, Berlin, 2001.

[2] Dmitri P. Bertsekas. The Auction Algorithm for Assignment and Other Network Flow Problems: A Tutorial. *Interfaces*, 20(4):133–149, July 1990.

[3] Sylvia Botelho and Rachid Alami. M+: a scheme for multi-robot cooperation through negotiated task allocation and achievement. In *Proc. of the IEEE Intl. Conf. on Robotics and Automation (ICRA)*, pages 1234–1239, Detroit, MI, May 1999.

[4] Peter Brucker. *Scheduling Algorithms*. Springer-Verlag, Berlin, 2nd edition, 1998.

[5] John L. Bruno, Edward G. Coffman, and Ravi Sethi. Scheduling Independent Tasks To Reduce Mean Finishing Time. *Communications of the ACM*, 17(7):382–387, July 1974.

[6] Luiz Chaimowicz, Mario F. M. Campos, and Vijay Kumar. Dynamic Role Assignment for Cooperative Robots. In *Proc. of the IEEE Intl. Conf. on Robotics and Automation (ICRA)*, pages 293–298, Washington, DC, May 2002.

[7] Thomas H. Cormen, Charles E. Leiserson, and Ronald L. Rivest. *Introduction to Algorithms*. MIT Press, Cambridge, Massachusetts, 1997.

[8] Randall Davis and Reid G. Smith. Negotiation as a Metaphor for Distributed Problem Solving. *Artificial Intelligence*, 20(1):63–109, 1983.

[9] Jean-Louis Deneubourg, Guy Theraulaz, and Ralph Beckers. Swarm-made architectures. In *Proc. of the European. Conf. on Artificial Life (ECAL)*, pages 123–133, Paris, 1991.

[10] Francis Y. Edgeworth. *Mathematical Psychics: An Essay on the Application of Mathematics to the Moral Sciences*. Augustus M. Kelley, New York, 1967. Originally published in 1881.

[11] David Gale. *The Theory of Linear Economic Models*. McGraw-Hill Book Company, Inc., New York, 1960.

[12] Brian P. Gerkey and Maja J Matarić. Sold!: Auction methods for multi-robot coordination. *IEEE Transactions on Robotics and Automation*, 18(5):758–768, October 2002.

[13] Harold W. Kuhn. The Hungarian Method for the Assignment Problem. *Naval Research Logistics Quarterly*, 2(1):83–97, 1955.

[14] Eyal Kushilevitz and Noam Nisan. *Communication Complexity*. Cambridge University Press, Cambridge, 1997.

[15] Maja J Matarić. Behavior-based control: Examples from navigation, learning, and group behavior. *J. of Experimental and Theoretical Artifical Intelligence*, 9(2–3):323–336, 1997.

[16] R. Preston McAfee and John McMillan. Auctions and Bidding. *J. of Economic Literature*, 25(2):699–738, June 1987.

[17] John Von Neumann and Oskar Morgenstern. *Theory of Games and Economic Behavior*. J. Wiley, New York, Third edition, 1964.

[18] Esben H. Østergård, Maja J Matarić, and Gaurav S Sukhatme. Distributed multi-robot task allocation for emergency handling. In *Proc. of the IEEE/RSJ Intl. Conf. on Intelligent Robots and Systems (IROS)*, pages 821–826, Wailea, Hawaii, October 2001.

[19] Lynne E. Parker. L-ALLIANCE: A Mechanism for Adaptive Action Selection in Heterogeneous Multi-Robot Teams. Technical Report ORNL/TM-13000, Oak Ridge National Laboratory, October 1995.

[20] Lynne E. Parker. ALLIANCE: An architecture for fault-tolerant multi-robot cooperation. *IEEE Transactions on Robotics and Automation*, 14(2):220–240, April 1998.

[21] David V. Pynadath and Milind Tambe. Multiagent Teamwork: Analyzing the Optimality and Complexity of Key Theories and Models. In *Proc. of Intl. Conf. on Autonomous Agents and Multi Agent Systems*, pages 873–880, Bologna, Italy, July 2002.

[22] Barry Brian Werger and Maja J Matarić. Broadcast of Local Eligibility for Multi-Target Observation. In Lynne E. Parker, George Bekey, and Jacob Barhen, editors, *Distributed Autonomous Robotic Systems 4*, pages 347–356. Springer-Verlag, 2000.

[23] Robert Zlot, Anthony Stentz, M. Bernardine Dias, and Scott Thayer. Multi-Robot Exploration Controlled by a Market Economy. In *Proc. of the IEEE Intl. Conf. on Robotics and Automation (ICRA)*, pages 3016–3023, Washington, DC, May 2002.

Explicit Communication in Designing Efficient Cooperative Mobile Robotic System

Y. K. Lam[1]   E. K. Wong[2]   C. K. Loo[3]

Centre for Robotics and Automation, Faculty of Engineering and Technology
Multimedia University (Malacca Campus), Jalan Ayer Keroh Lama, Bukit Beruang, 75450 Melaka, Malaysia
Tel: +60-6-252-3276, Fax: +60-6-231-6552
E-mail: [1]yklam@mmu.edu.my, [2]ekwong@mmu.edu.my, [3]ckloo@mmu.edu.my

Abstract – This paper presents the design of ant-inspired control strategies that mimic the ant colony foraging behavior. Inter-agent communication, in particular explicit communication is applied to create multiple cooperating mobile robots in order to accomplish the foraging task. Explicit communication can significantly multiply the capabilities and effectiveness of teams of robotic systems. However, the drawback is it introduces more interference among robots at the goal and home region in comparison to robotic system without the implementation of inter-agent communication. Thus, to investigate how well the explicit communication can be adopted in designing efficient cooperative mobile robotic system, experiments were carried out on the simulated robots. The efficiency of the strategies are measured in terms of three criteria: time, density of robots and interference.

Keywords – Explicit communication, foraging, cooperative mobile robots

I. INTRODUCTION

Foraging task is chosen to investigate the impact of communication to the robot societies in this paper. Foraging has a strong biological basis. Many ant species, for instance, perform the foraging task as they gather food. Foraging is also an important subject of research in the mobile robotics community; it relates to many real-world problems. Among other things, foraging robots may find potential use in mining operations, explosive ordnance disposal and waste or specimen collection in hazardous environments.

In this research, foraging task involves the collection of objects of interest (pucks) clustered or localized in the environment. Robots move away from home and wander around to look for pucks. Upon encountering a puck, the robot moves towards it and grasps it. After attachment, the robot returns the object to home base.

In fact, arranging more than one robot without the communication capability to perform foraging task is sufficient to resemble the ant colony behavior. Nevertheless, from robotic point of view, in order for an implementation to be considered efficient, the amount of time required for robots to accomplish task must be minimized. Therefore, many robot designers have introduced simple communication to the collective robotic systems. There is no doubt the efficiency is improved with the aid of inter-agent communication.

More recently, the collective robotics community focuses on two major types of inter-agent communication that are implicit and explicit communication. Explicit communication requires explicit signaling and reception of the communicated information [1]. As opposed to explicit, when robots do not require a deliberate act of transmission, it is described as implicit. Robots are communicating through the environment. Implementation of implicit communication is preferred in the collective robotics nowadays because of lower cost consumption. However, it is of crucial interest to understand what can be gained by the introduction of explicit communication in a team of autonomous robots in conjunction with the cost. Explicit communication can be divided into two categories: local and global communications, which are also known as point-to-point and broadcast communications respectively.

From the design point of view, motor schema-based architecture is chosen as the guide to design the control strategies. The overall method of schema-based robotics is to provide behavioral primitives that can act in a distributed, parallel manner to yield intelligent robotic action in response to environmental stimuli [2]. Primitive behaviors can be combined cooperatively to create a high-level behavior that is obtained by multiplying the vector response of each motor schema by a gain, then summing and normalizing the result. With the introduction of communication to robotic system, the control strategies are no longer purely based on reactive schema-based architecture, but the combination of reactive and deliberative architecture. Reactive architecture is defined as one which "tightly couples perception to action without the use of intervening abstract representations or time history. On the other hand, deliberative architectures rely on abstract representation of the world [2]. Architectures which extend purely systems with some memory capabilities lie between these two extremes [3]. This approach is imperative for robots to trace the visited goal position.

II. RELATED WORK

Parker presented the architecture of a group of heterogeneous robots, ALLIANCE, that successfully deal with real-world applications, including a laboratory version of hazardous waste cleanup and a cooperative box pushing demonstration. ALLIANCE uses one-way broadcast communication; no negotiation or two-way conversations are utilized. From the point of control mechanism, it uses the concept of behavior-based subsumption architecture introduced by Brook [4]. Mataric applies the same method to develop homing, aggregation, dispersion, following and wandering behaviors for robots in a foraging task. In addition, Mataric investigated the communication between foraging robots by introducing robot chains [5] and leaving landmarks in shared localization space [6], which are indirectly mimicking the ant colonies. Balch and Arkin have also investigated the social communication among mobile robots performing different tasks [1].

III. FORMS OF INTER-AGENT COMMUNICATION

This section is started by clarifying the control strategies without communication and with communication, dividing the latter into two aforementioned categories: local and global communication architectures. Local communication allows the robotic system to send message from one robot to another or to a supervisor with the help of a dedicated physical channel such as radio or infrared link. On the other hand, global communication differs significantly from the local communication in the way of passing message. Another characteristic difference of the control architectures presented in this paper is about the time history memory capability where the resulting robot behavior from local communication is determined not only by the reaction to a stimulus but also by some internal registers which keep track of the time history [3], while global communication keeps the time history in the host (workstation). The strategy without communication works as the baseline schemas for designing and evaluating strategies with inter-agent communication. Figure 1, 2 and 3 demonstrates three Finite State Automaton (FSA) implementing motor-schema based reactive and deliberative multi-agent systems.

A. Strategy 1 (S1): FSA for Robots Forage without Communication

Figure 1 has been shown in [7]. It is constructed by three simple motor schemas (behaviors): Wander, Acquire and Deliver. It is obvious that robots do not interact with each other. Nevertheless, these schemas are sufficient for low intelligent robots to forage like ant colonies in a cooperative manner.

B. Strategy 2 (S2): FSA for Robots with Private Crumbs Forage with Communication and Acknowledgement to Peer and Home Base

A slightly more intelligent robotic system is presented where the robots are able to recall the visited region (Figure 2). Once encountered a region with more than one puck, the robot will keep the region's position in its internal register. The position data in register is called private crumb [6] in this paper. Private crumb is retrieved after robot deposits the attached puck. An assumption is made concerning the private crumb does not "evaporate" as in the case of ant's pheromone but it disappears once robot acquires the last puck from the region or when it detects the region has no more pucks.

Additionally, robots are able to transmit private crumb to the nearest peer while delivering puck to home base if pucks are still available in the region. The peer robot will keep the received message, peer crumb as its private crumb. This method applies Goal Communication in [1]. It is also clear that robots have the capability to sense the number of pucks available in the region.

After depositing puck, robot will acknowledge with home base about the number of pucks left in the region. Robot will revisit the region if only number of pucks with respect to the particular home crumb recorded in home base greater than zero. Home crumb is the visited region position data set by the first robot exploring the region. In addition, robot carrying puck on the way back to home base will automatically less its buffer data of number pucks left with respect to private crumb for every transmission to peer robot. This is to avoid the waste of revisiting time after robot recruited peer robots, in other words, assigned the acquiring and delivering tasks to the peer robot.

C. Strategy 3 (S3): FSA for Robots Forage with Communication to Home Base

The major differences of this strategy with S2 are the time history accessibility and the way of inter-agent message passing. Robots keep the visited position data to and retrieve it from home base (host workstation). And, robots do not directly interact with peers but through the broadcast from home base. The first robot to visit a region with more than one puck will report to home base about the position. The position data is called global crumb in this paper. Home base then broadcasts to the rest of robots about the location. As a result, all the wandering robots will head towards the location simultaneously. It is then the responsibility for the robot that acquires the last puck to report to home base so that home base deletes the global crumb. The global crumb does not "evaporate" in simulation.

D. Discussion

It is obvious that S2 is using the architecture of point-to-point communication (local communication), while S3 uses broadcast communication (global communication) architecture.

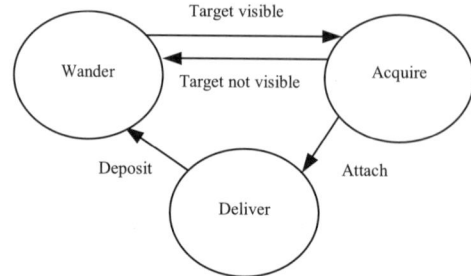

Fig. 1 FSA for Robots Forage without Communication (S1)

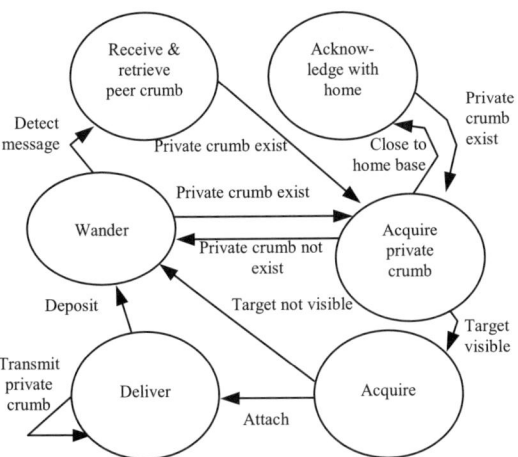

Fig. 2 FSA for Robots with Private Crumbs Forage with Communication and Acknowledgement to Peer and Home Base (S2)

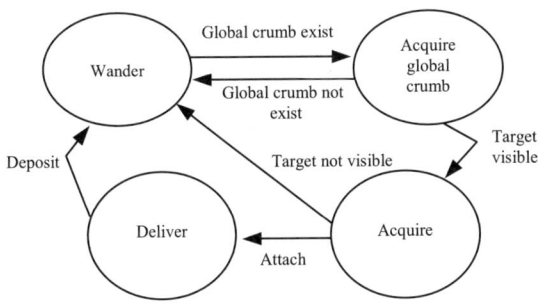

Fig. 3 FSA for Robots Forage with Communication to Home Base (S3)

IV. SIMULATION ENVIRONMENT

The simulation is carried out using Java-based Teambots Simulator [8]. The robotic agents with radius, 0.247 unit execute their task in a 6 x 6 unit environment with velocity, 0.02032 unit per step. The unit is dimensionless. All robots could only carry one puck. It is assumed that the mass of puck is fixed and every robot moved away from home simultaneously at the initial stage.

Clustered pucks are examined instead of distributed pucks for this research, as stated in [9], interaction did not improve performance for distributed pucks, but did for localized pucks. Therefore, the cluster size greater than one is evaluated to differentiate the distributed and localized pucks. And, the impact of communication upon the efficiency of cooperation can be investigated if there is more than one puck in a cluster.

Balch [7] states that within each task domain, an experimental space was explored by varying one or more independent variables and evaluating the resulting systems. The experiments are designed based on the parameters shown in Table I.

TABLE I
EXPERIMENTAL PARAMETER VALUES

Variables	Description
Cluster size (cs)	2-7 pucks
Number of robots (n)	1 to 5 for each strategy or FSA
World	One cluster of pucks
Obstacles	No
Number of trials	30 for each combination of cs and n
Steps constraint	10000

From the observation of a number of simulation runs, the robots could complete task (collect all pucks available in the environment) in average less than 8000 steps. In between, infinite loop or deadlock always happens when robots compete to acquire a target at the same territory. Consequently, to avoid lockups in infinite loops in the event the society is unable to complete the task for that particular world, simulation is allowed to continue for 10000 steps before starting a new trial. Since most runs complete in less than 8000 steps, it is highly likely that the system will never complete the task if it does not do so before failure is asserted. In order to ensure statistical significance, multiple worlds are created by randomly positioning the location of cluster of pucks for every simulation run.

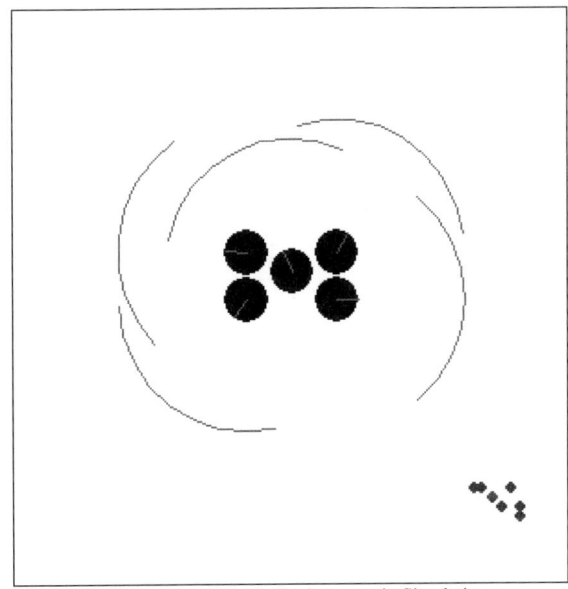

Fig. 4 Clustered Pucks Environment in Simulation

V. METHODOLOGY

Experiments were carried out in simulation environment. In general, design and modeling is an important stage prior to simulation. Therefore, FSA is designed initially to model and configure behavior-based agents as shown in Figure 1, 2 and 3. Data collection and analysis is subsequently done after every simulation experiment. The task accomplishment time in every trial of simulation is recorded.

A. Time and Cost

For this research, time to complete task was chosen as the primary performance metric. Strategy that yields minimum completion time is considered as the best architecture. The performance data is visualized as a 3-dimensional surface with the X axis reflecting the number of robots and the Y axis indicating the number of pucks. The Z or height axis shows the time to complete task for that combination of robots and pucks (Figure 5).

In order to quantify the difference between performances of three aforementioned strategies mentioned above, a performance ratio based on completion time is computed (Figure 6). For example, the performance time of Strategy 2 (S2) is divided by the performance time of Strategy 1 (S1). Result less than 1.0 indicates S2 has higher performance compared to S1.

In addition, the number of robots required for a system that can be directly equated to cost could also be estimated from the time performance comparison.

B. Interference

Interference refers to the situation where two robots attempt to occupy the same place at the same time; it is measured as the amount of time agents spend avoiding one another [7]. Interference reduces the rate at which agents can deliver pucks. In this research, interference is computed as the function of work, W (number of pucks collected over completion time) and density of robots, D (number of robots over environment area). By theoretical analysis, increasing the density of robots will yield higher interference, thereby defeats the work efficiency. In contrast, lower density of robots will have less interference that benefits the work efficiency. With this relation, the interference is modeled as such in this research:

$$Interference = f(W, D)$$
$$= \frac{D}{W}$$
$$= \frac{(n-1) \times (c+1) \times t}{p \times a} \quad (1)$$

where
n = number of robots
c = improvement of time performance
$$= \begin{cases} 0 \text{ for S1} \\ \frac{1}{r} \text{ for S2, S3} \end{cases}$$
t = completion time
p = number of pucks, for $p > 0$
a = Environment area, for $a > 0$ unit^2

Number of robots, n is less one to ensure zero interference at one robot. Time performance improvement, $c = r^{-1}$ for S2 and S3 because the improvement rate, r is inversely proportional to interference. Table II shows the average interference rate computed using (1) with the assumption that no specific cooperative strategies are applied in system.

VI. RESULTS AND ANALYSIS

This section shows the experimental comparison in assessing the impact of communication on performance.

A. Time and Cost

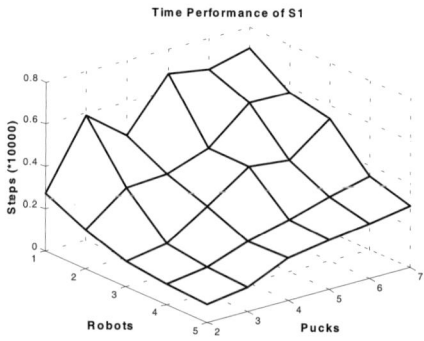

(a)

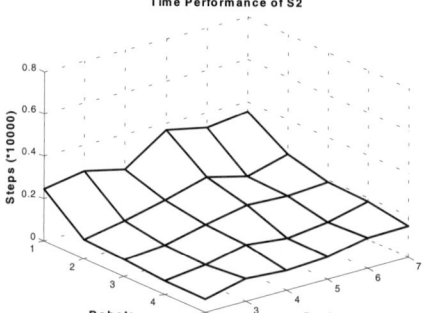

(b)

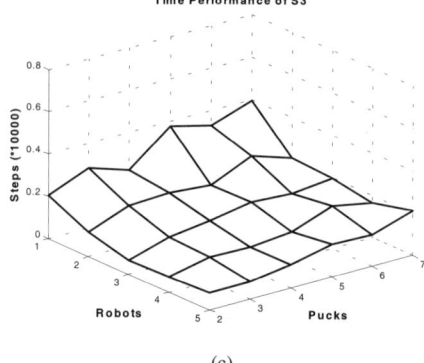

(c)

Fig. 5 Time Performance

Figure 5 illustrates time performance is significantly improved with the application of communication architecture. As shown in Figure 5a, robots are able to complete task in average less than 8000 steps without interact between each other. Once interaction is enforced, task completion time is ultimately reduced by almost half as shown in Figure 5b and 5c.

Precise improvement rate could be finalized from the computation of time performance ratio as illustrated in Figure 6. The result shows local communication (S2) and global communication (S3) improved performance rate, r in the forage task an average of 46% and 41% respectively.

In addition, with the implementation of communication architecture, the number of robots to be occupied in a system can be reduced, thereby lessen the cost consumption with the tolerance of time performance. As stated in [1], a system with two robots is generally best for three or more attractors (pucks) for a system without communication. This implies that n robots will likely collect more than or equal to $n+1$ pucks. Figure 5a has also shown for n number of pucks, an obvious time decreasing occurs from one to $n-1$ robots and saturates after $n-1$ robots. The saturation indicates increasing number of robots no longer improves the time performance. In conjunction with the relation of time performance ratio and number of robots, $0.54n$ ($n>1$) robots for local communication and $0.59n$ ($n>1$) robots for global communication are sufficient to collect $n+1$ or more pucks in the environment. If the environment contains only one or two pucks, one robot is the best choice for all the cases: no communication, local and global communication.

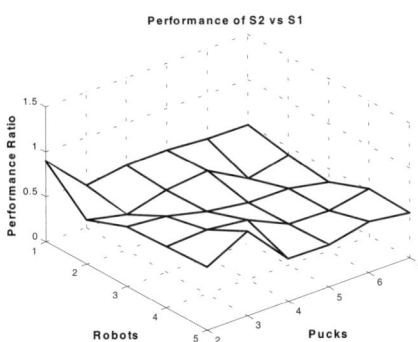

Fig. 6 Performance ratio of S2 over S1

From the hardware implementation point of view, one of the major drawbacks of global communication architecture is that a single radio channel with limited bandwidth must be shared by all the team members [3]. As a result, a global communication channel can quickly become a bottleneck, limiting team performance if the team size is not small and there is a large amount of information to be transmitted. Its cost consumption is also much higher than the cost of local communication architectures that commonly apply simple infrared transmitter and receiver. Consequently, many researchers have started to focus on local communication architectures in order to overcome the limitation of global schemes. Moreover, the performance ratio indicated local communication is 5% better than global communication on the average. Although the difference is small, local communication (S2) is evidently the best choice of communication architecture in terms of cost and hardware implementation.

B. Interference

TABLE II
INTERFERENCE

Strategy	Average Interference rate
S1	30.3391
S2	51.7858
S3	62.6513

The key drawback of introducing communication architecture to robotic system is the increasing of interference. By quantitatively analysis, S2 creates 70% more interference compared to S1 whilst S3 yields 107% more interference than S1. In general, both S2 and S3 have almost twofold multiplied the interference rate compared to S1.

Without the implementation of communication architecture, the existence of interference is due to competition for food and territory.

As for the case of S2, the competition for territory is raised when robots travel back and forth between home and the same pucks region. While on the way back to home, robots may have to avoid their peers, which are traveling to pucks region in the opposite direction at the same time. Basically, the increase of interference rate is dependent with the recruitment rate. As more robots are recruited, more competitions will happen at the pucks and home region as well as the path between home and pucks region.

The interference from S3 is highly dependent with the number of robots. Once a robot found the pucks region, it recruited all robots in the environment through the broadcasting from home base. As a result, robots from different parts of the environment headed towards the reported region concurrently. Competitions for pucks and territory certainly took place and concentrated at the same zone. That is the reason more robots yield excessive interference consequently defeat the performance.

In order to eliminate the occurrence of interference, cooperative strategies such as task allocation schemes or spatio-temporal schemes [10] shall be introduced. These schemes are designed to avoid condition where two robots attempt to occupy same territory concurrently.

VII. SUMMARY AND CONCLUSIONS

Extensive simulation studies have been performed to investigate the impact of communication on the efficiency of cooperative mobile robotic system. In this research, foraging task imitating ant colony behavior was devised to measure three significant performance metrics that include time, density of robots and interference. Density of robots can be equated to cost. The goal is to examine how well the communication schemes could optimize the performance metrics.

Generally, two types of communication strategies have been introduced in cooperative robotic research area. There are implicit and explicit communication strategies. This paper focuses on the latter, dividing it into local communication (as shown in control strategy 2, S2) and global communication (as illustrated in control strategy 3, S3). These two strategies were evaluated based on the fundamental control strategy, S1, which does not apply any communication scheme.

The principal results derived from experiments are:

- By introducing communication architectures to the fundamental control strategy S1, as demonstrated in S2 and S3, time to accomplish forage task is adequately reduced by almost half of S1's performance time. This is because communication schemes have effectively minimized searching time. Instead of searching pucks independently, robots could recruit peers through explicit communication once they found the pucks region.
- From the time performance point of view, the precise performance ratio of S2 and S3 are 0.54 and 0.59 with respect to S1 correspondingly. Lower performance ratio indicates the strategy can achieve better result.
- Foraging without communication needs at least n robots if the environment is expected to have $n+1$ or more pucks. On the other hand, foraging with communication that can be categorized to local and global communication will need at least $0.54n$ and $0.59n$ robots respectively with $n>1$. In other words, n robots are able to collect $1.46(n+1)$ or more pucks for the case of local communication and $1.41(n+1)$ or more pucks for the case of global communication without overwhelming the completion time. This relation is essential to determine the number of robots to be occupied in a system, thereby estimates the cost consumption.
- By utilizing explicit communication, number of robots can be reduced by about 43 percent of those occupied in no communication system.
- Interference definitely eliminates the performance rate. The result shows global communication yields the highest interference rate followed by local communication if compared to no communication foraging. Explicit communication most likely twofold multiplied the interference. This reflects that the performance rate will adequately increase twice if interference is not added in. One of the excellent ways to eradicate interference is to add in task allocation schemes or spatio-temporal schemes.

The result analysis implies that local communication is the best explicit communication architecture that potentially increases efficiency of cooperative mobile robotic system, which particularly performing search and get task.

In this paper, it has been shown that simulation approach can predict the efficiency of a cooperative robot algorithm by considering time, cost, number of robots and interference. Nevertheless, simulation serves only as a prediction tool. To demonstrate the simulation results, and to move towards a completely functional society, future work involves instantiate forage behavior on mobile robots adapted with infrared local communication and radio link global communication.

VIII. ACKNOWLEDGEMENTS

This work is fully supported by Centre for Robotics and Automation, Faculty of Engineering and Technology (FET), Multimedia University, Malaysia.

IX. REFERENCES

[1] Balch. T. and Arkin R. C., "Communication in Reactive Multiagent Robotic Systems", *Autonomous Robots*, vol. 1, no. 1, 1994, pp. 27-52.

[2] Arkin R. C., *Behavior-based Robotics*, The MIT Press: Cambridge, MA; 1998, pp. 143-213.

[3] Martinoli, *Swarm Intelligence in Autonomous Collective Robotics: From Tools to the Analysis and Synthesis of Distributed Collective Strategies*, Ph.D. Thesis Nr. 2069, Oct. 1999, DI-EPFL, Lausanne, Switzerland.

[4] R. A. Brooks, "A robust layered control system for a mobile robot", *IEEE Journal of Robotics and Automation*, vol. RA-2, no. 1, 1986, pp. 14-23.

[5] Werger, B. B., and M. J. Mataric, "Robotic "food" chains: Externalization of state and program for minimal-agent foraging", in *Proceedings of the Fourth International Conference on Simulation of Adaptive Behavior*, 1996, pp. 625-634.

[6] Richard T. Vaughan, Kasper Stoy, Gaurav S. Sukhatme and M. J. Mataric, "Whistling in the Dark: Cooperative Trail Following in Uncertain Localization Space", in *Proceeding of the International Conference Autonomous Agents*, 2000, Barcelona, Spain.

[7] Balch. T., *Behavioral Diversity in Learning Robot Teams*, Ph.D Thesis, Dec. 1998, College of Computing, Georgia Institute of Technology, Atlanta, GA.

[8] Balch T., 31 July 2001. Citing Internet sources URL http://www.teambots.org.

[9] Sugawara, K. and M. Sano, "Cooperative Acceleration of Task Performance: Foraging Behavior of Interacting Multi-Robots System", *PHYSICA D* (*Nonlinear Phenomena*) vol. 100, nos. 3, 4, Jan. 1997, pp. 343-354.

[10] D. Goldberg and M. J. Mataric, *Stay Outta My Way: Interference as a Guide for Designing Efficient Group Behavior*, Brandeis University Computer Science Technical Report CS-96-186, May 1996.

A Hybrid-Systems Approach to Potential Field Navigation for a Multi-Robot Team *

Jing Ren and Kenneth A. McIsaac
Dept. of Elec. and Comp. Eng.
University of Western Ontario
Elborn College, 1201 Western Rd
London ON
Canada N6G 1H1
jren2@uwo.ca, kmcisaac@engga.uwo.ca

Abstract

We consider potential field-based cooperative motion planning for a distributed team of semi-autonomous robots. We present a changing navigation function to allow the robots to incorporate new sensor data into their maps of the environment. We choose a Gaussian function to model attractors and a higher-order Gaussian-like function to model obstacles in order to avoid undesired local minima. Using arguments from hybrid systems theory, we show that this changing navigation function can be viewed as a mode-specific team Lyapunov function that stabilizes the system at all times. We have verified our approach in simulations of a robot team mapping and foraging in an initially unknown environment. The team is able to map the environment, noting the location of all obstacles and attractive objects, then retrieve the attractors and return them to a goal position. Potential field navigation succeeds in this task while avoiding collisions between robots and obstacles as well as collisions among team members.

1 Introduction

An area of growing interest in the field of robotics is the field of **hybrid systems** research. Loosely speaking, hybrid systems are systems with both continuous and discrete dynamics. One common place example would be rigid multi-body dynamics (e.g. - walking robots with intermittent ground contact) where the active set of differential equations switches based on the type (rolling, sliding, etc.) and number of contacts. Another would be an autonomous robot using logic based control.

Various mathematical definitions (an often cited work is [5]) and modeling formalisms such as hybrid automata, state machines, reactive systems, Petri nets, etc. have been proposed for modeling and analyzing such systems (see [1, 2, 9] for an overview of work in this rapidly growing research area). One important distinction is that of **hybrid dynamics** versus **hybrid control**. The rigid body contact problem is obviously a system with hybrid dynamics since the set of differential equations prescribing the time evolution of the system actually changes in a discontinuous fashion. Another important subclass of hybrid systems comprises systems with unchanging dynamics which are subject to logic based control. In this paper, we restrict our attention to this second class of systems. In particular, we explore techniques to extend the potential field approach to robot motion planning using concepts from hybrid systems theory.

The potential field method for motion planning has been studied extensively in the past decade [11, 7]. In Koditschek's basic formulation [10], artificial "hills" and "valleys" in the robot's world map lead naturally to a stable path towards a "low-energy" goal position. Researchers have since built on this concept to include collision avoidance [13] and multi-robot manipulation [12]. Of particular interest in this work is a recent paper by Esposito [6], in which motion plans generated using the potential field method were refined to satisfy unpredictable run-time constraints. We follow an analogous approach in this paper, and update our navigation function at run-time as new information about the environment and about other team members becomes available. By combining the technique of artificial potential fields and some ideas from hybrid systems theory, we show that we can find a **team Lyapunov function** to achieve a stable system.

2 Assumptions of Robot Capabilities

For the purposes of algorithm development, we assume a certain set of robot capabilities, all of which can be implemented by off-the-shelf components. The robots are assumed to have knowledge of their position, through a positioning system such as GPS. Robots are assumed to

---

*This work was supported by an NSERC grant.

have a sensor (ie: camera) capable of determining the relative position of obstacles and targets. Finally, we assume that robots have omnidirectional navigational capabilities. In future work, we will explore modifications to our control structure that allow us to relax some of these assumptions.

Finally, we assume that the robots communicate sensor and pose information to create a shared *global knowledge base*. We have designed our guidance software using an agent-based [8, 14] approach, as in Figure 2. In each robot, the *Sensing* agent is responsible for processing sensor information. New target and obstacle information, as well as periodic robot pose reports are broadcast to all team members through the *Sender* agent. (For the purposes of simulation, communication is performed over peer-to-peer socket connections.) The *KnowledgeBaseUpdate* agent is responsible for integrating local sensor information with remote information. The *DecisionMaking* agent performs motion planning based on the information stored in the *KnowledgeBase*, making no distinction between information from local and remote sources. In this way, the team shares a common world model.

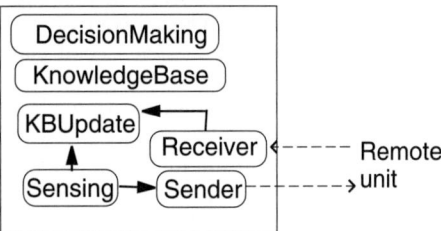

Figure 1: Agent based navigation software. Local and remote sensor information are integrated into the knowledge base, and no distinction is made between them during motion planning.

3 Problem Statement

We test our approach in simulations of a simple search-and-forage application, similar to those studied in [3]. We assume a team of Q robots is assigned to search an unknown field. (Figure 2 demonstrates a typical configuration.) The configuration q_i of each robot is given by the vector $q_i = (x_i, y_i)$ of the position of its center of mass. We also define $q = (q_1, q_2, ..., q_Q)$ as the state vector of the robot team. The area of the field is known to be $M \times N$ units, and the area is assumed to contain an unknown number of convex *obstacles* (solid in the figure) and *targets* (represented by the $ symbol). It is assumed that there is sufficient space between pairs of obstacles to allow a robot's passage. For the purposes of searching, the area is divided into an appropriate number of known and unknown *sectors*. The area of one sector corresponds to the area that can be surveyed in one sensor sweep. In our simulations, the team is assigned to first search the area, determining the locations of all obstacles and targets, then return the target objects to a home position. Throughout the task, collisions between team members and between robots and obstacles are forbidden.

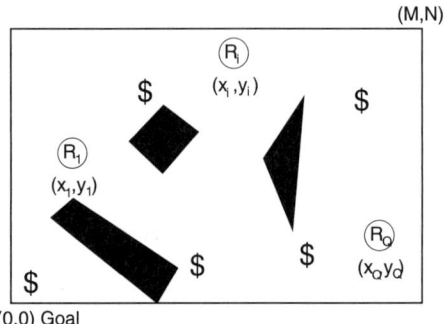

Figure 2: A simple search-and-forage application. Robots, R_i, are represented by circles located at $q_i = (x_i, y_i)$. Point targets are represented by the $ symbol, and obstacles are black polygons.

4 Potential Field Models and Navigation Functions

We base our potential fields on the two-dimensional Gaussian function. Attractive points, such as targets and the goal position, located at (a_x, a_y) are represented by the negative Gaussian *attractor* function (Figure 3):

$$f_A(x,y) = 1 - e^{-\frac{(x-a_x)^2+(y-a_y)^2}{2\sigma^2}}. \qquad (1)$$

Repulsive points, such as obstacles and other robots, located at (r_x, r_y) are modeled with the circular, two-dimensional Gaussian-like *repulsor* function :

$$f_R(x,y) = e^{-\frac{1}{2}\left(\frac{(x-r_x)^2+(y-r_y)^2}{\sigma^2}\right)^C}. \qquad (2)$$

For some positive integer C. The variance, σ, is a measure of the size of the obstacle. The variable C determines the *effective range (steepness)* of the obstacle (see Figure 4 for the effect of varying these parameters). We represent convex obstacle shapes with a superscribed circle.

These Gaussian-like attractor and repulsor functions are attractive to us for the following reasons:

- Adjustments of the size and position of attractors and repulsors can easily be accomplished by adjusting the mean (location (a_x, a_y) or (r_x, r_y)) and variance, σ.

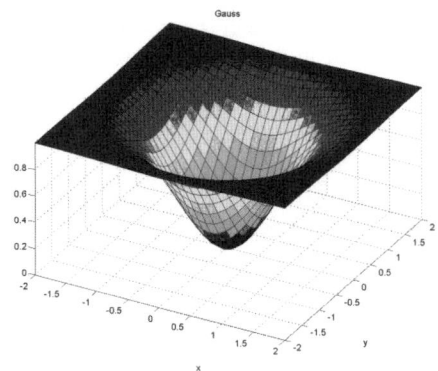

Figure 3: The negative Gaussian attractor function. The attractor in the figure is located at $(a_x, a_y) = (0, 0)$.

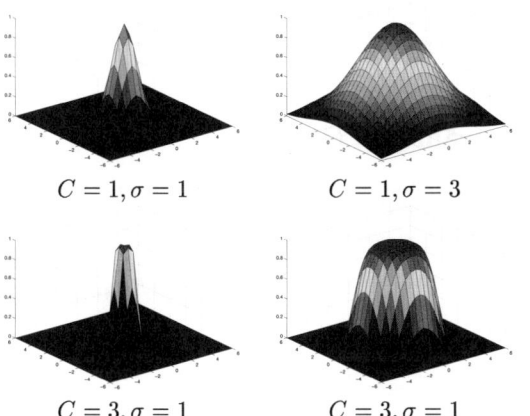

Figure 4: The effect of variance, σ, and effective range parameter, C on the repulsor function of Equation 2. The values of σ and C used to generate each repulsor function are given below the associated plot. In all the figures, repulsor is located at $(r_x, r_y) = (0, 0)$.

- Changing the variance, σ allows us to control the affecting area of the repulsor or attractor to correspond to the object's physical size.

- The effective range parameter C in our higher-order repulsor function means that the effect of repulsive obstacles is highly localized. Robots near a given obstacle are strongly repelled; robots distant from the obstacle essentially navigate independent of it.

- We can place effective upper bounds on the problem of local minima which often arises in artificial potential field-based navigation [7]. Since the affecting area of obstacles is so easily controlled, we can potentially specify a minimum clearance between obstacles to guarantee that local minima will not occur.

5 A Hybrid Systems Model of the Search and Forage Task

General hybrid systems [5], are concerned with the time evolutions of systems represented with a continuous state vector q and a discrete **mode** S, such that in mode S_i the dynamics of q are governed by the differential equation $\dot{q} = f_i(q)$. In general, a different differential equation is associated with each mode, and the system switches from mode S_i to S_j based on some external event or sensor input.

5.1 Definitions of Modes and Mode Switching

In our simulation of the search-and-forage task, robots switch between three generic modes, called **Search**, **Forage** and **Transport**. In the Search mode, robots are attracted to unexplored territory. In the Forage mode (entered when the entire area has been searched), robots are attracted to targets. In the Transport mode (entered after a robot has collected a target) robots are attracted to the goal. In all three modes, robots are repelled by obstacles and other team members.

However, based on these three basic modes, a robot can be in an indefinite (but finite) number of specific modes, based on sensor information, and specifically, on the number N_U of unsearched sectors, the number N_O of found obstacles, and on the number N_T of found, free targets. For convenience, we write the name of each mode with these three numbers appended, eg: Search_20_4_3 would indicate that 20 cells remain unknown, and that four obstacles and three free targets have been found. Mode switches occur based on one of the following events:

- Whenever the number N_U of unsearched sectors decreases during the search phase.

- Whenever the number N_O of found obstacles increases during the search phase,

- Whenever the number N_T of found targets increases during the search phase.

- Whenever a robot picks up a target for transport to the goal

- Whenever a robot has successfully returned a target to the goal and is beginning to forage.

To clarify this technique, we diagram in Figure 5 a simple state diagram for a single robot in a nine sector evironment with one obstacle and two targets. Mode switches occur on the arcs, and mode names are given in the circles.

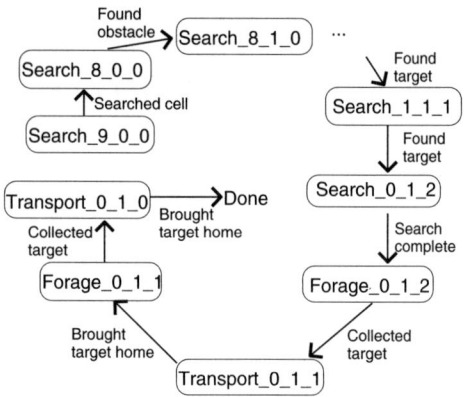

Figure 5: The mode switching diagram for a simple search-and-forage application. Mode switches occur on the arcs, for the reasons given by the text on the arc. The robot begins in the Search_9_0_0 mode. For convenience, most of the arcs corresponding to changes of state due decreases in N_U have been elided.

5.2 Modal Dynamics and Navigation Functions

Using the technique of artificial potential fields, we construct a **navigation function**, $V_i^X(q)$ for each robot i in mode X (where X can be any of the modes defined above). The (mode-dependent) dynamics of each robot are then given by:

$$\dot{q}_i = -\alpha \frac{\frac{\partial V_i^X}{\partial q_i}}{\left|\frac{\partial V_i^X}{\partial q_i}\right|} \quad (3)$$

In Equation 3, the operator $\frac{\partial V_i^X(q)}{\partial q_i}$ represents the gradient of $V_i^X(q)$ with respect to only q_i. Thus, we use a gradient descent method to generate a unit vector direction for $\dot{q}_i$, and the constant velocity parameter α to determine robot speed. We use a unit gradient because our Gaussian-like potential functions decay very rapidly and robot speed should not depend on position.

To construct the **navigation function** for a given mode, we use the three part formula:

$$V_i^X(q) = VA^X(q_i) + VO^X(q_i) + \sum_{i \neq j} VR(q_i, q_j) \quad (4)$$

In Equation 4, $VA^X(q_i)$ represents the sum of the effects on robot i of all the N_A attractors in the system during mode X. During Search phases, the unsearched cells are the attractors, and therefore $N_A = N_U$. During Forage phases, free targets are the attractors, so $N_A = N_T$. Finally, during the transport phase, there is only one attractor (the goal), so $N_A = 1$. In general, then:

$$VA^X(q_i) = \sum_{k=1}^{N_A} \left(1 - e^{-\frac{(x_i - (a_k)_x)^2 + (y_i - (a_k)_y)^2}{2\sigma^2}}\right). \quad (5)$$

$VO^X(q_i)$ represents the sum of the effects on robot i of all the known, fixed obstacles in the system during mode X. This value is always N_O, so we can state the general case as:

$$VO^X(q_i) = \sum_{k=1}^{N_O} \left(e^{-\frac{1}{2}\left(\frac{(x_i - (r_k)_x)^2 + (y_i - (r_k)_y)^2}{\sigma^2}\right)^C}\right). \quad (6)$$

Finally, the functions $VR(q_i, q_j)$ represent the repulsor functions between pairs of robots i and j. Note that VR is not mode dependent, since the number of robots is assumed to be constant. Thus, in general:

$$VR(q_i, q_j) = e^{-\frac{1}{2}\left(\frac{(x_i - x_j)^2 + (y_i - y_j)^2}{\sigma^2}\right)^C} \quad (7)$$

6 Stability Analysis

The definition of our modal navigation functions, and of our mode-switching rules, allow us to show that the system (entire robot team) is globally stable at all times and in all states.

6.1 Intramodal (Dynamic) Stability

Within each mode, we can define a mode-specific Lyapunov function of the form:

$$V^X(q) = \sum_{i=1}^{Q} V_i^X(q) - \sum_{i=1}^{Q} \sum_{j=i+1}^{Q} VR(q_i, q_j) \quad (8)$$

$$= \sum_{i=1}^{Q} \left(VA^X(q_i) + VO^X(q_i)\right) +$$

$$\sum_{i=1}^{Q} \sum_{j=i+1}^{Q} VR(q_i, q_j) \quad (9)$$

where the step from Equation 8 to Equation 9 is justified by the fact that $VR(q_i, q_j) = VR(q_j, q_i)$.

To show intra-modal Lyapunov stability, we are required to show $V^X(q) \geq 0 \;\forall q$ and $\dot{V}^X(q) < 0 \;\forall q, t$. $V^X(q) \geq 0$ follows naturally from the definition of $V^X(q)$ in Equation 9. To show that $V^X(q)$ is always decreasing, we begin using the form of Equation 8. For convenience, in the derivations that follow, we have replaced terms of the form $VR(q_i, q_j)$ with the short form VR_{ij}:

$$\dot{V}^X(q) = \sum_{i=1}^{Q} \frac{\partial V_i^X(q)}{\partial q_i} \dot{q}_i + \sum_{i=1}^{Q} \sum_{i \neq j} \frac{\partial V_i^X(q)}{\partial q_j} \dot{q}_j \quad (10)$$

$$- \sum_{i=1}^{Q} \sum_{j=i+1}^{Q} \left(\frac{\partial VR_{ij}}{\partial q_i} \dot{q}_i + \frac{\partial VR_{ij}}{\partial q_j} \dot{q}_j\right)$$

Now, observe that since q_j only appears in $V_i^X(q)$ through the VR_{ij} terms, the second term of Equation 10 can be rewritten as:

$$\sum_{i=1}^{Q} \sum_{i \neq j} \frac{\partial V_i^X}{\partial q_j} \dot{q}_j \quad (11)$$

$$= \sum_{i=1}^{Q} \sum_{i \neq j} \frac{\partial VR_{ij}}{\partial q_j} \dot{q}_j \quad (12)$$

$$= \sum_{i=1}^{Q} \sum_{j=i+1}^{Q} \frac{\partial VR_{ij}}{\partial q_j} \dot{q}_j + \sum_{i=1}^{Q} \sum_{j=1}^{i-1} \frac{\partial VR_{ij}}{\partial q_j} \dot{q}_j \quad (13)$$

$$= \sum_{i=1}^{Q} \sum_{j=i+1}^{Q} \frac{\partial VR_{ij}}{\partial q_j} \dot{q}_j + \sum_{p=1}^{Q} \sum_{s=p+1}^{Q} \frac{\partial VR_{ps}}{\partial q_p} \dot{q}_p \quad (14)$$

$$= \sum_{i=1}^{Q} \sum_{j=i+1}^{Q} \left(\frac{\partial VR_{ij}}{\partial q_i} \dot{q}_i + \frac{\partial VR_{ij}}{\partial q_j} \dot{q}_j \right) \quad (15)$$

Thus, the second and third terms in Equation 10 will cancel, and we will have:

$$\dot{V}^X(q) = \sum_{i=1}^{Q} \frac{\partial V_i^X(q)}{\partial q_i} \dot{q}_i \quad (16)$$

If we substitute for $\dot{q}_i$ using the dynamics defined in Section 5.2, we will have:

$$\dot{V}^X(q) = -\alpha \sum_{i=1}^{Q} \frac{\partial V_i^X(q)}{\partial q_i} \frac{\frac{\partial V_i^X}{\partial q_i}}{\left| \frac{\partial V_i^X}{\partial q_i} \right|} \quad (17)$$

$$= -\alpha \sum_{i=1}^{Q} \left| \frac{\partial V_i^X(q)}{\partial q_i} \right| \quad (18)$$

and we will have $\dot{V}^X(q) < 0 \; \forall t, q$ as required.

6.2 Intermodal (Hybrid) Stability

One of the central problems in hybrid control theory is that a hybrid system is not necessarily stable, even though it switches between stable modes [4]. With our switching technique, defined in Section 5.1, we can guarantee that this problem will not occur. Although the order of mode switches can not be predicted in advance, our switching technique guarantees that the mode-transition graph will be *acyclic*. The hybrid system must therefore always make progress towards the goal, since mode switches only occur when the team has made progress towards solving the overall task and there is never a situation when a given mode is re-entered. We therefore have stable switching, and stable team dynamics within each mode, so we can guarantee team stability for all time.

7 Simulation Results

In order to validate our control approach, we have performed simulations of robot teams of varying size in a

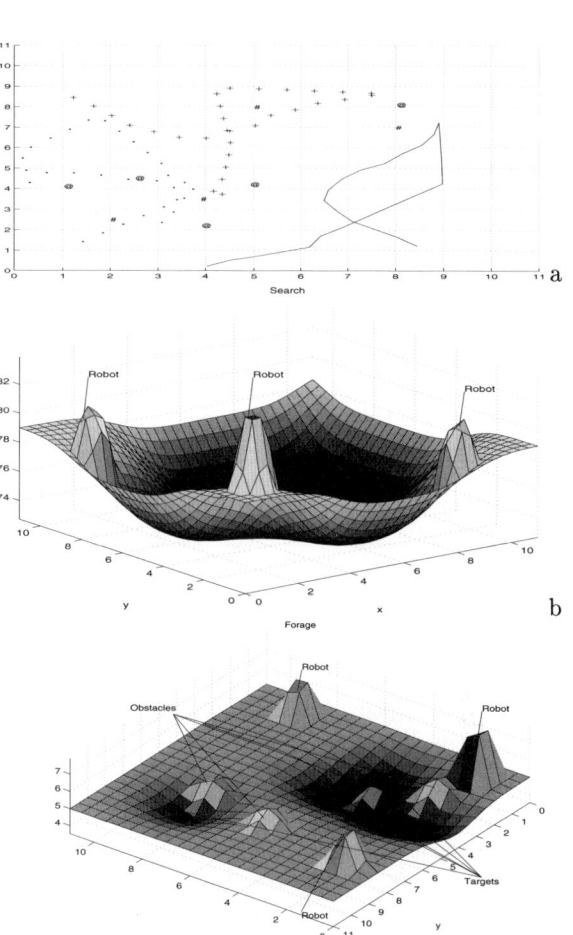

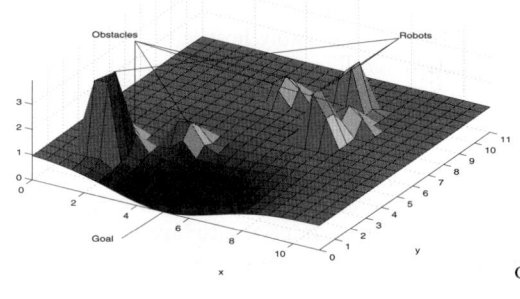

Figure 6: (a)The courses followed by the robots during the search phase (robot 1 solid, robot 2 dots, robot 3 '+' signs). (b)The modal Lyapunov function $V^{\text{Search\_144\_0\_0}}(q)$ at the beginning of the simulation. Attractors are unsearched cells and repulsors are other robots. (c)The modal Lyapunov function $V^{\text{Forage\_0\_5\_3}}(q)$ at the beginning of the forage phase (after all the cells have been searched) (d)The modal Lyapunov function $V^{\text{Transport\_0\_5\_0}}(q)$ after the last target has been collected for transport to the goal.

varying field of play. Figure 6.2 shows the results of a typical simulation, played in an 12 × 12 grid containing 4 obstacles (represented with the @ symbol) with a team of 3 robots trying to find 5 targets (represented by the $ symbol). Figure 6.2(a) shows the paths followed by the three robots during the search phase. Figure 6.2(b) shows the contour of the team Lyapunov function $V^{\text{Search-144-0-0}}(q)$ at the beginning of the simulation. Attractors (valleys) in this function are given by the unsearched cells. In Figure 6.2(c), we plot the contour of the team Lyapunov function $V^{\text{Forage-0-5-3}}(q)$ immediately after the search phase has been completed. Attractors (valleys) represent the location of target objects, and repulsors (hills) are the locations of found obstacles. Finally, Figure 6.2(c) shows the contour of the team Lyapunov function $V^{\text{Transport-0-5-0}}(q)$ immediately after the last target has been collected for transport to the goal. The only attractor in this case is the goal.

8 Conclusions and Future Work

In this paper, we have shown a technique for developing stable motion plans for communicating robot teams using Lyapunov theory and the basic tools of hybrid systems. We have proposed the use of Gaussian-like attractor and repulsor functions that allow us to control the affecting range of obstacles and attractors. Not only does this allow us to model attractors and repulsors of varying sizes, but it has the potential to yield predictions of the minimum obstacle separation required to guarantee the system will not halt in a local minimum state.

Our motion plans can guarantee that team members do not collide, but we have only considered the problem of stable obstacles. In future work, we will explore the problem of moving, possibly hostile obstacles. We have also assumed that all targets are of equal value and equal size. In future work we plan to model targets of varying configurations, and to require coordinated robot action to move large targets to the goal.

We have made a significant number of assumptions about robot capabilities. In future work we intend to relax some of these assumptions, to investigate the effects on the stability of our motion plans. For example, we will explore the effect of non-holomonic constraints on robot motion and the effect of sensor noise and sensor inaccuracies.

One of the fundamental assumptions we have made about our robots is that they can communicate enough sensor and pose information to create a "team shared memory", ensuring that every robot has access to the same store of information. While we feel that trends in communication and wireless networking will eventually make this capability commonplace, it is difficult to achieve with today's technology. As a result, we will explore the sensitivity to communication errors, such as lost packets and lost team members, and especially propagation delays in communication, which are sure to occur as the team size increases.

References

[1] R. Alur, T. Henzinger, and E. Sontag, editors. *Hybrid Systems III: Verification and Control. LNCS 1066.* Springer-Verlag, 1996.

[2] P. Antsaklis, W. Kohn, A. Nerode, and S. Sastry, editors. *Hybrid Systems II. LNCS 999.* Springer-Verlag, 1995.

[3] T. Balch and R. Arkin. Communication in reactive multiagent robotic systems. *Autonomous Robots*, 1(1):27–52, 1994.

[4] M. S. Branicky. Multiple lyapunov functions and other analysis tools for switched and hybrid systems. *IEEE Transactions on Automatic Control*, 43(4):475–482, April 1998.

[5] M. S. Branicky, V. S. Borkar, and S. K. Mitter. A unified framework for hybrid control. In *Proc. Int. Conf. on Decision and Control*, pages 4228–4234, Lake Buena Vista, FL, Dec 1994. IEEE.

[6] J. M. Esposito and V. Kumar. A method for modifying closed-loop motion plans to satisfy unpredictable dynamic constraints at run-time. In *Proc. IEEE Int. Conf. Robotics and Automation*, pages 1691–1696, Washington, May 2002.

[7] S. S. Ge and Y. J. Cui. New potential functions for mobile robot path planning. *IEEE Transactions on Robotics and Automation*, 16(5):615–620, October 2000.

[8] H. Ghenniwa and M. Kamel. Interaction devices for co-ordinating cooperative distributed systems. *Automation and Soft Computing*, 6(2):173–184, 2000.

[9] R. Grossman, A. Nerode, A. Ravn, and H. Rischel, editors. *Hybrid Systems. LNCS 736.* Springer-Verlag, 1993.

[10] D. E. Koditschek. Robot planning and control via potential functions. In O. Khatib, J. J. Craig, and T. Lozano-Perez, editors, *The Robotics Review 1*, pages 349–367, 1989.

[11] Y. Koren. Potential field methods and their inherent limitations for mobile robot navigation. In *Proc. IEEE Int. Conf. Robotics and Automation*, pages 1398–1404, April 1991.

[12] P. Song and V. Kumar. A potential field approach to multi-robot manipulation. In *Proc. IEEE Int. Conf. Robotics and Automation*, pages 1217–1222, Washington, May 2002.

[13] T. S. Wilkman, M. S. Branicky, and W. S. Newman. Reflexive collision avoidance: A generalized aproach. In *Proc. IEEE Int. Conf. Robotics and Automation*, pages 3:31–36, May 1993.

[14] M. F. Wood and S. DeLoach. An overview of the multiagent systems engineering methodology. In *AOSE*, pages 207–222, 2000.

Calculating Possible Local Displacement of Curve Objects using Improved Screw Theory

Jun Takamatsu
Department of Computer Science,
University of Tokyo
7-3-1 Hongo, Bunkyo-ku, Tokyo, Japan

Koichi Ogawara
Institute of Industrial Science,
University of Tokyo,
4-6-1 Komaba, Meguro-ku, Tokyo, Japan

Hiroshi Kimura
Grad. School of Information Systems,
Univ. of Electro-Communications
1-5-1 Chofugaoka, Chofu City, Tokyo, Japan

Katsushi Ikeuchi
Institute of Industrial Science,
University of Tokyo,
4-6-1 Komaba, Meguro-ku, Tokyo, Japan

Abstract— Various methods to recognize assembly tasks using possible local displacement of objects have been proposed. To calculate this displacement, the screw theory is employed. It is equivalent to the first order Taylor expansion of the displacement.

However, such methods can treat polyhedral objects only. Because the screw theory cannot treat curvature information of objects. In this paper, we propose a method to calculate possible local displacement of curve objects using improved screw theory, which is equivalent to the second order Taylor expansion of the displacement, and verify the validity of the proposed method.

I. INTRODUCTION

For recognizing assembly tasks and making automatically robot programming to execute such tasks, various methods have been proposed[1][2][3][4]. These methods require to calculate possible local displacement of objects for recognition.

To calculate this displacement, the *screw theory*[5] or tools with equivalent capabilities are usually employed. They are equivalent to the first order Taylor expansion of the displacement (referred to as *first order displacement*), and the displacement is formulated as simultaneous linear inequalities. That is a good characteristic because a powerful tool to calculate such inequalities, the *theory of the polyhedral convex cones*[6], has already been established.

However, the first order displacement cannot treat curvature information of objects. For example, every first order displacement of a cubic object as shown in Figure 1 is the same. That is much different from truth.

In this paper, we propose a method to formulate and calculate possible local displacement of curve objects using improved screw theory[7], which is equivalent to the second order Taylor expansion of the displacement (referred to as *second order displacement*). Then we verify the validity of a proposed method.

This paper is organized as follows: Section 2 describes preliminaries: the screw representation and the representation of curve lines and surfaces. Section 3 formulates the second order possible displacement of curve objects. Section 4 roughly introduces a method to calculate the second order displacement. Section 5 applies the proposed method to various contact relations between two curve objects. Section 6 concludes this paper.

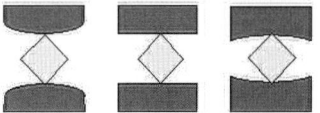

Fig. 1. Same possible local displacement?

II. PRELIMINARIES

A. Screw Representation

To represent local displacement, we employ the screw representation[5]. It represents the displacement as a combination of translation along a *screw axis* and rotation about the same axis.

The displacement can be uniquely decided by a direction $\mathbf{r}$ and a location $\mathbf{c}$ of the screw axis, the ratio s of the translation to the rotation, and a rotation angle $\Delta\theta(>0)$ in the screw representation. After the displacement, the location of any point $\mathbf{v}$ is calculated by Equation (1).

$$\mathbf{v}_m = R(\mathbf{v}-\mathbf{c}) + \mathbf{c} + s\mathbf{r} \qquad (1)$$

$$R = I + \sin\Delta\theta [\mathbf{r}]_\times + (1-\cos\Delta\theta)[\mathbf{r}]_\times^2$$

$$[\mathbf{r}]_\times = \begin{pmatrix} 0 & -r_z & r_y \\ r_z & 0 & -r_x \\ -r_y & r_x & 0 \end{pmatrix}$$

In this paper, we assume that $\mathbf{c}$, and s are functions with a variable $\Delta\theta^1$.

---
[1] We assume that $\mathbf{r}$ is fix, because rotation with a fixed axis direction is usually preferred.

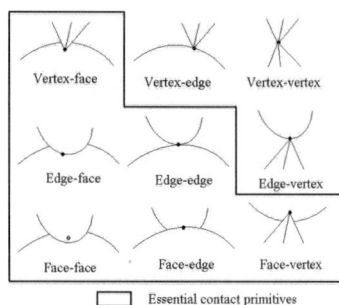

Fig. 2. Nine types of contact primitives

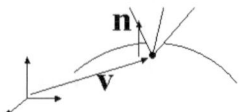

Fig. 3. Vertex-face contact

B. Representation of Curve Line and Surface

We assume that a curve line is parametrized by the length s from a fixed point on it as Equation (2).

$$\mathbf{x} = \mathbf{f}(s) \quad (2)$$

Applying the Taylor expansion to the equation near the point, $\mathbf{p} = \mathbf{f}(s_p)$, Equation (3) is obtained, where $\mathbf{R}_3(s)$ is the third order reminder term.

$$\mathbf{x} = \mathbf{p} + \mathbf{e}_1 \cdot (s - s_p) + \frac{1}{2}\kappa \mathbf{e}_2 \cdot (s - s_p)^2 + \mathbf{R}_3(s) \quad (3)$$

$\mathbf{e}_1$ is a tangent normal, $\mathbf{e}_2$ is a normal vector satisfied $\mathbf{e}_1 \cdot \mathbf{e}_2 = 0$, and κ is a curvature on the point $\mathbf{p}$.

Given the function $f(\mathbf{x})$ which calculates the length between a point $\mathbf{x}$ and a surface, such a surface is formulated by $f(\mathbf{x}) = 0$. Applying the Taylor expansion to the equation, Equation (4) is obtained.

$$f(\mathbf{x}) = \mathbf{n} \cdot (\mathbf{x} - \mathbf{p}) + \frac{1}{2}(\mathbf{x} - \mathbf{p})^T V (\mathbf{x} - \mathbf{p}) + R_3(\mathbf{x}) \quad (4)$$

$\mathbf{n}$ is a surface normal and V is a curvature matrix.

In this paper, we assume that every curve line or surface is represented as the second order approximation, that is, the equation which is obtained by removing third order reminder term R_3 from Equation (3) or (4).

III. SECOND ORDER DISPLACEMENT FOR CURVE LINES AND SURFACES

In this paper, we consider a contact relation between two objects: one is movable and the other is fixed. An arbitrary object is composed of object primitives: vertices, edges (curve lines), and faces (curve surfaces). Every contact relation between two objects can be represented as a combination of contact primitives. A contact primitive is composed of two contacting object primitives. Therefore, nine types of contact primitives exist as shown in Figure 2. A vertex-face contact means that a vertex in a movable object contacts a face in a fixed object.

A. Essential Contact Primitive

Hirukawa et. al. proposed the method to derive the first order possible displacement from any contact relations between two polyhedral objects[9]. They derived this displacement using a contacting point and a separating plane. A separating plane composes of a contacting plane or the plane which two contacting edges is on.

And they illustrated that the first order possible displacement is represented as one system of simultaneous linear inequalities, if and only if an unique separating plane exist on every contacting point, that is, a contact relation dose not include contact primitives (referred to as *singular contact primitives*) as follows: 1) A convex vertex contacts a convex edge. 2) A convex vertex contacts a convex vertex.

We define following six contact primitives as essential contact primitives: vertex-face, edge-face, face-face, edge-edge, edge-face, face-face contacts (Show the area surrounded by bold line in Figure 2). That reason is as follows: 1) These six contact primitives can compose a contacting point and a unique separating curve surface. 2) Another three contact primitives can be regarded as a combination of some essential contact primitives, if they are not singular contact primitives.

In this paper, we only treat contact relations not including singular contact primitives. Therefore, the second order possible displacement can be represented by one system of simultaneous inequalities.

B. Vertex-Face Contact

Consider the case that a vertex contacts a face in the point $\mathbf{v}$ as shown in Figure 3. After the displacement, the location of the point $\mathbf{v}$ is represented by Equation (5).

$$\mathbf{v}_m = R(\mathbf{v} - \mathbf{c}) + \mathbf{c} + s\mathbf{r} \quad (5)$$

When a curve surface is represented as Equation (4), the length Δ_{vf} between a vertex and a curve surface after the displacement represents Equation (6).

$$\Delta_{vf} = \mathbf{n} \cdot (\mathbf{v}_m - \mathbf{v}) + \frac{1}{2}(\mathbf{v}_m - \mathbf{v})^T V (\mathbf{v}_m - \mathbf{v}) \quad (6)$$

Applying the Taylor expansion to Equation (6), we obtain Equation (7), where $\mathbf{t}_1 = \mathbf{r}$, $\mathbf{t}_2 = \mathbf{c}(0) \times \mathbf{r} + s'(0)\mathbf{r}$,

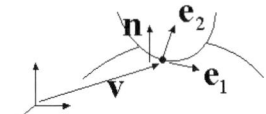

Fig. 4. Edge-face contact

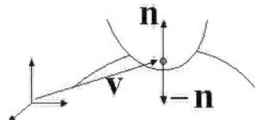

Fig. 5. Face-face contact

$\mathbf{t}_3 = 2\mathbf{c}'(0) \times \mathbf{r} + s''(0)\mathbf{r}$.

$$\begin{aligned}\Delta_{vf} &= \mathbf{n} \cdot (\mathbf{t}_1 \times \mathbf{v} + \mathbf{t}_2)\Delta\theta \\ &+ (\mathbf{n} \cdot \mathbf{t}_3 + (\mathbf{n} \times \mathbf{t}_1) \cdot (\mathbf{t}_1 \times \mathbf{v} + \mathbf{t}_2) \\ &+ (\mathbf{t}_1 \times \mathbf{v} + \mathbf{t}_2)^T V(\mathbf{t}_1 \times \mathbf{v} + \mathbf{t}_2))\frac{\Delta\theta^2}{2} \\ &+ R_3(\Delta\theta) \end{aligned} \quad (7)$$

The possible displacement is formulated by $\Delta_{vf} \geq 0$.

C. Edge-Face Contact

Consider the case that an edge contacts a face in the point $\mathbf{v}$ as shown in Figure 3. We assume that a curve line is represented by Equation (8).

$$\mathbf{e}(s) = \mathbf{v} + s\mathbf{e}_1 + \frac{1}{2}s^2 \kappa \mathbf{e}_2 \quad (8)$$

After the displacement, a curve line is represented by Equation (9).

$$\mathbf{e}_m(s) = \mathbf{v}_m + sR\mathbf{e}_1 + \frac{1}{2}s^2 \kappa R\mathbf{e}_2 \quad (9)$$

When a curve surface is represented as Equation (4), the length Δ_{ef} between a point on the edge, $\mathbf{e}_m(s_p)$, and the curve surface is represented by Equation (10).

$$\begin{aligned}\Delta_{ef} &= \mathbf{n} \cdot (\mathbf{e}_m(s_p) - \mathbf{v}) \\ &+ \frac{1}{2}(\mathbf{e}_m(s_p) - \mathbf{v})^T V(\mathbf{e}_m(s_p) - \mathbf{v})\end{aligned} \quad (10)$$

When the length is minimum, Equation (11) must be satisfied. Such minimum length is the length between an edge and a face.

$$\frac{\partial \Delta_{ef}}{\partial s_p} = 0 \quad (11)$$

And, Equation (12) must be satisfied, because an edge contacts a face without penetrating.

$$\mathbf{n} \cdot \mathbf{e}_1 = 0 \quad (12)$$

Applying the Taylor expansion to Equation (10) and using Equation (11) and (12), we obtain Equation (13).

$$\begin{aligned}\Delta_{ef} &= \mathbf{n} \cdot (\mathbf{t}_1 \times \mathbf{v} + \mathbf{t}_2)\Delta\theta \\ &+ (\mathbf{n} \cdot \mathbf{t}_3 + (\mathbf{n} \times \mathbf{t}_1) \cdot (\mathbf{t}_1 \times \mathbf{v} + \mathbf{t}_2) \\ &+ (\mathbf{t}_1 \times \mathbf{v} + \mathbf{t}_2)^T V(\mathbf{t}_1 \times \mathbf{v} + \mathbf{t}_2) \\ &- a^2(\kappa \mathbf{n} \cdot \mathbf{e}_2 + \mathbf{e}_1^T V \mathbf{e}_1))\frac{\Delta\theta^2}{2} \\ &+ R_3(\Delta\theta)\end{aligned} \quad (13)$$

$$a = -\frac{(\mathbf{e}_1 \times \mathbf{n}) \cdot \mathbf{t}_1 + \mathbf{e}_1^T V(\mathbf{t}_1 \times \mathbf{v} + \mathbf{t}_2)}{\kappa \mathbf{n} \cdot \mathbf{e}_2 + \mathbf{e}_1^T V \mathbf{e}_1}$$

The possible displacement is formulated by $\Delta_{ef} \geq 0$.

D. Face-Face Contact

Consider the case that one face contacts the other face in the point $\mathbf{v}$ as shown in Figure 5. We assume that two faces are represented by Equation (14).

$$\begin{aligned}f(\mathbf{x}) &= -\mathbf{n} \cdot (\mathbf{x} - \mathbf{v}) + \frac{1}{2}(\mathbf{x} - \mathbf{v})^T V_f(\mathbf{x} - \mathbf{v}) = 0 \\ g(\mathbf{x}) &= \mathbf{n} \cdot (\mathbf{x} - \mathbf{v}) + \frac{1}{2}(\mathbf{x} - \mathbf{v})^T V_g(\mathbf{x} - \mathbf{v}) = 0\end{aligned} \quad (14)$$

After the displacement, the curve surface, $f(\mathbf{x}) = 0$, is represented by Equation (15).

$$\begin{aligned}f_m(\mathbf{x}) &= -R\mathbf{n} \cdot (\mathbf{x} - \mathbf{v}) \\ &+ \frac{1}{2}(\mathbf{x} - \mathbf{v})^T R V_f R^T (\mathbf{x} - \mathbf{v}) = 0\end{aligned} \quad (15)$$

Let $\mathbf{p}$ be any point on the surface $g(\mathbf{x}) = 0$. We assume that $\mathbf{p}$ is represented as Equation (16), where $\mathbf{k}_1$ and $\mathbf{k}_2$ are two tangent vector of the surface with maximum and minimum curvature, and $\mathbf{k}_1 \times \mathbf{k}_2 = \mathbf{n}, \mathbf{k}_2 \times \mathbf{n} = \mathbf{k}_1, \mathbf{n} \times \mathbf{k}_1 = \mathbf{k}_2$ are satisfied.

$$\mathbf{s} = q_1 \mathbf{k}_1 + q_2 \mathbf{k}_2 + q_3 \mathbf{n} + \mathbf{v} \quad (16)$$

Because $\mathbf{p}$ is on the surface, Equation (17) must be satisfied, where κ_1 and κ_2 are maximum and minimum curvature.

$$q_3 = -\frac{1}{2}\kappa_1 q_1^2 - \frac{1}{2}\kappa_2 q_2^2 \quad (17)$$

The length between two face is equal to the minimum of Equation (18).

$$\Delta_{ff} = f_m(\mathbf{p}) \quad (18)$$

If Δ_{ff} is minimum, Equation (19) must be satisfied.

$$\frac{\partial \Delta_{ff}}{\partial q_1} = 0, \frac{\partial \Delta_{ff}}{\partial q_2} = 0 \quad (19)$$

Applying the Taylor expansion to Equation (18) and using Equation (17) and (19), Equation (20) is obtained, where $V_{diff} = V_f - V_g$ and $K = (\mathbf{k}_1 \ \mathbf{k}_2)$

$$\begin{aligned}\Delta_{ff} &= \mathbf{n} \cdot (\mathbf{t}_1 \times \mathbf{v} + \mathbf{t}_2)\Delta\theta \\ &+ (\mathbf{n} \cdot \mathbf{t}_3 - (\mathbf{n} \times \mathbf{t}_1) \cdot (\mathbf{t}_1 \times \mathbf{v} + \mathbf{t}_2) \\ &+ (\mathbf{t}_1 \times \mathbf{v} + \mathbf{t}_2)^T V_f (\mathbf{t}_1 \times \mathbf{v} + \mathbf{t}_2) \\ &- \mathbf{n} \cdot \mathbf{b})\frac{\Delta\theta^2}{2} + R_3(\Delta\theta)\end{aligned} \quad (20)$$

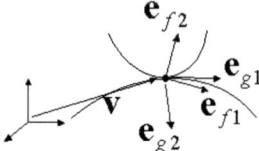

Fig. 6. Edge-edge contact

$$\mathbf{b} = \kappa_1 c_1^2 + \kappa_2 c_2^2$$
$$\begin{pmatrix} c_1 \\ c_2 \end{pmatrix} = M^- \mathbf{d}$$
$$M = K^T V_{diff} K$$
$$\mathbf{d} = \begin{pmatrix} \mathbf{k}_1^T V_f(\mathbf{t}_1 \times \mathbf{v} + \mathbf{t}_2) - \mathbf{k}_2 \cdot \mathbf{t}_1 \\ \mathbf{k}_2^T V_f(\mathbf{t}_1 \times \mathbf{v} + \mathbf{t}_2) + \mathbf{k}_1 \cdot \mathbf{t}_1 \end{pmatrix}$$

The possible displacement is formulated by $\Delta_{ff} \geq 0$.

E. Edge-Edge Contact

In this paper, we define the length between two edges as Equation (21), where $\mathbf{v}_1$ and $\mathbf{v}_2$ are points on edges of moving and fixed objects, $\mathbf{t}_1$ and $\mathbf{t}_2$ are tangent vectors on $\mathbf{v}_1$ and $\mathbf{v}_2$, and a direction of the vector $\mathbf{t}_1 \times \mathbf{t}_2 (\neq \mathbf{0})$ is outward to a fixed object.

$$\min_{\mathbf{v}_1,\mathbf{v}_2} (\mathbf{t}_1 \times \mathbf{t}_2) \cdot (\mathbf{v}_1 - \mathbf{v}_2) \qquad (21)$$

When the equation is equal to 0, two edges contact each other. And when the equation is less than 0, they penetrate each other.

Consider the case that one edge contacts the other edge in the point $\mathbf{v}$ as shown in Figure 6. We assume that two curve lines are represented by Equation (22).

$$\mathbf{f}(l) = \mathbf{v} + l\mathbf{e}_{f1} + \frac{1}{2}l^2 \kappa \mathbf{e}_{f2}$$
$$\mathbf{g}(l) = \mathbf{v} + l\mathbf{e}_{g1} + \frac{1}{2}l^2 \kappa \mathbf{e}_{g2} \qquad (22)$$

After the displacement, the curve line $\mathbf{f}(l)$ is represented by Equation (23).

$$\mathbf{f}_m(l) = \mathbf{v}_m + lR\mathbf{e}_{f1} + \frac{1}{2}l^2 \kappa R\mathbf{e}_{f2} \qquad (23)$$

First, we consider the case that two edges are not locally on the same plane. In this case, Equation (24) is satisfied.

$$\mathbf{e}_{f1} \times \mathbf{e}_{g1} \neq 0 \qquad (24)$$

The length between two edges is equal to the minimum of Equation (25).

$$\Delta_{ee} = \left(\frac{\partial \mathbf{f}_m(l_1)}{\partial l_1} \times \frac{\partial \mathbf{g}(l_2)}{\partial l_2} \right) \cdot (\mathbf{f}_m(l_1) - \mathbf{g}(l_2)) \qquad (25)$$

If Δ_{ee} is minimum, Equation (26) must be satisfied.

$$\frac{\partial \Delta_{ee}}{\partial l_1} = 0, \frac{\partial \Delta_{ee}}{\partial l_2} = 0 \qquad (26)$$

TABLE I
SECOND ORDER DISPLACEMENT

coefficient of $\Delta\theta$	coefficient of $\Delta\theta^2$	possible displacement
> 0 (first order repelling)	—	
= 0 (first order reciprocal)	> 0 (second order repelling)	possible
	= 0 (second order reciprocal)	
	< 0 (second order contrary)	
< 0 (first order contrary)	—	impossible

Applying the Taylor expansion to Equation (25) and using Equation (26), we obtain Equation (27).

$$\begin{aligned}
\Delta_{ee} &= \mathbf{n} \cdot \mathbf{t}_3 + (\mathbf{n} \times \mathbf{t}_1) \times (\mathbf{t}_1 \times \mathbf{v} + \mathbf{t}_2) \\
&+ 2((\mathbf{t}_1 \times \mathbf{e}_{f1}) \times \mathbf{e}_{g1}) \cdot (\mathbf{t}_1 \times \mathbf{v} + \mathbf{t}_2) \\
&+ \mathbf{n} \cdot (\kappa_f a_1^2 \mathbf{e}_{f2} - \kappa_g a_2^2 \mathbf{e}_{g2})
\end{aligned} \qquad (27)$$

$$a_1 = \frac{(\mathbf{e}_{f2} \times \mathbf{e}_{g1}) \cdot (\mathbf{t}_1 \times \mathbf{v} + \mathbf{t}_2)}{(\mathbf{e}_{f1} \times \mathbf{e}_{g1}) \cdot \mathbf{e}_{f2}}$$

$$a_2 = \frac{(\mathbf{e}_{g2} \times \mathbf{e}_{f1}) \cdot (\mathbf{t}_1 \times \mathbf{v} + \mathbf{t}_2)}{(\mathbf{e}_{f1} \times \mathbf{e}_{g1}) \cdot \mathbf{e}_{g2}}$$

The possible displacement is formulated by $\Delta_{ee} \geq 0$.

When Equation (24) is not satisfied, this contact primitive can be regarded as a combination of another contact primitives as the case of polyhedral objects.

F. Face-Vertex and Face-Edge Contact

Consider the case that a face contacts a vertex in one point. Viewing the displacement from the vertex, the face inversely displaces. From the definition, the inverse displacement can be represented as $-\mathbf{t}_1, -\mathbf{t}_2, -\mathbf{t}_3$. Therefore, the equation to represent the second order displacement is obtained by substituting $-\mathbf{t}_1, -\mathbf{t}_2, -\mathbf{t}_3$ to Equation (7). As the same way, we obtain the equation to represent the second order displacement in a face-edge contact case is obtained.

IV. CALCULATING THE SECOND ORDER DISPLACEMENT

The second order possible displacement can be calculated by investigating signs of coefficients of $\Delta\theta$ and $\Delta\theta^2$ as shown in Table I.

First, we calculate the first order displacement. Then, we calculate the second order displacement by classifying the first order reciprocal motion. For calculating the first order displacement, various methods have been

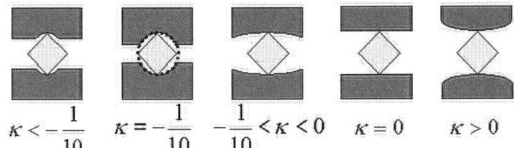

Fig. 7. Example

proposed[10][11]. We proposed the method to classify the first order reciprocal motion into the second order reciprocal motion or not[7]. However, it is so difficult to classify the first order reciprocal motion into the second order repelling, reciprocal, or contrary motions. That is an open problem. Therefore, we roughly introduce a method to classify the first order reciprocal motion into the second order reciprocal motion or not.

Because the first order possible displacement is represented as one system of simultaneous linear inequalities, the first order reciprocal motion is represented as a linear sum of some bases as shown in Equation (28), where a_i is any real number.

$$[\mathbf{t_1}, \mathbf{t_2}] = \sum a_i \mathbf{b_i} \quad (28)$$

Substituting the equation to coefficients of $\Delta \theta^2$, they are generally represented as Equation (29), where $h_i(a_1, \ldots, a_r)$ is the second order homogeneous equation.

$$\mathbf{n}_i \cdot \mathbf{t}_3 + h_i(a_1, \ldots, a_r) \quad (29)$$

To classify the first order reciprocal motion into the second order reciprocal motion or not, we need to solve Equation (30).

$$\bigcap \mathbf{n}_i \cdot \mathbf{t}_3 + h_i(a_1, \ldots, a_r) = 0 \quad (30)$$

Searching linear dependent combinations of $\{\mathbf{n}_i\}$, equations including h_i only are obtained. Such equations are naturally the second order homogeneous. Next we decompose these equations to linear equations: 1) If it is equivalent to $A_1^2 + \cdots + A_n^2 = 0$, it can be decomposed into Equation (31).

$$\bigcap_i^n A_i = 0 \quad (31)$$

2) If it is equivalent to $A_1^2 - A_2^2 = 0$, it can be decomposed into Equation (32).

$$A_1 + A_2 = 0 \bigcup A_1 - A_2 = 0 \quad (32)$$

V. EXAMPLE

We apply the proposed method to a contact relation consisting of four vertex-face contacts as shown in Figure 7. We assume that locations of four vertices are $(0, 20, 10)$, $(0, 20, -10)$, $(0, -20, 10)$, and $(0, -20, -10)$, surface normals of two faces are $(0, 0, 1)$ and $(0, 0, -1)$, curvatures of two faces along the y-axis are 0, and curvatures along the x-axis are κ.

First, we formulate the first order possible displacement as Equation (33).

$$\begin{pmatrix} 10 & 0 & 0 & 0 & 0 & -1 \\ -10 & 0 & 0 & 0 & 0 & 1 \\ -10 & 0 & 0 & 0 & 0 & -1 \\ 10 & 0 & 0 & 0 & 0 & 1 \end{pmatrix} \begin{pmatrix} \mathbf{t_1} \\ \mathbf{t_2} \end{pmatrix} \geq \mathbf{0} \quad (33)$$

We obtain Equation (34) by solving Equation (33). The solution means an object cannot translate along the z-axis and rotate about the x-axis, and other displacements are the first order reciprocal motions.

$$\begin{pmatrix} 10 & 0 & 0 & 0 & 0 & -1 \\ -10 & 0 & 0 & 0 & 0 & -1 \end{pmatrix} \begin{pmatrix} \mathbf{t_1} \\ \mathbf{t_2} \end{pmatrix} = \mathbf{0} \quad (34)$$

We classify all the first order reciprocal motions into the second order reciprocal motions or not. We select bases of the first order reciprocal motions as Equation (35)

$$\mathbf{s_1} = (0, 1, 0, 0, 0, 0), \mathbf{s_2} = (0, 0, 0, 1, 0, 0)$$
$$\mathbf{s_3} = (0, 0, 0, 0, 1, 0), \mathbf{s_4} = (0, 0, -1, 0, 0, 0) \quad (35)$$

We obtain Equation (36) which the second order reciprocal motion must be satisfied (See the next page).

We obtain Equation (37) to (40) by searching linear dependent combinations.

$$20(1 + 10\kappa)s_1^2 + 2\kappa s_2^2 + 800\kappa s_4^2 = 0 \quad (37)$$
$$40(1 + 20\kappa)s_1 s_4 + 80\kappa s_2 s_4 = 0 \quad (38)$$
$$20(1 + 10\kappa)s_1^2 + 40(1 + 20\kappa)s_1 s_4$$
$$+ 2\kappa s_2^2 + 80\kappa s_2 s_4 + 800\kappa s_4^2 = 0 \quad (39)$$
$$20(1 + 10\kappa)s_1^2 - 40(1 + 20\kappa)s_1 s_4$$
$$+ 2\kappa s_2^2 - 80\kappa s_2 s_4 + 800\kappa s_4^2 = 0 \quad (40)$$

When $\kappa > 0$, $s_1 = s_2 = s_4 = 0$ is obtained from Equation (37). That means only translation along the y-axis can maintain the contact relation.

When $\kappa = 0$, $s_1 = 0$ is obtained from Equation (37). That means rotation about the y-axis cannot maintain the contact relation.

When $-1/10 < \kappa < 0$, $s_4 = 0 \bigcup (1 + 20\kappa)s_1 + 2\kappa s_2 = 0$ is obtained from Equation (38). 1) When $s_4 = 0$, $\sqrt{10(1 + 10\kappa)}s_1 + \sqrt{\kappa}s_2 = 0 \bigcup \sqrt{10(1 + 10\kappa)}s_1 - \sqrt{\kappa}s_2 = 0$ is obtained form Equation (37). That means translation along x axis is necessary for rotation about the y-axis to maintain the contact relation. 2) When $(1 + 20\kappa)s_1 + 2\kappa s_2 = 0$, it is difficult to analyze what the result means. The result represents the case that rotation axis is on the xz-plane. In this case, one cannot solve the answer intuitively.

When $\kappa = -1/10$, $s_2 = s_4 = 0$ is obtained from Equation (37). That means translation along the x-axis and rotation about the z-axis cannot maintain the contact relation.

$$(0,0,-1)\cdot \mathbf{t_3} + 10(1+10\kappa)s_1^2 + (1+20\kappa)s_1s_2 + 20(1+20\kappa)s_1s_4 + \kappa s_2^2 + 40\kappa s_2s_4 + 400\kappa s_4^2 = 0$$
$$(0,0,1)\cdot \mathbf{t_3} + 10(1+10\kappa)s_1^2 - (1+20\kappa)s_1s_2 + 20(1+20\kappa)s_1s_4 + \kappa s_2^2 + 40\kappa s_2s_4 + 400\kappa s_4^2 = 0$$
$$(0,0,-1)\cdot \mathbf{t_3} + 10(1+10\kappa)s_1^2 + (1+20\kappa)s_1s_2 - 20(1+20\kappa)s_1s_4 + \kappa s_2^2 - 40\kappa s_2s_4 + 400\kappa s_4^2 = 0$$
$$(0,0,1)\cdot \mathbf{t_3} + 10(1+10\kappa)s_1^2 - (1+20\kappa)s_1s_2 - 20(1+20\kappa)s_1s_4 + \kappa s_2^2 - 40\kappa s_2s_4 + 400\kappa s_4^2 = 0 \quad (36)$$

When $\kappa < -1/10$, $s_1 = s_2 = s_4 = 0$ is obtained from Equation (37), the result is the same as $\kappa 0 > 0$.

VI. CONCLUSION

In this paper, we proposed a method to formulate and calculate the second order displacement. First, we formulate the second order displacement in each contact primitives. These formulas are very different from each other, however, they can be treated as the same way when calculating.

Then, we verified the proposed method by using the case that four vertex-face contacts exist. We illustrated that a solution of calculating the second order displacement is changed in concord with the change of the face curvature. We found that the solution is correct and the proposed method is superior to the original screw theory.

VII. ACKNOWLEDGMENTS

This work is supported in part by the Japan Science and Technology Corporation (JST) under the Ikeuchi CREST project, and in part by the Grant-in-Aid for Scientific Research on Priority Areas (C) 14019027 of the Ministry of Education, Culture, Sports, Science and Technology.

VIII. REFERENCES

[1] K. Ikeuchi and T. Suehiro : "Toward an assembly plan from observation part i: Task recognition with polyhedral objects," *IEEE Trans. on Robotics and Automation*, Vol. 10, No. 3, Jun. 1994.

[2] Y. Kuniyoshi, M. Inaba, and H. Inoue : "Learning by watching: Extracting reusable task knowledge from visual observation of human performance," *IEEE Trans. on Robotics and Automation*, Vol. 10, No. 6, Dec. 1994.

[3] M. Tsuda, T. Takahashi, and H. Ogata : "Generation of an assembly-task model analyzing human demonstration," *Journal of the Robotics Society of Japan*, Vol. 18, No. 4, pp. 535 – 544, 2000.

[4] H. Onda, T. Ogasawara, H. Hirukawa, K. Kitagaki, A. Nakamura, and H. Tsukune : "A telerobotics system using planning functions based on manipulation skills and teaching-by-demonstration technique in vr," *Journal of the Robotics Society of Japan*, Vol. 18, No. 7, pp. 979 – 994, 2000.

[5] M. S. Ohwovoriole and B. Roth : "An extension of screw theory," *Journal of Mechanical Design*, Vol. 103, pp. 725 – 735, Oct. 1981.

[6] H. W. Kuhn and A. W. Tucker : "Linear inequalities and related systems," *Annals. of Mathematics Studies*, Vol. 38, , 1956.

[7] J. Takamatsu, H. Kimura, and K. Ikeuchi : "Improved screw theory using second order terms," *IEEE Inter. Conf. on Intelligent Robots and Systems*, pp. 1614 – 1618, 2002.

[8] E. Rimon and J. Burdick : "Mobility of bodies in contact - i: A new 2nd order mobility index for multiple-finger grasps," *IEEE Inter. Conf. on Robotics and Automation*, pp. 2329 – 2355, 1994.

[9] H. Hirukawa, T. Matsui, and K. Takase : "A general algorithm for deriving constraint of contact between polyhedra from geometric model," *Journal of the Robotics Society of Japan*, Vol. 9, No. 4, pp. 415 – 426, 1991.

[10] H. Hirukawa, T. Matsui, and K. Takase : "A fast algorithm for the analysis of the constraint for motion of polyhedra in contact and its application to departure motion planning," *Journal of the Robotics Society of Japan*, Vol. 9, No. 7, pp. 841 – 848, 1991.

[11] S. Hirai : "Kinematics and statics of manipulation using the theory of polyhedral convex cones and their application to the planning of manipulative operations," *Journal of the Robotics Society of Japan*, Vol. 17, No. 1, pp. 68 – 83, 1999.

Knot Planning from Observation

Takuma Morita Jun Takamatsu Koichi Ogawara† Hiroshi Kimura†† Katsushi Ikeuchi

Institute of Industrial Science, University of Tokyo
†*Japan Science and Technology Cooperation*
††*Univ. of Electro-communications*
moritaku@cvl.iis.u-tokyo.ac.jp

Abstract

Learning from Observation (LFO) has been widely applied in various types of robot system. It helps reduce the work of the programmer. But the available systems have application limited to rigid objects. Deformable objects are not considered because: 1) it is difficult to describe their state and 2) too many operations are possible on them. In this paper, we choose the knot tying as case study for operating on nonrigid bodies, because a "knot theory" is available and the type of operations is limited. We describe the Knot Planning from Observation (KPO) paradigm, a KPO theory and a KPO system.

1 Introduction

The Learning from Observation (LFO) paradigm was proposed around 1990. In this paradigm a robot makes observations of human tasks, understands them and generates a program to reproduce the tasks. To understand tasks, information obtained from vision systems is automatically transformed into abstract information. Thus, LFO reduces the human effort in programming a robot to imitate the task. There is a lot of researches on LFO[1, 2, 3, 4].

However, these methods consider only the assembly of rigid objects. Deformable objects are not considered because their geometry is unpredictable. It is difficult to accurately express their position and their pose.

In this paper, the knot tying task is used to study the manipulation of deformable objects. We select this task, because of the following reasons.

1. Knot tying has restrictive conditions prohibiting deformation in which object parts move through each other. Compared to other deformable objects, knot tying has much less possible operations.

2. Knot tying already has an available mathematical model the "knot theory"[5, 6].

There is lots of related research on knot tying tasks. [7] dealt with rope handling by using hand-eye coordination. [8] proposed a language to describe a knot. [9] and [10] introduced a knot theory.

To perform a knot-tying task, we have been developing a Knot Planning from Observation (KPO) system. The KPO system makes observations of a human tying a knot through a vision system. The observations are evaluated and a set of commands is generated. By following this set of commands the robot is able to tie a similar knot.

This paper overviews the KPO paradigm in Section 2. Then from Section 3 to Section 6, we describe the theory the KPO system is based on. The principles of the knot theory are explained in Section 3. This is followed by the expression of states in Section 4. Construction of movement primitives is then presented in Section 5. Section 6 presents a method that extracts movement primitives from state transitions. Section 7 shows the currently implemented KPO system, and its experimental results. Section 8 summarizes the work and describes the remaining parts of KPO.

2 KPO paradigm

In a KPO system, an operator ties a knot in front of the robot's visual system (stereo camera system). A set of images from the camera is used as the observation. From the observation, the knot tying is recognized by the following modules.

1. *Configuration Recognition Module (CRM)*
   Extracting rope region from each image, performing image processing, recognizing a knot as arranged points and segments.

2. *State Recognition Module (SRM)*
   Recognizing knot states.

3. *Task Recognition Module (TRM)*
   Recognizing operations based on the result of the State Recognition Module

4. *Task Execution Module (TEM)*
   Reproducing a knot based on the result of the Task Recognition Module.

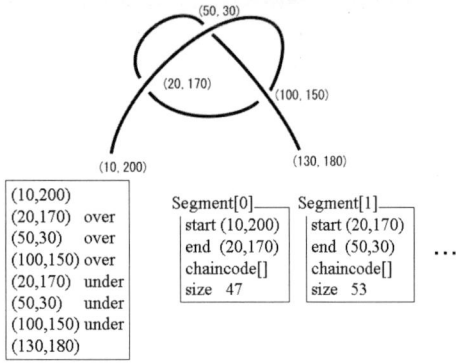

Figure 1: K-data

Figure 2: Knot

In a KPO system, the CRM, the SRM and the TRM are very important.

In CRM the position and pose of a rope are recognized by performing thinning, recognizing segment connectivity and judging the vertical position of intersections. The result of the CRM is expressed in form of K-data. An example of K-data is shown in Figure 1. K-data consists of a vector of intersections and a vector of segments. These vectors are arranged in order from one of the ends of a rope to another end. An intersection is expressed by 2D coordinates on an image and the relative vertical position (over or under). A segment is expressed by a starting intersection, an end intersection, a size and a chain code which connects these intersection.

The SRM transforms the K-data into P-data[11]. P-data is a data structure designed according to the knot theory. P-data contains less information than K-data. It has only the topological information (segments' connectivity, vertical position of an intersection). No 2D coordinate is provided.

The TRM compares two states (P-data) and recognizes a state transition. The KPO system is presupplied with a set of movement primitives and chooses an appropriate movement primitive corresponding to the state transition. Movement primitives are operations which serve as elements of knot tying. The knot-tying task is recognized as a sequence of movement primitives. A robot then collects the required parameters from the vision system to tie a similar knot.

3 The knot theory

In designing KPO systems, the following two problems need to be solved.

- How should a state be expressed?
- What kind of operations should be chosen as movement primitives?

Both problems are answered by the knot theory. In this section, the principles of the knot theory are described.

3.1 A knot and a projection

Figure 2 shows an example of a knot. The knotted ring in this figure is created by tying a rope and connecting both ends. This ring has no end and cannot be untied without cutting it. A ring like this is defined as a knot. Theoretically, the knot does not have thickness (i.e. the knot is a simple closed curve in 3D space).

Most of the methods for information transmission are two dimensional, so the knot is projected from R^3 to R^2 space. Examples of projections are shown in Figure 2. In projected images, the lower strand is drawn with gaps around the position where it crosses below the other strand.

3.2 Reidemeister moves

Two knots are equivalent if one can be deformed to the other without cutting the string.

If we manipulate a string in R^3 space, it is easy to create the equivalent knots by following this rule. But we need to be able to show these 'legal' moves in projected images.

Reidemeister[5] proved that any equivalent knot can be created from the current knot by three types of moves. These moves are known as the Reidemeister moves. The Reidemeister moves are shown in Figure 3.

Reidemeister move I adds or removes one crossing through a simple loop. Reidemeister move II adds or removes two crossings simultaneously from a knot. Reidemeister move III allows a strand to be moved to the other side of a crossing.

4 State expression

When we construct the state expression, the following 2 conditions are required.

1. The state expression is an abstract data structure that does not depend on parametric information.
   The expression must always be the same for one

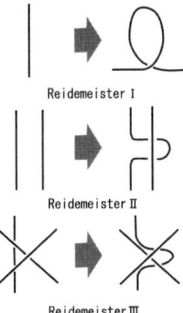

Figure 3: *Reidemeister moves*

kind of knot, even though the knot has undergone some motion or its shape/position is different.

2. The transformation to the state expression is reversible.
   The original projection can be restored from the state expression. For example, if we express knot states by connectivity matrices, we cannot restore the matrices to the original projections. That's because when two segments are between the same two intersections, we cannot determine which segment is outside/inside from the connectivity matrices.

K-data is not applicable for the SRM, because of its dependence on parametric information. We transform K-data into P-data. P-data[11] is chosen, because it is simple and satisfies both conditions. P-data is created through the following process.

1. At first we choose one end of a knot as a starting end.

2. From the starting end, we follow the knot to the other end. If we encounter an intersection, we number the intersection according to the order of encounter. Each intersection has two numbers.

3. We refollow the knot. When we encounter an intersection, we record the intersection number from the direction of encounter and the other one. We also determine a sign plus/minus and a relative vertical position (over/under) from the relation of the direction of the current and the other strand.

4. Finally, we code relative vertical positions and signs into a set of numbers as follows: 1: over/-, 2: under/-, 3: over/+, 4: under/+.

A sign is determined by the following equation.

$$sign = \frac{\vec{l}_{over} \times \vec{l}_{under}}{|\vec{l}_{over} \times \vec{l}_{under}|} \cdot \vec{e}_z$$

Where, $\vec{l}_{over}$ and $\vec{l}_{under}$ are the direction vectors of the upper and lower strand, respectively, at their intersection point. e_z is a unit vector parallel to the z axis.

If the strand that passes over the left hand side of the strand that passes under, the sign is plus. And if the strand that passes over is in the right hand side of the strand that passes under, the sign is minus.

From the above process, P-data for the knot in Figure 4 is obtained as follows.

$$\begin{array}{cccccc} 1 & 2 & 3 & 4 & 5 & 6 \\ 4 & 5 & 6 & 1 & 2 & 3 \\ 3 & 1 & 2 & 4 & 2 & 1 \end{array}$$

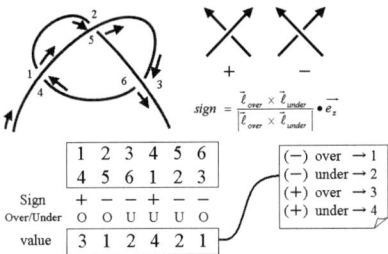

Figure 4: *P-data*

A theorem in in a graph theory guarantees that the projection can be restored from P-data. Refer to [11] for the restoration of the projected image from P-data.

5 Construction of movement primitives

A mathematical knot is a closed curve and does not have open ends. For closed curves, Reidemeister moves change the projection of a knot without changing the topology of a knot.

However in our research, we have to deal with open curves. Open curves are not mathematical knots. But if we virtually connect both ends, open curves can be considered as knots. So unlike closed curves, both projections and topology are changeable through operations.

Operations toward ropes can be categorized into two types: operations that change topology and the ones that do not.

To describe the knot-tying operations Cross and Reidemeister moves are defined as movement primitives. Cross is an operation in which the one end of the rope crosses any segment. It changes topology of knots. In the other hand, Reidemeister moves do not change the topology of knots. Furthermore, Cross is operated on both the segment and the open end of the rope, while Reidemeister moves change only the segment of the rope.

Every kind of knot tying can be expressed by a sequence of these four movement primitives (Cross, Reidemeister move I, Reidemeister move II and Reidemeister move III). However, for our current analysis (overhand knot, eight knot, bowline knot, harness hitch, bow tie, single loop bow, two-half knot and tautline hitch) the Reidemeister move III has not been used. We believed that more complex knot, it may be used.

Figure 5 shows the analysis of the bowline knot. R_I and C in the figure represent Reidemeister move I and Cross, respectively.

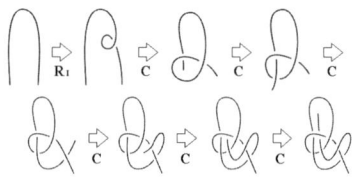

Figure 5: *Analysis of Moyai-musubi*

6 Recognition of operations

In this section, we describe a method to recognize the movement primitives by the change of P-data.

We also show how to recognize Cross, Reidemeister move I and Reidemeister move II. Reidemeister move III is omitted, since it is too complex.

In the following, P-data at time t is expressed as P_t. $n(P_t)$ is defined as the number of columns of P-data, i.e., twice the number of intersections on a projection. $intersection(l,m)$ is the intersection whose corresponding intersection numbers are l and m. $segment(l,m)$ is the segment whose corresponding intersection numbers are l and m.

1. Cross

    An example of Cross is shown in Figure 6. The P-data in the left side and the right side are P_{t-1} and P_t, respectively.

    If we remove the columns corresponding to $intersection(3,8)$ (these are surrounded by frames in Figure 6) from P_t and reorder the intersection numbers, i.e., subtract the intersection numbers 3 to 7 by one, P_t becomes equal to P_{t-1}.

    Such removing and reordering is defined as $C(P_t, 8)$.

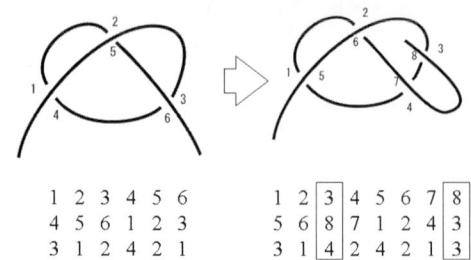

Figure 6: *Change of P-data under Cross*

Only $C(P_t, 1)$ and $C(P_t, n(P_t))$ are considered generally, because Cross is an operation involving ends.

If $C(P_t, 1) = P_{t-1}$, a Cross occurs at the end-1 side. If $C(P_t, n(P_t)) = P_{t-1}$, there is a Cross at the end-$n(P_t)$ side.

2. Reidemeister move I

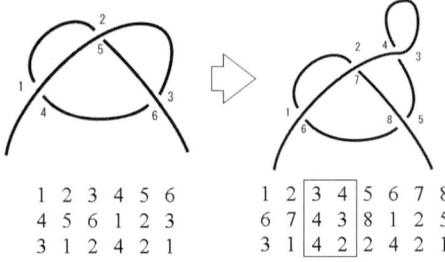

Figure 7: *Change of P-data under Reidemeister move I*

An example of Reidemeister move I is shown in Figure 7.

If we remove the columns corresponding to $intersection(3,4)$ (these are surrounded by frames in Figure 7.) from P_t and reorder the intersection numbers, i.e., subtract the intersection numbers 5 to 8 by two, P_t becomes equal to P_{t-1}.

Such removing and reordering is defined as $R_I(P_t, 3)$.

If $R_I(P_t, k) = P_{t-1}$, a Reidemeister move I occurs at $segment(k-1, k)$ at time $t-1$.

In the case of Figure 7, $R_I(P_t, 3) = P_{t-1}$ is fulfilled, so a Reidemeister move I occurs at the $segment(2,3)$ at time $t-1$.

3. Reidemeister move II

    An example of Reidemeister move II is shown in Figure 8.

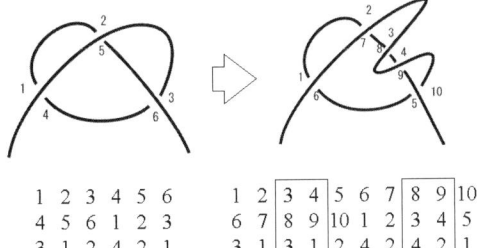

Figure 8: *Change of P-data under Reidemeister move II*

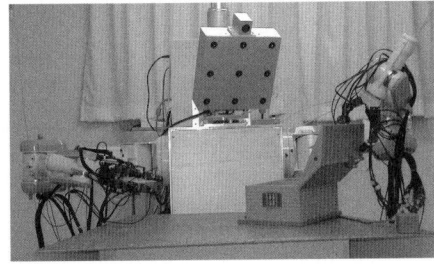

Figure 9: *Our robot*

If we remove the rows corresponding to $intersection(3,8)$ and $intersection(4,9)$ (these are surrounded by frames in Figure 8.) from P_t and reorder the intersection numbers, i.e., subtract the intersection numbers 5 to 7 by two and subtracting the intersection number 10 by four, P_t becomes equal to P_{t-1}.

Such an operation is defined as $R_{II}(P_t, 3)$.

We calculate $R_{II}(P_t, k)$ toward $k = 1, 2, \cdots, n(P_t)$ in order. If $R_{II}(P_t, k) = P_{t-1}$, the KPO system can recognize that Reidemeister move II occurs at $segment(k-1, k)$ and the $segment(b_{k-1}, b_k)$ at time $t-1$. Here, b_i is the value at the second line of P-data, at column i.

In the case of Figure 8, $R_{II}(P_t, 3) = P_{t-1}$ is fulfilled, so a Reidemeister move II occurs at $segment(2,3)$ and $segment(b(2), b(3))$ ($segment(5,6)$) at time $t-1$.

7 Implementation and experimental result

To test our proposed system, The CRM, the SRM and the TRM were implemented.

In this experiment, our platform is a humanoid robot (Figure 9). The robot has a 9-eye stereo camera, two arms and 4 fingers in each arm.

7.1 Configuration Recognition Module

In the CRM, K-data is generated from the set of images taken by the 9-eye stereo camera. The generation procedure is described here with the results of each process (Figure 10).

1. Extracting the region of a rope from images by subtraction. (Figure 10(a))

2. Thinning by Hilditch filter. (Figure 10(b))

3. Finding feature points (intersection and the end of the rope). (Figure 10(c))

4. Encoding segments between feature points into the chain-code in K-data.(Figure 10(d))

5. Removing short segments. (E.g. area inside a circle in Figure 10(b) and Figure 10)

6. Finding the intersections mistakenly separated by edge thinning.

7. Merging incorrect intersections to form the correct ones. (Figure 10(f))

8. Numbering segments from one end to the other end.

9. Calculating the vertical positions of crossings from the 9-eye stereo system.

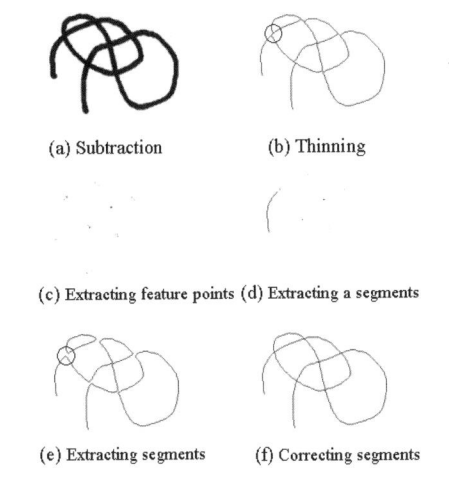

Figure 10: *Results of each stage in the Configuration Recognition Module*

7.2 State Recognition Module and Task Recognition Module

The SRM generates P-data from K-data. Then the TRM extracts movement primitives from P-data transitions. The method in Section 6 was applied to the actual knot. Because the knot in Figure 10 is complex and difficult to evaluate, a simpler knot is presented.

The experimental result is shown in Figure 11.

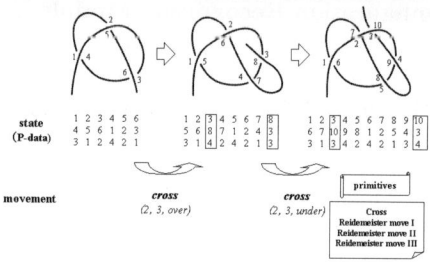

Figure 11: State & Result of SRM & TRM

7.3 Evaluation

A successful result of the CRM is shown in Figure 10. But at present, the rate of images succesfully analyzed is low (about 10%), because of unrobustness of obtaining the relative vertical positions from the vision system. To solve this problem, we are developing an algorithm.

On the other hand, the SRM and TRM run robustly.

8 Summary

8.1 Summary

In this paper, we propose a novel algorithm to recognize operations within a knot. The KPO system is created to make a robot able to tie a knot from observation. The system applied KPO theory, a kind of LFO algorithm. KPO theory applies the knot theory and represents the knot in form of a state expression (P-data) and movement primitives. Our recognition algorithm is applied P-data to reconstruct the knot. The Configuration Recognition Module, the State Recognition Module and the Task Recognition Module have been implemented and showed that our proposed recognition algorithm could correctly extract the operations made with a knot. The extracted operations can be applied by a robot to perform the knot-tying task.

8.2 Future works

Some of future research directions toward a complete KPO system include:

1. Configuration Recognition Module
   The current CRM cannot treat video. The CRM has to deal with occlusions and be robuster to observation noises, in order to treat video. In order to solve this problem, we are examining using improved Snakes[12].

2. Task Execution Module
   We are just beginning to implement the TEM. Before completion, there will be many problems; manipulation, grasping strategies..

Acknowledgments

This work is supported in part by the Japan Science and Technology Corporation (JST) under the Ikeuchi CREST project, and in part by the Grant-in-Aid for Scientific Research on Priority Areas (C) 14019027 of the Ministry of Education, Culture, Sports, Science and Technology. Special thanks to S. Auethavekiat and K. Bernardin for the english correction.

References

[1] K. Ikeuchi and T. Suehiro: "Toward an assembly plan from observation, Part I: Task recognition with polyhedral objects," IEEE Trans. Robot. Automat., **10**, 3, pp. 368–385 (1994).

[2] J. Takamatsu, H. Tominaga, K. Ogawara, H. Kimura and K. Ikeuchi: "Symbolic representation of trajectories for skill generation," International Conference on Robotics and Automation, **4**, pp. 4077–4082 (2000).

[3] Y. Kuniyoshi, M. Inaba and H. Inoue: "Learning by watching: extracting reusable task knowledge from visual observation of human performance," IEEE Trans. Robot. Automat., **10**, 6, pp. 799–822 (1994).

[4] K.Ogawara, S. Iba, T. Tanuki, H. Kimura and K. Ikeuchi: "Acquiring hand-action models by attention point analysis," International Conference Robotics and Automations, **4**, pp. 465–470 (2001).

[5] K. Reidemeister: "KNOT THEORY," BCS Associates (1983).

[6] C. C. Adams: "The Knot Book–an elementary introduction to the mathematical theory of knots–," W.H.FREEMAN AND COMPANY (1994).

[7] H. Inoue and M. Inaba: "Hand-eye coordination in rope handling," Proc. of ISRR; Published as Robotics Research, M. Brady and R. Paul, pp. 163–174, MIT PRESS (1984).

[8] J. E. Hopcroft, J. K. Kearney and D. B. Krafft: "A case study of flexible object manipulation," The International Journal of Robotics Research, pp. 41–50 (1991).

[9] M. Yamada, R. Budiarto, H. Seki and H. Itoh: "Topology of cat's cradle diagrams and its characterization using knot polynomials," Journal of Information Processing Society of Japan, **38**, pp. 1573–1582 (1997).

[10] T. M. A. Fink and Y. Mao: "THE 85 WAYS TO TIE A TIE–The Science and Aesthetics of Tie Knots–," Broadway Books (2000).

[11] M. Ochiai, S. Yamada and E. Toyoda: "COMPUTER AIDED KNOT THEORY," Makino Shoten (1996).

[12] M. Kass, A. Witkin and D. Terzopoulos: "Snakes: Active contour models," International Journal of Computer Vision, **1**, pp. 321–331 (1988).

Estimation of essential interactions from multiple demonstrations

Koichi Ogawara   Jun Takamatsu†   Hiroshi Kimura††   Katsushi Ikeuchi†

Japan Science and Technology Cooperation
†Institute of Industrial Science, University of Tokyo
††Univ. of Electro-Communications
ogawara@cvl.iis.u-tokyo.ac.jp

Abstract— To learn a new everyday task under the "Learning from Observation" framework, the system needs to detect which parts of the demonstration are essential to complete the task without task-dependent knowledge. In the previous research, we proposed a technique to estimate essential interactions in a task by integrating multiple demonstrations which represent virtually the same task. Although, the technique could automatically segment the essential interactions and determine the number of the interactions, the segmentation algorithm depends on some heuristics and only stationary interactions could be obtained.

In this paper, a novel technique is proposed, which overcomes this limitation and can estimate almost any types of interactions. In this approach, a demonstrator needs to give a explicit signal once during each essential interaction as a hint on the occurrence of the essential interaction. From visual information and these signals, the system automatically analyzes the essential parts of the task and their periods, and also detects which environmental objects are interacted with the manipulated object. These information is hard to be obtained from a single demonstration, because of the ambiguity in interpreting the interaction especially in cluttered environment. The proposed method is evaluated in a simulation and also in a real world by using a humanoid robot.

I. INTRODUCTION

In the past few years, many projects have been started to aim for supporting human beings in daily environment with the robotics technology[1]. One of the key obstacles to be solved is the lack of adaptability to cope with new situations. Unlike the robots in the factories, all the reactions to all the possible events cannot be programmed in advance for those who are expected to work in everyday environment. So the robots must have a way of learning and expanding their knowledge and ability.

As a part of learning and expanding the ability of a robot by itself, we have been studying how to obtain everyday manipulation tasks under the "Learning from Observation" framework[2], [3]. In this framework, a robot analyzes a demonstration of a new task by sensors such as vision or cyber-gloves, and constructs an abstract representation of the task which is used to reproduce the obtained task. Most of the past researches in this framework can be divided into 2 approaches; one is to record the trajectory, and the applied force if necessary, of the manipulated object and try to maintain the same motion at the time of reproduction[4], the other is to decompose a demonstration into a sequence of the pre-designed primitives, each of which represents a specific phenomenon appeared in a task. These primitives are designed in advance so that the finite set of these primitives can cover all the phenomena appeared in the subject task[2], [5], [6], [7].

However, neither of the approaches cannot be used to understand unknown tasks. The trajectory based approaches assume that the position of the objects with which the manipulated object interacts is not changed along the entire task and the obtained trajectory can be used directly to reproduce the demonstration, however this condition is not met for reproduction in different environment. On the other hand, the primitive based approaches assume that all the possible primitives are pre-designed so that any tasks can be represented by concatenating appropriate primitives. For example, typical assembly tasks are described as a process of adding constraints to the degree of freedom of the manipulated object and low-level primitives such as contact-state transitions are useful to represent assembly tasks, however they are not a suitable set of primitives for general everyday behavior such as pouring or wiping.

In general, it is hard to design a set of primitives that can cover all the possible everyday behaviors, so a practical way is to learn a behavior (primitive) at the time when the robot encounters a new one which cannot be understood from the already obtained knowledge as we do.

To realize this process, two steps are required, namely "differentiation" and "classification". A demonstrated task is first differentiated to a sequence of essential interactions and then each of which is recognized to check if it is already obtained or not. If not, it is added to the knowledge as a new primitive. In this paper, the former part of the process, differentiation, is described. The latter part is partly covered by [8].

To detect all the essential interactions without prior knowledge, we proposed a technique to integrate observations from multiple demonstrations which virtually represent the same task[3]. In that method, the sequence of interactions which are common to all the demonstrations are extracted, which are considered to be essential to complete that task.

However, because of the weakness of the segmentation algorithm used, only the relatively stationary interactions such as pouring can be detected by this method. To overcome this limitation, in this paper, we propose a new improved method which can detect almost any interactions, including cyclic behavior such as wiping and dynamic behavior such as hammering from multiple demonstrations.

In the following sections, we first describe the problem definitions in Section II. The limitation of the previous

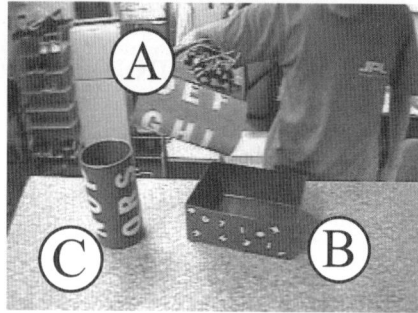

Fig. 1. Ambiguity in detecting an interaction.

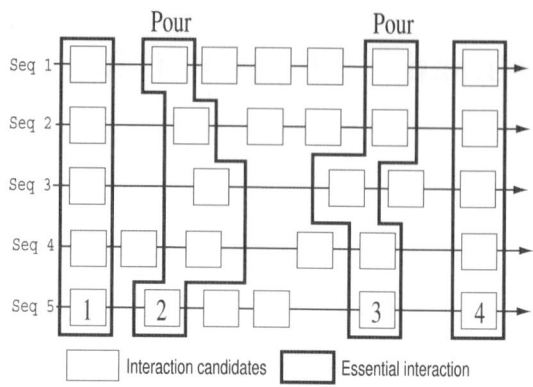

Fig. 2. Previous estimation of essential interactions.

method and the improvement in the proposed method is also described. Then the detail of our technique is presented in Section III. The technique is evaluated in simulation and in real-environment, and the result is presented in section IV. The results are compared with that of the previous study and are also applied to a humanoid robot to reproduce the same task to show the validity of the estimated interactions. Finally, we conclude in Section V.

II. PROBLEM DEFINITIONS

We assume that the target everyday tasks can be represented as a sequence of interactions between a manipulated object and an environmental object. To differentiate a demonstrated task, we need to extract essential interactions which are necessary to complete the task. An essential interaction is represented as a sequence of relative relationships, the trajectory of the manipulated object in the target object's coordinate frame, in a certain period of time.

In general, it is hard to detect essential interactions from one demonstration. If we know that contact-state transitions are key feature of the essential interactions, we may detect them by recognizing contact-states, but even so, it is a hard job to estimate exact contact-state transitions from vision or cyber-gloves.

Anyway, in general case, we can't count on the contact-state transitions or adjacency of the objects to detect essential everyday behavior, especially in cluttered environment where the task may be composed of a series of interactions with different objects.

Fig. 1 shows one such example. There are one manipulated object A, and 2 environmental objects B and C. We can easily explain the scene because we have knowledge about the function of each object and also knowledge about the situations where these objects are typically used. But, without those knowledge, there exists ambiguity in interpreting this scene namely; (1) A interacts with B, (2) A interacts with C, (3) This scene is not essential to complete the task. Even if we get to know that A in the scene is involved in an essential interaction, ambiguity about the period of the segment remains.

A. Limitation of the Previous Approach

To solve this problem, in the previous research[3], we proposed a method to estimate essential interactions by integrating multiple demonstrations. In that method, several demonstrations of virtually the same task are observed. The difference in demonstrations is in the arrangement of the objects and/or in the occurrence of non-essential motions.

The method follows 2 steps. At the first step, the essential interaction candidates are extracted for each demonstration as shown in the small boxes in Fig. 2. Each candidate keeps the information about the relative trajectory of the manipulated object in a certain period of time. At the second step, the likelihood of all the possible combinations of the candidates are evaluated and the optimal path which maximizes the sum of the evaluations of the combinations are calculated by multi dimensional DP matching technique. These combinations as shown in the large regions in Fig. 2 are considered to be essential.

The advantage of this method is that it detects not only the corresponding target object in cluttered environment but also the total number of the essential interactions and the bound of each interaction. However, to detect essential interaction candidates, the variance of the pose of the manipulated object in the target object's coordinate frame is evaluated along the time axis and those regions whose variance is lower than a certain threshold are detected as interaction candidates. The clear disadvantage of this segmentation algorithm is that only the slow interactions such as pouring and spooning can be detected.

B. Extension to Detect Any Interactions

To overcome this limitation, in this paper, we propose a new improved method which can detect almost any interactions, including cyclic behavior such as wiping and dynamic behavior such as hammering.

The main cause of the limitation comes from the unknown number of the essential interactions and also comes from the uncertainty about the periods corresponding to each essential interaction. This uncertainty makes it hard to estimate multiple essential interactions simultaneously only from raw visual information of demonstrations.

So we introduce extra information by a demonstrator, a signal corresponding to each essential interaction, to reduce this difficulty. This extra instruction can be realized without effort by giving a simple oral signal such as "Here" once during each interaction. Anyway, it is quite natural to give such a signal to teach ourselves.

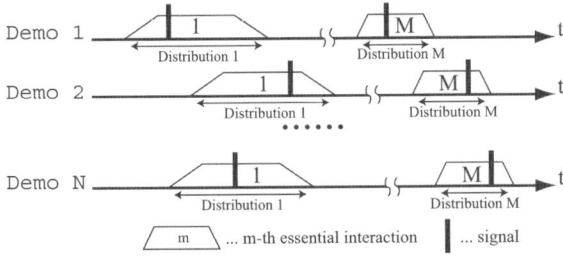

Fig. 3. Overview of the subject problem.

Because a signal is generated in a certain instant during each essential interaction, the total number of the essential interactions are known directly. So, for each essential interaction, the problems to be solved are; (1) Which environmental object is interacted with the manipulated object, and (2) What the start time and stop time of each interaction are.

III. ESTIMATION OF ESSENTIAL INTERACTIONS

Fig. 3 shows the overview of the subject problem. N demonstrations are performed and each demonstration contains M essential interactions. Each essential interaction is made of a distribution of motion, i.e. the middle part of an interaction are almost identical among all the demonstrations, while the both ends of an interaction are affected by the adjacent motion and are not the same. Broadly speaking, the width of the m-th distribution remains as same for each demonstration, and, even if the start time of each essential interaction may be different, the order of the interactions along the time axis is fixed. Further, a signal is accompanied with each essential interaction, and the time frame of the occurrence of each signal in each interaction is not known.

Each demonstration is recorded by tracking the manipulated object in 3D space. Before a demonstration begins, we assume that all the objects including the one which later be manipulated are on the table in front of the robot. At this time, the pose of all the objects are estimated by the localization technique introduced in [9].

Then, for each demonstration, the manipulated object are tracked in 3D space by a stereo vision system using the similar technique described in [9]. The stereo vision system produces color images and disparity images in 30 fps. If a 3D CAD model of the manipulated object and the pose of the object in the current time frame are known, the pose of the next time frame can be estimated by localizing the 3D model in the 3D space generated from the disparity image in the next time frame. The initial pose of the manipulated object is known from the above procedure, so, by applying this technique sequentially, the entire trajectory of the manipulated object in a demonstration is obtained.

When d-th demonstration has T^d samples, the pose of the manipulated object in the Object s's coordinate frame in d-th demonstration in time frame t is represented in the form $< p_s^d(t), q_s^d(t) >$; where $p(t)$ means 3 values position vector and $q(t)$ means 4 values quaternion vector.

Each essential interactions is represented as a trapezoid function with the height equals to 1. The height of a trapezoid at time frame t means the probability that the corresponding relative pose of the manipulated object at t being the nature of the interaction. The oblique side of a trapezoid means the end of the motion where the essential interaction and the adjacent motion are blended.

A trapezoid function of the m-th essential interaction in d-th demonstration is represented by 2 parameters; the mean μ_m^d of the trapezoid and the standard deviation σ^d as shown in Fig. 4 and Fig. 5. Note that, although μ_m^d is not dependent on each other, σ^d takes the same value among all the demonstrations. That means, for each m-th essential interaction, it must be performed in the same velocity among all the demonstrations.

In the proposed method, the following flow is repeatedly processed until the parameters converge.

1) Initial μ are set to be the time frame of each signal. Initial σ are set to be a constant value.
2) Estimate the valid pose corresponding to an essential interaction among all the sampled trajectories.
3) Update the parameters μ, σ by fitting each trapezoid.
4) Estimate the target object during each essential interaction.
5) If the change of the updated parameters are larger than a certain threshold, go to 2.

A. Estimation of Valid Sampled Pose

If only the trajectory of the manipulated object corresponding to an essential interaction is sampled, the parameters of the distributions can be solved by, for example, EM algorithm. But, in our case, the trajectory is sampled uniformly, so we first need to estimate the validness of each sampled pose.

The sampled pose is valid if all the corresponding relative pose $< p_s^d(t), q_s^d(t) >$ in a certain object s's coordinate frame among the demonstrations are similar during the entire interaction. So, based on the current estimation of the parameters, the validness of the pose in the n-th demonstration sampled in time t can be evaluated as $e^{-W_a W(t,n,m,s)}$ where $W(t,n,m,s)$ is calculated as shown in Eq.(1).

$$p\_mean(t) = \frac{\sum_{d=1}^N p_s^d(t)}{N}$$

$$q\_mean(t) = \frac{\sum_{d=1}^N q_s^d(t)}{N} \text{ by Slerp}$$

$$W(t,n,m,s) = \qquad\qquad\qquad\qquad (1)$$
$$W_p \sqrt{\frac{\sum_{d=1}^N |p_s^d(t - \mu_m^n + \mu_m^d) - p\_mean(t)|^2}{N}}$$
$$+ W_q \sqrt{\frac{\sum_{d=1}^N acos((q_s^d(t - \mu_m^n + \mu_m^d) \cdot q\_mean(t)))^2}{N}}$$

where the weight term W_p is set to be 0.002 and W_q is set to be $\frac{1}{\pi}$ to normalize $W(t,n,m,s)$. W_a is set to be 4.6 so that $e^{-W_a W(t,n,m,s)}$ reaches 0.01, i.e. almost 0, when $W(t,n,m,s)$ reaches 1.

Then, for each demonstration n and for each m-th essential interaction, the convolution of $e^{-W_a W(t,n,m,s)}$ and

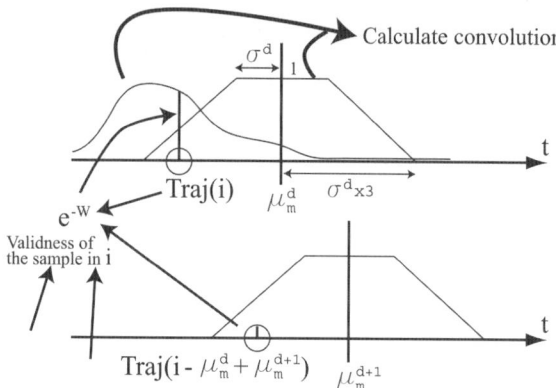

Fig. 4. Estimation of the validness of each sampled pose.

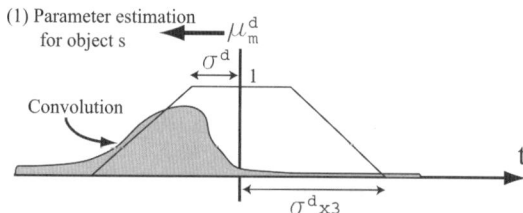

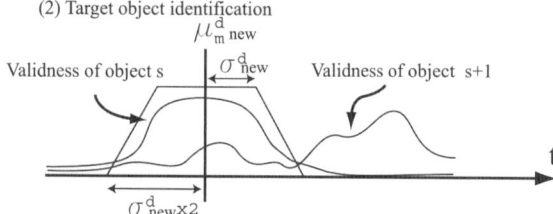

Fig. 5. Parameter estimation and target object identification.

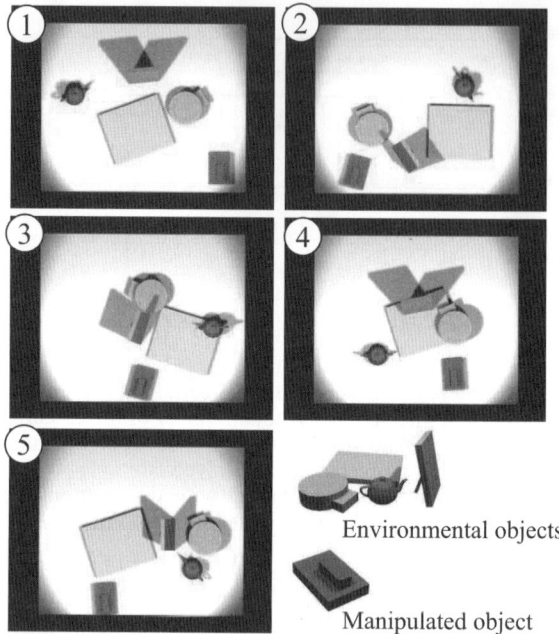

Fig. 6. Arrangement of the objects.

the trapezoid is calculated along t as shown in Fig. 4. The base length of the trapezoid is set to be $\sigma^n \times 6$, so that the area larger than the current trapezoid can be searched for when estimating σ_n in the next section. This convolution function $c_{n,m,s}(t)$ is considered as an index of the sample t being essential.

B. Estimation of the New Parameters

Then, the new parameters of the trapezoid, $\mu_m^d new$ and $\sigma_m new$ are calculated as the mean and the standard deviation of each convoluted distribution as shown in the top half of Fig. 5.

At this time, we keep the set of new parameters for each environmental object. So these new parameters are the candidates for the parameters in the next step and the parameters generated from the most appropriate object s are selected.

C. Estimation of Interacted Objects

Finally, the proper target object for each interaction is determined. First of all, the validness of each sampled pose is calculated by using the new parameters for each object s as shown in the bottom half of Fig. 5. Then, the integral of the convolution of the validness and the trapezoid is calculated for each object s and the one which results in the largest value is selected as a target object in the m-th essential interaction as shown in Eq.(2). Because this is the evaluation step, the base length of the trapezoid is set to be $\sigma^n \times 4$ to cover almost 80% of the distribution.

$$\text{target\_obj}_m = \operatorname{argmax}_s \sum_{d=1}^{N} \int_{1}^{T^d} \text{trapezoid}^d(t) \cdot \text{validness}^d(t) dt \quad (2)$$

In the case of Fig. 5, the object s results in better evaluation than that of the object $s+1$.

IV. EXPERIMENT

To evaluate the proposed method, 2 types of experiments are carried out. To show the variety of the possible interactions that can be handled, trajectories including cyclic interactions and dynamic interactions are generated in a virtual world and the generated data is evaluated by our method. The reason why the trajectories are generated in a virtual world, not in the real world, is due to the limitation of our 3D tracking technique. Those dynamic motions such as hammering are hard to be tracked stably by out current vision system, though we are tackling this problem now. So, in this study, to get dynamic and cyclic trajectories precisely, the motion is simulated in a virtual world.

The second experiment is made to show the validness of the proposed technique compared to the previous approach. For this purpose, the same data which were

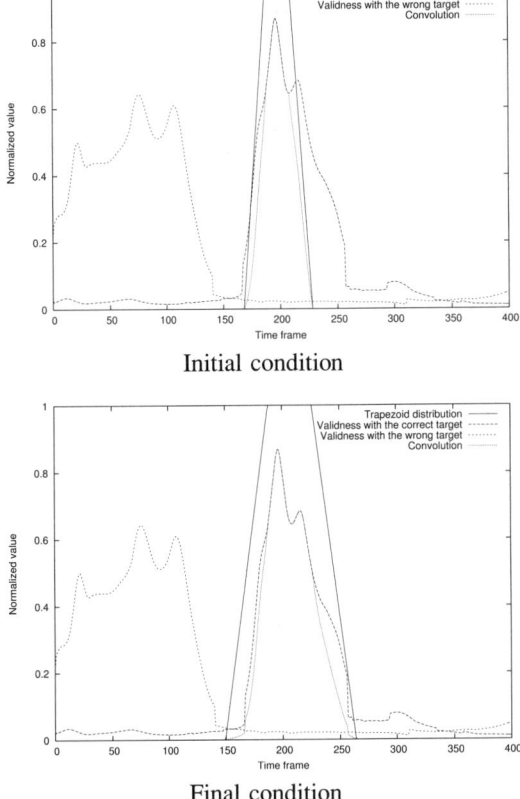

Fig. 8. Result of parameter estimation in a virtual world.

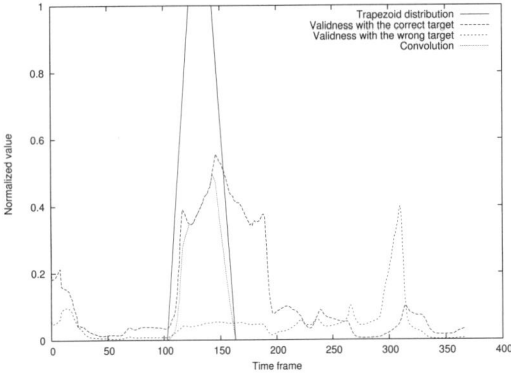

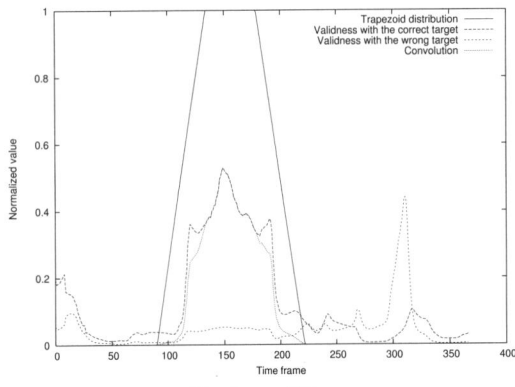

Fig. 9. Result of parameter estimation in a real world.

obtained previously in the real world is used; though all the motions included are relatively stationary.

A. Result in Simulation

5 demonstrations are performed in a virtual world and the initial arrangement of the objects are shown in Fig. 6. 4 environmental objects and 1 manipulated object are placed on the table.

Fig. 7 shows the trajectory of the manipulated object in 5th demonstration and 3 essential interactions included there. The 1st and the 2nd essential interactions are cyclic motions, while the 3rd one is a dynamic motion where a virtual demonstrator strikes the manipulated object on the stage. The trajectory is generated manually from a spline curve with velocity, and the initial pose of each object and the trajectory data is directly obtained. Signals are generated randomly during each essential interaction.

Fig. 8 shows the initial estimation and the final estimation of the 2nd essential interaction in Fig. 7. We can see that the trapezoid is correctly fitted to the correct validness distribution. The target objects for each essential interaction are also estimated correctly.

B. Result in Real Environment

The data obtained in the previous research (Fig. 2) is used again to show the generality of our method compared to the previous method which depends on some heuristics to detect essential interactions. The initial distribution is transformed to the estimated distribution, but in the case of the 1st essential interaction, the result is different between the proposed method and the previous method. This is caused from the illness of the heuristics used in the previous research and the estimated distribution of the proposed method is much more natural.

TABLE I
COMPARISON BETWEEN THE ESTIMATED ESSENTIAL PERIODS.

	1st essential int.		2nd essential int.	
	start	end	start	end
Initial	113	153	230	270
Estimated	112	200	215	313
Previous	68	184	202	309

TABLE I shows the comparison between the estimated essential periods from the proposed method and the previous result.

C. Reproduction by the robot

At the time of reproduction by a robot, the relative trajectory corresponding to each sequence of essential interactions are converted to the desired trajectory of the

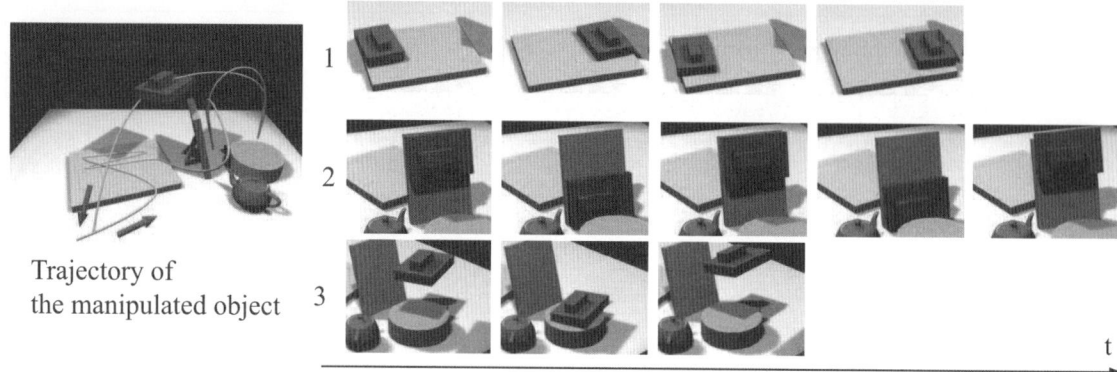

Fig. 7. Trajectory and essential interactions.

Fig. 10. Reproduction by a robot.

end-effector and are maintained to keep the same relative relationship in the target object's coordinate frame. Fig. 10 shows the result of the reproduction by a humanoid robot.

V. CONCLUSION

To understand an unknown task without prior task-dependent knowledge, the advantage of the use of multiple demonstrations to obtain a series of essential interactions is explained. Then the previous research on estimation of essential interactions from multiple demonstrations is described and the limitation on the variety of interactions is also pointed out.

To overcome this limitation, a novel method which can recognize almost any interactions including cyclic motions such as wiping and dynamic motions such as hammering is proposed. With a little extra effort on demonstrator's side, i.e. giving a signal for each essential interactions, the system automatically detects the target object of interaction and also estimates the period of each essential interaction. The proposed method is evaluated in virtual environment and also in the real world by using a humanoid robot.

VI. ACKNOWLEDGMENTS

This work is supported in part by the Japan Science and Technology Corporation (JST) under the Ikeuchi CREST project, and in part by the Grant-in-Aid for Scientific Research on Priority Areas (C) 14019027 of the Ministry of Education, Culture, Sports, Science and Technology.

VII. REFERENCES

[1] M. Graf, B. Hagele. Dependable interaction with an intelligent home care robot. *1st IARP/IEEE-RAS Joint Workshop*, 2001.

[2] K. Ikeuchi and T. Suehiro. Toward an assembly plan from observation part i: Task recognition with polyhedral objects. *IEEE Trans. Robotics and Automation*, 10(3):368–384, 1994.

[3] K. Ogawara, J. Takamatsu, H. Kimura, and K. Ikeuchi. Generation of a task model by integrating multiple observations of human demonstrationsacquiring hand-action models by attention point analysis. In *Int. Conference on Robotics and Automation*, pages 1545–1550, 2002.

[4] N. Delson and H. West. Robot programming by human demonstration: Adaptation and inconsistency in constrained motion. In *Int. conf. on Robotics and Automation*, pages 30–36, 1996.

[5] Y. Kuniyoshi, M. Inaba, and H. Inoue. Learning by watching. *IEEE Trans. Robotics and Automation*, 10(6):799–822, 1994.

[6] J. Takamatsu, H. Tominaga, K. Ogawara, H. Kimura, and K. Ikeuchi. Symbolic representation of trajectories for skill generation. In *Int. conf. on Robotics and Automation*, pages 4077–4082, 2000.

[7] B. Dufay and J. C. Latombe. An approach to automatic robot programming based on inductive learning. *Int. Journal of Robotics Research*, 3(4):3–20, 1984.

[8] K. Ogawara, J. Takamatsu, H. Kimura, and K. Ikeuchi. Modeling manipulation interactions by hidden markov models. In *Int. Conference on Intteligent Robot and Systems*, pages 1096–1101, 2002.

[9] K. Ogawara, J. Takamatsu, H. Kimura, and K. Ikeuchi. Extraction of fine motion through multiple observations of human demonstration by dp matching and combined template matching. In *10th IEEE Int. Workshop on Robot and Human Communication (ROMAN) 2001*, pages 8–13, 2001.

Synthesize Stylistic Human Motion from Examples

Atsushi Nakazawa , Shinichiro Nakaoka and Katsushi Ikeuchi
Institute of Industrial Science, University of Tokyo
Japan Science and Technology Corporation
nakazawa@cvl.iis.u-tokyo.ac.jp

Abstract— The human body motion synthesis is highly necessary for humanoid robots' motion planning and computer animations. In this paper, new method for generating human-like natural motions based on the motion database acquired by motion capture systems is described. On the analysis step, the acquired motions are divided into some motion segments, and then the characteristic poses and motions are archived as 'motion styles'. The motion style is a kind of the human skill, and it's unique to the motions' scenario, such as the different kinds of dances. On the synthesis step, users direct the key poses of human figures. The system generates the characteristic motions according to the user's directions and motion style database. The experiment result shows that this method can synthesize the realistic 'stylized' motions with this framework.

Keywords— human motion, dance motions, motion style, motion synthesis, humanoid robot

I. INTRODUCTION

Synthesizing realistic human motion is very necessary for humanoid robot's motion planning and computer animation. Compared to the industrial purposes, such applications highly need human-like reality for generated motions. Our group aims to develop the total techniques to imitate stylistic human motions such as the dance motions by humanoid robots, for the purpose of the digital human motion archive [1]. The overview of our project is shown in figure 1. The human motion is acquired by the motion capture systems, then the motion analysis methods are applied for the purpose of motion searching and editing. Finally, original or synthesized motions are displayed by using humanoid robots. Users can recognize the motion effectively through the realistic display.

In this paper, we describe the motion synthesis method that can generate realistic human motion based on the analysis of the captured human motions. On the motion analysis step, the original motion sequence is divided into the motion segments, and the fundamental motions are extracted through the motion analysis. According to the results, characteristic poses and body movements are retrieved. These poses and movements are unique to the motions' scenario, such as the different kinds of dances, personality or other factors. On the synthesis step, the user designs the motion by directing the end effector's positions of the key frames. The system generates the key poses and the transition motion between them, finally new motion sequence can be generated.

The study for human motion synthesis has been done in the robotics and computer animation fields; they are categorized in two types. The most basic researches are based on finding the pose or transition by optimizing the particular evaluation values, such as jerk of joint angles [3], integration of the joint torque [2]. They are mainly considering the arm movement and proposed models are evaluated by the simulation result and the actual human motion. For the whole body motion synthesis, Tak et. al proposed the method by keeping whole body balancing [4]. Their method uses the ZMP (Zero Moment Point) as the evaluation value, which indicates if they can keep the dynamic balance.

These methods are based on the finding optimal solution of pre-defined evaluation values, but they have other problems that generated motion always draws a single path. In reality, a human body motion has characteristics according to scenarios. For example, a person performing dance motions, the motion will not always follows the optimal trajectory.

The solution for this problem is using some example motions of actual humans'. The new motion can be synthesized by using them. Ijspeert et. al proposes the method that the robot acquire its control command by self learning method. The robot system learn its control by comparing its movement and the actual human motion [5]. Bland et.al has propose the idea of the "Style Machine" [6]. In this method, the same class motions with different style (such as walking slowly, rapidly, etc.) has described by Hidden Malkov Model(HMM), and the other stylistic motions can be generated by the analysis result of this HMM. Lee et,al proposed the model that the motions consists of the hierarchy [7]. The human body motion is described by the direction of body links, and they are also expressed by B-spline. Because this B-spline is described by hierarchical knot vectors, new motions can be synthesized by manipulating these vectors.

All these methods can not design the particular desired human poses and motions, but combine or deform slightly existing motion segments. Our method aims to solve this issue that enables the user to design the motion sequence much directly. The user can direct the end effectors' posi-

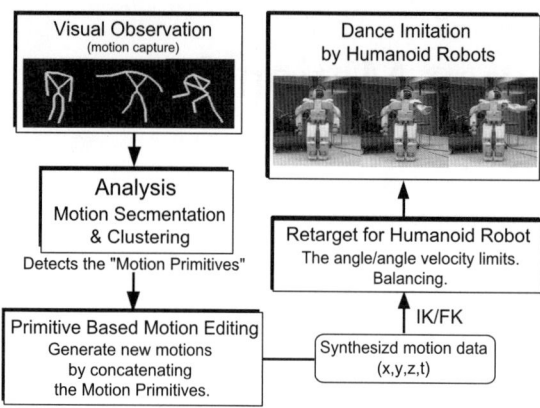

Fig. 1. Overview of our project.

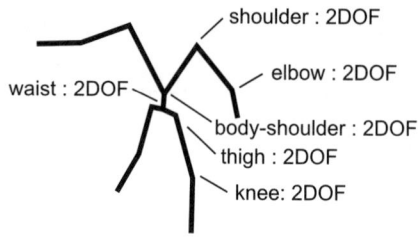

Fig. 2. The human body model.

tions of the key frames key positions, the resulting motion are synthesized as its naturally being in the desired scenario. Even if the user directed the same positions on the key frames, the generated motions are different if the different database has used for the synthesis.

II. THE BASIC IDEA

We think that the human motion sequence is consisting of some kinds of the motion elements, namely primitive motions. The primitive motions are defined on every body portions (arms and legs), and the same primitive motions appear more than a time on a motion sequence. Therefore, the whole motion sequence is also the sequence of the primitive motions:

$$HumanMotionSequence = PrimitiveMotionA \\ + PrimitiveMotionB + ...$$

Moreover, a primitive motion consists of a motion base and a motion style:

$$PrimitiveMotion = MotionBase \oplus MotionStyle$$

The motion base is calculated mathematically from the border conditions of the motion segments (start and end poses). The motion style is the displacement between the real human motion and the motion base. Accordingly, the original motions are recovered from the key poses and the motion styles. On the analysis stage, these key poses and motion styles are stored into the motion database.

On the synthesis step, users design the human motion by directing end-effectors positions of the human figure. The two border conditions - the first and last key poses of human figure - are obtained from the key pose database, and then the motion base is calculated. Finally, the system retrieves the most suitable motion style and applies it to the motion base, which is calculated by the key poses. Because the motion styles are unique to the kinds of the original human motions, the generated motions are different between the used motion styles. And they also keep the features of the original human motions.

III. MOTION ANALYSIS

We used 15 measurement points for whole body representation: hands (L,R), elbows (L,R), shoulders (L,R), head, hip, body center, waists (L,R), thighs (L,R) and feet (L,R), and 18 DOF for the whole body: waist to body (2DOF), body to the shoulders (2DOF x 2), shoulders to elbows (2DOF x 2), elbows to hands (2DOF x 2), waist to thighs (2DOF x 2), thighs to knees (2DOF x 2) and knees to feet (2DOF x 2) (Figure 2). On the motion analysis, the motion sequence is segmented and primitive motions are detected, then the key poses and motion styles are retrieved and stored as the motion database.

A. Segmentation

The aim of the motion segmentation is to find the start and stop frame of the end effectors' movements. On the top of the analysis, we define the body center coordinate system at each motion frames. This coordinate has the waist direction as the X-axis and the perpendicular direction as the Z-axis. The four end effector's positions (hands and feet) are mapped onto these coordinate systems, and their velocities are calculated.

The segmentation is done by detecting the local minimum of the velocity. To prevent the over segmentation, the gaussian filter has applied beforehand and these terms are checked for each segments.

1. The velocity of the segmented frames is less than the threshold.
2. The amplitude of the velocity inside of the segment is larger than the threshold.

The poses on the segmented frames are registered into the database as the key poses.

B. Detecting Primitive Motions

For detective primitive motion of the motion sequences, the correlations between the motion segments are evaluated by comparing the end effectors' trajectory. We use the DP matching distance for this purpose. Assume that the end effectors positions in the motion segment m, n are described as $V_m = \{vm_1, vm_2, ..., vm_{im} | vm_i \in R^3\}$, $V_n = \{vn_1, vn_2, ..., vn_{in} | vn_i \in R^3\}$, then the distance between these segments $D(m,n)$ can be calculated with following :

$$D(m,n) = S(V_m, V_n)$$
$$S(k,l) = d_{k,l} + min(S_{k,l-1}, S_{k-1,l-1}, S_{k-1,l})$$
$$d_{i,j} = |vm_i - vn_j|$$

After the correlations between all combinations of the segments are calculated, they are clustered by using the nearest neighbor algorithm. As the result, the segments in which the end effectors pass the similar trajectory are in the same cluster. Finally, the segments in the same clusters are equalized and registered as the primitive motions. Moreover, we can easily recognize the motion sequence is the repetition of the primitive motions, and also having some kinds of structures (see figure 4).

C. Extracting the motion base and the motion style

The motion bases and the motion styles are extracted from primitive motions. The motion base can be generated by interpolating the body links at two segment frames (the start and end frames). The human body link n on the motion primitive m is described as $X_{m,n}(t_m) = \{x_{m,n}(t_m), p_{m,n}(t_m) | x_{m,n} \in R^3, p_{m,n} \in R^3, 1 \le t_m \le T_m\}$, where x_n, p_n, T_m are the connected position to the parent body link, the direction of the link and the number of frames in the segment m, respectively. The basic motion is obtained by simply linear interpolation of the border conditions:

$$Xb_{m,n}(t_m) = \{(1 - \frac{t_m}{T_m})x_{m,n}(0) + \frac{t_m}{T_m}x_{m,n}(T_m),$$
$$(1 - \frac{t_m}{T_m})p_{m,n}(0) + \frac{t_m}{T_m}p_{m,n}(T_m)$$
$$|x_{m,n} \in R^3, p_{m,n} \in R^3, 1 \le t_m \le T_m\}$$

For extracting the motion style, we use the new coordinate system $Rs_n(t)$ which is set up on each body links (figure 3). On this coordinate systems, $x_{m,n}(t)$ is the origin, $p_{m,n}(0)$ is the x-coordinate and $(p_{m,n}(T_m) - p_{m,n}(0)) \times p_{m,n}(0)$ is the z-coordinate.

Accordingly, the movement of each body link is described by two parameters, z-axis rotation $rz(t)$ and y-axis rotation $ry(t)$. The motion base $\{rz^b_{m,n}(t), ry^b_{m,n}(t)\}$ is defined as linear z-axis rotation:

$$(rz^b_{m,n}(t), ry^b m, n(t)) = (d_z \frac{t}{T_m}, 0)$$

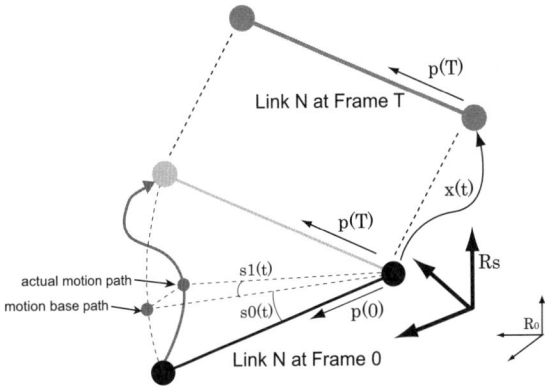

Fig. 3. The coordinate systems of the human body link and definition of the motion style parameters.

The motion style parameters $(s0_{m,n}(t), s1_{m,n}(t))$ is the displacement between the actual motion and the motion base.

$$s0_{m,n}(t) = rz_{m,n}(t) - rz^b_{m,n}(t)$$
$$s1_{m,n}(t) = ry_{m,n}(t)$$

Due to the definition of the coordinate system $Rs_n(t)$, the border conditions of the parameters are zero. These motion style parameters means that the $s0$ is the fluctuations of the motion and the $s1$ indicates the distortion from the motion base. The motion base is calculated mathematically from the border conditions, the resulting motion seems to be artificial and mechanical. By applying the motion styles onto the motion base, the result becomes similar to the human motion, because these parameters describe the distortion and fluctuation of the human motion. We also thinks the motion styles are characteristic between the scenario of the original motions. This means the motion style is a kind of the human skill.

IV. MOTION SYNTHESIS

With the acquired poses and motion style parameter database, new stylized motions can be synthesized. The user indicates the end effectors' positions and the durations of the motion segments. According to these values, the system can synthesize new motion by using the database.

A. Synthesize Key Poses

At the beginning, the first and the end poses of the motion segment is determined. The user indicates the end effectors' positions of each frame in body center coordinate. The system searches the pose that has the nearest end effector's position from the database. If some candidates are found, the user can design the desired poses by blending them. The result is used for the initial value for synthesizing poses. To 'fit' the initial pose to the user's desired one, Jacobian based inverse kinematics solver is employed.

The initial joint angles $\theta_0 \in R^N$ are given by the found pose, and the desired joint angles $\theta_{designed} \in R^N$ can be acquired by following algorithm (N is the number of DOF of the portions):

$\theta = \theta_0$
x = ForwardKinematics(θ)
while(| x - $x_{designed}$ | > thresh)
$\quad J^W = W^{-1}J^T(JW^{-1}J^T)^{-1}$
$\quad \dot{\theta} = J^W (x_{designed} - x)$
$\quad \theta = \theta + \alpha \cdot \dot{\theta}$
$\quad$ x = ForwardKinematics(θ)
end
$\theta_{designed} = \theta$

Where $x_{designed}$ is the user designed position, J is the Jacobian matrix of the target body portion and α is the constant (0.01 - 0.1). Because this algorithm minimize the equation $\dot{\theta}^T W \theta$, the matrix W is the diagonal matrix which indicate the weight of each joints. For the arm movement, we used smaller value for body-to-shoulder joints than other ones because of the observation of the motion. Using this algorithm, each portions' start and end poses of the motion segments can be obtained. Because we uses the pose database of original motion sequence as the initial poses, the resulting ones keep the characteristics of the original motion sequence.

B. Synthesize the Transition

We developed two-step approach for synthesizing transition motion. First, the motion base is calculated from the border conditions. After that, the the appropriate motion style is retrieved from the motion database and applied. The resulting motion has human-like reality due to the motion style.

The motion base is synthesized by using the algorithm described in the last section and the border conditions obtained by the key pose synthesis. Assume that the user designed the motion segment that the end effector is moving from $x(0)$ to $x(T_{sty})$ in the body center coordinate. And the stored transition motion in the motion database is the path from $y(0)$ to $y(T_{ref})'$, the distance between them are obtained by :

$$d = \sum_{i=1}^{N} ||\{(1-\frac{i}{N})x(0) + \frac{i}{N}x(T_{sty})\}$$
$$-\{(1-\frac{i}{N})y(0) + \frac{i}{N}y(T_{ref})\}||$$

According to the nearest motion primitive, the motion style is retrieved and applied the synthesized motion base.

The final synthesized result of the body link n is acquired by following:

$$x_{sty}(t) = {}^0 R_s R_z(\frac{\delta r_z}{T_{sty}}) \cdot R_z(s1(\frac{T_{ref}}{T_{sty}})t)$$
$$\cdot R_y(s2(\frac{T_{ref}}{T_{sty}})t) \cdot \{l_n \ 0 \ 0 \ 1\}^T$$

Where ${}^0R_s$, R_z and R_y are the 4x4 homogeneous matrix from local coordinate R_s to world coordinate, rotation matrix around z-axis and y-axis, l_n is the length of the link and δr_z is the displacement angle along z-axiz between the first and end frames. This algorithm is based on the idea that the body movement that draws the similar end effector's path would have the similar motion styles. In this case, we do not consider the actual transition path of the end effectors, because we would like to synthesize the motion's characteristics only from the border conditions.

V. EXPERIMENTS

We have captured three kinds of Japanese fork dances at 60Hz frequency as the motion database. The length of these ones are 1500(harukoma), 1517(nishi-monai-ondo) and 4500 frames(soran-bushi), totally 7517 frames. Figure 4 and table I show the detection results of the motion segments and primitive motions of these motions. We can recognize the number of the primitive motions is about 10% to 50% of the number of the original motion segments. Especially in Harukoma dance motion, all hands motions are described about 1/10 numbers of primitives.

Name	Segment	Primitive	Ratio
Soran-Bushi(left hand)	140	58	41.4%
Soran-Bushi(right hand)	150	61	40.7%
Soran-Bushi(left foot)	153	26	17.0%
Soran-Bushi(right foot)	153	30	19.6%
Harukoma(left hand)	86	9	10.4%
Harukoma(right hand)	89	8	9.0 %
Harukoma(left foot)	57	8	14.0%
Harukoma(right foot)	50	8	16.0%
Nishi-Monai-Ondo(left hand)	46	21	45.7%
Nishi-Monai-Ondo(right hand)	45	22	48.9%
Nishi-Monai-Ondo(left foot)	39	7	17.9%
Nishi-Monai-Ondo(right foot)	41	7	17.1%

TABLE I
THE ANALYSIS RESULT OF THE THREE DANCE MOTIONS (ON THE LEFT HAND MOVEMENT).

Figure 6 shows the key pose frames segmented by the left hand's velocity. As we can see, each poses has the characteristic body shape if the end effector is located at near positions. Figure 5 is the synthesized result by using these

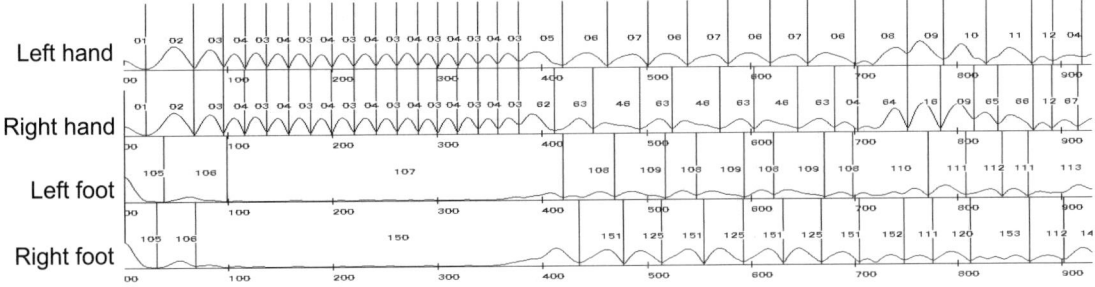

Fig. 4. The result of the segmentation and detection of primitive motions on Soran-Bushi motion sequence. The number of each segments indicate the ID of the primitive motions.

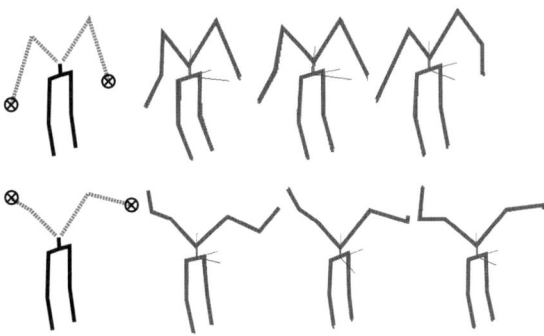

Fig. 5. The synthesized poses with different databases. Left: The end effecor position that user directed, Left Center, Right Center, Right: The synthesized result with the poses of Harukoma, Nishi-Monai, Soran-Bushi.

different three key pose databases. Although the same positions are directed for end effectors, the synthesized poses are different between them.

Figure 7 to figure 9 show the synthesis result of the stylized motion. These results are base on the synthesis result of the poses in the last experiment. We can recognize the difference between the motion base and stylized motions in the arm movement. Because the motion base is the linear interpolation of the border conditions, the resulting motions are always short-cut paths between them. And the transition angle speeds are constant, they seems to be a kind of mechanical motion. On the other hand, the stylized motions not only draw the shortest path between the start and end frames. For example, the stylized motion according to the Harukoma dance database, the left hand motion is far longer way compared to the motion base, and velocity of the right hand motion are very different. We can see the similar characteristics in other stylized dance motions and proposed stylized parameter can actually synthesize the feature of the characteristic poses and transition features.

VI. CONCLUSION

In this paper, we proposed the motion synthesis method that can generate realistic human motion according to the motion capture databases. This method is based on the idea that the human motion is consisting of the characteristic poses, motion base and the motion style. For synthesizing poses, original human motions are segmented to detect the stop motion frames. After the motion base has generated mathematically from the border conditions (synthesized start and end poses), the motion style is applied. The motion style can be obtained from the motion database according to the desired transition path. We can confirm this motion style has large effect for increasing the human-like reality through the experiment results. Moreover, the synthesized results are different according to the motion database, even if the user directs the same information for the synthesis. This feature can be useful for computer animation and imitating human skill by the robots. Now we can generate the motion which the user designed their end effectors positions under the constraint that all of the body portions have the stop frame at the same time. But in reality, this constraint cannot be always appropriate. For this issue, we are now trying to detect the relation between the stop frame of each portion, and use for the designing.

REFERENCES

[1] Nakazawa A., Nakaoka S., Kudoh S., Ikeuchi K. Yokoi K.: Imitating Human Dance Motions through Motion Structure Analysis, Proc. of International Conference on Intelligent Robots and Systems (IROS2002), 2002.
[2] Y.Uno, M.Kawato, R.Suzuki: Formation and Control of Optimal Trajectory in Human Multi-Joint Arm Movement-Minimun Torque Change Model, Biological Cybernetics 61, pp.89-101, 1989.
[3] T.Flash, H.Hogan: The coordination of Arm Movements, Journal of Neuroscience, pp.1688-1703, 1985.
[4] Tak S., Song. O., Ko H.: Motion Balance Filtering, Proc. of Eurographics 2000, Vol.19, No.3, pp.437-446, 2000.
[5] Ijspeert A.J., Nakanishi J., Schaal S.: Movement imitation with non-linear dynamical systems in humanoid robots, International Conference on Robotics and Automation (ICRA2002), pp 1398-1403, 2002.
[6] Brand, M.E.; Hertzmann, A.: Style Machines, Proc. of ACM SIGGRAPH2000, pp. 183-192, 2000.
[7] Lee J. and Shin S.Y. : A Hierarchical Approach to Interactive Motion Editing for Human-like Figures, Proc. of ACM SIGGRAPH'99, pp.39-48, 1999.

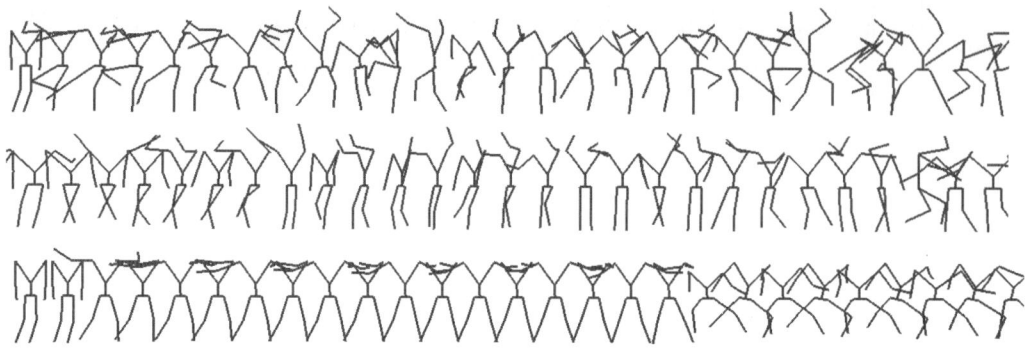

Fig. 6. The key poses of the three kinds of dances (segmented according to the left hand's movement).

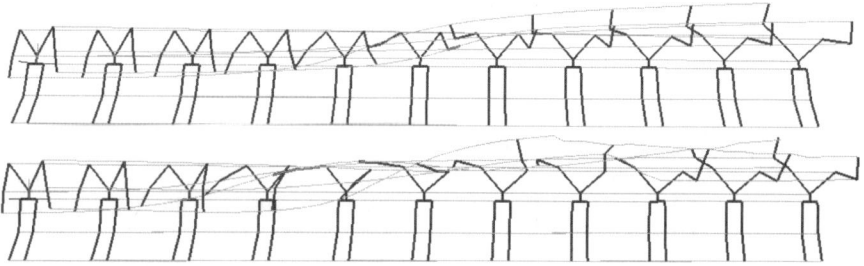

Fig. 7. The synthesized results with Harukoma motion database. Upper: Motion Base, Lower: Motion Base + Motion Style.

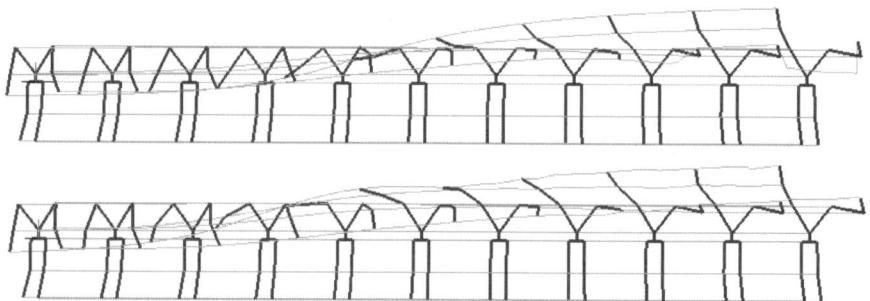

Fig. 8. The synthesized results with Nishi-Monai motion database. Upper: Motion Base, Lower: Motion Base + Motion Style.

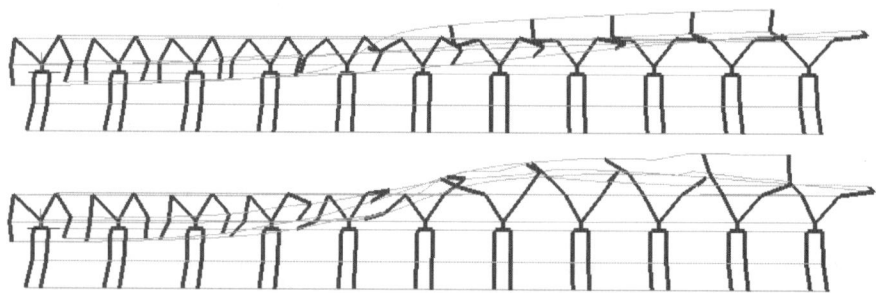

Fig. 9. The synthesized results with Soran-Bushi motion database. Upper: Motion Base, Lower: Motion Base + Motion Style.

Generating Whole Body Motions for a Biped Humanoid Robot from Captured Human Dances

Shinichiro Nakaoka   Atsushi Nakazawa   Kazuhito Yokoi†
Hirohisa Hirukawa †   Katsushi Ikeuchi
Institute of Industrial Science, University of Tokyo
4-6-1 Komaba, Meguro-ku, Tokyo, 153-8505, Japan
nakaoka@cvl.iis.u-tokyo.ac.jp
†Intelligent Systems Institute, National Institute of Advanced Industrial Science and Technology
Tsukuba Central 2, 1-1-1 Umezono, Tsukuba, Ibaraki 305-8568, Japan

Abstract— The goal of this study is a system for a robot to imitate human dances. This paper describes the process to generate whole body motions which can be performed by an actual biped humanoid robot. Human dance motions are acquired through a motion capturing system. We then extract symbolic representation which is made up of primitive motions: essential postures in arm motions and step primitives in leg motions. A joint angle sequence of the robot is generated according to these primitive motions. Then joint angles are modified to satisfy mechanical constraints of the robot. For balance control, the waist trajectory is moved to acquire dynamics consistency based on desired ZMP. The generated motion is tested on OpenHRP dynamics simulator. In our test, the Japanese folk dance, 'Jongara-bushi' was successfully performed by HRP-1S.

I. INTRODUCTION

Traditional dances are considered as intangible cultural assets. However, some of them are disappearing because of a lack of successors. We are attempting to preserve these dances through computer and robotic technology.

The simplest way to preserve the traditional dances is to record their motions through a motion capturing system. However, the recorded data is insufficient for preservation because we cannot watch the actual dances again. Furthermore, it is difficult to master the dances only from the recordings.

This has motivated us to develop a robot system to imitate human dances. Pollard et al.[8] have realized a dance by a humanoid based on captured human motion . But the motion is limited to the upper body and the waist is fixed on a stand. In our study, a robot performs dances including leg motions and is able to maintain its balance independently.

A. Imitation and Behavior Models

It is desirable that humanoid robots can master many kinds of tasks without a burden of complicated programming by a human. This requires the ability to imitate human behaviors just as humans learn from one another. Many researchers have defined an abstract model to recognize and reproduce human behavior. For example, Inamura et al.[2] proposed a general framework for imitating whole body motions.

We hypothesized that a specific model to dance motions is required and the model has two-level structure: *motion primitives* and *styles*. The motion primitives are the high level structure which constructs the motion overview like a musical score. They represent the intentions of the dancer in some sense. The styles express skill or characteristics of motion details. We proposed a method to extract the motion primitives by analyzing trajectories of limbs [5]. Representation based on our model enables various applications such as an adaptive performance to stage condition or creating new sequences of dance action. Also, the model in itself can contribute to dance preservation because extracted information on the model helps people learn the dances.

In this paper, we define *primitive steps* as motion primitives to generate feasible leg motions of the actual robot. The motion primitives of arms are used as *essential postures* to express the characteristics of the original dances.

B. Balance Control during Complex Motions

Recently the development of biped humanoid robots has been active and their walking ability has been advanced [1]. Kajita et al.[3] proposed the generation method of stable walking patterns.

In addition to simple walking, these robots must have the potential to perform more complex motions with the whole body. However, a complex motion such as dances is difficult to maintain balance.

To realize these motions, one reasonable approach is to create an initial motion and to transform it to keep balance [9][10]. This process should preserve the characteristics of the original motion particularly for dances. For this, the important factors of the characteristics in the dance motions must be clarified.

II. OVERVIEW OF THE PROPOSED SYSTEM

A robot motion is generated from human dance performance through a series of steps. Figure1 depicts the

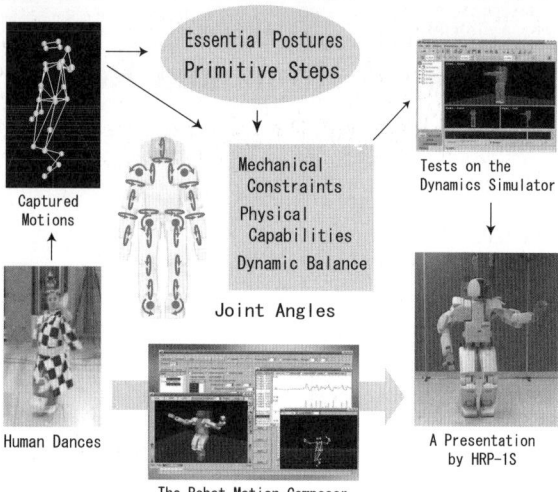

Fig. 1. Overview of the System

overview of our proposed system. We currently focus on Japanese folk dances, 'Jongara-bushi' (Fig.2).

Dance motions are acquired by an optical type motion capturing system made up of eight cameras. A dancer has 30 markers on his or her joints, and marker positions are recorded at the rate of 200 frames per second. Then marker positions are analyzed to extract primitive motions: the essential postures and the step primitives.

The robot is moved according to the joint angle trajectories. The trajectories are generated from marker positions and motion primitives. The generating process for arm and leg motions are different. Arm joint angles are mainly calculated by inverse kinematics of markers and leg joint angles are mainly generated as patterns from step primitives.

The generated angles are not under the constraints of the robot and they tend to have inconsistent balance for

Fig. 2. Jongara-Bushi

dynamic forces. Therefore the angle trajectories must be modified to solve such problems. Information of essential postures is used to express dance characteristics.

The motion is tested on OpenHRP dynamics simulator [4] to verify its validity. Last of all, the motion is carried out on a real humanoid robot HRP-1S [11].

We have developed the Robot Motion Composer, which can automate the generating process. Users can interactively change the motion parameters and create new motion sequence which can be verified by animation. Information of a target robot can be given in the same format as the OpenHRP model file.

III. GENERATING JOINT ANGLE TRAJECTORIES

A. Constraints of the Robot

Humanoid robots are supposed to have the body structure similar to that of human. For example, HRP-1S has 7-DOF arms, 6-DOF legs and a 1-DOF neck; this seems sufficient for human-like movement. However, the structure has considerable constraints for dance motions, because many dances include motions such as torso twists, swings of arms in wide arcs, etc. For some parts of motions, adequate DOF may not be available. Also the robot structure has singular points. For example, the shoulder joint is constructed of a three rotation sequence of pitch, yaw and roll. When the robot raises its arm horizontally, the DOF of the shoulder joint decreases and the robot cannot freely change the direction of the arm from that position.

One approach to avoid the problems of the mechanical structure is to increase the number of the joints. But it would be not easy to attach the new joint mechanism and actuator because of the lack of space within the body. Another approach is to develop a new mechanism. Okada et al. proposed the Cybernetic Shoulder which has no singular point on simple mechanism [7].

The capacity of actuators is another constraint. Dance motions often require quicker motions than walking or everyday tasks. Although HRP-1S has the most powerful actuators currently available, it is not sufficient for most dances. More powerful actuators or lighter weight body are necessary.

The hardware development to eliminate these constraints is important. However, we are now concentrating on realizing a seemingly good motion by using the current robots.

B. Arm Joint Angles

In arms, the initial value of the joint angle trajectories are calculated directly from position correlations with related joint markers. Angles and angular velocities of the initial value must be limited within possible ranges of the robot. The initial motion may imply posture which is in the neighborhood of singular points. At the posture

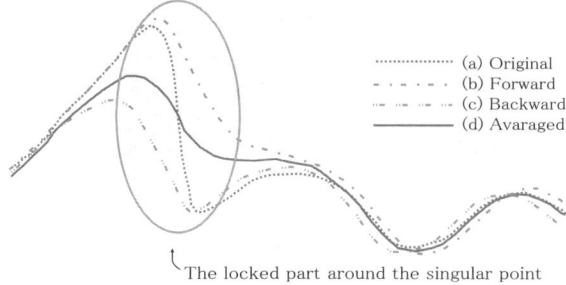

Fig. 3. The limiter of the joint angular velocity and a trajectory around singular point

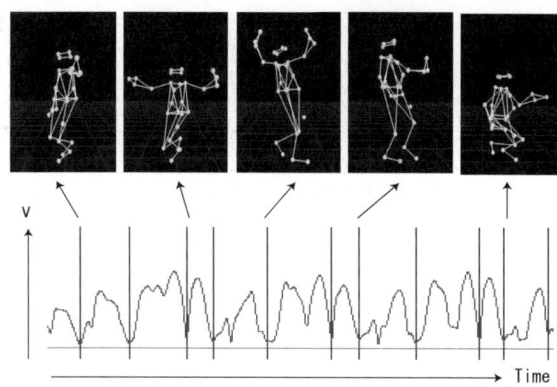

Fig. 4. Motion segments and stopping postures. The graph shows a speed of a hand movement. Vertical lines show segment boundaries.

near singular point, the valid moving patterns are limited and a movement may be locked. This problem can be considered as a problem of angular velocity because a non-continuous curve of velocity around the locked frame can be considered to be a high gradient curve on the discrete system.

Pollard et al. [8] proposed a method to limit angular velocities. In their method, the new angle trajectory is created by following the original trajectory under the velocity limits. This process is formulated as follows.

$$\dot{\theta}_i = \theta_i - \theta_{i-1}, \quad (1)$$
$$\ddot{\theta}'_{i+1} = 2\sqrt{K_s}(\dot{\theta}_i - \dot{\theta}'_i) + K_s(\theta_i - \theta'_i), \quad (2)$$
$$\dot{\theta}'_{i+1} = max(\dot{\theta}_L, min(\dot{\theta}_U, \dot{\theta}'_i + \ddot{\theta}'_{i+1})) \quad (3)$$
$$\theta'_{i+1} = \theta'_i + \dot{\theta}'_{i+1} \quad (4)$$

where θ_i is the original joint angle, θ'_i is new joint angle, i is a joint number, $\dot{\theta}_L$ and $\dot{\theta}_U$ are the lower and upper velocity limits.

This process generates a similar trajectory to the original one within the limit. Consequently, it must delay from the original one all through the motion (Fig.3-b). Then another trajectory is created by the inverse process of the above equations from the end to the start point, a result becomes the trajectory of the future instance compared to the original one (Fig.3-c). Finally, both trajectories are averaged to get a trajectory whose shape is overlapped with the original one. (Fig.3-d). This trajectory preserves the characteristics of the original one well. The locked parts around singular points in the initial motion is also changed into a smooth curve.

C. Clarifying the Essential Postures

We analyzed the trajectory of the hands and the arm motion is segmented at the frames in which the speed of hands is approximately zero [5]. At these frames, the dancer makes particular postures. These postures are the important, essential ones, because the set of these postures is the unique representation of a particular dance. Hence, the dancer pauses to show it clearly.

However, in the modified trajectories, the essential postures becomes ambiguous, because the adapting process separately changes the stopping time of each joint. As a result, the motion becomes obscured and it does not resemble to the original one. For a better representation, it is important to clearly stop the motions at the essential postures.

Figure 4 shows the joint angle trajectories of the original and the modified motion. To clarify the essential postures, the extreme points (where velocity is zero) of all the joint should be arranged at the segment boundary within the possible range (Fig.5). This process is as follows.

1) In each segment boundary, find the nearest extreme point.
2) For the point far from the boundary,
   - Slide the point horizontally to the boundary, if new gradient is under the velocity limit.
   - Otherwise, slide the point vertically to the level which is under the limit.
   - connect the trajectory to the new extreme point by stretching the original trajectory.

D. Leg Pattern Generation

The generating process of leg motions is different from that of the arm motions. First, symbols of the motion primitives are extracted from the original motion. Then all leg motions are created from patterns associated with the primitives.

One of the reasons for using primitives is the limitation of leg movement in robots. In the actual robots, the constraints of the leg motions in dances is more stricter than that of the arm motion. Humans often use toe support, which is incapable of our robot. The robot sole must be completely flat against the ground when it contacts. Support legs cannot expand sufficiently without toe support. Because of the articulatio coxa constraint, the robot cannot cross its legs adequately. These constraints

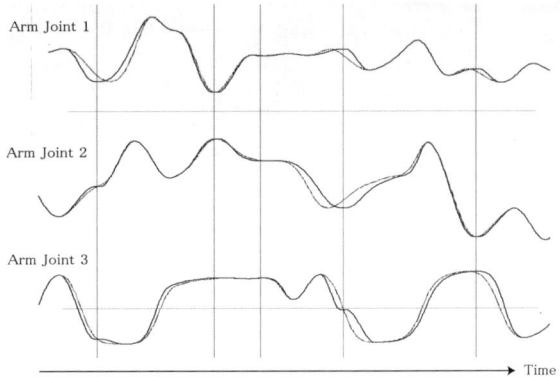

Fig. 5. Arrangement of the extreme points near by the segment boundaries (vertical lines). Dotted lines are the original trajectories and solid lines are arranged ones.

STAND SQUAT STEP

	Parameters of the primitive
STAND	A previous state is maintained
SQUAT	• The time and the waist height at the deepest posture • The period of the action
STEP	• Which is the swing foot ? • The time at the highest foot posture • The highest position and attitude of the swing foot • The position and the attitude of the swing foot when it lands on the ground. (The positions and the attitudes is relative to the waist coordinate.)

Fig. 6. Motion primitives of the legs

make it unsuitable to generate the leg motions directly from the captured ones.

Therefore, in order to satisfy the limitation of leg motions, a feasible motion is created from the motion primitives instead of the raw motion information. This approach generates proper motions according to the capacity of a robot. In addition, this makes it possible to perform the dance interactively, or adaptively in various stage conditions. Essential of this approach is to achieve a performance with intentions.

In Jongara-bushi, the movement of the legs is segmented into actions and the actions are classified into three primitive symbols; STAND, SQUAT and STEP. Figure 6 shows these primitives. Primitive symbols are extracted by analyzing sole and waist positions. The extraction process is as follows.

From the height of toe and heel markers, we can recognize whether the foot is supporting the body on the ground (*support state*) or floating for the next step (*swing state*). The motion is segmented at which the foot state is changed and each segment is classified. In each segment, if the body is supported by one leg, primitive is STEP. If the position of the waist is low in the segment, primitive is SQUAT. Otherwise, it is STAND. Primitive parameters are determined from the position and the attitude of the foot at the segment boundary, the highest position of a swing leg, the lowest waist position, etc.

A leg motion is initially created as a trajectory of foot position and a transition of foot attitude. The position and attitude are based on the waist coordinate. In the STEP primitive, the trajectory is created by interpolating through the initial, peak and the final state. The sole attitude is constrained to lie flat against the ground when the foot is on the ground. In the SQUAT primitive, the trajectory is created by interpolating the height of the waist position. Then joint angles of each frame are calculated with inverse kinematics of the foot trajectory.

IV. BALANCE CONTROL

The robot has the capacity to take postures along the motion generated through the above process. However, the robot is unable to keep its balance when it performs the motion standing by itself on the ground. Because the generated motion does not always satisfy the dynamic consistency in interaction with the ground. This section describes the method of motion modification for the balance control.

A. Dynamic Balance and Zero Moment Point

Leg motion is generated under the assume that all the area of a foot sole contacts with the floor when it is supporting the body. In other words, a sole does not rotate during that time. In terms of dynamics, this assume is satisfied when the point at which the moment to the robot body is zero exists in the area of the sole surface. In this time, the sole does not rotate. The point is called 'zero moment point (ZMP)' and the area is called *supporting area*. If a robot is supported by both feet, supporting area corresponds to the convex area which consists of both soles.

Given the physical model of a robot, a trajectory of ZMP can be calculated from motion data for the robot, under the assume that the supporting area is infinite. If ZMP moves out of an actual supporting area, the motion is impossible to perform because the actual motion must imply rotation of the supporting sole at that time, so that the sole moves away from the ground and the robot falls down.

Therefore the motion must be modified to keep ZMP in the supporting area. It is necessary to create a modified ZMP trajectory which is always in the supporting area and acquire knowledge of how to modify the original motion to realize the desired ZMP.

ZMP is calculated from the motion of all the robot body segments. Following equation shows the calculation on x-axis. (The calculation is separately performed on each axis.)

$$x_{zmp} = \frac{\sum m_i z_i \ddot{x}_i - \sum \{m_i(\ddot{z}_i + g)x_i + (0,1,0)^T \mathbf{I}_i \dot{\omega}_i\}}{-\sum m_i(\ddot{z}_i + g)}$$

where x_{zmp} is the position of ZMP on x axis, x_i, z_i are the position of each segment, m_i is the mass, I_i is the inertia tensor ω_i is the vector of angular velocity and g is gravitational constant.

Under this equation, we need a modification of segment motions from the modification of ZMP. But this problem is difficult just as it is. Nishiwaki et al. [6] proposed a method to solve this problem. On discrete system, supposing all the segments are restricted to be translated horizontally in the same distance, the following equation is acquired.

$$x_{zmp}^e(t_i) = \frac{-hx^e(t_{i+1}) + (2h + g\Delta t^2)x^e(t_i) - hx^e(t_{i-1})}{g\Delta t^2}$$

where x_{zmp}^e is a difference between the original ZMP and desired ZMP, x^e is a difference between the position of original segment and the modified position, t_i is the time at frame i, h is the height of the center of mass, Δt is time per one frame. x^e is calculated from x_{zmp}^e as tridiagonal simultaneous liner equations.

In practice, the proposed translation can be approximated to an upper body translation. This method is easily converge in iteration. Although a modification is limited to horizontal translation of upper body, this method is effectively control the balance.

B. Creating a ZMP Trajectory

It is fundamental that the desired ZMP is inside the supporting area. If a supporting area remains on one state, ZMP should remain a stable point in the area such as a center of the area or just below the ankle joint. In practice, supporting state changes with steps. A stability of the motion depends on a ZMP trajectory with state transitions. For a stable transition, the ZMP trajectory should be as smooth as possible. In this study, we applied the following criteria.

- In STEP period, ZMP must locate at center of a supporting sole.
- In STAND period, ZMP moves from a previous position to the next supporting position by third order polynomial equation. Initial velocity and accelerations and final ones are kept zero.
- If period of STAND is long, transition is separated into three steps: (1) from a previous position to center of supporting area, (2) stay there and (3) move to the next supporting position.
- If period of STAND state is short, ZMP movement speeds up and robot motion becomes unstable. Adequate transition time is required for stable motion. In this case, ZMP movements is expanded so that it starts in the previous state and extends into the next state. Acceleration and deceleration of ZMP is done in those states.

From above method, we generate a desired ZMP trajectory and modify the trajectory of the upper body position. Figure 7 shows a sequence of support state and ZMP trajectory.

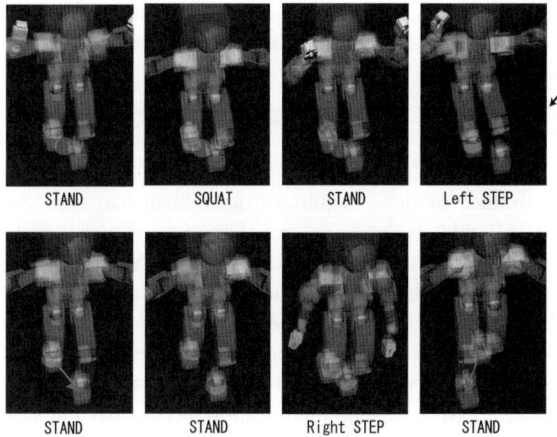

Fig. 7. A motion with support state transition. A markers on the feet shows ZMP.

C. Dynamic Stability of Arm Motions

Pollard method in Section 3 can also control angular acceleration by constant Ks in equation (2). The higher Ks is, the bigger acceleration average becomes. The response to the original motion is better with higher Ks. On the contrary, when Ks is smaller, response is worse but dynamics stability is better. When arms move in wide arc with high acceleration, the robot will be unstable. If the leg motion is in unstable state at that time, balance control is difficult. In this case, Ks around such arm actions should be small to restrict ZMP movement. On the contrary, when a leg action is in stable state, Ks can be increased for better imitation.

V. EXPERIMENTS

We tested the generated motions on the OpenHRP platform. On this platform, both a real and a virtual robot can be controlled by the same software. Since we currently focus on generating the feasible motion patterns in advance, we used the default controller which requires a sequence of joint angles and desired ZMP, and just plays that motion.

On simulation, the virtual robot can perform the motion keeping balance along the entire sequence. However, a support foot slides and the motion becomes unstable when the robot widely rotates the waist to turn. Balance modification does not consider horizontal (yaw) element of moment, which is not concerned with ZMP. Balance modification should be improved to solve this behavior.

Then we tested the motion on the real robot HRP-1S. As the first step, we prepared the motion in which leg action is restricted within squats. In an initial experiment, arms could not follow the given motion and became unstable. We had to reduce the speed of the motion as 2.5 times slow throughout the whole sequence. Then the robot could perform the whole motion standing on its legs in stable (Fig. 8).

VI. CONCLUSION

This paper described the system for robots to perform dance motions acquired from human dancers. Our system could generate the robot motions which satisfy the mechanical constraints and dynamics consistency. We realized the dance performance by the humanoid robot standing on its legs. It has been confirmed that the step primitives are valid for generating feasible leg motions from complex motions such as dances. However, we should improve balance modification to consider the yaw rotation so that more stable motion is acquired. Then we are going to test the motion including step actions on the actual robot.

VII. REFERENCES

[1] Kazuo Hirai, Masato Hirose, Yuji Haikawa, and Toru Takenaka. The development of honda humanoid robot. *Proceedings of International Conference on Robotics and Automation*, May 1998.

[2] Tetsunari Inamura, Iwaki Toshima, and Yoshihiko Nakamura. Acquisition and embodiment of motion elements in closed mimesis loop. *Proceedings of International Conference on Robotics and Automation*, 2002.

[3] Shuuji Kajita, Fumio Kanehiro, Kenji Kaneko, Kazuhito Yokoi, and Hirohisa Hirukawa. The 3d linear inverted pendulum mode: A simple modeling biped walking pattern generation. *Proceedings of International Conference on Intelligent Robots and Systems*, 2001.

[4] Fumio KANEHIRO, Kiyoshi FUJIWARA, Shuuji KAJITA, Kazuhito YOKOI, Kenji KANEKO, Hirohisa HIRUKAWA, Yoshihiko NAKAMURA, and Katsu YAMANE. Open architecture humanoid robotics platform. *Proceedings of International Conference on Robotics and Automation*, 2002.

[5] Atsushi Nakazawa, Shinichiro Nakaoka, and Katsushi Ikeuchi. Imitating human dance motions through motion structure analysis. *Proceedings of International Conference on Intelligent Robots and Systems*, 2002.

[6] Koichi NISHIWAKI, Satoshi KAGAMI, Yasuo Kuniyoshi, Masayuki INABA, and Hirochika INOUE. Online generation of humanoid walking motion based on a fast generation method of motion pattern that follows desired zmp. *Proceedings of International Conference on Intelligent Robots and Systems*, 2002.

[7] Masafumi OKADA, Yoshihiko NAKAMURA, and Shin ichiro HOSHINO.

[8] Nancy S. Pollard, Jessica K.Hodgins, Marcia J. Riley, and Christopher G. Atkeson. Adapting human motion for the control of a humanoid robot. *Proceedings of International Conference on Robotics and Automation*, 2002.

[9] Seyoon Tak, Oh young Song, and Hyeong-Seok Ko. Motion balance filtering. *Proceedings of EUROGRAPHICS*, 19(3), 2000.

[10] Katsu Yamane and Yoshihiko NAKAMURA. Dynamics filter - concept and implementation of on-line motion generator for human figures. *Proceedings of International Conference on Robotics and Automation*, 2000.

[11] Kazuhito Yokoi, Fumio Kanehiro, Kenji Kaneko, Kiyoshi Fujiwara, Shuji Kajita, and Hirohisa Hirukawa. A honda humanoid robot controlled by aist software. *Proceedings of the IEEE-RAS International Conference on Humanoid Robots*, 2001.

Fig. 8. A Performance of Jongara-bushi by HRP-1S

Improving camera displacement estimation in eye-in-hand visual servoing: a simple strategy

Graziano Chesi*, Koichi Hashimoto

Abstract— The problem of estimating the camera displacement in eye-in-hand visual servoing is considered, and a simple strategy based on the idea that the estimates accuracy can be improved if the fact that the point correspondences used throughout the visual servoing are relative to the same 3D points is taken into account is presented. In particular, an accurate scaled euclidean reconstruction of the object is built in the first steps of the visual servoing by suitably using existing linear methods, and from this reconstruction the camera displacement is suitably estimated. Extensive proves performed in random conditions have shown that the proposed approach provides significantly better results with respect to the existing linear methods actually used in visual servoing.

Index Terms— Visual servoing, Point correspondences, Camera displacement estimation.

I. INTRODUCTION

Eye-in-hand visual servoing systems consist of a feedback control based on the view of an object acquired by a vision system, usually one camera, mounted on the robot end effector. Although some standard visual servoing approaches require a priori the knowledge of the 3D model of the observed object, such a knowledge is not always available especially if the visual servoing has to be performed in a dynamic framework. For this reason model-free control algorithms have been developed which drive the camera just by comparing the current view of the object with its desired view, and are known as "teaching-by-showing" visual servoing approaches (see for example [1], [2]).

In most of these visual servoing approaches the control law is computed on the basis of the Camera Displacement (CD) existing between current and desired position. This is the case of position-based visual servoing (pure and not pure, see for example [3], [4], [5]) where the feedback error is defined directly in terms of the rotational and translational components of the displacement, of image-based visual servoing (pure and not pure, see for example [6], [7], [8], [9]) where the dispacement is needed to calculate the image Jacobian which defines the control law, of hybrid methods as [10], [11] where the dispacement is needed to calculate feedback error and other parameters of the control law, and of the recent developed method [12] invariant to camera intrinsic parameters where the knowledge of the CD is needed to compute the Jacobian and consequently the control law. The CD is computed from point correspondences that unavoidably are affected by measurement uncertainties. This means that the CD is not exactly known, and its estimation error obviously influences the stability and performance of the visual servoing algorithm.

In visual servoing the CD is computed from point correspondences in two main ways: through the essential matrix which is usually estimated through the eight point algorithm (see for example [13]), and through the homography matrix relative to a virtual plane which can be estimated with the methods proposed in [14], [15], [16]. These algorithms, called also linear methods, provide almost closed-form solutions for the CD estimate and can be executed at video rate, necessary requirement for visual servoing application. Other algorithms, called nonlinear methods, have been proposed in the computer vision area for obtaining more accurate CD estimates, see for example [17], [18], but they require the solution of nonlinear systems and/or some optimization procedures and, therefore, are not suitable to be executed at video rate.

In this paper, a simple strategy for obtaining more accurate CD estimates in visual servoing is presented. The basic idea is that the estimates accuracy can be improved if the fact that the point correspondences used throughout the visual servoing are relative to the same 3D points is taken into account. In order to exploit this structural constraint in a light computational way suitable for video rate applications, a two-phases algorithm is devised as follows. First, a Scaled Euclidean Reconstruction (SER) of the object is estimated using the essential matrix algorithm and the homography matrix algorithm above mentioned. In order to select at each step of the visual servoing the best reconstruction estimate among all the computed ones, an accuracy index based on euclidean congruency is introduced. This procedure is performed during the first steps of the visual servoing when the CD is the largest one and more accurate estimates can be obtained. Second, the CD is linearly estimated from the best reconstruction estimate as the rotation and translation which minimize the quadratic error of the projective equations.

Our approach can be applied in presence of any camera motion, included the pure rotation one in which the epipolar geometry degenerates, with the only assumption that the observed spatial points are not coplanar. Extensive proves performed with random initial camera dispacements, random points configurations and random noise in absence and in presence of image coordinates discretization have shown that the proposed method provides significantly better results with respect to the essential matrix algorithm and the homography matrix algorithm, which are the methods actually used in visual servoing.

(Corresponding author) G. Chesi is a visiting researcher from the Department of Information Engineering, University of Siena, Italy, at the Department of Information Physics and Computing, University of Tokyo, Japan, under a fellowship program of the European Commission. E-mail: chesi@k2.t.u-tokyo.ac.jp

K. Hashimoto is with the Department of Information Physics and Computing, University of Tokyo, Japan. E-mail: koichi@k2.t.u-tokyo.ac.jp

II. PRELIMINARIES

A. Notation

We indicate with $\mathbf{I}_n$ the identity matrix of dimension n, and with $\mathbf{0}_n$ the null vector $n \times 1$. Let $\mathcal{F}^*$ be the absolute reference frame and the desired camera frame. The 3D point $\mathbf{q} = [x, y, z, 1]^T$ (expressed in homogeneous coordinates) projects on $\mathcal{F}^*$ on the point $\mathbf{m}^*$ defined by $d^*\mathbf{m}^* = [\mathbf{I}_3, \mathbf{0}_3]\mathbf{q}$ where d^* is the depth with respect to $\mathcal{F}^*$. Let $\mathcal{F}$ be the current camera frame and let $\mathbf{R}$ and $\mathbf{t}$ be respectively the rotation and translation of $\mathcal{F}$ with respect to $\mathcal{F}^*$. The 3D point $\mathbf{q}$ projects on $\mathcal{F}$ on the point $\mathbf{m}$ defined by $d\mathbf{m} = [\mathbf{R}, \mathbf{t}]\mathbf{q}$ where d is the depth with respect to $\mathcal{F}$ (see Figure 1). The frame points $\mathbf{m}^*$ and $\mathbf{m}$ (expressed in normalized homogeneous coordinates) project on the camera image plane respectively on the points $\mathbf{Km}^*$ and $\mathbf{Km}$ (expressed in pixel homogeneous coordinates) where $\mathbf{K}$ is the intrinsic parameters matrix. Matrix $\mathbf{R}$ can be decomposed in the angle of rotation θ and in the normalized axis of rotation $\mathbf{u}$ through the relation $\mathbf{R} = e^{[\theta \mathbf{u}]_\times}$ where $[\theta \mathbf{u}]_\times$ is the skew symmetric matrix associated to vector $\theta \mathbf{u}$.

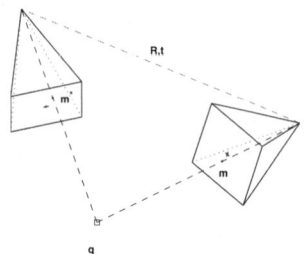

Fig. 1. 3D point $\mathbf{q}$ and corresponding projections $\mathbf{m}^*$ and $\mathbf{m}$.

B. Camera displacement estimation: existing linear methods

We briefly remind how the CD estimation is performed from point correspondences through the essential matrix and through the homography matrix which are the methods actually used in visual servoing (see for example [16] for more details). In the sequel we will compare the results provided by these methods with the ones provided by our approach.

1) Essential Matrix Algorithm (EMA): The first step consists of estimating the fundamental matrix $\mathbf{F}$ through the eight points algorithm from the points expressed in pixel coordinates and suitably normalized in order to avoid ill conditioned numerical problems (see for example [19], [13]). The essential matrix $\mathbf{E}$ is then computed from $\mathbf{F}$ and the intrinsic parameters matrix as $\mathbf{E} = \mathbf{K}^T \mathbf{F} \mathbf{K}$. The rotation $\mathbf{R}$ and the normalized translation $\mathbf{t}/\|\mathbf{t}\|$ are estimated from the relation $\mathbf{E} = [\mathbf{t}]_\times \mathbf{R}$ as described in [20] by minimizing the quadratic error with respect to all translations $\mathbf{t}$ and all rotations $\mathbf{R}$.

This method fails in the case of pure rotation motion since $\mathbf{E}$ vanishes for $\mathbf{t} = \mathbf{0}_3$ and $\mathbf{R}$ cannot be recovered.

2) Homography Matrix Algorithm (HMA): We refer to the algorithm proposed in [16] which presents some improvements with respect to the ones proposed in [14], [15]. The first step consists of estimating the collineation matrix $\mathbf{G}$ relative to a virtual plane defined by three non-collinear points in both images. In order to avoid ill conditioned numerical problems, the choice of these three points is done by selecting the ones maximizing the area of the corresponding triangle. In the coordinate system defined by the selected three points the collineation matrix is diagonal and its three entries are computed by calculating two singular value decompositions which solve a homogeneous system of degree three derived from epipolar constraints on the points not belonging to the virtual plane. The homography matrix $\mathbf{H}$ is then computed from $\mathbf{G}$ and the intrinsic parameters matrix as $\mathbf{H} = \mathbf{K}^{-1} \mathbf{G} \mathbf{K}$ and it is related to the CD by $\mathbf{H} = \mathbf{R} + \mathbf{t}\mathbf{n}^T/d$ where $\mathbf{n}$ is the normal to the virtual plane expressed in $\mathcal{F}^*$ and d is the distance between the virtual plane and the origin of $\mathcal{F}^*$. The rotation $\mathbf{R}$ and the normalized translation $\mathbf{t}/\|\mathbf{t}\|$ are estimated from $\mathbf{H}$ using the algorithms proposed in [21] or [22]. Since in the general case there are two solutions for such parameters, the procedure has to be repeated considering another virtual plane in order to eliminate the indeterminacy, choosing after the common solution between the two pairs.

In the case of pure rotation, $\mathbf{H}$ is still defined contrarily to $\mathbf{E}$ in the previous method. Nevertheless, the matrix used to estimate the diagonal entries of the collineation in the new coordinate system loses rank and these three unknowns have to be recovered with a different approach.

III. ESTIMATION STRATEGY

Our approach for obtaining better estimates of the CD with respect to the existing linear methods is based on the idea of exploiting the information that the point correspondences used throughout the visual servoing are relative to the same spatial points. The algorithm consists of two phases, one for estimating the SER of the object and one for estimating the CD from the available reconstruction estimate. We describe the algorithm firstly for the case of not null translation, i.e. generic motion, and secondly for the case of pure rotation motion.

A. Generic motion case

In the first phase of the algorithm we estimate the SER of the object in the desired camera frame. Our strategy consists of selecting, at a generic step of the visual servoing, the best reconstruction estimate among the best reconstruction estimate selected at the previous step and the reconstruction estimates obtained at the present step through the CD estimates provided by the EMA and the HMA. The selection is done through the introduction of a suitable index that measures the accuracy of a given reconstruction estimate. In order to define such an accuracy index let us firstly show how the SER can be estimated from an estimate of the CD. In absence of uncertainty, any observed point $\mathbf{q}_i$, $i = 1, \ldots, n$, can be computed up to the distance between current and desired camera centers from the relations

$$\begin{cases} d_i^* \mathbf{m}_i^* = [\mathbf{I}_3, \mathbf{0}_3] \mathbf{q}_i \\ d_i \mathbf{m}_i = [\mathbf{R}, \mathbf{t}/\|\mathbf{t}\|] \mathbf{q}_i. \end{cases} \quad (1)$$

Eliminating d_i^* and d_i we can rewrite the previous relations as

$$\mathbf{L}(\mathbf{m}_i^*, \mathbf{m}_i, \mathbf{R}, \mathbf{t}) \mathbf{q}_i = \mathbf{0}_4 \quad (2)$$

where $\mathbf{L}(\mathbf{m}_i^*, \mathbf{m}_i, \mathbf{R}, \mathbf{t}) \in \mathbb{R}^{4\times 4}$. In presence of uncertainty on the quantities $\mathbf{m}_i^*, \mathbf{m}_i, \mathbf{R}, \mathbf{t}$ the previous equality may not be satisfied, and a normalized estimate[1] $\hat{\mathbf{q}}_i$ of $\mathbf{q}_i$ can be obtained from the estimates $\hat{\mathbf{m}}_i^*, \hat{\mathbf{m}}_i, \hat{\mathbf{R}}, \hat{\mathbf{t}}$ by solving the trivial linear least-squares minimization

$$\hat{\mathbf{q}}_i \leftarrow \min_{(\mathbf{q}_i)_4 = 1} \left\| \mathbf{L}(\hat{\mathbf{m}}_i^*, \hat{\mathbf{m}}_i, \hat{\mathbf{R}}, \hat{\mathbf{t}}) \mathbf{q}_i \right\| \quad (3)$$

where $(\mathbf{q}_i)_4$ denotes the fourth entry of $\mathbf{q}_i$. Hence, the reconstruction estimate is given by

$$\hat{\mathcal{Q}} = \{\hat{\mathbf{q}}_i, \ i = 1, \ldots, n\}. \quad (4)$$

Let us define the normalized accuracy index $\mu(\hat{\mathcal{Q}})$ of $\hat{\mathcal{Q}}$ as follows:

$$\mu(\hat{\mathcal{Q}}) = \max_{i=1,\ldots,n} \frac{\sigma_m(\mathbf{L}_i(\hat{\mathcal{Q}}))}{\sigma_M(\mathbf{L}_i(\hat{\mathcal{Q}}))} \quad (5)$$

where $\mathbf{L}_i(\hat{\mathcal{Q}})$ is the matrix $\mathbf{L}(\cdot, \cdot, \cdot, \cdot)$ used in (3) to compute the i-th point $\hat{\mathbf{q}}_i$ of $\hat{\mathcal{Q}}$, and $\sigma_m(\cdot)$ and $\sigma_M(\cdot)$ denote respectively the minimum and maximum singular values. The quantity $\mu(\hat{\mathcal{Q}})$ measures the error occurred in the reconstruction process due to the presence of uncertainty and, hence, it is an accuracy index of the reconstruction estimate $\hat{\mathcal{Q}}$ ($\mu(\mathcal{Q})$ is in fact zero if the measurements are not corrupted, and it is strictly positive if the minimum in (3) is greater than zero for some i). By using the index $\mu(\cdot)$ we select, at a generic step of the visual servoing, the best reconstruction estimate $\hat{\mathcal{Q}}_{best}$ among the best one selected at the previous step and the reconstruction estimates $\hat{\mathcal{Q}}_E$ and $\hat{\mathcal{Q}}_H$ obtained from (3)-(4) using the rotation and normalized translation estimates provided respectively by the EMA and the HMA. We compute the reconstruction estimates only during the first steps of the visual servoing since in these steps the CD is the largest one and more accurate estimates can be obtained.

In the second phase of the algorithm we estimate, at each step of the visual servoing, the CD using the current best reconstruction estimate $\hat{\mathcal{Q}}_{best}$. Specifically, the system of equations

$$d_i \mathbf{m}_i = [\mathbf{R}, \mathbf{t}] \mathbf{q}_i, \ i = 1, \ldots, n \quad (6)$$

can be rewritten, eliminating d_i, as

$$\mathbf{M}(\mathbf{m}_1, \mathbf{q}_1, \ldots, \mathbf{m}_n, \mathbf{q}_n) \mathbf{s} = \mathbf{0}_{2n} \quad (7)$$

where $\mathbf{M}(\cdot, \ldots, \cdot) \in \mathbb{R}^{2n\times 12}$ and $\mathbf{s} \in \mathbb{R}^{12}$ contains the entries of the projection matrix $[\mathbf{R} \ \mathbf{t}]$. Let us indicate with $\mathbf{R}(\mathbf{s})$ and $\mathbf{t}(\mathbf{s})$ the rotation and translation parameterized by $\mathbf{s}$. Then, given the estimates $\hat{\mathbf{m}}_i$ of the image points and the estimates $\hat{\mathbf{q}}_i$ in $\hat{\mathcal{Q}}_{best}$ for all $i = 1, \ldots, n$, we compute the estimate $\hat{\mathbf{s}}$ of $\mathbf{s}$ by solving the trivial linear least-squares minimization

$$\hat{\mathbf{s}} \leftarrow \min_{\det(\mathbf{R}(\mathbf{s}))=1} \|\mathbf{M}(\hat{\mathbf{m}}_1, \hat{\mathbf{q}}_1, \ldots, \hat{\mathbf{m}}_n, \hat{\mathbf{q}}_n)\mathbf{s}\|. \quad (8)$$

Let us observe that, in presence of uncertainty, the so computed $\hat{\mathbf{s}}$ does not identify a true projection matrix in the general

---

At the k-th step of the visual servoing:
1) if $k < k_{lim}$ then:
   a) compute $\hat{\mathcal{Q}}_E$ and $\hat{\mathcal{Q}}_H$ from (3)-(4) using the rotation and normalized translation estimates provided respectively by the EMA and the HMA;
   b) if $\hat{\mathcal{Q}}_{best}$ has not been defined yet then $\hat{\mathcal{Q}}_{best} \leftarrow \min\{\mu(\hat{\mathcal{Q}}_E), \mu(\hat{\mathcal{Q}}_H)\}$, otherwise $\hat{\mathcal{Q}}_{best} \leftarrow \min\{\mu(\hat{\mathcal{Q}}_{best}), \mu(\hat{\mathcal{Q}}_E), \mu(\hat{\mathcal{Q}}_H)\}$;
2) compute $\hat{\mathbf{R}}_I$ and $\hat{\mathbf{t}}_I$ from (8), (9) and (10).

TABLE I

Camera displacement estimation algorithm in the generic motion case.

case since $\mathbf{R}(\hat{\mathbf{s}})$ may not be a rotation matrix. Therefore, we define our estimates $\hat{\mathbf{R}}_I$ and $\hat{\mathbf{t}}_I$ of the CD as

$$\hat{\mathbf{R}}_I \leftarrow \min_{\mathbf{R} \text{ is a rotation matrix}} \|\mathbf{R}(\hat{\mathbf{s}}) - \mathbf{R}\| \quad (9)$$

$$\hat{\mathbf{t}}_I = \mathbf{t}(\hat{\mathbf{s}}). \quad (10)$$

The constrained least-squares minimization in (9) defining $\hat{\mathbf{R}}_I$ admits a closed form solution that requires just a singular value decomposition (see [20] for details). Notice that the so defined $\hat{\mathbf{R}}_I$ is the closest rotation matrix to $\mathbf{R}(\hat{\mathbf{s}})$.

Algorithm in Table I summarizes the described algorithm for the estimation of the CD. The quantity k_{lim} has been introduced in order to compute the reconstruction estimate only in the first steps of the visual servoing.

B. Pure rotation motion case

In this case we cannot reconstruct the object in the euclidean space up to a scale factor as described Section III-A (i.e., one common scale factor for all points of the object) since the translation between current and desired camera center is null. We can however compute the direction in the euclidean space of each observed point. Specifically, in absence of uncertainty, we can write for any observed point $\mathbf{q}_i = [x_i, y_i, z_i, 1]^T$

$$\begin{cases} d_i^* \mathbf{m}_i^* = \mathbf{I_3}[x_i, y_i, z_i]^T \\ d_i \mathbf{m}_i = \mathbf{R}[x_i, y_i, z_i]^T. \end{cases} \quad (11)$$

Eliminating d_i^* and d_i we can rewrite the previous relations as

$$\tilde{\mathbf{L}}(\mathbf{m}_i^*, \mathbf{m}_i, \mathbf{R})[x_i, y_i, z_i]^T = \mathbf{0}_4 \quad (12)$$

where $\tilde{\mathbf{L}}(\mathbf{m}_i^*, \mathbf{m}_i, \mathbf{R}) \in \mathbb{R}^{4\times 3}$. In presence of uncertainty on the quantities $\mathbf{m}_i^*, \mathbf{m}_i, \mathbf{R}$ the previous equality may not be satisfied, and a normalized estimate[2] $[\hat{x}_i, \hat{y}_i, \hat{z}_i]^T$ of $[x_i, y_i, z_i]^T$ can be obtained from the estimates $\hat{\mathbf{m}}_i^*, \hat{\mathbf{m}}_i, \hat{\mathbf{R}}$ by solving the trivial linear least-squares minimization

$$\begin{bmatrix} \hat{x}_i \\ \hat{y}_i \\ \hat{z}_i \end{bmatrix} \leftarrow \min_{x_i^2+y_i^2+z_i^2=1} \left\| \tilde{\mathbf{L}}(\hat{\mathbf{m}}_i^*, \hat{\mathbf{m}}_i, \hat{\mathbf{R}}) \begin{bmatrix} x_i \\ y_i \\ z_i \end{bmatrix} \right\|. \quad (13)$$

Hence, the reconstruction estimate is given by

$$\hat{\mathcal{D}} = \{[\hat{x}_i, \hat{y}_i, \hat{z}_i]^T, \ i = 1, \ldots, n\}, \quad (14)$$

---

[1] Vector $\hat{\mathbf{q}}_i$ is normalized since estimated from $\mathbf{q}_i$ up to the distance between current and desired camera centers.

[2] Since only the direction of $[x_i, y_i, z_i]^T$ can be estimated, we define $[\hat{x}_i, \hat{y}_i, \hat{z}_i]^T$ as an unit vector.

> At the k-th step of the visual servoing:
> 1) if $k < k_{lim}$ then:
>   a) compute $\hat{\mathcal{D}}_H$ from (13)-(14) using the rotation estimate provided by the HMA;
>   b) if $\hat{\mathcal{D}}_{best}$ has not been defined yet then $\hat{\mathcal{D}}_{best} = \hat{\mathcal{D}}_H$, otherwise $\hat{\mathcal{D}}_{best} \leftarrow \min\{\tilde{\mu}(\hat{\mathcal{D}}_{best}), \tilde{\mu}(\hat{\mathcal{D}}_H)\}$;
> 2) compute $\hat{\mathbf{R}}_I$ from (18) and (19).

TABLE II

Camera displacement estimation algorithm in the pure rotation motion case.

and, analogously to Section III-A, the normalized accuracy index $\tilde{\mu}(\hat{\mathcal{D}})$ of $\hat{\mathcal{D}}$ is defined as:

$$\tilde{\mu}(\hat{\mathcal{D}}) = \max_{i=1,\ldots,n} \frac{\sigma_m(\tilde{\mathbf{L}}_i(\hat{\mathcal{D}}))}{\sigma_M(\tilde{\mathbf{L}}_i(\hat{\mathcal{D}}))} \quad (15)$$

where $\tilde{\mathbf{L}}_i(\hat{\mathcal{D}})$ is the matrix $\mathbf{L}(\cdot,\cdot,\cdot)$ used in (13) to compute the i-th point $\hat{\mathcal{D}}$. By using the index $\tilde{\mu}(\cdot)$ we select, at a generic step of the visual servoing, the best reconstruction estimate $\hat{\mathcal{D}}_{best}$ among the best one selected at the previous step and the reconstruction estimate $\hat{\mathcal{D}}_H$ obtained from (13)-(14) using the rotation estimate provided by the HMA (in fact, in the pure rotation motion case the essential matrix vanishes and hence the EMA cannot be used to estimate the rotation).

From the current best reconstruction estimate $\hat{\mathcal{D}}_{best}$ we estimate the rotation as follows. From the system of equations

$$d_i \mathbf{m}_i = \mathbf{R}[x_i, y_i, z_i]^T, \, i = 1, \ldots, n \quad (16)$$

eliminate d_i and obtain

$$\tilde{\mathbf{M}}(\mathbf{m}_1, [x_1, y_1, z_1]^T, \ldots)\mathbf{w} = \mathbf{0}_{2n} \quad (17)$$

where $\tilde{\mathbf{M}}(\cdot, \ldots, \cdot) \in \mathbb{R}^{2n \times 9}$ and $\mathbf{w} \in \mathbb{R}^9$ contains the entries of the $\mathbf{R}$. Let us indicate with $\mathbf{R}(\mathbf{w})$ the rotation parameterized by $\mathbf{w}$. Then, analogously to Section III-A, given the estimates $\hat{\mathbf{m}}_i$ of the image points and the estimates $[\hat{x}_i, \hat{y}_i, \hat{z}_i]^T$ in $\hat{\mathcal{D}}_{best}$, we compute the estimate $\hat{\mathbf{w}}$ of $\mathbf{w}$ by solving

$$\hat{\mathbf{w}} \leftarrow \min_{\det(\mathbf{R}(\mathbf{w}))=1} \left\| \tilde{\mathbf{M}}(\hat{\mathbf{m}}_1, [\hat{x}_1, \hat{y}_1, \hat{z}_1]^T, \ldots)\mathbf{w} \right\|, \quad (18)$$

and from this we calculate our estimate $\hat{\mathbf{R}}_I$ of $\mathbf{R}$ analogously to (9):

$$\hat{\mathbf{R}}_I \leftarrow \min_{\mathbf{R} \text{ is a rotation matrix}} \|\mathbf{R}(\hat{\mathbf{w}}) - \mathbf{R}\|. \quad (19)$$

Algorithm in Table II summarizes the described algorithm for the estimation of the CD.

C. Why should this strategy work better than the other linear methods?

The reason is that our approach takes into account the information that the point correspondences used throughout the visual servoing to estimate the CD are relative to the same spatial point, while the EMA and the HMA do not (in other words our approach exploit more information). In order to understand quantitatively the difference between these methods, let us consider for example the generic motion case. Once that a reconstruction estimate of the object has been obtained we are in position to estimate the CD using only 6 points (since matrix $\mathbf{M}(\cdot, \ldots, \cdot)$ has dimension $2n \times 12$) while the other linear methods require at least 8 points. Obviously this difference allows us to obtain more accurate estimates. Moreover, by introducing a suitable index that measures the accuracy of a given reconstruction estimate, we are able to single out an accurate one, of course relatively to the level of the present noise.

D. Summarizing remarks

First, our approach requires the same number n of observed points required by the EMA and the HMA since we exploit these algorithms to build a reconstruction estimate of the object and since we do not require a greater number for estimating the CD from the reconstruction estimate. In particular, at least $n = 8$ points in the generic motion case and $n = 4$ points in the pure rotation motion case are required.

Second, we exclude the special case in which the observed points are coplanar since the matrices $\mathbf{M}(\cdot, \ldots, \cdot)$ and $\tilde{\mathbf{M}}(\cdot, \ldots, \cdot)$ lose rank.

Third, the computational burden of the proposed approach is quite low. Indeed, let us consider the generic motion case. In the first phase of the algorithm we execute the EMA and the HMA plus $2n$ singular value decompositions of 4×4 matrices (for computing the index $\mu(\cdot)$). In the second phase we just compute one singular value decomposition with dimensions $2n \times 12$ (for computing $\hat{\mathbf{s}}$) and one with dimensions 4×4 (for computing $\hat{\mathbf{R}}_I$). Moreover, the first phase is executed only at the beginning of the visual servoing.

IV. SIMULATION RESULTS

In this section we compare the accuracy of the CD estimates provided by our approach with the ones provided by the EMA and the HMA summarized in Section II-B. In order to obtain a reliable comparison between these methods we have proceeded as follows.

First, we have generated 100 visual servoing camera trajectories in absence of noise for both generic motion and pure rotation motion cases by applying, respectively, image-based visual servoing and position-based visual servoing [3]. Each trajectory is characterized by a different observed object, each of them composed of a cloud of 12 points randomly generated in a cube with 30 cm edge. The desired camera center is located at 75 cm from the cube center and the orientation axes of the desired camera frame coincide with the cube ones. The image size is 320×200 and the intrinsic parameters matrix is chosen as $\mathbf{K} = [320, 0, 160; 0, 200, 100; 0, 0, 1]$ which ensures that the object lies in the field of view of the camera in the desired frame. The initial camera frame of each trajectory is then randomly selected with respect to the desired camera one under the constraints that the object has to lie also in the field of view of the camera in this frame and that the angle of

---

[3]Although it is intuitive that the kind of visual servoing does not affect the comparison we are going to do, we have selected two different algorithms for showing more general results. Observe that in the pure rotation motion case we cannot use image-based visual servoing since this control law generates a translation motion even if the translation between initial and desired frame is null.

rotation between initial and desired frame is not greater than 120 degrees. We have introduced this last constraint on the initial CD in order to minimize the probability that the object may get out from the field of view during the visual servoing (since image-based and position-based visual servoing present this drawback), although this does not constitute any problem to the convergence of the visual servoing in simulation stages.

Second, for each trajectory we have performed, for three different levels of image noise and for three different numbers of observed points, 50 times the CD estimation along the camera path corrupting each time the image point coordinates by random noise under the selected noise level. The noise levels, that is the maximum amplitudes of the random noise added to each image coordinate, are chosen as 0.25 pixels (noise level $N1$), 0.50 pixels ($N2$) and 0.50 pixels followed by coordinates discretization ($N3$). The numbers of observed points are chosen as 8, 10 and 12.

The indexes used to compare the CD estimation algorithms are the rotation error e_R and translation error e_t between, respectively, true rotation $\mathbf{R}$ and estimate $\hat{\mathbf{R}}$, and true translation $\mathbf{t}$ and estimate $\hat{\mathbf{t}}$, defined as follows:

- e_R: the distance between $\mathbf{R}$ and $\hat{\mathbf{R}}$, that is the length of the shortest geodesic starting at $\mathbf{R}$ and ending at $\hat{\mathbf{R}}$. This distance is given by the rotation angle of the matrix $\mathbf{R}\hat{\mathbf{R}}^T$;
- e_t: the angle between the normalized vectors $\mathbf{t}/\|\mathbf{t}\|$ and $\hat{\mathbf{t}}/\|\hat{\mathbf{t}}\|$.

We use the notations $e_R(N, n, c, s, i)$ and $e_t(N, n, c, s, i)$ to indicate that the errors depend on the noise level N ($N = N1, N2, N3$), on the number n of observed points ($n = 8, 10, 12$), on the trajectory c of the camera ($1 \leq c \leq 100$), on the step s along the trajectory c, and on the noise simulation i ($1 \leq i \leq 50$). In order to have compact representations of the results we define the following average errors for the rotation (the ones for the translation are analogous):

- $\bar{e}_R(N, n, c, s)$: average of $e_R(N, n, c, s, i)$ over the 50 noise simulations;
- $E_R(N, n, c)$: average of $\bar{e}_R(N, n, c, s)$ over the steps of the trajectory c;
- $\bar{E}_R(N, n)$: average of $E_R(N, n, c)$ over the 100 trajectories.

Finally:

- in our approach we have generically selected $k_{lim} = 10$ in the algorithms of Tables I and II;
- in the HMA we have used the algorithm proposed in [21] to recover the rotation and normalized translation from the homography matrix.

A. Generic motion case

Table IV-A shows the average errors $\bar{E}_R(N, n)$ and $\bar{E}_t(N, n)$ obtained for the chosen noise levels and numbers of observed points. As we can see, the errors provided by our approach are significantly smaller than the ones provided by the other linear methods. Observe again that $\bar{E}_R(N, n)$ and $\bar{E}_t(N, n)$, for each pair (N, n), are averages over a large number of situations (CDs and noise values), in particular over $50 \sum_{c=1}^{100} l(c) = 528700$ situations (for the generic motion

	$n = 8$	$n = 10$	$n = 12$
$N1$	**1.153 ; 18.78** 3.900 ; 36.60 3.239 ; 37.72	**0.454 ; 11.96** 1.454 ; 23.88 1.182 ; 24.41	**0.349 ; 9.708** 1.107 ; 20.84 0.792 ; 19.55
$N2$	**2.251 ; 28.12** 6.142 ; 46.25 5.007 ; 46.81	**0.935 ; 19.72** 2.607 ; 33.33 2.132 ; 34.28	**0.706 ; 16.73** 2.018 ; 30.19 1.472 ; 29.01
$N3$	**3.063 ; 33.65** 7.785 ; 50.80 6.145 ; 50.99	**1.329 ; 24.30** 3.488 ; 38.25 2.802 ; 38.93	**1.180 ; 21.35** 2.725 ; 35.25 1.973 ; 34.00

TABLE III

Average errors $\bar{E}_R(N, n)$ and $\bar{E}_t(N, n)$ (in degrees) for the generic motion case. Each row of the cells has the form $\bar{E}_R$; $\bar{E}_t$. The first row of each cell is relative to our approach (in bold characters), the second row to the HMA and the third row to the EMA.

	$n = 8$	$n = 10$	$n = 12$
$N1$	**1.399** 8.402	**0.6784** 5.965	**0.4720** 5.026
$N2$	**2.326** 13.16	**1.221** 10.05	**0.9037** 8.928
$N3$	**3.011** 15.99	**1.781** 12.69	**1.259** 11.69

TABLE IV

Average error $\bar{E}_R(N, n)$ (in degrees) for the pure rotation motion case. The first row of each cell is relative to our approach (in bold characters), the second row to the HMA.

case) being $l(c)$ the number of steps of the visual servoing along the trajectory c. This clearly suggests that, at least averagely, our approach provides more accurate estimates than the EMA and the HMA.

In order to show some other detail of the generated simulations, let us select for example the case $N = N1$ and $n = 12$. Figure 2 shows the average errors $E_R(N1, 12, c)$ and $E_t(N1, 12, c)$ versus the trajectory c. Figure 3 shows, for a generic trajectory $\bar{c}$, the rotation and translation components of the camera versus the step s of the visual servoing and the image trajectory of the points computed in absence of noise. Figure 4 shows, for the same trajectory $\bar{c}$, the average errors $\bar{e}_R(N1, 12, \bar{c}, s)$ and $\bar{e}_t(N1, 12, \bar{c}, s)$ versus s.

B. Pure rotation motion case

Table IV-B shows the average error $\bar{E}_R(N, n)$ obtained with our approach and with the HMA (as already said, the essential matrix vanishes in the pure rotation case and hence the EMA cannot be used). Also in this case the errors provided by our approach are significantly smaller than the other ones. Analogously to the generic motion case, error $\bar{E}_R(N, n)$ is an average over a large number of situations for each pair (N, n), in particular over $50 \sum_{c=1}^{100} l(c) = 1076100$ situations being $l(c)$ the number of steps of the visual servoing along the trajectory c.

V. CONCLUSION

A simple strategy for estimating the CD in visual servoing has been presented. The proposed approach exploits the fact that the point correspondences used throughout the visual servoing are relative to the same 3D points in order to obtain better estimates with respect to other linear algorithms. In particular, an accurate SER of the object is built in the first steps of the visual servoing by suitably using the EMA and the HMA, and from this reconstruction the CD is suitably estimated. Extensive proves performed with random initial camera dispacements, random points configurations and random noise in absence and in presence of image coordinates discretization have shown that the proposed method provides significantly better results with respect to the EMA and HMA.

REFERENCES

[1] K. Hashimoto, *Visual Servoing: Real-Time Control of Robot Manipulators Based on Visual Sensory Feedback*. Singapore: World Scientific, 1993.
[2] S. Hutchinson, G. Hager, and P. Corke, "A tutorial on visual servo control," *IEEE Trans. on Robotics and Automation*, vol. 12, no. 5, pp. 651–670, 1996.
[3] W. Wilson, C. Hulls, and G. Bell, "Relative end-effector control using cartesian position-based visual servoing," *IEEE Trans. on Robotics and Automation*, vol. 12, no. 5, pp. 684–696, 1996.
[4] C. Taylor and J. Ostrowski, "Robust vision-based pose control," in *Proc. IEEE Int. Conf. on Robotics and Automation*, (San Francisco, California), pp. 2734–2740, 2000.
[5] B. Thuilot, P. Martinet, L. Cordesses, and J. Gallice, "Position based visual servoing: keeping the object in the field of vision," in *Proc. IEEE Int. Conf. on Robotics and Automation*, (Washington, D.C.), pp. 1624–1629, 2002.
[6] K. Hashimoto, T. Kimoto, T. Ebine, and H. Kimura, "Manipulator control with image-based visual servo," in *Proc. IEEE Int. Conf. on Robotics and Automation*, pp. 2267–2272, 1991.
[7] B. Espiau, F. Chaumette, and P. Rives, "A new approach to visual servoing in robotics," *IEEE Trans. on Robotics and Automation*, vol. 8, no. 3, pp. 313–326, 1992.
[8] K. Deguchi, "Optimal motion control for image-based visual servoing by decoupling translation and rotation," in *Proc. Int. Conf. on Intelligent Robots and Systems*, pp. 705–711, 1998.
[9] P. Corke and S. Hutchinson, "A new partitioned approach to image-based visual servo control," *IEEE Trans. on Robotics and Automation*, vol. 17, no. 4, pp. 507–515, 2001.
[10] E. Malis, F. Chaumette, and S. Boudet, "2 1/2 D visual servoing," *IEEE Trans. on Robotics and Automation*, vol. 15, no. 2, pp. 238–250, 1999.
[11] F.-X. Espiau, E. Malis, and P. Rives, "Robust features tracking for robotic applications: Towards 2 1/2 D visual servoing with natural images," in *Proc. IEEE Int. Conf. on Robotics and Automation*, (Washington, D.C.), pp. 574–579, 2002.
[12] E. Malis, "Vision-based control invariant to camera intrinsic parameters: stability analysis and path tracking," in *Proc. IEEE Int. Conf. on Robotics and Automation*, (Washington, D.C.), pp. 217–222, 2002.
[13] O. Faugeras and Q.-T. Luong, *The Geometry of Multiple Images*. Cambridge (Mass.): MIT Press, 2001.
[14] B. Boufama and R. Mohr, "Epipole and fundamental matrix estimation using the virtual parallax property," in *IEEE Int. Conf. on Computer Vision*, (Cambridge, USA), pp. 1030–1036, 1995.
[15] B. Coupael and K. Bainian, "Stereo vision with the use of virtual plane in the space," *Chinese Journal of Electronics*, vol. 4, no. 2, pp. 32–39, 1995.
[16] E. Malis and F. Chaumette, "2 1/2 D visual servoing with respect to unknown objects through a new estimation scheme of camera displacement," *Int. Journal of Computer Vision*, vol. 37, no. 1, pp. 79–97, 2000.
[17] R. Deriche, Z. Zhang, Q.-T. Luong, and O. Faugeras, "Robust recovery of the epipolar geometry for an uncalibrated stereo rig," in *Proc. European Conf. on Computer Vision*, (Stockholm, Sweden), 1994.
[18] Z. Zhang, "Determining the epipolar geometry and its uncertainty - a review," *Int. Journal of Computer Vision*, vol. 27, no. 2, pp. 161–195, 1998.
[19] R. Hartley, "In defence of the 8-point algorithm," in *Proc. Int. Conf. on Computer Vision*, (Santa Margherita Ligure, Italy), pp. 1064–1070, 1995.
[20] J. Weng, T. Huang, and N. Ahuja, "Motion and structure from two perspective views: Algorithms, error analysis, and error estimation," *IEEE Trans. on Pattern Analysis and Machine Intelligence*, vol. 11, no. 5, pp. 451–476, 1989.
[21] O. Faugeras and F. Lustman, "Motion and structure from motion in a piecewise planar environment," *Int. Journal of Pattern Recognition and Artificial Intelligence*, vol. 2, no. 3, pp. 485–508, 1988.
[22] Z. Zhang and A. Hanson, "Scaled euclidean 3D reconstruction based on externally uncalibrated cameras," in *IEEE Symp. on Computer Vision*, (Coral Gables, Florida), 1995.

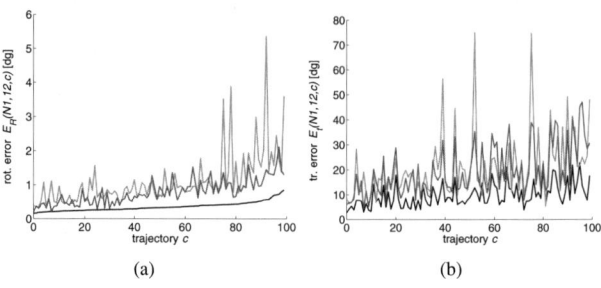

Fig. 2. Errors $E_R(N1,12,c)$ and $E_t(N1,12,c)$ (black: our approach; green: HMA; red: EMA). The trajectories c are ordered so that $E_R(N1,12,c)$ provided by our approach does not decrease with respect to c (in order to facilitate the understanding of the figure).

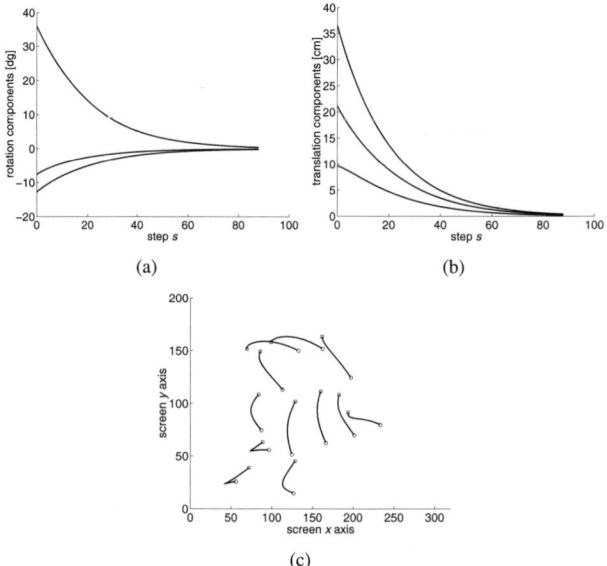

Fig. 3. Rotation components (a), translation components (b) and camera view (c) for trajectory $\bar{c}$.

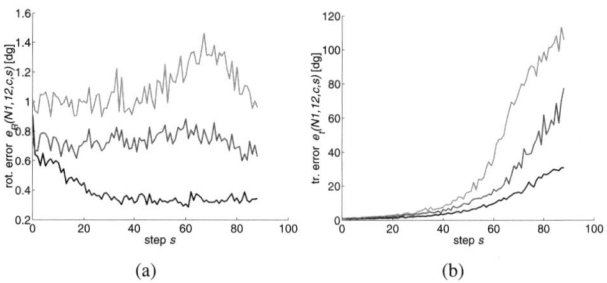

Fig. 4. Errors $\bar{e}_R(N1,12,\bar{c},s)$ and $\bar{e}_t(N1,12,\bar{c},s)$ (black: our approach; green: HMA; red: EMA).

Optimal Landmark Configuration for Vision-Based Control of Mobile Robots

Darius Burschka, Jeremy Geiman, and Gregory Hager
Computational Interaction and Robotics Laboratory
Johns Hopkins University
Baltimore, MD 21218
{burschka|jag|hager}@cs.jhu.edu

Abstract— We analyze the problem of finding the optimal placement of tracked primitives for robust vision-based control of a mobile robot. The analysis evaluates the properties of the Image Jacobian matrix, used for direct generation of the control signals from the error signal in the image, and the accuracy of the underlying sensor system. The analysis is then used to select optimal tracking primitives that ensure good observability and controllability of the mobile system for a variety of sensor system configurations.

The theoretical results are validated with our mobile robot for system configurations that use standard video cameras mounted on a pan-tilt head and catadioptric systems.

I. MOTIVATION

Localization is a fundamental task on mobile systems. Knowledge about the current position relative to landmarks in a local area is essential for planning and task specification. In many cases it is not necessary to use sophisticated models of the environment to specify the path of a mobile system. Many systems merely need to follow pre-specified paths learned in a *teaching* phase to fulfill their tasks, e.g. mail delivery robots, storage management robots, sentry robots, etc. This is in fact our goal: to develop a system that can be *walked* in a teaching phase through the environment. During this phase it learns the path based on position of identified tracking primitives, like color blobs, gray-scale patterns or disparity regions. It uses this knowledge later to repeat this path by generating the control signal directly from the error between the expected position in the image for the tracked primitives and their actual position during the *replay* step.

A standard localization method is to measure angles between landmark positions and to compute the pose from intersections of the circles that represent the possible positions from each possible pair of bearing measurements [14], [10], [6], [2]. We extend this idea by measuring both azimuth and elevation angles of a landmark to represent its position on a sphere instead of a circle (Fig. 1).

There have been a number of papers on the process of selecting useful features points or natural markers in image data [7], [8], [9], [11], [12], [13], [15]. These approaches select optimal landmarks based on their appearance in the image. In this paper we analyze the problem of finding the optimal placement of tracked landmarks based on their 3D position in the world.

In [3] we presented a system that allows vision-based control of mobile robots in a local path segment with pre-selected tracking primitives (*features*). The system directly uses the input from a sensor that provides measurements of the angular directions of the incident light rays to navigate in a local area (Fig. 1). This ideal sensor can be approximated with a variety of sensor configurations such as omnidirectional cameras or standard cameras mounted on pan-tilt heads, to give just a short list of possible configurations.

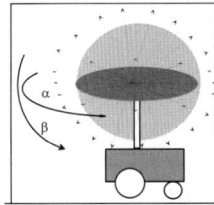

Fig. 1. Ideal generic sensor system measuring the horizontal α and vertical β angle of incidence.

Theoretical considerations presented in this paper are validated in multiple experiments. They show that varying landmark configurations do significantly influence the quality of the generated control signals. Finding optimal landmark configurations is essential for several areas of work with mobile systems. The correct selection depends partly on spatial resolution of the used sensor system. The results from this paper help to choose an optimal sensor system for a specific room structure with pre-specified artificial landmarks. In on-line landmark selection process during the initial phase of operation in a local path segment, the results of this paper help to choose landmarks allowing best control signals in a given segment.

Our goal is to investigate the properties of the Jacobian matrix used to map deviations in the 3D space onto changes in the perception of the ideal sensor (Fig. 1).

The remainder of this article is structured as follows. The next section describes the geometry of the vision-based control problem. Section III describes the evaluation of the system sensitivity to changes in the input parameters and the expected input sensitivities of the camera systems. Section IV presents numerical evaluation and experimental results from a real system. We close with a discussion of future work.

II. VISION-BASED CONTROL SYSTEM

As already described in [3] in more detail, the presented navigation system operates in two steps. In a *teaching* phase the user takes the robot for a *walk*. The robot saves the image positions of selected tracking primitives (e.g. color blobs, gray-scale patterns, etc.) to identify positions in the world that later are used to repeat this path autonomously in the *replay* phase. The control signals are generated directly from the error signal in the image using the Image Jacobian matrix described in Section II-B.

A. Spherical Image Projection

We assume a non-holonomic mobile system with unicycle kinematics throughout the article. The system operates in the (x, z, Θ) coordinates (Fig. 2).

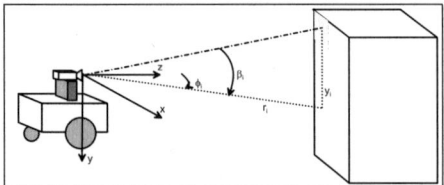

Fig. 2. Coordinate system used in the system.

We describe the imaging properties of the generic sensor system depicted in Fig. 1 in spherical coordinates (α - azimuth angle, β - elevation angle). The origin of the robot coordinate system is assumed to be at the center of rotation of the mobile system, the optical z axis points in the "forward" direction of the robot motion, and the x axis points to the "right" of the robot orientation (Fig. 2). A point in space relative to the robot can then be described by the triple (x_i, y_i, z_i).

We define the spherical coordinates (α_i, β_i) in the camera projection of an observed point P_i to

$$\alpha_i = \arctan \frac{x_i}{z_i} \wedge \beta_i = \arctan \frac{y_i}{\sqrt{x_i^2 + z_i^2}} \quad (1)$$

B. The Image Jacobian

Now, assuming holonomic motion in the plane, we can compute the following Image Jacobian that relates the change of angles in the image, (α_i, β_i), to changes in position in the (x, z)-plane from (1):

$$\mathcal{J}_i^t = \begin{pmatrix} \frac{\partial \alpha_i}{\partial x} & \frac{\partial \alpha_i}{\partial z} & \frac{\partial \alpha_i}{\partial \Theta} \\ \frac{\partial \beta_i}{\partial x} & \frac{\partial \beta_i}{\partial z} & \frac{\partial \beta_i}{\partial \Theta} \end{pmatrix} = \begin{pmatrix} \frac{z_i}{x_i^2 + z_i^2} & -\frac{x_i}{x_i^2 + z_i^2} & -1 \\ -\frac{x_i y_i}{(x_i^2 + y_i^2 + z_i^2)\sqrt{x_i^2 + z_i^2}} & -\frac{y_i z_i}{(x_i^2 + y_i^2 + z_i^2)\sqrt{x_i^2 + z_i^2}} & 0 \end{pmatrix} \quad (2)$$

The dependency on the unknown position (x_i, z_i) of the robot relative to the tracked landmark can be avoided considering the geometry of the system to:

$$\sqrt{x_i^2 + y_i^2 + z_i^2} = \frac{y_i}{\sin \beta_i}, \quad \sqrt{x_i^2 + z_i^2} = \frac{y_i}{\tan \beta_i},$$
$$x_i = \frac{y_i \cdot \sin \alpha_i}{\tan \beta_i}, \quad z_i = \frac{y_i \cdot \cos \alpha_i}{\tan \beta_i} \quad (3)$$
$$\mathcal{J}_i^t = \begin{pmatrix} \frac{\tan \beta_i \cdot \cos \alpha_i}{y_i} & -\frac{\tan \beta_i \cdot \sin \alpha_i}{y_i} & -1 \\ -\frac{\sin^2 \beta_i \cdot \sin \alpha_i}{y_i} & -\frac{\sin^2 \beta_i \cdot \cos \alpha_i}{y_i} & 0 \end{pmatrix}$$

Note in particular that the Image Jacobian is a function of only one unobserved parameter, y_i, the height of the observed point. Furthermore, this value is *constant* for motion in the plane. Thus, instead of estimating a time-changing quantity as is the case in most vision-based control, we only need to solve a simpler static estimation problem. We refer to [3] for a detailed description of the y_i-estimation.

In the following text we assume that the system already learned the positions of the tracked objects as columns in a matrix $\mathcal{M}_p$ in the *teaching* phase [3]. In the *replay phase* the values $\mathcal{M}_p[t][i]$ representing the stored positions for the tracker i at the time stamp t together with the estimations of y_i are used to calculate the position error of the robot $\Delta w^t = (dx, dz, d\Theta)^T$.

For each tracked landmark we can write the dependency of the observation error in the image Δe_i^t on the position error Δw^t using (2) to

$$\Delta e_i^t = \mathcal{J}_i^t \cdot \Delta w^t, \quad \text{with} \quad \Delta e_i^t = \mathcal{M}_p[t][i] - p_i^t. \quad (4)$$

Since all observations depend on the same position error Δw^t and we are interested in estimation of the position error Δw^t from the error in the camera image Δe_i^t, we need to invert the equation (4). It is not possible to estimate all three values of Δw^t from one landmark (α_i, β_i) in the image. We compute a "stacked" observation vector Δe^t using (4) to

$$\Delta e^t = \mathcal{J}^t \cdot \Delta w^t, \quad \text{with}$$
$$\mathcal{J}^t = (\mathcal{J}_1^t, \dots, \mathcal{J}_N^t)^T \wedge \Delta e^t = (\Delta e_1^t, \dots, \Delta e_N^t)^T \quad (5)$$

From (5) we can estimate Δw^t using the pseudo-inverse $(\mathcal{J}^t)^{-1}$ of the "stacked" Image Jacobian matrix from (2) to

$$\Delta w^t = (\mathcal{J}^t)^{-1} \cdot \Delta e^t, \quad \text{with} \quad (\mathcal{J}^t)^{-1} = (\mathcal{J}^{t^T} \mathcal{J}^t)^{-1} \mathcal{J}^{t^T} \tag{6}$$

The value Δw^t describes the error in the 3D position that we use to generate the control signals for the robot.

III. SYSTEM ANALYSIS

The relative error in the solution caused by perturbations of parameters can be estimated from the condition number of the Image Jacobian matrix J. The condition number is the ratio between the largest and the smallest singular value of the matrix J.

The condition number estimates the sensitivity of solution of a linear algebraic system to variations of parameters in matrix J and in the measurement vector b.

Consider the equation system with perturbations in matrix J and vector b:

$$(\mathcal{J} + \epsilon \delta \mathcal{J}) x_b = b + \epsilon \delta b \tag{7}$$

The relative error in the solution caused by perturbations of parameters can be estimated by the following inequality using the condition number κ calculated for J (see [5]):

$$\frac{\|x - x_b\|}{\|x\|} \leq \kappa \left(\epsilon \frac{\|\delta \mathcal{J}\|}{\|\mathcal{J}\|} + \epsilon \frac{\|\delta b\|}{\|b\|} \right) + \mathcal{O}(\epsilon^2) \tag{8}$$

Therefore, the relative error in solution x can be as large as condition number times the relative error in J and b. The condition number together with the singular values of the matrix J describe the sensitivity of the system to changes in the input parameters.

In the following subsections we investigate the observability and accuracy of the output parameters (x, y, z) from the input stream of the camera system (sec. III-A) and the influence of the real sensor on the achievable accuracy of the system (sec. III-B).

A. Optimal Landmark Configuration for the Image Jacobian Matrix

The singular values can be obtained as positive square roots of the eigenvalues of the matrix $J^T \cdot J$. With $y_{i \in \{1,...,N\}}$ as heights of the tracked objects, $\alpha_{i \in \{1,...,N\}}$ as azimuth angles to them and $\beta_{i \in \{1,...,N\}}$ as their elevation angles. The resulting matrix for N landmarks has the form shown in (9).

The system estimates three parameters $(dx, dy, d\Theta)$ from the image positions (u_i, v_i) of all tracked primitives (features) $i \in \{1, \ldots, N\}$ (4). Therefore, at least two *features* are necessary to estimate all 3 position parameters.

Each feature contributes a measurement of a distance Δr_i from the robot to the feature in the ground plane and an orientation $\Delta \Theta$ relative to it. The equation (3) can then be written in this case as:

$$\mathcal{J}_i^t = \begin{pmatrix} 0 & -1 \\ -\frac{\sin^2 \beta_i}{y_i} & 0 \end{pmatrix} = \begin{pmatrix} 0 & -1 \\ -\frac{1}{y_i \cdot \left(1 + \left(\frac{r_i}{y_i}\right)^2\right)} & 0 \end{pmatrix}$$

$$r_i = \sqrt{x_i^2 + z_i^2} \tag{10}$$

From the equation (10) we learn that an error $\Delta \Theta$ is directly forwarded to the output value α_i, while the value Δr, the error in the distance to the feature, is scaled with the value

$$\kappa_r = \left[y \cdot \left(1 + \left(\frac{r_i}{y_i} \right)^2 \right) \right]^{-1} \tag{11}$$

Since in our case the measurement error is in the image space, the resulting errors in the world are dependent on the reciprocal values.

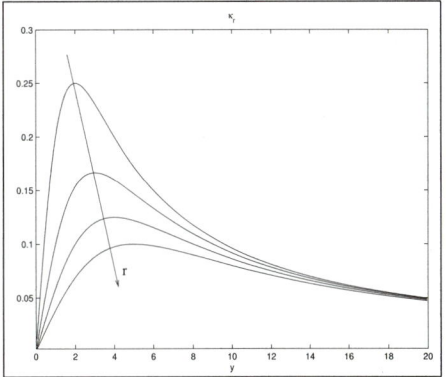

Fig. 3. Dependency of κ_r on y_i and r_i.

We deduce from the above equation that the optimum placement of the feature should maximize the above expression to allow good observability of the position error Δr. The optimal value can be estimated to

$$\frac{d\kappa_r}{dy_i} = -\frac{1}{x_i^2 \left(1 + \frac{r_i^2}{y_i^2}\right)} + \frac{2 \cdot r_i^2}{y_i^4 \left(1 + \frac{r_i^2}{y_i^2}\right)^2} = 0$$

$$\Rightarrow \quad y_i = \pm r_i \quad \Rightarrow \beta_i = \arctan \frac{y_i}{r_i} \tag{12}$$

that corresponds to an angle $|\beta_i| = 45°$.

It is shown that any linear system has at least one solution component whose sensitivity to perturbations is proportional to the condition number of the matrix, but there may exist many components that are much better conditioned [4]. The sensitivity for different components of the solution changes depending on the configuration of the landmarks and the relative position of the robot to them.

$$\mathcal{J}^T \cdot \mathcal{J} = \begin{pmatrix} \sum_{i=1}^N \left(\frac{\tan^2 \beta_i \cdot \cos^2 \alpha_i}{y_i^2} + \frac{\sin^4 \beta_i \cdot \sin^2 \alpha_i}{y_i^2} \right) & \sum_{i=1}^N \left(\frac{\sin^4 \beta_i \cdot \sin \alpha_i \cdot \cos \alpha_i}{y_i^2} - \frac{\tan^2 \beta_i \cdot \cos \alpha_i}{y_i^2} \right) & \sum_{i=1}^N \left(-\frac{\tan \beta_i \cdot \cos \alpha_i}{y_i} \right) \\ \sum_{i=1}^N \left(\frac{\sin^4 \beta_i \cdot \sin \alpha_i \cdot \cos \alpha_i}{y_i^2} - \frac{\tan^2 \beta_i \cdot \cos \alpha_i}{y_i^2} i \right) & \sum_{i=1}^N \left(\frac{\tan^2 \beta_i \cdot \sin^2 \alpha_i}{y_i^2} + \frac{\sin^4 \beta_i \cdot \cos^2 \alpha_i}{y_i^2} \right) & \sum_{i=1}^N \left(\frac{\tan \beta_i \cdot \sin \alpha_i}{y_i} \right) \\ \sum_{i=1}^N \left(-\frac{\tan \beta_i \cdot \cos \alpha_i}{y_i} \right) & \sum_{i=1}^N \left(\frac{\tan \beta_i \cdot \sin \alpha_i}{y_i} \right) & N \end{pmatrix} \tag{9}$$

Figure 4 shows that the sensitivity to perturbations in image coordinates is highest when close to the two landmarks selected for this evaluation. The two peaks in Figure 4 indicate this. The sensitivity to changes in the x-coordinate are slightly better (left) than the sensitivity to the changes in z direction (right in Figure 4).

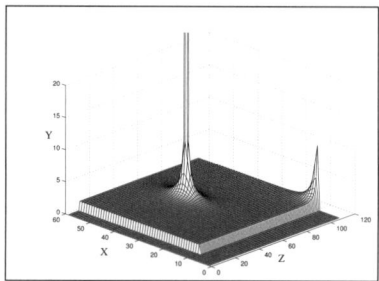

Fig. 5. Sensitivity (Y) to changes in Θ.

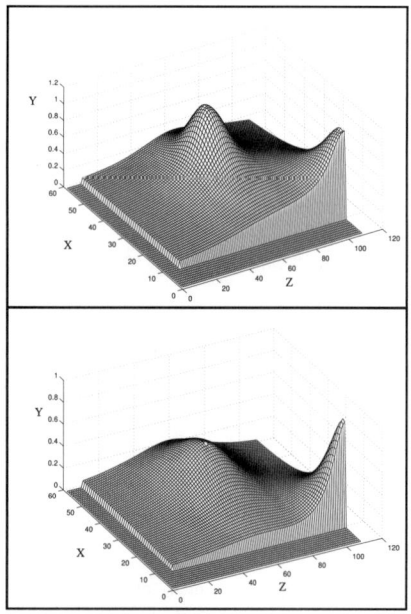

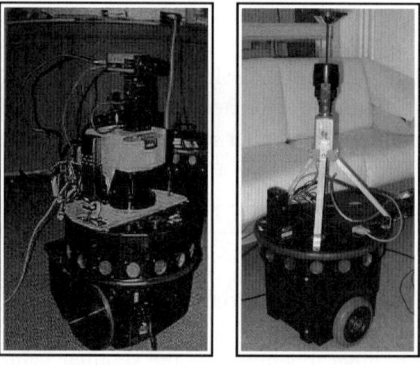

Fig. 6. Two implementations of the ideal sensor from Figure 1: (left) combination of a pan-tilt head with standard camera; (right) omnidirectional camera.

Fig. 4. Sensitivity (Y) to changes in x (top) and z (bottom) coordinates for two landmarks at (3,2,5) and (0,1,10).

This result could be validated in the result section with a real sensor system (section IV-B).

Figure 5 shows the expected uniform sensitivity to angular errors $d\Theta$ (10). The visible singularities are at the positions of the tracked landmarks, where a simultaneous measurement of the horizontal components of both landmarks is not possible.

B. Angular Resolution of the Sensor System

We assumed so far in this paper that the used sensor system meets the requirement from Figure 1 of uniform angular resolution in the α and β directions.

This ideal sensor can only be approximated with real, physical sensor systems. In our experiments we used two different sensor configurations that approximate this ideal sensor (Fig. 6).

The standard video camera has a limited field of view. A camera equipped with an 8mm lens has a field of view of approximately $50°$. The angular resolution of this camera varies depending on the distance from the optical center and is highest around the center. The resolution of this camera ϵ_p around center can be estimated to

$$\epsilon_p = \arctan \frac{p_x}{f}, \quad p_x - \text{pixelsize}, \quad f - \text{focal length} \tag{13}$$

This type of camera is very sensitive to coordinate changes in the world, especially around the optical center. However, the limited field of view restricts the possible configurations of the landmarks to just a narrow field in front of the camera. Considering the requirement for the landmarks to be visible over a long path segment, this kind of camera gives poor control results at the beginning of the segment, when the landmarks are still far away. Objects at larger distances away appear closer to the center where the resolution is higher than in the periphery. The characteristics of the angular resolution for this type of camera slightly compensates the decrease in the sensitivity of the Jacobian matrix for larger distances from the robot.

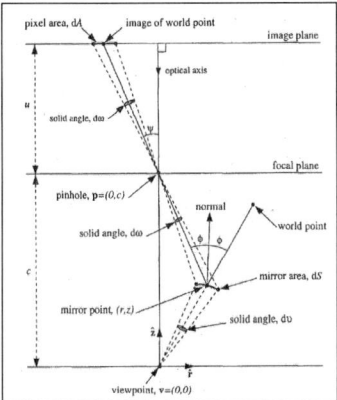

Fig. 7. The geometry used to derive the resolution of a catadioptric sensor (from [1]).

According to the derivation in [1] the resolution of a catadioptric camera (i.e. omnidirectional camera) $dA/d\nu$ with a hyperbolic mirror relates to the resolution of a standard camera $dA/d\omega$ as (see Fig. 7):

$$\frac{dA}{d\nu} = \frac{r^2 + z^2}{(c-z)^2 + r^2} \cdot \frac{dA}{d\omega} \quad (14)$$

where (r, z) is a point on the mirror being imaged and c is the distance between the viewpoint of the mirror-camera configuration and the focal point of the lens. Using simple properties of hyperboloids it follows that the factor in (14) increases with r. This system has the highest resolution around the periphery. In the configuration depicted in Figure 6 objects in the periphery are in the largest distance from the robot. Here again the imaging properties of the camera system compensate for the flaws in the sensitivity of the control system. The omnidirectional camera has the advantage to track features over a larger azimuth range of 360° compared to the 50° of a standard camera. This allows landmarks to be much closer to the robot during the initial landmark selection. Additionally, it allows them to stay visible over the same path length since the robot can pass between them without losing track of any of them. Since the sensitivity of the system decreases quadratically with the distance r_i to the landmark (11), reducing the initial distance results in a high sensitivity gain of the system observed on our mobile system.

IV. RESULTS

A. Sensitivity of the Jacobian Matrix

Based on the considerations in Section 3 we conclude that the system tries to minimize the distances to the tracked features to keep κ_r maximal (see (11) and Fig. 3).

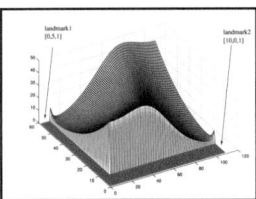 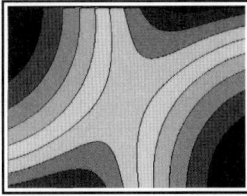

Fig. 8. Distribution of the condition number values for 2 similar landmarks: (left) surface plot; (right) contour plot.

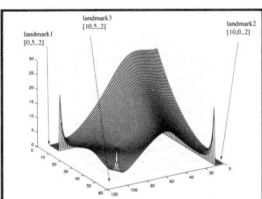 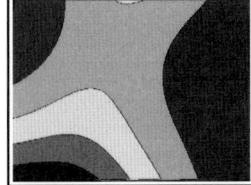

Fig. 9. Distribution of the condition number value for 3 similar landmarks: (left) surface plot; (right) contour plot.

The plotted examples for two and three landmarks in Figures 8 and 9 confirm this assumption. Each landmark results in a cavity in the plotted surface representing the condition number.

The influence of a single landmark on the condition number can be evaluated by choosing two landmarks with extreme different impacts on the shape of the surface. In the following example we have chosen two landmarks at the coordinates (3,4,5) and (0,0.1,10). We follow from equation (11) that the influence of the second landmark ($y_2 = 0.1$) is several magnitudes lower than that of the first landmark ($y_1 = 5$). This assumption is verified in Figure 10.

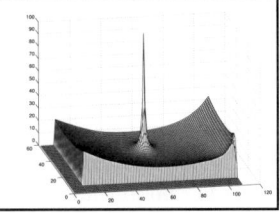 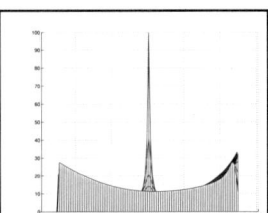

Fig. 10. Influence of a single landmark on the condition number: (left) perspective view; (right) projection from one side.

B. Overall Sensitivity of the System

As mentioned already in Section III, the error of the system output consists of the error in the system input projected by the Jacobian matrix onto the output data space. For the sensor configuration consisting of an omnidirectional sensor that measures the angles of the incident rays, we measured the input error in the image space (Fig. 11).

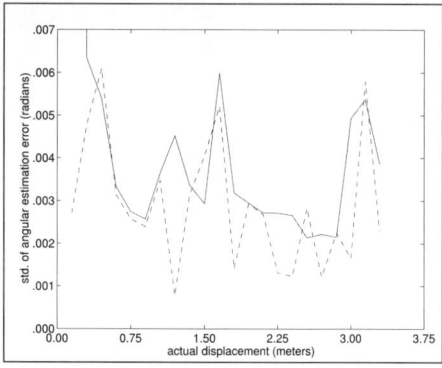

Fig. 11. Standard deviation of the error in the estimation of α (solid) and β (dashed) angles (in radians) from the tracking process vs. displacement as system moves away from tracked features.

As already predicted in Section III-A, the x-coordinate estimates better follow the changes in the image coordinates than the z-coordinate estimates (Fig. 12). The sensitivity for the x-coordinates is higher and allows for a higher accuracy of control in this direction.

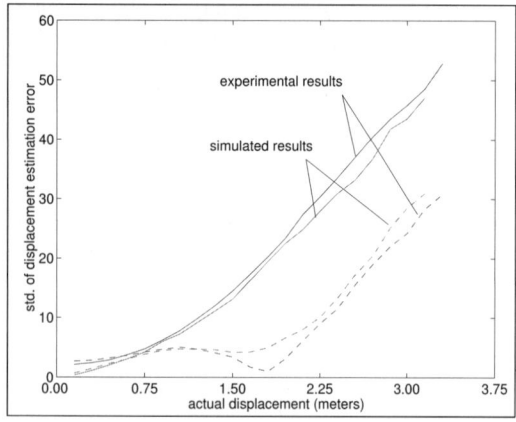

Fig. 12. Standard deviation of the error in the estimation of the displacement for x (solid) and z (dashed) as system moves away from tracked features. Experimental and simulated results shown.

Fig. 12 validates the correctness of the used system model. The predicted error for both coordinates (x, z) matches the actual measurements on our mobile system.

V. CONCLUSIONS

Our work is part of a system for sensor based navigation in which natural landmarks are used as markers or landmarks. In this system, a set of natural landmarks is used for navigation in a given path segment. The sets change each time any of the landmarks disappear due to occlusions or limited field of view of the conventional camera system. This defines the boundaries between the path segments. The landmarks can be selected manually by the user in a teaching phase or they are selected automatically by the system.

We have discussed the optimal selection of landmarks to obtain the best control results for a given environment based on the sensitivity of the sensor system used to model the ideal sensor with uniform angular resolution from Figure 1 and based on projection properties of the Image Jacobian matrix used to generate the control signals from the image data.

The presented results can be used for optimal configuration of catadioptric and standard camera systems. A correct selection of the imaging properties of the camera can compensate some of the insensitivities of the control system such as for landmarks at larger distances from the robot.

In the future, the system will be extended to automatically select new landmarks for omnidirectional navigation based on the results from this paper.

Acknowledgments

This work was supported by the MARS project, and by the NSF RHA program.

VI. REFERENCES

[1] S. Baker and S.K. Nayar. A Theory of Catadioptric Image Formation. In *Proc. of IEEE International Conference on Computer Vision*, 1998.

[2] M. Betke and L. Gurvits. Mobile Robot Localization Using Landmarks. *IEEE Trans. on Robotics and Automation*, 13:251–263, 1997.

[3] D. Burschka and G. Hager. Vision-based control of mobile robots. In *Proc. International Conference on Robotics and Automation*, pages 1707–1713, 2001.

[4] S. Chandrasekaran and I. C. F. Ipsen. On the Sensitivity of Solution Components in Linear Systems of Equations. *SIAM Journal on Matrix Analysis and Applications*, 16(1):93–112.

[5] J. Demmel. The componentwise distance to the nearest singular matrix. *SIAM Journal on Matrix Analysis and Applications*, 13:10–19, 1992.

[6] U.D. Hanebeck and G. Schmidt. Set theoretic localization of fast mobile robots using angle measurements. *IEEE Conf. on Robotics and Automation*, page 1387, 1996.

[7] S. Se, D. Lowe, and J. Little. Vision-based mobile robot localization and mapping using scale-invariant features. In *IEEE Conf. on Robotics and Automation*, pages 2051–2058, 2001.

[8] R. Sim and G. Dudek. Mobile robot localization from learned landmarks,. In *IROS*, 1998.

[9] Saul Simhon and Gregory Dudek. Selecting targets for local reference frames. In *Proc. IEEE Int. Conf. on Robotics and Automation*, pages 2840–2845, 1998.

[10] K.T. Sutherland and W.B. Thompson. Localizing in unstructured environments: Dealing with the errors. *IEEE Transactions on Robotics and Automation*, 10:740–754, 1994.

[11] Y. Takeuchi, P. Gros, M. Hebert, and K. Ikeuchi. Visual learning for landmark recognition. In *Proc. Image Understanding Workshop*, pages 1467–1473, 1997.

[12] Sebastian Thrun. Finding landmarks for mobile robot navigation. In *IEEE Conf. on Robotics and Automation*, pages 958–963, 1998.

[13] C. Tomasi and J. Shi. Good features to track. In *Proc. IEEE Conf. on Comp. Vision and Patt. Recog.*, pages 593–600, 1994.

[14] U. Wiklund, U. Anderson, and K. Hyypa. Agv navigation by angle measurements. *Proc. 6th Int. Conf. AGV Systems*, pages 199–212, 1988.

[15] E. Yeh and D. Kriegman. Toward selecting and recognizing natural landmarks. In *IEEE Int. Workshop on Intelligent Robots and Systems*, pages 47–53, 1995.

Visual Navigation Of An Autonomous Robot Using White Line Recognition

Huaming Li, Changhai Xu, Qionglin Xiao, Xinhe Xu

Institute of Artificial Intelligence & Robotics
Northeastern University
Wenhua Road 3-11, Shenyang
China

Abstract – This paper proposes a navigation system for an autonomous mobile robot that is designed to participate in the annual international robot competition, ABU (Asian Broadcast Union) Robocon 2002. The self-localization of the robot is done by detecting the position and orientation of vertically intersecting white guide lines laid on the game field, with a USB camera pointing to the ground perpendicularly.

Given the information of the playground, the robot can decide where to go autonomously by calculating its position and comparing it with the predetermined routine coordinate in the map. With the visual navigation system the vehicle can follow the line promptly and stop at any given intersection precisely.

1. INTRODUCTION

ABU (Asian Broadcast Union) Robocon 2002 is held in Tokyo, Japan. The annual contest will propose different themes every year. This year the aim is to design a robot competing for points in a game field, which is 12m long and 10.2m wide, with 30mm white guide lines contrasting with the green background marked on it. As shown in Fig. 1, the yellow, green and red circles on the cross represent silos of different height valued different marks where the robots should put in their balls in order to earn points. Since the robot that can enter the inner part of the game field must be autonomous, one of the key steps to win is to navigate the robot to desired position with the white guide lines.

Though there exist many methods that can be used to detect the white line, such as infrared (IR) [2], laser and optical line sensors [3] and they are easy in principia, considering the potential need of objects colour recognizing and more width to be seen with high resolution, we designed a wheeled vehicle equipped with an Acer notebook computer and a USB CCD camera from Logitech as the lines detector. The USB camera enables the system to obtain the digital images directly without using frame grabbers, thus making the system more economical and compact.

This digital image processing system has two advantages over a traditional analogy vision system, which are very helpful to the algorithm adopted in lines detection mentioned in part 4. First, in a USB camera the A/D conversion is performed close to the CCD sensor, thus

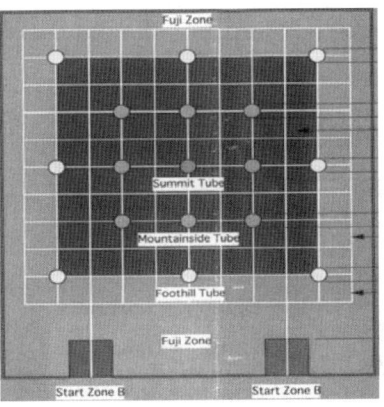

Fig. 1: Path layout of white guide line on the game field

confining the amount of electronic noise to an absolute minimum [4]. Second there is an advantage that with the notebook computer and a USB hub we can conveniently build a low-cost multi-camera sensing system by sharing several USB cameras with a single processor [5]. Also we can switch between the cameras with great flexibility by software programming, such as adopting the Microsoft DirectShow method, without any extra hardware expense.

A motion controller is designed to keep the robot walking along the guide line as well as possible by using a tangent method, which not only ensures the robot's correct speed and direction but offers a continuous and convenient adjusting procedure.

II. VISUAL NAVIGATION SYSTEM STRUCTURE

The visually navigated vehicle system discussed in this paper is composed of several processes as shown in Fig. 2. During the navigation process, the robot can recognize the white guide lines, calculate the orientation and adjust the wheels' speed to follow the lines precisely and smoothly. When a cross is detected, the robot updates its coordinate information and selects a path in the predetermined route library. Then the computer sends corresponding wheel velocity commands via RS-232C to motor controller where appropriate PWM wave is generated to drive the robot.

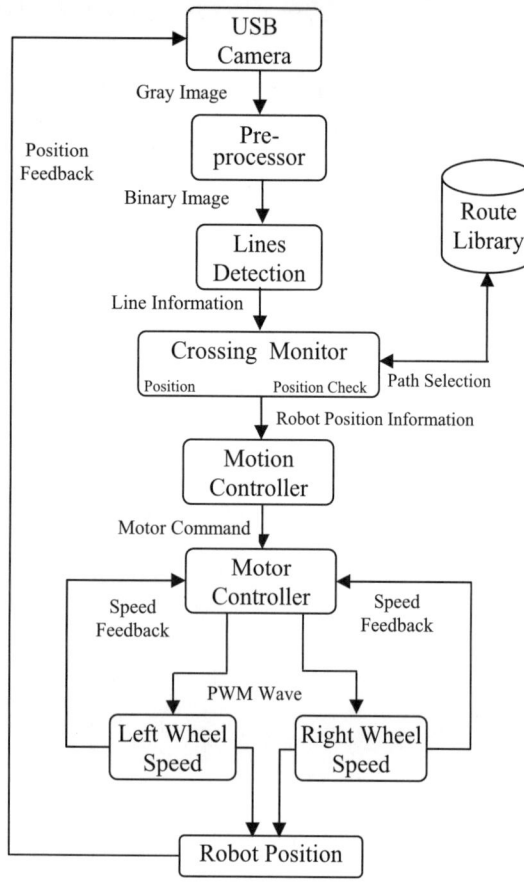

Fig. 2: Visually navigated vehicle
system using white line recognition

III. IMAGE CAPTURE AND PRE-PROCESSING

We adopt a Logitech QuickCam Pro 3000 as our USB CCD camera, pointing perpendicularly to the ground. QuickCam Pro can support a true resolution up to 640x480 pixels and a capture speed up to 30 fps. Here we sample the ground with a resolution of 320x240 at a speed of 30 fps. The driver of the camera can help us get a gray image directly. Introducing the Microsoft DirectShow technique and COM programming, we can locate a pointer to a block of memory where the image's RGB values are stored.

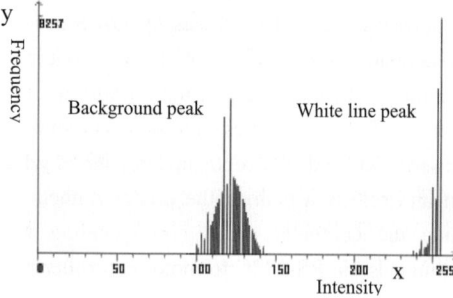

Fig. 3: Brightness histogram of the sample image

Taking account of the unfavourable practical field illumination condition and the intensely competitive process, we designed a particular mechanical structure equipped with independent light source to provide the optimal illumination. Also we introduced an adaptive segmentation method to enable the automatic adjustment of segmentation threshold according to dramatic environment change. In this way, the environment light disturbance is reduced to the lowest and the reliability of the visual navigation system is guaranteed.

Because of the simplification of game field ground, containing only two colours, background green and guide lines white, the binarizing process is relatively simple. As shown in Fig. 3, the histogram shape of a sample image illustrates a typical kind of bi-modal with right peaks made up of pixels of the white lines, and left ones made up of the dark green background.

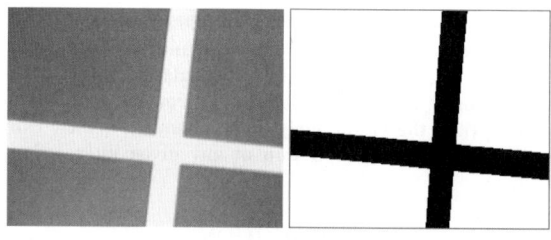

Fig. 4: (a) Sample image of the cross guide lines on the
ground, (b) binarized sample image with
a proper minimum error threshold

It makes intuitive sense for us to employ the criterion of minimum error to determine the threshold, the gray-level that has a minimum histogram value between the two maxima mentioned above. The binaries image of the sample is shown in Fig. 4.

IV. LINES DETECTION

In this part, we try to recognize the guide lines and calculate their equations in a rectangular system, from which we can get the information of the robot's position such as the orientation and distance error to the guide lines. Sobel operator can be used as simple line edge detector here, but it must process every pixel in the image, which comes with a wealth of computation work. Since clear binary images of the ground can be obtained by the pre-processing, in this paper we propose a fast and efficient way to recognize the guide lines, which is to detect the edge of the images but not all of them.

First step, we program to scan the pixels of the four edges of the binaries image and find out the "crucial points" whose gray values change greatly contrasting to the former or latter points. Then we calculate those crucial points' coordinates in the Cartesian coordinate system with the origin set at the center of the image. With this value we can obtain the white line equation.

Additional filtering is done on detection of the crucial points to remove ill-behaved dots due to some noise disturbance like the unexpectable dust reflection. Any change that continues less than three pixels is ignored, though this filter occasionally affects some particular lines detection when a guide line runs through an image angle and as it happens only less than 3 pixels are on one edge. However, this problem will be amended by a method talked about later.

Second step is to calculate the equation of the line adopting the two points method represented as follows:

$$\frac{y-y_1}{x-x_1} = \frac{y_2-y_1}{x_2-x_1} \quad (1)$$

Which can be transformed into:

$$y = \frac{y_2-y_1}{x_2-x_1}(x-x_1) + y_1$$
$$= \frac{y_2-y_1}{x_2-x_1}x + \frac{y_1 x_2 - x_1 y_2}{x_2-x_1} \quad (2)$$

where (x_1, y_1) and (x_2, y_2) are the two points' coordinates.

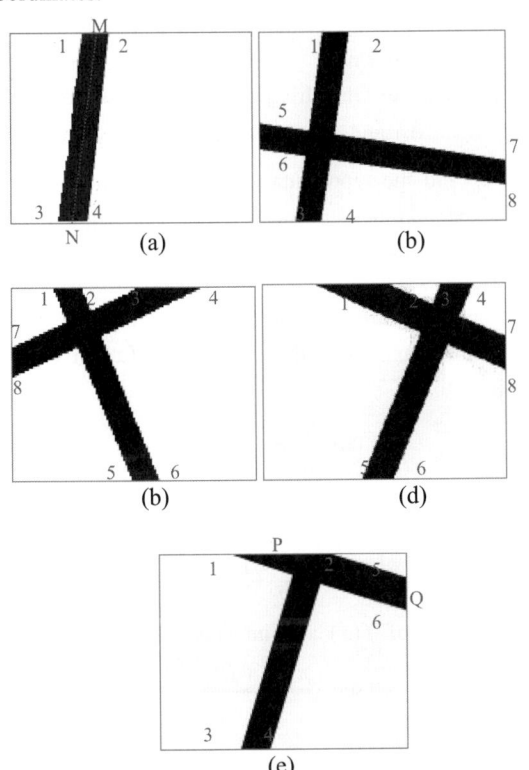

Fig. 5: Typical situations encountered in lines detection

However, how to determine the points used in this equation is a problem that needs more discussion. There are several typical situations demonstrated in Fig. 5, according to how many points are detected on the edges and how they distribute.

(a) is a simple but important fundamental situation, where two crucial points on top edge and two on bottom are detected, marked by number 1-4 in the image. This result means only one line is in the scene. Under the assumption that the white line equation we desire is represented by the center line marked with red dot dash in (a), the coordinate of white line center point M and N needed for the line equation can be calculated using the x value of point 1,2 and 3,4. The coordinate of point M can be represented as $(\frac{x_1+x_2}{2}, \frac{H}{2})$, and that of N is $(\frac{x_3+x_4}{2}, -\frac{H}{2})$, where H is the image height. Once the two points are settled, we can use (1) and (2) to get the white line equation shown in (3).

$$y = -\frac{2H}{x_3+x_4-x_1-x_2}x + \frac{H}{2}\frac{x_3+x_4+x_1+x_2}{x_3+x_4-x_1-x_2} \quad (3)$$

where $-\frac{2H}{x_3+x_4-x_1-x_2}$ is the slope.

We should notice a tricky situation that requires additional attention. When there exists a line that is very close to vertical, we must avoid the division by zero. If the x value difference between point 1 and 2 or 3 and 4 is very small (we use less than 5 pixels in our program), the line should be regarded as vertical and we just draw a line as

$$x = \frac{x_m + x_n}{2}.$$

(b) can be considered as an overlapped image made up of (a) and an image similar to (a) but with a single line crossing from the left to right. So we can use almost the same method as mentioned in situation (a) to get the two crossing lines' equation. In this case there are altogether 8 crucial points in the image, 2 points on each edge. It is a typical situation when the vehicle is passing a cross. Points located on opposite edges are grouped together to be substituted into(1) and (2).

(c) and (d) are two reciprocally horizontally flipped images. The number of crucial points is the same with (b). However, there are 4 points detected on the same top edge, which makes us faced with a problem that how to group those points to calculate the line equations. In case (c), points 1,2 and 5,6 should be in a group, while in (d) it is 1,2 and 7,8. To solve this problem, we take advantage of the priori knowledge that any two cross guide lines in the game field intersect vertically. Just choose any two center points on different edges as a group; then calculate the two lines' slopes to verify whether their product is near -1 which means the two lines are vertical. Thus we get the proper grouping result.

(e) is a troublesome one which we call a "T" pattern. This situation can be judged by its special 6 points distributing. There should be 4 crucial points on the top if we omit the line width; however, unfortunately the other two points are just immerged in the practical line. In our system, we solve this problem by estimating a center point between point 1 and 2 marked as P; then figure out one line equation with P and its neighbor center point Q. The other line can be calculated by point/slope method with the preknown vertical information. In fact, there are other "T" patterns, but all can be solved by similar ways. Fig. 5 shows the detection result of (e) grabbed from the GUI.

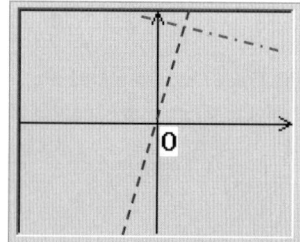

Fig. 6: "T" pattern recognizing result

Third step is to provide the orientation error information with the calculated slope and the distance error information from the origin to the guide line. Suppose the line equation is represented by $Ax + By + C = 0$ where A, B and C are three constants, the distance is

$$D = \frac{|C|}{\sqrt{A^2 + B^2}} \quad (4)$$

Considering the motion blurring and other disturbance, we verify every recognized result by comparing the slope product with -1 when there are two lines in the image. If a result is not accordant, the system uses the last correct detection result stored in advance as a substitution. In this way, the problem in the detection of the crucial points mentioned in the first step can be solved, also the continuity and stability of the system are guaranteed.

V. MOTION CONTROLLER

When the robot has understood its relationship with the guide lines, how to keep itself on the line and going forward in the meanwhile is immediately put on the agenda. To solve this problem we intend to let the vehicle travel along a tangent curve, as shown in Fig. 7. On one hand, this method ensures that the robot's main task is to get closer to the guide lines when it is at a longer distance to the line and to advance quickly when it is close to the guide line; on the other hand, this method realizes the smoothness of control, which will inevitably enhance control quality.

Here we just take it for example that the robot is on the left side of the guide line and their distance is between zero and L (a cautionary value). When the distance is equal to or above L, the robot will keep its centroid fixed, turn until it confronts the line with angle $\frac{\pi}{4}$ and head directly until the distance reduces to $(-L, 0)$.

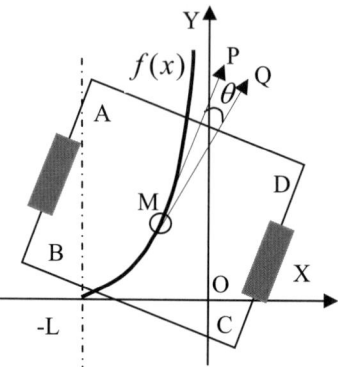

Fig. 7: Angle error control with tangent method

In Fig. 7, Y-coordinate is set in accordance with the objective line and rectangle $ABCD$ represents the autonomous robot with line segment AD as the front side and point M as the centroid. Here we suppose the mass distribution of the robot is even, thus the centroid happens to be its geometrical center.

Suppose the distance from point M to Y-coordinate is x_0 ($-L < x_0 < 0$) and the robot's direction is $\overrightarrow{MP}$. The distance x_0 and the representation of $f(x)$ can uniquely determine X-coordinate. Then we obtain the plane rectangular coordinates on which the following discussion is based.

Compared with the standard tangent curve with period π, the equation of the curve $y = f(x)$ in Fig. 7 should be

$$\frac{\pi}{2} \cdot \frac{y}{L} = \tan\left(\frac{\pi}{2} \cdot \frac{x + L}{L}\right) \quad (5)$$

that is

$$y = f(x) = \frac{2L}{\pi} \tan \frac{\pi(x + L)}{2L} \quad (6)$$

The tangent slope of $f(x)$ at point M is

$$\left.\frac{dy}{dx}\right|_{x=x_0} = \sec^2\left(\frac{\pi(x_0 + L)}{2L}\right) \quad (7)$$

therefore

$$\theta_d = \arctan \left.\frac{dy}{dx}\right|_{x=x_0}$$
$$= \arctan\left(\sec^2\left(\frac{\pi(x_0 + L)}{2L}\right)\right) \quad (8)$$

where θ_d is the angle we will use as the object direction to adjust the robot.

Now calculate the angle error, which will determine the output of the motion controller.

$$\theta_e = \theta_d - \theta_s \qquad (9)$$

where θ_s is the current angle of the robot.

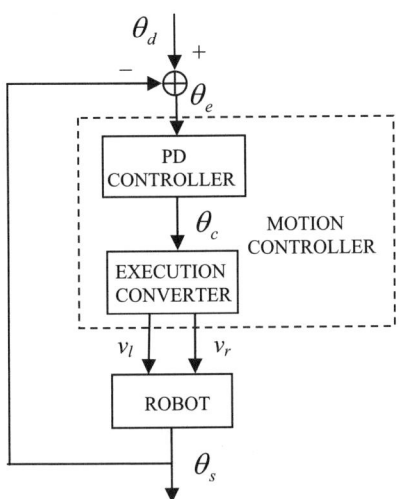

Fig. 8: Structure of motion controller

Next step is to adjust θ_e to promote the system's stability and response speed by using some algorithms. Here we introduce incremental type PD control algorithm due to its simple control structure and widely accepted performance and as a result we get the control value θ_c.

Furthermore, in order to make the controller flexible based on dynamic process, that is, adaptable to different conditions, the increment of θ_c is chosen as

$$\Delta\theta_c = -(\alpha\theta_e + (1-\alpha)\dot{\theta}_e) \qquad (10)$$

where α ($0 \leq \alpha \leq 1$) can be adjusted online and $\dot{\theta}_e$ is the derivative of θ_e.

The real-time adjustment of α will try to minimize the performance index

$$J = \int_0^\infty |\theta_e(t)| dt \qquad (11)$$

Suppose θ_e and $\dot{\theta}_e$ are obtained at sample time K, then we will set

$$\theta_c(K+1) = \theta_c(K) + \Delta\theta_c(K+1) \qquad (12)$$

and

$$\Delta\theta_c(K+1) = -\alpha\theta_e(K) - (1-\alpha)\dot{\theta}_e(K) \qquad (13)$$

The output angle of the robot will follow $\theta_c(K+1)$ but may delay m sample periods. The robot's actual angle θ_s at next sample time can be predicted approximately as

$$\hat{\theta}_s(K+1) = \theta_s(K) + \gamma(\theta_c(K+1) - \theta_s(K)) \qquad (14)$$

where $\gamma = 1/m$. So the error at time $K+1$ can be represented as

$$\hat{\theta}_e(K+1) = \hat{\theta}_d(K+1) - \hat{\theta}_s(K+1)$$

$$= \hat{\theta}_d(K+1) - \theta_s(K)$$
$$\quad - \gamma(\theta_c(K+1) - \theta_s(K)) \qquad (15)$$

where $\hat{\theta}_d(K+1)$ can be predicted by the representation of $f(x)$ and the current speeds of the two wheels. Substituting (12) and (13) for $\theta_c(K+1)$ results in

$$\hat{\theta}_e(K+1) = -\gamma\theta_c(K) + \alpha\gamma\theta_e(K) + (1-\alpha)\gamma\dot{\theta}_e(K)$$
$$\quad + \hat{\theta}_d(K+1) - (1-\gamma)\theta_s(K) \qquad (16)$$

and the desired α can be calculated through

$$|\hat{\theta}_e(K+1)| = 0 \qquad (17)$$

And then the control value θ_c we are pursuing is finally gained in conformity to (13).

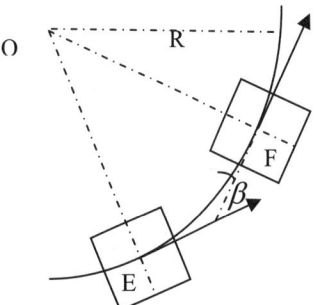

Fig. 9: Convertion from θ_c to wheel speeds

In this system we control the robot in the end through continuously changing the two wheels' speeds, so we have to associate θ_c with the speeds. This is realized in the execution converter.

Now consider the case that the robot walks from point E at time KT ($K=0,1,2,...$) to point F at time $(K+1)T$ with the left wheel's speed v_l and the right v_r. Obviously during this short period of time the route of the vehicle's centroid is a part of a circle.

If the radius of that circle is R and the width of the robot is W, then we can obtain three equations as follows

$$v_l T = (R - W/2)\beta, \qquad (18)$$

$$v_r T = (R + W/2)\beta, \qquad (19)$$

and

$$vT = R\beta \qquad (20)$$

where β is the angle that the robot has changed in time T. Also the centriod's speed v can be achieved as

$$v = (v_l + v_r)/2 \qquad (21)$$

If the centroid's speed v is set with a reference value v_{ref} according to actual condition and θ_c is considered, we can finally get the wheels' speed v_l and v_r respectively as

$$v_l = v_{ref} - \frac{W\theta_c}{2T} \qquad (22)$$

and

$$v_r = v_{ref} + \frac{W\theta_c}{2T} \qquad (23)$$

Fig. 10 shows the simulation result of the tangent method, where the yellow rectangle represents the centroid of the moving robot.

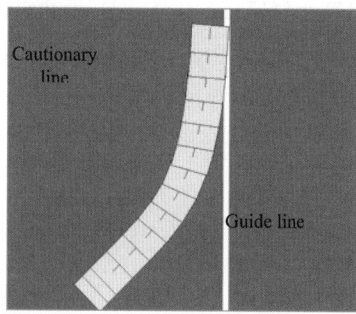

Fig. 10: Simulation result

VI. CONCLUSION AND FUTURE WORK

In this paper, we propose a visual navigation system for a robot participating in the Robocon 2002 contest. It has compact structure mainly composed of a notebook computer and a USB camera. With the fast line detection algorithm, the robot can move in the game field following the guide lines on any path at an approximate speed of 0.8m/s and stop at desired point to finish the precise shooting action.

With the flexible structure, we can easily compose a multi-camera sensing system by adding a USB hub to connect other cameras. In this way the collision avoidance and color object recognizing can be done. Also we can introduce the IEEE-1394 high performance serial bus (firewire) or new USB (Universal Serial Bus) 2.0 which can support up to 480Mbps bandwidth comparing with the current 12Mbps to obtain high-quality images at higher speed in future work.

VII. ACKNOWLEDGMENTS

Our special thanks go to Mr. Sun Lujun, Mr. Cong Dehong and the lovely members of our robot team in Northeastern University (NEU) for their great appreciated help.

VIII. REFERENCES

[1] Ishikawa, S., H. Kuwamoto, and S. Ozawa, "Visual navigation of an autonomous vehicle using white line recognition", IEEE *Transactions on Pattern Analysis and Machine Intelligence*, volume 10 (1988), number 5, pp. 743-749.

[2] http://www.robotstore.com/download/3-554_and_3-573_sensors.pdf

[3] J. Kaiser, P. Schaeffer: "A smart optical sensor for direct robot control", *in Proceedings of* IEEE International Conference on Mechatronics and Machine Vision in Practice 2001, Hong Kong, China, August 27-29, 2001.

[4] Iwan Ulrich and Illah Nourbakhsh, "Firewire Untethered: High-Quality Images for Notebook Computers", *Advanced Imaging Magazine*, January 2000, pp. 69-70.

[5] M. Saxena, V. Selvaraj, R. Dhareshwar, E. L. Hall, Univ. of Cincinnati "Video switching and sensor fusion for multicamera sensing systems", *Proceedings of SPIE* Vol. 4197 [4197-18].

[6] Betke, M., "Learning and Vision Algorithms for Robot Navigation", PhD Thesis, Department of Electrical Engineering and Computer Science, Massachusetts Institute of Technology. 1995.

[7] A. Clérentin, L. Delahoche, E. Brassart "Omnidirectional sensors cooperation for multi-target tracking", *IEEE Int. Conf. on Multisensor Fusion and Integration* MFI 2001, Aug. 01, Germany, pp. 335-340.

[8] J. Tanner and C. Mead. "An integrated analog optical motion sensor", *VLSI Signal Processing*, volume 2, New York, 1988. pages 59—87.

[9] Kluge, Karl, Yarf. "Simple Line Detection, An Open-Ended Framework for Robot Road Following". School of Computer Science, Carnegie Mellon, February 1993, Technical Report CMU-CS-93-104.

[10] Matthias O. Franz, Bernhard SCH Olkopf, "Learning View Graphs for Robot Navigation", *Autonomous Robots*, 1998, vol.5, pp.111–125.

A switching control law for keeping features in the field of view in eye-in-hand visual servoing

Graziano Chesi*, Koichi Hashimoto, Domenico Prattichizzo, Antonio Vicino

Abstract—In this paper, a visual servoing strategy for dealing with the problem of keeping the observed points in the camera field of view is proposed. The approach consists of a switching control law based on camera displacement estimation and regulated from the position of the points in the image. In absence of uncertainties on the intrinsic parameters and optical axis direction, global stability is achieved and all points are kept in the field of view. Moreover, the trajectory length is minimized in the rotational space and, for some cases, also minimized in the translational one. Robustness against uncertainties is also guaranteed.

Index Terms— Visual servoing, Point correspondences, Field of view, Switching control.

I. INTRODUCTION

In eye-in-hand visual servoing systems the control law is computed on the basis of the view of an object acquired by a vision system, usually one camera, mounted on the robot end effector (see, e.g., [1] and [2]). As a consequence, a fundamental requirement on visual servoing algorithms is to keep the observed object in the field of view during the robot control. However, this requirement is difficult to satisfy if the geometrical model of the object or the camera parameters are unknown.

In position-based visual servoing (PBVS) (see, e.g., [3] and [4]) it is possible that points leave the field of view since the camera is controlled in the 3D space and no control is performed in the image domain. Although in image-based visual servoing (IBVS) (see, e.g., [5],[6] and [7]) the camera is controlled in the image domain, this problem is still present and, moreover, the global convergence is not guaranteed and the resulting 3D trajectory can be very unsatisfactory (see [8]).

Recently, several approaches have been developed to deal with the image constraint. In [9] a hybrid method based on the use of PBVS to control the rotation and IBVS to control the position of a reference point is proposed. In [10] a hybrid method which controls the position of an ellipse in the image including all points is presented. In [11] and [12] partitioning methods are used to steer the camera through decoupled control laws. Approaches based on the use of potential fields for repelling points from the image boundary have been also

(Corresponding author) G. Chesi is a visiting researcher from the Department of Information Engineering, University of Siena, Italy, at the Department of Information Physics and Computing, University of Tokyo, Japan, under a fellowship program of the European Commission. E-mail: chesi@k2.t.u-tokyo.ac.jp

K. Hashimoto is with the Department of Information Physics and Computing, University of Tokyo, Japan. E-mail: koichi@k2.t.u-tokyo.ac.jp

D. Prattichizzo and A. Vicino are with the Department of Information Engineering, University of Siena, Italy. E-mail: {prattichizzo,vicino}@dii.unisi.it

proposed, as in the partitioning method [13] and as in the path-planning technique [14]. However, in most cases it is not guaranteed either that all points remain in the field of view or that the camera converges to the desired situation from any feasible initial one. More recently, a modified PBVS has been proposed to deal with the image constraint in [15], but only the object frame origin is guaranteed to remain in the field of view.

In this paper, a new visual servoing approach for dealing with the field of view problem is presented. The proposed strategy consists of a switching control law regulated by the position of the observed points in the image. Specifically, when all points lie in a pre-specified sub area of the screen the camera is steered by using a position-based control law, and when at least one point gets out from the sub area the camera is steered by using some rotational or translational control laws able to push the critical points inside the sub area and to guarantee the final convergence. Knowledge of the 3D model of the object is not required and a low computational burden is needed to the on-line implementation. Occlusions are also allowed if the number of visible and tracked points is however sufficient to estimate the camera displacement between current and desired situation.

This technique ensures, for known intrinsic parameters and optical axis direction, that the camera reaches the desired situation from any feasible initial one (global stability) and that all points are kept in the field of view. Moreover, for known intrinsic parameters, the camera trajectory length is minimized in the rotational space and, for some cases including all the ones in which the features do not get out from the image sub area, also minimized in the translational space. Robustness against uncertainty on the intrinsic parameters and optical axis direction is also guaranteed. In particular, simulation results show that stability is achieved also in presence of large uncertainties and image noise.

II. PRELIMINARIES

A. Notation and problem formulation

Let us introduce the notation used throughout this paper. We indicate with e_1, e_2, e_3 the column vectors of the 3×3 identity matrix I_3, with 0_3 the null vector 3×1, with $\|A\|$ the Frobenius norm of matrix A, and with $[v]_\times$ the skew-symmetric matrix of vector $v = [v_1, v_2, v_3]^T \in \mathbb{R}^3$ defined as $[v]_\times = [0, -v_3, v_2; v_3, 0, -v_1; -v_2, v_1, 0]$. Moreover:

- $c_a \in \mathbb{R}^3, a = i, c, d$: (initial, current and desired) camera centers with respect to the absolute frame;
- $R_a \in \mathbb{R}^{3 \times 3}, a = i, c, d$: camera rotation matrices with respect to the absolute frame;

- $\mathbf{m}_{k,a} \in \mathbb{R}^3, a = i, c, d$: image projections (in normalized homogeneous coordinates) of the k-th 3D point $\mathbf{q}_k \in \mathbb{R}^3$, according to

$$\mathbf{m}_{k,i} = \frac{\mathbf{R}_a^T(\mathbf{q}_k - \mathbf{c}_a)}{\mathbf{e}_3^T \mathbf{R}_a^T(\mathbf{q}_k - \mathbf{c}_a)}, \quad a = i, c, d; \quad (1)$$

- $\mathbf{p}_{k,a} \in \mathbb{R}^3, a = i, c, d$: image projections (in pixel homogeneous coordinates) of the k-th 3D point $\mathbf{q}_k \in \mathbb{R}^3$, according to $\mathbf{p}_{k,a} = \mathbf{K}\mathbf{m}_{k,a}, a = i, c, d$, where $\mathbf{K} \in \mathbb{R}^{3 \times 3}$ is the intrinsic parameters matrix $\mathbf{K} = [f_x, s, i_x; 0, f_y, i_y; 0, 0, 1]$ being f_x, f_y the focal lengths, i_x, i_y the principal points, and s the skew.

We consider the following problem. A set of n 3D points is observed, firstly, from the desired camera situation $\{\mathbf{c}_d, \mathbf{R}_d\}$, and, secondly, from the initial camera situation $\{\mathbf{c}_i, \mathbf{R}_i\}$ (the camera situation during the visual servoing is called current camera situation and it is indicated as $\{\mathbf{c}_c, \mathbf{R}_c\}$). The point correspondences between the two views are known. Then, our aim is to steer the camera from the current to the desired situation ensuring that all points are kept in the field of view of the camera during the visual servoing.

B. Camera displacement estimation and object reconstruction

The components of the camera displacement existing between current and desired camera situations are the rotation matrix $\mathbf{R} \in \mathbb{R}^{3 \times 3}$ and the translation vector $\mathbf{t} \in \mathbb{R}^3$ (expressed in the current camera frame) defined by $\mathbf{R}_d = \mathbf{R}_c \mathbf{R}$ and $\mathbf{c}_d = \mathbf{c}_c + \mathbf{R}_c \mathbf{t}$. Using point correspondences from two perspective views it is possible to recover $\mathbf{R}$ and the normalized translation $\mathbf{t}/\|\mathbf{t}\|$. This can be done, for example, through the essential matrix using the eight-point algorithm (see [16], [17] for details) or through the homography matrix relative to a virtual plane using the algorithm reported in [18]. From $\mathbf{R}$ and $\mathbf{t}/\|\mathbf{t}\|$ it is possible to reconstruct the object up to a positive scale factor (see Appendix for details).

III. SWITCHING VISUAL SERVOING ALGORITHM

A. Basic idea

The basic idea of our strategy consists of realizing a switching visual servoing where a position-based control law is selected when all points lie *inside*[1] the image, and some rotational or translational control laws are selected when at least one point lies on the image boundary in order to push such a point inside the image and to guarantee the final convergence. In particular, at any time instant in which all points lie inside the image, the camera velocities are selected according to the following standard PBVS:

$$\begin{cases} \mathbf{v}_r &= \lambda_r \mathbf{r}, \\ \mathbf{v}_t &= \lambda_t \mu \mathbf{t}/\|\mathbf{t}\|, \end{cases} \quad (2)$$

where $\mathbf{v}_r, \mathbf{v}_t \in \mathbb{R}^3$ are, respectively, the translational and rotational input velocities (expressed in the current camera frame), $\lambda_r, \lambda_t \in \mathbb{R}$ are positive gains, $\mathbf{r} \in \mathbb{R}^3$ is the vector containing the angular components of $\mathbf{R}$, i.e. $\mathbf{R} = e^{[\mathbf{r}]_\times}$, and

[1] We say that a point lies inside a region A if the point belongs to A but does not belong to the boundary of A.

$\mu \in \mathbb{R}$ is the image error ensuring the stop condition defined as

$$\mu = \sqrt{\frac{1}{n}\sum_{k=1}^{n} \|\mathbf{m}_{k,d} - \mathbf{m}_{k,c}\|^2}. \quad (3)$$

Then, at any time instant in which at least one point lies on the image boundary, the camera velocities are selected as follows:

- if the points lying on the image boundary will move inside the image by applying only the rotational control law in (2) [2], then

$$\begin{cases} \mathbf{v}_r &= \lambda_r \mathbf{r}, \\ \mathbf{v}_t &= \mathbf{0}_3; \end{cases} \quad (4)$$

- otherwise, if the points lying on the image boundary will move inside the image by applying only the translational control law in (2) [2], then

$$\begin{cases} \mathbf{v}_r &= \mathbf{0}_3, \\ \mathbf{v}_t &= \lambda_t \mu \mathbf{t}/\|\mathbf{t}\|; \end{cases} \quad (5)$$

- otherwise, the camera is sent away from the observed object through a backward translational motion along the the optical axis by selecting

$$\begin{cases} \mathbf{v}_r &= \mathbf{0}_3, \\ \mathbf{v}_t &= -\lambda_b \mu \mathbf{e}_3, \end{cases} \quad (6)$$

where $\lambda_b \in \mathbb{R}$ is a positive gain.

Observe that by selecting (6) we are always able to push inside the image any point that has reached the image boundary. Nevertheless, we try not to select (6) by checking, firstly, if it is possible to push such points inside the image by selecting (4) and, secondly, if it is possible by selecting (5) [3], in order to avoid the backward motion that produces a longer trajectory in the translational space. Observe also that the rotational velocity is always selected in two possible ways only: as $\mathbf{v}_r = \lambda_r \mathbf{r}$ or as $\mathbf{v}_r = \mathbf{0}_3$. This will allow us to achieve convergence and to follow the shortest path in the rotational space.

This basic idea, however, cannot be implemented in practice since visual servo systems are unavoidably discrete time systems and, hence, it is not possible to guarantee that a point escaping from the field of view will lie on the image boundary at any sampling instant before leaving the image. Moreover, the stability region in the space of the intrinsic parameters is restricted by the fact that the predicted direction along which the points move by selecting (4) or (5) depends on the available estimate of these parameters.

B. Practical implementation of the basic idea

First, in order to deal with the problem introduced by the time discretization we define a band along the image boundary for detecting points escaping from the field of view. In fact, while it is not possible to guarantee that an escaping point will lie on the image boundary at any sampling instant, it is

[2] We can predict the direction of the points image motion with a simple numerical test. This will be clarified in Section III-B and in the Appendix.
[3] The chosen priority between (4) and (5) is due to the fact that it is not possible to achieve the translational convergence with (5) without firstly achieving the rotational one with (4), while it is possible the opposite.

possible to guarantee that such point will lie on a sufficiently large band along the image boundary. Second, in order to achieve larger robustness against uncertainties on the intrinsic parameters we do not select control laws (4) and (5) if at least one point is too close to the image boundary. Hence, we define a second band surrounded by the previously defined one and select control law (6) if at least one point lies on the outer band, while we eventually select control laws (4) and (5) if no point lies on the outer band and at least one lies on the inner band. Figure (1) shows the image partitioned in the band A_2 (outer band), band A_1 (inner band), and remaining area A_0.

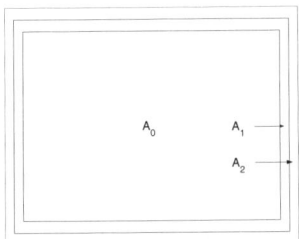

Fig. 1. Partition of the image in the bands A_2 and A_1, and in the area A_0.

In order to formally state the proposed strategy let us define the following conditions:

- *condition CR*: any point lying outside A_0 will initially move toward A_0 by selecting the camera velocities as in (4);
- *condition CT*: any point lying outside A_0 will initially move toward A_0 by selecting the camera velocities as in (5)

(conditions CR and CT can be easily checked from $\mathbf{R}$ and $\mathbf{t}/\|\mathbf{t}\|$ as shown in Appendix). Then, let us introduce the algorithm states characterized as follows:

- *state S1*: all points lie inside A_0;
- *state S2R*: (not all points lie inside A_0) & (all points lie inside $A_0 \cup A_1$) & (condition CR holds);
- *state S2T*: (not all points lie inside A_0) & (all points lie inside $A_0 \cup A_1$) & (condition CR does not hold) & (condition CT holds);
- *state S3*: any case not included in S1, S2 or S3.

Therefore, we implement the basic idea described in Section III-A by realizing a switching controller that, at any time instant, selects the camera velocities $\mathbf{v}_r$ and $\mathbf{v}_t$ in function of the current state as shown in Table I.

algorithm state	camera velocities
S1	select as in (2)
S2R	select as in (4)
S2T	select as in (5)
S3	select as in (6)

TABLE I

Switching visual servoing algorithm: camera velocities selection in function of the current state.

From the previous discussions it follows that the thickness of bands A_2 and A_1 has to be selected not less than the maximum shift of any point between two sampling instants (in order to guarantee that a point escaping from A_0 will lie at any sampling instant on A_1, and that one escaping from A_1 will lie at any sampling instant on A_2). Such a thickness depends on the camera speed and on the sampling rate of the visual servo system, and amounts of few pixels for standard applications.

IV. PROPERTIES OF THE PROPOSED STRATEGY

In this section we analyze stability, performance and robustness of the switching visual servoing proposed in Table I. Before introducing such an analysis, some basic assumptions have to be clarified:

- all points are assumed to lie inside A_0 in the initial and desired views, and the corresponding 3D points are supposed stationary during the visual servoing;
- occlusions are allowed if the number of points not occluded and tracked during the visual servoing is sufficient to estimate the camera displacement between current and desired situation;
- the estimation of the optical axis direction is assumed to be performed through robot joint angles and, hence, independent from the *actual* intrinsic parameters;
- no object is assumed along the camera trajectory.

The proofs of the following results can be found in [19].

Theorem 1: In absence of uncertainties on the intrinsic parameters and optical axis direction, the visual servoing is globally stable and the object is kept in the field of view.

Theorem 2: In absence of uncertainties on the intrinsic parameters, the camera follows the shortest path in the rotational space and, in some cases including all the ones in which no point gets out from A_0, also in the translational space.

Remark Although the obtained camera trajectory in the translational space may correspond to the shortest one for some cases (see Example 1 in the next section), it is possible that such trajectory is far from the shortest one for other cases due to the backward motion. This is the cost to pay in our approach for keeping all points in the field of view. However, we point out that most existing visual servoing algorithms (for example most algorithms completely or partially based on IBVS) not only do not guarantee to satisfy the visibility constraint and/or to achieve global convergence, but also do not try to minimize the camera trajectory length, contrarily to our approach.

Theorem 3: (Stability robustness) The visual servoing is robustly stable with respect to uncertainties on the intrinsic parameters and on the optical axis direction. In particular, stability is achieved for any uncertainty on the optical axis direction.

Theorem 4: (Visibility robustness) The visual servoing robustly keeps the object in the field of view with respect to uncertainties on the intrinsic parameters and on the optical axis direction. In particular, this visibility requirement is satisfied for any uncertainty on the intrinsic parameters.

V. SIMULATIONS RESULTS

In this section some simulation results of the proposed visual servoing strategy are presented. In all examples the desired camera situation is $\{\mathbf{c}_d, \mathbf{R}_d\} = \{\mathbf{0}_3, \mathbf{I}_3\}$, and the same configuration of $n = 8$ points is used to estimate the camera displacement through the essential matrix. The screen size is 320×200 pixels and the bands A_1 and A_2 of Figure 1 have thickness equal to 5 pixels. The intrinsic parameters matrix is $\mathbf{K} = [320, 0, 160; 0, 200, 100; 0, 0, 1]$.

A. Example 1

The initial camera situation is $\{\mathbf{c}_i, \mathbf{R}_i\} = \{[-100, 24, -10]^T cm, e^{[[\pi/12, \pi/6, -\pi/8]]_\times}\}$ as shown in Figure 2. Figure 3 shows the results obtained in absence of uncertainties (on the intrinsic parameters and optical axis direction). Since the intrinsic parameters are supposed known the trajectory length in the rotational space is minimized according to Theorem 2. Observe that, in this case, the trajectory length is minimized also in the translational space as indicated by Figure 3(d) which shows that the camera center follows a straight line (that is, the algorithm never reaches state S3 corresponding to the backward motion).

Figure 4 shows the results obtained in presence of large uncertainties. In particular, the wrong estimate of the intrinsic parameters matrix used to estimate the camera displacement is $\hat{\mathbf{K}} = [410, 0, 200; 0, 290, 130; 0, 0, 1]$. Uncertainties on the optical axis direction are not supposed since, again, the algorithm never reaches state S3 and hence the backward motion is not used.

B. Example 2

In this example a large initial camera displacement is considered, specifically $\{\mathbf{c}_i, \mathbf{R}_i\} = \{[-90, 20, 120]^T cm, e^{[[-\pi/2, \pi/2, -3\pi/10]]_\times}\}$ as shown in Figure 5. Figure 6 shows the results obtained in absence of uncertainties. Differently from Example 1, the algorithm this time reaches state S3 and hence the backward motion is used to keep points in the field of view.

Figure 7 shows the results obtained in presence of large uncertainties. In particular, the wrong estimate of the intrinsic parameters matrix is again selected as $\hat{\mathbf{K}} = [410, 0, 200; 0, 290, 130; 0, 0, 1]$, and the estimate of the optical axis direction is supposed to be corrupted along both perpendicular directions by an unknown noise belonging to the interval $[-25, 25]$ degrees (it can be verified that this noise cannot make points leaving the field of view for the chosen band A_2).

Figure 8 shows the results obtained in presence of the same uncertainties on the intrinsic parameters and optical axis direction and in presence of image noise. In particular, each point coordinate is, firstly, corrupted by an additive noise with maximum amplitude equal to 0.5 pixels and, secondly, rounded towards the nearest integer. It turns out that the points are still kept in the field of view and the camera converges toward the desired situation. However, due to image noise, the behaviour of PBVS close to the convergence is unsatisfactory (oscillations around the desired situation may be present) especially if few points are used to estimate the camera displacement as in this case. This problem can be overcome by switching to a standard IBVS in the last steps of the visual servoing when the camera is by now very close to the desired situation and hence points do not get out from the field of view by applying such a technique (in this example IBVS is applied from the vertical dashed line indicated in Figures 8(a)-(b)).

VI. ABOUT EXPERIMENTS AND SOME COMMENTS

At present we are not able to report real experiments in which the proposed approach is used. For this reason, in the simulations described in Section V we have tried to consider typical experimental conditions, as uncertainties on the intrinsic parameters, uncertainties on the optical axis direction and image noise, in order to show possible experimental results.

Other aspects regarding the implementation of the proposed approach are as follows. First, the simple control laws used in (2) can be substituted with PID ones in order to take into account constraints on the camera velocities, especially at the beginning of the visual servoing, and achieve better performance.

Second, whenever any point should become not visible due to occlusions, it is possible to try to still keep such point in the field of view *in order to be successively re-tracked* by computing its image projections from the last computed scaled euclidean reconstruction containing it. Obviously, an accurate enough estimate of the intrinsic parameters should be available.

Third, as it can be seen from the simulation results, chattering phenomenon is present when any point gets out from A_0, corresponding to discontinuous camera velocities. This is a characteristic of our strategy that, however, can be reduced by decreasing the control gains when some points are too close or outside A_0.

VII. CONCLUSION

A visual servoing strategy for dealing with the problem of keeping the observed points in the camera field of view has been proposed. The approach consists of a switching control law based on camera displacement estimation and regulated from the position of the points in the image. Stability, performance and robustness have been investigated. Specifically, it has been shown that, in absence of uncertainties on the intrinsic parameters and optical axis direction, global stability is achieved and all points are kept in the field of view. Moreover, the trajectory length is minimized in the rotational space and, for some cases, also minimized in the translational one. Robustness against uncertainties is also guaranteed. In particular, examples with a coarsely calibrated camera, large uncertainties on the optical axis direction and image noise have shown that the proposed strategy is quite robust.

ACKNOWLEDGEMENT

The authors would like to thank the reviewers of this paper for their helpful comments.

References

[1] K. Hashimoto, *Visual Servoing: Real-Time Control of Robot Manipulators Based on Visual Sensory Feedback*. Singapore: World Scientific, 1993.

[2] S. Hutchinson, G. Hager, and P. Corke, "A tutorial on visual servo control," *IEEE Trans. on Robotics and Automation*, vol. 12, no. 5, pp. 651–670, 1996.

[3] R. Basri, E. Rivlin, and I. Shimshoni, "Visual homing: surfing on the epipole," *Int. Journal of Computer Vision*, vol. 33, no. 2, pp. 22–39, 1999.

[4] C. Taylor and J. Ostrowski, "Robust vision-based pose control," in *Proc. IEEE Int. Conf. on Robotics and Automation*, (San Francisco, California), pp. 2734–2740, 2000.

[5] A. Sanderson, L. Weiss, and C. Neuman, "Dynamic sensor-based control of robots with visual feedback," *IEEE Trans. on Robotics and Automation*, vol. RA-3, pp. 404–417, 1987.

[6] J. Feddema and O. Mitchell, "Vision-guided servoing with feature-based trajectory generation," *IEEE Trans. on Robotics and Automation*, vol. 5, no. 5, pp. 691–700, 1989.

[7] K. Hashimoto, T. Kimoto, T. Ebine, and H. Kimura, "Manipulator control with image-based visual servo," in *Proc. IEEE Int. Conf. on Robotics and Automation*, pp. 2267–2272, 1991.

[8] F. Chaumette, "Potential problems of stability and convergence in image-based and position-based visual servoing," in *The confluence of vision and control* (G. H. D. Kriegman and A. Morse, eds.), pp. 66–78, Springer-Verlag, 1998.

[9] E. Malis, F. Chaumette, and S. Boudet, "2 1/2 D visual servoing," *IEEE Trans. on Robotics and Automation*, vol. 15, no. 2, pp. 238–250, 1999.

[10] G. Morel, T. Leibezeit, J. Szewczyk, S. Boudet, and J. Pot, "Explicit incorporation of 2D constraints in vision based control of robot manipulators," in *Experimental Robotics VI* (P. Corke and J. Trevelyan, eds.), Springer-Verlag, 2000.

[11] K. Deguchi, "Optimal motion control for image-based visual servoing by decoupling translation and rotation," in *Proc. Int. Conf. on Intelligent Robots and Systems*, pp. 705–711, 1998.

[12] P. Oh and P. Allen, "Visual servoing by partitioning degrees-of-freedom," *IEEE Trans. on Robotics and Automation*, vol. 17, no. 1, pp. 1–17, 2001.

[13] P. Corke and S. Hutchinson, "A new partitioned approach to image-based visual servo control," *IEEE Trans. on Robotics and Automation*, vol. 17, no. 4, pp. 507–515, 2001.

[14] Y. Mezouar and F. Chaumette, "Visual servoing by path planning," in *Proc. 6th European Control Conference*, (Porto, Portugal), pp. 2904–2909, 2001.

[15] B. Thuilot, P. Martinet, L. Cordesses, and J. Gallice, "Position based visual servoing: keeping the object in the field of vision," in *Proc. IEEE Int. Conf. on Robotics and Automation*, (Washington, D.C.), pp. 1624–1629, 2002.

[16] O. Faugeras and Q.-T. Luong, *The Geometry of Multiple Images*. Cambridge (Mass.): MIT Press, 2001.

[17] J. Weng, T. Huang, and N. Ahuja, "Motion and structure from two perspective views: Algorithms, error analysis, and error estimation," *IEEE Trans. on Pattern Analysis and Machine Intelligence*, vol. 11, no. 5, pp. 451–476, 1989.

[18] E. Malis and F. Chaumette, "2 1/2 D visual servoing with respect to unknown objects through a new estimation scheme of camera displacement," *Int. Journal of Computer Vision*, vol. 37, no. 1, pp. 79–97, 2000.

[19] G. Chesi, K. Hashimoto, D. Prattichizzo, and A. Vicino, "A switching control law for keeping features in the field of view in eye-in-hand visual servoing," technical report, 2003.

Appendix

In order to check condition CR and CT we first compute a scaled euclidean reconstruction of the object in the current camera frame as follows. Let us denote with $\bar{\mathbf{q}}_k \in \mathbb{R}^3$ the reconstruction of the 3D point $\mathbf{q}_k$ up to the scale factor $\|\mathbf{c}_d - \mathbf{c}_c\|$. Points $\bar{\mathbf{q}}_k$ have to satisfy the system of equations obtained by writing for all $k = 1, \ldots, n$ equation (1) with the substitutions $\mathbf{q}_k \to \bar{\mathbf{q}}_k$, $\mathbf{c}_c \to \mathbf{0}_3$, $\mathbf{c}_d \to \mathbf{t}/\|\mathbf{t}\|$, $\mathbf{R}_c \to \mathbf{I}_3$ and $\mathbf{R}_d \to \mathbf{R}$. From this system we compute $\bar{\mathbf{q}}_k$ through linear least-squares.

Now, let $\mathbf{p}_{k,c}$ be a point lying on A_1 and let us indicate $\bar{\mathbf{q}}_k$ as $\bar{\mathbf{q}}_k = [\bar{x}_k, \bar{y}_k, \bar{z}_k]^T$. Condition CR can be checked as follows. By selecting the camera velocities as in (4) the angular components move according to $\mathbf{r}_c \to \mathbf{r}_c + \alpha \mathbf{r}$ where $\alpha \in \mathbb{R}$ is positive, and consequently $\mathbf{p}_{k,c}$ moves according to

$$\mathbf{p}_{k,c} \to \tilde{\mathbf{p}}_{k,c} = \mathbf{K} \left[\frac{\bar{\mathbf{q}}_k^T \mathbf{R} \mathbf{e}_1}{\bar{\mathbf{q}}_k^T \mathbf{R} \mathbf{e}_3}, \frac{\bar{\mathbf{q}}_k^T \mathbf{R} \mathbf{e}_2}{\bar{\mathbf{q}}_k^T \mathbf{R} \mathbf{e}_3}, 1 \right]^T. \quad (7)$$

Then, for small α, $\mathbf{p}_{k,c}$ moves along the direction

$$\left. \frac{\partial \tilde{\mathbf{p}}_{k,c}}{\partial \alpha} \right|_{\alpha=0} = \frac{\mathbf{K}}{\bar{z}_k^2} \left[\begin{array}{c} \bar{z}_k \bar{\mathbf{q}}_k^T [r]_\times \mathbf{e}_1 - \bar{x}_k \bar{\mathbf{q}}_k^T [r]_\times \mathbf{e}_3 \\ \bar{z}_k \bar{\mathbf{q}}_k^T [r]_\times \mathbf{e}_2 - \bar{y}_k \bar{\mathbf{q}}_k^T [r]_\times \mathbf{e}_3 \\ 0 \end{array} \right]. \quad (8)$$

Hence, condition CR can be checked by calculating the direction in (8) and checking if point $\mathbf{p}_{k,c}$ along this direction moves toward A_0 or not. Condition CT can be analogously checked. Let us indicate $\mathbf{t}/\|\mathbf{t}\|$ as $\mathbf{t}/\|\mathbf{t}\| = \bar{\mathbf{t}} = [\bar{t}_x, \bar{t}_y, \bar{t}_z]^T$. By selecting the camera velocities as in (5) the camera center moves according to $\mathbf{c}_c \to \mathbf{c}_c + \alpha \bar{\mathbf{t}}$ where $\alpha \in \mathbb{R}$ is positive, and consequently $\mathbf{p}_{k,c}$ moves according to

$$\mathbf{p}_{k,c} \to \tilde{\mathbf{p}}_{k,c} = \mathbf{K} \left[\frac{\bar{x}_k - \alpha \bar{t}_x}{\bar{z}_k - \alpha \bar{t}_z}, \frac{\bar{y}_k - \alpha \bar{t}_y}{\bar{z}_k - \alpha \bar{t}_z}, 1 \right]^T. \quad (9)$$

Then, for small α, $\mathbf{p}_{k,c}$ moves along the direction

$$\left. \frac{\partial \tilde{\mathbf{p}}_{k,c}}{\partial \alpha} \right|_{\alpha=0} = \frac{\mathbf{K}}{\bar{z}_k^2} \left[\begin{array}{c} \bar{x}_k \bar{t}_z - \bar{z}_k \bar{t}_x \\ \bar{y}_k \bar{t}_z - \bar{z}_k \bar{t}_y \\ 0 \end{array} \right]. \quad (10)$$

Hence, condition CT can be checked by calculating the direction in (10) and checking if point $\mathbf{p}_{k,c}$ along this direction moves toward A_0 or not.

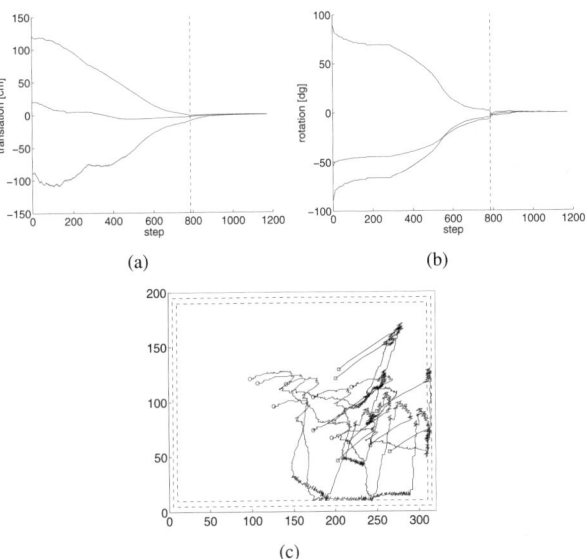

Fig. 8. Example 2 (presence of uncertainties and image noise). Translation (a), rotation (b) and camera view (c).

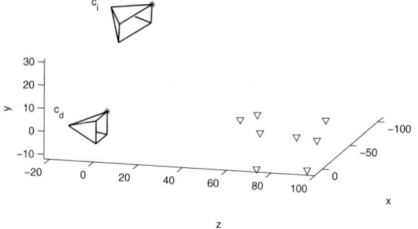

Fig. 2. Example 1: 3D points and camera in the initial and desired situations.

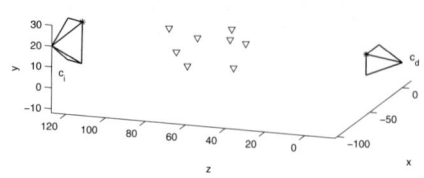

Fig. 5. Example 2: 3D points and camera in the initial and desired situations.

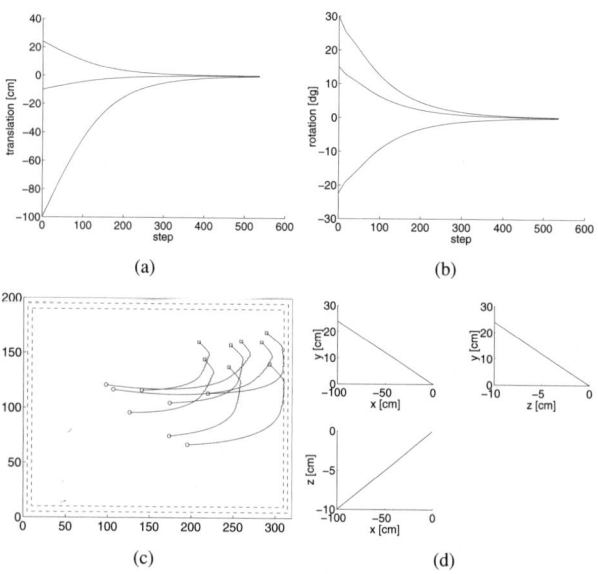

Fig. 3. Example 1 (absence of uncertainties). Translation (a), rotation (b), camera view (c) and camera center trajectory (d).

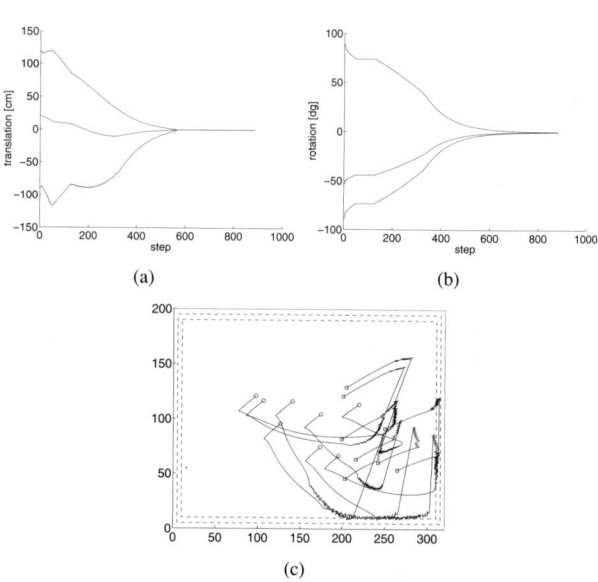

Fig. 6. Example 2 (absence of uncertainties). Translation (a), rotation (b), camera view (c).

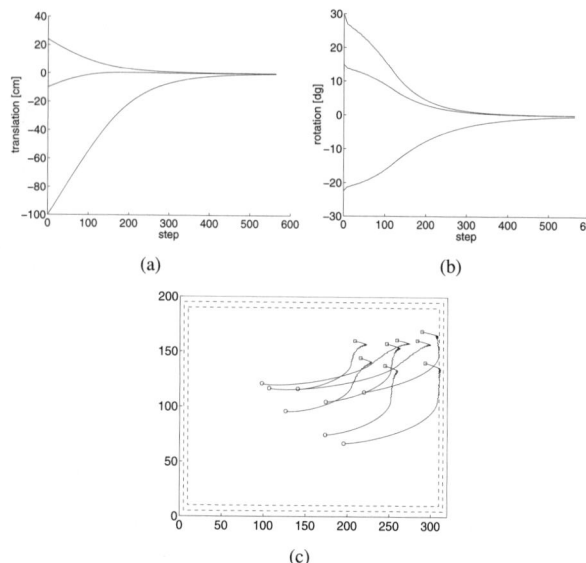

Fig. 4. Example 1 (presence of uncertainties). Translation (a), rotation (b) and camera view (c).

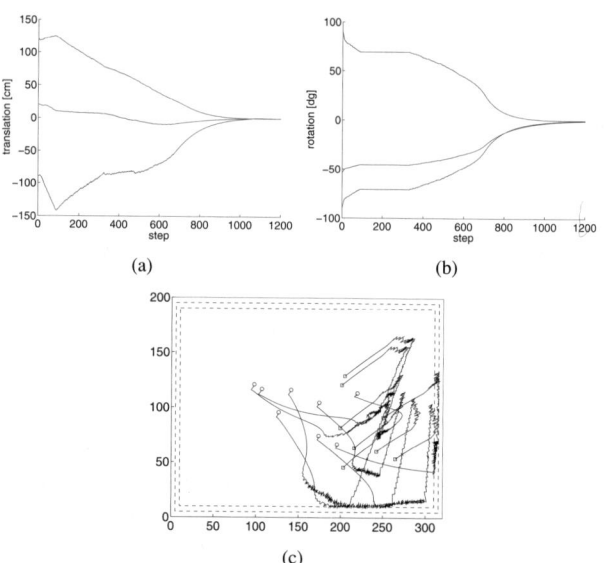

Fig. 7. Example 2 (presence of uncertainties). Translation (a), rotation (b) and camera view (c).

3934

Visual Registration and Navigation using Planar Features

Gabriel A. D. Lopes
EECS Department, College of Engineering
University of Michigan, Ann Arbor, MI
Email: glopes@umich.edu

Daniel E. Koditschek
EECS Department, College of Engineering
University of Michigan, Ann Arbor, MI
Email: kod@umich.edu

Abstract—This paper addresses the problem of registering the hexapedal robot, RHex, relative to a known set of beacons, by real-time visual servoing. A suitably constructed navigation function represents the task, in the sense that for a completely actuated machine in the horizontal plane, the gradient dynamics guarantee convergence to the visually cued goal without ever losing sight of the beacons that define it. Since the horizontal plane behavior of RHex can be represented as a unicycle, feeding back the navigation function gradient avoids loss of beacons, but does not yield an asymptotically stable goal. We address new problems arising from the configuration of the beacons and present preliminary experimental results that illustrate the discrepancies between the idealized and physical robot actuation capabilities.

I. INTRODUCTION

This paper reports on our progress in adapting the fixed camera, moving beacon visual servoing algorithms of Cowan et al [1] to the beacon "inside out" version of the problem — a moving camera reacting to a fixed beacon — that arises when attempting to register a mobile robot vehicle relative to some effective landmark in its visual field. Specifically, we are interested in applying these ideas to the hexapedal robot, RHex [2], [3] considered as operating in the (three degree of freedom) horizontal plane. The lower level controls presently operative in our legged machine result in horizontal plane behavior nicely modeled by a unicycle [4]. The reduced affordance of this nonholonomically constrained model precludes the possibility of point stabilization by any smooth feedback law [5], and the navigation function will eventually play the role of a control Lyapunov function [6] in this research. In the present paper, we illustrate the interplay between the navigation function (our task model), its realization in physical hardware, and the preliminary navigation results that we have obtained to date both in extensive simulation studies and on the physical RHex platform.

A. Background Literature

Several authors have dealt with the problem of vision based navigation [7], [8], almost exclusively, to date, in indoor environments. Ostrowski [9], [10] uses a blimp equipped with a camera that implements a diffeomorphism between image plane features and robot pose to maintain a constant distance from the beacon. Ezio and Chaumette [11] decouple the rotation and translation degrees of freedom to position a fully actuated camera arm in relation to a collection of features. Cowan [1], [12], [13] servos a 6 dof arm to a predefined pose, by introducing a Navigation function that guarantees the features stay in the field of view (FOV) of the camera at all times.

The problem to be solved in this paper entails navigation of an autonomous hexapod robot in an environment with known beacons using vision. The paper is an extension of Cowan's work [12] in the sense that it generalizes the configuration of the beacons (landmarks) in the planar version of the problem and implements the controller on a legged platform, maintaining the same emphasis on convergence to the goal with no FOV violations, modulo the reduced control affordance introduced by the kinematic constraints of the mobile platform.

The mobile platform of present interest is RHex, [2], [3], (illustrated in figure 1), a hexapedal machine with passive compliant legs that afford impressive mobility. Much of the theoretical inquiry into this machine has been confined to its behavior in the sagittal plane [3], [14], leaving a significant gap in the characterization of its operation in the horizontal plane. For present purposes, when only small accelerations of the body are required, we will find it acceptable to characterize RHex's horizontal plane mechanics via the standard quasi-static "unicycle" model — a nonholonomically constrained machine whose velocity can be commanded in the fore-aft and heading directions relative to the body. The efficacy of this highly simplified model for the present quasi-static operating regime is documented by a comparison of simulation and experimental results, below. No doubt, extending these techniques to the full dynamical regime of which RHex is capable will require a plant model far more accurately informed about its complex Lagrangian mechanics.

B. Organization of the paper

Section II describes analytically the generalization of the visual servoing algorithm developed by Cowan [13]. In section III it is shown that different configurations of beacons yield different pose measurement error for a given location and therefore, a proper choice of beacons may help improve the accuracy of the pose. Simulations of the globally convergent controller are made for a fully actuated rigid body and for a unicycle. Finally section IV describes the implementation on RHex.

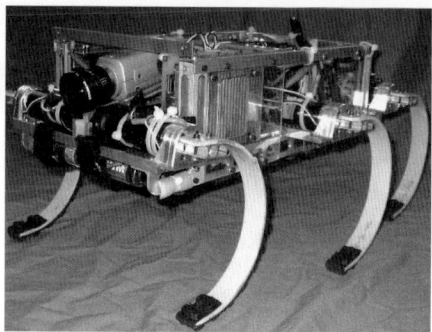

Fig. 1. RHex - Robot hexapod

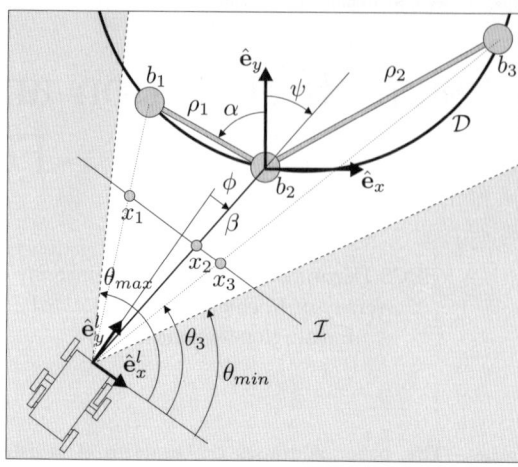

Fig. 2. The parameters $(\alpha, \rho_1, \rho_2) \in \mathcal{B}$ define the beacon configuration. The parameters (ϕ, ψ, β) define the coordinates in $\mathcal{Q}$, the robot's configuration space. The dashed lines represent the FOV boundary with parameters $(\theta_{min}, \theta_{max})$. The line $\mathcal{I}$ represents the camera projection line and the points (x_1, x_2, x_3) are the projection of each beacon b_i into $\mathcal{I}$.

II. 2D Visual Servoing

Throughout this paper we assume a perfectly calibrated 1 dimensional pinhole camera, and our robot model assumes motion in only the horizontal plane. We further assume that the geometry of the beacons is known and the correspondence problem is solved. The algorithm presented here uses 3 beacons to extract and regulate full relative pose. In [12] it is shown that an algebraic inverse of the camera map is obtained for 3 collinear points in $SE(2)$. Although this result may be sufficient for some applications, for real environments it will become important for the robot to be able to handle more general configurations of beacons. With appropriate image preprocessing one can assign signatures to natural elements like trees, stones, etc, but natural beacons are, in general, non-collinear. We now generalize the methods of [12], [13] to accommodate arbitrary beacon configurations.

A. Pose computation

The first step is to define a parameterization of the beacons and find the camera map that relates the projected beacon coordinates in the camera image line $\mathcal{I}$ to the robot pose in $SE(2)$ (the camera projection plane is reduced to a line in $SE(2)$). For any configuration of beacons in a plane, define the beacon parameter space $\mathcal{B} \subset \mathbb{R}^2 \times S$ by fixing the world frame so that the second point is at the origin and the remaining points lie in lines going through the origin with congruent angles. Figure 2 illustrates the configuration of the beacons with parameters (α, ρ_1, ρ_2).

$$\mathcal{B} := \{(\rho_1, \rho_2, \alpha) \in SE(2) \mid \rho_1 > 0, \rho_2 > 0, 0 \leq \alpha < \pi\}$$

The coordinates of each beacon b_i in the world frame are:

$$\begin{bmatrix} b_1 & b_2 & b_3 \end{bmatrix} = \begin{bmatrix} \rho_1 R_\alpha \hat{\mathbf{e}}_y & \mathbf{0} & \rho_2 R_\alpha^T \hat{\mathbf{e}}_y \end{bmatrix}$$

To build the camera map it is convenient to use polar coordinates, in effect passing to a new space $\mathcal{Q}$, diffeomorphic to the robot configuration space. $\mathcal{Q}$ expresses in a computationally tractable form the fact that the robot configuration space has the topology of a solid torus (after removing a disk enclosing the beacons from the robot's available workspace). The motivation to introduce such a coordinate system arises from the fact that in $\mathcal{Q}$ the set of self-occlusions appears as a literal (2 dimensional) torus, providing significant geometrical insight into the self-occlusion problem. Figure 2 illustrates the parameterization of the space $\mathcal{Q}$ with parameters (ϕ, ψ, β).

Having adopted a representation for the beacon configuration and the robot configuration space it is now necessary to determine for a given beacon the set of robot configurations for which occlusion-free servoing can be accomplished. Define the facing set $\mathcal{F}$ as the set of configurations for which the robot lies "in front" of the set of beacons, i.e. the beacons appear to face the robot sensor. Intuitively the beacons must keep a certain order in the camera projection line $\mathcal{I}$. Define the function f_i that returns a vector that goes through beacon b_i for a given configuration $\mathbf{q} = (\phi, \psi, \beta)$.

$$f_i(\mathbf{q}) := R_\phi R_\psi b_i + \beta R_\phi \hat{\mathbf{e}}_y$$

The facing set if then defined by (1) where J is a skew symmetric matrix:

$$\mathcal{F} := \{\mathbf{q} \in \mathcal{Q} \mid f_i(\mathbf{q})^T J f_j(\mathbf{q}) > 0; i < j\} \quad (1)$$

Define the visible set $\mathcal{V}$ as the set of configurations for which the beacons are in the FOV of the camera sensor, where $\theta_{min}, \theta_{max}$ are the FOV camera parameters illustrated in figure 2 and function "$\angle$" returns the angle of a vector.

$$\mathcal{V} := \{\mathbf{q} \in \mathcal{Q} \mid \theta_{min} < \angle(f_i(\mathbf{q})) < \theta_{max}, i = 1, 2, 3\}$$

The previous sets arise from geometrical insight, necessary for the vision implementation, but are not sufficient to fully characterize the set of configurations for which pose computation can be accomplished. In fact, as shown next, the camera map may not always be injective in $\mathcal{F} \cap \mathcal{V}$. This is due to the generalization of [13] by allowing any beacon configuration. It is shown here that the injectivity is lost at worst on the zero set of the function $\Theta_{\mathbf{b}}$ (a factor in the determinant of the jacobian of $c_{\mathbf{b}}$) in which set the inverse image can have

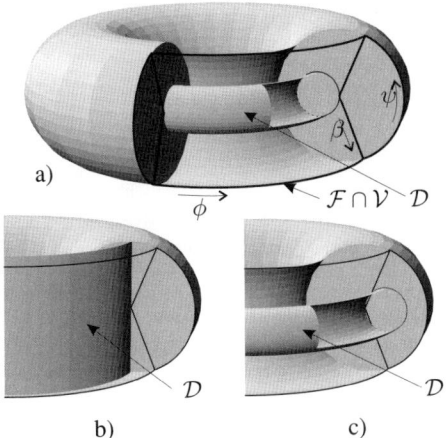

Fig. 3. Illustration of $\mathcal{Q}$. The thick black lines represent the intersection of the visible set and facing set slicing $\mathcal{Q}$. a) convex configuration. b) linear configuration. c) concave configuration. For the concave configuration c) $\mathcal{D}$ disconnects $\mathcal{F} \cap \mathcal{V}$.

cardinality 2. We then introduce the degenerate set $\mathcal{D}$ using the function $\Theta_{\mathbf{b}}$:

$$\Theta_{\mathbf{b}}(\mathbf{q}) \;:\; \mathcal{B} \times \mathcal{Q} \to \mathbb{R}$$
$$\mathbf{q} \mapsto \rho_1 \sin(\alpha - \psi) + \rho_2 \sin(\alpha + \psi) + \beta \sin(2\alpha)$$

$$\mathcal{D} := \{\mathbf{b} \in \mathcal{B}, \mathbf{q} \in \mathcal{Q} \mid \Theta_{\mathbf{b}}(\mathbf{q}) = 0\}$$

If the configuration space is understood topologically as a thickened torus, then the degenerate set will in general be a thin torus that disconnects $\mathcal{Q}$. If the three beacons are collinear then $\mathcal{D}$ becomes a cylinder. Figure 3 illustrates the solid torus sliced by the FOV of the camera sensor. One should notice that when the FOV slices the configuration space it may be the case that the degenerate set does not disconnect the facing set $\mathcal{F} \cap \mathcal{V}$ as in figures 3a) and 3b). In fact, only if the set of beacons is configured in a concave shape will $\mathcal{D}$ disconnect $\mathcal{F}$ (figure 3c)). Define the free configuration space $\mathcal{W} \subset \mathcal{Q}$ by:

$$\mathcal{W} := (\mathcal{F} \cap \mathcal{V}) - \mathcal{D}$$

Proposition 1. *There exists a smooth and smoothly invertible map from the free configuration space $\mathcal{W}$ into a subset of T^3.*

Proof: Consider the map:

$$c_{\mathbf{b}}(\mathbf{q}) \;:\; \mathcal{W} \to T^3 \qquad (2)$$
$$\mathbf{q} \mapsto \begin{bmatrix} \angle(f_1(\mathbf{q})) & \angle(f_2(\mathbf{q})) & \angle(f_3(\mathbf{q})) \end{bmatrix}^T$$

Smoothness is easily verifiable. It is sufficient to show that the map $c_{\mathbf{b}}$ is a local diffeomorphism in the neighborhood of a point in $\mathcal{W}$ and that the cardinality of the inverse image in the co-domain is unity. This is necessary since $\mathcal{D}$ disconnects $\mathcal{F} \cap \mathcal{V}$.

The determinant of the Jacobian of the camera map degenerates only on the degenerate set and outside the visible and facing sets. Therefore $c_{\mathbf{b}}$ is a local diffeomorphism in $\mathcal{W}$:

$$|D_{\mathbf{q}} c_{\mathbf{b}}| = \frac{\rho_1 \rho_2 \Theta_{\mathbf{b}}}{\|f_1\|^2 \|f_2\|^2}$$

Now suppose that there exist two configurations $\mathbf{q}'$ and $\mathbf{q}''$ such that $c_{\mathbf{b}}(\mathbf{q}') = c_{\mathbf{b}}(\mathbf{q}'')$. This is equivalent to saying that each of the points $f_i(\mathbf{q}'')$ and $f_i(\mathbf{q}'')$ are in a line that goes through the origin, i.e.:

$$f_i(\mathbf{q}'') \times f_i(\mathbf{q}'') = 0 \qquad (3)$$

For $i = 2$ we get $(\beta' R_{\phi'} \hat{\mathbf{e}}_y)^T J(\beta'' R_{\phi''} \hat{\mathbf{e}}_y) = 0$ which simplifies to $\beta' \beta'' \sin(\phi' - \phi'') = 0$. Since β' and β'' cannot be null this results in $\phi' = \phi'' + k\pi$ with $k \in \mathbb{N}$. For $i = 1, 3$ equation (3) simplifies to:

$$\begin{cases} \Theta_{\mathbf{b}}(\mathbf{q}'') \sin(\psi' + \alpha) - \Theta_{\mathbf{b}}(\mathbf{q}') \sin(\psi'' + \alpha) = 0 \\ \Theta_{\mathbf{b}}(\mathbf{q}'') \sin(\psi' - \alpha) - \Theta_{\mathbf{b}}(\mathbf{q}') \sin(\psi'' - \alpha) = 0 \end{cases} \qquad (4)$$

Eliminating $\Theta_{\mathbf{b}}(\mathbf{q}')$ and $\Theta_{\mathbf{b}}(\mathbf{q}'')$ from the previous equations we get:

$$\sin(2\alpha) \sin(\psi' - \psi'') = 0 \Rightarrow \psi' = \psi'' + k\pi, k \in \mathbb{N}$$

Finally using equation (4) with $\psi'' = \psi' = \psi$ completes the result:

$$\begin{cases} \rho_1(\beta' - \beta'') \sin(\alpha + \psi) = 0 \\ \rho_2(\beta' - \beta'') \sin(\alpha - \psi) = 0 \end{cases} \Rightarrow \beta' = \beta''$$

$\square$

Call $c_{\mathbf{b}}$ the camera map and define the set $\mathcal{Y} = c_{\mathbf{b}}(\mathcal{W})$. To find the inverse camera map the same constructive method is used as in [12]. Let the projection of the beacons in the camera projection line be $(\theta_1, \theta_2, \theta_3) \in T^3$ (i.e. $\theta_i = \arctan(x_i) + \pi/2$ as illustrated in figure 2) and let Y and Y' be:

$$Y = \begin{bmatrix} \cos(\theta_1) & \cos(\theta_2) & \cos(\theta_3) \\ \sin(\theta_1) & \sin(\theta_2) & \sin(\theta_3) \end{bmatrix}$$
$$Y' = \begin{bmatrix} \rho_1 \cos(\theta_1 - \alpha) & 0 & \rho_2 \cos(\theta_3 + \alpha) \\ \rho_1 \cos(\theta_1 - \alpha) & 0 & \rho_2 \cos(\theta_3 + \alpha) \end{bmatrix}$$

The robot's pose is computed by the following expressions, where $Y^\dagger$ is the pseudo-inverse of Y^T and $Y_\perp$ is the orthogonal complement of the subspace generated by the lines of $Y^\dagger$:

$$\begin{aligned} \phi &= \theta_2 + \frac{\pi}{2} \\ \psi &= \angle(\delta R_\phi^T J Y' Y_\perp) \qquad (5) \\ \beta &= \frac{\|Y^\dagger Y'^T J Y' Y_\perp\|}{\|Y' Y_\perp\|} \end{aligned}$$

Having an explicit closed form expression for the camera map and its inverse parameterized by the beacon configuration, one may now address the question: how does the beacon configuration affect the pose computation error? For convex and collinear beacons it is expected that the pose computation error grows with distance, but for concave beacon configuration a more complex error structure is expected, due to the degenerate set $\mathcal{D}$. Section III approaches this question through a numerical study.

For concave beacons the free configuration space $\mathcal{W}$ is disconnected. A new question arises: If the robot's initial and goal pose lie in distinct connected components of $\mathcal{W}$ can occlusion-free navigation be accomplished? In other words, is it possible to "puncture" the disconnecting degenerate set $\mathcal{D}$? In section III-B numerical simulation suggest that in general it is possible to accomplish global convergent occlusion-free navigation even in the presence of a disconnecting degenerate set $\mathcal{D}$.

B. Navigation function

Since the camera map is a diffeomorphism between the free configuration space $\mathcal{W}$ and the projected beacons on the camera projection line $\mathcal{I}$, one can build a potential function φ so that the system $\dot{\mathbf{y}} = -\nabla\varphi(\mathbf{y})$ is globally asymptotically stable in $\mathcal{Y}$. Next, we use the camera map $c_\mathbf{b}$ to pull back the velocities from a known globally convergent system into $\mathcal{Q}$. Let φ be a potential function:

$$\varphi(\theta) : T^3 \to [0,1]$$
$$\theta \mapsto \frac{\bar{\varphi}(\theta)^k}{\epsilon + \bar{\varphi}(\theta)^k} \quad (6)$$

$$\bar{\varphi}(\theta) := \frac{\left(\sum_{i=1}^{3}(\theta_i - \theta_i^*)^2\right)^m}{(\theta_{max} - \theta_1)(\theta_1 - \theta_2)(\theta_2 - \theta_3)(\theta_3 - \theta_{min})}$$

By construction the function φ equals unity on the boundary of $\mathcal{Y}$ and has a global minima at the goal configuration. φ is also continuous and differentiable and therefore it is a navigation function in $\mathcal{Y}$ as defined in [15]. The parameters ϵ, k and m shape the function φ to allow fine tuning of the resultant velocity vector field, $(\theta_1^*, \theta_2^*, \theta_3^*)$ represent the robot goal configuration in T^3 and $(\theta_{min}, \theta_{max})$ are the FOV parameters described in figure 2.

The final ingredient is to pullback the gradient vector field $\nabla\varphi$ into the world space. Two new maps are introduced to accomplish that: Φ maps coordinates in $\mathcal{Q}$ into $SE(2)$ in the local robot frame. Υ maps local robot coordinates into world coordinates. See the appendix for details on these maps. Define the full camera map $\bar{c}_\mathbf{b} = c_\mathbf{b} \circ \Phi \circ \Upsilon(x_w)$. Writing the gradient system in the world space we get 7, then apply the chain rule on $\bar{c}_\mathbf{b}$.

$$\dot{x}_w = \mathbf{u} = -\nabla\left(\varphi \circ \bar{c}_\mathbf{b}\right)(x_w) = -D\bar{c}_\mathbf{b}^T \cdot \nabla\varphi(x_w) \quad (7)$$

C. Unicycle model

In the previous sections it is assumed that the robot is fully actuated. In reality the dynamical model of RHex, used in the experiments, is not yet fully modeled in all gaits and terrains of interest. Several assumptions are made in order to implement the algorithms previously described. Most importantly, on the strength of empirical experience and the longer term theoretical perspective of [4], we adopt for RHex's horizontal plane behavior the model of a quasi-static unicycle. The motion control software written for RHex implements a tripod gate for a normal walk. At any time 3 legs always touch the ground. The "walk mode" used in the experiments has the

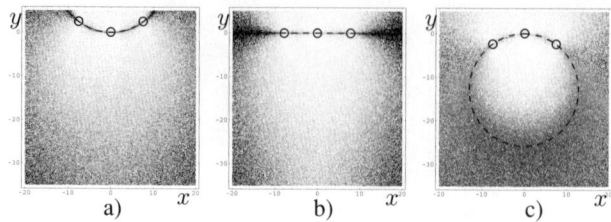

Fig. 4. Illustration of simulated pose computation error. **a)** Convex beacons; **b)** collinear beacons; **c)** Concave beacons. Darker values mean larger error. Beacons are represented by the small circles and the image of the degenerate set $\mathcal{D}$ is represented by the dashed curves.

control inputs of forward velocity and turn velocity, therefore the unicycle model is the natural choice to implement. The input velocity vector of the robot is computed by projecting the desired velocity vector given by equation (7) into the y axis of the local body frame ($\hat{\mathbf{e}}_y^l$). The turn velocity is applied directly. Let $[\mathbf{u}\ u_\theta]^T = -\nabla\left(\varphi \circ \bar{c}_\mathbf{b}\right)$. Then the unicycle model equation becomes:

$$\begin{bmatrix}\dot{x}_w \\ \dot{y}_w \\ \dot{\theta}_w\end{bmatrix} = \begin{bmatrix}-\sin(\theta_w) & 0 \\ \cos(\theta_w) & 0 \\ 0 & 1\end{bmatrix}\begin{bmatrix}\langle\mathbf{u},\hat{\mathbf{e}}_y^l\rangle \\ u_\theta\end{bmatrix} \quad (8)$$

Local minima are introduced when the projection of the velocity vector yields a zero vector and the turn velocity is null. Numerical simulations verify this fact in section III-C.

III. SIMULATIONS

A. Pose computation error

For concave beacon configurations the degenerate set disconnects the free configuration space $\mathcal{W}$. Therefore, it is expected that in a small neighborhood of $\mathcal{D}$ the camera map Jacobian is small potentially introducing large numerical errors. To visualize the extent and magnitude of this problem, pose computation error equation (9) was simulated for intervals of x_w and y_w with $\theta_w = 0$. A random noise vector δ with Gaussian distribution is added to the computation of the inverse camera map to simulate the noise from the camera. The simulated error is computed by:

$$\mathbf{e}_\delta = \|\mathbf{x}_w - \bar{c}_\mathbf{b}^{-1}(\delta + \bar{c}_\mathbf{b}(\mathbf{x}_w))\| \quad (9)$$

One can notice that the error increases with distance from the beacons as expected. In figure 4 c) the pose computation error increases when the robot is close to $\mathcal{D}$. This clearly suggests that concave beacon configurations are not desirable.

B. Fully actuated rigid body

Figure 5 illustrates the simulation of a fully actuated body using equation (7). The initial conditions range from $x \in [-5,5]$ meters, $y = -5$ meters and $\theta \in [0, \frac{\pi}{2}]$. The goal location is at $(0,-2,0)$. In section II-A the possibility of puncturing the degenerate set is contemplated. Figure 5 shows that in simulation, using a concave beacon configuration, the algorithm converges successfully in all the trials. This suggests that in theory it is safe to puncture the degenerate set.

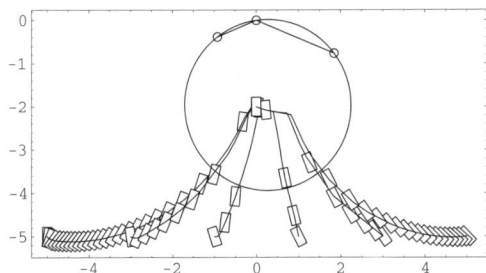

Fig. 5. Simulation of a fully actuated body with a concave beacon configuration. The degenerate set $\mathcal{D}$ is represented by the large circle. On top of the trajectories represented by the solid lines, the pose of the robot is plotted for fixed time intervals to give a crude idea of the robot's velocity.

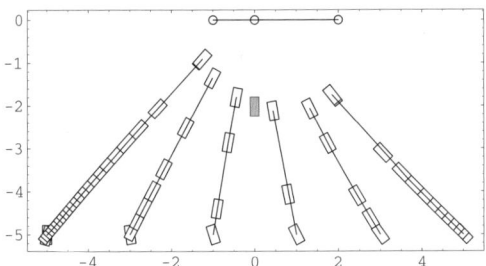

Fig. 6. Simulation of the non-holonomic constrained unicycle with a linear beacon configuration. The goal pose is represented by the gray rectangle.

Nevertheless, as shown previously, the pose computation error increases close to $\mathcal{D}$ and therefore, although puncturing is safe in theory, it should be avoided in practice.

C. Unicycle model

Figure 7 illustrates the same simulation of figure 5 but now using the unicycle model. Figure 6 illustrates a simulation for a linear beacon configuration. For both figures 6 and 7 the non-holonomic constraint predictably introduces local minima and, in general, as expected, the robot does not reach the goal. As figure 7 illustrates, the degenerate set $\mathcal{D}$ does not perturb the robot's motion.

IV. Experiments

This section describes the experiments performed with RHex in order to validate the algorithms developed in this

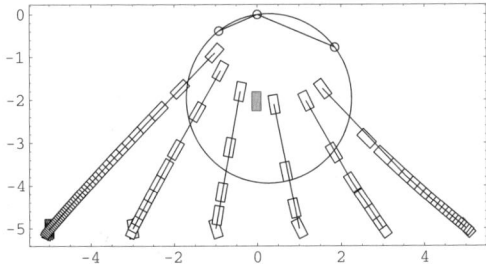

Fig. 7. Simulation of the non-holonomic constrained unicycle with a concave beacon configuration. The degenerate set $\mathcal{D}$ is represented by the large circle. The goal pose is represented by the gray rectangle.

paper. The robot's body frame contains 2 PC104 stacks, a motor drive board, a camera, accelerometers and a gyroscope. The first stack is equipped with i/o boards to read the motor encoder information, motor temperature, etc. It runs the robot's controller using a supervisor implemented with the RHexLib library [16], [17]. The 2nd stack, connected to a digital camera through a FireWire port, does all of the image processing. To implement the low level image processing functions a new fast vision library (SVision) was written. A discretized version of equation (7) is implemented. In order to accommodate for the input velocities allowed in the robot's walk mode the function w_i, defined in the appendix, is introduced. We then get the discretized equation of motion:

$$\xi_{k+1} = \xi_k + \Delta_k H_\xi \cdot W(H_\xi^T \cdot \mathbf{u}) \quad (10)$$

where $W(x_1, x_2) = (w_1(x_1), w_2(x_2))$ with w_i defined in the appendix, H_ξ is the "non-holonomic projection matrix", Δ_k is a gain factor and $\mathbf{u}$ is the input velocity vector obtained in equation (7).

$$H_\xi = \begin{bmatrix} -\sin(\xi_\theta) & 0 \\ \cos(\xi_\theta) & 0 \\ 0 & 1 \end{bmatrix}$$

The robot is positioned approximately 2 meters away from a set of beacons and a "snapshot" is taken. The location of the beacons recorded in the snapshot's image plane is fed into the navigation function as the goal pose. The robot is then moved into different initial conditions and it is released as represented by the triangles in the right side of figures 8 and 9. In general it is not expected that the robot will get back to goal point, only to the apparent curve of equilibrium points suggested in the numerical simulations. Due to the differences between the presumed quasi-static unicycle model and RHex's true locomotion behaviour, some failures occurred as reported in the following table. A trial is considered a failure if the beacons leave the FOV of the robot's camera.

experiment	failure rate
#1	5 out of 23
#2	0 out of 18

For experiment #1, illustrated in figure 8, a linear beacon configuration is used. Experiment #2 verifies the results obtained in simulation suggesting that it is safe to puncture the degenerate set $\mathcal{D}$: figure 9 shows that the robot successfully reaches a small neighborhood of the goal pose and it is not perturbed by the singularity $\mathcal{D}$ represented by the large circle. One can notice that in experiment #1 the number of failures is higher then experiment #2. This is due to the more careful selection of the scaling, saturation and dead zone parameters of function w_i, used in equation 10 and defined in the appendix.

V. Conclusions

The experiments in section IV reveal that navigation using visual servoing can be accomplished for the 3 beacon algorithm. It is verified experimentally that the algorithm is fairly robust to parameter uncertainty. By taking a snapshot of

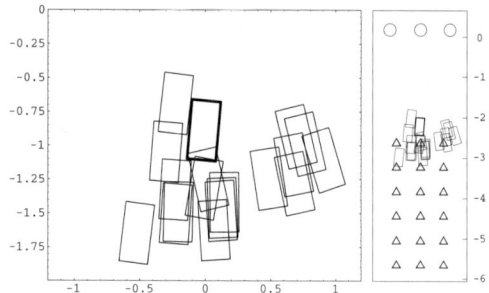

Fig. 8. Experiment #1. Linear beacon configuration. The picture on the right represents a view of the hall: the circles represent the beacons, the triangles represent initial conditions and the thick rectangle represents the goal pose. The picture on the left is a detailed view of the final pose for all the experiments. The units are in meters.

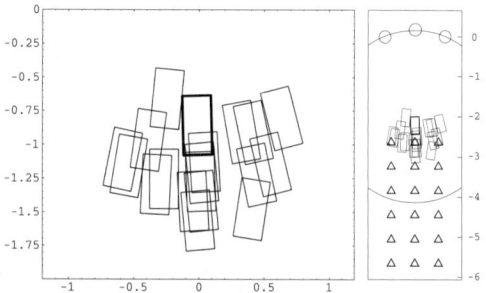

Fig. 9. Experiment #2. Linear beacon configuration. The picture on the right represents a view of the hall: the small circles represent the beacons, the large circle represents $\mathcal{D}$ the triangles represent initial conditions and the thick rectangle represents the goal pose. The picture on the left is a detailed view if the final pose for all the experiments. The units are in meters.

the goal location the parameter uncertainty error is implicitly subtracted during the trial run, as discussed at greater length in [1]. The simulations reported in section III qualitatively resemble the results obtained in the experiments and serve to validate the modeling assumptions used for the robot's dynamics. However, ultimately, the known limitations in the quasi-static model, built into the present version of these algorithms, limit the achievable accuracy.

A. Future work

Naturally, the next step is to modify the 3 beacon algorithm to account for the non-holonomic model of the robot. Following that, a better dynamical model for RHex should be developed. In the longer term, we seek to replace the current bright red beacons with landmarks derived from natural elements of the scene, allowing the robot to use trees, rocks, and other objects to navigate the real world.

VI. ACKNOWLEDGMENTS

We thank Clark Haynes for help with the experimental infrastructure and Noah Cowan for his advice and numerous helpful suggestions. This research was supported by DARPA/ONR N00014-98-1-0747 and the Fundação para a Ciência e Tecnologia - Portugal, with the fellowship PRAXIS XXI/BD/18148/98.

REFERENCES

[1] N. J. Cowan, J. D. Weingarten, and D. E. Koditschek, "Visual servoing via navigation functions," *Transactions on Robotics and Automation*, August 2002.
[2] U. Saranli, M. Buehler, and D. E. Koditschek, "Rhex: A simple and highly mobile hexapod robot," *The International Journal of Robotics Research*, vol. 20, no. 7, pp. 616–631, July 2001.
[3] R. Altendorfer, N. Moore, H. Komsuoglu, M. Buehler, H. Brown Jr., D. McMordie, U. Saranli, R. J. Full, and D. Koditschek, "Rhex: A biologically inspired hexapod runner," *Autonomous Robots*, vol. 11, pp. 207–213, 2001.
[4] U. Saranli and D. E. Koditschek, "Template based control of hexapedal running," in *IEEE Int. Conf. on Robotics and Automation*, May 2003.
[5] R. W. Brockett, *Asymptotic stability and feedback stabilization*, ser. Differential Geometric Control Theory. Birkhauser, Boston: R. W. Brockett, R. S. Millman, and H. J. Sussmann, 1983, pp. 181–191.
[6] E. D. Sontag, ""a lyapunov-like" characterization of asymptotic controllability," *SIAM J. Contr. Opt*, vol. 21, pp. 462–471, 1983.
[7] S. A. Hutchinson, G. D. Hager, and P. I. Corke, "A tutorial on visual servo control," *IEEE Trans. Robotics and Automation*, vol. 12, no. 5, pp. 651–670, Oct 1996.
[8] P. I. Corke, "Visual control of robot manipulators - a review," 1994.
[9] C. J. Taylor and J. P. Ostrowski, "Robust visual servoing based on relative orientation," in *Int. Conf. on Robotics and Automation*, 1998.
[10] H. Zhang and J. P. Ostrowski, "Visual servoing with dynamics: Control of an unmanned blimp," in *Int. Conf. on Robotics and Automation*, 1999.
[11] E. Malis, F. Chaumette, and S. Boudet, "2-1/2-d visual servoing," *IEEE Transactions on Robotics and Automation*, pp. 238–250, 1999.
[12] N. J. Cowan, "Vision-based control via navigation functions," Ph.D. dissertation, University of Michigan, 2001.
[13] N. J. Cowan, G. A. D. Lopes, and D. E. Koditschek, "Rigid body visual servoing using navigation functions," in *Conference on Decision and Control*. Sydney, Australia: IEEE, 2000, pp. 3920–3926.
[14] R. Ghigliazza, R. Altendorfer, P. Holmes, and D. Koditschek, "Passively stable conservative locomotion," submitted to *SIAM Journal of Applied Dynamical Systems*, 2002.
[15] D. E. Koditschek and E. Rimon, "Robot navigation functions on manifolds with boundary," *Advances in Applied Mathematics*, vol. 11, pp. 412–442, 1990.
[16] U. Saranli and E. Klavins, "Rhexlib programmer's reference manual," University of Michigan, Tech. Rep.
[17] U. Saranli, "Simsect hybrid dynamical simulation environment," University of Michigan, Tech. Rep., 2000.

APPENDIX

Define Φ as the map from local body coordinates in $SE(2)$ into $\mathcal{Q}$:

$$\Phi(x_b, y_b, \theta_b) \ : \ SE(2) \to \mathcal{Q}$$

$$(x_b, y_b, \theta_b) \mapsto \begin{bmatrix} \arctan(-x_b/y_b) \\ \theta_b - \arctan(-x_b/y_b) \\ \sqrt{x_b^2 + y_b^2} \end{bmatrix}$$

Define Υ as the map from world coordinates to body coordinates:

$$\Upsilon(x_w, y_w, \theta_w) \ : \ SE(2) \to SE(2)$$

$$(x_w, y_w, \theta_w) \mapsto - \begin{bmatrix} R_{\theta_w}^T & \mathbf{0} \\ \mathbf{0} & 1 \end{bmatrix} \begin{bmatrix} x_w \\ y_w \\ \theta_w \end{bmatrix}$$

Define the function $w_i(x)$ with saturation τ_i, scaling κ_i and dead zone v_i as:

$$w_i(x) := \begin{cases} 0 & if \ x \in [-v_i, v_i] \\ \max(\min(\kappa_i x, \tau_i), -\tau_i) & if \ x \notin [-v_i, v_i] \end{cases}$$

A Computer-Aided Probing Strategy for Workpiece Localization

Zhenhua Xiong, Michael Yu Wang and Zexiang Li *

Abstract

This paper presents an optimal planning problem for workpiece measurement. Two sequential optimization algorithms are introduced to find maximum determinant solutions. Then, based on a reliability analysis of workpiece localization and the sequential optimization algorithms, a computer-aided probing strategy is proposed. With this strategy, given the desired translation and orientation error bounds and desired confidence limit, we can experimentally find the least number of points needed to measure. Simulation results show the efficiency of the computer-aided probing strategy.

1 Introduction

In manufacturing literature, workpiece localization is a problem as follows: assuming a rigid workpiece is arbitrarily fixtured to a machine table, determine the position and orientation of the workpiece frame relative to a known machine frame from a set of coordinates measured on the workpiece [1].

Given a set of measurement points $y_i \in \mathbb{R}^3$ ($i = 1, \cdots, n$) sampled from the workpiece surfaces in the machine reference frame C_W and the corresponding surface descriptions S_i in (CAD) model frame C_M, the problem is formulated as a least squares problem with the objective function given by

$$\mathcal{E}(g, x_1, \cdots, x_n) = \sum_{i=1}^{n} \| y_i - g x_i \|^2, \quad (1)$$

where $g \in SE(3)$ is the Euclidean transformation transforming the CAD frame C_M of the workpiece to a known machine reference frame C_W, and $x_i \in \mathbb{R}^3$ is the home point of $y_i \in \mathbb{R}^3$ on the corresponding surface S_i.

---

*Z.H. Xiong is with the Robotics Institute of Shanghai Jiao Tong University(mexiong@sjtu.edu.cn). M. Y. Wang is with the Dept. of ACAE of the Chinese University of Hong Kong. Z.X. Li is with the Dept. of EEE of Hong Kong University Science & Technology.

When we get the point set y_i in the machine reference frame C_W using a computer-controlled coordinate measuring machine (CMM) or on-machine touch probe, there are two questions need to be answered:
1) In order to do a reliable workpiece localization, how many points should we probe on the workpiece surfaces?
2) If a point number is specified, how should we plan the measurement points on the CAD model?

For the first question, a reliability analysis method of workpiece localization [2] can be applied to check whether the localization result can be accurate enough with the measurement point set y_i. The second question is about measurement synthesis on the CAD model. We call it optimal planning problem. In this paper, a computer aided probing strategy is proposed. The strategy is a combination of reliability analysis of workpiece localization and optimal planning. With this probing strategy, a set of optimally planned points will be probed, while the transformation errors of the workpiece are within given error bounds.

The remainder of the paper is organized as follows. In Section 2, we review existing coordinate sampling strategies for reverse engineering or dimension inspection. In Section 3, based on an accuracy analysis model from fixture planning, we present two sequential optimization algorithms for optimal planning on workpiece surfaces. In Section 4, we present the computer-aided probing strategy for workpiece localization. Simulation results of optimal planning and probing strategy are given in Section 5. Finally, we draw conclusions in Section 6.

2 Literature review

Minimizing sampling time and cost while keeping accurate measurement is the key concern in coordinate metrology of workpiece surfaces. In [3], the authors reviewed error sources of sampling and different sampling strategies. It is noted that, in order to accurately measure part geometry, much higher sampling densities than those in the current practice must be incorporated.

Due to the limitations in the speed of most machines, minimal sampling is desired for industry practice. Many efforts have been devoted to developing a rapid and accurate sampling strategy. Woo [4] [5] investigated two deterministic sequences of number, namely, Hammersley sequence and Halton-Zaremba sequence, which give low level of discrepancy. After comparing with uniform sampling in 2D, it is found that the adoption of either sequence would give rather significant improvement in the accuracy or savings in sample size and hence in time. Later on, in [6] and [7], it is shown that, by modelling a machined surface as a Wiener process, the root-mean-square (RMS) error of measurement is equivalent to the L_2-discrepancy of the complement of the sampling points. The Zaremba sequence, which is optimal in terms of L_2-discrepancy, was shown to require quadratically fewer points than does the uniform or random sequence with the same order of accuracy in measurement.

In [8], a feature-based methodology which integrates the Hammersley sequence and a stratified sampling method was developed. The method can be used to derive the sampling strategy for multiple feature surfaces with multiple variances. The stratified Hammersley sampling was found to be more robust than the stratified random sampling and the stratified uniform sampling.

It is noted that the above sampling methods are mostly used in reverse engineering or dimensional inspection applications. The purpose is to reduce measurement error of sampled geometric features with real surfaces. For workpiece localization application, we want to accurately recover the position and orientation of the workpiece subject to sampling errors. So, we always need to consider the geometric relations among several surfaces. It is known that different set of sampling points will give different transformation results [9]. Since the sampling errors are inevitable, to ensure good recovery of Euclidean transformation g, a good planning of sampling points is very important.

In [9], an upper bound of the transformation error was estimated by the normalized sensitivity measure. This measure serves as an index that reflects the joint effect of both the number of measurement points and the geometric attributes of measurement locations. Although such an index has been proposed for a given part geometry, synthesis of the measurement points has not been done.

In [10], a simple example is used to show how the localization of probing points affect the possible displacement region of the object. A method using hitting sets and set covers was developed to obtain near-optimal probe placements for any known polygonal object. But for a sculptured workpiece, there is still a lack of sampling planning method.

In [11], an index was used for planning fixture locators based on the variance of resultant localization error. Nonlinear programming was used to minimize the variance. The method can only deal with one continuous surface. When there are many surfaces in consideration, there exists a combinatorial problem on point number assignment of different surfaces.

For the application of medical image registration in [12], a $3D$ model is constructed from images using a sensor such as a computed tomographic (CT) scanner. In the synthesis procedure, a noise amplification index is used to automatically generate near-optimal data configurations [13]. Four numerical methods were applied. It was shown that the planning of 10 to 75 optimal points may need up to several hundred minutes of computational time. Such a method is suitable only for off-line planning.

3 Sequential optimal probing

In this section, we first present a model for localization accuracy analysis originally developed for fixture planning. Then, two sequential optimization algorithms are given for optimal planning of the probing points.

3.1 Localization accuracy analysis

As described above, the position and orientation of the workpiece are determined from the measured coordinate data. At each probe point, there exists certain measurement error due to the accuracy of the probing device as well as the geometric and surface inaccuracy of the workpiece. Therefore, the positioning errors of the coordinate data will result in positional (translational and rotational) errors of the workpiece derived from the measurement data.

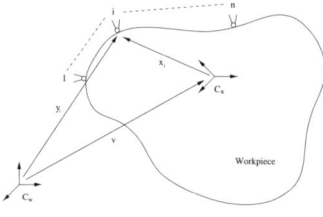

Figure 1: Workpiece localization with probing data

As shown in Figure 1, for each measured coordinate point y_i in the world frame C_W, its home point x_i is in the body frame C_B on the workpiece. From the kinematic analysis previously developed for workpiece fixturing ([14, 15]), we have the following linear model:

$$\delta y_i = -[(n_i)^T \; (n_i \times x_i)^T]\delta\xi = h_i^T \delta\xi, \quad (2)$$

where n_i is the unit normal at the ith home point, δy_i represents the projection of the positioning error the ith measured point along the normal direction, $\delta\xi^T = [\delta v^T \; \delta\omega^T]$ defines the small perturbation of the workpiece position in the twist coordinate [16].

For a collection of n measured points, we can combine the equations at all measurements to obtain

$$\delta y = G^T \delta\xi, \quad (3)$$

where $\delta y = [\delta y_1 \; \delta y_2 \cdots \delta y_n]^T$ and $G = [h_1 \; h_2 \cdots h_n]$. This establishes the linear relation between probing errors and the workpiece location errors.

3.2 Sequential optimal probing algorithms

Probe planning is to find the best set of positions on the workpiece for measurement among a set of virtually infinite feasible probing positions. The globally optimal solution is certainly difficult to find even for a polygonal workpiece [10]. Therefore, any optimal planning algorithm is usually developed for suboptimal solutions.

In this paper, we further restrict the feasible set of probing points in a discrete domain instead of continuous surfaces as in [15, 17]. The advantages of workpiece discretization are obvious. It makes the planning applicable for arbitrary geometric features. It is also noted that fine discretization should not significantly affect the planning results.

In [15], it is noticed that

$$\|\delta y\|^2 = \delta y^T \delta y = \delta\xi^T (GG^T)\delta\xi = \delta\xi^T M \delta\xi, \quad (4)$$

where the matrix M is called information matrix. Then $det(M)$ is used as the index of optimization. By maximizing $det(M)$, one can minimize the variance of the workpiece location error $\delta\xi$ in equation (3).

One advantage of using the determinant function in designing a sequential optimization algorithm is as follows. Given a discrete point set of N points, for an initial set of n candidate points, the information matrix is

$$M = GG^T = \sum_{i=1}^{n} h_i h_i^T. \quad (5)$$

Now, if we need to add (or delete) the jth point from the candidate set of $(N-n)$ points, the resulting information matrix $M_{(j)}$ and its inverse are given as

$$M_{(j)} = GG^T \pm h_j h_j^T \quad (6)$$

$$M_{(j)}^{-1} = M^{-1} \mp (M^{-1}h_j)(M^{-1}h_j)^T/(1 \pm p_{jj}) \quad (7)$$

where

$$p_{jj} = h_j^T M^{-1} h_j.$$

In fact, p_{jj} is a diagonal element of the so-called prediction matrix $G^T(GG^T)^{-1}G$, and it is easy to show that $0 \le p_{jj} \le 1$.

Furthermore,

$$det(M_{(j)}) = (1 \pm p_{jj}) det(M). \quad (8)$$

These recursive relations lead us to a sequential deletion algorithm. Starting from N points discrete point set, the candidate point j which minimizes p_{jj} is chosen and is subtracted from the point set. This procedure is repeated till a pre-specified number of measurement points remain. This algorithm represents a "top-down" approach and its efficiency is quite limited when the initial number N is large.

An alternative approach of "bottom-up" [17] is a sequential addition algorithm. In an initial step, we select six points out of the N discrete point set, such that matrix G is of full rank. Then, an interchange step is engaged to improve the initial planning till no further increase. After the six points, we can add additional point from the left $(N-6)$ points by maximizing p_{jj}. One by one, we will obtain a set of measurement points with the pre-specified number.

4 Computer-aided probing strategy

With the sequential optimization algorithms and a rough estimation of workpiece location, a touch probe (or CMM) can be used to measure a set of points y_i on the workpiece surfaces. The set of points will be used to localize the workpiece in the machine reference frame C_W with the existing localization algorithms. In general, for a least squares algorithm, more points give better result. To determine the quality of sampling and whether enough points are sampled, we first perform a reliability analysis of workpiece localization to evaluate the localization results.

4.1 Reliability analysis

Assume that we have determined a workpiece localization algorithm to use, such as the Hong-Tan algorithm [1]. Let $g^* = (R^*, p^*)$ and $x_i^*, i = 1, 2, \cdots n$ be the solution, and $\mathcal{E}_*$ the value of the objective function at (g^*, x_i^*). Because of inevitable errors in the measurement and a finite number of points are to be probed over certain regions of the workpiece, the computed transformation g^* will differ from the (unknown) actual (or true) transformation g^∞. Using the F-test [18] in statistical analysis, the translational error between the computed and the actual transformations is bounded, with a confidence limit $(1-\epsilon)$ [2], by

$$d = \sqrt{\delta p_x^2 + \delta p_y^2 + \delta p_z^2} \leq ((F_{\epsilon(l,l)} - 1)\mathcal{E}_*/\lambda_p)^{1/2} := \alpha_p^n, \tag{9}$$

where λ_p is the smallest eigenvalue of the translational error residue matrix J_p,

$$J_p = N_p^T N_p, \quad N_p = \begin{bmatrix} n_1^T \\ n_2^T \\ \vdots \\ n_n^T \end{bmatrix}, \tag{10}$$

n_i is the surface normal at x_i^*, and $F_{\epsilon(l,l)}$, $l = n - 6$, is the critical value at the ϵ-level in statistic analysis [18]. Similarly, we can get the estimated orientation error bound α_r^n [2].

4.2 Computer-aided probing strategy

Suppose that the desired accuracy requirement of a particular localization task has been obtained and is specified by (α_p^d, α_r^d). If we assume that the accuracy can be achieved by probing enough points on the workpiece surfaces, we then need to determine the minimal number of points so that the resulting estimation errors are bounded by α_p^d and α_r^d, respectively. The impact of the choice of measurement points on the accuracy of the computed transformation can not be underestimated.

On the basis of the sequential optimization method and reliability analysis of workpiece localization, the computer-aided probing strategy is described as follows:

(**Computer-aided probing strategy:**)

Input: (a) CAD model of the workpiece, with N' discretized points;
  (b) Surface finishing and sensor accuracy information;

Output: Estimated transformation that is within given error bounds;

Step 0: (a) Determine acceptable transformation error bounds (α_p^d, α_r^d) and confidence limit $(1-\epsilon)$;
  (b) Manually probe seven points (set $n = 7$);
  (c) Compute the transformation and transformation error. If $(\alpha_p^n \leq \alpha_p^d)$ and $(\alpha_r^n \leq \alpha_r^d)$, exit. Else continue;
  (d) Align the CAD model relative to the machine frame on the user screen, manually eliminate unaccessible regions and points of the CAD model. Let N be the candidate set available for measurement. Measure the optimal planned seven points $(n = 7)$;

Step 1: (a) Set k, a new set of points to be planned;
  (b) Use the sequential optimization algorithm to generate k points from the remaining set of $N - n$ points;
  (c) Set $n = n + k$;

Step 2: (a) Cluster and sequence the previous set of k points;
  (b) Generate a path to probe these points;
  (c) Command the touch probe to sample the points;

Step 3: (a) Compute the transformation using all n points;
  (b) Compute the translational and orientational errors. If they are within the error bound, exit. Else if $(n < N)$ go to Step 1, and if $(n \geq N)$, report that no satisfactory transformation can be found and exit.

Note that in the last case when no satisfactory transformation can be found, either the error bounds have to be relaxed or a finer discretization model has to be generated from the CAD model. In Step 1(b), we can use the reversed sequence by the sequential deletion method or directly use the sequential addition method.

5 Simulation results

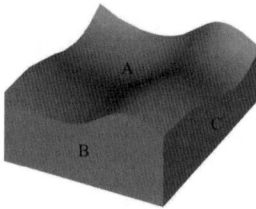

Figure 2: The simulation model for computer aided probing strategy

Here we show the simulation results using a model shown in Figure 2. Assume that we are interested in using surface A, B and C only. We first discretize the three surfaces. The discretized model is shown in Figure 3. For this simulation, we discretize surface A with 877 points, surface B with 365 points and surface C with 357 points. There are totally 1599 points, i.e. $N' = 1599$. The acceptable transformation error bounds are set at $\alpha_p^d = 0.1mm$ and $\alpha_r^d = 0.1$ degrees, and the confidence limit $(1 - \epsilon)$ is set at 95%.

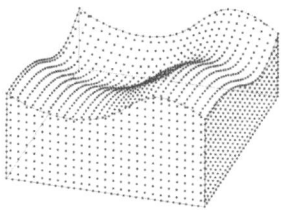

Figure 3: The discretized simulation model

Since regions near surface boundaries are not desirable for probing, we leave a $5mm$ margin from every boundary of the surfaces A, B and C. Thus, a total of 991 points remain, i.e. $N = 991$. These points are shown in Figure 4.

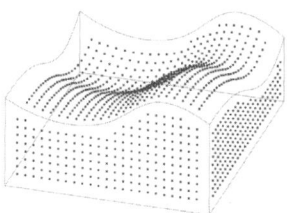

Figure 4: The discretized simulation model for computation

Then, the sequential deletion algorithm is applied to get an optimal six points as shown in Figure 5. They are generated in $69s$ on a PII 400 PC, suing MATLAB. In this case, $(detM) = 7.056 \times 10^{11}$. In the procedure, we also get a sequence of deleted points. The points, that are deleted earlier are less desirable to probe than the points deleted later.

Let $g_a = (R_a, p_a) \in SE(3)$ and $g_* = (R_*, p_*)$ be given and computed Euclidean transformations, respectively. The orientation and translation errors are defined as:

$$\mathcal{E}_R = |\theta|, \; where \; e^{\hat{\omega}\theta} = R_*^T R_a, \; \|\omega\| = 1$$

and $\mathcal{E}_p = \|p_* - p_a\|$. Measurement data are generated based on the 991 discretized points. A known Euclidean transformation g_a is applied to these points. Ran-

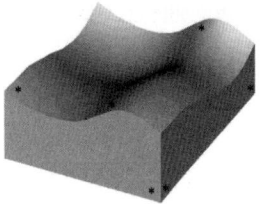

Figure 5: Optimal planning scheme with six points on the sculptured model

dom noise, which simulates measurement errors and dimensional errors, is added to these points. The random noise is assumed to be of normal distribution $N(\mu, \sigma^2)$, where the mean μ is measurement bias and the variance σ^2 depends on machining processes and measurement devices. Seven points with optimal planning scheme are first used to localize the workpiece. The Hong-Tan algorithm is used in this simulation.

After each localization, the translation and orientation errors are estimated. If both estimated errors are within the given error bounds α_p^d and α_r^d, we terminate the simulation and report success. Otherwise, several more points are added and the process continues. In the first simulation, we set $\mu = 0.01$ and $\sigma^2 = 0.01$. It turns out that when the number of points reaches 95, both error bounds are satisfied. The measurement scheme for 95 points is shown in Figure 6.

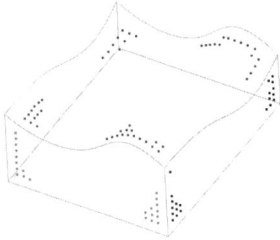

Figure 6: Optimal planning scheme with 95 points on the sculptured model

If we use the sequential addition algorithm, after several interchanges from a randomly generated six points, we then compare the six points with one generated by sequential deletion algorithm in Figure 7. The points planned by sequential addition are marked with triangles, while the points planned by sequential addition are marked with asterisks. In this case, we only need averagely $0.05s$ to get the solution using MATLAB, with a result of $(detM) = 9.180 \times 10^{11}$. Then, it takes averagely $0.066s$ to add an additional optimal point. From the time aspect, the sequential addition algorithm is more desirable, since we need not computer all the point sequence as in the sequential deletion algorithm.

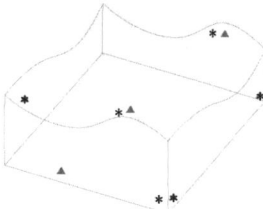

Figure 7: Optimal planning scheme comparison on the sculptured model

For the point sequence generated with sequential addition algorithm, we also set $\mu = 0.01$ and $\sigma^2 = 0.01$. It turns out that when the number of points reaches 225, both error bounds are satisfied. The measurement scheme for 225 points is shown in Figure 8.

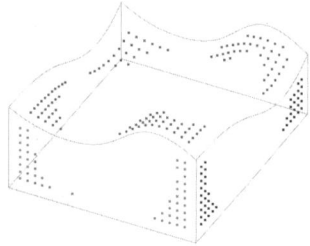

Figure 8: Optimal planning scheme with 225 points on the sculptured model

6 Conclusion

In order to do a reliable workpiece localization, we need a good planning of measurement points on workpiece surfaces. In this paper, first, with a model for localization accuracy analysis, we propose a maximum determinant planning strategy. Two sequential optimization algorithms are then introduced, which makes online planning of measurement points possible. After that, a computer-aided probing strategy is proposed, which is combined with the reliability analysis of workpiece localization. Simulation results show that, given the desired translation and orientation error bounds and desired confidence limit, we can find the least number of points needed to measure. The computer-aided probing strategy gives a feasible answer to the two proposed questions of workpiece measurement.

References

[1] Z.X. Li, J.B. Gou, and Y.X. Chu. Geometric algorithms for workpiece localization. *IEEE Transactions on Robotics and Automation*, 14(6):864–78, Dec. 1998.

[2] Y.X. Chu, J.B. Gou, and Z.X. Li. Workpiece localization algorithms: Performance evaluation and reliability analysis. *Journal of Manufacturing Systems*, 18(2):113–126, Feb. 1999.

[3] R.J. Hocken, J. Raja, and U. Babu. Sampling issues in coordinate metrology. *Manufacturing Review*, 6(4):282–94, Dec. 1993.

[4] T.C. Woo and R. Liang. Dimensional measurement of surfaces and their sampling. *Computer-Aided Design*, 25(4):233–239, 1993.

[5] T.C. Woo, R. Liang, C.C. Hsieh, and N.K. Lee. Efficient sampling for surface measurements. *Journal of Manufacturing Systems*, 14(5):345–54, 1995.

[6] R. Liang, T.C. Woo, and C.C. Hesieh. Accuracy and time in surface measurement, part 1: mathematical foundations. *Journal of Manufacturing Science and Engineering, Transactions of ASME*, 120:141–49, Feb. 1998.

[7] R. Liang, T.C. Woo, and C.C. Hesieh. Accuracy and time in surface measurement, part 2: optimal sampling sequence. *Journal of Manufacturing Science and Engineering, Transactions of ASME*, 120:150–55, Feb. 1998.

[8] G. Lee, J. Mou, and Y. Shen. Sampling strategy design for dimensional measurement of geometric features using coordinate measuring machine. *Int. J. Mach. Tools Manufact.*, 37(7):917–34, 1997.

[9] C.H. Menq, H. Yau, and G. Lai. Automated precision measurement of surface profile in CAD-directed inspection. *IEEE Trans. on Robotics and Automation*, 8(2):268–278, 1992.

[10] John Canny and Eric Paulos. Optimal probing strategies. *International Journal of Robotics Research*, 20(8):694–704, Aug. 2001.

[11] W. Cai, S.J. Hu, and J.X. Yuan. A variational method of robust fixture configuration design for 3-D workpieces. *Journal of Manufacturing Science and Engineering*, 119:593–602, November 1997.

[12] D.A. Simon. *Fast and Accurate Shape-Based Registration*. PhD thesis, Carnegie Mellon University, December 1996.

[13] A. Nahvi and J.M. Hollerbach. The noise amplification index for optimal pose selection in robot calibration. In *IEEE Intl. Conf. on Robotics and Automation*, volume 1, pages 647–56, 1996.

[14] H. Asada and A.B. By. Kinematic analysis of workpiece fixturing for flexible assembly with automatically reconfigurable fixtures. *IEEE Journal of Robotics and Automation*, RA-1(2):86–93, June 1985.

[15] M.Y. Wang. An optimum design for 3-D fixture synthesis in a point set domain. *IEEE Transactions on Robotics and Automation*, 16(6):939–46, December 2000.

[16] R. Murray, Z.X. Li, and S. Sastry. *A Mathematical Introduction to Robotic Manipulation*. CRC Press, 1994.

[17] M.Y. Wang and Diana M. Pelinescu. Optimizing fixture layout in a point-set domain. *IEEE Transactions on Robotics and Automation*, 17(3):312–23, June 2001.

[18] P.G. Hoel. *Introduction to mathematical statistics*. John Wiley & Sons, 5th edition, 1984.

Structured Product Coding System (SPCS) for Product Cost Evaluation in a CAE/CAD/CAM Product (C3P) Environment

Chi-haur Wu, Swee M. Mok* and Yujun Xie
Department of Electrical and Computer Engineering
Northwestern University, Evanston, IL 60208
*Motorola Inc, Schaumburg, IL 60196
chwu@ece.nwu.edu, swee.mok@motorola.com

Abstract

Virtual design and manufacturing has become a key technology in reducing both design errors and manufacturing cost. In order to have a realistic evaluation, the cost of assembling, disassembling, and manufacturing a product has to be realized. A product's design and its manufacturability will give an overall manufacturing efficiency and production cost. To eliminate future production failure and control the cost, management has to be able to foresee and evaluate the whole process from design to production, easily and wisely. In recognizing the importance of this capability, a CAE/CAD/CAM Product (C3P) management tool based on a structured product coding system (SPCS) is developed for the process of virtual design and manufacturing.

1. Introduction

Products that are difficult to assemble will raise manufacturing cost. Moreover, products that are difficult to disassemble will also increase cost in the near future as manufacturers are asked to recycle their products by a more environmentally conscious society. To overcome this problem, new products must be designed for easy assembly and disassembly, especially when automation is desired. Normally, the design is passed from the design department to the manufacturing department for production. This process partitions the cooperation between designing and manufacturing processes, hampers the ability of designers and manufacturing engineers to optimize the product as a whole, such as minimizing cost and production cycle time. In our previous work [2-4], we developed a design-to-manufacturing (DTM) process using a virtual assembly and disassembly (VIRAD) system. This paper will describe additional work that we have completed based on the VIRAD system in the product design and analysis section.

The developed virtual manufacturing VIRAD system shown in Fig. 1 will automate the process of analyzing parts for manufacturability and predict the cost of assembly and disassembly operations without physical prototypes. In our proposed DTM process, a new product is first designed using a solid modeling CAD system. The CAD system will have its own proprietary design database. From the CAD database, blue prints of production tools and parts for making the product and the assembly sequence can be obtained. The product design data is exported from the CAD system using a standardized format known as the ISO-10303 Standard for the Exchange of Product Model Data, or STEP [6]. Using STEP, products design data are imported into the VIRAD system to create generic assembly and disassembly (GENAD) trees [1].

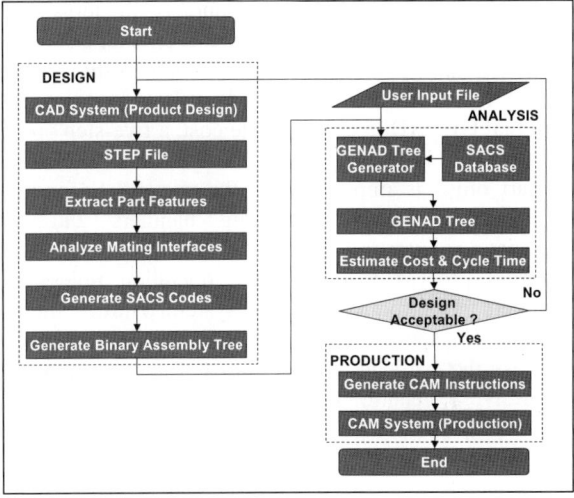

Figure 1. Virtual assembly/disassembly manufacturing system.

Inside VIRAD system, a part-feature extractor algorithm will read the parts' information including their principal axis of inertia, the part features vectors, and assembly sequence and store them in a part feature table. An expert system will be designed to work inside VIRAD to assist a product designer to select part features and to make design changes to achieve a better manufacturing index value. Finally, when a product has been designed to meet its operating and manufacturability index specifications, it is released for manufacturing. To efficiently design such a virtual manufacturing system in

---

* S. Mok will be receiving his Ph.D degree from the Dept. of ECE, Northwestern University, in June 2003.

The authors would like to acknowledge Dr. Iwona Turlik and Dr. Thomas Babin from the Motorola Advanced Technology Center, Motorola Labs, and the Northwestern University / Motorola Center for Communications Research for their support and research funding.

a C3P environment, a structured product coding system (SPCS) is developed to code essential product information for data tracking and evaluation.

2. C3P-SPCS System Design

Accurate modeling and simulation of product assembly and disassembly operations are key features of the proposed virtual manufacturing VIRAD system. The foundation of VIRAD is a newly developed hierarchical model named the generic assembly and disassembly tree-based work cell [2-4]. It models assembly and disassembly operations carried out inside a work cell using a developed structured assembly coding system (SACS) [5], which uses numerical codes to represent a set of well-defined assembly and disassembly procedures. By associating each SACS code with its operation cost, a product's total manufacturing cost can be estimated.

A generic assembly and disassembly (GENAD) tree [2] is a parts-merging binary tree that can represent a work cell environment including its parts, tools, fixtures and manipulators. Moreover, all objects in the work cell are grouped into parts and handlers. Parts are consumable items for making the product. Handlers are tools in the work cell such as: robots, feeders, end-effectors, and other non-consumables items, for manipulating the parts to assemble or disassemble the product. In order to generate a virtual GENAD tree to estimate cost, a five-step process illustrated in Fig. 2 is defined. Firstly, a binary tree (part-to-part only) is imported into the system. Secondly, a C3P-SPCS database containing handlers and their associating operating cost is defined – this database creation process is a one-time operation unless new handlers are introduced into the system. Thirdly, the user selects the appropriate handlers from the database to process the parts in the previously entered binary tree. Fourthly, a generic assembly and disassembly tree is generated; thus, cost can then be estimated using the GENAD tree.

2.1 Structures Product Coding System (SPCS)

To facilitate the data for evaluation and analysis, a structured product coding system is developed. This coding system is an extended SACS coding system [5] to store all data involved in designing and manufacturing a product. The original SACS coding system [5] was defined to handle the mating operations of parts and handlers. But to generate a generic assembly and disassembly GENAD tree, we need to extend SACS to include the assembly and disassembly operations between intermediate modules that contain a combination of parts and handlers. In this section, we will describe the new SPCS coding technique to address this problem. In addition, this SPCS coding structure is also defined to enable us to capture the relationships between the parent and child nodes in the GENAD tree.

As illustrated in Fig. 2, a binary assembly tree is first input to the system. In our previous results [2], a postfix expression was defined to provide two important pieces of information for assembling the product. They are the sequence for merging all the parts, and the types of operations used. Each assembly or disassembly operation is carried out by a moving part onto a stationary part, as defined by the structured assembly coding system [5]. Their assembly or disassembly operation is represented by 'OP' that can be set to the symbol '+' or '-', respectively. Each SACS code is a 4-digit numeric code that represents the type of constraint assembly operation between two parts. For example, "AD=A D + 1012" means that a part AD will be assembled by a stationary part A and a moving part D with a SACS code 1012. After having been assembled, the part AD can be either a final or intermediate part. If it is the latter, another assembly or disassembly operation will further operate on part AD to create more complex part.

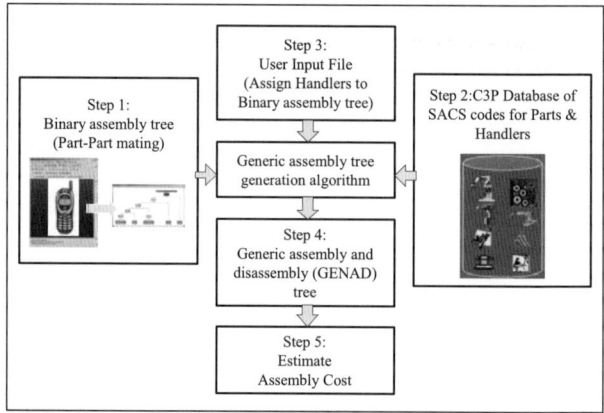

Figure 2. The virtual process of C3P-SPCS.

Before a part is ready to be merged with other parts, a sequence of operations must take place. For example, the part should be staged in a feeder first, and then a robot with a gripper can capture the part from the feeder. These sequence of operations between the parts and their specific handlers can be defined explicitly by a user and encoded into the GENAD tree. To generate data for SPCS code, we defined a specific product input file that defines various handlers and/or equipment required for executing part-to-part mating operations. The format of this product input file is defined in Figure 3., in which three fields Stationary Object, Moving Object, and Assembly Node Location (lower left of Fig 3.) are used to represent the merging operations of two parts, two handlers, or a part and a handler. All operational definitions for assembling two objects are placed between Begin-End blocks. The first two fields contain names of parts or handlers that will be operated on. One object is stationary while the other moves during an assembly operation, as defined by SACS [5]. The third field is divided into two sub fields named "Level" and "Index." The former represents the tree level at which the operation has to take place in the assembly tree. The latter is a unique integer to distinguish different assembled parts or handlers on the same tree level. For example, all leave nodes of an assembly tree will have level zero, and the assembled product will have the highest level. In other words, operations that can be executed at the same time will have the same level. The example in Fig. 3 shows that P_A (part named A) and H_G_A (handler: part gripper for part A) are attached to the assembly tree leave nodes at level 0 while the assembled part and handler are at level 1. In the physical world, part A is first put onto feeder A. At the same time, robot 1 grasps gripper A. Then robot 1 with gripper A will be used to pick up part A from the feeder.

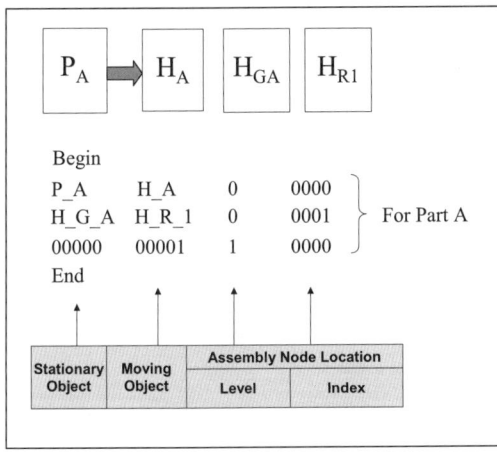

Figure 3. Data format of a product input file.

Parts and handlers that are on the same level in the assembly tree are candidates for simultaneous operation in a work cell. As a result, we can use this code as a guide for selecting parts for parallel manufacturing operations to speed up assembly operations. After reading data of parts and handlers from its product input file, a product database using SPCS coding technique can be generated for evaluating and manufacturing this product.

The format of SPCS code consists of eight fields shown as in Fig. 4. These eight fields are Stationary Part, Moving Part, New Part, SACS Code, SACS Operation, Assembly Tree Level, Assembly Cost, and Disassembly Cost (column combined with assembly cost in Fig. 4). Each SPCS code represented as a row in Fig. 4 describes the production property of each object, whereby each object can be a part, handler, or intermediate part. In Fig. 4, the first field in the SPCS database table, on the left most column, is named Name and it is a string expression to identify the part, handler, or intermediate part. This string is encoded using the following format:

P/H_ObjectName[_ObjectNumber]. It has three sub-fields divided by the symbol "_". The string expression itself is encoded with information to identify the object as a part or a handler or an intermediate module.

Name	Stationary_Part		Moving_Part		New_Part		SACS Code	SACS Operation	Assem. Tree Level	Assembly (Disassem-Bly) Cost
	Type	Index	Type	Index	Type	Index				
AGAINST_FIT	NA	NA	NA	NA	0000	0000	0023	NA	NA	A/D0023
SCREW_FIT	NA	NA	NA	NA	0000	0001	0023	NA	NA	A/D0023
Simple Part Mating Operations...	NA	NA	NA	NA	0000	...	...	NA	NA	A/D...
SPH_AGAINST	NA	NA	NA	NA	0000	0018	2300	NA	NA	A/D2300
P_0	NA	NA	NA	NA	0001	0000	NA	NA	NA	...
P_1	NA	NA	NA	NA	0001	0001	NA	NA	NA	...
P_N	NA	NA	NA	NA	0001	N	NA	NA	NA	...
...	NA	NA	NA	NA	...	...	...	NA	NA	...
H_G_0	NA	NA	NA	NA	0002	0000	3011	NA	NA	A/D3011
H_G_N	NA	NA	NA	NA	0002	N	...	NA	NA	A/D...
H_0	NA	NA	NA	NA	0003	0000	3010	NA	NA	A/D3010
H_N	NA	NA	NA	NA	0003	N	...	NA	NA	A/D...
H_R_0	NA	NA	NA	NA	0004	0000	3010	NA	NA	A/D3010
H_R_N	NA	NA	NA	NA	0004	N	...	NA	NA	A/D...
...	NA	NA	NA	NA	...	...	...	NA	NA	A/D...
P_0.H_0	0001	0000	0003	0000	0005	0000	3010	+	0	A3010
H_G_0.H_R_0	0002	0000	0004	0000	0005	0001	3011	+	0	A3011
H_0.P_0.H_G_0.H_R_0	0005	0000	0005	0001	0005	0002	3011	+	1	A3011
H_0 / P_0.H_G_0.H_R_0	0003	0000	0005	0002	0005	0003	3010	-	2	D3010
Additional modules...	...	...	...	...	0005	...	...	...	...	A/D....

Figure 4. A partial SPCS table representing parts, handlers, and intermediate parts.

For each SPCS code, the next three columns Stationary_Part, Moving_Part, and New_Part (column 2 to 4 from left) are sub-divided into two fields: Type and Index. Type is an integer for identifying the object in that row as a part, handler or mating operation. Moreover, an Index is used to identify different objects that are of the same Type. We currently have 6 Types defined. Type 0 is used for part mating operation while Type 1 is used for representing a simple part. Handlers would be labeled as Type 2, 3, or 4, depending if they are a gripper, a feeder or a robot, respectively. Finally, Type 5 represents an intermediate part. In future, we can add new labels when needed, say, when a new kind of handler is added to the system to represent leaves of a GENAD tree, two fields: Stationary_Part and Moving_Part (column 2 and 3 from left) in the SPCS database table are not used for the traditional SACS codes, parts, and handlers. Only an intermediate part (combination of parts and handlers) uses these two fields to describe objects in a GENAD tree. The fifth column is reserved for the SACS codes associated with either an assembly or disassembly operation. This is an important field for generating the correct assembly and disassembly instruction for manufacturing, and for calculating the final product assembly and disassembly cost. All traditional SACS operations will still use the developed SACS codes. Simple parts by themselves do not have a value and are filled with zeroes. Handlers have predefined operations and their corresponding SACS codes are entered here. This field is most interesting for intermediate parts that have a combination of parts and handlers as part of its assembly or disassembly process. An intermediate part, depending on its operation, can either be using a SACS code related to a part-part, part-handler or handler-handler assembly/disassembly operation. The sixth column contains either a symbol '+' or '-' for representing an assembly or disassembly operation. By definition, the part-part operations obtained from the product's binary assembly tree is always '+', while a handler's related operation can have both types. The seventh column is filled with an integer value representing the assembly tree level at which the part mating will take place. This value is useful for later optimization of the assembly tree and for optimizing factory assembly equipment layouts.

As for the last two columns, Assembly_Cost and Disassembly_Cost are used for cost estimation. Note that the cost associated with a SACS operation is a composite object. That is, this cost is actually made up of a set of variables relevant to characterizing the 'cost' of an operation. These variables include but are no limited to: 1) cycle time for executing the operations, 2) tooling cost for enabling the SACS code execution, 3) process yield over a large number of operations, 4) mean time between failure during a large number of operations, and 5) amount of manual intervention needed to operate the equipment. A nominal cost is calculated for each SACS operation by assigning a coefficient (based on their importance) to each of the measurable cost variables.

2.2 Generic assembly and disassembly (GENAD) tree

The SPCS coding system is designed in a way that a product's GENAD tree can be efficiently generated for estimating the total assembly and disassembly cost. To properly evaluate the cost, we must consider the cost of handlers (i.e. tooling, conveyors, robots) for building the product. As explained, SPCS database table will collect all those related information from product's binary assembly tree and product input file. As a result, from the data stored in the SPCS database table of a product, product's GENAD tree can model all the manufacturing processes for assembling and disassembling the parts either to build or disassemble the product. A key process in generating the GENAD tree is in creating the intermediate parts with parts and handlers, and mapping out their interactions in building the final product. An intermediate part may contain a combination of multiple parts and handlers.

2.3 Cost estimation

For a product that is fully reversible in terms of assembly and disassembly operations, the total assembly cost is obtained by adding all operations as we traverse the GENAD tree from its leaf nodes to root node. Similarly, total disassembly cost is obtained by traversing the same tree from the root to its leaf nodes. Moreover, the cost of assembling an intermediate part (i.e. a module) can also be calculated by summing the cost below it. As before, the disassembling cost of a module down to its constituent parts can be obtained by traversing the tree in the opposite direction.

3. Simulation

We will now present a simulation study using a telephone handset that consists of 18 parts [2], as shown in Fig. 5. To improve clarity, each part's name will be abbreviated using an alphabetic name: A through R, as shown in Fig. 5. Handlers for manipulating a specific part will also have the same alphabetic name as the part. The names of all parts and handlers will start off with P and H from their left hand side, respectively. A binary tree for assembling the telephone product is also shown in Fig. 5. It could be generated manually by the product designer, or by extracting it out from the CAD data file [2]. The assembly tree spans 11 levels, starting at level zero at the bottom of the diagram. Parts are progressively added to form the telephone at the top, represented by the following string: ADCEFBGHIJMLKOPQRN. At each level, a SACS code (e.g. 0023, 2111) is used to represent the required assembly operation.

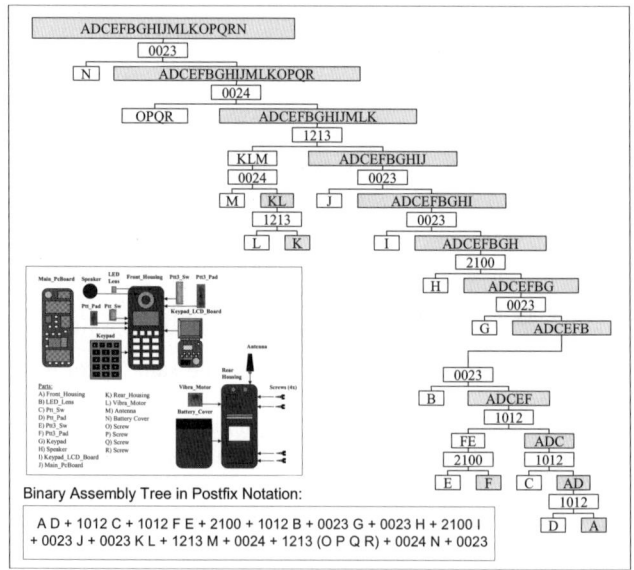

Figure 5. A binary assembly tree of the telephone.

3.1 Product Input File

An important step in assembling a product is in determining the correct tools or handlers for carrying out the needed operations. Following the steps defined in Fig. 4, a product input file for handling the 18 parts of this telephone set is defined in Fig. 6. The file assigned a handler to every part defined in the assembly tree's leaf nodes. For example, P_A (part) requires handler H_A (part feeder), while handler H_G_A (part gripper) requires another handler H_R_1 (robot) for manipulation.

Moreover, the robot with a gripper (H_R_1.H_G_A) will be used to pick part A from its feeder (H_A.P_A), to form H_R_1.H_G_A.P_A.H_A. This set of information is defined within a BEGIN-END block. Each row of information inside the block is defined as: Stationary_Part, Moving_Part, Level, and Index.

```
BEGIN
P_A       H_A       0   0
H_G_A     H_R_1     0   1
00        01        1   0
END
BEGIN
P_B       H_B       0   0
H_G_B     H_R_2     0   1
00        01        1   0
END
BEGIN
P_C       H_C       0   0
H_G_C     H_R_2     0   1
00        01        1   0
END
BEGIN
P_D       H_D       0   0
H_G_D     H_R_2     0   1
00        01        1   0
END
......
BEGIN
H_S_OPQR  H_R_2     0   0
P_P       00        1   0
END
BEGIN
H_S_OPQR  H_R_2     0   0
P_Q       00        1   0
END
BEGIN
H_S_OPQR  H_R_2     0   0
P_R       00        1   0
END
```

Figure 6. Product Input File for assembling the telephone.

3.2 Database of SPCS codes for handlers & parts

For this telephone simulation example, there are 18 parts, 14 grippers, 14 feeders, 1 customer-defined handler, and 3 robots. This is in addition to 24 simple part mating operations that has been defined for SACS. As described in section 2.1, SPCS codes are set up for the product. Part of SPCS table entries are shown in Fig. 7.

Figure 7. Partial table entries of SPCS codes for product.

3.3 Product's GENAD tree

Since SPCS database table of a product consists all initial information about parts and handlers, telephone's GENAD tree can be easily generated from its binary assembly tree and its SPCS database. Inside the GENAD tree, all intermediate parts will be generated. They will form new parts and be stored in newly created rows in the SPCS database table to complete the product information. Each node in a GENAD tree consists of five fields: New_Part#, New_Part_Name, SACS code for operation, Moving_part and Stationary_Part.

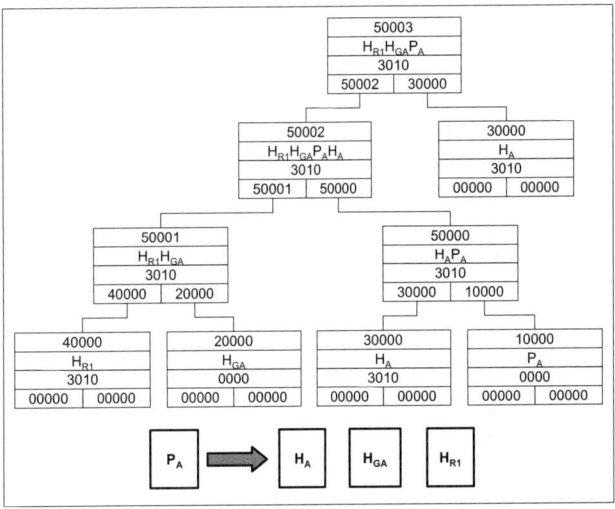

Figure 8. Part of GENAD tree showing part A being assembled using its associated handlers.

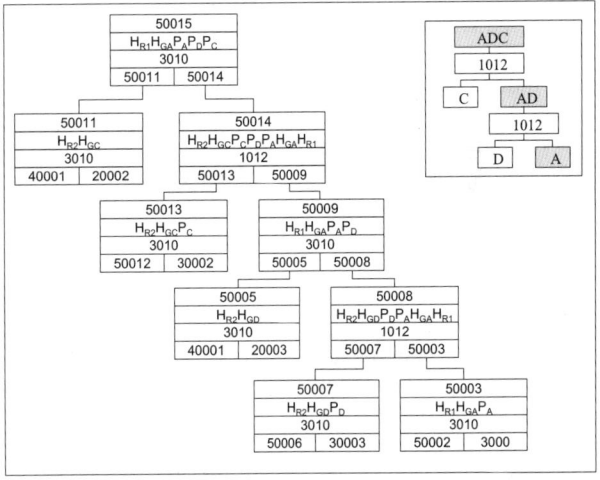

Figure 9. Portion of a GENAD tree for three parts.

In a GENAD tree, the objects at level 0 are all leave nodes and they do not have any child node. All leave entries in a GENAD tree were filled with data extracted from SPCS table. When an intermediate part is created in a GENAD tree, shown in Fig. 8, two children nodes, Stationary and Moving parts, are referenced. The children nodes' SPCS codes and names are obtained from the product input file whereby the parts and handlers are defined. For example, when handler H_A picks up part H_A to form an intermediate part $H_A P_A$, the fields of $H_A P_A$ were extracted from SPCS table and binary assembly tree. As shown in Fig. 8, two child fields are 30000 (combining 0003 and 0000) and 10000 (combining 0001 and 0000), representing H_A and P_A. The handler H_A uses a SACS code of 3010. The name of this new intermediate part is $H_A P_A$ and it is assigned a new part number 50000 (combining Type=0005 and Index=0000) on the product's input file. For bookkeeping, this new part will form a new row on the SPCS database table.

Repeating above process, Figure 9 displays a portion of GENAD tree for assembling three parts: P_A, P_D and P_C with required handlers and grippers. As shown in the binary assembly tree, at the top right corner of Fig. 9, part A and part D are first assembled, followed by part C. In the corresponding GENAD tree, note how the handlers of part D are disassembled before the new intermediate part# 50009 with P_A and P_D is formed. On the top of Fig. 9, the assembled part ADC was only held by the handlers of part A. Although it is not shown, subsequent steps will disassemble the handler from part ADC to complete the product assembly sequence.

After completing generation of a GENAD tree for a product, database of product's SPCS table will also be completely created and stored.

3.4 Cost estimation

In a GENAD tree, each SACS code of a SPCS code is associated with two operation costs related to assembly and disassembly, respectively. By summing up these individual costs for all SACS operations in an assembly tree, a total cost for product assembly and disassembly can be attained. Based on the defined SPCS table for the product, the estimated maximum assembly operation cost for assembling the simulated telephone is equal to (47*c3010) + (2*c2100) + (3*c1012) + (4*c0023) + (2*c1213) + (5*c0024) + (2*c3000) + (42*d3010).

In our example, SACS operation 3010 was used 47 times for assembling parts and was used 42 times for disassembling parts. In addition, the cost for the required handlers will be HA + HD + HC + HF + HE + HB + HG + HH + HI + HJ + HM + HL + HK + HSOPQR + HN + HGA + HGD + HGC + HGEF + HGF + HGB + HGG +HGH + HGI + HGJ + HGM + HGL + HGK + HGN + HR1 + HR2 + HR3. When calculating the disassembly cost, each assembly operation becomes a disassembly operation and vice-versa. In the simulation, every handler was assumed to be capable of performing both assembly and disassembly operations. The cost of disassembling this telephone can be evaluated by (47*d3010) + (2*d2100) + (3*d1012) + (4*d0023) + (2*d1213) + (5*d0024) + (2*d3000) + (42*c3010). In most practical situations, the product assembly and disassembly operations are not so easily interchangeable, and handlers are usually designed to only perform either assembly or disassembly operations. As a result, their assembly and disassembly operation cost equation will not be the same.

4. Conclusion

A special structured product coding system (SPCS) is developed for generating generic assembly and disassembly tree from products' binary assembly tree so that a product can be evaluated in a C3P environment. This structured coding system is designed to code handlers and intermediate parts in addition to product's parts. A SPCS database table will be created to store all information relating to a product's parts, handlers, and the intermediate parts that are created during an assembly or disassembly process. Based on this SPCS data table, product's GENAD tree can be created for product's evaluation on design, cost and efficiency. The system was simulated using a telephone with 18 parts, and a set of user defined handlers including feeders, parts grippers, and robots. The final assembly and disassembly cost was calculated by summing the operations costs, and the handlers used.

5. References

1. Mok, S, K. Ong, and C.H. Wu, "Automatic Generation of Assembly Instructions using STEP", IEEE ICRA2001, Seoul, Korea, May 2001.
2. Mok, S, C.H. Wu, and D.T. Lee, "Modeling Automatic Assembly and Disassembly Operations for Virtual Manufacturing," IEEE Trans. on Systems, Man, and Cybernetics, PART A: Systems and Humans, V. 31, No. 3, pp. 223-232, May 2001.
3. Mok, Swee, Chi-haur Wu, and D.T. Lee, "A System for Analyzing Automatic Assembly and Disassembly Operations," IEEE International Conference on Robotics and Automation, San Francisco, California, Apr. 27, 2000.
4. Mok, Swee, Chi-haur Wu, and D.T. Lee, "A Hierarchical Workcell Model for Intelligent Assembly and Disassembly," IEEE International Symposium on Computational Intelligence in Robotics and Automation (CIRA99), Monterey, California, Nov. 8-9, 1999.
5. Wu, Chi-haur, Myong Gi Kim, "Modeling of Part-Mating Strategies for Automating Assembly Operations for Robots," IEEE Trans. On Systems, Man, and Cybernetics, Vol. 24, No. 7, July 1994, pp. 1065-1074.
6. U.S. Product Data Association, "An American National Standard: Product Data Exchange using STEP (PDES), Part 1-ANS US PRO/IPO-200-001, Part 41-ANS US PRO/IPO-200-041-1994, Part 42-ANS US PRO/IPO-200-042-1994, Part 43-ANS US PRO/IPO-200-043-1994, Part 203-ANS US PRO/IPO-200-203-1994", Trident Research Center, N. Charleston, SC.

"Unilateral" Fixturing of Sheet Metal Parts Using Modular Jaws with Plane-Cone Contacts*

K. "Gopal" Gopalakrishnan[‡], Matthew Zaluzec[†], Rama Koganti[†], Patricia Deneszczuk[†] and Ken Goldberg[‡]

[‡]IEOR & EECS, U.C. Berkeley and the [†]Ford Motor Co.

Abstract - To fixture sheet metal parts for welding, we propose "unilateral fixtures" consisting of modular fixturing elements that lie almost completely on one side of the part. These are based on cylindrical jaws with conical grooves which provide the equivalent of 4 point contacts.

We propose a two-phase procedure for designing unilateral fixtures. The first phase is a geometric algorithm that assumes the part is rigid and computes vg-grips (vertex-groove grips). The vg-grip algorithm uses a fast sufficient test for immobility to generate a list of vg-grips and find bounds on jaw cone angles for each. The second phase is a fast heuristic procedure that uses FEM to arrange secondary contacts to reduce part deformation.

For a part described by n concave "virtual vertices", a list of vg-grips and minimum half cone angles for each vg-grip can be generated in $O(n^2)$ time. We also propose a quality metric based on the sensitivity of the part's orientation to an infinitesimal relaxation of the jaws that can be evaluated in constant time for a given fixture. For an FEM model with m nodes, the second phase takes $O(m^3 r)$ time to arrange r secondary contacts for each vg-grip.

I. INTRODUCTION

Sheet metal parts are created by stamping and bending planar sheets. To assemble industrial parts such as automotive bodies and large appliances, such sheet metal panels need to be accurately located and held in place by fixtures to permit welding. Existing fixturing methods are usually:
1. Bulky (often larger than the part they are holding).
2. Dedicated to each product model (and hence require a large investment in materials for each model).
3. Designed by human intuition (and hence subject to delays and errors in design).

Our goal is to develop a new approach to fixturing based on fixtures that are:
1. Compact, using a reduced set of contacts (so that the fixture and loading mechanism lie almost completely on one side of the part, to maximize access for welding and inspection).
2. Modular (so that the fixture hardware can be reconfigured quickly and reused for different product models).
3. Designed by CAD/CAM software that can analyze the sheet-metal geometry and automatically design the fixtures based on mathematical models.

This paper is a first step in this direction. We propose

---
* This research is supported in part by the Ford Motor Company and NSF Award DMI-0010069. For more information, contact gopal@ieor.berkeley.edu or goldberg@ieor.berkeley.edu.
Submitted to IEEE ICRA 2003.

fixturing with two types of modular jaws. One type consisting of cylindrical jaws with conical grooves formed by a pair of coaxial frustums joined at their narrow ends. These jaws contact part edges as illustrated. The second type provides point contacts that support the part from the interior. We refer to these as *"unilateral fixtures"*: the fixturing hardware lies almost entirely on one side of the sheet metal part to maximize access for operations such as welding and inspection.

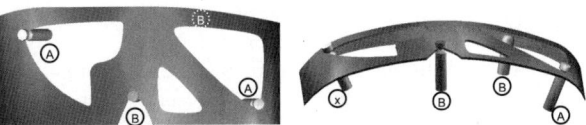

Fig. 1 Unilateral Fixture with 5 contacts, shown in two views. The sheet metal part is held with two primary jaws A and two secondary jaws B.

We present a two-phase design procedure for unilateral fixtures. The first phase is a geometric algorithm that assumes the part is rigid and locates one pair of primary jaws to immobilize the rigid part. This phase is purely geometric. The second phase adds additional secondary jaws using a heuristic Finite Element Method (FEM) procedure to reduce deformation to within specified tolerances. This two-phase approach allows us to reduce the number of FEM iterations in contrast to pure FEM methods such as [16] and [2].

Cylindrical jaws with v-grooves achieve plane-cone contacts. These facilitate frictionless immobilization of a part with just 2 jaws: each jaw is equivalent to 4 contacts because it makes contact with 2 edges and each edge is in contact with 2 faces of the groove. These grooved jaws achieve *"vg-grips"* (vertex-groove grips) that locate the part in the desired plane and in the desired position and orientation within the plane. Vg-grips are an example of minimalist robotics, or RISC, as presented in [3].

In vg-grips, each jaw groove makes contact with part edges at two distinct points. For curved part edges, this occurs not only at vertices, but also where the curvature of the edge is higher than the jaws' curvature. Below, we define "virtual vertices" at regions of high curvature. We develop a fast sufficient test to check for vg-grips that hold the part in form-closure. This test uses results from the planar analysis reported in [5]. A vg-grip is called *expanding* if the jaws move away from each other and *contracting* if they move towards each other. For sheet metal parts, only expanding vg-grips are used as compressive forces cause buckling.

II. RELATED WORK

Grasping and fixturing both involve holding a part in a way that limits its mobility. Bicchi and Kumar [1] and Mason [10] provide concise surveys of research on robot grasping. Grasps can be classified as force or form-closure. Form-closure occurs when any neighboring configuration of the part results in collision with an obstacle. Force-closure occurs if any external wrench can be resisted by applying suitable forces at the contacts [10, 19].

Gripper contacts can be modeled as frictional points, frictionless points or soft contacts [21]. [17] and [23] prove that 4 and 7 frictionless point contacts are necessary to establish form closure in the plane and in 3D respectively and [12] and [9] proved that 4 and 7 point contacts suffice.

Van der Stappen et al [24] describe an efficient algorithm to compute all placements of four frictionless point contacts on a polygonal part that ensure form-closure. Given a set of four edges, they show how to compute critical contact placements in constant time. The time complexity of their algorithm is bounded by the number of such sets.

Rimon and Burdick [Rimon96, 98] were the first to identify and introduce the notion of second order force closure. Immobility is defined to occur if any trajectory results in the decrease of distance between the part and at least one obstacle it is in contact with. First and Second orders of immobility arise due to the truncation of the Taylor expansions of the distances at the first and second order terms respectively. [18] shows that generic planar parts can be immobilized (second-order) with three frictionless contacts if they are placed with infinite precision. Ponce et al [15] give an algorithm to compute such configurations.

Rimon and Blake [20] give a method to find *caging grasps*, configurations of jaws that constrain parts in a bounded region of C-space such that actuating the gripper results in a unique final configuration. They consider the opening parameter of the jaws as a function of their positions and use stratified Morse theory to find caging grasps. In this paper, we look at the distance between the jaws and use the fact that they are at a strict extremum to show that the part is immobilized.

Wang [25] examines the errors in machined features in relation to the errors in locator position and locator surface geometric errors. The relation is expressed using a critical configuration matrix for the part. Wang suggests an optimal locator configuration based on error sensitivity of multiple features machined on the part.

In [5], we give a fast test for form-closure in 2D using cylindrical jaws. Such grips are called v-grips. We extended the results to 3D to hold parts such that only translation along the jaws is possible. We also suggest a metric to evaluate the sensitivity of 2D v-grips to small relaxations in the jaws' positions. We will make use of some results in [5] to prove our theorems, and extend the metric to evaluate vg-grips.

Plut and Bone [13,14] proposed inside-out and outside-in grips using two or more frictionless point contacts at linear or curved part edges. They show how to find such grips where the distance between contacts is at an extremum. They achieve form closure in 3D using horizontal V-shaped circumferential grooves (VCGs). Our unilateral model minimizes fixture profile on one part exterior and generalizes their analysis with an exact test for 3D form-closure, a new quality metric, and a method for locating secondary contacts based on FEM.

[6] pioneered the application of finite element models for fixturing sheet metal parts. Their analytic model determines the deformation, stresses and clamping forces for the workpiece, assuming Coulomb friction at the contacts but does not propose a method to design fixtures.

[11] determines the positions of the primary datum (the datum points needed to locate the part in the correct plane) for 3-2-1 fixturing to minimize deformation. They use a finite element model of the part to model the deformation, and optimize the fixture locations using the gradient of the objective function. Their work is extended by [16] and [2]. [16] designs a fixture for a sheet metal part by weighing the deformation and the number of fixtures in the objective function, using a remeshing algorithm, but without addressing properties specific to sheet metal parts. Cai et al describe an N-2-1 fixturing principle in [2]. This is used instead of the conventional 3-2-1 principle to reduce deformation of sheet-metal parts. They use N ($>=3$) locators for the primary datum, (i.e. they use N datum points to locate the sheet metal part in the correct plane) in their fixtures. They model the sheet metal parts using finite elements with quadratic interpolation, constraining nodes in contact with the primary datum to only in-plane motion. For a known force, linear static models are used to predict deformation. To make their algorithm faster, instead of remeshing the part for different locator positions, they express the constrained displacement at the locator by using a linear interpolation of displacement at the adjacent nodes. Care is taken to avoid buckling while placing fixture elements. In contrast, our two phase approach is a hybrid of geometric and FEM methods.

[22] considers the fixturing of a sheet metal workpiece using clamps and locators fixed on a base-plate with t-slots. The height of the fixture elements are variable, and are adjusted to fit the shape of the part. Determining the positions of the locators and clamps is formulated as a non-linear programming problem in terms of the part deformation.

Li et al [7] describe a procedure to design fixtures for two sheet metal parts that are to be welded to produce a good fit along the seam to be welded. The fixtures are designed using a finite element model to determine either an optimal fixture or a robust fixture.

The unilateral fixturing approach is inspired by Toyota's "Global Body Line" auto assembly system [4].

This modular system fixtures different auto models using a reduced set of hardware.

III. PROBLEM STATEMENT

To determine a suitable fixture for the part under consideration, we need as input a contiguous connected 2D surface with holes whose thickness is assumed to be small compared to the dimensions of the features on the part. It is defined by a CAD model that consists of a list of its edges: both external and internal (holes) in terms of spline curves. For each edge, the side of the edge on which the part lies is also specified. A FEM mesh discretizing the part as a surface embedded in 3D is also specified. This is a triangular or quadrilateral mesh (but other meshes can be used in general). The part's thickness and material properties are also specified. Primary jaws consist of 2 coaxial frustums of cones joined at their narrow ends which have equal radii (called the radius of the jaw). Secondary jaws may either be of the same shape as primary jaws, or may be point contacts supporting the interior of the part (away from edges). All jaws are assumed to be rigid and all contacts are assumed to be frictionless. The part is subjected to a set of known external wrenches specified as a list describing each wrench vector and the node of the part's mesh where it is applied. For each node, the direction of the part's interior, i.e. the direction in which the unilateral fixture may lie, is specified. A tolerance δ is specified. This is the maximum deformation (i.e. magnitude of displacement from original position) of any point on the part as a result of the applied wrenches.

<u>Input:</u> CAD model of part with FEM mesh (as specified above), Young's modulus and Poisson's ratio of the part, jaw radii, list of applied wrenches, and allowed tolerance δ.

<u>Output:</u> A list of unilateral fixtures that specify positions and orientations of each jaw within the given tolerances and bounds on the cone angles of each primary jaw, or a report that no solution exists.

IV. PHASE I: LOCATING PRIMARY JAWS

A. Problem Statement

In Phase I we assume the part is rigid and proceed to design and determine the locations of 2 jaws that can immobilize the part. These primary jaws are designed to engage the part at its concavities such that the intersections of the frustums in the jaws are seated in the plane of the sheet metal part.

For the part to contact the jaws on the plane of intersection of its frustums, the local radius of curvature of the part needs to be large compared to the jaws' radius. If this is not true, contact does not occur on the plane, but instead, on the surfaces of the individual cones. Therefore, at such candidate jaw locations, we assume local planarity of the part and linearity of the edges for first order analysis of immobility, since only local shape is of importance. We construct tangents at the points of contact. We call these tangents the part's "virtual edges", and the point of intersection of the edges, the corresponding "virtual vertex". If we approximate the part locally using the virtual edges and vertices, immobility of the approximation will be equivalent to the immobility of the original part up to the first order. The jaws' positions are described in terms of the virtual vertices. Virtual vertices are concave by definition. Given 2 virtual vertices v_a and v_b, we call the unordered pair $<v_a, v_b>$ a vg-grip if the part is held in form-closure when the jaws' grooves engage the part at the edges defining v_a and v_b.

The FEM portion of the input such as the mesh and the part's material properties are irrelevant for this phase. This phase can give an output of a list (possibly empty) of vg-grips for the part, with bounds on jaw cone angles for each listed vg-grip, sorted by the quality metric described below.

The following sufficient test for form-closure makes use of the results reported in [5].

B. Sufficient Test for Vg-grips

As shown in Figure 2, we define a coordinate system such that the direction of the x axis is taken from v_a to v_b. In a projection perpendicular to the x axis, the z-axis is defined as the bisector of the acute angle between the projection of jaw axes. When jaw axes projections are parallel, the z axis is defined at 45° to the jaws' axes. The y axis is perpendicular to the x and z axes using the right hand rule. Let the points of contact have position vectors r_{a1}, r_{a2}, r_{b1} and r_{b2}. Let the vectors a_a and a_b be the axes of the jaws with positive z components and the centers of the intersections of the cones be c_a and c_b. (The subscripts a and b denote the jaws near vertices v_a and v_b.) We define q_{a1} as: $e_x \times ((r_{a1}-v_a)-(r_{a1}-v_a.e_x)e_x) = e_x \times (r_{a1}-v_a)$, and similarly q_{b1}, q_{a2} and q_{b2}.

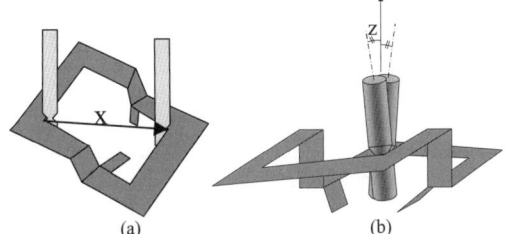

Fig. 2. (a) The x axis is chosen along the line connecting the vertices v_a and v_b. (b) In a projection perpendicular to the x-axis, the z-axis is chosen as the bisector of the acute angle between the jaws' axes' projections.

Theorem 1: Assuming that the part is rigid, immobility is achieved if all of the following are satisfied:
(a) The projection of the part and jaws on the x-y plane is an expanding 2D v-grip.
(b) The projection of the part and jaws on the x-z plane is an expanding 2D v-grip.

(c) The angle between $\mathbf{q}_{a1}$ and the inward normal to at least one of the cones at $\mathbf{r}_{a1}$ is less than 90°, and the angle between $-\mathbf{q}_{a1}$ and at least one of the inward normals at $\mathbf{r}_{a1}$ is less than 90°. And similarly for $\mathbf{q}_{a2}$, $\mathbf{q}_{b1}$, $\mathbf{q}_{b2}$.

Our 2D analysis reported in [5] must be extended since cross sections in the planes of contact project to ellipses and not circles on the coordinate planes. As a result, the transformed part (to adjust for non-zero jaw radius) is obtained by drawing edges parallel to the original part edges through the ellipses' centers.

C. Proof of Theorem 1

The distance between the jaws is defined as the x component of the distance between the centers of the cones' intersections. We will show that any small displacement of the part requires a decrease in distance between the jaws if one jaw is fixed and the other is allowed to translate. Hence, since the jaws are fixed, the part will be in form closure.

In showing this, we use the following result from [5]: an expanding v-grip is equivalent to a strict local maximum of the distance between the jaws as the jaws move along the perimeter of the part. Also, that the distance is a strict local maximum if the jaws are allowed to move anywhere on the plane without colliding with the part.

Consider any small displacement of the part. This can be denoted as the sum of 3 translations and 3 rotations (along and about the x, y and z axes). We show that as the part is subject to each of these components of displacement while keeping the distance between them at the local maximum of the possible distances, the distance between them decreases.

From condition (c) in Theorem 1, any rotation of the part about the x axis should result in a decrease of distance between the jaws. This is because the vectors $\mathbf{q}_{xi}$, x=a, b; i=1, 2, give the direction of the instantaneous velocities of each contact. Hence, if a jaw stays in the same position, it collides with the part. Hence, it has to move either towards or away from the vertex. If cannot move towards the vertex because of the following reason: if we scale down the part and the jaw about the vertex, such that the distance between the scaled jaw and the vertex is equal to the distance between the vertex and the jaw after the rotation, the scaled jaw would collide with the part after an identical rotation (since the conditions are scale-independent). Since a smaller jaw would collide with the part in such a position, the original bigger jaw will also collide with part, since the vertex and edges of the part do not change on scaling. Hence, each jaw is pushed away from the vertex.

First order form-closure is robust in the sense that immobility is guaranteed allowing for small changes in part geometry. Since none of the axes are perpendicular to the planes of intersections of each jaw's cones, conditions (a) and (b) of theorem 1 ensure that the projections of the part on the x-y and x-z planes are in form closure after an infinitesimal rotation of the part about the x-axis. We note that the distance between the vertices does not change as a result of rotation about the x-axis. Since the distance between the vertices remains the same due to such a rotation and since the edges are linear and the vertices concave, it follows from the results in [5] that the distance between the jaws decreases.

Condition (a) also implies that and translation along the x or y axes, and rotation about the z axis will result in further increase in the distance between the jaws. Condition (b) implies that any further translation along x or z axes and rotation about the y axis leads to another increase in distance. Thus, any displacement of the part results in a displacement of the jaws, hence proving that form-closure is achieved if the jaws are fixed.

V. PHASE I: CANDIDATE JAW LOCATIONS

As stated in section IV, while contact occurs near vertices for a part defined by linear edges, parts with curved edges have virtual vertices where the jaws engage the part. These virtual vertices are potential locations for jaws. A virtual vertex is called a candidate jaw location if a jaw engaging the part at the virtual vertex makes contact with the part at two points in the plane of intersection of the frustums. Candidate jaw locations are identified using the algorithm described below.

The algorithm uses the fact that jaws contact the part at 2 points only if there is a concave vertex between the points of contact or if part of the edge contained between the points of contact is concave and has higher curvature than the jaw.

Step 1: Set list l as list of the part's concave vertices. Set list l_c to an empty list.

Step 2: Traverse each edge of the part. For each edge, numerically identify concave stretches with radius of curvature less than jaw radius, and add the end points (with higher arc-length) to l.

Step 3: For each point i in l, traverse the edge starting from the point i in the direction of increasing arc-length, constructing discs tangential the edge till the disc touches the part at 2 points or the entire edge is traversed back up to the position of the current element of l.

   If the entire edge was not traversed and if the edge at the second point of contact is in plane with the disc, add the center to l_c. Replace the current element of l by the point of intersection of the tangents.

   Else, delete the current element of l.

Step 4: Traverse l_c for duplicates and eliminate them and the corresponding elements in l.

Step 5: Return the list l as the list of candidate locations and l_c as the list of centers.

VI. BOUNDS ON CONE ANGLES

Conditions (a) and (b) in theorem 1 are independent of the cone shapes for a given jaw radius. Hence, bounds on the cone angles that satisfy Theorem 1 are determined only by the condition (c). In the worst case, $\pm\mathbf{q}_{xi}$, x=a, b, are tangential to the cones for at least 1 value of i=1, 2. Hence, if we project $\pm\mathbf{q}_{xi}$ to the plane containing $\mathbf{r}_{xi}$ and $\mathbf{a}_x$, the acute angles between the projections and $\pm\mathbf{a}_x$ gives a candidate lower bound for the half cone angle for the upper cone. The lower bound is chosen as the higher of candidate bounds obtained from $\mathbf{q}_{x1}$ and $\mathbf{q}_{x2}$. For the example shown in figure 1, the bounds for the half cone angles for the 4 cones were 18°, 21°, 18°, and 26°.

VII. PHASE II: LOCATING SECONDARY JAWS

In Phase II, we assume two jaws hold the part as determined by Phase I. Part deformation is then modeled using FEM, based on a given part mesh. We use brick elements in our implementation. Rigid jaws constrain the positions of the nodes on which they lie. By the nature of the expanding vg-grip, the forces exerted by each jaw on the part are directed away from each other. Any jaw that is added to the fixture can constrain the part only if the jaw does not lose contact with the part as it deforms.

VIII. UNILATERAL FIXTURE DESIGN ALGORITHMS

Based on sections IV through VII, we describe below our two-phase unilateral fixture design algorithm for sheet metal parts:

(Phase I: Kinematic analysis)
Step 1: Set list of fixtures L_f to be empty.
Step 2: Generate a list of all virtual vertices. Store this as $(v_1, v_2, v_3, \ldots)$
Step 3: For each unordered pair of vertices $<v_i, v_j>$, apply Theorem 1 to determine if $<v_i, v_j>$ is a vg-grip.

(Phase II: Deformation model)
Step 4: For each vg-grip $<v_i, v_j>$, with primary jaws at v_i and v_j:
 i. Define E = the set of candidate edge nodes and F = set of candidate face nodes as all mesh nodes of the part's perimeter and interior respectively.
 ii. Traverse E and remove nodes j if either:
  a. Jaw at j and primary jaws collide, or
  b. Jaw at j and any primary jaw cause the part to buckle (the pair exerts a compressive force).
 iii. With the current set of jaws, compute the deformation of the part at each node.
 iv. Determine the maximum deformation d.
 v. If this $d < \delta$, add the current fixture to L_f and go to step 5.
 vi. If E and F are empty, proceed to next vg-grip from step i.
 vii. Let maximum deformation for all jaws in E and F occurs at node i. If i is an edge node, place a conical jaw at the candidate node. Else, if the deformation at i is towards the interior of the assembly, place a point contact at i. with maximum deformation normal to the part (call it node i).
 viii. If i is in E, then from E remove node i and all nodes j such that either:
  a. Jaws at i and j collide, or
  b. Jaws at i and j cause the part to buckle (they exert a compressive force).
  Else, remove i from F.
 ix. Go to step iv.
Step 5: For each fixture, compute bounds on cone angles and store them in L_f.
Step 6: Return L_f, the list of acceptable fixtures and bounds on cone angles.

The algorithm generates a natural actuation order. However, this need not be the best order in which to actuate each jaw.

Figure 3 shows an example of the first 2 iterations (figures 3(a) and 3(b)) for an example part. For a tolerance of 1mm, the fixture is shown in figure 3(c).

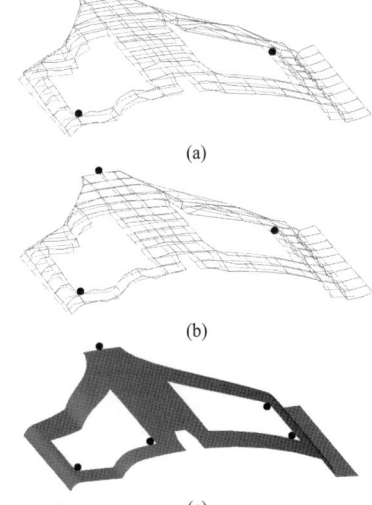

Fig. 3 Deformed and undeformed meshes for first 2 iterations of Phase II. Final fixture(c) required 4 iterations.

IX. QUALITY METRIC

We propose as an intuitive fixture quality metric the maximum change in orientation along any of the coordinate axes due to an infinitesimal relaxation of the jaws. This is based on the metric we proposed in [5] for 2D v-grips. It is quantified by $|d\theta/dl|$, l being the distance between the jaws, and θ the orientation. For the y and z components, this reduces to the metric obtained from the 2D v-grip metrics. For rotation about the x axis, this is not the case. We find an approximate value for $|d\theta/dl|$ by assuming that the contacts lie on the vertices of the v-groove in the projection of the jaws on a plane perpendicular the plane containing the contacts and the edges. Since the contacts on the jaw projection hold the jaw in a v-grip, we know that distance between the contacts increases by $q_a \Delta\theta$, where q_a is the quality metric for this v-grip. Hence, if the original distance between the center of jaw a and the vertex is d_a, the distance after rotation is $d_a(1+ q_a\Delta\theta/|r_{a1}-r_{a2}|)$. Thus, the metric for rotation about x-axis simplifies to $|r_{a1}-r_{a2}|/d_a \, q_a + |r_{b1}-r_{b2}|/d_b \, q_b$). The quality of the vg-grip is the maximum of the metrics for all 3 rotations.

Figure 4 shows an example part (a) and its front and top views, (b) and (c) respectively. For jaws with all half-cone angles $60°$, the vg-grip shown in (c) is evaluated as better than that shown in (d).

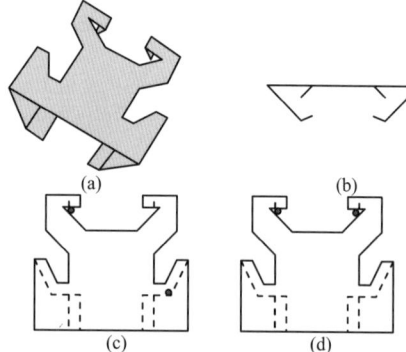

Fig. 4 Example Part: (a) oblique view, (b) front view. Vg-grip (c) is evaluated as better than vg-grip (d) by the quality metric.

X. COMPLEXITY

For a sheet metal part, given n virtual vertices, there are at most $O(n^2)$ pairs of candidate jaw locations. For each pair, the algorithm is applied in constant time. Hence phase I runs in $O(n^2)$ time. In phase II, if the FEM mesh has m nodes, at most m secondary jaws are needed as deformations are at maxima at the mesh nodes. Hence, for r jaws, the algorithm runs in $O(m^3r)$ time for each vg-grip.

X. EXTENSIONS AND FUTURE WORK

We will physically test unilateral fixtures generated by the two-phase procedure and design modular baseplate hardware that can facilitate unilateral fixturing. We also will incorporate other fixture elements such as vacuum cups that can apply suction forces as in [8].

X. ACKNOWLEDGEMENTS

We thank Prof. Gary Bone, Ron Alterovitz and Anthony Levandowski for feedback and advice.

X. REFERENCES

[1] A. Bicchi and Vijay Kumar, Robotic Grasping and Contact: A Review, Proceedings of IEEE International Conference on Robotics and Automation, pp348-353, 2000

[2] Cai W., Hu S.J., Yuan J.X., Deformable sheet metal fixturing: principles, algorithms, and simulations. Transactions of the ASME. Journal of Manufacturing Science and Engineering, vol.118, (no.3), ASME, Aug. 1996. p.318-24.

[3] J. F. Canny, K. Y. Goldberg, "RISC" industrial robotics: recent results and open problems, IEEE International Conference on Robotics and Automation, vol.3, pp.: 1951 -1958, 1994.

[4] http://toyota.irweb.jp/IRweb/corp_info/body_line/ Global Portal, Toyota.

[5] Gopalakrishnan K., Goldberg K., Gripping parts at concave vertices, Proceedings,IEEE International Conference on Robotics and Automation, 2002, Page(s): 1590 -1596, Volume: 2 , 2002.

[6] Lee J. D., and Haynes L. S., Finite Element Analysis of Flexible Fixturing Systems, ASME J. Eng. Ind., 109, pp. 134–139, 1987.

[7] Li B., Shiu B.W., Lau K.J., Fixture configuration design for sheet metal assembly with laser welding: a case study. International Journal of Advanced Manufacturing Technology, vol.19, (no.7), Springer-Verlag, 2002. p.501-9.

[8] Li H. F., Ceglarek D., Shi J., A Dexterous Part-Holding Model for Handling Compliant Sheet Metal Parts, ASME Transactions, Journal of Manufacturing Science and Engineering. Vol. 124, No. 1, pp. 109-118, 2002.

[9] X. Markenscoff, L. Ni and C. H. Papadimitriou, The Geometry of Grasping, International Journal of Robotics Research, Vol. 9, No. 1, pp 61-74, 1990.

[10] Mason M.T., Mechanics of Robotic Manipulation, The MIT Press, 2001.

[11] Menassa R., De Vries W., Optimization Methods Applied to Selecting Support Positions in Fixture Design, ASME Journal of Engineering for Industry, vol 113, pp. 412-418, 1991.

[12] B. Mishra, J. Schwarz, and M. Sharir, On the existence and Synthesis of Multifinger Positive Grips, Algorithmica 2, 1987.

[13] Plut, W.J., Bone, G.M., Limited mobility grasps for fixtureless assembly, Robotics and Automation, 1996. Proceedings., 1996 IEEE International Conference on , Volume: 2 , 1996, Page(s): 1465 -1470 vol.2.

[14] Plut, W.J., Bone, G.M., 3-D flexible fixturing using a multi-degree of freedom gripper for robotic fixtureless assembly, Robotics and Automation, 1997. Proceedings., 1997 IEEE International Conference on , Volume: 1 , 1997, Page(s): 379 -384 vol.1.

[15] Jean Ponce, Joel Burdick and Elon Rimon, Computing the Immobilizing Three-Finger Grasps of Planar Objects, Proceedings of the Worskshop on Computational Kinematics, 1995.

[16] Rearick M.R., Hu S.J., Wu S.M., Optimal Fixture Design for Deformable Sheet Metal Workpieces, Transactions of NAMRI/SME, vol. XXI, pp. 407-412.

[17] F. Reuleaux, The Kinematics of Machinery. New York: Macmillan 1876, republished by New York: Dover, 1963.

[18] Rimon E. and Burdick J., On force and form closure for multiple finger grasps, Proceedings of IEEE International Conference on Robotics and Automation, 1996, pp. 1795 -1800 vol.2.

[19] Elon Rimon and Joel Burdick, Mobility of bodies in contact - I, IEEE transactions on Robotics and Automation, 14(5): 696-708, 1998.

[20] Rimon, E. and Blake, A., Caging planar bodies by one-parameter two fingered gripping systems, International Journal of Robotics Research, v18, n3, March 1999, pp. 299-318.

[21] Salisbury, J.K. Kinematics and Force Analysis of Articulated Hands. Ph.D. Thesis, Stanford University, 1982

[22] Sela M.N., Gaudry O., Dombre E., Benhabib B., A reconfigurable modular fixturing system for thin- walled flexible objects, International Journal of Advanced Manufacturing Technology, 13:611-617, 1997.

[23] P. Somoff, Uber gebiete von schraubengeschwindigkeiten eines starren korpers bieverschiedener zahl von stuz achen, Zeitschrift fur Mathematic and Physik, vol. 45, pp. 245-306, 1900.

[24] Van der Stappen A.F., Wentink C., Overmars M.H., Computing form-closure configurations, Proceedings of IEEE International Conference on Robotics and Automation, vol.3, pp. 1837 -1842, 1999.

[25] M. Y. Wang, Tolerance analysis for fixture layout design, Assembly Automation, a special issue on Automated Fixturing, vol. 2, no. 2, pp. 153 – 162, 2002.

Realization of Fault Tolerant Manufacturing System and its Scheduling based on Hierarchical Petri Net Modeling

YoungWoo Kim*, Akio Inaba**, Tatsuya Suzuki* and Shigeru Okuma*

*Dept. of Electrical Engineering, Graduate School of Nagoya University
Furo-cho, Chikusa-ku, Nagoya, 464-8603, Japan
Phone:+81-52-789-2778, FAX:+81-52-789-3140,
Email:kim,suzuki,okuma@okuma.nuee.nagoya-u.ac.jp

**Gifu Prefectural Research Institute of Manufactural Information Technology,
4-179-1,Sue, Kagamihara, Gifu. Japan
Phone:+81-583-79-9300, FAX:+81-583-79-3301,
E-Mail:gifu-irtc@go.jp

Abstract

This paper presents a new hierarchical scheduling method for a large-scale production system based on a hierarchical Petri net model, which consists of FOHPN and TPN. The automobile production system equipped with 2 stand-by lines is focused as one of a typical large-scale system and these stand-by lines are controlled by binary signal. In a high level, the FOHPN model is used to represent continuous flow in production of an entire system, and MLD description is used to control the net dynamics of FOHPN. Also in a low level, TPN is used to represent production environment of each sub-line in a decentralized manner, and MCT algorithm is applied to find a feasible semi-optimal process sequences for each sub-line.

Keywords: *Manufacturing, Hybrid Petri Net, Mixed Logical Dynamical System, Scheduling, MIQP, RTA**,

1 Introduction

The model-based approaches have clear advantages in a scheduling of manufacturing system compared with other non model-based approaches. The reason of this can be stated as follows: (1) It is easy to consider the various practical constraints which come from production environments. (2) It is easy to monitor the current situation of the production system. (3) It is easy to combine the model with powerful search algorithms. Among several modeling tools, Petri Net (PN) is known as one of the promising tools due to its graphical understanding and algebraic manipulability. Various methods have been developed for the scheduling of the manufacturing systems based on Petri net model. For a large-scale system, however, the direct application of model-based approach often requires unreasonably large computational effort. One of the interesting ideas to overcome this problem is to build the model in decentralized and hierarchical manner. From this point of view, first of all, the decentralized and hierarchical modeling technique for the manufacturing system is proposed in this paper. In the proposed modeling, the macro behavior of the entire system is approximated by a continuous flow model and the micro behavior of each local sub-system is described by the conventional discrete model, such as Timed Petri Net (TPN) model. As for the continuous part, First-Order Hybrid Petri Net (FOHPN) proposed by F. Baluzzi et.al [2] is adopted. FOHPN is a kind of fluid approximation model is recognized a suitable model for the high-level decision-making.

In [2], the FOHPN model is combined with Linear Programming (LP), and the optimal process speed at each machine was found "myopically" by applying LP. Although this approach can provide systematic planning schemes for large-scale systems, the way of cooperation with low-level scheduling, which generates concrete feasible process sequences, has not been fully discussed yet. On the other hand, according to the increase of the demand for the reliability, the design of the fault tolerant manufacturing system is attracting great attention. The fault tolerant manufacturing system should have mechanism of the real-time detection and fault recovery. Therefore, this paper secondly tries to realize the fault tolerant and highly agile manufacturing system by introducing a stand-by sub-lines. The stand-by sub-lines are used in the case that some machine breaks down and/or unexpected heavy requirements are imposed on some sub-line due to an urgent interruptive manufacturing demand. From these considerations, this paper thirdly presents a new scheduling technique which consists of following two planning engines:

High level : For the FOHPN model, which represents the continuous production flow of the entire system, the Mixed Logical Dynamical System (MLDS) [3] description with certain sampling interval is introduced. In our problem setup, the binary (logical) signal, which takes a value of 0 or 1, is specified to enable or disable the use of the stand-by sub-lines. Then, the macro behavior of the entire system is planned by solving the Mixed Integer Quadratic Programming (MIQP). The obtained results are transmitted to low level scheduler for each sub-line as a production requirement.

Low Level : For the TPN model of each sub-line, the Minimization of Completion Time (MCT) algorithm [1], which is based on the reactive graph search, is applied to find a feasible and semi-optimal process sequences for each sub-production line. The goal of the low level planner is to realize the production requirement requested from high level scheduler.

Our proposed scheduling method macroscopically tries to find an optimal flow of process in entire system and generate microscopically the processing sequence taking into consideration the physical constraints which come from the real shop floor. The advantage of our scheduling method is its adaptability for unexpected change of production environment such as processing failure, conveyance delay and so on.

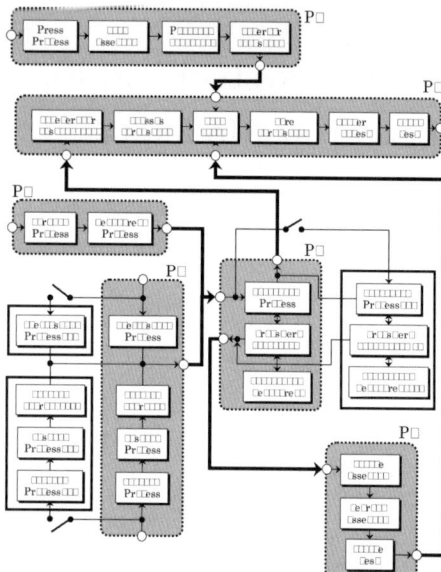

Figure 1: Automobile production system

The usefulness of the proposed idea is demonstrated for the automobile production system which is one of the typical large scale manufacturing system.

2 Hierarchical modeling based on FOHPN and TPN

In this section, the production environment of the automobile manufacturing system focused in this paper is explained and its hierarchical modeling method based on FOHPN and TPN is introduced.

2.1 Outline of automobile production system

The automobile production system is generally composed of following six sub-production lines : (1) Body assembly line (production line 1 : P1), (2) Main assembly line (P2), (3) Metalworking line 1 (P3), (4) Metalworking line 2 (P4), (5) Machining line (P5) and Engine assembly line (P6).

Fig.1 shows the system environment of automobile production system where metalworking line 1 (P3) and machining line (P5) have auxiliary stand-by line (enclosed by black box) for the case of unexpected breakdown or excessively heavy burden. The entire system includes confluence and diffluence of each production line and each line may include serial and parallel sequences.

2.2 Hierarchical modeling based on FOHPN and TPN

In a high level modeling, the macroscopic behavior of an automobile production system is described by FOHPN, which helps us to understand the process flow in an entire system. Fig.2 shows the FOHPN model for the automobile production system of Fig.1 and is defined as follows :

$$N_H = (P_H, T_H, v, I_H^-, I_H^+, M_H^0) \quad (1)$$

where its components $P_H, T_H, v, I_H^+, I_H^+$ and M_H^0 are given in the following:

- Set of places P_H

The set of places $P_H = P_{H_C} \cup P_{H_D}$ is partitioned into a set of continuous places P_{H_C} and a set of discrete places P_{H_D}. P_{H_C} is used to represent all load stations and unload station in entire systme. P_{H_D} is also used to represent the signal which enables or disables the auxiliary stand-by line to work. In Fig.2, variables l, j and s of continuous place $p_{C_{l,j,s}} \in P_{H_C}$ imply the index of production line, the index of job type to be processed and the index of part to be processed, respectively. $p_{D_{l,j}} \in P_{H_D}$ represents a 'control place' for the use of the auxiliary stand-by line. If the corresponding place has a token, then the corresponding line is allowed to work.

- Set of transitions T_H

The set of transitions T_H in a high level modeling represent the flow of the works from macroscopic point of view. There are two kinds of transitions as follows: (1) transition corresponding to the production line and (2) transition corresponding to the stand-by line.

The transition t of the FOHPN model is *enabled* at time τ if for both continuous and discrete places $p \in {}^\bullet t$, $m_h(p, \tau) \geq I_H^-(p, t)$. Here, m_h is the marking of the FOHPN model and is partitioned into continuous marking m_{h_c} and discrete marking m_{h_d} and I_H^- is forward incidence relationship discribed later. In Fig.2, indices l and j of continuous transition $t_{l,j}$ imply production line number and job number to be processed, respectively.

- Firing speed of continuous transition v

The function $v_{l,j}(\tau)$ specifies the firing speed assigned to continuous transition $t_{l,j}$ at time τ. $v_{l,j}(\tau)$ must satisfy

$$0 \leq v_{l,j}(\tau) \leq v_M(l,j) \quad (2)$$

where $v_M(l,j)$ indicates the maximum firing speed of continuous transition $t_{l,j}$. The firing speed represents the number of works to be processed in the corresponding line during specified unit time interval (see below).

- Functions I_H^- and I_H^+

The functions, $I_H^-(p,t)$ and $I_H^+(p,t)$ specify the backward and forward incidence relationship between transition t and place p which precedes or follows the transition, respectively. I_H^+ and I_H^- represent the number of works which flow along the corresponding arc during unit time.

- Initial marking M_H^0

M_H^0 is the initial marking.

- Virtual net dynamics

The net dynamics of FOHPN is supposed to be represented by first order differential equation as for each continuous place $p \in P_{H_C}$ as follows :

$$\frac{dm_{h_c}(p,\tau)}{d\tau} = \sum_{t_{l,j} \in I_e} C(p, t_{l,j}) \cdot v_{l,j}(\tau) \quad (3)$$

where $m_{h_c}(p, \tau)$ is the marking of continuous place p at time τ, $C(p,t)$ is incidence relationship given by $C(p,t) = I_C^+(p,t) - I_C^-(p,t)$, and I_e is the set of enabled continuous transitions $t_{l,j} \in p^\bullet \cup {}^\bullet p$. Also, $m_{h_c}(p,\tau)$ is supposed to have its maximum value $m_M(p)$.

The equation (3) is easily transformed to its discrete version (6) supposing that $v_{l,j}(\kappa T_s)$ is constant during two successive sampling instants.

$$m_{h_c}(p,(\kappa+1)T_s) = m_{h_c}(p,\kappa T_s) \\ + \sum_{l,j} C(p,t_{l,j}) \cdot v_{l,s}(\kappa T_s) \cdot T_s \quad (4)$$

where κ is a number of cycle and T_s is a sampling period. Note that this virtual net dynamics is used only for estimating overall behavior of an entire system and actual marking of the FOHPN model is decided by the evolution in TPN model at each sampling instant (See section 3).

In the low level modeling, the microscopic behavior in each production line is described by the TPN model. The TPN model enables us to take into consideration the various physical constraints which come from real shop floor and to introduce some powerful search engines to find (semi) optimal process sequences.

TPN model for line $l(l=1,\cdots,L)$ is defined as

$$N_{L_l} = (P_{L_l}, T_{L_l}, C_{L_l}, \theta_{L_l}, I^-_{L_l}, I^+_{L_l}, M^0_{L_l}) \quad (5)$$

where C_{L_l} is color information of token, θ_{L_l} is firing time and L is total line number. Each component of N_{L_l} is defined in similar way with N_H (See [1] for more detail).

Fig.3 shows the TPN model of production line 1 for the jobs listed in TABLE 1. A double and single circle in Fig.3 indicate a place of which capacity are more than one and only one, respectively. Also, white and black transitions are timed transitions of which firing time is more than one and only one, respectively. In Fig.3, Ln, Un, IBn, OBn, Mn and Pn indicate load station, unload station, input buffer, output buffer, machine and path, respectively. The TPN model represents the physical constraints and the behaviors explicitly, and enables us to analyze the status of the production line visually. The same formulation is applied to other production lines.

Table 1: Machine routings and processing times for production line 1

Operation Number	1	2	3	4	5
Job 1	M1 (4)	M4 (3)	M5 (2)	M6 (3)	M7 (3)
Job 2	M2 (3)				
Job 3	M3 (2)				

2.3 Interaction between FOHPN and TPN

The FOHPN model and the TPN models interact with each other at every sampling instant $\kappa \cdot T_s$ by means of "*marking converter*(MC)" and "*reference generator*(RG)". The *marking converter* generates the marking $m_h(p,\kappa T_s)$ $(p \in P_H)$ of the FOHPN model based on the markings of the TPN models,

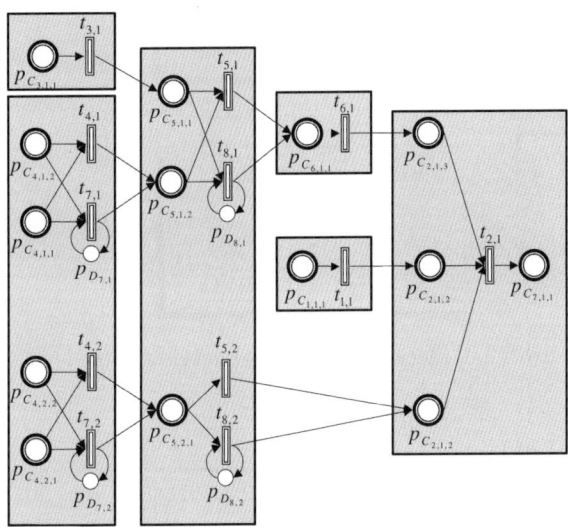

Figure 2: FOHPN model of automobile production system

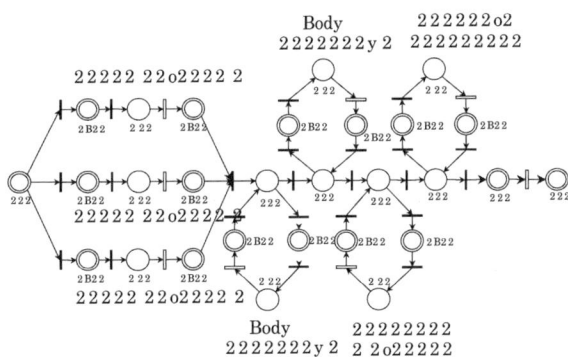

Figure 3: TPN model of production line 1

$m_l(p,\kappa T_s)$ $(p \in P_{L_l})$. Also, the *reference generator* calculates the reference number of works $\lfloor v_{l,j}(\kappa T_s) \cdot T_s \rfloor$ to be processed in corresponding TPN model with cutting off decimals in order to make this number integer. The interaction between the FOHPN model and the TPN models is depicted in Fig.4 and Fig.5.

3 Hierarchical scheduling based on hierarchical model

In this section, we develop a hierarchical scheduling method for "variable due-time problem" based on Petri net models shown in previous section. Here, variable due-time problem is the problem where the number of required products can vary for each sampling period.

Definition 1: T_s is the time horizon where the macro behavior is optimized based on the FOHPN model and the search engine such as branch-and-bound method. During T_s, the firing speed $v_{l,s}(\kappa T_s)$ of continuous transition $t_{l,s}$ is supposed to be constant.

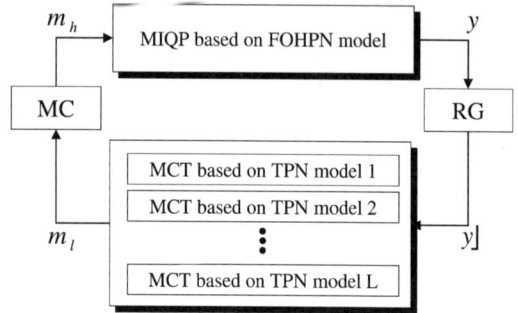

Figure 4: Interaction between high level and low level

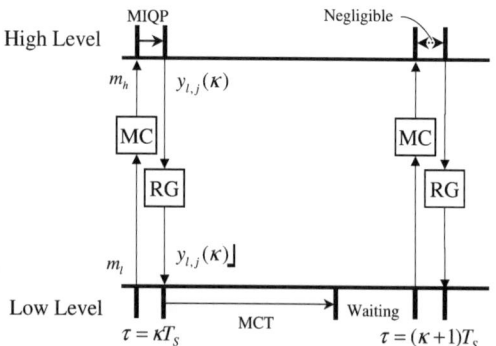

Figure 5: Relationship between high level and low level scheduler

Also T_s coincides with $N \cdot \Gamma$ in the TPN model. N is step number which is required to complete the process planned in a high level, and Γ is a unit time used to specify unit firing time for discrete transition in TPN model. □

3.1 Physical constraint and logical constraint

In order to control the behavior of the system shown in Fig.1, both of physical constraints which come from real shop floor and logical constraints which leads to an effective scheduling (e.g. efficient utilization of given machines or lines) must be considered.

In this work, the following constraints are introduced :

PC1) Physical constraint 1 : Each line l has its upper limit and lower limit to produce parts.

PC2) : All load stations, unload station, input buffer, output buffer, machine etc. have their upper and lower limit to keep parts.

LC1) Logical constraint 1 : The production in each stand-by line is allowed to only one kind of job between sampling instants.

LC2) : The production at the original line $f(l)$ as for corresponding stand-by line $g(l)$ is always carried out with its maximum production speed $v_M(f(l)) = min_{j=1}^{J_l}\{v_M(f(l),j)\}$ when operation for line $g(l)$ is allowed.

LC3) : The operation of stand-by line $g(l)$ requires always over the half speed of $v_M(g(l))$ in order to avoid wasteful operation of it.

Here $g(l)$ is the function which obtains the number of stand-by line which is controlled by u_l, $f(l)$ is the function which obtains the number of original line as for stand-by line $g(l)$ and J_l is a total job number to be processed at line l.

Since the proposed high level model includes both continuous (flow dynamics) and discrete (on-off control for stand-by line) aspects, some algebraic formulation which can handle both continuous and discrete aspects must be introduced. The MLD description has been developed so as to describe such system [3]. The MLD also enables us to easily consider some constraints given by inequalities and can be conbimed with powerful search engine such as Mixed Integer Quadratic Programming (MIQP) problem.

3.2 High level planning by means of MIQP

The net dynamics of the FOHPN model shown in Fig.2 can be described as for each place $p_{C_{l,j,s}}$ by the discrete version of a first order differential equation as follows

$$x_{l,j,s}(\kappa+1) = x_{l,j,s}(\kappa) + \sum_{t_{l,j} \in p^{\bullet}_{C_{l,j,s}} \cup {}^{\bullet}p_{C_{l,j,s}}} C(p_{C_{l,j,s}}, t_{l,j}) \cdot v_{l,j}(\kappa T_s) \cdot T_s, \quad (6)$$

where $x_{l,j,s}(\kappa)$ is the marking of continuous place $p_{C_{l,j,s}}$ at sampling index κ. The equation (6) can be formulated by the MLD description with the binary input $u \in \{0,1\}$ to control the firing of corresponding continuous transition.

$$x(\kappa+1) = Ax(\kappa) + Bu(\kappa) + b(\kappa)$$
$$y(\kappa) = Du(\kappa) + d(\kappa)$$
$$e(\kappa) \leq E_1 u(\kappa) + E_2 x(\kappa) \quad (7)$$

$$x(\kappa) = \begin{bmatrix} x_1(\kappa) \\ \vdots \\ x_{L+L_s}(\kappa) \end{bmatrix}, \begin{aligned} x_l(\kappa) &= [x_{l,1}(\kappa) \cdots x_{l,J_l}(\kappa)]^T, \\ x_{l,j}(\kappa) &= [x_{l,j,1}(\kappa) \cdots x_{l,j,S_{l,j}}(\kappa)]^T, \end{aligned}$$

$$y(\kappa) = \begin{bmatrix} y_1(\kappa) \\ \vdots \\ y_{L+L_s}(\kappa) \end{bmatrix}, y_l(\kappa) = [y_{l,1}(\kappa) \cdots y_{l,J_l}(\kappa)]^T,$$

$$u(\kappa) = \begin{bmatrix} u_{L+1}(\kappa) \\ \vdots \\ u_{L+L_s}(\kappa) \end{bmatrix}, u_{l'}(\kappa) = [u_{l',1}(\kappa) \cdots u_{l',J_{l'}}(\kappa)]^T,$$

subject to
$$x_{l,j,s}(\kappa) \in \mathcal{R}, \quad y_{l,j}(\kappa) \in \mathcal{R}, \quad u_{l',j}(\kappa) \in \{0,1\}$$
where,
$$1 \leq l \leq L \bullet \bullet L_s, \quad 1 \leq j \leq J_l,$$
$$1 \leq s \leq S_{l,j}, \quad 1 \leq l' \leq L_s,$$

Here, L is a total number of production line, $S_{l,j}$ is a total number of part to produce job j at line l and L_s is a total number of stand-by line. And the output $y_{l,j}(\kappa)$ is a number of token evolved from $p \in {}^\bullet t_{l,j}$ to $p \in t_{l,j}^\bullet$, input $u_{l,j}(\kappa)$ is a binary signal to control transition $t_{g(l),j}$. Also each matrix A, B, D, E_1, E_2 and vector $b(\kappa)$, $d(\kappa)$, $e(\kappa)$ have suitable dimensions. Note that since $u(\kappa)$ takes only binary signal, the state equation and the output equation of (7) include affine term as $b(\kappa)$ and $d(\kappa)$.

Definition 2: $x_{I_{l,j,s}}(\kappa)$ is *internal marking* of transition $t_{l,j} \in p^\bullet_{C_{l,j,s}}$. The internal marking amounts to the number of discrete tokens in all places of corresponding TPN model excluding load station $p \in P^1_{L_l}$. And $x_M(l,j,s)$ is maximum value of $x_{l,s}(\kappa)$ to keep tokens. □

By the introduction of internal marking, the FOHPN model has consistency with TPN model. All tokens evolved through transition $t_{l,j}$ during κth cycle can be observed in the form of $x_{l,j,s} + x_{I_{l,j,s}}$.

In FOHPN model, the constraint PC1) can be specified by introducing the maximum firing speed $v_M(l,j)$ and minimum firing speed $v_m(l,j) = 0$ as described in (8). The constraint PC2) also can be specified as (9).

$$\begin{aligned} y_{l,j}(\kappa) &\leq x_{I_{l,j,s}}(\kappa) + v_M(l,j) \cdot T_s \\ -y_{l,j}(\kappa) &\leq -x_{I_{l,j,s}}(\kappa) - v_m(l,j) \cdot T_s \end{aligned} \quad (8)$$

$$\begin{aligned} -x_{l,j,s}(\kappa) - v_{l_b,s_b}(\kappa) \cdot T_s + v_{l_a,s_a}(\kappa) \cdot T_s &\leq 0 \\ x_{l,j,s}(\kappa) + v_{l_b,s_b}(\kappa) \cdot T_s - v_{l_a,s_a}(\kappa) \cdot T_s &\\ \leq x_M(l,j,s) \end{aligned} \quad (9)$$

Here, v_{l_b,j_b} is the firing speed of $t_{l_b,j_b} \in {}^\bullet p$ ($p \in {}^\bullet t_{l,j}$) and v_{l_a,j_a} is the firing speed of $t_{l_a,j_a} \in p^\bullet$ ($t_{l,s} \in p^\bullet$).

The constraint LC1) can be represented by the following equation.

$$u_{l',j}(\kappa) = 1 \rightarrow \begin{array}{l} u_{l',1}(\kappa) = 0 \wedge \\ \vdots \\ u_{l',j-1}(\kappa) = 0 \wedge \\ u_{l',j+1}(\kappa) = 0 \wedge \\ \vdots \\ u_{l',J_l}(\kappa) = 0 \end{array} \quad (10)$$

Here, l' and j satisfy $1 \leq l' \leq L_s$ and $1 \leq j \leq J_{l'}$, respectively. Equation (10) can be transformed to the inequality as follows:

$$u_{l',1}(\kappa) + \cdots + u_{l',J_l}(\kappa) \leq 1. \quad (11)$$

The constraint LC2) can be represented by (12). This logical constraint can be transformed to the inequalities (13) and (14).

$$u_{l',1}(\kappa) = 1 \vee \cdots \vee u_{l',J_{l'}}(\kappa) = 1 \rightarrow \\ v_{f(l'),1}(\kappa) + \cdots + v_{f(l'),J_{l'}}(\kappa) = v_M(f(l')) \quad (12)$$

$$\begin{aligned} &-\{v_{f(l'),1}(\kappa) \cdots v_{f(l'),J_{l'}}(\kappa)\}T_s + \\ &u_{l',1}(\kappa)M_{l',j}(\kappa) \leq M_{l',j}(\kappa) - v_M(f(l'))T_s \\ &\qquad \vdots \\ &-\{v_{f(l'),1}(\kappa) \cdots v_{f(l'),J_{l'}}(\kappa)\}T_s + \\ &u_{l',J_{l'}}(\kappa)M_{l',j}(\kappa) \leq M_{l',j}(\kappa) - v_M(f(l'))T_s \end{aligned} \quad (13)$$

$$\begin{aligned} 0 &\leq v_{f(l'),1}(\kappa) \leq v_M(f(l'),1) \\ &\vdots \\ 0 &\leq v_{f(l'),J_{l'}}(\kappa) \leq v_M(f(l'),J_{l'}) \end{aligned} \quad (14)$$

Here, $M_{l,j}(\kappa) = \max_{l,j}\{v_{l,j}(\kappa)T_s\}$.

In a similar way, LC3) can be represented by (15) and then transformed to (16).

$$u_{l',1}(\kappa) = 1 \vee \cdots \vee u_{l',J_{l'}}(\kappa) = 1 \rightarrow \\ v_{g(l'),1}(\kappa) + \cdots + v_{g(l'),J_{l'}}(\kappa) \geq (1/2)v_M(g(l')) \quad (15)$$

$$\begin{aligned} &-\{v_{g(l'),1}(\kappa) \cdots v_{g(l'),J_{l'}}(\kappa)\}T_s + \\ &u_{l',1}(\kappa)M_{l',j}(\kappa) \leq M_{l',j}(\kappa) - (1/2)v_M(g(l'))T_s \\ &\qquad \vdots \\ &-\{v_{g(l'),1}(\kappa) \cdots v_{g(l'),J_{l'}}(\kappa)\}T_s + \\ &u_{l',J_{l'}}(\kappa)M_{l',j}(\kappa) \leq M_{l',j}(\kappa) - (1/2)v_M(g(l'),1)T_s \end{aligned} \quad (16)$$

Since the goal of our scheduling is to meet the variable due-time requierment, this can be achieved if the firing speed for the final line l_f, $v_{l_f,j}$ can trace the variable reference speed v_r (variable production requirement). The solution to this requirement can be found by minimizing the following objective function.

$$J(u(\kappa), x(\kappa), y(\kappa)) = \|y(\kappa) - V_r(\kappa) \cdot T_s\|_Q^2 \quad (17)$$

where, $V_r(\kappa)$ is the vector in which all elements take same value of reference speed for l_f with $L+L_s$ dimension (e.g. [5 5 5 5 5 $\cdots$]) in the case that reference speed of l_f is 5).

The basic form (7) of MLD with objective function (17) can be transformed to the canonical form of (0-1) MIQP(Mixed Integer Quadratic Programming) problem of (19) with introduction of $\chi(\kappa)$ of (18). The branch-and-bound method is well-known as a solver for this problem.

$$\chi(\kappa) = (x(\kappa), y(\kappa), v_{l,s}(\kappa), u(\kappa)) \quad (18)$$

Here, H is Hessian matrix and f is a first order cost coefficient vector.

$$\min_{\chi(\kappa)} \{\chi(\kappa)^T H \chi(\kappa) + f\chi(\kappa) \mid A_{ine}\chi(\kappa) \leq b_{ine}, \\ A_{eq}\chi(\kappa) = b_{eq}, \delta(\kappa) \subset \chi(\kappa) \in \{0,1\}\} \quad (19)$$

Table 2: $v_M \cdot T_s$ of continuous transition

Index of transition t	$t_{1,1}$	$t_{2,1}$	$t_{3,1}$	$t_{4,1}$	$t_{4,2}$	$t_{5,1}$
$v_M(t_i) \cdot T_s$	14	12	15	11	11	10
Index of transition t	$t_{5,2}$	$t_{6,1}$	$t_{7,1}$	$t_{7,2}$	$t_{8,1}$	$t_{8,2}$
$v_M(t_i) \cdot T_s$	14	15	11	11	10	14

3.3 Lower level planning by means of MCT algorithm

In a low level, the MCT algorithm is applied to the TPN models developed in section two, for each production line of automobile production system. The MCT algorithm is basically based on a RTA* algorithm and a supervisor. The RTA* algorithm was originally developed in the field of artificial intelligence and is one of the real-time graph search method. The RTA* algorithm is characterized by the great adaptation performance for the abrupt change of search space in given problem, whereas conventional graph search methods have some drawbacks in the fact that (1) the scheduling must be completed before the execution and (2) when an unexpected change of search environment such as machine trouble, occurs while the pre-planned scheduling solution is being executed, rescheduling for the remaining process is required.

Note that the MCT always finds time-optimal solution and does not try to control the production speed of each line explicitly. Once the low level scheduler for each line receives the reference production speed $v_{l,j}$ from the high level scheduler, then the low level scheduler runs the MCT and terminates it until the last transition (i.e. the transition which precedes the unload station) fires $v_{l,j} T_s \rfloor$ times or the sampling interval T_s elapses. If no works exist in line l, then the MCT is not executed until some works are imported from other line.

4 Numerical experiment

In order to confirm the usefulness of the proposed method, the numerical experiment was performed. We consider the manufacturing environments shown in Fig.2 where the capacities of $p_{C_{1,1,1}}$, $p_{C_{3,1,1}}$, $p_{C_{4,1,1}}$, $p_{C_{4,1,2}}$, $p_{C_{4,2,1}}$, $p_{C_{4,2,2}}$ in Fig.2 are 200 and the capacities of the other places of FOHPN model are 30. And the capacities of all input buffers and out buffers are 5. The maximum firing speed and conveyance speed of each line are listed in TABLE 2. These values can be easily found by applying MCT algorithm several times varying the number of works committed to each line.

4.1 Adaptability to the unexpected change of production requirement

The adaptability to the unexpected change of production requirement is investigated, assuming that reference speed is changed from 8 to 10 in the middle of production. Fig.6 shows the result in case that from fourth and fifth cycle, reference speed is changed to 10. Compared with the case of the third cycle

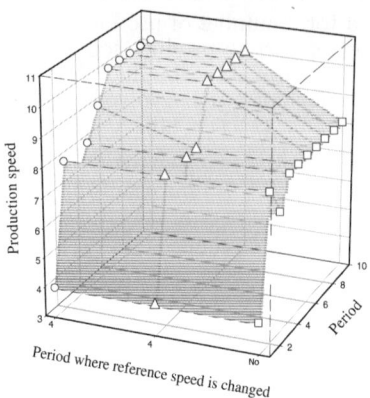

Figure 6: Adaptability to changed reference speed (from 8 to 10)

from which reference speed is changed, the middle one traced to 10 from sixth cycle without passing through transient state. Our proposed method shows great adaptation to unexpected change of requirement, system failure and so on by making use of "periodical planning scheme" in high level, and MCT algorithm which is reactive graph search algorithm in low level.

5 Conclusions

This paper presented a new scheduling method for a large-scale production system based on the hierarchical Petri net model, which consists of FOHPN and TPN. The automobile production system equipped with 2 stand-by lines is focused and these stand-by lines are controlled by binary signal. In a high level, the FOHPN has been used to represent continuous flow in production process of an entire system, and MLD form has been used to control the net dynamics of FOHPN. Also in a low level, TPN has been used to represent production environment of each sub-line in a decentralized manner, and MCT algorithm has been applied to find a feasible semi-optimal process sequences for each sub-line. The proposed hierarchical scheduling method has been confirmed very useful through numerical experiment.

References

[1] Y. Kim, A.Inaba, T. Suzuki and S. Okuma 'FMS scheduling based on timed petri net model and RTA* algorithm', Proc. of IEEE ICRA, pp. 848-853,Seoul, Korea, Mar, 2001

[2] F. Balduzzi, A. Giua and G. Menga 'First-Order Hybrid Petri Nets: A model for optimization and control', Trans. Robot. Automat., vol.16, no.4, pp. 382-399,2000

[3] A. Bemporad, M. Morari 'Control of systems integrating logic, dynamics, and constraints, Automatica, pp.407 427, 1997'

A New Concept of Modular Parallel Mechanism for Machining Applications

Damien Chablat and Philippe Wenger
Institut de Recherche en Communications et Cybernétique de Nantes[1]
1, rue de la Noë, 44321 Nantes, France
Damien.Chablat@irccyn.ec-nantes.fr

Abstract

The subject of this paper is the design of a new concept of modular parallel mechanisms for three, four or five-axis machining applications. Most parallel mechanisms are designed for three- or six-axis machining applications. In the last case, the position and the orientation of the tool are coupled and the shape of the workspace is complex. The aim of this paper is to use a simple parallel mechanism with two-degree-of-freedom (dof) for translation motions and to add one or two legs to add one or two-dofs for rotation motions. The kinematics and singular configurations are studied for each mechanism.

Key Words: Parallel Machine Tool, Isotropic Design, and Singularity.

1 Introduction

Parallel kinematic machines (PKM) are commonly claimed to offer several advantages over their serial counterparts, like high structural rigidity, high dynamic capacities and high accuracy [1]. Thus, PKM are interesting alternative designs for high-speed machining applications.

The first industrial application of PKMs was the Gough platform, designed in 1957 to test tyres [2]. PKMs have then been used for many years in flight simulators and robotic applications [3] because of their low moving mass and high dynamic performances [1]. This is why parallel kinematic machine tools attract the interest of most researchers and companies. Since the first prototype presented in 1994 during the IMTS in Chicago by Gidding&Lewis (the VARIAX), many other prototypes have appeared.

To design a parallel mechanism, two important problems must be solved. The first one is the singular configurations, which can be located inside the workspace. For a six-dof parallel mechanism, like the Gough-Stewart platform, the location of the singular configurations is very difficult to characterize and can change as a function of small variations of the design parameters [3]. The second problem is the non-homogeneity of its performance indices (condition number, stiffness...) throughout the workspace [1]. To the authors' knowledge, only one parallel mechanism is isotropic throughout the workspace [4, 5] but its stiffness is insufficient to be used in machining applications because its legs are subject to bending. Unfortunately, this concept is limited to three-dof mechanisms and cannot be extended to four or five-dof parallel mechanisms.

Numerous papers deal with the design of parallel mechanisms [4,6]. However, there is a lack of four- or five-dof parallel mechanisms, which are especially needed for machining applications [7].

To decrease the cost of industrialization of new PKM and to reduce the problems of design, a modular strategy can be applied. The translation and rotation motions can be divided into two separated parts to produce a mechanism where the direct kinematic problem is decoupled. This simplification yields also some simplifications in the definition of the singular configurations.

The organization of this paper is as follows. Next section presents design problems of parallel mechanisms. The kinematic description and singularity analysis of the parallel mechanism used, are reported in section 3. Section 4 is devoted to design of two new architectures of parallel mechanisms with one or two dofs of rotation.

---

[1] IRCCyN: UMR CNRS 6596, Ecole Centrale de Nantes, Université de Nantes, Ecole des Mines de Nantes

2 About parallel kinematic machines

2.1 General remarks

In a PKM, the tool is connected to the base through several kinematic chains or legs that are mounted in parallel. The legs are generally made of telescopic struts with fixed foot points (Figure 1a), or fixed length struts with moveable foot points (Figure 1b).

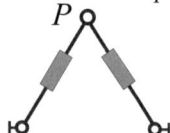
Figure 1a: A bipod PKM

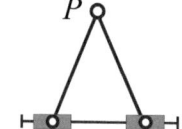
Figure 1b: A biglide PKM

For machining applications, the second architecture is more appropriate because the masses in motion are lower. The linear joints can be actuated by means of linear motors or by conventional rotary motors with ball screws. A classification of the legs suitable to produce motions for parallel kinematic machines is given by [7] with their degrees of freedom and constraints. The connection of identical or different kinematic legs permits the authors to define two-, three-, four- and five-dof parallel mechanisms. However, it is not possible to remove one leg from a four-dof to produce a three-dof mechanism because no modular approach is used.

2.2 Singularities

The singular configurations (also called singularities) of a PKM may appear inside the workspace or at its boundaries. There are two types of singularities [8]. A configuration where a finite tool velocity requires infinite joint rates is called a serial singularity. These configurations are located at the boundary of the workspace. A configuration where the tool cannot resist some effort and in turn, becomes uncontrollable is called a parallel singularity. Parallel singularities are particularly undesirable because they induce the following problems (i) a high increase in forces in joints and links, that may damage the structure, and (ii) a decrease of the mechanism stiffness that can lead to uncontrolled motions of the tool though actuated joints are locked.

Figures 2a and 2b show the singularities for the biglide mechanism of Fig. 1b. In Fig. 2a, we have a serial singularity. The velocity amplification factor along the vertical direction is null and the force amplification factor is infinite.

Figure 2b shows a parallel singularity. The velocity amplification factor is infinite along the vertical direction and the force amplification factor is close to zero. Note that a high velocity amplification factor is not necessarily desirable because the actuator encoder resolution is amplified and thus the accuracy is lower.

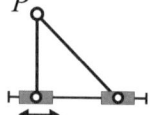
Figure 2a: A serial singularity

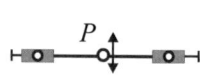
Figure 2b: A parallel singularity

The determination of the singular configurations for two-dof mechanisms is very simple; conversely, for a six-dof mechanism like Gough-Stewart platform, a mechanism with six-dof, the problem is very difficult [3]. With a modular architecture, when position and orientation of the mobile platform are decoupled, the determination of singularities is easier.

2.3 Kinetostatic performance of parallel mechanism

Various performance indices have been devised to assess the kinetostatic performances of serial and parallel mechanisms. The literature on performance indices is extremely rich to fit in the limits of this paper [10] (service angle, dexterous workspace and manipulability...). The main problem of these performance indices is that they do not take into account the location of the tool frame. However, the Jacobian determinant depends on this location [10] and this location depends on the tool used.

Another problem is that to the authors' knowledge there is no parallel mechanism, suitable for machining, for which the kinetostatic performance indices are constant throughout the workspace (like the condition number or the stiffness...). For a serial three-axis machine tool, a motion of an actuated joint yields the same motion of the tool (the transmission factors are equal to one). For a parallel machine, these motions are generally not equivalent. When the mechanism is close to a parallel singularity, a small joint rate can generate a large velocity of the tool. This means that the positioning accuracy of the tool is lower in some directions for some configurations close to parallel singularities because the encoder resolution is amplified. In addition, a high velocity amplification factor in one direction is equivalent to a loss of stiffness in this direction. The manipulability ellipsoid of the Jacobian matrix of robotic manipulators was defined two decades ago [9]. Unfortunately, this concept is quite difficult to apply when the tool frame can produce rotation and translation motions. Indeed, in this case, the Jacobian matrix is not homogeneous [10].

The first way to solve this problem is its normalization by computing its characteristic length [10]. The second one is to change the form of the Jacobian matrix. The first part of the mechanism for translational motion is optimized using homogeneous matrix. Then, the part dedicated to rotation motion can be optimized using the method introduced by [11].

3 Kinematics of mechanisms for translation motions.

Figure 3 shows a PKM with two dofs. The output body is connected to the linear joints through a set of two parallelograms of equal lengths $L = A_iB_i$, so that it can move only in translation.

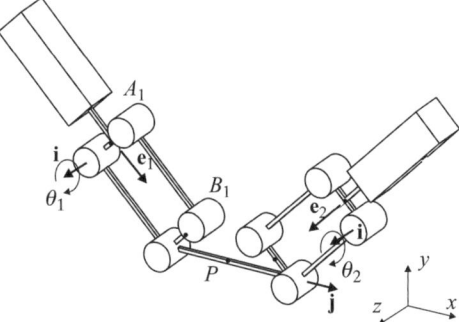

Figure 3: Parallel mechanism with two-dof

The two legs are PPa identical chains, where P and Pa stand for Prismatic and Parallelogram joints, respectively. This mechanism can be optimized to have a workspace whose shape is close to a square workspace and the velocity amplification factors are bounded [12].

The joint variables ρ_1 and ρ_2 are associated with the two prismatic joints. The output variables are the Cartesian coordinates of the tool center point $P = [x\ y]^T$. To control the orientation of the reference frame attached to P, two parallelograms can be used, which also increase the rigidity of the structure, Figure 3.

To produce the third translational motion, it is possible to place orthogonally a third prismatic joint. This one can be located as in the case of Figure 4. Another solution is to use the Orthoglide mechanism, an isotropic three-dof mechanism [13]. The choice between these two solutions depends on the main application of the milling machine. For example, for aeronautical pieces, the solution of fig. 4 is more appropriate because long and fine pieces are built. Conversely, for rapid prototyping of compact parts, we can choose the Orthoglide.

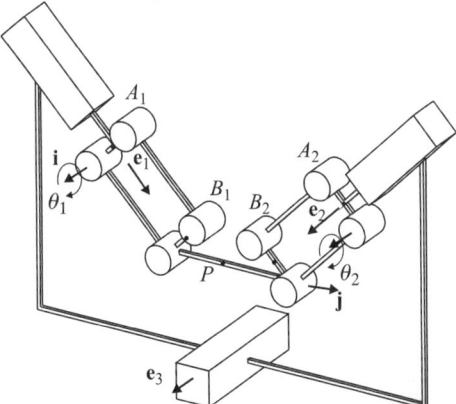

Figure 4: Hybrid mechanism with three-dof

The velocity $\dot{\mathbf{p}}$ of point P can be expressed in two different ways. By traversing the closed loop ($A_1B_1P - A_2B_2P$) in two possible directions, we obtain

$$\dot{\mathbf{p}} = \dot{\mathbf{a}}_1 + \dot{\theta}_1 \mathbf{i} \times (\mathbf{b}_1 - \mathbf{a}_1) \quad (1a)$$
$$\dot{\mathbf{p}} = \dot{\mathbf{a}}_2 + \dot{\theta}_2 \mathbf{i} \times (\mathbf{b}_2 - \mathbf{a}_2) \quad (1b)$$

where $\mathbf{a}_1$, $\mathbf{b}_1$, $\mathbf{a}_2$ and $\mathbf{b}_2$ represent the position vectors of the points A_1, B_1, A_2 and B_2, respectively. Moreover, the velocities $\dot{\mathbf{a}}_1$ and $\dot{\mathbf{a}}_2$ of A_1 and A_2 are given by $\dot{\mathbf{a}}_1 = \mathbf{e}_1\dot{\rho}_1$ and $\dot{\mathbf{a}}_2 = \mathbf{e}_2\dot{\rho}_2$, respectively.

For an isotropic configuration to exist where the velocity amplification factors are equal to one, we must have $\mathbf{e}_1.\mathbf{e}_2 = 0$ [12] (Figure 5). Two square useful workspaces can be used. The first one has horizontal and vertical sides. The second one has oblique sides but its size is higher.

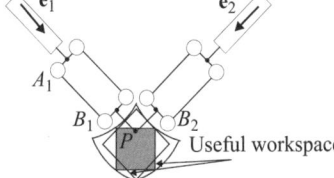

Figure 5: Cartesian workspace and isotropic configuration

We would like to eliminate the two passive joint rates $\dot{\theta}_1$ and $\dot{\theta}_2$ from Eqs. (1a-b), which we do upon dot-multiply the former by $(\mathbf{b}_1 - \mathbf{a}_1)^T$ and the latter by $(\mathbf{b}_2 - \mathbf{a}_2)^T$, thus obtaining

$$(\mathbf{b}_1 - \mathbf{a}_1)^T \dot{\mathbf{p}} = (\mathbf{b}_1 - \mathbf{a}_1)^T \mathbf{e}_1 \dot{\rho}_1 \quad (2a)$$

$$(\mathbf{b}_2 - \mathbf{a}_2)^T \dot{\mathbf{p}} = (\mathbf{b}_2 - \mathbf{a}_2)^T \mathbf{e}_2 \dot{\rho}_2 \quad (2b)$$

Equations (2a-b) can be cast in vector form, namely $\mathbf{A}\dot{\mathbf{p}} = \mathbf{B}\dot{\rho}$, with $\mathbf{A}$ and $\mathbf{B}$ denoted, respectively, as the parallel and serial Jacobian matrices,

$$\mathbf{A} \equiv \begin{bmatrix} (\mathbf{b}_1 - \mathbf{a}_1)^T \\ (\mathbf{b}_2 - \mathbf{a}_2)^T \end{bmatrix} \quad \mathbf{B} \equiv \begin{bmatrix} (\mathbf{b}_1 - \mathbf{a}_1)^T \mathbf{e}_1 & 0 \\ 0 & (\mathbf{b}_2 - \mathbf{a}_2)^T \mathbf{e}_2 \end{bmatrix}$$

where $\dot{\rho}$ is defined as the vector of actuated joint rates and $\dot{\mathbf{p}}$ is the velocity of point P, i.e.,

$$\dot{\rho} = \begin{bmatrix} \dot{\rho}_1 \\ \dot{\rho}_2 \end{bmatrix} \text{ and } \dot{\mathbf{p}} = \begin{bmatrix} \dot{x} \\ \dot{y} \end{bmatrix}$$

When $\mathbf{A}$ and $\mathbf{B}$ are not singular, we obtain the relations,

$\dot{\mathbf{p}} = \mathbf{J}\dot{\rho}$ with $\mathbf{J} = \mathbf{A}^{-1}\mathbf{B}$

Parallel singularities occur whenever the lines (A_1B_1) and (A_2B_2) are colinear, i.e. when $\theta_1 - \theta_2 = k\pi$, for $k = 1, 2, \ldots$. Serial singularities occur whenever $\mathbf{e}_1 \perp (\mathbf{b}_1 - \mathbf{a}_1)$ or $\mathbf{e}_2 \perp (\mathbf{b}_2 - \mathbf{a}_2)$. In [13], the range limits are defined to avoid these two singularities in using suitable bounds on the velocity factor amplification.

4 Kinematics of mechanisms for translation and rotation motions

The aim of this section is to define the kinematics of two mechanisms with one and two dofs of rotation, respectively. To be modular, the direct kinematic problem must be decoupled between position and orientation equations. A decoupled version of the Gough-Stewart Platform exists but it is very difficult to build because three spherical joints must coincide [14]. Thus, it cannot be used to perform milling applications.

The main idea of the proposed architecture is to attach a new body with the tool frame to the mobile platform of the two dofs defined in the previous section. The new joint admits one or two-dof according to the prescribed tasks.

4.1 Kinematics of a spatial parallel mechanism with one-dof of rotation

To add one-dof on the mechanism defined in section 3, we introduce one revolute joint between the previous mobile platform and the tool frame. Only one leg is necessary to hold the tool frame in position. Figure 6 shows the mechanism obtained with two translational dofs and one rotational dof.

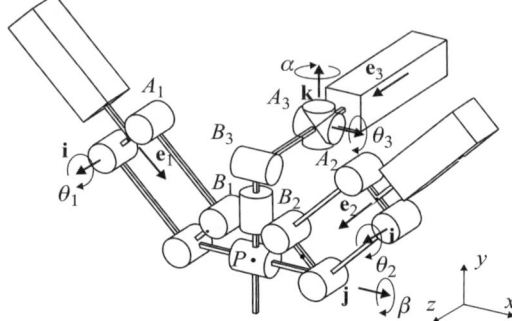

Figure 6: Parallel mechanism with two-dof of translation and one-dof of rotation

The architecture of the leg added is PUU where P and U stand for Prismatic and Universal joints, respectively [7]. The new prismatic joint is located orthogonaly to the first two prismatic joints. This location can be easily justified because on this configuration, i.e. when $(\mathbf{b}_1 - \mathbf{a}_1) \perp (\mathbf{b}_2 - \mathbf{a}_2)$ and $(\mathbf{b}_3 - \mathbf{a}_3) \perp (\mathbf{b}_3 - \mathbf{p})$, the third leg is far away from serial and parallel singularities.

Let $\dot{\rho}$ be referred to as the vector of actuated joint rates and $\dot{\mathbf{p}}$ as the velocity vector of point P,

$\dot{\rho} = [\dot{\rho}_1 \ \dot{\rho}_2 \ \dot{\rho}_2]^T$ and $\dot{\mathbf{p}} = [\dot{x} \ \dot{y}]^T$

Due to the architecture of the two-dof mechanism and the location of P, its velocity on the z-axis is equal to zero. $\dot{\mathbf{p}}$ can be written in three different ways by traversing the three chains A_iB_iP,

$$\dot{\mathbf{p}} = \dot{\mathbf{a}}_1 + \dot{\theta}_1 \mathbf{i} \times (\mathbf{b}_1 - \mathbf{a}_1) \quad (3a)$$

$$\dot{\mathbf{p}} = \dot{\mathbf{a}}_2 + \dot{\theta}_2 \mathbf{i} \times (\mathbf{b}_2 - \mathbf{a}_2) \quad (3b)$$

$$\dot{\mathbf{p}} = \dot{\mathbf{a}}_3 + (\dot{\theta}_3 \mathbf{j} + \dot{\alpha}\mathbf{k}) \times (\mathbf{b}_3 - \mathbf{a}_3) + \dot{\beta}\mathbf{j} \times (\mathbf{p} - \mathbf{b}_3) \quad (3c)$$

where $\mathbf{a}_i$ and $\mathbf{b}_i$ represent the position vectors of the points A_i and B_i for $i = 1, 2, 3$, respectively. Moreover, the velocities $\dot{\mathbf{a}}_1$, $\dot{\mathbf{a}}_2$ and $\dot{\mathbf{a}}_3$ of A_1, A_2 and A_3 are given by $\dot{\mathbf{a}}_1 = \mathbf{e}_1 \dot{\rho}_1$, $\dot{\mathbf{a}}_2 = \mathbf{e}_2 \dot{\rho}_2$ and $\dot{\mathbf{a}}_3 = \mathbf{e}_3 \dot{\rho}_3$, respectively.

We want to eliminate the passive joint rates $\dot{\theta}_i$ and $\dot{\alpha}$ from Eqs. (3a-c), which we do upon dot-multiplying Eqs. (3a-c) by $\mathbf{b}_i - \mathbf{a}_i$,

$$(\mathbf{b}_1 - \mathbf{a}_1)^T \dot{\mathbf{p}} = (\mathbf{b}_1 - \mathbf{a}_1)^T \mathbf{e}_1 \dot{\rho}_1 \quad (4a)$$

$$(\mathbf{b}_2 - \mathbf{a}_2)^T \dot{\mathbf{p}} = (\mathbf{b}_2 - \mathbf{a}_2)^T \mathbf{e}_2 \dot{\rho}_2 \quad (4b)$$

$$(\mathbf{b}_3 - \mathbf{a}_3)^T \dot{\mathbf{p}} = (\mathbf{b}_3 - \mathbf{a}_3)^T \mathbf{e}_3 \dot{\rho}_3 \\ + (\mathbf{b}_3 - \mathbf{a}_3)^T \dot{\beta}\mathbf{j} \times (\mathbf{p} - \mathbf{b}_3) \quad (4c)$$

Equations (4a-c) can be cast in vector form, namely,

$\mathbf{t} = \mathbf{J}\dot{\rho}$ with $\mathbf{J} = \mathbf{A}^{-1}\mathbf{B}$ and $\mathbf{t} = [\dot{x}\ \dot{y}\ \dot{\beta}]^T$

where $\mathbf{A}$ and $\mathbf{B}$ are the parallel and serial Jacobian matrices, respectively,

$$\mathbf{A} \equiv \begin{bmatrix} (\mathbf{b}_1 - \mathbf{a}_1)^T & 0 \\ (\mathbf{b}_2 - \mathbf{a}_2)^T & 0 \\ (\mathbf{b}_3 - \mathbf{a}_3)^T & -(\mathbf{b}_3 - \mathbf{a}_3)^T \mathbf{j} \times (\mathbf{p} - \mathbf{b}_3) \end{bmatrix}$$

$$\mathbf{B} \equiv \begin{bmatrix} (\mathbf{b}_1 - \mathbf{a}_1)^T \mathbf{e}_1 & 0 & 0 \\ 0 & (\mathbf{b}_2 - \mathbf{a}_2)^T \mathbf{e}_2 & 0 \\ 0 & 0 & (\mathbf{b}_3 - \mathbf{a}_3)^T \mathbf{e}_3 \end{bmatrix}$$

There are two new singularities when one leg is added. The first one is a parallel singularity when $(\mathbf{b}_3 - \mathbf{a}_3)^T \mathbf{j} \times (\mathbf{p} - \mathbf{b}_3) = 0$, i.e., when the lines $(A_3 B_3)$ and $(B_3 P)$ are colinear, and the second one is a serial singularity when $(\mathbf{b}_3 - \mathbf{a}_3)^T \mathbf{e}_3 = 0$, i.e., $(\mathbf{a}_3 - \mathbf{b}_3) \perp \mathbf{e}_3$. However, these singular configurations are simple and can be avoided by proper limits on the actuated joints.

4.2 Kinematics of a spatial mechanism with two-dof of rotation

To add two dofs on the mechanism defined in section 3, we introduce a universal joint, between the previous mobile platform and the tool frame. Two legs are necessary to hold the tool frame in position and we use the same archtitecture than for the previous mechanism, *i.e.* the *PUU* mechanism. Figure 7 depicts a parallel mechanism with two translational dofs and two rotational dofs.

Let $\dot{\rho}$ be referred to as the vector of actuated joint rates and $\dot{\mathbf{p}}$ as the velocity vector of point P,

$\dot{\rho} = [\dot{\rho}_1\ \dot{\rho}_2\ \dot{\rho}_3\ \dot{\rho}_4]^T$ and $\dot{\mathbf{p}} = [\dot{x}\ \dot{y}]^T$

$\dot{\mathbf{p}}$ can be written in four different ways by traversing the three chains $A_i B_i P$,

$\dot{\mathbf{p}} = \dot{\mathbf{a}}_1 + \dot{\theta}_1 \mathbf{i} \times (\mathbf{b}_1 - \mathbf{a}_1)$ (5a)

$\dot{\mathbf{p}} = \dot{\mathbf{a}}_2 + \dot{\theta}_2 \mathbf{i} \times (\mathbf{b}_2 - \mathbf{a}_2)$ (5b)

$\dot{\mathbf{p}} = \dot{\mathbf{a}}_3 + (\dot{\theta}_3 \mathbf{i}_3 + \dot{\alpha}_3 \mathbf{k}) \times (\mathbf{b}_3 - \mathbf{a}_3) + (\dot{\beta}\mathbf{j} + \dot{\gamma}\mathbf{i}) \times (\mathbf{p} - \mathbf{b}_3)$ (5c)

$\dot{\mathbf{p}} = \dot{\mathbf{a}}_4 + (\dot{\theta}_4 \mathbf{i}_4 + \dot{\alpha}_4 \mathbf{k}) \times (\mathbf{b}_4 - \mathbf{a}_4) + (\dot{\beta}\mathbf{j} + \dot{\gamma}\mathbf{i}) \times (\mathbf{p} - \mathbf{b}_4)$ (5d)

where $\mathbf{a}_i$ and $\mathbf{b}_i$ represent the position vectors of the points A_i and B_i, respectively, for $i=1,2,3,4$. Moreover, the velocities $\dot{\mathbf{a}}_1$, $\dot{\mathbf{a}}_2$, $\dot{\mathbf{a}}_3$ and $\dot{\mathbf{a}}_4$ of A_1, A_2, A_3 and A_4 are given by $\dot{\mathbf{a}}_1 = \mathbf{e}_1 \dot{\rho}_1$, $\dot{\mathbf{a}}_2 = \mathbf{e}_2 \dot{\rho}_2$, $\dot{\mathbf{a}}_3 = \mathbf{e}_3 \dot{\rho}_3$ and $\dot{\mathbf{a}}_4 = \mathbf{e}_4 \dot{\rho}_4$, respectively.

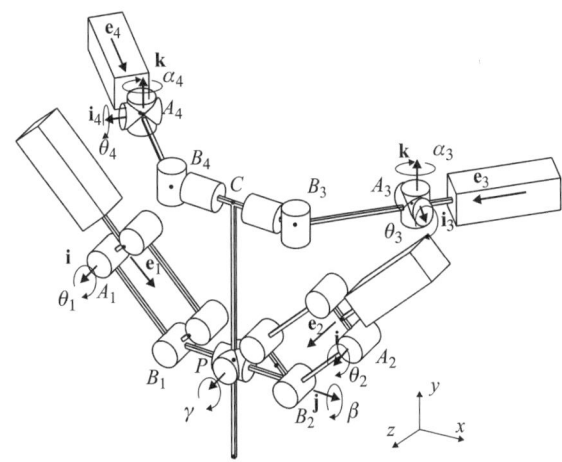

Figure 7: Parallel mechanism with two-dof of translation and two-dof of rotation

We want to eliminate the passive joint rates $\dot{\theta}_i$ and $\dot{\alpha}_i$ from Eqs. (6a-d), which we do upon dot-multiplying Eqs. (5a-d) by $\mathbf{b}_i - \mathbf{a}_i$,

$(\mathbf{b}_1 - \mathbf{a}_1)^T \dot{\mathbf{p}} = (\mathbf{b}_1 - \mathbf{a}_1)^T \mathbf{e}_1 \dot{\rho}_1$ (7a)

$(\mathbf{b}_2 - \mathbf{a}_2)^T \dot{\mathbf{p}} = (\mathbf{b}_2 - \mathbf{a}_2)^T \mathbf{e}_2 \dot{\rho}_2$ (7b)

$(\mathbf{b}_3 - \mathbf{a}_3)^T \dot{\mathbf{p}} = (\mathbf{b}_3 - \mathbf{a}_3)^T \mathbf{e}_3 \dot{\rho}_3$
$+ (\mathbf{b}_3 - \mathbf{a}_3)^T (\dot{\beta}\mathbf{j} + \dot{\gamma}\mathbf{i}) \times (\mathbf{p} - \mathbf{b}_3)$ (7c)

$(\mathbf{b}_4 - \mathbf{a}_4)^T \dot{\mathbf{p}} = (\mathbf{b}_4 - \mathbf{a}_4)^T \mathbf{e}_4 \dot{\rho}_4$
$+ (\mathbf{b}_4 - \mathbf{a}_4)^T (\dot{\beta}\mathbf{j} + \dot{\gamma}\mathbf{i}) \times (\mathbf{p} - \mathbf{b}_4)$ (7d)

Equations (8a-d) can be cast in vector form, namely,

$\mathbf{t} = \mathbf{J}\dot{\rho}$ with $\mathbf{J} = \mathbf{A}^{-1}\mathbf{B}$ and $\mathbf{t} = \begin{bmatrix} \dot{x} & \dot{y} & \dot{\beta} & \dot{\gamma} \end{bmatrix}^T$

where $\mathbf{A}$ and $\mathbf{B}$ are the parallel and serial Jacobian matrices, respectively,

$$\mathbf{A} \equiv \begin{bmatrix} (\mathbf{b}_1 - \mathbf{a}_1)^T & 0 & 0 \\ (\mathbf{b}_2 - \mathbf{a}_2)^T & 0 & 0 \\ (\mathbf{b}_3 - \mathbf{a}_3)^T & -(\mathbf{b}_3 - \mathbf{a}_3)^T \mathbf{j} \times (\mathbf{p} - \mathbf{b}_3) & -(\mathbf{b}_3 - \mathbf{a}_3)^T \mathbf{i} \times (\mathbf{p} - \mathbf{b}_3) \\ (\mathbf{b}_4 - \mathbf{a}_4)^T & -(\mathbf{b}_4 - \mathbf{a}_4)^T \mathbf{j} \times (\mathbf{p} - \mathbf{b}_4) & -(\mathbf{b}_4 - \mathbf{a}_4)^T \mathbf{i} \times (\mathbf{p} - \mathbf{b}_4) \end{bmatrix}$$

and

$$\mathbf{B} \equiv \begin{bmatrix} (\mathbf{b}_1 - \mathbf{a}_1)^T \mathbf{e}_1 & 0 & 0 & 0 \\ 0 & (\mathbf{b}_2 - \mathbf{a}_2)^T \mathbf{e}_2 & 0 & 0 \\ 0 & 0 & (\mathbf{b}_3 - \mathbf{a}_3)^T \mathbf{e}_3 & 0 \\ 0 & 0 & 0 & (\mathbf{b}_4 - \mathbf{a}_4)^T \mathbf{e}_4 \end{bmatrix}$$

When two legs are added, the same singular configurations as the previous mechanism occur, a parallel singularity occurs when the lines $(A_3 B_3)$, $(A_4 B_4)$ and (CP) are coplanar, and a serial singularity occurs when $(\mathbf{b}_3 - \mathbf{a}_3)^T \mathbf{e}_3 = 0$, i.e., $(\mathbf{a}_3 - \mathbf{b}_3) \perp \mathbf{e}_3$ or

$(\mathbf{b}_4 - \mathbf{a}_4)^T \mathbf{e}_4 = 0$, *i.e.*, $(\mathbf{a}_4 - \mathbf{b}_4) \perp \mathbf{e}_4$.

4.3 Discussion

The problem of the optimal design of the two mechanisms defined in the previous sections is not addressed in this paper. In future works, for both mechanisms, it is possible to define a configuration for which the Jacobian matrix is isotropic. This result permits us to define the location of the prismatic joints and to define the condition length to normalize the Jacobian matrix. A method to define the range limits is explained in [11], *via* a distance to the isotropic configuration.

5 Conclusions

In this paper, a new class of modular mechanisms is introduced with two, three, four and five dofs. All the actuated joints are prismatic joints, which can be actuated by means of linear motors or by conventional rotary motors with ball screws. The topology of the legs used to add one or two dof is the same. Only three types of joints are used, *i.e.*, prismatic, revolute and universal joints. All the singularities are characterized easily because position and orientation are decoupled for the direct kinematic problem and can be avoid by proper design. In the future, for the modular architecture, the lengths of the legs as well as their positions will be optimized, to take into account the velocity amplification factors.

6 Acknowledgments

This research was partially supported by the CNRS (Project ROBEA "Machine à Architecture compleXe"). The authors would like to thank Mr. Caro for his valuable remarks on this paper.

7 References

[1] Treib, T. and Zirn, O., "Similarity laws of serial and parallel manipulators for machine tools," Proc. Int. Seminar on Improving Machine Tool Performances, pp. 125-131, Vol. 1, 1998.

[2] Gough, V.E., "Contribution to discussion of papers on research in automobile stability, control an tyre performance," Proc. Auto Div. Inst. Eng., 1956-1957.

[3] Merlet, J-P, "The parallel robot," Parallel robots, Kluwer Academic Publ., Dordrecht, The Netherland, 2000.

[4] Kong, X. and Gosselin, C. M., "A Class of 3-DOF Translational Parallel Manipulators with Linear I-O Equations," Proc. of Workshop on Fundamental Issues and Future Research Directions for Parallel Mechanisms and Manipulators, Québec, Canada, 2002.

[5] Carricato, M. and Parenti-Castelli, V., "Singularity-Free Fully-Isotropic Translational Parallel Mechanisms," The International Journal of Robotics Research, Vol. 21, No. 2, pp. 161-174, February, 2002.

[6] Hervé, J.M. and Sparacino, F., 1991, "Structural Synthesis of Parallel Robots Generating Spatial Translation," Proc. of IEEE 5th Int. Conf. on Adv. Robotics, Vol. 1, pp. 808-813, 1991.

[7] Gao, F., Li, W., Zhao, X., Jin Z. and Zhao. H., " New kinematic structures for 2-, 3-, 4-, and 5-DOF parallel manipulator designs," Journal of Mechanism and Machine Theory, Vol. 37/11, pp. 1395-1411, 2002.

[8] Gosselin, C. and Angeles, J., "Singularity analysis of closed-loop kinematic chains," IEEE Transaction on Robotic and Automation, Vol. 6, No. 3, June 1990.

[9] Salisbury, J-K. and Craig, J-J., "Articulated Hands: Force Control and Kinematic Issues," The Int. J. Robotics Res., Vol. 1, No. 1, pp. 4-17, 1982.

[10] Angeles, J., *Fundamentals of Robotic Mechanical Systems*, Second Edition, Springer-Verlag, New York, 2002.

[11] Chablat, D. and Angeles, J., "On the Kinetostatic Optimization of Revolute-Coupled Planar Manipulators," Journal of Mechanism and Machine Theory, vol. 37,(4).pp. 351-374, 2002.

[12] Chablat, D., Wenger, Ph. and Angeles, J., "Conception Isotropique d'une morphologie parallèle: Application à l'usinage," 3rd International Conference On Integrated Design and Manufacturing in Mechanical Engineering, Montreal, Canada, May, 2000.

[13] Chablat, D. and Wenger, Ph, "Design of a Three-Axis Isotropic Parallel Manipulator for Machining Applications: The Orthoglide," Workshop on Fundamental Issues and Future Research Directions for Parallel Mechanisms and Manipulators, October 3-4, Québec, 2002.

[14] Khalil, W. and Murareci, D., "Kinematic Analysis and Singular configurations of a class of parallel robots," Mathematics and Computer in simulation, pp. 377-390, 1996.

An Error Restraining Method for Accurate Freeform Surface Cutting

A. Jaganathan and Y.J. Lin
Department of Mechanical Engineering
The University of Akron
Akron, OH 44325-3903
E-mail: ylin@uakron.edu

Abstract— Producing sculptured surfaces of special parts always poses challenges to machining industry. In this paper, an effective tool path error restraining method utilizing adaptive rules for cutting control is proposed for freeform surface machining. Based on the proposed approach an experimental verification is accomplished in milling a sample part with curved surfaces. Specifically, the parametric spatial curves representing the part's sculptured surface are approximated by sequences of connected line segments. From the current reference point of the cutting tool, the method has the capability of predicting the next reference point of the tool for a given feed. If the predicted position is not within the required tolerance, the algorithm will automatically adjust the position of the cutter so as to satisfy the tolerance requirements using feedback control philosophy. It is proved that the developed adaptive cutting laws are robust in achieving the desired freeform surface cutting with pre-specified tolerance requirements. The given tolerance is measured as the angular deviations to which the generated tool path deviates from the desired profile. Feedrate variations have been implemented in the investigation in the range between 5 *mm/second* and 30 *mm/second*. The tool paths generated with and without the adaptive mechanisms are compared. The experimental results demonstrate that the proposed tolerance feedback mechanism is very effective for producing parts having curved surfaces.

Keywords — Freeform surface cutting, Adaptive toolpath generation, Bezier curve interpolation, Cutting error restraining.

I. INTRODUCTION

CNC tools play an important role in modern automatic manufacturing systems. With increasing demands for better machining precision, methods for improving the performance of the CNC systems in contour machining are continuously being sought. One of the most important techniques manufacturing is to produce a part with free-form or parametric surfaces that are widely used in modern CAD/CAM systems.

In order to manufacture sculptured surfaces with an end-milling cutter, two questions must be answered. One is to set up a formula to generate the tool-center path, the other is to generate a few different tool paths for one sculptured surface to obtain surface finish. In this work, an adaptive tool path generation algorithm is proposed for machining a sculptured surface of a part using CNC machines. This algorithm has adaptive capability that it can steer the tool paths generated on the part such that the error between the generated path and the desired path based on a CAD model will always fall within an allowable machining tolerance based on precision requirements.

II. TOOL PATH GENERATION

In the past two decades, a number of effective CNC tool path generation methods have been proposed for CNC machining of various surface shapes [1-7]. Basically, these approaches were able to predict the step ahead positions of the cutter based on scientific calculations from the equations which are obtained from the mathematical integration of the velocity/acceleration components of the cutter. For linear path generation, the methods have been proved to be very good. However, when dealing with curved tool-paths generation, they demonstrate certain limitations in yielding smooth curvature gradients.

Aimed at eliminating or reducing the drawbacks of curved tool-path generation, many research works have been conducted in the past decades [10 -14]. Lin et al [13, 14] have developed an adaptive mechanism that is not only able to replicate the desired surface, but also able to control the machining tolerance within a given limit. This method is highly versatile as well as very robust. However, no experimental verifications of the proposed approach of precision machining has been reported.

In view of the limitations posed by some of the methods discussed above, a new cutting path generation algorithm for parametric surfaces and curves is proposed and implemented in this work. The algorithm possesses an adaptive nature. It is generic since it can equitably be applied to generate CNC tool paths for two-dimensional as well as three-dimensional curves and surfaces. In this method, the parametric surface curve on the three-dimensional sculptured surface is approximated by a sequence of linear segments. From the current reference point of the cutting tool, the method has the capability of predicting the next reference point of the tool for a given feedrate. If the predicted position is not within the allowed tolerances, the algorithm will automatically adjust the position of the cutter so as to satisfy the tolerance requirements, similar to a feedback control action.

III. THE PROPOSED CNC CUTTER CONTOUR INTERPOLATOR METHOD

Let's consider a generic curved surface. The direction of machining the surface is the horizontal direction. However, the other parameter v representing vertical/perpendicular direction of the curve is kept constant for each step of tool movement. In other words, the machining for each step can be represented by $r(u,c)$, where c is a constant according to the type of parametric curve being used and u varies from 0 to 1. To describe a typical cutting cycle with this scenario, let us first consider several requirements for an interpolator to be realized as follows. 1) The approximated cutting surface for the desired cutting actions must dictate the desired contour of the surface with negligible deviations. 2) The deviations of the cutting surface from the desired contour surface must meet a pre-specified tolerance δ_{all}. 3) The feedrate of the cutting must be nearly constant and be independent of the free-form surface variations.

It should also be kept in mind that the deviation of the cutting tool position between the actual proceeding position and the desired position must lie within the given tolerance δ_{all}.

The position vector of the cutting tool at every proceeding point is given by

$$\vec{P}_{i+1} = \vec{P}_i + d\vec{T} \qquad (1)$$

where $\vec{T}$ is a unit tangent of the surface curve, and $d = Vt$ mm. It is also equivalent to the step size of the cutter. When using a parametric curve, it is approximately equal to the arc length of the curve. The above equation can be used to generate the so-called Cutter Locations (CL) capable of replicating the contour of the desired surface, but it lacks a tolerance checking mechanism, i.e., if the cutting tool travels out of the allowable tolerance, the above mechanism does not have the capability of commanding the cutter back so that the position is within the allowable tolerance. This is one of the challenges while deriving the algorithm. In order to enhance the functionality of the algorithm, an adaptive mechanism is proposed and added to its capability.

A. The Adaptive Rules for Tool-Path Corrections

A scenario is suggested to make the desired geometric shape resemble the cutting surface as shown in Figure 1.

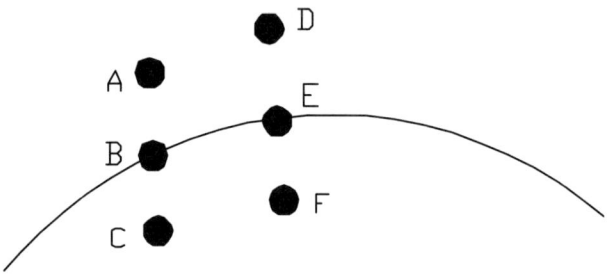

Fig. 1 Desired curved profile

Let us derive a general case that can be used for generic surface curves. The figure also indicates the next possible locations of the tool as it moves from its current position. There can be three possible positions where the cutter might pass from its present position. The proceeding position of the cutter can either be above the surface curve, below the curve or sometimes on the curve itself.

By simple vector algebra, the current position and next predicted position of the cutter can be calculated by the following equations respectively as

$$D_i = [\vec{P}_i - \vec{r}(u_i, c)] \bullet \vec{k}$$
$$D_{i+1} = [\vec{P}_{i+1} - \vec{r}(u_{i+1}, c)] \bullet \vec{k} \qquad (2)$$
$$\delta = \left| [\vec{r}(u_i, c) - \vec{r}(u_{i+1}, c)] \bullet \vec{k} \right|$$

To correct an exceeded position deviation, we incrementally rotate the step length about the current location by an angle $\Delta\theta_2$ until $\delta \leq \delta_{all}$ is satisfied. But in the first instance if the deviation is less than or equal to the allowable tolerance, then the algorithm predicts the proceeding tool position according to Equation (1) and the adaptive rule of it does not play any role.

If the predicted position of the cutter is at point E, then no correction is needed as the predicted point is exactly on the surface curve. The cutter is then driven to the position as predicted and the position of the cutter is updated for further cuts.

If the predicted position of the cutter is at point F and the deviation is less than or equal to the allowable tolerance, then the cut carries as predicted and the tool's position is updated for the following machining motion. However, if the deviation is greater than the allowable tolerance, the adaptive part of the algorithm comes into effect. But, instead of incrementally rotating the step size by $\Delta\theta_2$, we need to decrease the rotation by the same value till we get the deviation satisfies allowable conditions.

To cover all possibilities, it requires six other adaptive rules listed in Table 1. It is clear that only one rule applies at a time, depending on the underlying cutter's motion tendencies.

TABLE 1 ADAPTIVE RULES FOR CONTOURING ERROR CORRECTION

Deviation value of present position	Deviation value of next step position	Deviation condition of next step position	Required prediction rule
$D_i > 0$	$D_{i+1} < 0$	$\delta \leq \delta_{all}$	$\vec{P}_i = \vec{P}_{i+1}$
		$\delta > \delta_{all}$	$Increment\, by\, \Delta\theta_2$
	$D_{i+1} > 0$	$\delta \leq \delta_{all}$	$\vec{P}_i = \vec{P}_{i+1}$
		$\delta > \delta_{all}$	$Decrement\, by\, \Delta\theta_2$
	$D_{i+1} = 0$	$\delta = \delta_{all}$	$\vec{P}_i = \vec{P}_{i+1}$
$D_i = 0$	$D_{i+1} < 0$	$\delta \leq \delta_{all}$	$\vec{P}_i = \vec{P}_{i+1}$
		$\delta > \delta_{all}$	$Increment\, by\, \Delta\theta_2$
	$D_{i+1} > 0$	$\delta \leq \delta_{all}$	$\vec{P}_i = \vec{P}_{i+1}$
		$\delta > \delta_{all}$	$Decrement\, by\, \Delta\theta_2$
	$D_{i+1} = 0$	$\delta = \delta_{all}$	$\vec{P}_i = \vec{P}_{i+1}$
$D_i < 0$	$D_{i+1} < 0$	$\delta \leq \delta_{all}$	$\vec{P}_i = \vec{P}_{i+1}$
		$\delta > \delta_{all}$	$Increment\, by\, \Delta\theta_2$
	$D_{i+1} > 0$	$\delta \leq \delta_{all}$	$\vec{P}_i = \vec{P}_{i+1}$
		$\delta > \delta_{all}$	$Decrement\, by\, \Delta\theta_2$
	$D_{i+1} = 0$	$\delta = \delta_{all}$	$\vec{P}_i = \vec{P}_{i+1}$

B. New Step Position

Referring to Figure 2, if the step size of each cutter movement is allowed to revolve a full revolution about a specific point o, then it forms a space circle.

To obtain a Cartesian coordinates component mapping of the cutter location, they can be expressed as follows:

$$\Delta X_i = d \cos\theta_2 \sin\theta_1$$
$$\Delta Y_i = d \sin\theta_2 \quad\quad (3)$$
$$\Delta Z_i = d \cos\theta_2 \cos\theta_1$$

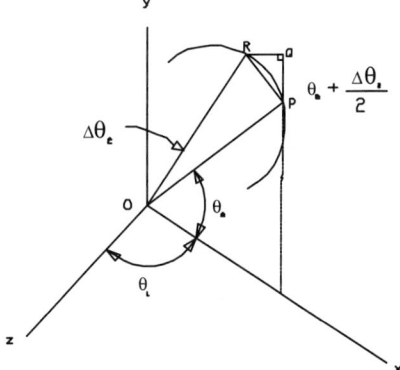

Fig. 2 Geometric relationship between current and next steps for cutter movement

Similarly, the components of the corrected position as a result of differential rotation of the step size are obtained as

$$X_{i+1}^c = d \cos(\theta_2 \pm \Delta\theta_2)\sin\theta_1 + X_i$$
$$Y_{i+1}^c = d \sin(\theta_2 \pm \Delta\theta_2) + Y_i \quad\quad (4)$$
$$Z_{i+1}^c = d \cos(\theta_2 \pm \Delta\theta_2)\cos\theta_1 + Z_i$$

where the superscript c indicates the coordinate of the cutter located at the corrected position.

It is our interest to find the parametric value u_{i+1}, that corresponds to the next step position $\vec{P}_{i+1}$ of the cutter in this subsection. Mathematically, it is known that the parametric curve on the surface can be calculated by a function of the arc length s along the underlying curve. By using Taylor's series the parametric value u_{i+1} can be approximated by the following equation as

$$u_{i+1} = u(s + s_0) = u(s_0) + s\frac{du}{ds}\bigg|_{s_0} + \frac{s^2}{2!}\frac{d^2u}{ds^2}\bigg|_{s_0} + O(s^3) +$$

where $u(s_0) = u_i$, and $O(s^4)$ stands for the fourth order (s^4) term. Therefore, if s is a very small value, then u_{i+1} can be approximated closely using the sum of the first four terms of the right-hand side of Eq. (5).

Since u on the parametric surface is a function of the arc length s, or $\vec{r}[u(s), c] = \vec{r}(s)$, the first derivative can be obtained by employing the vector differentiation to yield

$$\frac{du}{ds} = \frac{1}{\vec{r}_u \bullet \vec{T}} = \frac{1}{|\vec{r}_u|} \quad\quad (6)$$

where $\vec{r}_u$ is the first derivative of $\vec{r}$ with respect to the parametric variable u, $\vec{T}$ is the unit tangent of the underlying curve which is defined by

$$\vec{T} = \frac{\vec{r}_u}{|\vec{r}_u|} \qquad (7)$$

Similarly, the second and the third derivatives appeared in the third and fourth terms on the right-hand side of Equation (5) can be calculated respectively as follows:

$$\frac{d^2u}{ds^2} = \frac{d}{du}\left(\frac{du}{ds}\right)\frac{du}{ds} = -(\vec{r}_{uu} \cdot \vec{r}_u)(\vec{r}_u \cdot \vec{r}_u)^{-2} \qquad (8)$$

$$\frac{d^3u}{ds^3} = \frac{4(\vec{r}_{uu} \cdot \vec{r}_u) - |\vec{r}_u|^2 (\vec{r}_{uuu} \cdot \vec{r}_u + |\vec{r}_{uu}|^2)}{|\vec{r}_u|^7} \qquad (9)$$

where $\vec{r}_{uu}$ and $\vec{r}_{uuu}$ are the 2nd and 3rd derivatives of the cutter position vector with respect to the parametric variable u, respectively.

C. Computer Pseudo-Coding of the Algorithm

To implement the proposed automatic error checking and control mechanism for curved cutter path generation with computers, a pseudo-code is written based on the step-by-step cutting motion error checking and correction algorithm into the following eight steps.

1. Assign initial and current positions of the cutter and assume they are coincident. Also the cutter path is along the u-direction of the surface curve with $v = 0$.
2. Evaluate the next step position of the cutter using Equation (1).
3. Calculate the parametric value u_{i+1} for the next step position of the cutter using Equation (5).
4. Check the cutter error tolerance requirements using Equation (2) and determine the deviations between the actual position and the desired one of the cutter.
5. Examine whether the deviation obtained in Step (4) is less or equal to the allowed tolerance requirement or not. If yes, then go to the Step (7). If not, then refer to Table 1 derived previously with pertinent adaptive rules to determine the direction of cutter rotation.
6. Compute the minimum amount of differential rotation required to compensate based on the following equation.

$$\Delta\theta_2 \approx \frac{\delta - \delta_1}{s \bullet \cos\theta_2} \geq \frac{\delta - \delta_1}{s} \geq \frac{\delta - \delta_{all}}{s} \qquad (10)$$

Then calculate the compensated position using the following equation.

$$X_{i+1}^c = \Delta X_i \cos\Delta\theta_2 \pm \frac{\Delta Y_i \Delta X_i}{\sqrt{\Delta X_i^2 + \Delta Z_i^2}} \sin\Delta\theta_2 + X_i$$

$$Y_{i+1}^c = \Delta Y_i \cos\Delta\theta_2 \pm \sqrt{\Delta X_i^2 + \Delta Z_i^2} \sin\Delta\theta_2 + Y_i \qquad (11)$$

$$Z_{i+1}^c = \Delta Z_i \cos\Delta\theta_2 \pm \frac{\Delta Z_i \Delta Y_i}{\sqrt{\Delta X_i^2 + \Delta Z_i^2}} \sin\Delta\theta_2 + Z_i$$

where the superscript c represents corrected coordinates. Based on the compensated position, go back to Step (4) to continue the iteration process.

7. Check the parametric value u_{i+1}. If it is negative, then go to Step (2). Otherwise, continue until the value reaches 1.
8. Check the parametric value v. If it is less or equal to 1, then update v and go back to Step (1) for a new process cycle. Stop the process when v reaches 1.

IV. CNC Milling Implementation of the Proposed Algorithm

Having developed the algorithmic pseudo-code, the next step is to test the algorithm on a CNC machine. In the experimental phase, a metal cutting operation has to be selected. Since the algorithm developed is regarding precision surface machining, a CNC milling operation has been chosen to realize the metal cutting process with the proposed compensated algorithm.

A. The Experimental CNC Machine Specifications

The machine used for this experiment is a proLIGHT milling center. It is a three-axis milling machine controlled directly by a Pentium II-based computer. The proLIGHT control program, loaded on the computer can accept standard EIA RS-274D G&M codes that CNC machine tools recognize.

The sample workpieces were made of machinable solid wax material. The tools are various ball end milling cutters. In the experiment, we have to consider the difference between closed- and open-loop CNC controllers. Open-loop CNC controller is able to control the tool path during actual machining in real time. However, in the case of a closed-loop CNC architecture, the controller can only take in tool positions by reading the cutter location (CL) file from the controlled computer. To take advantage of the open architecture controller of the CNC, a program was written in *MATLAB*® to generate the CL files using the developed algorithm. The output of the file was then converted into a CNC format by the controller and used to run the experiment.

B. Cubic Bezier Curves

Bezier curves were used to represent the scenario of free-form contour surfaces cutter locations generation. The control points of the curves are the keys to generating the CL files. These curves and surfaces were adopted because Bezier curve representation is one that is utilized most frequently in CAD community. The curves are defined parametrically, which means that all the parameters have their geometric meanings. By varying the parameters we can construct specific points in three-dimensional space to represent free-form surfaces.

Given four control points, P_0, P_1, P_2, P_3, one can generate a curve $P(t)$, by letting

$$P_1^{(1)}(t) = tP_1 + (1-t)P_0$$
$$P_2^{(1)}(t) = tP_2 + (1-t)P_1$$
$$P_3^{(1)}(t) = tP_3 + (1-t)P_2 \quad (12)$$
$$P_2^{(2)}(t) = tP_2^{(1)}(t) + (1-t)P_1^{(1)}(t)$$
$$P_3^{(2)}(t) = tP_3^{(1)}(t) + (1-t)P_2^{(1)}(t)$$
$$P_3^{(3)}(t) = tP_3^{(2)}(t) + (1-t)P_2^{(2)}(t)$$

in which $P_3^{(3)}(t)$ is defined to be equal to $P(t)$.

This free-form shape construction is illustrated in Figure 3.

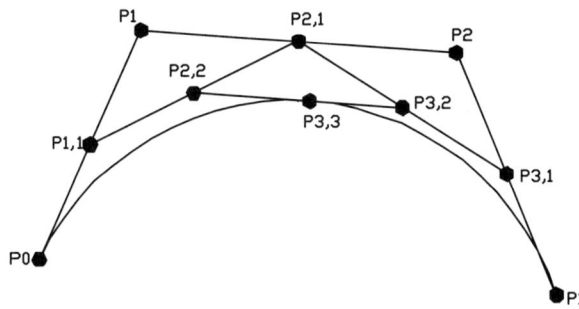

Fig. 3 Free-form shape construction example

Simplifying the above free-form curves construction expressions, we can rewrite them into the following form.

$$P(t) = P_3^{(3)}(t)$$
$$= t[tP_3^{(1)}(t) + (1-t)P_2^{(1)}(t)] + (1-t)[tP_2^{(1)}(t) + (1-t)P_1^{(1)}(t)]$$
$$= t^3 P_3 + 3t^2(1-t)P_2 + 3t(1-t)^2 P_1 + (1-t)^3 P_0 \quad (13)$$

The above expression is the analytical form of the curves simulated in the experiment.

To realize the Bezier curves geometrically, the following procedure is executed.

Given the control points, P_0, P_1, P_2, P_3, and a parametric value $t \in [0,1]$, the points on the Bézier curve can be generated by using

$$P(t) = P_3^{(3)}(t) \quad (14)$$

As for the analytical method, the control points, P_0, P_1, P_2, P_3 must also be given, then the Bézier curve can be defined to be

$$P(t) = \sum_{i=0}^{3} P_i B_{i,3}(t) \quad (15)$$

where

$$B_{0,3}(t) = (1-t)^3$$
$$B_{1,3}(t) = 3t(1-t)^2 \quad (16)$$
$$B_{1,3}(t) = 3t^2(1-t)$$
$$B_{3,3}(t) = t^3$$

are the so-called Bernstein polynomials of third degree.

C. Contour Machining Errors

The contour machining errors are the deviations of the actual tool path from the desired tool path. Since in a typical machining operation, the actual tool path does not coincide with the desired path, it consequently causes a contour error. Usually, the contour error is used to indicate the contour machining accuracy and therefore, is our main concern in cutting. The measurement of contour error is taken as the shortest distance between the desired and actual contours of cutter paths. Contouring accuracy of the cutter is often used as the major performance evaluation index of CNC machine tools and the algorithms that control them.

The factors (cutting conditions) that affect the surface finish produced in a milling process generally include cutting speed, feed, depth of cut, etc. For simplicity in this experimental work, the cutting speed is kept constant through the cutting cycle. However, the feed rates are varied to demonstrate the effectiveness of the proposed algorithm. The depth of cut is not considered as a parametric cutting condition in this work either.

V. Experimental Results and Performance Evaluation

Figure 4 shows the schematics of a desired general free-form surface to be produced. Figure 5 displays experimentally machined parts samples. The surface is used in the performance evaluation of the proposed algorithm. As mentioned previously, the surfaces are modeled as Bezier surfaces.

In the experiments, the cutter is to follow constant v parameter and varying u parameter. The surface curves are evaluated by using two methods. The first is the proposed algorithm that has the capability of adjusting the deviated offsets of the generated cutter contour in such a way that the measured offsets always fall within the pre-specified tolerances. The second method directly applies Equation (1) that does not have the adaptive correction capability. The containment for the algorithm is to generate a surface curve that not only has the contour of the desired curve, but also makes cutting offsets constrained within an allowable tolerance which can be assigned according to the precision requirements of the outcome.

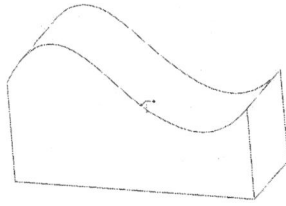

Fig. 4 Schematics of sample parts to be machined

Fig. 5 Machined parts samples produced by the proposed method

In the experiments, the cutter's feed rates were varied from *5mm/second* to *30 mm/second* in the performance evaluation of the algorithm. In all the cases, the sampling size was *0.005 seconds* and the allowable tolerance was preset to be *0.01mm*. The outcomes show that the adaptive capability of the algorithm is able to constrain the generated curve to move within the allowable tolerance so that we can cut as close to the desired curve as it can be according to the preset tolerances. Some samples of cut parts are shown in Figure 5.

Moreover, the intensity of adjustment of the cutting path is also measured for various feed rate in this experimental investigation, as listed in Table 1. These results demonstrate an interesting relationship between the intensity of adjustment and the feed rate of the cutter.

VI. Conclusion

In this paper a robust cutter path generation method was proposed for implementing freeform parts milling. The method is adaptive and capable of approximating the contour of the anticipated parts' surface by a number of steps. An adaptive algorithm was developed to generate the next cutter location continuously for a given feed rate. This generates the Cutter Location (CL) files constrained by preset tolerances that drive the CNC machine. Assisted by the tolerance checking mechanism of the algorithm, if the deviation between the next step cutter position and the desired location is greater than that is allowed, the tolerance checking mechanism adjusts the cutter location until it satisfies the desired tolerance.

To test the performance of the algorithm, a Bezier curve was formulated to model the surface patch representing that of freeform parts. A 3-axis CNC milling center with open-control architecture was used to implement the proposed cutting method. The algorithm was evaluated at feed rates ranging from *5 mm/second* to *30 mm/second* in a step increment of *5 mm/second*. The experimental results proved that the proposed algorithm was robust. The method was able to handle a cutting path tolerance up to *0.001mm* by measuring the intensity of adjustment for each different feed rate. Since the proposed algorithm can generate high precision and accurate freeform tool paths, it can be used as a guideline for developing an add-on adaptive mechanism to CNC tool path generators for wider applications.

Acknowledgement

This work was partially supported by the OBR Research Challenge Grant #R5629. Also Dr. T. Lee's technical consultation was gratefully acknowledged.

References

[1] Papaioannou, S.G., "A non-orthogonal interpolation algorithm for NC machine tools", *Computers in Industry*, **6**, 103-108, 1985.

[2] Qin, K., Bin, H., "Three-point recursion interpolation theory and algorithm of space circular arcs in CNC systems", *Computers in Industry*, **15**, 355-362, 1990.

[3] Shpitalni, M., Koren, Y., Lo, C.C., "Real-time curve interpolators", *Computer-Aided Design*, **26** (11), 832-838, November 1994.

[4] Yang, D.C.H., Kong, T., "Parametric interpolator versus linear interpolator for Precision CNC machining", *Computer-Aided Design*, **26** (3), 225-233, 1994.

[5] Loney, G.C., Ozsoy, T.M., "NC machining of free form surfaces", *Computer-Aided Design*, **19** (2), 85-90, 1987.

[6] Kim, K., Biegel, J.E., "A path generation method for sculptured surface manufacture", *Computers and Industry Engineering*, 14 (2), 95-101, 1988.

[7] Bedi, S., Quan, N., "Spline interpolation technique for NC machines", *Computer in Industry*, **18**, 307-313, 1992.

[8] Yeung, M.K., Walton, D.J., "Curve fitting with arc splines for NC tool path generation", *Computer-Aided Design*, **26** (11), 845-849, 1994.

[9] Qiu, H, Cheng, K., Li, Y., "Optimal circular arc interpolation for NC tool path generation in curve contour manufacturing", *Computer-Aided Design*, **29** (11), 751-760, 1997.

[10] Lo, C.C., "A new approach to CNC tool path generation", *Computer-Aided Design*, **30** (8), 649-655, 1998.

[11] Yeh, S.S., Hsu, P.L., "The speed-controlled interpolator for machining parametric curves", *Computer-Aided Design*, **31** (5), 349-357, 1999.

[12] Hwang, J.S., Chang, T.C., "Three-axis machining of Compound surfaces using flat and filleted end mills", *Computer-Aided Design*, **30** (8), 641-647, 1998.

[13] Lin, Y.J., Lee, T.S., "An adaptive tool path generation algorithm for precision surface machining", *Computer-Aided Design*, **31** (4), 237-247, 1999.

[14] Lee, T.S., "A unified approach for computer-aided precision machining of parts having sculptured surfaces", Ph.D. Dissertation, University of Akron, Akron, OH, 1998.

Robotic Metal Spinning
– Shear Spinning Using Force Feedback Control –

Hirohiko Arai

Intelligent Systems Institute
National Institute of Advanced Industrial Science and Technology
1-2 Namiki, Tsukuba, Ibaraki 305-8564, Japan   Email: h.arai@aist.go.jp

Abstract

Metal spinning is a plasticity forming process that forms a metal sheet or tube by forcing the metal onto a rotating mandrel using a roller tool. This is a study on metal spinning applying robot control techniques such as force feedback control with the aim to develop flexible and intelligent forming processes, and to expand a new application area for robot control. An experimental setup was developed for gathering basic data on the forming process. Some results of preliminary experiments are presented. The influence of the clearance between the roller and mandrel is also discussed. The author proposes applying hybrid position/force control for shear spinning, which is free from fine adjustment of the clearance. The effectiveness of the proposed method was experimentally verified.

1 Introduction

This study seeks to exploit robot control techniques such as force feedback control for *metal spinning*. We aim to develop flexible and intelligent forming processes, and to expand a new application area for robot control.

Metal spinning is a plasticity forming process that forms a metal sheet or tube by forcing the metal onto a rotating mandrel using a roller or a paddle tool (**Fig. 1**). It is widely used for producing round hollow metal parts and products, e.g. tableware, kitchenware, ornaments, lighting fixtures, parabola antennas, boilers, tanks, gas canisters, nozzles, engine parts, and tire wheels. This forming process is also known as a highly-skilled manufacturing craft by artisans that requires decades of experience. Even nose cones for H2 space rockets launched by NASDA in Japan are produced by such manual metal spinning.

Metal spinning has several merits over other metal forming processes as follows.
- It can create more complicated shapes than sheet metal stamping or deep drawing.
- As it needs only one mandrel, it is easier to set up for forming.
- Forming force is rather small and the forming apparatus can be compact.
- Material can be saved in comparison with cutting processes.
- It can provide precision products and good surface finishing.

These merits agree with recent trends in production technologies such as rapid prototyping and net shape manufacturing. Metal spinning can also be regarded as one of the incremental forming processes, which have been recently receiving attention in plasticity forming technologies.

However, progress in automated metal spinning is rather slower than in other areas of plasticity forming. Although numerically controlled spinning machines can achieve mass production of simple-shaped products, the programming of the machines depends greatly on skilled operators. Scientific research on metal spinning, mainly in production and material engineering, also does not seem very active. The mechanism of deformation is three-dimensional and too complicated for computer simulations. There are many control parameters and it requires the development of experimental setups. Consequently, the mechanics of the forming process have not been sufficiently clarified. In particular, forming procedures for *conventional spinning* have not been theoretically established. In fact, practical production depends on the experience of skilled workers.

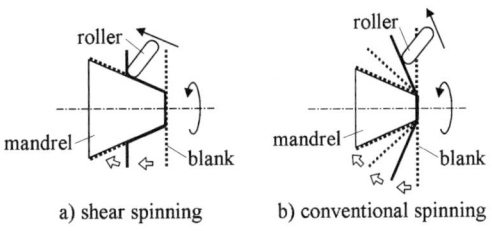

Fig. 1: Metal spinning

a) shear spinning   b) conventional spinning

Here we briefly review several previous studies on metal spinning. Hayama, who pioneered the earliest studies in this field, theoretically and experimentally investigated forming conditions of *shear spinning* [1, 2] and also studied the forming procedures of conventional spinning [3]. Kawai proposed building a database of forming examples and using it to select forming parameters [4]. Shima proposed flexible spinning, in which two rollers pinched the material and no mandrel was necessary [5]. Katupitiya proposed a metal spinning system that integrated human skills into an automated forming process [6].

In recent years, studies on industrial robots for manufacturing applications tend to be less and less active while most academic researchers are inclined towards non-manufacturing applications. In consequence, applications for industrial robots have not varied much from the conventional handling, assembly, welding and painting. Some researchers even consider industrial robots as mature or old-fashioned technology since they only take notice of such applications.

Current industrial robots are generally used for simple repetitive tasks of low added value, which substitute for unskilled factory workers. Such robots have value only if they are less expensive and used in mass production to achieve higher speed and yield. However, mass production is not the only form of operation in manufacturing industries. There are various types of manufacturing crafts that only experienced artisans can perform. Such crafts are usually of small quantity but can create high value added products. A new market for robot technologies might develop if robotics researchers were attracted to such areas, utilizing the potentials of accumulated techniques, e.g. sensory feedback control, to achieve valuable application tasks for which even expensive intelligent robots can be worthwhile.

Force feedback control of manipulators has been studied in robotics for many years. We now have a variety of theoretical and experimental knowledge on hybrid position/force control, impedance control, etc. However, these techniques have not been widely applied in practice except for only a few kind of tasks such as assembly and grinding. We must still seek effective applications for these control methods.

Metal spinning appears to be a suitable task for an industrial robot for several reasons. In manual metal spinning, the various senses of the worker, particularly force feeling via the tool, play an important role. Metal spinning needs much smaller forming force than other plasticity forming techniques, on the order of kilograms instead of tons, because it is based on local deformation. It involves many control parameters and needs dexterous motion with multiple degrees of freedom. It is suitable for limited production of a wide variety and is a process of high added value, which we can see from the fact that even manual production can be viable as a manufacturing business. Thus it is expected that the profitability of a force controlled industrial robot can be high.

In this research, we aim to make metal spinning more flexible and intelligent, by introducing robot control technologies, such as force control, into the forming process. The forming conditions are modified in real time based on feedback of the forming status to avoid forming defects and to obtain high-quality products.

Conventional robot tasks are mainly composed of *moving* an object. In contrast, this research encounters the novel aspects of *transforming* an object. We expect that challenging research subjects may develop from this research while utilizing the potentials of robot technologies developed so far.

The remainder of this paper is organized as follows. In Section 2, we describe an experimental setup for gathering basic forming data. The preliminary experimental results are shown in Section 3. We discuss the problem of clearance setting between the mandrel and roller in Section 4. In Section 5, we propose applying hybrid position/force control to metal spinning, and experimentally verify the effectiveness of our proposed control method.

2 Experimental setup

First, we developed an experimental setup for conducting basic experiments on metal spinning. Forming of different materials under various parameters can be done to gather forming data, e.g. success or failure of forming, precision of product, wall thickness, surface roughness, and forming force. In particular, generation of defects (wrinkles or breaks) can be analyzed by monitoring the conditions around their occurrence.

Figure 2 illustrates our experimental setup. The linear motion of x and y axes is driven by the ball-screws (2 mm/rev) and DC servo motors (60W). The

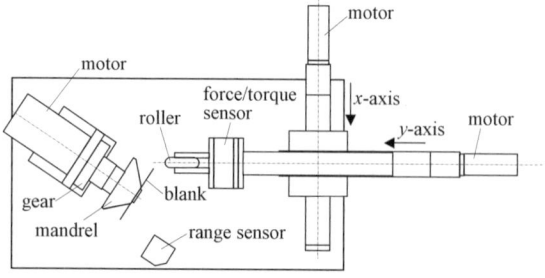

Fig. 2: Experimental setup

Table 1: Specifications of each axis

	x, y axis	θ axis
Rated Force/Torque	580 N	3.9 Nm
Rated Speed	100 mm/s	250 rpm
Resolution	0.5 μm/pulse	0.009 deg/pulse

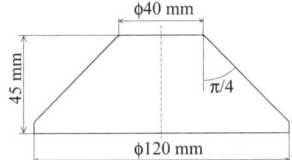

Fig. 3: Mandrel

Fig. 4: Forming experiment

Fig. 5: Blank and product

mandrel (θ axis) is rotated by a DC servo motor with a planetary gear (reduction rate: 1/10). The θ axis is slanted relative to the x axis by $\pi/3$ rad. Each motor has an incremental encoder (4000 p/r) for detecting the rotation angle. The specifications of each axis are listed in **Table 1**.

The diameter of the forming roller is 70 mm. The roundness of the edge is a 9.5 mm radius. The roller is made from alloy tool steel (AISI D2, quenched). A 6-axis force/torque sensor is equipped between the roller and y axis. The shape of the mandrel is illustrated in **Fig. 3**. The material of the mandrel is stainless steel (AISI 304). A laser range sensor measures wrinkles in the flange. The sensing resolution is 0.18 mm within a range of 40 to 120 mm.

A personal computer (Pentium, 233 MHz) receives the signals from the encoders, the force sensor and the range sensor via interface boards, and sends torque commands to the motor drivers via a D/A board. The sampling interval for the control is 1 ms.

3 Preliminary experiments

We conducted preliminary experiments to gather basic forming data using the setup described above. We first tried shear spinning, in which the roller was moved along the surface of the mandrel and the material was squeezed onto the mandrel. The blank was a round plate of pure aluminum (1100A-O, annealed) with a 120 mm diameter and 0.78 mm thickness.

In shear spinning, it is known that the wall thickness t of the product is represented as,

$$t = t_0 \sin \alpha \quad (1)$$

where t_0 is the thickness of the blank, and α is the angle between the side surface and the rotating axis of the mandrel. In our experiment, $t = 0.55$ mm since $t_0 = 0.78$ mm and $\alpha = \pi/4$ rad. The clearance between the mandrel and roller was set to 0.55 mm. The surface of the roller was lubricated with liquid lubricant (CRC5-56).

Each axis was controlled using simple high-gain PD feedback control. The x and y axes were controlled so that the roller moved along a straight line parallel to the surface of the mandrel at a constant velocity. The mandrel (θ axis) was also rotated at a constant angular velocity. The position and motor current of each axis, force at the roller, and range sensor signal were recorded during forming.

Figure 4 shows the forming experiment. **Figure 5** shows an example of the blank and the finished product. The product was formed to the shape of the mandrel.

Next, the forming force applied to the material by the roller was measured. The force component in line with the movement of the roller is defined as F_X, the normal force component against the surface of the mandrel is defined as F_Y, and the tangential component to the mandrel rotation is defined as F_Z (**Fig. 6**). F_X, F_Y and F_Z are plotted for the displacement X of the roller along the mandrel (**Fig. 7**). The velocity of the roller in the X direction was 0.1 mm/s, and the angular velocity of the θ axis was 120rpm (4π rad/s). The movement of the roller for one turn of the mandrel was 0.05 mm/rev.

F_Y, which forces the material onto the mandrel,

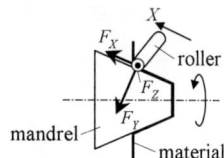

Fig. 6: Force components at roller

Fig. 8: Wrinkled flange

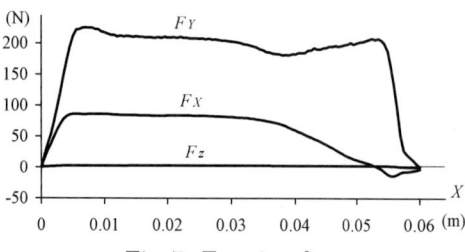

Fig. 7: Forming force

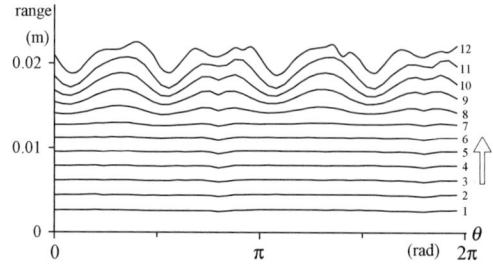

Fig. 9: Growth of wrinkles

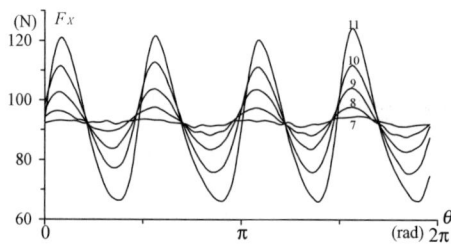

Fig. 10: Fluctuation of feeding force

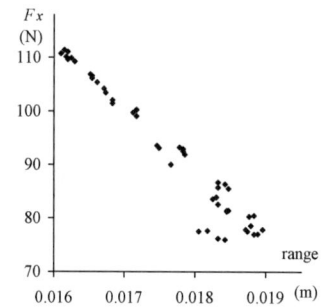

Fig. 11: Wrinkles and feeding force

was about 200N constantly. The feeding force of the roller, F_X, was about 80N at first, and later decreased gradually as the forming proceeded. As the width of the flange, i.e. the remaining unformed blank, became narrower, the stiffness of the flange reduced and the resistant force became weaker. The tangential force F_Z was about 3N maximum, and was much smaller than F_X and F_Y. We can see the forming force is so small that it can be applied by an industrial robot. It was also confirmed that the motor current was much smaller than the rated value and the torque of each motor had an adequate margin.

When the velocity of the forming roller is too large, wrinkles are generated at the flange due to buckling. The wrinkles disturb the uniform forming and degrade the quality of the product. Moreover, deep wrinkles might block the movement of the roller and make the forming impossible. **Figure 8** shows an example of the forming result when wrinkles occurred.

Growth of the wrinkles was experimentally investigated using a laser range sensor. **Figure 9** shows height of the flange at 5 mm from the outer edge. The velocity of the roller in X direction was 0.1mm/s and the angular velocity of the θ axis was 60rpm (2π rad/s). The roller moved 0.1 mm for one turn of the mandrel. The height of the flange from the attachment plane was measured during one turn for every 25 turns of the mandrel.

In the early phase of the forming (No. 1 to 6), the flange remained almost flat. As the forming progresses (No. 7 to 12), the shallow wrinkles slowly grew deeper. The stiffness of the narrow flange was insufficient to keep it flat.

The forming force was measured while the wrinkles were generated. **Figure 10** shows the feeding force F_X for the data of No. 7 to 11. As the wrinkles became deeper, the fluctuation of the feeding force increased. In **Figure 11**, F_X is plotted for the height of the flange during one turn (No. 10), where the phase of the θ axis was mutually shifted by π rad as the range sensor was located at the opposite side of the roller. The unevenness of the flange and the fluctuation of the feeding force are related almost linearly. Therefore, the growth of the wrinkles can be detected from the feeding force.

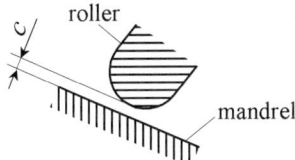

Fig. 12: Clearance between mandrel and roller

4 Clearance between mandrel and roller

Among the various forming parameters in metal spinning, the clearance between the mandrel and roller (**Fig. 12**) can be considered the most difficult to set, in view of controlling the forming machinery. In shear spinning, the wall thickness after the process is represented as Eq. (1), and the clearance should be exactly controlled equal to the thickness. When the clearance is too large, the precision is degraded because the material does not fully contact the mandrel. This upsets stable forming and wrinkles are likely to occur. Conversely, too small a clearance makes the forming force very large, and the flow of the material sometimes improperly deforms the product.

To obtain an appropriate clearance, the position of the roller relative to the mandrel must be strictly calibrated. The profile of the mandrel should also be exactly known. In addition, the roller should track the desired trajectory precisely under the forming force. When the stiffness of the machinery is insufficient, as is often the case with industrial robots, the set clearance cannot be maintained due to elastic distortion resulting from the forming force.

As the mandrel used in this study is a simple conical shape, the wall thickness of the product can be easily predicted from Eq. (1). However, it is difficult to know the exact distribution of the thickness if the shape of the mandrel is more complicated, e.g. when the inclination of the surface changes irregularly. In multi-pass conventional spinning, in which the material is formed in steps to fit the mandrel, the final thickness distribution is more difficult to estimate. The setting of the clearance fairly depends on the experience of the operators, and it should be adjusted after some forming trials.

We experimented on shear spinning under different clearance settings and compared the results. The theoretical wall thickness was $0.78 \sin \pi/4 = 0.55$[mm] from Eq. (1). The feeding velocity $\dot{X}$ of the roller was 0.1 mm/s and the angular velocity $\dot{\theta}$ of the mandrel was 60rpm, as in the example of the previous section. We tried four values for the clearance, $c = 0.25$ mm, 0.40 mm, 0.55 mm and 0.70 mm.

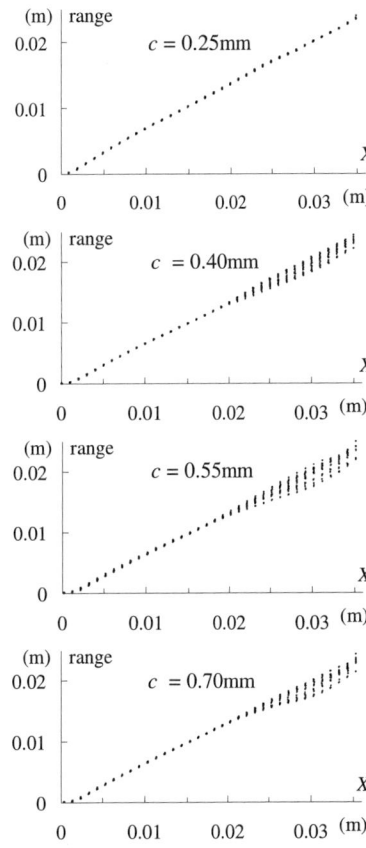

Fig. 13: Height of flange

Figure 13 shows the height of the flange measured by the laser range sensor. The data during one turn were plotted for every 10 turns of the mandrel. The forming was completed without wrinkles when $c = 0.25$ mm. At other clearance settings, the amplitude of the unevenness started to increase (the plot scattered) and wrinkles were generated.

The pushing force F_Y is compared in **Fig. 14**. The smaller the clearance was set, the larger F_Y became. The interior surface of the product was pressed onto the mandrel and appeared glossy when $c = 0.25$ mm. On the other hand, the surface finish of the blank sheet (fine scratches) remained when $c \geq 0.4$ mm. As the material separated from the mandrel, the precision of the product deteriorated.

Figure 15 shows the feeding force F_X of the roller. F_X oscillates when $c \geq 0.4$ mm and it also indicates the generation of wrinkles. When $c = 0.25$ mm, the feeding force is rather smaller than in other cases.

The clearance of 0.25 mm led to the best result, when the clearance is assumed to be too small. We also measured the actual wall thickness of the products using a micrometer. When $c = 0.25$ mm, the thickness was 0.55mm to 0.56 mm and nearly equaled to the

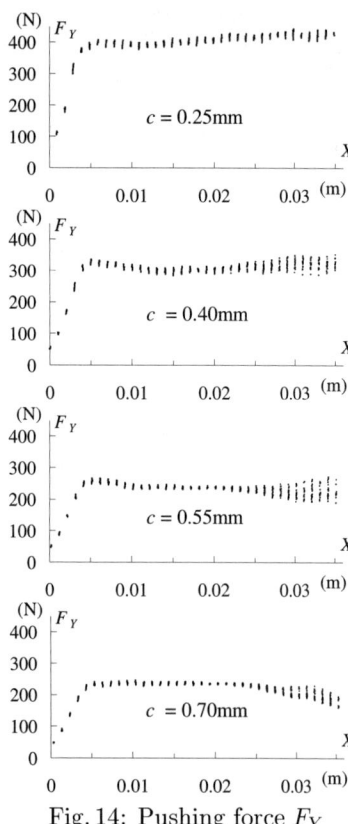

Fig. 14: Pushing force F_Y

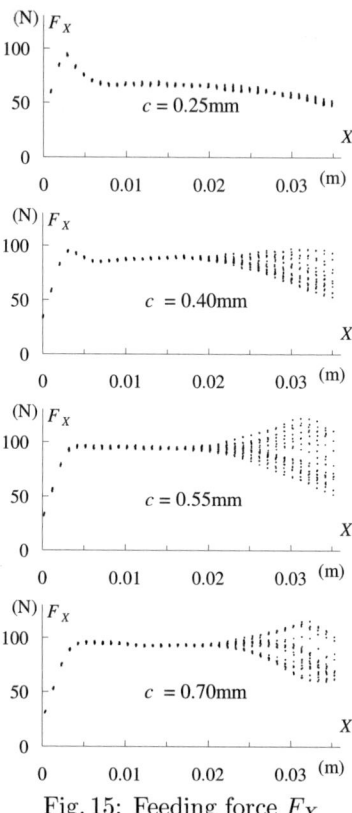

Fig. 15: Feeding force F_X

value from Eq. (1). It is supposed that the clearance of 0.25 mm was not actually achieved and the clearance enlarged nearly to the theoretical wall thickness because of the deformation arising from the forming force. As the deviation of the roller pass caused by servo error was 0.04mm maximum, the mechanical deformation was dominant. For $c = 0.55$ mm, which is nearest to the theoretical wall thickness, the deformation made the real clearance larger than the desired value and caused the wrinkles. The actual wall thickness was 0.59 mm to 0.62 mm near the flange and thicker than the theoretical value.

These results demonstrate that the clearance setting significantly affects the quality of the forming process. It might be preferable to set a smaller clearance than the theoretical value of Eq. (1) when the machinery does not have high stiffness. However, as the proper setting is dependent on the machinery and the formed object, some forming trials will be necessary. In contrast, when the machinery is very stiff, the clearance setting will be more critical and the roller should be controlled very precisely relative to the mandrel. In particular, when a thin sheet is spun, the admissible clearance will be severely restricted.

5 Force feedback control

The problems with the clearance setting discussed in the previous section come about because of the position-based control of the forming roller. The desired shape of the product can be obtained by fitting the material tightly to the mandrel. This can be accomplished by pressing the material onto the mandrel with an appropriate force, instead of leaving the clearance between the mandrel and the roller equal to the wall thickness. Hence we investigated the feasibility of force feedback control for metal spinning.

Hybrid position/force control [7] was applied in this study. As for the feeding direction X parallel to the mandrel surface, the forming roller was position-controlled in a constant velocity $\dot{X}_d$. The pushing force F_Y of the roller normal to the mandrel was controlled to be a constant value F_{Yd}. PD feedback and PI feedback were adopted for the position control and the force control, respectively. The control law is represented as,

$$\boldsymbol{f} = \begin{bmatrix} f_x \\ f_y \end{bmatrix} = \boldsymbol{f}_P + \boldsymbol{f}_F \qquad (2)$$

$$\boldsymbol{f}_P = \boldsymbol{M}\boldsymbol{J}^{-1} \begin{bmatrix} k_{vX}(\dot{X}_d - \dot{X}) + k_{pX}(X_d - X) \\ 0 \end{bmatrix}$$

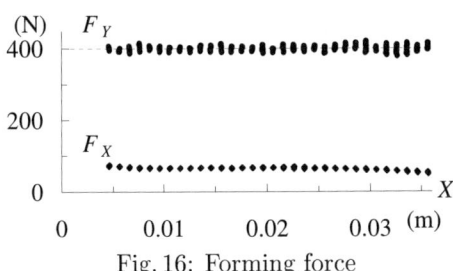

Fig. 16: Forming force

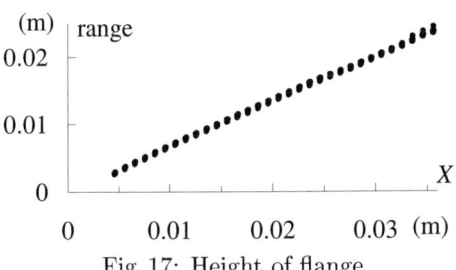

Fig. 17: Height of flange

$$f_F = J^T \begin{bmatrix} 0 \\ F_{Yd} + k_{pF}(F_{Yd} - F_Y) + k_{iF} \int (F_{Yd} - F_Y) \end{bmatrix}$$

where f_x, f_y are the force of the x and y axes, M is an inertia matrix, J is a Jacobian matrix of XY frame with regard to xy frame, and k_{vX}, k_{pX}, k_{pF} and k_{iF} are feedback gains.

This control law (2) was employed for the shear spinning experiment. The feeding velocity $\dot{X}$ was 0.1 mm/s and the mandrel rotation θ was 60 rpm, as in the previous section. The pushing force F_Y of the roller was controlled to 400N, based on the pushing force with 0.25 mm clearance, which led to a good result in the experiments of the previous section. At first, xy-position control was applied until the roller contacted the blank and the material was dented following the edge of the roller. Then the control was switched to the hybrid position/force control and the forming conducted.

Figure 16 shows the forming force F_X and F_Y. The pushing force F_Y of the roller was maintained about the desired value by the feedback control. **Figure 17** shows the height of the flange. The forming was achieved without generation of wrinkles. **Figure 18** shows an example of the finished product using this control method. The wall thickness was 0.56 mm to 0.57 mm, which almost agreed with the theoretical value of Eq. (1)

This method frees metal spinning from the setting of the clearance between the mandrel and roller. Fine positioning of the roller relative to the mandrel is unnecessary. The same value of the pushing force can be used regardless of the stiffness of the machinery. In addition, this method is effective not only for shear spinning but also for final phase of multi-pass conventional spinning when the material is forced onto the mandrel.

Fig. 18: Product by force control

6 Conclusions

We presented our study on metal spinning using robot control technology. First we discussed the background and purpose of the study. The experimental setup and preliminary experiments were described. In addition, we considered the effect of the clearance between the mandrel and the roller. We proposed metal spinning using hybrid position/force control, and verified its effectiveness experimentally.

References

[1] M. Hayama et al., "Experimental Study of Shear Spinning," Bulletin of Japan Society of Mechanical Engineers, Vol. 8, No. 31, pp. 453–460, 1965.

[2] M. Hayama et al., "Theoretical Study of Shear Spinning," Bulletin of Japan Society of Mechanical Engineers, Vol. 8, No. 31, pp. 460–467, 1965.

[3] M. Hayama, "Study of Pass Schedule in Conventional Simple Spinning," Bulletin of Japan Society of Mechanical Engineers, Vol. 13, No. 65, pp. 1358–1365, 1970.

[4] K. Kawai, S. Sawano and H. Ito, "A Trial Approach to Spinning Data Base," J. Japan Society for Technology of Plasticity, Vol. 30, No. 345, pp.1411–1415, 1989. (In Japanese)

[5] S. Shima, H. Kotera, H. Murakami and N. Nakamura, "Development of Flexible Spinning – A Fundamental Study," Advanced Technology of Plasticity, pp.557–560, 1996.

[6] J.Katupitiya, M.W.Yiu and D.Springford, "Automation of Metal Spinning Machines Using CNC Controllers and Human Skill Integration," Proc. The 4th Int. Conf. on Motion and Vibration Control (MOVIC'98), Zurich, Switzerland, 1998.

[7] T. Yoshikawa, "Dynamic hybrid position/force control of robot manipulators–description of hand constraints and calculation of joint driving force," IEEE J. Robitics and Automation, Vol. 3, No. 5, pp.386-392, 1987.

Robot Trajectory Integration for Painting Automotive Parts with Multiple Patches

Heping Chen, Ning Xi, Zhouhua Wei
Electrical and Computer Engineering Dept.
Michigan State University
East Lansing, MI

Yifan Chen, Jeffrey Dahl
Scientific Research Lab
Ford Motor Company
Dearborn, MI

Abstract— Automatic trajectory generation for spray painting is highly desirable for today's automotive manufacturing. Generating a trajectory for a surface with only one patch is widely studied to satisfy the paint thickness constraints. However, a complex surface has to be divided into several patches to satisfy the paint thickness and paint gun orientation constraints. Trajectory generation for a surface with multiple patches has not been addressed yet. In this paper, optimization processes are developed to optimize the paint thickness on a surface which consists of several patches. Optimization results are presented. Simulations are performed to verify the optimized parameters. Verification results is consistent with the optimization results.

I. INTRODUCTION

Spray painting is an important process in manufacturing many durable products, such as automobiles, furniture and appliances. The uniformity of paint thickness on a product can strongly influence the quality of the product. Paint gun trajectory planning is crucial to achieve the uniformity of paint thickness and it has been an active research area for many years. Currently, there are two trajectory generation methods: typical teaching method and automatic trajectory generation method.

Typical teaching method is complex, time-consuming, and the paint thickness is dependent on the operator's skill. Alternatively, some commercial software, such as *ROBCAD*^{TM}/Paint [1], can generate paint gun trajectories and simulate the painting process. However, the gun trajectories are obtained in an interactive way between the user and *ROBCAD*^{TM}/Paint, which is inefficient and error-prone. Automated generation of paint gun trajectories can be time-efficient and minimize paint waste and process time, but can also achieve optimal paint thickness.

Automated trajectory generation for a surface with a single patch has been widely studied. Suk *et al.* [2] presented an Automatic Trajectory Planning System (ATPS) for spray painting robots. But the paint gun model is quite simple. Antonio *et al.* [3] developed a framework for optimal trajectory planning to deal with the optimal paint thickness problem. The paint thickness is optimized. However, the paint gun path and the paint deposition rate must be specified. In practice, it is very difficult to get the paint deposition rate for a free-form surface. Chen *et al.* [4] presented an algorithm to generated a trajectory for a surface with one patch. The overlap distance and paint gun velocity are optimized. The paint thickness can satisfy the thickness constraints for a surface with only one patch. Although Choset [5] presented a method to generate paths for a surface with multiple patches, the method is to solve coverage problem.

To satisfy some constraints, such as thickness and gun orientation, a complex surface has to be divided into several patches [4]. Trajectory generation for a surface with multiple patches has not been studied yet due to the complexity of the intersection parts of multiple neighbor patches. Typically, the paint gun velocity is kept constant to optimize the paint thickness. However, to optimize the paint thickness for a surface with multiple patches, the paint gun velocity cannot be kept constant. In this paper, optimization processes are developed to optimize the paint thickness for a surface with multiple patches. The optimization results are presented. Simulations are performed for a surface with two flat patches to verify the optimized parameters. Simulation results is consistent with the optimized results

II. TOOL PROFILE AND THICKNESS REQUIREMENT

Different tool models have been used [2], [3], [6], [7], [8] in spray painting. Some models are quite simple [2] and some are quite complex [8]. Here a typical gun model [3], [6], [7] are adopted and shown in Figure 1(a). To generate a paint gun trajectory requires the knowledge of paint deposition rate. The paint deposition rate depends on many parameters, such as the tool standoff, the flow rate of material, the atomizing pressure and solvent concentration. Here we assume these parameters are fixed [9]. The typical profile of the paint deposition rate can be roughly approximated by parabolic curves [3], [7] as shown in Figure 1(b). The paint deposition rate G on a flat surface can be modelled as:

$$G = f(r, \theta) \quad (1)$$

where r is the distance from a point to the tool center inside a cone. θ is the fan angle. R is the spray cone radius. Goodman [10] presented a method to measure the paint deposition rate by covering a flat surface. Typically the tool standoff is kept constant [2], [3], [7], [8]. Therefore,

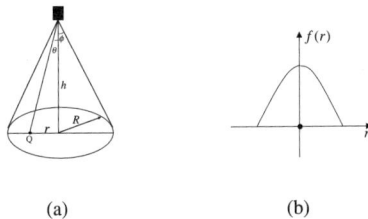

Fig. 1. (a) A tool model. (b) A tool profile

the paint deposition rate is only related to the distance r, i.e.

$$G = f(r) \quad (2)$$

To paint a surface, the paint thickness must satisfy some given constraints. Suppose the average thickness on a surface should be q_d. The paint thickness deviation from the average thickness is not discussed here because the parameter is not used in the optimization processes.

III. Optimization

A gun trajectory includes paint gun position, orientation and velocity. The paint gun position is related to the overlap distance between two paths. Since the optimization processes are performed on flat patches, the paint gun orientation is determined by the normals of the flat patches. Therefore, two parameters, the overlap distance and paint gun velocity need to be optimized.

The gun velocity and the overlap distance are determined by optimizing the painting process on the flat patch.

Theorem 3.1: Given a paint gun profile, the paint thickness on a flat patch is related to the paint gun velocity and the overlap distance. Moreover, the paint thickness is inversely proportional to the paint gun velocity.

The proof of the theorem is presented in [11]. This means:

$$\bar{q}(x,d,v) = \frac{1}{v}\rho(x,d) \quad (3)$$

where $\bar{q}(x,d,v)$ is the paint thickness on a plane; x the distance to the gun center; d the overlap distance; v the gun velocity; ρ is a function of x and d.

To find an optimal velocity v and overlap distance d, the mean square error of the thickness deviation from the required thickness q_d must be minimized, i.e.,

$$\min_{d\in[0,R],v} E_1(d,v) = \int_0^{2R-d}(q_d - \bar{q}(x,d,v))^2 dx \quad (4)$$

The maximum paint thickness and minimal paint thickness have to be optimized too because they will determine the paint thickness deviation from the average paint thickness.

$$\min_{d\in[0,R],v} E_2(d,v) = (q_{max} - q_d)^2 + (q_d - q_{min})^2 \quad (5)$$

From equation (4) and (5), we have:

$$\min_{d\in[0,R],v} E(d,v) = \frac{1}{2R-d}E_1(d,v) + E_2(d,v) \quad (6)$$

Corollary 3.2: The minimization of $E(d,v)$ is only related to the overlap distance d.

Proof: From Theorem 3.1, $\bar{q}(x,d,v)$ is inversely proportional to the paint gun velocity v. Then from equation (3), the maximum and minimum paint thickness can be expressed as:

$$q_{max} = \frac{1}{v}\rho_{max}(d), \qquad q_{min} = \frac{1}{v}\rho_{min}(d) \quad (7)$$

To find minimal $E(d,v)$, $\frac{\partial E(d,v)}{\partial v} = 0$. From equation (6), (7), and (3), we obtain:

$$v = \frac{\frac{1}{2R-d}\int_0^{2R-d}\rho^2(x,d)dx - \rho_{max}^2(d) - \rho_{min}^2(d)}{q_d[\frac{1}{2R-d}\int_0^{2R-d}\rho(x,d)dx + \rho_{max}(d) + \rho_{min}(d)]} \quad (8)$$

The paint gun velocity can be expressed as a function of the overlap distance d. Therefore, the minimization of $E(d,v)$ is only related to the overlap distance d. ∎

A golden section method [12] is adopted here to calculate the overlap distance and paint gun velocity by iteration.

A. Optimization Process for a Surface with Two Patches

The paint thickness optimization of two patches is much more complicated than that of one patch. The overlap distance and paint gun velocity should be kept the same as those in one patch except the intersection line between patches. According to the criteria that the main part of a paint gun path is parallel or perpendicular to the intersection line, different cases are studied: parallel-parallel case; parallel-perpendicular case; perpendicular-perpendicular case. Figure 2 shows the three different cases.

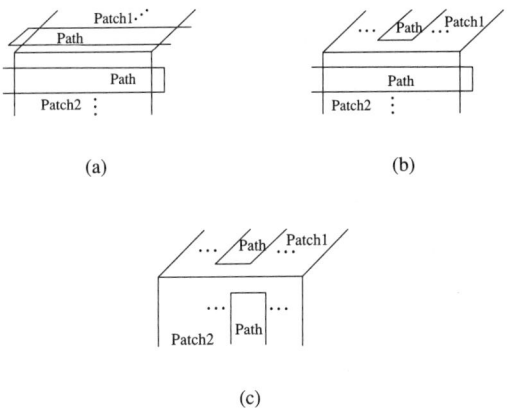

Fig. 2. (a) Case 1: parallel-parallel case. (b) Case 2: parallel-perpendicular case. (c) Case 3: perpendicular-perpendicular case.

The three different cases will be discussed individually. First we will discuss the parallel-parallel case as shown in Figure 3. In this case, we need to optimize the distance h

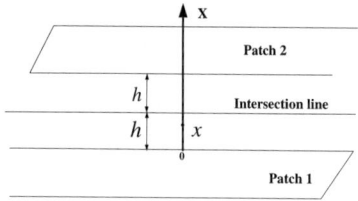

Fig. 3. Parallel-parallel case.

between the two paths. Because the two paths are symmetric, the distances of the two paths to the intersection line are the same. Suppose the angle between the two patches are α, then the paint thickness at the intersection part can be expressed as:

$$\bar{q}(x,h) = \begin{cases} \bar{q}_1(x,h) + \bar{q}_2(x,h)\cos\alpha & 0 \leq x \leq h \\ \bar{q}_1(x,h)\cos\alpha + \bar{q}_2(x,h) & h < x \leq 2h \end{cases} \quad (9)$$

Then, by optimizing equation (6), the optimized distance h can be obtained.

If a path is parallel to the intersection line between two patches, the paint thickness on the intersection part due to the parallel path will be evenly distributed. Therefore, before we discuss the optimization processes for parallel-perpendicular, the paint thickness on a patch in the perpendicular case is optimized. Figure 4(a) shows a path which is perpendicular to an intersection line.

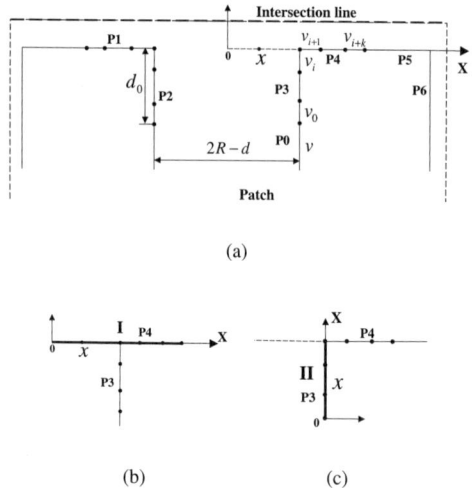

Fig. 4. Paint gun velocity optimization for a perpendicular path.

The paint gun velocity has to be optimized to optimize the paint thickness. Here we divide a path into segments as shown in Figure 4(a). d_0 is arbitrarily chosen. In each segment, the paint gun velocity is considered to be fixed.

In the figure, there are six pieces of path, P1,...,P6. P2, P3 and P6 are divided into $i+1$ segments respectively. The corresponding velocities are $v_0, \cdots, v_i$ respectively. P1, P4 and P5 are divided into k segments respectively. The corresponding velocities are $v_{i+1}, \cdots, v_{i+k}$ respectively. The minimum thickness and maximum thickness occur on Part I as shown in Figure 4(b). $x=0$ is the minimum paint thickness point and $x=(2R-d)$ is the maximum paint thickness point. Because the paint thickness is periodic along X axis, we only need to consider the paint thickness at $x \in [0, 2r-d]$. The paint thickness on Part I is calculated using the contribution of the six pieces of the path:

P1: $\quad q_{P_1}(x,j) = \frac{1}{v_j} \int_{\frac{2R-d}{2k}(j-1)}^{\frac{2R-d}{2k}(j-i)} f(|\frac{2R-d}{2}+y+x|)dy$

P2: $\quad q_{P_2}(x,j) = \frac{1}{v_j} \int_{\frac{j}{i+1}d_0}^{\frac{j+1}{i+1}d_0} f(\sqrt{(\frac{2R-d}{2}+x)^2 + (d_0-y)^2})dy$

P3: $\quad q_{P_3}(x,j) = \frac{1}{v_j} \int_{\frac{j}{i+1}d_0}^{\frac{j+1}{i+1}d_0} f(\sqrt{(\frac{2R-d}{2}-x)^2 + (d_0-y)^2})dy$

P4: $\quad q_{P_4}(x,j) = \frac{1}{v_j} \int_{\frac{2R-d}{2k}(j-1)}^{\frac{2R-d}{2k}(j-i)} f(|\frac{2R-d}{2}+y-x|)dy$

P5: $\quad q_{P_5}(x,j) = \frac{1}{v_j} \int_{\frac{2R-d}{2k}(j-1)}^{\frac{2R-d}{2k}(j-i)} f(|\frac{3(2R-d)}{2}-y-x|)dy$

P6: $\quad q_{P_6}(x,j) = \frac{1}{v_j} \int_{\frac{j}{i+1}d_0}^{\frac{j+1}{i+1}d_0} f(\sqrt{(\frac{3(2R-d)}{2}-x)^2 + (d_0-y)^2})dy$

$$(10)$$

Then the thickness on Part I is:

$$\begin{aligned} q_I(x) &= \Sigma_{j=0}^{i}(q_{P_2}(x) + q_{P_3}(x,j) + q_{P_6}(x,j)) \\ &+ \Sigma_{j=i+1}^{i+k}(q_{P_1}(x,j) + q_{P_4}(x) + q_{P_5}(x,j)) \end{aligned} \quad (11)$$

When optimizing the paint thickness on Part I, we should consider the thickness on Part II in Figure 4(c) also. The thickness on Part II can be expressed using P0, P3, P4 and P5:

P0: $\quad q'_{P_0}(x) = \frac{1}{v} \int_0^R f(R-y-x)dy$

P3: $\quad q'_{P_3}(x,j) = \frac{1}{v_j} \int_{\frac{j}{i+1}d_0}^{\frac{j+1}{i+1}d_0} f(|y-x|)dy$

P4: $\quad q'_{P_4}(x,j) = \frac{1}{v_j} \int_{\frac{2R-d}{2k}(j-1)}^{\frac{2R-d}{2k}(j-i)} f(\sqrt{(d_0-x)^2 + y^2})dy$

P5: $\quad q'_{P_5}(x,j) = \frac{1}{v_j} \int_{\frac{2R-d}{2k}(j-1)}^{\frac{2R-d}{2k}(j-i)} f(r)dy$

where: $\quad r = \sqrt{(d_0-x)^2 + (\frac{3(2R-d)}{2}-y)^2} \quad (12)$

Then the thickness are Part II is:

$$q_{II}(x) = q'_{P_0}(x) + \Sigma_{j=0}^{i} q'_{P_3}(x,j) + \Sigma_{j=i+1}^{i+k}(q'_{P_4}(x,j) + q'_{P_5}(x,j))$$

According to equations (4), (5), (6), we have:

$$\min_{v_j, j \in [0, i+k]} E(v_j) = \frac{1}{2R-d} E_I(v_j) + \frac{1}{d_0} E_{II}(v_j) + E_{III}(v_j) \quad (13)$$

where E_I, E_{II} and E_{III} are defined as:

$$\begin{aligned} E_I(v_j) &= \int_0^{2R-d}(q_d - q_I(x))^2 dx \\ E_{II}(v_j) &= \int_0^{d_0}(q_d - q_{II}(x))^2 dx \\ E_{III}(v_j) &= (q'_{max} - q_d)^2 + (q_d - q'_{min})^2 \end{aligned} \quad (14)$$

where q'_{max} and q'_{min} are the maximum and minimum thickness in $q_I(x)$ and $q_{II}(x)$.

The paint thickness is optimized using equation (13). This is a multi-variable unconstraint optimization problem. The steepest-descent algorithm [13] is adopted here to optimize equation (13). Then the optimized velocities are obtained such that the paint thickness is optimized. After the thickness is obtained, the maximum thickness point P$_{max}$ and the minimum thickness point P$_{min}$ can be found on Part I.

Figure 5 shows the parallel-perpendicular case. Suppose the angle between the two patches are α, the

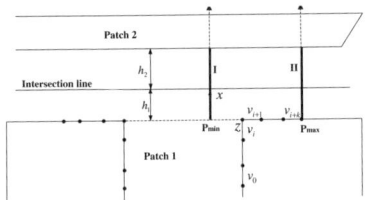

Fig. 5. Parallel-parallel case.

paint thickness on Part I and II can be calculated:

$$q_I(x) = \begin{cases} q_1(x) + q_2(x)\cos\alpha & 0 \le x \le h_1 \\ q_1(x)\cos\alpha + q_2(x) & h_1 < x \le h_1 + h_2 \end{cases}$$

$$q_{II}(x) = \begin{cases} q'_1(x) + q'_2(x)\cos\alpha & 0 \le x \le h_1 \\ q'_1(x)\cos\alpha + q'_2(x) & h_1 < x \le h_1 + h_2 \end{cases} \quad (15)$$

where $q_1(x)$ and $q_2(x)$ are the paint thickness on Part I due to the paths on patch 1 and patch 2 respectively. $q'_1(x)$ and $q'_2(x)$ are the paint thickness on Part II due to the paths on patch 1 and patch 2 respectively. Similar method as that in the perpendicular case (equation (10) and (11)) is used to calculate the paint thickness. Then the error function can be developed:

$$\min_{h_1,h_2} E(h_1,h_2) = \frac{1}{h_1+h_2}[E_I(h_1,h_2) + E_{II}(h_1,h_2)] + E_{III}(h_1,h_2) \quad (16)$$

where $E_I(h_1,h_2)$, $E_{II}(h_1,h_2)$ and $E_{III}(h_1,h_2)$ are defined as:

$$E_I(h_1,h_2) = \int_0^{h_1+h_2}(q_d - q_1(x))^2 dx$$
$$E_{II}(h_1,h_2) = \int_0^{h_1+h_2}(q_d - q_2(x))^2 dx$$
$$E_{III}(h_1,h_2) = (q'_{max} - q_d)^2 + (q_d - q'_{min})^2 \quad (17)$$

where q'_{max} and q'_{min} are the maximum and minimum thickness in $q_1(x)$ and $q_2(x)$. By optimizing equation (16), h_1 and h_2 can be obtained.

For perpendicular-perpendicular case, according to the relative position of the path in patch1 and that in patch2, the optimized paint thickness is different. To get the optimized paint thickness on a surface with two patches, the paint thickness due to one path should compensate the paint thickness due to the other. Figure 6 shows the perpendicular-perpendicular case. After calculating the

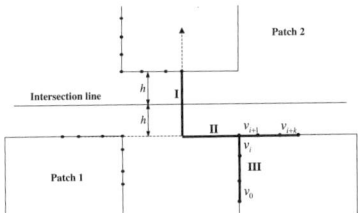

Fig. 6. Perpendicular-perpendicular case.

paint thickness on Part I, II and III using the similar method in the perpendicular case, the similar error function in parallel-perpendicular case can be formed. In this case, the distance h and the velocities have to be optimized together to get the optimized thickness. The steepest-descent algorithm [13] is adopted here to optimize the paint thickness.

B. Optimization Analysis for a Surface with Multiple Patches

Figure 7 shows a surface with three patches. The paint

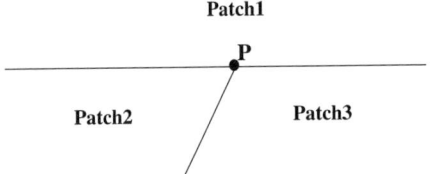

Fig. 7. A surface with three patches.

thickness between patch 1 and 2, patch 2 and 3, patch 1 and 3 is optimized. The only part we need to consider is the thickness on the area around intersection point P. Since the paint thickness between patch 2 and patch 3 is optimized, we can merge them into one patch: patch I. The paint thickness on patch I is optimized. Since the paint thickness between patch 1 and 3, patch 1 and 2 is optimized, the paint thickness between patch I and patch 1 should be optimized also. Therefore, the paint thickness on the area around the intersection point P is optimized.

This method can be applied to a surface with more than three patches which intersect at one point. Therefore, if the paint thickness between any two patches is optimized, the paint thickness on a surface which is consist of multiple patches should be optimized.

IV. IMPLEMENTATION AND VERIFICATION

Suppose the required average thickness is $q_d = 50 \ \mu m$. The spray radius $R = 50 \ mm$. The paint deposition rate is

$$f(r) = \frac{1}{10}(R^2 - r^2) \ \mu m/s \quad (18)$$

After optimization, the gun velocity and the overlap distance were calculated. $v = 323.3$ mm/s and $d = 39.2$ mm The maximum and minimum thickness are $\bar{q}_{max} = 52.02$ μm and $\bar{q}_{min} = 48.05$ μm. The optimized paint thickness is shown in Figure 8.

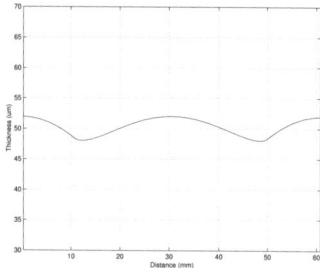

Fig. 8. The optimized paint thickness on a surface with one patch.

After performing the optimization process for PA-PA case, the optimized paint thickness for $\alpha = 30^o$ is shown in Figure 9(a). The minimum and maximum paint thickness are 48.8 and 51.5 μm respectively.

For the velocity optimization process for the perpendicular case, $i = 5$, $d_0 = 2R$ and $k = 6$ are chosen. The optimized paint gun velocities are:

$$v_0 = 272.2 mm/s, \quad v_1 = 333.1 mm/s, \quad v_2 = 459.2 mm/s$$
$$v_3 = 336.4 mm/s, \quad v_4 = 226.7 mm/s, \quad v_5 = 355.3 mm/s$$
$$v_6 = 547.2 mm/s, \quad v_7 = 690.8 mm/s \quad (19)$$

The optimized paint thickness is shown in Figure 9(b).

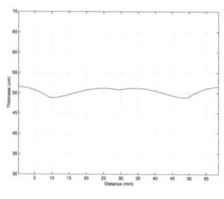 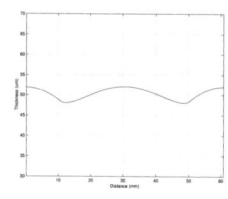

(a) (b)

Fig. 9. The optimized paint thickness (a) PA-PA case when $\alpha = 30^o$; (b) the perpendicular case.

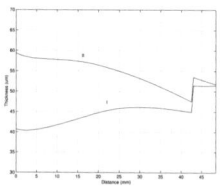 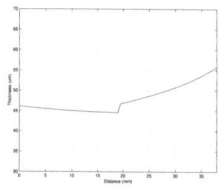

(a) (b)

Fig. 10. The optimized paint thickness: (a) PA-PE when $\alpha = 30^o$; (b) PE-PE when $\alpha = 30^o$.

For PA-PE case, the optimized paint thickness for $\alpha = 30^o$ is shown in Figure 10(a). The minimum and maximum paint thickness are 40.7 and 59.4 μm respectively.

For PE-PE case, the optimized paint gun velocities when $\alpha = 30^o$ are:

$$v_0 = 252.0 mm/s, \quad v_1 = 308.4.1 mm/s, \quad v_2 = 425.2 mm/s$$
$$v_3 = 311.5 mm/s, \quad v_4 = 209.9 mm/s, \quad v_5 = 329.0 mm/s$$
$$v_6 = 506.7 mm/s, \quad v_7 = 639.6 mm/s \quad (20)$$

The optimized paint thickness for $\alpha = 30^o$ is shown in Figure 10(b). The minimum and maximum paint thickness are 44.6 and 55.8 μm respectively.

The maximum and minimum paint thickness for the three cases when $\alpha = 30^o$ are shown in Table I. A surface

TABLE I
THE SIMULATION RESULTS

Case	Maximum thickness (μm)	Minimum thickness (μm)
PA-PA	48.8	51.5
PA-PE	40.7	59.4
PE-PE	44.6	55.8

with two flat patches are generated and rendered into triangles. The trajectory for each patch is generated and the optimized parameters are applied to calculate the paint thickness using a simulation model [4]. The part rendered into triangles is shown in Figure 11.

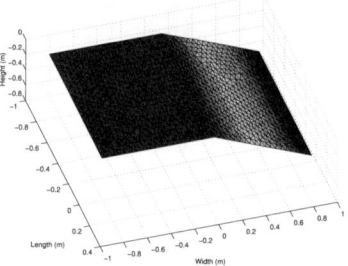

Fig. 11. The part with two flat patches when $\alpha = 30^o$.

The paths of the part are generated for the parallel-parallel, parallel-perpendicular and perpendicular-perpendicular cases. Figure 12 shows the path and paint thickness for parallel-parallel case. Figure 13 for parallel-perpendicular case. Figure 14 for perpendicular-perpendicular case.

The maximum and minimum paint thickness for the three cases when $\alpha = 30^o$ are shown in Table II.

The results shown in Table I and II are quite close. This means the optimized parameters can optimize the

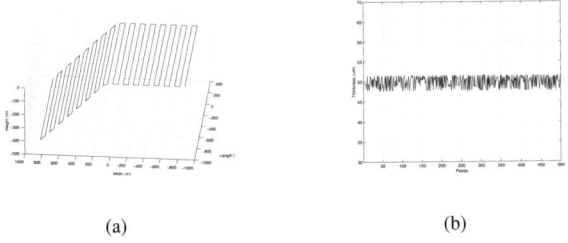

Fig. 12. Verification results for PA-PA case. (a) The Path. (b) The paint thickness.

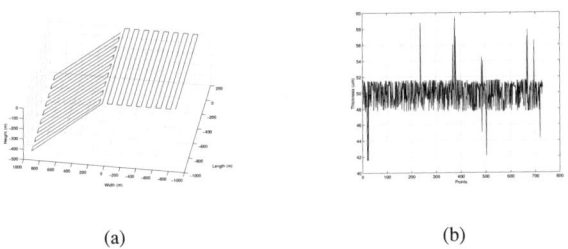

Fig. 13. Verification results for PA-PE case. (a) The Path. (b) The paint thickness.

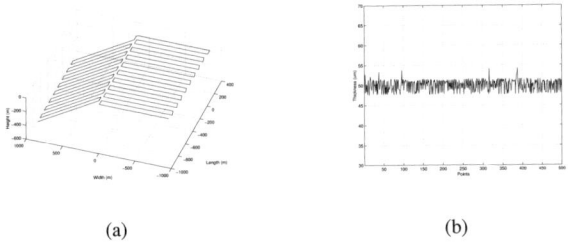

Fig. 14. Verification results for PE-PE case. (a) The Path. (b) The paint thickness.

TABLE II
THE SIMULATION RESULTS

Case	Maximum thickness (μm)	Minimum thickness (μm)
PA-PA	47.6	51.6
PA-PE	41,6	59.5
PE-PE	47.5	54.5

paint thickness in different cases. From the optimization results and the verification results, we can see that the optimized paint thickness for the parallel-parallel case is quite uniform. The paint thickness for the best case of perpendicular-perpendicular case is also uniform. The paint thickness deviation from the required thickness is about $5\mu m$. However, the paint thickness for the parallel-perpendicular case are quite large. The paint thickness deviation from the required thickness is about $10\mu m$. In path generation, the parallel-perpendicular case should be avoided.

V. CONCLUSION

Paint thickness optimization methods for a surface with multiple patches are developed. Optimization processes are presented for one flat surface and a surface with two patches. The paint thickness optimization for a surface with multiple patches is analyzed. Simulations are performed to verify the optimized parameters. The optimization results and the verification results show that the parallel-parallel case is the best and parallel-perpendicular case is the worst among the three cases. The paint thickness optimization method can also be applied in other material quantity optimization, such as spray forming.

VI. ACKNOWLEDGMENTS

Research is partially supported under NSF Grant IIS-9796300, IIS-9796287, EIA-9911077 and DMI 0115355. The authors would like to thank Tecnomatix Inc. for its providing us with the software ROBCAD.

VII. REFERENCES

[1] Tecnomatix. *ROBCAD/Paint Training*. Tecnomatix, Michigan, USA, 1999.
[2] S. Suh, I. Woo, and S. Noh. Automatic trajectory planning system (atps) for spray painting robots. *Journal of Manufacturing Systems*, 10(5):396–406, 1991.
[3] J. K. Antonio, R. Ramabhadran, and T. L. Ling. A framework for optimal trajectory planning for automated spray coating. *International Journal of Robotics and Automation*, 12(4):124–134, 1997.
[4] H. Chen, W. Sheng, N. Xi, M. Song, and Y. Chen. Automated robot trajectory planning for spray painting of free-form surfaces in automotive manufacturing. In *IEEE International Conference on Robotics and Automation*, volume 1, pages 450 –455, 2002.
[5] H. Choset. Coverage of known spaces: The boustrophedon cellupdar decomposition. *Autonomos Robots*, 9:247–253, 2000.
[6] E. Freund, D. Rokossa, and J. Rossmann. Process-orientated approach to an efficient off-line programming of industrial robots. In *Proceedings of the 24th Annual Conference of the IEEE Industrial Electronics Society, IECON '98.*, volume 1, pages 208 –213, 1998.
[7] W. Persoons and H. Van Brussel. Cad-based robotic coating with highly curved surfaces. In *International Symposium on Intelligent Robotics (ISIR'93)*, volume 14, pages 611–618, 1993.
[8] P. Hertling, L. Hog, R. Larsen, J. W. Perram, and H. G. Petersen. Task curve planning for painting robots. i. process modeling and calibration. *IEEE Transactions on Robotics and Automation*, 12(2):324 –330, April 1996.
[9] J. K. Antonio. Optimal trajectory planning for spray coating. In *IEEE International Conference on Robotics and Automation*, pages 2570 –2577, San Diego, California, May 1994.
[10] E. D. Goodman and L. T. W. Hoppensteradt. A method for accurate simulation of robotic spray application using empirical parameterization. In *IEEE International Conference on Robotics and Automation*, volume 2, pages 1357 –1368, Sacramento, California, April 1991.
[11] H. Chen, N. Xi, Y. Chen, and J. Dahl. Cad-guided spray gun trajectory planning of free-form surfaces in manufacturing. *Journal of Advanced Manufacturing Systems*, June 2003.
[12] S. S. Rao. *Optimization: Theory and Application*. John Wiley & Sons, Inc., New York, USA, 1983.
[13] P. E. Gill, W. Murray, and M. H. Wright. *Practical Optimization*. Academic Press, New York, NY, USA, 1981.

A Novel 2-DOF Parallel Mechanism Based Design of a New 5-Axis Hybrid Machine Tool

Xin-Jun Liu[*], Xiaoqiang Tang, and Jinsong Wang

Manufacturing Engineering Institute, Department of Precision Instruments, Tsinghua University
Beijing, 100084, P.R. of CHINA

Abstract

In this paper, the concept of a new 5-axis hybrid machine tool is proposed. The machine is a tool being with both parallel and serial structures, which is based on a novel 2-DOF parallel platform and serial orientations. The machine tool has advantages in terms of high stiffness, high dexterity, high speed, and being capable of long components manufacturing. The kinematics characteristics, such as inverse and forward kinematics, conditioning indices, of the novel 2-DOF parallel platform are studied. The dimensional synthesis based on the workspace and conditioning indices is presented. The results of the paper are very useful for the design of the hybrid machine tool.

1. Introduction

The parallel kinematics machine (PKM) is a new type of machine tool which was firstly showed at the 1994 Internatinoal Manufacturing Technology Show in Chicago by two American machine tool companies, Giddings & Lewis and Ingersoll. These machine tools, named Hexpod, were based on the paradigm of the spatial 6-DOF parallel mechanism. The parallel kinematics machine technology promises to offer manufacturers a number of advantages relative to conventional machine tools, such as a higher stiffness-to-mass ratio, higher speeds, higher accuracy, reduced installation requirements, mechanical simplicity, and high flexibility.

The six-DOF Stewart platform [1] is one PKM configuration that has been used in a number of new machine tool designs at the beginning of the born of PKMs. For machining applications, disadvantages of the Stewart platform include a complex workspace, limited orientation range of motion and a requirement of six actuators for a five degree-of-freedom task (milling, drilling, and similar operations). Moreover, there are some disadvantages for the parallel kinematics itself, e.g., the forward kinematics can not be described in closed-form, and the calibration is difficult, and so on. For such reasons, many researchers begin to pay their attentions to PKMs with less than 6 DOFs [2,3], especially, hybrid PKMs [4,5,6], such as the Tricept HPI (Neos), Hexam (Toyoda), PA35 (Hitachi Seiki), and Georg V (IFW-University of Hannover). PKMs with hybrid kinematics are always built as Tripod structures, for which all points within the workspace are reachable with high dynamics and high accuracy [5] through the used parallel mechanism. By means of the two-axis wrist joint the end-effector gets the desired orientation in the workspace. By this arrangement of the kinematics the dexterity of the system can be increased compared to fully parallel kinematics (Hexapod systems). Another advantage to design a machine tool as hybrid structure based on a 3- or 4-DOF parallel mechanism, to the author's knowledge, there is no hybrid machine tool is based on a 2-DOF parallel mechanism, is that the stiffness can be improved by increasing redundant constraints.

This paper proposes another design concept for the hybrid machine tool, which is a 5-axis hybrid gantry structure based on a novel 2-DOF parallel mechanism, a two-axis wrist joint and a long movable worktable. The machine tool has following advantages: (a) high stiffness, (b) high dexterity, (c) high speed, and (d) being capable of the manufacture for long components. The machine tool has been developed by Tsinghua University and the Second Machine Tool Works of QiQiHaer in China.

Firstly, a novel 2-DOF parallel mechanism is proposed. The output of the moving platform is planar translations of a rigid body but not a point. The kinematics design, such as the workspace, inverse and forward kinematics problems, the conditioning indices, and dimensional synthesis, is discussed, The results are very useful for the design of the machine tool.

2. Description of the novel 2-DOF parallel mechanism

Two-DOF parallel mechanisms are very important systems in the family of the parallel mechanism. The existing planar two-DOF parallel mechanisms [7,8] can only position a point not a rigid body in a plane.

The novel 2-DOF parallel mechanism proposed in this paper is shown in Fig.1. A schematic of the mechanism is shown in Fig.2, where the base is labeled 1 and the moving platform is labeled 2. The moving platform is connected to the base by two identical legs. Each leg consists of a planar four-bar parallelogram: links 2, 3, 4, and 5 for the first leg; 2, 6, 7, and 8 for the

---

[*] Corresponding Author: xinjunl@yahoo.com

second leg. In each planar four-bar parallelogram, the joints are all revolute pairs. Links 3 and 8 are actuated by prismatic actuators, respectively. Motions of the moving platform are achieved by the combination of movements of the links 3 and 8 that can be transmitted to the platform by the system of the two parallelograms. Due to the structure, one can see that the moving platform or the rigid body 2 has two pure translational degrees of freedom with respect to the base because of the planar four-bar parallelograms. What we should notice is that the system is an over-constraint one. To obtain two DOFs of a rigid body in this system, only one planar four-bar parallelogram is enough. The reason to use two planar four-bar parallelograms is to increase the system's stiffness and make the system symmetry.

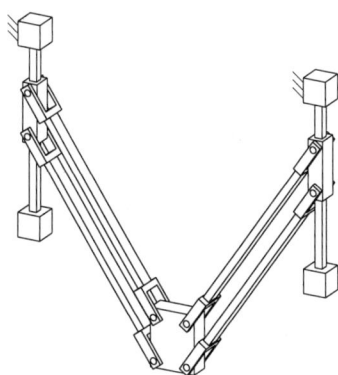

Figure 1: *A novel planar 2-DOF parallel mechanism*

3. Kinematics Analysis

As illustrated in Fig.2, a reference frame $\Re : O - xy$ is fixed to the base and a moving reference frame $\Re' : O' - x'y'$ is attached to the moving platform, where O' is the reference point on the moving platform. For the characteristic of a planar four-bar parallelogram, we can consider chains P_1B_1 and P_2B_2, as shown in Fig.2, to resolve the kinematics of the mechanism. And vectors $p_{i\Re'}$ and $p_{i\Re}$ ($i = 1, 2$) will be defined as the position vectors of points P_i in frames $\Re'$ and $\Re$, respectively, vectors $b_{i\Re}$ ($i = 1, 2$) as the position vectors of points B_i in frame $\Re$. The geometric parameters of the mechanism are $P_iB_i = L$ ($i = 1, 2$), the moving platform parameter r, and the distance between two guideways $2R$. And the position of point O' in the fixed frame $\Re$ is denoted as vector

$$c_\Re = (x, \ y)^T \quad (1)$$

Vectors of $b_{i\Re}$ in the fixed frame $\Re$ can be written as

$$b_{1\Re} = (R \ \ y_1)^T, \quad b_{2\Re} = (-R \ \ y_2)^T \quad (2)$$

where y_i are actuated inputs. And vectors $p_{i\Re}$ in the fixed frame $\Re$ can be written as

$$p_{i\Re} = p_{i\Re'} + c_\Re, \quad i = 1,2 \quad (3)$$

where

$$p_{1\Re'} = (r \ \ 0)^T, \quad p_{2\Re'} = (-r \ \ 0)^T \quad (4)$$

Then the inverse kinematics problem of the mechanism can be solved by writing following constraint equation

$$\|p_{i\Re} - b_{i\Re}\| = L \quad i = 1,2 \quad (5)$$

that is

$$y_1 = \pm\sqrt{L^2 - (r+x-R)^2} + y \quad (6)$$

$$y_2 = \pm\sqrt{L^2 - (x-r+R)^2} + y \quad (7)$$

from which we can see that there are four solutions for the inverse kinematics of the mechanism. Hence, for a given mechanism and for prescribed values of the position of the moving platform, the required actuated inputs can be directly computed from Eqs. (6) and (7). To obtain the configuration as shown in Fig.1, the "$\pm$" in Eqs.(6) and (7) should be "+".

The objective of the direct kinematics solution is to define a mapping from the known set of the actuated inputs to the unknown pose of the output platform. From Eqs.(6) and (7), the direct kinematics of the mechanism can be described as

$$x = ay + b \quad (8)$$

where

$$a = \frac{y_2 - y_1}{2(R-r)}, \quad b = \frac{y_1^2 - y_2^2}{4(R-r)} \quad (9)$$

and

$$y = \frac{-f \pm \sqrt{f^2 - 4eg}}{2e} \quad (10)$$

in which

$$e = a^2 + 1, \quad f = 2a(r+b-R) - 2y_1,$$
$$g = (r+b-R)^2 + y_1^2 - L^2 \quad (11)$$

To obtain the forward configuration as shown in Fig.1, the "$\pm$" in Eq.(10) should be "−".

From above equations, we can see that the inverse and direct kinematics problems of the mechanism are very easy and can be described as closed-forms.

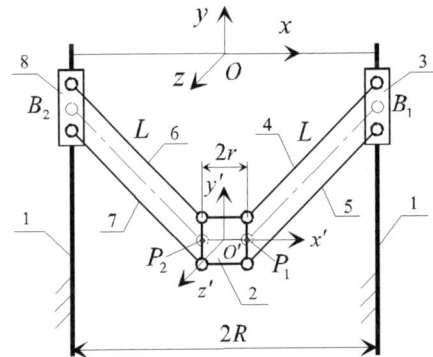

Figure 2: A schematic of the mechanism

4. Jacobian matrix and conditioning indices

Equation (5) can be differentiated with respect to time to obtain the velocity equations. This leads to an equation of the form

$$A \begin{pmatrix} \dot{y}_1 \\ \dot{y}_2 \end{pmatrix} = B \begin{pmatrix} \dot{x} \\ \dot{y} \end{pmatrix} \quad (12)$$

where A and B are, respectively, the 2×2 inverse and forward Jacobian matrices of the mechanism and can be expressed as

$$A = \begin{bmatrix} y - y_1 & 0 \\ 0 & y - y_2 \end{bmatrix}, \quad B = \begin{bmatrix} r + x - R & y - y_1 \\ x - r + R & y - y_2 \end{bmatrix} \quad (13)$$

If the matrix A is not singular, the Jacobian matrix of the mechanism can be obtained as

$$J = A^{-1} B = \begin{bmatrix} \dfrac{r + x - R}{y - y_1} & 1 \\ \dfrac{x - r + R}{y - y_2} & 1 \end{bmatrix} \quad (14)$$

Let

$$|E\lambda - J| = 0 \quad (15)$$

where $E = \begin{bmatrix} 1 & 0 \\ 0 & 1 \end{bmatrix}$ and there is

$$\lambda^2 + s\lambda + t = 0 \quad (16)$$

where

$$s = -\left(\frac{r + x - R}{y - y_1} + 1 \right), \quad t = \frac{r + x - R}{y - y_1} - \frac{x - r + R}{y - y_2} \quad (17)$$

Then the eigenvalue of Jacobian matrix J can be obtained,

$$\lambda = \frac{-s \pm \sqrt{s^2 - 4t}}{2} \quad (18)$$

As well known, the dexterity of a parallel mechanism can be evaluated by the conditioning index (CI) $1/\kappa$ [9], which is written as

$$1/\kappa = \lambda_2 / \lambda_1 \quad (19)$$

where λ_1 and λ_2 are the maximum and minimum eigenvalues of the Jacobian matrix, which can be obtained from Eq.(18), respectively. The corresponding global conditioning index (GCI) will be

$$\eta = \int_W 1/\kappa \, dW \Big/ \int_W dW \quad (20)$$

where W is the reachable workspace of the mechanism.

In the conditioning indices, the singular configurations should be avoided. From above analysis, one can see that the mechanism is very simple. And the singularity analysis will also be simple, which can be reached from Jacobian matrices A and B. When $|A| = 0$ and $|B| \neq 0$, from Eq.(13), there is $y = y_1$ or $y = y_2$, which means that the first or second leg is parallel to x axis. This corresponds to the first kind of singularity. If $|B| = 0$ and $|A| \neq 0$, the second kind of singularity occurs, i.e., $r + x = R$ for the first leg when x is positive and $R + x = r$ for the second leg when x is negative. In such case, the mechanism is in the configuration that four bars of the parallelogram in one of the two legs are parallel to each other. $|B| = 0$ and $|A| = 0$ lead to the third kind of singularity, in which the two legs are both parallel to x axis. The geometric parameter condition for this singularity is $r + R = L$.

5. Workspace of the mechanism

One of the most important issues in the process of design of the mechanism is its workspace. For parallel mechanisms, this issue may be more critical since parallel mechanisms will sometimes have a rather limited workspace.

The workspace of the planar 2-DOF parallel mechanism is often represented as a region of the plane. And the determination of the workspace is more simply, which can be obtained geometrically from the inverse kinematics equation. From Eq.(5), one obtains

$$(r + x - R)^2 + (y - y_1)^2 = L^2 \quad (21)$$
$$(x - r + R)^2 + (y - y_2)^2 = L^2 \quad (22)$$

which means that if y_i are specified, Eqs.(21) and (22) represent two circles centered at $(R - r, \; y_1)$ and $(r - R, \; y_2)$, respectively. Their radii are L. If $y_i \in [y_{i\min}, \; y_{i\max}]$, Eqs. (21) and (22) represent two enveloping surfaces, each of which is the locus of a circle (the radius is L), when the center is rolling on line segments $x = R - r$ and $x = r - R$ ($y \in [y_{i\min}, y_{i\max}]$), respectively. The intersection of the two enveloping surfaces is the workspace of the mechanism.

For example, the workspace of such a mechanism with $R = 4$, $r = 1$, $L = 6$ and $y_i \in [-2, \; 2]$ is shown in Fig.3.

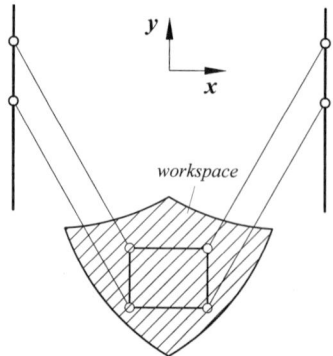

Figure 3: A workspace example

6. The dimensional synthesis

6.1 The dimensional synthesis based on the workspace

The objective of the dimensional synthesis is to determine geometric parameters of the mechanism for a desired workspace. The parameters are R, r, L and the input $|y_{i\min} - y_{i\max}|$. In this paper the desired workspace is assumed to be given by a rectangle $x_W \times y_W$.

Because the moving platform only has translational DOFs, the parameter r will not bring any effect to performances of the mechanism. The determination of parameter r is depended on the designer's demand, in this paper $r = 75$mm. One of advantages of the mechanism is that it gives a total freedom in the choice of the workspace in y direction, as shown in Fig.2. The workspace volume along $y-$axis, denoted as y_W, will not effect the parameters R and L but the input $|y_{i\min} - y_{i\max}|$, for which we can consider firstly the workspace volume along $x-$axis, denoted as x_W, to determine parameters R and L. When the moving platform reaches the boundary of the desired workspace, as shown in Fig.4, there are

$$\cos\alpha = \frac{d + x_W - r}{L}, \quad \sin\beta = \frac{d - r}{L} \quad (23)$$

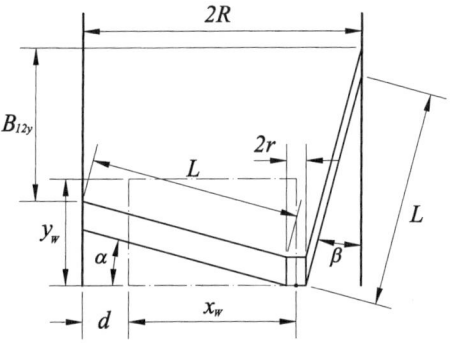

Figure 4: Dimensional synthesis of the mechanism

When the moving platform travels x_W along $x-$axis, the input, denoted as B_{12y}, of each leg should be

$$B_{12y} = L(\cos\beta - \sin\alpha) \quad (24)$$

from which we can see that if $\alpha = \beta$ there is $B_{12y} = x_W$, which means that the ratio between the input and the output is

$$B_{12y} : x_W = 1:1 \quad (25)$$

Actually, we hope that the effect of the input to output should be small, i.e., $B_{12y} : x_W > 1:1$. If the condition of Eq.(25) is considered in the design of the mechanism, which means $\alpha = \beta$, Eq.(23) can be rewritten as

$$d = \frac{(x_W - r)\tan\alpha + r}{1 - \tan\alpha} \quad (26)$$

from which one can obtain $d = 417.5$mm if $\alpha = 10°$. The parameter α can get other value, which will depend on the designer's demand. But if α is very small the configuration of the mechanism as shown in Fig.4 will be near the singularity. What is more, if $\alpha = 0$, it is not difficult to find out that the 2-DOF parallel mechanism proposed in this paper reaches singularity. Other parameters R, L, and $|y_{i\min} - y_{i\max}|$ will be obtained as

$$R = 2d + x_W, \quad L = \frac{d-r}{\sin\beta}, \quad (27)$$

and

$$|y_{i\max} - y_{i\min}| = x_W + y_W \quad (28)$$

Therefore, if $x_W = 1.6$m, $y_W = 1.0$m, $\alpha = 10°$, and $r = 75$mm, we will obtain

$$R = 1217.5\text{mm}, \quad L = 1972.5\text{mm},$$
$$|y_{i\max} - y_{i\min}| = 2600.0\text{mm} \quad (29)$$

6.2 The dimensional synthesis based on the CI

As shown by Strang [10], the condition number of a matrix is used in numerical analysis to estimate the error generated in the solution of a linear system of equations by the error on the data. When applied to the Jacobian matrix, the condition number will give a measure of the accuracy of the Cartesian velocity of the end effector and the static load acting on the end effector. It can also be used to evaluate the dexterity and stiffness of a manipulator [9,11]. After we reach the results of dimensional design based on workspace, we should consider the stiffness of the mechanism, which will make our design better.

In this Section, what we should do is to find out relationship between the conditioning indices and parameters of the mechanism. So that we can obtain the optimal parameters of the mechanism for the given workspace $1.6\text{m} \times 1\text{m}(x_W \times y_W)$. Firstly, the distribution of the conditioning index $1/\kappa$ in x_W of the workspace is plotted as shown in Fig.5, in which the parameter $L = 1972.5$mm and R is specified as $R \in [x_W/2, \; L + r - x_W/2]$. From Fig.5, we can see that

- The conditioning index is symmetric with respect to $x = 0$;
- The index reaches its maximum value when $x = 0$, for any value R.

In the process of dimensional synthesis, only x_W is considered. The GCI, Eq.(20), will be rewritten as

$$\eta = \int_{x_W} 1/\kappa \, dW \Big/ \int_{x_W} dW \quad (30)$$

The relationship between GCI and parameter R is illustrated as Fig.6, from which one obtains that if $L = 1972.5$mm, the GCI reaches its maximum value $\eta = 0.5616$ in $R \in [1203, 1210]$. And the

dimensional synthesis result will be:
$$R = 1203.0\text{mm}, \quad L = 1972.5\text{mm},$$
$$|y_{i\max} - y_{i\min}| = 2528.04\text{mm} \quad (31)$$

The index to evaluate the volume of the mechanism is defined roughly as
$$V = R \times |y_{i\max} - y_{i\min}| + L \times 4 \quad (32)$$

Then, there is $V = 3.05\text{m}^2$ for the design.

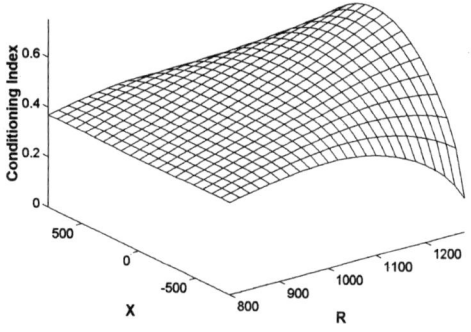

Figure 5: Distribution of the CI in workspace when L is specified

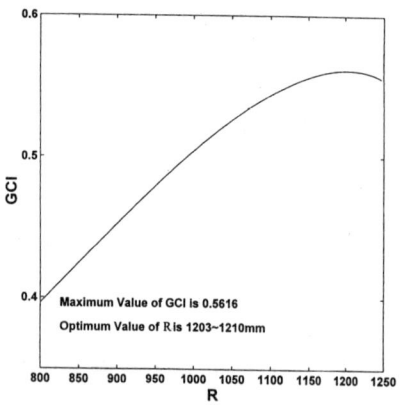

Figure 6: The relationship between the GCI and R

Similarly, when $R = 1217.5\text{mm}$ is specified, the corresponding maps are shown in Fig.7 and Fig.8, respectively, from which we can reach another dimensional synthesis result:
$$R = 1217.5\text{mm}, \quad L = 2061.5,$$
$$|y_{i\max} - y_{i\min}| = 2342.6\text{mm} \quad (33)$$

The maximum value of GCI will be $\eta = 0.5710$ in $L \in [2061.5, \ 2068.5]$, and there is $V = 2.86\text{m}^2$. It is clear that the workspace/volume ratio of the later design is less than that of the former design.

7. The design of a 5-axis hybrid machine tool

Recently, more and more hybrid machine tools [4,5,6] based on parallel mechanisms are proposed because of their high speed, high mobility, and high flexibility. According to demands of the Second Machine Tool Works of QiQiHaer in China, which are high speed, high mobility, high stiffness, and being capable of long components manufacturing, e.g., the vane, a 5-axis hybrid machine tool with gantry structure is proposed, as shown in Fig.9. The machine tool is based on the planar 2-DOF parallel mechanism proposed in this paper, which provides the machine tool with high stiffness and high speed. Especially, the upper and lower links of each of the two planar four-bar parallelograms are substituted by two plates, which can improve the system's stiffness. By means of the two-axis wrist joint the end-effector gets the desired orientation in the workspace, which provides it with high mobility. Only single-DOF joints are used in the machine tool, which can increase the accuracy. The worktable can move freely along the z-axis, which endows the machine tool with the capability of manufacturing long components.

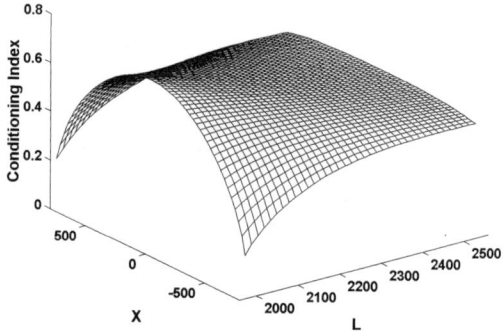

Figure 7: Distribution of the CI in workspace when R is specified

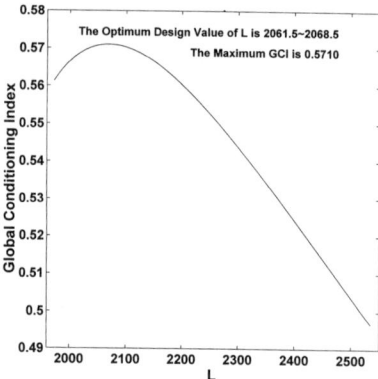

Figure 8: The relationship between the GCI and L

Additionally, in this design, width of each of the four plates is increasing from the bottom end to the top, which can also increase the system's stiffness. As shown in Fig.9, each leg consists of one four-bar parallelogram. Letting each leg comprise two or more four-bar parallelograms, i.e., increasing the system's redundant constraints, will improve the system's stiffness largely, at the same time the accuracy of manufacture is increased. In order to avoid the interference between the upper plate and the lower plate, the moving platform is design as trapezoidal profile, as shown in Fig.9.

One can see that, in this parallel mechanism design, all joints are single-DOF ones, which means that it is not difficult to fabricate all joints with high accuracy and fine tolerance.

The new hybrid machine tool is developed in associate with the Second Machine Tool Works of QiQiHaer as shown in Fig.10. Now the device is under test to machine a kind of vane. Other properties about this machine tool will be reported in the future work.

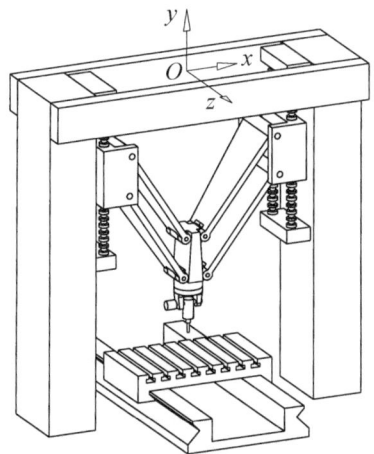

Figure 9: A new 5-axis hybrid machine tool

Figure 10: The developed 5-axis hybrid machine tool

8. Conclusion

Recently, the parallel kinematics machine with less than 6 DOFs becomes more attractive. In this paper, a novel planar 2-DOF parallel mechanism is proposed, the design theory is studied in detail. And a new 5-axis hybrid machine tool is presented based on the 2-DOF parallel mechanism, motivated by the application in the Second Machine Tool Works of QiQiHaer in China. The machine tool has many advantages, such as high speed, high stiffness, high mobility, high accuracy, and being capable of long components manufacturing. We believe that the machine tool will be more interesting in the manufacturing.

ACKNOWLEDGMENT

This research is sponsored by the State High-Technology Development Program of China (No. 2002AA421180).

References

[1] D. Stewart, A platform with six degree of freedom, In: *Proc. Inst. Mech. Engineering*, Vol.180, London, 1965, pp.371-386.

[2] T.Moriwaki, Survey of R&D Activities Related to Parallel Mechanisms in Japan, In: C.R. Boer, L. Molinari-Tosatti and K.S. Smith (editors), *Parallel Kinematic Machines*, Springer-Verlag London Limited, 1999, pp.431-440.

[3] Xin-Jun Liu, J. Wang, F. Gao and L.-P. Wang, On the analysis of a new spatial three degrees of freedom parallel manipulator, *IEEE Transactions on Robotics and Automation*, Vol.17, No.6, 2001, pp.959-968.

[4] Bruno Siciliano, The Tricept robot: inverse kinematics, manipulability analysis and closed-loop direct kinematics algorithm, *Robotica*, Vol.17, 1999, pp.437-445.

[5] H.K. Tonshoff, H. Grendel and R. Kaak, Structure and characteristics of the hybrid manipulator Georg V, In: C.R. Boer, L. Molinari-Tosatti and K.S. Smith (editors), *Parallel Kinematic Machines*, Springer-Verlag London Limited, 1999, pp.365-376.

[6] Lung-Wen Tsai, Multi-degree-of-freedom mechanisms for machine tools and the like, US Patent US5656905, 1997.

[7] D. McCloy, Some comparisons of serial-driven and parallel driven manipulators, *Robotica*, Vol.8, 1990, pp355-362.

[8] C.M. Gosselin, Kinematische und statische analysis eines ebenen parallelen manipulators mit dem freiheitgrad zwei, *Mechanisms and Machine Theory*, Vol.31, No.2, 1996, pp.149-160.

[9] C.M. Gosselin, The optimum design of robotic manipulators using dexterity indices, *Roboitcs and Autonomous Systems*, Vol.9, 1992, pp.213-226.

[10] J.K. Salisbury, J.J. Graig, Articulated hands: force control and kinematic issues, *The International Journal of Robotics Research*, Vol.1, No.1, 1982, pp.4-17.

[11] C. Gosselin, Stiffness Mapping for Parallel Manipulators, *IEEE Transactions on Robotics and Automation*, Vol.6, No.3, 1990, pp.377-382.

Psychological and Social Effects of Robot Assisted Activity to Elderly People who stay at a Health Service Facility for the Aged

Kazuyoshi Wada[*1,2], Takanori Shibata[*1,3] Tomoko Saito[*1], Kazuo Tanie[*1,2]

*1 Intelligent Systems Institute, AIST
1-1-1 Umezono, Tsukuba, Ibaraki, 305-8568 Japan
*2 Institute of Engineering Mechanics, University of Tsukuba
*3 PRESTO, JST
{k-wada, shibata-takanori, tomo-saito, tanie.k}@aist.go.jp

Abstract

We have been developing mental commit robots that provide psychological, physiological, and social effects to human beings through physical interaction. The appearances of these robots look like real animals such as cat and seal. The seal robot was developed especially for therapy. We have applied seal robots to assisting activity of elderly people at a health service facility for the aged. In order to investigate psychological and social effects of seal robots to the elderly people, we evaluated elderly people's moods by face scales (which express person's moods by illustration of person's faces) and Profile of Mood States (which measures person's moods by questionnaires). Seal robots were provided into the facility for three weeks. As the results, feelings of elderly people were improved by interaction with the seal robots.

1. Introduction

Due to improvement of our living environment, dietary life and progress of medical, we have obtained the longest life in our history [1]. However, in most advanced countries, the number of elderly people who need nursing because of dementia, bedridden, and so on, has been increasing. Then, there are many people who stay in an elderly institution for long time, until the end of their life. Moreover, nursing staff's body and mental poverty by manpower shortage and increasing of load is becoming a big problem. Especially, mental stress of nursing causes Burnout syndrome [2]. It makes nursing staff into irritation and losing sympathy to patients. Therefore, it is important to improve "quality of life (QOL)" of elderly people because this helps them to spend their life healthily and independently. It also saves social cost for elderly people.

It is said that interaction with animals heal human mind from many years ago. Its effects are applied to medical. Especially in the United States, animal assisted therapy and activity are becoming popular at hospital and nursing home [3]. A doctor or nurse makes a program for therapy. Following three effects are expected in animal assisted therapy and activity:

(1) Psychological effect (e.g. relaxation, motivation)
(2) Physiological effect (e.g. improvement of vital sign)
(3) Social effect (e.g. activation of communication among inpatients and caregivers)

In addition to these effects, animal assisted therapy at nursing homes brings effect of rehabilitation to elderly people who have decreased his moving ability, and offers laughter and enjoyment to a patient who has few remainders of his life [4]. Moreover, there are some cases that the therapy improved state of elderly people who were dementia.

However, most hospitals and nursing homes, especially in Japan, don't accept animals even though they admit effects of animal assisted therapy and activity. They are afraid of negative effects of animals to human beings such as allergy, infection, bite, and scratch.

We have been building animal type robots as examples of artificial emotional creatures [5-15]. The animal type robots have physical bodies and behave actively while generating goals and motivations by themselves. They interact with human beings physically. When we engage physically with an animal type robot, it stimulates our affection. Then we have positive emotions such as happiness and love, or negative emotions such as anger and fear. Through physical interaction, we develop attachment to the animal type robot while evaluating it as intelligent or stupid by our subjective measures. In this research, animal type robots that give mental value to human beings are referred to as "mental commit robot." We have developed cat robot and seal robot as the mental commit robot.

We have applied seal robots as substitution of real animals to therapy of children at a university hospital [12].

This was referred to as robot-assisted therapy (RAT). Moods of children were improved by interaction with the robot. Moreover, the robot encouraged children to communicate with each other and caregivers. In one striking instance, a young autistic patient recovered his appetite and his speech abilities during the weeks when the robot was at the hospital. In another case, nurses noted the rehabilitative benefits for a long-term patient, unable to leave her bed, who was willing to stroke and pet the animal.

In addition, we have applied seal robots to robot-assisted activity (RAA) for elderly people who used a day service center [13-15]. The day service center is an institution that aims to decrease nursing load of a family by keeping elderly people in daytime. The robots improved their moods and brought vigor to them. Moreover, nursing staff's mental poverty decreased because the elderly people spent their time by themselves with the robots.

In this paper, we applied seal robots to assist activity of elderly people at a health service facility for the aged, in order to investigate psychological and social effects of seal robots to the elderly people who stayed at the facility. Then, we compared with effects of the seal robot and those of a placebo seal robot that was changed its motion generation program.

Chapter 2 explains a seal robot and placebo seal robot that were used for RAA. Chapter 3 describes ways of experiments and explains the effects of RAA to elderly people. Chapter 4 discusses current results of RAA and future works. Finally, chapter 5 concludes this paper.

2. Seal Robot and Placebo Seal Robot

2.1. Specifications of Seal Robot

Seal robot, Paro was developed to have physical interaction with human beings (Fig.1). Paro's appearance is from a baby of harp seal, which has white fur for three weeks from its born. As for perception, Paro has tactile, vision, audition, and posture sensors beneath its soft white artificial fur. In order for Paro to consist of a soft body, a tactile sensor was developed and implemented. As for action, it has seven actuators; two for each eyelids, two for neck, one for each front fin, and one for two rear fins. Weight of Paro is about 3.0 [kg].

Paro has a behavior generation system that consists of hierarchical two layers of processes: proactive and reactive processes. These two layers generate three kinds of behaviors; proactive, reactive, and physiological behaviors:

(1) Proactive Behaviors: Paro has two layers to generate its proactive behaviors: behavior-planning layer and behavior-generation layer. Considering internal states, stimuli, desires, and a rhythm, Paro generates proactive behaviors.

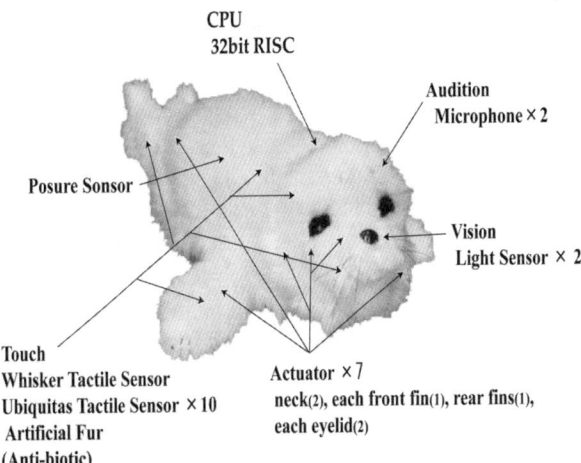

Fig.1 Seal Robot "Paro"

(a) Behavior-planning layer: This has a state transition network based on internal states of Paro and Paro's desire produced by its internal rhythm. Paro has internal states that can be named with words of emotions. Each state has numerical level and is changed by stimulation. The state decays by time. Interaction changes internal states and creates character of Paro. The behavior-planning layer sends basic behavioral patterns to behavior-generation layer. The basic behavioral patterns include some poses and some motions. Here, although "proactive" is referred, proactive behaviors are very primitive compared with those of human beings. We implemented similar behaviors of a real seal into Paro.

(b) Behavior generation layer: This layer generates control references for each actuator to perform the determined behavior. The control reference depends on strength of internal states and their variation. For example, parameters change speed of movement, and the number of the same behavior. Therefore, although the number of basic patterns is countable, the number of emerging behaviors is uncountable because numeral parameters are various. This creates living like behaviors. In addition, as for attention, the behavior-generation layer adjusts parameters of priority of reactive behaviors and proactive behaviors based on strength of internal states. This function contributes to situated behavior of Paro, and makes it difficult for a subject to predict Paro's action.

(c) Long-term memory: Paro has a function of reinforcement learning. It has positive value on preferable stimulation such as stroked. It also has negative value on undesirable stimulation such as beaten. Paro put values on relationship between stimulation and behaviors. Gradually, Paro can be shaped to preferable behaviors of its owner.

(2) Reactive behaviors: Paro reacts to sudden stimulation. For example, when it hears big sound suddenly, Paro pays attention to it and looks at the direction. There are some patterns of combination of stimulation and reaction. These patterns are assumed as conditioned and unconscious behaviors.

(3) Physiological behaviors: Paro has a rhythm of a day. It has some spontaneous desires such as sleep based on the rhythm.

2.2. Specifications of Placebo Seal Robot

We often experience that we lose interest in toys when we found its mechanism. Therefore, we consider following hypothesis:

The robots that execute only defined simple motions are predicted its motions by people, and they lose interest in the robots. Moreover, the robots also lose its effects to the people.

According to this hypothesis, we changed regular Paro's program, and made placebo Paro as follows.

Proactive behaviors: repetition of following five kinds of actions.
(1) Blink
(2) Swing rear fins to right and left
(3) Swing both front fins to forward and backward
(4) Swing head to right and left
(5) Cry → Return to (1)

Reactive behaviors: following simple reactions against stimuli.
(1) Cry (sound is different from proactive motion's cry)
(2) Raise head

3. Robot Assisted Activity for Elderly People

We applied Paro to robot-assisted activity for elderly people at a health service facility for aged in order to investigate its effects on elderly people. The health service facility for aged is an institution that provides several services, such as stay in the institution, day care and rehabilitation to elderly people. People who need nursing can stay in there during a certain period. In order to rehabilitate into society, they are provided daily care and trained to be able to spend their daily life independently during their staying at the institution. When we started experiment at the institution, about 100 elderly people were staying in there. Moreover, about 30 people of them were dementia. People who were not dementias stayed in A and B building. On the other hand, people who were dementia stayed in C building, and they were isolated from other people.

Before starting the robot-assisted activity, we explained the purposes and ways of the experiment to elderly people

Table 1 Basic Attribute of 23 Subjects

	A	B
Total number of people	12	11
Male	4	2
Female	8	9
Age(AV±SD)	84.6±7.0	85.5±5.4

Fig.2 Interaction between Elderly People and Paro

who stayed A and B building, and received their approval. Symptoms of the elderly people who approved the investigation were various with different reasons (no answer to questionnaires, bedridden, etc). Some people were impossible to be investigated. Then, a nursing staff that knew usual states of the elderly people well evaluated them, and decided who could be investigated. After the evaluation, the number of subjects was 23. 12 subjects stayed in A building, and 11 subjects were in B building. Their basic attributes are shown in Table 1.

3.1. Ways of activity

Regular Paro was provided to the subjects who stayed in B building, and placebo Paro was provided to the subjects who stayed in A building. In order to prevent that subjects of each group interact with other group's Paro, they interacted with each group's Paro in different place in the institution. Moreover, we kept the existence of two kinds of Paro secret from subjects. Each groups interacted with each Paro about one hour at a time, four days a week for three weeks. We prepared a desk to set Paro in the center of people, and the subjects were arranged up as shown Fig.2. However, all the subjects couldn't interact with Paro at the same time. Therefore, we moved Paro among subjects in turn, and we made each subject's interaction time with Paro to be same.

3.2. Ways of evaluation

In order to investigate elderly people's moods before and after introduction Paro to the institution, the following two kinds of data and extra information were collected.
(1) Face scale [16] (Fig.3)
(2) Profile of Mood States (POMS) [17]
(3) Comments of nursing staffs

The Face Scale contains 20 drawings of a single face, arranged in serial order by rows, with each face depicting a slightly different mood state. A graphic artist was consulted so that the faces would be portrayed as genderless and multiethnic. Subtle changes in the eyes, eyebrows, and mouth were used to represent slightly different levels of mood. They are arranged in decreasing order of mood and numbered from 1 to 20, with 1 representing the most positive mood and 20 representing the most negative mood. As the examiner pointed at the faces, the following instructions were given to each patient: "The faces below go from very happy at the top to very sad at the bottom. Check the face which best shows the way you have felt inside now."

POMS is one of popular questionnaires, which measures person's moods [17]. POMS is used in various research fields such as medical therapy and psychotherapy. It can measure six mood states at the same time: Tension-Anxiety, Depression-Defection, Anger-Hostility, Vigor, Fatigue, and Confusion. It has 65 items concerning moods. Each item was evaluated by five stages of 0-4: 0 = not at all, 1 = a little, 2 = moderately, 3 = quite a bit, and 4 = extremely. 58 of 65 items are classified into the six mood states, and we calculate total scores of each mood states. (Note: 7 items are dummy items) Then, we translate the total scores into standard scores by using special table.

Moreover, we investigated familiarity with Paro for once a week by questionnaires. The questionnaires have 3 items: I like Paro, I speak to Paro, and Paro is like a child or grandchild for me. These items were evaluated by five stages: 0 = not at all, 1 = a little, 2 = moderately, 3 = quite a bit, and 4 = extremely.

3.3. Results of evaluation

The face scale and POMS were applied to subjects, a week before introduction of Paro, 2nd and 3rd week after introduction.

As for face scale, we obtained data from 7 people of regular Paro group, and from 12 people of placebo Paro group. Fig.4 shows average face value. Average scores of regular Paro group decreased from about 9.0 (before introduction) to 7.0 (3rd week). Moreover, placebo Paro group's average scores also decreased from about 7.0 (before introduction) to 6.3 (3rd week). Therefore, interaction with regular and placebo Paro improved mood of subjects.

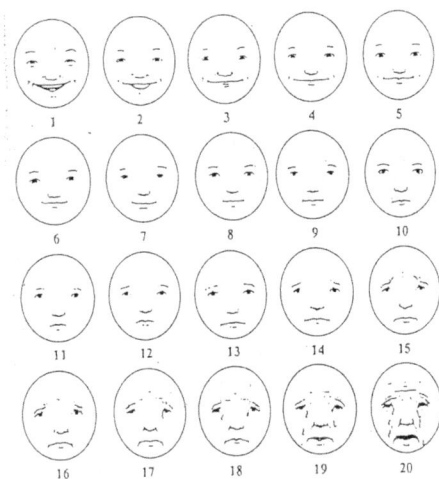

INSTRUCTIONS: The faces above go from very happy at the top to very sad at the bottom. Check the face which best shows the way you have felt inside now

Fig.3 Face Scale

As for POMS, we obtained data from 3 people of regular Paro group, and from 9 people of placebo Paro group. Fig.5 shows average standard scores of Depression-Dejection. Here, 50 standard points means average score of Depression-Dejection of over 60 years old Japanese people. Average standard scores of regular Paro group decreased from about 61 (before introduction) to 47 (3rd week). Moreover, placebo Paro group's average standard scores also decreased from about 58 (before introduction) to 51 (3rd week). Therefore, interaction with regular and placebo Paro improved depression and dejection of subjects.

As for other factors such as "Tension-Anxiety", "Anger-Hostility", "Fatigue" and "Confusion" also decreased. However, these scores didn't change as large as "Depression-Dejection". As for "Vigor", its scores decreased. We think that it means people relaxed and calmed down by interaction with Paro.

As for comments and observations of nursing staffs, both groups of subjects were waiting for Paro and participated interaction with Paro willingly. Paro increased their laughing, and encouraged subjects to communicate with each other and nursing staffs. In an interesting instance, an elderly man who was fastidious and difficult to communicate with other people, sang songs to Paro with big voice many times, and he made other people laughing. Another elderly made a song of Paro and sang it to Paro.

As for questionnaires concerning familiarity with Paro, We obtained data from 9 people of regular Paro group, and from 12 people of placebo Paro group. Fig.6 shows results of average scores of "I like Paro" that was one of the questionnaires' items. Average score of regular Paro group decreased from about 3.0 to 1.5. On the other hand, average score of placebo Paro group kept high value, about 3.0 for three weeks. As a statistic analysis, we applied Friedman's test to the change of score of each group. As a result, a significant change was seen in change of score of regular Paro group ($p < 0.05$). Therefore, subjects of placebo Paro group didn't lose interest in placebo Paro. On the other hand, regular Paro group's interest in their Paro decreased. As for other items such as "I speak to Paro" and "Paro is like a child or grandchild for me", regular Paro group's average scores of each item were 2.0, and placebo Paro group's average scores of those were 3.0 for 3 weeks. However, there were no significant differences and changes in scores of both groups.

4. Discussions

We investigated the effects of Paro on elderly people who were staying in health service facility for the aged. Then, we compared the effects by the regular Paro with those by a placebo Paro. Against our expectation, face scale scores of regular and placebo Paro groups improved, and their standard scores of Depression-Dejection of POMS decreased after introduction of Paro. From these results, regular and placebo Paro improved elderly people's moods. Especially, Paro was effective to their depression.

Moreover, we investigated familiarity with Paro by questionnaires. The results were interesting. As for a question item "I like Paro", average score of regular Paro group decreased. On the other hand, average score of placebo Paro group kept high value. Therefore, subjects of placebo Paro group didn't lose interest in placebo Paro, and regular Paro group's interest in their Paro decreased.

Before experiment, we expected that people would lose interest in placebo Paro, because its reaction was very simple. However, our expectation was wrong. Subjects of placebo Paro group kept interaction with their Paro, and they didn't notice that placebo Paro's reaction was simple.

From these results, following 3 questions occur to us:
Q1. Was placebo Paro really liked more than regular Paro by subjects?
Q2. Why didn't subjects lose interest in placebo Paro?
Q3. Why did regular Paro group's interest in their Paro decrease?

As for Q1, there were some differences between subjects of regular and placebo Paro groups, and we couldn't compare with regular and placebo Paro simply. For example, there was a man who made people excited, and placebo Paro group was more independent than regular Paro group. However, we couldn't randomize the subjects because of limitation of the institution.

As for Q2, we consider following 2 reasons:

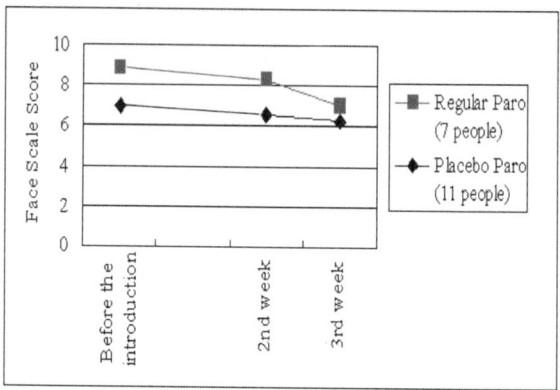

Fig.4 Average Face Scale Scores of Elderly People for 4 weeks

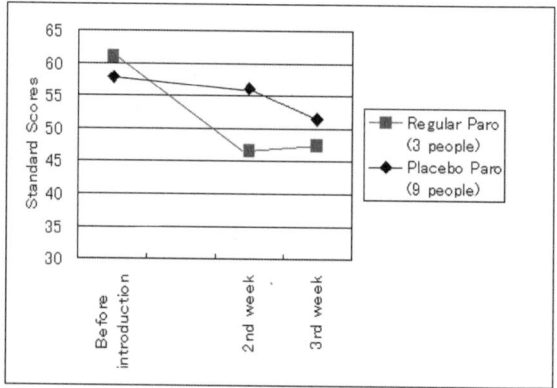

Fig.5 Average Standard Scores of "Depression-Dejection" of POMS of Elderly People for 4 weeks

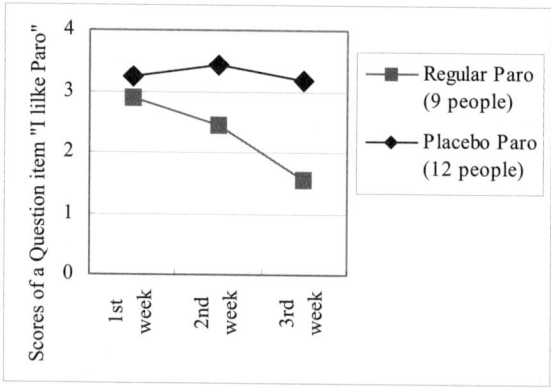

Fig.6 Average Scores of a Question item "I like Paro" of Elderly People for 3 weeks

(1) It was difficult for subjects to notice that placebo Paro's reaction was one pattern.

Subjects interacted with Paro in two or more people at the same time. Therefore, each subject's interaction time with Paro was not as long as they could notice that its reaction was one pattern.

(2) Reaction that cry and raise its head had special meanings.

Some subjects said "good boy" when Paro raised its head. They felt that Paro answered their calling.

As for Q3, we consider that people might felt that regular Paro was impolite, because it has too various reactions (include neglect) against stimuli.

In order to clarify these points, we will carry out experiments that use more number of Paro, and compare the effects of Paro, those of the placebo Paro and those of another placebo Paro (ex. swing its head to right and left against stimuli) to same subject.

In this research, we used questioners, POMS, because it can measure six mood states accurately. However, it had many items, and some subjects refused to answer it with passage of time. We will make more simple questioners that measure moods of elderly people.

5. Conclusions

We applied seal type mental commit robots, Paro to robot-assisted activity for elderly people at a health service facility for the aged. The experiment was carried out for 4 weeks in total. Then, we compared the effects by the regular Paro with those by a placebo Paro. The results show that interaction with regular and placebo Paro has psychological effect and social effect to elderly people.

Physiologically, we used urinary tests to find that robot-assisted activity decreased stress reaction in the elderly clients. The details are described in [18].

We will have further experiments and research in different conditions and situations. Moreover, we will investigate relationship between functions of a mental commit robot and its effects to elderly people in robot-assisted activity.

References

[1] UN, World Population Prospects: The 1996 Revision
[2] C. Maslach, Burned-out, Human Behavior, Vol.5, No.9, pp. 16-22, 1976.
[3] M. M. Baum, N. Bergstrom, N. F. Langston, L. Thoma, Physiological Effects of Human/Companion Animal Bonding, Nursing Research, Vol. 33. No. 3, pp. 126-129 (1984)
[4] J. Gammonley, J. Yates, Pet Projects Animal Assisted Therapy in Nursing Homes, Journal of Gerontological Nursing, Vol.17, No.1, pp. 12-15, 1991.
[5] T. Shibata, et al., Emotional Robot for Intelligent System - Artificial Emotional Creature Project, Proc. of 5th IEEE Int'l Workshop on ROMAN, pp. 466-471 (1996)
[6] T. Shibata and R. Irie, Artificial Emotional Creature for Human-Robot Interaction - A New Direction for Intelligent System, Proc. of the IEEE/ASME Int'l Conf. on AIM'97 (Jun. 1997) paper number 47 and 6 pages in CD-ROM Proc.
[7] T. Shibata, et al., Artificial Emotional Creature for Human-Machine Interaction, Proc. of the IEEE Int'l Conf. on SMC, pp. 2269-2274 (1997)
[8] T. Tashima, S. Saito, M. Osumi, T. Kudo and T. Shibata, Interactive Pet Robot with Emotion Model, Proc. of the 16th Annual Conf. of the RSJ, Vol. 1, pp. 11, 12 (1998)
[9] T. Shibata, T. Tashima, and K. Tanie, Emergence of Emotional Behavior through Physical Interaction between Human and Robot, Procs. of the 1999 IEEE Int'l Conf. on Robotics and Automation (1999)
[10] T. Shibata, T. Tashima, K. Tanie, Subjective Interpretation of Emotional Behavior through Physical Interaction between Human and Robot, Procs. of Systems, Man, and Cybernetics, pp. 1024-1029 (1999)
[11] T. Shibata, K. Tanie, Influence of A-Priori Knowledge in Subjective Interpretation and Evaluation by Short-Term Interaction with Mental Commit Robot, Proc. of the IEEE Int'l Conf. On Intelligent Robot and Systems (2000)
[12] T. Shibata, et al., Mental Commit Robot and its Application to Therapy of Children, Proc. of the IEEE/ASME Int'l Conf. on AIM'01 (July. 2001) paper number 182 and 6 pages in CD-ROM Proc.
[13] K. Wada, T. Shibata, T. Saito, K. Tanie, Robot Assisted Activity for Elderly People and Nurses at a Day Service Center, Proc. of the IEEE Int'l Conf. on Robotics and Automation pp.1416-1421, 2002.
[14] K. Wada, T. Shibata, T. Saito, K. Tanie, Analysis of Factors that Bring Mental Effects to Elderly People in Robot Assisted Activity, Proc. of the IEEE Int'l Conf. On Intelligent Robot and Systems (2002)
[15] T. Saito, T. Shibata, K. Wada, K. Tanie, Examination of Change of Stress Reaction by Urinary Tests of Elderly before and after Introduction of Mental Commit Robot to an Elderly Institution, Proc. of the 7th Int. Symp. on Artificial Life and Robotics Vol.1 pp.316-319, 2002.
[16] C. D. Lorish, R. Maisiak, The Face Scale: A Brief, Nonverbal Method for Assessing Patient Mood, Arthritis and Rheumatism, Vol. 29, No. 7, pp. 906-909 (1986)
[17] McNair DM, Lorr M, Droppleman LF, Profile of Mood States, San Diego: Educational and Industrial Testing Service, 1992.
[18] T. Saito, T. Shibata, K. Wada, K. Tanie, Change of Stress Reaction by Introduction of Mental Commit Robot to a Health Services Facility for the Aged, Proc. of Joint 1st International Conference on Soft Computing and Intelligent Systems and 3rd International Symposium on Advanced Intelligent Systems, paper number 23Q1-5, in CD-ROM Proc., 2002.

Therapy of Hemiparetic Walking by FES

Markus Weber, Friedrich Pfeiffer

Institute for Applied Mechanics, Technical University of Munich, Garching, Germany

e-mail: weber@amm.mw.tum.de

Abstract

Functional Electrical Stimulation (FES) is a suitable method for the therapy of hemiparetic walking. However, FES requires suitable human and muscle models in order to achieve fast improvements of the patient's gait. Furthermore, an intelligent control scheme is to be employed considering the patient's autonomous motor activity.

This paper presents the general therapeutic approach to therapy the hemiparetic gait. Further, human body and muscle models are presented and control algorithm are proposed. Concluding, results from simulations and experiments are presented and debated.

1 Introduction

Apoplexy is one of the common causes for disability, characterized by disturbance of memory and paralysis mainly on one side (hemiplegia). The associated social isolation and the lack of mobility imply research for suitable methods in therapy of hemiparetic walking.

Besides physiotherapy [11, 12], the artificial agitation of musculature by Functional Electrical Stimulation is more and more used for therapy.

The theory of FES and its application is already known in the field of paraplegia, where a desired movement of the patient (e.g. standing up) is achieved by stimulating the atrophied muscles using a prosthesis [9, 13]. In this respect, the necessity of appropriate models for the human body including the muscles could be shown. The emerging parameters are mostly determined by regression equations [5] or calculated by preceding measurements [2].

In terms of movement control several approaches are proposed. [8] gives an overview of different designs of feedback control. In [1] a model-based, adaptive control is used, whereas in [4] a combination of learning and PID control is presented. Furthermore, papers treating the application of neuronal networks [7] and finite state control [6] can be found in literature.

In this paper a therapy method for hemiparetic walking employing FES is described. The general approach is presented in sector 2. Modelling and controller design are described in sectors 3 and 4. Simulation and experimental results referring to the human-muscle model and control design are presented in sector 5. The paper concludes with a discussion about future work.

2 General conception

The main idea of FES-based gait therapy on hemiparetic walking is to enable the patient to activate his muscles at the right time of the gait cycle by giving him a haptic feedback. By the time, the patient relearns the natural gait. In order to achieve fast improvements, a combination of physiotherapy and muscle stimulation is used in order to restore the patients unassisted mobility within 4-5 weeks.

A patient who has suffered an apoplexy gets a physiotherapeutic treatment to restore the flexibility of the joints and the muscles (ca. 1-2 weeks). In the following, a stimulation of the muscles is exerted to the person. For therapy with FES, electrodes are fixed on the patient's skin. Computed stimulation parameters (pulse width, electric current, frequency) are transmitted to the patient, so that the generated joint torque assists the patient's gait. The resulting joint angles are measured by an optical system based on markers and sent back to the control. In the following, the control algorithm alters the stimulation parameters to improve the patient's gait (Figure 1).

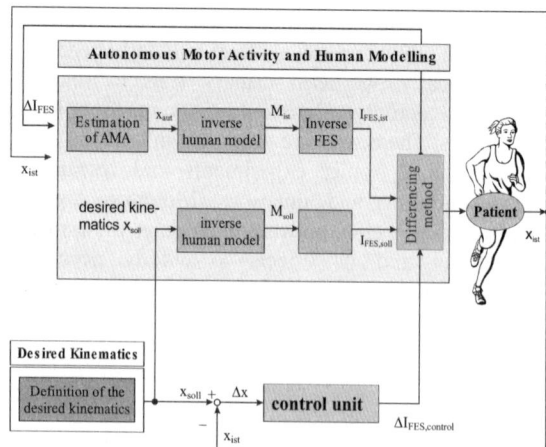

Figure 1: schematic structure for FES

In detail, the therapy of hemiparetic gait is divided in the estimation of the autonomous motor activity (AMA), the definition of the reference trajectory, the human and muscle models and the differencing method (Figure 1).

In contrast to paraplegia patients, who get no nerval feedback because of the lesion in the spinal column, the hemiparetic patient has a remain motor activity after the 2 weeks physiotherapy. This (AMA) has to be considered by computing the parameters for stimulation. In order to get the AMA, the patient's movement without stimulation is measured every day of therapy. These measurements also serve as a basis to gauge the patient's progress. The therapist can estimate the patient's AMA by regarding the interference with FES and the patient's state (e.g. patient's fear impairs the actual AMA). According to the computed joint angles, the parameters for stimulation can be calculated using inverse models of human body including the muscle.

On the other hand, a desired value of the state variables (joint angles) must be found. This part of the therapy plays an important role, because the joint angles must be computed to ensure a stable gait cycle for each day of therapy. Furthermore these defined joint angles must include the patient's improvements. In this context, the therapist defines the patient's "daily learning target", e.g. the knee joint angle should be increased from 30° to 50° in the initial-swing phases of gait cycle. On the basis from the gained experiences, an automated procedure must be developed in the future. This algorithm has also to ensure an optimal devolution of the therapy by using the therapist's experience.

By knowing the reference joint trajectories, the reference parameters of stimulation can be computed, using the inverse models of human body and muscle. These and the parameters of the AMA are compared and combined in the differencing method. This method ensures that only the deviation to the reference angles is stimulated.

Furthermore, a feedback control is applied in order to compensate disturbances or inaccuracies. These errors appear, as the patient's behaviour can only be estimated roughly (the patient's remain motor activity has a direct influence on the movement that can perturb the whole system behaviour). Also noise from the measuring system and other hardware components affect the system behaviour.

3 Autonomous Motor Activity and Human Modelling

3.1 Autonomous Motor Activity

The estimation of the actual AMA gives the therapist the possibility to adapt the measured AMA to the patient's constitution. Figure 2 gives more detail information about this part.

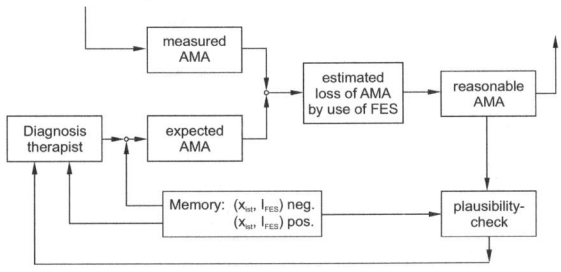

Figure 2: Design of the AMA-estimation

On the basis of the past patient's gait-therapy, the therapist estimates the actual AMA in reference to the measured AMA. In the next step, the resulting joint angles must be investigated concerning loss of AMA by the application of FES. The concluding plausibility check by the therapist ensures that mistakes in evaluating the actual AMA can be avoided.

3.2 Human Modelling

3.2.1 Dynamic Human Model

In order to achieve good results in planning the variables for stimulation, a suitable model of the human body is required. This model is used for transforming the state variables into joint torques (inverse dynamic model) and for simulating the patient's movement (direct dynamic model).

The used model is based on the theory described in [5]. The human body is described by 12 bodies (upper arm, forearm, head, Thorax-Abdomen, Pelvis, Femoral, shank, foot) with 42 DOF. [5] shows that the dynamics of head and arms are of smaller importance than the rest of the body. Additionally, these extremities are not planned to be stimulated, so that the model [5] was reduced by these bodies. The implemented human model comprises thus 8 bodies with 27 DOF.

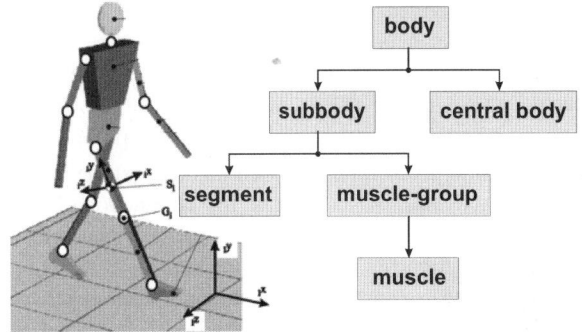

Figure 3: Human modelling [5]

According to Newton-Euler's mathematical approach, the equation of motion can be written as:

$$\sum_i (\mathbf{J}^T_{trans,i}(\dot{\mathbf{p}}_i - \mathbf{F}_i^e) + \mathbf{J}^T_{rot,i}(\dot{\mathbf{L}}_i - \mathbf{M}_i^e)) = 0 \quad (1)$$

By geometrical correlations and kinematical data, (1) can be transformed into the standard-equation of motion:

$$\mathbf{M}(\mathbf{q}) \cdot \ddot{\mathbf{q}} - \mathbf{h}(\mathbf{q},\dot{\mathbf{q}},t) = \mathbf{B} \cdot \mathbf{u} \quad (2)$$

Due to the application of object-orientated programming language, adjustments as restructuring or simplification/detailing the body system can be made easily. Future investigations will improve the model of the human body concerning the spinal column.

3.2.2 Muscle Modelling

The used muscle model is based on the models presented in [3, 9, 10]. It is divided into 3 functional parts:

the activation dynamics (AD), the contraction dynamics (CD) and the global balance (GB) (Figure 4). In order to achieve good results, the model should match real muscle properties.

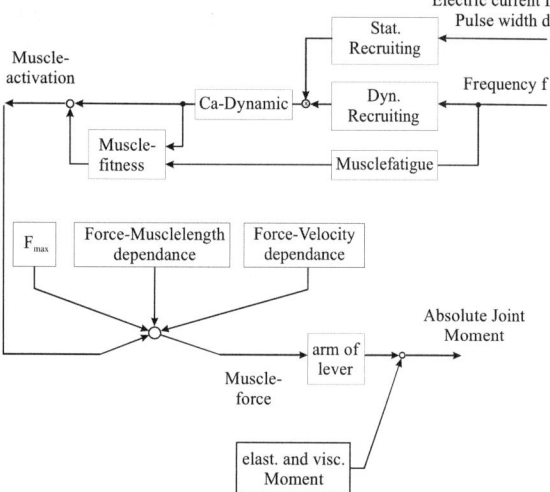

Figure 4: Muscle Model

The AD describes the relationship between the stimulation parameters (pulse width, electric current, frequency) and the activation of the muscle. This muscle-activation is a standardised value, which equals 1 when all muscle fibres are activated and thus the max. muscle force is produced. This max. force is dependant to the current angles. Analogous, it equals 0 when no fibres are activated.

The model is orientated on the physiological recruiting of the muscle. Recruiting means in this context the process that activates muscle fibres for contraction. Normally, the muscle consists of 2 types. Type I is characterized by a high barrier to release an action potential (the released pulse is transmitted by chemical messenger to the muscle fibre, which is contracting itself). Further it is characterized by a slow contraction velocity and only small fatigue. Type II has a low barrier to release an action potential, a high contraction velocity and a high fatigue.

At the natural recruiting, the muscle fibres of type I are recruited first. With an increasing force, more and more muscles of type II (Figure 5a) are recruited. Furthermore, the temporal summation (Figure 5b) enables another scaling of the complete muscle force. Thereby, the number of action potential sent to the muscle is varied.

Regarding artificial recruiting by FES, the charging level (I*d), which is sent to the stimulation electrodes, has only influence in a small region. Due to the lower barrier, only muscle fibres of type II are activated in the local area around the electrodes (Figure 5a). Moreover, it is not possible to activate different muscle fibres in the same cross-section at different times as in the natural activation (Figure 5b). Thus the frequency must be set higher for the same result. But in general, it can be shown that both natural and artificial recruiting are based

on the same logic. Thus, the larger the pulse width or the electric current is, the higher is the number of muscle fibres which are "activated" (areal summation). On the other hand a higher frequency produces a stronger activation of the muscle fibre itself (temporal summation). The order of changing pulse width, electric current and frequency is also orientated to natural procedure: The pulse width (areal recruiting) is altered firstly, secondly the electric current (areal recruiting), which is limited by the patient's threshold of pain and lastly the frequency (temporal summation), which has great influence on the fatigue.

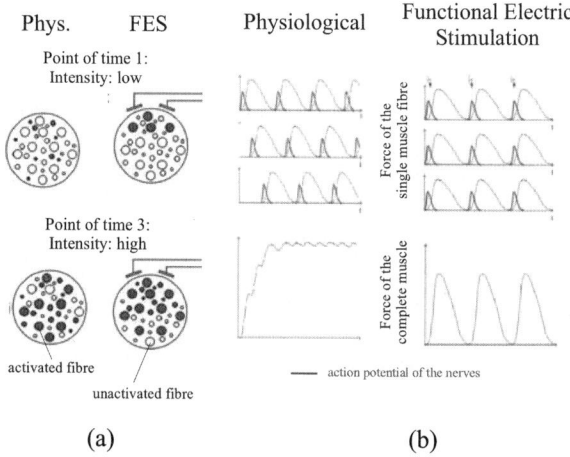

Figure 5: areal summation (a); temporal summation (b)

In the next step, the modelled Calcium-dynamics considers the natural time lag of the emission of the Ca-ionic, which initiate the muscle contraction. Furthermore, the model of the muscle-fatigue takes the decreasing muscle-fitness into account.

The CD takes the influence of joint angles and joint angular velocities on the muscle force into account. The coefficients are designed such that the max. isometric force is calculated, when the contraction velocity equals 0 and the muscle length is optimal.

In the GB, the muscle forces are mapped on the joint torques. Furthermore, elastic and viscose properties of the muscle are projected onto the joint and added to the joint torque. The elastic and viscose torques regarding each joint (e.g. knee) are given by equations [2], because not all muscles can be parameterised and a more exact torque can be computed in that way for the complete joint. After the summation of all torques concerning a joint, the resulting torque is used for computing the direct dynamics.

As both natural and artificial recruiting are based on the same theory, it is possible to compare the stimulation parameters computed from the AMA and from the reference joint angles. This comparison is made in the differencing method. It is necessary, because only the deviation of gait must be stimulated to assist the patient's gait. Principally there are two ways of differencing. Either to make the comparison on the level of the joint torques or on the level of the stimulation pulses. Considerations have shown that the differencing method on basis of the stimulation parameters gives better results. Thus the following approach could be found:

$$a * f * Q = f_{des} * Q_{des} - f_{act} * Q_{act} \quad (3)$$

Q denotes the charging level $I*d$, f the frequency. Factor a represents the factor that considers that there is an overlap regarding the muscle fibres, activated by natural and artificial stimulation.

4 Control Design for Hemiparetic Walking

In contrast to patients with paraplegia, hemiparetic persons have a remain motor activity, which has a great influence on the stimulation and thus on the movement. Additionally, there are uncertainties and inaccuracies in the human modelling. Therefore, the usage of appropriate control units is necessary. Thereby, a cyclic control that computes the whole gait cycle is realized. In the future, it is aimed to develop a real time control.

The control methods discussed are described schematically in Figure 6. Generally, the controller gets the information about the current and the desired joint angles. In the following, the control algorithm computes the joint moments to compensate the deviations from reference trajectory and measured motion. The inverse muscle-model generates the stimulation pulses, which are sent to the electrodes. The patient's motion is fed back to the control unit.

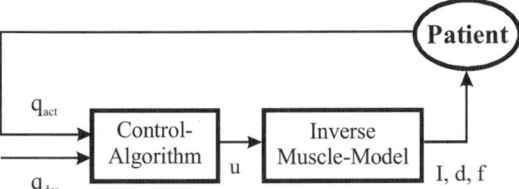

Figure 6: Schematic control design

The first, in this context discussed approaches are PID and feedback-linearisation for the whole body control. The decision for PID and feedback-linearisation is based on the fact that the system behaviour is non-linear and thus linear and non-linear control design can be compared. Additionally, whole body control means that each joint angle can be controlled by stimulation. Consequently, the joint moments to be stimulated are:

$$\mathbf{u} = c_i \left(\int (\mathbf{q}_{des} - \mathbf{q}_{act}) dt \right) + c_p (\mathbf{q}_{des} - \mathbf{q}_{act}) + \\ + c_d (\dot{\mathbf{q}}_{des} - \dot{\mathbf{q}}_{act}) \quad (4)$$

for the PID control, and

$$\mathbf{u} = \mathbf{B}^{-1}(\mathbf{M} \cdot \mathbf{v} - \mathbf{h}(\dot{\mathbf{q}}, \mathbf{q}, t)) \quad (5)$$
$$\mathbf{v} = \ddot{\mathbf{q}}_{des} + \lambda_1 (\dot{\mathbf{q}}_{act} - \dot{\mathbf{q}}_{des}) + \lambda_2 (\mathbf{q}_{act} - \mathbf{q}_{des}) \quad (6)$$

for the feedback-linearisation according to the human model described in (2). Matrix B has full rank, and therefore it is invertible. Considering vector $\vec{v}$ leads to a linear system behaviour. The influence of the muscle model according to this performance will be discussed in the following sector.

Due to the fact that not all joint moments can be stimulated (e.g. arms, thorax-abdomen), the realisation of a full-body-control is not possible. In principle, the control of the non-stimulated joint angles could be regarded as a kind of natural control, but the lack of proving experiments and the limited patient's capability to control the affected side require a further control approach.

Another in this paper described control concept is based on the consideration that non-stimulated joint angles are given as constraints. Consequently, the equation of motion becomes:

$$\mathbf{M} \cdot \ddot{\mathbf{q}} - \mathbf{h}(\dot{\mathbf{q}}, \mathbf{q}, t) + \mathbf{W} \cdot \lambda = \mathbf{B} \cdot \mathbf{u} \quad (7)$$
$$\mathbf{q} = \mathbf{J} \cdot \mathbf{q}_{min} + \mathbf{q}_{rest} \quad (8)$$

Thereby, $W^T \lambda$ is the joint moment that complies the constraint condition, u is the vector of the stimulated joint moments. Additionally, the reduced control Matrix $J^T B$ is invertible and the already presented control approaches (PID, Feedback Linearisation) can be used (4ff), replacing the actual and the reference state variables and its derivations by the stimulated (q_{min}) ones.

Further improvements of the control can be achieved by compensating the time lags. These time delays emerge due to the use of hardware components, which are connected by a can bus-system. Furthermore, simulation results show that the inverted Ca-Dynamics leads to instabilities in the muscle model (section 5). The neglect of the Ca-Dynamics produces a further time lag. Thus, the used estimator is defined by a combination of an extrapolation of the current joint angles and the desired state after the time lag. This combination is used, as better results can be achieved compared to using only the current joint angle. The estimated joint angles, which are computed by the estimation algorithm (9) are transferred to the control unit.

In terms of extrapolation, a quadratic approximation was used. Therefore the computed state variable is set to:

$$\mathbf{q}_{estimate} = a * (\mathbf{q}_{cur} + \dot{\mathbf{q}}_{cur} * t_{del} + 1/2 * \ddot{\mathbf{q}}_{cur} t_{del}^2) + \\ + (1-a) * \mathbf{q}_{ref, del} \quad (9)$$

In this context, t_{del} is the time lag, the indices *del* means delayed, *ref* means reference and *cur* means the current state. Factor a is dependant to the current deviation of joint angle.

Simulation results comparing the different control unit approaches are presented in the following sector.

5 Simulation and Experimental Results

5.1 Human modelling

Although the muscle models described in subsection 3.2.2 is proposed widely in literature, confirming experiments must be done. The used parameters for describing the muscle model are acquired by statistic measurements. Thus, there is an deviation in these parameters as different patients (age, paralysis) are stimulated. Therefore, a single joint stimulation at the knee flexion was performed as rectangle function (Figure 7). The desired motion was provided by 2 antagonistic groups of muscles: Vasti (Knee-Extension) and Hamstring (Knee-Flexion). Moreover, the stimulated subjects were normal persons without paralysis. The generated movement was measured by goniometers and the stimulation was open looped, as only human and muscle models had to be verified.

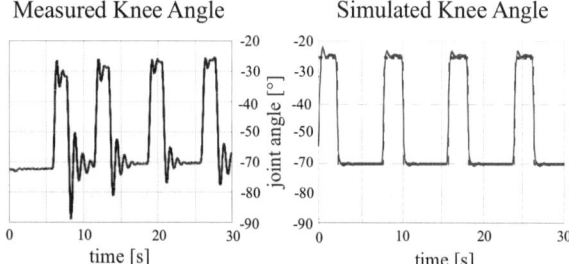

Figure 7: Comparison Simulation Measurements

Comparing measured and simulated movement, the amplitude and frequency of the movement were identical. Exceptions are the overshoots in the measured motion, which are not seen in simulation. This discrepancy cannot be justified in the muscle or human model, as the used stimulation pattern had a similar continuous devolution alike the joint angle devolution. The overshoots are caused by the proband's reaction on the changing stimulation parameters. An attenuation of the overshoots can be seen after some seconds. The proband got used to the stimulation pattern, so that the influence of the proband's motor activity on the knee movement decreased. In order to avoid these overshoots, an adequate control, regarding the patient's AMA must be used (Section 4).

5.2 Control

Before applying the control scheme, a stable system behaviour must be guaranteed to enlarge the patient's confidence in the therapy and thus aid the patient's recovery. Therefore the control algorithm is simulated with respect to the influence of the muscle model and the stability of the control. Furthermore, a comparison of the approaches presented in sector 4 is made. Concluding, the improvements due to the estimator are shown.

The feedback-linearisation (fl) enables to apply the methods of linear control design. Therefore, an exponential reduction of the gait-deviation can be achieved. As this fl doesn't consider the inverse muscle model, the deviation of the reference trajectory is reduced more slowly (Figure 8).

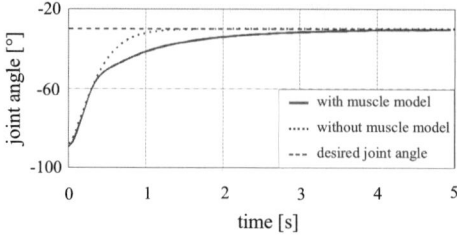

Figure 8: Influence of the Muscle Model

Different approaches in control design were illustrated in the previous sector to enable a comparison between linear and non-linear control. Therefore, a point-to-point control of the knee extension has been simulated (Figure 9). Thus, the system stability, the offset and the control response time can be derived from the devolution of the joint angles. Thereby, flexion- and extension group of muscles concerning femoral, lower-leg and foot-motion were used to generate the computed joint moments. The controller were designed to reach the desired reference angles as fast as possible. Constraints were only given by the maximal muscle force that can be produced by each muscle.

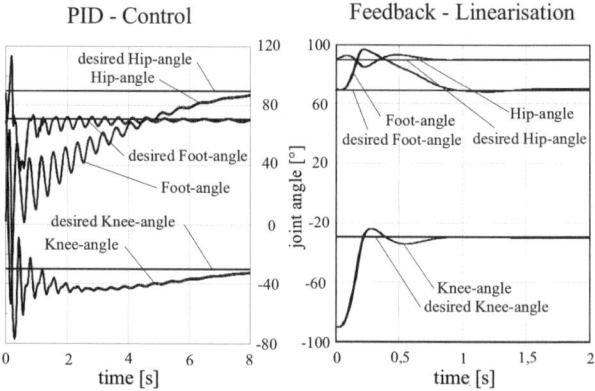

Figure 9: Comparison of the Control-Approaches

Regarding the two simulation diagrams, it is obvious that the PID control is slower than the fl control. On the other hand, the fl control is more affected by inaccuracies or modelling errors. Furthermore, the PID control is more suitable for a real-time application as it is less computational expensive.

Another aspect of control design deals with estimation. Simulation results revealed that the muscle models had to be simplified due to stability problems. Therefore, the estimator must consider the time-lag, which result from the neglected Ca-Dynamics.

In the following a comparison between the controlled variable with and without estimator is presented (Figure 10). A sinusoidal reference trajectory of the knee-joint angle is used in the simulation. Furthermore, fl control is applied and only the knee joint muscles are stimulated.

The devolution of the joint angles, which is shown in Figure 10, shows two aspects:

Firstly, if the time delay is known exactly, the estimator approximates the actual devolution of the joint angle after the time delay. Especially, the target control torque is, besides from the time delay, identical to the reference torque, if the deviation from the reference angle is negligibly.

On the other hand, the control energy is high without estimator. Oscillations occur as the controller tries to minimize the tracking error while getting the results after the time delay.

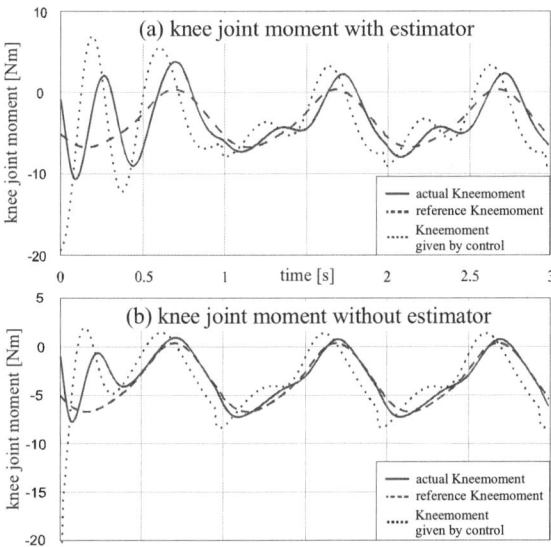

Figure 10: Control variable (a) with, (b) without estimator

6 Discussion

A method for therapy of hemiparetic gait is presented, giving more detailed information about human modelling (body, muscle) and control strategies. Moreover results from simulations and experiments are given.

Further investigations will concentrate on improving the model, the control and the gait planning. In particular, the spinal column will be detailed in the human body model. Concerning control, methods for calculation of time lag from the measured kinematics and kinetics must be found and approaches for determining the patient's autonomous motor activity must be developed.

In order to realize the gait stimulation by using the presented methods, the software must be able to compute the parameters for stimulation in real-time. Finally, a suitable tool that transforms the therapist's target specification into a gait cycle of human walking is to be developed.

7 References

[1] Sung-Nien Yu, May-Kuen Wong, "An Adaptive Control System for Electrical Stimulated Muscle: A Simulation Study", Biomedical Engineering – Applications, Basis & Communications, pp. 302 – 311, 1998

[2] T. Edrich; R. Riener; J. Quintern, " Analysis of passive elastic joint moment in paraplegics", IEEE Transactions on Biomedical Engineering, Vol. 47 (2000), No. 8, pp. 1058-1065

[3] T. Watanabe, et. al., "A Stimulus Frequency-Controlled Muscle-Model for FES Simulator", 18th Annual International Conference of the IEEE Engineering in Medicine and Biology Society, Amsterdam, pp. 551-552, 1996

[4] D.-C. Lim, S.-M. Baek, "Biped Walking by learning", Proceedings of the 32nd ISR (International Symposium on Robotics), 19-21 April 2001, pp. 1557 – 1562, 2001

[5] C. Lutzenberger, "Dynamik des menschlichen Ganges", Fortschritts-Berichte VDI, Vol. 17: Biotechnik/Medizintechnik, No. 218, 2001

[6] P.C. Sweeney, . M. Lyons, "Finite State Control of functional electrical stimulation for the rehabilitation of gait", Medical & Biological Engineering & Computing 2000, Vol. 38, pp. 121 – 126, 2000

[7] K. Y. Tong, M.H. Granat, "Gait Control System for functional electrical stimulation using neuronal network", Medical & Biological Engineering & Computing 1999, Vol. 37, pp. 35 – 41, 1999

[8] M. Ferrarin, R. Riener, J. Quintern, "Model-based Control of FES-Induced Single Joint Movements", IEEE Transactions on neuronal systems and rehabilitation engineering, Vol. 9, No. 3, pp. 245 – 257, 2001

[9] R. Riener, "Model-based development of neuroprostheses for paraplegic patients", Philosophical Transactions: Biological Sciences Vol. 354 No. 1385, pp. 877 – 894, 1999

[10] R. Riener, T. Fuhr, " Patient-Driven Control of FES-Supported Standing Up: A Simulation Study", IEEE Transactions on Rehabilitation Engineering, Vol. 6, No. 2, pp. 113-124, 1998

[11] E. Koenig, F: Müller, N. Mai, "Prinzipien der motorischen Rehabilitation und Frührehabilitation", in T. Brandt, J. Dichgans, H. C. Diener, "Therapie und Verlauf neurologischer Erkrankungen", 3. edition, pp. 941-959, Kohlhammer Verlag, 1998.

[12] K. Scheidmann, H. Brunner, F. Müller, „Sequenzeffekte in der Laufbandtherapie", Neurologie & Rehabilitation, Vol. 5, No. 4, pp. 198-202, 1999

[13] Yo-Luen Chen, Yen-chen Li, "The development of a closed-loop controlled functional electrical stimulation (FES) in gait training", Journal of medical Engineering & Technology, Vol. 25, No. 2, pp. 41 – 48, 2001

Assistance of Self-Transfer of Patients Using a Power-Assisting Device

Kiyoshi NAGAI, Isao NAKANISHI and Hideo HANAFUSA
Ritsumeikan University, 1-1-1, Noji-higashi, Kusatsu, Shiga ,JAPAN

Abstract

A power-assisting device is being developed to help older persons and people with disabilities move themselves from beds to wheelchairs. The device is intended to contribute to a more independent life for these people and to greater utilization of their latent abilities. Self-support assistance, i.e., the support provided by a power-assisting device to the person without the need for aid from a caregiver, and capability of the device to assist the movements of the people are discussed. A method for the production of power assistance and motion guidance is then proposed for the device. Finally, a test of the performance of the developed system is described.

Keywords: Self-Support, Power-Assisting Device, Transfer Motion, Power Assistance, Motion Guidance

1 Introduction

People who are unable to walk, whether because of age or other disability, often have problems when moving from a seated or reclining posture to another seated or reclining posture. Consequently, such people are often bedridden in the absence of a caregiver to help them. Furthermore, assisting such transfer motions imposes a heavy physical burden on caregivers and is liable to lead to back pain and injuries.

Many devices are already being manufactured [1] to assist caregivers in such tasks, including hoists and other lifting devices. These devices relieve caregivers of most of the physical burden involved in these tasks. However, the design of these devices still requires that people be almost constantly accompanied by caregivers. A different approach that emphasizes the people's independence has been tried in which a robot arm mounted on a rail attached to the ceiling is used to assist older persons in walking [2]. However, installing this device in existing houses is difficult because of the need to attach the rail to the ceiling. On the other hand, transfer support-equipment assisting independent transfer while moving on a floor has also been developed [3]. Although the starting and end positions of the equipment can be determined according to each user, the assisting method should be studied in more detail with the aim of providing better independent transfer assistance to its user.

We have developed a power-assisting device that can assist older persons or people with disabilities with the motions involved in transfer between posture positions [4], and have proposed a control law governing the functions of power assistance, which applies force to assist the user, and motion guidance, which leads the user through the required postures during the motion. However, the control law should be modified, since the proposed control law does not permit adequate adjustment of the functions.

In this paper, we discuss self-support assistance using a power-assisting device and describe the developed device. We propose a new method of assistance for the device to apply during the user's transfer motions. The new control law can adjust the functions of power assistance and motion guidance. We put the resulting device to the test, investigating basic functions required for the transfer motion.

2 Self-support assistance using power-assisting device

2.1 Guideline of assistance for self-support

The aim of the power-assisting device is to help older persons and people with disabilities to change positions without the need for assistance from a caregiver. These people will be termed the users of the device, and the device's forms of assistance are the assisting force, guiding motion, and keeping the user's balance. Utilizing the user's latent abilities as much as possible is important, and the device only provides assistance to the extent of the user's deficiency. Thus, the user needs to be capable of maneuvering the device. In this point, the device is different from hoists and other lifting devices that help caregivers and provide most of the force required to complete the transfer motions.

The self-support device needs to satisfy the following requirements simultaneously.

R1. The device must provide all the required assisting functions.

R2. Assisting functions should be easy for the user to use.

R1 is considered to be the primary requirement for achieving self-transfer, whereas R2 is a secondary requirement for the users to maneuver the device.

The required assisting functions are as follows:

- Power Assistance
- Motion Guidance

Power assistance means lending force to the user in motion, while motion guidance means leading the user through the required motions while helping the user to attain the appropriate postures.

The following conditions should be satisfied to ensure that the user can easily use the assisting functions.

1. The user should easily understand what to do during the motions.

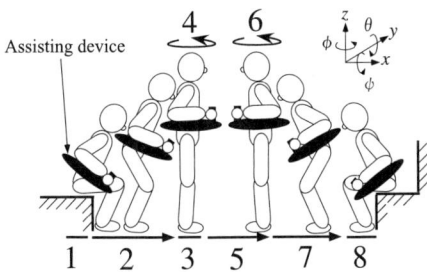

Fig. 1: Transfer motion

2. The user should be able to maneuver the device without difficulty.
3. The user should not find it easy to master the device's operation.

Meeting both the assisting functions and usability requirements simultaneously is important because the users need to maneuver the device by themselves.

2.2 Target users and transfer motions

The target users of the device fall into the following category: users who are unable to stand up by themselves, or who can only do so with great difficulty, but are able to walk with suitable support if they have some help in standing up. Our target users need help in moving from one seated posture to another seated posture (e.g., from the edge of a bed to a wheelchair). This target motion is indispensable in the daily lives of users who are unable to stand up by themselves.

The primitive motions executed when the user moves from a bed to a wheelchair are categorized in the ways shown in Fig. 1. The numbering of the items in the list below corresponds to the numbering in Fig. 1.

1. Adjustment of the initial position and posture of standing up
2. Motion of standing up
3. Standing motions and postures
4. Turning to face the target position
5. Walking to the target position
6. Turning to face away from the seat
7. Motion of sitting down
8. Adjustment of seated position and posture

In the case we are examining, steps two and six above are the foci of assistance, since both motions require the user to make significant muscular exertions with his/ her legs.

2.3 Assisting functions for self-transfer

The required assisting functions for the respective primitive motions in section 2.2 are as follows:

Standing up
1. A function through which the device can be moved by the user to a position that the user can conveniently receive the device's assisting force,
2. Power assistance that lends force to the user in motion, and motion guidance that leads the user through the required motions while keeping the user's balance,
3. Motion guidance that keeps the user's balance,

Walking
4–6. Motion guidance that leads the user through the required motions while keeping the user's balance,

Sitting down
7. A function to adjust the initial sitting position and posture, and motion guidance that reduces the downward speed of the user's body and keeps the user's balance,
8. Motion guidance that keeps the user's balance.

The methods of realizing the above functions are as follows: Power assistance is realized by a force-based control to lend a constant-direction force to the user, while motion guidance is realized by a position-based control to lead the user through the required motions while keeping the user's balance. The functions of the above list are realized with a control device operated by the user. Function 7 is realized by a position-based control to move the device that is operated by user command.

3 Developed power-assisting device

3.1 Overview

We have already designed and manufactured the power-assisting device and have described it in ref. [4]. Figures 2 and 3 show the structure and external appearance of the device, respectively.

The following specification applies to the design of the device.

A) The device must be capable of assisting the user through the sequences of primitive transfer motions.

B) The device must be suitable for installation in a four-and-a-half-mat Japanese room, which measures 2.73 [m] × 2.73 [m].

Condition A is the core requirement for the device. Condition B requires that the device be suitable for setting up in almost all existing Japanese dwellings. Thus, for maximum effectiveness, both conditions should apply.

We adopted a wire-driven mechanism capable of satisfying both conditions, since a mechanism of this type can be set on pillars in an existing room, thus avoiding the use of ceiling boards.

We adopted a structure that consists of an upper part, which is capable of motion in the horizontal plane, and a lower part, which is capable of motion in the vertical plane. The upper part mainly assists the user in walking and the lower part mainly assists the user in standing up and sitting down. This structure simplifies the design of the individual parts of the device.

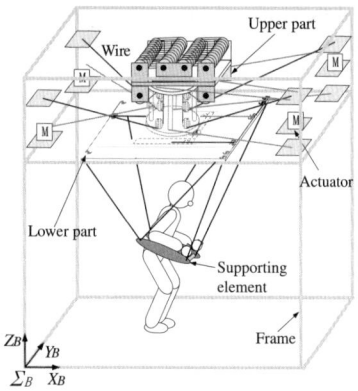
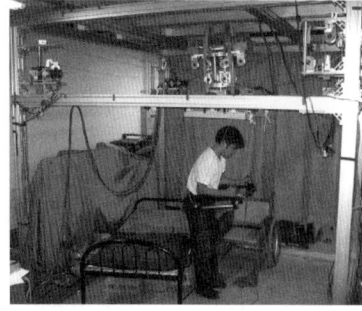
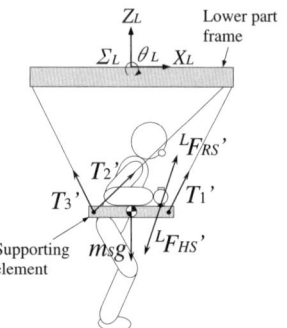

Fig. 2: A power-assisting device

Fig. 3: Photo of the power-assisting device

Fig. 4: Relationship between user and lower part of the device

To ensure that device would be suitable for setting up in existing houses, we imposed the restriction that the height of the whole device, other than the user support element, be kept in the range between 1.8 [m] and 2.4 [m]. Here, we had to design a new mechanism to avoid the placement of excessive tension on the wires because of the narrow margins. That is, we separated the upper part into a height-keeping mechanism, which stays at a fixed height, and a mechanism that moves in a horizontal plane. The latter mechanism obviates requirements for excessive tension on the wires because the height-keeping mechanism copes with the load applied to the upper part.

3.2 Motion range and feasible assisting force

In this section, we quantitatively represent the relationship between the motion range of the device and the feasible assisting force provided by the device. Figure 4 shows the relationship between the user and the lower part of the device in the vertical plane. The static dynamics relationship is given as follows:

$$^L F_{RS}' = W_S' T_L' - {}^L g_S' \quad (1)$$
$$= \tilde{W}_S' \tilde{T}' \quad (2)$$
$$\tilde{W}_S' = [W_S' W_g'] \in R^{3 \times 4} \quad (3)$$
$$\tilde{T}' = [T_{L1}', T_{L2}', T_{L3}', m_S g]^T \in R^4 \quad (4)$$
$$^L g_S' = m_S g W_g' \quad (5)$$
$$^L F_{HS}' = -{}^L F_{RS}' \quad (6)$$

where $^L F_{RS}'$, $^L F_{HS}'$, T_L' and $^L g_S'$ are the projection of the forces applied to the user by the supporting element, the forces applied to the supporting element by the user, the tensions of wires, and gravitational force onto the $X_L - Y_L$ plane, respectively. W_S' and W_g' are the transpose of the Jacobian matrices, respectively. m_S is the mass of the supporting element, and g is the acceleration due to gravity.

Here, the set of feasible assisting forces provided by the supporting element can be represented with the following equation, since the wire tensions are greater than or equal to zero and lower than or equal to the maximum tension.

$$F_{F_{RS}} = \{^L F_{RS}' | {}^L F_{RS}' = \tilde{W}_S' \tilde{T}',$$
$$0 \leq T_{Li} \leq T_{Limax} (i = 1-3)\} \quad (7)$$

The supporting element has two types of motion range, A_{r_S} and P_{r_S} shown in Fig. 5. The shaded portions in Fig. 5 are the sets of the feasible assisting forces provided by the supporting element for different postures. The supporting element can apply forces in every direction, and rank($\tilde{W}_S'$) = 3 and $T_{Li}' > 0 (i = 1 \sim 3)$ are satisfied in A_{r_S}. Thus, the posture of the supporting element can be kept by only $\tilde{T}'$. On the other hand, forces applied by the user are needed to keep the posture of the supporting element in P_{r_S}, since the supporting element can apply forces only in a specific direction. The user can easily apply this force by gripping the lever on the supporting element.

The necessary condition for self-transfer assistance is represented as follows:

$$R_{F_{RS}} \subseteq F_{F_{RS}} \quad (8)$$

where $R_{F_{RS}}$ is the set of the required assisting forces to be provided by the supporting element. Therefore, the supporting element can exist in P_{r_S} through sequences of primitive motions satisfying Eq. (8). If the supporting elements are A_{r_S} and P_{r_S}, a wide range of motion will be ensured, while avoiding a cumbersome lower part. This is in contrast with the other wire-driven mechanisms whose motion range is only A_{r_S}.

Figure 6 shows an example of the assistance process of standing up. The shaded portions in Fig. 6 are the set of feasible assisting forces provided by the supporting element in respective positions. First, the user moves the supporting element from state (1) to state (2) in Fig. 6. The forces applied by the user are

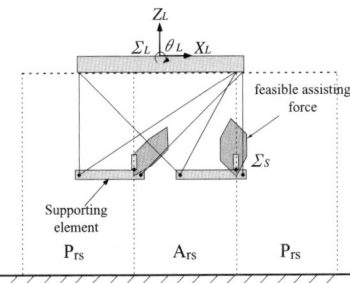
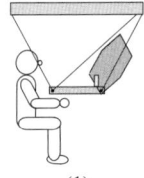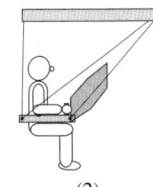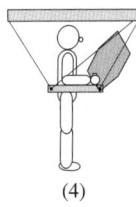

Fig. 5: Motion range of supporting element

Fig. 6: Example of the process of assistance to stand up

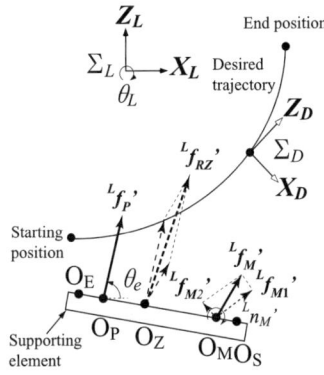

Fig. 7: Assisting forces

needed because the supporting element is moved from the posture in A_{rs} to the other posture in P_{rs}. The supporting element can be in either A_{rs} or P_{rs} in state (2), since the required assisting forces for starting to stand up are the sum of the upward forces and forward forces. Second, the user stands up while receiving assisting forces from state (2) to state (4). The required assisting forces, which are the sum of upward forces and forward forces of the motion, contain the set of feasible assisting forces. Considering the circumstances mention above, the device can assist users if the supporting element is in P_{rs} in the process of standing up. Assistance for sitting down can be represented in a similar way.

4 Assisting functions and usability

4.1 Control law for assisting functions

The device provides assisting forces to the forearm of the user with the supporting element shown in Fig. 4. In designing the control law for the device, we determined the productive mechanism of the assisting forces applied to the user by the device, and the control law is derived so as to produce the assisting forces without consideration of the dynamics of the user or the productive mechanism of muscular strength. The reasons for adopting the above approach are that modeling the dynamics of the user and the productive mechanism of muscular strength is difficult and we assume that the user should be able to maneuver the device without help.

The assisting force $^L\boldsymbol{F_{RS}}'$ provided to the user by the supporting element is the sum of the assisting forces for the two required functions mentioned in section 2.1 (Fig. 7). The forces are related as follows:

$$^L\boldsymbol{F_{RS}}' = {}^L\boldsymbol{F_{PS}}' + {}^L\boldsymbol{F_{MS}}'$$
$$= \boldsymbol{J_{PS}}'^{\mathrm{T}\,L}\boldsymbol{F_P}' + \boldsymbol{J_{MS}}'^{\mathrm{T}\,L}\boldsymbol{F_M}' \quad (9)$$

$$^L\boldsymbol{F_P}' = [{}^L\boldsymbol{f_P}'^{\mathrm{T}}, 0]^{\mathrm{T}}$$
$$= [f_P \cos\theta_e, f_P \sin\theta_e, 0]^{\mathrm{T}} \quad (10)$$

$$^L\boldsymbol{F_M}' = \begin{bmatrix} {}^L\boldsymbol{f_M}' \\ {}^Ln_M' \end{bmatrix} = \begin{bmatrix} {}^L\boldsymbol{f_{M1}}' + {}^L\boldsymbol{f_{M2}}' \\ {}^Ln_M' \end{bmatrix} \quad (11)$$

$$^L\boldsymbol{F_{RZ}}' = \boldsymbol{J_{ZS}}'^{\mathrm{T}\,L}\boldsymbol{F_{RS}}'$$
$$= \begin{bmatrix} {}^L\boldsymbol{f_{RZ}}' \\ 0 \end{bmatrix} = \begin{bmatrix} {}^L\boldsymbol{f_P}' + {}^L\boldsymbol{f_M}' \\ 0 \end{bmatrix} \quad (12)$$

where $^L\boldsymbol{F_P}'$ represents the power assisting forces, which is the translatory force from a fixed point on the supporting element O_P in the direction of angle of elevation θ_e. $^L\boldsymbol{F_M}'$ represents the motion guidance assisting forces from a fixed point on the supporting element O_M, which is the sum of the forces to lead the user through the required motions $^L\boldsymbol{f_{M1}}'$ and to keep the user's balance $^L\boldsymbol{f_{M2}}'$ and $^Ln_M'$. $^L\boldsymbol{f_{M1}}'$ is the force to track the desired trajectory in the forward direction, while $^L\boldsymbol{f_{M2}}'$ and $^Ln_M'$ are the restitutive forces, which are produced when the supporting element deviates from the desired trajectory. $^L\boldsymbol{F_{RZ}}'$ represents the assisting forces, which is the translatory force from a moving point on the supporting element O_Z. Therefore, $^L\boldsymbol{F_M}'$ and O_Z change according to the user's motion because $^L\boldsymbol{F_M}'$ is changed by the response of the supporting element.

Here, we derive the control law to produce the $^L\boldsymbol{F_{RS}}'$ applied to the user by the supporting element. The following control law can be derived using Eqs. (1) and (9), eliminating $^L\boldsymbol{F_{RS}}'$ and adding $\boldsymbol{T_{LE}}'$, which represent the tensions to prevent slack in the wires.

$$\boldsymbol{T_L}' = \boldsymbol{W_S}^{-1}({}^L\boldsymbol{F_M}' + {}^L\boldsymbol{F_{PS}}' + {}^L\boldsymbol{g_S}')$$
$$+ \boldsymbol{T_{LE}}' \quad (13)$$

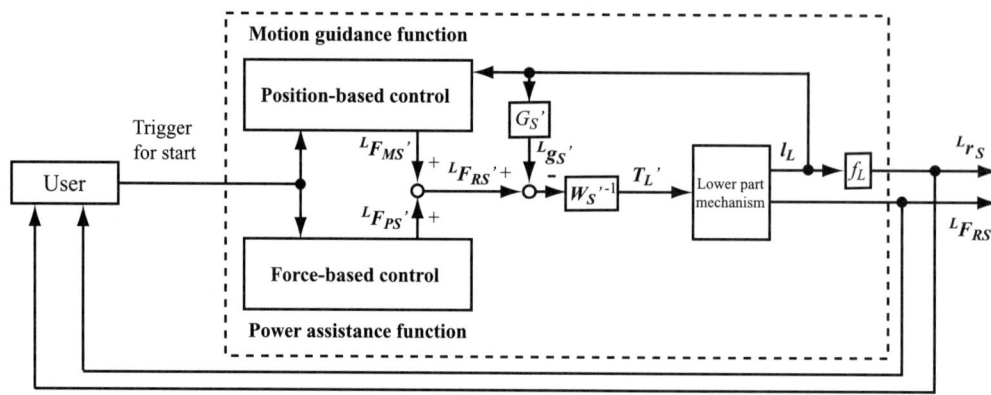

Fig. 8: Block diagram of control law

$$^LF_{MS}{'} = -D_d{'}{}^L\dot{r}_S{'} - K_d{'}(^Lr_S{'} - {}^Lr_{Sd}{'})$$
$$-K_I{'}\int_0^t(^Lr_S{'} - {}^Lr_{Sd}{'})dt \quad (14)$$

$$D_d{'} = {}^LR_D{'}{}^DD_d{}^LR_D{'}{}^T \quad (15)$$

$$K_d{'} = {}^LR_D{'}{}^DK_d{}^LR_D{'}{}^T \quad (16)$$

$$K_I{'} = {}^LR_D{'}{}^DK_I{}^LR_D{'}{}^T \quad (17)$$

$$^DD_d = \text{diag.}(^Dd_{dx}, {}^Dd_{dz}, {}^Dd_{d\theta}) \quad (18)$$

$$^DK_d = \text{diag.}(^Dk_{dx}, {}^Dk_{Iz}, {}^Dk_{d\theta}) \quad (19)$$

$$^DK_I = \text{diag.}(^Dk_{Ix}, {}^Dk_{Iz}, {}^Dk_{I\theta}) \quad (20)$$

where $D_d{'}$, $K_d{'}$ and $K_I{'}$ are the respective feedback gain matrices, and $^LR_D{'}$ is the rotation matrix between Σ_L, which is the coordinate system of the lower part of the device, and Σ_D, which is the coordinate system of the desired trajectory of the supporting element. The block diagram of the control law is shown in Fig. 8.

4.2 Usability

We introduce the following functions to ensure usability as described in section 2.1.

F1. The user can determine the progress of the motions.

F2. The user can adjust the tensions to prevent slack in the wires.

F3. The user can easily deal with the start timing of standing up.

F4. The user can adjust the initial position and facing when sitting down.

F1 and F2 are realized using two button-type on-off switches installed on the top of the right and the left levers, which are on the supporting element and gripped by the user. The right switch is used to move to the next phase of the motions, while the left switch is used for making adjustments, which are adjusting the tensions of the wires and the initial position and facing when sitting down. To adjust the tensions of the wires, the user holds down the left switch to increases the tensions. A confirmation sound is provided after these switches have been pressed so that the user can be alerted to the progress of the motions.

F3 works in that F_P in Eq. (10) is determined as shown in Fig. 9. Although the start timing of standing up t_0 is determined by pressing the right switch by the user, F_P between t_0 and t_{20} is determined as shown in Fig. 9 so that the user can predict the timing of increasing the assisting forces.

F4 is realized using a three-axes force sensor installed on the right lever. The process of adjusting the initial position and facing of sitting down is as follows: 1) The user presses the left switch before sitting down to move on to the adjustment phase of the initial position of sitting down. 2) The user applies the force on the right lever in the direction of the intended motion, and then the supporting element moves in this direction, because the upper part tracks the desired position of the upper part as indicated by the direction of the user's force on the right lever. After adjusting the initial position of sitting down, the user presses the right switch to move on to the phase of sitting down or he/she presses the left switch to adjust the initial facing of sitting down. 3) The user applies the force on the right lever (to the right or left to indicate the direction of turning), after which the supporting element turns in the indicated direction, because of the changing angle of the upper part. After adjusting the initial facing for sitting down, the user presses the right switch to progress to the next phase of sitting down or the left switch to adjust the initial position of sitting down.

These methods ensure good usability because the user can operate the device intuitively.

4.3 Adjusting parameters

4.3.1 Framework of adjusting parameters

To adjust the functions to the individual user, we introduced two sets of parameters, preset parameters and parameters that are continuously adjusted with

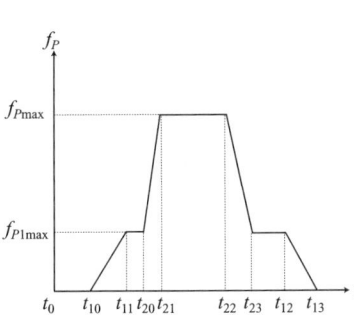

Fig. 9: Assisting force

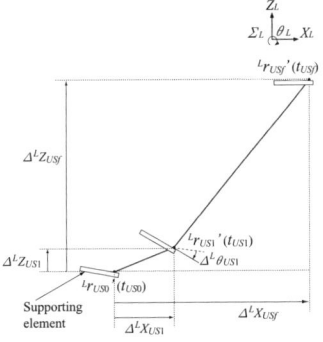

Fig. 10: Trajectory for standing up

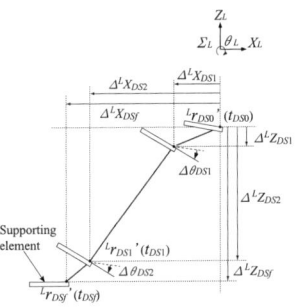

Fig. 11: Trajectory for sitting down

the user's motion. The preset parameters are used to fit the device's power assistance and motion guidance functions. The continuously adjusted parameters are used to adjust the tensions of the wires and the timing of starting the user's motion, the actual starting position, and posture while standing up and sitting down. The details of the adjusting parameters of each motion are described in the following sections.

4.3.2 Standing up

The power assisting force for standing up was determined as shown in Fig. 9, and the desired trajectory of the supporting element was determined as shown in Fig. 10. The preset parameters regarding power assistance are the times and the forces shown in Fig. 9 and an angle of elevation θ_e, while regarding motion guidance, they are the times t_{US1} and t_{USf} and the differences of each position and the posture of the supporting element Δx_{US1}, Δz_{US1}, $\Delta \theta_{US1}$, Δx_{USf} and Δz_{USf} and the posture of the supporting element θ_{USf} shown in Fig. 10 and feedback gain matrices D_d', K_d' and K_I'. The parameters that are adjusted during the process of helping a user to stand up are the user's initial position and posture, and the timing of the start of the motion t_{US0}.

The assisting device can assist the user to stand up when the above parameters are appropriately adjusted.

4.3.3 Walking

This device must support the user's motions while he or she is walking, because the target users are ones able to walk with suitable support. In the motions involved in walking and turning, the user operates the upper part to move to the position where he or she wants to sit down. Here, the parameters that are adjusted before the user starts to walk are the speeds of movement and turning, and the compliance of the supporting element. The parameters that are adjusted while the user is walking are the timing of the motion and the position of sitting down.

4.3.4 Sitting down

The trajectory of the supporting element shown in Fig. 11 is for adjusting the user's motion when sitting down. The preset parameters regarding power assistance are the same as for standing up, while regarding motion guidance, they are the times t_{DS1}, t_{DS2} and t_{DSf} and the differences of each position and posture of the supporting element Δx_{DS1}, Δz_{DS1}, $\Delta \theta_{DS1}$, Δx_{DS2}, Δz_{DS2}, $\Delta \theta_{DS2}$, Δx_{DSf}, Δz_{USf} and θ_{DSf} shown in Fig. 11 and feedback gain matrices D_d', K_d' and K_I'. The parameters that are adjusted during the process of sitting down are the user's initial position and posture, and the timing of the motion t_{DS0}.

Here, the essential difference between the motions of standing up and sitting down is that the gravitational force acts as an accelerant in the latter motion. Therefore, the assisting forces need to be applied as a brake that reduces the downward speed of the user's body.

5 Experiment

The experiment was to determine whether the power assistance and motion guidance functions could be adjusted according to the proposed control law.

The user in this experiment was Dr. Hideo Hanafusa, a 79-year-old man, standing 1.7 [m] and weighing 68 [kg]. His physical condition was normal on that day. His walking function has declined a little compared with that of a young person, but he is able to stand up, walk and sit down without help. The experiments of standing up were tested twice. The power assistance was strong and motion guidance level was low in the first test (Case 1), while power assistance was weak and motion guidance level was high in the second test (Case 2). The power assisting function was adjusted by f_{Pmax} in Fig. 9, while the motion guidance function was adjusted by the feedback gains K_d', D_d', K_I' in Eq. (14). The feedback gains were determined by the natural angular frequency of each direction in coordinate Σ_D so that the device would be a second-order system with critical damping in coordinate Σ_D if the user and assisting force ${}^L F_{RS}'$ would be absent.

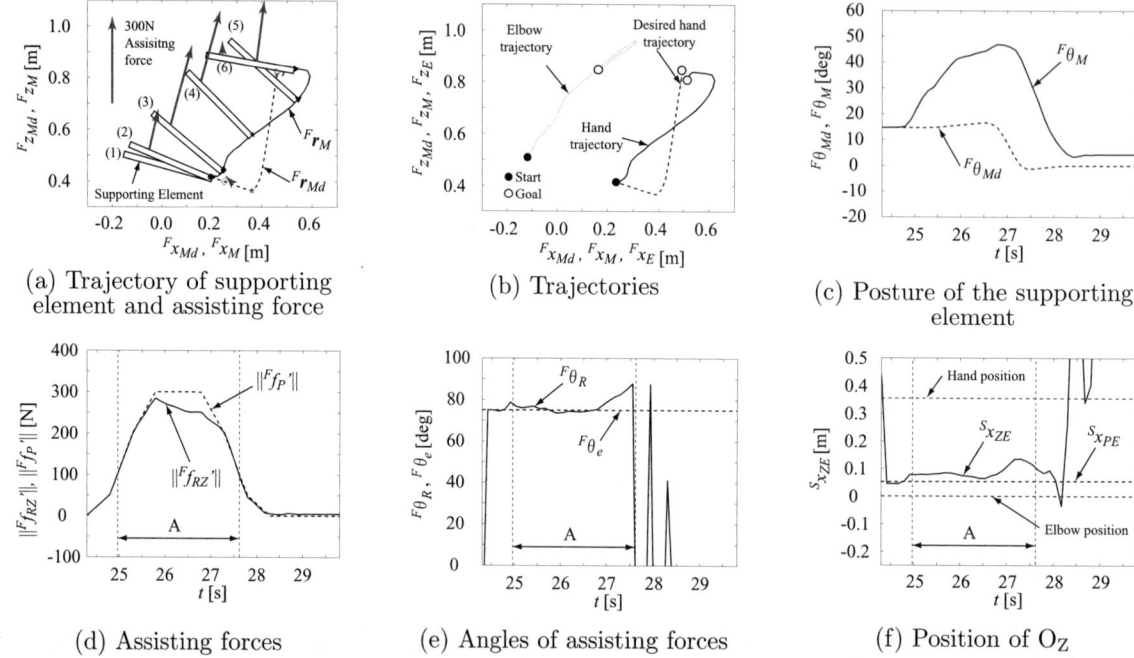

Fig. 12: Experimental results (Case 1)

Table 1 shows the parameters for the strength of power assistance and motion guidance in each case.

The acting point of power assisting force O_P was set up 0.055 m from the elbow position of the user in the direction of his hand, while the acting point of motion guidance force O_M was set up at the user's hand position O_S. The desired trajectory of the device in Case 1 was determined by the natural standing up motion of a graduate student of our laboratory, while the desired trajectory of the device in Case 2 was a little different from Case 1's because some parameters were adjusted according to the user's comments.

The experimental results of standing up are shown in Figs. 12 and 13, and an example of the user's posture in standing up is shown in Fig. 14. The parameters in Figs. 12 and 13 are represented in Σ_F setting on the floor, except the parameters in Figs. 12(f) and 13(f). The positions of the supporting element and assisting force given by the device in Figs. 12(a) and 13(a) were plotted every 0.75 [s]. Figures 12(b) and 13(b) show the desired hand trajectory and elbow and hand trajectories, while Figs. 12(c) and 13(c) show the posture of the supporting element. Figures 12(d) and 13(d) show the assisting force $\|{}^F\!f_{RZ}{}'\|$ and power-assisting force $\|{}^F\!f_P{}'\|$. A in Figs. 12(d) to (f) and B in Figs. 13(d) to (f) are the period of $\|{}^F\!f_{RZ}{}'\| \geq 100$ N. Figures 12(e) and 13(e) show the angles of the assisting forces, and Figs. 12(f) and 13(f) show the trajectory of the acting point of the assisting force O_Z. The trajectories in Fig. 12(e), (f) and 13(e), (f) are important in periods A and B.

In Case 1, the assisting force $\|{}^F\!f_{RZ}{}'\|$ was almost the same as the power-assisting force $\|{}^F\!f_P{}'\|$ because of the weak motion guidance function, and the angle of the assisting force ${}^F\theta_R$ and the acting position of the assisting force O_Z were almost the same as the angle of the power assisting force ${}^F\theta_P$ and the acting position of the power assisting force O_P in period A, respectively (Figs. 12(d), (e), (f)). The displacement between the desired position and the response was large (Figs. 12(a), (b), (c)). The user's comment was "I could easily stand up receiving assisting force by the device".

In Case 2, the supporting element tracked the desired position better than it did in Case 1 because of the strong motion guidance function (Figs. 13(a), (b), (c)). The angle of the assisting force ${}^F\theta_R$ was almost the same as the angle of the power assisting force ${}^F\theta_P$ in period B, but the acting point of the assisting force O_Z was almost the center of the forearm because of the large motion guidance force ($=\|{}^F\!f_{RZ}{}'\|-\|{}^F\!f_P{}'\|$). The user's comment was "I felt awkward in standing up and felt an uncomfortable constraint".

We believe the reasons for the user's positive comment in Case 1 are as follows: 1) the assisting force was large when the buttock of the user rose up from the bed, 2) the user could easily receive the assisting force because the acting point of the assisting force was near his elbow, and 3) the device imposed no constraint on the user. We believe the reasons for the unfavorable comment in Case 2 are as follows: 1) the desired trajectory was bad for the user, and the device imposed unnecessary constraint on him, 2) the user could not utilize the assistance very well because the timing of the assisting force was bad for his standing up motion, and 3) the user could not easily receive the assisting

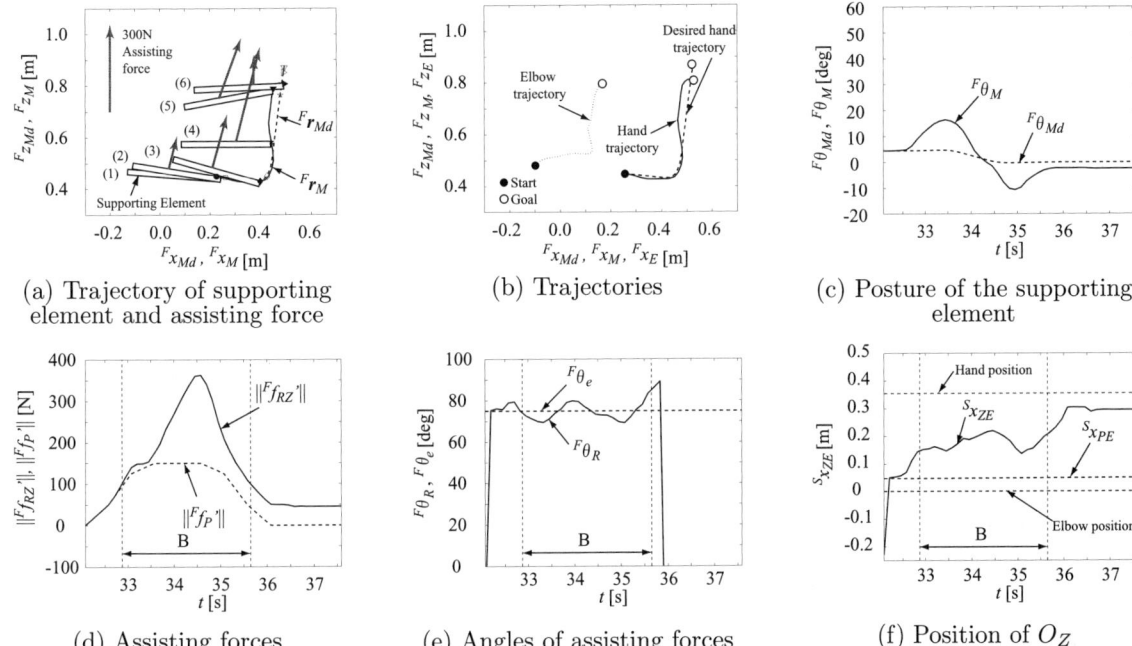

Fig. 13: Experimental results (Case 2)

Table 1: Parameters for the strength of power assistance and motion guidance

Case	PA	$f_{P\max}$	MG	$\omega_x, \omega_z, \omega_\theta$
1	strong	300[N]	weak	8, 8, 10 [rad/s]
2	weak	150[N]	strong	20, 20, 20 [rad/s]

force because the acting point of the assisting force was almost the center of the forearm. The user in this experiment liked the assistance of force only and disliked guidance of the uncomfortable motion because he can keep his own balance. That is, setting the power assisting function at a higher level than the motion guidance function is good for a user like him.

6 Conclusions

The major results that we obtained in the work described in this paper are summarized below.

1) The relationship between the device's range of motion and its feasible assisting force was found, thus determining its capability of assisting the user's own movements.

2) We proposed a control law governing the power assistance and motion guidance functions, along with an easily used interface.

3) The prototype device was tested to investigate its basic capability of assisting users with transfer motions. The functions of power assistance and motion guidance can be adjusted.

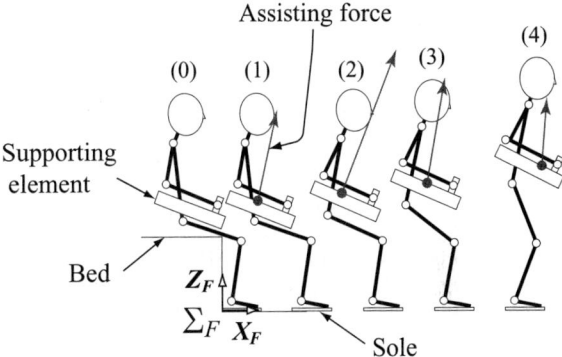

Fig. 14: Example of user's posture in standing up

References

[1] http://www.hcr.or.jp/english.html

[2] N. Suzuki, et al.: "System assisting walking and carrying daily necessities with an overhead robot arm for in-home elderlies", Proc. of the 22nd Annual Int'l Conf. of the IEEE Engineering in Medicine and Biology Society, vol.3, pp 2271–2274, 2000

[3] M. Egawa, et al.: "Standing and seating assistance by transfer support-equipment", Seimitsu Kogakkai Taikai Gakujutsu Koenkai Koen Ronbunshu, p. 401, 1998 (in Japanese)

[4] K. NAGAI, et al.: "Development of a Power Assistive Device for Self-Supported Transfer Motion", Proc. of 2002 IEEE/RSJ International Conference on Intelligent Robots and Systems, pp. 1433–1438, 2002.

Development of Rehabilitation System for the Upper Limbs in a NEDO Project

Ken'ichi Koyanagi*, Junji Furusho*, Ushio Ryu† and Akio Inoue ‡
* Graduate School of Engineering, Osaka University,
2-1 Yamadaoka, Suita, Osaka 565-0871, Japan.
Email: {koyanagi, furusho}@mech.eng.osaka-u.ac.jp
† System Equipments Div., Asahi Engineering Co., Ltd.,
4-1-8, Kohnan, Minato-ku, Tokyo 108-0075, Japan.
Email: ryu.ub@om.asahi-kasei.co.jp
‡ Electronics Materials & Devices Lab., Asahi Kasei Co.,
2-1, Samejima, Fuji, Shizuoka 416-8501, Japan.
E-mail: inoue.ag@om.asahi-kasei.co.jp

Abstract—Movements of the upper limbs are complicated, various and indispensable for daily activities. It therefore is important for the aged to exercise to keep their upper limb function. When something is wrong with the upper limb function because of disease or disorder, rehabilitation along with medical treatment is needed to recover function. Application of robotics and virtual reality technology makes possible for new training methods and exercises on upper limb rehabilitation and for quantitative evaluations to enhance the qualitative effect of training. However, rehabilitation systems applying training within a three-dimensional to upper limbs have not been in practical use. The authors have involved in a project managed by NEDO (New Energy and Industrial Technology Development Organization as a semi-governmental organization under the Ministry of Economy, Trade and Industry, Japan), "Rehabilitation System for the Upper Limbs and Lower Limbs," and developed a 3-DOF exercise machine for upper limb (EMUL). In this paper, the authors report on development of EMUL which can be use as a motion guide robot or a force display device. Particularly, mechanical structure safety kept by actuators using ER fluid is mentioned.

I. INTRODUCTION

The percentage of aged persons in society and their number are increasing, and their physical deterioration has become a social problem in many countries. Deterioration of body function has both direct influences like falls while walking, degeneration of motion ability and obstructions in daily activities. Additionally, it is mentioned that deterioration has indirect influences such as brain degeneration. Early detection of function deterioration and trying to maintain and/or recover functions is necessary, not only to decrease the numbers of aged who are bedridden or need nursing care, but also to enable the aged to take an active part in society. Movements of the upper limbs, such as for eating and operating appliances are complicated, various and indispensable for daily activities. It therefore is important for the aged to exercise to keep their upper limb function. When something is wrong with the upper limb function because of disease or disorder, rehabilitation along with medical treatment is needed to recover function [1].

Using apparatus that applies robotic technology of or virtual reality makes possible for new training methods and exercises to be introduced into upper limb rehabilitation. Feedback from the results of quantitative evaluations to patients by using computers can enhance the qualitative effect of training. Some rehabilitation systems for upper limbs have been developed but most of them apply training within a two-dimensional, horizontal plane, as does the MIT-MANUS [2] by MIT. Many movements, however, in daily activities need to move arms in a vertical direction. A system therefore that enables exercise in three-dimensions would seem to be more effective for such training. Although the MIME system [3] using PUMA-560 by VA and Stanford Univ. can give training in three-dimensions, the PUMA-560 is a robot originally developed for industrial use and may be not be sufficiently safe to train the aged and/or disabled.

This research has been involved in the development of a rehabilitation system for upper limbs in a NEDO (New Energy and Industrial Technology Development Organization) Project, "Rehabilitation System for the Upper Limbs and Lower Limbs" since 2001 [4]. Osaka University has already developed a 2-DOF (degrees of freedom) force display system using ER (Electrorheological) fluid and has carried out experiments in the basic properties, and clinical trials of rehabilitation for upper limbs [5], [6]. The purpose of this research was to develop a 3-DOF rehabilitation system for upper limbs based on this existing knowledge and to construct a rehabilitation system that included a quantitative evaluation of the training and a feedback system of the training results to the trainees.

This report proposes the application of an ER fluid

apparatus to a rehabilitation system that considers safety, and then it presents the 3-DOF upper limb support mechanism the authors have developed. The authors also describe the basic software to apply the system to practical rehabilitation training.

II. ER ACTUATORS FOR SAFETY

A. Safety Required in Rehabilitation

As mentioned above, a three-dimensional rehabilitation system has yet to be developed. Force display systems and medical robots, however, come very close; here we focus on the safety aspects of rehabilitation training and discuss such systems.

A rehabilitation system for upper limbs, different from systems for fingers or hands, needs to exercise a whole limb. The system therefore requires a large working space, a large display ability and a good back-drive ability. Such a rehabilitation system mechanism is equal to that of robot systems, but is different too from current industrial robots. The system is operated in constant contact with the trainee, and because the position of the end-effector, the control element, is close to the operator's face there is always the possibility of collision with and injury to the operator. Thereby it is necessary for operators to be able to use the system safely under any circumstances or conditions.

Passive-type systems [7], [8], which have no risk of a runaway, are from a safety viewpoint the best. However, training by passive exercise or active-assistive exercise needs active movements of a machine, so an active-type machine using an actuator is needed to enable such training.

Most active-type force display systems use servomotors as actuators. When the system using a servomotor gets out of control because of a bug in the software, computer trouble or the sensor, the end-effector could collide with the operator. Most of the systems, however, which have so far been developed are small, like PHANToM [9], and some studies of active-type force display systems emphasize their safety. As for studies of medical robot systems, systems with double sensors [10], [11] or which detect any unusual situations by the software handling the information from the sensors [12] have been proposed for safety. However, this makes the system complicated and does not necessarily improve reliability and maintainability of the system. Moreover they are expensive [13].

On the other hand, if the safety could be mechanically ensured, a safe system with high reliability and maintainability can be constructed relatively simply [14]. Thus, it is desirable that the following be realized mechanically.

- A speed limit
- A generative force limit
- Gravity compensation at an end point

B. ER Actuators

Although pneumatic actuators [15] and metal hydride actuators [16] have like servomotors been proposed as actuators for medical and welfare apparatus, their response speeds, generative force and mechanical rigidity are not sufficient. The authors consider an actuator using particle-type ER fluid [17] to be effective. ER fluid is a fluid whose rheological properties can be changed by applying a electrical field. Figure 1 shows the principle of an ER fluid actuator. The ER actuator is composed of an ER clutch and drive mechanisms such as a motor driving its input shaft and a reducer. The characteristics of the ER actuator [14] are as follows:

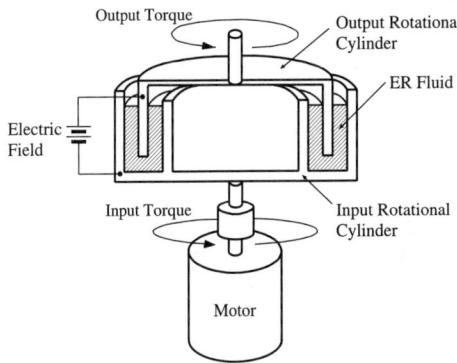

Fig. 1. Principle of ER actuator

- The maximum rotational speed of the output shaft is mechanically governed by the input shaft.
- It is not necessary to precisely control the rotational speed of the input shaft[14], [18].
- The output can be limited in a mechanically fashion by the performance of a high-voltage amplifier used in the system, because an applicable electrical field sets the maximum output torque.
- It can be used at low speed and high torque, so the reducing ratio of the speed reducer after the output shaft can be set low.
- It has high torque/inertia ratio and response is fast.
- As no contact-type clutch is used, there is no stick slip and the friction is low.
- In an emergency, cutting off the electrical field can quickly stop the output of the ER actuator.

Consequently, the advantages of using an ER actuator for a rehabilitation system would be:

- The speed at the end point can be mechanically limited and the risk of going out of control can be greatly reduced.
- The rotational speed of the input shaft does not need to be controlled precisely, so it is possible to drive the input shaft by a simple method.

- As it uses a clutch mechanism, slipping prevents giving too much load to the operator even though the output side of the actuator receives torque over the maximum output torque.
- The moment of inertia at the end point becomes low and it has good back-drive ability. A system with low inertia can reduce damage in a collision between the operator and the end-effector.
- It can output high-frequency torque. This is an advantage in the expression of collisions with hard objects.
- It is easy to make a mathematical model of the actuator and control the system.

III. THE DEVELOPMENT OF THE 3-DOF EXERCISE MACHINE FOR UPPER LIMB (EMUL)

A. Major Specifications of Whole System

This section describes the major specifications in the rehabilitation system for upper limbs that has been developed and consists of EMUL, computer, display, software, and so on.

Figure 2 shows a model of the developed system. A patient gets exercise, sitting on a chair by gripping the handle of the upper limb support machine with the right hand and thus training the right limb. For a patient who needs to train the left hand, another machine is used which has a mirror-image mechanism. When it is difficult for a patient to grip by him or herself, orthosis is used. The major target for this training are dystonic, ataxic or hemiplegic patients with stroke sequelae. The training is thought to include physical therapeutic exercises, such as passive and active exercises, occupational therapeutic exercise like eating movement and cognitive therapeutic exercises.

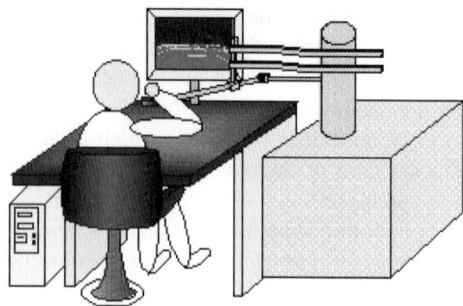

Fig. 2. Model of the developed rehabilitation system

B. The Development of a Prototype of an Upper Limb Support Mechanism

The authors have developed a 3-DOF upper limb support mechanism that has a performance suitable for rehabilitation systems for upper limbs, and can display force senses in three-dimensional space. They call the exercise machine for upper limb EMUL.

Fig. 3. The prototype 3-DOF upper limb support machine

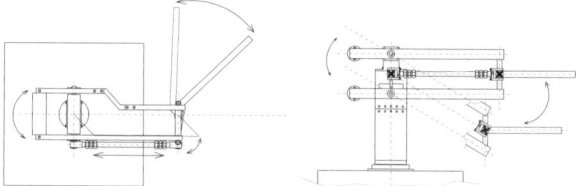

Fig. 4. Top view of the machine model

Fig. 5. Side view of the machine model

Figure 3 shows the picture of the first prototype made at Osaka University (proto-EMUL). Figures 4 and Fig. 5 show the motion of the end link part. The proto-EMUL was made mainly to verify the characteristics of the mechanism, which are as follows:

- The proto-EMUL has 2 DOF for horizontal rotation and 1 DOF for vertical rotation; the actuators drive them. This machine has three shafts: shaft 1 (horizontal rotation), shaft 2 (vertical rotation) and shaft 3 (horizontal rotation) from the foundation to the end. Link 1 is connected to shaft 1, link 2 to shaft 2 and link 3 to shaft 3. Thus, EMUL can assign only its end position in space.
- The end-effector has a 3 DOF handle, which does not impede 3 DOF at the end posture.
- The lengths of both link 2 and link 3 are 0.45 [m] and the height of the whole machine is about 1.0 [m].
- All the actuators are set on the base which does not move. A disadvantage in using ER actuators is the increase in the weight the actuators cause. Not moving the actuators themselves, however, can lower equivalent inertia of the control element and the necessary control force.
- Timing belts and pulleys mechanism and a bevel gear are used for the reducer between the actuator and the load-side shaft. Compared to a reduction gear mechanism or a wire and pulley mechanism, a timing belt mechanism has high mechanical rigidity and little backlash, and is suitable for a system to display comparatively large force senses.

- The load-side shaft has the triple shaft mechanism shown in Fig. 6 and transmits power to the upper link.
- The vertically operating part (link 2) has a parallel link mechanism, so constant gravitational torque acts on the vertical-rotation shaft irrespective of the posture of the end link part (link 3). This makes it easy to mechanically compensate for gravity because of the counterbalance, which leads to a lowering of the torque needed for the actuators and is effective both in saving space for the machine and in making the mechanism safer.
- The rotating part of link 3 has a spatial parallel link mechanism. Figure 7 shows the joint part. This joint has two orthogonal rotation shafts. This makes the friction loss low and the machine light. It also leads to a mechanism whose equivalent inertia as seen from the end-effector is low.
- The motion range is about $0.90W \times 0.54D \times 0.40H$ [m] at the maximum and the range where the manipulability is particularly good is about $0.30 \times 0.30 \times 0.30$ [m].
- DC servo motors are used as the actuators for this prototype. The generative force at the end is about 20 [N] in the direction where the force generates most from a basic position.

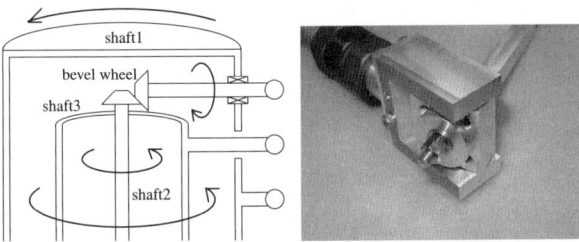

Fig. 6. Inner mechanism of shafts   Fig. 7. Spatial joint of EMUL

C. The Upper Limb Support Mechanism Using ER Actuators

An improved upper limb support mechanism using ER actuators has been developed in the Asahi Kasei Co. group. Figure 8 shows the 3-DOF exercise machine for upper limb using ER actuators (EMUL) and the whole rehabilitation system. The rotation of the ER actuators is unidirectional and the system has 6 ER actuators. This mechanism has the following characteristics besides the advantages mentioned in Sections II-B and III-B.

- The motion range has extended to about $0.90W \times 0.54D \times 0.50H$ [m] at maximum.
- The generative force at the end is about 23 [N] in each shaft direction within the horizontal plane, about 60 [N] in vertical directions, and has increased

Fig. 8. Picure of rehabilitation system with ER actuator

drastically as compared to the proto-EMUL using motors.

Table I shows a comparison of safety between the motors used in the first prototype and the ER actuators. The ER actuators are compared for situations where the clutch input shaft speed is set at 10 [rpm]. The authors reported in their previous paper [14] that even though the input shaft speed is set so low, there is no major problem with the basic performance of the force display. In this case, the end speed of the machine is limited to about 10.5[cm/s] at maximum. This is much lower than the standard value in ISO [19], 25[cm/s], so there is little risk of damaging a trainee in the case of a runaway and the system can be said to be more suitable for rehabilitation than other systems.

TABLE I
COMPARISON OF MOTOR AND ER ACTUATOR

	Motor	ER Actuator
Max. Torque	about 1.4 [N·m]	about 3.0 [N·m]
Mechanical Speed Limit	Impossible	Possible
Max. Speed	Over 2000 [rpm]	10 [rpm]
Max. Power	Over 340 [W]	about 3 [W]
When run away	Much Risk	Little Risk

IV. THE SOFTWARE FOR REHABILITAION TRAINING

A. The Kinematical Equation of EMUL

The kinematical equation of the end-effector has been determined for this machine. Figure 6 shows a schematic diagram of the inside structure of this mechanism. When shaft 1 rotates, the whole machine rotates and shafts 2 and shaft 3 inside look as if they rotated, relatively, in the opposite direction. Each shaft is connected to an ER actuator driving it through a belt and the pulley. All reducing ratios are 4.5, and link 2 that vertically moves has a bevel gear whose speed reducing ratio is 2.0.

Considering their influence, the kinematical equation of the end-effector can be determined by using the angle of the actuator, θ_{mi} as follows, where the directions both of coordinate axes, X-Y-Z and of the joint angle θ_i are set as shown in Fig. 9.

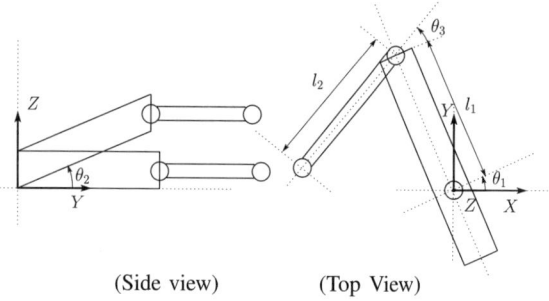

(Side view)   (Top View)

Fig. 9. Definition of coordinate axes and joint angles

$$\begin{cases} \theta_1 = \frac{2}{9}\theta_{m1} \\ \theta_2 = \frac{1}{9}(\theta_{m2} - \theta_{m1}) \\ \theta_3 = \frac{2}{9}(\theta_{m3} - \theta_{m1}) \end{cases}, \quad (i=1,2,3) \quad (1)$$

$$\begin{bmatrix} x \\ y \\ z \end{bmatrix} = \begin{bmatrix} -l_1 \sin\frac{2}{9}\theta_{m1} \cos\alpha - l_2 \sin\beta \\ l_1 \cos\frac{2}{9}\theta_{m1} \cos\alpha + l_2 \cos\beta \\ l_1 \sin\alpha \end{bmatrix} \quad (2)$$

here, $\alpha = \frac{1}{9}(\theta_{m2} - \theta_{m1}), \quad \beta = \frac{2}{9}\theta_{m3} + \frac{1}{2}\pi$

B. Trajectory Tracking Controls

Exercise therapy can be roughly classified into passive exercise, active-assistive exercise, active exercise and resistant exercise. The authors examine leading a patient's hand gripping the end-effector along a desired trajectory. The force that the system generates at the handle is classified into the force along the trajectory and the force to move the handle back to the original trajectory when it goes out of trajectory. When the force along the trajectory is strong, the system makes the patient's hand track the trajectory against the disturbance from the hand to perform passive exercise training. On the other hand, when the force along the trajectory is weak, the patient needs to generate the force along the trajectory and to exercise by himself to track the desired trajectory. The system can be used for active-assistive exercise training in this case.

Eating movements are indispensable to keep life and such disorders are serious. In this study, reaching movement training is examined as part of eating movement training. The control method mentioned above is considered to be able to be applied to passive and active-assistive exercise training in reaching training.

C. Compliance Controls

Compliance controls are introduced to perform the trajectory tracking controls mentioned in the preceding section; however, only the spring force by the displacement from the center of compliance is considered here.

The spring constants are determined as the following equations for each axis as shown in Fig. 9:

$$K = \begin{bmatrix} K_1 & 0 & 0 \\ 0 & K_2 & 0 \\ 0 & 0 & K_3 \end{bmatrix}, \quad K_i > 0 \ (i=1,2,3) \quad (3)$$

The elements of K are changed according to degree of impairment of patients.

Where $\boldsymbol{\theta_m} = [\theta_{m1}, \theta_{m2}, \theta_{m3}]^T$ is the vector indicating the rotation angle of an actuator and $\boldsymbol{x_H} = [x_H, y_H, z_H]^T$ is the end-effector's position vector in the space. Time-differentiating both sides of Eq. 2 gives the following:

$$\dot{\boldsymbol{x}}_H = J(\boldsymbol{\theta_m})\dot{\boldsymbol{\theta}}_m \quad (4)$$

where $J(\boldsymbol{\theta_m})$ is 3×3 Jacobian matrix.

The deviation between $\boldsymbol{x}_H$ and the desired point, $\boldsymbol{x_r}$, at some time is the displacement of the end-effector from the center of compliance, and then the force acting from the external to the end, $\boldsymbol{F}$, is written using the spring constants determined in Eq. 3 as follows:

$$\boldsymbol{F} = K(\boldsymbol{x}_H - \boldsymbol{x_r}) \quad (5)$$

The relationship between $\boldsymbol{F}$ and the torque an actuator should generate, $\boldsymbol{T}$, is expressed as follows:

$$\boldsymbol{T} = -J(\boldsymbol{\theta_m})^T \boldsymbol{F} \quad (6)$$

Bringing in the speed feedback into this equation to stabilize the control system gives the following:

$$\boldsymbol{T} = -J(\boldsymbol{\theta_m})^T \boldsymbol{F} - K_D \dot{\boldsymbol{\theta}}_m \quad (7)$$

where K_D is speed feedback gain.

D. Compliance Controls in an Arbitrary Direction

To make it easy to move the end-effector along the trajectory, it is recommended that the spring constant of the direction along the trajectory is set low and that of the direction vertical against the trajectory high. When the direction of the trajectory is on the X axis, lowering the spring constant of the direction of the axis forms operationality of the handle as shown in Fig. 10. This is called a compliance ellipsoid. Automatically setting the direction of the long axis of the ellipsoid to the direction of the desired trajectory at that time enables compliance controls according to an arbitrary desired trajectory.

The unit vector indicating the traveling direction of the desired trajectory, $\boldsymbol{x_c}$, is written as follows:

$$\boldsymbol{x_c} = \frac{1}{||\dot{\boldsymbol{x}}_r||}[\dot{x}_{r1}, \dot{x}_{r2}, \dot{x}_{r3}]^T \quad (8)$$

The rotation of the ellipsoid is performed by using rotation matrix, R, expressed with Euler angles, (ϕ, θ, ψ), and it is written as Eq. 9:

$$R = \begin{bmatrix} \cos\phi & -\sin\phi & 0 \\ \sin\phi & \cos\phi & 0 \\ 0 & 0 & 1 \end{bmatrix} \begin{bmatrix} \cos\theta & 0 & \sin\theta \\ 0 & 1 & 0 \\ -\sin\theta & 0 & \cos\theta \end{bmatrix} \quad (9)$$

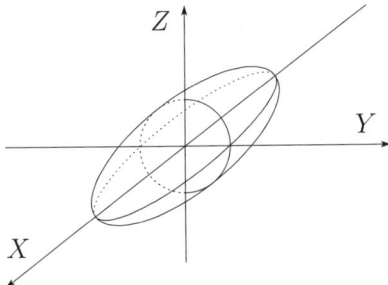

Fig. 10. Compliance ellipsoid

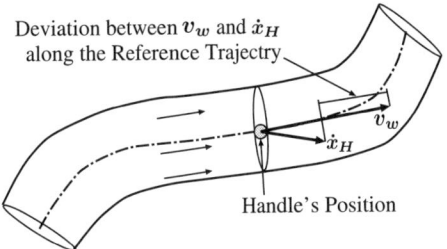

Fig. 11. Image view of free form curved virtual tube and assistive force as water flow

where the spring constants of the directions vertical against the long axis are uniform and $\psi = 0$. The Euler angles can be calculated by using Eq. 8 as follows:

$$\begin{cases} \cos\phi = \dfrac{\dot{x}_{r1}}{\sqrt{\dot{x}_{r1}^2 + \dot{x}_{r2}^2}} \\ \sin\phi = \dfrac{\dot{x}_{r2}}{\sqrt{\dot{x}_{r1}^2 + \dot{x}_{r2}^2}} \end{cases} \begin{cases} \cos\theta = \dfrac{\sqrt{\dot{x}_{r1}^2 + \dot{x}_{r2}^2}}{\|\dot{\boldsymbol{x}}_r\|} \\ \sin\theta = -\dfrac{\dot{x}_{r3}}{\|\dot{\boldsymbol{x}}_r\|} \end{cases} \quad (10)$$

The renewed spring constant, K', in the case of the long axis along the direction of the desired trajectory is given by Eqs. 9 and 3 as follows:

$$K' = R^T K R \quad (11)$$

E. Water Flow for Active-assistive Exercise Training

For another type of active-assistive exercise training, an assistive force like the force which the operator feels while his/her hand is in water is displayed in addition to the 3-D free form curved virtual tube (as Fig.11). To be concrete, some velocity along the virtual tube, or the velocity of flow, is set up, and then the force proportional to the deviation between the velocity of flow and the component of the handle velocity along the virtual tube is generated. When the direction of the velocity of flow set plus according to the direction of the teaching, the represent force assists the move of the end-effector. Therefore this software can be used as active-assistive exercise training or passive exercise training. On the other side, when the direction of the velocity of flow set plus against to the direction of the teaching, the represent force prevents the move of the end-effector. Therefore this software can be used as resistant exercise training. In this paper, the case of active-assistive exercise training is mentioned. Major difference from the Sec. IV-D is that operator's movements are not governed by time in this case.

The presentation of water flow is realized by compliance control as a compliance matrix is rotated along the segment on the target trajectory nearest to the position of the handle. The compliance matrix has only viscous characteristics, however, the force which should be presented is only generated in the direction of the target trajectory.

Here, the viscosity K_F is introduced as Eq. 12.

$$K_F = \begin{bmatrix} K_{FX} & 0 & 0 \\ 0 & 0 & 0 \\ 0 & 0 & 0 \end{bmatrix}, K_{FX} > 0 \quad (12)$$

This matrix is rotated to the direction of the 3-D free form tube. The new viscous matrix after rotation, K'_F, is written using a rotation matrix as follows:

$$K'_F = R^T K_F R \quad (13)$$

The direction of water flow is assumed as in the plus side of X-axis and its velocity is $\boldsymbol{v_w} = [v_{wx}, v_{wy}, v_{wz}]^T = [v_{wx}, 0, 0]^T$. $\boldsymbol{v_w}$ is rotated along the virtual tube to represent the water flow on the reference trajectory, then the force like the water flow is written as follows:

$$\boldsymbol{F} = K'_F (\dot{\boldsymbol{x}}_H - R\boldsymbol{v_w}) \quad (14)$$

The magnitude of $\boldsymbol{v_w}$ is changed according to degree of impairment of patients.

V. CONCLUSIONS

A 3-DOF motion exercise support machine for a rehabilitation system for upper limbs has been developed, and ER actuators have been installed to make the system safer. Major points are:

- Gravity compensation is mechanical.
- Using a parallel link mechanism keeps the machine light, and its friction loss low.
- The use of a timing belt mechanism heightens mechanical rigidity and decreases backlashes.
- The motion area and the generative force are large.
- Using ER actuators makes the system safer.

The authors are next developing basic software for using compliance controls. This will be applied to passive and active-assistive exercise training. They also have proposed a force-assist control, which generates force as water flow. This is for active-assistive exercise training.

In future work, the effect of our training system will be experimentally proved in medical facility.

ACKNOWLEDGMENTS

This study is supported financially by The National Research and Development Program for Medical and Welfare Apparatus (NEDO entrustment research), and the authors wish to express their gratitude.

VI. REFERENCES

[1] R. Nakamura, *Introduction to Rehabilitation Medicine*, Ishiyaku Publishers, Inc., 1998, (in Japanese).

[2] H. I. Krebs, B. T. Volpe, M. L. Aisen, and N. Hogan, "Increasing productivity and quality of care : Robot-aided neuro rehabilitation," *Jurnal of Rehabilitation Research and Development*, vol. 37, no. 6, pp. 639–652, 2000.

[3] C. G. Burgar, P. S. Lum, P. C. Shor, and H.F. Machiel Van der Loos, "Development of robots for rehabilitation therapy : The palo alto va/stanford experience," *Jurnal of Rehabilitation Research and Development*, vol. 37, no. 6, pp. 663–673, 2000.

[4] New Energy and Industrial Technology Development Organization (NEDO), *Transaction on Rehabilitation System for Upper Limb and Lower Limb in 2001 NEDO reports*, NEDO, 2002, http://www.nedo.go.jp/ (in press).

[5] J. Furusho and M. Sakaguchi, "New actuators using ER fluid and their applications to force display devices in virtual reality and medical treatments," *Int. J. of Modern Physics B*, vol. 13, no. 14,15&16, pp. 2151–2159, 1999.

[6] K. Koyanagi, T. Inoue, and J. Furusho, "Rehabilitation application of force display system using ER fluid," in *Proceedings of The 6th International Conference on Motion and Vibration Control (MOVIC2002)*, 2002, vol. 2, pp. 831–836.

[7] H. Davis and W. Book, "Torque control of a redundantly actuated passive manipulator," in *Proc. American Control Conf.*, 1997, pp. 959–963.

[8] M. Peshkin, J. E. Colgate, and C. Moore, "Passive robots and haptic displays based on nonholonomic elements," in *Proceedings of the IEEE 1996 International Conference on Robotics and Automation*, 1996, vol. 1, pp. 551–556.

[9] T. H. Massie, "Virtual touch through point interaction," in *Proceedings of the 6th International Conference on Artificial Reality and Tele-Existence (ICAT'96)*, 1996, pp. 19–38.

[10] M. Tanimoto, F. Arai, T. Fukuda, I. Takahashi, and M. Negoro, "Force display method for intravascular neurosurgery," in *Proc. 1999 IEEE Int. Conf. on Systems, Man, and Cybernetics*, 1999, vol. 4, pp. 1032–1037.

[11] D. Engel, J. Raczkowsky, and H. Wörn, "A safe robot system for craniofacial surgery," in *Proc. of the 2001 IEEE Int. Conf. on Robotics and Automation*, 2001, vol. 2, pp. 2020–2024.

[12] J. Zurada, A. L. Wright, and J. H. Graham, "A neuro-fuzzy approach for robot system safety," *IEEE Transactions on Systems, Man and Cybernetics*, vol. C-31, no. 1, pp. 49–64, 2001.

[13] G. C. Burdea, "Haptics issues in virtual environments," in *Proceedings of Computer Graphics International Conference 2000 (CGI2000)*, 2000, pp. 295–302.

[14] K. Koyanagi and J. Furusho, "Study on high safety actuator for force display," in *Proceedings of SICE Annual Conference 2002*, pp. 3118–3123.

[15] T. Noritsugu and T. Tanaka, "Application of rubber artificial muscle manipulator as a rehabilitation robot," *IEEE/ASME Trans. on Mechatronics*, vol. 2, no. 4, pp. 259–267, 1997.

[16] Y. Wakisaka, M. Muro, T. Kabutomori, H. Takeda, S. Shimizu, S. Ino, and T. Ifukube, "Application of hydrogen absorbing alloys to medical and rehabilitation equipment," *IEEE Trans. on Rehabilitation Engineering*, vol. 5, no. 2, pp. 148–157, 1997.

[17] G. Bossis ed., *Proceedings of the Eighth International Conference on Electrorheological Fluids and Magnetorheological Suspensions*, World Scientific, 2002.

[18] M. Sakaguchi and J. Furusho, "Force Display System Using Particle-Type Electrorheological Fluids," in *Proceedings of the 1998 IEEE International Conference on Robotics and Automation*, 5 1998, vol. 3, pp. 2586–2591.

[19] ISO10218, *Manipulating industrial robots -Safety*, 1992.

Implementation of a path planner to improve the usability of a robot dedicated to severely disabled people

M. Mokhtari [1,2] B. Abdulrazak [1,2], R. Rodriguez [1], B. Grandjean [2]

[1] HANDICOM Lab, Institut National des Télécomunications (INT), Evry, France.
[2] INSERM-U483, University Pierre & Marie Curie, Paris, France.
Email : Mounir.Mokhtari@int-evry.fr

Abstract -The design of robot dedicated to person with disabilities necessitate users implication in all steps of product development: design solution, prototyping the system, choice of users interfaces, and testing it with users in real conditions. However, before any design of any system, it is necessary to understand and meet the needs of the disabled users. In this paper, we describe our research activity on the integration of a robotic arm in the environment of disabled people who have lost the abilities to use their proper arms to perform daily living tasks and who are able to use an adapted robot to compensate, even partly, the problems of manipulation of objects in their environments. This paper presents our contribution for designing adaptable, configurable and personalized robot system based on the Manus robot. Some preliminary evaluation results are also presented.

1. INTRODUCTION

People who have lost the capabilities to use their proper arms to perform daily living task could use an adapted robot to compensate, even partly, the problems of object manipulation generated by their handicap. Several robotized systems have been performed this last decade to help people with severe motor disability. Many developments in the rehabilitation robotics field have lead to some encouraging robotic arms. They were developed to offer more independence to persons with severe disabilities and to perform tasks in their daily lives. The Manus arm is one of these robotic devices.

The scientific community in this field has been divided into two groups: the ones which thinks that the robotized system should be fully automated to perform complex tasks in an autonomous way without having the user in the control loop, and the ones which defend the idea of having the user in the command loop of the robot. The development done in the former case consisted on robotized workstations which have the advantages to perform task in an optimized way in term of trajectory and in term of time, but with the condition of having a structured environment which has been modeled. In the latter case, the research activity investigated the robotic arm, mounted on wheelchair or mobile base and controlled manually by the user. The main advantage of these systems is the possibility to perform task in an open and changing environment, but the optimization of the movement is based on the motivation of each user. Our team belongs mainly to the second group and in this paper we describe the current development in this field, which are supported by the European commission with the Commanus[1] project and the results obtained during the evaluation process. A new evaluation method, based on a quantitative analysis, has been developed to estimate exactly the usability of the system.

2. THE MANUS ROBOT

MANUS is a robot mounted on an electrical wheelchair (Fig 1). It is aimed to favor the independence of severally handicapped people who have lost their upper and lower limbs mobility, by increasing the potential activity and by compensating the prehension motor incapabilities. Manus is a robot with six degrees of freedom, with a gripper in the extremity of the arm which permits the capturing of objects (payload of 1,5 kg) in all directions, and a display. It is controlled by a 4x4 buttons keypad or by a joystick, and with the latest version, a mouse and a touch screen control was provided. A display unit gives the user current status of the MANUS.

Fig. 1 The Manus robot used in a supermarket

3. SOFTWARE COMMAND ARCHITECTURE

The Manus software architecture allows us to choose many modes in order to offer several possibilities of controlling the arm [12]. As shown in figure 2, the basic software

---

[1] COMMANUS project, EC DGXII Biomed-Craft program. Partners involved: Exact dynamics, TNO-TPD and RTD-HetDorp in the Netherlands, Oxime in UK,and INT, INSERM, AFM and A6R in France

architecture, called Manus modes, has three different control modes: The **Cartesian Mode**, which allows the user to control manually the arm and gripper motion in Cartesian space, the **Joint Mode** which allows a direct and separate control of the six arm joints, and the **Main Mode** which allows access to the above cited modes and allows the user to perform specific commands such us fold-in, fold-out, drink....

Regarding the evaluation results, we have developed new command architecture, called Commanus modes, and implemented several extra modes beside the Manus modes to meet the users needs. Four new control modes have been developed and integrated to the software architecture of Manus robot.

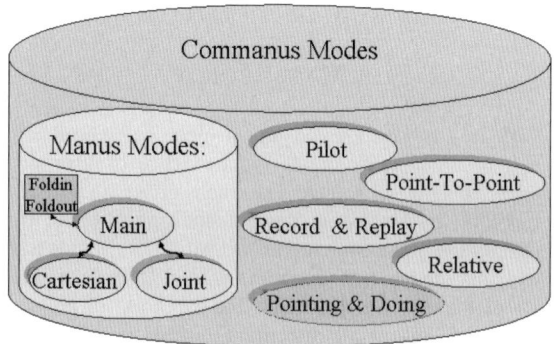

Figure 2 : Commanus command architecture

The first additional mode is called « **Record & Replay Mode** » which allows the user to record specific positions and movements, and when it is required to reach them later, only an automatic movement generated according to the recorded point will be required. When the gripper is in a specific position in a given space and a given configuration, the user will be able to record the coordinates of the point to be able to return directly to this point any time using only one action on the input device [4]. The second one is the "**Pilot mode**" which allow to handle Manus robot in the direction of the gripper following the main axis.

This mode has been developed mainly when using a 2D joystick to control the robot when practicing robot forward action in the direction of the target, which is similar to the human approaching movement to grip an object. The third mode is the **Relative mode** which allow movements of the gripper, usually where it is near the target, to perform small defined steps relatively to the target position and to the current position and orientation of the gripper. This mode is useful during task requiring high accuracy, such as inserting a video tape in the VCR.

During evaluation, we have noticed that, even with this modes, the user have to perform the same sequence of command action when processing repetitive tasks, such as eating or drinking with the Manus. This is natural for human movement, but heavy when using a robot The strategy we have followed was to try to identify the repetitive movements and implement them as automatic gestures available for the user. A gesture library was integrated in the software architecture and a **path planner** was developed to allow **point-to-point** and **pointing-and-doing** modes.

4. IMPLEMENTATION OF A PATH PLANNER

A. Gesture library

In the human physiology, any complete natural gesture is describe as being two-phased: an initial phase that transports the limb quickly towards the target location and a second long phase of controlled adjustment that allow to reach the target accurately. Those two phases are defined respectively as a transport component and a grasp component [8]. Each component is a spatio-temporal transformation between an initial state, and a final state of the arm. In our approach, we are interested in automating the first phase describe above. The second one require sensors (such as cameras and effort sensors) that are, from a conceptual point of view, impossible to be realised on the Manus.

The gesture library contains a set of generic global gestures that help disabled people to perform complex daily tasks. These gestures correspond to only a portion of any particular task. As shown below (Fig.3), each gesture (G_i) is characterised by an initial operational variable of the robot workspace (O_{ii}) corresponding to the initial robot arm configuration, and a final operational variable (O_{if}) corresponding to the final robot arm configuration.

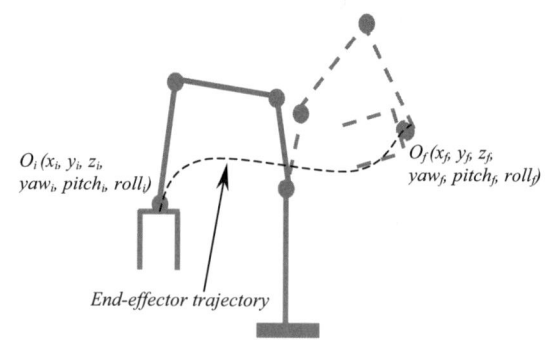

Figure 3: Robot configurations characterising a gesture

Each variable (O_i) is defined in the Cartesian space by the gripper position (x_i, y_i, z_i) and orientation (yaw_i, $pitch_i$, $roll_i$). The gestures generated by our system are linked only to the final operational variables. The **path planner** is able, from any initial arm configuration, to generate the appropriate trajectory to reach the final configurations.

We have pre-recorded twelve final operational variables which allow the user to record two others (Fig.4).

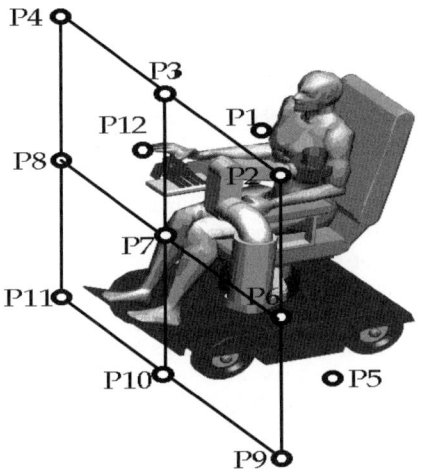

Figure 4: Point-to-Point control mode

B. Obstacles avoidance

To improve the point-to-point mode which perform movement in blind way without taking into account environmental obstacles, we have decided to design a new strategy based on the dynamic generation of 3D obstacle. One example of commonly performed task with the Manus is gripping a glass from a table as shown in figure 5.

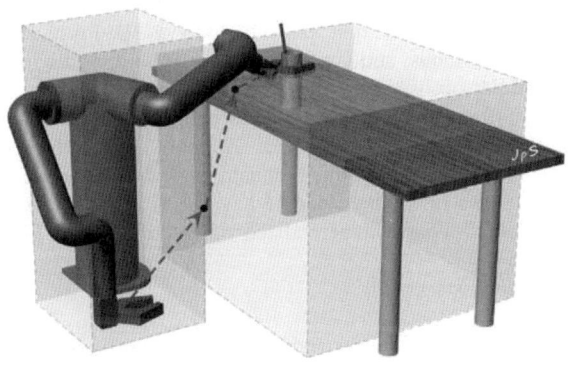

Figure 5 : Path planning process avoiding obstacle

The path planner takes into account obstacles located inside the working space of the robot between the initial and final configuration of the arm. Physical obstacles are virtually encapsulated in boxes playing the role of forbidden areas. In this case a first box represents the arm column which could not be crossed during the movement, and a second box representing the table obstacle. Intermediates points defining the robot trajectory are generated by the use of a developed avoidance algorithm based 3D geometrical calculation [2][3][10][11]. The path planner integrates the intermediate points when calculating an automatic gesture as defined above. Consequently, the control of the arm should be simplified for the user and this will offer gain in term of time and of control efficiency.

Actually, the 3D virtual boxes are statically defined to validate path planner functionalities. A dynamic definition of forbidden area is currently under design to allow the user defining obstacles according to his own changing environment.

This concept is complementary to the co-autonomy concept, described below, where, in case of a non defined obstacle, the user should always have the ability to modify the trajectory generated by the planner trajectory.

5. TOWARDS THE CO-AUTONOMY CONCEPT

The co-autonomy concept was recently introduced as a promising way to design assistive robots intended to meet the needs of disabled people[1]. This concept is based on the control charring between the human and the assistive robot. This approach was also proposed for obstacle avoidance in tele-robotic systems applications in hazardous environments [6].

Three types of situations were mentioned to define the co-autonomy concept.

1. the user is in total control.
2. the machine is in total control.
3. The user and the machine share the control.

The software command architecture is designed to fit this co-autonomy concept. In the first version of the command architecture, the first and the third type of the situations cited above can occur. Users are in total control when they are using the **Cartesian Mode** and share the control with an autonomous controller when they are using the **Point-to-Point Mode**. As describe in [4] the gesture in the Point-to-Point mode is controlled by the user by pressing, for example, the keypad button continuously until completion. The gesture stops if the button is released or continues otherwise (We can qualify such control as a pseudo-sharing control). This was designed to prevent collisions with the user, other persons, or obstacles.

Pressing a button of a keypad or pushing a stick of a joystick until completion of the gesture may sometimes be exhausting for some users with severe disability. To prevent from this fatigue, we thought to include the second type of situation of the co-autonomy concept in the command architecture and integrate the user in the autonomous control loop, ie. allowing him/her to intervene during the automated gestures. The user may then, during the progression of the arm towards the target, make gripper position adjustments. For example, it could occur that the path planner generates a trajectory that would go throw an obstacle. In this case, a collision of the arm with the obstacle will happen.

The user may then, act on the input device to ovoid this collision. Such a situation may be done, as shown in Figure 5, in three phases. An automatic phase, where the end-effector follows the trajectory processed by the path planner, a semiautomatic phase where the user intervenes to avoid the obstacle, and finally, another automatic phase, when the user stop intervening, and where a new trajectory towards the target is generated. Such control mode that we have called **Pointing-and-Doing** Control Mode which will complement the Point-to-Point Mode.

As shown in figure 6, the task is performed following different phases :

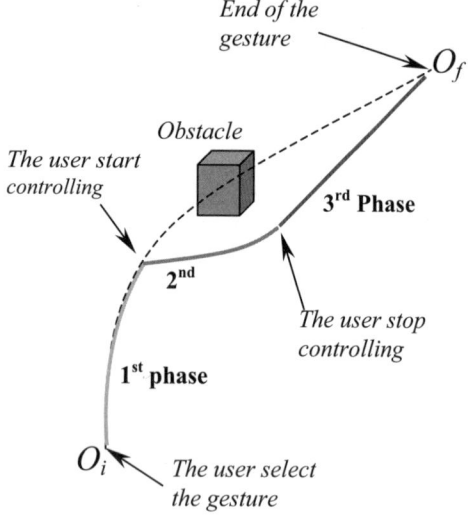

Figure 6: Pointing-and-Doing mode

1st phase: *autonomous phase*
The end-effector follows the trajectory processed by the path planner

2nd phase: *semi-autonomous phase*
The user intervenes during the autonomous phase to avoid the obstacle.

3rd phase: *autonomous phase*
The user stop intervening: A new trajectory towards the target is generated.

6. QUANTITATIVE USERS NEEDS ANALYSIS

A. Methodology

The aim was to develop original methods based on quantitative evaluation to analyze accurate data on the usability of the Manus robot and particularly the contribution of the new added modes. The idea was to record all the actions performed by the users on the input devices. The log file generated contains each command executed by user on device, the execution time of the robot, the corresponding mode, the robot gripper joint and position coordinates, and some other data. This method allow to see which time the user spends in each mode, how many actions he makes in each mode, and how many warnings and error messages has been generated.

The evaluation process is decomposed on tow phases: The **learning phase** and the **evaluation phase** of Manus to perform some specific tasks. During the learning phase, which could last from 10 minutes to an hours depending on the users; the users learn how to control Manus, how to use input device functionalities and how to swap between different control modes described above.

Below we present the results which correspond to eight months of evaluation recording with quadriplegic patient, mainly having spinal cord injuries and muscular dystrophies, at the rehabilitation hospital of Garches where our team installed an evaluation site.

The preliminary results presented correspond to eleven users from fifteen: Eight patients used a 3D joystick to control the Manus, and six users used a 16 buttons Keypad, and one person used a mouse scanning device. These eight months of recordings, mainly dedicated to the Commanus version, correspond to more than 37 hours of effective use Manus robot in institution. Evaluation outside the hospital and at homes of disabled people have been also performed, but with the Manus commercialized version of the robot [7].

B. Preliminary results

The first graph (Fig. 7) shows the time repartition during the whole evaluation duration (134.360.392ms =37 hours an 17 minutes):

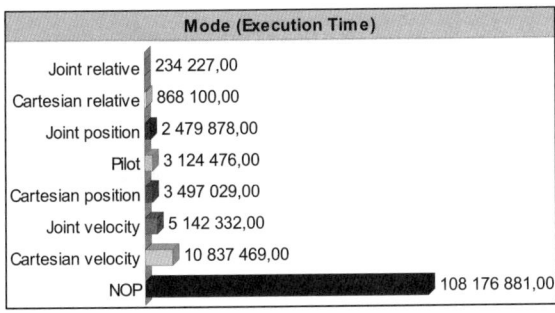

Fig. 7 Time repartition for control modes

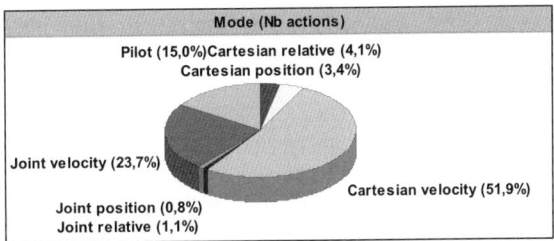

Fig. 9 number of action repartition for control modes

The "**NOP**" time (no actions or rest time) is considerable and represents **80,5%** of the total duration of the evaluation. But we have to make the distinction between four types of rest times:

- A no action time: when the user takes a real rest without switching off the Manus;
- A cognitive time: when the user is thinking on the sequence of action he plans;
- A physiological motor time: the physiological time necessary to execute a movement with hand or finger.

The "Cartesian velocity mode", corresponding to the **Cartesian Mode**, is the most frequently used (**8,1%** in time) and could be processed with three different speeds: slow, medium and high speed. As shown in figure 8, the Cartesian velocity mode speed medium (2) is the most frequently used (53% in time) where the "Cartesian mode" in high speed (3) is not really used (1,2%), only when users want to do large movement of the arm robot.

When a user wants to perform complex tasks, or when the gripper is close to the target, he usually chooses the "Cartesian mode" in low speed.

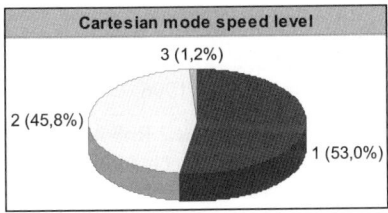

Fig. 8 Cartesian velocity speed repartition

The representation in term of events or actions performed on the input device is shown below (Fig. 9). The whole recordings time correspond to 9715 Events actions sent to the robot

Users manipulation on the input device generated 633 events without any robot activity (keypad buttons or joystick event without function). The Robot generated 92 warnings messages (robot in deadlocking configuration, limit of working space reached ...). To recover deadlocking configuration, the user have recourse to the **joint mode** to ovoid restarting the system.

During the evaluation users had to perform the same tasks, and to follow the same scenarios. Different parts have been performed according to the different control modes:

P-01: part using the basic modes (Cartesian and Joint)
P-02: P-01+Pilot mode
P-03: P-01+Point to point mode
P-04: P-01+Relative mode
P-09: free scenario (all modes)

Figure 10 shows parts repartition to some expert users (U-0i) to perform the all the tasks:

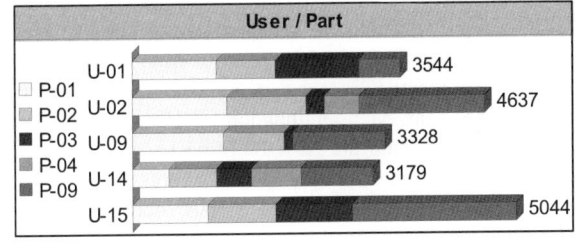

Fig. 10 Users-parts of scenarios repartition

We remark that within the P-02 and P-03 the users needed less number of actions, which means that the Pilot Mode and the Point-to-Point mode contribute in the reduction of the number of actions and then to the reduction of execution time.

7. CONCLUSION

In this paper we have tried to describe our design approach of an adaptable control system for Manus robot. The new architecture is designed to meet the disabled user needs in term of manipulation of the assistive robot Manus. these development are based on preliminary results obtained from quantitative and qualitative evaluation with the participation of disabled people [9].

This system is designed on one hand, to reduce manipulation problems that disabled users meet during complex tasks, and on the other hand, to solve the problems linked to the user-interface. With its new functions, we plan to reduce the task time and the number of commands necessary for complex task. For example, the task serve and drink a cup of water using Manus, and a keypad as an input device, takes about 5minutes and requires about 180 commands. As it was noticed in [7], the users spend more than 50% of the task duration seeking for the good strategies to reach the target or seeking for the appropriate button. The interest of the gesture library is to cop with these problems and propose a more intuitive control. For example, one command in the Point-to-Point mode is sufficient to perform the same results than 10 necessary commands with the Cartesian Mode.

The evaluation of the new architecture allowed us to bring some improvement to the system. The first trials with disabled patients showed their interest regarding the new added modes. The result obtained, are only preliminary results, and we can not yet pronounce on the real contribution of the new command architecture modes in the daily use of Manus at home and outside. More evaluations in real life conditions with the help of disabled people are necessary to test all the new functions offered by the proposed new system.

The actual development realised during this project leaded to a new software architecture for Manus which was integrated through the European Commanus project which has ended this year. The continuation of this research work will be insured through the AMOR[1] project which will start soon with the support of the European Commission. The aim is to propose a new generation of Manus robot taking into account the users requirements.

8. ACKNOWLEDGMENTS

The authors would like to thank the people who have participated actively in this presented research work, particularly J.P. Souteyrand from INSERM U.483 for graphical design, C. Dumas, ergotherapist, from the rehabilitation hospital of Garches, and C. Rose from the AFM (French Muscular dystrophies Association) for his support and for providing us two Manus robots.

---

[1] AMOR project EEC Growth program: Mechatronic upgrade & wheelchair integration of the Manus Arm manipulator. Partners involved: Exact dynamics, TNO-TPD and Koningh in the Netherlands, Ideasis and ExpertCam in Greece, Lund University in Sweden, HMC in Belgium, and INT and AFM in France

9. REFERENCES

[1] Chatila R., P. Moutarlier, N. Vigouroux, "Robotics for the Impaired and Elderly Persons", *IARP Workshop on Medical Robots*. Vienna, Austria. 1-2 Oct 1996.

[2] Coiffet P., «La Robotique, Principes et Applications», Ed. Hermes, Paris, 1992

[3] Coiffet P., «La Robotique, Principes et Applications», Ed. Hermes, Paris, 1992

[4] Didi N, Mokhtari M., Roby-Brami A., « preprogrammed gestures for robotic manipulators : An alternative to speed up task execution using Manus », In Proc. ICORR'99, Palo alto, California, July 1999

[5] Dombre E., Khalil W., «Modélisation et Commande des Robots», Ed. Hermes, Paris, 1988

[6] Guo C., T.J Tarn, N. Xi, A.K Bejczy," Fusion of Human and Machine Intelligence for Telerobotic systems", *the IEEE International Conference on Robotics and Automation.*, Nagoya, Japan, May 1995

[7] Heidmann J, Mokhtari M., Méthode d'évaluation quantitative et qualitative appliquée au développement d'une aide technique robotique pour la compensation du handicap moteur. Ed. Euopia Paris, RIHM (Revue d'Interaction Homme-Machine). Vol. 3, n°1, May 2002. P79-99.

[8] Jannerod M., "Intersegmental coordination during reaching at natural visual object". In J. Long & A. Baddeley (Eds.) *Attention and performance IX,* P153-169, Hillsdale, NJ: Lawrence Erlbaum Associates

[9] Mokhtari M., Didi N., Roby-Brami A., « A multidisciplinary approach in evaluating and facilitating the use of the Manus robot. IEEE-ICRA'99, Detroit, Michigan. May 1999.

[10] Pruski A., «Robotique Générale», Ed. Marketing, Collection Ellipses, Paris, 1988

[11] Pruski A., «Robotique Mobile, La Planification de Trajectoire», Ed. Hermes, Paris, 1996

[12] Vertut J., Coiffet.P., «Les Robots ». Tome 3a : Téléopération, Evaluation des technologies. Ed. Hermes. Paris, 1984.

Modeling the Kinematics and Dynamics of Compliant Contact

Vincent Duindam
(v.duindam@ieee.org)

Stefano Stramigioli
(s.stramigioli@ieee.org)

Drebbel Institute for Mechatronics
University of Twente, The Netherlands

Abstract

In this paper, we discuss the modeling of the kinematics and dynamics of compliant contact between bodies moving in Euclidean space. First, we derive the kinematic equations describing the motion of the contact point when two rigid bodies are rolling on each other. Secondly, we extend these results to describe the motion of the closest points between two rigid bodies moving freely in space. Then, we use these results to model compliant contact between bodies, using a spatial spring and a damper to model energy stored and dissipated during contact.

1 Introduction

In many models of walking machines (e.g. [1, 2, 3]), the contact between a foot and the ground is modeled as a discontinuous change of velocity of the foot, and instant dissipation of energy. This type of contact model can be used to accurately describe contact with a flat, hard surface. However, when the soil is something softer or more curved (such as sand, mud and rocks, typically encountered in outdoor and space applications), a more detailed contact model is needed to capture the compliance and the shape of the contact in order to be able to analyze the stability and robustness of the walking robot.

In this paper, we look at the modeling of this type of contact. More specifically, we look at the kinematic aspects of compliant contact. The dynamic aspects are also briefly discussed, but we refer to [4] for the details.

The kinematic analysis consists of two parts: First, *regular contact kinematics* describes the movement of the contact point between two rigid bodies as the bodies roll and slide over each other. In other words, it describes the velocity of two points (one on each body) that have zero distance. Second, *generalized contact kinematics* describes the movement of two points (one on each body) that have the smallest distance between them. Both regular and generalized contact kinematics are important in the modeling of compliant contact, as contact can be lost when two objects bounce off each other.

The kinematics of regular contact have been modeled before [5, 6], and recently these results have been extended to the generalized kinematics case [7]. Both approaches however rely on an orthogonal parameterization of the surfaces, and they need the definition of certain matrices and extra contact coordinates, which mystify the resulting equations and make it much harder to get a physical interpretation.

Contrary to the results mentioned before, we do not use a coordinate-based approach in this paper, so the results hold for any parameterization of the surfaces (not only orthogonal parameterizations), and we do not require the introduction of special extra coordinates and matrices. It thus gives a simpler, intuitive idea of the contact kinematics, and the equations are still physically interpretable and not just mathematical results. We used the results of this paper in a contact model based on the Port-Controlled Hamiltonian approach [8].

This paper is organized as follows: first, Section 2 provides the necessary mathematical preliminaries and notation about the dynamics of rigid bodies and about their surfaces. Then, Section 3 discusses the kinematics of rigid bodies in point contact, and Section 4 extends the results to the kinematics of rigid bodies moving freely in space. Section 5 briefly discusses the use of the kinematic equations in a lumped model of compliant contact between two bodies, and Section 6 shows simulations of this contact model. Finally, Section 7 presents the conclusions and an outlook on future research.

2 Preliminaries and Notation

2.1 Rigid Body Dynamics

In this paper, we deal with rigid bodies moving in the Euclidean space $\mathcal{E}$, which means we can describe the position and orientation of every body by an element of the special Euclidean group $SE(3)$, once a reference frame has been chosen. As shown for example in [9, 10], elements of this group can be represented by a homogeneous matrix of the form

$$H^i_j = \begin{bmatrix} R^i_j & p^i_j \\ 0 & 1 \end{bmatrix}$$

where R^i_j is a rotation matrix (element of the special orthonormal group $SO(3)$) and p^i_j is a vector in $\mathbb{R}^3$. H^i_j denotes the change of coordinates from a right-handed coordinate frame Ψ_j to another right-handed coordinate frame Ψ_i and can thus be used for example to describe the position and orientation of a body (with attached coordinate frame Ψ_j) relative to a reference (inertial) coordinate frame (Ψ_i).

The instantaneous velocity of a body i with frame Ψ_i relative to a body j with frame Ψ_j can be represented by a twist $T^{k,j}_i$, with

$$T^{k,j}_i = \begin{bmatrix} \omega^{k,j}_i \\ v^{k,j}_i \end{bmatrix}$$

where $\omega_i^{k,j}$ denotes the angular velocity of body i relative to body j expressed in coordinate frame Ψ_k, and $v_i^{k,j}$ denotes the instantaneous velocity (relative to frame Ψ_j) of the point fixed in frame Ψ_i that passes through the origin of frame Ψ_k. A twist can be regarded as the derivative of a homogeneous matrix in the following way, using what is called a right translation of a Lie group [11]:

$$\tilde{T}_i^{j,j} := \begin{bmatrix} \tilde{\omega}_i^{j,j} & v_i^{j,j} \\ 0 & 0 \end{bmatrix} = \dot{H}_i^j H_j^i \quad (1)$$

where $\tilde{\omega} = -\tilde{\omega}^T$ is the matrix equivalent to $(\omega \times \cdot)$.

We can also define a wrench W_i^k (the dual of a twist), which describes the generalized forces acting on body i and expressed in frame Ψ_k, as

$$W_i^k = \begin{bmatrix} \tau_i^k \\ F_i^k \end{bmatrix}$$

where F_i^k denotes the linear force and τ_i^k the momentum, acting on the point in the origin of frame Ψ_k. The dual product of a twist and a wrench (when expressed in the same coordinate frame) is equal to a power flow. More information on twists and wrenches can be found in [9, 10].

2.2 Surface Description

Consider a rigid body with a smooth, oriented surface $\mathcal{S}$, embedded in the Euclidean space $\mathcal{E}$. To this body, we rigidly attach a coordinate frame Ψ. In the frame Ψ, we can describe the surface (locally) as a bijective mapping $f : \mathcal{D} \to \mathcal{S}$, which assigns to each set of local coordinates $u \in \mathcal{D} \subset \mathbb{R}^2$ a point of the surface. The mapping f is a (local) parameterization of the surface, and we assume this parameterization to be well-defined, i.e. the derivative mapping $f_* = \frac{\partial f}{\partial u}$ is continuous and has kernel zero, i.e. the partial derivatives $\frac{\partial f}{\partial u}$ are independent at all points.

At each point of the surface, we can find the unit vector $n(p)$ normal to the surface (we can compute this for example by taking the cross product between the partial derivatives of f). We can identify these unit vectors with points on the unit sphere $\mathbb{S}^2$, if we think of the point on the sphere as the tip of the normal vector with its base point in the center of the sphere.

The Gauss mapping $g : \mathcal{S} \to \mathbb{S}^2$ is defined as the mapping which takes a point p on the surface and returns a point $g(p)$ on the sphere, corresponding to the unit normal at p. The smoothness and orientability of the surface ensure that the normal vector varies smoothly over the surface, and hence the mapping g is smooth. This means that we can also define the derivative mapping $g_* : T\mathcal{S} \to T\mathbb{S}^2$. This derivative can be interpreted as follows: if we move tangent to the surface at velocity $\zeta \in T\mathcal{S}$, then the normal vector changes with velocity $g_*\zeta \in T\mathbb{S}^2$. Since the vector $g(p)$ is perpendicular to the surface at p as well as to the sphere at $g(p)$, we can directly regard an element $g_*\zeta \in T_{g(p)}\mathbb{S}^2$ as an element $Pg_*\zeta \in T_p\mathcal{S}$, where P denotes the mapping from $T_{g(p)}\mathbb{S}^2$ to $T_p\mathcal{S}$.

The intuitive meaning of the differential g_* of the Gauss map is curvature: the vector $g_*\zeta$ for some $\zeta \in T\mathcal{S}$ describes the curvature of the surface when moving at velocity ζ. If $g_*(p)\zeta = 0$ for all $\zeta \in T_p\mathcal{S}$, then the surface is locally flat at p. If $\langle \zeta, Pg_*(p)\zeta \rangle > 0$ for all $\zeta \in T_p\mathcal{S}$, then we say that the surface is locally absolutely convex[1] at p.

Figure 1 shows the relations between the various mappings and spaces. It is important to note that f is a bijective mapping, hence it uniquely identifies coordinate-pairs to points on the surface and its derivative mapping f_* is invertible. This means that although the equations in the following sections do not contain local coordinates u or the surface parameterization f, we can always find coordinate expressions for these equations using f, f_* and their inverses.

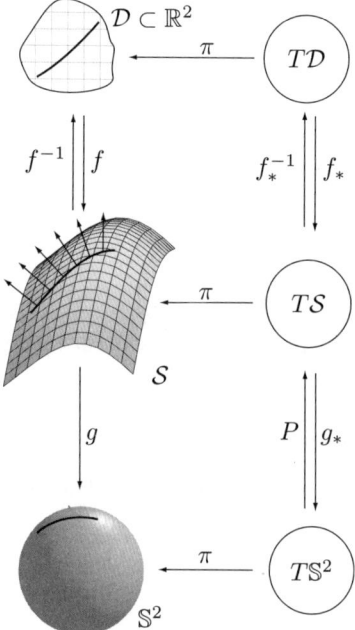

Figure 1: *Relation diagram showing the mappings between the coordinate patch $\mathcal{D}$, the surface $\mathcal{S}$, the unit sphere $\mathbb{S}^2$ and their tangent spaces. The canonical projection π is added for completeness; it takes an element (p, ζ) of the tangent bundle and returns its base point p.*

2.3 Numerical Notation

Since we want to use the formal results in numerical simulation, we need ways to represent the geometrical ideas in vectors and matrices. If we use the formats denoted in Table 1, then the geometric equations in the following sections can be implemented directly into numerical equations.

For example, say that we want to compute the mapping g_* and we only have a representation in local coordinates, i.e. a four-by-two matrix $\hat{g}_* : T\mathcal{D} \to T\mathbb{S}^2$ such that $\hat{g}_* = g_* f_*$. We can derive g_* from this, but then we need f_*^{-1}. Now f_* is represented by a four-by-two matrix (mapping the time derivative of local coordinates to the time derivative of a point in $\mathcal{E}$) and it cannot be inverted in the usual sense. However, if we take the definition given in

---
[1] We explicitly use the term 'absolutely convex' (which just means 'convex' in the usual sense) to distinguish it from the term 'relatively convex', which we define in Section 3.

the table (which is just the Moore-Penrose pseudo inverse), then the computation can be performed and $\hat{g}_* f_*^{-1}$ is a four-by-four matrix as required, mapping vectors $\dot{p} \in \mathbb{R}^4$ to vectors $g_* \dot{p} \in \mathbb{R}^4$ (of course, this mapping only has physical meaning for $\dot{p} \in T\mathcal{S}$).

Table 1: *Numerical implementation of the geometrical equations. A diamond ($\diamond$) denotes an arbitrary number.*

type	example	numerical format
local coordinates	u	$[\diamond \ \diamond]^T$
point in $\mathcal{E}$	p	$[\diamond \ \diamond \ \diamond \ 1]^T$
free vector in $\mathcal{E}$	$\dot{p}$	$[\diamond \ \diamond \ \diamond \ 0]^T$
surface parameterization	$f(u)$	$[\diamond \ \diamond \ \diamond \ 1]^T$
Gauss map	$g(p)$	$[\diamond \ \diamond \ \diamond \ 0]^T$
tangent mapping	f_*	$\begin{bmatrix}\diamond & \diamond & 0 \\ \diamond & \diamond & 0\end{bmatrix}^T$
inverse tangent mapping	f_*^{-1}	$(f_*^T f_*)^{-1} f_*^T$

3 Regular Contact Kinematics

We first consider the case of two rigid bodies in point contact, moving with a relative velocity represented by a twist $T_1^{2,2} = -T_2^{2,1}$ (this is exactly the case described by Montana [5]). We attach to each body i a coordinate frame Ψ_i and we assume to have a description f_i of the surface of this body, expressed in frame Ψ_i.

If we then express the location of the point of contact as two points p_1 (expressed in frame Ψ_1) and p_2 (expressed in frame Ψ_2), then if the two bodies are in point contact, we have

$$\begin{cases} p_1 = H_2^1 p_2 \\ g_1 = -H_2^1 g_2 \end{cases} \quad (2)$$

where we abbreviated $g_i := g_i(p_i)$. These equations just say that for point contact, the two contact points must be the same (when expressed in the same coordinate frame, in this case Ψ_1) and the normal vectors to the surfaces must be opposite.[2]

To obtain the kinematic equation relating the velocities of the contact points to the velocity of the bodies, we only need to take the time-derivative of (2) to obtain:

$$\begin{cases} \dot{p}_1 = \dot{H}_2^1 p_2 + H_2^1 \dot{p}_2 \\ \dot{g}_1 = -\dot{H}_2^1 g_2 - H_2^1 \dot{g}_2 \end{cases} \quad (3)$$

$$\begin{cases} \dot{p}_2 = H_1^2 \dot{p}_1 - H_1^2 \dot{H}_2^1 p_2 \\ g_{1*} \dot{p}_1 = \tilde{T}_2^{1,1} g_1 - H_2^1 g_{2*} \dot{p}_2 \end{cases} \quad (4)$$

where we pre-multiplied (3) by H_1^2 to obtain an expression for $\dot{p}_2$. If we now substitute this expression into the second

---

[2]Note that we use a homogeneous matrix H_2^1 to change coordinates for points (p_2) as well as for free vectors (g_2). Normally, free vectors only need to be rotated (using the rotation part of the coordinate transformation), but since we express these vectors numerically as a four-by-one matrix with its last element zero, we can just as well use multiplication by the full homogeneous matrix.

line of (4) and repeat the whole derivation with objects 1 and 2 switched, we obtain the desired kinematic equations:

$$\begin{aligned}(g_{1*} + H_2^1 g_{2*} H_1^2) \dot{p}_1 &= \tilde{T}_2^{1,1} g_1 - H_2^1 g_{2*} \tilde{T}_1^{2,2} p_2 \\ (g_{2*} + H_1^2 g_{1*} H_2^1) \dot{p}_2 &= \tilde{T}_1^{2,2} g_2 - H_1^2 g_{1*} \tilde{T}_2^{1,1} p_1 \end{aligned} \quad (5)$$

Let us now briefly discuss the conditions under which these equations have unique solutions $\dot{p}_1, \dot{p}_2$. Because of the symmetry, we only consider the first equation, i.e. the equation for $\dot{p}_1$. Because the four-by-four matrix $(g_{1*} + H_2^1 g_{2*} H_1^2)$ has a non-zero kernel, we cannot simply invert this matrix and always get a unique result. Instead, we must look at the equation from a geometrical point of view.

First of all, since we look at the motion of the contact point over the surface, we must have $\dot{p}_1 \in T_{p_1} \mathcal{S}_1$.

Secondly, since the domain of the mapping g_{2*} is $T_{p_2} \mathcal{S}_2$ and not all vectors in $\mathcal{E}$, we must have $\tilde{T}_1^{2,2} p_2 \in T_{p_2} \mathcal{S}_2$. This constraint means that the velocity of the instantaneous contact point ($\tilde{T}_1^{2,2} p_2$) can not have a component perpendicular to the surface, thus constraining the allowed relative motion to five degrees of freedom (which is clear from a physical point of view).

Finally, we need to ensure that a unique solution $\dot{p}_1$ exists for any twist satisfying the constraint above. Since both $\tilde{T}_2^{1,1} g_1$ and $H_2^1 g_{2*} \tilde{T}_1^{2,2} p_2$ are tangent to the surface, the co-domain of the matrix $(g_{1*} + H_2^1 g_{2*} H_1^2)$ must be the whole tangent plane to the surface, i.e. the matrix must have rank two. For physical reasons (no intersection of the surfaces) this means that the two surfaces must be *relatively convex*: the two non-zero eigenvalues of $(g_{1*} + H_2^1 g_{2*} H_1^2)$ must be larger than zero. Physically, relative convexity means that if one surface is concave, then the other body must be extra convex. Absolute convexity (as defined in Section 2) can be considered as a special case: an absolutely convex surface is relatively convex to a plane.

4 Generalized Contact Kinematics

In this section, we extend the results of Section 3 to the more general case as depicted in Figure 2: we do not consider just the kinematics of the point of contact between the two bodies, but we look at the kinematics of the points on the surfaces which have the shortest (in the Euclidean sense) distance between them. We call this problem the *generalized contact kinematics* problem.

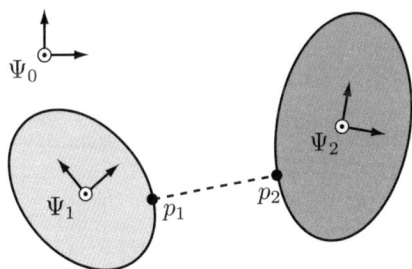

Figure 2: *Two rigid bodies and their generalized contact points p_1 and p_2.*

We use p_i, $i = 1, 2$ to denote the point on body i expressed in frame i such that the distance between p_1 and p_2

is the minimum distance between the bodies. This implies that the line connecting p_1 and p_2 must be perpendicular to both surfaces, which can be translated into the following equations:

$$\begin{cases} p_1 + \Delta g_1 = H_2^1 p_2 \\ g_1 = -H_2^1 g_2 \end{cases} \quad (6)$$

where $\Delta \in \mathbb{R}$ denotes the 'signed distance' between the generalized contact points:

$$\Delta = \langle g_1, H_2^1 p_2 - p_1 \rangle \quad (7)$$

i.e. $\Delta > 0$ means there is a distance $|\Delta|$ between the bodies, and $\Delta < 0$ means the bodies have a maximum penetration distance of $|\Delta|$. The use of this definition for distance (instead of the usual $\|H_2^1 p_2 - p_1\|$) turns out to be very useful in the modeling of contact dynamics in Section 5.

Theorem 1 *Given two rigid bodies and the generalized contact points as defined in (6). If the bodies are absolutely convex, then the velocity of the generalized contact points is uniquely determined by the following equations:*

$$\begin{cases} \left(g_{1*} + H_2^1 g_{2*} H_1^2 (I + \Delta g_{1*})\right) \dot{p}_2 = \\ \qquad \tilde{T}_2^{1,1} g_1 + H_2^1 g_{2*} (\dot{\Delta} g_2 - \tilde{T}_1^{2,2} p_2) \\ \left(g_{2*} + H_1^2 g_{1*} H_2^1 (I + \Delta g_{2*})\right) \dot{p}_1 = \\ \qquad \tilde{T}_1^{2,2} g_2 + H_1^2 g_{1*} (\dot{\Delta} g_1 - \tilde{T}_2^{1,1} p_1) \end{cases} \quad (8)$$

where $\tilde{T}_2^{1,1} = -H_2^1 \tilde{T}_1^{2,2} H_1^2$ can be any relative twist of the two bodies and $\Delta > \Delta_{min}$ for some $\Delta_{min} < 0$ depending on the surfaces and Δ is defined as in (7).

Proof We first compute the time derivative of Δ, e.g. the change of distance between the bodies.

$$\dot{\Delta} = \langle \dot{g}_1, H_2^1 p_2 - p_1 \rangle + \langle g_1, \dot{H}_2^1 p_2 + H_2^1 \dot{p}_2 - \dot{p}_1 \rangle$$
$$= \langle g_1, \dot{H}_2^1 p_2 \rangle \quad (9)$$
$$= \langle g_1, \tilde{T}_2^{1,1} H_2^1 p_2 \rangle \quad (10)$$

where (9) results since the normal vector g_1 is always perpendicular to the velocities ($\dot{p}_1$ and $H_2^1 \dot{p}_2$) of the contact points over the surface and since $\dot{g}_1$ is perpendicular to $H_2^1 p_2 - p_1$, and (10) results by applying (1).

Using this result for $\dot{\Delta}$, we can compute the time derivative of (6) to obtain the kinematics equation:

$$\begin{cases} \dot{p}_1 + \dot{\Delta} g_1 + \Delta g_{1*} \dot{p}_1 = \dot{H}_2^1 p_2 + H_2^1 \dot{p}_2 \\ g_{1*} \dot{p}_1 = -\dot{H}_2^1 g_2 - H_2^1 g_{2*} \dot{p}_2 \end{cases}$$
$$\begin{cases} \dot{p}_2 = H_1^2 (\dot{p}_1 + \dot{\Delta} g_1 + \Delta g_{1*} \dot{p}_1 - \dot{H}_2^1 p_2) \\ g_{1*} \dot{p}_1 = \tilde{T}_2^{1,1} g_1 - H_2^1 g_{2*} \dot{p}_2 \end{cases} \quad (11)$$

If we now substitute the first equation of (11) into the second, and repeat the whole derivation with objects 1 and 2

switched, we immediately obtain (8). Note that for $\Delta \equiv 0$, we recover the regular contact kinematics (5).

Now consider again the requirements for a unique solution $\dot{p}_1$. First we look at the term $\dot{\Delta} g_2 - \tilde{T}_1^{2,2} p_2$ in (8) and take the inner product with g_2:

$$\langle g_2, \dot{\Delta} g_2 - \tilde{T}_1^{2,2} p_2 \rangle = \dot{\Delta} - \langle H_2^1 g_2, H_2^1 \tilde{T}_1^{2,2} p_2 \rangle$$
$$= \dot{\Delta} - \langle g_1, \tilde{T}_2^{1,1} H_2^1 p_2 \rangle$$
$$= 0$$

where we used (6), (10), and the fact that a homogeneous transformation preserves the inner product. This shows that $\dot{\Delta} g_2 - \tilde{T}_1^{2,2} p_2$ is always tangent to the surface, so the right-hand side of (8) is well-defined for all twists $T_1^{2,2}$.

Whether $(g_{1*} + H_2^1 g_{2*} H_1^2 (I + \Delta g_{1*}))$ has rank two cannot be easily related to properties of the objects, since it also depends on the distance Δ. Even though an object may be relative convex (i.e. the contact points vary smoothly as the objects roll over each other), the contact points can jump when the objects are not in contact and move at a certain distance from each other. However, if the objects are absolutely convex, then invertibility is ensured for any $\Delta > \Delta_{min}$ for some $\Delta_{min} < 0$, i.e. for any positive distance, and for small enough penetrations, where Δ_{min} is the largest distance for which the matrix has rank less than two. $\square$

Although the kinematic equation (8) is similar to the results obtained in [7], the approach we used here does not depend on extra coordinates and orthogonal parameterization, and is therefore more transparent and easier to interpret and understand geometrically.

5 Generalized Contact Dynamics

In this section, we briefly review the concept of a lumped parameter model of compliant contact based on a spatial spring (as discussed in more detail in [4]) to show the modeling application of the kinematics as derived in the previous section. In the modeling process, we assume that the collision between the two objects is partially elastic; the energy stored during the collision is modeled as a spatial spring, and the energy dissipated during the collision is modeled as a damper. Since the damper is trivial to model (even if geometric), we only look at the spring in this section. Figures 2 and 3 illustrate the intuitive idea.

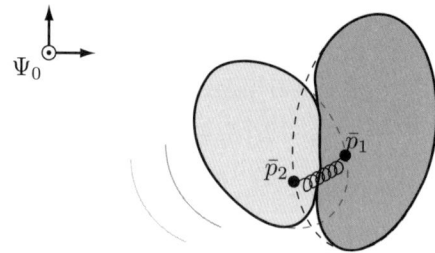

Figure 3: Schematic view of the spring representing the compliant contact as it is attached between the points $\bar{p}_1$ and $\bar{p}_2$ on the un-deformed bodies.

Starting from a no-contact situation, the position of the generalized contact points is monitored using the kinematic equations (7), (10), and (8). When at some time t_1 the bodies come into contact (Δ crosses zero), we connect a spatial spring [12] between two points $\bar{p}_1 = p(t_1)$ and $\bar{p}_2 = p(t_1)$ fixed with body 1 and body 2 respectively, in such a way that the initial center of stiffness is at the relative position and orientation of the bodies at impact.

After the impact, the objects continue to move, and the surfaces of the objects deform. We model this in the following way: the points $\bar{p}_1$ and $\bar{p}_2$ move with the objects as if the ideal, rigid shapes are maintained and penetrate each other. This means that $\bar{p}_1$ and $\bar{p}_2$ move away from each other, and the spring is charged.

Under the influence of external forces and contact (spring/damper) forces, the objects will start to slide and roll over each other. However, we do not want all movement of the bodies to charge the spring: pure rolling of two bodies over each other is not supposed to charge the spring, since no potential energy is stored in that case. On the other hand, when the bodies are sliding over each other (and in this way locally stretching the surfaces), we *do* want the spring to be charged, so that potential energy is stored.

The problem is that it is not so easy to distinguish between rolling and sliding if the bodies share an area of contact. In case of point contact, it would have been easy: rolling is when one body rotates relative to the other body around an axis in the tangent plane to the surfaces at the contact point, and sliding is in all other circumstances.

In the case of area contact, we cannot simply talk about *the* tangent plane to the surface. However, we can talk about the tangent planes at the points p_1 and p_2 of maximum penetration depth: these are completely specified by the normal vectors to the surfaces, i.e. by the Gauss maps. By (6) these tangent planes are parallel, and the 'average tangent plane' can be taken as a third parallel plane in between these two. The exact position of this plane depends on the relative stiffnesses of the two objects. We use this average tangent plane to decompose the twist in rolling and sliding, and then use the sliding component to define the spatial spring as attached between $\bar{p}_1$ and $\bar{p}_2$.

6 Simulations

We implemented the 3D kinematics and dynamics model in the simulation package 20sim [13] and simulated the dynamics of two ellipsoids bouncing on each other and on the floor under the influence of gravity. Since there are three objects, we need to have three copies of the contact model (one between each pair of objects) to be able to model all contact situations. Figure 4 shows a 2D schematic setup of the model. The sub-models are implemented using screw bond graphs [14, 10], which allow for easy modeling of the power ports to capture the energy balance of the system. We use relatively soft settings for the spring to be able to see more clearly what happens.

We drop the two ellipsoids at some distance right above each other, with zero initial velocity. Figures 5 and 6 show the results, indicating also the following time instants in the simulation:

(a) The two objects start from a certain height, with some distance between them, the largest distance is between the black ellipsoid and the ground.

(b) The grey ellipsoid hits the ground first and compresses a bit. When the black ellipsoid hits the grey, the grey

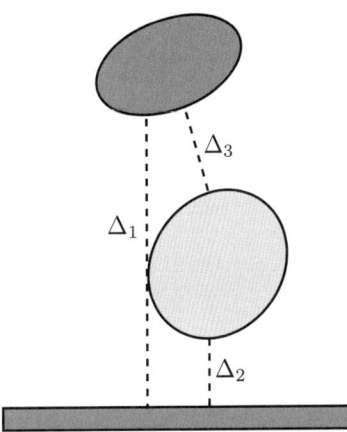

Figure 4: Schematic setup of the simulation model. We use three copies of the contact model to model all possible collisions between the three objects.

is penetrated more into the ground.

(c) The two ellipsoids start to roll over each other.

(d) As the black ellipsoid rolls over the grey, it approaches the ground fast.

(e) The black ellipsoid touches the ground.

(f) Both ellipsoids roll away, creating a distance between them.

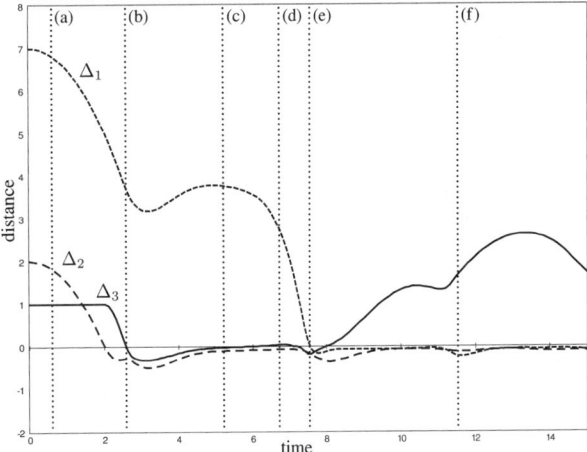

Figure 5: Time evolution of the distances between the three objects. The labels Δ_1, Δ_2, Δ_3 correspond to the labels in Figure 4, and the labels (a) through (f) correspond to the labels in Figure 6.

7 Conclusions and Future Work

In this paper, we derived the kinematic equations for regular and generalized contact. Starting from simple geometric equations that describe the properties of the contact points,

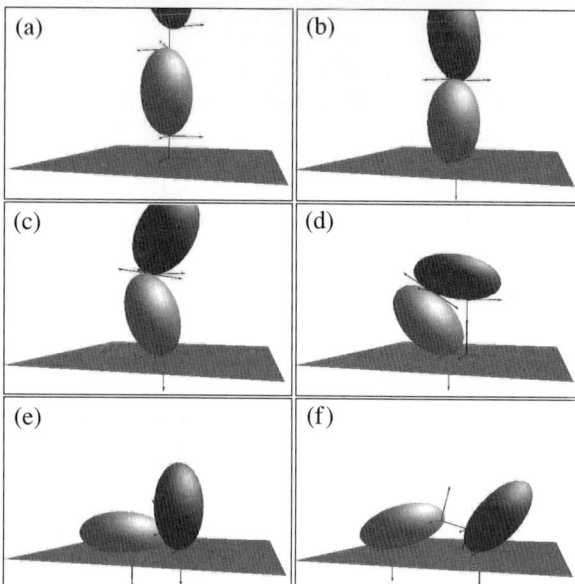

Figure 6: Snapshots of the simulation of two ellipsoids bouncing on each other and on the floor. The labels (a)-(f) correspond to the labels in Figure 5. The plots also show the contact frames, attached to the generalized contact points at each object.

we obtained the kinematic equations in very few steps. We briefly reviewed earlier work on the dynamics of contact that uses the kinematic results, and then showed a simulation of the contact model in action.

The model shows good behavior in terms of physical intuition: the bouncing of the objects on each other looks natural. However, the parameters that describe he compliance of the contact are still hand-picked. The next step in the modeling process would be to study practical results in contact modeling (e.g. Pacejka [15] is well-known for practical models of car tires) to obtain realistic parameter values that capture the behavior of real-life materials.

Another important aspect that needs to be added to the model is slip. In practice, the contact forces cannot grow infinitely large, but at a certain value the contact is broken and the objects slide freely over each other. This aspect is also very important for control, since it limits the forces the robot can exert on the ground and therefore restricts the set of suitable controllers for the robot in these circumstances; walking on sand may require different control than walking on tarmac.

Finally, we assumed in the kinematics analysis that the objects were (absolutely) convex, since this ensures that the contacts points move continuously over the surface. It would be interesting to see how the results could be applied to non-convex objects, for example by approximating these objects by a finite set of convex objects and comparing the distances of this set of possible contact points to obtain the *real* contact point as the minimum of this finite set.

The contact model described in this paper will be used in the 3D modeling of a walking robot to describe the contact of the feet with the soil. We hope it will help us in the analysis and design of controllers that stabilize the robot in walking and make it robust against variations in the soil.

Acknowledgment

This work has been done in the context of the European sponsored project GeoPlex with reference code IST-2001-34166. Further information is available at http://www.geoplex.cc.

References

[1] T. McGeer, "Passive bipedal running," in *Proc. R. Soc. Lond. B240*, 1990, pp. 107–134.

[2] D.E. Koditschek and M. Bühler, "Analysis of a Simplified Hopping Robot," *The International Journal of Robotics Research*, vol. 10, no. 6, pp. 587–605, December 1991.

[3] M. Garcia, A. Chatterjee, A. Ruina, and M. Coleman, "The Simplest Walking Model: Stability, Complexity, and Scaling," *ASME Journal of Biomechanical Engineering*, vol. 120, no. 2, pp. 281–288, April 1998.

[4] S. Stramigioli and V. Duindam, "A Novel Lumped Spatial Model of Tire Contact," in *Proceedings of the IEEE International Conference on Intelligent Transportation Systems*, 2002.

[5] David J. Montana, "The Kinematics of Contact and Grasp," *International Journal of Robotics Research 7(3)*, pp. 17–32, 1989.

[6] David J. Montana, "The Kinematics of Contact with Compliance," in *Proceedings of the IEEE Conference on Robotics and Automation*, 1989, pp. 770–774.

[7] M. Visser, S. Stramigioli, and C. Heemskerk, "Screw Bondgraph Contact Dynamics," in *Proceedings of the IEEE/RSJ International Conference on Intelligent Robots and Systems*, 2002.

[8] Arjan van der Schaft, L_2-*Gain and Passivity Techniques in Nonlinear Control*, Springer-Verlag, 2000.

[9] R.M. Murray, Z. Li, and S.S. Sastry, *A Mathematical Introduction to Robotic Manipulation*, CRC Press, 1994.

[10] S. Stramigioli, *Modeling and IPC Control of Interactive Mechanical Systems – A Coordinate-Free Approach*, Springer-Verlag, 2001.

[11] R. Gilmore, *Lie Groups, Lie Algebras, and Some of Their Applications*, John Wiley & Sons, 1974.

[12] S. Stramigioli and V. Duindam, "Variable Spatial Springs for Robot Control Applications," in *Proceedings of the IEEE/RSJ International Conference on Intelligent Robots and Systems*, 2001, pp. 1906–1911.

[13] Control Lab Products, "20sim version 3.2," http://www.20sim.com, 2002.

[14] H.M. Paynter, *Analysis and Design of Engineering Systems*, M.I.T. Press, 1961.

[15] H.B. Pacejka, *Tyre Factors and Vehicle Handling*, Delft University of Technology, 1978.

Inverse and Direct Dynamics of Constrained Multibody Systems Based on Orthogonal Decomposition of Generalized Force

Farhad Aghili

Canadian Space Agency, Saint-Hubert, Quebec, Canada, J3Y 8Y9
Email: farhad.aghili@space.gc.ca

Abstract— This paper presents a unified approach for inverse and direct dynamics of constrained multibody systems that can be served as a basis for analysis, simulation, and control. The compactness of the dynamics formulation can result in computational efficiency. Furthermore, the acceleration is explicitly related to the generalized force by an introduced "constraint inertia matrix" which is proved to be always invertible. Thus a simulation may proceed even with the presence of redundant constraints or singular configurations. The generalized forces are decomposed onto two orthogonal subspaces which are considered as control inputs for controlling position and constraint force. The motion controller scheme, remarkably, requires no force feedback, proves to be stable, and minimizes actuation force. Finally, numerical and experimental results obtained from dynamic simulation and control of constrained mechanical systems, based on the proposed inverse and direct dynamics formulations, are documented.

I. INTRODUCTION

Many robotic systems are formulated as multibody system with closed-loop topology, such as manipulators with end-effector constraints [10], [18], cooperative manipulators [9], robotic hands for grasping objects [5], parallel manipulators [14], humanoid robot and walking robots [16], and VR/Haptic applications [3]. In essence, simulation and control of such systems call for corresponding direct dynamics and inverse dynamics models, respectively. Mathematically, constraint mechanical systems are modelled by a set of n differential equations coupled with a set of m algebraic equations, i.e. Differential Algebraic Equations (DAE). Although computing the dynamics model is of interest for both simulation and control, the research works done in these two areas are rather divided. Surveys of the existing techniques for solving DAE may be found in [6], while model-based control of constrained manipulators can be found in [10], [7], [18].

The classical Lagrangian method performs poorly in the vicinity of singularities [12], because the augmented inertia matrix is invertible only with a full-rank Jacobian matrix. Model-based approaches for position/force [10] also rely on the assumption that the Cartesian constraints are linearly independent. The concept of coordinate partitioning is used for simulation [17], [15], [11], and control [13]. However, a fixed set of independent coordinates occasionally leads to ill-conditioned matrices [4], when the system changes its topology or the number of degrees of freedom. The augmented Lagrangian formulation proposed in [1], [2] can handles redundant constraints and singular situation. However, this formulation solves the equations of motion through an iterative process, and is not applicable for control.

In this work, we propose a unified formulation for direct and inverse dynamics of constrained mechanical systems which does not require the constraint equations be linearly independent. Also, in this method the vector of the generalized forces is decomposed onto two orthogonal subspaces considered as control inputs for a position and/or force control scheme. Remarkably, the motion control requires no force feedback and proves to be stable while minimizing the Euclidean norm of the generalized force.

This paper is organized as follows: The decomposition of acceleration obtained from the constraint equations is discussed in Section II. Using a projection operator, we derive models of inverse and direct dynamics in sections III and IV which are used as a bases for developing strategies for simulation and control of constrained mechanical systems in sections V and VI. Finally, sections VII and VIII report some simulation and experimental results.

II. Decomposition of Acceleration

The kinematics of a constraint mechanical system can be represented by a set of m nonlinear equations $\mathbf{\Phi}(\mathbf{q}) = \mathbf{0}$, where $\mathbf{q} \in \mathbb{R}^n$ is the vector of generalized coordinate, and $m \leq n$. Differentiating of the constraint equation gives

$$\mathbf{A}\dot{\mathbf{q}} = 0 \qquad (1)$$

where $\mathbf{A} = \partial \mathbf{\Phi}/\partial \mathbf{q}$ is the Jacobian of the constraint equation with respect to the generalized coordinate. Equation (1) specifies that any admissible velocity must belong to the null space of the Jacobian matrix, that is $\dot{\mathbf{q}} \in \mathcal{N}(\mathbf{A})$. Let $\mathbf{P} \in \mathbb{R}^{n \times n}$ be the orthogonal projection onto the null space, i.e. $\mathcal{R}(\mathbf{P}) = \mathcal{N}(\mathbf{A})$. Note that every orthogonal projection operator has these properties: $\mathbf{P}^2 = \mathbf{P}$ and $\mathbf{P}^T = \mathbf{P}$ [8]. Moreover, from the definition one can show that projector operator $(\mathbf{I}-\mathbf{P})$ projects onto the null space orthogonal $\mathcal{N}(\mathbf{A})^\perp$. As such, the velocity constraint equation (1) can be expressed by the notion of the projection operator, i.e.

$$(\mathbf{I} - \mathbf{P})\dot{\mathbf{q}} = 0. \qquad (2)$$

The time-derivative of the above equation gives

$$(\mathbf{I} - \mathbf{P})\ddot{\mathbf{q}} = \dot{\mathbf{P}}\dot{\mathbf{q}} = \mathbf{d}. \qquad (3)$$

The RHS of the above equation can be obtained as follows; Differentiation of $\mathbf{AP} = \mathbf{0}$ leads to $\mathbf{A}\dot{\mathbf{P}} = -\dot{\mathbf{A}}\mathbf{P}$. By multiplying both sides of the latter equation by $\dot{\mathbf{q}}$, we arrive at

$$\mathbf{Ad} = -\dot{\mathbf{A}}\dot{\mathbf{q}},$$

Consider $\mathbf{d}$ as an unknown variable. Then, in the above equation there are fewer independent equations than unknowns and hence a family of solutions exists. The theory of linear system of equations establishes [8] that the general solution can be expressed in terms of particular solution and homogeneous solution, i.e.

$$\mathbf{d} = -\mathbf{A}^+ \dot{\mathbf{A}}\dot{\mathbf{q}} + \mathbf{d}_n, \qquad (4)$$

where $\mathbf{A}^+$ is the *pseudo-inverse* of the Jacobian, and $\mathbf{d}_n \in \mathcal{N}(\mathbf{A})$ is the null space component. However, it is evident from (3) that $\mathbf{d} \in \mathcal{N}(\mathbf{A})^\perp$, and hence $\mathbf{d}_n = 0$. Thus

$$(\mathbf{I} - \mathbf{P})\ddot{\mathbf{q}} = -\mathbf{C}\dot{\mathbf{q}}, \qquad (5)$$

where $\mathbf{C} = \mathbf{A}^+ \dot{\mathbf{A}}$.

Note that the projection operator and the pseudo-inverse can be calculated by the Singular Value Decomposition (SVD) method [1] [8].

III. Projected Inverse Dynamics

Equations describing the DAE of multibody systems are formally written as

$$\mathbf{M}\ddot{\mathbf{q}} + \mathbf{h}(\mathbf{q},\dot{\mathbf{q}}) + \mathbf{A}^T \boldsymbol{\lambda} = \mathbf{f}, \qquad (6)$$

$$\mathbf{\Phi}(\mathbf{q}) = 0 \qquad (7)$$

here $\mathbf{M} \in \mathbb{R}^{n \times n}$ is the inertia matrix, $\mathbf{h}(\mathbf{q},\dot{\mathbf{q}}) \in \mathbb{R}^n$ contains Coriolis, centrifugal, gravitational terms, $\boldsymbol{\lambda} \in \mathbb{R}^m$ is the Lagrangian multiplier corresponding to the constraint force, and $\mathbf{f} \in \mathbb{R}^n$ is the vector of generalized force. In solving equations (6)-(7), it is typically assumed that: (i) the inertia matrix is positive definite and hence invertible (ii) the constraint equations are independent, i.e. the Jacobian matrix is not rank-deficient [6]. In this work, we solve DAEs without relying on the second assumption.

Let's define *generalized constraint force* as

$$\mathcal{F} = \mathbf{A}^T \boldsymbol{\lambda} \in \mathcal{R}(\mathbf{A}^T). \qquad (8)$$

The fundamental relationships between the range space and the null space associated with a linear operator is $\mathcal{R}(\mathbf{A}^T) = \mathcal{N}(\mathbf{A})^\perp$ [8]. Hence, one can say $\mathcal{F} \in \mathcal{N}^\perp(\mathbf{A})$. In other words, the projection operator $\mathbf{P}$ is an annihilator for the constraint force, i.e. $\mathbf{P}\mathcal{F} = 0$. Therefore, the constraint force can be eliminated from equation (6) if the equation is multiplied by $\mathbf{P}$, i.e.

$$\mathbf{P}(\mathbf{M}\ddot{\mathbf{q}} + \mathbf{h}) = \mathbf{P}\mathbf{f}. \qquad (9)$$

Equation (9) is called *projected inverse dynamics* of a constrained multibody system that is expressed in the so-called *descriptive form*. This is because matrix $\mathbf{PM}$ is singular and hence the acceleration can not be computed from the equation through the matrix inversion. Yet, equation (9) suffices for control applications as one can

---

[1]Assume that $r = \text{rank}(\mathbf{A})$. Then, there exist unitary matrices $\mathbf{U} = [\mathbf{U}_1 \quad \mathbf{U}_2]$ and $\mathbf{V} = [\mathbf{V}_1 \quad \mathbf{V}_2]$ so that $\mathbf{A} = \mathbf{U}\text{diag}\{\mathbf{\Sigma} \quad \mathbf{0}\}\mathbf{V}^T$, where $\mathbf{\Sigma} = \text{diag}(\sigma_1, \cdots, \sigma_r)$, and $\sigma_1 \geq \cdots \geq \sigma_r \geq 0$ are the singular values [8]. Then, the projection operator is $\mathbf{P} = \mathbf{V}_2\mathbf{V}_2^T$ and the *pseudo-inverse* of matrix $\mathbf{A}$ can be calculated by $\mathbf{A}^+ = \mathbf{V}_1\mathbf{\Sigma}^{-1}\mathbf{U}_1^T$.

compute the generalized force from position, velocity, and acceleration of the generalized coordinate.

IV. DIRECT DYNAMICS

As mentioned earlier, the acceleration cannot be determined uniquely from equation (9), because there are fewer independent equations than unknowns. Nevertheless; equation (5) compliments equation (9) so that a unique solution can be obtained by solving two equations together. To this end, we simply add both sides of the equations (5) and (9). Then, after factorization the resultant equation can be written concisely in the following form

$$\mathbf{M}_c \ddot{\mathbf{q}} = \mathbf{P}(\mathbf{f} - \mathbf{h}) - \mathbf{C}\dot{\mathbf{q}} \quad (10)$$

where $\mathbf{M}_c \in \mathbb{R}^{n \times n}$ is called *constraint inertia matrix* which is related to the unconstrained inertia matrix $\mathbf{M}$ by

$$\mathbf{M}_c = \mathbf{P}\mathbf{M} + (\mathbf{I} - \mathbf{P}), \quad (11)$$

Equation (10) constitutes the so-called *direct dynamics* of a constrained multibody system from which the acceleration can be solved. Note that the projection operator always exists and so does the constraint inertia matrix. But, in order to compute the acceleration from (10) requires that the constraint inertia matrix be invertible.

Theorem 1: If the unconstrained inertia matrix is full rank, then the constraint inertia matrix $\mathbf{M_c}$ is full rank too.

PROOF: We use the inversion lemma to prove the theorem. If matrix $\mathbf{M}_c$ is not full rank, then there must exist at least one non-zero vector $\boldsymbol{\xi} \neq 0$ lying in the matrix null space, that is $\mathbf{M}_c \boldsymbol{\xi} = 0$, or

$$\mathbf{P}\mathbf{M}\boldsymbol{\xi} + (\mathbf{I} - \mathbf{P})\boldsymbol{\xi} = 0 \quad (12)$$

The first term and the second term of the above equation are in two orthogonal subspaces and cannot cancel out each other. Hence, in order to satisfy the equation, both terms must be identically zero, i.e. $\mathbf{P}\mathbf{M}\boldsymbol{\xi} = 0$ and $(\mathbf{I} - \mathbf{P})\boldsymbol{\xi} = 0$. The former and the latter equations imply that $\mathbf{y} = \mathbf{M}\boldsymbol{\xi} \in \mathcal{N}^\perp$ and $\boldsymbol{\xi} \in \mathcal{N}$, respectively. Therefore, one can conclude that $\mathbf{y}$ is perpendicular to vector $\boldsymbol{\xi}$, i.e. $\boldsymbol{\xi}^T \mathbf{y} = \mathbf{0}$, or that

$$\boldsymbol{\xi}^T \mathbf{M} \boldsymbol{\xi} = 0. \quad (13)$$

But (13) is a contradiction because $\mathbf{M}$ is a positive definite matrix. Therefore, the null set of $\mathbf{M}_c$ is empty and the matrix is always invertible, and this completes the proof.

Theorem 1 is pivotal in showing the usefulness of the dynamics equation (10). This is because the theorem signifies the fact that constraint inertia matrix is always invertible irrespective of the constraint condition. This ensures that the acceleration can be always obtained from (10) even in case of redundant constraint equations or singular configurations.

A. Constraint Inertia Matrix

Despite of the conscience formulation of the dynamics by (10) and (11), there is an inconsistency in component of the inertia matrix $\mathbf{M}_c$ because different units are added in equation (11) – note that the projection operator $\mathbf{P}$ is unit-less. This may lead to a numerical pitfall. To overcome the unit conflict problem, we slightly modify the definition of the constraint inertia matrix as follows. Analogously to the procedure in the previous section, we first multiply equation (5) by $\mathbf{M}$ and then add both sides of the resultant equation with those of (9). That gives

$$\ddot{\mathbf{q}} = \mathbf{M}_c'^{-1}[\mathbf{P}(\mathbf{f} - \mathbf{h}) - \mathbf{C}'\dot{\mathbf{q}}], \quad (14)$$

where $\mathbf{C}' = \mathbf{M}\mathbf{C}$, and for a symmetric inertia matrix, i.e. $\mathbf{M}^T = \mathbf{M}$, the new constrained inertia matrix is given by

$$\mathbf{M}_c' = \mathbf{M} + \tilde{\mathbf{M}}, \text{ where } \tilde{\mathbf{M}} = \mathbf{P}\mathbf{M} - (\mathbf{P}\mathbf{M})^T \quad (15)$$

The unit conflict problem is thereby resolved with the new definition of the inertia matrix $\mathbf{M}_c'$.

It is also interesting to note that that $\tilde{\mathbf{M}}$ is a skew-symmetric matrix, i.e. $\tilde{\mathbf{M}}^T = -\tilde{\mathbf{M}}$, and thus for any vector $\mathbf{z} \in \mathbb{R}^n$ we can say $\mathbf{z}^T \tilde{\mathbf{M}} \mathbf{z} = 0$. Consequently, adding $\tilde{\mathbf{M}}$ to the inertia matrix in equation (15) preserves the positive definiteness property of the inertia matrix, or $\mathbf{M}$ is p.d. $\iff \mathbf{M}_c'$ is p.d..

B. Constraint Force and Lagrangian Multiplier

Equation (10) signifies the fact that only the null space projection of the generalized force contributes to motion of a constrained multibody system. This fact suggests that it is beneficiary to decompose the generalized force into two orthogonal subspaces

$$\mathbf{f} = \mathbf{f}_\| \oplus \mathbf{f}_\perp,$$

where $\mathbf{f}_\| = \mathbf{P}\mathbf{f} \in \mathcal{N}$ and $\mathbf{f}_\perp = (\mathbf{I} - \mathbf{P})\mathbf{f} \in \mathcal{N}^\perp$ are called *acting input force* (potent) and *passive input force* (impotent), respectively.

Now, by multiplying both sides of equation (6) by $\mathbf{I} - \mathbf{P}$ and then substituting the acceleration from equation (10) or (14), one can carry out the constraint force

$$\mathcal{F} = (\mathbf{f}_\perp - \mathbf{h}_\perp) - \boldsymbol{\mu}(\mathbf{f}_\| - \mathbf{h}_\|) + \boldsymbol{\mu}\mathbf{C}\dot{\mathbf{q}}, \quad (16)$$

where $\boldsymbol{\mu} = (\mathbf{I}-\mathbf{P})\mathbf{M}\mathbf{M}_c^{-1}$, and the nonlinear vector is decomposed analogously to the generalized forces. Having calculated the constraint force from (16), one can obtain the Lagrangian multiplier through pseudo-inversion, i.e. $\boldsymbol{\lambda} = \mathbf{A}^{+T}\mathcal{F}$ is the Jacobian matrix is full rank.

V. Simulation of Constrained Multibody Systems

Having computed acceleration from the equation, a simulation may proceed by integration of the acceleration to obtain the generalized coordinates. However, the integration inevitably leads to drift that eventually results in a large constraint error. In this section, we apply the Newton-Raphson method by using the pseudo-inverse for correcting the generalized coordinate in order to maintain the constraint condition precisely. It should be noted that using the pseudo-inverse herein does not impose any extra computation burden, because the pseudo-inverse should be obtained to compute the acceleration anyway.

Assume that the constraint condition is slightly violated, i.e. $\boldsymbol{\Phi}(\mathbf{q}_0) = \delta\boldsymbol{\Phi} \neq 0$. Therefore we seek a small compensation in the generalized coordinate $\delta\mathbf{q}$ such that the constraint condition is satisfied, i.e. $\boldsymbol{\Phi}(\mathbf{q}_0 + \delta\mathbf{q}) = 0$. This equation can be written by the first order approximation as

$$\boldsymbol{\Phi}(\mathbf{q}_0 + \delta\mathbf{q}) \approx \boldsymbol{\Phi}(\mathbf{q}_0) + \mathbf{A}\delta\mathbf{q}$$

The pseudo-inverse yields the minimum norm solution, i.e. $\|\delta\mathbf{q}\|$. Therefore, the following loop may be worked out iteratively until the error in the constraint falls into an acceptable tolerance, e.g. $\|\boldsymbol{\Phi}\| \leq \epsilon$.

$$\mathbf{q}_{k+1} = \mathbf{q}_k - \mathbf{A}^+ \boldsymbol{\Phi}(\mathbf{q}_k) \quad (17)$$

VI. Control

A. Projected inverse dynamics Control

Due to presence of r independent constraint, the actual number of degrees of freedom of the system is reduced to $k = n-r$. Thus, in principle, there must be k independent coordinates $\boldsymbol{\theta} \in \mathbb{R}^k$ from which the generalized coordinates can be derived, i.e. $\mathbf{q} = \boldsymbol{\psi}(\boldsymbol{\theta})$. Now, differentiation of the given function with respect to time gives

$$\dot{\mathbf{q}} = \boldsymbol{\Lambda}\dot{\boldsymbol{\theta}}, \quad \ddot{\mathbf{q}} = \boldsymbol{\Lambda}\ddot{\boldsymbol{\theta}} + \dot{\boldsymbol{\Lambda}}\dot{\boldsymbol{\theta}}, \quad (18)$$

where $\boldsymbol{\Lambda} = \partial\boldsymbol{\psi}/\partial\boldsymbol{\theta} \in \mathbb{R}^{n\times k}$. Since $\boldsymbol{\theta}^T = [\theta_1(\mathbf{q}), \cdots, \theta_\mathbf{k}(\mathbf{q})]$ constitutes a set of independent functions, the Jacobian matrix $\boldsymbol{\Lambda}$ must be full-rank. It is also important to note that any admissible function $\boldsymbol{\psi}(\cdot)$ must satisfy the constraint condition, that is

$$\boldsymbol{\Phi}(\boldsymbol{\psi}(\boldsymbol{\theta})) = 0 \quad \forall \boldsymbol{\theta}.$$

Using the chain-rule, one can obtain the time-derivative of the above equation $\mathbf{A}\boldsymbol{\Lambda}\dot{\boldsymbol{\theta}} = 0 \quad \forall \dot{\boldsymbol{\theta}}$. Since $\boldsymbol{\Lambda}$ is a full-rank matrix, the above equation is satisfied only if

$$\mathcal{R}(\boldsymbol{\Lambda}) = \mathcal{N}(\mathbf{A}) \quad (19)$$

Substituting the acceleration from (18) into the inverse dynamics equation (9) gives the dynamics in terms of the reduced-dimension coordinate, i.e.

$$\mathbf{P}\mathbf{M}(\boldsymbol{\Lambda}\ddot{\boldsymbol{\theta}} + \dot{\boldsymbol{\Lambda}}\dot{\boldsymbol{\theta}}) + \mathbf{P}\mathbf{h} = \mathbf{f}_\| \quad (20)$$

Suppose $\{\boldsymbol{\theta}_d(t), \dot{\boldsymbol{\theta}}_d(t), \ddot{\boldsymbol{\theta}}_d(t)\}$ denote desired trajectories of the new coordinates. Now, we propose the following control law

$$\mathbf{f}_\|^c = \mathbf{h}_\| + \mathbf{P}\mathbf{M}\boldsymbol{\Lambda}\mathbf{u}_p, \quad (21)$$

where $\mathbf{u}_p$ is an auxiliary control input as

$$\mathbf{u}_p = -\dot{\boldsymbol{\Lambda}}\dot{\boldsymbol{\theta}} + \boldsymbol{\Lambda}(\ddot{\boldsymbol{\theta}}_d + G_D\dot{\mathbf{e}} + G_P\mathbf{e}), \quad (22)$$

$\mathbf{e}_p = \boldsymbol{\theta}_d - \boldsymbol{\theta}$ is the position trajectory tracking error, and $G_P > 0$ and $G_D > 0$ are the PD feedback gains. In the sequel, superscript ^{c} is used for denoting control input. The control law in (21) is called *projected inverse dynamics control* (PIDC).

Theorem 2: While demanding minimum norm control input, the projected inverse dynamics control law (21)-(22) stabilizes the position tracking error, i.e. $\boldsymbol{\theta}(t) \to \boldsymbol{\theta}_d(t)$ as $t \to \infty$.

PROOF: First, we prove exponential stability of the position error. From equations (20), (22), and (21), one can conclude that the proposed control law leads to the following equation for tracking error

$$\mathbf{PM\Lambda}[\ddot{\mathbf{e}}_p + G_D \dot{\mathbf{e}}_p + G_P \mathbf{e_p}] = 0. \quad (23)$$

To show that the expression within the bracket is zero, we need to show that the matrix $\mathbf{PM\Lambda}$ is full-rank. In the following we will show that the matrix cannot have any null space and hence is full-rank. If the matrix has a null space, then $\exists \mathbf{x} \neq 0 \ni \mathbf{PM\Lambda x} = 0$. Let's define $\boldsymbol{\xi} = \boldsymbol{\Lambda} \mathbf{x}$. Recall that $\boldsymbol{\Lambda}$ is a full-rank matrix and that $\mathcal{R}(\boldsymbol{\Lambda}) = \mathcal{N}(\mathbf{A})$. Hence, $\boldsymbol{\xi} \neq 0$ and $\boldsymbol{\xi} \in \mathcal{N}$. On other hand, $\mathbf{PM}\boldsymbol{\xi} = 0$ implies that $\mathbf{M}\boldsymbol{\xi} \in \mathcal{N}^\perp$, and hence it is perpendicular to $\boldsymbol{\xi}$, i.e. $\boldsymbol{\xi}^T \mathbf{M} \boldsymbol{\xi} = 0$. But, this is a contradiction because $\mathbf{M}$ is a p.d. matrix. Consequently, $\mathcal{N}(\mathbf{PM\Lambda}) = \emptyset$, and it follows from (23) that

$$\ddot{\mathbf{e}}_p + G_D \dot{\mathbf{e}}_p + G_P \mathbf{e}_p = 0.$$

Hence the error dynamics can be stabilized by selecting adequate gains, that is $\boldsymbol{\theta} \to \boldsymbol{\theta}_d$ as $t \to \infty$. Moreover, due to orthogonality of the decomposed generalized force, we can say

$$\|\mathbf{f}^c\| = \|\mathbf{f}^c_\parallel\| + \|\mathbf{f}^c_\perp\|$$

where $\|\cdot\|$ denotes Euclidean norm of a vector. From the above norm relation, it is clear that $\mathbf{f}^c_\parallel$ is the minimum norm solution, since any other solution must have a component in $\mathbf{f}^c_\perp$ and this would increase the overall norm. Thus

$$\mathbf{f}^c = \mathbf{f}^c_\parallel \iff \min_{\theta(t) \to \theta_d(t)} \|\mathbf{f}^c\|.$$

B. Control of Constraint Force

Suppose that $\mathcal{F}_d$ represent desired constraint force. Then, considering $\mathbf{f}^c_\perp$ as control input, we propose the following control law

$$\mathbf{f}^c_\perp = \mathbf{h}_\perp + \boldsymbol{\mu}(\mathbf{f}_\parallel - \mathbf{h}_\parallel - \mathbf{C}\dot{\mathbf{q}}) + \mathbf{u}_{\mathcal{F}}, \quad (24)$$

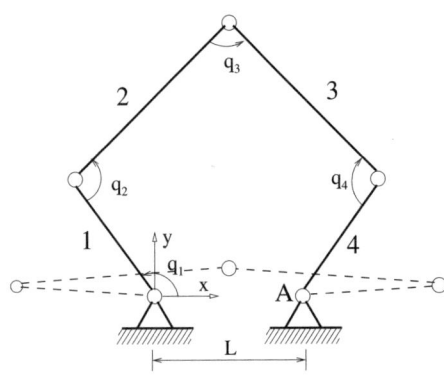

Fig. 1. The five-bar mechanism at initial configuration (solid), and at near the singular configuration (dashed).

where $\mathbf{u}_{\mathcal{F}}$ is the auxiliary control input which is traditionally chosen as

$$\mathbf{u}_{\mathcal{F}} = \mathcal{F}_\mathbf{d} + \mathbf{G_F e_f} + \mathbf{G_I} \int_0^t \mathbf{e_f} d\tau, \quad (25)$$

$\mathbf{e}_f = \mathcal{F}_d - \mathcal{F}$ is the force error, and $G_F > 0$ and $G_I > 0$ are the PI feedback gains. It should be pointed out that the integral term is not necessary but it improves the steady-state error. From (16), (24), and (25), one can obtain the error dynamics as

$$(G_F + 1)\dot{\mathbf{e}}_f + G_I \mathbf{e}_f = 0,$$

which is stable provided that the gains are positive definite, i.e. $\mathcal{F}(t) \to \mathcal{F}_d(t)$ as $t \to \infty$.

VII. SIMULATION RESULTS

In this section, we describe the results obtained from simulation of a typical five-bar mechanism, as shown in Fig. 1. The closed-loop is cut in the right hand support, i.e. point A in Fig.1, and translation motion of point A is prohibited by imposing constraint equations. The constraint error is controlled by the Newton-Raphson method where the error tolerance is $\epsilon = 10^{-14}$. Assuming that the gravity is the only applied force, the linkage falls from its initial condition at $\mathbf{q}_0 = [2.1\ 1.89\ 1.46\ 1.89]^T$ and $\dot{\mathbf{q}}_0 = 0$. Given $L = 1$, the Jacobian matrix is singular at position $\mathbf{q} = [\pi\ 0\ \pi\ 0]^T$.

Trajectories of joint angles are shown in Fig. 2A. As expected, the mechanism goes through the singular configuration that is clearly evident from the spike in the graph of the condition number of the Jacobian matrix in Fig. 3A. Nevertheless, as apparent from Fig 3B, $\mathbf{M}'_\mathbf{c}$ enjoys the

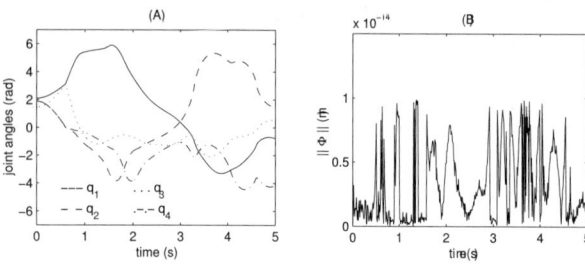

Fig. 2. Results from simulation of the five-bar mechanism.

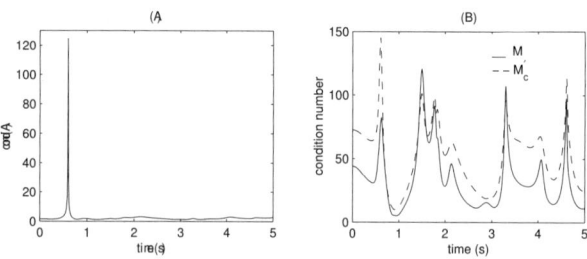

Fig. 3. The condition number of Jacobian (A). A comparison of the condition numbers of the inertia matrices $\mathbf{M}$ and $\mathbf{M'_c}$. (B)

same ill-conditioning property as that of the inertia matrix $\mathbf{M}$. As a result, the acceleration can be always computed and the simulation may proceed smoothly even if the system works in the vicinity of the singular position. This is evident form the plot of constraint error in Fig.2B showing that the proposed method is always able to maintain the constraint condition within the specified tolerance.

VIII. Experimental Results

In this section, we report comparative experimental results obtained from a constraint mechanical system shown in Fig. 4. The arm which was used for these experiments was a planar robot arm developed at CSA with three revolute joints which are driven by geared motors RH-8-6006, RH-11-3001, and RH-14-6002 from Hi-T Drive. The robot joints are equipped with optical encoders, while an ATI force sensor (gamma type) is installed in the robot wrist. The robot endpoint is connected to a slider by a hinge (global joint), as illustrated in Fig. 4, and the robot motion in Y-axis is thereby constrained.

Let the position and orientation of the robot endpoint be presented by $\{x, y, \theta\}$. Then from the topology of the kinematics, the constraint equation and the reduced-dimension coordinate can be specified by $\Phi = y(\mathbf{q}) - \mathbf{y_0} = \mathbf{0}$ — where $y_0 = -0.27m$ — and $\boldsymbol{\theta}^T = [x(\mathbf{q}), \theta(\mathbf{q})]$, respectively.

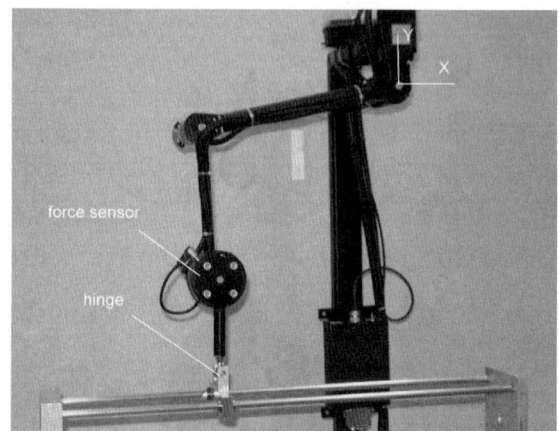

Fig. 4. The experimental setup.

In this experiment, the position feedback gains are $G_P = 480$ and $G_D = 45$, which corresponds to $3.5Hz$ bandwidth of the closed loop system. While the force feedback gains are $G_F = 3$, and $G_I = 0.1$. The desired position trajectory is specified as $x_d(t) = -0.075 + 0.225\sin(\pi t + \pi/9)$ and $\theta_d = 0$, while $\lambda_d = 0$.

Three different control schemes are implemented: the projected inverse dynamics control (PIDC) and the hybrid force/motion control described in Section VI as well as the standard inverse dynamics control (IDC). All controllers demonstrated a good motion tracking performance as illustrated in Fig. 5. Differences among the control schemes, though, is manifested in their force responses, as illustrated in Fig. 6. The trajectories of the contact force and those of the Euclidean norm of joint torque requested by the three controllers are plotted in Fig. 6A and Fig. 6B, respectively. It is clearly evident from Fig. 6B that the PIDC always requests minimum joint torque compared to the other controllers, albeit it doesn't yield zero constraint force, see Fig. 6A. The constraint force is regulated to the desired value zero when the force feedback law (24)-(25) is activated, but it gives rise to the requested joint torque, as can be clearly seen in Fig. 6A. The traditional IDC is also implemented that results in a large force. This is because, unlike PIDC which exhibits compliance in the constraint direction, the IDC tends to be stiff in all directions and that produces large force in case of position uncertainty.

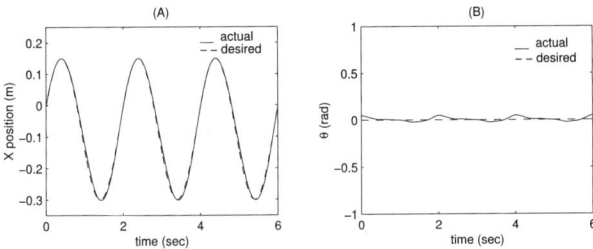

Fig. 5. The actual and desired trajectories of the position (A) and the orientation (B)

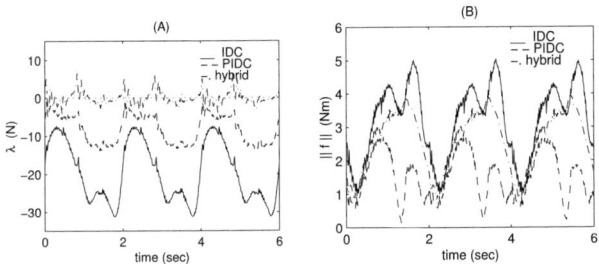

Fig. 6. Trajectories of the contact force (Lagrangian multiplier) (A), and those of the Euclidean norm of joint torque (B)

Conclusion

The unified approach for inverse and direct dynamics models of constrained multibody systems have been presented in a simple but comprehensive form. Subsequently, strategies for simulation and control of constrained multibody systems based on dynamics models have been developed without requiring the constraint equations be linearly independent. The equations of motion needs minimum arithmetic operations which can lead to computational efficiency in a simulation run. While the orthogonal decomposition of the generalized force gave control inputs required for controlling position and constraint force. The decomposition led to an optimal inverse dynamics control scheme which minimizes requested control input. The theoretical results have been partially demonstrated by simulation and experimental results.

References

[1] E. Bayo and J. Garcia de Jalon. A modified lagrangian formulation for the dynamic analysis of constrained mechanical systems. *Computer methods in applied mechanics and engineering*, 71:183–195, 1988.

[2] E. Bayo and R. Ledesma. Augmented lagrangian and mass-orthogonal projection methods for constrained multibody dynamics. *Nonlinear Dynamics*, 9:113–130, 1996.

[3] A. Bicchi, L. Pallottino, M. Bray, and P. Perdomi. Randomized parrallel simulation of constrained multibody systems for VR/Haptic applications. In *IEEE Int. Conference on Robotics and Automation*, pages 2319 – 2324, Seoul, Korea, 2001.

[4] W. Blajer, W. Schiehlen, and W. Schirm. A projective criterion to the coordinate partitioning method for multibody dynamics. *Applied Mechanics*, 64:86–98, 1994.

[5] A. Cole, J. Hauser, and S. Sastry. Kinematics and control of multifingered robot hand with rolling contact. *IEEE Trans. Robotics and Automation*, 34:398–404, 1989.

[6] J. Garcia de Jalon and Eduardo Bayo. *Kinematic and Dynamic Simulation of Multibody Systems*. Springer-Verlag, 1989.

[7] C. Cannudas de Wit, B. Siciliano, and G. Bastin (Eds). *Theorey of Robot Control*. Springer, London, Great Britain, 1996.

[8] G. H. Golub and C. F. Van Loan. *Matrix Computations*. The Johns Hopkins University Press, Baltimore and London, 1996.

[9] J. H. Jean and L. C. Fu. An adaptive control scheme for coordinated multimanipulator systems. *IEEE Trans. Robotics and Automation*, 9(2):226–231, 1993.

[10] O. Khatib. A unified approach for motion and force control of robot manipulators: The operational space formulation. *IEEE Transactions on Robotics and Automation*, RA-3(1):43–53, 1987.

[11] C. G. Liang and G. M. Lance. A differentiable null space method for constrained dynamic analysis. *Trans. ASME, JMTAD*, 109:450–411, 1987.

[12] S. T. Lin and M. C. Hong. Stabilization method for numerical integration of multibody mechanical systems. *Journal of Mechanical Design*, 120:565–572, 1998.

[13] N. H. McClamroch and D. Wang. Feedback stabilization and tracking in constrined robots. *IEEE Trans. on Automation Control*, 33:419–426, 1988.

[14] Y. Nakamura and M. Ghodoussi. Dynamics computation of closed-link robot mechanism with nonredundant actuators. *IEEE Trans. Robotics and Automation*, 5:294–302, 1989.

[15] P. E. Nikravesh and I. S. Chung. Application of euler parameters to the dynamic analysis of three dimensional constrained mechanical systems. *ASME J. Mech. Design*, 104:785–791, 1982.

[16] B. Perrin, C. Chevallereau, and C. Verdier. Calculation of the direct dynamics model of walking robots: Comparison between two methods. In *IEEE Int. Conf. Robotics and Automation*, pages 1088–1093, Apr. 1997.

[17] R. A. Wehage and E. J. Haug. Generalized coordinate partitioning of dimension reduction in analysis of constrained dynamic systems. *ASME J. Mech. Design*, 104:247–255, 1982.

[18] T. Yoshikawa. Dynamic hybrid position/force control of robot manipulators-description of hand constraints and calculation of joint driving force. *IEEE Transactions on Robotics and Automation*, RA-3(5):386–392, October 1987.

The 6×6 Stiffness Formulation and Transformation of Serial Manipulators via the CCT Theory

Shih-Feng Chen

Department of Mechanical Engineering, Lunghwa University of Science and Technology
Taoyuan, Taiwan 333, ROC, E-mail: dave2esc@mail.lhu.edu.tw

Abstract

This paper presents systematic methods to approach the conservative congruence transformation, CCT, via the geometrical analysis for the stiffness control transformation of serial manipulators. Through the strategy of changing basis, the 6×6 Cartesian stiffness of manipulators is shown to be basis dependent. The generalized formulation and symmetric property of the 6×6 Cartesian stiffness matrix are presented via the CCT theory in the presence of external loads. Examples of the serial manipulators are conducted to verify the conservative stiffness mapping.

1. Introduction

The stiffness transformation, from the joint stiffness of a manipulator to the Cartesian stiffness, is essential to the stiffness control algorithms. In 1980, Salisbury first derived the conventional congruence transformation to represent the mapping between the Cartesian and the joint stiffness matrices in robot manipulation [1]. In addition, Griffis, and Duffy [2] studied the global stiffness model of compliantly coupled systems with three elastic spring coupling between two planar rigid bodies, and discovered a 6×6 stiffness matrix that is in general asymetric. Ciblak and Lipkin [3] developed a formulation to represent the skew symmetric part of the 6×6 Cartesian stiffness matrix for conservative systems. Zefran and Kumar [4] used Lie groups to show the 6×6 stiffness matrix is asymetric in any conservative system subjected to a non-zero external load. Recently, there have been renewed interests in stiffness control which have posed challenges to the conventional understanding of the Cartesian stiffness matrix [5-14].

In 1999, Chen and Kao represented that the conventional congruence transformation derived by Salisbury in 1980 and well accepted by researchers until recently is not conservative [8]. The simulation results show there is a discrepancy between the work done in the joint space and that in the Cartesian space when applied to robotic stiffness control. It proves that the conventional formulation is only valid at unload position as well as the conservative Cartesian stiffness without the consideration of rotational effect in linear space is always symmetric. In this paper, we build upon the previous research results of the CCT and present a consistent methodology to formulate the 6×6 Cartesian stiffness matrix of the conservative system by using the geometrical methods. Examples of serial manipulators including the planar 3-dof manipulator and Stanford robot in [14] are utilized to verify that the skew-symmetric part of the 6×6 Cartesian stiffness matrix with respect to the manipulator (screw based) Jacobian is equal to negative one-half of the cross-product matrix formed by the external loads referenced to the fixed (inertial) frame.

From the point of view in basis mapping, this paper represents the geometrical interpretation of the CCT theory and establishes a systematic and complete stiffness transformation in robot systems. We must notice that the 6×6 Cartesian stiffness of manipulators is basis (coordinate) dependent when applied in robot stiffness control. Namely, the Cartesian stiffness of the conservative system could assume totally different physical forms even though they are representing the same system!

2. The CCT: A New Theory in Spatial Stiffness Mapping and Control

In 1999, Chen and Kao [8,9] proposed the conservative congruence transformation, CCT, which represents the relationship of stiffness control between the linear Cartesian space at the end-effector and the joint space at the joint space for a two-links planar manipulator in $\Re^{2\times 2}$. The CCT accounts for the changes of gasping geometry in the presence of external loads to correct the well-known conventional formulation derived by Salisbury in 1980.

However, the 6×6 Cartesian stiffness matrix of conservative systems including the linear and rotational motions in $SE(3)$ is quite different from the linear stiffness matrix. The difference is readily seen due to the well-known fact that the rotational components do not commute, i.e., the Cartesian coordinates are *not* on a *coordinate basis* [16,17]. Hence, the 6×6 Cartesian stiffness matrix is not expected to be symmetric, even for a conservative system.

In 2002, we performed the geometrical approach to the CCT for robotic spatial stiffness transformation and control [12], i.e.,

$$\mathbf{J}_\theta^T \mathbf{K} \mathbf{J}_\theta = \mathbf{K}_\theta - \mathbf{K}_g \quad (1)$$

where the Cartesian stiffness, $\mathbf{K} = d\mathbf{w}/d\mathbf{x}$, is a 6×6 matrix and the joint stiffness, $\mathbf{K}_\theta = d\boldsymbol{\tau}/d\boldsymbol{\theta}$, is a $n\times n$ symmetric matrix in general and the effect of change in geometry under external loads, $\mathbf{w}$, is captured by the $n\times n$ matrix $\mathbf{K}_g = [(\partial \mathbf{J}_\theta^T/\partial \theta_n)\mathbf{w}]$ with the i-th column element being $(\partial \mathbf{J}_\theta^T/\partial \theta_i)\mathbf{w}$. Here, the CCT represents the spatial stiffness mapping relationship between the coordinate basis in joint space and the non-coordinate basis in the Cartesian space.

Thus, the spatial Cartesian stiffness is given by the inverse of equation (1), i.e.,

$$\mathbf{K} = \mathbf{J}_\theta^{-T}(\mathbf{K}_\theta - \mathbf{K}_g)\mathbf{J}_\theta^{-1} \quad (2)$$

which shows that the property of the 6×6 spatial Cartesian stiffness, $\mathbf{K}$, is related to the joint stiffness, $\mathbf{K}_\theta$, as well as the changes in geometry under external forces is captured by the matrix $\mathbf{K}_g$. Because the joint stiffness is symmetric in general, the symmetric property of the 6×6 Cartesian stiffness is only related to the effective matrix, $\mathbf{K}_g = [(\partial \mathbf{J}_\theta^T/\partial \theta_n)\mathbf{w}]$.

3. Application of the CCT to Serial Manipulators

For robot manipulators, the Jacobian matrix is defined as the matrix that transforms the joint rates in the joint space to the velocity state in the Cartesian space. In this paper, the spatial Jacobian matrix is built upon the previous geometrical analysis and stiffness modeling via the CCT theory [12].

3.1. The Cartesian Velocity in Twist Basis

The rigid body motion describing the instantaneous position and orientation in three-dimensional space relevant to the fixed reference frame can be represented by a 4×4 homogeneous transformation matrix [15,18], i.e.,

$$\mathbf{T}(t) = \begin{bmatrix} \mathbf{R}(t) & \mathbf{p}(t) \\ \mathbf{0} & 1 \end{bmatrix} \quad (3)$$

where $\mathbf{R}(t)$ is a 3×3 rotation matrix with two key properties of $\mathbf{R}^T\mathbf{R} = \mathbf{R}\mathbf{R}^T = \mathbf{I}$ and $\det(\mathbf{R}) = +1$, $\mathbf{p}(t)$ is a 3×1 translation vector, and $\mathbf{0}$ is a 1×3 row vector of zeros. Thus, the tangent vector, $d\mathbf{T}/dt$, to this trajectory is given by

$$\frac{d\mathbf{T}(t)}{dt} = \begin{bmatrix} \dot{\mathbf{R}}(t) & \begin{matrix} v_{n,x} \\ v_{n,y} \\ v_{n,z} \end{matrix} \\ \mathbf{0} & 0 \end{bmatrix} \quad (4)$$

where the 3×1 column vector, $\mathbf{v}_n = [v_{n,x}, v_{n,y}, v_{n,z}]^T$, in the upper right corner represents the linear velocity of the origin of the end-effector coordinate frame. However, when we consider the right-invariant vector fields on this tangent group, i.e., $(d\mathbf{T}/dt)\mathbf{T}^{-1}$, it satisfies the following relation

$$\frac{d\mathbf{T}(t)}{dt}\mathbf{T}^{-1} = \begin{bmatrix} 0 & -\omega_z & \omega_y & v_{o,x} \\ \omega_z & 0 & -\omega_x & v_{o,y} \\ -\omega_y & \omega_x & 0 & v_{o,z} \\ 0 & 0 & 0 & 0 \end{bmatrix} \quad (5)$$

where the 3×3 sub-matrix in the upper left corner of equation (5), is a skew-symmetric matrix representing the angular velocity, and the 3×1 column vector, $\mathbf{v}_o = [v_{o,x}, v_{o,y}, v_{o,z}]^T$, in the upper right corner represents the linear velocity of a point in the end effector that is instantaneously coincident with the origin of the fixed frame, both viewed in the current fixed (inertial) frame. Therefore, $(d\mathbf{T}/dt)\mathbf{T}^{-1}$, a six-dimensional manifold, is the instantaneous twist in the fixed frame which belongs to *Lie Algebra*, *se(3)* [15]. On the other hand, the left-invariant vector fields on the tangent group, i.e., $\mathbf{T}^{-1}(d\mathbf{T}/dt)$, indicate the spatial velocity viewed in the body (moving) frame.

We can choose an ordered unit twist basis, $\mathbf{S} = \{\mathbf{S}_1; \mathbf{S}_2; \mathbf{S}_3; \mathbf{S}_4; \mathbf{S}_5; \mathbf{S}_6\}$, for the six-dimensional twist matrix group [16,17], i.e.,

$$\mathbf{S}_1 = \begin{bmatrix} 0&0&0&1\\0&0&0&0\\0&0&0&0\\0&0&0&0 \end{bmatrix} \quad \mathbf{S}_4 = \begin{bmatrix} 0&0&0&0\\0&0&-1&0\\0&1&0&0\\0&0&0&0 \end{bmatrix}$$
$$\mathbf{S}_2 = \begin{bmatrix} 0&0&0&0\\0&0&0&1\\0&0&0&0\\0&0&0&0 \end{bmatrix} \quad \mathbf{S}_5 = \begin{bmatrix} 0&0&1&0\\0&0&0&0\\-1&0&0&0\\0&0&0&0 \end{bmatrix} \quad (6)$$
$$\mathbf{S}_3 = \begin{bmatrix} 0&0&0&0\\0&0&0&0\\0&0&0&1\\0&0&0&0 \end{bmatrix} \quad \mathbf{S}_6 = \begin{bmatrix} 0&-1&0&0\\1&0&0&0\\0&0&0&0\\0&0&0&0 \end{bmatrix}$$

where the sub-basis $\{\mathbf{S}_1; \mathbf{S}_2; \mathbf{S}_3\}$ are related to the linear translation motion, which commute with one another; but $\{\mathbf{S}_4; \mathbf{S}_5; \mathbf{S}_6\}$ are related to the rotation motion of a rigid body and do not commute with one another. Note that the unit twist (Cartesian) basis is a *non-coordinate basis* that the Lie bracket of vector fields has to be non-zero [15]. Therefore, equation (5) can be rewritten in terms of the unit twist basis **S** to be

$$\dot{\mathbf{T}}\mathbf{T}^{-1} = v_{o,x}\mathbf{S}_1 + v_{o,y}\mathbf{S}_2 + v_{o,z}\mathbf{S}_3 + \omega_x\mathbf{S}_4 + \omega_y\mathbf{S}_5 + \omega_z\mathbf{S}_6 \quad (7)$$

where the twist components, $(v_{o,x}, v_{o,y}, v_{o,z}, \omega_x, \omega_y, \omega_z)$, are just the rigid body velocity in the Cartesian space. That is, $[v_{o,x}, v_{o,y}, v_{o,z}]^T$ is related to the translation motion along the X, Y, and Z axes, respectively, and $[\omega_x, \omega_y, \omega_z]^T$ is related to the consecutive rotation motion about the same axes.

Therefore, the 6×6 Cartesian stiffness matrix of a conservative system with the consideration of linear and rotational motions in terms of the unit twist basis is quite different from the 3×3 linear stiffness matrix. The difference can be readily seen due to the well-known fact that the rotational components in terms of the non-coordinate basis do not commute. This indicates that the spatial Cartesian coordinates are not on a coordinate basis. Thus, the 6×6 Cartesian stiffness matrix with respect to this *non-coordinate basis* is not expected to be symmetric, even for a conservative system. The non-symmetric property of the 6×6 Cartesian stiffness matrix will be discussed via the CCT in the following sections.

3.2. Derivation of the Jacobian Matrix via the Geometrical Methods

When we consider the motions of a serial manipulator parameterized by the joint variables, $\boldsymbol{\theta} = (\theta_1, \cdots, \theta_n)$, we can represent the end-effector velocity by the homogeneous transformation matrix, $\mathbf{T}(\theta)$. Let us consider a robot manipulator which consists of a series of links connected by joints. The $\mathbf{T}(\theta)$ matrix is simply a homogeneous transformation matrix describing the relative translation and rotation between link coordinate systems. Thus, the position and orientation of the end-effector of the manipulator with respect to the base frame can be described by the matrix product

$$\mathbf{T}(\theta) = {}^0_1\mathbf{T} \cdots {}^{i-1}_i\mathbf{T} \cdots {}^{n-1}_n\mathbf{T} \quad (8)$$

Where ${}^{i-1}_i\mathbf{T}$ indicates the position and orientation of the *i*-th link with respect to the (*i*-1)-th link. This can be systematically formulated using the Denavit-Hartenberg parameters.

For robot manipulators, the velocity state of the end-effector, **x**, can be expressed in several different ways. The most used definitions are the conventional Jacobian and manipulator (screw-based) Jacobian [14]. In a conventional Jacobian, the end-effector velocity is expressed as, $[v_{n,x}, v_{n,y}, v_{n,z}, \omega_x, \omega_y, \omega_z]^T$, that $\mathbf{v}_n = [v_{n,x}, v_{n,y}, v_{n,z}]^T$ is the linear velocity of the origin of the end-effector coordinate frame. Thus, the conventional Jacobian matrix is defined by $[v_{n,x}, v_{n,y}, v_{n,z}, \omega_x, \omega_y, \omega_z]^T = \mathbf{J}_\theta [d\theta_1/dt, \cdots, d\theta_n/dt]^T$. The linear velocity, $\mathbf{v}_n$, is obtained in equation (4), and the angular velocity is determined from $(d\mathbf{T}/dt)\mathbf{T}^{-1}$ in equation (5) viewed in the fixed frame. Obviously, we have to evaluate the linear velocity and the angular velocity of the end-effector separately. Thus, from the relationships $d\mathbf{T}/dt = v_{n,x}\mathbf{S}_1 + v_{n,y}\mathbf{S}_2 + v_{n,z}\mathbf{S}_3 + \ldots$ in equation (4), and the right-invariant vector field $(d\mathbf{T}/dt)\mathbf{T}^{-1} = v_{o,x}\mathbf{S}_1 + v_{o,y}\mathbf{S}_2 + v_{o,z}\mathbf{S}_3 + \omega_x\mathbf{S}_4 + \omega_y\mathbf{S}_5 + \omega_z\mathbf{S}_6$ in equation (7), we conclude

$$\sum_{i=1}^{6}\frac{\partial \mathbf{T}}{\partial \theta_i}\dot{\theta}_i = v_{n,x}\mathbf{S}_1 + v_{n,y}\mathbf{S}_2 + v_{n,z}\mathbf{S}_3 + \ldots$$
$$= \ldots + \omega_x{}^F\mathbf{S}_4 + \omega_y{}^F\mathbf{S}_5 + \omega_z{}^F\mathbf{S}_6 \quad (9)$$

That is, the conventional Jacobian matrix is obtained by the mapping relationship between bases $\{\mathbf{E}_i = \partial \mathbf{T}/\partial \theta_i\}$ and $\{\mathbf{S}_1; \mathbf{S}_2; \mathbf{S}_3; {}^F\mathbf{S}_4; {}^F\mathbf{S}_5; {}^F\mathbf{S}_6\}$, where $\{{}^F\mathbf{S}_i = \mathbf{S}_i\mathbf{T}\}$, as

$$\mathbf{E}_i = J_{\theta,1i}\mathbf{S}_1 + J_{\theta,2i}\mathbf{S}_2 + J_{\theta,3i}\mathbf{S}_3 \\ + J_{\theta,4i}{}^F\mathbf{S}_4 + J_{\theta,5i}{}^F\mathbf{S}_5 + J_{\theta,6i}{}^F\mathbf{S}_6 \quad (10)$$

In a manipulator (screw-based) Jacobian defined from $[v_{o,x}, v_{o,y}, v_{o,z}, \omega_x, \omega_y, \omega_z]^T = \mathbf{J}_\theta [d\theta_1/dt, \cdots, d\theta_n/dt]^T$, the linear velocity of a reference point, $\mathbf{v}_o = [v_{o,x}, v_{o,y}, v_{o,z}]^T$, in the end effector is instantaneously coincident with the origin of the fixed frame. When we consider the spatial velocity in the fixed (inertial) frame, i.e.,

$$\dot{\mathbf{T}}(t)\mathbf{T}^{-1} = (\mathbf{T}^{-1}\partial \mathbf{T}/\partial \theta_1)\dot{\theta}_1 + \cdots + (\mathbf{T}^{-1}\partial \mathbf{T}/\partial \theta_n)\dot{\theta}_n \quad (11)$$

Thus, the manipulator Jacobian matrix can be derived from equation (11), via the method of change of basis, i.e.,

$$\dot{\mathbf{T}}\mathbf{T}^{-1} = (J_{\theta,11}\mathbf{S}_1 + \cdots + J_{\theta,61}\mathbf{S}_6)\dot{\theta}_1 + \cdots \\ + (J_{\theta,1n}\mathbf{S}_1 + \cdots + J_{\theta,6n}\mathbf{S}_6)\dot{\theta}_n \quad (12)$$

where $J_{\theta,ij}$ denotes the components of the manipulator Jacobian matrix that relates the change of joint coordinates to the Cartesian (twist) coordinates of the end effector and *n* is the d.o.f. of a manipulator in joint space. The term $\mathbf{T}^{-1}\partial \mathbf{T}/\partial \theta_i = J_{\theta,1i}\mathbf{S}_1 + \cdots + J_{\theta,6i}\mathbf{S}_6$ in equation (11)

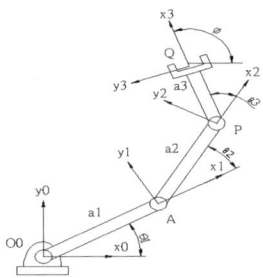

Fig. 1. Planar 3-dof manipulator

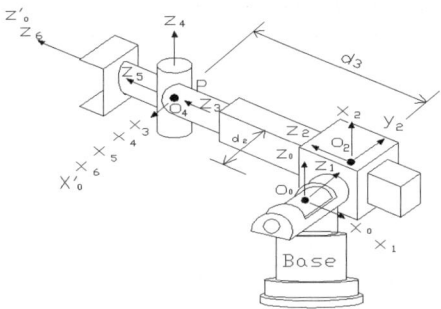

Fig. 2. Stanford manipulator.

determines the *i*-th column of the manipulator Jacobian matrix which indicates the *i*-th joint screw referenced to the moving frame. Thus, the manipulator (screw-based) Jacobian matrix referenced to the fixed (inertial) frame is derived from the relationship between the bases $\{\mathbf{E}_i = \partial \mathbf{T}/\partial \theta_i\}$ and $\{{}^F\mathbf{S}_i = \mathbf{S}_i \mathbf{T}\}$, i.e.,

$$\mathbf{E}_i = J_{\theta,1i}{}^F\mathbf{S}_1 + J_{\theta,2i}{}^F\mathbf{S}_2 + J_{\theta,3i}{}^F\mathbf{S}_3 \\ + J_{\theta,4i}{}^F\mathbf{S}_4 + J_{\theta,5i}{}^F\mathbf{S}_5 + J_{\theta,6i}{}^F\mathbf{S}_6 \quad (13)$$

In addition, the manipulator (screw-based) Jacobian matrix referenced to the body (moving) frame is derived from the relationship between the bases $\{\mathbf{E}_i = \partial \mathbf{T}/\partial \theta_i\}$ and $\{{}^B\mathbf{S}_i = \mathbf{T}\mathbf{S}_i\}$, i.e.

$$\mathbf{E}_i = J_{\theta,1i}{}^B\mathbf{S}_1 + J_{\theta,2i}{}^B\mathbf{S}_2 + J_{\theta,3i}{}^B\mathbf{S}_3 \\ + J_{\theta,4i}{}^B\mathbf{S}_4 + J_{\theta,5i}{}^B\mathbf{S}_5 + J_{\theta,6i}{}^B\mathbf{S}_6 \quad (14)$$

From equations (10), (13), and (14), we notice that the properties of the Jacobian matrices are determined by the basis mapping relationship. All of them describe the mapping from the coordinate basis, $\{\partial/\partial \theta_i\}$, in the joint space to the non-coordinate basis in Cartesian space. Thus, the 6×6 asymmetric Cartesian matrices of serial manipulators are expected!!

4. Examples

The formulation and transformation of the Cartesian matrices of the planar 3-dof manipulator and the Stanford manipulator are presented in this section [14].

4.1. Example I: A Planar Serial Manipulator

For the planar 3-dof manipulator that the coordinate system is attached to each link as shown in Fig. 1, the overall Denavit-Hartenberg transformation matrix is referred to [14].

Thus, the conventional Jacobian matrix is obtained from the mapping relationship between bases $\{\partial \mathbf{T}/\partial \theta_1; \partial \mathbf{T}/\partial \theta_2; \partial \mathbf{T}/\partial \theta_3\}$ and $\{\mathbf{S}_1; \mathbf{S}_2; {}^F\mathbf{S}_6\}$ as the following

$$\mathbf{J}_\theta = \begin{bmatrix} -a_1 s_1 - a_2 s_{12} - a_3 s_{123} & -a_2 s_{12} - a_3 s_{123} & -a_3 s_{123} \\ a_1 c_1 + a_2 c_{12} + a_3 c_{123} & a_2 c_{12} + a_3 c_{123} & a_3 c_{123} \\ 1 & 1 & 1 \end{bmatrix} \quad (15)$$

By assuming a diagonal joint stiffness matrix, $\mathbf{K}_\theta = diag(k_{\theta 11}, k_{\theta 22}, k_{\theta 33})$, and the external load, $\mathbf{w} = [f_x, f_y, m_z]^T$, applied at the end-effector, we find a symmetric 3×3 Cartesian stiffness matrix by the CCT. The result indicates that the CCT represents the transformation between coordinate bases, $\{\partial \mathbf{T}/\partial \theta_1; \partial \mathbf{T}/\partial \theta_2; \partial \mathbf{T}/\partial \theta_3\}$ in the joint space and, $\{\mathbf{S}_1; \mathbf{S}_2; {}^F\mathbf{S}_6\}$, in the Cartesian space.

On the other hand, the manipulator (screw-based) Jacobian matrix viewed in the fixed frame is given by the basis transformation relationship between $\{\partial \mathbf{T}/\partial \theta_1; \partial \mathbf{T}/\partial \theta_2; \partial \mathbf{T}/\partial \theta_3\}$ and $\{{}^F\mathbf{S}_1; {}^F\mathbf{S}_2; {}^F\mathbf{S}_6\}$, i.e.,

$$\mathbf{J}_\theta = \begin{bmatrix} 0 & a_1 s_1 & a_1 s_1 + a_2 s_{12} \\ 0 & -a_1 c_1 & -a_1 c_1 - a_2 c_{12} \\ 1 & 1 & 1 \end{bmatrix} \quad (16)$$

However, an asymmetric 3×3 Cartesian stiffness matrix is obtained with the following skew symmetric part.

$$\mathbf{K}_{skew} = 0.5 \begin{bmatrix} 0 & 0 & -f_y \\ 0 & 0 & f_x \\ f_y & -f_x & 0 \end{bmatrix} \quad (17)$$

In addition, the skew symmetric part of the Cartesian stiffness in terms of the manipulator Jacobian matrix viewed in the body frame is given by the negative matrix, $-\mathbf{K}_{skew}$. The result shows that the manipulator Jacobian matrix of the planar serial robot represents the mapping between the coordinate basis, $\{\partial \mathbf{T}/\partial \theta_1; \partial \mathbf{T}/\partial \theta_2; \partial \mathbf{T}/\partial \theta_3\}$, and the non-coordinate basis, $\{{}^F\mathbf{S}_1; {}^F\mathbf{S}_2; {}^F\mathbf{S}_6\}$ in the fixed frame or $\{{}^B\mathbf{S}_1; {}^B\mathbf{S}_2; {}^B\mathbf{S}_6\}$ in the body frame.

4.2. Example II: A Spatial Serial Manipulator

Fig. 2 shows a simplify diagram of Stanford arm that the origin of the fixed coordinate frame is located at the point of intersection of the first two joint axes, and the origin of the (x_6, y_6, z_6) frame is located at the point of intersection of the last three joint axes. Thus, the overall Denavit-Hartenberg transformation matrix is given by the matrix product, $\mathbf{T}(\theta_1, \theta_2, d_3, \theta_4, \theta_5, \theta_6) = {}_1^0\mathbf{T}\,{}_2^1\mathbf{T}\,{}_3^2\mathbf{T}\,{}_4^3\mathbf{T}\,{}_5^4\mathbf{T}\,{}_6^5\mathbf{T}$. From equation (10), we can obtain the conventional Jacobian matrix that represents the basis transformation between $\{\mathbf{E}_i = \partial \mathbf{T}/\partial \theta_i\}$ in the joint space and $\{\mathbf{S}_1; \mathbf{S}_2; \mathbf{S}_3; {}^F\mathbf{S}_4; {}^F\mathbf{S}_5; {}^F\mathbf{S}_6\}$ in the Cartesian space, i.e.,

$$\mathbf{E}_1 = -(d_3 s_1 s_2 + d_2 c_1)\mathbf{S}_1 + (d_3 c_1 s_2 - d_2 s_1)\mathbf{S}_2 + {}^F\mathbf{S}_6$$
$$\mathbf{E}_2 = d_3 c_1 c_2 \mathbf{S}_1 + d_3 s_1 c_2 \mathbf{S}_2 - d_3 s_2 \mathbf{S}_3 - s_1{}^F\mathbf{S}_4 + c_1{}^F\mathbf{S}_5$$
$$\mathbf{E}_3 = c_1 s_2 \mathbf{S}_1 + s_1 s_2 \mathbf{S}_2 + c_2 \mathbf{S}_3$$
$$\mathbf{E}_4 = c_1 s_2{}^F\mathbf{S}_4 + s_1 s_2{}^F\mathbf{S}_5 + c_2{}^F\mathbf{S}_6 \qquad (18)$$
$$\mathbf{E}_5 = (J_{\theta 45}){}^F\mathbf{S}_4 + (J_{\theta 55}){}^F\mathbf{S}_5 + (J_{\theta 65}){}^F\mathbf{S}_6$$
$$\mathbf{E}_6 = (J_{\theta 46}){}^F\mathbf{S}_4 + (J_{\theta 56}){}^F\mathbf{S}_5 + (J_{\theta 66}){}^F\mathbf{S}_6$$

where the terms of the 5-th and 6-th column of the manipulator Jacobian matrix are given by

$$J_{\theta,45} = -s_1 s_4 + c_1 c_2 c_4$$
$$J_{\theta,55} = c_1 s_4 + s_1 c_2 c_4$$
$$J_{\theta,65} = -s_2 c_4 \qquad (19)$$
$$J_{\theta,46} = s_5 s_1 c_4 + s_5 c_1 c_2 s_4 + c_1 s_2 c_5$$
$$J_{\theta,56} = -s_5 c_1 c_4 + s_5 s_1 c_2 s_4 + s_1 s_2 c_5$$
$$J_{\theta,66} = -s_2 s_4 s_5 + c_2 c_5$$

Through the CCT analysis, an asymmetric 6×6 Cartesian stiffness matrix in terms of the non-coordinate basis is obtained and the skew symmetric part is given by

$$\mathbf{K}_{skew} = 0.5 \begin{bmatrix} \mathbf{0} & \mathbf{0} \\ \mathbf{0} & (\mathbf{m}\times)^T \end{bmatrix} \qquad (20)$$

$$\mathbf{m}\times = \begin{bmatrix} 0 & -m_z & m_y \\ m_z & 0 & -m_x \\ -m_y & m_x & 0 \end{bmatrix} \qquad (21)$$

On the other hand, the manipulator (screw-based) Jacobian matrix viewed in the fixed frame is given by the bases transformation relationship between $\{\mathbf{E}_i = \partial \mathbf{T}/\partial \theta_i\}$ and $\{{}^F\mathbf{S}_i = \mathbf{S}_i \mathbf{T}\}$, i.e.,

$$\mathbf{E}_1 = {}^F\mathbf{S}_6;\ \mathbf{E}_2 = -s_1{}^F\mathbf{S}_4 + c_1{}^F\mathbf{S}_5;\ \mathbf{E}_3 = c_1 s_2{}^F\mathbf{S}_4 + s_1 s_2{}^F\mathbf{S}_5 + c_2{}^F\mathbf{S}_6$$
$$\mathbf{E}_4 = d_3 c_1 c_2{}^F\mathbf{S}_1 + d_3 s_1 c_2{}^F\mathbf{S}_2 - d_3 s_2{}^F\mathbf{S}_3 + c_1 s_2{}^F\mathbf{S}_4 + s_1 s_2{}^F\mathbf{S}_5 + c_2{}^F\mathbf{S}_6 \qquad (22)$$
$$\mathbf{E}_5 = (J_{\theta 15}){}^F\mathbf{S}_1 + (J_{\theta 25}){}^F\mathbf{S}_2 + (J_{\theta 35}){}^F\mathbf{S}_3 + (J_{\theta 45}){}^F\mathbf{S}_4 + (J_{\theta 55}){}^F\mathbf{S}_5 + (J_{\theta 65}){}^F\mathbf{S}_6$$
$$\mathbf{E}_6 = (J_{\theta 16}){}^F\mathbf{S}_1 + (J_{\theta 26}){}^F\mathbf{S}_2 + (J_{\theta 36}){}^F\mathbf{S}_3 + (J_{\theta 46}){}^F\mathbf{S}_4 + (J_{\theta 56}){}^F\mathbf{S}_5 + (J_{\theta 66}){}^F\mathbf{S}_6$$

Notice that the 4-th, 5-th, and 6-th rows of the manipulator (screw-based) Jacobian matrix in the fixed frame in equation (22), are same as those of the conventional Jacobian in equation (18), because both indicate the same rotation velocity in terms of the sub-basis $\{{}^F\mathbf{S}_4; {}^F\mathbf{S}_5; {}^F\mathbf{S}_6\}$. In addition, some terms of the 5-th and 6-th columns of the manipulator Jacobian matrix in equation (22), are given by

$$J_{\theta,15} = -c_4 s_1 d_3 - c_4 c_1 s_2 d_2 - c_1 c_2 s_4 d_3$$
$$J_{\theta,25} = c_4 c_1 d_3 - s_1 c_2 s_4 d_3 - s_1 s_2 d_2 c_4$$
$$J_{\theta,35} = -c_2 d_2 c_4 + s_2 s_4 d_3 \qquad (23)$$
$$J_{\theta,16} = c_2 d_3 s_5 c_1 c_4 - c_1 d_2 s_2 s_4 s_5 - s_1 d_3 s_5 s_4 + c_1 d_2 c_2 c_5$$
$$J_{\theta,26} = c_2 d_3 s_5 s_1 c_4 + c_1 d_3 s_5 s_4 - s_1 d_2 s_2 s_4 s_5 + s_1 d_2 c_2 c_5$$
$$J_{\theta,36} = -c_2 d_2 s_4 s_5 - s_2 c_5 d_2 - s_2 c_4 d_3 s_5$$

Once the manipulator Jacobian matrix in the fixed frame is obtained, the 6×6 Cartesian stiffness can be determined via the CCT. Thus, the skew symmetric part of the Cartesian stiffness matrix of the Stanford manipulator in the non-coordinate basis, $\{{}^F\mathbf{S}_i = \mathbf{S}_i \mathbf{T}\}$, is given by

$$\mathbf{K}_{skew} = 0.5 \begin{bmatrix} \mathbf{0} & (\mathbf{f}\times)^T \\ (\mathbf{f}\times)^T & (\mathbf{m}\times)^T \end{bmatrix} = 0.5(\mathbf{w}\times)^T = -0.5\mathbf{w}\times \qquad (24)$$

$$\mathbf{f}\times = \begin{bmatrix} 0 & -f_z & f_y \\ f_z & 0 & -f_x \\ -f_y & f_x & 0 \end{bmatrix} \qquad (25)$$

Through similar procedures, the skew symmetric part of the Cartesian stiffness matrix of the Stanford manipulator in the non-coordinate basis, $\{{}^B\mathbf{S}_i = \mathbf{T}\mathbf{S}_i\}$, is equal to one-half of the cross-product matrix formed by the external loads, i.e., $0.5\mathbf{w}\times$. Notice that the 6×6 Cartesian stiffness in the non-coordinate basis, $\{{}^B\mathbf{S}_i\}$, represents the transformation between the changes in Cartesian wrench and coordinates, i.e.,

$$[df_x, df_y, df_z, dm_x, dm_y, dm_z]^T = \mathbf{K}[dx, dy, dz, d\phi_x, d\phi_y, d\phi_z]^T$$

, both viewed in the body frame.

4.3. Discussions

The skew symmetric part of Stanford manipulator in terms of the manipulator (screw-based) Jacobian and the results shown in Section 3.4 are consistent with Lipkin's in [3]. In addition, the skew symmetric part of Stanford manipulator in equation (20), with respect to the conventional Jacobian is different from the form in equation (24), with respect to the manipulator Jacobian so as to the result of the planar 3-dof manipulator. No matter what the Cartesian stiffness matrices are defined, the CCT theory can represent the correct and valid stiffness control strategy!

In addition, the 6×6 Cartesian stiffness is shown to be an asymmetric matrix referenced to both fixed (inertial), $\{^F S_i\}$, and body (moving) frames, $\{^B S_i\}$. The Cartesian stiffness referenced to the fixed and moving frames at coincident points are transposes of each other.

5. Conclusion

In this paper, the conservative congruence transformation (CCT) is proven to be a valid and complete stiffness mapping relationship in robotic stiffness control. Through the geometrical methods, the 6×6 Cartesian stiffness of manipulators is shown to be basis (coordinate) dependent. Namely, the Cartesian stiffness of the conservative system could assume totally different physical forms even though they are representing the same system! In addition, the CCT theory shows that the properties of the stiffness matrices are related to the external loads and the differential Jacobian matrices which correct the erroneous conventional formulations used in the robotic stiffness control and accepted in textbooks!

Acknowledgment

This work was supported by the Taiwan National Science Council under Grant NSC91-2213-E-262-005.

References

[1] J. K. Salisbury, "Active stiffness control of a manipulator in Cartesian coordinates," in *Proc. 19th IEEE Conference on Decision and Control*, Albuquerque, NM, Dec. 1980, pp. 87-97.

[2] M. Griffis and J. Duffy, "Global stiffness modeling of a class of simple compliant couplings," *Mechanisms and Machine Theory* 28(2):207-224, 1993.

[3] N. Ciblak and H. Lipkin, "Asymmetric Cartesian stiffness for the modeling of compliant robotic system," in *ASME Conference-Robotics; Kinematics, and Control*, DE-Vol. 72, ASME, 1994, pp. 197-204.

[4] M. Zefran and V. Kumar, "Affine connections for the Cartesian stiffness matrix," in *Proc. of IEEE International Conference on Robotic and Automation*, 1997, pp. 1376-1381.

[5] Z. Lu and A. A. Goldenberg, "Robust impedance control and force regulation: Theory and experiments," *International Journal of Robotics Research*, 14(3):225-254, 1995.

[6] I. Kao, M. R. Cutkosky, and R. S. Johansson, "Robotic stiffness control and calibration as applied to human grasping tasks," *IEEE Transaction on Robotics and Automation* 13(4):557-566, 1997.

[7] S. Huang and J. M. Schimmels, "The bounds and realization of spatial stiffness achieved with simple springs connected in parallel," *IEEE Transaction on Robotics and Automation* 14(3):466-475, 1998.

[8] S.-F. Chen and I. Kao, "Conservative congruence transformation of stiffness control in robotic grasping and manipulation," in *the 9th International Symposium on Robotics Research*, Snowbird, Utah, USA, 1999, pp. 7-14.

[9] S.-F. Chen and I. Kao, "Conservative congruence transformation for joint and Cartesian stiffness matrices of robotic hands and fingers," *International Journal of Robotics Research*, 19(9): 835-847, 2000.

[10] S.-F. Chen, Y. Li, and I. Kao, "A new theory in stiffness control for dextrous manipulation," in *Proc. of IEEE Robotics and Automation Conference*, Seoul, Korea, May 2001, pp. 3047-3054.

[11] C. Huang and I. Kao, "Geometrical interpretations of conservative congruence transformation (CCT) for serial manipulators via screw theory," in *the 10th International Symposium on Robotics Research, Lorne, Victoria, Australia*, 2001.

[12] S.-F. Chen and I. Kao, "Geometrical approach to the conservative congruence transformation (CCT) for robotic stiffness control," in *Proc. of IEEE Robotics and Automation Conference*, Washington D.C., May 2002, pp. 544-549.

[13] Y. Li, S.-F. Chen, and I. Kao, "Stiffness control and transformation for robotic systems with coordinate and non-coordinate bases," in *Proc. of IEEE Robotics and Automation Conference*, Washington D.C., May 2002, pp. 550-555.

[14] L. W. Tsai, *Robot Analysis: the Mechanics of Serial and Parallel Manipulators.* John Wiley & Sons, New York, 1999.

[15] B. F. Schutz, *Geometrical Method of Mathematical Physics.* Cambridge University Press, 1980.

[16] J. M. Selig, *Geometrical Method in Robotics.* Springer-Verlag, 1st edition, 1996.

[17] R. M. Murray, Z. Li, and S. S. Sastry, *Robotic Manipulation. CRC Press.* 1st edition, 1994.

Dynamic Performance Analysis for Non-Redundant Robotic Manipulators in Contact

Alan Bowling
Dept. of Aerospace and Mechanical
Engineering, University of Notre Dame,
Notre Dame, IN 46556, USA

ChangHwan Kim
Dept. of Aerospace and Mechanical
Engineering, University of Notre Dame,
Notre Dame, IN 46556, USA

Abstract—This paper presents an analysis of the dynamic performance of a non-redundant robotic manipulator while in contact with a fixed environment, i.e. a rigid wall. This work extends an earlier performance characterization, the *Dynamic Capability Hypersurface*, which described a manipulator's ability to accelerate and apply forces at the end-effector, given the limitations on the manipulator's motor torques. The proposed extension includes an analysis of end-effector motion constraints and friction forces and moments associated with environmental contact. This analysis is applied to a PUMA 560 manipulator.

I. INTRODUCTION

Studies of dynamic performance usually focus on a manipulator's ability to accelerate its end-effector, [1], [2], [3], [4] for example. However, there has been a growing interest in considering the relationship between end-effector accelerations and forces/moments applied at the end-effector. This interest is sparked in part by the need to analyze tasks requiring end-effector motion while applying a contact force/moment, for instance buffing or sanding a surface. In this case of a flat surface, the end-effector must apply a normal force while also attempting to move in directions parallel to the surface. The ability to manipulate grasped and non-grasped objects is also determined by the acceleration and force/moment capabilities. This includes pushing and pulling objects as well as grasping and manipulating them.

In most performance analyses end-effector accelerations and forces are considered separately. Studies addressing only end-effector force/moments include [5], [6], [7]. A characterization which combines accelerations and forces/moments is the *Dynamic Capability Hypersurface* [8], [9]. However, that characterization did not consider the motion constraints and forces/moments, particularly friction, involved in contact with a rigid surface. This paper discusses the benefits from and difficulties with this analysis associated with the additional contact constraints.

In the following sections the constraints and contact forces/moments are discussed first. Then the capabilities for linear/angular accelerations, contact forces and moments, as well as linear/angular velocities, are analyzed subject to the additional constraints. This leads to the development of the Dynamic Capability Hypersurface for a contact configuration. The PUMA 560 manipulator is analyzed as an example using the proposed methodology.

II. CONTACT CONSTRAINTS

A. Linear Motions and Forces

The contact forces between the end-effector and the environment consist of a normal resistance force and tangential friction forces. Since the end-effector cannot penetrate into the environment, and assuming the surface does not move,

$$v_n = 0 \qquad \frac{dv_n}{dt} = \dot{v}_n = 0 \qquad (1)$$

where v_n is the component of the end-effector velocity in the direction normal to the surface, see Fig. 1. The normal force, the scalar f_n, preventing penetration can only have one direction since the surface cannot pull on the end-effector. With reference to Fig. 1,

$$f_n = |f_n|. \qquad (2)$$

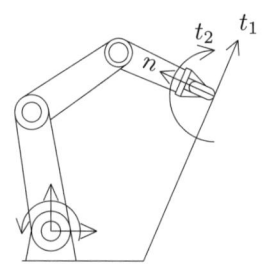

Fig. 1. Non-Redundant Planar Manipulator.

The friction forces are modeled using the Coulomb friction law. The motion subject to friction forces is typically classified into *sliding* and *sticking* phases as follows:

$$\begin{array}{ll} \boldsymbol{v}_t = 0, \ \dot{\boldsymbol{v}}_t = 0 & \text{sticking when} \quad \|\mathbf{f}_t\| \leq \mu_s |f_n| \\ \boldsymbol{v}_t \neq 0, \ \dot{\boldsymbol{v}}_t \neq 0 & \text{sliding when} \quad \|\mathbf{f}_t\| > \mu_s |f_n| \end{array} \qquad (3)$$

where $\boldsymbol{v}_t$ is the vector of end-effector velocities tangential to the surface, and μ_s is the static friction coefficient. The vector $\mathbf{f}_t$ represents the tangential contact forces. During the sliding phase, the following constraints are imposed

$$\|\mathbf{f}_t\| = \mu_d |f_n| \qquad \frac{\mathbf{f}_t}{\|\mathbf{f}_t\|} = -\frac{\boldsymbol{v}_t}{\|\boldsymbol{v}_t\|} \qquad (4)$$

where μ_d is the dynamic friction coefficient. A number of studies have used these constraints to analyze the contact situation, including [10], [11], [12], [13], [14], [15]. However, these studies focused on determining the next state of the manipulator, given the current acceleration, velocity, position/configuration, and the type of surface contact. In the proposed study, only the configuration and the type of surface contact are known, and an assessment of the acceleration and force/moment capabilities is sought.

B. Angular Motions and Moments

The angular contact constraints are more difficult to model than the analogous linear ones. This is partly because contact moments depend heavily on the type of surface contact involved. In [16] contacts were classified as point contact, soft finger contact, line contact, and plane contact. Each category has different angular motion constraints associated with it. Here the plane contact situation is considered, as shown in Fig. 1, where the surface of the end-effector is considered to be circular. The analysis of other types of contacts and surfaces can be obtained from this analysis. In plane contact the end-effector can only rotate about the axis normal to the plane,

$$\boldsymbol{\omega}_t = 0 \qquad \frac{d\boldsymbol{\omega}_t}{dt} = \dot{\boldsymbol{\omega}}_t = 0 \qquad (5)$$

where the vector $\boldsymbol{\omega}_t$ represents the end-effector's angular velocities about directions lying in the plane of the surface.

The frictional contact moment between a rotating circular cylinder and a planar surface is developed in [17] and [18]. They used the Coulomb friction law to develop a frictional shear stress on the surface. A uniform normal force across the surface is assumed, in order to obtain the shear stress distribution. This stress is integrated over the surface to obtain the friction moment m_n, used here to define the sticking and sliding phases for rotational motion as

$$\begin{aligned} \omega_n = 0, \; \dot{\omega}_n = 0 & \quad \text{sticking} & |m_n| &\leq \tfrac{2}{3} \, \text{R} \, \mu_s \, |f_n| \\ \omega_n \neq 0, \; \dot{\omega}_n \neq 0 & \quad \text{sliding} & |m_n| &> \tfrac{2}{3} \, \text{R} \, \mu_s \, |f_n| \end{aligned} \quad (6)$$

where ω_n and m_n are the components of the end-effector's angular velocity and contact moment about the normal axis and R is the radius of the contact surface. Other works have examined non-uniform stress distributions over the contact surface, for instance [19] who studied soft fingers. However, here the simple model in (6) will be used. During the sliding phase, the following constraints are imposed

$$|m_n| = \tfrac{2}{3} \, \text{R} \, \mu_d \, |f_n| \qquad \frac{m_n}{|m_n|} = -\frac{\omega_n}{|\omega_n|} . \quad (7)$$

III. Analysis of Dynamic Performance

In this section the analysis associated with the Dynamic Capability Hypersurface is developed considering the additional constraints discussed in Section II. The analysis begins by considering the equations of motion for a non-redundant manipulator in plane contact,

$$E \begin{bmatrix} \dot{\boldsymbol{v}} \\ \dot{\boldsymbol{\omega}} \end{bmatrix} + \mathcal{E} \begin{bmatrix} \mathcal{F} \\ \mathcal{M} \end{bmatrix} + \mathbf{h} + \mathbf{GRAV} = \boldsymbol{\Upsilon} \quad (8)$$

where E, $\mathcal{E}$, **GRAV**, $\boldsymbol{\Upsilon}$ and $\mathbf{h}$ contain the inertial properties, transpose of the manipulator Jacobian, gravity forces, actuator torques and the Coriolis and centrifugal forces as well as other velocity dependent terms. Also note that because of the constraints in (1) and (5),

$$\begin{bmatrix} \boldsymbol{v} \\ \boldsymbol{\omega} \end{bmatrix} = \begin{bmatrix} \boldsymbol{v}_t \\ \omega_n \end{bmatrix} \qquad \begin{bmatrix} \mathcal{F} \\ \mathcal{M} \end{bmatrix} = \begin{bmatrix} f_n \\ \mathbf{f}_t \\ m_n \\ \mathbf{m}_t \end{bmatrix} \quad (9)$$

where the vector $\mathbf{m}_t$ contains the contact moments about axes in the plane of the surface. The bounds on the actuator torques in (8) can be expressed as

$$\begin{aligned} \boldsymbol{\Upsilon}_{lower} \leq \; & E_v \, \dot{\boldsymbol{v}}_t + E_\omega \, \dot{\omega}_n + \\ & \mathcal{E}_\mathcal{F} \, \mathcal{F} + \mathcal{E}_\mathcal{M} \, \mathcal{M} + \mathbf{h} \; \leq \boldsymbol{\Upsilon}_{upper} \end{aligned} \quad (10)$$

where the matrices E and $\mathcal{E}$ are split into their linear and angular subspaces. $\boldsymbol{\Upsilon}_{lower}$ and $\boldsymbol{\Upsilon}_{upper}$ contain gravity forces and the bounds on the actuator torques. This model is developed in detail in [20], [8]. Equation (10) is considered row by row.

Each term in (10) represents an actuator torque, so the ith row can be expressed as

$$\Upsilon_{lower_i} \leq \Upsilon_{\dot{v}_i} + \Upsilon_{\dot{\omega}_i} + \Upsilon_{\mathcal{F}_i} + \Upsilon_{\mathcal{M}_i} + \Upsilon_{h_i} \leq \Upsilon_{upper_i}. \quad (11)$$

where

$$\Upsilon_{\dot{v}_i} = E_v^i \, \dot{\boldsymbol{v}}_t \quad (12)$$

and likewise for the other terms in (10) and (11). E_v^i is the ith row of E_v, and $i = \{1, \ldots, d\}$ where d is the number of *degrees-of-freedom* (DOF) of the end-effector, as well as the number of actuators.

The Dynamic Capability Hypersurface describes the worst-case performance of the manipulator. This description is obtained by examining the amount of torque required to perform all possible motions and apply all possible forces/moments at the end-effector. All possible accelerations are represented by a *balanced* acceleration sphere, circle or line segment, depending on the dimension of $\dot{\boldsymbol{v}}_t$,

$$\dot{\boldsymbol{v}}_t^T \dot{\boldsymbol{v}}_t = \|\dot{\boldsymbol{v}}_t\|^2 \quad (13)$$

and likewise for $\dot{\omega}_n$, $\boldsymbol{v}_t$, ω_n, f_n, $\mathbf{f}_t$, m_n and $\mathbf{m}_t$. In (13), the magnitude $\|\dot{\boldsymbol{v}}_t\|$ is a parameter having some fixed value, and likewise for the other terms. The worst-case performance is obtained by maximizing/minimizing the sum of the terms in (11), subject to the set of constraints represented by (13), for each relation in (10) [20], [8].

In the original hypersurface analysis, each term in (11) was maximized/minimized independently. This is disallowed by (4) and (7), which introduce a coupling between forces/moments and velocities. However, the analysis of the nonzero accelerations, recall (1) and (5), remains unchanged. These analyses, developed for the sticking and sliding cases, are discussed next.

A. End-Effector Sticking to Surface

In the sticking phase all velocities and accelerations are zero, thus (11) becomes

$$\Upsilon_{lower_i} \leq \Upsilon_{\mathcal{F}_i} + \Upsilon_{\mathcal{M}_i} \leq \Upsilon_{upper_i}. \quad (14)$$

The worst-case actuator torques are found as follows:

$$\begin{array}{l} \text{minimize/maximize} \\ \text{wrt } \{f_n, \mathbf{f}_t, m_n, \mathbf{m}_t\} \end{array} \quad \gamma_i = \Upsilon_{\mathcal{F}_i} + \Upsilon_{\mathcal{M}_i}$$

$$\begin{array}{ll}\text{subject to} & f_n = |f_n| \qquad m_n^2 = |m_n|^2 \\ & \mathbf{f}_t^T \mathbf{f}_t = \|\mathbf{f}_t\|^2 \qquad \mathbf{m}_t^T \mathbf{m}_t = \|\mathbf{m}_t\|^2 \\ & \|\mathbf{f}_t\| \leq \mu_s |f_n| \qquad |m_n| \leq \tfrac{2}{3} R \mu_s |f_n| \end{array} \quad (15)$$

where

$$\begin{aligned}\gamma_i &= \mathcal{E}_{\mathcal{F}}^i \begin{bmatrix} f_n \\ \mathbf{f}_t \end{bmatrix} + \mathcal{E}_{\mathcal{M}}^i \begin{bmatrix} m_n \\ \mathbf{m}_t \end{bmatrix} \\ &= \mathcal{E}_f^i f_n + \mathcal{E}_{\mathbf{f}}^i \mathbf{f}_t + \mathcal{E}_m^i m_n + \mathcal{E}_{\mathbf{m}}^i \mathbf{m}_t, \quad (16)\end{aligned}$$

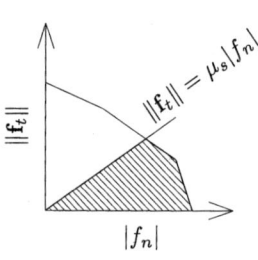

Fig. 2. Sticking Force and Moment Capabilities.

"wrt" means "with respect to," and $|f_n|$, $\|\mathbf{f}_t\|$, $|m_n|$ and $\|\mathbf{m}_t\|$ act as parameters in this optimization problem. The matrices $\mathcal{E}_{\mathcal{F}}^i$ and $\mathcal{E}_{\mathcal{M}}^i$ are split into their subspaces corresponding to the normal and tangential directions. Notice that the last two constraints in (15) do not involve the decision variables, and thus are applied later to the solution. Use of the Lagrange multiplier method yields,

$$\begin{aligned}(\gamma_i)_{max} &= \mathcal{E}_f^i |f_n| + \|\mathcal{E}_{\mathbf{f}}^i\| \|\mathbf{f}_t\| + \\ &\quad |\mathcal{E}_m^i| |m_n| + \|\mathcal{E}_{\mathbf{m}}^i\| \|\mathbf{m}_t\| \\ (\gamma_i)_{min} &= \mathcal{E}_f^i |f_n| - \|\mathcal{E}_{\mathbf{f}}^i\| \|\mathbf{f}_t\| - \\ &\quad |\mathcal{E}_m^i| |m_n| - \|\mathcal{E}_{\mathbf{m}}^i\| \|\mathbf{m}_t\|.\end{aligned} \quad (17)$$

The extremums for all relations in (10) can be expressed as

$$\mathbb{F} \begin{bmatrix} |f_n| & \|\mathbf{f}_t\| & |m_n| & \|\mathbf{m}_t\| \end{bmatrix}^T \leq \mathbb{T} \quad (18)$$

where the rows of the $4 \times 2d$ matrix $\mathbb{F}$ are defined as

$$\begin{aligned}\mathbb{F}^i &= \begin{bmatrix} \mathcal{E}_f^i & \|\mathcal{E}_{\mathbf{f}}^i\| & |\mathcal{E}_m^i| & \|\mathcal{E}_{\mathbf{m}}^i\| \end{bmatrix} \\ \mathbb{F}^{i+d} &= \begin{bmatrix} -\mathcal{E}_f^i & \|\mathcal{E}_{\mathbf{f}}^i\| & |\mathcal{E}_m^i| & \|\mathcal{E}_{\mathbf{m}}^i\| \end{bmatrix}\end{aligned} \quad (19)$$

and

$$\mathbb{T} = \begin{bmatrix} \Upsilon_{upper} \\ -\Upsilon_{lower} \end{bmatrix}. \quad (20)$$

Each relation in (18) represents a hyperplane in four-dimensional space. Sections of the hypersurface are taken in order to examine its characteristics. The section $|m_n| = \|\mathbf{m}_t\| = 0$ is associated with point contact. For this case, the relations in (18) are lines in $|f_n|$-$\|\mathbf{f}_t\|$ space. If these lines are overlaid on the same plot, the innermost envelope around the origin determines the worst-case performance, as shown in Fig. 2. This figure is similar to the well known *friction cone* which is flattened and halved when expressed in terms of force magnitudes. The sticking condition in (3), listed as one of the last constraints in (15), is enforced on the solution, as shown in Fig. 2. The hatched region describes the forces that can be applied to the surface without slipping/sliding. The sliding phase is discussed in the next section.

B. End-Effector Sliding on Surface

As mentioned earlier, the analysis of worst-case accelerations can be done independently of the other terms in (10). The procedure for the acceleration analysis is similar to the method discussed in Section III-A. This is not true for the velocity and force/moment analysis discussed in Section III-B.2.

1) Accelerations: The worst-case accelerations are found as follows:

$$\begin{array}{l}\text{minimize/maximize} \\ \text{wrt } \{\dot{\boldsymbol{v}}_t, \dot{\omega}_n\}\end{array} \quad \beta_i = \Upsilon_{\dot{v}_i} + \Upsilon_{\dot{\omega}_i} \quad (21)$$

$$\text{subject to} \quad \dot{\boldsymbol{v}}_t^T \dot{\boldsymbol{v}}_t = \|\dot{\boldsymbol{v}}_t\|^2 \quad \dot{\omega}_n^2 = |\dot{\omega}_n|^2.$$

The Lagrange multiplier method yields the solution

$$(\beta_i)_{max} = \|E_v^i\| \|\dot{\boldsymbol{v}}_t\| + |E_\omega^i| |\dot{\omega}_n| \quad (22)$$

where $(\beta_i)_{min} = -(\beta_i)_{max}$. The extremums for all of the relations in (10) can be expressed as

$$\mathbb{A} \begin{bmatrix} \|\dot{\boldsymbol{v}}_t\| & |\dot{\omega}_n| \end{bmatrix}^T \quad (23)$$

where the rows of the $2 \times 2d$ matrix $\mathbb{A}$ are defined as

$$\mathbb{A}^i = \mathbb{A}^{i+d} = \begin{bmatrix} \|E_v^i\| & |E_\omega^i| \end{bmatrix}. \quad (24)$$

2) Velocities and Contact Forces/Moments: The worst-case combinations of velocity and contact forces/moments are found as follows:

$$\begin{array}{l}\text{minimize/maximize} \\ \text{wrt } \{\boldsymbol{v}_t, \omega_n, \boldsymbol{\mathcal{F}}, \boldsymbol{\mathcal{M}}\}\end{array} \quad \alpha_i = \Upsilon_{\mathcal{F}_i} + \Upsilon_{\mathcal{M}_i} + \Upsilon_{h_i}$$

$$\begin{array}{ll}\text{subject to} & f_n = |f_n| \qquad m_n^2 = |m_n|^2 \\ & \mathbf{f}_t^T \mathbf{f}_t = \|\mathbf{f}_t\|^2 \qquad \mathbf{m}_t^T \mathbf{m}_t = \|\mathbf{m}_t\|^2 \\ & \boldsymbol{v}_t^T \boldsymbol{v}_t = \|\boldsymbol{v}_t\|^2 \qquad \omega_n^2 = |\omega_n|^2 \\ & \dfrac{\mathbf{f}_t}{\|\mathbf{f}_t\|} = -\dfrac{\boldsymbol{v}_t}{\|\boldsymbol{v}_t\|} \qquad \dfrac{m_n}{|m_n|} = -\dfrac{\omega_n}{|\omega_n|} \\ & \|\mathbf{f}_t\| = \mu_d |f_n| \qquad |m_n| = \tfrac{2}{3} R \mu_d |f_n|\end{array} \quad (25)$$

where

$$\alpha_i = \mathcal{E}_{\mathcal{F}}^i \boldsymbol{\mathcal{F}} + \mathcal{E}_{\mathcal{M}}^i \boldsymbol{\mathcal{M}} + \begin{bmatrix} \boldsymbol{v}_t \\ \omega_n \end{bmatrix}^T H_i \begin{bmatrix} \boldsymbol{v}_t \\ \omega_n \end{bmatrix} \quad (26)$$

and

$$\mathbf{h} = \begin{bmatrix} \begin{bmatrix} \boldsymbol{v}_t \\ \omega_n \end{bmatrix}^T H_1 \begin{bmatrix} \boldsymbol{v}_t \\ \omega_n \end{bmatrix} \\ \vdots \\ \begin{bmatrix} \boldsymbol{v}_t \\ \omega_n \end{bmatrix}^T H_d \begin{bmatrix} \boldsymbol{v}_t \\ \omega_n \end{bmatrix} \end{bmatrix}. \quad (27)$$

The last two constraints in (25) do not contain any decision variables and thus will be enforced on the solution. Substituting several of the constraints into the cost function eliminates some of the decision variables,

$$\alpha_i = \mathcal{E}_f^i |f_n| + \mathcal{E}_\mathbf{f}^i \mathbf{f}_t + \mathcal{E}_m^i m_n + \mathcal{E}_\mathbf{m}^i \mathbf{m}_t + \begin{bmatrix} \mathbf{f}_t \frac{\|\boldsymbol{v}_t\|}{\|\mathbf{f}_t\|} \\ m_n \frac{|\omega_n|}{|m_n|} \end{bmatrix}^T H_i \begin{bmatrix} \mathbf{f}_t \frac{\|\boldsymbol{v}_t\|}{\|\mathbf{f}_t\|} \\ m_n \frac{|\omega_n|}{|m_n|} \end{bmatrix} \quad (28)$$

which is maximized/minimized subject to

$$\mathbf{m}_t^T \mathbf{m}_t = \|\mathbf{m}_t\|^2 \quad \mathbf{f}_t^T \mathbf{f}_t = \|\mathbf{f}_t\|^2 \quad m_n^2 = |m_n|^2. \quad (29)$$

The combination of linear and quadratic terms in (28) and the coupling between forces, moments and velocities makes it very difficult to find a closed-form analytical solution to this problem. A numerical solution can be found easily if the magnitudes in the problem are specified. This fact is useful in solving the actuator selection problem [21], where the desired performance is specified. However here, an iterative search must be performed in order to obtain the velocity and force/moment information for the Dynamic Capability Hypersurface. This search will be discussed in a later publication. Here the solutions are expressed as,

$$\begin{aligned} (\alpha_i)_{max} &= \mathcal{E}_f^i |f_n| + \|\mathcal{E}_\mathbf{m}^i\| \|\mathbf{m}_t\| + \\ & \quad p(\|\mathbf{f}_t\|, |m_n|, \|\boldsymbol{v}_t\|, |\omega_n|) \\ (\alpha_i)_{min} &= \mathcal{E}_f^i |f_n| - \|\mathcal{E}_\mathbf{m}^i\| \|\mathbf{m}_t\| - \\ & \quad q(\|\mathbf{f}_t\|, |m_n|, \|\boldsymbol{v}_t\|, |\omega_n|) \end{aligned} \quad (30)$$

where p and q represent the unknown solutions. The extremums can be expressed as

$$\mathbb{F} \begin{bmatrix} |f_n| & \|\mathbf{m}_t\| \end{bmatrix}^T + \mathbb{C}(\|\mathbf{f}_t\|, |m_n|, \|\boldsymbol{v}_t\|, |\omega_n|) \quad (31)$$

where the rows of $\mathbb{F}$ and $\mathbb{C}$ are defined as

$$\begin{aligned} \mathbb{F}^i &= \begin{bmatrix} \mathcal{E}_f^i & \|\mathcal{E}_\mathbf{m}^i\| \end{bmatrix} \\ \mathbb{F}^{i+d} &= \begin{bmatrix} -\mathcal{E}_f^i & \|\mathcal{E}_\mathbf{m}^i\| \end{bmatrix} \\ \mathbb{C}^i &= p(\|\mathbf{f}_t\|, |m_n|, \|\boldsymbol{v}_t\|, |\omega_n|) \\ \mathbb{C}^{i+d} &= q(\|\mathbf{f}_t\|, |m_n|, \|\boldsymbol{v}_t\|, |\omega_n|) \end{aligned} \quad (32)$$

The situation when the velocity is zero and the end-effector is on the verge of motion, can be investigated analytically in closed-form. This analysis amounts to exploring the relationship between accelerations, forces and moments. For this case the force analysis results are the same as those discussed in Section III-A.

IV. DYNAMIC CAPABILITY HYPERSURFACE

In summary, the *Dynamic Capability Equations* which define the Dynamic Capability Hypersurface are:

End-Effector Sticking equation (18) when the sticking conditions in (3) and (6) are satisfied,

$$\begin{bmatrix} \mathbb{F} \\ \mu_s & -1 & 0 & 0 \\ \frac{2}{3} R \mu_s & 0 & -1 & 0 \end{bmatrix} \begin{bmatrix} |f_n| \\ \|\mathbf{f}_t\| \\ |m_n| \\ \|\mathbf{m}_t\| \end{bmatrix} \leq \begin{bmatrix} \mathbb{T} \\ 0 \\ 0 \end{bmatrix}. \quad (33)$$

As discussed earlier in Section III-A, the magnitude constraints are enforced on the solution.

End-Effector Sliding equations (23) and (31) when the sliding conditions in (3) and (6) are satisfied,

$$\begin{aligned} \mathbb{A} \begin{bmatrix} \|\dot{\boldsymbol{v}}_t\| \\ |\dot{\omega}_n| \end{bmatrix} + \mathbb{F} \begin{bmatrix} |f_n| \\ \|\mathbf{m}_t\| \end{bmatrix} + \mathbb{C}(\ldots) &\leq \mathbb{T} \\ \|\mathbf{f}_t\| - \mu_d |f_n| &= 0 \\ |m_n| - \tfrac{2}{3} R \mu_d |f_n| &= 0. \end{aligned} \quad (34)$$

The capabilities of the system on the verge of motion are determined from (18) and (23),

$$\begin{aligned} \mathbb{A} \begin{bmatrix} \|\dot{\boldsymbol{v}}_t\| \\ |\dot{\omega}_n| \end{bmatrix} + \mathbb{F} \begin{bmatrix} |f_n| & \|\mathbf{f}_t\| & |m_n| & \|\mathbf{m}_t\| \end{bmatrix}^T &\leq \mathbb{T} \\ \|\mathbf{f}_t\| - \mu_d |f_n| &= 0 \\ |m_n| - \tfrac{2}{3} R \mu_d |f_n| &= 0. \end{aligned} \quad (35)$$

Each of the relations in (35) represents a hypersurface in an six-dimensional magnitude space. The innermost envelope around the origin formed by overlaying these hypersurfaces in the same space determines the worst-case combinations of acceleration, force and moment described by the Dynamic Capability Hypersurface.

V. EXAMPLE: PUMA 560

One application of this analysis is to evaluate a manipulator's ability to polish or sand a surface. This involves applying normal forces to the surface while overcoming friction in the tangential directions so that the end-effector can move the buffing or sanding tool over the surface. This section examines the ability for the PUMA 560 to accomplish this task from the configuration shown in Fig. 3a. A portion of the surface to be worked is indicated by the parallelepiped at the end-effector. A friction coefficient of $\mu_s = \mu_d = 0.6$ was used; (it is for leather on steel.) A circular end-effector surface of radius $R = 1.5 in = 0.0381 m$ was also assumed. These results in Figs. 3b and 3c are generated using the dynamic model developed in [22].

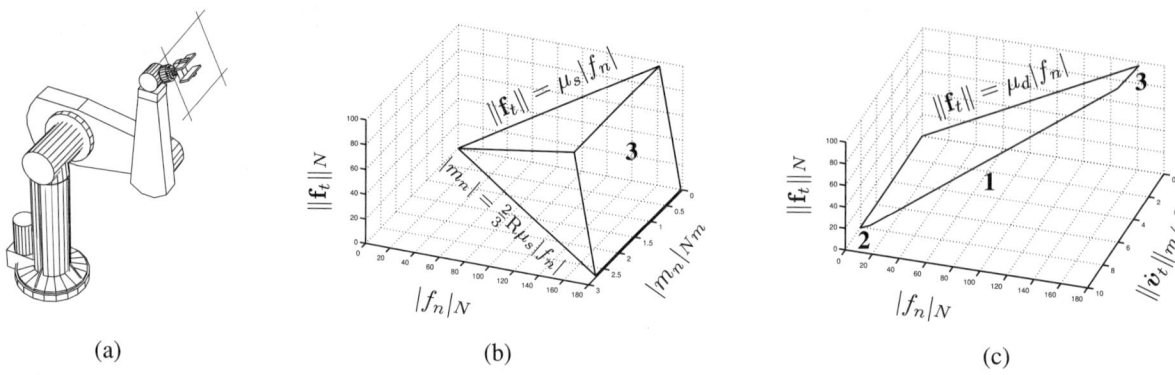

Fig. 3. PUMA 560 Dynamic Capability Hypersurface Sections.

Fig. 3b shows a section of the hypersurface defined by $\|\mathbf{m}_t\| = 0$, for the sticking condition described in (33). This surface describes the manipulators ability to apply contact forces and moments to a surface without slipping. However, in the context of buffing or sanding, this tells the amount of tangential force required to overcome tangential friction. The combinations of magnitudes of contact forces and moments lying in the interior of the surface can be applied to the surface in all directions included in f_n, $\mathbf{f}_t$ and m_n. Note that the surface patches without a numeric label, such as the "3" on the end surface, are defined by the no-slip conditions in (3) and (6). If the growth of contact force and moment follow a vector which breaks these surface patches, then the end-effector will begin to move/slide in translation and/or rotation.

However, the surface patch with the numeric label "3" is produced by a saturation of the third actuator of the PUMA 560; (the actuators are numbered starting from the base.) Saturation means the motor is supplying the maximum amount of torque/force it is capable of. The worst-case combinations of contact force and moment cannot grow such that they pass through this surface patch because the third actuator does not have sufficient torque capacity to allow it.

Fig. 3c shows a section of the hypersurface defined by (35) and $|\dot{\omega}_n| = |m_n| = \|\mathbf{m}_t\| = 0$, which describes the manipulator on the verge of motion. This describes the situation where static friction on the surface has been overcome and the end-effector is about to move. Notice that once the sticking condition is violated, the tangential and normal forces become dependent on one another. This dependency reduces what would have been a three-dimensional surface to a two-dimensional plane. The plane is bounded due to the saturation of actuators. The figure shows that the first, second and third actuators all saturate for different combinations of acceleration and applied forces.

The plane describes the acceleration capability while applying a contact force to the surface, the surface working scenario. Keep in mind that the plane describes the magnitudes of acceleration guaranteed to be achievable in all directions defined by $|f_n|$, $\|\mathbf{f}_t\|$ and $\|\dot{\mathbf{v}}_t\|$. Any acceleration-force-moment combination lying within the plane boundaries is achievable. The maximum acceleration can be achieved when the normal force is zero. As the normal force on the surface is increased, the amount of acceleration achievable in every direction reduces to zero.

VI. CONCLUSION

An extension of the previously developed Dynamic Capability Hypersurface to the case of contact with a rigid environment was presented. The extension involved incorporating kinematic constraints and contact friction forces/moments into the prior analysis. The capabilities of the mechanism were analyzed in terms of sticking and sliding phases of motion. The sticking phase was easily addressed by the Dynamic Capability Hypersurface analysis. Information about the acceleration and force/moment capabilities when the end-effector is on the verge of motion, just entering the sliding phase, was also easily obtainable. However a numerical/iterative solution is needed in order to study the effect of velocities and contact forces/moments on a mechanism's capabilities during the sliding phase. Key results found include a hypersurface description of the amount of torque required to overcome static friction, or the amount of force and moment that can be applied to a surface without slipping. A hypersurface description of the relationship between acceleration capability and applied contact forces at zero velocity was also found. This analysis was applied to the PUMA 560 robotic manipulator.

VII. REFERENCES

[1] Tsuneo Yoshikawa, "Dynamic manipulability of robot manipulators," in *Proceedings 1985 IEEE International Conference on Robotics and Automation*, 1985, pp. 1033–1038, St. Louis, Missouri.

[2] Oussama Khatib and Joel Burdick, "Optimization of dynamics in manipulator design: The operational space formulation," *The International Journal of Robotics and Automation*, vol. 2, no. 2, pp. 90–98, 1987.

[3] Timothy J. Graettinger and Bruce H. Krogh, "The acceleration radius: A global performance measure for robotic manipulators," *IEEE Journal of Robotics and Automation*, vol. 4, no. 1, pp. 60–69, February 1988.

[4] Yong-Yil Kim and Subhas Desa, "The definition, determination, and characterization of acceleration sets for spatial manipulators," in *The 21st Biennial Mechanisms Conference: Flexible Mechanism, Dynamics, and Robot Trajectories*, September 1990, pp. 199–205, Chicago, Illinois.

[5] J. Kenneth Salisbury and John J. Craig, "Articulated hands: Force control and kinematic issues," *The International Journal of Robotics Research*, vol. 1, no. 1, pp. 4–17, 1982.

[6] Pasquale Chiacchio, Yann Bouffard-Vercelli, and Francois Pierrot, "Evaluation of force capabilities for redundant manipulators," in *Proceedings IEEE International Conference on Robotics and Automation*, 1996, pp. 3520–3525, Kluwer Academic Publishers.

[7] Pasquale Chiacchio and Yann Bouffard-Vercelli, "Force polytope and force ellipsoid for redundant manipulators," *Journal of Robotic Systems*, vol. 14, no. 8, pp. 613–620, 1997.

[8] Alan Bowling, *Analysis of Robotic Manipulator Dynamic Performance: Acceleration and Force Capabilities*, Ph.D. thesis, Stanford University, June 1998.

[9] Alan Bowling and Oussama Khatib, "Modular redundant manipulator design for dynamic performance," in *Proceedings Thirteenth CISM-IFToMM Symposium on Theory and Practice of Robots and Manipulators (Romansy 13)*, July 1998, Paris, France.

[10] Yuan-Fang Zheng and H. Hemami, "Impact effects of biped contact with the environment," *IEEE Transactions on Systems, Man, and Cybernetics*, vol. SMC-14, no. 3, pp. 437–443, 1984.

[11] E.J. Haug, "Dynamics of mechanical systems with coulomb friction, stiction, impact and constraint addition-deletion-i: Theory," *Mechanism and Machine Theory*, vol. 21, no. 5, pp. 401–406, 1986.

[12] David Baraff, "Fast contact force computation for nonpenetration rigid bodies," in *Computer Graphics Proceedings, Annual Conference Series (SIGGRAPH 94)*, July 1994.

[13] Brian Mirtich and John Canny, "Impulse-based dynamic simulation," in *The Algorithmic Foundations of Robotics*, K. Goldberg, D. halperin, J.C. Latombe, and R. Wilson, Eds. 1995, A.K. Peters, Boston, MA, Proceedings from the workshop held in February, 1994.

[14] Brian Mirtich and John Canny, "Impulse-based simulation of rigid bodies," *In Symposium on Interactive 3D Graphics*, 1995.

[15] David E. Stewart, "Rigid-body dynamics with friction and impact," *SIAM Review*, vol. 42, no. 1, pp. 3–39, 2000.

[16] T. Mouri, T. Yamada, A. IWAI, N. Mimura, and Y. Funahashi, "Identification of contact conditions from contaminated data of contact force and moment," in *Proceedings IEEE International Conference on Robotics and Automation*, May 2001, vol. 1, pp. 597–603.

[17] Donald T. Greenwood, *Principles of Dynamics*, Prentice Hall, second edition, 1988.

[18] Ahmed A. Shabana, *Computational Dynamics*, John Wiley & Sons, first edition, 1988.

[19] Nicholas Xydas and Imin Kao, "Modeling of contact mechanics and friction limit surfaces for soft fingers in robotics, with experimental results," *The International Journal of Robotics Research*, vol. 18, no. 8, pp. 941–950, September 1999.

[20] Alan Bowling and Oussama Khatib, "The motion isotropy hypersurface: A characterization of acceleration capability," in *Proceedings IEEE/RSJ International Conference on Intelligent Robots and Systems*, October 1998, vol. 2, pp. 965–971, Victoria, British Columbia, Canada.

[21] Alan Bowling and Oussama Khatib, "Actuator selection for desired dynamic performance," in *Proceedings of the IEEE/RSJ International Conference on Intelligent Robots and Systems*, October 2002, vol. 2, pp. 1966–1973, Lausanne, Switzerland.

[22] Brian Armstrong, Oussama Khatib, and Joel Burdick, "The explicit model and inertial parameters of the puma 560 arm," in *Proceedings IEEE International Conference on Robotics and Automation*, 1986, vol. 1, pp. 510–518.

Nonholonomic Dynamic Rolling Control of Reconfigurable 5R Closed Kinematic Chain Robot with Passive Joints

Tasuku YAMAWAKI, Osamu MORI and Toru OMATA

Department of Precision Machinery Systems
Tokyo Institute of Technology
4259 Nagatsuta, Midoriku, Yokohama, Kanagawa, Japan
{m0130yamawaki, mori, omata}@pms.titech.ac.jp

Abstract We have developed a self-reconfigurable robot which can form a 5R closed kinematic chain, only two of whose joints are actuated. This paper discusses its dynamic rolling. When it rolls, it has one DOF of its absolute orientation besides two DOFs of its shape. We show that the absolute orientation is subject to an acceleration constraint, not a velocity constraint. Therefore, the dynamics of the rolling motion needs to be formulated to control it. This paper proposes a controller which can reduce its negative acceleration caused by gravity. The shape and orientation of the robot cannot be controlled simultaneously. This paper proposes a control strategy switching shape and orientation controllers. We verify the effectiveness of the control strategy by simulations and experiments.

1 Introduction

The purpose of this study is to develop a simple but useful robot. Such a robot is less expensive and more reliable than complex robots. A group of such robots could do more than a single complex robot.

As an example, we have developed a self-reconfigurable robot which can form 5R and 4R closed kinematic chains as shown in Fig. 1 by coupling the same two 2R open kinematic chains whose first joints are unactuated [1]. Fig. 2 shows a photo of the robot forming the 5R closed kinematic chain. We have proposed coupling of limbs with a passive joint(s) to form a parallel robot with the same number of actuators as its degrees of freedom (DOFs). Both the 5R and 4R closed kinematic chains can be used as a parallel robot with the same number of actuators as its DOFs. Our previous experiments reveal that the 5R closed kinematic chain can locomote by rolling as shown in Fig. 1 (B') [2].

Rolling motions of closed kinematic chains have been studied. Matsuo et al [4] studied learning of the rolling motion of a 6R closed kinematic chain using Genetic Algorithm. Yim et al [5] has developed a reconfigurable robot, which can form a closed kinematic chain. All joints of these robots are actuated. Lee and Sanderson[6] have developed the "Tetrobot" and studied its dynamic rolling motion.

In general a rolling motion is considered as the sequence of phases as shown in Fig. 3. In (A) a robot gains an initial velocity for rolling. In (B) through (D) it rolls about a point on the floor and in (E) it lands on the floor. It bounds on the floor in general as shown in (F), which can trigger the next step of rolling by saving some energy. Lee and Sanderson studied dynamic control to tip the Tetrobot for transition from (A) to (B) and to restore its shape in the phase of rolling about a point, (B) through (D). They also simulate the impact in the phase (E).

In this paper, we discuss dynamic rolling control of our 5R closed kinematic chain robot (*5R closed robot* or just *robot* for simplicity) in the phases of rolling about a point (B) through (D) and (F)) by focusing on its nonholonomic property. We show that the rolling velocity of the 5R closed robot can be accelerated/ decelerated even in these phases. Therefore the 5R closed robot can roll starting with a lower initial velocity.

The 5R closed robot has one DOF of its absolute orientation besides two DOFs of its shape. The absolute orientation is independent of the constraint for forming the closed kinematic chain. If its unactuated joints were actuated, the closed kinematic chain would be over-actuated but the absolute orientation could not be driven directly. Section 2 derives dynamic equations of motion and shows that the absolute orientation is subject to a second-order nonholonomic constraint. Since it is not subject to a velocity constraint, the dynamics of the rolling motion needs to be formulated to control it.

The shape and orientation of the robot cannot be controlled simultaneously. Section 3 proposes a control strategy switching shape and orientation controllers. Sections 4 and 5 show simulation and experimental results and the effectiveness of the proposed control strategy.

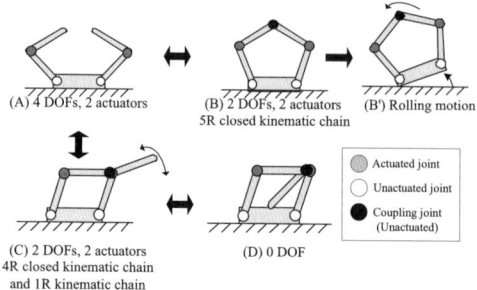

Figure 1: Self-reconfigurable robot by coupling

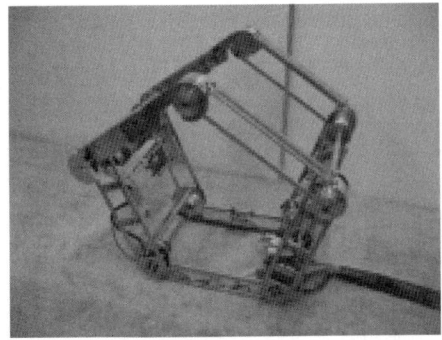

Figure 2: Our self-reconfigurable robot forming a 5R closed kinematic chain

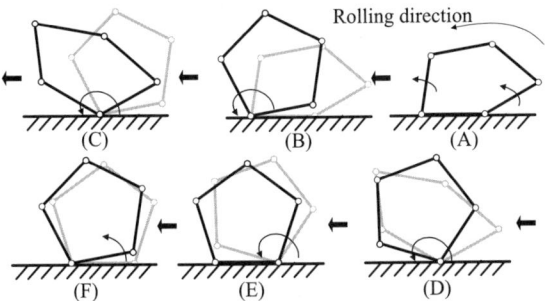

Figure 3: Phases of rolling motion

2 Dynamics for rolling motion

In the rolling motion, the orientation of this 5R closed kinematic chain can be classified into five cases as shown in Fig. 4 (A) to (E) according to the arrangement of actuators. If (E) is seen from the back of the sheet, the arrangement of actuators is seen as (E'), which is equivalent to (A). Therefore, their dynamic equations of motion are the same, while the rotation direction of (E) is opposite to that of (A). Similarly, (B) and (D) can be described as the same equations of motion. Therefore, we deal with only (A), (C) and (D).

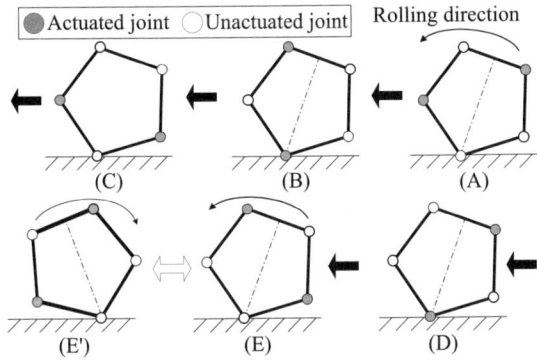

Figure 4: Arrangements of the actuators

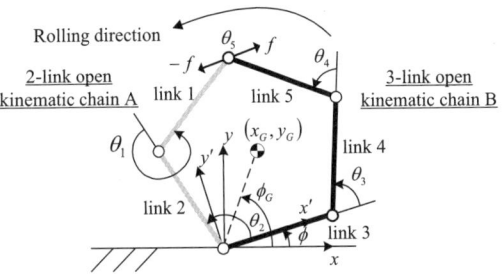

Figure 5: Common model of the rolling motion

Fig. 5 shows a common model for (A), (C) and (D). Let link3 be the link whose one end is leaving the floor and ϕ be its orientation angle from the floor. The joint angles θ_1 through θ_4 are defined as shown in the figure. Define generalized coordinates as $\mathbf{q} = (\theta_1\ \theta_2\ \theta_3\ \theta_4\ \phi)^T$. Corresponding to $\mathbf{q}$, the joint torque vector $\boldsymbol{\tau}$ is given by

$$\text{case (A)} \qquad \boldsymbol{\tau} = (\tau_1\ 0\ 0\ \tau_4\ 0)^T \qquad (1)$$
$$\text{case (C)} \qquad \boldsymbol{\tau} = (\tau_1\ 0\ \tau_3\ 0\ 0)^T \qquad (2)$$
$$\text{case (D)} \qquad \boldsymbol{\tau} = (0\ \tau_2\ 0\ \tau_4\ 0)^T \qquad (3)$$

Let (x_G, y_G) be the center of the gravity (COG) of the 5R closed kinematic chain and ϕ_G be the angle between the floor and COG, which we call the "COG angle". When $\phi_G < 90$ [deg], the angular velocity of the COG $\dot{\phi}_G$ is decelerated by gravity. If ϕ_G can reach $\phi_G > 90$ [deg], $\dot{\phi}_G$ is accelerated by gravity. The aim of this control for rolling is that $\phi_G > 90$ [deg]. Therefore ϕ_G is controlled actually instead of ϕ in Section 3.

This 5R closed kinematic chain is modeled as two open kinematic chains; one is the 2-link open kinematic chain 'A', consisting of links 1 and 2, the other is the 3-link open kinematic chain 'B', consisting of links 3 through 5. They are coupled at their end-points with the unactuated coupling joint θ_5. The constraint for forming the closed kinematic chain can be written with respect to the local coordinate system $x'y'$ attached to link 3 as

$$x' = l_1 c_{12} + l_2 c_2 = l_3 + l_5 c_{34} + l_4 c_3, \qquad (4)$$
$$y' = l_1 s_{12} + l_2 s_2 = l_3 + l_5 s_{34} + l_4 s_3, \qquad (5)$$

where s and c stand for sin and cos, and $s_{ij} = \sin(\theta_i + \theta_j)$ etc., and l_i is the length of the ith link. We have the velocity constraint by taking the time derivative of Eqs. (4) and (5):

$$J_A \begin{pmatrix} \dot{\theta}_1 \\ \dot{\theta}_2 \end{pmatrix} = J_B \begin{pmatrix} \dot{\theta}_3 \\ \dot{\theta}_4 \end{pmatrix}. \qquad (6)$$

The column vectors of $J_A \in R^{2\times 2}$ and $J_B \in R^{2\times 3}$, $\mathbf{j}_{Ai}$ and $\mathbf{j}_{Bi}$, are written as

$$\mathbf{j}_{A1} = \begin{pmatrix} -l_1 s_{12} \\ l_1 c_{12} \end{pmatrix}, \quad \mathbf{j}_{A2} = \begin{pmatrix} -l_1 s_{12} - l_2 s_2 \\ l_1 c_{12} + l_2 c_2 \end{pmatrix},$$

$$\mathbf{j}_{B1} = \begin{pmatrix} -l_5 s_{34} - l_4 s_3 \\ l_5 c_{34} + l_4 c_3 \end{pmatrix} \text{ and } \mathbf{j}_{B2} = \begin{pmatrix} -l_5 s_{34} \\ l_5 c_{34} \end{pmatrix}.$$

In Eq. (6), two variables from $\dot{\theta}_1$ through $\dot{\theta}_4$ can be selected as independent variables, since the number of the constraint equations is two. If its unactuated joints were actuated, the 5R closed kinematic chain would be over-actuated but ϕ could not be driven directly, since ϕ is independent of the constraint, Eq. (6). ϕ is subject to an acceleration constraint as described later.

Eq. (6) can be rewritten as

$$\Psi \dot{\mathbf{q}} = \mathbf{0} \in R^2 \quad (7)$$

where $\Psi = (\mathbf{j}_{A1} \; \mathbf{j}_{A2} \; \mathbf{j}_{B1} \; \mathbf{j}_{B2} \; \mathbf{0})$. By using Ψ in Eq. (7), the dynamic equation of motion of the 5R can be obtained in terms of the generalized coordinates $\mathbf{q}$:

$$M(\mathbf{q})\ddot{\mathbf{q}} + \mathbf{C}(\mathbf{q}, \dot{\mathbf{q}}) + \mathbf{G}(\mathbf{q}) = \boldsymbol{\tau} + \Psi^T \mathbf{f} \quad (8)$$

where $M(\mathbf{q}) \in R^{5 \times 5}$ is the inertia matrix, $\mathbf{C}(\mathbf{q}, \dot{\mathbf{q}}) \in R^5$ is the term of centrifugal force, $\mathbf{G}(\mathbf{q}) \in R^5$ is the gravitational term and $\mathbf{f} = (f_x \; f_y)$ is the constraint force acting on the end-points of the open kinematic chains A and B.

By examining the condition in the paper [6], we can prove that ϕ is subject to a second-order nonholonomic constraint. We chose ϕ and two joint angles from θ_1 to θ_4 as three independent variables in $\mathbf{q}$. The remaining two variables are dependent. Let $N \in R^{5 \times 3}$ be the matrix whose columns are the bases of the null space of Ψ and $\dot{\mathbf{p}} \in R^3$ be the vector consisting of the three independent variables. $\dot{\mathbf{q}}$ can be written as $\dot{\mathbf{q}} = N\dot{\mathbf{p}}$. Dot-multiplying both sides of Eq. (8) by N^T, we obtain

$$N^T M (N\ddot{\mathbf{p}} + \dot{N}\dot{\mathbf{p}}) + N^T (\mathbf{C} + \mathbf{G}) = N^T \boldsymbol{\tau} \quad (9)$$

since $\Psi N = \mathbf{0} \in R^{3 \times 3}$. Eq. (9) can be rewritten as

$$\tilde{M}\ddot{\mathbf{p}} + \tilde{\mathbf{C}} + \tilde{\mathbf{G}} = \tilde{\boldsymbol{\tau}} \quad (10)$$

where $N^T M N = \tilde{M} \in R^{3 \times 3}$, $N^T M \dot{N}\dot{\mathbf{p}} + N^T \mathbf{C} = \tilde{\mathbf{C}} \in R^3$, $N^T \mathbf{G} = \tilde{\mathbf{G}} \in R^3$ and $N^T \boldsymbol{\tau} = \tilde{\boldsymbol{\tau}} \in R^3$.

For instance, if $\dot{\theta}_1$ and $\dot{\theta}_4$ are chosen as independent variables, we have $\dot{\mathbf{p}} = (\dot{\theta}_1 \; \dot{\theta}_4 \; \dot{\phi})^T$ and

$$N^T = \begin{pmatrix} 1 & * & * & 0 & 0 \\ 0 & * & * & 1 & 0 \\ 0 & 0 & 0 & 0 & 1 \end{pmatrix}$$

where '*' denotes a certain real number. The third element of $\tilde{\boldsymbol{\tau}}$ is zero. So is it when other joint angles are chosen as independent variables. The third element of $\tilde{\mathbf{G}}$ is not constant. Therefore, the third equation is a second-order nonholonomic constraint [6]. Thus, ϕ is not subject to a velocity constraint.

3 Control design

3.1 Equations of motion including ϕ_G

We establish the dynamic equations of motion including ϕ_G. The COG of the system, (x_G, y_G), is given by

$$x_G = \frac{1}{m_0} \sum_{i=1}^{5} m_i x_i, \quad y_G = \frac{1}{m_0} \sum_{i=1}^{5} m_i y_i : \quad m_0 = \sum_{i=1}^{5} m_i$$

where m_i is the mass of the i th link, x_i and y_i are the COG of the i th link. ϕ_G can be written as

$$\tan \phi_G = \frac{y_G}{x_G}. \quad (11)$$

Taking the time derivative of Eq. (11), we obtain

$$L(\mathbf{q})\ddot{\mathbf{q}} - \ddot{\phi}_G = -K(\mathbf{q}, \dot{\mathbf{q}}) \quad (12)$$

where $L(\mathbf{q}) \in R^{1 \times 5}$ and $K(\mathbf{q}, \dot{\mathbf{q}}) \in R$. Taking the time derivative of Eq. (7), we obtain the constraints for $\ddot{\mathbf{q}}$,

$$\Psi \ddot{\mathbf{q}} + \dot{\Psi}\dot{\mathbf{q}} = \mathbf{0} \in R^2 \quad (13)$$

The dynamic equations of motion can be obtained from Eqs. (8), (12) and (13),

$$\begin{pmatrix} M(\mathbf{q}) & 0 & \Psi^T \\ L(\mathbf{q}) & -1 & 0 \\ \Psi & 0 & 0 \end{pmatrix} \begin{pmatrix} \ddot{\mathbf{q}} \\ \ddot{\phi}_G \\ -\mathbf{f} \end{pmatrix} = \begin{pmatrix} \boldsymbol{\tau} - \mathbf{C}(\mathbf{q}, \dot{\mathbf{q}}) - \mathbf{G}(\mathbf{q}) \\ -K(\mathbf{q}, \dot{\mathbf{q}}) \\ -\dot{\Psi}\dot{\mathbf{q}} \end{pmatrix} \quad (14)$$

In Eq. (14), the number of the constraints is eight and the number of the variables is ten, which are $\ddot{\theta}_1$ through $\ddot{\theta}_4$, $\ddot{\phi}$, $\ddot{\phi}_G$, $\mathbf{f}$ and $\boldsymbol{\tau}$. We can control two of them.

3.2 COG control mode

The COG control mode can control $\ddot{\phi}_G$ and the acceleration of one of the joints. Consider the case of Fig. 4 (A), for instance, and compute the joint torques τ_1 and τ_4. In Eq. (14), we express the matrix of the left-hand side as

$$\begin{pmatrix} M(\mathbf{q}) & 0 & \Psi^T \\ L(\mathbf{q}) & -1 & 0 \\ \Psi & 0 & 0 \end{pmatrix} = (\mathbf{w}_1 \; \cdots \; \mathbf{w}_8),$$

where $\mathbf{w}_i$ is the i th column vector of the matrix in the left-hand side. If we control $\ddot{\theta}_2$, Eq. (14) can be rewritten as

$$(\mathbf{w}_1 \; \mathbf{w}_3 \; \mathbf{w}_4 \; \mathbf{w}_5 \; \mathbf{w}_7 \; \mathbf{w}_8 \; \mathbf{z}_1 \; \mathbf{z}_4)(\ddot{\theta}_1 \; \ddot{\theta}_3 \; \ddot{\theta}_4 \; \ddot{\phi} \; -f_x \; -f_y \; \tau_1 \; \tau_4)^T$$
$$= -(\mathbf{w}_2 \; \mathbf{w}_6) \begin{pmatrix} \ddot{\theta}_2 \\ \ddot{\phi}_G \end{pmatrix} + \mathbf{h} \quad (15)$$

where $\mathbf{z}_1 = (-1 \; 0 \; 0 \; 0 \; 0 \; 0 \; 0)^T$, $\mathbf{z}_2 = (0 \; 0 \; 0 \; -1 \; 0 \; 0 \; 0)^T$ and

$$\mathbf{h} = \begin{pmatrix} \boldsymbol{\tau} - \mathbf{C}(\mathbf{q},\dot{\mathbf{q}}) - \mathbf{G}(\mathbf{q}) \\ -K(\mathbf{q},\dot{\mathbf{q}}) \\ -\ddot{\Psi}\dot{\mathbf{q}} \end{pmatrix}.$$

To obtain the joint torques, we rewrite Eq. (15) as follows

$$(\ddot{\theta}_1 \; \ddot{\theta}_3 \; \ddot{\theta}_4 \; \ddot{\phi} \; -f_x \; -f_x \; \tau_1 \; \tau_4)^T = \hat{M}_G^{-1}\hat{C}_G \quad (16)$$

where

$$\hat{M}_G = (\mathbf{w}_1 \; \mathbf{w}_3 \; \mathbf{w}_4 \; \mathbf{w}_5 \; \mathbf{w}_7 \; \mathbf{w}_8 \; \mathbf{z}_1 \; \mathbf{z}_4), \; \hat{C}_G = -(\mathbf{w}_2 \; \mathbf{w}_6)\begin{pmatrix} \ddot{\theta}_2 \\ \ddot{\phi}_G \end{pmatrix} + \mathbf{h}$$

and we also assume that $\hat{M}_G$ is non-singular. Section 4.3 discusses singularities of $\hat{M}_G$. $\ddot{\theta}_1$, $\ddot{\theta}_3$, $\ddot{\theta}_4$, $\ddot{\phi}$ and $\mathbf{f}$ are dependently determined by Eq. (16). Therefore the COG control mode cannot control the shape of the 5R closed kinematic chain which is determined by two of θ_1 through θ_4.

We discuss how to give a desired acceleration of $\ddot{\phi}_G$ for rolling. If ϕ_G is not controlled and the shape of the 5R closed robot is constant, $\dot{\phi}_G$ is decelerated by gravity and ϕ_G may not exceed 90 [deg]. Let the deceleration (negative acceleration) be $\ddot{\phi}_{grav}(<0)$. The COG control mode reduces the deceleration by setting the desired acceleration

$$\ddot{\phi}_G = \kappa \ddot{\phi}_{grav} \qquad 0 < \kappa < 1.0 \quad (17)$$

Note: If $\kappa > 1$ the COG mode decelerates $\dot{\phi}_G$ more than gravity alone. This is useful for the robot to descend a slope or to stop suddenly when it rolls about a point. If $\kappa < 0$, the COG mode absolutely accelerates $\dot{\phi}_G$ under gravity. However there is a tradeoff between reducing κ and maintaining an initial shape of the robot as discussed next.

$\ddot{\phi}_{grav}$ can be obtained as follows. If the shape is constant, the 5R closed robot can be regarded as a rigid-body. From the moment equilibrium condition, we have

$$\{I_G + m_0(x_G^2 + y_G^2)\}\ddot{\phi}_{grav} = -m_0 g x_G$$
$$\ddot{\phi}_{grav} = -\frac{m_0 g x_G}{I_G + m_0(x_G^2 + y_G^2)} \quad (18)$$

where I_G is the moment of inertia of the whole 5R closed robot, m_0 is its mass and g is the gravity acceleration.

The smaller κ is, the smaller the deceleration becomes, however if it is too small, the shape changes significantly. On the other hand, the bigger κ is, the more likely ϕ_G do not exceed 90 [deg]. In Section 4, we select an appropriate value of κ by simulation.

In the COG control mode, one of the joint angles can be controlled. The 5R closed robot can avoid singularities by controlling it as we discuss in Section 4.3.

3.3 Shape control mode

The shape control mode controls two of $\ddot{\theta}_1$ through $\ddot{\theta}_4$. If we chose $\ddot{\theta}_1$ and $\ddot{\theta}_4$, we can rewrite Eq. (14) by using $\mathbf{w}_i$,

$$(\mathbf{w}_2 \; \mathbf{w}_3 \; \mathbf{w}_5 \; \mathbf{w}_6 \; \mathbf{w}_7 \; \mathbf{w}_8 \; \mathbf{z}_1 \; \mathbf{z}_4)(\ddot{\theta}_2 \; \ddot{\theta}_3 \; \ddot{\phi} \; \ddot{\phi}_G \; -f_x \; -f_y \; \tau_1 \; \tau_4)^T$$
$$= -(\mathbf{w}_1 \; \mathbf{w}_4)\begin{pmatrix} \ddot{\theta}_1 \\ \ddot{\theta}_4 \end{pmatrix} + \mathbf{h} \quad (19)$$

We can also rewrite Eq. (17) as

$$(\ddot{\theta}_2 \; \ddot{\theta}_3 \; \ddot{\theta}_4 \; \ddot{\phi} \; -f_x \; -f_x \; \tau_1 \; \tau_4)^T = \hat{M}_S^{-1}\hat{C}_S \quad (20)$$

where

$$\hat{M}_S = (\mathbf{w}_2 \; \mathbf{w}_3 \; \mathbf{w}_5 \; \mathbf{w}_6 \; \mathbf{w}_7 \; \mathbf{w}_8 \; \mathbf{z}_1 \; \mathbf{z}_4), \; \hat{C}_G = -(\mathbf{w}_1 \; \mathbf{w}_4)\begin{pmatrix} \ddot{\theta}_1 \\ \ddot{\theta}_4 \end{pmatrix} + \mathbf{h}$$

and it is assumed that $\hat{M}_S$ is non-singular. Section 4.3 discusses the singularities of $\hat{M}_S$. The joint torques, τ_1 and τ_4, can be given by Eq. (20). Note that the shape control mode cannot control $\ddot{\phi}_G$ which is dependently obtained from Eq. (20).

A PD controller is applied to control two joint angles.

$$\ddot{\theta}_i = K_p(\theta_{ri} - \theta_i) - K_d\dot{\theta}_i \quad (21)$$

where θ_{ri} is a desired angle, and K_p and K_d are feedback gains.

3.4 Condition for switching control modes

The condition for switching the two control modes is as follows. When $\phi_G \leq 90$ [deg], we apply the COG control mode to achieve $\phi_G > 90$ [deg]. Once $\phi_G > 90$ [deg] is achieved, the shape control mode is applied. Until the robot lands on the floor, the shape can be controlled. It is our future work to obtain the optimal landing shape to stop or continue rolling. After switching to the shape control mode, ϕ_G may be less than 90 [deg] again because it is not controlled in the shape control mode. A solution is to delay the timing of switching the control modes.

4 Simulations

4.1 Physical parameters and initial conditions

Table 1 shows the length of the i th link, l_i, the mass, m_i, the length to the center of gravity, l_{ci}, and the inertia moment, I_i. The initial conditions are $\theta_1 = 288$ [deg], $\theta_2 = 108$ [deg], $\theta_3 = 72$ [deg], $\theta_4 = 72$ [deg], $\phi = 20$ [deg], $\dot{\theta}_1$ through $\dot{\theta}_4$ are 0 [deg/s], and $\dot{\phi} = \dot{\phi}_G = 68$ [deg/s]. Suppose that θ_1 and θ_4 are actuated joints as shown in Fig. 4 (A). The COG control mode controls $\ddot{\phi}_G$ and $\ddot{\theta}_2$

whose desired acceleration is 0 [deg/s^2]. Changing κ as 1.0, 0.8 and 0.1, we observe ϕ_G and the shape of the 5R closed kinematic chain. The shape control mode control $\ddot{\theta}_1$ and $\ddot{\theta}_4$ whose desired angles are $\theta_1 = 288$ [deg] and $\theta_4 = 72$ [deg].

Table 1 Physical parameters of each link

l_1	0.25 m	m_1	1.01 kg	l_{c1}	0.125 m	I_1	7.8×10^{-3} kg m^2
l_2	0.25 m	m_2	1.37 kg	l_{c2}	0.175 m	I_2	9.7×10^{-3} kg m^2
l_3	0.25 m	m_3	0.94 kg	l_{c3}	0.125 m	I_3	1.8×10^{-2} kg m^2
l_4	0.25 m	m_4	1.37 kg	l_{c4}	0.175 m	I_4	9.7×10^{-3} kg m^2
l_5	0.25 m	m_5	0.97 kg	l_{c5}	0.097 m	I_5	1.7×10^{-2} kg m^2

4.2 Simulation results

Figs. 6 and 7 show simulation results of ϕ_G and the configurations of the 5R closed robot, respectively. The configurations after that pointed by "a" are controlled by the shape control mode. When $\kappa = 1.0$ in Fig. 6, ϕ_G cannot reach 90 [deg] and return toward zero by gravity. When $\kappa = 0.8$ and $\kappa = 0.1$, ϕ_G can reach 90 [deg] and is accelerated by gravity after $\phi_G > 90$. In this simulation, the closer to one κ is, the smaller the change in the shape is. For other arrangements of actuators in Fig. 4 (C) and (D), the same results are obtained, since their torques are calculated to obtain the same desired $\ddot{\phi}_G$.

4.3 Singularity

We examine singular configurations at which $\hat{M}_G$ and $\hat{M}_S$ become singular in the workspace. These matrices are functions of three independent variables, ϕ and two variables from θ_1 through θ_4, however it turns out that ϕ is independent of the singular configurations in this example. When θ_2 and θ_4 are chosen as the independent variables, Fig. 8 shows the singular configurations. In the area (c), the 5R closed robot cannot actually form the closed shape geometrically. The condition number of $\hat{M}_G$ is more than 200 in the areas (a) and (b), that of $\hat{M}_S$ is more than 200 in the area (b). We define these areas as singular configurations, since the observed torques in these areas exceed the maximum limits of our actuators.

The dashed-dotted line is the simulation result when $\kappa = 0.1$ corresponding to Fig. 7 (B). The COG control mode is switched to the shape control mode at the point indicated as "×". The robot avoids the singular area (a), although it is not intended.

Using Fig.8, the robot can intendedly avoid singular configurations even in the COG control mode. When θ_2 is increased by the PD controller, the trajectory of configurations indicated by the dashed line is farther from the singular area (a) than that by the dashed-dotted line.

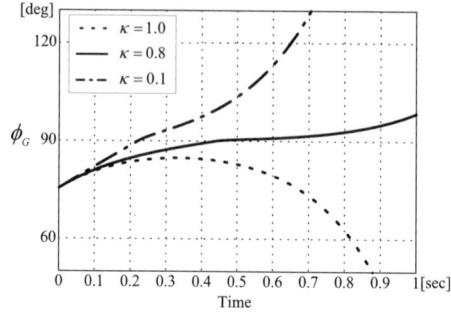

Figure 6: Angles of COG of the 5R closed kinematic chain

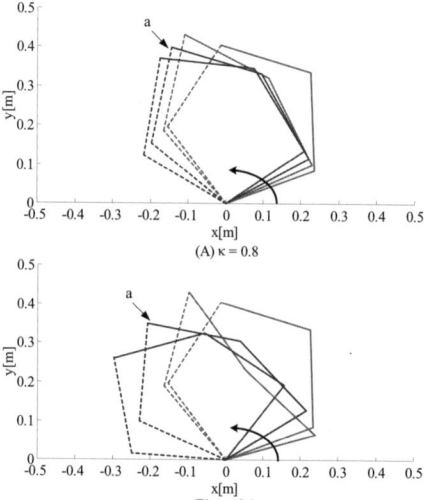

Figure 7: Simulation results

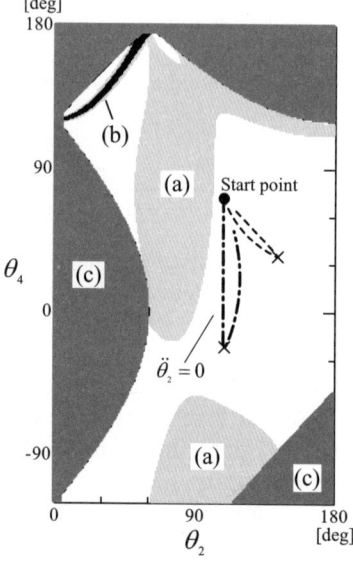

Figure 8: Singularities in the workspace of the simulation

5 Experiments

We conduct two experiments using the real robot in Fig. 2. The physical parameters of the robot have been shown in Table 1. In the first experiment, the switching control strategy proposed in Section 3.2 is applied with $\kappa = 0.8$. In the second experiment, only the shape control mode is used to maintain the initial shape of the robot. The two experiments are also conducted on a slope of 5 [deg].

The initial shape is $\theta_1 = 288$ [deg], $\theta_2 = 108$ [deg], $\theta_3 = 72$ [deg] and $\theta_4 = 72$ [deg]. θ_1 through θ_4 and $\dot{\theta}_1$ through $\dot{\theta}_4$ are measured by encoders, ϕ and $\dot{\phi}$ by a gyro sensor attached on link 3. ϕ_G and $\dot{\phi}_G$ are respectively obtained by Eq. (11) and taking the time derivative of Eq. (11).

Table 2 shows initial velocities of $\dot{\phi}_G$ in these experiments. Fig. 9 shows the experimental results of ϕ_G. (A) shows that the robot can achieve $\phi_G > 90$ by the switching control strategy (snapshots are shown in Fig. 10), but cannot when only the shape control mode is applied. The same results are obtained on the slope as shown in (B).

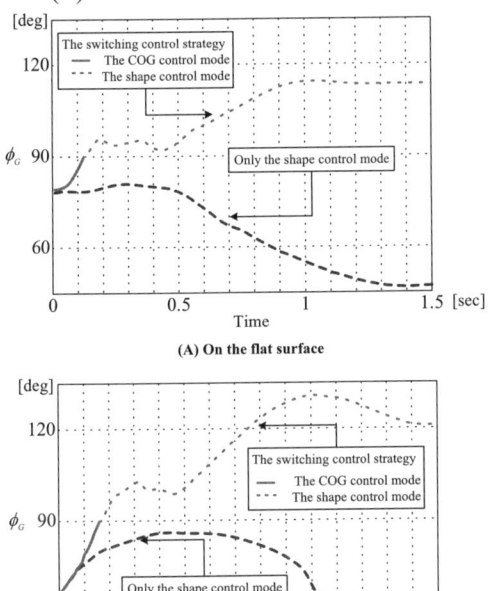

Figure 9: Experimental results

Table 2 Initial conditions

	Only the shape control mode	The switching control strategy
$\dot{\phi}_G$ on a flat surface	22.25 [deg/s]	21.21 [deg/s]
$\dot{\phi}_G$ on the slope	92.78 [deg/s]	89.02 [deg/s]

Figure 10: Snapshots of the rolling motion

6 Conclusion

For the dynamic rolling motion of the 5R closed kinematic chain robot, this paper has focused on its nonholonomic property and proposed a control strategy switching the two control modes: the COG control mode and the shape control mode. The COG control mode can reduce the negative acceleration caused by gravity which acts on the robot when it rolls about a point on the floor.

This robot we developed is self-reconfigurable. Since it can locomote by rolling when its shape is the 5R closed kinematic chain, it is useful if it can perform manipulative tasks after locomotion. The 5R closed kinematic chain can be used as a parallel manipulator and the 4R closed kinematic chain as shown in Fig. 1 may also be useful to perform some tasks. It is our future work to conduct experiments of manipulation.

Reference

[1] O. Mori and T. Omata, "Coupling of Two 2-Link Robots with a Passive Jointfor Reconfigurable Planar Parallel Robot", Proc. of the IEEE Int. Conf. on Robotics and Automation, pp. 4120-4125, 2002.

[2] O. Mori, T. Yamawaki and T. Omata, "Control of Self-Reconfigurable Parallel Robot by Coupling Open Kinematic Chains with Unactuated Joints," SICE Annual Conference 2002 CD-ROM, WM06-4, 2002.

[3] Y. Matsuo and V. Ampornaramveth, "A Wheel-Shaped Multi-Actuator System as a Test Bed for Autonomous Decentralized Motion Control," Trans. IEEE of Japan, 117-C(12), pp. 1840-1847, 1997.

[4] M. Yim, Y. Zhang and D. Duff : "Modular Robots," IEEE SPECTRUM, vol. 39, no. 2, pp. 30-34, The institute of Electrical and Electronics Engineers, 2002.

[5] W. H. Lee and A. C. Sanderson: "Dynamic Rolling Locomotion and Control of Modular Robots", IEEE Trans. on Robotics and Automation, vol. 18, no. 1, pp. 32-41, 2002.

[6] G. Oriolo and Y. Nakamura: "Control of Mechanical Systems with Second-Order Nonholonomic Constraints: Underactuated Manipulators," Proceedings of the 30th Conference on Decision and Control, pp.2398-2403, 1991.

Layered Multi Agent Architecture with Dynamic Reconfigurability

Eiichi Inohira, Atsushi Konno and Masaru Uchiyama
Department of Aeronautics and Space Engineering, Tohoku University
Aoba-yama 01, Sendai 980-8579, Japan
e-mail:{inohira, konno, uchiyama}@space.mech.tohoku.ac.jp

Abstract

This paper presents the software system architecture with the dynamic reconfigurability for highly autonomous systems. The proposed architecture is based on multi-agent model, which is suitable for autonomous, open and distributed systems. Since an agent is too sophisticated as the minimal component that we handle, we define a unit as a simple component of an agent. We present the advanced capability to dynamically reconfigure autonomous systems because their goals and environments vary frequently. We determine the specification of units and agents in order to analyze the behavior of overall system and to implement the dynamic reconfiguration as easy and simple as possible. In this paper, we also discuss the method to guarantee the system behavior during the dynamic reconfiguration in order to operate the robotic system without a hitch at any time, especially while reconfiguring.

1 Introduction

Enhancing autonomy of a robotic system is one of basic requests so that we would like to reduce the intervention in detail. In order to achieve the goal, an autonomous robotic system needs functions such as motion control, perception, planning, and learning. The software system becomes complex in structure since the above functions are implemented as the software. And so, it is very important that the architecture, which refers to specification of subsystems and underlying computational concepts[1].

Until now, many styles of the software system architecture for robotic systems have been proposed. Those are roughly classified into three types: hierarchical, behavioral and hybrid. Either of these architectures has some defects. The hierarchical architecture, which relies on a top-down approach, has poor flexibility. On the other hand, the behavioral architecture, which is based on a bottom-up approached, has low scalability. Although it is well known that both hierarchization and concurrent processing should be combined, it is not easy to actually design the hybrid architecture and to keep it simple. In addition to the above problem, there is another considerable problem due to the static configuration. The static configuration means that the software must be shut down and restarted every time it is modified. Furthermore, A robot must evacuate into safe positions and poses before a shutdown and restart of the software. In such a system, it takes much time to modify the goal and the environment. This is a serious problem in the case of autonomous mobile robots because their goals and environments vary frequently, even while operating. A solution to the problem is the dynamic reconfiguration, which is to change the system behaviors at runtime. The dynamic reconfigurability not only simplifies all the work regarding a robotic system, but also provides an advanced capability, that is, self-adaptation. For instance, a robot itself can modify the control algorithm according to a situation. The dynamic reconfigurability is the key technology since it can boost up the autonomy of autonomous robotic systems.

In this paper, we propose the software system architecture with dynamic reconfigurability. We also discuss the method to guarantee the system behavior while reconfiguring in order to operate autonomous robotic systems without a hitch at any time. The rest of this paper is organized as follows. Section 2 describes the layered multi agent architecture. Section 3 presents the underlying computational model. Section 4 discusses the method to guarantee of a system during the dynamic reconfiguration. We consider related work in Section 5, and conclude this paper in Section 6.

2 Layered multi agent architecture

Multi agent architecture is suitable for autonomous, open, distributed systems. It excels at modularity because a system consists of cooperative agents, which are independent concurrent processes. In multi-agent systems, we should consider the trade-off between the number of agents and the complexity of each agent[2]. However, this is hard work in the case of a large-scale system. The problem is that the conventional styles of multi agent architecture do not refer to the layered structure explicitly.

We define "the layered multi agent architecture" in order to solve the above problem and to target at a large-scale robotic system such as a full-featured humanoid robot. The basic idea is to incorporate multi agent architecture into a layered structure. In other words, a multi agent system is divided into an agent framework and components. In Fig. 1, the schematic view is shown. In our architecture, an agent consists of small and simple components in a uniform style. We define such components as "units." Although such an idea is very simple, there are the following benefits.

- It is easy to integrate a system.

- We can classify the components in detail for the purpose of reuse and portability.

- An agent can have the various and advanced faculties by combining units.

It is important to be separate into the system integration and the implementation of functions by an agent framework and units. The separation facilitates designing the architecture and realizing of the dynamic reconfiguration.

3 Computational model

3.1 Unit

A unit is a minimal element in our architecture and is a component of an agent. In order to separate units into platform-dependent and platform-independent components, the two types of a unit are defined: a normal unit and an interface unit. A normal unit consists of the following components.

- A data processing function

- An output buffer

Inputs of a unit function are output buffers of specified units as shown in Fig. 2. When a unit function is invoked, the output correspond to the inputs is calculated and then is written into the output buffer. In order to assume that the state of units is saved in only output buffers, we define that a unit must not access

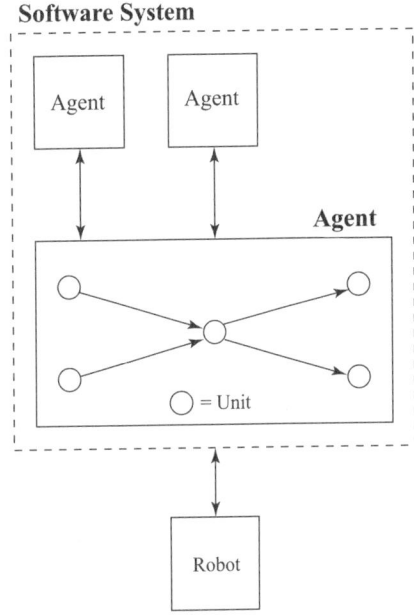

Figure 1: An generic diagram of the layer multi agent architecture

any memory area except output buffers. Also, we define that a normal unit does only calculation and is platform-independent.

An interface unit can interact with the following subsystems.

- Sensors and actuators

- File systems

- The other agents

For instance, an interface unit is allowed to read from a file, or write into a file as shown in Fig. 3. In this case of Fig. 4, the communication between agents is done via the interface unit that shares the output buffer with that of a unit of another agent. This is how the communication between existing normal units is realized without modification of them. An interface unit is platform-dependent so that it depends on the accessed subsystems.

Next, we discuss the unit structure with dynamic reconfigurability. In our architecture, we cannot decide the unit configuration at the stage of the design and development, that is, the static configuration. At this time, only data types of inputs and output of a unit can be determined. As shown in Fig. 5, we can first determine which unit the intended one is connected with after loading an agent with it. At this time, the agent must check the data types of the input

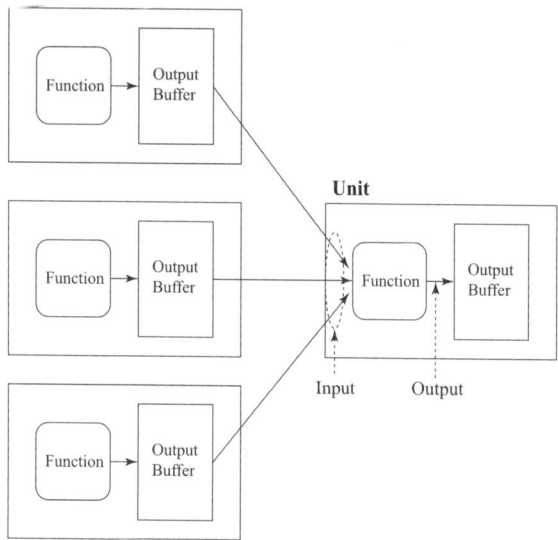

Figure 2: A generic diagram of the data flow between normal units

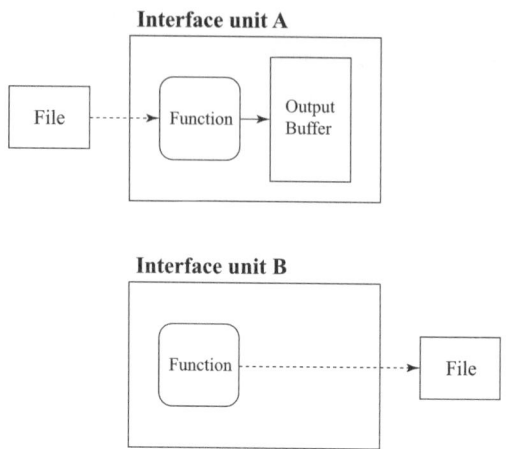

Figure 3: Examples of interface units

and output to prevent an error. Moreover, the following states of a unit are defined as shown in Fig. 6.

- Uninitialized
- Ready
- Terminated

Also, We should define the three functions correspond to the transition between these states.

3.2 Agent

In our architecture, the structure of an agent framework is shown in Fig. 7. The agent framework has the two faculties: the management of the units and execution of the unit functions. We define that an agent consists of the agent framework and units. An operator can interacts with an agent via a user interface and manage the unit configuration by a command at any time. The behavior of an agent depends on the unit configuration. Generally, an agent receives given inputs and then calculates an output according to a given algorithm. Each agent is independent and processed concurrently. But we define that unit functions of an agent are processed sequentially because of the simplification. The directed graph in Fig. 8 shows that the interrelation between the units. An agent invokes the unit functions in the user-defined sequence as the number in Fig. 8 indicates.

4 Dynamic reconfiguration

We present the method to implement the dynamic reconfiguration. In our architecture, there are the two cases of the reconfiguration: the agent reconfiguration and the unit reconfiguration. The former is to reconfigure a system by changing the agent configuration. The latter is to reconfigure an agent by changing the unit configuration. The system reconfiguration means all the above. As mentioned in Section 3.2, an agent only invokes the unit functions, and there is no direct effect of the agent configuration on the system. An agent gives us the distributed computing environment. Therefore, a system behavior depends only on the unit configurations.

Now, we discuss the problem regarding the dynamic reconfiguration. It is that the system guarantees the correct response of an autonomous robotic system at any time, especially during the dynamic reconfiguration. The correctness of the system behavior before the reconfiguration is responsible for an operator, who decides the system configuration. It applies to the correctness after the reconfiguration. However, it does not apply to the correctness during the reconfiguration because the operator cannot control it at runtime. For example, the process of the reconfiguration may preempt that of each agent. It means that the motion control of the robot is lost until the reconfiguration is completed. The system behavior cannot be guaranteed only by implementing the dynamic reconfiguration of the software. Therefore, we should design the architecture to prevent the process of the reconfiguration from interrupting that of each agent at least.

The simple solution is to process the unit recon-

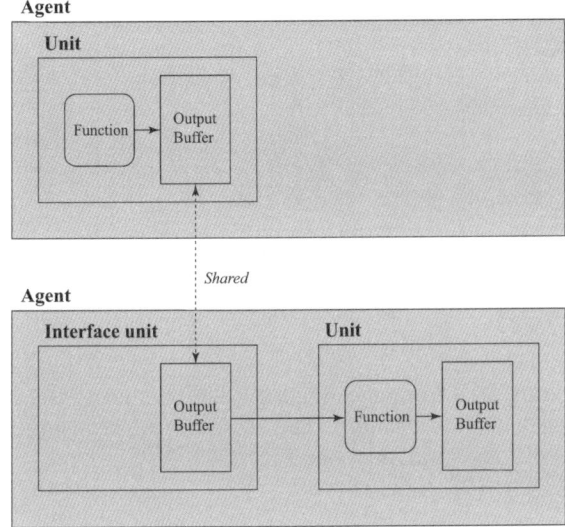

Figure 4: Communication between agents by using an interface unit

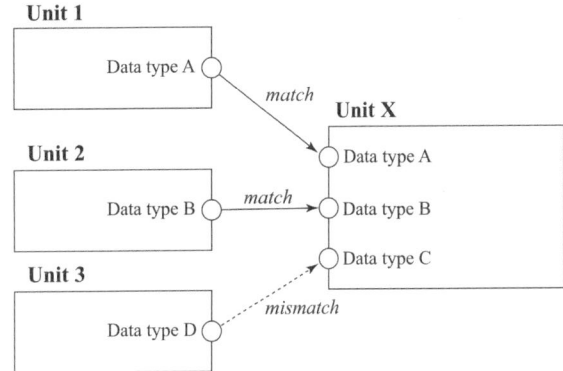

Figure 5: Type checking between units

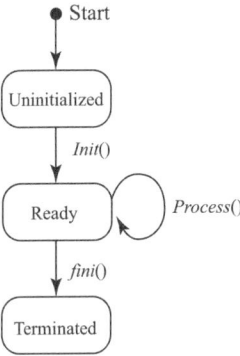

Figure 6: The state transition diagram of a unit

figuration in the background of the agents, that is, with a lower priority than theirs. The process of the reconfiguration is decomposed into the two stages as follows.

1. Calculate the new unit configuration from a command.

2. Switch to the new unit configuration from the old one.

The first stage may take much time when the number of the involved units is large. Thus there must be such a process in the background. This is no problem as long as the new configuration is not replaced with the old one. On the other hand, in the second stage, an agent must modify the unit configuration by itself before the unit functions are invoked. In this case, there is no trouble because it takes very short time to switch it. In our architecture, the aforementioned solutions can be easily implemented. The reason is that an agent has the two threads and invokes the unit functions sequentially as described in Section 3.2. For instance, the task transition during the unit reconfiguration is shown in Fig. 9. It may take much time to complete the reconfiguration in the above method. However, the process of an agent is not interrupted and the latency derived from switching is very short. Thus we can ignore the side effect of the dynamic reconfiguration of the software.

5 Related work

The robot architecture from some aspects is discussed in [1]. Here, we focus attention on particular types of the architecture that is targeted at robotic systems and has the dynamic reconfigurability.

First, D. B. Stewart et al. propose the dynamically reconfigurable real-time software framework [3] that based on the Chimera Methodology [4] . The software framework is a spin-off of a project to develop reconfigurable robots. The methodology and the RTOS mechanisms necessary to realize it are described in [3]. The dynamic reconfiguration of distributed robot controllers is also shown. Moreover, T. Q. Pham et al. present the self-adaptive control software systems [5] following the above research. It is based on a multi-agent model and also implemented mobile code by Java^{TM}.

Second, OPEN–R^{TM}, which is used for AIBO^{TM}, SDR–3X[6], etc., is presented in [7]. OPEN–R^{TM} is the standard architecture and interface for a re-

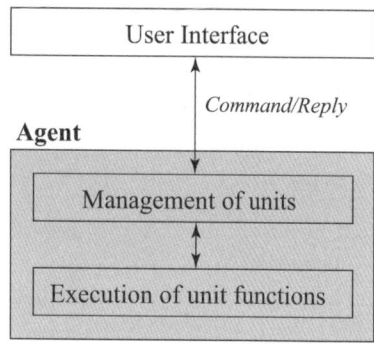

Figure 7: A block diagram of an agent framework

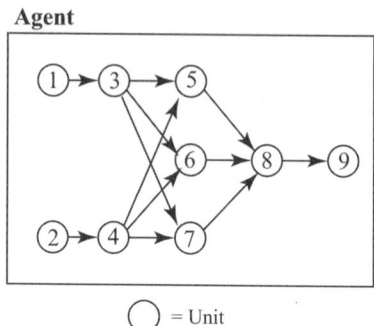

Figure 8: The unit configuration of an agent: the interconnection between the units and the sequence to invoke the unit functions

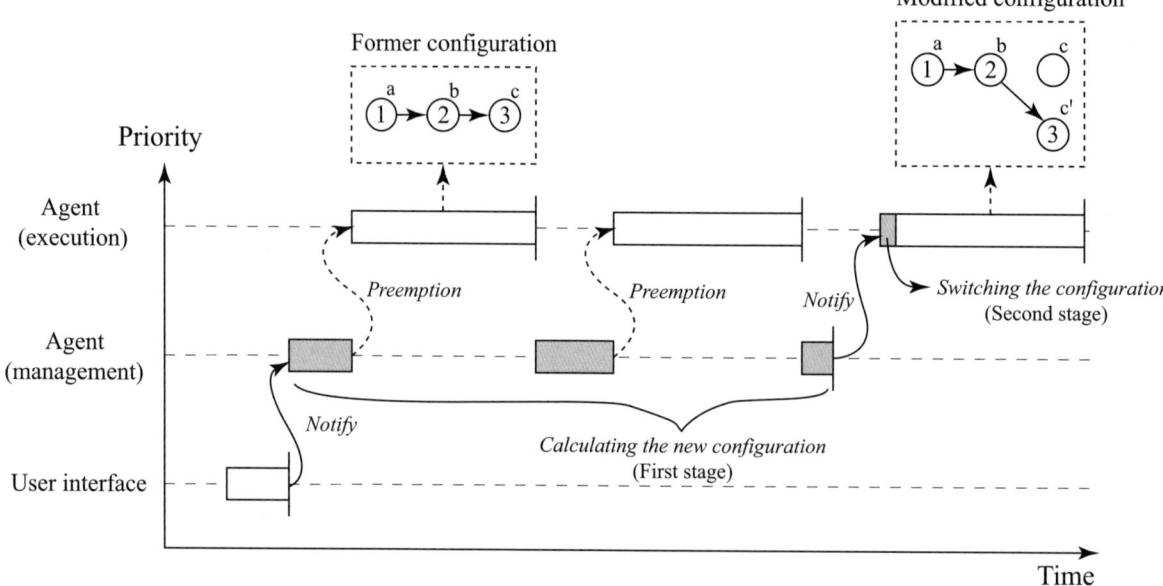

Figure 9: The task transition during the dynamic reconfiguration

configurable robot platform. It employs Aperios[TM] (a.k.a. Apertos) as the embedded operating system. Aperios[TM] supports dynamic updating of overall software. But there is little detail information about it.

Finally, S. Wang and K. G. Shin describe the reconfigurable software based on a Nested Finite State Machine[8] (NFSM) in [9]. The NFSM supports control logic reconfiguration at runtime and hierarchical composition of FSM's, which are the basic components. The dynamic reconfiguration of the motion control system for a milling machine is demonstrated.

The above researches refer to the software reconfiguration, but do not refer to the system behavior while reconfiguring at runtime. As mentioned in Section 4, only implementing the software reconfiguration cannot guarantee the system behavior during the reconfiguration. It is necessary to control the system behavior at that time. And so, this paper presents the method to solve the above problem.

There are the other recent studies on the robot architecture as follows: BeNet[10], Plugin architecture[11] and PredN[12]. But either study does not refer to the dynamic reconfigurability.

6 Conclusions

We proposed the layered multi agent architecture with dynamic reconfigurability. Our architecture is based on the multi-agent model, which is suitable to model dynamically reconfigurable software. A unit, which is a simple and small component of an agent, was defined since an agent is sophisticated. Thus our architecture consists of an agent framework and units, and combines openness with a layer structure.

We presented the underlying computational model. In our architecture, the roles of a unit and an agent are explicitly distinguished. A unit is a dynamically reconfigurable component to provide a function, and an agent sets up the distributed computing environment. Each agent only invokes the unit function sequentially. This policy facilitates analyzing the behavior of overall system and implementing the dynamic reconfiguration.

We discussed a system behavior during the dynamic reconfiguration. It is insufficient to implement the dynamic reconfiguration of the software in regard to guarantee the system behavior at any time. The reason is that the process of the reconfiguration may disturb the system behavior unless there is a particular mechanism to avoid such a situation. Thus we decomposed the process of the reconfiguration into the two stages. The new configuration is calculated in the background of agents, and an agent switches to the new one not while processing but before doing it.

As remarked above, the proposed architecture in this paper targets at autonomous robots, supports the dynamic reconfiguration, and guarantees the system behavior even during the reconfiguration. Now we apply this architecture to the software system of a humanoid robot.

References

[1] Ève Coste-Manière and Reid Simmons: Architecture, the Backbone of Robotic Systems, *Proceedings of the 2000 IEEE International Conference on Robotics and Automation*, pp. 67–72, 2000.

[2] Michael Wooldridge and Nicholas R. Jennings: Pitfalls of Agent-Oriented Development, *Proceedings of the 2nd International Conference on Autonomous Agents*, pp. 385–391, 1998.

[3] David B. Stewart, Richard A. Volpe, and Pradeep K. Khosla: Design of Dynamically Reconfigurable Real-Time Software Using Port-based Objects. *IEEE Transactions on Software Engineering*, Vol. 23, No. 12, pp. 759–776, 1997.

[4] David B. Stewart and Pradeep K. Khosla: The Chimera Methodology:Designing Dynamically Reconfigurable and Reusable Real-Time Software Using Port-based Objects. *International Journal of Software Engineering and Knowledge Engineering*, Vol. 6, No. 2, pp. 249–277, 1996.

[5] Theodore Q. Pham, Kevin R. Dixon, Jonathan R. Jackson, and Pradeep K. Khosla: Software Systems Facilitating Self-Adaptive Control Software, *Proceedings of the 2000 IEEE/RSJ International Conference on Intelligent Robots and Systems*, pp. 1094–1100, 2000.

[6] Tatsuzo Ishida, Yoshihiro Kuroki, Jinichi Yamaguchi, Masahiro Fujita, and Toshi T. Doi, Motion Entertainment by a Small Humanoid Robot Based on OPEN-R. *Proceedings of the 2001 IEEE/RSJ International Conference on Intelligent Robots and Systems*, pp. 1079–1086, 2001.

[7] Masahiro Fujita, Hiroaki Kitano, and Koji Kageyama: A reconfigurable robot platform, *Robotics and Autonomous Systems*, Vol. 29, pp. 199–132, 1999.

[8] Orest Storoshchuk, Shige Wang, and Kang G. Shin: Modelling Manufacturing Control Software, *Proceedings of the 2001 IEEE International Conference on Robotics and Automation*, pp. 4072–4077, 2001.

[9] Shige Wang and Kang G. Shin: Reconfigurable Software for Open Architecture Controller, *Proceedings of the 2001 IEEE International Conference on Robotics and Automation*, pp. 4090–4095, 2001.

[10] Tetsushi Oka, Junya Tashiro, and Kunikatsu Takase: Object-Oriented BeNet Programming for Data-Focused Bottom-Up Design of Autonomous Agents. *Robotics and Autonomous Systems*, Vol. 28, pp. 127–138, 1999.

[11] Fumio Kanehiro, Masayuki Inaba, and Hirochika Inoue: Developmental Software Environment that is applicable to Small-Size Humanoids and Life-Size Humanoids. *Proceedings of the 2001 IEEE International Conference on Robotics and Automation*, pp. 4084–4089, 2001.

[12] Oliver Stasse and Tasuo Kuniyoshi: PredN : Achiving efficiency and code re-usability in a programming system for complex robotic applications. *Proceedings of the 2000 IEEE International Conference on Robotics and Automation*, pp. 81–87, 2000.

Coordinating the Motions of Multiple Robots with Kinodynamic Constraints

Jufeng Peng
Department of Mathematical Sciences
Rensselaer Polytechnic Institute
Troy, New York 12180
pengj@rpi.edu

Srinivas Akella
Department of Computer Science
Rensselaer Polytechnic Institute
Troy, New York 12180
sakella@cs.rpi.edu

Abstract

This paper focuses on the coordination of multiple robots with kinodynamic constraints along specified paths. The presented approach generates continuous velocity profiles that avoid collisions and minimize the completion time for the robots. The approach identifies collision segments along each robot's path and then optimizes the motions of the robots along their collision and collision-free segments. For each path segment for each robot, the minimum and maximum possible traversal times that satisfy the dynamics constraints are computed by solving the corresponding two-point boundary value problems. Then the collision avoidance constraints for pairs of robots can be combined to formulate a mixed integer nonlinear programming (MINLP) problem. Since this nonconvex MINLP model is difficult to solve, we describe two related mixed integer linear programming (MILP) formulations that provide schedules that are lower and upper bounds on the optimum; the upper bound schedule is a continuous velocity schedule. The approach is illustrated with robots modeled as double integrators subject to velocity and acceleration constraints. An implementation that coordinates 12 nonholonomic car-like robots is described.

1 Introduction

Coordinating multiple robots with kinodynamic constraints, i.e. simultaneous kinematic and dynamics constraints ([8]), in a shared workspace without collisions has applications in manufacturing cells ([28]), AGV coordination in harbors and airports ([2]), and air traffic control ([24]). The general problem requires finding the trajectory (path and velocity profile) of each robot such that a specified objective, such as the task completion time, total time, or energy consumption, of the system is minimized.

We present here an approach to generate continuous velocity profiles for multiple robots with specified paths and dynamics constraints so their motions are collision-free and minimize the task completion time. This is in contrast to prior work that mostly addressed either the collision-free path or trajectory coordination of several robots without considering dynamics constraints ([23],[20],[36],[1]), or the search for time-optimal motions for a single robot ([5],[34]). An example application is the coordination of the motions of large numbers of AGVs along specified paths in harbors and airports ([2]). We must satisfy kinematic constraints, such as avoiding collisions between robots and with moving obstacles, and dynamics constraints, such as velocity and acceleration bounds, on the robot motions. By identifying the collision segments along a robot's path and when it can enter and exit its collision segments, we can combine the collision avoidance constraints for pairs of robots to formulate a mixed integer nonlinear programming (MINLP) problem. Since the resulting nonconvex MINLP formulation is difficult to solve, we use two related mixed integer linear programming (MILP) formulations, the *improved instantaneous* and *setpoint* formulations, that provide schedules that are lower and upper bounds on the optimal solution. We illustrate the approach using robots modeled as double integrators, and demonstrate its application to nonholonomic car-like robots with dynamics constraints.

1.1 Related Work

Multiple Robot Coordination: The problem of motion planning for multiple robots is to have each robot move from its initial to its goal configuration, while avoiding collisions with obstacles or other robots ([18]). This problem is highly underconstrained, and Hopcroft, Schwartz, and Sharir [12] showed that even a simplified two-dimensional case of the problem is PSPACE-hard. Recent efforts have focused on probabilistic approaches. A potential field randomized path planner was applied to multiple robot planning ([3]), and probabilistic roadmap planners have been developed for multiple car-like robots ([38]) and manipulators ([30]).

A slightly more constrained version of the problem is obtained when all but one of the robots have specified trajectories. This is the problem of planning a path and velocity for a single robot among moving obstacles ([27], [14]). To plan the motions of multiple robots, Erdmann and Lozano-Perez [9] assign priorities to robots and sequentially search for collision-free paths for the robots, in order of priority, in the configuration-time space.

If the problem is further constrained so that the paths of

the robots are specified, one obtains a path coordination problem. O'Donnell and Lozano-Perez [23] developed a method for path coordination of two robots. LaValle and Hutchinson [20] addressed a similar problem where each robot was constrained to a specified configuration space roadmap. The work most closely related to ours is that of Simeon, Leroy, and Laumond [36]. They perform path coordination for a very large number of car-like robots in the plane, where robots with intersecting paths can be partitioned into smaller sets. A more constrained version of this problem is the trajectory coordination problem where the trajectory (path and velocity) of each robot is specified. Previous work on trajectory coordination has focused almost exclusively on dual robot systems ([4], [7], [35]). Akella and Hutchinson [1] recently developed an MILP formulation to coordinate large numbers of robots with specified trajectories by changing only robot start times.

Trajectory Planning: There is a large body of work on the time optimal control of a single manipulator. Bobrow, Dubowsky, and Gibson [5] and Shin and McKay [34] developed algorithms to generate the time-optimal velocity profile of a manipulator along a specified path. Algorithms for minimum-time trajectory generation for a manipulator with dynamics and actuator constraints have also been developed ([29], [33]). Trajectory planning directly in the $2n$-dimensional state space that considers both kinematic and dynamic constraints is called *kinodynamic planning*. Donald et al. [8] developed a polynomial time approximation algorithm for kinodynamic planning for a single robot to generate near time-optimal trajectories. Fraichard [11] describes a trajectory planner for a car-like robot with dynamics constraints moving along a given path. Recent work on randomized kinodynamic planning includes the use of rapidly exploring random trees ([21]) and probabilistic roadmaps ([15]).

Air Traffic Control: Conflict resolution among multiple aircraft in a shared airspace ([37], [32], [24]) is closely related to multiple robot coordination. Tomlin, Pappas, and Sastry [37] synthesized safe conflict resolution maneuvers for two aircraft using speed and heading changes. Kosecka et al. [16] use potential field planners to generate conflict resolution maneuvers. Schouwenaars et al. [32] developed an MILP formulation for fuel-optimal path planning of multiple vehicles by using a discretized system model. Pallottino, Feron, and Bicchi [24] generate optimal conflict-free paths to minimize the total flight time and solve cases when either instantaneous velocity changes or heading angle changes are allowed.

2 Problem Overview

Given a set of n robots $\mathcal{A}_1, \ldots, \mathcal{A}_n$ with specified paths, the goal is to find the control inputs along the specified paths

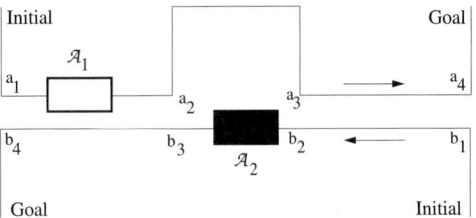

Figure 1: Example with two translating robots with two collision zones.

so that the completion time of the set of robots is minimized and their motions are collision free and satisfy their dynamics constraints. We assume that the start and goal configurations of each robot are collision-free, and that the specified paths for the robots are free of static obstacles. We further assume that each robot moves forward along its path without retracing its path.

2.1 Paths and Collision Zones

Each robot $\mathcal{A}_i$ is given a path γ_i, which is a continuous mapping $[0, 1] \to \mathcal{C}_i^{free}$. Let $\mathcal{S}_i = [0, 1]$ denote the set of parameter values s_i that place the robot along the path γ_i. The *coordination space* for n robots is defined as $\mathcal{S} = \mathcal{S}_1 \times \mathcal{S}_2 \times \cdots \times \mathcal{S}_n$. A feasible coordination is a schedule $\psi(t) : \mathcal{R}^+ \to \mathcal{S}$ in which $s_{init} = (0, 0, \ldots, 0)$ and $s_{goal} = (1, 1, \ldots, 1)$ and the robots do not collide. Note that there is a 1-to-1 mapping between s and the path length.

A *collision pair* $\mathcal{CP}_{ij}(s_i, s_j)$, where $s_i, s_j \in [0, 1]$ is defined as a pair of configurations $(\gamma_i(s_i), \gamma_j(s_j))$ where robot $\mathcal{A}_i$ and robot $\mathcal{A}_j$ collide, i.e., $\mathcal{A}_i(\gamma_i(s_i)) \cap \mathcal{A}_j(\gamma_j(s_j)) \neq \emptyset$. A *collision segment* for robot $\mathcal{A}_i$ is a contiguous interval $[s_i^{start}, s_i^{end}]$ over which $\mathcal{A}_i$ collides with some $\mathcal{A}_j$. That is, $\forall s_i \in [s_i^{start}, s_i^{end}], \exists s_j$ such that $\mathcal{A}(\gamma_i(s_i)) \cap \mathcal{A}(\gamma_j(s_j)) \neq \emptyset$.

An ordered pair of maximal contiguous intervals $([s_i^{start}, s_i^{end}], [s_j^{start}, s_j^{end}])$ in the coordination space $\mathcal{S}$ constitute a *collision zone* $\mathcal{CZ}_{ij}$ if and only if any point in one interval results in a collision with at least one point in the other interval (Figure 1). That is, $\forall s_i \in [s_i^{start}, s_i^{end}], \exists s_j \in [s_j^{start}, s_j^{end}]$ such that $\mathcal{A}(\gamma_i(s_i)) \cap \mathcal{A}(\gamma_j(s_j)) \neq \emptyset$, and $\forall s_j \in [s_j^{start}, s_j^{end}], \exists s_i \in [s_i^{start}, s_i^{end}]$ such that $\mathcal{A}(\gamma_i(s_i)) \cap \mathcal{A}(\gamma_j(s_j)) \neq \emptyset$.

In Figure 1, the collision zones are $([a_1, a_2], [b_3, b_4])$, and $([a_3, a_4], [b_1, b_2])$. A maximal interval that is not within any collision zone is called a *collision-free segment*. Each robot's path is decomposed into one or more collision segments and collision-free segments.

2.2 Optimal Control Problem For A Single Robot

Consider a robot $\mathcal{A}$ moving along a path segment. Let $\mathbf{x}(t)$ represent its state, $\mathbf{u}(t)$ be the control, γ be the path of $\mathcal{A}$, $J(\mathbf{x}, \mathbf{u})$ be the objective function, and $g(\mathbf{x})$ and $q(\mathbf{u})$ be the inequality constraints on the state variables and controls respectively. Then the optimal control problem, to compute the minimum and maximum segment traversal time for the robot subject to its dynamics and path constraints, can be written as:

$$\text{Minimize} \quad J(\mathbf{x}, \mathbf{u})$$
subject to:
$$\dot{\mathbf{x}} = f(\mathbf{x}, \mathbf{u})$$
$$g(\mathbf{x}) \leq 0$$
$$q(\mathbf{u}) \leq 0$$
$$\mathbf{x}(0) = \mathbf{x}_{start}$$
$$\mathbf{x}(\Delta T) = \mathbf{x}_{end}$$
$$\mathbf{x} \in \gamma$$

The minimum time control problem has $J(\mathbf{x}, \mathbf{u}) = \Delta T$, and the maximum time control problem has $J(\mathbf{x}, \mathbf{u}) = -\Delta T$ where $\Delta T = \int_0^{\Delta T} 1 dt$ is the time to traverse the segment. Feasible robot motions that give a minimum and a maximum of the objective over each segment are obtained by solving two TPBVPs (two-point boundary value problems) for each segment.

2.3 Coordinating Multiple Robots

Now consider the multiple robot system in which each robot has a specified path and dynamics constraints. The goal is to coordinate these robots to minimize a specified objective; in this paper it is the global completion time. The path of each robot is decomposed into collision segments and collision-free segments. The coordination of multiple robots can then be modeled as a mixed integer nonlinear programming (MINLP) problem, with each robot satisfying the traversal time constraints and collision avoidance constraints over each of its segments. Since this MINLP problem with nonconvex constraints is difficult to solve, we obtain schedules that provide a lower bound and an upper bound on the optimal solution by solving two related mixed integer linear programming (MILP) problems. We illustrate this approach using the double integrator model from optimal control ([6]).

This approach easily incorporates multiple moving obstacles with known trajectories. Each moving obstacle is treated like a robot with a known velocity profile whose collision constraints are included in the MILP formulations.

3 Instantaneous Model

We first consider a simplified model, the *instantaneous model*, where each robot moves only at its highest speed v_{max}, and can instantaneously start or stop with infinite acceleration. The discontinuous velocity schedule provided by the instantaneous model is a lower bound to the optimal continuous velocity schedule.

3.1 MILP Formulation

We now present a mixed integer linear programming (MILP) formulation for the instantaneous model. Let t_{ik} be the time when robot $\mathcal{A}_i$ begins moving along its kth segment and τ_{ik} be the traversal time for $\mathcal{A}_i$ to pass through segment k. Let ΔT_{ik}^{min} and ΔT_{ik}^{max} represent the minimum and maximum traversal time for $\mathcal{A}_i$ between the start point of segment k and the start point of segment $k + 1$. For the instantaneous model, $\Delta T_{ik}^{max} = \infty$. The minimum time for $\mathcal{A}_i$ to traverse a segment of length S_{ik} at its maximum velocity $v_{i,max}$ is $\Delta T_{ik}^{min} = S_{ik}/v_{i,max}$. The completion time C_{max} for the set of robots is greater than or equal to the completion time of each robot. Consider robots $\mathcal{A}_i$ and $\mathcal{A}_j$ with a shared collision zone where k and h are their respective collision segments. A sufficient condition for collision avoidance is that $\mathcal{A}_i$ and $\mathcal{A}_j$ are not simultaneously in their shared collision zone. That is, $t_{jh} \geq t_{i(k+1)}$ (when $\mathcal{A}_i$ exits segment k before $\mathcal{A}_j$ enters segment h) or $t_{ik} \geq t_{j(h+1)}$ (when $\mathcal{A}_j$ exits segment h before $\mathcal{A}_i$ enters segment k). These disjunctive constraints are converted to standard form ([22]) by introducing δ_{ijkh}, a binary variable that is 1 if robot $\mathcal{A}_i$ goes first along its kth segment and 0 if robot $\mathcal{A}_j$ goes first along its hth segment, and M, a large positive number. The resulting collision avoidance constraints to ensure the two robots $\mathcal{A}_i$ and $\mathcal{A}_j$ are not simultaneously in their shared collision zone are:

$$t_{jh} - t_{i(k+1)} + M(1 - \delta_{ijkh}) \geq 0$$
$$t_{ik} - t_{j(h+1)} + M\delta_{ijkh} \geq 0$$

The constraints for all robots are combined to form the instantaneous MILP formulation:

Minimize C_{max}
subject to:
$C_{max} \geq t_{i,last} + \tau_{i,last}$ for $i = 1, \ldots, n$
$t_{i(k+1)} = t_{ik} + \tau_{ik}$
$\Delta T_{ik}^{max} \geq \tau_{ik} \geq \Delta T_{ik}^{min}$
$t_{jh} - t_{i(k+1)} + M(1 - \delta_{ijkh}) \geq 0$
$t_{ik} - t_{j(h+1)} + M\delta_{ijkh} \geq 0$
$t_{ik} \geq 0$
$\delta_{ijkh} \in \{0, 1\}$

The collision avoidance constraints are conservative in not allowing two robots to simultaneously be in their collision zone, and in some cases lead to solutions that are not truly optimal. When a robot has overlapping collision zones with more than one robot, we subdivide its overlapping collision segments into several subsegments. The relevant pairs of subdivided collision zones are used to generate collision avoidance constraints.

The instantaneous model for multiple robot coordination can be viewed as a *job shop scheduling problem*, which is NP-hard ([26]). By reduction, the instantaneous model for robot coordination is NP-hard.

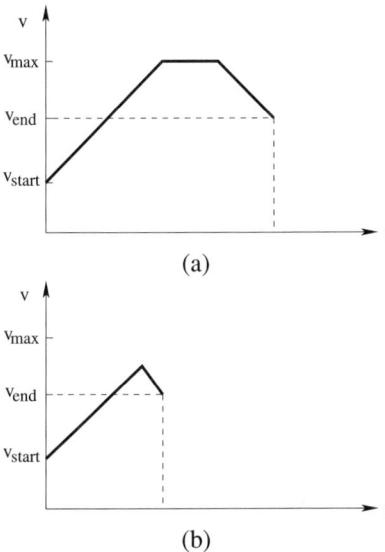

Figure 2: Minimum ΔT. Case (a): Velocity reaches v_{max}. Case (b): Velocity cannot reach v_{max}.

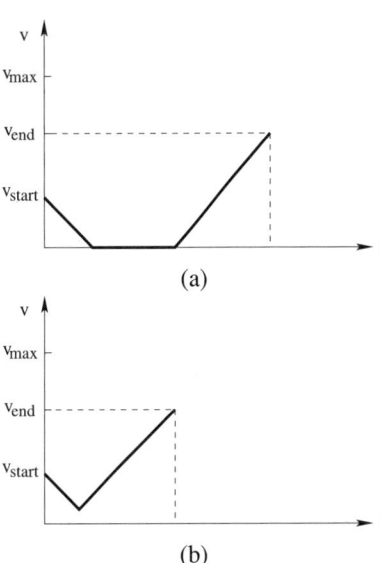

Figure 3: Maximum ΔT. Case (a): Velocity can decrease to zero. Case (b): Velocity cannot decrease to zero.

4 Continuous Velocity Model

We now consider generating a schedule with continuous velocity profiles for the robots consistent with their maximum velocity and acceleration bounds. To find the minimum and maximum times taken by a robot to traverse a segment, we solve two TPBVPs over the segment. We illustrate this procedure using the *double integrator* model from classical optimal control ([6]).

4.1 Single Robot on a Segment

A single robot moving along a path segment can be modeled as a double integrator with inequality constraints on the control input (acceleration) and the velocity state variable. The minimum time control of the double integrator model is well known ([6]) and we have extended this to obtain the maximum time control. Basically the solutions to these TPBVPs have a bang-bang or bang-off-bang control structure. Let S be the length of the segment, ΔT be the time taken to traverse the segment, and v_{start} and v_{end} be the velocities at the segment endpoints. The minimum ΔT and maximum ΔT each have two different cases, depending on whether S is sufficiently long for the robot to reach v_{max} (zero) for the minimum (maximum) time case. Note that if $\frac{|v_{end}^2 - v_{start}^2|}{2a_{max}} > S$, there is no feasible solution since the distance is too short for a feasible velocity profile.

1. Minimum ΔT (Figure 2):

   (a) If $S \geq \frac{v_{max}^2 - v_{start}^2 + v_{max}^2 - v_{end}^2}{2a_{max}}$,

   $$\Delta T^{min} = -\frac{(v_{max}^2 - v_{start}^2 + v_{max}^2 - v_{end}^2)}{2a_{max} \cdot v_{max}} + \frac{S}{v_{max}} + \frac{v_{max} - v_{start}}{a_{max}} + \frac{v_{max} - v_{end}}{a_{max}}$$

   (b) If $\frac{2v_{max}^2 - v_{start}^2 - v_{end}^2}{2a_{max}} > S \geq \frac{|v_{end}^2 - v_{start}^2|}{2a_{max}}$,

   $$\Delta T^{min} = \frac{v_{middle} - v_{start}}{a_{max}} + \frac{v_{middle} - v_{end}}{a_{max}}$$
   where $v_{middle} = \frac{1}{2}(2v_{start}^2 + 2v_{end}^2 + 4Sa_{max})^{\frac{1}{2}}$

2. Maximum ΔT (Figure 3):

   (a) If $S \geq \frac{(v_{start}^2 + v_{end}^2)}{2a_{max}}$, $\Delta T^{max} = \infty$.

   (b) If $\frac{1}{2}\frac{(v_{start}^2 + v_{end}^2)}{a_{max}} > S \geq \frac{1}{2}\frac{|(v_{end}^2 - v_{start}^2)|}{a_{max}}$,

   $$\Delta T^{max} = \frac{(v_{start} - v_{middle})}{a_{max}} + \frac{(v_{end} - v_{middle})}{a_{max}}$$
   where $v_{middle} = \frac{1}{2}(2v_{start}^2 + 2v_{end}^2 - 4Sa_{max})^{\frac{1}{2}}$

4.2 Continuous Velocity MINLP Formulation

Since the robot velocities are variables in the minimum and maximum time control for a robot over a segment, they introduce nonlinear constraints. We therefore formulate a mixed integer nonlinear programming (MINLP) model for generating a minimum time continuous velocity schedule. We have the usual completion time and collision avoidance constraints. The traversal time constraints are more complicated. Let $a_{i,max}$ and $v_{i,max}$ be the maximum acceleration and velocity of robot $\mathcal{A}_i$. Let v_{ik} represent the velocity of robot $\mathcal{A}_i$ at the start of segment k. Let ΔT_{ik}^{min} and ΔT_{ik}^{max} be the minimum and maximum traversal times for robot $\mathcal{A}_i$ along segment k. Let $\Delta T_{ik,1}^{min}(\Delta T_{ik,1}^{max})$ and $\Delta T_{ik,2}^{min}(\Delta T_{ik,2}^{max})$ represent the two possible minimum (maximum) values (Section 4.1). The binary variables $y_{ik,1}$ and $y_{ik,2}$ ($z_{ik,1}$ and $z_{ik,2}$) depend on whether or not the values of v_{ik} and $v_{i(k+1)}$ permit the robot to reach $v_{i,max}$ (zero) in S_{ik} and are used to select the feasible value of $\Delta T_{ik}^{min}(\Delta T_{ik}^{max})$. The MINLP formulation for the optimal continuous velocity schedule is:

Minimize C_{max}
subject to:
$C_{max} \geq t_{i,last} + \tau_{i,last}$ for $i = 1, \ldots, n$
$t_{i(k+1)} = t_{ik} + \tau_{ik}$
$\Delta T_{ik}^{max} \geq \tau_{ik} \geq \Delta T_{ik}^{min}$
$t_{jh} - t_{i(k+1)} + M(1 - \delta_{ijkh}) \geq 0$
$t_{ik} - t_{j(h+1)} + M\delta_{ijkh} \geq 0$
$t_{ik} \geq 0$
$\delta_{ijkh} \in \{0, 1\}$
$S_{ik} \geq \dfrac{(v_{i(k+1)}^2 - v_{ik}^2)}{2a_{i,max}} \geq -S_{ik}$
$S_{ik} - \dfrac{(v_{i,max}^2 - v_{ik}^2 + v_{i,max}^2 - v_{i(k+1)}^2)}{2a_{i,max}} - My_{ik,1} \leq 0$
$S_{ik} - \dfrac{(v_{i,max}^2 - v_{ik}^2 + v_{i,max}^2 - v_{i(k+1)}^2)}{2a_{i,max}} + My_{ik,2} \geq 0$
$\Delta T_{ik,1}^{min} = \dfrac{S_{ik}}{v_{i,max}} - \dfrac{(v_{i,max}^2 - v_{ik}^2 + v_{i,max}^2 - v_{i(k+1)}^2)}{2a_{i,max}v_{i,max}}$
$\qquad + \dfrac{v_{i,max} - v_{ik}}{a_{i,max}} + \dfrac{v_{i,max} - v_{i(k+1)}}{a_{i,max}}$
$\Delta T_{ik,2}^{min} = \dfrac{(v_{middle,ik}^{min} - v_{ik})}{a_{i,max}} + \dfrac{(v_{middle,ik}^{min} - v_{i(k+1)})}{a_{i,max}}$
$(v_{middle,ik}^{min})^2 = \dfrac{1}{4}(2v_{ik}^2 + 2v_{i(k+1)}^2 + 4S_{ik}a_{i,max})$
$\Delta T_{ik}^{min} = y_{ik,1} \cdot \Delta T_{ik,1}^{min} + y_{ik,2} \cdot \Delta T_{ik,2}^{min}$
$y_{ik,1} + y_{ik,2} = 1 \quad y_{ik,1}, y_{ik,2} \in \{0, 1\}$
$(S_{ik} - \dfrac{v_{ik}^2 + v_{i(k+1)}^2}{2a_{i,max}}) - Mz_{ik,1} \leq 0$
$(S_{ik} - \dfrac{v_{ik}^2 + v_{i(k+1)}^2}{2a_{i,max}}) + Mz_{ik,2} \geq 0$
$\Delta T_{ik,1}^{max} = \infty$
$\Delta T_{ik,2}^{max} = \dfrac{(v_{ik} - v_{middle,ik}^{max})}{a_{i,max}} + \dfrac{(v_{i(k+1)} - v_{middle,ik}^{max})}{a_{i,max}}$
$(v_{middle,ik}^{max})^2 = \dfrac{1}{4}(2v_{ik}^2 + 2v_{i(k+1)}^2 - 4S_{ik}a_{i,max})$
$\Delta T_{ik}^{max} = z_{ik,1} \cdot \Delta T_{ik,1}^{max} + z_{ik,2} \cdot \Delta T_{ik,2}^{max}$
$z_{ik,1} + z_{ik,2} = 1 \quad z_{ik,1}, z_{ik,2} \in \{0, 1\}$
$v_{i,max} \geq v_{ik} \geq 0$
$v_{i,initial} = v_{i,goal} = 0$

This MINLP problem has very difficult nonconvex constraints. Existing optimization techniques to solve MINLPs either require convexity or are not guaranteed to find the optimal solution for large problem sizes. Hence we solve two MILPs that differ only in their ΔT^{max} values to obtain good lower and upper bounds on the optimal solution; the bounds have been very close in our experiments. Assume for simplicity that the first and last segments are sufficiently long for each robot to go from zero to v_{max} and vice versa. (This assumption can be relaxed as discussed in [25].)

1. Lower bound MILP: A lower bound for the MINLP problem can clearly be obtained by solving the MILP for the instantaneous model with infinite acceleration. We obtain a tighter lower bound by formulating an *improved instantaneous model* that considers the acceleration and deceleration time over the first and last segments for each robot. The minimum traversal times for the first and last segments are then $\Delta T^{min} = S/v_{max} + v_{max}/2a_{max}$. Solving the resulting MILP gives a lower bound for the MINLP problem.

2. Upper bound MILP: The MINLP is transformed into an MILP problem by setting the velocities at the endpoints of each segment (except the initial and goal velocities) to the maximum feasible velocity. Solving this setpoint MILP problem (see next section) gives a feasible continuous velocity schedule, which is therefore an upper bound for the MINLP problem.

5 Setpoint Model

The *setpoint model* is used to generate a continuous velocity schedule. Since any continuous velocity schedule is an upper bound on the optimal continuous velocity schedule, the setpoint model is guaranteed to provide an upper bound on the MINLP problem. Here each robot's velocity is set to its maximum feasible velocity at its collision zones endpoints, thereby biasing the robots to move through their collision zones in the shortest possible time. Setting the velocity v_{ik} of each robot at the endpoints of its segments to $v_{i,max}$ transforms the MINLP formulation to an MILP formulation with ΔT^{min} and ΔT^{max} as follows:

$$\Delta T^{min} = \begin{cases} \dfrac{S}{v_{max}} & \text{if interior segment} \\ \dfrac{v_{max}}{2a_{max}} + \dfrac{S}{v_{max}} & \text{if first or last segment} \end{cases}$$

$$\Delta T^{max} = \begin{cases} \infty & \text{if } S \geq \dfrac{v_{max}^2}{a_{max}} \\ \dfrac{2v_{max} - 2(v_{max}^2 - a_{max}S)^{\frac{1}{2}}}{a_{max}} & \text{if } S < \dfrac{v_{max}^2}{a_{max}} \end{cases}$$

5.1 MILP Formulation

The MILP formulation for the setpoint model is identical to the formulation for the improved instantaneous model, and differs only in the ΔT^{max} parameter values. When the segment traversal time τ_{ik} generated by the MILP does not correspond to either a minimum time or maximum time trajectory over the segment, we have a simple algorithm to generate a feasible velocity profile for the double integrator.

6 Car-like Mobile Robots

We now illustrate our coordination approach on nonholonomic car-like robots with dynamics constraints. Paths that satisfy the nonholonomic constraints (Laumond [19]) typically require the robot to stop when there is a discontinuity in curvature (to change the steering direction) or when there is a cusp point (to reverse the robot motion direction). Therefore we use simple continuous curvature paths for a forward moving robot (Scheuer and Fraichard [31]).

6.1 Car-like Robot Model

The configuration of a robot is given by (x, y, θ, κ) where (x, y) represents the robot reference point at the midpoint of the rear axle, θ is the robot orientation, and κ is the signed path curvature. v is the robot velocity at its reference point.

We model a car-like robot of mass m moving on a plane with a friction coefficient μ as subject to the following dynamics constraints (Fraichard [11]):

1. Tangential acceleration constraints:

    (a) Acceleration constraints due to the engine force F are: $\frac{F_{min}}{m} \leq a \leq \frac{F_{max}}{m}$.

    (b) Sliding constraints to prevent slipping are:
    $-\sqrt{\mu^2 g^2 - \kappa^2 v^4} \leq a \leq \sqrt{\mu^2 g^2 - \kappa^2 v^4}$.

    Thus the (state dependent) acceleration constraints are:
    $a \geq \max(\frac{F_{min}}{m}, -\sqrt{\mu^2 g^2 - \kappa^2 v^4})$ and
    $a \leq \min(\frac{F_{max}}{m}, \sqrt{\mu^2 g^2 - \kappa^2 v^4})$.

2. Velocity constraints: In addition to the magnitude constraints $0 \leq v \leq v_{max}$, to ensure that $\mu^2 g^2 - \kappa^2 v^4 \geq 0$ we have the constraint $-\sqrt{\frac{\mu g}{|\kappa|}} \leq v \leq \sqrt{\frac{\mu g}{|\kappa|}}$.

    Thus the (state dependent) velocity constraints are:
    $0 \leq v \leq \min\left(v_{max}, \sqrt{\frac{\mu g}{|\kappa|}}\right)$.

6.2 Paths

The specified paths are chosen to be *simple continuous curvature paths* (SCC paths) (Scheuer and Fraichard [31]). Each path is C^2 continuous, so the path has continuous curvature and no cusps. Since the robot can follow the path without having to stop or reverse direction, we assume the robot moves forward monotonically along its path. The curvature κ of a path is upper bounded by κ_{max}, that is, the steering radius $\rho \geq \rho_{min} = 1/\kappa_{max}$. There is an upper bound on the time derivative of curvature, $\dot{\kappa}$.

We additionally assume $\mu^2 g^2 - \kappa^2 v^4 \geq (\frac{F_{max}}{m})^2$, which is true for typical values of the variables. This constraint can be expressed as a minimum steering radius constraint $\rho_{min} \geq v_{max}^2 / \sqrt{\mu^2 g^2 - (\frac{F_{max}}{m})^2}$ during path generation. This also implies that the maximum robot velocity is v_{max}.

6.3 Coordinating Multiple Car-like Robots

Consider a single car-like robot moving along a path segment with x and v representing its position and velocity respectively. The optimal control problem is:

$$\text{Min or Max } \Delta T = \int_0^{\Delta T} 1 \, dt$$

subject to:

$$\begin{pmatrix} \dot{x} \\ \dot{v} \end{pmatrix} = \begin{pmatrix} 0 & 1 \\ 0 & 0 \end{pmatrix} \begin{pmatrix} x \\ v \end{pmatrix} + \begin{pmatrix} 0 \\ 1 \end{pmatrix} a(t)$$

$$x(0) = -S \quad x(\Delta T) = 0$$
$$v(0) = v_{start} \quad v(\Delta T) = v_{end}$$
$$0 \leq v \leq v_{Max}(x)$$
$$-a_{Min}(x, v) \leq a(t) \leq a_{Max}(x, v)$$

where $v_{Max}(x) = \min\left(v_{max}, \sqrt{\frac{\mu g}{|\kappa|}}\right)$, $-a_{Min}(x, v) = \max(\frac{F_{min}}{m}, -\sqrt{\mu^2 g^2 - \kappa^2 v^4})$, and $a_{Max}(x, v) = \min(\frac{F_{max}}{m}, \sqrt{\mu^2 g^2 - \kappa^2 v^4})$. This TPBVP is difficult to solve because of the complex constraints on the state and control variables.

The minimum steering radius constraint $\rho_{min} \geq v_{max}^2 / \sqrt{\mu^2 g^2 - (\frac{F_{max}}{m})^2}$ makes $v_{Max}(x)$, $a_{Min}(x, v)$, and $a_{Max}(x, v)$ state independent constants. Therefore the double integrator formulation of Section 4.1 applies to the car-like robots above. Given a set of n car-like robots $\mathcal{A}_1, \ldots, \mathcal{A}_n$ with specified SCC paths that satisfy the above minimum steering radius constraints, we can generate collision-free continuous velocity profiles along the specified paths that minimize the completion time using the MILP formulations described earlier.

7 Implementation

We have implemented software in C++ to coordinate the motions of polyhedral robots with specified paths (Figure 4). We compute the collision zones using the PQP collision detection package (Larsen et al. [17]) by sampling uniformly along each robot's path. We generate the MILP formulations from the collision zones and solve them using the AMPL [10] and CPLEX [13] optimization packages. Since the setpoint formulation with its tighter constraints is solved much faster than the improved instantaneous formulation, we use the objective function value from the setpoint solution as an upper bound constraint in the improved instantaneous formulation. See Table 1 for running times measured on a Sun Ultra 60. The problem complexity depends primarily on the number of collision zones, and to a lesser extent on the number of robots. For a particularly difficult problem (for example, the radial case with a bottleneck at the center) or for a sufficiently large number of collision zones, the MILP time dominates the running time.

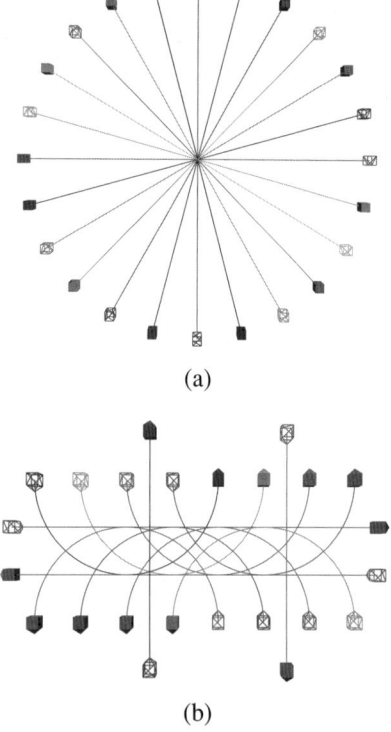

(a)

(b)

Figure 4: Overhead view of example paths for 12 robots: (a) Radial paths, with a bottleneck at the center (b) Simple continuous curvature paths. Goal configurations are indicated by solid polyhedra.

In our experiments, an optimal solution, indicated by a zero gap between the objective function values computed by the improved instantaneous and setpoint formulations, was found in almost all cases; the maximum gap observed was 8.84%. Example animations can be seen at www.cs.rpi.edu/~sakella/multikino/.

8 Conclusion

By combining techniques from optimal control and mathematical programming, we developed an MINLP formulation for minimum time collision-free coordination of multiple robots with kinodynamic constraints along specified paths. We then developed two related MILP formulations that give upper and lower bounds on the optimal solution. Although the MILP formulations for coordination of multiple robots are NP-hard, the availability of efficient collision detection software and integer programming solvers makes this approach practical for reasonable problem sizes.

There are several directions for future work. We have recently developed formulations for coordination of robots where each robot can move along a set of possible paths, and are investigating their computational feasibility. Analytically characterizing the gap between the improved in-

Num. of robots	Num. of collision zones	Collision detection time (secs)	MILP-S time (secs)	MILP-I time (secs)
5	13	18.67	0.04	0
8	42	55.67	0.13	0.08
10	71	88.26	0.53	0.17
12	82	115.81	0.61	0.25
8radial	29	30.53	3.87	0.095
12radial	86	70.53	160	60.67
12scc	154	65.62	12	1.167

Table 1: Sample run times for setpoint formulation (MILP-S) and improved instantaneous formulation (MILP-I). (The MILP-I formulation used the MILP-S solutions as upper bounds.) Collision checks were performed at 200 points along each path. AMPL presolve times are not included.

stantaneous model and setpoint model solutions, and developing heuristic algorithms for closing the gap is important. Extending the approach to systems with more complex dynamics, including aircraft and articulated robots, appears to be an attainable next step. Another interesting direction is online coordination of multiple robots using sensor estimates of robot positions and velocities.

Acknowledgments

This work was supported in part by RPI and by NSF under CAREER Award No. IIS-0093233. Andrew Andkjar implemented animation software that interfaced with the PQP software from the University of North Carolina and helped generate examples. Thanks to Prasad Akella for suggesting the problem, Seth Hutchinson for early discussions, and John Mitchell for advice.

References

[1] S. Akella and S. Hutchinson. Coordinating the motions of multiple robots with specified trajectories. In *IEEE International Conference on Robotics and Automation*, pages 624–631, Washington, DC, May 2002.

[2] R. Alami, S. Fleury, M. Herrb, F. Ingrand, and F. Robert. Multi-robot cooperation in the MARTHA project. *IEEE Robotics and Automation Magazine*, 5(1):36–47, Mar. 1998.

[3] J. Barraquand, B. Langlois, and J.-C. Latombe. Numerical potential field techniques for robot path planning. *IEEE Transactions on Systems, Man, and Cybernetics*, 22(2):224–241, 1992.

[4] Z. Bien and J. Lee. A minimum time trajectory planning method for two robots. *IEEE Transactions on Robotics and Automation*, 8(3):414–418, June 1992.

[5] J. E. Bobrow, S. Dubowsky, and J. S. Gibson. Time-optimal control of robotic manipulators along specified paths. *International Journal of Robotics Research*, 4(3):3–17, Fall 1985.

[6] A. E. Bryson and Y.-C. Ho. *Applied Optimal Control*. Hemisphere, Washington D.C., 1975.

[7] C. Chang, M. J. Chung, and B. H. Lee. Collision avoidance of two robot manipulators by minimum delay time. *IEEE Transactions on Systems, Man, and Cybernetics*, 24(3):517–522, Mar. 1994.

[8] B. R. Donald, P. Xavier, J. Canny, and J. Reif. Kinodynamic motion planning. *Journal of the ACM*, 40(5):1048–1066, Nov. 1993.

[9] M. Erdmann and T. Lozano-Perez. On multiple moving objects. *Algorithmica*, 2(4):477–521, 1987.

[10] R. Fourer, D. M. Gay, and B. W. Kernighan. *AMPL: A Modeling Language for Mathematical Programming*. Duxbury Press, 1993.

[11] T. Fraichard. Trajectory planning in dynamic workspace: a 'state-time space' approach. *Advanced Robotics*, 13(1):75–94, 1999.

[12] J. E. Hopcroft, J. T. Schwartz, and M. Sharir. On the complexity of motion planning for multiple independent objects: PSPACE-hardness of the "warehouseman's problem". *International Journal of Robotics Research*, 3(4):76–88, 1984.

[13] ILOG, Inc., Incline Village, NV. *CPLEX 6.6 User's Manual*, 1999.

[14] K. Kant and S. W. Zucker. Toward efficient trajectory planning: The path-velocity decomposition. *International Journal of Robotics Research*, 5(3):72–89, Fall 1986.

[15] R. Kindel, D. Hsu, J. Latombe, and S. Rock. Kinodynamic motion planning amidst moving obstacles. In *IEEE International Conference on Robotics and Automation*, pages 537–543, San Francisco, Apr. 2000.

[16] J. Kosecka, C. Tomlin, G. Pappas, and S. Sastry. Generation of conflict resolution maneuvers for air traffic management. In *IEEE/RSJ International Conference on Robots and Systems*, pages 1598–1603, Grenoble, France, 1997.

[17] E. Larsen, S. Gottschalk, M. Lin, and D. Manocha. Fast distance queries using rectangular swept sphere volumes. In *IEEE International Conference on Robotics and Automation*, pages 3719–3726, San Francisco, CA, Apr. 2000.

[18] J.-C. Latombe. *Robot Motion Planning*. Kluwer Academic Publishers, Norwell, MA, 1991.

[19] J.-P. Laumond, editor. *Robot Motion Planning and Control*. Lecture Notes in Control and Information Sciences, No. 229. Springer, 1998.

[20] S. M. LaValle and S. A. Hutchinson. Optimal motion planning for multiple robots having independent goals. *IEEE Transactions on Robotics and Automation*, 14(6):912–925, Dec. 1998.

[21] S. M. LaValle and J. J. Kuffner. Randomized kinodynamic planning. *International Journal of Robotics Research*, 20(5):378–401, May 2001.

[22] G. L. Nemhauser and L. A. Wolsey. *Integer and Combinatorial Optimization*. John Wiley and Sons, New York, 1988.

[23] P. A. O'Donnell and T. Lozano-Perez. Deadlock-free and collision-free coordination of two robot manipulators. In *IEEE International Conference on Robotics and Automation*, pages 484–489, Scottsdale, AZ, May 1989.

[24] L. Pallottino, E. Feron, and A. Bicchi. Conflict resolution problems for air traffic management systems solved with mixed integer programming. *IEEE Transactions on Intelligent Transportation Systems*, 3(1):3–11, Mar. 2002.

[25] J. Peng and S. Akella. Coordinating multiple robots with kinodynamic constraints along specified paths. In *Fifth Workshop on the Algorithmic Foundations of Robotics*, Nice, France, Dec. 2002.

[26] M. Pinedo. *Scheduling: Theory, Algorithms, and Systems*. Prentice-Hall, Englewood Cliffs, NJ, 1995.

[27] J. Reif and M. Sharir. Motion planning in the presence of moving obstacles. In *Proceedings of the 26th Annual Symposium on the Foundations of Computer Science*, pages 144–154, Portland, Oregon, Oct. 1985.

[28] A. A. Rizzi, J. Gowdy, and R. L. Hollis. Distributed coordination in modular precision assembly systems. *International Journal of Robotics Research*, 20(10):819–838, Oct. 2001.

[29] G. Sahar and J. M. Hollerbach. Planning of minimum-time trajectories for robot arms. *International Journal of Robotics Research*, 5(3):97–140, Fall 1986.

[30] G. Sanchez and J.-C. Latombe. Using a PRM planner to compare centralized and decoupled multi-robot systems. In *IEEE International Conference on Robotics and Automation*, pages 2112–2119, Washington, D.C., May 2002.

[31] A. Scheuer and T. Fraichard. Continuous-curvature path planning for car-like vehicles. In *Proc. of the IEEE-RSJ Int. Conf. on Intelligent Robots and Systems, volume 2*, pages 997–1003, Grenoble, France, Sept. 1997.

[32] T. Schouwenaars, B. D. Moor, E. Feron, and J. How. Mixed integer programming for multi-vehicle path planning. In *European Control Conference 2001*, Porto, Portugal, 2001.

[33] Z. Shiller and S. Dubowsky. On computing the global time-optimal motion of robotic manipulators in the presence of obstacles. *IEEE Transactions on Robotics and Automation*, 7(6):785–797, Dec. 1991.

[34] K. G. Shin and N. D. McKay. Minimum-time control of robotic manipulators with geometric path constraints. *IEEE Transactions on Automatic Control*, AC-30(6):531–541, June 1985.

[35] K. G. Shin and Q. Zheng. Minimum-time collision-free trajectory planning for dual robot systems. *IEEE Transactions on Robotics and Automation*, 8(5):641–644, Oct. 1992.

[36] T. Simeon, S. Leroy, and J.-P. Laumond. Path coordination for multiple mobile robots: A resolution-complete algorithm. *IEEE Transactions on Robotics and Automation*, 18(1):42–49, Feb. 2002.

[37] C. Tomlin, G. J. Pappas, and S. Sastry. Conflict resolution for air traffic management: A study in multi-agent hybrid systems. *IEEE Transactions on Automatic Control*, 43(4):509–521, Apr. 1998.

[38] P. Švestka and M. H. Overmars. Coordinated motion planning for multiple car-like robots using probabilistic roadmaps. In *IEEE International Conference on Robotics and Automation*, pages 1631–1636, Nagoya, Japan, May 1995.

Scalability and Schedulability in Large, Coordinated, Distributed Robot Systems

John D. Sweeney, Huan Li, Roderic A. Grupen, and Krithi Ramamritham

Laboratory for Perceptual Robotics
Department of Computer Science
University of Massachusetts, Amherst
{sweeney, lihuan, grupen, krithi}@cs.umass.edu

Abstract—Multiple, independent robot platforms promise significant advantage with respect to robustness and flexibility. However, coordination between otherwise independent robots requires the exchange of information; either implicitly (as in gestural communication), or explicitly (as in message passing in a communication network.) In either case, control processes resident on all coordinated peers must participate in the collective behavior. This paper evaluates the potential to scale such a coupled control framework to many participating individuals, where scalability is evaluated in terms of the schedulability of coupled, distributed control processes.

We examine how schedulability affects the scalability of a robot system, and discuss an algorithm used for off-line schedulability analysis of a distributed task model. We present a distributed coordinated search task and analyze the schedulability of the designed task structure. We are able to analyze communication delays in the system that put upper bounds on the size of the robot teams. We show that hierarchical methods can be used to overcome the scalability problem. We propose that schedulability analysis should be an integrated part of a multi-robot team design process.

Keywords: *multi-robot systems, schedulability analysis, coordinated control*

I. INTRODUCTION

As robotic technology becomes more mature, implementing distributed robot systems with a large number of robots will be possible. The expense and mean-time to failure of component hardware limit the size of fieldable teams to tens of robots [3]. However, in the future we can expect that developing technology will allow, and applications will require, teams with an order of magnitude more robots. While swarms of robots are an attractive idea, they are also frequently assumed to be composed of independent platforms and control processes. To do useful work, we may have to coordinate the activity of several robots, which introduces processing and communication constraints among the team. In this paper, we propose a coordination model and evaluate the bounds on coordinated robots teams that arise due to those constraints.

We present a distributed, coupled control framework applied to a leader/follower search task. The distributed controller can address multiple, concurrent objectives while maintaining global behavior constraints. Robust, closed-loop controllers are defined as primitives within the control architecture. Multiple controllers are combined via the nullspace projection operator: subordinate control actions are projected onto the nullspace of superior controllers, so that incorrect interactions between controllers are avoided. This allows "best-effort" guarantees to be made about global behavior. In the task of leader/follower search, the leader must concurrently search while maintaining connectivity by remaining within line-of-sight (LOS) of the follower.

In any robot system, interaction with the world imposes a real-time constraint on computation, whose logical correctness depends on the correctness of its outputs as well as their timeliness. The robot must be able to process sensor input, respond to dynamic environments, and send messages to other robots. Real-time specifications for such systems are derived from time constraints in the control processes and in the environment. If the system is unable to perform distributed control tasks in a timely manner, then overall performance suffers.

In this paper, we look at the issue of scalability from the perspective of real-time schedulability. A distributed multi-robot system is viewed as a collection of homogeneous processors. Each robot has a set of tasks that run periodically, with data flow between tasks on a single robot and between tasks on different robots. In the context of a real-time multi-robot system, schedulability analysis determines whether all tasks in the system can be scheduled to some period and deadline. We propose that schedulability analysis should be an integral part of the multi-robot system design process.

The paper is organized as follows. First we briefly present related work, then we give an overview of the distributed controller for concurrent, multi-objective tasks presented in [13]. This distributed controller is used to perform a leader/follower search task. Then we examine the controller using the algorithm developed in [8] for off-line schedulability analysis. We finish with conclusions and future work.

II. RELATED WORK

The control framework described in this paper is based on a bottom-up approach to control, similar to approaches such as the subsumption architecture [2], where robust,

This work was supported in part by NSF CDA-9703217, DARPA/IPTO DABT63-99-1-0022 and DABT63-99-1-0004.

low-level control primitives are combined to produce high-level behaviors. Individual controllers are constructed using the control basis approach [5], [13].

There is a lot of work in the literature on cooperative multi-robot teams, such as [3], which presents an overview of cooperative robotic techniques. However, the issue of the scalability of the coordination scheme is not fully addressed. Carpin et al. [4] have presented an approach for the leader/follower application. Their system is designed to allow teams of any size, however, they do not address the effects of using broadcast communications on the effective team size. Yoshida et al. [14] have examined how information propagation within a team and team performance were affected by using a shared communication channel and team size, but they did not consider real-time computing constraints.

Real-time operating systems for robotics that allow communication among distributed resources have been developed [1], [12]. There are also tools for designing controllers for real-time robotic systems such as [10]. Schedulability analysis for distributed real-time systems has also received a lot of attention in recent years [9]. For tasks with temporal constraints, researchers have focused on generating task attributes (e.g., period, deadline and phase) with the objective of minimizing the utilization and/or maximizing system schedulability while satisfying all temporal constraints. However, schedulability is clearly affected by both temporal characteristics and allocation of real-time tasks. A more comprehensive approach that takes into consideration task temporal characteristics and allocations, in conjunction with schedulability analysis, is required.

III. DISTRIBUTED, COORDINATED LEADER/FOLLOWER CONTROLLERS

We first give a brief overview of the architecture for reactive, coordinated controllers that address multiple, concurrent objectives in a mobile robot team, described in [13]. An example application of such a system is a multi-robot search task. Each robot is equipped with sensors specific to the search goal, in this case IR proximity sensors, and wireless communication. A team of robots R must search an unknown environment, while maintaining wireless connectivity throughout the team. The limited range of the wireless transmitters imposes a path constraint on the members of the team in that any pair of robots must ensure they are within line-of-sight (LOS) and within range specifications for the desired QoS or bandwidth in order to guarantee connectivity.

The control basis approach constructs a controller $\phi_\mathcal{E}^S$ by associating a state estimator, $\mathcal{S}$, and effectors, $\mathcal{E}$, with an objective function, or artificial potential, ϕ. In this paper, our artificial potentials are harmonic functions represented by discrete occupancy maps [6]. For example, the controller that enforces the LOS constraint is $\phi_j^{LOS_i}, i, j \in R$, where robot i generates the LOS region (computed from its position and an obstacle map), and robot j tries to stay within the LOS region by descending the potential ϕ. The search controller is ϕ_i^S, where robot i achieves search goal states S by greedy action on ϕ. The leader's search task implicitly avoids obstacles since it computes trajectories that move the leader away from obstacles and toward unexplored space.

A. The "Pull" Controller

A pairwise, concurrent, coordinated controller (denoted a "pull" controller) is constructed that allows the leader to search as long as the follower is within the LOS region:

$$\phi_i^S \triangleleft \phi_j^{LOS_i}, \quad (1)$$

where i is the leader and j is the follower. The "subject-to" operator ($\triangleleft$) allows concurrency by projecting the trajectory from ϕ_i^S onto the nullspace of $\phi_j^{LOS_i}$ (using the Moore-Penrose pseudoinverse, for example); ensuring that the leader's search task does not interact destructively with the LOS task. Here, the nullspace of $\phi_j^{LOS_i}$ refers to the nullspace of the Jacobian that maps changes in wheel displacements of robot j onto changes in the value of the artificial potential defined by ϕ^{LOS_i}. In general, planar mobile robots are not redundant with respect to their configuration space. However, a planar mobile robot may be redundant with respect to some objective function.

In addition to two robots, multiple robots can form a serial, kinematic chain by combining pull controllers. The robot at the head of the chain executes the controller in equation (1), while a robot k within the chain is involved in a pairwise pull controller with its neighbors:

$$\phi_k^{LOS_{k+1}} \triangleleft \phi_{k-1}^{LOS_k}. \quad (2)$$

The robot at the base of the chain is assumed to be a stationary communications hub for the team. This pull chain allows the leader to explore a great distance from the hub.

The leader/follower pull controller described above is implemented with two of our UMASS UBot mobile robots, each one using a 206 MHz StrongARM CPU with the K-Team Kameleon motor driver board. The controllers are implemented using the Player/Stage robot control system [7]. In the next section we describe our method of real-time schedulability analysis in a distributed control system, and analyze the schedulability of the pull controller.

IV. OUR APPROACH FOR REAL-TIME SCHEDULABILITY ANALYSIS

Equation (1) describes a coordinated controller that involves several processes: sensor processing to determine S and LOS_i, motor Jacobians that generate wheel velocities on platforms i and j, and processes that descend potential functions. From the real-time systems view, the scalability of such a scheme involves the ability to find feasible schedules as the task and processor sets increase, i.e., as the number of robots increases. The real-time constraint imposes a hard deadline on the amount of processing that can be completed in a given period of time. If robots are independent and do not collaborate in a coordinated control scheme, then scalability is not an interesting problem, since it becomes one of scheduling on an individual robot, which has been widely studied [11]. However, if the team members are cooperative, then, in addition to task constraints such as periods or deadlines, system level constraints are also introduced.

For instance, in order to achieve a common goal, robots may exchange messages. Therefore, communication costs and precedence constraints must be considered. The difficulty is that scheduling tasks with precedence constraints and individual deadlines for a multiprocessor system is an NP-complete problem even for unit processing time of each task. We propose a heuristic algorithm that takes into account all types of constraints to predict the scalability and schedulability for a large, recursive robot system [8]. In this context, recursive means that the task model of the system has a symmetric structure that can be easily generalized to accommodate additional robots. The pull controller is an example of a recursive, distributed system.

In the following sections, we will briefly review the system model and our approach to do real-time schedulability analysis for distributed coordinated robotics, which is discussed in [8].

A. System Model

The distributed robot system consists of m identical uni-processor sites. In this paper, we use *site* and *robot* interchangeably. The sites are connected by a shared communication medium from one site to another. Communications must be scheduled at specific times to assure that no contention for the channel occurs at run time.

Tasks we study here are real-time tasks that have the following characteristics:

- **Period**. This defines the inter-release times of instances of the task. One instance of the task should be executed every period.
- **Relative deadline**. This specifies the time at which each task instance must be completed.
- **Computation time**. This is the worst case execution time of any instance of the task.
- **Precedence relationships**. These constrain the execution order of the tasks and the production and consumption relationships of the data flow.
- **Locality constraints**. These relationships are based on the nature of the environment required for tasks to execute. For example, actuator control tasks must run on the processor that connects to the actuator. In this distributed task model, tasks without locality constraints can be assigned to any available processor.
- **Communication constraints**. Communication between tasks that are on different sites requires a communication medium and time to send or receive the message. Messages sent between tasks must be scheduled with the communication medium as a resource requirement. If the communication is modeled as a special task, then this task must satisfy the

$IR_{1,2}$	$POS_{1,2}$	$M_{1,2}$	ϕ_2^S	LOS_2	$\triangleleft$	$\phi_1^{LOS_2}$
20	20	20	25	1	10	25

TABLE I

WORST-CASE EXECUTION TIME (IN MILLISECONDS) FOR THE TASKS IN FIGURE 1.

precedence constraints with the two communicating tasks separately.

B. Algorithm Overview

In order to help understand the method, we now present a brief overview of the schedulability algorithm described in [8]. The first step of the algorithm assigns unallocated tasks to sites. A heuristic, which takes into account the trade-off between communication cost and processor workload, is used to assign tasks to sites. The basic idea of the heuristic is to cluster tasks with a high communication cost together on the same site while minimizing the utilization of each processor.

Next, we construct an extended task graph that includes communications represented as tasks. The algorithm uses communication tasks to model the communication cost and channel contention that occurs if the tasks are allocated to different sites. Then we build a comprehensive graph containing all instances of all tasks including communication tasks that will execute within the least common multiple (LCM) of all task periods, and preprocess precedence relations of tasks by setting up the relative earliest start time of consumers. Finally, a search is used to find a *feasible* schedule, if possible, mapping starting times to all tasks including communication, to determine if they can start and complete execution before their deadlines.

C. Pull Controller Analysis

The task model for the pull controller with two robots is shown in Figure 1. Every robot must execute IR obstacle detection tasks, denoted by IR_i, odometry tasks, POS_i, and motor control tasks, M_i. The three tasks IR_i, POS_i, and M_i, which are drawn with solid ellipses, are specific to the hardware of each robot, so they are all preallocated to run locally on each robot. The control tasks ϕ_2^S and $\phi_1^{LOS_2}$, the nullspace projection $\triangleleft_2$, and the sensor processing task LOS_2 may reside on a single robot, or be distributed between the pair, if necessary, to optimize processor utilization or communication costs. They are denoted with dotted ellipses. The functionality of the team is not affected by altering the allocations of the control tasks. However, a good allocation strategy does improve schedulability.

The communication cost between tasks, if they are distributed, is given in milliseconds on the corresponding arc. The computation times of the tasks are given in Table I. Computation times and communication costs were determined experimentally on the platform.

The sensor and motor tasks IR_i, POS_i, and M_i are designed to be updated periodically, and the control tasks

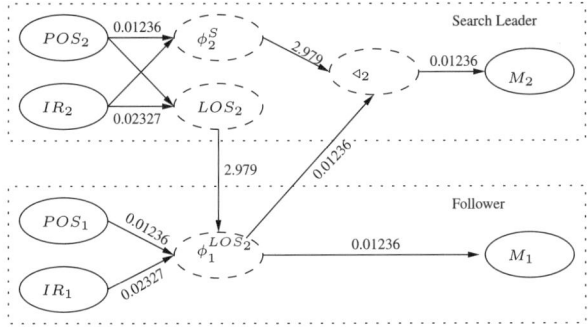

Fig. 1. The pull controller task model for two robots. Preallocated tasks are in solid ellipses, and dotted ellipses are tasks that may be distributed across the pair. The communication cost between tasks (in milliseconds) is shown on the arcs.

must execute periodically in order to consume the new sensor data and give new motor commands. Consequently, the periods of the control tasks should be based on the periods of the sensor and motor tasks. We determined experimentally that to achieve satisfactory performance, the sensor and motor tasks should run at least every 200 ms. The period of execution of all the tasks in the pull model determines the robot's responsiveness, and also affects the scalability of the system. After applying the scheduling algorithm using the "aggressive" allocation heuristic described in [8], we found that ϕ_2^S, LOS_2, and $\triangleleft_2$ are assigned to site 2, and $\phi_1^{LOS_2}$ is assigned to site 1. With this allocation of tasks, the only communications that require the wireless channel are $LOS_2 \to \phi_1^{LOS_2}$ and $\phi_1^{LOS_2} \to \triangleleft_2$.

The generalizability of the pull task model allows a robot to join the chain with only a minimal change in communication structure. Only the leader and its immediate follower execute the controller from equation (1), while the rest of the robots execute the controller in equation (2). A combined task model for a chain of n robots is shown in Figure 2. Since the controllers are designed to have the same real-time task characteristics and communication patterns within the chain, and our approach captures such recursive properties in advance, the result is a predictable allocation for the additional tasks. A larger team can be scheduled simply by generalizing the results from the smaller team. For example, if the nth robot joins the head of a chain with $n-1$ members, tasks ϕ_n^S, LOS_n, and $\triangleleft_n$ are known to be allocated to robot n. Also, schedulability can be ensured by checking if the laxity time is larger than the sum of the additional communication cost and computation time introduced by the n member.

With the allocation of tasks shown in Figure 2, we see that the only communication tasks that require the wireless channel are $LOS_i \to \phi_{i-1}^{LOS_i}$ and $\phi_{i-1}^{LOS_i} \to \triangleleft_i$. Thus, during every 200 ms period, both of those communication tasks must be scheduled on the wireless channel for every pair of robots in the team. This channel contention, along with the length of the period and precedence constraints,

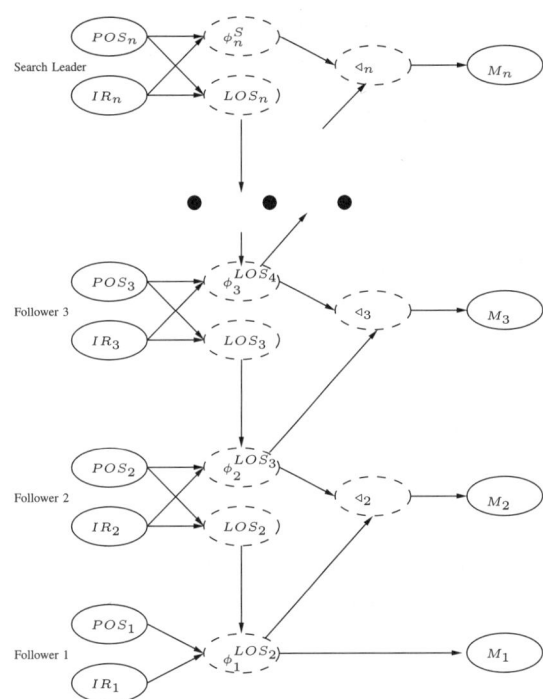

Fig. 2. The task model for a chain of pull controllers for n robots.

imposes a limit on the total number of robots that can be involved in a team.

How can the shared communication channel solution scale for larger teams? This question is answered by the analysis of the largest computation time within the team, which is the sum of execution times of any tasks, including communication tasks, along any path from an input task, or set of input tasks (tasks without incoming edges), to an output task (tasks without outgoing edges). We assume that communication occurring within a site can be ignored.

For a team with n robots as in our pull model, with the addition of a new robot, two communications need to be considered: $LOS_i \to \phi_{i-1}^{LOS_i}$ and $\phi_{i-1}^{LOS_i} \to \triangleleft_i$, $i \geq 2$. Because the robots are autonomous except for communication constraints between different members, the tasks scheduled locally need only to satisfy precedence constraints. For instance, POS_i and IR_i must be scheduled before $\phi_i^{LOS_{i+1}}$ (ϕ_i^S for the leader) and $\triangleleft_i$ can only be scheduled to start after the completion of $\phi_i^{LOS_{i+1}}$ (ϕ_i^S for the leader) and $\phi_{i-1}^{LOS_i} \to \triangleleft_i$.

Now let us consider the schedule of a leader/follower team. Initially, we start with a pair of robots. By using the earliest deadline plus the earliest start time first strategy [8], the leader has the longest execution time, since it must 1) transmit data from LOS_2 to $\phi_1^{LOS_2}$ using the communication channel, and 2) wait for data from $\phi_1^{LOS_2}$. Because of the parallel task execution on two robots, the total execution time of the leader is $T_L + C_{L \to F}$, where T_L is the sum of execution times of tasks allocated to the leader and $C_{L \to F}$ is the total communication cost. T_L and

$C_{L \to F}$ are computed as:

$$T_L = POS_2 + IR_2 + LOS_2 + \phi_2^S + \triangleleft_2 + M_2$$
$$= 20 + 20 + 1 + 25 + 10 + 20 = 96 \text{ ms}$$
$$C_{L \to F} = (LOS_2 \to \phi_1^{LOS_2}) + (\phi_1^{LOS_2} \to \triangleleft_2)$$
$$= 2.979 + 0.01236 \approx 2.99 \text{ ms.}$$

The computation time of $\phi_1^{LOS_2}$ is not included since it is equal to that of ϕ_2^S; otherwise, if it is larger, it needs to be taken into account. Thus the total execution time of the leader is $T_L + C_{L \to F} = 96 + 2.99 = 98.99$ ms.

The laxity within one period is $200 - 98.99 = 101.01$ ms. If a third robot joins the group and keeps the same control pattern, then even though the other tasks are running concurrently on different processors, the communication channel must be shared. Hence, the communication delay for each new member i comes from the accumulation of $LOS_i \to \phi_{i-1}^{LOS_i}$ and $\phi_{i-1}^{LOS_i} \to \triangleleft_i$, a delay of $2.979 + 0.01236 \approx 2.99$ ms. For a chain of size n, the leader will always have the longest execution time of

$$96 + 2.99(n-1) \text{ ms.}$$

Based on a period of 200 ms, the bound on the size of the pull chain is

$$200 = 96 + 2.99(n-1)$$
$$n = \lfloor \frac{200-96}{2.99} + 1 \rfloor = 35.$$

For the maximal chain size, the total execution time required for the 35th robot is $96 + 2.99 \cdot 34 = 197.66$ ms. Based on this analysis, we know that every 200 ms, each robot can successfully complete their work and achieve coordination. If more than 35 robots are involved, then there is no guarantee that messages will be be able to reach their destination before the task executes. This results in tasks executing with old data, and the system performance drops due to this time lag. A time-line showing the schedule for a team of four robots is shown in Figure 3.

If, during the design phase, we find that that we need to coordinate more robots than the upper bound, we can split the team into small groups geographically at run-time, where groups communicate with each other through one specialized element of the team to share information and/or decisions. Even though the communication resource is limited, we can predict in advance the available resources for a small group given the pre-analysis done by our algorithm. At run time a dedicated hierarchical communication model can be built just by looking up the grouping of the robots. Another solution is to redesign the system to reduce resource contention. This could entail reducing the amount of communication needed between members, by making the messages smaller, or eliminating communication by restructuring the flow of messages.

The task model for the pull controller shown in Figure 1 is one possible way of designing the controller. We can design multiple task models for a given controller and evaluate them using the schedulability analysis algorithm to determine how well they scale. Although many task

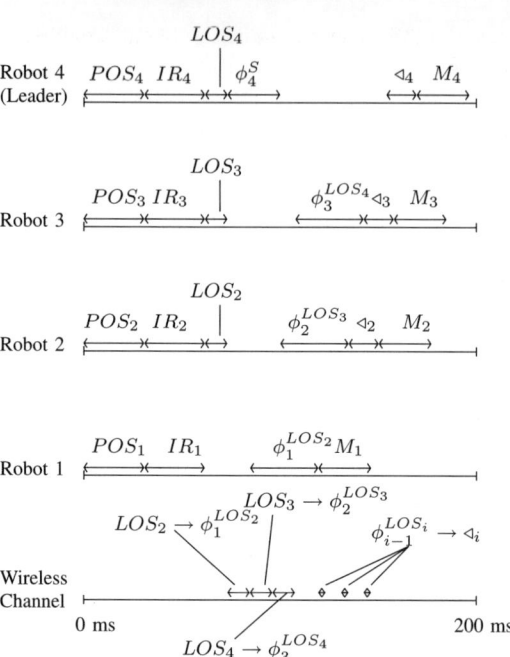

Fig. 3. A feasible schedule for a four-robot pull chain. Note that the schedule is not drawn to scale.

models achieve the same behavior, some are preferred because they allow more robots to coordinate or they may preserve more run-time flexibility when new or additional tasks are introduced.

V. CONCLUSIONS AND FUTURE WORK

We have presented a distributed implementation of a coordinated controller for leader/follower behavior maintaining a line-of-sight constraint. We have analyzed the task model for the controller and derived an upper bound on the size of a feasibly scheduled, coordinated team. In general, a distributed, real-time, multi-robot system has an inherent scale that is a function of the hardware limitations of the robots and the higher-level design of the system. Resource limitations such as a shared communication channel and limited bandwidth create an upper bound on the number of robots that are able to coordinate within the team with a feasible schedule. Schedulability of the overall system should be taken into account so that the value of coordination outweighs the costs. When designing a robot system, a large group can be split into several smaller teams at run-time, whose size is based on the results of the schedulability analysis.

In the future, we plan on extending this work to teams using a mixture of different controllers. The pull controller has a symmetric pair controller (denoted a "push" controller) that allows the follower to specify the LOS region to the leader. In addition, push and pull controllers can be combined along a chain in various combinations to achieve a goal. We want to analyze the combinations of controllers that a robot could have, so that at run-time the robot may lookup the schedule from a precomputed table when it joins a push/pull chain.

The work in this paper examines only one small part of the structure that makes a scalable robot team. A complete scalability analysis of a robot system would examine many factors other than schedulability; reliability and ease of maintenance are hardware-related factors that are particularly important in robotic systems. Other software design issues that affect scalability include interface usability; as the team grows, the operator must be able to effectively monitor or control the team's progress. All of these issues should be addressed during the design process of a robot team.

VI. REFERENCES

[1] J. Albus, R. Lumia, and A. Wavering. NASREM: The NASA/NBS standard reference model for telerobot control system architecture. In *Proceedings of the 20th Internation Symposium on Industrial Robots*, October 1989.

[2] R. Brooks. A robust layered control system for a mobile robot. *IEEE Journal of Robotics and Automation*, 2(1):14–23, March 1986.

[3] Y. U. Cao, A. S. Fukunaga, and A. B. Kahng. Cooperative mobile robotics: Antecedents and directions. *Autonomuos Robots*, 4:1–23, 1997.

[4] S. Carpin and L. E. Parker. Cooperative leader following in a distributed multi-robot system. In *Proceedings of IEEE International Conference on Robotics and Automation*. IEEE, 2002.

[5] J. Coelho and R. Grupen. A control basis for learning multifingered grasps. *Journal of Robotic Systems*, 14(7):545–557, 1997.

[6] C. Connolly and R. Grupen. On the applications of harmonic functions to robotics. *Journal of Robotics Systems*, 10(7):931–946, 1993.

[7] B. P. Gerkey, R. T. Vaughan, K. Stoy, A. Howard, M. J. Mataric, and G. S. Sukhatme. Most valuable player: A robot device server for distributed control. In *Proceedings of IEEE/RSJ International Conference on Intelligent Robots and Systems (IROS)*, pages 1226–1231. IEEE/RSJ, 2001.

[8] H. Li, J. Sweeney, K. Ramamritham, R. Grupen, and P. Shenoy. Real-time support for mobile robotics. Technical Report 03–08, University of Massachusetts, Amherst, 2003.

[9] J. C. Palencia and M. G. Harbour. Exploiting preceding relations in the schedulability analysis of distributed real-time systems. In *Proceedings of the 20th IEEE Real-Time Systems Symposium*, December 1999.

[10] S. Schneider, V. Chen, J. Steele, and G. Pardo-Castellote. The ControlShell component-based real-time programming system, and its application to the Marsokhod Martian Rover. In *Proceedings of the ACM SIGPLAN 1995 workshop on Languages, compilers, & tools for real-time systems*, pages 146–155. ACM, 1995.

[11] J. A. Stankovic and K. Ramamritham. *Advances in Real-time systems*. IEEE Computer Society, 1993.

[12] D. B. Stewart, D. E. Schmitz, and P. K. Khosla. Implementing real-time robotics systems using CHIMERA II. In *Proceedings of IEEE International Conference on Systems Engineering*. IEEE, 1990.

[13] J. Sweeney, T. Brunette, Y. Yang, and R. A. Grupen. Coordinated teams of reactive mobile platforms. In *Proceedings of IEEE International Conference on Robotics and Automation*. IEEE, 2002.

[14] E. Yoshida, T. Arai, J. Ota, and T. Miki. Effect of grouping in local communication system of multiple mobile robots. In *Proceedings of IEEE/RSJ International Conference on Intelligent Robots and Systems (IROS)*, pages 808–815. IEEE/RSJ, 1994.

Real-time Path Planning with Deadlock Avoidance of Multiple Cleaning Robots

Chaomin Luo & Simon X. Yang
Advanced Robotics and Intelligent Systems (ARIS) Lab
School of Engineering, University of Guelph
Guelph, Ontario N1G 2W1, Canada

Deborah A. Stacey
Dept. of Computing and Information Science
University of Guelph
Guelph, Ontario N1G 2W1, Canada

Abstract - In this paper, a cooperative sweeping strategy with deadlock avoidance of complete coverage path planning for multiple cleaning robots in a changing and unstructured environment is proposed, using biologically inspired neural networks. Cleaning tasks require a special kind of trajectory being able to cover every unoccupied area in specified cleaning environments, which is an essential issue for cleaning robots and many other robotic applications. Multiple robots can improve the work capacity, share the cleaning tasks, and reduce the time to complete sweeping tasks. In the proposed model, the dynamics of each neuron in the topologically organized neural network is characterized by a shunting neural equation. Each cleaning robot treats the other robots as moving obstacles. The robot path is autonomously generated from the dynamic activity landscape of the neural network, the previous robot location and the other robot locations. The proposed model algorithm is computationally efficient. The feasibility is validated by simulation studies on three cases of two cooperating cleaning robots. The multiple cleaning robots sweeping will not be trapped in deadlock situations.

I. INTRODUCTION

Nowadays cooperation of multiple robots becomes very important, since effectively cooperative sweeping of multiple robots can improve the work capacity of every single cleaning robot, share the cleaning tasks, and reduce the time to complete sweeping tasks. In some cleaning applications, the space needs to be cleaned in a limited short period. The path planning for cleaning robots is a special type of trajectory generation in 2-dimensional (2D) environment, which requires the robot path to pass through the whole areas in the workspace. Thus, it is commonly called complete coverage path planning (CCPP) or region filling. In addition to the cleaning robots, many other robotic applications also require complete coverage path planning such as vacuum robots, painter robots, land mine

This work was supported by Natural Sciences and Engineering Research Council (NSERC) and Materials and Manufacturing Ontario (MMO) of Canada.
Corresponding author: Simon X. Yang, syang@uoguelph.ca.

detectors, lawn mowers, and windows cleaners.

There are many research regarding multiple robots in coverage path planning (*e.g.*, [1], [2], [3], [4], [5], [6], [7], [8], [9]). Arai *et al.* [1] and Kurabayashi *et al.* [3] produced floor cleaning path planning algorithms for cooperative sweeping with movable obstacles. Those approaches are able to calculate appropriate distribution of path so that complete coverage path can be planed. Those approaches require the path, number, size and location of every movable obstacle. The time-varying environment information is difficult to be incorporated dynamically. It is hard to plan paths in unstructured environments with irregular obstacles. Kurabayashi *et al.* [4] proposed an off-line planning algorithm for multiple cleaning robots using a Voronoi diagram-like and boustrophedon approach, where a cost function is defined to obtain a near-optimal solution of the collective coverage task. The algorithm needs to consider in advance the knowledge of the areas with known obstacles. Thus it cannot deal with on-line cooperative sweeping. Butler *et al.* [2] suggested a distributed cooperative coverage algorithm DCR (distributed coverage of rectilinear environment). DCR performs independently on each robot in a team where the individual robot does not know the initial locations of their peers and applies to systems of robots operating in a rectilinear environment. The algorithm employs only intrinsic contact sensing to determine the boundaries of the environment. However, there is unavoidable path overlapping. Rekleitis *et al.* [6] employed a graph-like decomposition of space to resolve cooperative robot problem. Each robot is regarded as a beacon of other so as to explore the environment information and carry out the cooperative task. Wagner *et al.* [9] presented an approximate cellular decomposition approach for multi-robot sweeping. They employ dirt grid on the floor for communication among robots. The robots communicate each other by leaving traces. Tao and How [8] proposed a decentralized cooperative sweeping approach based on an on-line goal selection algorithm. They organized the multiple robots in a market-like structure to deal with cooperation issue. They combined cooperation and autonomous behaviors by using the hybrid system architecture. Singh and Fujimura [7] presented a map-making approach for cooperation and communication among the multiple robots by using an occupancy grid and the robots sensors to sweep all pixels of the grid in free space. Ota et al. [5] proposed an algorithm for acquiring and utilizing the motion skills, which is accomplished by employing hierarchical neural network. This is a learning-based approach.

In this paper, a novel neural network approach is proposed for path planning with deadlock avoidance of multiple cleaning robots in an arbitrarily varying environment. The robots can avoid collisions with obstacles and other robots and cooperatively work to improve cleaning productivity. The proposed model can deal with deadlock situations for multi-robot cooperation. The real-time path is generated employing a neural network, needlessly either any prior knowledge of environment, or any pre-defined map information. No learning procedures are required in the model. The advantage of the proposed neural networks is that the neurons are not repeatedly visited. It is more computationally efficient, more flexible and simpler for the proposed approach to implement autonomous CCPP comparing with other approaches. The dynamics of each neuron is characterized by a shunting equation derived from Hodgkin and Huxley's [10] membrane model for a biological neural system. There are only local lateral connections among neurons. Thus the computational complexity depends linearly on the neural network size. The varying environment is represented by the dynamic activity landscape of the neural network.

II. THE PROPOSED MODEL

In this section, the originality of the proposed neural network approach to real-time complete coverage path planning for multiple cleaning robots will be briefly introduced. Then the fundamental concept and model algorithm of the proposed approach will be presented.

A. Biological Inspiration

In 1952 Hodgkin and Huxley [10] proposed a computational model for a patch of membrane in a biological neural system using electrical circuit elements. In this model, the dynamics of voltage across the membrane, V_m, is described using state equation technique as

$$C_m \frac{dV_m}{dt} = -(E_p + V_m)g_p + (E_{Na} - V_m)g_{Na} \\ - (E_K + V_m)g_K, \quad (1)$$

where C_m is the membrane capacitance, E_K, E_{Na} and E_p are the Nernst potentials (saturation potentials) for potassium ions, sodium ions and the passive leak current in the membrane, respectively. Parameters g_K, g_{Na} and g_p represent the conductances of potassium, sodium and passive channels, respectively. This model provided the foundation of the shunting model and led to a lot of model variations and applications [11].

By setting $C_m = 1$ and substituting $x_i = E_p + V_m$, $A = g_p$, $B = E_{Na} + E_p$, $D = E_k - E_p$, $S_i^e = g_{Na}$ and $S_i^i = g_K$ in (1), a shunting equation is obtained

$$\frac{dx_i}{dt} = -Ax_i + (B - x_i)S_i^e(t) - (D + x_i)S_i^i(t), \quad (2)$$

where x_i is the neural activity (membrane potential) of the ith neuron. Parameters A, B and D are nonnegative constants representing the passive decay rate, the upper and lower bounds of the neural activity, respectively. Variables S_i^e and S_i^i are the excitatory and inhibitory inputs to the neuron. This shunting model was first proposed by Grossberg to understand the real-time adaptive behavior of individuals to complex and dynamic environmental contingencies, and has a lot of applications in visual perception, sensory motor control, and many other areas [11].

B. Model Algorithm

The fundamental concept of the proposed model is to develop a neural network architecture, whose dynamic neural activity landscape represents the dynamically varying environment. By properly defining the external inputs from the varying environment and internal neural connections, the unclean areas and obstacles are guaranteed to stay at the peak and the valley of the activity landscape of the neural network, respectively. The unclean areas globally attract the robot in the whole state space through neural activity propagation, while the obstacles have only local effect in a small region to avoid collisions. The real-time collision-free robot motion is planned based on the dynamic activity landscape of the neural network, the previous robot location and the other robot locations, to guarantee all areas to be cleaned and the robot to travel a smooth, continuous path with less turning and obstacles avoidance.

The proposed topologically organized model is expressed in a 2D Cartesian workspace $\mathcal{W}$ of the cleaning robots. The location of the ith neuron in the state space $\mathcal{S}$ of the neural network, denoted by a vector $q_i \in R^2$, *uniquely* represents a position in $\mathcal{W}$. In the proposed model, the excitatory input results from the unclean areas and the lateral neural connections, while the inhibitory input results from the obstacles only. Each neuron has local lateral connections to its neighboring neurons that constitute a subset $\mathcal{R}_i$ in $\mathcal{S}$. The subset $\mathcal{R}_i$ is called the receptive field of the ith neuron in neurophysiology. The neuron responds only to the stimulus within its receptive field. Thus, the dynamics of the ith neuron in the neuron network is characterized by a shunting equation as

$$\frac{dx_i}{dt} = -Ax_i + (B - x_i)\left([I_i]^+ + \sum_{j=1}^{k} w_{ij}[x_j]^+\right) \\ -(D + x_i)[I_i]^-, \quad (3)$$

where k is the number of neural connections of the ith neuron to its neighboring neurons within the receptive field $\mathcal{R}_i$. The external input I_i to the ith neuron is defined as $I_i = E$, if it is an unclean area; $I_i = -E$, if it is an obstacle area; $I_i = 0$, if it is a cleaned area, where $E \gg B$ is a very large positive constant. The terms $[I_i]^+ + \sum_{j=1}^{n} w_{ij}[x_j]^+$ and $[I_i]^-$ are the excitatory and inhibitory inputs, S_i^e and S_i^i in (2), respectively. Function $[a]^+$ is a linear-above-threshold function defined as, $[a]^+ = \max\{a, 0\}$, and the nonlinear function $[a]^-$ is defined as $[a]^- = \max\{-a, 0\}$. The connection weight w_{ij} from the ith neuron to the jth neuron is given by $w_{ij} = f(|q_i - q_j|)$, where $|q_i - q_j|$ represents the Euclidean distance between vectors q_i and q_j in the state space, and $f(a)$ is a monotonically decreasing function, such as a function defined as $f(a) = \mu/a$, if $0 \leq a < r_0$; $f(a) = 0$, if $a \geq r_0$, where μ and r_0 are positive constants. Therefore each neuron has only local lateral connections in a small region $[0, r_0]$. It is obvious that the weight w_{ij} is symmetric, i.e., $w_{ij} = w_{ji}$. A schematic diagram of the neural network in 2D is shown in Fig. 1, where r_0 is chosen as $r_0 = 2$. The receptive field of the ith neuron is represented by a circle with a radius of r_0. The 2D Cartesian workspace in the proposed approach is discretized into squares. The diagonal length of each discrete area is equal to the robot sweeping radius that is the size of robot effector or footprint. A topologically organized discrete map is used to represent the workspace. Each location uses a number to represent its environmental information. The neurons are placed uniformly on the space to represent cleaned locations, unclean locations and obstacles.

The proposed network characterized by Eqn. (3) guarantees that the positive neural activity can propagate to all the state space, but the negative activity only stays locally. Therefore, the unclean areas globally attract the robot, while the obstacles only locally

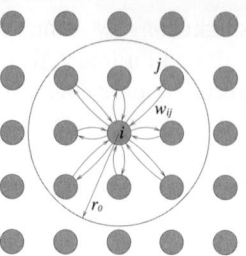

Fig. 1. Schematic diagram of the neural network for robot path planning when the state space is 2D. The ith neuron has only eight lateral connections to its neighboring neurons that are within its receptive field.

push the robot away to avoid collisions. The locations of the unclean areas and obstacles may vary with time, e.g., there are moving obstacles; the cleaned areas become unclean again. The activity landscape of the neural network dynamically changes due to the varying external inputs from the unclean areas and obstacles and the internal activity propagation among neurons. For energy and time efficiency, the robot should travel a shortest path (with least re-visit clean areas) and make least turning of moving directions. The robot path is generated from the dynamic activity landscape *and* the previous robot location to avoid least navigation direction changes. For a given current robot location in $\mathcal{S}$ (i.e., a location in $\mathcal{W}$), denoted by p_c, the next robot location p_n (also called "command location") is obtained by

$$p_n \Leftarrow x_{p_n} = \max\{x_j + cy_j, j = 1, 2, \cdots, k\}, \quad (4)$$

where c is a positive constant, k is the number of *neighboring neurons* of the p_cth neuron, i.e., all the possible next locations of the current location p_c. Variable x_j is the neural activity of the jth neuron, y_j is a monotonically increasing function of the difference between the current to next robot moving directions, which can be defined as a function of the previous location p_p, the current location p_c, and the possible next location p_j, e.g., a function defined as:

$$y_j = 1 - \frac{\Delta \theta_j}{\pi}, \quad (5)$$

where $\Delta \theta_j \in [0, \pi]$ is the turning angle between the current moving direction and next moving direction, e.g., if the robot goes straight, $\Delta \theta_j = 0$; if goes backward, $\Delta \theta_j = \pi$. Thus $\Delta \theta_j$ can be given as: $\Delta \theta_j = |\theta_j - \theta_c| = |\text{atan2}(y_{p_j} - y_{p_c}, x_{p_j} - x_{p_c}) - \text{atan2}(y_{p_c} - y_{p_p}, x_{p_c} - x_{p_p})|$. After the current location reaches its next location, the next location becomes a new current location (if the found next location is the same as the current location, the robot stays there without any movement). The current robot location *adaptively* changes according to the varying environment.

In a multi-robot system, if there exist two robots that work together to sweep in a workspace, the dynamics of the neurons in the neuron network are characterized by two shunting equations. Therefore, the dynamics of the ith neuron in the neuron network with regard to *one* robot is decided by a shunting equation as

$$\frac{dx_i}{dt} = -Ax_i + (B - x_i)\left([I_i]^+ + \sum_{j=1}^{k} w_{ij}[x_j]^+\right) - (D + x_i)[I_i]^-. \quad (6)$$

The external input I_i to the ith neuron is defined as $I_i = E$, if it is an unclean location; $I_i = 0$, if it is a cleaned area; $I_i = -E$, if it is an obstacle or another robot location. Similarly, the dynamics of the ith neuron in the neuron network with regard to *every other* robot is represented by a shunting equation as Eqn. (6). In a multi-robot system, each robot treats the other robots as moving obstacles so that they can avoid collisions and cooperatively work together.

The proposed neural network system is stable. The neural activity is bounded in a finite interval $[-D, B]$. In addition, the stability and convergence of the proposed model can be rigorously proved using a Lyapunov stability theory by rewriting it into Grossberg's general equation in [11] and proving it satisfy all the three required stability conditions.

It is inevitable that multiple cleaning robots have to deal with a deadlock situations in real world applications. When a cleaning robot arrives in a deadlock situation, i.e., all the neighboring locations are either obstacle or cleaned locations, all the neural activities of its neighboring locations are *not* larger than the activity at the current location, because its neighboring locations receive either negative external input (obstacles) or no external input (cleaned locations), and all the cleaned neighboring locations passed a longer decay time as they were cleaned earlier than the current location [see Eqn. (3)]. In the proposed model, the neural activity at the deadlock location will quickly decay to zero due to the passive decay term $-Ax_i$ in Eqn. (3). Meanwhile, due to the lateral excitatory connections among neurons, the positive neural activity from the unclean locations in the workspace will propagate toward the current robot location through neural activity propagation. Therefore, the robot is able to find a smooth path from the current deadlock location directly to an unclean location. The robot continues its cleaning task until all the areas in the workspace become cleaned. Thus the proposed model is capable of achieving complete coverage path planning with deadlock avoidance. The multiple cleaning robots will not be trapped in deadlock situations. The neural network model in this paper is in the sense that it is a biologically inspired neural dynamic model, other than a learning based neural network model.

III. SIMULATION STUDIES

The proposed neural network approach is capable of planning complete coverage path for multiple cleaning robots, autonomously without any human operation. In this section, the proposed approach is first applied to a two-deadlock situation to show the robot can escape from one or two deadlock situations. Then, cooperative sweeping in an indoor environment with deadlock situation is studied. Finally, it is applied to cooperative sweeping by four cleaning robots in an indoor environment. The multiple cleaning robots will not be trapped in deadlock situations.

A. CCPP in a Two-deadlock Situation

The proposed neural network model is first applied to a more complicated case in which there are two C-shaped deadlocks. The robot starts to sweep from original position at S (1,1) in Fig. 2A. After the mobile robot escapes from the first deadlock, it should go into the second deadlock to do cleaning work. Once the mobile robot reaches the end of the second deadlock, it can move to a designated location. The neural network has 30×30 discretely and topologically organized neurons, where all the neural activities are initialized to zero. In this case, the model parameters are set as follows: $A = 18$, $B = 1$, $D = 1$ for the shunting equation; $\mu = 0.7$, and $r_0 = 2$ for the lateral connections; and $E = 50$ for the external inputs. The robot starts to sweep from original position at S (1,1) and reaches $F1$, the end of the first deadlock (see Fig. 2A). Although all of the areas surround location $F1$ have been marked as cleaned, the robot is able to escape from the C-shaped deadlock. In the present model, the robot is able to autonomously identify the deadlock by using dynamic neural neighborhood analysis method and the deadlock avoidance algorithm, move out from the first deadlock, and move to the second deadlock with obstacles avoidance since the second deadlock has not been visited yet. After reaching $F2$, the end of the second deadlock, the robot searches to final designated location, the exit at location F (3,29) (see Fig. 2B) denoted by a square. The simulation result is shown in Fig. 2B. Initially, the external inputs I_i to all of the ith neurons are defined as E since the whole workspace are unclean, except, $I_i = -E$ in obstacles location. Every location is marked as $I_i = 0$ once the robot has cleaned the point. When the robot reaches the end of the deadlock, i.e., the robot finishes the cleaning work to the left of point F1 of the deadlock, the μ is set as 0.70, which determines connection weight W. The robot follows a continuous, smooth path to achieve the target position at F (3,29) by classic path planning shown using empty circles. The generated point-to-point path is illustrated in empty circles in Fig. 2B. The activity landscape of the neural network right after the robot reaches $F1$ in the first deadlock is shown in Fig. 3. F represents the final designated point that globally attracts the cleaning robot.

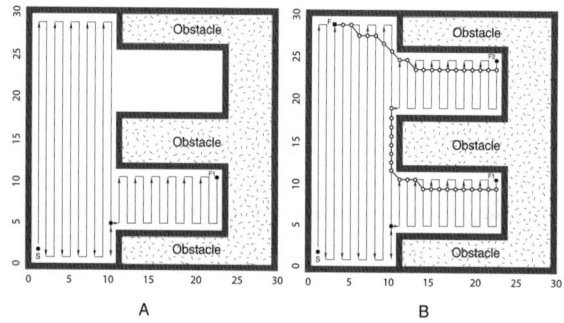

Fig. 2. A: The path when the robot reaches F1; B: The whole planned path.

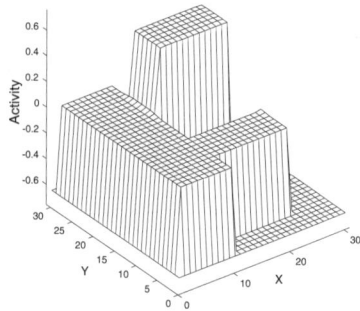

Fig. 3. The neural activity landscape of the neural network for the case in Fig. 2.

B. Cooperative Sweeping in an Indoor Environment with Deadlock Situations

The proposed model is then applied to a cooperative sweeping situation with obstacle avoidance in an indoor environment. In the simulation for multi-robot system, the neural network has 30×30 topologically organized neurons with zero initial neural activities. The model parameters are set as: $A = 50$, $B = 1$ and $D = 1$ for the shunting equation; $\mu = 0.7$ and $r_0 = 2$ for the lateral connections; and $E = 50$ for the external inputs. In Fig. 4A, one robot represented using solid dot starts to sweep from the lower-left corner $S1$ (1,1), the other exhibited using empty circle sweeps from the upper-right corner $S2$ (29, 29). After two robot sweep five columns, they encounter walls and then they start to clean in a narrow area like an aisle. Obviously, Robot 1 and Robot 2 reach the central corridor at the same time. They don't have any collision in the central area shown in Fig. 4A. For each robot in the central area, all the neighboring locations are either obstacle or cleaned locations. These two robot encounter deadlock situation in the central area. From the central area, the two robots are able to search point-to-point paths to move to unclean areas since unclean areas globally attract them. In this simulation shown in Fig. 4B. The robots follow continuous, smooth paths to achieve (5,9) point and (25,21) point, respectively. Then they generate complete coverage path planning in the two unclean areas and sweep those two unclean areas (see Fig. 4B).

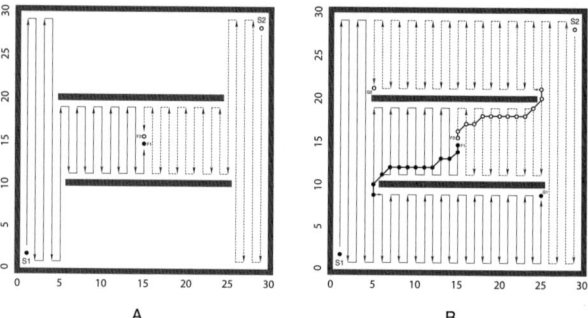

Fig. 4. CCPP of multi-robot in an unstructured environment. A: When the robots meet at the center; B: The generated entire paths.

C. Cooperative Sweeping by Four Cleaning Robots in an Indoor Environment

The proposed model is also applied to four cleaning robots cooperative sweeping in an indoor environment, where there exist four neural networks system and the four cleaning robots share common external input signal from sensory data representing environmental information. Each neural network has 19×19 topologically organized neurons with zero initial neural activities. The model parameters are set as: $A = 50$, $B = 1$ and $D = 1$ for the shunting equation; $\mu = 0.7$ and $r_0 = 2$ for the lateral connections; and $E = 50$ for the external inputs. In Fig. 5A, use of different lines is to distinguish the robots' generated paths. Robot 1 whose paths are represented by solid lines starts to move from the lower-left corner $S1$ at (1,1). Robot 2 whose paths are represented by dashed lines sweeps from the upper-left corner $S2$ at (1,18). Robot 3 whose paths are represented by dash-dotted lines starts to move from the upper-right corner $S3$ at (18,18). Robot 4 whose paths are represented by dash-dot-dot lines sweeps from the lower-right corner $S4$ at (18,1). In the simulation, the planned robot paths are shown in Fig. 5A, where the four robots search snake-trail complete coverage paths and meet in the central area. Because for each neural network, the positive neural activity can propagate to the whole state space of the neural network, each robot can plan a complete coverage path. If one area is cleaned by one robot, it will be marked cleaned by external input signal ($I_i = 0$), another robot will know that the area has cleaned. Thus when the four robots meet, they don't have any collision. It shows that these four cleaning robots are able to autonomously sweep the whole workspace. Not only can they sweep along zigzag coverage paths, but also are able to avoid collisions with each other. After they meet in the central area, where there is a deadlock situation, i.e., Robot 1 is at $F1$ (9,9); Robot 2 is at $F2$ (9,10); Robot 3 is at $F3$ (10,10); Robot 4 is at $F4$ (10,9), the robots are able to search point-to-point paths to move to any pre-defined targets. In this simulation shown in Fig. 5B, Robot 1 moves back to its original point $G1$ (1,1). Robot 2 goes back to its original point $G2$ (1,18). Robot 3 travels back to its original point $G3$ (18,18). Robot 4 moves back to its original point $G4$ (18,1). They can be pre-defined to move to any points. The targets can globally attract the robots in the whole workspace through neural activity propagation. This case has potential applications in sport fields such as basketball or volleyball contests. The four cleaning robots can be assigned to clean fields together and then go back their original points during sports contest interval.

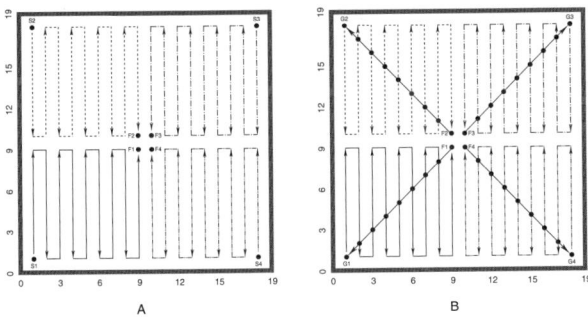

Fig. 5. *CCPP of four cleaning robots in an indoor environment. A: The robot paths when these four robots meet at the center; B: The entire robot paths after each robot returns its home position.*

IV. CONCLUSION

In this paper, a novel biologically inspired neural network approach to autonomous cooperative coverage path planning with deadlock avoidance of multiple cleaning robots is proposed. The developed approach is capable of autonomously planning collision-free path for multi-robot in unstructured, nonstationary environments. The effectiveness of the proposed paradigm is already been discussed and illustrated through simulation studies. The proposed model can deal with deadlock situations for multi-robot cooperation.

The model algorithm is computationally efficient. The robot path is is generated *without* explicitly optimizing any *global* cost functions, *without* any *prior* knowledge of the dynamic environment, and *without* any learning procedures.

References

[1] T. Arai, D. Kurabayashi, J. Ota, and S. Ichikawa, "Motion planning for cooperative sweeping with relocating obstacles," in *Proc. of IEEE Intl. Conf. on Systems, Man and Cybernetics*, Beijng, China, 1996, pp. 1513–1518.

[2] Z. J. Butler, A. A. Rizzi, and R. L. Hollis, "Cooperative coverage of rectilinear environments," in *Proc. of IEEE Intl. Conf. on Robotics and Automation*, San Francisco, USA, 2000, pp. 2722–2727.

[3] D. Kurabayashi, J. Ota, T. Arai, S. Ichikawa, S. Koga, H. Asama, and I. Endo, "Cooperative sweeping by multiple mobile robots with relocating portable obstacles," in *Proc. of IEEE/RSJ Intl. Conf. on Intelligent Robots and Systems*, Osaka, Japan, 1996, pp. 1472–1477.

[4] D. Kurabayashi, J. Ota, T. Arai, and E. Yoshida, "Cooperative sweeping by multiple mobile robots," in *Proc. of IEEE Intl. Conf. ON Robotics and Automation*, Minneapolis, USA, 1996, pp. 1744–1749.

[5] J. Ota, T. Arai, E. Yoshida, D. Kurabayashi, and J. Sasaki, "Motion skills in multiple mobile robot system," *Robotics and Autonomous Systems*, vol. 19, no. 1, pp. 57–65, 1996.

[6] I. M. Rekleitis, D. Dudek, and E. E. Milios, "Multi-robot exploration of an unknown environment, efficiently reducing the odometry error," in *Proc. of the 15th IEEE Intl. Joint Conf. on Artificial Intelligence*, Nagoya, Japan, 2000, pp. 1340–1345.

[7] K. Singh and K. Fujimura, "A navigation strategy for cooperative multiple mobile robots," in *Proc. of IEEE Intl. Conf. on Intelligent Robots and Systems*, Yokohama, Japan, 1993, pp. 283–288.

[8] W. M. Tao and K. Y. How, "A decentralized approach for cooperative sweeping by multiple mobile robots," in *Proc. of IEEE/RSJ Intl. Conf. on Intelligent Robots and Systems*, Victoria, Canada, 1998, pp. 380–385.

[9] I. A. Wagner and A. M. Bruckstein, "Distributed covering by ant-robots using evaporating traces," *IEEE Trans. on Robotics and Automation*, vol. 15, no. 5, pp. 918–933, 1999.

[10] A. L. Hodgkin and A. F. Huxley, "A quantitative description of membrane current and its application to conduction and excitation in nerve," *J. of Physiology (London)*, vol. 117, pp. 500–544, 1952.

[11] S. Grossberg, "Nonlinear neural networks: principles, mechanisms, and architecture," *Neural Networks*, vol. 1, pp. 17–61, 1988.

Hybrid Systems Modeling of Cooperative Robots

Luiz Chaimowicz[1,2], Mario F. M. Campos[1], and Vijay Kumar[2]

[1]DCC – Universidade Federal de Minas Gerais, Belo Horizonte, MG, Brasil, 31270-010
[2]GRASP Laboratory – University of Pennsylvania, Philadelphia, PA, USA, 19104
{chaimo, kumar}@grasp.cis.upenn.edu, mario@dcc.ufmg.br

Abstract

This paper proposes a methodology that uses hybrid systems to model multiple robots in the execution of cooperative tasks. Basically, each robot is represented by a hybrid automaton and the cooperative task execution is modeled by the composition of several automata. We describe in details our approach to perform the composition of these automata and demonstrate the effectiveness of the proposed methodology modeling a cooperative manipulation task.

1 Introduction

Cooperative robotics has been an active research field in recent years. Fundamentally, it consists in using a group of robots working cooperatively to execute various types of tasks, trying to increase the robustness and efficiency of task execution. An important aspect of cooperative robotics that has not received much attention so far is how to model the execution of tasks by multiple robots. In general, cooperative robotics systems normally require the representation of both continuous and discrete dynamics, together with synchronization, communication, etc. The modeling technique should address all these requirements and also provide ways of obtaining some formal results about the cooperative system.

In previous works [6, 7], we gave the first steps into modeling cooperative robotics using hybrid systems. In this paper, we extend and formalize this methodology, showing how the composition of several hybrid automaton, one for each robot, can be used to model the execution of cooperative tasks.

We have chosen hybrid systems in order to represent cooperative robotics for two main reasons. Firstly, hybrid systems provide a simple, structured, and formal way of reasoning about, representing, and implementing multi-robot cooperation. Also, as will be shown in this paper, the mapping from cooperative robotics to a hybrid automaton is relatively simple. The second main reason is that hybrid systems have a powerful theory in its background, which may allow the development of formal proofs about some aspects of the cooperative task execution. This aspects include stability, reachability of undesirable or goal states, etc. In this paper we focus on the cooperative task modeling while formal verification and other types of analysis are left as future work.

In spite of the large number of works dealing with coordination mechanisms and approaches to execute cooperative tasks, very few works have considered aspects of modeling and formal verification of cooperative robotics. A framework that uses a more formal approach to implement cooperative robotics has been developed in the GRASP Laboratory [2]. It uses the Charon language [4] for the description of multiple agents and their behavioral structure. Charon is a high level modeling language for the specification and simulation of hybrid systems, and can be used for modeling cooperative robotic systems. An example of this framework in the control of multi-robot formations can be found in [9]. Another work that tries to formalize the execution of cooperative robotics is described in [11], in which finite state machines are used to encapsulate behaviors and discrete transitions are controlled by binary sensing predicates. Finite state machines do not model continuous aspects of the system but can help in modeling the discrete part of the cooperation. Petri Nets, that are commonly used for modeling multi-process systems, have also been used to model robotics tasks [12, 13].

An interesting discussion about the specification and formal verification of robotic tasks can be found in [8]. One of the conclusions of that paper is that "all the area of hybrid systems (modeling, programming, formal verification) is a key research domain for the future, the results of which will find particularly relevant applications in robotics". Our hybrid systems modeling of cooperative robotics presented in this paper is a contribution in this direction.

2 Hybrid Systems

A hybrid system is a dynamical system composed by discrete and continuous states. The execution of a hybrid system can be defined by a sequence of steps: in each step, the system state evolves continuously according to a dynamical law until a discrete transition occurs. Discrete transitions are instantaneous state changes that separate continuous state evolutions [1].

Normally a hybrid system can be represented using a Hybrid Automaton. A hybrid automaton is a finite automaton augmented with a finite number of variables that can change continuously, as specified by differential equations, or discretely, according to specific assignments. Discrete states (control locations) contain evolution laws and the values of the variables change according to these laws while the system is in a specific discrete state. The discrete transitions of the system are labeled with a set of guards and assignments. A transition is enabled when the logical condition of its guard is satisfied. When a transition occurs, the assignment associated with that transition is executed, possibly modifying the values of the variables. Additionally, each control location has an invariant condition that must hold whenever the system is executing in that location. More formally, a hybrid automaton H can be defined as:

$$H = \langle Q, V, E, f, Inv, Init \rangle,$$

where:

- $Q = \{q_1, q_2, \ldots, q_n\}$ is the set of discrete states of the system, also called *control modes* or *control locations*.

- V is a finite set containing the variables of the system and can be composed by discrete (V_d) and continuous (V_c) variables: $V = V_d \cup V_c$. Each variable $x \in V$ has a value that is given by a function $\nu(x)$. This is called *valuation* (ν) of the variables. Thus, at any moment, the state of the system is given by a pair (q, ν), composed by the discrete state and the valuation of the variables.

- Discrete transitions between pairs of control modes are specified by control switches represented by E (also called *edges*). Each transition has an associated predicate g that is called a guard or jump condition. A discrete transition can only be taken (transition is enabled) if its predicate g is satisfied. Each transition may also have a reset statement r that changes the value of some variable or perform some action during a discrete transition. Finally, control switches can be tagged with a label $l \in \Sigma$. This label can be also called *event* and is important for synchronization issues in the composition of multiple automata. Thus, each discrete transition $e \in E$ can be represented by a tuple $\langle q_i, q_j, g, r, l \rangle$ where $q_i \in Q$ and $q_j \in Q$ are the source and destination states, g is the guard, r is the reset statement and $l \in \Sigma$ is the label.

- The dynamics of the continuous variables are determined by the flows f, generally described as differential equations inside each control mode.

- Invariants (Inv) are predicates related to the control modes. The system can stay in a certain control mode q while its invariant is satisfied and must leave the mode when it becomes invalid. If the invariant becomes false and there are no transitions enabled for the current control mode, the system is considered to be in a deadlock.

- $Init$ is the set of initial states of the system. Each initial state is composed by a pair $(q, \nu(X))$, where $q \in Q$, and $X \subseteq V$.

Other representations of hybrid automata may exist depending on the author and the context in which the definition is applied (for example [1] and [10]).

3 Modeling of Cooperative Robots

The behavior of a robot can be modeled using a hybrid automaton. Basically, it can be in one of several control modes, being controlled by different continuous equations in each mode. Information within each mode (such as the continuous states, sensor readings, etc.) can be described by continuous and discrete variables, and updated according to the equations inside each control mode and reset statements of each discrete transition. Coordination is done using the role assignment mechanism [6], in which the robots dynamically change their behavior during task execution. This can be modeled through the discrete transitions and is controlled by the guards and invariants.

Explicit communication can be represented considering that there are communication channels between agents and using a message passing mechanism. The basic actions are *send(channel, value)* to send a message containing a certain value in a channel and *receive(channel, variable)* to receive a message and put its value in a local variable. Messages are sent and received during discrete transitions. It is possible to use a *self-transition*, *i.e.*, a transition that does not change the discrete state, to receive and send messages. In this way, a robot can stay in one discrete mode and continuously send or receive messages.

The execution of a cooperative task by multiple robots can be modeled using a parallel composition of several automata, one for each robot. We built our approach to formally describe the composition of hybrid automata based on the composition of transition systems[1] [3], extending this concept for the composition of hybrid automata.

Let $H_1 = \langle Q_1, V_1, E_1, f_1, Inv_1, Init_1 \rangle$ and $H_2 = \langle Q_2, V_2, E_2, f_2, Inv_2, Init_2 \rangle$ be two hybrid automata. Some considerations have to be made before defining the composition. Firstly, we assume that the sets of

---
[1] A transition system can be considered the discrete part of a hybrid automata, *i.e*, the graph (Q, E) of the automata without considering variables, invariants, etc.

variables V_1 and V_2 are disjoint. This is a reasonable assumption since there are no shared variables. The two automata can share information, but this information will be stored in different variables and gathered in different ways. Another consideration is that two transitions from E_1 and E_2 labeled with the same label l are taken at the same time in both automata. This is a way of synchronizing the execution of both automata and can be done, for example, using communication. With these considerations, the parallel composition $H_1 \| H_2$ of H_1 and H_2 is the hybrid automaton:

$$H_1 \| H_2 = \langle Q_1 \times Q_2, V_1 \cup V_2, E', f_1 \cup f_2, Inv', Init' \rangle.$$

The control modes of the compound automaton are pairs (q_1, q_2), where $q_1 \in Q_1$ and $q_2 \in Q_2$, and the set of variables of $H_1 \| H_2$ are the union of the sets V_1 and V_2 from both automata. The discrete transitions (E') are defined based on the transitions of both automata:

1. for $l \in \Sigma_1 \setminus \Sigma_2$, for each transition $\langle q_1, q_1', g_1, r_1, l \rangle$ in E_1 and every $q \in Q_2$, E' contains the transitions $\langle (q_1, q), (q_1', q), g_1, r_1, l \rangle \forall q \in Q_2$;

2. for $l \in \Sigma_2 \setminus \Sigma_1$, for each transition $\langle q_2, q_2', g_2, r_2, l \rangle$ in E_2 and every $q \in Q_1$, E' contains the transition $\langle (q, q_2), (q, q_2'), g_2, r_2, l \rangle \forall q \in Q_1$;

3. if there is a transition labeled l on both automata ($l \in \Sigma_1 \cap \Sigma_2$), for every transition $\langle q_1, q_1', g_1, r_1, l \rangle$ in E_1 and $\langle q_2, q_2', g_2, r_2, l \rangle$ in E_2, E' contains the transition $\langle (q_1, q_2), (q_1', q_2'), g_1 \wedge g_2, r_1 \cup r_2, l \rangle$.

The flows f' of $H_1 \| H_2$ are simply the union of the flows from H_1 and H_2: $f' = f_1 \cup f_2$. Since the variable sets V_1 and V_2 are disjoint, the continuous update in each compound control mode (q_1, q_2) is simply the union of the continuous updates from both q_1 and q_2. In the same way, the invariant of each compound mode (q_1, q_2) is the conjunction (logical and) of the invariants of both q_1 and q_2: $Inv'(s_1, s_2) = Inv_1(s_1) \wedge Inv_2(s_2)$. This means that the execution of the automaton $H_1 \| H_2$ can stay in control mode (q_1, q_2) while the invariants of both q_1 and q_2 are true. Finally, the initial state set $Init'$ of $H_1 \| H_2$ is a composition of $Init_1$ and $Init_2$.

Theoretically, the composition of hybrid automata is not very complex and can be implemented by a computer program. But it is important to mention that the modeling of very large systems may require the use of very complex automata and the composition of several automata may cause an exponential explosion of discrete states. Basically, if we have n robots with m control modes each, the number of control modes in the compound automaton will be m^n, thus, the number of control modes of the compound automaton increases exponentially with the number of robots. To minimize these problems we may try to reduce the total number of states to be analyzed, trying to divide the modes of compound automaton. We will do this in the example of the next section, considering only the control modes in the compound automaton that are reachable from the initial state.

4 Example: Modeling Manipulation

In order to demonstrate the composition of hybrid automata in the modeling of cooperative tasks we have modeled a cooperative manipulation task in which two robots coordinate themselves to transport an object (box). We consider that the transportation is performed in one dimension and each robot i has an estimate of its position (x_i) and the position of the box (x_{bi}). In fact, they need only to know their distance to the box, which can be obtained using some kind of range sensor. We also consider that the robots can exchange messages using explicit communication. The robots use a very simple kinematic model, in which the only input is the linear velocity ($\dot{x} = u$). Figure 1 shows a diagram of this task.

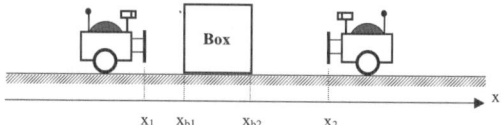

Figure 1: Diagram of two robots transporting a box.

Before formally describing the behavior of the robots using hybrid automata, let us give a brief overview of the execution. The robots can be in one of four discrete states (modes): Approach, Wait, Transport, and Lost. Each robot starts in the Approach mode, where it uses a proportional controller in order to get closer to the box. When the box is close enough, the robot switches to the Wait mode and sends a message to the other robot informing that its approach phase is complete. When both robots have approached the box, they synchronously start the Transport mode, in which they cooperate to move the box. In this mode, one of the robots (robot 1) is the leader and has a plan that sets its velocity. The follower (robot 2) estimates the leader's velocity (v_l) (through sensing or communication, for example) and sets its velocity to the estimated value. If one of the robots loses contact with the box, it sends a message to the other robot and switches to the Lost mode. The other robot will receive this message, send an acknowledgment and both robots will synchronize and switch to the Approach mode, restarting the cycle.

The execution of each robot can be formally modeled using a hybrid automaton. The automaton $H_1 = \langle Q_1, V_1, E_1, f_1, Inv_1, Init_1 \rangle$ for robot 1 is depicted in Figure 2. The variables of H_1 includes the position of the robot and the box, and three boolean conditions that indicate if the other robot has approached, lost the box, or sent an acknowledgment message. Some self-transitions are used to set these conditions according to the messages received from the other robot. One example is the transition $\langle A_1, A_1, \text{MsgAvailable}(C_{21}), \text{Recv}(C_{21}, \text{DockOk}_2), \phi \rangle$, that sets the variable DockOk_2 when a message from robot 2 is available in channel C_{21}. In Figure 2, it is possible to see the guards, actions, and labels of each transition: the guards are shown in normal font, the actions in bold and the labels in bold/italic. Note that some of the transitions may not have all of these components. Flows and invariants are set according to the behavior explained in the previous paragraph, and the execution starts in the Approach mode with the boolean conditions set to false.

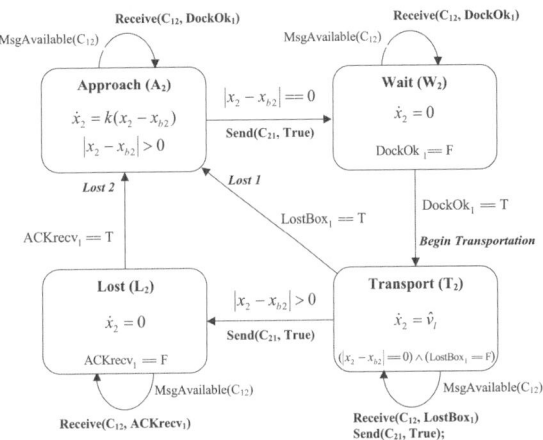

Figure 3: Hybrid automaton for robot 2.

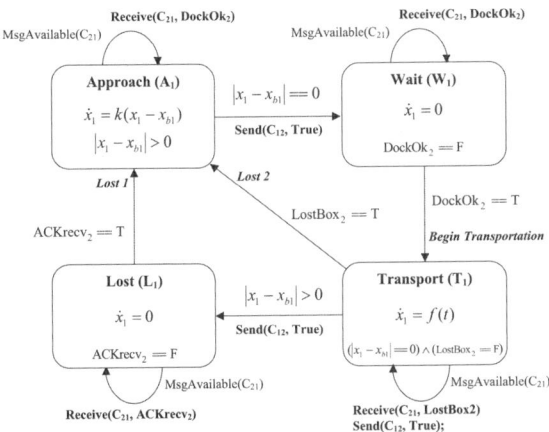

Figure 2: Hybrid automaton for robot 1.

The automaton $H_2 = \langle Q_2, V_2, E_2, f_2, Inv_2, Init_2 \rangle$ for robot 2 is very similar to H_1 and is shown in Figure 3. It is basically a copy of H_1 with the variables and communication channels renamed (from 1 to 2 and 2 to 1). The basic difference is in the dynamic equation of the transport mode. As mentioned, robot 2 is the follower and sets its velocity to be equal to the leader's velocity: $\dot{x}_2 = \hat{v}_l$.

The composition of H_1 and H_2 represents the cooperative execution of the task by the two robots. The hybrid automaton $H_1 \parallel H_2$ is built following the rules explained in the previous section. Figure 4 shows part of this automaton. In fact, $H_1 \parallel H_2$ is composed by 16 control modes, the product of the modes of the two automata ($Q_1 \times Q_2$), but, to simplify the figure, we only show the 8 control modes that are reachable from the initial state of the system.

The composition of control modes, invariants and flows is relatively simple, as explained in the previous section. The major challenge is to define the discrete transitions of the new automaton. According to the composition rules, there are three situations: (i) when there is a transition labeled l in E_1 but not in E_2, (ii) when there is a transition labeled l in E_2 but not in E_1, and (iii) when there is a transition labeled l in both E_1 and E_2. For situation (i), a transition $q_i \to q_j$ from mode q_i to mode q_j in E_1 generates a transition $(q_i, q) \to (q_j, q)$ in the compound automaton for each mode $q \in Q_2$. To exemplify, let us consider the transition $A_1 \to W_1 \in E_1$. In our modeling, we label only the transitions that are part of both automata. Transitions without labels are unique to H_1 or H_2. In the compound automaton, this transition should appear four times connecting the modes: AA $\to$ WA, AW $\to$ WW, AT $\to$ WT, AL $\to$ WL [2]. Since we are showing only the reachable modes, this transition appears twice in the automaton of Figure 4. The same thing happens for situation (ii) when there is a transition in H_2 that is not present in H_1. For example, the transition $T_2 \to L_2 \in E_2$ turns into the transitions AT $\to$ AL, WT $\to$ WL, TT $\to$ TL, LT $\to$ LL in the automaton $H_1 \parallel H_2$.

The other possible situation (iii) is when there is a transition labeled l in both H_1 and H_2. In this case, as explained in the previous section, a single transition is generated in the compound automaton. For example, the transitions $W_1 \to T_1 \in E_1$ and $W_2 \to T_2 \in E_2$ have the same label *Begin Transportation*. Their composition results in the transition $\langle \text{WW}, \text{TT}, \text{DockOk}_1 == \text{T} \wedge \text{DockOk}_2 == \text{T}, \phi, \text{Begin Transportation} \rangle$ in the compound automaton. As mentioned, these labeled transitions are used for synchronization. In the modeling, we consider that transitions that have the same label in both automata are taken at the same time. This is the case, for example, of the *Begin Transportation*

---

[2] To simplify the notation, we use only two letters for the names of the compound modes. The first letter is the mode of Q_1 and the second the mode of Q_2. For example the mode AW in $Q_1 \times Q_2$ is in fact the mode composed by A_1 and W_2.

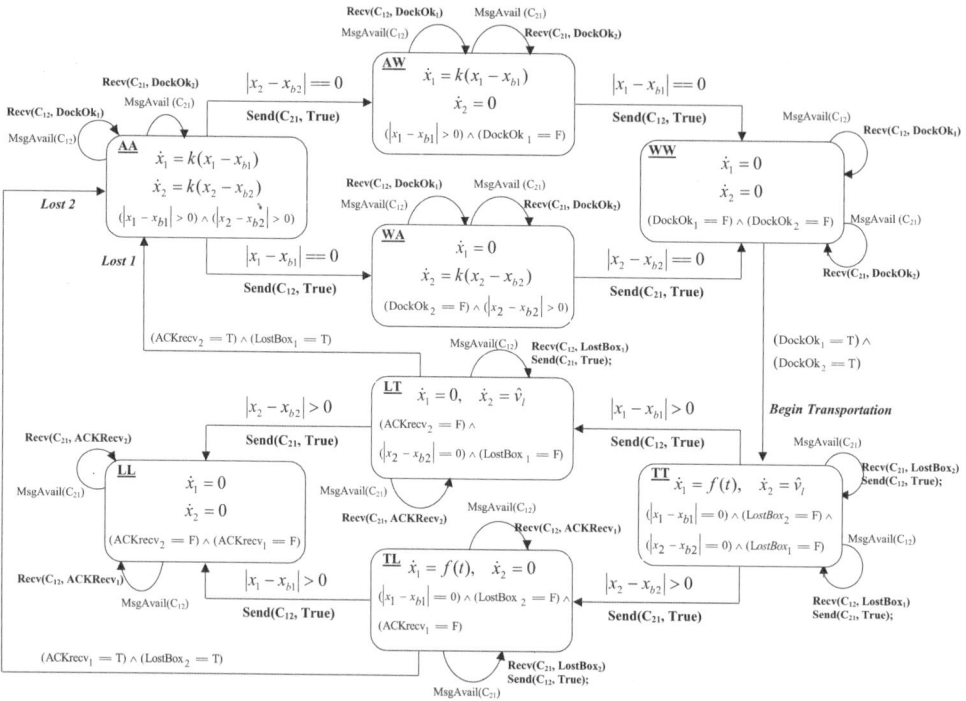

Figure 4: Compound automaton.

transition that synchronize the start of the transportation phase. Another example is the *Lost 1* transition, in which both robots go to the Approach mode when robot 2 receives the information that robot 1 have lost the box and robot 1 receives the acknowledgment.

In real implementations, synchronization is obtained using explicit communication. When a robot receives a certain message or acknowledgment, it sets some of its variables and switches to another discrete state. In the model presented in this example, we consider that a command to send a message is automatically followed by one to receive the message, before any continuous update of the automaton. If this condition is not followed, some undesirable situations such as deadlocks may happen. An example of deadlock is the mode LL of the compound automaton. The mode LL is a sink, since there are no transitions leaving from it. In normal situations, the system does not go to this mode from modes LT or TL (the transitions *lost 1* and *lost 2* to the AA mode are taken instead), but if there is some delay between the sending of a message and the receiving of an acknowledgment and both robots lose contact with the box in this interval this state can be reached. In fact, these deadlock situations can be avoided in the modeling by introducing some extra modes with the objective of waiting for specific messages and synchronizing the automata. A more detailed discussion regarding deadlock situations in this modeling can be found in [5].

5 Simulations

To show the effectiveness of the hybrid systems modeling we performed some simulations using MuRoS [5], a multi-robot simulator that we have developed for the execution of cooperative tasks. In these simulations we used a group of robots for the transportation of an object in an environment containing obstacles. The hybrid automaton for each robot is similar the ones shown in the previous section (Figures 2 and 3). The main difference is in the definition of the flows, because instead of using a leader-follower controller the robots use potential field controllers.

Initially, the robots are in the Approach mode and are attracted by the object. At the same time, they are repelled by each other, being able to distribute themselves along the object and prepare for the transportation. When one robot senses that it is close enough to the object, it goes to the Wait mode, and broadcasts a message communicating that it is ready. When all robots are ready they change to the Transport mode, and there is a controller switch so that each robot becomes attracted by the goal. If for some reason one robot losses contact with the object, it will change to the Lost mode, broadcast a message and stop moving. If all robots lose contact, they regroup and start docking again.

Figure 5 shows some snapshots of the simulator during the manipulation of a round object by ten holonomic robots in an environment with three obstacles. The goal position is marked with an x. Snapshot

(a) shows the robots in the Approach mode, starting to move in the direction of the object (light gray circle). Snapshot (b) shows nine robots in the Wait mode while the last one is still finishing the approach phase. In (c) eight of the robots have lost contact with the object because of a collision with an obstacle. It is important to note that the robots lose contact when the object is outside their sensor range, so two robots (showed in black) are still in contact. As these two robots move towards the goal, they will also lose contact with the object. When this happens, they regroup, grab the object again and resume the transport finishing the task (snapshot (d)).

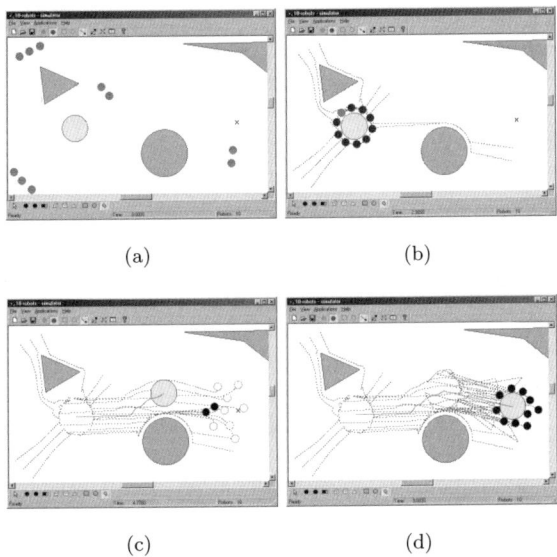

Figure 5: Snapshots of the manipulation task

6 Conclusion

In this paper, we modeled multi-robot cooperation under a hybrid systems framework, using hybrid automata to specify the behavior of each robot and the parallel composition of automata to model the cooperative task as a whole. This has allowed us to better formalize the execution of cooperative tasks by multiple robots providing a framework that can be used for developing formal proofs about the cooperation.

Future work is direct towards the use of the modeling presented in this paper in order to perform formal verification and analysis on cooperative task execution. As mentioned, hybrid systems have a powerful theory in its background and can be used to study stability and reachability of mixed continuous and discrete systems. Thus, we can use this theory in cooperative robotics to detect deadlocks, test reachability of undesired or target states and study stability of the cooperative system. For this, it may be necessary to abstract our representation trying to obtain simpler models (for example a linear automata) from our general hybrid automata, in order to use the tools already available for the analysis of these classes of hybrid systems.

References

[1] R. Alur, C. Coucoubetis, T. A. Henzinger, P.-H. Ho, X. Nicollin, A. Olivero, J. Sifakis, and S. Yovine. The algorithmic analysis of hybrid systems. *Theoretical Computer Science*, 138:3–34, 1995.

[2] R. Alur, A. Das, J. Esposito, R. Fierro, G. Grudic, Y. Hur, V. Kumar, I. Lee, J. Ostrowski, G. Pappas, B. Southall, J. Spletzer, and C. Taylor. A framework and architecture for multirobot coordination. In D. Rus and S. Singh, editors, *Experimental Robotics VII, LNCIS 271*. Springer Verlag, 2001.

[3] R. Alur and D. Dill. A theory of timed automata. *Theoretical Computer Science*, 126:183–235, 1994.

[4] R. Alur, R. Grosu, Y. Hur, V. Kumar, and I. Lee. Modular specification of hybrid systems in charon. In *Proceedings of the 3rd International Workshop on Hybrid Systems: Computation and Control*, 2000.

[5] L. Chaimowicz. *Dynamic Coordination of Cooperative Robots: A Hybrid Systems Approach*. PhD thesis, Univ. Federal de Minas Gerais - Brazil, June 2002. http://www.cis.upenn.edu/~chaimo/thesis.pdf.

[6] L. Chaimowicz, M. Campos, and V. Kumar. Dynamic role assignment for cooperative robots. In *Proceedings of the 2002 IEEE International Conference on Robotics and Automation*, pages 292–298, 2002.

[7] L. Chaimowicz, T. Sugar, V. Kumar, and M. Campos. An architecture for tightly coupled multi-robot cooperation. In *Proceedings of the 2001 IEEE International Conference on Robotics and Automation*, pages 2292–2297, 2001.

[8] B. Espiau, K. Kapellos, M. Jourdan, and D. Simon. On the validation of robotics control systems. part i: High level specification and formal verification. Technical Report 2719, INRIA - Rhône-Alpes, 1995.

[9] R. Fierro, A. Das, V. Kumar, and J. Ostrowski. Hybrid control of formations of robots. In *Proceedings of the 2001 IEEE International Conference on Robotics and Automation*, pages 3672–3677, 2001.

[10] T. Henzinger. The theory of hybrid automata. In *Proceedings of the 11th Annual Symposium on Logic in Computer Science*, pages 278–292, 1996.

[11] R. C. Kube and H. Zhang. Task modeling in collective robotics. *Autonomous Robots*, 4:53–72, 1997.

[12] D. Milutinovic and P. Lima. Petri net models of robotic tasks. In *Proceedings of the 2002 IEEE International Conference on Robotics and Automation*, pages 4059–4064, 2002.

[13] F. Wang and G. Saridis. Task translation in integration specification in intelligent machines. *IEEE Transactions on Robotics and Automation*, 9(3):257–271, 1993.

INVERSE DYNAMICS AND SIMULATION OF A 3-DOF SPATIAL PARALLEL MANIPULATOR

Yu-Wen Li, Jin-Song Wang, Li-Ping Wang, Xin-Jun Liu*

Department of Precision Instruments and Mechanology, Tsinghua University, Beijing, 100084, P. R. China
*School of Mechanical and Aerospace Engineering, Seoul National University, Seoul, Republic of Korea
Email: liyw00@mails.tsinghua.edu.cn (Yu-Wen Li)

Abstract: Recently the parallel manipulators with less DOF have attracted the researchers, but works on their dynamics are relative few. In this paper, an inverse dynamic formulation is presented by the Newton-Euler approach for a spatial parallel manipulator, which has two translational degrees of freedom and one rotational degree of freedom. The inverse kinematics analysis is firstly performed in closed form. Then the force and moment equilibrium equations for the manipulator are presented. According to the kinematic constraints of the legs and the platform, some joint constraint forces are eliminated and an algorithm to solve the actuator forces is given. In addition, ADAMS is used to perform the kinematic and dynamic simulation for the manipulator. The simulation results are compared to those derived from algebraic formulae and the comparison shows the validity of the mathematical model.

Keywords: Parallel manipulator, Inverse dynamics, Simulation

1 INTRODUCTION

There has been great amount of research on the application of parallel manipulators, such as machine tools [1] and industrial robots [2]. When good dynamic performance and precise positioning under high load are required, the dynamic model is important for their control. Because of the close-loop structure, their dynamic model is quite complicated. A lot of works have focused on the dynamics of Stewart platform, a 6-DOF parallel manipulator. Geng [3] developed Lagrangian equations of motion under some simplifying assumptions regarding the geometry and inertia distribution of the manipulator. Dasgupta and Mruthyunjaya [4,5] used the Newton-Euler approach to develop closed-form dynamic equations of Stewart platform, considering all dynamic and gravity effects as well as the viscous friction at joints. They observed that the application of the Newton-Euler approach is more economical in the case of parallel or hybrid manipulators than in serial manipulators.

In the last few years the parallel manipulators with less DOF have attracted the researchers and some of them have been used in the structure design of robotic manipulators. For example, Pierrot and Company [6] presented a new family of parallel robots with 4-DOFs. Wang and Gosselin [7] discussed the static balancing of the 3-RRS parallel manipulators using counter-weights and springs. Although the parallel manipulators with less DOF have been investigated to some extent, works on their dynamics are relatively few.

In this paper, the Newton-Euler approach is adopted to derive the inverse dynamic equations of a 3-DOF spatial parallel manipulator, which has two translational degrees of freedom and one rotational degree of freedom [8]. The mechanism and the coordinate systems of the manipulator are described in the next section. Then for inverse kinemaitcs, the position analysis is performed and the velocity and acceleration formulae are derived in closed form. Based on the force and moment equilibriums of the legs, movable platform and sliders, the inverse dynamics of the manipulator is developed in the fourth section. A numerical example is presented in the fifth section, concerning the nature of the actuator forces required to track some planned trajectories. In addition, ADAMS is used to perform the kinematic and dynamic simulation for the manipulator. The simulation results are compared to those derived from algebraic formulae.

2 DESCRIPTION OF THE MANIPULATOR

As shown in Figure 1, the parallel manipulator under consideration contains a triangular plate referred to as the movable platform. The platform is connected to a base platform, which consists of three guideways (2), (7) and (9), through the legs (1), (8) and (12). The legs (1) and (12)

have identical chains, consisting of a constant link which is connected to a universal joint (or two revolute joints) at the bottom end and a passive revolute joint at the other. The revolute joint is then attached to an active slider, which is mounted on the guideway (2) or (9). The third leg (8) consists of a constant link, a planar four-bar parallelogram, which is connected to a revolute joint at the bottom end and a passive revolute joint at the other. The revolute joint is attached to an active slider, which is mounted on the guideway (7). The motion of the movable platform is accomplished by the slide of three sliders on the guideways.

The parallel mechanism can be described by three parameters, shown in Figure 2. Parameter R denotes the size of the base platform, where $Ob_i = R$ ($i = 1, 2, 3$). The movable platform is an isosceles triangle, which is described by r, where $O'P_i = r$ ($i = 1, 2, 3$). And the length of each leg is denoted as L. The motion capability of the manipulator has been discussed by Liu [8]. It has two translational degrees of freedom in the Oyz plane and one rotational degree of freedom about the y axis. The purpose of this paper is to derive its dynamic model.

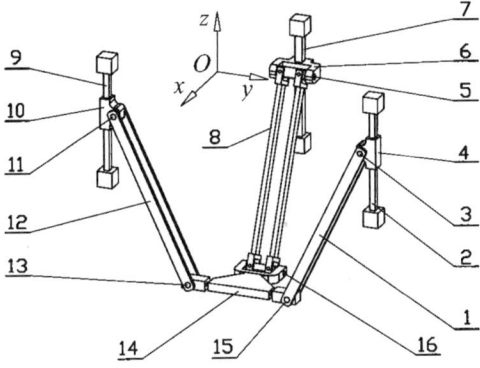

Figure 1 The spatial 3-DOF parallel manipulator

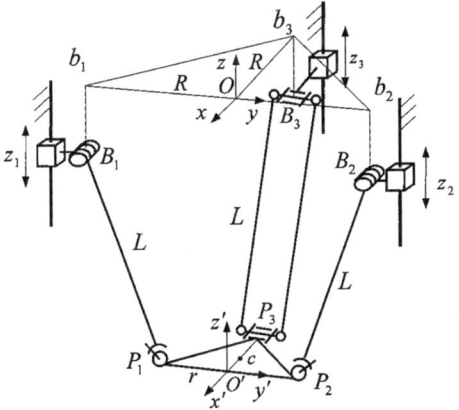

Figure 2 The geometric parameters of the special 3-DOF parallel manipulator

3 INVERSE KINEMATIC ANALYSIS

3.1 Position Analysis

A global frame $\{Oxyz\}$ is located at the center of the side b_1b_2 with the z axis normal to the base platform and the y axis along b_1b_2, shown in Figure 2. So the position vector of B_i and O' in the global frame are given as

$$B_1 = [0, -R, z_1]^T \quad B_2 = [0, R, z_2]^T \quad B_3 = [-R, 0, z_3]^T \quad (1)$$

$$O' = [0, y, z]^T \quad (2)$$

Therefore the pose of the platform can be described by vector $[y, z, \theta]^T$, where the angle θ is the rotational degree of the platform about the y axis. A platform frame $\{O'x'y'z'\}$ is established at the center of the side P_1P_2. The z' axis is perpendicular to the movable platform and the y' axis is along the side P_1P_2. The position vector of P_i in the platform frame can be written as

$$P_1' = [0, -r, 0]^T \quad P_2' = [0, r, 0]^T \quad P_3' = [-r, 0, 0]^T \quad (3)$$

And these position vectors can be transformed to the global frame as

$$P_1 = [0, y - r, z]^T \quad P_2 = [0, y + r, z]^T \quad (4)$$

$$P_3 = [-r\cos\theta, y, z + r\sin\theta]^T \quad (5)$$

Then the kinematic constraint equation of the manipulator is expressed as

$$\|P_i - B_i\| = L \quad (i = 1, 2, 3) \quad (6)$$

Substitute (1) (4), and (5) into (6) and the inverse kinematics solution can be computed as

$$z_1 = \sqrt{L^2 - (y - r + R)^2} + z \quad (7)$$

$$z_2 = \sqrt{L^2 - (y + r - R)^2} + z \quad (8)$$

$$z_3 = \sqrt{L^2 - y^2 - (R - r\cos\theta)^2} + r\sin\theta + z \quad (9)$$

3.2 Velocity Analysis

Because the legs (1) and (12) have one rotational degree of freedom about the x axis, the angular velocity ω and angular acceleration α of the leg can be written as

$$\omega_i = [\omega_i, 0, 0]^T \quad \alpha_i = [\alpha_i, 0, 0]^T \quad (i = 1, 2) \quad (10)$$

On the other hand, leg (8) has two rotational degrees of freedom. It can rotate about the y axis and unit vector k_3 where

$$k_3 = l_3 \times j \quad (11)$$

and l_3 is the unit vector along leg (8), which can be obtained as

$$l_3 = \frac{[R - r\cos\theta, y, z + r\sin\theta - z_3]^T}{L} \quad (12)$$

4093

So the angular velocity and angular acceleration of leg (8) can be written as

$$\boldsymbol{\omega}_3 = \left[\frac{-n}{L}\omega', \omega'', \frac{m}{L}\omega'\right]^T \quad \boldsymbol{a}_3 = \left[\frac{-n}{L}\alpha', \alpha'', \frac{m}{L}\alpha'\right]^T \quad (13)$$

where

$$m = R - r\cos\theta \quad n = z + r\sin\theta - z_3 \quad (14)$$

The velocity of point P_i can be written as

$$\dot{\boldsymbol{P}}_i = \dot{\boldsymbol{B}}_i + \boldsymbol{\omega}_i \times (\boldsymbol{P}_i - \boldsymbol{B}_i) \quad (15)$$

where $\dot{\boldsymbol{P}}_i$ and $\dot{\boldsymbol{B}}_i$ can be obtained from the time derivative of Equation (4), (5) and (1) respectively. Substitute $\dot{\boldsymbol{P}}_i$, $\dot{\boldsymbol{B}}_i$, and $\boldsymbol{\omega}_i$ into (15) and the angular velocity of the legs and velocity of the sliders can be computed as

$$\omega_i = -\frac{\dot{y}}{z - z_i} \quad (i = 1, 2) \quad (16)$$

$$\omega' = \frac{L\dot{y}}{m^2 + n^2} \quad \omega'' = \frac{\dot{m}}{n} + \frac{my\dot{y}}{n(m^2 + n^2)} \quad (17)$$

$$\dot{z}_1 = \dot{z} + \frac{y - r + R}{z - z_1}\dot{y} \quad \dot{z}_2 = \dot{z} + \frac{y + r - R}{z - z_2}\dot{y} \quad (18)$$

$$\dot{z}_3 = \dot{z} + r\cos\theta\dot{\theta} + \frac{m\dot{m} + y\dot{y}}{n} \quad (19)$$

where

$$\dot{m} = r\sin\theta\dot{\theta} \quad (20)$$

Equations (16) to (19) give the closed-form formulation for the velocity analysis of this parallel manipulator.

3.3 Acceleration Analysis

Again the acceleration of point P_i can be written as

$$\ddot{\boldsymbol{P}}_i = \ddot{\boldsymbol{B}}_i + \boldsymbol{a}_i \times (\boldsymbol{P}_i - \boldsymbol{B}_i) + \boldsymbol{\omega}_i \times (\dot{\boldsymbol{P}}_i - \dot{\boldsymbol{B}}_i) \quad (21)$$

where $\ddot{\boldsymbol{P}}_i$ and $\ddot{\boldsymbol{B}}_i$ can be obtained from the time derivative of $\dot{\boldsymbol{P}}_i$ and $\dot{\boldsymbol{B}}_i$ respectively. Substitute $\ddot{\boldsymbol{P}}_i$, $\ddot{\boldsymbol{B}}_i$, and $\boldsymbol{a}_i$ into (21) and the angular acceleration of the legs and acceleration of the sliders can also be computed in closed form as

$$\alpha_i = -\frac{\ddot{y}}{z - z_1} - \frac{(\dot{z} - \dot{z}_1)\dot{\omega}_i}{z - z_1} \quad (i = 1, 2) \quad (22)$$

$$\alpha' = \frac{L}{m^2 + n^2}\ddot{y} - \frac{(m\dot{m} + n\dot{n})L}{(m^2 + n^2)^2}\dot{y} \quad (23)$$

$$\alpha'' = \frac{\ddot{m}}{n} + \frac{m\dot{y}^2}{n(m^2 + n^2)} - \frac{\dot{m}\dot{n}}{n^2} - \frac{m\dot{n}y\dot{y}}{n^2(m^2 + n^2)} + \frac{my}{Ln}\alpha' \quad (24)$$

$$\ddot{z}_1 = \ddot{z} - (y - r + R)\alpha_1 - \omega_1\dot{y} \quad (25)$$

$$\ddot{z}_2 = \ddot{z} - (y + r - R)\alpha_2 - \omega_2\dot{y} \quad (26)$$

$$\ddot{z}_3 = \ddot{z} + r\cos\theta\ddot{\theta} - r\sin\theta\dot{\theta}^2 + A \quad (27)$$

where

$$\ddot{m} = r\sin\theta\ddot{\theta} + r\cos\theta\dot{\theta}^2 \quad \dot{n} = \dot{z} + r\cos\theta\dot{\theta} - \dot{z}_3 \quad (28)$$

$$A = \frac{n\dot{y}^2}{m^2 + n^2} + \frac{\dot{m}^2}{n} + \frac{m\dot{n}y\dot{y}}{n(m^2 + n^2)} + \frac{ny\alpha'}{L} + m\alpha'' \quad (29)$$

4 INVERSE DYNAMIC ANALYSIS

4.1 Dynamic equations for the legs

A local leg frame $\{B_i x_i y_i z_i\}$ ($i = 1,2,3$), shown in Figure 3, is attached to the leg with its origin at point B_i, the x_i axis along the leg and the y_i axis along the rotating axis of the revolute joint. The kinematic and dynamic parameters of the leg are transformed to a fixed leg frame (not shown separately in Figure 3) at point B_i, which is parallel to the global frame. The rotation matrix from the local leg frame to the fixed leg frame is

$$T_i = [\boldsymbol{l}_i \quad \boldsymbol{i} \quad \boldsymbol{l}_i \times \boldsymbol{i}] \quad (i = 1, 2) \quad T_3 = [\boldsymbol{l}_3 \quad \boldsymbol{j} \quad \boldsymbol{l}_3 \times \boldsymbol{j}] \quad (30)$$

where $\boldsymbol{l}_i$ is the unit vector along the ith leg and

$$\boldsymbol{l}_1 = \frac{[0, y - r + R, z - z_1]^T}{L} \quad \boldsymbol{l}_2 = \frac{[0, y + r - R, z - z_2]^T}{L} \quad (31)$$

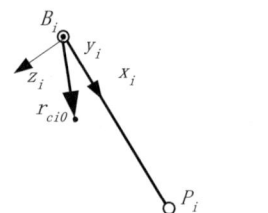

Figure 3 The ith local leg frame

If $\boldsymbol{r}_{ci0}$ ($i = 1,2,3$) denotes the position vectors of the centers of gravity of the legs in the respective local frame $\{B_i x_i y_i z_i\}$, then it can be transformed to the fixed leg frame as

$$\boldsymbol{r}_{ci} = T_i \boldsymbol{r}_{ci0} \quad (32)$$

The acceleration of the centers of gravity of the legs with respect to the global frame can be obtained as

$$\ddot{\boldsymbol{r}}_{ci} = \ddot{\boldsymbol{B}}_i + \boldsymbol{a}_i \times \boldsymbol{r}_{ci} + \boldsymbol{\omega}_i \times (\boldsymbol{\omega}_i \times \boldsymbol{r}_{ci}) \quad (33)$$

The moment of inertia $\boldsymbol{J}_i$ of the ith leg in the fixed leg frame can be obtained from its moment of inertia $\boldsymbol{J}_{i0}$ in its local frame by the rotation transformation

$$\boldsymbol{J}_i = T_i \boldsymbol{J}_{i0} T_i^T \quad (34)$$

Considering the moments acting on the legs with respect to point B_i, according to theorems of moment of momentum, we have

$$\boldsymbol{J}_i \boldsymbol{a}_i + m_{li} \boldsymbol{r}_{ci} \times \ddot{\boldsymbol{B}}_i - (\boldsymbol{P}_i - \boldsymbol{B}_i) \times \boldsymbol{F}_{pi} - m_{li} \boldsymbol{r}_{ci} \times \boldsymbol{g} - \boldsymbol{M}_{pi} = 0$$
$$(35)$$

where $\boldsymbol{F}_{pi}$ and $\boldsymbol{M}_{pi}$ is the constraint force and moment on the leg exerted by the platform, m_{li} is the mass of the ith leg and $\mathbf{g}$ is acceleration due to gravity. Because the legs (1) and (12) have one rotational degree of freedom about the x axis and leg (8) has two rotational degrees about the y axis and the unit vector $\mathbf{k}_3$, the quantity of $\mathbf{M}_{pi}$ along these directions is 0. Hence some constraint forces can be eliminated from Equation (35) as

$$-(z-z_1)F_{p1y}+(y-r+R)F_{p1z}=\mathbf{Q}_1\cdot\mathbf{i} \quad (36)$$

$$-(z-z_2)F_{p2y}+(y+r-R)F_{p2z}=\mathbf{Q}_2\cdot\mathbf{i} \quad (37)$$

$$nF_{p3x}-mF_{p3z}=\mathbf{Q}_3\cdot\mathbf{j} \quad (38)$$

$$-\frac{my}{L}F_{p3x}+\frac{(m^2+n^2)}{L}F_{p3y}-\frac{ny}{L}F_{p3z}=\mathbf{Q}_3\cdot\mathbf{k}_3 \quad (39)$$

where

$$\mathbf{Q}_i = \mathbf{J}_i\mathbf{\alpha}_i + m_{li}\mathbf{r}_{ci}\times\ddot{\mathbf{B}}_i - m_{li}\mathbf{r}_{ci}\times\mathbf{g} \quad (i=1,2,3) \quad (40)$$

4.2 Dynamic equations for the movable platform

If $\mathbf{r}_{cp0}$ denotes the position vectors of the centers of gravity of the platform in the platform frame $\{O'x'y'z'\}$, then in order to transformed the vector to the local fixed frame, which is parallel to the global frame, we have

$$\mathbf{r}_{cp} = \mathbf{R}\mathbf{r}_{cp0} \quad (41)$$

where $\mathbf{R}$ is the rotation matrix and

$$\mathbf{R}=\begin{bmatrix}\cos\theta & 0 & \sin\theta \\ 0 & 1 & 0 \\ -\sin\theta & 0 & \cos\theta\end{bmatrix} \quad (42)$$

The moment of inertia $\mathbf{J}_p$ of the platform in the local fixed frame can be obtained from its moment of inertia $\mathbf{J}_{p0}$ in the platform frame as

$$\mathbf{J}_p = \mathbf{R}\mathbf{J}_{p0}\mathbf{R}^T \quad (43)$$

Considering the moments with respect to point O' acting on the platform, theorems of moment of momentum lead to

$$\mathbf{J}_p\begin{bmatrix}0\\ \alpha\\ 0\end{bmatrix}+m_p\mathbf{r}_{cp}\times\begin{bmatrix}0\\ \ddot{y}\\ \ddot{z}\end{bmatrix}+\sum_{i=1}^{3}(\mathbf{P}_i-\begin{bmatrix}0\\ y\\ z\end{bmatrix})\times\mathbf{F}_{pi}+\sum_{i=1}^{3}\mathbf{M}_{pi}-m_p\mathbf{r}_{cp}\times\mathbf{g}-\mathbf{M}_e=0 \quad (44)$$

where m_p is the mass of the platform and $\mathbf{M}_e$ is the external moment on the platform. In order to eliminate $\mathbf{M}_{pi}$, dot multiply both sides of (44) with $\mathbf{j}$ and we have

$$F_{p3x}r\sin\theta + F_{p3z}r\cos\theta = \mathbf{U}\cdot\mathbf{j} \quad (45)$$

where

$$\mathbf{U} = -\mathbf{J}_p\begin{bmatrix}0\\ \alpha\\ 0\end{bmatrix}-m_p\mathbf{r}_{cp}\times\begin{bmatrix}0\\ \ddot{y}\\ \ddot{z}\end{bmatrix}+m_p\mathbf{r}_{cp}\times\mathbf{g}+\mathbf{M}_e \quad (46)$$

If $\mathbf{F}_e$ denotes the external force exerted on the platform at point O', considering the forces acting on the platform along the y and z direction in the global frame, Newton's equation gives

$$\sum_{i=1}^{3}F_{piy} = -m_p\ddot{y}+(m_p\mathbf{g}+\mathbf{F}_e)\cdot\mathbf{j} \quad (47)$$

$$\sum_{i=1}^{3}F_{piz} = -m_p\ddot{z}+(m_p\mathbf{g}+\mathbf{F}_e)\cdot\mathbf{k} \quad (48)$$

In equations (45), (47), (48), and (36) to (39), there are 7 unknowns, so the constraint forces at point P_i can be obtained as

$$[F_{p3x},F_{p1y},F_{p2y},F_{p3y},F_{p1z},F_{p2z},F_{p3z}]^T = \mathbf{B}^{-1}\mathbf{c} \quad (49)$$

where

$$\mathbf{B}=\begin{bmatrix}0 & -(z-z_1) & 0 & 0 & y-r+R & 0 & 0\\ 0 & 0 & -(z-z_2) & 0 & 0 & y+r-R & 0\\ n & 0 & 0 & 0 & 0 & 0 & -m\\ -\frac{my}{L} & 0 & 0 & \frac{m^2+n^2}{L} & 0 & 0 & -\frac{ny}{L}\\ r\sin\theta & 0 & 0 & 0 & 0 & 0 & r\cos\theta\\ 0 & 1 & 1 & 1 & 0 & 0 & 0\\ 0 & 0 & 0 & 0 & 1 & 1 & 1\end{bmatrix} \quad (50)$$

$$\mathbf{c} = [(\mathbf{Q}_1\cdot\mathbf{i}),(\mathbf{Q}_2\cdot\mathbf{i}),(\mathbf{Q}_3\cdot\mathbf{j}),(\mathbf{Q}_3\cdot\mathbf{k}_3),(\mathbf{U}\cdot\mathbf{j}),S_y,S_z]^T \quad (51)$$

where

$$S_y = -m_p\ddot{y}+(m_p\mathbf{g}+\mathbf{F}_e)\cdot\mathbf{j}$$

$$S_z = -m_p\ddot{z}+(m_p\mathbf{g}+\mathbf{F}_e)\cdot\mathbf{k}$$

4.3 Actuator forces

Considering the forces acting on the ith leg along the z direction, we obtain the force exerted on the leg by the slider along this direction as

$$F_{biz} = -F_{piz}+m_{li}(\ddot{\mathbf{r}}_{ci}-\mathbf{g})\cdot\mathbf{k} \quad (i=1,2,3) \quad (52)$$

Again, considering the forces acting on the ith slider along the z direction, we have the actuator force as

$$f_i = F_{biz}+m_{si}(\ddot{z}_i-\mathbf{g}\cdot\mathbf{k}) \quad (i=1,2,3) \quad (53)$$

where m_{si} is the mass of the ith silder.

5 SIMULATION RESULTS

5.1 Numerical example

Firstly the kinematic parameters of the parallel mechanism are given as

$r=100$ (mm) $R=250$ (mm) $L=400$ (mm)

And the dynamic parameters of the platform are

$\mathbf{r}_{cp0} = [-30.9653,0,0]^T$ (mm) $m_p=8.1573$ (kg)

$$\boldsymbol{J}_{p0} = \begin{bmatrix} 26895.0386 & 0 & 0 \\ 0 & 19036.6467 & 0 \\ 0 & 0 & 34712.5762 \end{bmatrix} (kg \times mm^2)$$

The dynamic parameters of the legs are

$$\boldsymbol{r}_{ci0} = [200, 0, 0]^T \text{ (mm)} \quad (i = 1, 2, 3)$$

$$m_{l1} = m_{l2} = 0.9803 \text{ (kg)} \quad m_{l3} = 1.9606 \text{ (kg)}$$

$$\boldsymbol{J}_{10} = \boldsymbol{J}_{20} = \begin{bmatrix} 49.0151 & 0 & 0 \\ 0 & 52374.1104 & 0 \\ 0 & 0 & 52374.1104 \end{bmatrix} (kg \times mm^2)$$

$$\boldsymbol{J}_{30} = \begin{bmatrix} 1225.3775 & 0 & 0 \\ 0 & 104748.2208 & 0 \\ 0 & 0 & 104748.2208 \end{bmatrix} (kg \times mm^2)$$

The mass of the sliders and acceleration due to gravity are

$$m_{si} = 0.975125 \text{ (kg)} \quad \boldsymbol{g} = [0, 0, -9806.65]^T \text{ (mm/s}^2)$$

In this numerical example, the platform moves for one second and the trajectory of the motion of the movable platform is given as

$$y = -1000t^4 + 1300t^3 - 300t^2 \text{ (mm)} \quad (54)$$
$$z = 400t^4 - 200t^3 - 200t^2 \text{ (mm)} \quad (55)$$
$$\theta = 0.4\sin(4\pi t) \text{ (rad)} \quad (56)$$

So the acceleration quantities are made continuous.

For the inverse kinematic analysis, the time histories of the displacement, velocities and acceleration of the sliders are computed and shown in Figure 4 to 6. Assuming that there is no external force and moment acting on the movable platform, the actuator forces are shown in Figure 7 to 9.

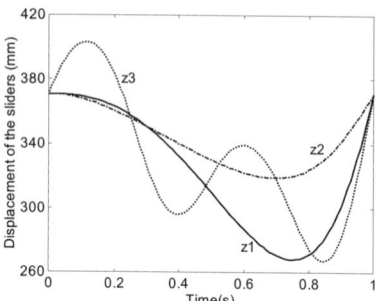

Figure 4   The displacement of the sliders

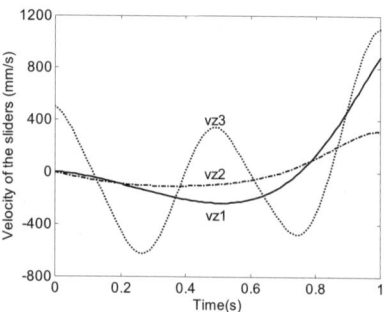

Figure 5   The velocity of the sliders

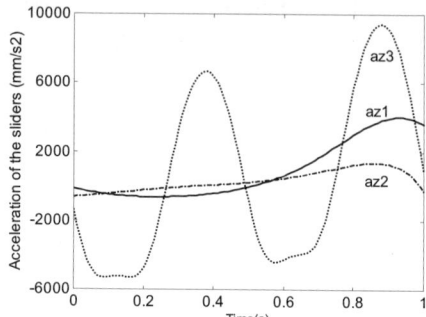

Figure 6   The acceleration of the sliders

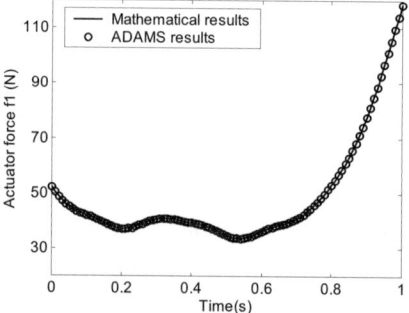

Figure 7   The Actuator force of the slider at B_1

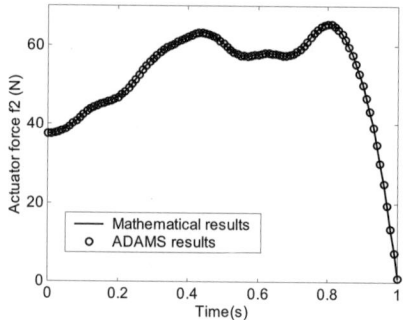

Figure 8   The Actuator force of the slider at B_2

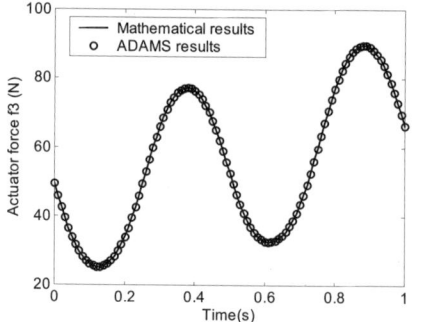

Figure 9   The Actuator force of the slider at B_3

5.2 Simulation in ADAMS

ADAMS is used to perform the kinematic and dynamic simulation for this manipulator. The kinematic and dynamic parameters of the legs, platform and sliders are the same as the numerical example. Some revolute, universal and prismatic JOINTs are defined to connect the rigid bodies according to Section 2. In ADAMS, the motion of a rigid body is driven by MOTION at a JOINT. Because the inverse kinematics can be presented in closed form, the MOTIONs at three prismatic JOINTs between the sliders and the ground are defined and input according to Equation (7) to (9). So the platform can move along the trajectory described as Equation (54) to (56). The simulation model in ADAMS is shown in Figure 10.

The simulation results of the actuator forces are also shown in Figure 7 to 9. It can be shown that the simulation results are very close to the mathematical results. Moreover, other kinematic and dynamic analysis, such as velocity, acceleration and kinetic energy, can also be output by ADAMS. However, it is difficult for ADAMS to give the inherent dynamic performance of the mechanism and optimize the structure of the manipulator. And the simulation results can hardly be used in the real-time control model. Therefore a mathematical model for its dynamics has to be established.

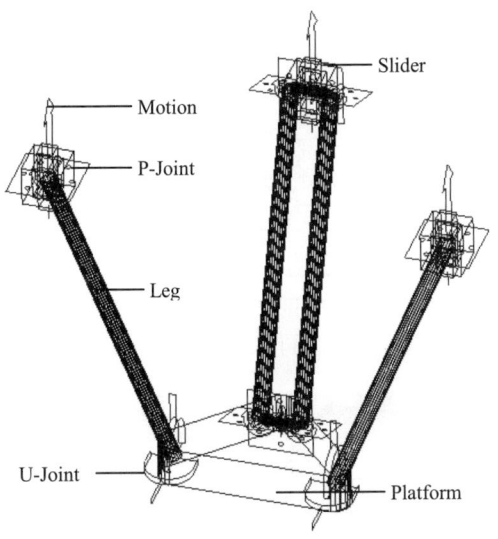

Figure 10   Simulation model in ADAMS

6 CONCLUSIONS

In this paper, the inverse kinematic and dynamic analysis of a spatial 3-DOF parallel manipulator is presented in closed form. The Newton-Euler approach is adopted to establish the dynamic equations. According to the different kinematic constraints of the three legs and the movable platform, some joint constraint forces are eliminated and the actuator forces are derived. The validity of this approach is shown by the comparison between the mathematical simulation results and ADAMS simulation results. The study presented in this paper provides a framework for the future researches such as the control model and structure optimization design of the spatial 3-DOF parallel manipulators. Moreover the presented method and process can also be applied to other parallel manipulators with less DOF.

ACKNOWLEDGEMENTS

The research has been supported by the National 863 Program of China (G2002AA424011) and the National Natural Science Foundation of China (G50275084).

RERERENCES

[1] J. Tlusty, J. Ziegert and S. Ridgeway, "Fundamental comparison of the use of serial and parallel kinematics for machine tools", Annals of CIRP, v49, 1999: 351-356

[2] K. Cleary and T. Brooks, "Kinematics analysis of a novel 6-DOF parallel manipulator", Proceeding of the IEEE Int. Conf. On Robotics and Automation, 1993: 708-713

[3] Z. Geng, L. S. Haynes, J. D. Lee and R. L. Carroll, "On the dynamic model and kinematic analysis of a class of Stewart platforms, Robotics and Autonomous system", v9, 1992: 237-254

[4] B Dasgupta and T. S. Mruthyunjaya, "A Newton-Euler formulation for the inverse dynamics of the Stewart platform manipulator", Mechanism and Machine Theory, v33, 1998: 1135-1152

[5] B Dasgupta and P. Choudhury, "A general strategy based on the Newton-Euler approach for the dynamic formation of parallel manipulators, Mechanism and Machine Theory, v34, 1999: 801-824

[6] F. Pierrot and O. Company, "H4: A new family of 4-dof parallel robots", Proceeding of the IEEE/ASME Int. Conf. On Advanced Intelligent Mechatronics, 1999: 508-513

[7] J. Wang and C. M. Gosselin, "Static balancing of spatial Three-Degree-of-Freedom parallel mechanism", Mechanism and Machine Theory, v34, 1999: 437-452

[8] Xin-Jun Liu, Jinsong Wang and Li-Ping Wang, "On the analysis of a new spatial Three-Degrees-of-Freedom parallel manipulator", IEEE Transaction on Robotics and Automation, v17(6), 2001: 959-968

Development of Force Displaying Device Using Pneumatic Parallel Manipulator and Application to Palpation Motion

Masahiro Takaiwa
Department of Systems Engineering,
Okayama University, 3-1-1 Tsushimanaka,
Okayama, 700-8530, Japan

Toshiro Noritsugu
Department of Systems Engineering,
Okayama University, 3-1-1 Tsushimanaka,
Okayama, 700-8530, Japan

Abstract—The goal of this study is to develop a mechanical system which display elastic characteristic like stiffness on the surface of human body aiming at applying to palpation simulator. Pneumatic parallel manipulator is employed as a driving mechanism, consequently, it brings capability of minute force displaying property owing to the air compressibility. Compliance control system without using force/moment sensor is constructed by introducing a disturbance observer and a compliance display scheme is proposed. The validity of the proposed scheme is verified experimentally.

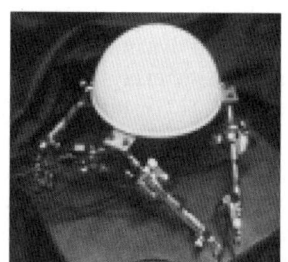 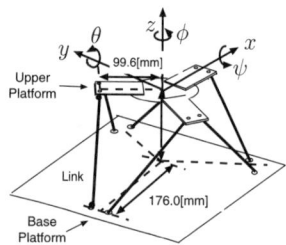

(a) Parallel Link Mechanism

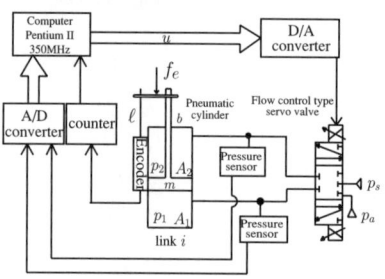

(b) Pneumatic Driving Circuit

Fig. 1. Developed Pneumatic Parallel Manipulator

I. INTRODUCTION

A palpation motion is one of the most important medical examination to find a human disease like a cancer of the breast. There is a great need for the development of a mechanical system as a palpation training simulator[1] or a displaying device for remote diagnosis[2]. In this study we aim at developing a mechanical system which can display compliance characteristic of human skin, such as, "stiffness" for the operator's palpation motion(pushing at a point or slide on the surface) as shown in Fig.2.

A pneumatic parallel manipulator is employed as our displaying device since it can drive a multiple d.o.f. for its compactness due to the parallel link mechanism and it can execute minute force regulation owing to the air compressibility of pneumatic actuator which is indispensable feature for the palpation motion. The air compressibility simultaneously implies a feature of safety and softness, which are indispensable for the mechanical system contact with human directly.

In order to display a concrete compliance feeling to an operator, a control strategy is proposed where the applied force from an operator is estimated with no use of force/moment sensor and the contact point on which an operator is touching is detected based on that estimated force and finally realize a corresponding compliance by constructing a compliance control system[3]. By regulating the reference compliance value according to the movement of a contact point(motion of the finger), a palpation action is realized. The validities of the proposed control systems are confirmed through some experiments and analysis.

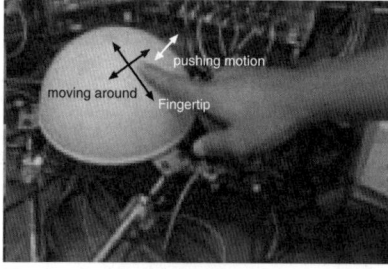

Fig. 2. Concept image of compliance display

II. OUTLINE OF PNEUMATIC PARALLEL MANIPULATOR

Fig. 1 (a) shows the developed pneumatic parallel manipulator. 6 pneumatic cylinders are employed to form so called Stewart type platform[4].

The position/orientation of the upper platform is expressed by a hand vector $h = [x, y, z, \phi, \theta, \psi]^T$ using roll-pitch-yaw angle notation. The origin of hand coordinate frame h is set at a center point of upper platform when manipulator stands in a standard posture.

TABLE I
SYSTEMS PARAMETERS

T_p	Time constant of pressure response
T_{pn}	Nominal time constant of pressure response
k_p	Steady gain of pressure response
k_{pn}	Nominal steady gain of pressure response
k_v	Steady gain between piston velocity and pressure
m	Equivalent mass for one cylinder
b	Viscous coefficient
f_e	External force applied on a link
A_1, A_2	cross sectional area of cylinder chamber
p_1, p_2	air pressure in chamber
ℓ	displacement of piston rod
J	Jacobian matrix
T_q, T_{pq}	Time constant of filter

Similarly a link vector is defined as $\boldsymbol{\ell} = [\ell_a, ..., \ell_f]^T$ with an element of a displacement of each piston rod. Force/moment vector works at an origin of $\boldsymbol{h}$ is defined as $\boldsymbol{f_m} = [f_x, f_y, f_z, \tau_\phi, \tau_\theta, \tau_\psi]^T$. The equivalent force vector acts on piston rod is denoted with $\boldsymbol{f_e}$ which satisfy the following relation.

$$\boldsymbol{f_m} = \boldsymbol{J}^T \boldsymbol{f_e} \quad (1)$$

, where $\boldsymbol{J}$ is Jacobian matrix and it forms the next relation in a parallel manipulator mechanism.

$$\frac{d\boldsymbol{\ell}}{dt} = \boldsymbol{J} \frac{d\boldsymbol{h}}{dt} \quad (2)$$

In the mean while, Fig.1(b) shows the pneumatic driving circuit of one cylinder. Low friction type pneumatic cylinder is employed (Airpel Co. Ltd., 9.3mm in internal diameter, 50mm in rod stroke). Pressure in each cylinder's chamber, p_1, p_2 are detected by pressure sensors and the displacement of piston rod ℓ is measured by wire type linear encoder. The A/D converter is of 12 bit resolution.

A control signal u calculated every sampling period(10 ms) in a computer corresponds to an input voltage of a servo valve (FESTO, 50 ℓ/min) through D/A converter(resolution of 12 bit), which regulates the difference pressure of each cylinder. Supply pressure p_s is set to be 400 kPa. Table I shows the control parameters.

The linearized state equations of pressure in cylinder's chamber are described by the following equation[5].

$$T_p \frac{dp_1}{dt} = -p_1 + k_p u - k_v \frac{d\ell}{dt} \quad (3.\text{a})$$

$$T_p \frac{dp_2}{dt} = -p_2 - k_p u + k_v \frac{d\ell}{dt} \quad (3.\text{b})$$

Equation of motion of piston rod is expressed by Eq.(4).

$$p_1 A_1 - p_2 A_2 = f_g = m \frac{d^2 \ell}{dt^2} + b \frac{d\ell}{dt} + f_e \quad (4)$$

III. RECOGNITION OF ELASTIC CHARACTERISTIC

A. Conceptional image

Fig.2 shows the concept image of compliance display. Human touches at an any point on the surface of an manipulator with their fingertip and implements a palpation motion by applying force for a various direction. The manipulator displays a corresponding force for the pushing motion of a fingertip by regulating compliance of the manipulator itself based on the displacement of fingertip and applied force. In order to realize such an action, a manipulator should have a function to detect which point a fingertip is pushing at and how much force is being applied. In the next section the strategy of compliance display including these detecting function is described.

B. Compliance control system

Fig.3 shows the proposed position based compliance control system[6]. The inner position control system is designed in order that the closed loop transfer function may follow the 3rd order system shown in Eq.(5).

$$\frac{\boldsymbol{H}}{\boldsymbol{H_r}} = \boldsymbol{G_r} = diag\left\{\frac{C}{s^3 + As^2 + Bs + C}\right\} \quad (5)$$

The inner block with a doublet represents a control system of generating force F_g as shown in Fig.4, which works to lower the influence of piston rod velocity that acts as disturbance on pressure response as shown in Eq.(3) as well as to make F_g to follow to the reference value with time constant T_{pn}[6].

First of all, the applied external force which works on a link equivalently is estimated by introducing a disturbance observer[7] for the transfer part $P_k(s)$, instead of measuring by installing a force/moment sensor which may loose a feature of compactness. The estimated disturbance $D(s)(= -F_e(s))$ is transfered to the hand coordinate force/moment vector $\boldsymbol{f_m}$ through a transpose of Jacobian matrix J^T and then fed back by being multiplied with a compliance matrix $\boldsymbol{K}^{-1} = diag\{K_x^{-1}, K_y^{-1}, K_z^{-1}, K_\phi^{-1}, K_\theta^{-1}, K_\psi^{-1}\}^T$.

C. Detecting contact force and contact point

Fig.5 shows a geometrical model where contact force vector $\boldsymbol{f}$ is applying at a contact point represented by position vector $\boldsymbol{R} = [x_0, y_0, z_0]^T$. So the first purpose of this study is to detect these vector $\boldsymbol{f}$ and $\boldsymbol{R}$ based on the estimated force/moment vector $\boldsymbol{f_m}$.

Here we consider $\boldsymbol{f_m} = [\boldsymbol{f_t}^T, \boldsymbol{\tau}^T]^T$ with transient force vector $\boldsymbol{f_t}$ and moment one $\boldsymbol{\tau}$. As you see that, force vector $\boldsymbol{f}$ is simply derived from the balance of translational force around the origin as

$$\boldsymbol{f} = \boldsymbol{f_t} \quad (6)$$

In the mean while, if the equation of manipulator's surface is known as Eq.(7), then the contact point can be derived based on the balance of moment shown by Eq.(8) in the following manner[8].

$$g(x_0, y_0, z_0) = 0 \quad (7)$$

$$\boldsymbol{R} \times \boldsymbol{f} = \boldsymbol{\tau} \quad (8)$$

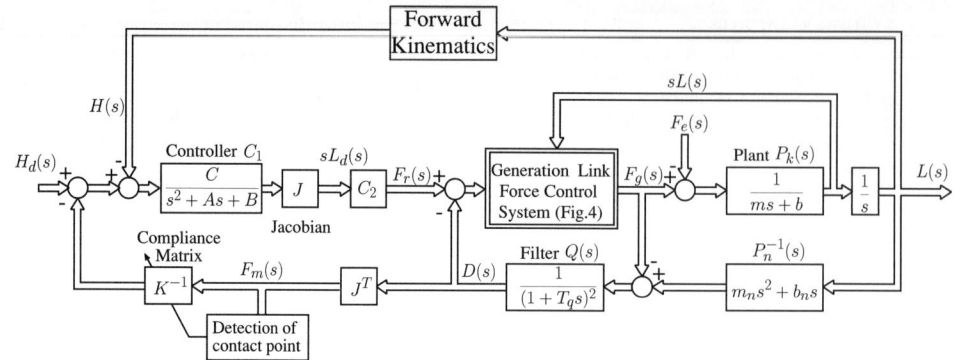

Fig. 3. Proposed compliance control system

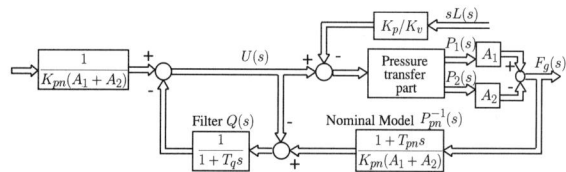

Fig. 4. Generation force control system

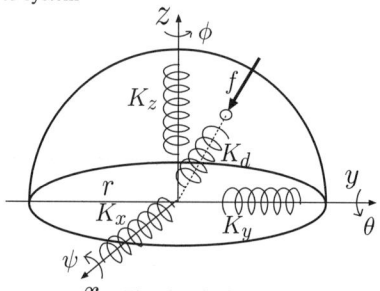

Fig. 6. Spring model

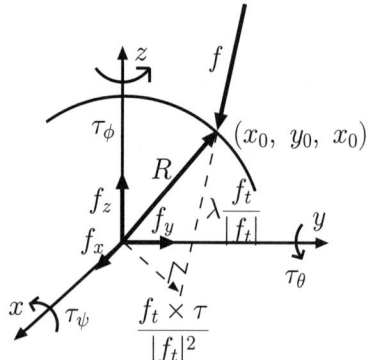

Fig. 5. Geometrical model

From Fig.5, position vector $\boldsymbol{R}$ can be described using a parameter λ as

$$\boldsymbol{R} = \frac{\boldsymbol{f_t} \times \boldsymbol{\tau}}{|\boldsymbol{f_t}|^2} + \lambda \frac{\boldsymbol{f_t}}{|\boldsymbol{f_t}|} \quad (9)$$

And $\boldsymbol{R}$ is obtained by substituting Eq.(9) into Eq.(7) to determine λ.

In our manipulator, as shown in Fig.1(a), a hemispherical shell, which is made of plaster, is introduced as the outer shape, whose center matches to the origin of $\boldsymbol{h}$.

Therefore giving an equation of hemispherical shell as Eq.(10) with radius r, contact point can be represented by Eq.(11), where $[x_1, y_1, z_1]^T$ is a first term of right hand side of Eq.(9).

$$x_0^2 + y_0^2 + z_0^2 = r^2 \quad (10)$$

$$x_0 = x_1 + f_x \sqrt{\frac{r^2 - (x_1^2 + y_1^2 + z_1^2)}{f_x^2 + f_y^2 + f_z^2}} \quad (11.a)$$

$$y_0 = y_1 + f_y \sqrt{\frac{r^2 - (x_1^2 + y_1^2 + z_1^2)}{f_x^2 + f_y^2 + f_z^2}} \quad (11.b)$$

$$z_0 = z_1 + f_z \sqrt{\frac{r^2 - (x_1^2 + y_1^2 + z_1^2)}{f_x^2 + f_y^2 + f_z^2}} \quad (11.c)$$

In the next, displaying corresponding compliance to the contact point is considered. Fig.6 shows a geometrical compliance model, where K_d is a desired stiffness along with the direction of the contact point. Hence the remaining problem is how much each element K_x, K_y, K_z should be set to realize the desired stiffness K_d for any contact point.

From a balance of translational force at the origin of $\boldsymbol{h}$, the desired stiffness K_d for the direction of the contact point (x_0, y_0, z_0) and the stiffness for each axis K_x, K_y, K_z satisfy the relation represented by Eq.(12), where $[\Delta x, \Delta y, \Delta z]^T$ is a displacement vector generated by an applied contact force. The left hand side corresponds to the desired force converted to the direction of contact point vector and the right one do the resultant force of each axis.

$$\left(K_d \frac{\Delta x \; x_0 + \Delta y \; y_0 + \Delta z \; z_0}{\sqrt{x_0^2 + y_0^2 + z_0^2}} \right)^2 = (K_x \Delta x)^2 + (K_y \Delta y)^2 + (K_z \Delta z)^2 \quad (12)$$

In order that K_x, K_y and K_z may satisfy Eq.(12), we introduce a constraint condition which makes the compliance control characteristic be normalized for each direction in the following manner. The closed loop relation of the control system shown in Fig.3 is described as

$$F_m = I_{mp}J^T JC_1 G_r^{-1}(I+I_{mp}J^T JC_1 K^{-1})^{-1}(H_d G_r - H) \quad (13)$$

, where I_{mp} corresponds to a mechanical impedance in the velocity control loop(namely satisfying $sL = sL_d + I_{mp}F_e$ in Fig.3). In our manipulator, it is designed so that $J^T J$ may become almost diagonal at the origin of hand coordinate frame h, which means that the relation between F_m and $(H_d G_r - H)$ can be considered to be diagonal. The value of $J^T J$ at origin of h is represented by Eq.(15), where non-diagonal elements are denoted as 0 since they are thoroughly small compared to the diagonal one.

$$J^T J = \begin{bmatrix} 1.5 & 0 & 0 & 0 & 0 & 0 \\ 0 & 1.5 & 0 & 0 & 0 & 0 \\ 0 & 0 & 2.9 & 0 & 0 & 0 \\ 0 & 0 & 0 & 3.0 \times 10^3 & 0 & 0 \\ 0 & 0 & 0 & 0 & 1.4 \times 10^3 & 0 \\ 0 & 0 & 0 & 0 & 0 & 1.4 \times 10^3 \end{bmatrix} \quad (14)$$

Seeing from Eq.(15), diagonal element corresponding to the z axis in $J^T J$ is almost 2 times as much as that for x and y axis. Therefore by setting each element of K according to the ratio in Eq.(15), the frequency characteristic between F_m and $H_d G_r - H$ for x, y, z direction becomes equivalent except for the influence of a static gain.

$$K_x : K_y : K_z = 1 : 1 : 2 \quad (15)$$

Substituting Eq.(15) into Eq.(12), desired stiffness for each axis is given as

$$K_x = \frac{K_d(\Delta x\, x_0 + \Delta y\, y_0 + \Delta z\, z_0)}{r\sqrt{\Delta x^2 + \Delta y^2 + 4\Delta z^2}} \quad (16.\text{a})$$

$$K_y = \frac{K_d(\Delta x\, x_0 + \Delta y\, y_0 + \Delta z\, z_0)}{r\sqrt{\Delta x^2 + \Delta y^2 + 4\Delta z^2}} \quad (16.\text{b})$$

$$K_z = \frac{2K_d(\Delta x\, x_0 + \Delta y\, y_0 + \Delta z\, z_0)}{r\sqrt{\Delta x^2 + \Delta y^2 + 4\Delta z^2}} \quad (16.\text{c})$$

In this study, the stiffness for the rotational direction K_ϕ, K_θ, K_ψ are all set to be 0, which means positioning control is implemented for the rotational direction.

Fig.7 shows the frequency characteristic of the coefficient of $H_d G_r - H$ in Eq.(13). By normalizing the frequency characteristic of the compliance control performance for each axis, the frequency characteristic of the desired compliance (admittance) can be also prescribed by the same frequency characteristic, which is useful in evaluating the realization of the desired compliance (admittance) in a frequency domain.

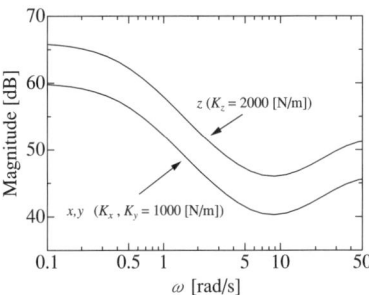

Fig. 7. Frequency characteristic

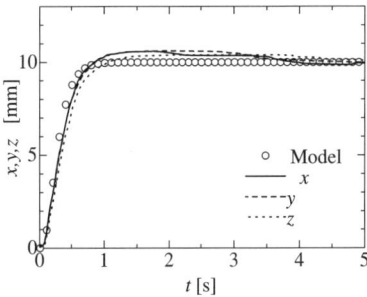

(a) Horizontal direction

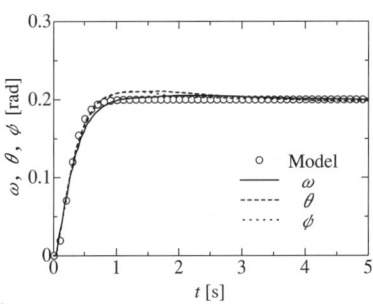

(b) Rotational direction

Fig. 8. Position control performance

IV. EXPERIMENTS AND DISCUSSION

A. Position control performances

Fig.8 shows the positioning step response, where (a) and (b) corresponds to the horizontal and rotational direction, respectively. A small circle ○ indicates the response of a model shown in Eq.(5), where parameters are chosen as $A = 38, B = 410, C = 1400$ in order that the step response may be almost the same with that of 2nd order system with $\omega_n = 8.0$ rad/s and $\zeta = 1.0$.

In the both figures, a little overshoot are confirmed but the obtained response for each direction is almost the same with that of the desired model, which proves an effectiveness of a proposed position control system.

B. Detection of contact force vector and contact point

Fig.9 shows the estimation performances of the contact force and contact point. Contact force is applied through a force sensor continuously 3 times for the same point (x_0, y_0, z_0)=(-21.4, 37.0, 74.0) [mm] as shown in the

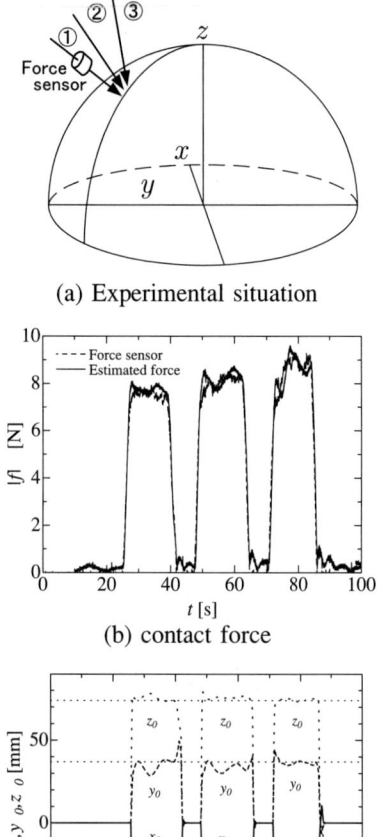

(a) Experimental situation

(b) contact force

(c) contact point

Fig. 9. Estimation performance of contact force and contact point

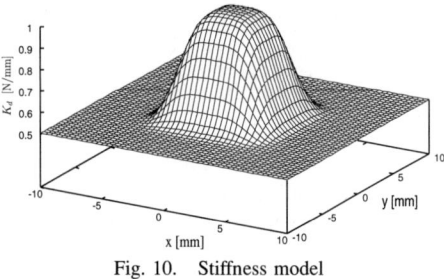

Fig. 10. Stiffness model

figure (a). The estimation performance of contact force and contact point are shown in (b) and (c), respectively. In spite that force is applied from various direction, both of the contact force and contact point can be confirmed to be estimated well, which proves the effectiveness of the proposed detection scheme.

C. Compliance display performance

Compliance displaying performances are verified. We introduce a geometrical model of a stiffness on a human skin as Eq.(17) apploximately, which means reference

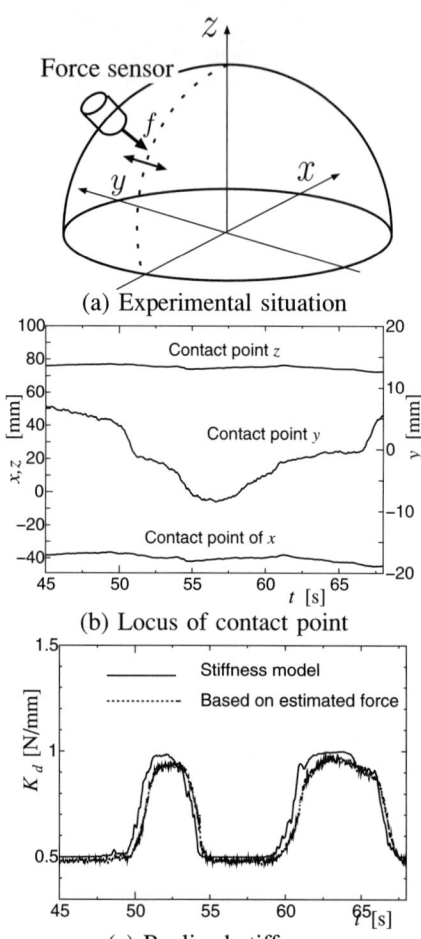

(a) Experimental situation

(b) Locus of contact point

(c) Realized stiffness

Fig. 11. Compliance display performance

stiffness K_d displaying for a fingertip has a maximum value of K_{max} at a point $[x_p, y_p, z_p]$ and according to going away from that point K_d closes to the minimum one K_{min}.

$$K_d = (K_{max} - K_{min})$$
$$\times \exp\{(-\frac{(x_0 - x_p)^4 + (y_0 - y_p)^4 + (z_0 - z_p)^4}{s_p})\}$$
$$+ K_{min} \qquad (17)$$

Fig.10 shows the geometrical image of Eq.(17), where restriction of $z_0 = 0$ is introduced to make possible it to be shown in the actual 3-d space, where $(x_p, y_p, z_p)=(0,0,0)$, $K_{max} = 1.0$[N/mm], $K_{min}=0.5$[N/mm], and spreading parameter $s_p = 200$.

Fig.11 shows the experimental result of compliance display by palpation motion. As shown in figure (a), human holds a force sensor and execute a round trip motion in order that contact point may go over the most rigid point with applying a force for normal direction continuously. The most rigid point is set to be $(x_p, y_p, z_p)=($-

even small variation of stiffness of 0.1 [N/mm] can be displayed, which is owing to the air compressibility and it is the advantage of employing a pneumatic driving system.

V. CONCLUSION

In this study, we developed a mechanical system using a pneumatic parallel manipulator, aiming at displaying a compliance characteristic on a human skin.

In order to realize such a motion, we proposed a compliance displaying scheme, where a contact point(which point on the surface of the device an operator is touching at) and contact force(how much force an operator is applying) are detected using an estimated force/moment with no use of general force/moment sensor and then display the target compliance determined according to the contact point by constructing a compliance control system.

Through some experiments, almost satisfactory control performances can be confirmed both in estimating contact force and contact point and in displaying the reference stiffness. The small variation of stiffness of 0.1 [N/mm] can be displayed, which is owing to the air compressibility of pneumatic actuator.

In addition to the further improvement of compliance displaying performance, the concrete recognition using not only sense of force feeling reported here but that of vision by constructing the deseased part in a computer using graphics image is under the current investigation.

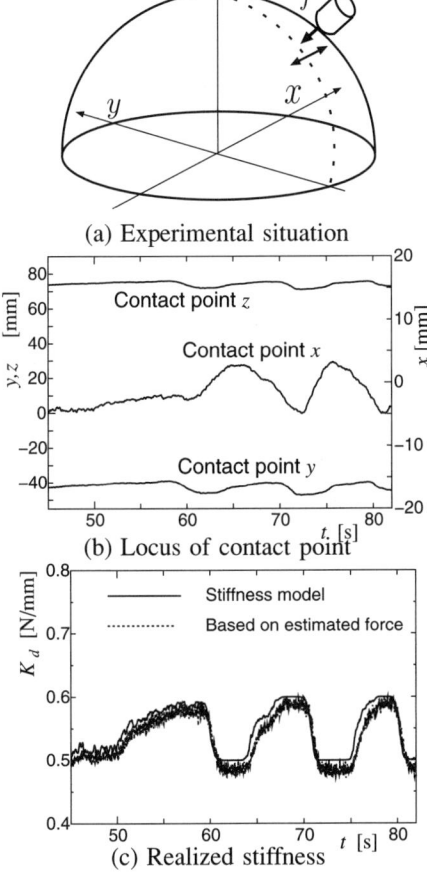

Fig. 12. Compliance display performance

40.0, 0.0, 75.5)[mm] and the stiffness is determined as $K_{max} = 1.0$ [N/mm], $K_{min} = 0.5$ [N/mm] based on the prior measurement stiffness data of a forearm.

Fig.11 (b) shows the locus of the contact point. Seeing from the figure, round trip motion is confirmed to be done for the direction along with y axis.

Fig.11 (c) shows the realized stiffness, where solid line corresponds to the value obtained from a models shown in Eq.(17), while dotted one indicates the actual realized stiffness obtained by a calculation of $|f_t|/\sqrt{(\Delta x)^2 + (\Delta y)^2 + (\Delta z)^2}$.

A response lag can be confirmed, which is considered to be resulted from dynamics lag of inner position control system, but almost satisfactory displaying property can be obtained, which proves that the proposed displaying scheme works properly. Further improvement of the displaying performance is the matter to be settled at present.

Fig.12 shows the same experimental results with Fig.11 except that the difference stiffness $K_{max} - K_{min}$ is quite small of 0.1 [N/mm] for the contact point of (x_p, y_p, z_p)=(0.0, -40.0, 75.5)[mm]. It is confirmed that

VI. REFERENCES

[1] J.Kim,S.De,Slinivasan, "Computationally efficient techniques for real time surgical simulation with force feedback", *Proc. 10th Symp. on Haptic Interface for Virtual Environment and Teleoperator Systems, HAPTICS 2002*, 2002, pp.51-57

[2] S.Majima and K.Matsushima, "Fuzzy Evaluation of Stiffness of Tissue by means of Micro-manipulator", *Proc. SPIE- Int. Soc. Opt. Eng. (USA)*, Vol.2101, No.1, 1993, pp. 521-526,

[3] M. Takaiwa and T. Noritsugu, "Development of Pneumatic Human Interface and Its Application to Compliance Display", *J. of Robotics and Mechatronics*, Vol.13, No.5,2001,pp.472-478

[4] D.Stewart, "A platform with Six Degrees of Freedom", *Proc. Inst. Mechanical Engineers*,180-15,1965,pp.371-386

[5] D.McCloy and H.R.Martin, "Control of Fluid Power: Analysis and Design", *2nd (Revised) Edition, (1980), pp.339*, John Wiley& Sons

[6] T. Noritsugu and M. Takaiwa, "Motion Control of Pneumatic Parallel Manipulator Using Disturbance observer", *Japan-U.S.A. Flexible Automations*, 1986

[7] T. Murakami and K. Ohnishi, "Advanced Control Technique in Motion Control", *The Nikkan Kogyo Shinbun Ltd., Japan*, 1990

[8] J.K.Salisbury, Jr., "Interpretation of Contact Geometries from Force Measurements", *Proc.of 1st Int. Symp. on Robotics Research(MIT Pewss)*,pp.565-577, 1984

Workspace and Dexterity Analyses of Hexaslide Machine Tools

A.B.Koteswara Rao
Dept. of Mechanical Engineering
Indian Institute of Technology Delhi
Hauz Khas, New Delhi–110 016
India
abkr_iitd@yahoo.co.in

P.V.M.Rao*
Dept. of Mechanical Engineering
Indian Institute of Technology Delhi
Hauz Khas, New Delhi–110 016
India
pvmrao@mech.iitd.ernet.in

S.K.Saha
Dept. of Mechanical Engineering
Indian Institute of Technology Delhi
Hauz Khas, New Delhi–110 016
India
saha@mech.iitd.ernet.in

Abstract - This paper presents kinematic analysis of a class of parallel manipulators, namely, Hexaslides, for machine tool applications. Hexaslides have constant-length legs. The inverse and direct kinematics solutions, to study the workspace properties of hexaslides are presented. Various kinematic performance indices, namely, workspace volume, workspace volume index i.e., the ratio of workspace volume to machine size and global dexterity index of hexaslides having same foot print size are used to study different rail-arrangements. Secondly, effect of two important parameters related to configuration, i.e., gap between adjacent rails in case of hexaglide configuration and inclination of rails incase of slanted configuration, on the performance measures is presented.

1. INTRODUCTION

Machine tools with parallel mechanisms, in which every axis is a direct link between the tool or mobile platform and fixed platform of the machine, unlike that in case of conventional serial machine tools where there exists a serial arrangement of feed axes and each axis has to carry load and moves the weight of all the following axes, meet the requirement of high dynamic performance, i.e., high stiffness constructions with little moving masses. Stewart platform [1] and the tyre test machine [2] are initial works in the area of parallel kinematic machines (PKMs). The machine tools based on parallel kinematic structures are reported in [3-6].

Parallel kinematic structures are made-up of one or more closed kinematic chains, whose end effector represents tool platform with several degree of freedom (d.o.f.) with respect to fixed platform. Guide chains coupling the two platforms can be moved independent to each other. One end of each guide chain, called strut, is coupled to the fixed platform while the other end is coupled to the tool platform by means of a suitable joint. Each joint allows several d.o.f. As there is no bending of legs, all axes simply have traction or compression forces. They are reconfigurable and can be built with relatively low investment as many of its components consist of standard machine elements.

Most common versions of PKMs, with six legs offering six-d.o.f., are Hexapods and Hexaslides. The hexaslides, based on the design chosen for feed drives, consist of six constant- length legs. The positions of the base joints, of the legs, on their respective rail -axes control the posture of the mobile platform in space. As the main moving parts, i.e., legs, can be made light but stiff, these find applications in machining, measuring, and handling etc. with high accuracy and precision. The Hexaslide machine tools (HSMs) require better workspace properties. Complex workspace is one of the major drawbacks of these HSMs. Moreover, the Jacobian matrix, *J*, which relates the joint rates to the output velocities, is not constant and not isotropic; the performances vary considerably for different points in the workspace and for different directions at one given point. This is a serious drawback for machining applications [6-8]. Hence in the design of HSMs, shape and size of the workspace, workspace volume index i.e., the ratio of workspace volume to machine size, dexterity, etc., are considered as the most important performance indices.

Out of the parameters that influence the performance of HSMs such as rail-arrangement, actuator stroke, leg-lengths, ranges of tool platform joints and double revolute joints; effect of rail-arrangement is the main concern in this work.

2. HEXASLIDE MACHINE TOOLS

A general hexaslide based machine tool consists of six distinct rails as shown in Fig. 1. The sliders move along their rails, whereas the legs of constant length are connected to the sliders through Revolute-Revolute joints. Other end of each leg is connected to the tool or mobile platform through spherical joints. Actuation of the sliders on their respective rails drives the tool platform in space.

There are primarily three machine tools based on hexaslides, namely, 1) Hexaglide [3], developed at ETH

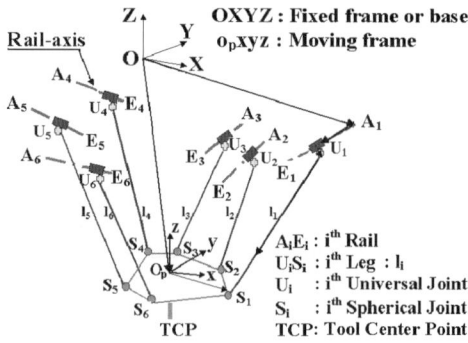

Fig.1 The general hexaslide machine tool

* Corresponding author

Zurich, consisting of coplanar and parallel rails as shown in Fig 2(a); 2) HexaM [4] developed by Toyoda consisting of slanted rails as shown in Fig 2(b) and 3) Linapod [5] developed at University of Stuttgart, which has the rails arranged in vertical direction as shown in Fig 2(c). These are all based on simple scissor drives [5] and the only difference is their rail-arrangement. The TIARA hexapod [9] with constant-length struts, consist of its rails resembling a tiara.

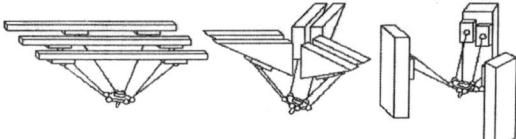

Fig. 2(a) Hexaglide   Fig. 2(b) HexaM   Fig. 2(c) Linapod
(courtesy: http://wwwrobot.gmc.ulaval.ca/~bonev/)

The workspace analysis of a general HSM based on vertex space concept [10] was proposed. Adopting a geometric algorithm, the volume and shape of the workspace at a given orientation, and their variation with orientation were discussed. Alternatively, a parameterized model [11] that applies to hexapods with fixed-length legs was developed to find the workspace and dexterity. However, there is no effort made to find the optimum rail angle. Moreover, the dexterity analysis was based on the variation in the singular values of the corresponding Jacobian matrix. The three different versions of HSMs, considered for comparison, consist of rails parallel by pairs only.

In regards to the indices of manipulator dexterity, the condition number ρ, given by $\rho = \sigma_{max} / \sigma_{min}$ where σ_{max} and σ_{min} are the largest and smallest singular values of the J, was used in [12]. Geometrically, the associated Jacobian matrix J describes a hyperellipsoid having lengths defined by its singular values. The determinant of this Jacobian matrix, $det(J)$ is proportional to the volume of hyperellipsoid. The condition number represents the sphericity of the hyperellipsoid. The manipulability measure [13] w, given by $w = \sigma_1\sigma_2\sigma_3\sigma_4\sigma_5\sigma_6 = \sqrt{det(JJ^T)}$ was defined to describe the ability of manipulator to change its position and direction in its workspace. The conditioning of J and the manipulability ellipsoid associated with J were used to optimize the workspace shape and performance uniformity of the Orthoglide [14]. Alternatively, global performance indices, that consider the dexterity of the manipulator over the entire workspace, were used in [15], [16].

For any practical purposes, prior to the development of any HSM with constant-length struts, it is necessary to identify a suitable arrangement of rails i.e., (i) either paired or unpaired rails, and (ii) orientation of the rails that fulfill the desired performances. Since no such data is readily available in the literature, a comparison among HSMs with different rail-arrangements related to various kinematic performance indices, namely, workspace volume, workspace volume index, and global dexterity index of the hexaslides having same foot print size but with different rail-arrangements, has been carried-out in this work. Effect of gap between adjacent rails and inclination angle of the rails on workspace and dexterity are also reported.

3. KINEMATIC ANALYSIS

The degree-of-freedom for a generalized HSM can be found, using the Kutz-bach criterion [17], as 6. For the kinematic analysis of the generic model of HSM, consider the loop $OA_1U_1S_1O_pO$ in Fig. 3, in which

O-XYZ : Fixed frame of reference attached to the base,
O_p-xyz : Moving frame attached to the tool platform,
$\boldsymbol{p} = \boldsymbol{OO_p} = [p_x, p_y, p_z]^T$, the position of center of the moving platform O_p in fixed frame and
$[R]$ is the Rotation matrix representing the orientation of the moving frame, O_p-xyz, with reference to O-XYZ.

for $i = 1$ to 6,
$\boldsymbol{u}_i$ is the unit vector along the ith rail in fixed frame; d_i is the distance of the ith actuator from A_i; $\boldsymbol{a}_i = \boldsymbol{OA}_i$, in fixed frame; $\boldsymbol{d}_i = \boldsymbol{A}_i\boldsymbol{U}_i$, in fixed frame; $\boldsymbol{l}_i = \boldsymbol{U}_i\boldsymbol{S}_i$, in fixed frame; $\boldsymbol{s}_i = \boldsymbol{A}_i\boldsymbol{S}_i$, in fixed frame; $\boldsymbol{b}_i = \boldsymbol{O}_p\boldsymbol{S}_i$, in moving frame.

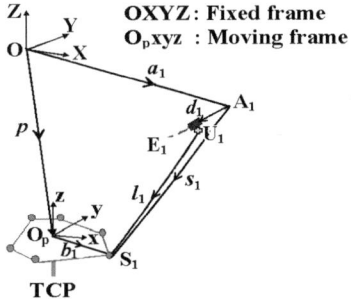

Fig. 3 Kinematic chain of HSM

3.1 Position Analysis

From the geometry, s_i can be written as,

$$\boldsymbol{s}_i = \boldsymbol{p} + [R]\boldsymbol{b}_i - \boldsymbol{a}_i = \boldsymbol{d}_i + \boldsymbol{l}_i, \text{ for } i=1 \text{ to } 6 \quad (1)$$

Since $\boldsymbol{d}_i = d_i\boldsymbol{u}_i$, (1) can be written as,

$$d_i\boldsymbol{u}_i = \boldsymbol{p} + [R]\boldsymbol{b}_i - \boldsymbol{a}_i - \boldsymbol{l}_i \quad (2)$$

or $(\boldsymbol{p}+[R]\boldsymbol{b}_i-\boldsymbol{a}_i-d_i\boldsymbol{u}_i)^T((\boldsymbol{p}+[R]\boldsymbol{b}_i-\boldsymbol{a}_i-d_i\boldsymbol{u}_i) = l_i^2$ (3)

The *inverse kinematic problem* can then be solved to find d_i using (3), while $\boldsymbol{p}$ and $[R]$ are given. Equation (3) is quadratic in d_i. There exists two solutions $d_i^{(1)}$, $d_i^{(2)}$. The true solution can be found based on the motion continuity. Alternatively, it can be shown that d_i is given by

$$d_i = \boldsymbol{u}_i \cdot \boldsymbol{s}_i - \sqrt{l_i^2 - [|\boldsymbol{s}_i|^2 - \{\boldsymbol{u}_i \cdot \boldsymbol{s}_i\}^2]}, \text{ for } i=1 \text{ to } 6 \quad (4)$$

Equation (4) also offers two values $d_i^{(1)}$, $d_i^{(2)}$. The true solution can be identified based on the motion continuity. The given pose is said to be achievable by the tool platform if the values of d_i satisfy the constraint

$$0 \leq d_i \leq r_i \quad (5)$$

along with the constraints imposed due to range of motion allowed by universal and spherical joints for all $i=1$ to 6, where r_i is the length of the ith rail.

To solve the *forward kinematics*, i.e., to find p and $[R]$ for given d_i, consider (3) in the following form:

$$f_i(x) = (q_i)^T (q_i) - l_i^2 = 0 \quad (6)$$

where $q_i = p + [R]b_i - a_i - d_i$. $\quad (7)$

For the HSM with the known geometry, i.e., a_i, and b_i, and given the values of d_i, (6) gives six scalar non-linear, simultaneous equations interms of the unknown vector $x = [p_x, p_y, p_z, \phi, \theta, \psi]^T$. As p_x, p_y and p_z are the three cartesian coordinates of the tool, and $\phi, \theta,$ and ψ are the *ZYZ* Euler angles [18], x represents *pose* vector of the tool platform. As (6) will not give closed-form solution, Newton-Raphson method [9], can be used.

3.2 Velocity Analysis

The time derivative of (2) yields,

$$\dot{d}_i u_i = v + \omega \times [R] b_i - \omega_i \times l_i \quad (8)$$

where $\dot{d}_i$ is the vector of actuator speeds, v and ω are the linear and angular velocities of the moving platform, and ω_i is the angular velocity of the ith leg. Taking the dot product of l_i on the both sides of (8), we get

$$\dot{d}_i u_i \cdot l_i = [v + \omega \times [R] b_i] \cdot l_i \quad (9)$$

Equation (9) may be represented in the matrix form as,

$$J_a \dot{d} = J_t \dot{x} \quad (10)$$

where $\dot{x} = [v^T \ \omega^T]^T$, vector of the end effector velocity; $\dot{d} = [\dot{d}_1 \ldots \dot{d}_6]^T$, the vector of actuator rates or speeds, $J_a = diag(l_1^T u_1 \ldots l_6^T u_6)$, which is a 6x6 matrix and J_t is given by,

$$J_t = \begin{bmatrix} l_1^T & ([R]b_1 \times l_1)^T \\ \vdots & \vdots \\ l_6^T & ([R]b_6 \times l_6)^T \end{bmatrix}_{6 \times 6}$$

Equation (10) may be re-written as

$$\dot{d} = J \dot{x} \quad (11)$$

where J is the Jacobian matrix of the tool platform and is given by, $J = J_a^{-1} J_t$. If J_a becomes singular for a certain position and orientation of the tool platform within the workspace of HSM, i.e., when $l_i^T u_i = 0$, for any $i=1$ to 6; it corresponds to the *stationary singularity* i.e., the leg standing perpendicular to the rail. In this case of singularity, the HSM loses one or more d.o.f. When J_t becomes singular, which corresponds to the *uncertainity singularity*, the HSM gains one or more d.o.f. Taking the time derivative of (8), acceleration analysis may also be done. As this analysis is not required for the present work, it is not presented.

4. PERFORMANCE MEASURES

In the design of HSMs, much concern is given to the shape and size of the workspace, ratio of workspace volume to machine size, dexterity etc., which are presented next.

4.1 Workspace Volume (WSV)

The complete or total workspace of a HSM is a six dimensional space for which complete graphical representation is very difficult to obtain. Out of different types of subsets of the complete workspace [19], the most commonly determined workspace is the constant-orientation workspace, which is a three-dimensional space or volume reachable by center of the tool platform (or the TCP) while orientation is constant. For the workspace evaluation, a search method [20], based on the inverse kinematics, is applied. Search proceeds by defining a bounding box covering a maximum possible reachable space of HSM, and then slicing the bounding box into a number of layers with each layer is being discretized into points. For each of these points the distance, d_i in Fig.3 is calculated and the constraints are checked. If the constraints are not violated, the point under computation is considered within the workspace, otherwise considered outside of the workspace. The workspace volume is defined as,

$$WSV = \sum A_m \Delta z \quad (12)$$

where A_m is the reachable area in the mth layer, and Δz is the layer interval.

4.2 Workspace Volume Index (WVI)

The workspace volume index is the ratio of workspace volume to the size of HSM. Size of HSM is taken as the product of area of the fixed platform and the maximum reach of the TCP in Z-direction.

4.3 Global Dexterity Index (GDI)

Dexterity is a measure of kinematic performance of HSMs. It depicts the ability to arbitrarily change its position and orientation, or apply forces and torques in arbitrary directions during machining. The Global Dexterity Index [15] is given by

$$GDI = \frac{\int_W \left(\frac{1}{\kappa}\right) dW}{\int dW} \quad (13)$$

where dW is an infinitesimal small element representing one of the workspace points and κ is the condition number of the Jacobian at that point. It represents the uniformity of manipulatability within the entire workspace. In the present work, for the comparison of HSMs, GDI using (13) is used with, $\kappa = \|J\| \|J^{-1}\|$, where $\|.\|$ refers to the 2- norm.

5. COMPARISON OF HSMs

A program has been developed in MATLAB to model a generic HSM configuration and to determine shape and size of the workspace, range of moving platform, workspace volume index and the global dexterity index. The results obtained from the program are verified by finding the workspace volume for certain known cases. One such validation involves finding workspace of a *Test machine*, shown in Fig. 4(a), with following details.

5.1 Test Machine and its Workspace

Fixed platform : Regular hexagon of side 500 units
Moving platform : Regular hexagon of side 100 units
Legs : 6 Nos of each 400 units length
Rails : 6 Nos of each 250 units length

Rails are arranged vertically through the vertices of fixed platform. For this test machine, the workspace would be just a vertical line as the tool tip traces a straight-line and all these points would be singular points suffering from the stationary singularity. The workspace plot, obtained from the program is shown in Fig. 4(b).

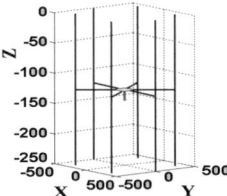
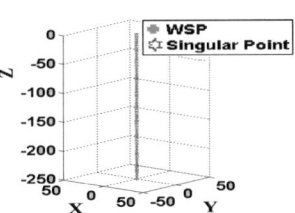

Fig. 4(a) Test machine   Fig. 4(b) Workspace and singular points

5.2 Configurations of HSM considered for Comparison

The performance indices of HSMs having different rail-arrangements have been evaluated. Results for some typical cases, which also include most of the rail-arrangements proposed by researchers in the past, have been presented. These configurations shown in Fig. 5(a) to Fig. 5(g) are as follows:

HSM with Radial rails: All the rail-axes are radial and are symmetric with Z-axis as shown in Fig. 5(a)
HSM with Radial rails parallel by pairs: The rail-axes are parallel by pairs, the pairs being radial and symmetric with Z-axis as shown in Fig. 5(b)
HSM with Slanted rails: All the rail-axes are slanted and symmetric with Z-axis as shown in Fig. 5(c)
HSM with Slanted rails parallel by pairs: The rail-axes are slanted, parallel by pairs, the pairs being symmetric with Z-axis as shown in Fig. 5(d)
HSM with Vertical rails: All the rail-axes are vertical and symmetric with Z-axis as shown in Fig. 5(e)
HSM with Vertical rails parallel by pairs: The rail-axes are vertical, parallel by pairs, the pairs being symmetric with Z-axis as shown in Fig. 5(f)
HSM with rails parallel, coplanar: Hexaglide - Fig. 5(g).

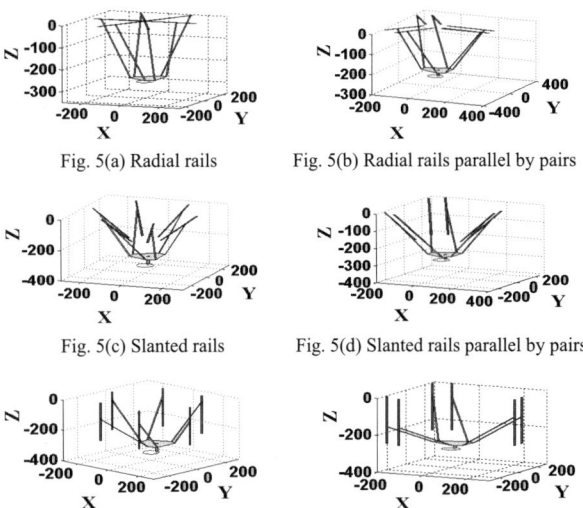

Fig. 5(a) Radial rails   Fig. 5(b) Radial rails parallel by pairs

Fig. 5(c) Slanted rails   Fig. 5(d) Slanted rails parallel by pairs

Fig. 5(e) Vertical rails   Fig. 5(f) Vertical rails parallel by pairs

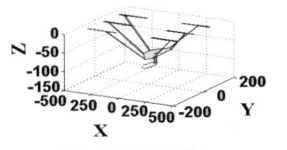

Fig. 5(g) Hexaglide

Fig 5(a)-(g) HSMs with different rail-arrangements

For the purpose of comparison, same sized fixed platform, same sized moving platform and same actuator stroke i.e., lengths of rails are used for all configurations. Fixed platform is a regular hexagon of side 0.255m for the HSM shown in Fig. 5(a), (c) and (e). It is a hexagon, for the HSM having rails parallel by pairs, with long and short sides equal to 0.5m and 0.08m, respectively. For the hexaglide configuration shown in Fig. 5(g), it is a rectangle. The gap between adjacent rails (G) and position of rails are considered such that the area of this rectangle is same as that of fixed platform in other cases. Tool platform is taken as a hexagon, for all HSMs, with long and short sides equal to 0.141m and 0.052m, respectively. Length of each rail is taken as 0.250m and length of each leg is taken as 0.3m for all cases of study.

As the first three columns of Jacobian matrix i.e., J in (11) are dimensionless while the last three columns have units of length, in order to analyze the global dexterity index independent of the physical size, the last three columns are divided by the distance between center of the tool platform and the center of the spherical joint on the tool platform.

The performance indices, workspace volume, workspace volume index, global dexterity index and range of TCP in X-direction (R_X), in Y-direction (R_Y) and in Z-direction (R_Z), for HSMs with different rail-arrangements, considering the HSM with its tool platform at constant-orientation i.e., horizontal only, are presented in Table 1. It can be observed from Table 1 that maximum workspace volume is offered by the HSM having slanted configuration with paired rails.

Comparison of the performance indices in case of hexaglide configuration for different gaps between adjacent rails is given in Table 2. It may be noted that Hexaglides 1, 2, 3, and 4 have same area of the fixed platform, which is a rectangle.

TABLE 1
PERFORMANCE INDICES FOR DIFFERENT VERSIONS OF HSMs

R_X, m	R_Y, m	R_Z, m	WSV, m^3	WVI	GDI
HSM with Radial rails					
0.14	0.14	0.04	0.00028	0.006	0.057
HSM with Slanted rails (48^0)					
0.44	0.44	0.22	0.01624	0.205	0.069
HSM with Vertical rails					
0.22	0.21	0.22	0.00385	0.049	0.083
HSM with Radial rails parallel by pairs, G = 0.080m					
0.32	0.36	0.11	0.00214	0.044	0.215
HSM with Slanted rails parallel by pairs (43^0), G=0.080m					
0.45	0.44	0.28	0.01797	0.231	0.257
HSM with Vertical rails parallel by pairs, G = 0.080m					
0.13	0.14	0.24	0.00183	0.026	0.291
Hexaglide, G = 0.102m (which offers maximum WSV)					
0.17	0.40	0.24	0.00931	0.230	0.034

TABLE 2
VARIATION IN PERFORMANCE INDICES WITH THE GAP BETWEEN ADJACENT RAILS FOR HEXAGLIDE

	WSV, m^3	WVI	GDI
Hexaglide-1: G = 0.080m	0.00049	0.029	0.018
Hexaglide-2: G = 0.085m	0.00264	0.098	0.025
Hexaglide-3: G = 0.102m	0.00931	0.230	0.034
Hexaglide-4: G = 0.168m	0.00011	0.002	0.021

Variation of performance indices of HSM with slanted configuration, having paired rails, and having unpaired rails, with the inclination of the rails with horizontal (β) have been computed and compared in Tables 3 and 4 respectively. Variation of WSV and GDI with the inclination of the rails is shown in Fig 6(a) and Fig. 6(b) respectively. Due to the space limitation, Shapes of workspaces of some salient HSMs having rails parallel by pairs only are shown in Fig. 7(a-d). In these plots, WSP and ICP correspond to the Workspace Points and Ill-Conditioned Points respectively. The reachable points, for which the condition number of the Jacobian matrix is more than 50, are considered as Ill-conditioned points.

TABLE 3
PERFORMANCE INDICES FOR DIFFERENT RAIL INCLINATION
(RAILS PARALLEL BY PAIRS)

β, deg	R_X, m	R_Y, m	R_Z, m	WSV, m^3	GDI
0	0.32	0.36	0.11	0.00214	0.215
22.5	0.46	0.48	0.21	0.01188	0.183
43	0.45	0.44	0.28	0.01797	0.257
67.5	0.32	0.30	0.31	0.01097	0.335
90	0.13	0.14	0.24	0.00183	0.291

TABLE 4
PERFORMANCE INDICES FOR DIFFERENT RAIL INCLINATION
(RAILS UNPAIRED)

β, deg	R_X, m	R_Y, m	R_Z, m	WSV, m^3	GDI
0	0.14	0.14	0.04	0.00028	0.057
22.5	0.33	0.33	0.12	0.00394	0.042
48	0.44	0.44	0.22	0.01624	0.069
67.5	0.36	0.37	0.25	0.01302	0.101
90	0.22	0.21	0.22	0.00385	0.083

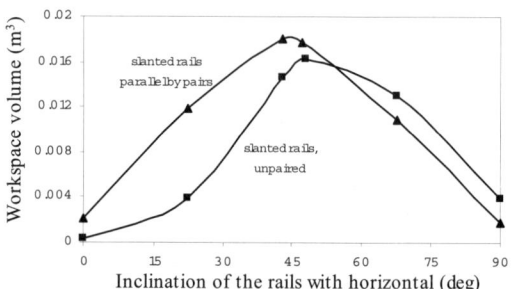

Fig. 6(a) Variation of workspace with the rail inclination

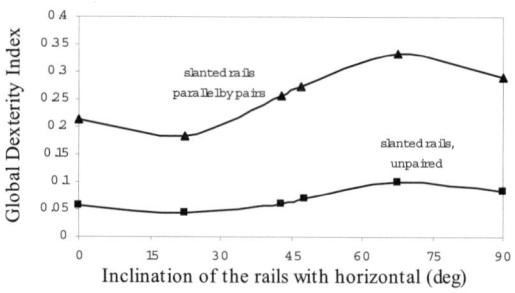

Fig. 6(b) Variation of Dexterity with the rail inclination

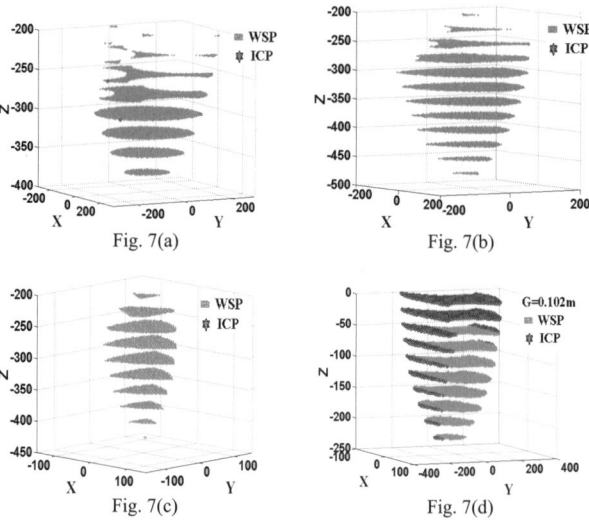

Fig. 7(a)-(d) Workspace plots of different HSMs
(a) for HSM-paired rails, β =22.5^0, (b) for HSM-paired rails, β =67.5^0, (c) for HSM-paired rails, β =90.0^0 and (d) for Hexaglide configuration

6. RESULTS AND DISCUSSION

In this work, workspace and dexterity analyses of a generic hexaslide manipulator was carried out. This generic model covers most of the specific configurations proposed by the researchers so far. Configurations include HSMs with both paired rails and unpaired rails. The results indicate that the workspace and dexterity strongly depend on the arrangement of rails. Effect of two important parameters related to configuration, gap between adjacent rails in case of hexaglides and inclination of rails incase of HSMs having slanted rails on workspace and dexterity was studied.

It was found that the HSM having slanted rails parallel by pairs offer maximum workspace. The optimum slanting angle was found to be in the vicinity of 43^0. A maximum value of work volume index, found to be 0.231 i.e., workspace would be 23.1% of the machine size, and was offered by the same HSM having slanted rails parallel by pairs with a slant angle of 43^0. But the dexterity offered by this configuration, however, is not optimum. The global dexterity index, in case of these HSMs with slanted rails parallel by pairs, ranges from 0.183 to 0.335 corresponding to 22.5^0 and 67.5^0 rail-inclinations respectively. In case of hexaglide configuration, maximum workspace volume is offered when the gap between adjacent rails is 0.102m. The results indicate that the workspace and dexterity strongly depend on the arrangement of rails i.e., gap between adjacent rails in case of hexaglide configuration and inclination of rails in case of HSMs with slanted rails has been established.

7. CONCLUSIONS

In this work, kinematic analyses of a class of parallel manipulators, namely, Hexaslides was carried-out. Various kinematic performance indices, namely, workspace volume, workspace volume index and global dexterity index of hexaslides having same foot print size were used to study different rail-arrangements. Influence of the gap between adjacent rails in case of hexaglide configuration and inclination of rails incase of slanted configuration, on the performance measures was also presented. Strong dependence of workspace and dexterity on gap between adjacent rails and inclination of rails has been established.

8. FUTURE WORK

The present work can be extended to a generic HSM configuration and workspace and dexterity analyses for all possible tool orientations. There is also a need to study the influence of other design parameters such as leg-length, rail-length etc. to arrive at an optimum design configuration. However, one main concern, which needs to be addressed in extending the work, is to improve the computational efficiency in evaluating the performance indices.

REFERENCES

[1] D.Stewart, "A platform with six degrees of freedom", *Proceedings Institution of Mechanical Engineers*, (Part-I), vol. 180, no. 15, 1965, pp. 371-386.

[2] V.E.Gough and S.G.Whitehall, "Universal Tyre test machine", in *Proceedings of the 9th International Technical Congress*, F.I.S.I.T.A., Institution of Mechanical Engineers, vol.. 117, 1962, pp.117-135.

[3] M.Honegger, A.Codourey and E.Burdet, "Adaptive control of the Hexaglide, a 6 DOF parallel manipulator", in *Proceedings of IEEE International Conference on Robotics and Automation*, Albuquerque, 1997, pp.543-548.

[4] M.Suzuki, K.Watanabe, T.Shibukawa, T.Tooyama and K.Hattori, "Development of milling machine with parallel mechanism", *Toyota Technical Review*, vol. 47, no. 1, 1997, pp.125-130.

[5] G.Pritschow and K.H.Wurst, "Systematic design of Hexapods and parallel link systems", *Annals of the CIRP*, vol. 46, no. 1, 1997, pp.291-295.

[6] J.Kim, F.C.Park and J.M.Lee, "A new parallel mechanism Machine tool capable of five-face machining", *Annals of the CIRP*, vol.48, no.1, 1999, pp.337-340.

[7] P.Wenger, C.Gosselin and B.Maille,"A Comparative study of serial and parallel mechanism topologies for Machine tools", in *Proceedings of PKM'99, Milano*, 1999, pp. 23-32.

[8] T.Treib and O.Zirn, "Similarity laws of serial and parallel manipulators for Machine Tools", in *Proceedings of The Fourth International Conference on Motion and Vibration Control - MOVIC'98* Zurich, August 25-28. (www.ifr.mavt.ethz.ch/movic98/proceedings/1c.pdf)

[9] D.N.Centea, H.Lacheray, F.Audren, R.Teltz and M.A.Elbestawi, "Development of TIARA Hexapod-a Machine tool based on parallel kinematic structures", T*ransactions of ASME Journal of Manufacturing Science and Engineering*, vol. 10, 1999, pp.869-878.

[10] I.A.Bonev and J.Ryu, "Workspace analysis of 6-PRRS parallel manipulators based on the vertex space concept", in *Proceedings of the ASME Design Engineering Technical Conferences*, September 12-15, 1999, *DETC 99 / DAC* -8647.

[11] F.Xi, "A comparison study on Hexapods with fixed-length legs", *International Journal of Machine Tools & Manufacture*, vol. 41, 2001, pp.1735-1748.

[12] J.Salisbury and J.Craig,"Articulated Hands:Force control and kinematic issues", *International Journal of Robotic Research*, vol.1, no. 1, 1982, pp.4-17.

[13] T.Yoshikawa, "*Foundations of robotics: analysis and control*", PHI, MIT Press, 1998, pp.131.

[14] P.Wenger and D.Chablat, "Design of a three-axis Isotropic parallel manipulator for machining applications: The Orthoglide", in *Proceedings of the Workshop on Fundamental Issues and Future Research Directions for Parallel Mechanisms and Manipulators*, October 3-4, Quebec, Canada, 2002, pp.16-24.

[15] C.Gosselin and J.Angeles, "A global performance index for the kinematic optimization of robotic manipulators", *ASME International Journal of Mechanical Design*, vol. 113(3), 1991, pp.220-226.

[16] Xin-Jun Liu, Zhen-Lin Jin and Feng Gao, "Optimum design of 3-DOF spherical parallel manipulators with respect to the conditioning and stiffness indices", *Mechanism and Machine Theory*, vol. 35, 2000, pp.1257-1267.

[17] G.N.Sandor, and A.G.Erdman, "*Advanced Mechanism Design: Analysis and Synthesis*", Prentice Hall, vol.2, 1988, pp.550-551.

[18] Lung-Wen Tsai, "*Robot Analysis*", John Wiley & Sons, Inc., 1999, pp 31-45.

[19] J.P.Merlet, "Determination of 6D workspaces of Gough-Type parallel manipulator and comparison between different geometries", *The International Journal of Robotics & Research*, vol.18, no.9, 1999, pp.902-916.

[20] O.Masory and J.Wang, "Workspace evaluation of Stewart platforms", *Advanced Robotics*, vol.9, no.4, 1995, pp.443-461.

Task Teaching to a Force-Controlled High-Speed Parallel Robot

Daisuke Sato, Takeshi Shitashimizu and Masaru Uchiyama
Department of Aeronautics and Space Engineering,
Graduate School of Engineering, Tohoku University,
Aoba-yama 01, Sendai 980-8579, Japan.
E-mail: {sato, shitashi, uchiyama}@space.mech.tohoku.ac.jp

Abstract

This paper purposes to implement fast and complicated tasks by a high-speed parallel robot, HEXA, for industrial applications and the others. Although this parallel robot does not use any force/torque sensors, it has sufficient capabilities to achieve various tasks by applying an integrated control of motion, force and compliance, which takes advantage of the back-driveability and friction compensation of the actuators. A key point is how simply and easily operators can bring out the performance. We develop an off-line task teaching system to realize the motion planning suitable for tasks. Some tasks are demonstrated by using this teaching and control system and we experimentally confirm the effectiveness of our system.

1 Introduction

Parallel robots have been developed and studied by many researchers because of serial robot deficiencies [1]. They are closed-loop mechanisms presenting great performances in terms of accuracy, rigidity and ability to manipulate large loads. For these characteristics, they have been used in a lot of applications ranging from medical science to amusement industry, and are becoming popular in machine tool industry.

In proportion to increases in the field of application, new requirements have arisen for these robots, for example, fast motion, compliance and force control. Clavel proposed the fast parallel manipulator DELTA with 3-DOF in translation [2]. And we developed a 6-DOF high-speed parallel robot HEXA [3, 4]. The overview is shown in Figure 1. This robot consists of the HEXA mechanism using direct-drive motors and possesses high speed, accuracy, rigidity and large load performances [5].

Moreover, HEXA is equipped with an integrated control of motion, force and compliance without using any force/torque sensors [6, 7]. In this control, the force between the end-effector and the external environment is detected by taking advantage of the

Figure 1: An overview of a high-speed parallel robot HEXA.

back-driveability and a friction compensation of the actuators. Therefore, because the weight of the moving parts remains light, we can keep the high-speed performance and cut expenses of the sensor.

Using this practical integrated control and usual direct teaching, the sufficient capabilities to realize various tasks have verified experimentally, for instance, sequential peg-in-hole insertion, inclined crank-turning, deburring and composite tasks [8]. However, to realize the motion planning suitable for tasks, operators must spend much time to iteratively adjust many technical parameters, which are set to smoothly switch the control modes. Therefore, a key point at the next stage is how simply and easily operators can handle HEXA.

There are some methods to provide a robot with tasks. Researchers make a lot of studies about teaching technologies, which is one of them, and it can be divided into four types, direct teach, remote teach, indirect teach and off-line teach. Recently, various off-line teaching methods have been developed, for example, to teach the robots through the CAD simulation data, to use the visual feedback and to make use of the human skills [9, 10, 11].

In view of the capabilities and the usability of the HEXA robot with these research backgrounds, especially for the high-speed performance and force control capability without using any force/torque sensors, it is very important to develop an effective teaching system that can utilize all the advantages.

To achieve this goal, we propose a 3D graphics-based off-line teaching system. The teaching approach of this system is characterized in that it creates and modifies the operating commands in units of primitive motions. Concretely, the primitive motions consist of the teaching points data and the control parameters of the robot. By dealing with the whole motion of implemented task as the units of small motions, it becomes simple and easy to teach the force and compliance elements and include the operator's ideas in the command data. In addition, the teaching work and time is saved because the primitive motions are reusable.

Following the experimental results of two tasks, peg-in-hole insertion and crank-turning, we show the effectiveness of our teaching system and demonstrate that it can bring out the capabilities of HEXA.

2 Off-line Teaching of Constrained Tasks

Although the parallel robot HEXA does not use any force/torque sensors, it has sufficient capabilities to achieve various tasks by applying a motion, force and compliance control to exploit the motor back-driveability. The details of this integrated control, please refer to Appendix A. The key point is how simply and easily operators can handle the robot with the integrated control. Particularly, it is important to deal with the teaching of constrained tasks because of the industrial applications. But it is very difficult for the operators to manage the parameters of the force control and to teach constrained tasks to the robot. Therefore, the new teaching system is nessesarry for the HEXA robot to bring out the performance and implement fast and compliant tasks.

2.1 Off-line Task Teaching System

The setup of the off-line task teaching system is composed of two subsystems: the HEXA robot real-time control subsystem and a task teaching subsystem with 3D graphics model and Graphical User Interface (GUI) shown in Figure 2.

The teaching subsystem runs the graphic control panels and the 3D graphics model, and generates the teaching data depending on the planned task. Moreover, computing the kinematics, it draws and simulates the graphics to actually understand how the real system will behave with the current task planning. It also analyzes the experimental data utiliz-

Figure 2: 3D graphics model and GUI environment.

ing MATLAB^{TM} software. On the other hand, the robot control system acquires the planning data from the teaching subsystem and executes the desired tasks with the integrated control.

2.2 Task Teaching Method

The procedure of the proposed off-line task teaching method is divided into four steps. And the task teaching goes by the three types of coordinate systems, the HEXA base coordinate system, the end-effector coordinate system and the task coordinate system. Here, the task coordinate system can arbitrarily be defined by the operator. Using only the end-effector coordinate system, it is difficult for the operator implement suitable strategies for each task. Especially, it is very hard to set the parameters for the force control. The task coordinate system can solve this problem and reflects the operator's ideas into the teaching.

Step 1: Teaching

Using the GUI and the graphics simulator, the following parameters can be selected/set for each teaching point or point-to-point interval to generate the teaching data for the robot:

- the friendly teaching coordinate system and the simulator's viewing point according to the task,
- the position and orientation of the end-effector,
- the control mode between the current and the next teaching point along with the reference force and stiffness of the end-effector,
- the time interval to execute a task between the current and the next teaching point,
- the interpolation methods (5th order, 4-1-4, or spline interpolation) separately for each of the reference position, orientation, force and stiffness, depending on how well a method defines an individual path,
- the gravity and friction compensation methods for an individual point-to-point interval,

- the control gains and servo sampling time for each task interval.

For spline interpolation, the MATLAB™ Spline Toolbox is used to reduce the program development efforts. Due to the guaranteed smooth velocity and acceleration in the interpolation, it is possible to arbitrarily select the teaching data and even the execution time intervals without reducing the accuracy.

Moreover, this teaching system can group several teaching points into a unit of primitive motion. And the system can edit and save these units utilizing the following functions: the parallel translation, the scaling translation, the rotational transformation, the splitting and attachment, the extraction and insertion, the selection of control parameters.

Increase of the number of these parameters raises the complication of the teaching procedure. However, these teaching functions are made user-friendly to create and modify motion data by treating them in units of the small motions and employing variable task coordinate systems.

Step 2: Planning

Based on the above teaching parameters and task conditions, the following reference data at each sampling time interval are generated by the task planning program and saved to the data files (refer to the symbol of each parameter described in the equation (4) of Appendix A):

- the end-effector position and orientation data in the HEXA base coordinate system,
- the rotation angle θ_{com} and angular velocity for each motor $\dot{\theta}_d$,
- the forces and torques in the end-effector coordinate system F_d,
- the end-effector stiffness matrix K,
- the control modes, gravity and friction compensation modes at a reference point.

Step 3: Verification

Using these teaching data files, the operator runs the 3D simulations to verify whether the generated teaching data are feasible to realize the desired task(s) or not. This step is omissible if it is not necessary to verify the motions. However, in the case of complex tasks, if some problems are noticed in the generated data, all the previous steps are to be repeated to generate another set of teaching data. Accordingly, this step is very helpful to recognize the robot motions beforehand and to correct these parameters.

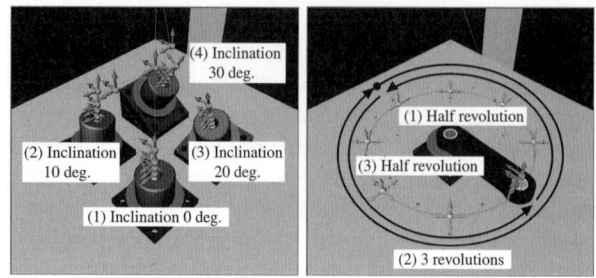

(a) Peg-in-hole insertion.   (b) Crank-turning.

Figure 3: Two task teaching experiments.

Step 4: Execution

Finally, once the feasibility of the taught task is verified, the control algorithm accesses these reference data files and executes the desired task(s) on the HEXA robot.

Therefore, using these steps and the system functions, the operator can manipulate the whole system in the way he/she feels convenient to teach the tasks. Moreover, it is possible to achieve fast and compliant execution of the tasks by the HEXA robot.

3 Task Teaching Experiments

To verify the functionality of the proposed off-line task teaching system, we have experimentally implemented the planning and teaching of two tasks: a peg-in-hole insertion task and a crank-turning task.

In the peg-in-hole insertion task illustrated in Figure 3 (a), four sockets are fixed on the xy plane with different inclination angles of 0, 10, 20, and 30 degrees. The length and diameter of the peg are 0.030 m and 0.016 m respectively. The clearance between the peg and the hole is 0.05×10^{-3} m. The robot inserts the peg, attached to the end-effector, into these holes in a sequential order.

In the crank-turning task illustrated in Figure 3 (b), a crank fixed on the xy plane is rotated with a peg attached to the end-effector of HEXA. The length, diameter, clearance of peg, and the radius of crank are 0.015 m, 0.017 m, 0.05×10^{-3} m, and 0.120 m, respectively. After inserting the peg into a hole at the end of the crank, which is located at 0.10 m from the center, the crank is turned 180 degrees clockwise, then three complete revolutions in the counter-clockwise direction, and then 180 degrees in the clockwise direction illustrated.

3.1 Teaching the Tasks to HEXA

The teaching for the peg-in-hole insertion task is done in five steps, illustrated in Figure 4.

At the beginning, the operator creates a path of the

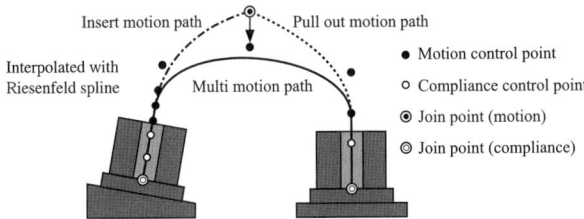

Figure 4: The teaching points and the switching points of the control modes for peg-in-hole insertion task.

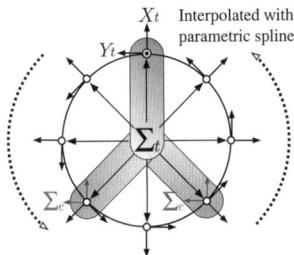

Figure 5: The task coordinate systems for crank-turning task.

peg inserting motion for one hole and saves this motion data as a unit of primitive motions. Next, using the extracting function, the operator makes the created inserting motion of a new path to pull the peg out of the hole. In detail, the extracting function gets the teaching data, which consist of the created inserting motion, and replaces them in reverse order.

This modified data is saved for the retraction motion of the peg. Moreover, the operator integrates the retraction motion with the insertion using the attachment function. Hereby, a peg-in-hole motion for one hole is created as a unit of new basic motion. Next, using the edit functions of this teaching system, for example, the parallel translation, the rotational transformation, and the attachment function, the operator creates another peg-in-hole motion which is different as the hole is inclined. And he/she combines all of them to make another unit. This makes the almost-complete required motions.

However, only simple attachment of four peg-in-hole motions cannot achieve the effective results. The operator reconstructs the four motions as one unit and finally, selects and adjusts the control modes and parameters for each unit. The teaching points in Figure 4 represent the switching among the control modes. Customizing these control parameters, the operator can realize more effective motions for tasks.

In the same way, the crank-turning task are created from the teaching data in the units of primitive motions. In the crank-turning task, in addition to the above insertion and retraction motions, the oper-

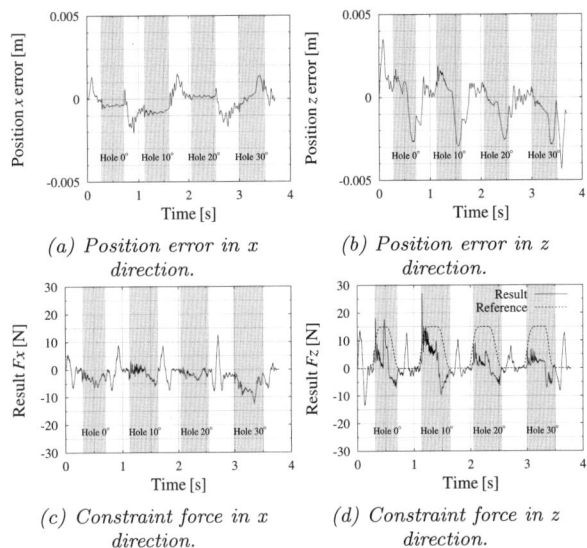

Figure 6: The results of peg-in-hole insertion.

ator creates a new motion that is to turn the crank one revolution in the clockwise direction, illustrated in Figure 5.

Splitting and extracting this turning motion by the edit functions, three turning motions, a clockwise half revolution, a counterclockwise half revolution, three revolutions clockwise can be readily prepared. After the operator attaches these motions to a single task, the control parameters and modes are adjusted.

Moreover, for these tasks, we have tried to reduce the execution time as much as possible.

3.2 Experimental Results and Discussion

The experimental results showing the features of the two tasks are shown in Figures 6, 7. Plots (a), (b) in each set show the position errors between the reference and the real data, and plots (c), (d) show the constraint forces on the end-effector.

In accordance with the above mentioned teaching approach, we have created the teaching data and executed the two experiments. At several times, the task execution failed because of the geometrical errors between the graphics and the real system, and the inappropriate control parameters. However, it is possible to know these errors to some extent by using the data analysis function of the system. Therefore, they are immediately known, and the parameters can be edited. Even the control parameters can also be edited in detail after the easy debugging of the tasks. In this way, as shown in these figures, smooth task motions with high accuracy and small force errors have been realized. Moreover, all tasks could be realized

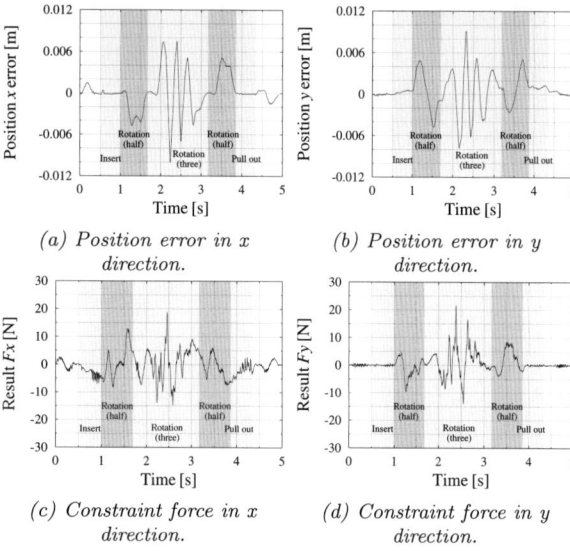

Figure 7: The results of crank-turning.

with sufficiently high velocities.

Besides these geometrical and control parameters' errors, some problems occurred due to the excessive friction between the peg and the hole. This friction causes a decrease in the driving force of the robot and hence the task cannot be completed properly. Accordingly, this problem could also be corrected by adding to the reference forces shown in Figure 6 (d).

4 Conclusions

In this paper, we have proposed a graphics based off-line task teaching system and implemented it for a high-speed parallel robot HEXA. To demonstrate its performance, two tasks were taught and executed. In the task teaching, using the units of primitive motions, the task coordinate system and user-friendly teaching system functions, the operator can create and modify the motion data of the tasks and the control parameters which are controlled for the position, force, and compliance. From the results obtained, it could be concluded that it is possible to exploit the very specific characteristics of the robot with a speedy task teaching. Consequently, we have confirmed the effectiveness of this off-line task teaching system and demonstrated that it can bring out the capabilities of HEXA for various applications.

References

[1] J. -P. Merlet, "Parallel Robots," *Kluwer Academic Publishers*, 2000.

[2] R. Clavel, "DELTA, A Fast Robot with Parallel Geometry," *Proc. of 18th Int. Symp. on Industrial Robots*, pp. 91–100, 1988.

[3] F. Pierrot, M. Uchiyama, P. Dauchez and A. Fournier, "A New Design of a 6-DOF Parallel Robot," *J. of Robotics and Mechatronics*, Vol. 2, No. 4, pp. 308–315, 1990.

[4] M. Uchiyama, K. Iimura, F. Pierrot, K. Unno and O. Toyama, "Design and Control of a Very Fast 6-DOF Parallel Robot," *Proc. of the IMACS/SICE Int. Symp. on Robotics, Mechatronics and Manufacturing Systems*, pp. 473–478, 1992.

[5] M. Uchiyama, K. Masukawa and T. Sadotomo, "Experiment on Dynamic Control of a HEXA-Type Parallel Robot," *Proc. of 1st World Automation Congress (WAC'94)*, pp. 281–286, 1994.

[6] M. Uchiyama, T. Miwa and D. N. Nenchev, "A Very Fast Parallel Robot to Be Applied to Dexterous Motion," *Proc. of the World Automation Congress (WAC '96)*, Vol. 3, pp. 753–758, 1996.

[7] M. Uchiyama, D. Kim and B. Porapukham, "Integrated Control of Motion, Force and Compliance of a Robot without Using Any Force/Torque Sensor," *Proc. 2nd IMACS Int. Multiconference, Computational Engineering in Systems Applications (CESA '98)*, Vol. 4, pp. 833–838, 1998.

[8] D. Kim and M. Uchiyama, "A Force/Torque Sensor-less Realization of Fast and Dexterous Tasks with a Parallel Robot," *2000 IEEE Int. Conf. on Industrial Electronics Control and Instrumentation (IECON-2000)*, pp. 223–228, 2000.

[9] Y. Kuniyoshi, M. Inada and H. Inoue, "Teaching by Showing, generating Robot Programs by Visual Observation of Human Performance," *Proc. 20th Int. Symp. on Industrial Robotics*, pp. 119–126, 1989.

[10] S. Liu and H. Asada, "Teaching Human Motion/Force Skills to Robots," *J. of the Robotics Society of Japan*, Vol. 13, No. 5, pp. 592–598, 1995.

[11] H. Onda, H. Hirukawa, F. Tomita, T. Suehiro and K. Takase, "Assembly Motion Teaching System using Position/Force Simulator – Generating Control Program –," *Proc. of the 1997 IEEE/RSJ Int. on Intelligent Robots and Systems*, Vol. 2, pp. 938–945, 1997.

Appendix A: Motion, Force and Compliance Control to Exploit Motor Back-Driveability

HEXA enjoys the advantages of high speed, accuracy, rigidity and large load performances. For this parallel robot, we have proposed the integrated control of motion, force and compliance.

The approach combines several control algorithms. It realizes position control in free space and compliance control in contact phase. When reference force command is given, it carries out force control utilizing the back-driveability of the direct-drive motors in place of force/torque sensors. In addition to combine with the friction compensation of the motors and the gravity compensation of the moving parts, smoothly switching among the control modes is achieved [6, 7, 8].

A.1 Friction compensation

The friction compensation model for the i-th motor is shown in Figure 8, where $\dot{\theta}_i$ is the motor velocity, $\tau_{fr,i}$ is the frictional torque, $R_{si(\pm)}$ is the maximum static friction, $R_{i(\pm)}$ is the dynamic friction, and $\pm\epsilon$ is a threshold of the velocity to decide the state of motion. And we consider the following two ways of the friction compensation.

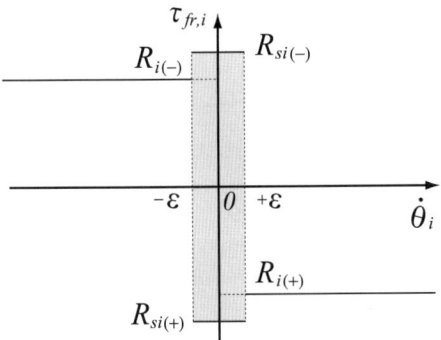

Figure 8: The friction compensation model.

A.1.1 Feedforward compensation

The friction compensation torque $\tau_{fr\_ff,i}$ of the i-th motor depends on the desired motor velocity $\dot{\theta}_{d,i}$, which is calculated from the desired trajectory. The compensation torque is calculated by

$$\tau_{fr\_ff,i} = \begin{cases} R_{i(+)} & \epsilon < \dot{\theta}_{d,i} \\ R_{si(+)} & 0 < \dot{\theta}_{d,i} \leq \epsilon \\ R_{si(-)} & -\epsilon \leq \dot{\theta}_{d,i} < 0 \\ R_{i(-)} & \dot{\theta}_{d,i} < -\epsilon \end{cases} \quad (1)$$

In the case when $\dot{\theta}_{d,i} = 0$,

$$\tau_{fr\_ff,i} = \begin{cases} R_{si(+)} & \tau_{d,i} \geq 0 \\ R_{si(-)} & \tau_{d,i} < 0 \end{cases} \quad (2)$$

where $\tau_{d,i}$ denotes the desired torque.

A.1.2 Feedback compensation

The friction compensation torque $\tau_{fr\_fb,i}$ depends on the current motor velocity $\dot{\theta}_i$. The compensation torque is given by

$$\tau_{fr\_fb,i} = \begin{cases} R_{i(+)} & \epsilon < \dot{\theta}_i \\ R_{si(-)} \Leftrightarrow R_{si(+)} & -\epsilon \leq \dot{\theta}_i \leq \epsilon \\ R_{i(-)} & \dot{\theta}_i < -\epsilon \end{cases} \quad (3)$$

where $\Leftrightarrow$ denotes the alternation of the maximum static friction torques $R_{si(\pm)}$ at each sampling time. This causes a *dither* effect.

A.2 Integrated Control of Motion, Force and Compliance

The integrated control is given as follows:

$$\dot{\boldsymbol{\theta}}_{com} = \dot{\boldsymbol{\theta}}_d + W_{pid}\left(\boldsymbol{K}_p \delta\boldsymbol{\theta} + \boldsymbol{K}_i \int \delta\boldsymbol{\theta} dt + \boldsymbol{K}_d \delta\dot{\boldsymbol{\theta}}\right)$$
$$+ W_{frc} \boldsymbol{K}_v^{-1} \boldsymbol{J}^T \boldsymbol{F}_d + W_{cmp} \boldsymbol{K}_v^{-1} \boldsymbol{J}^T \boldsymbol{K} \boldsymbol{J} \delta\boldsymbol{\theta}$$
$$+ \boldsymbol{K}_v^{-1} \left\{ S_g \boldsymbol{\tau}_g - (1 - S_{fr}) \boldsymbol{\tau}_{fr_f f} - S_{fr} \boldsymbol{\tau}_{fr_f b} \right\} \quad (4)$$

where $\dot{\boldsymbol{\theta}}_{com}, \dot{\boldsymbol{\theta}}_d \in \boldsymbol{R}^6$ denotes a vector of the motor command and a vector of the feedforward term of the desired velocities, respectively. $\delta\boldsymbol{\theta}, \delta\dot{\boldsymbol{\theta}}$ are the motor angular errors and the angular velocity errors. $\boldsymbol{K}_p, \boldsymbol{K}_i, \boldsymbol{K}_d, \boldsymbol{K}_v \in \boldsymbol{R}^{6\times6}$ stand for the proportional gains, the integral gains, the derivative gains and the velocity feedback gains of the motor drivers, respectively. $\boldsymbol{F}_d \in \boldsymbol{R}^6$ is a vector of the end-effector reference forces and torques, $\boldsymbol{J}, \boldsymbol{K} \in \boldsymbol{R}^{6\times6}$ are a Jacobian matrix and a stiffness matrix o the end-effector. $\boldsymbol{\tau}_g, \boldsymbol{\tau}_{fr_ff}, \boldsymbol{\tau}_{fr_fb} \in \boldsymbol{R}^6$ are vectors of the joint torques for the gravity compensation, the feedforward friction compensation and the feedback friction compensation, respectively.

$W_{pid}, W_{frc}, W_{cmp}$ $(0 \leq W \leq 1)$ are scalar weighting factors for the position control, the force control, the compliance control, respectively. And the position control can not run along with the compliance control, and hence their weighting factors are bound with $0 \leq W_{pid} + W_{cmp} \leq 1$. S_g, S_{fr} denote scalar numbers (0 or 1) to select the gravity compensation and two types of the friction compensation according to the control mode.

DYNAMIC ANALYSIS OF CLAVEL'S DELTA PARALLEL ROBOT

Staicu St. & Carp-Ciocardia D. C.

Department of Mechanics, University "Politehnica" of Bucharest, Romania

Some iterative matrix relations for the geometric, kinematic and dynamic analysis of a Delta parallel robot are established in this paper. The prototype of this manipulator is a three degree of freedom spatial mechanism, which consists of a system of parallel chains. Supposing that the position and the translation motion of the platform are known, an inverse dynamic problem is solved using the virtual powers method. Finally, some recursive matrix relations and some graphs for the moments and the powers of the three active couples are determined.

Key words: robotics, manipulator, platform, matrix, dynamics.

1. Introduction

The parallel robots are spatial mechanical structures that consist of kinematic closed chains. Generally, a parallel manipulator, have two platforms. One of them is attached to the fix reference frame. The other one can have arbitrary motions in its workspace. Three mobile legs, made up as serial robots, connect the effector, which is attached to the moving platform, to the fixed platform. The elements of the robot are connected one to the other by spherical joints, revolute joints or prismatic joints.

The parallel manipulators have some special characteristics with respect to the serial robots such as: more rigid structure, high orientation accuracy, stabile functioning, control on the limits of velocities and accelerations, suitable position of the acting systems and a good positional repetitivity. The parallel robots are equipped with hydraulic or pneumatic actuators. They have a robust construction and they can move bodies of considerable masses and dimensions with high speeds. This is why the mechanisms, which produce a translation or spherical motion to a platform, are based on the concept of parallel manipulator.

The most known application is the flight simulator with six degree of freedom, which is in fact the Gough-Stewart platform [Stewart, 1965; Merlet, 1997]. The parallel manipulator Star [Hervé and Sparacino, 1992; Tremblay and Baron, 1999] and the parallel Delta robot [Clavel, 1988; Zsombor-Murray, 2001] equipped with three engines, which have a parallel setting, train on the effector in a three degree of freedom general translation motion, used in quick operations of *pick and place*. Angeles (1997), Wang and Gosselin (2001) developed the direct kinematic analysis of a prototype of spherical manipulator Agile Wrist, which has three concurrent rotations.

This paper establishes some recursive matrix relations used for positional, kinematic and dynamic analysis for a three degree of freedom Delta robot. In 1988, R. Clavel developed the prototype of this robot at the Lausanne Federal Polytechnic Institute.

2. Inverse geometric model

The following elements are the elements of the topological structure of one of the three kinematic closed chains of the manipulator, respectively: an engine, an active revolute joint, an intermediary mechanism with four revolute links that connect four bars, which are parallel two and two, and finally a passive revolute link connected to the moving platform (fig.1).

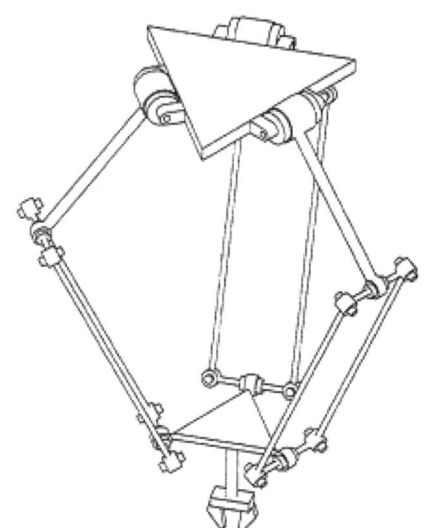

Fig. 1. The Clavel's Delta Robot

Let $Ox_0y_0z_0$ (T_0) be a fix cartesian frame. A three degrees of freedom Delta manipulator is moving with respect to this reference frame. The manipulator has three legs. The elements of these legs have known dimensions and masses. One of the three active elements of the robot is the first body of the leg A. This is a homogenous crank, which rotates about the axis $A_1 z_1^A$ with the angular velocity ω_{10}^A and the angular acceleration ε_{10}^A. It has the length $A_1 A_2 = l_1^A$, the mass m_1^A and the tensor of inertia $\hat{J}_1^A$. The transmission bar $A_3 A_6 = l_2^A$ is connected to the $A_2 x_2^A y_2^A z_2^A$ (T_2^A) frame and it has a relative rotation with the angle φ_{21}^A, so that $\omega_{21}^A = \dot{\varphi}_{21}^A$ and $\varepsilon_{21}^A = \ddot{\varphi}_{21}^A$. It has the mass m_2^A and the tensor of inertia $\hat{J}_2^A$.

Further on, two identical and parallel bars with same length $l_3^A = l_6^A$, rotate about the T_2^A frame with the angle $\varphi_{32}^A = \varphi_{62}^A$. They have also the same mass $m_3^A = m_6^A$ and the same tensor of inertia $\hat{J}_3^A = \hat{J}_6^A$. The parallelogram is closed by an element T_4^A, which has the same length and mass with T_2^A. Its tensor of inertia is $\hat{J}_4^A$. This element rotates with the relative angle $\varphi_{43}^A = \varphi_{32}^A$.

The platform of the robot is an equilateral triangle. The relation $l = \sqrt{3}\left(l_0^A - l_3^A \sin \beta_A\right)$ gives the side dimension of this triangle, which has the mass m_5^A. Let us denote with $\omega_{54}^A = \dot{\varphi}_{54}^A$ (fig.2), the angular velocity of the platform with respect to the nearby body T_4^A. The following angles give the initial position of the manipulator:

$$\alpha_A = \frac{\pi}{3}, \alpha_B = \pi, \alpha_C = -\frac{\pi}{3}, \beta_A = \beta_B = \beta_C = \frac{\pi}{6}. \quad (1)$$

Let us consider the rotation angles φ_{10}^A, φ_{10}^B, φ_{10}^C, of the three actuators A_1, B_1, C_1, the parameters which give the position of the mechanism. In the inverse geometric problem, one can consider that the coordinates of the mass centre of the platform, x_0^G, y_0^G, z_0^G, give the position of the mechanism.

Pursuing the leg A in the $OA_1A_2A_3A_4A_5$ way, one obtains the following passing matrices:

$$a_{10} = a_{10}^\varphi \theta_1 \theta_2 a_{\alpha_A}, \quad a_{21} = a_{21}^\varphi a_{\beta_A} \theta_3, \quad a_{32} = a_{32}^\varphi \theta_1 \theta_2, \quad (2)$$
$$a_{43} = a_{43}^\varphi \theta_3, \quad a_{54} = a_{54}^\varphi a_{\beta_A} \theta_1 \theta_4, \quad a_{62} = a_{32},$$

where one denoted [Staicu, 1998]:

$$\theta_1 = \begin{bmatrix} 0 & 0 & -1 \\ 0 & 1 & 0 \\ 1 & 0 & 0 \end{bmatrix}, \theta_2 = \begin{bmatrix} 0 & 1 & 0 \\ -1 & 0 & 0 \\ 0 & 0 & 1 \end{bmatrix}, \theta_3 = \begin{bmatrix} -1 & 0 & 0 \\ 0 & 1 & 0 \\ 0 & 0 & -1 \end{bmatrix},$$

$$\theta_4 = \begin{bmatrix} -1 & 0 & 0 \\ 0 & -1 & 0 \\ 0 & 0 & 1 \end{bmatrix}, a_{\alpha_A} = \begin{bmatrix} \cos \alpha_A & \sin \alpha_A & 0 \\ -\sin \alpha_A & \cos \alpha_A & 0 \\ 0 & 0 & 1 \end{bmatrix},$$

$$a_{\beta_A} = \begin{bmatrix} \cos \beta_A & \sin \beta_A & 0 \\ -\sin \beta_A & \cos \beta_A & 0 \\ 0 & 0 & 1 \end{bmatrix}, \quad (3)$$

$$a_{k,k-1}^\varphi = \begin{bmatrix} \cos \varphi_{k,k-1}^A & \sin \varphi_{k,k-1}^A & 0 \\ -\sin \varphi_{k,k-1}^A & \cos \varphi_{k,k-1}^A & 0 \\ 0 & 0 & 1 \end{bmatrix},$$

$$a_{k0} = \prod_{j=1}^{k} a_{k-j+1,k-j} \quad (k=1,2,\ldots,5).$$

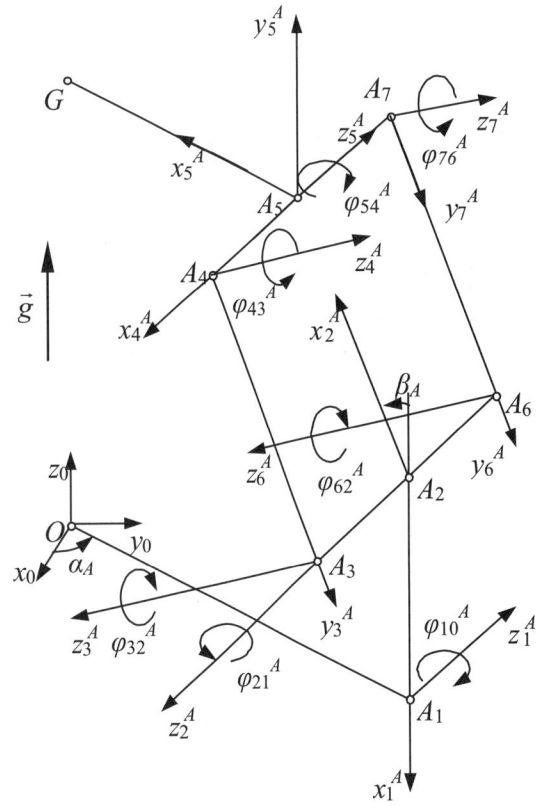

Fig. 2. The kinematic schema of the mechanism

If the other legs, B and C, of the mechanism are pursued, some analogous relations can be written.

The translation conditions of the platform are given by the following identities

$$a_{50}^{\circ T} a_{50} = b_{50}^{\circ T} b_{50} = c_{50}^{\circ T} c_{50} = I \quad (4)$$

and by the following matrices

$$a_{50}^\circ = \frac{1}{2}\begin{bmatrix} -1 & -\sqrt{3} & 0 \\ 0 & 0 & 2 \\ -\sqrt{3} & 1 & 0 \end{bmatrix}, b_{50}^\circ = \begin{bmatrix} 1 & 0 & 0 \\ 0 & 0 & 1 \\ 0 & -1 & 0 \end{bmatrix}, \quad (5)$$

$$c_{50}^\circ = \frac{1}{2}\begin{bmatrix} -1 & \sqrt{3} & 0 \\ 0 & 0 & 2 \\ \sqrt{3} & 1 & 0 \end{bmatrix}.$$

From this relations, one obtains the following relations between angles:

$$\varphi_{54}^A = -\varphi_{20}^A = \varphi_{21}^A - \varphi_{10}^A,$$
$$\varphi_{54}^B = -\varphi_{20}^B = \varphi_{21}^B - \varphi_{10}^B, \quad (6)$$
$$\varphi_{54}^C = -\varphi_{20}^C = \varphi_{21}^C - \varphi_{10}^C.$$

Supposing that the motion of the mass centre of the platform along an ellipses is given by the relations

$$\vec{r}_0^G = \begin{bmatrix} x_0^G & y_0^G & z_0^G \end{bmatrix}^T$$
$$x_0^G(t) = x_0^{G*} \sin\left(\frac{\pi}{3}t\right)$$
$$y_0^G(t) = y_0^{G*} \left[1 - \cos\left(\frac{\pi}{3}t\right)\right] \quad (7)$$
$$z_0^G(t) = l_1^A + l_3^A \cos\beta_A - z_0^{G*}\left[1 - \cos\left(\frac{\pi}{3}t\right)\right],$$

the angles φ_{10}^A, φ_{21}^A, φ_{32}^A, φ_{10}^B, φ_{21}^B, φ_{32}^B, φ_{10}^C, φ_{21}^C, φ_{32}^C are given by the following geometric conditions:

$$\vec{r}_{10}^A + \sum_{k=1}^{4} a_{k0}^T \vec{r}_{k+1,k}^A + a_{50}^T \vec{r}_5^{GA} =$$
$$= \vec{r}_{10}^B + \sum_{k=1}^{4} b_{k0}^T \vec{r}_{k+1,k}^B + b_{50}^T \vec{r}_5^{GB} = \quad (8)$$
$$= \vec{r}_{10}^C + \sum_{k=1}^{4} c_{k0}^T \vec{r}_{k+1,k}^C + c_{50}^T \vec{r}_5^{GC} = \vec{r}_0^G,$$

where one denoted:

$$\vec{u}_1 = \begin{bmatrix} 1 \\ 0 \\ 0 \end{bmatrix}, \vec{u}_2 = \begin{bmatrix} 0 \\ 1 \\ 0 \end{bmatrix}, \vec{u}_3 = \begin{bmatrix} 0 \\ 0 \\ 1 \end{bmatrix}, \tilde{u}_3 = \begin{bmatrix} 0 & -1 & 0 \\ 1 & 0 & 0 \\ 0 & 0 & 0 \end{bmatrix},$$
$$\vec{r}_{10}^A = l_0^A a_{\alpha_A}^T \vec{u}_1, \vec{r}_{21}^A = -l_1^A \vec{u}_1 \quad (9)$$
$$\vec{r}_{32}^A = \frac{l_2^A}{2}\vec{u}_3, \vec{r}_{43}^A = -l_3^A \vec{u}_2$$
$$\vec{r}_{54}^A = -\frac{l_2^A}{2}\vec{u}_1, \vec{r}_5^{GA} = \left(l_0^A - l_3^A \sin\beta_A\right)\vec{u}_1.$$

3. Velocities and accelerations

The motions of the component elements of each leg (for example the leg A) are characterised by the following skew symmetric matrices [Staicu and Carp-Ciocardia, 2001]

$$\tilde{\omega}_{k0}^A = a_{k,k-1}\tilde{\omega}_{k-1,0}^A a_{k,k-1}^T + \omega_{k,k-1}^A \tilde{u}_3, \quad (10)$$

associated to the absolute angular velocities given by the recurrence relations

$$\vec{\omega}_{k0}^A = a_{k,k-1}\vec{\omega}_{k-1,0}^A + \omega_{k,k-1}^A \vec{u}_3, \omega_{k,k-1}^A = \dot{\varphi}_{k,k-1}^A. \quad (11)$$

The velocity $\vec{v}_{k0}^A$ of the joint A_k is given by the relation

$$\vec{v}_{k0}^A = a_{k,k-1}\{\vec{v}_{k-1,0}^A + \tilde{\omega}_{k-1,0}^A \vec{r}_{k,k-1}^A\},$$
$$\vec{v}_{k,k-1} = \vec{0}, (k=1,2,...,5). \quad (12)$$

The following matrix relations give the kinematic constraints:

$$\omega_{10}^A \vec{u}_i^T a_{10}^T \vec{u}_3 + \omega_{21}^A \vec{u}_i^T a_{20}^T \vec{u}_3 + \omega_{54}^A \vec{u}_i^T a_{50}^T \vec{u}_3 = 0$$
$$\omega_{10}^A \left(l_1^A \vec{u}_i^T a_{10}^T \tilde{u}_3 \vec{u}_1 + l_3^A \vec{u}_i^T a_{10}^T \tilde{u}_3 a_{21}^T a_{32}^T \vec{u}_2\right) +$$
$$+ \omega_{21}^A l_3^A \vec{u}_i^T a_{20}^T \tilde{u}_3 a_{32}^T \vec{u}_2 + \omega_{32}^A l_3^A \vec{u}_i^T a_{30}^T \tilde{u}_3 \vec{u}_2 = \quad (13)$$
$$= -\vec{u}_i^T \dot{\vec{r}}_0^G, \quad (i=1,2,3).$$

The relations (13) give the Jacobi matrix of the mechanism. This matrix is an essential element for the analysis of the robot workspace.

Also, the relations (13) represent *the connectivity conditions of the relative angular velocities*. These relations give the angular velocities ω_{10}^A, ω_{21}^A, ω_{32}^A, $\omega_{54}^A = \omega_{21}^A - \omega_{10}^A$ as a function of the translation velocity of the platform.

Let us assume that the robot has a virtual motion determined by the angular velocities $\omega_{10a}^{Av} = 1$, $\omega_{10a}^{Bv} = 0$, $\omega_{10a}^{Cv} = 0$. The characteristic virtual velocities expressed as functions of the position of the robot are given by the connectivity conditions of the relative velocities of the loops A-B and B-C:

$$\vec{u}_i^T a_{50}^T \vec{v}_{50a}^{Av} = \vec{u}_i^T b_{50}^T \vec{v}_{50a}^{Bv} = \vec{u}_i^T c_{50}^T \vec{v}_{50a}^{Cv}, (i=1,2,3) \quad (14)$$
$$\omega_{21}^{Av} = 1 + \omega_{54}^{Av}, \omega_{21}^{Bv} = \omega_{54}^{Bv}, \omega_{21}^{Cv} = \omega_{54}^{Cv}.$$

Some other compatibility relations can be obtained if one considers successively that $\omega_{10a}^{Bv} = 1$ and $\omega_{10a}^{Cv} = 1$.

The angular accelerations, ε_{10}^A, ε_{21}^A, ε_{32}^A, ε_{54}^A, of the elements of the robot are given by some new *connectivity conditions*, obtained by deriving the relations (13). The following relations result:

$$\varepsilon_{10}^A \vec{u}_i^T a_{10}^T \vec{u}_3 + \varepsilon_{21}^A \vec{u}_i^T a_{20}^T \vec{u}_3 + \varepsilon_{54}^A \vec{u}_i^T a_{50}^T \vec{u}_3 = 0$$
$$\varepsilon_{10}^A \left(l_1^A \vec{u}_i^T a_{10}^T \tilde{u}_3 \vec{u}_1 + l_3^A \vec{u}_i^T a_{10}^T \tilde{u}_3 a_{21}^T a_{32}^T \vec{u}_2\right) +$$
$$+ \varepsilon_{21}^A l_3^A \vec{u}_i^T a_{20}^T \tilde{u}_3 a_{32}^T \vec{u}_2 + \varepsilon_{32}^A l_3^A \vec{u}_i^T a_{30}^T \tilde{u}_3 \vec{u}_2 =$$
$$= -\vec{u}_i^T \ddot{\vec{r}}_0^G - \omega_{10}^A \omega_{10}^A \left(l_1^A \vec{u}_i^T a_{10}^T \tilde{u}_3 \tilde{u}_3 \vec{u}_1 + \right.$$
$$\left. + l_3^A \vec{u}_i^T a_{10}^T \tilde{u}_3 \tilde{u}_3 a_{21}^T a_{32}^T \vec{u}_2\right) - \quad (15)$$
$$- \omega_{21}^A \omega_{21}^A l_3^A \vec{u}_i^T a_{20}^T \tilde{u}_3 \tilde{u}_3 a_{32}^T \vec{u}_2 -$$
$$- \omega_{32}^A \omega_{32}^A l_3^A \vec{u}_i^T a_{30}^T \tilde{u}_3 \tilde{u}_3 \vec{u}_2 -$$
$$- 2\omega_{10}^A \omega_{21}^A l_3^A \vec{u}_i^T a_{10}^T \tilde{u}_3 a_{21}^T \tilde{u}_3 a_{32}^T \vec{u}_2 -$$

$$-2\omega_{10}^A \omega_{32}^A l_3^A \vec{u}_i^T a_{10}^T \tilde{u}_3 a_{21}^T a_{32}^T \tilde{u}_3 \vec{u}_2 -$$
$$-2\omega_{21}^A \omega_{32}^A l_3^A \vec{u}_i^T a_{20}^T \tilde{u}_3 a_{32}^T \tilde{u}_3 \vec{u}_2 \text{ , } (i=1,2,3).$$

If the other two kinematic chains of the manipulator are pursued analogous relations can be easily obtained.

The following recurrence relations give the angular accelerations $\vec{\varepsilon}_{k0}^A$ and the accelerations $\vec{\gamma}_{k0}^A$ of joints

$$\vec{\varepsilon}_{k0}^A = a_{k,k-1} \vec{\varepsilon}_{k-1,0}^A + \varepsilon_{k,k-1}^A \vec{u}_3 +$$
$$+ \omega_{k,k-1}^A a_{k,k-1} \tilde{\omega}_{k-1,0}^A a_{k,k-1}^T \vec{u}_3$$
$$\tilde{\omega}_{k0}^A \tilde{\omega}_{k0}^A + \tilde{\varepsilon}_{k0}^A =$$
$$= a_{k,k-1} \left(\tilde{\omega}_{k-1,0}^A \tilde{\omega}_{k-1,0}^A + \tilde{\varepsilon}_{k-1,0}^A \right) a_{k,k-1}^T + \qquad (16)$$
$$+ \omega_{k,k-1}^A \omega_{k,k-1}^A \tilde{u}_3 \tilde{u}_3 + \varepsilon_{k,k-1}^A \tilde{u}_3 +$$
$$+ 2\omega_{k,k-1}^A a_{k,k-1} \tilde{\omega}_{k-1,0}^A a_{k,k-1}^T \tilde{u}_3$$
$$\vec{\gamma}_{k0}^A = a_{k,k-1} \left[\vec{\gamma}_{k-1,0}^A + \left(\tilde{\omega}_{k-1,0}^A \tilde{\omega}_{k-1,0}^A + \tilde{\varepsilon}_{k-1,0}^A \right) \vec{r}_{k,k-1}^A \right].$$

The relations (13), (15) represent the *inverse kinematic model* of the Delta robot.

4. Equations of motion

Three electric engines, A, B, C, that generate three couples of moments $\vec{m}_{10}^A = m_{10}^A \vec{u}_3$, $\vec{m}_{10}^B = m_{10}^B \vec{u}_3$, and $\vec{m}_{10}^C = m_{10}^C \vec{u}_3$, which have the directions of the axes $A_1 z_1^A$, $B_1 z_1^B$, $C_1 z_1^C$, control the motion of the legs of the manipulator. The force of inertia and the resultant moment of the forces of inertia of the rigid body T_k are determined with respect to the centre of the joint O_k. On the other hand, the characteristic vectors $\vec{f}_k^*$ and $\vec{m}_k^*$ evaluate the influence of the action of the weight $m_k \vec{g}$ and of other external and internal forces applied to the same element T_k of the robot.

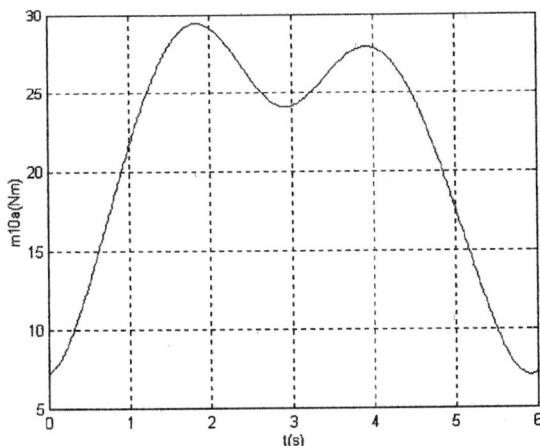

Fig. 3. The moment m_{10}^A

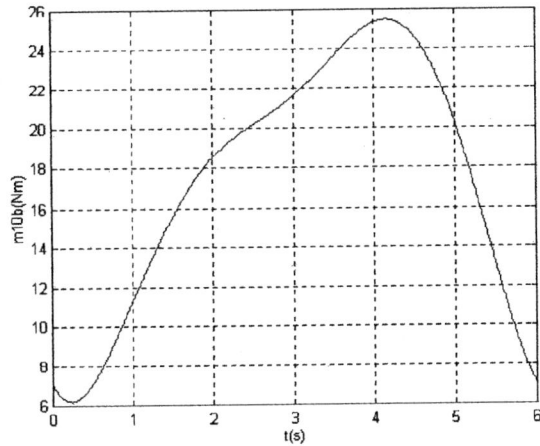

Fig. 4. The moment m_{10}^B

Let us consider that the motion of the platform is known. In these conditions, one determines first the position, the velocity and the acceleration of each joint. Then the forces and the moment that are acting each body are determined. Finally, one calculates the moments of the active couples. There are three methods, which can provide the same results concerning these moments. The first one is using the Newton-Euler classic procedure, the second one applies the Lagrange' equations and multipliers formalism and the third one is based on the virtual work principle.

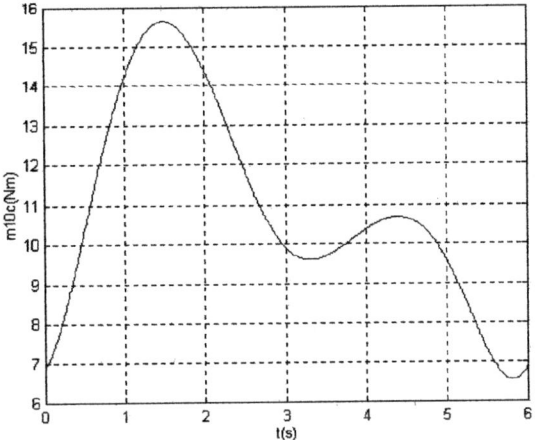

Fig. 5. The moment m_{10}^C

Kane and Levinson (1985) obtained some vectorial recursive relations concerning the equilibrium of the generalized forces that are applied to a serial robot arm.

In the inverse dynamic problem, in this paper, one applies the virtual powers method in order to establish some recursive matrix relations for the moments and the powers of the three active couples. Some graphs of these moments and powers are also obtained.

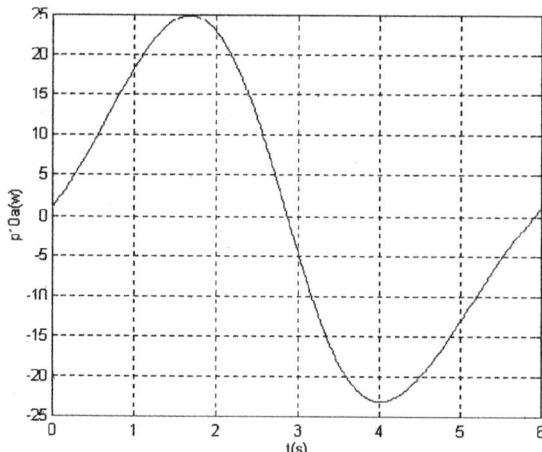

Fig. 6. The power of the first actuator

As the virtual velocities method shows, the dynamic equilibrium condition of the mechanism is that the virtual power of the external, internal and inertia forces, which is developed during a general virtual displacement, must be null. Applying *the fundamental equations of the parallel robots dynamics* obtained by St. Staicu (2000), the following matrix relation results

$$m_{10}^A = \vec{u}_3^T \left[\vec{M}_1^A + \omega_{54a}^{Av} \vec{M}_5^A + \right.$$
$$+ \omega_{21a}^{Av} \vec{M}_2^A + \omega_{32a}^{Av} \left(\vec{M}_3^A + \vec{M}_4^A + \vec{M}_6^A \right) +$$
$$+ \omega_{21a}^{Bv} \vec{M}_2^B + \omega_{32a}^{Bv} \left(\vec{M}_3^B + \vec{M}_4^B + \vec{M}_6^B \right) + \quad (17)$$
$$\left. + \omega_{21a}^{Cv} \vec{M}_2^C + \omega_{32a}^{Cv} \left(\vec{M}_3^C + \vec{M}_4^C + \vec{M}_6^C \right) \right],$$

where one denoted:

$$\vec{F}_{k0}^A = m_k^A \left[\vec{\gamma}_{k0}^A + \left(\widetilde{\omega}_{k0}^A \widetilde{\omega}_{k0}^A + \widetilde{\varepsilon}_{k0}^A \right) \vec{r}_k^{CA} \right] - \vec{f}_k^{*A}$$
$$\vec{M}_{k0}^A = m_k^A \widetilde{r}_k^{CA} \vec{\gamma}_{k0}^A + \hat{J}_k^A \vec{\varepsilon}_{k0}^A + \widetilde{\omega}_{k0}^A \hat{J}_k^A \vec{\omega}_{k0}^A - \vec{m}_k^{*A}$$
$$\vec{F}_k^A = \vec{F}_{k0}^A + a_{k+1,k}^T \vec{F}_{k+1} \quad (18)$$
$$\vec{M}_k^A = \vec{M}_{k0}^A + a_{k+1,k}^T \vec{M}_{k+1} + \widetilde{r}_{k+1,k}^T a_{k+1,k}^T \vec{F}_{k+1},$$
$$(k = 1, 2, ..., 6).$$

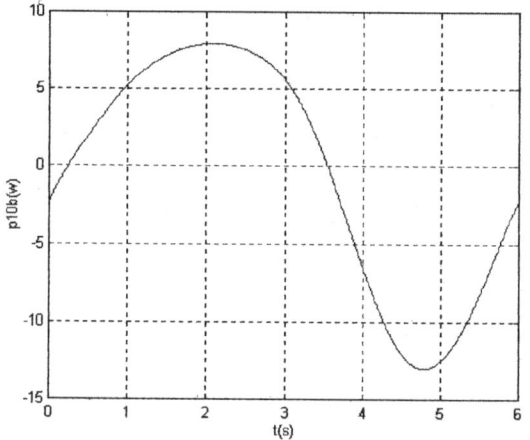

Fig. 7. The power of the second actuator

The relations (17) and (18) represent the *inverse dynamic model* of the parallel Delta robot.

As application let us consider a robot which has the following characteristics:

$x_0^* = 0.03m$, $y_0^* = 0.05m$, $z_0^* = 0.07m$,

$l_0^A = l_3^A = 0.35m$, $l_1^A = 0.1m$, $l_2^A = 0.05m$,

$m_1^A = 2.5kg$, $m_2^A = m_4^A = 2kg$

$m_3^A = m_6^A = 5kg$, $m_5^A = 15kg$

$$\hat{J}_1^A = \begin{bmatrix} 0.05 & & \\ & 0.1 & \\ & & 0.1 \end{bmatrix}, \hat{J}_2^A = \begin{bmatrix} 0.02 & & \\ & 0.02 & \\ & & 0.01 \end{bmatrix}$$

$$\hat{J}_3^A = \begin{bmatrix} 0.4 & & \\ & 0.2 & \\ & & 0.4 \end{bmatrix}, \hat{J}_4^A = \begin{bmatrix} 0.04 & & \\ & 0.08 & \\ & & 0.08 \end{bmatrix}.$$

Finally, one obtains the graphs of the moments m_{10}^A (fig.3), m_{10}^B (fig.4), m_{10}^C (fig.5) and of the powers p_{10}^A (fig.6), p_{10}^B (fig.7), p_{10}^C (fig.8) given by the active couples of the three actuators.

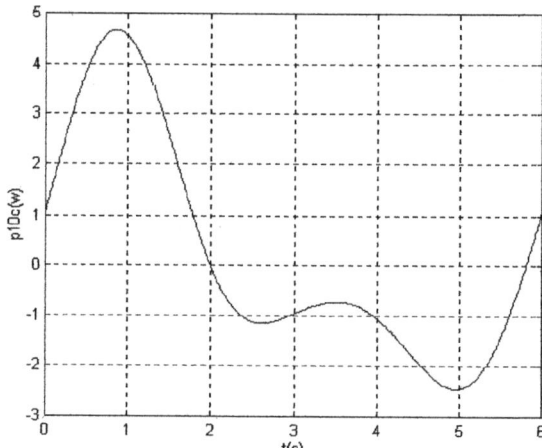

Fig. 8. The power of the third actuator

5. Conclusions

1. Within the inverse positional analysis some exact relations that give in real time the position, the velocity and the acceleration of each element of the parallel robot have been established.

2. Using the Newton-Euler classic method, which takes into account each separate body of the mechanism, an one hundred and five equations system, that must be solved, would result. Finally, the moments of the active couples could be obtained.

The analytical calculi involved in the Lagrange's equations and multipliers formalism are too long and they have risk of making errors. Also, the time for numerical calculus grows with the number of the bodies of the mechanism.

3. The new approach based on the virtual work principle establishes a direct recursive determination of the variation in real time of the moments and the powers of the active couples. The iterative matrix relations, (17) and (18), of the theoretical model of dynamic simulation, can be transformed in a model for the automatic command of the parallel Delta robot.

6. References

[1] Stewart, D. (1965), A Platform with Six Degrees of Freedom, *Proceedings of the Inst. Mech. Engrs.,* 180, 15.
[2] Merlet, J.-P. (1997), *Les robots paralléles,* Hermes.
[3] Hervé, J. M., Sparacino, F. (1992), Star. A New Concept in Robotics, *Proceedings of the Third Int. Workshop on Advances in Robot Kinematics,* Ferrara.
[4] Tremblay, A., Baron, L. (1999), Geometrical Synthesis of Parallel Manipulators of Star-Like Topology with a Genetic Algorithm, *IEEE Int. Conf. on Robotics and Automation,* Detroit, Michigan.
[5] Clavel, R. (1988), Delta: a Fast Robot with Parallel Geometry, *Proceedings of the 18th Int. Symposium on Industrial Robot,* Lausanne.
[6] Zsombor-Murray, P.J. (2001), Kinematic Analysis of Clavel's "Delta" Robot, *CCToMM Symposium on Mechanisms, Machines and Mechatronics,* Saint-Hubert (Montréal).
[7] Angeles, J. (1997), *Fundamentals of Robotic Mechanical Systems. Theory, Methods and Algorithms,* Springer-Verlag, New York.
[8] Wang, J., Gosselin, C. (2001), Representation of the Singularity Loci of a Special Class of Spherical 3-dof Parallel Manipulator with Revolute Actuators, *CCToMM Symposium on Mechanisms, Machines and Mechatronics,* Saint-Hubert (Montréal).
[9] Staicu, St. (1998), *Theoretical Mechanics,* Edit. Didactica & Pedagogica, Bucharest.
[10] Staicu, St. (2000), Méthodes matricielles en dynamique des mécanismes, *Scientific Bulletin, Series D, Mechanical Engineering,* University "Politehnica" of Bucharest, 62, 3.
[11] Staicu, St., Carp-Ciocardia, D.C. (2001), On an Inverse Dynamic Problem in Robotics, *Proceedings of the 12th Int. DAAAM Symposium,* Vienna.
[12] Kane, T.R., Levinson, D.A. (1985), *Dynamics. Theory and Applications,* Mc Graw Hill

STAICU Stefan, Professor, Dr.
E-mail: staicu@cat.mec.pub.ro
313 Splaiul Independentei, Bucharest, Romania.
CARP-CIOCARDIA Daniela Craita, Reader, Dr.
E-mail: carp@cat.mec.pub.ro

Learning Implicit Models during Target Pursuit

Chris Gaskett[‡], Peter Brown[†1], Gordon Cheng[‡], and Alexander Zelinsky[†]

[†]Robotic Systems Laboratory, Department of Systems Engineering, RSISE,
The Australian National University, Canberra, ACT 0200 Australia
{pfb,alex}@syseng.anu.edu.au, http://www.syseng.anu.edu.au/rsl/

[‡]Department of Humanoid Robotics and Computational Neuroscience,
ATR Computational Neuroscience Laboratories, Kyoto, Japan
{cgaskett,gordon}@atr.co.jp, http://www.cns.atr.co.jp/hrcn/

Abstract— Smooth control using an active vision head's verge-axis joint is performed through continuous state and action reinforcement learning. The system learns to perform visual servoing based on rewards given relative to tracking performance. The learned controller compensates for the velocity of the target and performs lag-free pursuit of a swinging target. By comparing controllers exposed to different environments we show that the controller is predicting the motion of the target by forming an implicit model of the target's motion. Experimental results are presented that demonstrate the advantages and disadvantages of implicit modelling.

I. INTRODUCTION

This research demonstrates learned control using one joint of a 4DOF active vision head. The task is to control the position of a target object to the centre of the robot's field of view by reacting to the target's movement, as shown in Fig. 1. The controller performs *visual servoing*—closed-loop control based on visual information—to fixate on both static and moving targets [1].

Although visual servoing does not necessarily require learning, a learning system can reduce the amount of required knowledge about the system to be controlled and reduce the reliance on calibration. A visual servoing system with learning capabilities could learn to predict the movements of the target. Investigation of human vision has shown that adaption and learning improves the efficiency of gaze control [2, 3] and compensates for imperfections or changes in the eyes [4–6]. Eye movement skills may develop incrementally during infancy [7].

Reinforcement learning is a suitable approach for learning visual servoing. Reinforcement learning systems do not require an explicit dynamic model of the system to be controlled or a teacher to present ideal behaviour. Behaviour can be optimised over time, where the optimisation criteria is set through a reward function [8]. Our earlier research successfully applied a reinforcement learning algorithm to mobile robot tasks [9]. Although we were able to show that the

[†]This research was performed at the Robotic Systems Laboratory, ANU. For a video demonstrating the results or for further information please contact Chris Gaskett.
[1]Department of Foreign Affairs and Trade, Canberra, Australia.

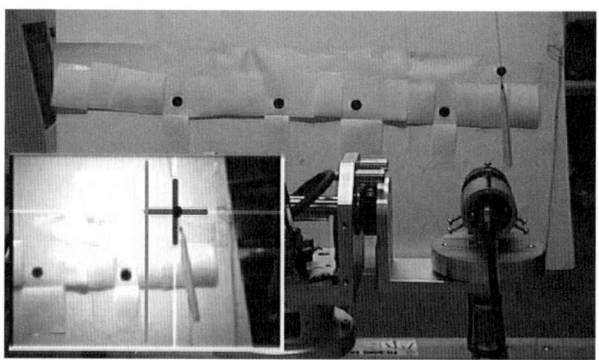

Fig. 1. The current target-object's position in the robot's view (inset lower left) is marked with a red cross. The desired position is shown by a green line to the left of the cross.

robot moved smoothly, we could not confirm that the algorithm was predicting target movement since the robot moved slowly. Further, in the uncontrolled environment it was difficult to perform comparative experiments. Performing experiments on an active head allowed repeated, safe experiments at speeds high enough to investigate the dynamic qualities of a learned controller.

A controller with good dynamic properties would pursue moving targets smoothly with the minimum possible lag. Smooth, low-lag pursuit reduces blurring and reduces the probability of losing the target from view. Lag is introduced by delays in sensing, processing, and actuation. To minimise lag, the controller should compensate for the velocity of the target, predict the target's movement, and be capable of producing smoothly varying actions.

II. THE LEARNING ALGORITHM

In reinforcement learning tasks the learning system must discover by trial-and-error which *actions*, u, are most valuable in particular *states*, x [8]. In reinforcement learning nomenclature the *state* is a representation of the current situation of the learning system's environment. The *action* is an output from the learning system that can influence its environment. The learning system's choice of actions in response to states is called its *policy*.

Evaluative feedback is provided in the form of a scalar *reward* signal, r, that may be delayed. The reward signal is defined in relation to the task to be achieved; reward is given when the system is successfully achieving the task.

Q-Learning is a method for solving reinforcement learning problems. Q-Learning stores the *expected value*, $Q(x, u)$, of performing each action in each state, assuming that the actions with the highest expected values will be performed thereafter:

$$Q(x_t, u_t) = E\left[r_t(x_t, u_t, X_{t+1}) \right.$$
$$+ \gamma r_{t+1}\left(X_{t+1}, \arg\max_{u_{t+1}} Q(X_{t+1}, u_{t+1}), X_{t+2}\right)$$
$$\left. + \gamma^2 r_{t+2}\left(X_{t+2}, \arg\max_{u_{t+2}} Q(X_{t+2}, u_{t+2}), X_{t+3}\right) + \ldots\right]$$

where probabilistic variables are capitalised; and γ is the discount factor, between 0 and 1, that makes rewards that are earned later exponentially less valuable. The action-values are updated through the one-step Q-update equation [10]:

$$Q(x_t, u_t) \xleftarrow{\alpha} r(x_t, u_t, x_{t+1}) + \gamma \max_{u_{t+1}} Q(x_{t+1}, u_{t+1})$$

where α is a learning rate (or step size), between 0 and 1, that controls convergence.

Accurate pursuit of a moving target requires continuously variable actuator commands, and the ability to respond to smooth changes in state. However, the world of discourse for most reinforcement learning algorithms is a symbolic representation. They treat continuous variables, for example speeds or positions, as discretised values. Discretisation does not allow smooth control and disregards important sensed information.

We avoided the limitations of discrete state and action reinforcement learning by applying our continuous state and action Q-learning method: wire fitted neural network Q-learning, or WFNN. The WFNN method is based on an idea from Baird and Klopf [11]. It combines a feedforward neural network with a moving least squares approximator to implement Q-learning (see Fig. 2). All of the learned parameters are stored in the neural network; the interpolator assists by generalising between similar actions and performing structural credit assignment. The action, not only the action's expected value, is an output from the neural network. Thus the action can vary smoothly in response to smooth changes in the state. The algorithm is capable of learning and selecting actions in real-time. It also supports off-policy learning, allowing learning from observation of other controllers.

The WFNN algorithm is described in detail in [12], which also includes a description of several other continuous state and action Q-learning algorithms. Further methods are described in [13–16]. The purpose of this paper is not redescription of our learning algorithm; rather, it is to discuss issues that affect all reinforcement learning systems.

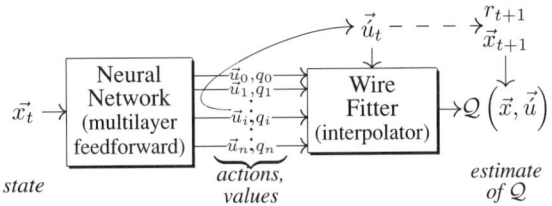

Fig. 2. Architecture of the WFNN learning system

III. EXPERIMENTAL PLATFORM: HYDRA

The experiments were performed on the HyDrA binocular active vision platform shown in Fig. 3 [17]. HyDrA is equipped with a pair of colour NTSC cameras and has four mechanical degrees of freedom: pan, tilt, left verge-axis, and right verge-axis. The experiments reported here used only the right verge-axis and the corresponding camera. The same approach could be used for the other joints, either using one MDOF controller, or using several independent controllers. The right verge-axis controller could be transferred directly for the left verge-axis joint.

Fig. 3. The HyDrA binocular active vision head

IV. EXPERIMENTAL TASK

The task was to control HyDrA to fixate on static and moving targets. The dynamics of the system and the camera parameters were unknown to the controller. The tracking targets were a lead weight suspended with string, a target mounted on a stick, and fixed targets attached to a backing-board. Target tracking was performed through grey-scale template matching. The target with the highest correlation with a stored template image was chosen automatically as the *current target*. This can be regarded as a random selection. Once a target had been selected, the target tracking process searched for the current target in a small window around the current target's last location. Therefore, it remained fixated on one target, rather than switching randomly between the visible targets. If the

State	θ: joint angle
	$\dot{\theta}$: joint angular velocity
	x: pixels error to target
	Δx: pixels velocity of target
Action	$\dot{\theta}'$: desired joint angular velocity
Reward $\sum$	$-\left|\dot{\theta}-\dot{\theta}'\right|$: smooth joint motion
	$-\left|x\right|$: movement to target
	$-\left|\Delta x\right|$: keep target still

Fig. 4. Task representation. All components are scaled by hand-crafted weighting factors.

current target was lost, the tracking process searched again over the entire view for the best matching target and selected a new current target. The learning system measured the state and produced actions at a rate of 15Hz. The low control rate exacerbates the problem of lag and, consequently, facilitates comparison of lag reduction methods.

To approach the visual servoing problem through reinforcement learning a state, action, and reward formulation must be chosen. Figure 4 describes the representation and reward function for the fixation task. The output or action of the controller is the desired joint angular velocity. The state representation is composed of: joint position and velocity; and target position and velocity from the vision system. Joint angular velocity as returned by the HyDrA system software is a copy of the last velocity command sent; it is not measured. The joint position was included since the joint behaves differently at the extremes of its range of rotation. Image velocity was estimated from the position of the target in two consecutive frames.

The reward is a weighted sum of negative components that punish for coarse joint motion, error in the target's position, and target movement. Minimising the error in the target's position is the main learning task. The punishment terms are included to improve the quality of the solution by encouraging smooth motion.

V. RESULTS

Initially, the robot's motions were random and the target was lost several times. User intervention was required to bring the target back into view. Within a few minutes the robot's movements pursued the target. However, learning a controller that was competent at both pursuing moving targets and fixating on static targets was difficult. The controller was always competent at some parts of the task, but not others. More consistent results were obtained by dividing the task between two controllers: the first specialising in step movement to a target (the static-trained controller), the second specialising in smooth pursuit (the swing-trained controller). Evaluating both of these controllers with both tasks helped us to understand why achieving competence at both tasks simultaneously was difficult.

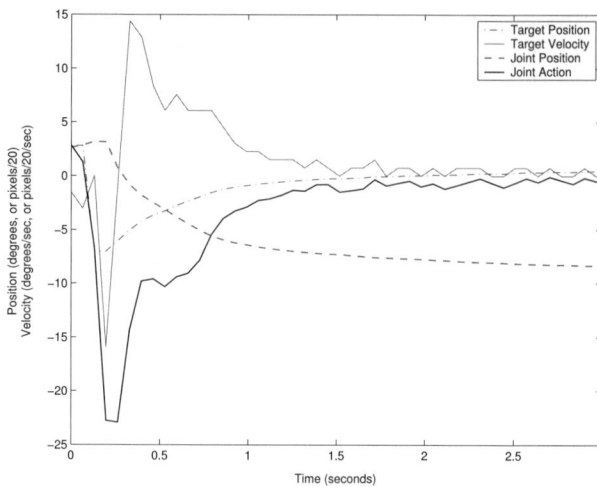

Fig. 5. Results from the static-trained controller pursuing a static target. The joint action is the un-filtered output from the learned controller. The target data is the error in the target object's position and velocity based on image data in pixels, approximately scaled to the same degrees and degrees per second units as the joint data.

A. Training with Static Targets—
Evaluating with Static Targets

The static targets were the hand-held target and the targets fixed to the backing-board. The controller learnt to reliably fixate to the static targets in both directions within a few minutes. A graph of the response showing fixation to a newly selected target is shown in Fig. 5.

Smoothness of joint motion was varied by adjusting the weighting of the *smooth joint motion* and *pixels velocity of target* coarseness penalties in the reward function (see Fig. 4).

With appropriate coarseness penalties, the commanded action changed smoothly in response to the step change in target position. This is unlike the behaviour of a simple PID family controller and shows that the learned controller took the cost of acceleration into account. The learned controller accelerated more quickly than it decelerated, unlike HyDrA's existing trapezoidal profile motion (TPM) controller [17].

The TPM controller accelerated with fixed acceleration to a maximum ceiling velocity, coasted at that velocity, then decelerated at a fixed rate equal to the acceleration rate. The learned controller's strategy of accelerating more quickly seems practical since final positioning accuracy is more dependent on the deceleration phase than the acceleration phase. Large saccadic motions in humans also accelerate more quickly than they decelerate [6].

B. Training with Static Targets—
Evaluating with Swinging Targets

Figures 6 and 7 show the performance of the controller trained with static targets when pursuing a swinging target. Learning was manually disabled during these experiments so that the controller must perform the task based only on its experience with static targets.

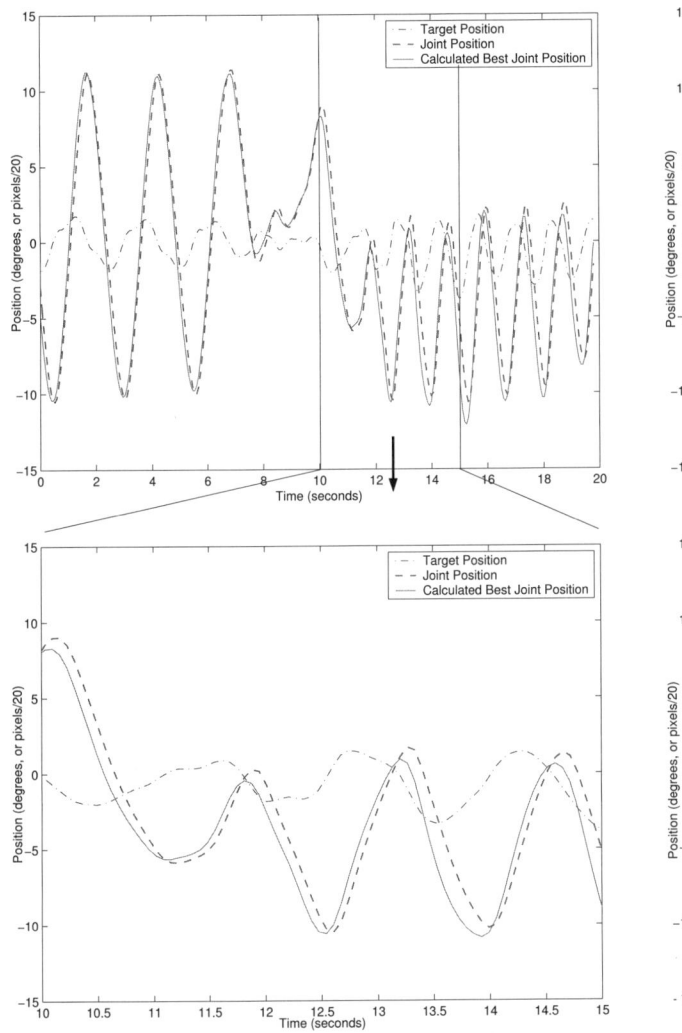

Fig. 6. Position data from the static-trained controller following a swinging target. The ideal target position is 0, the centre of the field of view.

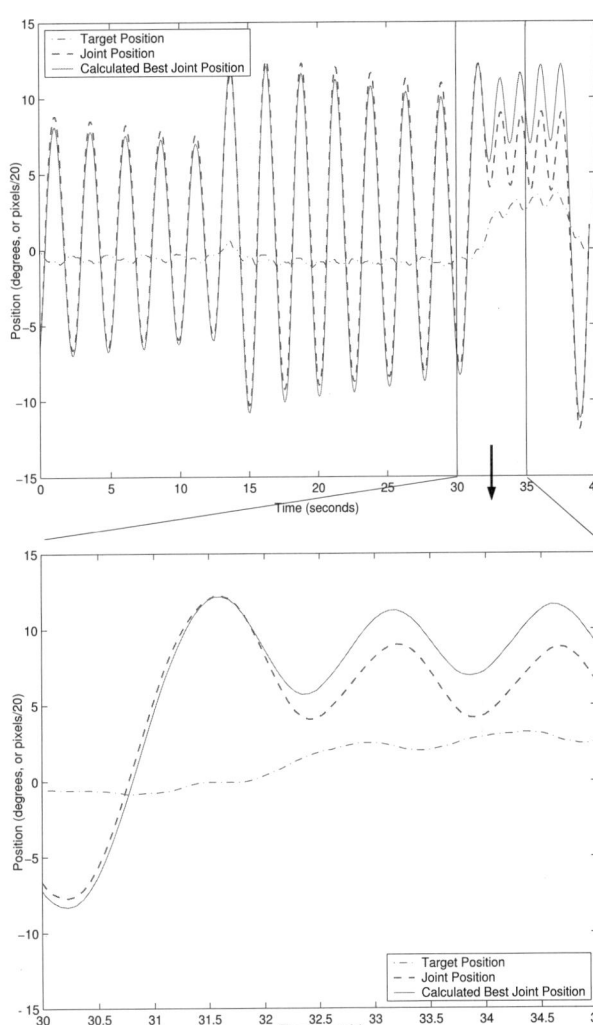

Fig. 8. Position data from the swing-trained controller following a swinging target

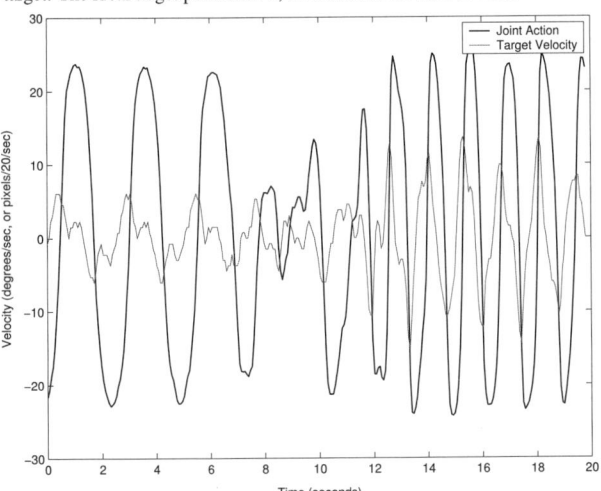

Fig. 7. Velocity data from the static-trained controller following a swinging target. The ideal target velocity is 0, i.e. the target's position is stationary in the field of view.

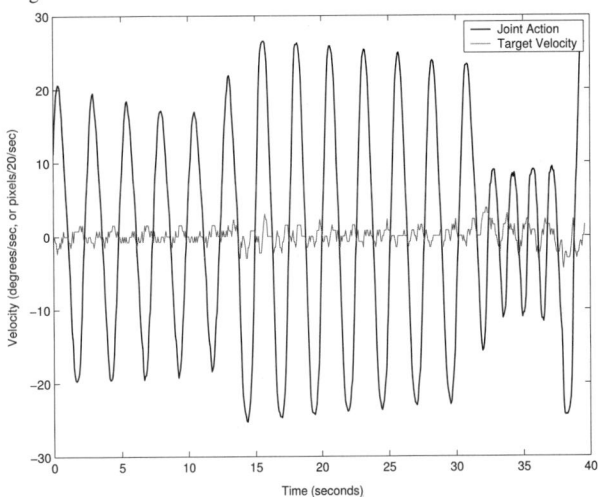

Fig. 9. Velocity data from the swing-trained controller following a swinging target

When attempting to follow smoothly moving targets there was some lag. Lag is inevitable due to sensing and actuation delays unless the controller can predict the movement of the target. Figure 6 shows that the positioning error is highest at the bottom of the swinging motion, when the target's velocity is highest. Ideally the target position and velocity should be maintained at zero. Using our controller there was some variance in the target position and velocity but the bias was small.

C. Training with Swinging Targets—
Evaluating with Swinging Targets

When trained only with swinging targets the controller learnt to smoothly follow the target with excellent performance within a few minutes. Figures 8 and 9 show that the swing-trained controller eliminated the pursuit lag that was seen with the static-trained controller. The swing-trained controller had small target velocity bias, target velocity variance, and target position variance. However, there was some bias in the target position; the bias was largest when the target was swung at a higher than normal velocity in a different position. At all times the target velocity bias and variance remained small.

D. Training with Swinging Targets—
Evaluating with Static Targets

Figure 10 shows the swing-trained controller's behaviour when exposed to a static target. Learning was again disabled so that the controller must perform the task based only its experience with swinging targets. The target was held static to the left for 17 seconds. During that time the active head failed to purse the target and was nearly static. When the swinging target was released the controller successfully pursued the target. The behaviour was repeatable across training sessions.

The controller rotated the camera towards the right during the first swing, matching the target velocity rather than immediately eliminating the target position bias. The target position bias was eliminated when the target was about to swing back. The graph shows that the joint was wiggling slightly while the target was static. The large downward spike in target velocity was due to the release of the target. The lower (zoomed) portion of the graph shows that the controller was not responding to the change in target position as the target was released; it was responding to the velocity of the target. The upward spike in target velocity was due to an overshoot in the joint velocity as the controller matched the velocity of the target.

Another unexpected behaviour sometimes occurred when a pursued swinging target was seized near the extreme of its swing. The controller did not fixate the camera consistently on the now still target: it moved the joint back a few degrees towards the central position, stopped, moved to fixate on the target again, stopped, then repeated the process. Sometimes the controller made several of the ticking motions then

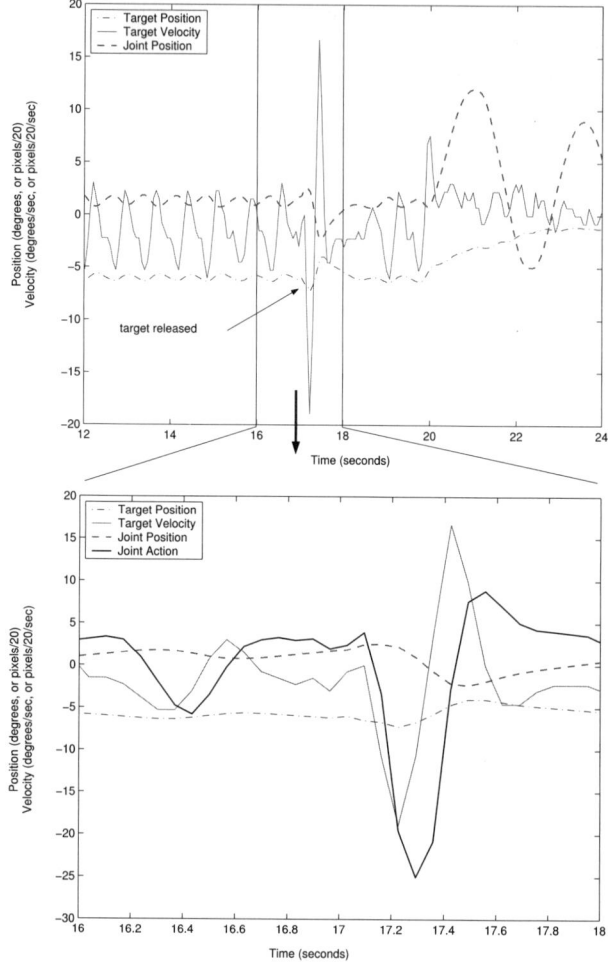

Fig. 10. Swing-trained controller failing to pursue a stationary target

stopped, fixated on the target. On other occasions the ticking motion was repeated, interspersed by pauses of variable length, until the target was released.

VI. DISCUSSION

Although the static-trained and swing-trained controllers had the same initial conditions, the learned controllers differed because the controllers were exposed to different experiences. This resulted in qualitatively different behaviour. The difference was emphasised when the controllers were applied to tasks other than their speciality.

The static-trained controller did not accurately compensate for the velocity of swinging targets and lagged somewhat. Further, the bias in the target position was small, indicating that the static-trained controller responds strongly to the target position.

The behaviour of the swing-trained controller when exposed to static targets is strong evidence that the controller is predicting target behaviour. When faced with a static target the controller acted as if it was expecting swinging behaviour—waiting for the target to swing towards the middle, and moving in expectation that the target will swing towards the middle. Prediction of target behaviour also explains the zero-lag tracking performed by the swing-trained controller when exposed to swinging targets.

To verify the theory, we compared the actions of the control system to the dynamics of a pendulum model. The pendulum model was developed based on measurement of the weight of the target hanging on the string, the length of the string, and the gravitational constant. Incorporating the approximate relationship between the robot's joint angle and the angle of the string gives the required joint acceleration to match target acceleration due to gravity[2]:

$$\ddot{\beta} = \frac{\beta g}{l}\sqrt{1 - \left(\frac{\beta d}{l}\right)^2} \qquad (1)$$

Comparison of the model to the behaviour of the controller does not necessarily require analysis of experimental data gathered using the robot. It is not possible to investigate the controller's behaviour over *multiple* time steps without a dynamic model of the environment, but the controller's behaviour over a *single* time step can be extracted directly from the controller itself by providing an appropriate state input. In this case the state vector has joint velocity, target position, and target velocity of zero. The joint position is the independent variable; the dependent variable is the joint acceleration, which is calculated from the initial joint velocity and output joint velocity. Figure 11 shows joint acceleration for various joint positions. It compares the output of the swing-trained and static-trained controllers with the pendulum model and a static object model ($\ddot{\beta} = 0$ radians/s^2). The output from the swing-trained controller resembles the pendulum model, while the output from the static-trained controller is similar to the static model. This evidence supports the theory that the controllers are predicting the behaviour of the target.

We validated the theory further by investigating the situation in which a new, static target has just appeared. In this case the state vector has joint velocity, joint position, and target velocity set to zero. The target position is the independent variable. As before, the dependent variable is the joint acceleration. Figure 12 shows the output from the static-trained and swing-trained controllers. The static-trained controller moves in the direction of the target. The swing-trained controller moves inconsistently: if the target is to the right it moves towards the target, but if the target is to the left it moves away! The result shows that the swing-trained controller was specialised towards pursuing swinging targets and was not capable of fixating to static targets.

[2]Equation (1) is an approximation, and based on rough measurements of the pendulum string length and the distance from the camera to the target at rest. The variables are $\ddot{\beta}$ radians/s^2, the joint acceleration to match target acceleration; and β radians, the joint angle to match pendulum angle. The constants are $g = 9.81\text{m/s}^2$, the acceleration due to gravity; $l = 1.6\text{m}$, the length of the pendulum string; and $d = 0.64\text{m}$, the distance from the camera to target at rest.

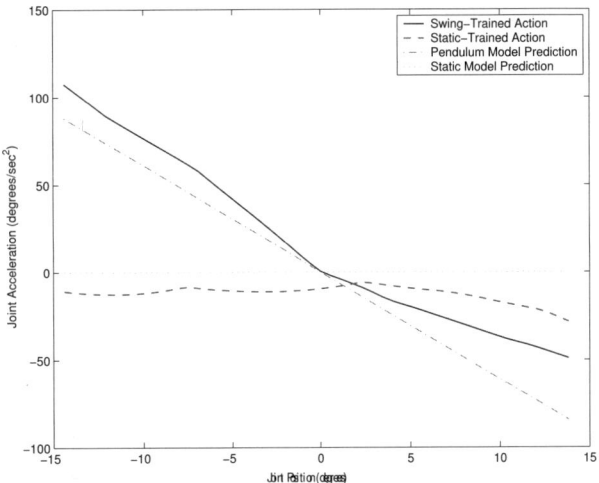

Fig. 11. Actual output and model-predicted joint acceleration for both the static-trained and swing-trained controllers. Joint velocity, target position, and target velocity are zero.

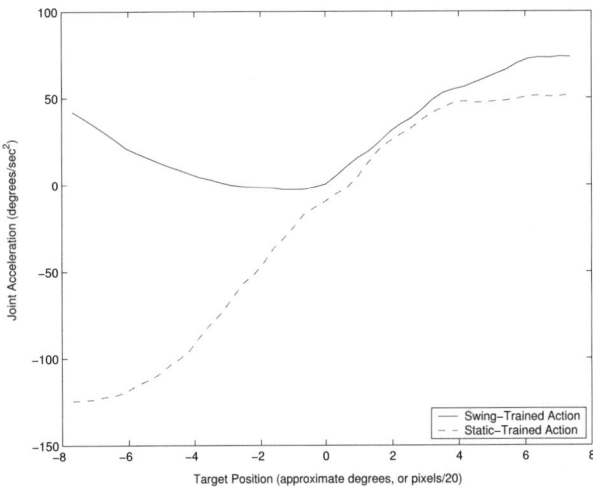

Fig. 12. Joint acceleration for both the static-trained and swing-trained controllers when exposed to a static change in target position. Joint position, joint velocity, and target velocity are zero.

The WFNN algorithm does not represent the target's behaviour or the dynamics of the active-head mechanism explicitly. An explicit model would be a specific part of the system that can be identified as a model of the target's behaviour, or a model of the active-head dynamics. Rather, Q-learning systems represent behaviour and dynamics *implicitly* through the stored action-values; reinforcement learning systems are known as model-free.

The state representation shown in Fig. 4 makes no distinction between state variables of the robot, such as joint velocity, and state variables from the environment, such as target position. There is no explicit connection between the actions and the related state variables. The learning algorithm is interacting with a meta-environment consisting of the robot and its environment, in which much of the relevant state information is unmeasured and not represented by the state vector. However, the Q-learning algorithm assumes interaction with a Markov environment, that is, the probability distribution of the next-state is affected only by the execution of the action in the current, measured state.

Consequently, the swing-trained controller learnt to predict the acceleration of the swinging target based on the joint position (Fig. 11). Inclusion of the joint position in the state information was necessary since the joint behaves differently at the extremes of its rotation range. Nevertheless, predicting target behaviour based on joint position is obviously flawed since the model is no longer applicable if the position of the pendulum's anchor changes (i.e. the position at rest changes). Learning a controller is competent both issuing swinging targets and static targets was difficult because the learning system confounds the movement of the target with its joint movement. This illustrates a flaw in the model-free learning system paradigm: failing to separate controllable mechanisms from uncontrollable environment can lead to learning a controller that is fragile with respect to the behaviour of the environment.

The results could be dismissed as merely another example of over-fitting, except that the type of over-fitting is highly specific, and occurs due to confounding controllable mechanisms with the uncontrollable environment. Avoiding the problem requires a method of specifying, or learning, the distinction.

VII. RELATED WORK

Several other controllers for active heads have included explicit mechanisms for coping with delays by modelling their effect and attempting to compensate [18–21]. Shibata and Schaal's [22] active head controller had an *explicit* module for learning target behaviour that was capable of adapting to patterns in target movement in only a few seconds.

Human eye movements demonstrate a range of mechanisms for minimising lag. Basic human eye movements include *saccades, smooth pursuit, vergence*, and the *vestibular-ocular reflexes* [23]. In general, the purpose of these movements is to allow the gaze to *fixate* on a particular object. Saccades are jerky eye movements that are generated by a change of focus of attention [6]. During these fast eye movements the target is not visible, making saccades an *open-loop* mechanism. Saccades are based on both position and velocity so that the saccade can compensate for the expected movement of the target. Estimating the target velocity before the saccade is generated also allows smooth pursuit to commence immediately at approximately the correct velocity [24]. Murray et al. [21] demonstrated this capability for a non-learning active head.

Smooth pursuit movements follow moving objects and keep them in the centre of the field of view [25]. The eye movement is driven by the velocity of the object, not its position [26]. Positional errors are corrected through saccades. Smooth pursuit is closed-loop, predictive and adaptive [2].

Velocity of the target is an important part of the human gaze control system and non-learning mechanical active and controllers. Yet, many learning active head controllers do not use velocity information—they can not make a different decision based on whether the target is swinging towards the centre of the view or away.

For example, Berthouze et al. [27] smooth pursuit controller, based on Feedback Error Learning (FEL) [28], did not use velocity of the target. Also, current artificial ocular-motor map techniques for saccadic motion have not considered target velocity [29–32]. Researchers at the LiraLab developed an integrated system, combining various biologically inspired eye and head movement mechanisms [33]. Target velocity was not considered, except when compensating for induced pan movements [34].

Shibata and Schaal's [18] controller included velocity components as well as position components, based on a model of the human vestibulo-ocular reflex (VOR) and optokinetic reflex (OKR). The VOR model included a learning component, using FEL with an eligibility trace mechanism [35].

Reinforcement learning was applied to active head control by Piater et al. [36]. The system controlled one degree of freedom, vergence movements, in which both eyes turn inward or outward to look at an object at a particular distance [37]. Five discrete actions were available: changes in vergence angle between 0.1 and 5 degrees. The direction of the change was hard-wired. States were also represented discretely, including the positioning error. The discrete state and action representation without generalisation is ill-suited for this task. Vergence motions were made using discrete actions in a few steps; in contrast, human vergence motions are smooth and closed-loop [37]. The lack of generalisation in the state and action representations requires exploration of every possible representable state and action. The pure-delayed reward signal was the negative of the final positioning error. Given that this error signal is available at all times a delayed reward statement of the problem would probably result in faster learning. Piater et al.'s work appears to be the only application of reinforcement learning to active head control.

VIII. CONCLUSION

Continuous state and action reinforcement learning was successfully applied to control of an active head. The learned controller generated precise, smoothly varying actions. Further, the system considered the velocity of the target and performed lag-free tracking of a swinging target. This was possible through implicitly predicting the target's behaviour. Reinforcement learning's ability to optimise behaviour over time helps to compensate for sensing delays.

Extracting the implicit models through synthetic input was a valuable technique that allowed us to gain a deeper understanding of the controller's behaviour. The controller developed qualitatively different behaviours depending on the learning environment. Although the lag-free tracking performance was excellent, the controller's solution was somewhat fragile with respect to changes in the target's behaviour. The model-free approach is the source of the fragility: it makes no distinction between controllable mechanisms and uncontrollable environment.

ACKNOWLEDGEMENTS

We thank the anonymous reviewers for their comments, Dr. Thomas Brinsmead for assistance with the pendulum model, and Leanne Matuszyk and Orson Sutherland for introducing us to HyDrA.

REFERENCES

[1] S. Hutchinson, G. D. Hager, and P. I. Corke, "A tutorial on visual servo control," *IEEE Transactions on Robotics and Automation*, vol. 12(5):pp. 651–670, 1996.

[2] R. Jürgens, A. W. KornHuber, and W. Becker, "Prediction and strategy in human smooth pursuit eye movements," in G. Lüer, U. Lass, and J. Shallo-Hoffman, eds., *Eye Movement Research: Physiological and Psychological Aspects*, C. J. Hogrefe, Göttingen, Germany, 1988.

[3] D. L. Zhao, A. G. Lasker, and D. A. Robinson, "Interactions of simultaneous saccadic and pursuit prediction," in *Proc. of Contemporary Ocular Motor and Vestibular Research: A Tribute to David A. Robinson*, pp. 171–180, 1993.

[4] R. J. Leigh and D. S. Zee, "Oculomotor disorders," in [23].

[5] H. Collewijn, A. J. Martins, and R. M. Steinman, "Compensatory eye movements during active and passive head movements: Fast adaption to changes in visual magnification," *Journal of Physiology*, vol. 340:pp. 259–286, 1983.

[6] W. Becker, "Saccades," in [23].

[7] R. N. Aslin, "Development of smooth pursuit in human infants," in *Proc. of the Last Whole Earth Eye Movement Conference*, Florida, 1981.

[8] R. S. Sutton and A. G. Barto, *Reinforcement Learning: An Introduction*, Bradford Books, MIT, 1998.

[9] C. Gaskett, L. Fletcher, and A. Zelinsky, "Reinforcement learning for a vision based mobile robot," in *Proc. of the IEEE/RSJ International Conference on Intelligent Robots and Systems (IROS2000)*, Takamatsu, Japan, 2000.

[10] C. J. C. H. Watkins and P. Dayan, "Technical note: Q learning," *Machine Learning*, vol. 8(3/4):pp. 279–292, 1992.

[11] L. C. Baird and A. H. Klopf, "Reinforcement learning with high-dimensional, continuous actions," Tech. Rep. WL-TR-93-1147, Wright Laboratory, 1993.

[12] C. Gaskett, D. Wettergreen, and A. Zelinsky, "Q-learning in continuous state and action spaces," in *Proc. of the 12th Australian Joint Conference on Artificial Intelligence*, Sydney, Australia, 1999.

[13] F. Saito and T. Fukuda, "Learning architecture for real robot systems—extension of connectionist Q-learning for continuous robot control domain," in *Proc. of the International Conference on Robotics and Automation (IROS'94)*, pp. 27–32, 1994.

[14] Y. Takahashi, M. Takeda, and M. Asada, "Continuous valued Q-learning for vision-guided behavior," in *Proc. of the IEEE/SICE/RSJ International Conference on Multisensor Fusion and Integration for Intelligent Systems*, 1999.

[15] F. Maire, "Bicephal reinforcement learning," in *Proc. of the 7th International Conference on Neural Information Processing (ICONIP-2000)*, Taejon, Korea, 2000.

[16] W. D. Smart and L. P. Kaelbling, "Practical reinforcement learning in continuous spaces," in *Proc. of the 17th International Conference on Machine Learning*, 2000.

[17] O. Sutherland, S. Rougeaux, S. Abdallah, and A. Zelinsky, "Tracking with hybrid-drive active vision," in *Proc. of the Australian Conference on Robotics and Automation (ACRA2000)*, Melbourne, Australia, 2000.

[18] T. Shibata and S. Schaal, "Biomimetic gaze stabilization based on feedback-error learning with nonparametric regression networks," *Neural Networks*, vol. 14(2), 2001.

[19] C. Brown, "Gaze controls with interactions and delays," *IEEE Transactions on Systems, Man, and Cybernetics*, vol. 20(1):pp. 518–527, 1990.

[20] D. W. Murray, F. Du, P. F. McLauchlan, I. D. Reid, P. M. Sharkey, and J. M. Brady, "Design of stereo heads," in A. Blake and A. Yuille, eds., *Active Vision*, MIT Press, 1992.

[21] D. W. Murray, K. J. Bradshaw, P. F. McLauchlan, I. D. Reid, and P. M. Sharkey, "Driving saccade to pursuit using image motion," *International Journal of Computer Vision*, vol. 16(3):pp. 205–228, 1995.

[22] T. Shibata and S. Schaal, "Biomimetic smooth pursuit based on fast learning of the target dynamics," in *Proc. of the IEEE International Conference on Intelligent Robots and Systems (IROS2001)*, 2001.

[23] R. H. S. Carpenter, ed., *Eye Movements*, vol. 8 of *Vision and Visual Dysfunction*, Macmillan, 1991.

[24] A. L. Yarbus, *Eye Movements and Vision*, Plenum Press, New York, 1967.

[25] J. Pola and H. J. Wyatt, "Smooth pursuit: Response characteristics, stimuli and mechanisms," in [23].

[26] C. Rashbass, "The relationship between saccadic and smooth tracking eye movements," *Journal of Physiology*, vol. 159:pp. 338–362, 1961.

[27] L. Berthouze, S. Rougeaux, Y. Kuniyoshi, and F. Chavand, "A learning stereo-head control system," in *Proc. of the World Automation Congress/International Symposium on Robotics and Manufacturing*, France, 1996.

[28] M. Kawato, K. Furawaka, and R. Suzuki, "A hierarchical neural network model for the control and learning of voluntary movements," *Biological Cybernetics*, 1987.

[29] H. Ritter, T. Martinetz, and K. Schulten, *Neural Computation and Self-Organizing Maps: An Introduction*, Addison Wesley, 1992.

[30] R. P. Rao and D. H. Ballard, "Learning saccadic eye movements using multiscale spatial filters," in *Proc. of Advances in Neural Information Processing Systems 7 (NIPS94)*, 1994.

[31] M. Marjanović, B. Scassellati, and M. Williamson, "Self-taught visually guided pointing for a humanoid robot," in *Proc. of the Fourth International Conference on Simulation of Adaptive Behavior*, MA, 1996.

[32] M. Pagel, E. Maël, and C. von der Malsburg, "Self calibration of the fixation movement of a stereo camera head," *Machine Learning*, vol. 31(1-3):pp. 169–186, 1998.

[33] G. Metta, F. Panerai, R. Manzotti, and G. Sandini, "Babybot: an artificial developing robotic agent," in *Proc. of From Animals to Animats: Sixth International Conference on the Simulation of Adaptive Behavior (SAB 2000)*, Paris, 2000.

[34] F. Panerai, G. Metta, and G.Sandini, "Learning VOR-like stabilization reflexes in robots," in *Proc. of the 8th European Symposium on Artificial Neural Networks (ESANN 2000)*, Bruges, Belgium, 2000.

[35] A. G. Barto, R. S. Sutton, and C. W. Anderson, "Neuronlike adaptive elements that can solve difficult learning control problems," *IEEE Transactions on systems, man and cybernetics*, vol. SMC-13:pp. 834–846, 1983.

[36] J. H. Piater, R. A. Grupen, and K. Ramamritham, "Learning real-time stereo vergence control," in *Proc. of the 14th International Symposium on Intelligent Control (ISIC '99)*, 1999.

[37] S. J. Judge, "Vergence," in [23].

A New Algorithm of Adaptive Iterative Learning Control for Uncertain Robotic Systems

Chun-Te Hsu, Chiang-Ju Chien, and Chia-Yu Yao
Department of Electronic Engineering, Huafan University, Taipei County, Taiwan
E-mail : cjc@huafan.hfu.edu.tw, Fax : 886-2-86606675

Abstract—In this paper, we propose a new adaptive iterative learning control (AILC) scheme for a class of parametric uncertain robotic systems with disturbances. The main feature of the proposed AILC scheme is that all the estimated parameters are updated by a new adaptive law which combines time-domain and iteration-domain adaptation. This new adaptive law is designed without using projection or deadzone mechanism and can be applied to system with non-periodic or non-repeatable disturbance. Via a rigorous technical analysis, it is shown that all adjustable parameters as well as the internal signals remain bounded in the time-domain for each iteration and the tracking error can be driven to zero in the iteration-domain. Finally, the learning performance will be demonstrated by a simulation example.

I. INTRODUCTION

Iterative learning control (ILC) is one of the most effective control approaches in dealing with the tasks of repeated tracking control or periodic disturbance rejection. Basically, the traditional ILC updates its current input by adding the tracking error and/or its time derivative to the previous input with a suitable linear combination mechanism [1]-[5]. It can easily improve the control performance by these simple self-tuning processes without using accurate system models. However, due to some difficulties or limitations to apply traditional ILC for certain applications, other new types of ILC algorithms have been widely studied in recent years. One of the most interesting and important developments is the so-called adaptive iterative learning control (AILC). Substantial efforts in the area of AILC have been reported in [6]–[13] for past decade. The main feature of AILC is to estimate the uncertain parameters, which are in turn used to generate the current control input. Because of the iteration based control problem, the adaptive learning laws for estimation of the unknown parameters are mostly designed in iteration-domain [6]–[10]. In general, projection or deadzone mechanism is necessary to construct the iteration-domain based adaptive laws in order to guarantee the stability and convergence of the learning system. In [11], the control parameter is updated in the time-domain similar to the traditional adaptive control approach, but without considering the effect of disturbance. In [12], the AILC is designed based on an estimation procedure using a Kalman filter and an optimization of a quadratic criterion. Recently, an AILC scheme with both time-domain and iteration-domain adaptations for uncertain robotic system was presented in [13]. But the adaptive laws are designed separately. A time-domain adaptive law estimates the robotic system parameters so that the upper bounds on these parameters are not necessary. However, another iteration-domain learning law which learns the desired input and repeatable disturbance still needs the bound for estimation.

In this paper, a new adaptive AILC scheme is proposed for a class of parametric uncertain robotic systems with disturbances. The new adaptive law which combines time-domain and iteration-domain adaptation is proposed to update the control parameters so that the upper bound for any parameter is no longer needed. The adaptive law can become a pure time-domain learning law or iteration-domain learning law if a weighting parameter is suitably chosen. Using this new design, the disturbance can be totally rejected even it is not periodic or not repeatable. A rigorous proof via the technique of Lyapunov-like analysis is given to guarantee the stability and convergence of the closed-loop learning system. It is shown that all adjustable parameters as well as internal signals are bounded in time domain for each iteration and the tracking error will asymptotically converge to zero in iteration domain if iteration number is large enough.

Similar to [13], we use the concept of $L_{pe}[0,T]$ in the subsequent discussions to denote the set of Lebesgue measurable (or piecewise continuous) real valued (vector) functions with L_{pe} norm

$$|\cdot|_{pe} = \begin{cases} \left(\int_0^T |\cdot|^p dt\right)^{\frac{1}{p}} & \text{if } p \in [1, \infty) \\ \sup_{0 \leq t \leq T} |\cdot| & \text{if } p = \infty \end{cases}$$

where $|\cdot|$ denotes the absolute value of a scalar function, or the Euclidean or any other consistent norm of a vector.

II. PLANT DESCRIPTION AND CONTROL OBJECTIVE

In this paper, we consider a robot system with n rigid bodies which can perform a given task repeatedly over a finite time interval $[0, T]$ as follows:

$$D(q^j(t))\ddot{q}^j(t) + B(q^j(t), \dot{q}^j(t))\dot{q}^j(t) + f(q^j(t), \dot{q}^j(t)) + d^j(t) = u^j(t) \quad (1)$$

where j denotes the index of iteration number, $q^j(t) \in R^n$ is the generalized joint coordinate vector, $D(q^j(t)) \in R^{n \times n}$ is the inertia matrix, $B(q^j(t), \dot{q}^j(t)) \in R^n$ is the centripetal plus Coriolis force vector, $f(q^j(t), \dot{q}^j(t)) \in R^n$ is the gravitational plus frictional forces, $d^j(t) \in R^n$ is an unknown disturbance vector, and $u^j(t) \in R^n$ is the joint torque vector. The symmetric inertia matrix $D(q^j(t))$ is assumed to be positive definite and bounded for all $t \in [0, T]$ and iteration $j \geq 1$ as $0 < \lambda_1 I \leq D(q^j(t)) \leq \lambda_2 I$ where $\lambda_1, \lambda_2 > 0$ and I is an $n \times n$ identity matrix. The matrix $\dot{D}(q^j(t)) - 2B(q^j(t), \dot{q}^j(t))$ is assumed to be skew-symmetric, that is, $x^\top (\dot{D}(q^j(t)) - 2B(q^j(t), \dot{q}^j(t)))x = 0$

for all $x \in R^n$ and $x \neq 0$. Let $q_d(t)$, which is independent of iteration j, be the desired trajectory to be tracked during the time interval $[0, T]$. Furthermore, we assume that the system parameters of the robot manipulator is not completely known. The control objective is to find an adaptive iterative learning controller $u^j(t)$ such that when iteration number j is large enough, the joint position $q^j(t)$ will converge to $q_d(t)$ as close as possible even there exists some non-repeatable disturbance $d^j(t)$. In this paper, the desired joint position, velocity, acceleration and control input vectors are denoted as $q_d(t)$, $\dot{q}_d(t)$, $\ddot{q}_d(t)$ and $u_d(t)$, respectively. In other words, without considering the effect of disturbance $d^j(t)$, the following equation will hold,

$$D(q_d(t))\ddot{q}_d(t) + B(q_d(t), \dot{q}_d(t))\dot{q}_d(t) \\ + f(q_d(t), \dot{q}_d(t)) = u_d(t) \quad (2)$$

The assumptions required to achieve the control objective are summarized as follows:

(A1) The desired control input $u_d(t)$ and disturbance $d^j(t)$ are bounded such that $|u_d(t) + d^j(t)| \leq \psi$ for some existed but unknown positive constant ψ and for all $t \in [0, T]$, $j \geq 1$.

(A2) $q_d(t)$ and $\dot{q}_d(t)$ are measurable and bounded.

(A3) The initial resetting condition is satisfied, i.e., $q^j(0) = q_d(0)$, $\dot{q}^j(0) = \dot{q}_d(0)$, for all $j \geq 1$.

In the next section, the time argument t will be omitted for the notation brevity.

III. DESIGN OF THE ADAPTIVE ITERATIVE LEARNING CONTROLLER

To achieve the control objective, the learning control input $u^j(t)$ at jth iteration is designed as:

$$u^j = u_f^j + u_{a\ell}^j \quad (3)$$

where u_f^j is the feedback control input and $u_{a\ell}^j$ is the adaptive learning control input that compensates for the unknown system parameters and disturbance. As the most control algorithms for robot manipulator, the feedback control input u_f^j is responsible for the stabilization of the closed-loop system, which is designed as the following traditional PD-type controller:

$$u_f^j = \beta L \left(\dot{e}^j + a e^j \right) = \beta L z^j \quad (4)$$

where $z^j \equiv \dot{e}^j + a e^j$, $e^j \equiv q_d - q^j$, β is a positive feedback gain, L is a symmetric positive define matrix and a is a positive constant. Since the parameters of robot system are not completely known and the disturbance is unknown and non-repeatable, the desired compensated control input is not available before controller design. Similar to [6], [7], [13], the algorithm $u_{a\ell}^j$ is designed based on the estimated parameters as follows:

$$u_{a\ell}^j = \widehat{D}_e(q^j)\ddot{q}_d + \widehat{B}_e(q^j, \dot{q}^j)\dot{q}_d + \widehat{f}_e(q^j, \dot{q}^j) \\ + a\left(\widehat{D}(q^j)\dot{e}^j + \widehat{B}(q^j, \dot{q}^j)e^j\right) + sign(z^j)\widehat{\psi}^j \quad (5)$$

where $\widehat{D}_e(q^j) = \widehat{D}(q^j) - \widehat{D}(q_d)$, $\widehat{B}_e(q^j, \dot{q}^j) = \widehat{B}(q^j, \dot{q}^j) - \widehat{B}(q_d, \dot{q}_d)$, $\widehat{f}_e(q^j, \dot{q}^j) = \widehat{f}(q^j, \dot{q}^j) - \widehat{f}(q_d, \dot{q}_d)$, and $sign(z^j) = [sign(z_1^j), \cdots, sign(z_n^j)]^\top$. The final term of (5) is an additional control force which is used to compensate for the unknown bounded desired input u_d and disturbance d^j. In general, this is a combination of computed torque input and variable structure input based learning scheme. By using the technique of linear parameterization for robot systems, we can rearrange $u_{a\ell}^j$ with a suitably chosen system parameter vector as:

$$u_{a\ell}^j = Y_e^j \widehat{\theta}^j + sign(z^j) \widehat{\psi}^j \quad (6)$$

where $Y_e^j \equiv Y(q^j, \dot{q}^j, q_d, \dot{q}_d, \ddot{q}_d) \in R^{n \times p}$ is the regression matrix and $\widehat{\theta}^j \in R^p$ is the estimated parameter vector with p being a suitably defined positive integer. Now, substituting the control input u^j into the robot system (1), we have

$$D(q^j)\ddot{q}^j + B(q^j, \dot{q}^j)\dot{q}^j + f(q^j, \dot{q}^j) + d^j \\ = \beta L z^j + Y_e^j \widehat{\theta}^j + sign(z^j)\widehat{\psi}^j \quad (7)$$

If we define $D_e(q^j) = D(q^j) - D(q_d)$, $B_e(q^j, \dot{q}^j) = B(q^j, \dot{q}^j) - B(q_d, \dot{q}_d)$, and $f_e(q^j, \dot{q}^j) = f(q^j, \dot{q}^j) - f(q_d, \dot{q}_d)$, we can rearrange (7), by adding $-a(D(q^j)\dot{e}^j + B(q^j, \dot{q}^j)e^j)$ on both sides of (7), as follows

$$\begin{aligned} &D(q^j)\dot{z}^j + B(q^j, \dot{q}^j)z^j + \beta L z^j \\ &= D(q^j)\ddot{q}_d + B(q^j, \dot{q}^j)\dot{q}_d + f(q^j, \dot{q}^j) \\ &\quad + a(D(q^j)\dot{e}^j + B(q^j, \dot{q}^j)e^j) \\ &\quad + d^j - Y_e^j \widehat{\theta}^j - sign(z^j)\widehat{\psi}^j \\ &= D_e(q^j)\ddot{q}_d + B_e(q^j, \dot{q}^j)\dot{q}_d + f_e(q^j, \dot{q}^j) \\ &\quad + a\left(D(q^j)\dot{e}^j + B(q^j, \dot{q}^j)e^j\right) \\ &\quad D(q_d)\ddot{q}_d + B(q_d, \dot{q}_d)\dot{q}_d + f(q_d, \dot{q}_d) \\ &\quad + d^j - Y_e^j \widehat{\theta}^j - sign(z^j)\widehat{\psi}^j \\ &= Y_e^j \theta - Y_e^j \widehat{\theta}^j + u_d + d^j - sign(z^j)\widehat{\psi}^j \quad (8) \end{aligned}$$

If the right-hand side of (8) is zero, it is well known that the system would be asymptotically stable [14], [15]. Hence, it is now clear from (8) that the learning control forces $Y_e^j \widehat{\theta}^j$ and $sign(z^j)\widehat{\psi}^j$ are used to compensate for the unknown system parameter vector θ and unknown desired control torque u_d and disturbance d^j, respectively. Suitable adaptation laws will be deduced to estimate θ and the bound ψ of $u_d + d^j$. In order to guarantee that the signals are bounded in time domain and error converges in iteration domain, we propose the following adaptation laws which combine time domain and iteration domain as follows:

$$(1 - \gamma_1)\dot{\widehat{\theta}}^j = -\gamma_1 \widehat{\theta}^j + \gamma_1 \widehat{\theta}^{j-1} + \beta_1 Y_e^{j\top} z^j \quad (9)$$

$$(1 - \gamma_2)\dot{\widehat{\psi}}^j = -\gamma_2 \widehat{\psi}^j + \gamma_2 \widehat{\psi}^{j-1} + \beta_2 |z^j| \quad (10)$$

where $\widehat{\theta}^j(0) = \widehat{\theta}^{j-1}(T), \widehat{\psi}^j(0) = \widehat{\psi}^{j-1}(T)$ for all $j \geq 1$ and $0 < \gamma_1, \gamma_2 < 1, \beta_1, \beta_2 > 0$. For the first iteration, $\widehat{\theta}^0(t)$ is set to be any constant vector θ^0 or zero vector, and $\widehat{\psi}^0(t)$ is set to be a small positive number $\psi^0, \forall t \in [0,T]$. The positive parameters β_1, β_2 and γ_1, γ_2 are defined as the adaptation gains and weighting gains, respectively. The weighting gains decide the weight between time-domain and iteration-domain adaptation for these two laws. In general, (9) and (10) will become pure time-domain adaptation laws if $\gamma_1 = \gamma_2 = 0$, or pure iteration-domain adaptation laws if $\gamma_1 = \gamma_2 = 1$.

IV. ANALYSIS OF STABILITY AND CONVERGENCE

In this section, stability and convergence of the proposed adaptive iterative learning control system are considered. To begin with, the boundedness of internal signals at first iteration will be first established.

Lemma 1: Consider the uncertain robot system (1) which satisfies assumption (A1). If the control task repeats over a finite time interval $[0,T]$ with a given desired position and velocity trajectories q_d and $\dot{q}^j$ satisfying assumption (A2) and (A3), then the proposed adaptive iterative learning controller (3) and adaptation laws (9), (10), will ensure that all the internal signals at first iteration are bounded, i.e., $e^1, z^1, \widehat{\theta}^1, \widehat{\psi}^1, u^1, \dot{\widehat{\theta}}^1, \dot{\widehat{\psi}}^1 \in L_{\infty e}[0,T]$.

Proof: Define the estimation parameter error as $\widetilde{\theta}^j = \widehat{\theta}^j - \theta$ and $\widetilde{\psi}^j = \widehat{\psi}^j - \psi$ and choose a Lyapunov function candidate as

$$V_a^j = \frac{1}{2} z^{jT} D(q^j) z^j + \frac{(1-\gamma_1)}{2\beta_1} \widetilde{\theta}^{jT} \widetilde{\theta}^j + \frac{(1-\gamma_2)}{2\beta_2} (\widetilde{\psi}^j)^2 \quad (11)$$

By using the fact that $\dot{D}(q^j) - 2B(q^j, \dot{q}^j)$ is skew-symmetric and L is a symmetric positive define matrix, the derivative of V_a^j can be computed as

$$\begin{aligned}
\dot{V}_a^j &= z^{jT} D(q^j) \dot{z}^j + \frac{1}{2} z^{jT} \dot{D}(q^j) z^j \\
&\quad + \frac{(1-\gamma_1)}{\beta_1} \widetilde{\theta}^{jT} \dot{\widetilde{\theta}}^j + \frac{(1-\gamma_2)}{\beta_2} \widetilde{\psi}^j \dot{\widetilde{\psi}}^j \\
&= z^{jT} \left[-B(q^j, \dot{q}^j) z^j - \beta L z^j - Y_e^j \widetilde{\theta}^j + u_d + d^j \right. \\
&\quad \left. - sign(z^j) \widehat{\psi}^j \right] + \frac{1}{2} z^{jT} \dot{D}(q^j) z^j \\
&\quad + \frac{(1-\gamma_1)}{\beta_1} \widetilde{\theta}^{jT} \dot{\widetilde{\theta}}^j + \frac{(1-\gamma_2)}{\beta_2} \widetilde{\psi}^j \dot{\widetilde{\psi}}^j \\
&= z^{jT} \left[-\beta L z^j - Y_e^j \widetilde{\theta}^j + u_d + d^j - sign(z^j) \widehat{\psi}^j \right] \\
&\quad + \frac{\widetilde{\theta}^{jT}}{\beta_1} \left[(1-\gamma_1) \dot{\widetilde{\theta}}^j \right] + \frac{\widetilde{\psi}^j}{\beta_2} \left[(1-\gamma_2) \dot{\widetilde{\psi}}^j \right]
\end{aligned}$$

Substituting the adaptation laws (9), (10) into above equation, we have

$$\begin{aligned}
\dot{V}_a^j &= -\beta z^{jT} L z^j - z^{jT} Y_e^j \widetilde{\theta}^j \\
&\quad + z^{jT} (u_d + d^j) - z^{jT} sign(z^j) \widehat{\psi}^j \\
&\quad + \frac{\widetilde{\theta}^{jT}}{\beta_1} \left[-\gamma_1 (\widehat{\theta}^j - \theta) + \gamma_1 (\widehat{\theta}^{j-1} - \theta) + \beta_1 Y_e^{jT} z^j \right] \\
&\quad + \frac{\widetilde{\psi}^j}{\beta_2} \left[-\gamma_2 (\widehat{\psi}^j - \psi) + \gamma_2 (\widehat{\psi}^{j-1} - \psi) + \beta_2 |z^j| \right] (12)
\end{aligned}$$

Now we check the term $z^{jT}(u_d + d^j) - z^{jT} sign(z^j) \widehat{\psi}^j + \widetilde{\psi}^j |z^j|$ in above equation

$$\begin{aligned}
z^{jT}(u_d + d^j) &- z^{jT} sign(z^j) \widehat{\psi}^j + \widetilde{\psi}^j |z^j| \\
&\leq |z^j||u_d + d^j| - |z^j|\widehat{\psi}^j + \widetilde{\psi}^j|z^j| \\
&\leq |z^j|\psi - |z^j|\widehat{\psi}^j + \widetilde{\psi}^j|z^j| \\
&= -|z^j|\widetilde{\psi}^j + \widetilde{\psi}^j|z^j| \\
&= 0 \quad (13)
\end{aligned}$$

Substituting the result of (13) into (12) yields

$$\dot{V}_a^j \leq -\beta z^{jT} L z^j - \frac{\gamma_1}{\beta_1} \widetilde{\theta}^{jT} \widetilde{\theta}^j + \frac{\gamma_1}{\beta_1} \widetilde{\theta}^{jT} \widetilde{\theta}^{j-1} - \frac{\gamma_2}{\beta_2} (\widetilde{\psi}^j)^2 + \frac{\gamma_2}{\beta_2} \widetilde{\psi}^j \widetilde{\psi}^{j-1} \quad (14)$$

For the first iteration $j=1$, it is note that $\widetilde{\theta}^0(t) = \widehat{\theta}^0(t) - \theta = \theta^0 - \theta \equiv \bar{\theta}^0$ and $\widetilde{\psi}^0(t) = \widehat{\psi}^0(t) - \psi = \psi^0 - \psi \equiv \bar{\psi}^0$ are bounded for all $t \in [0,T]$ so that (14) can be rewritten as

$$\begin{aligned}
\dot{V}_a^1 &\leq -\beta z^{1T} L z^1 - \frac{\gamma_1}{\beta_1} \widetilde{\theta}^{1T} \widetilde{\theta}^1 + \frac{\gamma_1}{\beta_1} \widetilde{\theta}^{1T} \bar{\theta}^0 \\
&\quad - \frac{\gamma_2}{\beta_2} (\widetilde{\psi}^1)^2 + \frac{\gamma_2}{\beta_2} \widetilde{\psi}^1 \bar{\psi}^0 \\
&= -\beta z^{1T} L z^1 - \frac{\gamma_1}{2\beta_1} \widetilde{\theta}^{1T} \widetilde{\theta}^1 - \frac{\gamma_2}{2\beta_2} (\widetilde{\psi}^1)^2 \\
&\quad - \frac{\gamma_1}{2\beta_1} \widetilde{\theta}^{1T} \widetilde{\theta}^1 + \frac{\gamma_1}{\beta_1} \widetilde{\theta}^{1T} \bar{\theta}^0 - \frac{\gamma_1}{2\beta_1} \bar{\theta}^{0T} \bar{\theta}^0 + \frac{\gamma_1}{2\beta_1} \bar{\theta}^{0T} \bar{\theta}^0 \\
&\quad - \frac{\gamma_2}{2\beta_2} (\widetilde{\psi}^1)^2 + \frac{\gamma_2}{\beta_2} \widetilde{\psi}^1 \bar{\psi}^0 - \frac{\gamma_2}{2\beta_2} (\bar{\psi}^0)^2 + \frac{\gamma_2}{2\beta_2} (\bar{\psi}^0)^2 \\
&= -\beta z^{1T} L z^1 - \frac{\gamma_1}{2\beta_1} \widetilde{\theta}^{1T}(t) \widetilde{\theta}^1 - \frac{\gamma_2}{2\beta_2} (\widetilde{\psi}^1(t))^2 \\
&\quad - \frac{\gamma_1}{2\beta_1} (\widetilde{\theta}^{1T} - \bar{\theta}^{0T})(\widetilde{\theta}^{1T} - \bar{\theta}^0) - \frac{\gamma_2}{2\beta_2} (\widetilde{\psi}^1 - \bar{\psi}^0)^2 \\
&\quad + \frac{\gamma_1}{2\beta_1} \bar{\theta}^{0T} \bar{\theta}^0 + \frac{\gamma_2}{2\beta_2} (\bar{\psi}^0)^2 \\
&\leq -kV_a^1 + \bar{k}^0 \quad (15)
\end{aligned}$$

where $k = \min\{\frac{1}{2\beta}, \frac{\gamma_1}{1-\gamma_1}, \frac{\gamma_2}{1-\gamma_2}\}, \bar{k}^0 = \frac{\gamma_1}{2\beta_1} \bar{\theta}^{0T} \bar{\theta}^0 + \frac{\gamma_2}{2\beta_2} (\bar{\psi}^0)^2$. Note that the initial value $V_1^1(0)$ is bounded since $z^1(0) = 0, \widetilde{\theta}^1(0) = \widehat{\theta}^1(0) - \theta = \widehat{\theta}^0(T) - \theta = \bar{\theta}^0$, and $\widetilde{\psi}^1(0) = \widehat{\psi}^1(0) - \psi = \widehat{\psi}^0(T) - \psi = \bar{\psi}^0$. Together with the result from (15), it implies $V_a^1, z^1, \tilde{\theta}^1, \tilde{\psi}^1 \in L_{\infty e}[0,T]$ and hence, $e^1, \dot{e}^1$ (by the definition of $z^1 = \dot{e}^1 + ae^1, a > 0$), $\hat{\dot{\theta}}^1$ (by 9), $\hat{\dot{\psi}}^1$ (by 10) $\in L_{\infty e}[0,T]$. Since $e^1, \dot{e}^1, z^1 \in L_{\infty e}[0,T]$, we have $Y_e^1 \in L_{\infty e}[0,T]$ from lemma 1 in [13]. This implies $u^1 = Y_e^1 \hat{\theta}^1 - sign(z^1)\hat{\psi}^1 \in L_{\infty e}[0,T]$. Q.E.D.

Lemma 2: The proposed AILC can ensure that
(L1) $\lim_{j \to \infty} \tilde{\theta}^{jT}(T)\tilde{\theta}^j(T) = \theta_T^T \theta_T$, and $\lim_{j \to \infty}(\tilde{\psi}^j(T))^2 = (\psi_T)^2$ for some constants $\theta_T^T \theta_T$ and $(\psi_T)^2$.
(L2) $\tilde{\theta}^{jT}(T)\tilde{\theta}^j(T)$, $(\tilde{\psi}^j(T))^2$ and $z^{jT}(T)D(q^j(T))z^j(T)$ are bounded for all $j \geq 1$.
(L3) $\lim_{j \to \infty} z^{jT}(T)D(q^j(T))z^j(T) = 0$, and
$\lim_{j \to \infty} \int_0^T z^{jT}(t)Lz^j(t)dt = 0$.

Proof: Choose a positive function $V^j(T)$ as
$$V^j(T) = \int_0^T \left[\frac{\gamma_1}{2\beta_1}\tilde{\theta}^{jT}\tilde{\theta}^j + \frac{\gamma_2}{2\beta_2}(\tilde{\psi}^j)^2\right]dt$$
$$+ \frac{1-\gamma_1}{2\beta_1}\tilde{\theta}^{jT}(T)\tilde{\theta}^j(T) + \frac{(1-\gamma_2)}{2\beta_2}(\tilde{\psi}^j(T))^2$$

The difference between $V^j(T)$ and $V^{j-1}(T)$ can be derived by using integration by parts as follows

$V^j(T) - V^{j-1}(T)$
$= \int_0^T \left[\frac{\gamma_1}{2\beta_1}\left(\tilde{\theta}^{jT}\tilde{\theta}^j - \tilde{\theta}^{j-1T}\tilde{\theta}^{j-1}\right)\right.$
$\left. + \frac{\gamma_2}{2\beta_2}((\tilde{\psi}^j)^2 - (\tilde{\psi}^{j-1})^2)\right]dt + \frac{(1-\gamma_1)}{\beta_1}\int_0^T \tilde{\theta}^{jT}\dot{\tilde{\theta}}^j dt$
$+ \frac{(1-\gamma_1)}{2\beta_1}\tilde{\theta}^{jT}(0)\tilde{\theta}^j(0) - \frac{(1-\gamma_1)}{2\beta_1}\tilde{\theta}^{j-1T}(T)\tilde{\theta}^{j-1}(T)$
$+ \frac{(1-\gamma_2)}{\beta_2}\int_0^T \tilde{\psi}^j \dot{\tilde{\psi}}^j dt$
$+ \frac{(1-\gamma_2)}{2\beta_2}(\tilde{\psi}^j(0))^2 - \frac{(1-\gamma_2)}{2\beta_2}(\tilde{\psi}^{j-1}(T))^2$
$= \int_0^T \left[\frac{\gamma_1}{2\beta_1}\left(\tilde{\theta}^{jT}\tilde{\theta}^j - \tilde{\theta}^{j-1T}\tilde{\theta}^{j-1}\right)\right.$
$\left. + \frac{\gamma_2}{2\beta_2}((\tilde{\psi}^j)^2 - (\tilde{\psi}^{j-1})^2)\right]dt$
$+ \frac{1}{\beta_1}\int_0^T \tilde{\theta}^{jT}\left[-\gamma_1 \tilde{\theta}^j + \gamma_1 \tilde{\theta}^{j-1} + \beta_1 Y_e^{jT} z^j\right]dt$
$+ \frac{1}{\beta_2}\int_0^T \tilde{\psi}^j\left[-\gamma_2 \tilde{\psi}^j + \gamma_2 \tilde{\psi}^{j-1} + \beta_2|z^j|\right]dt$
$= \int_0^T \left[-\frac{\gamma_1}{2\beta_1}(\tilde{\theta}^j - \tilde{\theta}^{j-1})^T(\tilde{\theta}^j - \tilde{\theta}^{j-1})\right.$
$\left. -\frac{\gamma_2}{2\beta_2}(\tilde{\psi} - \tilde{\psi}^{j-1})^2 + \tilde{\theta}^{jT}Y_e^{jT}z^j + \tilde{\psi}^j|z^j|\right]dt$
$\leq \int_0^T \left[\tilde{\theta}^{jT}Y_e^{jT}z^j + \tilde{\psi}^j|z^j|\right]dt \quad (16)$

where we use the facts that $\tilde{\theta}^j(0) = \tilde{\theta}^{j-1}(T)$ and $\tilde{\psi}^j(0) = \tilde{\psi}^{j-1}(T)$. If we define another Lyapunov function $V_b^j = \frac{1}{2}z^{jT}D(q^j)z^j$, we can use similar method in lemma 1 and easily derive the following inequality:

$$\dot{V}_b^j \leq -\beta z^{jT}Lz^j - z^{jT}Y_e^j\tilde{\theta}^j - \tilde{\psi}^j|z^j| \quad (17)$$

Integrating both side of (17) from 0 to T, we have

$V_b^j(T) - V_b^j(0)$
$\leq \int_0^T \left[-\beta z^{jT}Lz^j - z^{jT}Y_e^j\tilde{\theta}^j - \tilde{\psi}^j|z^j|\right]dt \quad (18)$

or equivalently,

$\int_0^T \left[\tilde{\theta}^{jT}Y_e^{jT}z^j + \tilde{\psi}^j|z^j|\right]dt$
$\leq -V_b^j(T) - \int_0^T \beta z^{jT}Lz^j dt \quad (19)$

where we use the property of $V_b^j(0) = \frac{1}{2}z^{jT}(0)D(q^j(0))z^j(0) = 0$ due to assumption (A3). Substituting (19) into (16), it yields

$V^j(T) - V^{j-1}(T) \leq -\frac{1}{2}z^{jT}(T)D(q^j(T))z^j(T)$
$- \int_0^T \beta z^{jT}Lz^j dt \quad (20)$

Since $V^1(T)$ is bounded by lemma 1, and $V^j(T)$ is positive and monotonically decreasing by (20), we conclude that $V^j(T)$ is bounded for all $j \geq 1$ and will converge to some limit value $V(T)$ (independent of j) as j approaches infinity. Hence, all terms in $V^j(T)$ converge and (L1) of lemma 2 follows. On the other hand, (20) implies that

$$\frac{1}{2}z^{jT}(T)D(q^j(T))z^j(T) \leq V^{j-1}(T) - V^j(T) \leq V^1(T) \quad (21)$$

and

$$\beta \int_0^T z^{jT}(t)Lz^j(t)dt \leq V^{j-1}(T) - V^j(T) \leq V^1(T) \quad (22)$$

for all $j \geq 1$. From (21), the boundness of $z^{jT}(T)D(q^j(T))z^j(T)$ is established for all iteration, and hence (L2) of lemma 2 follows. Finally, as $\lim_{j \to \infty} V^{j-1}(T) - V^j(T) = 0$, (L3) of lemma 2 is achieved from (21), and (22). This completes the proof. Q.E.D.

Until now, we have shown that all the internal signals for the first iteration are bounded, and $\tilde{\Theta}^j(T), \tilde{\psi}^j(T)$ or equivalently $\tilde{\Theta}^j(0), \tilde{\psi}^j(0)$) are bounded for all $j \geq 1$ too. In the following theorem, the boundedness of all internal signals at each iteration and the convergence of $z^j(t)$ will be established.

Theorem 1: Consider the problem set-up in lemma 1 again. The proposed AILC will guarantee the tracking performance and system stability as follows :

(T1) $z^j, e^j, \dot{e}^j, \widehat{\theta}^j, \dot{\widehat{\theta}}^j, \widehat{\psi}^j, \dot{\widehat{\psi}}^j, u^j \in L_{\infty e}[0,T]$, for all $j \geq 1$.

(T2) $\lim_{j \to \infty} z^{j\top} L z^j = 0$, for all $t \in [0,T]$.

Proof:

(T1) Since $z^1, \widetilde{\theta}^1, \widetilde{\psi}^1 \in L_{\infty e}[0,T]$ as shown in lemma 1, if we assume $z^{j-1}, \widetilde{\theta}^{j-1}, \widetilde{\psi}^{j-1} \in L_{\infty e}[0,T]$, $\dot{V}_a^j$ in (14) can be rewritten as

$$\dot{V}_a^j \leq -\beta z^{j\top} L z^j - \frac{\gamma_1}{\beta_1}\widetilde{\theta}^{j\top}\widetilde{\theta}^j - \frac{\gamma_2}{\beta_2}(\widetilde{\psi}^j)^2$$
$$+ \frac{\gamma_1}{\beta_1}\widetilde{\theta}^{j\top}(t)\bar{\theta}^{j-1} + \frac{\gamma_2}{\beta_2}\widetilde{\psi}^j(t)\bar{\psi}^{j-1}$$
$$\leq -k V_a^j + \bar{k}^{j-1} \quad (23)$$

where $k = \min\{\frac{1}{2\beta}, \frac{\gamma_1}{1-\gamma_1}, \frac{\gamma_2}{1-\gamma_2}\}$, $\bar{k}^{j-1} = \frac{\gamma_1}{2\beta_1}\bar{\theta}^{j-1\top}\bar{\theta}^{j-1} + \frac{\gamma_2}{2\beta_2}(\bar{\psi}^{j-1})^2$ and $\bar{\theta}^{j-1}, \bar{\psi}^{j-1}$ are the upper bounds on $|\widetilde{\theta}^{j-1}(t)|$ and $|\widetilde{\psi}^{j-1}(t)|$, respectively. Since the initial condition of V_a^j is bounded for all $j \geq 1$ due to (L2) of lemma 2, we conclude from (23) that $z^j, \widetilde{\theta}^j, \dot{\widehat{\theta}}^j, \widehat{\psi}^j, \dot{\widehat{\psi}}^j$, and $e^j, \dot{e}^j, u^j \in L_{\infty e}[0,T]$ by the same argument given in lemma 1. Hence, (T1) of theorem 1 is achieved by using mathematical induction.

(T2) By using V_b^j in lemma 2 and (T1) in this theorem, we have $\dot{V}_b^j$ and $V_b^j \in L_{\infty e}[0,T]$. On other hand, $\lim_{j \to \infty} \int_0^T z^{j\top}(t) L z^j(t) dt = 0$, or equivalently, $\lim_{j \to \infty} \int_0^T V_b^j dt = 0$ due to (L3) of lemma 2. We can finally conclude that $\lim_{j \to \infty} z^{j\top}(t) L z^j(t) = \lim_{j \to \infty} V_b^j(t) = 0$, for all $t \in [0,T]$ by using similar argument of Barbalat's lemma [16]. Q.E.D.

V. A SIMULATION EXAMPLE

In this section, a computer simulation is conducted to demonstrate the learning effect of the proposed AILC. Here we consider a two-link planar robotic system [14] as follows:

$$\begin{bmatrix} H_{11} & H_{12} \\ H_{21} & H_{22} \end{bmatrix} \begin{bmatrix} \ddot{q}_1^j \\ \ddot{q}_2^j \end{bmatrix} + \begin{bmatrix} -h\dot{q}_2^j & -h(\dot{q}_1^j + \dot{q}_2^j) \\ h\dot{q}_1^j & 0 \end{bmatrix} \begin{bmatrix} \dot{q}_1^j \\ \dot{q}_2^j \end{bmatrix} + \begin{bmatrix} d_1^j \\ d_2^j \end{bmatrix} = \begin{bmatrix} u_1^j \\ u_2^j \end{bmatrix} \quad (24)$$

where $H_{11} = m_1 l_{c1}^2 + m_2(l_1^2 + l_{c2}^2 + 2l_1 l_{c2}\cos(q_2^j)) + I_1 + I_2$, $H_{12} = H_{21} = m_2 l_1 l_{c2}\cos(q_2^j) + m_2 l_{c2}^2 + I_2$, $H_{22} = m_2 l_{c2}^2 + I_2$, $h = m_2 l_1 l_{c2}\sin(q_2^j)$. Here, m_i, I_i, l_i and l_{c_i} represent mass, inertia, length of link i, and the distance from the previous joint to the center of mass of link i, respectively. The technique of linear parameterization is applied to form the term of $Y_e^j \theta$. With six suitably chosen parameters $\theta_1, \cdots, \theta_6$, the regression matrix Y_e^j will be in the form of

$$Y_e^j = \begin{bmatrix} y_{e11}^j & y_{e12}^j & y_{e13}^j & y_{e14}^j & y_{e15}^j & y_{e16}^j \\ y_{e21}^j & y_{e22}^j & y_{e23}^j & y_{e24}^j & y_{e25}^j & y_{e26}^j \end{bmatrix}$$

with $y_{e11}^j = a\dot{e}_1^j$, $y_{e12}^j = y_{e11}^j$, $y_{e13}^j = a(\dot{e}_1^j + \dot{e}_2^j)$, $y_{e14}^j = (\cos(q_2^j) - \cos(q_{d2}))(2\ddot{q}_{d1} + \ddot{q}_{d2}) + \cos(q_2^j)(y_{e11}^j + y_{e13}^j) - \sin(q_2^j)(\dot{q}_2^j\dot{q}_{d1} + \dot{q}_1^j\dot{q}_{d2}) + 2\sin(q_{d2})\dot{q}_{d1}\dot{q}_{d2} - 2\sin(q_2^j)\dot{q}_2^j\dot{q}_{d2} + \sin(q_{d2})\dot{q}_{d2}^2 - a\sin(q_2^j)(\dot{q}_2^j e_1^j + (\dot{q}_1^j + \dot{q}_2^j)e_2^j)$, $y_{e15}^j = y_{e11}^j$, $y_{e16}^j = y_{e13}^j$, $y_{e21}^j = 0$, $y_{e22}^j = 0$, $y_{e23}^j = y_{e13}^j$, $y_{e24}^j = (\cos(q_2^j) - \cos(q_{d2}))\ddot{q}_{d1} + \cos(q_2^j)(a\dot{e}_1^j) + \sin(q_2^j)\dot{q}_1^j(\dot{q}_{d1} + ae_1^j) - \sin(q_{d2})\dot{q}_{d1}^2$, $y_{e25}^j = 0$, $y_{e26}^j = y_{e16}^j$.

In the simulation example, the physical parameters are specified as $m_1 = 10$Kg, $m_2 = 5$Kg, $l_1 = 1$m, $l_2 = 0.5$m, $l_{c1} = 0.5$m, $l_{c2} = 0.25$m, $I_1 = 0.83$Kg m^2 and $I_2 = 0.3$Kg m^2. Here the control objective is to let $q^j = [q_1^j, q_2^j]^\top$ track the desired trajectory $q_d = [q_{d1}, q_{d2}]^\top = [\sin(3t), \cos(3t)]^\top$ as close as possible over a finite time interval $[0, 5]$. The disturbance is set to be $d^j = [d_1^j, d_2^j]^\top = [h_1^j \sin(w_1^j t), h_2^j \cos(w_2^j t)]^\top$ where $h_1^j \in \{-0.3, 0.3\}, h_2^j \in \{-0.1, 0.1\}, w_1^j \in \{0, 3\}, w_2^j \in \{0, 3\}$. To achieve this control objective, the proposed AILC (3), (4), (6), (9) and (10) is applied with the following design parameters, feedback gains $a = 1$, $\beta = 30$, adaptation gains $\beta_1 = \beta_2 = 10$, weighting gains $\gamma_1 = \gamma_2 = 0.5$, and the symmetric gain matrix $L = \begin{bmatrix} 20 & 0 \\ 0 & 20 \end{bmatrix}$. The sampling period of the simulation is set to 0.001(sec). At the first iteration, $\widehat{\theta}^0(0)$ is set to be a zero vector, and $\widehat{\psi}^0(0)$ is set to be 0.1.

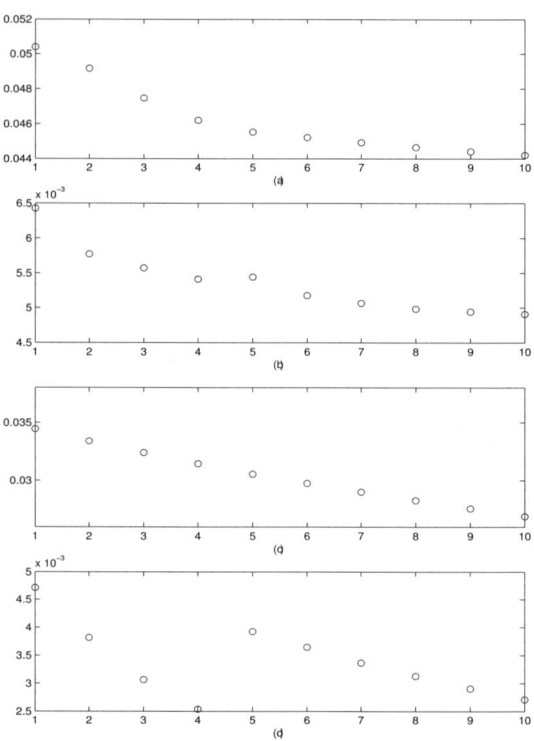

Figure 1 : $L_{\infty e}$ norms of joint position errors e_1^j (a) and e_2^j (b) versus iteration j, and RMSE of joint position errors e_1^j (c) and e_2^j (d) versus iteration j.

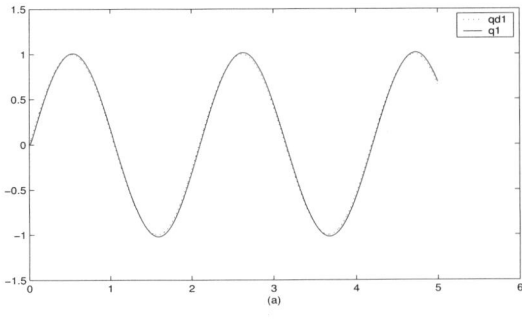

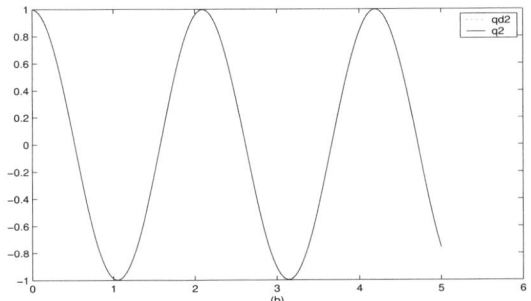

Figure 2 : Responses of joint position q_1^j, q_2^j versus time t at 10th iteration.

Figure 1(a) and Figure 1(b) show the $L_{\infty e}$ norm of tracking errors versus iteration j for $e_1^j = q_{d1} - q_1^j$ and $e_2^j = q_{d2} - q_2^j$. On the other hand, the root mean square errors of e_1^j and e_2^j versus iteration j are shown in Figure 1(c) and Figure 1(d), respectively. As iteration increases, the asymptotic convergence of the tracking error is clearly found from the simulation results. Figure 2 shows the learning performance at 10th iteration. Once the iteration number is large enough, the disturbance can be rejected and the tracking errors converge to an acceptable residual.

VI. CONCLUSIONS

A new algorithm for adaptive iterative learning control of parametric uncertain robotic system is proposed in this paper. Without the upper bounds on the desired parameters, the adaptive law combining time-domain and iteration-domain adaptation can generate the control parameters for the current iteration to guarantee the stability and convergence of the ILC system. By estimating the upper bound of input disturbance, the control scheme can totally reject the disturbance even it is not periodic or not repeatable. We show that the control parameters and internal signals are bounded along the time axis for all iterations and the tracking error will asymptotically converge to zero if the iteration number is large enough. A two-link planar robotic system is used as a simulation example to demonstrate the effect and performance of the proposed AILC system.

VII. ACKNOWLEDGMENTS

This work was supported by National Science Council, R.O.C., under Grant NSC90-2213-E-211-002

VIII. REFERENCES

[1] S. Arimoto, S. Kawamura and F. Miyazaki, "Bettering operation of robots by learning," *J. of Robot. Syst.*, vol. 1, no. 2, pp. 123-140, 1984.

[2] G. Heinzinger, D. Fenwick, B. Paden and F. Miyazaki, "Stability of learning control with disturbances and uncertain initial conditions," *IEEE Transactions on Automatic Control*, vol. 37, no. 1, pp. 110–114, 1992.

[3] T.Y. Kuc, J.S. Lee and K. Nam, "An iterative learning control theory for a class of nonlinear dynamic systems," *Automatica*, vol. 28, no. 6, pp. 1215-1221, 1992.

[4] C.J. Chien and J.S. Liu, "A P-type iterative learning controller for robust output tracking of nonlinear time-varying systems," *Int. J. Control*, vol. 64, no. 2, pp. 319-334, 1996.

[5] H.S. Lee and Z. Bien, "Study on robustness of iterative control with non-zero initial error," *Int. J. Control*, vol. 64, no. 3, pp. 345-359, 1996.

[6] B.H. Park, T.Y. Kuc and J.S. Lee, "Adaptive learning of uncertain robotic systems," *Int. J. Control*, vol. 65, no. 5, pp. 725-744, 1996.

[7] B.H. Park, J.S. Lee and T.Y. Kuc, "Adaptive learning control of robotic systems and its extension to a class of nonlinear systems," Chapter 13 of *Iterative Learning Control - Analysis, Design Integration and Application*, Kluwer Academic Publishers, pp. 239-259, 1998.

[8] T.Y. Kuc and W.G. Han, "An adaptive PID learning control of robot manipulators," *Automatica*, vol. 36, pp. 717-725, 2000.

[9] J.X. Xu, and B. Wiswanathan, "Adaptive robust iterative learning control with dead zone scheme," *Automatica*, vol. 36, pp. 91-99, 2000.

[10] C.J. Chien and L.C. Fu, "An iterative learning control of nonlinear systems using neural network design," *Asian Jornal of Control*, vol. 4, no. 1, pp. 21-29, 2002.

[11] M. French and E. Rogers, "Nonlinear iterative learning control for a class of nonlinear dynamic systems", *Proceedings of the 37th conference on Decision and Control*, pp.175-180, Florida, USA, 1998.

[12] M. Norrlöf, "An adaptive iterative learning control algorithm with experiments on an industrial robot," *IEEE Transtractions on Robotics and Automation*, vol. 18, no. 2, pp. 245-251, 2002.

[13] J.Y. Choi and J.S. Lee, "Adaptive iterative learning control of uncertain robotic systems," *IEE Proc. D, Control Theory Application*, vol. 147, no. 2, pp. 217-223, 2000.

[14] J.J.E. Slotine and W. Li, *Applied Nonlinear Control*, Prentice-Hall, New Jersey, 1991.

[15] M.W. Spong and M. Vidyasagar, *Robot dynamics and control*, Wiley, New York, 1989.

[16] P.A. Ioannou and J. Sun, *Robust Adaptive Control*, Englewood Cliffs, NJ, Prentice-Hall, 1996.

Learning to Optimize Mobile Robot Navigation Based on HTN Plans

Thorsten Belker
Dept. of Computer Science
Univ. of Bonn, Germany
belker@cs.uni-bonn.de

Martin Hammel
Dept. of Computer Science
Univ. of Bonn, Germany
hammel@cs.uni-bonn.de

Joachim Hertzberg
Fraunhofer AIS
Sankt Augustin, Germany
hertzberg@ais.fraunhofer.de

Abstract

High-level symbolic representations of actions to control the working of autonomous robots are used in all hybrid (reactive and deliberative) robot control architectures. Abstract action representations serve several purposes, such as structuring the control code, optimizing the robot performance, and providing a basis for reasoning about future robot action.

The paper presents results about re-designing the RHINO navigation system by introducing an HTN plan layer. Besides yielding a more structured robot control software, this layer is used as a basis for optimizing the navigation performance by plan transformations. We show how a robot can learn to select plan transformations based on projections of its intended behavior. Our experimental evaluation shows that the overall robot navigation performance is increased by almost 42 % when using learned projective models to select plan transformations.

1 Introduction

Abstract plan or task layers have been used in robot control since SHAKEY's times [17], and they are essential in hybrid robot control architectures [15, 12]. In McDermott's terminology [14], a plan is that part of a robot control program, which the robot cannot only execute, but also reason about and manipulate. According to that broad view, a plan may serve many purposes in a robot control system: As [1] has it, "the use of plans enables these robots to flexibly interleave complex and interacting tasks, exploit opportunities, quickly plan their courses of action, and, if necessary, revise their intended activities".

In this paper, we describe a technique for using a plan layer in robot control for optimizing the performance in robot navigation by learning from past navigation experience. We start with the well-known RHINO software [22], building on top of it abstract navigation tasks from which plans are generated and executed. We use HTNs [16] as the plan format, as it allows standard plans to be generated very efficiently and makes it easy for the control system programmer to express prior knowledge about priorities and preferred decompositions of navigation tasks under certain circumstances. Moreover, the hierarchical nature of HTN plans makes them a handy substrate for dealing with execution failure by jumping to higher levels of abstraction within the current plan and pick alternative task expansion strategies. This has been one of the reasons for developing it for the archetype of HTN planners, NOAH [19].

Based on the declarative representation of robot navigation in the HTN format, the expected action of the robot on a given task can be projected into the future, allowing its performance to be estimated and, wherever possible, to be improved if alternative courses of action are available that can be chosen instead. This idea of plan transformation is inspired by the work of Beetz and McDermott [2]. As past experience from navigation tasks can of course be accumulated, we end up in a life-long learning framework in which the robot is able, based on the navigation plan format, to improve its navigation performance. Our results show performance improvements of 42 % on average.

The remainder of this paper is organized as follows. In the next section, we briefly recapitulate the RHINO navigation system and discuss its major shortcomings. Then we introduce – mostly by way of example – HTN planning and discuss how HTN planning can help to improve the RHINO system. After that, the technical core of the paper describes the procedure of optimizing navigation performance using learning techniques. We finally present empirical results of running the procedure, and conclude.

2 The RHINO Navigation System

The RHINO navigation system is well known for two reasons. First, its robust performance in two early tourguide projects in the Deutsches Museum in Bonn [6] and the Smithsonian Institute in Washington D.C [21] and second, the consequent application of probabilistic algorithms to both map learning [22] and localization [8, 10].

Figure 1 depicts the main components of the RHINO sys-

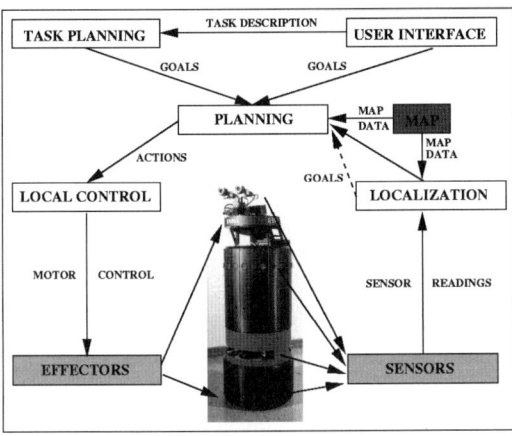

Figure 1: Components of the RHINO system.

tem and their interaction. The localization component tries to fit the sensor readings of a mobile robot into a model (a grid map) of the robot's working environment to estimate the robot's current position or pose within the environment and reduce uncertainty caused by unreliable dead-reckoning. A robust self-localization can be achieved using Bayes' rule to integrate noisy odometry information and noisy sensor information into a probability distribution of the robot's current position or pose within a known environment [8, 10]. The occupancy grids of the environment needed for localization can be learned [22]. While position tracking and navigation planning can be done concurrently, sensing actions might become necessary if the position uncertainty of the robot becomes too high. Burgard et al. suggest an entropy-based approach to compute sensing actions for active localization [7].

The navigation planning component receives goal points from either a user interface, a task planner or – in the case of active localization – from a localization component. It generates a sequence of actions that can be executed by a local reactive control component. In most systems, navigation planning is considered as an instance of path planning [13]. However, RHINO's navigation planner computes navigation policies rather than paths. Navigation policies are functions that assign a navigation action to each (discrete) position or pose of the robot. From a navigation policy, a path to the goal can be computed efficiently for every state of the robot. This is advantageous when the robot often deviates from its pre-planned path, e.g. due to drift or unexpected obstacles. Thrun et al. propose to use a deterministic value iteration to compute a navigation policy for a fine-grained two-dimensional occupancy grid map of the robot's environment [22].

The local control component ensures a safe execution of high-level navigation actions. It translates these actions into sequences of motor control commands that can be directly executed by the mobile robot. It guarantees a safe navigation by reacting to unforeseen and dynamic obstacles. The dynamic window approach to collision avoidance [9] generates trajectories for local navigation tasks based on simple models of the robot dynamics and chooses the trajectories with the highest utility with respect to some given evaluation function. A recent extension of this approach [4] combines the dynamic window approach with local planning to compute a better evaluation function.

As the local navigation component can fail to execute a navigation action due to inaccurate motor control, unexpected obstacles, and inherent limitations of the local control component, the planning component should be able to handle failures at execution time and plan for their avoidance. We will introduce a symbolic navigation planner which can account for execution failures on both levels. By projecting the execution costs of alternative plans, the planning component can optimize the robot's average performance using plan transformations.

3 HTN Plans for Robot Navigation

The basis for the improvement of the robot's performance reported in the rest of this paper is the representation of abstract navigation actions in terms of HTN (Hierarchical Transition Network) plans for robot navigation. By introducing a plan layer into the robot control, we continue the rich tradition of hybrid robot control architectures [15, 12].

HTN planning originates from early work by Sacerdoti [19]. Since then, it has been used as a technique in several domain-independent and special-purpose planners. Milestones include the SIPE-2 system [23] and its various applications, as well as BridgeBaron [20], the winner of the 1997 computer bridge world championship. SHOP [16] is a modern, domain-independent HTN planner incorporating the BridgeBaron design knowledge.

HTN planning specifies a planning problem as a task network, i.e., a set of tasks together with constraints on the order in which they can be performed and restrictions on how variables may be bound. Tasks may be *elementary*, i.e., executable by the robot (in our case), or *compound*, i.e., to be expanded into a task network. The expansion, in general, is not unique. Planning is performed by expanding compound tasks until only elementary tasks remain. The resulting network of elementary tasks is a solution plan for the problem. In this paper, we use only total-order task networks and ground instances of tasks, handling linear propositional plans all the time. This simplification does, of course, not apply to HTN planning in general.

To introduce some terminology first, we call *plan stub* an HTN with an elementary task as the first task in its ordering. The strategy of stopping to reduce elementary tasks in an HTN as soon as a plan stub has been found, is called the *lazy expansion principle*. This principle is used in our case as the robot may start navigating even before a complete solution plan has been found. The rationale is that we cannot assume that no unforeseen events occur during plan execution (which would make plan suffixes unexecutable or irrelevant) and that plans may include sensing actions, the results of which may not be known at planning time. The lazy expansion principle reduces waste of planning time in these cases.

To present an example, let us introduce the operator inventory for our navigation planner. We have the following schemata for elementary tasks:

SetTarget(x, y, d) sets the target point (x, y) for the low-level routine for collision-free drive control, which has to be approached up to a precision of d cm.

TurnTo(x, y) causes the robot to rotate on the spot until heading towards the point (x, y).

TurnToFree$()$ causes the robot to rotate on the spot until heading towards some free space.

MoveForward(d) causes the robot to move by d cm straight forward.

MoveBackward(d) causes the robot to move by d cm straight backward.

All elementary tasks must have an implementation in terms of low-level control routines of the robot, so that executing an elementary task means calling the respective routine. That does, of course, not guarantee that each and every elementary task instance, or its corresponding control routine, respectively, can be successfully executed. Failure is possible, as usual. This issue will be addressed below.

We have two types of compound tasks, the schemata of which are:

ApproachPoint(x, y, d) drives the robot to position (x, y) within an error radius of d. In terms of the expansion hierarchy, **ApproachPoint** is a middle-level task that serves for dealing with self-generated intermediate target points.

MDPgoto(x, y) drives the robot to the user-specified target point (x, y). It is the highest task in the expansion hierarchy. Over the time, it gets expanded into a sequence of **ApproachPoint** operators.

In the experiments, we consider the following expansions. The task **MDPgoto**(x, y) can be expanded into **ApproachPoint**(t_x, t_y, c) with the target point (t_x, t_y) either 2 m (default), 1 m or 4 m ahead on the optimal path to the goal point (x, y) and with $c=1$ m, $c=0.5$ m, or $c=2$ m respectively. **ApproachPoint**(x, y, d) is either expanded into **SetTarget**(x, y, d) (default), into the sequence **TurnTo**(x, y), **SetTarget**(x, y, d), into the sequence **MoveBackward**(30), **SetTarget**(x, y, d), or into a sequence of **TurnToFree**, **MoveForward**, **TurnTo**, and **SetTarget**.

Within a plan, an *instance* of any task has all arguments fully instantiated, as we are dealing with purely propositional plans here. Moreover, all task instances have an additional argument saying whether they are PENDING, i.e., not yet executed (for elementary tasks) or expanded (for compound tasks), or EXPANDED in the case of compound tasks. Executed tasks are simply deleted from the current plan.

Representing a plan as a stack, here is an example. Assume the navigation target is the position $(1521.31, 1563.8)$ on some given floor map. This would be transformed into the one-task plan

MDPgoto$(1, \text{PENDING}, 1521.31, 1563.8)$

where the first two arguments are the task instance ID and the status, resp., and the following ones are like in the task schema descriptions given above. (This pattern will re-appear in all other task instances to follow.)

Dealing with the top task of the stack means expanding it, in this case, since it is compound. Using the default expansion, this yields

ApproachPoint$(2, \text{PENDING}, 1434.38, 1009.38, 100)$
MDPgoto$(1, \text{EXPANDED}, 1521.31, 1563.8)$

and, expanding the **ApproachPoint** task,

SetTarget$(3, \text{PENDING}, 1434.38, 1009.38, 100)$
ApproachPoint$(2, \text{EXPANDED}, 1434.38, 1009.38, 100)$
MDPgoto$(1, \text{EXPANDED}, 1521.31, 1563.8)$

As the topmost task is elementary, this is a plan stub, and according to the lazy expansion principle, this operator gets immediately executed by the robot, causing a physical robot drive action.

Assuming that all goes well, the control routine implementing the **SetTarget** task terminates successfully. This task pops out, and so does **ApproachPoint** in consequence. The **MDPgoto** task, however, is not yet finished, as its target point is not yet reached according to the robot's self-localization. In consequence, it gets re-expanded, yielding

ApproachPoint$(4, \text{PENDING}, 1476.88, 1158.12, 100)$
MDPgoto$(1, \text{EXPANDED}, 1521.31, 1563.8)$

the topmost task of which would get expanded into the respective **SetTarget** task and executed as before. If all keeps going well, this cycle of expand-execute-pop is repeated until the final target point is reached and the **MDPgoto** task pops out.

Failures to execute an elementary task are reported by the low-level control routines by raising exceptions of different types. Assume that executing the top task in the plan

SetTarget(5, PENDING, 1476.88, 1158.12, 100)
ApproachPoint(4, EXPANDED, 1476.88, 1158.12, 100)
MDPgoto(1, EXPANDED, 1521.31, 1563.8)

results in an exception of type NO-ADMISSIBLE-TRAJECTORY, i.e., the low-level execution cannot find an unoccluded local path from the current position to the point (1476.88, 1158.12), based on the recent sensor readings. As a result, the failed task would pop out and the next expansion alternative for the **ApproachPoint** task would be patched in, resulting in

MoveBackward(6, PENDING, 30)
SetTarget(7, PENDING, 1476.88, 1158.12, 100)
ApproachPoint(4, EXPANDED, 1476.88, 1158.12, 100)
MDPgoto(1, EXPANDED, 1521.31, 1563.8)

If this should fail again, there are more ways of expanding **ApproachPoint**. Only if all expansion alternatives for some compound task have been exhausted, there will be backtracking in the tree of possible expansions, or, if no more backtracking is possible, the execution of the main task fails and permanent failure is reported.

The two examples demonstrate that the HTN framework nicely supports the fast generation of default plans and the handling of exceptions caused by unexpected obstacles or inaccurate effectors. In addition, HTNs can be used to optimize the robot's navigation performance, that is, to minimize the expected execution time. In some situations, for example, it might be advantageous for the robot to turn to the target point before trying to approach it. In other situations, e.g. in a wide corridor, it might improve the robot's performance to have a target point that is more distant and allows the robot to drive faster. All these different courses of action can be represented as different expansions of the **ApproachPoint** or **MDPgoto** schema. As soon as an opportunity of improving the robot's navigation plan has been detected, the planner can backtrack in the expansion tree and select another expansion that seems to be more suitable for the current situation.

4 Learning to Optimize Navigation Performance

In the previous section, we have argued that HTNs support the optimization of navigation performance by opportunistic plan transformations. However, the detection of opportunities for plan improvements is often difficult and the specification of a good detector requires much insight in the robot's operation. In this section, we therefore propose a method that projects different plan stubs and selects the one which causes the lowest expected execution costs (time). Due to the declarativity of the HTNs, the execution of different tasks can be monitored and a prediction of the execution costs can be learned from past experience.

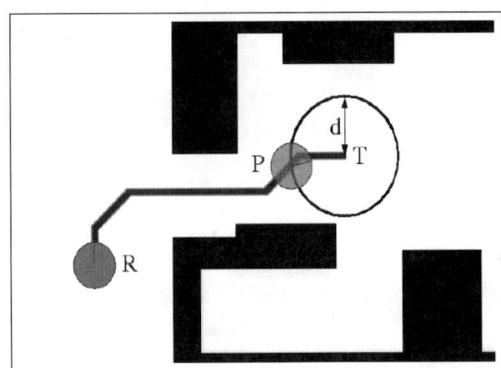

Figure 2: Projection of the position of a robot after successful completion of a navigation task.

The projection of the position and orientation of the robot after successfully executing **SetTarget**(x, y, d) is done by computing the optimal path to the target $T = (x, y)$ and following the path to the first position $P = (x', y')$ on the path that is at most d cm away from T. The robot is supposed to face T. The time the robot needs to arrive at position P has to be estimated based on previous experience of executing this kind of task. As the time the robot needs to execute the navigation task depends on the shape and length of the path to the projected state P, we use a set of features derived from the path. Besides the path length, the path curvature (the ratio of the euclidian distance to the target and the path length) and the initial angle of the robot towards the path, the feature set contains features derived from the main axis clearance histogram of the environment map [3]. The feature *crossesDoor*, for example, is true if the path crosses a region with low main axis clearance. The other primitive tasks are straight forward to project as the execution costs are either constant or only depend on the angle the robot has to turn. Compound tasks are projected by projecting their (default) expansion.

For the learning itself we have applied model trees [11, 18], an extension of regression trees [5] where the tree leafs contain linear predictions rather than constant predictions. Model trees implement piecewise linear regression models with main-axis parallel boundaries. We have decided to use tree-based induction methods for the function approximation as they provide in addition to a value prediction an explanation of the results, which can be translated into symbolic rules and is thus well accessible for human inspection. Figure 3 shows a rule learned for the task of predicting the durations of a navigation task.

```
IF      (pathCurvature < 1.05)    AND
        (NOT crossesDoor)         AND
        (pathLength ≥ 110.00)     AND
        (pathLength < 130.00)
THEN    duration = 1/23.99 * pathLength
```

Figure 3: One of the rules learned for the prediction task.

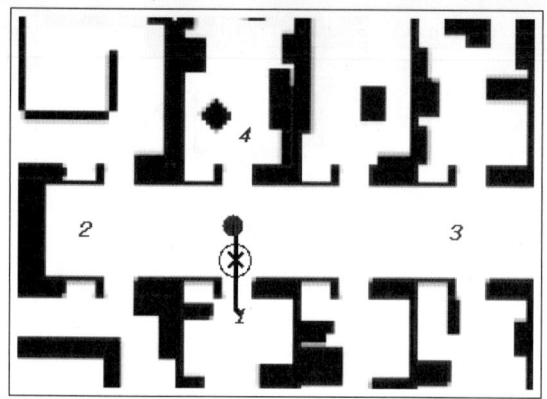

Figure 4: The four goal points for the real world experiments.

Figure 5: The Pioneer II platform

5 Experimental Results

In this section, two experiments will be presented that demonstrate the use of HTN plans for improving mobile robot navigation. The first experiment shows how plan transformations can be used to optimize navigation performance. While in this experiment hand-coded transformation rules are used, the second experiment shows how machine learning techniques can help to disburden the human programmer from specifying these rules.

5.1 Plan Transformations

In this experiment, the following three hand-coded transformation rules are used to improve the robot's navigation performance.

1. If the robot has its back to the target point and its clearance is low, then the robot turns towards the target point before approaching it.

2. If the robot's clearance gets small along the path fast, then the target point is set only one meter ahead.

3. If the robot's path has a low curvature and the clearance stays high along the path, then the target point is set four meters ahead.

These transformations are implemented using the alternative expansions of the **ApproachPoint** task (rule 1) and the **MDPgoto** task (rules 2,3). During execution of the plan, the robot starts to execute the default expansion of a given task immediately and does not wait until all alternative expansions have been considered. During execution the transformation rules are evaluated and applied if possible.

To evaluate the use of these transformations, the robot executes the sequence of four navigation tasks shown in Figure 4 ten times, both with and without transformations. The tasks are chosen such that the behavior of the robot in four different situations can be analyzed. Entering an office, leaving an office, navigation in the hallway, and navigation from one office to another. By applying the above mentioned transformation rules, the navigation performance of a PIONEER II robot with laser range finder (Figure 5) is on average improved by 30.88 %.

5.2 Learning to Predict Navigation Performance

Instead of specifying transformation rules by hand, the robot can transform plans based on projections of its behavior and a learned prediction of the time it takes the robot to execute a navigation task. The second experiment shows that this results in a navigation performance which is comparable to the behavior achieved with the transformation rules described above which we consider as expert knowledge.

To generate the data to learn the prediction function, the robot repeatedly executes a sequence of 10 navigation tasks in simulation by randomly selecting a possible reduction of the **MDPgoto** and **ApproachPoint** tasks.

Using a greedy error reduction splitting criterion, linear prediction functions, a depth-limit stopping criterion with a limit of 7, and a reduced-error post-pruning criterion, the model tree learner generates a set of 14 rules with one to six preconditions from about 7000 training examples.

The learned rules are used to predict the performance of the robot. If an alternative reduction of the current task is projected to result in a better performance of the robot than the default reduction, the plan is transformed accordingly. This results in a navigation performance of the PIONEER II which is 41.88 % better than without using plan transformations and 8.4 % better than with hand-coded transformation rules. Both performance improvements are statistically significant with respect to a significance level of 0.95.

However, the performance gain differed considerably for the four tasks. While there was no significant performance gain for the first task (entering the office), the performance gain for the second task (leaving the office) was 51.37 %, for the third task (navigation in the hallway) 50.68 %. In the last task the robot on average performed even 65.97 % better. The results show that the transformations are especially useful for the task of leaving an office and fast navigation in the hallway.

Conclusion

Symbolic navigation plans allow control knowledge to be expressed and used in the robot controller in a structured way—that is the rationale behind all hybrid robot control architectures. The HTN framework allows for compact and efficient representation of such plans and alleviates to deal with execution failures in the lower level control routines in a transparent way.

The contribution of this paper is, first, to demonstrate the advantage of using an abstract HTN layer of navigation plans in a robot navigation system that has been used successfully over an extended period of time on various kinds of robots. Besides a transparent execution failure handling, such a layer facilitates opportunistic plan transformations. In the experiments, the navigation performance could be improved by 30.88 % using three hand-coded transformation rules. Second, we have shown the way for using machine learning techniques to predict the robot's performance in different situations. The learned models are essential for projecting the robot's behavior when executing a given plan. Plan projections can be used to trigger plan transformations whenever they promise an improvement of the robot's performance. Combing planning and learning in this way, the robot could improve its performance by 41.88 % on average. The learned rules thus outperform the hand-coded rules by 8.4%.

Future work will include the learning of transformation rules. The learning will be based on previously acquired projective model as described in this paper and therefore not require a human trainer.

Acknowledgments

T. Belker is partly supported by the German Research Foundation (DFG) under contract number BE 2200/3-1. J. Hertzberg is partly supported by the German Federal Ministry of Research (BMBF) by the project AgenTec (01AK905B).

References

[1] M. Beetz, J. Hertzberg, M. Ghallab, and M. Pollack. Preface. In M. Beetz, J. Hertzberg, M. Ghallab, and M. Pollack, editors, *Advances in Plan-Based Control of Robotic Agents*. Springer, LNAI 2466, 2002.

[2] M. Beetz and D. McDermott. Expressing transformations of structured reactive plans. In *Recent Advances in AI Planning. Proceedings of the 4th European Conference on Planning*, 1997.

[3] T. Belker, M. Beetz, and A. B. Cremers. Learning of plan execution policies for indoor navigation. *AI Communications*, 15(1):3–16, 2002.

[4] T. Belker and D. Schulz. Local action planning for mobile robot collision avoidance. In *IROS*, 2002.

[5] L. Breiman, J.H. Friedman, R.A. Olshen, and C.J. Stone. *Classification and Regression Trees*. Wadsworth, Inc., Belmont, CA, 1984.

[6] W. Burgard, A.B. Cremers, D. Fox, D. Hähnel, G. Laemeyer, D. Schulz, W. Steiner, and S. Thrun. Experiences with an interactive museum tour-guide robot. *Artificial Intelligence*, 114(1-2), 1999.

[7] W. Burgard, D. Fox, and S. Thrun. Active mobile robot localization. In *Proceedings of the Fifteenth International Joint Conference on Artificial Intelligence*, 1997.

[8] D. Fox, W. Burgard, F. Dellaert, and S. Thrun. Monte carlo localization: Efficient position estimation for mobile robots. In *Proceedings of the Sixteenth National Conference on Artificial Intelligence*, Orlando, FL, 1999.

[9] D. Fox, W. Burgard, and S. Thrun. The dynamic window approach to collision avoidance. *IEEE Robotics and Automation Magazine*, 4(1), 1997.

[10] D. Fox, W. Burgard, and S. Thrun. Markov localization for mobile robots in dynamic environments. *Journal of Artificial Intelligence Research*, 11, 1999.

[11] A. Karalic. Linear regression in regression tree leaves. In *Proceedings of ISSEK '92 (International School for Synthesis of Expert Knowledge)*, 1992.

[12] D. Kortenkamp, R.P. Bonasso, and R. Murphy, editors. *Artificial Intelligence and Mobile Robots: Case studies of successful robot systems*. MIT Press, Cambridge, MA, 1998.

[13] J.-C. Latombe. *Robot Motion Planning*. Kluwer Academic Publishers, 1991.

[14] D. McDermott. Robot planning. *AI Magazine*, 13(2):55–79, 1992.

[15] R. Murphy. *Introduction to AI Robotics*. MIT Press, 2000.

[16] D.S. Nau, Y. Cao, A. Lotem, and H. Munoz-Avila. Shop: Simple hierarchical ordered planner. In *Proc. IJCAI-99*, 1999.

[17] N.J. Nilsson. Shakey the roboot. Technical Report 323, SRI International, Menlo Park, California, 1984.

[18] J. Quinlan. Learning with contionous classes. In *Proceedings of the 5th Australien Joint Conference on Artificial Intelligence*, 1992.

[19] E.D. Sacerdoti. *A Structure for Plans and Behavior*. Elsevier/North Holland, 1977.

[20] S.J.J. Smith, D.S. Nau, and T. Throop. Success in spades: Using ai planning techniques to win the world championship of computer bridge. In *Proc. AAAI-98/IAAI-98*, pages 1079–1086, 1998.

[21] S. Thrun, M. Bennewitz, W. Burgard, A.B. Cremers, F. Dellaert, D. Fox, D. Haehnel, C. Rosenberg, N. Roy, J. Schulte, and D. Schulz. Minerva: A second generation mobile tour-guide robot. In *Proceedings of the IEEE International Conference on Robotics and Automation (ICRA'99)*, 1999.

[22] S. Thrun, A. Buecken, W. Burgard, D. Fox, T. Fröhlinghaus, D. Hennig, T. Hofmann, M. Krell, and T. Schmidt. Map learning and high-speed navigation in rhino. In [12], 1998.

[23] D. Wilkins. *Practical Planning. Extending the Classical AI Planning Paradigm*. Morgan Kaufmann, San Mateo, CA, 1988.

Design and Implementation of a Behavior-Based Control and Learning Architecture for Mobile Robots

Il Hong Suh[1], Sanghoon Lee[1], Bong Oh Kim[1], Byung Ju Yi[1], and Sang Rok Oh[2]

[1]Department of Electrical Engineering and Computer Science
Hanyang University, Seoul, KOREA
Tel : +82-2-2290-0392, E-mail : ihsuh@hanyang.ac.kr

[2]Intelligent System Control Research Center, Korea Institute of Science and Technology,
P. O. Box 131, Cheongryang, Seoul 130-650, KOREA

Abstract – A behavior-based control and learning architecture is proposed, where reinforcement learning is applied to learn proper associations between stimulus and response by using two types of memory called as short Term Memory and Long Term Memory.
In particular, to cope with delayed-reward problem, a knowledge-propagation (KP) method is proposed, where well-designed or well-trained S-R(stimulus-response) associations for low-level sensors are utilized to learn new S-R associations for high-level sensors, in case that those S-R associations require same objective such as obstacle avoidance. To show the validity of our proposed KP method, comparative experiments are performed for the cases that (i) only a delayed reward is used, (ii) some of S-R pairs are preprogrammed, (iii) immediate reward is possible, and (iv) our KP method is applied

I. INTRODUCTION

For implementation of autonomous and intelligent system, a lot of research works have been done in many areas including cognition, reasoning, and learning. When compared with level of human intelligence, behaviors of low-level animals could not be considered as intelligent one. But, recently, it has been understood that those behaviors should be counted as sufficiently intelligent when considering the environment of the low-level animals. Some robot control systems involving behavior-based architectures of such low-level animals have been proposed [1]-[4].

A distinct feature of behavior-based control architecture can be described as follows: an action is selected without use of cognitive models, reasoning, and planning, among possible actions of a robot associated with a stimulus (or state) when an environmental state is given to a robot. Here, there are two representative behavior-based control architectures; Subsumption [4] and Schema [3].

In the Subsumption architecture, an action with higher priority can always subsume other possible actions associated with the state. And in schema-based architectures, all actions associated with the state would be combined. In those architectures, if associations between set of behaviors and set of stimulus are fixed, then the robot shows same behaviors under same environment. But, for the adaptation on uncertain and dynamic environment, it is necessary to change associations; strengthening or weakening current connections between sensor state and actions, and linking a new sensor state with proper actions (or behaviors). For such a purpose, reinforcement learning techniques have been employed [5]-[8].

When designing an intelligent robot by using a behavior-based control architecture involving reinforcement learning, several factors should be taken into account. For example, all necessary state-action pairs (or stimulus-response behaviors), and their relations should be defined and designed, where relations could be setup in a hierarchical or a parallel fashion [2].

On the other hand, it usually takes a long time to learn some necessary associations between stimuli and behaviors by reinforcement learning technique [11][12]. There are two types of rewards; immediate reward and delayed reward.

Delayed rewards have to be used to evaluate the suitability of past course of action of the robot. While this process is theoretically possible, it tends to be unacceptably inefficient: feedback information is often too rare and episodic for an effective learning process to take place in realistic robotic applications. A way to bypass this problem is to use a trainer to continuously monitor the behavior of a robot and provide immediate reinforcements. To produce an immediate reinforcement, the trainer must be able to judge how well each single robot action fits into the desired behavior pattern.

In this paper, a behavior-based control and learning architecture is proposed, where a behavior is selected among behaviors associated with a given sensor state by considering internal desires, and also reinforcement learning is applied to learn proper associations between sensor states and behaviors. In our design, two types of memory are employed for the learning. One is short term memory (STM) in which stimulus-response (SR) pairs are recorded along the time. The other is long term memory (LTM) to which stimulus-response pairs are moved from STM together with their reliability, when a reward is received. And in particular, to solve delayed-reward problem, a knowledge-propagation (KP) method is suggested, where well-designed or well-trained S-R associations for low-level sensors such as ultra sonic sensors are utilized to learn new S-R associations for high-level sensors such as CCD camera, in case that those S-R associations could require same objective such as obstacle avoidance.

To show the validity of our proposed KP method, comparative experiments are performed for the cases that (i) only a delayed reward is used, (ii) some of S-R pairs are preprogrammed, (iii) immediate reward is possible, and (iv) our KP method is applied. From such experiments, we will show that KP method could be more effective in learning S-R behaviors in delayed reword environment than other methods.

II. A Behavior-based Control and Learning Architecture

A. General Architecture

Block diagram of our proposed behavior-based control and learning architecture is represented as in Fig. 1 In our architecture, there are 5 modules; Sensor Module, Perception Module, Memory Module, Motor Module, and Behavior Selection Module. Sensor Module is composed of physical sensor and logical sensor. Physical sensor is attached into the robot, and transforms physical quantity into electrical signals. And, logical sensor can be considered as processing part of the physical sensor signals to detect stimulus from the environment. Explicit use of the logical sensor can enhance the openness of our behavior-based architecture in the sense that any physical sensors can be easily attached into our software system thru our logical sensors by simply defining necessary information into pre-specified DB protocols. Perception Module (PM) plays a role of transferring output of logical sensors to the behavior modules, where each logical sensor information is properly weighted by stimulus recognition filters. Memory module consists of short-term memory (STM) and long-term memory (LTM).

STM is used to record stimulus-response pair at every tick (sampling time) until a new reward or reinforcement is enforced. Then, the stored information is transferred to LTM with reliability index, where S-R pairs are analyzed to learn current association between stimulus and response.

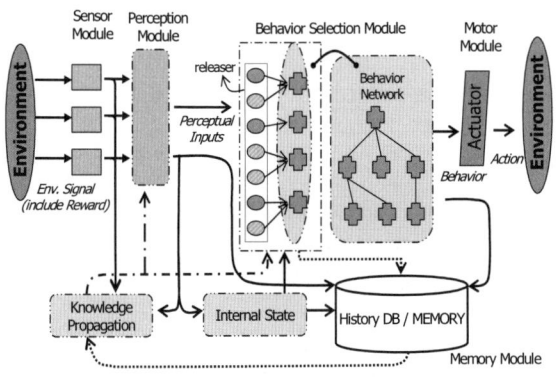

Fig. 1 Behavior-based control architecture.

Motor module plays a role of controlling actuators to perform selected behaviors. Behavior Selection Module includes releaser which activates behaviors, behavior network which explains relations among behaviors, and associations between releasers and behaviors. Output of PM is transferred to all releasers, and every releaser computes how much current stimulus coming from PM is related or associated with its own behavior, where one releaser handling one stimulus is connected with one behavior to form an S-R behavior [2]. Behavior value is then determined by using only releaser output, or by using releaser output as well as internal state values [5]-[8]. Behavior selection is done in such a way that maximum behavior value is chosen [14], or some behaviors [15], where values are greater than a pre-specified threshold value, are competed by means of a competition network. Maximum value method lets the system perform the action right after action selection. But, competition method requires competition among behaviors with values above a threshold, which relatively cause complexity problem [16]. On the other hand, priority-based behavior selection can be also applied. All such behavior selection techniques do not employ cognitive reasoning and/or planning to select a behavior responding to incoming stimulus. In this sense, we call such techniques as behavior-based technique.

In our work, a competition-based action-selection is utilized. And, to learn appropriate association between stimulus and its possible responding behaviors, reinforcement learning will be applied. Specifically, our architecture is designed such that new S-R behaviors can be inserted into the behavior network whenever strength of association between a new stimulus and some behavior gets higher than a threshold.

B. Memory for Learning

Behavior exploration is executed to find a proper behavior for a new stimulus (or environmental state), and to learn optimal behavior which is better than current behavior.

To find a proper behavior for a new stimulus, an arbitrary behavior is performed. And rewards may be received or not. Such an arbitrary behavior to respond to a new stimulus is recorded in STM along the time regardless of reward signal. If a reward is received, then past history of S-R behaviors are transferred from STM to LTM. In this process, if some S-R behaviors are already registered in LTM, reliability of those S-R behaviors are updated by

$$V_{ij} = V_{ij(t-1)} + \eta \frac{1 - V_{ij(t-1)}}{d_k} \qquad (1).$$

where

η : learning rate,

V_{ij} : reliability of between stimulus and behavior at time t,

d : time difference on ST memory,

i : index of stimulus,

j : index of behavior,

k : index of ST memory.

And, new S-R behaviors whose reliability value is above a threshold are registered in the behavior network. Exploration for optimal behavior can be performed by ε-greedy policy [17] or by skill learning [18] which generates a new behavior by changing motor command parameters. Here, ε-greedy policy is employed.

In Figs. 2 and 3, operational procedure and its pseudo codes of STM and LTM are summarized.

It is remarked that our learning as explained above can be considered as an associative learning between stimulus and behaviors using STM and LTM. There are two types of associative learning; one is classical conditioning and the other is operant conditioning. Classical conditioning is an association forming process by which a stimulus that previously did not elicit a response comes to elicit a response. Operant conditioning is a process through which the consequences of a response increase or decrease the likelihood that the response will occur again. In our case, operant conditioning is employed for learning.

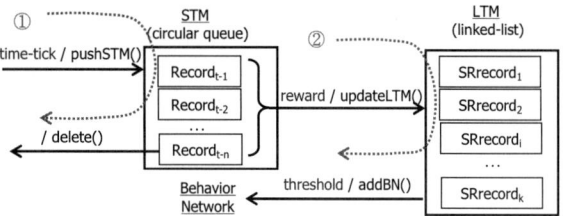

Fig. 2 STM and LTM operations.

① if received<timetick> then
   Begin thread
     if(query(stimulus,behavior))) // to PM and BSM
     {
       makeRecord()   // stimulus and behavior data
       pushToSTM(data);
     }
   End thread

② if received<reward> then
   Begin thread
     do
     {
       popFromSTM()   // pop the data from STM
       if(find(stimulus, behavior)) // in LTM
         updateReliability() // update reliability in LTM using Eq. (1)
       else
       {
         insertData()   // insert data to LTM
         updateReliability()
       }
     }while(isEmptySTM())
   End thread

Fig. 3 Pseudo codes for STM and LTM operations.

III. Knowledge Propagation

A. Behavior Learning by Delayed Reward

In general, it is difficult for the robot to get rewards or reinforcement immediately just after a behavior in applying reinforcement learning technique. This is due to a difficulty that a robot cannot evaluate how well its behavior is fitted for the goal or for the satisfaction of its own motivation. Such as evaluation may be possible, if a trainer is available. In this case, the trainer should watch out the robot until it completes necessary reinforcement learning which might require a long time operation. Thus, it is practically difficult to apply. After all, delayed-rewards have to be received. Delayed reward greatly slow down the learning process, because reward signal is often too rare and episodic for an effective learning process to take place in realistic applications. Several techniques can be considered to bypass such a delayed reward problem [19].

First, robot is driven to completely learn necessary S-R behaviors. In this case, it cannot be known when learning is completed. In fact, a complete learning may not be guaranteed due to insufficient number of episodes.

Second, reinforcement program (RP) can be constructed to replace a human trainer. This artificial trainer could provide the robot with immediate reward, and thus robot can be effectively trained by reinforcement learning. However, it is not clear how we can construct such a reinforcement program, since we should know all possible stimuli coming from robot sensors, and should understand what to do for those stimuli. Unfortunately, such a perfect understanding may be unrealistic. If wrong RP is applied, incorrect S-R behaviors may be generated, which we would like to avoid in the design of robot actions.

Third, direct programming (DP) can be applied, if we exactly know what behavior should be matched with a given stimulus. DP can reduce the number of S-R behaviors to be learned. But, all S-R behaviors cannot be directly preprogrammed, since there could be a lot of unexpected stimuli to the robot. Thus, even in DP case, learning is still required as in other cases. It is expected that DP can enhance speed of reinforcement learning owing to reduction of number of S-R pairs to be learned.

In this work, to partially cope with delayed-reward problem of reinforcement learning, a novel type of knowledge propagation (KP) is proposed. To be specific, suppose that a robot is perfectly trained to respond to a stimulus from a low level of robot sensor, which let the robot achieve an internal desire or motivation. Also suppose that a robot gets a new stimulus from a different or high level of robot sensors, and robot should respond to the new stimuli to achieve the same type of internal desire or motivation of existing S-R behaviors made by the low level of robot sensors. Then, new S-R behavior made by the high level sensor can be expected to be learned by using the existing S-R behaviors made by low-level sensor.

For example, suppose that obstacle avoidance behavior by sonar sensors (low-level sensors) be innately programmed or pre-learned in a mobile robot. Then if a new vision sensor (high-level sensor) is mounted to the mobile robot, an obstacle can be detected as a new stimulus from the vision sensor, while the robot could avoid the obstacle by the obstacle avoidance behavior learned by sonar sensors. For this case, a new obstacle-avoidance behavior can be effectively learned to respond to the obstacle stimulus coming from the vision sensor by employing such an existing obstacle-avoidance behavior of the robot. We will call this type of learning as knowledge propagation (KP). Fig. 4 shows the conceptual diagram of our proposed KP module.

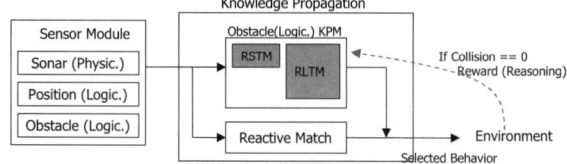

Fig. 4 Knowledge Propagation Module.

B. Knowledge Propagation Algorithm

As shown in Fig. 4, two types of memories, RSTM and RLTM, are included in the module. Recall that in STM, every selected behavior and its associated stimulus is recorded at every time tick. But, in RSTM, a stimulus-response for a sensor (e.g., sonar sensor) is recorded, only if the stimulus for the sensor is considered to be associated with other stimulus from a different sensor (e.g., vision sensor). When a reward is given, history of S-R behaviors for the sensor (usually, relatively low-level sensor) is transformed to RLTM together with reliability index computed by the similar equation of Eq. (1). Detailed memory operation for RSTM and RLTM is omitted since it is similar with that for STM and LTM in Sec2.

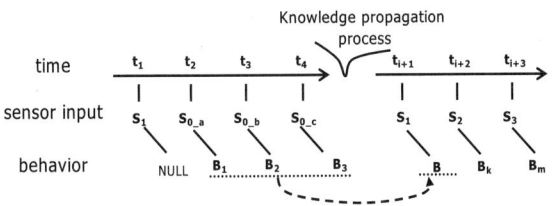

Fig. 5 Knowledge Propagation Process.

To be more specific, consider the timing chart in Fig. 5, when at time t1, a new stimulus S1 (obstacle) from high-level sensor like CCD camera is applied to the robot. And at time t2, t3, and t4, the robot receives stimulus S0_a, S0_b, and S0_c, respectively, from a low-level sensor like sonar sensor, and thus executes innately preprogrammed avoidance behaviors B1, B2, and B3, respectively. If objectives of S1-R and S0-R behavior are known as the same, then KP process begins to analyze B1, B2, and B3 to make a new behavior B for the stimulus S1. Here, B is made by vector sum of B1, B2, and B3, for the robot to rapidly avoid obstacle when receiving the stimulus S1. After all, a new S1-B pair is made by the knowledge propagated from S0_a-B1, S0_b-B2, and S0_c-B3 pairs, and is then available to the robot. This process is done in reasoningKPM() of the pseudo code in Fig. 6.

```
sensorModuleUpdate()
perceptionModuleUpdate()
reactiveMatch()
selectBehavior()

if( existKPMPercept() )
    saveKPMMemory( KPMPercept, currBehvaior )

if( receivedReward() )
{
    reasoningKPM()         // start reasoning
    updateKPMMemory()      // update reliability in RLTM
}
```

Fig. 6 Pseudo Code of Knowledge Propagation.

It is remarked that KP method would be better than DP and KP methods owing to the use of RSTM. To be specific, recall that each logical sensor is connected with an RSTM. Then, we can understand that when delayed reward is received, KP process gets to know what behaviors should be rewarded by searching for RSTM or by analyzing RSTM. This process can be considered as replacing delayed reward with immediate reward. Thus, KP can show enhanced learning speed. And also, a behavior is chosen generated by referring to those rewarded behaviors, which reduces exploration trials. But, RP and DP methods have to explore behaviors until a robot receives a positive reward for the behavior on each stimulus.

IV. EXPERIMENTS

A. AmigoBot System

AmigoBot of ActivMedia Robotics Company [22] is employed for our experiments as shown in Fig. 7. The robot has 8 sonar sensors and one CCD camera. Pentium PC(233MHz) is used to control the robot, where 900 MHz RF modem is utilized to communicate with robot, and 2.2 GHz A/V receiver is used to get the video data of CCD camera.

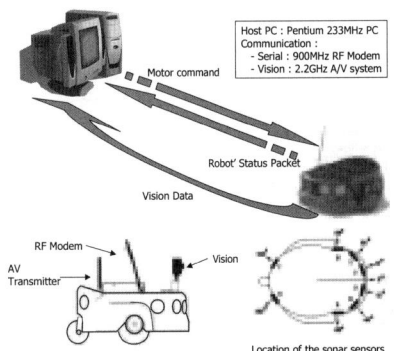

Fig. 7 Control System and Robot.

B. Learning Experiments

To show the validity of our proposed architecture and KP algorithm, experimental set-up is organized as in Fig. 8, where successful robot task is defined as running along the circular track without violation of traffic signal and without collision to the walls as well as an obstacle.

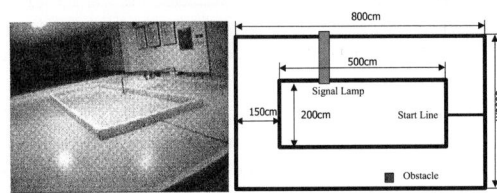

Fig. 8 Environment.

When a run is considered successful, the robot will get a reward. Here, it is assumed that physical properties of the track such as, length and width should not be changed during whole experiments. But a traffic signal in the track will be arbitrarily changed from red to green or from green to red, and an obstacle is given at any place in the track.

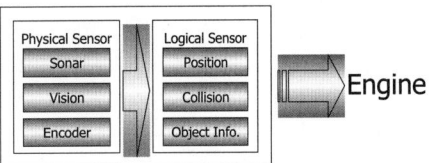

Fig. 9 Sensor Module

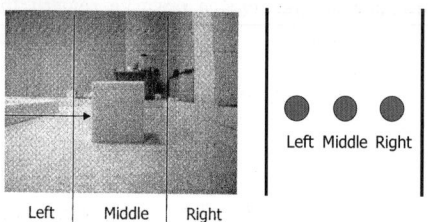

(a) Vision Physical Sensor   (b) Position Logical Sensor

Fig. 10 Obstacle Logical Sensor

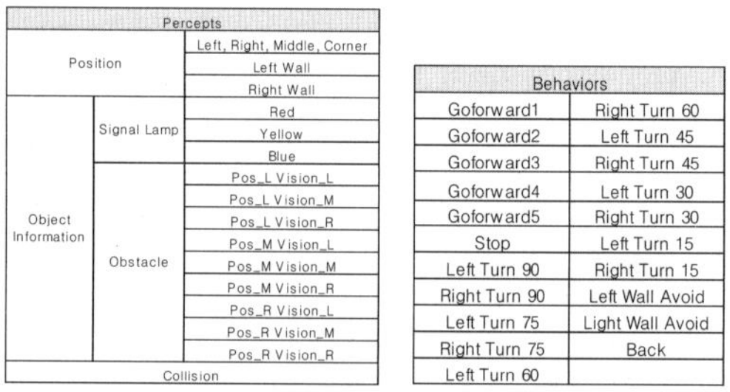

(a) Percepts

(b) Behaviors

Fig. 11 Percepts and Behaviors

TABLE I
Innate Knowledge

Stimulus	Behavior
Corner	LeftTurn90
Left Wall	LeftWallAvoid
Right Wall	RightWallAvoid
Collision	Back

TABLE II
Experiment Results in case Robot has only innate knowledge

	Simulation	Experiment
Success	5	2
Episodes	250	100

Sensor Module of the architecture in Fig. 1 is designed as in Fig. 9. And 9 different stimuli for obstacle detection are employed as in Fig. 10, and percepts and behaviors are designed as in Fig. 11. Initial stimulus-response pairs are given in TABLE I as innate knowledge. TABLE II shows how the robot could behave in the track only with innate knowledge in TABLE I. As shown in TABLE II, only 2% of 100 real episodic robot runs could be counted successful. This implies that no further learning is possible by means of reinforcement learning due to shortage of number of successful episodes which cause delayed-reward problem.

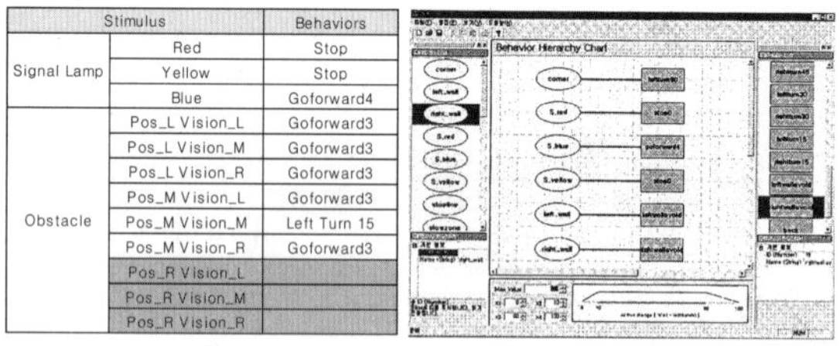

Fig. 12 S-R editor tool for DP method.

Now, to show the effectiveness of KP method, same experiment as above is performed for the DP, RP, and KP methods. In the DP, correct responses to the traffic signal, and correct behaviors to same obstacle-stimuli are directly programmed by using our developed S-R editor tool as shown in Fig. 12. Here, remaining 3 S-R behaviors are to be learned by reinforcement learning technique. For the RP, a computer program is written and a human trainer is employed to give proper rewards to the responses for the traffic signals and an obstacle. And finally, for the KP, S-R behaviors for the sonar sensors and position logical sensors are designed. Fig. 13 shows an exemplar design of such S-R behaviors.

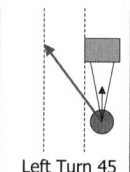

Left Turn 45

Fig. 13 An Example of S-R Behaviors for sonar sensor and logical position sensor.

TABLE III
Reliability of LTM

Behavior ID	M-M			M-R			R-L			R-M			R-R		
	DP	RP	KP	DP	RP	KP	DP	RP	KP	DP	RP	KP	DP	RP	KP
goforward1	0	0	0	0	0.13	0	0	0	0	0	0	0	0	0	0
goforward2	0	0	0	0	0.17	0	0	0	0	0	0	0	0	0	0
goforward3	0	0	0	1	0.18	0.73	0	0	0	0	0	0	0	0	0
goforward4	0	0	0	0	0.33	0	0	0	0	0	0	0	0	0	0
goforward5	0	0	0	0	0.13	0	0	0	0	0	0	0	0	0	0
stop	0	0	0	0	0	0	0	0	0	0	0	0	0	0	0
left turn 90	0	0	0	0	0	0	0	0	0	0	0	0	0	0	0
right turn 90	0	0	0	0	0	0	0	0	0	0	0	0	0	0	0
left turn 75	0	0	0	0	0	0	0	0	0	0	0	0	0	0	0
right turn 75	0	0	0	0	0	0	0	0	0	0	0	0	0	0	0
left turn 60	0	0	0	0	0	0	0	0	0	0	0	0	0	0	0
right turn 60	0	0	0	0	0	0	0	0	0	0	0	0	0	0	0
left turn 45	0	0	0	0	0	0	0.08	0.1	0	0.12	0.14	0	0	0	0
right turn 45	0	0	0	0	0	0	0	0	0	0	0	0	0	0	0
left turn 30	0	0.12	0	0	0	0	0.29	0.42	0.67	0.17	0.36	0.48	0.04	0.29	0.33
right turn 30	0	0	0	0	0	0	0	0	0	0	0	0	0	0	0
left turn 15	0	0	0	0	0	0	0	0	0	0	0	0	0	0	0
right turn 15	1	0.18	0.27	0	0	0	0	0	0	0	0	0	0	0	0
left wall avoid	0	0	0	0	0	0	0	0	0	0	0	0	0	0	0
right wall avoid	0	0	0	0	0	0	0	0	0	0	0	0	0	0	0
back	0	0	0	0	0	0	0	0	0	0	0	0	0	0	0

50 episodic trials are performed for each method. TABLE III shows the reliability index values for S-R behaviors of each method. Here, we omitted S-R behaviors for traffic signals and other stimuli. Reliability index values show how much a stimulus is strongly connected to a behavior. It is here observed from TABLE III that RP showed 5 behaviors such as goforward1, goforward2, goforward3, goforward4, and goforward5 for the case of M-R. M-R is the case from Fig. 10 that robot is in the middle and obstacle is in the right. On the other hand, for M-R, KP method showed an appropriate behavior, goforward3, with the reliability value higher than RP. It can be also observed from TABLE III that reliability values of S-R behaviors of M-M, R-L, R-M, and R-R cases for KP method are higher than those for DP and RP methods. This implies that for the same episodic trial numbers, KP method enable the robot to learn necessary S-R behaviors more reliable than other two methods in delayed reward environment.

It is remarked that fast learning and high reliability of KP method would come from the use of RSTM which is connected with a logical sensor. Specifically, when a delayed reward is received, KP process can immediately know what behaviors should be rewarded by searching for

RSTM or by analyzing RSTM. That is, KP process can be considered as replacing delayed reward with immediate reward. Thus, KP can show an enhanced learning speed. But, RP and DP methods have to explore behaviors until a robot receives a positive reward for the behavior on each stimulus. Thus, for RP and DP methods, it may take a relatively long time to learn correct S-R behaviors.

V. CONCLUSIONS

In this paper, a behavior-based control and learning architecture was proposed, where reinforcement learning was applied to learn proper associations between sensor states and behaviors. To solve delayed-reward problem, a knowledge-propagation (KP) method was proposed. And to show the validity of our proposed KP method, comparative experiments were performed for the cases that (i) only a delayed reward was used, (ii) some of S-R pairs were preprogrammed, (iii) immediate reward was possible, and (iv) our KP method was applied. From the experiments, we showed that KP method enabled the robot to learn necessary S-R behaviors faster and more reliable than RP and DP methods.

Acknowledgement

This work has been supported in part by Next-Generation New Technology Development Program entitled as Control/Recognition Technology for Personal Robots.

VI. REFERENCES

[1] R.C. Arkin, Behavior-Based Robotics, MIT Press, 1998.
[2] R.R. Murphy, Introduction to AI Robotics, MIT Press, 2000.
[3] R.C. Arkin, "Towards cosmopolitan robots: Intelligent navigation in extended man-made environments," Ph.D. Thesis, COINS Tech, Rpt., 97-80, Univ. of Massachusetts, Dept. of Computer and Information Science, pp. 143-177, 1987.
[4] R.A. Brooks, "A robust layered control system for a mobile robot," IEEE J. Robotics and Automation, vol. RA-2, no. 1, pp. 14-23, 1986.
[5] S. D. Touretzky, and L.M. Saksida, "Skinnerbots," Proc. Int. Conf. Simulation of Adaptive Behavior (SAB96), pp. 285 – 294, 1996.
[6] B. Blumberg, "Old Tricks, New Dogs: Ethology and Interactive Creatures," Ph.D. Thesis, The Media Lab, MIT, Cambridge, 1996.
[7] S.Y. Yoon, "Affective Synthetic Characters," Ph.D. Thesis, The Media Lab, MIT, Cambridge, 2000.
[8] J. Pauls, "Pigs and People," Project Report, Division of Information, University of Edinburgh, 2001.
[9] P. Maes, "The dynamics of action selection," Proc. Int. Joint Conf. Artificial Intelligence, Detroit, MI, pp. 991-997, 1989.
[10] A. Saffiotti, K. Konolige, and E. Ruspini, "A multivalued logic approach to integrating planning and control," Artificial Intelligence 76, pp. 481-526, 1995.
[11] A.F.R. Araujo, and A.P.S. Braga, "Reward-Penalty Reinforcement Learning Schema for Planning and Reactive Behavior," Proc. IEEE Int. Conf. System, Man, and Cybernetics, vol. 2, pp. 1485-1490, 1998.
[12] R. Genov, S. Madhavapeddi, and G. Cauwengerghs, "Learning to Navigate from Limited Sensory Input: Experiments with the Khepera Microrobot," Proc. Int. Conf. Neural Networks, vol. 3, pp. 2061-2064, 1999.
[13] C.J.C.H. Watkins, "Learning from delayed rewards," Ph.D. Thesis, Cambridge University, Cambridge, England, 1989.
[14] K. Lorenz, "The comparative method in studying innate behavior patterns," Symposia of the Society for Experimental Biology, 4, pp. 221-268, 1950.
[15] P. Maes, "How to do the right thing," Connection Science, 1, 291-323, 1989.
[16] A. Ludlow, "The evolution and simulation of a decision maker," In: Analysis of Motivational Process, Academic Press, 1980.
[17] R.S. Sutton, and A.G. Barto, Reinforcement Learning, MIT Press, 1988.
[18] L.S. Crawford, "Learning Control of Complex Skills," Ph.D. Thesis, Biophysics, University of California, Berkeley, September 1998.
[19] M. Dorigo, and M. Colombetti, Robot Shaping: An Experiment in Behavior Engineering, MIT Press, 1998.
[20] I.P. Pavlov, Selected works, Foreign Languages Publishing House, Moscow, 1950.
[21] B.F. Skinner, The behavior of organisms: An experimental analysis, Englewood Cliffs, NJ: Prentice Hall, 1938.
[22] AmigoBot User's Guide, 2000.
[23] R.C. Arkin, and J. Diaz, "Line-of-sight constrained exploration for reactive multiagent robotic teams," 7th Int. Workshop on Advanced Motion Control, pp. 455-461, 2002.
[24] M. Likhachev, M. Kaess, and R.C. Arkin, "Learning behavioral parameterization using spatio-temporal case-based reasoning," Proc. Int. Conf. Robotics and Automation, vol. 2, pp. 1282-1289, 2002.
[25] J.B. Lee, M. Likhachev, and R.C. Arkin, "Selection of behavioral parameters: integration of discontinuous switching via case-based reasoning with continuous adaptation via learning momentum," Proc. Int. Conf. Robotics and Automation, vol. 2, pp. 1275-1281, 2002.
[26] M. Likhachev, and R.C. Arkin, "Spatio-temporal case-based reasoning for behavioral selection," Proc. Int. Conf. Robotics and Automation, vol. 2, pp. 1627-1634, 2001.
[27] J.B. Lee, and R.C. Arkin, "Learning momentum: integration and experimentation," Proc. Int. Conf. Robotics and Automation, vol. 2, pp. 1975-1980, 2001.

On Learning Control with Limited Training Data

Yongsheng Ou and Yangsheng Xu
Department of Automation and Computer-Aided Engineering
The Chinese University of Hong Kong

Abstract

In this paper, we study the interpolation approach in reducing the problem of small training sample sizes severely affecting the learning control performance of artificial neural networks when the dimension of the input variables is high. We use the local polynomial fitting approach to individually rebuild the time-variant functions of system states. Based on these functions, we can effectively produce new unlabelled training samples. We show that by using additional unlabelled samples, the learning control performance can be improved and, therefore, the overfitting phenomenon can be mitigated. Furthermore, experimental results verified these claims.

1 Introduction

Learning control from a human expert demonstration can be considered as a process of building a mapping between system states and control inputs, i.e., a function regression problem. An important problem in learning control is the effect of small training sample size in estimation performance. It is well known that when the ratio of the number of training samples to the VC (*Vapnik-Chervonenkis*) dimension of the function is small, the estimates of the regression function are not accurate and, therefore, the learning control results may not be satisfactory. Moreover, the learning processes of most artificial neural networks have the local minima problem and small training samples can make the local minima much deeper. Meanwhile, the real-time sensor data always have random noise. Both of them have bad effect on the learning control performance. Thus, we need large sets of data to overcome these problems. Since most control processes are short, a large number of training samples can be very expensive and time consuming to acquire. In this work, our main aim is to produce more new training samples but without increasing the cost (called unlabelled sample, here) and to enforce the learning effect, so as to improve the learning control.

The main problem in statistical pattern recognition is to design a classifier. A considerable amount of effort has been devoted to design a classifier in small training sample size situations [2]-[4]. Many methods and theoretical analysis have focused on the nearest neighbor re-sample or bootstrap re-sample. however, the major problem in learning control is function approximation. There is limited research in function regression under conditions of sparse data. Janet and Alice's work in [5] examined the three resampling methods (cross validation, jackknife and bootstrap) for function estimation.

In this paper, we use the local polynomial fitting approach to individually rebuild the time-variant functions of system states. Then, through interpolation in a smaller sampling rate, we can rebuild any number of unlabelled samples (new samples).

The organization of the paper is as follows. In Section 2, the problem of limited sample size is provided. The *overfitting* phenomenon, which is the loss of function regression that is observed when the dimensionality of the neural network (ANN) structure is large while the training sample size is small, is briefly discussed in Section 3. In Section 4, the basic idea of the local polynomial fitting approach is described mathematically. Our proposed re-sampling method is addressed and the effect of new training samples is discussed in Section 5. In Section 6, a dynamically stable system, which is our experimental system, is introduced. In Section 7, some experimental results, which demonstrate the validity of proposed approach, are provided. The final remarks are presented in Section 8.

2 Problem Statement

Our learning control problem can be considered as building a map between the system states X and the control inputs Y. Both $X = (x_1, x_2, ..., x_m)$ and Y may be continuous time-various vectors, where x_i is one of the system states. Furthermore, without the loss of generality, we restrict Y to be a scalar for purposes of simplifying the discussion.

$$Y = F(X) + \epsilon, \qquad (1)$$

where $\epsilon \in R$ is the total sensor errors from real-time measurement. We assume that ϵ has the Gaussian distribution with $E(\epsilon) = 0$ and variance $\sigma^2 > 0$.

Here we focus our attention on the estimation of regression function F. We show that for a fairly broad

---

*This work is supported in part by Hong Kong Research Grant Council under the grants CUHK 4403/99E, CUHK 4228/01E and Hong Kong SAR Government under grand ITS/140/01.

class of sampling schemes t_i with fixed and identical sampling rate $\Delta > 0$. Assume that $(\hat{F})$ is an estimation of Y from a random sample T of size n. A training pair $T_i = (X_i, Y_i)$ $(i = 1, 2, ..., n)$ consists of a system state vector X_i and the control input Y_i. For any possible system state vector X_o, define the estimation error as

$$Err_o = Y_o - \hat{F}(X_o).$$

The target of the estimation is to make the mean of Err_o equal zero and the variance of Err_o as small as possible.

3 The Overfitting Phenomenon

Scholkopf [6] pointed out that the *actual risk* $R(w)$ of the learning machine is expressed as:

$$R(w) = \int \frac{1}{2}\|f_w(\mathbf{x}) - y\| dP(\mathbf{x}, y). \quad (2)$$

The problem is that $R(w)$ is unknown, since $P(\mathbf{x}, y)$ is unknown.

The straightforward approach to minimize the *empirical risk*,

$$R_{emp}(w) = \frac{1}{l}\sum_{i=1}^{l} \frac{1}{2}\|f_w(\mathbf{x})_i - y_i\|,$$

turns out not to guarantee a small actual risk $R(w)$, if the number l of training examples is limited. In other words, a small error on the training set does not necessarily imply a high *generalization* ability (i.e., a small error on an independent test set). This phenomenon is often referred to as *overfitting*. For the learning problem, the *Structure Risk Minimization* (SRM) principle is based on the fact that for any $w \in \Lambda$ and $l > h$, with a probability of at least 1-η, the bound

$$R(w) \leq R_{emp}(w) + \Phi(\frac{h}{n}, \frac{log(\eta)}{n}) \quad (3)$$

holds, where the *confidence term* Φ is defined as

$$\Phi(\frac{h}{n}, \frac{log(\eta)}{n}) = \sqrt{\frac{h(log\frac{2n}{h} + 1) - log(\eta/4)}{n}}.$$

The parameter h is called the VC (*Vapnik-Chervonenkis*) dimension of a set of functions, which describes the capacity of a set of functions. Usually, to decrease $R_{emp}(w)$ to a certain bound, most ANNs with complex mathematical structure have a very high value of h. It is noted that when n/h is small (for example, less than 20, when the training sample is small in size), Φ has a large value. When this occurs, the performance poorly represents $R(w)$ with $R_{emp}(w)$. As a result, according to the SRM principle, a large training sample size is required to acquire a satisfactory learning machine.

4 Local Polynomial Fitting

Suppose that we are given by noisy samples of a signal $x(t)$,

$$z_s = x(t_s) + \epsilon_s, \ s = 1, 2, 3, ..., N, \quad (4)$$

where ϵ_s i.i.d., $E(\epsilon_s) = 0$, $E(\epsilon_s^2) = \sigma^2$. It is assumed that $x(t)$ belongs to the nonparametric class of piecewise continuous m-differentiable functions

$$\mathcal{F} = |x^{(m)}(t)| \leq L_m.$$

Our goal is to estimate $x_s = x(t_s)$ depending on observation $(z_s)_{s=1}^N$ with a point-wise mean squared error (MSE) risk that is as small as possible.

Suppose that locally the regression function $x(t)$ can be approximated by

$$x(t_s) \approx \sum_{j=0}^{m} \frac{x^{(j)}(t)}{j!}(t_s - t)^j \equiv \sum_{j=0}^{p} \beta_j(t_s - t)^j, \quad (5)$$

for t_s in a neighborhood of t, by using Taylor's expansion. From a statistical modelling viewpoint, Equation 5 models $x(t_s)$ locally by a simple polynomial model. This suggests the use of a locally weighted polynomial regression

$$J_h(t) = \sum_{i=1}^{n}\{z_s - \sum_{j=0}^{p}\beta_j(t_s - t)^j\}^2 K_h(t_s - t), \quad (6)$$

where $K(\cdot)$ denotes a kernel function and h is a window size or a bandwidth. If $\hat{\beta} = (\hat{\beta}_0, ..., \hat{\beta}_m)$ denotes the solution to the above weighted least squares problem, then by Equation 5, $j!\hat{\beta}_j(t)$ estimates $x^{(j)}(t), j = 0, ..., m$. Minimizing Equation 6 leads to the following set of equations. Let

$$K_h(t) = K(t/h)/h,$$

$$u_{n,j} = \frac{1}{n}\sum_{i=1}^{n}(\frac{t_s - t}{h})^j K_h(t_s - t),$$

$$a_{n,j} = \frac{1}{n}\sum_{i=1}^{n}(\frac{t_s - t}{h})^j K_h(t_s - t)Z_s.$$

Putting

$$U_n = \begin{pmatrix} u_{n,0} & \cdots & u_{n,m} \\ \vdots & \ddots & \vdots \\ u_{n,m} & \cdots & u_{n,2m} \end{pmatrix}, A_n = \begin{pmatrix} a_{n,0} \\ \vdots \\ a_{n,m} \end{pmatrix}$$

the solution to Equation 6 can be expressed as

$$\hat{\beta} = diag(1, h^{-1}, ..., h^{-p})U_n^{-1}A_n. \quad (7)$$

5 Resampling Approach

The purpose here is to improve the function estimation performance of ANN learning controllers by using the polynomial fitting approach interpolation samples. Let $x_i^T = \{x_i(t_1), x_i(t_2), ...x_i(t_n)\}$ be an original training sample set of one system state, by the same sampling rate $\Delta > 0$ and $T_j = (X_j, Y_j)$, where $X_j = \{x_1(t_j), x_2(t_j), ...x_m(t_j)\}$ is an original training sample point. An unlabelled sample set $\bar{T} = \{\bar{T}_1, \bar{T}_2, ...\bar{T}_N\}$ which has the size of N can be generated by the following.

Step 1 Using the original training sample set x_1^T to produce the segment of local polynomial estimation $\hat{x}_1(t)$ of $x_1(t)$ with respect to time $t \in (t_1, t_n)$. Let

$$x_1 = x(t) + \epsilon_1$$

where $x_1, x, t \in R$

Step 2 Repeat step 1 to produce the local polynomial estimation $\hat{x}_i(t)$ ($i = 2, 3, ..., m$) and $\hat{Y}(t)$ of $x_i(t)$ ($i = 2, 3, ..., m$) and $Y(t)$ with respect to time $t \in (t_1, t_n)$.

Step 3 Divide the sampling rate $\Delta > 0$ by $k = N/n + 1$ to produce new sampling rate $\Delta_k = \Delta/k$. We produce new unlabelled samples $\bar{T}$ by combination of interpolating the polynomials $\hat{x}_i(t)$ ($i = 1, 2, ..., m$) and $\hat{Y}(t)$ in the new sampling rate.

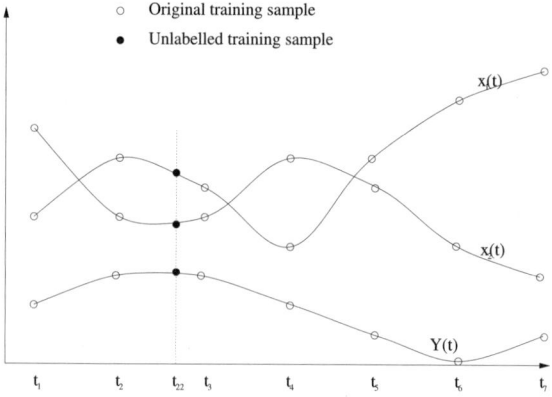

Figure 1: Examples of the interpolation sample generation, when $k = 3$.

Figure 1 shows an example of one unlabelled training sample generation process. In our method, unlabelled training samples can be generated in any number.

Barron [12] has studied the way in which the residual sum-of-squares error decrease as the number of parameters in a model is increased. For neural networks he showed that this error falls as $O(1/M')$, where M' is the number of hidden nodes in an one hidden layer network. By contrast, the error only decreases as $O(1/M^{2/d})$, where d is the dimensionality of the input space, for polynomials or indeed any other series expansion in which it is the coefficients of linear combinations of fixed functions which are adapted. However, from the former analysis, the number of hidden nodes M' cannot be chose any large. For the practitioner, his theory provides the guidance to choose the number of variables d, the number of network nodes M', and the sample size n, such that both $1/M'$ and $(M'd/n)\log n$ are small. In particular, with $M' \sim (n/(d \log n))^{1/2}$, the bound on the mean squared error is a constant times of $(d \log n/n)^{1/2}$.

For the unlabelled training sample data can be produced by any large number by interpolation, if we choose some suitable large number of nodes (M'), we can have obtained a very high learning precise. At this time, the learning error comes from the polynomial fitting, instead of neural network learning. However, we have improved the learning by these original limited training samples, for the learning problem can be considered turn the inputs from $x \in R^d$ to $t \in R^1$.

Especially, when $d = 1$, $3M' \approx M$, where M is the order of the polynomial. Thus, if the training sample size is little be large and M can be chose larger than three. For this one dimension input to one dimension output mapping problem, to choose polynomial curve fitting is better than network learning.

Using this approach, we have turned a multivariate function estimation problem to be a univariate function estimation problem (i.e. one-by-one mapping). For dealing with univariate function estimation problems, there have already been a number of research works, for example, local polynomial fitting approach.

6 Experimental System

The single-wheel gyroscopically-stabilized robot takes advantage of the dynamic stability of a single wheel. For such a system, the research issue on small training sample data becomes more crucial. Figure 2 shows a photograph of the third Gyrover prototype.

Gyrover is a sharp-edged wheel with an actuation mechanism fitted inside the rim. The actuation mechanism consists of three separate actuators: (1) a spin motor, which spins a suspended flywheel at a high rate and imparts dynamic stability to the robot; (2) a tilt motor, which steers the Gyrover; and (3) a drive motor, which causes forward and/or backward acceleration by driving the single wheel directly.

The Gyrover is a single-wheel mobile robot that is dynamically stabilizable but statically unstable. As a mobile robot, it has inherent nonholonomic constraints [7]. First-order nonholonomic constraints include constraints at joint velocities and Cartesian space velocities. Because no actuator can be used directly for stabilization in the lateral direction, it is an underactuated nonlinear system. This gives rise to a second-order nonholonomic constraint as a consequence of dynamic constraints introduced by accelerative forces on passive joints.

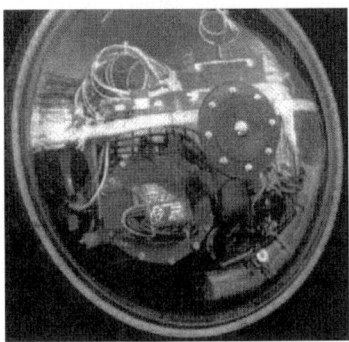

Figure 2: Gyrover: A single-wheel robot.

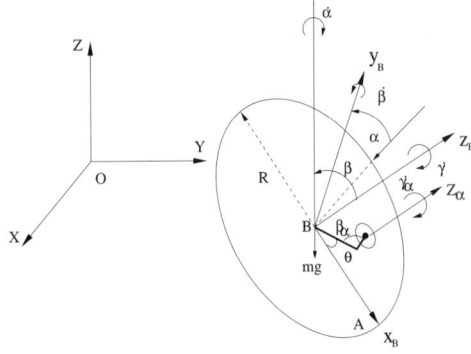

Figure 3: Definition of the Gyrover's system parameters.

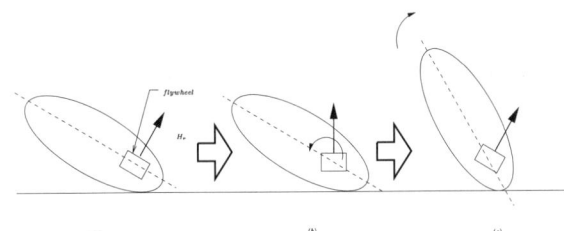

Figure 4: Mechanism of tilt-up motion for the robot.

In this effort, we are, however, faced with a number of difficult challenges that are absent or much less daunting in more traditional quasi-static systems. Since system dynamics plays an integral role in the behavior of dynamically stable systems, an accurate analytic model is necessary in order to apply classical control techniques, yet such a model is often difficult to arrive at in practice. In general, these systems exhibit dynamics that are highly coupled, nonlinear, and vary substantially depending on precise configuration of the systems; moreover, friction and other difficult-to-model physical parameters often impact dynamically stable systems more severely than conventional quasi-static systems (e.g. four-wheel vehicles). We have certainly observed this to be so in our on-going experiments with Gyrover. Despite these challenges, we do, however, hold one important card – namely, the fact that humans are typically capable of demonstrating precisely those desired control maneuvers that are so challenging for mechanical systems to master. Therefore, we believe that capturing and transferring human skill to these systems is an ideal and fitting paradigm for autonomously controlling dynamically stable systems.

To represent the dynamics of the Gyrover, we need to define the coordinate frames: three for position (X, Y, Z), and three for the single-wheel orientation (α, β, γ). The Euler angles (α, β, γ) represent the precession, lean and rolling angles of the wheel respectively. (β_a, γ_a) represent the lean and rolling angles of the flywheel respectively. They are illustrated in Figure 3.

7 Experimental Study

The aim of this experiment is to illustrate how to use the proposed local polynomial fitting interpolation re-sampling approach and validate the approach.

The control problem consists of tilting up Gyrover from the horizontal position. It is an important process in the automatic control of Gyrover. Moreover, one of Gyrover's advantages compared with multi-wheel vehicles is that it can right itself after falling over. According to the conservation of angular momentum, the robot is able to recover from falls by tilting the flywheel. The tilting direction of flywheel opposes the leaning direction of the robot. Figure 4 shows the mechanism of the tilt-up motion for the single wheel robot.

We built up an ANN learning controller based on learning imparted from expert human demonstrations. Here, the ANN is a cascade neural network architecture with node-decoupled extended Kalman filtering learner (CNN), which can decrease the local minima problem for its adjustable hidden neural number.

In the experimental system, there are two major control inputs: U_0 controlling the rolling speed of the single wheel $\dot{\gamma}$, and U_1 controlling the angular position of the flywheel β_a. For the manual-model (i.e., controlled by a human), U_0 and U_1 are input by joysticks, and in auto-model they are derived from the software controller. During all experiments, we only focus on the value of U_1 and fix the value of U_0 to zero. Among the system state variables, only β, β_a are used during the training process, as the trained model inputs, and its output is U_1.

The tilt-up motion, which last about 6 seconds on average, includes two processes: one involves Gyrover righting itself, and the other entails it maintaining a vertical state so as to remain upright for several seconds. First, a human expert demonstrated the tilt-up motion where the sampling interval time was $0.062s (62ms)$ and produced 91 original training sample data. Table 1 displays some raw sensor data from the human expert control process.

Input		Output
β_a	β	U_1
4.0930	157.1000	180.0000
0.1427	149.7800	216.0000
-30.5816	160.7600	234.0000
-55.2331	171.0080	203.0000
-20.3402	149.2920	162.0000

Table 1: Sample human control data.

Based on these original training sample points, the CNN learner had poor learning performance. Using this original learning controller, Gyrover could not be tilted up. After the human expert righted Gyrover several times, we combined these all sets of sample points. As a result, the performance of the CNN learner was more effective and the experiment was successful. Thus, the reason for the failure of the single set of sample training process is due to the small training samples.

In this work, we used the approach outlined above to solve the small sample problem relying on the first set of original training sample data. First, we did the function estimation (curve fitting) for lean angle β with regard to time t. From Equation 7, we can collect the polynomial coefficients $\hat{\beta}$. An easier way is to use the MATLAB function "$lpolyreg$".[1] We could also use MATLAB function "$interp1$" to obtain the interpolation of lean angle. The MATLAB functions use a similar algorithm to the local polynomial fitting approach. Figure 5 shows the local polynomial fitting for lean angle β with the polynomial degree of $m = 15$.

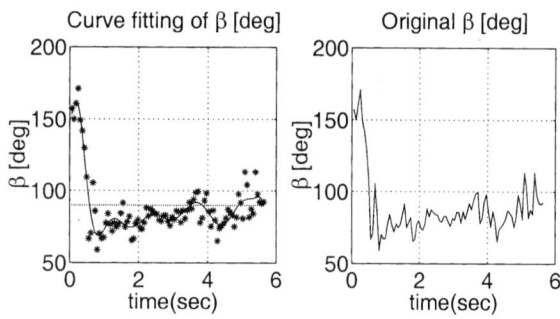

Figure 5: Local polynomial fitting for lean angle β.

---

[1] Readers may download an implementation of the Direct Plug-In (DPI) method of Ruppert, Sheather and Wand (JASA, 1995) for local polynomial regression in http://www.acae.cuhk.edu.hk/ you/lpolyreg.tar.gz .

Second, we can perform the function estimation for both β_a and U_1 as for lean angle β. Third, we produced the new unlabelled training sample by interpolation in the time interval $0.010s$ $(10ms)$ and produced 564 unlabelled training sample data. Table 2 displays some unlabelled samples from the interpolation process.

Input		Output
β_a	β	U_1
0.2105	154.5492	200.2818
-5.8567	155.4685	208.9759
-11.2569	156.4426	215.6724
-16.0087	157.4303	220.6101
-20.1324	158.3957	220.6101

Table 2: Unlabel sample data.

By putting these 564 unlabelled sample data into a new CNN learning model, we have a new CNN-new neural network model. We can, therefore, compare it with the old CNN model CNN-old, which is produced by the original single set training data. We use β and β_a of another set of human demonstration training data as inputs to compare the outputs of these two neural network models. Figure 6 shows the comparison of the human control, CNN-new model learning control and CNN-old model learning control. From the figure shows, the CNN-new model has smaller error variance than that of the CNN-old, i.e., it may have better learning control performance. Practical experiments confirm this.

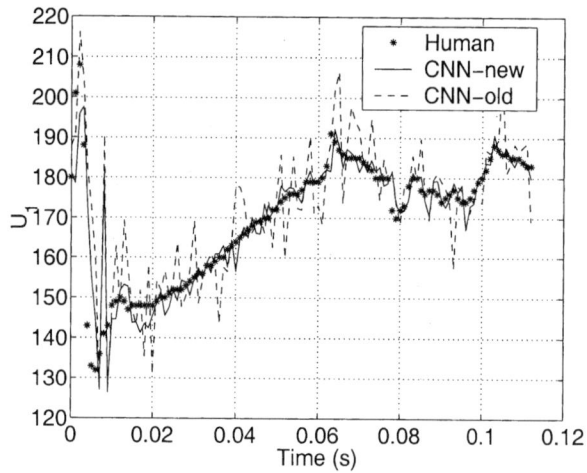

Figure 6: Comparison of U_1 in another set of training data.

Subsequently, we use the CNN-new model as a learning controller to right Gyrover automatically in real-time. This experiment was also successful. Figures 7 and 8 show the control state variables lean angle

β and control command U_1 of a set of human and the learning controls.

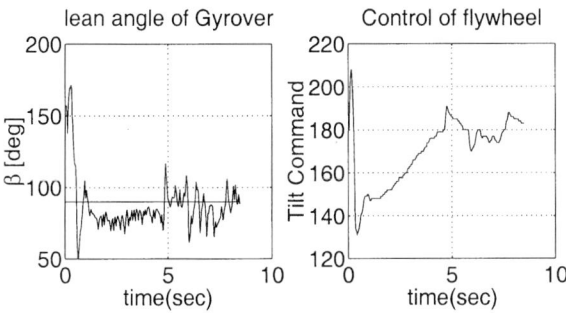

Figure 7: A set of human control.

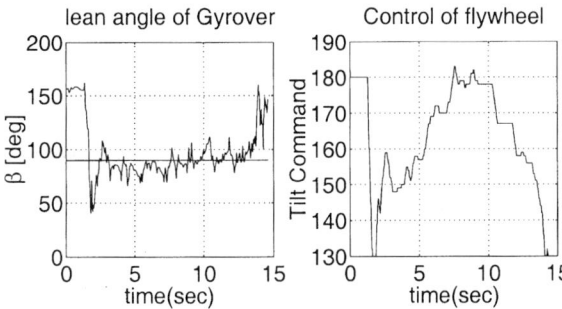

Figure 8: The CNN-new model learning control.

8 Discussions and Concluding Remarks

In this paper, a method of adding unlabelled training samples to enhance the learning control performance was studied. It is observed that by incorporating unlabelled samples into the estimation process, the overfitting phenomenon might be mitigated. The mitigation of the overfitting phenomenon is important in a learning control process, especially of a dynamically stable system like Gyrover, for the following reasons. First, the initial condition may differ from case to case. Second, the controllers and the system states are combined to form an integrated system and interact with each other. Third, if the learning model does not set up well the mapping between the system states and the control inputs effectively, the learning controllers from model will be rigid and fail to make the control process converge to the targets.

An important practical point that needs to be kept in mind is that the points near the boundary of the local polynomial fitting curve may have large errors. Thus, in practice, we need avoid using the head and the tail of the fitting curve in the interpolation process. Although in theory the additional unlabelled training samples should always improve performance, this might not always be the case in practice. This is because our method is based on an assumption that the original sample size is not too small, but enough to make any system state estimation achieve some degree of precision.

In this paper, a new method of adding new training samples without increasing costs is proposed. We use the local polynomial fitting approach to individually rebuild the time-variant functions of system states. Through interpolating, we can effectively produce any number of additional training samples. A number of experiments based on a dynamically stable system – Gyrover – show that by using additional unlabelled samples, the learning control performance can be improved, and therefore the overfitting phenomenon can be mitigated.

References

[1] Y. Xu, H. B. Brown and K. W. Au, "Dynamics mobility with single-wheel configuration," *Int. J. of Rob. Res.*, vol. 18, no. 7, pp. 728-738, 1999

[2] S. J. raudys and A. K. Jain, "Small sample size effects in statistical pattern recognition : Recommendations for practitioners," *IEEE Trans. PAMI-13*, no. 3, pp. 252-264, 1991.

[3] Y. Hamamoto, S. Uchimura, T. Kanaoka and S. Tomita, "Evaluation of artificial neural network classifiers in small sample size situations," *Proc. Int. Conf. Neural Networks*, pp. 1731-1735, 1993.

[4] Y. Hamamoto, Y. Mitani and S. Tomita, "On the effect of the noise injection in small training sample size situations," *Proc. Int. Conf. Neural Information Processing*, vol. 1, pp. 626-628, 1994.

[5] J. M. Twomey and A. E. Smith, "Bias and Variance of validaton methods for function approximation neral networks uner conditions of sparse data ", *IEEE Transaction on SMC-Part C: Application and Review*, vol. 28, no. 3, pp. 417-430, 1998.

[6] S. Bernhard, C. J. C. Burges and A. Smola, "Advanced in kernel methods support vector learning," *Cambridge, MA, MIT Press*, 1998.

[7] Y. Ou and Y. Xu, "Stabilization and line tracking of the gyroscopically stabilized robot," *Proc. IEEE Int. Conf. on Robotics and Automation*, vol. 2, pp. 1753-1758, 2002.

[8] J. Fan and I. Gijbels, *"Local polynomial modelling and its application"*. London: Chapman and Hall, 1996.

[9] Bellman, R., *"Adaptive Control Processes: A Guided Tour"*. New Jersey: Princeton University Press, 1961.

[10] D. Ruppert, S. J. Sheather and M. P. Wand, "An Effective Bandwidth Selection for Local Least Squares Regression," *Journal of the American Association*, vol. 90, no. 432, pp. 1257-1270, 1995.

[11] David Ruppert, "Empirical-Bias Bandwidths for Local Polynomial Nonparametric Regression and density estimation ," *Journal of the American Association*, vol. 92, no. 439, pp. 1049-1062, 1997.

[12] Andrew R. Barron, "Universal approximation Bounds for Superpositions of a Sigmoidal Function," *IEEE Trans. on Information Theory*, vol. 39, no. 3, pp. 930-945, 1993.

Ada – Intelligent Space: An artificial creature for the Swiss Expo.02

Kynan Eng, Andreas Bäbler, Ulysses Bernardet, Mark Blanchard, Marcio Costa[1], Tobi Delbrück, Rodney J Douglas, Klaus Hepp, David Klein, Jonatas Manzolli[1], Matti Mintz[2], Fabian Roth, Ueli Rutishauser, Klaus Wassermann, Adrian M Whatley, Aaron Wittmann, Reto Wyss, Paul F M J Verschure

Institute of Neuroinformatics, University/ETH Zurich, Switzerland
[1]*Nucleo Interdisciplinar de Comunicacao Sonora, University of Campinas, Brazil*
[2]*Psychobiology Research Unit, Tel-Aviv University, Tel-Aviv, Israel*

Abstract

Ada is an entertainment exhibit that is able to interact with many people simultaneously, using a language of light and sound. "She" received 553,700 visitors over 5 months during the Swiss Expo.02 in 2002. In this paper we present the broad motivations, design and technologies behind Ada, and a first overview of the outcomes of the exhibit.

1. The Ada Project

Ada is an interactive space developed for the Swiss national exhibition Expo.02 located in Neuchâtel. Conceptually, "she" can be seen as an inside-out robot with visual, audio and tactile input, and non-contact light and sound effectors. Visitors to Ada are immersed in an environment where their only sensory stimulation comes from Ada herself (and other visitors). Like an organism, Ada's output is designed to have a certain level of coherence and convey an impression of a basic unitary sentience to her visitors. She can communicate with them collectively by using global lighting and background music to express overall internal states, or on an individual basis through the use of local light and sound effects.

To realise such a space, several simultaneous lines of research and development were pursued. Topics under investigation include:

- Audio processing and localisation
- Multi-modal tracking
- Automatic music composition
- Real-time neuromorphic control systems
- Human-machine interaction via whole-body locomotion

From a more application-oriented perspective, Ada was used to gain practical experience in handling large-scale behavioural integration issues in autonomous systems. Certain components of the final system are also being evaluated for technology transfer projects.

Development of Ada commenced in late 1998 and ramped up to a maximum team size of about 25 people. A total of over 100 people were directly involved in the construction and running of Ada. The exhibit ran continuously for up to 12 hours a day over 5 months from 15 May to 20 October 2002. During this period 550,000 visitors entered the space.

Diversity was a defining characteristic of the project. The technical development team came from many different nations and disciplines, ranging from biological sciences through engineering to musical composition. On top of this was a team of architects, artists, publicists, scenographers, on-site managers and guides for handling the production and operation of the exhibit. The financial stakeholders represented most of the main sectors of Swiss society: government (via the Expo.02 organisation), education (ETH Zurich), private industry (Manor, a department store chain) and two private foundations. Correspondingly, the project goals had to address many issues simultaneously. These issues included: contractually defined requirements for system uptime, visitor numbers and standard of entertainment experience, special events for sponsors and media, and encouragement of discussion of the societal impact of future autonomous technologies. Positive publicity for all stakeholders also had to be assured as far as possible. Their requirements had to be balanced against the primary research goals of the project.

This paper presents an overview of Ada's design: her sensors and effectors, system architecture and behaviours. An overview of the operational results is also given, followed by a brief description of the data collection strategies.

2. Sensors, Effectors and Core Services

In total (including auxiliary exhibition areas), Ada has 15 video inputs, 367x3 tactile inputs, 9 audio input

Table 1: Descriptions of functionality and types of software found at different levels in Ada

Level	Functionality	Software
4: Behavioural modulation	Goal function evaluation, behaviour mode selection, emotional model	Simulated neurons (IQR421 [2])
3: Behavioural modules	Coordinated high-level interactions	Simulated neurons, software agents
2: Sensorimotor processes	Filtering of raw input data	Procedural or object-oriented code
1: Device I/O drivers	Interface to hardware	Procedural or object-oriented code
0: Hardware devices	Motor control, sound production, sensor reading, light setting	On-device logic

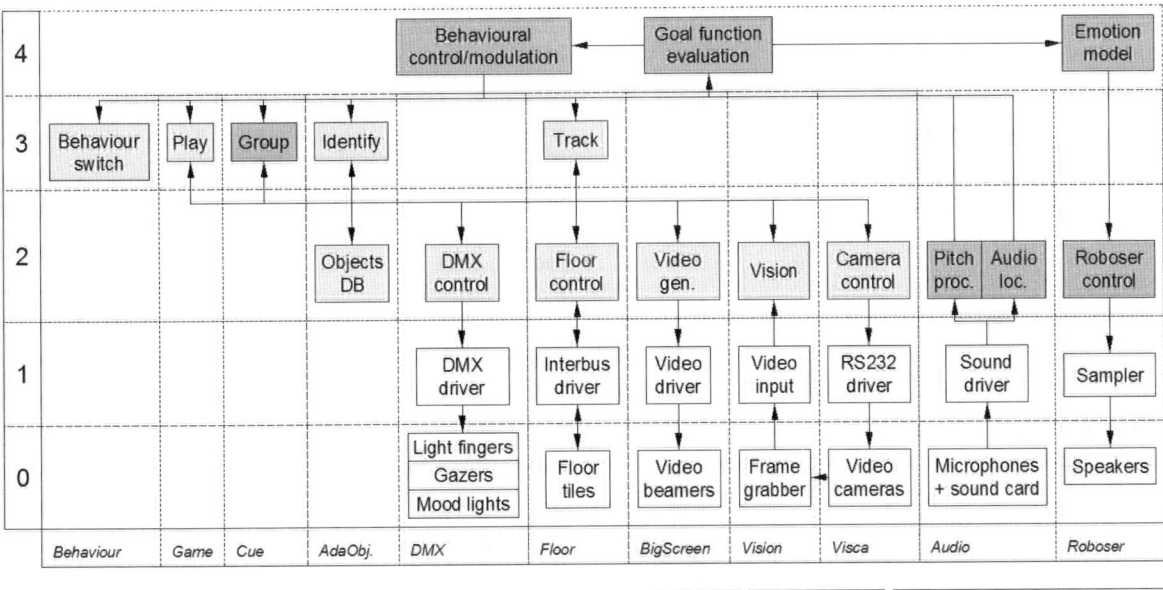

Figure 1: Overview of Ada system architecture, organised into conceptual layers.

channels, 46 mechanical degrees of freedom, 17 output audio channels, 367x3 floor tile lights, 30 ambient lights and 20 full-screen video outputs. All of these inputs and outputs can be addressed independently, giving a rich array of sensory modalities and output possibilities. Ada has the following sensory capabilities:

- **Vision:** Pan-tilt cameras called *gazers* are available to Ada for focused interactions with specific visitors. The cameras have zoom and digital filtering capabilities that are controlled on-line.

- **Hearing:** There are clusters of three fixed microphones each in the ceiling plane, with which Ada is able to localise sound sources by triangulation. Some basic forms of sound and word recognition and pitch extraction are available.

- **Touch:** Ada has a "skin" of 0.66 m wide hexagonal pressure-sensitive floor tiles [3] that can detect the presence of visitors by their weight. Each contains a microcontroller and sits on a serial bus running an industrial automation protocol called Interbus.

As well as sensing, Ada can also express herself and act upon her environment in the following ways:

- **Visual:** Ada uses a 360° ring of 12 LCD projectors to express her internal states and visitor interaction dynamics. These projectors collectively show a single, unified display of 3D objects covering multiple screens in real-time, as well as live video that can move with smooth transitions between screens. There is also a ring of ambient lights for setting the overall visual tone of the space. Local visual effects can be created using the red, green and blue neon lights in each floor tile in Ada's skin.

- **Audio:** Ada is able to generate a wide range of sound effects. These sounds can be distributed across the entire space or localised using a matrix mixer. She expresses herself using sound and music composed in real-time on the basis of her internal states and sensory input. She can also change the pitch of her output depending on what she hears from her visitors. The composition is generated using a system called Roboser [13].

- **Touch:** Ada has twenty 16-bit pan-tilt *light fingers* for pointing at visitors or indicating different locations in the space. They are standard theatre lights on a serial bus called DMX, which is also used to control the ambient lights and the gazers.

The core services of Ada support her higher-level functions. A *tracking system* uses information from the floor tile pressure sensors to determine the location, speed, direction and weight of visitors. The limited resolution of the tiles means that it is not always possible to distinguish individual paths, so in some cases Ada only knows about the presence of groups of people at certain locations. To obtain more information about individual visitors, a *vision system* deploys gazers to collect images of people who have been localised on the floor. The *audio system* localises and recognises basic sounds (e.g. the word "Ada", pitch, note and key) to help identify salient individuals. On the output side, the *Roboser* system composes real-time music and sound effects, a *video server* allows the visualisation of saved and live images, and a *DMX server* controls the light fingers, gazers and ambient lights.

3. Behaviours and Interactions

The degree of success with which visitors can be convinced that Ada is an artificial organism depends on the nature of their interactions. The operation of the space needs to be coherent, real-time, and reliable enough to work for extended periods of time. As well as this, it must be understandable to visitors and sufficiently rich in the depth of interactions so that visitors feel the presence of a basic unitary intelligence.

To provide for a natural progression in visitor interaction, Ada incorporates at least four basic behavioural functions. First, she can *track* individual visitors or groups of visitors, possibly (but not necessarily) giving them an indication that they are being tracked. At the same time, she can *identify* those visitors who are more "interesting" than others because of their responsiveness to simple cues that Ada uses to probe their reactions. These people are encouraged to form a *group* in part of the space through the use of various light and sound cues. When the conditions are appropriate, Ada rewards a group of visitors by *playing* one of a number of games with them. She continuously evaluates the results of her actions and expresses emotional states accordingly, and tries to regulate the distribution and flow of visitors. These four behavioural functions are decomposed into smaller behaviours that call on the core services as needed.

4. System Architecture

Ada's architecture can be roughly sketched out as a series of levels (see Figure 1), with a gradient of decreasing biological plausibility as the proportion of traditional procedural code increases. The types of data stored or "learned" at each level also follows a similar gradient of biological plausibility. Each level contains modules that communicate with other modules in the same layer, as well as with modules in adjacent layers. The metaphor used is that of distributed brain-like computation, characterised by tight coupling within individual modules and loose coupling between modules. The underlying software is a mixture of simulated neural networks, agent-based systems and conventional procedural or object-oriented software. By far the largest amount of technical effort in the project was expended on developing this hybrid system and tuning it so that the different computing paradigms used could co-exist and co-operate usefully. The types of computations performed at each of the different levels are summarised in Table 1.

Different communication protocols are used to connect the components of the system: a socket-based protocol for the simulated neural networks, and an asynchronous message-based middleware for data transfer between agents. Data exchanged in this way includes floor data (input), visitor tracking data (internal), behavioural states (internal) and DMX device control (output).

5. System Goals and Action Selection

Conceptually, Ada is an artificial organism that tries to maximise its own goal functions, which we interpret as her "happiness". This means that the system as a whole must implicitly or explicitly compute its level of happiness, which can then be used to determine if certain actions contribute to this goal. As a first approximation we can write:

$$H = f(g_s, g_r, g_i)$$

H = overall goal or "happiness"
g_s = survival
g_r = recognition
g_i = interaction

Survival is a measure of how well Ada satisfies her basic requirements, which are to maintain a certain flow of visitors over time and to keep these people moving with a certain average speed. *Recognition* quantifies how well Ada has been able to track and collect data about people, as a pre-condition for more advanced interactions. This process can be seen as Ada "carving" objects out of the world of her sensory data, which is implemented as a progressive filtering of the sensory data and the creation of objects in an internal database once certain criteria of persistence and coherence have been satisfied. *Interaction* measures the number of successful human interactions that Ada has been involved in, with more complex interactions such as games being weighted more highly.

As a system, Ada has the goal of maximising the value of H. There are multiple strategies for achieving this: for example, Ada could encourage high visitor throughput, but in doing so have very few possibilities for recognition and interaction (g_s high, g_r and g_i low). Alternatively, Ada could also achieve an equivalent value of H with only a few visitors in the space, but with high recognition and interaction with each visitor (g_s low, g_r and g_i high). The actual computation of H occurs over multiple levels: an explicit top-level calculation is done using simulated neurons, and in parallel individual behaviours also calculate their own contributions to the parameters for H.

The results of the H calculation are combined with other high-level inputs and the state history to select the most appropriate behavioural state for Ada. Behaviour selection occurs at multiple levels – for example, the floor tiles display colours that depend on the local effects in use as well as the overall state of the space. At the top level, a neural modulation scheme activates and inhibits the underlying behaviours. This modulation can take a variety of forms, including a "hard" winner-take-all (WTA) scheme, a "soft" multiple-winner WTA, or a scheme where the behaviours run freely. The extent to which the behaviour modulation needs to be "hard" depends on the subjective evaluation of how the behaviours interact and/or interfere with each other. In practices, because of the constraints imposed by a high visitor flow rate, the behavioural control is run in a "hard" mode.

6. Computational Infrastructure

Ada runs on a 100 Mbit network of 31 PCs (AMD Athlon XP 1800+, 1.0 Gb RAM, Linux). Driver cards are used for DMX and Interbus communications. In addition, 40 frame grabbers, 4 sound cards and 11 dual-headed accelerated graphics cards are installed. Laptops on a wireless LAN enable system testing and tuning to occur while walking around in the main space.

7. Operational Issues

Due to the extremely large number of visitors that wanted to see Ada, it was necessary to control their flow very rigidly to avoid problems with overcrowding. This was necessary both for safety reasons and to ensure that each visitor had a certain minimum amount of space with which to interact with Ada. Table 2 summarises a typical visitor experience in the space and Figure 2 shows a typical scene in the main space.

During normal operation, the main space received about 25 visitors at a time, giving a nominal capacity of about 300 visitors per hour and an instantaneous occupancy of 125 visitors at any one time. The tracking system worked as reliably for wheelchairs and children weighing more than about 20 kg as it did for adults.

Figure 2: *A typical live user interaction scene within Ada. Visible are floor tiles, a visitor being highlighted by a light finger (centre left – the light finger itself is in the ceiling frame), a dynamic 3D visualisation (top) and a live gazer video on the screens (top left).*

Table 2: *Typical visitor experience in Ada exhibit*

Region	Visitor experience	Time (min)
Queue (outside)	"Brainworkers" neuroscience video on big screen (10 min)	0-90
Conditioning tunnel	Sequential introduction to individual sensors and effectors	5
Voyeur corridor	View group of visitors interacting with Ada, listen to guide's explanation	5
Main space	Interact with Ada	5
Brainarium	View "control room" screens and look back into main space	5
Explanatorium	Art by H R Giger depicting future technology, guest book, videos with statements by scientists, credits poster	5

Ada ran for over 1700 hours on 159 consecutive days with an uptime of better than 98.3%, where uptime was defined as having a system that functioned well enough to enable a normal flow of visitors through the exhibit. Discounting outages due to deficiencies in building services that were beyond the control of the project team, the overall system uptime was over 99.1%. Nine stable versions of the Ada software were released during the Expo, incorporating incremental improvements in user functionality and data logging facilities. On any given day, either the latest development version or the stable version could be run, depending on the demands for testing and experimentation.

The public reaction to the exhibit was overwhelmingly positive. Surveys conducted by the Expo.02 organisation indicated that Ada was one of the 5 most popular attractions out of over 60 at the Expo. An online poll [8] also indicated that Ada was the most popular of the IT-related exhibits at the Expo. The visitor queue length of >30 minutes outside the exhibit for the entire duration of the Expo indicated that the final attendance of 553,700 could have been even higher if the capacity of the space had been larger.

The satisfactory operational result was partly the result of accumulated experience during the development process. From 1998 onwards, ten increasingly large public tests were run to evaluate the feasibility and scalability of the underlying technologies, gauge visitor impressions, and test different interaction scenarios. The two key issues that stood out from the results of the tests were the need for effective visitor flow control, and the importance of communicating Ada's intentions clearly through the use of effective cues and visitor pre-conditioning sequences. One direct consequence of this experience was the decision to employ guides to actively inform Ada's visitors as much as possible about what they would see in the exhibit.

8. Data Logging and Experiments

Data can be logged at several different parts of the system simultaneously. A brief description of the types of data logged is given in Table 3.

Table 3: Types of data logged in Ada

Description	Type
Floor raw load data: sensor values and calibrated RGB neon output colour	ASCII Text
Floor server data: positions of all currently loaded floor tiles at every time step	ASCII Text
Tracking data: Onset, path and endpoint of all visitor tracks on the floor	ASCII Text
Neural network internal states for behaviour selection	ASCII Text
Camera view (up to 4 gazers or overhead cameras simultaneously)	DV
Gazer view of tracked visitors	MPEG
Visitor questionnaires gauging their responses to their experience in Ada	ASCII Text
Ada automatically generated music	MIDI
Sound in space	WAV

Data is accumulated in multiple locations at ~5 Gb per hour during active logging, not including DV or WAV data. The data is automatically sent to a central repository each evening for backup on to DVD-R. A timeserver keeps all timestamps across the cluster synchronised to within a few tenths of a second, which is sufficient for most types of analysis.

As of this writing, some preliminary analysis of the ~150 Gb of collected data had been performed. Development, experiments and analysis have been focussing on the following areas:

- Verification of correct system operation with regard to its internal world-view representation
- Automatic calibration of gazers and multi-modal visual/tactile tracking
- Audio localisation and recognition in noisy spaces
- Assessment of visitor reactions to the exhibit based on demographic measures, and an investigation of the effects of various manipulations of the functionality of the space on visitor perception
- Automatic neural behaviour and action selection
- Automatic music generation based on an internal emotional model
- The ability of the space to actively affect the speed and position of visitors, and the effect of boundary conditions (entry/exit placement, pre-conditioning sequences) on their distribution in the space

9. Related Projects

Ada's closest relative from a project perspective was also her direct physical neighbour – the EPFL Robotics exhibit at the Expo.02. This project dealt with different technical content to Ada (autonomous cooperating museum guide robots), but both projects were of similar size and operated under similar conditions.

Several research projects deal with issues related to home automation and "intelligent rooms", and many companies offer commercial home automation systems such as the GE Smart series from GE Industrial Systems [7]. This system offers a substrate for connecting electrical devices and home network services with a common software interface. The control system software is based on rule sets or driven directly by end users, either within the building or via remote links. In this sort of system, the design emphasis is on ease of end-user installation, operation and customisation, rather than advanced behavioural functionality.

More advanced control systems exist in projects such as the Intelligent Room at MIT [10]. The Intelligent Room project aims to develop systems that support human activities in a seamless, flexible way. To date, work has been done on context-aware speech and gesture recognition, flexible resource allocation [6] and an agent-based extension to Java called MetaGlue. Ada has a similar set of functionalities, but with three main differences. Firstly, Ada is a completed product and is much larger than the Intelligent Room, in terms of physical size, number of components and degree of behavioural integration. Secondly, the design of the user interaction with the space is immersive rather than invisible – the building does not serve its users' needs in the background, but is an active participant in their experiences. Thirdly, the space actively tries to achieve its own goals by engaging its users.

A similar project, also named the Intelligent Space, is being pursued at the University of Tokyo [1]. The general approach is to design a platform to facilitate communication between the entities that inhabit it – whether they be humans, robots, or components of the space itself. The concept of a Distributed Intelligent Network Device (DIND) is proposed for connecting devices in the space. Each DIND has sensors, processing and communications components. In this way the space is seen not as an explicit entity like Ada, but as a common networking medium in a physical area. Another group at the University of Tokyo has developed a system for accumulating human behaviour in a small prototypical apartment [12] using mainly tactile sensors. Two noteworthy developments are a pressure-sensitive bed and a high-resolution pressure-sensitive floor [11] for use in the invalid care industry where 24-hour monitoring of patients is desirable.

An animal-like analogue to Ada is the *Mutant* dog robot [5] and its commercially available successor *Aibo* from Sony. Ada and Aibo are both complete systems designed to interact with the general public, and both integrate visual, audio and tactile information to produce behaviour. They both have an internal emotional model and layered system architectures: Aibo's architecture is agent-based, while Ada has a hybrid of simulated neural networks and agent-based software components. Sony has formalised its system architecture in the OPENR model for building robots [4]. The main differences between Aibo and Ada are the obvious ones of appearance and size. By looking like a dog, Aibo has an inherent advantage over Ada for human interactions. A decision made in designing Ada was to explore the limits of human interactions that could be supported without the use of pre-existing metaphors, and to discourage visitors from anthropomorphising the system. On the engineering front, Aibo has the dual challenges of miniaturisation and minimising power consumption, whereas Ada faces power consumption constraints on a much larger scale.

10. Outlook

Ada is one of the largest-scale artificial organisms yet created. During five months of successful active operation, Ada successfully entertained over half a million visitors, while also serving as a platform for research into several different topics. While definite plans were not available at the time of writing, it is expected that Ada will be reassembled in a new location to allow an ongoing mix of research and public performance.

11. Acknowledgments

Ada is supported by: ETH/University Zurich, Expo.02, Manor AG, Velux Stiftung and Gebert Rüf Stiftung.

12. References

[1] Appenzeller, G, Lee, J-H and Hashimoto, H, "Building Topological Maps by Looking at People: An Example of Cooperation between Intelligent Spaces and Robots", *Proceedings of the International Conference on Intelligent Robots and Systems (IROS 1997)*

[2] Bernardet, U, Blanchard, M and Verschure, PFMJ, "IQR: a distributed system for real-time real-world neuronal simulation", *Neurocomputing*, 2002, 44-46: 1043-1048

[3] Delbrück, T, Douglas, R J, Marchal, P, Verschure, P and Whatley, A M, "A device for controlling a physical system", EU patent application 99120136.9-2215 (1999)

[4] Fujita, M and Kageyama, K, "An Open Architecture for Robot Entertainment" (1997), *Proceedings of the First International Conference on Autonomous Agents*, pp 435-442, ACM Press

[5] Fujita, M and Kitano, H, "Development of an Autonomous Quadruped Robot for Robot Entertainment" (1998), *Autonomous Robots* 5: 7-20

[6] Gajos, K, Weisman, L and Shrobe, H, "Design Principles For Resource Management For Intelligent Spaces", *Proceedings of the Second International Workshop on Self-Adaptive Software (IWSAS'01)*

[7] GE Industrial Systems, GE Smart web page, www.ge-smart.com

[8] Infoweek online poll, Sep 2002, www.infoweek.ch

[9] Lee, J-H & Hashimoto, H; "Intelligent Space – Its concept and contents" (2002), *Advanced Robotics* Vol. 16 No. 3: 265-280

[10] MIT Intelligent Room project web page, www.ai.mit.edu/projects/iroom

[11] Morishita, H, Fukui, R, Sato, T, "High Resolution Pressure Sensor Distributed Floor for Future Human-Robot Symbiosis Environments", *Proceedings of the International Conference on Intelligent Robots and Systems (IROS 2002)*

[12] Noguchi, H, Mori, T, Sato, T, "Construction of Accumulation System for Human Behavior Information in Room", *Proceedings of the International Conference on Intelligent Robots and Systems (IROS 2002)*

[13] Wasserman, K C, Blanchard, M, Bernadet, U, Manzolli, J M and Verschure, P F M J (2000), "Roboser: An Autonomous Interactive Composition System", In: I Zannos (Ed.), *Proceedings of the International Computer Music Conference (ICMC)*, pp. 531-534. San Francisco, CA, USA: The International Computer Music Association.

Human Behavior Interpretation System based on View and Motion-based Aspect Models

Masayuki Furukawa
Dept. of Adaptive Machine Systems, Grad.
Sch. of Eng. Osaka Univ.,
2-1 Yamada-oka, Suita 565-0871, Osaka, Japan

Yoshio Kanbara
Fuculty of Systems Eng., Wakayama Univ.,
930 Sakaedani, Wakayama 640-8510, Japan

Takashi Minato
Dept. of Adaptive Machine Systems, Grad.
Sch. of Eng. Osaka Univ.,
2-1 Yamada-oka, Suita 565-0871, Osaka, Japan

Hiroshi Ishiguro
Dept. of Adaptive Machine Systems, Grad.
Sch. of Eng. Osaka Univ.,
2-1 Yamada-oka, Suita 565-0871, Osaka, Japan

Abstract—This paper proposes an interpretation system for recognizing human motion behaviors and constructing the behavior rules called Behavior Grammar. The system recognizes human motion behaviors based on the gestures, locations, directions, and distances by using a distributed omnidirectional vision system (DOVS). The DOVS consisting of multiple omnidirectional cameras is a prototype of a perceptual information infrastructure for monitoring and recognizing the real world. The sequences of interpreted behaviors are represented as a behavior graph to extract behavior rules. This paper shows how the system realizes robust and real-time visual recognition based on View and Motion based Aspect Models (VAMBAM) and the resultant behavior graph.

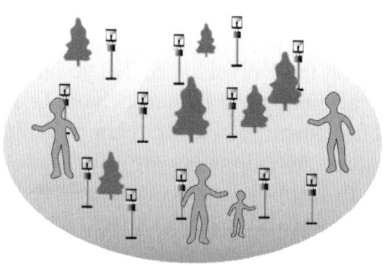

Fig. 1. Distributed omnidirectional vision system

I. INTRODUCTION

We have developed a *Distributed Omnidirectional Vision System* (DOVS) shown in Fig. 1 as a prototype of *Perceptual Information Infrastructure* (PII) [7] that supports activities of agents such as human and robots in an environment. In this system, it is important to recognize human behaviors.

In the past research, there are many recognition systems that recognize human gestures (for example [3], [12], [2], [5], [4], [1]). But these systems deal with mere gestures and don't use other information such as human positions and the time of day. To recognize human activities of daily living, the system needs the positions, moving distances of humans, moving directions, contexts of human behaviors, etc. This paper proposes an extended system that interprets human behaviors using not only gestures but also the positional and contextual information.

In our system, the recognition target is a human walking in a room monitored by a DOVS. Therefore, we need a location-free and rotation-free gesture recognition system. As such a system, we have proposed the gesture recognition system using *View and Motion-based Aspect Models* (VAMBAM) [8]. VAMBAM is an omnidirectional aspect model of human gestures.

The system proposed in this paper is realized by integrating VAMBAM and a real-time human tracking system developed so far [14], [15], and it acquires rich information on human behavior such as gestures, positions, moving distances, etc. Further, the system interprets the human behavior according to interpretation rules.

The rule is defined by the acquired detected gestures and other information. By observing human activities of daily life, we have acquired the rules in detail as much as possible.

The interpretation rules are defined by carefully observing given by a designer in a top-down way. However, our observations and the rich rules make the interpretation natural. We build a recognition system that can analyze human activities by this interpretation function.

To analyze human activities, we propose a Behavior Grammar consisting of human behavior rules. In the field of speech recognition, a language grammar plays an important rule for extracting meanings of sentences. Similarly, it is considered that the grammar of human behaviors is necessary for extracting the meanings and intentions of human behaviors in human behavior recognition.

To construct Behavior Grammar, we represent the sequences of behaviors as a behavior graph of which nodes are the behaviors interpreted by our system and arcs

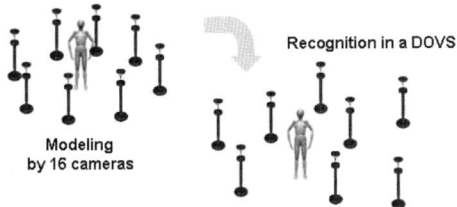

Fig. 2. Modeling and recognition

Fig. 3. 16 feature vectors

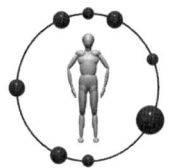

Fig. 4. Weighted 16 feature vectors

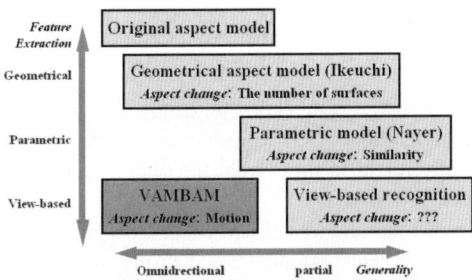

Fig. 5. Several approach for visual modeling

are the behavior transitions. From this behavior graph, we extract the behavior rules according to a method of inferring a grammar from a finite automaton [16]. In this paper, we present an example behavior graph as the first step.

In the following sections, section II and III explain the details of VAMBAM and the Real-time Human Tracking system. Next, section IV describes the method to interpret a human behavior. Finally, section V reports experimental results and discussions are given in section VI.

II. VAMBAM

A VAMBAM has two features as follows:
1) Omnidirectional model of a gesture
2) Motion-based segmentation of the gesture

As shown in Fig.2, the multiple cameras observe a human from various viewing point to build the VAMBAM. Then, the system recognizes human gestures in the distributed vision system that consists of multiple cameras distributed in the environment. A VAMBAM maintains the visual information in a form of view-based aspect model. In the modeling, a gesture is observed as image sequences. Then, the image sequence is transformed to a feature vector that represents an aspect of the gesture based on the motion information. Thus, VAMBAM maintains omnidirectional and motion-based gesture information. This gesture recognition not using 3-D models has not been developed so far [13].

A. Omnidirectional Aspect Model

The system has two phases for recognizing human gestures: modeling phase and recognition phase (see Fig.2). In the modeling phase, the system acquires VAMBAMs by using 16 cameras surrounding a human (we can change the number of cameras). The human makes several gestures and the system memorizes each gesture as an individual VAMBAM. In the recognition phase, the system consisting of multiple omnidirectional cameras finds a human, tracks, and then matches acquired images from several cameras with the stored VAMBAMs. This section explains the modeling phase. An ideal model for object recognition is to have all visual information from all directions like a complete 3-D model. If there are an infinite number of cameras surrounding an object, we can directly acquire such an ideal view-based model. VAMBAM is an approximation of such an ideal model. A problem is how to reduce the number of cameras surrounding a modeling target in the modeling phase. It depends on complexity of the target. Since the purpose of the DOVS is to recognize coarse human gestures, we have decided the number as 16 based on a preparatory experiment.

By using 16 cameras, the system acquires 16 view sequences and transforms them to 16 feature vectors. In Fig.3, the sphere shows a feature vector. Here, we suppose all of the cameras are located at the same height. Of course, we can extend the camera arrangement, but it is not needed for our current gesture recognition.

All 16 cameras do not have the same priority. For recognizing a particular gesture, for example breathing gesture, the front camera is more important than the side cameras. Therefore, each feature vector has a weight estimated from the motion information. If the camera observes more moving pixels by background subtraction than the others, it has a bigger weight. Fig.4 shows the conceptual figure of the weight vectors. The weights define aspect changes

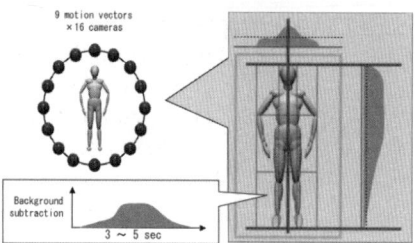

Fig. 6. Acquisition of a feature vector

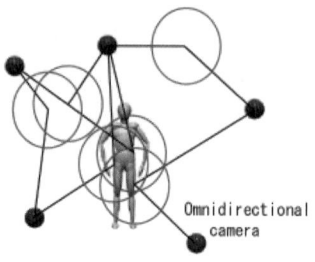

Fig. 7. *N*-ocular stereo by a DOVS

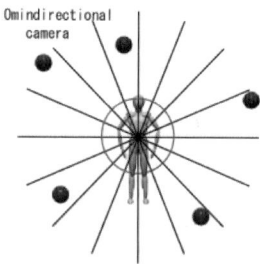

Fig. 8. Matching with a VAMBAM

in VAMBAM. Fig.4 shows 9 feature vectors that have positive weights, and it represents 9 aspects. The system efficiently refers to the weight for searching the target in the recognition phase.

Here, let us state VAMBAM as an aspect model. Aspect models represent omnidirectional visual information surrounding the target. Koenderink [11] proposed the original concept based on 3-D structure of the object. Ikeuchi [6] modified for passive vision systems and defined the aspect change by the number of observable surfaces of the object. In their works, the important point is that the aspect model maintains omnidirectional visual information for the generality. On the other hand, view/appearance-based methods and parametric methods are also a kind of aspect model. The view-based method defines the aspect change as a difference among views by template matching or feature-based comparison. The parametric method defines the aspect change as a distance in the principle component parameter space. They, however, do not maintain the omnidirectional visual information in the previous works. VAMBAM is a view-based approach as described in the next subsection, but it is also an aspect model that maintains omnidirectional visual information. Fig.5 shows relationships among the existing methods and our method.

B. Motion-based Segmentation

Another important feature of VAMBAM is motion-based segmentation of visual information. As mentioned before, the motion-based segmentation reduces the amount of memory and enables us to utilize omnidirectional view-based aspect models for practical systems.

In Fig.3, each feature vector is given as shown in Fig.6. The camera finds a moving object by background subtraction and tracks it as a candidate of a human. Then, it detects the center of gravity by making histograms along vertical and horizontal axes. The histograms give the height and width of the human. Based on the parameters, the system defines 9 regions that cover the human as shown in the left of Fig.6, tracks the human fitting the regions. In each region, the system recodes the moving pixels detected by background subtraction and acquires 9 motion vectors in 3–5 sec. observation. Here, the observation length is adjusted according to the gesture. The acquired 9 motion vectors define a feature vector as represented with a sphere in Fig.6. As mentioned before, several feature vectors do not contain motion information because of the viewing direction. By referring to the average and variance of the 9 motion vectors, the system assigns a weight to the feature vector.

The motion-based segmentation by background subtraction is one of the most robust and stable feature extractions and many practical systems use background subtraction. If we consider that "motion" represents relationships between static objects and events and gives meaning to the static objects, it is natural to build a recognition system based on the motion-based feature extraction.

Further, this representation of a gesture is compact. Suppose to record a 5 sec. gesture. A motion vector consists of integers and 16 feature vectors have integers and the quarter is selectively memorized by the weights. If we represent an integer with 8 bits, the necessary memory size will be 4000 bits and it is not large for recognizing a dozen of gestures in real time.

III. Two Techniques to Support VAMBAM

A. Real-time Tracking of Walking Human

For realizing the location-free and direction-free gesture recognition, the system needs two more functions: one is tracking of walking humans in real time, the other is a robust matching method with gesture models.

By utilizing redundant visual information of the DOVS, we can build a robust and real-time multiple camera stereo system. This stereo, called N-ocular stereo, is an extension of trinocular stereo [10]. The basic process of N-ocular stereo is as follows:

1) Detect moving regions on the images by background subtraction.
2) For all combination of two moving regions on different omnidirectional cameras, estimate the distance to the region and apply a circle as a human model (the circles in Fig.7).
3) Check the overlapping of the circles and determine human locations.

As described above, we implement the N-ocular stereo as a model-based stereo method. Generally, the model-based stereo is more stable than the feature-based stereo in the case where the target is known. However, appearance of a human body frequently deforms on the image. Therefore, we have approximated the human body with a circle (the diameter is 60 cm) on the horizontal plane that is parallel to the floor. When three or more circles overlap each other, the system decides a human exists in the center of gravity of the circles [14], [15].

This N-ocular stereo for tracking humans in a DOVS is executed in real-time with a standard computer and it robustly tracks up to 3 humans simultaneously with four omnidirectional cameras. The reason of the robustness is in the redundant observation by multiple omnidirectional vision sensors. The method based on background subtraction is, generally speaking, noisy against change of lighting condition. However, the multiple observations from different viewing angles suppress the noise.

Another issue in the N-ocular stereo is precise localization and identification of the omnidirectional cameras. For this issue, we have already developed a method that automatically performs the localization and identification [9].

B. 3-D Dynamic Programming for Matching

The recognition system needs to handle multiple gesture models simultaneously. Therefore, a robust and real-time matching technique is required.

The basic idea is to use conventional continuous dynamic programming (CDP). We have extended this CDP to interpolate discrete observation by 16 cameras, which is called *Spatio-Temporal Continuous Dynamic Programming* (STCDP, see Fig.9). In the recognition phase, a stored model does not exactly match to the observations by the DOVS as shown in Fig.8. The STCDP searches the best match by swiveling the model within 360/16 degrees. In addition to this, the STCDP finds the direction

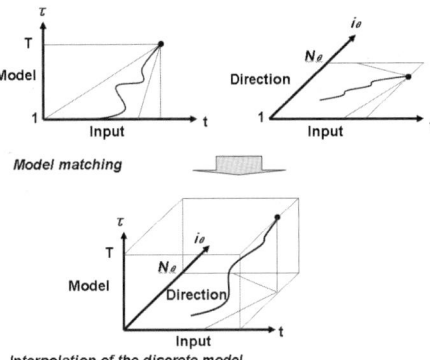

Fig. 9. STCDP

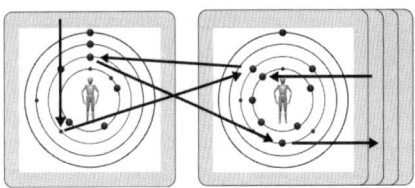

Fig. 10. Attention control in the matching

of target by rotating the models 360 degrees. Thus, this 3-D matching realizes the rotation-free gesture recognition by using the discrete model.

In the current implementation, the STCDP sequentially matches the input with 10 stored models. However, the STCDP needs to be modified to handle a large number of models. The weights assigned to the VAMBAM reduce the computational time drastically. By referring to the weights, the STCDP can dynamically select possible feature vectors and ignore redundant search for impossible models in the early stage as shown in Fig.10. In other words, we can see attention control in the recognition phase using VAMBAM.

IV. INTERPRETATION OF HUMAN BEHAVIORS

A. Interpretation rules

To interpret human behaviors based on the sensory data, we define interpretation rules. The rules are classified into the following six types.

- Position interpretation rules
- Moving distance interpretation rules
- Gesture constraint interpretation rules
- Human action interpretation rules
- Action constraint interpretation rules
- Behavior interpretation rules

The position interpretation rules (position IRs) define relative human positions to objects such as "beside of chair". The moving distance IRs define human motions such as "shortly", "from a desk to a bed" and so on. The gesture constraint IRs define constraints among continuous human gestures. For example, we do not cook on a bed and we must take a seat before rising from the seat. These three types of IRs are applied before the human action IRs.

The human action IRs define human actions based on the following action information in a form of if-then rule.

- Gesture
- Position
- Moving distance
- Moving direction
- Human appearance
- Human disappearance
- Number of people close to the target human

For example, if the recognized gesture and detected position are "stand up" and "chair" respectively, the system interprets them as the human action is "rise from a chair". The action constraint IRs define constraints among continuous actions like the gesture constraint IRs. The behavior IRs interpret human behaviors according to the recognized actions and its duration. For example, if a person does not move for a long time, the system interprets the behavior as "he/she is thoughtful."

B. Procedure for Interpreting Human Behaviors

The procedures for interpreting human behaviors are as follows:

1) Collect data for the six features, i.e. gesture, position, moving distance, moving direction, appearance and disappearance, by using the gesture recognition system that tracks humans in real time and recognizes their gestures with VAMBAMs. These data are interpreted with the gesture, position and moving distance IRs.
2) Acquire candidates of human actions based on the human action IRs. Here, we have rather loose tolerances in the matching with the six features by adjusting conditions of the if-then rule and pick up candidates according to the matching scores.
3) Select a human action that has the highest score from the candidates that satisfy the action constraints IRs.
4) Compute the duration of the acquired action.
5) Acquire an interpretation of human behaviors according to the behavior IRs.

V. EXPERIMENT

A. Experimental Conditions

We have evaluated the interpretation system in a laboratory environment shown in Fig.11. The circles in this figure show 16 omnidirectional sensors distributed in the laboratory room. Before experiments, we built VAMBAMs and defined the interpretation rules by carefully observing daily human activities in this environment. The number

Fig. 11. Experimental environment

TABLE I

RECOGNITION RATIO

Episode	Time [sec]	Interpretation times	Error	Correct ratio [%]
1	212	1784	242	86
2	223	1401	89	94
3	201	1740	146	92
4	198	1467	114	92
5	221	1551	92	94
6	210	1448	122	92
7	214	1754	129	93
8	204	1567	91	94
Total	1683	12712	1025	92

of VAMBAMs (i.e. the number of gestures) is 13 and behavior IRs is 1551.

They are "take a seat", "walking towards a door", for example. In order to verify the system performance, we have prepared several episodes. All episodes consist of an almost identical behavior sequence, but the subjects (recognition targets) are different among the episodes. The duration of each episode is about 210 seconds. The system interprets human behaviors ten times per second.

B. Experimental Results

Table I shows recognition ratios. To calculate the recognition ratios, we gave the ground truth by ourselves. Each episode includes the behaviors such as "exiting a room" and "entering a room". The system did not interpret when a subject disappeared from the laboratory room. There is little or no difference of the recognition ratios among subjects. Fig.12 shows a part of resultant interpretations. The average of the recognition ratios was 92%. We consider this is sufficient as a general system for interpreting human behaviors.

Further, we made the behavior graph. An example of the graph is shown in Fig.13. The nodes of this graph are the interpreted behaviors and the arrows are their transitions. The thickness of each arrow is proportional

Standing beside a bookshelf
Taking a book from a bookshelf
Standing beside a bookshelf
Walking beside a bookshelf
Walking beside the desk A
Standing beside a sofa
Sitting
Still Sitting
Standing
Walking beside a sofa
Walking beside the desk B
Walking beside a meeting room
Walking toward a door
Standing beside the desk C
Going out by a door

Fig. 12. Example of interpretation

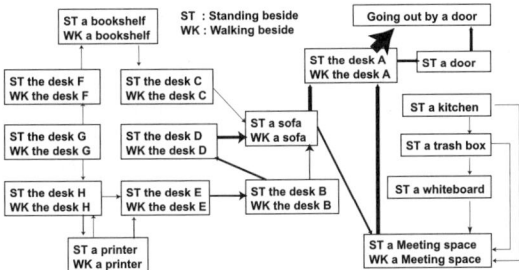

Fig. 13. An example of behavior graph

to the frequency of the corresponding transition. In this case, the topology of the nods shows a structure of human behaviors according to the environmental structure. We consider it is possible to extract characteristics of human behaviors depends on the environment as a behavior grammar by using a method for inferring a grammar from a finite automaton [16]. This grammar extraction is our ongoing study.

VI. CONCLUSION

This paper has proposed an interpretation system for recognizing human motion behaviors and describing the behaviors. By defining the interpretation rules and applying them to the real time gesture recognition system, we could realize the robust behavior-interpretation system.

To support human daily activities, the Perceptual Information Infrastructure requires a robust recognition function. We consider it is important to find general patterns in human behaviors. They are expected to be formalized as a behavior grammar. Our interpretation system will be powerful tool to construct the behavior grammar.

VII. REFERENCES

[1] A. Bobick and A. Wilson. A state-based technique for the summarization and recognition of gesture. In *Proc. ICCV*, pages 382–388, 1995.

[2] L. W. Campbell, D. A. Becker, A. Azarbayejani, A. Bobick, and A. Pentland. Invariant features for 3-d gesture recognition. In *Proc. AFGR*, pages 157–162, 1996.

[3] T. Darrell and A. Pentland. Space-time gestures. In *Proc. CVPR*, pages 335–340, 1993.

[4] J. Davis and M. Shah. Recognizing hand gestures. In *Proc. ECCV*, pages 331–340, 1994.

[5] Y. Guo, G. Xu, and S. Tsuji. Understanding human motion patterns. In *Proc. ICPR*, pages 325–329, 1994.

[6] K. Ikeuchi. Automatic generation of object recognition program. *Proc. of IEEE*, 71(8), 1988.

[7] H. Ishiguro. Distributed vision system: A perceptual information infrastructure for robot navigation. In *Proc. IJCAI*, pages 36–41, 1997.

[8] H. Ishiguro and T. Nishimura. Vambam: View and motion-based aspect models for distributed omni-directional vision systems. In *Proc. IJCAI*, pages 1375–1380, 2001.

[9] K. Kato, H. Ishiguro, and M. Barth. Identifying and localizing robots in a multi-robot system environment. In *Proc. IROS*, 1999.

[10] Y. Kitamura and M. Yachida. Three-dimensional data acquisition by trinocular vision. *Advanced Robotics*, 4(1):29–42, 1990.

[11] J. J. Koenderink and A. J. van Doorn. The internal representation of sold shape with respect to vision. *Biological Cybernetics*, 32:211–216, 1797.

[12] S. Nagaya, S. Seki, and R. Oka. A theoretical consideration of pattern space trajectory for gesture spotting recognition. In *Proc. AFGR*, pages 72–77, 1996.

[13] V. I. Pavlovic, R. Sharma, and T. S. Huang. Visual interpretation of hand gestures for human-computer interaction: A review. *IEEE Trans. PAMI*, 19(7):677–695, 1997.

[14] T. Sogo, H. Ishiguro, and M. Trivedi. Real-time target localization and tracking by n-ocular stereo. In *IEEE Workshop on Omni. Vision*, pages 153–160, 2000.

[15] T. Sogo, H. Ishiguro, and M. Trivedi. *Panoramic Vision: Sensors, Theory and Applications*. Springer-Verlag, 2001.

[16] F. Thollard, P. Dupont, and C. de la Higuera. Probabilistic dfa inference using kullback-leibler divergence and minimality. In *Proc. ICML*, pages 975–982, 2000.

Collaborative Capturing of Experiences with Ubiquitous Sensors and Communication Robots

Norihiro Hagita[1,2], Kiyoshi Kogure[2], Kenji Mase[1,2], and Yasuyuki Sumi[1]

[1]ATR Media Information Science Laboratories
[2]ATR Intelligent Robotics and Communication Laboratories
E-mail: hagita@atr.co.jp

Abstract - We propose an intelligent environment for "co-experience communication" consisting of stationary and wearable sensors, and communication robots. The environment captures many events collaboratively from multiple viewpoints using a combination of stationary sensors that track people's behaviors and interactions, and wearable sensors that observe the viewpoints and areas of interest of individual persons. Humanoid communication robots also play a role as assistants to help visitors pay attention to significant events. First of all, the basic framework of the environment is described. Next, our developing exhibition room is introduced as an example of an intelligent environment.

I. INTRODUCTION

The history of media usage strongly suggests that we human beings have an inherent motivation to disseminate our feelings and experiences to each other using any and all kinds of media. The Internet, a recent example, allows person-to-person, person-to-community, and community-to-community communication. It is natural to ask: what will the next generation medium be and what forms of interaction will it allow? We predict that a next-generation ubiquitous network is going to allow us to communicate with anybody, anywhere, anytime, in any way. According to Mark Weiser, ubiquitous computing is a method of enhancing computer use by making many computers available throughout the physical environment, but making them effectively invisible to users [1]. The recent boom in ubiquitous interfaces increases the possibility of new business coming from our daily experience of interaction, such as casual conversations, casual attitudes, missed events, incidental/accidental events in a hospital, etc.

Let's consider an exhibition as a typical interaction situation. It generates numerous conversations between presenters and visitors. Such a huge collection of interactions is a valuable information source for both the sponsors and the visitors. However, typical outcomes of the event are usually summarized data, such as the number of visitors, the number of sponsors, and a stochastic summarization of questionnaire sheets. If the interactions and behaviors of individuals could be observed by using stationary sensors (stationary cameras, microphones, etc.) and wearable sensors (portable cameras and microphones, physiological sensors, etc.), we might gain more significant information on the exhibition's effects and the visitors' responses. That is, the room would become an intelligent environment where individual events are collected using these sensors, then visualized and exchanged. Assistants would also play important roles to introduce visitors to a specified booth and help them to gain additional information besides that given by presenters. In previous related work, we have already proposed the notion of facilitating encounters and knowledge sharing among people who have shared interests and experiences in conferences using a hand-held guidance system, exhibit displays, an information kiosk, web-based off-site services and a visualized community network [2]. Our recent study [3] has shown that humanoid communication robots, such as the "robovie" system, help humans pay attention to a specified direction, and create an opportunity to communicate, when they speak and provide humans with greetings, and playful words.

The present paper proposes an intelligent environment that combines stationary and wearable sensors with communication robots. First, we describe the basic framework of our project The project focuses on a combination of stationary sensors that capture people's interactions, and wearable sensors that observe personal interests and experiences in place of hand-held guidance systems, a community ware system that collects, visualize, and exchanges records of events in the community, and humanoid communication robots that serve as assistants in the environment. Next, we describe a prototype exhibition room with these sensors, as an example of an intelligent environment.

II. BASIC FRAMEWORK

A. Basic Idea

Figure 1 illustrates the concept of an intelligent environment as used in our project. It includes stationary sensors (stationary cameras, microphones, LED ID sensors) and wearable sensors (portable cameras and microphones, biometric sensors, LED ID sensors), and communication robots (autonomous humanoid robots [3]

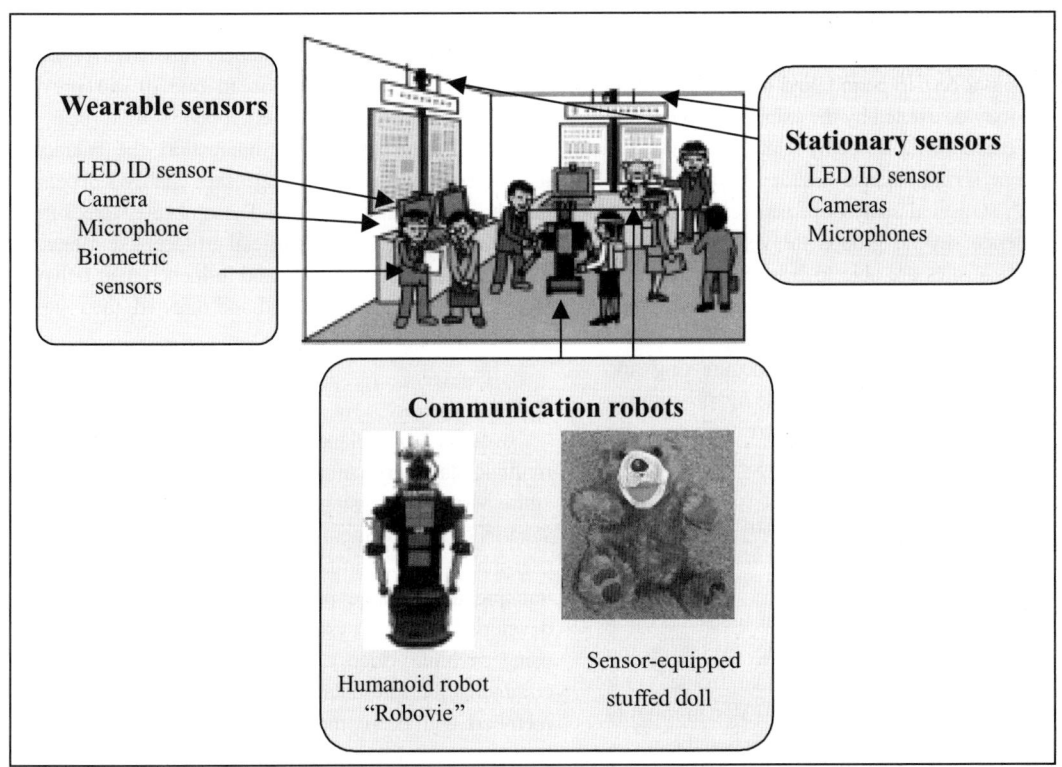

Figure 1 Concept of the intelligent environment being developed at ATR.

and sensor-equipped stuffed dolls [12]).

The first issue to consider in building an intelligent environment is to constantly capture the interaction and behavior of people in the room. The hand-held guidance system in C-MAP (Context-aware Mobile Assistant Project) [2] is only able to automatically capture the initial interaction to each exhibit, since it still requires the manual entry of experiences, areas of interest and emotional responses according to guidance menus. In order to solve this problem, we adopt a combination of stationary and wearable sensors in place of the hand-held guidance system.

These sensors allow us to effortlessly capture day-to-day experiences. The combination of stationary cameras and LED ID sensors is utilized to identify each person and the location in the room. That is, stationary and portable cameras collaboratively detect LED ID tags to identify persons within their field of vision as well as track people who pass through. Audio and video streams are collected using stationary cameras and microphones, as well as portable cameras and microphones. The combination of stationary cameras and microphones, portable cameras, headset microphones and biometric sensors is utilized to detect areas of personal interest, temporal and spatial situations, and emotional status. We, therefore, assume that each person wears an LED ID sensor, a head-mounted camera, a microphone and biometric sensors for the environment to capture personal information. Each object, such as a PC monitor and a communication robot, also wears an LED ID sensor. Moreover, the communication robots, called "Robovie", can capture the sensory data using their own omni-directional camera, stereo cameras, microphone, tactile, and ultrasonic sensors autonomously[3]. A sensor-equipped stuffed doll can also capture events by using its own camera, microphone, bend sensors, accelerometers, tactile sensors, proximity sensors, and temperature sensors [12].

The second issue is to select the significant parts of people's interaction and behavior out of a huge collection of sensor data. It is impossible to index these manually, since we estimate that more than 100 GB of data per day is collected in the case of our ubiquitous sensor room. Instead of indexing data manually, we focus on the ID tags. When stationary and/or portable cameras detect multiple ID tags, the persons corresponding to the ID numbers may be communicating with each other during the period. The ID tags may thus be used to automatically select that period as a significant part of the corresponding persons' interaction and behavior out of the collected audio and video streams. As a result, we may automatically index an interaction database with the information that these persons were talking during that period.

The third is to share personal experiences among people. Automatic indexing of personal interaction data will facilitate knowledge communications. For example, when ID tags detect that exhibitor E1 interacted with visitor V1 during a certain period, and biometric sensors detect that V1 may be excited by the exhibition of E1, and the interest of exhibitor E2 is deeply related to that of E1, the environment should induce visitor V1 to visit the booth of E2. A statistical analysis of people's interaction data may generate several groups for visitors' areas of interest. These statistics are able to help people navigate exhibition routes, and introduce exhibitors and other visitors who share a new visitor's interest.

The fourth issue is to facilitate encountering and introducing visitors. Since personal guide agents running on PalmGuide and AgentSalon [9] are effective for facilitating these types of interaction in C-MAP [2], we will also use personal guide agents to facilitate communication among people. We also focus on the possibility of communication robots as assistants in the room as well as personal guide agents.

In an exhibition or meeting situation, assistants often help visitors pay attention to specific events in an exhibition, and suggest an appropriate route. Our autonomous humanoid robot, "Robovie" [3], can roughly recognize human faces using an omni-directional camera, maintain eye-contact with a specific person, recognize about fifty Japanese or English words, and speech-synthesize about three hundred sentences. It has sensor-action units for representing friendly interactive behaviors such as greeting, nodding, kissing, singing, hugging, etc. For example, 'joint attention', in which a robot gazes at a human and then points at an object is a significant behavior for a robot to serve as an assistant in the room. In joint attention, humans tend to glance in the direction of the object pointed at by the robots as shown in Fig. 2. This behavior may help visitors pay attention to specific objects or events. Various arm movements will help visitors decide the appropriate route or the path to a booth of interest, as determined using the shared interest data.

Fig. 2 An example of the robot's 'joint attention' behaviors

Stuffed toys have a strong influence on human behavior, simply by their intimate appearance, although they are not active like a humanoid robot. In one of our previous studies [12] with a sensor-equipped stuffed bear that generates sound with tactile interaction, the balance of utterances between conversing users changed depending on who was holding the bear. These observations support our belief that expressive artificial artifacts like robots and toys can enter people's lives and influence their behavior. When appropriately controlled and directed, these can be good assistants for humans.

B. Related works

In current surveillance-based ubiquitous environments, multiple stationary cameras allow a representation of the entire visual field using the information from the video streams of these cameras. The Smart Room[4] provides a user-interface to a virtual environment using vision-based tracking and a large projection screen. Chiu et al. [5] developed a capture system that records meeting activities using multiple video cameras and collects memos and comments on presentation slides written by audience members with pen-computers. The system also provides a browsing tool for integrating, recording, and collecting data. Similar systems have been explored in educational environments, such as Classroom2000 [6] or eClass, and for ubiquitous computing environments in the home [7].

Microsoft's Easy Living [8] system is a recent project to build intelligent environments that facilitate unencumbered person-to-person interaction, as well as interaction with computers and appliances. It utilizes vision-tracking of user interaction, multiple sensors, a geometric model of the world, device-independent communication, data protocols, etc. It also focuses on a shared representation of the world.

At ATR, we have already developed C-MAP, which provides users with personalized guidance according to temporal and spatial situations as well as individual areas of interest [2]. We proposed the notion of facilitating encounters and sharing knowledge among people having a common interest and common experiences at a conference using a hand-held guidance system, exhibit displays, an information kiosk, web-based off-site services and a visualized community network [2]. The ATR Semantic Map is a visual interface for visually exploring community information accessible via the Internet and on information kiosks located in exhibition sites. AgentSalon [9] also facilitates face-to-face knowledge exchange and discussion between users by inviting them to chat via their personal agents, which maintain personalized data about their areas of interest and experiences. Here, in order to make this system easier for users, we focus on a combination of stationary and wearable sensors in place of a hand-held guidance system such as PalmGuide. Our basic framework for intelligent environments at ATR comes from these works.

Wearable sensors are also important for capturing personal interests and emotions. Many kinds of wearable sensors have been developed. Intelligent wearable camera systems have been developed that allow us to effortlessly capture day-to-day experiences [10, 11]. Also, RFID (radio frequency identification) tags are promising wearable devices for detecting the location of each person.

These sensors are utilized to analyze different opinions, areas of interest and exhibitions of particular interest to individual visitors at an exhibition. The collection of sensor data may provide important feedback for improving presentations, display and booth design, and for more effective tour planning for visitors.

III. INTELLIGENT ENVIRONMENT at ATR

Based on the basic framework in Section 2, we have been building an intelligent environment in an exhibition booth at our labs. The entire system consists of stationary and portable (or wearable) data capturing clients, captured data servers, application servers, and communication robots, as shown in Fig. 3. A stationary data capture client consists of a desktop PC with an Ethernet connection to a captured data server and an audio sensor, a visual sensor, and an

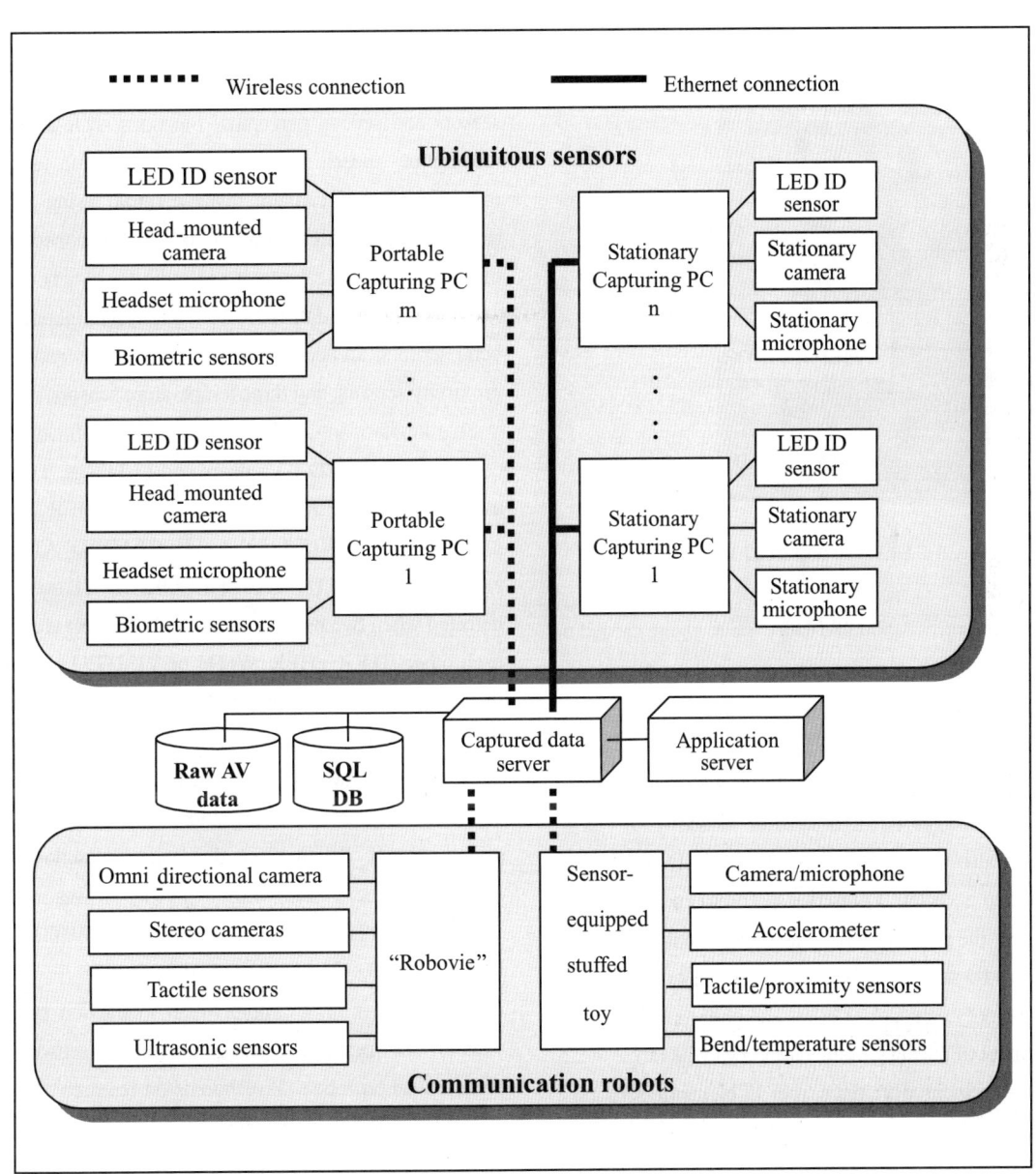

Fig. 3 The entire sensor system of the intelligent environment at ATR

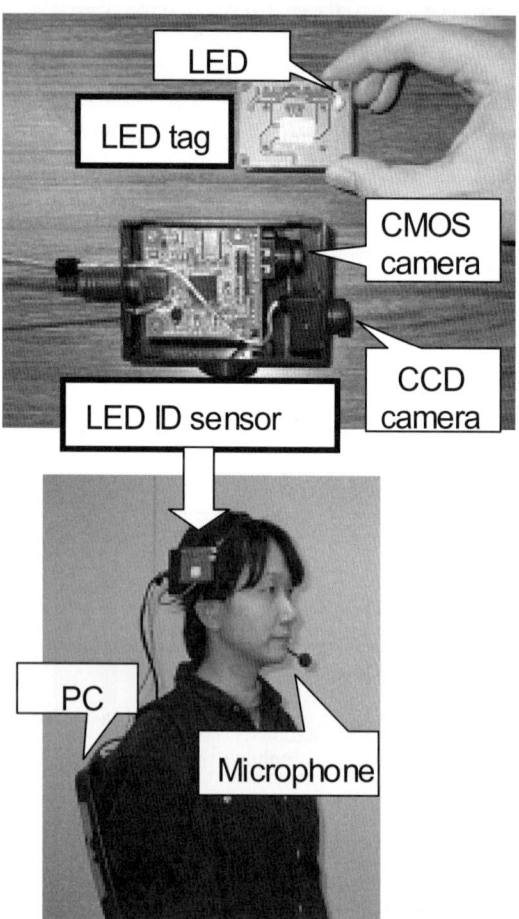

Fig. 4 LED tag and LED ID sensor

Fig. 5 An example of an interaction scene in the exhibition booth at ATR

LED ID sensor. A portable data capture client, carried or worn by a person, consists of a lightweight notebook PC with an IEEE802.11a wireless connection to a captured data server (56 Mbps), and an audio sensor (headset microphone), a visual sensor (CCD camera) and an LED ID sensor (CMOS camera), as shown in Fig. 4. Figure 4 also shows an infrared LED (LED tag) that is attached to a visitor, a presenter, a communication robot, or each of the posters and displays at the exhibition booth. The LED ID sensors make it possible to detect LED tags, that is identify persons or objects and observe the X and Y coordinates of the tag, within their view in the range of 2.5 m. A few persons wore three types of biometric sensors: a pulse physiology sensor, a skin conductance sensor, and a temperature sensor on their finger. The autonomous humanoid robot, Robovie, has a CPU based on RT-Linux, weighs 50 kg, and is with 114 cm tall. It has ultrasonic sensors and several perceptual functions such as speech recognition, speech synthesis, person detection using an omni-directional camera, gaze-control using stereo cameras, and tactile sensation. It allows connection to other servers or the Internet via radio-LAN. The sensor –equipped stuffed doll has a camera, a microphone, four bend sensors, an accelerometer, five tactile sensors, two proximity sensors, and three temperature sensors.

Five kinds of interaction primitives are defined by the combination of LED ID sensors and LED tags. They are as follows: TALKED_WITH, TOGETHER_WITH, LOOKED_AT, VISITED, and STARED_AT. For example, TALKED_WITH or LOOKED_AT events are detected when two users or a user and an object are facing each other. TOGETHER_WITH or VISITED events are detected when two users or a user and an object are detected by the same LED ID sensor over a preset interval. STARED_AT events are detected when a user is capturing another user or an object for at least twice the preset interval. When more than one interaction event occurred, the following priority was used: TALKED_WITH > TOGETHER_WITH > LOOKED_AT > VISITED > STARED_AT.

Figure 5 shows an example of an interaction between visitors and presenters using ubiquitous sensors and communication robots. Five booths on research activities were set up. 16 presenters, 63 visitors and a communication robot utilized the wearable sensors. A total of 540 GB of audio-visual input data and over 380,000 ID sensing data were obtained from the

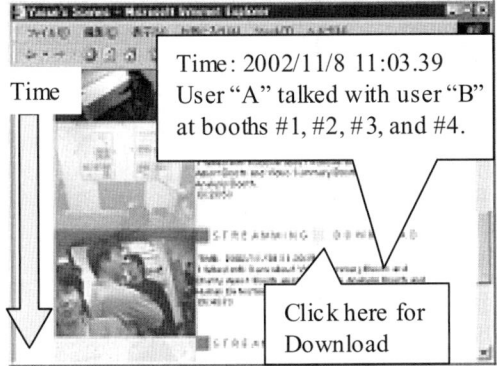

Fig. 6 HTML based video summary for a user automatically generated from interaction data.

experiments of a two-day exhibition. The output was automatically annotated by the interaction events defined above. Each thumbnail corresponds to a video clip and was highlighted or faded depending on the interaction priority. Robovie also wore an LED ID sensor and captured audio-visual data just as humans did. Figure 6 shows an example of the HTML output of a video scene for user "A".

IV. CONCLUSION

This paper proposed an intelligent environment consisting of ubiquitous sensors and communication robots. The basic framework of the project was first described. The basic idea comes from our works on C-MAP (Context-aware Mobile Assistant Project) for collecting, visualizing, and exchanging users' events in a community, and the friendly interaction capability of autonomous humanoid robots. We have chosen an exhibition room at research laboratories as an example of an intelligent environment. The aim of the project is to build an intelligent environment in which communication robots serve visitors as assistants, and where all of the visitors' experiences can be captured using ubiquitous sensors, and shared with other visitors and exhibitors. We introduced the entire sensor system of an exhibition room and the user's video summary that automatically generated from the interaction data.

V. ACKNOWLEDGEMENT

We would like to acknowledge Dr. Akira Utsumi, Dr. Sidney Fles, Dr. Tetsuya Matsuguchi, Mr. Sadanori Ito, Ms. Noriko Suzuki, Mr. Atsushi Nakahara, Mr. Tetsushi Yamamoto and Ms. Keiko Nakao for discussion and implementation of the exhibition. We are indebted to Dr. Hiroshi Ishiguro, Dr. Takahiro Miyashita, and Mr. Takayuki Kanda for giving a few snaps and comments on communication robots. We are also grateful to members of our labs for several discussions. This research was supported by the Telecommunications Advancement Organization of Japan.

VI. REFERENCES

[1] Weiser, Mark, "The Computer for the Twenty-First Century," *Scientific American*, pp. 94-104, September 1991.

[2] Y. Sumi, and K. Mase, "Supporting the awareness of shared interests and experiences in communities," *International Journal of Human-Computer Studies*, Vol.56, No.1, pp.127-146, 2002.

[3] T. Kanda, H. Ishiguro, M. Imai, T. Ono and K. Mase, "A constructive approach for developing interactive humanoid robots," *IEEE/RSJ International Conference on Intelligent Robots and Systems*, Lausane, 2002.

[4] A. Pentland, "Smart rooms," *Scientific American*, pp.54-62, April 1996.

[5] P. Chiu, A Kapuskar, S. Reitmeier, and L. Wilcox, "Room with a rear view: Meeting capture in a multimedia conferenceroom," *IEEE Multimedia*, Vol. 7, No.4, 2000.

[6] G. D. Abowd, "Classroom 2000: An experiment with the instrumentation of a living educational environment," *IBM Systems Journal*, Vol.38, No.4, pp.508-530, 1999.

[7] C. D. Kidd, R. Orr, G. D. Abowd, C. G. Atkeson, I. A. Essa, B. MacIntyre, E. Mynatt, T.-E. Startner, and W. Newstetter, "The aware home: A living laboratory for ubiquitous computing research," *In Proceedings of CoBuild'99 (Springer LNCS1670)*, pp.190-197, 1999.

[8] B. Brumitt, B. Meyers, J. Krumm, A. Kern and S. Shafer, "EasyLiving: Technologies for Intelligent Environments," in Peter Thomas and Hans W. Gellersen editors, *Handheld and Ubiquitous Computing, 2000, Lecture Notes in Computer Science*, Vol. 1927, pp.12-29, Springer, 2000.

[9] Y. Sumi, and K. Mase, "AgentSalon: facilitating face-to-face knowledge exchange through conversations among personal agents," *Proc. of Agents 2001*, pp.393-400, 2001.

[10] S. Mann, "Humanistic intelligence: wearcomp as a new framework for intelligent signal processing," *Proc. of the IEEE*, Vol. 86, No.11, pp.2123-2151, 1998.

[11] T. Kawamura, Y. Kono and M. Kidode, "A novel video retrieval method to support a user's recollection of past events for wearable information playing", *In Proc. of the Second IEEE Pacific-Rim Conference on Multimedia*, pp.24-31, 2001.

[12] T. Yonezawa and K. Mase: Musically Expressive Doll in Face-to-face Communication, in Proc. on International Conference on Multimodal Interfaces, pp. 417-422, Pittsburgh, Oct. 2002.

Self-Identification of Distributed Intelligent Networked Device in Intelligent Space

Hideki HASHIMOTO[1,2], Joo-Ho LEE[1] and Noriaki ANDO[1]

[1]*Institute of Industrial Science, University of Tokyo*
Komaba 4-6-1, Meguro-ku, Tokyo, Japan 153-8505
[2]*PRESTO, JST*
{h.hashimoto,leejooho,n-ando}@ieee.org

Abstract— The Intelligent Space is a space where we can easily interact with computers and robots, and get useful service from them. To achieve such a space, Distributed Intelligent Networked Device(DIND) has been proposed. Many DINDs are installed in a space and they cooperate each other to make the space, Intelligent Space. In this paper, optimal DIND placement, self-calibration of DIND and handover protocol for cooperation among DINDs will be described.

I. INTRODUCTION

'Intelligent Space' has been proposed by Hashimoto lab. in university of Tokyo[1]. Intelligent Space is an environmental system, which is able to support human in informative and physical ways. Most of intelligent system interacts with human in a passive space, but in Intelligent Space, a space, which contains human and artificial systems, is an intelligent system itself. Human and artificial systems become clients of Intelligent Space and simultaneously the artificial systems become agents of Intelligent Space. Since the whole space is an intelligent system, Intelligent Space, a spatial system, is able to monitor and to provide services to clients easily. Specific tasks, which cannot be achieved only by Intelligent Space, are accomplished by utilizing its clients. For examples, Intelligent Space utilizes computer monitors to provide information to the human, and robots are utilized to provide physical services to the human as physical agents. Robot as well as human is supported by Intelligent Space if it is necessary. When a robot is lacking of sensors to navigate around in Intelligent Space, the robot is treated as a client of Intelligent Space and lacking information is provided to the robot by Intelligent Space.

The ultimate goal of Intelligent Space project is to accomplish an environment that comprehends human's intentions and satisfies them. It seems that such a system is hardly achieved, since a huge number of functions should be prepared and human-like intelligence is required. Even though such a complete system cannot be achieved immediately, it is convinced that a useful system can be achieved with current technology by proper system integration.

The purpose of this paper is to propose an architecture of Intelligent Space based on distributed intelligent sensors. Since Intelligent Space is a spatial system, its size and shape are not determined and restricted. Distributed intelligent networked device (DIND) is developed as a proper element of building such kinds of system[2], [3]. Ordinary space is changed into Intelligent Space by installing many DINDs aroud the space. However, when DINDs are installed in a space, it should be considered that where to place DINDs and how many DINDs are required. After installing DINDs, DINDs should be calibrated and their roles should be defined. Protocol, which makes DINDs cooperate is also required.

It will be described that a space is watched by many distributed intelligent sensors and the sensors cooperate with each other in Intelligent Space. By cooperation of intelligent sensors, the service of Intelligent Space is provided in a space globally and seamlessly. An intelligent sensor in Intelligent Space watches a fixed local area and it provides position based data continuously for high-level and complex functions.

II. SELF-IDENTIFICATION

In the Intelligent Space, the most basic element is DIND [3]. To make Intelligent Space with multiple DINDs, following functions are required.

A. Optimal placement of DIND

Since current DIND is based on sensor of CCD camera, its placement becomes important. If the DINDs are located badly, it may cause big error in localization and proper service may not be afforded. DIND placement based on features of CCD camera and functional requirements of Intelligent Space should be considered.

B. Calibration of DIND and role allocation

From optimal placement of DINDs, roughly DINDs can be located around a space. However to perform proper functions, DINDs should be calibrated carefully and their role assignment depending on position should be performed. After this process, DINDs are able to work as elements of Intelligent Space.

C. Handover protocol

Since Intelligent Space is a spatial system and its size and shape are not fixed. To afford service with DINDs in various spaces as Intelligent Space, DIND should cooperate each other so that the whole space becomes a system. Seamless and continuous service from Intelligent Space only become possible, if there is a protocol that can make DINDs cooperate each other.

Above three functions are necessary ingredients for achieving Intelligent Space in real environment with

DINDs.

III. Determination of Optimal Placement of DINDs

In general, optimal arrangement of cameras depends mainly on tasks. It is difficult to consider all possible cases because the number of system design parameters becomes huge according to features of cameras used in the system, and the size and shape of target workspace. Furthermore, it is not so easy to deal with three or more cameras at a time. Thus, in this paper, only the case of two cameras is considered and arrangement of more than two cameras is left for future work. A pinhole camera model is considered as a CCD camera model and camera simulator, which calculates the field of view from two cameras depending on objects' size and workspace is explained.

A. Evaluation function

In consideration of five factors (*Maximization of Area, Stereo Matching and Baseline, The Number of Pixels, Symmetric Arrangement*), camera parameters (Figure 1), θ_{p1}, θ_{p2}, θ_{t1}, θ_{t2}, h_{c1}, h_{c2}, bl will be determined for optimal arrangement of cameras in this section.

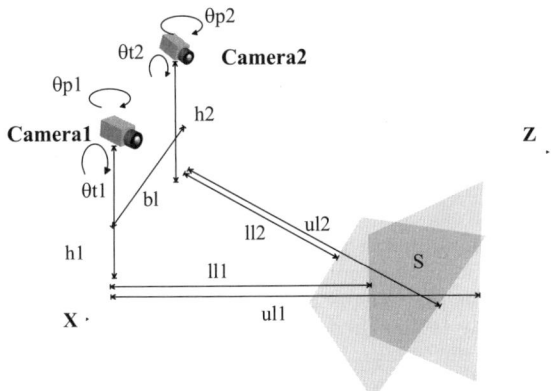

Fig. 1. Camera Parameters

To determine optimal parameter values of these factors, performance indexes are utilized. In general, performance indexes changes according to the aims of tasks and needs from users. Thus we define the performance index function that can evaluate four factors; area S, average of quantitative error e, number of pixels on target image pn, ratio between base line bl and distance of center of area d. Performance index is defined as (1).

$$J = k_S S_{std} - k_e e_{std} + k_{pn} pn_{std} - k_{sm}(bl/d)_{std} \quad (1)$$

Since the performance index depends on the task condition and positioning systems, weight coefficients, $k_{\{S,e,pn,sm\}}$, are tuned for the target task. In the index function, X_{std} represents standardized value.

We are interested in making index value J as big as possible. There are a lot of methods of maximizing functions such as genetic algorithm[4], Powell's method and conjugate gradient method[5]. Among these methods, we chose downhill simplex method[5]. The downhill simplex method was originally developed by Nelder and Mead in 1965. The downhill simplex method requires only function evaluations, not derivatives. Since derivatives of our performance index cannot be calculated, we use this method for optimization.

Using this method, we simulated under several restrictions given by the task and the performance of image processing. That is,

$$\begin{cases} e_{qmax} \leq e_{allow} \\ pn \geq MP \end{cases} \quad (2)$$

where e_{qmax} is maximum quantization error, e_{allow} is allowable error in the task, pn is the amount of pixels of the object in the image and MP is minimum amount of pixels processed in the image processing. To keep symmetric characteristic, we assumed $\theta_{p1} = -\theta_{p2}$, $\theta_{t1} = \theta_{t2}$, $h_{c1} = h_{c2}$.

B. Current Arrangement

Before performing optimization, we simulated current system with the current parameters of camera sensors under the restrictions shown at (2). Camera parameters of our current system are shown in Table I(Before). Figure 2 is the result of simulation. Cameras are positioned at $X = -1.950$ and $X = 1.950$ respectively. Z-axis represents quantization error. Figure 2 shows that the stereovision area is $9.04\,\mathrm{m}^2$ and mean of error is $10.89\,\mathrm{mm}$.

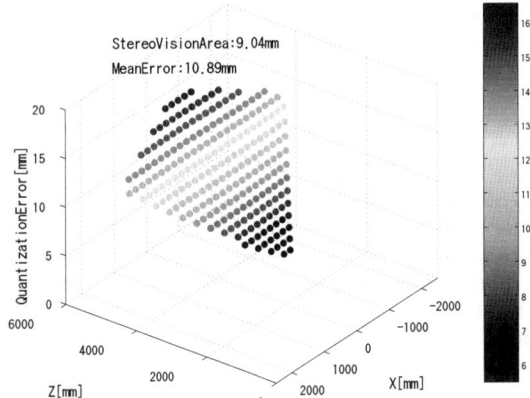

Fig. 2. Before Optimization

C. After Optimization

After simulation of our current arrangement, optimal arrangement of cameras was performed under the same restrictions shown at (2). Figure 3 is the result of simulation and Table I(After) shows optimal camera parameters. Cameras are positioned at $X = -2.250$ and $X = 2.250$ respectively. Figure 3 shows that quantization error is smaller than allowable error $2\,\mathrm{cm}$. Stereovision area is $11.28\,\mathrm{m}^2$ and mean of error is $8.48\,\mathrm{mm}$. Compared with Figure 2, stereovision area become larger and mean of error become smaller. In this optimization, k_S=8.5, k_e=7.5, k_{pn}=5.5 and k_{sm}=4.5.

TABLE I
Optimal Parameters and Size of Area

	θ_p[deg]	θ_t[deg]	h_c[m]
Before	20.0	20.0	2.4
After	38.0	15.0	2.2
	bl[m]	S[m^2]	e_{qmean}[cm]
Before	3.9	9.04	10.89
After	4.5	11.28	8.48

Although the cameras in simulator are not perfectly modeled to real cameras, this result shows better performance is obtained after optimization.

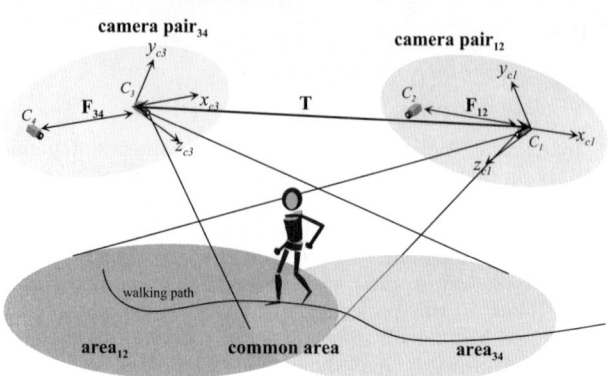

Fig. 4. Multi camera calibration by using pair stereovision camera.

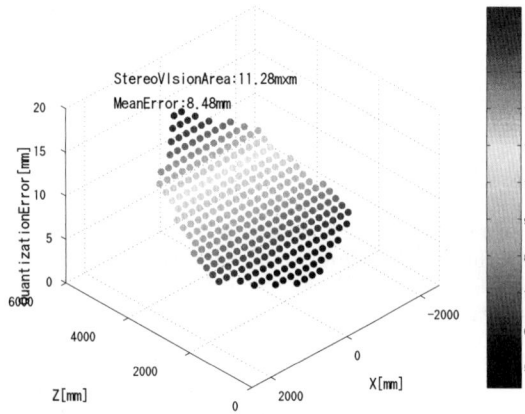

Fig. 3. After Optimization

IV. Multiple Camera Calibration

In this section, we describe multiple camera calibration in Intelligent Space.

To realize Intelligent Space as a space-scalable system using multiple cameras, automatic camera calibration function is indispensable. A lot of camera type DINDs in Intelligent Space are connected to network and they detect each other on the network. However, these DINDs cannot know a geometric relation mutually only by connecting with a network. To realize seamless space, which can measure the position of objects by using cameras in the space, it is necessary that each cameras know their geometric relation i.e. Fundamental matrix (**F**-matrix) [6]. Moreover camera-type DINDs should know their supporting area and neighbor DINDs of its area. Because each DIND have to catch behavior of object and DINDs have to control robots in supporting area.

The aim of the multiple camera calibration is to reconstruct the geometrical relation among many cameras in Intelligent Space.

A. Making a pair stereo vision system

Camera calibration is the process of relating the ideal model of the camera to the actual physical device and of determining the position and orientation of the camera with respect to a world reference system. Depending on the model used, there are different parameters to be determined. The pinhole camera model is broadly used and the parameters to be calibrated are classified in two groups:
1. Internal (or intrinsic) parameters. Internal geometric and optical characteristics of the lenses and the imaging device.
2. External (or extrinsic) parameters. Position and orientation of the camera in a world reference system.

In our system, Tsai's camera calibration algorithm is utilized for making a pair stereovision system.

B. Making seamless measurable space

Camera calibration also requires the establishment of correspondences between image points and world coordinates. A traditional multi-camera system might include two to five cameras. In this case, manual methods are tolerable. One approach is to carefully measure image coordinates using a graphical user interface. However, as the number of cameras increases, such methods become increasingly tedious and error prone.

We are developing "automatic DIND cell construction system". Basic idea is the following. In Figure 4, two cameras: C_1 and C_2 make a pair, which works as a stereovision system. Other two cameras C_3 and C_4 also make a pair. These pairs have a narrow common area, and when some objects pass the common area, two camera pairs can detect these position based on their own coordinate. Each camera pair save object's position with timestamp and send data to other pair camera via network. Using the data, coordinate transformation matrix **T** can be calculated.

n points get from camera sets C_1, C_2 and C_3, C_4 are $\boldsymbol{x}_{Ai}, \boldsymbol{x}_{bi} : i = 1 \sim n$ and we can get,

$$\boldsymbol{b} = \begin{bmatrix} \boldsymbol{x}_{A1}^T \\ \vdots \\ \boldsymbol{x}_{An}^T \end{bmatrix}, \boldsymbol{A} = \begin{bmatrix} \boldsymbol{x}_{B1}^T & 1 \\ \vdots & \vdots \\ \boldsymbol{x}_{Bn}^T & 1 \end{bmatrix}, \boldsymbol{x} = \begin{bmatrix} \boldsymbol{R}^T \\ \boldsymbol{t}^T \end{bmatrix} \quad (3)$$

$$\boldsymbol{b} = \boldsymbol{A}\boldsymbol{x} \quad (4)$$

$\boldsymbol{T}(\boldsymbol{R}, \boldsymbol{t})$ are derived from

$$\boldsymbol{x} = (\boldsymbol{A}^T \boldsymbol{A})^{-1} \boldsymbol{A} \boldsymbol{b}.$$

A pair stereovision supports an area, which is called DIND cell, and a lot of cells cover target space. As shown in Figure 5, DIND cells administrate object tracking and

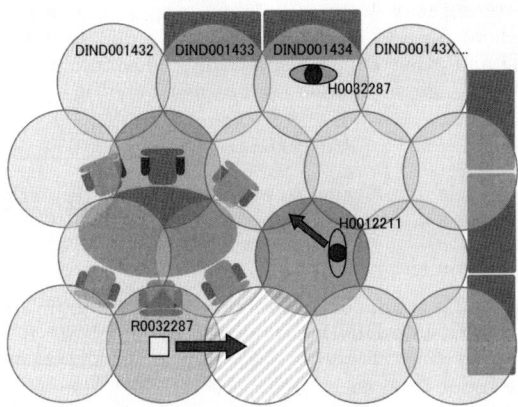

Fig. 5. Camera-type DINDs create *DIND cells*.

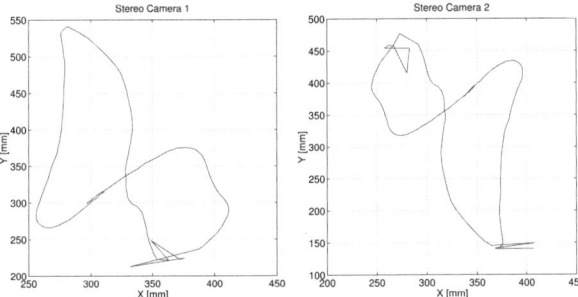

Fig. 6. Tracking 3D trajectory of stereo camera1 and camera2.

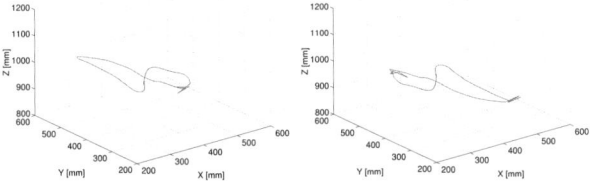

Fig. 7. Tracking 2D trajectory of stereo camera1 and camera2.

robot agent control in their supporting area. Each DIND's supporting area is narrow because of camera resolution constraints and processing power. Therefore to realize seamless tracking, recognition and control, a lot of DINDs have to collaborate each other.

Each DIND cell have narrow common area is laid on to perform calibration camera pairs. This area is also important for handing over of object tracking and robot control authority.

C. Experiment

An experiment to obtain transformation between two stereo camera, was performed. To simplize the experiment, these two cameras was installed adjacently so that they have rectangle observation area and 30 cm width common area on the floor. Figure 6, 7 show trajectory of target object, which was observed two tracking vision system.

The trajectory data of the tracking, which was obtained from two stereo cameras, are stored with timestamp. The translation between two stereo pair was calculated from these data. Rotation $\boldsymbol{R}$[rad] and translation $\boldsymbol{T}$[m] are given as the following.

$$\begin{bmatrix} R_x \\ R_y \\ R_z \end{bmatrix} = \begin{bmatrix} -0.0030 \\ -0.1013 \\ -3.0744 \end{bmatrix} \text{[rad]}, \begin{bmatrix} T_x \\ T_y \\ T_z \end{bmatrix} = \begin{bmatrix} -0.1633 \\ -0.8100 \\ 0.0280 \end{bmatrix} \text{[m]} \quad (5)$$

By iterating this process, DIND's supporting areas are networked geometrically.

V. SCALABILITY OF INTELLIGENT SPACE

Even though one DIND is able to provide services by its function modules, cooperation with other DINDs is necessary to accomplish Intelligent Space. Since one DIND monitors a restricted area, to provide services seamlessly and continuously through entire space, DINDs should cooperate with each other. Moreover, in some functions, information from other DINDs is required so that cooperation among function modules should be prepared.

A. Localization of Human

To localize human in a space, at least two vision sensors are required. This means that a function module of localizing human requires information from other DIND. To reconstruct 3D position of human in Intelligent Space, the human localization function module matches local clusters and clusters from other DINDs. During the cluster matching process, noises are filtered out in consideration of size, position, epipolar conditions, etc. On the basis of calibration data of each DIND including absolute position of DINDs in Intelligent Space, 3D reconstruction is performed.

B. Localization of Robot

Unlike the case of localizing human, to localize a mobile robot in the Intelligent Space, only one DIND is required. Because the heights (z-axis) of the targets are known, only one camera is needed for precise 3D reconstruction. The information on heights of the robots and their identification color-codes is stored in database of DIND in Intelligent Space. Based on height of robot from database, 3D positions of color-codes are calculated. According to the database of the DIND, each color-code, placed around a mobile robot, is checked. If a color-code, which is not shown in the database, is detected, it is removed from the candidates. From this process the DIND is able to recognize which robot it is and localize the robot by referring database. To get rid of errors, 2D euclidean distances between detected color-codes are checked. If the distance is too long or too short, the detected color-codes are neglected. The geometrical relations, stored in the database, are compared between the color-codes and the robot is localized from geometrical relations of color-codes.

C. Control of Robot

Mobile robots in the Intelligent Space need neither range sensors to detect obstacles nor positioning sensors to localize themselves. The robots only require passive color bars-codes and a communication device to communicate with DINDs. Due to communication between robots and DINDs, robots are able to use sensors in DINDs freely. It is a kind of resource sharing and this feature leads robots to low cost, since expensive sensors, such as gyroscope sensor, laser range finder, etc are not needed. In Intelligent Space, robots are both client and agent of the space. As explained above, to provide physical service in Intelligent Space, a robot as an agent of Intelligent Space is used.

C.1 Control loop

Mobile robot control function module produces control input for the mobile robots. A DIND senses mobile robots, and robot localization function module localizes it. According to desired path and estimated position, control input for the mobile robots are generated and transferred to the mobile robots through wireless LAN. The flow of information makes closed loop.

C.2 Reliability rank

DIND is an independent device, and its functions, including localization, control of mobile robots, are performed completely within it. Thus, if a mobile robot is moving in the area, at which one DIND is monitoring, the robot is guided without any difficulty. However, to guide robots in wide area, which only one DIND cannot cover, we define DIND that has control authority of a robot as dominant DIND for the robot. When a robot moves from an area to a different area, dominant DIND for the robot should be changed automatically. It is called handing over of control authority. Dominant DIND has control authority of the robots and only one dominant DIND exists to a robot at once. Therefore, the control authority should be handed over to the next DIND smoothly at the proper time and the proper location. To solve this problem, the reliability rank is devised. High reliability stands for that DIND can guide a robot robustly and precisely. Generally, near area from DIND and center area of an image, which DIND captures, have highest reliability rank, and boundary of the image and far area from DIND have lowest reliability, since vision camera is adopted as a sensor of a DIND. Figure 8 introduces how the reliability rank is determined. Figure 8(a) is an area monitored by a DIND and it is divided into 4 parts according to distance from DIND. In Figure 8(b), the image is divided into 3 areas. Even though a good calibration algorithm is adopted, outskirts of the image distort big since a vision sensor with lens is used. Based on these two partition methods, actual reliability ranks in monitored area are determined as shown in Figure 8(c).

C.3 Hand over protocol

Figure 10 describes the protocol of handing over of control authority of a robot. At the first, DIND(n) is dominant DIND to the robot. DIND(n) requests other DINDs about reliability rank of the robot. The other DINDs replies their reliability rank on the robot and the current dominant DIND, DIND(n), compares those values with own rank. If DIND(n+1) has higher rank than rank of DIND(n), then current dominant DIND(n) transfers authority of control to the DIND(n+1) which has higher rank on the robot. Then the new dominant DIND, DIND(n+1), controls the robot. However, if authority of control is shifted to a new DIND with the condition,

$$R_d > R_{other} \qquad (6)$$

, chattering may occur in handing over of control authority, where R_d and R_{other} are reliability rank of dominant DIND and other DIND respectively. Therefore instead of (6), following condition is used to avoid chattering.

$$R_d > R_{other} + \alpha \qquad (7)$$

, where α is offset value to avoid chattering. Empirically, α is set to 1 in actual system.

C.4 Experimental result

Figure 9 is results of controlling a mobile robot in Intelligent Space. Two DINDs cooperated to complete a navigation of the robot. Figure 9(a) and (b) are localized outputs of the robot from each DIND. Even though they failed in localization in some area, handing over of control authority was performed properly and the robot was controlled seamlessly and continuously. Figure 9(c) shows that which DIND controlled the robot. '+' indicates that the robot was controlled by one DIND at that position and 'x' is by other DIND. The trajectory is discontinuous at the position where hand over occurs. However, due to (7), chattering does not occur and the robot is controlled smoothly.

VI. CONCLUSION AND FUTURE WORK

In this paper, early results on self-identification of DIND are introduced. Required functions, such as optimal displacement of DIND, calibration of DIND and handover protocol of DINDs, are described. However, currently the functions are not complete and they are not combined into a DIND yet. By the deadline of camera ready paper of ICRA2003, authors will complete the system and further research results will be shown in the paper.

REFERENCES

[1] H. Hashimoto, "Intelligent Interactive Space based for Robots", Proceedings of the 2000 International Symposium on Mechatronics and Intelligent Mechanical System for 21 Century, pp26, 2000.
[2] Hashimoto Hashimoto, "Intelligent Space - How to Make Spaces Intelligent by using DIND? -", Proceedings of the IEEE International Conference on Systems, Man and Cybernetics (SMC'02), 2002.10
[3] G. Appenzeller, J. H. Lee and H. Hashimoto, "Building Topological Maps by Looking at People: An Example of Cooperation between Intelligent Space and Robots", Proceedings of IROS97, 00, pp1326-1333, September, 1997.
[4] J. Koza, "Genetic Programming", MIT Press, Cambridge, MA, 1992.
[5] William H. Press, Saul A. Teukolsky, William T. Vetterling, Brian P. Flannery, "Numerical Recipes in C", Second Edition, 1992.
[6] R. J. Tsai, "A Versatile Camera Calibration Technique for High Accuracy 3D Machine Vision Metrology Using Off-the-Shelf TV Cameras and Lenses", IEEE Journal of Robotics and Automation, Vol. 3, No. 4,pp.323-344,1987.
[7] B.Brumitt, B.Meyers, J.Krumm, A.Kern and S.Shafer, "EasyLiving: Technologies for Intelligent Environments", Proceedings of the International Conference on Handheld and Ubiquitous Computing, September 2000.
[8] A. Ward, A. Jones, A. Hopper, "A New Location Technique for the Active Office", IEEE Personal Communications, vol4, No5, pp42-47, October, 1997.
[9] S. Blostein, T. Huang, "Error analysis in stereo determination of 3-D point positions", IEEE Transactions on Pattern Analysis and Machine Intelligence, 9(6), pp752-765, November, 1987.
[10] G. Wei and S. Ma, "Implicit and explicit camera calibration: Theory and experiments", IEEE Transactions on Pattern Analysis and Machine Intelligence, 16(5), pp469-480, 1994.
[11] J. A. Nelder, R. Mead, "A Simplex Method for Function Minimization", Computer Journal, vol. 7, pp. 308-313, 1965.

(a) Based on distance from DIND (b) Based on captured image (c) Combined reliability map

Fig. 8. Reliability rank map

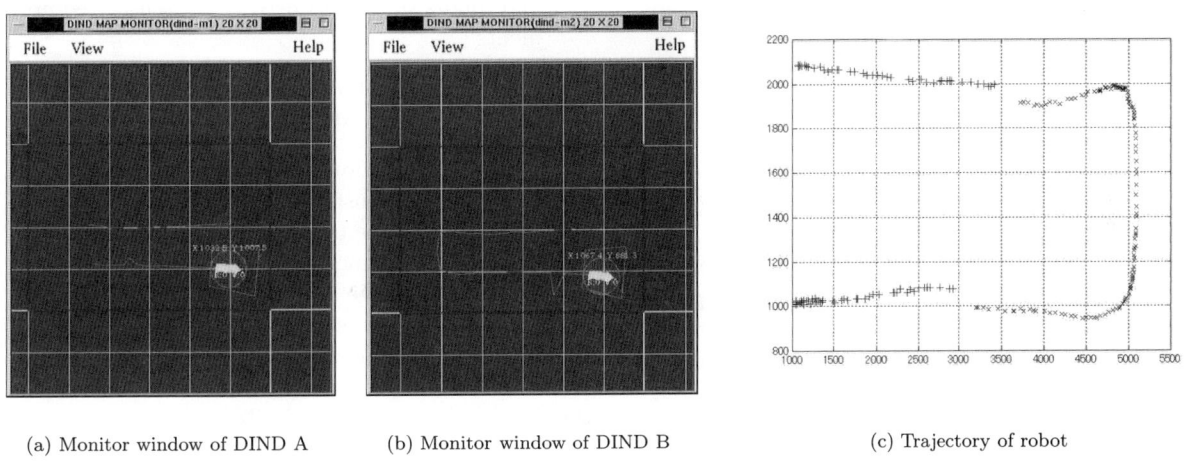

(a) Monitor window of DIND A (b) Monitor window of DIND B (c) Trajectory of robot

Fig. 9. Control of robot with multiple DINDs

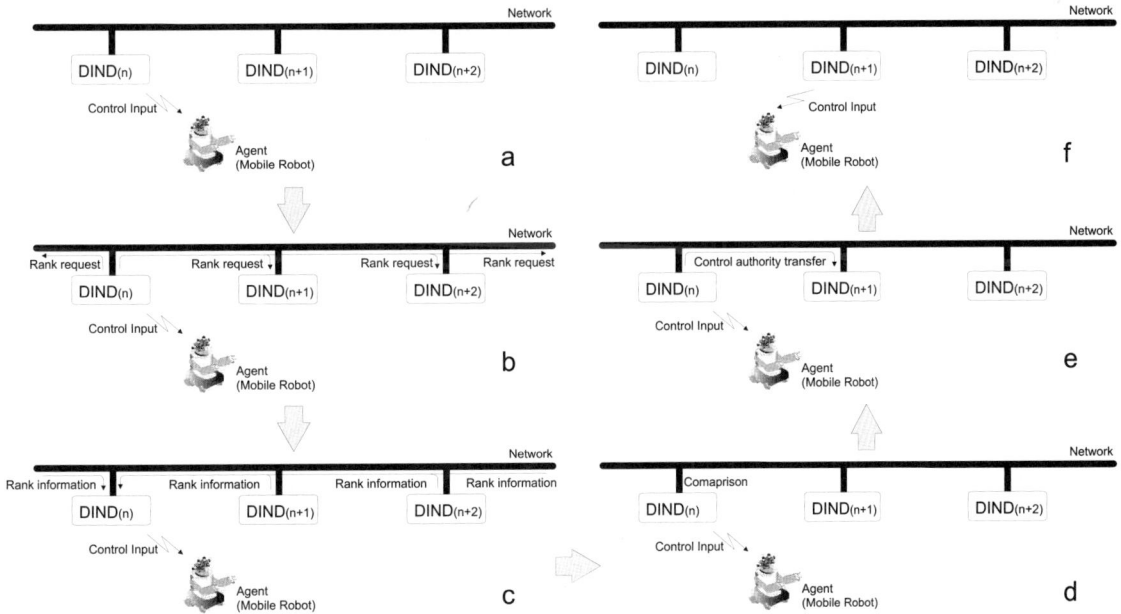

Fig. 10. Procedure of handover

4177

Expression Method of Human Locomotion Records for Path Planning and Control of Human-symbiotic Robot System based on Spacial Existence Probability Model of Humans

Rui FUKUI[1], Hiroshi MORISHITA[2] and Tomomasa SATO[1]

[1]*University of Tokyo*, [2]*HMI Corp.*
E-mail {*fukui,hiroshi,tomo*} *@ics.t.u-tokyo.ac.jp*

Abstract

In this paper, a novel describing method of human locomotion trajectory record was proposed, and methods of its application to the path planning and control of human-symbiotic mobile robots were indicated. The novel describing method, named as Existence Record Probability Map (abbreviated ERPM hereafter), was generated through the following three steps: 1) measuring human locomotion path using pressure distribution sensor floor, 2) applying a probability potential model of human body to the measured path, and 3) integrating such probability potential maps for some time period. From an ERPM, a) an existence probability value, b) ridge lines of the potential, and c) Q-value at a point on a ridge line, were elicited. As a usage of these information, methods of a) finding a position where a collision probability of a mobile robot and a human is high, b) finding major trajectories of human locomotion, and c) evaluation of path width to avoid collision with a human, were indicated.

1 Introduction

A number of researches have been conducted on mobile robot system, and path planning to avoid collision has become one of the major areas. In the case of factory, space, and nuclear plant application, path planning to avoid collision with an object is emphasized. On the other hand, development of humanoids and welfare robots are quite active, and in the near future, a human-robot symbiosis environment, where robots assist humans in ordinary houses, may become feasible. In this case, not only collision prevention but also more sophisticated control like keeping some distance with humans which is regarded as being comfortable for the humans, will become essential. To realize such human-robot symbiosis environment, the following two techniques are important:

- Technique to measure the positions of humans and robots simultaneously by the sensory equipments

- Technique to control the position and action of robots based on the measured human position

As for the first technique, there seems to be two approaches: One is to install sensory equipments in the environment, and the other is to install them in the robots and humans such as wearable position transmitter and so on. The authors think the former one more feasible because sensory equipments in the environment do not restrict the ordinary movement of humans and robots. Intelligent Room of MIT [1], the Aware Home of Georgia Tech [2], Neural Network House of University of Colorado [3] are the examples sensory environment system to measure the positions of humans and robots. In the Intelligent Room, human position is measured by vision sensor, and floor type sensor is used for personal identification in the Aware Home. The Neural Network House is equipped with sensors to monitor motion, temperature, light and sound. However, these intelligent environments are developed mainly for informative support to humans, namely they are not intended to create human-robot symbiosis environments for the purpose of mechanical support to humans. In the near future, not only informative support but also mechanical support such as fetching things out of the reach of a human, will become important. To realize this mechanical support, it is essential to develop the techniques to measure and accumulate the locomotion of humans correctly, and to process the information to the shape suitable for path and motion planning. From this viewpoint, this paper firstly proposes the method to create a potential map named ERPM, which describes the major path of human locomotion as a potential, and secondly proposes methods to utilize the characteristic values elicited from ERPM for path planning of human symbiosis mobile robots. In chapter 2, methods to measure the locomotion record of humans are explained. In chapter 3, the ERPM, a novel expression of human locomotion trajectory records and the characteristic values which can be elicited from it are discussed. In chapter 4, methods of applying the ERPM and the characteristic values made from actual human locomotion data to the path planning and control of human symbiosis mobile robots are shown. Chapter 5 is the conclusion.

2 Measurement of human locomotion by the sensor floor

Before explaining a novel expression method for human locomotion records, this section describes measurement method of human movement by sensor floor which can measure the pressure distribution as a bitmap image.

2.1 Outline of the sensor floor

To measure pressure distribution, sensor floor[4] shown Figure 1 was used for detecting human locomotion trajectory. This sensor floor covers 2m×2m area with 256×256=65,536 pressure measurement points. The sensor floor consists of 16 floor sensor

Figure 1: *Pressure Sensor Distributed Floor to Obtain Footprints of Humans*

units. Figure 2 shows the details of floor sensor unit. As showed in Figure 2(right), each floor sensor unit is composed of upper sheet and bottom pattern. When pressure is applied on the upper sheet, it deforms and touches the bottom pattern to make the short-circuit. This short-circuit is detected by the controller. The distribution of switch ON and switch OFF information are converted to a bitmap image and is put on the serial bus line as the output. Table 1 shows the specification of the sensor floor.

Figure 2: *Floor Sensor Unit*

Table 1: *Sensor Floor Specification*

Sensing Area	2000 × 2000 mm
Sensor Pitch	7 mm
Output Data	ON-OFF(1bit)
Transition Pressure	25 kPa
Sampling Frequency	Approx. 16 Hz
Data Transfer Rate	115200 bps
Data Type	Compressed Character

2.2 Data obtained from the sensor floor

Figure 3 is the typical data obtained from the sensor floor. The figure displays the data of human walking on bare foot. While human is walking on the sensor floor, immediately after the human foot touches the sensor floor, small blob image comes out. Then the size of blobs becomes larger and reaches to its maximum when all foot backside touches the floor, and becomes smaller as his foot separates from the upper sheet. This dynamic transition is captured as a series of bitmap images.

3 Novel expression method of human locomotion record (ERPM) and its characteristic values

This section describes the calculation algorithm to obtain ERPM, and the characteristic values elicited from the ERPM. Figure 4 shows the flow of data processing. There are five processing steps in the flow. The ERPM is obtained through the first three steps, and the characteristic values are elicited in the latter two steps.

1. Interpolation of human footprints by B-spline curve.

2. Expansion of B-spline curve width.

3. Accumulating potential maps.

4. Extraction of ridge lines from ERPM.

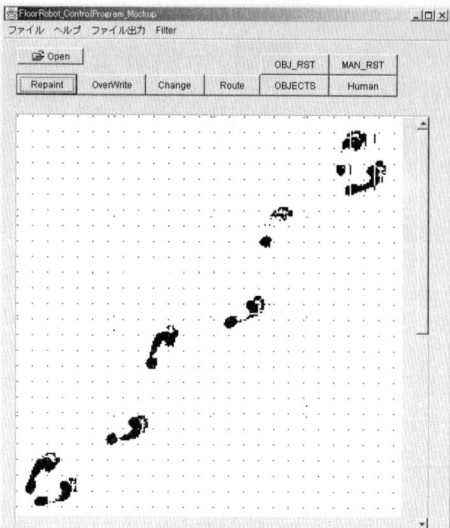

Figure 3: Pressure Data Obtained when Walking on Bare Foot

5. Calculation of Q-value at each points on a ridge line of ERPM.

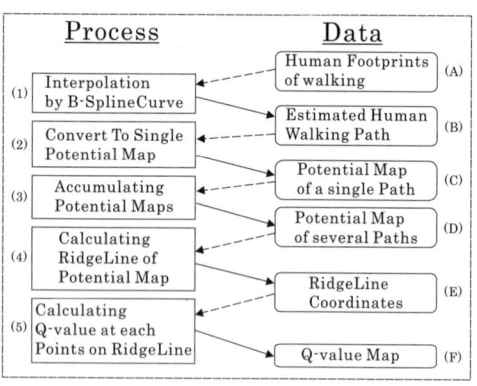

Figure 4: Data Processing Flow

3.1 Interpolation of human footprints by B-spline curve

As described before, discrete footprint image can be obtained from the sensor floor. An image of one foot doesn't always become a single blob. In case when one footprint consist of two blobs, one is an image of tiptoe and the other is an image of heel. In these cases human foot position is estimated by calculating the gravity center of two or more blobs. By repeating this step, touch down points of each foot are obtained and a B-spline curve that interpolate those points is calculated. This B-spline curve is defined as human walking path. More research should be done to determine which is the best among various interpolation methods, but in this paper we use B-spline curve for convenience.

3.2 Expansion of B-spline curve width

In the previous step, we got the estimated human walking path. In this step, we need to express human body width. There may be two kinds of human body width. One is the breadth of shoulders that occupies the space consistently. The other is the region where hands and feet pass by beyond the breadth of shoulders while walking. The latter region should be treated by a probability method. To realize this, probability model shown Figure 5(A) should be arranged along the B-spline curve. Figure 5(A) shows that the probability becomes lower in the outside of the cylinder. The parameters of this conic cylinder model need to be determined by experimental method. In this paper, we used cone and cylinder like Figure 5(B),(C) for simplicity, and obtained qualitatively proper results.

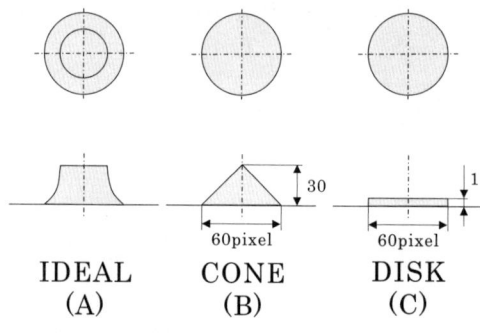

Figure 5: Models Used to Expand B-spline Line

3.3 Accumulating Potential Maps

When a human walks in a room, even if start and end point are fixed, the routes he takes varies according to the subject and to the cases. However a typical route emerges after many trips. To express such situation, human walking data that have same start area and same goal area are processed by step(1),(2), and ERPM is made by accumulating the output.

3.4 Extraction of ridge lines from ERPM

There can be found some ridge lines in ERPM, and they can be reagarded as typical human locomotion routes because because these ridge lines connect the peaks of ERPM. Usually there shows middle size ridge lines in addition to the typical route. Sorting the extracted ridge lines by its height makes it possible to detect a typical route.

3.5 Calculation of Q-value at each points on a ridge line

Even on the same typical ridge line, the potential value is not constant. For example, 1) If there is narrow area in the path where only one person can pass the area at the same time, all people must walk on that point. Therefore the potential of the point becomes high and the shape of potential map becomes sharp. In the opposite case, 2) If the width of the road is enough for several people passing through at the same time, people can select their route freely, and the potential becomes low and dull. To treat this numerically, a virtual cross section perpendicular to the typical route is assumed. The shape of this cross section is generally a hill shape. Firstly h is defined as the height of the hill, and w as the width of hill at the half height. Secondly a value $Q = h/w$ is adopted, and Q is called as "Q-value". 1)This Q-value becomes big when the ERPM is sharp at the point, and 2) becomes small when the potential is dull. This value can be used for indicating concentration degree of human walking to the typical route.

4 Experiments and Discussion

4.1 Experiments

Figure 6 shows the setup of experiments. 1m×1m Goal and Start areas are prepared at the downleft and upright corners of 2m square floor. Two chairs are used as obstacles that restrict human walking path. In the case of experimental setup 1, subjects are supposed to walk between the two chairs. In the case of experimental setup 2, subjects are supposed to walk around the two chairs. Figure 7 is the footprints image obtained in the experimental setup 1.
In the Figure 8, "◯"marks denote the gravity center of footprints image, and the line is the B-spline curve that interpolates these points. 100 cones shown in Figure 5(B) are arranged on the B-spline curve in Figure 8, and by calculating enveloping surface of cones the potential of single path shown Figure 9 is obtained. Figure 10 displays the potential of single path made by arranging disks shown Figure 5(C) instead of cones. Comparing Figure 10 with Figure 9, Figure 9 potential seems to reflect the human locomotion in the sense that probability of existence becomes lower, getting apart from the curve. Figure 11 shows the potential calculated by accumulating cone shape potentials shown in Figure 5(B) instead of calculating enveloping surface. However this calculation seems to be inadequate for single path potential because the potential values at the center and the ends on the same route are different.

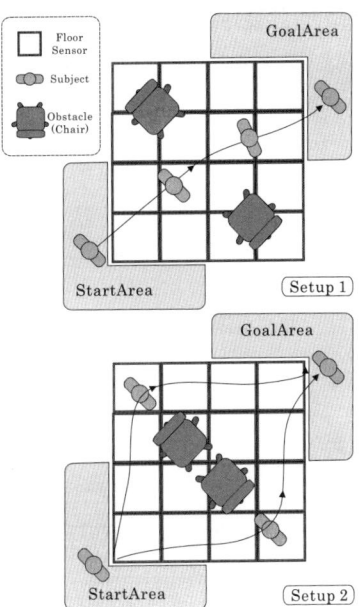

Figure 6: *Experimental Setup*

Under the experimental setup 1 of Figure 6, subjects were asked to walk from start area to goal area freely for 120 times. These 120 footprints images are interpolated by B-spline curve and transformed to single path potential maps by arranging cones on the curve. Figure 12 shows the accumulated potential map(ERPM) of 120 single path potential maps. At the center area the potential values are high, because the walking path is restricted by the two chairs. Start area and goal area have certain width, and there are many choices of routes for subjects. Therefore the potential spread broad. Similarly, experiment under setup 2 of Figure 6 was carried out for 120 times. Figure 13 shows the potential map(ERPM). In this experiment setup, there are obstacles at the center of floor, and two routes are available for subjects, right and left routes. These two routes are clearly shown in the potential map.

Figure 14, Figure 15 is the ridge lines extracted from Figure 12, Figure 13 potential maps respectively. In Figure 14, there are main ridge line running from downleft to upright, and some extra ridge lines around start area and goal area. This result means that some subjects didn't walk through the main typical route, but took other routes. For example, a subject seems to have entered the floor from left floor edge and exited from right floor edge. In the experimental setup 2 of Figure 15, some extra

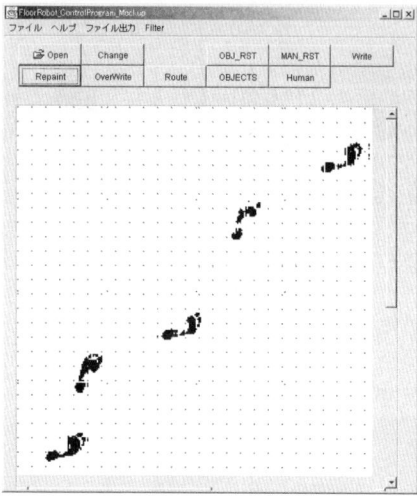

Figure 7: An Example of Pressure Data Obtained in Experimental Setup 1

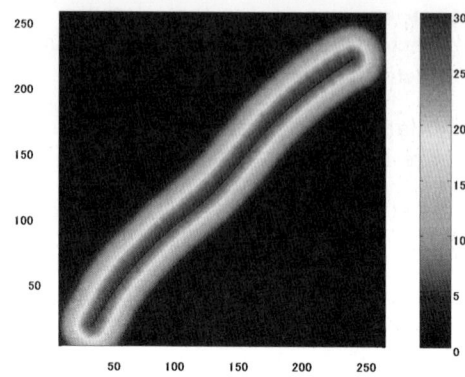

Figure 9: Expansion of Walking Path by Enveloping Cone-shaped Potentials

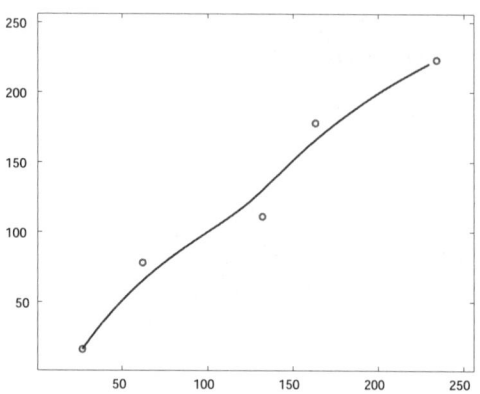

Figure 8: Interpolation of Footprints by B-spline Curve

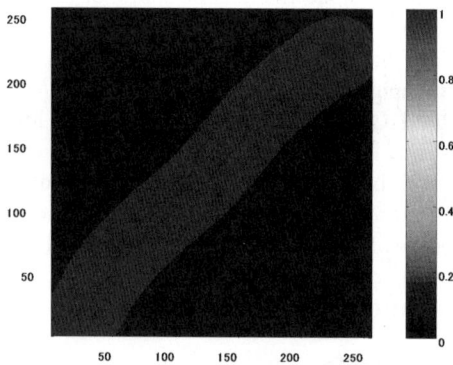

Figure 10: Expansion of Walking Path by Enveloping Disk-shaped Potentials

routes can be seen near the floor edges in addition to the main routes. Among many ridge lines of Figure 14, one ridge line was selected as a shortest course from downleft start area to upright goal area. Figure 16 shows the extracted ridge line. Figure 17 is the graph of Q-value change along the selected ridge line. Left edge of graph corresponds to the donwleft corner of Figure 17 map, and right edge to the upright corner. Q-value is quite high at the center of floor and relatively low at the corner.

4.2 Discussion

Adoption of temporal weight. In the experiments of this paper, each single path potential are accumulated with the same weight. However if the passage of time from data sampling moment to the present is considered in the way that weight parameters are changed depending on the passage of time, dynamic potential map can be obtained. In such a dynamic potential map, typical route and Q-values will be changing continuously. That is to say, if the reciprocal number of the passage of time is used for weight parameter, past walking data is treated lightly and latest walking data is emphasized.

Application for robot path planning. In a human symbiotic robot system, robot and human live together. In such a system, path planning for robot movement is significant in the following two points 1) Robot must fit into human daily life, 2) Robot shouldn't disturb human life. In the former case, when mobile robot fits into human life, robot can moves on the typical route of human walking route. In this case robot should be controlled as moving along the ridge line of ERPM. In the latter case, robot is thought not to disturb human life. If robot moves along the ridge line, robot may collide with human, and such situation is thought to be inadequate. In such case, robot should check the Q-value at each

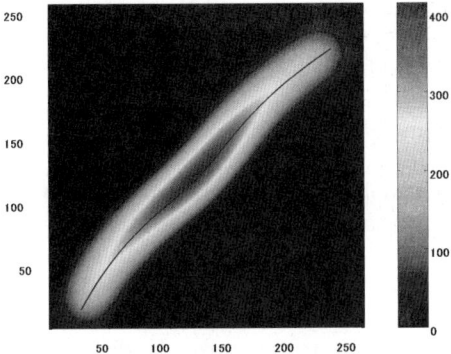

Figure 11: Expansion of Walking Path by Accumulating Cone-shaped Potentials

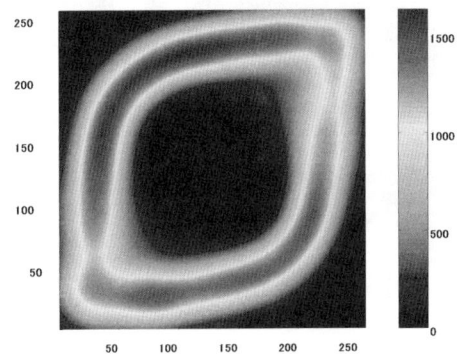

Figure 13: Potential Map made by Accumulating Several Paths (Experimental Setup 2)

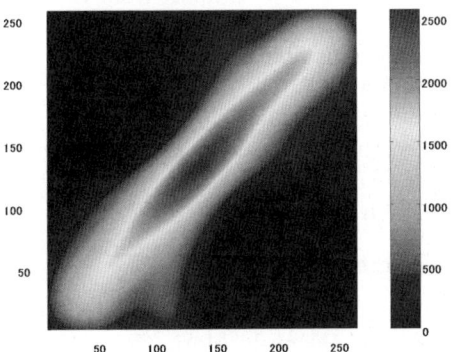

Figure 12: Potential Map made by Accumulating Several Paths (Experimental Setup 1)

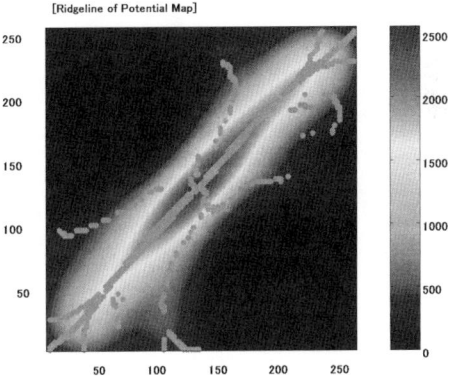

Figure 14: Ridge Lines Extracted from the Potential Map (Experimental Setup 1)

points of the ridge line. Low Q-value indicates that the route along the ridge line have extra space for robot, and the robot can go through this extra space. Conversely high Q-value means the route don't have so much extra space and people can't walk away from the ridge line. A robot may come across people at such point because people must walk tightly along the ridge line, so the robot must confirm the vacancy on the route before going into such point and pass there as quick as possible.

5 Conclusion

In this paper, a novel expression method of human locomotion records was proposed, and basis for its application to the path planning and control of human symbiosis mobile robots was shown. In order to get an expression, the following two procedures are necessary: 1) measuring the human locomotion path by using pressure distribution sensor floor, and 2) producing 2D human existence probability map by applying the human existence probability model to the path and by making integration of them for some

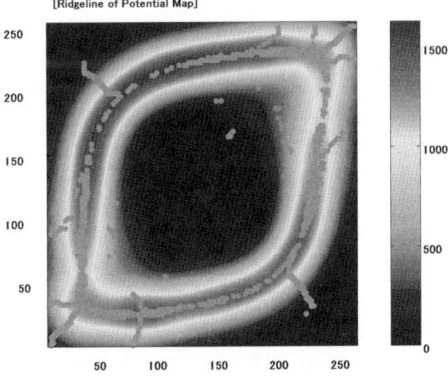

Figure 15: Ridge Lines Extracted from the Potential Map (Experimental Setup 2)

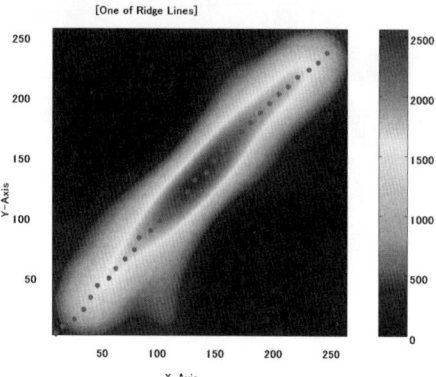

Figure 16: One Ridge Line of Potential Map (Experiment 1)

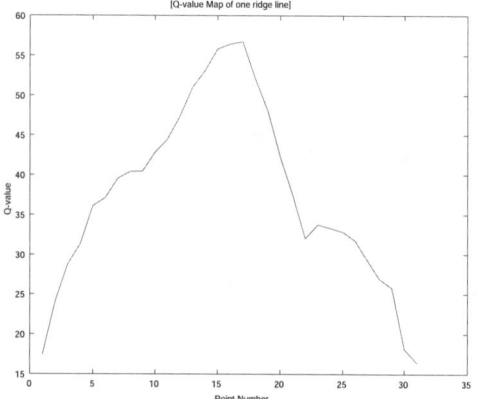

Figure 17: Q-value Transition One Ridge Line(Experiment 1)

time period. This map is the probability expression of human locomotion records, and the authors named it as Existence Record Probability Map (ERPM). In the experiments, 2m×2m pressure distribution sensor floor was used to measure the footprints generated during the human locomotion. Two types of path settings were experimented by setting two obstacles on the floor. 120 trips in each setting were recorded and ERPMs were calculated respectively. From these ERPMs, ridge lines were extracted, and Q-values were calculated. In this paper, characteristic values extracted from ERPM and its application method to the path planning and control of human symbiosis robots were discussed. This application method is summarized as follows: a) Direct usage of ERPM value: probability of a human at a certain point is calculated by dividing the ERPM by total number of trips. Therefore when a mobile robot goes into such point that ERPM value is high, special care should be taken to avoid collision with a human. b) Usage of ridge line of ERPM: the ridge lines of ERPM show the typical locomotion path of humans. Therefore, when one wants to control the mobile robot to follow the same path as humans travel, this ridge line can be used as a path for the mobile robot. c) Usage of Q-value on the ridge line: the Q-value which is obtained from the cross section figure perpendicular to the ridge line, shows the allowance for the mobile robot path to the typical human locomotion path. Therefore in case the mobile robot should be programmed to take the same path as humans take and also to minimize the disturbance to the humans, this Q-value plays a important role. If Q-value is low, the mobile robot can take a little apart path from the typical human locomotion path. When the Q-value is high, the path is narrow and it is impossible to take a path apart from the typical human path. In this case, the mobile robot is reqeuested to proceed after confirming that no one is on the typical path and to pass that area as quick as possible. Introducing the temporal weight onto the ERPM will emphasize the newly recorded paths, and this time-considered ERPM will be useful to describe the changing environments.

Acknowledgments

The authors would like to express their great thanks to our colleagues Taketoshi Mori, Tatsuya Harada, and Masamichi Simosaka for the fruitful discussion.

References

[1] Nicholas Hanssens, Ajay Kulkarni, Rattapoom Tuchinda, and Tyler Horton "Building Agent-Based Intelligent Workspaces" *The International Workshop on Agents for Business Automation. Las Vegas, NV, 2002*

[2] Kidd, Cory D., Robert J. Orr, Gregory D. Abowd, Christopher G. Atkeson, Irfan A. Essa, Blair MacIntyre, Elizabeth Mynatt, Thad E. Starner and Wendy Newstetter. "The Aware Home: A Living Laboratory for Ubiquitous Computing Research" *the Second International Workshop on Cooperative Buildings, October 1999.*

[3] Mozer, M. C. "The neural network house: An environment that adapts to its inhabitants." *the American Association for Artificial Intelligence Spring Symposium on Intelligent Environments, pp.110-114, Menlo Park, CA: AAAI Press. 1998.*

[4] H.Morishita, Rui Fukui and Tomomasa Sato "High Resolution Pressure Sensor Distributed Floor for Future Human-Robot Symbiosis Environments", *IEEE/RSJ IROS 2002 Proceedings, pp.1246-1251, EPFL, Lausanne, Switzerland, September 2002.*

Safe Path Planning in an Uncertain-Configuration Space

Alain Lambert
IEF, UMR CNRS 8622
Université Paris-Sud
91405 Orsay, France
Alain.Lambert@ief.u-psud.fr

Dominique Gruyer
LIVIC, INRETS/LCPC
Route de la Minière-Satory
78000 Versailles, France
Dominique.Gruyer@inrets.fr

Abstract

The objective of this paper is to bring an effective response to the safe path planning problem which should be solved in an uncertain-configuration space. Firstly, a path planning method dealing with localization uncertainties is proposed, where the uncertainties in both position and orientation of a non-holonomic mobile robot are considered. The safety of this method is due to the mixing of the planning phase and the navigation phase using the same process of localization (the Kalman filter). Secondly, while previous works planned safe paths in the configuration space, we show that it is necessary to plan safe paths in an uncertain-configuration space. Then, we introduce the novel concept of "towers of uncertainties" and show the effectiveness of this concept with some examples.

1 Introduction

In most cases, navigating a vehicle implies (a) planning a path, i.e a continuous sequence of configurations (position and orientation), that takes the robot to its goal, and then (b) following this path. Unfortunately planning a path taking into account geometric constraints [7] is not sufficient. Practical experiments on path following show us that such paths cannot be followed, because nothing is as perfect in reality as assumed during the planning (dead reckoning is not perfect in the real world). So there is a strong need to take into account the uncertainties during the planning phase.

Consequently, path planning with uncertainty has motivated a number of research works [1, 3, 8, 9, 10, 11, 14]. The goal of these works is to plan safe paths: paths which can be followed safely (with accuracy and without collision with mapped objects) in real world. Unfortunately, those works firstly consider simplifying assumptions (point robot performing straight motion, perfect exteroceptive sensing ...) and secondly odometry and sensors are used in a very simple way. Furthermore, at navigation time, the paths generated by these methods are not very satisfying because the localization procedure used is not the same as in the planning phase. So, the information given by the localization procedures is different and plans must be adapted conforming to these data. To cope with this problem, recent works propose that planning and navigation use the same localization scheme and robot model [5, 6].

However, in all the papers where the robot's uncertainty is explicitly represented, none of the authors explain the necessity of planning in an uncertain-configuration space. We are going to show in this paper that planning in a configuration space with a collision test taking into account uncertainties is not sufficient and can lead to a poor or no solution (although a solution might exist). To solve this problem, we propose a new planner in a 4D space (configuration+uncertainty space) which uses the novel concept of towers of uncertainties.

This paper is organized as follows. First the localization of the vehicle in its environment is presented. Next, we show a framework for computing a safe path. Section 4 demonstrates the shortcoming of this framework and introduces the concept of towers of uncertainties (which allows the path planner to work in an uncertain-configuration space). Then, before drawing conclusions, section 5 shows experimental results which demonstrate the soundness of our approach.

2 The vehicle and its environment

Our vehicle is equipped with telemetric and odometric sensors. The data from these different sensors are combined with an Extended Kalman Filter in a classical localization process. The localization, which relies on an ideal 2D world map of the environment,

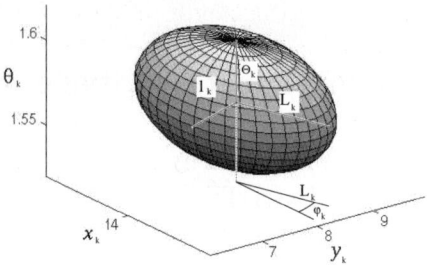

Figure 1: Representation of the uncertainties on the robot's configuration

has already been explained in [5]. Both planning and navigation use the same map where wall and obstacles are represented as polygonal lines (see fig. 2(a)). Bold lines represent detectable walls (landmarks) and dashed lines represent undetectable obstacles (in reference to the remote on-board sensors used). The mapping of the environment has three purposes:
-localization in real world
-detection of possible collisions during path planning
-simulation of sensors during path planning
The two last points should be further discussed. If we want to plan a safe path, we need an estimation of the uncertainty of the vehicle when it follows our planned path in the real world. The estimation of the uncertainty in the real world is done by a two phase process. Firstly a sensor simulation is done; this is quite easy as we know the robot configuration, the map and the sensors. Next, the sensors measurements are brought to an EKF which realizes the simulated localization. The result of this process is a covariance matrix which can be visualized as an ellipsoid [13] which represents the set of possible configurations of the vehicle (according to a probability which is commonly fixed to 90%). We believe that if we plan a path that avoid collisions during path planning, we will avoid these collisions during navigation. Collision avoidance is done during path planning by enlarging the vehicle with its uncertainty ellipsoid and by testing the result with the obstacle's location.

3 Path planning in the configuration space

3.1 Building of the roadmap

Our path planner relies upon a roadmap, i.e a graph in the configuration space whose edges are feasible and

```
01 CLOSE←∅
02 OPEN←node_start
03 while (OPEN≠∅)
04   node←head(OPEN) //a OPEN node with the smallest f value
05   CLOSE←CLOSE + node
06   OPEN←OPEN - node
07   if (node = node_goal) then exit(success)
08   NEW_NODES←successors(node)
09   for each new_node of NEW_NODES do
10     if ( (configuration(new_node) ∉ configurations(CLOSE))
          and (new_node is collision_free) ) then
11       g(new_node) ← g(node) + d_dubins(node, new_node)
12       f(new_node) ←g(new_node) +d(new_node, node_goal)
13       if configuration(new_node) ∉ configurations(OPEN) then
14         OPEN←OPEN + new_node
15         parent(new_node)←node
16       else n_node←extract(OPEN, configuration(new_node))
17         if g(new_node) < g(n_node) then
18           OPEN←OPEN - n_node + new_node
19           parent(new_node)←node
20         endif
21       endif
22     endif
23   endfor
24 endwhile
25 exit(no_solution)
```
Algorithm 1: A* without uncertainty management

collision-free (but not safe) paths for the mobile robot. The origin of the roadmap is not relevant. It may be automatically generated or manually built, but it must capture the connectivity of the collision-free configuration space. Fraichard has shown in [3] how to use the Probabilistic Path Planner in order to compute roadmaps that satisfy the requirements above in the case of a car-like robot subject to nonholonomic constraints. We compute our roadmap in the same way.

3.2 Finding the optimal path

The planner is based upon an A* algorithm [4] (an A algorithm combined with a heuristic function which speeds up the computing) which generates a path that minimizes a cost function. Such an algorithm ensures the optimality of the produced path (according to the cost function). Globally, it explores a tree where each node is a specific configuration (position and orientation) and computes a cost function 'g' along a path P. 'g' is an iterated combination of the sum of the distance d_{dubins} [2] between two configurations:

$$g(P) = \sum_{i=1}^{n} d_{dubins}(q_i, q_{i-1}) \quad (1)$$

The A* algorithm is described in Algorithm 1 where each node is uniquely described by its configuration. There are 3 lists : OPEN (contains nodes that will probably be developed), CLOSE (nodes that have already been developed) and NEW_NODES (the successors of node 'node').

Starting from node=node_start (line 2), the algorithm finds the successors (NEW_NODES) of 'node' (line 08) and puts the "good" successors (line 10 and 17) in the OPEN list (line 14 or 18). OPEN is ordered with increasing 'f' value. 'node' is put on the CLOSE list (line 05). The test "collision_free", at line 10, is not mandatory because our roadmap is already collision free (nevertheless, it will be mandatory in §3.3). Next (next loop, line 24 and 03), a node of OPEN with the minimum 'f' value is selected (extracted) as the 'node' that will be developed (this node is put in CLOSE and its successors are put in OPEN)...

The algorithm continues until the node selected from OPEN is node_goal (line 07) then the solution is given from node_goal to node_start by following the parents. If no solution is found (the OPEN list is empty), the algorithm stops on line 25.

3.3 Computing of safe paths

During the graph expansion, the planner develops node after node, and needs the uncertainties on the robot's localization in each of these nodes in order to choose a safe path and to avoid possible collisions. The prediction step of Kalman filtering is done along the path joining two nodes (as done in navigation). The estimation step is done at each node [5].

It is possible to detect possible collisions between the robot and the environment by computing the robot's uncertainties ellipsoid and testing the intersection of the enlarged robot with the obstacles. If a collision occurs, the current node is deleted (the node will not be included in the graph: see line 10 of the algorithm 1 where "collision_free" is now done with an enlarged robot). In this way, no path colliding with obstacles will be generated.

Thus, we have shown how we can plan a safe path in a configuration space with computed uncertainties during the graph expansion. Unfortunately the collision test which stops the progression of the graph search on some nodes of the graph could forbid the planner to find a safe path as explained in the next part.

4 Path planning in the uncertain-configuration space

4.1 On the necessity of planning in an uncertain-configuration space

Planning minimal length paths was done by the oldest planners in the configuration space. Without telemetric sensor and assuming tiny uncertainties (the initial configuration is assumed to be known with accuracy and the dead reckoning is assumed to be perfect), our planner can produce such paths from a Start to a Goal configuration under non holonomic constraints (see figure 2(a) where the uncertainties on the robot location are represented by ellipses[1]). Taking into account more important uncertainties such as the odometry being no longer accurate or the initial location is poorly defined (identical to the real world uncertainties), our planner cannot find a solution due to the obstacles (fig. 2(a)). Good behaviour of the planner is shown in figure 2(b), where P is the nearest eligible location to the goal. Unfortunately, our planner cannot find a solution to the Goal for a vehicle equipped with short range sensors[2].

The question is now: why does the planner fail although a solution exists (fig. 2(c))? The cause is the graph's exploration which develops node after node and forbids new exploration of previously explored nodes. The algorithm develops nodes which have an increasing value of the cost function [4]. If the algorithm finds a previously explored node it discards the possibility of reexploring this node (because the cost function of the latter path is bigger then the cost function of the previously explored path). In other words, if the algorithm has found the smallest path to go toward a configuration it keeps this smallest path and does not search to find a longer path (that is why we choose this algorithm: to minimize our cost function which is distance dependent!).

We can now come back to the example of figure 2(b). Based on the collision test the successor of P node has been discarded by the planner. Unfortunately, following the described algorithm, a problem now occurs at point P (see fig. 2(b)) during the graph exploration because this point (node) has already been explored by a straight path. Consequently it cannot be explored anymore and no one path going through this point can be generated. Hence no one solution can be found.

4.2 Introducing the concept of towers of uncertainties

In the previous section we pointed out the failure of the algorithm which cannot explore previously ex-

---
[1] The uncertainty on the robot orientation is not represented on the figure for legibility purpose.

[2] By introducing a new parameter (the volume of the uncertainties ellipsoid) in the cost function, we can find a solution (by a good tuning of the parameters). Nevertheless we think that it is not a good choice because: (a) the resulting path could become uncessarily excessively long and (b) in some case no solution could be found.

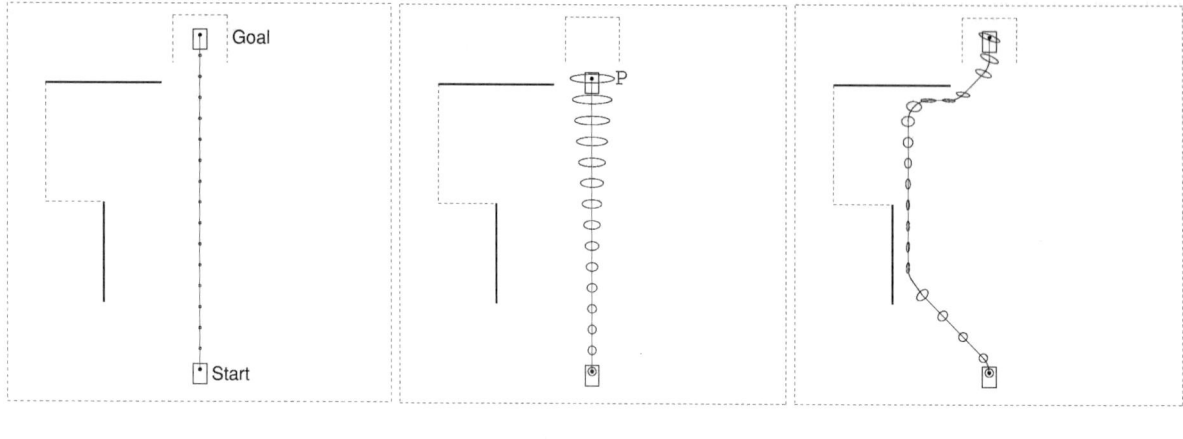

(a) Minimal lenght path

(b) Problematic case in the configuration space

(c) Path planning in the uncertain-configuration space

Figure 2: Planning should be done in the uncertain-configuration space

plored nodes. We could solve this problem by allowing a previously explored node to be reexplored when the algorithm reaches it with a smaller uncertainty. Then in order to avoid break in the graph we should remove all the already visited successors of this node. This is not problematic because they will be revisited. The shortcoming is that the described strategy will produce an uncessarily excessively long path (but extremely safe path).

As our guideline is to plan the shortest safe path, we look for another modification of the A* algorithm. It seems that the only way to solve our problem is to plan paths in an extended configuration space: at each node in the graph should correspond not only a configuration but equally the associated uncertainty. In this way the drawback mentioned on the previous section will be overriden. As the resulting dimension of the uncertain-configuration space is high (4+3), we proposed the concept of uncertainties towers to reduce the search space. The dimension of an uncertainties tower is "one" and, at each level, a node with included uncertainties is associated. The resulting graph search is consequently done in a 4D space which is easily trackable. Figure 3 exhibits three towers of our graph. Each tower has a position (x, y, θ) and each level of each tower correspond to a particular uncertainty (L, l, φ, Θ)(see figure 1). For the graph search, we define each node as $(x, y, \theta, L, l, \varphi, \Theta)$. Lines represents topological links between two configurations (a Dubins' path [2]; see [12] for a faster algorithm than [2]) and dashed lines represents links (edges) between

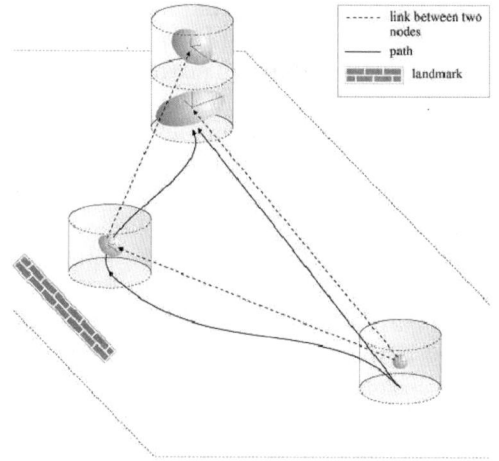

Figure 3: Towers of uncertainties

two nodes. The topological links could be computed in advance but the towers should be built during the graph expansion: the more localisation areas there are, the higher the towers are. It seems impossible to compute the towers in advance as their height depends on the initial configuration and its associated uncertainty. In figure 4, the upper tower has two levels because two possible paths go to this tower:
-the smallest path with the biggest uncertainty
-the biggest path with the smallest uncertainty
Each of these two levels has one associated node that will be developed by the graph search.

```
01 CLOSE←∅
02 OPEN←node_start
03 while (OPEN≠∅)
04   node←head(OPEN)
05   CLOSE←CLOSE + node
06   OPEN←OPEN - node
07   if (node = node_goal) then exit(success)
08   NEW_NODES←successors(node)
09   for each new_node of NEW_NODES do
10     if configuration(new_node)∉configurations(OPEN,CLOSE)
         and (new_node is collision_free) then
11       g(new_node)← g(node) + d_dubins(node, new_node)
12       f(new_node)←g(new_node)+d(new_node,node_goal)
13       build(NEW_TOWER, configuration(new_node))
14       add_level(NEW_TOWER, new_node)
15       OPEN←OPEN + new_node
16       parent(new_node)←node
17     else
18       TOWER←extract_tower( configuration(new_node))
19       level← - 1
20       do
21         add_level←false
22         level←level + 1
23         n_node←extract_node(TOWER,level)
24         if ( ( g(new_node) ≥ g(n_node) and
               ellipsoid(n_node) ⊄ ellipsoid(new_node))
             or ( g(new_node) < g(n_node)) )
             then add_level←true
25       while( level≠top_level(TOWER) and add_level=true)
26       if (add_level=true)
27         level←insert(new_node, TOWER)
28         OPEN←OPEN + new_node
29         parent(new_node)←node
30         U_NODES←nodes(upper_levels(TOWER, level))
31         for each u_node of U_NODES do
32           if (ellipsoid(new_node) ⊆ellipsoid(u_node) ) then
33             remove(TOWER,u_node)
34             OPEN←OPEN - u_node
35           endif
36         endfor
37       endif
38     endif
39   endfor
40 endwhile
41 exit(no_solution)
```

Algorithm 2: A* with towers of uncertainties

4.3 Modification of the A* algorithm

One point should be further elaborated: when should we add up a new level on a tower? The simplest case occurs when no tower exists at the node configuration. Then, we can build a tower with its first level (see algorithm 2, line 10 to 16). Each node is described by its configuration and its uncertainties. Although not mentioned, the uncertainties are computed at line 08 : each time that a new node is discovered, its associated uncertainty is computed.

Assuming that the lastest computed path is longer than all the previous ones, we should add a new level only if the new uncertainty is "better" then all the previously computed uncertainties. Consequently, we should compare the new uncertainties' ellipsoid with all the previous ones. The only way to decide if ellipsoid1 is "better" (not worse) than ellipsoid2, is to verify that ellipsoid1 does not include ellipsoid2. So, if the new ellipsoid does not include any of the ellipsoids located in the lower levels, then we choose to add a new level (see line 24 of algorithm 2).

Assuming that the last computed path is shorter than all the previous ones, we can add a new level to the tower (line 24). But in this case, we should verify that the upper levels (higher the levels, higher the 'g' cost) are still valid; their ellipsoids should be good enough to be kept (line 30 to 36). Unfortunately, in most of the cases, the path is longer than some other paths and shorter than the others. Consequently, the algorithm is a mixture of the two cases described above (line 28 to 38).

We now come back to the problematic example exhibited in the previous section. In figure 2(c) the vehicle first gets close to the nearest wall and follows it in order to get localization information. Next it can cross the horizontal road in order to reach the next wall and reduce its uncertainty which grows during this open loop motion. Having a small uncertainty at its location, the vehicle can then safely enter the corridor.

5 Examples

The following examples aim to show how our planner can adapt its strategies according to different initial and goal uncertain configurations. We consider a robot equipped with sensors which has a longer range than in the previous examples. On the first example (fig. 4(a)), the final uncertainties are unbounded. The resulting path exhibits minimal length characteristics. If we reduce the final allowed uncertainties then the robot should enter the upper corridor in order to relocalize itself before going to the goal (fig. 4(b)). By increasing the initial uncertainties (fig. 4(c)), the robot cannot enter the first corridor (due to the cumulative uncertainties). Consequently, it turns to the right to reduce its uncertainties (using the two lower landmarks) before going to the goal.

6 Conclusion

In this paper we have presented a new planner which works in an uncertain-configuration space and uses a localization method widely used in navigation.

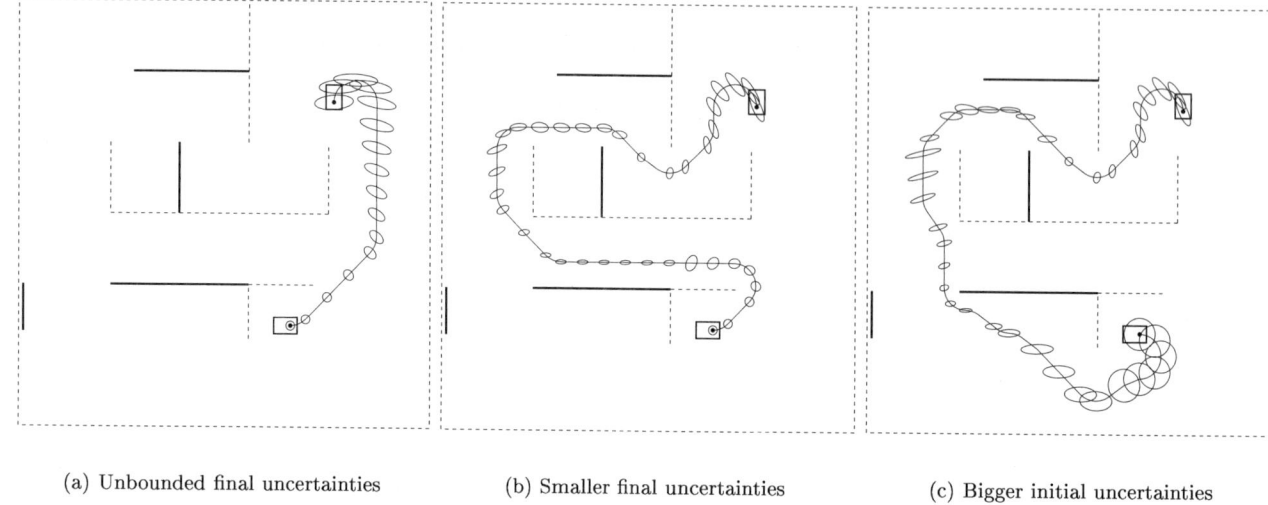

(a) Unbounded final uncertainties   (b) Smaller final uncertainties   (c) Bigger initial uncertainties

Figure 4: Different initial and final uncertainties lead to different planned paths

The concept of towers of uncertainties allows us to plan shortest safe paths which were far to be possible before. Further works will concern new results with deep analyzing and proofs of our algorithm.

References

[1] B. Bouilly, T. Siméon and R. Alami. *A numerical technique for planning motion strategies of a mobile robot in presence of uncertainty*, IEEE Int. Conf. on Robotics and Automation, pp. 1327-1332, Nagoya, May 1995.

[2] L. Dubins. *On curves of minimal lenght with a constraint on a average curvature and with a prescribed initial and terminal positions and tangents*, American Journal of Mathematics, no79, pp. 497-516, 1957.

[3] Th. Fraichard and R. Mermond. *Path planning with uncertainty for car-like robots*, IEEE Int. Conf. on Robotics and Automation, pp 27-32, May 1998.

[4] P. Hart, N. J. Nilsson and B. Raphael. *A formal basis for the heuristic determination of minimum cost paths*, IEEE Transaction on System Sciences and Cybernetics, Vol. 4, pp. 100-107, 1968.

[5] A. Lambert and N. Le Fort-Piat. *Safe task planning integrating uncertainties and local maps federation*, International Journal of Robotics Research, 19(6), pp. 597-611, June 2000.

[6] A. Lambert and Th. Fraichard. *Landmark-based safe path planning for car-like robots*, IEEE Int. Conf. on Robotics and Automation, pp 2046-2051, San Francisco, April 2000.

[7] J.C. Latombe. *Robot motion planning*, Kluwer Academic publishers, 1991.

[8] A. Lazanas and J.C Latombe. *Motion Planning with uncertainty: a landmark approach*, Artificial intelligence. 76(1-2), pp. 287-317, 1995.

[9] N. Le Fort-Piat, I. Collin, D. Meizel. *Planning robust displacement missions by means of robot-tasks and local maps*, pp. 99-114, vol. 20, Robotics and Autonomous Systems, 1997.

[10] L.A. Page and A.C Sanderson. *Robot motion planning for sensor-based control with uncertainties*, IEEE International Conference on Robotics and Automation, pp. 1333-1340, Nagoya, May 1995.

[11] L. Shen, V.J. Lumelsky, A.M. Shkel. *Hazard and safety regions for paths with constrained curvature*, Mathematical Methods in Applied sciences, Vol. 21, pp. 1655-1679, 1998.

[12] A.M. Shkel, V. J. Lumelsky. *Classification of the Dubins set*, Robotics and Autonomous Systems, 34(4), pp. 179-202, 2001.

[13] R.C. Smith and P. Cheeseman. *On the representation and estimation of spatial uncertainty*, International Journal of Robotics Research, pp.56-68, 5(4), 1986.

[14] H. Takeda, C. Facchinetti and J.C. Latombe. *Planning the motions of a mobile robot in a sensory uncertainty field*, IEEE Transactions on PAMI, 16(10), pp. 1002-1017, October 94.

On-Line Safe Path Planning in Unknown Environments

Chen Weidong, Fan Changhong and Xi Yugeng

Institute of Automation
Shanghai Jiao Tong University
Shanghai, 200030, P.R. China
changhongfan@sjtu.edu.cn, wdchen@sjtu.edu.cn, ygxi@sjtu.edu.cn

Abstracts - For the on-line safe path planning of a mobile robot in unknown environments, the paper proposes a simple Hopfield Neural Network (HNN) planner. Without learning process, the HNN plans a safe path with consideration of "too close" or "too far". For obstacles of arbitrary shape, we prove that the HNN has no unexpected local attractive point and can find a steepest climbing path, if a feasible path(s) exists. To effectively simulate the HNN on sequential processor, we discuss algorithms with $O(N)$ time complexity, and propose the constrained distance transformation-based Gauss-Seidel iteration method to solve the HNN. Simulations and experiments demonstrate the method has high real-time ability and adaptability to complex environments.

I. Introduction

Planning a collision-free path to navigate a mobile robot through an unknown environment has received considerable attentions. The path planning methods [5,6] includes the graph-search, the roadmap and the potential field methods *etc*. The A* algorithm [5] is a typical graph-search method for path re-planning, and its time complexity is $O(N\log N)$ for an $\sqrt{N}\times\sqrt{N}$ grid map. The roadmap method [6] has fewer nodes than the grid map and can plan path very quickly. But the online reconstruction of the roadmap is time-consuming. The artificial potential field (APF) method, proposed by Kathib [8], is computationally advantageous, but introduces unexpected local attractive point.

Recently, Glasius [4] and Yang [13-15] respectively used the numerical potential fields (NPF) of generalized HNN to plan collision-free path in a dynamic, unknown environment. These HNN take parallel time complexities $O(N)$ to find feasible paths. But there're some problems to be analyzed for their method. 1) The completeness: [4,13-15] paid more attention to the stability of their HNN, and deemed the stability of the HNN would insure the finding of a feasible path if it exists, but ignore the analysis of the completeness. Unfortunately, we found HNN of [13-15] existed unexpected attractive point sometimes. 2) The effective application on sequential processors: because the outputs of those HNN nodes are very small (such as 10^{-100} Voltage), those HNN are still hard to be precisely realized by nowadays hardware technology, and can only be simulated on computers. A single sequential processor can do but pseudo-parallel simulation of the HNN. So in simulations on a sequential processor, continuous HNN of [4,13-15] become discrete HNN similar to Kassim's WENN [7], and also need $O(N^2)$

time to form feasible paths. Lagoudakis [6] used the "Raster Scan" (RS), a sequential scan, to accelerate the propagation of the NPF of Glasius' HNN. But the NPF's propagation with RS is constrained by the obstacle nodes, and there is still much useless computing.

The safety of the planned path is nontrivial for navigating a mobile robot. Some methods for path planning try to find the shortest path. But the shortest path always clips the corners of obstacles and runs along the edges of obstacles, and this forms the "too close" problem [16]. Some other methods, such as the Voronoi diagram (VD) [5], plan path along the middle axis of the free configuration space. But the VD method always produces path that is far away from the shortest path, and forms the "too far" problem. For a static and known environments, Zlinsky [16] increased the cost of area nearing obstacles and used a distance transformation for the A* algorithm to search for the safe path, but the time complexity is still $O(N\log N)$. Muniz [10] used a feed-forward neural network to learning the planning of safe path, but the learning process is time-consuming. Yang [14] extended his HNN method to find a safe path in unknown environments. But as above discussion, method in [14] needs $O(N^2)$ time on a sequential processor, and the stability of HNN in [14] can't guarantee the elimination of unexpected attractive points.

We propose a simple HNN for safe path planning. This model has the following characters: 1) The neural nodes are locally connected; 2) The safety of the collision-free path is considered in the weight design of the HNN, and then no learning process is needed to keep the robot suitable distance away from obstacles. We give the condition for the existing of the feasible path, and then we prove the HNN can find a steepest climbing path if there is feasible path(s) and the HNN don't have unexpected attractive points, that means the method is complete. For effective application on a sequential processor, we use the constrained distance transformation (CDT) [11] information to guide the asynchronous iteration of the HNN and can plan a safe path in $O(N)$ time.

Section II describes the HNN. The stability of HNN and the condition for the existing of feasible path(s) are analyzed in Section III; Section IV analyzes the safe path planning. The effective solving of the HNN is given in Section V, and Section VI provides simulations and experiment. We summarize conclusions in Section VII.

II. The HNN model

The 2-dimensional environment may be unknown initially, and obstacles are not assumed to be convex. The robot R has only local sense ability and can incrementally construct the map of the environment. The configuration space [9] of the mobile robot R is C; obstacles are represented by O_i, $i=1,2,..,m$, which are sets of the unreachable points in C; $R(q)$ represents the robot R at the point $q \in C$. So the unreachable set is $\cup O_i' = \cup \{q \in C \mid R(q) \cap O_i \neq \Phi\}$, and the free configuration space is $C_{free} = C \setminus \cup O_i'$.

Fig.1 illustrates Net, the locally connected neural network similar to that in [7]. Each neural node in Net represents a configuration point in the discrete configuration space C. According to the connectivity between the neighboring points in C, each node of Net at most is connected with $2d$ neighboring nodes. If $d=2$, the nodes are 4-connected; and if $d=4$, the nodes are 8-connected. We select $d=4$ in following simulations and experiments. The robot makes decisions at discrete time, $T=1,2,..$, and moves along the steepest climbing path, if it exists, to reach the target.

We assume the number of nodes in Net is N for a $\sqrt{N} \times \sqrt{N}$ grid map. The output of the ith node is $x_i(t)$ for $1 \le i \le N$, and the 8-connected neighboring set of the ith node is NE_i. The target node is $Tset$, $1 \le Tset \le N$. And only the $Tset$ node has external constant input $I>0$.

The dynamic equation of the ith node is

$$\dot{x}_i(t) = \begin{cases} -Ax_i(t) + D_i(T)y_i(t), & \text{if } i \neq Tset \\ -Ax_i(t) + D_i(T)y_i(t) + I, & \text{otherwise} \end{cases} \quad (1)$$

$$y_i(t) = m \sum_{j \in NE_i} \omega_{ij} x_j(t) \quad (2)$$

Equations (1) and (2) form a linear system. In (1), $A>0$ is the negative feedback gain. $D_i(T)$ represents the accumulated information about the environment at discrete time T. If the ith node is occupied by obstacle, then $D_i(T)=0$; otherwise $D_i(T)=1$.

In (2), m is a positive gain coefficient. The connected weight from the node j to node i is $\omega_{ij} = \omega_{ij0} \omega_{ijs} \in (0,1]$. ω_{ij0} represents the factor of the Euclidean distance between nodes i and j, and we select $\omega_{ij0} = (2md/A)^{0.414} < 1$ (in following theorems we asks $2md/A<1$) for the diagonal 8-connected neighboring nodes, while $\omega_{ij0}=1$ for the 4-connected neighboring nodes. Additionally, $0 < \omega_{ijs} \le 1$ represents the safety factor from the jth node to the ith one, and is given in Section IV. Because the safety factor from j to i may be different that from i to j, so we don't assume the connected weights are symmetrical, and this is different from [4,13-15].

The system equation of Net is rewritten in

$$\dot{x}(t) = -Ax(t) + G(T)x(t) + U \quad (3)$$

where only the element of U corresponding to $Tset$ is I, and the other elements of U are zeros.

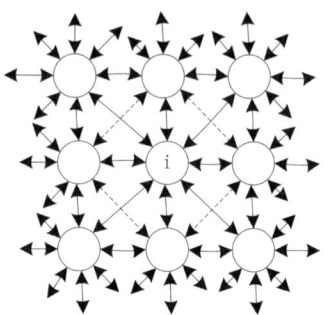

Fig.1 8-connected HNN of the 2-dimensional environment

III. Stability and non-negative of the HNN

Glasius [4] and Yang [13-15] respectively used the Liapunov method to prove the stabilities of their models. For the simple HNN model (3), The M-matrix [2] is introduced to elegantly prove the global exponential stability (GES) of (3) and the non-negative property of the HNN' equilibrium point which describes whether one node has feasible path or not.

Definition 1 [2]: The $n \times n$ matrix $Q=\{q_{ij}\}$ is a Z_n-matix, if $q_{ii}>0$ and q_{ij} is not larger than 0 for $i \neq j$.

Lemma 1 [2]: the Z_n-matrix Q is a nonsingular M-matrix, is equivalent to:

1) There exists positive diagonal matrix D, such that he matrix QD is a strictly diagonal dominant matrix and the diagonal elements of QD is larger than zero;
2) The real part of each eigenvalue of Q is positive;
3) Q is nonsingular and $Q^{-1} \ge 0$;

Theorem 1: If $A>2md$, then the system (3) has only one equilibrium point x_e and is GES.

Proof. The system matrix of the linear system (3) is $(-AE+G(T))$, where E is the $N \times N$ unit matrix. It can be easily known that the sum of elements in every row (column) of $G(T)$ is smaller than $2md$. By the definition 1, if D is an unit matrix and $A>2md$, $(AE-G(T))D = (AE-G(T))$ is a strictly diagonal dominant Z_n-matrix. By 1) of *Lemma* 1, $(AE-G(T))$ is nonsingular M-matrix, and then the real part of every eigenvalue of $(AE-G(T))$ is larger than 0 by 2) of *Lemma* 1, that is the real part of every eigenvalue of $(-AE+G(T))$ is smaller than 0. According to the linear system theory, system (3) has only one equilibrium point x_e and is GES. The proof is complete.

Theorem 2: If $A>2md$, then (a) the equilibrium point satisfies $x_e \ge 0$; (b) $x_{ei}=0$ for any node i that don't have feasible path to the target node $Tset$; and $x_{ei}>0$ for any node i that has feasible path to the target node $Tset$.

Proof. (a) Because the equilibrium point x_e of the system (3) satisfies $\dot{x}_e = 0$, we have $(AE-G(T))x_e=U$. By theorem 1, if $A>2md$, then $(AE-G(T))$ is nonsingular M-matrix of Z_n, and $x_e=[AE-G(T)]^{-1}U$. Then by 3) of *Lemma* 1 and $U \ge 0$, we have $[AE-G(T)]^{-1} \ge 0$ and $x_e=[AE-G(T)]^{-1}U \ge 0$.

(b). If there is no feasible path between the ith node and $Tset$, we separate all the nodes that have feasible paths to the ith node from the system (3), and a new linear system forms. By theorem 1, this new system is also GES. And more, it can be easily known that this new linear system has no external input, and so $x_{ei}=0$.

Otherwise, we suppose that $Tset$ has a feasible path which passes the non-obstacle nodes i_1, i_2, i_3,...to reach the ith node. Firstly, from (a), $Tset$ gives a positive excitation to the node i_1, and more all other excitations to i_1 are not smaller than 0, so it is immediate that the stable state of the node i_1 is larger than zero. Analogically, all nodes on the feasible path between $Tset$ and the ith node have stable states that larger than zero. The proof is complete.

Form Theorem 2, we have the corollary:
Corollary 1: $x_{ei}>0$ is equivalent to that there is feasible path between the ith node and $Tset$.

In following, $Net(T)$ represents the set of nodes that have feasible paths to $Tset$ at time T. The stable state of the nodes in $Net(T)$ are noted as $\bar{x}_e$, and $\bar{x}_e$ satisfies $\bar{x}_e > 0$.

IV. Path planning

This section firstly proves the NPF of the HNN doesn't have unexpected attractive point. So if there is feasible path(s), the NPF of the HNN must form a steepest climbing path from the starting node to the target.

1 The steepest climbing path

Similar to Glasius' and Yang's models, our model also uses the steepest climbing path of the NPF of the HNN to navigate the robot to the target, which needs that the NPF can't have local maximal attractive point at any node but the target node. In the following theorem, we analyze the local maximal property of the NPF formed by $\bar{x}_e > 0$, and show that the NPF has only one local maximal point that's just at the target point.

Theorem 3: If $A>2md$, the global NPF formed by $\bar{x}_e > 0$ has only one local maximal point, and this local maximal point must be at the target node.

Proof by contradiction. Assume that $\bar{x}_e$ has a local maximal point at the ith node, $i \in Net(T)$ and $i \neq Tset$. By (1), (2) and $\dot{x}_e = 0$, we know that

$$m \sum_{j \in NE_i} \omega_{ij} x_{e_j} = A\bar{x}_{e_i} > 0$$

$$0 < m \sum_{j \in NE_i} \omega_{ij} x_{e_j} \leq 2md \max_{j \in NE_i} \omega_{ij} x_{e_j}$$

then

$$2md \max_{j \in NE_i} \omega_{ij} x_{e_j} \geq A\bar{x}_{e_i} > 0$$

But by the precondition that $A>2md$, we know $\max_{j \in NE_i} x_{e_j} \geq \max_{j \in NE_i} \omega_{ij} x_{e_j} > \bar{x}_{e_i} > 0$, which means that there is at least one node in the neighborhood of the ith node has a larger output than the ith node. So this is in opposition to the assumption, and means that the NPF formed by $\bar{x}_e$ impossible has local maximal point at non-target node.

All elements of $\bar{x}_e$ are larger than 0 and the number of the elements of $\bar{x}_e$ is enumerable, so $\bar{x}_e$ has at least one global (and local at the same time) maximal point.

From the above two outcomes, it is immediately known that the global NPF formed by $\bar{x}_e > 0$ has only one local maximal point, and this local maximal point must be at the target node. The proof is complete.

Theorem 3 indicates that in the neighborhood of any non-target node I in $Net(T)$, there is a node j having the largest output in node i's neighborhood. We call the node j is the steepest climbing node of the node i. if $j \neq Tset$, node j also has a new steepest climbing node. If the steepest climbing nodes are selected one by one, a steepest climbing nodes sequence from the node i to the node $Tset$ is formed. By contraries, the outputs of neighboring nodes of any node not in $Net(T)$ are zeroes, so there doesn't exist the steepest climbing path for any node not in $Net(T)$.

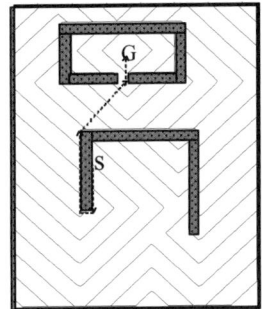 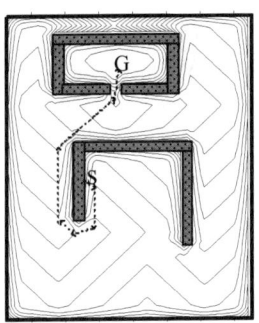

(1) (2)

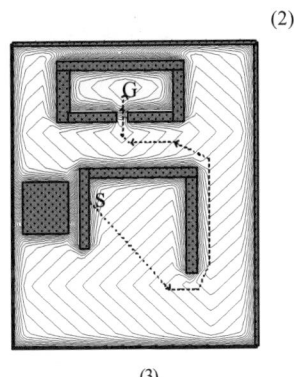

(3)

Fig.2 ▨ obstacle on; S: starting point; G: target point; (1) The contour of the NPF of the HNN without safety consideration and the shortest path; (2) The contour of the NPF of the HNN with safety consideration and the safe-path; (3) The new contour of the NPF of the HNN with safety consideration and the corresponding new safe-path

Notes: Model (3) can be extended to Yang' HNN [13] whose GES condition is $A>0$. But in simulations, if $A>0$ and

$A<2md$, the extended model has unexpected attractive points sometimes. So our analysis of theorem 3 is necessary.

2 Safe path planning

If $\omega_{ijs}=1$, the HNN (3) will also plan the shortest feasible path similar to [4,13,15]. This means that the HNN (3) doesn't consider the safety factor and implicitly optimize the length of the feasible path when $\omega_{ijs}=1$. Fig.2(1) illustrates the contour mapping of the equipotential surfaces of the NPF formed by x_e when $\omega_{ijs}=1$. The steepest climbing path search forms the shortest path from the starting node S to the target node G in Fig.2(1). Obviously the shortest path always clips the corners of obstacles and runs along the edges of obstacles.

From the theorem 3, if $\omega_{ijs}>0$, the NPF of $\bar{x}_e$ will forms the steepest climbing paths from any nodes that have feasible path to the target node. This means, that so long as we select $\omega_{ijs}>0$ then the NPF of $\bar{x}_e$ will not have unexpected maximal point. To consider the safe factor of the path, we decrease the weights of nodes that near obstacles to reduce the output of these nodes. A node is more near obstacles, the outputs of this node is smaller. When the robot is close to obstacles, the small weights ω_{ijs} will form repulsive potential filed to impulse the robot away from the obstacles. This kind of design of ω_{ijs} will not have unexpected local maximal point.

First, we compute the Euclidean distance (ED) of every no-obstacle nodes' to the nearest obstacle node by the Euclidean distance transformations (EDT) [12] in $O(N)$ time. For example, the ith nodes' ED is d_i. Then according to the sizes of the grid and the mobile robot, the safe distance limit D_{safe} is selected by the users: if $d_i<D_{safe}$, the safe factor is considered in the weight ω_{ijs}; otherwise, not considered. We select the following function to consider the safe factor:
$\omega_{ijs} = (2md/A)^{fs(d_i)}$, where the safe function $fs(d_i)$ satisfies that $fs(d_i)>0$ and $fs(d_i)$ is a non-increasing function with d_i, that is

$$fs(d_i) = \begin{cases} ks/d_i, & \text{if } 0 < d_i < D_{safe} \\ 0, & otherwise \end{cases} \quad (4)$$

where the parameter $ks>0$ can be adjusted.

By the above selection of ω_{ijs}, ω_{ij} will decrease very quickly while the node i is closer to obstacles, and as illustrated by Fig.2(2) the contour lines of the NPF of x_e near the obstacle is dense, so the robot will be repulsed away from obstacles while the robot still traces the gradient of the NPF to reach the target. Obviously, the safe-path in Fig.2(2) is longer than that in Fig2(1), but the safe-path is advantageous for the robot' safety and forwarding speed.

The safe-path in Fig.2(2) is still similar to that in Fig.2(1). In Fig.2(3), a new obstacle is added to the map of Fig.2(1), the shortest path is still feasible, but the shortest path must pass a strait channel, but the HNN plans a new safe-path in Fig.2(3), which is quite different from that in Fig.2(1) and Fig.2(2).

V. The effective solving of the HNN

Discussing the effective solving of the HNN is important when the HNN is simulated on a sequential processor. A single processor has only sequential computational ability, and can only do pseudo-parallel simulation of HNN. In simulations on a sequential processor, continuous HNN in [4,13-15] become discrete HNN, and need $O(N^2)$ time [7] to form feasible paths. Lagoudakis used the "Raster Scan" (RS), a sequential scan, to accelerate the simulation of HNN of [4], and this formed asynchronous HNN. But the NPFs' propagation with RS is constrained by the obstacle nodes, and there are still many useless computing. We discuss how to effectively solve the equation $[AE-G(T)]x_e=U$ of HNN (3).

1) The direct method

Because each nodes of the HNN has only local connections to nodes in its neighborhood, the $N \times N$ matrix $AE-G(T)$ has at most $9N$ non-zero elements. So by the sparse matrix technology [17], $[AE-G(T)]x_e=U$ can be solved in $O(N)$ time. The direct method is advantageous for a known map. But the programming for solving sparse matrix is complicated, and the solving of the new equilibrium can't use the last equilibrium when new obstacles are added to the map.

2) The indirect method——The iterative method

Lagoudakis's "raster scan" [6] simulation turned the parallel HNN of [4] into a sequential or asynchronous HNN. The RS method could be considered a small integration step Gauss-Seidel iteration method (GSIM) [3], which is an efficient indirect method for solving linear system of equations. Compared the pseudo-parallel simulation on sequential processor, the GSIM accelerates the propagation of NPF quickly. But constrained by the obstacle nodes, after the GSIM is used once, the RS method can't propagate the NPF over all nodes that have feasible path to the target node, and there is still much useless computing.

We use the constrained distance transformation (CDT) [11] to decide the propagation order of the NPF, and then use the order to refresh the GSIM. The gist of this CDT-based GSIM (CDTGSIM) is that: firstly, the nodes are classified into different classes by their constrained connected distances (CCD) to the target node; and then the smaller CCD a class has, the priority the nodes in this class are calculated by the GSIM. For example, the node in the class with CCD=0 (that's just the target node itself) is firstly calculated by the GSIM; secondly the nodes in the class with CCD=1 are calculated by the GSIM; thirdly the nodes in the class with CCD=2 are calculated by the GSIM …, until all nodes that have feasible paths are calculated once. If the starting node can't be expanded by the CDT, then there is no feasible path; otherwise repeating the CDTGSIM several times to approximate to x_e of the HNN (3).

The on-line CDTGSIM is detailed as following:

(1) The 4-connected constrained distances between all non-obstacle nodes and *Tset* are simply computed by the Breadth-First Search [18] in $O(N)$ time and then are classified into different classes by their distances to *Tset* in $O(N)$ time. Assume the largest 4-connected distance N_0, and the number of the nodes (include *Tset*) that have feasible paths to the target node is N_1. Set $j_0=0$, *repnum*=1.

If the starting node is not expanded by CDT, go to (5).

Merge all nodes in the N_0 number of classes into an increasing sequence $\tilde{x}$ according to their CDT values, and we notes the *i*th node in this sequence is $\tilde{x}^i$.

(2) GSIM is used for each node whose 4-connected distance to the target is j_0, e.g. for the *i*th node $\tilde{x}^i$ of $\tilde{x}$:

$$\tilde{x}^i(repnum+1) = \frac{m}{A}(\sum_{j<i \cap j\in NE_i} \omega_{ij}\tilde{x}^j(repnum+1) + \sum_{j>i \cap j\in NE_i} \omega_{ij}\tilde{x}^j(repnum))$$

The above iteration formula means that the states of the nodes that have been iterated will be directly used by the successive nodes, and forms a asynchronous iteration.

(3) $j_0+1 \rightarrow j_0$. If $j_0>N_0$, go to (4); else go to (2)

(4) $j_0=1$, *repnum*+1→*repnum*. If *repnum* is smaller than pre-set iteration number, go to (2); else go to (6).

(5) There is no feasible path, so stop and go to (7).

(6) Search the steepest climbing path, and go to (7).

(7) In the next decision processing, if new obstacle information is obtained, then go to (1) to recalculate the iterative order of the CDTGSIM; else the iterative order of the CDTGSIM is unchanged and directly go to (2).

The time complexity from (2) to (4) is $O(N)$. So it is obviously that the whole process from (1) to (6) for a path re-planning in every decision period is just $O(N)$.

VI. Simulations and Experiments

In simulations, we assume: 1) the robots' perception radius is 10 times of the length of the grid; 2) the HNN is computed twice by the CDTGSIM every discrete time with the new obstacle information.

A Pioneer 2 mobile robot whose decision period is 0.1 second is used in experiments. The robot has 16 sonars and the effective detect ranges of these sonars are limited to 1.5 meters. The environment of 10×10 square meters is divided into a grid map with 100×100 grids. We use the histogamic in-motion mapping method [2] to decide whether a grid is occupied by obstacle or not. The obstacle distance transformation [12] is used to cut down the very narrow pass that can't be passed by the robot.

The processor for simulations and experiments is a 300M CPU with Linux OS. For a large grid map with 500×500 grids, the processor can do 30 times CDTGSIM per second.

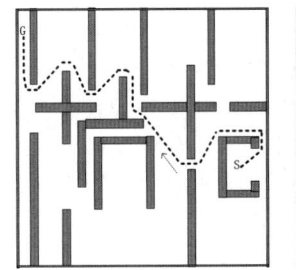 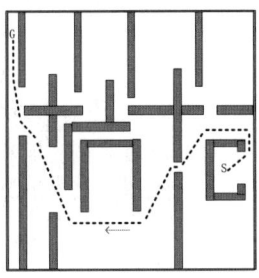

(1) safe path with $ks=5$ (2) safe path with $ks=10$

Fig 3 The CDTGSIM for the HNN to plan safe-path

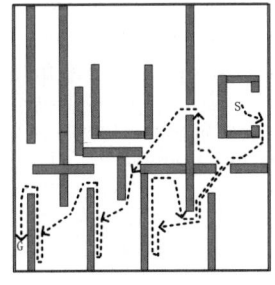 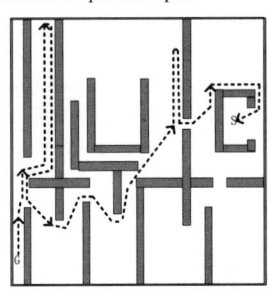

(1) From S to G   (2) From G back to S

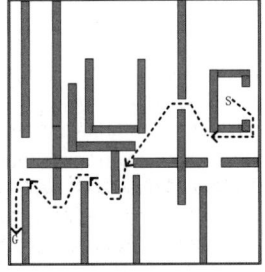

(3) From S to G again

Fig.4 The on-line path planning from S to G

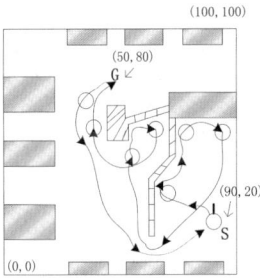

Fig.5 Experiment in the unknown environment: the real line gives the trajectories from S to G and from G back to S

1 The fast speed of the CDTGSIM

Within our experience, the CDTGSIM is very effective to solve the HNN for safe-path planning. In this simulation, the initial map is known and it has 500×500 grids. We select $A=8$, $m=1$, $ks=5$ and $D_{safe}=5$. After twice CDTGSIM, the

HNN plans a safe-path in Fig.3(1). The CPU time for this safe-path is less than 0.4s. A new safe-path is also planned for a larger $ks=10$ after twice CDTGSIM in Fig.3(2). The new safe-path is longer than that in the Fig.3(1), but the new safe-path is more smooth. The CPU time for the new safe-path is less than 0.4s.

The CPU time reported here is very faster than the results reported in [4,6,13-15], and shows that the CDTGSIM has high real-time ability for path planning.

2 Simulations under the unknown environment

Initially the robot doesn't know any obstacle information about the environment. After 283 times CDTGSIM, the robot reaches G the first time in Fig.4(1), and takes 312 times CDTGSIM back to S in Fig.4(2). And then the robot moves from S to G again with 133 times CDTGSIM in Fig.4(3). In all the round-trip, the robot doesn't clip the corners of obstacles or run along the edges of obstacles, and isn't trapped in unexpected local attractive point.

3 Experiment under the unknown environment

The forward speed of the real robot is proportional to the distances of forward obstacles, and is limited under 250 mm/s. Without initial information about the obstacles, the robots' spent much time to explore the enviro nment, and used 85 seconds on reaching G. Then with the grid map of the detected obstacles, the robot used less time, 30 seconds, to back to S.

VII. Conclusions

A HNN is used to plan safe path on-line. Without learning processing, the HNN can plan a safe-path that compromises between the "too close" and "too far" paths. For environments of arbitrary shape, the NPF of the HNN has no unexpected attractive points. Effective solving of the NPF of the HNN based on distance transformation is given. Simulations and experiments demonstrated the method has high real-time ability and adaptability to complex environments.

VIII. Acknowledgements

This work is supported by National high-tech research and development plan under grant 2001AA422140, by National Science Foundation under grant 60105005.

Reference

[1] Berman, A. *Nonegative matrices in the mathematical science*. 2nd ed. New York Academic, 1994

[2] Borenstein, J., Y. Koren, *The vector field Histogram --- fast obstacle avoidance for mobile robots*. IEEE RA, 1991,Vol.7(3):278-288

[3] Connolly, C.I. et al, *Path planning using Laplace's Equation*. ICRA, 1990,Vol.3:2102-2106

[4] Glasius, R., A. Komoda, *Neural network dynamics for path planning and obstacle avoidance*. Neural Networks, 1995, Vol.8(1):125-133

[5] Huang, Y., Ahuju, N., *Gross-motion planning*, ACM Computing Surveys, 1992, Vol.24(3): 220-291

[6] Lagoudakis, M.G., Maida, A.S. Neural maps for mobile robot navigation. 1999 Int. Joint Conf. on Neural Networks, 1999, Vol.3: 2011-2016

[7] Kassim, A., V. Kumar. *Path planners based on wave expansion neural network*. Robotics and Autonomous Systems, 1999, Vol.26:1-22

[8] Kathib, O., *Real-time obstacle avoidance for manipulators and mobile robots*. Int. J. Rob. Res. 1986,Vol.5:90-98

[9] Latombe, J.C., *Robot motion planning*. Kluwer Academic Publishers, 1991

[10] Muniz, F., et al, *Neural controller for a mobile robot in nonstationary environment*. the 2nd IFAC on Intelligent Autonomous Vehicles, Helsinki, Finland, 1995: 275-284

[11] J. Piper, E. Granum, *Computing distance transformations in convex and non-convex domains*. pattern Recognition, 1987, Vol.20, pp: 599-615

[12] Shih, F.Y., Liu, J.J., *Size-invariant four-scan Euclidean distance transformation*, Pattern Recognition, 1998,Vol.31(11):1761-1766

[13] Yang, S.X., M. Max, *Neural Network Approaches to Dynamic Collision-Free Trajectory Generation*. IEEE SMC Part B. 2001, Vol.31(3):302-318

[14] Yang, S.X., M. Max, *An efficient neural network method for real-time motion planning with safety consideration*, Robotics and Autonomous Systems, 2000, Vol.32: 115-128

[15] Yang, S.X., M. Max, *An efficient neural network approach to dynamic robot motion planning*, Neural Networks, 2000, Vol.13: 143-148

[16] Zelinsky, A., *Using path transforms to guide the search for findpath in 2D*. The Int J. of Robotics Research, 1994, Vol.13(4):315-325

[17] S. Pissanetzky, *Sparse matrix Technology*. New York: Academic, 1984

[18] Shaffer, C.A. *Data structures and algorithm analysis*. Prentice Hall Publishers, 1996

Robot Motion Decision-Making System in Unknown Environments

S. Boonphoapichart*, S. Komada*, T. Hori**, and W. A. Gruver***

*Dept. of Electric and Electronic Eng., Mie University
1515 Kamihama-cho, Tsu-shi, Mie-ken, 514-8507 Japan
somchai@hori.elec.mie-u.ac.jp, komada@elec.mie-u.ac.jp

**Dept. of Electronics & Information Eng., Aichi University of Technology
50-2 Manori, Nishihazama-cho, Gamagori-shi, Aichi-ken, 443-0047 Japan
hori@aut.ac.jp

***Intelligent Robotics and Manufacturing Systems Laboratory, School of Engineering Science
Simon Fraser University, Burnaby, BC V5A1S6 Canada
gruver@cs.sfu.ca

Abstract – A fuzzy decision-making system for multi-criteria robot motion planning and control is described for a constant speed omni-directional mobile robot with active stereo cameras capable of panoramic viewing. The robot can be commanded to move to a target point through an unknown environment while satisfying multiple criteria. Advantages of the proposed method are confirmed by simulation. A multi-agent approach is also described for path planning with multiple robots.

I. INTRODUCTION

Robot motion planning can occur in known or unknown environments. For known environments, the robots acquire all information about the environment and then generate a path to the target. Methods based on optimization and computational intelligence have been developed to deal with this issue [1-2]. For unknown environments, in which the robot does not have a prior knowledge of its environment or the environment is rapidly changing, some methods have been proposed based on grid-based control, sensor based control, and robot learning [3-5].

Previous proposed methods treat robot path planning with only one sub-objective such as shortest distance or collision [6-7]. In this research, we describe a method for robot path planning, called *Fuzzy Decision-Making System*, that accommodates multiple criteria in unknown environments [8-9]. The effectiveness of the approach is confirmed by a method to deal with multiple objectives called *Multi-factorial Evaluation*.

II. FUZZY DECISION-MAKING

A. Multi-Criteria Fuzzy Decision-Making

Fuzzy sets provide a basis for representing uncertainty. Fuzzy decision-making [10] can be applied to obtain decisions that satisfy specified criteria. Ordered weighted aggregation operators provide a method for quantifiable-guided aggregation. Let us assume $\{\sigma_1(x), \sigma_2(x), ..., \sigma_i(x), ..., \sigma_m(x)\}$ is a fuzzy set consisting of goals or constraints (criteria) and x is an object such that for any criterion rule σ_i we have

$$D(x) = f(\sigma_1(x), \sigma_2(x), \sigma_3(x), ..., \sigma_m(x)) \quad (1)$$
$$X^* = arg\{g(D(x))\} \quad (2)$$

where $\sigma_i(x) \in [0,1]$ is the fuzzy value of ith (criterion)
$D(x)$ is a fuzzy set defined over alternatives X
$f(x), g(x)$ are the decision algorithms
X^* is the argument of $g(D(x))$

To determine the degree to which x satisfies all criteria, we determine $f(x)$ to be a function such that

$$D(x) = \min_{x \in X}\{\sigma_1(x), \sigma_2(x), \sigma_3(x), ..., \sigma_m(x)\} \quad (3)$$

Thus, x must satisfy $\{\sigma_1$ and σ_2 and $\sigma_3...$ and $\sigma_m\}$.

To determine the degree to which x satisfies at least one of the criteria, $g(x)$ is a function such that

$$X^* = arg\{\max_{x \in X}(D(x))\} \quad (4)$$

In this case x must satisfy the condition $\{\sigma_1$ or σ_2 or $\sigma_3...$ or $\sigma_m\}$.

Since the minimizing function is chosen to consider all criteria for each alternative in (3), we choose weightings defined by

$$D(x) = \min(\sigma_1^{\alpha_1}(x), \sigma_2^{\alpha_2}(x), ..., \sigma_n^{\alpha_n}(x)) \quad (5)$$

where α_i is the weighting factor of ith criterion and $\sum \alpha_i = n$.

B. Multi-Factorial Evaluation

Multi-factorial evaluation is a special case of multiple objectives decision-making. Its purpose is to evaluate an object relative to an objective in a fuzzy decision environment [11].

Let u be a set of objects for evaluation, let $\pi = \{f_1, f_2, ..., f_m\}$ be a set of factors in the evaluation, let $E = \{e_1, e_2, ..., e_p\}$ be a set of grades or qualitative classes used in the evaluation, and let $\varphi_j^{(k)}$ be the objective function rule j in qualitative class k.

Let $r_{ij}(u) = \varphi_j^{(k)}(f_j(u))$. Then the $m \times p$ multi-factorial evaluation matrix can be written as

$$R^{(u)} = \begin{bmatrix} r_{11}(u) & r_{12}(u) & \ldots & r_{1p}(u) \\ r_{21}(u) & r_{22}(u) & \ldots & r_{2p}(u) \\ \vdots & \vdots & \ldots & \vdots \\ r_{m1}(u) & r_{m2}(u) & \ldots & r_{mp}(u) \end{bmatrix} \quad (6)$$

Let $W = [w_1\ w_2\ \ldots\ w_m]$ be a constant matrix of weighting factors and $w_1+\ldots+w_m = 1$. A multi-factorial evaluation model is shown in Fig. 1.

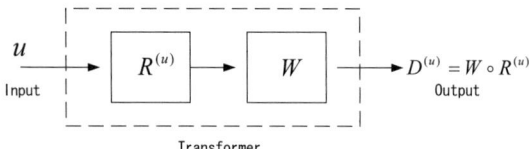

We use the multi-factorial evaluation model to show that our fuzzy decision-making system can give a better result than that obtained with one objective.

III. SYSTEM ARCHITECTURE

A. Autonomous System Architecture

An *autonomous system* is one that has the ability to plan and execute its own actions. A control system with high autonomy should have the ability for self-governance [12]. The functional level architecture of an autonomous controller is shown in Fig. 2.

The management and organization level separates the user command into tasks to be coordinated. The coordination level considers the tasks, and determines the subtasks for execution level, where algorithms in software and hardware will be used to determine control commands. Subsequently, autonomous machines execute the subtasks. When subtasks are finished, they will send responses to the upper level and consider the next subtask. If the subtasks are impossible, the upper level will be asked to make a new decision.

B. Process Diagram by Using Multi-Criteria Fuzzy Decision-Making

We consider a system consisting of a mobile robot or a group of mobile robots that are commanded to move to a target point in an unknown environment. The system processes information from sensors to satisfy multiple criteria. The position of the obstacles is obtained by the use of vision sensors on each mobile robot. Finally, the sensor data is used for multi-criteria fuzzy decision-making.

As shown in Fig.3, we specify the task decision "rough path (direction decision)" and the subtask "precise path (trajectory decision)". The mobile robot must determine a suitable initial direction and a path to follow.

From (1) and (2) fuzzy sets are generated from alternative paths and fuzzy rules involving multiple

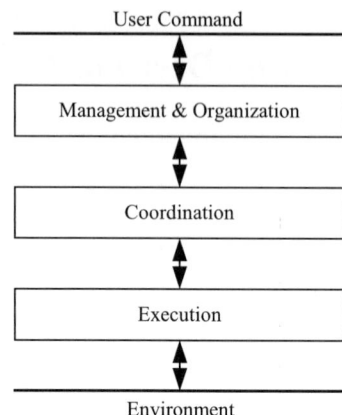

Fig. 2 Functional architecture

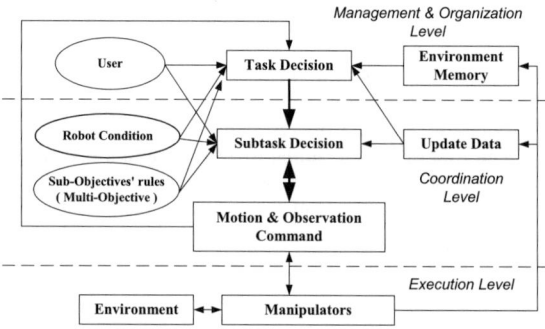

Fig. 3 Proposed process diagram

criteria. Data from the visual sensors will be transformed to fuzzy values for multi-criteria fuzzy decision-making using requirements for geometries of safety, shortest path, past direction, and visual path. After the multi-criteria fuzzy decision-making phase has been completed, an appropriate direction will be obtained.

The approach for obtaining a suitable trajectory is similar. The robot continues until the subtasks are finished, or the subtasks cannot be executed because a new obstacle is sensed or new information is obtained. Then, the process repeats.

C. Alternative Path Detection Method

The visual sensors provide environmental data for the mobile robot system. Raw data from the visual sensors are used to generate alternative paths by

$$|dist(ang) - dist(ang-1)| \geq k \quad (7)$$

in which *dist(ang)* is the distance from visual sensors at the angular position, *ang* (deg); *k* is constant equal to (*robot size*+(*n*)*(*safety compensation*)); $n \in \{1,2,\ldots\}$.

D. Criteria Fuzzy Rules

The sensor data is transformed to fuzzy values by using the rules shown in Fig. 4(a), (b), (c), and (d). At task decision level (direction decision level), *x* is a

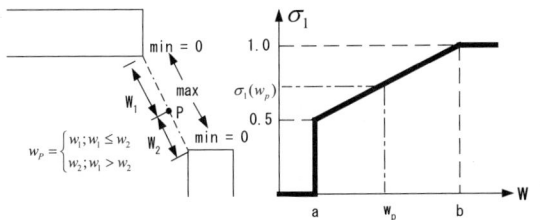

$$\sigma_1 := Safety\_rule(w_P) = \begin{cases} 0; w_P \leq Robot\_size + Safety\_compensation = a \\ \dfrac{0.5 \times (w_P + b) - a}{b - a}; a \leq w_P < 5 \times a = b \\ 1; w_P \geq b \end{cases}$$

(a) Safety rule (σ_1)

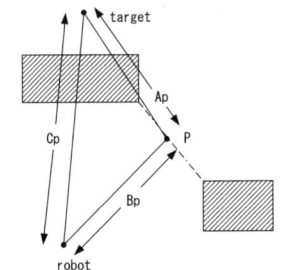

$$\sigma_2 := Shortest\_part\_rule_P = \dfrac{C_P}{A_P + B_P}$$

(b) Shortest rule (σ_2)

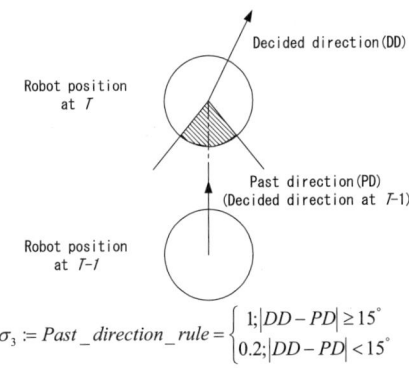

$$\sigma_3 := Past\_direction\_rule = \begin{cases} 1; |DD - PD| \geq 15° \\ 0.2; |DD - PD| < 15° \end{cases}$$

(c) Past direction rule (σ_3)

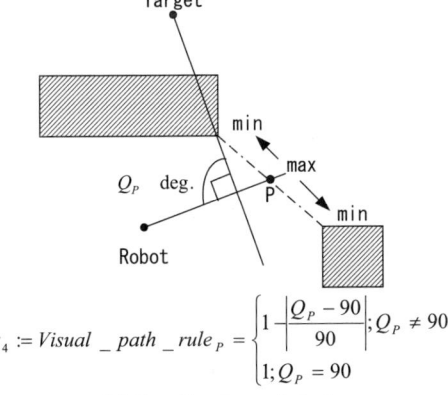

$$\sigma_4 := Visual\_path\_rule_P = \begin{cases} 1 - \left|\dfrac{Q_P - 90}{90}\right|; Q_P \neq 90 \\ 1; Q_P = 90 \end{cases}$$

(d) Past direction rule (σ_4)

Fig. 3 Geometries of criteria fuzzy rules

direction obtained using fuzzy rules for the constraint. For example, if the width of alternative path ($w_1 + w_2$) is w_P, we apply the rule in Fig. 4(a). The shortest path rule in Fig. 4(b) results in a trajectory for which the best fuzzy value is the shortest path. The past direction rule in Fig. 4 (c) is a conditional rule that is only used when the robot must move backwards to avoid previous directions.

At subtask decision level (trajectory decision level), the process is similar to task decision level by using the rules in Fig. 4(a), 4(b) and 4(d). The visual path rule (Fig. 4(d)) results in the possibility that the target and change their plan.

IV. SIMULATION EXPERIMENTS

A. Simulation Parameters

The parameters of the robot system are defined by Fig. 5. Since the radius is 1 unit, and the safety compensation factor is 1 unit, the total safety factor of 4 units is equal to k in (7). Table I summarizes the results.

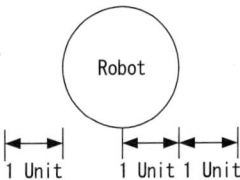

Fig. 5 Definition of robot parameters

TABLE I
SIMULATION PARAMETERS.

Robot shape	Circle
Robot size	1 unit
n (compensation times)	2
Safety compensation	1 unit

B. Simulation Results

Simulation results are displayed in a top view, and referenced to user frame. Fig. 6 shows the robot trajectory when an initial plan is successful, and Fig. 7 shows the robot trajectory when the initial plan is changed because of a new obstacle. The robot trajectories and safety fuzzy values are shown in Table II. A trajectory for a more complex environment is shown in Fig. 8.

C. Evaluation

We use the multi-factorial evaluation method to assess the quality of the robot trajectory. The values for total distance and average safety are f_1 and f_2, respectively. The evaluation is defined by e_1 = poor, e_2 = fair, e_3 = good, and e_4 = excellent, and $E = \{e_1, e_2, e_3, e_4\}$. The quality standards of the robot trajectories in Fig. 6 and 7 are shown in Tables III and IV, respectively. For this

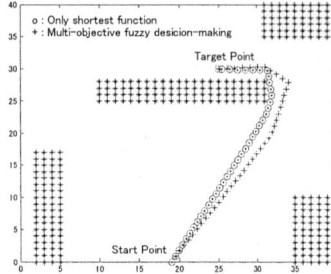

Fig. 6 Robot trajectory when the initial plan is successful

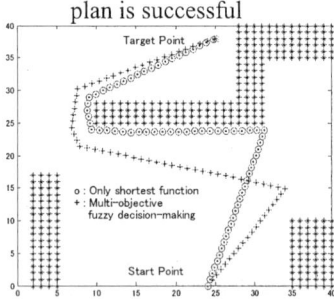

Fig. 7 Robot trajectory when the plan is changed after finding new obstacle

TABLE II
SUMMARY OF ROBOT TRAJECTORY AND AVERAGE SAFETY FUZZY VALUES

	Sub – Objective	Initial Plan			
		Successful (Fig.6)		Changed (Fig.7)	
		Length (unit)	Avg. safety (fuzzy value)	Length (unit)	Avg. safety (fuzzy value)
o	Only shortest path	45	0.159	70	0
+	Multi-criteria fuzzy decision-making	49	0.759	72	0.583

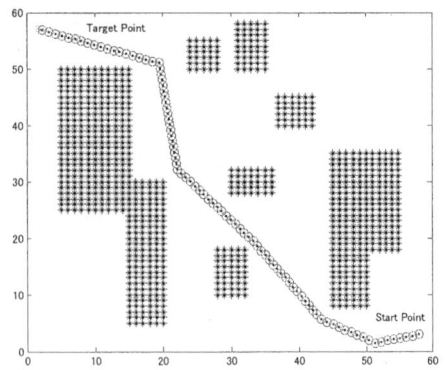

Fig. 8 Robot trajectory for a complex environment

purpose, the weighting matrix W is defined as [0.5 0.5], i.e. the shortest path rule and safety rule are equally important.

The membership functions for the qualification rules in Table III, IV are used in the distance evaluation fuzzy rule and safety factor evaluation fuzzy rule, as shown in

TABLE III
QUALITY OF ROBOT TRAJECTORY FOR FIG. 6

Class/Factor	Distance (length)	Safety (fuzzy value)
e_1: Excellent	42 - 50	0.85 - 1.00
e_2: Good	50 - 63	0.70 - 0.85
e_3: Fair	63 - 76	0.50 - 0.70
e_4: Poor	> 76	< 0.50

TABLE IV
QUALITY OF ROBOT TRAJECTORY FOR FIG. 7

Class/ Factor	Distance (length)	Safety (fuzzy value)
e_1: Excellent	68 - 82	0.85 - 1.0
e_2: Good	82 - 102	0.70 - 0.85
e_3: Fair	102 - 122	0.50 - 0.70
e_4: Poor	> 122	< 0.50

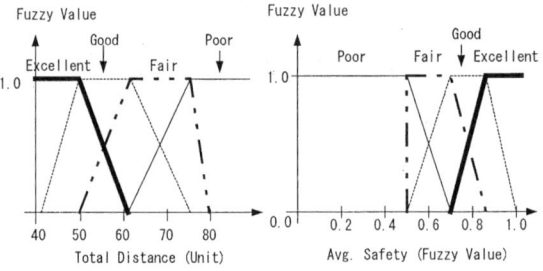

Fig. 9 Membership functions for Fig. 6

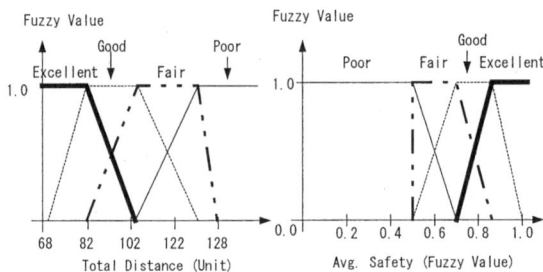

Fig. 10 Membership functions for Fig. 7

Fig. 9 and Fig. 10, respectively. The distance evaluation fuzzy rule is defined in terms of the shortest distance (= 42 in Fig. 6, = 68 in Fig. 7) for 1 time unit, 1.2 time units, 1.5 time units, and 1.8 time units. The average safety fuzzy value is an average value of each safety fuzzy value at the trajectory level from the first to the last decision, as shown in Table II.

From Table II, Table III, Table IV, Fig. 9, and Fig. 10, we obtain the shortest trajectory and multi-criteria fuzzy decision-making trajectory as follows:

$$R^{(u)}_{only\_shortest} = \begin{bmatrix} 0 & 0 & 0.38 & 1 \\ 1 & 0 & 0 & 0 \end{bmatrix} \quad (8)$$

$$R^{(u)}_{multi-criteria} = \begin{bmatrix} 0 & 0 & 0.88 & 1 \\ 0.59 & 1 & 0.43 & 0 \end{bmatrix} \quad (9)$$

and applying the weighting matrix W yields

$$D^{(u)}_{only\_shortest} = [0.5 \quad 0 \quad 0.19 \quad 0.5] \quad (10)$$

$$D^{(u)}_{multi-criteria} = [0.29 \quad 0.5 \quad 0.65 \quad 0.5] \quad (11)$$

From (10), we obtain that the path is excellent relative to shortest distance but poor relative to safety. However, the multi-criteria fuzzy decision-making trajectory from (11) is good. For Fig. 7 in which the plan is changed after finding a new obstacle, we obtain

$$D^{(u)}_{only\_shortest} = [0.5 \quad 0 \quad 0.06 \quad 0.5] \quad (12)$$

$$D^{(u)}_{multi-criteria} = [0 \quad 0.31 \quad 0.65 \quad 0.70] \quad (13)$$

Equations (10) and (12) cannot be used to conclude the quality of the shortest path trajectory. However, it can be concluded that the trajectory determined by multi-criteria fuzzy decision-making is excellent. Further research is being conducted to determine the class of unknown environments for which fuzzy decision-making system is suitable.

V. PATH PLANNING WITH MULTIPLE ROBOTS

Based on simulations that were conducted, the proposed fuzzy decision-making system can determine a path for a single autonomous mobile robot moving in an unknown environment. The fuzzy decision-making system approach is not only suitable for one robot, but it can also be applied to path planning with multiple robots.

A. Multi-Agent Path Planning

It is assumed that the robots, which have the same constant speed, are commanded to move through a narrow opening that is only large enough for one robot. Multi-criteria fuzzy decision-making can also be applied to decide the order of the robots so that each robot can plans its trajectory. When the trajectories have been determined, the current position of each robot relative to its start point is sent to host computer to initiate a motion decision based on multi-criteria fuzzy decision-making. The multi-agent system is shown in Fig. 11.

During the process of commanding the robot by the host computer, the criteria in Fig. 12 are used as rules for multi-criteria fuzzy decision-making so the higher moving order (HMO) robot moves before the lower moving order (LMO) robot.

The nearest target distance rule considers the robot operation time to reach the decided point relative to the distance from the current point to the target point. The turning angle rule considers the time lost by the robot when it prepares to change its direction of motion.

B. Collision Protection

After the host computer receives data from the robots, it determines the moving order. For this process, the host computer can predict the operation time T used until the

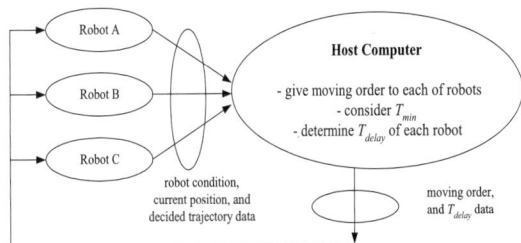

Fig. 11 Multi-agent robot system

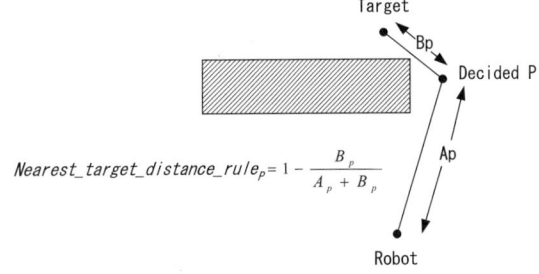

(a) Nearest target distance rule

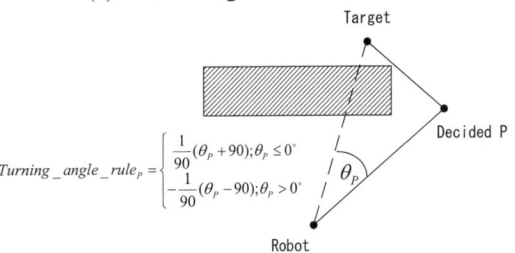

(b) Turning angle rule

Fig. 12 Geometries of the main criteria rules

trajectory segment has been traversed. These operation times of each robot are used for collision protection until time T_{delay} to avoid collision among the robots.

Let T_i be a subset of the predicted robot operation time T. Then T_{min} in (14) is the minimum time that the robots can move from their current positions to finish their trajectory segments. T_{min} is chosen to predict the position of each robot to check safety among robots (Fig. 13).

$$T_{min} = \min_{A,B,C \in i}(T_A, T_B, T_C) \quad (14)$$

As a consequence, the LMO robot can avoid collision with the HMO robots at $T_{cur}+T_{min}$, the host computer compares safety ranges (Fig. 14) between each robot and all HMO robots, thereby determining T_{delay} in (15). In addition, T_{delay} for HMO robots must not be exceeded the time for LMO robots to avoid collision. The LMO robot must wait for T_{delay} until the HMO robots can move. Thus

$$T_{delay,i} = \max\left(\frac{D_{HMO,i}}{v}\right) \quad (15)$$

where D_{HMO} are collision distances (Fig. 14) to the other HMO robots related to robot-i at $T_{cur}+T_{min}$, and v is the speed of robot.

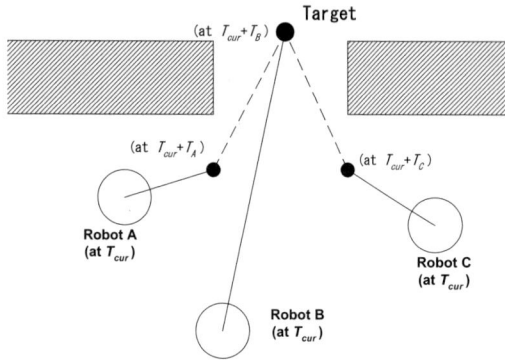

Fig.13 Robot positions at T_{cur} and $T_{cur}+T_i$

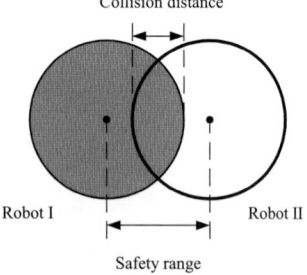

Fig. 14 Safety range and collision distance at $T_{cur}+T_{min}$

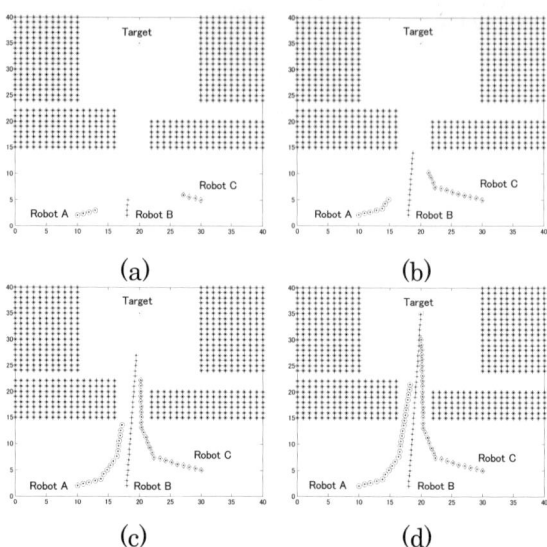

Fig. 15 Simulation results for multi-agent system

Simulation results for this scenario are shown in Fig. 15.

VI. CONCLUSIONS

A fuzzy decision-making system is proposed for multi-criteria robot motion planning. The method has decision levels for determining a rough direction and a precise trajectory. Advantages for operation in unknown environments are confirmed by comparison with a shortest path trajectory obtained by multi-factorial evaluation. Finally, a multi-agent approach is described for multi-criteria fuzzy decision-making with multiple robots.

The results of this research indicate that the proposed fuzzy decision-making system can perform robot path planning with multiple criteria and it can be implemented in a multi-agent system to accommodate multiple robots moving in unknown environments.

VII. REFERENCES

[1] L. Kavraki and J.C. Latombe, "Randomized Preprocessing of Configuration Space for Fast Path Planning," in *Proc. of the 1994 IEEE Intl. Conf. on Robotics and Automation*, CA, pp. 2138-2145.

[2] T. Danner and L. E. Kavraki, "Randomized Planning for Short Inspection Paths," in *Proc. of the 2000 IEEE Intl. Conf. on Robotics and Automation*, San Francisco, pp. 971-976.

[3] R. J. Szczerba, D. Z. Chen, and J. J. Uhran Jr., "A Grid-Based Approach for Finding Conditional Shortest Paths in an Unknown Environment," *Technical Report #94-34*, Dept. of Computer Science and Engineering, University of Notre Dame, November 1994.

[4] G. Yasuda and H. Takai, "Sensor-Based Path Planning and Intelligent Steering Control of Nonholonomic Mobile Robots," in *Proc. of the 27th Annual Conference of the 2001 IEEE Industrial Electronics Society*, Denver, CO, pp. 317-322.

[5] T. Fukuda and N. Kubota, "An Intelligent Robotic System Based on a Fuzzy Approach," in *Proc. of the IEEE*, Vol. 87, No. 9, September 1999, pp. 1148-1470.

[6] J. Sellen, "Direction Weighted Shortest Path Planning," in *Proc. of the 1995 IEEE Intl. Conf. on Robotics and Automation*, Nagoya, pp. 1970-1975.

[7] B. Freisleben and T. Kunkelmann, "Combining Fuzzy Logic and Neural Network to Control an Autonomous Vehicle," in *Proc. of the 1993 IEEE Intl. Conf. on Fuzzy Systems*, pp. 321-326.

[8] S. Boonphoapichart, S. Komada, and T. Hori, "Robot's Motion Decision-Making System in Unknown Environment," in *Proc. of the 2002 Japan Industry Applications Society Conference*, Kagoshima, Japan, pp. 705-708.

[9] S. Boonphoapichart, S. Komada, and T. Hori, "Robot's Motion Decision-Making System in Unknown Environment and its Application to a Mobile Robot," in *Proc. of 2002 IEEE Intl. Conf. on Industrial Technology*, Thailand, pp. 18-23.

[10] C. Carkssib and R. Fuller, *Fuzzy Reasoning in Decision Making and Optimization*, Physica-Verlag, 2002, pp. 65-99.

[11] H. Xing Li and V. C. Yen, *Fuzzy Sets and Fuzzy Decision-Making*, CRC Press, 1995, pp. 161-187.

[12] P. J. Antsaklis and K. M. Passino, *An Introduction to Intelligent and Autonomous Control*, Kluwer Academic Publishers, 1992, pp. 1-26.

Probability of success and Uncertainty analysis in Path Planning

Dapena Eladio
Department of System Engineering,
Los Andes University,
Av. Tulio Febres. 5101, Merida, Venezuela

Moreno Luis
Dept. System Engineering and Automation,
Carlos III University,
Butarque 15 Leganes, Spain

Abstract— This paper shows the robustness path planning for mobile robots in indoors environments. The planning includes an accurate uncertainty in position analysis and the probability of success calculation as a path quality measurement. A topological and metric environment model, for the environment representation, is proposed. This model includes localization and risk transit zones. During the planning process, all the paths to complete a mission, are generated. Obtaining parameters such as: goal uncertainty, path feasibility, path distance and path uncertainty. These parameters allow the most adequate paths selection to fulfil a mission.

I. INTRODUCTION

During the past two decades, researchers in mobile robotics have dealt with different path planning methods. In most cases, the methods goal is to find feasible free collision paths, which will meet the initial and final configurations, to complete a mission. Some researchers propose methods where the robots configuration is perfectly known, at each instant, during the planning and navigation stages. This is not always possible. It is essential to deal with uncertainty, in the planning stage, when the position errors approach values close to the missions permitted thresholds. Frequent use plans, based on geometrical models, assuming null uncertainty, are clearly insufficient. Thus, the use of schedulers, which not explicit deal with uncertainty, is limited to simple situations where the errors are less than the allowed thresholds to complete the missions. In general, the proposed paths are not always robust enough to guarantee the missions success. The planning methods, with or without uncertainty, are conditioned by the environment characteristics and the way of representing it.

Recently, different researchers have started to work on planning methods that take into consideration the uncertainty [6], [5], [1], [4]. In this work, the paths selection combining parameters such as: probability of success, maximum uncertainty reached in the path, goal uncertainty and path distance, is proposed. The path planning presented in this paper, uses an enriched environment model which includes geometrical information, localization zones, restricted zones, walls, obstacles, etc. During the planning process, all paths for a mission are generated, obtaining parameters such as: goal uncertainty, path feasibility, path distance, path uncertainty. These parameters, permit the most adequate path selection to accomplish a mission.

II. ENVIRONMENT MODEL

The proposed model generates an undirect graph, from the geometric map. It combines simple geometric information, symbolic information, connectivity between geometrical and symbolic elements, connections feasibility, etc., in a single model named Extended Topo-Geometric Model, ETGM [3]. The ETGM, is a graph with geometrical information, which allows the obtention of optimum routes using, path distance, path feasibility, goal uncertainty and maximum path uncertainty criteria. The workspace representation can be either obtained using classical techniques, as Voronoi graph, visibility graph, cell decomposition, etc., on a simple geometric map or in exploration missions. The nodes correspond to the (x,y) map positions, which provide some useful information to the path planning. Sometimes, the set made up of the node and the simple map, is enriched with symbolic information, related to special points of interest or to other geometrical elements. In this paper it is proposed that the information, provided by the nodes, must be of the following type: **geometrical** a single position in the x,y Cartesian plane; **perceptive** when the node shows a zone where known elements, for the observation, exist; and **approximation** when the node represents the entrance to a zone considered of risk. The workspace as it is defined, does not allow to carry out planning tasks on its own. Obviously, the establishment of an adjacent relationship between nodes is required. The segment set S, transitable or not, is defined as the group made up of the Cartesian product $Q \times Q$, where the pairs $(q_i = q_j)$ are eliminated, see equation 1. The number of segments is given by the expression $Q*(Q-1)$, where Q is the number of nodes.

$$S : \left\{ \forall (q_i, q_j) = S_{i,j} \; / \; \left\{ (q_i, q_j) \in Q \right\} \wedge (q_i \neq q_j) \right\} \quad (1)$$

The segments map generated, does not fulfil this paper requirements of obtaining feasible paths for the missions. It contains segments which cross some workspace elements such as: walls, obstacles, etc.

In this environment representation stage, all those segments non suitable for navigation are rejected, used the follow criteria:

- Those which intersect environment elements.
- All those segments which cross a restriction zone, when its nodes do not match with the corresponding approximation nodes.

III. THE KINEMATIC MODEL

The robots position will be represented by the vector of its spatial variables $X(k)$, as a point in the Cartesian plane, with $x(k)$ and $y(k)$ coordinates and an orientation $\theta(k)$, shown in equation 2.

$$X = \begin{bmatrix} x & y & \theta \end{bmatrix}' \quad (2)$$

The simplified kinematic model proposed in [8], describes how the robots position changes, in time, in relation to a initial position, in response to a $u(k)$ control input made up of a $T(k)$ translation followed by a $\triangle \theta(k)$ rotation: $u(k) = [T(k), \theta(k)]'$. The state, for a given instant, is obtained from the state transition function $F(X(k), u(k))$, represented in equation 3.

$$F(X(k), u(k)) = \begin{vmatrix} x(k) + T(k) * \cos(\theta(k)) \\ y(k) + T(k) * \sin(\theta(k)) \\ \theta(k) + \triangle \theta(k) \end{vmatrix} \quad (3)$$

The simplified plant model is reduce to equation 4.

$$X(k+1) = F(X(k), u(k)) + v(k) \quad (4)$$

where: $F(X(k), u(k))$, is a non-linear state transition function, v(k) is a noise source assumed to be a zero-mean Gaussian with covariance $Q(k) \rightarrow N(0, Q(k))$ and finally $u(k)$ is the control input.

The equation allows to obtain the mean vector estimation in the k+1 position. It is now necessary, to estimate the covariances matrix P in the same position. The first two moments, **the mean** and **the covariance** of the distribution function, which follow the spatial position relationship, must be determined. The covariances matrix related to the prediction, in the non-linear spatial relationship case, is obtained from the Taylor series expansion. The $P_{k+1/k}$ covariances matrix equation is shown in equation 5.

$$P(k+1/k) = \nabla F * P(k/k) \nabla F' + Q(k) \quad (5)$$

where: ∇F is the state transition function jacobian, obtained as the linearizing result around the estimated state. The state transition function jacobian is described in equation 6.

$$\nabla F = \begin{vmatrix} 1 & 0 & -T(k) * \sin(\widehat{\theta}(k/k)) \\ 0 & 1 & T(k) * \cos(\widehat{\theta}(k/k)) \\ 0 & 0 & 1 \end{vmatrix} \quad (6)$$

IV. DEALING WITH UNCERTAINTIES

To obtain the uncertainty representation in the robots nominal position, in relation to an initial position, the covariances matrix calculated in the estimated position, is used. It is assumed that the probability distribution function of the robots position is a multi-variable Gaussian (x, y, q), shown in equation 7.

$$\rho(X) = \frac{1}{\sqrt{(2*\pi)^n} * \det |P|} \times \exp^{(-\frac{1}{2})*((X-\widehat{X})*P^{-1}*(X-\widehat{X}))} \quad (7)$$

where n, is the number of dimensions, P is the covariances matrix, $\widehat{X}$ is the mean nominal vector, X is a vector representing a point in particular.

The constant probability contour, of this distribution, has the ellipsoids shape in the n-dimensional space (ellipses for two dimension), centered in the means vector $\widehat{X}$. The present work, is only focused on the two dimensional position error. This error generates an ellipse in the plane. This simplification reduces the covariances matrix P, to a 2x2 matrix, only using the terms related to the x, y coordinates, obtaining the probability distribution represented in equation 8.

$$\rho(x, y) = \frac{1}{2*\pi \sqrt{\sigma_x * \sigma_y (1-\rho^2)}} \times e^{\left(-\left(\frac{1}{2*(1-\rho^2)}\right)*\left(\frac{x^2}{\sigma_x^2} + \frac{2*\rho_{xy}}{\sigma_x \sigma_y} + \frac{y^2}{\sigma_y^2}\right)\right)} \quad (8)$$

where ρ is the correlation coefficient between the x, y variables. The ellipse equation is:

$$\left(X - \widehat{X}\right)' * P_x^{-1} * \left(X - \widehat{X}\right) = K^2 \quad (9)$$

where K, is a constant selected for the desired confidence range. In [9] the equations to extract the ellipses parameters from the covariances matrix, are found.

A. Uncertainty reduction

Most of mobile robots have one or more perception systems, which permit to carry out localizations and therefore, the uncertainty reduction in position. In general, the traditional path planning does not use this fact to obtain off-line paths [2]. This is because the localization processes need the data supplied by the sensors, which is not available during the plans generation. Despite this fact, as it will be later described, a covariances matrix adjustment is possible in the estimated position, without the need of physical data from the sensors. On the other hand, the perception system varies considerably from one platform to another. The position adjustment, varies in a similar way, depending on the system type; being different a laser localization than the one carried out by a vision system or by ultrasound sensors. Thus, compromising the planning system to a specific perception system will limit its use. Two techniques are proposed to confront the uncertainty reduction [3].

1) Static: This technic assumes that when the vehicle enters a localization mark influence zone its covariance matrix can be reduced, independently of the sensors and the robots actual position. In this case, two reduction types are proposed: direct an linear.

- *Linear reduction.* The actual covariances matrix is linearly reduced, in all its components, by a reduction factor K_r, statistically obtained.
- *Direct Reduction.* The covariances matrix is directly substituted by a matrix of valid constants, on the localization zone. This last approach is similar to the one used by *Lazanas* and *Latombe* in [7] in what they call *perfection isles*.

2) Dynamic: This technic uses a reduction uncertainty model similar to the one used in the navigation stage. The static reduction has the disadvantage of not including, in the reduction, the sensors, the position, etc. information. The **E**xtended **K**alman **F**ilter **EKF** is used, in this paper, to adjust the covariances matrix. The EKF is used because the state transition and the observation functions are not linear.

It is very frequent to find navigation systems with localization cycles based on the *EKF*. The filter includes different sensors measurements and it adjusts the state vector position and its covariances matrix. A localization cycle for mobile robot, according to [8], is made up of the following stages: position prediction, observation, observation prediction, matching and estimation. During the off-line planning, it is not worth considering the observation stage, because the measurements obtained from the sensors are not available. This fact does not allow carrying out adjustments in the robots position. Despite this fact, the same does not apply for the covariances matrix adjustment. The equation which defines the covariances matrix adjustment is observation independent. For this reason the proposed cycle for the uncertainty reduction will be formed by the following stages: Position prediction, observation prediction, matching and estimation.

The linear uncertainty reduction is used, to obtain the present work results, due to its simplicity. During the tests that were carried out the three reduction models were implemented. It can be then stated that the linear and dynamic uncertainty reduction values do not differ very significantly.

B. Feasibility

In this paper, a new concept of feasibility is proposed as a quality measurement within the segments. The feasibility, despite of not being defined in all the path segments, shows a great interest in those segments which cross the zones considered of risk or of restricted transit. Let suppose the path, which joins nodes A, B, C, D, shown in figure 1. In this figure the ellipses uncertainty and the approximations and restricted transit zones, are represented. On the other hand, during the training environment stage, adapts the segment feasibility to the unitary value, where the segment not across a restricted zone. The $t(S_{i,j})$ function permits, in the planning stage, the obtention of the probability of crossing the segment.

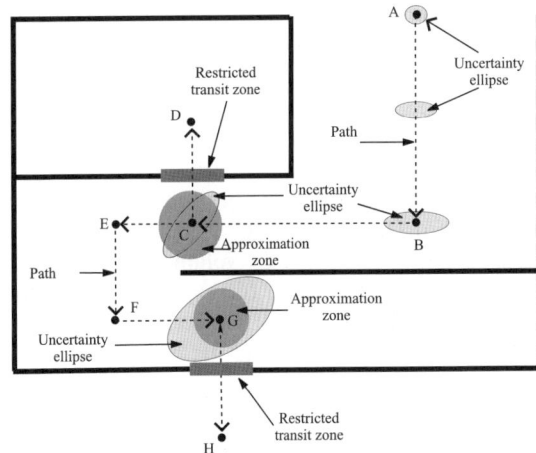

Fig. 1. Path which joins nodes A, B, C, D

The feasibility concept, defined before, are individually applied to all segments, obtaining the following expressions:

- $\overline{AB}$ Segment   Feasibility 1,
- $\overline{BC}$ Segment   Feasibility 1,
- $\overline{CD}$ Segment $t(S_{CD})$.

where $t(S_{CD})$, is the function which determines the probability of reaching node D, from node C. The calculation is independent from the probability of reaching node C. As it was mentioned previously, the probability of overcoming the segment is calculated individually. Being of special interest, those segments, which cross restricted transit zones. The restricted transit zones have associated an approximation zone, that is, *To overcame a restricted transit zone, the robot* **must be placed** *inside the approximation zone*. This is also applicable to the localization marks and its respective zones of influence, *To observe a localization mark, the robot* **must be placed** *inside the mark influence zone*. The statement **must be placed**, leads to the position uncertainty model, presented in previous sections. Therefore, the probability of overcoming the restricted transit zone or *feasibility*, is closely related to the position uncertainty, within the approximation node. This is described in equation 10.

$$t(S_{i,j}) = \iint_{ZA} \rho(X) \delta X$$

$$0 \leq t(S_{i,j}) \leq 1 \quad (10)$$

where $\rho(X)$, is the probability distribution function of the position vector X, defined in equation 8, $S_{i,j}$, is the segment which joins nodes i and j, with direction from i to j and ZA is the approximation area related to node i. The function $t(S_{i,j})$, is **not commutative**, and depends on the approximation node uncertainty. At the same time, the approximation node uncertainty depends on the distance travelled to that node. To obtain the feasibility calculation the robot position normal bi-variated distribution function, centered in the approximation node and bounded by the approximation area, must be integrated.

V. PATH PLANNING CRITERIA

A. Uncertainty criteria

The uncertainty measurement in position is the Generalized Standard Deviation **GSD**, considered as the areas scalar measurement in R^2, or the volume in R^3, where the data is distributed. The GSD, for any node, is obtained as the generalized variance **GV** square root, see equation 11. The GV is the variances and the covariances matrix determinant, P.

$$GV = \det(P)$$
$$GSD = \sqrt{GV} \quad (11)$$

1) Goal uncertainty: This term is referred as the robots orientation and position uncertainty, once the zone (of interest) to accomplish the mission, is reached. The goal uncertainty limitations will be given by the mission. The uncertainty measurement in position, is given by the state variables (x, y, θ) covariances matrix, in the final position. This matrix represents the gathered error, during the robot movements from q_0 to q_f and the uncertainty reductions carried out in the localization zones.

2) Uncertainty Profiles: When the uncertainty profile is used, the goal uncertainty is not only of interest. The uncertainty behavior, along all the route, is of great importance too. In the mission, a threshold for the uncertainty limits, must be considered. This will allow to reject those paths (*where the robot could get lost*), which will exceed the threshold. The best path, in this case, is the path with minimum value, following the maximum path uncertainty criterion.

B. Path feasibility

The path feasibility $T(\tau_j)$, is defined as **The robots probability of reaching the goal mission M, following path τ_j**. The feasibility is a path quality measurement and it is obtained by equation 12.

$$T_{\tau_j} = T_{q_0} * (\prod_{i}^{k} T_{q_i}) * T_{q_f} \quad ; \quad \{\forall \ q_i \in \tau_j\} \quad (12)$$

where T_{q_i}, is either the probability of reaching node i, from its predecessor, or the segment feasibility which joins node i with its predecessor in the node i direction.

C. Path distance travelled

Another important parameter when selecting paths, during the decision-taking process, is the distance travelled $D(\tau_j)$ shown in the equation below.

$$D(\tau_j) = \sum_{i=1}^{n}(\overline{S_i}) \quad \therefore \quad (S_i \in \tau_j) \quad (13)$$

VI. PATH PLANNING

Let the mission M_i, be a mission commanded to a mobile robot in an environment represented by the $ETGM^{M_i}$. The path planning is defined as: *Finding the best evaluated paths, individually or grouped, following the goal uncertainty, path uncertainty, path feasibility and path distance criteria.*

The path selection will depend on each mission conditions. For example: goal position accuracy, dependability, low charged batteries, travelled distance time, environment, etc. These are factors which condition the decision-taking process. Two additional functions, which relate the basic parameters described in previous sections, are proposed.

A. A cost function

A cost function, which relates the distance and feasibility parameters $C(D(\tau_j), T(\tau_j))$. The cost function proposed is the distance and feasibility ratio. The function tries to favour those paths with larger feasibility, with regard to those paths whose distance travelled is alike, but the path feasibility is smaller.

B. Parametric function

The second function proposed, is the normalized parametric function. This function assigns a weight to the different path planning parameters, see equation 14.

$$P_i(D_i, T_i, IO_i, IC_i) = P_1 * \frac{D_i}{D_{Min}} + P_2 * \frac{1}{T_i} + P_3 * \frac{IO_i}{IO_{Min}} + P_4 * \frac{IC_i}{IC_{Min}} \quad (14)$$

where D_i is path i travelled distance, D_{Min} is the minimum travelled distance for the τ paths set, T_i path i feasibility, IO_i is the goal GSD for the i path, IO_{Min} is the minimum goal GSD for the τ paths set, IC_i is the maximum path GSD reached for the i path and IC_{Min} is the smallest value of the maximums path GSD reached for the τ paths set.

TABLE I
PATHS PARAMETERS TABLE

	Paths		
	38	46	24
Distance m.	**16.87**	27.72	24.37
Feasibility	0.10	**0.85**	0.76
Goal Uncert. m^2	0.64	**0.36**	0.36
Max. Uncert. m^2	3.24	**1.10**	1.20
Cost function	0.17	0.04	**0.03**
Parametric function	5.71	6.13	**5.62**

VII. RESULTS

The *ETGM* for mission M_1 see figure 2, is used during the parameters calculation, for all **72** possible paths. Localization processes are simulated in those nodes which correspond. The sensors type, used for localization, is an ultrasound ring. The doors are considered as risk and localization zones.

The planning results will choose the path labelled **38**, as the path of less distance travelled, the path labelled **46**, as the best path evaluated for the feasibility, the goal uncertainty and maximum path uncertainty parameters, and the path labelled **24** as the best path evaluated for the cost and parametric functions. The values for the planning criteria, for the best evaluated paths **38, 46, 24**: Path distance, probability of success, goal uncertainty, maximum path uncertainty and the cost and parametric functions, are shown in table I.

The path labelled as **38**, best evaluated in distance but with probability of success (feasibility) 0.10, gives little possibilities of reaching the goal, and it must be rejected. Despite of the uncertainty reductions, the maximum path uncertainty reached ($3.54\,m$), is very large. This is a clear example where, a path optimization, based exclusively in minimizing the distance travelled, is inadequate considering the environment characteristics.

For paths **46** and **24**, the feasibility, the goal uncertainty and the maximum path uncertainty criteria values are very similar. Both, the cost and parametric functions, agree in choosing the path labelled as **24**. The distance travelled for this path decreases in a few meters, with a slight feasibility deterioration. In this case, path **24** feasibility reduction, in regard to path **46**, does not compromised the mission success. Therefore, the cost and parametric functions generate the optimum path to fulfil the mission. Finally, the decision of selecting the best path concerns the planners experience and each missions individual conditions.

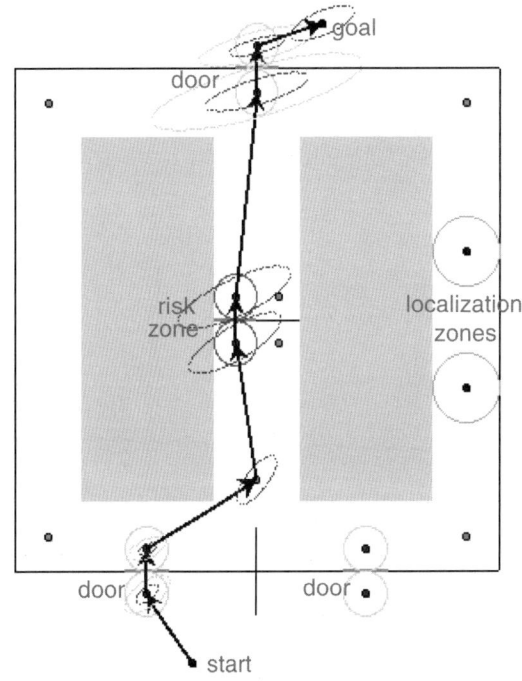

Fig. 2. Path 38

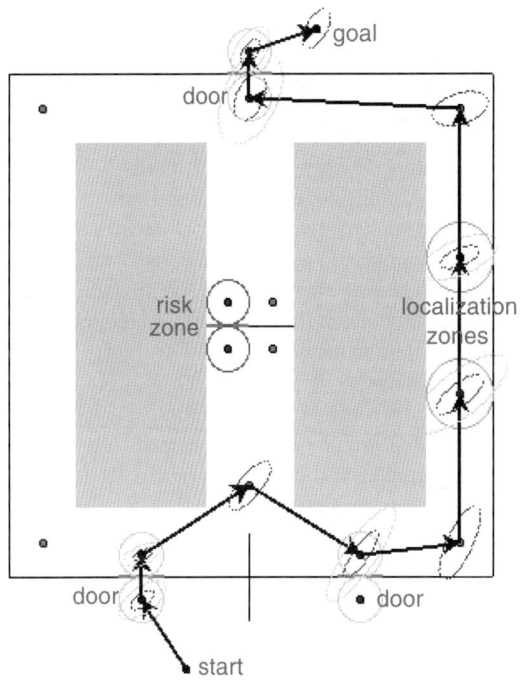

Fig. 3. Path 46

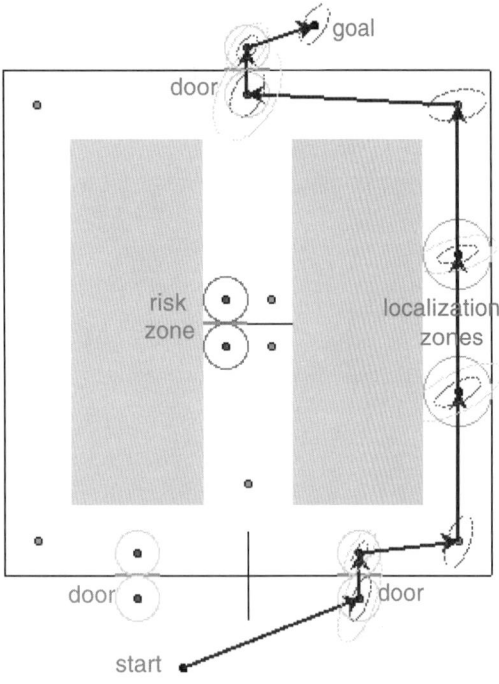

Fig. 4. Path 24

VIII. Conclusion

Dealing with uncertainty, in off-line path planning stages, is very important when trying to £nd the best paths to accomplish the missions. The contribution of this paper is the following: a new concept of probability of success (feasibility) for the paths, an evolution model of the mobile robot uncertainty and an environment model. An algorithm which generates all paths, for any possible mission, was developed. This algorithm, also calculates parameters such as: feasibility, goal uncertainty, maximum path uncertainty and path distance, used for a robust path planning. The path planning incorporates the mobile robot kinematic model and the perception system characteristics, for the off-line paths generation. This fact, allows that the generated plans, will approach the real situation accomplished by the robot, in the navigation stage. The proposed path planning method has been successfully implemented and tested under real conditions in a B-21 mobile robot.

IX. Acknowledgments

The authors gratefully acknowledge the contribution of Angela Nombela.

X. References

[1] Bouilly, B. and Siméon, T. A sensor-based motion planner for mobile robot navigation with uncertainty. In *International Workshop, RUR'95 Proceedings*, pages 235–247, Berlin, Germany 1996, 1996. Springer-Verlag.

[2] Dapena G. Eladio. Path planning with uncertainty. In *18th International Conference on CAD/CAM, Robotics and Factories of the Future*, volume 02, pages 491–496. CARs-FOF2002, July 2002. CARs-FOF2002 : 18th International Conference on CAD/CAM, Robotics and Factories of the Future 3, 4 and 5 July 2002. PORTO, PORTUGAL.

[3] Dapena G. Eladio. *Plani£cación de Caminos con Incertidumbre*. PhD thesis, Universidad Carlos III de Madrid, Mayo 2002.

[4] Fraichard, Th. and Mermond, R. Integrating uncertainty and landmarks in path planning for car-like robots. In Elsevier, editor, *Intelligent Autonomous Vehicle 1998 (IAV'98)*, pages 397–402, Kidlington, UK; 1998, 1998. Elsevier.

[5] Khatib, M., Bouilly, B., Simeon, T., and Chatila, R. Indoor navigation with uncertainty using sensor-based motions. In *Proceedings of the 1997 IEEE International Conference on Robotics and Automation*, pages 3379–3384, Albuquerque. New Mexico. USA, April 1997.

[6] Lambert, Alain and Le Fort-Piat, Nadine. Safe task planning integrating uncertainties and local maps federations. *The International Journal of Robotics Research*, 19(6):597–611, June 2000.

[7] Lazanas, Anthony and Latombe, Jean-Calude. Motion planning with uncertainty a landmark approach. *Arti£cial Intelligence*, pages 287–317, 1995.

[8] Leonard, John J. and Durranr-Whyte, Hugh F. *Directed Sonar Sensing for Mobile Robot Navigation*. Kluwer Academic Publishers, 1992.

[9] Smith, Randall C. and Cheeseman, Peter. On representation and estimation of spatial uncertainty. *The International Journal of Robotics Research*, 5(4):56–67, 1986.

A Neural Network Model that Calculates Dynamic Distance Transform for Path Planning and Exploration in a Changing Environment

Dmitry V. Lebedev, Jochen J. Steil, Helge Ritter

AG Neuroinformatik, Faculty of Technology,
University of Bielefeld,
P.O.-Box 10 01 31, 33501 Bielefeld,
Germany

Abstract – In this paper, we present a neural network model that realizes a dynamic version of the distance transform algorithm (used for path planning in a stationary domain). The novel version is capable of performing path generation for highly dynamic environments. The neural network has discrete-time dynamics, is locally connected, and, hence, computationally efficient. No preliminary information about the world status is required for the planning process. Path generation is performed via the neural-activity landscape, which forms a dynamically-updating potential field over a distributed representation of the configuration space of a robot. The network dynamics guarantees local adaptations and includes a set of strict rules for determining the next step in the path for a robot. According to these rules, planned paths tend to be optimal in a L_1 metric. Simulation results in a series of experiments for various dynamical situations prove the effectiveness of the proposed model.

I. INTRODUCTION

One of the most important attributes of a robotic system is its ability to plan the paths and to navigate autonomously. At the same time, an "intelligent" navigation is characterized by the capability of adapting a route dynamically in the case of sudden appearance of other objects, or obstacles. There exists a lot of research on path planning (see, e.g., [1]-[5]). A number of neural network approaches has also been proposed to solve this problem ([6]-[17]). The capability of a multilayer perceptron to learn successfully the navigation task in a maze-like environment has been demonstrated in [6]. In [7], a self-organizing Kohonen net with nodes of two types has been used. In [8], a description of a network with oscillating behavior, that solves the problem of path planning for an object with two degrees of freedom (DOFs), formulated as a dynamic programming task, is given. The algorithm, proposed in [9], uses a set of intermediate points, connected by elastic strings. Gradient forces of the potential field, generated by a multilayer neural network, minimize the length of the strings, forcing them at the same time to round the obstacles. In [10], a multilayer feed-forward network for performing real-time path planning was applied. The neural network for path finding described in [11] has three layers of neurons with recurrent connections in the local neighborhoods. The dynamics of the network emulates the diffusion process.

Most of the known neural network approaches require, however, full knowledge of environment and can be applied only for a stationary domain. Besides that, optimality of the path is often left out of consideration.

For a non-stationary domain, a topologically-organized Hopfield-like neural network for dynamic trajectory generation has been proposed in [15] and improved recently in [16], but some additional efforts are required for tuning the network parameters.

In this paper we present a novel neural network dynamics for finding a path in a dynamic world. The model is based in interweaving manner on three paradigms, (a) the notation of configuration space as a framework for a flexible object representation [18]; (b) potential field building, which is a generic and elegant method for formation/reconstruction of a path; and, (c) a wave expansion mechanism, that guarantees an efficient construction of the potential field. In comparison with our previous results in [19], the network dynamics has been reconsidered, simplified, and yields now shorter and smoother paths.

The model has been simulated and tested in the context of autonomous path planning and exploration for various types of dynamical changes in the environment, and has demonstrated efficient and effective path generation capabilities.

The paper is organized as follows. In section 2, we describe the general idea of the proposed algorithm and give a formalization of the problem. Section 3 contains the description of the neural network model. We illustrate simulation results in section 4 and conclude with a discussion in section 5.

II. THE PROPOSED ALGORITHM

A. The General Idea

The original version of the distance transform algorithm was presented first in [20] and then exploited extensively for path planning and navigation tasks in a stationary domain (see in [21]-[23]). In the essence of this algorithm lies the idea of distance propagation in the workspace around the goal position, such that the value of a cell after application of the algorithm corresponds to the cost of the path to the goal (for more details see [20], [21]). The path itself is found by following the steepest descent with respect to the calculated distance values.

One could notice, that the idea of distance transform is very similar to the generation of discretized numerical potential fields (see, e.g., [17] and [24]). So, in [17] the model of a wave expansion neural network has been proposed, that computes a distance transform (named "grid potentials") over a discretized representation of the configuration space of a robot. The methods [17], [21] and [24] can be characterized as "one-time-go-through". This means that the procedure of potential field generation is performed only one time and is finished usually as soon as each position in the whole workspace (or configuration space) a numerical value has been assigned. Thus, these methods can not be applied effectively for a dynamic domain. In [23], e.g., a replanning of the whole path is initiated each time, when the robot encounters an obstacle. However, in an environment populated densely by obstacles or other agents, the replanning of the whole path can become a restricting factor for real-time navigation capabilities of a robot.

For planning a path in a time-varying world, we propose a novel neural network model that calculates dynamic distance transform, or dynamic grid potentials. To form the desired grid potentials, a wave-expansion mechanism is used. During the process of potential generation, the activity is spread around the source of excitation, and the minimum value of the generated potential field stays always at the excitory point, which in turn attracts the robot.

The most important challenge, distinguishing our model from the approach described in [17], is the proposed neural network dynamics, which makes an effective combination of (1) repetitive wave expansions, and (2) rules to cope with dynamically-generated waves of neural activity. These rules are based on a set of threshold-like functions and are included into the network dynamics to ensure the proper formation of a dynamic distance transform and to guarantee that a robot will move only along a safe route.

To provide the repetitive wave expansions, a regular excitation source is fed at the target neuron. This results in origination at each time step of a new wave of neural activity in the network field. Neural activity, therefore, changes locally, while propagating through the network field, and adapts to the dynamical status of environment. Since in the case of a stationary environment, wave fronts yield paths, which are optimal in a L_1 metric (see [17]), dynamical paths, which are generated by the proposed model, also tend to keep such optimality. Consequently, longer paths to the target are cut off automatically by the algorithm.

B. Formalization of the Problem

Without loss of generality, we can define the configuration space $C \subset \Re^d$ to be a regularly discretized hypercube, where d is the number of DOFs of a robot. For a robot in C the starting and the final configurations, denoted accordingly S and T, are defined. Suppose at the time t_k, there is a number N_k of obstacles (i.e. of forbidden configurations) in C. At that moment of time, positions of all obstacles in C form the obstacle region $O_k = \{O_i^k\} = \{(X_{i_1}^k, ..., X_{i_d}^k)\}$, where the obstacle coordinates in C are denoted by vectors $(X_{i_1}^k, ..., X_{i_d}^k)$, $1 \leq i \leq N_k$. Let $\tau(t_k) = (p_1(t_k), .., p_d(t_k))$ define the configuration of a robot in C at the time t_k. The task is to find a safe (i.e. a collision-free) path τ, that satisfies the conditions: $\tau(t_s) = S$,

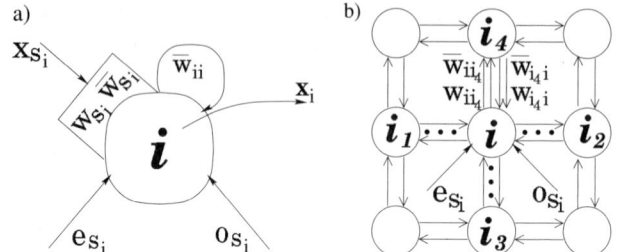

Fig. 1. a) The model of the neuron; b) Network neighborhood in $2D$.

$\tau(t_g) = T$, $\tau(t_k) \cap O_k = \emptyset$, $t_s \leq t_k \leq t_g$, where t_s and t_g are the time of start and the time of reaching the goal, respectively.

III. THE NEURAL NETWORK MODEL

A. The Network Architecture

The neural network has a parallel locally connected structure of cellular type. Depending on the dimensionality of C the network field may consist either of a single layer (for a 2D configuration space), or a set of layers with locally connected neurons. The arrangement of the neurons coincides with the discretized representation of C, i.e. each discrete position in C is associated with a neuron in the network field. Fig. 1a shows the neuron model.

The i-th neuron is connected with $n = 2d$ immediate neighbors, where d is the dimensionality of C. We will denote the set of neighbors of the i-th neuron as s_i, i.e. $s_i = \{i_1, ..., i_n\}$, and an enumeration of neurons in the local neighborhoods is fixed. A neuron neighborhood for an example 2D configuration space is depicted in Fig. 1b.

The network can be viewed as a discrete-time dynamical system, which can be fully described by a set of neuron state vectors $X_i = [x_i, \widehat{W}_{s_i}] \subset \Re^{2n+2}$. The first element x_i of vector X_i is the activity level, or the output of the i-th neuron, which is a real scalar quantity. The second element, vector $\widehat{W}_{s_i} = [w_{i_1 i}, ..., w_{i_n i}, \bar{w}_{ii}, \bar{w}_{i_1 i}, ..., \bar{w}_{i_n i}]$ consists of two sets of connection weights, defining the synaptic strengths of the connections between neuron i and its immediate neighbors. Notice, that w_{ji}, $\bar{w}_{ji}$, w_{ij} and $\bar{w}_{ij}$ are *four* connection weights between neurons i and j and that weights w_{ji} and $\bar{w}_{ji}$ follow different update rules.

Activity levels of the neurons in the neighborhood of the neuron i comprise the vector $X_{s_i} = [x_{i_1}, x_{i_2}, ..., x_{i_n}]$. The neuron i is active if $x_i > 0$ and inactive otherwise.

Additionally, the activity of the i-th neuron can be influenced by the excitory and inhibitory inputs from the neighboring neurons, which form accordingly vectors $e_{s_i} = [e_{i_1}, ..., e_{i_n}, e_i]$ and $o_{s_i} = [o_{i_1}, ..., o_{i_n}, o_i]$, whose elements can be of value zero or one only.

B. Path Planning Process

In this part, we present the underlying idea of the dynamic generation of distance potentials and its formal description.

State equations (1)-(3) contain the rules providing the activity evolution of the neuron i and of the connection weights, associated with the latter.

$$x_i(t+1) = e_i \cdot (x_i(t) + 1) + (1 - e_i) \times \\ r(\sum_{j \in s_i} w_{ji}(t) \cdot (x_j(t) + 2)), \quad (1)$$

$$w_{ji}(t+1) = U(P_k(X_{s_i}(t))), \quad j \in s_i, \quad (2)$$

$$\bar{w}_{ji}(t+1) = x_j(t), \quad j \in \{s_i \cup \{i\}\}. \quad (3)$$

The functions $r(.)$ and $U(.)$ are defined as

$$r(\tau) = \begin{cases} [\tau], & \tau > 0 \\ 0, & \tau \leq 0 \end{cases}, \quad U(\tau) = \begin{cases} 1, & \tau > 0 \\ 0, & \tau \leq 0 \end{cases},$$

where $[\tau]$ denotes the integral part of τ.

Initially, activity levels of all neurons and all connection weights are zero. Available information about the current status of the environment is applied to the external neuron inputs. The neurons corresponding to stationary or dynamical obstacles in C are inhibited by inputs $o_i = 1$. The neuron associated with the target position in C is excited by its input $e_i = 1$ and initiates the propagation of activity in the neural field. As can be seen from (1), the activity level of the target neuron is incremented by one at each time step and remains the global minimum of the potential field during the network evolution. The activity of other neurons depends on the current state of the neurons in the local neighborhoods. Equations (2) guarantee that among the incoming connections from the local neighbors only one will be positive. This neighboring neuron is chosen by the function $P_k(x_i, X_{s_i}, \widehat{W}_{s_i}, o_{s_i})$:

$$P_k(.) = \begin{cases} \Omega_1, & k = 1 \\ \Omega_k + I(\sum_{l=1}^{k-1} \Omega_l), & 1 < k \leq n \end{cases}. \quad (4)$$

This function queries the neighboring neurons according to the preassigned enumeration, and selects the first neuron k, which corresponds to a positive value of the expression:

$$\Omega_k = F(o_k) \cdot U(x_k(t)) \cdot E(x_k(t), \bar{w}_{ki}(t)) \times \\ F(g(x_i(t), \bar{w}_{ii}(t))) \times \quad (5) \\ (U(x_i(t)) \cdot D(x_i(t), x_k(t)) + F(x_i(t))).$$

Here $g(x_i(t), \bar{w}_{ii}(t)) = F(x_i(t)) \cdot U(\bar{w}_{ii}(t))$, and functions $I(.), E(.), D(.)$ and $F(.)$ are given by

$$I(\tau) = \begin{cases} -c, & \tau > 0 \\ 0, & \tau \leq 0 \end{cases} \text{ [17], where } c \gg 1,$$

$$E(\tau_1, \tau_2) = \begin{cases} 1, & \tau_1 \neq \tau_2 \\ 0, & \tau_1 = \tau_2 \end{cases},$$

$$D(\tau_1, \tau_2) = \begin{cases} 1, & \tau_1 - \tau_2 \geq 0 \\ 0, & \tau_1 - \tau_2 < 0 \end{cases}, \text{ and } F(\tau) = 1 - U(\tau).$$

From the formulas (5), (4), and (2) follows that the connection weight w_{ji} is assigned the value one if and only if the following conditions are fulfilled:
a) the neuron j has changed its activity level at the previous time step; this means that the neuron j can be considered as a candidate, from which the connection weight w_{ji} can receive a positive value;
b) it has zero on its inhibitory input; i.e., the neuron j does not correspond to an obstacle at the current time step;
c) the i-th neuron had already a non-zero activation at the previous time step and the activity level of the neuron j is not larger than that of the i-th neuron; i.e., at the next time step the activity of neuron i will be larger than the activity of neuron j, what provides a proper gradient formation of the grid potentials;
d) the current activity of the i-th neuron is not zero and the connection weight $\bar{w}_{ii}$ is not positive simultaneously; i.e., if this condition is false, all connection weights w_{ji} will receive zero value, that results in zero activity value for the i-th neuron (see (1)). This is done to prohibit activity oscillations in the local neighborhoods, and to guarantee, therefore, a correct generation of grid potentials.
e) the neuron j is active; i.e., the conditions a), b), c), d) and e) are fulfilled also for this neuron.

If one of these conditions is false, the corresponding weight w_{ji} becomes zero. Connection weights $\bar{w}_{ji}$ (3) are responsible for storing the activity values of the neighboring neurons, calculated at the previous time step.

Permanent excitation of the target neuron via its external input e_i leads at each time step to generation of a new wave of neural activity in the network field. These waves carry updated information about the environment status. Therefore, the dynamical activity landscape accounts for changes in the environment and adapts to them.

A robot starts moving as soon as the first wave of neural activity has reached its initial position. Due to the strict rules, which are included into the network state equations and provide a proper potential field building, the next step rule for a robot becomes rather simple: it should move in the direction of the positive weight from the neighboring neuron, or, formally, $\tau(t_s + n) = \{p_j : w_{ji} > 0, j \in s_i\}$, where p_j is the next position of a robot in the workspace (or the next configuration in C), associated with the neuron j, t_s is the starting time, and $n > 0$ denotes the n-th discrete time step. This rule ensures that a robot always moves along a safe and shortest path.

IV. SIMULATION EXAMPLES

In this section we illustrate simulation results for various types of dynamical changes in the environment. To demonstrate the dynamic nature of the proposed algorithm more clearly, all the experiments have been done for a point robot in a 2D workspace, that, however, does not restrict general applicability of the model.

Stationary obstacles in the workspace are shown on 2d-plots in a light-gray color and dynamical obstacles are colored black. Paths of the robot are represented by continuous curves. Black squares, settling spontaneously the workspace, denote randomly appearing obstacles. SP and TP stand for Start and Target Position.

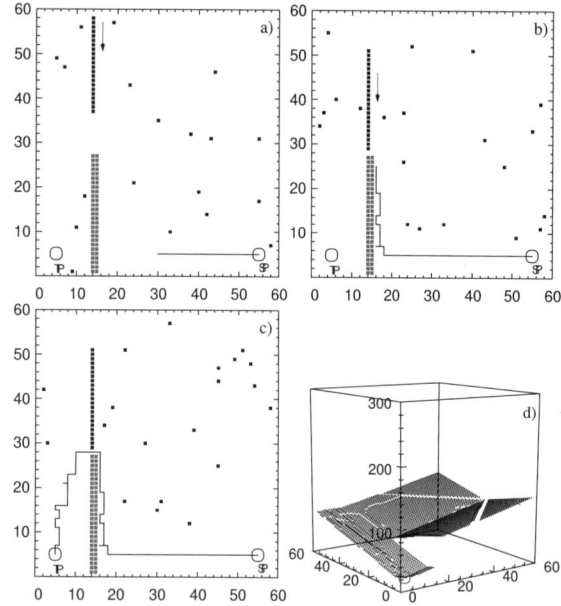

Fig. 2. The "open gate" situation: a),b) - two intermediate stages; c) The whole path; d) Activity landscape at the moment of reaching the target.

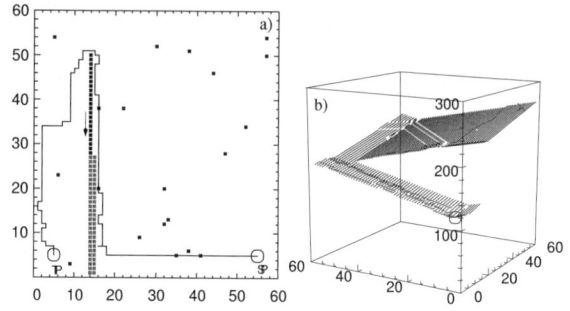

Fig. 3. The "closed gate" situation: a) The robot follows dynamically another route; b) Activity landscape at the moment of arriving at the goal.

We used for our experiments a network field, consisting of 3721 (61 × 61) neurons over the discretized workspace representation. The borders of the workspace were treated as obstacles. For test simulations we chose the neighboring neurons labeled as shown in Fig. 1b. Therefore, whenever possible, the robot prefers first to move horizontally.

A. The "Open Gate" Situation

This model situation is shown in Fig. 2. The workspace is cluttered with static obstacles. After 50 path steps of the robot, the dynamic obstacle starts to move in the direction shown by the arrow, see Fig. 2a. It stops at the position shown in Fig. 2c, leaving a small gate open. The robot traverses through the gate while avoiding 20 random obstacles, which appear in the workspace at each time step. The resulting path of the robot and the potential field, corresponding to the time of reaching the goal, are illustrated in Fig. 2c and 2d, accordingly.

B. The "Closed Gate" Situation

The initial setup for this test example is very similar to the "open gate" situation. The robot travels through the same intermediate stages, as shown in Fig. 2a and 2b. But in the new situation, the moving obstacle closes the gate before the robot can pass through it, see Fig. 3a Then the robot reacts dynamically to the change in the environment. The final path and the activity landscape at the moment of reaching the target are depicted in Fig. 3a and 3b, respectively.

C. "Freezing up" Dynamic Obstacles

In this model situation the dynamic obstacles start to move in the direction of the arrows, when the robot makes its first step (see Fig. 4a). After 20 path steps of the robot, the obstacles are frozen as shown in Fig. 4c. The activity landscape then adapts quickly to the status of the environment. The resulting activity landscape, reflecting the structure of the stationary workspace at the moment of arrival at the goal, is shown in Fig. 4d. The path of the robot is depicted in Fig. 4c.

D. "Warming up" Dynamic Obstacles

In this example situation the start and the target positions for the robot are as in the previous example. The obstacles appear in the workspace in the positions shown in Fig. 5a. The robot starts moving and after 5 path steps the obstacles drift in the directions of the arrows (Fig. 5c). The robot adapts dynamically its path in this complex situation and approaches the goal successfully (see Fig. 5c and 5d).

E. Occupation by Random Obstacles

We performed with our model a series of experiments with randomly appearing obstacles. Some examples for 20, 150, and 250 obstacles are illustrated in Fig. 6. These model examples demonstrate clearly the tendency of the path to be optimal in a L_1 metric. The path length increases gradually with the number of obstacles.

F. Autonomous Workspace Exploration

In this example the task is to reach a goal in the environment with stationary obstacles. It is assumed that the robot can detect an obstacle only if the latter lies immediately in front of it. The given workspace with obstacles is shown in Fig. 7a. During the path planning process the positions of obstacles are treated as free. Unlike the algorithm, reported in [23], where the path planning is done from scratch each time, as a new obstacle is detected by the robot, in the proposed approach the discovered obstacles are integrated dynamically into the path generation process. This allows the robot to move more continuously in real time, and not to wait until the whole path is

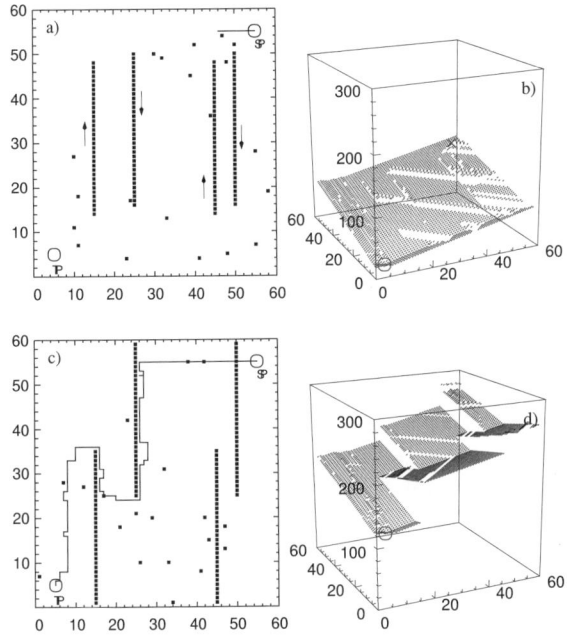

Fig. 4. "Freezing up" dynamic obstacles: a), b) Moving obstacles and the activity landscape at an initial stage of navigation; c) The planned path; d) Activity landscape at the moment of reaching the target.

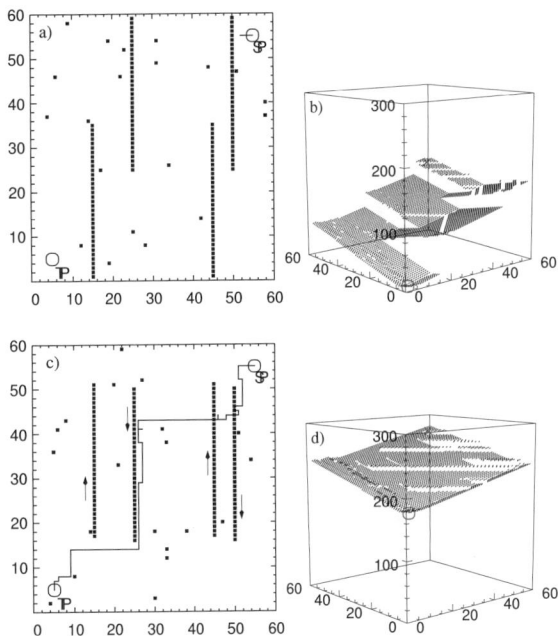

Fig. 5. "Warming up" dynamic obstacles: a), b) Arrangement of the obstacles and activity landscape at the start of the traverse; c) Reaching the goal; d) Activity landscape at the moment of arrival at the target.

replanned. The resulting path and the obstacles, detected by the robot during the traverse, are illustrated in Fig. 7c.

V. CONCLUSIONS

We have presented a novel model of a neural network, which is capable of calculating a dynamic distance transform (or dynamic grid potentials), useful for route planning in a changing environment. The proposed neural network dynamics combines in an efficient manner a wave expansion mechanism with a set of rules for detection of the next candidate path step for a robot. Due to local interactions between neurons and a regular excitation at the target neuron, the neural activity landscape adapts and accounts for environmental changes. This provides a proper formation of the grid potentials. The target point stays always at the minimum value of the potential field and attracts the robot to the goal. Since the network has a highly parallel and locally connected structure (all the neurons in the network field update their states simultaneously), the generation of grid potentials is an extremely fast process. The main properties of the proposed neural network model are:

- no *a priori* knowledge of the environment is needed;
- no learning process is required;
- the network is locally connected and highly parallel;
- computational complexity grows linearly in the number of neurons in the field;
- tendency to gain optimal paths in a L_1 metric;
- fast activity propagation makes possible real-time planning.

The presented neural network dynamics has been tested in the context of autonomous navigation and exploration on various types of complex dynamical changes in the environment, including appearance, disappearance, and drift of obstacles, avoidance of random obstacles, occupying the workspace. It has shown both the capabilities of fast adaptation to dynamical changes and a fast activity stabilization in the absence of the latter. The planned paths are safe and have the tendency to be optimal in a L_1 metric.

Due to the fast dynamical updating of potential field, robot navigates actively without waiting until the environment presents "good-traversal opportunities". Hence, one can consider the proposed approach as a compromise between tendency to path optimality, and active and mobile reaction to environmental changes. The described approach could be ap-

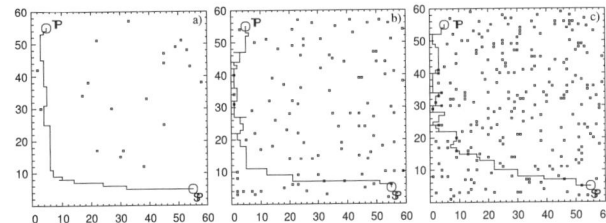

Fig. 6. Planning paths, avoiding randomly appearing obstacles: a) With 20 obstacles (the path length is 108 steps); b) With 150 obstacles (the path length is 130 steps); c) With 250 obstacles (the path length is 150 steps).

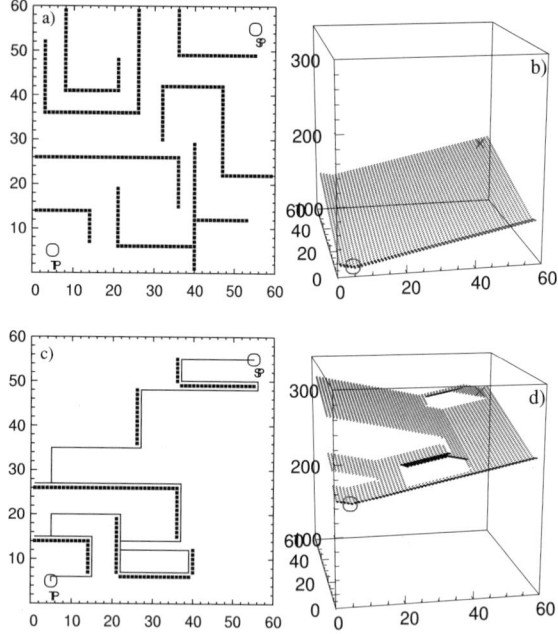

Fig. 7. Autonomous task-oriented exploration: a) The view of the given maze-like workspace; b) Potential field at the beginning of the navigation; c) The found path and the obstacles, discovered during the traverse; d) Activity landscape at the moment of coming to the goal.

plied for path planning of both mobile autonomous systems and robotic manipulators.

VI. ACKNOWLEDGMENTS

The work of the first author has been done with financial support from the DFG (German Research Council), the grant GRK256-2. Dmitry Lebedev would like to thank also Prof. V.V. Mayorov for the discussion and supporting of the ideas, presented in this paper.

VII. REFERENCES

[1] J.C. Latombe, *Robot Motion Planning*, Kluwer Acad. Publ., 1991.

[2] Y.K. Hwang and N. Ahuja, Gross Motion Planning - A Survey, *ACM Computing Surveys*, vol. 24(3), 1992, pp. 219-291.

[3] D. Henrich, Fast Motion Planning by Parallel Processing - A Review, *J. of Intel. and Robotic Syst.*, vol. 20, 1997, pp. 45-69.

[4] Jacob T. Schwartz and Micha Sharir, Algorithmic Motion Planning in Robotics, *Handbook of theoretical computer science* (ed. J. van Leeuwen), Elsevier, 1990, pp. 391-430.

[5] B. Kroese and J. van Dam, Neural Vehicles, *Neural Systems for Robotics* (eds. O. Omidvar, P. van der Smagt), Academic Press, 1997.

[6] D.C. Dracopoulos, Neural Robot Path Planning: The Maze Problem, *Neural Computing & Applications*, vol. 7, 1998, pp. 115-120.

[7] Jules M. Vleugels, Joost N. Kok and Mark H. Overmars, Motion Planning Using a Colored Kohonen Network, *Technical Reports*, Dep. of Computer Science, Utrecht Univ., 1993, Report RUU-CS-93-38.

[8] M. Lemmon, 2-Degree-of-freedom Robot Path Planning using Cooperative Neural Fields, *Neural Comput.*, vol. 3, 1991, pp. 350-362.

[9] S. Lee and G. Kardaras, Collision-Free Path Planning with Neural Networks, in *Proc. 1997 Int. Conf. on Robotics and Automation*, pp. 3565-3570.

[10] J. Park and S. Lee, Neural Computation For Collision-free Path Planning, in *Proc. 1990 IEEE Conf. on Neural Networks*, pp. 229-232.

[11] Th. Kindermann, H. Cruse and K. Dautenhahn, A fast, three-layer neural network for path finding, *Network: Computation in Neural Systems*, vol. 7, 1996, pp. 423-436.

[12] D.A. Ageev and A.Yu. Istratov, Neural Network Implementation for the Optimal Path Problem, *J. of Computer and Systems Sciences Int.*, vol. 37(1), 1998, pp. 118-125.

[13] G. Bugmann, J.G. Taylor and M. Denham, Route Finding by Neural Nets, *Neural Networks* (ed. J.G. Taylor), Alfred Waller Ltd, 1995, pp. 217-230.

[14] F.A. Kolushev and A.A. Bogdanov, Neural Algorithms of Path Planning for Mobile Robots in Transport Systems, in *Proc. 1999 IEEE Int. Joint Conf. on Neural Networks*.

[15] R. Glasius, A. Komoda and S. Gielen, Neural Network Dynamics for Path Planning and Obstacle Avoidance, *Neural Networks*, vol. 8(1), 1995, pp. 125-133.

[16] Simon X. Yang and Max Meng, An efficient neural network approach to dynamic robot motion planning, *Neural Networks*, vol. 13(2), 2000, pp. 143-148.

[17] Ashraf A. Kassim and B.V.K. Vijaya Kumar, Path Planning for Autonomous Robots Using Neural Networks, *Journal of Intelligent Systems*, vol. 7(1-2), 1997, pp. 33-56.

[18] T. Lozano-Perez, Spatial planning: a configuration space approach, *IEEE Transactions on Computers*, 1983, pp. C-32:108-120.

[19] D.V. Lebedev, J.J. Steil and H. Ritter, A New Wave Neural Network Dynamics for Planning Safe Paths of Autonomous Objects in a Dynamically Changing World, in *Proc. 2002 Int. Conf. on Neural Networks and Applications*, pp. 4171-4176.

[20] R.A. Jarvis and J.C. Byrne, Robot Navigation: Touching, Seeing and Knowing, in *Proc. 1986 1st Austral. Conf. on Artificial Intelligence*.

[21] A. Zelinsky, Using path transforms to guide the search for find-path in 2D, *Int. J. of Robotic Research*, vol. 13(4), 1994, pp. 315-325.

[22] Y.T. Chin, H. Wang, L. P. Tay and Y. C. Soh, Vision Guided AGV Using Distance Transform, in *Proc. 2001 32nd Int. Symp. on Robotics (ISR'2001)*, pp. 1416-1421.

[23] A. Zelinsky, A Mobile Robot Navigation Exploration Algorithm, *IEEE Trans. of Robotics and Autom.*, vol. 8(6), 1992, pp. 707-717.

[24] J. Barraquand and J.C. Latombe, Robot motion planning: A distributed representation approach, *Int. J. of Robotics Research*, vol. 10(6), 1991, pp. 628-649.

Motion Planning for a Crowd of Robots

Tsai-Yen Li
Computer Science Department
National Chengchi University,
Taipei, Taiwan, R.O.C.

Hsu-Chi Chou
Computer Science Department
National Chengchi University,
Taipei, Taiwan, R.O.C.

Abstract - Moving a crowd of robots or avatars from their current configurations to some destination area without causing collisions is a challenging motion-planning problem because the high degrees of freedom involved. Two approaches are often used for this type of problems: *decoupled* and *centralized*. The tradeoff of these two approaches is that the decoupled approach is considered faster while the centralized approach has the advantage of being complete. In this paper, we propose an efficient centralized planner that is much faster than the traditional randomized planning approaches. This planner uses a hierarchical sphere tree structure to group robots dynamically. By taking advantage of the problem characteristics on independently moving robots, we are able to design a practical planner with the centralized approach when the number of robots is rather large. We use several simulation examples to demonstrate the efficiency and effectiveness of the planner.

I. INTRODUCTION

The problem of directing a fleet of robots or moving a crowd of avatars are often raised in the context of robot contest, computer animation, and simulation for urban planning. The problem is challenging because the high degrees of freedom involved when the number of robots becomes large. The curse of dimensionality makes the problem difficult to solve [16]. Generally speaking, there are two main approaches to the planning problem for multiple robots: *decoupled approach* and *centralized approach*. The tradeoffs between these two approaches lie on efficiency and completeness. The decoupled approach is typically faster but lacks completeness while the centralized approach can be made complete but might need a large amount of planning time and storage space.

When the degrees of freedom in a system are rather independent in nature, the decoupled approach might be a good solution since the planning time for each decomposed problem could be rather short. The algorithm developed for solving a simple subproblem can also be complete. However, when the planner for the decomposed subproblem fails, there are usually no good algorithms that can backtrack and systematically try alternative decomposition. If we choose to use a centralized approach to solve the problem, a complete method can be developed to search the composite configuration space systematically. However, since the size of the composite configuration space is overwhelming, a systematic search dooms to be impractical. Therefore, most planners with the centralized approach use a randomized algorithm to achieve probabilistic completeness.

Although randomized algorithms have been shown to be a practical approach to solve motion-planning problems in high dimensional configuration space, we found that they may fail to find a feasible path when the decoupled degrees of freedom are actually interfering with each other such as in the case of robot crowds. In this paper, we propose a novel centralized planning approach that moves the robots in groups formed dynamically with a sphere-tree structure. We use several examples to demonstrate that the traditional randomized path planners fall short when the number of robots becomes large. With the new approach, on the other hand, we can plan for a larger number of robots in a more efficient way. We have also implemented a decoupled planner to demonstrate that the centralized planner could be a better choice in terms of completeness and efficiency.

The rest of the paper will be organized as follows. We will review the related work on the planning problem for multiple robots in the next section. In the third section, we will describe the basic problem and present an implemented planner with a decoupled approach. Then, we will propose our new centralized approach in Section IV. In Section V, we will use some experimental data to demonstrate the effectiveness of our approach. Finally, we will conclude the paper in the last section.

II. RELATED WORK

Surveys of motion planning algorithms can be found in [11] and [7]. According to [7], path planning can be viewed as either *centralized* or *distributed*. The centralized planning typically considers all robots and their degrees of freedom altogether and therefore usually entails a high dimensional composite search space. In a distributed approach, each individual robot plans and adjusts its paths in parallel with other robots until feasible paths for all robots are found.

In [11], the taxonomy about planning for multiple robots is somewhat different. The approaches are classified into two categories: *centralized* and *decoupled*. The decoupled planning is different from the distributed planning on that the robots are planned sequentially in the decoupled approach. Two variations exist in decoupled planning: (1) *prioritized planning* that considers one robot at a time under the constraint of previously planned paths for other robots, (2) the *path coordination method* that schedules the execution of individually planned paths to avoid interference. The work of [6] is an example of decoupled approach where all robots are prioritized and planned with respect to only higher-priority robots. A similar approach has also been proposed in [14] to generate the motions of two manipulator robots in an on-line manner. In [13], a decoupled approach has been used to generate motions for avatars in a virtual environment.

Most methods originally developed for single-robot systems can be applied in centralized planning. However, due to the high dimensionality of such a system, a complete planner is usually intractable. In the past decade, the randomized planning approach [1] has attracted much attention and been successfully demonstrated in many applications with difficult problems [5][10]. Early randomized planners use artificial potential fields built in the workspace to guide the search in C-space and use random walks to escape local minima. A typical planner with this approach is the RPP (Randomized Path Planner) [2]. Most of the randomized planners developed in the last few years use the PRM (Probabilistic Roadmap Method) approach [9]. In such an approach, we build a random roadmap in the C-space in a preprocessing step and try to answer planning queries at a later time as quickly as possible. A common feature of the randomized planners is that they are probabilistic complete, which means that if there exists a feasible path and there is no time limits, then the planner will be able to find it eventually.

The research on generating motions for crowds of agents can be found in the literature of Robotics, Artificial Life, and Computer Animation. A good survey of cooperative robotics can be found in [3]. In [8], a flocking model was used for a crowd of robots to follow a leader robot. A similar approach has been adopted in [12] to simulate a crowd of avatars led by a leader capable of generating collision-free motions. Realistic flocking behaviors for virtual creatures such as birds or fishes have been successfully simulated with artificial forces [17]. In [3], roadmap consisting of medial axes is used to guide the simulation for a flock of avatars. However, a common weakness of these approaches is that they cannot guarantee that a feasible motion plan can be generated for the whole system even if such a plan exists.

III. THE DECOUPILED PLANNING APPROACH

In this section, we will first give a general description of the path-planning problem for multiple robots. The problem definition, in fact, might be different for different applications at various situations. However, we will focus on the problem suitable for the decoupled approach in this section and briefly describe a planner implemented with this approach. Examples generated with this approach will be given at the end.

A. Considerations for Different Applications

Depending on the applications, one can define the planning problem for multiple robots slightly differently. For example, depending on the time when the problem is raised, a planner may need to plan for all robots at a time or it may be called sequentially for each robot when their paths are needed. For the first case, the problem can be solved with either a decoupled or a centralized approach. However, for the second case, a decoupled planning is more appropriate since the path for each robot is generated at different time.

Another application attribute that might affect the choice of the planning approach is the specification of the goal configuration. If each robot can be given a definite goal configuration at run time, we can use either approach to solve the problem. However, specifying the goal configuration for a large number of robots is a tedious task. If we must generate the motions for all robots at a time, we are more likely to specify a rough destination region for the robots instead of individual goals. If the destination region is not very large, the decoupled approach may not be a good choice because the robots that reach the region earlier are likely to block the entrance and prevent later robots to reach the region. In the next subsections, we will assume that the planning requests are issued at run time while the motions of other robots are being executed, and each robot will be given a specific goal configuration.

B. Motion-Planning Problem for Multiple Robots

Assume that we are given a geometric description of the robot and polygonal obstacles in a 2D workspace. We assume that each robot can be represented by an enclosing circle of radius r. Due to geometric symmetry, we can use only two parameters (x, y) to describe the configuration q^i for a robot i. Suppose that there exist n robots ($n>1$) in the workspace. We denote the individual configuration space (C-space) for robot i by C^i. Then the composite configuration space for the multi-robot system is defined as $C = C^1 \times C^2 \times ... \times C^n$, where a configuration in C is denoted by q. Each robot has to satisfy the geometric constraint that they cannot collide with each other or with obstacles. In addition, each robot must move under a velocity limit constraint.

In this section, we will assume that the planning requests are issued at different times for different robots while the motions of other robots are being executed. Each request defines a planning problem for a robot from its current configuration to a specified goal configuration. Although not necessary, it is usually desirable not to disturb the current motion plans of other robots when we try to find a feasible motion for a robot. Therefore, a decoupled approach is more appropriate for this case.

C. Decoupled Planning Approach

Assume that we are given a motion-planning problem for multiple robots as described in the previous subsection. The path of the ith robot (denoted by τ^i, $i = 1$ to n) is known as a function of time t, including when it is static. In our decoupled approach, for the kth robot under consideration we augment its C-space by the time dimension to form the so-called *Configuration-Time Space* (*CT-space*). A conceptual example is depicted in Fig. 1. There are two types of

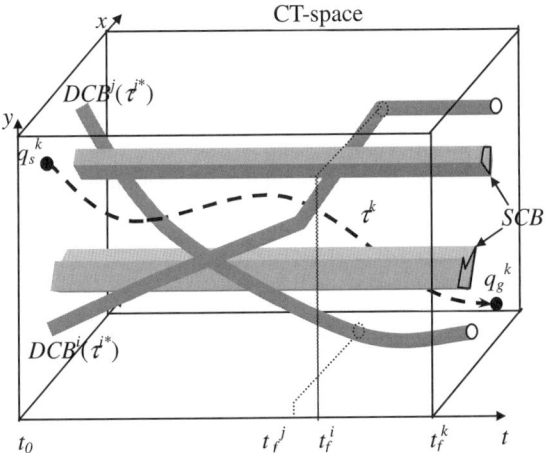

Fig. 1. Searching for a feasible path amongst obstacle regions in the CT-space.

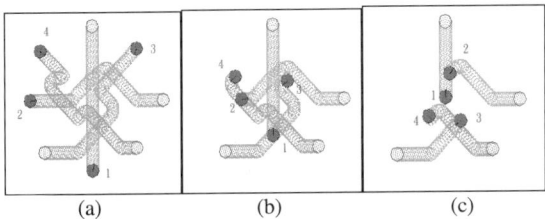

Fig. 2. An example of decoupled planning with coordinated crossing motions for four robots

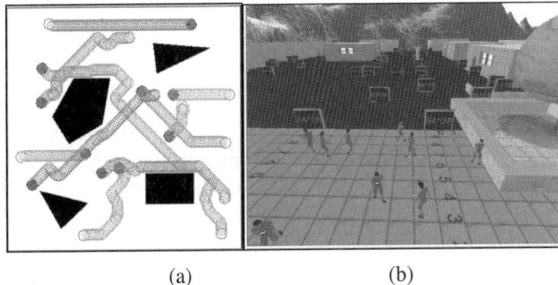

Fig. 3: An example of decoupled planning for a crowd of avatars moving independently in a virtual world

forbidden regions in the CT-space representing obstacle regions that the robot should avoid entering. One (denoted by SCB) is due to the static obstacles while the other (denoted by DCB) is due to other moving robots. Note that SCB is axis-parallel extrusion of 2D obstacles in time while DCB's are curve extrusions of the obstacle regions imposed by other moving robots. When the ith robot finishes its motion at time t_f^i, we assume that it will stay there unless otherwise instructed. Equivalently, we are extending the path for the ith robot to infinity and this extended path is denoted by τ^{i*}. The objective of the path planner is to find a collision-free path τ^k for the kth leader in the CT-space that can connect the current (q_s^k) configuration at the current time (t_0) to the specified goal configuration (q_g^k) at some time (t_f) in the future. Because of the velocity constraint the slope at any point along a legal path in this CT-space must be positive (because time is not reversible) and less than some user-specified value (maximal velocity). A Best-First planning algorithm can be adopted to search for a feasible path in such a CT-space.

D. Planning Examples

Fig. 2 shows an example of decoupled planning for four robots moving across each other. The trace of their paths is shown, and the planning order (priority) is depicted beside the robots. Note that the first robot chooses a straight-line path since it has the highest priority. The later a robot is planned, the more detoured its path usually will be. Another example of multiple robots moving independently without colliding with each other is shown in Fig. 3(a). In Fig. 3(b), we show a snapshot of how the planner has been used to simulate a human crowd in a virtual environment.

In the example of Fig. 2, the average planning time for each robot is about 105ms on a regular PC with 650MHz CPU. When the number of robots increases, one can expect that the planning time will have a quadratic growth because a robot has to check collisions with other $n-1$ robots. Although the growth is not as fast as the exponential growth in the centralized approach, the advantage that the decoupled approach can be used in an on-line manner for interactive applications may not be valid when n increases to some large value.

IV. THE CENTRALIZED PLANNING APPROACH

A. Revising Problem Definition

When the number of robots increases to some large value, say 100, it becomes impractical to specify the goal configuration for each individual robot interactively. In this case, it is more desirable to specify a rough destination region for the crowd of robots to move to. We assume that the region is a circle of radius R centered at (x_g, y_g), specified by the user. The goal is reached if all robots can enter the region enclosed by the circle. We assume that the order of entering and their relative positions are not important. Although a decoupled approach can be used to solve the problem but such an approach often fails to find a path simply because the robots that arrive early may prevent later robots to enter the region. Therefore, a centralized approach is preferred. However, one has to face the curse of dimensionality as the number of robots increases.

Among the randomized path planning algorithms proposed in the literature, the RPP (Randomized Path Planner) and PRM (Probabilistic Roadmap Method) are the two mainstream methods. However, both methods share some

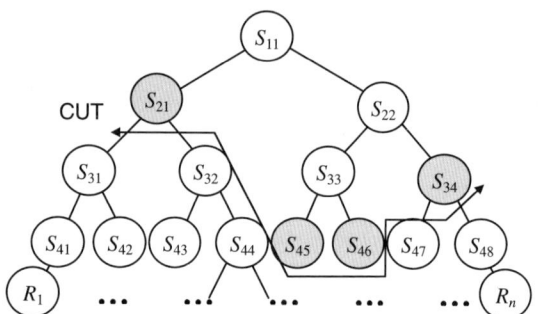

Fig. 4. A sphere tree and a profile cut separating spheres with and without collisions

```
procedure Down_Motion_with_Grouping()
1.  SUCCESS : = false
2.  Append( qi, τ )    { τ is the path for down motion }
3.  nStep : = 0    { nStep is number of legal moves }
4.  CUT : = root    { a profile cut list of the sphere tree }
5.  while ¬ SUCCESS
6.      nTrial : = 0    { nb of trials for lower legal neighbors }
7.      nTotalTrial : = 0    { nb of trials for local minima }
8.      while nTrial < nMaxTrial
9.          nTrial : = nTrial + 1
10.         nTotalTrial : = nTotalTrial + 1
11.         for all spheres o in CUT    { o is a sphere }
12.             o' : = SelectLegalNeighbor(o)    { random select }
13.             if o' = NULL then Split(o, CUT) { split o }
14.         q_new : = Conf(CUT)    { find corresponding conf }
15.         if Legal(q_new) then    { collision-free or not }
16.             nTrial : = nTrial + 1
17.             if U(q_new) < U_min then    { lower potential }
18.                 Append(q_new, τ )
19.                 nStep : = nStep + 1
20.                 break    { while }
21.             else if Crowded() then RebuildST()
22.         if nTrial >= nMaxTrial then SplitLargestSphere(CUT)
23.         if nStep mod nPeriod = 0 then RebuildST()
24.     if U_min = 0 then SUCCESS : = true
25.     if nTotalTrial > nMaxTotalTrial then break {local min.}
```

Fig. 5: The Down_Motion_with_Grouping algorithm

common characteristics and have their own pathological cases [1]. For example, the PRM planners typically have difficulties in connecting two roadmap components through long narrow passages. The RPP planner could be better in solving difficult planning problems but it typically takes a longer time when the heuristic potential fields used in the planner are misleading.

Although the motion-planning problem for multiple robots can be solved with either centralized approach, we think pathological cases often occur in the traditional planners as the number of robots increases. When the number of robots is large, it is more likely that the robots are rather crowded at the initial and goal configurations. In this case, the probability of finding a legal neighboring configuration becomes rather low since the robots all move independently. As long as one robot is in collision, the overall system configuration becomes illegal. Therefore, the size of solution space is rather small compared to the whole problem space when we allow every robot to have its full degrees of freedom. The set of legal configurations would be limited to those that move the robots at the periphery outwards first. However, the probability of choosing such a configuration is rather low in a random process. Therefore, we need to improve the traditional planner by accounting for this problem characteristic.

B. Grouping Robots with a Hierarchical Sphere Tree

As described in the previous subsection, when we plan for multiple independent robots, the pathological case happens because we are giving the robots too much freedom. Allowing only a few robots to move at a time may be a good idea but the planner may not be probabilistic complete any more. In addition, one still has to determine who to move first. Therefore, we propose to organize the crowd of robots into a hierarchical sphere tree structure and move the robots as a set of robot groups whenever possible. A sphere tree is a binary tree, whose leaf nodes represent geometric primitives, such as a circle or a sphere, composing the shape of a robot. Each internal node represents a sphere whose size is large enough to enclose its children spheres. This type of sphere tree structure is commonly used to reduce the number of calls to expensive collision detection routines [15]. As long as the bounding volume of a node at a higher level does not cause collisions, further examination below the node becomes unnecessary.

We build a sphere tree for the robots at their initial configuration. The robots are organized in a hierarchical structure where each leaf node represents a robot, as shown in Fig. 4. Since each leaf node belongs to a list of ancestor sphere nodes of various sizes, robots can be grouped and moved with different levels of grouping. When we move an internal sphere node, all robots under the node also move for the same amount. When a sphere node moves, the ancestor spheres up to the root must update their radius accordingly in order to enclose their children nodes. This is somewhat different from the case of pure collision detection applications where the relative positions between spheres in a sphere tree do not change because most applications assume that the robot is a rigid body.

C. RPP with Hierarchical Grouping

The low probability of moving to a legal neighbor makes the planning problem a pathological case for path planners, especially for PRM-based planner. Therefore, we have chosen to improve the potential field based planners (RPP) by incorporating a hierarchical grouping strategy to increase the chance of finding a legal neighbor. The RPP algorithm consists of alternative calls to the Down_Motion and Brownian_Motion procedures. The modifications that we

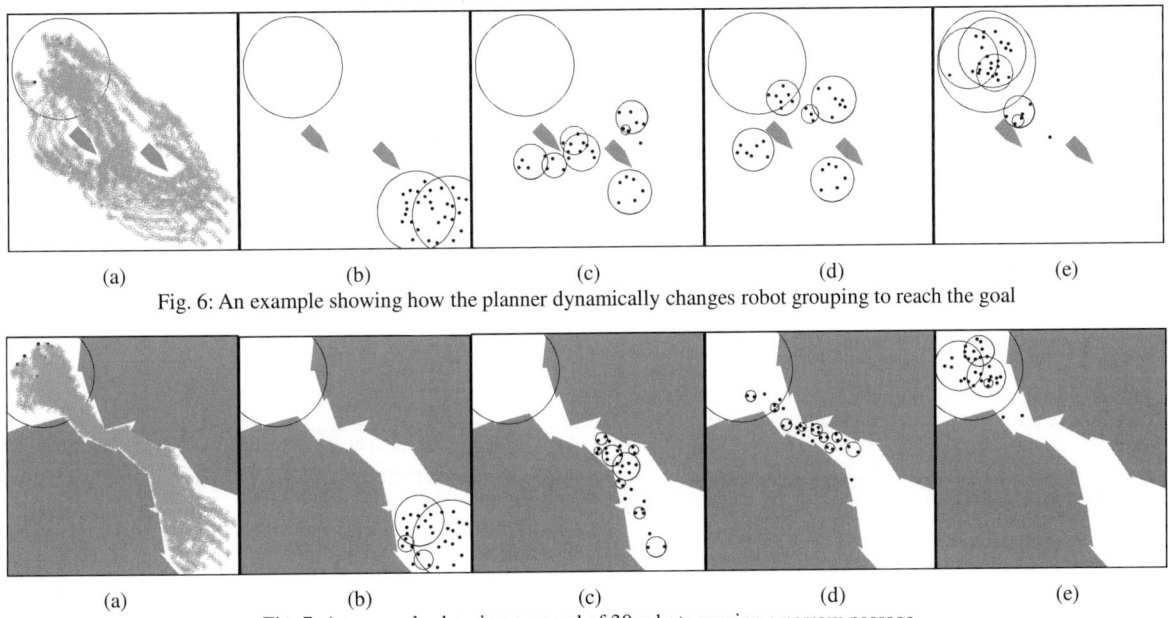

Fig. 6: An example showing how the planner dynamically changes robot grouping to reach the goal

Fig. 7: An example showing a crowd of 30 robots passing a narrow passage

have made are mainly on using the grouping strategy to generate legal neighbors in the Down_Motion procedure.

In Fig. 5, we show the modified procedure, called Down_Motion_with_Grouping. The idea is to freeze the relative positions between robots as much as possible by grouping them with hierarchical spheres as described in the previous subsection. The grouping attempts start from the root of the sphere tree and walk down toward the leaf nodes when the current grouping sphere collides with obstacles. When testing the possibility of grouping robots at an internal node, we randomly try a few neighboring configurations for the sphere (line 12) until a legal (collision-free with obstacles) configuration has been found or all trials fail. When the attempt fails, we will recursively walk down to the next level and attempt to move its two children spheres independently. When a legal configuration for the whole system has been found, we will have a list of grouping spheres (depicted in grey in Fig. 4) at various levels that forms a profile cut on the sphere tree. We will record this "CUT" location and start the next trial from it. In addition to considering the collisions with obstacles, we also have to check the inter-collision between robots (line 15). If the new system configuration is legal, then we check if the new configuration has a smaller potential value than the current one. If so, we will move the robots to the new configuration, and the search for the next legal configuration with a lower potential starts over again.

We set a limit on the number of trials for moving the robot system to a configuration with a lower potential value. If the number is reached, we further lower the CUT by splitting the largest sphere (line 22). This step enables the CUT to move to the lowest level (leaves) and restores the freedom of each robot. However, if we can only lower the CUT, the advantage of grouping the robots will disappear eventually when all robots retain their freedom. Therefore, we also have to consider moving the CUT upward as well. However, according to [12], attempting to merge nodes in every step from bottom up might not be more efficient than updating the list from the root node down after a few steps. In addition, when two sibling spheres move away from each other, the radii of their parent spheres may increase to a degree that rebuilding the sphere tree is desirable. Therefore, we periodically rebuild the sphere tree (line 23) and update the CUT from the top down to maintain a more representative sphere tree for future uses. In addition, when we detect that the robots are away from the obstacles but the inter-collision between robots are severe (tested in the Crowded function in line 21), we also choose to reorganize their relative positions by rebuilding the sphere tree.

We build a numerical potential field [2] in the C-space of a robot to guide the search. Since the goal is a destination region instead of a single configuration, we set the potential values of all configurations in the region to zero. The overall potential $U(q)$ for a system configuration q is the sum of the potentials for each individual robot. When all robots enter the region, U will become zero. If the system has made a given number (nMaxTotalTrial) of trials to move to a lower potential without success, we will assume that a local minimum has been reached and the procedure will return its current available path found so far. A random walk will be used to escape the local minimum as in the traditional RPP.

V. EXPERIMENTS

We have implemented the RPP planner with dynamic hi-

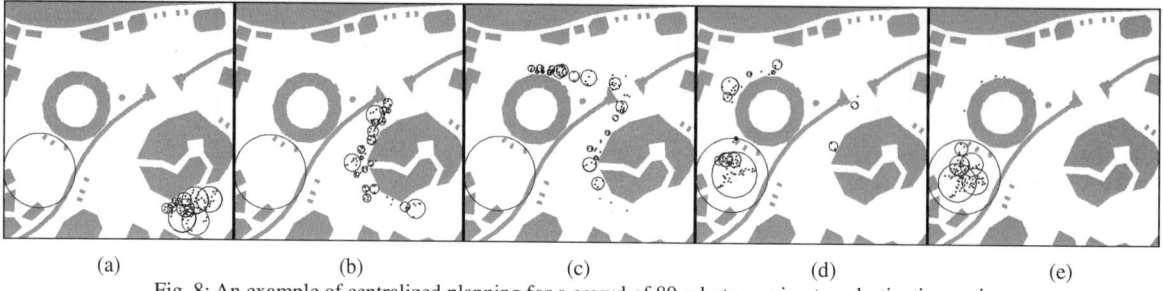

Fig. 8: An example of centralized planning for a crowd of 80 robots moving to a destination region

Table 1. Comparisons of the planning times (in seconds) for moving different number of robots in three different planners

N	Decoupled	Centralized RPP w/o Sphere Tree	Centralized RPP w/ Sphere Tree
10	6.67	0.93	0.28
20	15.29	2.85	0.72
30	23.60	6.76	1.21
40	29.52	13.45	1.51
50	43.03	26.49	3.36
60	62.49	47.84	3.71
70	89.10	70.28	4.46
80	122.87	110.39	13.40
90	127.44	175.04	17.73
100	204.683	239.59	27.99
120	275.21	447.09	56.84
140	526.91	810.07	71.15
160	2224.05	1570.83	170.89
180	---	2359.68	262.44
200	---	---	309.91
220	---	---	784.93
240	---	---	1594.19
260	---	---	1365.34
280	---	---	2502.34

Table 2. The planning times (seconds) for using different methods to rebuild the sphere tree

Methods	(A) no rebuild	(B) when crowded	(C) periodic	(D) periodic and crowded
Planning time (sec.)	93.25	53.84	5.47	4.88

erarchical grouping in the Java language. We have conducted extensive experiments to demonstrate the efficiency of the planner. All experimental data reported in this section were measured on a regular PC with 1.2GHz CPU.

A. Planning Examples

In Fig. 6 and Fig. 7, we show the example paths generated by the planner for two different workspaces. In both examples, there exist thirty robots trying to move from their configurations to a destination region depicted in circle. The first subfigure(a) in both examples shows the traces of the paths executed by the robots. Note that in the example of Fig. 6, the robots move together as large groups until they encounter obstacles. In this case, the robots are organized into smaller groups and resume more degrees of freedom. The sphere tree may be rebuilt at run time so that we can see the robots are grouped differently in Fig. 6(d). As some robots reach the destination region, they still have some degrees of freedom to move inside the area so that they do not block the entrance where they entered. Fig. 7 shows an example in another workspace where there exists a narrow passage that forces the robots to move individually as they pass the passage. This is a pathological case for the planner since the advantage of moving robots in groups disappears and we still have to pay the cost of maintaining the sphere tree. However, since we rebuild the sphere tree periodically, the planner can resume moving in groups as soon as they pass the passage. We found that the planner with dynamic grouping is still more efficient than the traditional RPP in this example, which implies that the overhead of maintaining the sphere tree is rather low.

B. Performance Comparisons

We have conducted extensive experiments to compare the performances of the decoupled planner, the traditional RPP planner and the new planner. The results are shown in Table 1. We run the three planners (decoupled, centralized with and without using sphere tree to group robots) for the workspace shown in Fig. 8. Snapshots along an example path generated for 80 robots are shown in Fig. 8. The number of robots ranges from 10 to 300 in the experiment. All planners are forced to terminate when the planning time exceeds one hour (marked with '---' in Table 1). Note that the new planner outperforms the decoupled and the traditional planners in all cases and the more the number of robots, the more improvement that we can observe. Although the traditional RPP planner is probabilistic complete, it is terminated after one hour of trial when the number of robots reaches 200. On the other hand, the new planner can still find a path when the number of robots reaches 280 (or 560 DOF).

We also have conducted experiments to study the effects of rebuilding the sphere tree that keeps its size small. In two ways, we rebuild the sphere tree. One is by detecting the situation that most collisions occur between robots instead of between robots and obstacles while the other is by periodic updates. The results are shown Table 2. The cases (B) and (C) correspond to the two methods above. Note that timely rebuild of the sphere tree can improve the overall performance and the effect of periodic updates seems to be more significant than the other case.

C. Discussions

The improvement of the new planner over the traditional RPP planner is quite significant. Although we are not attempting to deal with the curse of dimensionality, we have significantly lowered the constant of the exponent to make the planner practical when the number of robots is large. The centralized planner outperforms the decoupled planner in most cases when the number of robots is not large. Besides, the centralized planner has the advantage of being probabilistic complete that the decoupled planner does not have. Detail data from our experiments show that the new planner with dynamic grouping is more efficient mainly because the number of inter-robot collisions has been significantly reduced. This observation reveals that the original idea of using a hierarchical sphere tree to group robots dynamically in order to reduce inter-robot collisions is quite effective.

VI. CONCLUSIONS

The problem of path planning for multiple robots is getting more attentions in Robotics and Computer Animation. However, the traditional planners do not seem to be able to solve the problem as efficiently as in other cases. In this paper, we reviewed the different approaches proposed in the literature, and implemented the planners with these approaches for comparisons. We also have proposed a new planner based on the RPP planner to improve the planning performance for a large number of robots. Experiments show that this new method can significantly reduce inter-robot collisions and therefore is more effective for this type of multiple-robot planning problem.

VII. ACKNOWLEDGMENTS

This work was partially supported by National Science Council under contract NSC 91-2213-E-004-005.

VIII. REFERENCES

[1] J. Barraquand, L. Kavraki, J.C. Latombe, T.Y. Li, and P. Raghavan, "A Random Sampling Scheme for Path Planning," *Intl. J. of Robotics Research*, 16(6), pp.759-774, Dec. 1997.

[2] J. Barraquand and J. Latombe, "Robot Motion Planning: A Distributed Representation Approach," *Intl J. of Robotics Research*, 10:628-649, 1991.

[3] O. B. Bayazit, J.M. Lien, N. M. Amato, "Simulating Flocking Behaviors in Complex Environments," *Proc. of the Pacific Conf. on Computer Graphics and Applications*, 2002.

[4] Y.U. Cao, A.S. Fukunaga, A.B. Kahng, and F. Meng, "Cooperative Mobile Robotics: Antecedents and Directions," in *IEEE/TSJ Intl. Conf. on Intelligent Robots and Systems*, pp.226-234, 1995.

[5] H.S. Chang and T.Y. Li, "Assembly Maintainability Study with Motion Planning," *Proc. of 1995 IEEE Intl. Conf. on Robotics and Automation*, Nagoya, Japan, May 1995.

[6] M. Erdmann and T. Lozano-Perez, "On Multiple Moving Objects," AI Memo No. 883, Artificial Intelligence Laboratory, MIT, 1986.

[7] K. Fujimura, *Motion Planning in Dynamic Environments*, Springer-Verlag, New York, 1991.

[8] V. Gervasi and G. Prencipe, "Flocking by a Set of Autonomous Robots," Technical Report: TR-01-24, Department of Information, University of Di Pisa, Italy, 2001.

[9] L. Kavraki, P.Svestka, J. Latombe, and M. Overmars, "Probabilistic Roadmaps for Fast Path Planning in High-Dimensional Configuration Spaces," *IEEE Trans. on Robotics and Automation*, 12:566-580, 1996.

[10] Y. Koga, K. Kondo, J. Kuffner, and J.C. Latombe, "Planning Motions with Intentions," *Computer Graphics (SIGGRAPH'94)*, pp.395-408, 1994.

[11] J. Latombe, *Robot Motion Planning*, Kluwer, Boston, MA, 1991.

[12] T.Y. Li and J.S. Chen, "Incremental 3D Collision Detection with Hierarchical Data Structures," in *Proc. of ACM Symp. on Virtual Reality Software and Technology, (VRST'98)*, pp.139-144, Taipei, Taiwan, 1998.

[13] T.Y. Li Y.J Jeng, and S.I Chang, "Simulating Virtual Human Crowds with a Leader-Follower Model," *Proc. of the 2001 Computer Animation Conf.*, Korea, 2001.

[14] T.Y. Li and J.C. Latombe, "Online Manipulation Planning for Two Robot Arms in a Dynamic Environment," *Intl. J. of Robotics Research*, 16(2):144-167, 1997.

[15] S. Quinlan, "Efficient Distance Computation between Non-Convex Objects," *Proc. of Intl. Conf. on Robotics and Automation*, pp.3324-3329, San Diego, CA, 1994.

[16] J.H. Reif, "Complexity of the Mover's Problem and Generalizations," *Proc. of the 20th IEEE Symp. on Foundations of Computer Science*, pp. 421-427, 1979.

[17] C. Reynolds, "Steering Behaviors For Autonomous Characters," *Proc. of Game Developers Conf.*, 1999.

Motion Planning for Multiple Mobile Robots using Dynamic Networks

Christopher M. Clark & Stephen M. Rock
Aerospace Robotics Lab
Department of Aeronautics & Astronautics
Stanford University
{chrisc, rock}@sun-valley.stanford.edu

Jean-Claude Latombe
Department of Computer Science
Stanford Universty
latombe@cs.stanford.edu

Abstract - A new motion planning framework is presented that enables multiple mobile robots with limited ranges of sensing and communication to maneuver and achieve goals safely in dynamic environments. To combine the respective advantages of centralized and de-centralized planning, this framework is based on the concept of centralized planning within dynamic robot networks. As the robots move in their environment, localized robot groups form networks, within which world models and robot goals can be shared. Whenever a network is formed, new information then becomes available to all robots in this network. With this new information, each robot uses a fast, centralized planner to compute new coordinated trajectories on the fly. Planning over several robot networks is decentralized and distributed. Both simulated and real-robot experiments have validated the approach.

I. INTRODUCTION

When many robots operate in the same environment, high-level motion planning is required for the robots to reach their goals while avoiding collisions among themselves and with static and moving obstacles. In unknown or partially known environments, it is unlikely that a system of sensors can provide global knowledge. In addition, continuous inter-robot communication is usually not feasible. Instead, only robots that are sufficiently close to each other can exchange information, e.g., share their goals and local world models.

This paper introduces a new planning framework that exploits the changing communication links between robots, as the robots move, to combine the respective advantages of centralized and decentralized planning. More precisely, our approach is based on *dynamic robot networks* that are capable of: 1) forming dynamically whenever communication and sensing capabilities permit; 2) sharing world models and robot goals within each network; and 3) constructing "on the fly" coordinated trajectories for all robots in each network using a fast centralized motion planner.

An overview of this approach is presented in Section II. A background review (Section III) justifies the choices made in our approach. We then describe aspects of our framework in more detail, namely the representation of partial world models (Section IV) and the planning technique (Section V). Section VI presents the test-platform used for simulations and robot experiments. Section VII gives some experimental results.

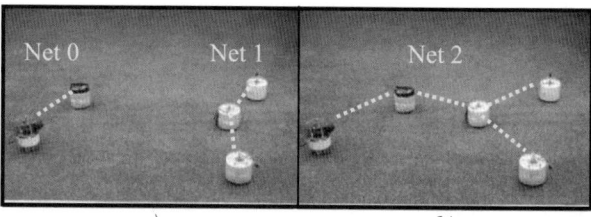

Fig. 1 Example with 5 robots. Dashed lines between robots depict communication links. In a) the robots form two distinct networks Net0 and Net1. In b), two robots have moved, and the two networks in a) have merged into Net2.

II. PLANNING IN DYNAMIC NETWORKS

A. Network Formation

When any two robots are within communication range of each other, they establish a communication link. Define G to be the graph whose nodes are the robots and edges are the communication links. A *network* of robots is any group of $k \geq 1$ robots forming a maximal connected component of G. So, any two robots in a network can communicate through one or several communication links, but two robots from different networks can not.

Fig. 1a shows an environment with 5 robots, where 2 networks have formed. In Net1, the top and bottom robots can exchange information via their communication links with the middle robot. Because robots are moving to achieve their goal locations, the networks are dynamic. Robots may leave networks and/or form new networks (see Fig. 1b). An application level protocol ensures that at any time robots in each network can access the local sensing information of all other robots in the same network, and hence share a common world model.

B. Planning Process

Motion planning in a network N is triggered by any one of the following events:

- N just got formed, i.e., two robots from different networks entered one another's communication range.

- A significant change in the world model occurs, e.g., a robot in N senses a new obstacle.

- A new goal location is requested for one or several robots in N.

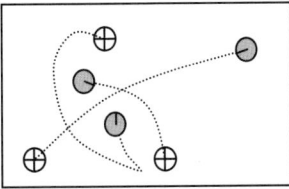 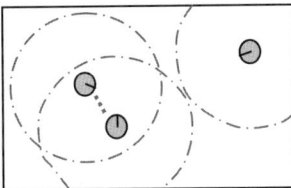

a) All three robots (grey circles) are at their initial locations. The two left robots are in communication range and form a network. Their centralized planners create coordinated collision-free trajectories that lead toward the goals (cross-hairs). The right robot forms a network by itself, and its trajectory is planned independently from the other two. The robots start moving along these trajectories.

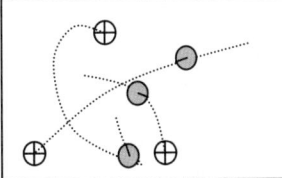 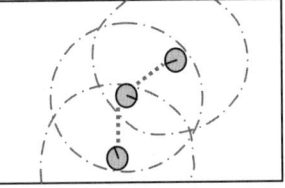

b) As the robots move along their trajectories, the middle robot and the right robot enter communication range with each other, and all three robots now form a larger network.

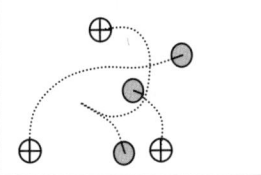 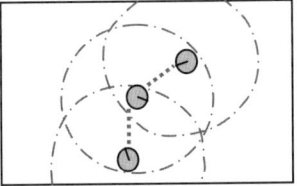

c) A new plan is made for all three robots in the network. This plan consists of collision-free trajectories for all three robots.

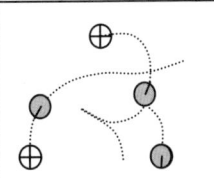 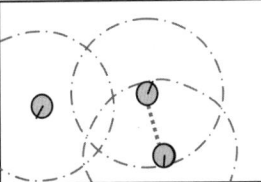

d) As robots move along their new trajectories, they leave communication range of each other and some network links are broken. They continue to follow the planned trajectories.

Fig. 2 Top-down view of a planning example with three robots. In each of the fours snapshots, the illustration on the left shows the robots following their trajectories to their respective goals (cross-hairs). The diagram on the right depicts the communication range of each robot and the existing communication links.

When such a triggering event occurs, data is exchanged between the robots in N, so that each one gets an updated world model that combines the local world model and goal of every robot. Once robots have shared this information, each robot runs its own copy of a centralized motion planner to construct coordinated trajectories for all robots in the network. When the planner terminates, each robot broadcasts its plan to all other robots in the network. Each robot selects the same best plan and immediately starts executing its trajectory in this plan. The planner is a single-query probabilistic-roadmap (PRM) planner similar to the one presented in [12] (see Section V).

This process is illustrated in Fig. 2 on a simple example involving 3 robots, with no obstacles. A triggering event automatically occurs at the start of the process, as the first networks get formed.

Since robots also have limited sensing, the world model shared through a network is partial. Planning is done using this model. As robots move, their sensors may detect previously unknown obstacles or a change in the trajectory followed by a known obstacle. Such an event triggers a re-planning operation within the network where the new obstacle or change of trajectory was detected.

III. BACKGROUND REVIEW

Most previous work on multi-robot motion planning can be grouped into *centralized* and *decentralized* planning [2,23]. While centralized planning considers all robots together as if they were forming a single multi-body robot [4,6,17,22,26,27], a decentralized planner plans for each robot separately before coordinating the individual plans by tuning the robot velocities along their respective paths [1,3,9,14,15,19,21,25]. A variant of decentralized, called *prioritizing planning*, plans for one robot at a time, in some sequence, considering the robots whose trajectories have already been planned as moving obstacles [5,10].

Centralized planners can be advantageous because they allow the possibility of completeness and global optimization. For example, it was shown in [23] that a centralized planner based on PRM techniques can reliably solve problems requiring the tight coordination of multiple articulated arms, while decentralized planners based on similar PRM techniques fail often. On the other hand, centralized planning may take more time due to the high dimensionality of the configuration spaces that are searched. A worse drawback is that they require all information (partial world models and robot goals) to be centralized in one single place, which is only possible if the robots have unlimited communication capabilities. This is not the case in many practical settings.

A major advantage of decentralized planning is that it allows for distributed planning. Each robot can then plan its own trajectory using its own partial model of the environment. If two robots eventually get close to one another and risk colliding, simple velocity-tuning techniques or reactive techniques can be used to locally coordinate their motions. However, a fully distributed approach fails to exploit the fact that localized groups of robots can exchange information to improve planning.

While decentralized planning is potentially less computationally intensive because it searches several configuration spaces of smaller dimensionality, it cannot offer any completeness or optimality guarantee. Various attempts have been made to improve the outcome of decentralized planners (e.g., [3,5,11]). In particular, a negotiation scheme between localized groups of robots is used in [3] to assign priority orders to robots, which allow the decentralized planner to compute trajectories of reduced lengths. This negotiation scheme demonstrates the benefits of localized inter-robot communication, and is the technique most closely related to the robot network planning framework presented in this paper. But decentralized planning remains intrinsically incomplete.

The planning approach presented in this paper exploits the respective advantages of centralized and decentralized planning. In each robot network, it uses a centralized single-query PRM planner to increase completeness and still provide fast on-the-fly planning. However, planning is distributed over the various networks – hence, planning over multiple networks is decentralized – to accommodate the fact that robots from different networks cannot share information. The triggering event caused by the merging of two previously distinct networks into a single network leads the robots in this new network to take advantage of the information they now share by centrally re-planning their coordinated trajectories.

Planning with incomplete world models and on-the-fly re-planning when a sensor detects the presence of a still unknown obstacle or a change in an obstacle's trajectory have previously been described in [12, 16] for a single robot. We use similar techniques, but extend them to multiple robot networks.

IV. WORLD MODEL

Describing the world model in a concise but useful form is necessary to allow for information sharing between robots in the same network. In the experimental system that we have built, world models simply consist of a list of robots and their descriptions, and a list of obstacles and their descriptions. The following table outlines the information stored in each list:

World Model Description

---
1) List of Robot Descriptions
   - State (position and velocity)
   - Size (Radius)
   - Most Recent Update Time
   - Information Source
   - Goal position
   - Current Trajectory

2) List of Obstacle Descriptions
   - State (position and velocity)
   - Size (Radius)
   - Most Recent Update Time
   - Information Source

Robots report their own size and state, while obstacle sizes and states are estimated by robot sensors. The most recent update time is useful when updating world models with information received from other robots. The information source is a robot ID that indicates which robot sensed (or communicated with) the object. It is used to keep track of which robots are currently in the network. Several assumptions were made to allow such a concise world model:

- Each robot has access to its own state relative to a global coordinate system (e.g., GPS).

- Each object is approximated as a circular object to allow its geometry to be described by a single parameter, its radius.

- Each obstacle has constant linear velocity estimated by a robot's sensor. As in [12], if at any later time its trajectory is found to diverge by more than some threshold from the predicted trajectory (either because the obstacle did not move at constant velocity, or because the error in the velocity estimate was too high), then the robot that detects this divergence calls for the construction of a new plan within its network. The planner "grows" the obstacles (and the robots) to allow for some errors in predicted trajectories of the objects.

- All objects in the environment are easily identifiable by robot sensors, which can also precisely estimate their positions and velocities. Any discrepancy between two local world models can be easily resolved.

The second assumption is rather easy to eliminate, as it has been shown before that PRM planners can efficiently deal with geometrically complex robots and obstacles (e.g., [22]). In [12], the third assumption has been shown to be quite reasonable, even when obstacle velocities change frequently, provided that (re-)planning is fast enough. The last assumption is more crucial. In our experimental system, it is enforced by engineering the vision system appropriately (Section VI). In the future, it will be important to relax this assumption by using more general sensing systems and data fusion techniques [20].

V. MOTION PLANNING ALGORITHM

As indicated earlier, motion planning within a robot network is done using a centralized single-query PRM planner (more precisely, several copies of this planner running in parallel). This planner searches the joint state×time space C of the k robots in this network. The state of each robot is defined by the two coordinates of its center and two velocity parameters, so C has $4k+1$ dimensions. This representation can easily be extended to other robots. For instance, we have implemented a version of the planner for robots in three-dimensional space [8]. The planner searches C for a collision-free trajectory from the initial state of the robots to their goal state. The resulting trajectory defines the coordinated motions of the robots to their respective goals.

Our planner searches C by incrementally building a tree of milestones (the roadmap), as described in [12,13,17]. At each iteration, it selects a milestone m in the current roadmap, generates a collision-free state m' at random in a neighborhood of m in C and, if the path from m to m' tests collision-free, installs m' as a new milestone in the roadmap. The search terminates when m' falls into an "endgame" region around the goal. See [12] for details.

As in [12,24], our planner satisfies kinodynamic constraints as follows: to generate each new milestone m', it picks a control input at random and integrates the equations of motion of the robots over a short duration.

We name our planner Kinodynamic Randomized Motion Planning - KRMP. As shown in [12], under reasonable assumptions on the free space, the probability of not finding a plan when one exists decreases exponentially to 0 as the number of milestones increases. This is a major advantage over our previous work in [7,9], which used a decentralized prioritized planning approach. Note, however, that the fact that the planner is probabilistically complete does not imply that the entire system is also probabilistically complete. The robots use partial world models and thus need to re-plan their trajectories when they encounter discrepancies in their model, (e.g. new obstacles). Since there is no guarantee that a series of complete plans is itself a complete plan, the robots are not guaranteed to find a global plan if one exists. While it is unclear to what extent the notion of completeness applies when planning for global goals with only partial knowledge of the environment, it is still desirable to achieve completeness in the system's components whenever this is possible.

The work in [12] also demonstrated empirically that the above techniques successfully compute trajectories for a single robot with kinodynamic motion constraints, in real-time. To enable motion planning within robot networks, KRMP extends this previous work to accommodate multiple robots. Modified techniques are needed to 1) select milestones for expansion, 2) generate new milestones, and 3) define the endgame region. Below we present the technique used to generate new milestones.

When planning for multiple robots, one may generate m' using the following "parallel" approach: first, pick the control inputs for all the robots at random; next, integrate the motions of all the robots concurrently; if no collision is detected, then record the endpoint as a new milestone, otherwise pick another set of control inputs. We found that this technique yields a high rejection rate, especially in tight space. This led us to develop the following "sequential" approach: consider the robots in some order, pick the control inputs one robot at a time and integrate their motion (considering the previous robots as moving obstacles); if the motion collides, pick new control inputs or change the motion of a previous robot. Experiments show that this sequential approach makes it possible to get each new milestone much faster, without affecting the probabilistic completeness of the overall planner.

Finally, we take advantage of the various processors available in a robot network by concurrently running a separate copy of KRMP on each robot of the network. Each copy uses a different seed of the random number generator, hence constructs different roadmaps. We set the same timeout constraint (typically, a small fraction of a second) on every robot. Each robot then returns a plan or its failure to generate one. The same best plan is selected by the robots and each robot immediately executes its new trajectory. This is made possible because we use a PRM planning approach.

VI. EXPERIMENTAL TEST-PLATFORM

A. Micro-Autonomous RoverS Test-Platform

Located in the Aerospace Robotics Lab at Stanford University, the Micro-Autonomous RoverS (MARS) test-platform is used to model mobile robots in a two-dimensional workspace. The platform consists of a large 3m x 2m flat, granite table with six autonomous robots that move about the table's surface. The robots are cylindrical in shape and use two independently driven wheels that allow them to rotate on the spot, but inhibit lateral movement (nonholonomic constraint). Each robot is equipped with its own planner (copy of KRMP) and controller that are located off-board.

B. Sensors

An overhead vision system is used to track the states of all objects on the table. The vision system processor calculates these states and publishes them to all applications that subscribe (see Section VI). This makes global state information available to all robots. To simulate the limited sensing range that would occur when sensors are mounted on robots, the object states are filtered such that robots only receive state information regarding objects within some preset range of the robot.

C. Network Communication

Fig. 3 shows the computer/network architecture of the MARS test-platform. All the processing is done off-board. Two processors are assigned to each robot, respectively for planning and control. These computers are connected through a LAN. All communication within the LAN is accomplished with Real Time Innovation's Network Data Delivery Service (NDDS) software. Because a LAN is used for inter-robot communication instead of a wireless medium, there are no physical barriers to limit the range of communication. Hence the communication barrier is simulated.

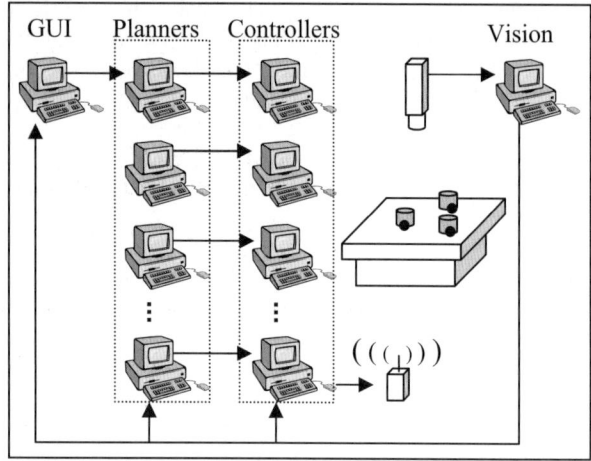

Fig. 3 Network architecture of MARS test-platform

NDDS is based on a publish/subscribe architecture. To broadcast messages by flooding a robot network, the sender will publish a message to which all robots subscribe. Before robots can receive their subscriptions, the messages are filtered so that only robots within some predetermined range of the sender will receive the message. This effectively simulates a discrete physical communication range.

VII. EXPERIMENTS

A. Physical Experiments

To illustrate the applicability of the planner to a physical system, real robot experiments with up to 5 robots have been carried out. One example of such an experiment is illustrated in Fig. 4. The left photos are screen-shots of the GUI taken throughout the experiment. The right photos show the physical hardware, and were taken at the same time as the corresponding GUI screen-shots. In the GUI, robots are depicted as small circles and obstacles are depicted as larger circles. Robot goal locations are indicated by cross-hairs, and lines leading to the goals depict the trajectories. When robots form a network as described in Section II, it is indicated by a color change. Hence robots within a network have a common color, and this color will differ between networks.

All five robots are initially located at the close end of the table (i.e. bottom of the GUI screen). Communication and sensing ranges were limited to 0.75 m. Robot colors indicate that 2 networks have formed on startup, one in the bottom left and one in the bottom right. As the experiment progresses, the robots follow their trajectories to reach their goal locations at the far end of the table.

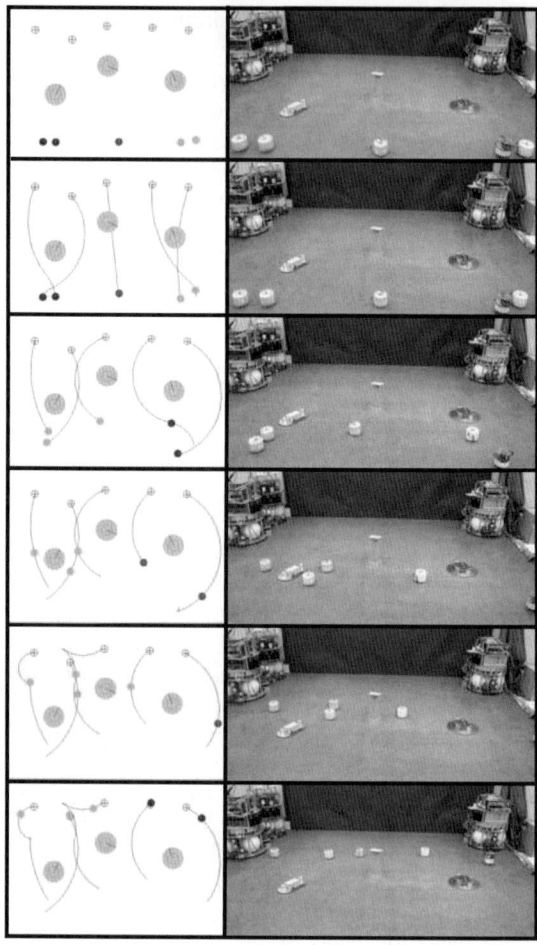

Fig. 4 Example experiment on the MARS test-platform involving 5 robots and 3 obstacles.

Along the way, networks are continually changing as illustrated by the robots changing colors between frames. A result of this is real-time re-planning. This is illustrated by the fact that trajectories change between frames. Throughout the experiment, robots planned an average of 3.4 times, and planning times were an average of 9 ms.

B. Simulations

The physical experiments shown above validate the planner's ability to function on real robots. However, the limited number of robots and obstacles available prevent us from performing experiments that demonstrate the planner's ability to handle more complex scenarios.

A single scenario was simulated that incorporates 12 robots, 6 static obstacles and 6 moving obstacles. The workspace was given dimensions 4m x 6m while robots and obstacles had diameters of 0.14m.

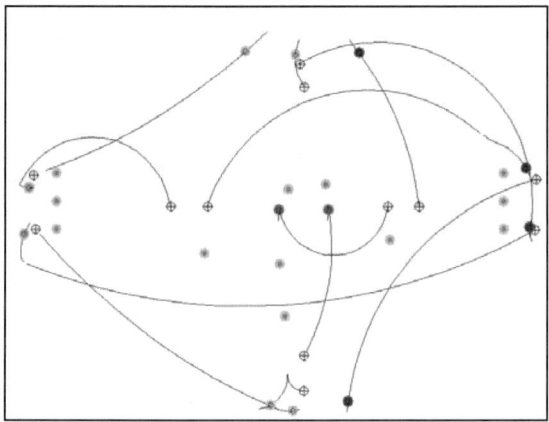

Fig. 5 Screen-shot of the test scenario.

To add complexity to the scenario, 4 of the moving obstacles were directed towards a network of 2 robots with little room to maneuver, (see middle of Fig. 5). Also, 2 networks of 2 robots were placed between a row of 3 obstacles and a workspace boundary.

The scenario was run 25 times with different initial random seeds. Despite the apparent difficulty of the scenario, the planner demonstrated fast planning times (an average of 15.8 ms), while planning for up to 5 robots in a network. To provide an idea of the level of complexity, robots formed on average 49 different networks throughout simulations that lasted several minutes.

Table 1 Simulation data for test scenario

Avg. number of robots per plan	2.12
Avg. planning time (ms)	17.3
Avg. number of plans per robot per simulation	5.07
Avg. number of networks formed per simulation	49.4

VIII. CONCLUSIONS

The motion planning framework presented has demonstrated its effectiveness in planning for multiple mobile robots within a bounded workspace. It plans with a high probability of success in environments involving robots, stationary obstacles and moving obstacles. Planning times of less than 100 ms allowed the robots to plan on-the-fly and react to changes in the environment.

Future work includes incorporating more sophisticated methods of modeling the environment into the communication system. Another future direction will be to investigate the effects of varying the ratio between sensor range and communication range.

REFERENCES

[1] R. Alami, F. Robert, F. Ingrand, & S.Suzuki. Multi-Robot Cooperation Through Incremental Plan-Merging, *Proc. IEEE Int. Conf. on Robotics and Automation*, p. 2573-2678, 1995.
[2] T. Arai & J. Ota. Motion Planning of multiple mobile robots. *Proc. IEEE/RSJ Int. Conf. on Intelligent Robots and Syst.*, p. 1761-1768, 1992.
[3] K. Azarm & G. Schmidt. Conflict-Free Motion of Multiple Mobile Robots Based on Decentralized Motion Planning and Negotiation, *Proc. IEEE Int. Conf. on Robotics and Automation*, p. 3526-3533, 1997.
[4] J. Barraquand, B. Langlois, & J.C. Latombe. Numerical Potential Field techniques for Robot Path Planning, *IEEE Tr. On Syst., Man, and Cyb.*, 22(2):224-241, 1992.
[5] M. Bennewitz, W. Burgard & S. Thrun. Optimizing Schedules for Prioritized Path Planning of Multi-Robot Systems, *Proc. Int. Conf. on Robotics and Automation*, 2001.
[6] S.J. Buckley. Fast Motion Planning for Multiple Moving Robots. *Proc. IEEE Int. Conf. on Robotics and Autom.*, p. 1419-1424, 1989.
[7] C. Clark & S. Rock. Randomized Motion Planning for Groups of Nonholonomic Robots, *Proc. Int. Symp. of Artificial Intelligence, Robotics and Automation in Space*, 2001.
[8] C. Clark, S. Rock & J. C. Latombe. Dynamic Networks for Motion Planning in Multi-Robot Space Systems, *Proc. Int. Symp. of Artificial Intelligence, Robotics and Automation in Space*, 2003.
[9] C. Clark, T. Bretl, & S. Rock. Kinodynamic Randomized Motion Planning for Multi-Robot Space Systems, *Proc. of IEEE Aerospace Conf.*, 2002.
[10] M. Erdmann & T. Lozano-Perez. On Multiple Moving Objects, *Proc. IEEE Int. Conf. on Robotics and Automation*, p. 1419-1424, 1986.
[11] Y. Guo & L. E. Parker. A Distributed and Optimal Motion Planning Approach for Multiple Mobile Robots, *Proc. IEEE Int. Conf. on Robotics and Automation*, p. 2612-2619, 2002.
[12] D. Hsu, R. Kindel, J.C. Latombe, & S. Rock. Randomized Kinodynamic Motion Planning with Moving Obstacles, *Int. J. of Robotics Research*, 21(3):233-255, March 2002.
[13] D. Hsu, J.C. Latombe, & R. Motwani. Path planning in expansive configuration spaces, *Proc. IEEE Int. Conf. on Robotics and Automation*, p. 2719-2726, 1997.
[14] K. Kant & S. Zucker. Toward efficient Trajectory Planning: The path-velocity decomposition, *Int. J. of Robotics Research*, 5(3):72-89,1986.
[15] S. Kato, S. Nishiyama, & J. Takeno. Coordinating mobile robots by applying traffic rules, *Proc. IEEE/RSH Int. Conf. on Intelligent Robots and Systems*, p. 1535-1541, 1992.
[16] J.J. Kuffner. *Autonomous Agents for Real-Time Animation*. PhD Thesis, Computer Science Dept., Stanford U., 1999.
[17] S.M. LaValle & S.A. Hutchinson. Optimal Motion Planning for Multiple Robots Having Independent Goals, *IEEE Tr. on Robotics and Automation*, 14:912-925, 1998.
[18] S.M. LaValle & J.J. Kufner. Randomized Kinodynamic Planning," *Int. J. of Robotics Research*, 20(5):278-300, 2001.
[19] V.J. Lumelsky & K.R. Harinarayan. Decentralized Motion Planning for Multiple Mobile Robots: The Cocktail Party Model, *Autonomous Robots J.*, 4:121-135, 1997.
[20] P. Moutarlier & R. Chatila. Stochastic Multisensory Data Fusion for Mobile Robot Location and Environment Modelling. *Proc. Int. Symp. on Robotics Research, Tokyo*, 1989.
[21] D. Parsons & J. Canny. A Motion Planner for Multiple Mobile Robots, *Proc. IEEE Int. Conf. on Robotics and Autom.*, p. 8-13, 1992
[22] G. Sánchez & J.C. Latombe. On Delaying Collision Checking in PRM Planning : Application to Multi-Robot Coordination, *Int. J. of Robotics Research*, 21(1):5-26, Jan. 2002.
[23] G. Sánchez-Ante & J.C. Latombe. Using a PRM Planner to Compare Centralized and Decoupled Planning for Multi-Robot Systems, *Proc. IEEE Int. Conf. on Robotics and Autom..*, 2002.
[24] S. Sekhavat, P. Svetska, J.P. Laumond, & M.H. Overmars. Multilevel Path Planning for Nonholonomic Robots Using Semiholonomic Subsystems. *Int. J. of Robotics Res.*, 17:840-857, 1998.
[25] T. Siméon, S. Leroy, & J.P. Laumond. Path Coordination for Multiple Mobile Robots: a Geometric Algorithm, *Proc. Int. Joint Conf. on Artificial Intelligence*, 1999.
[26] P. Svestka, & M.H. Overmars. Coordinated Motion Planning for Multiple Car-Like Robots Using Probabilistic Roadmaps, *Proc. IEEE Int. Conf. on Robotics and Autom., p. 1631-1636, 1995*.
[27] C.W. Warren. Multiple Path Coordination using Artificial Potential Fields, *Proc. IEEE In. Conf. on Robotics and Autom.*, p. 500-505, 1990.

Reduced Order Motion Planning for Nonlinear Symmetric Distributed Robotic Systems

M. Brett McMickell
mmcmicke@nd.edu

Bill Goodwine
goodwine@controls.ame.nd.edu

Aerospace and Mechanical Engineering
University of Notre Dame
Notre Dame, Indiana 46556

Abstract— This paper develops a motion planning algorithm which exploits symmetry in distributed systems to reduce complexity and motion planning design time. The motion planning computations are carried out on a reduced order system, then extended to larger-order equivalent systems in such a way that the objectives of the larger system are satisfied and collision avoidance is guaranteed. The algorithm maintains a rigid body formation as a group robots follows a specified trajectory at the beginning and end of the trajectory. At this point, our algorithm is open loop. A simulation of four robots maintaining a square formation is presented to demonstrate the utility of the algorithm.

I. INTRODUCTION

This paper considers the motion planning problem for symmetric distributed robotic systems which consist of, perhaps many, robots working together to perform a specified task. As the number of robots increases, so does the overall dimension and complexity of the system. There have been many efforts exploring high level planning and coordination between groups of robots [15], [4], [7], [8], [13]; however, none of these attempted to directly exploit any of the symmetry properties that distributed systems may possess. The aim of this work is to consider discrete symmetries to "reduce" the order of complexity of large-scale distributed systems. In this paper, we consider nonlinear robotic systems with equations of motion of the following, general form

$$\dot{x} = \sum_{i=1}^{m} g_i(x) u_i, \quad (1)$$

where the $g_i(x)$ are smooth analytic vector fields and the u_i's are admissible control inputs.

The problem is to find an algorithm which produces a set of inputs that steers a group of robots from a given initial position and orientation to a final position and orientation while maintaining a rigid body formation at the beginning and end of the trajectory. While rigid body formation control for systems of mobile robots clearly has been achieved before [2], [3], [5], [11], [12], [16], [17], the main contribution of this paper is the development of a motion planning algorithm which exploits symmetry of a distributed robotic system to simplify the design process and reduce the time necessary to develop motion plans for large, complex systems. This paper considers motion planning for symmetric distributed robotic systems in which there is no state information communicated between individual robots. Extending these results to an entire equivalence class of systems, as we have done for controllability in [9], will be the subject of a future paper.

The type of symmetry we consider is when certain robots of the overall system can be interchanged without affecting the dynamics of the overall system. The general idea is that a distributed system is comprised of sets of multiple, repeated instances of identical that can be interchanged. Motion planning is designed considering only one of these robots and then is mapped to the other robots in such a way that the total computational burden is far less than if the motion were planned for each robot separately. Furthermore, collision avoidance is guaranteed.

The remainder of this paper is organized as follows. A description of symmetric distributed systems is given in Section II. This is followed by the development of an equivalence relation between vector fields, which leads to the definitions of symmetric systems. A brief summary of piecewise constant motion planning is given in Section III. This leads to the motion planning algorithm for symmetric distributed systems. Section IV presents simulation results demonstrating the utility of the motion planning algorithm on a system of four mobile robots maintaining a square formation.

II. DRIFTLESS SYMMETRIC DISTRIBUTED SYSTEMS

This section outlines background material from the authors' previous work [9], [10] necessary to formulate our motion planning algorithm for symmetric nonlinear distributed systems. In this section, we provide the definitions and notation related to our representation of a driftless distributed system, define a symmetric nonlinear distributed system, and define an equivalence relation between different symmetric nonlinear distributed systems. The equivalence relation naturally leads to an equivalence class of control systems.

A. Driftless nonlinear distributed systems

We will consider smooth analytic driftless systems of the form

$$\Sigma: \quad \dot{x} = g_{1,1}(x)u_{1,1} + g_{1,2}(x)u_{1,2} + \cdots \quad (2)$$
$$+ g_{2,1}(x)u_{2,1} + g_{2,2}(x)u_{2,2} + \cdots$$
$$\vdots$$
$$+ g_{n,1}(x)u_{n,1} + g_{n,2}(x)u_{n,2} + \cdots,$$

for all $x \in M$ where M is a smooth manifold, $g_{i,j}$ are smooth analytic vector fields on M.

Since we are considering distributed systems, the system is assumed to be organized into individual robots, corresponding to which are certain vector fields and control inputs. In Equation (2), the first subscript on the g's and u's identifies the robot to which the vector field and control input corresponds, and the second subscript indexes different vector fields and inputs within that subsystem. Let $\tilde{g}_i(x)$, then it represents the *ordered set* of vector fields associated with the ith robot, i.e., $\tilde{g}_i(x) = \{g_{i,1}, g_{i,2}, \ldots\}$.

We assume that M is partitioned into a set of m regular submanifolds, M_i, such that M is the Cartesian product of the M_i, i.e., $M = \prod_{i=1}^m M_i$. Each submanifold M_i represents a *subsystem* or *robot* of the distributed system. In this paper, each M_i would represent the configuration space for robot i in the system and $\{u_{i,1}, u_{i,2}, \ldots\}$ would be the control inputs for that robot.

B. Symmetric nonlinear distributed systems

Now we will consider what it means for a nonlinear distributed system to be symmetric. Recall from the introduction that the motivating idea is that there is a subset of individual robots that can be interchanged without changing the dynamics of the overall team of robots. Mathematically this will be represented by the fact that vector fields from various robots will, in some sense, be equivalent. Since the vector fields from different robots are defined on different spaces, we need a definition of equivalence which is more than just requiring that they be 'identical'.

DEFINITION II.1 *Two vector fields, g_1 And g_2 are equivalent, denoted $g_1 \sim g_2$, if there exists a diffeomorphism, $\psi: M \mapsto M$, such that*

$$\psi_* g_1 = g_2,$$

where ψ_* is the push forward of ψ,

$$\psi_* g(x) = T\psi \circ g \circ \psi^{-1}(x),$$

and T is the usual tangent operation (see [1]).

The definition of vector field equivalence applies to general submanifolds without any assumptions regarding the relationship between the coordinate systems defined on different robots; however, often each robot is parameterized identically so that the diffeomorphism, ψ, in Definition II.1 is a simple permutation of states.

Since typically equivalence is determined by a permutation of coordinates, we first review the symmetric group and it's action on a set. Recall that the symmetric group of order $p!$, denoted S_p, is the group of permutations of p objects. A permutation of a set $X = \{1, \ldots, p\}$ is a one-to-one mapping of X onto itself. Such a permutation ρ is written,

$$\rho = \begin{pmatrix} 1 & 2 & \cdots & p \\ k_1 & k_2 & \cdots & k_p \end{pmatrix},$$

which represents that 1 is mapped to k_1, 2 is mapped to k_2, etc. The following example further illustrates vector field equivalence.

EXAMPLE II.2 Consider a system of five robots where each robot is parameterized by one state and let g_2 and g_3 be given by,

$$g_2(x) = \begin{bmatrix} x_1 \\ \cos x_2 \\ x_2 + 1 \\ 0 \\ x_2 x_5 \end{bmatrix}, \quad g_3(x) = \begin{bmatrix} x_1 \\ x_3 x_2 \\ \cos x_3 \\ x_3 + 1 \\ 0 \end{bmatrix}$$

and,

$$\psi: M \mapsto M$$

defined by

$$\psi(x_1, x_2, x_3, x_4, x_5) = (x_1, x_5, x_2, x_3, x_4),$$

which corresponds to interchanging robot 2 with node 3. Also, ψ is related to $\rho \in S_4$ where

$$\rho = \begin{pmatrix} 1 & 2 & 3 & 4 & 5 \\ 1 & 3 & 4 & 5 & 2 \end{pmatrix},$$

where the x_1 coordinate is fixed and coordinates $x_2, \ldots, x_5$ are mapped according to ρ. The inverse mapping is

$$\psi^{-1}(x_1, x_2, x_3, x_4, x_5) = (x_1, x_3, x_4, x_5, x_2).$$

Invariance of a system with respect to interchanging robot 2 and 3 requires that

$$\psi_* g_2 = g_3 \quad \text{and} \quad \psi_*^{-1} g_3 = g_2.$$

In detail,

$$\psi_* g_2(x_1, x_2, x_3, x_4, x_5)$$
$$= T\psi \circ g_2 \circ \psi^{-1}(x_1, x_2, x_3, x_4, x_5)$$
$$= T\psi \circ g_2(x_1, x_3, x_4, x_5, x_2)$$
$$= T\psi \circ \begin{bmatrix} x_1 \\ \cos x_3 \\ x_3 + 1 \\ 0 \\ x_3 x_2 \end{bmatrix}$$
$$= \begin{bmatrix} 1 & 0 & 0 & 0 & 0 \\ 0 & 0 & 0 & 0 & 1 \\ 0 & 1 & 0 & 0 & 0 \\ 0 & 0 & 1 & 0 & 0 \\ 0 & 0 & 0 & 1 & 0 \end{bmatrix} \begin{bmatrix} x_1 \\ \cos x_3 \\ x_3 + 1 \\ 0 \\ x_3 x_2 \end{bmatrix}$$
$$= \begin{bmatrix} x_1 \\ x_3 x_2 \\ \cos x_3 \\ x_3 + 1 \\ 0 \end{bmatrix}$$
$$= g_3(x).$$

A similarly straight-forward computation shows that $\psi_*^{-1} g_3(x) = g_2(x)$.

Given an equivalence relation among vector fields, we now define a symmetric nonlinear distributed system.

DEFINITION II.3 *Let a symmetry orbit,* **O**, *be subset of p robots in Σ, let* **F** *be the subset of Σ containing $n - p$ fixed robots, and let $\rho \in S_p$. The system Σ is a symmetric nonlinear distributed system if*

$$\tilde{g}_i \sim \tilde{g}_{\rho(j)} \quad \forall i \in \{1, \ldots, p\} \text{ and } \forall \rho \in S_p,$$

where $\tilde{g}_i$ is the ordered set of driftless vector fields corresponding to robot i in **O**.

III. NONLINEAR MOTION PLANNING FOR SYMMETRIC SYSTEMS

The motion planning algorithm developed in this paper is an extension of piecewise constant motion planning algorithm from [6]. A complete description of this motion planning algorithm is beyond the scope of this paper, so only an outline will be provide in this section. This section also provides a method for ensuring that there are no collisions between robots.

A. Piecewise motion planning

Piecewise constant motion planning works exactly for systems whose controllability Lie algebra is nilpotent, *i.e.*, there exists a $k > 0$, such that $[g_{i_1}, \ldots, [g_{i_{p-1}}, g_i], \ldots] = 0, \forall p > k$ and for all vector fields g_i. For systems that are not nilpotent, the method provides only an approximate solution and explicit error bounds on the resulting error are given in [6].

The basic idea of piecewise continuous motion planning is to decompose the desired trajectory into multiple subtrajectories along vector fields which, when evaluated at a point, form a basis for the tangent space of the configuration space. For underactuated systems, the basis will contain motion in the directions of a Lie brackets. Recall, a Lie bracket in coordinates is given by

$$[g_1, g_2] = \frac{\partial g_2}{\partial x_i} g_1(x) - \frac{\partial g_1}{\partial x_i} g_2(x).$$

Motion in a Lie bracket direction can be approximated using the following four segment flow,

$$\phi^t_{[g_1, g_2]}(x_0) = \phi^{\sqrt{t}}_{-g_2} \circ \phi^{\sqrt{t}}_{-g_1} \circ \phi^{\sqrt{t}}_{g_2} \circ \phi^{\sqrt{t}}_{g_1}(x_0) \quad (3)$$

where $\phi^t_g(x_0)$ represents the flow along the vector field g for time t starting at point x_0. Using basis vector fields, any smooth trajectory can be represented by the *Chen-Fleiss series*,

$$S_t(q) = e^{h_s(t) B_s} e^{h_{s-1}(t) B_{s-1}} \cdots e^{h_2(t) B_2} e^{h_1(t) B_1}, \quad (4)$$

where $[h_1, \ldots, h_s]$ are functions know as *Phillip Hall coordinates*, $B_1, \ldots, B_s$ are *Phillip Hall basis elements*, and S is the series representation of a given trajectory. Phillip hall basis elements are related to the original system and their Lie brackets and are chosen such that the the basis elements are full rank over the desired trajectory.

The Chen-Fliess series satisfies the following differential equation, referred to as the *formal differential extended system*,

$$\dot{S}(t) = S(t)(B_1 v_1 + \cdots + B_s v_s), \quad (5)$$

where $v_1, \ldots, v_s$ are fictitious inputs. The inputs are referred to as fictitious because they are inputs associated with basis vector fields and may not be available to the actual system.

By differentiating Equation 4 and equating it to Equation 5, we can solve for the $\dot{h}$'s in terms of the fictitious inputs, which results in a ordinary differential equation,

$$\dot{h} = Q(h)v, \quad h(0) = 0, \quad (6)$$

where Q(h) is a coefficient matrix in terms of the Phillip Hall coordinates. The solution of this differential equation represents the evolution of the system in response to the fictitious inputs.

B. Motion planning for symmetric distributed systems

Motion planning for a nonlinear symmetric system of p robots in $\mathbb{R}^{m \times p}$ using rigid body formation is accomplished as follows. Let each robot consist of n states, where $n \geq m$. First, determine a rotation matrix $R(wt) \in SO(m)$ such that $R(wT)$ produces the desired final orientation of the rigid formation, where w is the rotational velocity of the rigid formation. Next, choose a trajectory, $q(t) \in C^1$, connecting the initial center of the formation to a desired final center for a given $t \in [0, T]$. This can be done using any C^1 function. Note that $q(t) \in \mathbb{R}^m$. Therefore, the trajectory of a robot i in the rigid body is given by,

$$p_i(t) = R(wt)P_i + q(t), \quad (7)$$

where $P \in \mathbb{R}^n$ is the initial starting position of robot i relative to the center of the formation. The rigid formation uniquely determines the position of each robot, but it does not constrain the robot's orientation. Let $r(t) \in C^1$ describe the changing orientation of the robots. The rigid body trajectory for robot i can be written as,

$$\gamma_i(t) = A(wt)\hat{P}_i + \hat{q}(t) + \hat{r}(t),$$

where $\hat{P}$ is the robot i's initial state (position and orientation) with respect to the center of the rigid body, $A(wt)$ is an augmented rotation given by,

$$A(wt) = \begin{bmatrix} R(wt) & 0 \\ 0 & 1 \end{bmatrix},$$

$\hat{q}(t)$ is an augmented trajectory given by,

$$\hat{q}(t) = \begin{bmatrix} q(t) \\ 0 \end{bmatrix},$$

and $\hat{r}(t)$ is given by,

$$\hat{r}(t) = \begin{bmatrix} 0 \\ r(t) \end{bmatrix}.$$

Taking the derivative of the trajectory, we find

$$\dot{\gamma}_i(t) = \dot{A}(wt)\hat{P}_i + \dot{\hat{q}}(t) + \dot{\hat{r}}(t).$$

Select s linearly independent Phillip Hall basis elements, $\{B_1, \ldots, B_s\}$, and determine the corresponding fictitious inputs. To do this, define an ordered matrix $\tilde{C}$ composed of all the linearly independent vector fields,

$$\tilde{C}(\gamma_i(t)) = [g_1(\gamma_i(t)), g_2(\gamma_i(t)), \ldots, g_s(\gamma_i(t))].$$

Recall that some of the g_i's will be Lie brackets between vector fields in the original system. The basis elements were chosen so they have full rank over the entire trajectory. Therefore, $\tilde{C}$ is invertible. The fictitious inputs for robot i are given by,

$$v_i = \tilde{C}^{-1}(\gamma_i(t))\dot{\gamma}_i(t).$$

where $v = [v_1, \ldots, v_n]^T$. The Phillip Hall coordinates corresponding to the fictitious inputs are determined by solving Equation 6 and the initial condition h(0)=0. This gives the control inputs for the extended system. Inputs in the extended system that are associated with motion in a Lie bracket direction are approximated using Equation 3.

The motion plan for developed for this robot can now be extended to other equivalent robots using the following Proposition.

PROPOSITION III.1 *Let Σ be a robotic system containing p robots, such that, $\Sigma : \dot{x} = g_1(x)u_1 + \cdots + g_n(x)u_n$. If Σ is symmetric distributed system, then the Phillip hall coordinates of any two robots, i and j, in the symmetry orbit of the system with a desired trajectory given by Equation 7, are related by,*

$$\dot{h}_i = Q(h)v_i = Q(h)(\psi_* \tilde{C}_j(\psi(\gamma_j(t))))^{-1} \psi_* \dot{\gamma}_j(t). \quad (8)$$

Proof: Consider the trajectory given by Equation 7 for two robots, i and j, which are in the symmetry orbit. Let $\Sigma_i : \dot{x}_i = g_1(x_i)u_1 + \cdots + g_m(x_i)u_n$ and $\Sigma_j : \dot{x}_j = f_1(x_j)u_1 + \cdots + f_m(x_j)u_n$ trajectories of the two system are related by the diffeomorphism, ψ, such that

$$\gamma_i(t) = \psi(\gamma_j(t)) \implies \dot{\gamma}_i(t) = \psi_* \dot{\gamma}_j(t).$$

Let $\tilde{C}_i$ be the ordered matrix of vector fields corresponding with system Σ_i. The ordered matrices of vector fields, $\tilde{C}$, are also related by the diffeomorphism,

$$\begin{aligned}\tilde{C}_i(\gamma_i(t)) &= \psi_* \tilde{C}_j(\psi(\gamma_j(t))) \\ &= [\psi_* f_{k_1}(\psi(\gamma_j(t))), \ldots, \psi_* f_{k_m}(\psi(\gamma_j(t)))],\end{aligned}$$

which implies that $\tilde{C}_i^{-1}(\gamma_i(t)) = (\psi_* \tilde{C}_j)^{-1}(\psi(\gamma_j(t)))$. Therefore, the fictitious inputs, v_i, are also related through the diffeomorphism,

$$\begin{aligned}v_i &= \tilde{C}_i(\gamma_i(t))^{-1}\dot{\gamma}_i(t) \\ &= (\psi_* \tilde{C}_j(\psi(\gamma_j(t))))^{-1}\psi_* \dot{\gamma}_j(t).\end{aligned}$$

Diffeomorphisms are natural with respect to Lie brackets, *i.e,*

$$[\psi_* f, \psi_* g] = \psi_*[f, g],$$

which implies that if system Σ_i is nilpotent of order k, then Σ_j is also nilpotent of order k. Therefore, $Q_i(h) = Q_j(h) = Q(h)$. Therefore,

$$\dot{h}_i = Q(h)v_i = Q(h)(\psi_* \tilde{C}_j(\psi(\gamma_j(t))))^{-1}\psi_* \dot{\gamma}_j(t).$$

∎

C. Collision avoidance

The overall approach is to decompose the complete trajectory into subtrajectories that are small enough to ensure there is no collision in the system. Since we are considering small motions, we will consider the system locally in $\mathbb{R}^n$. For the trajectory, γ, given in Equation 7 for $t \in [0, T]$, let $\mathcal{R}_i = \min_{t \in [0,T]} \|\gamma_i(t) - \gamma_j(t)\|$, such that $i \neq j$, *i.e.* the closest any robot gets to robot i while following the trajectory. Also, let $\Delta_i = \|\gamma_i(T) - \gamma_i(0)\|$. Consider a linear trajectory $\Gamma_i(t) = \gamma_i(0) + t(\gamma_i(T) - \gamma_i(0))$ connection the initial position to the final position. Recall, the fictitious inputs are calculated by solving $\dot{\gamma}_i(t) = [g_1(\gamma_i(t)), \ldots, g_s(\gamma_i(t))]v_i$. Applying this to the linear trajectory, $\Gamma_i(t)$, we find

$$\|\dot{\Gamma}_i\| = \|\gamma_i(T) - \gamma_i(0)\| < \|[g_1(\gamma_i(t)), \ldots, g_s(\gamma_i(t))]\|\|v_i\|.$$

From this equation, we find that the fictitious inputs are bounded by a constant, α_i, *i.e.*, $\|v_i\| < \alpha_i \|\dot{\Gamma}_i\| = \alpha_i \Delta_i$.

By construction of the real inputs from the fictitious inputs, $\|u\| < \alpha_i \Delta^{1/k}$ where k is the order of the highest Lie bracket needed to make $\tilde{C}$ full rank. Let $x_{i,\max} = \max_{t \in [0,T]} \|x_i(t) - \gamma_i(0)\|$ denote the flow that is maximally distant from the starting point. Note, this is not necessarily $\gamma_i(T)$. Now, pick a ball, $\mathcal{B}_i$ of radius $\mathcal{R}_i$ centered at the initial point. Let η_i be the maximum norm of all the first order vector fields for all points in the ball $\mathcal{B}_i$. The distance, $\|x_{i,\max} - \gamma_i(0)\|$ is necessarily bounded by the sum of the norms of each individual flow associated with one real control input, $u_{i,j}^l$. That is,

$$\|x_{i,\max} - \gamma_i(0)\| \leq \sum_l \sum_j \|\int_0^1 g_l u_{i,j}^l \, dt\|.$$

We know, $\|u_i^l\| \leq \alpha_i \Delta_i^{1/k}$ and $\|g_l(x)\| \leq \eta_i$ for all x_i. Therefore,

$$\|x_{i,\max} - \gamma_i(0)\| \leq \sum_l \sum_j \eta_i \alpha_i \Delta^{1/k},$$

and since $\Delta_i = \|\gamma_i(T) - \gamma_i(0)\|$, by choosing the desired final point close enough to the stating point, the robots will not collide. Because Δ_i is raised to the power of $1/k$, if k is large, then Δ_i may need to be exceedingly small. This approach is very conservative and the appropriate step length may best be identified experimentally.

IV. EXAMPLE

Consider a group of four simple robotic unicycles each described by [14],

$$\begin{bmatrix} \dot{x} \\ \dot{y} \\ \dot{\theta} \end{bmatrix} = \begin{bmatrix} \cos\theta \\ \sin\theta \\ 0 \end{bmatrix} u_1 + \begin{bmatrix} 0 \\ 0 \\ 1 \end{bmatrix} u_2 \quad (9)$$

where u_1 is the linear velocity input and u_2 is the angular velocity input. All robots are identically parameterized, so the diffeomorphism, ψ, is simply a translation mapping.

The robots are initially in a square formation centered about the origin a distance of unity apart. The robots are to follow a linear path, $q(t) = [t, t, 0]^T$ for a time $t \in [0, 1]$ with the orientation of the square rotating by an angle π. The initial and final points are illustrated in Figure 1 (a).

Motion planning is done on one robot and then extended to the other robots. Equation 9 describing the mobile robots is not nilpotent; however, it is nilpotentizable (see [6]). Using inputs,

$$u_1 = \frac{1}{\cos(\theta)} w_1 \quad (10)$$
$$u_2 = \cos^2(\theta) w_2,$$

the system becomes

$$\begin{bmatrix} \dot{x} \\ \dot{y} \\ \dot{\theta} \end{bmatrix} = \begin{bmatrix} 1 \\ \tan\theta \\ 0 \end{bmatrix} w_1 + \begin{bmatrix} 0 \\ 0 \\ \cos^2\theta \end{bmatrix} w_2,$$

which is nilpotent of order 2. The motion plan for the transformed systems using piecewise constant inputs will be exact. Solving equation 6 for the desired motion, we find that the Phillip Hall coordinates for the ith robot are,

$$\begin{bmatrix} h_{1i}^k \\ h_{2i}^k \\ h_{3i}^k \end{bmatrix} = \begin{bmatrix} \cos(\pi t) P_{1i}^0 + \sin(\pi t) P_{2i}^0 + t - P_{1i}^k \\ 0 \\ -\sin(\pi t) P_{1i}^0 + \cos(\pi t) P_{2i}^0 + t - P_{2i}^k \end{bmatrix},$$

where $P_1^0 i$ and $P_2^0 i$ are the initial x and y positions of the ith robot, respectively. The desired motion is greater than the maximum collision bound, so the motion must be divided into segments. Each segment, i, of robot i has different h values denoted h_i^k. Since the state of the robots changes with each step, we denote the starting state at the kth segment of the motion as P_i^k. For example, the initial state of robot 1 is $P_1^0 = [-1, 1, 0]^T$, so the corresponding Phillip Hall coordinates for the initial step are,

$$\begin{bmatrix} h_{11}^0 \\ h_{21}^0 \\ h_{31}^0 \end{bmatrix} = \begin{bmatrix} -\cos(\pi t) + \sin(\pi t) + t + 1 \\ 0 \\ \sin(\pi t) + \cos(\pi t) + t - 1 \end{bmatrix}.$$

After determining the motion plan for a robot j, the motion plan is extended to the other robots using Proposition III.1. The resulting Phillip hall coordinates for the ith robot are,

$$\begin{bmatrix} h_{1i}^k \\ h_{2i}^k \\ h_{3i}^k \end{bmatrix} = \begin{bmatrix} \cos(\pi t)\psi(P_{1j}^0) + \sin(\pi t)\psi(P_{2j}^0) + t - \psi(P_{1j}^k) \\ 0 \\ -\sin(\pi t)\psi(P_{1j}^0) + \cos(\pi t)\psi(P_{2j}^0) + t - \psi(P_{2j}^k) \end{bmatrix}$$

Figure 1 displays a simulation of the four robots. Figure 1 (a) displays the robots initial and final positions shown as 'o' and 'x', respectively. Figure 1 (b)-(e) show the motion during each of the four subtrajectories necessary to move to final position without a collision. The complete motion plan for the mobile robotic system is given in Figure 1 (f). Note, the maximum step size determined by the collision bound given in Section III-C is conservative. The step size for this example was computed experimentally.

V. CONCLUSIONS AND FUTURE WORK

A motion planning algorithm for nonlinear symmetric systems that exploits the symmetry of a system has been developed. The algorithm is based on piecewise constant inputs [6], which is exact for nilpotent systems. A bound on the maximum step size is provide that ensures the motion is collision free. A simulation of a group of mobile robots was used to demonstrate the utility of this algorithm.

The motion plan developed in this paper is for noninteracting symmetric robots. Future work is directed toward removing this restriction and developing a motion planning algorithm for symmetric distributed systems that can be designed on a "reduced order" equivalent system. We are also exploring a more general formation control which would consider time-varying formations. Furthermore, we are experimentally testing motion control algorithms on a

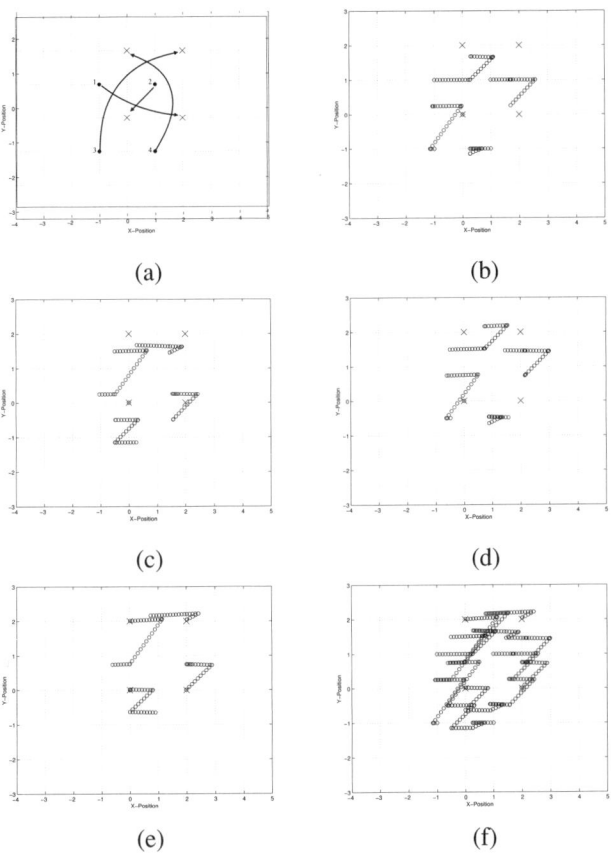

Fig. 1. Plot (a) displays the initial and final configurations shown as "o" and "x", respectively. Plots (b) - (e) displays the motion plan in four steps to avoid collisions. Plot (f) displays the combined motion plan of all four steps.

VI. ACKNOWLEGEMENT

The partial support of the National Science Foundation grant NSF CCR01-13131 is gratefully acknowledged.

VII. REFERENCES

[1] R. Abraham, J. E. Marsden, and T. Ratiu. *Manifolds, Tensor Analysis, and Applications.* Springer–Verlag, second edition, 1988.

[2] M. Egerstedt and X. Hu. Formation constrained multi-agent control. In *IEEE International Conference on Robotics and Automation*, pages 3961–3966. IEEE, 2001.

[3] J. Alexander Fax and Richard M. Murrray. Information flow and cooperative control of vehicle formations. In *IFAC World Congress*, 2002. to appear.

[4] S. Hirose, R. Damoto, and R. Kawakami. Study of super-mecho-colony (concept and basic experimental setup). In *IEEE/RSJ International Conference on Intelligent Robots and Systems*, volume 3, pages 1664–1669, 2000.

[5] Eric Klavins. Automatic synthesis of controllers for distributed assembly and formation forming. In *IEEE International Conference on Robotics and Automation*, pages 3296–3302, May 2002.

[6] G. Lafferriere and Hector J. Sussmann. A differential geometric approach to motion planning. In X. Li and J. F. Canny, editors, *Nonholonomic Motion Planning*, pages 235–270. Kluwer, 1993.

[7] Xu Liying, S. Zien-Sabatto, and A. Sekmen. Development of intelligent behaviors for mobile robot. In *Proceedings of the 33rd Southeastern Symposium on System Theory*, pages 383–386, 2001.

[8] R. Logan and S. Theodoropoulos. The distributed simulation of multiagent systems. In *Proceedings of IEEE*, volume 89, pages 174–185, 2001.

[9] M. Brett McMickell and Bill Goodwine. Reduction and nonlinear controllability of symmetric distributed systems. In *IEEE/RSJ International Conference on Intelligent Robots and Systems*, pages 1232–1237, 2001.

[10] M. Brett McMickell and Bill Goodwine. Reduction and nonlinear controllability of symmetric distributed systems. *International Journal of Control*, 2002. Submitted for review.

[11] Mark B. Milam, Nicolas Petit, and Richard M. Murray. Constrained trajectory generation for microsatellite formation flying. In *AIAA Guidance, Navigation, and Control Conference*, 2001.

[12] Reza Olfati-Saber and Richard M. Murrray. Distributed cooperative control of multiple vehicle formations using structural potential functions. In *IFAC World Congress*, 2002. to appear.

[13] M. Quinn. A comparison of approaches to the evolution of homogenous multi-robot teams. In *Proceedings of the 2001 Conference on Evolutionary Comutations*, volume 1, 2001.

[14] Shankar Sastry. *Nonlinear Systems: Analysis, Stability, and Control*, chapter 11. Springer, 1999.

[15] T. Sugar, J.P. Desai, V. Kumar, and J.P. Ostrowski. Coordination of multiple mobile manipulators. In *IEEE International Conference on Robotics and Automation*, pages 3022–3027, 2001.

[16] Ichiro Suzuki and Masafumi Yamashita. Distributed autonymous mobile robts: Formation of geometric patterns. *SIAM J. Comput.*, 28(4):1347–1363, 1999.

[17] Hiroaki Yamaguchi and Joel Burdick. Time-varying feedback control for nonholonomic mobile robots forming group formations. In *Proceedings of the 37th IEEE Conference on Decision and Control*, pages 4156–4163, 1998.

Evasion of Multiple, Intelligent Pursuers in a Stationary, Cluttered Environment Using a Poisson Potential Field

Ahmad A. Masoud

Electrical Engineering Department, KFUPM, P.O. Box 287, Dhahran 31261, Saudia Arabia
Tel.: 03-860-3740, E-mail: masoud@kfupm.edu.sa

Abstract

In this paper a new potential field approach is suggested for the evasive navigation of an agent that is engaging multiple pursuers in a stationary environment. Here, the gradient of a potential field that is generated by solving the Poisson equation subject to a set of mixed boundary conditions is used to generate a sequence of directions to guide the motion of an evader so that it will escape a group of pursuers while avoiding a set of forbidden regions (clutter). The focus here is on continuous evasion where the agent does not have the benefit of a target zone (e.g., a shelter) which up on reaching it can discontinue engaging the pursuers. The capabilities of the approach are demonstrated using simulation experiments.

I. Introduction

Evasive navigation is an important tactical aid that is needed to enhance survivability of an agent operating in an adversarial environment [1]. It is also a non-determinate game in which an agent (evader) has to move from an initial location to a final one while avoiding a number of pursuers in an environment that may be populated by forbidden regions. The presence of clutter complicates the evasion strategy which is usually studied for one pursuer in an open environment [2]. Clutter (figure-1) excludes simple solutions to the evasion problem such as running along a straight line toward infinity using the highest possible speed. It also excludes commonly used maneuvers such as protean behavior [3] in which an evader turns in an unpredictable manner to confuse a faster, but less maneuverable, pursuer. In this case, it is highly likely that simple reflexive control will not work. A high level controller is needed to fuse the context in which the actors are operating, the strategy and intentions of the pursuers with the decision making process used to guide the evader's actions.

Figure-1: A cluttered environment.

Many aspects of pursuit-evasion have been investigated. In [4] the problem of a group of pursuers searching a cluttered environment for an intruder using flash lights is investigated. On the other hand, in [5] the behavior of an agent trying to hide from a pursuer is evaluated in terms of the amount of protection provided by the environment. In [6] an algorithm is suggested for a group of robots to capture a fugitive agent moving on a grid. In [7] an intelligent controller is suggested for intercepting a known, well-informed target that is intelligently maneuvering in a cluttered environment to evade capture. An intensive literature survey on pursuit evasion may be found in [8].

The focus here is on evasive navigation of an agent that is being tracked by multiple pursuers in a cluttered environment for the case where the evader does not have a target point present. A target point is equivalent to a shelter which when reached the evader no longer have to engage the pursuers. The situation necessitates that the evader continuously engage the pursuers. This presents the evader with a considerable intellectual burden, especially when facing intelligent pursuers who may be cooperating and have the ability to learn regularities or patterns in the evader's behavior and evolve a capture strategy.

The approach suggested in this paper for tackling the above problem is an alternative to a recent approach suggested by the author [13] that utilizes a modified version of the harmonic potential field (HPF) approach to behavior synthesis [9,10,11] for constructing an intelligent controller to guide an evader in a situation such as the one described above. While the approach in [13] utilizes a vector boundary value problem (VBVP) to solve for the phase field of the harmonic potential (the solution of the magnitude field of the HPF is bypassed), the approach in this paper generates the field by solving a standard Poisson equation subject to an appropriate set of mixed boundary conditions. The use of Poisson equation is motivated by the fact that solution of this type of BVPs is much simpler than that of a VBVP. Highly stable, off-the-shelf numerical packages exist for solving the Poisson BVP (e.g., PDE tool-box of MATLAB), which is not the case for VBVPs. Also the properties of the solution of the Poisson equation are well-understood, and are thoroughly documented in the literature.

The paper is organized as follows: in section II the evasion problem is formulated. Section III gives a brief background of the harmonic potential field approach, the difficulties in its application to the evasion problem, and a recent attempt by the author to modify this approach so it can be used for continuously evading multiple pursuers. In section IV the Poisson approach is introduced and the BVP that generates the evasion field is provided. Simulation results are provided in section V, and conclusions are stated in section VI.

II. Problem Formulation

The pursuers and evader are assumed to be operating in a multidimensional environment (R^N) that is populated by stationary forbidden regions (O, $\Gamma = \partial O$). All actors are required to restrict their activities to the subset, Ω, of the multidimensional space known as the workspace ($\Omega = R^N - O$). The location of the i'th pursuer is xp_i. A group of L pursuers is assumed to be operating in Ω. The location of the group is described using the vector $XP = [xp_1\ xp_2\ ...\ xp_L]^t$.

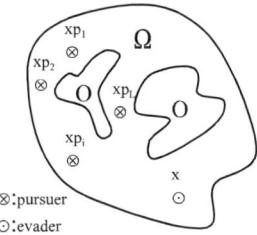

Figure-2: the pursuit-evasion environment

The evader has full knowledge of the environment and the location of the pursuers. Likewise, the pursuers, who may be communicating, are assumed to have full knowledge of the environment and the location of the evader, figure-3.

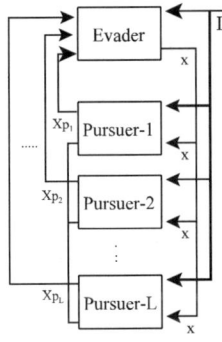

Figure-3: interaction between the evader and the pursuers

The high-level guidance mechanism which the evader is using aggregates the data about the environment (Γ), and the locations of the pursuers (XP) in order to advise the evader regarding the direction it needs to head along if it is to be safe and escape capture. Although the controller yields only a reference trajectory for the evader to follow, many techniques exist for translating a trajectory marked by the gradient field of an HPF into a control signal. A summery of some of these techniques may be found in [12]. Mathematically speaking, constructing the evasion control requires the construction of the gradient dynamical system:

$$\dot{x} = -\nabla V(x, XP(t), \Gamma) \qquad (1)$$

such that: $\qquad x(t) \in \Omega \qquad \forall t$

and $\qquad \sum_{i=1}^{L} |xp_i - x| > 0 \qquad \forall t,$

where ∇ is the gradient operator, and V is the HPF.

III. Background

Difficulties with the HPF approach:

In the HPF approach, the navigation field is synthesized using the BVP: $\quad \nabla \cdot \nabla V(x) = \nabla^2 V(x) \equiv 0 \quad \forall x \in \Omega \qquad (2)$

$$V(x) = 1\big|_{x=\Gamma}, \quad V(x) = 1\big|_{x=xp_i(t), i=1,...,L}, \quad V(x) = 0\big|_{x=x_t}$$

where x_t is the target point, the potential, V(x), is valid for only the time instant t. The control at time t is derived from the negative gradient of V(x):

$$u = -\nabla V(x). \qquad (3)$$

Harmonic functions assume their minima and maxima on their boundaries (here Γ, XP(t), and x_t). There are no stagnating points in Ω where ∇V vanishes. Therefore, the highest potential, V=1, will be at $x = \Gamma$, and $x=xp_i(t)$, i=1,...,L; while the lowest potential, V=0, will be at $x=x_t$. Figure-4a shows a rectangular forbidden region confining the motion of three pursuers and an evader whose goal is to reach the target without running into the pursuers or the walls of the workspace. Figuers-4b, and c show the HPF, and its negative gradient field respectively.

Removing the goal point, x_t, where the potential is fixed to zero, from the BVP in (2) makes the maxima and minima of

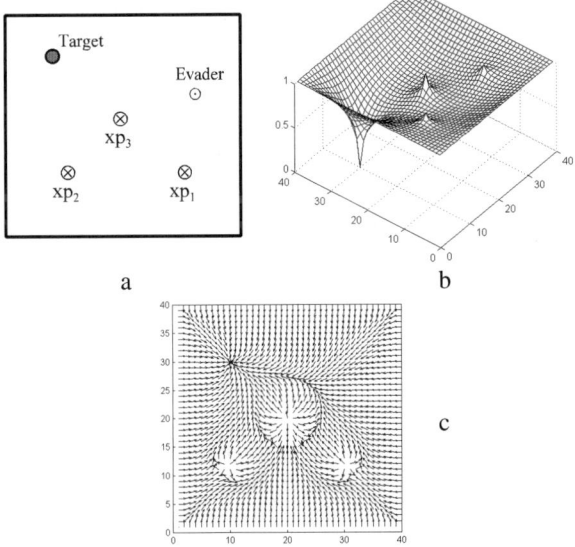

Figure-4: a: environment, b: potential, c: negative gradient

V(x) equal to 1. In other words the value of V(x) is a constant equal to 1 for all points in Ω. This causes the gradient field to degenerate every where in Ω ($\nabla V(x) \equiv 0, \forall x \in \Omega$) making it impossible to construct the evasion field, figures-5a,b.

The Modified HPF Approach:

The removal of the goal point, x_t, from (2) causes the potential to become flat and the gradient field to degenerate. While the magnitude field of ∇V, $a(x)$, degenerates, the phase field, $G(x)$, remains stable and computable. This makes it possible to adapt the generating BVP to work for the case where no target point is explicitly specified.

$$\nabla V(x) = a(x)G(x) . \quad (4)$$

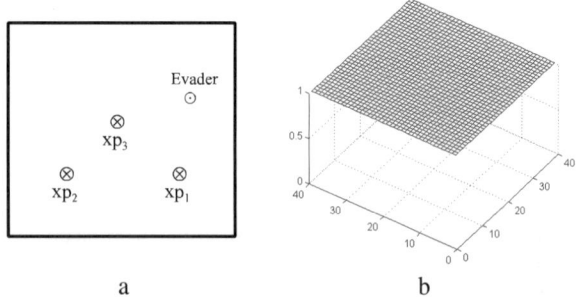

Figure-5: a: the environment, b: the potential field

The BVP is: solve $\quad \nabla \cdot G(x) \equiv 0 \quad X \in \Omega \quad (5)$
subject to: $\quad G(x) = n\Gamma \big|_{x=\Gamma}, \quad G(x) = n\Gamma i \big|_{x=\Gamma i}$,
and $\quad |G(x)| = 1, \quad\quad\quad i=1,...L$

Where $n\Gamma_i$ is a unit vector field orthogonal to Γi,

$$\Gamma_i = \{x : |x - xp_i| = \delta, \delta > 0\}, \quad (6)$$

and $n\Gamma$ is a unit vector field orthogonal to Γ. The above vector BVP (VBVP) is solved for the environment shown in figure-6a. The corresponding evasion field is shown in figure-6b.

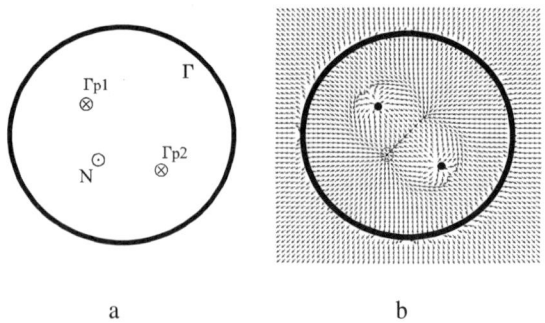

Figure-6: a. environment, b. evasion field

Although a target point was not specified in the modified BVP, a stable equilibrium point, N (i.e. a target point), spontaneously emerged in the synthesized field (a north pole, figure-6b). Unlike (2) where target location is a priori specified, in the modified VBVP, the target location is free to move in a manner dependant on the environment and the locations of the pursuers. The target keeps adapting its location positioning itself as far as possible from the pursuers

and the forbidden regions. The intelligent, high-level controller suggested here for continuously steering the evader away from harm, accepts Γ and XP(t) as inputs, and generates N(t) as an output. The generated time sequence of locations , N(t), is the one the evader has to follow in order to avoid harm and capture, figure-7.

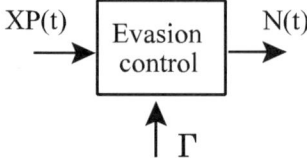

Figure-7: suggested evasion controller

IV. The Suggested Approach

When the potential field was first suggested, the most serious problem facing it was deadlock, or local minima at which ∇V vanishes trapping motion short of reaching the target. The HPF approach to motion planning solved the deadlock problem by forcing ∇V to satisfy Laplace equation for each point inside Ω,

$$\nabla^2 V(x) = \nabla \cdot \nabla V(x) \equiv 0 \quad x \in \Omega \quad (7)$$

By satisfying the Laplace equation, the divergence of the gradient of the potential is forced to zero inside Ω. Physically speaking, the divergence of a vector field is defined as the outflow of the flux generated by the field when the volume of the closed area the field is passing through shrinks to zero. In other words, the flux that goes inside that close area must leave (figure-8a). This prevents stagnation of the flux and in turn prevents deadlock.

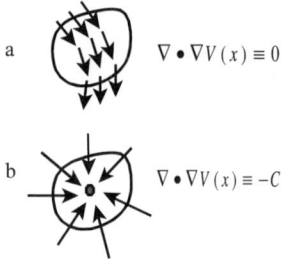

Figure-8: Physical interpretation of a divergence,
a: Laplace equation, b: Poisson equation

Unfortunately when motion is to be planned in order to continuously avoid multiple pursuers, the target point of the evader cannot be *a priori* specified. A target point which is accounted for using a point source is responsible for inducing a field that fills Ω. For motion to be actuated there has to be a vector field everywhere in Ω. Also, for a field to exist a source must be present. The Poisson equation offers an alternative to the Laplace equation in this regard. The Poisson equation:

$$\nabla^2 V(x) = \nabla \cdot \nabla V(x) \equiv -C , \quad (8)$$

constrains the divergence of ∇V to a negative constant value. This amounts to densely covering the workspace with point sources (figure-9) with a field sinking in each (figure-8b). Under the influence of the proper boundary conditions the fields from the micro-sources aggregate to yield a global field pattern. Although not *a priori* specified, a context-dependant, equilibrium point emerges marking the goal point which the evader should move toward.

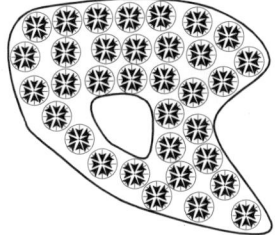

Figure-9: A workspace filled with point field sources

The Generating BVP:

The BVP used for generating the potential whose gradient constitutes the evasion field is: solve

$$\nabla^2 V(x) \equiv -C \qquad x \in \Omega$$

subject to: $\dfrac{\partial V(x)}{\partial n} + q \cdot V(x) = g \Big|_{x \in \Gamma \cup \Gamma_i} \quad i=1,..,L \quad (9)$

Where C, q, and g are positive constants.

V. Results

The capabilities of the Poisson potential field approach are demonstrated by simulation. In figure-10 an agent, whose location is marked by a **+** sign, is attempting to evade capture by two pursuers (marked by circles). Although the stationary environment, and the locations of the pursuers are known, their tactics, future moves, and any coalitions are not *a priori* known to the evader. As can be seen, the evasion, gradient field from the Poisson potential keeps adapting to the movements of the pursuers in a manner that accounts for the contents of the environment. The adaptation takes place so that the stable equilibrium point of the field is situated as far as possible away from the pursuers and the forbidden regions.

Cooperative surround and block pursue:

Here (Figure-11), the pursuing agents attempt to form a ring around the evader and gradually reduce its radius. Moreover, they monitor the direction along which the evader is heading and attempt to group along that direction to block its escape route

$$\dot{\rho}_i = -C \cdot en_i \qquad (10)$$
$$\dot{\theta}_i = K \cdot SGN(\dot{X}^t \, et_i)$$

where ρ_i is the distance between the evader and the i'th pursuer, θ_i the angle of the i'th pursuer, en_i and et_i are the unit vectors normal and tangent to the i'th pursuer, and SGN(x) = [+1 for x>0, 0 for x=0, -1 for x<0].

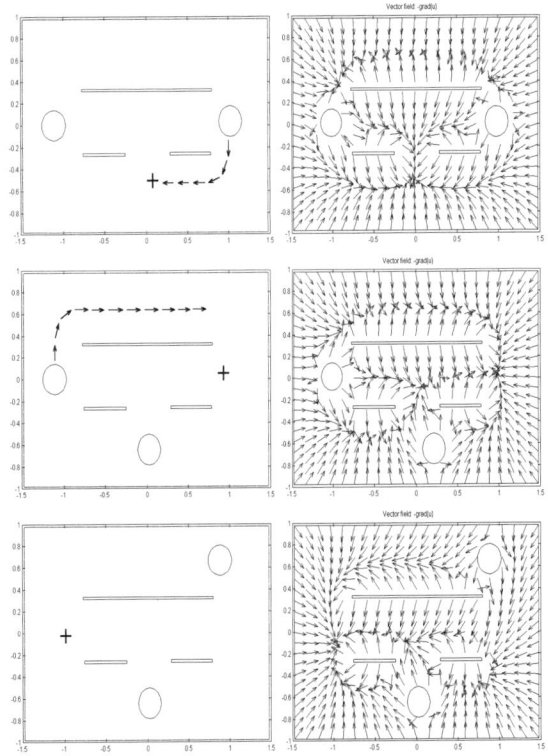

Figure-10: Movements of the evader and pursuers and the corresponding gradient evasion field.

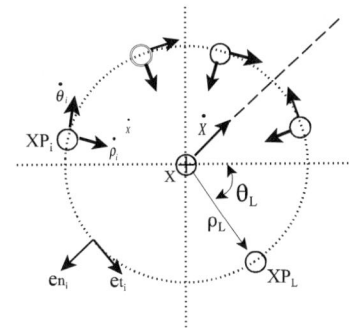

Figure-11: cooperative surround and block pursue

The distance between the evader and the i'th pursuer is:
$$D_i(t) = |X - XP_i| \qquad i = 1,...., L . \quad (11)$$
The distance of the evader to the nearest pursuer (DM(t)) is taken to be the measure of safety:
$$DM(t) = \min|_i D_i(t) \qquad i = 1,...., L. \quad (12)$$
Success of the pursuers is indicated by their ability to drive DM(t) close to zero.

It is a widely accepted belief that if an evader moves in a purely random manner, its chance of escaping capture is improved. Figure-12 shows snapshots of an agent utilizing random walk to evade capture by eight pursuers utilizing the block and surround strategy. A uniformly distributed random number generator is used to generate, at each time step, independent increments along the x and y directions. As can be seen, the pursuers manage to close in on the evader and surround it frequently reducing the radius of the ring to a very small value. The trajectory of the evader is shown in figure-13, and the corresponding DM(t) is shown in figure-14.

In figure-15 the evader replaces the random walk approach with the gradient, evasion field from the Poisson potential. Despite being initially surrounded by a large number of pursuers, the evader manages to outmaneuver them, break the ring, and escape keeping a steady distance away from the pursuers. Moreover, the evader manages to strip the pursuers of their advantage as a group that is capable of utilizing many patterns of behavior for capture. The manner in which the evader maneuvers to escape capture causes the formation of the pursuers to clump in effect reducing the group action into the action of a single agent that is lagging behind the evader. The trajectory of the evader is shown in figure-16, and the corresponding DM(t) is shown in figure-17.

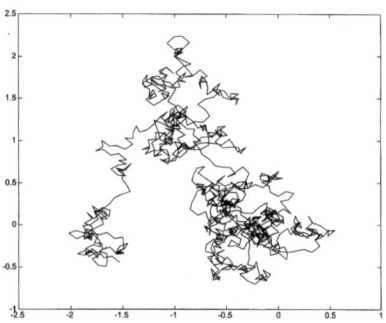

Figure-13: Trajectory of the evader

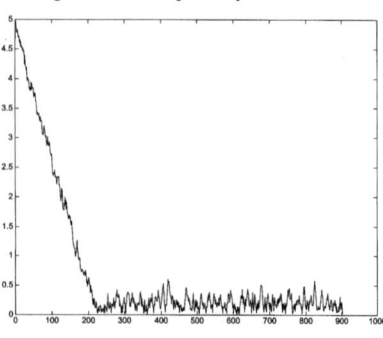

Figure-14: DM(t)

VI. Conclusions

In this paper the problem of continuously evading multi-pursuers in a stationary, cluttered environment is addressed. The high-level controller sensitizing the evader to the contents of its environment is constructed from the gradient of a potential field that satisfy the Poisson equation (also known as the Laplace-poisson equation). The gradient field is supposed to guide the actions of the evader in an attempt to escape capture by the pursuers. In addition to providing tactical

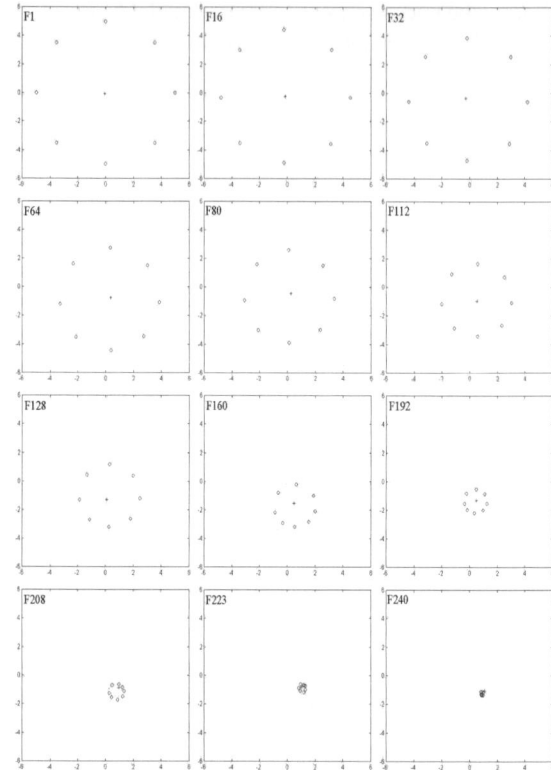

Figure-12: Evasion using random walk.

capabilities for an agent attempting to survive in an adversarial environment, the suggested high-level controller may help to shed light on the origin of purposive behavior. It is believed that any mechanism concerned with the generation of goal-oriented behavior must be supplied with an *a priori* specified goal around which the guidance field is constructed. The suggested controller is a proof that goal-oriented, purposive behavior can be synthesized without having to *a priori* specify a goal. While the preliminary results regarding the performance of the suggested evasion controller are encouraging, in-depth mathematical analysis and simulation experiments remain to be done.

Acknowledgment:

The author would like to thank King Fahd University of Petroleum and minerals for its support of this work.

References:

[1] S. M. Amin, E. Y. Rodin, A. Garcia-Ortiz, "Evasive Adaptive Navigation and Control Against Multiple Pursuers", Proceedings of the American control Conference, Albuquerque, New Mexico, June 1997, pp. 1453-1457.

[2] M. Wahde, M. G. Nordahl, "Coevolving Pursuit-Evasion Strategies in Open and Confined Regions", in Adami et al., Artificial Life VI, MIT Press, pages 472-476, 1998.

[3] G. F. Miller and D Cliff. "Protean behavior in dynamic games: Arguments for the co-evolution of pursuit-evasion

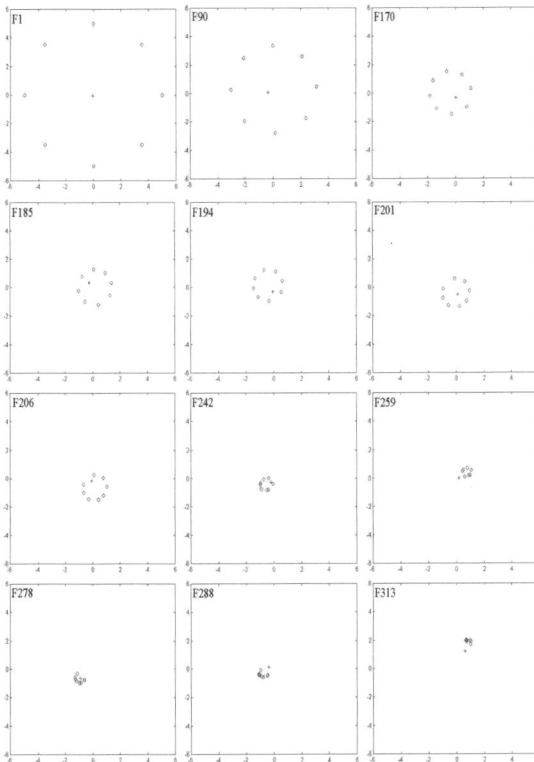

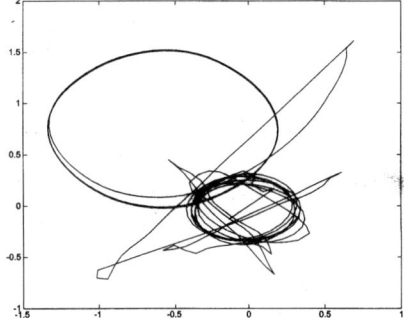

Figure-16: Trajectory of the evader

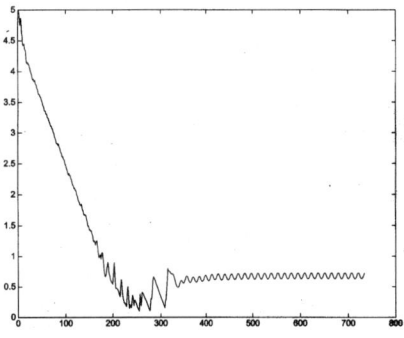

Figure-17: DM(t)

Figure-15: evader utilizing the gradient of a Poisson potential for escape

tactics". In J-A Meyer D Cliff, P Husbands and S Wilson, editors, From Animals to Animats 3, Proc. of the 3rd Int. Conf. on Simulation of Adaptive Behavior, pages 411--420. MIT Press/Bradford Books, 1994.

[4] S. M. Lavalle, J. E. Hinrichsen, "Visibility-Based Pursuit-Evasion: The Case of Curved Environemnts", IEEE Transactions on Robotics and Automation, Vol. 17, No. 2, April 2001, pp. 196-202.

[5] T. C. Ly, S. Venkatesh, D. Kieronska, "Agents in Adversarial Domains-Modeling Environments in Parallel", Proceedings of the fourth annual conference on AI simulation, and planning in high autonomy systems, 20-22 September 1993, Tuscon, AZ, USA, pp. 81-87.

[6] K. Sugihara, I. Suzuki, "On a Pursuit-Evasion Problem Related to Motion Coordination of Mobile Robots", proceeding of the twenty first annual Hawaii International Conference on System Sciences, 1988, Vol. IV, 5-8 January 1988, Kailuakona, HI, USA, pp. 218-226.

[7] A. A. Masoud, "A Boundary Value Problem Formulation of Pursuit-Evasion in a Known Stationary Environment: A Potential Field Approach", IEEE International Conference on Robotics and Automation, May 21-27, Nagoya, Japan pp. 2734-2739.

[8] E. Y. Rodin, "A Pursuit-Evasion Bibiliography-version 2", Computers and Mathematics with Applications, Vol. 18, No. 1-3, 1989, pp. 245-320.

[9] A A. Masoud, S. A. Masoud, "Motion Planning in the Presence of Directional and Obstacle Avoidance Constraints Using Nonlinear Anisotropic, Harmonic Potential Fields", the IEEE International Conference on Robotics and Automation, San Francisco, CA, April 24-28, 2000, pp. 2944-2951.

[10] A. A. Masoud, S. A. Masoud, "A Self-Organizing, Hybrid, PDE-ODE Structure for Motion Control in Informationally-deprived Situations", The 37th IEEE Conference on Decision and Control, Tampa Florida, Dec. 16-18, 1998, pp. 2535-2540.

[11] A. A. Masoud, S. A. Masoud, "Robot Navigation Using a Pressure Generated Mechanical Stress Field, The Biharmonic Potential Approach", The 1994 IEEE International Conference on Robotics and Automation, May 8-13, 1994 San Diego, California, pp. 124-129.

[12] S. A. Masoud, A. A. Masoud, "Constrained Motion Control Using Vector Potential Fields", The IEEE Transactions on Systems, Man, and Cybernetics, Part A: Systems and Humans. May 2000, Vol. 30, No.3, pp.251.

[13] A. Masoud, "Evasion of Multiple Pursuers in a Stationary, cluttered, Environment: A Harmonic Potential Field Approach", 2002 IEEE International Symposium on Intelligent Control (ISIC), 27-30 October 2002, Vancouver, BC, Canada.

Closed Loop Navigation for Multiple Non - Holonomic Vehicles [*]

Savvas G. Loizou   Kostas J. Kyriakopoulos

Control Systems Laboratory, Mechanical Eng. Dept
National Technical University of Athens, Greece
{sloizou, kkyria}@central.ntua.gr

Abstract

In this paper we incorporate dipolar potential fields used for nonholonomic navigation into a novel potential function designed for multi – robot navigation. The derived navigation function is suitable for navigation of multiple nonholonomic vehicles. A properly designed discontinuous feedback control law is applied to steer the nonholonomic vehicles. The derived closed form control scheme provides robust navigation with guaranteed collision avoidance and global convergence properties, as well as fast feedback, rendering the methodology particularly suitable for real time implementation. Collision avoidance and global convergence properties are verified through non - trivial computer simulations.

1 Introduction

Multiple robot navigation is a research area with an increasing research interest over the last decade [19, 11, 8, 18, 7]. In the last few years multi - robot navigation for Non - Holonomic vehicles is gaining increasing attention [14, 2, 4, 5].

Our main interest is to deduce global convergent control schemes with collision avoidance, suitable for real time implementation. Many researchers consider the local stabilization issues [20, 2, 3] without any deadlock resolution mechanism. There are also several attempts to attack the problem with neural nets [21, 5] and with fuzzy logic controllers [4]. In [14] a global convergent algorithm is presented for nonholonomic path planning, based on probabilistic roadmaps, but the me thodology cannot be used for real time implementation due to its complexity.

Nonholonomic stabilization has attracted the attention of the control community over the years, due to the fact that nonholonomic systems do not satisfy the Brockett's necessary smooth feedback stabilization condition [1]. In this paper we address the problem of multiple nonholonomic robot navigation by constructing a potential function that can handle both multiple robot situations and provide feasible nonholonomic trajectories due to its dipolar structure.

The rest of the paper is organized as follows: Section 2 introduces the motivating problem. Section 3 outlines the concept of multiple robot navigation functions. Section 4 presents the discontinuous feedback control scheme. Section 5 presents simulation results for a number of non - trivial navigational tasks. Finally, section 6 summarizes the conclusions and indicates our current research directions.

2 Problem Statement

Consider the following system of m nonholonomic vehicles:

$$\begin{aligned} \dot{x}_i &= u_i \cdot \cos(\theta_i) \\ \dot{y}_i &= u_i \cdot \sin(\theta_i) \\ \dot{\theta}_i &= w_i \end{aligned} \quad (1)$$

with $i \in \{1\ldots m\}$. (x_i, y_i, θ_i) are the position and orientation of each robot, u_i and w_i are the translational and rotational velocities respectively.

---
[*] The authors want to acknowledge the contribution of the European Commission through contract IST-2001-33567-MICRON

The problem can be now stated as follows: *"Given the nonholonomic system (1), derive a feedback kinematic control law that steers the system from any initial configuration to the goal configuration avoiding collisions. The environment is assumed perfectly known and stationary, while each robot acts as a potential obstacle to the others."*

3 Multi-Robot Navigation Functions

In a previous work [9] the authors presented an extension to the navigation function methodology with applications to multiple robot navigation. In this section we present how this novel class of potential functions can be enhanced with a dipolar structure [15] to provide trajectories suitable for nonholonomic navigation.

As it was shown in [9] the function: $\varphi = \frac{\gamma_d^k}{(\gamma_d^k + G)^{1/k}}$ proposed by [6] for single robot navigation, with a proper selection of G can be used for multiple robot navigation and can be made a navigation function by an appropriate choice of k. Our assumption that we have spherical robots and spherical obstacles does not constrain the generality of this work since it has been proven [6] that navigation properties are invariant under diffeomorphisms. Methods for constructing analytic diffeomorphisms are discussed in [13, 12] for point robots and in [16, 17] for rigid body robots.

Let us assume the following situation: We have m mobile robots, and their workspace $W \subset R^2$. Each robot R_i, $i = 1 \ldots m$ occupies a disk in the workspace: $R_i = \{q \in R^2 : \|q - q_i\| \leq r_i\}$ where $q_i \in R^2$ is the center of the disk and r_i is the radius of the robot. The position vector of the robots is represented by $q = [q_1 \ldots q_m]$. The orientation vector of the robots is represented by $\theta = [\theta_1 \ldots \theta_m]$ where θ_i represents the orientation of each robot. The configuration of each robot is then represented by $p_i = \begin{bmatrix} q_i & \theta_i \end{bmatrix} \in R^2 \times (-\pi, \pi]$ and the configuration space C is spanned by $p = \begin{bmatrix} q_1^T \ldots q_m^T & \theta_1 \ldots \theta_m \end{bmatrix}^T$.

3.1 Mathematical Tools - Terminology

The robot proximity functions, a measure for the distance between two robots i and j, are defined by: $\beta_{i,j}(q) = q^T D_{ij} q - (r_i + r_j)^2$, where r_i is the radius of the i'th robot and D_{ij} is defined in [9]. We will use the term '**relation**' to describe the possible collision schemes that can be defined in a multi robot - obstacles scene. The '**set of relations**' between the members of a set can be defined as the set of all possible collision schemes between the members. A **binary relation** is a relation between two robots. Any relation can be expressed as a set of binary relations. A '**relation tree**' is the set of robots-obstacles that form a linked team. Each *relation* may consist of more than one tree (figure 1). We will call the number of binary relations in a relation, the '**relation level**'.

 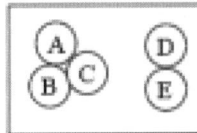

Figure 1 : (a) One – tree relation, (b) Two tree relation

A **relation proximity function (RPF)** provides a measure of the distance between the robots involved in a relation. Each relation has it's own *RPF*. An *RPF* assumes the value of zero whenever the related robots collide and increases wrt the distance of the related robots: $b_R = q^T \cdot P_R \cdot q - \sum_{\{i,j\} \in R} (r_i + r_j)^2$ where R is the set of binary relations (e.g. for the relation in figure (1.a) $R = \{\{A,B\},\{A,C\},\{B,C\},\{D,E\}\}$) and $P_R = \sum_{\{i,j\} \in R} D_{i,j}$ is the **relation matrix** of *RPF*. The gradient and Hessian of the *RPF* are: $\nabla b_R = 2 P_R \cdot q$ and $\nabla^2 b_R = 2 P_R$.

A **Relation Verification Function (RVF)** is defined by:

$$g_{R_j}\left(b_{R_j}, B_{R_j^C}\right) = b_{R_j} + \lambda \cdot b_{R_j} \Big/ \left(b_{R_j} + B_{R_j^C}^{1/h}\right) \quad (2)$$

where $\lambda, h > 0$, R_j^C is the complementary to R_j set of *relations* in the same level, j is an index number defining the relation in the level and $B_{R_j^C} = \prod_{k \in R_j^C} b_k$. An *RVF* is zero if a relation holds while no other relation from the same level holds and has the properties: (a) $\lim_{x \to 0} \lim_{y \to 0} g_x(x,y) = \lambda$, (b) $\lim_{y \to 0} \lim_{x \to 0} g_x(x,y) = 0$.

Based on the above properties, in a robot proximity situation, one can verify that: if $(g_{R_j})_k = 0$ at some *level* k then $(g_{R_i})_h \neq 0$ for any *level* h and $i \neq j$ in level k. It should be noted hereby that since in the highest relation level only one relation exists, there will be no complementary relations and the RVF will be identical to the RPF e.g. $\lambda = 0$ for this relation.

We can now define $G = \prod_{L=1}^{n_L} \prod_{j=1}^{n_{R,L}} (g_{R_j})_L$, with n_L the number of *levels* and n_R, L the number of *relations* in level L. Figure (2) demonstrates several types of relations of a four – member team.

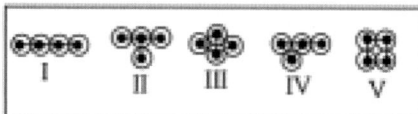

Figure 2: I, II are level 3; IV, V are level 4 and III is a level 5 relation

3.2 Dipolar Navigation Functions

To be able to produce a dipolar potential field, φ must be modified as follows:

$$\varphi = \frac{\gamma_d^k}{\left(\gamma_d^k + H_{nh} \cdot G\right)^{1/k}} \quad (3)$$

where H_{nh} has the form of a pseudo - obstacle. A possible selection of H_{nh} would be: $H_{nh} = \varepsilon_{nh} + \left(\prod_{i=1}^{m} \eta_{nh_i}\right)^{\mu}$ with $\eta_{nh_i} = \|(q - q_d) \cdot \mathbf{n}_{d_i}\|^2$, where $\mathbf{n}_{d,i} = \begin{bmatrix} O_{1 \times 2(i-1)} & \cos(\theta_{d,i}) & \sin(\theta_{d,i}) & O_{1 \times 2(m-i)} \end{bmatrix}^T$ and μ a tuning parameter. Subscript d denotes destination. Moreover $\gamma_d = \|p - p_d\|^2$, i.e. the angle is incorporated in the distance to the destination metric. The proposed modifications of the potential function does not affect its navigation properties [10], as long as the workspace is bounded and $\varepsilon_{nh} > \varepsilon(k)$.

4 Non - Holonomic Control

In the following analysis we will use V for denoting the navigation function instead of φ for notational consistency.

Define $M = \{1, \ldots, m\}$ and $\Omega = P(M)$ where P denotes the power set operator. Assuming that Ω is an ordered set, let N_j denote the j'th element of Ω where $j \in \{1, \ldots, 2^m\}$. Then $N_j \subseteq M$ with $N_1 = \{\emptyset\}$ and $N_{2^m} = M$. We can now define: $\Delta_j = K_\theta \cdot \sum_{i \in \{M \setminus N_j\}} (V_{\theta_i} \cdot (\theta_{nh_i} - \theta_i)) - K_u \sum_{i=1}^{m} (|V_{x_i} \cdot \cos(\theta_i) + V_{y_i} \cdot \sin(\theta_i)| \cdot Z_i) - K_\theta \cdot \sum_{i \in N_j} V_{\theta_i}^2$ with $Z_i = K_u \cdot (V_{x_i}^2 + V_{y_i}^2) + K_z \left((x_i - x_{d_i})^2 + (y_i - y_{d_i})^2\right)$ where V_q denotes the derivative $\frac{\partial V}{\partial q}$ of V along q. Define $H = \{j : \Delta_j < 0\}$ and $\rho = \left\{j : \Delta_j = \max_{i \in H}(\Delta_i)\right\}$. We can now state the following:

Proposition 1. *The system (1) under the control law:*

$$\omega_i = K_\theta \cdot (\theta_{d_i} - \theta_i), \, i \in M \quad \Delta_1 \leq 0$$
$$\omega_l = K_\theta \cdot (\theta_{d_l} - \theta_l), \, l \in \{N_p\}, \, \Delta_1 > 0$$
$$\omega_j = -K_\theta \cdot V_{\theta_j}, \, j \in \{M \setminus N_p\}, \, \Delta_1 > 0$$

$$u_i = -\operatorname{sgn}\left(V_{x_i} \cdot \cos(\theta_i) + V_{y_i} \cdot \sin(\theta_i)\right) \cdot Z_i, \quad i \in M$$

is globally asymptotically stable.

Proof. The navigation function V studied in the previous section serves as a Lyapunov function candidate. We will now examine the derivative of V along the trajectories of (1): $\dot{V} = \frac{\partial V}{\partial t} + \nabla V \cdot \dot{\mathbf{x}} = \nabla V \cdot \dot{\mathbf{x}}$ since $V = V(x)$ with $\dot{\mathbf{x}} = \begin{bmatrix} \dot{x}_1 & \dot{y}_1 & \dot{\theta}_1 & \ldots & \dot{x}_m & \dot{y}_m & \dot{\theta}_m \end{bmatrix}^T$ and $\nabla V = \begin{bmatrix} \frac{\partial V}{\partial x_1} & \frac{\partial V}{\partial y_1} & \frac{\partial V}{\partial \theta_1} & \ldots & \frac{\partial V}{\partial x_m} & \frac{\partial V}{\partial y_m} & \frac{\partial V}{\partial \theta_m} \end{bmatrix}^T$. Substituting we get:

$$\dot{V} = \sum_{i=1}^{m} \left(\frac{\partial V}{\partial x_i} \dot{x}_i + \frac{\partial V}{\partial y_i} \dot{y}_i + \frac{\partial V}{\partial \theta_i} \dot{\theta}_i\right) =$$
$$\sum_{i=1}^{m} \left(u_i \left(V_{x_i} \cdot \cos(\theta_i) + V_{y_i} \cdot \sin(\theta_i)\right) + \dot{\theta}_i V_{\theta_i}\right)$$

We are interested in establishing that $\dot{V} < 0$ almost everywhere, and the sets of points where $\dot{V} = 0$ except from the destination are not invariant. Applying the proposed controls, we get:

For $\Delta_1 \leq 0$ we have:

$$\omega_i = K_\theta \cdot (\theta_{d_i} - \theta_i), \; i \in M$$

$$u_i = -\text{sgn}(V_{x_i} \cdot \cos(\theta_i) + V_{y_i} \cdot \sin(\theta_i)) \cdot Z, \\ i \in M$$

Then $\dot{V} = \Delta_1 \leq 0$. To proceed with the proof we will need the following lemma:

Lemma 1. *If $\Delta_1 > 0$ then $\exists i \in \{1, \ldots, 2^m\} : \Delta_i < 0$*

Proof. If $\Delta_1 > 0$ then since: $-K_u \sum_{i=1}^{m} (|V_{x_i} \cdot \cos(\theta_i) + V_{y_i} \cdot \sin(\theta_i)| \cdot Z_i) \leq 0$ It must be $K_\theta \cdot \sum_{i=1}^{m} (V_{\theta_i} \cdot (\theta_{nh_i} - \theta_i)) > 0$ which means that there exists at least one k for which $V_{\theta_k} \neq 0$ and the term $-K_\theta \cdot \sum_{i \in N_j} V_{\theta_i}^2$ of some Δ_i will be negative definite. For the worst case scenario, $\Delta_{2^m} < 0$ since $N_{2^m} = M$. □

For $\Delta_1 > 0$ then there is at least one j for which $\Delta_j < 0$ as we deduced from (Lemma 1) and thus $\rho \neq \{\emptyset\}$. We choose $j = \rho$ because we want the maximum possible number of robots to follow the dipole generated Non-Holonomic trajectories. The rest will be doing a conflict avoidance manoeuver. The controls in those cases take the form:

$$\omega_l = K_\theta \cdot (\theta_{d_l} - \theta_l), \; l \in \{N_p\}, \; \Delta_1 > 0$$
$$\omega_j = -K_\theta \cdot V_{\theta_j}, \; j \in \{M \setminus N_p\}, \; \Delta_1 > 0$$

$$u_i = -\text{sgn}(V_{x_i} \cdot \cos(\theta_i) + V_{y_i} \cdot \sin(\theta_i)) \cdot Z_i, \\ i \in M$$

Then $\dot{V} = \Delta_\rho \leq 0$

Now let $E = \{\mathbf{x} : \dot{V}(\mathbf{x}) = 0\}$ and $E \supset S = \{x : \omega_i = u_i = 0, \forall i \in M\}$ is an invariant set. From the proposed control law, it can be seen that $u_i = 0, \forall i \in M$ only at the destination, and for all other configurations the controller provides a direction of movement. According to LaSalle's invariance principle, the trajectories of the system converge asymptotically to the largest invariant set, which is the destination configuration □

5 Simulations

To verify the navigation properties of the methodology, we set up a simulation with four nonholonomic unicycles that are about to navigate from an initial to a final configuration, without hitting each other. The robots are placed at several initial configurations and the paths travelled are recorded and depicted in the figures that follow. The chosen configurations constitute non - trivial setups, since the straight paths connecting initial and final positions are obstructed by other robots.

In the first case (figure 5) the four robots were equally sized and positioned at: $[q_1^T \ldots q_4^T] = [\,0.1732 \; -0.1 \; -0.1732 \; -0.1 \; 0.0 \; 0.2 \; 0.0 \; 0.0\,]$ with angles $[\theta_1 \ldots \theta_4] = [\,\pi/2 \; \pi \; 0 \; -\pi\,]$ and their destination configuration was set at: $[{}^dq_1^T \ldots {}^dq_4^T] = [\,-0.1732 \; 0.1 \; 0.1732 \; 0.1 \; 0.0 \; -0.2 \; 0.0 \; 0.0\,]$ with $[{}^d\theta_1 \ldots {}^d\theta_4] = [\,0 \; 0 \; 0 \; 0\,]$. Figure (5a) denotes the initial (R1...R4) and target (T1...T4) configurations of the four robots. Figures (5b-5d) depict the trajectories of the robots. As can be seen, the multirobot navigation function successfully resolves all the proximity situations and the nonholonomic controller successfully steers the system to its destination.

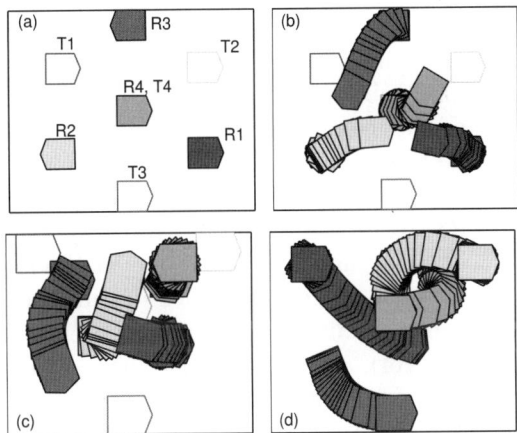

Figure 3 : *(a) Initial Conf., (b,c) Intermediate Conf., (d) Intermediate and Final Configurations*

In the next simulation, robots (R1...R3) were equally sized and robot R4 had half the radius of the rest. In this scenario, robots (R1...R3) are placed at their target configurations (figure 4a), obstructing robot (R4) to achieve its destination. As can be seen in this simulation (figures 4b-4e), the robots (R1...R3), exhibit a cooperative behavior, departing momentarily from their destinations to allow robot R4 to manoeuver to its destination.

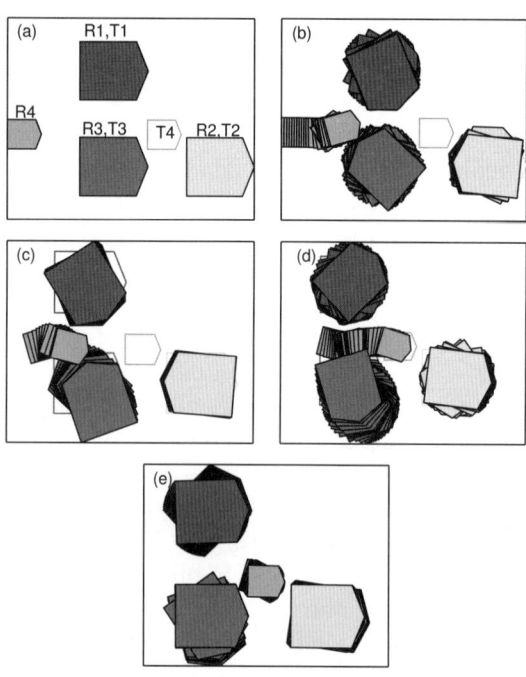

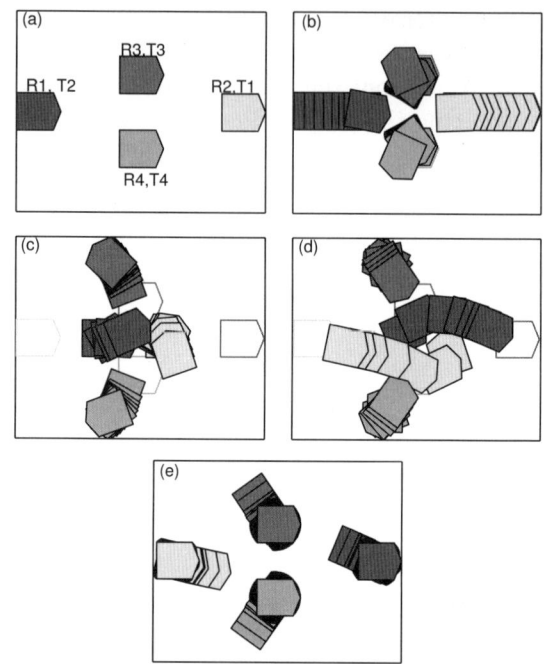

Figure 4 : (a) Initial Conf., (b,c,d) Intermediate Conf., (e) Intermediate and Final Configurations

Figure 5 : (a) Initial Conf., (b,c,d) Intermediate Conf., (e) Intermediate and Final Configurations

In the last simulation (figure 5a), we have again equally sized robots, but the two of them (R3, R4) were placed at their destination configurations (T3, T4), while the other two (R1, R2) were placed at the destinations of each other (T2, T1). Again robots (R3, R4) are obstructing (R1, R2). As can be seen and in this simulation, the methodology succeeds to steer the robots to their destination and resolves the proximity situations encountered. The robots (R3, R4), in a cooperative manner depart momentarily from their destination configurations to allow (R1, R2) to reach their targets. In all simulations, after all robots reach their targets, the system remains stable to the destination configuration.

6 Conclusions - Issues for further research

In this paper we successfully merged two powerful concepts: Dipolar Potential Fields (DPF) for nonholonomic navigation and Multirobot Navigation Functions (MNF). The derived Dipolar Multirobot Navigation Function (DMNF), along with the specially designed discontinuous feedback control law, provides guaranteed global convergence of the system. The methodology due its closed loop nature provides a robust navigation scheme with guaranteed collision avoidance and it's global convergence properties guarantee that a solution will be found if one exists. The closed form control law and the analytic expression of the potential function and its derivatives, provides fast feedback and makes the methodology particularly suitable for real time implementation. The methodology can be easily applied to a three dimensional workspace and through proper transformations to arbitrarily shaped robots.

Current research directions are towards decentralized multiple robot navigation with limited workspace knowledge, limited vision capability, cooperation between mobile robots, formation control, as well as locomotion issues.

References

[1] R. W. Brockett. Control theory and singular riemannian geometry. In *New Directions in Appl. Math.*, pages 11–27. Springer, 1981.

[2] J. P. Desai and V. Kumar. Nonholonomic motion planning for multiple mobile manipulators. *Proc. of IEEE Int. Conf. on Robotics and Automation*, pages 3409–3414, 1997.

[3] J. P. Desai, C. Wang, M. Zefran, and V. Kumar. Motion planning for multiple mobile manipulators. *Proc. of IEEE Int. Conf. on Rob. and Automation*, pages 2073–2078, 1996.

[4] B. J. Driessen, J. D. Feddema, and K. S. Kwok. Decentralized fuzzy control of multiple nonholonomic vehicles. *Proc. of the American Control Conference*, pages 404–410, 1998.

[5] E. Hu, S. Yang, and D. Chiu. A non-time based tracking controller for multiple nonholonomic mobile robots. *Proc. of IEEE Int. Conf. on Rob. and Autom.*, pages 3954–3959, 2002.

[6] D. E. Koditschek and E. Rimon. Robot navigation functions on manifolds with boundary. *Advances Appl. Math.*, 11:412–442, 1990.

[7] J. C. Latombe. *Robot Motion Planning*. Kluwer Academic Publishers, 1991.

[8] Y.H. Liu et al. A practical algorithm for planning collision free coordinated motion of multiple mobile robots. *Proc of IEEE Int. Conf. on Robotics and Autom.*, pages 1427–1432, 1989.

[9] S. G. Loizou and K. J. Kyriakopoulos. Closed loop navigation for multiple holonomic vehicles. *To Appear, Proc. of IEEE/RSJ Int. Conf. on Intelligent Robots and Systems*, 2002.

[10] S. G. Loizou and K. J. Kyriakopoulos. Closed loop navigation for multiple non-holonomic vehicles. Tech. report, NTUA, http://users.ntua.gr/sloizou/academics/TechReports/TR0202.pdf, 2002.

[11] V. J. Lumelsky and K. R. Harinarayan. Decentralized motion planning for multiple mobile robots: The cocktail party model. *Journal of Autonomous Robots*, 4:121–135, 1997.

[12] E. Rimon and D. E. Koditschek. The construction of analytic diffeomorphisms for exact robot navigation on star worlds. *Trans. of the American Mathematical Society*, 327(1):71–115, September 1991.

[13] E. Rimon and D. E. Koditschek. Exact robot navigation using artificial potential functions. *IEEE Trans. on Robotics and Automation*, 8(5):501–518, 1992.

[14] P. Svestka and M. H. Overmars. Coordinated motion planning for multiple car-like robots using probabilistic roadmaps. *Proc. of IEEE Int. Conf. on Robotics and Automation*, pages 1631–1636, 1995.

[15] H. G. Tanner and K. J. Kyriakopoulos. Nonholonomic motion planning for mobile manipulators. *Proc of IEEE Int. Conf. on Robotics and Automation*, pages 1233–1238, 2000.

[16] H. G. Tanner, S. G. Loizou, and K. J. Kyriakopoulos. Nonholonomic stabilization with collision avoidance for mobile robots. *Proc. of IEEE/RSJ Int. Conf. on Intelligent Robots and Systems*, pages 1220–1225, 2001.

[17] H. G. Tanner, S. G. Loizou, and K. J. Kyriakopoulos. Nonholonomic navigation and control of cooperating mobile manipulators. *Accepted, IEEE Trans. on Robotics and Automation*, 2002.

[18] E. Todt, G. Raush, and R. Suárez. Analysis and classification of multiple robot coordination methods. *Proc. of IEEE Int. Conf. on Rob. and Autom.*, pages 3158–3163, 2000.

[19] P. Tournassoud. A strategy for obstacle avoidance and its applications to multi - robot systems. *Proc. of IEEE Int. Conf. on Robotics and Automation*, pages 1224–1229, 1986.

[20] H. Yamaguchi and J. W. Burdick. Time-varying feedback control for nonholonomic mobile robots forming group formations. *Proc. of IEEE Int. Conf. on Decision and Control*, pages 4156–4163, 1998.

[21] X. Yang and M. Meng. Real-time motion planning of car-like robots. *Proc. of IEEE/RSJ Int. Conf. on Intelligent Robots and Systems*, pages 1298–1303, 1999.

Designing a Secure and Robust Mobile Interacting Robot for the Long Term

N. Tomatis[†‡], G. Terrien[†], R. Piguet[†], D. Burnier[‡], S. Bouabdallah[‡], Kai O. Arras[‡], R. Siegwart[‡]

[†]BlueBotics SA
PSE-C
CH-1015 Lausanne
n.tomatis@ieee.org

[‡]Autonomous Systems Lab, EPFL
Swiss Federal Institute of Technology Lausanne
CH-1015 Lausanne
r.siegwart@ieee.org

Abstract

This paper presents the genesis of RoboX. This tour guide robot has been built from the scratch based on the experience of the Autonomous Systems Lab. The production of 11 of those machines has been realized by a spin-off of the lab: BlueBotics SA. The goal was to maximize the autonomy and interactivity of the mobile platform while ensuring high robustness, security and performance. The result is an interactive moving machine which can operate in human environments and interacts by seeing humans, talking to and looking at them, showing icons and asking them to answer its questions. The complete design of mechanics, electronics and software is presented in the first part. Then, as extraordinary test bed, the Robotics exhibition at Expo.02 (Swiss National Exhibition) permits to establish meaningful statistics over 5 months (from May 15 to October 20, 2002) with up to 11 robots operating at the same time.

1. Introduction

The task of a tour guide robot is to be able to move around autonomously in the environment, to acquire the attention of the visitors and to interact with them efficiently in order to fulfill its main goal: give the visitors a pre-defined tour. The environment is known and accessible, but a general approach requiring no environmental changes is better suited for a commercial purpose. For the same reason a fully-autonomous and self-contained robot is preferable. Furthermore such a machine is required to have a long live cycle and a high mean time between failure (MTBF), which minimizes the need of human supervision and guarantees a good credibility of the machine with respect to the visitors.

Within the Expo.02, the Swiss National Exhibition, the *Robotics* exhibition takes place in Neuchâtel, where the main thematic is *nature and artifice*. *Robotics* is intended to show the increasing proximity between man and machine. The visitors interact with up to 11 autonomous, freely navigating tour guide robots, which present the exhibit going from industrial robotics to cyborgs on a surface of 320 m^2.

2. Related Work

The tour-guide robot task can be subdivided in two separate issues, which are navigation and interaction.

Navigation: A limited number of researchers have demonstrated autonomous navigation in exhibitions or museums [4], [11], [14], [7] and [15]. Most of these systems have still some limitations in their navigation approaches. For instance *Rhino* [4] and *Minerva* [14] have shown their strengths in museums for one week (19 kilometers) and two weeks (44 kilometers) respectively. However, their navigation has two major drawbacks: it relies on off-board resources, and due to the use of raw range data for localization and mapping it is sensible to environmental dynamics. *Sage* [11], *Chips, Sweetlips, Joe* and *Adam* [15], use a completely different approach for permanent installations in museums: the environment is changed by adding artificial landmarks to localize the robot. This approach performed well, as shown with a total of more than half a year of operation and 323 kilometers for *Sage* [11] and a total of more than 3 years and 600 kilometers for *Chips, Sweetlips, Joe* and *Adam* [15]. However their movements, but for *Adam*, are limited to a predefined set of unidirectional safe routes in order to simplify both localization and path-planning. Another permanent installation which is operating since March 2000 is presented in [7]. Three self-contained mobile robots navigate in a restricted and very well structured area. Localization uses segment features and a heuristic scheme for matching and pose estimation.

Interaction: Human-centered and social interactive robotics is a comparatively young field in mobile robotic research. However, several experiences where untrained people and robots meet are available. The analysis of the first public space experience with *Rhino* [4] underlines the importance of improving human-robot interfaces in order to ease the acceptance of robots by the visitors. In [14] *Minerva* attracted visitors and gave tours in a museum. It was equipped with a face and used an emotional state machine with four states to improve interaction. The *Mobot Museum Robot Series* [11] and [15] focused on the interaction. Robustness and reliability were identified as an important point for the credibility of a public robot. The permanent installation at the *Deutsches Museum für Kommunikation* in Berlin [7], uses three robots which have the task to welcome visitors, offer them exhibition-related information and to entertain them.

The system presented here is designed to offer enhanced interactivity with complete autonomous navigation in a completely self-contained robot and without requiring changes of the environment. Furthermore it is intended to work permanently with minimal supervision.

3. Design

The typical environment of an exhibit, which is highly dynamic, and the visitor experience expected with such a robot impose various constraints on the design and control. This leads to the following specification of the mobile platform:

- Highly reliable and fully autonomous navigation in unmodified environments crowded with hundreds of humans.
- Bidirectional multi-modal interaction based on speech (English, German, French and Italian), facial expressions and face tracking, icons (LED matrix), input buttons, and robot motion.
- Safety for humans, objects, and the robot itself all the time.
- Minimal human intervention and simple supervision.

The esthetic of the robot has been designed in collaboration with artists, industrial designers, and scenographers. The result of the design of both hardware and software is RoboX: a mobile robot platform ready for the real world (figure 1).

Given the above mentioned specifications, the mechanical, electronic, and software design are now presented.

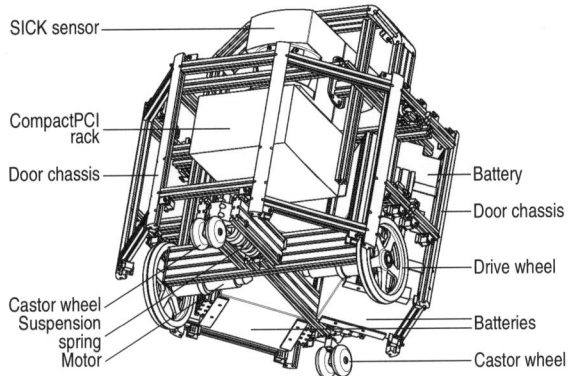

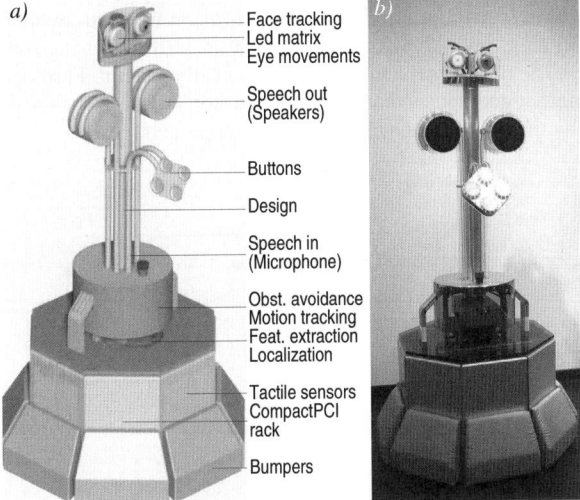

Figure 1: a) Functionality of the tour guide robot RoboX. b) An image of RoboX 9.

3.1 Mechanical Design

The navigation base (lower part of the robot) consists mainly in a CompactPCI rack with two control computers, two laser range sensors (SICKs LMS-200), the batteries, eight bumpers and the differential drive actuators with harmonic drive gears. The base (figure 2) has an octagonal shape with two actuated wheels on a central axis and two castor wheels. In order to guarantee good ground contact of the drive wheels, one of the castor wheels is mounted on a spring suspension. This gives an excellent manoeuvrability and stability to the 1.65 m high robot.

Figure 2: Mechanical design of the RoboX base.

The upper part of the robot incorporates the interaction modules. The face includes two eyes with two independently actuated pan-tilt units and two mechanically coupled eyebrows. The left eye is equipped with a color camera, which is used for face tracking. The right eye integrates a LED matrix for displaying symbols and icons. The eyebrows further underline eye expressions by means of a rotational movement. Behind the face, a gray scale camera pointing to the ceiling is mounted for localization purpose.

The main input device for establishing a bidirectional communication with the humans are four buttons which allow the visitors to reply to questions the robot asks. The robot can further be equipped with a directional microphone matrix for speech recognition even though this remains challenging in the very noisy environment of an exhibition.

3.2 Electronic Design

The control system (figure 3) has been designed very carefully by keeping in mind that the safety of the humans and the robot has to be guaranteed all the time. It is composed of a CompactPCI rack containing an Intel Pentium III card and a Motorola PowerPC 750 card. The latter is connected by the PCI backplane to an analog/digital I/O card, a Bt848-based frame grabber, an encoder IP module and a high bandwidth RS-422 IP module. Furthermore a Microchip PIC processor is used as redundant security system for the PowerPC card.

The navigation software runs on the hard real-time operating system XO/2 [3] installed on the PowerPC. This processor has direct access to the camera looking at the ceiling, the two SICK sensors, the tactile plates and the main drive motors. It communicates with the interaction PC through Ethernet via an on-board hub.

The interaction software is running under Windows 2000 on an industrial PC. This allows using commercial off-the-shelf (COTS) software for speech synthesis and recognition, and makes scenario development easier. The PC has direct access to the eye camera, the eyes and eyebrows controller, the input buttons, the two loudspeakers, and the microphone.

The robot (both CPUs) is connected by a radio Ethernet to an external computer for supervision only, in order to track its status at any time on a graphical interface.

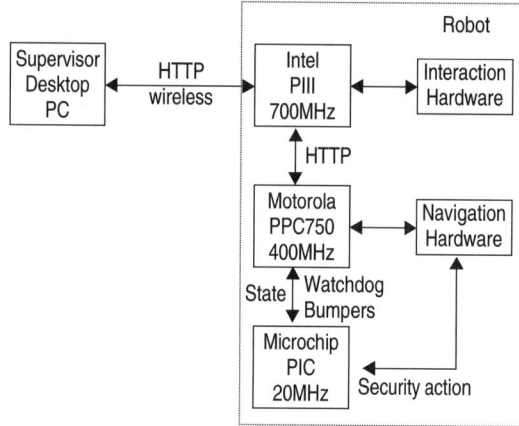

Figure 3: Simple scheme of the electrical design

3.3 Software Design

As explained in the section above, the robot is composed of both an Intel Pentium and a Motorola PowerPC system. The software has been designed without taking into account this fact based on the functionality which was to be developed. However, as soon as the implementation started, the objects have been assigned to one of the two distributed embedded systems. For hardware related objects (mainly sensor drivers) the choice was obvious. For the others, their relevance to safety has been evaluated: due to the hard real-time characteristics of XO/2, all the time-critical objects in relation with the security have been implemented on the PowerPC. Objects requiring COTS components have been implemented on the Windows machine because of their wider availability (f.e. MBrola for speech out, small FireWire camera in the eye for face tracking, etc.).

The resulting object distribution is represented in figure 4. In the following part of this section each component of figure 4 is briefly presented starting with the interaction system followed by the navigation. A complete description of the interaction of RoboX can be found in [9]. Its navigation system is presented in [2].

Interaction

Scenario Controller: It is the central object of the interaction subsystem, which accesses all the other objects. A *scenario* is a sequence of tasks of all modalities (speech, face expression, motion, LED matrix, etc.). A sophisticated tour-guide scenario consists of several small scenarios which are played by the scenario controller.

People detection: It permits to detect movements of objects around the robot by means of the laser scanners. By assuming a static environment, these moving objects are either humans or other robots. The moving objects are then tracked by means of *Kalman Filters*.

Speech Out: By using software permitting either text-to-phonemes-to-speech or directly text-to-speech, this object permits the robot to talk in four languages (English, German, French, and Italian). Furthermore, files of format .wav and .mp3 can be played.

Buttons Controller: This controls the main input device for the interaction between the robot and the humans. Four capacitive buttons with different colored lights are used in combination with questions from the speech out to close the interaction loop.

LED Matrix: The LED matrix is in the right eye. Its controller permits to show icons and animations.

Eyes Controller: The eyes can be moved independently. The controller has a set of predefined expressions, which can be directly played.

Face Tracking: The color camera in the left eye is used to track skin colored regions. The approach is based on [8]. In combination with the eyes controller, this permits to track a face on the image and with the movement of the eyes.

Navigation

Odometry Driver: Calculates the position and uncertainty of the robot based on the wheel rotations.

Speed Controller: Regulates the speed defined by the obstacle avoidance with a PID controller accessing the encoders and updates the odometry.

Localization: Uses a new approach [1] based on an *Extended Kalman Filter* [5] to correct the odometry with exteroceptive sensors (laser scanners, CCD camera).

Obstacle Avoidance: Calculates a collision free path by initializing the path with a *NF1* function [10] and using the *Elastic Band* approach [12] to dynamically adapt it. Furthermore it guarantees that the robot can stop before collision at any time with the *Dynamic Window* approach [6].

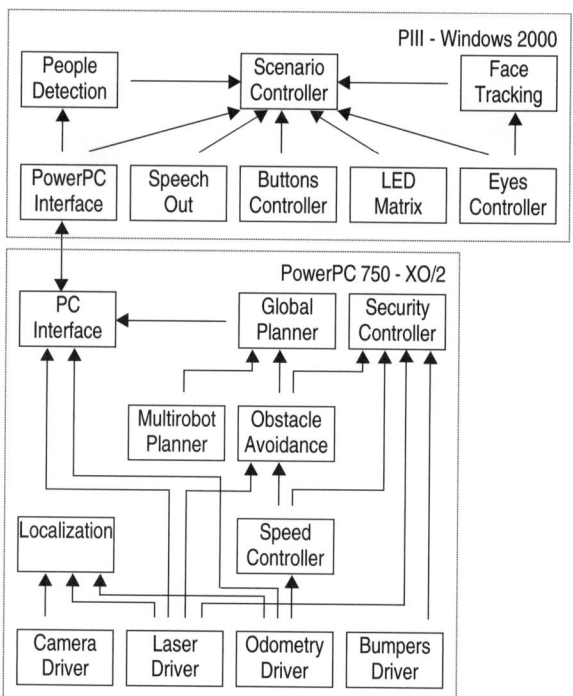

Figure 4: Object distribution of the software on the distributed embedded system.

Multirobot Planner: Synchronizes the movement of the robots to avoid having many robots going to the same place.

Global Planner: Plans the navigation of the robot on the a priori map level, by defining via points which permit to reach the goal point within the graph representing the map.

Security Controller: Guarantees that the robot cannot become dangerous even in case of failure, by supervising the safety-critical software and sensors. Due to the importance of this issue for a robot sharing the environment with humans, the next section presents the security system in details.

4. Security

In this section the involvement of the security issue in the design of the whole system is pointed out in more details.

All the software which relates to the movement of the robot is defined as safety critical. In order to guarantee the security of both the users and the robot itself, safety is on three levels: the operating system, the software implementation and the redundancy of the hardware.

4.1 Operating System

All the navigation software is implemented on the PowerPC which is operated by XO/2, a deadline driven hard real-time operating system [3]. Due to its characteristics XO/2 helps for all the components [13] of safety, which are:
- *Safety*: nothing bad happens.
- *Progress*: the right thing do (eventually) happen.
- *Security*: things happens under proper supervision.

Static safety is guaranteed by the strong-typing characteristic of Oberon-2, the language used under XO/2. Many errors are already found at compile-time instead of run-time. Furthermore, index-checks, dynamic type systems and especially the real-time compatible garbage collector guarantee dynamic safety by forbidding almost any memory-management related errors.

The deadline driven scheduler is in charge of progress: it guarantees that each task is executed within the predefined deadline. Of course this is possible only if the constellation of the tasks running on the PowerPC requires less than 100% of the CPU. For this, the duration of each tasks has to be known. Admission tests are performed at each installation of a new real-time task to guarantee their feasibility. As soon as the progress of all real-time tasks is guaranteed, the CPU is scheduled between the non-real-time tasks depending on their priorities.

Each error causes a *system trap* which is under complete control of the operating system. The system knows exactly where the error took place, who called this part of the code up to the task currently running (stack trace). This is very helpful for debugging, but it is even more important for security because for each task an *exception handler* can be defined. The actions which have to take place in such a case can therefore be properly defined.

4.2 Software Security

Tasks whose failure could cause injuries to people or damage objects required special attention during design. Software watchdogs are therefore implemented for the speed controller, the obstacle avoidance, the laser driver and the bumpers driver (figure 4). Failure of one of these tasks is detected by the security controller which then either restarts the failed task or stops the robot, turns on the alarm blinker and sends an e-mail to the maintenance. This permit to centralize the control of the security and to refer to a single object if a problem occurs. Furthermore, the security controller generates a watchdog signal on a digital output permitting to know if both the operating system and the security controller are still running.

4.3 Hardware Redundancy

The above mentioned software permits to have a consistent control system running on the PowerPC. However, this isn't enough to guarantee the security of the robot and its surrounding. Even in case of failure of the electronics or problems on the operating system of the PowerPC, the robot must remain un-dangerous. For this, the robot has a third processor: a Microchip PIC (figure 3). The software running on it checks the watchdog generated by the security controller, awaits acknowledgements from the security for each bumper contact and controls that the pre-defined maximal speed is never exceeded. If one of these conditions is not respected the redundant security software running on the PIC safely stops the robot (it shortcuts the phases of the motors) and puts it in emergency mode (acoustic alarm).

5. Experiments

The whole operational period, 159 days from May 14 to September 17, is available for statistics. Each day from six to eleven freely navigating tour-guide robots have given tours from 9:30am to 8:00pm until August, then from 9:00am to 8:00pm in September and from 9:00am to 9:00pm in October on the surface of the exhibit which is approximately 320 m^2.

5.1 Definitions

Failure: A failure is any kind of problem which requires human intervention. The only exceptions are the emergency button, which can be pressed and released also by visitors, and the situations where the robot remains blocked because it is to near too an object. In the latter case, the staff can displace the robot by a switch which disconnects the motors from the amplifiers and allows to move the 115 kilograms robot easily.

Uncritical: Uncritical failures are those which do not stop the task of the robot. For example, a failure consisting in a robot which stops sending an image to the supervisor is not critical for the tour the robot is giving to the visitors.

Critical: Critical failures stop the robot until human intervention is performed. An example is the failure of the scenario controller or of the obstacle avoidance.

Reboot: Critical failures requiring a reboot of either the Pentium or the PowerPC are treated separately because they require more time before the robot is again operational.

5.2 Results

During the 159 days of operation the robots served more than 680'000 visitors for a total of 13'313 hours of operational time. In order to perform their job, they travelled 3'316 kilometers for a total moving time of more than 9'415 hours meaning that the mean displacement speed is 0.098 meters per second. As it can be seen in table 1, the uncritical failures represent only a small portion of the total amount of failures (6.7%). Furthermore they do not disturb the operation of the robot. They are therefore not treated in the following analysis which will focus on the critical and reboot failures of the whole robot first and then of the PowerPC.

Run time	13'313 h
Movement time	9'415 h
Travelled distance	3'316 km
Speed (average / max)	0.098 / 0.6 m/s
Failures (total / critical / uncritical)	4'378 / 4'086 / 292
Critical failures (PC / PPC / HW)	3'216 / 694 / 176
Visitors	686'405

Table 1: Five months of operation. More than 13'000 hours of work, where the RoboXes have travelled 3'300 kilometers and served more than 680'000 visitors.

As it can be seen in figure 5, the beginning of the exposition in the middle of May showed that some work was still to be done. The software running on the PC was very unstable due especially to errors in treating the list of the tasks running in the scenario controller.

The mean time between failure (MTBF) of the whole robot (PC, PowerPC and hardware) during the first three weeks was 1.41 hours. This has improved to 4.61 hours from week four to the end, which means that during one day with 10 robots, the staff had to perform a mean of 25 interventions. The type of interventions goes from the simple double-click to restart an application (typical intervention on the PC) to the change of a motor amplifier (very rare, it happened five times, two of them due to a motor defect). After the first three weeks, the MTBF already doubled.

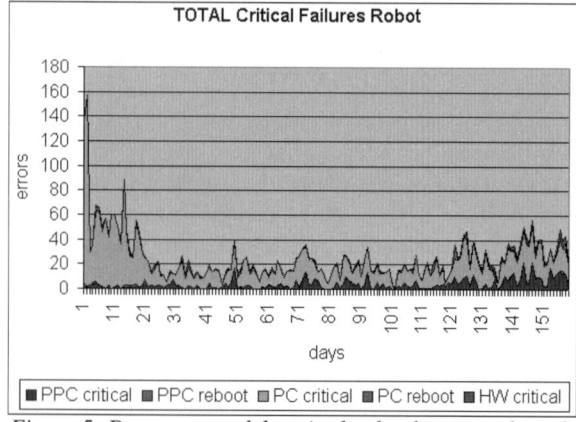

Figure 5: Due to many delays in the development, the software was still in the test phase at the beginning of the exposition. The first three weeks represent a huge improvement in the stability of the software, especially on the PC side.

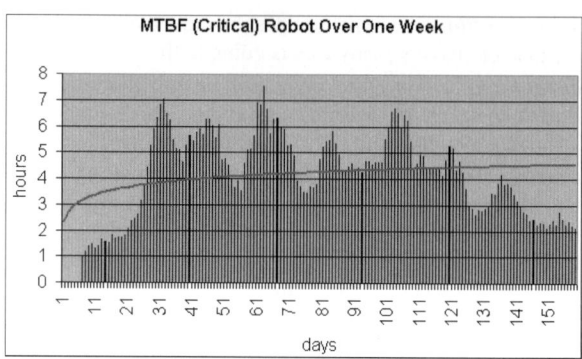

Figure 6: The mean time between critical failure of any kind (PC, PowerPC, hardware). The improvement has been constant exponential during the first four weeks, where the most important errors have been found.

In figure 7 all the critical failures coming from the navigation software (PowerPC) are shown. During the first three weeks, errors in the safety-critical tasks were treated by the security controller, but could sometimes require a reboot in order to restart the trapped task. This has been partly corrected allowing for much faster intervention in case of failure. Critical failures in figure 7 also contain errors which have not directly to do with the software: situations where the robot went lost. The main reason for lost-situations are visitors or untrained staff members who handle the robots without using the switch to disconnect the motors from the amplifiers during a manual intervention. This causes unmodeled odometry errors of such an extent that the robot went lost (robot kidnapping). Note that this type represents 73% of the critical failures of the PowerPC (504 failures) and that they are not software failures, but situations which the localization system cannot handle since underlying assumptions are violated (more details in [2]).

The MTBF for the PowerPC (figure 8 (a)) was between 20 and 80 hours already at the beginning of the exposition. By taking into account only the software errors (figure 8 (b)), the MTBF over the whole period is 70.1 hours.

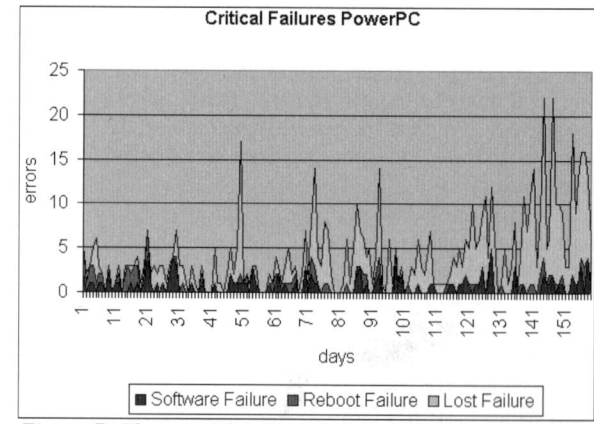

Figure 7: The critical failures of the PowerPC (navigation system). Some of the critical errors require the reboot of the PowerPC. Lost failures are not software errors (they are not "bugs").

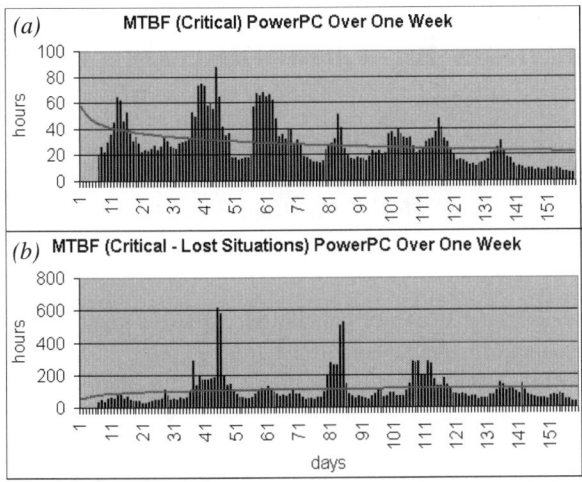

Figure 8: The MTBF (critical) with (a) and without (b) lost situations. By not taking into account the lost situations (b) the MTBF is very high (mean 70.1 hours).

Hardware failures (figure 9) are due to some uncritical design errors at the beginning (robot doors), some motor-amplifier problems and to the high temperature between day 33 and day 40 causing some component failures.

5.3 Lessons to be Learned

The characteristics of this project give an extraordinary chance of learning by experience. Thousands of hours of operation permit to improve the software and hardware to a level which is simply not achievable in smaller projects. This was shown during the exploitation, where some errors were found after few days of operation while others appeared for the first time after one or two months. The best example are the failures of the laser scanners on week 5 due to the temperature in the exhibit. This failure wasn't taken into account by the security causing the obstacle avoidance to permanently receive the last available scan and the robot to collide with the next encountered object. This problem is since then under supervision of the Security Controller (figure 4).

Another interesting point is the difference in the software reliability implemented under PC and PowerPC. The better result of the PowerPC is due to the real-time XO/2 operating system which has been developed for embedded systems focusing on robustness and safety [3], and also to the longer experience in navigation at the Autonomous Systems Lab in contrast to the new interaction software which has been developed only for this application starting in late year 2000.

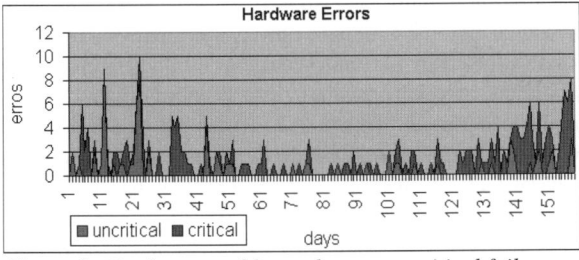

Figure 9: Hardware problems also cause critical failures.

6. Conclusion

This project represents a milestone in the field of mobile robotics: for the first time interactive mobile robots are produced (11 robots) and used for a long time (five months) as real products instead of prototypes as in former projects. The paper presents their characteristics first, then goes into details about the mechanical, electrical and software design. The security issue is faced seriously for ensuring security of the humans and the robot itself all the time. In the experiments section the results of the whole project (159 days of operation) of the *Robotics* exposition are presented and analyzed focusing on the amount and type of robot failures.

References

[1] Arras, K. O., J. A. Castellanos, and R. Siegwart (2002). <u>Feature-Based Multi-Hypothesis Localization and Tracking for Mobile Robots Using Geometric Constraints</u>. IEEE International Conference on Robotics and Automation, Washington DC, USA.

[2] Arras, K. O., R. Philippsen, N. Tomatis, M. De Battista, M. Schilt, and R. Siegwart (2002). <u>A Navigation Framework for Multiple Mobile Robots and its Application at the Expo.02 Exhibition</u>. IEEE International Conference on Robotics and Automation, Taipei, Taiwan.

[3] Brega, R., N. Tomatis, K. Arras, and R. Siegwart (2000). <u>The Need for Autonomy and Real-Time in Mobile Robotics: A Case Study of XO/2 and Pygmalion</u>. IEEE/RSJ International Conference on Intelligent Robots and Systems, Takamatsu, Japan.

[4] Burgard, W., A. B. Cremers, et al. (1999). "Experiences with a Interactive Museum Tour-Guide Robot." <u>Artificial Intelligence</u> 00(1999): 1-53.

[5] Crowley, J. L. (1989). <u>World Modeling and Position Estimation for a Mobile Robot Using Ultrasonic Ranging</u>. IEEE International Conference on Robotics and Automation, Scottsdale, AZ.

[6] Fox, D., W. Burgard, et al. (1997). "The Dynamic Window Approach to Collision Avoidance." <u>IEEE Robotics & Automation Magazine</u>: 23-33.

[7] Graf, B., R. D. Schraft, et al. (2000). <u>A Mobile Robot Platform for Assistance and Entertainment</u>. International Symposium on Robotics, Montreal, Canada.

[8] Hilti, A., I. Nourbakhsh, B. Jensen, and R. Siegwart (2001). <u>Narrative-level Visual Interpretation of Human Motion for Human-robot Interaction</u>. IEEE/RSJ International Conference on Intelligent Robots and Systems, Maui, Hawaii.

[9] Jensen, B., G. Froidevaux, X. Greppin, A. Lorotte, L. Mayor, M. Meisser, G. Ramel and R. Siegwart (2002). <u>The interactive autonomous mobile system RoboX</u>. IEEE/RSJ International Conference on Intelligent Robots and Systems, Lausanne, Switzerland.

[10] Latombe, J.-C. (1991). <u>Robot motion planning</u>. Dordrecht, Netherlands, Kluwer Academic Publishers.

[11] Nourbakhsh, I., J. Bodenage, et al. (1999). "An Effective Mobile Robot Educator with a Full-Time Job." <u>Artificial Intelligence</u> 114(1-2): 95-124.

[12] Quinlan, S. and O. Khatib (1993). <u>Elastic bands: connecting path planning and control</u>. IEEE International Conference on Robotics and Automation.

[13] Szyperski, C. and J. Gough (1995). <u>The role of programming languages in the life-cycle of safe systems</u>. Second International Conference on Safety Through Quality (STQ'95), Kennedy Space Center, Cape Canaveral, Florida, USA.

[14] Thrun, S., M. Beetz, et al. (2000). "Probabilistic Algorithms and the Interactive Museum Tour-Guide Robot Minerva." <u>International Journal of Robotics Research</u> 19(11): 972-99.

[15] Willeke, T., C. Kunz, et al. (2001). <u>The History of the Mobot Museum Robot Series: An Evolutionary Study</u>. Florida Artificial Intelligence Research Society (FLAIRS), Florida.

The Current Opinion on the use of Robots for Landmine Detection

S.Rajasekharan
Department of Computer Science, University of Hull, United Kingdom, e-mail: S.Rajasekharan@dcs.hull.ac.uk

C.Kambhampati
Department of Computer Science, University of Hull, United Kingdom

Abstract— Anti-Personal landmines are a significant barrier to economic and social development in a number of countries. Several sensors have been developed but each one will probably have to find, if it exists, a specific area of applicability, determined by technological as well as economical or even social factors, and possibly other sensors to work with using some form of sensor fusion. A significant issue concerns the safety of the deminer. For every 5000 mines removed, one deminer is killed. The design of an accurate sensor may reduce the amount of time needed to determine whether a landmine exists, but does not increase the safety of the deminer. Since the safety issues during the eradication process are of great concern, the use and integration of cheap and simple robots and sensors in humanitarian demining has been raised. This removes the deminer from close contact with mines. This paper is concerned mainly with the work that has been done in the area of robotics and landmine detection, the problems involved, the issues that have been overlooked and the future of robots in the field.

I. INTRODUCTION

Anti-Personal landmines are a significant barrier to economic and social development in a number of countries, especially Vietnam, Cambodia, Afghanistan and Angola [1]. Since the 1940's, metal detectors have been used to detect buried landmines and this method remains unchanged today. Unfortunately, this can be an extremely tedious and slow process [2]. Metal detectors currently used by demining teams cannot differentiate a mine from metallic debris, which leads to 100-1000 false alarms for each real mine in minefields where the soil is contaminated by large quantities of shrapnel, metal scraps and cartridge cases [4]. Increasing metal detector sensitivity only results in the discovery of even smaller debris in the ground. The only current alternative is to prod the soil at a shallow angle using rigid sticks of metal to determine the shape of an object:this is an intrinsically dangerous operation.

It must be noted that the solutions developed by the military are generally *for* the military and cannot be used for the purposes of humanitarian demining. For the military, mine detection rates of 80% are accepted since all the military need are a quick breach in a minefield. For humanitarian mine clearing it is obvious that the system must have a detection rate approaching the perfection of 99.6% [1].

All efforts have been directed towards an improved mine detector but what is essentially needed is a detector/sensor that will reliably confirm that the ground being tested does not contain an explosive device, with a reliability approaching 100% [3].

One would assume, therefore, that the problem lies within the sensor technology. Radar (ground penetrating (GPR), wideband, arrays, synthetic aperture radar), infrared and microwave radiometry, explosive vapour sensors, acoustic sensors, electromagnetic induction, magnetometers, and electrical impedance tomography are some of the techniques which have been tried [4]. None of these technologies presented seems in fact capable of reaching, in a very large number of situations, good enough detection while maintaining a low false alarm rate. Rather, each one will probably have to find, if it exists, a specific area of applicability, determined by technological as well as economical or even social factors, and possibly other sensors to work with using some form of sensor fusion.

Robots have been suggested for the problem of landmine detection but has been met with controversy [5]. There are two main reasons for this. The first is that the experts in the field believe that the solution lies in building a better sensor. Assume then that a sensor is developed that is capable of producing a 99% accuracy rate and it is then fixed to a hand held device as is done with the metal detector. The most significant accomplishment then is the speed needed to determine whether a mine exists or not. Unfortunately, it is still a dangerous operation for the deminer that uses the hand held sensor. For every 5000 mines detected and removed, one deminer is killed. Thus it would be fair to say that having an unmanned platform to carry the sensor and use it would be ideal for this problem. Alternatively, a team of robots employed to detect mines in a specific terrain using intelligent on board sensors will be better equipped than deminers provided the sensors can actually detect the mines and the robots search the area thoroughly [6]. Robots do not tire like humans, provided they have enough power, and their performance is not affected by psychological tension and trauma. So if this is the case why haven't there been significant results in the last decade with robots detecting landmines? So we come to the second controversial issue: awareness. A general lack of awareness of the problem has led in the majority of the cases to inappropriate solutions that never make it to the minefield. This paper is concerned mainly with the work that has been done in the area of robotics and landmine detection, the problems involved, the issues that have been overlooked and the future of robots in the field. It must be noted that these robots have been designed solely for the purpose of detection rather than demining or clearance. Several mine clearance vehicles exist and detailed work on them my be found in [7], [8], [9].

A. Existing Methods

Apart from hand-held metal detectors which was mentioned in the introduction, the hand-probing technique of the deminer is the most reliable method of mine clearance. A probe is manually inserted into the soil at a 30 degree angle, approximately every five centimeters. When an object is detected with higher stiffness compared to the environment stiffness, more examinations are conducted to identify the shape and size of the object. If the object is determined to be a potential mine, a mine clearing team is called to uncover or detonate the object. A Manual method for investigating a suspect metal object in hard ground conditions in Afghanistan is shown in Fig 1.

Dogs can and are used to smell explosive vapours and/or traces, similarly to what is done at airports and in other security application. Well run dog programs are nowadays generally accepted by most humanitarian demining organisations for area verification and minefield delineation (area reduction). The use of dogs for individual mine detection is slightly more controversial since there is a lack of coherent and universal testing protocols for dogs. Other limiting factors are unfavorable climate conditions and the fact that dogs can tend to tire. Further information on mine sniffing dogs may be found at [10].

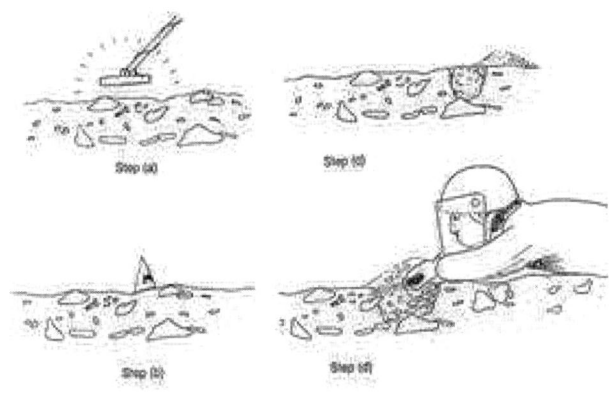

Fig. 1. Manual method for investigating a suspect metal object in hard ground conditions in Afghanistan. Step (a) - object located with metal detector (typically 1 object per 3 sq metres on average, but up to 30 objects per square metre, or as few as one object for 50 sq metres). Step(b) - object location marked. Step(c) - deminer scrapes surface carefully to see if fragment is lying on the ground. Otherwise, he digs a trench 30cm behind the location. Step(d) - deminer works forward, dismantling the ground piece by piece with a prodder (usually a bayonet) until the object is found. This diagram has been taken from http://www.mech.uwa.edu.au/jpt/demining/countries/ minefields.html)

B. Robots: The story so far

In the last decade since robot research in landmine detection began various projects have started, some of these have been finished and very few have made it to the minefield. A few are described in the following sections.

1) Automated Probing:: Several robots have been designed to imitate the human deminer [11]. Fig ?? shows an idea conceived by Antoniae and Ratkoviae [12] for an automated probe. The EOD (Explosive Ordanance Disposal) robot was designed by the Croatian Ministry of the Interior. The terrain is first search with a five centimeter raster. If an obstacle is detected, the terrain is searched in the close vicinity of that point. If the obstacle is detected at one or more neighbouring points, there is a possibility of a landmine presence. Object recognition algorithms are then used to analyse collected data, determining the size and shape of the buried object. If the object is a mine, a mine removal procedure is administered. The robot then moves five centimeters ahead and repeats the procedure. Shahri and Naghdy [13] have tried to automate the hand-probing mine detection by a mechatronics device. The device inserts a bayonet into the soil. It is controlled by modelling the dynamics of the manipulator and environment and adapting for variation in the stiffness sensed by the bayonet when it comes in contact with the mine or any other object in the soil. A neuro-fuzzy impact controller is used to control the impact control while dealing with uncertainties and changes in environment parameters. The performance of the controller was validated through computer simulation and experimental results. These results show that the device can maintain its stability after a change in the environment stiffness coefficient.

Franklin et al [14] proposed a new theory of geometric sensing and probing in the mine detection context. Rather than use a single sensor into probing, they look at multiple sensors in optimal probing in a given region. This was implemented as a simulation.

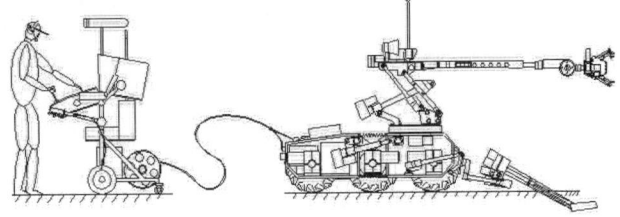

Fig. 2. Tele-operated probe in action

2) Wheeled Robots:: Several wheeled robots [15], [16], [17] have been designed, most notably the Pemex BE [18]. The Pemex BE is a light weight 2-wheels robot with sensors located inside a half-sphere which acts as a third supporting point(See Fig 3). It weighs less than 16kg and can easily dismantled and carried out as hand luggage(Fig 4). It is battery operated with an autonomy of 60 minutes and can move at a speed of up to 6km/h. The robot performs better as a carrier of sensors than a landmine detecting robot. Work still needs to be done on the sensor side of the research to find a more reliable, cheap and sensitive sensor for the robot.

Fig. 3. Pemex BE climbing a pile of rocks

Fig. 4. Pemex BE packed and ready to be carried

Like the Pemex BE, the TRIDEM 1 [19] is a wheeled mine detecting robot built by the Royal Military Academy in Belgium. The vehicle is equipped with three independent drive/steer wheels connected by a triangular frame. Each wheel has two electrical motors and the frame supports the control electronics.
IRobot's Fetch program offered new approach to counter mine problem: a team of low cost robotic mine hunters that will provide rapid and complete coverage of the mine field [20]. The first FETCH demonstrated the aspects of UXO clearance. Operating autonomously, fetch was able to navigate to a given point, perform a local search, find and pickup a UXO, transport the munitions to the disposal area, and place the UXO ont he ground. In addition to autonomous operation, high level supervision and direct operator control were demonstrated. Fetch 2 is a test bed on which to address the questions that arise when multiple mine clearing robots are employed to sweep an area.

3) Walking Robots: Several walking robots have been developed in this time. COMET I and COMET II [21] are six-legged robots designed for humanitarian demining. The total weight of the robot is around 120kg, the width 1300mm and body height is around 1000mm. It is equipped with two computers and its high level motion control and external recognition is assigned to a host computer based on tele-operation. The task for a walking robot in a minefield is extremely complex: the robot must ensure that each leg avoids the mines precisely while keeping a stable attitude. Since force sensors cannot be used to control the walking gait of the robot on a minefield, an attitude control method to realize stable gait on uneven terrain was studied. The walking robots have radar and metal detectors attached to them for the detection purpose. COMET I can walk slowly at a speed of 12m-20m per hour with precise detection mode using six metal detectors. It can also move in day and night using IR cameras so that the area detected is approximately from $250m^2$ to $500m^2$ a day. COMET II is faster and is also equipped with two manipulators and a grass cutter.
Legged and tracked vehicles were also being developed by RMA[19]. The Hunter is an automated electrical tracked platform. When the metal detector detects something, the robot stops and an alarm is reported to the operator.

4) Algorithms and simulations for searches: Research on demining includes many different aspects and in particular the design of efficient and intelligent strategies for (1) demining regions of interest using a variety of sensors, (2) detecting and classifying mines, and (3) searching for mines by autonomous agents. Strategies for directing autonomous agents or robots to search for mines in a pre-selected area or minefield based on spatio-temporal distributions are considered in [22]. The initial model for search is static and is only modified by the information that is collected by the robot (presence or absence of mines at locations visited). Prior information about the minefield is provided by means of a probability distribution of the presence of mines. The authors show that a random search is no different to a sweeping search since neither uses information about the minefield.

Zhang et al [23] look at probabilistic methods for robotic landmine search by directing the search based on the spatial distribution of the minefield. The authors attempt to efficiently extract the pattern of the spatial distribution of the minefield during the beginning of the search process. The extracted probability distribution for the configuration of the minefield is then used to search for more mines.

A very basic algorithm is used by Bauer et al [24] using the model dynamic equations of the robot and non-linear control theory such as Kalman filtering.

Further coverage methods have been studied [25] for a team of independent robots that can cooperatively cover their shared environment. It comprises a reactive coverage algorithm which operates without explicit knowledge of cooperation and tries to maintain cooperative relationships with other robots while increasing efficiency of the team. The authors show that any team of square robots with intrinsic contact sensing can successfully cover a finite rectilinear environment efficiently.

Cassinis et al [6] looked at strategies for navigation of robot swarms to be used in landmine detection. Simulations were performed to compare different search strategies such as random movements and relay clustering based on animal behaviour. In relay clustering, when one robot finds a mine in a random search, the position is relayed back to other robots. Other robots then head towards this particular robot.

Researchers at the University of Hull are currently looking at object detection for surface landmine detection using wheeled robots [26]. The search is performed in a grid based manner. With all the robots beginning the search at the same point, each robot moves to the corner of its designated area. Once, in its respective grid, the grid is divided in half due to the ultrasonic map of the robot. The robot then performs a search of one half and when completed, moves to the next half of the grid. When an object is detected, it is relayed back to the server and the remaining area is split into two again and the search is resumed.

5) Airborne Systems: Researches have often stressed that given improved sensor technology, airborne systems may be the most appropriate carriers for landmine detection. A solution from Schiebel Technology, the CAMCOPTER [9], is an unmanned remote controlled autonomous vertical take off and landing aerial platform. For the purpose of landmine detection, the UAV employs an inertial navigation system coupled with a differential GPS (Global Positioning System) for precise positioning, and an electro-optical/infrared sensor suite attached to the universal payload mounting base. Several sensors are currently being developed for integration with the CAMCOPTER such as GPR and vapor sample collector. Apart from being used for detection, the CAMCOPTER can also provide deminers with adequate and up-to-date maps from which to conduct their operations. The large format mapping camera takes aerial photographs of what the minefield looks like today, which can then be tailored to meet current needs. Work on subsurface imaging for site characterization for airborne and vehicle GPR are also being studied [27].

C. Issues

Although several robot solutions for landmine detection have been suggested and designed very few have come out of the workshops and onto the minefields. There are several reasons for this. Most robotic equipment thus far produced, with the exception of the Pemex BE have been extremely expensive: expensive to build, expensive to run and expensive to maintain. Thus they remain in the laboratory they were built in because the end-user cannot afford to use such equipment. Another reason for lack of successful solutions is the complexity of the machine. Technical knowledge in the 3rd World is low when it comes to robotics and hence it can be difficult to make end users enthusiastic. A more serious problem is to do with the application and task. A lot of work has been a by-product or after thought and thus ends up being unsuitable for the actual detection problem. Most of this research is an extension of mobile robot research - designing robots that have been originally used within the confines of a laboratory.

A problem often overlooked by robot scientists is vegetation and environment. Minefields often have large amounts of thick vegetation and terrain is not always flat or suitable for common wall-following robots to traverse. In many cases, preliminary work is ignored before design and implementation. Environment and especially vegetation affects performance of robots in an adverse way. Thus the reality of the situation usually hampers any decision to send an expensive robot to the field.

D. Suggestions

As can be seen from the previous sections, the problem has not been the technology but rather the delivery of technology. The most important lesson to be learnt from the research is the lack of communication with the end-user - in this case the demining group. If the demining group is directly involved with the design of equipment from the early stages it is less likely that there will be a reluctance on their part in accepting new technology. The following sections describe aspects of a robot that has been widely accepted by researchers who agree that robots can have a possible part to play in humanitarian demining.

1) Low cost, lightweight and high mobility: Cost and weight are extremely important for a robot deminer. The possibility of creating cheaper and smaller robots is not fiction anymore. At the University of Reading, researchers have built simple robot insects with a budget less than ε 50 [28]. The autonomous robots are capable of avoiding obstacles, wall following and switching to semi-autonomous behaviour.

2) Control and Communication: The principle requirement is that such a demining vehicle should operate in a remote control mode, or at least in semi-autonomous mode. The idea of using cooperative or distributed robot teams has been put forward by several researchers [20], [6]. The advantage in using a team of robots lies mainly in speed and effectiveness. In many complex tasks that can also be tedious for humans, a robot team can be more beneficial that a single robot. In the landmines case, this is not so straight forward. As the number of robots increase, so does the cost, weight and certainly the complexity.

Verification methods have to be more stringent as to see that no areas are overlooked by the robots. Also the movement of these robots must be studied thoroughly so that the most effective formation on the minefield is chosen: NOT the quickest. Inexpensive robots may well tend to use randomized search strategies [17] rather than coordinated search strategies for two reasons: (1) the increase in effectiveness provided by a coordinated search strategy decreases as the capability of the search sensor decreases; and (2) the cost of implementing the navigation capabilities necessary to support a coordinated search strategy may be prohibitive, relative to the cost of a less capable search element. However, careful analysis is required to select and implement an appropriate strategy that efficiently provides the required area coverage.

Within the context of cooperation, several control questions arise. Should a centralised method or decentralised be chosen? A centralised system would be slow (depending on the number of robots) and not very fault tolerant. On the other hand a decentralised system can be faster and more computationally efficient [29]-[31]. However, having a decentralised control systems means communication between the robots. This may require detection to flow between sensors and coordinator or the deminer controlling the robots. This will undoubtedly add noise in the data which should not be an extra problem in a situation where sensors are not a 100% accurate.

3) Sensor integration: The sensor is the most important aspect of a robot used in the detecting process. If the sensor fails or performs inadequately then there is no point to the robot being there. Hence great care must be taken to make sure that the sensors performance is not adversely affected by its carrier. Immediately there are several contradictions. The most popular form of advanced sensor in use is the GPR which can weigh around 20kg [32]. Therefore a robot carrying a GPR will not meet the light weight requirement.

An additional issue is the performance of the sensor in relation to the robots movement. In the case where several robots are working together, will it be possible to accurately calculate the exact area that has been tested, the area that has been overlapped by another robot and the area overlooked? In overlapping cases, how does this affect results? Will some sort of sensor fusion be required here? With these issues in mind, it may be easier to design both robot and sensor together.

4) All terrain robot: Although vital to the mine infested areas, this is an extremely difficult objective to accomplish while satisfying cost. The existing all-terrain robots like the MARS rover [33] are expensive. However, the Pemex BE [18] has contradicted this.

5) Simplicity, fault tolerance and easy maintenance: Simplicity is essential to be included for a robot on the minefield. Whether, semi-automatic, tele-operated or autonomous, operating the robot must not be difficult. Essentially the robot should be an extension or the deminer's metal detector or sniffer dog rather than a robot that needs extensive training to use. Fault tolerance and robustness are issue that contradict cost and weight. A robust robot is usually heavy and able to withstand the fragments of an exploding mine but this contradicts earlier weight requirement. Hence these issues need further investigation and consideration.

II. SUMMARY

This paper has surveyed the robots and search methods for landmine detection over the last decade. It has also stressed certain guidelines for future robot design. Most of the projects mentioned have not met these guidelines and hence a majority of them have not made it to the minefields for testing. It is not the technology we should blame: it is the delivery mechanism. Detection is an immensely difficult problem to which there is no quick, easy solution. The basic step - examining what the users need and what you have the ability to provide - is often overlooked in the eagerness to leap right into an exciting development program.

Landmines were recognised as a major humanitarian problem, and there was a build-up of funds and interest to combat them. Research organizations, many of which were already involved in related research, jumped on this wave of public opinion and found easy access to funds and thus vast amounts have been spent on technology development. A major part has also been wasted on unrealistic schemes where the lack of results could have been predicted and avoided. Thus they have not been accepted by the demining groups. Acceptance of new technologies are not only due to expense: reluctance to try a new technology is also a reason. The prospect of increased efficiency are not always enough to ensure the purchase and use of equipment. Hence, demining groups have to be brought in on the demining process right from the start.

So it remains that the most common and preferred detection method is manual probing, sniffer dogs and hand held detectors. Research in robot and landmine research has decreased in the last two years but ideas and projects have occasionally appeared. Although, these are not in strict adherence to the points provided here, they may shed light on other aspects of detection process. Several projects are being pursued by Navy and Military organisations [] but these as explained before cannot be used for humanitarian demining. However, this does not mean that robotics research for humanitarian demining should be abandoned - on the contrary - it should be continued with renewed effort. We may not find the miracle solution with a single robot or a group of robots but we will learn more along the way that will certainly help deminers in ridding the world of landmines.

REFERENCES

[1] "Anti-personal Mines: An Overview 1996", *International Commitee of the Red Cross*, Geneva, 1996.

[2] K.Eblagh " Practical Problems in Demining and their Solutions" , *Proceedings Eurel International Conference on The Detection of Abandoned Landmines*, Edinburgh, 1996, 1-5.

[3] K.Langer " A guide to Sensor Design for LandMine Detection" , in *Proceedings Eurel International Conference on The Detection of Abandoned Landmines*, Edinburgh, 1996, 30-32.

[4] B.Gros and C.Bruschini " Sensor Technologies for the Detection of Antipersonnel Mines: A Survey of Current Research and System Developments" , *Proceedings 6th International Symposium on Measurement and Control in Robotics*, Brussels, 1996, 564-569.

[5] J.Trevelyan , "Robots and Landmines", *Industrial Robot*, 24(2), 1997, pp114-125.

[6] R.Cassinis, G.Bianco, A.Cavagini, P.Ransenigo, " Landmines detection methods using swarms of simple robots" , in *6th International Conference on Intelligent Autonomous (IAS-6)*, JUL 25-27, 2000 INTELLIGENT AUTONOMOUS SYSTEMS 6, 212-218, 2000

[7] L.Dyck, " Claim and reality: mechanically assisted demining" , *Journal of Mine Action* 3(2), Summer 1999.

[8] R.Hess " Mechanical Assistance Systems for Humanitarian Mine & UXO clearance" , *Journal of Mine Action* 3(2), Summer 1999.

[9] M.K.Habib " Mine Clearance Techniques and Technologies for Effective Humanitarian Demining" , *Journal of Mine Action* 3(2), Summer 1999.

[10] C.Harwood, B.Howell, R.Keely, J.Richardier (Eds) " The use of Dogs for operations related to humanitarian mine clearance" , *Handicap International*, 1998.

[11] T.Williams, K.Dawson-Howe " Automated Force Sensed Probing for Buried Landmine Detection" , *(Technical Report, Dept. Computer Science, Trinity College, Dublin, Ireland, 1997*

[12] D.Antoni, I.Ratkovi, " Ground probing sensor for automated mine detection, (to be published)" , *KoREMA*, Opatija 1996.

[13] A.M.Shahri, F.Naghdy " Adaptive Fuzzy Force Control of An Anti-Personnel (AP) Mine Detector Robot" , in *Canadian Conference on Electrical and Computer Engineering*, 2001, Vol I & II p99-104.

[14] D.E.Franklin, A.B.Kahng, M.A.Lewis " Distributed sensing and probing with multiple search agents: toward system level landmine detection solutions" , in *Detection Technologies for Mines and Minelike Targets, Proceedings of SPIE*, Vol.2496.

[15] E.Colon, P.Alexandre, J.Weemals, I.Doroftei " Development of a high mobility wheeled robot for humanitarian mine clearance", in *SPIE Conference on Robotic and Semi-Robotic Ground Vehicle Technology*, Orlando, Florida, April 1998, SPIE Vol 3366.

[16] S.H.Salter, CNG Gibson " Map-Driven Platforms for Moving Sensors in Minefields" , *Journal of Mine Action* 3(2), Summer 1999.

[17] D.Gage " Many Robot MCM Search Systems" , *Proceedings of the Autonomous Vehicles in Mine Countermeasures Symposium*, Monterey CA, 4-7 April 1995.

[18] JD. Nicoud and P.Machler, " Demining Robots", *Intelligent Autonomous Systems*, 1995.

[19] Y.Baudoin, M.Acheroy, M.Piette, J.P.Salmon " Focus on Machine Assisted Demining", *Humanitarian Demining*, Summer 1999, Vol.3, No.2.

[20] P.K.Pook, S.J.Finney, K.Barrett, G.Whittinghill " Control of the FETCH team of robots", in *Mobile Robots XIII and Intelligent Transportation Systems PROCEEDINGS OF THE SOCIETY OF PHOTO-OPTICAL INSTRUMENTATION ENGINEERS (SPIE)*, 3525 25-31, 1998.

[21] H Uchida, K.Nonami " Quasi force control of mine detection six-legged robot COMET1 using attitude sensor" , *Climbing and Walking Robots*, 4th Int Conf. 2001, p979-988

[22] E.Gelenbe, Y.Cao " Autonomous Search for Mines" , in *SPIE Proceedings* Vol 3079, 691 - 703.

[23] Y.Zhang, M.J.Schervish, E.Acar, H.Choset " Probabilistic methods for robotic landmine search" , in *Mobile robots and telemanipulator and telepresence technologies*, VII 2001 p8-19

[24] B.Bauer, G.Cook " Algorithm and simulation development of autonomous robotics search" , in *Proceedings of the 23rd International Conference on Industrial Electronics, Control, and Instrumentation*, Vols. 1-4, 1278-1283, 1997

[25] Z.Butler, A.Rizzi, R.Hollis " Proc. Complete Rectilinear Environments" , *Algorithmic and computational robotics, new directions*, 2001, p51-61

[26] X.Yang, C.Kambhampati " Object Detection for Surface Landmine Detection" , in *IEEE International Conference on Systems, Man and Cybernetics*, SMC 2003, Tunisia.

[27] P.Fiorini, A.Fijany and A.Bejczy " Hierarchical Subsurface Imaging for Site Characterization Using Airborne and Vehicle Mounted Ground Penetrating Radar" , from *www.nasa.gov/people/fiorini*

[28] I.D.Kelly, D.A.Keating " Faster learning of control parameters through sharing experiences of autonomous mobile robots" , *International Journal of Systems Science* 29 (7):783-793 1998

[29] C.Kambhampati and S.Rajasekharan " Decomposed Modelling and Control of Multi-Robot systems" , *International Journal of Robotics*, 16 (4), 2001, 162-171.

[30] C.Kambhampati and S.Rajasekharan, " Multiple manipulator control from a human motor control perspective", *To be published in IEEE Transactions on Robotics and Automation*, June 2003.

[31] C.Kambhampati and S.Rajasekharan, " Multiple manipulator modelling from a human motor control perspective", *To be published in Biological Cybernetics*.

[32] J.W.Brooks " Applications of GPR Technology to Humanitarian Demining Operations in Cambodia: Some Lessons Learned" , *diwww.epfl.ch/lami/detec*, 1997.

[33] L. Matthies, E. Gat, R. Harrison, B. Wilcox, R. Volpe, and T. Litwin. " Mars microrover navigation: Performance evaluation and enhancement" , *Autonomous Robots*, 2(4):291–311, 1995.

Sensor-based motion planning for car-like mobile robots in unknown environments

Claudio Lanzoni
LAR-DEIS
University of Bologna
Italy
{cla_73@libero.it}

Abraham Sánchez and René Zapata
LIRMM - University Montpellier II
161, rue Ada 34392, Montpellier - France
{asanchez, zapata@lirmm.fr}

Abstract— This work deals with the sensor-based motion planning problem for car-like robots. Sensor-based versions of Lazy DRM and Lazy LRM are used to exploit the information obtained from sensors and to compute a feasible collision-free path. The algorithm tries to reach the goal, executing the local method in the known free region. If it succeeds, a path to the goal is found and the algorithm finishes. Otherwise, the algorithm executes more scans to extend its free space, an so on.

We have performed some simulations that show the promise of our approach.

I. INTRODUCTION

Various methods have been developed in both motion planning (MP) and sensor-based motion planning (SBMP) in robotics. Nevertheless, most researches have addressed these problems separately. Motion planning methods assume complete knowledge of both the robot and the environment. These methods build some paths (set of sub-goals) which are free of obstacles. Its main advantage is to prove the existence of a solution which permits the robot to reach its destination and to generate collision-free map-making. However, they have some well-known drawbacks. For example, an exact model of the environment is needed which unfortunately cannot be defined in most applications. Also, it is difficult to handle correctly dynamic modifications of the environment due to the addition of objects and to the presence of obstacles in motion.

Sensor-based motion planning incorporates sensors information, reflecting the current state of the environment, into a robot's planning process, as opposed to classical motion planning. Local methods are mainly used in these approaches. They could be called *reactive strategies* and are completely based on sensory information. Therefore, an absolute localization is not requisite and only the relative interactions between the robot and the environment have to be assessed. In these circumstances, a structural modelling of the environment is unnecessary, but the robot has to acquire through its sensory inputs a set of stimulus- response mechanisms. In this scheme, the robot is generally expected to carry out simple tasks. They do not guarantee a solution for the mission because of the occurrence of deadlock problems. For more efficiency and safety, perception tools have to be increased (several types of sensors, including, e.g., cameras) to get more pertinent information about the environment. But then it is not easy process the data under real time constraints. These constraints often lead to a degradation of the accuracy and the richness of the information.

We propose an approach for solving the motion planning problem for car-like robots operating in an unknown environment containing obstacles of arbitrary shapes. The framework for our planning approach is inspired by a recent variant of PRM, which solves motion planning problems, called Lazy PRM (Lazy Probabilistic Roadmap Method) [4]. The Lazy PRM approach builds a roadmap, where all nodes and edges are assumed to be collision-free and searches it at hand for a shortest path between robot's initial and goal configurations. Experiments show that in many cases, only a very small fraction must be explored to find a feasible path.

In a recent work [11], experimental and theoretical results prove that the use of deterministic sampling offers advantages in comparison to the traditional random sampling in PRM context. In [17], we have extended this approach for car-like robots. Experimental results show a better coverage of configuration space for both versions, multiple query and single query. Additionally, we have proposed a version of lazy LRM (using Sukharev grid); with this version we obtained the smaller preprocessing time compared to other lazy versions [17]. We strongly suggest the use of low discrepancy point sets to build the lazy roadmap (e.g., Halton/Hammersley, Faure, Sobol, Sukharev grid).

Since lazy PRM approach does not consider information about the obstacles, we suppose that the start and goal configurations are known to be collision-free. Using an A^* strategy and sensor information, the algorithm computes a feasible path to allow the robot to reach the goal configuration. Planning as it senses, the planner needs from 5 to 30 scans (depending on the scene complexity), while avoiding collisions with the obstacles.

II. RELATED WORK

A. Model-based motion planning

Motion planning for high dimensional configuration space has been very successful with randomized approaches [13], [7], [2], [6], [10]. This set of algorithms capture the connectivity of robot's free C-space in a finite graph structure, such as Ariadne's Clew Algorithm (ACA) [3], Probabilistic Roadmap Methods (PRM) [13], [7] and Rapidly-Exploring Random

Trees (RRT) [10]. In the graph structure, each node represents a free robot configuration, and an edge between two nodes represents that a simple local method can find a collision-free path between the corresponding robot configurations. These algorithms adopt different strategies for placing configurations (nodes) and for connecting them (for example, OBPRM [2], Hsu's algorithm [6]).

Computational complexity of motion planning is known to be exponential to the degrees of freedom of the robot. The graph (or tree) in these roadmap-based approaches is constructed without explicitly representing the obstacles of the configuration space, and is therefore particularly useful as a representation for high dimensional configuration spaces.

B. Sensor-based motion planning

Sensor-based navigation enables a robot to explore an unknown environment, with an assumption of simple and weak sensors. The algorithm proposed by Lumelsky ("Bug algorithm") is one of the most famous [12]. Choset and Burdick present in [5] the hierarchical generalized Voronoï graph (HGVG) which is a roadmap that serves as a basis for sensor based robot motion planning. Authors only focus in the roadmap construction.

An another roadmap approach proposed by Yu and Gupta [21], is used for solving sensor-based collision-free motion planning for articulated robots arms. Their approach incrementally builds a roadmap that represents the connectivity of free C-space, as it senses the physical environment.

Another work presents a variant of generalized Voronoï graph (GVG) [14], authors propose a local smooth path planning algorithm for car-like robots. Basically, an initial path is generated by conventional path planning algorithm using GVG theory, and it is deformed smoothly to enable car-like robot's. Unfortunately, its solution is not simple and the computational cost becomes much higher than the conventional GVG approach.

An other sensor-based approach has assumed abstract sensors that provide distances in C-space is presented in [16].

A sensor-based version of the Ariadne's Clew Algorithm is proposed in [1] to incrementally search for the free space and compute a path to a goal configuration to a 2-link planar robot. The authors assume that the knowledge of workspace is partially known and that it can be increased by using a laser-like sensor.

Roadmaps have several properties: *connectivity*, *accessibility*, and *departability*. These properties imply that the planner can construct a path between any two points in a connected component of the robot's free space by first finding a collision-free path onto the roadmap (accesibility), traversing the roadmap to the vicinity of the goal (connectivity), and then constructing a collision-free path from a point on the roadmap to the goal (departability).

III. THE SB-LAZY DRM ALGORITHM

The robot, equipped with one or more sensors (assume distance or range sensor), is required to plan and execute collision-free motions in an environment initially unknown to the robot. This lack of a priori knowledge (at least complete knowledge) about the environment is the fundamental difference between sensor-based motion planning and classical motion planning. The first approach is, therefore, not an off-line process as in the case in classical motion planning problem. Instead, motion is generated step by step while more and more knowledge about the environment is accumulated incrementally. This changes the nature of the problem in a fundamental way. Notice that the planning space (the C-space) and the sensor space (physical environment) are very different. This is a crucial distinction from the assumption in most of the current approaches to sensor-based planning for mobile robots; in order to simplify the planning problem, they all treated the mobile robot as a single point (or a circular cross section) in 2D or 3D physical space. The configuration spaces for these mobile robots are basically the same as the physical space.

A. Definitions and notation

Configuration space, $\mathcal{CS}$, is the n-dimensional space which parameterizes the robot's degrees of freedom. The set of configurations at which the robot intersects some obstacle in the workspace is called $\mathcal{CS}_{obs}$. Let q be a point in C-space. Correspondingly, $\mathcal{CS}_{free}$ is the free C-space and $\mathcal{P}_{free}$ is the free physical space. $\mathcal{A}$ represents the robot. $\mathcal{A}(q) \subset \mathcal{P}$ denotes the physical space occupied by the robot at configuration q.

We can assume that a telemetric rotary range sensor, mounted on the robot, provides the distances information when required by the motion planner (see Figure 1). The additional increment in free physi-cal space due to a scan of physical space is noted $\triangle \mathcal{P}_{free}$. We assume that a routine ENVIRONMENT_SCAN returns this $\triangle \mathcal{P}_{free}$. Corresponding to $\triangle \mathcal{P}_{free}$, $\mathcal{CS}_{free}$ is augmented by $\triangle \mathcal{CS}_{free}$. Formally, let $\mathcal{CS}_{free} = \mathcal{M}(\mathcal{P}_{free})$, where $\mathcal{M}$ is the mapping from $\mathcal{P}_{free}$ to $\mathcal{CS}_{free}$, then $\triangle \mathcal{CS}_{free} = \mathcal{M}(\mathcal{P}_{free} \cup \triangle \mathcal{P}_{free}) - \mathcal{M}(\mathcal{P}_{free})$, where $-$ denotes set difference.

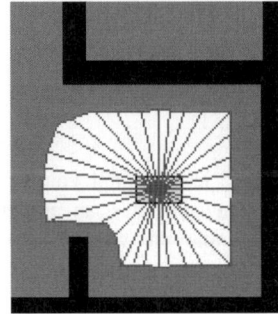

Fig. 1. An illustration of a scanned region.

B. Planning process

The initial and final robot configurations are nodes in the roadmap. $\mathcal{P}_{free}$ is initially assumed to be a small region that surrounds the robot for initial movement. $\mathcal{CS}_{free}$ is $\mathcal{M}(\mathcal{P}_{free})$, although we do not explicitly calculate it. At each iteration, the robot takes a set of measured distances. Distance to obstacles is performed with the algorithm proposed in [19], [20].

The planner attempts to reach the goal, executing the local planner (Reeds & Shepp paths [15]) in the known free region. If it succeeds, a path to the goal is found and the algorithm finishes. Otherwise, the algorithm executes more scans to extend its known space, and so on. The entire process is repeated until either the goal is found or no path can be found with current constraints. It is not necessary to set a maximum amount of time to avoid loops because, after exploring all the reachable positions in the roadmap without success, the robot gets back to start configuration. The process is shown in Figure 2.

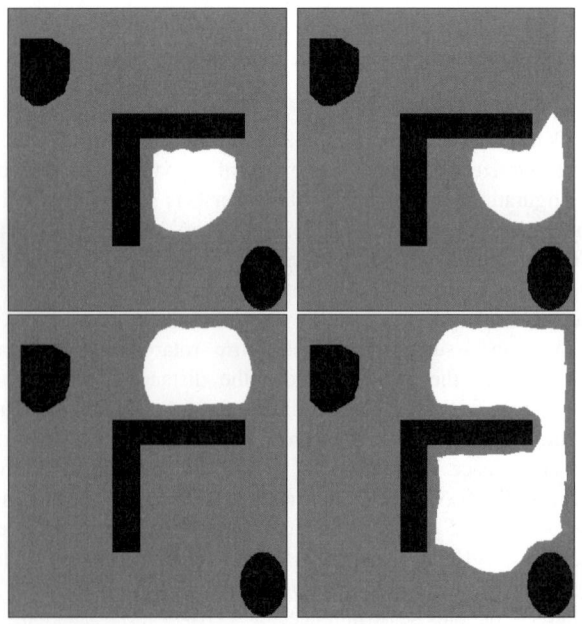

Fig. 2. A sequence of snapshots showing sensor-based planning task. Black regions correspond to obstacles, gray region represents the unknown space and the white one is the scanned region.

C. Algorithm

The robot starts off in an initial configuration ($q_{start} \in \mathcal{P}_{free}$). The algorithm generates an initial lazy roadmap in the attempt to sample the workspace via the function ROADMAP_INIT. Deterministic sampling is strongly recommended as it always gave best results in terms of workspace coverage [17]. The key step is to guide the research in the graph via the sensor information, using an A^* strategy and step by step update the graph by "eliminating edges" that would lead to a colliding configuration (function ROADMAP_UPDATE). At any given iteration i, the planner attempts to reach the closest node to goal configuration via a path generated by a local planner. To do so, an ENVIRONMENT_SCAN must be performed to ensure that both chosen node and path are on a local collision-free area; the condition assures that the chosen configuration can be reached in safety, without the risk to collide with the obs-tacles. The function ROBOT_MOVE generates control data to make the robot follow the generated path. We denote $\mathcal{CS}_{local\_free}$, the local collision-free area and the function that generates it from sensor information in current robot configuration q_i is ENVIRONMENT_SCAN. The function ROBOT_REACH_GOAL(q_i) returns TRUE if the current robot's configuration is the goal configuration. A key feature of the algorithm is the function ROBOT_STEP_BACK(q_i): it returns the configuration to reach in case no further progress toward the goal is allowed from the current one. That gives to the algorithm the powerful capability to exit from a local minimum, making it possible to find another path from one of previously visited nodes.

The algorithm for sensor-based is described as follows:

SB-LAZY-DRM
$\mathcal{G} \leftarrow$ ROADMAP_INIT
$q_0 \leftarrow q_{start}$
$i \leftarrow 1$
While ROBOT_REACH_GOAL = FALSE **do**
  $\mathcal{G}_c(q_i) \leftarrow$ set of connected configurations to q_i, from $\mathcal{G}$
  $\mathcal{CS}_{local\_free}(i) \leftarrow$ ENVIRONMENT_SCAN(q_i)
  $\mathcal{G}_{c\_free}(i) \leftarrow$ set of configurations $q_j \in \mathcal{CS}_{local\_free}(i)$, from $\mathcal{G}_c(q_i)$
  $\mathcal{G} \leftarrow$ ROADMAP_UPDATE
  if $\mathcal{G}_{c\_free}(i) = \emptyset \wedge q_i = q_{start}$ **then**
    return **FAILURE** and exit
  else
    if $\mathcal{G}_{c\_free}(i) = \emptyset \wedge q_i <> q_{start}$ **then**
      $q \leftarrow$ ROBOT_STEP_BACK(q_i)
    else
      $q \leftarrow$ closest configuration to q_{goal} from $\mathcal{G}_{c\_free}(i)$
  $q_i \leftarrow q$
  ROBOT_MOVE(i)
  $i \leftarrow i + 1$
End

IV. IMPLEMENTING THE ALGORITHM

The algorithm was thought to be as general as possible, with the intent to maintain the highest level of flexibility. There are no limits to the type, the number or the range of the sensors mounted on the robot. However, simulations were done supposing a 360-degrees rotational telemetric sensor. A robot with less powerful sensing capabilities could require more scans to find a solution.

The two most powerful features of our approach are the capabilities to escape local minima and to exit with failure if no path can be found with current constraints and they both rely on the function ROBOT_STEP_BACK. For it to properly work the path should be stored in a stack-like data structure, allowing the algorithm to find the correct node in the roadmap to move to in case no further progress toward goal configuration is allowed. To avoid the function to be called in case of narrow passages, it can be added an euristic to perform a node-enhancement procedure to the roadmap in some cases and add its call inside of the procedure ROADMAP_UPDATE.

It is important that, after a ROBOT_STEP_BACK call, the roadmap's edge swept by the robot should be added to a prohibited edges' list. This will prevent the algorithm to loop in the attempt to reach the most promising node in the graph while allowing the robot to step back to the initial configuration after the whole graph has been explored. These features gave, basically, the reasons to implement a roadmap-based sensor-based motion planning algorithm.

V. EXPERIMENTAL RESULTS

In order to illustrate our approach and to show the robustness of the method, we apply it to several problems. Figure 3 illustrates the scenes used in our experiments. We have used Hammersley/Halton (H/H), Faure (F), Sobol (S) and Sukharev grid (SG) points as inputs to lazy DRM and lazy LRM, respectively.

Figure 4 shows a scene that contains obstacles of arbitrary shape, in the left picture the scanned region is shown and the right one the computed path. Table I-III summarizes the results with different deterministic sampling strategies for the different scenes. In figure 5, the paths computed for scenes 2 and 3 are shown.

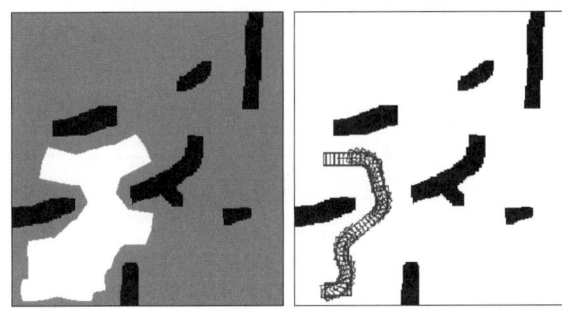

Fig. 4. Scene 1 illustrates the scanned region and the computed path.

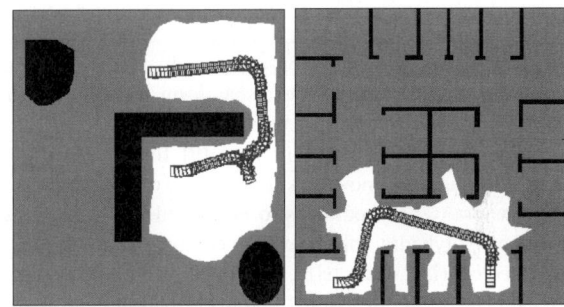

Fig. 5. Paths computed for scenes 2 and 3.

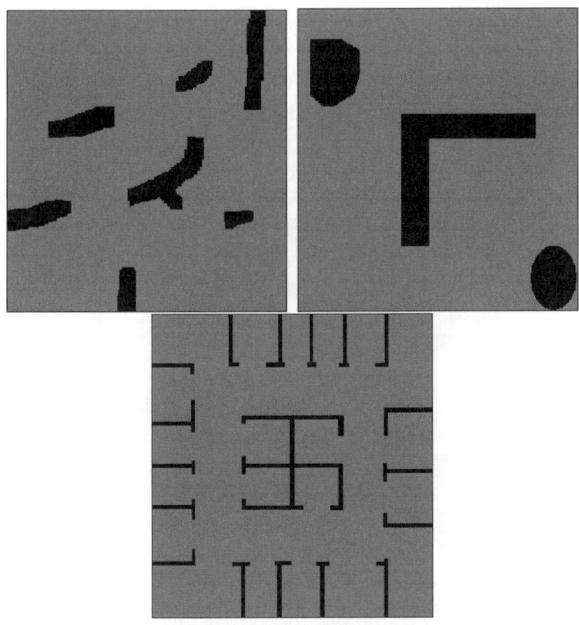

Fig. 3. Three different environments used to test our approach.

A series of experiments showed that the SB-Lazy-DRM algorithm works efficiently and is able to find a path in fairly complex unknown environments. The run time for one iteration varies, depending on the number of generated nodes and on the complexity of the physical space. The algorithm is executed on a Pentium III 866 computer. Planner was implemented in Matlab, using partially C.

Several examples are shown in Figure 6, each path and its corresponding scanned region are included.

VI. CHARACTERIZING OUR ALGORITHM

A practical robotic application involves moving a robot amidst obstacles, without colliding with them. This problem

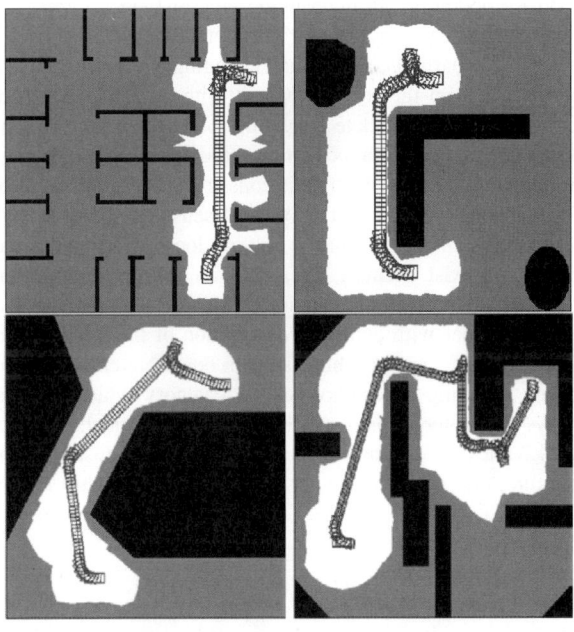

Fig. 6. We present different examples that prove the robustness of our approach. The figures at top correspond to scenes 1 and 2, whereas the one lower contains a large corridor.

TABLE I
PERFORMANCE DATA FOR THE SCENE 1.

Sampling method	number of scans	run time	number of nodes
SG	7	23.21	81
H/H	17	43.24	100
F	8	25.00	100
S	6	17.26	120

TABLE II
PERFORMANCE DATA FOR THE SCENE 2.

Sampling method	number of scans	run time	number of nodes
SG	20	75.70	64
H/H	27	89.96	100
F	26	86.84	100
S	13	39.70	130

of obstacle avoidance has been extensively studied and has generated a vast literature. The research on motion planning in the presence of obstacles can be classified into two large categories. In the first category (see [9] for a survey), for example, popularly known as Piano Mover's problem, an object (generally polyhedral) is to be moved in a scene filled with other polyhedral obstacles. A complete knowledge of the obstacles (i.e., their size, position, orientation, etc) is assumed. The scene, or a suitable transformation of it, is then divided into allowed and forbidden regions for the object's movement. A path is generated through a graph search, in a graph formed with these regions. This one-time, off line and computationally very intensive procedure is most suitable for repetitive operations in unchanging environment.

The second category deals with robot applications involving unknown (or partially known) and changing environments. The sensors fitted on the robot supply it with information on the obstacles in the immediate neighborhood of the robot. This information can only be a local or a partial description of robot's environment. The robot's path is cons-tructed on-line, point by point, using at any instant only local sensory information of the environment. A path based only on local information may not always exist because the global requirement of moving a robot between two given points generally requires a global motion planner with complete description of the environment (i.e., the first category). This results in one more classification. A path planning scheme of second category which can not always guarantee a path can be called, heuristical. Most of the schemes of second category has been proposed in the literature, one of most popular is presented by Khatib in [8].

TABLE III
PERFORMANCE DATA FOR THE SCENE 3.

Sampling method	number of scans	run time	number of nodes
SG	8	24.99	100
H/H	11	37.47	200
F	9	27.09	200
S	8	25.83	220

A motion planning scheme of second category which can always guarantee a path, when there exists one, can be called non-heuristical or provable.

The algorithm proposed by Lumelsky and Stepanov [12] is non-heuristical. This algorithm is based on the Jordan Curve Theorem, and cannot generally be applied to environments of dimension greater than two.

Roughly speaking, the reasons why sensor-based algorithms can keep the convergence but the artificial potential algorithm cannot maintain it are as follows:

1) The "going" and "avoiding" behaviors are alternatively switched in the sensor-based approach, on the other hand, they are mixed each other in the artificial potential algorithm,
2) A mobile robot always take notice of only one obstacle to avoid in the sensor-based approach, on the other hand, a mobile robot pays attention to multiple neighbor obstacles simultaneously to avoid the artificial potential algorithm.

Due to these differences, although the sensor-based approach obtains the convergence, the artificial potential loses it. In the following Figure 7 we showed an example of an environment that contains local minima, our algorithm solves the problem; however, the use of an artificial potential strategy would require some heuristic to escape local minima.

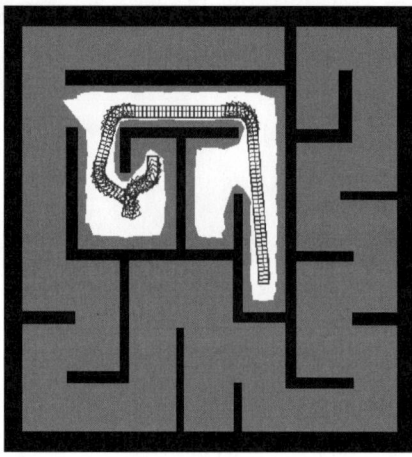

Fig. 7. The scene contains local minima (or narrow passages), our approach solves the problem correctly. We observed, that the algorithm executed more scannings.

VII. DISCUSSION

An important goal of robotics research is the development of systems which encompass versatile, "intelligent" robot behavior. Obviously, such systems require extensive knowledge of the robot's environment. For example, planning of collision-free paths - a rather fundamental task - needs information about obstacles' size and position. In complicated, cluttered environments it is often impossible, or, at least, extremely difficult, to provide a world model *a priori*. Thus, the robot should create the world model automatically performing sensing actions.

Assuming that the environment does not vary with time and that initially the worlds are completely unknown, robot motions and sensor measurements are planned to iteratively explore the surroundings. A representation of the workspace is created incrementally, integrating the collected sensor data.

We characterize two types of robot movement in sensor-based motion planning. The first types, called *Pmove*, corresponds to a physical motion of the robot from one configuration to another one and it is executed to carry the robot to a new scanning position. Pmove, therefore, expands the known physical space and C-spaces. The cost of a Pmove is not only the graph search but also the physical movement of the robot.

The second type is called *CSmove* and is executed by the planner to search free C-space. Whenever the local planner is called, it is as if the planner was simulating a robot movement. The cost of CSmove is essentially the computational cost of the local planner which is mainly the cost of collision detection for a sequence of configurations. Normally, the cost of CSmove would be higher in computation time. An objective for a sensor-based planner could, for example, be to reduce the number of CSmove.

VIII. CONCLUSIONS AND FUTURE WORK

We presented and implemented an approach to sensor-based motion planning for car-like robots. It consists in incrementally search into the roadmap as the sensor sense new free region in physical space. A series of experiments showed that SB-Lazy-DRM algorithm works efficiently and that it is able to find a path even in complex environments. The necessary and sufficient criteria for completeness of SB-Lazy-DRM remains an open problem. Our conjecture, that SB-Lazy-DRM is resolution complete, needs to be proven.

Like all randomized motion planning algorithms, the algorithm presented in [21] is probabilistically complete. Nevertheless, as more configurations are generated, the probability of finding path increases. For instance, SB-LAZY-DRM planner presented in this paper, chooses the next scan in somewhat heuristic manner. Future work includes the development of theoretical framework to choose the next scans. Our work also has some similarities with the work presented in [18]. We intend to apply the approach to a real robot.

We will extend the approach proposed in [21]. Initially, a probabilistic roadmap of the known portion of workspace can be built in the standard manner. Then, guided by some heuristics, a node on this roadmap can be selected to which, subsequently, the robot moves. Then, again, the robot can sense the environment and extend the probabilistic roadmap in those portions of its environment that were previously unknown to it. This process can be repeated until the robot's environment is well mapped.

ACKNOWLEDGMENT

Claudio Lanzoni is supported by LAR-DEIS, University of Bologna, Italy. Abraham Sánchez is on leave from Computer Science Dept. University of Puebla-México, and is supported by PROMEP-BUAP -2000-19 (PROMEP/BUAP-447). René Zapata is partially supported by Action de Collaboration France-México (ECOS-Nord, Action M01 M01).

REFERENCES

[1] J. M. Ahuactzin and A. Portilla. A basic algorithm and data structures for sensor-based path planning in unknown environments, *Proc. of the IEEE International Conference on Intelligent Robots and Systems*, 2000.

[2] N. M. Amato, O. Bayazit, L. K. Dale, C. Jones, and D. Vallejo. OBPRM: An obstacle-based PRM for 3D workspaces, *Workshop on the Algorithmic Foundations of Robotics*, pp. 155-168, 1998.

[3] P. Bessière, E. Mazer, and J. M. Ahuactzin. The Ariadne's clew algorithm: Global planning with local methods, *Workshop on the Algorithmic Foundations of Robotics*, 1994.

[4] R. Bohlin and L. Kavraki. Path planning using lazy PRM, *Proc. of the IEEE Robotics and Automation Conference*, pp. 521-528, April 2000.

[5] H. Choset and J. Burdick. Sensor based motion planning: The hierarchical generalized voronoï graph, *Workshop on Algorithmic Foundations of Robotics*, J-P. Laumond and M. Overmars (editors), A. K. Peters, 1996.

[6] D. Hsu, L. E. Kavraki, J-C. Latombe, R. Motwani, and S. Sorkin. On finding narrow passages with probabilistic roadmap planners, *Workshop on the Algorithmic Foundations of Robotics*, pp. 141-154, 1998.

[7] L. E. Kavraki, P. Švetska, J-C. Latombe, and M. H. Overmars. Probabilistic roadmaps for path planning in high-dimensional configuration spaces, *IEEE Transactions on Robotics and Automation*, Vol. 12, No. 4, pp. 566-580, August 1996.

[8] O. Khatib. Real-time obstacle avoidance for manipulators and mobile robots, *The International Journal of Robotics Research*, Vol. 5, No. 1, pp. 90-98, 1986.

[9] J. C. Latombe. *Robot motion planning*, Kluwer Academic Publishers, 1991.

[10] S. M. LaValle and J. J. Kuffner. Randomized kinodynamic planning, *Proc. of the IEEE Robotics and Automation Conference*, pp. 473-479, 1999.

[11] S. M. LaValle and M. S. Branicky. On the relationship between classical grid search and probabilistic roadmaps. *Proc. Workshop on the Algorithmic Foundations of Robotics*, december 2002.

[12] V. Lumelsky and A. Stepanov. Path planning strategies for a point mobile automaton moving amongst unknown obstacles of arbitrary shape, *Algorithmica*, Vol 3., No. 4, pp. 403-430, 1987.

[13] M. H. Overmars and P. Švestka. A probabilistic learning approach to motion planniing, *Algorithmic Foundations of Robotics*, A. K. Peters, Ltd., pp. 19-37, 1995.

[14] K. Nagatani, I. Iwai and Y. Tanaka. Sensor based navigation for car-like mobile robots using generalized voronoï graph, *Proc. of the IEEE International Conference on Intelligent Robots and Systems*, 2001.

[15] J. A. Reeds and R. A. Shepp. Optimal paths for a car that goes both forward and backwards, *Pacific Journal of Mathematics*, Vol. 145, No. 2, pp. 367-393, 1990.

[16] E. Rimon and J. Canny. Construction of c-space roadmaps from local sensory data. What should the sensors look for?, *Proc. of the IEEE Robotics and Automation Conference*, pp. 117-123, 1994.

[17] A. Sánchez, R. Zapata and C. Lanzoni. On the use of low discrepancy sequences in nonholonomic motion planning, *To Appear in ICRA 2003*.

[18] A. Stentz. The focussed D^* algorithm for real-time replanning, *Proc. of the International Joint Conference on Artificial Intelligence*, August 1995.

[19] M. Venditelli and J-P. Laumond. Visible positions for a car-like robot amidst obstacles, *Workshop on Algorithmic Foundations of Robotics*, J-P. Laumond and M. Overmars (editors), A. K. Peters, pp. 213-227, 1996.

[20] M. Venditelli, J-P. Laumond and C. Nissoux. Obstacle distance for car-like robot, *IEEE Transactions on Robotics and Automation*, Vol. 15, No. 4, pp. 678-691, 1999.

[21] Y. Yu and K. Gupta. Sensor-based probabilistic roadmaps: Experiments with and eye-in-hand system, *Journal of Advanced Robotics*, 2000.

ns of the 2003 IEEE
Heterogeneous Implementation of an Adaptive Robotic Sensing Team[*]

Bradley Kratochvil[†], Ian T. Burt[‡], Andrew Drenner[†], Derek Goerke[†], Bennett Jackson[†],
Colin McMillen[†], Christopher Olson[†], Nikolaos Papanikolopoulos[†§], Adam Pfeifer[†],
Sascha A. Stoeter[†], Kristen Stubbs[†], David Waletzko[‡]

Center for Distributed Robotics, Department of Computer Science and Engineering
University of Minnesota, Minneapolis, U.S.A.

Abstract: *When designing a mobile robotic team, an engineer is faced with many design choices. This paper discusses the design of a team consisting of two different models of robots with significantly different sensing and control capabilities intended to accomplish a similar task. Two new robotic platforms, the COTS Scout and the MegaScout are described along with their respective design considerations.*

1 Introduction

In 2001, the University of Minnesota developed its second-generation Scout robots [4]. These small 40 mm × 110 mm robots are intended to function as an information gathering team in situations potentially dangerous for humans. The Scouts contain sensing equipment such as cameras and small low-power microprocessors for control. These robots are designed to fit a very small form factor while containing as much sensing equipment and computational power as possible. This led to a complex and almost entirely custom design, which creates difficulties when trying to upgrade or mass-produce the devices. Due to the Scouts' small size and computing power, another agent is intended to accompany them into the field for deployment and supervision. Initially, this role was filled by a modified ATRV-JR known as the Ranger. Although the Ranger is a platform intended to transport the Scouts into position, it is not human transportable itself. In an effort to address these issues in a robotic team, two new robotic platforms have been developed.

---

[*]This material is based upon work supported by National Science Foundation through grant #EIA-0224363, the Defense Advanced Research Projects Agency, Microsystems Technology Office (Distributed Robotics), ARPA Order No. G155, Program Code No. 8H20, issued by DARPA/CMD under Contract #MDA972-98-C-0008, and Microsoft Corporation.
[†]Center for Distributed Robotics and Dept. of Computer Science and Engineering, University of Minnesota.
[‡]Dept. of Mechanical Engineering, University of Minnesota.
[§]Corresponding Author.

The Commercial-Off-The-Shelf (COTS) Scout, detailed in Section 2, has been developed to address the issues of small form factor and ease of deployment. The main focus of this platform is to reduce the cost and complexity of the device by limiting it to only a mobile camera platform with no processing power. This has led to a low-cost teleoperable mobile robotic platform. Since this platform is limited to teleoperation, another entity, human or robotic, must accompany it during deployment. For the case of a robotic command-and-control center, the MegaScout platform has been developed. The MegaScout, described in Section 3, is a larger, more powerful sensor platform that can either perform missions on its own or in teams consisting of COTS Scouts or other MegaScout robots. The platform is intended to combine the command and control nature of the Ranger with the portability and compact nature of the Scout. Although these platforms are still in their final fabrication stages, Section 5 describes future enhancements that expand the platforms' mobility, durability, and sensing.

2 Related Work

One of the problems when applying a distributed robotic team to a variety of tasks in hazardous environments is maintaining effective communication links which keep operators aware of the situation the robot is in. Projects such as [5] have worked to create a relay network in which each robot serves both as a node of the network and is capable of performing local sensing tasks. Such networks are crucial for tasks in Urban Search And Rescue (USAR) environments where relocating an operator to regain wireless communication is not permitted [1].

Designing robust robotic platforms for operation outside of the laboratory is a difficult task. In order to function in a variety of environments, researchers have looked into making easily reconfigurable or self-reconfiguring robots such as the PolyBot [9] and the

Figure 1: A prototype COTS Scout in an outdoor setting.

CONRO [2] for increased locomotion capability. Supporting the reconfiguration of sensors [6] also allows for more useful real world applications. Another example of creating hybrid heterogeneous robotic teams to accommodate problems associated with limited locomotion and sensing capabilities of a single platform are the Packbot and Throwbot discussed in [7].

3 COTS Scout

As its name implies, the COTS Scout, shown in Figure 1, is designed to incorporate as much standard hardware as possible. This approach is motivated by the high cost to reproduce the original Scout robots. While the original Scout prototypes cost on the order of thousands of dollars, the COTS Scout prototypes cost under $300. This has largely been made possible by using a significantly different approach to the robot design. The original Scout is based around a digital system. It has digital command communications and micro-controllers to handle its operation. Although this allows for a more sophisticated robot and more possible uses, it adds a great deal of complexity. The COTS Scout is designed to be as simple as possible while meeting its design objectives that are:

- Low construction cost,
- Large communication range,
- Ease of maintenance,
- Low operating cost.

To meet these objectives, the COTS Scout is designed around equipment available for radio controlled vehicles. The main shell of the COTS Scout is a 44 mm × 97 mm long cylinder. All supporting hardware is mounted to this shell for structural support. A jumping foot has been omitted from this design because of the increase in production costs, but multiple wheel sizes allow the COTS Scout to traverse a variety of terrains. The current wheels consist of 57 mm, 64 mm, and 76 mm discs. Table 1 shows experimental data taken from the COTS Scouts operating on a variety of terrains.

The main command receiver for the COTS Scout has the capability of using any of 50 different crystals enabling an equal number of command frequencies. The command radio is able to receive commands from approximately 45 m indoors and 90 m line-of-sight. The video transmission system is based around a FM modulator that is capable of three different channels. Since the COTS Scouts are capable of using many more control channels, hardware was added to allow each Scout to turn its video transmitter off or on. This allows multiple COTS Scouts to be used in the same area. The video transmitter can broadcast approximately 15 m indoor and 18 m line-of-sight using standard antennas. Due to the rugged construction of the COTS Scout, it can withstand the impact of being thrown more than 15 m horizontally onto a hard surface.

4 MegaScout

The MegaScout, Figure 2, is designed to support a Scout robotic team as well as perform missions independently. When supporting a Scout mission, the MegaScout can be equipped with command transmitters and video receivers to communicate with a host of Scouts. It can then relay this information back to a base station, effectively increasing the range of a Scout team. If communication back to the base station should fail, the MegaScout is also equipped with sufficient computing power to control the Scout team. When supporting a Scout mission, the MegaScout can be equipped with an array of sensors that would otherwise be too large to be deployed on the Scouts. In order to accomplish this, work has focused on developing a robust robotic platform that can easily be modified

Terrain	*Wheel diameter*		
	57 mm	64 mm	76 mm
5 cm tall grass	0 m/s	0.03 m/s	0.04 m/s
Small debris	0 m/s	0.17 m/s	0.40 m/s
Concrete	0.39 m/s	0.42 m/s	0.50 m/s
Max. incline	35°	35°	28°

Table 1: COTS Scout performance characteristics.

and will operate in a variety of environments.

During the initial design phases of the MegaScout several capabilities were specified, and the robot was designed around them:

- Software Design: A robust, stable architecture would be needed to support the platform. The architecture should provide the developer with a sufficiently abstract interface, so that hardware can be changed or added with little or no change to the high level functions.

- Electronic Design: The electronics would need to be designed to support a variety of different sensors. The device needs a 32-bit central processor. The systems would be robust enough that software errors should not damage them.

- Mechanical Design: The robot would need to be able to withstand the impact of being thrown or dropped. It would need to operate in a wide variety of harsh terrains and be modular to support future upgrades.

The resulting robot is the MegaScout. Discussion of this platform will be divided into the above three areas: software, electronics, and mechanical hardware.

4.1 Software

Since the MegaScouts are intended to operate as part of a heterogeneous robot team, a very generic software API was defined. This allows command and control software to treat all individual robots as uniformly as possible. In the context of this API, a robot is modeled as a hierarchical collection of sensor and actuator objects, whose interfaces represent general types of robotic parts (motors, encoders, cameras, range finders, etc.) and whose implementations are left undefined. Concrete packages have been written to fill in all the implementation details necessary for the operation of the equipment on each MegaScout. A programmer writing an implementation for another type of robot can pick and choose from these packages or write new packages whose objects comply with the same interfaces.

The exact composition of each robot is defined by an XML file that enumerates its sensory equipment and its mechanical actuators along with any configuration data needed for each part. A robot object is then assembled dynamically at runtime by parsing the XML description into a hierarchy of objects drawn from the selected component packages. This structure makes it easy to support the robot's field-swappable sensory equipment without the need to recompile any code,

Figure 2: A prototype standard equipped MegaScout.

and allows each robot to maintain separate calibration information to account for variations between sensors.

The MegaScout runs White Dwarf embedded Linux on its main processor. Using Linux simplifies building the robot from relatively low cost commodity hardware while at the same time providing easy remote access to the system via SSH over a wireless network. It also makes it possible to customize the operating system very precisely to fit the limited disk and CPU resources available onboard. The control software is written in C++, which provides a rich object-oriented environment and fast, native executables. A large selection of compatible utility libraries is available.

4.2 Electronics

As mentioned earlier, much of the design for the MegaScout robot revolved around the electronics component selection. Early in the design process, a PC104+ stack was selected as the main processing module. The advantages of using a standard central processing board are the ease of upgradeability, little need for hardware customization, relatively small form factor, and industry standard parts.

By limiting the amount of electronics that our engineers create, the entire design of the platform can be accomplished more rapidly allowing more time to be spent on developing applications. The PC104+ stack chosen for the MegaScout has a 166 MHz Pentium processor, 128 MB of RAM, and a 128 MB Flash disk.

Although the Pentium chip offers a high amount of processing power, it is known not to be as power efficient as some of the other embedded processors (StrongARM, Crusoe, Geode, etc.). These new processors

Figure 3: The MegaScout equipped with balloon tires.

could offer potential savings in battery power, but were not chosen for this design due to their relatively recent introduction into the market and reduced selection of board features. Along with the main processing module, the PC104+ stack contains a 4-channel frame grabber and a PC Card adapter.

Due to the rapid increases in communications technology, it was deemed necessary for the robots communication hardware to be easily upgradeable. The PC Card module contains an IEEE 802.11b card, which serves as the main communication module for the robot. A PC Card was chosen as an interface to allow the robot to easily use different communication hardware such as CDPD cellular cards. With the current wireless hardware in the robot, it is able to communicate approximately to 100 m indoors and 600 m outdoors. The robot is equipped with a switched diversity gain antenna array for the wireless communication module, which improves the communications of the system by reducing fading regions [8]. For analog video, the initial prototypes of the MegaScouts are equipped with the same FM 900 MHz transmitters as the COTS Scout. This transmitter is included to provide a control station with high quality video to be used when the robot is moving at high speeds.

The included framegrabber allows the MegaScout to process analog video onboard as opposed to the original Scout system where all video was processed remotely. This greatly increases the robot's ability to perform local behaviors as well as reduces the amount of wireless communication traffic. By using digital video, the user has enhanced control of the quality and quantity of video being sent back to a central control area. The MegaScout is also equipped with a USB port, enabling USB cameras to be utilized. The main advantage of the USB camera is that it allows more processing to be done off the main CPU board.

The BrainStem by Acroname facilitates the communication between the main CPU stack and the rest of the electronics. Communication with the BrainStem is handled through a serial interface. Each BrainStem has 5 analog, 5 digital, and 5 servo I/O ports to communicate with the robot's sensors. The MegaScout has a BrainStem in each of the sensor bays as well as one specialized unit for motor control. This enables each robot to communicate with a variety of different types of sensors and manipulators. An interface layer between the main CPU and the other electronics is useful when the main CPU is upgraded or a large number of sensors change. This minimizes the effect such changes have on the rest of the systems.

Initially, there are only two different types of sensor payloads for the MegaScouts. One payload contains a camera with a tilting mechanism, an infrared range finder, and an analog video transmitter. The other contains a USB camera with an Eltec heat detection sensor. The Eltec sensor is used to detect motion in the infrared spectrum emitted by humans. A third sensor payload is under development, and will have a video receiver and command transmitter for the COTS Scout. The sensor bays are designed such that they can be replaced with a minimal impact on the rest of the system.

The electronics are powered by two 118 Wh Lithium Polymer batteries. With these packs in place, the robot has approximately 20 hours of standby time and 4 to 5 hours of average runtime. The Lithium Polymer packs were chosen due to their high energy density and ability to source large amounts of current. These particular battery packs are equipped with monitoring circuitry to watch battery health as well as provide battery statistic gathering capabilities.

4.3 Mechanical Hardware

The MegaScout was designed to operate in the difficult environments of urban warfare, urban search and rescue, and general field use. As a result, a great deal of engineering went into the mechanical design of the platform. Specifically, the design allows for use over a variety of different terrains including mud, dirt, and debris. This is accomplished through the minimization of entry points and the use of weather resistant features such as gaskets and shaft seals. Due to the sealed nature of the shell, all sensing to the outside world occurs through polycarbonate and other sensor transparent materials.

The MegaScout robotic platform is not only designed to tolerate environmental hazards, but also severe mechanical shock loading. A grade 5 tita-

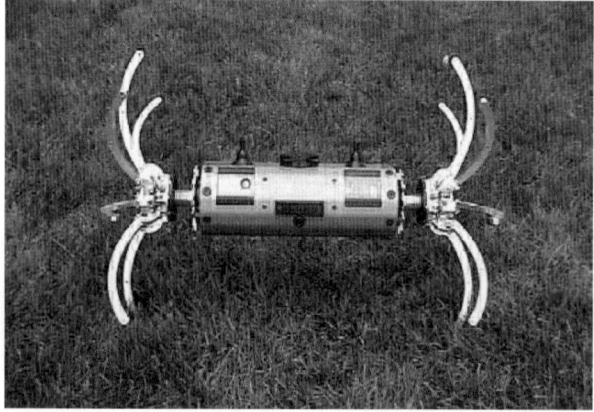

Figure 4: The actuated-wheel MegaScout shown with its wheels retracted and expanded.

nium hull provides a tough and lightweight exoskeleton that resists local deformation under impact conditions. Lightweight Fortal aluminum gearboxes and bulkheads are used to provide a rigid support structure. The wheels are mounted on hardened chrome alloy output shafts which are designed to take severe impacts. This allows deployment under less than ideal conditions with a minimal risk of damage to the platform. A modular approach to the drive system allows for several types of wheels to be used and swapped easily in the field, while allowing provisions for future enhancements.

Since the form factor of the platform is of a two-wheeled robot, a tail capable of withstanding rugged use is included in addition to careful placement of the axle with respect to the center of gravity. This arrangement counters the high torques generated by accelerations. The MegaScout platform is designed to incorporate two 150 W output motors. These are run at 42 V to avoid resistive losses associated with high power low voltage systems. In addition, robust electronic control is facilitated through the use of low currents, allowing a total drive system efficiency near 90 %. These high torque motors are used to accommodate actuated wheels, and provide robust acceleration. The standard platform is designed for a top speed of around 6 m/s.

There are currently three types of wheels developed for the MegaScout. Each of them provides advantages under different circumstances. The standard wheel, shown in Figure 2, for the MegaScout consists of a billet Fortal aluminum hub with a 25 mm thick urethane tread. This not only encompasses the perimeter of the wheel but also extends onto the outer facing hub surface. Shore A durometer 60 urethane was chosen to balance traction, general wear, and shock travel. This allows for a lightweight wheel that can take moderate radial and axial impacts. A diameter of 230 mm provides environmental isolation for external parts by providing sufficient ground clearance.

Since the inertial energy stored in the mass of the robot is essentially constant for a given velocity, the larger the distance the system traverses under deceleration, the lower the resulting body forces required to transfer inertial loads. Thus, for severe duty, large foam filled rubber tires are used as shown in Figure 3. The 300 mm outer diameter with a hub diameter similar to the MegaScout hull allows for over 60 mm of radial travel. This design allows the platform to withstand deployment far above the point of interest, as well as providing general dampening of shock loading. The large diameter of the wheel also allows the platform to surmount rougher terrain than standard wheels.

With increased wheel diameter comes enhanced environmental mobility [3]. However, large wheels make the portability of the platform difficult. In order to keep a small robot with the benefits of higher ground clearance, variable-sized actuated wheels have been developed as shown in Figure 4. Through the use of eight radially symmetric, four-bar linkages, the diameter of the wheel may be altered to any point within mechanical limits. This is accomplished by using a solenoid to selectively fix the rotation of a threaded shaft with respect to the reference frame of the platform. The inner linkage pivots are fixed to a ring with threads that mate to the threaded shaft. By driving normally with the solenoid engaged, the ring is forced to slide axially and actuate the linkages. Friction with respect to the drive train keeps the wheel from actuating while not engaged with the solenoid. Careful pivot placement allowed for optimal transmission angles to be obtained throughout linkage travel. To minimize user encumbrance while transporting the platform and to facilitate a compact form, the wheels wrap around the outer hull of the robot in the closed position.

5 Future Work

The COTS Scout platform is currently in its third revision, and work is focusing on increasing the durability of the platform while reducing the cost. Initial results from this work imply that the prototyping cost for the COTS Scout can be reduced to below $150.

Although the MegaScout is still in its final prototyping stages, work has begun on expanding the platform's capabilities. Efforts are being focused on developing new deployment options, mobility modes, and sensor payloads. Future work will enhance the actuated wheels by replacing the rigid outer wheel segments with flexible segments, providing a suspension and a smoother riding profile. Another concept under development is to captivate a coiled spring to all eight ends of the wheel to provide a round surface at all stages of actuation. A flexible membrane is then stretched over the entire wheel to provide traction and prohibit foreign matter from interfering with the mechanism. Both concepts will be blended to provide passive suspension while maintaining a round wheel throughout actuation.

Currently, designs are in progress for a magnetic wheel system which, combined with improved weatherproofing, could operate in environments such as the sides of buildings or ships. During the early design phases, the idea of joining MegaScouts together was proposed and provisions for interconnectivity were included in the design. Combining the MegaScouts will allow the traversal of obstacles such as stairs and transportion of objects too large for a single robot. Due to the modularity of the sensor bays, new sensors will be some of the simplest additions to the MegaScout platform. Additions that are currently being investigated are such devices as GPS modules, Geiger counters, and infrared cameras.

6 Conclusions

The COTS Scout and the MegaScout are a team of new robotic platforms. The COTS Scout is a portable durable, and inexpensive platform with limited capabilities that can be deployed in many hard to reach locations. The expanded capabilities of the MegaScout enable it to be used in a wider array of situations by sacrificing a small portion of its portability. These systems are intentionally designed with a high degree of standardization to decrease the amount of time needed for prototyping and allow greater amounts of time to be spent researching applications of these new technologies.

References

[1] J. Casper and R. R. Murphy. Workflow study on human-robot interaction in USAR. In *Proc. of the IEEE Int'l Conf. on Robotics and Automation*, May 2002.

[2] A. Castaño, W.-M. Shen, and P. Will. CONRO: Towards deployable robots with inter-robot metamorphic capabilities. *Autonomous Robots*, 8(3):309–324, 2000.

[3] A. Drenner, I. Burt, B. Kratochvil, B. J. Nelson, N. Papanikolopoulos, and K. B. Yeşin. Communication and mobility enhancements to the scout robot. In *Proc. of the IEEE/RSJ Int'l Conf. on Intelligent Robots and Systems*, Lausanne, Switzerland, Oct. 2002.

[4] D. F. Hougen, J. C. Bonney, J. R. Budenske, M. Dvorak, M. Gini, D. G. Krantz, F. Malver, B. Nelson, N. Papanikolopoulos, P. E. Rybski, S. A. Stoeter, R. Voyles, and K. B. Yesin. Reconfigurable robots for distributed robotics. In *Government Microcircuit Applications Conf.*, pages 72–75, Anaheim, CA, Mar. 2000.

[5] H. G. Nguyen, H. Everett, N. Manouk, and A. Verma. Autonomous mobile communication relays. In *Proceedings of the SPIE conference on AeroSense/Unmanned Ground Vehicle Technology IV*, Orlando, FL, 2002.

[6] D. Rus and K. Kotay. Versatility for unknown worlds: Mobile sensors and self-reconfiguring robots. In A. Zelinsky, editor, *Field and Service Robotics*. 1998.

[7] J. Spofford, D. Anhalt, J. Herron, and B. Lapin. Collaborative robotic team design and integration. In *Proceedings of SPIE Unmanned Ground Vehicle Technology II*, Apr. 2000.

[8] G. Stuber. *Principles of Mobile Communication, Second Edition*. Kluwer Academic Publishers, Boston, 2001.

[9] M. Yim, D. G. Duff, and K. D. Roufas. Polybot: a modular reconfigurable robot. In *Proc. of the IEEE Int'l Conf. on Robotics and Automation*, 2000.

A System for Volumetric Robotic Mapping of Abandoned Mines

Sebastian Thrun[†], Dirk Hähnel[†‡], David Ferguson[†], Michael Montemerlo[†], Rudolph Triebel[†], Wolfram Burgard[‡], Christopher Baker[†], Zachary Omohundro[†], Scott Thayer[†], William Whittaker[†]

[†] School of Computer Science
Carnegie Mellon University
Pittsburgh, PA, USA

[‡] Department of Computer Science
University of Freiburg
Freiburg, Germany

Abstract—This paper describes two robotic systems developed for acquiring accurate volumetric maps of underground mines. One system is based on a cart instrumented by laser range finders, pushed through a mine by people. Another is a remotely controlled mobile robot equipped with laser range finders. To build consistent maps of large mines with many cycles, we describe an algorithm for estimating global correspondences and aligning robot paths. This algorithm enables us to recover consistent maps several hundreds of meters in diameter, without odometric information. We report results obtained in two mines, a research mine in Bruceton, PA, and an abandoned coal mine in Burgettstown, PA.

I. INTRODUCTION

The lack of accurate maps of inactive, underground mines poses a serious threat to public safety. According to a recent article [3], "tens of thousands, perhaps even hundreds of thousands, of abandoned mines exist today in the United States. Not even the U.S. Bureau of Mines knows the exact number, because federal recording of mining claims was not required until 1976." The lack of accurate mine maps frequently causes accidents, such as a recent near-fatal accident in Quecreek, PA [18]. Even when accurate maps exist, they provide information only in 2-D, which is usually insufficient to assess the structural soundness of abandoned mines.

Hazardous operating conditions and difficult access routes suggest that robotic exploration and mapping of abandoned mines may be a viable option. The idea of mapping mines with robots is not new. Past research has predominantly focused on acquiring maps for autonomous robot navigation in active mines. For example, Corke and colleagues [8] have built vehicles that acquire and utilize accurate 2-D maps of mines. Similarly, Baily [1] reports 2-D mapping results of an underground area using advanced mapping techniques. None of these techniques generate volumetric maps of mines.

In general, the mine mapping problem is made challenging by the lack of global position information underground. As a result, mine mapping must be approached as a *simultaneous localization and mapping*, or SLAM, problem [10], [15], [20]. In SLAM, the robot acquires a map of its environment while simultaneously estimating its own position relative to this map. The SLAM problem is known to be particularly difficult when the environment possesses cyclic structure [5], [6], [13], [21]. This is because cycles pose hard correspondence problems that arise due to the (relatively) large position error accrued by a vehicle when closing cycles. Mines often contain a large number of cycles, hence the ability to handle cycles is essential for successful approaches to mapping mines.

This paper describes a SLAM algorithm for acquiring 3-D models of underground mines that can accommodate multiple cycles. Our algorithm uses a scan matching algorithms for constructing 2-D mine maps described in [14]. To close cycles, however, it utilizes an iterative correspondence algorithm based on the *iterative closest point* algorithm (ICP) [4], adapted to the problem of establishing correspondence in cyclic maps. 3-D maps are generated by applying scan matching to 3-D measurements after the 2-D mapping is complete.

Our algorithm has successfully enabled two robotic systems to acquire 3-D maps of mines. The first such system consists of an instrumented cart, which is pushed manually through a mine. This system is a low-cost solution to the mine mapping problem, but it can only be brought to bear in environments accessible to people. Our second system consists of a rugged robotic platform equipped with laser range sensors. Abandoned mines, when dry, are often subject to low oxygen levels, poisonous gases, and they may be structurally unstable. Since bringing humans into such mines exposes them to a serious danger of life, the employment of autonomous robotic systems appears to be natural solution. This paper provides results obtained in two different mines, both located in Pennsylvania, USA. One of these mines is a research mine, accessible to people. Another is a former deep mine turned into a strip mine, inaccessible to people but accessible to robotic vehicles.

II. THE ROBOT SYSTEMS

Figure 1 shows the two robotic systems used in our research. On the left is a cart, equipped with four 2-D laser range finders. The laser range finders provide information about the mine cross section ahead of the vehicle, and the ground and ceiling structure. The center panel in Figure 1 shows the Groundhog robot, a tele-operated device constructed from the chassis of two ATVs [2]. The robot is equipped with two 2-D laser range finders, one pointed forward for 2-D mapping and one pointed towards the ceiling for 3-D mapping. The right panel of this figure shows Groundhog's

Fig. 1. From left to right: Mine mapping cart with four laser range finders, pushed manually through a mine. Groundhog robot used for breaching difficult mine environments. Strip mine in Burgettstown, PA. None of the vehicles provide any odometry information.

descent into an abandoned mine in Burgettstown, PA. Unfortunately, neither of these systems possess odometers or inertial sensors. Thus, the location of the vehicles relative to their points of entry can only be recovered from the range scan data.

III. MINE MAPPING ALGORITHM

A. 2-D Scan Matching

In a first processing stage, our approach applies the scan registration technique described in [14] to recover locally consistent pose estimates, which is reminiscent of prior work in [4], [13], [17]. This algorithm aligns scans by iteratively identifying nearby points in pairs of consecutive range scans, and then calculating the relative displacement and orientation of these scans by minimizing the quadratic distance of these pairs of points. The result of registering scans in this way is a relative displacement and orientation between two consecutive scans:

$$\delta_t = \begin{pmatrix} \Delta x_t & \Delta y_t & \Delta \theta_t \end{pmatrix}^T \quad (1)$$

This relative information makes it possible to recover an estimate of the global coordinates at which a scan was acquired. We will denote such global coordinates by

$$\xi_t = \begin{pmatrix} x_t & y_t & \theta_t \end{pmatrix}^T \quad (2)$$

where x_t and y_t are Cartesian coordinates in 2-D, and θ_t is the robot's orientation relative to the global coordinate system at time t. The global coordinates are recovered by applying the following recursive estimation equation:

$$\xi_t = f(\xi_{t-1}, \delta_t) \quad (3)$$

with

$$f(\xi_{t-1}, \delta_t) = \begin{pmatrix} x_{t-1} + \Delta x_t \cos\theta_t + \Delta y_t \sin\theta_t \\ y_{t-1} - \Delta x_t \sin\theta_t + \Delta y_t \cos\theta_t \\ \theta_{t-1} + \Delta\theta_t \end{pmatrix} \quad (4)$$

Unfortunately, the pairwise scan registration technique is unable to recover the *global* structure of an environment. This is specifically problematic in environments that contain cyclic structure. Figure 3a shows an occupancy grid map [11] after executing the local ICP scan registration.

While this map is consistent at the local level, it is inconsistent at the global level due to inconsistencies that arise form the accumulation of small errors in the ICP scan matching procedure. The remaining problem is one of correspondence. To acquire globally consistent maps, we need to know the points in time the robot traversed the same mine segment. This problem is generally considered one of the most challenging problems in robotics, and has been addressed by several researchers [5], [6], [13], [21].

B. Building Consistent Maps With Many Cycles

Our approach uses a modified version of the iterative closest point algorithm (ICP) to estimate the correspondence between robot poses at different points in time. To obtain a globally consistent map, our approach iterates a step in which correspondences are identified, and a step in which a path is recovered from the hypothesized correspondences. This iterative optimization procedure is familiar from the literature on ICP [4], the expectation maximization [9], and the RANSAC [12] algorithm in computer vision (see also [22]). The iteration of both steps leads to a sequence of poses $\xi_t^{[0]}, \xi_t^{[1]}, \ldots$ of increasing global consistency.

The initial poses are obtained from the local scan matcher described in the previous section: $\xi_t^{[0]} := \xi_t$. Figure 2a shows the sequence of poses, subsampled in five-meters intervals for computational efficiency. In a first step, possible correspondences are identified. Our algorithm identifies pairs of poses $\xi_{s_i}^{[n]}$ and $\xi_{t_i}^{[n]}$, indexed by s_i and t_i, which fulfill multiple criteria: they have to be nearby; they have to lie on approximately parallel path segments; and the line connecting them has to be approximately orthogonal to their respective paths. Figure 2b shows the pose pairs identified by our algorithm in the first iteration.

Next, a new set of poses is calculated that matches these correspondences. To calculate such poses in closed form, our approach transforms the relative pose information δ_t into quadratic constraints between adjacent poses. More specifically, our approach applies the following Taylor expansion

$$\xi_t \approx \hat{\xi}_t^{[n]} + F_{t-1}^{[n]} (\xi_{t-1} - \xi_{t-1}^{[n]}) \quad (5)$$

Here $\hat{\xi}_t^{[n]}$ denotes the pose "prediction" $f(\xi_{t-1}^{[n]}, \delta_t)$, and $F_{t-1}^{[n]}$

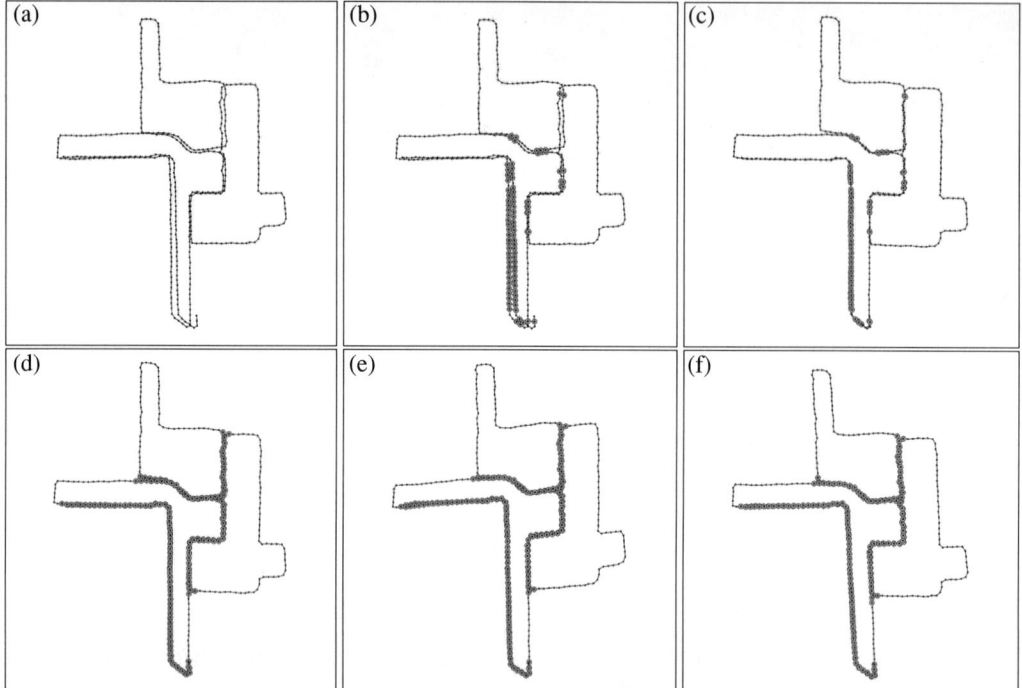

Fig. 2. Global correspondence: (a) Path of the robot, with a node placed every five meters; (b) initial set of correspondences; (c) path obtained under these correspondences; (d) new set of correspondences obtained using the new path; (e) optimal path under these new correspondences; (f) final path and correspondences after three full iterations of the algorithm.

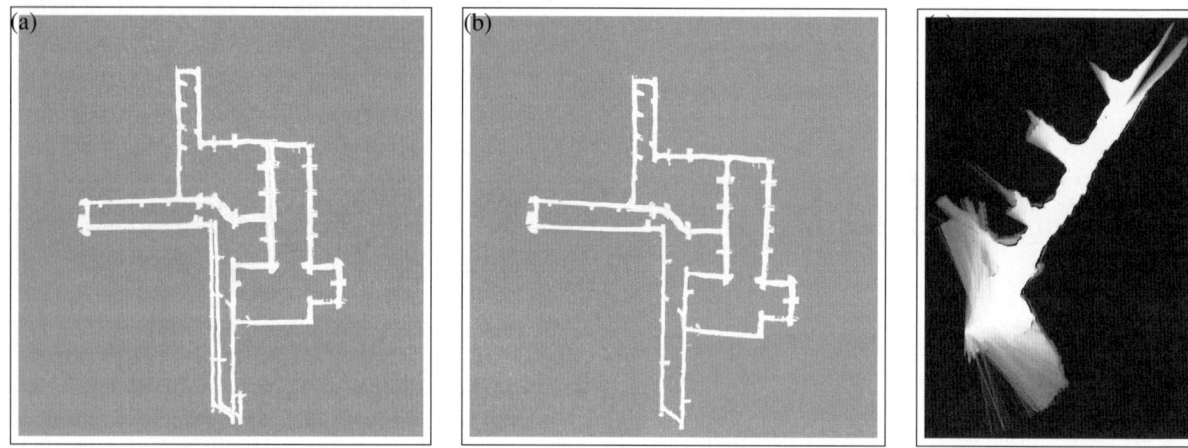

Fig. 3. (a) Map of the coal mine in Bruceton, PA, based on scan matching. This map is obtained by incremental scan matching, and the resulting poses form the input to our loop closing algorithm. (b) Map obtained using our loop closing routines. This map measures approximately 250 by 200 meters in size and contains three large loops. (c) 2-D map of a mine in Burgettstown, PA.

is the tangent to the function f at $\xi_{t-1}^{[n]}$:

$$F_{t-1}^{[n]} = \frac{\partial f(\xi_{t-1}^{[n]}, \delta_t)}{\partial \xi_{t-1}} \quad (6)$$

$$= \begin{pmatrix} 1 & 0 & -\Delta x_t \sin \theta_t^{[n]} + \Delta y_t \cos \theta_t^{[n]} \\ 0 & 1 & -\Delta x_t \cos \theta_t^{[n]} - \Delta y_t \sin \theta_t^{[n]} \\ 0 & 0 & 1 \end{pmatrix}$$

Both $\hat{\xi}_t^{[n]}$ and $F_{t-1}^{[n]}$ are constants in the optimization to follow. The goal of the optimization is to identify poses ξ_t that minimize the quadratic distance to the approximation in (5). This is achieved by minimizing the following quadratic error:

$$\sum_t (\xi_t - \hat{\xi}_t^{[n]} - F_{t-1}^{[n]} (\xi_{t-1} - \xi_{t-1}^{[n]}))^T H_t (\xi_t - \hat{\xi}_t^{[n]} - F_{t-1}^{[n]} (\xi_{t-1} - \xi_{t-1}^{[n]})) \quad (7)$$

Turning the exact calculation in (4) into an optimization problem enables us to "bend" the path of the robot. The matrix H_t measures the penalty associated with bending the path. Mathematically, H_t characterizes the negative log-likelihood of a Gaussian noise model of the ICP scan matcher. Ideally, H_t should be extracted by analyzing the curvature of the ICP target function under translation and rotation. In our software, we simply use a fixed diagonal matrix for the penalty H_t.

The correspondences are incorporated into the optimization through an additional quadratic penalty function. Each pair (t_i, s_i) in the set of pairwise correspondences is mapped into a quadratic cost function of the type:

$$(\xi_{t_i} - \xi_{s_i})^T Z (\xi_{t_i} - \xi_{s_i}) \qquad (8)$$

Here Z is a diagonal penalty matrix. Technically, our approach does not enforce $\xi_{t_i} = \xi_{s_i}$; instead, it minimizes the quadratic distance between these poses, with the penalty z. The total cost function of incorporating all correspondences is given by

$$\sum_i (\xi_{t_i} - \xi_{s_i})^T Z (\xi_{t_i} - \xi_{s_i}) \qquad (9)$$

To solve the coupled quadratic optimization problem, we now conveniently reorder the terms in (7) and (9). All terms linear in ξ_t in (7) are collected in a large matrix $A^{[n]}$, and all remaining constants into the vector $c^{[n]}$. Similarly, all linear terms in (9) are subsumed in a matrix $B^{[n]}$. The sum of (7) and (9) is then of the following quadratic form:

$$J^{[n]} = (A^{[n]}\xi - c^{[n]})^T \mathcal{H} (A^{[n]}\xi - c^{[n]}) + (B^{[n]}\xi)^T \mathcal{Z} B^{[n]}\xi$$

Here $\xi = \xi_1, \xi_2, \ldots$ is the vector of all poses, and $\mathcal{H}$ and $\mathcal{Z}$ are high-dimensional versions of H and Z, respectively. Minimizing this quadratic expression is now straightforward. In particular, we calculate its first derivative

$$\begin{aligned} \frac{\partial J}{\partial \xi_t} &= A^{[n]T} \mathcal{H} (A^{[n]}\xi - c^{[n]}) + B^{[n]T} \mathcal{Z} B^{[n]}\xi \\ &= \left[A^{[n]T} \mathcal{H} A^{[n]} + B^{[n]T} \mathcal{Z} B^{[n]} \right] \xi \\ &\quad - A^{[n]T} \mathcal{H} c^{[n]} \end{aligned} \qquad (10)$$

Setting this expression to zero gives us the new set of poses $\xi^{[n+1]}$:

$$\xi^{[n+1]} = \left[A^{[n]T} \mathcal{H} A^{[n]} + B^{[n]T} \mathcal{Z} B^{[n]} \right]^{-1} A^{[n]T} \mathcal{H} c^{[n]}$$

This calculation involves multiplying and inverting matrices whose dimensions is are linear in the number of robot poses. These matrices are sparse; however, they can still be humongous. Our software therefore subsamples the set of all poses: As indicated in Figure 2, only a single pose is included for every five meters of robot motion. In the specific data set shown in Figure 2, this reduces the number of pose variables form 13,116 to 381, a dimension that is easily handled by efficient linear algebra libraries. Adjusted poses for those poses not included in the optimization are easily recovered through linear interpolation. Finally, we note that the linearization is only an approximation, and multiple iterations of the minimization may be required. In our experiments, we always obtained good results in the first two iterations of the optimization.

Figure 2c shows the resulting alignment for the previously calculated correspondences. While the path is now globally consistent in the area where correspondences were identified, it is still inconsistent in other areas. Iterating the basic algorithm leads to the remaining panels in Figure 2. As is easily seen, our approach succeeds in recovering a globally consistent map. The algorithm converges when the correspondences are identical to the ones estimated in the previous iteration.

C. Globally Consistent 2-D Occupancy Maps

Based on the pose estimates obtained in the previous step, our approach extracts an occupancy grid map from the results of the path alignment. It does so by applying once again the scan matching algorithm used to establish the initial relative pose estimates [14], but this time using the poses $\xi_t^{[n]}$ obtained in the global alignment step as an additional constraint. As above, this constraint is represented by a quadratic penalty function, which is easily incorporated into the classical scan matching algorithm (which also optimizes a quadratic function).

Figure 3b shows the map obtained from data acquired in the Bruceton research mine. The map measures 250 by 200 meters in size, and has been constructed without any odometry information.

D. Volumetric 3-D Mapping

In a final step, our approach recovers a 3-D map of the mine. This map is obtained by utilizing the upward pointed 2-D laser and (in the case of the instrumented cart) the downward pointed 2-D laser. Good initial maps are obtained by using the 2-D pose information to construct a 3-D map, via the obvious geometric projections, as described in [16]. Unfortunately, such a reconstruction is only valid for planar environments; in non-planar environments, both volumetric lasers may be tilted, and estimating the tilt is essential for the accuracy of the resulting maps.

Our approach utilizes a forward-pointing vertical laser, presently only available on the robotic cart, which provides a vertical cross-section of the mine as the robot moves. This cross-section enables the robot to register its ceiling and ground scans while simultaneously recovering its pitch (the roll cannot presently recovered). This estimation is performed using a 3-D variant of the scan matching technique describe in [14], using the results of the 2-D pose estimation as a starting point.

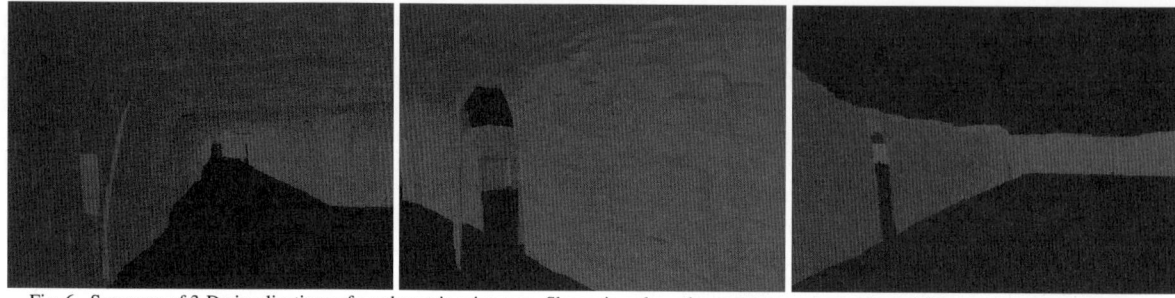

Fig. 6. Sequence of 3-D visualizations of a volumetric mine map. Shown in red are the sensor measurements used for generating the mine map.

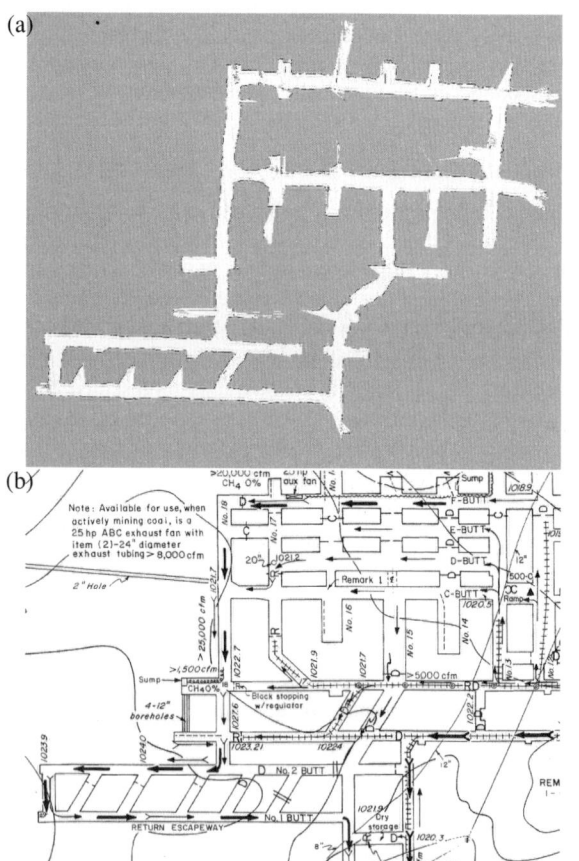

Fig. 4. (a) 2D Mine map acquired by the robotic cart; (b) a hand-drawn map of the corresponding mine segment for comparison.

Fig. 5. (a) A section of the 3-D map produced using the mine-cart and our 2-D mapping algorithms; (b) similar section, using 3-D scan matching for post-processing the mine maps.

IV. RESULTS

We obtained all our data in segments of two abandoned coal mines in Pennsylvania. The Bruceton Mine is geographically close to the Quecreek Mine in Somerset County. It is operated as a research mine by the U.S. Bureau of Mines, enabling us to enter robotic equipment without the need for explosion-proof certification. The Burgettstown Mine is an abandoned mine in a dangerously unstable state. Human access is prohibited and the floor of the mine is covered in a thick toxic sludge known as "yellow boy." The entrance to this mine was discovered only days before the robotic mapping expedition, at which point it was fully submersed. In preparation for the robot mapping experiment, water was pumped out of the mine. Mines of this type pose threats to people due to the low oxygen levels and the danger of collapse.

We already discussed example 2-D maps acquired in both mines, and shown in Figure 3. A second map is shown in Figure 4, along with a hand-drawn map of the corresponding mine section. This map was acquired by the robotic cart. Figure 6 shows example views of a 3-D volumetric maps obtained using this system. The lower hemispheres of the maps are missing because our robot has no downward pointing laser. Views of a full 3-D map acquired by the robotic cart is shown in Figure 5. Here we illustrate the effect of the final scan registration step in the full 3-D model—a step

Fig. 7. A section of the 3-D map produced by the Groundhog robot in the Burgettstown mine. This robot possesses no downward pointed laser; hence the map only shows the ceiling and upper side walls of the mine.

that requires a total of four laser range finders. From this map, the total volume of the mine is easily calculated; information that is typically not available from existing mine maps.

Maps of the Burgettstown mine are shown in Figures 3c and 7. These maps are much smaller than those of the Bruceton mine. However, their significance lies in the fact that they have been acquired in an environment inaccessible to people. The entrance of the mine is shown in Figure 1c. Figure 7 shows a view of a 3-D map acquired by our Groundhog robot. As before, only the upper half of the mine has been mapped, since the robot possesses no downward pointed sensor.

V. CONCLUSION

We have presented systems and algorithms for robotic mapping of underground mines. Both of our systems are equipped with laser range finders to recover ego-motion and to build accurate maps. Our approach relies on 2-D scan matching to recover a locally consistent map, and on a 2-D global alignment algorithm for generating globally consistent maps. The resulting maps and robot paths form the basis for integrating the 3-D information, acquired by additional scanners pointed at the ceiling and the floor of a mine. A final optimization step further improves the spatial consistency of the resulting 3-D mine map.

While we find that in the mines explored so far, our approach consistently produces accurate maps, the greedy nature of this algorithm makes it possible to get stuck in local minima. Algorithms such as RANSAC [12] are applicable to reduce the danger of getting stuck in a local minimum, at the expense of increased computational complexity.

We believe that existing techniques for mobile robot exploration [7], [19], [23] can be adapted for the purpose of autonomously exploring mines. Such an extension would overcome a crucial limitation of the present approach, namely its reliance on human tele-operation.

ACKNOWLEDGMENTS

This research was motivated and supported by an ongoing subterranean mapping course at CMU. Pictures of the mine-cart were taken by Tom Stepleton. We thank Ben Wegbreit, John Leonard and Michael Bosse and the CMU Mine Mapping Class for their input on our algorithm design. The research has been sponsored by DARPA's MARS Program (contract N66001-01-C-6018 and contract NBCH1020014), which is gratefully acknowledged.

REFERENCES

[1] T. Bailey. *Mobile Robot Localisation and Mapping in Extensive Outdoor Environments*. PhD thesis, University of Sydney, 2002.

[2] C. Baker, Z. Omohundro, S. Thayer, W. Whittaker, M. Montemerlo, and S. Thrun. A case study in robotic mapping of abandoned mines. Submitted to FSR'03.

[3] J.J. Belwood and R.J. Waugh. Bats and mines: Abandoned does not always mean empty. *Bats*, 9(3), 1991.

[4] P. Besl and N. McKay. A method for registration of 3d shapes. *Trans. Pattern Analysis and Machine Intelligence*, 14(2):239–256, 1992.

[5] M. Bosse, J. Leonard, and S. Teller. Large-scale CML using a network of multiple local maps. In [15].

[6] M. Bosse, P. Newman, M. Soika, W. Feiten, J. Leonard, and S. Teller. An Atlas framework for scalable mapping. *ICRA-03*.

[7] W. Burgard, D. Fox, M. Moors, R. Simmons, and S. Thrun. Collaborative multi-robot exploration. *ICRA-00*.

[8] P. Corke, J. Cunningham, D. Dekker, , and H. Durrant-Whyte. Autonomous underground vehicles. *CMTE Mining Technology Conf.-96*.

[9] A.P. Dempster, A.N. Laird, and D.B. Rubin. Maximum likelihood from incomplete data via the EM algorithm. *Journal of the Royal Statistical Society, Series B*, 39(1):1–38, 1977.

[10] G. Dissanayake, H. Durrant-Whyte, and T. Bailey. A computationally efficient solution to the simultaneous localisation and map building (SLAM) problem. In [15].

[11] A. Elfes. *Occupancy Grids: A Probabilistic Framework for Robot Perception and Navigation*. PhD thesis, CMU, 1989.

[12] M. A. Fischler and R. C. Bolles. Random sample consensus: A paradigm for model fitting with applications to image analysis and automated cartography. *Communications of the ACM*, 24:381–395, 1981.

[13] J.-S. Gutmann and K. Konolige. Incremental mapping of large cyclic environments. *CIRA-00*.

[14] D. Hähnel, D. Schulz, and W. Burgard. Map building with mobile robots in populated environments. *IROS-02*.

[15] J. Leonard, J.D. Tardós, S. Thrun, and H. Choset, editors. *Notes of the ICRA Workshop on SLAM*, ICRA, 2002.

[16] Y. Liu, R. Emery, D. Chakrabarti, W. Burgard, and S. Thrun. Using EM to learn 3D models with mobile robots. *ICML-01*.

[17] F. Lu and E. Milios. Globally consistent range scan alignment for environment mapping. *Autonomous Robots*, 4:333–349, 1997.

[18] E. Pauley, T. Shumaker, and B. Cole. Preliminary report of investigation: Underground bituminous coal mine, non-injury mine inundation accident (entrapment), July 24, 2002, Quecreek, Pennsylvania, 2002. Black Wolf Coal Company.

[19] R. Simmons, D. Apfelbaum, W. Burgard, M. Fox, D. an Moors, S. Thrun, and H. Younes. Coordination for multi-robot exploration and mapping. *AAAI-00*.

[20] R. Smith, M. Self, and P. Cheeseman. Estimating uncertain spatial relationships in robotics. In *Autonomous Robot Vehicles*: 167–193. Springer, 1990.

[21] S. Thrun. A probabilistic online mapping algorithm for teams of mobile robots. *International Journal of Robotics Research*, 20(5):335–363, 2001.

[22] B. Triggs, P. McLauchlan, R. Hartley, and A. Fitzgibbon. Bundle adjustment–A modern synthesis. In *Vision Algorithms: Theory and Practice*, Springer, 2000.

[23] B. Yamauchi, P. Langley, A.C. Schultz, J. Grefenstette, and W. Adams. Magellan: An integrated adaptive architecture for mobile robots. TR 98-2, ISLE, Palo Alto, CA.

Application of Moment Invariants to Visual Servoing

Omar Tahri
IRISA/INRIA Rennes
Campus de Beaulieu
35 042 Rennes-cedex, France
E-mail: otahri@irisa.fr

François Chaumette
IRISA/INRIA Rennes
Campus de Beaulieu
35 042 Rennes-cedex, France
E-mail: chaumett@irisa.fr

Abstract— In this paper, we present how moment invariants can be used to design a decoupled 2D visual servoing scheme and to minimize the nonlinearity of the interaction matrix related to the selected visual features. Experimental results using a 6 dof eye-in-hand system to position a camera parallel to planar objects of complex shape are presented to demonstrate the efficiency of the proposed method.

I. INTRODUCTION

In 2D visual servoing [8], [10], the control of the robot motion is performed using data directly extracted from the images acquired by one (or several) camera(s). Let $\mathbf{s}$ be a set of k features that characterize the image of the considered object. Once $\mathbf{s}$ is given by a differentiable function, the time derivative of $\mathbf{s}$ and the relative motion between the camera and the object can be related by the classical equation:

$$\dot{\mathbf{s}} = \mathbf{L_s}\mathbf{T} \tag{1}$$

where $\mathbf{T}$ is the relative velocity between the camera and the object. The matrix $\mathbf{L_s}$ is called the features Jacobian or the interaction matrix related to $\mathbf{s}$. The control scheme consists in canceling the task function:

$$\mathbf{e} = \mathbf{C}(\mathbf{s} - \mathbf{s}^*) \tag{2}$$

where $\mathbf{s}$ is the current state and $\mathbf{s}^*$ the desired state. To control the 6 dof of the system, we usually choose $\mathbf{C} = \mathbf{I_6}$ if $k = 6$. If we specify an exponentially decoupled decrease of the task function:

$$\dot{\mathbf{e}} = -\lambda\,\mathbf{e} \tag{3}$$

where λ is a proportional gain, the control law for an eye-in-hand system observing a static object is given by:

$$\mathbf{T_c} = -\lambda\,\widehat{\mathbf{L_s}}^+\mathbf{e} \tag{4}$$

where $\mathbf{T_c} = (V_X, V_Y, V_Z, \Omega_X, \Omega_Y, \Omega_Z)^T$ is the camera velocity sent to the low-level robot controller, $\widehat{\mathbf{L_s}}$ is a model or an approximation of $\mathbf{L_s}$, and $\widehat{\mathbf{L_s}}^+$ the pseudo-inverse of $\widehat{\mathbf{L_s}}$. Several kind of image features $\mathbf{s}$ were proposed in the past. Most works were concerned with known and simple objects. They assume that the objects in the scene can be expressed with simple features such as points, straight lines, ellipses and more. The group of the objects that these methods can be applied to is thus limited. These methods have also the basic requirement of feature matching between initial and desired images, which is generally not easy to obtain. Other methods try to surmount the problems mentioned above, by using for example the Eigen space method [7] or the polar signature of an object contour [5]. Recently, a new method was proposed using image moments [4]. In this paper, we propose significant improvements to this method.

To date, an appropriate question in the visual servoing field is to determine the visual features to use in the control scheme in order to obtain an optimal behavior of the system. A first necessary condition of the convergence is that the interaction matrix must be not singular. Hence, a good choice of the features must allow to obtain a large domain where the matrix $\mathbf{L_s}$ has full rank 6. A good way to ensure this condition is to design a decoupled control scheme, i.e. to try to associate each camera dof with only one visual feature. Such control would make easy the determination of the potential singularities of the considered task, as well as the choice of $\widehat{\mathbf{L_s}}$. Unfortunately, a such totally decoupled control is ideal and seems impossible to reach. It is however possible to decouple the translational motion from the rotational one. In practice, it can be obtained using 3D visual servoing [19], but this approach requires the knowledge of a 3D CAD model of the object. It can also be obtained using $2\,1/2D$ visual servoing [12], where the knowledge of a 3D model is not required. However, in this case, a homography matrix must be computed at each iteration of the control scheme. In 2D visual servoing, first attempts have been recently proposed in [6], [4]. In this paper, we present a more efficient method using moment invariants.

In 2D visual servoing, the behavior of the features in the image is generally satisfactory. On the other hand, the robot trajectory in 3D space is quite unpredictable and may be really unsatisfactory for large rotational displacements [3]. In fact, the difference of behaviors in image space and 3D space is due to the non linearities in the interaction matrix. To explain that, let us consider the basic interaction matrix related to the coordinates (x, y) of an image point:

$$\mathbf{L_x} = \begin{pmatrix} -1/Z & 0 & x/Z & xy & -1-x^2 & y \\ 0 & -1/Z & y/Z & 1+y^2 & -xy & -x \end{pmatrix} \quad (5)$$

We can see that the dynamic of $\dot{x}$ and $\dot{y}$ with respect to the camera velocity components are really not the same: some are inversely proportional to the depth Z of the point, some are linearly dependent to the image coordinates, while others depend on them at second order. Even if we are able to design a control scheme such that the error in the image has an exponential decoupled decrease, the robot dynamics will be unlikely very far from such an exponential decoupled decrease, because of the strong non linearities in $\mathbf{L_x}$. The robot trajectory will thus be very far from the optimal one (typically, a straight line as for the translation and a geodesic as for the rotation). An important problem is thus to determine visual features such that they minimize the non linearities in the related interaction matrix. In this paper, following the recent work described in [11], three new visual features are given to control the translational dof. They are such that the related interaction matrix is diagonal and constant with respect to these dof.

In the remainder of the paper, we first briefly recall some definitions and important properties of moments. We then determine in Section 3 six visual features to control the six robot dof. The obtained control scheme is finally validated in Section 4 trough experimental results.

II. MOMENT INVARIANTS

A. Definitions

The 2D moments m_{pq} of order $p+q$ of the density function $f(x,y)$ are defined by:

$$m_{pq} \equiv \int_{-\infty}^{+\infty} \int_{-\infty}^{+\infty} x^p y^q f(x,y) dx dy \quad (6)$$

The centred moments μ_{pq} are computed with respect to the object centroid (x_g, y_g). They are defined by:

$$\mu_{pq} = \int_{-\infty}^{+\infty} \int_{-\infty}^{+\infty} (x-x_g)^p (y-y_g)^q f(x,y) dx dy \quad (7)$$

where $x_g = \frac{m_{10}}{a}$ and $y_g = \frac{m_{01}}{a}$, $a = m_{00}$ being the object area.

The moments of a density function f exist if f is piecewise continuous and has nonzero values only in a finite region of the space. The centred moments are known to be invariant to 2D translational motion. The moment invariants to rotations are generally given in a polynomial form. In the literature, several works propose various methods to derive moment invariants. Reddi [16] obtained moment invariants to rotation using radial and angular moments, Teague [17], Belkassim [2], Walin and Kübler [18] derived Zernike moments invariant to rotation, Abu-Mustapha and Psaltis [1], Flusser [9] obtained invariants to rotation from complex moments. Finally, several formula have been proposed for invariants to scale, such as for example [13]:

$$I = \frac{m_{pq}}{m_{00}^{(p+q+2)/2}} \quad (8)$$

This normalization will be used in Section 3.1 to decouple the features involved in the control of the translational dof. More details about moment invariants can be found in Mukundan [14] and Prokop [15]. We just present some invariants to 2D rotations, to scale, and to 2D translations. They will be used in the design of the features involved to control the rotational velocities Ω_X and Ω_Y.

$$R_1 = \frac{I_{n_1}}{I_{n_2}}, \ R_2 = \frac{I_{n_3}}{I_{n_2}}, \ R_3 = \frac{I_{n_4}}{I_{n_6}}, \ R_4 = \frac{I_{n_5}}{I_{n_6}} \quad (9)$$

where

$$\begin{aligned} I_{n_1} = &\mu_{30}^4 + 6\mu_{30}^3 \mu_{12} + 9\mu_{30}^2 \mu_{12}^2 + 6\mu_{30}^2 \mu_{21} \mu_{03} \\ &+ 2\mu_{30}^2 \mu_{03}^2 + 18\mu_{30}\mu_{21}\mu_{12}\mu_{03} \\ &+ 6\mu_{30}\mu_{12}\mu_{03}^2 + 9\mu_{21}^2\mu_{03}^2 + 6\mu_{21}\mu_{03}^3 + \mu_{03}^4 \end{aligned} \quad (10)$$

$$\begin{aligned} I_{n_2} = &3\mu_{30}^2 \mu_{12}^2 + 2\mu_{30}^2 \mu_{03}^2 - 6\mu_{30}\mu_{21}^2\mu_{12} \\ &- 6\mu_{30}\mu_{21}\mu_{12}\mu_{03} + 2\mu_{30}\mu_{12}^3 + 3\mu_{21}^4 \\ &+ 2\mu_{21}^3\mu_{03} + 3\mu_{21}^2\mu_{03}^2 - 6\mu_{21}\mu_{12}^2\mu_{03} + 3\mu_{12}^4 \end{aligned} \quad (11)$$

$$\begin{aligned} I_{n_3} = &-\mu_{30}^2\mu_{03}^2 + 6\mu_{30}\mu_{21}\mu_{12}\mu_{03} - 4\mu_{30}\mu_{12}^3 \\ &- 4\mu_{21}^3\mu_{03} + 3\mu_{21}^2\mu_{12}^2 \end{aligned} \quad (12)$$

$$I_{n_4} = (\mu_{50} + 2\mu_{32} + \mu_{14})^2 + (\mu_{05} + 2\mu_{23} + \mu_{41})^2 \quad (13)$$

$$I_{n_5} = (\mu_{50} - 2\mu_{32} - 3\mu_{14})^2 + (\mu_{05} - 2\mu_{23} - 3\mu_{41})^2 \quad (14)$$

$$I_{n_6} = (\mu_{50} - 10\mu_{32} + 5\mu_{14})^2 + (\mu_{05} - 10\mu_{23} + 5\mu_{41})^2 \quad (15)$$

B. Interaction matrix of 2D moments

In this paragraph, we recall from [4] the interaction matrix of the 2D moments. In the following we assume that the object belongs to a plane whose equation is given by:

$$\frac{1}{Z} = Ax + By + C$$

We also assume that the image is binary or that the grey level does not change when the camera moves (i.e. $\frac{df(x,y)}{dt} = 0$). In that case, the interaction matrix $\mathbf{L}_{m_{ij}}$ related to m_{ij} can be determined:

$$\mathbf{L}_{m_{ij}} = \begin{pmatrix} m_{vx} & m_{vy} & m_{vz} & m_{wx} & m_{wy} & m_{wz} \end{pmatrix} \quad (16)$$

where:

$$\begin{cases} m_{vx} = -i(Am_{ij} + Bm_{i-1,j+1} + Cm_{i-1,j}) - Am_{ij} \\ m_{vy} = -j(Am_{i+1,j-1} + Bm_{ij} + Cm_{i,j-1}) - Bm_{ij} \\ m_{vz} = (i+j+3)(Am_{i+1,j} + Bm_{i,j+1} + Cm_{ij}) - Cm_{ij} \\ m_{wx} = (i+j+3)m_{i,j+1} + jm_{i,j-1} \\ m_{wy} = -(i+j+3)m_{i+1,j} - im_{i-1,j} \\ m_{wz} = im_{i-1,j+1} - jm_{i+1,j-1} \cdots \end{cases}$$

Similarly, the interaction matrix related to the centred moments μ_{ij} is given by:

$$\mathbf{L}_{\mu_{ij}} = \begin{pmatrix} \mu_{vx} & \mu_{vy} & \mu_{vz} & \mu_{wx} & \mu_{wy} & \mu_{wz} \end{pmatrix} \quad (17)$$

with:

$$\begin{aligned}
\mu_{vx} &= -(i+1)A\mu_{ij} - iB\mu_{i-1,j+1} \\
\mu_{vy} &= -jA\mu_{i+1,j-1} - (j+1)B\mu_{ij} \\
\mu_{vz} &= -A\mu_{wy} + B\mu_{wx} + (i+j+2)C\mu_{ij} \\
\mu_{wx} &= (i+j+3)\mu_{i,j+1} + ix_g\mu_{i-1,j+1} \\
&\quad +(i+2j+3)y_g\mu_{ij} - in_{11}\mu_{i-1,j} - jn_{02}\mu_{i,j-1} \\
\mu_{wy} &= -(i+j+3)\mu_{i+1,j} - (2i+j+3)x_g\mu_{ij} \\
&\quad -jy_g\mu_{i+1,j-1} + in_{20}\mu_{i-1,j} + jn_{11}\mu_{i,j-1} \\
\mu_{wz} &= i\mu_{i-1,j+1} - j\mu_{i+1,j-1}
\end{aligned}$$

where $n_{ij} = 4\mu_{ij}/m_{00}$. For the positions where the object is parallel to the image plane (i.e. $A = B = 0$), we can check from the first two components of $\mathbf{L}_{\mu_{ij}}$ that the variation of the centred moments with respect to V_X and V_Y vanishes, which proves that these moments are invariant to 2D translations parallel to the image plane when (and only when) the object is also parallel to the image plane. For the same positions, it is easy to prove that the variation of the scale moment invariants with respect to V_Z vanishes. Even if the invariance to translation is local (i.e. only valid when $A = B = 0$), these features depend mainly on the rotation. We will thus use them to control the rotational dof.

III. CHOICE OF THE FEATURE VECTOR

In this section, we select from the previous theoretical results six features to control the six dof of the robot. Our objective is to obtain a sparse interaction matrix that changes slowly around the desired position of the camera. We will see that the solution we present is such that the interaction matrix is triangular when the object is parallel to the image plane. Furthermore, we will see that, for the same positions, the elements corresponding to translational motions form a constant diagonal block, which is independent of depth. In [4], this last interesting property was not satisfied.

We now assume that the desired position of the image plane and the object is parallel (i.e. $A = B = 0$) and we denote $\mathbf{L}_s^\parallel$ the interaction matrix for such positions. In the following, we will only be concerned with $\mathbf{L}_s^\parallel$ since it will be used to build the model $\widehat{\mathbf{L}}_s$ of $\mathbf{L}_s$ in the control scheme (4).

A. Features to control the translational dof

In [6], [4], the three visual features used to control the translational dof have been selected to be the coordinates x_g, y_g of the center of gravity and the area a of the object in the image. In that case, we obtain from (16):

$$\begin{aligned}
\mathbf{L}_{x_g}^\parallel &= \begin{pmatrix} -C & 0 & Cx_g & \epsilon_1 & -(1+\epsilon_2) & y_g \end{pmatrix} \\
\mathbf{L}_{y_g}^\parallel &= \begin{pmatrix} 0 & -C & Cy_g & 1+\epsilon_3 & -\epsilon_1 & -x_g \end{pmatrix} \quad (18) \\
\mathbf{L}_a^\parallel &= \begin{pmatrix} 0 & 0 & 2aC & 3ay_g & -3ax_g & 0 \end{pmatrix}
\end{aligned}$$

with $\epsilon_1 = n_{11} + x_g y_g$, $\epsilon_2 = n_{20} + x_g^2$ and $\epsilon_3 = n_{02} + y_g^2$. Even if the above matrix is triangular, we can note that its elements are strongly non linear. Moreover, the features do not have the same dynamic with respect to each translational dof.

Our choice is based on these intuitive features, but adding an adequate normalization. More precisely, we define:

$$a_n = Z^*\sqrt{\frac{a^*}{a}}, \quad x_n = a_n x_g, \quad y_n = a_n y_g \quad (19)$$

where a^* is the desired area of the object in the image, and Z^* the desired depth between the camera and the object. The interaction matrices related to these normalized features can be easily determined from (18). Noting that $Z^*\sqrt{a^*} = Z\sqrt{a} = \sqrt{S}$ where S is the area of the 3D object, we obtain:

$$\begin{aligned}
\mathbf{L}_{x_n}^\parallel &= \begin{pmatrix} -1 & 0 & 0 & a_n\epsilon_{11} & -a_n(1+\epsilon_{12}) & y_n \end{pmatrix} \\
\mathbf{L}_{y_n}^\parallel &= \begin{pmatrix} 0 & -1 & 0 & a_n(1+\epsilon_{21}) & -a_n\epsilon_{11} & -x_n \end{pmatrix} \quad (20) \\
\mathbf{L}_{a_n}^\parallel &= \begin{pmatrix} 0 & 0 & -1 & -3y_n/2 & 3x_n/2 & 0 \end{pmatrix}
\end{aligned}$$

with $\epsilon_{11} = n_{11} - x_g y_g/2$, $\epsilon_{12} = n_{20} - x_g^2/2$, and $\epsilon_{21} = n_{02} - y_g^2/2$. Since a_n is inversely proportional to $\sqrt{a}$, we find again the recent result given in [11] stating that the variation of such features depends linearly of the depth (note the constant term in the third element of $\mathbf{L}_{a_n}^\parallel$). The normalization by $Z^*\sqrt{a^*}$ has just be chosen so that this constant term is equal to -1. Furthermore, the design of x_n and y_n allows us to completely decouple the three selected features with respect to the translational dof. This property was expected from (8). We also obtain the same dynamics for the three features and the three translational dof (note the diagonal block equal to $-\mathbf{I}_3$ in (20)). This very nice property will allow us to obtain an adequate robot translational trajectory.

Finally, we can notice from the analytical and the numerical values of $\mathbf{L}_{x_n}^\parallel$ and $\mathbf{L}_{y_n}^\parallel$ (see (24)) the classical coupling between V_X and Ω_Y, and between V_Y and Ω_X. In fact, this natural coupling allows the object to remain as much as possible in the camera field of view.

B. Features to control the rotational dof

As in [6], [4], we use $\theta = \frac{1}{2}\arctan(\frac{2\mu_{11}}{\mu_{20}-\mu_{02}})$ defined as the orientation angle of the principal axis of inertia with the X-axis of the image frame. From Figure 1, we can notice that there are two solutions for θ: θ and $\theta + \pi$. However, the third order moments can be used to solve this ambiguity since a rotation of an object by π changes the sign of its third order moments.

We also use two moment invariants R_i and R_j chosen in (9). The related interaction matrices can be obtained from (17). We obtain (after tedious developments):

$$\begin{aligned}
\mathbf{L}_{R_i}^\parallel &= \begin{pmatrix} 0 & 0 & 0 & R_{i_{wx}} & R_{i_{wy}} & 0 \end{pmatrix} \\
\mathbf{L}_{R_j}^\parallel &= \begin{pmatrix} 0 & 0 & 0 & R_{j_{wx}} & R_{j_{wy}} & 0 \end{pmatrix} \quad (21) \\
\mathbf{L}_\theta^\parallel &= \begin{pmatrix} 0 & 0 & 0 & \theta_{wx} & \theta_{wy} & -1 \end{pmatrix}
\end{aligned}$$

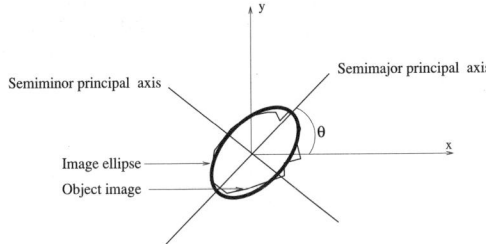

Fig. 1. Image ellipse

where the analytical form of the elements corresponding to Ω_X and Ω_Y can not be given here by lack of place (if interested, θ_{wx} and θ_{wy} can be found in [4]). As expected, we can notice the invariance of the selected features with respect to any 3D translational motion (remember that we consider here that $A = B = 0$), and the invariance of R_i and R_j with respect to Ω_Z. Finally, we can see from (24) that θ depends essentially on the rotation Ω_Z around the optical axis. As for R_i and R_j, they are chosen such that $\mathbf{L}_{R_i}^{\|}$ and $\mathbf{L}_{R_j}^{\|}$ are as orthogonal as possible. In the next section, we present experimental results where R_3 and R_4 have been selected. For the considered object, this choice has given the best results.

IV. EXPERIMENTAL RESULTS

This section presents some experimental results obtained with a six dof eye-in-hand system. The moments are computed at video rate after a simple binarisation of the aquired image, without any spatial segmentation. As already explained, we have used as visual features vector:

$$\mathbf{s} = \begin{pmatrix} x_n & y_n & a_n & R_3 & R_4 & \theta \end{pmatrix}^T \quad (22)$$

In our experiments, the parameters of the object plane in the camera frame are given approximately for the desired position ($A = B = 0$, $C = 2$, which corresponds to $Z^* = 0.5$ m). They are not estimated at each step. For the two first experiments, a correct value of the camera intrinsic parameters has been used. The desired value $\mathbf{s}^*$ is given by:

$$\mathbf{s}^* = \begin{pmatrix} \widehat{Z}^* x_g^* & \widehat{Z}^* y_g^* & \widehat{Z}^* & R_3^* & R_4^* & \theta^* \end{pmatrix}^T \quad (23)$$

where $x_g^*, y_g^*, R_3^*, R_4^*$ and θ^* are computed directly from the desired image (acquired during an off-line learning step), and where $\widehat{Z}^*$ has been set to 0.5 m. We can note from (23), (22) and (19) that using a wrong value $\widehat{Z}^*$ for Z^* has no influence on the convergence of the system ($\mathbf{s} = \mathbf{s}^*$ only for the desired position whatever the setting of value $\widehat{Z}^*$). It will just induce the same gain effect (with value $\widehat{Z}^*/Z^*$) for the decreasing of the three first features. An esperiment with a wrong setting of $\widehat{Z}^*$ is described in Section IV-C.

A. Pure translational motion

We first compare the results obtained with our features and those obtained using the centroid coordinates (x_g, y_g)

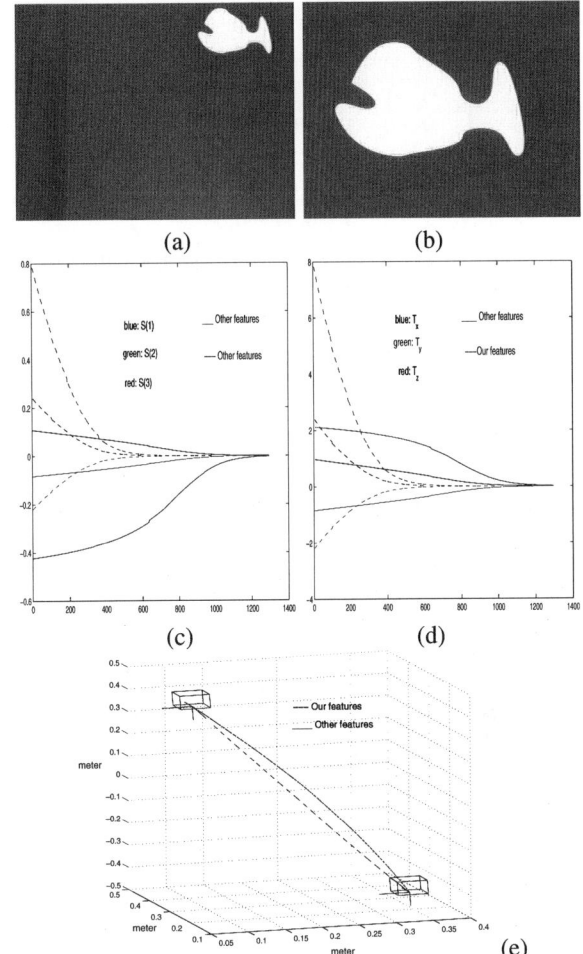

Fig. 2. Results for a pure translational motion: (a) initial image, (b) desired image, (c) visual features ($\mathbf{s} - \mathbf{s}^*$), (d) camera velocity $\mathbf{T_c}$, (e) camera 3D trajectory

and the area a for a pure translational motion between the initial and desired images (given on Figure 2.a and 2.b). For both schemes, we have used $\widehat{\mathbf{L}_s} = \mathbf{L}_{s|s=s^*} = \mathbf{L}_{s|s=s^*}^{\|}$ in the control scheme (4) and gain λ has been set to 0.1.

We can see on Figure 2 the improvements brought by the proposed features (in dashed lines) since they allow to obtain the same exponential decoupled decrease for the visual features and for the components of the camera velocity. As expected, the camera 3D trajectory is thus a pure straight line using the proposed features, while it is not using the other ones.

B. Complex motion

We now test our scheme for a displacement involving very large translation and rotation to realize between the initial and desired images (see Figures 3.a and 3.b). The interaction matrix computed at the desired position has the following form:

$$\mathbf{L}^{\|}_{\mathbf{s}|\mathbf{s}=\mathbf{s}^*} = \begin{pmatrix} -1 & 0 & 0 & 0.01 & -0.52 & 0.01 \\ 0 & -1 & 0 & 0.51 & -0.01 & 0.01 \\ 0 & 0 & -1 & -0.02 & -0.01 & 0 \\ 0 & 0 & 0 & -0.33 & 0.62 & 0 \\ 0 & 0 & 0 & -0.61 & 0.09 & 0 \\ 0 & 0 & 0 & -0.04 & -0.08 & -1 \end{pmatrix} \quad (24)$$

We can note that this matrix is block triangular with main terms around the diagonal. The value of its condition number (equal to 2.60) is also very satisfactory. Finally, we have used the following model of $\mathbf{L_s}$ in the control scheme (4):

$$\widehat{\mathbf{L}}_\mathbf{s} = \frac{1}{2}(\mathbf{L}^{\|}_\mathbf{s} + \mathbf{L}^{\|}_{\mathbf{s}|\mathbf{s}=\mathbf{s}^*}) \quad (25)$$

This choice has given the best experimental results. The obtained results are given on Figure 3. They show the good behavior of the control law. First, we can note the fast convergence towards the desired position (while the system does not converge for the six visual features proposed in [4]). Then, there is no oscillation in the decrease of the visual features (see Figure 3.c), and there is only one small oscillation for only two components of the camera velocity (see Figure 3.d). Finally, even if the rotation to realize between the initial and the desired positions is very large, the obtained camera 3D trajectory is satisfactory (see Figure 3.e), while it was an important drawback for classical 2D visual servoing.

C. Results with a bad camera calibration and object occultation

We now test the robustness of our approach with respect to a bad calibration of the system. In this experiment, errors have been added to camera intrinsic parameters (25% on the focal length and 20 pixels on the coordinates of the principal point) and to the object plane parameters ($\widehat{Z}^* = 0.8m$ instead of $Z^* = 0.5m$). We can also notice that the lighting conditions for the desired and the initial positions given on Figure 4.a and 4.b are different. Furthermore, an occultation has been generated since the object is not completely in the camera field of view at the begining of the servo. The obtained results are given in Figure 4. We can notice that the system converges despite the worse conditions of experimentations and, as soon as the occultation ends (after iteration 30), the behavior of the system is similar to those of the previous experiment, which validates the robustness of our scheme with respect to modeling errors.

V. Conclusion

In this paper, we have proposed a new visual servoing scheme based on the moments of an object. Our approach presents several advantages: there is no constraint on the object shape and the model of the object is also not required. Moment invariants have been used to decouple the camera dof, which allows the system to have a large convergence domain and a good behavior for the

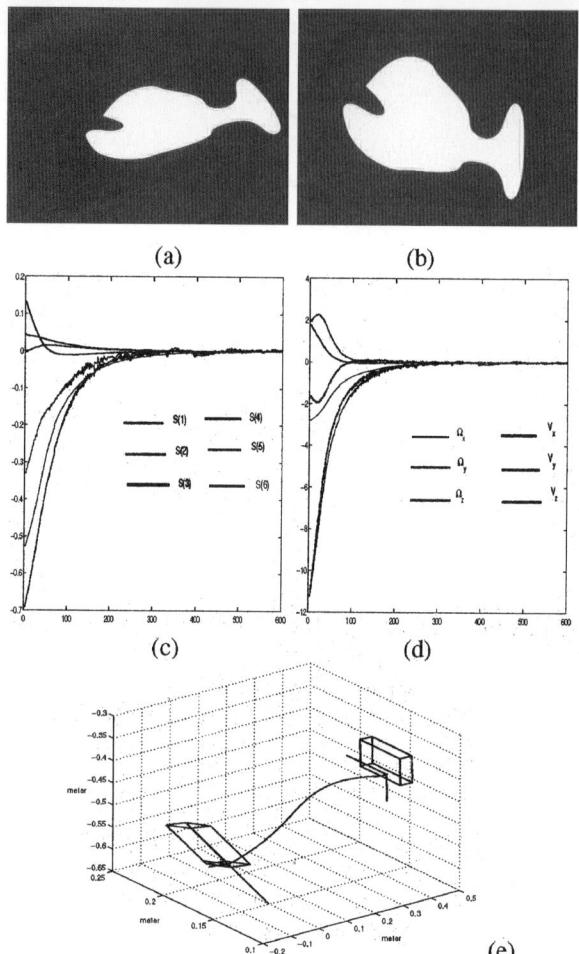

Fig. 3. Results for a complex motion: (a) initial image, (b) desired image, (c) visual features $(\mathbf{s} - \mathbf{s}^*)$, (d) camera velocity $\mathbf{T_c}$, (e) camera 3D trajectory

features in the image and for the camera trajectory. The experimental results show the validity of the approach and its robustness with respect to calibration errors. To improve further the obtained results, future works will be devoted to the development of an estimation scheme of the pose between the object and the camera.

VI. REFERENCES

[1] Y. S. Abu-Mustapha and D. Psaltis. Image normalisation by complex moments. *IEEE Trans. on PAMI*, 7(1):46–55, 1985.

[2] S. O. Belkassim, M. Shridhar, and M. Ahmadi. Shape-contour recognition using moment invariants. In *10th Int. Conf. on Pattern Recognition*, pages 649–651, Atlantic City, NJ, USA, June 1990.

[3] F. Chaumette. Potential problems of stability and convergence in image-based and position-based visual servoing. In A.S. Morse D. Kriegman, G. Hager, editor, *The Confluence of Vision and Control*, number 237, pages 66–78. Springer-Verlag, 1998. LNCIS.

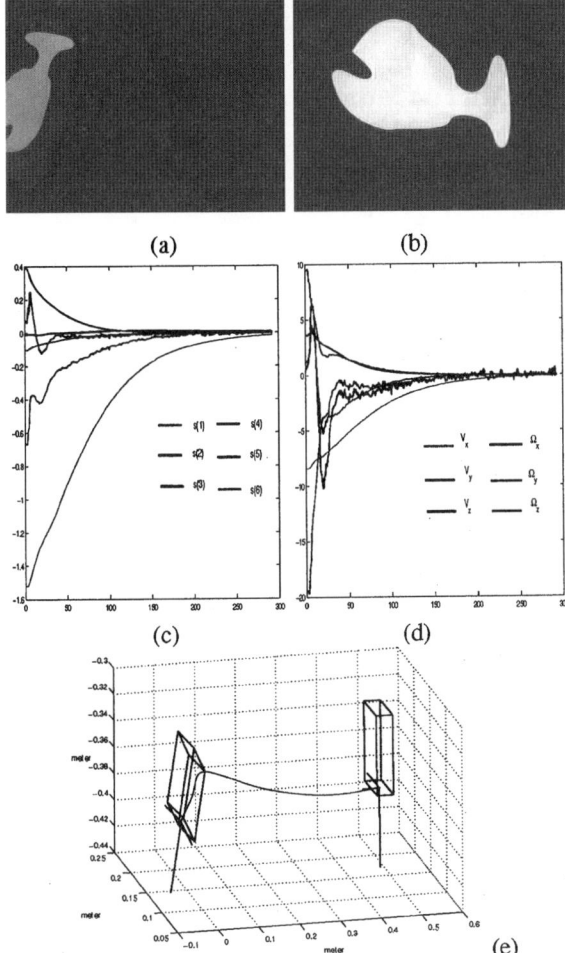

Fig. 4. Results using a bad camera calibration: (a) initial image, (b) desired image (c) visual features $(s - s^*)$, (d) camera velocity $\mathbf{T_c}$, (e) camera 3D trajectory

[4] F. Chaumette. A first step toward visual servoing using image moments. In *IEEE/RSJ IROS'02*, pages 378–383, Lausanne, Switzerland, Oct. 2002.

[5] C. Collewet and F. Chaumette. A contour approach for image-based control of objects with complex shape. In *IEEE/RSJ IROS'00*, pages 751–756, Takamatsu, Japan, Nov. 2000.

[6] P. I. Corke and S. A. Hutchinson. A new partitioned approach to image-based visual servo control. *IEEE Trans. on Robotics and Automation*, 17(4):507–515, Aug. 2001.

[7] K. Deguchi. A direct interpretation of dynamic images with camera and object motions for vision guided robot control. *Int. Journal of Computer Vision*, 37(1):7–20, 2000.

[8] B. Espiau, F. Chaumette, and P. Rives. A new approach to visual servoing in robotics. In *IEEE Trans. on Robotics and Automation*, volume 8, pages 313–326, June 1992.

[9] J. Flusser. On the independance of rotation moment invariants. *Pattern Recognition*, 33:1405–1410, 2000.

[10] S. Hutchinson, G. Hager, and P. Corke. A tutorial on visual servo control. *IEEE Trans. on Robotics and Automation*, 12(5):651–670, Oct. 1996.

[11] R. Mahony, P. Corke, and F. Chaumette. Choice of image features for depth-axis control in image-based visual servo control. In *IEEE/RSJ IROS'02*, pages 390–395, Lausanne, Switzerland, Oct. 2002.

[12] E. Malis, F. Chaumette, and S. Boudet. 2 1/2 d visual servoing. *IEEE Trans. on Robotics and Automation*, 15(2):238–250, Apr. 1999.

[13] A. G. Mamistvalov. n-dimensial moment invariants and conceptual theory of recognition n-dimensional solids. *IEEE Trans. on PAMI*, 20(8):819–831, Aug. 1998.

[14] R. Mukundan and K. R. Ramakrishnan. *Moment Functions in Image Analysis Theory and Application*. World Scientific Publishing Co.Pte.Ltd, 1998.

[15] R.J. Prokop and A. P. Reeves. A survey of moments based techniques for unoccluded object representation. *Graphical models and Image Processing*, 54(5):438–460, Sep. 1992.

[16] S. S. Reddi. Radial and angular moment invariants for image identification. *IEEE Trans. on PAMI*, 3(2):240–242, 1981.

[17] M.R. Teague. Image analysis via the general theory of moments. *Journal of Opt. Soc. of America*, 70:920–930, Aug. 1980.

[18] A. Walin and O. Kübler. Complete sets of complex zernike moments invariants and the role of the pseudo-invariants. *IEEE Trans. on PAMI*, 17(11):1106–1110, Nov. 1995.

[19] W. Wilson, C. Hulls, and G. Bell. Relative end-effector control using cartesian position-based visual servoing. *IEEE Trans. on Robotics and Automation*, 12(5):684–696, Oct. 1996.

Modeling and Vision-based Control of a Micro Catheter Head for Teleoperated In-Pipe Inspection

Saliha Boudjabi[1], Antoine Ferreira[1], Alexandre Krupa[2]

[1] Laboratoire Vision et Robotique (LVR), ENSIB- University of Orléans
10, Bd. Lahitolle, 18020 Bourges, France
[2] Laboratoire des Sciences de l'Image, de l'Informatique et de la Télédétection (LSIIT), ULP
Strasbourg, France
antoine.ferreira@ensi-bourges.fr

Abstract

The objective of this project, named MESIA (Multifunctional Micro-Endoscope Head System for Industrial Applications) is to develop an integrated multi-sensor system and to create intelligent sensor fusion for detection of micro-cracks. This paper is focused on the modeling of a 2-DOF pan-tilt platform actuated by Ni-Ti shape memory alloy (SMA) wires and antagonistic mechanical springs in order to control visually the CCD camera motion. A Preisach model is used and experimentally identified. The derived model is then exploited to design a position controller which compensates for the hysteretic nonlinearity. Finally, by controlling the catheter head orientation through a head motion tracker, in-pipe visual tracking has been experimentally verified from a remote site.

1. INTRODUCTION

The needs for inspection and diagnosis of industrial intra-tubular structures of millimeter sized diameter, i.e., vapor generator, electric turbines, nuclear stations, requires the development of active catheter devices with multifunctional abilities able to detect, measures and intervene in flooded pipes. The objective of this project (Fig.1b), named MESIA (Multifunctional Micro-Endoscope Head System for Industrial Applications) is to develop multi-sensor system and to create intelligent sensor fusion for detection of micro-cracks. It is envisaged that a CCD micro-camera would measure the surface geometry of the drained part of a sewer, while an ultrasonic sonar scanner measures the flooded part. Up to now, various technologies have been investigated in the actuation of micro endoscope head systems, such as electrostatic [1], pneumatic [2] and shape memory alloy (SMA) [3] microactuators. It appears that considering its interesting characteristics for low-powered and miniaturized micro head endoscopes with coarse and fine motion capabilities, SMA actuation is the mostly employed. As illustration of this technology for real in-pipe industrial inspection, a 2-DOF pan-tilt platform prototype actuated by two antagonistic Ni-Ti shape memory alloy wires has been developed (Fig.1a). Its goal is to control visually the CCD camera motion. Practically, the extreme environmental operating conditions of flooded pipes (inert gas, variable temperature, high pressure), lead to strong dynamic nonlinearities of SMA actuators, i.e. hysteresis, mechanical stress and thermal behavior. For

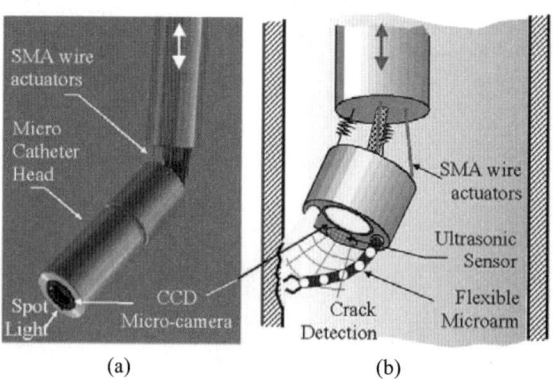

Figure 1. a) Current developed prototype and (b) schematic representation of active catheter head end-effector with multifunctional possibilities, e.g., vision, ultrasonic sonar, light, micro-tool.

accurate motion control, linearization based inverse model control methods should be employed. In [4], thermomechanical models with a limited number of physical parameters are provided, and used for closed loop control. A number of authors also use Preisach [5] or KP[6] modeling as input-output static mapping. These models do not account for the physics of the system but provides a generic way of modeling hysteretic systems. They use simple identification methods, and can be exploited in a control loop. Also, a number of convenient properties for control design, such as positivity, passivity, etc.., have been formally proven from Preisach models in [7].

In the first part, we briefly present the 2-DOF pan-tilt platform actuated by two antagonistic Ni-Ti shape memory alloy (SMA) wires. In Sections 3 and 4 we show how generalized Preisach model can be used for modeling and for vision-based control of the micro catheter head prototype. Then, Section 5 shows experiments of visual inspection in a non-flooded pipe made by a remote operator through a teleoperation system assisted by a graphical user interface (GUI) before to conclude in Section 6.

2. SYSTEM DESIGN

The micro catheter head end-effector is a 12 mm diameter system which includes (Fig.1):
- a two degree of freedom pan-tilt device, each degree of freedom being actuated by one Ni-Ti

shape memory alloy (SMA) wire and an antagonistic mechanical spring;
- a miniature CCD camera sensor integrating an achromatic lens,
- a piezoelectric ultrasonic sensor of 5 mm diameter for micro-crack characterization.
- a micro-tool acting as a marquer and/or a pointer (e.g.,micro-drilling tool) [8].

Once heated through Joule effect, an SMA wire changes from martensite phase to austenite phase. It is thus contracted, which generates a rotation of the mobile plate around the extremity of the fixed joint. Once cooled (through natural convection) it can recover its initial length (back rotation) if a force is exerted by the antagonistic spring. The electrical connection of the CCD sensor is possible through micro holes that have been drilled through the in-pipe base and the mobile part. The CCD sensor is connected to 7×25mm board which pre-amplifies the CCD signal. The ultrasonic sensor is connected to a signal conditioner board integrated into the catheter. The maximum deviation angle is approximately 60 degrees for a maximum input current of 400mA.

3. APPLICATION OF THE PREISACH MODEL TO SMA ACTUATORS

3.1. Generalized Preisach Model

The Preisach model is a phenomenological model which has recently been applied for modeling hysteresis in smart actuators. The SMA wires are current-controlled through the linear relation $i(t)=u(t)/R$ with $R=20\Omega$. In the following model, we assume that the electrical input is given by the applied voltage $u(t)$. Figure 2 shows experimental hysteresis curves describing the nonlinear angular motion versus applied voltage behavior of the SMA-driven platform. Not modeling hysteresis in SMA material can lead to inaccuracy in open-loop control, and can generate amplitude-dependent phase shifts and harmonic distortion that reduce the effectiveness of feedback control. In order to implement an efficient visual tracking control, hysteresis have to be taken into account. The classical Preisach model equation relating the shape memory alloy deformation $f(t)$ to the electric current $u(t)$ is [9]:

$$f(t) = \iint_P \mu(\alpha,\beta) \cdot [\gamma_{\alpha\beta}(u(t))] \cdot d\alpha \cdot d\beta \quad (1)$$

where $\gamma_{\alpha\beta}[.]$ are elementary hysteresis operators with switching values α and β and whose values are determined by the input current signal $u(t)$. The output of these operators is multiplied by the corresponding Preisach functions $\mu(\alpha,\beta)$, and then summed continuously over all possible values of α and β. The function $\mu(\alpha,\beta)$ is a weighting function estimated from the measured data and is defined as :

$$\mu(\alpha,\beta) = \frac{1}{2} \cdot \frac{\partial^2 f(\alpha,\beta)}{\partial\alpha\partial\beta}$$

From Fig.2, it is clear that the SMA actuators hysteresis loop is only defined in the first quadrant of the $u(t)$-$f(t)$ plane. So the value of the hysteresis operator $\gamma_{\alpha\beta}[u(t)]$ are selected as switching between 0 and 1. Knowing what the maximum and minimum values of the input are, the condition $\alpha \geq \beta$ leads to a limiting triangle T on the α–β plane (Fig.3) which is defined such that the function $\mu(\alpha,\beta)$ is equal to zero outside T. On the α–β half-plane, there is a one-to-one correspondence between operators $\gamma_{\alpha\beta}[u(t)]$ and points (α,β) which implies at each pair of values defines a unique operator $\gamma_{\alpha\beta}[.]$ with switching values α and β [9].

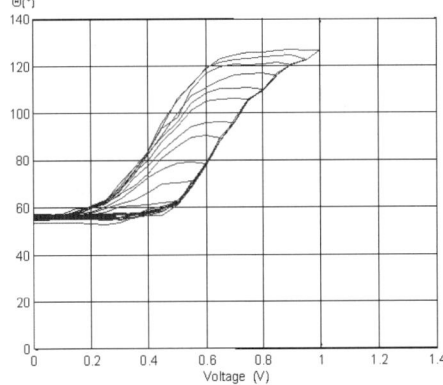

Figure 2. Angular position versus voltage curves showing hysteresis behavior of SMA wire actuators during martinsite and austenite phases.

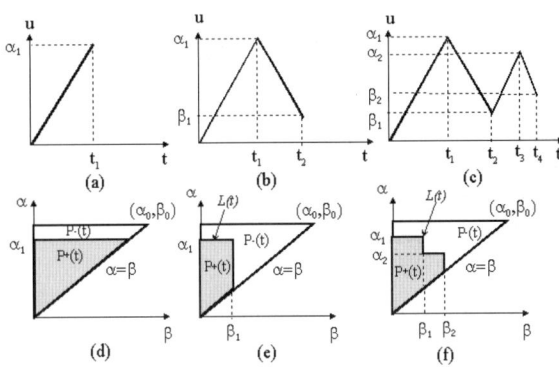

Figure 3. Mechanism of storage of input history in the Preisach plane. Graphs (a)-(c) depict the time trajectory of voltage input and graphs (d)-(f) show the corresponding subdivisions of the triangle T and the formation of the interface $L(t)$.

At each instant of time and as a result of applying an input $u(t_i)$, the limiting triangle in the half-plane can be divided into two areas, $P^+(t)$ and $P^-(t)$ (Fig.3a). All operators that belong to $P^+(t)$ are equal to 1 and those that belong $P^-(t)$ are equal to 0. For detailed explanations, the reader may refer to [9]. As the input starts to monotonically increase, the hysteresis operators $\gamma_{\alpha\beta}[u(t)]$ with switching values less than the current input $u(t_1)$ switch to "+1" position (P^+ region). The demarcation between the two sets is a horizontal line given by the equation $\alpha = \alpha_1 = u(t_1)$, and shown in Fig.3(d). Next, when the input starts to monotonically decrease from $u(t_1)$ to $u(t_2)$ all the hysteresis operator $\gamma_{\alpha\beta}[u(t)]$ with switching values larger than the current input value $u(t_2)$ switch to the "0" position (P. region). Geometrically, this means that the previous subdivision of the limiting triangle T is changed, and the interface link $L(t)$, which express the boundary between the regions $P^+(t)$ and $P^-(t)$, moves from right to left (Fig.3b-e). Based on the above, the latter two displacements are denoted $F(u_{t1}, u_{t2})$ and are defined as:

$$F(u_{t1}, u_{t2}) = f_{u_{t1}} - f_{u_{t1}u_{t2}} = \iint_P \mu(\alpha, \beta) \cdot d\alpha \cdot d\beta \quad (2)$$

The function $F(u_{t1}, u_{t2})$ is a double integral of the weighting function $\mu(\alpha, \beta)$ over the region $P_+(u_{t1}, u_{t2})$. In the third step, voltage is again increased to a value α_2 at time t_3, which is less than α_1. This increase the input results in the formation of a new horizontal link in $L(t)$ at $\alpha = \alpha_2 = u(t_3)$. Next, as voltage is decreased again to β_2 greater than β_1, a new vertical link appears on the interface function $L(t)$. The figure of the final two links in $L(t)$ is shown in Fig.3(f).

3.2. Identification and Numerical Implementation

The Preisach model can be implemented by using Eq.(2) to compute the output $f(t)$. However, the computation of the weighting function $\mu(\alpha, \beta)$ requires double differentiation which amplifies errors always present in experimentation. In general, the expression of the SMA expansion, in the case of a monotonically increasing input, is:

$$f(t) = \sum_{k=1}^{n-1} [F(M_k, m_{k-1}) - F(M_k, m_k)] + F[u(t), m_{n-1}] \quad (3)$$

where $f(t)$ stands for the SMA expansion estimated using the classical Preisach model, (n-1) represents the number of maxima/minima stored (not wiped-out) and the pairs $\{(M_k, m_k)\}$ represent the sequence of maximum and minimum values of the input signal constituting the coordinates interface $L(t)$. The corresponding triangles are depicted in Fig.4(a-b). By considering

$$F(M_k, m_k) = f_{M_k} - f_{M_k m_k} \quad (4)$$
$$F(M_k, m_{k-1}) = f_{M_k} - f_{M_k m_{k-1}}$$

Eq.(3) can be rewritten as (Fig.4a):

$$\dot{u}(t) > 0: \quad f(t) = \sum_{k=1}^{n-1} [f_{M_k m_k} - f_{M_k m_{k-1}}] + f_{u(t)} - f_{u(t) m_{k-1}} \quad (5)$$

and similarly, when the input is monotonically decreasing (Fig.4b):

$$\dot{u}(t) < 0: \quad f(t) = \sum_{k=1}^{n-1} [f_{M_k m_k} - f_{M_k m_{k-1}}] + f_{M_n u(t)} - f_{M_n m_{k-1}} \quad (6)$$

From this, it is possible to explicit expressions that can be used to calculate the response $f(t)$ of a SMA actuator

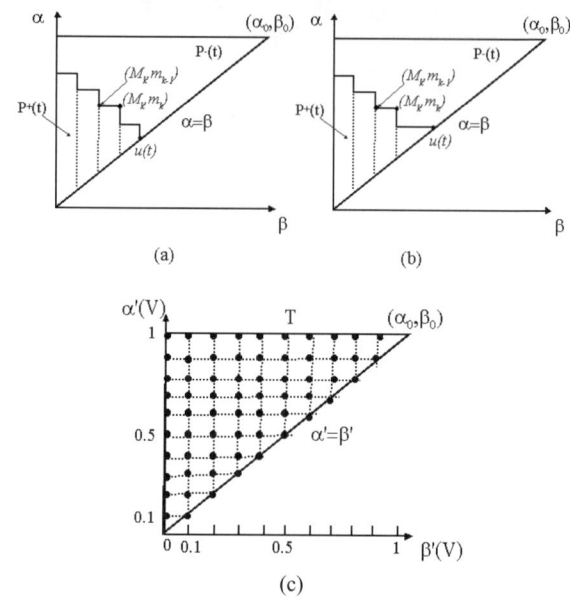

Figure 4. Triangle T for numerical implementation of Preisach model when (a) input is decreasing, (b) input is increasing and (c) square mesh in α–β plane.

subject to a known arbitrary input voltage sequence $u(t)$. Numerical implementation of the above form of the Preisach model requires the experimental determination of $F(\alpha, \beta)$ at a finite number of points within the triangle T. For this purpose, a square mesh covering the triangle T is created, divided into a number of squares and triangles, and the corresponding values of $f(\alpha, \beta)$ are experimentally determined (Fig.4c). The parameters of the Preisach model $F(\alpha, \beta)$ are calculated using Eq.(5) and (6). The alternating series of dominant extrema (M_k, m_k) are calculated from the time history of the input voltage, to determine the vertices of the interface $L(t)$. This series is continuously updated at each instant of time. Finally, the output $f(t)$ of the hysteresis nonlinearity is calculated using the Preisach model parameters $\{F_{\alpha,\beta}\}$, in (5) and (6). In the experiments reported in this article, $\alpha_0 = \beta_0 = 1.2V$. A grid of points with $\Delta\alpha = \Delta\beta = 0.1V$ in the complete range of variation of α and β on the triangle T is created. Then, $f_{\alpha,\beta}$ are experimentally determined for these grid points, and the Preisach model parameters model are calculated. For points of the (α, β) plane lying within of the squares, the interpolation Eq. (7) is used, and for points lying within any of the triangles, the linear interpolation Eq.(8) is used.

$$f_{\alpha'\beta'} = c_0 + c_1\alpha' + c_2\beta' + c_3\alpha'\beta' \quad (7)$$
$$f_{\alpha'\beta'} = c_0 + c_1\alpha' + c_2\beta' \quad (8)$$

The c_i-coefficients in (7)-(8) are found for each square and triangle using the experimental values $f_{\alpha'\beta'}$ at the corners of the squares and triangles, respectively. In order to verify the validity of the identified model, an arbitrary sequence was applied to the experimental system. Figure 5 shows the good agreement between the real and simulated outputs.

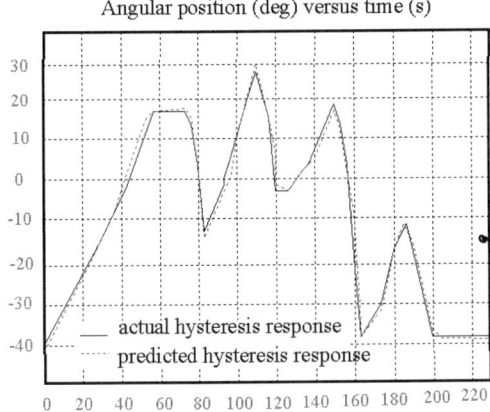

Figure 5. Actual and predicted hysteresis response under an arbitrary input excitation.

4. VISION-BASED CONTROL

4.1. Inversion of the Preisach Model Hysteresis

As explained in the previous section, the modeling of hysteresis allows to design controllers that correct these hysteretic effects and improve accuracy of the orientation of the micro catheter head. Essential to the synthesis of such controller architectures is the development of the inverse Preisach model. Given the Preisach model parameters $\{F_{\alpha',\beta'}\}$ and the associated interface $L(t)$ for the triangle T, the inverse Preisach model determines the voltage $u(t+\Delta t)$ that will result in a desired orientation angle $\theta_d(t+\Delta t)$ that at the next instant. In the formulation of the inverse Preisach model, two distinct cases corresponding to decreasing $\theta_d(t+\Delta t) < \theta_d(t)$ and increasing $\theta_d(t+\Delta t) > \theta_d(t)$ need to be considered. When $\theta_d(t+\Delta t) = \theta_d(t)$, then $u(t+\Delta t) = u(t)$ as shown in Fig. 6. The terms θ_d, θ_m and u are the signal entry of the Preisach model, the output of the Preisach model and the inverse Preisach model, respectively. Consequently, it is possible to pursue model-based compensation of hysteresis nonlinearity in these actuators using an open-loop compensation strategy depicted in Fig.7. The final inverse model consists of two blocks: an inverse linear transfer function which approximates the heating process occurred on SMA wires and an inverse nonlinear Preisach block. The performance of the compensation scheme is experimentally investigated for variations of angular orientation (Fig.8). The experimental angular position trajectory follows accurately the first rising ramp for slowly varying inputs which is not the case at the end of the input profile due to strong dynamical effects and variable dynamic behavior of the SMA heating process. A significant error exists between the reference and the actual signal after a long period of operation.

4.2. 2-D Visual Servoing Approach

We define as visual features the projection of the center of

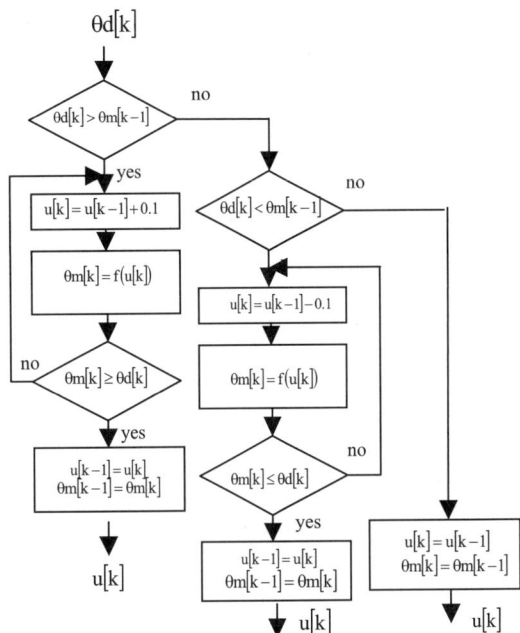

Figure 6. Algorithm of Preisach model inversion.

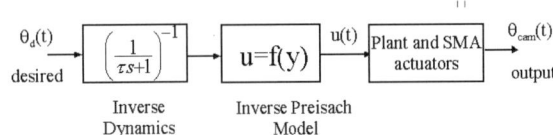

Figure 7. Model for open-loop compensation of hysteresis.

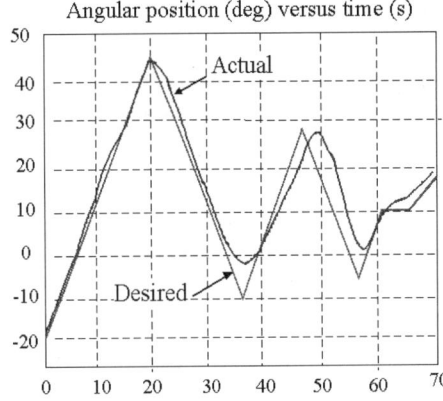

Figure 8. Tracking control using open-loop control.

gravity of the target: $\mathbf{P}=(X,Y)^T$ and we control the camera in order to see it centered in the image : $\mathbf{P_d}=(X_d,Y_d)^T = (0,0)^T$. The image Jacobian is given by [10]:

$$\mathbf{L} = \begin{pmatrix} XY & -(1+X^2) \\ 1+Y^2 & -XY \end{pmatrix} \quad (9)$$

and the resulting control law is simply estimated from :

$$\begin{pmatrix} \theta_X^{act} \\ \theta_Y^{act} \end{pmatrix} = -\lambda \begin{pmatrix} \dfrac{Y}{1+X^2+Y^2} \\ -\dfrac{X}{1+X^2+Y^2} \end{pmatrix} + \mu I, \qquad (10)$$

where I, λ and μ are an integral term to attenuate the tracking errors, and constant parameters, respectively.

4.3. Internal-based Model Control

Different closed-loop controllers have been developed for control of SMA actuators based on inverse Preisach model $u=f^{-1}(y)$. In [11], the inverse Preisach model is used in a feedforward compensation, in addition to a PI feedback control loop. In, [12], a model reference adaptive control (MRAC) using the inverse Preisach model as inverse compensator is used. These methods can be poorly applied on our prototype taking into account the variability of the in-pipe operating conditions. A more robust approach based on feedback linearization with an internal model has been studied (Fig.9).

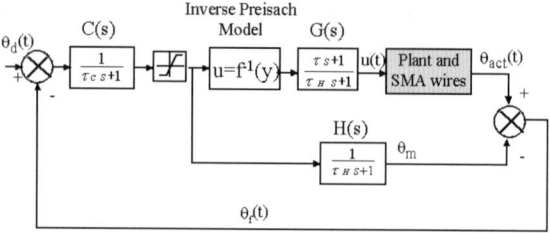

Figure 9. Scheme of the internal model based control.

The desired input and output position vectors are defined by $\theta_d = [\theta^d_x, \theta^d_y]$ and $\theta_{act} = [\theta^{act}_x, \theta^{act}_y]$, respectively. The vector position $\theta_m = [\theta^m_x, \theta^m_y]$ defines the internal model output. The function G(s) is used to compensate the phase-lag due to dynamics of the plant by placing a slow pole ($-1/\tau_H$). The main characteristics of the internal model based control is to guarantee the closed-loop stability by adjusting a simple corrector C(s) and to ensure the robustness against perturbations. The filter H(s) describes the ideal transfer function between the actual angular position and the input of the inverse Preisach model. In order to verify the efficiency of the proposed approach, Fig.10 shows the tracking results for different desired positions by considering only one degree of freedom of the micro catheter head. It is shown that the closed loop behaviour using the inverse Preisach model linearizes greatly the system dynamics. Figure11 shows the a two-dimensional closed-loop tracking of a circular trajectory. From this experiment, it can be seen that the vision-based closed tracking confirms that the generalized inverse Preisach model linearizes the system dynamics. We can conclude that off-line identification can reduce the computational overhead for real time control implementation.

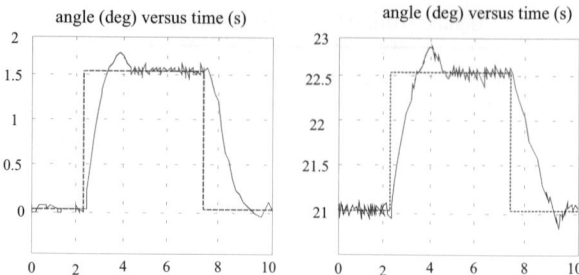

Figure 10. Closed loop response with Preisach model based controller.

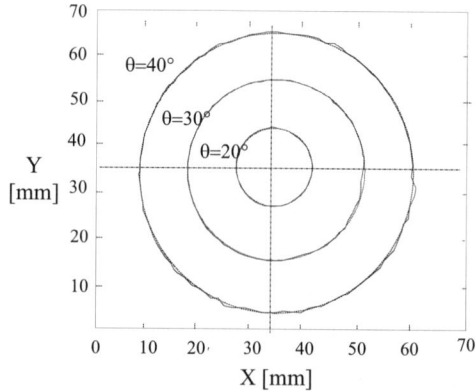

Figure 11. Vision-based tracking closed loop response of a circular trajectory.

5. TELEOPERATED VISUAL IN-PIPE INSPECTION

5.1. Head Motion Tracking System

Figure 12 shows the input and output devices for teleoperated in-pipe inspection. A joystick is used as an input device for manoeuvering of the microcatheter locomotion, whereas the pan and tilt motions of the camera are controlled by the output of a motion tracking system (Flick of Birds, *Ascension Technology Corp.*) attached to an head mounted display (*Seam Team Corp.*).

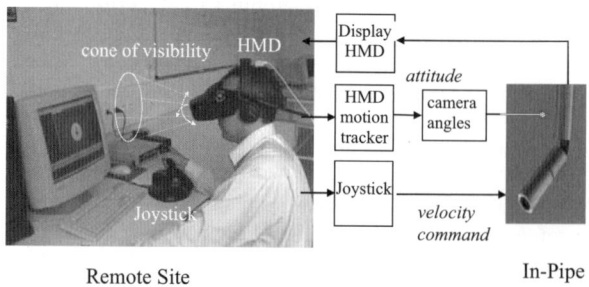

Figure 12. Vision-based head motion tracking system for in-pipe inspection.

The camera control commands, which are calculated on the basis of the head attitude and direction detected by the head motion tracking sensor, are sent to the camera angle controller. The latter calculates the scaled operator attitude in order to be within the *cone of visibility*. This cone corresponds to the scaled orientation limits of the catheter head system. Similarly, the motion commands that are calculated on the basis of the joystick position are sent. The joystick has 2-DOF for forward-backward and left-right, and a button which is pushed when the operator wants to select a reference image point. The real image captured by micro-vision camera is displayed on the HMD.

5.2. Strategy for Visual Inspection

The main strategy for in-pipe inspection process flow is described as follows:

Step1: The catheter head is inserted in the pipe and visual inspection is performed through the operator head motion attitude.

Step2: The supervisor guides a cursor on a monitor (using a computer mouse) which the visual tracking system accepts as a control input to a visual servoing marquer.

Step3: Focus on the detected crack.

Step4: A gradient method is applied in order to extract effect regions from the pipe surface image. The main interest is to identify the points where the variation in intensity levels are high, regardless of the lighting conditions.

Step5: Ultrasonic characterization of the crack is performed under visual tracking due to instabilities of the catheter head through antagonistic effects (pressure, temperature, liquid flow).

Step6: Reconstruction of a 3D-graphic representation of the failure (2D location and depth) through sensor fusion at the pixel level.

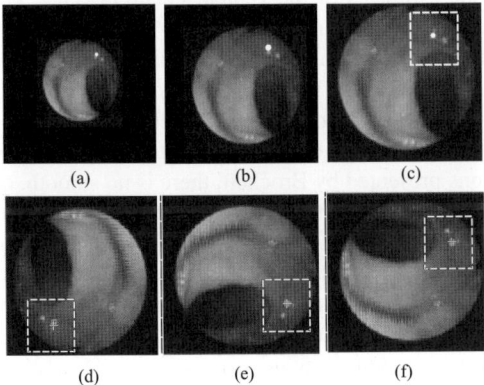

Figure 14. Images of a pipe with a 1mm hole. The location of the hole is marked with a small square. Images from (a) to (c) are acquired with a standard lighting system when the catheter head is inserted. The crack can be easily identified (white spot). Note that the spot next to the white spot is created by light reflection and it is not a defect. Images from (d) to (f) shows the efficiency of visual tracking. The white spot is efficiently tracked during attitude motion of the HMD system for further visual inspection.

5.3. Experiments

Some experiments have been performed in an non-flooded metallic pipe (20mm diameter) to test the efficiency and robustness of the remote visual tracking system in real operating conditions (see Fig.14).

5. CONCLUSION

This paper has shown the first implementation of an integrated design of a multifunctional micro catheter head for in-pipe industrial inspection. The first step of integration of modeling and vision-based control issues have addressed. An extended Preisach methodology has proven to be very efficient for both modeling and control aspects. However, further work will be carried out in order to implement adaptive control of hysteretic systems in order to match with the in-pipe variation of operating conditions. Visual tracking experiments made in a non-flooded pipe has shown that it was possible to decouple attitude head motion of the operator when the visual tracker is switch on. The next-term objective (Step 5 and 6) of this research is to assess partially flooded millimeter-sized pipes using the described multi-sensor system.

6. REFERENCES

[1]: A. Sadamoto *et al.* "Wireless micromachine for in-pipe visual inspection and the possibility of biomedical applications", *32nd Int. Symposium on Robotics*,19-21 April, Seoul, 2001, pp.433-438.

[2]: S. Kumar, I. M Kassim, V.K. Asari, "Design of a vision-based microrobotic colonoscopy system", *Advanced Robotics*, Vol.14, No.2, 2000, pp.87-104.

[3]: K. Ikuta, M. Tsukamoto, S. Hirose, "Mathematical Model and Experimental Verification of Shape Memory Alloy for Designing Microactuator", *IEEE Int. Conf. on Rob. and Automation*, 1988.

[4]: H.Benzaoui, "Modélisation thermomécanique et commande d'actionneurs en alliage à mémoire de forme pour la microrobotique", PhD dissertation of university of Franche-Comté, 1998.

[5]: R.B. Gorbet, D.W.L. Wang, K.A. Morris, "Preisach Model Identification of a Two-Wire SMA Actuator", *IEEE Int. Conf. On Robotics and Automation*, 1998.

[6]: G.V.Webb, D.C. Lagoudas, A. Kurdial "Hysteresis Modeling of SMA Actuators for Control Applications", *Jour. of Intelligent Material Systems and Structures*, Vol.9, 1998, pp.432-448.

[7]: R.B. Gorbet, "Control of Hysteretic Systems with Preisach Representation", PhD thesis, University of Waterloo, 1997.

[8]: J. Forêt, A. Ferreira, J-G. Fontaine, "GA-based control of a binary-continuous joints of a multi-link micromanipulator using SMA actuators", *IEEE Conference on Robots and Intelligent Systems*, Oct. 1-4, Lausanne, Switzerland, 2002, pp.1736-1741.

[9]: I.D. Mayergoyz, "Mathematical models of hysteresis", Springer-Verlag, New-York, 1991.

[10] : B. Espiau, F. Chaumette, and P. Rives, "A new approach to visual servoing in robotics", *IEEE Trans. on Robotics and Automation*, 8(3) June 1992, pp. 313-326.

[11]: P. Ge, M. Jouaneh, "Generalized Preisach model for hysteresis nonlinearity of piezoceramic actuators", *Precision Engineering*, Vol.20, 1997, pp.91-111.

[12]: D. Song, C.J. Li, "Modeling of piezoactuator's nonlinear and frequency dynamics", *Mechatronics*, Vol.9, pp.391-410, 1999.

Visual Servoing of a Car-like Vehicle – an Application of Omnidirectional Vision

Kane Usher, Peter Ridley
School of Mechanical, Medical and
Manufacturing Engineering,
Queensland University of Technology,
2 George Street, Brisbane 4001, Australia

Peter Corke
CSIRO Manufacturing and Infrastructure
Technology,
P.O. Box 883, Kenmore 4069, Australia

Abstract—In this paper, we develop the switching controller presented by Lee et al. [14] for the pose control of a car-like vehicle, to allow the use of an omnidirectional vision sensor. To this end we incorporate an extension to a hypothesis on the navigation behaviour of the desert ant, *cataglyphis bicolor*, which leads to a correspondence free landmark based vision technique. The method we present allows positioning to a learnt location based on feature bearing angle and range discrepancies between the robot's current view of the environment, and that at a learnt location. We present simulations and experimental results, the latter obtained using our outdoor mobile platform.

I. INTRODUCTION

A key skill for the autonomous navigation of mobile robots is the ability to find particular locations in a workspace. Further to this, in order to perform useful tasks, a mobile robot requires the ability to servo to particular poses in the environment. For the nonholonomic, car-like vehicle used in these experiments, Brockett [3], showed that there is no smooth, continuous control law which can locally stabilise such systems.

Insects in general display amazing navigation abilities, traversing distances far surpassing the best of our mobile robots on a relative scale. To do this, evolution has provided insects with many 'shortcuts' enabling the achievement of relatively complex tasks with a minimum of resources in terms of processing power and sensors [21]. In particular, the high ground temperatures encountered by the desert ant, *cataglyphis bicolor*, eliminates pheremones as a potential navigation aid, as is used by ants in cooler climates [12]. The desert ant navigates using a combination of path integration and visual homing.

Visual homing, also known as visual piloting, is the process of matching an agent's current view of a location in a distinctive locale to a (pre-stored) view at some target position. Discrepancies between the two views are used to generate a command that drives the agent closer to the target position. The process enables the agent to 'find' positions in distinctive locales. These distinctive locales can then be linked to generate paths through an environment [9], [11], eliminating the need for complex map-like representations, instead embedding this knowledge in terms of what the agent's sensors can 'see'.

When applied to nonholonomic mobile robots, the constraints of Brockett's theorem prevent the insect-based strategies from completely resolving the pose stabilisation problem; they enable servoing to a position but cannot guarantee a particular orientation (see e.g. [12], [21] – it should be noted that these vehicles could spin on their own axis, and thus orientation errors on reaching 'home' could be resolved with a simple switching strategy – the problem is more complex for a car-like vehicle).

In the control community, the problem of stabilising a mobile robot to a specific pose has generally been approached from two directions; the open-loop and closed-loop strategies. Open loop strategies seek to find a bounded sequence of control inputs, driving the vehicle from an initial position to some arbitrary position, usually working in conjunction with a motion planner (e.g. [13], [18]). Finding this sequence of control inputs is difficult for the nonholonomic case and, in the event of disturbances, a new plan has to be formulated.

The closed-loop strategies consist of designing a feedback loop using proprioceptive and exteroceptive sensors to provide estimates of the vehicle's state (see e.g. [16]). Feedback control systems are generally more robust to uncertainty and disturbances when compared to their open-loop counterparts. All real mobile robots and sensors are subject to noise and uncertainty — feedback control would thus seem essential. However, due to the well-known limitations presented by Brockett, there is no smooth, continuous control law which can locally stabilise closed-loop nonholonomic systems to a pose [3]. These limitations can be side-stepped by either relaxing the constraints on desired pose (i.e. stabilising to a point without a guarantee on orientation, as for the insect inspired approaches), using discontinuous control techniques (see e.g. [1], [2], [5], [14]), or by using time-varying control (see e.g. [20]). Much of the literature does not address what has been found in this study to be a significant limitation of many control algorithms for the pose stabilisation of car-like vehicles — input saturation.

For the task at hand, we argue the case for the feedback control methods, with vision used as the primary sensor. However, it is clear that a measure of open-loop planning is also usually required in order to prevent deadlock situations from occurring. Here, we are concerned only with the feedback component for the task of servoing to a position and orientation. The use of visual feedback is finding increasing application in the solution to this problem (see e.g. [5], [6], [8], [19]), but the use of omnidirectional camera systems has not yet been fully explored.

The remainder of this paper is arranged as follows: Section II details the switching control technique and some of the adaptations we have made to cope with noisy vision data and to ensure stability; Section III describes our experimental system and briefly outlines some preliminary results; Section IV concludes the paper and presents some directions of future interest.

II. CONTROL STRATEGY

In this section, the switching control strategy presented by Lee et al. [14] is developed. We then show how a derivative of the insect inspired Average Landmark Vector model of navigation presented by Lambrinos et al. [12] can be used to provide the required quantities to the switching controller.

A. Kinematics

Our vehicle is (approximately) car-like in its kinematics. Referring to Fig. 1, the kinematics in Cartesian space of our experimental vehicle are:

$$\begin{aligned} \dot{x} &= v\cos\theta \\ \dot{y} &= v\sin\theta \\ \dot{\theta} &= v\frac{\tan\phi}{L} \end{aligned} \quad (1)$$

where v is the vehicle's forward velocity (measured at the centre axle of the rear wheels), L is the vehicle's length, ϕ is the steering angle, and the point (x,y) refers to the centre of the front axle. (It is more usual to define the coordinates of a rear-wheel driven vehicle with respect to the centre of the rear axle. However, on our vehicel, the sensors which provide position information are mounted over the front axle and thus it makes more sense to use this as the reference point — it is a simple transformation between the front and rear points on the vehicle.)

B. Control Law Development

The technique presented by Lee et al. [14] operates in Cartesian space and sidesteps Brockett's theorem via a discontinuity in the control law. The control law operates in two distinct stages. Without loss of generality, consider the goal pose to be $(x,y,\theta) = (0,0,0)$. The initial stage involves minimising y and θ; i.e. the vehicle converges to the x-axis with an orientation of 0. The second stage then moves the vehicle along the x-axis to the desired point. Each controller is described in the following sections.

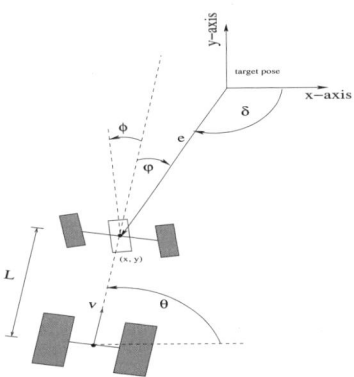

Fig. 1. The vehicle and the coordinate system used. All angles are counter-clockwise positive.

1) Control Law — First Stage: The first stage of the control law takes the vehicle to the x-axis with the correct orientation. This is achieved via the minimisation of the Lyapunov-like function [14]

$$V = \frac{1}{2}k_1 y^2 + \frac{1}{2}\theta^2$$

This function is radially unbounded and positive semi-definite. Its derivative is:

$$\begin{aligned} \dot{V} &= k_1 y \dot{y} + \theta \dot{\theta} \\ &= k_1 y v \sin(\theta) + \theta v \frac{\tan(\phi)}{L} \end{aligned}$$

With the following choice for the steering angle,

$$\phi = \arctan\left[-\frac{L}{v}\left(k_2\theta + k_1 v \frac{\sin(\theta)}{\theta}y\right)\right]$$

Lee et al. [14] showed that y and θ converge to zero.

The velocity of the vehicle is chosen according to its initial orientation with respect to the goal. If the vehicle is facing the goal, it is given a constant positive velocity, else it is given a constant negative velocity, i.e.

$$v = \begin{cases} k_3 & \cos\varphi_{initial} < 0 \\ -k_3 & \cos\varphi_{initial} \geq 0 \end{cases}$$

where φ is the initial orientation of the goal relative to the vehicle.

The velocity v could be switched at any time, depending on obstacle layout etc.. Lee addressed local minima problems via the use of 'intermediate' points. At this stage in our work, we assume there are no obstacles, and we simply set the velocity equal to a constant, k_3, with the sign assigned as explained above. However, we set a limitation on the distance the vehicle can be from the goal, and reverse the velocity if this distance is exceeded.

2) Control Law — Second Stage: On reaching the y-axis and $\theta = 0$, the following control law is invoked:

$$v = -k_3 x$$
$$\phi = \arctan\left[-\frac{L}{v}\left(k_2\theta + k_1 v \frac{\sin(\theta)}{\theta} y\right)\right]$$

bringing the vehicle to the desired pose of $(x, y, \theta) = (0, 0, 0)$. In the original work of Lee et al., the steering angle was set to zero in the second stage of control. We have found that in practice, due to the imperfections of 'real' systems, it is necessary to control the steering angle in the second stage of control. An analysis of the Lyapunov function for this case shows that convergence is still guaranteed with the above choice of steering angle.

In combination, these control laws stabilise the vehicle to the desired pose from any initial condition. Again in practice, the imperfections of 'real' systems requires that switching to the second stage of control occurs when the y-axis distance and the error in orientation (θ) drop below pre-specified thresholds.

The control technique presented above assumes that the distance and angles to the target location can be measured, from which the quantities required for the controller can be calculated. In the next section, we detail a landmark-based vision technique which yields this information without explicit knowledge of the robot's pose; sensing is provided by both an omnidirectional camera and a compass.

C. Ant Navigation

The desert ant, cataglyphis bicolor, is unable to use pheremones to navigate due to the high ground temperatures found in its habitat. Thus, it relies on a combination of visual piloting and path integration, enabling it to find nest openings of less than a few millimetres after foraging journeys of several hundred metres [12], [22]; see Fig. 2 for an example of a foraging journey. On this journey, the ant has travelled a round trip distance of over two hundred metres, returning to a nest with an opening of less than 5 mm, equating to a drift rate of less than 0.0025% in its navigation system. Comparing this to high-end commercial inertial navigation units for land based navigation, which have drift rates of the order 0.1% of distance travelled, demonstrates the effectiveness of this insects navigation system.

Here however, we are interested in the visual piloting component of the ants navigation system. When visual piloting, ants take a rather unprocessed view of the target location and match it with a current view, using the discrepancies to derive a direction of movement. Matching characteristics used are the differences in landmark bearings, apparent size and apparent height [12].

An elegant, correspondence free, homing method developed from hypotheses on how these ants might use visual piloting is the Average Landmark Vector model. An

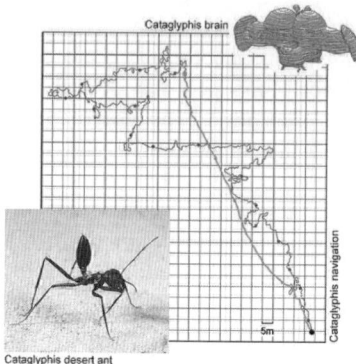

Fig. 2. An example of the amazing navigation feats of the humble ant. Diagram courtesy www.neuroscience.unizh.ch/e/groups/wehner00.htm, Rudiger Wehner.

ALV for any particular position in the workspace is found by summing unit vectors towards all currently visible landmarks, and dividing by the number of landmarks. By matching the current ALV with a pre-stored ALV of the target location, a homing vector can be formed which drives the agent (robot) towards the target location [12]. In order to consistently add the vectors in the ALV model, an absolute reference direction is required, and, unless apparent size information is incorporated, a minimum of three landmarks is needed.

The nature of our sensor, an omnidirectional vision system, led us to investigate improvements to the ALV method. In its original form, the ALV method required the bearings to landmarks only. Range information could be incorporated by including landmark apparent size, and slight improvements to the performance could be made. However, we have found that by including range information directly, a significant improvement is made and in fact, the distance and angle to the goal are yielded directly. In addition, the minimum required landmarks is reduced to one. An example of the IALV method is shown in Fig. 3. In essence, the IALV method is equivalent to finding a position relative to the centroid of the landmarks in the workspace.

As with the ALV method, the IALV method is purely sensor-based. Landmark bearings are readily ascertained with an omnidirectional camera. If a flat-earth assumption is made, range information can be derived from an omnidirectional camera image through the geometry of the camera/mirror optics. Alternatively, optic flow techniques could be used to determine landmark range [4].

One of the advantages of the ALV, and hence the IALV method, is that knowledge of a target location is contained within a single quantity. This reduces the need for complex map-like representations of the environment and is well suited for a topological navigation method, (see e.g. [7], [11], [15], [17]). Additionally, landmarks need not be unique, and the need for landmark correspondence is also

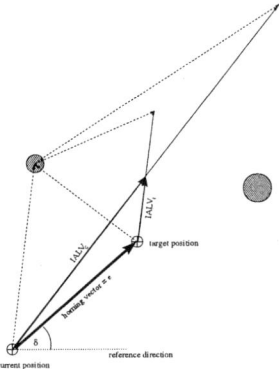

Fig. 3. Illustration of the the IALV method for two landmarks in a workspace. The IALV's are found by adding the vectors to the individual landmarks, and dividing the resulting vector by the number of landmarks. The home vector is then calculated by subtracting the target IALV from the currently IALV.

bypassed. Many of the other homing algorithms require that the landmarks in the current image be matched with those at the target location, usually by minimisation of the sum of the bearing differences (see e.g. [21]). If landmarks are occluded or missing, these methods can fail. Of course, like all sensor-based techniques, this method has a finite catchment area, limited by the omnidirectional sensor's range and, in addition, has the potential to suffer from perceptual aliasing, or in a similar sense, the local minima problem.

The homing vector provided by the IALV method can be used to drive the agent towards home but does not provide a means of guaranteeing a final orientation. However, the quantities derived from the IALV can easily be converted to the states required by the switching controller developed earlier.

D. Simulations

In this simulation, landmarks are modelled as points, consistent with the requirements of the IALV model which merely requires the range and bearing to one or more non-unique landmarks in the workspace. An IALV is taken at a target location, nominally the origin, with this target IALV being matched with the agent's current IALV throughout the vehicle's journey.

The vehicle model used is simply the kinematic equations (as given in equation 1). Also included in the model are input saturation and first-order delays in the steering and velocity loops. A simulation of the omndirectional sensor (with random noise) provides the IALV, from which the required states are derived and fed to the switching controller. Fig. 4 shows the path generated, pose, and control inputs for a starting pose of $(x, y, \theta) = \left(0, -5, \frac{\pi}{2}\right)$ and a goal pose of $(0,0,0)$. Gains were set to $(k_1, k_2, k_3) = (0.35, 0.1, 0.1)$ for these simulations, based on an analysis

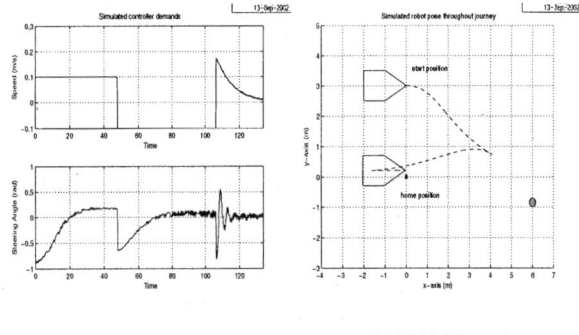

(a) Demanded inputs. (b) Vehicle's path.

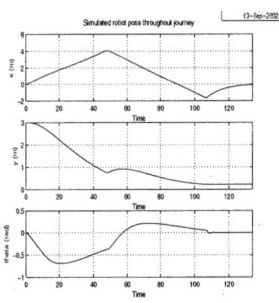

(c) Pose throughout journey.

Fig. 4. Results of the simulation for a starting pose of $(x, y, \theta) = (0, 3, 0)$.

of the linearised system. The method works for all starting and goal poses.

The method has been extensively tested in simulation and found to be quite robust to input saturation and noise. The next section presents some experimental results which validate our simulations.

III. EXPERIMENTS

Experimental validation of this visual servoing technique was conducted on our outdoor research testbed. In these experiments, artificial landmarks in the form of red traffic cones (witches hats) were used. The following sections give a brief overview of the vehicle and outline the image processing used to extract the landmarks.

A. Robotic Testbed

The experimental platform is a Toro ride-on mower which has been retro-fitted with actuators, a control system, and a computer, enabling control over the vehicle's operations. All control and computing occurs on-board. The vehicle is fitted with an array of sensors including odometry, GPS, a magnetometer, a laser range-finder (for collision avoidance only) and an omnidirectional camera (see Fig. 5 for a photograph of the vehicle). For the experiments cited here, the primary sensor used is the

Fig. 5. The experimental platform. Note the omnidirectional camera mounted over the front wheels and the box at the rear which houses the control and computer system.

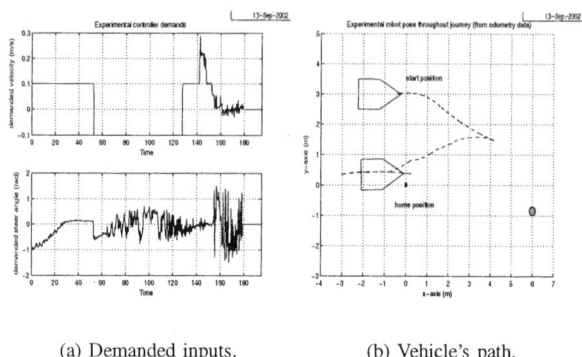

(a) Demanded inputs.   (b) Vehicle's path.

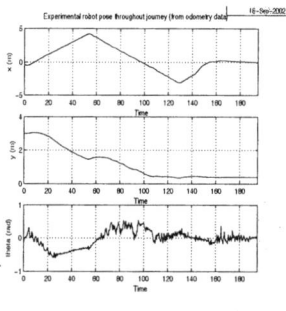

(c) Pose throughout journey.

Fig. 6. Experimental results for a starting pose of $(x, y, \theta) = (0, 3, 0)$. Note the similarity to the results achieved with the simulation plotted in Figure 4.

omnidirectional camera with the magnetometer providing an absolute reference direction.

B. Image Processing

The distinguishing feature of the landmarks used in this experiment is their colour. Hence, we use colour segmentation to track them; not a trivial task in an outdoor environment with no control over lighting conditions. Our frame-grabber provides a YCrCb signal, and, to reduce processing time, we work directly in this colour space.

We use a bivariant histogram on the Cr and Cb of the objects we wish to track, creating a two-dimensional lookup table on colour. As each image is acquired, pixels that fall within the histogram are flagged as belonging to a landmark. Blob extraction is then performed, the objects sorted by size, and very small and very large objects eliminated. The image coordinates of the centroid of each blob are then converted to a polar form, giving the radial distance and bearing of each blob with respect to the centre of the image. At no stage do we attempt to 'unwrap' the image; we believe this to be a waste of valuable processing time and instead work directly with the radial image.

The relative landmark bearings are then combined with the robot's orientation obtained from the magnetometer to give bearings with respect to the reference direction. Landmark range however is not directly available; we use a flat-earth assumption and knowledge of the camera/mirror optics to derive an estimate of range given a landmarks radial pixel distance from the centre of the image [6], [10].

C. Results and Discussion

The testing arena used for these experiments is a large shed. The vehicle was placed in the middle of the shed defining the goal pose $(x, y, \theta) = (0, 0, 0)$. A 'landmark' was then placed at $(x, y) = (5.85, -1)$, and the target IALV was found. The vehicle was then manually moved to $(x, y, \theta) = (0, 3, 0)$, and the control system was activated. Fig. 6 depicts the results of the experiment, showing the vehicle servoing to the goal pose based on vision and compass data alone. Fig. 6 (c) shows the results of the IALV calculation which gives the state fed into the switching controller. Although this data is extremely noisy, the system seems quite robust to this.

The similarity to the simulation results plotted in Fig. 4 is quite clear, although the experimental system did take longer to stabilise to the desired pose. This could be due to the quite coarse onboard velocity measurements. Also, the rather noisy data provided by the vision system could be in effect here.

IV. CONCLUSION

This paper has described a method of stabilising a car-like vehicle to a target pose based on the discrepancies between a target view of the landmarks in a workspace and the robot's current view.

The robot's view of the workspace is summarised by a single quantity, the Improved Average Landmark Vector, which augments the original formulation of Lambrinos et al. [12] with range information. At each instant, the robot compares the current IALV with that at the target location which yields directly the distance and orientation to the target position. At no stage is there a requirement for landmark correspondence; this represents a key advantage over many other homing methods. This information is then fed into switching controller which stabilises the vehicle to the desired pose based on the information provided by the vision system. Furthermore, the method uses polar representations of the landmarks, and thus, the need for processor intensive image unwarping is circumvented.

We have presented both simulation and real expereimental results showing the validity of our approach even in the face of input saturation and noise. Future work includes extensive experimental validation of the method, along with a deeper understanding of the impact of dynamic elements in the control loops. Finally, it is clear that this technique is ideally suited to topological navigation, and this too represents a direction for further research.

V. ACKNOWLEDGMENTS

The technical support of the CMIT Tractor Team — Peter Hynes, Stuart Wolfe, Stephen Brosnan, Graeme Winstanley, Pavan Sikka, Elliot Duff, Les Overs, Craig Worthington and Steven Hogan — is gratefully acknowledged, while Jonathon O'Brien at UNSW is gratefully thanked for the loan of the Toro ride-on mower.

The first author gratefully acknowledges the funding and technical support provided by CMIT, the funding provided by an APA grant through QUT and would also like to thank Peter Corke and Peter Ridley for their guidance and support throughout this course of research.

VI. REFERENCES

[1] Michele Aicardi, Giuseppe Casalino, Antonio Bichi, and Aldo Balestrino. Closed loop steering of unicycle-like vehicles via Lyapunov techniques. *IEEE Robotics and Automation Magazine*, pages 27–35, March 1995.

[2] A. Astolfi. Exponential stabilization of a car-like vehicle. In *Proceedings of the International Conference on Robotics and Automation*, pages 1391–1396, Nagoya, Japan, 1995. IEEE.

[3] R.W. Brockett. Asymptotic stability and feedback stabilization. In R. W. Brockett, R. S. Millman, and H. J. Sussman, editors, *Differential Geometric Control Theory*, pages 181–191. Birkhauser, Boston, USA, 1983.

[4] J. S. Chahl and M. V. Srinivasan. Range estimation with a panoramic visual sensor. *Journal of the optical society of America*, 14(9), September 1997.

[5] Fabio Conticelli, Bendetto Allota, and Pradeep K. Khosla. Image-based visual servoing of nonholonomic mobile robots. In *Proceedings of the 38rd Conference on Decision and Control*, pages 3496–3501. IEEE, Phoenix, Arizona, USA, December 1999.

[6] A.K. Das, R. Fierro, V. Kumar, B. Southall, J. Spletzer, and C.J. Taylor. Real-time vision based control of a nonholonomic mobile robot. In *International Conference on Robotics and Automation*, pages 1714–1719, Seoul, Korea, May 2001. IEEE.

[7] Jose Gaspar, Niall Winters, and Jose Santos-Victor. Vision-based navigation and environmental representations with an omnidirectional camera. *IEEE Transactions on Robotics and Automation*, 16(6):890–898, December 2000.

[8] Koichi Hashimoto and Toshiro Noritsugo. Visual servoing of nonholonomic cart. In *International Conference on Robotics and Automation*, pages 1719–1724, Albuqueque, New Mexico, USA, April 1997. IEEE.

[9] Jiawei Hong, Xiaonan Tan, Brian Pinette, Richard Weiss, and Edward M. Riseman. Image-based homing. In *International Conference on Robotics and Automation*, pages 620–625, Sacramento, California, USA, April 1991. IEEE.

[10] I. Horswill. Polly: A vision-based arificial agent. In *Proceedings of the eleventh national conference on artificial intelligence (AAAI'93)*, Washington DC, USA, July 1993. MIT Press.

[11] Benjamin Kuipers and Yung-Tai Byun. A robot exploration and mapping strategy based on a semantic hierarchy of spatial representations. *Robotics and Autonomous Systems*, 8:47–63, 1991.

[12] Dimitrios Lambrinos, Ralf Moller, Thomas Labhart, Rolf Pfiefer, and Rudiger Wehner. A mobile robot employing insect strategies for navigation. *Robotics and Autonomous Systems*, 30:39–64, 2000.

[13] Jean-Claude Latombe. *Robot Motion Planning*. Kluwer Academic, 1991.

[14] Sungon Lee, Manchul Kim, Youngil Youm, and Wankyun Chung. Control of a car-like mobile robot for parking problem. In *International Conference on Robotics and Automation*, pages 1–6, Detroit, Michigan, 1999. IEEE.

[15] Tod S. Levitt and Daryl T. Lawnton. Qualitative navigation for mobile robots. *Artificial Intelligence*, 44:305–360, 1990.

[16] Alessandro De Luca, Giuseppe Oriolo, and Claude Samson. Feedback control of a nonholonomic car-like robot. In J.-P. Laumond, editor, *Planning Robot Motion*, chapter 4. Springer Verlag, 1997.

[17] Maja J. Mataric. Integration of representation into goal-driven behaviour-based robots. *IEEE Transactions on Robotics and Automation*, 8(3):304–312, June 1992.

[18] Richard M. Murray and S. Shankar Sastry. Nonholonomic motion planning: Steering using sinusoids. *IEEE Transactions on Automatic Control*, 38(5):700–716, May 1993.

[19] Pierpaolo Murrieri, Daniele Fontanelli, and Antonio Bicci. Visual-servoed parking with limited view angle. In *Preprints of the International Symposium on Experimental Robotics*, Sant'Angelo d'Ischia, Italy, 2002.

[20] C. Samson and K. Ait-Abderrahim. Feedback control of a non-holonomic wheeled cart in cartesian space. In *International Conference on Robotics and Automation*, pages 1136–1141, Sacramento, California, USA, April 1991. IEEE.

[21] Keven Weber, Svetha Venkatesh, and Mandyam Srinivasan. Insect-inspired robotic homing. *Adaptive Behavior*, 7(1):65–97, 1999.

[22] R. Wehner, B. Michel, and P. Antonsen. Visual navigation in insects: Coupling of egocentric and geocentric information. *Journal of Experimental Biology*, 199:129–140, 1996.

Quadrotor Control Using Dual Camera Visual Feedback

Erdinç Altuğ, James P. Ostrowski, Camillo J. Taylor
GRASP Lab. University of Pennsylvania, Philadelphia, PA 19104, USA
E-mail: {erdinc, jpo, cjtaylor}@grasp.cis.upenn.edu

Abstract—In this paper, a vision-based stabilization and output tracking control method for a four-rotor helicopter has been proposed. A novel 2 camera method has been described for estimating the full 6 DOF pose of the helicopter. This two camera system is consisting of a pan-tilt ground camera and an onboard camera. The pose estimation algorithm is compared in simulation to other methods (such as four point method, and a stereo method) and is shown to be less sensitive to feature detection errors on the image plane. The proposed pose estimation algorithm and non-linear control techniques have been implemented on a remote controlled quadrotor helicopter.

I. INTRODUCTION

The purpose of this study is to explore control methodologies and pose estimation algorithms that will make an unmanned aerial vehicle (UAV) autonomous. An autonomous UAV will be suitable for applications like search and rescue, surveillance and remote inspection. Rotary wing aerial vehicles have distinct advantages over conventional fixed wing aircrafts on surveillance and inspection tasks, since they can take-off/land in limited spaces and easily hover above the target. A *quadrotor* is a four rotor helicopter. Recent work in quadrotor design and control includes quadrotor [1], X4-Flyer [2], mesicopter [3] and hoverbot [4]. Also, related models for controlling the VTOL aircraft are studied by Hauser et al [5].

Quadrotor is an under-actuated, dynamic vehicle with four input forces and six output coordinates. Unlike regular helicopters that have variable pitch angle rotors, a quadrotor helicopter has four fixed pitch angle rotors. Advantages of using a multi-rotor helicopter are the increased payload capacity and high maneuverability. Disadvantages are the increased helicopter weight and increased energy consumption due to the extra motors. The basic motions of a quadrotor are generated by varying the rotor speeds of all four rotors, thereby changing the lift forces. The helicopter tilts towards the direction of slow spinning rotor, which enables acceleration along that direction. Therefore control of the tilt angles and the motion of the helicopter are closely related and estimation of orientation (roll and pitch) is critical. Spinning directions of the rotors are set to balance the moments, therefore eliminating the need for a tail rotor. This is also used to produce the desired yaw motions. A good controller should properly arrange the rotor speeds so that only the desired states change.

In order to create an autonomous UAV, precise knowledge of the helicopter position and orientation is needed. This info can be used to stabilize, hover the helicopter or for tracking an object. The pose estimation of a 3D robot has also been studied by [9], [10], [11], [12]. But in these papers, a single onboard camera has been used and the estimates were obtained by combining image data with readings from the inertial navigation systems, GPS or gyros. Our primary goal is to investigate the possibility of a purely vision-based controller on the quadrotor. Limited payload capacity does not permit the use of heavy navigation systems or GPS. Moreover the GPS does not work at indoor environments. One can still setup an indoor GPS system or use small navigation systems but, cost limits the use of these systems. This study utilizes a two camera system for pose estimation. Unlike previous work that either utilizes monocular views or stereo pairs, our two cameras are set to see each other. A ground camera that has pan-tilt capability, and an onboard camera are used to get accurate pose information. The proposed pose estimation algorithm is compared in simulation with other methods like a four-point algorithm [13], a state estimation algorithm [14] and a direct method that uses the area estimations of the blobs. A backstepping like controller [7], [8] shown in [1] has been implemented and shown effective in simulations of the dynamical quadrotor model. The proposed pose estimation algorithm and the control techniques have been implemented on a remote controlled, battery powered model helicopter.

II. HELICOPTER MODEL

The quadrotor helicopter model is shown in Figure 1. A body-fixed frame (**B**) is assumed to be at the center of gravity of the quadrotor, where the z-axis is pointing upwards. This body axis is related to the inertial frame (**O**) by a position vector p=(x,y,z) ∈ **O** and a rotation matrix $R: O \rightarrow B$, where $R \in \mathbf{SO}(3)$. A ZYX Euler angle representation has been chosen to represent the rotations. It is composed of 3 Euler angles, (ϕ, θ, ψ), representing yaw, roll (rotation around y-axis) and pitch (rotation around x-axis), respectively.

A spinning rotor produces moment as well as thrust. Let F_i be the thrust and M_i be the moment generated by rotor i, that is spinning with rotational speed of w_i.

Let $V_b \in \mathbf{B}$ be the linear velocity in body-fixed frame and $w_b \in \mathbf{B}$ the angular velocity. Therefore the velocities will be

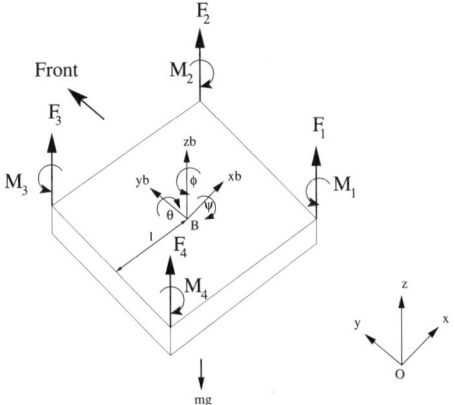

Fig. 1. 3D Quadrotor Model.

$$V_b = R^T \dot{p} \quad (1)$$
$$skew(w_b) = R^T \dot{R} \quad (2)$$

where $skew(w) \in \mathbf{so}(3)$ is the skew symmetric matrix of w. To represent the dynamics of the quadrotor, we can write the Newton-Euler equations as follows

$$m\dot{V}_b = F_{ext} - w_b \times mV_b \quad (3)$$
$$I_b \dot{w}_b = M_{ext} - w_b \times I_b w_b. \quad (4)$$

F_{ext} and M_{ext} are the external forces and moments on the body-fixed frame. I_b is the inertia matrix, and m is the mass of the helicopter.

Drag on a moving object [17] is given by $Drag = \frac{1}{2} C_d \rho v^2 A$, in which ρ is the density of air, A is the frontal area, C_d is the drag coefficient, and V is the velocity. Assuming constant ρ, the constants at the above equation can be combined to form C, which simplifies drag to $Drag = Cv^2$.

The force generated by a rotor [17] which is spinning with rotational velocity of w is given by $F = bL = \frac{\rho}{4} w^2 R^3 abc(\theta_t - \phi_t)$, where b is the number of blades on a rotor, θ_t is the pitch at the blade tip, ϕ_t is the inflow angle at the tip. By combining the constant terms as constant variable D, this equation simplifies to $F_i = Dw_i^2$.

Therefore F_{ext} and M_{ext} will be

$$F_{ext} = -C_x \dot{x}^2 \hat{i} - C_y \dot{y}^2 \hat{j} + (T - C_z \dot{z}^2)\hat{k} - R \cdot mg\hat{k} \quad (5)$$
$$M_{ext} = M_x \hat{i} + M_y \hat{j} + M_z \hat{k} \quad (6)$$

where C_x, C_y, C_z are the drag coefficients along x, y and z axes, respectively. T is the total thrust and M_x, M_y, M_z are the moments generated by the rotors. The relation of thrust and moments to the rotational velocities of rotors is given as follows

$$\begin{pmatrix} T \\ M_x \\ M_y \\ M_z \end{pmatrix} = \begin{pmatrix} D & D & D & D \\ -Dl & Dl & Dl & -Dl \\ -Dl & -Dl & Dl & Dl \\ CD & -CD & CD & -CD \end{pmatrix} \begin{pmatrix} w_1^2 \\ w_2^2 \\ w_3^2 \\ w_4^2 \end{pmatrix}. \quad (7)$$

The above matrix $M \in \mathbb{R}^{4 \times 4}$ is full rank for $l, C, D \neq 0$. The rotational velocity of rotor i (w_i), can be related to the torque of motor i (τ_i) as

$$\tau_i = I_r \dot{w}_i + K w_i^2 \quad (8)$$

where I_r is the rotational inertia of rotor i, K is the reactive torque due to the drag terms.

Motor torques τ_i should be selected to produce the desired rotor velocities (w_i) in Equation 8, which will change the external forces and moments in Equations 5 and 6. This will produce the desired body velocities and accelerations in Equations 3 and 4.

III. CONTROL OF A QUADROTOR HELICOPTER

A controller should pick suitable rotor speeds w_i for the desired body accelerations. Let's define the control inputs to be

$$\begin{aligned} u_1 &= (F_1 + F_2 + F_3 + F_4) \\ u_2 &= l(-F_1 + F_2 + F_3 - F_4) \\ u_3 &= l(-F_1 - F_2 + F_3 + F_4) \\ u_4 &= C(F_1 - F_2 + F_3 - F_4). \end{aligned} \quad (9)$$

C is the force-to-moment scaling factor. The u_1 represents a total thrust on the body in the z-axis, u_2 and u_3 are the pitch and roll inputs and u_4 is a yawing moment. Backstepping controllers [7] are useful when some states are controlled through other states. Since motions along the x and y axes are related to tilt angles θ and ψ respectively, backstepping controllers given in [1] can be used to control tilt angles enabling the precise control of the x and y motions (inputs u_2 and u_3). The altitude and the yaw, can be controlled by PD controllers

$$\begin{aligned} u_1 &= \frac{g + K_{p1}(z_d - z) + K_{d1}(\dot{z}_d - \dot{z})}{\cos\theta \cos\psi} \\ u_4 &= K_{p2}(\phi_d - \phi) + K_{d2}(\dot{\phi}_d - \dot{\phi}). \end{aligned} \quad (10)$$

The proposed model and the controllers were tested in simulations. In Figure 2, the quadrotor moves from (30,40,150) to the origin with an initial yaw of 20 degrees and zero tilt angles. Note the tilt-up motions of the quadrotor towards the end of the simulation, performed to slow down and reach the origin with zero velocity. Figure 3 shows the motion of the quadrotor during this simulation.

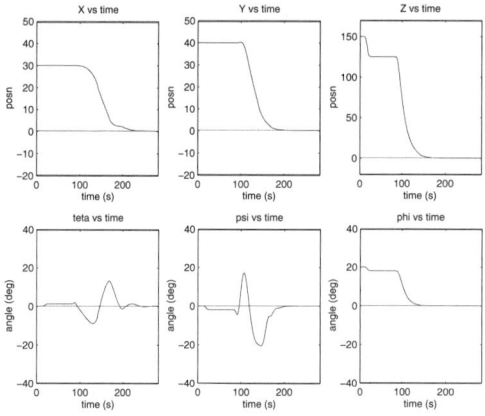

Fig. 2. Quadrotor simulation results.

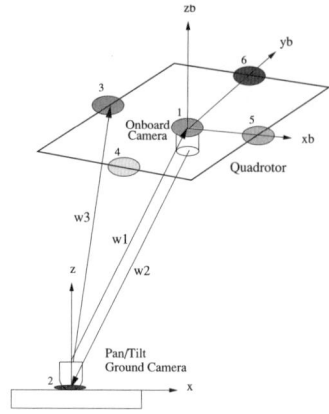

Fig. 4. Two camera pose estimation method.

Fig. 5. Quadrotor tracking with a camera.

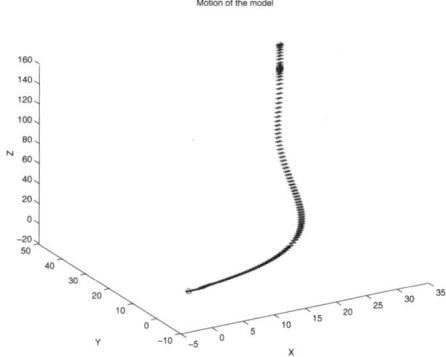

Fig. 3. The path of the quadrotor.

IV. POSE ESTIMATION

For autonomous helicopters, estimation of motion (relative position, orientation and velocities) is important for surveillance and remote inspection tasks or taking off / landing from a site. For our project, the goal is to obtain the pose from vision rather than complex and heavy navigation systems or GPS. For this purpose, a pair of color cameras track multi-color blobs located under the helicopter and on the ground. A blob tracking algorithm is used to determine the positions and areas of the blobs on the image planes. Therefore the purpose of the pose estimation algorithm is to obtain (x,y,z) positions, tilt angles (θ, ψ), the yaw angle (ϕ) and the velocities of the helicopter, in real-time relative to the ground camera frame. These can be represented as the Rotation Matrix, $R \in \mathbf{SO(3)}$, defining the body-fixed frame of the helicopter with respect to the ground camera frame, where $R^T R = I$, $det(R) = 1$, and the relative position vector $p \in \mathbf{R}^3$, which is the position of the helicopter with respect to the ground camera.

A. Two Camera Pose Estimation Method

The two camera pose estimation method uses a pan-tilt ground camera and an onboard camera. Previous work on vision-based pose estimation utilizes monocular views or stereo pairs. Our two camera pose estimation method involves the use of two cameras that are set to see each other as shown in Figure 4. This method is especially useful for autonomous taking off or landing. Colored blobs of 2.5 cm radius are attached to the bottom of the quadrotor and to the ground camera as shown in Figure 5. Tracking two blobs on the quadrotor image plane and one blob on the ground image frame is found to be enough for accurate pose estimation. To minimize the error as much as possible, five blobs are placed on the quadrotor and a single blob is located on the ground camera. The blob tracking algorithm tracks the blobs and returns image values (u_i, v_i) for $i = 1 \cdots 6$.

The cameras have matrices of intrinsic parameters, A_1 and A_2. The unit vector $w_i \in \mathbf{R}^3$ from each camera to the blobs can be found as

$$w_i = inv(A_1).[u_i \quad v_i \quad 1]', \quad w_i = w_i/norm(w_i)$$
$$\text{for} \quad i = 1,3,4,5,6 \qquad (11)$$
$$w_2 = inv(A_2).[u_2 \quad v_2 \quad 1]', \quad w_2 = w_2/norm(w_2)$$

Let L_a be the vector pointing from blob-1 to blob-3 in Figure 4. Vectors w_1 and w_3 are related by

$$\lambda_3 w_3 = \lambda_1 w_1 + RL_a \quad (12)$$

where λ_1 and λ_3 are unknown scalars. Taking the cross product with w_3 gives

$$\lambda_1(w_3 \times w_1) = RL_a \times w_3. \quad (13)$$

This can be rewritten as

$$(w_3 \times w_1) \times (RL_a \times w_3) = 0. \quad (14)$$

Let the rotation matrix R be composed of two rotations: the rotation of θ degrees around the vector formed by the cross product of w_1 and w_2 and the rotation of α degrees around w_1. In other words:

$$R = Rot(w_1 \times w_2, \theta) \cdot Rot(w_1, \alpha) \quad (15)$$

where $Rot(a,b)$ means the rotation of b degrees around the unit vector a. The value of θ can be found from the dot product of vectors w_1 and w_2.

$$\theta = acos(w_1 \cdot w_2) \quad (16)$$

The only unknown is the angle α. Let M be a matrix described as:

$$M = (w_3 \times w_1) \times (w_3 \times (R(w_1 \times w_2, \theta))). \quad (17)$$

Using Rodrigues' formula [15], Equation 14 can be simplified to

$$M \cdot L_a + \sin\alpha \cdot M\widehat{w_1} \cdot L_a + (1 - \cos\alpha) \cdot M \cdot (\widehat{w_1})^2 \cdot L_a = 0. \quad (18)$$

This is a set of three equations in the form of $A\cos\alpha + B\sin\alpha = C$, which can be solved by

$$\alpha = \arcsin \frac{B \cdot C \pm \sqrt{(B^2 \cdot C^2 - (A^2 + B^2) \cdot (C^2 - A^2))}}{A^2 + B^2} \quad (19)$$

One problem here is that $\alpha \in [\pi/2, -\pi/2]$, because of the arcsin function. Therefore one must check the unit vector formed by two blobs to find the heading, and pick the correct α value.

Thus, the estimated rotation matrix will be $R = Rot(w_1 \times w_2, \theta) \cdot Rot(w_1, \alpha)$. Euler angles (ϕ, θ, ψ) defining the orientation of the quadrotor can be obtained from rotation matrix, R.

In order to find the relative position of the helicopter with respect to the inertial frame located at the ground camera frame, we need to find scalars λ_i, for $i = 1 \cdots 6$. λ_1 can be found using Equation 12. The other λ_i values $(\lambda_2, \lambda_3, \lambda_4, \lambda_5, \lambda_6)$ can be found from

$$\lambda_i w_i = \lambda_1 w_1 + RL_i. \quad (20)$$

L_i is the position vector of ith blob in body-fixed frame. To reduce the errors, λ_i values are normalized using the blob separation, L.

The center of the quadrotor will be

$$\begin{aligned} X &= (\lambda_3 w_3(1) + \lambda_4 w_4(1) + \lambda_5 w_5(1) + \lambda_6 w_6(1))/4 \\ Y &= (\lambda_3 w_3(2) + \lambda_4 w_4(2) + \lambda_5 w_5(2) + \lambda_6 w_6(2))/4 \quad (21) \\ Z &= (\lambda_3 w_3(3) + \lambda_4 w_4(3) + \lambda_5 w_5(3) + \lambda_6 w_6(3))/4. \end{aligned}$$

B. Comparing the Pose Estimation Methods

The proposed two camera pose estimation method is compared to other methods using a Matlab simulation. Other methods used were a four-point algorithm [13], a state estimation algorithm [14], a direct method that uses the area estimations of the blobs, and a stereo pose estimation method that uses two ground cameras that are separated by a distance d.

The errors are calculated using angular and positional distances, given as

$$\begin{aligned} e_{ang} &= \| log(R^{-1} \cdot R^{est}) \| \\ e_{pos} &= \| p - p^{est} \|. \end{aligned} \quad (22)$$

R^{est} and p^{est} are the estimated rotational matrix and the position vector. Angular error is the amount of rotation about a unit vector that transfers R to R^{est}.

Figures 6, 7 and 8 show the motion of the quadrotor and the pose estimation errors. Quadrotor moves from the point (22, 22, 104) to (60, 60, 180) cm., while (θ, ψ, ϕ) changes from (0.7, 0.9, 2) to (14, 18, 40) degrees. A random error up to 5 pixels were added on image values. The blob areas were also added a random error of magnitude ± 2. The comparison of the pose estimation methods and the average angular and positional errors are given on Table 1.

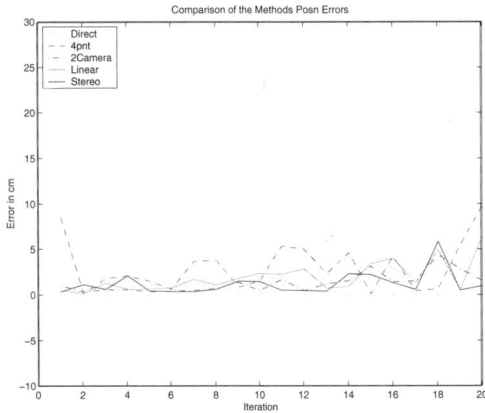

Fig. 6. Comparison of the Position Estimation Errors.

It can be seen from the plots and Table 1 that, the estimation of orientation is more sensitive to errors on the image plane than the position estimation. The poor orientation estimate of the direct method is because of the blob areas which are subject to the random noise. For the stereo method, the value of the baseline is important for

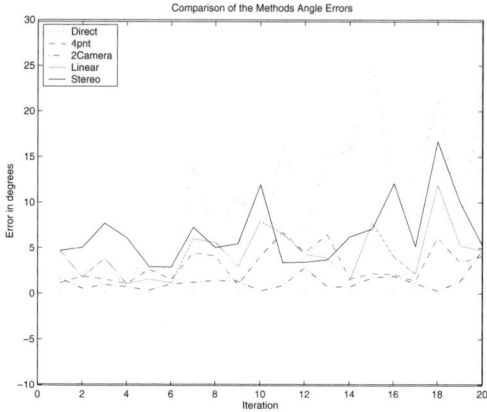

Fig. 7. Comparison of the Orientation Estimation Errors.

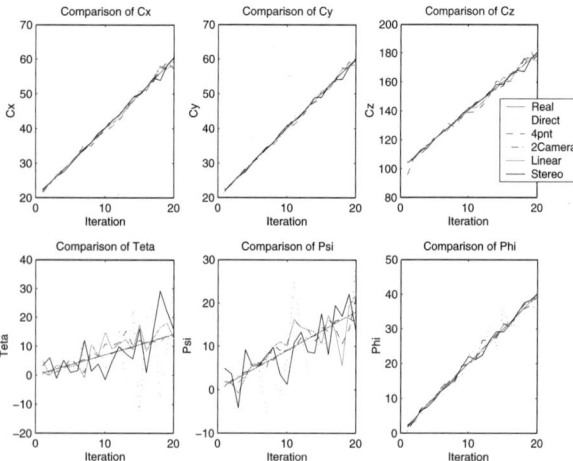

Fig. 8. Estimated helicopter positions and orientation angles.

TABLE I
COMPARISON OF THE POSE ESTIMATION METHODS.

Method	Angular E. (deg.)	Positional E. (cm.)
Direct M.	10.2166	1.5575
4 Pnt. M.	3.0429	3.0807
2 Camera M.	1.2232	1.2668
Linear M.	4.3700	1.8731
Stereo M.	6.5467	1.1681

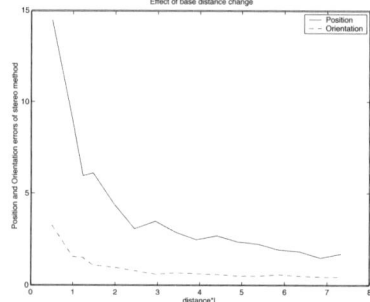

Fig. 9. Effects of baseline distance change on pose estimation at the Stereo Method.

pose estimation. Figure 9 shows the effects of the baseline distance change on the estimation. As the baseline distance approaches 3 times the distance between blobs (6L), the stereo method appears to be giving good estimates. The need for a large baseline for stereo pairs is the drawback of the stereo method. The average error of the two camera method is 1.22 degrees for angular estimation and 1.26 cm. for positional estimation, so we can conclude that the two camera method is more effective comparing to the other methods. The two camera method has other advantages, such as its ability to perform even when some blobs are lost. But for other methods the loss of a single blob will result bad pose estimates.

V. EXPERIMENTS

The proposed controllers and the pose estimation algorithms have been implemented on a remote-controlled battery-powered helicopter shown in Figure 10a. It is a commercially available model helicopter called HMX-4. It is about 0.7 kg, 76 cm long between rotor tips and has about 3 minutes flight time. This helicopter has three gyros on board to stabilize itself. An experimental setup shown in Figure 10b was prepared to prevent the helicopter from moving too much on the x-y plane, while enabling it to turn and ascend/descend freely. Vision based stabilization experiments were performed using the two camera pose estimation method. In these experiments two separate computers were used. Each camera was connected to separate computers which were responsible for performing blob tracking. PC-1 did the computation for the onboard camera and sent the information to PC-2 via the network. PC-2 was responsible for the ground camera and for the calculation of the control signals. These signals were then sent to the helicopter with a remote control device that uses the parallel port.

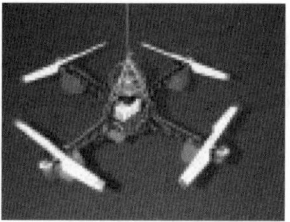 

Fig. 10. a) Quadrotor Helicopter, b) Experimental Setup.

Controllers described in [1] were implemented for the experiment. Figure 11 shows the results of this experiment using the two camera pose estimation method, where height, x, y and yaw angle are being controlled. The mean and standard deviation are found to be 129 cm and 13.4 cm for z, 5.86 degrees and 17.2 degrees for ϕ respectively. The results from the plots show that the proposed controllers do an acceptable job despite the pose estimation errors.

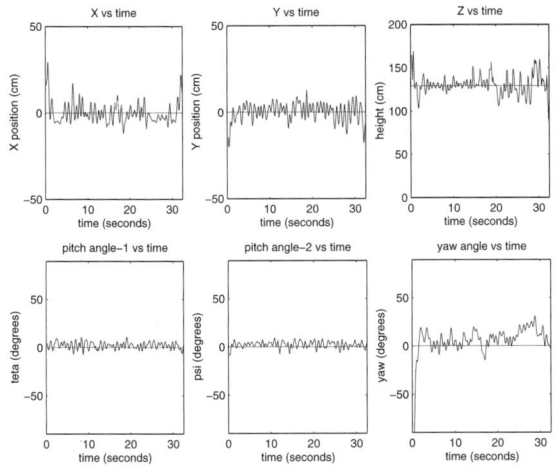

Fig. 11. The results of the height x, y and yaw control experiment with two camera pose estimation method.

VI. CONCLUSIONS AND FUTURE WORK

We have presented a novel two camera method for pose estimation. The method has been compared to other pose estimation algorithms and shown to be more effective especially when there are errors on the image plane. Backstepping controllers have been used to stabilize and perform output tracking control. Simulations performed on Matlab Simulink show the ability of the controller to perform output tracking control even when there are errors on state estimates. The proposed controllers and the pose estimation method have been implemented on a remote-control, battery-powered model helicopter. Initial experiments on a tethered system showed that the vision-based control is effective in controlling the helicopter. Our future work will include placing the ground camera on top of a mobile robot and enabling take-off/landing from a mobile robot. Such functionalities will be useful for inspection, chase and other ground-air cooperation tasks.

VII. ACKNOWLEDGMENTS

We gratefully acknowledge support from the Republic of Turkey Ministry of National Education, NSF and DARPA.

VIII. REFERENCES

[1] E. Altuğ, J. P. Ostrowski, R. Mahony, *Control of a Quadrotor Helicopter using Visual Feedback*, Proceedings of the IEEE International Conference on Robotics and Automation, Washington, D.C., May 2002, pp. 72-77.

[2] T. Hamel, R. Mahony, A. Chriette, *Visual Servo trajectory tracking for a four rotor VTOL aerial vehicle*, Proceedings of the 2002 IEEE International Conference on Robotics and Automation, Washington, D.C., May 2002, pp. 2781-2786.

[3] I. Kroo, F. Printz, *Mesicopter Project*, Stanford University, http://aero.stanford.edu/mesicopter.

[4] J. Borenstein, *Hoverbot Project*, Univ. of Michigan, www-personal.engin.umich.edu/~johannb/hoverbot.htm

[5] J. Hauser, S. Sastry, G. Meyer, *Nonlinear Control Design for Slightly non-minimum Phase Systems: Application to V/STOL Aircraft*, Automatica, vol 28, No:4, 1992, pp. 665-679.

[6] P. Martin, S. Devasia, B. Paden . *A different look at output tracking: Control of a VTOL aircraft*, Automatica No:1, 32, 1996, pp. 101-107.

[7] S. Sastry, *Nonlinear Systems; Analysis, Stability and Control*, Springer-Verlag, 1999.

[8] T. Hamel, R. Mahony, *Visual servoing of a class of under-actuated dynamic rigid-body systems*, Proceedings of the 39th IEEE Conference on Decision and Control, 2000.

[9] O. Amidi, *An Autonomous Vision Guided Helicopter*, Ph.D. Thesis, Carnegie Mellon University, August 1996.

[10] H. Zhang, *Motion Control for Dynamic mobile robots*, Ph.D. Thesis, University of Pennsylvania, June 2000.

[11] H. Shim, *Hierarchical Flight Control System Synthesis for Rotorcraft-based Unmanned Aerial Vehicles*, Ph.D. Thesis, University of California, Berkeley, Fall 2000.

[12] Y. Ma, J. Kosecka, S. Sastry, *Optimal Motion from image Sequences: A Riemannian viewpoint*, Research Report, University of California Berkeley, UCB/ER/M98/37.

[13] A. Ansar, D. Rodrigues, J. Desai, K. Daniilidis, V. Kumar, M. Campos, *Visual and Haptic Collaborative Tele-presence*, Computer and Graphics, Special Issue on Mixed Realities Beyond Convention, Oct 2001.

[14] C. S. Sharp, O. Shakernia, S. S. Sastry, *A Vision System for Landing an Unmanned Aerial Vehicle*, IEEE Conference on Robotics and Automation, 2001.

[15] R. Murray, Z. Li, S. Sastry, *A Mathematical Introduction to Robotic Manipulation*, CRC Press, 1994.

[16] E. Trucco, A. Verri, *Introductory Techniques for 3-D Computer Vision*, Prentice-Hall, 1998, pp. 322-325.

[17] R. Prouty, *Helicopter Performance, Stability, and Control*, Krieger Publishing Company, June 1995.

Production Cycle-Time Analysis Based on Sensor-Based Stage Petri Nets for Automated Manufacturing Systems

ShihSen Peng
Department of Mechanical Engineering
Chinese Military Academy
Taiwan
shihsen@yahoo.com

MengChu Zhou
Department of Electrical and Computer Engineering, New Jersey Institute of Technology, Newark, NJ 07102, USA and Institute of Automation, CAS, Beijing 100080, China
zhou@njit.edu

Abstract

Production cycle time reduction in their discrete-event control systems (DECS) helps increase the productivity of automated manufacturing systems (AMS). Methods developed to evaluate the production cycle time are usually based on either the Design for Manufacture (DFM) or Design for Production (DFP) scheduling techniques. To evaluate the real cycle time at the programming level of controllers such as the ladder logic design of programmable logic controller (PLC) in DECS, this paper discusses a method to analyze the production cycle time based on the sensor-based stage Petri nets technique. The production time can be estimated at each stage directly from all the I/O sensors that are represented by the extended Petri nets----sensor-based stage Petri net (SBSPN). The production cycle time required to complete each product is marked on the individual stage transition through the real timers in the SBSPN model. For the production of multiple products, different production cycles times are estimated through the stage-by-stage real timers of controller program. These production cycle times are able to evaluate the bottleneck of integrated manufacturing systems. An example is used to illustrate the approach.

I. Introduction

In discrete-event control systems (DECS), programmable logic controller (PLC) is one of the basic lower level controllers. Ladder diagram (LD) is the popular language for PLC implementation. Design of LD normally consists of input sensors, output actuators, timers, and the counters, which are integrated into rungs to match the required control logic. In a complicate LD design, a stage-programming technique can be used to prevent the interlocking problem [3-4]. However, it is not easy to analyze the designed LD for production cycle time. An SBSPN technique in this study intends to represent all the I/O sensors statuses of the designed LD for analysis by eliminating the need to consider a interlocking problem in complicate LD logic. An SBSPN is an extended Petri net developed as a tool to maintain and analyze the ladder logic design for the DECS. This paper discusses the way to evaluate the production cycle time directly from a PLC program based on the SBSPN technique. For the multiple products produced in a flexible machining cell, a comparison of all the estimated production cycles times shows a clear picture that whether the designed I/O sensors allocation and LD developed for the PLC may cause bottleneck in the controlled process of an integrated system. Thus, the result can be used to improve the PLC implementation.

II. Sensor-Based Stage Petri Net

A sensor-based stage Petri net (SBSPN) [1-2] is the extension of real-timed Petri net (RTPN) by adding a self-loop, priority stage place in the stage control net (SCN) to the corresponding timed transitions at that stage and follow the behavior of a CPU scanning process in PLC. This scanning process can be represented by a token flow cyclically through the simultaneous stage places and transitions of SCN. The flow of tokens is used to drive the associated elements of RTPN at each stage. A SBSPN is a union of SCN and RTPN that can be formally defined as follows:

$$SBSPN = (SCN) \cup (RTPN)$$
$$= (P_{Si}, T_{Si}) \cup (P, T, I, O, m_0, D, X, Y)$$

where:
- $\{P, T, I, O, m_0, D, X, Y\}$ is RTPN as defined before [5-6]; and
- $SCN = \{P_{Si}, T_{Si}\}$

where:
P_{Si} is the set of stage places and T_{Si} is the set of stage transitions in the SCN
$\forall\ SCN \in RTPN;$
Let $O\ (P_{Si}, T_{Si}) = I\ (P_{Si}, T_{Si}) = 1$

Since SCN is established as a stage-enabled net simply to monitor the sequential status of RTPN, there is only one token allowed to flow sequentially and cyclically through SCN. The I/O status of RTPN at each stage is checked by the priority places of SCN first once its token moves up to that stage. The feedback token in the priority place then moves back to its transition for further stage scanning processes. A double directed arc in the SCN represents the priority choice, self-loop characteristic of a stage token to enable the stage I/O combination logic. The execution of SBSPN is thus controlled by the token flow of SCN through the priority place at each stage. If no SCN token presents at the stage, no action in that stage is enabled to take place. On the other hand, even if the scanning token presents at the priority place, no action of RTPN is executed if any input sensor is not simultaneously activated to achieve the combinational logic at the stage. This is the reason we can call SCN as a stage-control net to monitor and also integrate the design of RTPN for PLC stage programming. The proposed SCN in this study can simplify the ladder program design process. Once the control variables for a discrete event process are identified, a function between inputs/outputs at each stage can then be easily defined. The individual logic relationship at each stage is thus expressed as the Boolean equation of the stage without considering the interlocking problem from other variables at different stages.

III. SBSPN of a Flexible Machining Cell

Consider a flexible machining cell (FMC) consisting of an articulated robot, a lathe, a vertical machine center (VMC), a pull control buffer, and a PLC as shown in Figure 1. The production flexibility of FMC is dependent on the ladder logic (LD) designed for PLC. In this study, ten input sensors and thirteen output actuators are linked to construct the ladder logic of PLC to process three different parts (A, B, and C). All the I/O sensors are defined and interpreted as places and transitions in Tables 1-2 to establish the SBSPN model for the PLC integrating process. The production cycle time analysis based on the SBSPN modeling stage-by-stage is used to validate whether the designed control logic of the integration process may cause any bottleneck problems for the FMC implementation. Thus, using this method to update or redesign the ladder logic of PLC integrating technique would be more efficiently.

Table 1. Sensor Places and Interpretations for SBSPN Model

Input Sensor Places	Interpretations
S_{LO}	Lathe door opens
S_{LA}	Part A installed at lathe spindle
S_{CA}	Part A counter at output conveyor
S_{MO}	Machine center door opens
S_{MB}	Part B installed at machine spindle
S_{CB}	Part B counter at output conveyor
S_{LC}	Part C installed at lathe spindle
S_{MC}	Part C installed at machine spindle
S_{CC}	Part C counter at output conveyor
S_{bC}	Buffer full sensor

Table 2. Transitions and Interpretations for SBSPN Model

Output Transitions	Interpretations
t_{RAL}	Robot loads part A to lathe
t_{RBM}	Robot loads part B to VMC
t_{RCL}	Robot loads part C to lathe
t_{RCM}	Robot loads part C from lathe to VMC
t_{RAC}	Robot unloads part A to output conveyor
t_{RBC}	Robot unloads part B to output conveyor
t_{RCLB}	Robot unloads part C from lathe to buffer
t_{RCBM}	Robot loads part C from buffer to VMC
t_{RCC}	Robot unloads C from VMC to conveyor
t_{LA}	Lathe machines part A
t_{MB}	VMC machines part B
t_{LC}	Lathe machines part C
t_{MC}	VMC machines part C

To simplify the FMC model configuration, several conditions are assumed as follows:
1) A pallet with universal fixture for parts A, B, and C is designed and ready for robot to pick and place at each machine.
2) Robot has a universal gripper to pick parts A, B, and C at the same spot with unlimited supply.
3) All robot programs are ready in the controller for PLC implementation. Each concluded operation moves the robot back to home position automatically.
4) All NC programs for lathe and VMC to machine parts A, B, and C are ready in the controller for PLC implementation. Each NC program concludes with the spindle stops and door open automatically.
5) A buffer between lathe and VMC is designed to accumulate three C parts for deadlock avoidance. A sensor (S_{bC}) is used as a feedback of full buffer for the production pull control. The designed mechanism (a decoupler) allows robot to pick and place part C at the same spot of buffer.
6) Three sensors (S_{CA}, S_{CB}, and S_{CC}) are installed at output conveyor to count the number of finished parts automatically. The ordered lot sizes m, n, and k for parts A, B, and C are controllable as well.

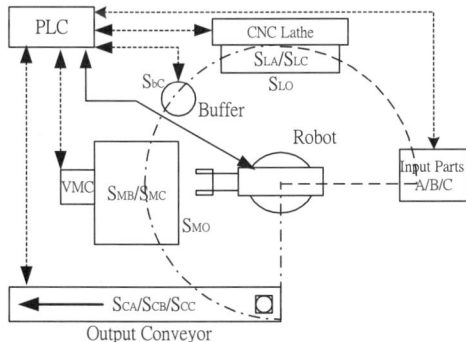

Figure 1. A Flexible Machining Cell Model

Using ten sensor places and thirteen timed transitions defined above, the logic combinations for FMC operations can be expressed into thirteen stages as follow:

$t_{RAL} = S_{LO} * \overline{S_{AL}} * \overline{S_{CA}}$ --------------------Stage 1

$t_{LA} = \overline{S_{LO}} * \overline{S_{AL}}$ --------------------Stage 2

$t_{RBM} = S_{MO} * \overline{S_{BM}} * \overline{S_{CB}}$ --------------------Stage 3

$t_{MB} = \overline{S_{MO}} * \overline{S_{BM}}$ --------------------Stage 4

$t_{RAC} = S_{LO} * \overline{S_{AL}}$ --------------------Stage 5

$t_{RBC} = S_{MO} * \overline{S_{CB}} * \overline{S_{BM}}$ --------------------Stage 6

$t_{RCL} = S_{LO} * S_{CL}$ --------------------Stage 7

$t_{LC} = \overline{S_{LO}} * S_{CL}$ --------------------Stage 8

$t_{RCM} = S_{LO} * S_{MO} * \overline{S_{CL}} * S_{CM}$ ----------Stage 9

$t_{RCLB} = S_{LO} * S_{bC} * \overline{S_{CL}}$ --------------------Stage 10

$t_{RCBM} = S_{MO} * \overline{S_{bC}} * S_{CM}$ --------------------Stage 11

$t_{MC} = \overline{S_{MO}} * S_{CM}$ --------------------Stage 12

$t_{RCC} = S_{MO} * \overline{S_{CC}} * \overline{S_{CM}}$ --------------------Stage 13

All the logic combinations expressed above can be converted into a complete SBSPN representation as shown in Figure 2. The SBSPN representation depicts the ladder logic of a PLC to integrate all the I/O sensors for the FMC to produce parts A, B, and C without considering the complex interlocking problems of the LD design. Since the PLC interacts with a real-time control process, the input sensors (except timers and counters) are marked as simultaneous places and all the output actuators are represented by the timed transitions. The input sensor places are safe due to its ON/OFF binary status. The timed duration of each output transition represents the output actuator of PLC to activate the robot transportation programs and the NC codes for different types of machining operations. These programs are preloaded and ready in the robot or NC machine controllers for LD programming. For safety consideration, all the robot operations must start and conclude the operation program at home position and the NC codes for each machine must start running from a stopped spindle, confirmed door closed of NC machine, and conclude at the stopped spindle and then door open for next operation. A buffer locating between lathe and VMC is used to store up to three C parts temporally and avoid the system deadlock from the production bottlenecking conflict.

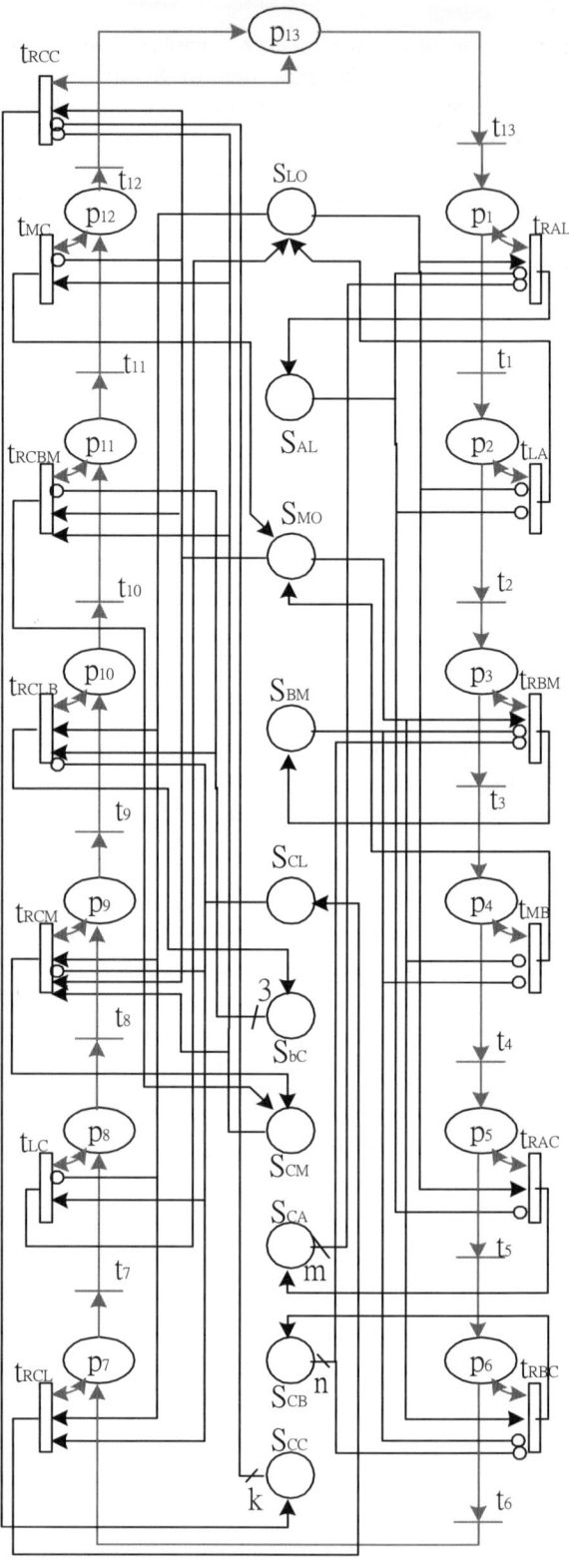

Figure 2. A complete SBSPN representation for the FMC model integration

Counter places S_{CA}, S_{CB}, and S_{CC} are installed on the output conveyor to count the number of finished part and feedback the counted number back to the SBSPN controller to increase the number of tokens in the lot size sensors S_{CA}, S_{CB}, and S_{CC}. For instance, once the number of tokens in S_{CA} equals to m, the part A stops supplying automatically. It causes the sensors places, S_{AL} and S_{CA}, to have no more token shown. The timed transitions, t_{RAL}, t_{LA}, and t_{RAC}, then stop to fire for part A production process.

The CPU scanning property and stage programming of the PLC programming representation are depicted clearly in the complete SBSPN representation for the FMC model as shown in Figure 2. There are thirteen stages in the SBSPN graph. The combination logic expression at each stage is enabled by several input sensors to activate the output actuator as a time transition. The timed duration of each transition is dependent on the logic of designed rungs for LD or program length of NC codes for both the robot manipulations and NC machine processes.

IV. Production Cycle-Time Analysis of SBSPN

The CPU scanning process on the LD of PLC is used to monitor the input sensor statuses, and update output actuator statuses of ladder logic stage by stage. Then, it integrates all the stages together into cyclic operations. The real-timed actuator transitions are activated as timed ones once all the input sensors have right values. The real-time durations of all the timed transitions are accumulated as a production cycle time for the integrated FMC and then can be evaluated from the complete SBSPN representation model. Since different lot sizes of parts are ordered, different production schedules may cause different production cycle times to complete all the lot size required to produce in the designed FMC. By using the SBSPN technique and the possible production schedule combinations in the designed FMC model, the real time production cycle can be evaluated from the I/O logic combination in terms of time unit.

Suppose that all the timed transitions consume fixed time units as shown in Table 3, and all production cycle times based on the SBSPN model representation can be evaluated from the scheduled combinations for FMC as follows:

(A) Total production time of each part:
1) Part A:
$T_{RAL}(5)$ → $T_{LA}(10)$ → $T_{RAC}(5)$ = 20 time units
2) Part B:
$T_{RBM}(5)$ → $T_{MB}(20)$ → $T_{RBC}(5)$ = 30 time units
3) Part C:
$T_{RCL}(5)$ → $T_{LC}(15)$ → $T_{RCM}(5)$ → $T_{MC}(25)$ → $T_{RCC}(5)$
=55 time units

Table 3. Machine operation and robot transportation duration time units required in the FMC

Machine Operation/Robot Transport		Time Units
T_{RAL}	Robot transports part A to lathe	5
T_{LA}	Lathe machines part A	10
T_{RAC}	Robot transports part A from lathe to output	5
T_{RBM}	Robot transports part B to VMC	5
T_{MB}	VMC machines part B	20
T_{RBC}	Robot transports part B from VMC to output	5
T_{RCL}	Robot transports part C to lathe	5
T_{LC}	Lathe machines part C	15
T_{RCM}	Robot transports part C to VMC	5
T_{MC}	VMC machines part C	25
T_{RCC}	Robot transports part C to output	5
T_{RCLB}	Robot transports part C from lathe to buffer	5
T_{RCBM}	Robot transports part C from buffer to VMC	5

(B) Production cycle-time of scheduled combinations:
1) The production cycle time of parts A→B→C sequence (Figure 3);
2) The production cycle-time of parts B → C sequence only (part A is completed) (Figure 4);
3) The production cycle-time of part A → C sequence only (part B is completed) (Figure 5); and
4) The production cycle-time of part A→ B sequence only (part C is completed) (Figure 6).

The above estimation shows that the total production time of part A is 20 time units, part B is 30 time units, and part C is 55 time units. Without the ladder logic integration in FMC, the total production time to produce each part once in a strict sequence requires 105 time units (20 + 30 + 55). This is the production real time required based on the RTPN technique assuming no breakdown of the robot or machines. For the SBSPN technique, both SCN and RTPN are combined to integrate the LD of PLC for FMC implementation. The production cycle-time is evaluated based on the stage-by-stage programming of PLC ladder logic. Production cycle-times of four possible schedule combinations are evaluated using Gantt chart for comparison. The system deadlock and bottleneck problems, if existing, can then be detected directly through the SBSPN representation.

Figures 3-6 show different schedule combinations for the FMC implementation that is depicted clearly in the SBSPN model. The production cycle-time of each combination in the FMC is the maximum time span required to complete the scheduled production combination without any further interrupted delay. Using the SBSPN elements and the time units required, the production cycle-time of FMC to produce parts A, B, and C in sequence is 75 time units as shown in Figure 3. Compare the total production time to the maximum time span, which is the production cycle-time of the integrated FMC, a total production time saving for this sequential combination is about 28.57% [(105-75)/105]. Since the lot size ordered for each part may be different, after one of the part required is finished, all the machines in the FMC can be used to manufacture only two-left part combinations. Figures 4-6 show the production cycle times of other three combinations, respectively. They indicate that the one in Figure 4 has 23.52%[(85-65)/85] production time saving, Figure 5 has 20% [(75-60)/75] production time saving and Figure 6 has 30% [(50-35)/50] production time saving. In Figures 3 and 4, a buffer is not needed. This means that buffer is not required after part B or C finished individually. Figure 3 shows three different time spans to complete part A, B, and C in sequence. Normally, the production cycle-time is the largest in this case.

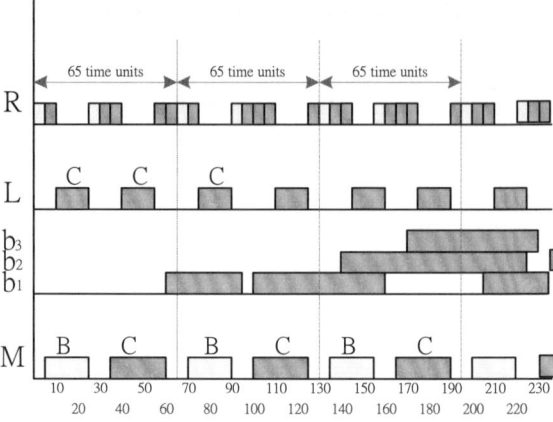

Figure 4. The production cycle time of FMC to manufacture part B → C in sequence only

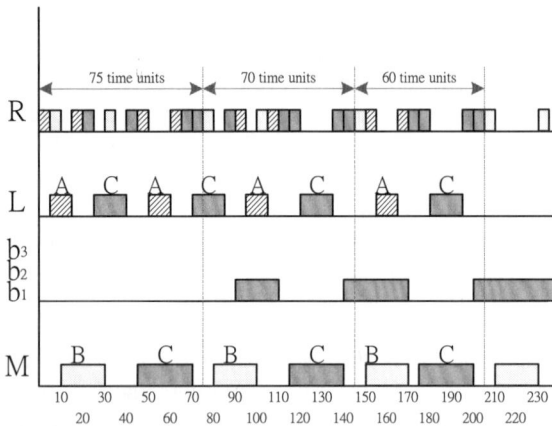

Figure 3. The production cycle time of FMC to manufacture parts A→B→C in sequence

After part A lot size finished, since both part B and C need the job on VMC, as shown in Figure 4, three spaces of the buffer are almost overlapped within time period 210 to 230 time units. This means that the buffer is full in this period. If the first part C not moving out shortly, a deadlock could happen once the part C job finished at lathe. To solve this problem, add more buffer space may avoid the system deadlock. Based on the SBSPN analysis for the illustrated example, the FMC configuration does confirm the deadlock free operations in all the sequential combinations of the FMC implementation. The bottleneck resource is the one that is the busiest in a complete cycle. For example, a Robot is bottleneck in Figure 3 and thus its performance improvement can produce the positive effect on the system productivity.

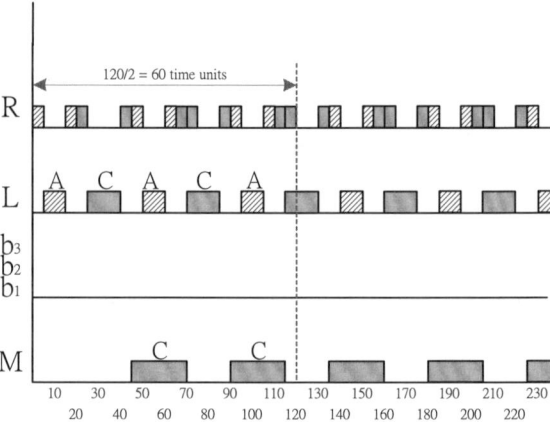

Figure 5. The production cycle time of FMC to manufacture part A → C in sequence only

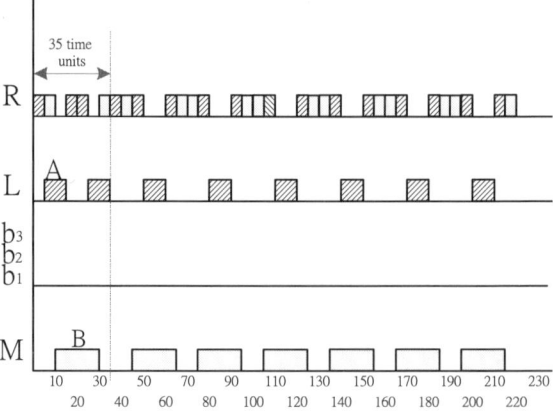

Figure 6. The production cycle time of FMC to manufacture part A → B in sequence only

V. Conclusion and Future Work

This paper discusses a method to analyze the production cycle-time based on the SBSPN technique. The approach is dealing with the real time span of different production sequences through the ladder logic of a PLC controller in an automated manufacturing system using the sensor- based stage programming Petri net representation. After the analysis of the illustrated example, several results are concluded as follows:

1) The production cycle time analysis based on the SBSPN for the automated manufacturing system is a real time evaluation that can be estimated through the PLC stage programming.
2) With the CPU scanning process control net linking into the real-time control net, the SBSPN representation is a good picture to depict the different production sequences through the ladder logic of PLC implementation.
3) A three-space buffer designed for Part C is pull control mechanism used to avoid deadlock during FMC implementation. It is very useful to reduce production cycle-time, evaluate the bottleneck problem, and improve the design performance of the ladder logic of PLC integration for FMC model.
4) To control lot size of each part, counters can be installed individually on the output conveyor to detect the finished part passing through. In the SBSPN model, these counters are represented as sensor places with m, n, and k directed arcs connecting to the timed transitions with inhibitor. Whenever the lot sizes are required to change, one may just update the directed arc number of counter places to control the time transition for firing. Such update does not affect the SBSPN modeling structures as well as its characteristics.
5) As shown in Figure 3, the bottleneck resource in a complete cycle of FMC implementation is the robot that carries the busiest operation. Its performance improvement can produce the positive effect on the system productivity.

Performance evaluation using the sensor-based stage Petri net (SBSPN) technique for more complex automated manufacturing systems is our further work in this field of study.

Acknowledgements

The second author is partially supported by the National Outstanding Young Scientist Research Award (Class B) from the National Natural Science Foundation of China.

References

1. Peng, S. S., and Zhou, M. C., "Sensor-Based Stage Petri Net Modeling of PLC Logic Programs for Discrete-Event Control Design," *International Journal of Production Research,* 41(3), 629-644, Feb. 2003.
2. Peng, S. S., and Zhou, M. C., "Sensor-based stage Petri net modeling of PLC logic programs for discrete-event control design," in *Proc. of 2002 IEEE Int. Conf. on Robotic & Automation*, May 2002, Washington DC, USA, pp. 1907-1912.
3. Peng, S. S., and Zhou, M. C., "Petri net based PLC stage programming for discrete-event control design," in *Proc. of 2001 IEEE Int. Conf. on Systems, Man, and Cybernetics,* October 2001, Tucson, AZ, USA, pp. 2706-2710.
4. Peng, S. S., and Zhou, M. C., "Conversion between ladder logic diagrams and Petri nets in discrete-event control design---a survey," in *Proc. of 2001 IEEE Int. Conf. On Systems, Man, and Cybernetics,* October 2001, Tucson, AZ, USA, pp.2682-2687.
5. Zhou, M. C., and Venkatesh, K., *Modeling, Simulation, and Control of Flexible Manufacturing Systems – A Petri Net Approach*. World Scientific Publishers, Singapore, 1998.
6. Zhou, M. C., and Twiss, E., "Design of industrial automated systems via relay logic programming and Petri nets," *IEEE Tran. On Systems, Man, and Cybernetics*. Vol.28, no.1, pp.137-150, Feb. 1998.

A Colored Timed Petri Net Model to Manage Resources in Complex Automated Manufacturing Systems

Maria Pia Fanti
Dept. of Electrical and Electronic Engineering
Polytechnic of Bari
Via Re David 200 - 70125 BARI
Italy

Abstract – Automated Manufacturing Systems (AMSs) can process different parts according to operation sequences sharing a finite number of resources. In these systems deadlock situations can occur so that the flow of parts is permanently inhibited and the processing of jobs is partially or completely blocked. This paper proposes a control strategy to manage resources in complex systems where multiple resource acquisitions are allowed to complete a working operation (Conjunctive Resource Service System - CRSS). The AMS structure and dynamics is described by a Colored Timed Petri net model, suitable for following resource changes and working procedure updating. Moreover, on the basis of the deadlock characterization obtained by digraph tools, an event-based controller is defined to avoid deadlock in CRSSs.

I. INTRODUCTION

Automated Manufacturing Systems (AMSs) can produce different parts and presents frequent reconfigurations of production resources. To describe AMS it is necessary to use a flexible model able to easily follow the structure modification. In addition, the software control must be flexible and reconfigurable to satisfy the demand of a changeable production. Moreover, in-process jobs compete for a finite number of resources, so that blocking, conflict and deadlocks can arise. Therefore, in recent years great attention is given to the deadlock resolution problem in AMS.

Most of technical literature uses Petri nets (PN) or Colored Petri nets (CPN) to model AMS and to derive deadlock prevention and avoidance algorithms [1], [3], [4], [9], [15]-[17]. In addition, direct digraphs represent an alternative approach to model the interactions between jobs and resources [2], [5]. Finally, in [12], [14] the AMS is modeled as a finite state automaton and some efficient and scalable deadlock avoidance algorithms are developed. However, all the mentioned approaches address deadlock in systems where each job requires at each step of its process a single unit of a single resource (SU-RAS). Recently, Park and Reveliotis [13] propose deadlock avoidance policies for systems with multiple resource acquisition and flexible routing (CD-RAS). More precisely, CD-RAS are systems where every process stage of jobs can acquire an arbitrary number of units from an arbitrary resource set.

In order to analyze and characterize deadlock conditions in complex resource allocation systems, this paper considers AMS, named Conjunctive Resource Service Systems (CRSS), where at each stage a part can require a single unit of an arbitrary number of resource types. Even if this situation is less general than the sequential allocation system analyzed in [13], it represents a more common situation in AMS where for example a single robot can cooperate with a single unit of a machine to process a part [18].

The objective of the paper is two folds. First, it introduces a Colored Timed Petri Net (CTPN) model [10] to describe the AMS structure and to implement the control strategies working on the basis of the knowledge of the system state. Following the Ezpeleta and Colom [4] idea, this more complex system is modeled considering separately its structure and its dynamics characterization.

Second, the paper proposes an efficient deadlock avoidance policy, suitable to manage resource acquisition in CRSS. The control strategy is based on the necessary and sufficient deadlock conditions proved in [7] and it constituted of two main actions. The first one checks whether a job can acquire next set of resources on the basis of logical conditions referred to the knowledge of next marking. The second action allows a new job entering the system if it does not determine deadlock or restricted deadlock in the future. In particular, restricted deadlock is not a deadlock condition but it is a situation of permanent blocking caused by the control inhibition. Moreover, the second control action bases the decision on the result of a simulation run performed under a deadlock avoidance algorithm and a priority law ruling concurrent job selection.

The paper is organized as follows: Section II describes the AMS. Moreover, Sections III and IV build the Petri net model structure and the CTPN modeling the AMS dynamics, respectively. Furthermore, Section V analyzes and characterizes a deadlock marking in the CTPN. Finally, Section VI defines the deadlock avoidance strategy and Section VII draws the conclusions.

II. THE SYSTEM DESCRIPTION

The AMS is described by the resource set $R=\{r_m, m=1,2,\ldots,R\}$ where r_m with $m=1,2,\ldots,R-1$ may represent a generic resource and r_R is a fictitious resource that jobs acquire as they leave the system. The capacity of r_m is an integer, say $C(r_m)$, indicating the maximum number of jobs that such a resource can simultaneously hold. Since the fictitious resource is always available, we assume $C(r_R)=|J|$, where J is the set of jobs to produce and $|X|$ refers to the cardinality of the set X. Now, processing each element from J requires a sequence of operations, named working procedure and indicated by $w_k=(o^k_1,\ldots,o^k_{L_k})$ where L_k is the number of operations necessary to complete the job. Each operation o_i with $i=1,2,\ldots,L_k$ is represented by the couple $o^k_i=(\rho^k_i,\tau^k_i)$ where the set $\rho^k_i \subset R$ collects the resources required in the i-th operation and $\tau^k_i \in \Re+$ represents the deterministic processing time of the corresponding operation, with $\Re+$ the set of non-negative real numbers.

Moreover, we assume that each set ρ^k_i is partitioned in two disjoint subsets: $\rho^k_i=\alpha^k_i\cup\beta^k_i$ with $\alpha^k_i\cap\beta^k_i=\emptyset$. While the resources from $\alpha^k_i\neq\emptyset$ are only released when the job acquires the resources for the next operation, the ones in β^k_i are released as soon as the operation goes to an end. Obviously, if ρ^k_i contains only one machine, then it is $\rho^k_i=\alpha^k_i$ and $\beta^k_i=\emptyset$. In the following, we use the statement "resource set $\gamma\subset R$ belongs to operation $o^k_i=(\rho^k_i,\tau^k_i)$" if and only if (iff) $\gamma\subseteq\rho^k_i$. We remark that the last subset of each working procedure is $\rho^k_{Lk}=\{r_R\}$ to indicate that the job leaves the system and the corresponding processing time is $\tau^k_{Lk}=0$. In the following, assuming that the total number of the working procedures necessary to process the jobs from J is W, the set containing all the working procedures is denoted by $W=\{w_1, w_2,\ldots,w_W\}$.

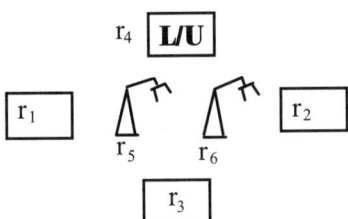

Fig.1 The robotic assembly system for Example 1

Example 1. To illustrate the previous definitions, we consider the system of Fig. 1 similar to the robotic assembly system described by Zhou and Di Cesare [18]. The AMS consists of three workstations (r_1, r_2, r_3), a loading/unloading station (r_4) and two robot manipulators (r_5, r_6). Moreover, r_7 represents the system output and J is the set of job to produce. When either r_1 or r_2 is ready to execute the assembly task, it requests robots r_5 and r_6 and acquires them. The assembled job releases the robots and detaining either r_1 or r_2, requires an operation on r_3. Finally, station r_4 unloads the finished job.

TABLE I
WORKING PROCEDURES FOR EXAMPLE 1

w_k	$\alpha^k_1\cup\beta^k_1$	$\alpha^k_2\cup\beta^k_2$	$\alpha^k_3\cup\beta^k_3$	$\alpha^k_4\cup\beta^k_4$	$\alpha^k_5\cup\beta^k_5$
k=1	$\{r_4\}\cup\emptyset$	$\{r_1\}\cup\{r_5,r_6\}$	$\{r_3\}\cup\emptyset$	$\{r_4\}\cup\emptyset$	$\{r_7\}\cup\emptyset$
k=2	$\{r_4\}\cup\emptyset$	$\{r_2\}\cup\{r_5,r_6\}$	$\{r_3\}\cup\emptyset$	$\{r_4\}\cup\emptyset$	$\{r_7\}\cup\emptyset$

The resource capacities are $C(r_m)=1$ for $m=1,2,4,5$ and 6, $C(r_3)=2$ and $C(r_7)=|J|$. Now, the two working procedures are described by the introduced formalism: $w_k=(o^k_1,o^k_2,o^k_3,o^k_4,o^k_5)=((\rho^k_1,\tau^k_1),(\rho^k_2,\tau^k_2),(\rho^k_3,\tau^k_3),(\rho^k_4,\tau^k_4),(\rho^k_5,0))$ with $k=1,2$. The sets α^k_i and β^k_i are defined in Table 1. More precisely, operations o^1_2 and o^2_2 require the conjunctive service of r_1, r_5, r_6 and r_2, r_5, r_6 respectively. Since, robots r_5 and r_6 are released as soon as the operation goes to an end, the jobs detain r_1 and r_2, respectively, after the process.

III. THE PETRI NET MODELING THE AMS

Since the controller decisions are based on the system state knowledge, it is necessary to formally describe the AMS and to model its dynamics. This section proposes a method to build in a modular way the Petri net modeling the system layout.

An ordinary (marked) Petri net is a bipartite digraph $PN=(P, T, F, m)$ where P is a set of places and T is a set of transitions [18]. The set of arcs $F\subset(P\times T)\cup(T\times P)$ is the flow relation. Given a PN and a node $x\in P\cup T$, the set $\bullet x=\{y\in P\cup T : (y,x)\in F\}$ is the preset of x while $x\bullet=\{y\in P\cup T : (x,y)\in F\}$ is the post-set of x. The state of a PN is given by its current marking that is a mapping m: $P\to\overline{N}$ where m is described by a $|P|$-vector and $\overline{N}$ is the set of non-negative integers. Moreover, the i-th component of m, indicated with m(p) represents the number of tokens in the i-th state place $p\in P$.

In our model the ordinary Petri net $PN_k=(P,T^k,F^k,m^k_0)$ describes the working procedure $w_k\in W$. A place $r_m\in P$ denotes the resource $r_m\in R$ and a token in r_m represents a job detaining an item of resource r_m. The set T^k is the transition set and the set of arcs $F^k\subset(P\times T^k)\cup(T^k\times P)$ is built by using the following procedure. Let us consider the generic working procedure $w_k=(o^k_1,\ldots,o^k_i,o^k_{i+1},\ldots,o^k_{Lk})$ with $o^k_i=(\rho^k_i,\tau^k_i)$ and $o^k_{i+1}=(\rho^k_{i+1},\tau^k_{i+1})$ and let us suppose that $\rho^k_i=(\alpha^k_i,\beta^k_i)$ and $\rho^k_{i+1}=(\alpha^k_{i+1},\beta^k_{i+1})$. The transition $t^k_0\in T^k$ models the loading of a job in the system that has to follow w_k and it is such that $t^k_0\in\bullet r_m$ for each $r_m\in\rho^k_1$ and $\bullet t^k_0=\emptyset$. Now, each transition $t^k_{i,i+1}\in T^k$ models the release of each resource $r_n\in\alpha^k_i$ and the acquisition of resources $r_m\in\rho^k_{i+1}$ i.e., $t^k_{i,i+1}\in\bullet r_m$ for each $r_m\in\rho^k_{i+1}$ and $t^k_{i,i+1}\in r_n\bullet$ for each $r_n\in\alpha^k_i$. If the set $\beta^k_i\neq\emptyset$, the resources of β^k_i are released as soon as the operation o^k_i goes to an end. To model this situation a transition $t^k_i\in T^k$ is introduced such that $t^k_i\in r_m\bullet$ for each $r_m\in\beta^k_i$ and $t^k_i\bullet=\emptyset$. Moreover, the initial marking m^k_0 is defined $m^k_0(r_m)=0$ for each $r_m\in R$.

The model of the whole AMS describing all possible working procedures in W is obtained by merging the defined Petri nets PN_k for $k=1,\ldots,W$. Therefore, the merged $PN=(P,T,F,Inh,m_0)$ is given by: $P=\cup_k P^k$, $T=\cup_k T^k$, $F=\cup_k F^k$, $m_0(r_m)=0$ for each $r_m\in P$. The obtained transition set T can be partitioned in three subsets: T_W is the set of transitions t^k_0 with $k=1,2,\ldots,W$ modeling the loading of jobs in the system according to the working procedure $w_k\in W$; T_F is the set of transitions $t^k_{i,i+1}$ modeling the flow of parts through the system; T_R is the set of transitions t^k_i modeling the release of the cooperating resource set β^k_i that does not detain the job after the executed operation. Finally, inhibitor arcs are introduced to model the finite capacities of each resource. More precisely, for each $t\in T_W\cup T_F$ and for each $r_m\in t\bullet$ with $m\neq R$, there exists an inhibitor arc between r_m and t labeled $C(r_m)$, i.e. $Inh(r_m,t)=C(r_m)$. Hence, an inhibitor arc labeled $C(r_m)\in\overline{N}$ implies that as soon as there are $C(r_m)$ tokens in r_m, the arc inhibits the firing of t.

IV. THE PETRI NET MODELING THE AMS DYNAMIC

A Colored Timed PN integrates the PN modeling resources and working procedures with the model of the parts flowing through the system and describes the AMS behavior. Moreover, the time attribute allows various time-based measures to be added in the system model [13], [14] and the implementation of the discrete-event control strategy to manage the AMS.

A. The Colored Timed Petri Net Model

A CTPN=$(P, T, Co, Inh, \mathbf{C}^+, \mathbf{C}^-, \Omega, M_0)$ is a 8-tuple where P, T and Inh are respectively the sets of places, transitions and inhibitor weight function previously defined. Indeed, the introduced ordinary PN represents the skeleton of the CTPN.

Now, denoted by J_M the set of jobs in process at marking M of the CTPN, we indicate with WP(j) the working procedure associated with any job $j \in J_M$ and with RWP(j) the residual working procedure, i.e., the resource sets necessary for each $j \in J_M$ to complete its processing. For sake of clarity, some auxiliary notation is used. The symbols FR(j) and SR(j) respectively identify the first and the second resource sets in RWP(j) while the set HR(j)$\subseteq$FR(j) represents the resource set currently held by j and released when such a job acquires the resources for the next operation.

Co is a color function defined from $P \cup T$ to a set of finite and not empty sets of colors [10]. Co maps each place $r_m \in P$ to a set of possible token colors $Co(r_m)$ and each transition $t \in T$ to a set of possible occurrence colors $Co(t)$. In our model, each job $j \in J_M$ is modeled by a set of colored token and each token color is the couple <RWP(j),j>. Hence, the marking M of the CTPN represents the state of the AMS and it is a mapping defined over P so that $M(r_m)$ is a set of elements of $Co(r_m)$, also with repeated elements (i.e., a multiset [10]) corresponding to token colors in the place p. In particular, if <RWP(j),j>$\in M(r_m)$, then $j \in J_M$ detains an item of resource r_m and RWP(j) is the sequence of operations that j has to perform starting from the current marking. Consequently, the color domain of place $r_m \in P$ is: $Co(r_m)=\{$<RWP,j> such that $j \in J$ and RWP is a subsequence of some $w_k \in W$ starting with an operation containing $r_m\}$.

Considering that at the initial marking M_0 no job is in process, we set $M_0(r_n)$=<0> for each $r_n \in P$.

Moreover, Co associates with each transition $t^k_{i,i+1} \in T_F$ a set of possible colors. In particular, let us suppose $t^k_{i,i+1} \in r_n \bullet$ and $t^k_{i,i+1} \in \bullet r_m$ for each $r_n \in \alpha^k_i$ and for each $r_m \in \rho^k_{i+1}$, the corresponding occurrence color set is $Co(t^k_{i,i+1})=\{$<RWP,j> such that $j \in J$, RWP is a subsequence of some $w_k \in W$, α^k_i belongs to the first operation of RWP and ρ^k_{i+1} belongs to the second operation of RWP$\}$.

Analogously, Co associates with each transition $t^k_i \in T_R$ such that $t^k_i \in r_n \bullet$ for each $r_n \in \beta^k_i$ the following set of possible occurrence colors:
$Co(t^k_i)=\{$<RWP,j> such that $j \in J$, RWP is a subsequence of some $w_k \in W$ and β^k_i belongs to the first operation of RWP$\}$.

In addition, the $|P| \times |T|$ matrices $\mathbf{C}^+$ and $\mathbf{C}^-$ are the post-incidence and the pre-incidence respectively. In particular, $\mathbf{C}^+(r_m,t)$ associates to each set of color of $Co(t)$ a set of color of $Co(r_m)$. Moreover, $\mathbf{C}^+(r_m,t)$ ($\mathbf{C}^-(r_m,t)$) is represented by means of an arc from t to r_m (from r_m to t) labeled with the function $\mathbf{C}^+(r_m,t)$ ($\mathbf{C}^-(r_m,t)$). The pre- and the post-incidence matrices $\mathbf{C}^-$ and $\mathbf{C}^+$, respectively, are:

1. for each $(r_n,t) \in F$ $\mathbf{C}^-(r_n,t)=I_d$ where I_d stands for "the function makes no transformation in the elements", otherwise $\mathbf{C}^-(r_n,t)=0$. This definition means that each token leaving a resource $r_n \in P$ is not modified;
2. for each $(t,r_m) \in F$, $\mathbf{C}^+(r_m,t)=U$ where U is a function that updates the color <RWP> with the color <RWP'>, otherwise $\mathbf{C}^+(r_m,t)=0$. More precisely, RWP' is the residual working procedure obtained from RWP by cutting its first element o_i.

The set Ω is defined by $\Omega=\{Co(x): x \in P \cup T\}$.

A transition $t \in T$ is enabled at a marking M with respect to a color $c \in Co(t)$ iff for each $r_m \in \bullet t$, $M(r_m) \geq \mathbf{C}^-(r_m,t)(c)$. When fired, this gives a new marking M': this will be denoted as M[t(c)>M'. A sequence $M[t_1(c_1)>M_1[t_2(c_2)>M_3...M_{n-1}[t_n(c_n)>M'$ is denoted by M[δ>M', where $\delta=t_1(c_1)t_2(c_2)...t_n(c_n)$ is a firing sequence. In such a case we say that M' is reachable from M. The symbol $Reach(M)$ denotes the set of reachable markings from M.

In our model, each transition $t^k_{i,i+1} \in T_F$ is enabled if the following conditions are verified for each $r_m \in t^k_{i,i+1} \bullet$ and for each $r_n \in \bullet t^k_{i,i+1}$:
C1) $C(r_m) \geq |M(r_m)|-1$,
C2) $M(r_n) \geq \mathbf{C}^-(r_n, t^k_{i,i+1})($<RWP(j),j>$)$, i.e., <RWP(j),j>$\in M(r_n)$.

Condition C1) refers to the condition imposed by the inhibitor arcs, the second one represents the enabling condition of the CTPN at marking M. If transition $t^k_{i,i+1}$ satisfies C1) or C2), it is said resource or color enabled, respectively.

Now, to investigate the performance of the system it is convenient to extend the CPN with the time concept [10], [11]. To do this we introduce a global clock, i.e., the clock values $\tau \in \Re^+$ represent the model continuous time. Moreover, the temporization of Petri nets can be achieved by attaching a time stamp to each token. The time stamp describes the earliest delay after which the token becomes available and can be removed by an enabled transition. In addition, it is described by the function $s: Co \to \Re+$, where $s(c)$ describes the time elapsed from the arrival of a token of color $c \in Co$ to the actual place. Hence, $s(c)$ is reset to zero as soon as the c-color token arrives to the place. Moreover, we specify the time delay by means of the color of the token. When $s(c)$ is equal or larger than the time specified in the color token c, the transition enabled by the considered token is *ready* for the execution.

Hence, in the defined CTPN each token is characterized by its color <RWP(j),j>=$(\rho^k_I, \rho^k_{I+1}, ..., \rho^k_{LK})$ and its stamp $s($<RWP(j),j>$)$. So, $t^k_{i,i+1} \in T_F$ can occur at time τ if it is color-resource enabled and ready ($s($<RWP(j),j>$) \geq \tau^k_i$). Moreover, each transition $t^k_0 \in T_W$ ($t^k_i \in T_R$) can fire if it is resource enabled (color enabled and ready).

Example 2. Let us consider the AMS described in Example 1. Fig. 2 depicts the CTPN at marking M with $J_M=\{j_1\}$ and $M(r_1)=M(r_5)=M(r_6)=$<RWP(j_1),j_1>$==$<(o^1_2,o^1_3,o^1_4,o^1_5),j_1> with $o^1_2=(\{r_1\} \cup \{r_5,r_6\}, \tau^1_2)$, $M(r_3)=M(r_4)=M(r_2)=$<0>.

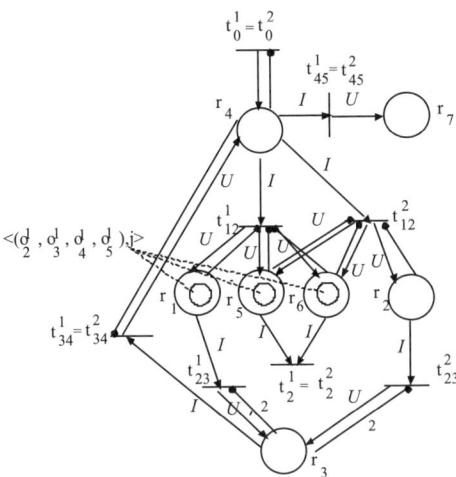

Fig.2. The CTPN modeling the AMS for Example 2

B. The Events

The marking of the CTPN can change on the occurrence of events involving resource releasing and acquiring. In particular, when a job completes an operation, it releases the resources of the set FR(j)-HR(j) without delay and the job holds the resources in HR(j) as long as all the resources necessary for next operation are not available. Moreover, a job entering the system is a further event modifying the marking. Hence, the events relevant for the deadlock analysis are of three types:

type 1 event: a new job enters the system. Each of these events is identified by a label $\sigma_1=(j,w_k)$, where $j \in J$ denotes the job entering the system and $w_k \in W$ is the working procedure the job has to follow;

type 2 event: a job $j \in J_M$ progresses from the resource set it currently holds (say α_i=HR(j)) to next resource set (ρ^k_{i+1}=SR(j)). Such an event can be identified by the triple $\sigma_2=(j, \alpha^k_i, \rho^k_{i+1})$. Moreover, if $\rho^k_{i+1}=\{r_R\}$ then j leaves the system.:

type 3 event: a job $j \in J_M$ completes an operation and releases the resource set involved in the co-operation (say $\beta^k_i \neq \emptyset$). The type 3 event is described by the pair $\sigma_3=(j,\beta^k_i)$.

The CTPN models the $\sigma_1=(j,w_k)$ event occurrence by the firing of t^k_0. Moreover, event $\sigma_3=(j,\beta^k_i)$ can occur if transition $t^k_i \in r_n \bullet$ for each $r_n \in \beta^k_i$ is color enabled and ready. After a type 3 event occurrence, the corresponding type 2 event $\sigma_2=(j, \alpha^k_i, \rho^k_{i+1})$ can happen without delay if $t^k_{i,i+1}$ is resource enabled. Let Σ_1, Σ_2 and Σ_3 respectively indicate the sets of events of type 1, 2 and 3 allowed for in the considered AMS.

Before closing this section, an important point is emphasized. To avoid ambiguity, a specification of the mechanism establishing priority for concurrent jobs is necessary. Typical concurrent problems are the selection of a job, among those blocked for receiving service from a resource that becomes idle or the selection of a job among those simultaneously completing an operation. To specify the above selection mechanism, the priority laws are defined as a set of functions $\Pi=\{\pi\}$ where $\pi: M \to J_M$ and M is the set of admissible markings of the CTPN.

V. DEADLOCK CHARACTERIZATION

This section recalls some results characterizing deadlock in CRSS [7]. The *transition digraph*, indicated by $D_T(M)=[N,E_T(M)]$, describes at the current marking M, in a concise way, the interactions between jobs and resources. While the vertex set N represents the resource set and is fixed ($N=R=P$), the edge set changes as the marking is updated: an edge e_{nm} is in $E_T(M)$ iff there exists a job $j \in J_M$ such that $r_n \in$ HR(j) in the marking M and r_m is one of the next resources, i.e. $r_m \in$ SR(j). An edge $e_{nm} \in E_T(M)$ may represent one or more jobs detaining r_n and requesting r_m. On the other hand, more edges can represent only one job requiring the conjunctive service of a set of resources. To take into account these situations, we associate with each edge $e_{nm} \in E_T(M)$ the weight expressed by the following job set: $J_M(e_{nm})=\{j \in J_M: r_n \in$ HR(j) and $r_m \in$ SR(j)$\}$. Obviously, $J_M(e_{nm})=\emptyset$ if $e_{nm} \notin E_T(M)$.

At this point we establish the relation between the described CTPN model and the introduced digraph tool to detect deadlocks. To this aim, we remind that a digraph containing N nodes is completely characterized by its (N×N) adjacency matrix [8]. We assume that the (m,n)-entry of the adjacency matrix is equal to one iff the digraph contains an arc from the n-th node to the m-th one.

Starting from the CTPN, it is possible to obtain the adjacency matrix $\mathbf{A}_M$ of $D_T(M)$ at each marking M by using the pre- and post-incidence matrices of the CTPN. Finally, the sets $J_M(e_{nm})$ associated with each non zero (m,n)-entry of $\mathbf{A_M}$, can be easily obtained from marking M: $J_M(e_{nm})=\{j \in J_M:$ $<$RWP(j),j$> \in M(r_n)$ and r_m belongs to second operation of RWP(j)$\}$

As mentioned before, a deadlock state is a condition in which each member in a job set waits indefinitely for other jobs in the same set to release resources.

From the Petri net point of view, a deadlock means that once some marking has been reached, some color-enabled transition cannot be fired anymore because it can not be resource enabled. The consequence in our model is that the flow of some jobs is interrupted and the system can not produce all the jobs from *J*. We said *Deadlock Marking* the marking that corresponds to a deadlock condition.

As for systems with single unit resource service [5,6], we use digraphs to characterize deadlock. Hence, we relate deadlock conditions to some particular strong subdigraphs [8] of $D_T(M)$, characterized by the following definition.

Definition 1. Let $\Gamma=(N_\Gamma,E_\Gamma)$ be a strong subdigraph of $D_T(M)$. We call Γ a *"Busy Conditioned Strong Subdigraph"* of $D_T(M)$ (BCSS for brevity) if the following properties hold true:

D1a) *Busy:* all the resources from N_Γ are busy;
D1b) *Conditioned:* for each job j holding resources from N_Γ (i.e., HR(j)$\cap N_\Gamma \neq \emptyset$), there exists an edge $e_{hm} \in E_\Gamma$ such that $j \in J_M(e_{hm})$.

The following result proved in [7] shows the relation between a BCSS and a deadlock marking.

Proposition 1. M is a deadlock marking for the CTPN iff there exists at least one BCSS in $D_T(M)$.

Finally, if the system does not contain conjunctive resource service, the notion of BCSS reduces to the one of Maximal-weigh Zero-outdegree Strong Component (MZSC) introduced in [6].

VI. DEADLOCK AVOIDANCE POLICY

An AMS real-time control avoids deadlock by ruling the allocation of resources at each event occurrence, on the basis of the current system state knowledge. In particular, the controller applies a policy that decides whether events of type 1 and 2 can occur (*enabled*) or if it is necessary to inhibit them (*inhibited*). Since a type 3 event does not involve resource acquisition, it can not cause a deadlock.

In term of CTPN modeling, a transition $t \in T$ is said to be controlled if its firing is determined by a control policy (CP) when t is ready and enabled according to conditions C1 and C2. Moreover, if a set of transitions is controlled and can fire according to a CP, then the CTPN is said "*controlled under CP*". More formally, in our model a CP is a mapping that associates with each event $\sigma \in \Sigma_1 \cup \Sigma_2$ and with each marking M a control action that enables or inhibits event σ.

To avoid deadlock it is sufficient to guarantee that each in process job can go on so that can reach under the CP the output of the system. Hence, we define *task marking* the marking associated with the system state in which all the in process jobs have been produced and have reached the system output. More formally, we introduce the following definitions.

Definition 2: Marking M* is called *task marking* of the CTPN if $M^*(r_R) \neq <0>$ and $M^*(r_n) = <0>$ for each r_n with $r_n \neq r_R$.

Definition 3: A CTPN controlled under CP is said to be deadlock free at marking $M \in Reach(M_0)$, if there exists a controlled firing sequence ω such that $M[\omega>M^*$.

Hence, the deadlock avoidance controller synthesis problem that we consider is finding a CP to keep the CTPN deadlock-free at each marking $M \in Reach(M_0)$, i.e., able to reach M* after the complete job production.

Starting from Proposition 1, an intuitive deadlock avoidance policy is defined, based on a look-ahead procedure of only one step. More precisely, this policy (named CP1) updates next marking M', builds the new transition digraph $D_T(M')$ and inhibits the event iff such a digraph contains a BCSS.
CP1
for each $M \in \mathcal{M}$ and for each $\sigma_i \in \Sigma_i$ with i=1,2:

$f_i(\sigma_i, M) = 0$ if the transition digraph $D_T(M')$, exhibits a BCSS.
$f_i(\sigma_i, M) = 1$ otherwise.

Although CP1 prevents transitions immediately leading to a deadlock, it might lead the system to some situations called *restricted deadlock*. In such a condition, the system is not in a deadlock state, but it inevitably incurs a permanent blocking caused by the control inhibition. We showed in [7] that it is necessary to modify CP1 in order to obtain a CP under which the CTPN behavior is deadlock and restricted deadlock free in any case.

Definition 4: The marking M* is said to be reachable from marking $M \in Reach(M_0)$ under f_2 and the priority rule Π, if there exists a controlled firing sequence $\delta = t_1(c_1)t_2(c_2)\ldots t_n(c_n)$ with $t_i(c_i) \in T_F \cup T_R$ for i=1,...,n such that $M[\delta>M^*$.

We remark that the firing sequence δ controlled by f_2 is a succession of events of type 2 and 3. However, ambiguity and concurrency may occur among ready and enabled transitions. Nevertheless, concurrency is solved by the priority rule Π that selects only one event to fire, without ambiguity. Hence, if M* is reachable under f_2 and Π, then the sequence δ is univocally determined. Considering previous considerations, CP1 is modified as follows:
CP2

Let the CTPN be at time τ and at marking $M \in Reach(M_0)$ and denote with M' the marking obtained from M after or $\sigma_1 \in \Sigma_1$ either $\sigma_2 \in \Sigma_2$ event occurrence:

$f_1(\sigma_1, M) = 1$ if M* is reachable from M' under f_2 and the priority rule Π.
$f_1(\sigma_1, M) = 0$ otherwise.
$f_2(\sigma_2, M) = 0$ if the transition digraph associated to M', exhibits a BCSS.
$f_2(\sigma_2, M) = 1$ otherwise

The following proposition proves that the system controlled by CP2 is deadlock and restricted deadlock free.

Proposition 2: Let the CTPN be at time τ and at marking $M \in Reach(M_0)$. If M* is reachable from marking M under f_2 and the priority rule Π, then the CTPN controlled under CP2 and the priority rule Π is deadlock free.

Proof. Let us suppose that the CTPN is at time τ and at marking $M \in Reach(M_0)$. If M* is reachable from marking M under f_2 and the priority rule Π, then there exists only one controlled firing sequence δ so that $M[\delta>M^*$ under f_2 * and the priority rule Π. Hence, the evolution of the CTPN controlled under CP2 follows the firing sequence δ till a type 1 event σ_1 has to occur. Moreover, let us suppose that the CTPN reaches marking M_1 such that $M[\delta_1>M_1$ where δ_1 is a subsequence of δ. Called M' the marking reached after the occurrence of σ_1, if there exists a new controlled firing sequence δ_2 under f_2 and the priority rule Π, so that $M'_1[\delta_2>M^*$, then f_1 enables σ_1. After the occurrence of the type 1 event, the AMS performance follows the event sequence δ_2 as long as a new type 1 event has to occur. Concluding, the system deadlock freeness is guaranteed.

A. Computational Complexity and comparison with existing policies

An algorithm working in polynomial time [6], [7] and requiring $O[|E_T(M)| \times R^2]$ operations updates the transition digraph and verifies whether it contains a BCSS when a type

2 event has to occur. On the other hand, the controller checks whether the system after a type 1 event can reach the final marking by executing a simulation run. It is easy to show that applying function f_1 of CP2 requires $O\{L \times |J_M|^2 \times |E_T(M)| \times R^2\}$ operations.

We remark that the obtained deadlock avoidance policy is realized by two control actions. The first one, defined by f_2, rules the flow of jobs in the AMS and performs a logical check based on the knowledge of next CTPN marking. The second control action, defined by f_1, governs the system job inputs and bases the decisions on the future controlled evolution of the system. More precisely, f_1 performs a simulation run and checks whether M* is reachable from the actual marking under f_2 and the priority rule Π, considering that no new job enters the system. In addition, the model time parameter does not influence the correctness of the proposed deadlock avoidance policy. It allows us to obtain an event sequence that is close to the real succession of the events determined by the operation times. Comparing the approach with the method proposed by Hsieh and Chang [9], we point out that here the simulation text is used just to avoid restricted deadlock states.

Moreover, the proposed CP can be applied to a complex class of sequential production processes. This type of resource allocation systems is a subclass of the CD/RAS that has been studied in [13]. Since, the deadlock avoidance policies proposed in [13] are not maximally permissive, the problem of obtaining a polynomial and not restrictive deadlock avoidance policy is open. On the other hand, f_2 avoids immediate deadlocks by means of a one-step look-ahead strategy. Consequently, CP2 results maximally permissive if any RDs cannot occur in the system. In addition, while the results obtained in [13] are strictly linked with the used PN model structure, CP2 does not need software reconfiguration whether the AMS layout and the working procedures change.

VII. CONCLUSIONS

This paper proposes a strategy managing the acquisition and the release of resources in Automated Manufacturing Systems (AMSs) in order to avoid deadlock. A large class of AMS is considered in which operations can require simultaneous service of a set of resources (Conjunctive Resource Service Systems- CRSS). A Colored Timed Petri Net (CTPN) that is suitable both for modeling relations and structural interactions among resources and for measuring and managing system performances describes the AMS dynamics. Moreover, an important peculiarity of the proposed model consists in the modularity of the CTPN building to ensure an easy real time update of the model. In addition, applying some results obtained by using digraphs to model interaction between jobs and resources, the paper defines a deadlock avoidance policy that rules the resource acquisition and the job inputs. A basic characteristic of the proposed deadlock avoidance strategy is that it does not need software reconfiguration and updating when the AMS layout and the working procedures change. Hence, the CTPN model and the proposed control strategy are suitable for following the frequent changes of the production system.

REFERENCES

[1] Z. A. Banaszak and B. H. Krogh, "Deadlock Avoidance in Flexible Manufacturing Systems with Concurrently Competing Process Flows", *IEEE Trans. on Robotics and Automation*, vol. 6, no. 6, pp. 724-734, 1990.

[2] H. Cho, T.K. Kumaran, and R.A. Wysk, "Graph-Theoretic Deadlock Detection and Resolution for Flexible Manufacturing Systems", *IEEE Trans. on Robotics and Automation*, vol. 11, 3, pp. 413-421, 1995.

[3] F. Chu and X. Xie "Deadlock Analysis of Petri Nets Using Siphons and Mathematical Programming" *IEEE Trans. On Robotics and Automation*, vol 13, no. 6, December 1997..

[4] J. Ezpeleta and J. M. Colom, "Automatic synthesis of colored Petri net for the control of FMS", *IEEE Transactions on Robotics and Automation*, vol. 13, no. 2, 327-337, 1997.

[5] M.P. Fanti, B. Maione, S. Mascolo, B. Turchiano, "Event Based Feedback Control for Deadlock Avoidance in Flexible Production Systems," *IEEE Trans. on Robotics and Automation*, vol. 13, no. 3, June 1997, pp. 347-363.

[6] M.P. Fanti, B. Maione, and B. Turchiano, "Deadlock Avoidance in Flexible Production Systems with multiple Capacity Resources," *Studies in Informatics and Control*, **7**, 4, pp 343-364, 1998.

[7] M.P. Fanti, B. Turchiano, "Deadlock Analysis in Automated Manufacturing Systems with conjunctive resource Service", *IEEE Int. Conference on Robotics and Automation,* May 11-15, Washington, U.S.A., 2002.

[8] Harary F., Norman R.Z., and Cartwright D., *Structural Models: An Introduction to the Theory of Directed Graphs*. John Wiley & Sons, Inc. New York, 1965.

[9] F. Hsieh and S. Chang, "Dispatching-Driven Deadlock Avoidance Controller Synthesis for Flexible Manufacturing System," *IEEE Trans. on Robotics and Automation,* vol. 10, no. 2, pp. 196-209, (1994).

[10] K. Jensen, Colored Petri nets: basic concepts, analysis methods and practical use. Vol. 1, New York Springer, 1992.

[11] Z. Jiang, M.J. Zho, R. Y. K. Fung, and P. Y. Tu, "Temporized Coloured Petri nets with changeable structure (CPN-CS) for performance modelling of dynamic production systems", *Int. J. Production Research*, vol. 38, no. 8, 1917-1945, 2000.

[12] M. A. Lawley, S. A. Reveliotis, and P. M. Ferreira, "A Correct and scalable deadlock avoidance policy for Flexible Manufacturing Systems", *IEEE Trans. on Automatic Control*, vol. 14, no. 5, pp. 796-809, 1998.

[13] J. Park and S.A. Reveliotis "Deadlock Avoidance in Sequential resource Allocation Systems with Multiple Resource Acquisitions and Flexible Routings", *IEEE Trans. on Automatic Control*, vol. 46, no. 10, 1572-1583, 2001.

[14] S. A. Reveliotis and P. M. Ferreira, "Deadlock Avoidance Policies for Automated Manufacturing Cells", *IEEE Trans. on Robotics and Automation,* **12**, 6, pp. 845-857, (1996).

[15] N. Viswanadham, Y. Narahari, and T.L. Johnson, "Deadlock Prevention and Deadlock Avoidance in Flexible Manufacturing Systems Using Petri Net Models," *IEEE Trans. on Robotics and Automation,* vol. 6, no. 6, pp. 713-723, 1990.

[16] N. Wu, "Necessary and sufficient Conditions for deadlock-free operation in Flexible Manufacturing Systems using Colored Petri Net model." *IEEE Transaction on Systems, Man, and Cybernetics – Part C: Applications and Reviews*, vol. 29, no. 2, pp. 192-204, 1999.

[17] N. Wu, and M.C. Zhou, "Avoiding deadlock and reducing starvation and blocking in automated manufacturing systems", *IEEE Transaction on Robotics and Automation*, vol. 17, no. 5, pp. 657-668, 2001.

[18] M.C. Zhou, and F. DiCesare, Petri Net Synthesis for Discrete Event Control of Manufacturing Systems. Kluwer Academic Publishers, Boston, 1993.

Controller Synthesis via Mapping Task Sequence to Petri Nets in Multi-Agent Collaboration Applications

Wenbiao Han and Mohsen A. Jafari

Industrial & Systems Engineering, Rutgers University
96 Frelinghuysen Road, Piscataway, NJ 08854, USA
wenbiao@alumni.rutgers.edu, jafari@rci.rutgers.edu

Abstract - A systematic multi-agent controller synthesis approach is presented in this paper. An abstract format for multi-agent task sequence is defined first. The typical elementary task sequences in multi-agent collaboration applications and the corresponding Petri Nets (PN) models are presented. This correspondence is used as templates to map a task sequence to a synthesized multi-agent PN controller model. We demonstrate the proposed approach using a two-agent controller synthesis problem in a layered manufacturing application.

I. INTRODUCTION

A number of agents can be employed to complete a common task in a collaboration environment. For example, one robot agent delivers a machined part, while another visual agent inspects the part and informs the flaw information (if any) to robot, which invokes appropriate correcting actions. This procedure may be repeated for each part. Like a single agent controller, a multi-agent controller needs to meet control specifications and achieve desirable system behaviors such as liveness (non-deadlock) and cyclicness. However, due to interactive nature of multi-agent collaboration, controller design complexity is increased, which may lead to longer design cycle and larger cost. An efficient and cost-effective multi-agent controller synthesis approach, which ensures desirable system behaviors (mentioned above), is demanded. To this end, each agent controller may be first modeled using Petri Nets (PN) and a multi-agent controller model may be derived via synthesizing the individual agent controller models. The problem is how to formalize this process.

Object-Oriented PN (OOPN) has been used to model manufacturing cell controller in [1] [2] and floor controller in [3]. The analytical power of PN formalism is employed to conduct system analysis (e.g., reachability analysis). However, these modeling schemes need to manually construct synthesized controller models. This paper proposes a systematic multi-agent controller synthesis approach, which may lead to automated controller synthesis design. Our approach consists of following steps: (1) find a multi-agent task sequence for completing a common task (Note that each unit task in the task sequence will be conducted by one agent); (2) systematically map the task sequence to synthesized PN using templates; (3) verify if the system behaviors are satisfactory. To the authors' knowledge, this is the first work to propose the mapping from elementary tasks to the corresponding PN models and perform multi-agent controller synthesis based on the derived elementary PN models.

The rest of this paper is organized as follows. Section 2 defines the format of task sequence. Section 3 proposes the PN modeling scheme for agents. A systematic approach to map a task sequence to a PN model is presented in section 4. Section 5 demonstrates the proposed approach with an example of two-agent controller synthesis in layered manufacturing. This paper is concluded in section 6.

II. DEFINITION OF TASK SEQUENCE

A task consists of a sequence of subtasks or actions. Each action may be executed once or repeatedly. For example, a robot may move an object from one location to another via a sequence of subtasks such as moving, picking-up, and putting-down. Formally, for a task T that involves n actions in one cycle (run), we have

$$T = \{ \alpha_i \mid \alpha_{i-1} \prec \alpha_i \prec \alpha_{i+1}, \ \alpha_i \in AC_k \in A_k, \ i \in [i,n] \}$$

where α_i is a subtask or an action of index i; $\alpha_i \prec \alpha_{i+1}$ means α_{i+1} starts after α_i completes; AC_k is the action set of the agent A_k (k is an index). α_i can be a null action, which is represented as ε. In general, if α_i is a subtask, it may be in one of two forms, an *AND* form and an *OR* form. The actions in the *AND* form will be executed in parallel by two or more agents, while the actions in the *OR* form will be executed optionally by the same or different agents, depending on the branch conditions. We use symbol & to represent *AND* and symbol / to represent *OR*. According to different combinations of $\prec$, & and /, we list in Table 1 the typical cases that may exist in a task sequence. The term inside [] represents the branching conditions. For example, the following expression stands for a task sequence,

$$T_e := \alpha_1 \prec \alpha_2 \ \& \ \beta_1 \prec \alpha_3 \prec [c_1]\beta_2/[c_2]\beta_3 \prec \alpha_4$$

where $\alpha_i \in AC_1$, $i \in [1,4]$, and $\beta_j \in AC_2$, $j \in [1,3]$. α_2

TABLE 1. THE TYPICAL ELEMENTARY TASK FORMATS IN A TASK SEQUENCE

Cases		Subtask Format	Explanations
simple sequence	1	$\alpha_1 \prec \alpha_2$	Actor does not change.
	2	$\alpha_1 \prec \beta_1$	An actor switches from one to another.
conditional branches	3	$\alpha_1 \prec [c_1]\alpha_2 \mid [c_2]\alpha_3$	Actor does not change.
	4	$\alpha_1 \prec [c_1]\alpha_2 \mid [c_2]\beta_1$	Two optional actions are performed by two different actors depending on the switching conditions.
	5	$\alpha_1 \prec [c_1]\beta_1 \mid [c_2]\beta_2$	Actor changes. Two optional actions are performed by the same actors.
concurrency	6	$\alpha_1 \prec \alpha_2 \ \& \ \beta_1$	Two agents perform two actions concurrently. α_2 and β_1 are synchronized, i.e., the agent that completes its action first needs to wait for another agent to complete before starting next action.

and β_1 occur concurrently while β_2 and β_3 are optionally executed depending on the switching conditions, c_1 and c_2. For an action sequence $\gamma_1 \prec \gamma_2$, if the actors (agents) of γ_1 and γ_2 are different, i.e., $\gamma_1 \in A_i$, $\gamma_2 \in A_j$, $i \neq j$, then A_i needs to send A_j a message after it completes γ_1 in order to inform A_j to execute γ_2.

Note that the internal behavior of each unit task is usually well defined, while the task sequence of an agent in completing a complete task is application-dependent. One way to derive the task sequence is by intelligent planning approach in artificial intelligence (AI) [4]. That is, by defining an initial world state, a goal world state, and action operators of an agent in a formal language (e.g., predicate logic), one may obtain a task sequence that leads the world from the initial state to the goal state using an intelligent task planner.

III. PETRI NETS MODELING OF AGENTS

The second step of our proposed approach is to systematically map a task sequence to a multi-agent PN model. Before this step, we present the formalism to use PN in modeling multi-agent controller.

In a manufacturing environment, an agent (e.g., robot) is an entity that consists of sensors, actuators, a communication unit, and a controller. An agent obtains environment information via its sensors, receives messages (commands or information) from peer agents via its communication unit, determines the response to these inputs via its controller, and performs appropriate actions via its actuators.

To deal with message passing between agents, we introduce "message place" to our modeling paradigm as also suggested in [2] and [5]. We introduce two places for message passing into our formalism, *mo* (message out) and *mi* (message in). An example of the PN model for two collaborating agents is

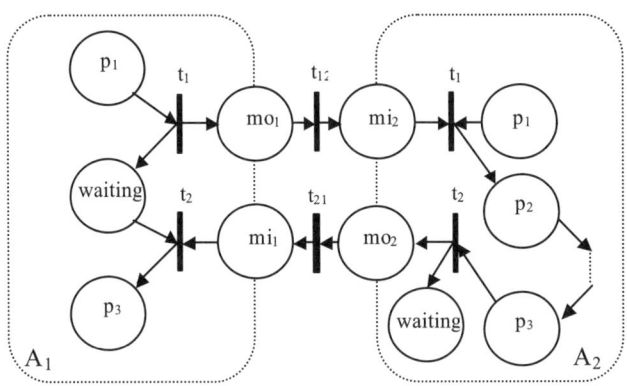

Fig. 1 A PN Model of Two Collaborating Agents

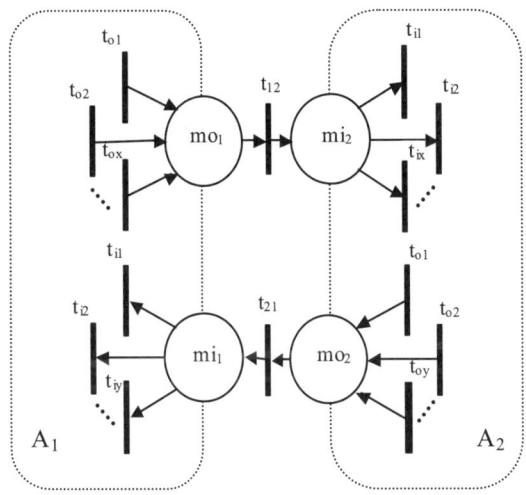

Fig. 2 Multiple Messages Between Two Agents

presented in Fig. 1. Note that there are only one message-input place (*mi*) and one message-output place (*mo*) for the PN model of each agent, while usually there are a number of transitions that are directed from the message-input place and a number of transitions that are linked to the message-output place (see Fig. 2). Both the message-input place and the message-output place are macro places, in which the message tokens are input to or output from a queue (resource place) of finite size (see Fig. 3). The function of the queue is to buffer the incoming or outgoing messages when multiple agents communicate with one another. In Fig. 3, the input transitions of the place *enqueue* in m_1 and the output transitions of the place *dequeue* in m_2 have one-to-one corresponding relationships and the same meanings.

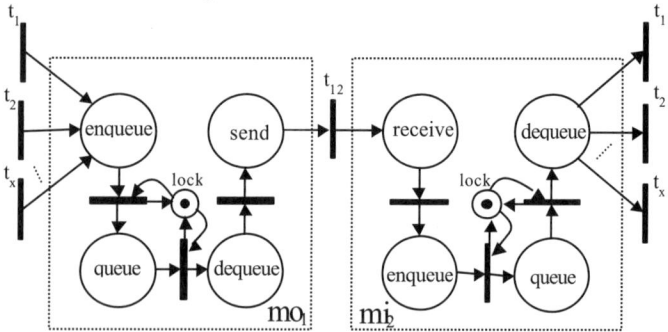

Fig. 3 Internal Representations of Macro Places *mo* and *mi*

IV. MAPPING TASK SEQUENCE TO PETRI NETS

Assuming a task sequence has been obtained (e.g., through an intelligent task planner), we want to transform it into the corresponding PN. To this end, we apply task-mapping templates to convert elementary tasks into the equivalent PN. Fig. 4 presents the mapping templates for those elementary tasks listed in Table 1. Whenever the task involves more than one agent, message passing must happen. Therefore, except for the first template (mapping $\alpha_1 \prec \alpha_2$) and the third template (mapping $\alpha_1 \prec [c_1]\alpha_2 \mid [c_2]\alpha_3$), all other four templates have used message passing. It should be noted that task $\alpha_1 \prec \alpha_2 \ \& \ \beta_1$ requires two agents to synchronize the completion of their concurrent subtask, α_2 and β_1, as shown in its corresponding PN model (see Fig. 4-6). With these mapping templates, a multi-agent PN controller can be systematically synthesized out of each agent's PN model.

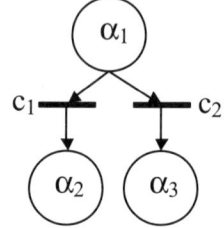

(4-3) Mapping for $\alpha_1 \prec [c_1] \alpha_2 \mid [c_2] \alpha_3$

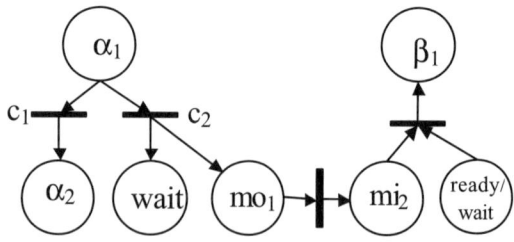

(4-4) Mapping for $\alpha_1 \prec [c_1] \alpha_2 \mid [c_2] \beta_1$

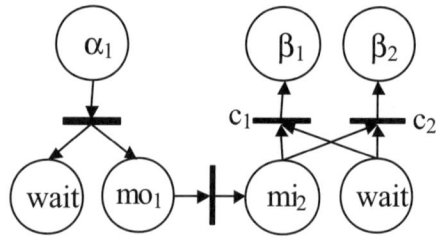

(4-5) Mapping for $\alpha_1 \prec [c_1] \beta_1 \mid [c_2] \beta_2$

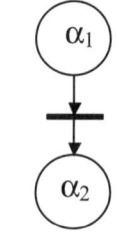

(4-1) Mapping for $\alpha_1 \prec \alpha_2$

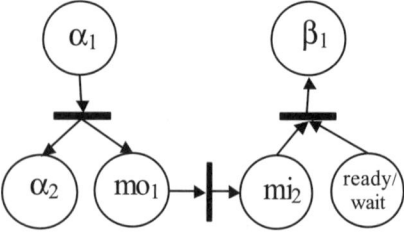

(4-2) Mapping for $\alpha_1 \prec \beta_1$

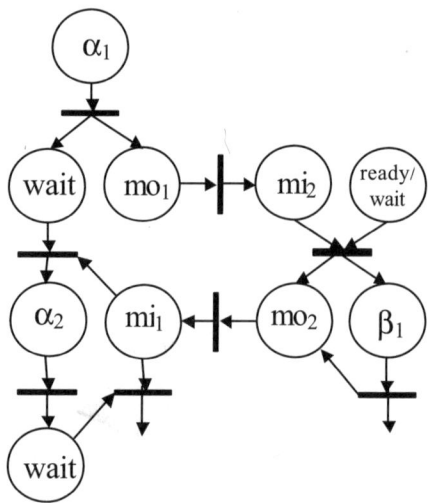

(4-6) Mapping for $\alpha_1 \prec \alpha_2 \ \& \ \beta_1$

Fig. 4 Templates for Mapping Task Sequence to PN Model

V. APPLICATION OF MULTI-AGENT CONTROLLER SYNTHESIS APPROACH IN LAYERED MANUFACTURING SYSTEM

Layered manufacturing (LM) refers to the technologies that build objects by adding materials in 2.5D layers according to a tool path plan. Fused deposition-based LM is being used to fabricate functional components (such as sensors and actuators) by depositing heated semi-solid materials (such as ceramics, metals) layer by layer [6]. In order to build parts with acceptable layer quality, we need to identify overfills and underfills in each layer.

We employ two subsystems to accomplish this task, Process Monitoring Subsystem (PMS) and Machine Vision Subsystem (MVS) [7]. The PMS and the MVS are two agents that collaborate to build a part and remove errors in each layer. The major steps of manufacturing process for making a layer are as follows: (1) The MVS computes the positions where the camera needs to take images. Note that these positions could be different from layer to layer; (2) The PMS controls and monitors material deposition following a predefined tool path; (3) Upon the completion of a layer, the PMS moves the camera to a set of points obtained in the first step, where the MVS will take partial images of a layer; (4) the MVS assembles a full image of the layer using the obtained partial images and identify overfills and/or overfills according to the full image; (5) the MVS notifies the PMS of the layer status (e.g., no error, overfill and/or underfill errors) and the PMS will take the corresponding correcting actions (such as filling in underfills, polishing overfills) if there are errors. As an example, Fig. 5 shows a camera movement trajectory when a square part is built. In each layer of this part, the camera will take an image at each numbered points. Note that each square around a number indicates the image size. These 9 partial images will be assembled to obtain a full image of a layer.

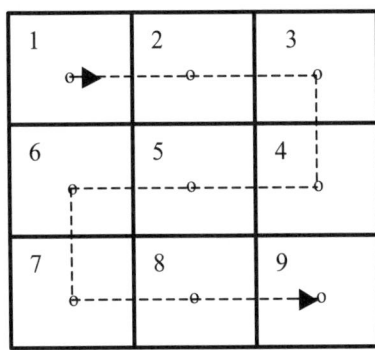

Fig. 5 An Example for A Sequence of Points Where A Camera May Take Images

The action sets of the PMS and the MVS and their functions are presented in Table 2. Note that an action consisting of a number of sub-actions can be represented as a sub-net. If we use $[i=n_1:n_2:n_3]\{task\ sequence\ A\}$ to represent a repeated task sequence A with i (which starts from n_1 and ends in n_3) increased by n_2 at each loop, the collaboration task sequence of the PMS and the MVS can be described as follows:

$GetCameraPosition() \prec BuildLayer() \prec [i = 0:1:n-1]$
$\{MoveCamera(P_i, P_{i+1}) \prec TakeImage(P_{i+1})\} \prec$
$MosaicImage() \prec ProcessImage() \prec [Error\ Found]$
$CompensateError() \mid [No\ Error\ Found]\ \varepsilon$

TABLE 3. Subtasks in PMS and MVS Collaboration

Task Format	Subtask Instance
$\alpha_1 \prec \alpha_2$	$BuildLayer() \prec MoveCamera(P_0, P_1)$
	$TakeImage(P_n) \prec mosaicImage()$
	$MosaicImage() \prec ProcessImage()$
$\alpha_1 \prec \beta_1$	$GetCameraPosition() \prec BuildLayer()$
	$MoveCamera(P_i, P_{i+1}) \prec$ $TakeImage(P_{i+1})$, $i \in [0, n-1]$
	$TakeImage(P_{i+1}) \prec$ $MoveCamera(P_i, P_{i+1})$, $i \in [0, n-2]$, $n>1$
$\alpha_1 \prec$ $[c_1]\beta_1 \mid [c_2]\beta_2$	$ProcessImage() \prec [Error\ Found]$ $CompensateError() \mid [No\ Error\ Found]\ \varepsilon$

TABLE 2. DEFINITIONS OF FUNCTIONAL LOGIC FOR PMS AND MVS

Agent	Action	Function Description
PMS	BuildLayer()	To deposit materials in a layer according predefined tool path and deposition plan.
	MoveCamera(x, y)	To move a camera from position x to position y.
	CompensateError()	To remove overfills and underfills in a layer.
MVS	GetCameraPosition()	To obtain the positions where the camera will take images.
	TakeImage(x)	To take an image at position x.
	MosaicImage()	To assemble images taken in various positions.
	ProcessImage()	To process an assembled image and identify overfills and underfills.

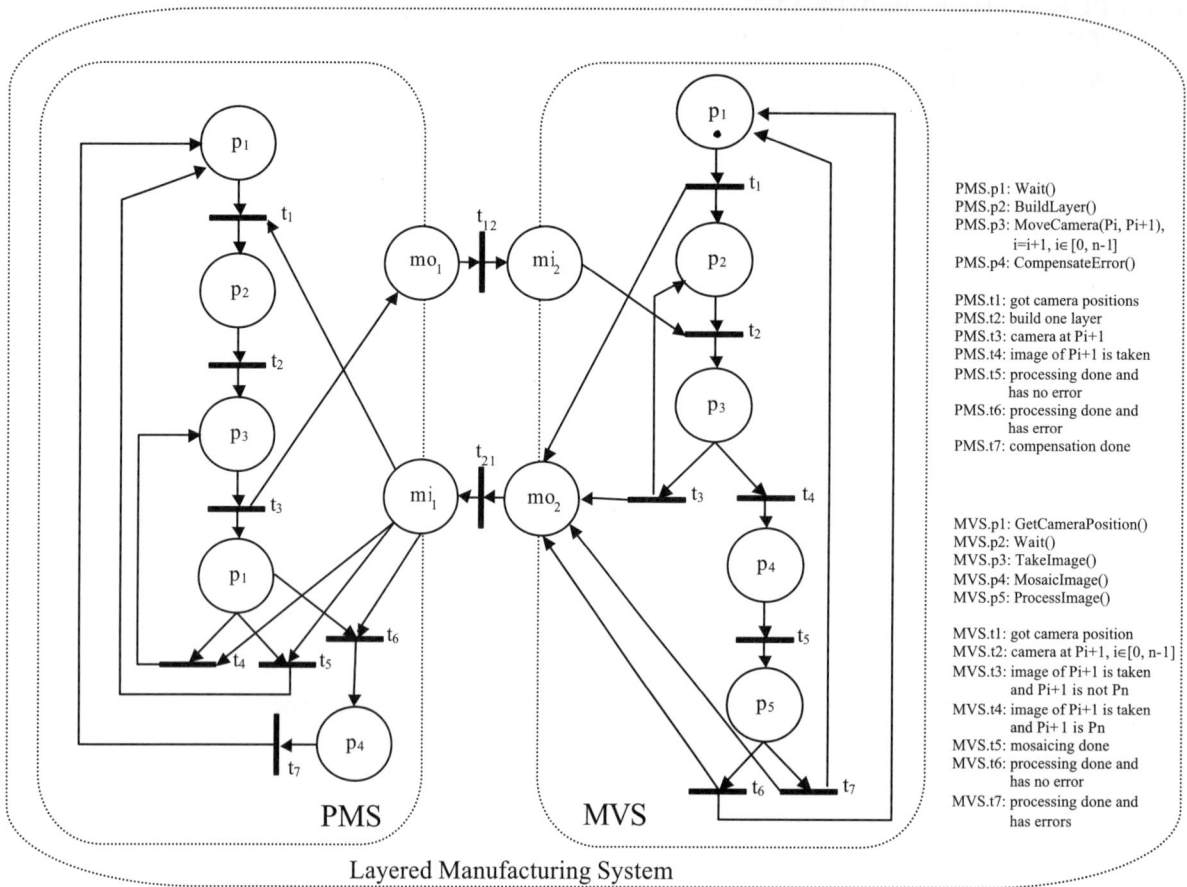

Fig. 6 PN Model of Two Collaborating Agents: PMS and MVS

Table 3 lists the subtasks in this task sequence and their corresponding generic formats. After transforming this task sequence using the mapping templates presented in Fig. 4, we obtain the PN controller shown in Fig. 6. It can be seen that: (1) the PN models of the PMS and the MVS are live; (2) the synthesized controller model is live; (3) the synthesized controller model is cyclic. We are working on the theory that can be used to identify liveness and cyclicness of a synthesized PN based multi-agent controller model.

VI. CONCLUSION

A multi-agent controller needs to meet control specifications and to ensure desirable system behaviors such as liveness and cyclicness. A cost-effective approach is desirable in multi-agent controller designs. To achieve this goal, we propose a systematic multi-agent controller synthesis approach, which may lead to automated multi-agent controller synthesis. We define task sequence format first and present six typical elementary task sequences. PN is used to model multi-agent controller because of its power in analyzing system liveness and cyclicness. We map a task sequence to a synthesized PN controller model using predefined templates. The proposed approach is demonstrated using a two-agent controller synthesis problem in a layered manufacturing application. This procedure may be automated via a Computer Aided Software Engineering (CASE) tool. The future works are: (1) to implement such a CASE tool; (2) to develop a theory that can be used to verify the system behaviors of a multi-agent controller model according to each individual agent controller model.

VII. REFERENCES

[1] Y. K. Lee and S. J. Park (1993), "OPNets: An Object-Oriented High-Level Petri Net Model for Real-Time System Modeling", *Journal of Systems Software*, vol. 20, pp. 69-86.

[2] L. Wang (1996), "Object-oriented Petri Nets for Modeling and Analysis of Automated Manufacturing Systems", *Computer Integrated Manufacturing Systems*, vol. 25, no. 2, pp. 111-125.

[3] K. Venkatesh and M. Zhou (1998), "Object-Oriented Design of FMS Control Software Based On Object Modeling Technique Diagrams and Petri Nets", *Journal of Manufacturing Systems*, vol. 17, no. 2, pp. 118-136.

[4] D. S. Weld (1999), "Recent Advances in AI Planning", *AI Magazine*, Summer 1999, pp. 93-123.

[5] E. Y. Lin and C. Zhou (1999), "Modeling and Analysis of Message Passing in Distributed Manufacturing Systems", *IEEE Transactions on Systems, Man, and Cybernetics – Part C: Applications and Reviews*, vol. 29, no. 2, May 1999.

[6] W. Han, M. A. Jafari, S. C. Danforth, and A. Safari (2002), "Tool Path-Based Deposition Planning in Fused Deposition Processes", *Transactions of The ASME, Journal of Manufacturing Science and Engineering*, vol. 124, issue 2, pp. 462-472.

[7] M. A. Jafari, W. Han, F. Mohammadi, A. Safari, S. C. Danforth, and N. A. Langrana (2000), "A Novel System for Fused Deposition of Advanced Multiple Ceramics", *Rapid Prototyping Journal*, vol. 6, no. 3, pp. 161-174.

Fuzzy Petri nets for monitoring and recovery

Daniel RACOCEANU**, Eugenia MINCA*, Noureddine ZERHOUNI**

*Electrical Faculty of Valahia University, 18, Bd. Unirii, 0200, Targoviste, Romania, minca@valahia.ro
**Laboratoire d'Automatique de Besançon UMR CNRS 6596, 25, rue Alain Savary, 25000 Besançon, France,
daniel.racoceanu@ens2m.fr, zerhouni@ens2m.fr

Abstract - In this paper, we propose a unitary tool for modeling and analysis of discrete event systems monitoring. Uncertain knowledge of such tasks asks specific reasoning and adapted fuzzy logic modeling and analysis methods. In this context, we propose a new fuzzy Petri net called Fuzzy Reasoning Petri Net: the FRPN. The modeling consists in a set of two collaborative FRPN. The first is used for the fault dynamic state of the system by temporal spectrum of the marking. A monitoring fuzzy Petri net (MFPN) represents the fault tree. The second model, the recovery fuzzy Petri net (RFPN) corresponds to recovering activities. The two proposed models form a dynamic loop for production system monitoring and recovery. Production system is supposed to be modelized using temporal Petri nets, with the assumption that the primary fault symptoms are detected. These symptoms are considered in the MFPN and evolve according to all other derived faults of the system. Synchronizing signals, corresponding to different warning levels and alarms, form the interface between these two tools.

Keywords: Discrete event system, Fuzzy logic, Fuzzy Petri nets, Monitoring, Recovery, Diagnosis, Fault tree.

1. FUZZY LOGIC APPROACH FOR MONITORING AND RECOVERY

The complex analysis of re-configurable discrete event systems requires specific flexible and fast reasoning modeling and analysis methods. The field of the artificial intelligence seems very appropriate for monitoring and recovery applications. In the class of modeling tools, fuzzy Petri nets [4, 6, 7] are very interesting tools for discrete event systems characterized by an imprecise knowledge. J. Cardoso, R. Valette and D. Dubois have published in [3] a very complete overview of the existing fuzzy Petri nets (FPN) approaches.

Chen [4] and Gao, Zhou, & all [8] have some important research contributions in fuzzy reasoning Petri nets. For monitoring tasks modeling, we complete this approach by defining an extension of the FPN able to integrate, by a fuzzy temporal approach, the instant of occurrence of a default or degradation of a discrete event system. The recovery is the consequence of a reasoning based on a fuzzy logical rule basis, modelized by another FPN tool with a similar typology. This FPN recovery model has a double interface, one with the monitoring (detection and diagnosis) system and the second one with the studied (supervised) discrete event system. In theses interfaces, information is transmitted by an emission/reception protocol, using a representation inspired by the synchronized Petri nets [11], adapted to the fuzzy variables transmission.

In the next paragraph, we define the fuzzy Petri net for monitoring. Following, the recovery fuzzy Petri net is introduced. In order to illustrate the interest of the proposed modeling and analysis algorithms, an example of industrial system study is presented at the end.

2. FUZZY PETRI NET FOR MONITORING

2.1. Fuzzy monitoring

Fuzzy Petri nets are used for fuzzy reasoning modeling based on the static logic rules [4, 6, 7] according to their evolution in time. In this sense, we propose a fuzzy Petri net for monitoring called monitoring fuzzy Petri net – MFPN. By a fuzzy temporal approach, this tool permits us to model the dynamic behavior of a monitoring system. The MFPN treats a fuzzy logic rule base obtained from the logic expression of the supervised system fault tree. Inadequate for linear logical reasoning [8], the MFPN doesn't apply to the resources state modeling [2, 3, 12]. MFPN represents the union or the intersection of logical reasoning, while respecting the specific concepts of the fuzzy logic [1, 4, 6, 7]. Analysis possibilities of such a tool offers us refined information of every basic and derived defect. We also show the impact of critical cases and strategies (represented by the critical path in fault tree) to the strategy of the prognosis function.

2.2. Monitoring fuzzy Petri nets definition

In the first part of our study, we treat the problem of a discrete event system monitoring. We suppose that the surveyed system is modelized by a temporal Petri net, able to include the fault symptom detection, using appropriate devices like watching dogs [5].

The monitoring fuzzy Petri net - MFPN is defined as being the n-uplet: MFPN = < **P**, **T**,**D**, I, O, f, F, **?S**, **!R**, α, β, λ > , with:

- **P** = { p_1, p_2, ... p_n} - the finite set of places modeling possible faults, identified at the discrete event system level. Two types of faults characterize this fault set: basic faults and derived faults. The considered faults can be as well transient that persistent;
- **T** = { t_1, t_2, ... t_m } - the finite set of transitions, representing the fault evolution, corresponding to the set of logical fuzzy rules R. Every transition is associated to a fuzzy rule;
- **D** = { d_1, d_2, ... d_n } - the finite set of logical propositions that defines the rule basis R;
- I : **T**→ **P** - the input function of places;
- O: **P** → **T** - the output function of places;
- f : **T**→ **F** - the function that associates to the every rule modelized by a transition, a function F describing the credibility degree $\mu = F(t)$ of the rule. The instant t corresponds to the detection of a fault symptom in the surveyed discrete event system;
- **?S** ={ s_1, s_2,..., .s_l} - the set of fuzzy symptoms (signals) received by the monitoring system from the surveyed discrete event system;
- **!R** ={ r_1, r_2,..., .r_l} - the set of fuzzy recovery information (signals) emitted by the monitoring system. The recovery tool will use these signals;
- α: **P** → [0,1] - the function giving a fuzzy value α_j of credibility for each place p_i corresponding to the logic proposition $d_i \in D$. This parameter represents the occurrence possibility of the corresponding fault;
- β : **P** → **D** - the bijective function that associates a logic proposition d_i to each place $p_i \in P$;
- λ : **P** → [0,1] - the function that associates an acceptance/permissiveness warning threshold λ_i of the fault corresponding to each $p_i \in P$ of the critical path of the fault tree. These thresholds represent the starting point of the recovery strategies.
- M_0 - the basic faults places initial marking. Every token of the marking M_0 is associated to the fuzzy number 1 that means the certainty of the basic fault occurrence. By convention, places associated to the basic faults are not represented in the global model of the MFPN.

Each transition of the MFPN represents a fuzzy logic elementary proposition: $d_i \rightarrow d_j$. The transition is associated to a function F(t) describing the degree of credibility of the corresponding proposition at the time t (firing possibility at time t). The function $\mu_i = F(t)$ represents the membership function of the fuzzy variable t to the fuzzy set defined by the linguistic variable: "occurrence of the fault d_j" (fig.1). Being variable in time, the value of credibility μ gives to every rule, a *dynamic credibility* character. The interval [0 ΔT] represents the total studied period. A degradation is recorded if the firing time belongs to [t_0, t_{max}) and the fault occurs if the firing time belongs to [t_{max}, ΔT].

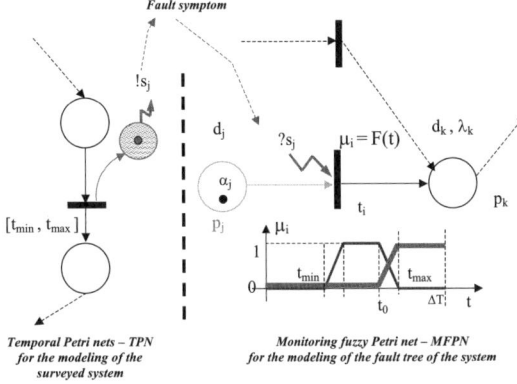

Fig. 1. Fault propagation modeling using MFPN.

Every place of the MFPN is related to a (basic or derived) fault of the supervised system. The marking of a place is associated to a fuzzy credibility value α ∈[0, 1]. For the MFPN represented in the figure 1, the place p_j is associated to the observed fault (new experience) and the place p_k to the derived fault (conclusion). The fuzzy value α_k associated to the marking of the conclusion, will be calculated using the generalized modus ponens [1]. In our approach, we consider the operators: T(u,v)=min(u,v), ⊥(u,v)=max(u,v) and the generalized modus-ponens operator **T**$_{probabilistic}$(u, v)=u·v. The value α_k of credibility of the conclusion will be calculated using the next formula [4, 6, 7]:

$$\alpha_k = \alpha_j \cdot \mu_i \Rightarrow \mu_i = \frac{\alpha_k}{\alpha_j} \qquad (1)$$

We suppose [6] that a transition firing consumes copies of tokens of the places of the M_0. This hypothesis is justified by the respect of the logical reasoning principle: if a hypothesis is true, it always remains true.

Proposition 1: *The degree of truth μ corresponding to a transition associated to the proposition*

$$(D_1^x \wedge D_2^x \wedge \ldots \wedge D_k^x) \to D^y$$

which receive in a synchronous way the signals $\{?s_1, ?s_2, \ldots, ?s_k\}$, is:

$$\mu = \frac{\min(\alpha_1 \cdot \mu_1, \alpha_2 \cdot \mu_2, \ldots, \alpha_m \cdot \mu_k)}{\min(\alpha_1, \alpha_2, \ldots, \alpha_k)}$$

if the transition receives temporal synchronized signals for the input positions and
$\mu = 0$,
if the transition does not receive any signal $?s_i$.

We specify that signals $\{?s_1, ?s_2, \ldots, ?s_k\}$ represent instants of elementary implication validation:

$$D_1^x \to D^y, \quad D_2^x \to D^y, \quad \ldots \quad D_k^x \to D^y$$

Proposition 2: *If a fuzzy implication transition doesn't receive its corresponding fuzzy symptom signal, the transition representing the elementary logical proposition $D_i^x \to D^y$ has a degree of truth $\mu = 0$.*

The synchronization information between the supervised discrete event system model (temporal Petri net – TPN) and the fault propagation corresponding model (monitoring fuzzy Petri net - MFPN) are illustrated in figure 1 using a fuzzy signal representation inspired by the synchronized Petri nets [11].

3. FUZZY PETRI NETS FOR RECOVERY

3.1. Recovery problem in monitoring

In a fuzzy monitoring system, the recovery requests a the use of a tool able to integrate temporal synchronous fuzzy information of the monitoring system, related to a base of fuzzy logic rules. This detailed information can describe the critical faults associated to the critical path of the MFPN model. Thus, we propose a second tool called recovery fuzzy Petri net (RFPN), representing a fuzzy expert system.

Using the fuzzyfication of the embedded variable (the marking value of the place that receives the signal), the fuzzy signals represent input variables coming from an inference mechanism to a base of fuzzy logic rules (fig.2). This base **R** is designed according to the strategy adopted for an optimal recovery. The so obtained RFPN model represents a variant of a fuzzy controller for discrete event systems. The transitions of this model materialize the generalized modus ponens operator, by the composition of input variables and the base of rules **R**. The outputs of the RFPN are also synchronized fuzzy temporal signals. They are correction controls for the supervised system, either fuzzy recovery signals for the subsystems of the next decision level.

3.2. RFPN definition

The recovery fuzzy Petri net - RFPN is defined as being the n-uplet :

$$RdPFR = \langle P, T, x, y, D, X_k, Y_q, R,$$
$$?!S, M_0, \partial, I, O, f, \lambda, \dagger_x, \dagger_y \rangle$$

with:

$\mathbf{P} = \mathbf{P}^k \cup \mathbf{P}^q$ - the finite set $\{p_1, p_2, \ldots p_{k \times q}\}$ of input $\mathbf{P}^k$ and output $\mathbf{P}^q$ places;

$\mathbf{T} = \{tf_1, tf_2, \ldots tf_n\}$ - the finite set of transitions specialized in inference/aggregation operations of logic rules and fuzzy variables defuzzyfication;

$\mathbf{x} = \{x_1, x_2, \ldots x_k\}$ - the finite set of markings of $p_i \in \mathbf{P}^k$ places;

$\mathbf{y} = \{y_1, y_2, \ldots y_q\}$ - the finite set of output normalized variables, associated to the markings of $p_j \in \mathbf{P}^q$ places;

$\mathbf{D} = \mathbf{D}^x \cup \mathbf{D}^y$ - the finite set of the logic variables. $\mathbf{D}^x, \mathbf{D}^y$ are subsets of variables that are respectively in the antecedence and in the consequence of the base of rules $\mathbf{R}$;

$\mathbf{R} = \bigcup_{w=1}^{r} R^w, \quad R^w : \mathbf{D}^x \to \mathbf{D}^y$ - the fuzzy logic rules set. We supposes that the set $\mathbf{R}$ of rules can be incomplete from the point of view of the possibilities of exhaustive combination of the logical variables associated to the $\mathbf{D}^x$ and respectively $\mathbf{D}^y$;

$\mathbf{X}^k = \{X_{11}, X_{12}, \ldots X_{k1}, X_{k2}, \ldots X_{ki}\}$ - the finite set of membership functions, defined on the universe [0,1] of the variables $\{x_1, x_2, \ldots x_k\}$ associated to the logic variables of $\mathbf{D}^x$. k represents the number of input variables;

$\mathbf{Y}^q = \{Y_{11}, Y_{12}, \ldots Y_{q1}, Y_{q2}, \ldots Y_{qj}\}$ - the finite set of membership functions, definite on the universe of variable $\{y_1, y_2, \ldots y_q\}$, partners to the logical variable $\{y_1, y_2, \ldots y_q\}$ of $\mathbf{D}^y$. q is the number of output variables;

$?! \ \mathbf{S} = \{r_1, r_2, \ldots r_{k \times q}\}$ - the set of received/emitted (fuzzy) signals for recovery/control. The signal r_i induces a vector of fuzzy numbers in the input places;

$M_0 : P^k \to \{v_i = \langle 0, 0, \ldots 0 \rangle\}_{i=1:k}$ - the initial marking of the input places $p_i \in P^k$;

$\partial : P \to \{v_1, v_2, \ldots v_{k \times q}\}$ - function that associates a dimension v_i to every input place. For an input place, v_i is the number of membership functions defined on the universe of the variable x_i. For an output places, v_i will be equal to 1;

$I : T_f \to P$ and $O : T_f \to P$ - (places) input and output functions;

$f : P \to \bigcup_{w=1}^{k \times q} F^w$ - function that associates to every place p_i, the membership function v_i corresponding to the fuzzy description F^i of the variable x_i;

$\lambda : P^k \to [0,1]$ - function that associates to a place $p_i \in P^k$ a threshold value of acceptance/permissiveness warning λ_i of the corresponding fault, from the point of view of the recovery. The parameter λ_i is associated to places belonging to the critical path of the fault tree;

$\dagger_x : X^k \to D^x$ - a bijective function that associates a fuzzy set X_{ij} to a logic proposition D_{ij}^x being in the antecedence of a logic implication;

$\dagger_y : Y^q \to D^y$ - a bijective function that associates a fuzzy set Y_{ij} to a logic proposition D_{ij}^y being in the consequence of a logic implication.

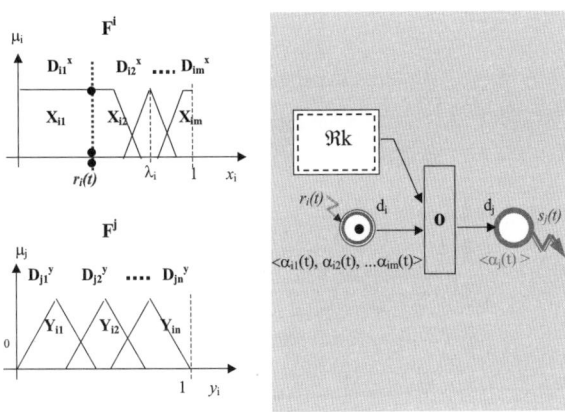

Fig. 2. Recovery modeling using RFPN.

For the complex systems, the recovery action permits a modular approach (fig.3).

Proposition 3: *If a fuzzy signal represents the input of several modular recovery subsystems, it will be multiplied by a simple transition.*

Remark: *The RFPN model is open and can have input interface places with elaborated information obtained with other intelligent sensors. The presence of these places permits us to develop, by the different techniques of the artificial intelligence, a distributed intelligent monitoring platform.*

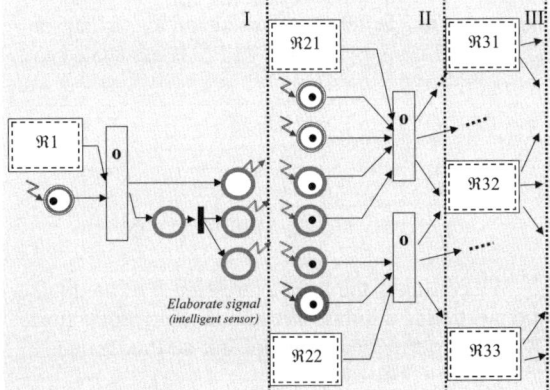

Fig. 3. Modular recovery system.

Each input or output place of the RFPN is associated to a fuzzy description. For the input places, we describe the marking variable of the place, whereas for the output places we describe the normalized control.

For each subsystem modelized by RFPN, we identify a transition specialized in a complex operation of input variables composition. Each base of logic rules R represents the fuzzy implications describing the knowledge base of the expert. Each implication respects the "if-then" model represents the logical dependence of linguistic variable $\{D^x \bigcup D^y\}$ associated to the fuzzy sets $\{X^k, Y^q\}$.

The composition of K rules requires an aggregation mechanism using intersection or means. Two methods of approximate reasoning exist: the AI (the aggregation of rules followed by the inference) method and the IA (the inference of the every rule followed by the aggregation) method, that give near similar results. We chose the first approach $\mathbf{R} = \bigcap_{k=1}^{K} {}_{T,\Sigma} R^k$, where $\bigcap_{T,\Sigma}$ means the operation of aggregation using the T - triangular norm or Σ - mean.

Proposition 3: *In order to obtain the fuzzy variable associated to the rule base R, we apply the ZMA approach (Zadeh-Mamdani-Assilian) [1]. In this approach, a fuzzy rule R^u is generally interpreted like a superposition of simultaneously true logical propositions:*

$R^u \Leftrightarrow$

$\Omega(X_{1u}, \Omega(X_{2u}, \Omega(\ldots, \Omega(X_{k-1,u}, X_{ku})))) \otimes (Y_{1u}) \Rightarrow$

$\mathbf{R} = \mathbf{D}^x \left(\dagger_x \circ \dagger_y \circ \otimes \right) \mathbf{D}^y$

For an arbitrary input variable, modelized by a fuzzy set X', while applying the concept of generalized modus ponens, we obtain a fuzzy exit set Y', that will be obtained by the composition between fuzzy sets **R** and X'.

$$Y' = R \circ X' \qquad Y' = \sup_{x \in X} \mathbf{T}(X', R),$$

where **T** is the generalized modus ponens operator.

A defuzzyfication operation f^{-d} will be necessary to get the exact value of **Y**. In fact, a specialized transition of the RFPN is associated to the operations:

$$\mathbf{Y} = f^{-d}\left(\sup_{x \in X} \mathbf{T}(X', \bigcap_{k=1}^{K}{}_{T,\Sigma} R^k)\right) \Rightarrow \mathbf{Y} = f^{-d}\left(\sup_{x \in X} \mathbf{T}(X', \mathbf{R})\right)$$

4. INDUSTRIAL APPLICATION

We considers the logic expression F of the fault propagation in a failure tree of a flexible production system of our partner, the "Institut de Productique" of Besançon [10] :

$$F = [(a+b+c+d) * e] + b + c$$

where + and * operators represent the union or the intersection of the logic variables {a, b, c, d, e}. This expression corresponds to the previous failure tree with the set of associates of rules (fig.4). The critical path indicates places that send warning signals in the recovery model (fig.6). Output variables u_1 and u_2 represent controls of the system. The logic reasoning between the MFPN model and the RFPN model, constitute the base of the fuzzy logic rules (fig.5).

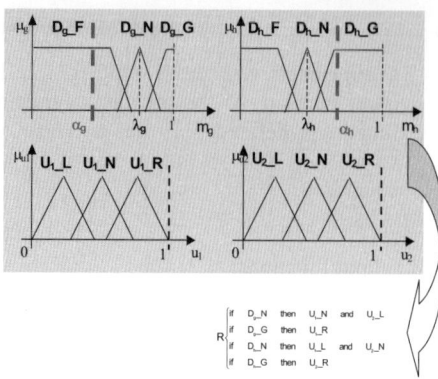

Fig. 5. Fuzzy rule base for the recovery.

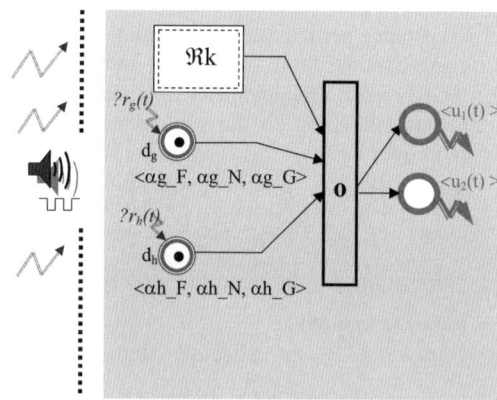

Fig. 6. Modeling of the real-time recovery system.

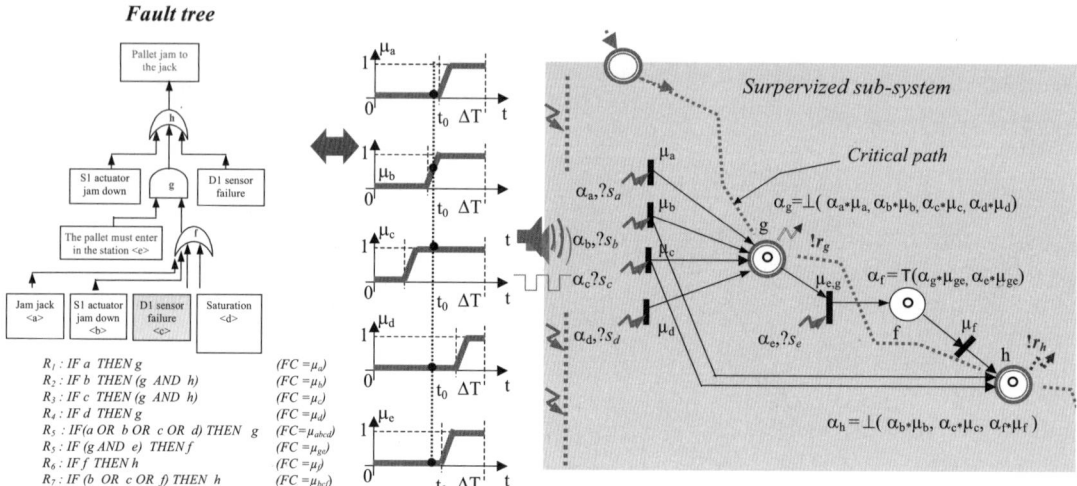

Fig. 4. Modeling of the function F using the MFPN.

5. CONCLUSIONS

In this paper, we have proposed two fuzzy Petri net tools, MFPN - monitoring fuzzy Petri nets and RFPN - recovery fuzzy Petri nets, able to modeling and analyze monitoring and recovery tasks of a discrete event system, using a temporal fuzzy approach. The discrete event system and the monitoring and recovery models communicate using fuzzy synchronous information. This approach permits us to give a finer spectrum of the discrete event system failures. The use of the fuzzy logic and of the associated degree of credibility for a failure event/evolution includes a prognosis point of view in the monitoring fuzzy tool.

For complex systems monitoring, our results permit the use of a modular approach. The proposed tools form an open monitoring system and give also the possibility to integrate some others monitoring modules using other diagnosis and prognosis techniques. The open architecture can thus integrate a more complete maintenance platform.

In order to improve our fuzzy Petri net tool, a future research is represented by the extension including the negation and other logical operations like suggested in [8], by keeping our temporal specification.

REFERENCES

[1] B. Bouchon-Meunier: La logique floue. Collection Que sais-je, n.2702 Ed. Presses Universitaires de France, Paris, 1994, France.

[2] J. Cardoso, R. Valette, D. Dubois : Petri nets with uncertaine markings, LNCS Advances in Petri nets 1990, Vol. 483, G. Rozenberg, Ed.Springer Verlang, 1991, pp. 64 - 78.

[3] J. Cardoso, R. Valette, D. Dubois, Fuzzy Petri Nets : An Overview, 13th Word Congress of IFAC, Vol. I : Identification II, Discrete Event Systems, San Francisco, CA, USA, June 30 - July 5, 1996, pp. 443-448.

[4] S.-M. Chen, J.-S. Ke, J.-F. Chang : Knowledge Representation Using Fuzzy Petri Nets, IEEE Transaction on knowledge and data engineering, Vol. 2, No.3, September 1990.

[5] M. Combacau : Commande et Surveillance des Systèmes a Evènements Discrets Complexes : Application aux Ateliers Flexibles. PhD, Université Paul Sabatier, Toulouse, December 1991, France.

[6] C.G. Looney : Fuzzy Petri nets for rule-based Decision making IEEE Trans. on Systems Man and Cybernetics, January-February 1988 Vol 18, N 1, pp 178-183.

[7] M.L. Garg and S.I. Ahson and P.V. Gupta: A Fuzzy Petri net for knowledge representation and reasoning, Information Processing Letters, V39, 1991, p.165-171.

[8] M. Gao, M. Zhou, X. Huang, Z. Wu, Fuzzy Reasoning Petri Nets, to be published.

[9] B. Pradin Chézalviel, R. Valette: Petri nets and Linear logic for process oriented diagnosis, IEEE/SMC Int. Conf. on Systems, Man and Cybernetics: Systems Engineering in the Service of Humans, Le Touquet, October 17-20, 1993, France

[10] S. Proos, Analyse des defaillances des systemes industriels, Rapport de stage DEA Informatique, Automatique et Productique, Laboratoire d'Automatique de Besançon, September, 2001, France.

[11] D. Racoceanu, N. Zerhouni, N. Addouche, Modular modeling and analysis of a distributed production system with distant specialized maintenance, Proc. of the 2002 IEEE International Conference on Robotics and Automation, on CD ROM, 7 pages, pp. 4046-4052, May 11-15, 2002, Washington, USA.

[12] R. Valette, J. Cardoso, D. Dubois : Monitoring manufacturing System by means of Petri Nets with Imprecise Markings, IEEE International Symposium on Intelligent Control 1989, September 25-26, Albany N.Y., USA.

Information Systems as a Tool for Specification of Concurrent Systems

Zbigniew Suraj

Abstract—The paper presents an approach to the synthesis problem. The approach is based on the rough set philosophy, Boolean reasoning and Petri nets. Information systems as a tool for specification of concurrent systems are treated. However colored Petri nets as a model for concurrency are used. The proposed approach in the $ROSECON$ computer system has been implemented. It can be used to deal with various problems arising from the process design domain.

Index Terms— Information systems, minimal rules, knowledge discovery, concurrent systems, colored Petri nets.

I. INTRODUCTION

In the paper rough set methods for the synthesis problem are considered. The research is motivated by the problems coming from the domains such as, for example, knowledge discovery systems, data mining, control design, decomposition of information systems, object identification in real-time (see e.g. [11], [13], [14], [21], [23]).

The synthesis problem is related to synthesizing a concurrent system model from observations or the specification of processes running in a given concurrent system. The problem has been discussed in the literature for various types of formalisms [1], [4]. Our approach is based on the rough set philosophy [10], Boolean reasoning [2] and colored Petri nets [5].

The rough set theory is dealing with incomplete or imprecise data. In general, in order to understand used data it is necessary to derive underlying knowledge about the data, i.e., what it represents. Such knowledge can be represented in many forms. The (production) rules are the most common form of knowledge representation in the rough set theory. In the paper, the rules will be represented in the following condition-action form: IF(condition) THEN (action), where "condition" is a list of conditions linked by logical operators (AND, OR) and "action" is an action to be taken if the "condition" evaluates to true.

However it is assumed that knowledge encoded in data table represented by an information system is represented by rules automatically extracted from the system. The approach presented here uses a new method for generating the minimal rules in order to construct a solution of the considered synthesis problem. The rules describe all dependencies between the local states of processes in the system. This paper provides an algorithm for constructing a model of a given concurrent system in the form of a net. The net construction consists of two stages. In the first stage, all dependencies represented by means the minimal rules between local states of processes in the system are extracted. In the second stage, a colored Petri net corresponding to these dependencies is built.

This paper is organized as follows. Section II presents the basic concepts underlying the information systems and colored Petri nets and fixes notations. The synthesis problem is formulated in section III. The algorithm for constructing a concurrent model in the form of a *CP*-net corresponding to a given information system on the base of extracted rules from the information system is described in Section IV. The solution properties are presented in Section V. Section VI deals with the concluding remarks.

II. PRELIMINARIES

In this section we recall basic notions and notation of rough set theory as well as colored Petri nets.

A. Information Systems

Information systems (sometimes called data tables, knowledge representation systems etc.) are used for representing knowledge. The notion of an information system presented here is due to Z. Pawlak [10].

An *information system* is a pair $S = (U, A)$, where U is a non-empty, finite set of *objects*, called the *universe*, A is a non-empty, finite set of *attributes*, i.e., $a : U \to V_a$ for $a \in A$, where V_a is called the *value set* of a. The set $V = \bigcup_{a \in A} V_a$ is said to be the domain of A.

In the paper attributes are interpreted as processes of a given concurrent system, the values of attributes - as local states of processes, and objects - as global states of the system.

For $S = (U, A)$, a system $S' = (U', A')$ such that $U \subseteq U'$, $A' = \{a' : a \in A\}$, $a'(u) = a(u)$ for $u \in U$ and $V_a = V_{a'}$ for $a \in A$ will be called an $U' - extension$ of S (or an extension of S, in short). S is then called a *restriction* of S'. If $S = (U, A)$ then $S' = (U, B)$ such that $A \subseteq B$ will be referred to as a $B - extension$ of S. S is also called a *subsystem of S'*.

Example II.1. Consider an information system $S = (U, A)$ with $U = \{u_1, u_2, u_3, u_4\}$, $A = \{a, b\}$ and the values of the attributes as in Table I.

This information system can be treated as a specification of system behavior concerning a toy communication system presented in Fig. 1. The communication system consists of two communicating devices a and b. A communication between

Z. Suraj is with the Institute of Mathematics, Rzeszów University, Poland, and the Chair of Computer Science Foundations, University of Information Technology and Management, Rzeszów, Poland. E-mail: zsuraj@wenus.wsiz.rzeszow.pl

TABLE I
AN INFORMATION SYSTEM S.

U/A	a	b
u_1	0	1
u_2	1	0
u_3	0	2
u_4	2	0

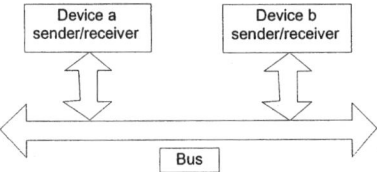

Fig. 1. A scheme of a communication system.

TABLE II
THE DISCERNIBILITY MATRIX $M(S)$ FOR THE INFORMATION SYSTEM S
FROM EXAMPLE II.1.

U	u_1	u_2	u_3	u_4
u_1				
u_2	a,b			
u_3	b	a,b		
u_4	a,b	a	a,b	

these devices is realized by means a common bus. We assume that the devices a and b can work as information senders as well as information receivers. These devices can be in three local states: sending (0), receiving (1) or disconnecting from bus (2). Entries of the table we interpret as local states of devices (attributes) a and b whereas objects u_1, u_2, u_3, u_4 - as global states of the communication system. In the table we have four global states. For instance, in the first global state (u_1) of the system the device a sends information whereas the device b receives information. In a similar way we can interpret the remaining global states of the system. In the following we shall use this example as a specification for constructing a concurrent model of the given communication system in the form of a colored Petri net.

In a given information system, in general, we are not able to distinguish all single objects (using attributes of the system). Namely, different objects can have the same values on considered attributes. Hence, any set of attributes divides the universe U into some classes which establish a partition [10] of the set of all objects U. It is defined in the following way.

Let $S = (U, A)$ be an information system. With any subset of attributes $B \subseteq A$ we associate a binary relation $ind(B)$, called an *indiscernibility relation*, which is defined by $ind(B) = \{(u, u') \in U \times U$ for every $a \in B, a(u) = a(u')\}$.

Notice that $ind(B)$ is an equivalence relation.

If $u\ ind(B)\ u'$, then we say that the objects u and u' are indiscernible with respect to attributes from B. In other words, we cannot distinguish u from u' in terms of attributes in B.

Any information system $S = (U, A)$ determines an *information function* $Inf_A : U \to 2^{A \times V}$ defined by $Inf_A(u) = \{(a, a(u)) : a \in A\}$. The values of an information function will be represented by vectors of the form $(v_1, \ldots, v_m)$, $v_i \in V_{a_i}, i = 1, \ldots, m$, where $m = card(A)$. Such vectors are called *information vectors* (over A and V). The set $\{Inf_A(u) : u \in U\}$ is denoted by $INF(S)$.

Hence, $u\ ind(A)\ u'$ if and only if $Inf_A(u) = Inf_A(u')$.

A *decision table* is any information system of the form $S = (U, A \cup \{d\})$, where $d \notin A$ is a distinguished attribute called *decision*. The elements of A are called *conditional attributes*.

Let $S = (U, A)$ be an information system, where $A = \{a_1, \ldots, a_m\}$ and V is the domain of A. Pairs (a, v), where $a \in A, v \in V$ are called *descriptors* over A and V. Instead of (a, v) we write $a = v$.

B. Discernibility Matrix

Frequently discernibility of objects is more interesting than the specific values of attributes. In these situations an information system may be represented as a discernibility matrix. Skowron and Rauszer [18] have introduced two notions, namely the *discernibility matrix* and the *discernibility function*, which will help to compute minimal forms of rules with respect to the number of attributes on the left hand side of the rules. With these two notions, we can store the differences between the attributes of each pair of objects in a matrix called *discernibility matrix*.

Let $S = (U, A)$ be an information system, and let us assume that $U = \{u_1, \ldots, u_n\}$, and $A = \{a_1, \ldots, a_m\}$. By $M(S)$ we denote an $n \times n$ matrix (c_{ij}), called the *discernibility matrix* of S, such that $c_{ij} = \{a \in A : a(u_i) \neq a(u_j)\}$ for $i, j = 1, \ldots, n$.

Intuitively an entry c_{ij} consists of all the attributes discerning objects u_i and u_j. Since $M(S)$ is symmetric and $c_{ii} = \emptyset$ for $i = 1, \ldots, n$, $M(S)$ can be represented using only elements in the lower triangular part of $M(S)$, i.e., for $1 \leq j < i \leq n$.

With every discernibility matrix $M(S)$ one can uniquely associate a *discernibility function* $f_{M(S)}$, defined as follows.

A *discernibility function* $f_{M(S)}$ for an information system S is a Boolean function of m propositional variables $a_1^*, \ldots, a_m^*$ (where $a_i \in A$ for $i = 1, \ldots, m$) defined as the conjunction (AND) of all expressions $\bigvee c_{ij}^*$, where $\bigvee c_{ij}^*$ is the disjunction (OR) of all elements of $c_{ij}^* = \{a^* : a \in c_{ij}\}$, where $1 \leq j < i \leq n$ and $c_{ij} \neq \emptyset$. In the sequel we write a instead of a^*.

Example II.2. For the information system from Example II.1, we obtain the following discernibility matrix $M(S)$ presented in Table II and discernibility function $f_{M(S)}$ as follows: $f_{M(S)} = (a$ OR $b)$ AND a AND b. After simplification (using the absorption laws) we get the following minimal disjunctive normal form of the discernibility function: $f_{M(S)} = a$ AND b.

C. Rules in Information Systems

Rules express some of the relationships between values of

the attributes described in the information systems.

Let $S = (U, A)$ be an information system and let $B \subset A$. For every $a \notin B$ we define a function $d_a^B : U \to P(V_a)$ such that $d_a^B(u) = \{v \in V_a : \text{there exists } u' \in U \; u' \, ind(B) \, u \text{ and } a(u') = v\}$, where $P(V_a)$ denotes the powerset of V_a. Hence, $d_a^B(u)$ is the set of all the values of the attribute a on objects indiscernible with u by attributes from B. If the set $d_a^B(u)$ has only one element, this means that the value $a(u)$ is uniquely defined by the values of attributes from B on u.

Let $S = (U, A)$ be an information system and V be the domain of A.

A *rule* of an information system S is any expression of the following form:

(1) IF $(a_{i_1} = v_{i_1})$ AND ... AND $(a_{i_r} = v_{i_r})$ THEN $(a_p = v_p)$, where $a_p, a_{i_j} \in A$, $v_p \in V_{a_p}$, $v_{i_j} \in V_{a_{i_j}}$ for $j = 1, \ldots, r$.

We say that the rule (1) is *true* in S if and only if a set of all objects from S for which the dependency expressed by the rule is not empty.

By $D(S)$ we denote the set of all rules true in S.

Let $R \subseteq D(S)$. An information vector $\mathbf{v} = (\mathbf{v}_1, ..., \mathbf{v}_m)$ is *consistent* with R if and only if for any rule IF $(a_{i_1} = v_{i_1})$ AND ... AND $(a_{i_r} = v_{i_r})$ THEN $(a_p = v_p)$ in R if $\mathbf{v}_{i_j} = v_{i_j}$ for $j = 1, ..., r$ then $v_p = \mathbf{v_p}$.

Let $S' = (U', A')$ be a U'-extension of $S = (U, A)$. We say that S' is a *consistent extension* of S if and only if $D(S) \subseteq D(S')$. S' is a *maximal consistent extension* of S if and only if S' is a consistent extension of S and for any consistent extension S'' of S' we have $D(S'') = D(S')$.

In the paper we apply the Boolean reasoning approach to the rule generation [17]. The Boolean reasoning approach, due to G. Boole, is a general problem solving method consisting of the following steps: (i) construction of a Boolean function corresponding to a given problem; (ii) computation of prime implicants *[27] of the Boolean function; (iii) interpretation of prime implicants leading to the solution of the problem.

D. Minimal Rules in Information Systems

Now we present a method for generating the minimal form of rules (i.e., rules with a minimal number of descriptors on the left hand side) in information systems. The method is based on the idea of Boolean reasoning applied to discernibility matrices defined in subsection B and modified here for our purposes. In fact, this subsection presents the first stage in the construction of a concurrent model of knowledge embedded in a given information system.

Let $S = (U, A)$ be an information system and $a \in A$. We are looking for all minimal rules in S of the form: IF $(a_{i_1} = v_{i_1})$ AND ... AND $(a_{i_r} = v_{i_r})$ THEN (a, v), where $a_{i_j} \in A$, $v_{i_j} \in V_{a_{i_j}}$, $a \in A$, $v \in V_a$ for $j = 1, \ldots, r$.

The above rules express functional dependencies between the values of the attributes of S. These rules are computed

---

*An implicant of a Boolean function f is any conjunction of literals (variables or their negations) such that if the values of these literals are true under an arbitrary valuation v of variables then the value of the function f under v is also true. A prime implicant is a minimal implicant. Here we are interested in implicants of monotone Boolean functions only, i.e., functions constructed without negation.

TABLE III
THE INFORMATION SYSTEM $(U, B \cup \{b\})$ WITH THE FUNCTION d_b^B WHERE $B = \{a\}$

U/B	a	b	d_b^B
u_1	0	1	$\{1,2\}$
u_2	1	0	$\{0\}$
u_3	0	2	$\{1,2\}$
u_4	2	0	$\{0\}$

from systems of the form $S' = (U, B \cup \{a\})$ where $B \subset A$ and $a \in A - B$.

First, for every $v \in V_a, u_l \in U$ such that $d_a^B(u_l) = \{v\}$ a modification $\boldsymbol{M}(S'; a, v, u_l)$ of the discernibility matrix is computed from $M(S')$.

By $\boldsymbol{M}(S'; a, v, u_l) = (c_{ij}^*)$ (or $\boldsymbol{M}$, in short) we denote the matrix obtained from $M(S')$ in the following way:

if $i = l$ then $c_{ij}^* = \emptyset$;
if $c_{lj} \neq \emptyset$ and $d_a^B(u_j) \neq \{v\}$ then $c_{lj}^* = c_{lj} \cap B$
else $c_{lj}^* = \emptyset$.

Next, we compute the discernibility function f_M and the prime implicants of f_M taking into account the non-empty entries of the matrix $\boldsymbol{M}$ (when all entries c_{ij}^* are empty we assume f_M to be always true).

Finally, every prime implicant a_{i_1} AND ... AND a_{i_r} of f_M determines a rule IF $(a_{i_1} = v_{i_1})$ AND ... AND $(a_{i_r} = v_{i_r})$ THEN $(a = v)$, where $a_{i_j}(u_l) = v_{i_j}$ for $j = 1, \ldots, r$, $a(u_l) = v$.

The set of all rules constructed in the above way for any $a \in A$ is denoted by OPT(S, a).

We put OPT$(S) = \bigcup \{$ OPT$(S, a) : a \in A \}$.

The next example illustrates how to find all non-trivial dependencies between the values of attributes in a given information system.

Example II.3. Let us consider the information system S presented in Table I. Let us start by computing the rules corresponding to the set of attributes $B = \{a\}$ and the distinguished attribute b.

We have the following subsystem $S_1 = (U, B \cup \{b\})$ from which we compute a part of searched minimal rules.

In the table the values of the function d_b^B are also given. The discernibility matrix $\boldsymbol{M}(S_1; b, v, u_l)$ where $v \in V_b$, $u_l \in U$, $l = 1, 2, 3, 4$, obtained from $M(S)$ in the above way is presented in Table IV.

The discernibility functions corresponding to the values of the function d_b^B are the following:

Case 1. For $d_b^B(u_2) = \{0\} : a \wedge a = a$.

We consider non-empty entries of the column labelled by u_2 (see: Table IV), i.e., a and a; next a is treated as Boolean variable and the disjunction a and a is constructed from these entries; finally, we take the conjunction of all the computed

TABLE IV
THE DISCERNIBILITY MATRIX $M(S_1; b, v, u_l)$ FOR THE MATRIX $M(S)$ PRESENTED IN TABLE II.

U	u_1	u_2	u_3	u_4
u_1				
u_2	a		a	
u_3				
u_4	a		a	

disjunctions to obtain the discernibility function corresponding to $M(S_1; b, v, u_l)$.

Case 2. For $d_b^B(u_4) = \{0\} : a \wedge a = a$.

Hence we obtain the following minimal rules: IF $(a = 1)$ THEN $(b = 0)$, IF $(a = 2)$ THEN $(b = 0)$.

In a similar way we can compute the minimal rules from the information system $S_2 = (U, B \cup \{a\})$ where $B = \{b\}$. After computation we get the following minimal rules: IF $(b = 1)$ THEN $(a = 0)$, IF $(b = 2)$ THEN $(a = 0)$.

Eventually, we obtain the set $OPT(S)$ of minimal rules corresponding to all functional dependencies in the considered information system S: IF $(a = 1)$ THEN $(b = 0)$, IF $(a = 2)$ THEN $(b = 0)$, IF $(b = 1)$ THEN $(a = 0)$, IF $(b = 2)$ THEN $(a = 0)$.

E. Colored Petri Nets

In what follows, it is assumed that the reader is familiar with colored Petri nets in K. Jensen [5]. The colored Petri nets provide a basis for modelling, simulating and analyzing concurrent systems.

A *Colored Petri Net* (*CP-net*) is a tuple $CPN = (\Sigma, P, T, A, N, C, G, E, I)$ satisfying the requirements below: Σ is a nonempty, finite set of types which are called *color sets*, P is a finite set of *places*, T is a finite set of *transitions*, A is a finite set of *arcs*, $N : A \to (P \times T) \cup (T \times P)$ is a *node function*, $C : P \to \Sigma$ is a *color function*, G is a *guard function*, E is an *arc expression function*, I is an *initialization function*.

Colored Petri nets provide data typing (color sets) and sets of values of a specified type for each place. The expression $E(p, t)$ specifies the input associated with the arc from input place p to transition t, and the expression $E(t, p')$ specifies a transformation (activity) performed by transition t on its inputs $E(p, t)$ to produce an output for place p'. A guard $G(t)$ is an enabling condition associated with transition t. The multi-set in a place is called the *marking* of the place. A multi-set is a set which can have multiple appearances of the same element. The initial distribution of multi-sets among the places is called *initial marking* and is denoted by $M0$. A marking M' is said to be *reachable* from a marking M if and only if there exists a finite occurrence sequence starting in M and ending in M'. The set of markings which are reachable from M is denoted by $[M_0 >$. A marking is *reachable* if and only if it belongs to $[M_0 >$. A CP-net is said to be *live* if, no matter what marking has been reached from the initial marking, it is possible to ultimately fire any transition of the net by progressing through some further finite occurrence sequence starting from the current marking.

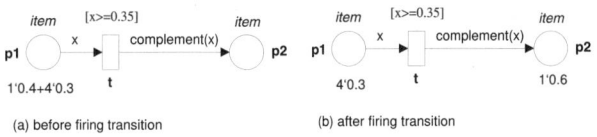

(a) before firing transition   (b) after firing transition

Fig. 2. Sample *CP*-net.

Example II.4. A sample *CP*-net is given in Figure 2, where place $p1$ supplies a value of x of type *item* to transition t, which outputs $complement(x) = 1 - x$ to place $p2$ whenever the token x in place $p1$ satisfies the guard $[x >= 0.35]$. The notation $1`0.4 + 4`0.3$ specifies that a multi-set contains one element with the value 0.4 and four elements with value 0.3. The prefix 1 indicates the number of tokens in the multi-set (in this case, one token in place $p1$), and the suffix indicates that x has been assigned an x-value. Whenever transition t fires, it augments the multi-set associated with place $p2$.

Further an *CP*-net corresponding to an information system S will be denoted by CPN_S.

III. THE SYNTHESIS PROBLEM

The aim of this section is to formulate the synthesis problem by using the rough set formalism. The idea of concurrent system representation by information systems is due to Professor Z. Pawlak [10]. Some relationships of information systems and the rough set theory with the synthesis problem have been recently discussed in ([6]-[9], [13], [14], [19]-[20], [21]-[23], [25]).

Let $S = (U, A)$ be an information system, where U is a set of global states (objects) of the system, A is a set of local processes (attributes). With every local process $a \in A$ is associated a finite set V_a of its internal states (values of attributes). We assume that the behavior specification of a modelled system is presented in the form of such data table. Each row in the table includes record of local states of processes from A, and each record is labelled by an element from the set U of global states (objects) of the system. The columns in the table are labelled by the names of processes.

The problem is: Construct for a given information system S its concurrent model in the form of a colored Petri net (*CP*-net) CPN_S with the following property: the reachability set of markings $[M_0 >$ of CPN_S defines an extension S' of S created by adding to S all new global states corresponding to markings from $[M_0 >$ where M_0 denotes an initial marking of the net. Moreover S' is the largest extension of S [24], [16] with that property. The initial marking of CPN_S corresponds to any global state of S.

IV. THE SOLUTION

This section presents an algorithm for constructing a concurrent model in the form of *CP*-net on the base minimal rules extracted from a given information system S. That is the second stage in the construction of a concurrent model of a given information system.

Let $S = (U, A)$ be an information system and let OPT(S) be a set of all minimal rules in S.

ALGORITHM for constructing a concurrent model CPN_S of S:

Input: An information system S.

Output: CPN_S - the concurrent model of S in the form of a CP-net.

Step 1. Extract the minimal rules from an information system S by using a method described in section II.

Step 2. Construct the net representing all attributes in a given information system S. Each place p_a of CPN_S corresponds to an attribute $a \in A$ of S. The color sets of places in the net are labelled by means the names of attributes of S. For each place the color set of place consists of colors labelled by means the names of attribute values of a given attribute. There is only one transition t in the constructed net. The transition t represents the global state changes. The initial marking M_0 of the net corresponds to an object $u \in U$ of S chosen in an arbitrary way.

Step 3. The net obtained in *Step* 2 is extended by adding a guard expression to the transition t. The guard expression is determined using the minimal rules in S.

In order to realize the *Step* 3 of the above algorithm we can perform the following procedure.

PROCEDURE for computing a guard expression:

Input: A set OPT(S) of all minimal rules in S.

Output: A guard expression corresponding to OPT(S).

Step 1. Rewrite each rule from OPT(S) to the disjunctive normal form of Boolean formula.

Step 2. Construct the conjunction of formulas obtained in *Step* 1.

Step 3. Use Boolean algebra laws for simplification of the formula obtained in *Step* 2 in order to get its minimal disjunctive normal form. The resulting formula is the guard expression corresponding to OPT(S).

Example IV.1. After execution of the above procedure with the set OPT(S) from Example II.3 we obtain the following Boolean expression: $(a = 0)$ AND $(b = 0)$ OR $(a = 0)$ AND $(b = 1)$ OR $(a = 0)$ AND $(b = 2)$ OR $(a = 1)$ AND $(b = 0)$ OR $(a = 2)$ AND $(b = 0)$. The guard expression corresponding to OPT(S) is as follows: $(ya = (a = 0))$ AND $(yb = (b = 0))$ OR $(ya = (a = 0))$ AND $(yb = (b = 1))$ OR $(ya = (a = 0))$ AND $(yb = (b = 2))$ OR $(ya = (a = 1))$ AND $(yb = (b = 0))$ OR $(ya = (a = 2))$ AND $(yb = (b = 0))$.

The concurrent model of S in the form of CP-net constructed by using the above algorithm is shown in Fig. 1. The guard expression form associated with the transition t (see Fig. 1) differs slightly from the presented above. It follows from the syntax of the $CPN\ ML$ language accessible into the $Design/CPN$ system [28].

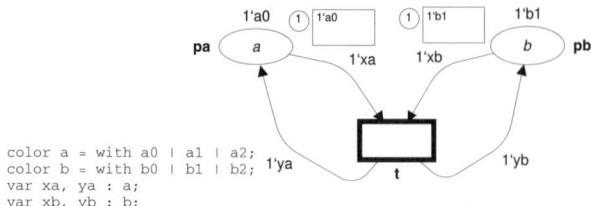

```
color a = with a0 | a1 | a2;
color b = with b0 | b1 | b2;
var xa, ya : a;
var xb, yb : b;
```

[(ya=a0 andalso yb=b0) orelse (ya=a0 andalso yb=b1) orelse (ya=a0 andalso yb=b2) orelse (ya=a1 andalso yb=b0) orelse (ya=a2 andalso yb=b0)]

Fig. 3. CP-net as a concurrent model of S.

TABLE V
THE REACHABILITY SET OF MARKINGS $[M_0 >$ OF CP-NET FROM FIG. 3.

Marking/Place	p_a	p_b
M_1	1'(a=0)	1'(b=1)
M_2	1'(a=1)	1'(b=0)
M_3	1'(a=0)	1'(b=2)
M_4	1'(a=2)	1'(b=0)
M_5	1'(a=0)	1'(b=0)

The following markings of the constructed CP-net by admissible bindings in the net can be generated:

The set of markings of the constructed CP-net corresponds to all global states consistent with all rules valid in information system S. Hence a maximal consistent extension S' of S has the form presented in Table VI. It is worth to observe that the object corresponding to the marking M_5 of the constructed CP-net is new. Moreover, the information vector $(0,0)$ determined by the object u_5 is consistent with the rules from OPT(S). The set of all information vectors INF(S') = $\{(0,1), (1,0), (0,2), (2,0), (0,0)\}$.

V. THE SOLUTION PROPERTIES

The algorithm described in section IV has the following two properties.

Proposition 1: Let S be an information system and S' its maximal consistent extension. Moreover, let CPN_S be a CP-net constructed by using our algorithm presented in section IV. Then the reachability set $[M_0 >$ of CPN_S is equal to the set of all information vectors INF(S'), i.e., $[M_0 > =$ INF(S').

Proposition 2: Let S be an information system and CPN_S its concurrent model constructed by using our algorithm

TABLE VI
THE LARGEST EXTENSION S' OF S FROM TABLE 1.

U'/A	a	b
u_1	0	1
u_2	1	0
u_3	0	2
u_4	2	0
u_5	0	0

presented in section IV. Then the net CPN_S is live at the initial marking M_0.

VI. CONCLUDING REMARKS

In the paper an approach to the construction of a concurrent model from a given information system has been demonstrated. The approach allows to generate automatically from an arbitrary information system its concurrent model in the form of a CP-net.

A construction of a concurrent model represented by low-level Petri nets based on the minimal rules as well as the inhibitor rules [7] extracted from a given information system is more complicated than a construction of such model represented by CP-nets by using the same kind of rules. In the case of CP-nets it is possible to construct coherent, readable and simple concurrent model on the base of the minimal rules as well as the inhibitor rules. For CP-nets the both models are similar w.r.t. the structure and the dynamics of the resulting net [8]. The difference between them concerns the form of guard expressions assigned to the transitions in the constructed nets. In fact, the both guard expressions are equivalent in the logical sense [25].

Applications of concurrent model obtained from a given information system in the form of a classical Petri net [15] has been discussed in [19], [21]-[23].

The method proposed in the paper has been implemented according to a new programming technology in the $ROSECON$ system [6] running on $IBM\ PC$ computers under Windows operating system. The $ROSECON$ system is being developed in the Chair of Computer Science Foundations at the University of Information Technology and Management in Rzeszów. At present the $ROSECON$ system makes possible computing reducts [10], rules, inhibitor rules, components and coverings of information systems specified by data tables [21], and creating their concurrent models in the form nets with inhibitor expressions [7] or CP-nets [8], [25]. This system is permanently evolved.

In the forthcoming paper [9] we study an application of the presented methodology for the synthesis of concurrent systems specified by the dynamic information systems [22]. Moreover, in the next paper we shall consider the synthesis problem discussed here into the context of process languages introduced by L. Czaja in [3].

VII. ACKNOWLEDGMENT

I am grateful to the anonymous referees for helpful comments.

REFERENCES

[1] E. Badouel and Ph. Darondeau, "Theory of Regions", in W. Reisig, G. Rozenberg (eds.), *Lectures on Petri Nets I: Basic Models. Advances in Petri Nets*, Springer, 1998, pp. 529-586.
[2] E.M. Brown, *Boolean Reasoning*, Kluwer, 1990.
[3] L. Czaja, "On the Analysis of Petri Nets and their Synthesis from Process Languages" (to appear).
[4] J. Desel and W. Reisig, "The synthesis problem of Petri nets", *Acta Informatica*, vol. 33, no. 4, 1996, pp. 297-315.
[5] K. Jensen, *Coloured Petri Nets. Basic Concepts, Analysis Methods and Practical Use*, vol. 1, Springer, 1992.
[6] K. Pancerz and Z. Suraj, "ROSECON: A system for automatic discovering of concurrent models from data tables", in *Proceedings of the IX Environmental Conference on Mathematics and Informatics*, Korytnica, June 12-16, 2002, Poland (in Polish), p. 34.
[7] K. Pancerz and Z. Suraj, "From Data to Nets with Inhibitor Expressions: A Rough Set Approach", in [26], pp. 102-106.
[8] K. Pancerz and Z. Suraj, "The Synthesis of Concurrent Systems Specified by Information Systems with Using the Coloured Petri Nets", in H.-D. Burkhard, L. Czaja, G. Lindemann, A. Skowron, P. Starke (eds.), *Proceedings of the Concurrency, Specification and Programming (CS&P'2002). Workshop*, vol. 2, Berlin, October 7-9, 2002, Germany, pp. 255-266.
[9] K. Pancerz and Z. Suraj, "Modelling Concurrent Systems Specified by Dynamic Information Systems: A Rough Set Approach", in *Proceedings of the International Workshop on Rough Sets in Knowledge Discovery and Soft Computing*, Warsaw, April 12-13, 2003 (to appear).
[10] Z. Pawlak, *Rough Sets Theoretical Aspects of Reasoning About Data*, Kluwer, 1991.
[11] Z. Pawlak, "Concurrent versus sequential the rough sets perspective", *Bulletin of the EATCS*, vol. 48, 1992, pp. 178-190.
[12] Z. Pawlak, "Some Remarks on Explanation of Data and Specification of Processes Concurrent", *Bulletin of International Rough Set Society*, vol. 1, no. 1, 1997, pp. 1-4.
[13] J.F. Peters, A. Skowron, Z. Suraj, W. Pedrycz, and S. Ramanna, "Approximate Real-Time Decision Making: Concepts and Rough Fuzzy Petri Net Models", *International Journal of Intelligent Systems*, vol. 14, no. 4, 1998, pp. 4-37.
[14] J.F. Peters, A. Skowron, and Z. Suraj, "An Application of Rough Set Methods in Control Design", *Fundamenta Informaticae*, vol. 43, nos. 1-4, 2000, pp. 269-290.
[15] W. Reisig, *Petri Nets. An Introduction*, Springer, 1985.
[16] W. Rzasa and Z. Suraj, "A new method for determining of extension and restriction of information system", in J.J. Alpigini, J.F. Peters, A. Skowron, N. Zhong (eds.), *Proceedings of the Third International Conference a Rough Sets and Current Trends in Computing*, October 14-16, 2002, Malvern, PA, USA, Lecture Notes in Artificial Intelligence, vol. 2475, Springer, 2002, pp. 197-204.
[17] A. Skowron, "A Synthesis of Decision Rules: Applications of Discernibility Matrices", in *Proceedings of a Workshop on Intelligent Information Systems, Practical Aspects of AI II*, Augustw, Poland, pp. 30-46.
[18] A. Skowron and C. Rauszer, "The discernibility matrices and functions in information systems", in R. Slowinski (ed.), *Intelligent Decision Support: Handbook of Applications and Advances of Rough Sets Theory*, Kluwer, 1992, pp. 331-362.
[19] A. Skowron and Z. Suraj, "Rough Sets and Concurrency", *Bulletin of the Polish Academy of Sciences*, vol. 41, no. 3, 1993, pp. 237-254.
[20] A. Skowron and Z. Suraj, "Parallel Algorithm for Real-Time Decision Making: A Rough Set Approach", *Journal of Intelligent Information Systems*, vol. 7, Kluwer, 1996, pp. 5-28.
[21] Z. Suraj, "Discovery of Concurrent Data Models from Experimental Tables: A Rough Set Approach", *Fundamenta Informaticae*, vol. 28, nos. 3-4, December 1996, pp. 353 - 376.
[22] Z. Suraj, "The Synthesis Problem of Concurrent Systems Specified by Dynamic Information Systems", in L. Polkowski and A. Skowron (eds.), *Rough Sets in Knowledge Discovery*, vol. 2, Physica-Verlag, 1998, pp. 418-448.
[23] Z. Suraj, "Rough Set Methods for the Synthesis and Analysis of Concurrent Processes", in L. Polkowski, S. Tsumoto, T.Y. Lin (eds.), *Rough Set Methods and Applications*, Springer, 2000, pp. 379-488.
[24] Z. Suraj, "Some Remarks on Extensions and Restrictions of Information Systems", in W. Ziarko, Y. Yao (eds.), *Rough Sets and Current Trends in Computing*, Lecture Notes in Artificial Intelligence, vol. 2005, Springer, 2001, pp. 204-211.
[25] Z. Suraj, "Information Systems as a Tool for Specification of Concurrent Systems", in V. Tchaban (ed.), *Proceedings of the 7-th International Modelling School of AMSE-UAPL*, 12-17 September, 2002, Alushta, Ukraine, pp. 47-50.
[26] Z. Suraj (ed.), *Proceedings of the Sixth International Conference on Soft Computing and Distributed Processing*, Rzeszów, June 24-25, 2002, Poland.
[27] I. Wegener, *The complexity of Boolean functions*, Wiley and B.G. Teubner, Stuttgart 1987.
[28] http://www.daimi.au.dk/designCPN/

Dynamic modeling of a parallel robot. Application to a surgical simulator.

N. LEROY[1], A.M. KÖKÖSY[1,2], W. PERRUQUETTI[1]
1 LAIL UMR 8021 CNRS,
Ecole Centrale de LILLE, BP 48,
59 651 VILLENEUVE D'ASCQ CEDEX - FRANCE.
e-mail : Nicolas.Leroy@ec-lille.fr Wilfrid.Perruquetti@ec-lille.fr
phone : (33) 3 20 33 54 50
2 Institut Supérieur d'Electronique du Nord (ISEN)
41 Bd VAUBAN, 59 046 LILLE CEDEX - FRANCE.
e-mail : Annemarie.Kokosy@isen.fr, phone : (33) 3 20 30 40 23

Abstract— This paper presents a closed form solution for the dynamic model of a parallel robot. This robot is utilized like a haptic interface for a surgical simulator used in the amnicentesis operation. The dynamic model is obtained by using the Lagrange formulation applied to parallel robot. It is proved that for a large class of parallel robots which has some properties, it is possible to avoid the explicit calculation of Lagrange multipliers. The aim is to simplify the calculation for a real-time application.

I. Introduction

Surgical gestures require a great dexterity, and robots could be used to improve the skill of doctors, or to teach this gesture to students. The surgical simulators have the advantage of a full time availability and of a high range of different simulations of medical problems. From this perspective, a simulator of amniocentesis was made to provide a means of training for this difficult gesture. An amniocentesis is a technique to take amniotic liquid. The operation is done under a scanning control, and a syringe is used to take the liquid. The more the baby's birth is close, the more the operation is tactful. The robot simulates the behavior of the womb. The user interacts with a syringe. An important request from the doctors is that the syringe is not tied up with the robot. The user hooks firstly the syringe on a skullcap. After that, the haptic interface gives a force feedback to the user and a screen gives a visual feedback.

This paper presents the structure of the haptic interface which the user interacts with, and its dynamic modeling. A previous study [1] shows the necessity of a dynamic model of the robot, with a view to controlling with force feedback. Dynamic modeling of series robots are commonly made with Euler-Lagrange equations [2], [3], [4]. A systematic way to obtain these equations is the use of the Denavit-Hatenberg method. Nevertheless, the domain of parallel robotics is less known and the problems are more complex [5]. The dynamic equations of a parallel robot are classically obtained by virtually opening the closed loops [6], [7]. The robot becomes a series robot whose equations are obtained by Euler-Lagrange method. Afterwards, the connections put on side are reintroduced by the use of Lagrange multipliers. If the parallel robot belongs to a class having some useful properties for the Lagrangian expression and the constraints, the explicit calculation of the Lagrange multipliers can be avoided. In this way, the calculation of the dynamic model is simpli...ed. The dynamic equations of the haptic interface which will be obtained are based on the expression of the geometric model and provides a reduced expression to decrease the calculations.

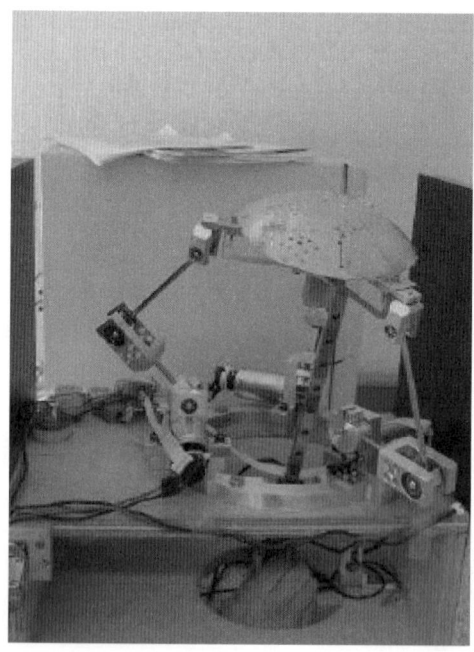

Fig. 1. The haptic interface

The reported results were obtained within the framework of project "Alcove" (LIFL UMR 8022 CNRS).

II. ROBOT DESCRIPTION

A. Global description

The haptic interface is a parallel robot with 3 d.o.f. The platform, called skullcap (dashed zone in figure 2), moves on a spherical surface (θ et φ) with variable ray (l) (figure 2). This skullcap is supported by three legs distributed regularly. Each leg is made of two mobile segments. The skullcap is also maintained by a connection on the level of point O, to avoid its rotation around the axis (OO_2). The robot has 3 active joints (on A_0, B_0, C_0), and 18 passive joints.

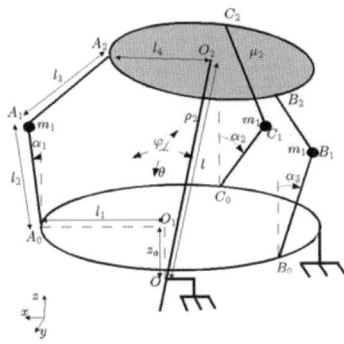

Fig. 2. General outline of the parallel robot

Geometrical constants of the manipulator are (see figure 2):

$$\begin{aligned} z_a &= OO_1 \\ l_1 &= O_1A_0 = O_1B_0 = O_1C_0 \\ l_2 &= A_0A_1 = B_0B_1 = C_0C_1 \\ l_3 &= A_1A_2 = B_1B_2 = C_1C_2 \\ l_4 &= O_2A_2 = O_2B_2 = O_2C_2. \end{aligned}$$

B. Mechanical links

The system is composed of seven mobile bodies, which are dependent between them by the following connections (figures 2):

in A_0, B_0, C_0: connections driving actuators (3×1 d.o.f.)

in A_1, B_1, C_1: connections kneecaps (3×3 d.o.f.)

in A_3, B_3, C_3: kneecaps with two degrees of freedom (type cardan joint) (3×2 d.o.f.)

in O: a connection with 3 d.o.f (2 rotations θ, φ and 1 translation l).

The position in the space of the robot is defined by $\mathbf{q}$ and $\boldsymbol{\alpha}$:

$$\begin{aligned} \mathbf{q} &= \begin{pmatrix} \theta & \varphi & l \end{pmatrix}^\mathbf{T} \\ \boldsymbol{\alpha} &= \begin{pmatrix} \alpha_1 & \alpha_2 & \alpha_3 \end{pmatrix}^\mathbf{T}. \end{aligned} \quad (1)$$

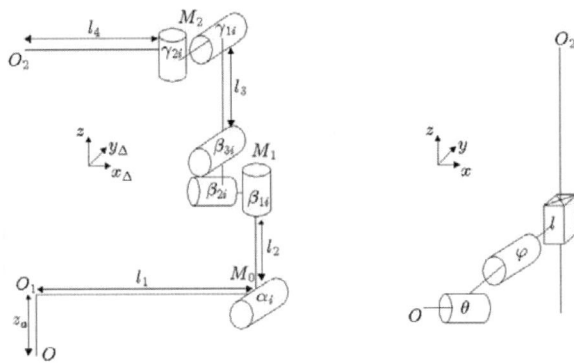

Fig. 3. Schema of the mechanical links for each coresponding coordinates (angles or translation)

$\mathbf{q}$ and $\boldsymbol{\alpha}$ represent the position of the skullcap and of the motor, respectively. They are linked by geometrical constraints: the knowledge of $\mathbf{q}$ involves the knowledge of $\boldsymbol{\alpha}$, and vice versa. This constraints are obtained by the geometrical inverted model which gives the value of $\boldsymbol{\alpha}$ in function of $\mathbf{q}$.

The three legs are identical up to a rotation (z, Δ_i). In order to establish the geometric model we consider a general leg, where the points (M_0, M_1, M_2) are respectively the points (A_0, A_1, A_2), (B_0, B_1, B_2), or (C_0, C_1, C_2). It is defined in the reference (x_Δ, y_Δ, z) (figure 3) with

$$\begin{aligned} x_\Delta &= x\cos\Delta_i + y\sin\Delta_i \\ y_\Delta &= -x\sin\Delta_i + y\cos\Delta_i \end{aligned}, \quad i \in \{1,2,3\}. \quad (2)$$

The angle value Δ_i depends on the leg we consider. The three legs are regularly distributed, then

$$\Delta_1 = 0, \quad \Delta_2 = \frac{2\pi}{3}, \quad \Delta_3 = -\frac{2\pi}{3}. \quad (3)$$

C. Mass distribution

The system is composed of seven mobile bodies, but they have not all a significant mass. The heaviest body is the skullcap. Because of its weight and its size, we consider its complete dynamic. Under the Denavit-Hartenberg method, the dynamic of this body is described by its inertial matrix. This one depends on the mass distribution which is not homogeneous. At point O_2 in the skullcap reference, this matrix is:

$$J_c = \begin{pmatrix} I_1 & 0 & I_4 & I_5 \\ 0 & I_2 & 0 & 0 \\ I_4 & 0 & I_3 & I_6 \\ I_5 & 0 & I_6 & I_7 \end{pmatrix}. \quad (4)$$

The zeros in the inertial matrix are due to a plane of symmetry $(O_2, \vec{x}, \vec{z})$. By definition of the inertial matrix, we know that:

$$I_5 = m_2 x_g, \quad I_6 = m_2 z_g, \quad I_7 = m_2, \quad (5)$$

where m_2 is the mass of the skullcap, and $\begin{pmatrix} x_g & 0 & z_g \end{pmatrix}^T$ is the position of the gravity centrum. In proportion, other bodies have not a significant mass, they are only the thin stem, except for the joints in (A_1, B_1, C_1). These joints are quite heavy, so in these points we consider some identical punctual masses, noted m_1. Finally, we obtain the following expression for the inertia matrix J_c by adding the different contributions of each leg (see figure 3):

$$J_c = \pi l_4^2 \mu_2 \begin{pmatrix} \frac{l_4^2}{4} & 0 & 0 & 0 \\ 0 & \frac{l_4^2}{4} & 0 & 0 \\ 0 & 0 & 0 & 0 \\ 0 & 0 & 0 & 1 \end{pmatrix}$$

$$+ L\rho_2 \begin{pmatrix} \delta^2 & 0 & \frac{L\delta}{2} & -\delta \\ 0 & 0 & 0 & 0 \\ \frac{L\delta}{2} & 0 & \frac{L^2}{3} & -\frac{L}{2} \\ -\delta & 0 & -\frac{L}{2} & 1 \end{pmatrix}. \quad (6)$$

III. GEOMETRIC MODEL

In the next section IV, we will establish the dynamic equation of the robot by using the Lagrange equation. The calculation of the dynamic model of the robot requires the knowledge of its inverse geometric model.

The inverse geometric model gives the position of the actuators joints (denoted by $\boldsymbol{\alpha}$) according to the position of the skullcap (denoted by $\mathbf{q}$):

$$\boldsymbol{\alpha} = \boldsymbol{\Omega}(\mathbf{q}), \, \boldsymbol{\alpha} = \begin{pmatrix} \alpha_1 & \alpha_2 & \alpha_3 \end{pmatrix}^T, \mathbf{q} = \begin{pmatrix} \theta & \varphi & l \end{pmatrix}^T. \quad (7)$$

For this, we use the constraint $\|M_1 M_2\| = l_3$ (for each leg). Since the position of M_2 depends on $\mathbf{q}$ (see figure 3) and the position of the point M_1 depends on $\boldsymbol{\alpha}$ we obtain the desired relation (7). The details are given by the following relations:

$$OM_2 = rot(\theta, x) \, rot(\varphi, y) \, trans(l, z)$$
$$rot(\Delta_i, z) \, trans(l_4, x) \begin{pmatrix} 0 \\ 0 \\ 0 \\ 1 \end{pmatrix}, \quad (8)$$

$$OM_1 = trans(z_a, z) \, rot(\Delta_i, z) \, trans(l_1, x)$$
$$rot(\alpha_i, y) \, trans(l_2, z) \begin{pmatrix} 0 \\ 0 \\ 0 \\ 1 \end{pmatrix}, \quad (9)$$

$$l_3^2 = \|M_1 M_2\|^2 = (OM_2 - OM_1)^2. \quad (10)$$

By using the Denavit-Hartenberg method, equation (10) can be expressed in the form:

$$V_{ci} \cos(\alpha_i) + V_{si} \sin(\alpha_i) + V_i = 0 \text{ avec } i \in \{1, 2, 3\} \quad (11)$$

with

$$\begin{aligned}
V_{ci} =&\, l_2^2 + 2(z_a + (\cos\theta\sin\varphi\cos\Delta_i - \sin\theta\sin\Delta_i)l_4 - l\cos\theta\cos\varphi)l_2 \\
V_{si} =&\, l_2^2 + 2l_2((\cos\Delta_i l_1 - \sin\varphi l)\cos\Delta_i \\
&+ (\sin\Delta_i l_1 + \sin\theta\cos\varphi l)\sin\Delta_i) \\
&+ 2l_2 l_4(-\cos\Delta_i^2 \cos\varphi - \sin\Delta_i(\sin\theta\sin\varphi\cos\Delta_i + \cos\theta\sin\Delta_i)) \\
V_i =&\, l^2 - l_3^2 + l_1^2 + za^2 + l_4^2 + 2((\sin\Delta_i\sin\theta\cos\varphi - \sin\varphi\cos\Delta_i)l_1 \\
&+ ((\cos\Delta_i^2\cos\theta - \sin\Delta_i\sin\theta\sin\varphi\cos\Delta_i - \cos\Delta_i^2\cos\varphi - \cos\theta)l_1 \\
&+ (\cos\theta\sin\varphi\cos\Delta_i - \sin\theta\sin\Delta_i)za)l_4 - za\cos\theta\cos\varphi)l
\end{aligned} \quad (12)$$

for each $i \in \{1, 2, 3\}$ (for each leg). This equation has two solutions corresponding to the intersection of the sphere of centre M_2 radius l_3 and the circle of centre M_0 radius l_2 in the plane (x_Δ, z). One of these solutions is reachable by the system and the other one is not. The reachable solution is:

$$\begin{aligned} \alpha_i &= 2\arctan(T_i), \\ T_i &= \frac{-V_{si} + \sqrt{V_{ci}^2 + V_{si}^2 - V_i^2}}{V_i - V_{ci}}, \, i \in \{1, 2, 3\}. \end{aligned} \quad (13)$$

The restricted inverse geometric model is thus given by equation (13).

Remark 1: For computation purpose, the function $\boldsymbol{\Omega}^{-1}$ is interpolated by the following second order polynomial:

$$\hat{\boldsymbol{\Omega}}^{-1} = \mathbf{gain} \times \mathbf{h}(\boldsymbol{\alpha}) \quad (14)$$

with

$$\mathbf{h}(\boldsymbol{\alpha}) = \begin{pmatrix} 1 \\ \alpha_1 \\ \alpha_2 \\ \alpha_3 \\ \alpha_1\alpha_1 \\ \alpha_1\alpha_2 \\ \alpha_1\alpha_3 \\ \alpha_2\alpha_2 \\ \alpha_2\alpha_3 \\ \alpha_3\alpha_3 \end{pmatrix}, \quad (15)$$

$$\mathbf{gain} = \begin{pmatrix} 0 & 0.0021 & 0.4136 \\ 0 & 0.4048 & -0.0654 \\ -0.3560 & -0.2036 & -0.0654 \\ 0.3560 & -0.2036 & -0.0654 \\ 0 & -0.0571 & 0.0020 \\ 0.0066 & -0.0036 & 0.0094 \\ -0.0066 & -0.0036 & 0.0094 \\ 0.0505 & 0.0289 & 0.0020 \\ 0 & 0.0071 & 0.0094 \\ -0.0505 & 0.0289 & 0.0020 \end{pmatrix}^T. \quad (16)$$

The maximum error between the interpolation and the function $\boldsymbol{\Omega}^{-1}(\boldsymbol{\alpha})$ is $\Delta \mathbf{q}_{\max} = \begin{pmatrix} 0.4° & 0.4° & 2mm \end{pmatrix}^T$, and this error is on the envelope of the workspace defined by the construction of the robot ($\alpha_1, \alpha_2, \alpha_3 \in [50°; 120°]$). The function $\hat{\boldsymbol{\Omega}}^{-1}$ thus obtained has a low calculating time. The matrix $\frac{\partial \hat{\boldsymbol{\Omega}}^{-1}}{\partial \mathbf{q}}$ can then be calculated very easily by

$$\mathbf{gain} \begin{pmatrix} 0 & 1 & 0 & 0 & 2\alpha_1 & \alpha_2 & \alpha_3 & 0 & 0 & 0 \\ 0 & 0 & 1 & 0 & 0 & \alpha_1 & 0 & 2\alpha_2 & \alpha_3 & 0 \\ 0 & 0 & 0 & 1 & 0 & 0 & \alpha_1 & 0 & \alpha_2 & 2\alpha_3 \end{pmatrix}^T \quad (17)$$

like for $\frac{d}{dt}\left(\frac{\partial \hat{\Omega}}{\partial \mathbf{q}}^{-1}\right)$:

$$\mathbf{gain}\begin{pmatrix} 0 & 1 & 0 & 0 & 2\dot{\alpha}_1 & \dot{\alpha}_2 & \dot{\alpha}_3 & 0 & 0 & 0 \\ 0 & 0 & 1 & 0 & 0 & \dot{\alpha}_1 & 0 & 2\dot{\alpha}_2 & \dot{\alpha}_3 & 0 \\ 0 & 0 & 0 & 0 & 0 & 0 & \dot{\alpha}_1 & 0 & \dot{\alpha}_2 & 2\dot{\alpha}_3 \end{pmatrix}^T. \quad (18)$$

IV. Dynamic model

Newton-Euler or Lagrange formalism are generally used to establish the dynamic model of a manipulator. The Lagrange formalism is better in our case to obtain the state model. The Lagrangian L is defined as $L = K - P$, where K is the kinetic energy and P is the potential energy.

The dynamic equations of the system can be formulated in terms of the Lagrangian function as follows [8]:

$$\frac{d}{dt}\frac{\partial L}{\partial \dot{\mathbf{q}}} - \frac{\partial L}{\partial \mathbf{q}} = \mathbf{F}^T, \quad (19)$$

with $\mathbf{F}$ the generalized force of the outside actions (syringe, engine and friction). The generalized force $\mathbf{\Gamma}_s$, $\mathbf{\Gamma}_a$, $\mathbf{\Gamma}_f$ are the different contributions to $\mathbf{F}$. $\mathbf{\Gamma}_s$ is the action of the syringe. We suppose that $\mathbf{\Gamma}_s$ acts directly on $\mathbf{q}$. This assumption gives a more simple expression. $\mathbf{\Gamma}_a$ is the torque from the actuators which act on $\boldsymbol{\alpha}$. $\mathbf{\Gamma}_f$ is the friction force of the actuators. Friction is a complex nonlinear force that is difficult to model accurately. We consider two distinct components of friction: viscous friction and Coulomb friction:

$$\mathbf{\Gamma}_{fi} = \mathbf{K}_{vi}\dot{\alpha}_i + \mathbf{K}_{ci}\,\mathbf{signe}(\dot{\alpha}_i) \text{ for } i \in \{1, ..., 3\}, \quad (20)$$

with K_{vi} the gain of the viscous friction of the ith actuator and K_{ci} the gain of the Coulomb friction of the ith actuator.

$\mathbf{q}$ has all the independent coordinates which give the position of the system. But we also can use $\boldsymbol{\alpha}$, or any set of coordinates which gives the position of the system.

In general, for systems with geometrical constraints, Lagrange multipliers are used to introduce the constraints in the dynamic equations [6].

In order to obtain the dynamic model of the robot, each term of the next equation must be expressed [6]:

$$\frac{d}{dt}\frac{\partial L}{\partial \dot{\mathbf{x}}} - \frac{\partial L}{\partial \mathbf{x}} + \lambda\frac{\partial \mathbf{C}}{\partial \mathbf{x}} = \mathbf{0}, \quad (21)$$

where λ represents the row vector of Lagrange multipliers and $\mathbf{C}$ is the column vector of the constraints. So λ and $\mathbf{C}$ must be calculated. However, under certain conditions it is possible to avoid these hard calculations. In fact, if it possible to decompose the Lagrangian in two parts, the first depending only on $\mathbf{q}$ and the second depending only on $\boldsymbol{\alpha}$, calculation of the dynamic model of the parallel robot by using equation (21) becomes very easy. This is shown as follows.

Theorem 1: If the Lagrangian of a system is $L = L_1(\mathbf{q}, \dot{\mathbf{q}}) + L_2(\boldsymbol{\alpha}, \dot{\boldsymbol{\alpha}})$, and the system has the constraints $\boldsymbol{\alpha} = \boldsymbol{\Omega}(\mathbf{q})$ with $\boldsymbol{\Omega}$ a sufficient differential function, then the equations of the system are given by

$$\left(\frac{d}{dt}\frac{\partial L_1}{\partial \dot{\mathbf{q}}} - \frac{\partial L_1}{\partial \mathbf{q}}\right) + \left(\frac{d}{dt}\frac{\partial L_2}{\partial \dot{\boldsymbol{\alpha}}} - \frac{\partial L_2}{\partial \boldsymbol{\alpha}}\right)\frac{\partial \boldsymbol{\Omega}}{\partial \mathbf{q}} = 0. \quad (22)$$

Proof: ▶ The Lagrangian L of the system depends on $(\mathbf{x}, \dot{\mathbf{x}})$, with $\mathbf{x} = \begin{pmatrix} \mathbf{q}^T & \boldsymbol{\alpha}^T \end{pmatrix}^T$. The system has the constraints $\mathbf{C} = \boldsymbol{\Omega}(\mathbf{q}) - \boldsymbol{\alpha} = \mathbf{0}$. The equations of the system are:

$$\frac{d}{dt}\frac{\partial L}{\partial \dot{\mathbf{x}}} - \frac{\partial L}{\partial \mathbf{x}} + \lambda\frac{\partial \mathbf{C}}{\partial \mathbf{x}} = \mathbf{0}, \quad (23)$$

or, if we replace $\mathbf{x}$ by its elements:

$$\begin{cases} \frac{d}{dt}\frac{\partial L}{\partial \dot{\mathbf{q}}} - \frac{\partial L}{\partial \mathbf{q}} + \lambda\frac{\partial \mathbf{C}}{\partial \mathbf{q}} = \mathbf{0} \\ \frac{d}{dt}\frac{\partial L}{\partial \dot{\boldsymbol{\alpha}}} - \frac{\partial L}{\partial \boldsymbol{\alpha}} + \lambda\frac{\partial \mathbf{C}}{\partial \boldsymbol{\alpha}} = \mathbf{0} \end{cases}. \quad (24)$$

By deriving the constraints $\mathbf{C}$ with respect to $\mathbf{q}$ and $\boldsymbol{\alpha}$ we have

$$\begin{aligned} \frac{\partial \mathbf{C}}{\partial \mathbf{q}} &= \frac{\partial \boldsymbol{\Omega}}{\partial \mathbf{q}} \\ \frac{\partial \mathbf{C}}{\partial \boldsymbol{\alpha}} &= -\mathbf{Id}. \end{aligned} \quad (25)$$

Using relations (25) in equation (24) and eliminating λ, we obtain the following equation:

$$\frac{d}{dt}\frac{\partial L}{\partial \dot{\mathbf{q}}} - \frac{\partial L}{\partial \mathbf{q}} + \left(\frac{d}{dt}\frac{\partial L}{\partial \dot{\boldsymbol{\alpha}}} - \frac{\partial L}{\partial \boldsymbol{\alpha}}\right)\frac{\partial \boldsymbol{\Omega}}{\partial \mathbf{q}} = 0. \quad (26)$$

Therefore, using relation $L = L_1(\mathbf{q}, \dot{\mathbf{q}}) + L_2(\boldsymbol{\alpha}, \dot{\boldsymbol{\alpha}})$ in this equation, it is easy to deduce equation (22), which proves the theorem.
◀

A. Lagrangian calculation

For our robot, the Lagrangian can be expressed as:

$$L = \left(\sum_{i=1}^{3} K_i - \sum_{i=1}^{3} P_i\right) + (K_c - P_c). \quad (27)$$

where K_i and P_i ($i \in \{1, 2; 3\}$) are the kinetic and potential energies of the ith mass m_1 on the ith leg. These energies only depend on $\boldsymbol{\alpha}$ the position of the actuators. K_c and P_c are the kinetic and potential energies of the skullcap, and only depend on the position of the skullcap $\mathbf{q}$.

The kinetic energy K_i is:

$$\sum_{i=1}^{3} K_i = \frac{1}{2}m_1 l_2^2 \dot{\boldsymbol{\alpha}}^T \dot{\boldsymbol{\alpha}}. \quad (28)$$

with m_1 the mass at the end of the body M_0M_1 and l_2 the length of the body (see figures 2 and 3). The potential energy P_i is:

$$\sum_{i=1}^{3} P_i = m_1 g l_2 \begin{pmatrix} 1 & 1 & 1 \end{pmatrix} \cos\boldsymbol{\alpha},$$

$$\cos\boldsymbol{\alpha} = \begin{pmatrix} \cos\alpha_1 & \cos\alpha_2 & \cos\alpha_3 \end{pmatrix}^T. \quad (29)$$

By using the Denavit-Hartenberg method [2], we can obtain the expression of the kinetic energy K_c of the skullcap:

$$K_c = \frac{1}{2}\sum_{j=1}^{3}\sum_{k=1}^{3} Trace\left(\frac{\partial T_c}{\partial q_j} J_c \frac{\partial T_c^T}{\partial q_k}\right) \dot{q}_j \dot{q}_k \quad (30)$$

with $J_c = \begin{pmatrix} I_1 & 0 & I_4 & I_5 \\ 0 & I_2 & 0 & 0 \\ I_4 & 0 & I_3 & I_6 \\ I_5 & 0 & I_6 & I_7 \end{pmatrix}$, $\mathbf{q} = \begin{pmatrix} \theta & \varphi & l \end{pmatrix}^T$, and $T_c = rot(\theta, x)rot(\varphi, y)trans(l, z)$.

Thus:

$$K_c = \mathbf{\dot{q}^T M \dot{q}}, \quad \mathbf{M} = \begin{pmatrix} ft & 0 & 0 \\ 0 & fp & -I_5 \\ 0 & -I_5 & I_7 \end{pmatrix} \quad (31)$$

with

$$\begin{aligned} ft &= I_2 + I_1 + (l^2 I_7 + 2l I_6 + I_3 - I_1)\cos^2\varphi \\ &\quad - 2(I_4 + l I_5)\sin\varphi\cos\varphi, \\ fp &= l^2 I_7 + 2l I_6 + I_3 + I_1. \end{aligned}$$

The potential energy P_c of the skullcap is:

$$P_c = g\cos\theta\left(-I_5 \sin\varphi + (l I_7 + I_6)\cos\varphi\right). \quad (32)$$

B. Dynamic Equations

Using Theorem 1 we obtain:

$$\left(\frac{\partial \mathbf{\Omega}^T}{\partial \mathbf{q}}\right)^{-1}\left(M\ddot{\mathbf{q}} + \mathbf{M_k} + \frac{\partial \mathbf{P_c}^T}{\partial \mathbf{q}} - \mathbf{\Gamma_s}\right) = \mathbf{\Gamma_a} + g m_1 l_2 \sin\alpha - m_1 l_2^2 \ddot{\alpha} \quad (33)$$

with

$$\mathbf{M_k} = \begin{pmatrix} \dot{\theta}\dot{\varphi} & \dot{\theta}\dot{l} & 0 \\ -\dot{\theta}^2/2 & 0 & \dot{\varphi}\dot{l} \\ 0 & -\dot{\theta}^2/2 & -\dot{\varphi}^2/2 \end{pmatrix}\begin{pmatrix} \frac{\partial ft}{\partial \varphi} \\ \frac{\partial ft}{\partial l} \\ \frac{\partial fp}{\partial l} \end{pmatrix}$$

and

$$\frac{\partial \mathbf{P_c}^T}{\partial \mathbf{q}} = g\begin{pmatrix} -((I_6 + l I_7)\cos\varphi - I_5 \sin\varphi)\sin\theta \\ -((I_6 + l I_7)\sin\varphi + I_5 \cos\varphi)\cos\theta \\ I_7 \cos\varphi\cos\theta \end{pmatrix} \quad (34)$$

The state model expresses the acceleration according to positions, velocities and controls. But in the model, we have $\ddot{\mathbf{q}}$ and $\ddot{\alpha}$, so we need the expression of $\ddot{\alpha}$ according to $\ddot{\mathbf{q}}$. That can be done using the expression of constraints $\alpha = \Omega(\mathbf{q})$ and derives this expression twice:

$$\dot{\alpha} = \frac{\partial \Omega}{\partial q}\dot{\mathbf{q}} \quad (35)$$

$$\ddot{\alpha} = \frac{d}{dt}\left(\frac{\partial \Omega}{\partial \mathbf{q}}\right)\dot{\mathbf{q}} + \frac{\partial \Omega}{\partial \mathbf{q}}\ddot{\mathbf{q}}. \quad (36)$$

Thus, by using the equation (33), it results:

$$\left(M + m_1 l_2^2 \frac{\partial \Omega^T}{\partial \mathbf{q}}\frac{\partial \Omega}{\partial \mathbf{q}}\right)\ddot{\mathbf{q}} + \mathbf{M_k} + \frac{\partial \mathbf{P_c}^T}{\partial \mathbf{q}} - \mathbf{\Gamma_s} = $$
$$-\frac{\partial \Omega^T}{\partial \mathbf{q}}\left(m_1 l_2^2 \frac{d}{dt}\left(\frac{\partial \Omega}{\partial \mathbf{q}}\right)\dot{\mathbf{q}} - g m_1 l_2 \sin\Omega(\mathbf{q}) - \mathbf{\Gamma_a}\right). \quad (37)$$

We can also express the model using $\boldsymbol{\alpha}$:

$$\left(\left(\frac{\partial \Omega^{-1}}{\partial \mathbf{q}}\right)^T M \frac{\partial \Omega^{-1}}{\partial \mathbf{q}} + m_1 l_2^2 \mathrm{Id}\right)\ddot{\boldsymbol{\alpha}} - g m_1 l_2 \sin\alpha - \mathbf{\Gamma_a} = $$
$$-\left(\frac{\partial \Omega^{-1}}{\partial \mathbf{q}}\right)^T\left(M\frac{d}{dt}\left(\frac{\partial \Omega^{-1}}{\partial \mathbf{q}}\right)\dot{\boldsymbol{\alpha}} + \mathbf{M_k} + \frac{\partial \mathbf{P_c}^T}{\partial \mathbf{q}} - \mathbf{\Gamma_s}\right) \quad (38)$$

The form of this equation is the same as the form of the dynamic model of the robot manipulator:

$$M1(\boldsymbol{\alpha})\ddot{\boldsymbol{\alpha}} + M2(\boldsymbol{\alpha}, \dot{\boldsymbol{\alpha}}, \mathbf{\Gamma_s}) = \mathbf{\Gamma_a}. \quad (39)$$

Remark 2: It is interesting to note that the inertia matrix $M1(\boldsymbol{\alpha})$ has the same property as the inertia matrix of the robot manipulators. Hence, it is symmetric and positve definite. These properties are very useful for the control synthesis.

Proof of the symmetry. The matrix M, (31) is symmetric, so the matrix $M1(\boldsymbol{\alpha})$ is also symmetric.

Proof of the positive definiteness. The matrix M is positive definite if ft and the determinant of the matrix $\begin{pmatrix} fp & -I_5 \\ -I_5 & I_7 \end{pmatrix}$ are positive. In order to prove that ft is positive,

$$\begin{aligned} ft &= (l^2 I_7 + 2l I_6 + I_3 - I_1)\cos^2\varphi \\ &\quad - 2(I_4 + l I_5)\sin\varphi\cos\varphi + I_2 + I_1, \quad (40) \end{aligned}$$

we use the trigonometric relations

$$\sin\varphi = \frac{2t}{1+t^2},\ \cos\varphi = \frac{1-t^2}{1+t^2}, t = \tan(\frac{\varphi}{2}).$$

By replacing in equation (40) $\sin\varphi$ and $\cos\varphi$ by the above expressions, we obtain:

$$ft = (t^2 - 1)[A(l)t^2 + 4tB(l) - A(l)] + 4t^2(I_2 + I_1), \quad (41)$$

where

$$A(l) = l^2 I_7 + 2l I_6 + I_3 + I_2, \quad B(l) = I_5 l + I_4.$$

$A(l)$ is positive for all $l \in \mathbb{R}$. If $\varphi \in]-\frac{\pi}{2}, \frac{\pi}{2}[$ then $t^2 - 1 < 0$. Expression $[A(l)t^2 + 4tB(l) - A(l)] < 0$ for all $l, \varphi \in \mathbb{R}$. Hence, $ft > 0$ for all $l \in \mathbb{R}$ and for $\varphi \in]-\frac{\pi}{2}, \frac{\pi}{2}[$.

The principal minor $\begin{pmatrix} fp & -I_5 \\ -I_5 & I_7 \end{pmatrix}$ is positive for all $l, \varphi \in \mathbb{R}$. Hence, the inertia matrix $M1(\boldsymbol{\alpha})$ is positive definite.

V. CONCLUSION

Surgical robots are economic stakes, they need to be controlled with a good precision and at a high speed rate to provide good performance. Parallel robots are a good solution to have a stiffer system but their models are more complicated and involve a high range of calculation. This paper proposes a simplified method to calculate the dynamic model of the parallel model. In fact, if the parallel robot belongs to a class having some useful properties for the Lagrangian expression and the constraints, the explicit calculation of the Lagrange multipliers can be avoided. It is also proved that the inertia matrix of the haptic interface has the same properties as the matrix inertia of the robot manipulators. These properties will be useful for the control synthesis.

REFERENCES

[1] C. Duriez, "Mise en oeuvre et commande d'un robot parallèle comme interface haptique d'un simulateur d'amniocentèse," Master's thesis, DEA de Réalité Virtuelle et Maîtrise des Systèmes Complexes (Paris), 2001.

[2] R. Paul, *Robot Manipulators: Mathematics, Programming, and Control.* The MIT press, 1981.

[3] R. Schilling, *Fundamentals of Robotics : Analysis and Control.* Prentice-Hall International, 1990.

[4] C. Canudas de Wit, B. Siciliano, and G. Bastin, *Theory of Robot Control.* Springer, 1996.

[5] J. Merlet, *Les Robots Parallèles.* HERMES, 1997.

[6] M. Ait-Ahmed, *Contribution À la Modélisation Géométrique et Dynamique Des Robots Parallèles.* PhD thesis, LAAS, Toulouse, 1993.

[7] W. Khalil and S. Guegan, "A novel solution for the dynamic modeling of gough-stewart manipulators," (Washington, DC), pp. 817–822, IEEE International Conference on Robotics and Automation, May 2002.

[8] Landau and Lifshitz, *Theorical Physics*, vol. 1. Mechanics. Pergamon Press, 1960.

Path Trackability and Verification for Parallel Manipulators

C. K. Kevin Jui and Qiao Sun*
Department of Mechanical and Manufacturing Engineering
University of Calgary
Calgary, AB T2N 1N4, Canada
*qsun@ucalgary.ca

Abstract—Even though parallel manipulators are uncontrollable at force singular positions, our analysis reveals that motion through force singularities may be possible using a path tracking type of approach. This article establishes the condition for trackability of paths in the presence of force singularities, and presents the tools required to verify path trackability. Simulation results of a 2 DOF planar parallel manipulator are shown.

I. Introduction

Parallel manipulators encounter difficulties in both forward kinematics and inverse dynamics at force singularities. Existing research on the remedies to these difficulties generally focuses either on the use redundancies to eliminate singularities ([3] and [4]), or on singularity avoidance ([1], [5] and [14]). Very little attention has been paid to the possibility of operating parallel manipulators in the presence of force singularities. The few research papers on this possibility ([12] and [13]) have only addressed the difficulties in kinematics but have not adequately addressed the dynamic aspect of force singularities.

Our analysis is restricted to the class of nonredundant parallel manipulators for which the configuration of the manipulator can be uniquely specified by the position of the end-effector in the task space. Many popular parallel manipulators belong to this class, including the Stewart Platform. The dynamics of an n DOF parallel manipulator of this class can be expressed as [6]:

$$\mathbf{M}(\mathbf{x})\ddot{\mathbf{x}} + \mathbf{b}(\mathbf{x},\dot{\mathbf{x}}) = \mathbf{J}(\mathbf{x})\boldsymbol{\tau}, \quad (1)$$

where $\mathbf{x} \in \mathbf{R}^n$ represents the end-effector position in task space, $\mathbf{M} \in \mathbf{R}^{n \times n}$ is the inertia matrix, $\boldsymbol{\tau} \in \mathbf{R}^n$ represents the actuator forces, and $\mathbf{b} \in \mathbf{R}^n$ contains the centrifugal, Coriolis, and gravitational terms.

At or near force singularities, the inverse dynamics solution of (1) tends to demand an unfeasibly large $\boldsymbol{\tau}$. This results in actuators being saturated and the manipulator moving in an uncontrolled fashion.

It is to be shown that the problem of unbounded inverse dynamics solutions at force singularities may be resolved by relaxing the rigid constraint on timing caused by trajectory parameterization with respect to time. We consider path as a geometrically constrained one dimensional entity in the task space parameterized by an artificial parameter that is not time. Trajectory is defined as path with timing specified along the path. There is a trend in recent years to change timing instead of path in order to move serial robots through kinematic singularities ([7], [8], [9], and [11]). Our analysis follows this trend except we focus on the dynamic aspect of force singularities. We will refer to force singularities simply as singularities in the remainder of this article. Our method for path verification is developed based upon the techniques of minimum time path tracking ([2], [15], [16], and [17]).

II. Dynamics Under Path Parameterization

Path parameterization is incorporated into dynamics here to simplify the dynamics equation. This also enables us to analyze the motion feasibility and constraints experienced by a parallel manipulator in the neighbourhood of singularities.

A. Path Parameterization and Dynamics

Let us parameterize a path $\mathbf{x} = \mathbf{r}(s)$, s.t. $\mathbf{r}'(s) \neq \mathbf{0}$, for $s \in [s_0, s_f]$, where $\mathbf{r} : \mathbf{R}^1 \to \mathbf{R}^n$ is a C^1 vector-valued function, $'$ denotes the derivative with respect to s, and the parameter s is not time. The velocity and acceleration along the path in the task space are related to $\dot{s}$ and $\ddot{s}$ by $\dot{\mathbf{x}} = \mathbf{r}'(s)\dot{s}$ and $\ddot{\mathbf{x}} = \mathbf{r}'(s)\ddot{s} + \mathbf{r}''(s)\dot{s}^2$; which can be substituted into (1) to result in:

$$\mathbf{M}(s)\mathbf{r}'(s)\ddot{s} + \mathbf{M}(s)\mathbf{r}''(s)\dot{s}^2 + \mathbf{b}(s,\dot{s}) = \mathbf{J}(s)\boldsymbol{\tau}. \quad (2)$$

At singularities, $\mathbf{J}$ is rank deficient; the numerical solution of (2) produces an unbounded inverse dynamics solution of $\boldsymbol{\tau}$ for an arbitrarily defined finite acceleration. However, a necessary and sufficient condition for the existence of inverse dynamics solution is [10]

$$rank\left(\begin{bmatrix} \mathbf{Mr}'\ddot{s} + \mathbf{Mr}''\dot{s}^2 + \mathbf{b} & \mathbf{J} \end{bmatrix}\right) = rank(\mathbf{J}). \quad (3)$$

This gives us hope that even when $\mathbf{J}$ is rank deficient by one, we can still find a bounded $\boldsymbol{\tau}$ if we restrict the choice of $\ddot{s}$ in (2). When $\mathbf{J}$ is rank deficient by more than one, however, it is generally impossible to satisfy (3). Fortunately, the most commonly encountered singularities in practice have $\mathbf{J}$ being rank deficient by one and our analysis is restricted to such singularities.

The problem of unbounded inverse dynamics solution is, however, not just restricted to the situation where $\mathbf{J}$ is rank deficient. Practically, actuators can only produce limited forces and the solution of (2) tends to demand unfeasibly large forces for an arbitrarily defined acceleration in the neighbourhood of singularities. If we were to move the manipulator through singular regions, then we should determine what motion is feasible and only ask the manipulator to execute a feasible motion.

Parameterization of the path simplifies the dynamics equation (1) of $2n$ states in $\mathbf{x}$ and $\dot{\mathbf{x}}$ to (2) of two states in s and $\dot{s}$ [16]. This enables us to determine the feasible motion along the path. Bounds of the actuator forces impose constraints on the feasible $\dot{s}$ and $\ddot{s}$ along the path. By determining the feasible $\dot{s}$ and $\ddot{s}$ along the path, we can determine the feasible motion for which the manipulator can carry out along the path. Analysis of motion feasibility plays a particularly important role for path tracking in the presence of singularities.

B. Minimum and Maximum Accelerations

For the purpose of path verification, we need a way to determine the minimum and maximum accelerations $\ddot{s}_{\min}$ and $\ddot{s}_{\max}$ that can be achieved at a given position and velocity $(s, \dot{s})$. Rewrite (2) in the following form:

$$\mathbf{a}_c \ddot{s} + \mathbf{b}_c = \mathbf{J} \boldsymbol{\tau}, \quad (4)$$

where $\mathbf{a}_c = \mathbf{M}\mathbf{r}'$ and $\mathbf{b}_c = \mathbf{M}\mathbf{r}''\dot{s}^2 + \mathbf{b}$.

Let us assume constant bounds for $\boldsymbol{\tau}$ such that $\tau_i \in [\tau_{i\min}, \tau_{i\max}]$ for the i-th actuator. The bounded region of $\boldsymbol{\tau}$ in the joint space can be transformed by $\mathbf{J}\boldsymbol{\tau} - \mathbf{b}_c$ to an $(n-d)$ dimensional parallelepiped in terms of generalized forces in the task space, where d is the co-dimension of $\mathbf{J}$. Fig. 1 shows a two dimensional example of this transformation. If we are to obtain a feasible inverse dynamics solution, then the line directed by $\mathbf{a}_c$ must intersect this transformed parallelepiped and $\mathbf{a}_c \ddot{s}$ must be constrained within the intersection such that $\ddot{s}_{\min} \leq \ddot{s} \leq \ddot{s}_{\max}$. Because a parallelepiped is convex, the intersection can be determined by checking the $2n$ facets of the parallelepiped.[1]

The actuator force $\boldsymbol{\tau}$ is bounded inside of a right parallelepiped in terms of forces in the joint space. The basis $\mathbf{t}$ of this right parallelepiped is defined as:

$$\mathbf{t} = diag\{\tau_{1\max} - \tau_{1\min}, \cdots, \tau_{n\max} - \tau_{n\min}\}.$$

Also define a vector

$$\boldsymbol{\tau}_{\min} = \begin{bmatrix} \tau_{1\min} & \cdots & \tau_{n\min} \end{bmatrix}^T.$$

[1]While the transformed parallelepiped becomes degenerate at singularity, a facet only becomes degenerate when $null(\mathbf{J})$ falls into the $(n-1)$ basis of the corresponding facet in the joint space. We restrict our analysis to cases where $\mathbf{a}_c$ only passes through non-degenerate facets, which are the typical situations encountered in practice.

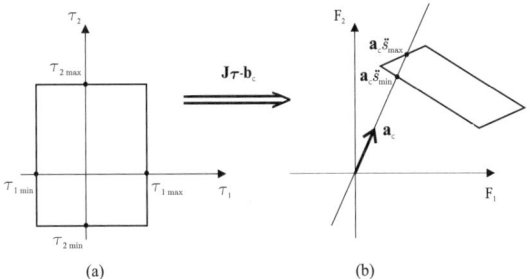

Fig. 1. Bounded region of $\boldsymbol{\tau}$ in the joint space (a), and the transformed region in task space (b).

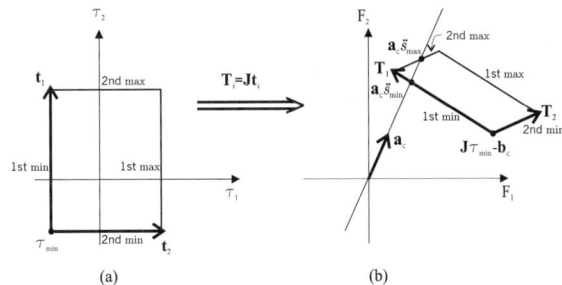

Fig. 2. Facets of the bounded region of $\boldsymbol{\tau}$ (a), and facets of the transformed region (b).

The feasible $\boldsymbol{\tau}$ is constrained inside of a right parallelepiped which can be expressed as:

$$\boldsymbol{\tau} = \boldsymbol{\tau}_{\min} + \mathbf{t}\mathbf{h}, \mathbf{h} = \begin{bmatrix} h_1 & h_2 & \cdots & h_n \end{bmatrix}^T, h_i \in [0,1], \forall i. \quad (5)$$

Pairs of parallel facets exist and the i-th pair shares the same basis $\mathbf{t}_i$ with $h_i = 0$ on one facet and $h_i = 1$ on the other (Fig. 2 (a)). The $n \times (n-1)$ matrix $\mathbf{t}_i$ is simply $\mathbf{t}$ without its i-th column.

Recall that the dynamics equation (4) is expressed in terms of generalized forces in the task space. We can transform basis $\mathbf{t}$ of the bounded region of $\boldsymbol{\tau}$ to basis $\mathbf{T}$ of the transformed region by $\mathbf{T} = \mathbf{J}\mathbf{t}$. Subsequently, $\mathbf{T}_i = \mathbf{J}\mathbf{t}_i$ defines the basis of the i-th pair of facets of the transformed region (Fig. 2(b)). Rearrange (4) to include actuator force bounds of (5):

$$\mathbf{a}_c \ddot{s} = \mathbf{T}\mathbf{h} + (\mathbf{J}\boldsymbol{\tau}_{\min} - \mathbf{b}_c). \quad (6)$$

The intersection between the line directed by $\mathbf{a}_c$ and the $(n-1)$ dimensional hyperplane containing the i-th min facet in terms of generalized forces in the task space can be found by solving:[2]

$$\begin{bmatrix} \mathbf{a}_c & -\mathbf{T}_i \end{bmatrix} \mathbf{h}_c = \mathbf{J}\boldsymbol{\tau}_{\min} - \mathbf{b}_c, \quad (7)$$

[2]If $abs\left(det\left(\begin{bmatrix} \mathbf{a}_c & -\mathbf{T}_i \end{bmatrix}\right)\right) < e_l$ where e_l is a small positive real number to be specified, the i-th pair of facets are considered either degenerate or being parallel to $\mathbf{a}_c$.

where $\mathbf{h}_c = \begin{bmatrix} \ddot{s} & h_{c2} & \cdots & h_{cn} \end{bmatrix}^T$ and it is *not* the same as the $\mathbf{h}$ of (6); $\ddot{s}$ is potentially an extremal parameter acceleration and $\mathbf{a}_c \ddot{s}$ lies on the $(n-1)$ dimensional hyperplane containing the i-th min facet; $h_{c2} \cdots h_{cn}$ have the same meaning as elements of $\mathbf{h}$ in (5). The line directed by $\mathbf{a}_c$ passes through this facet if $h_{cj} \in [0,1], \forall j = 2, \cdots, n$. Similarly, the intersection between the line directed by $\mathbf{a}_c$ and the i-th max facet in terms of generalized forces in the task space can be found by solving:

$$\begin{bmatrix} \mathbf{a}_c & -\mathbf{T}_i \end{bmatrix} \mathbf{h}_c = \mathbf{J}(\boldsymbol{\tau}_{\min} + \boldsymbol{\tau}_{bi}) - \mathbf{b}_c \quad (8)$$

where $\boldsymbol{\tau}_{bi}$ is the i-th column of $\mathbf{t}$, and the line directed by $\mathbf{a}_c$ intersects this facet if $h_{cj} \in [0,1], \forall j = 2, \cdots, n$. Solve (7) for each min facet and (8) for each max facet; there should be either one pair of intersections between the line directed by $\mathbf{a}_c$ and the $2n$ facets of the transformed parallelepiped or no intersections at all. If a pair of intersections are found, then the corresponding values of $\ddot{s}$ are compared and assigned to $\ddot{s}_{\min}$ and $\ddot{s}_{\max}$. Otherwise the line directed by $\mathbf{a}_c$ does not intersect the transformed parallelepiped and no feasible acceleration exists; therefore, $\ddot{s}_{\min}$ and $\ddot{s}_{\max}$ are not assigned by this algorithm.

Because the transformed parallelepiped degenerates at singularities, the feasible range of acceleration typically degenerates to $\ddot{s}_{\min} = \ddot{s}_{\max}$ as well.[3] Execution of an arbitrarily defined acceleration at singularities is therefore impossible and results in an unfeasible inverse dynamics solution.

C. Minimum and Maximum Velocities

At a given position s along the path, the range of feasible acceleration $[\ddot{s}_{\min}, \ddot{s}_{\max}]$ depends on the velocity $\dot{s}$. A velocity $\dot{s}$ is said to be *admissible* if the range of feasible acceleration $[\ddot{s}_{\min}, \ddot{s}_{\max}]$ is not null ([2] and [17]). The admissibility of a given $\dot{s}$ can be determined using the algorithm proposed in section II-B for $\ddot{s}_{\min}$ and $\ddot{s}_{\max}$. If valid $\ddot{s}_{\min}$ and $\ddot{s}_{\max}$ are successfully determined, then $\dot{s}$ is considered admissible; otherwise it is considered inadmissible. Let us assume the manipulator only moves in the direction of non-negative $\dot{s}$,[4] therefore narrowing the set to only positive admissible velocities. Let us also assume that frictions are ignored in the dynamics model. This eliminates the existence of gaps in the set of admissible velocities [17]. These assumptions therefore restrict the admissible velocities to $[\dot{s}_{\min}, \dot{s}_{\max}]$, such that $\dot{s}_{\max} \geq \dot{s}_{\min} \geq 0$. If $\dot{s} = 0$ is an admissible velocity, then we consider $\dot{s}_{\min} = 0$.

---

[3]At singularity, $\ddot{s}_{\min} \neq \ddot{s}_{\max}$ iff $\mathbf{a}_c \in range(\mathbf{J})$ and the line directed by $\mathbf{a}_c$ passes through the transformed parallelepiped, which is already degenerate. This situation rarely occurs in practice.

[4]Because it is not meaningful to have the manipulator moving backward in path tracking.

III. PATH VERIFICATION AND BOUNDARY CURVES IN THE PHASE SPACE

In order to verify the trackability of a path, we need to examine its characteristics in the $s - \dot{s}$ phase space. Analysis in the phase space reveals how the manipulator's velocity is constrained along a path and this in turn verifies the trackability of a path. The results can be conveniently presented in a two dimensional graph.

A. Velocity Limit Curves

The curve connecting all $\dot{s}_{\max}, \forall s \in [s_0, s_f]$ in the phase space, is referred to as the Upper Velocity Limit Curve in [16], and we use UVC as its acronym here; the curve connecting all $\dot{s}_{\min}$ where $\dot{s}_{\min} > 0, \forall s \in [s_0, s_f]$, is referred to as the Lower Velocity Limit Curve (LVC) here. Admissible velocity in the phase space is bounded by the UVC from above and by the LVC and the s-axis from below. On the velocity limit curves, at least one actuator of the manipulator is operating at its limit [2]. A trajectory that crosses the UVC or the LVC acquires an inadmissible velocity and causes the manipulator to stray from the path due to actuator saturation. Because $\dot{s}_{\min} > 0$ implies $\dot{s} = 0$ is an inadmissible velocity, the manipulator cannot pause (not even instantaneously) on the path at positions under the LVC. The presence of LVC indicates the occurrence of actuator saturation at low velocities. LVC is unique to parallel manipulators and only occurs in the neighbourhood of force singularities.[5]

B. Slope of Trajectory

The slope of a trajectory $\dot{s}'$ at a point $(s, \dot{s})$ is related to $\dot{s}$ and $\ddot{s}$ by [17]

$$\dot{s}' = \frac{d\dot{s}}{ds} = \frac{d\dot{s}}{dt}\frac{dt}{ds} = \frac{\ddot{s}}{\dot{s}}. \quad (9)$$

Therefore the minimum and maximum slopes are $\dot{s}'_{\min} = \ddot{s}_{\min}/\dot{s}$ and $\dot{s}'_{\max} = \ddot{s}_{\max}/\dot{s}$, respectively, and $\dot{s}'$ is feasible if $\dot{s}' \in [\dot{s}'_{\min}, \dot{s}'_{\max}]$. The slope of feasible trajectory entering and leaving a point $(s, \dot{s})$ is bounded inside of the set $[\dot{s}'_{\min}, \dot{s}'_{\max}]$, and this set varies with s and $\dot{s}$.

At a point that is away from singular positions, and away from the UVC and LVC, $\ddot{s}_{\min} < \ddot{s}_{\max}$ holds and a trajectory has a range of slope to choose in $[\dot{s}'_{\min}, \dot{s}'_{\max}]$. However, this range shrinks in the singular neighborhood and near the UVC and LVC. At singular positions, and on the UVC and the LVC, the range of acceleration generally degenerates to $\ddot{s} = \ddot{s}_{\min} = \ddot{s}_{\max}$; therefore constraints the range of trajectory slope to $\dot{s}' = \dot{s}'_{\min} = \dot{s}'_{\max}$. In order for a manipulator to track a path, its trajectory must conform to the rigid constraint on slope in the phase space.

---

[5]It is generally safe to assume that a manipulator's actuators are powerful enough so that it can pause at positions away from force singularities.

On the UVC, a point is said to be a trajectory sink [15] if the *maximum* acceleration trajectory at this point moves towards an inadmissible velocity region [16], and a point is said to be a trajectory source [15] if the *minimum* acceleration trajectory at this point moves towards an admissible velocity region [16]. Similarly, on the LVC, we consider a point to be a trajectory sink if the *minimum* acceleration trajectory at this point moves towards an inadmissible velocity region, and a trajectory source if the *maximum* acceleration trajectory at this point moves towards an admissible velocity region.

C. Boundary Curves

Because the range of trajectory slope degenerates in the singular neighbourhood, and near the UVC and LVC, there exist regions in the phase space that are inaccessible to trajectories, regions where trajectories are forced into crossing the UVC or the LVC to acquire inadmissible velocities, and regions where trajectories are forced into crossing the s-axis. A trajectory that acquires an inadmissible velocity causes the manipulator to stray from the path. A trajectory that is forced into crossing the s-axis attains a negative velocity and starts to move backward; therefore not being able to complete path tracking. Because of the existence of these off-limit regions, the UVC and the LVC are not the true boundaries of feasible trajectory, and these off-limit regions must also be avoided.

1) Switching Points: To identify the boundaries of the aforementioned off-limit regions, the first step is to identify *switching points* where the velocity limit curves switch from sink to source [15]. Switching points are the only places where a trajectory can touch the velocity limit curves without becoming inadmissible [16].

To the right of a switching point, the velocity limit curve is a trajectory source, and there exists a region inaccessible to feasible trajectories because trajectories in this region must have been inadmissible prior to appearing at source points on the velocity limit curve. To the left of a switching point, the velocity limit curve is a trajectory sink, and there exists a region where trajectories are forced into crossing the velocity limit curve. The boundary curves of such regions can then be obtained by integrating extremal acceleration trajectories from the switching points.

2) Upper Boundary Curves: Four types of upper boundary curves of feasible trajectories exist in the phase space, which are identified in [2] and [17]: (a) Forward maximum acceleration trajectory from $(s_0, 0)$. (b) Backward minimum acceleration trajectory from $(s_f, 0)$. (c) Forward maximum acceleration trajectory from an upper switching point. (d) Backward minimum acceleration trajectory from an upper switching point. The boundary curves terminates upon reaching the velocity limit curves, s_0, s_f, or the s-axis.

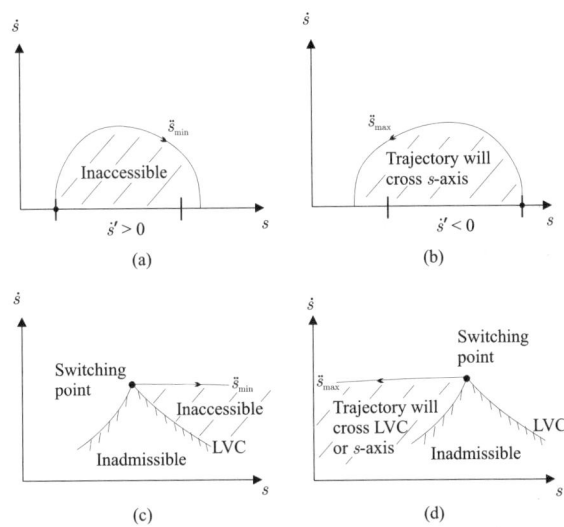

Fig. 3. Types of lower boundary curves.

3) Lower Boundary Curves: Away from singularities, $\ddot{s}$ can take on both positive and negative values at low velocities. In the neighborhood of singularities, $\ddot{s}$ may become strictly positive or strictly negative on the s-axis, which means the trajectory slope is strictly positive or strictly negative in such segments on the s-axis.

Fig. 3 illustrates the four types of lower boundary curves of feasible trajectories in the phase space. The first two types are related the segments just mentioned: (a) Forward minimum acceleration trajectory from the leftmost point of a segment on the s-axis where $\dot{s}'$ is strictly positive; the region below this trajectory is inaccessible to forward moving trajectories as trajectories in this region must have originated from under the s-axis. (b) Backward maximum acceleration trajectory from the right-most point of a segment on the s-axis where $\dot{s}'$ is strictly negative; trajectories that traverse below this trajectory will cross the s-axis, acquire a negative velocity and become unable to complete path tracking.

The remaining two types of upper boundary curves exist due to the degenerate range of trajectory slope close to and on the LVC: (c) Forward minimum acceleration trajectory from a lower switching point; the region below this trajectory is inaccessible because trajectories in this region are inadmissible prior to appearing at sources on the LVC. (d) Backward maximum acceleration trajectory from a switching point; trajectories that traverse below this trajectory cannot speed up fast enough to avoid crossing the LVC or to avoid crossing the s-axis at a point where $\dot{s}'$ is strictly negative.

Similar to the upper boundary curves, the lower boundary curves terminates upon reaching the velocity limit curves, s_0, s_f, or the s-axis. Like the LVC, the lower boundary curves are unique to parallel manipulators and only occur in the neighbourhood of force singularities. Feasible trajectories connecting from $(s_0, 0)$ to $(s_f, 0)$ are bounded from below by these four types of lower boundary curves and the s-axis.

D. Path Trackability

We have discussed the admissibility of velocity, constraints on acceleration and trajectory slope, and boundary curves. A trajectory connecting two points in the phase space is considered feasible if it conforms to the constraints on trajectory slope and remains admissible. The existence of feasible trajectory is the idea behind the following theorem on path trackability.

Theorem 1: A path is trackable iff there exists a feasible trajectory that connects from $(s_0, 0)$ to $(s_f, 0)$ in the phase space.

Proof: Tracking a feasible trajectory only demands feasible bounded actuator forces, which means a feasible trajectory can be executed along the path. Tracking an unfeasible trajectory demands unfeasible actuator forces, which causes actuator saturation and leads to deviation from the desired path. Therefore if a feasible trajectory exists between $(s_0, 0)$ and $(s_f, 0)$, then it can be used for path tracking, otherwise it is not possible to track the path due to actuator saturation.

Remark 1: Additionally, we also require that $\ddot{s}_{\min} < 0$ and $\ddot{s}_{\max} > 0$ at both $(s_0, 0)$ and $(s_f, 0)$ to consider a path to be trackable. These are required for practical reasons so that the manipulator can pause at both s_0 and s_f.

Theorem 2: A path is trackable iff the range of admissible velocity $[\dot{s}_{\min}, \dot{s}_{\max}]$ does not diminish, and none of the upper boundary curves hit the LVC or the s-axis.

Proof: If the admissible velocity diminishes, then no feasible trajectory can connect $(s_0, 0)$ and $(s_f, 0)$. If the upper boundary curves hit the LVC or the s-axis, then forward trajectories from $(s_0, 0)$ and backward trajectories from $(s_f, 0)$ will either become inadmissible or will cross the s-axis. Therefore upper boundary curves that hit the LVC or the s-axis separates the phase space and indicates the non-existence of feasible trajectory between $(s_0, 0)$ and $(s_f, 0)$.

If no upper boundary curves hit the LVC or the s-axis, then the upper boundary curves can be connected to form the minimum time trajectory along the path ([2][15] and [17]). The minimum time trajectory is feasible, therefore, indicates the path is trackable.

Remark 2: If the type (a) and type (b) upper boundary curves cross each other, then they can be connected to form the minimal time trajectory and the path is trackable.

IV. SIMULATION RESULTS

Let us consider an example using the two DOF manipulator depicted in Fig. 4. The pair of legs of this manipulator

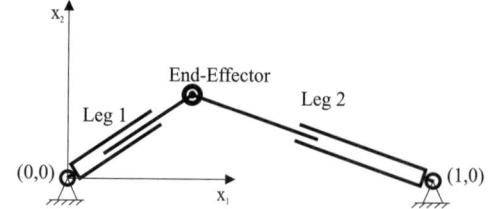

Fig. 4. A planar two DOF parallel manipulator.

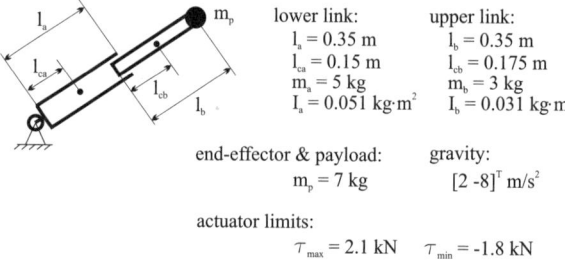

lower link:
$l_a = 0.35$ m
$l_{ca} = 0.15$ m
$m_a = 5$ kg
$I_a = 0.051$ kg·m^2

upper link:
$l_b = 0.35$ m
$l_{cb} = 0.175$ m
$m_b = 3$ kg
$I_b = 0.031$ kg·m^2

end-effector & payload:
$m_p = 7$ kg

gravity:
$[2\ -8]^T$ m/s^2

actuator limits:
$\tau_{\max} = 2.1$ kN $\tau_{\min} = -1.8$ kN

Fig. 5. Link and payload properties of the two DOF manipulator.

is identical and the end-effector is holding a point mass payload. Properties of the manipulator links and payload are given in Fig. 5. This manipulator is being asked to track the two parameterized paths given in Table I, which are also shown in Fig. 6. These paths are parameterized between $s_0 = -2$ and $s_f = 5$.

The UVC, LVC, and boundary curves of these two paths are shown in Fig. 7. Since the upper boundary curves of path 1 hit the s-axis, which indicates path 1 is untrackable. The upper boundary curves of path 2 do not hit the LVC or the s-axis. Therefore, path 2 is trackable, even through it passes through singularities.

V. SUMMARY AND DISCUSSION

Our analysis reveals that parallel manipulator may carry out path tracking through singular positions. We have also established the condition for path trackability, and the tools required for verifying path trackability. The simulation carried out for the two example paths determines that one of the paths is in fact trackable, even though it passes through force singularities.

Since our approach attempts to move through force singularities instead of avoiding them, the full workspace of a parallel manipulator could be utilized. Singularity avoidance algorithms, on the other hand, tend to greatly

TABLE I

PATH PARAMETERIZATION.

#	Parameterization r =
1	$[0.47 + 0.085\cos(s), 0.05 + 0.085\cos(s)]^T$
2	$[0.47 + 0.085\cos(s), -0.073 + 0.085\sin(s)]^T$

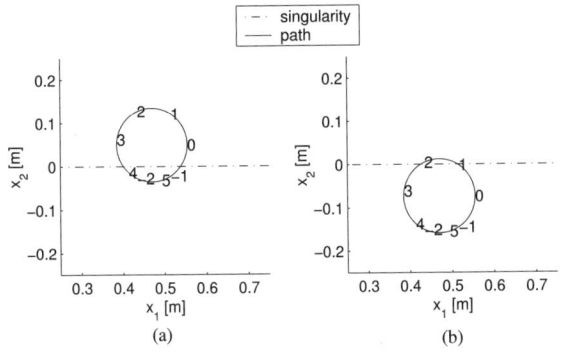

Fig. 6. Parameterized paths: (a) path 1, (b) path 2.

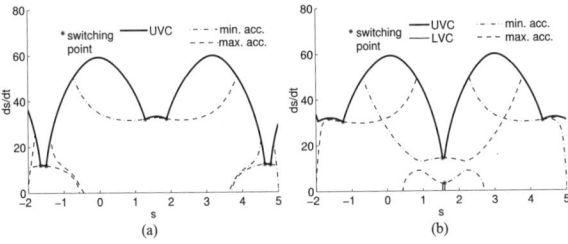

Fig. 7. Velocity limit curves and boundary curves: (a) path 1 - untrackable, (b) path 2 - trackable.

reduce the effective workspace of a parallel manipulator. Our approach also does not require the increased cost and added design complexity that comes with redundancies. Therefore, our approach should serve as a good alternative remedy to singularity avoidance and redundancies.

VI. REFERENCES

[1] S. Bhattacharya, H. Hatwal, and A. Ghosh. "Comparison of an exact and an approximate method of singularity avoidance in platform type parallel manipulators", *Mechanism and Machine Theory*, 33(7):965–974, 1998.

[2] J. Bobrow, S. Dubowsky, and J. Gibson. "Time-optimal control of robotic manipulators along specified paths", *The International Journal of Robotics Research*, 4(3):3–17, 1985.

[3] C. L. Collins and G. L. Long. "The singularity analysis of an in-parallel hand controller for force-reflected teleoperation", *IEEE Transactions on Robotics and Automation*, 11(5):661–669, 1995.

[4] B. Dasgupta and T. Mruthyunjaya. "Force redundancy in parallel manipulators: Theoretical and practical issues", *Mechanism and Machine Theory*, 33(6):727–742, 1998.

[5] B. Dasgupta and T. Mruthyunjaya. "Singularity-free path planning for the stewart platform manipulator", *Mechanism and Machine Theory*, 33(6):711–725, 1998.

[6] Z. Geng, L. Haynes, J. Lee, and R. Carroll. "On the dynamic model and kinematic analysis of a class of stewart platforms", *Robotics and Autonomous Systems*, 9:237–254, 1992.

[7] J. Kieffer. "Manipulator inverse kinematics for untimed end-effector trajectories with ordinary singularities", *The International Journal of Robotics Research*, 11(3):225–237, 1992.

[8] J. Kieffer. "Differential analysis of bifurcations and isolated singularities for robots and mechanisms", *IEEE Transactions on Robotics and Automation*, 10(1):1–10, 1994.

[9] J. E. Lloyd and V. Hayward. "Generating robust trajectories in the presence of ordinary and linear-self-motion singularities", in *Proceedings of the 1998 IEEE International Conference on Robotics and Automation*, pages 3228–3234, Leuven, Belgium, 1998.

[10] Y. Nakamura. *Advanced Robotics: Redundancy and Optimization*. Addison-Wesley Publishing Company, 1991.

[11] D. N. Nenchev. "Tracking manipulator trajectories with ordinary singularities: A null space-based approach", *The International Journal of Robotics Research*, 14(4):399–404, 1995.

[12] D. N. Nenchev, S. Bhattacharya, and M. Uchiyama. "Dynamic analysis of parallel manipulators under the singularity-consistent parameterization", *Robotica*, 15:375–384, 1997.

[13] D. N. Nenchev and M. Uchiyama. "Singularity-consistent path planning and motion control through instantaneous self-motion singularities of parallel-link manipulators", *Journal of Robotic Systems*, 14(1):27–36, 1997.

[14] M.-H. Perng and L. Hsiao. "Inverse kinematic solutions for a fully parallel robot with singularity robustness", *The International Journal of Robotics Research*, 18(6):575–583, 1999.

[15] F. Pfeiffer and R. Johanni. "A concept for manipulator trajectory planning", *IEEE Transactions on Robotics and Automation*, RA-3(2):115–123, 1987.

[16] Z. Shiller and H.-H. Lu. "Computation of path constrained time optimal motions with dynamic singularities", *Journal of Dynamic Systems, Measurement, and Control*, 104:33–40, 1992.

[17] K. Shin and N. McKay. "Minimum-time control of robotic manipulators with geometric path constraints", *IEEE Transactions on Automatic Control*, AC-30(6):531–541, 1985.

Control and Experiments of a Multi-purpose Bipedal Locomotor with Parallel Mechanism

Yusuke Sugahara[1], Tatsuro Endo[1], Hun-ok Lim[2,3] and Atsuo Takanishi[1,3]

[1] Graduate School of Science and Engineering, Waseda University, Tokyo, Japan
[2] Department of System Design Engineering, Kanagawa Institute of Technology, Kanagawa, Japan
[3] Humanoid Robotics Institute, Waseda University, Tokyo, Japan

#59-308, 3-4-1 Ookubo, Shinjuku-ku, Tokyo, 169-8555 Japan
Tel: +81-3-5286-3257, Fax: +81-3-5273-2209
sugahara@suou.waseda.jp, http://www.takanishi.mech.waseda.ac.jp/

Abstract

This paper describes a control and a structure of a battery driven bipedal robot, WL-15(Waseda Leg-15), which has a 6-DOF parallel mechanism. WL-15 has been designed as a multi-purpose locomotor of robotic systems. This robot is controlled by the QNX real-time operating system. Using Math Works' SIMULINK and OPAL-RT's RT-LAB executed on Windows NT and QNX target systems, control software and pattern generator are developed. A walking control capable of dealing with various dynamic walking is proposed and tested indoors and outdoors using the WL-15. Dynamic biped walking is realized with the maximum step length of 200[mm], the maximum walking speed of 0.8[s/step] and the minimum speed of 1.92[s/step]. Also, the WL-15 is confirmed to be able to turn 90[deg] in two steps and carry a load of 18[kg].

1. Introduction

To date, most of the humanoid robots have employed wheels and crawlers for their locomotion mechanism. The reason is that a biped walking is essentially unstable, and it is difficult to treat it as an independent locomotion mechanism like the wheels or crawlers.

From the standpoint that the expected locomotion form of a humanoid is ultimately a biped walking, we have studied the mechanism and control of bipedal robots about 35 years [1, 2, 3, 4]. In our research, we have set one of the final targets as the development of a bipedal robot as a leg module that is sufficient for practical use as a multi-purpose locomotor of robotic systems.

There are a few researches on an application of the bipedal robot as a locomotor. ParaWalker was developed, which has 6-DOF that is the minimum of required DOF preserving the characteristics of walking robots. This robot has been developed for tasks such as leveling concrete [5]. Furthermore, a walking type wheelchair was studied, and a parallel mechanism was adopted for its leg mechanism [6]. TITRUS-II was developed for various tasks [7]. However, these robots are difficult to be employed as stable locomotive systems in human living and working environments because of the lack of DOF.

Therefore, we have developed a biped robot, WL-15 (Waseda Leg - No.15) having 6-DOF parallel mechanisms [8]. This robot is composed of two legs and a waist, is small and light, and walks independently using battery (Figure 1). In this paper, we present a control strategy for stable walking, without the design part of the WL-15 proposed already. The control method bases on a model-based walking control that is used in controlling WABIAN [9]. Using this control method, the

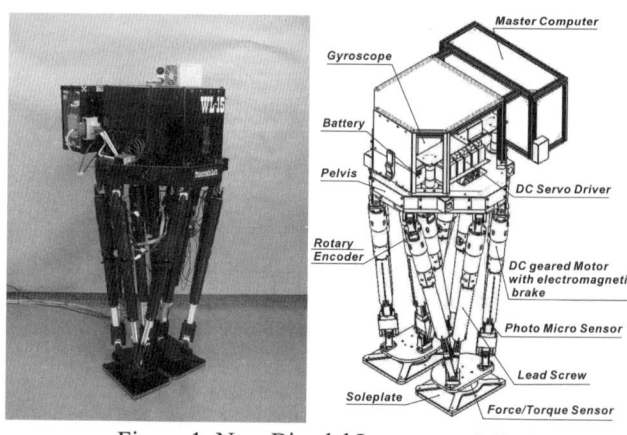

Figure 1: New Bipedal Locomotor WL-15

moments produced by the locomotion are cancelled. Especially, we discuss in detail how to generate the compensation motion in this paper.

This paper is organized as follows. In Section 2, we describe a design concept of WL-15. Section 3 describes a walking control method. Section 4 illustrates experimental results. Finally, Section 5 provides conclusions.

2. Design Concept and Hardware of WL-15

As stated previously, this biped machine was developed as a leg module that is sufficient for practical use as a multi-purpose locomotor of the robot system. Therefore, many features were required, as to accomplish a high independence, specifically it had to be battery driven, and it had to be able to carry a payload, and so on.

In order to satisfy these functional requirements, we have focused on the 6-DOF parallel mechanisms. As for the bipedal robots studied so far, most of them use serial linkage mechanisms for their legs. Paying attention to an easy inverse kinematics, an equalized position error, a high rigidity, a high output power and so on, the 6-DOF parallel mechanism has been adopted as the leg of our machine, and we have planned fulfilling those many requirements.

Since this machine was developed so that those features that were described previously will be fulfilled, the applications by the following methods are expected.

An application of the locomotion mechanism of a humanoid robot in which wheels and crawlers are used is conceivable first, and applications as a locomotion mechanism of a walk support machine, a wheelchair or for a medical-treatment field are also conceivable. The applications to new intelligent mobility or a new entertainment vehicle that people can drive are also interesting.

WL-15 is a battery powered bipedal robot, which introduces the 6-DOF parallel mechanism as the mechanism used for its legs to eventually develop a leg module that is sufficient for practical use as a multi-purpose locomotor. Taking advantage of the mechanical rigidity of the parallel mechanism, we have succeeded in the weight reduction by means of using polyacetal resin in many of the components instead of metals, and by cutting down on the weight of gears, motors, drivers and other elements. By using electromagnetic brakes, it was made possible to keep the robot in the same posture with less energy consumption.

WL-15's gross weight is 64 kg including a 7 kg battery weight, and a height of about 1.2 m.

3. Control Method of Bipedal Walking

The walking control method of WL-15 is a simplified version of WABIAN's model based walking control. For the high robustness of this algorithm, it was possible to have utilized effectively after adding the minimum change also in walking control of WL-15. Since WL-15 has no trunk, and Legs are constituted as parallel mechanisms, the two points changed by applying this algorithm to WL-15 are the following:
- Approximation model of the robot
- Inverse kinematics of the legs

This algorithm consists of the following four main parts.
1. Modeling of the robot
2. Derivation of the ZMP equations
3. Computation of approximate waist motion
4. Computation of strict waist motion by iteratively computing the approximate waist motion

The other component of the control method is a program control for walking using preset walking patterns transformed from the motion of the lower-limbs.

In this section, we describe the algorithm for computing the compensatory waist motion.

3.1 Modeling of the Robot
Let the walking system be assumed as follows:
(1) The robot is a system of particles.
(2) The floor for walking is solid and not moved by any force or moment.
(3) A Cartesian coordinate system is determined as shown in Figure 2. Here, the X-axis and Y-axis form a plane which is the same as that of the floor.
(4) The contact region between the foot and the floor is a set of contact points.
(5) The coefficient of friction for rotation around the X, Y and Z-axes is zero at the contact point.

3.2 Derivation of the ZMP Equations and Computation of Waist Motion
By assuming that the other masses surrounding the waist are parts of the waist, we could first define an approximation model of the trunk and the position vectors as shown in Figure 2. The moment balance around point P on the floor can be expressed as below:

$$\sum_{i}^{all\_particles} m_i(\mathbf{r}_i - \mathbf{r}_p) \times (\ddot{\mathbf{r}}_i + \mathbf{G}) + \mathbf{T} = \mathbf{0} \cdot \quad (1)$$

Point P is defined as ZMP, so $\mathbf{T} = \mathbf{0}$, and we denote the position vector of P as $P_{zmp}(x_{zmp}, y_{zmp}, 0)$. To consider the relative motion of each part, a moving coordinate $\overline{W}\text{-}\overline{XYZ}$ is established on the waist of the robot on a parallel with the fixed coordinate O-XYZ (shown in Figure 2). $Q(x_q, y_q, z_q)$ is the position vector of the origin of $\overline{W}$ from the origin of O. Using the moving coordinate frame, equation (1) can be modified as follows:

$$\sum_{i}^{all\_particles} m_i(\bar{\mathbf{r}}_i - \bar{\mathbf{r}}_{zmp}) \times \{\ddot{\bar{\mathbf{r}}}_i + \ddot{\mathbf{Q}} - \overline{\mathbf{G}} \\ + \dot{\overline{\omega}} \times \bar{\mathbf{r}}_i + 2\overline{\omega} \times \dot{\bar{\mathbf{r}}}_i + \overline{\omega} \times (\overline{\omega} \times \bar{\mathbf{r}}_i)\} = \mathbf{0} \quad (2)$$

where $\bar{r}_{zmp}$ is the position vector of ZMP with respect to the O. $\overline{\omega}$ is the angular velocity vector of the origin of $\overline{W}\text{-}\overline{XYZ}$.

Assuming that a moving coordinate does not rotate, this equation is expanded into equations (3) and (4) by putting the terms representing the motion of the upper-limb particles on the left-hand side as unknown variables, and the terms representing the moment generated by the lower-limb particles on the right-hand side as known parameters, named $M(M_x, M_y)$ respectively.

$$m_T(\bar{z}_T - \bar{z}_{zmp})(\ddot{\bar{x}}_T + \ddot{\bar{x}}_q - g_x) - m_T(\bar{x}_T - \bar{x}_{zmp})(\ddot{\bar{z}}_T + \ddot{\bar{z}}_q - g_z) \\ m_W(\bar{z}_W - \bar{z}_{zmp})(\ddot{\bar{x}}_W + \ddot{\bar{x}}_q - g_x) - m_W(\bar{x}_W - \bar{x}_{zmp})(\ddot{\bar{z}}_W + \ddot{\bar{z}}_q - g_z) \\ = -M_y \quad (3)$$

$$m_T(\bar{y}_T - \bar{y}_{zmp})(\ddot{\bar{z}}_T + \ddot{\bar{z}}_q - g_z) - m_T(\bar{z}_T - \bar{z}_{zmp})(\ddot{\bar{y}}_T + \ddot{\bar{y}}_q - g_y) \\ m_W(\bar{y}_W - \bar{y}_{zmp})(\ddot{\bar{z}}_W + \ddot{\bar{z}}_q - g_z) - m_W(\bar{z}_W - \bar{z}_{zmp})(\ddot{\bar{y}}_W + \ddot{\bar{y}}_q - g_y) \\ = -M_x \quad (4)$$

However, these equations are interferential and non-linear, because each equation has the same variable z_T, and therefore it is difficult to derive analytic solutions from them. Thus, the other stage of approximation is needed. By assuming that neither the waist nor the trunk particles move vertically, i.e., the trunk moves on the horizontal plane only, the equations can be decoupled and linearized. The linearized equations (5) and (6) are thereby obtained.

$$m_T(\bar{z}_T - \bar{z}_{ZMP})(\ddot{\bar{x}}_T + \ddot{\bar{x}}_q) - m_T(\bar{x}_T - \bar{x}_{ZMP})(-\bar{g}_z) \\ + m_W(\bar{z}_W - \bar{z}_{ZMP})(\ddot{\bar{x}}_W + \ddot{\bar{x}}_q) - m_W(\bar{x}_W - \bar{x}_{ZMP})(-\bar{g}_z) \\ = -M_y \quad (5)$$

$$m_T(\bar{y}_T - \bar{y}_{ZMP})(-\bar{g}_z) - m_T(\bar{z}_T - \bar{z}_{ZMP})(\ddot{\bar{y}}_T + \ddot{\bar{y}}_q) \\ + m_W(\bar{y}_W - \bar{y}_{ZMP})(-\bar{g}_z) - m_W(\bar{z}_W - \bar{z}_{ZMP})(\ddot{\bar{y}}_W + \ddot{\bar{y}}_q) \\ = -M_x \quad (6)$$

In equations (5) and (6), the known clause of the left side is moved to the right-hand side and the right-hand side is anew replaced with $-\mathbf{M}^* = -[M^*_x, M^*_y]^T$, the next will be obtained.

$$m_T(\bar{z}_T - \bar{z}_{ZMP})\ddot{\bar{x}}_T - m_T(-\bar{g}_z)\bar{x}_T \\ + m_W(\bar{z}_W - \bar{z}_{ZMP})\ddot{\bar{x}}_W - m_W(-\bar{g}_z)\bar{x}_W = -M^*_y \quad (7)$$

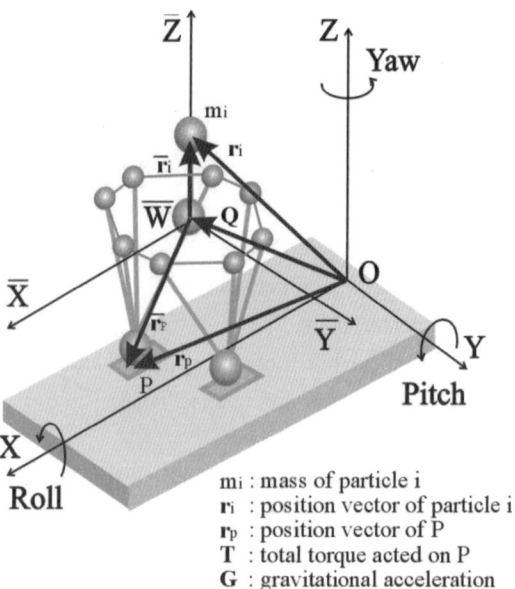

m_i : mass of particle i
r_i : position vector of particle i
r_p : position vector of P
T : total torque acted on P
G : gravitational acceleration

O-XYZ : fixed coordinate system
$\overline{W}\text{-}\overline{XYZ}$: moving coordinate system

Figure 2: Definition of Coordinate and Vector

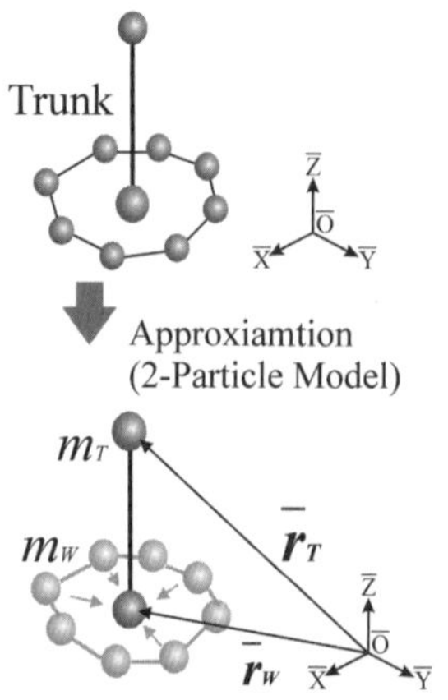

Figure 3: Approximation Model of Trunk and Waist

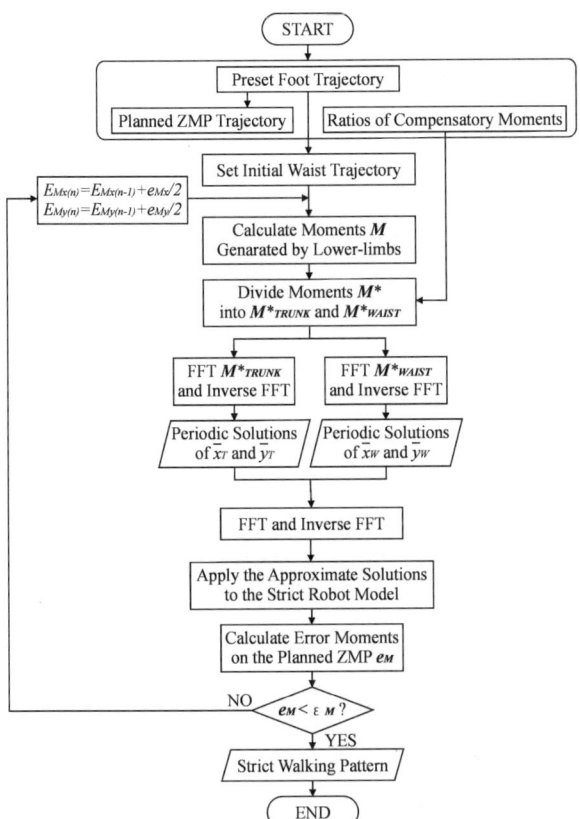

Figure 4: Flowchart to Compute Trunk Motion

$$-m_T(\bar{z}_T - \bar{z}_{ZMP})\ddot{\bar{y}}_T + m_T(-\bar{g}_z)\bar{y}_T \\ -m_W(\bar{z}_W - \bar{z}_{ZMP})\ddot{\bar{y}}_W + m_W(-\bar{g}_z)\bar{y}_W = -M^*_x \quad (8)$$

Here, since the total of four trajectories as $\bar{x}$ and $\bar{y}$ of each trunk and waist are unknown and sufficient equations do not exist, the approximation solution of the moment compensation trajectories of the pitch and roll axis cannot be specified.

Then, the following four differential equations are built into four strange trajectories $\bar{x}_T$, $\bar{x}_W$, $\bar{y}_T$ and $\bar{y}_W$ by distributing the compensated moment in an approximation model to the waist and a trunk, respectively. It enables us to specify an approximation solution in only one.

$$m_T(\bar{z}_T - \bar{z}_{ZMP})\ddot{\bar{x}}_T - m_T(-\bar{g}_z)\bar{x}_T = -M^*_{yTRUNK} \\ + m_W(\bar{z}_W - \bar{z}_{ZMP})\ddot{\bar{x}}_W - m_W(-\bar{g}_z)\bar{x}_W = -M^*_{yWAIST} \quad (9) \\ M^*_{yTRUNK} + M^*_{yWAIST} = M^*_y$$

$$-m_T(\bar{z}_T - \bar{z}_{ZMP})\ddot{\bar{y}}_T + m_T(-\bar{g}_z)\bar{y}_T = -M^*_{xTRUNK} \\ -m_W(\bar{z}_W - \bar{z}_{ZMP})\ddot{\bar{y}}_W + m_W(-\bar{g}_z)\bar{y}_W = -M^*_{xWAIST} \quad (10) \\ M^*_{xTRUNK} + M^*_{xWAIST} = M^*_x$$

However, since WL-15 now does not have the DOF of the trunk, the ratio of M^*_{yTRUNK} and M^*_{xTRUNK} has been determined for the trunk and the waist to take the same trajectories.

In these equations, M^*_y and M^*_x are known, because they are derived from the lower-limb's motion and the time trajectory of ZMP. In the case of steady walking, M^*_y and M^*_x are periodic functions because each particle of the lower-limbs and the time trajectory of ZMP move periodically for the moving coordinate $\overline{W}\text{-}\overline{XYZ}$. Thus, each equation can be represented as a Fourier series. By comparing the Fourier transform coefficients from both sides of each equation, we can easily acquire the approximate periodic solution for trunk motion.

The above computation is applicable not only to steady walking, but also to complete walking. That is, by regarding complete walking as one walking cycle, and making static standing states before and after walking long enough, we could apply the algorithm to it.

Further, in order to compute strict solutions, we proposed an algorithm that computes the approximate solutions iteratively. The flowchart of the algorithm is shown in Figure 4. $\varepsilon(\varepsilon_{My}, \varepsilon_{Mx})$ determines a specific tolerance level of the moment error.

4. Software Environment

We adopted a real-time OS called QNX as an OS of WL-15. By using QNX and a software called RT-LAB which OPAL-RT developed, the time and effort of the development can be saved considerably. RT-LAB is software which can perform a real-time simulation or hardware control by the hardware-in-the-loop system using MATLAB/Simulink of Mathworks. By using MATLAB/Simulink, a development period can be sharply shortened by using a block diagram as compared with the case of using the C language.

By using this system, QNX which is a server that was

accessed from Windows NT which is the Host PC, control with a control cycle of 1 ms was performed. Specifically, Host PC controls actuators as the generated pattern described previously. PI control which detects the rotation angle of motors from encoders and feeds back a speed reference value to motor drivers is performed.

5. Walking Experiment

5.1 Experiment method

The following experiments were conducted in order to evaluate the performance of the developed WL-15:

(a) Standard walking experiment: The walking speed is 0.96 s/step, and the step length is configured as 100 mm, we checked the time trajectories of angles of WL-15's pelvis.

(b) Validation experiment of the maximum step length: We changed the step length between 100 mm and 250 mm, and confirmed the maximum step length which WL-15 can walk to the front and the side.

(c) Validation experiment of the maximum walking speed: We changed the walking speed between 1.92 and 0.80 s/step, and confirmed the maximum walking speed which WL-15 can walk to the front and the side.

(d) Validation experiment of the maximum turn angle: We changed the turn angle and confirmed the maximum turn angle which WL-15 can walk.

(e) Validation experiment of the payload: The maximum weight of payload on its waist which WL-15 can walk with at a step length of 100 mm and walk speed of 0.96 s is confirmed.

(f) Walking experiment done outside: By battery drive and command transmission using a wireless LAN, we tried a perfect wireless operation in WL-15. To confirm it, the walking experiment was done outside.

(g) Validation experiment of posture-holding without Batteries: To confirm the posture-holding function through negative operation electromagnetic brakes attached in the actuators after cutting the power supply, the power supply was disconnected after walking and the battery was taken out.

5.2 Experiment Results

The following results were obtained by experiments. Maximum step length is 200 mm (Forward/backward, right/left, Figure 5, 6) and maximum walking speed: 0.80

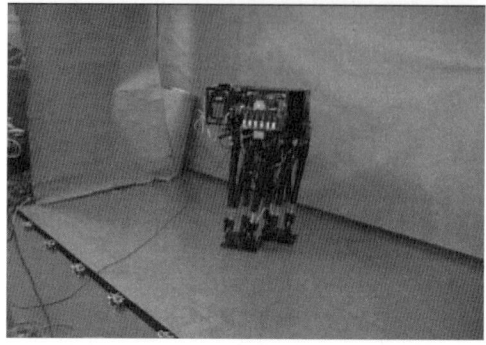

Figure 5: Walking with 200 mm Step Length (Forward)

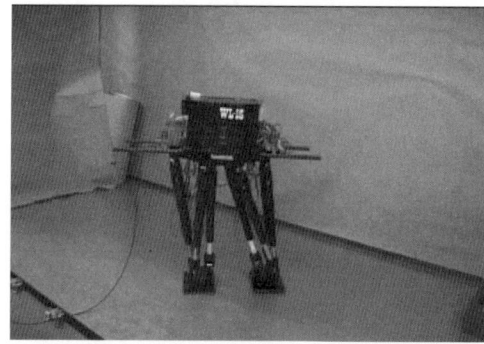

Figure 6: Walking with 200 mm Step Length (Side)

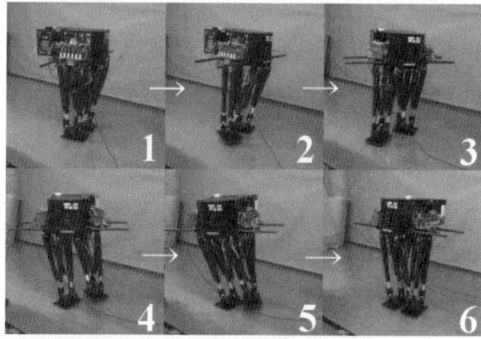

Figure 7: 90 deg Turn in 2 Steps

Figure 8: Validation of the Payload

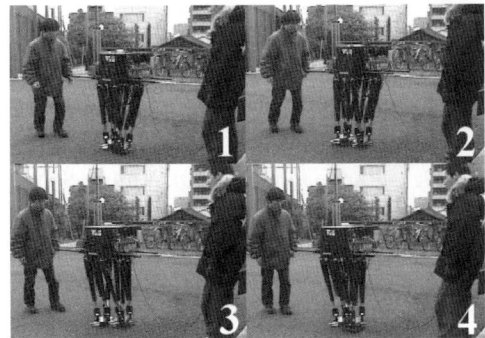

Figure 9: Walking Outside

s/step, minimum walking speed: 1.92 s/step. The maximum turn angle in 2 Steps is 90 deg (Figure 7). The payload of WL-15 is 18 kg (Figure 8).

The image of the walking experiment in the outdoors is shown in Figure 9.

The result of the validation experiment of the posture-holding after battery extraction, it is validated that WL-15 can hold posture without battery.

6. Conclusions and Future Works

We have developed a battery powered bipedal robot WL-15 having 6-DOF parallel mechanisms as the mechanisms for the legs to develop a leg module that is sufficient for practical use as a multi-purpose locomotor. The walking control method is a simplified version of WABIAN's model based walking control. For the high robustness of this algorithm, it was possible to have effectively utilized after adding the minimum change also in the walking control of WL-15.

Through various walking experiments, the performance of WL-15 is confirmed. The maximum step length is 200 mm, the maximum walking speed is 0.80 s/step, and the minimum is 1.92 s/step. The maximum turn angle is 90 deg/2 steps. Concerning the payload, it was confirmed that WL-15 can walk loading 18 kg, and also, WL-15 can hold posture without battery.

Now that we have succeeded in the basic walking experiments using WL-15, our next goal is to walk in the human-living environments. We will also continue to study more walking experiments to achieve walking at higher speeds, with heavier loads, and etc.

Acknowledgment

This study has been conducted as a part of the humanoid project at the Humanoid Robotics Institute, Waseda University. This research was supported by TMSUK Co., Ltd., SANYO Electric Soft Energy Co., Ltd., and Neorium Technology Co., Ltd. The authors would like to express sincere thanks to them for their financial supports. Also the authors would like to thank Takuya Hosobata, Yutaka Mikuriya and Hiroyuki Sunazuka for help us in developing WL-15.

References

[1] I. Kato, S. Ohteru, H. Kobayashi, K. Shirai, and A. Uchiyama, "Information-power machine with senses and limbs," Proc. CISM-IFToMM Symp. Theory and Practice of Robots and Manipulators, Udine, Italy, Sep.1973, pp.12-24.

[2] A. Takanishi, M. Ishida, Y. Yamazaki, and I. Kato, "The realization of dynamic walking by the biped walking robot," Proc. ICRA'85, pp.459-466.

[3] A. Takanishi, T. Takeya, H. Karaki, M. Kumeta, and I. Kato, "A control method for dynamic walking under unknown external force," Proc. IROS'90, pp.795-801.

[4] H. O. Lim, A. Ishii, and A. Takanishi, "Motion pattern generation for emotion expression," Proc. Int. Symp. Humanoid Robots, 1999, pp.36-41.

[5] Y. Ota, Y. Inagaki, K. Yoneda and S. Hirose, "Research on a Six-Legged Walking Robot with Parallel Mechanism", Proc. IROS'98, pp.241-248

[6] Y. Takeda, M. Higuchi, and H. Funabashi, "Development of a walking chair (Fundamental investigations for realizing a practical walking chair)", Proc. CLAWAR2001, pp.1037-1044.

[7] K. Takita, R. Hodoshima, and S. Hirose, "Fundamental Mechanism of Dinosaur-like Robot TITRUS-II Utilizing Coupled Drive", Proc. IROS 2000.

[8] Y. Sugahara, T. Endo, H. O. Lim and A. Takanishi, "Design of a Battery-powered Multi-purpose Bipedal Locomotor with Parallel Mechanism", Proc. IROS 2002.

[9] S. A. Setiawan, S. H. Hyon, J. Yamaguchi and A. Takanishi, "Physical Interaction between Human And a Bipedal Humanoid Robot -Realization of Human-follow Walking-", Proc. ICRA 99.

Design of a Redundantly Actuated Leg Mechanism

So, Byung Rok[*], Yi, Byung-Ju[*], Kim, Wheekuk[**], Oh, Sang-Rok[***] Park, Jongil[*], Kim, Young Soo[****]

* School of Electrical Engineering and Computer Science, Hanyang University, Korea
** Department of Control and Instrumentation Engineering, Korea University, Korea
*** Intelligent System Control Research Center, KIST, Korea
**** Department of Medicine, Hanyang University, Korea
bj@hanyang.ac.kr

Abstract – In humanoid robot system, many human-body motions such as walking, running, jumping, etc require large power. To achieve a high power-to-weight ratio, this paper proposes a new design of the leg mechanism using parallel kinematic chains involving redundant actuators. The kinematics for the leg mechanism is derived and a kinematic index to measure force transmission ratio are introduced. It is demonstrated through simulation that incorporation of redundant actuator into the leg mechanism enhances the power of the mechanism approximately 4 times of the minimum actuation. The leg mechanism is developed and has been integrated into the biomimetic system for the purpose of payload enhancement.

1. Introduction

In humanoid robot system, many human-like motions such walking, running, jumping, etc require large power. So far, most of the biped robots developed have been designed as the serial-type mechanisms. Even though the advantage of serial structure is the wide workspace, the power-to-weight ratio is the major drawback. Therefore, it is necessary to propose an alternative design to cope with this problem.

To date, we can find many alternative designs using parallel mechanisms embedded into leg structures. Morisawa, et al [1-2] proposed a 3D biped robot in which each leg is composed of a 6-DOF parallel mechanism. Also, they compared a serial mechanism with parallel mechanism in biped robot [3]. Sugahara, et al [4] introduced a biped robot with a total of 12 active linear actuators at each leg. Takita, et al [5] proposed a linear actuator with a ball screw and a coupled drive mechanism to develop dinosaur-like biped walking robot. Their mechanism is driven by one linear actuator and a coupled drive consisting of two linear actuators to achieve 3-DOF motion of the dinosaur-like biped walking robot. Also, Sellaouti, et al [6] also proposed a 3-DOF parallel mechanism to drive a biped robot with linear actuators. David, et al [7-8] introduced a linear actuator, called a Series Elastic Actuator, with a ball screw to achieve a high force/mass and power/mass ratio and developed a M2–3D bipedal walking robot with linear actuation.

In this paper, we also develop a linear actuator using a ball-screw mechanism for linear motion and propose a new design of leg mechanism using parallel structure. Typically, parallel structures have many potential input locations where actuators can be placed. It has been reported that redundant actuation mode has many merits in terms of singularity avoidance, payload enhancement, and utilization of many subtasks. Special feature of this work is the utilization of redundant actuators to enhance the force transmission ratio of humanoid. For this, the kinematics for the leg mechanism and a kinematic index to measure actuator power are introduced. It is demonstrated through simulation that incorporation of redundant actuator into the hip module enhances the power of the mechanism approximately 4 times of the minimum actuation. The leg mechanism is developed and has been integrated into humanoid.

2. Mechanism Architecture

Fig. 1 shows a 3-DOF leg mechanism. The yaw motion (θ_{h1}) is directly achieved by one rotary actuator, and the roll, pitch and knee motions are achieved by the embedded actuators inside the parallel structure. The leg mechanism is composed of three chains. Each of the left and right chains has one active linear joint and five passive revolute joints (upper universal joint and lower spherical joint). Also, the middle chain has two passive revolute joints in which additional, redundant actuators will be placed in the proposed design.

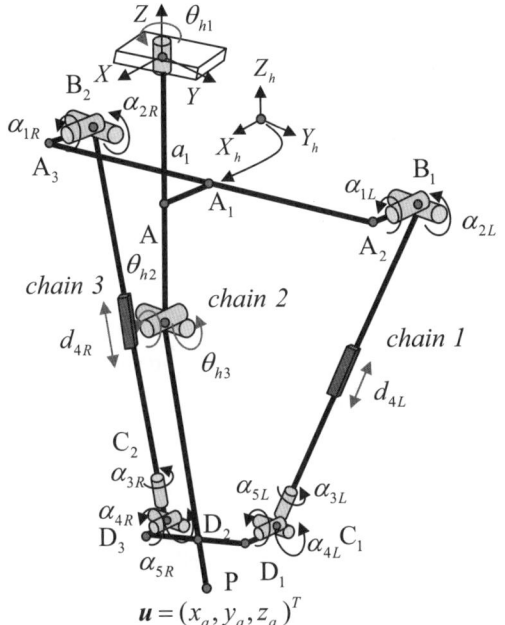

Fig. 1. Leg Mechanism

3. Kinematic Modeling for Leg Mechanism

The modeling methodology integrates the *Generalized Principle of D'Alembert* with the method of kinematic influence coefficients (*KIC*) resulting in closed form vector expressions. The reader is referred to Freeman and Tesar [9] for a more detailed description of the following scheme. In the following, the letter G stands for 1st order *KIC* matrix, and superscribed quantities are considered and indicate dependent parameters with subscripts denoting independent parameters.

3.1 Open-chain Kinematics

Note that the four open-chains of the leg mechanism have a common kinematic constraint at the knee position as shown in Fig. 1. Let the knee position be $u = (x_a, y_a, z_a)^T$.

The Jacobians of the four chains are expressed as

$$[_1G_\phi^u] = [S_{11} \ S_{12} \ S_{13} \ S_{14} \ S_{15} \ S_{16}],$$

$$[_2G_\phi^u] = [S_{21} \ S_{22}],$$

$$[_3G_\phi^u] = [S_{31} \ S_{32} \ S_{33} \ S_{34} \ S_{35} \ S_{36}],$$

where

$$S_{rn} = \begin{cases} _rS_n \times (P - _rR_n) & : \ n^{th} \text{ revolute joint} \\ _rS_n & : \ n^{th} \text{ prismatic joint} \end{cases}.$$

$_rS_n$ denotes the unit vector along the nth joint axis in the rth chain, expressed in terms of the hip reference frame, $_rR_n$ the origin of the nth coordinate frame with respect to the hip reference frame($X_h Y_h Z_h$), and P the position vector of the ankle position with respect to the reference frame.

3.2 Internal Kinematics

Excluding the yaw motion of the hip module, the mobility of this mechanism is two. Thus, at least two actuators are required to control the mechanism. There are several choices in the selection of the independent joints (i.e. actuator location). An internal kinematic relationship between the dependent joints and independent joints is derived from [10]

$$\dot{u} = [_1G_\phi^u]_1\dot{\phi} = [_2G_\phi^u]_2\dot{\phi} = [_3G_\phi^u]_3\dot{\phi}. \quad (1)$$

Choosing d_{4L} and d_{4R} as the independent joints (ϕ_a) and the other joints as the dependent joints (ϕ_p). Eq. (1) can be rearranged as the following form

$$[A]\dot{\phi}_p = [B]\dot{\phi}_a, \quad (2)$$

where

$$[A] = \begin{pmatrix} [_1G_a^u] & 0 \\ [_1G_a^u] & -[_3G_a^u] \end{pmatrix}, \quad (3)$$

$$[B] = \begin{pmatrix} -[_1G_p^u] & [_2G_p^u] & 0 \\ -[_1G_p^u] & 0 & [_3G_p^u] \end{pmatrix}, \quad (4)$$

$$\dot{\phi}_a = (\dot{d}_{4L} \ \dot{d}_{4R})^T, \quad (5)$$

and

$$\dot{\phi}_p = (\dot{\alpha}_{1L} \ \dot{\alpha}_{2L} \ \dot{\alpha}_{3L} \ \dot{\alpha}_{4L} \ \dot{\alpha}_{5L} \ \dot{\theta}_{h2} \ \dot{\theta}_{h3} \ \dot{\alpha}_{1R} \ \dot{\alpha}_{2R} \ \dot{\alpha}_{3R} \ \dot{\alpha}_{4R} \ \dot{\alpha}_{5R})^T. \quad (6)$$

Now, premultiplying the inverse matrix of $[A]$ to both sides of Eq. (2) yields

$$\dot{\phi}_p = [G_a^p]\dot{\phi}_a = [A]^{-1}[B]\dot{\phi}_a, \quad (7)$$

where $[G_a^p]$ denotes the first-order KIC relation between $\dot{\phi}_p$ and $\dot{\phi}_a$.

According to the duality existing between the velocity

vector and force vector, the force relation between the independent joints and the dependent joints is described by

$$T_a = [G_a^p]^T T_p. \quad (8)$$

When any additional actuators are attached on the joint θ_{h2} or θ_{h3} of the middle chain, the system becomes a redundantly actuated system. In this case, the effective load referenced to the independent joints can be written by

$$T_a^* = T_a + [G_a^p]^T T_p = [G_a^A]^T T_A, \quad (9)$$

where

$$[G_a^A] = \begin{bmatrix} I \\ [G_a^p] \end{bmatrix}, \quad (10)$$

$$T_a = (T_{4L}\ T_{4R})^T, \quad (11)$$

$$T_A = (T_{4L}\ T_{4R}\ T_{h3})^T.$$

In Eq. (9), T_A and $[G_a^A]$ denote a force vector for activated joints and the velocity relationship between the independent joints and the actuator joints. Likewise, when more redundant actuators are employed, $[G_a^A]$ can be reformulated by collecting the corresponding rows from $[G_a^p]$. In this case, T_A is composed of three joint torques.

3.3 Forward Kinematics

Since the joints ($_r\phi$) of the rth chain is composed of some of the independent and dependent joints, $_r\dot{\phi}$ can be expressed in terms of the independent joints by

$$_r\dot{\phi} = [^r G_a^\phi]\dot{\phi}_a, \quad (12)$$

where the rows of the matrix $[^r G_a^\phi]$ is formed with rows corresponding to $_r\dot{\phi}$ by collecting rows from $[G_a^p]$ and by augmenting the other rows with a unity in the ith row and j^{th} column and with zeros in all other elements of the ith row if $_r\phi_i = \phi_{a_i}$.

Thus, the forward kinematics for the common object is obtained by embedding the first-order internal KIC into one of the rth pseudo open-chain kinematic expression as follows :

$$\dot{u} = [_r G_\phi^u]_r\dot{\phi} = [G_a^u]\dot{\phi}_a, \quad (13)$$

where the forward Jacobian is determined by

$$[G_a^u] = [_r G_\phi^u][_r G_a^\phi]. \quad (14)$$

According to Eq. (13), the force relation between the independent joints and the operational force vector is described by

$$T_a = [G_a^u]^T T_u = [G_a^A]^T T_A, \quad (15)$$

where T_u denotes the operational load vector.

The force relationship between T_A and T_u is then given by

$$T_A = \{([G_a^A]^T)^+ [G_a^u]^T\} T_u = [G_A^u]^T T_u. \quad (16)$$

where $([G_a^A]^T)^+$ denotes the pseudo-inverse of $[G_a^A]^T$.

3.4 Kinematic Index

Based on the effective force relationship between the operational force vector and the input force vector, the 2-norm ratio of the output load to the input load can be expressed as

$$\frac{\|T_A\|}{\|T_u\|} = \sqrt{\frac{T_u^T [G_A^u][G_A^u]^T T_u}{T_u^T T_u}}, \quad (17)$$

where $\|T_A\|$ and $\|T_u\|$ are defined as

$$\|T_A\|^2 = T_A^T T_A, \quad (18)$$

$$\|T_u\|^2 = T_u^T T_u. \quad (19)$$

Based on the Rayleigh quotient, the output bound with respect to the input load is given as

$$\lambda_{min} \|T_u\| \le \|T_A\| \le \lambda_{max} \|T_u\|, \quad (20)$$

where λ_{min} and λ_{max} are the square root of minimum and maximum singular values of $[G_A^u][G_A^u]^T$, respectively. These singular values are used in determining the bounds of the force transmission ratio. Specifically, λ_{max} is defined as the maximum force transmission ratios for a unit operational load of $\|T_u\|$. Smaller λ_{max} implies that the total input power becomes smaller for the given operational load.

4. Simulation Work

Fig. 2 shows the workspace of the leg mechanism expressed with respect to the roll and pitch angles. A maximum value of the rolling motion is 25 degrees when fixing the pitch and knee angles as zero, a maximum value of the pitching motion is 50 degrees when fixing the roll and knee angles as zero. The motion ranges of the joints are relatively smaller that those of serially connected open-chain. Thus, the proposed leg mechanism may be inadequate to humanoid system, but can be useful in the design of some biomimetic systems resembling two or four legged animals. Takita, et al [5] proposed a similar leg mechanism with small motion range, but it was successfully applied to the design of a Dinosaur-like Robot.

Fig. 3 represents the maximum force transmission ratios for several cases. Fig. 3(a) is the case of a 2-DOF serial mechanism. This structure denotes the middle chain excluding parallel structure of Fig. 1. The others are the cases of proposed mechanism including parallel structure. The result shows that the force transmission ratio of the 2-DOF serial-type structure is larger than the others. It means that a serial structure requires more power than proposed parallel structure. Fig. 3(b) is the case of non-redundant actuation in the parallel structure. Also, Fig. 3(c) and Fig. 3(d) are the cases of redundant actuation. It can be observed from the simulation results that when one redundant actuator is placed on the joint θ_{h2} or θ_{h3}, the force transmission ratios become smaller. To be specific, when one redundant actuator is placed on the joint θ_{h3}, the maximum force transmission ratio is one fourth of that of minimum actuation. In other words, the power of the system can be saved as one fourth of minimum actuation just by employing one additional actuator. Thus, the proposed leg mechanism involving redundant actuators in its parallel kinematic chain can be beneficially employed as a module of Humanoid or legged mechanism to increase the load carrying capacity of the system.

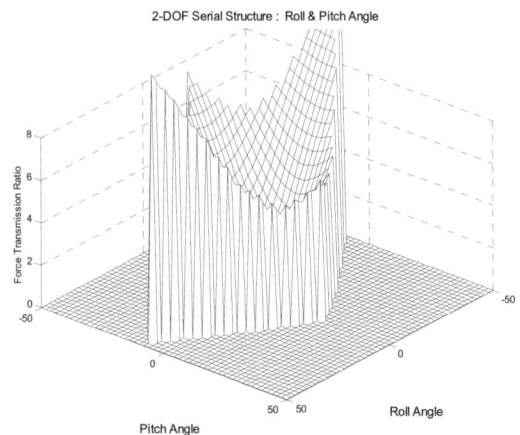

(a) 2-DOF Serial Structure

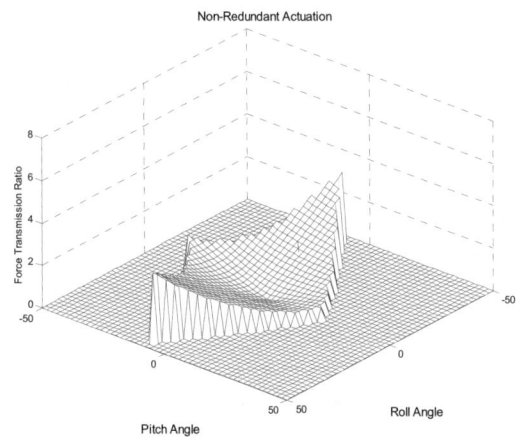

(b) Non-Redundant Actuation

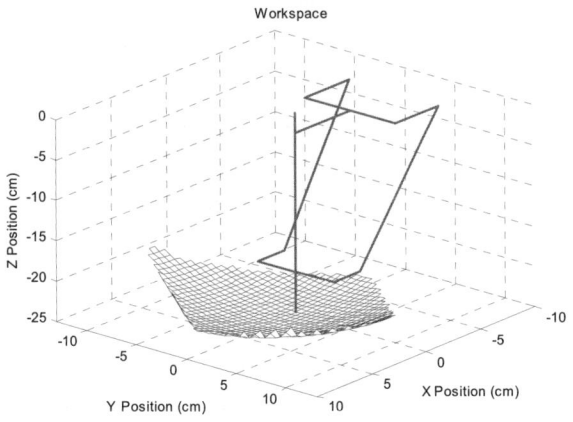

Fig. 2. Workspace of the Hip Module

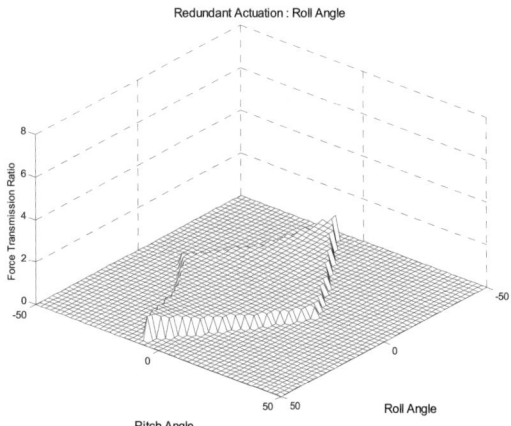

(c) Redundant Actuation at Roll Joint θ_{h2}

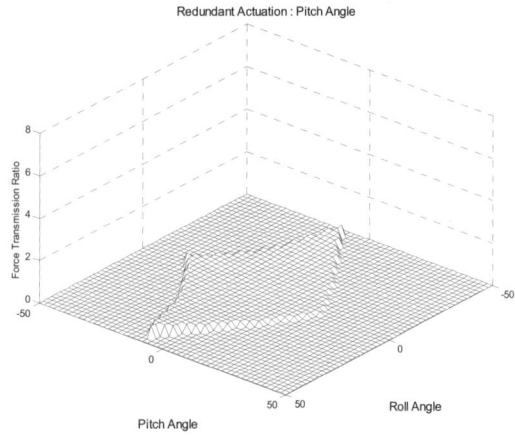

(d) Redundant Actuation at Pitch Joint θ_{h3}

Fig. 3. Maximum Force Transmission Ratios

5. Implementation

Fig. 4 shows a linear actuator consisting of a ball screw and a DC motor. This module has been developed by referencing the series elastic actuator of MIT Leg Lab [7-8].

Fig. 5 shows the prototype of the leg mechanism for humanoid. Two linear actuators and one rotary actuator drive a 2 DOF hip module. Fig. 6 shows the biped robot employing the proposed parallel mechanism. Each leg has six linear actuators. The first and third chains of the hip module have universal and spherical joint at the upper and the lower part of the chain, respectively. And, the second chain has two revolute joints with encoders and redundant actuators integrated to them. For more powerful action, the knee module is driven by dual linear actuators. Also, the ankle module is denoted using parallel structure driven linear actuators.

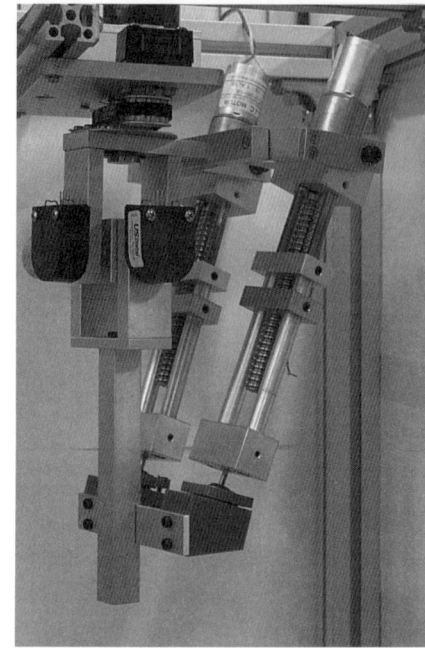

Fig. 5. The Prototype of Hip Module

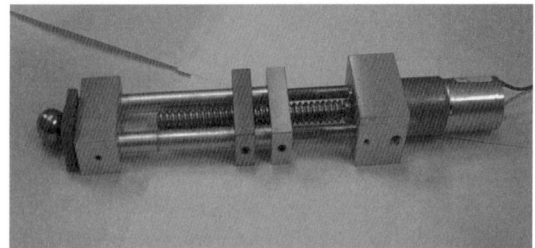

Fig. 4. Linear Actuator Module

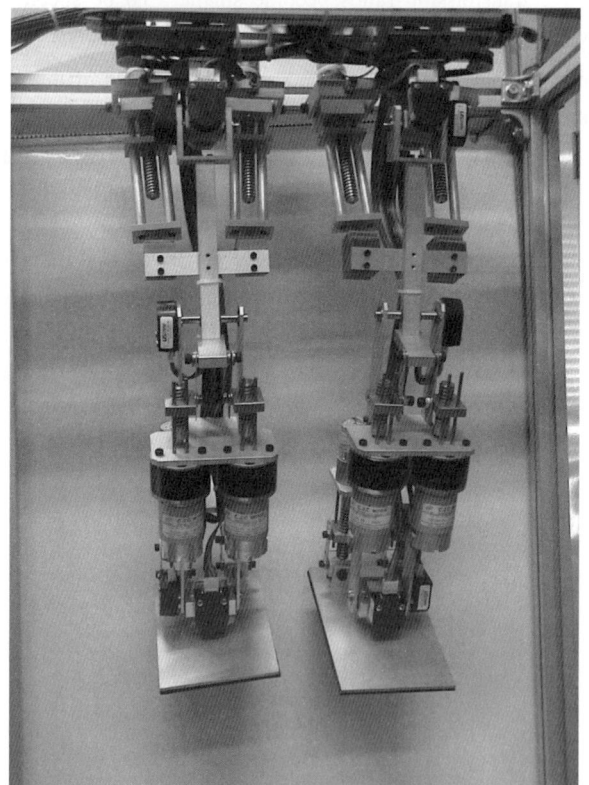

Fig. 6. Biped Robot

6. Conclusions

In this paper, we propose a leg mechanism employing parallel structure. A linear actuator mechanism using a ball-screw is self-developed to drive the linear joints of the hip module. As a special feature, redundant actuators are incorporated into the parallel structure to enhance the force transmission capability. It is shown through simulation that the parallel structure exhibits improved force transmission on characteristic as compared to the serial structure use and that the force transmission performance of the redundantly actuated case in parallel structure is 4 times superior to the original minimum actuation. The developed hip mechanism has been implemented to a small size biped robot. Future work is the integration of the leg module to the full-scale humanoid system or some biomimetic systems resembling two or four legged animals.

References

[1] M. Morisawa, T. Yakoh, T. Murakami and K. Ohnishi, "An approach to biped robot with parallel mechanism", *Proceedings 6th International Workshop on Advanced Motion Control*, pp. 537-541, 2000.

[2] M. Morisawa, Y. Fujimoto, T. Murakami and K. Ohnishi, "A walking pattern generation for biped robot with parallel mechanism by considering contact force", *Proceedings on IEEE 27th Annual Conference, IECON '01, Industrial Electronics Society*, pp 2184-2189, 2001.

[3] M. Morisawa, T. Yakoh, T. Murakami and K. Ohnishi, "A Comparison Study between Parallel and Serial Linked Structure in Biped Robot System"", *Proc. of the IEEE International Conference on Industrial Electronics, Control and Instrumentation*, Nagoya, Japan, pp. 537-541, 2000.

[4] Y. Sugahara, Endo Tatsuro, Lim. Hun-ok and A. Takanishi, "Design of a battery-powered multi-purpose bipedal locomotor with parallel mechanism", *Proc. of the IEEE/RSJ International Conference on Intelligent Robots and System*, Lausanne, Switzerland, pp. 2658–2663, 2002.

[5] K. Takita, R. Hodoshima, and S. Hirose, "Fundamental Mechanism of Dinosaur-like Robot TITRUS-II Utilizing Coupled Drive," *Proc. of the IEEE International Conference on Intelligent Robots and Systems*, Takamatsu, Japan, pp. 1670-1675, 2000.

[6] R. Sellaouti, A. Konno, and F.B. Ouezdou, "Design of a 3DOFs Parallel Actuated Mechanism for a Biped Hip Joint," *Proc. of the IEEE International Conference on Intelligent Robotics and Automation*, Washington, DC, United States, pp. 1161-1166, 2002.

[7] D.W. Robinson, J.E. Pratt, D.J. Paluska and G.A. Pratt, "Series Elastic Actuator Development for a Biomimetic Walking Robot," *Proc. of the IEEE/ASME International Conference on Advanced Intelligent Mechatronics*, Atlanta, GA, United States, pp. 561-568, 1999.

[8] G.A. Pratt, and M.W. Williamson, "Series Elastic Actuators," *Proc. of the IEEE International Conference on Intelligent Robots and Systems*, Vol. 1, pp. 399-406, 1995.

[9] R.A. Freeman and D. Tesar, "Dynamic modeling of serial and parallel mechanisms/robotic systems, Part I-Methodology, Part II-Applications," *Proceedings on 20th ASME Biennial Mechanisms Conference, Trends and Development in Mechanisms, Machines, and Robotics*, Orlando, FL, DE-Vol. 15-3, pp. 7-27, 1988.

[10] So. B.R., Yi. B.-J., Oh. S.-R. and Suh. I.H., "An Independent Joint-Based Compliance Control Method for a Five-bar Finger Mechanism Via Redundant Actuators," *Proc. of the IEEE International Conference on Intelligent Robotics and Automation,* pp. 2140-2146, 1999.

Probabilistic Motion Planning for Parallel Mechanisms

J. Cortés, T. Siméon
LAAS-CNRS
7, avenue du Colonel-Roche
31077 Toulouse - France
{jcortes,nic}@laas.fr

Abstract

Despite the increasing interest on parallel mechanisms during the last years, few researchers have addressed the motion planning problem for such systems. The few existing techniques lie onto a representation of the workspace of the mechanism (or its boundary). However, obtaining this representation is generally too difficult, only partial solutions exist for particular cases. In this paper we propose a general approach based onto probabilistic motion planning techniques. This approach does not need any modeling of the robot's workspace. It combines random sampling techniques with simple but general geometric algorithms that guide the sampling toward feasible configurations satisfying the closure constraints of the parallel mechanism. The efficiency and the generality of the method are demonstrated onto several complex mechanisms made up with serial or parallel associations of Stewart platforms, or created with several redundant robots manipulating an object.

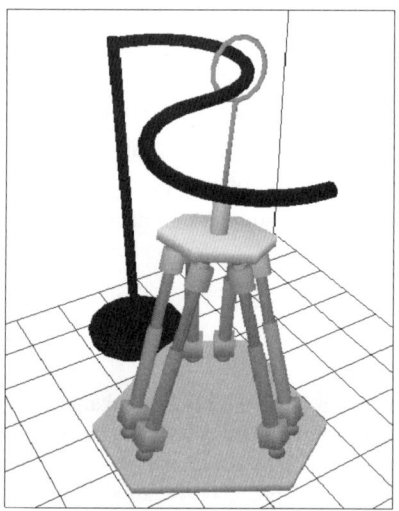

Fig. 1. Example of motion planning problem for a Stewart platform.

I. INTRODUCTION

A parallel manipulator is a mechanism in which the end-effector is connected to the base by at least two independent kinematic chains [19]. The most representative parallel manipulator is the six-degrees-of-freedom (d.o.f.) mechanism known as Stewart platform [26], [8]. This definition can be also applied to more complex multi-loop mechanisms formed by several manipulators handling an object. In this paper we address the problem of planning collision-free motions for such general parallel mechanisms.

The few existing techniques for trajectory validation and motion planning of parallel mechanisms [22] use a representation of the workspace of the mobile platform. The difficulty to compute such representations limits the generality of these approaches, like with deterministic motion planning techniques [16] that first relied on an exact model of the collision-free configuration-space CS_{free} of the mobile system. Probabilistic motion planning techniques [12] do not need to build a model of the space where they are applied. This property was the key of their success during the last decade. However, the closure constraints of parallel mechanisms (i.e. multiple loops) remain an important challenge for probabilistic motion planning methods and the few existing approaches [17], [11], [6] are mostly limited to single-loop mechanisms. We present an extension of our work on motion planning for closed kinematic chains [6] using the PRM framework, to efficiently deal with multi-loop mechanisms such as general parallel robots.

A first application of the approach is to capture the self-collision-free workspace of the parallel mechanism into a small data structure (a random visibility roadmap [25]). Once computed for a given mechanism, this data structure can be used to generate in real-time valid motions avoiding self-collisions between the links of the mechanism. In presence of obstacles, the proposed approach also allows us to solve motion planning problems like the one illustrated in Figure 1 where the path to extract the ring mounted onto a Stewart platform from the "s-shaped" obstacle is computed in only a few seconds.

Section II first gives a brief overview of probabilistic motion planning techniques. In this same section, we discuss about the extension of these techniques to handle closed kinematic chains. Parallel mechanisms are presented in Section III. Our approach for sampling random configurations of such systems is explained in Sections IV and V. Results in Section VI show the generality of the method through different applications for various kinds of systems.

II. PROBABILISTIC MOTION PLANNING

Probabilistic motion planning techniques appeared in the last decade as an alternative to deterministic approaches. In particular, *Probabilistic RoadMap (PRM)* methods [12] have been mostly developed. These techniques have demonstrated to be efficient and general tools for motion computing.

PRM Principle: The general PRM principle is to construct a graph (roadmap) that captures the topology of CS_{free}. The nodes are randomly sampled configurations satisfying intrinsic conditions in this space (e.g. collision-free). The edges are short feasible paths (*local paths*) linking "nearby" nodes.

PRM Variants: Several algorithms [1], [3], [27], [25], [2] have been proposed sharing this basic idea. These methods mostly differ from their sampling strategies. In particular, the *visibility-PRM* approach [25] is used in our solution. The algorithm building the graph only keeps the sampled configurations in two cases: when they link several connected components of the roadmap or when they can not be connected to any of these components. The main advantage is to compute a smaller roadmap which significantly decreases the number of calls to the local planner (the most expensive step of the roadmap construction) compared to other approaches. Figure 2 shows two roadmaps for the same 2D environment. The left one has been computed by a *basic*-PRM algorithm that keeps every valid sampled configuration. The right one, obtained by the visibility approach, encodes the same information in a much smaller structure.

The mentioned PRM techniques are called *multiple-query*. Once the roadmap is computed, motion planning queries are solved by connecting the start and goal configurations to the graph and searching a path in it. Other algorithms dedicated to solve *simple planning queries* have been developed from the same principles than PRM (e.g. *RRT* [15], *SBL-PRM* [23]).

All techniques above require the generation of random configurations of the mechanism. This is a trivial process in the case of open kinematic chains. On the contrary, when the mechanism contains loops, samples must be generated into a variety instead of a configuration-space. The difficulty to compute (and to connect) such configurations remains a challenge for the application of probabilistic motion planners.

PRM and Closure Constraints: Only a few works extending the PRM framework to deal with closed-chain mechanisms can be found in the literature [17], [11], [6]. The approach in [6] demonstrates good performance onto complex 3D closed chains involving tenths of d.o.f.. Each single-loop in the mechanism is broken (as initially proposed in [11]) into two chains (*passive* and *active*). The random node generation combines a sampling technique called *Random Loop Generator (RLG)* with forward kinematics for the active chain and inverse kinematics for the remaining (passive) part of the loop in order to force the closure. When computing the edges, the local planner is limited to act onto the active joints. The passive part of each loop follows the motion of the rest of the chain using point to point inverse kinematics.

The main interest of RLG is that it produces random samples for the active chain that have a high probability to be reachable by the passive part. The algorithm in [6] performs well on independent single-loops and was also applied to some cases of multi-loops. However, this approach requires an extension to efficiently handle more general closed-chain mechanisms. Parallel mechanisms are a more complex instance that presents particular interest.

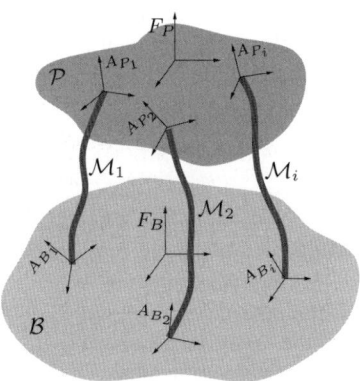

Fig. 3. General description of a parallel mechanism.

III. PARALLEL MECHANISMS

Description: A parallel mechanism is composed of a base $\mathcal{B}$, a platform $\mathcal{P}$ and n kinematic chains $\mathcal{M}_i$ linking them. We call A_{B_i} and A_{P_i} the frames corresponding to the connections of each $\mathcal{M}_i$ to $\mathcal{B}$ and $\mathcal{P}$ respectively. F_B and F_P are the frames associated with $\mathcal{B}$ and $\mathcal{P}$ (see Figure 3).

The spatial situation of $\mathcal{P}$ (usually called *pose*) is defined by a vector $q_\mathcal{P}=\{x,y,z,\theta_x,\theta_y,\theta_z\}$. The three firts elements represent the position of F_P with respect to F_B. The orientation is given by three consecutive rotations

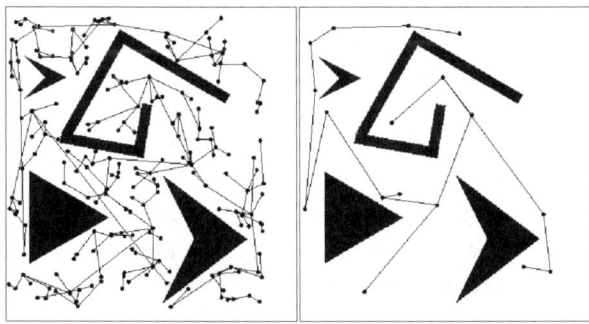

Fig. 2. Basic-PRM and visibility-PRM in the same 2D environment.

around the axes of F_P [1]. The platform is considered to be the end-effector of a parallel mechanism. Hence, an equivalence can be established between poses of $\mathcal{P}$ and points of the workspace of the system.

Workspace: The workspace WS_P of a parallel mechanism is usually computed from the workspaces WS_{M_i} of the chains $\mathcal{M}_i$ and the dimensions related to $\mathcal{P}$. The difficulty is that WS_P can not be decoupled into two three-dimensional (graphically representable) sub-spaces because of the dependence between position and orientation of the end-effector. Therefore, only sub-sets of the workspace may be represented. Most of the existing works are limited to the determination of some particular sections of the positional workspace with constant orientation of the platform [10], [18]. Other techniques compute the feasible rotations of the platform around a fixed point [21].

Configuration: The configuration of a parallel mechanism is defined by the joint values of the chains $\mathcal{M}_i$ (and the pose of $\mathcal{P}$). Configurations satisfying the closure constraints could be easily computed from a model of its workspace. Hence, a feasible pose $q_\mathcal{P}$ could be directly obtained from this representation. The spatial situation of the connection-frames A_{P_i} w.r.t. the A_{B_i} would be given by $q_\mathcal{P}$, and the configuration of the chains $\mathcal{M}_i$ linking these frames could be then computed by inverse kinematics techniques.

However, modeling the workspace of a general parallel mechanism remains an open problem [20]. In next section, we describe an algorithm that generates random configurations of a general parallel mechanism without requiring the explicit computation of WS_P.

IV. RANDOM CONFIGURATION SAMPLING FOR PARALLEL MECHANISMS

We propose a general approach that combines random sampling techniques with simple geometric operations for generating random configurations of parallel mechanisms. *Spherical shells* approximating the WS_{M_i} are used to progressively compute the pose $q_\mathcal{P}$ of $\mathcal{P}$. The algorithm first generates the position parameters of $q_\mathcal{P}$ and then it computes the rotation parameters. Such obtained pose of $\mathcal{P}$ correspond to random samples in a conservative approximation of WS_P. Then, the existence of a feasible configuration is checked for each chain $\mathcal{M}_i$ linking A_{B_i} and A_{P_i}. The configuration of the parallel mechanism is kept when all the $\mathcal{M}_i$ connect the base and the platform, else the process is iterated. Next paragraphs detail the main features of the approach.

WS_{M_i} **Approximation:** In a similar way than in [6], spherical shells bounding the reachable workspace (only in position) of the chains $\mathcal{M}_i$ are used. A trade-off between accuracy and computing time justifies this choice. A spherical shell is defined by the intersection

[1]The approach is valid for other representations of the orientation (e.g. Euler angles).

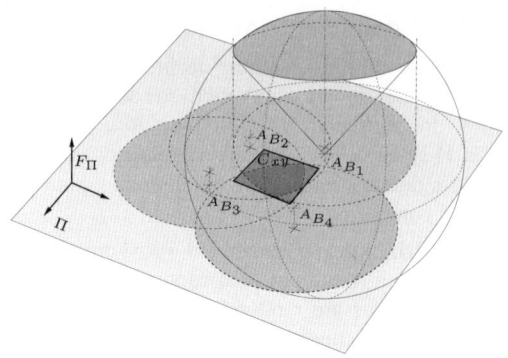

Fig. 4. Illustration of the computation of C_{xy}.

of the volume between two concentric spheres and a cone whose vertex coincides with their common center. Spherical shells are used in collision detection as boundary volumes of objects because of the fast computation of their intersections [14]. Parameters characterizing the spherical shell are derived from the features of the chains $\mathcal{M}_i$. The center is the origin of the frame A_{B_i}. The external and internal radii correspond respectively to the maximum and minimum extension of the chain. The axis of the cone cutting the full shell is a vector associated with A_{B_i} (normally its z-axis). The half-opening angle is the maximum angle between this axis and the vector passing through the origins of A_{B_i} and A_{P_i}.

Platform Position: Given the fixed spatial situation of the A_{B_i} in F_B, a plane Π can be computed by interpolating the position of the frame origins (when there are more than two). We next explain the method to generate the position of the platform w.r.t. the frame F_Π associated with this plane.

First, a rectangle C_{xy} approximating the orthogonal projection of WS_P on Π is computed as follows. The spherical shell approximating each WS_{M_i} is augmented by the distance from A_{P_i} to F_P. The projection of the external portion of sphere on Π corresponds in general to an ellipse. C_{xy} is the rectangle bounding the intersection of these ellipses. Figure 4 illustrates this process in the case of four chains $\mathcal{M}_i$. For clarity purpose, we have only represented the external surface of the augmented shell of the chain $\mathcal{M}_1$.

The generation of a pose of $\mathcal{P}$ begins by randomly sampling a point p_{xy} in C_{xy}. Then, the intersection of the line perpendicular to Π passing through p_{xy} with each one of the augmented spherical shells is computed. (when one or several volumes are not intersected, a new point p_{xy} must be sampled). The result of this operation are one or several intervals in z (relative to Π) for each $\mathcal{M}_i$. The intersection of such intervals represents a conservative approximation of the set of reachable positions of the platform for a given p_{xy}. The z coordinate of the origin of F_P is generated by randomly sampling in this set.

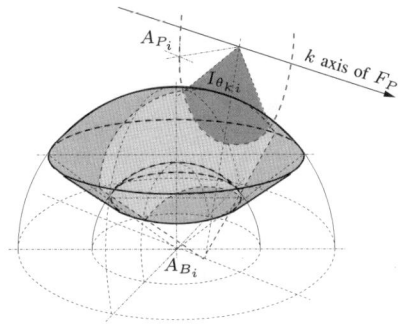

Fig. 5. Illustration of the computation of $I_{\theta_{k_i}}$.

Platform Orientation: For a given position of $\mathcal{P}$, its orientation is generated by progressively computing the three elementary rotations. We next explain the process for one rotation.

The rotation of $\mathcal{P}$ around an axis k produces a circular motion of each A_{P_i}. The intersection of the spherical shell approximating each WS_{M_i} with the circle generated by its corresponding A_{P_i} is computed. The result is the set $I_{\theta_{k_i}}$ of values of the rotation parameter making A_{P_i} reachable (in position) by $\mathcal{M}_i$ (considering our approximation). Figure 5 illustrates this operation. When the circle intersects all the shells, the value of the rotation θ_k is obtained by randomly sampling in the interval(s) resulting of the intersection of the $I_{\theta_{k_i}}$ sets. The process is iterated when any of the computed intersections is null.

$\mathcal{M}_i$ Configuration & Validation: The conservativeness of the approach is essential in order to guarantee that $WS_\mathcal{P}$ will be completely sampled. However, the validity of each configuration must be tested. A configuration is valid when the platform's pose induces feasible joint values of all the chains $\mathcal{M}_i$.

For parallel manipulators (Stewart platform type) obtaining the configuration of the $\mathcal{M}_i$ for a given pose of $\mathcal{P}$ is straightforward due to the simple nature of these chains. In the general context this process is more complex. Each triplet $\{A_{B_i}, A_{P_i}, \mathcal{M}_i\}$ is treated as a closed kinematic chain. When $\mathcal{M}_i$ is a non-redundant system, its configuration can be directly obtained by inverse kinematics. In case of redundancy, we use the RLG algorithm presented in [6] to generate it.

Note that this approach also allows to handle particular cases such as mechanisms where the position of the platform is fixed w.r.t. the base and/or the rotational mobility is limited (i.e. rotating only around one or two axes).

V. Complex Mechanisms

The presented approach has been extended to handle more complex systems obtained by the associations of parallel mechanisms. It has been also adapted to a particular case of highly-redundant chains $\mathcal{M}_i$.

Associations: When n parallel mechanisms are connected in *series* (see left image in Figure 8), each platform $\mathcal{P}_i$, $i=1..n-1$, becomes the base for the next platform. The process of generation and validation of the configuration is progressively achieved for each platform starting from the base ($\mathcal{P}_1$) to the top ($\mathcal{P}_n$). When the sampled pose of a given $\mathcal{P}_i$ is not valid, the process does not re-start completely. It is only iterated from $\mathcal{P}_{i-1}$. When several parallel mechanisms are disposed in *parallel* (see right image in Figure 8) they form a "main" parallel system. Each mechanism can be considered as a chain $\mathcal{M}_i$ of the main system. Therefore, their platforms become passive elements of the whole mechanism.

Mobile $\mathcal{M}_i$ bases: Such case occurs for example when the $\mathcal{M}_i$ chains correspond to mobile manipulators (i.e. articulated system composed of an arm mounted on a mobile base [13], see Figure 10). Spherical shapes are suitable to bound the reachable workspace of the arm, but not for the whole mobile manipulator. We have extended the approach to the case where the A_{B_i} can freely move on parallel planes. While computing $q_\mathcal{P}$, these frames are considered to be placed at the position that maximizes the variation of each parameter. Then, feasible random configurations of the chains $\mathcal{M}_i$ are computed by RLG.

VI. Results

The approach has been implemented into the motion planning software *Move3D* [24]. In this section we comment some of the obtained results for very different parallel mechanisms. Numerical results correspond to tests performed with a Sun Blade 100 workstation.

Self-collision-free motions: The first experiment aims at demonstrating the performance of the approach to compute self-collision-free motions of the Stewart platform. The roadmap computed for this mechanism can be used to generate such motions in real-time. The left image in Figure 6 shows an example of the self-collision configurations to be avoided. The graph illustrated in the other image of this figure was computed by the visibility-PRM [25] approach in 22 seconds. It only contains one connected component made up with 11 configurations. This small graph covers more than the 99.99% of the robot workspace. The roadmap construction required the generation of 17442 configurations of the mechanism, of which 2328 were found to be collision-free. Using our sampling strategy, 30840 platform's poses were tested for the generation of valid configurations (more than 50% of success). With similar tests performed using standard random sampling techniques to generate the platform's pose [2], less than a 2% of the samples produced valid configurations of the mechanism. This illustrates the important gain (about 25 times faster) using the proposed sampling approach.

[2]Samples are taken in a six-dimensional-box bounding the space of feasible poses of $\mathcal{P}$.

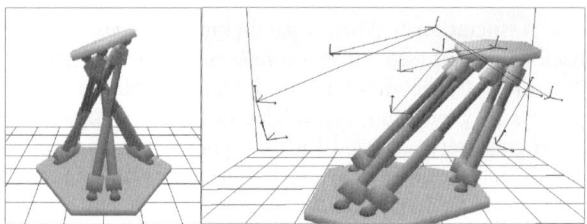

Fig. 6. Illustration of the self-collision-free visibility-PRM computed for a Stewart platform.

Motion planning for Stewart platforms: The second example illustrated by Figure 7 shows a constrained motion planning problem for the Stewart platform. The figure shows the start and goal configurations and the trace of the solution path. Note that the motion requires extreme deformations of the mechanism. A graph containing this solution was computed in 60 seconds. Once computed, it allows to process motion planning queries in some hundredths of second. The two images in Figure 8 correspond to motion planning problems involving associations of parallel mechanisms. The manipulator of the left image is a model of the Logabex-LX4 [5]. The arm is composed of four Stewart platforms connected in series. Motion planning queries solving problems where the manipulator with the grasped bar changes from one to another opening of the bridge were computed by RRT algorithm [15] in a few seconds. The right image illustrates an example where two sets of three Stewart platforms cooperate in a assembly task. This type of association as been proposed in [4] for the manipulation of large objects. The motion to assemble the two puzzle-like parts was computed in only 15 seconds.

Parallel systems including manipulator arms: The two last examples show the generality of the approach. In both cases, the mechanism consists of several robotic arms grasping an object. The problem illustrated in Figure 9, where four *6R* manipulators have to unhook an object and to insert it into the cylindrical axis, was solved using RRT in less than 1 second. The last example (see Figure 10)

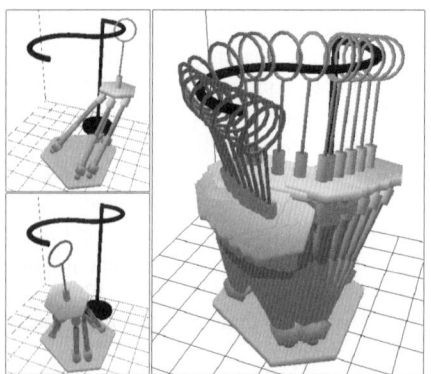

Fig. 7. Sequence of the solution motion for the s-bar problem.

Fig. 8. Examples of associations of Stewart platforms.

combines two types of difficulty. First, the system composed by the three holonomic mobile manipulators and the piano is a very complex parallel mechanism (9 d.o.f. for each $\mathcal{M}_i$ chain). Also, the complexity of the scene makes the validation of collision-free configurations and local paths harder. A graph that permits to rapidly compute any feasible motion in this scene was computed using the visibility-PRM approach in about 5 minutes. In this example, the redundancy of the manipulators ($\mathcal{M}_i$ chains) is treated by the RLG algorithm as explained in Section V.

VII. CONCLUSIONS

The proposed approach allows to extend the PRM framework to efficiently handle complex mechanisms with multiple loops. Our aim is to reach the highest level of generality. The approach can deal with the most general definition of parallel mechanisms and its efficacy was demonstrated onto complex examples (e.g. serial/parallel associations of Stewart platforms, parallel system with redundant chains). A possible improvement of this approach could be to integrate constraints for avoiding singular configurations along the trajectory [7], [20].

We are currently investigating the application of our closed-chain PRM approach to highly articulated mechanisms encountered in molecular models. Hence, tools for analyzing the motion of loops in protein structures [9] should help biologists to better understand the important processes such as protein-ligand or protein-protein interactions and protein folding.

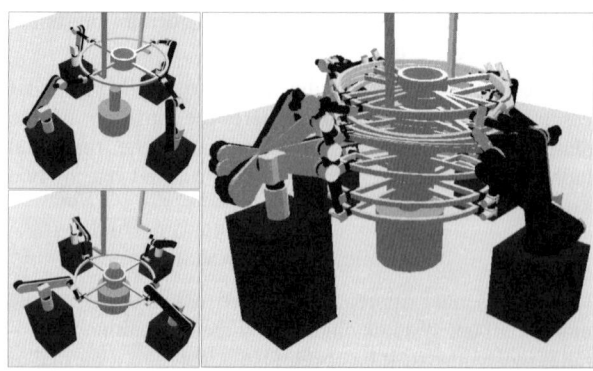

Fig. 9. Four robotic arms manipulate an object.

Fig. 10. A piano mover problem with three cooperating mobile arms.

Acknowledgments: This paper has greatly benefited from discussions with Jean-Pierre Merlet and Lluís Ros.

VIII. REFERENCES

[1] N. Amato, O.B. Bayazit, L. Dale, C. Jones and D. Vallejo. OBPRM: An Obstacle-Based PRM for 3D Workspaces. In *Workshop on Algorithmic Foundations of Robotics*, 1998.

[2] R. Bohlin and L. Kavraki. Path Planning using Lazy PRM. In *IEEE Int. Conf. on Robotics and Automation*, 2000.

[3] V. Boor, M. Overmars and A. van der Stappen. The Gaussian Sampling Strategy for Probabilistic Roadmap Planners. In *IEEE Int. Conf. on Robotics and Automation*, 1999.

[4] K.S. Chai and K. Young. Designing a Stewart Platform-based Cooperative System for Large Component Assembly. In *IEEE Int. Conf. on Methods and Models in Automation and Robotics*, 2001.

[5] S. Charentus. *Modélisation et commande d'un robot manipulateur redondant composé de plusieurs plateformes de Stewart*. PhD thesis. Rapport LAAS n.90146, 1990.

[6] J. Cortés, T. Siméon and J.P. Laumond. A Random Loop Generator for Planning the Motions of Closed Kinematic Chains using PRM Methods. In *IEEE Int. Conf. on Robotics and Automation*, 2002.

[7] B. Dasgupta and T.S. Mruthyunjaya. Singularity-Free Path Planning for the Stewart Platform Manipulator. In *Mechanism and Machine Theory*, vol. 33(6), pp. 711-725, 1998.

[8] B. Dasgupta and T.S. Mruthyunjaya. The Stewart Platform Manipulator: A Review. In *Mechanism and Machine Theory*, vol. 35(1), pp. 15-40, 2000.

[9] A. Fiser, R.K. Do, and A. Šali. Modeling of Loops in Protein Structures. In *Protein Science* 9, 2000.

[10] C. Gosselin. Stiffness Mapping for Parallel Manipulators. In *IEEE Transaction on Robotics and Automation*, vol. 6(3), pp. 377-382, 1990.

[11] L. Han and N. Amato. A Kinematics-Based Probabilistic Roadmap Method for Closed Kinematic Chains. In *Workshop on the Algorithmic Foundations of Robotics*, 2000.

[12] L. Kavraki, P. Švestka, J.C. Latombe and M. Overmars. Probabilistic Roadmaps for Path Planning in High-Dimensional Configuration Spaces. In *IEEE Transaction on Robotics and Automation*, vol. 12(4), pp. 566-580, 1996.

[13] O. Khatib, K. Yokoi, K. Chang, D. Ruspini, R. Holmberg and A. Casal. Coordination and Decentralized Cooperation of Multiple Mobile Manipulators. In *Journal of Robotic Systems*, vol 13, pp. 755-64, 1996.

[14] S. Krishnan, A. Pattekar, M. Lin and D. Manocha. Spherical Shell: A Higher Order Bounding Volume for Fast Proximity Queries. In *Workshop on the Algorithmic Foundations of Robotics*, 1998.

[15] J. Kuffner and S. Lavalle. RRT-Connect: An Efficient Approach to Single-Query Path Planning. In *IEEE Int. Conf. on Robotics and Automation*, 2000.

[16] J.C. Latombe. *Robot Motion Planning*. Kluwer Academic Publishers, 1991.

[17] S. LaValle, J.H. Yakey and L. Kavraki. A Probabilistic Roadmap Approach for Systems with Closed Kinematic Chains. In *IEEE Int. Conf. on Robotics and Automation*, 1999.

[18] O. Masory and J. Wang. Workspace Evaluation of Stewart Platforms. In *Advanced Robotics Journal*, vol. 9(4), pp. 443-461, 1995.

[19] J.P. Merlet. *Parallel Robots*. Kluwer, Dordrecht, 2000.

[20] J.P. Merlet. Parallel Robots: Open Problem. In *9th Int. Symp. of Robotics Research*, 1999.

[21] J.P. Merlet. Determination of the Orientation Workspace of Parallel Manipulators. In *Journal of Intelligent and Robotic Systems*, vol. 13, pp. 143-160, 1995.

[22] J.P. Merlet. *Manipulateurs parallèles, 7eme partie: Vérification et planification de trajectoire dans l'espace de travail*. Rapport de Recherche 1940, INRIA, 1993.

[23] G. Sánchez and J.C. Latombe. A Single-Query Bi-Directional Probabilistic Roadmap Planner with Lazy Collision Checking. In *Int. Symp. on Robotics Research*, 2001.

[24] T. Siméon, J.P. Laumond and F. Lamiraux. Move3D: A Generic Platform for Path Planning. In *4th Int. Symp. on Assembly and Task Planning*, 2001.

[25] T. Siméon, J.P. Laumond and C. Nissoux. Visibility-Based Probabilistic Roadmaps for Motion Planning. In *Advanced Robotics Journal*, vol 14(6), 2000.

[26] D. Stewart. A Platform with Six Degrees of Freedom. In *Proc. of the Institution of Mechanical Engineers*, 180 (Part I, 15), pp. 371-386, 1965.

[27] S. Wilmarth, N.Amato and P.Stiller. MAPRM: A Probabilistic Roadmap Planner with Sampling on the Medial Axis of the Free Space. In *IEEE Int. Conf. on Robotics and Automation*, 1999.

ROBUST TASK-SPACE CONTROL OF HYDRAULIC ROBOTS

O. Becker[1], I. Pietsch[2] and J. Hesselbach[3]

Institute for Machine Tools and Production Engineering
Technical University of Braunschweig, Germany
E-mail: [1]o.becker@tu-bs.de [2]i.pietsch@tu-bs.de [3]j.hesselbach@tu-bs.de

Abstract

This paper presents a model-based robust controller for hydraulically driven robot manipulators. The approach guarantees that the output error of the plant remains within a prescibed bound despite of model uncertainties. Manipulator dynamics and actuator dynamics including the dynamics of the valves are taken into account. The stability proof is based on the Lyapunov method and passivity arguments. It is shown that the stability proof is not restricted to the proposed control law of the hydraulic actuators but also holds for arbitrary hydraulic force control laws as long as certain requirements are fulfilled.

1 INTRODUCTION

Although electrically driven robots are widely used in an increasing number of applications, there are still many industrial tasks where hydraulic actuators can be used advantageously. Machinery in construction industry and forming industry as well as large flight-simulators take advantage of the high power-to-weight ratio, the stiffness and the short response time of hydraulic drives. On the other hand, hydraulic actuators suffer from their strong nonlinear behavior combined with large model uncertainties, which makes modelling and design of appropriate controllers a challenging task.

Many industrial applications are equipped with simple proportional or PD controllers, using the fact that hydraulic actuators resemble rather velocity sources than force sources [19]. Some of these controllers are augmented by a disturbance rejection strategy. Since the eigenfrequency of a hydraulic actuator changes with the actuator stroke and the nonlinearities are not negligible, the standard fix-gain-controllers often result in conservative system performance [18]. The corresponding dynamic behavior is even worse if the actuators are strongly coupled, e.g. in the case of parallel robots.

Model-based control algorithms have successfully been applied to electrically driven robots, resulting in a strong improvement of position and tracking performance compared to classical approaches [16]. However, one crucial assumption in model-based control schemes is the instantaneous and precise generation of the desired force or torque according to the control signal. This requirement is not fulfilled for hydraulic actuators, since their nonlinear behavior has to be taken into account. This problem can be alleviated, if an additional inner force control-loop for the hydraulic actuators is used. Thus, the overall performance depends strongly on the performance of this inner loop.

In the last years the main focus in the field of force control of hydraulic actuators was the control of single hydraulic rods. Heintze [10] compares a cascaded pressure controller using variable gains with a controller based on the inverse dynamics of the hydraulics. Alleyne [4] concludes that a PID force controller is inherently limited in its force tracking ability. In [2] he presents an adaptive, Lyapunov-based control algorithm for force tracking control of a symmetric hydraulic actuator. Sohl [21] introduces a Lyapunov-based controller for a differential hydraulic actuator without considering the valve dynamics. Nguyen [13] presents a sliding mode force control algorithm, neglecting valve dynamics as well. Sirouspour [18] uses the backstepping design for an adaptive Lyapunov-based controller. The valve dynamics in this approach are modeled as a linear second-order system. Niksefat [14], [15] proposes an explicit force controller which is based on the quantitative feedback theory (QFT).

In contrast to this well explored field of pure hydraulic actuator control, there are only a few papers studying the control of hydraulically driven manipulators. Early investigations are based on a linearized model [11], [17], therefore lacking global stability proofs. Stability proofs are also missing in decentralized control laws that are presented by Edge [8] and Heinrichs [9]. Both, Sirouspour [19] and Bu [5] present a model-based adaptive position tracking controller for hydraulically driven manipulators without taking the valve dynamics into account.

In this paper a novel model-based sliding-mode controller for position tracking of hydraulically driven manipulators is developed. The valve dynamics are taken into account, which is crucial for high bandwidth tracking [2]. In opposite to adaptive control algorithms, the presented robust control law does not need persistent excitation. Even if a continuous approximation of the switching control law is used, the stability of the approach and the boundedness of the errors can be proven. Furthermore, it will be shown that the whole system remains stable as long as the errors of the force control loops are bounded. Therefore, the proposed robust hydraulic force control law can be replaced by any algorithm that guarantees boundedness of the force error. The effectiveness of the approach is shown by simulations.

2 ROBOT MODEL INCLUDING ACTUATOR DYNAMICS

The dynamic equations of motion for a robot manipulator with n d.o.f. including joint stick-slip friction can be written as [16]

$$M(x)\ddot{x} + C(x,\dot{x})\dot{x} + g(x) + f_f(\dot{x}) = f, \quad (1)$$

where $f \in \mathbb{R}^n$ is the vector of the actuator forces, $x \in \mathbb{R}^n$ is the vector describing end-effector motion, $M = M^T \in \mathbb{R}^{n \times n}$ is the positive definite inertia matrix, $C \in \mathbb{R}^{n \times n}$ is the matrix of Coriolis and centrifugal terms, $g \in \mathbb{R}^n$ is the vector of gravity forces and $f_f \in \mathbb{R}^n$ describes the actuator friction. Generally all system matrices and vectors are only known with a certain accuracy. Therefore, we define the errors between the estimated terms and the real terms as

$$(\tilde{\bullet}) = (\hat{\bullet}) - (\bullet). \quad (2)$$

The errors are assumed to be upper bounded according to

$$|\tilde{m}_{ij}| < b_{\tilde{m}_{ij}}, |\tilde{c}_{ij}| < b_{\tilde{c}_{ij}} |\dot{x}_j|,$$
$$|\tilde{g}_i| < b_{\tilde{g}i}, \left|\tilde{f}_{fi}\right| < b_{\tilde{f}_{fi}} \quad (3)$$

where $\tilde{m}_{ij}$ and $\tilde{c}_{ij}$ are the elements of the matrices $\tilde{M}$ and $\tilde{C}$, respectively.
The dynamic behavior of the hydraulic actuators can be described as

$$\dot{\tau} = a_h + c_h z_h x_v \quad (4)$$

where a_h, c_h and z_h are chosen as in [13] and x_v is the spool displacement of the valve. The parameters of (4) are also assumed not to be known exactly but obtain upper bounded errors:

$$c_{h,\min} < c_{hi} < c_{h,\max}$$
$$\beta = \sqrt{\frac{c_{h,\max}}{c_{h,\min}}} > 1$$
$$|a_{hi} - \hat{a}_{hi}| \leq \alpha \quad (5)$$

3 ROBUST POSITION CONTROL

In the following section a new approach for robust position control based on sliding-mode control is derived which is shown to stabilize both, the robot dynamics and the dynamics of the hydraulic actuators. The design is broken down into two steps: First, we use a sliding-mode approach that guarantees global asymptotic stability of the robot. In a second step the Lyapunov function of the first step is enhanced in order to take the actuator dynamics into account. Conditions for a second switching term are derived resulting in a globally stable overall system.

3.1 SLIDING-MODE-CONTROL

Step 1: Choose

$$V_1 = \frac{1}{2}\left[s^T M s\right] + \frac{1}{2}\left[\tilde{x}^T K_P \tilde{x}\right] \quad (6)$$

as a Lyapunov function candidate [20], where $\tilde{x} = x - x_d$ is the error of the endeffector pose.

$$s = \dot{\tilde{x}} + \Lambda \tilde{x} = \dot{x} - \dot{x}_r \quad (7)$$

denotes a deviation vector,

$$\dot{x}_r = \dot{x}_d - \Lambda \tilde{x} \quad (8)$$

can be considered as reference endeffector velocity. With (1) and (7), the time derivative of (6) along the system trajectories yields

$$\dot{V}_1 = s^T\left[f - M\ddot{x}_r - C\dot{x}_r - g - f_f\right] + \tilde{x}^T K_P \dot{\tilde{x}}. \quad (9)$$

A virtual control law for the robot is chosen as

$$f_d = \hat{f} - k_c \mathrm{sgn}(s), \quad (10)$$

with

$$\hat{f} = \hat{M}(x)\ddot{x}_r + \hat{C}(x,\dot{x})\dot{x}_r + \hat{g}(x) \quad (11)$$
$$+ \hat{f}_f - K_P \tilde{x} - K_D s,$$

and

$$k_c\, \mathrm{sgn}(s) = \begin{pmatrix} \vdots \\ k_{ci}\mathrm{sgn}(s_i) \\ \vdots \end{pmatrix}, \quad i = 1\ldots n. \quad (12)$$

Λ, K_P, K_D are positive definite diagonal matrices. After defining the force error $\tilde{f} = f - f_d$, where $\tilde{f}$ results from the dynamics of the nonideal actuators, and using (10-12), $\dot{V}_1$ becomes

$$\dot{V}_1 = -s^T K_D s - \tilde{x}^T \Lambda K_P \tilde{x} + s^T \tilde{f} + \quad (13)$$
$$s^T\left[\tilde{M}\ddot{x}_r + \tilde{C}\dot{x}_r + \tilde{g} + \tilde{f}_f\right] - \sum_{i=1}^n k_{ci}|s_i|.$$

If each switching parameter k_{ci} is chosen to verify[1]

$$k_{ci} \geq \sum_j b_{\tilde{m}_{ij}} |\ddot{x}_{rj}| + \sum_j b_{\tilde{c}_{ij}} |\dot{x}_j||\dot{x}_{rj}|$$
$$+ b_{\tilde{g}i} + b_{\tilde{f}_{fi}} + \eta_{ci}, \quad (14)$$

with $\eta_{ci} > 0$, it follows that

$$k_{ci} \geq \left|\left[\tilde{M}(x)\ddot{x}_r + \tilde{C}_q(x,\dot{x})\dot{x}_r + \tilde{g}(x) + \tilde{f}_f\right]_i\right| + \eta_{ci} \quad (15)$$

(where $[\bullet]_i$ denotes the i-th row of vector $(\bullet)$) and therefore

$$\dot{V}_1 \leq -s^T K_D s - \tilde{x}^T \Lambda K_P \tilde{x} + s^T \tilde{f} - \sum_{i=1}^n \eta_{ci} |s_i|. \quad (16)$$

Note that $\dot{V}_1 < 0$ for $\tilde{f} = 0$.

Step 2: For writing convenience, define

$$W = -s^T K_D s - \tilde{x}^T \Lambda K_P \tilde{x} \quad (17)$$
$$+ s^T \left[\tilde{M}(x)\ddot{x}_r + \tilde{C}_q(x,\dot{x})\dot{x}_r\right.$$
$$\left. + \tilde{g}(x) + \tilde{f}_f\right] - \sum_{i=1}^n k_{ci}|s_i| < 0,$$

where k_{ci} is chosen according to (14). Use the Lyapunov function candidate

$$V_2 = V_1 + \frac{1}{2}\tilde{\tau}^T \Gamma \tilde{\tau} \quad (18)$$

which now includes the actuator dynamics. $\tilde{\tau} = \tau - \tau_d$ denotes the actuator force error. The time derivative of V_2 is

$$\dot{V}_2 = W + \tilde{\tau}^T \left(J^{-1}s + \Gamma \dot{\tilde{\tau}}\right), \quad (19)$$

where $\tilde{f} = J^{-T}\tilde{\tau}$ has been used. Regarding (17) and (19),

$$\tilde{\tau}^T \left(J^{-1}s + \Gamma \dot{\tilde{\tau}}\right) \stackrel{!}{<} 0 \quad (20)$$

has to hold in order to make V_2 a Lyapunov function. Let $s_{qi} = \sum_{j=1}^n (J^{-1})_{ij} s_j$. Thus, (20) holds, if

$$\tilde{\tau}_i \left(s_{qi} + \Gamma_{ii}\dot{\tilde{\tau}}_i\right) \leq -\eta_{x_{vi}} |\tilde{\tau}_i| < 0, \quad (21)$$

with $\eta_{x_{vi}} > 0$. Condition (21) can be met choosing each component of the control law as

$$x_{vi} = \hat{x}_{vi} - \hat{c}_h^{-1} z_h^{-1} k_{x_{vi}} sgn(\tilde{\tau}_i), \quad (22)$$

---
[1] Note that in contrast to sliding-mode approaches in the joint-space (e.g. [6]), k_{ci} cannot be estimated using the maximum singular values of $\tilde{M}, \tilde{C}$, since the first three components of $\tilde{M}(x)\ddot{x}_r + \tilde{C}_q(x,\dot{x})\dot{x}_r$ represent a force whereas the last three components represent a torque. Using the singular values of $\tilde{M}, \tilde{C}$ and the norm of $x_{(r)}$ would result in an inconsistency of units [7]. Instead, (14) can be used to determine a lower bound for k_{ci}.

with

$$\hat{x}_{vi} = \hat{c}_h^{-1} z_h^{-1} \left(\dot{\tau}_{di} - \hat{a}_h - \Gamma_{ii}^{-1} s_{qi}\right). \quad (23)$$

$k_{x_{vi}}$ is calculated by inserting (4) and (22), (23) in condition (21):

$$-\eta_{x_{vi}} |\tilde{\tau}_i| \geq \tilde{\tau}_i \Gamma_{ii} \left(\Gamma_{ii}^{-1} s_{qi} + a_h + c_h z_h \hat{x}_{vi} \right.$$
$$\left. - c_h \hat{c}_h^{-1} k_{x_{vi}} sgn(\tilde{\tau}_i) - \dot{\tau}_{di}\right). \quad (24)$$

It follows [13]:

$$-\eta_{x_{vi}} \geq \Gamma_{ii} \left(\Gamma_{ii}^{-1} s_{qi} + a_h + c_h z_h \hat{x}_{vi} \right. \quad (25)$$
$$\left. - c_h \hat{c}_h^{-1} k_{x_{vi}} sgn(\tilde{\tau}_i) - \dot{\tau}_{di}\right) sgn(\tilde{\tau}_i)$$

$$c_h \hat{c}_h^{-1} k_{x_{vi}} \geq \Gamma_{ii}^{-1} \eta_{x_{vi}} + (a_h - \hat{a}_h \quad (26)$$
$$+ \underbrace{\hat{a}_h - \dot{\tau}_{di} + \Gamma_{ii}^{-1} s_{qi}}_{-\hat{c}_h z_h \hat{x}_{vi}} + c_h z_h \hat{x}_{vi}) sgn(\tilde{\tau}_i)$$

$$k_{x_{vi}} \geq c_h^{-1} \hat{c}_h \Gamma_{ii}^{-1} \eta_{x_{vi}} + c_h^{-1} \hat{c}_h |a_h - \hat{a}_h|$$
$$+ \hat{c}_h z_h \left|c_h^{-1} \hat{c}_h - 1\right| |\hat{x}_{vi}|. \quad (27)$$

By including the error bounds (5) we arrive at the switching term k_{vi}:

$$k_{x_{vi}} \geq \beta \left(\Gamma_{ii}^{-1} \eta_{x_{vi}} + \alpha\right) + \hat{c}_h z_h (\beta - 1) |\hat{x}_{vi}|. \quad (28)$$

The choice of the switching terms according to (14) and (28) results in global asymptotic stability of the robot with hydraulic actuators resulting in vanishing force error $\tilde{\tau} \to 0$ and endeffector position and velocity error since $s \to 0$.

3.2 CONTINUOUS APPROXIMATION

High-frequency switching of the sliding-mode controller provokes undesirable chattering when the switching surfaces are reached. This effect can be avoided by smoothing control laws (10) and (22): The $sgn(\bullet)$ functions are replaced with $sat(\bullet)$ functions:

$$f_d = \hat{f} - k_c sat\left(\frac{s}{\phi}\right), \quad \phi > 0 \quad (29)$$

$$x_{vi} = \hat{c}_h^{-1} z_h^{-1} \left(\dot{\tau}_{di} - \hat{a}_h - k_{x_{vi}} sat\left(\frac{\tilde{\tau}_i}{\phi_\tau}\right)\right), (30)$$

$$\phi_\tau > 0, \quad (31)$$

with

$$sat(x) = \begin{cases} x & \forall |x| \leq 1 \\ sgn(x) & \forall |x| > 1 \end{cases}. \quad (32)$$

The stability proof is inspired by [1] and [3]: First we show that $\tilde{\tau}$ decreases to a boundery layer

$$|\tilde{\tau}_i| \leq \phi_\tau, \quad (33)$$

by choosing

$$V_3 = \frac{1}{2}\tilde{\tau}^T \Gamma \tilde{\tau} \quad (34)$$

as Lyapunov function candidate. It is easy to show that
$$\dot{V}_3 < 0 \tag{35}$$
as long as $|\tilde{\tau}_i| > \phi_\tau$ by using (24-27), where the term $\Gamma_{ii}^{-1} s_{qi}$ is omitted. Thus, $\tilde{\tau}_i \to \phi_\tau$ until $|\tilde{\tau}_i| \leq \phi_\tau$. In order to be stable, s has to stay bounded even if $\tilde{\tau}$ has not yet reached its boundary layer. This can be proven by regarding (6) as a storage function. Rearranging its time derivative (16) and using (15), where $|s_i| > \phi$ is assumed, i.e. control law (29) degenerates to (10), yields a relationship between the force error $\tilde{f}$ as input and the error variable s as output

$$s^T \tilde{f} \geq \dot{V}_1 + s^T K_D s + \tilde{x}^T \Lambda K_P \tilde{x} + \sum_{i=1}^{n} \eta_{ci} |s_i|. \tag{36}$$

The mapping from $\tilde{f} \to s^T$ is strictly passive [12] and therefore bounded-input bounded-output (BIBO) stable. Since $\tilde{f} = J^{-T} \tilde{\tau}$, s is also bounded, given a bounded $\tilde{\tau}$ (assuming that J^{-1} is not singular).

In a last step we show that s_i is bounded by ϕ as long as $\tilde{\tau}_i$ stays inside its boundary layer: As described before, for $|s_i| > \phi$ control law (29) degenerates to (10) and (16). Redefining η_{ci} in (16) as

$$\eta_{ci} = \eta_{ci}^* + \phi_\tau \sum_{j=1}^{n} \left|\left(J^{-T}\right)_{ij}\right|, \quad \eta_{ci}^* > 0, \tag{37}$$

leads to $\dot{V}_1 < 0$, again for $|s_i| \geq \phi$:

$$\dot{V}_1 \leq -s^T K_D s - \tilde{x}^T \Lambda K_P \tilde{x} \tag{38}$$

$$+ \sum_{i=1}^{n} \left(\underbrace{\sum_{j=1}^{n} \left(s_i \left(J^{-T}\right)_{ij} \tilde{\tau}_j - |s_i| \left|\left(J^{-T}\right)_{ij}\right| \phi_\tau \right)}_{\leq 0 \, \forall \, |\tilde{\tau}_i| \leq \phi_\tau} \right.$$

$$\left. - \eta_{ci}^* |s_i| \right), \tag{39}$$

where (37) and (16) have been used. Thus, $|s_i|$ decreases to ϕ and reaches this boundary layer after a finite time t_s.

It is important to note that the proof is not dependent on the choice of the control law (30) as long as $|\tilde{\tau}_i| \leq \phi_\tau$ can be guaranteed. The robust sliding-mode approach can, for example, be substituted by an adaptive approach [2].

3.3 VALVE DYNAMICS

In case of non negligible valve dynamics, we propose to apply one more sliding control law, thus guaranteeing $x_{vi} = x_{vdi}$, where x_{vdi} is the desired spool position of valve i. Assume that the dynamic behavior of a fast servo valve can be approximated by the following linear first-order differential equation

$$\dot{x}_{vi} = \underbrace{-\frac{1}{T_v} x_{vi}}_{f_{vi}} + \underbrace{\frac{1}{T_v}}_{b_v} u_i \tag{40}$$

where u_i is the voltage that is applied to the valve and T_v is the valve time constant. f_{vi} and b_v are assumed not to be known exactly but to be upper and lower bounded by

$$\beta_v = \sqrt{\frac{b_{v,\max}}{b_{v,\min}}} \tag{41}$$

and

$$\left|\hat{f}_{vi} - f_{vi}\right| < F_v. \tag{42}$$

The spool position error amounts $\tilde{x}_{vi} = x_{vi} - x_{vdi}$. In order to obtain exact system tracking $x_{v,i} = x_{vdi}$, we define a sliding surface $s_v = 0$ as

$$s_{vi} = \tilde{x}_{vi}. \tag{43}$$

The choice of

$$u_i = \hat{b}^{-1} [\hat{u}_i - k_{vi} \, sgn(s_{vi})], \tag{44}$$

with

$$\hat{u}_i = -\hat{f}_{vi} + \dot{x}_{vd} \tag{45}$$

as control law for the valve dynamics, where k_{vi} is chosen as

$$k_{vi} > \beta_v (F_v + \eta_v) + (\beta_v - 1) |\hat{u}_i|, \tag{46}$$

satisfies the sliding condition $s\dot{s} \leq -\eta_\nu |s_{\nu i}|$ [20][2].

The sliding-mode control guarantees a vanishing tracking error under the assumption of zero switching time of the amplifier. In reality the switching time is finite, possibly causing a small tracking error. It can be shown that after a slight modification of (28), the system stays stable in presence of small deviations, as long as $|\tilde{x}_{vi}| = |x_{vi} - x_{vdi}| < \phi_{x_v}$, which can be guaranteed using (22) as control law:
(35) holds, if

$$\tilde{\tau}_i \Gamma_{ii} \dot{\tilde{\tau}}_i \leq -\eta_{x_{vi}} |\tilde{\tau}_i| < 0. \tag{47}$$

Inserting (22) in (47), referring to (4) and using $x_{vi} = x_{vdi} + \tilde{x}_{vi}$ (x_{vi} in (22) is substituted by x_{vdi}) yields

$$-\eta_{x_{vi}} |\tilde{\tau}_i| \geq \tilde{\tau}_i \Gamma_{ii} (a_h + c_h z_h \hat{x}_{vi} \tag{48}$$
$$- c_h \hat{c}_h^{-1} k_{x_{vi}} sgn(\tilde{\tau}_i) + c_h z_h \tilde{x}_{vi} - \dot{\tau}_{di}).$$

---
[2] Regarding nonlinear valve dynamics, Slotine [20] shows that the above approach can be enhanced easily for nonlinear systems in the form $x_{vi}^{(n)} = f_{vi} + b_{vi} u_i$.

Following the same procedure as in (25)-(27) leads to

$$k_{x_{vi}} \geq \beta \left(\Gamma_{ii}^{-1} \eta_{x_{vi}} + \alpha \right) + \hat{c}_h z_h \left(\beta - 1 \right) |\hat{x}_{vi}| + \hat{c}_h z_h \phi_{x_v}, \quad (49)$$

where the term $\hat{c}_h z_h \phi_{x_v}$ ensures that $|\tilde{\tau}_i| \leq \phi_\tau$ as long as $|\tilde{x}_{vi}| < \phi_{x_v}$ after replacing (28) by (49). Furthermore, the stability proof of the continuous approximated control law (30) still holds.

4 SIMULATION

Simulations of the dynamical behavior of a hexapod are performed with the proposed smoothed control algorithm. The hexapod is modeled as a rigid 13-body system. Realistic physical parameters for the masses, moments of inertia etc. are used. The hydraulic model is based on the commercial simulation software "hyvos 6.0" from the Bosch-Rexroth company. In order to verify the robustness properties of the proposed approach, the parameters used for the controllers are varied up to 20 % compared to the model parameters. During the simulation, the desired path of the tool center point is a circle, parallel to the base plane, with a diameter of 75 mm. The desired path velocity amounts roughly to 50 mm/s. Additionally the platform rotates around the pitch axes and the roll axes with sin- or cos-characteristics, respectively, resulting in a kind of tumbling motion with a maximum angle of 15° and 1.5 Hz (the trajectory is chosen for a special metal forming process, for which the hexapod is designed). Due to the rotation, the maximum actuator velocity reaches $\approx$300 mm/s. After 1.5 s a force of 50 kN perpendicular to the platform and a force of 5 kN parallel to the roll-axis are applied to the tool center point. Fig. 1 shows that the actual force tracks the desired force very well.

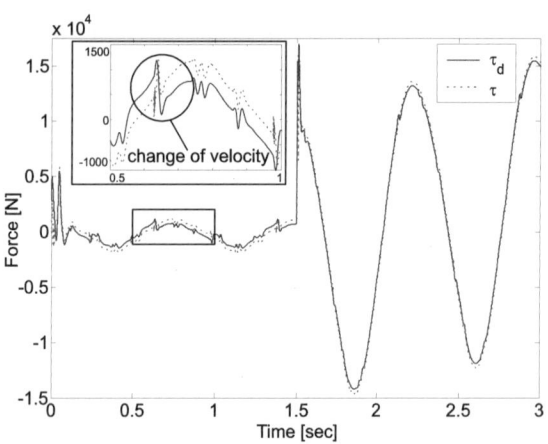

Figure 1: Desired force, actual force

In Fig. 2 the proposed algorithm is compared to a well tuned PID-position controller, which is augmented by a feedforward control and a disturbance rejection. As can be seen, the model based control law achieves a significant improvement over the enhanced PID algorithm in presence of large process forces.

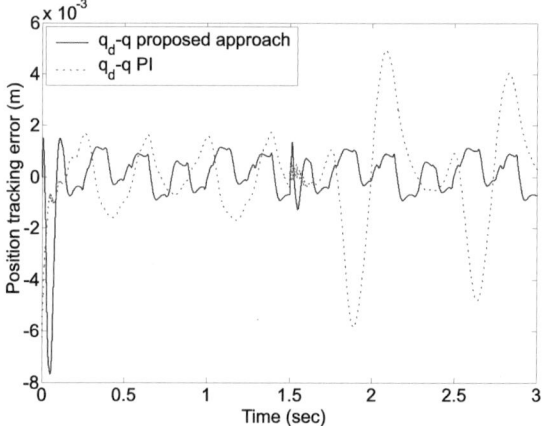

Figure 2: Position tracking error

A drawback of the backstepping-like control law is the sensitivity to sensor-noise, therefore high-quality measurements are necessary. One possibility to overcome this drawback is to introduce a heuristic limitation of the time-derivatives in the control law. This method, reduces the influence of sensor noise in the simulations significantly, leading concurrently only to a slight reduction of the control performance in comparison to perfect sensor signals.

5 CONCLUSION

A novel robust model-based position control scheme for hydraulically driven manipulators has been presented. The global asymptotic stability of the switching control law has been proven despite model uncertainties of both manipulator and actuators. After smoothing this control law to avoid actuator chattering, the output error of the plant is proven to stay within a prescribed bound. Another advantage of the proposed scheme is its modular structure which simplifies the tuning of the controller parameters significantly.

ACKNOWLEDGEMENT

This work was supported by the German Research Foundation DFG.

References

[1] A. Alleyne. Multiple surface sliding control. In: *ASME Int. Mechanical Engineering Congress and Exposition*, pp. 93–99, 1994.

[2] A. Alleyne and R. Liu. A simplified approach to force control for electro-hydraulic systems. *Control Engineering Practice*, 8(12):1347–1356, 2000.

[3] A. Alleyne and R. Liu. Systematic control of a class of nonlinear systems with application to electro-hydraulic cylinder pressure control. *IEEE Trans. on Control Systems Technology*, pp. 623–634, 2000.

[4] A. Alleyne, R. Liu, and H. Wright. On the limitations of force tracking control for hydraulic active suspensions. In: *Proc. Of the American Control Conf.*, pp. 43–47, June 1998.

[5] F. Bu and B. Yao. Observer based coordinated adaptive robust control of robot manipulators driven by single-rod hydraulic actuators. In: *Proc. IEEE Int. Conf. on Robotics and Automation*, pp. 3034–3039, 2000.

[6] L. Cai and G. Song. Robust postion / force control of robot manipulators during contact tasks. In: *Proc. of American Control Conf. (ACC)*, pp. 216–220, 1994.

[7] J. Duffy. The fallacy of modern hybrid control theory that is based on orthogonal complements of twist and wrench spaces. *Journal of Robotic Systems*, pp. 139–144, 1990.

[8] K. A. Edge and F. G. de Almeida. Decentralized adaptive control of a directly driven hydraulic manipulator. In: *Journal of Systems and Control Engineering*, pp. 197–205, 1995.

[9] B. Heinrichs, N. Sepehri, and A. B. Thorton-Trump. Position-based impedance control of an industrial hydraulic manipulator. In: *Proc. IEEE Int. Conf. on Robotics and Automation*, pp. 284–290, 1996.

[10] J. Heintze and A. J. J. van der Weiden. Inner-loop design and analysis for hydraulic actuators, with an application to impedance control. *Control Engineering Practice*, pp. 1323–1330, 1995.

[11] R. Hoffman. Dynamics and control of a flight simulator motion system. In: *Canadian Conf. on Automatic Control*, Montreal, Canada, 23.-24. Mai 1979.

[12] H. K. Khalil. *Nonlinear Systems*. Prentice-Hall, Upper Saddle River, NJ, 2. edition, 1996.

[13] Q. H. Nguyen, Q. P. Ha, D. C. Rye, and H. F. Durrant-Whyte. Force/position tracking for electrohydraulic systems of a robotic excavator. In: *Proc. 39th IEEE Conf. on Decision and Control*, pp. 5224–5229, 2000.

[14] N. Niksefat and N. Sepehri. Design and experimental evaluation of a robust force controller for an electro-hydraulic actuator via quantitative feedback theory. *Control Engineering Practice*, pp. 1335–1345, 2000.

[15] N. Niksefat and N. Sepehri. Designing robust force control of hydraulic actuators. *IEEE Control Systems Magazine*, pp. 66–77, April 2001.

[16] L. Sciavicco and B. Siciliano. *Modelling and Control of Robot Manipulators*. McGraw-Hill, London, 1996.

[17] N. Sepehri, G. A. M. Dumont, P. D. Lawrence, and F. Sassani. Cascade control of hydraulically actuated manipulators. *Robotica*, 8:207–216, 1990.

[18] M. R. Sirouspour and S. E. Salcudean. On the nonlinear control of hydraulic servo-systems. In: *Proc. of the IEEE Int. Conf. on Robotics and Automation*, 2000.

[19] M. R. Sirouspour and S. E. Salcudean. Nonlinear control of hydraulic robots. *IEEE Trans. on Robotics and Automation*, 17(2):173–182, 2001.

[20] J.-J. E. Slotine and W. Li. *Applied Nonlinear Control*. Prentice-Hall, 1991.

[21] G. A. Sohl and J. E. Bobrow. Experiments and simulations on the nonlinear control of a hydraulic servosystem. In: *Proc. of the American Control Conference (ACC)*, pp. 631–635, 1997.

Passivity Monitor and Software Limiter which Guarantee Asymptotic Stability of Robot Control Systems

Katsuya KANAOKA

Department of Robotics
Ritsumeikan University
Kusatsu, Shiga, 525-8577, Japan
kanaoka@se.ritsumei.ac.jp

Tsuneo YOSHIKAWA

Department of Mechanical Engineering
Kyoto University
Kyoto, 606-8501, Japan
yoshi@mech.kyoto-u.ac.jp

Abstract

In this paper, we propose a passivity monitor which evaluates stability of a system based on the concept of passivity, and a software limiter which turns the system stable if it is originally unstable. It is shown that asymptotic stability will be guaranteed for arbitrary robots with arbitrary control schemes and environments, even if their properties are unknown. The validity of the passivity monitor and the software limiter is verified by numerical simulations and experiments performed on various unstable systems.

1 Introduction

Conventionally, robot systems were only in isolated environments. In recent years, however, working environments for humans and robots become overlapping. A robot system which *never becomes unstable in any situation* is required to coexist with humans. One basic solution is to add hardware limiters on such a system. But, the distinction between stable and unstable is rather rough. The indices of stability in such hardware limiters are just the restrictions of the range of joint angles or the maximum electric current to actuators, and so on. Moreover, once the hardware limiters are activated, the subsequent operation of the system will be uncontrolled and some secondary disaster may be caused. Therefore, only the hardware limiters are insufficient to guarantee safety. An intelligent software limiter can provide a better solution.

Passivity of mechanical robot systems [1] is an important property. Many researchers are considering the passivity to solve some problems on analysis and control. Systems which restrict the total amount of stored energy have already proposed, for example, the energy balance monitor [2] and the passivity observer/passivity controller [3, 4]. However, these systems monitor the passivity in transmission lines of a master slave system, or in a virtual space of a haptic system. So they do not necessarily guarantee stability of arbitrary robot systems. The *passivity monitor* proposed here gives a stability index of robot control systems, including arbitrary hardwares, arbitrary control schemes, and arbitrary environments. The intelligent *software limiter*, constructed by using the passivity monitor, guarantees asymptotic stability of the output from the robot control systems.

2 Passivity of Robot Systems

In this section, the concept of the passivity is described. First, its definition is given [1].

When an input u and an output y of a system have the same dimension n, the system is called passive if the following inequality is always satisfied,

$$\int_0^t y^T(\tau)u(\tau)d\tau \geq -\gamma_0^2 \qquad (^\forall t > 0) \qquad (1)$$

where γ_0 is a certain constant value.

The equation of motion of an n-dof robot can be written as

$$u = M(q)\ddot{q} + \frac{1}{2}\dot{M}(q)\dot{q} + S(q,\dot{q})\dot{q} + g(q) + d(\dot{q}) \quad (2)$$

where $q \in \Re^n$ is the joint displacement vector, $u \in \Re^n$ is the joint driving force/torque vector. The first term in the right side is the inertia term, and $M(q) \in \Re^{n \times n}$ is positive definite symmetric. The second and third terms are the nonlinear terms. $S(q,\dot{q}) \in \Re^{n \times n}$ is a skew-symmetric matrix. $g(q) \in \Re^n$ is the potential term. $d(\dot{q}) \in \Re^n$ is the dissipation term expressing damping or friction. Each element of $d(\dot{q})$ always has the same sign as the corresponding element of $\dot{q}$. The following relations are satisfied,

$$K(\boldsymbol{q}, \dot{\boldsymbol{q}}) = (1/2)\dot{\boldsymbol{q}}^T \boldsymbol{M}(\boldsymbol{q})\dot{\boldsymbol{q}} \tag{3}$$

$$\boldsymbol{g}(\boldsymbol{q}) = (\partial P(\boldsymbol{q})/\partial \boldsymbol{q}^T)^T \tag{4}$$

where $K(\boldsymbol{q}, \dot{\boldsymbol{q}})$ and $P(\boldsymbol{q})$ are the kinetic and potential energy, respectively. The total internal energy E of the robot can be written as

$$E(\boldsymbol{q}, \dot{\boldsymbol{q}}) = K(\boldsymbol{q}, \dot{\boldsymbol{q}}) + P(\boldsymbol{q}) \tag{5}$$

It is known that the robot systems are passive with respect to the velocity output $\boldsymbol{y} = \dot{\boldsymbol{q}}$. Both sides of (2) are multiplied by $\dot{\boldsymbol{q}}^T$ and integrated,

$$\begin{aligned}\int_0^t \dot{\boldsymbol{q}}^T \boldsymbol{u}\, d\tau &= \int_0^t \dot{\boldsymbol{q}}^T \left\{ \boldsymbol{M}\ddot{\boldsymbol{q}} + \frac{1}{2}\dot{\boldsymbol{M}}\dot{\boldsymbol{q}} + \boldsymbol{S}\dot{\boldsymbol{q}} + \boldsymbol{g} + \boldsymbol{d} \right\} d\tau \\ &= E(\boldsymbol{q}(t), \dot{\boldsymbol{q}}(t)) - E(\boldsymbol{q}(0), \dot{\boldsymbol{q}}(0)) + \int_0^t \dot{\boldsymbol{q}}^T \boldsymbol{d}\, d\tau \\ &\geq -E(\boldsymbol{q}(0), \dot{\boldsymbol{q}}(0)) = -\gamma_0^2\end{aligned} \tag{6}$$

therefore (1) is always satisfied for robot systems on the conditions that $\dot{\boldsymbol{q}}^T \boldsymbol{S} \dot{\boldsymbol{q}} = 0$ (skew-symmetric property) and $\dot{\boldsymbol{q}}^T \boldsymbol{d} \geq 0$. It is also found that γ_0^2 is the initial value of the internal energy E.

An output of a passive system will be asymptotically stable when a linear time-invariant and strictly positive real feedback input is given. For example, consider the joint PD control

$$\boldsymbol{u}_{pd} = -\boldsymbol{K}_p(\boldsymbol{q} - \boldsymbol{q}_d) - \boldsymbol{K}_d \dot{\boldsymbol{q}} \tag{7}$$

as shown in **Figure 1**. $\boldsymbol{q}_d \in \Re^n$ is the desired joint displacement. $\boldsymbol{K}_p$ and $\boldsymbol{K}_d \in \Re^{n \times n}$ are the proportional and derivative gain matrices, and they are positive diagonal. In the case that the robot expressed by (2) is given the control input $\boldsymbol{u} = \boldsymbol{u}_{pd}$,

$$\lim_{t \to \infty} \boldsymbol{q} = \boldsymbol{q}_0, \quad \lim_{t \to \infty} \dot{\boldsymbol{q}} = \boldsymbol{0} \tag{8}$$

is realized, because the *monitored system* in **Figure 1** is passive. $\boldsymbol{q}_0$ is an equilibrium point which depends on (2) and (7).

3 Passivity Monitor

As shown in the preceding section, robot systems are passive and asymptotically stable by joint PD control. However, the system of **Figure 2** is not necessarily stable. Namely, some new control input $\boldsymbol{u}_x$ is added for a certain advanced task, or a non-negligible disturbance $\boldsymbol{u}_{dis}$ enters the system.

The feature that *the joint PD control (7) makes a passive system asymptotically stable* is used to evaluate the stability of robot systems. The system of **Figure 2** is asymptotically stable if the monitored system is passive. The monitored system is not passive if the system is not asymptotically stable. Therefore, the passivity of the monitored system to the joint PD control can be used as a quantitative evaluation index of the stability of the system.

3.1 Configuration of Passivity Monitor

The control systems, for which the passivity monitor can be used, have the following input

$$\boldsymbol{u} = \boldsymbol{u}_{pd} + \boldsymbol{u}_x + \boldsymbol{u}_{dis} \tag{9}$$

where $\boldsymbol{u}_{pd}$ is the joint PD control input (7), $\boldsymbol{u}_x$ is an arbitrary control input, and $\boldsymbol{u}_{dis}$ is a disturbance input from an arbitrary environment (See **Figure 2**).

The energy stored in the monitored system serves as the passivity monitor.

$$E_{pd} = \int_0^t \dot{\boldsymbol{q}}^T \boldsymbol{u}_{pd}\, d\tau \tag{10}$$

The passivity of the monitored system is evaluated by this E_{pd}. Namely, after setting up the initial internal energy E_0 of the monitored system at a suitable value, if the stored energy restriction

$$E_{pd} \geq -E_0 \tag{11}$$

is satisfied, the monitored system can be said to be passive and the output $\dot{\boldsymbol{q}}$ is asymptotically stable. If it is not satisfied, the monitored system is not passive and the output $\dot{\boldsymbol{q}}$ can get unstable.

3.2 Physical Meanings of Passivity Monitor

In this section, the meanings of the passivity monitor restriction (11) is considered with respect to the

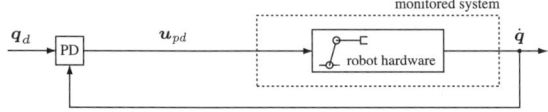

Figure 1: Robot system with PD control

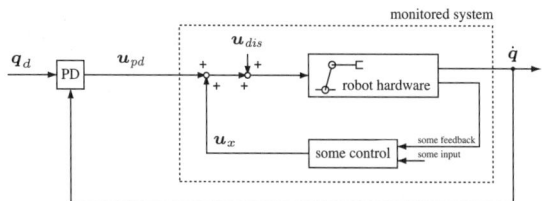

Figure 2: Robot system with PD, some control, and disturbance

joint PD control. From (7) and (10), the energy entering the PD control system ($-E_{pd}$) is obtained as

$$-E_{pd} = \int_0^t \dot{\boldsymbol{q}}_e^T \boldsymbol{K}_p \boldsymbol{q}_e d\tau - \int_0^t \dot{\boldsymbol{q}}_d^T \boldsymbol{K}_p \boldsymbol{q}_e d\tau + \int_0^t \dot{\boldsymbol{q}}^T \boldsymbol{K}_d \dot{\boldsymbol{q}} d\tau \quad (12)$$

where the joint displacement error vector $\boldsymbol{q}_e = \boldsymbol{q}_d - \boldsymbol{q}$. For convenience, the joint PD control is considered as springs and dampers attached to the joints. So the potential energy of the springs is

$$E_p(t) = (1/2) \, \boldsymbol{q}_e^T(t) \boldsymbol{K}_p \boldsymbol{q}_e(t) \quad (13)$$

From (12) and (13), the restriction (11) becomes equivalent to the following inequality,

$$E_p + \int_0^t \dot{\boldsymbol{q}}^T \boldsymbol{K}_d \dot{\boldsymbol{q}} d\tau \\ \leq E_{p0} + E_0 + \int_0^t \dot{\boldsymbol{q}}_d^T (\boldsymbol{K}_p \boldsymbol{q}_e) d\tau \quad (14)$$

where $E_{p0} = E_p(0)$. The first term on the left side of (14) is the potential energy of the P control springs. The second term expresses the total amount of the energy dissipated in the D control dampers. The first and second terms on the right side are constant, and they are the initial internal energy of the P control springs and the monitored system, respectively. The third term is the total amount of the work which the equilibrium position (target value) change $\dot{\boldsymbol{q}}_d$ has performed against the force/torque of the P control springs ($-\boldsymbol{K}_p \boldsymbol{q}_e$).

In the case that the equilibrium position change $\dot{\boldsymbol{q}}_d$ satisfies the following inequality

$$\int_0^t \dot{\boldsymbol{q}}_d^T (\boldsymbol{K}_p \boldsymbol{q}_e) d\tau \leq E_{pmax} - E_{p0} - E_0 \quad (15)$$

where E_{pmax} is an arbitrary bounded constant, (14) can be transformed into

$$E_p + \int_0^t \dot{\boldsymbol{q}}^T \boldsymbol{K}_d \dot{\boldsymbol{q}} d\tau \leq E_{pmax} \quad (16)$$

Therefore, (11) of the passivity monitor means the potential energy restriction in the P control springs. The second term on the left side is positive and never decreases, it can make the potential energy restriction more conservative.

Although the equilibrium position input $\boldsymbol{q}_d$ is required to satisfy (15) to guarantee the passivity of the monitored system, $\boldsymbol{q}_d$ can be time-variant and the task to achieve by $\boldsymbol{q}_d(t)$ of the joint PD control (7) can depart from that of *some control* in the figures.

4 Software Limiter

When the passivity monitor detects that the restriction (11) is not satisfied, it is necessary to recover the passivity of the monitored system by some method. Although it is possible to shut down the control input, the robot system can cause some secondary disaster.

By using the passivity monitor, an intelligent software limiter is constructed, which recovers the passivity of the monitored system without shutting down. The whole configuration is shown in **Figure 3**. The new input to the robot hardware is

$$\boldsymbol{u} = w_{pd} \boldsymbol{u}_{pd} + w_x \boldsymbol{u}_x + w_v \boldsymbol{u}_v + \boldsymbol{u}_{dis} \quad (17)$$
$$\boldsymbol{u}_v = -\boldsymbol{K}_v \dot{\boldsymbol{q}} \quad (18)$$

where $w_x(t)$ ($0 \leq w_x \leq 1$) is the weight of a generated energy decreasing limiter, $w_v(t)$ ($w_v \geq 0$) is the weight of a dissipated energy increasing limiter, and $w_{pd}(t)$ ($0 \leq w_{pd} \leq 1$) is the weight of a joint PD control limiter. These are all explained in detail in the following sections.

4.1 Generated Energy Decreasing Limiter

The first software limiter makes the stored energy E_{pd} increase, by decreasing the generated energy in the monitored system. It makes the passivity of the monitored system recover. The weight w_x is decreased until the restriction (11) gets satisfied. However, in the case that the monitored system is non-passive due to the disturbance $\boldsymbol{u}_{dis}$, the restriction (11) is not necessarily satisfied even if the limiter weight w_x is decreased.

Once the generated energy decreasing limiter is configured, trial and error tests to determine the suitable control gains for $\boldsymbol{u}_x$ become unnecessary. When w_x is fixed, $\boldsymbol{u}_x$ can be replaced by $w_x \boldsymbol{u}_x$. Then, the gains for $\boldsymbol{u}_x$ can be calculated, which make the output $\dot{\boldsymbol{q}}$ asymptotically stable.

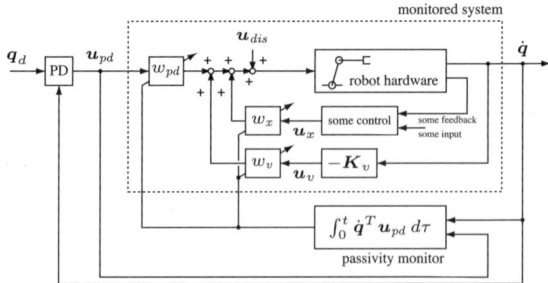

Figure 3: System with passivity monitor and software limiter

For example, the weight w_x can be set to

$$w_x = \begin{cases} 1 & (E_u < E_{pd}) \\ \frac{1}{2}\left\{1 - \cos\left(\frac{E_{pd} - E_l}{E_u - E_l}\right)\pi\right\} & (E_l \leq E_{pd} \leq E_u) \\ 0 & (E_{pd} < E_l) \end{cases} \quad (19)$$

where E_u and E_l ($E_u \geq E_l \geq -E_0$) are suitable constants.

4.2 Dissipated Energy Increasing Limiter

The second software limiter makes the stored energy E_{pd} increase, by increasing the dissipated energy in the monitored system. The weight w_v is increased until the restriction (11) gets satisfied. This limiter makes the passivity recover, even if the unknown disturbance u_{dis} lies at the origin of the non-passivity of the monitored system.

For example, the weight w_v can be set to be

$$w_v = \begin{cases} 0 & (E_{pd} \geq E_l) \\ (E_l - E_{pd})^m & (E_{pd} < E_l) \end{cases} \quad (20)$$

where E_l ($E_l \geq -E_0$) and m are suitable constants.

4.3 Joint PD Control Limiter

An arbitrary control scheme is not necessarily expressed as a combination with the joint PD control u_{pd} like (9). On the other hand, the roles of the joint PD control u_{pd} are, (a) an input to the passivity monitor, (b) an alternative control scheme in the case that the monitored system gets non-passive. Namely, if u_x can maintain the passivity of the monitored system, the joint PD control u_{pd} does not necessarily need to be input into the robot hardware.

Therefore, if the control scheme u_x does not include the joint PD control and if the monitored system has been passive, the role of the joint PD control can be restricted only as the input to the passivity monitor. The function of the passivity monitor is not lost even if $w_{pd} \neq 1$, because (16) is still satisfied. In the case that the monitored system becomes non-passive, the software limiter(s) should be activated to make the output $\dot{q}$ asymptotically stable and w_{pd} is regulated toward 1 as the alternative control scheme.

For example, the limiter weight w_{pd} can be set as follows.

$$w_{pd} = 1 - w_x \quad (21)$$

In addition, although w_x, w_v, and w_{pd} are considered to be scalars for simplicity, we can define them as diagonal matrices $\boldsymbol{w}_x$, $\boldsymbol{w}_v$, and $\boldsymbol{w}_{pd}$ with differently weighted diagonal elements, to set up the weights individually.

4.4 Proof of Asymptotic Stability

In this section, it is proven that the asymptotic stability can be guaranteed by using the passivity monitor and the software limiter.

From (10), (17), (18), and (21), the stored energy restriction (11) of the passivity monitor is equivalent to the following inequality:

$$\int_0^t \dot{\boldsymbol{q}}^T w_x(\boldsymbol{u}_x - \boldsymbol{u}_{pd})\,d\tau + \int_0^t \dot{\boldsymbol{q}}^T \boldsymbol{u}_{dis}\,d\tau$$
$$\leq E_0 + \int_0^t \dot{\boldsymbol{q}}^T \boldsymbol{u}\,d\tau + \int_0^t \dot{\boldsymbol{q}}^T w_v \boldsymbol{K}_v \dot{\boldsymbol{q}}\,d\tau \quad (22)$$

Theorem *For an arbitrary robot, with an arbitrary control scheme $\boldsymbol{u}_x$ and with an arbitrary disturbance $\boldsymbol{u}_{dis}$, a passivity monitor and a software limiter are configured as **Figure 3**. The robot control system of **Figure 3** can be made asymptotically stable by defining suitable values for w_x and w_v.*

Assumptions *The robot hardware is expressed as (2). An arbitrary driving force/torque can be applied to each joint of the robot system. Both a sensor and a actuator are collocated in each joint. The values except time t are all bounded.*

Proof *As described in Section 2, if the joint PD control input (7) is applied to a passive system, the output $\boldsymbol{y} = \dot{\boldsymbol{q}}$ is asymptotically stable. In other words, if the passivity of the monitored system in **Figure 3** is proven, the asymptotic stability of the output $\boldsymbol{y} = \dot{\boldsymbol{q}}$ is guaranteed. The necessary and sufficient condition for the monitored system to be passive is that the restriction (11) is satisfied. Since (11) is equivalent to (22), the values except time t are all bounded, and the lower limit of the second term on the right side of (22) is restricted by $-\gamma_0^2$, there always exist $w_x(t)$ and $w_v(t)$ satisfying (11) in any state of the system. The monitored system becomes passive by setting these $w_x(t)$ and $w_v(t)$ as the limiter weights in (17). Therefore, the asymptotic stability of the output $\boldsymbol{y}$ has been proven.*

5 Numerical Simulations

In this section, dynamics simulations are carried out when a robot control system turns unstable because of a sensor delay or a disturbance.

Table 1: Parameters for numerical simulations

link & joint i	length l_i [m]	mass center l_{gi} [m]	mass m_i [kg]	inertia I_i [kg·m^2]	damping coefficient d_i [N·m·s/rad]
1	0.5000	0.2500	3.0000	0.24586	0.1000
2	0.5000	0.3125	4.0000	0.29273	0.1000

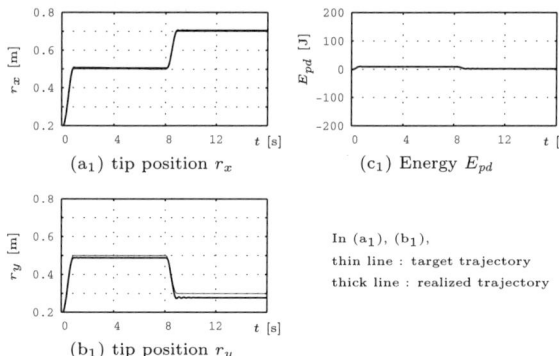

Figure 4: Simulation results (joint PD without delay)

The controlled robot is a planar 2-dof serial link manipulator (See **Figure 5**). The parameters are shown in **Table 1**. The armature inertia (converted to the output axes) of the joint 1 and 2 is set to 1.0 [kg·m^2]. The joint displacement vector $q = [\theta_1, \theta_2]^T$ and the tip position vector $r = [r_x, r_y]^T$ from Σ_R satisfy the kinematic relation $r = f(q)$.

The sampling period for the control is set to 1.0 [ms]. The 4th-order Runge-Kutta method is used for the numerical integration, with the integration step of 0.1 [ms].

5.1 Instability due to sensor delay

Trajectory control on a tip desired trajectory r_d is considered. One possible solution is to change r_d into a joint desired trajectory q_d by calculating the inverse kinematics $q_d = f^{-1}(r_d)$ and applying the joint PD control (7) as $u = u_{pd}$. This system will be asymptotically stable, however, some positioning offset may arise in this joint PD control, because of gravity.

The PD feedback gains are set to $K_p = \text{diag}(1000$ [N·m/rad], 1000 [N·m/rad]) and $K_d = \text{diag}(10$ [N·m·s/rad], 10 [N·m·s/rad]). r_d is expressed as a 5th-order polynomial with respect to time. The desired trajectory is from $r_0 = [0.2, 0.2]^T$ to $r_1 = [0.5, 0.5]^T$ at time 0.0 [s] to 1.0 [s], then from r_1 to

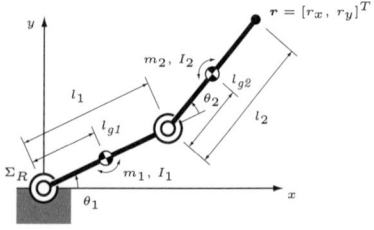

Figure 5: Numerical model of 2-dof manipulator

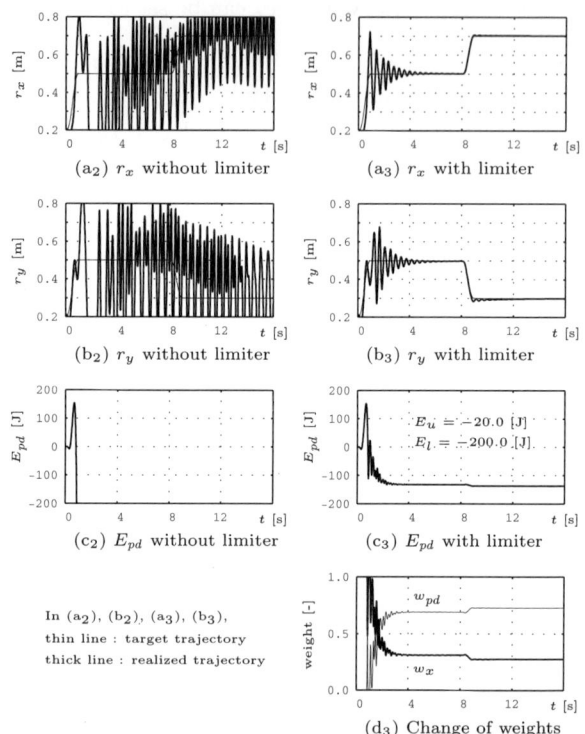

Figure 6: Simulation results (joint PD and tip position PID with sensor delay)

$r_2 = [0.7, 0.3]^T$ at time 8.0 [s] to 9.0 [s]. The inverse kinematics is solved on the condition that $\theta_2 \leq 0$. The gravitational acceleration works downward along the y-axis.

The simulation results are **Figure 4** (a$_1$) and (b$_1$). The offset has arisen because of gravity. The stored energy E_{pd} is shown in (c$_1$). These results mean this control is not accurate although it is stable.

Next, another control scheme with directly fed back the tip position error is applied.

$$u_x = J^T \left(K_{rp} r_e + K_{ri} \int_0^t r_e d\tau + K_{rd} \dot{r}_e \right) \quad (23)$$

J is the Jacobian matrix $J = \partial r / \partial q^T$, r_e is the tip position error vector $r_e = r_d - r$. The feedback gains are set to $K_{rp} = \text{diag}(500$ [N/m], 500 [N/m]), $K_{ri} = \text{diag}(1000$ [N/m·s], 1000 [N/m·s]) and $K_{rd} = \text{diag}(20$ [N·s/m], 20 [N·s/m]). When the control input $u = u_x$ is applied to the manipulator of **Figure 5**, no positioning offset can be realized. However, the following hardware restriction has to be taken into account. The tip position r is sensed by a vision sensor of which output r_{vis} is updated only at 30 [Hz]. The tip position r fed back to the control (23) is given

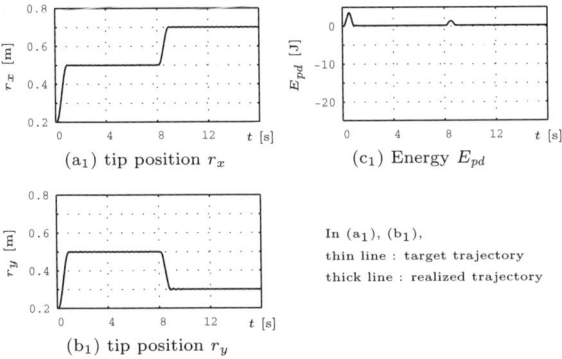

Figure 7: Simulation results (joint PD without disturbance)

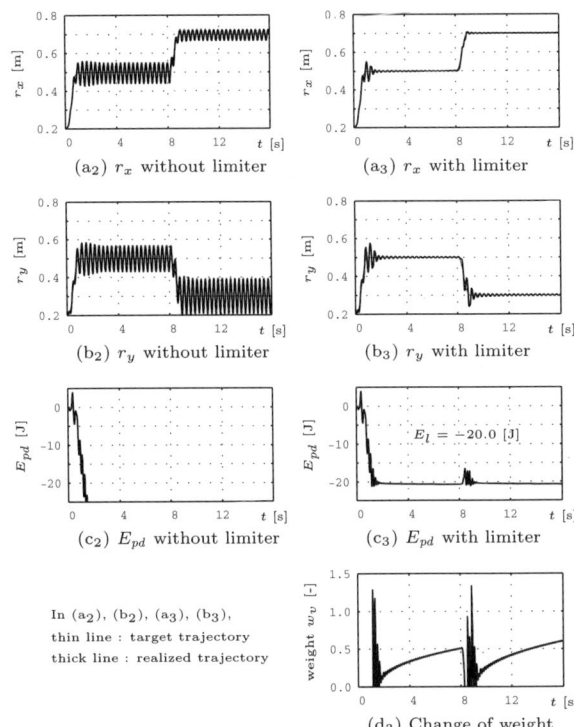

Figure 8: Simulation results (joint PD with disturbance)

as the following equation with the low pass filter (the cutoff frequency $\omega_c = 30\pi$ [rad/s]):

$$r(s) = \frac{\omega_c}{s + \omega_c} r_{vis}(s) \qquad (24)$$

If the control system is configured as mentioned above and the control is performed on the same desired tip trajectory as **Figure 6** (a_1) and (b_1), it becomes unstable as shown in **Figure 6** (a_2) and (b_2).

The passivity monitor (10) is added to this system. The control input is (17) on the condition that $w_{pd} \equiv 0$, $w_x \equiv 1$, $\boldsymbol{u}_v \equiv \boldsymbol{u}_{dis} \equiv \boldsymbol{0}$, (7), and (23). The stored energy E_{pd} in the monitored system, observed by the passivity monitor, is shown in **Figure 6** (c_2). Its rapid decreasing means the monitored system is getting non-passive.

Then, the generated energy decreasing limiter is added to the unstable control system. The control input is (17), where $\boldsymbol{u}_v \equiv \boldsymbol{u}_{dis} \equiv \boldsymbol{0}$. The limiter weights are (19) and (21) where $E_u = -20.0$ [J] and $E_l = -200.0$ [J]. The simulation results on the same desired trajectory are shown in **Figure 6** (a_3) and (b_3). The unstable system turns into stable. The realized trajectory follows the target position change between 8.0 [s] and 9.0 [s]. The stored energy in the monitored system and the change of the limiter weights are shown in **Figure 6** (c_3) and (d_3). The generated energy in the monitored system is suppressed appropriately by the software limiter. Moreover, since $w_x > 0$, even if the limiter is on, the effect of the tip PID feedback (23) has not been lost. Therefore, no positioning offset due to gravity occurs, as shown in (a_3) and (b_3).

5.2 Resonance with Disturbance

Consider a periodic disturbance $\boldsymbol{f}_{dis}$ to the tip, acting with a frequency equal to the natural frequency of the system. The joint PD control (7) is applied to the manipulator of **Figure 5**. The joint PD gains $\boldsymbol{K}_p$ and $\boldsymbol{K}_d$ are set to the same values as the preceding section. The first mode natural frequency is $\omega_e = 19.04$ [rad/s], at the tip position $\boldsymbol{r}_1$ for $\boldsymbol{q}_1 = [\pi/2, -\pi/2]^T$. Then, the disturbance $\boldsymbol{f}_{dis}$ [N] applied to the tip and the equivalent disturbance $\boldsymbol{u}_{dis}$ [Nm] at the joints are:

$$\boldsymbol{u}_{dis} = \boldsymbol{J}^T \boldsymbol{f}_{dis} = \boldsymbol{J}^T \begin{bmatrix} 0 \\ 50 \sin(\omega_e t) \end{bmatrix} \qquad (25)$$

Figure 7 (a_1) and (b_1) shows the results when only the joint PD control (7) is used and without disturbance. Since this simulation is carried out under no gravity, unlike the preceding section, the desired tip trajectory is realized precisely. The stored energy in the monitored system E_{pd} is shown in **Figure 7** (c_1).

Under the assumption that $w_{pd} \equiv 1$ and $\boldsymbol{u}_v \equiv \boldsymbol{u}_x \equiv \boldsymbol{0}$ for the control input (17), the joint PD control (7) and the disturbance (25) are applied to the manipulator of **Figure 5**. This system will become unstable by resonance as shown in **Figure 8** (a_2) and (b_2). The stored energy E_{pd} in the monitored system is **Figure 8** (c_2). This means that the monitored system is getting non-passive due to the disturbance (25).

Next, the dissipated energy increasing limiter is

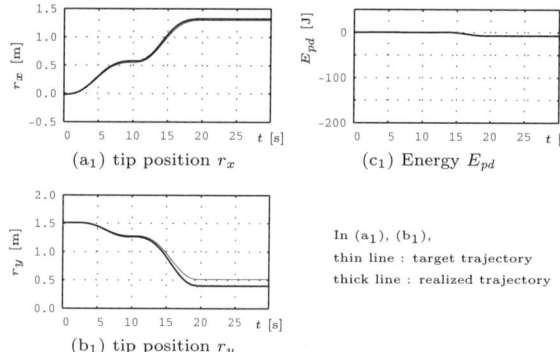

Figure 9: Experimental results (joint PD with flexibility)

added to the unstable system. Considering (7) and (18) as the control schemes, with the control input (17) under the condition that $w_{pd} \equiv 1$ and $\boldsymbol{u}_x \equiv \boldsymbol{0}$. The disturbance is given by (25). The limiter weight w_v is (20) where $E_l = -20.0$ [J] and $m = 2$. The simulation results on the same desired trajectory $\boldsymbol{r}_d$ are shown in **Figure 8** (a_3) and (b_3). The unstable system turns into stable. The realized trajectory follows the target position change between 8.0 [s] and 9.0 [s]. The stored energy in the monitored system and the change of the limiter weights are shown in **Figure 8** (c_3) and (d_3). It can be seen that the dissipated energy in the monitored system is enhanced appropriately by the software limiter.

6 Experiments

In this section, an example of the passivity monitor and the software limiter is given to decide control gains which make a flexible manipulator asymptotically stable.

6.1 Experimental System

An overview of the experimental system is shown in **Figure 10**. This is a 2-dof flexible manipulator

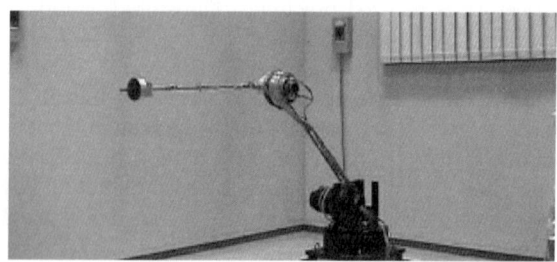

Figure 10: Experimental system

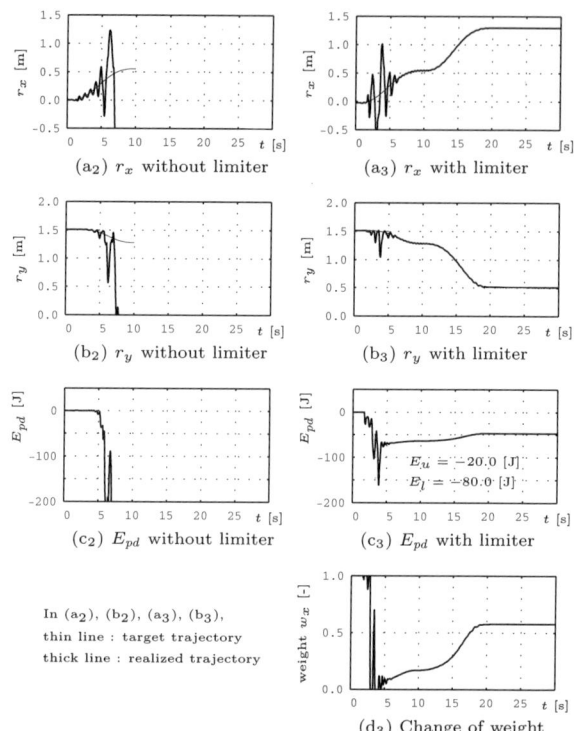

Figure 11: Experimental results (joint PD and compensation with flexibility)

in a vertical plane, with the same joint composition as **Figure 5**. The link lengths are $l_1 = 0.728$ [m] and $l_2 = 0.7838$ [m]. The flexible parts are round bars made of S45C carbon steel, the diameters are $d_1 = 0.013$ [m] and $d_2 = 0.008$ [m]. The elastic displacements are measured by 8 sets of strain gauges attached on the flexible links.

6.2 Instability due to flexibility

First, positioning control of this flexible manipulator by the joint PD control (7) is considered. The feedback gains are $\boldsymbol{K}_p = \text{diag}(1000$ [N·m/rad], 1000 [N·m/rad]) and $\boldsymbol{K}_d = \text{diag}(10$ [N·m·s/rad], 10 [N·m·s/rad]). A desired tip trajectory $\boldsymbol{r}_d$ is expressed as a 5th-order polynomial with respect to time. It goes from $\boldsymbol{r}_3 = [0.0000, 1.5118]^T$ to $\boldsymbol{r}_4 = [0.5542, 1.2822]^T$ at time 0.0 [s] to 10.0 [s], then goes from $\boldsymbol{r}_4$ to $\boldsymbol{r}_5 = [1.2986, 0.5148]^T$ at time 10.0 [s] to 20.0 [s]. The inverse kinematics is solved on the condition that $\theta_2 \leq 0$. The realized tip trajectory $\boldsymbol{r}$ is shown in **Figure 9** (a_1) and (b_1). It is found that the offset is caused by the flexibility and gravity. The stored energy E_{pd} is shown in (c_1).

Next, quasi-static compensation control is applied to the experimental system. This control scheme realizes precise tip positioning by compensating the elastic displacement e of the flexible manipulator. The control scheme uses the following integration term for the compensation.

$$u_x = K_i \int_0^t (q_d + \Delta q - q) dt \quad (26)$$

$$\Delta q = -J_q^+ J_e \Delta e \quad (27)$$

where Δq is the quasi-static compensation vector to compensate the tip error Δr due to Δe. Here, J_q and J_e are the Jacobian matrices: $J_q = \partial r / \partial q^T$ and $J_e = \partial r / \partial e^T$. J_q^+ is the pseudo inverse matrix of J_q.

The passivity monitor (10) is added to this system according to **Figure 3** with the control input (17), on the condition that $w_{pd} \equiv w_x \equiv 1$, $u_v \equiv u_{dis} \equiv 0$, (7), and (26). The joint PD gains K_p and K_d are all the same as in the preceding experiment of **Figure 9**. The integral gains are set to $K_i = \text{diag}(5000\ [\text{N·m/rad·s}], 5000\ [\text{N·m/rad·s}])$. If the quasi-static compensation control is performed on the same tip target trajectory as **Figure 9** (a_1) and (b_1), it becomes unstable as shown in **Figure 11** (a_2) and (b_2). The stored energy E_{pd} in the monitored system, observed by the passivity monitor, is shown in **Figure 11** (c_2). Its rapid decreasing means that the monitored system is getting non-passive.

In this case, appropriate gains could be determined by trial and error. But, these trials may be very dangerous because of the instability. It is possible to decide appropriate gains, while guaranteeing asymptotic stability, if the software limiter is applied.

The generated energy decreasing limiter is added to the unstable control system. The control input is (17) on the condition that $w_{pd} \equiv 1$, $u_v \equiv u_{dis} \equiv 0$, (7), and (26). The limiter weights are (19) and (21) with $E_u = -20.0$ [J] and $E_l = -80.0$ [J]. When the software limiter is configured like this, the experimental results on the same target trajectory as the last one are shown in **Figure 11** (a_3) and (b_3). The unstable system turns into stable. The stored energy in the monitored system and the change of the limiter weights are shown in **Figure 11** (c_3) and (d_3). The generated energy in the monitored system is suppressed appropriately by the software limiter. Moreover, since $w_x > 0$, even if the limiter is on, the effect of the integral feedback (26) of the quasi-static compensation control has not been lost. Therefore, no positioning offset occurs. This is shown in (a_3) and (b_3).

7 Conclusion

In this paper, the passivity monitor and the software limiter have been proposed. They guarantee the asymptotic stability of arbitrary robot control systems, with arbitrary control schemes and arbitrary environments. The passivity monitor and the software limiter can be widely applied, not only to the robot systems, but also to general machine systems. They are comparatively easy to implement and use. In the future, the validity of the passivity monitor and the software limiter for space robots, bipedal robots, etc. will be examined.

References

[1] S. Arimoto, "Control Theory of Non-linear Mechanical Systems — A Passivity-based and Circuit-theoretic Approach," *Oxford Science Publications*, 1996.

[2] Y. Yokokohji, T. Imaida, and T. Yoshikawa, "Bilateral Control with Energy Balance Monitoring Under Time-Varying Communication Delay," *ICRA 2000*, pp. 2684–2689, 2000.

[3] B. Hannaford and J.H. Ryu, "Time Domain Passivity Control of Haptic Interfaces," *ICRA 2001*, pp. 1863–1869, 2001.

[4] J.H. Ryu, D.S. Kwon, and B. Hannaford, "Stable Teleoperation with Time Domain Passivity Control," *ICRA 2002*, pp. 3260–3265, 2002.

Hierarchical Velocity Field Control for Robot Manipulators*

Javier Moreno

Centro de Investigación y Desarrollo
de Tecnología Digital, CITEDI–IPN
Ave. del Parque 1310, Mesa de Otay,
Tijuana, B.C., 22510, MEXICO
email: moreno@citedi.mx
fax: +52 (646) 1 75 05 54

Rafael Kelly

División de Física Aplicada, CICESE
Apdo. Postal 2615, Adm. 1
Ensenada, B.C. 22800, MEXICO
email: rkelly@cicese.mx
fax: +52 (664) 6231388

Abstract

This paper concerns the velocity field control in operational space of robot manipulators. Velocity field control is a recent control formulation in robotics. A velocity field defines the desired robot velocity in the operational space as a function of its current position, thus the robot performs the desired motions. In this paper, a controller is proposed for operational space velocity field control. The proposed controller is based on a hierarchical structure that result of using the kinematic control concept and a joint velocity controller. Experimental results on a two degrees–of–freedom direct–drive robot arm illustrate the viability of the proposed scheme.

Keywords: *Velocity field, Robot control, Stability, Direct–drive arm.*

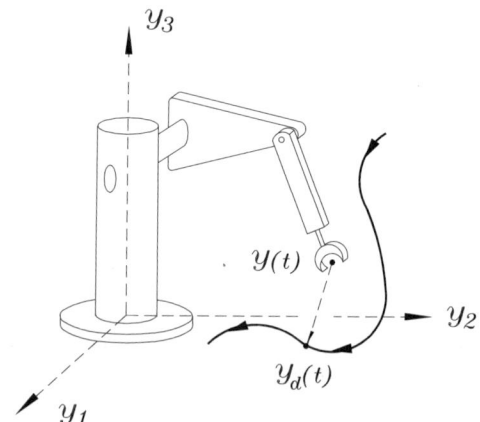

Figure 1: Trajectory tracking control in Cartesian space

1 Introduction

The dynamics in joint space of a serial–chain n-link robot manipulator considering the presence of friction at the robot joints can be written as [1],[2]

$$M(q)\ddot{q} + C(q,\dot{q})\dot{q} + g(q) + f(\dot{q}) = \tau \qquad (1)$$

where q is the $n \times 1$ vector of joint displacements, $\dot{q}$ is the $n \times 1$ vector of joint velocities, τ is the $n \times 1$ vector of applied torque inputs, $M(q)$ is the $n \times n$ symmetric positive definite manipulator inertia matrix, $C(q,\dot{q})\dot{q}$ is the $n \times 1$ vector of centripetal and Coriolis torques, $g(q)$ is the $n \times 1$ vector of gravitational torques, and $f(\dot{q})$ is the vector of forces and torques due to friction, such as viscous and Coulomb friction, present at the robot joints.

The following property is satisfied by the dynamic model (1) [3]:

$$x^T \left[\frac{1}{2}\dot{M}(q) - C(q,\dot{q})\right] x = 0 \quad \forall\ x, q, \dot{q} \in \mathbb{R}^n, \qquad (2)$$

*Work partially supported by CONACyT, and CYTED.

where the matrix $C(q,\dot{q})$ is written using Christoffel symbols and $\dot{M}(q)$ stands for the time derivative of the inertia matrix $M(q)$.

Denoting $h(q) : \mathbb{R}^n \to \mathbb{R}^m$ the robot direct kinematics, then the position and orientation $y \in \mathbb{R}^m$ of the end–effector is given by

$$y = h(q). \qquad (3)$$

The time derivative of the direct kinematic model (3) yields the differential kinematics

$$\dot{y} = \frac{d}{dt}h(q) = \frac{\partial h}{\partial q}\dot{q} = J(q)\dot{q} \qquad (4)$$

where $J(q)$ is the so–called analytical Jacobian matrix [3]. In this paper the analytical Jacobian $J(q)$ is assumed of full–rank and bounded.

Specifications of many tasks to be executed by robot manipulators are traditionally established by means of a desired timed trajectory $y_d(t)$ for the end–effector pose y. Once the specifications are given in terms of the desired pose trajectory $y_d(t)$, then the

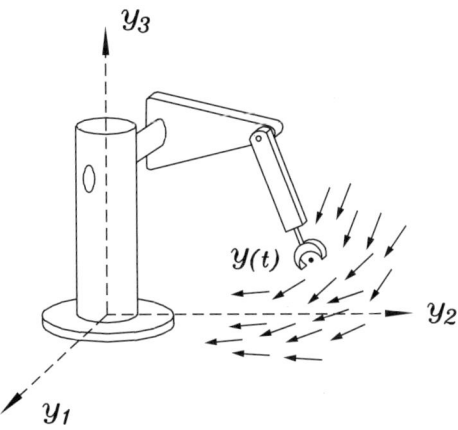

Figure 2: Desired velocity field in Cartesian space

motion control aim is to achieve asymptotic tracking, that is, to ensure

$$\lim_{t\to\infty} [\boldsymbol{y}_d(t) - \boldsymbol{y}(t)] = \boldsymbol{0}.$$

This approach corresponds to the well known motion control —or trajectory tracking control— of robot manipulators in operational space which has been widely addressed in the literature. See [2],[3], for surveys on operational space tracking control. Figure 1 illustrates the concept of trajectory tracking control.

On the other hand, passive velocity field control has been recently introduced by [4],[5], which attempts to be an alternative to motion control. In this control philosophy, the task to be accomplished by the robot is coded by means of a smooth desired *velocity vector field* defined in the operational configuration space $\mathcal{G}$ and denoted as a map

$$\boldsymbol{v}(\boldsymbol{y}): \begin{array}{ccc} \mathcal{G} & \to & T\mathcal{G} \\ \boldsymbol{y} & \mapsto & \boldsymbol{v}(\boldsymbol{y}) \end{array}$$

where $T\mathcal{G}$ denotes the tangent bundle of $\mathcal{G}$ [6].

A velocity field defines a tangent vector (the desired end-effector velocity $\dot{\boldsymbol{y}}_d$) at every point of the robot operational configuration space. Figure 2 illustrates the specification of motion by means of a velocity field. This Figure depicts a velocity field defined in the three dimensional Cartesian space of a robot arm which assigns a desired velocity vector (arrow) to each point in the operational space.

Although the idea of motion specification independent of time for a robotic task seems at first sight illogical, in many application the timing of the desired trajectory is unimportant compared to the coordination and synchronization requirements between the various degrees of freedom. In this way, velocity field control approach is particularly well suited to contour following tasks for machining operations such as cutting, milling and deburring [8].

The velocity field control is reformulated in this paper without regard of closed–loop passivity requirements [7]:

The <u>velocity field control</u> objective in operational <u>space</u> is established as the design of the torques input $\boldsymbol{\tau}$ so that

$$\lim_{t\to\infty} [\boldsymbol{v}(\boldsymbol{y}(t)) - \dot{\boldsymbol{y}}(t)] = \boldsymbol{0}, \quad (5)$$

where the difference between the desired velocity field $\boldsymbol{v}(\boldsymbol{y})$ and the manipulator end–effector velocity $\dot{\boldsymbol{y}}$ defines the velocity field error.

In this situation the desired velocity field $\boldsymbol{v}(\boldsymbol{y})$ is defined so that if the velocity $\dot{\boldsymbol{y}}$ of the output matches the velocity field $\boldsymbol{v}(\boldsymbol{y})$, that is $\dot{\boldsymbol{y}} = \boldsymbol{v}(\boldsymbol{y})$, then the robot output is guided in the operational space for completing the desired robot motions. Thus, instead of requiring the arm end-efector tip to be at specific location at each instant time as it is imposed in trajectory tracking control, in velocity field control the arm tip will match with the flow lines of the desired velocity field, as it can be seen in Figure 2.

A control strategy based on a hierarchical structure is proposed. This structure consists of using the kinematic control concept for joint velocity resolution and an asymptotically stable joint velocity controller.

This paper is organized as follows. Section 2 deals with velocity field control for robot manipulators. In Sections 3, experimental results carried out on a direct–drive robot are shown. Finally, some concluding remarks are drawn in Section 4.

2 Velocity field control: Kinematic control

The kinematic control approach considers that the system input is the joint velocity $\dot{\boldsymbol{q}}$ [2],[3],[9]. The differential kinematic model (4) is at the origin of the kinematic control concept. It is implicitly assumed that the robot at hand is equipped with an ideal joint velocity control system guaranteeing that the actual joint velocity $\dot{\boldsymbol{q}}$ matches the desired joint velocity $\boldsymbol{\omega}_d$, i.e.,

$$\dot{\boldsymbol{q}}(t) \equiv \boldsymbol{\omega}_d(t). \quad (6)$$

Because the analytical robot Jacobian $J(\boldsymbol{q})$ is assumed full–rank, and inspired from the resolved motion rate control philosophy [10], we propose the fol-

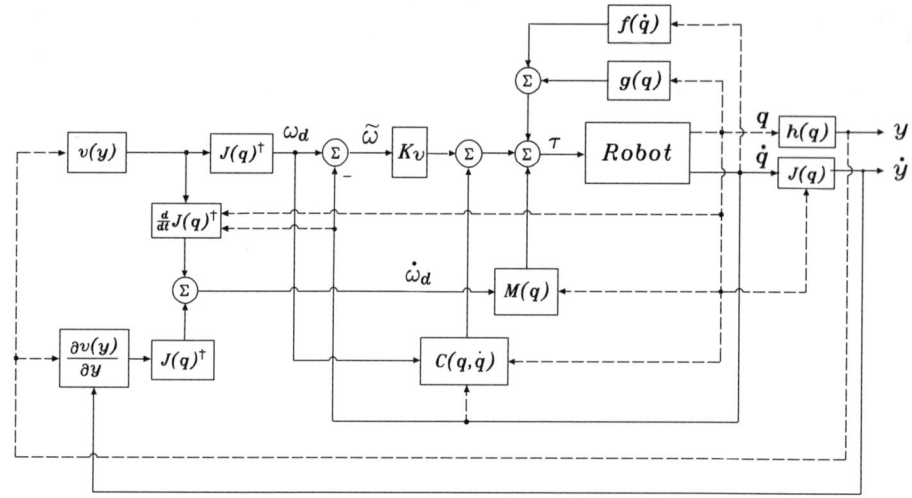

Figure 3: Block diagram of the hierarchical velocity field control

lowing control law to generate the desired joint velocity $\boldsymbol{\omega}_d$

$$\boldsymbol{\omega}_d = J(\boldsymbol{q})^\dagger \boldsymbol{v}(\boldsymbol{y}), \quad (7)$$

where $J(\boldsymbol{q})^\dagger = J(\boldsymbol{q})^T \left[J(\boldsymbol{q}) J(\boldsymbol{q})^T \right]^{-1}$ is the Jacobian right pseudoinverse. Under the assumption of exact joint velocity tracking (6) and substituting (7) into (4) we get

$$\dot{\boldsymbol{y}} = \boldsymbol{v}(\boldsymbol{y}),$$

satisfiying directly the velocity field control objective (5).

The control law (7) will guarantee the velocity field control objective as long as the velocity matching assumption (6) holds. However, most of velocity controllers used in practice assure —in the best case— only asymptotic velocity tracking instead of exact tracking. The consequence of this fact is that the velocity matching assumption (6) may fail in practice depending on the manipulator servo system and the range of requested desired joint velocity $\boldsymbol{\omega}_d$. Let us define the joint velocity error as

$$\tilde{\boldsymbol{\omega}} = \boldsymbol{\omega}_d - \dot{\boldsymbol{q}}. \quad (8)$$

In this way, from the differential kinematic (4) and the definition of the joint velocity error (8) we have

$$\dot{\boldsymbol{y}} = \boldsymbol{v}(\boldsymbol{y}) - J(\boldsymbol{q})\tilde{\boldsymbol{\omega}}. \quad (9)$$

In equation (9), we can see that the operational space robot velocity $\dot{\boldsymbol{y}}$ is disturbed by the term $J(\boldsymbol{q})\tilde{\boldsymbol{\omega}}$. Thus, as soon as the joint velocity error $\tilde{\boldsymbol{\omega}}$ vanishes, the operational space robot velocity $\dot{\boldsymbol{y}}$ will match the desired velocity field $\boldsymbol{v}(\boldsymbol{y})$.

2.1 Velocity field control: Dynamic control via an hierarchical approach

It is possible to achieve the velocity field control objective using the kinematic control law (7), and a control law defined in the joint space for generating suitable torques.

A joint velocity controller can be derived from the PD+ control philosophy [11] giving

$$\boldsymbol{\tau} = M(\boldsymbol{q})\dot{\boldsymbol{\omega}}_d + C(\boldsymbol{q},\dot{\boldsymbol{q}})\boldsymbol{\omega}_d + \boldsymbol{g}(\boldsymbol{q}) + K_v \tilde{\boldsymbol{\omega}} + \boldsymbol{f}(\dot{\boldsymbol{q}}), \quad (10)$$

where K_v is a symmetric positive definite matrix, and $\dot{\boldsymbol{\omega}}_d(t)$ stands for the desired joint acceleration. This controller is an asymptotic velocity controller in the sense that it produces a closed–loop system which is globally asymptotically stable and the joint velocity error $\tilde{\boldsymbol{\omega}}(t)$ vanishes as $t \to \infty$.

If the desired joint velocity $\boldsymbol{\omega}_d$ is chosen according to the kinematic control law (7) and the desired acceleration as

$$\dot{\boldsymbol{\omega}}_d = J(\boldsymbol{q})^\dagger \frac{\partial \boldsymbol{v}(\boldsymbol{y})}{\partial \boldsymbol{y}} \dot{\boldsymbol{y}} + \left[\frac{d}{dt} J(\boldsymbol{q})^\dagger \right] \boldsymbol{v}(\boldsymbol{y}), \quad (11)$$

then the control system can be seen as a hierarchical control scheme having an outer loop defined in the operational space, and an inner loop defined the joint space. Figure 3 depicts the hierarchical controller (7), (10), and (11) in block diagram form.

After some manipulations, using equations (1) and (10), the closed–loop system is given by

$$M(\boldsymbol{q})\dot{\tilde{\boldsymbol{\omega}}} + C(\boldsymbol{q},\dot{\boldsymbol{q}})\tilde{\boldsymbol{\omega}} + K_v \tilde{\boldsymbol{\omega}} = \boldsymbol{0}. \quad (12)$$

Observe that $\tilde{\boldsymbol{\omega}} = \boldsymbol{0}$ is the unique equilibrimum point of (12).

We propose the following Lyapunov function candidate:

$$V(\widetilde{\boldsymbol{\omega}}) = \frac{1}{2}\widetilde{\boldsymbol{\omega}}^T M(\boldsymbol{q})\widetilde{\boldsymbol{\omega}},$$

whose time derivative along the closed–loop system trajectories (12) is given by

$$\dot{V}(\widetilde{\boldsymbol{\omega}}) = -\widetilde{\boldsymbol{\omega}}^T K_v \widetilde{\boldsymbol{\omega}},$$

where the property (2) was used. The time derivative of the Lyapunov function is globally negative definite. Thus, invoking the Lyapunov direct method [12], we have that $\widetilde{\boldsymbol{\omega}} = \mathbf{0}$ is a globally asymptotically stable equilibrimum point, therefore

$$\lim_{t\to\infty} \widetilde{\boldsymbol{\omega}}(t) = \mathbf{0}.$$

We have assumed that the robot Jacobian $J(\boldsymbol{q})$ is full rank and bounded. This allows the conclusion from equation (9) that

$$\lim_{t\to\infty} \dot{\boldsymbol{y}}(t) = \boldsymbol{v}(\boldsymbol{y}(t)).$$

Thus, the velocity field control objective (5) is satisfied with controller (7), and (10).

3 Experimental results

Velocity field control experiments have been carried out in order to assess the performance of the proposed velocity field controller.

We have built at CICESE Research Center a two degrees–of–freedom experimental direct–drive robot arm to enable real–time control experiments. The arm moves in the vertical plane as shown in Figure 4. The motors used in the robot arm are the DM1015–B and DM1004–C models from Parker Compumotor for the shoulder and elbow joints respectively. The motors are operated in torque mode, so they act as torque source and accept an analog voltage as a reference of torque signal. The control algorithm is executed at 1 [msec] sampling period in a PC host computer equipped with the data adquisition board MFIO–3A from Precision MicroDynamics. The origin of the Cartesian frame is attached at the axis of rotation of the first joint while y_1 and y_2 denote the horizontal and vertical axes respectively. With this convention, the direct kinematics is given by

$$\boldsymbol{h}(\boldsymbol{q}) = \begin{bmatrix} l_1 \sin(q_1) + l_2 \sin(q_1 + q_2) \\ -l_2 \cos(q_1) - l_2 \cos(q_1 + q_2) \end{bmatrix},$$

which leads to the following Jacobian matrix

$$J(\boldsymbol{q}) = \begin{bmatrix} l_1 \cos(q_1) + l_2 \cos(q_1 + q_2) & l_2 \cos(q_1 + q_2) \\ l_2 \sin(q_1) + l_2 \sin(q_1 + q_2) & l_2 \sin(q_1 + q_2) \end{bmatrix}.$$

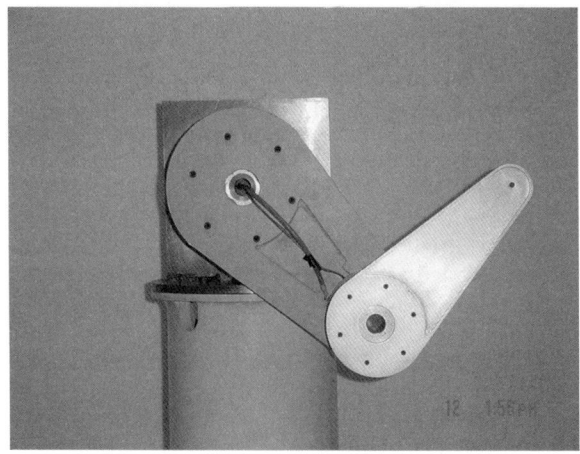

Figure 4: Experimental arm

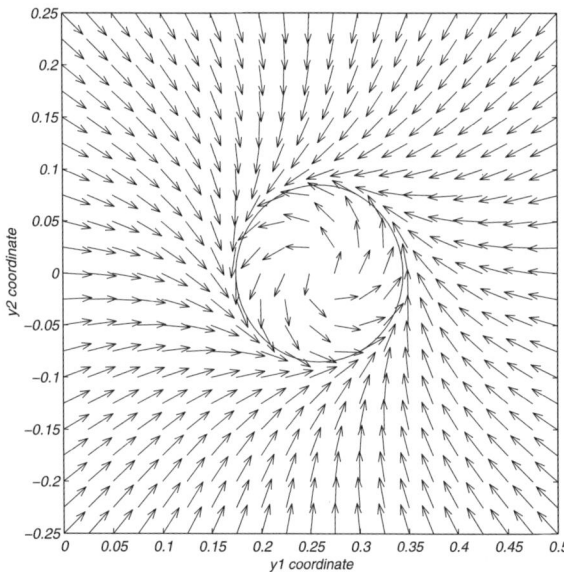

Figure 5: Desired velocity field used in experiments

For the experimental system $l_1=l_2=0.26$ [m]. The dynamics of the experimental robot arm is described in the Appendix.

3.1 Desired velocity field

The experiments of velocity field control have considered a velocity field that encode the task of tracing a circle in the operational space. This means that the flow lines of the velocity field converges to a circular contour. The proposed velocity field $\boldsymbol{v}(\boldsymbol{y})$ to generate the flow lines convergent to the circle in the y_1–y_2 plane is shown in Figure 5. The desired speed at the circle is $v_0 = 0.65$ [m/sec].

4377

The mathematical expression of the desired velocity field $v(y)$ is given by

$$v(y) = -k(y) f(y) \begin{bmatrix} 2[y_1 - y_{c1}] \\ 2[y_2 - y_{c2}] \end{bmatrix} + c(y) \begin{bmatrix} -2[y_2 - y_{c2}] \\ 2[y_1 - x_{c1}] \end{bmatrix}$$

where $f(y) = [y_1 - y_{c1}]^2 + [y_2 - y_{c2}]^2 - r_0^2$ with $y_{c1} = 0.26$ [m], $y_{c2} = 0.0$ [m], and $r_0 = 0.085$ [m]. Denoting $\nabla f(y)$ the gradient of $f(y)$, the functions $k(y)$ and $c(y)$ are defined as

$$k(y) = \frac{k_0}{|f(y)| \, \|\nabla f(y)\| + \epsilon}, \quad (13)$$

$$c(y) = \frac{v_0 \exp^{-\alpha |f(y)|}}{\|\nabla f(y)\|}, \quad (14)$$

where $k_0 = 0.65$ [m/s], $v_0 = 0.65$ [m/s], $\alpha = 50$ [m^{-2}], and $\epsilon = 0.00075$ [m^3].

3.2 Experiments

The hierarchical control scheme described by equations (7), and (10) was tested. We set the gains related with this controller as

$$K_v = \mathrm{diag}\{6.0, 5.0\} \; [\mathrm{Nm} \cdot \mathrm{sec/rad}]$$

The compenents of the velocity field error are shown in Figures 6 and 7, respectively. As expected, the velocity field error remains around zero. Good performance is observed from Figure 8 where the path of the arm tip is depicted. The velocity field errors present maximum peaks in steady state of 0.075 [m/sec], which gives a relative error of 11.5 % with respect to the desired speed at the circular contour $v_0 = 0.65$ [m/sec]. With the proposed controller the desired velocity field is followed with good performance since the arm tip is guided towards the circular contour as shown is Figure 8. One important observation is that no fast motion transient was presented at the beginning of the motion. One reason for this is that the robot attained the desired velocity specified by the velocity field at the current configuration in a fast way with measured control effort.

4 Conclusions

In this paper we have presented a solution to the velocity field control for robot manipulators. The proposed solution consists of a hierarchical control structure based on kinematic control for the resolution of desired joint velocity, and an asymptotically joint velocity controller. Experiments were carried out on a two degrees–of–freedom direct–drive arm, showing good performance in the tracking of the desired velocity field.

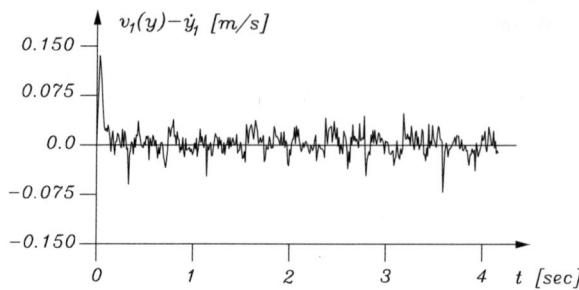

Figure 6: Component of the velocity field error: $v_1(y) - \dot{y}_1$

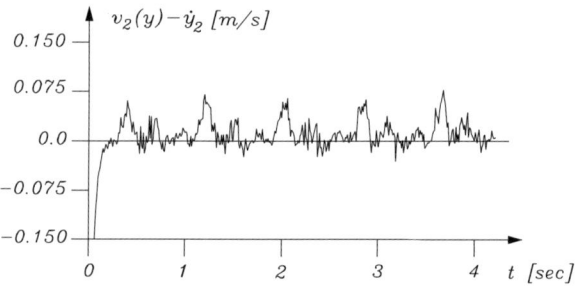

Figure 7: Component of the velocity field error: $v_2(y) - \dot{y}_2$

Appendix

We present below the entries of the robot dynamics. The elements $M_{ij}(q)$ $(i,j = 1, 2)$ of the inertia matrix are:

$$\begin{aligned}
M_{11}(q) &= 0.3353 + 0.02436\cos(q_2), \\
M_{12}(q) &= 0.01267 + 0.01218\cos(q_2), \\
M_{21}(q) &= 0.01267 + 0.01218\cos(q_2), \\
M_{22}(q) &= 0.01267.
\end{aligned}$$

The elements $C_{ij}(q, \dot{q})$ $(i, j = 1, 2)$ of the centrifugal and Coriolis matrix are:

$$\begin{aligned}
C_{11}(q, \dot{q}) &= -0.01218\sin(q_2)\dot{q}_2, \\
C_{12}(q, \dot{q}) &= -0.01218\sin(q_2)\dot{q}_1 - 0.01218\sin(q_2)\dot{q}_2, \\
C_{21}(q, \dot{q}) &= 0.01218\sin(q_2)\dot{q}_1, \\
C_{22}(q, \dot{q}) &= 0.0.
\end{aligned}$$

The entries of the gravitational torque vector $g(q)$ are given by:

$$\begin{aligned}
g_1(q) &= g[1.1731\sin(q_1) + 0.04685\sin(q_1 + q_2))], \\
g_1(q) &= g[0.04685\sin(q_1 + q_2)],
\end{aligned}$$

where $g = 9.81$ [m/sec^2].

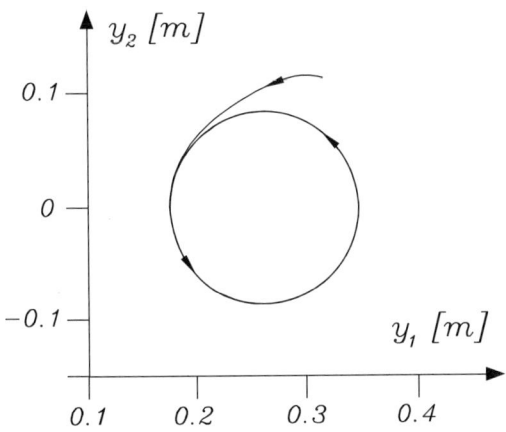

Figure 8: Path of the arm tip

Friction torques have been modeled by

$$f_1(\dot{q}) = f_{v_1}\dot{q}_1 + f_{C_1}\text{sgn}(\dot{q}_1),$$
$$f_2(\dot{q}) = f_{v_2}\dot{q}_2 + f_{C_2}\text{sgn}(\dot{q}_2),$$

where

$$\text{sgn}(x) = \begin{cases} 1 & x > 0, \\ -1 & x < 0. \end{cases}$$

Following the procedures exposed in [13], and references therein, we have estimated the parameters of the viscous and Columb model of friction for the experimental arm.

For joint 1: f_{v_1}=0.2741 [Nm sec/rad], and f_{C_1}=1.29 [Nm].

For joint 2: f_{v_2}=0.1713 [Nm sec/rad], and f_{C_2}=0.965 [Nm].

References

[1] M.W. Spong and M. Vidyasagar, *Robot dynamics and control*, John Wiley and Sons, New York, 1989.

[2] L. Sciavicco and B. Siciliano, *Modeling and control of robot manipulators*, Springer, London, 2000.

[3] C. Canudas de Wit, B. Siciliano and G. Bastin (Eds.), *Theory of robot control*, Springer-Verlag, London, 1996.

[4] P. Y. Li and R. Horowitz, "Passive velocity field control of mechanical manipulators" *IEEE Trans. on Robotics and Automation*, Vol. 15, No. 4, pp. 751–763, 1999.

[5] P.Y. Li and R. and Horowitz, "Passive velocity field control (PVFC): Part I–Geometry and robustness" *IEEE Trans. on Automatic Control*, Vol. 46, No. 9, pp. 1346–1359, 2001.

[6] R. M. Murray, Z. Li, and S. S. Sastry, *A mathematical introduction to robotic manipulation*, CRC Press, Boca Raton, Florida, 1994.

[7] J. Moreno and R. Kelly, "On manipulator control via velocity fields", *Proc. 15th IFAC World Congress*, Barcelona, Spain, July, 2002.

[8] P. Y. Li, "Coordinated contour following control for machining operations – A survey", *Proc. of the American Control Conference*, San Diego, CA., pp. 4543–4547, June 1999.

[9] B. Siciliano, "Kinematic control of redundant robot manipulators", *Journal of Intelligent and Robotic Systems*, vol. 3, pp. 201–212, 1990.

[10] D. E. Whitney, "Resolved motion rate control of manipulators and human prostheses", *IEEE Trans. on Man–Machine Systems*, vol. MMS–10, No. 2, pp. 47–53, June 1969.

[11] D. Koditschek, "Natural motion for robot arms", *Proc. of the 23rd IEEE Conference on Decision and Control*, Las Vegas, NV, pp. 733–735, 1984.

[12] M. Vidyasagar, *Nonlinear Systems Analysis*, Prentice–Hall, Englewood Cliffs, 1993.

[13] Kelly R., J. Llamas and R. Campa, "Measurement procedure for viscous and Coulomb friction", *IEEE Trans. Instrum. Meas.*, Vol. 49, No. 4, pp. 857–861, August 2000.

Better Robot Tracking Accuracy with Phase Lead Compensated ILC

Yongqiang Ye and Danwei Wang*
School of Electrical and Electronic Engineering
Nanyang Technological University
Singapore 639798

Abstract

In this paper, a learning control scheme is proposed to improve robot tracking accuracy. Through the analysis in frequency domain, it is shown that phase lead compensation can broad the learnable frequency band of a learning control system. The phase lead compensation is realized by phase lead filtering the error of last repetition. In theory a filter whose phase difference to the system is within $\pm 90^\circ$ can be a candidate for the phase lead compensation process. Experimental results on an industrial robot system show that the proposed scheme is both effective and robust against dynamic modeling errors.

1 Introduction

Most industrial robot systems are used for repetitive operations. In consideration of the periodicity of the tasks, a control method called *Iterative Learning Control* (ILC) has been proposed for such cases [1]. Knowledge of the dynamics is often imprecise and, further, the information about the system is prone to possible changes during operation. One of the important features of the learning control is that it requires less a priori knowledge about the controlled system in the system design phase. Learning control aims to produce zero tracking error during the whole period of a process operation including the transient part using minimum knowledge of the system. This is accomplished by using the past experience with the same task in order to improve the performance in the future. The input signal is updated based on the recorded data from previous trials to make the output converge to the desired output.

*Email: edwwang@ntu.edu.sg Phone:+65-67905376
Fax: +65-67920415

In [1], the input update utilized the derivative signals of the previous error signal. Thus the learning law is termed D-type ILC. A drawback of D-type ILC is that it introduces high frequency noises. This problem is overcome by P-type ILC [2] which learns from the tracking error itself as follows

$$u_j(t) = u_{j-1}(t) + L(\cdot)e_{j-1}(t) \quad (1)$$

When P-type ILC is used together with D-type ILC, so called PD-type ILC, this has been shown to be effective in ensuring the convergence of the tracking errors [5].

An alternative approach for ILC is to phase lead compensate the error of last repetition $e_{j-1}(t)$ in (1) and the ILC law has the following form

$$u_j(t) = u_{j-1}(t) + L(\cdot)e'_{j-1}(t) \quad (2)$$

where $e'_{j-1}(t)$ is a phase lead filtered version of $e_{j-1}(t)$. One benefit of the phase lead compensation is a significantly broader learnable frequency band [10] which will result a higher precision. Generation of a proper $e'_{j-1}(t)$ in (2) can be an analysis and design problem in frequency domain. In this paper, the necessity of phase lead compensation is shown in frequency domain. As a realization of the phase lead compensation, a filter is used to phase lead filter the error $e_{j-1}(t)$, resulting in a $e'_{j-1}(t)$. Phase bound of the candidate phase lead filter is derived.

2 ILC Controlled Robot System

Consider a robot joint system modeled by the transfer function

$$\frac{Y(s)}{U(s)} = G_p(s) \quad (3)$$

where the $Y(s)$ is the output joint angle and $U(s)$ is the control input torque. It is assumed that any feedback control, if implemented, has been included in the transfer function in (3). The frequency characteristics of (3) is given as follows

$$G_p(j\omega) = N_p(\omega)\exp(j\theta_p(\omega)) \qquad (4)$$

where $N_p(\omega)$ and $\theta_p(\omega)$ are the magnitude characteristics and phase characteristics, respectively.

The output Laplace transform for the jth operation cycle is

$$Y_j(s) = G_p(s)U_j(s) \qquad (5)$$

Suppose $Y_d(t)$ is a desired joint angle trajectory defined on a finite time operation interval $[0,T]$. The jth operation cycle produces a tracking error $E_j(s) = Y_d(s) - Y_j(s)$.

Let the Laplace transform of the learning law be

$$U_j(s) = U_{j-1}(s) + k\Phi_c(s)E_{j-1}(s) \qquad (6)$$

where k is the scalar learning gain and $\Phi_c(s)$ is the learning compensator in Laplace form whose DC gain is 1, $\lim_{s\to 0}\Phi_c(s) = 1$. The learning compensator $\Phi_c(s)$ has frequency characteristics $\Phi_c(j\omega) = N_c(\omega)\exp(j\theta_c(\omega))$ with $N_c(\omega)$ and $\theta_c(\omega)$ being its magnitude characteristics and phase characteristics, respectively.

Use (5) and (6) and we get

$$E_j(s) = [1 - kG_p(s)\Phi_c(s)]E_{j-1}(s) \qquad (7)$$

$[1 - kG_p(s)\Phi_c(s)]$ can be viewed as a transfer function from the tracking error at operation cycle $(j-1)$ to the tracking error at operation cycle j. If we want the tracking error decay monotonically every operation cycle for all frequencies, the condition for tracking error contraction at steady state is

$$|1 - kG_p(j\omega)\Phi_c(j\omega)| < 1 \qquad (8)$$

A cutoff frequency is introduced to stop learning for the error components with frequencies at which this condition is violated.

Using the characteristics of $G_p(j\omega)$ and $\Phi_c(j\omega)$, (8) leads to

$$kN_p(\omega)N_c(\omega) < 2\cos(\theta_p(\omega) + \theta_c(\omega)) \qquad (9)$$

If $k > 0$, (9) necessarily requires

$$-90^\circ < \theta_p(\omega) + \theta_c(\omega) < 90^\circ \qquad (10)$$

Condition (10) is vital because if (10) is satisfied we can always find a learning gain k small enough to satisfy (9). Therefore the selection of learning compensator $\Phi_c(j\omega)$ is based on (10). Of course, $\theta_c(\omega) + \theta_p(\omega) = 0$, is the ideal case [11]. But full phase cancellation is impossible as an exact system model is unavailable.

3 Phase Lead Compensated ILC

3.1 Phase Lead Filtering

Because stable robotic systems are minimum phase systems and $\theta_p(\omega)$ is negative, learning compensator $\theta_c(\omega)$ should generally introduce phase lead (a positive $\theta_c(\omega)$). Several phase lead filtering methods have been proposed in the existing learning control literatures. System inversion [7] is one simple solution and the phase lead filter is the inverse system model. Inverting a model will introduce substantial noises. In [3], the phase lead filter is the discretized inverse of the resonance mode in a system. This can be regarded as a partial system inversion. In [4, 6], two different learning gain matrices are used to phase lead filter the error 'supervector' for discrete-time systems. The first one is the transpose of the system's Markov parameter matrix. The second one comprises a set of gains that are computed from the inverse of the plant's phase characteristics using Inverse Discrete Fourier Transform (IDFT).

Half of the zero-phase filtering process [9] (also in [8]) is firstly utilized in [11], termed *reverse sequence method*, to generate a desired phase lead to cancel the phase lag of the system. The phase lead filtering process in [11] is restated here. Suppose we have the error signal $e(t)$, $t \in [0,T]$ after one operation cycle. Firstly, this time sequence $e(t)$ is reversed to produce another series $e1(t)$, i.e., $e1(t) = e(T-t)$. Secondly, filter $e1(t)$ by a filter $F(s)$ (with $N_f(\omega)$ and $\theta_f(\omega)$ being its magnitude characteristics and phase characteristics, respectively) and the output is recorded as $e2(t)$. Finally, the time sequence of $e2(t)$ is reversed again to get $e3(t)$. Then the signal $e3(t)$ is leading the original error signal $e(t)$ by a phase $-\theta_f(\omega)$, i.e.,

$$E3(j\omega) = N_f(\omega)e^{-j\theta_f(\omega)}E(j\omega) \qquad (11)$$

where $E(j\omega)$ is the Fourier transform of $e(t)$ and $E3(j\omega)$ is the Fourier transform of $e3(t)$. When a filter $F(s)$ is used in the above process, it is termed a phase lead filter.

3.2 Learning Law

The phase lead compensated learning control law in Fourier form is concluded as

$$\begin{cases} E1(j\omega) = e^{jT\omega} E_{j-1}^*(j\omega) \\ E2(j\omega) = F(j\omega) E1(j\omega) \\ E3(j\omega) = e^{jT\omega} E2^*(j\omega) \\ U_j(j\omega) = U_{j-1}(j\omega) + kE3(j\omega) \end{cases} \quad (12)$$

where $*$ denotes complex conjugate. According to (11), $E3(j\omega) = F^*(j\omega) E_{j-1}(j\omega)$ and (12) can be reduced to

$$U_j(j\omega) = U_{j-1}(j\omega) + kF^*(j\omega) E_{j-1}(j\omega) \quad (13)$$

3.3 Phase Bound of Candidate Filters

Replace $\Phi_c(j\omega)$ by $N_f(\omega) e^{-j\theta_f(\omega)}$ in (8), and (9) and (10) become

$$kN_p(\omega) N_f(\omega) < 2\cos(\theta_p(\omega) - \theta_f(\omega)) \quad (14)$$
$$-90^o < \theta_p(\omega) - \theta_f(\omega) < 90^o \quad (15)$$

Condition (15) shows that a filter whose phase difference to the system is within a $\pm 90^o$ bound can be a potential phase lead filter.

4 Robot Performance and Experiments

The proposed scheme is tested on an industrial robot, SEIKO TT3000, which is a SCARA type robotic manipulator, as shown in Fig. 1. Fig.

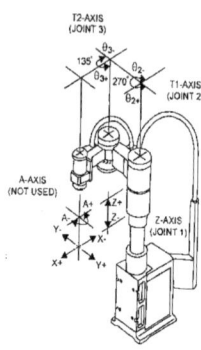

Figure 1: Experimental Robot Arm

2 shows the hardware setup of the robot control system.

Joints 2 and 3 control the two links moving in a horizontal plane. The dynamics of these two

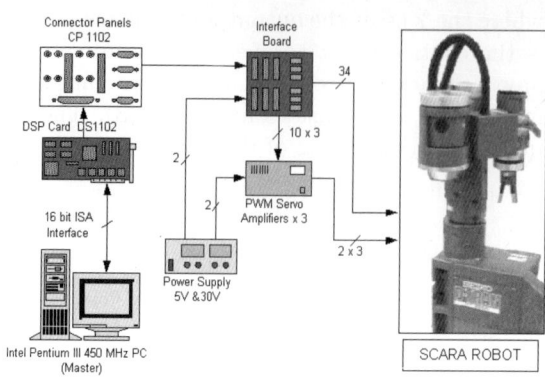

Figure 2: Robot Control System

joints possess the non-linear and coupling characteristics and are given as

$$\begin{bmatrix} (m_2+m_3)a_2^2 + m_3 a_3^2 + 2m_3 a_2 a_3 \cos\theta_3 & m_3 a_3^2 + m_3 a_2 a_3 \cos\theta_3 \\ m_3 a_3^2 + m_3 a_2 a_3 \cos\theta_3 & m_3 a_3^2 \end{bmatrix} \begin{bmatrix} \ddot{\theta}_2 \\ \ddot{\theta}_3 \end{bmatrix}$$
$$+ \begin{bmatrix} -m_3 a_2 a_3 (2\dot{\theta}_2 \dot{\theta}_3 + \dot{\theta}_3^2)\sin\theta_3 \\ m_3 a_2 a_3 \dot{\theta}_2^2 \sin\theta_3 \end{bmatrix} + \begin{bmatrix} f_{v2}\dot{\theta}_2 \\ f_{v3}\dot{\theta}_3 \end{bmatrix} + \begin{bmatrix} F_{c2} \\ F_{c3} \end{bmatrix} = \begin{bmatrix} \tau_2 \\ \tau_3 \end{bmatrix} \quad (16)$$

where θ_j, f_{vj}, F_{cj} and τ_j, $j = 2,3$ are the joint angles, viscous frictions, Coulomb frictions and control torques of joints 2 and 3, respectively. m_j and a_j, $j = 2,3$, are the masses and mass centers of links 2 and 3, respectively. Joints 2 and 3 are firstly stabilized by decentralized feedback P controllers with $k_{p2} = k_{p3} = 0.1$, Fig. 3. Two SISO

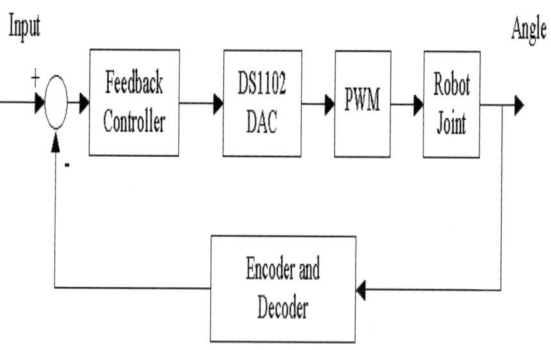

Figure 3: System Structure of Feedback Loop

learning controllers work for the two closed-loop systems in a decentralized way. The nonlinear and coupling characteristics in these two joints are good test on the robustness of the proposed learning law.

The desired joint angle trajectories for joints 2

and 3 are the same as in [10],

$$y_d(t) = \sum_{n=1}^{51} a_n[1-\cos(\omega_n t)] \text{ degree}; \ 0 \le t \le 10 \text{ second}$$

where ω_n are $0.1\pi, 2\pi, 4\pi, 6\pi, ..., 100\pi$, and the amplitude are $a_n = 80e^{-\omega_n}$. The sampling time is 0.01second. The learning gain k is 1 and no cutoff is employed in all experiments.

To illustrate the proposed learning law, the following three filters with different orders are used as phase lead filters, respectively.

$$F_1(s) = \frac{25}{s+25} \quad (17)$$

$$F_2(s) = \frac{625}{s^2 + 25s + 625} \quad (18)$$

$$F_3(s) = \frac{15625}{s^3 + 50s^2 + 1250s + 15625} \quad (19)$$

Fig. 4 and 6 show the phase lead filtering effect on the errors of joints 2 and 3 at repetition 0 under the three phase lead filters. (The input is $y_d(t)$ for both joints at repetition 0). Note that the phase lead filtered errors are shifted forward with respect to the original errors. Fig. 5 and 7 are amplified parts of Fig. 4 and 6. Fig. 8

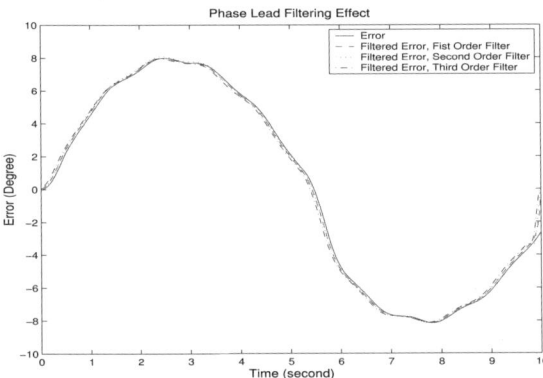

Figure 4: Phase Lead Filtering Effect on Error of Joint 2

and 9 illustrate the RMS error histories of joints 2 and 3 under the three different phase lead filters. Though our robot joints are second order systems, the forementioned three filters all make ILC converge. This implies that the learning law is robust against the modeling error of the robot joint dynamics.

Two other filters whose order differences to the robot joints are more than one are also tested in the experiments,

$$F_0(s) = 1 \quad (20)$$

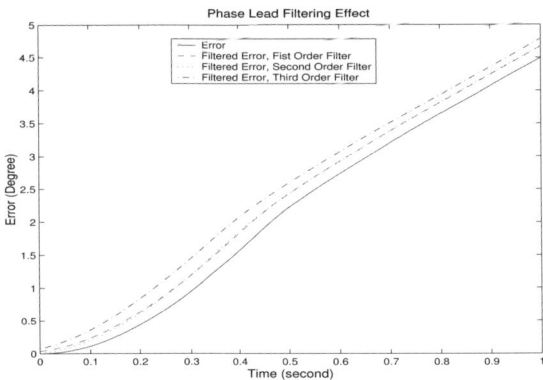

Figure 5: Phase Lead Filtering Effect on Error of Joint 2 (1st Second)

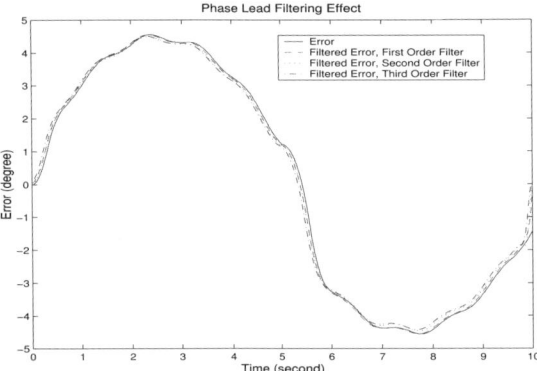

Figure 6: Phase Lead Filtering Effect on Error of Joint 3

$$F_4(s) = \frac{390625}{s^4 + 50s^3 + 1875s^2 + 31250s + 390625} \quad (21)$$

Note that $F_0(s)$ is constant 1 which means there is no phase lead compensation and the learning law is simply a kind of P-type law,

$$u_j(t) = u_{j-1}(t) + e_{j-1}(t) \quad (22)$$

The two learnings both exhibit the behavior of first convergence followed by divergence because (15) is violated at high frequencies, Fig. 10 and 11. In the case of ILC with no filter, with the increasing of the errors, the two joints hit the hardware limits more and more fiercely and we have to stop the experiment after repetition 136 for fearing of damaging the robot.

For comparison, three system inversion ILC experiments have also been carried out. Their learning compensators are the inverse of the three

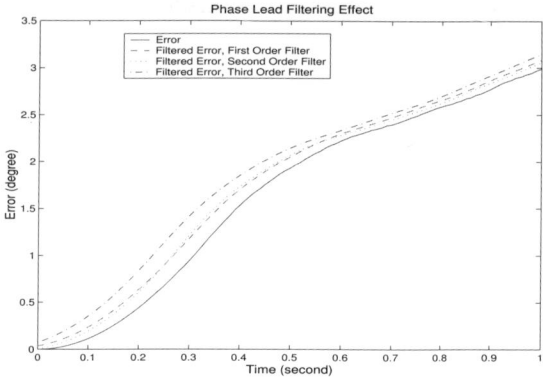

Figure 7: Phase Lead Filtering Effect on Error of Joint 3 (1st Second)

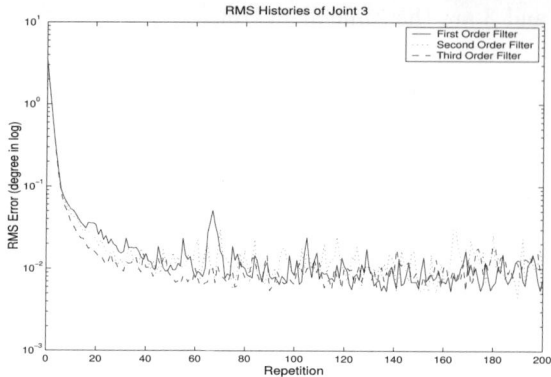

Figure 9: RMS Error Histories of Joint 3

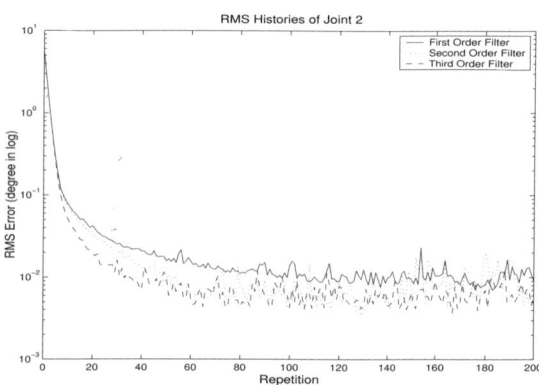

Figure 8: RMS Error Histories of Joint 2

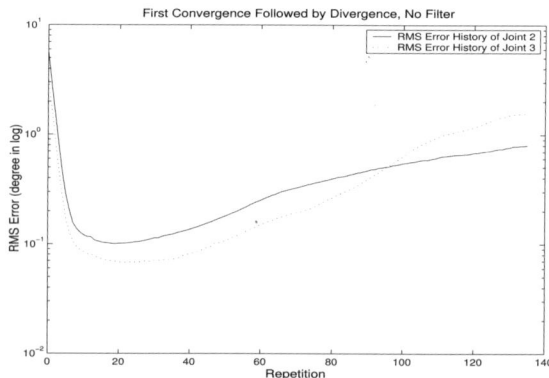

Figure 10: Failed Learning, No Phase Compensation

filters, $F_1(s)$, $F_2(s)$ and $F_3(s)$, respectively. In all the three cases, the RMS errors decrease at repetition 1, then diverge quickly, Fig. 12 and 13. The properly phase lead compensated ILC is more robust than the system inversion ILC.

5 Conclusion

The phase lead compensated learning control uses the time reverse and filtering techniques to phase lead filter the error. The robustness of the phase lead compensated learning control law is rooted in the fact that a filter whose phase difference to the system is within a $\pm 90^o$ bound can be a potential phase lead filter. Analysis and experiments show that robot performance can be improved in better tracking accuracy by the SISO ILC law in the presence of the nonlinear and coupling effects. As illustrated by the design of phase lead filters (17), (18) and (19), the proposed scheme is very robust against the modeling errors of the robot dynamics, including the case where the orders of the filter and the real system are different by one, be it higher or lower.

References

[1] S. Arimoto, S. Kawamura and F. Miyazaki, "Bettering operation of robots by learning," *Journal of Robotic System*, vol. 1, 1984, pp. 123-140.

[2] S. Arimoto, "Learning control theory for robotic motion," *International Journal of Adaptive Control and Signal Processing*, vol. 4, 1990, pp. 543-564.

[3] H. Elci, R. W. Longman, M. Phan, J.-N. Juang and R. Ugoletti, "Discrete frequency based learning control for precision motion control," *Proceeding of the 1994 IEEE Inter-*

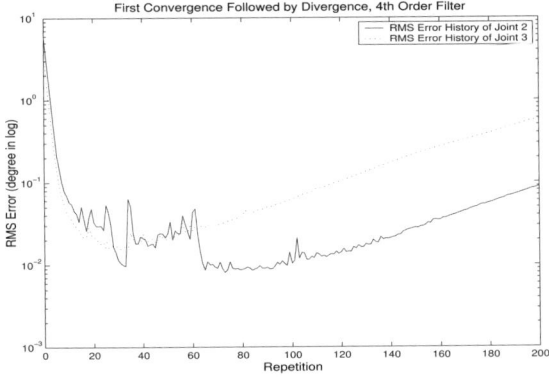

Figure 11: Failed Learning, 4th Order Filter

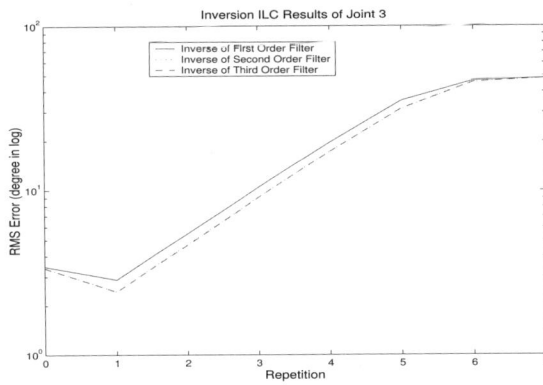

Figure 13: System Inversion ILC, Joint 3

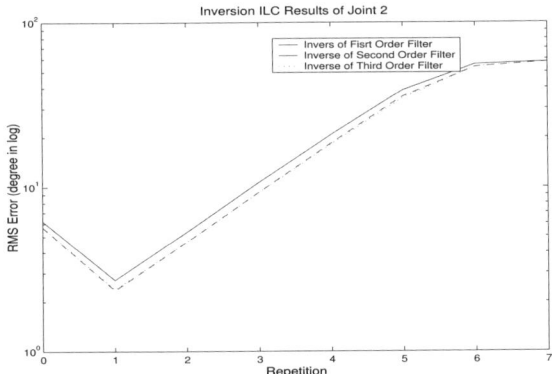

Figure 12: System Inversion ILC, Joint 2

national Conference on Systems, Man, and Cybernetics, San Antonio, USA, vol. 3, 1994, pp. 2767-2773.

[4] H. Elci, R. W. Longman, M. Phan, J.-N. Juang, and R. Ugoletti, "Automated learning control through model updating for precision motion control," ASME Adaptive Structures and Composite Materials: Analysis and Application, vol. 54, 1994, pp. 299-314.

[5] G. Heinzinger, D. Fenwick, B. Paden and F. Miyazaki, "Stability of learning control with disturbance and uncertain initial conditions," IEEE Transaction on Automatic Control, vol. 37, 1992, pp. 110-114.

[6] H. S. Jang and R. W. Longman, "An update on a monotonic learning control law and some fuzzy logic learning gain adjustment techniques," Advances in Astronautical Sciences, vol. 90, 1996, pp. 301-318.

[7] G. Lee-Glauser, J.-N. Juang and R. W. Longman, "Comparison and combination of learning controllers: computational enhancement and experiments," Journal of Guidance, Control and Dynamics, vol. 19, 1996, pp. 1116-1123.

[8] R. W. Longman, "Iterative learning control and repetitive control for engineering practice," International Journal of Control, vol. 73, 2000, pp. 930-954.

[9] R. W. Longman, "Practical design methods for iterative learning control and repetitive control," Tutorial Notes of the Sixth International Conference on Control, Automation, Robotics and Vision, 2000, Singapore.

[10] R. W. Longman and T. Songchon, "Tradeoffs in designing learning/repetitive controller using zero-phase filter for long term stabilization," Advances in the Astronautical Sciences, vol. 99, 1999, pp. 673-692.

[11] Y. Ye and D. Wang, "Phase Cancellation Learning Control," Submitted to ASME Journal of Dynamic Systems, Measurement and Control.

Forcefree Control with Independent Compensation for Inertia, Friction and Gravity of Industrial Articulated Robot Arm

Satoru GOTO and Masatoshi NAKAMURA

Department of Advanced Systems Control Engineering,

Saga University,

Honjomachi, Saga 840-8502,

Japan

Nobuhiro KYURA

Department of Electrical and Computer Engineering,

Kinki University in Kyushu

11-6 Kayanomori, Iizuka 820-8555

Japan

Abstract— Forcefree control can realize the flexible motion of industrial articulated robot arms without any change of the built-in controller. In this paper, the forcefree control was extended to realize flexible motion emulating operational circumstance free of inertia, friction and gravity through the independent compensation of inertia, friction and gravity. The property of the forcefree control with independent compensation was also investigated by experimental study of an actual industrial articulated robot arm, where the external force was measured with a force sensor which was attached to the tip of the robot arm.

Keywords— Forcefree control, independent compensation, industrial articulated robot arm

I. INTRODUCTION

A lot of industrial robot arms are worked in industry. Some works of industrial robot arms, such as pulling-out of products made by die casting, require flexible motion according to external force. A number of methods for flexible motion realization of the robot arms such as impedance control[1], [2], [3] and compliance control[4], [5] are proposed. However, these methods are difficult to be used in industrial robot arms because of the complexity of algorithms and the necessity to change the built-in controller of the industrial robot arms. On the other hand, servo float method can realize the flexible motion of the industrial robot arms and it has already used in industry[6]. The servo float method, however, requires mode change of the controller for the realization of the flexible motion and the change is not convenient for applications.

The authors have proposed the forcefree control[7] and it can be used to realize the flexible motion without change of the built-in controller. The forcefree control realizes a motion of the industrial robot arm as if it were in circumstance of non-gravity and non-friction condition.

In this paper, the previously proposed forcefree control[7] was extended to realize flexible motion emulating operational the circumstance free of inertia, friction and gravity through the independent compensation of inertia, friction and gravity. The property of the forcefree control with independent compensation was also investigated by experimental study of an actual industrial robot arm.

II. FORCEFREE CONTROL WITH INDEPENDENT COMPENSATION

A. Concept of Forcefree Control of Industrial Robot Arm

When a person pushes the tip of an industrial robot arm, the robot arm does not move because the robot arm is actuated by the servo motor connected to the links with a high ratio gear. Servo controller of an industrial robot arm keeps position of the robot arm until the input signal responsible for motion is applied. The torque generated by the external force is compensated by the servo controller.

The forcefree control can realize the motion of the industrial robot arm under non-friction and non-gravity condition[7]. By use of the forcefree control, the robot arm moves by the external force directly as if it were under the condition of non-friction and non-gravity through appropriate control of servo motors. In this paper, the forcefree control is extended to realize flexible motion without experiencing the effect of inertia, friction and gravity for operation.

B. Dynamics of Industrial Articulated Robot Arm

Dynamic model of an industrial articulated robot arm is required for the construction of the forcefree control with independent compensation. Dynamics of an articulated robot arm is expressed by

$$H(q)\ddot{q} + D\dot{q} + \mu\mathrm{sgn}(\dot{q}) + h(q\,\dot{q}) + g(q) = \tau_s + \tau_f \quad (1)$$

where $H(q)$ is inertia matrix, $D\dot{q}$ and $\mu\mathrm{sgn}(\dot{q})$ are friction terms, $h(q\,\dot{q})$ is coupling nonlinear term, $g(q)$ is gravity term, q is output of joint angle, τ_s is the torque input to the robot arm and τ_f is the torque caused by external force. In industrial robot arms, servo controller (P and PI type control) is used to control the motion of robot arm and the control loop of the servo controller is shown in right hand side of Fig. 1, where K_p, K_v and K_τ

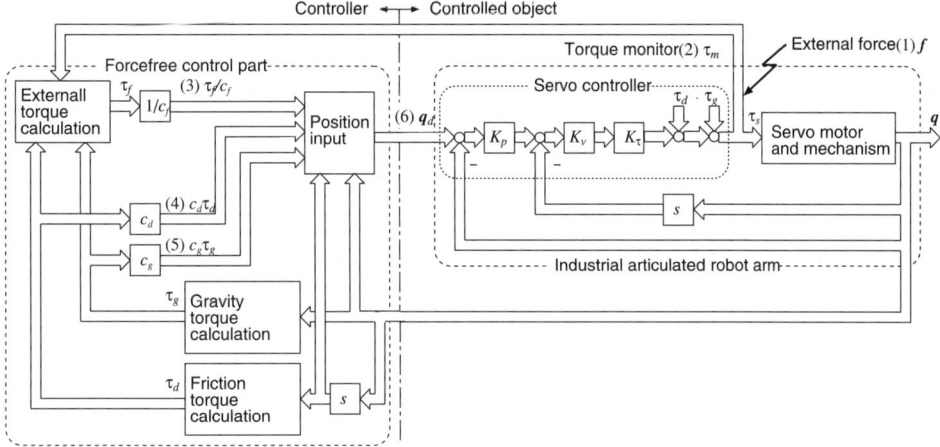

Fig. 1. Block diagram of forcefree control by independent compensation.

are position loop gain, velocity loop gain and torque constant, respectively[8], [9], [10]. The servo controller generates the torque input to the robot arm as

$$\tau_s = K_\tau(K_v(K_p(q_d - q) - \dot{q})) + \tau_d + \tau_g - \tau_f \quad (2)$$

where q_d is the input of joint angle, τ_d is the friction compensation torque and τ_g is the gravity compensation torque. As expressed in (2), the servo controller includes the friction compensation and the gravity compensation through integral action of PI controller in the servo controller. Here, we assume that the friction and the gravity are ideally compensated by the servo controller as

$$\tau_d = D\dot{q} + \mu \text{sgn}(\dot{q}) \quad (3)$$

$$\tau_g = g(q). \quad (4)$$

The torque caused by external force τ_f is also compensated by the servo controller because the servo controller in the industrial robot arm is designed such that the stiffness of the robot arm is enough high and the external force never move the robot arm.

The total dynamic equation of the industrial articulated robot arm including the servo controller is given by substituting (2), (3) and (4) for (1) as[7], [11]

$$H(q)\ddot{q} + h(q\,\dot{q}) = K_\tau(K_v(K_p(q_d - q) - \dot{q})). \quad (5)$$

C. Algorithm of Forcefree Control with Independent Compensation

Forcefree control with independent compensation for inertia, friction and gravity means that the effect of inertia, friction and gravity to the robot arm motion can be assigned arbitrarily. The dynamics of forcefree control with independent compensation is described by

$$H(q)\ddot{q} + h(q\,\dot{q}) = (1/c_f)\tau_f - c_d\tau_d - c_g\tau_g. \quad (6)$$

Here, the coefficients c_f, c_d and c_g can be adjusted to vary the effect of the inertia, friction and gravity, independently. For instance, $c_f = 1$, $c_d = 0$ and $c_g = 0$, corresponds to the original forcefree control[7] and $c_f = c_d = c_g = 0$ corresponds to the perfect compensation of the inertia, friction and gravity.

Block diagram of the forcefree control with independent compensation is shown in Fig. 1. The inputs of joint angle (q_d) for the forcefree control is obtained by substituting (6) for (5) and solving for q_d as

$$q_d = K_p^{-1}(K_v^{-1}K_\tau^{-1}((1/c_f)\tau_f - c_d\tau_d - c_g\tau_g) + \dot{q}) + q \quad (7)$$

where τ_f is the joint torque corresponding to the external force f on the tip of robot arm and it is obtained by

$$\tau_f = -(\tau_m - \tau_d - \tau_g). \quad (8)$$

Here, τ_m is the output value of torque monitor and it includes the friction torque τ_d and the gravity torque τ_g. The torque monitor is usually attached to the servo controller of the industrial robot arm and is used to check the value of torque.

The algorithm of the forcefree control with independent compensation is explained briefly (see Fig. 2). 1) External force (f) is added to the robot arm. 2) Torque monitor detects the external force (f). 3) The friction torque (τ_d) is calculated by eq. (3), 4) The gravity torque (τ_g) is calculated by eq. (4). 5) External torque (τ_f) is calculated by eq. (8). 6) Torque of inertia compensation (($1/c_f)\tau_f$) is calculated. 7) The friction compensation torque ($c_d\tau_d$) is calculated. 8) The gravity compensation torque ($c_g\tau_g$) is calculated. 9) The position input (q_d) is generated by eq. (7). 10) Finally, the position input (q_d) is given

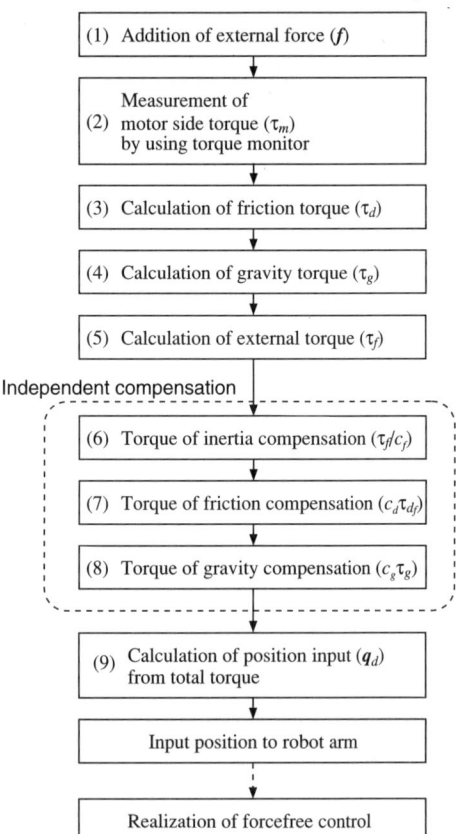

Fig. 2. Flowchart of forcefree control.

to the servo controller. According to the above algorithm, the forcefree control with independent compensation is realized.

The key advantage of the proposed forcefree control is the realization of the forcefree control by inputting appropriate position to the servo controller without modification of the servo controller. Besides, the robot arm would realize the optimal motion by the external force.

III. VERIFICATION OF FORCEFREE CONTROL WITH INDEPENDENT COMPENSATION

A. Condition

An industrial articulated robot arm (Performer-MK3S, YA-HATA Electric Machinery Mfg., Co., Ltd) was used for the experiment of the forcefree control with independent compensation. A structure of the experimental equipment is shown in Fig. 3. Two links of Performer-MK3S was used for the experiment.

The link lengths of the robot arm were $l_1 = 0.25$[m], $l_2 = 0.215$[m], and masses of the links were $m_1 = 2.86$[kg], $m_2 = 2.19$[kg], respectively. Position loop gain was $K_p = \mathrm{diag}\{25, 25\}$[1/s], velocity loop gain was $K_v = \mathrm{diag}\{150, 150\}$[1/s], torque constant was $K_\tau = \mathrm{diag}\{0.017426, 0.036952\}$[Nm/(rad/s^2)]. The parameter K_τ was estimated experimentally using the ramp input of velocity without position loop. The coefficient of viscous damping was $D = \mathrm{diag}\{4.68, 2.72\}$[Nms/rad], coefficient of coulomb friction was $\mu = \mathrm{diag}\{0.5, 0.5\}$[Nm]. Coefficient of viscous damping and that of coulomb friction were determined by the trial and error manner. Various simulations were carried out with different coefficients of viscous damping and coulomb friction. The coefficients were determined when the simulation result was close to the experimental result. The experimental condition is as follows. Coefficients corresponding to the inertia, friction and gravity were $c_f = 1$, $c_d = 1$ and $c_g = 0$, sampling interval was $\Delta t = 0.005$[s], experimental time was 5[s]. The initial position of the robot arm was (0.3, 0.3)[m].

In realization of forcefree control, a force sensor which measures the value of the external force is not required because the torque monitor is used to estimate the torque caused by the external force. However, the force sensor was used to measure the value of the external force. The property of the proposed forcefree control was verified by the comparison between the experimental result and and the simulation result using measured external force.

B. Simulation and Experimental Result

Simulation and experimental result of the forcefree control with independent compensation are shown in Fig. 4. Fig. 4(a) shows the external force of the tip of the robot arm along X-axis and (b) shows that along Z-axis, which were measured by attached force sensor. Fig. 4(c) shows the torque output of link 1 caused by the external force and (d) shows that of link 2. Fig. 4(e) shows the position trajectory of link 1 and (f) shows that of link 2. Fig. 4(g) shows the velocity trajectory of link 1 and (h) shows that of link 2. Fig. 4(i) shows the locus of the tip position of the robot arm. In Fig. 4(c)-(i), the bold lines show the experimental result and the dotted lines show the simulation results.

As shown in Fig. 4, the robot arm was moved by an external force of around -120 [N] whereas the masses of the links were $m_1 = 2.86$[kg] and $m_2 = 2.19$[kg]. The absolute value of the external force is not so large and the result shows that the forcefree control with independent compensation was realized with actual industrial robot arm. The experimental result and the simulation result have the similar tendency, however, the absolute values are slightly different. The difference is caused by the modeling error of the robot arm dynamics, especially coefficients of viscous damping and coulomb friction.

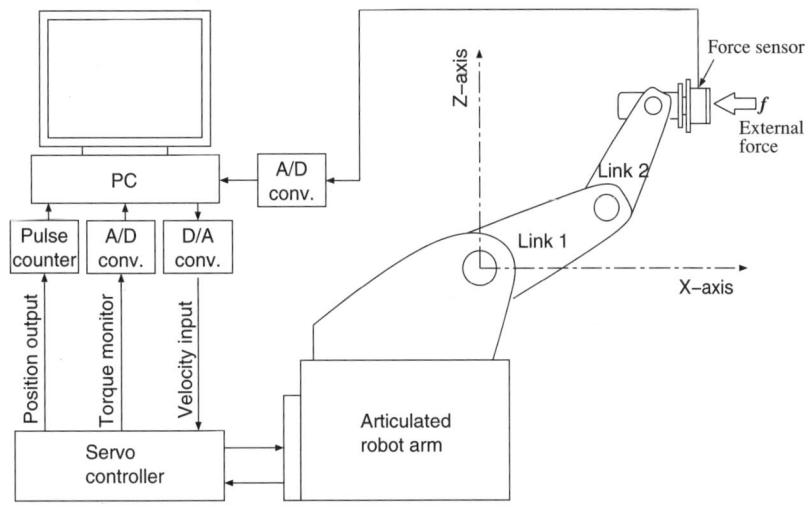

Fig. 3. Experimental equipment.

IV. DISCUSSION

A. Comparison with Other Methods

The proposed forcefree control with independent compensation is compared with the other methods of force control. Compliance control is famous and applicable to articulated robot arms. The compliance control[4], [5] is that the desired mechanical impedance of mass-damper-spring system is introduced between the tip of the robot arm and the environment. The robot arm is controlled such that the desired mechanical impedance can be achieved. When the spring constant of the mechanical impedance is set to zero, the compliance control, especially position command type one, is closed to the proposed forcefree control. The position command type compliance control, however, requires the measurement of external force, and it does not consider the servo controller and the torque monitor of the industrial articulated robot arms. Moreover, the robot arm must follow the position command without delay. The proposed forcefree control is specified for the industrial articulated robot arms, and the characteristics of the servo controller and the torque monitor are effectively used.

Characteristics of the forcefree control and the servo float method[6] is compared. (i) The contact force between the tip of the robot arm and the environment tends to be zero for the forcefree control, and it is set by the designer for the servo float method. Hence, the servo float method requires tuning of set contact force. In other words, the role of the servo float method is not the control of the flexibility but the control of the contact force. (ii) The forcefree control is sensitive to the external force. The servo float method actuates robot arm when the external force exceeds the set contact force. (iii) The forcefree control has flexibility, if external force is applied to every part of the robot arm. The servo float method, however, depends on the set contact force to the tip of the robot arm. From such point of view, the property of the forcefree control is more suitable for flexible motion control compared with the servo float method.

B. Applications of Forcefree Control with Independent Compensation

In industry, robot arms are used in various applications, and some applications require the forcefree control, for instance, the realization of direct-teaching for teaching playback type robot arms. Generally, teaching of industrial articulated robot arms is carried out by using an operational equipment called a teach pendant. Smooth teaching can be achieved if the direct-teaching is realized. Here, in the direct-teaching the robot arm is manually moved by the human operator. Non-gravity and non-friction condition is desirable for the implementation of direct-teaching. From such point of view, specialty, the forcefree control is applicable for the direct-teaching. The proposed forcefree control with independent compensation is also applicable to pull-out-work. The pull-out-work is operated as follows; a) the hand of the robot arm grasps the workpiece, b) the workpiece is pushed out by the push-rod, c) the workpiece is released by the force from the cast. The motion of the robot arm requires flexibility in order to follow the pushed workpiece.

Further, each joint could be monitored for unexpected torque deviation from the desired torque profile as a result of unplanned circumstances such as accidental collision with an object or human being. Under such circumstances, forcefree control mode can be invoked and thereby assure the avoidance of

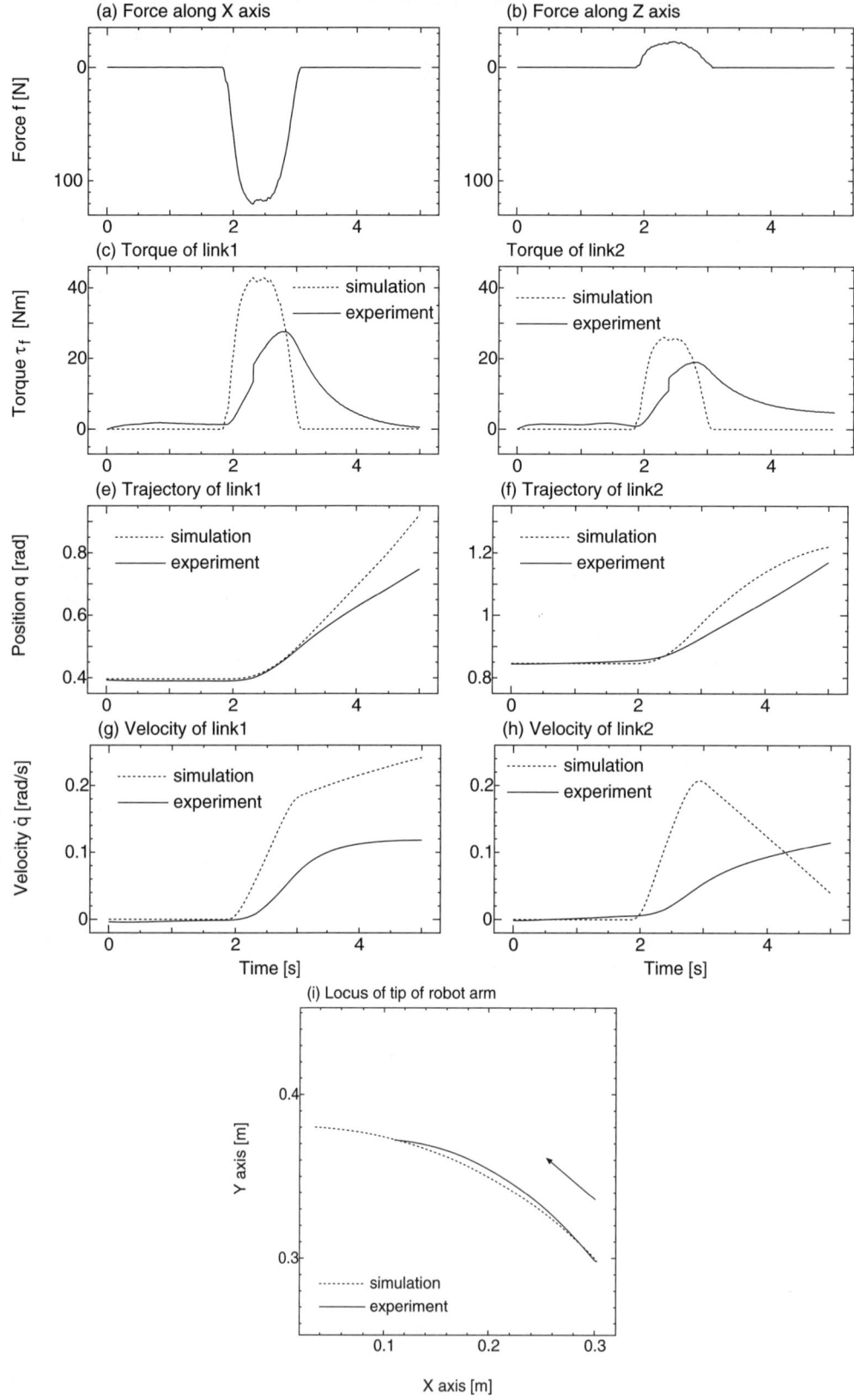

Fig. 4. Simulation and experimental result of forcefree control with independent compensation in the addition of external force input

damages. Hence, the forcefree control can also improves the safety of works with human operator.

V. CONCLUSIONS

The forcefree control with independent compensation for inertia, friction and gravity of industrial articulated robot arms was proposed. The corrective measures for inertia, friction and gravity of the robot arm was adjusted by selecting the appropriate coefficients of the respective compensation terms in the forcefree control. An experimental result of an actual industrial robot arm was successfully realized by the proposed method. The proposed method requires no change of the hardware of the robot arm and easily acceptable to industrial applications.

ACKNOWLEDGMENTS

The authors would like to thank Dr. D. Kushida, Mr. Y. Matsuyama and Mr. Y. Ishida, Department of Advanced Systems Control Engineering, Graduate School of Science and Engineering, Saga University, Japan for their valuable suggestions and experimental studies.

REFERENCES

[1] N. Hogan "Impedance Control; An Approach to Manipulation: Part I~III", *Trans. of the ASME Journal of Dynamic System, Measurement, and Control*, vol. 107, pp. 1-24, 1985.

[2] H. A. Chae, G. A. Christpher and M. H. John, *Model-Based Control of a Robot Manipulator*, The MIT Press, Cambridge, 1988.

[3] L. Sciavicco and B. Siciliano, *Modelling and Control of Robot Manipulators*, pp. 271-280, Springer, London, 2000.

[4] M. T. Mason "Compliance and Force Control for Computer Controlled Manipulators", *IEEE Trans. on Systems, Man, and Cybernetics,*, vol. 11, No. 6, pp. 418-432, 1981.

[5] B. Michael, M. H. John, L. J. Timothy, L. P. Tomas and T. M. Matthew, *Robot Motion: Planning and Control*, The MIT Press, Cambridge, 1982.

[6] H. Nagata, Y. Inoue, K. Yasuda "Sensorless Flexible Control for Industrial Robot", *16th Conf. of RSJ*, vol. 3, pp. 1533-1534, 1998 (in Japanese).

[7] D. Kushida, M. Nakamura, S. Goto, N. Kyura "Human Direct Teaching of Industrial Articulated Robot Arms Based on Forceless Control", *5th International Symposium on AROB*, vol. 1, pp. 383-386, 2000.

[8] M. Nakamura, S. Goto, N. Kyura "Control of Mechatronic Servo System", *Morikita Shuppan*, 1998 (in Japanese).

[9] N. Kyura "The Development of a Controller for Mechatronics Equipment", *IEEE Trans. on Industrial Electronics*, vol. 43, No. 1, pp. 30-37, 1996.

[10] N. Kyura and H. Ono "Mechatronics — An Industrial Perspective", *IEEE/ASME Trans. on Mechatronics*, vol. 1, No. 1, pp. 10-15, 1996.

[11] K. S. Fu, R. C. Sonzalez and C. S. G. Lee, *Robotics Control, Sensing, Vision, and Intelligence*, McGraw-Hill, Inc., Singapore, pp. 82-144, 1987.

CHALLENGES IN VR-BASED ROBOT TELEOPERATION

Cheng-Peng Kuan and *Kuu-young Young
Computer & Communications Research Laboratories
Industrial Technology Research Institute, Chutung, Taiwan
*Department of Electrical and Control Engineering
National Chiao-Tung University, Hsinchu, Taiwan
*email: kyoung@cc.nctu.edu.tw

ABSTRACT

Teleoperation techniques unite the human operator as the supervisor and the machine as the manipulator. Because both the human operator and machine are involved in the control loop and they are connected via the network instead of a direct link, the development of a telerobotic system poses challenges different from systems involving machines alone. And these challenges become more severe when the telerobotic system is used for compliance task, in which simultaneous control of both position and force are demanded and inevitable contact with the environment is encountered. In this paper, we discuss the challenges we may face in designing a telerobotic system. We then describe the networked VR-based telerobotic system developed in our laboratory and report the results by using the developed telerobotic system to execute several different kinds of compliance tasks.

Key words: Teleoperation, Network, Compliance Task, Virtual Reality (VR).

1 Introduction

Teleoperation technologies have been applied to hazardous or uncertain environments such as nuclear plants, outer space, or deep oceans, and also to highly automated systems that are not necessarily hazardous but which demand human intervention for detecting and monitoring abnormalities, such as aviation. One impressive example is the historic mission of Mars landing of the Prospector spacecraft on July 4, 1997, which sent the space robot, Sojourner, on the surface of the Mars first ever, marking a milestone for the field of space robotics and teleoperation [1]. Along with the development of network, teleoperation has gained much publicity. The IEEE Robotics and Automation Magazine, March 2000, devotes an entire issue to discuss global remote control through internet teleoperation [2]. The progress of nanotechnologies brings teleoperation into the world of micro- and nano-robots. And the teleoperation techniques also receive much attention from the fields of education and medicine, among others [3].

Through teleoperation, the human operators can enter remote environments with scales or physical laws much different from those in the normal world, while they are actually not there. This specialty, on the other hand, makes the development of a teleoperation system demand creation of virtual environments that make the operators feel actually present at the remote sites. The generation of telepresence becomes much more challenging, however, when the teleoperation system is used to execute compliance tasks, because simultaneous control of both position and force is demanded, and because contact with environments is inevitable. Furthermore, compliance tasks of the same kind may even evoke different interactions with the environment. For instance, burrs vary in size and distribution on various target objects in deburring compliance tasks. In addition, teleoperation indicates that the human operator is cooperating with the control mechanism equipped on the remote manipulator during task execution, and they form a control loop connected via the network. Consequently, due to the transmission time delay and the incompatibility between the manipulative devices used by the operator and the manipulator in the remote site, the operator usually experiences unnatural and ineffective manipulation [4, 5]. These issues highly increase the complexity in teleoperating a compliance task.

To tackle the complexity aforementioned, especially when the task becomes more complex and delicate, such as telesurgery and remote assembly, many schemes propose to incorporate more intelligence into the telerobotic system, for instance, the concepts of shared autonomy control, tele-sensor-programming, virtual mechanism, and predictive telemanipulation control [6, 7, 8]. To achieve a higher degree of autonomy in dealing with uncertain environments and various tasks, the human operator and remote manipulator may need to enhance their own capabilities and also their cooperation. In this paper, we focus on the challenges in designing a telerobotic system for compliance tasks. We then describe the telerobotic system developed in our laboratory. This system is developed in a virtual environment due to recent gains in the capabilities and popularity of virtual reality (VR) to generate realistic telepresence. The system provides both haptic and visual information [4], and equips an intelligence controller on the remote robot manipulator [9]. We use this networked VR-based telerobotic system to execute several different kinds of compliance tasks and analyze how the human operator and the remote intelligence controller can cooperate and also the performances.

2 The Challenges

Figure 1 shows the block diagram of a general networked VR-based telerobotic system. The system consists of VR I/O devices and the VR engine in the operator site and the robot and sensors in remote site. The VR I/O devices provide human-computer interaction for the system. Through these devices, the operator inputs commands into the computer and receives feedback, such as visual, auditory, and haptic feedback, that yields the feeling of immersion. The VR engine can be viewed as the simulation manager in a VR-based telerobotic system. It formulates the virtual environment, renders the scene, and manages object behaviors in the simulation loop, in addition to processing sensor data from the real world and sending commands to the robot manipulator. A successful implementation of the VR engine has an essential influence on the performance of the telerobotic system by providing a more realistic modeling of the real world and better communication between the virtual and real worlds. The robot is located at the remote site of the telerobotic system. Thus, sensors, such as position and force sensors, are needed to provide information about the situation at the remote site. The robot receives commands from the operator via the drive unit, and sensor data are sent to the VR engine via an intermediate transmission line, both of which may induce transmission time delay. According to Figure 1, we list some major challenges in designing a telerobotic system for compliance tasks:

- Telepresence of remote environments, including visual, auditory, and haptic senses;

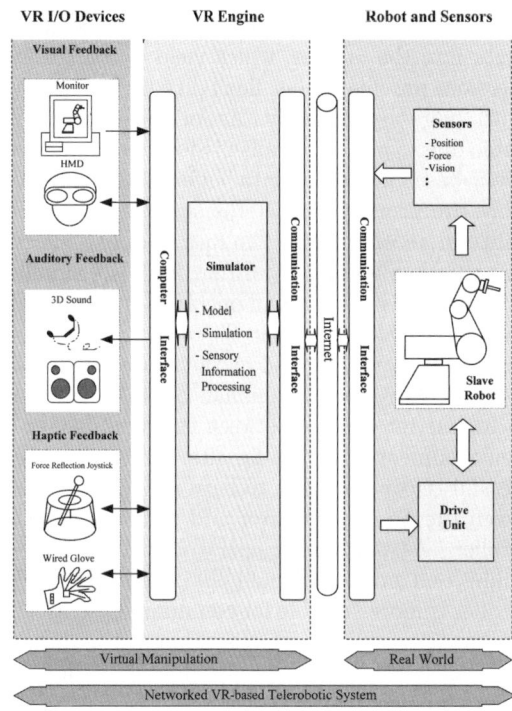

Figure 1: A networked VR-based telerobotic system.

- Design of VR I/O devices, such as head mount display, 3D glasses, force-reflection joystick, and data glove;

- Incompatibility between the manipulative device and the remote manipulator;

- Time delay due to network transmission;

- Supporting tools helpful for human operators in manipulation;

- Remote intelligence controller dealing with interaction between the manipulator and the environment;

- Cooperation between the human operator and remote intelligence controller.

Researches have been actively dedicated to tackle these challenges. For generation of telepresence of remote environments, the virtual environment should be constructed as similar as the real one in not only the outlooks and locations of the objects in the environment, but also their physical properties, such as mass, damping, and stiffness. For this purpose, fine vision systems and force data processors are built up to provide the visual and tactile information necessary.

Meanwhile, new generations of VR I/O devices are brought into the market, which yield better feeling of immersion, provide higher flexibility in manipulation, and are more user-friendly. As for the time delay inevitable due to network transmission, techniques, such as event-based planning, data buffering, and predictive display, are proposed to let the system still maintain stability in accomplishing the task. Supporting tools, such as path planning algorithm and VR simulator, are used to assist the human operator in manipulation and decision-making. And in the remote site, the intelligence controller, equipped on the robot manipulator, is designed to relieve the human operator from dealing with the interaction between the manipulator and the environment from the far site. One key issue for successful teleoperation is to have salient cooperation between the human operator and remote intelligence controller. As the human operator may be better at planning and navigation and the remote intelligence controller is more suitable for command execution and force management, their cooperation can be based on their specialties and the specific characteristics of the given compliance tasks.

3 Developed Networked VR-Based Telerobotic System

According to the discussions in Sec. 2, we have developed a networked VR-based telerobotic system, as shown in Figure 2. At the operator site, the impulse engine, developed by the Immersion Corporation (U.S.A.), is used as the force-reflection joystick. This joystick has five degrees of freedom in motion, with three of them equipped with force reflection, and its maximal output force is about 8.9 N. Because the impulse engine is quite expensive and generates a small maximal output force, we thus developed a wheel-typed force-refection joystick, as shown in Figure 3. By bringing in the feature of a steering wheel into the commonly used joystick, we designed the joystick to be two degrees of freedom, with its X axis rotating in a full 360° range and the Y axis moving perpendicular to the X axis in a 40° range. Both of the axes are equipped with force reflection, and generate the maximal output force at 43 N and 60 N, respectively. At the remote site, a Mitsubishi RV-M2 type five-axis robot manipulator equipped with a JR3 force-moment sensor is used. Two personal computers (PC1 and PC2) with Pentium CPUs are located at the operator and remote site, respectively. PC1 is mainly used for developing the VR simulator and communicating with the force-reflection joystick, with PC2 for developing the remote intelligence controller and communicating with the RV-M2 drive unit and JR3 force-moment sensor. For implementing the VR simulator, the WorldToolKit (WTK) (Sense8 Corporation, U.S.A.) is used to develop the simulation manager and TrueSpace2 (Caligari Corporation, U.S.A.) is the 3D modeling software.

As Figure 2(a) shows, the human operator sends in motion command M_h to move the VR robot manipulator in the virtual environment using the force-reflection joystick; this modifies M_h into M_j. Via PC1 and then PC2, motion command M_j is sent to the RV-M2 drive unit which in turn generates torques T_a to move the real robot manipulator. Actual robot positions P_R are fed back to the VR simulator residing at PC1 via the drive unit and PC2 to synchronize the motions of the VR and real robot manipulators. Contact forces F_c induced when the robot manipulator interacts with remote objects are measured using the JR3 force-moment sensor mounted on the robot manipulator. The measured contact forces F_c are first processed by the JR3 interface and then sent to the remote intelligence controller residing at PC2. The remote intelligence controller can then use P_R and F_c to perform position and force control, if necessary. F_c is also sent to the VR simulator to generate realistic VR object deformation and to the force-reflection joystick to generate haptic feeling H_f [9].

To tackle the complexity exhibited by the compliance task during teleoperation, the human operator and remote intelligence controller should possess salient capabilities and proper coordination in between. We equip several supporting tools to assist the human operator in manipulation and decision-making, discussed in Sec. 3.1. We install the compliance control strategy into the remote intelligence controller, and it is designed to be able to deal with the interaction between the robot manipulator and the environment autonomously, discussed in Sec. 3.2. For different kinds of compliance tasks, the load distribution in planning and control for the human operator and remote intelligence controller may also be different. For instance, for a maze-passing task, the path planning may rely mainly on the human operator, and the force requirement may not be that crucial. For a surface-tracking task, it may demand a higher accuracy in position than in force. And for a peg-in-hole task, both the requirements on position and force need to be well taken care of. Therefore, the human operator and remote intelligence controller may coordinate with each other in task execution. For instance, when the remote intelligence controller is responsible for force management, the compliance control strategy installed in it will be used; on the other hand, when the human op-

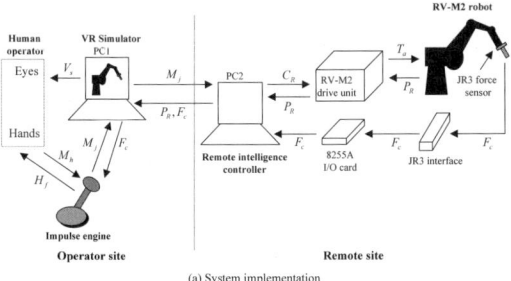

(a) System implementation

(b) System view

Figure 2: The developed telerobotic system: (a) system implementation and (b) system view.

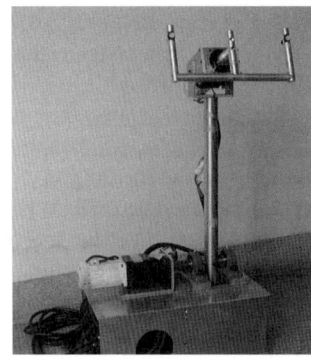

Figure 3: The developed wheel-typed force-reflection joystick.

erator takes charge, the remote intelligence controller only needs to provide force feedback.

3.1 Human operator

The human operator can usually take a global view on the events, and quickly assess changes in the environment and proceed with necessary modification [10]. Thus, she (he) may be in a better position to plan and guide the motion during task execution. To assist the human operator in planning and guidance, the proposed VR-based telerobotic system provides several supporting tools, including modules for generating haptic and visual information and a collision-free path planning algorithm.

A. Haptic and visual information

When the human operator is responsible for force management, the force feedback from the environment becomes crucial. To let the human operator be able to feel and also visualize the influences of the interactive forces on the environment, the proposed VR-based telerobotic system provides both haptic and visual information by using the techniques of force reflection and object deformation, respectively. Via the force-reflection joystick, the system furnishes the operator with haptic feeling using a VR force-reflection strategy, rather than receiving contact forces measured by the force sensor mounted on the remote robot manipulator directly [5, 9]. The basic idea for this VR force-reflection strategy is to generate the VR reflected force using the estimated stiffness of the remote object derived from the measured position and force data. The least-square linear regression method is used for estimating real-time stiffness K_o through processing a series of continuously measured position and force data, P_R and F_c. With the estimated object stiffness K_o, the VR reflected force F_r can then be derived as

$$F_r = K_o \cdot (P_R - P_{cs}) \quad (1)$$

where is P_{cs} the location of the contact surface. In addition to the consideration of time delay, the use of the estimated stiffness for VR force generation can also avoid the sensitivity problem usually encountered in using the sensed force directly. As for the visual information V_s, we used a simple spring model, similar to that described in Eq.(1), to generate VR object deformation due to the forces induced when the robot manipulator interacts with remote objects. For the proposed VR force-reflection strategy and the object deformation generation, a more general mass-spring-damper model can also be used at the expense of processing time.

B. Collision-free path planning algorithm

As the environment may be filled with obstacles, it is of concern whether the human operator is able to guide the remote robot manipulator to pass through the obstacles in the way to reach the goal. The human operator is considered to excel at navigating a mobile robot in a complex environment due to her (his) global vision and intelligence to interpret the environment. However, when it comes to a multi-joint robot manipulator where every point of the manipulator body is subject to potential collision, the human operator may not outperform a computer algorithm in obstacle avoidance [10]. In addition, the human operator is not accurate in teleoperating a remote robot manipulator. Therefore, a collision-free path planning algorithm is used to assist the human operator in manipulation [11]. The human operator only needs to provide the goal location and the direction to move, and the path planning algorithm will then guide the robot manipulator to avoid the obstacles and reach the goal autonomously.

3.2 Remote intelligence controller

Because the intelligence controller is located at the remote site, it is more suitable to deal with the uncertainties and the interaction with the environment. We design the remote intelligence controller to tackle the interacting robot manipulator and environment autonomously and cooperate with the human operator for compliance task execution. We thus develop, in the remote intelligence controller, a compliance control scheme that can adapt to environmental variations and ensure system stability in the presence of modeling uncertainties and external disturbances. The proposed compliance control scheme is basically the type of hybrid control and briefly described below [9]. The scheme first receives the motion command D_j, which represents the hand movement of the operator and provides the direction for the robot manipulator to follow. With D_j and the position and force information sent from the sensors equipped on the robot manipulator, the scheme derives the command C_R sent to the robot manipulator for execution. To determine a proper C_R when the stiffness of the environment and the shape of the environmental surface are not known in advance, a learning control algorithm is developed for force regulation. When the robot manipulator moves to contact with the environment from the free space, D_j is usually not along the direction of the environmental surface due to the imprecise manipulation of the operator. D_j can generally be divided into two portions: one is along and the other perpendicular to the environmental surface. D_s is taken just as the projection of D_j on the environmental surface. However, since the shape of the environmental surface is not known exactly, D_j cannot be projected onto the surface directly. Instead, the measured contact force F_c, providing the directional information for the environmental surface, is used for deriving this projection. On the other hand, D_f, perpendicular to D_s, is not the projection of D_j normal to the environmental surface. The use of D_f is to let the measured contact force F_c approach a desired force F_d, and it is derived using a learning process [9]. With D_s and D_f, the compliance control scheme can achieve surface tracking with force regulation.

4 Experiments

Experiments based on the maze-passing, surface-tracking, and peg-in-hole compliance tasks, shown in Figures 4(a)-(c), were performed to investigate how the human operator and the remote intelligence controller can cooperate in planning and control when teleoperating compliance tasks. For the maze-passing task shown in Figure 4(a), the human operator was asked to move the peg equipped on the robot end-effector to pass through the maze. The human operator performed all the planning and control on position and force, and the remote intelligence controller just executed the commands. The experimental results show that the human operator accomplished the task successfully in the assistance of both the visual information and haptic feeling provided by the VR simulator and the force-reflection joystick, respectively. In comparison, we let the human operator perform the same task without force reflection. Although the task could still be accomplished, the contact force became much larger and some large bounces occurred.

For the surface-tracking task shown in Figure 4(b), the human operator was asked to manipulate the robot to let the peg equipped on the end-effector move along an unknown environmental surface, while maintaining a desired contact force. We used two approaches to execute the task: first, we let the human operator perform all the planning and control and the remote intelligence controller simply execute the commands; second, we let the human operator proceed with the planning on both position and force and the remote intelligence controller execute the position and force control. The experimental results show that both approaches could achieve surface tracking, while the first approach led to oscillating force responses and could not maintain the desired contact force. By contrast, the second approach yielded stable force responses close to the desired value, which could be attributed to the effect of the remote intelligence controller.

Finally, the human operator was asked to perform the peg-in-hole task shown in Figure 4(c). In this task, the operator first moved the peg toward the carton until it contacted the surface, moved the peg along the surface with a desired contact force until the hole was reached, and then inserted the peg into the hole until the peg pressed on the bottom of the hole with a desired contact force. The human operator acted as

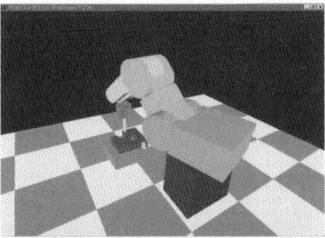

(a) The maze-passing compliance task.

(b) The surface-tracking compliance task.

(c) The peg-in-hole compliance task.

Figure 4: The compliance tasks for the experiments: (a) the maze-passing compliance task, (b) the surface-tracking compliance task, and (c) the peg-in-hole compliance task.

a navigator and guided the motion, and the remote intelligence controller proceeded with the planning on force and executed both the position and force control. The results demonstrate that the coordination of the human operator and the remote intelligence controller in this way successfully accomplished the peg-in-hole task, which exhibited many typical situations usually encountered during compliance task execution.

5 Conclusion

In this paper, we have discussed some major challenges in robot teleoperation. According to them, we have developed a networked VR-based telerobotic system. Experiments based on several different types of compliance tasks demonstrate the necessity and effectiveness of proper coordination between the human operator and the remote intelligence controller. Due to the variety of the compliance tasks and the uncertainties exhibited by the environments, as one of the future works, the remote intelligence controller will be incorporated with higher autonomy and more intelligence. Better manipulative devices and supporting tools also deserve developing. Another worthwhile future work is to investigate the feasibility of real-time closed-loop control for teleoperating the compliance task. Thus, the robot may not only just walk around the garden, but also possibly move on to perform telesurgery [3].

This work was supported in part by the National Science Council, Taiwan, under grant NSC 90-2213-E-009-093.

References

[1] Bekey, G.: On Space Robotics and Sojourner Truth, *IEEE Robotics and Automation Magazine* **4**(3) (1997), 3-4.

[2] Robots on the Web, *IEEE Robotics and Automation Magazine* **7**(1) (2001).

[3] Goldberg, K. (ed.): *The Robot in the Garden*, MIT Press, 2000.

[4] Kazerooni, H. and Her, M.G.: The Dynamics and Control of a Haptic Interface Device, *IEEE Trans. Robotics and Automation* **10**(4) (1994), 453-464.

[5] Kim, W.S., Hannaford, B., and Bejczy, A.K.: Force-Reflection and Shared Compliant Control in Operating Telemanipulators with Time Delay, *IEEE Trans. Robotics and Automation* **8**(2) (1992), 176-185.

[6] Hirzinger, G., Brunner, B., Dietrich, J., and Heindl, J.: Sensor-Based Space Robotics — ROTEX and Its Telerobotic Features, *IEEE Trans. Robotics and Automation* **9**(5) (1993), 649-663.

[7] Joly, L. and Andriot, C.: Imposing Motion Constraints to a Force Reflecting Telerobot Through Real-Time Simulation of Virtual Mechanisms, in *IEEE Int. Conf. on Robotics and Automation*, (1995), 357-362.

[8] Kotoku, T.: A Predictive Display with Force Feedback and Its Application to Remote Manipulation System with Transmission Time Delay, in: *IEEE Int. Conf. on Intelligent Robots and Systems*, (1992), 239-246.

[9] Kuan, C.P. and Young, K.Y.: VR-based Teleoperation for Robot Compliance Control, *J. Intelligent and Robotic Systems* **30**(4) (2001), 377-398.

[10] Ivanisevic, I. and Lumelsky, V.J.: Configuration Space as a Means for Augmenting Human Performance in Teleoperation Tasks, *IEEE Trans. Systems, Man, and Cybernetics, Part B: Cybernetics* **30**(3) (2000), 471-484.

[11] Sheu, C.H. and Young, K.Y.: A Heuristic Approach to Robot Path Planning Based on Task Requirements Using a Genetic Algorithm, *J. Intelligent and Robotic Systems* **16**(1) (1996), 65-88.

Adaptive Fusion of Sensor Signals based on Mutual Information Maximization

Tetsushi Ikeda   Hiroshi Ishiguro   Minoru Asada

Dept. of Adaptive Machine Systems,
Graduate School of Engineering, Osaka University

ikeda@er.ams.eng.osaka-u.ac.jp   {ishiguro,asada}@ams.eng.osaka-u.ac.jp

Abstract—The research approaches utilizing ubiquitous sensors to support human activities have become of major interest lately. Sensor fusion is one of the fundamental issues to develop such intelligent environments. The sensor fusion in previous works is performed in the task-level layer through individual representations of the sensors. Therefore, it does not provide new information by fusing sensors. This paper proposes another method that fuses sensory signals based on mutual information maximization in the signal-level layer. The fused signal provides us new information that cannot be obtained from individual sensors. As an example, this paper also shows experimental results in an audio-visual fusion task.

I. INTRODUCTION

As the Internet develops, the information systems supporting human daily activities is becoming more important. At the same time, several new research issues have become clear. The major problem of the current information infrastructure is in the perceptual function. In order to support human activities, the system needs to recognize them through various sensors.

Recent research activities named "intelligent room"[3],"smart room"[9], "robotic room"[8] and "perceptual information infrastructure"[7] are working on the recognition with ubiquitous sensors embedded in the environment.

One of the features of the ubiquitous sensor system is to use various sensors, such as cameras, microphones, touch sensors, and so on. Therefore, it is important to integrate/fuse such sensors and generate more robust and more task-oriented information.

There are two approaches in the sensor fusion: one is to use each sensor in a limited situation and the other is to fuse sensory signals directly. Let us consider a task to recognize a speaking person in an environment monitored by a ubiquitous sensor system with cameras and microphones. The former approach individually uses the cameras and microphones. The system detects positions of humans by analyzing visual data taken by the cameras in the environment, and simultaneously it receives the voice with the microphones and detects the location of the sound source. Then the system identifies a speaking person by fusing voice and audition based on the representations of the positions or directions given by both of the sensors.

The latter approach solves the task more simply. Humans can recognize a speaking person by voice and lip motions. That is, a human can identify the talking person in vision and audition by directly fusing the sensory information. The latter approach is similar to such the human ability. That is the system directly finds corresponding information between different sensor signals and provides new information that directly gives the solution of the task. This paper proposes a novel method following the latter approach, which is signal-level sensor.

Before proposing our idea, let us briefly review previous works. Several studies on signal-level sensor fusion have been reported so far.

Becker and Hinton[1][2] proposed to train neural networks by using a criterion that maximize mutual information among output from the networks. They showed that the networks extract the common information shared by inputs of the networks. Cutler and Davis[4] localized a speaker who utters a specific word by fusing visual and audio data with a TDNN (Time-Delay Neural Network). In this work, the networks need to be trained for each word, since the relation between video and audio signals is different for each word. Hershey et al.[6] and Fisher et al.[5] also localized a speaker in the image by computing mutual information between video and audio signals.

A problem of these approaches is that the statistical model for sensor signal analysis is not adaptive to dynamical changes of the scene. That is, they assume that the sound source does not move. In order to apply these sensor fusion methods to a dynamic environment, the system needs to deal with temporal changes of the model. As a method to solve the problem, this paper proposes a simple method to track the sound source prior to computation of mutual information between video and audio.

II. ACQUISITION OF RELATION BETWEEN SENSORS BASED ON MUTUAL INFORMATION

A. Mutual information between sensors

Let $x(t)$ be the time sequence of sensory data from sensor X, and $y(t)$ from sensor Y, respectively. Mutual information between $x(t)$ and $y(t)$ is represented as

$$I(x;y) = H(x) + H(y) - H(x,y) \qquad (1)$$

where $H(x)$ is entropy of $x(t)$, and $H(x,y)$ is mutual entropy between $x(t)$ and $y(t)$. They are defined as:

$$H(x) = -\sum_{t} p(x(t))\ \log\ p(x(t))$$

$$H(x,y) = -\sum_{t_x,t_y} p(x(t_x),y(t_y))\ \log\ p(x(t_x),y(t_y))$$

Here, mutual information I is computed with a fixed-length time window whose length is T.

Now, let us assume that $x(t)$ and $y(t)$ are jointly Gaussian[6]. The mutual information can be replaced with

$$\frac{1}{2}\log\frac{1}{1-\rho(x,y)^2} \qquad (2)$$

where $\rho(x,y)$ is correlation function between $x(t)$ and $y(t)$.

B. Computing mutual information in dynamically changing environments

If the relation between sensors do not change, the signals from sensors can be fused by the above method. Let us assume $x(t)$ and $y(t)$ are video and audio signal respectively. In previous approaches[5][6], $x(t)$ is intensity of a pixel or a region in the imaging sensor. It is assumed that the target is at a fixed position in the observed image. The mutual information among sensors is acquired with a long time interval and the location of speaker is detected in the image.

In general, however, the relation among sensors change dynamically. Suppose a person walking in an environment where a sensor network observes. The target that each sensor observes often change as the person moves. Since the above assumption does not hold, localization fails in such a situation(Fig. 1).

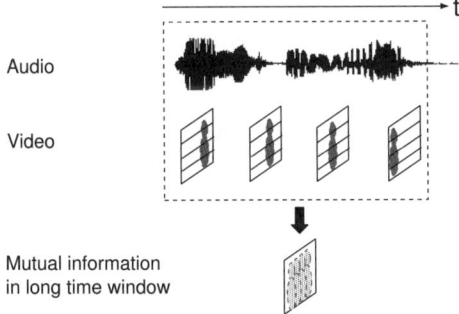

Fig. 1. Previous methods

C. Gaze control for observing moving targets

Ideally, it is required to estimate movements of all parts of targets to adapt to the changes of relation between sensors. By computing mutual information between the motion vector of each part and audio signal, the sound source is located based on this perfect tracking.

However, it is hard to estimate movements of targets precisely. For solving this problem, our approach proposed in this paper is to track the center of the person, not all movements of parts of the body. As shown in the next section, we detect the objects in the environment by background subtraction. Prior to fusing video and audio, the trajectory of each detected object is acquired. By computing mutual information between video and audio along the acquired trajectory, we obtain stable relation among sensors (Fig. 2).

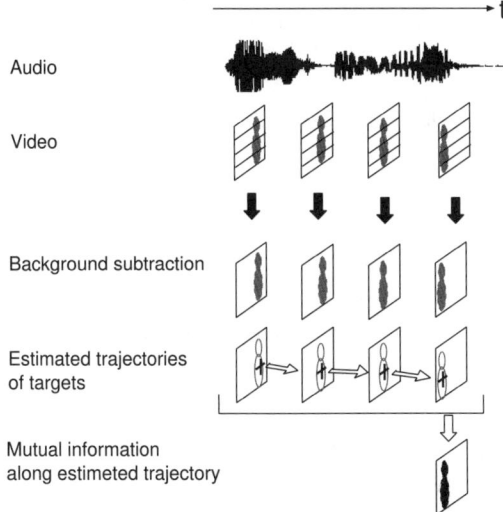

Fig. 2. Process flow of the proposed method

III. FUSION OF VISION AND AUDITION

As an example, we focus on sensor fusion between vision and audition and consider locating a walking person in video images as a task. One camera and one microphone are used. The camera observes moving legs and the microphone monitors the sound of footsteps. The relation between video and audio signals is obtained by computing mutual information between these signals.

The video signal is sampled at 30 frames/second, the image size is 160x120, and the intensity of each pixel is referred as $x(t)$ in equation (2). The audio signal is sampled at 16 kHz, and the average energy in each video frame is computed with a Hanning window and it is referred as $y(t)$. Fig. 3 and Fig. 4 shows samples of the video and audio signals, respectively.

A. The computing process

Step 1. Background subtraction and extraction of targets

First, we perform background subtraction and detect moving regions. Then perform a dilation operation two times. Each pixel in the image is substituted with the highest value in 4 nearest neighbors. Then, we perform binalization for the acquired image and extract connected regions and their centroids by labeling.

Figs. 5 shows experimental results of this step. The cross mark in Fig. 5(b) indicates the centroid of the detected region.

(a) frame 232

(b) frame 511

(c) background

Fig. 3. Example images

(a) Result of background subtraction

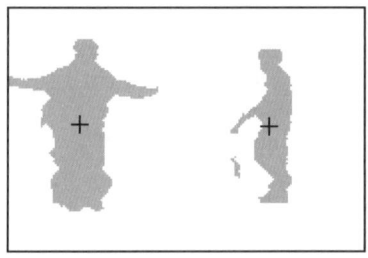

(b) Extracted regions and their centroids

Fig. 5. Processing results at step1

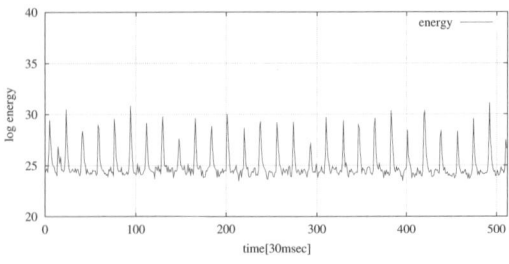

Fig. 4. Example audio signal

Step 2. Find correspondences between regions detected in consecutive images

In general, a few regions are extracted in step 1. These regions correspond to targets in environment that includes a sound source. To extract each sequence of the target, this step finds correspondence of the centroids of regions between consecutive images.

As a result, sequences of regions are extracted. Each sequence indicates a trajectory of a target moving in the environment.

Step 3. Estimate the mutual information

Along each sequence of regions acquired in previous step, we compute mutual information along the detected trajectory in a constant time window (this window length is described as T). All detected regions in each target are aligned to overlap all of the centroid. The mutual information can be used for verifying the results of the step 2. It is possible to feed this result back to the determination of the window length. This is one of our future works.

B. Results

Figs. 6 shows experimental results without tracking. Video signal is a sequence of observed intensity at a fixed position. The darker pixels indicate higher mutual information in the figure. The time window T was set to 256[frames] in all experiments. Figs. 6 (a) and (b) show the result at frame 64 and 256, respectively. The results have not included any remarkably darker regions. That means the sound source localization is failed.

Figs. 7 shows results by proposed method. Video signal is a sequence of observed intensity in a moving region. By computing mutual information along the extracted trajectory, regions that correspond to walker's leg have not been clearly detected.

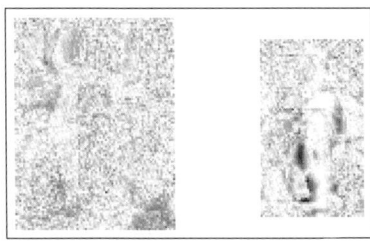

(a) at frame 64

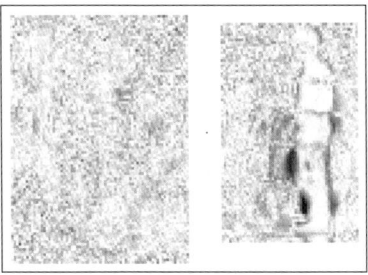

(b) frame 256

Fig. 7. Result (with tracking)

(a) at frame 64

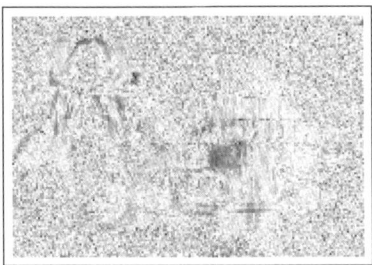

(b) frame 256

Fig. 6. Result (without tracking)

Fig. 8 shows samples of the video signals along trajectories. Fig. 8(a) is a video signal in the regions for the walker, and Fig. 8(b) is a video signal in the region for the other person. It is clear the signal in Fig. 8(a) is highly correlated with audio signal (Fig. 9).

IV. CONCLUSION

This paper has proposed a method that fuses sensory signals at the signal level based on mutual information maximization. The proposed method adapts to the dynamical changes of the relation between sensors by tracking centroids of the regions detected by background subtraction.

To confirm the effectiveness of the proposed method, we have applied it to track a walking person. The experimental result has shown that the proposed method can fuse audio signal and video signals and localize the signal source under a condition where the relation between sensors is dynamically changing.

In this paper, we have focused on audio and vision. But, it is possible to apply this method to other sensors used in a perceptual information infrastructure. Our next step is to deal with other sensors and to extend the method for applying to more complicated environments where many talking persons exist.

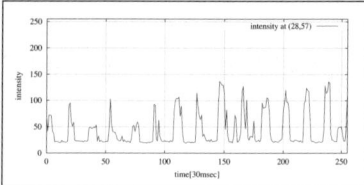

(a) In the region of the walker

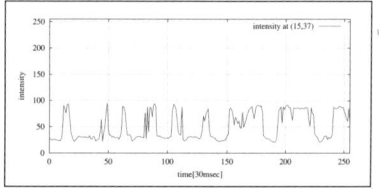

(b) In the region of the other person

Fig. 8. Sample of the video signal

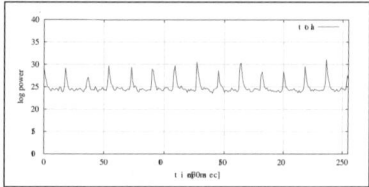

Fig. 9. Audio signal

V. REFERENCES

[1] S. Becker. Mutual information maxmization: Models of cortical self-organization. *Network: Computation in Neural Systems*, 7(1), 1996.

[2] S. Becker and G. E. Hinton. A self-organizing neural network that discovers surfaces in random-dot stereograms. *Nature*, 355(9):161–163, 1992.

[3] R. A. Brooks. Intelligent room project. In *Proc. of the Second International Cognitive Technology Conference*, 1997.

[4] R. Cutler and L. Davis. Look who's talking: Speaker detection using video and audio correlation. In *IEEE International Conference on Multimedia and Expo (ICME)*, 2000.

[5] J. W. Fisher III, T. Darrell, W. T. Freeman, and P. Viola. Learning joint statistical models for audio-visual fusion and segregation. In *Advances in Neural Information Processing Systems*, 2000.

[6] J. Hershey, H. Ishiguro, and J. R. Movellan. Audio vision: Using audio-visual synchrony to locate sounds. In *Proc. of Neural Information Processing Systems (NIPS'99)*, 1999.

[7] H. Ishiguro. Distributed vision system: A perceptual information infrastructure for robot navigation. In *Proc. Int. Joint Conf. Artificial Intelligence*, pp. 36–41, 1997.

[8] T. Mori and T. Sato. Robotic room: Its concept and realization. *Robotics and Autonomous Systems*, 28(9):141–148, 1999.

[9] A. Pentland. Smart rooms. *Scientific American*, 274(4):68–76, 1996.

Reinforcement Learning Congestion Controller for Multimedia Surveillance System

Ming-Chang Hsiao, Kao-Shing Hwang, Shun-Wen Tan and Cheng-Shong Wu
Department of Electrical Engineering
National Chung Cheng University
160, Ming-Hsiung, Chia-Yi 621, Taiwan

Abstract - The use of reinforcement learning scheme for congestion control in factory surveillance network is presented in this paper. Traditional methods perform congestion control by means of monitoring the queue length. When the queue length is greater than a predefined threshold, the source rate is decreased at a fixed rate. However, the determination of the congested threshold and sending rate is difficult for these methods. We adopted a simple reinforcement learning method, called Adaptive Heuristic Critic (AHC), to solve the problem. The AHC controller maintains an expectation of reward and takes the best policy to control source flow. By way of learning and then taking right actions, simulation results have shown that the approach can promote the system utilization and decrease packet loss.

Keywords: Congestion Control, Reinforcement Learning, Adaptive Heuristic Critic (AHC) method.

I. INTRODUCTION

In the study of high-speed network supporting multimedia service for factory automation, an interesting problem arises in the control of traffic source that do not reserve network resource, and hence the network may be congested. Congestion is a result of a mismatch between the network resources available for these connections and the amount of the traffic admitted for transmission. Therefore, a multimedia high-speed network must have an appropriate flow control scheme [1] not only to guarantee QoS for existing links but also to achieve high system utilization.

Congestion control is difficult owing to the uncertainties and the highly time-varying of the different traffic pattern. Recently, neural fuzzy network [2]-[4] have been employed in congestion control. Tarraf [2] proposes a learning neural network (NN) controller. It provides a cost function to tune the weights of the backward propagating neural network, which is used to generate the control signal. The major advantage of NN controller is the reduction of network complex statistic behavior by measuring the queue length only. Cheng [3] proposes a neural fuzzy approach for connection admission controller for multimedia high-speed networks.

In this paper, a simple reinforcement learning method, called adaptive heuristic critic (AHC) [5], is adopted to solve this problem. AHC is an actor-critic method of reinforcement learning, which is consist of a critic and an actor. Temporal-Difference (TD) [6] learning is used in our AHC congestion controller to learn directly from raw experience without a model of the environment's dynamics. Simulation results have shown that by way of learning and then taking right actions, our method can promote the utilization of the available bandwidth and decrease packet loss.

II. SYSTEM MODEL

Our system model is shown in Fig. 1. There are several voice/video sources sending packets to the multiplexer at various rates. For simplicity, we assume that these packets are all with a fixed length. We also assume that the multiplexer has a buffer of fixed size and outputs at a constant rate.

The congestion controller senses the state of the multiplexer and makes decision on which rate the sources should use to reduce the packet loss rate and to increase the utilization of the multiplexer's output bandwidth.

The decisions made by the congestion controller are fed back to the traffic sources as control signals.

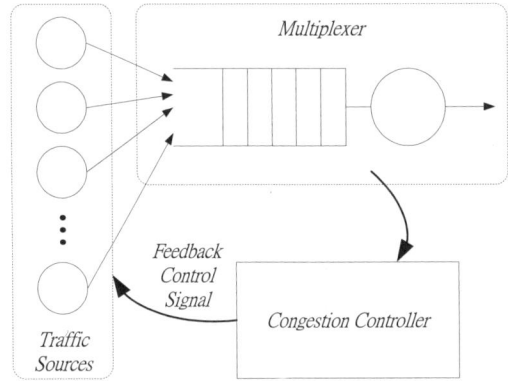

Fig. 1 The framework of the feedback congestion control system.

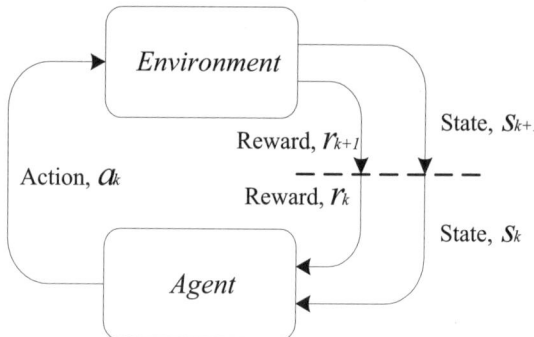

Fig. 2. The interaction between the reinforcement learning agent and the environment.

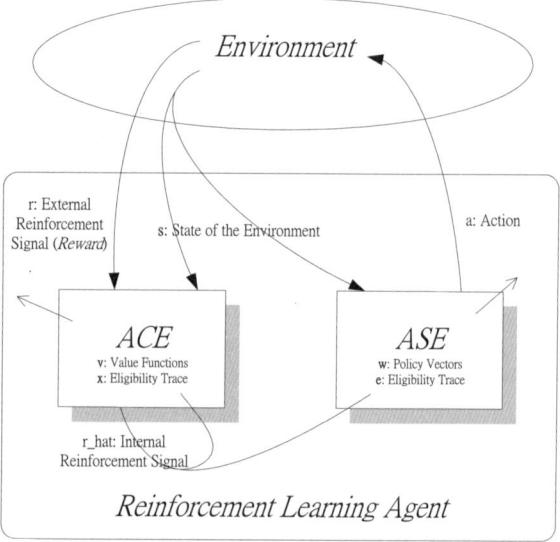

Fig. 3. Adaptive Heuristic Critic Reinforcement Learning Agent.

III. REINFORCEMENT LEARNING CONGESTION CONTROLLER

A. Reinforcement Learning Problems

Recently, reinforcement learning methods are frequently applied to the domain of control and are suitable to solve the problem that is too complicated to have an explicit model. Reinforcement learning problems are the problems of achieving a goal by learning from interactions between an agent and its environment. The learner and the decision maker is called the agent, and everything outside the agent make up the environment, which includes anything that cannot be changed arbitrarily by the agent.

Fig. 2 depicts the interaction between the agent and the environment. For step k of learning, the actions, a_k, are the decisions that the agent is learning to make, and the state, s_k, is the combination of anything the agent can know, which might be useful to make these decisions. The reward given by the environment to the agent, r_k, is a special mathematical value that the agent tries to maximize over time.

The agent and the environment interact continually: At the beginning of each step of learning, the agent senses the state of the environment and gets the reward from the environment. The agent then selects an action according to its policy, which is the mapping from states to probabilities of selecting each possible action. The agent learns to change its policy every step to maximize the total amount of reward it receives over the long run, and reinforcement learning methods specify how the agent changes its policy as a result of its experience.

After the agent selects an action, the environment responds to the action by changing its state to s_{k+1} and giving a new reward, r_{k+1}, to the agent. Then the agent senses the state of the environment again and starts the next step of learning.

B. Adaptive Heuristic Critic Method

We adopt an AHC method, one of reinforcement learning methods, to design our congestion controller. Fig. 3 shows the structure of an AHC reinforcement learning agent. It is composed of two elements: the adaptive critic element (ACE) and the associative search element (ASE).

The ASE maintains the policy vectors, for each state, which are composed of weights used to calculate the probabilities of selecting one of the actions under condition of the given state. On the other hand, the ACE maintains value functions, which are defined as the expected amount of reward that can be received when the agent enters a given state. On every step of learning, the ACE uses TD method to calculate the internal reinforcement signal, which is used to update the value functions in ACE and the policy vectors in ASE.

Let S be the set of all possible states, and A be the set of all possible actions. $v[i]$ is the value function for state i, and $w[i][j]$ is the weight for selecting action j given the state i. Let $x[i]$ and $e[i][j]$ be the eligibility traces used when updating $v[i]$ and $w[i][j]$, respectively. They account for the correlation between the received reward and the state entered several steps ago. The value of $x[i]$ increases in the steps that state i is entered, and decays with the discounting rate in the other steps. It is similar to $e[i][j]$. Let α be the learning rate of $w[i][j]$ and β be the learning rate of $v[i]$, η be the discounting rate of $e[i][j]$ and λ be the discounting rate of $x[i]$. The algorithm of an AHC agent is described in Fig. 4.

C. AHC Congestion Controller

We take the multiplexer and the traffic sources shown in Fig. 1 as the environment and the congestion controller as the agent. The time interval of a learning step is 1ms. Let n_k, n_{k-1} be the buffer utilization at the beginning of k, $k-1$ step of learning, respectively, and u_k, u_{k-1} be the sources' packet generating rate of k, $k-1$ cycle, respectively. We choose the set of the environment's states to be the set of all possible combinations of (n_k, n_{k-1}, u_k, u_{k-1}).

We assume that the encoding schemes used in the sources are variable, and there are three values of output rate that can be used: $u(k)=1\times R_t$, $u(k)=0.75\times R_t$, $u(k)=0.5\times R_t$, and $u(k)=0.25\times R_t$, where R_t is the highest output rate of the source. So there are three possible actions to select for our AHC congestion controller.

When the state of the environment becomes preferable, for example, the occupied buffer size decreases thus the packet loss rate reduces, the reward given by the environment is set to be 1, otherwise the reward is set to be 0. The AHC congestion controller learns to gather more and more reward by applying the algorithm depicted in Fig. 4.

IV. SIMULATION

For simulations, we assume that the packets are all with a fixed length, 100 bytes, and that the buffer size of the multiplexer is fixed to 20 packets. The constant output rate of the multiplexer is 100 Mbps. There are two types of traffic sources: voice and video. For voice sources we use ON/OFF model to model its behavior, with the probability for on, $p_{ON}=0.35$, and the probability for off, $p_{OFF}=0.65$. During the ON period the source sends packets to the multiplexer with a constant rate. For video sources we use a first order autoregressive Markov process $X(n)$ to model the sending rate during nth frame. $X(n)$ is defined with:

$$X(k) = m \times X(k-1) + n \times G(k), \quad (1)$$

where m and n are constants, and $G(k)$ is a sequence of independent Gaussian random variables.

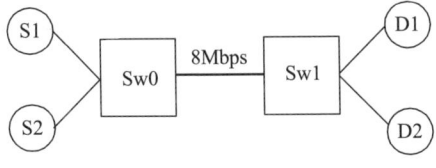

Fig. 5. A model with two ATM switches.

Initialization:
  For all $i \in S$:
    $v[i] \leftarrow 0$, $x[i] \leftarrow 0$.
  For all $i \in S$ *and all* $j \in A$:
    $w[i][j] \leftarrow 0$, $e[i][j] \leftarrow 0$.

For each step k of learning:
  Obtain the state of the environment, s_k.
  Receive the reward, r, from the environment.
  Calculate the internal reinforcement signal, $\hat{r}$, using $\hat{r} \leftarrow r + \gamma \times v[s_k] - v[s_{k-1}]$.
  For all $i \in S$:
    Update **v** *and* **x** *with*
    $$v[i] \leftarrow v[i] + \beta \times \hat{r} \times x[i]$$
    and
    $$x[i] \leftarrow \begin{cases} \lambda \times x[i] + (1-\lambda), & \text{if } s_k = i \\ \lambda \times x[i], & \text{otherwise} \end{cases}.$$
  For all $i \in S$ *and all* $j \in A$:
    Update **w** *and* **e** *with*
    $$w[i][j] \leftarrow w[i][j] + \alpha \times \hat{r} \times e[i][j]$$
    and
    $$e[i][j] \leftarrow \begin{cases} \lambda \times e[i][j] + (1-\lambda), & \text{if } s_k = i, a_{k-1} = j \\ \lambda \times e[i][j], & \text{otherwise} \end{cases}$$
  Select the action, a, from all $j \in A$ *with probabilities*
  $$\text{Prob}\{a_k = j \mid s_k = i\} = \frac{\exp(w[i][j])}{\sum_{n \in A} \exp(w[i][n])}.$$
  Apply the action, a_k, to the environment and starts next step of learning.

Fig. 4. The algorithm of the AHC reinforcement learning agent.

Fig. 5 shows the configuration of model in which two ATM switches, Sw0 and Sw1 are cascaded. In addition, two sources, S1 and S2, are connected to Sw0, and two destinations, D1 and D2, are connected to Sw1. We have done simulations for three scenarios: 1) traffic with best effort, 2) with enhanced proportional rate-control algorithm (EPRCA) [7] and 3) with AHC congestion controller.

V. RESULTS

The Results of our simulation are shown in Fig. 6-10. Fig. 6 shows the probability mass function (PMF) of the buffer utilization of three scenarios while the total traffic load is 80%. Obviously, the buffer utilization is almost identical. In contrast, Fig. 7 and Fig. 8 shows the PMF of the buffer utilization for three scenarios with load 100% and 120%, respectively. We can see that with AHC congestion controller, the buffer utilization is kept low and thus the probability of buffer overflow is reduced.

From Fig. 9 we can see that the cell loss rate (CLR) with AHC congestion controller is greatly reduced, while the amount of packet delivered is kept high, which is shown in Fig. 10. This indicates that the output bandwidth is nearly fully utilized.

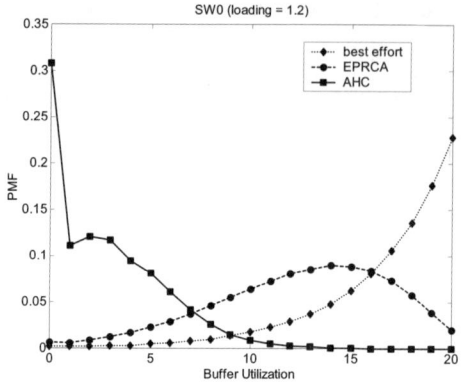

Fig. 8. PMF of buffer utilization under loading = 1.2

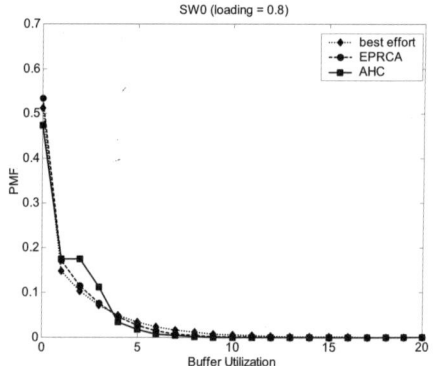

Fig 6. PMF of buffer utilization under loading = 0.8

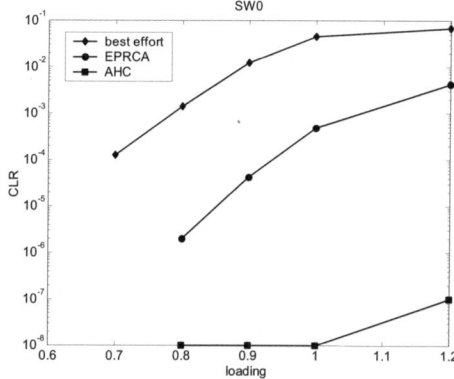

Fig. 9. CLR vs. various loading.

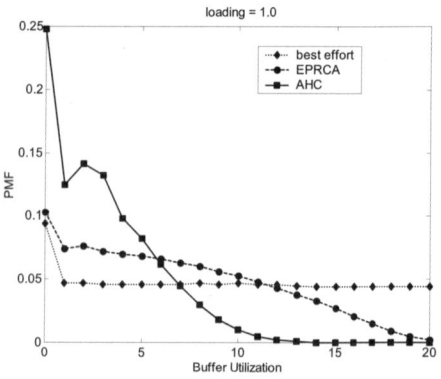

Fig. 7. PMF of buffer utilization under loading = 1.0

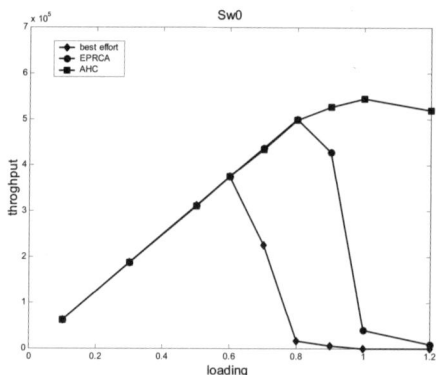

Fig. 10. Throughput behavior vs. various loading.

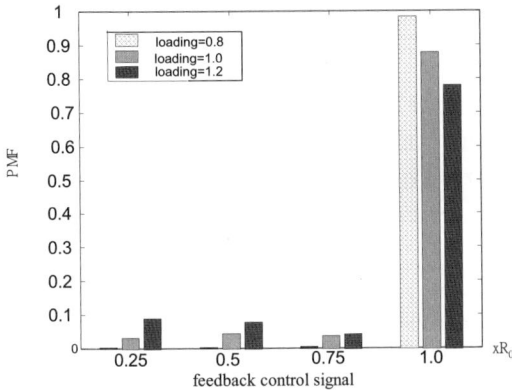

Fig. 11. PMF of feedback control signal (sending rate) under various loading

Fig. 11 shows the PMF of the sending rate selected by the AHC congestion controller with different traffic load. Most of the time the AHC is operating at a speed of the maximum sending rate to transmit packets form sources to various destinations, and only a small amount of low rates are used to accommodate traffic to avoid cell loss, so that the throughput is definitely high. When the traffic load is heavy, the increase of the PMF of lower sending rates is for the purpose of keeping the level of the quality. The cost of CLR damaging video/voice signal is much more than that of the reduction of the sending rate, so the quality of the video/voice traffic will keep high.

VI. Conclusion

In this paper we proposed a reinforcement learning congestion controller implemented by adopting an AHC reinforcement learning agent into the controller. We described the reason why we need a congestion controller for factory surveillance system. The detail of the AHC congestion controller and its algorithm are given. Simulation results shows that our AHC congestion controller improves the cell loss rate performance and thus providing better quality and reliability of service to the multimedia surveillance system.

Acknowledgement

This work was financially supported by National Science Council Taiwan, under Grant No. NSC 91-2213-E-194-004

References

[1] Panos Gevros, Jon Crowcoft, Peter Kirstein and Saleem Bhatti, "congestion control Mechanisms and the Best Effort Service Model," IEEE network, May/June 2001, pp 16-26.

[2] A. A. Tarraf, I. W. Habib, and T. N. Saadawi, "Reinforcement Learning Based Neural Network Congestion Control for ATM Networks", *Proceeding of MILCOM 1995*, page 668-672.

[3] Ray-Guang Cheng, Chung-Ju Chang, and Li-Fong Lin, "A QoS-Provisioning Neural Fuzzy Connection Admission Controller for Multimedia High-Speed Networks ", *IEEE/ACM Trans. on Networking*, vol. 7, no. 1, Feb. 1999.

[4] S. J. Lee and C. L. hou, "A Neural-Fuzzy System for Congestion Control in ATM Networks", *IEEE Transactions on System, Man. and Cybernetics*, Vol. 30, pp. 2-9, 2000.

[5] R. S. Sutton and A. G. Barto, "*Reinforcement Learning An introduction*", Cambridge, Mass., MIT Press, 1998.

[6] R. S. Sutton, "Learning to Predict by the Methods of Temporal Differences", *Machine Learning*, Vol. 3, pp. 9-44, 1988.

[7] Robots L. "Enhanced PRCA (proportional rate-control algorithm)". ATM Forum Contribution 94-0735R1, 1994.

Student Performance Evaluation in Web Based Access to Robot Supported Laboratories

H. E. Motuk
ASELSAN Inc.
e-mail: e109293@metu.edu.tr

A. M. Erkmen

I. Erkmen

Middle East Technical University
Dept. of Electrical Engineering
06531 Ankara Turkey

e-mail: aydan@metu.edu.tr

e-mail: erkmen@metu.edu.tr

Abstract:

Our aim in this work is to generate an intelligent interface for profiling and evaluating each student remotely accessing a robot assisted lab setup. The interface provides the student with the "hands on" experience by necessary visual feedback, giving this user as much freedom as possible to control the experiment. Besides, the system evaluates the user's performance, adapts the context to the level of user's acquired knowledge and skill, and thus intelligently coaches the user to successfully do the experiment. The system's ability is demonstrated on an example and the behavior of each module in the proposed system is discussed. The major contribution of this paper is taking the student behavior into consideration while providing online robotics experimentation, evaluating the user's performance and coaching him/her towards the successful achievement of the tasks.

1. Introduction

Distant access of laboratories through the Internet is an important and new issue in distance education. It also defines new pathways for internet telerobotics and teleoperation. Since classical laboratory setups normally open to local students have to be shared by online students, the setups should be assisted by robotic devices doing real time, the necessary online manipulation. Sharing of the robot assisted experimental resources between different institutes worldwide can also be enabled and students outside campus can be provided with real lab experimentation online. There are numerous applications of online lab access for educational purposes over the Internet. Germany's DERIVE project develops a mechactronics learning environment where on-site and remote components merge into a cooperative learning process [1]. Another German project VVL [2] focuses on research, development, testing and evaluation of virtual laboratories in engineering and computer sciences. The experiments are both telecontrolled and teleobserved in real laboratories. PEARL project [3] at Open University UK aims at delivering practical experimentation, where students work together over the internet, much as they would do in a teaching laboratory.

A general framework for implementing and deploying remote experimentation solutions is presented in [4], where the motivation for remote experimentation is discussed and best practices for the selection of the physical systems in automatic control education, the client-server architecture and the user interface are suggested for an online access to an inverted pendulum at EPFL. Rohrig et al [5] presents a platform independent approach to remote experimentation. Problems occurring in conjunction with platform-free remote laboratory experiments such as scheduling time, security and authentication, and analysis and visualization of the experiment data are addressed and the architecture of the remote laboratory for control theory at the University of Hagen is presented.

Reviews of adaptive and intelligent technologies in the context of web-based distance education, kinds of technologies available, their implementations on the web and the place of these technologies in a large scale web based education are done in [6, 7]. Besides those reviews on web oriented adaptive learning technologies, reviews are also conducted [8, 9] on generic, ANN based, adaptive user modeling systems. Capuano et al [10] presents a student model, based on student cognitive states and learning preferences and use it to generate automatic curriculum for different students to monitor student knowledge by testing them and assign recovery material if necessary, generating an "Adaptive Web Based Tutoring".

Our aim in this work is to generate an intelligent interface for profiling and evaluating each student remotely accessing a robot assisted lab setup. The lab apparatus to be used can either be a robot or a device that is connected to the internet. The interface should provide the student with the "hands on" experience of experimentation by necessary visual feedback, giving this user as much freedom as possible to control the experiment. Besides, the system should evaluate the user's performance, adapt the context to the level of user's acquired knowledge and skill, and thus intelligently coach the user to successfully do the experiment and get the most useful experience out of the experimentation. The concepts and tools borrowed from fields such as web-based intelligent tutoring, human-computer interaction, user-adapted interaction and internet telerobotics are necessary for a successful accomplishment of our requirements in education oriented lab access through the Internet.

The main objective of this study is developing an intelligent interface that can be used for the remote access of robot-supported laboratory and for experimentation through the Internet. The main differences from works in the literature are taking the student behavior into consideration while providing online robotics experimentation, evaluating the user's performance and coaching him/her towards the successful achievement of the tasks while authenticating the user. Thus, the proposed approach is a combination of concepts borrowed from intelligent tutoring and internet robotics.

From the nature of the internet, the system will serve to a diverse number of students each having different knowledge and skill levels. The system is thus designed as be adaptive to these different levels and provide each student with enough assistance for accomplishing the desired experiment and getting the necessary knowledge and experience. Students are then introduced with experiments having different complexity levels according to their past and the present performances. The system grades the students according to their performances and keeps the grades and the student profiles in a database. The system also has an authentication part to ensure security and recall a previous user from the database.

2. The System

The block diagram of the proposed system is given in the Figure-1. The system is composed of an error quantization module, a classifier of error types based on an artificial neural network and a decision process. There is also an authentication subsystem for the security of the application which denies unidentified user access with bizarre error patterns. The whole system takes errors done during the task and outputs informative messages sent to the experiment interface for the student, such as new task assignments or repeats, and the user grade values to be added to the user record.

2.1 Error Quantization Module

For each experiment and its levels, the correct sequence of actions and success limits are determined a priori by the laboratory experimentation manager. The error evaluation module outputs automatically either of the two values "–1" or "1" as a measure of performance for each user action. The value of "1" means that the user action is inside the success limits in the correct context of the experiment and therefore the action is correctly executed. A "-1" value means that the action is outside of the limits of success in the related context, and therefore the action is wrong. "-1" and "1" values are collected in the form of an input vector for the classifier until a specified number actions are completed.

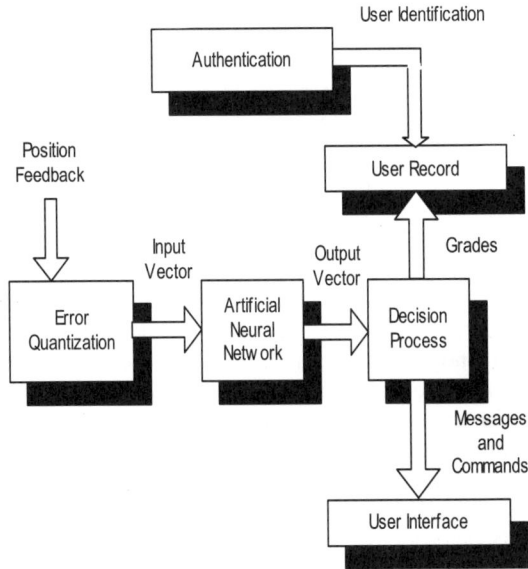

Figure-1: The block diagram of the system

This specified number of actions determines the evaluation interval. An experiment consists of many evaluation intervals which are not transparent to the user. Thus, the student is not aware of his/her evaluation intervals and this parameter does not hinder his/her performance. For the implementation given in this paper, the evaluation interval takes a value among three preset values of 8, 12, and 16. The evaluation interval value is determined for a specific user considering the user's performance while carrying out a certain experiment. It can either increase or decrease in the course of the experiment. By default, the value is taken as 8. During the experiment, this value increases one step in the case of three successive successful intervals with zero experiment repeat value in the user record or two successive full successful intervals with a repeat value smaller than two. This value is decreased one step in the case of successive failures or full failures. (The notions of full-success, success, full-failure and failure are the outputs of the decision process which will be presented in section 2.3). The evaluation interval value determines somewhat the freedom given to the user. In the case of a large interval, the performance is evaluated more rarely and evaluation results are more in the favor of the user: the user may make minor errors but still considered as successful.

2.2 Error Classification Using ANN

The input vector to this error classifier module represents a model of the user in the evaluation interval in terms of the quantized error performance. After each completed evaluation period, the resulting bipolar vector is input to an ANN based classifier that typifies the error made. The neural network is a feed-forward network having one hidden layer. The number of input nodes of the network is determined by the evaluation interval value and has the value of 8, 12 or 16; hence there exist three different neural networks in the case of our illustrative example. The number of hidden nodes is the number of input nodes-5. The sensitivity of our proposed system to this constant (5) is currently being considered and the detailed sensitivity analysis of the system to each parameter involved is incorporated in an upcoming journal paper. The number of the output nodes is 3 in all cases, defining the error classes to be used in the decision process for grades and task reassignments. The back propagation algorithm is used for training the neural networks which are each trained by default by 8 input vectors of 8, 12 or 16 dimension, representing different typical user behaviors for the different evaluation intervals, 8, 12 and 16. The number of training vectors represents the classes of user behaviors which are considered as having 8 main user characteristics which are:

- no errors in all steps
- small number of errors in the first steps
- small number of errors in the middle steps or distributed unevenly
- small number of errors in the last steps
- errors in all steps
- large number of errors in the first steps
- large number of errors in the middle steps or distributed unevenly
- large number of errors in the last steps

2.3 Decision Process

The decision process is a rule-based system which uses fuzzy control principles to make decisions about the performance of the user according to the error classes made within the respective interval. Each of the three outputs of the neural network is fuzzified into 4 sets such that we recover a continuum out of unbiased quantized errors and classes of user behaviors. These sets are labeled as LOW, MED1, MED2 and HIGH and triangular membership functions are used. They cover up the interval between –1 and 1 which are the outmost margins of each network output. The names of the sets imply the range of an output node value falling between the margins. Figure 2 shows the fuzzification of each node's output. Fuzzy rules are constructed to select among 8 levels of success. The 8 levels are labeled as f, f1, f2, f3, s3, s2, s1, and s respectively.

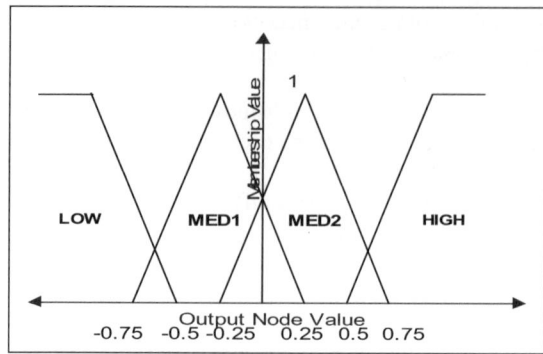

Figure-2: Fuzzification functions

The labels represent the following characteristics.

- F – Full failure
- F1 – Failure with errors mostly in first part of the interval.
- F2 – Failure with errors mostly in the middle part or distributed unevenly on the interval.
- F3 – Failure with errors mostly in the last part of the interval.
- S3 – Success with small number of errors in the first part of the interval.
- S2 – Success with small number of errors in the middle part or distributed unevenly.
- S1 – Success with small number of the errors in the last part of the interval.
- S – Full success in the interval.

The online student performance evaluation decision is finalized by selecting the output set which has the maximum membership value. According to this selection, a corresponding message informing the user of his/her performance is shown on the user interface. The necessary grade is assigned to the user at the end of the experiment by taking the arithmetic mean of the grades of each evaluation interval. For the full success case, highest grade is assigned. In other success cases the assigned grade is lowered, for S1 the least and S3 the most. The system requires the user to repeat the experiment from the beginning in case of all failures.

3. Application

The system is applied to a Tele-robotics system in the BILTIR CAD/CAM Robotics Center of the Middle East Technical University [11]. An ABB IRB 2000 6 DOF industrial robot is controlled by an intelligent interface through the Internet with a remote experimentation scenario. The robot control interface is written in Microsoft Visual basic 6.0. The proposed system is added as a module to the robot control interface, which is called as a function upon completion of each user action. The experiments consist of drawing various shapes and letters by the pen, which is attached to a mechanical gripper held by the robot, on a plain paper, ranging from easier tasks to much harder ones. Figure-4 shows robot's gripper and the pen from two different angles.

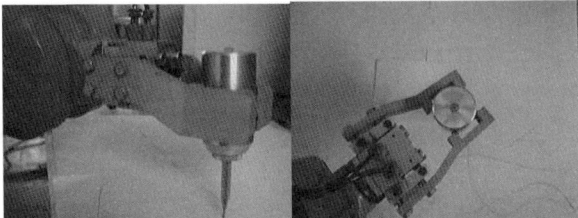

Figure-4: Gripper and pen holder side view and top view [11]

In each call of the evaluation system, the current position of the tip of the pen that robot holds is passed as a parameter to the evaluation system which consists of x and y coordinate values. The graphical user interface for the robot control and experimentation can be seen in figure-4. The robot is manipulated by clicking and moving the red spot on the right. The big white area represents a simulated version of the actual paper that is used during the experiment, and the position of the tip of the pen that the robot holds is shown in this area by a small dot that is green when the pen is up and red when the pen touches the paper. Thus, the resulting drawing done by manipulating the robot is drawn in this area as well.

The system evaluates the user performance, gives grade for each experiment, assists the user by informing of its performance and directs the user to repeat an experiment in the case of failure. In the context of this experimentation, for the evaluation interval number 8, at most 2 errors in an interval is considered as success, where at most 4 and 5 errors are considered as success in the cases of 12 and 16 respectively. The system also has an access to the user database containing user information where updates to the user record fields are made. Stored user data helps the system to recall a user that has previously done experiments. Moreover, the number of the experiment level to start from and the initial evaluation interval value are determined using this information.

4. Illustrative Example

4.1. Results and discussion of each module in the system

To illustrate the system in operation, we will consider results from each system module for a given experimental input vector and discuss each result within its own module.

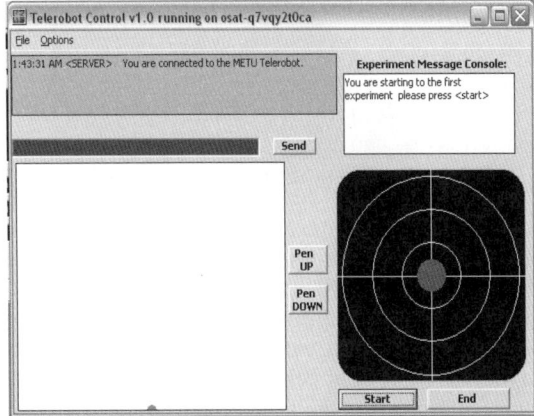

Figure-4: The user interface

For this particular example, error quantization module takes position coordinate values that result from actual operation and compare them to the desired values. If the position error is outside the performance limits then "-1" value is assigned automatically to the action performed, otherwise "1" value is assigned which indicates an acceptable move.

Table-1 shows the error values and corresponding outputs for the case of evaluation interval value 12 and its circle drawing experiment and the error made on a portion of that circle. The desired absolute error values are obtained from the circle equation. The graphical representation of this situation in the interface is shown in figure-5.

Action No.	Error	Output
1	5	1
2	10	1
3	12	-1
4	25	-1
5	5	1
6	4	1
7	0	1
8	2	1
9	9	1
10	14	-1
11	35	-1
12	22	-1

Table-1: Results from error quantization module (threshold value =10)

Figure-5: Graphical representation for Table-1

The sensitivity of this module is directly dependent on the quantization threshold. Quantization unbiases the errors initially from the past of the user which makes the experiment more independent of initial student fatigue, instant loss of attention, etc.

The resulting unbiased behavior vector, which is the last column of table-1, is the actual input to the ANN. ANN outputs that correspond to different input vectors are shown in table-2. At this stage, the ANN's output just indicates an error class about user performance while the fuzzy decision making part gives necessary meaning to the ANN's output through its evaluation.

Input Vector	Output Vector
[1 1 1 1 1 1 1 1 1 1 1 1]	[1 1 1]
[1 -1 -1 -1 1 1 1 1 1 1 1 1]	[-0.5 1 1]
[1 1 -1 1 1 -1 1 -1 -1 1 1 -1]	[0.5 -2.7e-4 -6.9e-4]
[1 1 1 1 1 1 1 -1 -1 -1 -1 -1]	[1 0.5 -1]
[1 1 1 1 -1 -1 1 1 1 1 1 1]	[1 9.1e-4 1]

Table-2: Results from the ANN

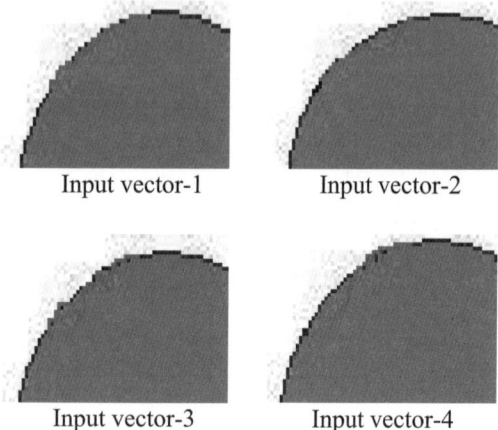

Input vector-1   Input vector-2

Input vector-3   Input vector-4

Input vector-5

Figure-6: Graphical representations of input vectors on Table-2

The graphical representations for the situations represented by the input vectors in table-2 are shown in figure-6.

The fuzzy decision process provides evaluation meaning to the ANN outputs by outputting the class that a particular user behavior falls in and the user grade for that interval. In the final evaluation output, a message is issued according to the performance class on the interface and the grade for the interval is temporarily stored to calculate the overall grade for the experiment. Table-3 shows various inputs and outputs of the fuzzy decision process.

Input Vector	Resulting Class	Resulting Grade
[1 1 1]	S	100
[-0.5 1 1]	S3	70
[0.5 -2.7e-4 -6.9e-4]	F2	- (Repeat)
[1 0.5 -1]	F3	- (Repeat)
[1 9.1e-4 1]	S2	80

Table-3: Results from fuzzy decision part

4.2. System Performance

The system's sensitivity to the different input vectors, or in other words different user behaviors which are exhibited during experiment depends on several parameters, the most important of which are neural network's training set and the error to be converged, and the number and topologies of fuzzy sets. Our main focus is on a detailed analysis of our system to these parameters. However, a primary analysis can be made by testing our system with randomly generated input vectors which are user task errors.

During test runs, the main confusion of the decision module is found to be between the "F2" and the "S2" cases, which is due to the input vector of the ANN exceeding the number of errors to be considered as success, which is for our example, 4 errors maximum for success when the evaluation interval value is 12. In such cases the system may decide success in the task instead of failure. This happens when the errors are distributed unevenly throughout the evaluation interval, that is the input vector to the ANN has "-1" values distributed unevenly among "1" values. As an example, the input vector [-1 1 1 -1 1 1 1 -1 1 1 -1 -1] is classified as "S2", although the correct class should be "F2". Consequently, our system exhibits a poor evaluation performance when the user does not show an effort of learning during the task by making random errors. Random errors and no effort of learning during experimentation are naturally not within the aim of our study. Our system focuses on evaluating students during an experimentation and learning from it. When we test our system with randomly generated input vectors to be used as user task errors meaning that there are no learning efforts during experiments, the evaluation system generates wrong decisions and grades but, still with low enough percentages:

- evaluation interval value=8   error rate=4%
- evaluation interval value=12 error rate=6%
- evaluation interval value=16 error rate=10%

The other lack of sensitivity of the system to randomly created inputs that became apparent after testing is that, the "F2" decision dominates much of the user behaviors. For example in the case the evaluation interval value is 16, the inputs, with number of errors ("-1" values in the input vector to the ANN) ranging between 6 to 12, result in the "F2" decision category and even if there are less consecutive errors say 5, in the first 5 actions, the remaining errors distributed unevenly over the other actions render the performance as "F2". This means any user (student) that does not exhibit an improving effort in the experimentation will be doomed by our system to continuously repeat the same task with failure grades.

Needless to say that, under actual user operation for distance education with lab access, input error vectors will not at all be random since the user will make learning efforts and, our system confusion in consistently pointing to failure will dramatically decrease and the user behaviors will fall into correct categorizations with much better success.

5. Conclusion

The aim of this work is to add intelligence to the experiment control interface for distance education through laboratory works in order to evaluate the performances of different users, and to coach them for better achievement of the experiment goals.

The proposed system is called an "evaluation subsystem", which computes its error inputs from the online robot supported application, classifies the user experimental behavior and generates outputs to the user interface in the form of messages such as repeat commands or new experimental assignments and to the user record database in the form of grades. Our system is actually in use at the site: http://www.me.metu.edu.tr

6. Acknowledgement:

This work has been applied on the control interface of an ABB robot through the Internet, developed by A. Boyaci and I. Konukseven in METU Mechanical Eng. Dept.

7. References

[1] Dr. rer. Nat. Ulrich Karras, Dipl.-Inform. Hauke Ernst, "DERIVE, Distributed Real and Virtual Learning Environment for Mechatronics and Tele-service" WS 2001, International Workshop on Tele-Education in Mechatronics Based on Virtual Laboratories, 18 – 21 July 2001, Weingarten, Germany.

[2] http://www.vvl.de

[3] http://kmi.open.ac.uk/projects/pearl/index.html

[4] Xavier Vialta, Denis Gillet and Christophe Salzman, "Contribution to the Definition of Best Practices for the Implementation of Remote Experimentation Solutions" IFAC Workshop on Internet Based Control, IFAC Workshop on Internet Based Control Education, IBCE'01, Madrid, Spain, December 12-14, 2001

[5] Christoph Rohrig, Andreas Jochheim, "Java Based Framework for Remote Access to Laboratory Experiments" In Proc. IFAC/IEEE Symposium on Advances in Control Education, Gold Coast, Australia, 2000

[6] Peter Brusilovsky, "Adaptive Educational Systems on the World-Wide-Web: A Review of Available Technologies" Workshop "WWW-Based Tutoring" In: Proceedings of at 4th International Conference on Intelligent Tutoring Systems (ITS'98), San Antonio, TX, August 16-19, 1998.

[7] Peter Brusilovsky, "Adaptive Hypermedia", User Modeling and User-Adapted Interaction 11: 87-110, 2001. Kluwer Academic Publishers.

[8] Alfred Kobsa "Generic User Modeling Systems"; User Modeling and User-Adapted Interaction. 11(1-2), 49-63. (2001)

[9] R. Yasdi, "A Literature Survey on Applications of Neural Networks for Human-Computer Interaction", Neural Comput & Applic (2000)9:245–258 2000 Springer-Verlag London Limited

[10] Nicola Capuano, Marco Marsella, Saverio Salerno, "ABITS: An Agent Based Intelligent Tutoring System for Distance Learning", Proceedings of the International Workshop on Adaptive and Intelligent Web-based Educational Systems held in Conjunction with ITS 2000 Montreal, Canada.

[11] Ali O. Boyaci, Internet and Socket Control of an ABB Robot Arm" MSc. Thesis Middle East Technical University, Mech. Eng. Dept., Ankara, Turkey September 2002.

Co-operative Control of Internet Based Multi-robot Systems with Force Reflection

Wang-tai Lo, Yun-Hui Liu
Dept. of Automation and Computer
Aided Engineering
The Chinese University of Hong Kong

Imad Elhajj, Ning Xi
Dept. of Electrical and Computer
Engingeering
Michigan State University

Yinghai Shi, Yuechao Wang
Shenyang Institute of Automation,
Chinese Academy of Science

Abstract

With the rapid development of information technology, Internet has evolved from a simple data-sharing media to an amazing information world where people can enjoy different kinds of services. Recently, the use of the Internet has been expanded to the field of automation, i.e. using the Internet as a tool to control equipment located at remote sites. This paper presents a cooperative robot system consisting of a robot hand and a mobile robot carrying a stereo vision, which can be tele-operated by operators at different sites via the Internet. To overcome the instability and reliability problem caused by the random time delay of the Internet communication, we adopt an event as the reference for controller design of the system. A vision-based method is adopted to maintain interactions among the operations. Results obtained in teleoperation experiments among Hong Kong, the mainland China, and USA will be demonstrated to confirm the usefulness and effectiveness of the developed method and system.

1. Introduction

With the rapid development of information technology, Internet is applied in different fields nowadays. Recently, the use of the Internet has been extended to the field of automation by using it as a command transmission media for controlling the machine remotely. With this remote control technique, the operator is able to control the robot everywhere. It is convenient for several different operators at different location to collaborate concurrently in real time.

Collaborative teleoperation is an interesting area in engineering. Today, many works have been reported in the cooperative control of telerobot over the Internet. Goldberg et al [1] set up a collaborative teleoperated system. Through the developed client's Internet browser, several users can play the well known Ouija board game together. Elhajj et al [2] developed a multi-site Internet-based teleoperation system which allows operators from Hong Kong and Japan control the mobile manipulator located at USA cooperatively in real time. Chong et al. [3] built a tele-manipulation test bed in which one local operator and one remote operator control the robot with a local on-line graphics simulator to tackle the time delay. Kheddar et al [4] developed a long distance multi-robot teleoperation system between Japan and France using an intermediate functional representation of the real remote world by the means of virtual reality. Suzuki et al [5] [6] designed a human interface system to control multi-robot using the World Wide Web. Each robot in the system has its own ID number, and the operator is able to operate all of them by using the developed interface system. And, the system is extended to the use of cooperative inspection.

With the characteristics of wide spread, high speed and low cost, Internet seems to be an ideal communication media for teleoperation. However, control via Internet is still facing some difficulties, such as time delay. Due to the unpredictable network congestion and varying data transmission routine, the time delay of the Internet is therefore unpredictable and varying. Although the random time delay may not affect the data transmission, it may result in the instability of the system and greatly affect the reliability and performance of the system. Many researchers have proposed different ways to solve this critical issue [7]-[10]. One of the effective approaches is to adopt a non time-based controller [11] [12].

In this paper, a telerobotics system consisting of one multi-fingered robot hand and one mobile robot is presented and the cooperation between a mobile robot and a fixed robot arm is being concerned. Different from others in which the robots are working on similar task, in our system the mobile robot and the multi-fingered robot hand are responsible for different task. The mobile robot with camera installed is used to explore the working environment, while the robot hand is used to perform the grasping task based on the images captured by the mobile robot. Although the two robots are controlled independently, they are linked up by force interaction. A visual tracking module is introduced to determine interactions between two robots by monitoring the distance between the mulitfingered robot hand and the mobile robot, and thus monitoring the robots' motion.

The developed system can be operated remotely and collaboratively by multi operators at different location in real time via Internet with force reflecting. And, the problem of random time is overcome by applying an event-based controller. Compared with other event-based systems which must wait for the completion of all the cycles, in our system each robot is independent form the from the others, offering advantages of being simple, yielding better performance, being less affected by time delay. The feasibility of the system is verified successfully by several experiments conducted between China, U.S.A, and Hong Kong.

This project is supported in part by the Hong Kong Research Grant Council and the National Science Foundation of China under grants CUHK4166/98E, CUHK4173/00E and N_CUHK404/01.
Email : yhliu@acae.cuhk.edu.hk

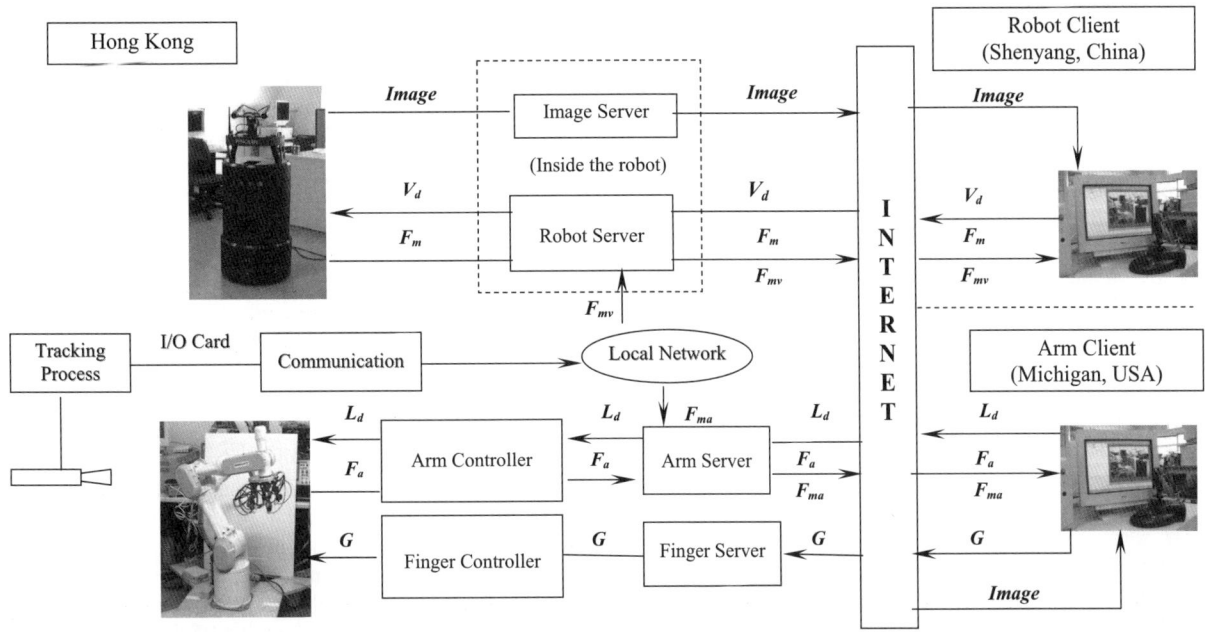

Figure 1. The architecture of the internet based cooperative telerobotics system

2. The Internet-based cooperative robots system

A real-time force reflecting cooperative telerobotics system is developed. The system can be divided into four main subsystems which are client module, mobile robot module, multi-fingered robot hand module, and visual tracking module. All these modules are connected together through the Internet with TCP/IP communication protocol to form an Internet-based cooperative multi-robot system. The architecture of the system and the variable used are shown in Fig. 1.

2.1 Client Module

The input device is a Microsoft SideWinder Force Feedback Pro. programmable force feedback joystick, which has 3 degrees of freedom for data input. Moreover, it can generate feedback force in conventional x and y directions and simulate different feelings of vibration.

The communication between the computer and the joystick is done by the client interface programmed in Visual C++ with the DirectX technology. The client program takes the position of the joystick and the buttons' status and sends these information to the server as motion command. Besides, the program is also used for commanding the joystick to render the reflected forces received from the server.

2.2 Mobile Robot Module

The mobile robot is the B21 indoor mobile robot manufactured by Real World interface. It is driven by a four wheel synchronous mechanism which allows the robot moving freely in translation and rotation. On the top of the robot, a pan tilt head with two colour CCD cameras are installed. To keep the robot online, a wireless Ethernet bridge is incorporated to connect the robot to the Internet.

Inside the robot, there are two computers. One is the robot server, another is the image server. The robot server is responsible for the communication and the motion control of the mobile robot. It receives motion commands from the operator and collects the workspace information through the installed sensors. The image server is running VIC for image transmission. VIC is a video conferencing application software package developed by the Network Research Group at the Lawrence Berkeley National Laboratory in collaboration with the University of California, Berkeley.

2.3 Multi-fingered Robot Hand Module

The multi-fingered robot hand is composed of one robot arm and a five-fingered robot hand. The robot arm itself is the PA-10 portable general-purpose intelligent arm from Mitsubishi, which has 7 joints in all with symmetrical operation range and no offset from the centre. For the end effector, there are total 5 fingers installed. They are the product of Yaskawa Electric each of which has 3 degrees of freedom and can be controlled independently.

A SUN SparcStation 4, running the arm server and the finger server, is used for the communication between the remote and local site. Through the serial port, the arm server forwards the desired offset and orientation of the hand accepted from the operator to the arm controller, and gets the position of the hand from it. For the finger server, it is responsible for sending the predefined joint angles of each finger in grasping or ungrasping gesture to the finger controller depended on the received grasping signal.

2.4 Visual Tracking Module

The module is composed by two computers which are connected through an I/O card. The tracking process is carried out by the computer with a Matrox pulsar frame grabber card installed. And, the communication part is done by another computer. The real time images of the arm's workspace are taken from the CCD camera which is hanged above the floor level about 2.5m. The image of the workspace is shown in Fig. 2. Two makers in black colour are placed on the mobile robot and the robot hand respectively. The use of the marker is to make the robot easier to be recognized and reduce the noise from the background.

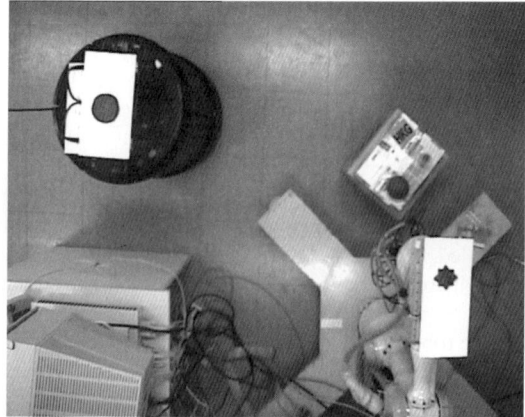

Figure 2. The image of the workspace

3. Event-based Control for Force Reflecting Teleoperation

Time delay in teleoperation greatly affects the performance, stability of the system and action synchronization. This problem is more critical when the control commands and the feedback information are transmitted through the Internet in real time, as the time delay over the Internet is varying and unpredictable. To illustrate the randomness of the time delay, Fig. 3 and Fig. 4 show the round trip delay between Hong Kong and USA, and that between Hong Kong and China.

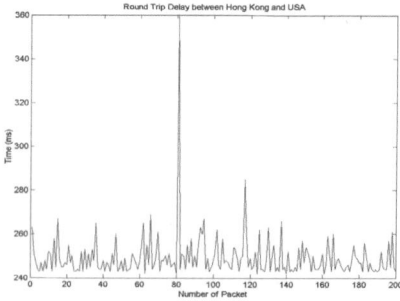

Figure 3. The round trip delay between HK and USA

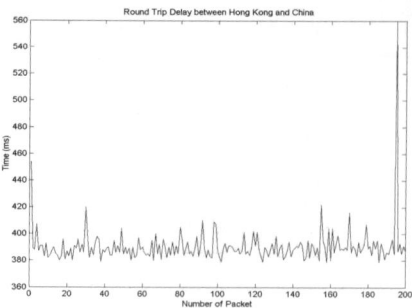

Figure 4. The round trip delay between HK and China

To solve the time delay problem, the concept of event-based control law is applied to the developed telerobotics system. Event-based controller is a non-time based controller, which takes an event, instead of time, as the action reference parameter. The event parameter is the physical output of the system, however, it is not necessary to have a physical meaning. After applying the event-based controller, the model of all the components in the developed system will be a function of the event. As the system is now no longer related to the time, it can be independent from the random time delay problem.

3.1 Modeling

3.1.1 Model of Joystick
Assuming the joystick is a spring-mass model, the dynamics of the joystick is:

$$M_j \ddot{X}_j(s_j) + K_j X_j(s_j) = F_p(s_j) + F_j(s_j) \quad (1)$$

$$F_j(s_j) = F_s(s_j) + F_v(s_j) \quad (2)$$

where M_j is the mass of the joystick, K_j is the spring constant of the joystick, and $X_j(s_j)$ is the displacement of the joystick from the centre. $F_p(s_j)$ is the force applied by the operator and $F_j(s_j)$ is force generated by the installed motor. $F_s(s_j)$ is the reflected obtained based on the environment information and $F_v(s_j)$ is the interactive force between two robots obtained by visual tracking module. S_j is the reference event of the joystick.

As the operator will move the joystick to a new position according to the forces he felt, the input data can be described as:

$$P(s_j + 1) = K_p \times F_j(s_j) \quad (3)$$

where P is the position value of the joystick, and K_p is the scaling constant of the joystick. $F_j(s)$ is the reflected force described in equation (2). Once the position of the joystick has been obtained, the motion command will be derived by following equation:

$$C(s_j) = K_i \times P(s_j) \quad (4)$$

where K_i is the positive scaling constant and C is the motion command of the robots.

3.1.2 Model of Mobile Robot
Due to the obstacle algorithm, the robot may not move exactly with the desired velocity. In fact, the robot will move at $V_a(s_m)$, the actual velocity of the robot. Therefore, the dynamic of the mobile base is written as following:

$$M_m \times \dot{V}_a(s_m) = F_m(s_m) + T_m(s_m) \quad (5)$$

where M_m is the mass of the mobile robot. T_m is the driving force of the motor, and F_m is the reflected forces obtained based on the environmental information. S_m is the reference event of the mobile robot module.

3.1.3 Model of Multi-fingered Robot Hand

The end effector of the robot hand will move to a new position once the desired position offset is received. Each joint will displace for a certain offset to achieve the desired position. The dynamics equation of the robot hand can be described as following:

$$H(q)\ddot{q} + (\frac{1}{2}\dot{H}(q) + S(q,\dot{q}))\dot{q} + G(q) = T_a(s) + J^T F_a(s_a) \quad (6)$$

where q is the generalized coordinates of the robot hand, and $H(q)$ is a symmetric and positive definite inertial matrix. $S(q, \dot{q})$, $G(q)$, and J are the skew-symmetric matrix, the gravity force, and the Jacobian matrix of the manipulator respectively. T_a is the torque applied by the actuators and F_a is the feedback forces obtained based on the shortest distance between the end effector of the arm and the boundaries of the workspace. S_a is the event reference of the multi-fingered robot hand module

3.2 Control

To prove the developed system is stable after applying the event-based controller, we first assume that the dynamic of the robots is stable with reference time t. As stated in [13], the event s should be a non-decreasing function of time. The selection of the event s, in our case, is the number of the executed cycle. As the number of executed cycle is always increasing with time once the connection is established, the system stability can be guaranteed regardless of the time-delay by choosing it as a reference parameter. As consequence, S_j, S_m and S_a are the cycle number of the joystick, mobile robot module, and multi-fingered robot hand module respectively.

As the mobile robot and the multi-fingered robot hand are controlled independently and no needs to wait for other side to complete the cycle, therefore the cycle number of the mobile robot module S_m may not equal to that of the multi-fingered robot hand module S_a. However, the joystick controlling the mobile robot has the same cycle number as S_m, while another joystick controlling the robot arm has the same cycle number as S_a.

3.3 Force Feedback

Besides the visual information, force feedback also plays an important role in teleoperation. With the force feedback, the operator can have a better understanding on how the robot interacts with the remote environment, and therefore increase the dexterity of teleoperation.

In developed system, force feedback is applied on three different aspects: obstacle avoidance, boundary avoidance, and rendering the interaction between the mobile robot and the robot hand.

3.3.1 Obstacle avoidance

The sonar sensors around the mobile robot are used to detect the existence of obstacles and get the obstacle distance. If the obstacle is 50cm away from the mobile robot, no force will be generated and the robot can be controlled to move freely. However, when there is a detected obstacle closer than 50cm, the robot server will send the feedback force to the operator to alert him the existence of the object. And, at the same time, the speed of the robot will be slowed down by half of the desired velocity to allow the operator to have more time to response. Furthermore, if the distance between the robot and the object detected is less than 30cm, the robot will be stopped to prevent from damaging itself and the working environment.

The direction of the feedback force will tell the operator the position of the object relative to the robot, and the magnitude of the force is related to the desired velocity of the robot. The relation between the desired translation velocity and the reflected force can be written by following equations,

$$F_{mx}(s) = V_{dy}(s) \times \cos\theta \times r \times K_m \quad (7)$$

$$F_{my}(s) = V_{dy}(s) \times \sin\theta \times r \times K_m \quad (8)$$

F_{mx} and F_{my} are the force feedback in the direction of x and y respectively. V_{dy} is the desired velocity in translation and θ is the orientation of the obstacle relative to the heading of the mobile base. K_m is the gain. D is the distance between the robot and the obstacle in cm, and r is a variable depended on D, where

$$r = \begin{cases} 0 & D > 50 \\ 0.5 & \text{for } 30 < D < 50 \\ 1 & D < 30 \end{cases} \quad (9)$$

3.3.2 Boundary Avoidance

In the multi-fingered robot hand module, a workspace in the shape of rectangular box is defined to avoid the arm reaching the singular position. When the end effector is away the boundaries 5cm or more, the operator will not feel any force, and he can control the arm to move in all directions. But, if the shortest distance between the arm and the boundaries is less than 5cm, the operator will receive the feedback force from the arm server to prevent him from driving the arm to the workspace limit. And, the robot hand will stop moving if it reaches the boundary of the workspace.

The direction of the reflected force is opposite to the motion of the arm, and so the operator will know which direction he should move to keep the arm away from the boundary. The Forces can be represented as follows,

$$|F_{ax}(s)| = |F_{ay}(s)| = |F_{az}(s)| = \begin{cases} K_a & \text{for } D_a < 5 \\ 0 & \text{otherwise} \end{cases} \quad (10)$$

where K_a is the force constant, D_a the shortest distance between the end effector and the workspace boundaries in cm. F_{ax}, F_{ay} are the force feedbacks in the direction of x and y respectively. As the joystick can only generate forces in x and y direction, F_{az}, the force generated when the arm is reached the upper or lower limit, will be a vibration force.

3.3.3 Interaction Rendering

In order to explore the working environment with a better view point, the mobile robot sometimes may need to enter the workspace of the robot arm. The robot hand is different from stationary obstacle, since it can move in both x, y, z direction. Therefore, the end effector may not be detected by the sonar sensor and collie with the mobile robot. To prevent the collision, the visual tracking module is introduced to find out the interaction between the end effector and the mobile robot.

In the real time image, the mobile robot and the robot hand is being tracked, and the position and the intermediate distance of the robots are found. The operators will not feel the feedback force until the distance between two robots is less than 60cm. And, the force will increase linearly when the mobile robot approaches to the robot hand. The reflected force is only used to alert the operators and will not affect the motion of the robots.

Both operators receive the feedback force with the same magnitude but in different direction. The mobile robot is assumed that it is always pointing to the arm's workspace and keeps away from the end effector by moving backward, so its operator will get the backward reflected force. For the robot hand module, with the end effector as the centre, the operator can receive the feedback force in 4 directions depended on the quadrant of the mobile robot located. The interactive force can be described as follows,

$$|F_{vm}(s)| = |F_{va}(s)| = \begin{cases} K_v \times (60 - D_v) & \text{for } D_v < 60 \\ 0 & \text{otherwise} \end{cases} \quad (11)$$

where K_v is the force constant and D_v is the distance between the mobile robot and the end effector. F_{vm} and F_{va} are the interactive force sent to the operator of mobile robot and that of the robot arm respectively.

4. Experimental Results

To evaluate the developed system, several experiments were carried out among the Mainland China, Hong Kong, USA to verify its performances. Described here is one of the experiments in which the mobile robot and the mutifingered robot hand are located in Hong Kong, and the operators from China and USA are cooperated to complete the specific task. The mobile robot module is controlled by the operator in Shenyany, China, and the multi-fingered arm module is operated by the operator in Michigan State, USA.

In this experiment, the operator at China controls the mobile robot to help the operator at USA in grasping the object by providing a better point of view of the robot hand's workspace. Next to the target object, there are two boxes. After the object is grasped by the robot hand successfully, the operator at China moves the mobile robot to view at either one of the boxes telling the operator at USA which box he should put the object.

During the experiment, operators will not only receive the feedback force for obstacle avoidance and boundary avoidance, they will also obtain the interaction between the robot arm and the mobile robot. The interactive force can help the operators to keep a distance between two robots, and therefore prevent the collision between the robots even the mobile robot has entered to the robot hand's workspace.

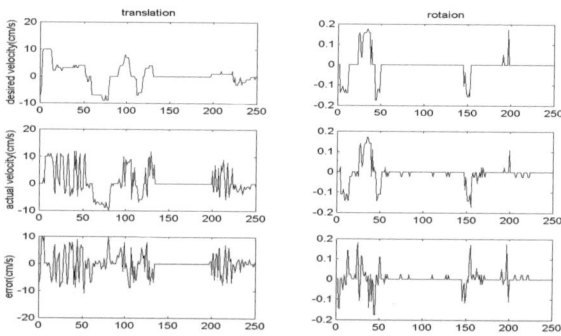

Figure 5. Experimental result of the mobile robot.

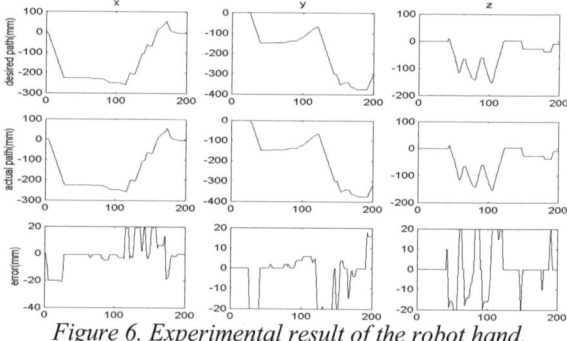

Figure 6. Experimental result of the robot hand.

In Fig. 5, the plots of the actual velocity, the desired velocity and the velocity error of the mobile robot in translation and rotation are shown in the left column and the right column respectively. In the graph of the translation velocity, the inconsistence between the actual and desired velocity is due to the poor performance of the controller at slow speed. When the mobile robot enter the robot arm workspace, the controller fails to move the mobile robot smoothly as the robot is only allowed to move at half of the desired velocity if it is near an object. However, the rotation of the robot does not affect by the obstacle avoidance algorithm, so the actual rotation velocity is at the same pattern as the desired one.

And, the first column, the second column, and the third column of Fig. 6 are showing the desired path, the actual path and the position error of the arm in x, y, z respectively. From the graphs, the robot arm are moved corresponding to the motion command and the response of the robot hand is fast and synchronized.

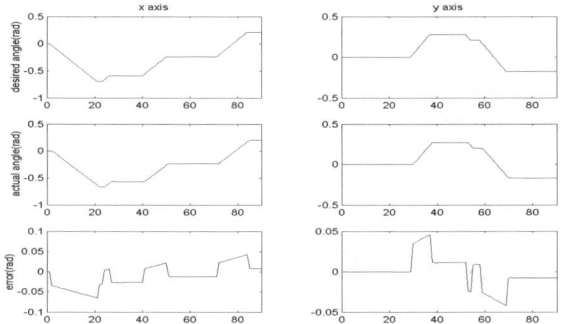

Figure 7. Experimental result of orientation control

Another experiment is carried out to verify the performance of orientation control of the robot hand. The orientation of the end effector about x and y axis is controlled. As the fingers are connected with a number of cables, due to the safety reason, the orientation about the z axis is unable to be changed. The results are shown in Fig. 7. The left column shows the desired orientation, the actual orientation and the orientation error about x axis and the right column shows the same things about y axis. The plot shows that the system is stable and the motion of the hand follows the commands sent by the operator.

5. Conclusion

Presented in this paper is a cooperative control of multi robots for Internet-based teleoperation with force reflecting. The proposed approach is demonstrated by a cooperative telerobotic system consisting of a mobile robot and a multi-fingered robot hand. In addition, a visual tracking module is introduced to find out the interaction between two robots. To overcome the time delay problem, an event based controller is adopted. And, the performance of the presented system is verified by several teleoperation experiments carried out among China, Hong Kong, and USA.

With such remote control technique, different operators at different location can collaborate concurrently all over the world via the Internet. It is hoping that such technique can be expanded to the general application such as intelligent home and 'e-hospital' in the future.

Acknowledgment

The authors would like to acknowledge Mr. Martin Leung and Mr. William Chen for their helps in developing the systems, and Mr. Wai Keung Fung for his helps in conducting the experiments.

References

[1] K. Goldberg, B. Chen, R. Solomon, S. Bui, B. Farzin, J. Heitler, D. Poon, G. Smith, "Collaborative Teleoperation via the Internet", Proceedings of the IEEE International Conference on Robotics & Automation, April 2000, pp. 2019-2024.

[2] I. Elhajj, N. Xi, W. K. Fung, Y. H. Liu, Y. Hasegawa, T. Fukuda, "Modeling and Control of Internet Based Cooperative Teleoperation", Proceedings of the IEEE International Conference on Robotics & Automation, 2001, pp. 662-667.

[3] N.Y. Chong, T. Kotoku, K. Ohba, K. Komoriya, F. Ozaki, H. Hashimoto, J. Oaki, K. Maeda, N. Matsuhira, K. Tanie, "Development of a Multi-telerobot System for Remote Collaboration", Proceedings of the IEEE/RSJ International Conference on Intelligent Robots and Systems, 2000, pp. 1002-1007.

[4] A. Kheddar, P. Coiffet, T. Kotoku, K. Tanie, "Multi-Robots Teleoperation – Analysis and Prognosis", IEEE International Workshop on Robot and Human Communication, 1997, pp.166-171.

[5] T. Suzuki, T. Fujii, K. Yokota, H. Asama, H. Kaetsu, I. Endo, "Teleoperation of Multiple Robots through the Internet", IEEE International Workshop on Robot and Human Communication, 1996, pp. 84-89.

[6] T. Suzuki, T. Sekine, T. Fujii, H. Asama, I. Endo, "Cooperative Formation among Multiple Mobile Robot Teleoperation in Inspecton Task", Proceedings of the IEEE International Conference on Decision and Control, 2000, pp. 358-363.

[7] K. Brady, T.J. Tarn, "Internet-Based Teleoperation", Proceedings of the IEEE International Conference on Robotics & Automation, May 2001, pp. 644-649.

[8] A. Sano, H. Fujimoto, T. Takai, "Network-Based Force-Reflecting Teleoperation", Proceedings of the IEEE International Conference on Robotics & Automation, April 2000, pp. 3126-3131.

[9] G. Niemeyer, J.J. E. Slotine, "Towards Force-Reflecting Teleoperation Over the Internet", Proceedings of the IEEE International Conference on Robotics & Automation, May 1998, pp. 1909-1915.

[10] R. C. Luo, T. M. Chen, C. C. Yih, "Intelligent Autonomous Mobile Robot Control Through the Internet", Proceedings of the IEEE International Symposium on Industrial Electronics, 2000, pp. PL6-PL11.

[11] Y. Liu, C. Chen, M. Meng, "A Study on the Teleoperation of Robot Systems via WWW", Canadian Conference on Electrical and Computer Engineering, 2000, pp. 836-840.

[12] I. Elhajj, N. Xi, Y. H. Liu, "Real-Time Control of the Internet Based Teleoperation with Force Reflection", Proceedings of the IEEE International Conference on Robotics & Automation, April 2000, pp. 3284-3289.

[13] N. Xi, T.J. Tarn, "Stability analysis of non-time referenced Internet-based telerobotic systems", Robotic and Autonomous Systems, 2000, Vol. 32, pp 173-178

The Bridge Test for Sampling Narrow Passages with Probabilistic Roadmap Planners

David Hsu*,[1]   Tingting Jiang[†]   John Reif[†]   Zheng Sun[†]

*Department of Computer Science
National University of Singapore
Singapore, 117543, Singapore
dyhsu@comp.nus.edu.sg

[†]Department of Computer Science
Duke University
Durham, NC 27708, USA
{ruxu, reif, sunz}@cs.duke.edu

Abstract— Probabilistic roadmap (PRM) planners have been successful in path planning of robots with many degrees of freedom, but narrow passages in a robot's configuration space create significant difficulty for PRM planners. This paper presents a hybrid sampling strategy in the PRM framework for finding paths through narrow passages. A key ingredient of the new strategy is the *bridge test*, which boosts the sampling density inside narrow passages. The bridge test relies on simple tests of local geometry and can be implemented efficiently in high-dimensional configuration spaces. The strengths of the bridge test and uniform sampling complement each other naturally and are combined to generate the final hybrid sampling strategy. Our planner was tested on point robots and articulated robots in planar workspaces. Preliminary experiments show that the hybrid sampling strategy enables relatively small roadmaps to reliably capture the connectivity of configuration spaces with difficult narrow passages.

I. INTRODUCTION

During the past decade, probabilistic roadmap (PRM) planners [ABD+98], [BK00], [BOvdS99], [HLM99], [KŠLO96], [NSL99], [LK01] have emerged as a powerful framework for path planning of robots with many degrees of freedom (dofs). The main idea of a classic PRM planner [KŠLO96] is to sample at random a robot's configuration space to construct a network, called a *roadmap*, that captures the connectivity of the free space. PRM planners are both simple to implement and efficient, and thus have found many applications, including robotics, virtual prototyping, computer animation, and computational biology (see, *e.g.*, [ABG+02], [ADS02], [HLM99], [KL00], [LK01], [SLvGC01], [SLB99]).

Despite the success of PRM planners, path planning for many-dof robots is difficult. Several instances of the problem have been proven to be PSPACE-hard [HJW84], [Rei79], [SS83]. It is unlikely that random sampling, the key idea behind PRM planners, can overcome such difficulty entirely. Indeed, narrow passages in a robot's configuration space pose significant difficulty for PRM planners. Intuitively a narrow passage is a small region critical to the

[1] Part of the work was completed while the author was at the University of North Carolina at Chapel Hill.

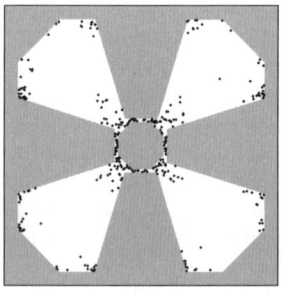

Fig. 1. An example of samples generated with the bridge test. In this and all later figures, black dots indicate sampled milestones, and shaded regions indicate obstacles.

connectivity of the free space. We can also give formal characterizations [BKL+97], [HLM99] using the notion of *visibility sets*. To capture the connectivity of the free space accurately, a PRM planner must sample configurations in the narrow passages. This is difficult, because narrow passages have small volumes, and the probability of drawing random samples from small sets is low.

In this paper, we propose a hybrid sampling strategy in the PRM framework in order to find paths through narrow passages efficiently. Key to our new strategy is the *bridge test*, which boosts the sampling density inside narrow passages and thus improves the connectivity of roadmaps. In a bridge test, we check for collision at three sampled configurations: the two endpoints and the midpoint of a short line segment s. We accept the midpoint as a new node in the roadmap graph being constructed, if the two endpoints are in collision and the midpoint is collision-free. We call this a bridge test, because the line segment s resembles a bridge: the endpoints of s, located inside obstacles, act as piers, and the midpoint hovers over the free space. Inside narrow passages, building *short* bridges is easy, due to the geometry of narrow passages; in wide-open free space, doing so is much more difficult. By favoring short bridges, we increase the chance of accepting configurations inside narrow passages (Fig. 1).

The bridge test uses only collision checking as a primitive operation and does not require complex geometric processing in the configuration space. It can be easily generalized to high-dimensional configuration spaces. It is also simple to implement and runs efficiently.

While being very effective in boosting the sampling density inside narrow passages, the bridge test severely reduces the sampling density in wide-open collision-free regions. This may be undesirable, because nodes in the roadmap need to cover the free space adequately [BKL+97]. Interestingly the difficulty encountered by the bridge test can be overcome by the uniform sampling strategy, which tends to place many samples in wide-open free space. The strengths of these two strategies complement each other naturally, and are combined with suitable weights to produce a hybrid sampling strategy to achieve better results. Our approach is related to the stratification methods for Monte Carlo integration [KW86].

The difficulty posed by narrow passages and its importance were noted in early work on PRM planners (see, e.g., [KŠLO96]) and were later articulated in [HKL+98]. Several sophisticated sampling strategies can alleviate this difficulty, but a satisfactory answer remains elusive. One possibility is to sample more densely near obstacle boundaries [ABD+98], [BOvdS99], because configurations inside narrow passages lie close to obstacles. This approach admits a simple, efficient algorithm, the Gaussian sampler [BOvdS99]. However, many configurations near obstacle boundaries lie outside of narrow passages and do not help in improving the connectivity of roadmaps. So despite the improvement, sampling near obstacle boundaries may waste many samples in uninteresting regions. See Fig. 3 for a comparison with samples generated with the bridge test. In some special cases, the Gaussian sampler can be extended to reduce the number of wasted samples by paying a higher computational cost [BOvdS99]. Other approaches for sampling narrow passages include dilating the free space [HKL+98] and retracting to the medial axis of the free space [WAS99]. Both require complex geometric operations that are difficult to implement in high-dimensional configuration spaces. The visibility roadmap [NSL99] is related to the narrow passage problem. It tries to reduce the number of unnecessary samples by checking their visibility.

The rest of the paper is organized as follows. Section II gives an overview of our planner. Sections III and IV describe and analyze the bridge test, and show how to combine it with uniform sampling to produce the hybrid sampling strategy. Section V reports experimental results. Section VI discusses alternatives to some choices made in our current planner. Section VII summarizes the main results and points out direction for future research.

II. OVERVIEW OF THE PLANNER

A classic multi-query PRM planner proceeds in two stages. In the first stage, it tries to construct a roadmap

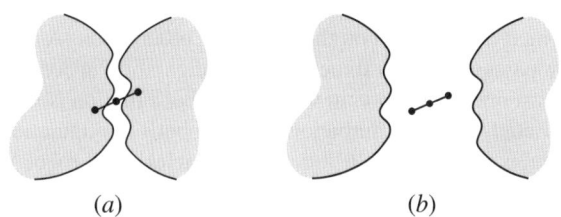

Fig. 2. Building short bridges is much easier in narrow passages than in wide-open free space.

graph G that captures the connectivity of the free space $\mathcal{F}$. The nodes of G are randomly sampled points from $\mathcal{F}$, called *milestones*. There is an edge between two milestones if they can be connected via collision-free canonical paths, typically, straight-line segments. A good roadmap G has two properties. First, the set of milestones in G covers the free space well. In other words, for every point $p \in \mathcal{F}$, there is a collision-free straight-line segment between p and a milestone in G with high probability. Second, there is an edge in G between two milestones q and q', if and only if q and q' lie in the same connected component of $\mathcal{F}$. After constructing the roadmap, the planner searches it for a collision-free path between two given query configurations in the second stage. In this paper, we address only the first stage, roadmap construction. Methods for the second stage are well-known [KŠLO96], [ABD+98].

Our goal is to build a good roadmap by sampling a small number of well-placed milestones. To obtain milestones in narrow passages, we pay a higher cost per milestone than simpler strategies such as uniform sampling; however, our roadmap size is often much smaller, thus saving lots of time in checking whether collision-free straight-line paths exist between pairs of milestones. The trade-off is well worthwhile as shown by our experiments (see Section V).

The sampling distribution that we use is a weighted mixture of π_B, the distribution generated by the bridge test, and π_U, the uniform distribution. We describe how to construct π_B and combine the two distributions in the next two sections. After generating the milestones, for every pair of milestones close to each other, we check whether a collision-free straight-line segment exists between them. If so, we insert an edge between them into the roadmap.

III. THE BRIDGE TEST

Narrow passages in a free $\mathcal{F}$ are small regions critical in preserving the connectivity of a roadmap built in $\mathcal{F}$. It is difficult to sample in narrow passages because of their small volumes. Any sampling distribution based on the volumes is likely to fail. In particular, the uniform distribution does not work well. Furthermore, when dealing with many-dof robots, we do not have an explicit representation of configuration space $\mathcal{C}$ and cannot locate narrow passages directly by processing the global geometry of $\mathcal{C}$.

The bridge test is designed to boost the sampling density inside narrow passages using only simple tests of local geometry. It is based on the following observation. A narrow passage in an n-dimensional configuration space has at least one direction v, in which the robot's motion is very restricted. Small perturbation of the robot's configuration along v results in collision with obstacles. The robot is free to move only in those directions perpendicular to the restricted ones. Therefore, for a collision-free point p in a narrow passage, it is easy to sample at random a short line segment s through p such that the endpoints of s lie in obstacles in $\mathcal{C}$ (Fig. 2a). The line segment s is called a *bridge*, because it resembles a bridge across the narrow passage, with the endpoints of s acting as piers and the point p hovering over the free space. We say that a point $p \in \mathcal{F}$ passes the bridge test, if we succeed in building a bridge through p. Clearly building *short* bridges is much easier in narrow passages than in wide-open free space (Fig. 2). By favoring short bridges over longer ones, we increase the chance of accepting points in narrow passages.

Sampling milestones. To sample a new milestone using the bridge test, we pick a line segment s from $\mathcal{C}$ at random by choosing its endpoints and determine whether s passes the bridge test. If so, we insert the midpoint of s into the roadmap G as a new milestone. The details are shown in Algorithm 1, which is called Randomized Bridge Builder (RBB). RBB calls the function CLEARANCE to determine whether a point in $\mathcal{C}$ is collision-free.

Algorithm 1 Randomized Bridge Builder (RBB).
---
1. **repeat**
2.   Pick a point x from $\mathcal{C}$ uniformly at random.
3.   **if** CLEARANCE(x) returns FALSE **then**
4.   Pick a point x' in the neighborhood of x according to a suitable probability density λ_x.
5.   **if** CLEARANCE(x') returns FALSE **then**
6.   Set p to be the midpoint of line segment $\overline{xx'}$.
7.   **if** CLEARANCE(p) returns TRUE **then**
8.   Insert p into G as a new milestone.

To perform the bridge test, RBB uses only one geometric primitive, CLEARANCE, which can be implemented very efficiently using a collision detection algorithm (see, *e.g.*, [Qui94], [GLM96]). The bridge test is purely local and does not require processing the global geometry of $\mathcal{C}$.

RBB pays a higher cost to obtain a milestone than simpler strategies such as uniform sampling, because it accepts a milestone only if a sampled point passes the bridge test, which makes three calls to CLEARANCE each. However, RBB generates milestones in narrow passages critical in capturing the connectivity of the free space, resulting in much smaller roadmaps; lots of computation time is saved in checking whether collision-free paths exist between pairs of milestones, a much more expensive operation than the simple collision check CLEARANCE.

Choosing the probability density λ. The density function λ determines how frequently a particular bridge is chosen for a test. Short bridges are preferred over longer ones in order to increase the probability of sampling in narrow passages. We choose λ_x to be a Gaussian with its center at x and the same small standard deviation σ for each dimension of $\mathcal{C}$. If we have *a priori* information on the narrow passages, then there may be other distributions more suitable than this radially symmetric Gaussian. See Section VI for further discussion.

Analysis of the sampling distribution. One may wonder: what does π_B, the probability density created by RBB, look like? To calculate π_B, let us first define X and X' to be two random variables, representing respectively the two endpoints of a bridge. The first endpoint X is distributed uniformly over the set of configuration-space obstacles $\mathcal{B} = \mathcal{C} \backslash \mathcal{F}$. So the density $f_X(x)$ is non-zero if and only if x lies in $\mathcal{B}$. Assume, without loss of generality, that $\mathcal{B}$ has volume 1. Then $f_X(x)$ is 1 if $x \in \mathcal{B}$ and 0 otherwise. Given $X = x$, we choose the other endpoint X' according to the density λ_x. The point X' is accepted only if it lies in $\mathcal{B}$. Let I be a binary function such that for any point $p \in \mathcal{C}$, $I(p) = 1$ if $p \in \mathcal{B}$ and 0 otherwise. The conditional density of X' given X is given by

$$f_{X'|X}(x' \mid x) = \lambda_x(x')I(x')/Z_x,$$

where $Z_x = \int_\mathcal{C} \lambda_x(x')I(x')\,dx'$ is a normalizing constant. To calculate π_B at a point $p \in \mathcal{F}$, we condition on X:

$$\pi_\text{B}(p) = \int_\mathcal{C} f_{X'|X}(x' \mid x) f_X(x)\,dx. \quad (1)$$

Note that p is the midpoint of the line segment $\overline{xx'}$ and so $x' = 2p - x$. Substituting the expressions for f_X, $f_{X'|X}$, and x' into (1), we have

$$\pi_\text{B}(p) = \int_\mathcal{B} \lambda_x(2p - x)I(2p - x)/Z_x\,dx. \quad (2)$$

Recall that λ_x is a Gaussian with its center at x and a small standard deviation. The density λ_x is large if $x' = 2p - x$ lies close to x. Furthermore, the integrand in (1) is non-zero only if $I(2p - x) = 1$, *i.e.*, $x' \in \mathcal{B}$. For a point p in a narrow passage, both conditions are more likely satisfied, resulting in a large value for π_B at p.

Comparison with sampling near obstacle boundaries. RBB is related to the Gaussian sampler [BOvdS99]. Both use one simple geometric primitive CLEARANCE to create favorable distributions. Their objectives, however, are quite different. RBB increases the sampling density inside narrow passages; the Gaussian sampler increases the sampling density near obstacle boundaries. RBB is slightly more expensive: it makes one more call to CLEARANCE per sample than the Gaussian sampler. However, by focusing on narrow passages, RBB gains efficiency by avoiding sampling uninteresting obstacle boundaries that do not contribute to improving the connectivity of roadmaps.

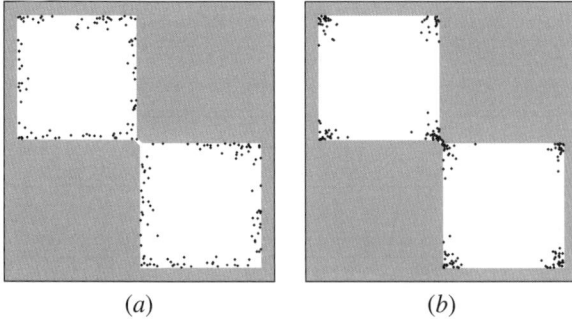

Fig. 3. The samples generated by (a) the Gaussian sampler and (b) RBB.

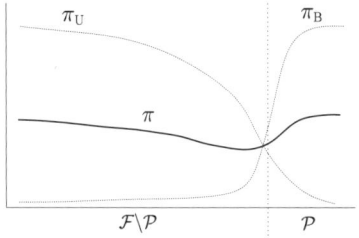

Fig. 4. The hybrid sampling distribution π. The distributions π_B and π_U perform well on $\mathcal{P}$ and $\mathcal{F}\backslash\mathcal{P}$, respectively. Combining them with suitable weights leads to good performance over the entire sampling domain.

See Fig. 3 for a comparison between these two sampling strategies. In some special cases, an extension of the Gaussian sampler can reduce the number of wasted samples by picking a triple instead of a pair of points and checking that two of the three picked points lie in *different* obstacles [BOvdS99]. It is unclear how well this extension works in general, *e.g.*, in a narrow passage formed by one single non-convex obstacle.

The idea of sampling near obstacle boundaries fails when the boundaries are uninteresting. The bridge test may fail, too, though less often. This happens when $\mathcal{F}$ contains sharp corners, because near the tip of a corner, it is easy to build short bridges. In Fig. 3(b), RBB generated a number of milestones in the six corners of $\mathcal{F}$. These samples are unhelpful. Nonetheless, our experiments suggest that the benefits gained by sampling in narrow passages outweigh the computation time wasted in sampling near sharp corners (see Section V).

IV. COMBINING COMPLEMENTARY SAMPLING DISTRIBUTIONS

We have seen that the bridge test is effective in boosting the sampling density in $\mathcal{P}$, the subset of $\mathcal{F}$ occupied by narrow passages. The density π_B is heavily biased towards $\mathcal{P}$. At the same time, π_B penalizes wide-open collision-free regions: few points are sampled in $\mathcal{F}\backslash\mathcal{P}$. This may be undesirable, because a good roadmap must cover the entire free space adequately.

Interestingly we can make up the deficiency of π_B with the uniform distribution π_U, which samples $\mathcal{F}$ with probability proportional to the volumes of subsets in $\mathcal{F}$. For π_U, most samples fall into $\mathcal{F}\backslash\mathcal{P}$. The two sampling distributions complement each other: π_U provides good coverage of $\mathcal{F}\backslash\mathcal{P}$, and π_B samples more densely in $\mathcal{P}$ and thus improves the connectivity of the roadmap. They are combined to produce a hybrid sampling distribution:

$$\pi = (1-w) \cdot \pi_B + w \cdot \pi_U, \quad (3)$$

where w is a weight, with $0 \leq w \leq 1$. The choice of w depends on the difficulty of sampling in narrow passages and the number of milestones needed to cover $\mathcal{F}$. The best choice depends on the specific problem. Currently we set w manually to favor π_B, because we assume that $\mathcal{F}$ contains at least some difficult narrow passages. We intend to conduct more experiments to determine the range of w that works well for typical problems.

One fruitful way of thinking about this hybrid distribution π is to divide the free space $\mathcal{F}$ into two subsets, the narrow passages $\mathcal{P}$ and its complement $\mathcal{F}\backslash\mathcal{P}$. We use a different sampling strategy tailored to each subset to achieve good performance over the entire sampling domain. See Fig. 4 for an illustration. This approach is related to the stratification methods for Monte Carlo integration [KW86] and the multiple-importance sampling for ray-tracing photo-realistic images [VG95].

The significance of a hybrid distribution is not about putting together two distributions, but rather about identifying distributions complementary in their strengths and combining them so that their individual strengths are preserved. Our approach differs from the previous work (*e.g.*, [DA01]) in that the two sampling distributions π_B and π_U naturally complement each other. No computation is necessary to explicitly decompose the sampling domain.

To implement the hybrid distribution, we can certainly generate new random points from π_U, but actually we can get at least some of these points "for free" by reusing the points rejected in line 3 of Algorithm 1.

V. IMPLEMENTATION AND EXPERIMENTS

To test the hybrid sampling strategy, we applied it to both a point robot and articulated robots in planar environments. Preliminary experiments indicate that our planner is able to efficiently capture the connectivity of free spaces containing difficult narrow passages.

Implementation details. Two parameters need to be chosen for our hybrid sampling strategy. First, for RBB, we chose the density function λ to be an independent Gaussian for each dof of a robot, with a small standard deviation σ to bias towards sampling short bridges. In our experiments, we set σ to be roughly 10% of the smallest allowable range of motion among all dofs. Making σ too small may adversely impact the performance of the planner. The reason is that if a bridge is too short, the second endpoint

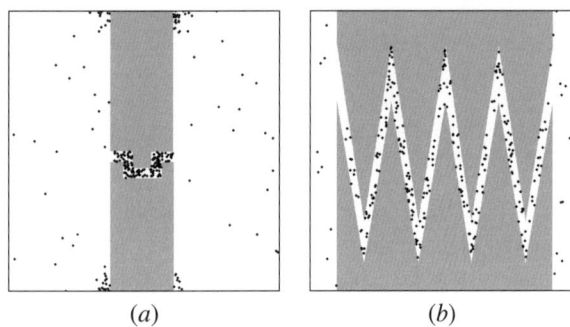

Fig. 5. Environments used for testing our planner.

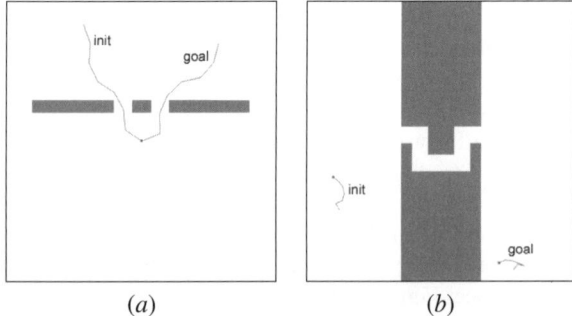

Fig. 6. Experiments with (a) a 7-dof articulated robot with a fixed base and (b) an 8-dof articulated robot with a mobile base.

TABLE I

PERFORMANCE STATISTICS OF DIFFERENT SAMPLING STRATEGIES.

Env.	Sampler	N_{mil}	N_{clear}	N_{con}	Time (sec.)
A	uniform	773	1859	3184	0.44
	RBB	46	4071	266	0.08
	hybrid	63	4543	384	0.09
B	uniform	675	1710	2685	0.81
	RBB	16	3040	46	0.03
	hybrid	22	3836	75	0.04
C	uniform	566	963	2556	0.19
	RBB	359	47372	1674	0.44
	hybrid	69	6680	515	0.06
D	uniform	111	315	839	0.03
	RBB	101	2255	873	0.04
	hybrid	98	1776	951	0.04
E	uniform	3200	4677	37270	325
	RBB	1092	735056	20206	148
	hybrid	1164	638421	19303	174
F	uniform	13699	24005	83170	1431
	RBB	1091	806759	23290	193
	hybrid	384	231305	3146	38

N_{mil} : number of milestones in the resulting roadmap
N_{clear}: number of calls to CLEARANCE
N_{con} : number of calls to check collision-free connection between two milestones

of the bridge would be unable to cross the narrow passage and fall in $\mathcal{F}$, causing many potentially useful points to fail the bridge test. The second parameter is the weight for combining π_{B} and π_{U}. We use the ratio 5:1 in favor of π_{B}, meaning that for every five milestones generated from π_{B}, we pick one milestone from π_{U}.

Our program was implemented in Java, and the results reported below were acquired from a PC with a 1.8GHz Pentium 4 processor.

Experimental results. We first tested our planner on a point robot in planar environments, containing different kinds of narrow passages. Environment A (Fig. 1) consists of four chambers connected by multiple narrow corridors. To go from one chamber to the diagonally opposite one, the point robot must pass through two narrow corridors. Environment B (Fig. 3) contains a very narrow and short corridor that connects two large square chambers. The corridor in environment C (Fig. 5a) is longer and has multiple turns. So each milestone in the corridor has low visibility and covers only a small portion of the free space. Environment D (Fig. 5b) contains a very long narrow corridor. It illustrates an interesting scenario in which RBB and the Gaussian sampler behave similarly, because almost every point in the narrow passages lies close to obstacle boundaries and *vice versa*.

We also performed preliminary experiments with planar articulated robots. Environment E contains a seven-dof articulated robot with a fixed base (Fig. 6a). At both the initial and goal configurations, the robot is trapped in narrow openings and must execute difficult maneuvers in order to find a path. Fig. 6b shows environment F. The workspace is similar to that in environment C, but the robot is an articulated robot with six links and a mobile base, eight dofs in total.

For each of the environments A–D, we generated 30 random queries that require the point robot to go through narrow passages. For environment E and F, we handpicked the queries. We then performed 10 independent runs for each query. We terminated the planner as soon as a path was found between query configurations, and recorded the running times and other statistics. For comparison, we performed the same experiments with both pure RBB (without mixing with uniform sampling) and hybrid sampling. We also used uniform sampling as a way to calibrate the relative difficulty of queries. The results for each environment and sampling strategy are averaged and reported in Table I. Note that the running times were acquired from a Java implementation. So the relative performance is more important than the absolute values of running times.

Table I shows consistent results from experiments on different robots and environments. Hybrid sampling is usually the best performer in terms of running times. Although it rejects more samples and makes more calls to CLEARANCE, it produces a roadmap with a smaller number of milestones and thus greatly reduces the time in checking that pairs of milestones can be connected via collision-free straight-line segments. In general, such a connection check is much more expensive than a call to CLEARANCE. So hybrid sampling was able to achieve good overall performance.

For some queries, pure RBB performs as well as hybrid sampling. However, when the initial and goal configurations lie in wide-open free space (environment C and F), pure RBB performs much worse, because it places almost all samples inside narrow passages and does not cover the free space well. In such cases, pure RBB may need to use a larger σ to cover the free space.

The only exception to our general observations is environment D. As expected, all three sampling strategies generated roadmaps of comparable sizes to answer the queries. However, since hybrid sampling and RBB made more calls to CLEARANCE, they took slightly longer.

VI. DISCUSSION

In designing the bridge test, the choice of the density function λ is important. So far we have assumed that λ is a radially symmetric Gaussian, which works well if a robot's dofs are all symmetric, *e.g.*, a point robot. The symmetry breaks down on free-flying rigid-bodies and articulated robots, for which each dof must be scaled to reflect its influence on the global movement of the robot. To deal with the problem, we assign a different Gaussian standard deviation to each dof. This simple extension has already been implemented for planar articulated robots in our experiments.

Furthermore, one may question whether there are other density functions better than the Gaussian. In practice, we may have some estimates on the width of narrow passages. For instance, if a robot is stuck in a narrow corridor, it has a rough idea of how much room there is to maneuver based on its knowledge of the environment. Let us assume that the width of the narrow passage is roughly c. Then it is useless to sample bridges much shorter than c, because the bridge test is bound to fail, as explained in Section V. So λ should have a shape that peaks at roughly c and decreases quickly to 0 as the length of the bridge gets much shorter or longer.

VII. CONCLUSION AND FUTURE WORK

We have presented a new sampling strategy in the PRM framework for finding paths through narrow passages. A key ingredient of the new strategy is the bridge test, which boosts the sampling density inside narrow passages. The bridge test makes use of only one geometric primitive, which checks whether a configuration is collision-free. The bridge test is purely local and can be implemented efficiently in high-dimensional configuration spaces. The strengths of the bridge test and the uniform sampling are complementary. We combine them to obtain a hybrid sampling strategy that generates small roadmaps that cover the free space well and have good connectivity. In our preliminary tests on a point robot and articulated robots with up to eight dofs, our planner was able to reliably capture the connectivity of free spaces with difficult narrow passages.

Several interesting issues regarding the bridge test and the hybrid sampling strategy require further exploration.

We are conducting additional experiments with high-dof planar articulated robots to better understand the strength and weakness of our planner. We are also implementing the planner for free-flying rigid bodies in 3-D workspaces. Based on our experience with PRM planners and that of other researchers, we are confident that the new planner will perform well in 3-D workspaces.

We would also like to further develop the hybrid sampling strategy. The important issues here are to identify sampling distributions that are naturally complementary and do not require explicitly decomposing the sampling domain, and to find a systematic weight assignment method that preserves the strengths of individual distributions.

VIII. ACKNOWLEDGEMENTS

This work is partially supported by DARPA/AFSOR contract F30602-01-2-0561 and NSF grants EIA-0218376, EIA-0218359, 11S-01-94604, EIA-0086015. Hsu was supported by the NSF ITR grant CCR-0086013.

IX. REFERENCES

[ABD+98] N.M. Amato, O.B. Bayazit, L.K. Dale, C. Jones, and D. Vallejo. OBPRM: An obstacle-based PRM for 3D workspaces. In P.K. Agarwal et al., editors, *Robotics: The Algorithmic Perspective: 1998 Workshop on the Algorithmic Foundations of Robotics*, pages 155–168. A. K. Peters, Wellesley, MA, 1998.

[ABG+02] M.S. Apaydin, D.L. Bruglag, C. Guestrin, D. Hsu, and J.C. Latombe. Stochastic roadmap simulation: An efficient representation and algorithm for analyzing molecular motion. In *Proc. ACM Int. Conf. on Computational Biology (RECOMB)*, pages 12–21, 2002.

[ADS02] N.M. Amato, K.A. Dill, and G. Song. Using motion planning to map protein folding landscapes and analyze folding kinetics of known native structures. In *Proc. ACM Int. Conf. on Computational Biology (RECOMB)*, pages 2–11, 2002.

[BK00] R. Bohlin and L.E. Kavraki. Path planning using lazy PRM. In *Proc. IEEE Int. Conf. on Robotics & Automation*, pages 521–528, 2000.

[BKL+97] J. Barraquand, L.E. Kavraki, J.C. Latombe, T.-Y. Li, R. Motwani, and P. Raghavan. A random sampling scheme for path planning. *Int. J. Robotics Research*, 16(6):759–774, 1997.

[BOvdS99] V. Boor, M.H. Overmars, and F. van der Stappen. The Gaussian sampling strategy for probabilistic roadmap planners. In *Proc. IEEE Int. Conf. on Robotics & Automation*, pages 1018–1023, 1999.

[DA01] L.K. Dale and N.M. Amato. Probabilistic roadmaps—putting it all together. In *Proc. IEEE Int. Conf. on Robotics & Automation*, pages 1940–1947, 2001.

[GLM96]　S. Gottschalk, M. Lin, and D. Manocha. OBB-Tree: A hierarchical structure for rapid interference detection. In *SIGGRAPH 96 Conference Proceedings*, pages 171–180, 1996.

[HJW84]　J.E. Hopcroft, D.A. Joseph, and S.H. Whitesides. Movement problems for 2-dimensional linkages. *SIAM J. on Computing*, 13(3):610–629, 1984.

[HKL+98]　D. Hsu, L.E. Kavraki, J.C. Latombe, R. Motwani, and S. Sorkin. On finding narrow passages with probabilistic roadmap planners. In P.K. Agarwal et al., editors, *Robotics: The Algorithmic Perspective: 1998 Workshop on the Algorithmic Foundations of Robotics*, pages 141–154. A. K. Peters, Wellesley, MA, 1998.

[HLM99]　D. Hsu, J.C. Latombe, and R. Motwani. Path planning in expansive configuration spaces. *Int. J. Computational Geometry & Applications*, 9(4 & 5):495–512, 1999.

[KL00]　J.J. Kuffner and J.C. Latombe. Interactive manipulation planning for animated characters. In *Proc. Pacific Graphics*, 2000.

[KŠLO96]　L.E. Kavraki, P. Švestka, J.C. Latombe, and M.H. Overmars. Probabilistic roadmaps for path planning in high-dimensional configuration space. *IEEE Trans. on Robotics & Automation*, 12(4):566–580, 1996.

[KW86]　M.H. Kalos and P.A. Whitlock. *Monte Carlo Methods*, volume 1. John Wiley & Son, New York, 1986.

[LK01]　S.M. LaValle and J.J. Kuffner. Randomized kinodynamic planning. *Int. J. Robotics Research*, 20(5):278–400, 2001.

[NSL99]　C. Nissoux, T. Siméon, and J.-P. Laumond. Visibility based probabilistic roadmaps. In *Proc. IEEE/RSJ Int. Conf. on Intelligent Robots and Systems*, pages 1316–1321, 1999.

[Qui94]　S. Quinlan. Efficient distance computation between non-convex objects. In *Proc. IEEE Int. Conf. on Robotics and Automation*, pages 3324–3329, 1994.

[Rei79]　J.H. Reif. Complexity of the mover's problem and generalizations. In *Proc. IEEE Symp. on Foundations of Computer Science*, pages 421–427, 1979.

[SLB99]　A.P. Singh, J.C. Latombe, and D.L. Brutlag. A motion planning approach to flexible ligand binding. In *Proc. Int. Conf. on Intelligent Systems for Molecular Biology*, pages 252–261, 1999.

[SLvGC01]　T. Siméon, J.P. Laumond, C. van Geem, and J. Cortes. Computer aided motion: Move3D within MOLOG. In *Proc. IEEE Int. Conf. on Robotics & Automation*, pages 1494–1499, 2001.

[SS83]　J.T. Schwartz and M. Sharir. On the 'piano movers' problem: II. General techniques for computing topological properties of real algebraic manifolds. *Advances in Applied Mathematics*, 4:298–351, 1983.

[VG95]　E. Veach and L.J. Guibas. Optimally combining sampling techniques for Monte Carlo rendering. In *SIGGRAPH 95 Conference Proceedings*, pages 419–428, 1995.

[WAS99]　S.A. Wilmarth, N.M. Amato, and P.F. Stiller. MAPRM: A probabilistic roadmap planner with sampling on the medial axis of the free space. In *Proc. IEEE Int. Conf. on Robotics & Automation*, pages 1024–1031, 1999.

Improving the Connectivity of PRM Roadmaps[1]

Marco Morales   Samuel Rodríguez   Nancy M. Amato

PARASOL Lab, Department of Computer Science, Texas A&M University

{marcom,sor8786,amato}@cs.tamu.edu

Abstract

In this paper we investigate how the coverage and connectedness of PRM roadmaps can be improved by adding a connected component (CC) connection step to the general PRM framework. We provide experimental results establishing that significant roadmap improvements can be obtained relatively efficiently by utilizing a suite of CC connection methods, which include variants of existing methods such as RRT and a new ray tracing based method. The coordinated application of these techniques is enabled by methods for selecting and scheduling pairs of nodes in different CCs for connection attempts. In addition to identifying important and/or promising regions of C-space for exploration, these methods also provide a mechanism for controlling the cost of the connection attempts. In our experiments, the time required by the improvement phase was on the same order as the time used to generate the initial roadmap.

I. INTRODUCTION

Motion planning has applications in robotics, computer animation, computer-aided design, bioinformatics, etc. The motion planning problem is known to be PSPACE-hard and is likely to require time that is exponential in the number of degrees of freedom of the robot [9].

Some of the most successful approaches to motion planning are the *probabilistic roadmap methods* (PRMs)[7]. PRMs map C-space by producing a graph whose nodes are valid configurations and whose edges represent valid paths between the nodes they connect. This graph, or roadmap, can be used to solve queries by connecting the start and goal to the roadmap and finding a path between them in the roadmap. PRMs are probabilistically complete, that is, the probability of finding a path using a PRM, if one exists, approaches one as the running time grows to infinity [2].

The goal of a roadmap is to represent the connectivity and topology of the free C-Space (C-free.) However, PRMs trade completeness for speed. For certain problems, this may result in roadmaps that fail to represent C-free well, e.g., zero or multiple roadmap components in one region of C-free (see Fig. 1).

Several PRM variants have been proposed to improve the quality of roadmaps. Some of these variants attempt to solve the connectivity problem by generating nodes in hard to map regions of C-free such as narrow passages and

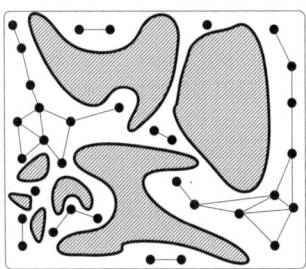

Fig. 1. Multiple components in one region of C-free.

cluttered areas (e.g., OBPRM [1], GaussianPRM [3], and MAPRM [10]). Other methods address this problem in the connection stage with more sophisticated connection techniques.

In this paper we concentrate on the connection phase. We propose to add a new step to the general PRM framework that is applied after the roadmap is constructed. Our goal is to improve the roadmap quality by connecting and growing the existing connected components of the roadmap.

Although an enhancement step for improving roadmaps has been proposed before, it was focused on placing more nodes in the roadmap. The techniques we consider here focus on exploring C-free in a systematic way while trying to find paths between pairs of nodes in different connected components. We propose new techniques and variants of existing methods and perform a comparative evaluation to study the relative strengths and weaknesses of the various methods.

II. THE PRM FRAMEWORK

PRM-based motion planning is a two step process. First, during preprocessing, a roadmap is constructed. In the query phase, pairs of start and goal configurations are connected to the roadmap and then paths between them are extracted using graph search techniques. Sometimes, attempts are made to improve the roadmap.

<u>PRM FRAMEWORK</u>
1. Roadmap Construction
1.1 Node generation
1.2 Node connection
2. Roadmap Improvement
3. Roadmap Query
3.1 Connect start and goal to the roadmap
3.2 Search for a path on the roadmap

---

[1]This research supported in part by NSF Grants ACI-9872126, EIA-9975018, EIA-0103742, EIA-9805823, ACR-0113971, CCR-0113974, EIA-9810937, EIA-0079874, and by the Texas Higher Education Coordinating Board grant ATP-000512-0261-2001. Morales supported in part by a Fulbright/Garcia Robles (CONACYT) Fellowship.

A. Roadmap construction

Roadmap construction has generally been thought of as having two steps: Node Generation and Node Connection.

PRM nodes are randomly generated in C-space. Each generated node is evaluated, and if valid, is added to the roadmap. The original PRMs generated nodes by uniform sampling in C-space [7]. Various techniques have been developed that bias the generation of nodes toward specific areas of C-free to improve the likelihood that the roadmap represents the connectivity of C-free. Notable examples include OBPRM [1], Gaussian Sampling [3], and MAPRM [10].

After the nodes have been generated, they are connected. Local planners are used to attempt to connect nearby nodes. If the local planner determines that there is a valid path connecting the two nodes, then an edge is added to the roadmap. Many local planners have been used in PRMS [1], but the most popular is the straight line in C-space method, which validates the intermediate configurations on the straight line in C-space connecting the two nodes.

B. Roadmap query

Once the roadmap has been constructed, it can be used to solve arbitrary motion planning queries. First we attempt to connect the start and goal configuration to the same connected component of the roadmap. If we cannot, then a valid path is not contained in the current roadmap. However, one might be found if the connected components to which the start and the goal are connected can be connected to each other, either directly or by connecting to other connected components.

C. Roadmap improvement

A roadmap that is to be used for multiple, arbitrary queries should cover C-space as completely as possible. Since PRMs often trade completeness for the speed, the connectivity of the roadmap might not represent the connectivity of the free C-space. In these cases, the roadmap needs to be expanded to improve its coverage and connectivity. This problem has usually been addressed with techniques focused on placing more nodes in particular areas which are not well represented in the roadmap. This process has been called roadmap expansion [7].

In this work we propose adding a new step focused on finding paths between existing CCs of the roadmap (see Fig. 2). This step attempts to improve the connectivity of the roadmap by connecting previously generated connected components. We propose using methods that are suitable for exploring areas of C-free between pairs of CCs. The difference from previous approaches is that we focus on exploring instead of on placing nodes, although we will place more nodes as a byproduct of the attempts to connect pairs of CCs.

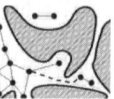

Fig. 2. The dashed line shows a path between two CCs.

III. RELATED WORK

Researchers have approached the problem of improving roadmap quality in ways ranging from purely random approaches to biasing node generation to particularly interesting areas of C-free. In general, the probability of generating a node in a region of C-free decreases as the volume of that region gets smaller and as the complexity of that region increases. There are many problems in which such areas form bridges between larger open spaces. The lack of nodes in such areas will result in roadmaps that do not reflect the connectivity of C-free.

Random reflections. Horsch et al. [6] proposed a method referred to as *random reflections at C-Space obstacles*. The method shoots a ray from one subgraph (connected component) in a random direction. Every time the ray hits an obstacle, the last free configuration is added to the roadmap and the ray is reflected in a random direction. Attempts are made to connect the new node to the k-closest nodes of the "target" subgraph, where k is some small constant. They assume that C-free is connected, and claim that the rays will eventually connect the subgraphs.

A similar technique can be found in [7] in the expansion step of roadmap construction. They generate additional nodes from existing nodes in "difficult" regions. To expand a node, a short random-bounce walk starting from that node is performed, 'bouncing' in random directions when collisions occur. When the process terminates, they add to the roadmap the final node and edges which include the path computed.

Biased sampling strategies. Amato et al. [1] used a three-staged connection strategy that includes two stages to attempt connection of CCs and to grow CCs. Attempts are made to connect smaller connected components to larger ones. The paths explored during failed connection attempts are retained and used to expand the connected components. Small connected components are grown because they might be small because they are in difficult areas of C-free.

RRT Connect. Rapidly-Exploring Random Trees (RRT) is a technique that has been used to solve motion planning queries [8]. In particular, the RRT properties can be applied to grow two RRTs, one rooted at the start and one at the goal, toward each other. This is done by generating a random configuration and extending the nearest neighbor from one RRT toward that configuration. If there is no collision, then a new node and edge are added to the RRT. This new node will be the configuration that the other RRT will try to extend toward. A greedy heuristic could be used here to extend the nearest neighbor of the second RRT as far as possible or completely toward the new configuration.

IV. CONNECTING ROADMAP COMPONENTS

The roadmap improvement stage that we propose to add to the PRM framework focuses on finding paths between existing connected components. We divide this stage into three steps: preprocessing, choosing pairs of CCs to connect, and attempting connection between chosen CC pairs.

In section IV-A we describe the algorithm in general terms. Criteria such as size or distance, can be used to choose pairs of CCs and to schedule the order for attempting connections. The methods we have implemented are described in section IV-A.

We studied three connection strategies for improving the connectivity of the roadmap: an RRT-based method (Section IV-B), a ray-tracing based method (Section IV-C), and a simple method which attempts to connect smaller CCs to larger CCs (Section III). The goal of this connection process is to produce a roadmap which represents the connectivity of C-free better than the initial roadmap.

A. Algorithm

Our strategy for connecting CCs is described in the algorithm below. First, the roadmap is preprocessed to expand or grow existing connected components according to the method that will be used to connect them. Then, in step 3, a schedule consisting of an ordered list of pairs of CCs to attempt to connect is created. Next each pair of connected components (cc_1, cc_2) is processed. First, a set of representative vertices (cc_1^s, cc_2^s) of each connected component (cc_1, cc_2) is selected (steps 5 and 6). Then a connection method is used to attempt to connect cc_1 to cc_2. Parameters specific to the connection method and a target roadmap to store the subgraph obtained are also provided. The final step is to merge the target roadmap into the roadmap. The process iterates for each pair of CCs.

Preprocessing. The preprocessing of the roadmap corresponds to the roadmap enhancement step of previous researchers. It consists of growing the connected components around some vertices. The techniques used may depend on the CC connection method.

Scheduling. The schedule is an ordered list of at most max_sched_size pairs of connected components between which connections will be attempted. In many cases the number of CCs is small enough that it is feasible to try

CONNECTCCS(RDMP,METHOD,PARAMETERS)
1.   preprocessing(rdmp,method,parameters)
2.   ccs = rdmp.GetCCs()
3.   schd_pairs = schedule(ccs,$schd\_mode$,max_schd_size)
4.   for each pair(cc_1,cc_2) in schd_pairs
5.   if (not SameCC(cc_1,cc_2))
6. cc_1^s = sampleVertices(cc_1,max_smpl_size)
7. cc_2^s = sampleVertices(cc_2,max_smpl_size)
8.   method.ConnectCCs(cc_1, cc_1^s,cc_2,cc_2^s,
                parameters,$target\_rdmp$)
9.   rdmp.MergeRdmp($target\_rdmp$)
11.   endif
11. endfor

Fig. 3. General framework used to attempt connections between connected components.

to connect every pair, however, it may be desirable to try some pairs first. The schedule can be created in different ways ($schd\_mode$). The schedule is constructed by ordering the CCs and forming pairs consisting of adjacent CCs in the ordered list. Different permutations can be obtained by ordering the connected components based on the distance between them, the number of vertices in them, the density of nodes in them, etc.

The criteria we chose for ordering CCs is the distance between them. In our implementation, the distance between CCs is the distance between their centers of mass, where the center of mass of a connected component is simply the average of all the configurations in it. Even though the straight line distance between a pair of CCs is not necessarily related to the distance that should be traveled in order to connect them, it is often a good indicator.

Sampling connected components. The number of vertices in each connected component can vary greatly. Some CCs could be large, while others might be composed of only a few or even one vertex. Some connection methods might use configurations belonging to the CCs in order to start connection attempts, as is the case of the ray tracing method described in Section IV-C.

To reduce the number of configurations used by the connection methods, we select a sample of the configurations of each connected component. When the connected component is smaller than a predefined limit (max_sampl_size), the sample is every configuration in the CC. Otherwise, we take an approximation of the boundary of the connected component by selecting the n_s nodes that are the farthest from the center of mass of the connected component (see Fig. 4 Left).

Attempting connections and updating the roadmap. After selecting a pair of connected components and their sampled configurations, a connection method is called. If it manages to connect the pair of CCs, then it puts the resulting subgraph in a target roadmap ($target\_rdmp$)

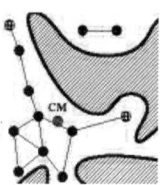 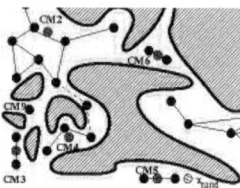

Fig. 4. **Left:** the hatched circles show the two farthest configurations from the center of mass(CM). **Right:** the dashed line shows a connection between CCs CM2 and CM4 using RRT-based connection. CM2 grows toward the random configuration x_{rand}.

which will be merged into the original roadmap. In the next sections we describe the connection methods we studied.

B. RRT-based connection

The basic RRT can be used to grow any two graphs toward one another. Although previous research was geared towards connecting two RRTs to each other, in our work we adapt this technique to bias the growth of any two CCs toward one another (Fig. 4 Right).

In the RRTconnect-components algorithm (Fig. 5), we sample a random configuration x_{rand}, find the nearest configuration x_{near} to it in cc_1, and attempt to grow x_{near} toward x_{rand} as shown in Steps 2–4. If a new node x_{new} and edge can be created, cc_2 will be grown toward x_{new} from its nearest neighbor x_{near2} in cc_2 (steps 8 - 11). If the node is free and an edge can be added, then a new node x_{new2} and edge are added to cc_2.

If a connection is made between the two connected components or if the maximum number of iterations, n_{iter}, is exceeded then the connection process stops.

RRTCONNECTCCS(cc_1,*,cc_2,*,K,STEP,ROADMAP,*)
1. while ((cc_1 is not connected to cc_2) and (i < n_{iter}))
2. x_{rand} = random configuration
3. x_{near1} = nearest neighbor in cc_1 to x_{rand}
4. x_{new1} = extend x_{near1} toward x_{rand} for step length
5.   if ((x_{new1} can connect to cc_1) and
6.   (x_{new1} is not in collision))
7.   Add node and edge to roadmap
8.   if (new node and edge added) x_{rand} = x_{new1}
9.   else x_{rand} = random configuration
10. x_{near2} = nearest neighbor in cc_2 to x_{rand}
11. x_{new2} = extend x_{near2} toward x_{rand} for step length
12.   if ((x_{new2} can connect to cc_2) and
13.   (x_{new2} is not in collision))
14.   Add node and edge to roadmap
15.   i++
16. endwhile

Fig. 5. Connecting CCs in a roadmap using RRT.

C. Ray tracing based connection

Ray tracing is a technique that has been used in computer graphics to render very high quality pictures of three-dimensional environments. Images are produced by following the paths of light rays shot through a *view window* as they bounce around the environment while changing their color properties. When a ray hits an object it bounces, and the material it hits becomes a new source of illumination from which more rays are shot in an effect known as radiosity [4], [5].

Ray tracing can also be viewed as a search technique that maps the environment as the rays travel through it. Of particular interest to us is the ability of ray tracing techniques to cover the environment. However, this technique is computationally intensive.

Random reflections, a technique similar to ray tracing, has been used to improve roadmaps in PRMs, as discussed in Section III. But the use of random reflections has been limited to bouncing. In this paper, we propose to exploit ray tracing techniques to guide our search for connections between CCs.

Ray Tracing for connecting connected components. Our ray tracing technique (Fig. 6) consists of shooting a ray in C-space from one connected component toward another, tracing the ray, and connecting it to the CCs along its route. For each CC in our schedule of CC pairs, we select sample (source) configurations as described in section IV-A. The rays are shot towards the other connected component and traced to see if any of the ray configurations is reachable from other CCs. The rays that manage to connect a pair of CCs are incorporated into the roadmap.

RAYTRACINGCONNECTCCS(cc_i,cc_i^s,cc_j,cc_j^s,
  TARGET_RDMP,MAX_RAYS,MAX_BOUNCES,MAX_LENGTH)
1. shootCfgs = cc_i^s
2. targetCfgs = cc_j^s
3. for (each shootCfgs[i] in shootCfgs)
4.   rdmpRays[i] = traceRay (shootCfgs[i], targetCfgs,
      target_rdmp, max_rays, max_bounces, max_length)
5. endfor
6. if (path found)
7.   target_rdmp.MergeRdmp(rdmpRays)
8. endif

Fig. 6. Ray-tracing based CC connection.

Ray tracer description. We use a simple ray tracer shown below. A ray is shot from the source configuration towards the first target configuration along a straight line in C-space. This ray continues until it collides with a C-obstacle. The last free configuration on the ray is tested to see if it can be reached from any target configuration. If so, a path has been found. Otherwise the ray bounces away from the C-obstacle and the process iterates. The process stops when a path has been found or when the ray has reached a maximum limit in either length (*max_length*) or number of bounces (*max_bounces*). *max_rays* rays are shot to attempt connections.

TRACERAY(SOURCE, TARGETCFGS, ROADMAP,
   MAX_RAYS, MAX_BOUNCES, MAX_LENGTH)
1. `while` ((not path found) and (rays shot ¡ max_rays))
2.   set initial direction toward targetCfgs[0]
3.   `while` ((not path found) and (ray is within limits))
4.   `if` (ray is reachable from a cfg in targetCfgs)
5.   path found, finish ray
6.   `else if` (ray collides)
7.   bounce ray, increase length and bounces
8.   `end if`
9.   `endwhile`
10.   `if` (path found)
11.   add ray to roadmap
12.   `endif`
13. `endwhile`

V. Experimental Results

Our experimental results were obtained on an Intel Pentium 4 1.8 GHz CPU with 512 MB of RAM and 256 KB of cache memory, running the Linux 2.4.9-31 operating system. Our code was written in C++ and compiled with gcc 2.96. It was implemented within our group's motion planning library which includes many PRM variants.

We tested our techniques for connecting CCs on environments of varying difficulty. Tables I and II show some PRM roadmap improvement results for two different environments: stairs-block and walls-serial.

A. Experimental setup and design

Before applying our techniques for connecting CCs, we created a roadmap for each environment. The same roadmap was used as input for all experiments in that environment. First, we applied each technique separately to enable us to compare their individual performances. Next, we applied them in combination, one after another. By cascading the techniques, each method makes progress in exploring areas of C-free and discovering regions that can be used by other methods to make more progress in their turn. This exploits the unique strengths of the different techniques.

The stairs-block environment (Fig. 7 left) has a 6-DOF robot that can travel through two narrow passages. An initial roadmap with 104 nodes and 15 CCs was generated using OBPRM in 4.45 seconds.

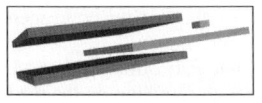 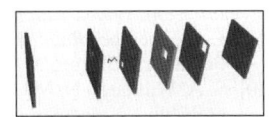

Fig. 7. The stairs-block environment (left) and the walls-serial environment (right).

The walls-serial environment (Fig. 7 right) consists of five chambers with small openings connecting them, and a 4-link revolute-articulated 9-DOF robot. An initial roadmap with 400 nodes and 5 CCs was generated for this environment using a basic (uniform sampling) PRM in 13.59 seconds.

B. Results

Table I shows the results obtained for the stairs-block environment. The RayTracing technique was the best performer, connecting all CCs into a single CC in the shortest time. RRT also did quite well, obtaining a roadmap with only two CCs. ConnectCCs was not able to reduce the number of CCs even though we modified its parameters so that it attempted all possible inter-CC pairwise connections.

TABLE I
RESULTS FOR STAIRS-BLOCK ENVIRONMENT.

Initial Roadmap OBPRM, 15 CCs (sizes 46,23,13,10,2, 10 of size 1)						
Connection Method	Initial Roadmap			Improved Roadmap		
	CCs	Nodes	Time[s]	CCs	Nodes	Time[s]
RRT	15	104	4.45	2	419	5.27
RayTracing	15	104	4.45	1	145	4.57
ConnectCCs	15	104	4.45	15	104	0.82
Combination of methods in the specified order						
RayTracing	15	104	4.45	6	119	0.70
RRT	6	119	5.15	1	437	1.09
ConnectCCs	1	437	6.24	1	437	0.00
total improvement time:						1.79

Using a combination of methods resulted in the best performance. The number of CCs was reduced to one in two-fifths of the time used to generate the roadmap. RayTracing reduced the number of CCs from 15 to 6, and then RRT took advantage of areas discovered by RayTracing and was quickly able to combine the six remaining CCs to form one CC.

Table II shows the results obtained for the walls-serial environment. RRT reduced the number of CCs to two in 35.23 seconds while RayTracing was able to connect all the initial connected components in 245.22 seconds. ConnectCCs managed to reduce the number of CCs to three in 73.54 seconds, and to two after 661.34 seconds, much longer than it took RayTracing to obtain one CC.

When using a combination of methods in the walls-serial environment, we obtained a roadmap with a single CC in only 22.66 seconds when the methods were applied in the order RRT, ConnectCCs, and RayTracing. After RRT was able to expand the roadmap and reduce the number of CCs from five to four, ConnectCCs was able to benefit from the already improved roadmap and quickly reduce the number of CCs in the roadmap from four to two. Ray Tracing finished the connection of the CCs. When the order was RayTracing, RRT, and ConnectCCs,

TABLE II
RESULTS FOR WALLS-SERIAL ENVIRONMENT.

Initial Roadmap BasicPRM, 5 CCs (sizes 129, 100, 62, 58, 51)						
Connection Method	Initial Roadmap			Improved Roadmap		
	CCs	Nodes	Time[s]	CCs	Nodes	Time[s]
RRT	5	400	13.59	2	795	35.23
RayTracing	5	400	13.59	1	594	245.22
ConnectCCs$_1$	5	400	13.59	3	400	73.54
ConnectCCs$_2$	5	400	13.59	2	400	661.34
Combination of methods in the specified order						
RRT	5	400	13.59	4	556	9.34
ConnectCCs	4	556	22.93	2	556	1.35
RayTracing	2	556	24.28	1	621	11.97
total improvement time:						22.66
Combination of methods in the specified order						
RayTracing	5	400	13.59	3	428	26.26
RRT	3	428	39.85	3	1034	17.28
ConnectCCs	3	1034	57.13	1	1034	4.73
total improvement time:						48.27

we also obtained a single CC. This took slightly longer than in the first combination but was still significantly faster than using any method alone.

None of the methods alone is clearly superior. Although RayTracing managed to merge all the CCs into only one, it was very expensive in the walls-serial environment. On the other hand, although RRT did not manage to merge all the CCs into one, it did make good progress.

Cascading the methods is clearly the way to go. In the three cases, the cascaded methods took much less time than the time spent by any of the methods used independently. It seems that by cascading the methods we take advantage of their individual strengths. Each method discovers regions for the following methods to exploit. However, it is not clear which ordering is better. It does seem that ConnectCCs performs well and at minimal expense when used after the other methods.

VI. CONCLUSIONS

In this paper we have investigated how the coverage and connectedness of PRM roadmaps can be improved by adding a connected component (CC) connection step to the general PRM framework. Our study of several CC connection strategies establishes that significant roadmap improvements can be obtained relatively efficiently. We have also seen that different methods have different strengths, and, moreover, in the situations we studied, the best strategy was in fact to utilize multiple CC connection techniques.

Our proposed addition to the PRM framework for cascading the application of different techniques enables each technique to take advantage of the increasingly improved roadmap. We propose methods for selecting and scheduling pairs of nodes in different CCs for connection attempts. In addition to identifying important and/or promising regions of C-space for exploration, these methods also provide a mechanism for controlling the cost of the connection attempts. In our experiments, the time required by the improvement phase was on the same order as the time used to generate the initial roadmap.

We proposed a new ray-tracing based connection technique which appears to facilitate rapid discovery of new regions in C-free. It is a good complement to our RRT-based connection variant which incorporates the good coverage properties of the RRT methods. Both these methods are complemented by a third method, ConnectCCs, which has been used in traditional PRMs to quickly connect easily connectable CCs.

There are many areas of improvement for this work. In the general framework, we plan to investigate additional methods for selecting and scheduling representative nodes in different CCs for connection attempts and an automatable process for determining the order, and manner, in which to apply the various connection methods. We also plan to explore improvements in the ray tracing based connection technique.

VII. ACKNOWLEDGMENTS

We would like to thank the PARASOL Lab at Texas A&M University.

VIII. REFERENCES

[1] N. M. Amato, O. B. Bayazit, L. K. Dale, C. V. Jones, and D. Vallejo. OBPRM: An obstacle-based PRM for 3D workspaces. In *Proc. Int. Workshop on Algorithmic Foundations of Robotics (WAFR)*, pages 155–168, 1998.

[2] J. Barraquand, L.E. Kavrakiand J.C. Latombe, T.Y. Li, R. Motwani, and P. Raghavan. A random sampling scheme for path planning. *Int. J. of Rob. Res*, 16(6):759–774, 1997.

[3] V. Boor, M. H. Overmars, and A. F. van der Stappen. The Gaussian sampling strategy for probabilistic roadmap planners. In *Proc. IEEE Int. Conf. Robot. Autom. (ICRA)*, pages 1018–1023, 1999.

[4] M. Cohen and J. Wallace. *Radiosity and Realistic Image Synthesis*. Academic Press, 1993.

[5] A. S. Glassner. *An Introduction to Ray Tracing*. Academic Press, 1989.

[6] T. Horsch, F. Schwarz, and H. Tolle. Motion planning for many degrees of freedom – random reflections at c-space obstacles. In *Proc. IEEE Int. Conf. Robot. Autom. (ICRA)*, pages 3318–3323, 1994.

[7] L. Kavraki, P. Svestka, J. C. Latombe, and M. Overmars. Probabilistic roadmaps for path planning in high-dimensional configuration spaces. *IEEE Trans. Robot. Automat.*, 12(4):566–580, August 1996.

[8] J. J. Kuffner and S. M. LaValle. RRT-Connect: An Efficient Approach to Single-Query Path Planning. In *Proc. IEEE Int. Conf. Robot. Autom. (ICRA)*, pages 995–1001, 2000.

[9] J. C. Latombe. *Robot Motion Planning*. Kluwer Academic Publishers, Boston, MA, 1991.

[10] S. A. Wilmarth, N. M. Amato, and P. F. Stiller. MAPRM: A probabilistic roadmap planner with sampling on the medial axis of the free space. In *Proc. IEEE Int. Conf. Robot. Autom. (ICRA)*, pages 1024–1031, 1999.

HPRM: A Hierarchical PRM

Anne D. Collins
Department of Mathematics,
Stanford University
collins@math.stanford.edu

Pankaj K. Agarwal
Department of Computer Science,
Duke University
pankaj@cs.duke.edu

John L. Harer
Department of Mathematics,
Duke University
john.harer@duke.edu

Abstract— We introduce a hierarchical variant of the probabilistic roadmap method for motion planning. By recursively refining an initially sparse sampling in neighborhoods of the $\mathcal{C}$-obstacle boundary, our algorithm generates a smaller roadmap that is more likely to find narrow passages than uniform sampling. We analyze the failure probability and computation time, relating them to path length, path clearance, roadmap size, recursion depth, and a local property of the free space. The approach is general, and can be tailored to any variety of robots. In particular, we describe algorithmic details for a planar articulated arm.

I. INTRODUCTION

One of the central problems in robotics is the *motion planning problem*: Given a robot $\mathcal{R}$ and a workspace $\mathcal{W}$ containing a set $\mathcal{O}$ of obstacles, determine a collision-free motion between specified initial and final configurations of $\mathcal{R}$ [16].

The set of configurations of a robot $\mathcal{R}$ with d degrees of freedom can be represented by a d-dimensional *configuration space* $\mathcal{C}$. The *free space* $\mathcal{F} \subseteq \mathcal{C}$ is defined to be the set of configurations in which $\mathcal{R}$ does not intersect any obstacle. The motion planning problem can be formulated as computing a path in $\mathcal{F}$ between two given configurations. The best known algorithm for motion planning, by Canny [4], takes time $n^{O(d)}$, where n measures the complexity of the obstacles, and the problem is known to be PSPACE-Hard [18]. Canny's algorithm and many other approaches compute a one-dimensional *roadmap* $\mathcal{R}$: a graph embedded in $\mathcal{C}$ that correctly captures the connectivity of $\mathcal{F}$, in the sense that any path in $\mathcal{F}$ is homotopic to a path in $\mathcal{R}$. See [10] for a survey on motion planning.

To get faster algorithms, we must sacrifice completeness, by allowing the planner to occasionally fail to return a collision-free path even though one exists. Recent research has focused on Monte Carlo approaches to motion planning, where the best we can hope for is "probabilistic completeness," an assurance that we find a solution with high probability. Our approach is a variant of the *probabilistic roadmap method* (PRM) [15]. The basic PRM constructs a roadmap $\mathcal{R}$ by sampling the free space uniformly and connecting a pair of sample configurations if a free path joining them can be easily generated by a *simple planner*, which tests, for example, whether the straight line joining them is free. A motion-planning query is then answered by connecting the initial and final configurations to nearby sample points, and searching in $\mathcal{R}$ for a path between them. This approach is particularly useful in situations in which a large number of paths are to be planned in the same environment. PRM variants seek to enhance the roadmap in narrow corridors that uniform sampling is likely to miss [3], [11], [10]. Although these planners perform well in practice, little work has been done on analyzing their performance [14], [13], [12].

PREVIOUS WORK. Kavraki et al. [14] were the first to analyze the performance of PRM. They introduce the notion of an ε-good free space, one for which the simple planner connects any $x \in \mathcal{F}$ to at least a fraction ε of $\mathcal{F}$, and present a probabilistically complete planner for such spaces. But their algorithm relies on calls to an expensive complete planner. For ε-good free spaces that satisfy some extra assumptions, the so-called *expansive* free spaces of [12], a complete planner is no longer needed. However, the extra conditions are not very natural, and the algorithm is specifically tailored to "single-shot" planning, where only one query is asked. Answering multiple queries may require building a new roadmap each time.

An alternative approach is taken in [13], where the probability that two points are connected is related to the length and clearance of a path between them, and the number of points sampled. Since they sample points uniformly at a density determined by the narrowest part of $\mathcal{F}$, it oversamples high clearance regions. Hence the efficiency could be adversely affected by a tiny small-clearance region.

OUR RESULTS. We present a hierarchical variant of the probabilistic roadmap method for motion planning. Our approach combines a local property similar to that in [14] and the clearance results of [13], with a hierarchical sampling scheme that samples only as densely as the path clearance dictates, and does not rely on calls to a deterministic planner. The roadmap we generate is not only typically smaller than that obtained by uniform sampling, but also has a higher probability of success. The approach is general and can be tailored to any variety of robots. In particular, we provide algorithmic details for a planar articulated arm in Section V.

II. Preliminaries

A roadmap of $\mathcal{F}$ is a graph $\mathcal{R} = (\mathcal{M}, \mathcal{E})$, where $\mathcal{M}$ is a set of sampled configurations, called *milestones*, of $\mathcal{F}$. Assume that we have a *simple planner*, which quickly constructs a free path of "simple shape" between two milestones if there exists one; for example, a common approach is to check whether the straight segment joining two milestones is free. We add an edge $(x,y) \in \mathcal{E}$ if the simple planer produces a free path from x to y. A simple planner thus trades efficiency with accuracy — it may not find a free path between two configurations even if there exists one. For technical reasons we assume that whenever there exists a free ball containing two configurations, the simple planner successfully finds a path between them.

In the following, we denote the ball in $\mathcal{C}$ of radius r centered at x by $B(r,x)$, and denote the volume of a subset $S \subseteq \mathcal{C}$ by $|S|$. The *clearance* $cl(x)$ of $x \in \mathcal{F}$ is the distance from x to the nearest point on $\partial \mathcal{F}$; the clearance of a path is the minimum clearance over all points on the path.

We next introduce a local property of the free space, which is similar to a localization of the ε-good property of [14], except that ours is not defined in terms of visibility.

DEFINITION 2.1: $\mathcal{F}$ is (δ, ε)-*free* if for all $\delta' \leq \delta$ and $x \in \mathcal{F}$, at least a fraction ε of the points within δ' of x are free, that is,
$$|B(\delta', x) \cap \mathcal{F}| \geq \varepsilon \cdot |B(\delta', x)|.$$

For example, a square S of side length 1 is $(1, 1/4)$-free since $|B(\delta, x) \cap S| \geq |B(\delta, x)|/4$ for any $x \in S$, $\delta \leq 1$. Intuitively, spaces with narrow passages can still have good values for (δ, ε), provided that the narrow regions are relatively "short", surrounded only by "thin" obstacles, or narrow in only a few dimensions.

Finally, we review the result of [13]: If configurations a and b can be connected by a path $\Gamma : [0, L] \to \mathcal{F}$ that has clearance at least r, then by uniformly sampling $\mathcal{F}$ at an appropriate density, we are guaranteed to connect a to b in the resulting roadmap with high probability. The argument proceeds by covering Γ with balls of radius $r/2$ whose centers are no more than $r/2$ apart. Milestones in consecutive balls will always be connected, since there exists a free ball of radius r which contains both $r/2$-balls. Therefore, if we sample $\mathcal{F}$ so densely that there is a milestone in each of these balls, the roadmap generated will connect a to b.

III. Hierarchical PRM

Let $\Gamma : [0, L] \to \mathcal{F}$ be a free path joining configurations a and b. If only a small portion of Γ has small clearance, then uniformly sampling $\mathcal{F}$ at the density determined by the minimum clearance is unnecessary; it suffices to sample so densely only in regions of small clearance. As

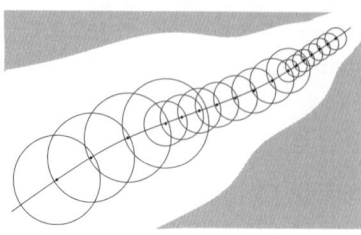

Fig. 1. Higher clearance parts of Γ are covered by larger balls.

in [13], we imagine covering Γ with balls, but we now let the *local* clearance of Γ dictate the radius of each ball; see Figure 1. We then sample $\mathcal{F}$ so that each ball contains a milestone.

Our sampling scheme proceeds as follows: Let
$$\mathcal{F}(r) = \{x \in \mathcal{F} \mid cl(x) \leq r\}$$
be the free r-neighborhood of $\partial \mathcal{F}$. We choose two parameters r_1 and N appropriately, and set $r_i = r_{i-1}/2$ for $i > 1$. We select the milestones in phases. We perform two steps in the ith phase of the algorithm.

(S1) We randomly sample a collection $\mathcal{M}_i$ of points, called i-milestones, in $\mathcal{F}(5r_i/2)$ at a uniform density $N/|B(3r_i, 0)|$.

(S2) For each i-milestone x, we use the simple planner to decide which milestones in $\mathcal{M}_{i-1} \cup \mathcal{M}_i$ that lie within a distance $3r_i/2$ of x should be connected to x.

UNIFORMLY SAMPLING $\mathcal{F}(r)$. In order to sample a neighborhood of the $\mathcal{C}$-obstacles without explicitly computing them, assume that we have available a *clearance oracle* that determines, given $x \in \mathcal{F}$ and a radius r, whether $cl(x) \leq r$. Note that we only need a bound on $cl(x)$, not the exact value. We describe such an oracle for a robot arm in Section V.

Our approach is based on the following straightforward idea, which extends to multiple sets in the obvious way.

LEMMA 3.1: *The union $A \cup B$ of two sets A and B can be uniformly sampled by first uniformly sampling A at the desired density, then uniformly sampling B at the same density, but discarding those points that are also in A.*

Suppose that we have ordered the $(i-1)$-milestones, and let $X_j = \langle x_1, \ldots, x_j \rangle \subseteq \mathcal{M}_{i-1}$; note that any order will do, as it is for bookkeeping purposes only. Then step (S1) of HPRM proceeds as follows:

(S1a) For $x_j \in \mathcal{M}_{i-1}$, we uniformly sample a collection Y_j of N points in $B(3r_i, x_j)$.

(S1b) For each $y \in Y_j$, we add y to $\mathcal{M}_i$ if $y \in \mathcal{F}(5r_i/2)$ and if $B(3r_i, y) \cap X_{j-1} = \emptyset$.

That is, we sample N potential i-milestones around each $(i-1)$-milestone, but discard those which lie in the balls already covered by previously treated $(i-1)$-milestones. Lemma 3.1 implies that $\mathcal{M}_i$ is a uniform $(N/|B(3r_i,0)|)$-dense sampling of $\mathcal{F}(5r_i/2)$, provided that the balls $B(3r_i,x_j)$ do indeed cover $\mathcal{F}(5r_i/2)$.

IV. ANALYSIS

In this section, we sketch a proof of the probabilistic completeness of HPRM. A number of details have been omitted from this abstract; see [7] for a complete proof.

Let r_1 be the initial radius parameter from Section III, and suppose that $\mathcal{F}$ is (r_1,ε)-free for some ε. Then, by Def. 2.1, for any $x \in \mathcal{F}$ and for any i, at least a fraction ε of the points within distance $r_i = r_1/2^{i-1}$ of x are free. The next lemma dictates the parameter N.

LEMMA 4.1: *Let $x \in \mathcal{M}_{i-1}$, let Y be a set of N points selected uniformly from $B(3r_i,x)$, and let $S = Y \cap \mathcal{F}$. Choosing*

$$N \geq \max\left\{6^d \ln\left(\frac{1}{\gamma}\right), \frac{3^d}{\varepsilon} \ln\left(\frac{1}{\gamma}\right)\right\},$$

we have

(a) $z \in \mathcal{F} \cap B(2r_i,x) \implies \Pr[B(r_i,z) \cap S = \emptyset] \leq \gamma.$

(b) $z \in \mathcal{F} \cap B(2r_i,x)$ *and* $cl(z) \geq r_i/2$
$\implies \Pr[B(r_i/2,z) \cap S = \emptyset] \leq \gamma.$

The proof of Lemma 4.1 is straightforward. Roughly speaking, it says that for any $z \in \mathcal{F}$ within distance r_{i-1} of an $(i-1)$-milestone, the following two conditions hold with probabibility at least $1-\gamma$: (a) there is an i-milestone within r_i of z, regardless of the clearance of z; and (b) if, in addition, z has clearance at least $r_i/2$, then there is an i-milestone within $r_i/2$ of z.

Since the probability that a given ball contains an i-milestone depends upon the presence of a nearby $(i-1)$-milestone, which in turn depends on the presence of a nearby $(i-2)$-milestone, and so on, we can prove the following lemma.

LEMMA 4.2: *If $cl(x) \geq r_i/2$, then the probability that we fail to sample an i-milestone in $B(r_i/2,x)$ is at most $i \cdot \gamma$.*

Let Γ be a collision-free path between two configurations a and b. Partition Γ into pieces, where Γ_i is the portion of Γ whose clearance is between r_i and r_{i-1}. Cover Γ_i with balls of radius $r_i/2$ whose centers lie on Γ_i and are no more than $r_i/2$ apart. If there is an i-milestone in each of the $2|\Gamma_i|/r_i$ balls of radius $r_i/2$ covering Γ_i, then our assumption that the simple planner connects two milestones contained in a free ball ensures that the resulting roadmap will connect a to b. Thus the probability that we fail is bounded by the probability that one of the balls covering Γ does not contain a milestone. Setting $L_i = |\Gamma_i|$, Lemma 4.2 implies that:

THEOREM 4.3: *Let $\Gamma : [0,L] \to \mathcal{F}$ be a free path from a to b which has clearance at least r_k. Then the probability that HPRM fails to connect a and b is*

$$\Pr[\text{fail}] \leq 2\gamma \sum_{i=1}^{k} \frac{iL_i}{r_i}.$$

We next bound the time to generate the roadmap. Let τ (resp. σ) be the time taken by the clearance oracle (resp. simple planner). Let m_i be the number of i-milestones, and consider the ith phase of HPRM. For each of the $m_{i-1}N$ points x sampled in step (S1a), step (S1b) takes time τ to determine whether $x \in \mathcal{F}(5r_i/2)$ and another $O(m_{i-1})$ time to determine whether x lies in any of the previously sampled balls. In step (S2), the simple planner tests each i-milestone for connection with $O(m_{i-1}+m_i)$ $(i-1)$- and i-milestones.

THEOREM 4.4: *Suppose that $\mathcal{F}$ is (r_1,ε)-free. Then the time spent on the ith phase of HPRM is*

$$O(\tau m_{i-1}N + m_{i-1}^2 N + \sigma m_i(m_{i-1}+m_i)),$$

where m_i is the number of i-milestones in $\mathcal{R}$, N is as in Lemma 4.1, and τ (resp. σ) is the time taken by the clearance oracle (resp. simple planner).

REMARK 4.5: The running time can be improved in certain cases. Suppose we use the L_∞-norm on $\mathcal{C}$, as in Section V. Then balls are axis-parallel hypercubes, and the problem of finding the i-milestones in a given ball can be formulated as answering an *orthogonal range query*. Using a data structure known as a *kd-tree* [2], we can reduce some factors of m to $m^{1-1/d}$; see [7] for details.

Since they are stated in full generality, Theorems 4.3 and 4.4 are rather cumbersome. In order to compare HPRM to the uniform sampling scheme of [13], let us suppose that both L_i and $|\mathcal{F}(5r_i/2)|$ decrease as i increases. In particular, we assume that there exist two constants $0 < c_1, c_2 < 1$ such that:

(A1) $L_{i+1} \leq c_1 L_i$ for all i, and

(A2) $|\mathcal{F}(5r_{i+1}/2)| \leq c_2 |\mathcal{F}(5r_i/2)|$

Corollaries 4.6 and 4.7 bound the failure probability and roadmap size, respectively, subject to these assumptions. Proofs can be found in [7].

COROLLARY 4.6: *Let $\Gamma : [0,L] \to \mathcal{F}$ be a path from a to b with clearance at least r_k that satisfies (A1) with $c_1 \leq 1/2$. Then the probability that we fail to connect a to b in $\mathcal{R}$ is*

$$\Pr[\text{fail}] \leq (\gamma L/r_1) \cdot k^2.$$

For comparison, if we sample $\mathcal{F}$ at the k-milestone density, the argument in [13] gives $\Pr[\text{fail}] \leq (\gamma L/r_1) \cdot 2^k$; thus, our algorithm is more likely to find a path.

The value of c_1 in assumption (A1) which ensures that our hierarchical roadmap has a smaller failure probability than uniform sampling depends upon k, the recursion depth. In fact, it seems likely that the appropriate bound in Lemma 4.6 is $c_1 \leq 1/\sqrt[k-1]{4}$. Lemma 4.7 shows that a similar bound holds for the constant c_2.

COROLLARY 4.7: *Suppose that the free space $\mathcal{F}$ satisfies (A2) for some $c_2 \leq 1/\sqrt[k]{2}$. Then the total number of milestones in $\mathcal{R}$, to level k, is $|\mathcal{M}| = \sum m_i < u$, where u is the number of points needed to uniformly sample all of $\mathcal{F}$ at the k-milestone density.*

Thus HPRM generates a *smaller* roadmap for any free space that satisfies (A2). Note that in the limit as $i \to \infty$, we have $|\mathcal{F}(5r_{i+1}/2)|/|\mathcal{F}(5r_i/2)| \to 1/2$, so this is not a particularly strong condition; in fact, if $|\mathcal{F}(5r_{i+1}/2)|/|\mathcal{F}(5r_i/2)|$ is roughly $1/2$ for all i, then $|\mathcal{M}| \leq u/2^{k-1}$, and our roadmap is significantly smaller.

V. CLEARANCE ORACLE FOR A PLANAR ARM

In this section, we discuss how to implement the clearance oracle for a d-link planar articulated arm $\mathfrak{R}$ with revolute joints, which has one of its endpoints anchored at the origin of the 2-dimensional workspace $\mathcal{W}$. $\mathfrak{R}$ has one degree of freedom for each link, and therefore a total of d degrees of freedom. Let $\phi = (\phi_1, \ldots, \phi_d) \in \mathcal{C}$ correspond to the configuration of the arm in $\mathcal{W}$ where ϕ_i is the absolute orientation of link i, as measured with respect to the x-axis of $\mathcal{W}$. We permit $\mathfrak{R}$ to intersect itself. Note that $\mathfrak{R}$ can only reach points in the component of $\mathcal{W} - \mathcal{O}$ that contains the base, so any hole in a $\mathcal{W}$-obstacle can be ignored. We therefore assume that $\mathcal{O}$ is a collection of pairwise disjoint simple polygons. We assume that $\mathcal{W}$ is bounded by a rectangle. An obstacle $O \in \mathcal{O}$ maps to a $\mathcal{C}$-obstacle, corresponding to the set of configurations for which $\mathfrak{R}$ intersects O.

A key idea underlying our algorithm is to project a ball or path in $\mathcal{C}$ to the 2-dimensional workspace $\mathcal{W}$, and determine whether the image intersects the $\mathcal{W}$-obstacles $\mathcal{O}$. Although the details of how to compute the image in $\mathcal{W}$ of a $\mathcal{C}$-space ball or path depends upon the robot in question, the method we present for intersecting this $\mathcal{W}$-image with $\mathcal{O}$ is more general.

A configuration ϕ maps to $\mathcal{W}$ as follows: Parametrize link i by t, so that $t = 0$ at joint $(i-1)$ and $t = 1$ at joint i. If the length of link i is ℓ_i, then the location of the point t along link i when $\phi = (\phi_1, \ldots, \phi_d)$ is given by the map $\Pi_i : \mathcal{C} \times [0,1] \to \mathcal{W}$, defined recursively by

$$\Pi_i(\phi, t) = \Pi_{i-1}(\phi, 1) + (t \cdot \ell_i \cos \phi_i, t \cdot \ell_i \sin \phi_i),$$

where $\Pi_0(\phi, t) = (0, 0)$. The portion of $\mathcal{W}$ occupied by link i in configuration ϕ is $\Pi_i(\phi, [0, 1])$.

To determine whether $cl(\phi^0) \leq \rho$, we compute for each $i \leq d$,

$$\Lambda_i = \Pi_i(B(\rho, \phi^0), [0, 1]),$$

the set of all points in $\mathcal{W}$ that link i will occupy for some $\phi \in B(\rho, \phi^0)$. Then $cl(\phi^0) \leq \rho$ if and only if one of these *link regions* Λ_i intersects $\mathcal{O}$.

Before proceeding, we note that some care must be taken when working in this non-Euclidean configuration space. In particular, the distance between two points on $\mathbb{S}^1$ is the length of the shorter of the two paths around the circle, and is therefore always less than π. For example, $\text{dist}(5\pi/3, \pi/3) = 2\pi/3$, and not $4\pi/3$. We use the L_∞ norm on $\mathcal{C}$, where the distance between ϕ and ϕ^0 is $|\phi - \phi^0| = \max_i\{\text{dist}(\phi_i - \phi_i^0)\}$.

COMPUTING THE LINK REGIONS Λ_i. Note that Π_i is independent of ϕ_j for $j > i$, so we can view Λ_i as the image under Π_i of the product of an i-dimensional ball with the interval $[0, 1]$.

Λ_1 is easy to compute. $\Pi_1(\phi, t) = (t\ell_1 \cos \phi_1, t\ell_1 \sin \phi_1)$, and the image of $[\phi_1^0 - \rho, \phi_1^0 + \rho] \times [0, 1]$ is a cone.

The projection for link 2 is $\Pi_2(\phi, t) = (x, y) \in \mathcal{W}$, where

$$\begin{aligned} x &= \ell_1 \cos \phi_1 + t\ell_2 \cos \phi_2, \\ y &= \ell_1 \sin \phi_1 + t\ell_2 \sin \phi_2. \end{aligned}$$

The link region Λ_2 is the image under Π_2 of the region

$$U = [\phi_1^0 - \rho, \phi_1^0 + \rho] \times [\phi_2^0 - \rho, \phi_2^0 + \rho] \times [0, 1],$$

a subset of $\mathbb{S}^1 \times \mathbb{S}^1 \times [0, 1]$. Clearly, the image of each edge of this cube is a candidate for a boundary curve of Λ_2. Unfortunately, these curves are not always sufficient.

We recall the following result from differential topology; see any standard text, for example [9].

THEOREM 5.1: *Let M and N be two manifolds without boundary, and let $f : M \to N$ be a map from M to N. Suppose that f is a submersion at $p \in M$; that is, the derivative map $df_p : T_p M \to T_{f(p)} N$ has rank $n = dim(N)$. Then there exists a neighborhood of $f(p)$ in $f(M)$ that is diffeomorphic to $\mathbb{R}^n$.*

If f is not a submersion at p, we say that p is a *critical point* of f, and $f(p)$ is a *critical value*.

Let V be an open face of U. Since V is a manifold without boundary, Theorem 5.1 applies to the restriction of Π_2 to V. Now, a point on $\partial \Lambda_2$ does not have a neighborhood in Λ_2 which is diffeomorphic to $\mathbb{R}^2$. Therefore, $\Pi_2|_V$ must fail to be a submersion at any $(\phi, t) \in V$ for which $\Pi_2(\phi, t) \in \partial \Lambda_2$.

To illustrate, consider the open face V of U with $t = 1$: $V = (\phi_1^0 - \rho, \phi_1^0 + \rho) \times (\phi_2^0 - \rho, \phi_2^0 + \rho) \times \{1\}$. The

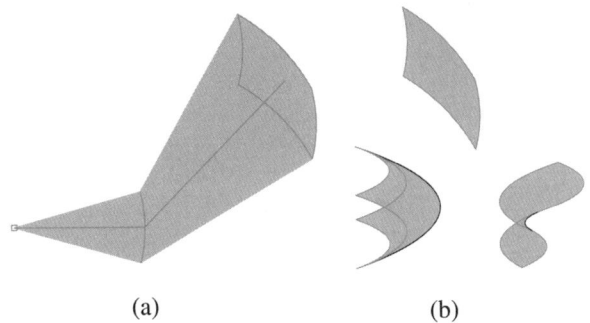

Fig. 2. (a) $\mathcal{W}$-region $\Lambda_1 \cup \Lambda_2$ swept out by a 2-link arm if $\phi^0 = (0, \pi/4)$, $\rho = \pi/12$. (b) The region swept out by the end of link 2, for various ϕ^0. Critical curves, where interior points map to the image boundary, are drawn in black.

restricted map $\Pi_2|_V = \Pi_2(\cdot, 1) : V \to \mathcal{W}$ has derivative

$$d\Pi_2(\cdot, 1)_\phi = \begin{pmatrix} -\ell_1 \sin \phi_1 & -\ell_2 \sin \phi_2 \\ \ell_1 \cos \phi_1 & \ell_2 \cos \phi_2 \end{pmatrix}.$$

Its rank is less than $dim(\mathcal{W}) = 2$ if and only if the determinant vanishes:

$$\ell_1 \ell_2 (\cos \phi_1 \sin \phi_2 - \cos \phi_2 \sin \phi_1) = \ell_1 \ell_2 \sin(\phi_2 - \phi_1) = 0.$$

If $\sin(\phi_2 - \phi_1) \neq 0$ for any point in V, then the image of V is a nice embedding of itself; however, if V contains critical points, its image appears folded, or twisted. See Figure 2.

Let Σ_i be the collection of critical curves obtained by projecting to $\mathcal{W}$ the locus of critical points in $B(\phi^0, \rho) \times [0, 1]$ where Π_i, restricted to one of its open faces, is not a submersion. Then $\partial \Lambda_i \subseteq \Sigma_i$. Note that Σ_i always contains at least the images of the edges of $B(\phi^0, \rho) \times [0, 1]$, since any map from a 1-dimensional manifold to the plane has rank at most 1. We omit the proof of the following lemma.

LEMMA 5.2: *Each critical curve in Σ_i is a line segment or a circular arc. Moreover, if $\rho < \pi/4$, then $|\Sigma_i| = \Theta(3^i)$.*

INTERSECTING Λ_i WITH $\mathcal{O}$. Finally, we test each link region Λ_i for intersection with the obstacles. If none of the link regions intersect any obstacle, then $cl(\phi^0) > \rho$.

We describe how to detect an intersection between Λ_i and $\mathcal{O}$. Without loss of generality assume that $i = d$. Λ_d intersects $\mathcal{O}$ if and only if a vertex of $\mathcal{O}$ lies inside Λ_d, a vertex of Σ_d lies inside $\mathcal{O}$, or an edge of Σ_d intersects an edge of $\mathcal{O}$. We first test whether any vertex of Σ_d lies in $\mathcal{O}$. We preprocess $\mathcal{W} - \mathcal{O}$ in $O(n \log n)$ time for planar point-location queries [8] and query it with each vertex of Σ_d in $O(\log n)$ time per vertex. If any vertex lies inside $\mathcal{O}$, then $cl(\phi^0) < \rho$ and we are done.

Otherwise, we walk along each arc of Σ_d and test for intersection with $\partial \mathcal{O}$. This can be accomplished by constructing a data structure for the so-called *circular-arc-shooting problem*, studied in [1], [6], in which one wishes to preprocess a set of polygons so that the first intersection of a query circular arc and the polygons can be determined as one follows the circular arc. Cheng et al. [6] have shown that a simple polygon can be preprocessed, in time $O(n \log^2 n)$, into a data structure of size $O(n \log n)$ so that a circular-arc-shooting query can be answered in $O(\log^2 n)$ time. However, since $\mathcal{W} - \mathcal{O}$ is not simply connected, we cannot directly use the algorithm of [6]. Instead, we proceed as follows.

If $\mathcal{O}$ has w (simply) connected components, we introduce w artificial edges to $\mathcal{W}$, $w - 1$ of which join the components of $\mathcal{O}$ into one simply connected obstacle and the final edge extends to the bounding box of $\mathcal{W}$. Let $\tilde{\mathcal{O}}$ denote the union of $\mathcal{O}$ with these auxiliary edges. If we regard each edge as a thin rectangle, $\mathcal{W} - \tilde{\mathcal{O}}$ is a simple polygon and we preprocess it using the algorithm by Cheng et al. Let $\mathcal{H}$ be the resulting data structure.

Given a curve $\alpha \in \Sigma_d$ that starts in $\mathcal{W} - \mathcal{O}$, we determine the first intersection point ξ of α with $\mathcal{W} - \tilde{\mathcal{O}}$ (along α). If ξ lies on an edge of $\mathcal{O}$, then α intersects $\mathcal{O}$. We thus conclude that $cl(\phi^0) < \rho$, and we are done. Otherwise, α intersects one of the artificial edges at ξ, and we begin a new search in $\mathcal{H}$ to detect whether the remaining portion of the arc intersects $\mathcal{O}$.

By choosing the artificial edges carefully, we can limit the number of searches performed. Chazelle and Welzl [5] showed that for any set of w points in the plane, one can construct a spanning path with the property that every circle intersects $O(\sqrt{w})$ of its edges. Such a path is said to have *low stabbing number*. The algorithm described in [5] takes $O(w^3)$ time, but by generalizing the algorithm described in [17], the running time can be improved to $O(w^{3/2} \log w)$. We choose one representative point from each component of $\mathcal{O}$. Let W be the set of w representative points. We compute the above spanning path π on W and choose our artificial edges from among the $O(n\sqrt{w})$ pieces of π in $\mathcal{W} - \mathcal{O}$. After having computed π, the last step can be accomplished in $O(n \log n)$ time by a sweep-line algorithm [8]. Since α crosses $O(\sqrt{w})$ auxiliary edges, we spend $O(\sqrt{w} \log^2 n)$ time determining whether an edge of Σ_d intersects $\mathcal{O}$.

Finally, suppose that no curve of Σ_d intersects $\mathcal{O}$. Then Λ_d intersects $\mathcal{O}$ only if one of the components of $\mathcal{O}$ lies entirely within a face of the arrangement Σ_d. In this case, it suffices to determine whether any representative point in W lies inside Λ_d. We partition Λ_d into $2^{O(d)}$ pseudo-trapezoids, each of which is bounded by two vertical edges and two arcs (circular arcs or line segments) of $\partial \Lambda_d$, e.g., by computing the vertical decomposition of Λ_d [19]. We then determine whether any of these pseudo-trapezoids Δ contain a point of W. This can be done by computing $O(\sqrt{w})$ intersection points of π and $\partial \Delta$ in a total of $O(\sqrt{w} \log w)$ time. If any segment of π intersects $\partial \Delta$ an odd number of times, then a point of W lies inside

Δ. Hence, we can test whether any component of $\mathcal{O}$ is contained in Λ_d in time $O(2^{O(d)}\sqrt{w}\log w)$.

In summary, the clearance oracle proceeds as follows: (i) preprocess $\mathcal{W} - \mathcal{O}$ in $O(n\log n)$ time for point-location queries; (ii) choose a set W of w representative points, one from each polygon of $\mathcal{O}$, construct a spanning path π of W so that a circular arc intersects only $O(\sqrt{w})$ edges of π, and preprocess π for computing intersections between π and a circular arc; and (iii) preprocess $\mathcal{O}$ for circular-arc-shooting queries in $O(n\log^2 n)$ time using π and the data strucure of Cheng et al. [6]. A query can be answered in $O(\sqrt{w}\log^2 n)$ time. Now, given a configuration ϕ^0 and a radius ρ, and for each $i = 1, \ldots, d$, we compute the $2^{O(i)}$ curves Σ_i that define the link region Λ_i, and test whether Λ_i intersects $\mathcal{O}$ in time $O(\sqrt{w}\log^2 n \, 2^{O(i)})$. Since $\sum_{i=1}^{d} 2^{O(i)} = 2^{O(d)}$, we have the following theorem:

THEOREM 5.3: *For a planar articulated arm with d links, fixed base, and revolute joints, moving among w disjoint simple polygonal obstacles with a total of n vertices, we can determine whether the clearance of a configuration is within a specified bound in time* $\tau = 2^{O(d)}\sqrt{w}\log^2 n$ *after spending* $O(n\log^2 n + w^{3/2}\log w)$ *time in preprocessing the obstacles in* $\mathcal{W}$.

Note that the dependence on d and n is completely decoupled, and that τ is sublinear in n.

REMARK 5.4: The projection technique desribed above can also be used as a simple planner for a d-link planar arm. Given a pair of configurations u and v, it either reports that it cannot find a collision-free path between u and v, or it returns a collision-free path consisting of at most d line segments, each of which is parallel to an axis (i.e., only one ϕ_i changes along each edge). The running time is $O(d^2\sqrt{w}\log^2 n)$. We omit the details from this extended abstract. Unlike the most commonly used simple planner, which determines whether a path is free by sampling points along the path and checking whether all the sampled points lie in the free space, this planner never returns a path that is not collision free.

VI. REFERENCES

[1] P. K. Agarwal and M. Sharir. Circle shooting in a simple polygon. *J. Algorithms*, 14(1):69–87, 1993.

[2] P.K. Agarwal and J. Erickson. Geometric range searching and its relatives. Advances in Discrete and Comput. Geom. (B. Chazelle, J. Goodman, and R. Pollack, eds.), AMS, Providence, 1998.

[3] N.M. Amato, O.B. Bayazit, L.K. Dale, C. Jones, and D. Vallejo. OBPRM: An obstacle-based prm for 3d workspaces. In *Proc. of the Workshop on Algorithmic Foundations of Robotics*, pages 155–168, March 1998.

[4] J.F. Canny. *The Complexity of Robot Motion Planning*. M. I. T. Press, Cambridge, 1988.

[5] B. Chazelle and E. Welzl. Quasi-optimal range searching in spaces of finite VC-dimension. *Discrete and Computational Geometry*, 4:467–490, 1989.

[6] S.-W. Cheng, H. Everett, O. Cheong, and R. van Oostrum. Hierarchical vertical decompositions, ray shooting, and circular arc queries in simple polygons. In *Proc. 15th Symp. Comp. Geom.*, pages 227–236, 1999.

[7] A.D. Collins. *Configuration Spaces in Robotic Manipulation and Motion Planning*. PhD thesis, Duke University, 2002.

[8] Mark de Berg, Marc van Kreveld, Mark Overmars, and Otfried Schwarzkopf. *Computational Geometry: Algorithms and Applications*. Springer-Verlag, Berlin, 1997.

[9] V. Guillemin and A. Pollack. *Differential Topology*. Prentice-Hall, 1974.

[10] D. Halperin, L. E. Kavraki, and J.-C. Latombe. Robotics. In Jacob E. Goodman and Joseph O'Rourke, editors, *Handbook of Discrete and Computational Geometry*, chapter 41, pages 755–778. CRC Press LLC, Boca Raton, FL, 1997.

[11] D. Hsu, L. Kavraki, J.-C. Latombe, R. Motwani, and S. Sorkin. On finding narrow passages with probabilistic roadmap planners. In *3rd Workshop on the Algorithmic Foundations of Robotics*, 1998.

[12] D. Hsu, J.-C. Latombe, and R. Motwani. Path planning in expansive configuration spaces. *Int. J. of Comp. Geom. and Apps*, 9(4 & 5):495–512, 1999.

[13] L. Kavraki, M.N. Kolountzakis, and J.-C. Latombe. Analysis of probabilistic roadmaps for path planning. *IEEE Trans. Rob. Aut.*, 14(1):166–171, 1998.

[14] L. Kavraki, J.-C. Latombe, R. Motwani, and P. Raghavan. Randomized query processing in robot motion planning. In *Proc. 27th ACM Symp. on Theory of Computing*, pages 353–362, Las Vegas, NV, 1995.

[15] L. Kavraki, P. Svestka, J.-C. Latombe, and M. Overmars. Probabilistic roadmaps for path planning in high dimensional configuration spaces. *IEEE Trans. on Robotics and Automation*, 12(4):566–580, 1996.

[16] J.-C. Latombe. *Robot Motion Planning*. Kluwer Academic Publishers, Boston, 1991.

[17] J. Matousek. More on cutting arrangements and spanning trees with low stabbing number. Technical Report B-90-2, Freie Universitat Berlin, 1990.

[18] J.H. Reif. Complexity of the mover's problem and generalizations. In *Proc. of the 20th IEEE Symp. on Foundations of Comp. Sci.*, pages 421–427, 1979.

[19] M. Sharir and P.K. Agarwal. *Davenport-Schinzel Sequences and Their Geometric Applications*. Cambridge University Press, 1995.

A General Framework for Sampling on the Medial Axis of the Free Space

Jyh-Ming Lien   Shawna L. Thomas   Nancy M. Amato

Department of Computer Science
Texas A&M University
{neilien,sthomas,amato}@cs.tamu.edu

Abstract. *We propose a general framework for sampling the configuration space in which randomly generated configurations, free or not, are retracted onto the medial axis of the free space. Generalizing our previous work, this framework provides a template encompassing all possible retraction approaches. It also removes the requirement of exactly computing distance metrics thereby enabling application to more realistic high dimensional problems. In particular, our framework supports methods that retract a given configuration* exactly *or* approximately *onto the medial axis. As in our previous work,* exact *methods provide fast and accurate retraction in low (2 or 3) dimensional space. We also propose new* approximate *methods that can be applied to high dimensional problems, such as many DOF articulated robots. Theoretical and experimental results show improved performance on problems requiring traversal of narrow passages. We also study tradeoffs between accuracy and efficiency for different levels of approximation, and how the level of approximation effects the quality of the resulting roadmap.*

1 Introduction

Due to the computational infeasibility of complete motion planning algorithms, recent attention has focused on probabilistic methods which sacrifice completeness for computational feasibility. In particular, several algorithms, known collectively as *probabilistic roadmap methods* (PRMs), have been shown to perform well in a number of practical situations, see, e.g., [9]. The idea behind these methods is to create a graph (or roadmap) of randomly generated collision-free configurations. Connections between these nodes are made by a simple and fast local planning method. Actual global planning is then carried out on the roadmap. These methods run quickly and are easy to implement. Unfortunately, simple situations exist in which they perform poorly, e.g., when paths are required to pass through narrow passages in configuration space.

The *medial axis*, or *generalized Voronoi diagram*, has a long history in motion planning, see [2, 4, 5, 6, 11, 15]. This is because the medial axis $MA(C_{free})$ of the free space C_{free} (the set of all collision-free configurations) has lower dimension than C_{free} but is still a complete representation for motion planning purposes. Paths on the medial axis have appealing properties such as large clearance from obstacles. However, the medial axis is difficult and expensive to compute explicitly, particularly in higher dimensions. The Medial Axis PRM (MAPRM) [16, 17] combines these two approaches by generating random networks whose nodes lie on the medial axis of C_{free} which yields improved performance on problems requiring traversal of narrow passages.

Previous work developed MAPRM for two dimensional C-spaces [16] and rigid, convex bodies in three dimensional space [17]. In this paper, we present a general MAPRM framework. Our generalized framework extends MAPRM to arbitrary bodies and high DOF robots. The framework enables sampling on the medial axis in high (> 6) dimensional configuration space through the use of *approximate* methods for computing clearance and penetration depth.

2 Related Work

PRMs are easy to implement, run quickly, and are applicable to a wide variety of robots. Various sampling schemes and local planners have been used, see [8, 9, 14]. A shortcoming of these methods is their poor performance on problems requiring paths through narrow passages in the free space. This is a direct consequence of how the nodes are sampled from C_{free}. For example, uniform sampling over C_{free}, is unlikely to provide any samples in small volume corridors. Intuitively, such narrow corridors may be characterized by their large surface area to volume ratio. Several techniques have been proposed to increase the number of nodes sampled in such narrow corridors [1, 3, 7, ?, 17].

A PRM variant, MAPRM, was proposed in [16, 17]. MAPRM generates random networks whose nodes lie on the medial axis of the free C-space. It is difficult and expensive to compute the medial axis explicitly, particularly in higher dimensions. As shown in [16, 17] for low dimensional C-space, it is possible, however, to efficiently retract any sampled configuration, free or not, onto the medial axis of the free space without having to compute the medial axis. Sampling and retracting in this way has been shown to give improved performance on problems requiring traversal of narrow passages for rigid bodies in two or three dimensions. Even for 6D C-space, MAPRM uses an inefficient brute force method to find penetrations between polyhedral objects. The requirement for exact computation of clearance and penetration depth in C-space is the primary rea-

---

[1]This research supported in part by NSF Grants ACI-9872126, EIA-9975018, EIA-0103742, EIA-9805823, ACR-0113971, CCR-0113974, EIA-9810937, EIA-0079874, and by the Texas Higher Education Coordinating Board grant ATP-000512-0261-2001. Thomas is supported in part by an NSF Graduate Research Fellowship.

son MAPRM can not be applied in arbitrary dimensions. In the following section, we introduce a general framework to cope with these difficulties while still maintaining the good properties of MAPRM.

3 Generalized MAPRM Framework

In this section we present a general framework for MAPRM. Let C be the C-space, C_{obst} be the C-obstacle, and C_{free} be $C \smallsetminus C_{obst}$. We begin by sketching the MAPRM strategy. To retract any sampled configuration onto the medial axis of C_{free}, all that is needed is the closest point on ∂C_{obst}. If the sampled configuration $p \in C_{free}$, MAPRM computes the closest point, q, on ∂C_{obst} to p, and pushes p away from q until the nearest point on ∂C_{obst} is different. If $p \in C_{obst}$, MAPRM first pushes the configuration to C_{free} and then retracts it to the medial axis as before. Figure 1 shows the extended retraction map, $r(p)$, for 2D C-space.

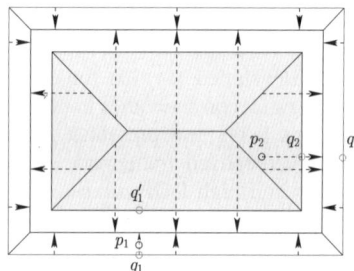

Figure 1: The extended retraction map, $r(p)$. $p_1 \in C_{free}$. q_1 is the witness for p_1's clearance, and q'_1 is the second witness when p_1 reaches the medial axis. $p_2 \in C_{obst}$. q_2 is p_2's witness for penetration depth. When p_2 is pushed into free space, q_2 will be clearance witness of p_2 and q'_2 is p_2's second clearance witness when p_2 reaches the medial axis.

Algorithm 3.1 Basic MAPRM Framework
Preprocessing:
Input. N, the number of nodes to generate.
Output. N nodes in C_{free} connected into a roadmap.
 1: **repeat**
 2:    Sample a configuration p from C-space.
 3:    **if** $p \in C_{free}$ **then**
 4:      q = **NearestContactCfg_Clearance**(p)
 5:      Set retraction direction $\vec{v} = \overrightarrow{qp}$ and start point $s = p$
 6:    **else**
 7:      q = **NearestContactCfg_Penetration**(p)
 8:      Set retraction direction $\vec{v} = \overrightarrow{pq}$ and start point $s = q$
 9:    **end if**
10:    Starting at configuration s, move robot in direction $\vec{v}$ until it has two nearest boundary points (i.e., is on medial axis).
11: **until** N nodes have been generated
12: Build connections between nodes using local planners.

The general MAPRM framework for roadmap construction is sketched in Algorithm 3.1. It involves uniform sampling in C, followed by application of the extended retraction map. If the initial sample is free, we compute a configuration q witnessing the minimum clearance using the *NearestContactCfg_Clearance* subroutine in Line 4. If the initial sample is in collision, a witness q to the minimum penetration depth is computed by the *NearestContactCfg_Penetration* subroutine in Line 7. Let $\ell_r(p)$ denote the reaction distance of $r(p)$. The cost of retracting p to the $\text{MA}(C_{free})$ is $\lceil \frac{\ell_r(p)}{d} \rceil \times \text{NC}(p)$, where d is the resolution of the workspace and $\text{NC}(p)$ is the cost of finding the nearest contact configuration.

Note that Algorithm 3.1 makes no assumption about the dimension of C, distance metrics, the robot, or the environment (e.g., convexity, articulation). In fact, only the clearance and penetration depth computations depend on these properties. That is, the only steps of Algorithm 3.1 that depend on the problem instance are *NearestContactCfg_Clearance* and *NearestContactCfg_Penetration*.

For convex rigid bodies in two and three dimensions, there exist algorithms and libraries [10, 13] to efficiently compute clearance and penetration depth. Thus, in this case, the clearance and penetration depth, and configurations realizing them, can be computed exactly. Such methods were discussed in [16, 17].

Unfortunately, computing clearance and penetration depth for higher DOF robots is hard because it is difficult to define good distance metrics. For example, the previous approach [17] of defining the shortest distance between two configurations as purely translation is not appropriate for articulated robots. Indeed, to the best of our knowledge, no efficient algorithms exist for finding penetration depth for non-convex or articulated bodies. In this paper, we propose the use of approximate methods for computing clearance and penetration for high dimensional C.

The MAPRM algorithms proposed in this paper are classified in Table 1 according to their level of approximation. MAPRM [16, 17] uses exact computation for clearance and penetration depth. MAPRM$^\sim$ uses exact computation for clearance but an approximate approach for penetration depth and can thus be applied to environments containing non-convex rigid bodies[1]. MAPRM$^\approx$ uses an approximate method for both clearance and penetration depth and can be applied to arbitrarily high DOF robots.

Algorithm	Clearance Computation	Penetration Computation
MAPRM	exact	exact
MAPRM$^\sim$	exact	approximate
MAPRM$^\approx$	approximate	approximate

Table 1: MAPRM algorithms for various approximation levels.

4 MAPRM

MAPRM was first developed for a point robot in the plane

---
[1] MAPRM$^\sim$ uses the approximate method only if it is necessary. Thus in some situations, e.g., the robot is entirely contained in an obstacle, penetration depth can be calculated just like clearance.

[16] and then extended to convex rigid bodies in 3D [17].

4.1 MAPRM for a Point Robot in 2D

When the environment is composed of polygonal objects, clearance or penetration depth for a given configuration is just the shortest distance from this point to the boundary of the polygons. This algorithm plays the roles of both NearestContactCfg_Clearance and NearestContactCfg_Penetration in the framework (Algorithm 3.1). For 2D C-space, MAPRM has been shown, theoretically and empirically, to increase sampling in narrow corridors [16].

Algorithm 4.1 NearestContactCfg in 2D

Preprocessing:
Input. A configuration p.
Output. The nearest contact configuration q from p.
1: Find nearest configuration q on ∂C_{free} to p by computing the distance from p to ∂C_{free}.
2: **return** q.

4.2 MAPRM for a Rigid Body in 3D

In MAPRM for convex rigid bodies in three dimensions [17], the objective is to compute the nearest contact configuration in ∂C_{obst} without explicitly computing C_{obst}.

Let p be a sampled configuration. If $p \in C_{free}$, then the nearest contact point is a witness realizing the clearance (Algorithm 4.2), which is provided by several algorithms and collision detection packages [10, 13].

Algorithm 4.2 NearestContactCfg_Clearance in 3D

Input. A collision-free configuration p.
Output. The nearest contact configuration q from p or **failure**.
1: Find a point, b, on B and a point, r, on Robot such that $dist(b, r)$ is smallest.
2: **return** $p + \overrightarrow{rb}$.

If $p \in C_{obst}$, we compute the nearest contact point by computing its penetration depth (Algorithm 4.3). If all objects in the environment are convex, the penetration can be computed efficiently [12, 13]. Otherwise, a brute force method which tests all feature pairs is applied. Clearly, the brute force approach is not practical for large complex environments. MAPRM for a convex, rigid body in 3D has been shown to increase sampling in narrow corridors [17].

Algorithm 4.3 NearestContactCfg_Penetration in 3D

Input. An in-collision configuration p.
Output. The nearest contact configuration q from p or **failure**.
1: **if** Robot and Obstacles are all convex objects **then**
2:    Use Lin-Canny closest features algorithm [12].
3: **else**
4:    Use brute force method [16, 17].
5: **end if**
---

5 Approximate Variants of MAPRM

In this section, we present an approximate approach which enables us to apply the MAPRM philosophy in more general situations, such as high DOF robots. There are several complications with extending the strategy for 2D and 3D MAPRM to high DOF robots. For instance, finding the nearest contact configuration requires the ability to compute witnesses for clearance and penetration depth. Unfortunately, the exact closest contact configuration in C cannot be computed without computing the ∂C_{obst}, a computationally expensive process which PRM methods are designed to avoid.

We are able, however, to approximate the C-space clearance (or penetration) of a configuration without computing ∂C_{obst}. Let $Cl(p, \vec{v})$ be the C-space clearance of configuration p in direction $\vec{v}$. Then the approximate clearance is defined as: $Cl(p) = \min(\{Cl(p, \vec{v_i}) : i = 1 \ldots N\})$. To compute $Cl(p, \vec{v})$, we walk out from p towards $\vec{v}$ until the collision state changes. As we increase N, the approximate clearance approaches the actual clearance. Penetration is defined in the same way; i.e., $Pt(p) = \min(\{Pt(p, \vec{v_i}) : i = 1 \ldots N\})$. Algorithm 5.1 describes the process. It can be used to implement the *NearestContactCfg_Clearance* or *NearestContactCfg_Penetration* subroutines in Algorithm 3.1. The time complexity of Algorithm 5.1 is $O\left(N \times \lceil \frac{\ell}{d} \rceil\right) \times T(CD)$, where ℓ is $Cl(p)$ or $Pt(p)$, d is the resolution of the workspace, and $T(CD)$ is the cost for collision detection. The accuracy of the approximation depends on N and d. However, while N has a more profound effect on accuracy, $Cl(p)$ and $Pt(p)$ using large d can be refined by the bisection search.

MAPRM$^\sim$ approximates the penetration depth in the basic MAPRM framework (Algorithm 3.1) while clearances are computed explicitly. This allows the extension to non-convex rigid bodies.

MAPRM$^\approx$ approximates both clearance and penetration depth in the basic MAPRM framework. By approximating both clearance and penetration, we can apply the MAPRM sampling strategy to arbitrary robots with high DOF.

Algorithm 5.1 NearestContactCfg in for general C-space.

Input. A configuration c.
Output. An approximately closest contact configuration c'.
1: Let *CollStat()* be the collision detection function.
2: Let $c_i = c$, $i = 1$ to N.
3: Randomly create N normalized vectors, $\vec{v_1}$ to $\vec{v_N}$.
4: **while** true **do**
5:   **for** $i = 1$ to N **do**
6:     $c_i = c_i + \vec{v_i}$.
7:     **if** *CollStat*$(c_i) \neq$ *CollStat*(c) **then**
8:       **if** $(c \in C_{obst})$ **then** $c_i = c_i - \vec{v_i}$.
9:       **return** c_i.
10:     **end if**
11:   **end for**
12: **end while**

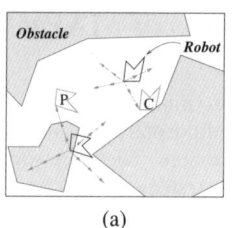 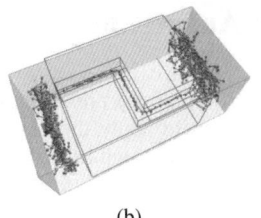 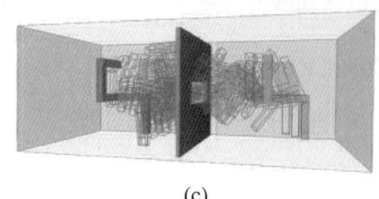 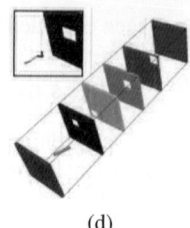

(a) (b) (c) (d)

Figure 2: (a) We approximate a configuration's C-space clearance (penetration) by finding its clearance (penetration) in several random directions. The configuration marked as "c" ("p") is an approximate contact configuration witnessing the clearance (penetration). (b) **S-Tunnel** is contained in a $6.5 \times 6.5 \times 14$ bounding box. The robot is a cube with side lengths of 0.6; the obstacle is a solid box of size $6 \times 6 \times 9$ with the indicated corridor cut through it. The corridor has a 1×1 cross section. The obstacle is in wireframe with the indicated corridor cut through it. An example of a roadmap generated by MAPRM$^\sim$ is shown. (c) **Hook** contains a hook-like rigid robot and an obstacle ($200 \times 150 \times 5$) with a hole ($100 \times 30 \times 5$) inside. The robot is composed of 5 blocks ($10 \times 10 \times 50$) with different orientations. The swept volume illustrates a solution path. (d) **Walls** contains six parallel walls ($4 \times 4 \times 0.25$). Each of the four walls in the middle has a (1×1) hole. The robot is between the two leftmost walls. The goal is placed between the two rightmost walls.

6 Experimental Results

In this section, we show how the MAPRM variants perform in practice. PQP [10] and V-Clip [13] are used to provide collision detection and exact closest pair calculations. Options for sampling and connection are controlled under the same condition unless stated otherwise. The execution of each method was terminated when the roadmap contained a component that reached from one mouth of the corridor to the other. Experiments were carried out on an Intel Pentium 4 processor at 1.80 GHz.

We tested examples in 3D environments with a variety of robots (Section 6.1). We also studied the tradeoffs between accuracy and efficiency of different approximate levels (Section 6.2).

6.1 3D Environments

We tested three environments: s-tunnel, hook with a hook-like rigid robot, and walls with both a stick robot and an articulated robot; see Figure 2 (b), (c), and (d). In Table 2, We present experimental results for MAPRM, MAPRM$^\sim$, and MAPRM$^\approx$, and compare them with results from uniformly sampled PRM. We averaged the results over 10 runs with different seeds.

6.1.1 S-Tunnel environment

In Figure 2(b), the obstacle is composed of 6 convex pieces which enables us to use V-Clip [13] to find exact contact configurations. The robot must pass through the tunnel to solve the query.

Uniform random sampling in Table 2 was unable to solve the query with a roadmap size of 64000 nodes and 11 hours of execution time. In contrast, all MAPRM variations were able to produce a valid solution path. MAPRM$^\sim$ takes slightly longer than MAPRM, because the approximation calculation in MAPRM$^\sim$ requires more time than the exact calculation. This can be seen in the longer average node generation time of 749.83ms versus 41.01ms for

EXPERIMENTAL RESULTS FOR MAPRM ALGORITHMS

Env.	Method	Samp. time(s)	Connect time(s)	Total time(s)	Solved
S-Tunnel	Uniform	155	35570	39744	N
	MAPRM	14	9	22	Y
	MAPRM$^\sim$	19	5	26	Y
	MAPRM$^\approx$	45	53	213	Y
Hook	Uniform	686	97014	99869	N
	MAPRM$^\sim$	33	74	120	Y
	MAPRM$^\approx$	44	45	105	Y
Walls (Stick)	Uniform	2	161	162	Y
	MAPRM$^\sim$	11	30	40	Y
	MAPRM$^\approx$	25	35	61	Y
Walls (Articulated)	Uniform	2	119	121	Y
	MAPRM$^\approx$	47	51	99	Y

Table 2: For the S-Tunnel environment, MAPRM$^\sim$ uses $N = 4$ and MAPRM$^\approx$ uses $N = 100$ for penetration and $N = 4$ for clearance. For the Hook environment, MAPRM$^\sim$ uses $N = 20$ and MAPRM$^\approx$ uses $N = 4$ rays for penetration and 20 rays for clearance. For the Walls environment with the stick robot, MAPRM$^\sim$ uses $N = 4$ and MAPRM$^\approx$ uses $N = 20$ for clearance and penetration. For the Walls environment with an articulated robot, MAPRM$^\approx$ uses $N = 4$ for clearance and penetration.

MAPRM. MAPRM$^\approx$ is the slowest of the MAPRM algorithms because both clearance and penetration calculations are approximated, requiring more time during node generation.

6.1.2 Hook Environment

In the hook environment (Figure 2(c)), the robot starts from the left side of the environment and the unique solution path requires it to twist through the hole to reach the goal in the right side of the environment.

In Table 2, uniform random sampling was unable to solve the query with a roadmap size of 128000 nodes and 28 hours of execution time. MAPRM could not be used because the environment contains non-convex objects. In contrast, MAPRM$^\sim$ and MAPRM$^\approx$ were able to solve the query with only 2815 nodes in just a couple minutes. For environments like this, it is critical that nodes are generated in the narrow corridor. The results demonstrate MAPRM's ability to increase sampling in narrow corridors, even when clearances and penetrations are only approximated.

6.1.3 Walls environment

In our third example (Figure 2(d)), the robot must pass through the holes in the walls to reach the goal at the other end of the environment. We tested both a rigid robot and a 4-link articulated robot.

In Table 2, MAPRM could not be used because the environment contains non-convex objects. Both MAPRM$^\sim$ and MAPRM$^\approx$ out-perform uniform sampling. Although MAPRM$^\sim$ finds a solution faster than MAPRM$^\approx$ the difference is not as pronounced as in the s-tunnel experiment. This is because constraints on rotation decrease the performance of MAPRM$^\sim$.

The articulated robot has high DOF, so only MAPRM$^\approx$ could be applied. MAPRM$^\approx$ again beat uniform random sampling by solving the query with a roadmap half the size. The speedup is not as great as the roadmap size reduction because MAPRM$^\approx$ requires more time to generate a node (30.69ms on average) than uniform random sampling (0.72ms on average). Note that the time MAPRM$^\approx$ lost during node generation was more than made up for during node connection. This shows MAPRM$^\approx$'s nodes to be of higher quality than those from uniform random sampling.

6.2 Approximation Study

The efficiency of the approximate methods, MAPRM$^\sim$ and MAPRM$^\approx$ depends on the efficiency of computing the approximate clearance and penetration; e.g., $N \times \lceil \frac{Cl(p)}{d} \rceil$ for C-space clearance. In this section we study how N relates to the minimum time required to find a solution path and how this varies between environments. We are interested in questions such as what environment properties indicate a need for greater clearance accuracy and greater penetration accuracy. To investigate these issues, we varied the value of N for both clearance and penetration calculations.

6.2.1 Accuracy and Computation Time

First, we studied accuracy and computation time by varying N for both clearance and penetration depth. Accuracy is based on the normalized distance between the exact and approximate contact configurations. Larger distances indicate a less accurate result. The computation time is measured based on mean time to generate one contact configuration.

Figure 3(a) shows the computation time for different combinations of N for penetration and clearance. Since the s-tunnel environment has little free space, the computation for clearance is faster than that for penetration depth. It is clear that the computation time grows linearly with N.

Figure 3(b) shows the error rates introduced by approximation. Let error rates for clearance and penetration depth be $ErrorRate = \frac{\sum_{i=1}^{n} dist(cfg_i^e, cfg_i^a)}{\sum_{i=1}^{n} dist(cfg_i^e, cfg_i^o)}$. Then accuracy is defined as the reciprocal of the $ErrorRate$. Here cfg_i^o is the i-th randomly sampled configuration, cfg_i^e is the exact nearest contact configuration for cfg_i^o, and cfg_i^a is an estimated contact configuration. Since $dist(cfg^e, cfg^a)$ depends on properties of the environment, such as the size of the obstacles and the volume of the free space, we normalize the distance by dividing by the exact retraction distance, $dist(cfg^e, cfg^o)$. In the s-tunnel environment, large obstacles will produce a large mean distance between the exact nearest contact configuration and the configuration in collision, and small free space will have small mean exact clearance. From Figure 3(b), one can notice that the error rates drop significantly in the interval of 4 to 100, and little improvement is shown for > 100.

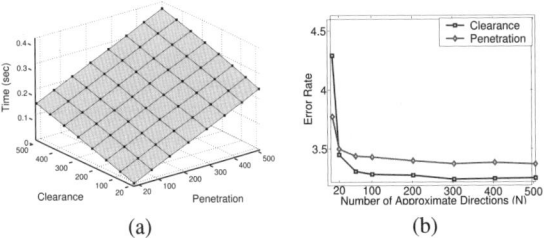

(a) (b)

Figure 3: Approximation study using S-Tunnel environment. Average over 55,000 samples. (a) The computation time for finding one nearest contact configuration for different N. (b) The error rate for finding a nearest contact configuration for different N.

6.2.2 S-Tunnel

Figure 4 shows the results of the approximation study for the s-tunnel environment (Figure 2). The larger N is, the more accurate the contact configuration will be and the more computation time it will take.

For MAPRM$^\sim$, the best value for N is 4. Due to the small volume of the tunnel relative to the obstacle, the accuracy of the penetration depth is important and larger values of N should solve the problem faster. However, the robot is so small that most of samples are contained inside the obstacle completely. This is a case in which penetration depth can be calculated exactly. Thus, the probability of using an approximation is relatively small. This is why $N = 4$ is enough to generate a good roadmap.

For MAPRM$^\approx$, the best combination was $N = 4$ for clearance and $N = 100$ for penetration. A close second was $N = 20$ for clearance and $N = 100$ for penetration. Because the corridor is small compared to the surrounding obstacle, accurate penetration calculations are critical to the algorithm's success. This is reflected in the fact that the top two clearance/penetration combinations had $N = 100$ for penetration. Planning in the space outside the corridor is trivial and the robot is small compared to the available space. This is reflected in the coarser clearance approximation ($N = 4$ or 20).

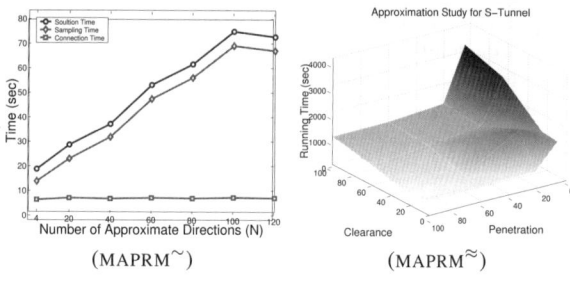

Figure 4: S-Tunnel approximation study.

6.2.3 Walls

Figure 5 shows the results of the approximation study for the walls environment with a stick robot (Figure 2(a)).

For MAPRM$^\sim$, the best value for penetration was $N = 20$, and for MAPRM$^\approx$, the best clearance/penetration combination was $N = 20$ for clearance and $N = 20$ for penetration. In this environment, the walls are thinner than in the s-tunnel, so penetration calculations do not need to be approximated at as fine a detail. Also, the spaces between the walls are not cluttered. Planning here is easy to moderate, so only $N = 4$ or 20 is needed for clearance calculations.

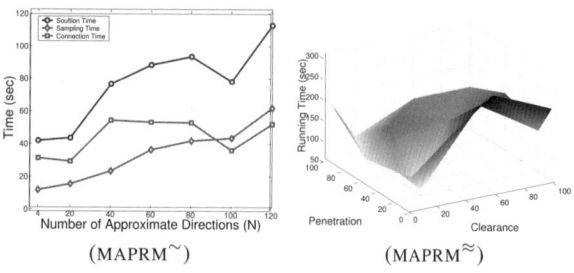

Figure 5: Walls with stick robot approximation study.

7 Conclusion

We have described a general framework for sampling configurations on the medial axis of the free C-space. It enables users to "plug in" appropriate retraction functions according to the properties of given problem. In particular, we propose using exact and approximate methods to implement the retraction functions. Exact methods, MAPRM, for convex rigid bodies in two and three dimensions, and approximate methods, MAPRM$^\sim$ and MAPRM$^\approx$, for arbitrary configuration space are proposed. Both MAPRM$^\sim$ and MAPRM$^\approx$ out performs uniform sampling in all studied environments. MAPRM$^\approx$ out performs MAPRM$^\sim$ when rotation is significant for rigid robots. Only MAPRM$^\approx$ can be used for high DOF robots. We observed the relationship between accuracy, i.e. number of approximate directions, and properties of environments, e.g. clearance and penetration depth.

Acknowledgment

Some of the models studied were provided by Jean-Paul Laumond's and Lydia Kavraki's groups.

References

[1] N. Amato, O. B. Bayazit, L. K. Dale, C. Jones, and D. Vallejo. OBPRM: An obstacle-based PRM for 3D workspaces. In P. K. Agarwal, L. E. Kavraki, and M. Mason, editors, *Proc. Workshop Algorithmic Found. Robot.* A. K. Peters, Wellesley, MA, 1998.

[2] F. Aurenhammer. Voronoi diagrams: A survey of a fundamental geometric data structure. *ACM Comput. Surv.*, 23:345–405, 1991.

[3] V. Boor, M. H. Overmars, and A. F. van der Stappen. The Gaussian sampling strategy for probabilistic roadmap planners. In *Proc. IEEE Int. Conf. Robot. Autom. (ICRA)*, pages 1018–1023, 1999.

[4] M. Foskey, M. Garber, M. Lin, and D. Manocha. A voronoi-based hybrid motion planner. In *Proc. IEEE/RSJ International Conf. on Intelligent Robots and Systems (IROS 2001)*, 2001.

[5] K. Hoff, T. Culver, J. Keyser, M. Lin, and D. Manocha. Interactive motion planning using hardware-accelerated computation of generalized voronoi diagrams. In *Proceedings of the 2000 International Conference on Robotics and Automation (ICRA 2000)*, 2000.

[6] C. Holleman and L. E. Kavraki. A framework for using the workspace medial axis in prm planners. In *Proceedings of the 2000 International Conference on Robotics and Automation (ICRA 2000)*, pages 1408–1413, San Fransisco, CA, 2000. IEEE Press.

[7] D. Hsu, L. E. Kavraki, J.-C. Latombe, R. Motwani, and S. Sorkin. On finding narrow passages with probabilistic roadmap planners. In *Proc. 1998 Workshop Algorithmic Found. Robot.*, Wellesley, MA, 1998. A. K. Peters.

[8] L. Kavraki. *Random Networks in Configuration Space for Fast Path Planning*. PhD thesis, Stanford Univ., Stanford, CA, 1995.

[9] L. E. Kavraki, P. Švestka, J.-C. Latombe, and M. H. Overmars. Probabilistic roadmaps for path planning in high dimensional configuration spaces. *IEEE Trans. Robot. Autom.*, 12:566–580, 1996.

[10] E. Larsen, S. Gottschalk, M. C. Lin, and D. Manocha. Distance queries with rectangular swept sphere volumes. In *Proc. of IEEE Int. Conference on Robotics and Automation*, 2000.

[11] J.-C. Latombe. *Robot Motion Planning*. Kluwer Academic Publishers, Boston, 1991.

[12] M. Lin and J. Canny. A fast algorithm for incremental distance calculation. In *Proceedings of International Conference on Robotics and Automation*, pages 1008–1014, 1991.

[13] B. Mirtich. V-Clip: Fast and robust polyhedral collision detection. Technical Report TR97-05, MERL, 201 Broadway, Cambridge, MA 02139, USA, July 1997.

[14] M. Overmars and P. Svestka. A probabilistic learning approach to motion planning. In *Proc. Workshop on Algorithmic Foundations of Robotics*, pages 19–37, 1994.

[15] C. Pisula, K. Hoff, M. Lin, and D. Manocha. Randomized path planning for a rigid body based on hardware accelerated voronoi sampling. In *Proc. Workshop on Algorithmic Foundations of Robotics (WAFR 2000)*.

[16] S. A. Wilmarth, N. M. Amato, and P. F. Stiller. MAPRM: A probabilistic roadmap planner with sampling on the medial axis of the free space. In *Proc. IEEE Int. Conf. Robot. Autom. (ICRA)*, pages 1024–1031, 1999.

[17] S. A. Wilmarth, N. M. Amato, and P. F. Stiller. Motion planning for a rigid body using random networks on the medial axis of the free space. In *Proc. ACM Symp. on Computational Geometry (SoCG)*, pages 173–180, 1999.

A General Framework for PRM Motion Planning*

Guang Song   Shawna Thomas   Nancy M. Amato

Department of Computer Science
Texas A&M University
College Station, TX 77843-3112
{gsong,sthomas,amato}@cs.tamu.edu

Abstract

An important property of PRM roadmaps is that they provide a good approximation of the connectivity of the free C-space. We present a general framework for building and querying probabilistic roadmaps that includes all previous PRM variants as special cases. In particular, it supports no, complete, or partial node and edge validation and various evaluation schedules for path validation, and it enables path customization for variable, adaptive query requirements. While each of the above features is present in some PRM variant, the general framework proposed here is the only one to include them all. Our framework enables users to choose the best approximation level for their problem. Our experimental evidence shows this can result in significant performance gains.

1 Introduction

Probabilistic Roadmap (PRM) motion planning methods have been the subject of much recent work. These methods create a graph of randomly generated collision-free configurations which are connected by a simple and fast local planning method. Actual global planning (queries) is carried out on this graph. The initial PRMs were shown to be very successful in solving difficult problems in high-dimensional configuration spaces (C-space) that had previously resisted efficient solution [16]. These successes motivated the application of PRMs to challenging problems arising in a variety of fields including robotics (e.g., closed-chain systems [14, 19]), CAD (e.g., maintainability [6, 11]), deformable objects [2, 5, 15]), and even computational Biology and Chemistry (e.g., ligand docking [7, 22], protein folding [3, 23, 24]). Indeed, it can be argued that the PRM framework was instrumental in this broadening of the range of applicability of motion planning, as many of these problems had never before been considered candidates for automatic methods.

The strength of PRMs comes from their efficiency and effectiveness in approximately representing the connectivity of the free C-space. However, if the queries only utilize a small portion of the roadmap, then the time spent constructing the unused portion of the roadmap is wasted. This consideration motivated researchers to propose PRM variants, such as Lazy PRM [9], Fuzzy PRM [21], and Customizable PRM [25], that postpone some roadmap construction operations until query time.

While PRMs are very good at finding *a* path, they do not support applications which might impose particular, variable path requirements, e.g., maintaining a particular clearance or minimizing the robot's rotation. This issue has not received as much attention as, e.g., the narrow passage problem, because simply finding any path was considered a necessary first step. Some probabilistic methods have been developed to enable one to specify fixed [8, 18] and variable [25] path requirements.

In this paper, we present a general framework for building and querying probabilistic roadmaps. This framework includes all previous PRM variants as special cases. In particular,

- it encompasses variable levels of validation, ranging from no validation to complete validation, of nodes and edges during roadmap construction,
- it provides for various evaluation schedules for the edges when validating paths extracted from the roadmap, and
- it enables customization of paths to satisfy variable, adaptive query requirements.

While each of the above features is present in some PRM variant, the general framework proposed in this paper is the only one to include them all.

2 Related Work

In this section we briefly mention other approaches targeted at either improving the efficiency of PRMs or in supporting requirements on the query path other than the usual collision-free requirement.

---

*This research supported in part by NSF Grants ACI-9872126, EIA-9975018, EIA-0103742, EIA-9805823, ACR-0081510, ACR-0113971, CCR-0113974, EIA-9810937, EIA-0079874, and by the Texas Higher Education Coordinating Board grant ATP-000512-0261-2001. Song supported in part by an IBM TJ Watson PhD Fellowship and Thomas supported in part by an NSF Graduate Research Fellowship.

General PRM framework				
PRM Method	Validation			Query Prefs
	Node	Edge	Path	
Basic PRM	complete	complete	in order	fixed
Lazy PRM	none	none	in order	fixed
Fuzzy PRM	complete	none	priority	fixed
C-PRM	tunable	tunable	in order	adaptable
General	tunable	tunable	tunable	adaptable

Table 1: Special cases of the General PRM framework.

Lazy PRM. Lazy PRM [9] initially assumes all nodes and edges to be collision-free during roadmap construction, resulting in fast roadmap construction. During the query, the nodes (first) and edges (second) along the path are checked for collision. If a node or edge is found to be in-collision, it is removed from the roadmap, and a new shortest path is extracted. This process repeats until a collision-free path is found or a path no longer exists between the start and goal.

Fuzzy PRM. A similar approach called Fuzzy PRM was proposed in [21]. Fuzzy PRM postpones edge validation but validates all nodes during roadmap construction. Its priority-based evaluation scheme for edges is described in Section 5.

C-PRM. Customizable PRM (C-PRM) also allows for incomplete validation of nodes and edges during roadmap construction and is the first PRM approach to enable multiple, variable path requirements or preferences [25]. C-PRM first builds a coarse roadmap by performing partial validation of roadmap nodes and/or edges. In the query phase, the roadmap is validated and refined only in the area of interest for the query and is customized in accordance with any specified path preferences. This approach, like Lazy and Fuzzy PRM, postpones validation checks until the query phase to yield more efficient roadmap construction. In addition, it gives one the ability to customize the same roadmap for different path preferences.

3 General PRM Framework

In this section, we present a general PRM framework that includes all previous PRM variants as special cases. Table 1 provides an overview of the relation of our framework to several well known PRM variants.

Our framework consists of two phases: roadmap construction and query. Ultimately, all nodes and edges on the solution path must be fully validated. This validation can be performed during roadmap construction, as in the original PRMs [16], completely postponed until a path is extracted in the query phase, as in Lazy PRM [9], or partially performed during roadmap construction and completed during the query phase, as in Fuzzy PRM [21] or C-PRM [25]. A less studied issue is the order in which incompletely validated edges are checked when validating a path; most PRM variants simply consider the edges in order, but

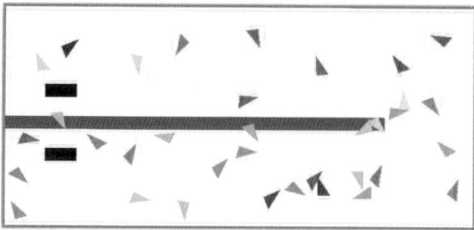

Figure 1: An environment where partial validation is important in estimating the free C-space and its connectivity. The rectangular robot is shown at the start and goal.

Fuzzy PRM uses a priority-based evaluation scheme. Our general PRM formulation encompasses all such options.

A strength of PRMs is that they provide a good approximation of the connectivity of the free C-space. While eliminating all validation from the roadmap construction phase sacrifices this desirable property, useful approximations can be computed by partial validation of nodes and/or edges. This can be illustrated in the simple environment shown in Figure 1 in which both partial node and edge validation would be useful in better approximating the connectivity of the free C-space which would improve query efficiency.

3.1 Roadmap Construction

Node generation. During node generation, N nodes are sampled. Traditional PRMs sample nodes uniformly at random. Other techniques have been developed to increase the number of sampled nodes in narrow passages [1, 26]. Nodes are then added to the roadmap if they pass a validation test.

Node evaluation ranges from no validation (as in Lazy PRM) to complete validation (as in traditional PRMs). Complete validation checks that the node meets all the given requirements. Methods for partially validating nodes are discussed in Section 4.1.

Node connection. During node connection, an attempt is made to connect each node with its k closest neighbors, for some small constant k, via some local planner (e.g., a straight line in C-space). If there exists a valid path, an edge between the two nodes is added to the roadmap. (Other connection techniques have also been developed [13].)

Edge evaluation also ranges from no validation to complete validation. Complete validation ensures that the entire edge (i.e., all the nodes along that edge) meets the given requirements. An edge can be partially validated if only a subset of the requirements are completely validated or if a requirement is only approximately validated. Methods for approximate edge validation will be discussed in Section 4.2.

3.2 Query Phase

During the query, the start and the goal are connected to the roadmap. Then, an algorithm (e.g., Dijkstra's [12]) is used to search the roadmap for a shortest path.

Path evaluation. If the roadmap was not completely validated during construction (i.e., checked that nodes and edges meet all requirements), then the nodes and edges along the path must be checked. If a portion of the path is invalid, it is removed from the roadmap, and a new path is extracted. This repeats until a valid path is found or the start and goal are no longer connected. This strategy is followed in Lazy PRM [9], Fuzzy PRM [21], and C-PRM [25]. Identifying invalid portions of the path quickly will reduce the total query time.

Path requirements. Usually there are other desirable properties for a path in addition to the basic collision-free requirement. Table 2 lists some common requirements. C-PRM [25] is the only PRM variant to support such variable query requirements.

Common Path Requirements	
Application	Requirement
CAD	clearance
Mobile robots	clearance, smoothness
Manipulators	singularity, clearance
Ligand binding	low potential
Protein folding	low potential

Table 2: Common path requirements.

Enforcing path preferences. It is not practical to store all the information needed to verify such properties in the roadmap. Moreover, in some cases it is difficult, or even impossible, to associate such information with roadmap nodes or edges because it relates to global properties of the solution path. For example, one might ask for a path for a mobile robot that makes at most k sharp turns. Since this is not a local property and the start and goal are not known in advance, one cannot prune the roadmap so that it contains only valid paths. In this example, and indeed in many interesting queries, the only way to enforce such requirements is during the query stage.

In many cases, one would like to request a path meeting several path preferences but with different priorities. Our general PRM framework can support such queries by iteratively refining, or customizing, the roadmap according to the prioritized requirements. At each stage, invalid nodes and edges are removed as they are discovered, and the process iterates. Finally, a path that meets the preferences to the largest degree possible is found, and in the process, the roadmap is customized to these preferences.

4 Building Approximate Roadmaps

Evaluation during roadmap construction covers a spectrum from no validation (e.g., Lazy PRM and Fuzzy PRM) to complete validation (e.g., traditional PRM). In this section we present techniques that approximately validate nodes and edges.

4.1 Approximate Node Evaluation

Complete node validation consists of checking that the node satisfies all the given requirements. Approximate node evaluation makes an educated guess about whether the node satisfies a given requirement. The most common motion planning requirement is that the robot is collision-free. Collision detection can be time consuming, especially if the robot or the environment is complex. Below, we present both conservative and aggressive approximate validation methods to reduce roadmap construction time.

Bounding volume approximation. One way to simplify collision detection calculations is replace the robot's and/or the obstacles' complex geometry with simplified geometries. Here we present three approaches based on bounding volumes.

Bounding box approximation. An object's geometry is replaced by its bounding box. Although this is conservative in that it may discard valid nodes, it dramatically reduces the cost of collision detection.

Bounding sphere approximation. An object's geometry is replaced by a bounding sphere. While this approximation allows for very efficient collision checks, it is even more conservative than the bounding box.

Convex hull approximation. An object's geometry is replaced by its convex hull. This approximation is less conservative than the others since it provides a tighter fit to the original object. Collision detection packages can then take advantage of techniques for convex objects. The time saved in collision checking compensates for the $O(n \log n)$ time required to compute the convex hull [4].

Center of mass approximation. A node's collision status can also be approximated by checking if the robot's center of mass lies inside any obstacle. Unlike the previous methods, this is an aggressive technique and can add invalid nodes to the roadmap.

Grid-based approximation. Here, an approximate value for each cell in a grid-based decomposition of C-space. During roadmap construction, the value for each configuration is determined by a simple, fast table lookup. Of course, due to the high-dimensionality of most interesting C-spaces, this method is infeasible in many cases. Nevertheless, there exist some applications where useful approximations can be computed on a coarse grid (e.g., potential energy calculations for ligand binding [25]). Also, sometimes reasonable C-space approximations can be provided from a grid-based decomposition of the *workspace*.

4.2 Approximate Edge Evaluation

Our goal is to perform fast, approximate validation of the edges. Many invalid edges can be quickly discovered, resulting in roadmaps that reflect the connectivity of the free C-space better than roadmaps built without edge validation.

Binary resolution approximation. The strategy is to validate the intermediate configurations on an edge (e.g., a straight-line in C-space) according to a binary partitioning strategy. We first validate the midpoint, then the midpoints of the two resulting subsegments, etc. Our experience shows this strategy discovers many invalid edges quickly. A similar idea of increasing resolution checks is used in Fuzzy PRM [21] when attempting to increase edge probabilities during the query stage.

Overlapping spheres approximation. Another way to estimate whether an edge is collision-free is to use a test based on the C-space clearance of its two endpoints. If the sum of the two C-space clearances is greater than the distance between the endpoints, then there exists a collision-free path connecting them. (A path could still exist even if the spheres do not intersect.) Since we cannot compute exact clearances in C-space, we compute an approximation by selecting n directions at random, finding the C-space clearance in those directions, and using the minimum as the C-space clearance approximation (see Figure 2). Clearly, the accuracy of the approximation is very sensitive to the number n, and it is possible that invalid edges are added to the roadmap. In practice, values as small as $n = 3$ work well.

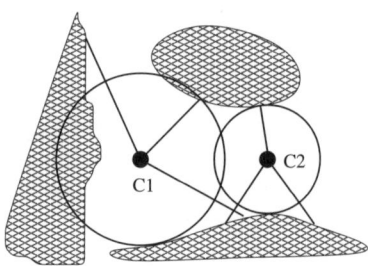

Figure 2: Overlapping spheres. The (possibly invalid) edge $(C1, C2)$ is added to the roadmap if the C-space clearance approximation spheres $C1$ and $C2$ overlap.

5 Path Evaluation

To validate a particular path, every node and edge along that path must be checked to meet all the given requirements. If invalid nodes/edges are identified quickly, the time required to validate the path is reduced. Thus, the order in which these are checked effects the total query time.

Sequential path evaluation. The approach employed by most PRM variants checks the edges in the path sequentially. First the nodes are checked in order from the start to the goal. If they all meet the given requirements, the edges are checked in order from the start to the goal.

Priority-based path evaluation [21]. While sequential path evaluation is straightforward to implement, a more efficient approach would identify which edges are most likely invalid and check these first. Each edge is assigned a weight according to the probability that it meets the given requirement. To validate a path, each edge along the path is inserted into a priority queue. The edge with the lowest probability is removed from the queue, checked at a higher resolution, and assigned a new probability. This repeats until all edges meet the requirement (probability of 1) or an edge is found invalid (probability of 0), removed, and a new path is extracted.

Fuzzy PRM implements such a strategy for the collision-free requirement using the probability:

$$P(e) = \begin{cases} [cosh(\lambda \frac{d(c_1,c_2)}{2^l})]^{-2^l} & \text{if } l < l_{\max} \\ 1 & \text{if } l \geq l_{\max} \end{cases}$$

where $d(c_1, c_2)$ is the length of the edge, l is the level at which the edge has been checked, and $l_{\max}$ is the highest resolution required. (See [21] for more details.) This approach can be extended to other path requirements by redefining the probability function.

6 Experimental Results

All experiments were performed on a 1.8 GHz PC using our C++ OBPRM library, which includes implementations of many PRM variants.

6.1 Roadmap Construction Techniques

In this section, we study how different roadmap construction techniques effect the algorithm's performance, as measured in terms of roadmap size and running time.

Node evaluation. We study various levels and techniques for node validation using the house environment containing a table and a piano shown in Figure 3. The goal is to move the piano out of the house. We built several roadmaps with different node evaluation techniques. All edges are added to roadmaps without any validation. The results are shown in Table 3. Because nodes are generated uniformly at random, we ran each experiment 10 times and averaged the results.

As seen in Table 3, it is useful to perform some level of node validation. If no validation is performed, the size of the corresponding minimal roadmap to solve the query is the largest among all cases, and also the most expensive to build. Moreover, the query time is significantly longer as the roadmap contains many invalid nodes that are encountered and discarded during the query process. While complete node validation performs much better, it can be seen that some partial node validation further reduces running

Validation Method	Approx. Type	Objects Applied	Nodes	Edges	Setup Time	Build Time	Query (s)		Total (s)	
							Seq.	Pri.	Seq.	Pri.
None	n/a	all	209	2328	0.0	1.61	6.25	2.15	7.86	3.76
Complete	n/a	all	64	704	0.0	0.25	13.8	0.50	14.05	0.75
Approximate	bounding box	all	179	2003	0.006	1.21	5.34	0.77	6.55	1.98
Approximate	convex hull	all	102	1156	0.026	0.46	7.81	0.74	8.30	1.23
Approximate	bounding box	robot only	52	565	0.001	0.14	8.89	0.39	9.03	0.53
Approximate	convex hull	robot only	68	753	0.01	0.28	14.16	0.51	14.45	0.80

Table 3: Node Validation Comparison. The table shows the minimal roadmap required to solve the query shown in Figure 3. Edges were not validated. Results are averaged over 10 runs. Setup time is the time spent preparing the bounding volume. Build time is the time spent constructing the roadmap.

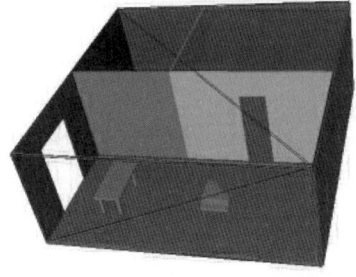

Figure 3: The piano mover's problem.

time. In this particular environment, the best performance is achieved when the robot, a complex piano model, is approximated by its bounding box. The gain can be attributed to the simpler bounding box, which makes collision detection much faster. The more complex convex hull bounding volume is slightly slower overall. There are cases, however, where the more accurate convex hull bounding volume has proven useful [20]. Brock and Kavraki [10] used a similar strategy of approximating the robot by a simpler structure in their decomposition-based planning work.

Edge evaluation. Figure 1 shows an environment where approximate edge validation may be advantageous. Approximate edge validation can coarsely represent the connectivity of the C-space while saving time over complete validation. Table 4 compares several levels of approximate edge validation against no validation and complete validation.

The results show that approximate edge validation greatly reduces roadmap construction time while eliminating most invalid edges. For example, approximate edge validation at the coarsest level eliminates 74% of the invalid edges while only slightly increasing roadmap construction time over no edge validation. The benefit of eliminating these invalid edges is clearly seen in the query time. The query time for a roadmap with no edge validation is much longer because many edges must be checked and removed before a valid path is found. Roadmap construction time is reduced by an order of magnitude over complete edge validation while only slightly increasing the query time.

Table 4 also shows that there is a tradeoff between saving time during roadmap construction and eliminating invalid edges. The optimal balance between these will vary according to the environment and the difficulty of the query. In this case, the best performance is achieved with 0.05 resolution, i.e., 20 times lower than the original resolution.

6.2 Path Evaluation Order

We study the two evaluation orders for path validation described in Section 5: sequential and priority-based. We compare their running times, shown in the last four colums of Tables 3 and 4, for the two environments given in Figures 1 and 3.

For the relatively uncluttered house environment (Figure 3), the results in Table 3 indicate that priority-based path validation performs significantly better than its sequential counterpart; recall that no edge validation is performed during roadmap construction. For the cluttered environment (Figure 1), nodes are completely validated and edges are approximately validated. Table 4 shows that the priority-based validation is only slightly better in most cases. This is because sequential path validation can identify an invalid edge nearly as quickly as priority-based path validation in cluttered environments.

7 Conclusion

In this paper, we present a general framework for building and querying probabilistic roadmaps that includes all previous PRM variants as special cases. In particular, it supports no, complete, or partial node and edge validation and various path validation schedules. Partial node and edge validation also facilitates the imposition of different or additional requirements at query time. We provide experimental evidence that using appropriate techniques for approximately validating nodes and edges during roadmap construction enables one to build roadmaps at a significantly reduced cost while still approximately capturing the connectivity of the free C-space. We also find that priority-

Validation Method	Resolution Level	Nodes	Edges	Build Time (s)	Query Time		Total Time (s)	
					Sequential	Priority	Sequential	Priority
Complete	1.0	300	2304	11.50	1.85	1.32	13.35	12.82
Approximate	0.500	300	2315	6.62	1.76	1.39	8.38	8.01
Approximate	0.250	300	2327	4.29	1.80	1.46	6.09	5.75
Approximate	0.100	300	2367	2.87	2.09	1.45	4.96	4.32
Approximate	0.050	300	2423	2.39	2.23	1.47	4.62	3.86
Approximate	0.025	300	2589	2.03	3.05	2.53	5.08	4.56
None	0.0	300	3398	1.56	15.64	18.6	17.20	20.16

Table 4: Edge Validation Comparison. The table shows the minimal roadmap required to solve the query for the environment shown in Figure 1. Nodes were completely validated. Resolution level is the level at which each edge is validated. It ranges from 0 (no validation) to 1 (complete validation). 0.5 means that half of the nodes along the edge were validated (i.e., every other one).

based path validation proves superior to the sequential evaluation order typically used to validate roadmap paths. Future work involves developing automatic methods for determining appropriate, problem-dependent, approximation levels.

Acknowledgments

We would like to thank the robotics group at Texas A&M. Some of the models studied were provided by Jean-Paul Laumond's and Lydia Kavraki's groups.

References

[1] N. M. Amato, O. B. Bayazit, L. K. Dale, C. V. Jones, and D. Vallejo. OBPRM: An obstacle-based PRM for 3D workspaces. In *Proc. Int. Workshop on Algorithmic Foundations of Robotics (WAFR)*, pages 155–168, 1998.

[2] E. Anshelevich, S. Owens, F. Lamiraux, and L. Kavraki. Deformable volumes in path planning applications. In *Proc. IEEE Int. Conf. Robot. Autom. (ICRA)*, 2000.

[3] M.S. Apaydin, A.P. Singh, D.L. Brutlag, and J.-C. Latombe. Capturing molecular energy landscapes with probabilistic conformational roadmaps. In *Proc. IEEE Int. Conf. Robot. Autom. (ICRA)*, pages 932–939, 2001.

[4] C. Bradford Barber, David P. Dobkin, and Hannu Huhdanpaa. The quickhull algorithm for convex hulls. *ACM Trans. Math. Softw.*, 22(4):469–483, December 1996.

[5] O. B. Bayazit, J.-M. Lien, and N. M. Amato. Probabilistic roadmap motion planning for deformable objects. In *Proc. IEEE Int. Conf. Robot. Autom. (ICRA)*, pages 2126–2133, 2002.

[6] O. B. Bayazit, G. Song, and N. M. Amato. Enhancing randomized motion planners: Exploring with haptic hints. In *Proc. IEEE Int. Conf. Robot. Autom. (ICRA)*, pages 529–536, 2000.

[7] O. B. Bayazit, G. Song, and N. M. Amato. Ligand binding with OBPRM and haptic user input: Enhancing automatic motion planning with virtual touch. In *Proc. IEEE Int. Conf. Robot. Autom. (ICRA)*, pages 954–959, 2001. This work was also presented as a poster at RECOMB 2001.

[8] P. Bessiere, J. M. Ahuactzin, E.-G. Talbi, and E. Mazer. The ariadne's clew algorithm: Global planning with local methods. In *Proc. IEEE Conf. Intel. Rob. Syst. (IROS)*, volume 2, pages 1373–1380, 1993.

[9] R. Bohlin and L. E. Kavraki. Path planning using Lazy PRM. In *Proc. IEEE Int. Conf. Robot. Autom. (ICRA)*, pages 521–528, 2000.

[10] O. Brock and L.E. Kavraki. Decomposition-based motion planning: A framework for real-time motion planning in high-dimensional configuration places. In *Proc. IEEE Int. Conf. Robot. Autom. (ICRA)*, 2001.

[11] H. Chang and T. Y. Li. Assembly maintainability study with motion planning. In *Proc. IEEE Int. Conf. Robot. Autom. (ICRA)*, pages 1012–1019, 1995.

[12] T. H. Cormen, C. E. Leiserson, and R. L. Rivest. *Introduction to algorithms*. MIT Press and McGraw-Hill Book Company, 6th edition, 1992.

[13] L. K. Dale, G. Song, and N. M. Amato. Faster, more effective connection for probabilistic roadmaps. Technical report, Dept. of Computer Science, Texas A&M University, Dec 1999.

[14] L. Han and N. M. Amato. A kinematics-based probabilistic roadmap method for closed chain systems. In *Algorithmic and Computational Robotics – New Directions (WAFR 2000)*, pages 233–246, 2000.

[15] L. Kavraki, F. Lamiraux, and C. Holleman. Towards planning for elastic objects. In *Proc. Int. Workshop on Algorithmic Foundations of Robotics (WAFR)*, 1998.

[16] L. Kavraki, P. Svestka, J. C. Latombe, and M. Overmars. Probabilistic roadmaps for path planning in high-dimensional configuration spaces. *IEEE Trans. Robot. Automat.*, 12(4):566–580, August 1996.

[17] J. C. Latombe. *Robot Motion Planning*. Kluwer Academic Publishers, Boston, MA, 1991.

[18] S. M. LaValle and J. J. Kuffner. Randomized kinodynamic planning. In *Proc. IEEE Int. Conf. Robot. Autom. (ICRA)*, pages 473–479, 1999.

[19] S.M. LaValle, J.H. Yakey, and L.E. Kavraki. A probabilistic roadmap approach for systems with closed kinematic chains. In *Proc. IEEE Int. Conf. Robot. Autom. (ICRA)*, 1999.

[20] J.-M. Lien, Marco Morales, and Nancy M. Amato. Neuron prm: A framework for constructing cortical networks. Technical Report 01-002, PARASOL Lab, Dept. of Computer Science, Texas A&M University, Oct 2001.

[21] C. L. Nielsen and L. E. Kavraki. A two lovel fuzzy prm for manipulation planning. *IEEE/RSJ International Conference on Intelligent Robotics and Systems*, 2000.

[22] A.P. Singh, J.C. Latombe, and D.L. Brutlag. A motion planning approach to flexible ligand binding. In *7th Int. Conf. on Intelligent Systems for Molecular Biology (ISMB)*, pages 252–261, 1999.

[23] G. Song and N. M. Amato. A motion planning approach to folding: From paper craft to protein folding. In *Proc. IEEE Int. Conf. Robot. Autom. (ICRA)*, pages 948–953, 2001.

[24] G. Song and N. M. Amato. Using motion planning to study protein folding pathways. In *Proc. Int. Conf. Comput. Molecular Biology (RECOMB)*, pages 287–296, 2001.

[25] G. Song, S. L. Miller, and N. M. Amato. Customizing PRM roadmaps at query time. In *Proc. IEEE Int. Conf. Robot. Autom. (ICRA)*, pages 1500–1505, 2001.

[26] S. A. Wilmarth, N. M. Amato, and P. F. Stiller. MAPRM: A probabilistic roadmap planner with sampling on the medial axis of the free space. In *Proc. IEEE Int. Conf. Robot. Autom. (ICRA)*, pages 1024–1031, 1999.

Authors' Index

A

Aakchi, Kazuhiko	1633
Abadie, Joël	3219
Abbott, Jake J.	2798
Abdallah, Muhammad	109
Abdulrazak, B.	4023
Acosta, Leopoldo	1588
Agarwal, Pankaj K.	4433
Aghili, Farhad	4035
Agrawal, Sunil K.	1209, 3023, 3029, 3225
Aiyama, Yasumichi	2031
Akella, Srinivas	4066
Alami, R.	2914
Albiston, Brian W.	83
Albu-Schäffer, Alin	3101, 3704
Alegre, Fernando	2344
Alhaj, Ali	3055
Alici, Gürsel	3666
Allebach, Jan P.	2622
Allen, Peter K.	145, 1582, 1824
Almeida, Luis	139
Alsina, Pablo J.	3752
Altendorfer, Richard	37
Alterovitz, Ron	1793
Althaus, Philipp	1551
Althoefer, Kaspar	103, 121, 2555, 2561
Altuğ, Erdinç	4294
Alvarez, Diego	3347
Alvarez, Juan C.	3347
Alves, João	139
Amat, J.	2824
Amato, Nancy M.	741, 2424, 2854, 4427, 4439, 4445
Ambrose, R.O.	2543
Amin-Naseri, M.R.	2928
Ando, Masakazu	3855
Ando, Noriaki	4172
Ando, Shingo	2871
Andrade-Cetto, Juan	1576
Andreff, Nicolas	1191
Ang, Marcelo H. Jr.	3428
Ang, Wei Tech	1781
Angeles, Jorge	773, 3120, 3379
Antonelli, Gianluca	1464
Aoki, Shigeru	2323
Apostoloff, Nicholas	2097
Arai, Fumihito	250, 300, 306, 3624
Arai, Hirohiko	3977
Arai, Tamio	995, 2269, 2356, 2448, 2586, 3770
Arai, Tatsuo	652
Aranda, Joan	386
Arata, Jumpei	2663
Araújo, Rui	1312
Arimoto, S.	2336
Arkin, Ronald C.	109, 2727
Armingol, J.M.	1324
Arras, Kai O.	1992, 4246
Asada, H. Harry	282, 646, 1711, 2224
Asada, Minoru	4398
Asaka, Kinji	1830
Asama, Hajime	2269, 2448, 3849
Asamura, Naoya	3207
Ascari, Luca	2657
Atkeson, Christopher G.	2368
Au, Samuel	646
Avadhanula, S.	1842

B

Bäbler, Andreas	4154
Babu, Kartik	3613
Babvey, Sharareh	957
Bachmann, Eric R.	1171
Bae, J.-H.	2336
Bailey, Tim	1966
Baille, G.	2430
Bajcsy, Ruzena	1694
Baker, Christopher	4270
Bakkum, Douglas	109
Ban, Shigeki	491
Bandlow, Jonathan A.	3613
Barbagli, Federico	809, 1259
Barberá, H. Martínez	2147
Barth, Eric J.	188, 628
Basile, Francesco	1440
Basmajian, Arin	2224
Batalin, Maxim A.	2714
Baur, Charles	3716
Bayle, B.	69
Becker, O.	4360
Belker, Thorsten	4136
Belli, Rossella	2212
Belta, Calin	2498
Bengtsson, Johan	3491
Benhabib, Beno	169, 1527
Bennewitz, Maren	2000
Benoit, Michel	1875
Benson, Eric	1209
Bergamasco, Massimo	3260
Bernardet, Ulysses	4154

Bessière, P.	2104	Burdick, Joel W.	1304, 1482, 1817, 2579, 3600, 3619	Cavallaro, Ettore	2212
Bétemps, M.	658			Çavuşoğlu, M. Cenk	2818
Bi, Z.M.	2317			Ceccarelli, Marco	355, 3654
Biagiotti, L.	3187	Burgard, Wolfram	1557, 2000, 4270	Chablat, Damien	3965
Bianco, Giovanni	3467			Chai, Mong-Lu	2549
Bicchi, Antonio	2412, 3175	Burgess, J.	3600	Chaillet, Nicolas	2960
Biegelbauer, Georg	151	Burnier, D.	4246	Chaimowicz, Luiz	4086
Birglen, Lionel	1139	Burschka, Darius	875, 3917	Chan, Ho-Yin	288
Bizdoaca, Nicu	2079	Burt, Ian T.	90, 4264	Chan, M.L.	2611
Blaer, Paul	1582	Buskey, Gregg	546	Chang, Chih-Fu	1283
Blanchard, Mark	4154	Buss, Martin	1356	Chang, Fan-Ren	3804
Bleuler, H.	670	Butterfass, J.	684, 3164	Chang, Jau-Lung	3517
Bluethmann, William J.	2543, 2806	Byun, Kyung-Seok	503	Chang, Pyung-Hun	3692
				Chang, Shi-Chung	163
Bonvilain, Agnès	2960	**C**		Chang, Shih-Jie	3776
Boonphoapichart, S.	4197			Chang, Wen-Chung	2549
Borst, Ch.	702	Cabido-Lopes, Manuel	2375	Chang, Wen-Jer	2330, 2616
Bosscher, Paul	336, 834	Caccia, Massimo	977	Chatila, Raja	803
Bosse, Michael	1234, 1899	Caffaz, Andrea	2786	Chaumette, François	3055, 4276
Botturi, Debora	3303	Cai, H.G.	3164, 3249	Chen, Cheng-Lun	2622
Bouabdallah, S.	4246	Campbell, Christina L.	2806	Chen, Chih-Chieh	983
Boudjabi, Saliha	4282	Campbell, D.	1380	Chen, Chuanyu	181
Boukallel, Mehdi	3219	Campolo, Domenico	3339	Chen, Chun-Hung	163
Bouloubasis, Antonios K.	2287	Campos, Mario F.M.	4086	Chen, Chu-Song	1677
Bowling, Alan	4048	Cao, Zhiqiang	735	Chen, Haoxun	1743
Bowling, Michael	2281	Cappiello, G.	2230	Chen, Heping	3504, 3984
Bretl, Timothy	2946	Carbajo, J.	1330	Chen, I-Ming	761, 773
Brock, Oliver	3385	Carbone, Ciro	1440	Chen, Jeng-Shi	3266
Brogliato, Bernard	2218	Carbone, Giuseppe	3654	Chen, Jiun-Hung	1677
Brown, H. Benjamin	3834	Cariou, C.	115	Chen, Longhui	1881
Brown, Peter	4122	Carlson, Jennifer	274	Chen, Mu-Chen	3554
Browning, Brett	2281	Carp-Ciocardia, D.C.	4116	Chen, Pei-Feng	3421
Bruce, James	1277, 2281	Carpin, Stefano	3809	Chen, Peng	610
Brunn, D.	1545	Carreras, M.	971, 989	Chen, Qiang	696
Bruynickx, Herman	2599, 2766	Carrozza, M.C.	2230	Chen, S.Y.	2129
Bualat, M.	2535	Casalino, Giuseppe	2786	Chen, Shih-Feng	4042
Buehler, Martin	1368, 1380, 1386	Casals, A.	2824	Chen, Weidong	4191
		Castano, Andres	1	Chen, Xing	1001
Bunschoten, Roland	577	Castellanos, José A.	427	Chen, Yifan	3504, 3984
Burdet, E.	670	Causey, Greg	2772	Chen, Yi-Xiang	3776

Chen, Yong-Sheng	1677	
Cheng, Fan-Tien	596, 1723	
Cheng, Gordon	4122	
Chern, Ming-Yang	2085, 2110	
Chesi, Graziano	3911, 3929	
Chetwynd, Derek G.	1863	
Cheung, T.Y.	1446	
Chew, Chee-Meng	45	
Chiacchio, Pasquale	1440	
Chiang, Tsung-Che	1434	
Chiaverini, Stefano	1464	
Chien, Chiang-Ju	4130	
Chinellato, Eris	1133	
Chinzei, Kiyoyuki	652	
Chio, Tien-Sung	2701	
Chirikjian, Gregory S.	1594	
Chiu, Forng-Chen	983	
Chiu, George T.C.	2622	
Cho, Changhyun	3243	
Cho, Kyu-Jin	646	
Choi, H.B.	1185	
Choi, H.R.	1857	
Choi, Joonhyuk	2135	
Choi, Yi-King	349	
Choi, Yoon Ho	2006	
Chomchana, Trevai	2448	
Choset, Howie	1606	
Chou, Hsu-Chi	4215	
Christensen, Henrik I.	419, 1545, 1551, 1824, 2262, 3485	
Chu, Chengbin	1743	
Chuang, Jen-Hui	3353, 3365	
Chugo, Daisuke	3849	
Chung, Huan-Yuan	2330	
Chung, Jae Heon	521	
Chung, Ronald	1688	
Chung, Sheng-Luen	1033, 1050	
Chung, Wan Kyun	1606	
Chung, Woojin	2792, 2830	
Civita, Marco La	552	
Clark, Christopher M.	4222	
Cleghorn, William L.	3193	
Cline, Michael B.	3744	
Cobzas, Dana	1570, 2812	
Cole, G.R.	2024	
Collarini, Diego	2212	
Collewet, Christophe	3055	
Collins, Anne D.	4433	
Company, Olivier	1185, 1875	
Confente, Mirko	3303, 3467	
Conti, François	3716	
Corke, Peter I.	546, 4288	
Corso, Jason	875	
Cortés, J.	4354	
Cosma, Claudio	3303	
Costa, Marcio	4154	
Coué, C.	2104	
Cowling, P.I.	175	
Crinier, S.	1545	
Cufí, X.	989	
Culbert, C.J.	2543	
Cutkosky, Mark R.	1836	

D

Da Rosa, Vagner Santos	622	
Daescu, Ovidiu	3542	
Dahl, Jeffrey	3504, 3984	
Dahl, Torbjørn S.	2293	
Daifu, Shinichi	1098	
Daniilidis, Kostas	1913	
Dario, Paolo	232, 1086, 1092, 1836, 2212, 2230, 2657, 3576	
Dash, Anjan Kumar	761	
Davis, James	1001	
De Battista, Marc	1992	
De Gersem, Gudrun	2651	
de la Escalera, A.	1324	
De Luca, Alessandro	634	
De Rossi, Danilo	2412	
De Schutter, Joris	2599	
Dechev, Nikolai	3193	
del Pobil, Ángel P.	1133	
Delaplace, S.	1509	
Delbrück, Tobi	4154	
Dellaert, Frank	1960, 2344	
Deneszczuk, Patricia	3953	
Deng, Hongbin	2472	
Deng, Xinyan	1152	
Denk, J.	1343	
DeSouza, Guilherme N.	3473	
Devengenzo, Roman	809	
Devy, M.	1330	
Dias, Jorge	139	
Diftler, M.A.	2543	
Dissanayake, Gamini	944	
Divoux, Jean-Louis	2218	
Doh, Nakju	1606	
Dohring, Mark	3710	
Dombre, Etienne	2218	
Donamukkala, Raghavendra	25	
Dong, Lixin	300, 3624	
Douglas, Rodney J	4154	
Drenner, Andrew	4264	
Du, L.B.	3249	
Dubey, Rajiv	1247	
Duckett, Tom	434	
Dudek, Gregory	1907, 2907, 3434	
Duindam, Vincent	4029	
Dunbar-Jacobs, Jacqueline	25	
Duran, Olga	2561	
Durand, Jérôme	1682	
Durrant-Whyte, Hugh F.	944, 1521	

E

Eade, Ethan	2194
Ebert-Uphoff, Imme	336, 834
Efrat, Alon	3789
Eino, Jyun-ichi	899
Eladio, Dapena	4203
Elber, Gershon	1021
Elhajj, Imad	1646, 4414
Ellis, Matthew	3213
Elnagar, A.	2442
Emura, Takashi	57
Endo, Hisashi	362
Endo, Ken	1362
Endo, Tatsuro	4342
Eng, Kynan	4154
Enoki, Ryo	3729
Erdmann, Michael A.	3391
Erickson, Jeff	2242
Erkmen, A.M.	4408
Erkmen, I.	4408
Erni, J.	670
Espinosa-Romero, A.	3048
Estrin, Deborah	19

F

Fagg, Andrew H.	2677
Faloutsos, Petros	917
Fan, Changhong	4191
Fanti, Maria Pia	4306
Fantuzzi, C.	3290
Fattah, Abbas	3225
Fearing, Ronald S.	1146, 1164, 1842, 3339
Feiten, Wendelin	1899
Feng, Garry	3005
Feng, Gong	1086
Feng, Maria Q.	2362
Feng, Weidong	1749
Ferguson, David	4270
Fernández, Josep	386
Ferreira, Antoine	3076, 4282
Fierro-Rojas, J.D.	3042, 3048
Finkemeyer, Bernd	3069
Fiorini, Paolo	3303, 3467
Fischer, M.	684, 702
Fisher, Robert B.	1133
Fisher, William	2153, 2895
Fitzpatrick, Paul	3140
Folkesson, John	419
Fourquet, J.-Y.	69
Fox, Dieter	2836
Fraichard, Th.	2104
Fraisse, Philippe	2218
Fregene, Kingsley	2707
Frese, Udo	3704
Freund, E.	1705
Frew, Eric W.	3479
Frey, Chr. W.	2634
Frigola, Manel	386
Frisoli, Antonio	3260
Froidevaux, G.	2388
Fu, Li-Chen	1434, 1458
Fu, Long-Ming	3636
Fujimoto, Daisuke	816
Fujimoto, Hideo	2478
Fujita, Masahiro	471
Fujita, Yusuke	3359
Fujiwara, Kiyoshi	1620, 1633, 1640
Fukase, Takeshi	2356
Fukase, Yutaro	2985
Fukazawa, Yusuke	2269, 2448
Fukuchi, Masaki	262
Fukuda, Toshio	250, 300, 306, 785, 1646, 1811, 2206, 3624
Fukui, Rui	4178
Fukuoka, Y.	2037, 2043
Funabashi, Hiroaki	749
Fung, Wai Keung	3642
Furukawa, Masayuki	4160
Furukawa, Tomonari	944
Furuno, Seiji	3403
Furusho, Junji	202, 214, 4016

G

Gans, Nicholas R.	3061
Gao, Meimei	3548
Gao, X.H.	3249
Gaonkar, Roshan	1762
García, Rafael	133, 971, 989
Garg, Dinesh	1737
Gaskett, Chris	4122
Gassert, R.	670
Gau, X.H.	3164
Gautier, Maxime	3278
Ge, S.S.	1972
Geiman, Jeremy	3917
Geisberger, Aaron	1470
Geng, Juhong	2472
Genovese, Vincenzo	1318
Gerkey, Brian P.	3862
Gienger, M.	484
Gini, Maria	850
Giorgi, Fabio	2786
Goerke, Derek	4264
Gogola, Michael A.	188, 628
Göktoğan, Ali Haydar	2720
Goldberg, Ken	1793, 3953
Goldfarb, Michael	188, 628
Goldgeier, Michael	2510
Golmakani, Hamid R.	169
Gomez, Francisco	2555
González, Rafael C.	3347

González-Baños, Héctor H.	3789	
Goodwine, Bill	1600, 4228	
Gopalakrishnan, K.	3953	
Goradia, Amit	374, 1646	
Gosselin, Clément M.	1139	
Goto, Satoru	4386	
Goto, Takayuki	1946	
Gotoh, Taeko	2573	
Gotoh, Tatsuya	491	
Gouveia, Gonçalo	1312	
Grandjean, B.	4023	
Graves, Robert J.	3535	
Gravot, F.	2914	
Grebenstein, M.	684, 3164	
Green, Bill	1404	
Greenberg, Sam	3371	
Greppin, X.	2388	
Griswold, Karen	2895	
Grocholsky, Ben	1521	
Grupen, Roderic A.	2677, 3385, 4074	
Gruver, W.A.	2317, 4197	
Gruyer, Dominique	4185	
Gu, Guochang	1215	
Gu, Jason	1659	
Guan, F.	1972	
Guegan, Sylvain	3272	
Guglielmelli, Eugenio	232, 2212, 3576	
Guiraud, David	2218	
Guivant, Jose	412	
Guo, Jenhwa	983	
Guo, Shuxiang	1830	
Gupta, Kamal	2406	
Gutmann, Jens-Steffen	262	

H

Hada, Y.	2037
Hager, Gregory D.	727, 875, 1954, 3917
Hagita, Norihiro	4166
Hähnel, Dirk	1557, 4270
Haidacher, S.	684, 702, 1805
Haishi, Atsuo	368
Hakozaki, Mitsuhiro	3207
Hamel, William R.	208
Hammel, Martin	4136
Han, Wenbiao	4312
Hanafusa, Hideo	4008
Handa, Hiroyuki	2985
Hannaford, Blake	822
Hannan, Michael	3449
Hao, Yongxing	1209
Hara, Kei	1933, 2741
Harada, Kensuke	1620, 1627, 1633, 1640
Hardt, Michael	1356
Harer, John L.	4433
Hariti, Mohamed	1700
Hartono, Pitoyo	3571
Hasegawa, Takahiko	1098
Hasegawa, Tsutomu	708, 925, 2848
Hasegawa, Yasuhisa	250, 785, 1811
Hashimoto, Hideki	4172
Hashimoto, Koichi	3911, 3929
Hashimoto, Minoru	2055
Hashimoto, Shuji	3571
Hashizume, Hiroyuki	676
Hashizume, Makoto	2663
Hashizume, Takumi	899
Hashtrudi-Zaad, K.	3296
Hasunuma, Hitoshi	2998
Hatakeyama, Takuro	2305
Hattori, Shizuko	2979
Haung, Jian-Feng	3517
Hayet, J.B.	1330
Haynes, L.	604
Hayward, Vincent	3722
He, P.	3164
Hebbar, Ravi	3309
Hebert, Martial	3497
Heidemann, John	19
Henaff, P.	1509
Hepp, Klaus	4154
Hermosillo, J.	2430
Hernández, Sergio	1588
Hertzberg, Joachim	4136
Hesselbach, J.	4360
Hibi, Hitoshi	2865
Higashimori, Mitsuru	1115
Higashiura, Masaki	1811
Hildebrandt, A.	2182
Hirai, Hiroaki	226
Hirai, Shin'ichi	3169, 3441, 3737
Hiramatsu, Yuji	3582
Hirata, Yasuhisa	938, 2275
Hiroki, Takeuchi	2350
Hirose, Kazuya	749
Hirose, Shigeo	63, 368, 477, 2466
Hirukawa, Hirohisa	1620, 1627, 1633, 1640, 2985, 3905
Hirzinger, Gerd	684, 702, 1805, 3029, 3101, 3164, 3704
Ho, Chih-Ming	3630
Ho, Yueh Sheng	983
Hodgins, Jessica K.	3834
Hollis, R.L.	1253
Holmes, Philip	37
Homyk, A.	3600
Hong, Wei	1013
Hori, T.	4197
Hoshino, Hiroshi	2979
Hosoe, S.	2884

Hou, Ping-Cheng	2110
Howard, Andrew	868
Howard, Ayanna	2012
Hrabar, Stefan	558
Hsia, T.C. Steve	3321
Hsiao, Ming-Chang	4403
Hsieh, Bo-Wei	163
Hsu, Chun-Te	4130
Hsu, David	4420
Hsu, I-Chow	1793
Hsu, Jui-Cheng	1283
Hsu, Liu	458
Hsu, M.S.	3397
Hsu, Stephen	2194
Hsu, Su-Hau	1458
Hsu, Wen-Jing	181, 3510
Hsu, Y.C.	145
Hu, Chao	1659
Hu, H.Y.	3249
Hu, Tiemin	13
Huang, Chien-Sheng	616
Huang, H.J.	1446
Huang, Han-Pang	220, 1497
Huang, Pei-Zhi	3421
Huang, Qiang	2472
Huang, Shell-Ying	181
Huang, Shuguang	3082, 3095
Huang, Tian	330, 1863
Huang, Wesley H.	3613
Huang, Zhen	755, 1179, 1203, 1881, 1887
Huber, Eric	2806
Hung, Min-Hsiung	596
Hung, Yi-Ping	1677
Hussein, A.M.	2442
Hutchinson, Seth A.	1682, 3061
Hwang, Kao-Shing	4403
Hwang, Yoonkwon	31
Hygounenc, Emmanuel	540
Hyon, Sang-Ho	57

I

Ichige, Yukiko	2979
Ichikawa, Akihiko	306
Ichikawa, Hironobu	2640
Ichikawa, Kazuki	749
Ifju, Peter G.	1422
Igarashi, Hiroshi	2049
Iida, Michihisa	96
Ikeda, Atsutoshi	640
Ikeda, Tetsushi	4398
Ikemata, Yoshito	2478
Ikeuchi, Katsushi	3881, 3887, 3893, 3899, 3905
Ikuta, Koji	1098, 1103, 2640
Imai, Jun	3855
Imai, Kousuke	564
Inaba, Akio	3959
Inaba, Masayuki	911, 932, 1940, 2141, 2299, 3132
Inohira, Eiichi	31, 4060
Inoue, Akio	4016
Inoue, Hirochika	911, 932, 1940, 2141, 2299, 3132
Inoue, Kousuke	2067, 2073
Inoue, Takahiro	244
Ioka, Hayato	785
Ishida, Tatsuzo	471
Ishiguro, Hiroshi	1946, 4160, 4398
Ishihara, Hidenori	640
Ishii, Tomoyuki	368
Ishikawa, Masatoshi	1115, 1264, 2400
Ishikawa, Seiji	1410
Ishitsuka, Takashi	1850
Isler, Volkan	1913
Isoda, Shuzo	3582
Isozumi, Takakatsu	1633, 2985
Isto, Pekka	2934
Ito, Kazuyuki	791
Ito, Koji	3146
Ito, Tomotaka	362
Itoh, Kazuko	3588
Ivanescu, Mircea	2079
Iwata, Satoru	2663
Iyengar, S. Sitharama	1653
Izawa, Hidemitsu	2586
Izawa, Jun	3146
Izquierdo, M. Zamora	2147
Izumi, Kiyotaka	515

J

Jackson, Bennett	4264
Jacubasch, A.	2634
Jafari, Mohsen A.	4312
Jafry, H.	828
Jaganathan, A.	3971
Jagersand, Martin	1570, 2812
Janét, Jason	2194
Jeng, MuDer	1033, 1050
Jensen, B.	2388
Jensen, Björn	1503
Jeon, J.W.	1857
Jeon, Jinwoo	816
Jeong, Younkoo	1092, 2940
Jezernik, Karel	1068
Jiang, L.	3164
Jiang, Tingting	4420
Jiao, L.	1446
Jin, Ming-He	1800, 3164, 3249
Joehl, K.	684
Johansson, Rolf	3491, 3686
Johnson, K.W.	828
Jones, Chris	721
Jones, Erik D.	3321

Jonnalagadda, K.	2116	
Juang, Jyh-Ching	3266	
Jui, C.K. Kevin	4336	
Jung, K.M.	1857	
Jung, Myung-Jin	250	
Jüngel, Matthias	856	

K

Kaetsu, Hayato	3849
Kagami, Satoshi	911, 932, 2141
Kagami, Yoshiharu	2299
Kahlil, Wisama	3272
Kajita, Shuuji	1336, 1613, 1620, 1627, 1633, 1640
Kak, Avinash C.	1397, 3473
Kakikura, Masayoshi	2049
Kambhampati, C.	4252
Kamegawa, Tetsushi	791
Kamimura, Akiya	714
Kanada, Kensaku	785
Kanade, Takeo	552
Kanaoka, Katsuya	816, 4366
Kanazawa, Hiroyuki	2973
Kanbara, Yoshio	4160
Kanehiro, Fumio	1613, 1620, 1633, 1640, 2985
Kaneko, Kenji	1336, 1620, 1627, 1633, 1640, 2985
Kaneko, Makoto	664, 1115
Kang, Byoung Hun	1470, 3213
Kannan, Sampath	1913
Kao, Imin	3698
Kapuria, Anuj	25
Kashiwagi, Eiichi	380
Kassim, Irwan	1086
Katayama, Toshio	2466

Katsuki, Rie	995, 3770
Katsuragi, Tohoru	306
Kawabata, Kuniaki	3849
Kawahara, Tomohiro	664
Kawai, Yoshihiro	2985
Kawakami, Atsushi	63
Kawamura, Sadao	816
Kawamura, Tomoyuki	1533
Kawauchi, Naoto	2860, 2973
Kelly, Rafael	1062, 4374
Kennedy, Diane	2707
Khadem, S. Esmaeilzadeh	2928
Khamis, A.	3284
Khatib, Oussama	3716
Khoshnevis, Behrokh	2311
Khoshnevis, Berok	2516
Khosla, Pradeep K.	1781
Khosravi, M.A.	3108
Khuri-Yakub, Butrus T.	2892
Kiguchi, Kazuo	515, 779, 2206
Kikuchi, Haruka	2586
Kikuchi, Takehito	214
Kikuuwe, Ryo	1539
Kim, Bong Oh	4142
Kim, Byoung-Ho	3169
Kim, Byungkyu	1092, 2940
Kim, Byungmok	2940
Kim, ChangHwan	4048
Kim, Clay	2901
Kim, G.W.	3315
Kim, Gunhee	2792
Kim, H.M.	1857
Kim, Incheol	2123
Kim, Ja-Hee	1039
Kim, Jinsuck	2424, 2854
Kim, Jong-Hyuk	406
Kim, Jongwon	767
Kim, Jongwoo	2200
Kim, Kab-Il	2461

Kim, Kyunghoon	1515
Kim, Munsang	2135, 2792, 2830, 3243, 3315, 3415
Kim, Myung-Soo	349
Kim, Sewoong	208
Kim, Sungho	2123
Kim, Wheekuk	521, 690, 4348
Kim, Yoon Sang	822
Kim, Young Soo	4348
Kim, YoungWoo	3959
Kimura, Hiroshi	2037, 2043, 3881, 3887, 3893
Kimura, Ken'ichiro	3678
Kimura, Masafumi	3737
Kit, Chow Man	1646
Kitano, Hiroaki	392, 398, 1362
Kito, Tomomi	995, 3770
Kiyota, Yuuki	202
Klatzky, R.L.	1253
Klein, David	4154
Knoop, Steffen	1824
Kobayashi, H.	3126
Kobayashi, Masami	2998
Kobayashi, Yuichi	2356
Koburov, Stephen G.	3789
Koditschek, Daniel E.	37, 1374, 1391, 3935
Koenig, Sven	75, 3371
Koganti, Rama	3953
Kogure, Kiyoshi	4166
Koizumi, Norihiro	676
Kokaji, Shigeru	714
Kökösy, A.M.	4330
Komada, S.	4197
Komoriya, Kiyoshi	194
Komsuoğlu, Haldun	1391
Kondo, Eiji	2848
Kondo, Toshiyuki	3146

Koninckx, Bob	2766	
Koninckx, Philippe R.	2646	
Konishi, Masami	3855	
Konno, Atsushi	31, 1185, 2529, 4060	
Koo, T. John	3758	
Koseki, Yoshihiko	652	
Kosuge, Kazuhiro	938, 1080, 2275	
Koukam, Abderrafiaa	1700	
Koyachi, Noriho	652	
Koyama, Takeshi	2362	
Koyanagi, Ken'ichi	4016	
Koyasu, Hiroshi	893	
Kozlowski, K.	1068	
Kragic, D.	1545, 3485	
Krasny, Darren P.	3842	
Kratochvil, Bardley	4264	
Krishnaprasad, P.S.	2510	
Kröger, Torsten	3069	
Kröse, Ben	577, 2842	
Krosuri, Satya P.	312	
Krupa, Alexandre	4282	
Krut, Sébastien	1875	
Kuan, Cheng-Peng	4392	
Kuffner, James J.	911, 932, 2141	
Kugi, Andreas	3101	
Kumar, Vijay	2275, 2498, 4086	
Kume, Youhei	938	
Kunii, Yasuharu	2573	
Kunimoto, Masanari	1264	
Kuniyoshi, Yasuo	2299, 3132	
Kuntze, H.-B.	2634	
Kunz, C.	2535	
Kunze, K.	684	
Kuo, Chung-Hsien	616	
Kuo, Po-Ting	3804	
Kurazume, Ryo	925	
Kurokawa, Haruhisa	714	
Kuroki, Yoshihiro	471	
Kushihama, Kiyotaka	3441	
Kusumoto, Yukihiro	2171	
Kwak, J.W.	1857	
Kwan, C.	604	
Kweon, Inso	2123	
Kwok, Cody	2836	
Kwon, Ohung	1350	
Kwon, Paul G.R.	306	
Kyberd, P.J.	3231	
Kyriakopoulos, Kostas J.	7, 4240	
Kyura, Nobuhiro	4386	

L

la Torre, B.	1259	
Lai, Chih-Chiun	881	
Lam, Alan H. F.	3181	
Lam, C.P.	2024	
Lam, Raymond H.W.	1768, 3181	
Lam, Y.K.	3869	
Lambert, Alain	4185	
Lamon, Pierre	440	
Lampariello, R.	3029	
Lanzoni, Claudio	3764, 4258	
Larson, Bert	2901	
Laschi, Cecilia	232	
Latombe, Jean-Claude	2946, 3797, 4222	
Laugier, C.	1109, 2430	
LaValle, Steven M.	464, 2920	
Lawrence, P.D.	1297	
Laxton, Benjamin	1209	
Lazzarini, R.	2230	
Lebedev, Dmitry V.	4209	
Lee, B.H.	3315, 3415	
Lee, Chongwon	2792	
Lee, Choon-Young	1074	
Lee, Dongheui	2830	
Lee, Edward	1209
Lee, Eunjeong	3692
Lee, Gwo-Bin	3636
Lee, Hogil	521
Lee, Joo-Ho	4172
Lee, Ju-Jang	1074
Lee, Ka Keung	1671, 2567
Lee, Kok-Meng	3089
Lee, Rong Shean	1731
Lee, S.	2535
Lee, S.W.	1857
Lee, Sanghoon	4142
Lee, Tae-Eog	1039
Lee, Ti-Chung	905
Lee, Wen-Yo	1787
Lee, Woo Ho	3213
Lefebvre, Tine	2599
Lemoine, Philippe	3272
Lenain, R.	115
Lenser, Scott	1416
Leonard, J.	1234, 1921
Leonard, John	1899
Leonard, John J.	963
Leonard, Naomi Ehrich	2492
Leordeanu, M.	145
Leow, Wee Kheng	3428
Leow, Y.P.	3237
Lerasle, F.	1330
Leroy, N.	4330
Lett, Jean-François	850
Leung, Martin Y.Y.	3181
Li, Bin	2067, 2073
Li, Chuan	773
Li, Gary	2418
Li, Guangyong	3642
Li, Huaming	3923
Li, Huan	4074
Li, Jia-Wei	1800
Li, Kejie	2472
Li, Meng	330

Li, Ming	1954	Liu, G.F.	1869, 2683, 2695	Luo, Ren C.	1717, 2394
Li, Qinchuan	755, 1179, 1203, 1887	Liu, Hong	702, 1800, 3164, 3249	Luo, Z.W.	2884
Li, Qingguo	3594, 3607	Liu, Li-Wei	1497		
Li, T.Q.	3249	Liu, Peter Xiaoping	13, 1659		

M

Li, Tsai-Yen	3421, 4215
Li, Tzuu-Hseng S.	3776
Li, Wen J.	288, 1768, 3181, 3648
Li, Wujeng	590
Li, Y.F.	127, 2129
Li, Yangmin	3254
Li, Yanmei	3698
Li, Yanwen	1881
Li, Yu-Wen	4092
Li, Z.X.	1869, 2683, 2695, 3660
Li, Zexiang	3941
Li, Zhanxian	330
Li, ZhiWu	1452
Lian, Feng-Li	2504
Liang, Yongqiang	2579
Liao, Da-Yin	1027
Lien, Jyh-Ming	4439
Lim, Hun-ok	3654, 4342
Lim, Hyun-young	1092
Lim, Seungchul	2461
Lin, Chia-Ping	3554
Lin, Chien-Chou	3353, 3365
Lin, Ching-Po	2091
Lin, Hung-Hsing	1283
Lin, Jui Cheng	590
Lin, Pei-Chun	1391
Lin, R.X.	3397
Lin, Shih-Schön	1694
Lin, Shun-Yu	1434
Lin, Y.J.	3971
Lindemann, Stephen R.	2920
Lippiello, Vincenzo	3333
Liu, Pou	3624
Liu, Xiaoping	3254
Liu, Xin-Jun	767, 3990, 4092
Liu, Y.W.	3164
Liu, Yi-Hung	220, 1497
Liu, Yufeng	1227
Liu, Yugang	3254
Liu, Yunhui	1646, 2236, 3181, 4414
Liu, Z.G.	127
Lizarralde, Fernando	458
Lo, Wang-tai	4414
Lobo, Jorge	139
Löffler, K.	484
Loh, A.P.	1972
Loh, Chee Kit	181
Loh, W.K.	3237
Loizou, Savvas G.	4240
Loo, C.K.	3869
Lopes, Gabriel A.D.	3935
Loría, Antonio	1062
Lorotte, A.	2388
Lotti, F.	3187
Lou, Y.J.	1869
Low, Kian Hsiang	324, 3237, 3428
Lowe, D.G.	1297
Luewirawong, Thanaphon	1121
Luh, Peter B.	1749, 1756
Luis, Moreno	4203
Lumia, R.	2116
Lumsdaine, Arnold	208
Luo, Chaomin	4080
Luo, Qi	2605
Ma, Hongyu	3327
Ma, Shugen	2067, 2073
Ma, Yi	1013
Mabuchi, Kunihiko	1264
MacDonald, Bruce A.	1564
Maeda, Manabu	3582
Maeda, Yusuke	2586
Maeder, W.	670
Maekawa, Hitoshi	194
Maeno, Takashi	1362, 1533
Maeyama, Shoichi	1978
Magid, Evgeny	1021
Mahapatra, Saurabh	1241
Mahvash, Mohsen	3722
Mailey, Chris	2194
Maini, Eliseo S.	1318
Makarenko, Alexei	1521
Makino, Ryota	1264
Makksoud, Hassan El	2218
Malis, Ezio	1056
Manabe, Go	2305
Manickam, Arul	1653
Mäntylä, Martti	2934
Manzolli, Jonatas	4154
Marayong, Panadda	1954
Marcovich, Renzo	2212
Marhefka, Duane W.	3821
Marigo, Alessia	3175
Marquet, Frédéric	1191, 3278
Martinelli, Bruno	2212
Martinet, Philippe	115, 1191
Martinson, Eric Beowulf	1960, 2344, 2727
Maru, Noriaki	3036
Mascaro, Stephen A.	282

Mase, Kenji	4166	
Masehian, Ellips	2928	
Mason, Matthew T.	3391	
Masoud, Ahmad A.	4234	
Masoud, Osama	850	
Masuhara, Ban	3678	
Mata, M.	1324	
Matabosch, Carles	133	
Matarić, Maja J.	721, 868, 1665, 2293, 3862	
Mati, Yazid	157	
Matsubara, Takashi	380	
Matsumaru, Takafumi	362	
Matsumoto, Masaharu	3678	
Matsumoto, Tsutomu	797	
Matsunaga, Satoshi	664	
Matsuno, Fumitoshi	791, 2061	
Matsuoka, Yoky	238	
Matthews, Judith T.	25	
Matthies, Larry	1	
Mattone, Raffaella	634	
Mayor, L.	2388	
Mazer, E.	2104	
McBeath, Michael K.	3461	
McGhee, Robert B.	1171	
McIsaac, Kenneth A.	3875	
McKee, Gerard T	2287	
McKinney, Doug	1171	
McMickell, M. Brett	1600, 4228	
McMillen, Colin	4264	
McMordie, Dave	1386	
Medeiros, Adelardo A.D.	3752	
Meilă, Marina	2836	
Meisser, M.	2388	
Melchiorri, C.	3187	
Menciassi, Arianna	1092, 2657	
Mendoza, C.	1109	
Meng, J.	3660	
Meng, Max Q-H	13, 1659	
Menon, M.	1842	

Merlet, J-P.	1197	
Merrell, Roy	452	
Merrill, Ernest	109	
Messner, William C.	552	
Metta, Giorgio	3140	
Meybodi, Mohammad R.	957	
Micera, Silvestro	2212	
Mifune, Fumisato	2998	
Milios, Evangelos	1907	
Miller, Andrew T.	1824, 2262	
Miller, David P.	2436	
Mills, James K.	169, 2418, 3193	
Minato, Takashi	4160	
Minca, Eugenia	4318	
Minor, Mark A.	83, 312, 452	
Mintz, Matti	4154	
Mishima, Taketoshi	3849	
Misra, A.K.	3120	
Mitsuishi, Mamoru	676, 2663	
Miura, Jun	893	
Miwa, Hiroyasu	232, 3588	
Miyabe, Tomohiro	2529	
Miyamoto, Manabu	2663	
Miyazaki, Fumio	226	
Mizuta, Takahisa	995, 3770	
Mizuuchi, Ikuo	1940	
Mobasser, F.	3296	
Mochiyama, Hiromi	3672	
Mohri, Akira	497, 3403	
Mok, Swee M.	3947	
Mokhtari, M.	4023	
Momtahan, Omid	957	
Monacelli, E.	1509	
Montemerlo, Michael	1985, 4270	
Montestruque, Luis Antonio	1600	
Montiel, J.M.M.	1007	
Moon, Inhyuk	1515	
Morales, Antonio	1133	
Morales, Carlos A.	1588	
Morales, Marco	4427	

Moreno, Javier	4374	
Moreno, Wilfrido A.	1247	
Mori, Naoki	2171	
Mori, Osamu	4054	
Morinaga, Shinya	1080	
Morishita, Hiroshi	4178	
Morita, Takuma	3887	
Morizono, Tetsuya	1490, 2323	
Morris, Aaron	25	
Moser, Bryan	1756	
Moser, R.	670	
Mosse, Charles A.	1086	
Motomura, Kazuhiro	63	
Motoo, Kouhei	306	
Motuk, H.E.	4408	
Mun, Museong	1515	
Mundhra, Keshav	3461	
Muñoz, L.	2824	
Murakami, Kouji	708	
Murata, Satoshi	714	
Murphey, T.D.	3600, 3619	
Murphy, Robin R.	274	
Murray, Richard M.	2484, 2504	
Murrieta, Rafael	464	

N

Nagai, Kiyoshi	3678, 4008	
Nagasaka, Ken'ichiro	471	
Nagasaki, Takashi	1336	
Nagata, Fusaomi	2171	
Nahar, Dhiraj R.	318	
Nakadai, Kazuhiro	392, 398	
Nakajima, Masahiro	300, 3624	
Nakajima, Toshiya	2992	
Nakamura, Masatoshi	4386	
Nakamura, Shinya	2992	
Nakamura, Yoshihiko	51, 491, 1927, 3359	

Nakanishi, Isao	4008
Nakano, Yoshio	244
Nakaoka, Shinichiro	3899, 3905
Nakashima, Katsumi	2998
Nakauchi, Yasushi	380
Nakayama, Yuichiro	816
Nakazawa, Atsushi	3899, 3905
Nam, J.D.	1857
Namba, Kyota	3036
Namiki, Akio	1264, 2400
Naphattalung, Piyawat	380
Narahari, Y.	1737
Narimatsu, Katsumi	1749
Natale, Lorenzo	3140
Nebot, Eduardo	412
Nechyba, Michael C.	1422
Neira, José	427
Nejat, Goldie	1527
Nelson, Bradley J.	294, 3200
Nenche, Dragomir N.	2760
Nettleton, Eric	2720
Newman, Paul	1234, 1899, 1921
Newman, Wyatt S.	3309, 3710
Ng, Wan S.	1086
Ni, Ming	1756
Nickl, M.	684
Nicosevici, T.	971
Nieto, Juan	412
Nishi, Tatsushi	3855
Nishimoto, Shohei	3729
Nishiwaki, Koichi	911, 932, 2141
Nishiyama, Takashi	2979
Nishiyama, Tsuyoshi	995, 3770
Noborio, Hiroshi	1946, 3729
Noda, Kuniaki	3565
Nolin, J.T.	828
Noritsugu, Toshiro	2188, 4098
Nunes, Eduardo V.L.	458
Nuttin, M.	268

O

Oda, Kunihiko	214
Ogata, Tetsuya	3565
Ogawa, Hiroki	2586
Ogawa, Hironori	1264
Ogawara, Koichi	3881, 3887, 3893
Ögren, Petter	2492
Ogura, Yu	3582
Oh, Joon Seop	2006
Oh, Kun-Ku	767
Oh, Paul Y.	1404
Oh, Sang Rok	4142, 3158, 4348
Oh, So-Ryeok	3023
Oh, Sung-Nam	2461
Oh, Young Seok	3213
Ohmameuda, Yoshihiro	2067
Ohnishi, Masatoshi	244
Ohno, Kazunori	1978
Ohta, K.	2884
Ohtsuki, Kayoko	244
Ohya, Akihisa	256
Ohya, Kazuhisa	2998
Okada, Kei	2141
Okada, Masafumi	491
Okada, Nobuhiro	2848, 3497
Okamura, Allison M.	828, 1774, 1954, 2798
Okano, Yuzo	797
Okawa, Kazuya	3409
Okino, Akihisa	244
Okuchi, Tetsuya	3588
Okuma, Shigeru	3959
Okuno, Hiroshi G.	392, 398
O'Leary, M.D.	1774
Olson, Christopher	4264
Olsson, Tomas	3491
Omata, Toru	1817, 4054
Omohundro, Zachary	4270
Oomichi, T.	2860
Orin, David E.	3821, 3842
Oriolo, Giuseppe	3175
Ortín, D.	1007
Ostrowski, James P.	2200, 4294
Osuka, Koichi	96
Osumi, Hisashi	2031
Ota, Hiromichi	1875
Ota, Jun	995, 2269, 2448, 3770
Ota, Yusuke	477
Otake, Mihoko	2299
Ott, Christian	3101, 3704
Ottaviano, Erika	355
Ou, Yongsheng	3455, 4148
Ouelhadj, D.	175
Ozawa, R.	3126

P

Pagès, Jordi	133
Pai, Dinesh K.	3744
Palaniappan, Lingeshwaran	3789
Palmer, Luther R.	3821
Pámanes G., J. Alfonso	2753
Pan, Chi-Chun	3365
Pan, Feng	2592
Pana, Deniela	2079
Papadopoulos, Evangelos	1368
Papageorgiou, George	552
Papanikolopoulos, Nikolaos	90, 850, 1158, 4264
Paquier, Williams	803
Park, Edward J.	2418
Park, F.C.	2135
Park, Gwi-Tae	3158
Park, J.B.	3415
Park, Jae Byung	1397

Park, Jin Bae	2006
Park, Jong Hyeon	1350
Park, Jongil	4348
Park, Jong-Oh	1092, 2940
Park, Juyi	3692
Park, Kwi-Ho	3169
Parra-Vega, V.	3042, 3048
Parsa, K.	3120
Paulo, Sérgio	3529
Payandeh, Shahram	3594, 3607
Pearce, Janice L.	1158
Pearce, Roger A.	2424, 2854
Pedersen, L.	2535
Pedrosa, Diogo P.F.	3752
Peirs, Jan	2651
Pembeci, Izzet	727
Peng, Jufeng	4066
Peng, S.	2024
Peng, ShihSen	4300
Peng, Zhaoyang	3254
Pensky, D.H.	1705
Perçin, Gökhan	2892
Pernalete, Norali	1247
Perruquetti, W.	4330
Peters II, Richard Alan	2806
Petersson, Lars	2097
Petrovic, S.	175
Pfeifer, Adam	4264
Pfeiffer, Friedrich	484, 2954, 4002
Pfister, Samuel T.	1304
Phee, Soo J.	1086
Philippsen, Roland	446, 1503, 1992
Phoka, Thanathorn	2671
Piat, Emmanuel	3219
Pichler, Andreas	151
Pierrot, François	1185, 1875, 2218, 3278
Pietsch, I.	4360
Piguet, R.	4246
Pillonetto, Gianluigi	3809
Pinheiro Gomes, Sebastião Cícero	622
Pires, J. Norberto	3529
Piyabongkarn, D.	294
Platt, Robert Jr.	2543, 2677
Plietsch, R.	2634
Poignet, Philippe	2218, 3278
Ponce, Jean	2242
Pons, J.L.	3231
Popa, Dan O.	1470
Porta, Josep Maria	342, 2842
Potasek, D. P.	294
Poulakakis, Ioannis	1368
Pouliot, Jean	1793
Pradalier, C.	2430
Prahacs, Chris	1386
Prasser, David	1291
Pratt, Gill A.	45
Prattichizzo, Domenico	1259, 3929

Q

Qiao, Hong	2248
Quiñonero, J.P. Canovas	2147

R

Rabischong, Pierre	2218, 2657
Racoceanu, Daniel	4318
Rahimi, Mohammad	19
Rahman, M. Masudur	1410
Rajamani, R.	294
Rajasekharan, S.	4252
Ramamritham, Krithi	4074
Ramel, G.	2388
Ramirez, Gabriel	509
Rani, Pramila	2382
Rao, A.B. Koteswara	4104
Rao, Nageswara S.V.	1653
Rao, P.V.M.	4104
Rao, Sajit	3140
Redarce, T.	658
Reif, John	3782, 4420
Rekleitis, Ioannis	1907
Ren, Jing	3875
Renaud, M.	69
Renaud, Pierre	1191
Reveliotis, Spyros	1045
Reynaerts, Dominiek	2646, 2651
Ridao, P.	971, 989
Ridley, Matthew	2720
Ridley, Peter	4288
Rikoski, Richard J.	963
Riley, Marcia	2368
Rimon, Elon	1817, 2579, 2966
Ritter, Helge	4209
Rivero, D.M.	3284
Rives, Patrick	1056
Riviere, Cameron N.	1781
Rivlin, Ehud	1021
Roberts, Jonathan M.	546
Roberts, Randy S.	3321
Robertsson, Anders	3491, 3686
Roccella, S.	2230
Roche, Allen	3504
Rock, Stephen M.	2946, 3479, 4222
Rodriguez, R.	4023
Rodríguez, F.	3284
Rodríguez, Samuel	4427
Röfer, Thomas	856
Ros, Lluís	342, 355
Roth, Fabian	4154
Rothganger, Fred	2242

Roumeliotis, Stergios I.	1304
Roy, Binayak	2224
Rudas, Imre J.	1068
Ruichek, Yassine	1700
Ruo, Chi-Wei	3523
Rutishauser, Ueli	4154
Rybski, Paul E.	850, 1158
Ryu, Dongseok	3243
Ryu, Jaewook	2940
Ryu, Jee-Hwan	822
Ryu, Jeicheong	1515
Ryu, Ushio	4016

S

Sabatini, Angelo M.	1318
Sabe, Kohtaro	262
Sache, L.	670
Sadahiro, Teruyoshi	3017
Saeedi, P.	1297
Saeki, Masami	2628
Saffiotti, Alessandro	2780
Saha, Mitul	3797
Saha, S.K.	4104
Sahai, Ranjana	3339
Sahoo, Sambit	2657
Saida, Takao	3815
Saito, Takashi	1264
Saito, Tomoko	3996
Sakai, Satoru	96
Sakakibara, Shinsuke	2878
Sakamoto, Atsushi	1933
Salas, Joaquín	1893
Salcudean, S.E.	3296
Salerno, Alessio	3379
Salichs, M.A.	1324, 3284
Salisbry, Kenneth Jr.	809
Salvi, Joaquim	133
Sameshima, M.	2860
Sánchez, Abraham	3764, 4258
Sánchez-Ante, Gildardo	3797
Sanderson, Arthur C.	3213, 3535
Sandini, Giulio	3140
Sanfeliu, Alberto	1576
Sano, Akihito	2478
Santin, Chiara	2212
Santos, Nuno	1312
Santos-Victor, José	2375
Saranli, Uluc	1374
Sargent, R.	2535
Sarkar, Nilanjan	2382
Sasaki, Daisuke	2188
Sasaki, Keiji	1103
Sasaki, Taku	2973
Sastry, Shankar S.	571, 584, 1152, 3758
Sato, Daisuke	1940, 4110
Sato, Tomomasa	4178
Sawada, Kazuya	2979
Sawasaki, Naoyuki	2992
Sawodny, O.	2182
Scalzo, Alessandro	2018
Schenato, Luca	1146, 1152
Schenker, Paul S.	2287
Schilt, Martin	1992
Schimmels, Joseph M.	2592, 3082, 3095
Schmidt, Günther	887, 1343
Schmiedeler, James P.	3821
Schneider, K.	2182
Schrader, Cheryl B.	3692
Scilingo, Enzo Pasquale	2412
Seara, Javier F.	887
Sebastiani, F.	2230
Secchi, C.	3290
Seitz, N.	684, 3164
Sekhavat, S.	2430
Senda, Kei	797
Seneviratne, Lakmal D	121, 103, 2555, 2561
Seraji, Homayoun	2012
Sgambelluri, Nicola	2412
Sgorbissa, Antonio	2018
Shah, Hardik	19
Shakernia, Omid	571, 584
Shapiro, Amir	2966
Shen, Wei-Min	2311, 2516, 3828
Shen, Yi-Shiuan	1434
Sheng, Weihua	374, 3504
Shi, Yinghai	4414
Shibata, Takanori	3152, 3996
Shibukawa, T.	1185
Shigematsu, Bunji	1978
Shih, Ching-Long	1787, 3523
Shiina, Yoshikazu	2689
Shikata, Ritsu	1946
Shiller, Zvi	3359
Shimojo, Makoto	1264
Shinoda, Hiroyuki	3207
Shinohara, Tetsuya	491
Shinomiya, Hirotatsu	2979
Shiotani, Shigetoshi	2973
Shirai, Yoshiaki	893
Shiraishi, Atsushi	2992
Shirinzadeh, Bijan	3666
Shitashimizu, Takeshi	4110
Shukuya, Yuichirou	951
Shumway, Chris	3089
Shyr, Bor-Yeu	2085
Sian, Neo Ee	1613
Siciliano, Bruno	3333, 3576
Siegwart, Roland	440, 446, 1503, 1992, 2388, 4246
Sim, Robert	3434
Siméon, T.	4354
Simhon, Saul	2907

Simone, C.	1774
Sitti, Metin	1164
Skarmeta, A. Gomez	2147
Skidmore, George	1470, 3213
Skounakis, Nikos	7
Sloten, Jos Vander	2646
Smith, B.	145
Smith, Craig A.	2382
So, Byung Rok	4348
Soeder, Derek	3542
Soetens, Peter	2766
Soika, Martin	1899
Soldea, Octavian	1021
Song, Guang	4445
Song, Jae-Bok	503, 3243
Song, Kai-Tai	905, 2091
Sonohara, Yukitaka	1490
Souères, Philippe	540
Srinivasa, Siddhartha S.	3391
Stacey, Deborah A.	4080
Staicu, Stefan	4116
Stamos, Ioannis	145
Starr, G.	2116
Stefanini, Cesare	1836, 2657
Steil, Jochen J.	4209
Steinfeld, Aaron	25
Stephanou, Harry E.	1470, 3213
Stoeter, Sascha A.	90, 1158, 4262
Støy, K.	3828
Stramigioli, Stefano	3290, 4029
Strobl, Klaus H.	887
Stubbs, Kristen	4264
Su, Chan-Hung	1677
Su, Jianbo	3327
Su, Kuo L.	2394
Su, Yu-Chuan	1723
Sudsang, Attawith	1121, 2671
Suenaga, Kentaro	2061
Suga, Yuki	3565
Sugahara, Yusuke	4342
Sugano, Shigeki	3565
Sugar, Thomas G.	318, 3461
Sugawara, Yusuke	2992
Sugi, Masao	2586
Sugihara, Tomomichi	51
Sugiyama, Yuuta	3737
Suh, Il Hong	4142
Sukhatme, Gaurav S.	19, 558, 868, 1665, 2293, 2714
Sukkarieh, Salah	406, 2720
Sumi, Yasuyuki	4166
Sun, Chein-Chung	2330
Sun, Dong	528, 534, 3005
Sun, Guo-Ji	2176
Sun, Qiao	4336
Sun, Yu	294
Sun, Z.Q.	2522
Sun, Zenqi	2455
Sun, Zheng	3782, 4420
Sun, Zhenguo	696
Suraj, Zbigniew	4324
Surazhsky, Tatiana	1021
Suzuki, Katsuya	2640
Suzuki, Kenji	3571
Suzuki, Mototaka	3565
Suzuki, Takafumi	1264
Suzuki, Takahiro	3672
Suzuki, Tatsuya	3959
Suzumori, Koichi	2735
Svennebring, Jonas	75
Svinin, M.M.	2884
Sweeney, John D.	4074
Swindell, Scott	2436

T

Tabe, Keishiro	3571
Tadokoro, Naoki	2067, 2073
Tagawa, Takashi	2031
Taghirad, H.D.	3108
Tahara, K.	2336
Tahri, Omar	4276
Tai, Jen-Chao	2091
Tak, Younghun	2940
Takahashi, Katsumi	2305
Takahashi, Takeshi	380
Takahashi, Yoshihiko	2305
Takaiwa, Masahiro	2188, 4098
Takamatsu, Jun	3881, 3887, 3893
Takanishi, Atsuo	232, 244, 2979, 3582, 3588, 3654, 4342
Takanobu, Hideaki	244, 3582, 3588
Takase, K.	2037
Takashi, Toshinobu	899
Takata, Masanori	2735
Takeda, Yukio	749
Takesue, Naoyuki	202
Takeuchi, Ikuo	2979
Takiguchi, Jyun-ichi	899
Takita, Kensuke	2466
Tamaki, Tatsuya	477
Tan, Choopar	121
Tan, Jindong	374
Tan, Li	2436
Tan, Min	735
Tan, Shun-Wen	4403
Tan, Tieniu	1271
Tanaka, Hiromi T.	3441
Tanaka, Kanji	2848
Tanaka, Katsuya	2663
Tanaka, Kazuo	2362
Tanaka, Shinji	664
Tanaka, Takakazu	2206
Tanaka, Takayuki	2362
Tanemura, Takumi	3729

Tang, Hsiao-Wei	2646, 2651
Tang, Xiaoqiang	3990
Tang, Zhe	2455
Tanie, Kazuo	1336, 1613, 3152, 3996
Taniguchi, Masatoshi	610
Taniguchi, Yasuaki	951
Tao, C.W.	2611
Tao, WeiMin	2901
Tar, József K.	1068
Tardós, Juan D.	427
Tarn, Tzyh-Jong	1476, 2153, 2701, 2895
Taschereau, Richard	1793
Tatani, Koji	1927
Taylor, Camillo J.	4294
Teel, Andrew R.	1062
Teller, Seth	1899
Tendick, Frank	2818
Terrien, G.	4246
Terwijn, B.	2842
Terzopoulos, Demetri	917
Tews, Ashley D.	1665
Thayer, Scott	4270
Theingi	773
Thieffry, R.	1509
Thite, Shripad	2242
Thomann, G.	658
Thomas, Federico	342, 355
Thomas, Shawna	4445
Thomas, Shawna L.	4439
Thomas, Ulrike	3069
Thorpe, Charles	842
Thrun, Sebastian	25, 412, 842, 1227, 1557, 1985, 2000, 4270
Thuilot, B.	115
Todorovic, Sinisa	1422
Toh, Ah Cheong	181
Tokunaga, Yoshihiko	2979
Tomatis, N.	4246
Tomatis, Nicola	1992
Tomita, Fumiaki	2985
Tomita, Kohji	714
Tomizawa, Tetsuo	256
Tomokuni, Seiji	3737
Tomono, Masahiro	862
Torras, Carme	342
Torres, Jesús M.	1588
Tovar, Benjamín	464
Tovey, Craig	3371
Toyota, Toshio	610
Trevai, Chomchana	2269
Triebel, Rudolph	1557, 4270
Troccoli, A.	145
Tsai, Ching-Chih	1283
Tsai, Chi-Yi	905
Tsai, Elizabeth M.	741
Tsai, Jo Peng	1731
Tsai, Ming J.	3517
Tsai, Wen-Hsiang	881
Tsao, Ken-Jui	3804
Tsay, T.I. James	3397
Tsinarakis, G.J.	3559
Tso, S.K.	2522
Tsourveloudis, N.C.	3559
Tsubouchi, Takashi	1978
Tsuchiya, Naofumi	3565
Tsuda, Kunihiro	2171
Tsuji, Hiroshi	2973
Tsuji, Toshio	664
Tsujio, Showzow	1850
Tsumaki, Yuichi	2760
Tsumugiwa, Toru	1933, 2741
Tu, Jianping	1221
Tu, Yan	1749
Tuominen, Juha	2934
Turetta, Alessio	2786
Tzou, Jyh Hwa	1717

U

Uchiyama, Masaru	31, 1185, 2529, 4060, 4110
Ude, Ales	2368
Ueda, Jun	3011
Ueda, Naoki	3441
Ueda, Ryuichi	2356
Ueno, Takao	2998
Ueyama, Tsuyoshi	995, 3770
Uma, R.N.	3542
Umeda, Mikio	96
Umetani, Yoji	1490, 2323
Unger, B.J.	1253
Usher, Kane	4288
Ushimi, Nobuhiro	497

V

Valavanis, K.P.	3559
Van Brussel, Hendrik	268, 2646, 2651
van de Panne, Michiel	917
Vassura, G.	3187
Vecchi, F.	2230
Vela, Patricio A.	1482
Veloso, Manuela	1277, 1416, 2281
Vendittelli, Marilena	3175
Verschure, Paul F M J	4154
Vicino, Antonio	1259, 3929
Vidal, René	571, 584
Villani, Luigi	3333
Vincze, Markus	151
Viswanadham, N.	1737, 1762
Vivas, Andrès	3278

von Stryk, Oskar　1356

W

Waarsing, B.J.W.　268
Wada, Kazuyoshi　3152, 3996
Wada, Nobutaka　2628
Wade, Eric　1711
Wade, Keegan　2368
Wahl, Friedrich M.　3069
Wakabayashi, Kiyoshi　2992
Wakamatsu, Kunimitsu　3582
Wakimoto, Shuichi　2735
Waldron, Kenneth J.　109, 3821
Waletzko, David　4264
Walker, Ian　3449
Walter, Jennifer E.　741
Wang, Chia-Nan　1027
Wang, Chieh-Chih　842
Wang, Chin-Hui　1723
Wang, Danwei　4380
Wang, David W.L.　2707, 3114
Wang, Fei-Yue　3548
Wang, Guang　2472
Wang, H.　3249
Wang, Jinsong　1863, 3990, 4092
Wang, Jun　2747
Wang, Li-Ping　4092
Wang, Li-Sheng　3804
Wang, Michael Yu　2236, 3941
Wang, Pengpeng　2406
Wang, Shuo　735
Wang, Tza-Huei　3630
Wang, W.Y.　2611
Wang, Wenping　349
Wang, Xiao-hao　534
Wang, Xinyu　3207
Wang, Yin　2159
Wang, Yuechao　4414
Wang, ZhiDong　938, 2275
Warisawa, Shin'ichi　676, 2663
Washington, R.　2535
Washio, T.　1774
Wasik, Zbigniew　2780
Wassermann, Klaus　4154
Watanabe, Keigo　515, 779, 2171, 2206
Watanabe, Tetsuyou　1127
Waydo, Stephen　2484
Webb, Peter　2153
Weber, Markus　4002
Webster III, R.J.　828
Weghe, Michael Vande　238
Wehrmeyer, Joseph A.　188
Wei, R.　3164
Wei, Yucheng　1271
Wei, Zhouhua　3984
Wen, John T.　458, 1470
Wenger, Philippe　3965
Werger, Barry　2012
Whatley, Adrian M　4154
Whitehouse, David J.　1863
Whittaker, William　4270
Wieber, Pierre-Brice　2218
Wilkinson, David D.　238
Will, Peter　2311, 2516, 3828
Wittmann, Aaron　4154
Wollherr, Dirk　1356
Wong, Chun-Shin　220, 1497
Wong, E.K.　3869
Wong, Victor T.S.　3648
Wood, J.　2116
Wood, Robert J.　1146, 1842
Woodtli, H.R.　670
Wright, A.　2535
Wu, Cheng-Shong　4403
Wu, Chi-haur　3947
Wu, Hsiao-Pin　3554
Wu, Huai-yu　528, 534
Wu, Naiqi　1428
Wu, Qishi　1653
Wu, Sheng-Ming　2616
Wu, Wei-Chung　1146
Wu, Zhiming　2159, 2165
Wyeth, Gordon　1291
Wyss, Reto　4154

X

Xi, Ning　374, 1646, 1768, 3504, 3642, 3984, 4414
Xi, Yugeng　4191
Xiao, Jing　2605
Xiao, Qionglin　3923
Xie, Liangjun　735
Xie, Xiaolan　157, 1033
Xie, Yujun　3947
Xie, Z.W.　3164
Xiong, Shen-shu　534
Xiong, Zhenhua　3941
Xu, Changhai　3923
Xu, Gang　2165
Xu, J.J.　2683, 2695
Xu, R.　604
Xu, W.L.　2522
Xu, Xinhe　3923
Xu, Yangsheng　1671, 2567, 3455, 4148
Xue, Feng　3535

Y

Yachida, Masahiko　564
Yagi, Yasushi　564
Yamada, Takaaki　779

Yamada, Takeshi	2055	
Yamada, Yoji	1490, 2323	
Yamaguchi, Hideya	2255, 2689	
Yamaguchi, Jin'ichi	471	
Yamakita, Masaki	951, 2055, 3017	
Yamamoto, Keiichi	1103	
Yamamoto, Motoji	497, 3403	
Yamamoto, Takahiro	2640	
Yamamoto, Takahisa	2323	
Yamane, Katsu	3834	
Yamawaki, Tasuku	4054	
Yanagihara, Yoshitaka	2998	
Yang, Aiqiang	324	
Yang, Allen Y.	1013	
Yang, Ge	3200	
Yang, Guilin	761	
Yang, L.	3164	
Yang, Simon X.	13, 1221, 1659, 4080	
Yang, T.W.	2522	
Yang, Xin	515	
Yang, Yuandong	3385	
Yao, Chia-Yu	4130	
Yaralioglu, Göksenin G.	2892	
Yashima, Masahito	2255, 2689	
Yasuda, Kiminori	2171	
Ye, Yongqiang	4380	
Yeh, Sze-Chien	596	
Yen, Chiaming	590	
Yeo, Song Huat	761	
Yerex, Keith	2812	
Yi, Byung-Ju	521, 690, 4142, 4348	
Yi, Dong	690	
Yiu, Y.K.	3660	
Yokogawa, Ryuichi	1933, 2741	
Yokoi, Kazuhito	1336, 1613, 1620, 1633, 1640, 2998, 3905	

Yokokohji, Yasuyoshi	3815	
Yokoyama, Kazuhiko	2171, 2985	
Yoneda, Kan	477, 925	
Yoneda, Mitsunori	2979	
Yoon, Do-Young	3158	
Yoon, Youngrock	3473	
Yorozu, Yasuaki	3132	
Yoshida, Eiichi	714	
Yoshida, Shigenori	1940	
Yoshida, Tetsuji	2323	
Yoshidome, Takumi	2663	
Yoshikai, Tomoaki	1940	
Yoshikawa, Tsuneo	1127, 1539, 3011, 3815, 4366	
Yoshinaka, K.	1774	
You, Bum Jae	3158	
You, Song	2605	
Young, Kuu-young	4392	
Yu, Mengmeng	3642	
Yu, Wentao	1247	
Yu, Yong	1850	
Yuan, Ding	1688	
Yuan, Xiaobu	13	
Yuasa, Hideo	2269, 2448	
Yuen, David C.K.	1564	
Yukawa, Kimihito	640	
Yun, Xiaoping	1171	
Yuta, Shin'ichi	256, 862, 1978, 3409	

Z

Zaccaria, Renato	2018
Zaccone, Franco	2212
Zacharias, Franziska	850
Zagler, Andreas	2954
Zaluzec, Matthew	3953
Zapata, René	3764, 4258
Zapata D., JoséLuis	2753

Zecca, M.	2230
Žefran, Miloš	1241
Zeghloul, Saïd	509
Zelinski, Shannon	3758
Zelinsky, Alexander	2097, 4122
Zeng, Jianyang	3510
Zerhouni, Noureddine	4318
Zha, Hongbin	2848
Zhang, Bin	735
Zhang, Fumin	2510
Zhang, Hong	1570
Zhang, Mingjun	1476, 2153, 2895
Zhang, Ping	1671
Zhang, Rubo	1215
Zhang, W.J.	2317
Zhang, Wenzeng	696
Zhang, X.	604
Zhang, Ya-Chong	2176
Zhang, Ya-Jun	2176
Zhang, Yunong	2747
Zhang, Z.G.	2043
Zhao, Dongbin	696
Zhao, J.D.	3249
Zheng, Yuan F.	2006
Zhong, Yu	1215
Zhou, Changjiu	2455
Zhou, Chao	1271
Zhou, MengChu	1428, 1452, 3548, 4300
Zhou, Renbin	208
Zhou, Yu	1594
Zhou, Zhao-ying	528, 534
Ziaei, Kamyar	3114
Zisserman, A.	1007
Zollo, L.	3576
Zweiri, Yahya H.	103, 121
Zyda, Michael J.	1171

Categories' Index

Architecture and Programming

WP12	Control Architectures	2766
ThP4	Robot Programming through Visual Observation and Model-Based Knowledge	3881

Assembly Systems, Assembly Planning

WP6	Assembly Systems Design and Planning (I)	2579
ThA6	Assembly Systems Design and Planning (II)	3069

Biologically-Inspired Robots and Systems

WA3	Humanoid Robotics Software Platform: OpenHRP	1613
WM3	Snake-like Robots	2055
ThA3	Humanoid Robotics Project of METI	2973

Computer Aided Production Planning, Scheduling, and Control

TuA6	Computer Aided Scheduling	157
TuP6	Semiconductor Factory Automation	1027
ThM6	Computer Aided Production Planning (I)	3504
ThP6	Computer Aided Production Planning (II)	3941

Computer and Robot Vision

TuA5	3D Vision (I)	127
TuM3	Omnidirectional Vehicles	497
TuM5	Omnidirectional Vision	558
TuP1	Vision-Based Navigation (I)	875
TuP5	3D Vision (II)	995
TuE1	Vision-Based Navigation (II)	1304
TuE5	Visual Sensing and Application (I)	1397
WA2	Mobile Robot Design and Localization	1582
WA5	Stereo Vision and Visual Tracking	1677
WM5	3D Vision (III)	2116
WP5	Visual Sensing and Application (II)	2549
WP14	Mobile Robot Localization	2830
ThA5	Visual Servoing (I)	3036

ThM5	Visual Tracking	3473
ThP4	Robot Programming through Visual Observation and Model-Based Knowledge	3881
ThP5	Visual Servoing (II)	3911
ThP13	Intelligent Environment	4154
ThE5	Vision Based Control	4276

Computational Intelligence

TuM12	Computational Intelligence (I)	779
TuP12	Computational Intelligence (II)	1209
WA12	Computational Intelligence (III)	1893

Distributed, Cellular, and Multi Robots

TuM10	Distributed Robotic Systems	714
TuP3	Multi-Mobile Robot System (I)	938
WM10	Multiple Robots Coordination	2269
WP3	Cooperative Control of Multi-Vehicle Systems	2484
WP10	Multi-Robot Systems (I)	2701
ThP3	Multi-Mobile Robot System (II)	3849
ThP10	Multi-Robot Systems (II)	4060
ThE1	Multi-Robot Motion Planning	4215

Dynamics, Motion Control, Force/Impedance Control

TuA2	Control of Biped Robot (I)	31
TuM2	Control of Biped Robot (II)	471
TuM9	Dexterous Hand and Control	684
TuP2	Control of Biped Robot (III)	905
TuP7	Adaptive Control	1056
TuE7	Control Applications (I)	1458
WM2	Control of Quadruped Walking Robot	2024
WM7	Control Applications (II)	2176
WP2	Control of Biped Robot (IV)	2455
WP7	Control Applications (III)	2611
WP12	Control Architectures	2766
ThA7	Flexible Manipulator Control and Estimation	3101
ThA12	Identification and Control	3254
ThM2	Mobile Robot Control (I)	3379

ThM3	Mobile Robot Control (II)	3409
ThM12	Impedance Control	3686
ThP12	Learning Control	4122
ThE5	Vision Based Control	4276
ThE12	Robot Control	4360

Enterprise-Level Modeling, Analysis, and Supply Chain Coordination

| WA7 | Supply Chain Design, Analysis and Optimization | 1737 |
| ThM7 | Enterprise-Level Modeling and Analysis | 3535 |

Geometric Modeling

| TuA12 | Geometry Issues in Robotics | 342 |
| ThP9 | Contact | 4029 |

Human-Robot Interfaces, Haptics

TuA13	Human Robot Interaction (I)	374
TuM13	Haptic Interface (I)	809
TuP13	Haptic Interface (II)	1241
TuE8	Human Robot Interaction (II)	1490
WA13	Human Robot Interaction (III)	1927
WM13	Human Robot Interaction (IV)	2362

Intelligent/Flexible Manufacturing Systems

| TuP6 | Semiconductor Factory Automation | 1027 |
| ThP7 | Intelligent/Flexible Machine Control | 3965 |

Intelligent Transportation Systems

| WM4 | Intelligent Transportation Systems | 2085 |

Kinematics, Mechanics, and Mechanism Design

TuA3	Mobile Robot Wheel Mechanisms	63
TuA7	Actuator Design	188
TuA11	Mechanism Design	312
TuM7	Actuators and Drivers	622
TuM11	Parallel Robot (I)	749
TuP11	Parallel Robot (II)	1179

WA2	Mobile Robot Design and Localization	1582
WA11	Parallel Robot (III)	1863
WM11	Reconfigurable Robot and Special Robot	2299
WP11	Redundant Robots	2735
ThA4	Complex Robotic Systems	3005
ThA11	Robot Design	3225
ThM4	New Robotics	3441
ThM11	Robot Design and Analysis	3654
ThP11	Parallel Robot (IV)	4092
ThE11	Parallel Robot (V)	4330

Legged Robots

TuA2	Control of Biped Robot (I)	31
TuM2	Control of Biped Robot (II)	471
TuP2	Control of Biped Robot (III)	905
TuE2	New Challenges in Biped Locomotion	1336
TuE3	Dynamic Control of Multi-Legged Robot	1368
WM2	Control of Quadruped Walking Robot	2024
WP2	Control of Biped Robot (IV)	2455
ThA2	Micro and Pipe Crawler Walking Robot	2940
ThP2	Control of Multi-Legged and Multi-Joint Robot	3815

Manipulation and Grasping

TuA9	Mobility and Manipulation	250
TuM9	Dexterous Hand and Control	684
TuP9	Grasping Analysis	1115
WA9	Grasping and Manipulation (I)	1800
WM9	Grasping and Manipulation (II)	2236
WP9	Grasping and Manipulation (III)	2671
ThA9	Grasping and Manipulation (IV)	3164
ThM9	Manipulation	3594
ThP9	Contact	4029

Manufacturing System Architecture, Design, and Performance Evaluation

TuM6	Diagnostics and Networked Manufacturing Systems	590
TuP6	Semiconductor Factory Automation	1027

TuE6	Petri Nets in Automated Systems Design (I)	1428
WA6	Manufacturing Systems Architecture and Design (I)	1705
WM6	Manufacturing Systems Architecture and Design (II)	2147
ThE7	Petri Nets in Automated Systems Design (II)	4300

Mathematical Optimization

WM12	Mathematical Optimization	2330

Medical Robots and Systems

TuA8	Bio-Robotics	220
TuM8	Medical Diagnostic Robotics	652
TuP8	Endoluminal Surgery-Microendoscopy (I)	1086
WA8	Robotic Surgery	1768
WM8	Rehabilitation Robotics (I)	2206
WP8	Endoluminal Surgery-Microendoscopy (II)	2640
ThA8	Neuro-Robotics (I)	3132
ThM8	Neuro-Robotics (II)	3565
ThP8	Rehabilitation Robotics (II)	3996

Micro/Nano, Electro-Mechanical Sensor Systems and Robots

TuA7	Actuator Design	188
TuA10	Micro Robotics (I)	282
TuM7	Actuators and Drivers	622
TuP10	Micro Systems	1146
WA10	Micro Robotics (II)	1830
ThA2	Micro and Pipe Crawler Walking Robot	2940
ThA10	Micro Robotics (III)	3193
ThM10	Nano Robots and Manipulations	3624

Motion and Path Planning

TuA1	Mobile Robot Navigation (I)	1
TuM1	Mobile Robot Navigation (II)	440
TuP1	Vision-Based Navigation (I)	875
TuE1	Vision-Based Navigation (II)	1304
WA1	Map Building	1551
WM1	Mobile Robot Navigation (III)	1992

WP1	Mobile Robot Path Planning	2424
ThA1	Motion Planning (I)	2907
ThM1	Motion Planning (II)	3347
ThM14	Nonholonomic Path Planning	3752
ThP1	Motion Planning (III)	3782
ThP14	Path Planning with Uncertainty	4185
ThE1	Multi-Robot Motion Planning	4215
ThE14	Probabilistic Roadmap	4420

Petri Nets

TuE6	Petri Nets in Automated Systems Design (I)	1428
ThE7	Petri Nets in Automated Systems Design (II)	4300

Robotics and Automation in Agriculture and Construction

TuA4	Agriculture and Off-Road Robotics	96

Sensor Fusion, Sensor Based Robotics

TuA1	Mobile Robot Navigation (I)	1
TuA14	SLAM	406
TuM1	Mobile Robot Navigation (II)	440
TuM14	Localization (I)	842
TuP1	Vision-Based Navigation (I)	875
TuP14	Localization (II)	1271
TuE1	Vision-Based Navigation (II)	1304
TuE13	Sensor Application	1521
WA2	Mobile Robot Design and Localization	1582
WA14	Sensor Localization and Mapping	1960
WM1	Mobile Robot Navigation (III)	1992
WM14	Sensor-Based Robotics (I)	2394
WP14	Mobile Robot Localization	2830
ThA14	Sensor-Based Robotics (II)	3315
ThP13	Intelligent Environment	4154

Space Robots and Systems

TuM4	Helicopter/Air Vehicle	528
WP4	Space Robots	2516

Teleoperation, Telerobotics, and Network Robotics

WA4	Network Robotics	1646
WP13	Telerobotics	2798
ThA13	Teleoperation	3284
ThE13	Remote Robotics	4392

Total Quality Management, Maintenance, and Diagnostics

TuM6	Diagnostics and Networked Manufacturing Systems	590

Underwater Robotics

TuP4	Underwater Robotics	963

Virtual Reality

ThM13	Virtual Reality	3716

Wheeled Mobile Robots

TuA1	Mobile Robot Navigation (I)	1
TuA3	Mobile Robot Wheel Mechanisms	63
TuM1	Mobile Robot Navigation (II)	440
TuM3	Omnidirectional Vehicles	497
TuP1	Vision-Based Navigation (I)	875
TuP3	Multi-Mobile Robot System (I)	938
TuE1	Vision-Based Navigation (II)	1304
WA2	Mobile Robot Design and Localization	1582
WM1	Mobile Robot Navigation (III)	1992
WP1	Mobile Robot Path Planning	2424
WP14	Mobile Robot Localization	2830
ThM2	Mobile Robot Control (I)	3379
ThM3	Mobile Robot Control (II)	3409
ThP3	Multi-Mobile Robot System (II)	3849
ThE2	Mobile Robot Systems	4246

Industry Session

WE2	Advanced Industrial Robot Systems	2860
WE3	Robotics and Automation in Biotechnology	2884